Contents

ASM Handbook®

Formerly Ninth Edition, Metals Handbook

Volume 13
Corrosion

Prepared under the direction of the
ASM INTERNATIONAL Handbook Committee

Lawrence J. Korb, Co-Chairman
David L. Olson, Co-Chairman

Joseph R. Davis, Senior Editor
James D. Destefani, Technical Editor
Heather J. Frissell, Editorial Supervisor
George M. Crankovic, Assistant Editor
Diane M. Jenkins, Word Processing Specialist

Robert L. Stedfeld, Director of Reference Publications
Kathleen M. Mills, Manager of Editorial Operations

Editorial Assistance
J. Harold Johnson
Robert T. Kiepura
Dorene A. Humphries

The Materials
Information Society

First printing, September 1987
Second printing, September 1989
Third printing, December 1990
Fourth printing, December 1992
Fifth printing, March 1996

ASM Handbook is a collective effort involving thousands of technical specialists. It brings together in one book a wealth of information from world-wide sources to help scientists, engineers, and technicians solve current and long-range problems.

Great care is taken in the compilation and production of this volume, but it should be made clear that no warranties, express or implied, are given in connection with the accuracy or completeness of this publication, and no responsibility can be taken for any claims that may arise.

Nothing contained in the ASM Handbook shall be construed as a grant of any right of manufacture, sale, use, or reproduction, in connection with any method, process, apparatus, product, composition, or system, whether or not covered by letters patent, copyright, or trademark, and nothing contained in the ASM Handbook shall be construed as a defense against any alleged infringement of letters patent, copyright, or trademark, or as a defense against liability for such infringement.

Comments, criticisms, and suggestions are invited, and should be forwarded to ASM INTERNATIONAL.

Library of Congress Cataloging-in-Publication Data

ASM INTERNATIONAL

Metals handbook.

Includes bibliographies and indexes.
Contents: v. 1. Properties and selection—[etc.]—
v. 9. Metallography and microstructures—[etc.]—
v. 13. Corrosion.

1. Metals—Handbooks, manuals, etc. I. ASM
INTERNATIONAL. Handbook Committee.
TA459.M43 1978 669 78-14934
ISBN 0-87170-007-7 (v. 1)
SAN 204-7586

Printed in the United States of America

Foreword to the Fourth Printing

With the fourth printing of this Volume, it takes its place in the new *ASM Handbook* series. The *ASM Handbook* was established to build upon the proud tradition of the *Metals Handbook* and to position the series to meet the needs of future engineers, researchers, technicians, and students. The *ASM Handbook* series will encompass volumes from both the 9th and 10th Editions of *Metals Handbook*—as well as new and revised Volumes as they are released—in order to establish one comprehensive set of reference materials. This will allow a much more flexible approach to updating information: technological advances will be the impetus behind Volume revisions or the addition of new Volumes to the series. The title of the new *ASM Handbook* series reflects the increasingly interrelated nature of materials technology and emphasizes the position of ASM International as the premier source of authoritative materials information.

The Editors

Foreword to the First through Third Printings

Advances in the field of materials science have resulted in the improvement of existing materials and the development of new materials—along with a variety of new applications. Most major branches of engineering, particularly those involved in the design and construction of new mechanical or structural elements, depend on the results of mechanical tests for measurements of material properties.

For accurate analysis and use of test data, it is important that engineers, even those not engaged in actual testing work, possess a general knowledge of common test methods and an understanding of the applicability, advantages, and limitations of a given test procedure. This latest addition to the distinguished *Metals Handbook* series is intended to provide such an understanding; in it is detailed the maturing technical development of mechanical testing.

The production of this Handbook is a major accomplishment for all those involved. Although the 1948 and 1985 single-volume editions of *Metals Handbook* each include a section on mechanical testing, and Volumes 10 and 11 of the 8th edition series touch on various aspects of the subject, this marks the first time an entire Handbook volume has been devoted exclusively to this important technology.

The Society owes a debt of gratitude to the more than 200 authors and reviewers who volunteered their time and expertise to this undertaking. Under the guidance of John R. Newby of Armco Inc. and the ASM Handbook Committee, and in collaboration with the Handbook's editorial staff, they have produced for the technical community an authoritative reference work.

M. Brian Ives
President

Edward L. Langer
Managing Director

**The Ninth Edition of Metals Handbook
is dedicated to the memory of
TAYLOR LYMAN, A.B. (Eng.), S.M., Ph.D.
(1917-1973)
Editor, Metals Handbook, 1945-1973**

Preface

The cost of corrosion to U.S. industries and the American public is currently estimated at $170 billion per year. Although corrosion is only nature's method of recycling, or of returning a metal to its lowest energy form, it is an insidious enemy that destroys our cars, our plumbing, our buildings, our bridges, our engines, and our factories. Corrosion can often be predictable, such as the uniform corrosion of steel ship hulls or tanks, or it can be totally unpredictable and catastrophic, such as the hydrogen embrittlement or stress corrosion of critical structural members and pressure vessels in the aerospace and chemical processing industries. While corrosion obeys well-known laws of electrochemistry and thermodynamics, the many variables that influence the behavior of a metal in its environment can result in accelerated corrosion or failure in one case and complete protection in another similar case. We can no longer think of materials and environments as monolithic. It makes no sense to ask whether stainless steel is compatible with sulfuric acid. Rather, the question we must ask is which alloy of stainless steel, with which microstructure, with which design detail, is compatible with which sulfuric acid. What is the acid's temperature, concentration, pH, impurity level, types of trace species, degree of aeration, flow velocity, etc.?

Avoiding detrimental corrosion requires the interdisciplinary approach of the designer, the metallurgist, and the chemist. Sooner or later, nearly everyone in these fields will be faced with major corrosion issues. It is necessary to learn to recognize the forms of corrosion and the parameters that must be controlled to avoid or mitigate corrosion. This Handbook was written with these three engineering disciplines in mind. We have attempted to put together a reference book that is well rounded and complete in its coverage—for we want this to be the first book you select when researching a corrosion problem. Each article is indexed to other appropriate sections of the Handbook, and each provides a road map to the thousands of individual bibliographical references that were used to compile the information.

The Handbook is organized into eight major Sections. The first is a Glossary of metallurgical and corrosion terms used throughout the Volume. Nearly 600 terms are defined, selected from more than 20 sources. Of course, one of the most difficult terms to get corrosion experts to agree upon is a definition for "corrosion" itself, for where does one draw the line? Is not the hydride, which precipitates in a stressed titanium weld, a form of corrosion just as the hydrogen embrittlement of steel? And where does corrosion stop—with a metal, or is the environmental reaction of a ceramic or polymer also a form of corrosion? In this Handbook we have limited our discussion of corrosion to metals, by and large, but have included reactions with external environments which may diffuse inside a metal, leading to its destruction as an "internal environment."

The second Section covers the theory of corrosion from the thermodynamic and kinetic points of view. It covers the principles of electrochemistry, diffusion, and dissolution as they apply to aqueous corrosion and high-temperature corrosion in salts, liquid metals, and gases. The effects of both metallurgical and environmental variables on corrosion in aqueous solutions are discussed in detail.

The third Section describes the various forms of corrosion, how to recognize them, and the driving conditions or parameters that influence each form of corrosion, for it is the control of these parameters which can minimize or eliminate corrosion. For convenience, this Section is divided into articles on general corrosion, localized corrosion, metallurgically influenced corrosion, mechanically assisted degradation, and environmentally induced cracking. More than 20 distinct corrosion mechanisms are discussed.

In the fourth Section, methods of corrosion testing and evaluation in the laboratory as well as in-place corrosion monitoring are discussed. For each major form of corrosion (pitting, stress-corrosion cracking, etc.), the existing techniques used in their evaluation are discussed along with the advantages and limitations of each particular test and the quality of the test data generated.

The fifth Section looks at corrosion from the design standpoint. Which materials and design details minimize corrosion? What are the corrosion problems with weldments and how can they be addressed? Finally, how do you place an economic value on your selection of alternate materials or coatings?

The next Section reviews the various methods used for corrosion protection. These include surface conversion coatings, anodizing, ceramic coatings, organic coatings, metallic coatings (both as barrier metals and as sacrificial coatings), thermal spray coatings, CVD/PVD coatings, and other methods of surface modification. It also discusses the principles of and the approaches to anodic and cathodic protection. Finally, the various types and uses of corrosion inhibitors are thoroughly discussed.

The seventh Section covers the corrosion of 27 different metal systems, including all major structural alloy systems and precious metals, and relates the latest information on such topics as powder metals, cemented carbides, amorphous metals, metal matrix composites, hard chromium plating, brazing alloys, and clad metals. In many areas, complete articles have been written where only a few paragraphs were available in existing corrosion texts. For each metal system, the authors discuss the alloys available, the nature of the corrosion resistance film that forms on the metal, and the mechanisms of corrosion, including the metallurgical factors or elements that inhibit or accelerate corrosion. Various forms of corrosion are discussed as well as various environmental effects. The behavior of these metal systems in atmospheres (rural, marine, industrial), in waters (fresh water and seawater), and in alkalies, acids, salts, organic chemicals, and gases is discussed. Methods of corrosion protection most applicable to each metal system are reviewed.

The final Section of the Handbook is where all of this knowledge is put into practice. It vividly illustrates how far we've come in understanding and combating corrosion, and how far we have yet to go. The corrosion experiences of experts from 20 major industries are covered in detail—from fossil fuels to nuclear power, from the chemical processing to the marine industries, from prosthetic devices to the space shuttle, from pharmaceuticals to electronics, from petroleum production and refining to heavy construction. The authors describe the corrosion problems they encounter, tell how they solve them, and present illustrated case histories.

We think you will find this Handbook a broad-based approach to understanding corrosion, with sufficient data and examples to solve many problems directly, and references to key literature for further research into highly complex corrosion issues. There is no cookbook for corrosion avoidance! We hope this Volume with its road map of references will lead you to a better understanding of your corrosion problems and assist you in their solutions.

This Handbook would not have been possible without the generous contributions of the nearly 500 leading corrosion experts who donated their expertise as authors and reviewers. They represent many of the leading industries and educational institutions in this country and abroad. The articles in this Handbook represent tremendous individual efforts. We are also grateful to the Handbook staff at ASM INTERNATIONAL and for the extremely valuable contributions of several technical societies and industrial associations, including the National Association of Corrosion Engineers, the American Society for Testing and Materials, the Electric Power Research Institute, the Pulp and Paper Research Institute of Canada, the Tin Research Institute, the Institute of Paper Chemistry, the American Hot Dip Galvanizers Association, and the Lead Industries Association. In addition, we particularly appreciate the efforts of those who took responsibility for coordinating authors and papers for many articles or entire Sections of this Volume: Dr. Miroslav Marek, Dr. Bruce Craig, Dr. Steven Pohlman, Mr. Donald Sprowls, Mr. James Lackey, Dr. Herbert Townsend, Dr. Thomas Cape, Mr. Kenneth Tator, Dr. Ralph Davison, Dr. Aziz Asphahani, Mr. R. Terrence Webster, Mr. Robert Charlton, Mr. James Hanck, and Mr. Fred Meyer, Jr.

This has truly been a collective venture of the technical community. We thank those who willingly have shared their knowledge with all of us.

L. J. Korb
Co-Chairman

D. L. Olson
Co-Chairman

Policy on Units of Measure

By a resolution of its Board of Trustees, ASM INTERNATIONAL has adopted the practice of publishing data in both metric and customary U.S. units of measure. In preparing this Handbook, the editors have attempted to present data in metric units based primarily on Système International d'Unités (SI), with secondary mention of the corresponding values in customary U.S. units. The decision to use SI as the primary system of units was based on the aforementioned resolution of the Board of Trustees and the widespread use of metric units throughout the world.

For the most part, numerical engineering data in the text and in tables are presented in SI-based units with the customary U.S. equivalents in parentheses (text) or adjoining columns (tables). For example, pressure, stress, and strength are shown both in SI units, which are pascals (Pa) with a suitable prefix, and in customary U.S. units, which are pounds per square inch (psi). To save space, large values of psi have been converted to kips per square inch (ksi), where 1 ksi = 1000 psi. Some strictly scientific data are presented in SI units only. For example, fatigue crack growth rates have been given only in millimeters per cycle (mm/cycle).

To clarify some illustrations, only SI-based units are presented on artwork. References in the accompanying text to data in the illustrations are presented in both SI-based and customary U.S. units.

On graphs and charts, grids corresponding to SI-based units appear along the left and bottom edges; where appropriate, corresponding customary U.S. units appear along the top and right edges. In some instances, only SI-based or customary U.S. units have been used. An example of this is in the reporting of corrosion weight losses; only milligrams per square centimeter (mg/cm^2) or grams per square meter (g/m^2) are used.

Data pertaining to a specification published by a specification-writing group may be given in only the units used in that specification or in dual units, depending on the nature of the data. For example, the typical yield strength of aluminum sheet made to a specification written in customary U.S. units would be presented in dual units, but the thickness specified in that specification might be presented only in inches.

Data obtained according to standardized test methods for which the standard recommends a particular system of units are presented in the units of that system. Wherever feasible, equivalent units are also presented.

Conversions and rounding have been done in accordance with ASTM Standard E 380, with attention given to the number of significant digits in the original data. For example, an annealing temperature of 1570 °F contains three significant digits. In this case, the equivalent temperature would be given as 855 °C; the exact conversion to 854.44 °C would not be appropriate. For an invariant physical phenomenon that occurs at a precise temperature (such as the melting of pure silver), it would be appropriate to report the temperature as 961.93 °C or 1763.5 °F. In many instances (especially in tables and data compilations), temperature values in °C and °F are alternatives rather than conversions.

The policy on units of measure in this Handbook contains several exceptions to strict conformance to ASTM E 380; in each instance, the exception has been made in an effort to improve the clarity of the Handbook. The most notable exception is the use of $MPa\sqrt{m}$ rather than $MN \cdot m^{-3/2}$ or $MPa \cdot m^{0.5}$ as the SI unit of measure for fracture toughness. Other examples of such exceptions are the use of "L" rather than "l" as the abbreviation for liter and the use of g/cm^3 rather than kg/m^3 as the unit of measure for density (mass per unit volume).

SI practice requires that only one virgule (diagonal) appear in units formed by combination of several basic units. Therefore, all of the units preceding the virgule are in the numerator and all units following the virgule are in the denominator of the expression; no parentheses are required to prevent ambiguity.

Members of the ASM Handbook Committee (1986-1987)

Dennis D. Huffman
(Chairman 1986-; Member 1983-)
The Timken Company

Roger J. Austin (1984-)
Materials Engineering Consultant

Peter Beardmore (1986-)
Ford Motor Company

Deane I. Biehler (1984-)
Caterpillar Tractor Company

Robert D. Caligiuri (1986-)
SRI International

Richard S. Cremisio (1986-)
Rescorp International Inc.

Thomas A. Freitag (1985-)
The Aerospace Corporation

Charles David Himmelblau (1985-)
Lockheed Missiles & Space Company, Inc.

John D. Hubbard (1984-)
Hinderliter Heat Treating

L.E. Roy Meade (1986-)
Lockheed-Georgia Company

Merrill I. Minges (1986-)
Air Force Wright Aeronautical Laboratories

David V. Neff (1986-)
Metaullics Systems

David LeRoy Olson (1982-)
Colorado School of Mines

Paul E. Rempes (1986-)
Champion Spark Plug Company

Ronald J. Ries (1983-)
The Timken Company

E. Scala (1986-)
Cortland Cable Company, Inc.

David A. Thomas (1986-)
Lehigh University

Peter A. Tomblin (1985-)
De Havilland Aircraft of Canada Ltd.

Leonard A. Weston (1982-)
Lehigh Testing Laboratories, Inc.

Previous Chairmen of the ASM Handbook Committee

R.S. Archer
(1940-1942) (Member, 1937-1942)

L.B. Case
(1931-1933) (Member, 1927-1933)

T.D. Cooper
(1984-1986) (Member, 1981-1986)

E.O. Dixon
(1952-1954) (Member, 1947-1955)

R.L. Dowdell
(1938-1939) (Member, 1935-1939)

J.P. Gill
(1937) (Member, 1934-1937)

J.D. Graham
(1966-1968) (Member, 1961-1970)

J.F. Harper
(1923-1926) (Member, 1923-1926)

C.H. Herty, Jr.
(1934-1936) (Member, 1930-1936)

J.B. Johnson
(1948-1951) (Member, 1944-1951)

L.J. Korb
(1983) (Member, 1978-1983)

R.W.E. Leiter
(1962-1963) (Member, 1955-1958, 1960-1964)

G.V. Luerssen
(1943-1947) (Member, 1942-1947)

G.N. Maniar
(1979-1980) (Member, 1974-1980)

J.L. McCall
(1982) (Member, 1977-1982)

W.J. Merten
(1927-1930) (Member, 1923-1933)

N.E. Promisel
(1955-1961) (Member, 1954-1963)

G.J. Shubat
(1973-1975) (Member, 1966-1975)

W.A. Stadtler
(1969-1972) (Member, 1962-1972)

R. Ward
(1976-1978) (Member, 1972-1978)

M.G.H. Wells
(1981) (Member, 1976-1981)

D.J. Wright
(1964-1965) (Member, 1959-1967)

Authors and Reviewers

H. Ackerman
Edco Products, Inc.
Donald R. Adolphson
Sandia Laboratories
D.C. Agarwal
Haynes International, Inc.
V.S. Agarwala
Naval Air Development Center
John D. Alkire
Amoco Corporation
John R. Ambrose
University of Florida
Albert A. Anctil
Department of the Army
Phillip J. Andersen
Zimmer
D.B. Anderson
National Bureau of Standards
Peter L. Andresen
General Electric Research and Development Center
Dennis M. Anliker
Champion International Corporation
Frank J. Ansuini
Consulting Engineer
A.J. Armini
Surface Alloys Corporation
William G. Ashbaugh
Cortest Engineering Services
Aziz I. Asphahani
Haynes International, Inc.
Terje Kr. Aune
Norsk Hydro (Norway)
Denise M. Aylor
David Taylor Naval Ship Research & Development Center
Robert Baboian
Texas Instruments, Inc.
C. Bagnall
Westinghouse Electric Corporation
V. Baltazar
Noranda Research Centre (Canada)
Edward N. Balko
Englehard Corporation
Calvin H. Baloun
Ohio University
R.C. Bates
Westinghouse Electric Corporation
Michael L. Bauccio
The Boeing Company
Charles Baumgartner
General Electric Company
Richard Baxter
Sealand Corrosion Control, Ltd.

R.P. Beatson
Pulp and Paper Research Institute of Canada
John A. Beavers
Battelle Columbus Division
T.R. Beck
Electrochemical Technology, Inc.
S. Belisle
Noranda Inc. (Canada)
Robert J. Bell
Heat Exchanger Systems, Inc.
B.W. Bennett
Bell Communications Research
David C. Bennett
Champion International Corporation
E.L. Bereczky
Unocal Corporation
Carl A. Bergmann
Westinghouse Electric Corporation
I.M. Bernstein
Carnegie-Mellon University
A.K. Bhambri
Morton Thiokol Inc.
Robert C. Bill
Lewis Research Center
National Aeronautics & Space Administration
C.R. Bird
Stainless Foundry & Engineering, Inc.
Neil Birks
University of Pittsburgh
R. Ross Blackwood
Tenaxol, Inc.
Malcolm Blair
Delray Steel Casting, Inc.
A.J. Blazewicz
Babcock & Wilcox
J. Blough
Foster Wheeler Development Corporation
Michael E. Blum
FMC Corporation
Bennett P. Boffardi
Calgon Corporation
P.W. Bolmer
Kaiser Aluminum & Chemical Corporation
Rodney R. Boyer
Boeing Commercial Airplane Company
Samuel A. Bradford
University of Alberta (Canada)
Robert W. Bradshaw
Sandia National Laboratories

J.W. Braithwaite
Sandia National Laboratories
W.F. Brehm
Westinghouse Hanford Company
P. Bro
Technical Consultant
R. Brock
Teledyne CAE
Alan P. Brown
Argonne National Laboratory
M. Browning
Technical Consultant
S.K. Brubaker
E.I. Du Pont de Nemours & Company, Inc.
John C. Bruno
J&L Specialty Products Corporation
James H. Bryson
Inland Steel Company
R.J. Bucci
Alcoa Laboratories
Charles D. Bulla
ICI Americas Inc.
Donald S. Burns
Spraymetal, Inc.
H.E. Bush
Corrosion Consultant
Dwight A. Burford
Colorado School of Mines
J. Butler
Platt Brothers & Company
W.S. Butterfield
Beloit Corporation
L.E. Cadle
Texas Eastern Products Pipeline Company
John Campbell
Quality Carbide, Inc.
L.W. Campbell
General Magnaplate Corporation
Thomas W. Cape
Chemfil Corporation
Bernie Carpenter
Colorado School of Mines
Allan P. Castillo
Sandusky Foundry & Machine Company
Victor Chaker
The Port Authority of New York and New Jersey
George D. Chappell
Nalco Chemical Company

Robert S. Charlton
B.H. Levelton & Associates, Ltd. (Canada)
G. Dale Cheever
General Motors Research Laboratories
Newton Chessin
Martin Marietta Aerospace
Robert John Chironna
Croll-Reynolds Company, Inc.
Omesh K. Chopra
Argonne National Laboratory
Wendy R. Cieslak
Sandia National Laboratories
Ken Clark
Fansteel—Wellman Dynamics
Clive R. Clayton
State University of New York at Stony Brook
S.K. Coburn
Corrosion Consultants, Inc.
Robert Coe
Public Service Company of Colorado
B. Cohen
Air Force Wright Aeronautical Laboratories
Roland L. Coit
Technical Consultant
L. Coker
Exxon Chemical Company
N.C. Cole
Combustion Engineering Inc.
E.L. Colvin
Aluminum Company of America
J.B. Condon
Martin Marietta Energy Systems, Inc.
B. Cooley
Hoffman Silo Inc.
Richard A. Corbett
Corrosion Testing Laboratories, Inc.
B. Cox
Atomic Energy of Canada Ltd.
W.M. Cox
Corrosion and Protection Centre
University of Manchester (England)
Bruce Craig
Metallurgical Consultants, Inc.
K.R. Craig
Combustion Engineering Inc.
William R. Cress
Allegheny Power Service Corporation
Paul Crook
Haynes International, Inc.
Thomas W. Crooker
Naval Research Laboratory
Ronald D. Crooks
Hercules, Inc.
Carl E. Cross
Colorado School of Mines
Robert Crowe
Naval Research Laboratory
J.R. Crum
Inco Alloys International, Inc.
Daniel Cubicciotti
Electric Power Research Institute
William J. Curren
Cortronics, Inc.
Michael J. Cusick
Colorado School of Mines
Carl J. Czajkowski
Brookhaven National Laboratory
Brian Damkroger
Colorado School of Mines
P.L. Daniel
Babcock & Wilcox
Joseph C. Danko
American Welding Institute
Vani K. Dantam
General Motors Corporation
C.V. Darragh
The Timken Company
Ralph M. Davison
Avesta Stainless, Inc.
Sheldon W. Dean
Air Products and Chemicals, Inc.
Terry DeBold
Carpenter Technology Corporation
Thomas F. Degnan
Consultant
James E. Delargey
Detroit Edison
Stephen C. Dexter
University of Delaware
Ronald B. Diegle
Sandia National Laboratories
J.J. Dillon
Martin Marietta Energy Systems, Inc.
Bill Dobbs
Air Force Wright Aeronautical Laboratories
R.F. Doelling
The Witt Company
James E. Donham
Consultant
R.B. Dooley
Electric Power Research Institute
D.L. Douglass
University of California at Los Angeles
Donald E. Drake
Mobil Corporation
L.E. Drake
Stauffer Chemical Company
Carl W. Dralle
Ampco Metal
Edgar W. Dreyman
PCA Engineering, Inc.
Barry P. Dugan
St. Joe Resources Company
Arthur K. Dunlop
Corrosion Control Consultant
Walter B. Ebner
Honeywell Inc.
G.B. Elder
Union Carbide Corporation
Peter Elliott
Cortest Engineering Services Inc.
Edward Escalante
National Bureau of Standards
Charles L.L. Faust
Consultant
R. Fekete
Ford Motor Company
Ron Fiore
Sikorsky Aircraft
S. Fishman
Office of Naval Research
W.D. Fletcher
Westinghouse Electric Corporation
Mars G. Fontana
Materials Technology Institute
F. Peter Ford
General Electric Research & Development Center
Robert Foreman
Park Chemical Company
L.D. Fox
Tennessee Valley Authority
Anna C. Fraker
National Bureau of Standards
David Franklin
Electric Power Research Institute
Douglas B. Franklin
George C. Marshall Space Flight Center
National Aeronautics & Space Administration
David N. French
David French Inc.
R.A. French
BASF Corporation
R.E. Frishmuth
Cortest Laboratories
Allan Froats
Chromasco/Timminco, Ltd. (Canada)
P. Fulford
Florida Power and Light Company
J.M. Galbraith
Arco Alaska Inc.
J.W. Gambrell
American Hot Dip Galvanizers Association
S. Ganesh
General Electric Company
Richard P. Gangloff
University of Virginia
Thomas W. Gardega
National Thermal Spray Company
Warren Gardner
Department of the Air Force
Andrew Garner
Pulp and Paper Research Institute of Canada
D. Gearey
Corrosion and Protection Centre
University of Manchester (England)
George A. Gehring, Jr.
Ocean City Research Corporation
Floyd Gelhaus
Electric Power Research Institute
Randall M. German
Rensselaer Polytechnic Institute
William J. Gilbert
Croll-Reynolds Company, Inc.
Paul S. Gilman
Allied-Signal

William Glaeser
Battelle Columbus Division
Samuel V. Glorioso
Lyndon B. Johnson Space Center
National Aeronautics & Space Administration
Claus G. Goetzel
Stanford University
Michael Gold
Babcock & Wilcox
Barry M. Gordon
General Electric Company
Gerald M. Gordon
General Electric Company
Andrew John Gowarty
Department of the Army
Robert Graf
United Technologies Research Center
Richard D. Granata
Lehigh University
Stanley J. Green
Electric Power Research Institute
C.D. Griffin
Carbomedics, Inc.
Richard B. Griffin
Texas A&M University
John Grocki
Haynes International, Inc.
Earl C. Groshart
Boeing Aerospace Company
V.E. Guernsey
Electroplating Consultants International
Ronald D. Gundry
Buckeye Pipe Line Company
S.Wm. Gunther
Mangel, Scheuermann & Oeters, Inc.
Jack D. Guttenplan
Rockwell International
H. Guttman
Noranda Research Centre (Canada)
J. Gutzeit
Amoco Corporation
Charles E. Guzi
Procter and Gamble Company
Harvey P. Hack
David Taylor Naval Ship Research & Development Center
J.D. Haff
E.I. Du Pont de Nemours & Company, Inc.
Christopher Hahin
Materials Protection Associates
William B. Hampshire
Tin Research Institute, Inc.
James A. Hanck
Pacific Gas & Electric Company
Paul R. Handt
Dow Chemical Company
Michael Haroun
Oklahoma State University
Charles A. Harper
Westinghouse Electric Corporation
J.A. Hasson
E.F. Houghton & Company
David Hawke
Amax Magnesium
Gardner Haynes
Texas Instruments, Inc.
F.H. Haynie
Environmental Protection Agency
Robert H. Heidersbach
California Polytechnic State University
C. Heiple
Rockwell International
Lawrence E. Helwig
USX Corporation
James B. Hill
Allegheny Ludlum Corporation
James Hillis
Dow Chemical Company
John P. Hirth
Ohio State University
Norris S. Hirota
Electric Power Research Institute
N.J. Hoffman
Rockwell International
E.H. Hollingsworth
Aluminum Company of America (retired)
A. Craig Hood
ACH Technologies
R.L. Horst
Aluminum Company of America
J.B. Horton
J.B. Horton Company
K. Houghton
Wollaston Alloy Inc.
Louis E. Huber, Jr.
Technical Consultant
F.J. Hunkeler
NRC Inc.
H.Y. Hunsicker
Aluminum Company of America (retired)
J.R. Hunter
Pfizer Inc.
Carl A. Hutchinson
Federal Aviation Administration
S. Ibarra
Amoco Corporation
N. Inoue
Kubota America Corporation
R.I. Jaffee
Electric Power Research Institute
J.F. Jenkins
Naval Civil Engineering Laboratory
James W. Johnson
WKM—Joy Division
Mark J. Johnson
Allegheny Ludlum Corporation
Philip C. Johnson
Materials Development Corporation
Otakar Jonas
Consultant
Allen R. Jones
M&T Chemicals, Inc.
L. Jones
ERT, A Resource Engineering Company
R.H. Jones
Battelle Pacific Northwest Laboratories
R.M. Kain
LaQue Center for Corrosion Technology, Inc.
Herbert S. Kalish
Adamas Carbide Corporation
M.H. Kamdar
Benet Weapons Laboratory
Russell D. Kane
Cortest Laboratories
A. Kay
Akron Sand Blast & Metallizing Company
T.M. Kazmierczak
UGI Corporation
J.R. Kearns
Allegheny Ludlum Corporation
Victor Kelly
NDT International
G.D. Kent
Parker Chemical Company
H. Kernberger
Bohler Chemical Plant Equipment (Austria)
George E. Kerns
E.I. Du Pont de Nemours & Company, Inc.
R.J. Kessler
Department of Transportation
Bureau of Materials Research
Yong-Wu Kim
Inland Steel Company
Fraser King
Whiteshell Nuclear Research Establishment (Canada)
J.H. King
Chrysler Corporation
Thomas J. Kinstler
Metalplate Galvanizing, Inc.
W.W. Kirk
LaQue Center for Corrosion Technology, Inc.
Samuel Dwight Kiser
Inco Alloys International, Inc.
Erhard Klar
SCM Metal Products
D.L. Klarstrom
Haynes International, Inc.
D.T. Klodt
Manville Corporation
Gregory Kobrin
E.I. Du Pont de Nemours & Company, Inc.
G.H. Koch
Battelle Columbus Division
John W. Koger
Martin Marietta Energy Systems, Inc.
Thomas G. Kollie
Martin Marietta Energy Systems, Inc.
Juri Kolts
Conoco Inc.
Karl-Heintz Kopietz
Henry E. Sanson & Sons, Inc.

Karl A. Korinek
Parker Chemical Company
Curt W. Kovach
Crucible Materials Corporation
Peter Krag
Colorado School of Mines
H.H. Krause
Battelle Columbus Division
William D. Krippes
J.M.E. Chemicals
A.S. Krisher
ASK Associates
Clyde Krummel
Morton Thiokol, Inc.
Kenneth F. Krysiak
Hercules, Inc.
Paul Labine
Petrolite Research & Development
J.Q. Lackey
E.I. Du Pont de Nemours & Company, Inc.
G.Y. Lai
Haynes International, Inc.
F.K. Lampson
Marquordt Corporation
E.A. Lange
Technical Consultant
Bruce Lanning
Colorado School of Mines
John Larson
Ingersoll-Rand Company
S. Larson
Sundstrand Aviation
David S. Lashmore
National Bureau of Standards
R.M. Latanison
Massachusetts Institute of Technology
J.A. Laverick
The Timken Company
Herbert H. Lawson
Armco, Inc.
Harvey H. Lee
Inland Steel Company
T.S. Lee
National Association of Corrosion Engineers
Henry Leidheiser, Jr.
Center for Surface and Coating Research
Lehigh University
G.L. Leithauser
General Motors Corporation
Jack E. Lemons
University of Alabama School of Dentistry
G.G. Levy
Chrysler Corporation
Richard O. Lewis
University of Florida
Barry D. Lichter
Vanderbilt University
E.L. Liening
Dow Chemical Company
Bernard W. Lifka
Aluminum Company of America
Stephen Liu
Pennsylvania State University
Carl E. Locke
University of Kansas
A.W. Loginow
Consulting Engineer
F.D. Lordi
General Electric Company
C. Lundin
University of Tennessee
R.W. Lutey
Buckman Laboratories, Inc.
Fred F. Lyle, Jr.
Southwest Research Institute
Richard F. Lynch
Zinc Institute Inc.
A.J. Machiels
Electric Power Research Institute
J. Lee Magnon
Dixie Testing & Products Inc.
Gregory D. Maloney
Saureisen Cements Company
Paul E. Manning
Haynes International, Inc.
Miroslav I. Marek
Georgia Institute of Technology
Christopher Martenson
Sandvik Steel Company
J.A. Mathews
Duke Power Company
S.J. Matthews
Haynes International, Inc.
D. Mattox
Sandia National Laboratories
Daniel J. Maykuth
Tin Research Institute, Inc.
Joseph Mazia
Mazia Tech-Com Services, Inc.
M.M. McDonald
Rockwell International
J.E. McLaughlin
Exxon Research & Engineering Company
David H. Meacham
Duke Power Company
David N. Meendering
Colorado School of Mines
Jay Mehta
J&L Specialty Products Corporation
R.D. Merrik
Exxon Research & Engineering Company
Thomas Metz
Naval Air Propulsion Center
Fred H. Meyer, Jr.
Air Force Wright Aeronautical Laboratories
K. Miles
Pulp & Paper Research Institute of Canada
G.A. Minick
A.R. Wilfley & Sons, Inc.
K.L. Money
LaQue Center for Corrosion Technology, Inc.
B.J. Moniz
E.I. Du Pont de Nemours & Company, Inc.
Raymond W. Monroe
Maynard Steel Casting Company
Jean A. Montemarano
David Taylor Naval Ship Research & Development Center
J.F. Montle
Carboline Company
P.G. Moore
Naval Research Laboratory
Robert E. Moore
United Engineers and Constructors
Hugh Morrow
Zinc Institute Inc.
Robert E. Moser
Electric Power Research Institute
Max D. Moskal
Stone Container Corporation
Herbert J. Mueller
Corrosion Consultant
John J. Mueller
Battelle Columbus Division
S.K. Murarka
Abitibi-Price Inc. (Canada)
Charles A. Natalie
Colorado School of Mines
J. Lawrence Nelson
Electric Power Research Institute
James K. Nelson
PPG Industries, Inc.
R.J. Neville
Dofasco Inc. (Canada)
Dale C.H. Nevison
Zinc Information Center, Ltd.
R.A. Nichting
Colorado School of Mines
R.R. Noe
Public Service Electric and Gas Company
Peter Norberg
AB Sandvik Steel Company (Sweden)
W.J. O'Donnell
Public Service Electric and Gas Company
Thomas G. Oakwood
Inland Steel Research Laboratories
D.L. Olson
Colorado School of Mines
William W. Paden
Oklahoma State University
T.O. Passell
Electric Power Research Institute
C.R. Patriarca
Haynes International, Inc.
David H. Patrick
ARCO Resources Technology
Steven J. Pawel
University of Tennessee
G. Peck
Cities Service Oil & Gas Corporation
Bruno M. Perfetti
USX Corporation

Sam F. Pensabene
General Electric Company
Jeff Pernick
International Hardcoat, Inc.
William L. Phillips
E.I. Du Pont de Nemours & Company, Inc.
Joseph R. Pickens
Martin Marietta Laboratories
Hugh O. Pierson
Ultramet
D.L. Piron
École Polytechnique de Montreal (Canada)
Patrick Pizzo
San Jose State University
M.C. Place, Jr.
Shell Oil Company
Frederick J. Pocock
Babcock & Wilcox
Ortrun Pohler
Institut Straumann AG (Switzerland)
Steven L. Pohlman
Kennecott Corporation
Charles Pokross
Fansteel Inc.
Ned W. Polan
Olin Corporation
D.H. Pope
Rensselaer Polytechnic University
A.G. Preban
Inland Steel Company
R.B. Priory
Duke Power Company
R.B. Puyear
Monsanto Company
M. Quintana
General Dynamics Electric
Christopher Ramsey
Colorado School of Mines
Robert A. Rapp
Ohio State University
Louis Raymond
L. Raymond & Associates
George W. Read, Jr.
Technical Consultant
J.J. Reilly
McDonnell Douglas Corporation
Roger H. Richman
Daedalus Associates, Inc.
R.E. Ricker
National Bureau of Standards
O.L. Riggs, Jr.
Kerr McGee Corporation
Blaine W. Roberts
Combustion Engineering, Inc.
J.T. Adrian Roberts
Battelle Pacific Northwest Laboratories
Charles A. Robertson
Sun Refining & Marketing Company
H.S. Rosenberg
Battelle Columbus Division
Philip N. Ross, Jr.
Lawrence Berkeley Laboratory

Gene Rundell
Rolled Alloys
S. Sadovsky
Public Service Electric and Gas Company
William Safranek
American Electroplaters and Surface Finishers Society Headquarters
Brian J. Saldanha
Corrosion Testing Laboratories, Inc.
William Scarborough
Vickers, Inc.
Glenn L. Scattergood
Nalco Chemical Company
L.R. Scharfstein
Mobil Research and Development Company
S.T. Scheirer
Westinghouse Electric Corporation
John H. Schemel
Sandvik Specialty Metals Corporation
George Schick
Bell Communications Research
Mortimer Schussler
Fansteel Inc.
(retired)
Ronald W. Schutz
TIMET Corporation
B.J. Scialabba
JME Chemicals
John R. Scully
David Taylor Naval Ship Research & Development Center
J.J. Sebesta
Consultant
M. Sedlack
Technicon Enterprises Inc.
Ellen G. Segan
Department of the Army
R. Serenius
Western Forest Products Ltd. (Canada)
I.S. Shaffer
Department of the Navy
Sandeep R. Shah
Vanderbilt University
W.B.A. Sharp
Westvaco Research Center
C.R. Shastry
Bethlehem Steel Corporation
Barbara A. Shaw
David Taylor Naval Ship Research & Development Center
Robert A. Shaw
Electric Power Research Institute
Gene P. Sheldon
Olin Corporation
R.D. Shelton
Champion Chemicals, Inc.
T.S. Shilliday
Battelle Columbus Division
D.W. Shoesmith
Atomic Energy of Canada Ltd.
C.G. Siegfried
Ebasco Services, Inc.

W.L. Silence
Haynes International, Inc.
D.C. Silverman
Monsanto Company
G. Simard
Reid Inc. (Canada)
J.R. Simmons
Martin Marietta Corporation
Harold J. Singletary
Lockheed-Georgia Company
John E. Slater
Invetech, Inc.
J. Slaughter
Southern Alloy Corporation
George Slenski
Air Force Wright Aeronautical Laboratories
J.S. Smart III
Amoco Production Company
Albert H. Smith
Charlotte Pipe and Foundry Company
Dale L. Smith
Argonne National Laboratory
F.N. Smith
Alcan International Ltd. (Canada)
Gaylord D. Smith
Inco Alloys International, Inc.
Jerome F. Smith
Lead Industries Association, Inc.
Carlo B. Sonnino
Emerson Electric Company
Peter Soo
Brookhaven National Laboratory
N. Robert Sorenson
Sandia National Laboratories
C. Spangler
Westinghouse Electric Corporation
T.C. Spence
The Duriron Company, Inc.
Donald O. Sprowls
Consultant
Narasi Sridhar
Haynes International, Inc.
Stephen W. Stafford
University of Texas at El Paso
J.R. Stanford
Nalco Chemical Company
(retired)
E.E. Stansbury
University of Tennessee
T.M. Stastny
Amoco Corporation
A.J. Stavros
Union Carbide Corporation
T. Steffans
Anhauser-Busch Brewing Company, Inc.
Robert Stiegerwald
Bechtel National, Inc.
Donald R. Stickle
The Duriron Company, Inc.
T.J. Stiebler
Houston Light & Power Company
John G. Stoecker III
Monsanto Company

Paul J. Stone
Chevron U.S.A.
M.A. Streicher
University of Delaware
John Stringer
Electric Power Research Institute
T.J. Summerson
Kaiser Aluminum & Chemical Corporation
M.D. Swintosky
The Timken Company
W.R. Sylvester
Combustion Engineering, Inc.
Barry C. Syrett
Electric Power Research Institute
Robert E. Tatnall
E.I. Du Pont de Nemours & Company, Inc.
Kenneth B. Tator
KTA-Tator, Inc.
George J. Theus
Babcock & Wilcox
David E. Thomas
RMI Company
C.B. Thompson
Pulp & Paper Research Institute of Canada
Norman B. Tipton
The Singleton Corporation
P.F. Tortorelli
Oak Ridge National Laboratory
Herbert E. Townsend
Bethlehem Steel Corporation
K.L. Tryon
The Timken Company
R. Tunder
General Electric Company
Arthur H. Tuthill
Tuthill Associates Inc.
John A. Ulam
Clad Metals, Inc.
Robert H. Unger
TAFA Inc.
William Unsworth
Magnesium Elektron, Ltd. (England)
T.K. Vaidyanathan
N.Y.U. Dental Center
Ralph J. Valentine
VAL-CORR
J.H. VanSciver
Allied-Signal Corporation
Ellis D. Verink, Jr.
University of Florida
R. Viswanathan
Electric Power Research Institute
Ray Wainwright
Technical Consultant
James Walker
Federal Aviation Administration
Donald Warren
E.I. Du Pont de Nemours & Company, Inc.
Ray Watts
Quaker Petroleum Chemicals Company
William P. Webb
Failure Analysis Associates
R.T. Webster
Teledyne Wah Chang Albany
John R. Weeks
Brookhaven National Laboratory
Lawrence J. Weirick
Sandia National Laboratories
Donald A. Wensley
MacMillan Bloedel Research (Canada)
R.E. Westerman
Pacific Northwest Laboratory
Eddie White
Air Force Wright Aeronautical Laboratories
William E. White
Petro-Canada Resources
D. Whiting
Portland Cement Association
Ron Williams
Air Force Wright Aeronautical Laboratories
E.L. Williamson
Southern Company Services
G.G. Wilson
Stora Forest Industries (Canada)
G.C. Wood
Corrosion and Protection Centre
University of Manchester (England)
Ian G. Wright
Battelle Columbus Division
T.E. Wright
Alcan International Ltd. (Canada) (retired)
B.A. Wrobel
Northern Indiana Public Service Company
B.S. Yaffe
Diversey Wyandotte Metals
T.L. Yau
Teledyne Wah Chang Albany
Ronald A. Yeske
The Institute of Paper Chemistry
Edward Zysk
Englehard Corporation

Contents

Glossary of Terms

A

absorption. A process in which fluid molecules are taken up by a liquid or solid and distributed throughout the body of that liquid or solid. Compare with *adsorption*.

accelerated corrosion test. Method designed to approximate, in a short time, the deteriorating effect under normal long-term service conditions.

acid. A chemical substance that yields hydrogen ions (H^+) when dissolved in water. Compare with *base*.

acid embrittlement. A form of *hydrogen embrittlement* that may be induced in some metals by acid.

acid rain. Atmospheric precipitation with a pH below 5.6 to 5.7. Burning of fossil fuels for heat and power is the major factor in the generation of oxides of nitrogen and sulfur, which are converted into nitric and sulfuric acids washed down in the rain. See also *atmospheric corrosion*.

acicular ferrite. A highly substructured nonequiaxed *ferrite* formed upon continuous cooling by a mixed diffusion and shear mode of transformation that begins at a temperature slightly higher than the transformation temperature range for upper bainite. It is distinguished from *bainite* in that it has a limited amount of carbon available; thus, there is only a small amount of carbide present.

acrylic. Resin polymerized from acrylic acid, methacrylic acid, esters of these acids, or acrylonitrile.

activation. The changing of a passive surface of a metal to a chemically active state. Contrast with *passivation*.

active. The negative direction of *electrode potential*. Also used to describe corrosion and its associated potential range when an electrode potential is more negative than an adjacent depressed corrosion rate (passive) range.

active metal. A metal ready to corrode, or being corroded.

active potential. The *potential* of a corroding material.

activity. A measure of the *chemical potential* of a substance, where chemical potential is not equal to concentration, that allows mathematical relations equivalent to those for ideal systems to be used to correlate changes in an experimentally measured quantity with changes in chemical potential.

activity (ion). The ion concentration corrected for deviations from ideal behavior. Concentration multiplied by activity coefficient.

activity coefficient. A characteristic of a quantity expressing the deviation of a solution from ideal thermodynamic behavior; often used in connection with electrolytes.

addition agent. A substance added to a solution for the purpose of altering or controlling a process. Examples include wetting agents in acid pickles, brighteners or antipitting agents in plating solutions, and inhibitors.

adsorption. The surface retention of solid, liquid, or gas molecules, atoms, or ions by a solid or liquid. Compare with *absorption*.

aeration. (1) Exposing to the action of air. (2) Causing air to bubble through. (3) Introducing air into a solution by spraying, stirring, or a similar method. (4) Supplying or infusing with air, as in sand or soil.

aeration cell (oxygen cell). See *differential aeration cell*.

age hardening. Hardening by *aging*, usually after rapid cooling or cold working.

aging. A change in the properties of certain metals and alloys that occurs at ambient or moderately elevated temperatures after hot working or a heat treatment (quench aging in ferrous alloys, natural or artificial aging in ferrous and nonferrous alloys) or after a cold-working operation (strain aging). The change in properties is often, but not always, due to a phase change (precipitation), but never involves a change in chemical composition of the metal or alloy. See also *age hardening, artificial aging, natural aging, overaging, precipitation hardening, precipitation heat treatment, quench aging,* and *strain aging*.

alclad. Composite wrought product comprised of an aluminum alloy core having on one or both surfaces a metallurgically bonded aluminum or aluminum alloy coating that is anodic to the core and thus electrochemically protects the core against corrosion.

alkali metal. A metal in group IA of the periodic system—namely, lithium, sodium, potassium, rubidium, cesium, and francium. They form strongly alkaline hydroxides, hence the name.

alkaline. (1) Having properties of an alkali. (2) Having a pH greater than 7.

alkaline cleaner. A material blended from alkali hydroxides and such alkaline salts as borates, carbonates, phosphates, or silicates. The cleaning action may be enhanced by the addition of surface-active agents and special solvents.

alkyd. Resin used in coatings. Reaction products of polyhydric alcohols and polybasic acids.

alkylation. (1) A chemical process in which an alkyl radical is introduced into an organic compound by substitution or addition. (2) A refinery process for chemically combining isoparaffin with olefin hydrocarbons.

alligatoring. (1) Pronounced wide cracking over the entire surface of a coating having the appearance of alligator hide. (2) The longitudinal splitting of flat slabs in a plane parallel to the rolled surface. Also called fish-mouthing.

alloy plating. The codeposition of two or more metallic elements.

alpha ferrite. See *ferrite*.

alpha iron. The body-centered cubic form of pure iron, stable below 910 °C (1670 °F).

alternate-immersion test. A corrosion test in which the specimens are intermittently exposed to a liquid medium at definite time intervals.

aluminizing. Forming of an aluminum or aluminum alloy coating on a metal by hot dipping, hot spraying, or diffusion.

amalgam. An alloy of mercury with one or more other metals.

ammeter. An instrument for measuring the magnitude of electric current flow.

amorphous solid. A rigid material whose structure lacks crystalline periodicity; that is, the pattern of its constituent atoms or molecules does not repeat periodically in three dimensions. See also *metallic glass*.

amphoteric. A term applied to oxides and hydroxides which can act basic toward strong acids and acidic toward strong alkalis. Substances which can dissociate electrolytically to produce hydrogen or hydroxyl ions according to conditions.

anchorite. A zinc-iron phosphate coating for iron and steel.

anaerobic. Free of air or uncombined oxygen.

anion. A negatively charged ion that migrates through the electrolyte toward the *anode* under the influence of a potential gradient. See also *cation* and *ion*.

annealing. A generic term denoting a treatment, consisting of heating to and holding at a suitable temperature, followed by cooling at a suitable rate, used primarily to soften metallic materials, but also to simultaneously produce desired changes in other properties or in microstructure. The purpose of such changes may be, but is not confined to, improvement of machinability, facilitation of cold work, improvement of mechanical or electrical properties, and/or increase in stability of dimensions. When the term is used by itself, full annealing is implied. When applied only for the relief of stress, the process is properly called stress relieving or stress-relief annealing.

anode. The electrode of an electrolyte cell at which oxidation occurs. Electrons flow away from the anode in the external circuit. It is usually at the electrode that corrosion occurs and metal ions enter solution. Contrast with *cathode*.

anode corrosion. The dissolution of a metal acting as an *anode*.

anode corrosion efficiency. The ratio of the actual corrosion (weight loss) of an *anode* to the theoretical corrosion (weight loss) calculated by *Faraday's law* from the quantity of electricity that has passed.

anode effect. The effect produced by polarization of the *anode* in electrolysis. It is characterized by a sudden increase in voltage and a corresponding decrease in amperage due to the anode becoming virtually separated from the electrolyte by a gas film.

anode efficiency. Current efficiency at the *anode*.

anode film. (1) The portion of solution in immediate contact with the *anode*, especially if the concentration gradient is steep. (2) The outer layer of the anode itself.

anode polarization. See *polarization*.

anodic cleaning. Electrolytic cleaning in which the work is the anode. Also called reverse-current cleaning.

anodic coating. A film on a metal surface resulting from an electrolytic treatment at the *anode*.

anodic inhibitor. A chemical substance or mixture that prevents or reduces the rate of the anodic or oxidation reaction. See also *inhibitor*.

anodic polarization. The change of the electrode potential in the noble (positive) direction due to current flow. See also *polarization*.

anodic protection. (1) A technique to reduce the corrosion rate of a metal by polarizing it into its passive region, where dissolution rates are low. (2) Imposing an external electrical potential to protect a metal from corrosive attack. (Applicable only to metals that show active-passive behavior.) Contrast with *cathodic protection*.

anodic reaction. Electrode reaction equivalent to a transfer of positive charge from the electronic to the ionic conductor. An anodic reaction is an oxidation process. An example common in corrosion is: $Me \rightarrow Me^{n+} + ne^-$.

anodizing. Forming a *conversion coating* on a metal surface by anodic oxidation; most frequently applied to aluminum.

anolyte. The electrolyte adjacent to the *anode* in an *electrolytic cell*.

anti-fouling. Intended to prevent fouling of underwater structures, such as the bottoms of ships.

antipitting agent. An addition agent for electroplating solutions to prevent the formation of pits or large pores in the electrodeposit.

aqueous. Pertaining to water; an aqueous solution is made by using water as a solvent.

artificial aging. Aging above room temperature. See also *aging*. Compare with *natural aging*.

atmospheric corrosion. The gradual degradation or alteration of a material by contact with substances present in the atmosphere, such as oxygen, carbon dioxide, water vapor, and sulfur and chlorine compounds.

austenite. A solid solution of one or more elements in face-centered cubic iron. Unless otherwise designated (such as nickel austenite), the solute is generally assumed to be carbon.

austenitizing. Forming austenite by heating a ferrous alloy into the transformation range (partial austenitizing) or above the transformation range (complete austenitizing). When used without qualification, the term implies complete austenitizing.

auxiliary anode. In electroplating, a supplementary *anode* positioned so as to raise the current density on a certain area of the *cathode* and thus obtain better distribution of plating.

auxiliary electrode. An *electrode* commonly used in polarization studies to pass current to or from a test electrode. It is usually made from a noncorroding material.

B

backfill. Material placed in a drilled hole to fill space around anodes, vent pipe, and buried components of a cathodic protection system.

bainite. A metastable aggregate of *ferrite* and *cementite* resulting from the transformation of *austenite* at temperatures below the *pearlite* range but above M_s, the martensite start temperature. Bainite formed in the upper part of the bainite transformation range has a feathery appearance; bainite formed in the lower part of the range has an acicular appearance resembling that of tempered martensite.

banded structure. A segregated structure consisting of alternating nearly parallel bands of different composition, typically aligned in the direction of primary hot working.

base. A chemical substance that yields hydroxyl ions (OH^-) when dissolved in water. Compare with *acid*.

base metal. (1) The metal present in the largest proportion in an alloy; brass, for example, is a copper-base alloy. (2) An *active metal* that readily oxidizes, or that dissolves to form ions. (3) The metal to be brazed, cut, soldered, or welded. (4) After welding, that part of the metal which was not melted.

beach marks. Macroscopic progression marks on a fatigue fracture or stress-corrosion cracking surface that indicate successive positions of the advancing crack front. The classic appearance is of irregular elliptical or semielliptical rings, radiating outward from one or more origins. Beach marks (also known as clamshell marks or arrest marks) are typically found on service fractures where the part is loaded randomly, intermittently, or with periodic variations in mean stress or alternating stress. See also *striation*.

biaxial stress. See *principal stress (normal)*.

biological corrosion. Deterioration of metals as a result of the metabolic activity of microorganisms.

bipolar electrode. An *electrode* in an *electrolytic cell* that is not mechanically connected to the power supply, but is so placed in the electrolyte, between the *anode* and *cathode*, that the part nearer the anode becomes cathodic and the part nearer the cathode becomes anodic. Also called intermediate electrode.

bituminous coating. Coal tar or asphalt-based coating.

black liquor. The liquid material remaining from pulpwood cooking in the soda or sulfate papermaking process.

black oxide. A black finish on a metal produced by immersing it in hot oxidizing salts or salt solutions.

blister. A raised area, often dome shaped, resulting from (1) loss of adhesion between a coating or deposit and the base metal or (2) delamination under the pressure of expanding gas trapped in a metal in a near-subsurface zone. Very small blisters may be called pinhead blisters or pepper blisters.

blow down. (1) Injection of air or water under high pressure through a tube to the anode area for the purpose of purging the annular space and possibly correcting high resistance caused by gas blocking. (2) In connection with boilers or cooling towers, the process of discharging a significant portion of the aqueous solution in order to remove accumulated salts, deposits, and other impurities.

blue brittleness. Brittleness exhibited by some steels after being heated to a temperature within the range of about 200 to 370 °C (400 to 700 °F), particularly if the steel is worked at the elevated temperature.

blushing. Whitening and loss of gloss of a usually organic coating caused by moisture. Also called blooming.

brackish water. (1) Water having salinity values ranging from approximately 0.5 to 17 parts per thousand. (2) Water having less salt than seawater, but undrinkable.

breakdown potential. The least noble potential where *pitting* or *crevice corrosion*, or both, will initiate and propagate.

brightener. An agent or combination of agents added to an electroplating bath to produce a smooth, lustrous deposit.

brine. Seawater containing a higher concentration of dissolved salt than that of the ordinary ocean.

brittle fracture. Separation of a solid accompanied by little or no macroscopic plastic deformation. Typically, brittle fracture occurs by rapid crack propagation with less expenditure of energy than for *ductile fracture*.

burning. (1) Permanently damaging a metal or alloy by heating to cause either incipient melting or intergranular oxidation. See also *overheating*. (2) In grinding, getting the work hot enough to cause discoloration or to change the microstructure by tempering or hardening.

C

calcareous coating or deposit. A layer consisting of a mixture of calcium carbonate and magnesium hydroxide deposited on surfaces being cathodically protected because of the increased pH adjacent to the protected surface.

calomel electrode. An *electrode* widely used as a reference electrode of known potential in electrometric measurement of acidity and alkalinity, corrosion studies, voltammetry, and measurement of the potentials of other electrodes. See also *electrode potential, reference electrode,* and *saturated calomel electrode*.

calorizing. Imparting resistance to oxidation to an iron or steel surface by heating in aluminum powder at 800 to 1000 °C (1470 to 1830 °F).

carbonitriding. A *case hardening* process in which a suitable ferrous material is heated above the lower transformation temperature in a gaseous atmosphere of such composition as to cause simultaneous absorption of carbon and nitrogen by the surface and, by diffusion, create a concentration gradient. The process is completed by cooling at a rate that produces the desired properties in the workpiece.

carburizing. Absorption and diffusion of carbon into solid ferrous alloys by heating, to a temperature usually above Ac_3, in contact with a suitable carbonaceous material. A form of *case hardening* that produces a carbon gradient extending inward from the surface, enabling the surface layer to be hardened either by quenching directly from the carburizing temperature or by cooling to room temperature, then reaustenitizing and quenching.

case hardening. A generic term covering several processes applicable to steel that change the chemical composition of the surface layer by absorption of carbon, nitrogen, or a mixture of the two and, by diffusion, create a concentration gradient. The outer portion, or case, is made substantially harder than the inner portion, or core. The processes commonly used are carburizing and quench hardening; cyaniding; nitriding; and carbonitriding. The use of the applicable specific process name is preferred.

CASS test. See *copper-accelerated salt-spray test*.

cathode. The *electrode* of an *electrolytic cell* at which reduction is the principal reaction. (Electrons flow toward the cathode in the external circuit.) Typical cathodic processes are cations taking up electrons and being discharged, oxygen being reduced, and the reduction of an element or group of elements from a higher to a lower valence state. Contrast with *anode*.

cathode efficiency. Current efficiency at the *cathode*.

cathode film. The portion of solution in immediate contact with the *cathode* during *electrolysis*.

cathodic cleaning. *Electrolytic cleaning* in which the work is the *cathode*.

cathodic corrosion. Corrosion resulting from a cathodic condition of a structure usually caused by the reaction of an amphoteric metal with the alkaline products of *electrolysis*.

cathodic disbondment. The destruction of adhesion between a coating and its substrate by products of a *cathodic reaction*.

cathodic inhibitor. A chemical substance or mixture that prevents or reduces the rate of the cathodic or reduction reaction.

cathodic pickling. Electrolytic pickling in which the work is the *cathode*.

cathodic polarization. The change of the *electrode potential* in the active (negative) direction due to current flow. See also *polarization*.

cathodic protection. (1) Reduction of corrosion rate by shifting the *corrosion potential* of the electrode toward a less oxidizing potential by applying an external *electromotive force*. (2) Partial or complete protection of a metal from corrosion by making it a *cathode*, using either a galvanic or an impressed current. Contrast with *anodic protection*.

cathodic reaction. Electrode reaction equivalent to a transfer of negative charge from the electronic to the ionic conductor. A cathodic reaction is a reduction process. An example common in corrosion is: $Ox + ne^- \rightarrow Red$.

catholyte. The *electrolyte* adjacent to the cathode of an electrolytic cell.

cation. A positively charged ion that migrates through the electrolyte toward the *cathode* under the influence of a potential gradient. See also *anion* and *ion*.

caustic. (1) Burning or corrosive. (2) A hydroxide of a light metal, such as sodium hydroxide or potassium hydroxide.

caustic dip. A strongly alkaline solution into which metal is immersed for etching, for neutralizing acid, or for removing organic materials such as greases or paints.

caustic embrittlement. An obsolete historical term denoting a form of *stress-corrosion cracking* most frequently encountered in carbon steels or iron-chromium-nickel alloys that are exposed to concentrated hydroxide solutions at temperatures of 200 to 250 °C (400 to 480 °F).

cavitation. The formation and instantaneous collapse of innumerable tiny voids or cavities within a liquid subjected to rapid and intense pressure changes. Cavitation produced by ultrasonic radiation is sometimes used to effect violent localized agitation. Cavitation caused by severe turbulent flow often leads to *cavitation damage*.

cavitation corrosion. A process involving conjoint *corrosion* and *cavitation*.

cavitation damage. The degradation of a solid body resulting from its exposure to *cavitation*. This may include loss of material, surface deformation, or changes in properties or appearance.

cavitation-erosion. Progressive loss of original material from a solid surface due to continuing exposure to *cavitation*.

cell. Electrochemical system consisting of an *anode* and a *cathode* immersed in an *electrolyte*. The anode and cathode may be separate metals or dissimilar areas on the same metal. The cell includes the external circuit, which permits the flow of electrons from the anode toward the cathode. See also *electrochemical cell*.

cementite. A compound of iron and carbon, known chemically as iron carbide and having the approximate chemical formula Fe_3C. It is characterized by an orthorhombic crystal structure. When it occurs as a phase in steel, the chemical composition will be altered by the presence of manganese and other carbide-forming elements.

chalking. The development of loose removable powder at the surface of an organic coating usually caused by weathering.

checking. The development of slight breaks in a coating that do not penetrate to the underlying surface.

checks. Numerous, very fine cracks in a coating or at the surface of a metal part. Checks may appear during processing or during service and are most often associated with thermal treatment or thermal cycling. Also called check marks, *checking*, or *heat checks*.

chelate. (1) A molecular structure in which a heterocyclic ring can be formed by the unshared electrons of neighboring atoms. (2) A *coordination compound* in which a heterocyclic ring is formed by a metal bound to two atoms of the associated *ligand*. See also *complexation*.

chelating agent. (1) An organic compound in which atoms form more than one coordinate bond with metals in solution. (2) A substance used in metal finishing to control or eliminate certain metallic ions present in undesirable quantities.

chelation. A chemical process involving formation of a heterocyclic ring compound that contains at least one metal cation or hydrogen ion in the ring.

chemical conversion coating. A protective or decorative nonmetallic coating produced *in situ* by chemical reaction of a metal with a chosen environment. It is often used to prepare the surface prior to the application of an organic coating.

chemical potential. In a thermodynamic system of several constituents, the rate of change of the Gibbs function of the system with respect to the change in the number of moles of a particular constituent.

chemical vapor deposition. A coating process, similar to gas carburizing and carbonitriding, whereby a reactant atmosphere gas is fed into a processing chamber where it decomposes at the surface of the workpiece, liberating one material for either absorption by, or accumulation on, the workpiece. A second material is liberated in gas form and is removed from the processing chamber, along with excess atmosphere gas.

chemisorption. The binding of an adsorbate to the surface of a solid by forces whose energy levels approximate those of a chemical bond. Contrast with *physisorption*.

chevron pattern. A fractographic pattern of radial marks (shear ledges) that look like nested letters "V"; sometimes called a herringbone pattern. Chevron patterns are typically found on *brittle fracture* surfaces in parts whose widths are considerably greater than their thicknesses. The points of the chevrons can be traced back to the fracture origin.

chromadizing. Improving paint adhesion on aluminum or aluminum alloys, mainly aircraft skins, by treatment with a solution of chromic acid. Also called chromodizing or chromatizing. Not to be confused with chromating or chromizing.

chromate treatment. A treatment of metal in a solution of a hexavalent chromium compound to produce a *conversion coating* consisting of trivalent and hexavalent chromium compounds.

chromating. Performing a *chromate treatment*.

chrome pickle. (1) Producing a chromate *conversion coating* on magnesium for temporary protection or for a paint base. (2) The solution that produces the conversion coating.

chromizing. A surface treatment at elevated temperature, generally carried out in pack, vapor, or salt bath, in which an alloy is formed by the inward diffusion of chromium into the base metal.

clad metal. A composite metal containing two or more layers that have been bonded together. The bonding may have been accomplished by co-rolling, co-extrusion, welding, diffusion bonding, casting, heavy chemical deposition, or heavy electroplating.

cleavage. Splitting (fracture) of a crystal on a crystallographic plane of low index.

cleavage fracture. A fracture, usually of a polycrystalline metal, in which most of the grains have failed by *cleavage*, resulting in bright reflecting facets. It is associated with low-energy *brittle fracture*.

cold cracking. A type of weld cracking that usually occurs below 205 °C (400 °F). Cracking may occur during or after cooling to room temperature, sometimes with a considerable time delay. Three factors combine to produce cold cracks: stress (for example, from thermal expansion and contraction), hydrogen (from hydrogen-containing welding consumables), and a susceptible microstructure (plate martensite is most susceptible to cracking, ferritic and bainitic structures least susceptible). See also *hot cracking*, *lamellar tearing*, and *stress-relief cracking*.

cold working. Deforming metal plastically under conditions of temperature and strain rate that induce strain hardening. Usually, but not necessarily, conducted at room temperature. Contrast with *hot working*.

combined carbon. The part of the total carbon in steel or cast iron that is present as other than *free carbon*.

complexation. The formation of complex chemical species by the coordination of groups of atoms termed ligands to a central ion, commonly a metal ion. Generally, the ligand coordinates by providing a pair of electrons that forms an ionic or covalent bond to the central ion. See also *chelate*, *coordination compound*, and *ligand*.

compressive. Pertaining to forces on a body or part of a body that tend to crush, or compress, the body.

compressive strength. The maximum *compressive stress* a material is capable of developing. With a brittle material that fails in compression by fracturing, the compressive strength has a definite value. In the case of ductile, malleable, or semiviscous materials (which do not fail in compression by a shattering fracture), the value obtained for compressive strength is an arbitrary value dependent on the degree of distortion that is regarded as effective failure of the material.

compressive stress. A stress that causes an elastic body to deform (shorten) in the direction of the applied load. Contrast with *tensile stress*.

concentration cell. An *electrolytic cell*, the *electromotive force* of which is caused by a difference in concentration of some component in the electrolyte. This difference leads to the formation of discrete *cathode* and *anode* regions.

concentration polarization. That portion of the *polarization* of a cell produced by concentration changes resulting from passage of current through the electrolyte.

conductivity. The ratio of the electric current density to the electric field in a material. Also called electrical conductivity or specific conductance.

contact corrosion. A term primarily used in Europe to describe *galvanic corrosion* between dissimilar metals.

contact plating. A metal plating process wherein the plating current is provided by galvanic action between the work metal and a second metal, without the use of an external source of current.

contact potential. The potential difference at the junction of two dissimilar substances.

continuity bond. A metallic connection that provides electrical continuity between metal structures.

conversion coating. A coating consisting of a compound of the surface metal, produced by chemical or electrochemical treatments of the metal. Examples include chromate coatings on zinc, cadmium, magnesium, and aluminum, and oxide and phosphate coatings on steel. See also *chromate treatment* and *phosphating*.

coordination compound. A compound with a central atom or ion bound to a group of ions or molecules surrounding it. Also called coordination complex. See also *chelate, complexation*, and *ligand*.

copper-accelerated salt-spray (CASS) test. An *accelerated corrosion test* for some electrodeposits and for anodic coatings on aluminum.

corrodkote test. An *accelerated corrosion test* for electrodeposits.

corrosion. The chemical or electrochemical reaction between a material, usually a metal, and its environment that produces a deterioration of the material and its properties.

corrosion effect. A change in any part of the *corrosion system* caused by *corrosion*.

corrosion embrittlement. The severe loss of ductility of a metal resulting from corrosive attack, usually *intergranular* and often not visually apparent.

corrosion-erosion. See *erosion-corrosion*.

corrosion fatigue. The process in which a metal fractures prematurely under conditions of simultaneous corrosion and repeated cyclic loading at lower stress levels or fewer cycles than would be required in the absence of the corrosive environment.

corrosion fatigue strength. The maximum repeated stress that can be endured by a metal without failure under definite conditions of corrosion and fatigue and for a specific number of stress cycles and a specified period of time.

corrosion inhibitor. See *inhibitor*.

corrosion potential (E_{corr}). The *potential* of a corroding surface in an electrolyte, relative to a *reference electrode*. Also called rest potential, open-circuit potential, or freely corroding potential.

corrosion product. Substance formed as a result of *corrosion*.

corrosion protection. Modification of a *corrosion system* so that corrosion damage is mitigated.

corrosion rate. *Corrosion effect* on a metal per unit of time. The type of corrosion rate used depends on the technical system and on the type of corrosion effect. Thus, corrosion rate may be expressed as an increase in corrosion depth per unit of time (penetration rate, for example, mils/yr) or the mass of metal turned into corrosion products per unit area of surface per unit of time (weight loss, for example, g/m^2/yr). The corrosion effect may vary with time and may not be the same at all points of the corroding surface. Therefore, reports of corrosion rates should be accompanied by information on the type, time dependency, and location of the corrosion effect.

corrosion resistance. Ability of a metal to withstand *corrosion* in a given *corrosion system*.

corrosion system. System consisting of one or more metals and all parts of the environment that influence *corrosion*.

corrosivity. Tendency of an environment to cause *corrosion* in a given *corrosion system*.

counterelectrode. See *auxiliary electrode*.

couple. See *galvanic corrosion*.

covering power. The ability of a solution to give satisfactory plating at very low current densities, a condition that exists in recesses and pits. This term suggests an ability to cover, but not necessarily to build up, a uniform coating, whereas *throwing power* suggests the ability to obtain a coating of uniform thickness of an irregularly shaped object.

cracking (of coating). Breaks in a coating that extend through to the underlying surface.

crazing. A network of checks or cracks appearing on the surface.

creep. Time-dependent strain occurring under stress. The creep strain occurring at a diminishing rate is called primary creep; that occurring at a minimum and almost constant rate, secondary creep; and that occurring at an accelerating rate, tertiary creep.

creep-rupture embrittlement. *Embrittlement* under creep conditions of, for example, aluminum alloys and steels that results in abnormally low rupture ductility. In aluminum alloys, iron in amounts above the solubility limit is known to cause such embrittlement; in steels, the phenomenon is related to the amount of impurities (for example, phosphorus, sulfur, copper, arsenic, antimony, and tin) present. In either case, failure occurs by *intergranular cracking* of the embrittled material.

creep-rupture strength. The stress that will cause fracture in a creep test at a given time in a specified constant environment. Also called stress-rupture strength.

crevice corrosion. *Localized corrosion* of a metal surface at, or immediately adjacent to, an area that is shielded from full exposure to the environment because of close proximity between the metal and the surface of another material.

critical anodic current density. The maximum anodic current density observed in the active region for a metal or alloy electrode that exhibits active-passive behavior in an environment.

critical flaw size. The size of a flaw (defect) in a structure that will cause failure at a particular stress level.

critical humidity. The *relative humidity* above which the atmospheric corrosion rate of some metals increases sharply.

critical pitting potential (E_{cp}, E_p, E_{pp}). The lowest value of oxidizing potential at which pits nucleate and grow. It is dependent on the test method used.

current. The net transfer of electric charge per unit time. Also called electric current. See also *current density*.

current density. The current flowing to or from a unit area of an electrode surface.

current efficiency. The ratio of the electrochemical equivalent current density for a specific reaction to the total applied current density.

D

deactivation. The process of prior removal of the active corrosive constituents, usually oxygen, from a corrosive liquid by controlled corrosion of expendable metal or by other chemical means, thereby making the liquid less corrosive.

dealloying. The selective corrosion of one or more components of a solid solution alloy. Also called parting or selective leaching. See also *decarburization, decobaltification, denickelification, dezincification*, and *graphitic corrosion*.

decarburization. Loss of carbon from the surface layer of a carbon-containing alloy due to reaction with one or more chemical substances in a medium that contacts the surface. See also *dealloying*.

decobaltification. Corrosion in which cobalt is selectively leached from cobalt-base alloys, such as Stellite, or from cemented carbides. See also *dealloying* and *selective leaching*.

decomposition potential (or voltage). The *potential* of a metal surface necessary to decompose the electrolyte of a cell or a component thereof.

deep groundbed. One or more *anodes* installed vertically at a nominal depth of 15 m (50 ft) or more below the earth's surface in a drilled hole for the purpose of supplying *cathodic protection* for an underground or submerged metallic structure. See also *groundbed*.

delta ferrite. See *ferrite*.

dendrite. A crystal that has a treelike branching pattern, being most evident in cast metals slowly cooled through the solidification range.

denickelification. Corrosion in which nickel is selectively leached from nickel-containing alloys. Most commonly observed in copper-nickel alloys after extended service in fresh water. See also *dealloying* and *selective leaching*.

density (of gases). The mass of a unit volume of gas at a stated temperature and pressure.

density (of solids and liquids). The mass of unit volume of a material at a specified temperature.

deoxidizing. (1) The removal of oxygen from molten metals by use of suitable deoxidizers. (2) Sometimes refers to the removal of undesirable elements other than oxygen by the introduction of elements or compounds that readily react with them. (3) In metal finishing, the removal of oxide films from metal surfaces by chemical or electrochemical reaction.

depolarization. A decrease in the *polarization* of an electrode.

depolarizer. A substance that produces *depolarization*.

deposit corrosion. Corrosion occurring under or around a discontinuous deposit on a metallic surface. Also called poultice corrosion.

descaling. Removing the thick layer of oxides formed on some metals at elevated temperatures.

dezincification. Corrosion in which zinc is selectively leached from zinc-containing alloys. Most commonly found in copper-zinc alloys containing less than 85% copper after extended service in water containing dissolved oxygen. See also *dealloying* and *selective leaching*.

dichromate treatment. A chromate *conversion coating* produced on magnesium alloys in a boiling solution of sodium dichromate.

dielectric shield. In a *cathodic protection* system, an electrically nonconductive material, such as a coating, plastic sheet, or pipe, that is placed between an *anode* and an adjacent *cathode* to avoid current wastage and to improve current distribution, usually on the cathode.

differential aeration cell. An *electrolytic cell*, the *electromotive force* of which is due to a difference in air (oxygen) concentration at one electrode as compared with that at another electrode of the same material. See also *concentration cell*.

diffusion. (1) Spreading of a constituent in a gas, liquid, or solid, tending to make the composition of all parts uniform. (2) The spontaneous movement of atoms or molecules to new sites within a material.

diffusion coating. Any process whereby a base metal or alloy is either (1) coated with another metal or alloy and heated to a sufficient temperature in a suitable environment or (2) exposed to a gaseous or liquid medium containing the other metal or alloy, thus causing *diffusion* of the coating or of the other metal or alloy into the base metal with resultant changes in the composition and properties of its surface.

diffusion coefficient. A factor of proportionality representing the amount of substance diffusing across a unit area through a unit concentration gradient in unit time.

diffusion-limited current density. The *current density*, often referred to as *limiting current density*, that corresponds to the maximum transfer rate that a particular species can sustain because of the limitation of diffusion.

dimple rupture. A fractographic term describing *ductile fracture* that occurs through the formation and coalescence of microvoids along the fracture path. The fracture surface of such a ductile fracture appears dimpled when observed at high magnification and usually is most clearly resolved when viewed in a scanning electron microscope.

disbondment. The destruction of adhesion between a coating and the surface coated.

discontinuity. Any interruption in the normal physical structure or configuration of a part, such as cracks, laps, seams, inclusions, or porosity. A discontinuity may or may not affect the usefulness of the part.

dislocation. A linear imperfection in a crystalline array of atoms. Two basic types are recognized: (1) an edge dislocation corresponds to the row of mismatched atoms along the edge formed by an extra, partial plane of atoms within the body of a crystal; (2) a screw dislocation corresponds to the axis of a spiral structure in a crystal, characterized by a distortion that joins normally parallel planes together to form a continuous helical ramp (with a pitch of one interplanar distance) winding about the dislocation. Most prevalent is the so-called mixed dislocation, which is any combination of an edge dislocation and a screw dislocation.

double layer. The interface between an *electrode* or a suspended particle and an *electrolyte* created by charge-charge interaction leading to an alignment of oppositely charged ions at the surface of the electrode or particle. The simplest model is represented by a parallel plate condensor.

drainage. Conduction of electric current from an underground metallic structure by means of a metallic conductor. Forced drainage is that applied to underground metallic structures by means of an applied electromotive force or sacrificial anode. Natural drainage is that from an underground structure to a more negative (more anodic) structure, such as the negative bus of a trolley substation.

dry corrosion. See *gaseous corrosion*.

drying oil. An oil capable of conversion from a liquid to a solid by slow reaction with oxygen in the air.

ductile fracture. Fracture characterized by tearing of metal accompanied by appreciable gross plastic deformation and expenditure of considerable energy. Contrast with *brittle fracture*.

ductility. The ability of a material to deform plastically without fracturing, measured by elongation or reduction of area in a tensile test, by height of cupping in an Erichsen test, or by other means.

dummy cathode. (1) A *cathode*, usually corrugated to give variable current densities, that is plated at low current densities to preferentially remove impurities from a plating solution. (2) A substitute cathode that is used during adjustment of operating conditions.

dummying. Plating with *dummy cathodes*.

E

885-°F (475-°C) embrittlement. *Embrittlement* of stainless steels upon extended exposure to temperatures between 400 and 510 °C (750 and 950 °F). This type of embrittlement is caused by fine, chromium-rich precipitates that segregate at grain boundaries; time at temperature directly influences the amount of segregation. Grain-boundary segregation of the chromium-rich precipitates increases strength and hardness, decreases ductility and toughness, and changes corrosion resistance. This type of embrittlement can be reversed by heating above the precipitation range.

elastic deformation. A change in dimensions directly proportional to and in phase with an increase or decrease in applied force.

elasticity. The property of a material by virtue of which deformation caused by stress disappears upon removal of the stress. A perfectly elastic body completely recovers its original shape and dimensions after release of stress.

elastic limit. The maximum stress that a material is capable of sustaining without any permanent strain (deformation) remaining upon complete release of the stress.

elastomer. A natural or synthetic polymer, which at room temperature can be stretched repeatedly to at least twice its original length, and which after removal of the tensile load will immediately and forcibly return to approximately its original length.

electrical conductivity. See *conductivity*.

electrical isolation. The condition of being electrically separated from other metallic structures or the environment.

electrical resistivity. The electrical resistance offered by a material to the flow of current, times the cross-sectional area of current flow and per unit length of current path; the reciprocal of the conductivity. Also called resistivity or specific resistance.

electrochemical admittance. The inverse of *electrochemical impedance*.

electrochemical cell. An electrochemical system consisting of an *anode* and a *cathode* in metallic contact and immersed in an *electrolyte*. (The anode and cathode may be different metals or dissimilar areas on the same metal surface.)

electrochemical corrosion. Corrosion that is accompanied by a flow of electrons between cathodic and anodic areas on metallic surfaces.

electrochemical equivalent. The weight of an element or group of elements oxidized or reduced at 100% efficiency by the passage of a unit quantity of electricity. Usually expressed as grams per coulomb.

electrochemical impedance. The frequency-dependent complex-valued proportionality factor ($\Delta E/\Delta i$) between the applied potential or current and the response signal. This factor is the total opposition (Ω or $\Omega \cdot cm^2$) of an electrochemical system to the passage of charge. The value is related to the *corrosion rate* under certain circumstances.

electrochemical potential. The partial derivative of the total electrochemical free energy of a constituent with respect to the number of moles of this constituent where all factors are kept constant. It is analogous to the *chemical potential* of a constituent except that it includes the electric as well as chemical contributions to the free energy. The *potential* of an electrode in an electrolyte relative to a *reference electrode* measured under open circuit conditions.

electrochemical series. Same as *electromotive force series*.

electrode. (1) An electronic conductor used to establish electrical contact with an electrolytic part of a circuit. (2) An electronic conductor in contact with an ionic conductor.

electrode polarization. Change of *electrode potential* with respect to a reference value. Often the *free corrosion potential* is used as the reference value. The change may be caused, for example, by the application of an external electrical current or by the addition of an oxidant or reductant.

electrodeposition. The deposition of a substance on an *electrode* by passing electric current through an *electrolyte*.

electrode potential. The *potential* of an *electrode* in an *electrolyte* as measured against a *reference electrode*. The electrode potential does not include any resistance losses in potential in either the solution or external circuit. It represents the reversible work to move a unit charge from the electrode surface through the solution to the reference electrode.

electrode reaction. Interfacial reaction equivalent to a transfer of charge between electronic and ionic conductors. See also *anodic reaction* and *cathodic reaction*.

electrogalvanizing. The *electroplating* of zinc upon iron or steel.

electrokinetic potential. This *potential*, sometimes called zeta potential, is a potential difference in the solution caused by residual, unbalanced charge distribution in the adjoining solution, producing a double layer. The electrokinetic potential is different from the *electrode potential* in that it occurs exclusively in the solution phase; that is, it represents the reversible work necessary to bring a unit charge from infinity in the solution up to the interface in question but not through the interface.

electroless plating. A process in which metal ions in a dilute aqueous solution are plated out on a substrate by means of autocatalytic chemical reduction.

electrolysis. Production of chemical changes of the *electrolyte* by the passage of current through an *electrochemical cell*.

electrolyte. (1) A chemical substance or mixture, usually liquid, containing ions that migrate in an electric field. (2) A chemical compound or mixture of compounds which when molten or in solution will conduct an electric current.

electrolytic cell. An assembly, consisting of a vessel, electrodes, and an electrolyte, in which *electrolysis* can be carried out.

electrolytic cleaning. A process of removing soil, scale, or corrosion products from a metal surface by subjecting it as an *electrode* to an electric current in an electrolytic bath.

electrolytic protection. See *cathodic protection*.

electromotive force. Electrical potential; voltage.

electromotive force series (emf series). A list of elements arranged according to their standard *electrode potentials*, with "noble" metals such as gold being positive and "active" metals such as zinc being negative.

electron flow. A movement of electrons in an external circuit connecting an *anode* and *cathode* in a corrosion cell; the current flow is arbitrarily considered to be in an opposite direction to the electron flow.

electroplating. Electrodepositing a metal or alloy in an adherent form on an object serving as a *cathode*.

electropolishing. A technique commonly used to prepare metallographic specimens, in which a high polish is produced by making the specimen the *anode* in an *electrolytic cell*, where preferential dissolution at high points smooths the surface.

electrotinning. *Electroplating* tin on an object.

embrittlement. The severe loss of *ductility* or *toughness* or both, of a material, usually a metal or alloy. Many forms of embrittlement can lead to *brittle fracture*. Many forms can occur during thermal treatment or elevated-temperature service (thermally induced embrittlement). Some of these forms of embrittlement, which affect steels, include *blue brittleness, 885 °F (475 °C) embrittlement, quench-age embrittlement, sigma-phase embrittlement, strain-age embrittlement, temper embrittlement, tempered martensite embrittlement,* and *thermal embrittlement*. In addition, steels and other metals and alloys can be embrittled by environmental conditions (environmentally assisted embrittlement). The forms of environmental embrittlement include *acid embrittlement, caustic embrittlement, corrosion embrittlement, creep-rupture embrittlement, hydrogen embrittlement, liquid metal embrittlement, neutron embrittlement, solder embrittlement, solid metal embrittlement,* and *stress-corrosion cracking*.

endurance limit. The maximum stress that a material can withstand for an infinitely large number of fatigue cycles. See also *fatigue strength*.

environment. The surroundings or conditions (physical, chemical, mechanical) in which a material exists.

environmental cracking. *Brittle fracture* of a normally ductile material in which the corrosive effect of the environment is a causative factor. Environmental cracking is a general term that includes *corrosion fatigue, high-temperature hydrogen attack, hydrogen blistering, hydrogen embrittlement, liquid metal embrittlement, solid metal embrittlement, stress-corrosion cracking,* and *sulfide stress cracking*. The following terms have been used in the past in connection with environmental cracking, but are becoming obsolete: caustic embrittlement, delayed fracture, season cracking, static fatigue, stepwise cracking, sulfide corrosion cracking, and sulfide stress-corrosion cracking. See also *embrittlement*.

environmentally assisted embrittlement. See *embrittlement*.

epoxy. Resin formed by the reaction of bisphenol and epichlorohydrin.

equilibrium (reversible) potential. The *potential* of an electrode in an electrolytic solution when the forward rate of a given reaction is exactly equal to the reverse rate. The equilibrium potential can only be defined with respect to a specific electrochemical reaction.

erosion. Destruction of metals or other materials by the abrasive action of moving fluids, usually accelerated by the presence of solid particles or matter in suspension. When corrosion occurs simultaneously, the term *erosion-corrosion* is often used.

erosion-corrosion. A conjoint action involving *corrosion* and *erosion* in the presence of a moving corrosive fluid, leading to the accelerated loss of material.

eutectic. (1) An isothermal reversible reaction in which a liquid solution is converted into two or more intimately mixed solids on cooling, the number of solids formed being the same as the number of components in the system. (2) An alloy having the composition indicated by the eutectic point on an equilibrium diagram. (3) An alloy structure of intermixed solid constituents formed by a eutectic reaction.

eutectoid. (1) An isothermal reversible reaction in which a solid solution is converted into two or more intimately mixed solids on cooling, the number of solids formed being the same as the number of components in the system. (2) An alloy having the composition indicated by the eutectoid point on an equilibrium diagram. (3) An alloy structure of intermixed solid constituents formed by a eutectoid reaction.

exchange current. When an electrode reaches dynamic equilibrium in a solution, the rate of anodic dissolution balances the rate of cathodic plating. The rate at which either positive or negative charges are entering or leaving the surface at this point is known as the exchange current.

exchange current density. The rate of charge transfer per unit area when an electrode reaches dynamic equilibrium (at its reversible potential) in a solution; that is, the rate of anodic charge transfer (oxidation) balances the rate of cathodic charge transfer (reduction).

exfoliation. Corrosion that proceeds laterally from the sites of initiation along planes parallel to the surface, generally at grain boundaries, forming corrosion products that force metal away from the body of the material, giving rise to a layered appearance.

external circuit. The wires, connectors, measuring devices, current sources, etc., that are used to bring about or measure the desired electrical conditions within the test cell. It is this portion of the cell through which electrons travel.

F

failure. A general term used to imply that a part in service (1) has become completely inoperable, (2) is still operable but is incapable of satisfactorily performing its intended function, or (3) has deteriorated seriously, to the point that it has become unreliable or unsafe for continued use.

Faraday's law. (1) The amount of any substance dissolved or deposited in electrolysis is proportional to the total electric charge passed. (2) The amounts of different substances dissolved or deposited by the passage of the same electric charge are proportional to their equivalent weights.

fatigue. The phenomenon leading to fracture under repeated or fluctuating stresses having a maximum value less than the tensile strength of the material. Fatigue fractures are progressive and grow under the action of the fluctuating stress.

fatigue crack growth rate, da/dN. The rate of crack extension caused by constant-amplitude fatigue loading, expressed in terms of crack extension per cycle of load application.

fatigue life. The number of cycles of stress that can be sustained prior to failure under a stated test condition.

fatigue limit. The maximum stress that presumably leads to fatigue fracture in a specified number of stress cycles. If the stress is not completely reversed, the value of the mean stress, the minimum stress, or the stress ratio should also be stated. Compare with *endurance limit*.

fatigue strength. The maximum stress that can be sustained for a specified number of cycles without failure, the stress being completely reversed within each cycle unless otherwise stated.

ferrite. (1) A solid solution of one or more elements in body-centered cubic iron. Unless otherwise designated (for instance, as chromium ferrite), the solute is generally assumed to be carbon. On some equilibrium diagrams, there are two ferrite regions separated by an austenite area. The lower area is alpha ferrite; the upper, delta ferrite. If there is no designation, alpha ferrite is assumed. (2) In the field of magnetics, substances having the general formula: $M^{2+}O \cdot M_2^{3+}O_3$, the trivalent metal often being iron.

filiform corrosion. Corrosion that occurs under some coatings in the form of randomly distributed threadlike filaments.

film. A thin, not necessarily visible, layer of material.

fish eyes. Areas on a steel fracture surface having a characteristic white crystalline appearance.

flakes. Short, discontinuous internal fissures in wrought metals attributed to stresses produced by localized transformation and decreased solubility of hydrogen during cooling after hot working. In a fracture surface, flakes appear as bright silvery areas; on an etched surface, they appear as short, discontinuous cracks. Also called shatter cracks or snowflakes.

flame spraying. *Thermal spraying* in which a coating material is fed into an oxyfuel gas flame, where it is melted. Compressed gas may or may not be used to atomize the coating material and propel it onto the substrate.

foreign structure. Any metallic structure that is not intended as part of a *cathodic protection* system of interest.

fouling. An accumulation of deposits. This term includes accumulation and growth of marine organisms on a submerged metal surface and also includes the accumulation of deposits (usually inorganic) on heat exchanger tubing.

fouling organism. Any aquatic organism with a sessile adult stage that attaches to and fouls underwater structures of ships.

fractography. Descriptive treatment of fracture, especially in metals, with specific reference to photographs of the fracture surface. Macrofractography involves photographs at low magnification (<25×); microfractography, photographs at high magnification (>25×).

fracture mechanics. A quantitative analysis for evaluating structural behavior in terms of applied stress, crack length, and specimen or machine component geometry. See also *linear elastic fracture mechanics*.

fracture toughness. A generic term for measures of resistance to extension of a crack. The term is sometimes restricted to results of *fracture mechanics* tests, which are directly applicable in fracture control. However, the term commonly includes results from simple tests of notched or precracked specimens not based on fracture mechanics analysis. Results from test of the latter type are often useful for fracture control, based on either service experience or empirical correlations with fracture mechanics tests. See also *stress-intensity factor*.

free carbon. The part of the *total carbon* in steel or cast iron that is present in elemental form as graphite or temper carbon. Contrast with *combined carbon*.

free corrosion potential. *Corrosion potential* in the absence of net electrical current flowing to or from the metal surface.

free ferrite. *Ferrite* that is formed directly from the decomposition of hypoeutectoid austenite during cooling, without the simultaneous formation of cementite. Also called proeutectoid ferrite.

free machining. Pertains to the machining characteristics of an alloy to which one or more ingredients have been introduced to give small broken chips, lower power consumption, better surface finish, and longer tool life; among such additions are sulfur or lead to steel, lead to brass, lead and bismuth to aluminum, and sulfur or selenium to stainless steel.

fretting. A type of wear that occurs between tight-fitting surfaces subjected to cyclic relative motion of extremely small amplitude. Usually, fretting is accompanied by corrosion, especially of the very fine wear debris.

fretting corrosion. The accelerated deterioration at the interface between contacting surfaces as the result of corrosion and slight oscillatory movement between the two surfaces.

furan. Resin formed from reactions involving furfuryl alcohol alone or in combination with other constituents.

G

galvanic anode. A metal which, because of its relative position in the galvanic series, provides *sacrificial protection* to metals that are more noble in the series, when coupled in an electrolyte.

galvanic cell. A cell in which chemical change is the source of electrical energy. It usually consists of two dissimilar conductors in contact with each other and with an electrolyte, or of two similar conductors in contact with each other and with dissimilar electrolytes.

galvanic corrosion. Accelerated corrosion of a metal because of an electrical contact with a more noble metal or nonmetallic conductor in a corrosive electrolyte.

galvanic couple. A pair of dissimilar conductors, commonly metals, in electrical contact. See also *galvanic corrosion*.

galvanic couple potential. See *mixed potential*.

galvanic current. The electric current that flows between metals or conductive nonmetals in a *galvanic couple*.

galvanic series. A list of metals and alloys arranged according to their relative corrosion potentials in a given environment. Compare with *electromotive force series*.

galvanize. To coat a metal surface with zinc using any of various processes.

galvanneal. To produce a zinc-iron alloy coating on iron or steel by keeping the coating molten after hot dip galvanizing until the zinc alloys completely with the base metal.

galvanometer. An instrument for indicating or measuring a small electric current by means of a mechanical motion derived from electromagnetic or electrodynamic forces produced by the current.

galvanostatic. An experimental technique whereby an *electrode* is maintained at a constant current in an *electrolyte*.

gaseous corrosion. Corrosion with gas as the only corrosive agent and without any aqueous phase on the surface of the metal. Also called dry corrosion.

gamma iron. The face-centered cubic form of pure iron, stable from 910 to 1400 °C (1670 to 2550 °F).

general corrosion. See *uniform corrosion*.

Gibbs free energy. The thermodynamic function $\Delta G = \Delta H - T\Delta S$, where H is enthalpy, T is absolute temperature, and S is entropy. Also called free energy, free enthalpy, or Gibbs function.

glass electrode. A glass membrane *electrode* used to measure pH or hydrogen-ion activity.

grain. An individual crystal in a polycrystalline metal or alloy; it may or may not contain twinned regions and subgrains.

grain boundary. A narrow zone in a metal corresponding to the transition from one crystallographic orientation to another, thus separating one *grain* from another; the atoms in each grain are arranged in an orderly pattern.

grain-boundary corrosion. Same as intergranular corrosion. See also *interdendritic corrosion*.

graphitic corrosion. Deterioration of gray cast iron in which the metallic constituents are selectively leached or converted to corrosion products leaving the graphite intact. The term *graphitization* is commonly used to identify this form of corrosion, but is not recommended because of its use in metallurgy for the decomposition of carbide to graphite. See also *dealloying* and *selective leaching*.

graphitization. A metallurgical term describing the formation of graphite in iron or steel, usually from decomposition of iron carbide at elevated temperatures. Not recommended as a term to describe *graphitic corrosion*.

green liquor. The liquor resulting from dissolving molten smelt from the kraft recovery furnace in water. See also *kraft process* and *smelt*.

groundbed. A buried item, such as junk steel or graphite rods, that serves as the *anode* for the *cathodic protection* of pipelines or other buried structures. See also *deep groundbed*.

H

half cell. An *electrode* immersed in a suitable *electrolyte*, designed for measurements of *electrode potential*.

halogen. Any of the elements of the halogen family, consisting of fluorine, chlorine, bromine, iodine, and astatine.

hard chromium. Chromium plated for engineering rather than decorative applications.

hardenability. The relative ability of a ferrous alloy to form martensite when quenched from a temperature above the upper critical temperature. Hardenability is commonly measured as the distance below a quenched surface at which the metal exhibits a specific hardness (50 HRC, for example) or a specific percentage of martensite in the microstructure.

hardfacing. Depositing filler metal on a surface by welding, spraying, or braze welding to increase resistance to abrasion, erosion, wear, galling, impact, or cavitation damage.

hard water. Water that contains certain salts, such as those of calcium or magnesium, which form insoluble deposits in boilers and form precipitates with soap.

heat-affected zone. That portion of the base metal that was not melted during brazing, cutting, or welding, but whose microstructure and mechanical properties were altered by the heat.

heat check. A pattern of parallel surface cracks that are formed by alternate rapid heating and cooling of the extreme surface metal, sometimes found on forging dies and piercing punches. There may be two sets of parallel cracks, one set perpendicular to the other.

hematite. (1) An iron mineral crystallizing in the rhombohedral system; the most important ore of iron. (2) An iron oxide, Fe_2O_3, corresponding to an iron content of approximately 70%.

high-temperature hydrogen attack. A loss of strength and ductility of steel by high-temperature reaction of absorbed hydrogen with carbides in the steel resulting in *decarburization* and internal fissuring.

holidays. Discontinuities in a coating (such as porosity, cracks, gaps, and similar flaws) that allow areas of base metal to be exposed to any corrosive environment that contacts the coated surface.

hot corrosion. An accelerated corrosion of metal surfaces that results from the combined effect of oxidation and reactions with sulfur compounds and other contaminants, such as chlorides, to form a molten salt on a metal surface that fluxes, destroys, or disrupts the normal protective oxide. See also *gaseous corrosion*.

hot cracking. Also called solidification cracking, hot cracking of weldments is caused by the segregation at grain boundaries of low-melting constituents in the weld metal. This can result in grain-boundary tearing under thermal contraction stresses. Hot cracking can be minimized by the use of low-impurity welding materials and proper joint design. See also *cold cracking, lamellar tearing,* and *stress-relief cracking.*

hot working. Deforming metal plastically at such a temperature and strain rate that recrystallization takes place simultaneously with the deformation, thus avoiding any strain hardening. Contrast with *cold working.*

hot dip coating. A metallic coating obtained by dipping the base metal into a molten metal.

hot shortness. A tendency for some alloys to separate along grain boundaries when stressed or deformed at temperatures near the melting point. Hot shortness is caused by a low-melting constituent, often present only in minute amounts, that is segregated at grain boundaries.

humidity test. A corrosion test involving exposure of specimens at controlled levels of humidity and temperature. Contrast with *salt-fog test.*

hydrogen-assisted cracking (HAC). See *hydrogen embrittlement.*

hydrogen-assisted stress-corrosion cracking (HSCC). See *hydrogen embrittlement.*

hydrogen blistering. The formation of blisters on or below a metal surface from excessive internal hydrogen pressure. Hydrogen may be formed during cleaning, plating, corrosion, and so forth.

hydrogen damage. A general term for the embrittlement, cracking, blistering, and hydride formation that can occur when hydrogen is present in some metals.

hydrogen embrittlement. A process resulting in a decrease of the *toughness* or *ductility* of a metal due to the presence of atomic hydrogen. Hydrogen embrittlement has been recognized classically as being of two types. The first, known as internal hydrogen embrittlement, occurs when the hydrogen enters molten metal which becomes supersaturated with hydrogen immediately after solidification. The second type, environmental hydrogen embrittlement, results from hydrogen being absorbed by solid metals. This can occur during elevated-temperature thermal treatments and in service during electroplating, contact with maintenance chemicals, corrosion reactions, cathodic protection, and operating in high-pressure hydrogen. In the absence of residual stress or external loading, environmental hydrogen embrittlement is manifested in various forms, such as blistering, internal cracking, hydride formation, and reduced ductility. With a tensile stress or stress-intensity factor exceeding a specific threshold, the atomic hydrogen interacts with the metal to induce subcritical crack growth leading to fracture. In the absence of a corrosion reaction (polarized cathodically), the usual term used is hydrogen-assisted cracking (HAC) or hydrogen stress cracking (HSC). In the presence of active corrosion, usually as pits or crevices (polarized anodically), the cracking is generally called *stress-corrosion cracking* (SCC), but should more properly be called hydrogen-assisted stress-corrosion cracking (HSCC). Thus, HSC and electrochemically anodic SCC can operate separately or in combination (HSCC). In some metals, such as high-strength steels, the mechanism is believed to be all, or nearly all, HSC. The participating mechanism of HSC is not always recognized and may be evaluated under the generic heading of SCC.

hydrogen-induced cracking (HIC). Same as *hydrogen embrittlement.*

hydrogen overvoltage. *Overvoltage* associated with the liberation of hydrogen gas.

hydrogen stress cracking (HSC). See *hydrogen embrittlement.*

hydrolysis. (1) Decomposition or alteration of a chemical substance by water. (2) In aqueous solutions of electrolytes, the reactions of cations with water to produce a weak base or of anions to produce a weak acid.

hydrophilic. Having an affinity for water. Contrast with *hydrophobic.*

hydrophobic. Lacking an affinity for, repelling, or failing to absorb or adsorb water. Contrast with *hydrophilic.*

hygroscopic. (1) Possessing a marked ability to accelerate the condensation of water vapor; applied to condensation nuclei composed of salts that yield aqueous solutions of a very low equilibrium vapor pressure compared with that of pure water at the same temperature. (2) Pertaining to a substance whose physical characteristics are appreciably altered by effects of water vapor. (3) Pertaining to water absorbed by dry soil minerals from the atmosphere; the amounts depend on the physicochemical character of the surfaces, and increase with rising relative humidity.

I

immersion plating. Depositing a metallic coating on a metal immersed in a liquid solution, without the aid of an external electric current. Also called dip plating.

immunity. A state of resistance to corrosion or anodic dissolution of a metal caused by thermodynamic stability of the metal.

impingement corrosion. A form of *erosion-corrosion* generally associated with the local impingement of a high-velocity, flowing fluid against a solid surface.

impressed current. Direct current supplied by a device employing a power source external to the electrode system of a *cathodic protection* installation.

inclusions. Particles of foreign material in a metallic matrix. The particles are usually compounds (such as oxides, sulfides, or silicates), but may be of any substance that is foreign to (and essentially insoluble in) the matrix.

incubation period. A period prior to the detection of corrosion while the metal is in contact with a corrodent.

industrial atmosphere. An atmosphere in an area of heavy industry with soot, fly ash, and sulfur compounds as the principal constituents.

inert anode. An *anode* that is insoluble in the *electrolyte* under the conditions prevailing in the *electrolysis.*

inhibitor. A chemical substance or combination of substances that, when present in the environment, prevents or reduces corrosion without significant reaction with the components of the environment.

inorganic. Being or composed of matter other than hydrocarbons and their derivatives, or matter that is not of plant or animal origin. Contrast with *organic.*

inorganic zinc-rich paint. Coating containing a zinc powder pigment in an *inorganic* vehicle.

intensiostatic. See *galvanostatic.*

intercrystalline corrosion. See *intergranular corrosion.*

intercrystalline cracking. See *intergranular cracking.*

interdendritic corrosion. Corrosive attack that progresses preferentially along interdendritic paths. This type of attack results from local differences in composition, such as coring commonly encountered in alloy castings.

intergranular. Between crystals or grains. Also called intercrystalline. Contrast with *transgranular.*

intergranular corrosion. Corrosion occurring preferentially at grain boundaries, usually with slight or negligible attack on the adjacent grains. Also called intercrystalline corrosion.

intergranular cracking. Cracking or fracturing that occurs between the grains or crystals in a polycrystalline aggregate. Also called intercrystalline cracking. Contrast with *transgranular cracking.*

intergranular fracture. Brittle fracture of a metal in which the fracture is between the grains, or crystals, that form the metal. Also called intercrystalline fracture. Contrast with *transgranular fracture.*

intergranular stress-corrosion cracking (IGSCC). *Stress-corrosion cracking* in which the cracking occurs along grain boundaries.

intermediate electrode. Same as *bipolar electrode.*

internal oxidation. The formation of isolated particles of corrosion products beneath the metal surface. This occurs as the result of preferential oxidation of certain alloy constituents by inward diffusion of oxygen, nitrogen, sulfur, and so forth.

intumescence. The swelling or bubbling of a coating usually because of heating (term currently used in space and fire protection applications).

ion. An atom, or group of atoms, that has gained or lost one or more outer electrons and thus carries an electric charge. Positive ions, or *cations,* are deficient in outer electrons. Negative ions, or *anions,* have an excess of outer electrons.

ion exchange. The reversible interchange of ions between a liquid and solid, with no substantial structural changes in the solid.

iron rot. Deterioration of wood in contact with iron-based alloys.

isocorrosion diagram. A graph or chart that shows constant corrosion behavior with changing solution (environment) composition and temperature.

K

K_{ISCC}. Abbreviation for the critical value of the plane strain *stress-intensity factor* that will produce crack propagation by *stress-corrosion cracking* of a given material in a given environment.

knife-line attack. *Intergranular corrosion* of an alloy, usually stabilized stainless steel, along a line adjoining or in contact with a weld after heating into the sensitization temperature range.

kraft process. A wood-pulping process in which sodium sulfate is used in the caustic soda pulp-digestion liquor. Also called kraft pulping or sulfate pulping.

L

lamellar corrosion. See *exfoliation corrosion.*

lamellar tearing. Occurs in the base metal adjacent to weldments due to high through-thickness strains introduced by weld metal shrinkage in highly restrained joints. Tearing occurs by decohesion and linking along the working direction of the base metal; cracks usually run roughly parallel to the fusion line and are steplike in appearance. Lamellar tearing can be minimized by designing joints to minimize weld shrinkage stresses and joint restraint. See also *cold cracking*, *hot cracking*, and *stress-relief cracking*.

Langelier saturation index. An index calculated from total dissolved solids, calcium concentration, total alkalinity, pH, and solution temperature that shows the tendency of a water solution to precipitate or dissolve calcium carbonate.

ledeburite. The eutectic of the iron-carbon system, the constituents of which are *austenite* and *cementite*. The austenite decomposes into *ferrite* and cementite on cooling below Ar_1, the temperature at which transformation of austenite to ferrite or ferrite plus cementite is completed during cooling.

ligand. The molecule, ion, or group bound to the central atom in a *chelate* or a *coordination compound*.

limiting current density. The maximum current density that can be used to obtain a desired electrode reaction without undue interference such as from *polarization*.

linear elastic fracture mechanics. A method of fracture analysis that can determine the stress (or load) required to induce fracture instability in a structure containing a cracklike flaw of known size and shape. See also *fracture mechanics* and *stress-intensity factor*.

lipophilic. Having an affinity for oil. See also *hydrophilic* and *hydrophobic*.

liquid metal embrittlement. Catastrophic brittle failure of a normally ductile metal when in contact with a liquid metal and subsequently stressed in tension.

local action. Corrosion due to the action of "local cells," that is, galvanic cells resulting from inhomogeneities between adjacent areas on a metal surface exposed to an *electrolyte*.

local cell. A *galvanic cell* resulting from inhomogeneities between areas on a metal surface in an *electrolyte*. The inhomogeneities may be of physical or chemical nature in either the metal or its environment.

localized corrosion. Corrosion at discrete sites, for example, *crevice corrosion*, *pitting*, and *stress-corrosion cracking*.

long-line current. Current that flows through the earth from an anodic to a cathodic area of a continuous metallic structure. Usually used only where the areas are separated by considerable distance and where the current results from concentration-cell action.

luggin probe. A small tube or capillary filled with electrolyte, terminating close to the metal surface under study, and used to provide an ionically conducting path without diffusion between an *electrode* under study and a *reference electrode*.

M

macroscopic. Visible at magnifications to 25×.

macrostructure. The structure of metals as revealed by macroscopic examination of the etched surface of a polished specimen.

magnetite. Naturally occurring magnetic oxide of iron (Fe_3O_4).

martensite. A generic term for microstructures formed by diffusionless phase transformation in which the parent and product phases have a specific crystallographic relationship. Martensite is characterized by an acicular pattern in the microstructure in both ferrous and nonferrous alloys. In alloys where the solute atoms occupy interstitial positions in the martensitic lattice (such as carbon in iron), the structure is hard and highly strained; but where the solute atoms occupy substitutional positions (such as nickel in iron), the martensite is soft and ductile. The amount of high-temperature phase that transforms to martensite on cooling depends to a large extent on the lowest temperature attained, there being a rather distinct beginning temperature (M_s) and a temperature at which the transformation is essentially complete (M_f).

mechanical plating. Plating wherein fine metal powders are peened onto the work by tumbling or other means.

metal dusting. Accelerated deterioration of metals in carbonaceous gases at elevated temperatures to form a dustlike corrosion product.

metallic glass. An alloy having an amorphous or glassy structure. See also *amorphous solid*.

metallizing. (1) The application of an electrically conductive metallic layer to the surface of nonconductors. (2) The application of metallic coatings by nonelectrolytic procedures such as spraying of molten metal and deposition from the vapor phase.

microbial corrosion. See *biological corrosion*.

microscopic. Visible at magnifications above 25×.

microstructure. The structure of a prepared surface of a metal as revealed by a microscope at a magnification exceeding 25×.

mill scale. The heavy oxide layer formed during hot fabrication or heat treatment of metals.

mixed potential. The *potential* of a specimen (or specimens in a *galvanic couple*) when two or more electrochemical reactions are occurring. Also called galvanic couple potential.

molal solution. Concentration of a solution expressed in moles of solute divided by 1000 g of solvent.

molar solution. Aqueous solution that contains 1 mole (gram-molecular weight) of solute in 1 L of the solution.

mole. One mole is the mass numerically equal (in grams) to the relative molecular mass of a substance. It is the amount of substance of a system that contains as many elementary units (6.02×10^{23}) as there are atoms of carbon in 0.012 kg of the pure nuclide ^{12}C; the elementary unit must be specified and may be an atom, molecule, ion, electron, photon, or even a specified group of such units.

monomer. A molecule, usually an organic compound, having the ability to join with a number of identical molecules to form a *polymer*.

N

natural aging. Spontaneous aging of a supersaturated solid solution at room temperature. See also *aging*. Compare with *artificial aging*.

Nernst equation. An equation that expresses the exact *electromotive force* of a cell in terms of the activities of products and reactants of the cell.

Nernst layer, Nernst thickness. The diffusion layer or the hypothetical thickness of this layer as given by the theory of Nernst. It is defined by: $i_d = nFD[(C_o - C)/\delta]$, where i_d is the diffusion-limited current density, D is the diffusion coefficient, C_o is the concentration at the electrode surface, and δ is the Nernst thickness (0.5 mm in many cases of unstirred aqueous electrolytes).

neutron embrittlement. *Embrittlement* resulting from bombardment with neutrons, usually encountered in metals that have been exposed to a neutron flux in the core of a reactor. In steels, neutron embrittlement is evidenced by a rise in the ductile-to-brittle transition temperature.

nitriding. Introducing nitrogen into the surface layer of a solid ferrous alloy by holding at a suitable temperature (below Ac_1 for ferritic steels) in contact with a nitrogenous material, usually ammonia or molten cyanide of appropriate composition. Quenching is not required to produce a hard case.

nitrocarburizing. Any of several processes in which both nitrogen and carbon are absorbed into the surface layers of a ferrous material at temperatures below the lower critical temperature and, by diffusion, create a concentration gradient. Nitrocarburizing is performed primarily to provide an antiscuffing surface layer and to improve fatigue resistance. Compare with *carbonitriding*.

noble. The positive direction of *electrode potential*, thus resembling noble metals such as gold and platinum.

noble metal. (1) A metal whose *potential* is highly positive relative to the hydrogen electrode. (2) A metal with marked resistance to chemical reaction, particularly to oxidation and to solution by inorganic acids. The term as often used is synonymous with *precious metal*.

noble potential. A *potential* more cathodic (positive) than the standard hydrogen potential.

normalizing. Heating a ferrous alloy to a suitable temperature above the transformation range and then cooling in air to a temperature substantially below the transformation range.

normal solution. An aqueous solution containing one gram equivalent of the active reagent in 1 L of the solution.

normal stress. The stress component perpendicular to a plane on which forces act. Normal stress may be either tensile or compressive.

O

open-circuit potential. The *potential* of an electrode measured with respect to a reference electrode or another electrode when no current flows to or from it.

organic. Being or composed of hydrocarbons or their derivatives, or matter of plant or animal origin. Contrast with *inorganic*.

organic acid. A chemical compound with one or more carboxyl radicals (COOH) in its structure; examples are butyric acid, $CH_3(CH_2)_2COOH$; maleic acid, HOOCCH-CHCOOH; and benzoic acid, C_6H_5COOH.

organic zinc-rich paint. Coating containing zinc powder pigment and an *organic* resin.

overaging. *Aging* under conditions of time and temperature greater than those required to obtain maximum change in a certain property, so that the property is altered in the direction of the initial value.

overheating. Heating a metal or alloy to such a high temperature that its properties are impaired. When the original properties cannot be restored by further heat treating, by mechanical working, or by a combination of working and heat treating, the overheating is known as *burning*.

overvoltage. The difference between the actual electrode potential when appreciable electrolysis begins and the reversible electrode potential.

oxidation. (1) A reaction in which there is an increase in valence resulting from a loss of electrons. Contrast with *reduction*. (2) A corrosion reaction in which the corroded metal forms an oxide; usually applied to reaction with a gas containing elemental oxygen, such as air.

oxidized surface (on steel). Surface having a thin, tightly adhering, oxidized skin (from straw to blue in color), extending in from the edge of a coil or sheet.

oxidizing agent. A compound that causes *oxidation*, thereby itself being reduced.

oxygen concentration cell. See *differential aeration cell*.

ozone. A powerfully oxidizing allotropic form of the element oxygen. The ozone molecule contains three atoms (O_3). Ozone gas is decidedly blue, and both liquid and solid ozone are an opaque blue-black color, similar to that of ink.

P

partial annealing. An imprecise term used to denote a treatment given cold-worked material to reduce its strength to a controlled level or to effect stress relief. To be meaningful, the type of material, the degree of cold work, and the time-temperature schedule must be stated.

parting. See *dealloying*.

parts per billion. A measure of proportion by weight, equivalent to one unit weight of a material per billion (10^9) unit weights of compound. One part per billion is equivalent to 1 mg/g.

parts per million. A measure of proportion by weight, equivalent to one unit weight of a material per million (10^6) unit weights of compound. One part per million is equivalent to 1 mg/g.

passivation. (1) A reduction of the anodic reaction rate of an electrode involved in corrosion. (2) The process in metal corrosion by which metals become *passive*. (3) The changing of a chemically active surface of a metal to a much less reactive state. Contrast with *activation*.

passivator. A type of *inhibitor* that appreciably changes the potential of a metal to a more noble (positive) value.

passive. (1) A metal corroding under the control of a surface reaction product. (2) The state of the metal surface characterized by low corrosion rates in a potential region that is strongly oxidizing for the metal.

passive-active cell. A corrosion cell in which the *anode* is a metal in the *active* state and the *cathode* is the same metal in the *passive* state.

passivity. A condition in which a piece of metal, because of an impervious covering of oxide or other compound, has a *potential* much more positive than that of the metal in the active state.

patina. The coating, usually green, that forms on the surface of metals such as copper and copper alloys exposed to the atmosphere. Also used to describe the appearance of a weathered surface of any metal.

pearlite. A metastable lamellar aggregate of *ferrite* and *cementite* resulting from the transformation of *austenite* at temperatures above the *bainite* range.

phosphating. Forming an adherent phosphate coating on a metal by immersion in a suitable aqueous phosphate solution. Also called phosphatizing. See also *conversion coating*.

pH. The negative logarithm of the hydrogen-ion activity; it denotes the degree of acidity or basicity of a solution. At 25 °C (77 °F), 7.0 is the neutral value. Decreasing values below 7.0 indicate increasing acidity; increasing values above 7.0, increasing basicity.

physical vapor deposition. A coating process whereby the cleaned and masked component to be coated is heated and rotated on a spindle above the streaming vapor generated by melting and evaporating a coating material source bar with a focused electron beam in an evacuated chamber.

physisorption. The binding of an adsorbate to the surface of a solid by forces whose energy levels approximate those of condensation. Contrast with *chemisorption*.

pickle. A solution or process used to loosen or remove corrosion products such as scale or tarnish.

pickling. Removing surface oxides from metals by chemical or electrochemical reaction.

pitting. *Localized corrosion* of a metal surface, confined to a point or small area, that takes the form of cavities.

pitting factor. Ratio of the depth of the deepest pit resulting from corrosion divided by the average penetration as calculated from weight loss.

plane strain. The stress condition in *linear elastic fracture mechanics* in which there is zero strain in a direction normal to both the axis of applied tensile stress and the direction of crack growth (that is, parallel to the crack front); most nearly achieved in loading thick plates along a direction parallel to the plate surface. Under plane-strain conditions, the plane of fracture instability is normal to the axis of the principal tensile stress.

plane stress. The stress condition in *linear elastic fracture mechanics* in which the stress in the thickness direction is zero; most nearly achieved in loading very thin sheet along a direction parallel to the surface of the sheet. Under plane-stress conditions, the plane of fracture instability is inclined 45° to the axis of the principal tensile stress.

plasma spraying. A *thermal spraying* process in which the coating material is melted with heat from a plasma torch that generates a nontransferred arc; molten coating material is propelled against the base metal by the hot, ionized gas issuing from the torch.

plastic deformation. The permanent (inelastic) distortion of metals under applied stresses that strain the material beyond its *elastic limit*.

plasticity. The property that enables a material to undergo permanent deformation without rupture.

polarization. (1) The change from the open-circuit electrode potential as the result of the passage of current. (2) A change in the *potential* of an electrode during electrolysis, such that the potential of an *anode* becomes more noble, and that of a *cathode* more active, than their respective reversible potentials. Often accomplished by formation of a film on the electrode surface.

polarization admittance. The reciprocal of *polarization resistance* (di/dE).

polarization curve. A plot of *current density* versus *electrode potential* for a specific electrode-electrolyte combination.

polarization resistance. The slope (dE/di) at the *corrosion potential* of a potential (E)/current density (i) curve. Also used to describe the method of measuring corrosion rates using this slope.

polyester. Resin formed by condensation of polybasic and monobasic acids with polyhydric alcohols.

polymer. A chain of organic molecules produced by the joining of primary units called *monomers*.

potential. Any of various functions from which intensity or velocity at any point in a field may be calculated. The driving influence of an electrochemical reaction. See also *active potential, chemical potential, corrosion potential, critical pitting potential, decomposition potential, electrochemical potential, electrode potential, electrokinetic potential, equilibrium (reversible) potential, free corrosion potential, noble potential, open-circuit potential, protective potential, redox potential,* and *standard electrode potential*.

potential-pH diagram. See *Pourbaix (potential-pH) diagram*.

potentiodynamic (potentiokinetic). The technique for varying the *potential* of an electrode in a continuous manner at a preset rate.

potentiostat. An instrument for automatically maintaining an electrode in an electrolyte at a constant potential or controlled potentials with respect to a suitable reference electrode.

potentiostatic. The technique for maintaining a constant *electrode potential*.

poultice corrosion. A term used in the automotive industry to describe the corrosion of vehicle body parts due to the collection of road salts and debris on ledges and in pockets that are kept moist by weather and washing. Also called deposit corrosion or attack.

Pourbaix (potential-pH) diagram. A plot of the *redox potential* of a corroding system versus the pH of the system, compiled using thermodynamic data and the *Nernst equation*. The diagram shows regions within which the metal itself or some of its compounds are stable.

powder metallurgy. The art of producing metal powders and utilizing metal powders for production of massive materials and shaped objects.

precious metal. One of the relatively scarce and valuable metals: gold, silver, and the platinum-group metals. Also called *noble metal(s)*.

precipitation hardening. Hardening caused by the precipitation of a constituent from a supersaturated solid solution. See also *age hardening* and *aging*.

precipitation heat treatment. *Artificial aging* in which a constituent precipitates from a supersaturated solid solution.

precracked specimen. A specimen that is notched and subjected to alternating stresses until a crack has developed at the root of the notch.

primary current distribution. The current distribution in an *electrolytic cell* that is free of *polarization*.

primary passive potential (passivation potential). The potential corresponding to the maximum active current density (critical anodic current density) of an electrode that exhibits active-passive corrosion behavior.

primer. The first coat of paint applied to a surface. Formulated to have good bonding and wetting characteristics; may or may not contain inhibiting pigments.

principal stress (normal). The maximum or minimum value of the *normal stress* at a point in a plane considered with respect to all possible orientations of the considered plane. On such principal planes the shear stress is zero. There are three principal stresses on three mutually perpendicular planes. The state of stress at a point may be (1) uniaxial, a state of stress in which two of the three principal stresses are zero, (2) biaxial, a state of stress in which only one of the three principal stresses is zero, and (3) triaxial, a state of stress in which none of the principal stresses is zero. Multiaxial stress refers to either biaxial or triaxial stress.

profile. Anchor pattern on a surface produced by abrasive blasting or acid treatment.

protective potential. The threshold value of the *corrosion potential* that has to be reached to enter a *protective potential range*.

protective potential range. A range of *corrosion potential* values in which an acceptable corrosion resistance is achieved for a particular purpose.

Q

quench-age embrittlement. *Embrittlement* of low-carbon steels resulting from precipitation of solute carbon at existing dislocations and from precipitation hardening of the steel caused by differences in the solid solubility of carbon in ferrite at different temperatures. Quench-age embrittlement usually is caused by rapid cooling of the steel from temperatures slightly below Ac_1 (the temperature at which austenite begins to form), and can be minimized by quenching from lower temperatures.

quench aging. *Aging* induced by rapid cooling after solution heat treatment.

quench cracking. Fracture of a metal during quenching from elevated temperature. Most frequently observed in hardened carbon steel, alloy steel, or tool steel parts of high hardness and low toughness. Cracks often emanate from fillets, holes, corners, or other stress raisers and result from high stresses due to the volume changes accompanying transformation to martensite.

quench hardening. (1) Hardening suitable α-β alloys (most often certain copper or titanium alloys) by solution treating and quenching to develop a martensite-like structure. (2) In ferrous alloys, hardening by austenitizing and then cooling at a rate such that a substantial amount of austenite transforms to martensite.

quenching. Rapid cooling of metals (often steels) from a suitable elevated temperature. This generally is accomplished by immersion in water, oil, polymer solution, or salt, although forced air is sometimes used.

R

radiation damage. A general term for the alteration of properties of a material arising from exposure to ionizing radiation (penetrating radiation), such as x-rays, gamma rays, neutrons, heavy-particle radiation, or fission fragments in nuclear fuel material.

rare earth metal. One of the group of 15 chemically similar metals with atomic numbers 57 through 71, commonly referred to as the lanthanides.

reactive metal. A metal that readily combines with oxygen at elevated temperatures to form very stable oxides, for example, titanium, zirconium, and beryllium. Reactive metals may also become embrittled by the interstitial absorption of oxygen, hydrogen, and nitrogen.

recrystallization. (1) Formation of a new, strain-free grain structure from that existing in cold-worked metal, usually accomplished by heating. (2) The change from one crystal structure to another, as occurs on heating or cooling through a critical temperature.

redox potential. The *potential* of a reversible oxidation-reduction electrode measured with respect to a *reference electrode*, corrected to the hydrogen electrode, in a given *electrolyte*.

reducing agent. A compound that causes *reduction*, thereby itself becoming oxidized.

reduction. A reaction in which there is a decrease in valence resulting from a gain in electrons. Contrast with *oxidation*.

reference electrode. A nonpolarizable *electrode* with a known and highly reproducible *potential* used for potentiometric and voltammetric analyses. See also *calomel electrode*.

refractory metal. A metal having an extremely high melting point, for example, tungsten, molybdenum, tantalum, niobium, chromium, vanadium, and rhenium. In the broad sense, this term refers to metals having melting points above the range for iron, cobalt, and nickel.

relative humidity. The ratio, expressed as a percentage, of the amount of water vapor present in a given volume of air at a given temperature to the amount required to saturate the air at that temperature.

residual stress. Stresses that remain within a body as a result of *plastic deformation*.

resistance. The opposition that a device or material offers to the flow of direct current, equal to the voltage drop across the element divided by the current through the element. Also called electrical resistance.

resistivity. See *electrical resistivity*.

rest potential. See *corrosion potential* and *open-circuit potential*.

riser. (1) That section of pipeline extending from the ocean floor up the platform. Also, the vertical tube in a steam generator convection bank that circulates water and steam upward. (2) A reservoir of molten metal connected to a casting to provide additional metal to the casting, required as the result of shrinkage before and during solidification.

rust. A visible corrosion product consisting of hydrated oxides of iron. Applied only to ferrous alloys. See also *white rust*.

S

sacrificial protection. Reduction of corrosion of a metal in an *electrolyte* by galvanically coupling it to a more anodic metal; a form of *cathodic protection*.

salt fog test. An *accelerated corrosion test* in which specimens are exposed to a fine mist of a solution usually containing sodium chloride, but sometimes modified with other chemicals.

salt spray test. See *salt fog test*.

saturated calomel electrode. A *reference electrode* composed of mercury, mercurous chloride (calomel), and a saturated aqueous chloride solution.

scaling. (1) The formation at high temperatures of thick corrosion product layers on a metal surface. (2) The deposition of water-insoluble constituents on a metal surface.

season cracking. An obsolete historical term usually applied to *stress-corrosion cracking* of brass.

selective leaching. Corrosion in which one element is preferentially removed from an alloy, leaving a residue (often porous) of the elements that are more resistant to the particular environment. Also called *dealloying* or parting. See also *decarburization*, *decobaltification*, *denickelification*, *dezincification*, and *graphitic corrosion*.

sensitizing heat treatment. A heat treatment, whether accidental, intentional, or incidental (as during welding), that causes precipitation of constituents at grain boundaries, often causing the alloy to become susceptible to *intergranular corrosion* or *intergranular stress-corrosion cracking*. See also *sensitization*.

sensitization. In austenitic stainless steels, the precipitation of chromium carbides, usually at grain boundaries, on exposure to temperatures of about 550 to 850 °C (about 1000 to 1550 °F), leaving the grain boundaries depleted of chromium and therefore susceptible to preferential attack by a corroding (oxidizing) medium.

shear. That type of force that causes or tends to cause two contiguous parts of the same body to slide relative to each other in a direction parallel to their plane of contact.

shear strength. The stress required to produce fracture in the plane of cross section, the conditions of loading being such that the directions of force and of resistance are parallel and opposite although their paths are offset a specified minimum amount. The maximum load divided by the original cross-sectional area of a section separated by shear.

sigma phase. A hard, brittle, nonmagnetic intermediate phase with a tetragonal crystal structure, containing 30 atoms per unit cell, space group $P4_2/mnm$, occurring in many binary and ternary alloys of the transition elements. The composition of this phase in the various systems is not the same and the phase usually exhibits a wide range in homogeneity. Alloying with a third transition element usually enlarges the field of homogeneity and extends it deep into the ternary section.

sigma-phase embrittlement. *Embrittlement* of iron-chromium alloys (most notably austenitic stainless steels) caused by precipitation at grain boundaries of the hard, brittle intermetallic *sigma phase* during long periods of exposure to temperatures between approximately 560 and 980 °C (1050 and 1800 °F). Sigma-phase embrittlement results in severe loss in *toughness* and *ductility*, and can make the embrittled material susceptible to *intergranular corrosion*. See also *sensitization*.

slip. *Plastic deformation* by the irreversible shear displacement (translation) of one part of a crystal relative to another in a definite crystallographic direction and usually on a specific crystallographic plane. Sometimes called glide.

slow strain rate technique. An experimental technique for evaluating susceptibility to *stress-corrosion cracking*. It involves pulling the specimen to failure in uniaxial tension at a controlled slow strain rate while the specimen is in the test environment and examining the specimen for evidence of stress-corrosion cracking.

slushing compound. An obsolete term describing oil or grease coatings used to provide temporary protection against *atmospheric corrosion*.

smelt. Molten slag; in the pulp and paper industry, the cooking chemicals tapped from the recovery boiler as molten material and dissolved in the smelt tank as *green liquor*.

S-N diagram. A plot showing the relationship of stress, S, and the number of cycles, N, before fracture in fatigue testing.

soft water. Water that is free of magnesium or calcium salts.

solder embrittlement. Reduction in mechanical properties of a metal as a result of local penetration of solder along grain boundaries.

solid-metal embrittlement. The occurrence of *embrittlement* in a material below the melting point of the embrittling species. See also *liquid-metal embrittlement*.

solid solution. A single, solid, homogeneous crystalline phase containing two or more chemical species.

solute. The component of either a liquid or solid solution that is present to a lesser or minor extent; the component that is dissolved in the *solvent*.

solution. In chemistry, a homogeneous dispersion of two or more kinds of molecular or ionic species. Solution may be composed of any combination of liquids, solids, or gases, but they always consist of a single phase.

solution heat treatment. Heating an alloy to a suitable temperature, holding at that temperature long enough to cause one or more constituents to enter into *solid solution*, and then cooling rapidly enough to hold these constituents in *solution*.

solution potential. *Electrode potential* where half-cell reaction involves only the metal electrode and its ion.

solvent. The component of either a liquid or *solid solution* that is present to a greater or major extent; the component that dissolves the *solute*.

sour gas. A gaseous environment containing hydrogen sulfide and carbon dioxide in hydrocarbon reservoirs. Prolonged exposure to sour gas can lead to *hydrogen damage, sulfide-stress cracking,* and/or *stress-corrosion cracking* in ferrous alloys.

sour water. Waste waters containing fetid materials, usually sulfur compounds.

spalling. The spontaneous chipping, fragmentation, or separation of a surface or surface coating.

spheroidite. An aggregate of iron or alloy carbides of essentially spherical shape dispersed throughout a matrix of *ferrite*.

sputtering. A coating process whereby thermally emitted electrons collide with inert gas atoms, which accelerate toward and impact a negatively charged electrode that is a target of the coating material. The impacting ions dislodge atoms of the target material, which are in turn projected to and deposited on the substrate to form the coating.

stabilizing treatment. (1) Before finishing to final dimensions, repeatedly heating a ferrous or nonferrous part to or slightly above its normal operating temperature and then cooling to room temperature to ensure dimensional stability in service. (2) Transforming retained austenite in quenched hardenable steels, usually by cold treatment. (3) Heating a solution-treated stabilized grade of austenitic stainless steel to 870 to 900 °C (1600 to 1650 °F) to precipitate all carbon as TiC, NbC, or TaC so that *sensitization* is avoided on subsequent exposure to elevated temperature.

standard electrode potential. The reversible potential for an electrode process when all products and reactions are at unit activity on a scale in which the potential for the standard hydrogen half-cell is zero.

strain. The unit of change in the size or shape of a body due to force. Also known as *nominal strain*.

strain-age embrittlement. A loss in *ductility* accompanied by an increase in hardness and strength that occurs when low-carbon steel (especially rimmed or capped steel) is aged following *plastic deformation*. The degree of *embrittlement* is a function of aging time and temperature, occurring in a matter of minutes at about 200 °C (400 °F), but requiring a few hours to a year at room temperature.

strain aging. *Aging* induced by cold working.

strain hardening. An increase in hardness and strength caused by *plastic deformation* at temperatures below the recrystallization range.

strain rate. The time rate of straining for the usual tensile test. Strain as measured directly on the specimen gage length is used for determining strain rate. Because strain is dimensionless, the units of strain rate are reciprocal time.

stray current. Current flowing through paths other than the intended circuit.

stray-current corrosion. Corrosion resulting from direct current flow through paths other than the intended circuit. For example, by an extraneous current in the earth.

stress. The intensity of the internally distributed forces or components of forces that resist a change in the volume or shape of a material that is or has been subjected to external forces. Stress is expressed in force per unit area and is calculated on the basis of the original dimensions of the cross section of the specimen. Stress can be either direct (tension or compression) or shear. See also *residual stress*.

stress concentration factor (K_t). A multiplying factor for applied stress that allows for the presence of a structural discontinuity such as a notch or hole; K_t equals the ratio of the greatest stress in the region of the discontinuity to the nominal stress for the entire section. Also called theoretical stress concentration factor.

stress-corrosion cracking (SCC). A cracking process that requires the simultaneous action of a corrodent and sustained tensile stress. This excludes corrosion-reduced sections that fail by fast fracture. It also excludes intercrystalline or transcrystalline corrosion, which can disintegrate an alloy without applied or residual stress. Stress-corrosion cracking may occur in combination with *hydrogen embrittlement*.

stress-intensity factor. A scaling factor, usually denoted by the symbol K, used in *linear-elastic fracture mechanics* to describe the intensification of applied stress at the tip of a crack of known size and shape. At the onset of rapid crack propagation in any structure containing a crack, the factor is called the critical stress-intensity factor, or the *fracture toughness*. Various subscripts are used to denote different loading conditions or fracture toughnesses:

K_c. Plane-stress fracture toughness. The value of stress intensity at which crack propagation becomes rapid in sections thinner than those in which plane-strain conditions prevail.

K_I. Stress-intensity factor for a loading condition that displaces the crack faces in a direction normal to the crack plane (also known as the opening mode of deformation).

K_{Ic}. Plane-strain fracture toughness. The minimum value of K_c for any given material and condition, which is attained when rapid crack propagation in the opening mode is governed by plane-strain conditions.

K_{Id}. Dynamic fracture toughness. The fracture toughness determined under dynamic loading conditions; it is used as an approximation of K_{Ic} for very tough materials.

K_{ISCC}. Threshold stress-intensity factor for stress-corrosion cracking. The critical plane-strain stress intensity at the onset of stress-corrosion cracking under specified conditions.

K_Q. Provisional value for plane-strain fracture toughness.

K_{th}. Threshold stress intensity for stress-corrosion cracking. The critical stress intensity at the onset of stress-corrosion cracking under specified conditions.

ΔK. The range of the stress-intensity factor during a fatigue cycle.

stress-intensity factor range, ΔK. In fatigue, the variation in the *stress-intensity factor* in a cycle, that is, $K_{max} - K_{min}$.

stress raisers. Changes in contour or discontinuities in structure that cause local increases in stress.

stress ratio, A or R. The algebraic ratio of two specified stress values in a stress cycle. Two commonly used stress ratios are: (1) the ratio of the alternating stress amplitude to the mean stress, $A = S_a/S_m$; and (2) the ratio of the minimum stress to the maximum stress, $R = S_{min}/S_{max}$.

stress-relief cracking. Also called postweld heat treatment cracking, stress-relief cracking occurs when susceptible alloys are subjected to thermal stress relief after welding to reduce *residual stresses* and improve *toughness*. Stress-relief cracking occurs only in metals that can precipitation-harden during such elevated-temperature exposure; it usually occurs at *stress raisers*, is *intergranular* in nature, and is generally observed in the coarse-grained region of the weld *heat-affected zone*. See also *cold cracking, hot cracking,* and *lamellar tearing*.

striation. A fatigue fracture feature, often observed in electron micrographs, that indicates the position of the crack front after each succeeding cycle of stress. The distance between striations indicates the advance of the crack front across that crystal during one stress cycle, and a line normal to the striations indicates the direction of local crack propagation. See also *beach marks*.

subsurface corrosion. Formation of isolated particles of corrosion products beneath a metal surface. This results from the preferential reactions of certain alloy constituents to inward diffusion of oxygen, nitrogen, or sulfur.

sulfidation. The reaction of a metal or alloy with a sulfur-containing species to produce a sulfur compound that forms on or beneath the surface on the metal or alloy.

sulfide stress cracking (SSC). Brittle failure by cracking under the combined action of *tensile stress* and *corrosion* in the presence of water and hydrogen sulfide. See also *environmental cracking*.

surfactant. A surface-active agent; usually an *organic* compound whose molecules contain a *hydrophilic* group at one end and a *lipophilic* group at the other.

T

Tafel line, Tafel slope, Tafel diagram. When an electrode is polarized, it frequently will yield a current/potential relationship over a region that can be approximated by: $\eta = \pm B \log (i/i_o)$, where η is the change in open-circuit potential, i is the current density, and B and i_o are constants. The constant B is also known as the Tafel slope. If this behavior is observed, a plot on semilogarithmic coordinates is known as the Tafel line and the overall diagram is termed a Tafel diagram.

tarnish. Surface discoloration of a metal caused by formation of a thin film of corrosion product.

temper. (1) In heat treatment, to reheat hardened steel or hardened cast iron to some temperature below the eutectoid temperature for the purpose of decreasing hardness and increasing toughness. The process is also sometimes applied to normalized steel. (2) In tool steels, temper is sometimes inadvisably used to denote carbon content. (3) In nonferrous alloys and in some ferrous alloys (steels that cannot be hardened by heat treatment), the hardness and strength produced by mechanical or thermal treatment, or both, and characterized by a certain structure, mechanical properties, or reduction of area during cold working.

temper color. A thin, tightly adhering oxide skin (only a few molecules thick) that forms when steel is tempered at a low temperature, or for a short time, in air or a mildly oxidizing atmosphere. The color, which ranges from straw to blue depending on the thickness of the oxide skin, varies with both tempering time and temperature.

tempered martensite embrittlement. *Embrittlement* of ultrahigh-strength steels caused by tempering in the temperature range of 205 to 400 °C (400 to 750 °F); also called 350 °C or 500 °F embrittlement. Tempered martensite embrittlement is thought to result from the combined effects of cementite precipitation on prior-austenite grain boundaries or interlath boundaries and the segregation of impurities at prior-austenite grain boundaries.

temper embrittlement. *Embrittlement* of alloy steels caused by holding within or cooling slowly through a temperature range just below the transformation range. Embrittlement is the result of the segregation at grain boundaries of impurities such as arsenic, antimony, phosphorus, and tin; it is usually manifested as an upward shift in ductile-to-brittle transition temperature. Temper embrittlement can be reversed by retempering above the critical temperature range, then cooling rapidly.

tensile strength. In tensile testing, the ratio of maximum load to original cross-sectional area. Also called ultimate tensile strength.

tensile stress. A stress that causes two parts of an elastic body, on either side of a typical stress plane, to pull apart. Contrast with *compressive stress*.

tension. The force or load that produces elongation.

terne. An alloy of lead containing 3 to 15% Sn, used as a *hot dip coating* for steel sheet or plate. Terne coatings, which are smooth and dull in appearance, give the steel better corrosion resistance and enhance its ability to be formed, soldered, or painted.

thermal electromotive force. The *electromotive force* generated in a circuit containing two dissimilar metals when one junction is at a temperature different from that of the other. See also *thermocouple*.

thermal embrittlement. *Intergranular fracture* of maraging steels with decreased toughness resulting from improper processing after hot working. Thermal embrittlement occurs upon heating above 1095 °C (2000 °F) and then slow cooling through the temperature range of 815 to 980 °C (1500 to 1800 °F), and has been attributed to precipitation of titanium carbides and titanium carbonitrides at austenite grain boundaries during cooling through the critical temperature range.

thermally induced embrittlement. See *embrittlement*.

thermal spraying. A group of coating or welding processes in which finely divided metallic or nonmetallic materials are deposited in a molten or semimolten condition to form a coating. The coating material may be in the form of powder, ceramic rod, wire, or molten materials. See also *flame spraying* and *plasma spraying*.

thermocouple. A device for measuring temperatures, consisting of lengths of two dissimilar metals or alloys that are electrically joined at one end and connected to a voltage-measuring instrument at the other end. When one junction is hotter than the other, a *thermal electromotive force* is produced that is roughly proportional to the difference in temperature between the hot and cold junctions.

thermogalvanic corrosion. Corrosion resulting from an *electrochemical cell* caused by a thermal gradient.

threshold stress. Threshold stress for *stress-corrosion cracking*. The critical gross section stress at the onset of stress-corrosion cracking under specified conditions.

throwing power. (1) The relationship between the *current density* at a point on a surface and its distance from the *counterelectrode*. The greater the ratio of the surface resistivity shown by the electrode reaction to the volume resistivity of the electrolyte, the better is the throwing power of the process. (2) The ability of a plating solution to produce a uniform metal distribution on an irregularly shaped *cathode*. Compare with *covering power*.

tinning. Coating metal with a very thin layer of molten solder or brazing filler metal.

torsion. A twisting deformation of a solid body about an axis in which lines that were initially parallel to the axis become helices.

torsional stress. The shear stress on a transverse cross section resulting from a twisting action.

total carbon. The sum of the *free carbon* and *combined carbon* (including carbon in solution) in a ferrous alloy.

toughness. The ability of a metal to absorb energy and deform plastically before fracturing.

transcrystalline. See *transgranular*.

transcrystalline cracking. See *transgranular cracking*.

transference. The movement of ions through the *electrolyte* associated with the passage of the electric current. Also called transport or migration.

transgranular. Through or across crystals or grains. Also called intracrystalline or transcrystalline.

transgranular cracking. Cracking or fracturing that occurs through or across a crystal or grain. Also called transcrystalline cracking. Contrast with *intergranular cracking*.

transgranular fracture. Fracture through or across the crystals or grains of a metal. Also called transcrystalline fracture or intracrystalline fracture. Contrast with *intergranular fracture*.

transition metal. A metal in which the available electron energy levels are occupied in such a way that the *d*-band contains less than its maximum number of ten electrons per atom, for example, iron, cobalt, nickel, and tungsten. The distinctive properties of the transition metals result from the incompletely filled *d*-levels.

transition temperature. (1) An arbitrarily defined temperature that lies within the temperature range in which metal fracture characteristics (as usually determined by tests of notched specimens) change rapidly, such as from primarily fibrous (shear) to primarily crystalline (cleavage) fracture. (2) Sometimes used to denote an arbitrarily defined temperature within a range in which the ductility changes rapidly with temperature.

transpassive region. The region of an *anodic polarization* curve, noble to and above the passive *potential* range, in which there is a significant increase in current density (increased metal dissolution) as the potential becomes more positive (noble).

transpassive state. (1) State of anodically passivated metal characterized by a considerable increase of the corrosion current, in the absence of pitting, when the *potential* is increased. (2) The noble region of potential where an electrode exhibits a higher than passive current density.

triaxial stress. See *principal stress (normal)*.

tuberculation. The formation of *localized corrosion* products scattered over the surface in the form of knoblike mounds called tubercles.

U

ultimate strength. The maximum stress (tensile, compressive, or shear) a material can sustain without fracture, determined by dividing maximum load by the original cross-sectional area of the specimen. Also called nominal strength or maximum strength.

underfilm corrosion. Corrosion that occurs under organic films in the form of randomly distributed threadlike filaments or spots. In many cases this is identical to *filiform corrosion*.

uniaxial stress. See *principal stress (normal)*.

uniform corrosion. (1) A type of corrosion attack (deterioration) uniformly distributed over a metal surface. (2) Corrosion that proceeds at approximately the same rate over a metal surface. Also called general corrosion.

V

vacuum deposition. Condensation of thin metal coatings on the cool surface of work in a vacuum.

valence. A positive number that characterizes the combining power of an element for other elements, as measured by the number of bonds to other atoms that one atom of the given element forms upon chemical combination; hydrogen is assigned valence 1, and the valence is the number of hydrogen atoms, or their equivalent, with which an atom of the given element combines.

vapor deposition. See *chemical vapor deposition*, *physical vapor deposition*, and *sputtering*.

vapor plating. Deposition of a metal or compound on a heated surface by reduction or decomposition of a volatile compound at a temperature below the melting points of the deposit and the base material. The reduction is usually accomplished by a gaseous reducing agent such as hydrogen. The decomposition process may involve thermal dissociation or reaction with the base material. Occasionally used to designate deposition on cold surfaces by vacuum evaporation. See also *vacuum deposition*.

voids. A term generally applied to paints to describe *holidays*, holes, and skips in a *film*. Also used to describe shrinkage in castings and welds.

W

wash primer. A thin, inhibiting paint, usually chromate pigmented with a polyvinyl butyrate binder.

weld cracking. Cracking that occurs in the weld metal. See also *cold cracking, hot cracking, lamellar tearing,* and *stress-relief cracking*.

weld decay. *Intergranular corrosion*, usually of stainless steels or certain nickel-base alloys, that occurs as the result of *sensitization* in the *heat-affected zone* during the welding operation.

wetting. A condition in which the interfacial tension between a liquid and a solid is such that the contact angle is 0° to 90°.

wetting agent. A substance that reduces the surface tension of a liquid, thereby causing it to spread more readily on a solid surface.

white liquor. Cooking liquor from the kraft pulping process produced by recausticizing *green liquor* with lime.

white rust. Zinc oxide; the powdery product of corrosion of zinc or zinc-coated surfaces.

work hardening. Same as *strain hardening*.

working electrode. The test or specimen electrode in an *electrochemical cell*.

Y

yield. Evidence of *plastic deformation* in structural materials. Also called plastic flow or creep. See also *flow*.

yield point. The first stress in a material, usually less than the maximum attainable stress, at which an increase in strain occurs without an increase in stress. Only certain metals—those that exhibit a localized, heterogeneous type of transition from *elastic deformation* to *plastic deformation*—produce a yield point. If there is a decrease in stress after yielding, a distinction may be made between upper and lower yield points. The load at which a sudden drop in the flow curve occurs is called the upper yield point. The constant load shown on the flow curve is the lower yield point.

yield strength. The stress at which a material exhibits a specified deviation from proportionality of stress and strain. An offset of 0.2% is used for many metals.

yield stress. The stress level in a material at or above the *yield strength* but below the *ultimate strength*, i.e., a stress in the plastic range.

Z

zeta potential. See *electrokinetic potential*.

REFERENCES

- *Compilation of ASTM Standard Definitions*, 5th ed., American Society for Testing and Materials, 1982
- *Concise Encyclopedia of Science and Technology*, McGraw-Hill, 1984
- "Corrosion of Metals and Alloys—Terms and Definitions," ISO 8044, International Organization for Standardization, 1986 (available from the American National Standards Institute)
- *Dictionary of Scientific and Technical Terms*, 2nd ed., McGraw-Hill, 1978
- *Electroplating Engineering Handbook*, 3rd ed., A.K. Grahm, Ed., Van Nostrand Reinhold, 1971, p ix-xviii
- A.D. Merriman, *A Dictionary of Metallurgy*, Pitman Publishing, 1958
- *Metals Handbook Desk Edition*, American Society for Metals, 1985, p 1.1-1.42
- *Metals Handbook*, Vol 5, 9th ed., *Surface Cleaning, Finishing, and Coating*, American Society for Metals, 1982, p 379
- *Metals Handbook*, Vol 6, 9th ed., *Welding, Brazing, and Soldering*, American Society for Metals, 1983, p 1-20
- *Metals Handbook*, Vol 8, 9th ed., *Mechanical Testing*, American Society for Metals, 1985, p 1-15
- *Metals Handbook*, Vol 10, 9th ed., *Materials Characterization*, American Society for Metals, 1986, p 668-684
- *Metals Handbook*, Vol 11, 9th ed., *Failure Analysis and Prevention*, ASM International, 1986, p 1-11
- *Military Standardization Handbook, Corrosion and Corrosion Prevention of Metals*, MIL-HDBK-729, Section 10, Department of Defense
- "NACE Glossary of Corrosion Related Terms," National Association of Corrosion Engineers, 1985
- *The Nalco Water Handbook*, Nalco Chemical Company, F.N. Kemmer and J. McCallion, Ed., McGraw-Hill, 1979, p G1-G9
- *Science and Technology of Surface Coating*, B.N. Chapman and J.C. Anderson, Ed., Academic Press, 1974, p 435-445
- "Standard Definitions of Terms Relating to Corrosion and Corrosion Testing," G 15, *Annual Book of ASTM Standards*, Vol 03.02, American Society for Testing and Materials
- "Standard Definitions of Terms Relating to Electroplating," B 374, *Annual Book of ASTM Standards*, Vol 02.05, American Society for Testing and Materials
- "Standard Definitions of Terms Relating to Fatigue Testing and the Statistical Analysis of Fatigue Data," E 206, *Annual Book of ASTM Standards*, Vol 03.01, American Society for Testing and Materials
- "Standard Definitions of Terms Relating to Methods of Mechanical Testing," E 6, *Annual Book of ASTM Standards*, Vol 03.01, American Society for Testing and Materials
- "Standard Terminology Relating to Erosion and Wear," G 40, *Annual Book of ASTM Standards*, Vol 03.02, American Society for Testing and Materials
- "Standard Terminology Relating to Fracture Testing," E 616, *Annual Book of ASTM Standards*, Vol 03.01, American Society for Testing and Materials

Fundamentals of Corrosion

Section Chairman:
Miroslav I. Marek, Georgia Institute of Technology

Introduction

Miroslav I. Marek, School of Materials Engineering, Georgia Institute of Technology

PERHAPS THE MOST STRIKING FEATURE of corrosion is the immense variety of conditions under which it occurs and the large number of forms in which it appears. Numerous handbooks of corrosion data have been compiled that list the corrosion effects of specific material/environment combinations; still, the data cover only a small fraction of the possible situations and only for specific values of, for example, the temperature and composition of the substances involved. To prevent corrosion, to interpret corrosion phenomena, or to predict the outcome of a corrosion situation for conditions other than those for which an exact description can be found, the engineer must be able to apply the knowledge of corrosion fundamentals. These fundamentals include the mechanisms of the various forms of corrosion, applicable thermodynamic conditions and kinetic laws, and the effects of the major variables. Even with all of the available generalized knowledge of the principles, corrosion is in most cases a very complex process in which the interactions among many different reactions, conditions, and synergistic effects must be carefully considered.

All corrosion processes show some common features. Thermodynamic principles can be applied to determine which processes can occur and how strong the tendency is for the changes to take place. Kinetic laws then describe the rates of the reactions. There are, however, substantial differences in the fundamentals of corrosion in such environments as aqueous solutions, nonaqueous liquids, and gases that warrant a separate treatment in this Section.

Corrosion in Aqueous Solutions. Although atmospheric air is the most common environment, aqueous solutions, including natural waters, atmospheric moisture, and rain, as well as man-made solutions, are the environments most frequently associated with corrosion problems. Because of the ionic conductivity of the environment, corrosion is due to electrochemical reactions and is strongly affected by such factors as the electrode potential and acidity of the solution. As described in the article "Thermodynamics of Aqueous Corrosion," thermodynamic factors determine under what conditions the reactions are at an electrochemical equilibrium and, if there is a departure from equilibrium, in what directions the reactions can proceed and how strong the driving force is. The kinetic laws of the reactions are fundamentally related to the activation energies of the electrode processes, mass transport, and basic properties of the metal/environment interface, such as the resistance of the surface films (see the article "Kinetics of Aqueous Corrosion").

The fundamental kinetics of aqueous corrosion have been thoroughly studied. The simultaneous occurrences of several electrochemical reactions responsible for corrosion have been analyzed on the basis of the mixed potential theory, which provides a general method of intepreting or predicting the corrosion potential and reaction rates. The actual corrosion rates are then strongly affected by the environmental and metallurgical variables, as discussed in the articles "Effects of Environmental Variables on Aqueous Corrosion" and "Effects of Metallurgical Variables on Aqueous Corrosion," respectively. Special conditions exist in natural order and some industrial systems where biological organisms are present in the environment and attach themselves to the structure. Corrosion is expected by the presence of the organisms and the biological films they produce as well as the products of their metabolism, as described in the Appendix "Biological Effects" to the aforementioned article on environmental variables.

Corrosion in Molten Salts and Liquid Metals. These are more narrow but important areas of corrosion in liquid environments. Both have been strongly associated with the nuclear industry, for which much of the research has been performed, but there are numerous nonnuclear applications as well. In molten-salt corrosion, described in the article "Fundamentals of High-Temperature Corrosion in Molten Salts," the mechanisms of deterioration are more varied than in aqueous corrosion, but there are many similarities and some interesting parallels, such as the use of the $E - pO^{2-}$ diagrams similar to the E − pH (Pourbaix) diagrams in aqueous corrosion. Preferential dissolution plays a stronger role in molten-salt corrosion than in aqueous corrosion. Corrosion testing presents special problems and is much more involved than the familiar aqueous testing, usually requiring expensive circulation loops and purification of the salts. Although the literature on molten-salt corrosion is substantial, relatively few fundamental thermodynamic and kinetic data are available.

Liquid-metal corrosion, discussed in the article "Fundamentals of High-Temperature Corrosion in Liquid Metals," is of great interest in the design of fast fission nuclear reactors as well as of future fusion reactors, but is also industrially important in other areas, such as metal recovery, heat pipes, and various special cooling designs. Liquid-metal corrosion differs fundamentally from aqueous and molten-salt corrosion in that the medium, except for impurities, is in a nonionized state. The solubilities of the alloy components and their variation with temperature then play a dominant role in the process, and preferential dissolution is a major form of degradation. Mass transfer is another frequent consequence of the dissolution process. At the same time, the corrosion is strongly affected by the presence of nonmetallic impurities in both the alloys and the liquid metals.

Corrosion in Gases. In gaseous corrosion, the environment is nonconductive, and the ionic processes are restricted to the surface of the metal and the corrosion product layers (see the article "Fundamentals of Corrosion in Gases"). Because the reaction rates of industrial metals with common gases are low at room temperature, gaseous corrosion, generically called oxidation, is usually an industrial problem only at high temperatures when diffusion processes are dominant. Thermodynamic factors play the usual role of determining the driving force for the reactions, and free energy-temperature diagrams are commonly used to show the equilibria in simple systems, while equilibria in more complex environments as a function of compositional variables can be examined by using isothermal stability diagrams.

In the mechanism and kinetics of oxidation, the oxide/metal volume ratio gives some guidance of the likelihood that a protective film will be formed, but the major role belongs to conductivity and transport processes, which are strongly affected by the impurities and defect structures of the compounds. Together with conditions of surface film stability, the transport processes determine the reaction rates that are described in general form by the several kinetic rate laws, such as linear, logarithmic, and parabolic.

The most obvious result of oxidation at high temperatures is the formation of oxide scales. The properties of the scales and development of stresses determine whether the scale provides a continuous oxidation protection. In some cases of oxidation of alloys, however, reactions occur within the metal structure in the form of internal oxidation. Like corrosion in liquids, selective or preferential oxidation is frequently observed in alloys containing components of substantially different thermodynamic stability.

Thermodynamics of Aqueous Corrosion

Introduction

Miroslav I. Marek
Georgia Institute of Technology

CORROSION OF METALS in aqueous environments is almost always electrochemical in nature. It occurs when two or more electrochemical reactions take place on a metal surface. As a result, some of the elements of the metal or alloy change from a metallic state into a nonmetallic state. The products of corrosion may be dissolved species or solid corrosion products; in either case, the energy of the system is lowered as the metal converts to a lower-energy form. Rusting of steel is the best known example of conversion of a metal (iron) into a nonmetallic corrosion product (rust). The change in the energy of the system is the driving force for the corrosion process and is a subject of thermodynamics. Thermodynamics examines and quantifies the tendency for corrosion and its partial processes to occur; it does not predict if the changes actually will occur and at what rate. Thermodynamics can predict, however, under what conditions the metal is stable and corrosion cannot occur.

The electrochemical reactions occur uniformly or nonuniformly on the surface of the metal, which is called an electrode. The ionically conducting liquid is called an electrolyte. As a result of the reaction, the electrode/electrolyte interface acquires a special structure, in which such factors as the separation of charges between electrons in the metal and ions in the solution, interaction of ions with water molecules, adsorption of ions on the electrode, and diffusion of species all play important roles. The structure of this so-called double layer at the electrified interface, as related to corrosion reactions, will be described in the section "Electrode Processes" in this article.

One of the important features of the electrified interface between the electrode and the electrolyte is the appearance of a potential difference across the double layer, which allows the definition of the electrode potential. The electrode potential becomes one of the most important parameters in both the thermodynamics and the kinetics of corrosion. The fundamentals will be discussed in the section "Electrode Potentials," and some examples of the calculations of the potentials from thermodynamic data are shown in the section "Potential Versus pH (Pourbaix) Diagrams."

The electrode potentials are used in corrosion calculations and are measured both in the laboratory and in the field. In actual measurements, standard reference electrodes are extensively used to provide fixed reference points on the scale of relative potential values. The use of suitable reference electrodes and appropriate methods of measurement will be discussed in the section "Potential Measurements With Reference Electrodes."

One of the most important steps in the science of electrochemical corrosion was the development of diagrams showing thermodynamic conditions as a function of electrode potential and concentration of hydrogen ions. These potential versus pH diagrams, often called Pourbaix diagrams, graphically express the thermodynamic relationships in metal/water systems and show at a glance the regions of the thermodynamic stability of the various phases that can exist in the system. Their construction and application in corrosion, as well as their limitations, will be discussed in the section "Potential Versus pH (Pourbaix) Diagrams."

Electrode Processes

Charles A. Natalie
Department of Metallurgical Engineering
Colorado School of Mines

In the discussion of chemical reactions and valence, the topic of electrochemical reactions is usually treated as a special case. Electrochemical reactions are usually discussed in terms of the change in valence that occurs between the reacting elements, that is, oxidation and reduction. Oxidation and reduction are commonly defined as follows. Oxidation is the removal of electrons from atoms or groups of atoms, resulting in an increase in valence, and reduction is the addition of electrons to an atom or group of atoms, resulting in the decrease in valence (Ref 1).

Because electrochemical reactions or oxidation-reduction reactions can be represented in terms of an electrochemical cell with oxidation reactions occurring at one electrode and reduction occurring at the other electrode, electrochemical reactions are often further defined as cathodic reactions and anodic reactions. By definition, cathodic reactions are those types of reactions that result in reduction, such as:

$$M(aq)^{2+} + 2e^- \rightarrow M(s) \qquad \text{(Eq 1)}$$

Anodic reactions are those types of reactions that result in oxidation, such as:

$$M(s) \rightarrow M(aq)^{2+} + 2e^- \qquad \text{(Eq 2)}$$

Because of the production of electrons during oxidation and the consumption of electrons during reduction, oxidation and reduction are coupled events. If the ability to store large amounts of electrons does not exist, equivalent processes of oxidation and reduction will occur together during the course of normal electrochemical reactions. The oxidized species provide the electrons for the reduced species.

The example stated above, like many aqueous corrosion situations, involves the reaction of aqueous metal species at a metal electrode surface. This metal/aqueous interface is complex, as is the mechanism by which the reactions take place across the interface. Because the reduction-oxidation reactions involve species in the electrolyte reacting at or near the metal interface, the electrode surface is charged relative to the solution, and the reactions are associated with specific electrode potentials.

The charged interface results in an electric field that extends into the solution. This electric field has a dramatic effect on the solution near the metal.

A solution that contains water as the primary solvent is affected by an electrical field because of its structure. The primary solvent—water—is polar and can be visualized as dipolar molecules that have a positive side (hydrogen atoms) and a negative side (oxygen atoms). In the electric field caused by the charged interface, the water molecules act as small dipoles and align themselves in the direction of the electric field.

Ions that are present in the solution are also charged because of the loss or gain of electrons. The positive charged ions (cations) and negative charged ions (anions) also have an electric field associated with them. The solvent (water) molecules act as small dipoles; therefore, they are also attracted to the charged ions and align themselves in the electric field established by the charge of the ion. Because the electric field is strongest close to the ion, some water molecules reside very close to an ionic species in solution. The attraction is great enough that these water molecules travel with the ion as it moves through the solvent. The tightly bound water molecules are referred to as the primary water sheath of the ion. The electric field is weaker at distances outside the primary water sheath, but it still disturbs the polar water molecules as the ion passes through the solution. The water molecules

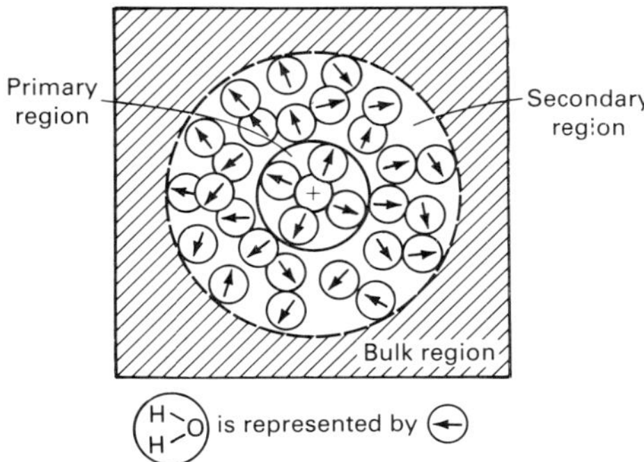

Fig. 1 Schematic of the primary and secondary solvent molecules for a cation in water

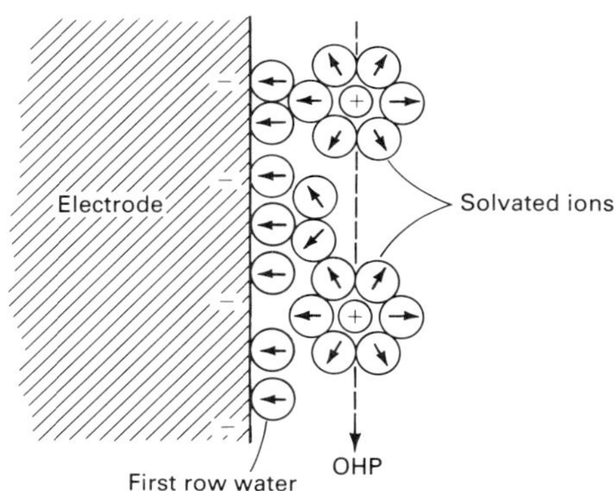

Fig. 2 Schematic of a charged interface and the locations of cations at the electrode surface

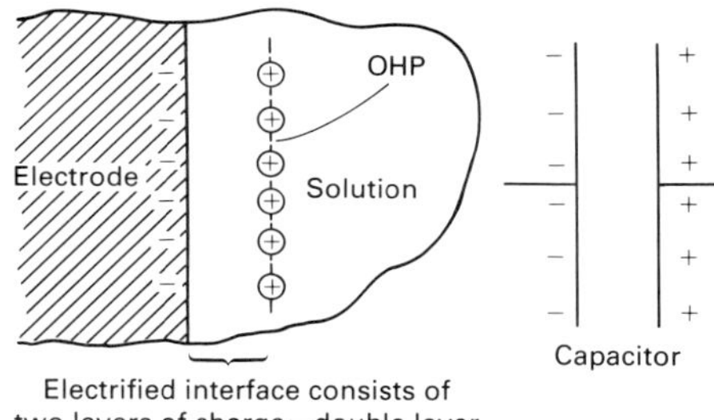

Fig. 3 Simplified double layer at a metal aqueous interface

that are disturbed as the ion passes, but do not move with the ion, are usually referred to as the secondary water sheath. Figure 1 shows a representation of the primary and secondary solvent molecules for a cation in water. Because of their smaller size relative to anions, cations have a stronger electric field close to the ion, and more water molecules are associated in their primary water sheath. However, anodic species have few, if any, primary water molecules. A detailed description of the hydration of ions in solution is given in Ref 2.

Because of the potential and charge established at the metal/aqueous interface of an electrode, ions and polar water molecules are also attracted to the interface because of the strong electric field close to the interface. Water molecules form a first row at the metal/aqueous interface. This row of water molecules limits the distance that hydrated ions can approach the interface. Figure 2 shows a schematic diagram of a charged interface and the locations of cations at the surface. Also, the primary water molecules associated with the ionic species limit the distance the ions can approach. For example, the plane of positive charge of the cations that reside near the surface of a negatively charged interface is a fixed distance from the metal due to the water molecules that are between the surface and the ions. This plane of charge is referred to as the Outer-Heimholz Plane (OHP).

Because of the structure of the charged interface described above, it is often represented (Ref 2) as a charged capacitor (Fig. 3). The potential drop across the interface is also often simplified as a linear change in potential from the metal surface to the OHP.

The significance of the electronic double layer is that it provides a barrier to the transfer of electrons. If there were no difficulty in the transfer of electrons across the interface, the only resistance to electron flow would be the diffusion of aqueous species to and from the electrode. The surface would be nonpolarizable, and the potential would not be changed until the solution was deficient in electron acceptors and/or donors.

This is of particular interest when dealing with the kinetics at the interface (see the article "Kinetics of Aqueous Corrosion" in this Volume). The double layer results in an energy barrier that must be overcome. Thus, reactions at the interface are often dominated by activated processes, and activation polarization plays a significant role in corrosion. The key to controlling corrosion usually consists of minimizing the kinetics; this slows the reaction rates sufficiently that corrosion appears to be stopped.

Electrode Potentials

Charles A. Natalie
Department of Metallurgical Engineering
Colorado School of Mines

The object of chemical thermodynamics is to develop a mathematical treatment of the chemical equilibrium and the driving forces behind chemical reactions. The desire is to catalog known quantitative data concerning equilibrium that can be later used to predict equilibria (perhaps even equilibria that has never been investigated by experimentation).

The driving force for chemical reactions has been expressed in thermodynamic treatments as the balance between the effect of energy (enthalpy) and the effect of probability. The thermodynamic property that relates to probability is called entropy.

The idea of entropy has been expressed as thermodynamic probability and is defined as the number of ways in which microscopic particles can be distributed among states accessible to them (Ref 3). The thermodynamic probability is an extensive quantity and is not the mathematical probability that ranges between 0 and 1.

Free Energy

The driving force for chemical reactions depends not only on chemical formulas of species involved but also on the activities of the reactants and products. Free energy is the thermodynamic property that has been assigned to express the resultant enthalpy of a substance and its inherent probability. At constant temperature, free energy can be expressed as:

$$\Delta G = \Delta H - T\Delta S \quad \text{(Eq 3)}$$

where ΔG is the change in free energy (Gibbs free energy), ΔH is the change in enthalpy, T is the absolute temperature, and ΔS is the change in entropy.

When reactions are at equilibrium and there is no apparent tendency for a reaction to proceed either forward or backward, it has been shown that (Ref 4):

$$\Delta G^\circ = -RT \ln K_{eq} \quad \text{(Eq 4)}$$

where ΔG° is the free energy change under the special conditions when all reactants and products are in a preselected standard state, R is the gas constant, and K_{eq} is the equilibrium constant. The standard free energy of formations for an extensive number of compounds as a function of temperature have been cataloged; this allows the prediction of equilibrium constants over a wide range of conditions. It is necessary only to determine the standard free energy change for a reaction (ΔG°, Eq 4) by subtracting the sum of the free energy of formations of the products at constant temperature.

If an electrochemical cell is constructed that can operate under thermodynamic reversible conditions (the concept of reversibility is described in more detail in the section "Potential Measurements With Reference Electrodes" in this article and in Ref 4) and if the extent of reaction is small enough not to change the activities of reactants and products, the potential remains constant, and the energy dissipated by an infinitesimal passage of charge is given by:

$$|\Delta G| = \text{charge passed} \cdot \text{potential difference}$$

or

$$|\Delta G| = nF \cdot |E| \quad \text{(Eq 5)}$$

where n is the number of electrons per atom of the species involved in the reaction, F is the charge of 1 mol of electrons, and E is the cell potential. Because free energy has a sign that denotes the direction of the reaction, a thermodynamic sign convention must be selected. The common U.S. convention is to associate a positive potential with spontaneous reactions; thus, the reaction becomes:

$$\Delta G = -nFE \quad \text{(Eq 6)}$$

If the reaction occurs under conditions in which the reactants and products are in their standard states, the equation becomes:

$$\Delta G^\circ = -nFE^\circ \quad \text{(Eq 7)}$$

Combination with Eq 4 results in:

$$\ln K_{eq} = \frac{-nFE^\circ}{RT} \quad \text{(Eq 8)}$$

thus allowing the prediction of equilibrium data for electrochemical reactions.

Table 1 Electromotive force series

See also Fig. 4, which shows a schematic of an electrochemical cell used to determine the potential difference between copper and zinc electrodes.

Electrode reaction	Standard potential at 25 °C (77 °F), volts versus SHE
$Au^{3+} + 3e^- \rightarrow Au$	1.50
$Pd^{2+} + 2e^- \rightarrow Pd$	0.987
$Hg^{2+} + 2e^- \rightarrow Hg$	0.854
$Ag^+ + e^- \rightarrow Ag$	0.800
$Hg_2^{2+} + 2e^- \rightarrow 2Hg$	0.789
$Cu^+ + e^- \rightarrow Cu$	0.521
$Cu^{2+} + 2e^- \rightarrow Cu$	0.337
$2H^+ + 2e^- \rightarrow H_2$	(Reference) 0.000
$Pb^{2+} + 2e^- \rightarrow Pb$	−0.126
$Sn_2 + 2e^- \rightarrow Sn$	−0.136
$Ni^{2+} + 2e^- \rightarrow Ni$	−0.250
$Co^{2+} + 2e^- \rightarrow Ni$	−0.277
$Tl^+ + e^- \rightarrow Tl$	−0.336
$In^{3+} + 3e^- \rightarrow In$	−0.342
$Cd^{2+} + 2e^- \rightarrow Cd$	−0.403
$Fe^{2+} + 2e^- \rightarrow Fe$	−0.440
$Ga^{3+} + 3e^- \rightarrow Ga$	−0.53
$Cr^{3+} + 3e^- \rightarrow Cr$	−0.74
$Cr^{2+} + 2e^- \rightarrow Cr$	−0.91
$Zn^{2+} + 2e^- \rightarrow Zn$	−0.763
$Mn^{2+} + 2e^- \rightarrow Mn$	−1.18
$Zr^{4+} + 4e^- \rightarrow Zr$	−1.53
$Ti^{2+} + 2e^- \rightarrow Ti$	−1.63
$Al^{3+} + 3e^- \rightarrow Al$	−1.66
$Hf^{4+} + 4e^- \rightarrow Hf$	−1.70
$U^{3+} + 3e^- \rightarrow U$	−1.80
$Be^{2+} + 2e^- \rightarrow Be$	−1.85
$Mg^{2+} + 2e^- \rightarrow Mg$	−2.37
$Na^+ + e^- \rightarrow Na$	−2.71
$Ca^{2+} + 2e^- \rightarrow Ca$	−2.87
$K^+ + e^- \rightarrow K$	−2.93
$Li^+ + e^- \rightarrow Li$	−3.05

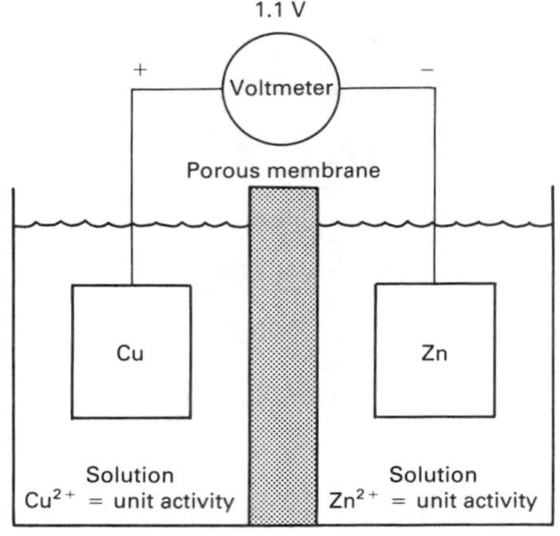

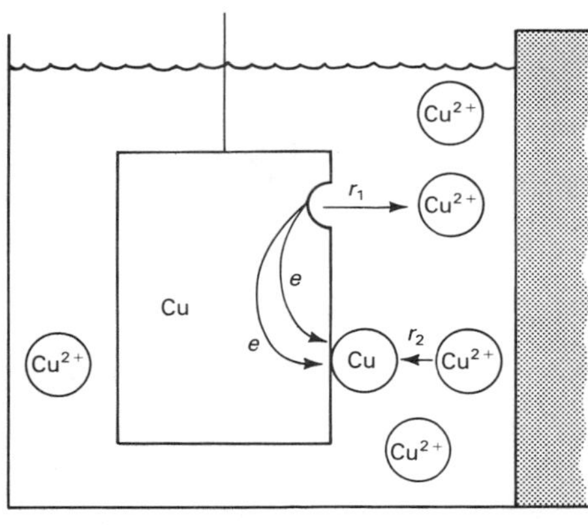

Fig. 4 Typical electrochemical cell (a) used to study the free energy change that accompanies electrochemical or corrosion reactions. In this example, the cell contains copper and zinc electrodes in equilibrium, with their ions separated by a porous membrane to mitigate mixing. For purposes of simplicity, the concentration of metal ions is maintained at unit activity; that is, each solution contains about 1 g atomic weight of metal ion per liter. The reactions taking place on each side of the cell are represented by Eq 10 and 11, and the half-cell reactions for copper and zinc electrodes are given in Table 1. The rates of metal dissolution and deposition must be the same as shown in (b), which illustrates copper atoms being oxidized to cupric ions and, at other areas, cupric ions being reduced to metallic copper. Equilibrium conditions dictate that the reaction rates r_1 and r_2 be equal. Source: Ref 5

Cell Potentials and the Electromotive Force Series

If a strip of zinc metal is placed in water, some zinc ions will be converted to aqueous zinc ions because of the relatively large tendency for zinc to oxidize. Because of the electrons remaining in the metal, the positively charged zinc ions will remain very close to the negatively charged zinc strip and thus will establish a double layer, as described in the section "Electrode Potentials" in this article (Fig. 1). The potential difference established between the solution and the zinc is of the order of 1 V, but because the double layer is very small, the potential gradient (change in potential with respect to distance) can be very high.

A negative electrode potential (with respect to the standard hydrogen electrode discussed below) exists for a zinc electrode in a solution of zinc ions. However, if a copper strip is placed in a solution containing copper ions, a positive potential is established between the more noble copper strip and the solution.

If, however, a metal is placed in a solution containing metal ions of a different nature, the first metal may dissolve, while the second metal deposits from its ions. A common example of this is the metal displacement reaction between zinc metal and copper ions, for which the complete oxidation-reduction reaction is:

$$Zn(s) + Cu(aq)^{2+} \rightarrow Zn(aq)^{2+} + Cu(s) \quad \text{(Eq 9)}$$

If the reverse procedure is tried, that is, copper metal placed in a solution containing zinc ions, no reaction will take place to any measurable extent. For example, if the solution containing zinc ions has no copper ions present initially, the reaction will occur to a very small extent, with the reaction stopping when a certain very small concentration of copper ions has been produced. In the opposite case, zinc metal will react with copper ions almost to completion; the reaction will stop only when the concentration of copper ions is very small.

The above experiment can be repeated with many combinations of metals, and the ability of one metal to replace another ion from solution can be used as a basis for tabulating the metals in a series. The table formed would show the abilities of metals to reduce other metal ions from solution. This electromotive force (emf) series for some common metals is shown in Table 1. The potentials listed in Table 1 are measured values, which will be described below as well as in the section "Potential Measurements With Reference Electrodes" in this article.

The reactions described in establishing an emf series are referred to as electrochemical reactions. Electrochemical reactions are those reactions that involve oxidation (increase in valence) and reduction (decrease in valence), as described in the section "Electrode Processes" in this article.

For the example of copper metal deposition using zinc metal, the oxidation reaction for producing electrons is:

$$Zn(s) = Zn(aq)^{2+} + 2e^- \quad \text{(Eq 10)}$$

Electrons are consumed by copper ion according to the following reduction reaction:

$$Cu(aq)^{2+} + 2e^- \rightarrow Cu(s) \quad \text{(Eq 11)}$$

To study the reactions discussed above (Eq 9 to 11), an electrochemical cell, such as the one shown and described in Fig. 4, can be constructed by using a copper electrode in a solution of copper sulfate as one electrode and a zinc electrode in a solution of zinc sulfate as the other electrode. If the external conduction path is short circuited, electrons will flow from the zinc electrode (anode) as zinc dissolves to the copper electrode (cathode); this causes the deposition of copper metal. This type of arrangement would demonstrate how some electrochemical reactions can take place with the reactants and products physically separated and how the overall process can be visualized as two separate reactions that occur together.

The two reactions listed in Eq 10 and 11 and shown schematically in Fig. 4 are often referred to as half-cell reactions. This nomenclature is due to the requirement that oxidation and reduction occur simultaneously under equilibrium conditions. Therefore, the reaction given in Eq 10 is defined as an oxidation half-cell reaction, and the reaction given in Eq 11 is a reduction half-cell reaction. The reaction in Eq 9 can be referred to as the overall electrochemical reaction and is the sum of the half-cell reactions given in Eq 10 and 11.

Because specific, or absolute, potentials of electrodes cannot be measured directly, an arbitrary half-cell reaction is used as a reference by defining its potential as 0. All other half-cell potentials can then be calculated with respect to this zero reference. As described in the following section "Potential Measurements With Reference Electrodes," the hydrogen ion reaction $2H^+ + 2e \rightarrow H_2$ (Table 1) is used as the standard reference point. It is not possible to make an electrode from hydrogen gas; therefore, the standard hydrogen electrode (SHE) potential is measured by using an inert electrode, such as platinum, immersed in a solution saturated hydrogen gas at 1 atm. All values of electrode potential, therefore, are with reference to SHE.

The potentials given in Table 1 are specifically the potentials measured relative to an SHE at 25 °C (77 °F) when all concentrations of ions are 1 molal, gases are at 1 atm of pressure, and solid phases are pure. This specific electrode potential is referred to as the standard electrode potential and is denoted by $E°$. The standard electrode potential for zinc—the accepted value for which is −0.763 (Table 1)—can be calculated by measuring the emf of a cell made up, for example, of a zinc and a hydrogen electrode in a zinc salt solution of known activity Zn^{2+} and H^+ (Fig. 5). This procedure could be repeated by exchanging the zinc electrode with any other metal and by assigning the half-cell electrode potentials measured for the electrochemical cells to the proper reactions in Table 1. Changes in concentration, temperature, and partial pressure will change the electrode potentials and the position of a particular metal in the emf series. In particular, the change in electrode potential as a function of concentration is given by the Nernst equation:

$$E = E° - \frac{RT}{nF} \ln \frac{(ox)}{(red)} \qquad \text{(Eq 12)}$$

where E is the electrode potential, $E°$ is the standard electrode potential, R is the gas constant (1.987 cal/K mol), T is the absolute temperature (in degrees Kelvin), n is the number of moles of electrons transferred in the half-cell reaction, F is the Faraday constant (F = 23 060 cal/volt equivalent), and (ox) and (red) are the activities of the oxidized and reduced species, respectively.

Electrode potentials, as described above, are always measured when zero current is flowing between the electrode and the SHE. The potential is thus a reversible measurement of the maximum potential that exists and an indication of the tendency for the particular reaction to occur. For example, metals listed in Table 1 above molecular hydrogen are more noble and less resistant to oxidation than the metals listed below hydrogen when standard-state conditions exist. This tendency is a thermodynamic quantity and does not take into account the kinetic factors that may limit a reaction because of such physical factors as protection by corrosion product layers.

Care should be taken when using an emf series such as that shown in Table 1. These values are for a very specific condition (standard state) and may not apply to a specific corrosion environment. More complete emf series (Ref 6, 7) and potentials in other environments (Ref 8) are available.

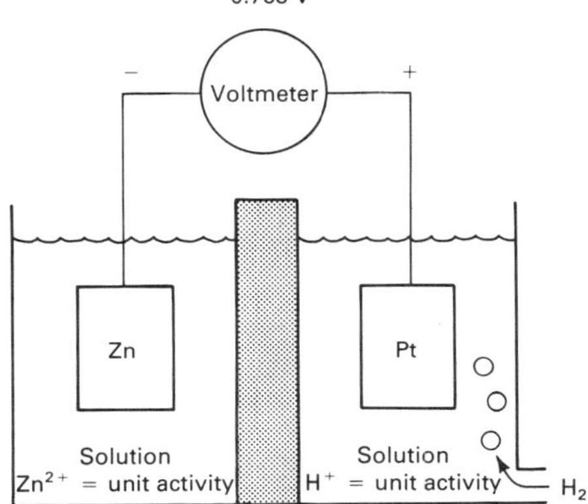

Fig. 5 Electrochemical cell containing a zinc electrode and hydrogen electrode

Returning to the example of an electrochemical cell with copper and zinc electrodes, it is apparent that the chemical energy that exists between the copper and zinc electrodes can be converted to electrical energy (as occurs in a battery). However, the external circuit can be replaced with a direct current (dc) power supply, which can be used to force electrons to go in a direction opposite to the direction they tend to go naturally. Both concepts are useful when dealing with corrosion because the oxidation of a metal will always be coupled to a cathodic reaction and because corrosion reactions are similar to the galvanic-type cell. Also, application of external potentials can be used to protect metals, as in cathodic protection (see the article "Cathodic Protection" in this Volume).

Corrosion processes are often viewed as the partial processes of oxidation and reduction previously described. The oxidation reaction (anodic reaction) constitutes the corrosion of the metallic phase, and the reduction reaction (cathodic reaction) is the result of the environment. Several different cathodic reactions are encountered in metallic corrosion in aqueous systems. The most common are:

$2H^+ + 2e^- \rightarrow H_2$

Hydrogen ion reduction

$O_2 + 4H^+ + 4e^- \rightarrow 2H_2O$

Reduction of dissolved oxygen (acid media)

$O_2 + 2H_2O + 4e^- \rightarrow 4OH^-$

Reduction of dissolved oxygen (basic media)

$M^{3+} + e^- \rightarrow M^{2+}$

Metal ion reduction

$M^{2+} + 2e^- \rightarrow M$

Metal deposition

Hydrogen ion reduction is very common because acidic media is so often encountered, and oxygen reduction is very common because of the fact that aqueous solutions in contact with air will contain significant amounts of dissolved oxygen. Metal ion reduction and metal deposition are less common and are encountered most often in chemical process streams (Ref 9). All of the above reactions, however, share one attribute: They consume electrons.

Potential Measurements With Reference Electrodes

D.L. Piron
Department of Metallurgical Engineering
École Polytechnique de Montreal

Electrode potential measurement is an important aspect of corrosion prevention. It includes determination of the corrosion rate of metals and alloys in various environments and control of the potential in cathodic and anodic protection. Many errors and problems can be avoided by intelligently applying electrochemical principles in the use of reference electrodes. Among the problems are the selection of the best reference for a specific case and selection of an adequate method of obtaining meaningful results.

It is important to note that many different reference electrodes are available, and others can be designed by the users themselves for particular problems. Each electrode has its characteristic rest potential value, which can be used to convert the results obtained into numbers expressed with respect to other references. These conversions are frequently required for comparison and discussion, and this involves use of E-pH (Pourbaix) diagrams, which will be discussed later in this article. The electrode selected must then be properly used, taking into account the stability of its potential value and the problem of resistance (IR) drop.

Electrode Potential Conventions

The use of reference electrodes is based on two fundamental conventions. One of these conventions sets a zero reference point in the potential scale, and the other gives a meaningful sign to potential values.

The Zero Convention. The potential of an electrode can be determined only with respect to another electrode, the reference electrode. As discussed previously in the section "Electrode Potentials" in this article, only the potential difference between two electrodes, each with its own specific potential, is measured.

The absolute value of the potential of a particular electrode cannot be obtained experimentally. One electrode, therefore, must be selected as 0 in the potential scale. By convention, the standard hydrogen electrode (SHE) was chosen, that is, the standard electrode potential for the reaction $2H^+ + 2e^- \rightarrow H_2$ is made to equal 0. This zero convention makes it possible to assign numbers to electrode potentials on the scale of electrode potentials. The SHE is arbitrarily fixed as the zero level, and all other potentials are expressed with respect to this reference. Practical measurements are performed with various reference electrodes having known values with respect to the SHE. For example, the saturated calomel electrode potential is +240 mV versus SHE, and the copper sulfate/copper ($CuSO_4$/Cu) electrode potential is +310 mV versus SHE.

The Sign Convention (The Reduction or Stockholm Convention). Electrode reactions may proceed in two opposite directions. For example, the Fe^{2+}/Fe system may undergo oxidation ($Fe \rightarrow Fe^{2+} + 2e^-$) or reduction ($Fe^{2+} + 2e^- \rightarrow Fe$).

The potential of this iron electrode is expressed with respect to the SHE = 0. The coupling of these two systems (Fe^{2+}/Fe and H^+/H_2), however, brings about the spontaneous oxidation of iron. The situation is entirely different with a Cu^{2+}/Cu system. In this case, the reduction is spontaneous in an electrochemical cell with a hydrogen electrode.

This difference in the spontaneous reaction direction with respect to hydrogen can be represented by a sign. This sign is also very useful in computing cell potentials from single electrode values.

The choice of a conventional direction for the reaction imposes a sign to the free energy. For the oxidation of Fe^{2+}/Fe, the ΔG_{ox}^{Fe} is negative, because a spontaneous reaction liberates energy. The ΔG_{red}^{Fe} would be positive for the reduction reaction. In the case of copper, however, the

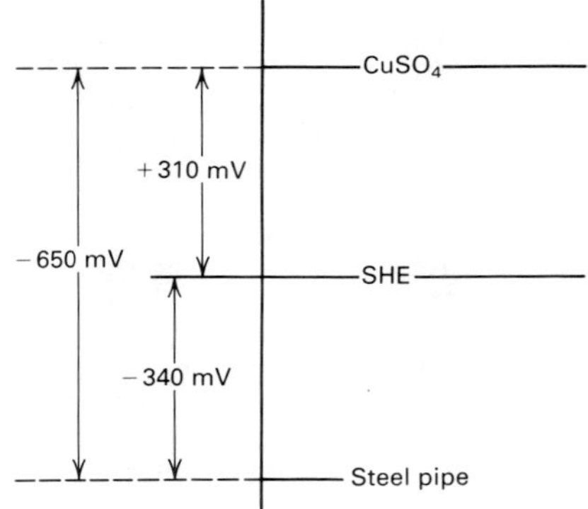

Fig. 6 Electrode potential conversion diagram

reduction of Cu^{2+}/Cu is spontaneous, and ΔG^{Cu}_{red} is therefore negative.

At the International Union of Pure and Applied Chemistry meeting held in Stockholm in 1953, it was decided to choose as the conventional direction the reduction reaction:

$$\text{ox} + ne^- \rightarrow \text{red}$$

where ox represents the oxidized species, n is the number of electrons e^-, and red is the reduced species.

The sign of the electrode can be determined by using the following reaction, which was discussed previously (see Eq 6):

$$\Delta G = -nFE$$

As a result, the Fe^{2+}/Fe system has a negative sign, and Cu^{2+}/Cu has a positive sign.

In this reduction convention, a negative sign indicates a trend toward corrosion in the presence of H^+ ions. The ferrous cations have a greater tendency to exist in aqueous solution than the H^+ cations. A positive sign indicates, on the contrary, that the H^+ ion is more stable than Cu^{2+}, for example.

The reduction convention selects a conventional direction reduction for electrochemical reactions. It is because this conventional direction is not necessarily the natural spontaneous direction that a sign can be given to the electrode potential.

Example of Potential Conversion. The need to be consistent in expressing electrode potentials versus references in a specific problem (regardless of the actual reference used in the measurement) is illustrated in the following example.

The electrode potential of a buried steel pipe is measured with respect to a $CuSO_4/Cu$ electrode, and the value is 650 mV for a pH 4 environment. If that value is mistakenly placed in the iron E-pH diagram, it could be concluded that corrosion is not going to take place. This conclusion would, however, be incorrect, because the E-pH diagrams are always computed with respect to the SHE. It is then necessary to express the result of the measurement with respect to that electrode before consulting the E-pH diagram. The measured electrode potential then has to be expressed with respect to the SHE.

Because the $CuSO_4/Cu$ electrode potential is +310 mV versus SHE, the number that expresses the measured potential is 310 mV higher with the $CuSO_4/Cu$ electrode than with the SHE. As a result, V_{SHE} should be −340 mV. The principle of this conversion is illustrated in Fig. 6 in an electrode potential reference conversion schematic.

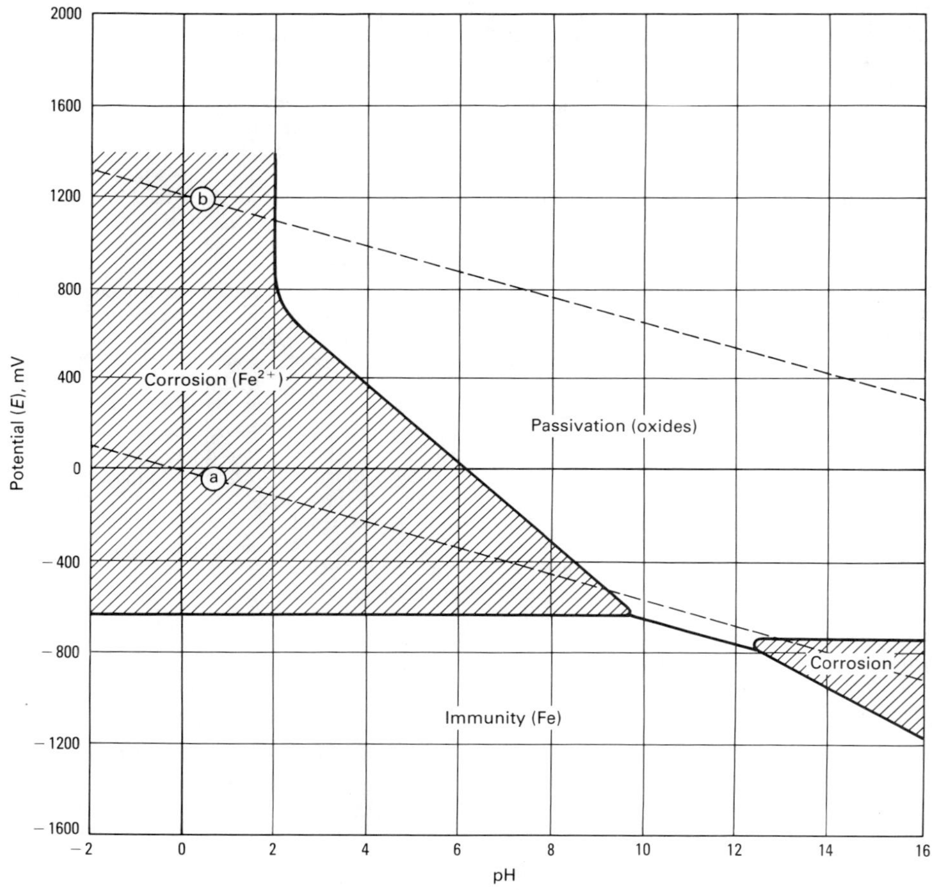

Fig. 7 Iron E-pH diagram. Dashed lines a and b are explained in Fig. 18 and in the corresponding text.

The value of −340 mV placed in the E-pH diagram at a pH 4 clearly indicates a corrosion region for iron (Fig. 7). It would then be definitely necessary to consider the cost benefit of a protection system for the steel pipe.

The Three-Electrode System

When a system is at rest and no significant current is flowing, the use of only one other electrode as a reference is sufficient to measure the test electrode potential. When a current is flowing spontaneously in a galvanic cell or is impressed to an electrolytic cell, reactions at both electrodes are not at equilibrium, and there is consequently an overpotential on each of them. The potential difference measured between these two electrodes then includes the value of the two overpotentials. The potential of only the test electrode cannot be determined from this measurement.

To obtain this value, a third electrode, the auxiliary electrode, must be used (Fig. 8). In this way, the current flows only between the test and the auxiliary electrodes. A high-impedance voltmeter placed between the test and the reference prevents any significant current flow through the reference electrode, which then does not show any overpotential. Its potential remains at its rest value. The test electrode potential and its changes under electric current flow can then be measured with respect to a fixed reference potential (most references are not made to be polarized by a current flow). The three-electrode system is widely used in the laboratory and in field potential measurement.

Electrode Selection Characteristics

Stable and Reproducible Potential. Electrodes used as references should offer an acceptably stable and reproducible potential that is free of significant fluctuations. To obtain these characteristics, it is advantageous, whenever possible, to use reversible electrodes, which can be easily made.

The $CuSO_4/Cu$ electrode is an excellent example of a good reversible electrochemical system; it is widely used as a reference electrode in the corrosion field. It can be easily made by immersing a copper rod in a saturated $CuSO_4$ aqueous solution, as shown in Fig. 9.

This electrode is reversible, because a small cathodic current produces the reduction reaction ($Cu^{2+} + 2e^- \rightarrow Cu$), while an anodic current brings about the oxidation reaction ($Cu \rightarrow Cu^{2+} + 2e^-$). This is not a corroding system like that of immersed iron, which dissolves anodically into

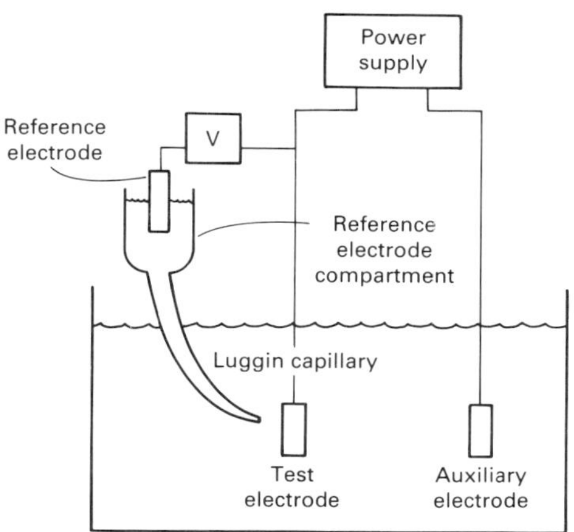

Fig. 8 Potential measurement with a luggin capillary. V, voltmeter

Fe^{2+}, because the immersed copper system produces the hydrogen evolution reaction under a cathodic current.

In the case of the $CuSO_4$/Cu electrode, the rest potential is the equilibrium potential that can be computed by the Nernst equation (Eq 12). At 25 °C (77 °F), it would be:

$$E_{Cu^{2+}/Cu} = +0.34 + \frac{0.059}{2} \log a_{Cu^{2+}}$$

where $E_{Cu^{2+}/Cu}$ is the Cu^{2+}/Cu equilibrium potential, 0.34 is the standard potential, and $a_{Cu^{2+}}$ is the activity of Cu^{2+} in the aqueous solution. This system then provides a well-defined reversible system that is reliable and easy to build.

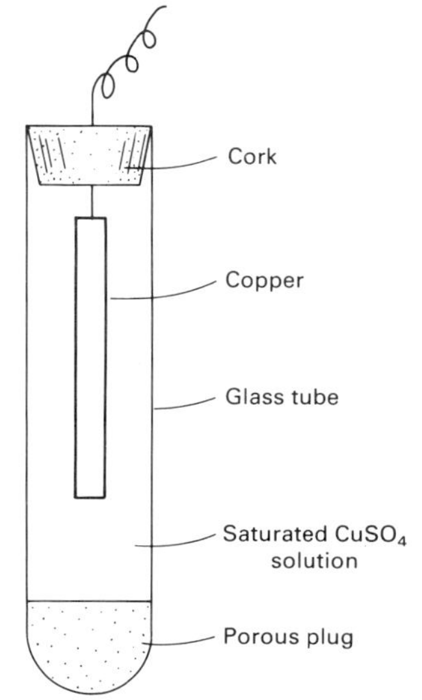

Fig. 9 Schematic of a $CuSO_4$/Cu reference electrode

In some practical cases, however, nonreversible electrodes are used. Although not as well defined, their potential stability in a particular environment is considered sufficient in certain applications. In the selection of reference electrodes, their durability, life expectancy, and price must also be considered.

Low Polarizability. The polarization of reference electrodes introduces an error in the potential measurement. The potential versus current density response, called a polarization curve, should show a low overpotential and a high exchange current, i_{ex}, as can be seen in Fig. 10 (line a). More detailed information on polarization curves can be found in the article "Kinetics of Aqueous Corrosion" and in the section "Electrochemical Methods of Corrosion Testing" of the article "Laboratory Testing" in this Volume.

A poor polarization characteristic for a reference electrode is represented by the dashed line (b) in Fig. 10. In this case, a small current density i_1 produces a significant potential change from E_{rev} to E_c^b. This results in a large change in the overpotential η_b. The electrode represented by line (a) offers much better polarization characteristics. Under the same current density i_1, the observed electrode potential E_c^a remains very close to the reversible value. The resulting overpotential η_a is then negligible.

It is a question of judgment as to how polarizable the reference electrode can be. The answer depends on the precision required and on the impedance of the voltmeter used. A high-impedance voltmeter may provide acceptable results with a more polarizable electrode than a less expensive measuring instrument.

The Liquid Junction Potential. Reference electrodes are usually made of metal immersed in a well-defined electrolyte. In the case of $CuSO_4$/Cu electrode, the electrolyte is a saturated $CuSO_4$ aqueous solution; for the saturated calomel electrode, it is a saturated potassium chloride (KCl) solution. This electrolyte that characterizes the reference electrode comes into contact with the liquid environment of the test electrode (Fig. 11). There is then direct contact between different aqueous media. The difference in chemical composition produces a phenomenon of interdiffusion. In this process, except for a few cases such as KCl, the cations and anions move at different speeds. However, for hydrogen chloride (HCl) solution in contact with another media, the H^+ ions move faster than the Cl^- ions. As a result, a charge separation appears at the limit between the two liquids (the liquid junction); this produces a potential difference called the liquid junction potential. This liquid junction potential is included in the measured potential, as expressed in:

$$V = V_T - V_R + V_{LJP}$$

where V_T is the unknown voltage to be measured, V_R is the reference electrode potential, and V_{LJP} is the unknown liquid junction potential.

In order to determine V_T, the liquid junction potential has to be eliminated or minimized. The

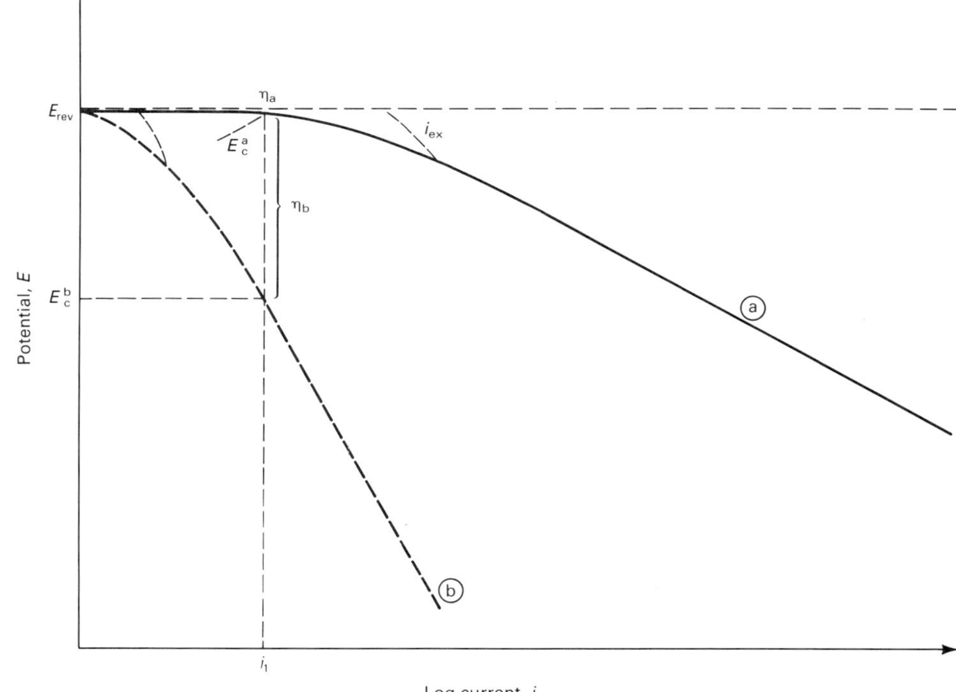

Fig. 10 Polarization curve for a good reference electrode (line a) and a poor reference electrode (line b)

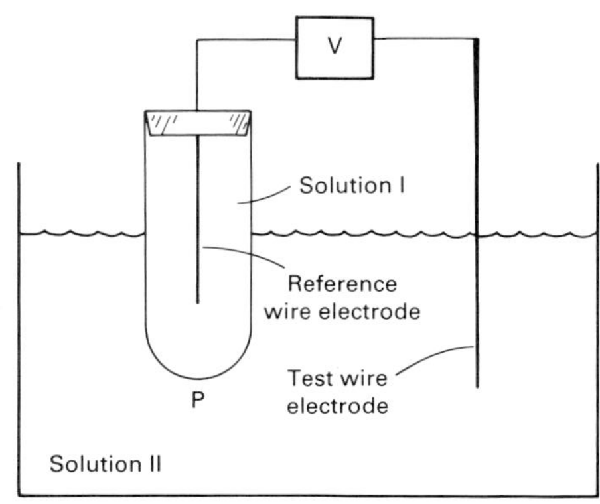

Fig. 11 Schematic of an electrochemical cell with liquid junction potential. P, interface

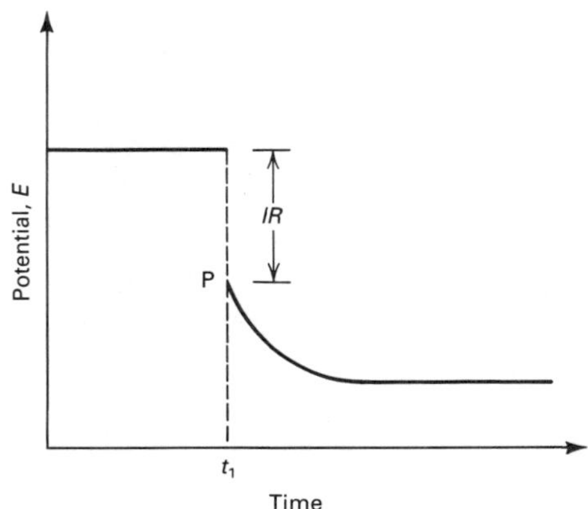

Fig. 12 The potential decay at current interruption. See text for discussion.

best way, when it is possible, is to design a reference electrode using electrolyte (Fig. 11) identical to the solution in which the test electrode is immersed. This can be done in some cases, for example, in overpotential measurement in a copper electrowinning cell. The reference electrode can be a copper wire in a glass tube simply immersed in $CuSO_4$ cell electrolyte. A simple $CuSO_4$/Cu reference electrode can be made in this way.

Most of the time, however, this ideal solution is not possible, and the best approach is to minimize the liquid junction potential by using a reference electrolyte with a chemical composition as close as possible to the corrosion environment. In some cases, the use of a solution of KCl (such as in the calomel electrode) offers a partial answer. The diffusion rates of potassium (K^+) and chloride (Cl^-) ions are similar. In contact with another electrolyte, a KCl solution does not produce much charge separation and, consequently, no significant liquid junction potential. The ions present in the other solution, however, also diffuse, and they may do so at different rates, thus producing some separation of charge at the interface P (Fig. 11).

The remaining liquid potential, after minimization, constitutes an error that is frequently accepted in electrode potential measurements, especially when compared with results determined under similar experimental conditions. Liquid junction potentials have to be minimized as much as possible. There is no general solution for this; each individual case has to be well thought out.

Operating Conditions for Reference Electrodes

When a reference has been selected for a particular application, its proper use requires caution, as well as measurement methods based on the same electrochemical principles. In a measurement of the potential of a polarized electrode, it is important not to polarize the reference electrode and to keep its reference value.

Very Low Current Density. It is important to use a reference electrode that operates at its known open-circuit potential and to avoid any significant overpotential. This is achieved by using a high-impedance voltmeter that has a negligible input current and, for polarized test electrodes, by using an auxiliary electrode in a three-electrode system.

The requirement is shown in Fig. 10 on curve a. The current must be maintained lower than i_1 to avoid a significant overpotential η_a. The value tolerated for η_a is a matter of judgment that depends on the accepted magnitude of error in the particular case under investigation. The use of an electrometer or a high-impedance voltmeter usually fulfills this requirement. The existence of an overpotential η_a could result, however, when less expensive equipment is used, and in this case, electrodes similar to electrode b in Fig. 10 should be avoided.

The *IR* Drop and Its Mitigation. The *IR* drop is an ohmic voltage that results from the electric current flow in ionic solutions. Electrolytes have an ohmic resistance, and when a current passes through them, an *IR* voltage can be observed between two distinct points.

When the reference electrode is immersed at some distance from a working test electrode, it is in the electric field somewhere along the current line. An electrolyte resistance exists along the line between the test and the reference electrode. Because a current flows through that resistance, an *IR* voltage appears in the potential measurement according to:

$$V = V_T - V_R + IR$$

where V_T is the test potential to be measured, V_R is the reference electrode potential, and *IR* is the ohmic drop. In this case, the liquid junction potential has been neglected. The *IR* drop constitutes a second unknown value in a single equation. It must be eliminated or minimized.

The Luggin capillary is a tube, usually made of glass, that has been narrowed by elongation at one end. The narrow end is placed as close as possible to the test electrode surface (Fig. 8), and the other end of the tube goes to the reference electrode compartment.

The Luggin capillary is filled with cell electrolyte, which provides an electric link between the reference and the test electrode. The use of a high-impedance voltmeter prevents the current from flowing into the reference electrode and consequently into the capillary tube between the test electrode and the reference electrode department (Fig. 8). This absence of current eliminates the *IR* drop, and the measurement of V_T is then possible. A residual *IR* drop may, however, exist between the tip of the Luggin capillary and the test. This is usually negligible, however, especially in high-conductivity media.

The remote electrode technique can be used only for measurement in an electrolyte with very low resistivity, usually in the laboratory. It is applicable, for example, in a molten salt solution, in which the ohmic resistance R is very small. In such a case, the reference electrode can be placed a few centimeters away from the test electrode because the product *IR* remains negligible. In other electrolytes (for example, in measurements in soils) the ohmic resistance is rather large, and the *IR* drop cannot be eliminated in this manner.

The Current Interruption Technique. In this case, when the current is flowing the *IR* drop is included in the measurement. A recording of the potential is shown in Fig. 12. At time t_1, the current is interrupted so that $I = 0$ and $IR = 0$.

At the moment of the interruption, however, the electrode is still polarized, as can be seen at point P in Fig. 12. The progressive capacitance discharge and depolarization of the test electrode takes some time. The potential measured at the instant of interruption then represents the test electrode potential corrected for the *IR* drop. Precise measurements of the potential of P are obtained with an oscilloscope.

Potential Versus pH (Pourbaix) Diagrams

D.L. Piron
Department of Metallurgical Engineering
École Polytechnique de Montreal

Potential-pH diagrams are graphical representations of the domain of stability of metal ions, oxides, and other species in solution. The lines that show the limits between two domains express the value of the equilibrium potential between two species as a function of pH. They are computed from thermodynamic data, such as standard chemical potentials, by using the Nernst equation (Eq 12). Potential-pH diagrams then provide a graphical expression of Nernst's law.

These diagrams also give the equilibrium of acid-base reactions independent of the potentials. These equilibria are represented by vertical lines at specific pHs. Potential-pH diagrams organize many important types of information that are useful in corrosion and in other fields of practice. They make it possible to discern at a glance the stable species for specific conditions of potential and pH.

When applied to a metal, the equilibrium potential line gives the limit between the domains of stability of the metal and its ions. For conditions of potential and pH corresponding to metal stability, corrosion cannot take place, and the system is in a region of immunity. However, when the potential and pH correspond to the stability of ions, such as Fe^{2+}, the metal is not stable, and it tends to oxidize into Fe^{2+}. The system is then in a corrosion region of the diagram.

In the case of iron, corrosion in deaerated water is expressed by the electrochemical reaction $Fe \rightarrow Fe^{2+} + 2e^-$, and the species Fe^{2+} and Fe are considered. The Nernst equation (Eq 12) makes it possible to compute the equilibrium potential for the system Fe^{2+}/Fe:

$$E_{eq}^{Fe} = E° + \frac{RT}{nF} \ln (Fe^{2+})$$

This equilibrium potential can be represented as a horizontal line in a partial *E*-pH diagram (see, for example, Fig. 7).

The line indicates the potential at which Fe and Fe^{2+} of a given concentration are in equilibrium and can coexist with no net tendency to transform into the other. Above the line is a domain of stability for Fe^{2+}; iron metal is not stable at these potentials and tends to dissolve as Fe^{2+} and thus increase the Fe^{2+} concentration. Below the equilibrium line, the stability of the metallic iron increases, and the equilibrium concentration of Fe^{2+} decreases; that is, the metal becomes immune. When the reaction of the metal with water produces an oxide that protects it, the metal is said to be passivated.

Diagrams such as Fig. 7 were first made by M. Pourbaix (Ref 10, 11) and have proved to be very useful in corrosion as well as in many other fields, such as industrial electrolysis, plating, electrowinning and electrorefining of metals, primary and secondary electric cells, water treatment, and hydrometallurgy. It is very important to emphasize that these diagrams are based on thermodynamic computations for a number of selected chemical species and the possible equilibria between them. It is then possible to predict from an E-pH diagram if a metal will corrode or not. It is, however, not possible to determine from these diagrams alone how long a metal will resist perforation.

Pourbaix diagrams offer a framework for kinetic interpretation, but they do not provide precise information on corrosion rates (Ref 12). Moreover, they are not a substitute for kinetic studies. Because each diagram is computed for a selected number of chemical species, the addition of one or more species to the system will introduce several new equilibria. Their representation in the E-pH diagram will produce a new diagram that is different from the previous one. For example, the simple diagram of gold in water does not show any possible solubility for that metal. The addition of cyanide ions to the system, however, makes possible the formation of a gold complex soluble in water. Gold that does not corrode in water can dissolve in the presence of cyanide. This property is the basis of gold plating and of the hydrometallurgy of that metal.

Computation and Construction of *E*-pH Diagrams

As discussed in the introduction to this article, E-pH diagrams are based on thermodynamic computations. The equilibrium potentials and the pH lines that set the limits between the various stability domains are determined from the chemical equilibria between the chemical species considered.

It is interesting and practical to realize here that there are three types of reactions to be considered.

- Electrochemical reactions of pure charge transfer
- Electrochemical reactions involving both electrons and H^+
- Pure acid-base reactions

As will be shown, graphical expressions at the Nernst equation (Eq 12) can be constructed from each of these reactions.

Reactions of Pure Charge Transfer. These electrochemical reactions involve only electrons and the reduced and oxidized species. They do not have protons (H^+) as reacting particles; consequently, they are not influenced by pH. An example of a reaction of this type is:

$$Ni^{2+} + 2e^- \rightarrow Ni$$

The equilibrium potential is given by the Nernst equation (Eq 12). In the case of the nickel reaction given above, it can be written:

$$E = E^\circ + \frac{RT}{nF} \ln (Ni^{2+}) \qquad \text{(Eq 13)}$$

where E is the equilibrium potential for Ni^{2+}/Ni; E° is the standard potential for Ni^{2+}/Ni; R, T, F, and n are defined in Eq 12; and (Ni^{2+}) is the Ni^{2+} activity in the solution.

The value of the potential obtained depends on the Fe^{2+} activity, but not on H^+ ions, which do not participate in the electrochemical reaction. The result is then independent of the pH, and it can be represented by the horizontal line in an E-pH diagram.

In order to obtain this result, it is necessary to compute the value of the standard potential, given by:

$$E^\circ = \frac{\Delta G^\circ}{nF} = \frac{\Sigma \nu_{ox}\, \mu^\circ_{ox} - \Sigma \nu_{red}\, \mu^\circ_{red}}{2F} \qquad \text{(Eq 14)}$$

where ν is the stoichiometric coefficients of oxidizing and reducing species and μ is given below. In this case under consideration:

$$E^\circ = \frac{\mu_{Ni^{2+}} - \mu_{Ni}}{2F}$$

where $\mu^\circ_{Ni^{2+}}$ and μ°_{Ni} are standard chemical potentials. By convention, the standard potential of a chemical element is 0. This gives $\mu^\circ_{Ni} = 0$, and simplifies the above equation:

$$E^\circ = \frac{\mu^\circ_{Ni^{2+}}}{2F}$$

The value of $\mu^\circ_{Ni^{2+}}$, which can be found in the *Atlas of Electrochemical Equilibria* (Ref 13), is 11 530 cal. Consequently:

$$E^\circ = \frac{11\ 530 \text{ cal}}{2 \text{ equiv.} \times 23\ 060 \text{ cal/V equiv.}} = -0.25 \text{ V}$$

This result can be introduced into Eq 13 as follows:

$$E_{Ni^{2+}/Ni} = -0.25 \text{ V} + \frac{2.3\ RT}{2F} \log (Ni^{2+})$$

where 2.3 R = 4.57 at a temperature of 25 °C (77 °F) or 298 °K. If 2.3 RT/F = 0.059 V, the equilibrium potential for Ni^{2+}/Ni will be:

$$E = -0.25 + 0.03 \log (Ni^{2+})$$

As previously stated for this case, the potential depends only on the activity of (Ni^{2+}), not on the pH. It is customary here to select four activities: 1 or 10^0, 10^{-2}, 10^{-4}, and 10^{-6} g ion/L. This will provide four horizontal lines, as shown in Fig. 13:

- At a concentration of 10^0 g ion/L, $E_{Ni^{2+}/Ni} = E^\circ = -0.25$ V
- With 10^{-2} g ion/L, $E_{Ni^{2+}/Ni} = -0.25 + 0.03 \log 10^{-2} = -0.31$ V
- With 10^{-4} g ion/L, $E_{Ni^{2+}/Ni} = -0.37$ V
- With 10^{-6} g ion/L, $E_{Ni^{2+}/Ni} = -0.43$ V

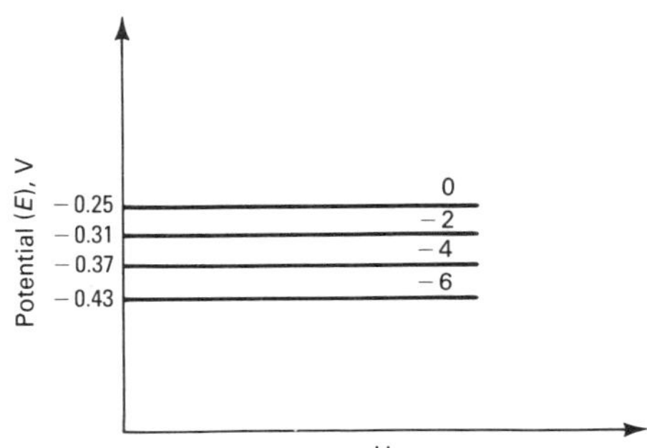

Fig. 13 Partial E-pH diagram for the $Ni^{2+} + 2e^- \rightarrow Ni$ reaction

For any activity of Ni^{2+} in the solution, a horizontal line represents the equilibrium potential, that is, the potential at which Ni^{2+} ions and Ni metal can coexist. Above the line is the region of stability of Ni^{2+} ions; nickel metal at these potentials will tend to corrode and produce Ni^{2+}, the stable species. Below the line, metallic nickel is stable, and nickel in these conditions will not corrode.

Reactions Involving Both Electrons and H^+. Nickel can also react with water to form an oxide, according to the electrochemical reaction:

$$Ni + H_2O \rightarrow NiO + 2H^+ + 2e^-$$

The standard potential E° is given by:

$$E^\circ = \frac{\mu^\circ_{NiO} - \mu^\circ_{Ni} - \mu^\circ_{H_2O}}{2F} = \frac{-51\ 610 - 0 + 56\ 690}{2 \times 23\ 060} = 0.11 \text{ V}$$

The Nernst equation (Eq 12) can in this case be written as follows:

$$E_{eq} = +0.11 + \frac{0.06}{2} \log \frac{(NiO)(H^+)^2}{(Ni)(H_2O)} \qquad \text{(Eq 15)}$$

The NiO and Ni are solid phases, and they are considered to be pure; their activity is therefore 1. The activity of water in aqueous solutions is also assumed to be 1. Equation 15 can then be simplified to:

$$E_{eq} = +0.11 + 0.03 \log (H^+)^2$$

Because pH = $-\log (H^+)$, it is possible to write:

$$E^{NiO}_{eq} = 0.11 - 0.06 \text{ pH} \qquad \text{(Eq 16)}$$

In this case, the equilibrium potential is a decreasing function of pH, as represented in a partial E-pH diagram (Fig. 14).

The diagonal line in Fig. 14 gives the value of the equilibrium potential for Ni and NiO at all pH values. Above the line, NiO is stable, and below it, nickel metal is stable.

Potential-pH diagrams are very general and can also be applied to electrochemical reactions involving nonmetallic chemicals. An example involving the reduction of nitrite (NO_2^-) in ammonia (NH_4^+) will be given here. In this case, the metal of the electrode supports the reaction by giving or taking away electrons as follows:

$$NO_2^- + 8H^+ + 6e^- \rightarrow NH_4^+ + 2H_2O$$

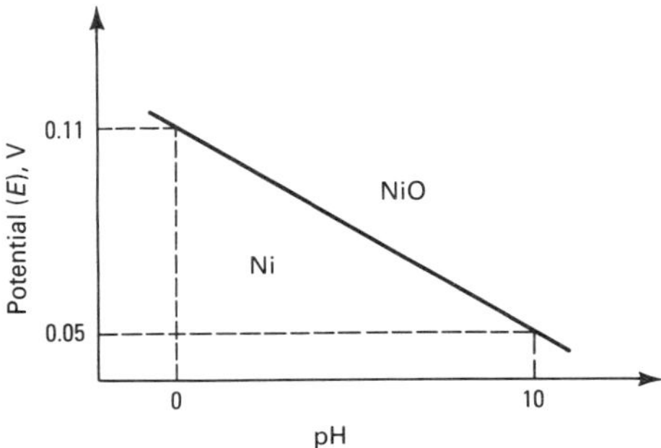

Fig. 14 Partial E-pH diagram for the $Ni + H_2O \rightarrow NiO + 2H^+ + 2e^-$ reaction

It is only a supporting electrode. The Nernst equation (Eq 12) can then be written:

$$E = E^\circ + \frac{RT}{6F} \ln \frac{(NO_2^-)(H^+)^8}{(NH_4^+)(H_2O)^2} \quad \text{(Eq 17)}$$

where $(H_2O) = 1$.

The equation for the standard potential is:

$$E^\circ = \frac{\mu^\circ_{NiO_2} + 8\mu^\circ_{H^+} - \mu^\circ_{NH_4^+} - 2\mu^\circ_{H_2O}}{2F}$$

$$= \frac{-8250 - (-19\,000) - 2(-56\,690)}{6 \times 23\,060}$$

$$= +0.897$$

The equilibrium potential is then given by:

$$E = 0.897 + 0.01 \log \frac{(NO_2)}{(NH_4^+)} - 0.08 \text{ pH}$$

It can be represented in an E-pH diagram for equal activity in NO_2^- and NH_4^+ by a decreasing line in Fig. 15.

Above the line, there is a region in which NO_2^- is predominantly stable but in equilibrium with smaller activities of NH_4^+. Below the line, NH_4^+ is predominantly stable, with smaller quantities of NO_2^-.

Pure Acid-Base Reactions. The previous computations showed that there are possible equilibria between the metal and its ions (such as Ni^{2+}/Ni) and between the metal and its oxide (NiO/Ni). In the case of cobalt it is possible, as shown in Fig. 16, to determine the equilibria for Co^{2+}/Co and for CoO/Co. The two equilibrium potential lines meet at some point P, and above them are two domains of stability for Co^{2+} and CoO. These two species are submitted to an acid-base chemical reaction:

$$Co^{2+} + H_2O \rightleftarrows CoO + 2H^+ \quad \text{(Eq 18)}$$

which does not involve electrons. It does not depend, then, on the potential, and it will be represented by a vertical line. Point P in Fig. 16 is one point on that line located at a pH 6.3 for an activity of 1 in Co^{2+} ions.

The pH value of that line can also be computed from the chemical equilibrium, with the general equation:

$$\Delta G^\circ = -RT \ln K$$

or

$$\log K = \frac{\Sigma \nu_R\, \mu^\circ_R - \Sigma \nu_P\, \mu^\circ_P}{2.3\, RT} \quad \text{(Eq 19)}$$

where K is a constant, ν_R is the stoichiometric coefficient of the reactants, ν_P is the stoichiometric coefficient of the product, μ°_R is the standard chemical potential of the reactant, and μ°_P is the standard chemical potential of the product. In this case, the equilibrium as given in Eq 18 can be written:

$$\log \frac{(CoO)(H^+)^2}{(Co^{2+})(H_2O)} = \frac{\mu^\circ_{Co^{2+}} + \mu^\circ_{H_2O} - \mu^\circ_{CoO}}{2.3\, RT}$$

By assuming that (CoO) and (H_2O) both have activities of 1 and by replacing the standard chemical potentials by their values given in the *Atlas of Electrochemical Equilibria* (Ref 13), it is possible to write:

$$\log \frac{(H^+)^2}{(Co^{2+})} = \frac{-12\,800 - 56\,690 - (-52\,310)}{2.3 \times 1.987 \times 298}$$

$$= -12.6 \quad \text{(Eq 20)}$$

and finally:

$$\log (Co^{2+}) = 12.6 - 2 \text{ pH} \quad \text{(Eq 21)}$$

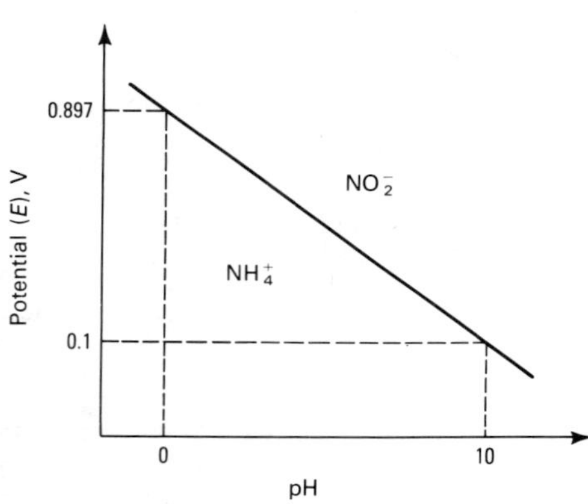

Fig. 15 Partial E-pH diagram for the NO_2^- + $8H^+$ + $6e^-$ → NH_4^+ + $2H_2O$ reaction

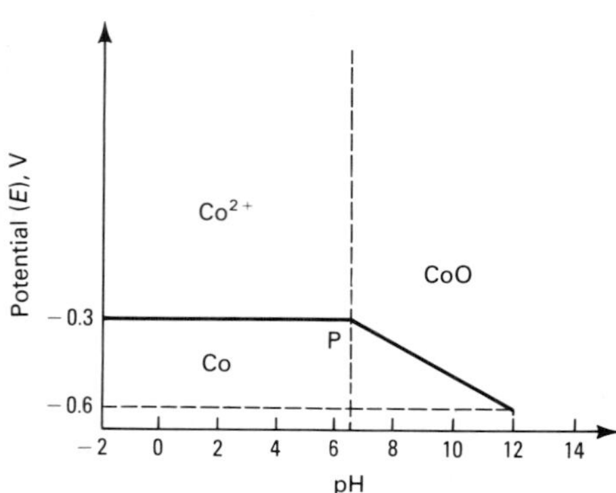

Fig. 16 Partial E-pH diagram for Co^{2+}/Co and CoO/Co

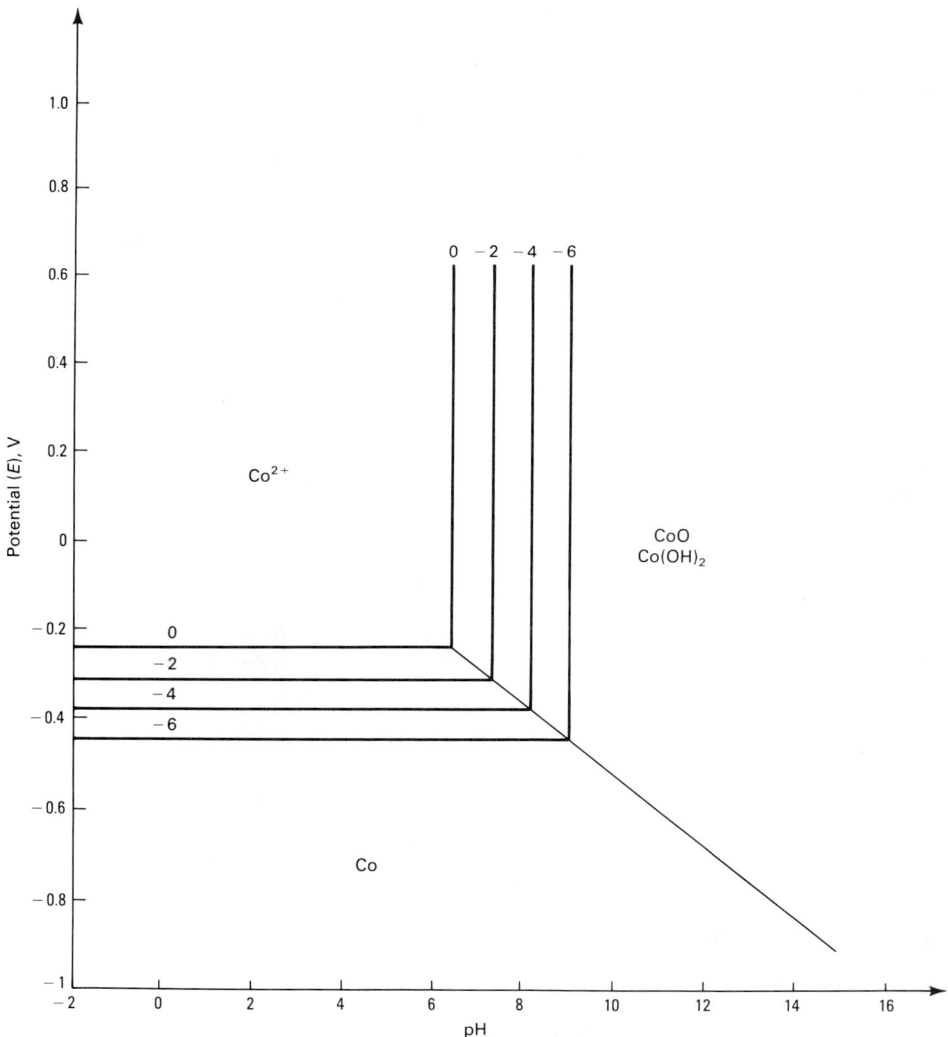

Fig. 17(a) Partial E-pH diagram for cobalt

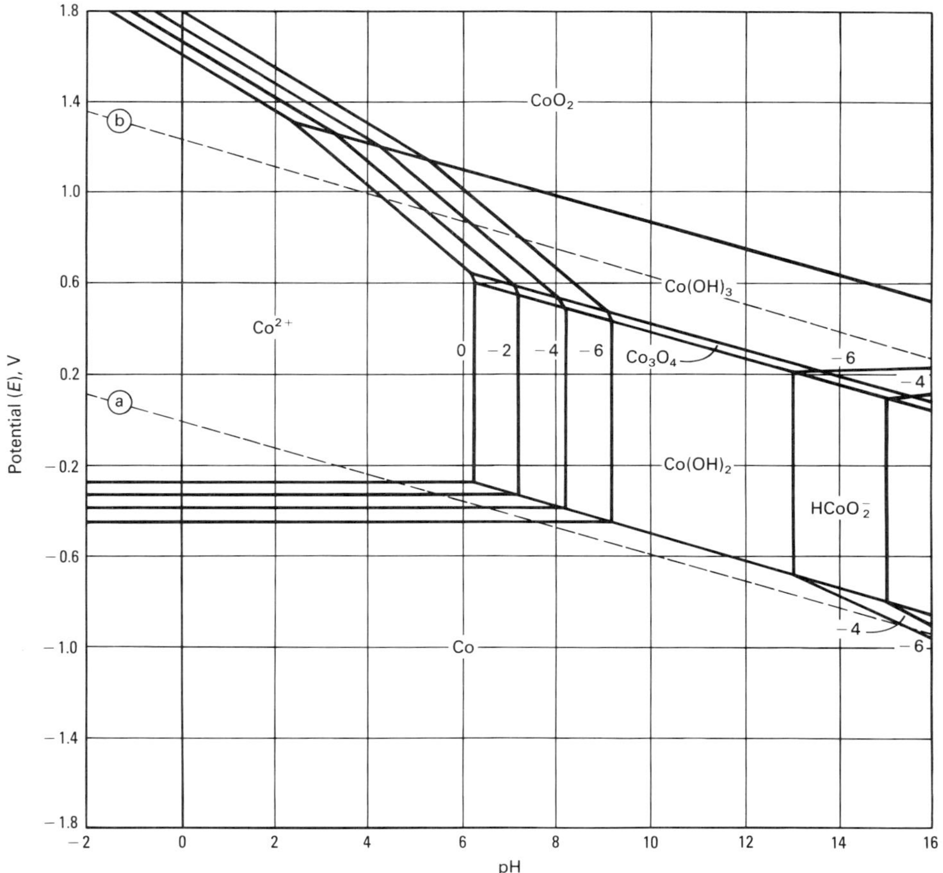

Fig. 17(b) *E*-pH diagram for cobalt

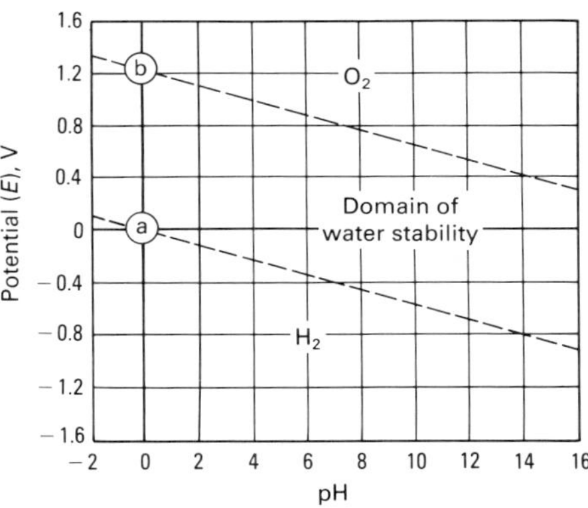

Fig. 18 The water *E*-pH diagram at 1 atm

for (Co^{2+}) = 1 or pH = 6.3. This verifies the value obtained by tracing the two equilibrium lines.

Figure 17(a) shows a partial *E*-pH diagram in which only three chemical species—Co, Co^{2+}, and CoO—are considered. There are, however, other possible chemical species, such as CoO_2 and $HCoO_2^-$, that must be considered. This introduces new equilibria that modify the diagram to give Fig. 17(b).

The Water *E*-pH Diagram. Pourbaix diagrams are traced for equilibrium reactions taking place in water; consequently, the water *E*-pH diagram always must be considered at the same time as the system under investigation. Water can be decomposed into oxygen and hydrogen, according to the following reactions:

$$2H^+ + 2e^- \rightarrow H_2$$

and

$$H_2O \rightarrow \tfrac{1}{2}O_2 + 2H^+ + 2e^-$$

There are then two possible electrochemical equilibria for which the equilibrium potential can be determined by using the Nernst equation (Eq 12). For hydrogen:

$$E_{H^+/H_2} = E^{\circ}_{H^+/H_2} + \frac{0.059}{2} \log \frac{(H^+)^2}{p_{H_2}}$$

where (H^+) is the activity of H^+ in water, and p_{H^2} is the pressure of hydrogen near the electrode. Because, by convention, $E^{\circ}_{H^+/H^2} = 0$, the above equation can be rewritten as follows:

$$E_{H^+/H_2} = -0.059 \text{ pH} - \frac{0.059}{2} \log p_{H_2} \quad \text{(Eq 22)}$$

The equilibrium potential for the system H_2/H_2O can be represented in Fig. 18 by line a, which decreases with the pH.

The equilibrium potential for the oxygen/water reaction is given by the Nernst equation:

$$E^{O2}_{eq} = E^{\circ}_{O_2} + \frac{0.059}{2} \log \frac{(p_{O_2})^{1/2}\,(H^+)^2}{(H_2O)}$$

where p_{O_2} equals the pressure of O_2 near the electrode. The activity is, as usual, assumed to be 1, and the standard potential for $0_2/H_2O$ is computed to be 1.23 V. The following can then be written:

$$E^{O2}_{eq} = 1.23 - 0.059 \text{ pH} + \frac{0.059}{4} \log p_{O_2} \quad \text{(Eq 23)}$$

Equation 23 is represented under 1 atm pressure by line b in Fig. 18.

It is interesting to note that the pressures of hydrogen and oxygen in the vicinity of the electrode are usually identical and nearly equal to the pressure that exists in the electrochemical cell. To be rigorous, the water vapor pressure should be taken into account, but it is frequently neglected as not being very significant.

When the pressure increases, line b in Fig. 18 is displaced upward in the diagram, and line a is lowered. The result is that the domain of water stability increases with increasing pressure. The water diagram is so important for a good understanding of the corrosion behavior of a metal that it is usually represented by dotted lines in all Pourbaix diagrams (Ref 13).

Practical Use of *E*-pH Diagrams

The *E*-pH diagram is an important tool for understanding electrochemical phenomena. It provides much useful thermodynamic information in a simple figure. A few cases are presented here to illustrate its practical use in corrosion.

Acid Corrosion of Nickel. A rod of nickel is immersed in an aqueous deaerated acid solution with a pH of 1 that contains 10^{-4} g ion/L of Ni^{2+} ions. The system is under 1 atm pressure. These conditions make it possible to simplify the *E*-pH diagram, as shown in Fig. 19.

At the metallic nickel/water interface, two electrochemical reactions are possible, and their equilibrium potentials can be computed:

$$Ni \rightarrow Ni^{2+} + 2e^-$$

with $E_{Ni} = -0.25 + 0.03 \log a_{Ni^{2+}}$, which for concentration $a = 10^{-4}$ gives $E_{Ni} = -0.37$ V (see Fig. 13). It follows that:

$$2H^+ + 2e^- \rightarrow H_2$$

with $E_H = -0.06$ pH. At pH = 1, $E_H = -0.06$ V.

The nickel equilibrium potential is then more active than that of hydrogen, and electrons tend to flow from the negative nickel to the more positive hydrogen. Because both reactions occur on the same electrode surface, the electrons can go directly from the nickel to the hydrogen. The two reactions then tend to proceed under a common electrode potential or mixed potential, with a value somewhere between the nickel and hydrogen equilibrium potentials.

The mixed potential E_M is then above the Ni^{2+}/Ni equilibrium potential in the region of Ni^{2+} stability (Fig. 19). Nickel is then not stable at low pH in water, and it tends to oxidize or corrode, producing Ni^{2+}, according to the reaction:

$$Ni \rightarrow Ni^{2+} + 2e^-$$

This charge separation could stop the ionization reaction if there were not another chemical reaction—the reduction of H^+. The mixed potential E_M is located below the H^+/H_2 equilibrium potential in a region where H_2 is stable (Fig. 19). As a result, H^+ can accept the electrons and is reduced according to $2H^+ + 2e^- \rightarrow H_2$, producing H_2 gas.

The Pourbaix diagram explains the tendency for nickel to corrode in strong acid solutions. It does not indicate the rate of corrosion, however. This important information has to be obtained from a kinetic experiment, for example, by measuring the corrosion current in a polarization experiment.

The Pourbaix diagram can also show that when the pH increases the difference between the nickel and the hydrogen equilibrium potential decreases in magnitude and that, consequently, the corrosion tendency becomes less important. For pHs between 6 and 8, Fig. 19 shows that hydrogen is more active than nickel. Under this condition, H^+/H^2 can no longer accept the electrons from nickel. Moreover, the potential of the system is in this case below the equilibrium potential of nickel in the region of metal immunity. In pure water at room temperature, nickel does not corrode for pHs between 6 and 8. Moreover, an increase in pressure according to Eq 22 lowers the equilibrium line of H^+/H_2 and does not change the equilibrium line of nickel. As a result, an increase in pressure favors the corrosion resistance of nickel. This behavior of nickel makes the metal slightly noble, and it is expected from the diagram to resist corrosion better than iron or zinc.

The presence of elements, such as chloride, not considered in Fig. 19 may increase the corrosion tendency of nickel. For pHs higher than 8, NiO, $Ni(OH)_2$, or Ni_3O_4 can form, as can be seen in Fig. 19. These oxides may in some cases protect the metal by forming a protective layer that prevents or mitigates further corrosion. This phenomenon is called passivation (passivation is described in detail in the article "Kinetics of Aqueous Corrosion" in this Volume). The presence of chloride is dangerous here, because it may attack the protective layer and then favor corrosion. Figure 19 also illustrates that for very strong alkaline solutions nickel may corrode as $HNiO_2^-$ when the potential is made anodic.

Corrosion of Copper. Observation of the copper *E*-pH diagram in Fig. 20 immediately reveals that the corrosion of copper immersed in deaerated acid water is not likely to occur. The H^+/H_2 equilibrium potential represented by line a is always more active than the Cu^{2+}/Cu equilibrium potential. The H^+ ions are then always in contact with immune copper metal that cannot corrode.

The presence of dissolved oxygen in nondeaerated solutions introduces another possible reaction—O_2/H_2O reduction, with an equilibrium potential more noble than that of Cu^{2+}/Cu. The O_2/H_2O system is then a good acceptor for the electrons abandoned by copper oxidation. The two electrochemical reactions:

$$\frac{1}{2}O_2 + 2H^+ + 2e^- \rightarrow H_2O$$

and

$$Cu \rightarrow Cu^{2+} + 2e^-$$

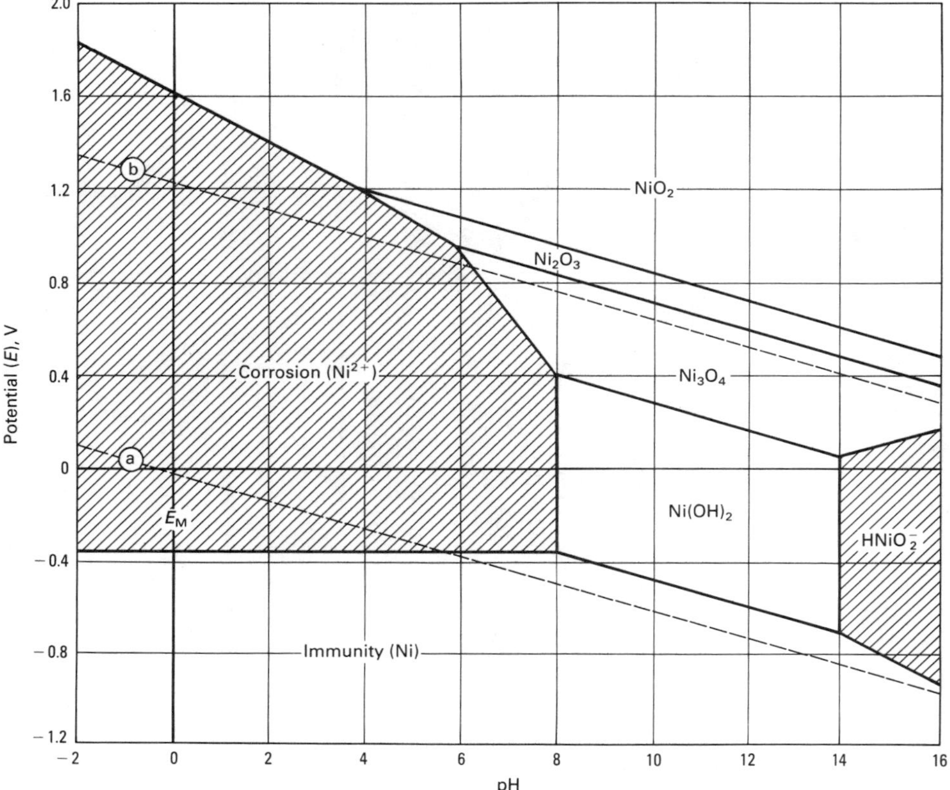

Fig. 19 *E*-pH diagram for nickel

take place at the same metal/solution interface at a common mixed potential.

This discussion assumes that the solution does not contain chloride or other compounds capable of forming soluble complexes with copper. In the presence of such impurities, another diagram must be traced for copper that in some conditions reveals different corrosion behavior. The diagram gives valuable information if all the substances present in the actual system under investigation are taken into account when it is traced.

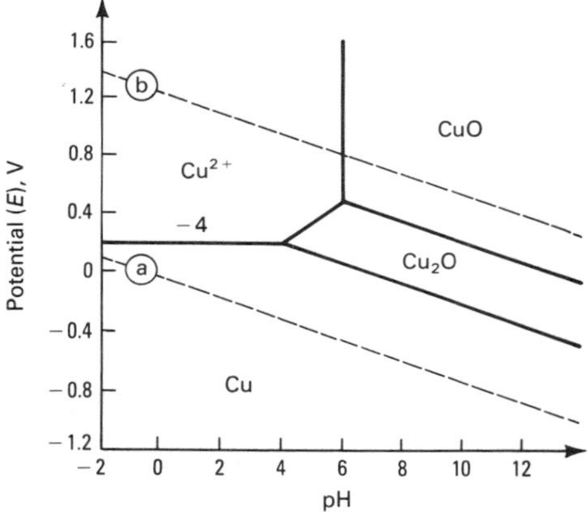

Fig. 20 Partial *E*-pH diagram for copper

REFERENCES

1. L. Pauling, *General Chemistry*, W.H. Freeman, 1964, p 338-360
2. J. O'M. Bokris and A.K.N. Reddy, *Modern Electrochemistry*, Vol 1, Plenum Press, 1977
3. J.M. Smith and H.C. Van Hess, *Introduction to Chemical Engineering Thermodynamics*, McGraw-Hill, 1975, p 159-162
4. K. Denbigh, *Principles of Chemical Equilibrium*, 2nd ed., Cambridge Press, 1981, p 133-186
5. M.G. Fontana, *Corrosion Engineering*, 2nd ed., McGraw-Hill, 1978, p 297-303
6. A.J. Bard, R. Parsons, and J. Jordan, *Standard Potentials in Aqueous Solutions*, Marcel Dekker, 1985
7. W.M. Latimer, *Oxidation Potentials*, Prentice Hall, 1964
8. F.L. La Que, *Corrosion Handbook*, H.H. Uhlig, Ed., John Wiley & Sons, 1948, p 416
9. M.G. Fontana, *Corrosion Engineering*, 2nd ed., McGraw-Hill, 1978, p 12
10. Thermodynamique des Solutions Aqueuses Diluées, Potentiel D'oxydo-Réduction (résumé de conférence), *Bull. Soc. Chim. Belgique*, Vol 48, Dec 1938
11. M. Pourbaix, *Thermodynamics of Dilute Aqueous Solutions*, Arnold Publications, 1949
12. R.W. Staehle, Marcel J.N. Pourbaix—Palladium Award Medalist, *J. Electrochem. Soc.*, Vol 123, 1976, p 23C
13. M. Pourbaix, *Atlas of the Electrochemical Equilibria*, NACE, 1974

Kinetics of Aqueous Corrosion

D.W. Shoesmith, Fuel Waste Technology Branch, Atomic Energy of Canada Ltd.

THE AQUEOUS CORROSION of metal is an electrochemical reaction. For metal corrosion to occur, an oxidation reaction (generally a metal dissolution or oxide formation) and a cathodic reduction (such as proton or oxygen reduction) must proceed simultaneously. For example, the corrosion of iron in acid solutions is expressed as follows:

Oxidation (anodic) $Fe \rightarrow Fe^{2+} + 2e^-$ (Eq 1)

Reduction (cathodic) $2H^+ + 2e \rightarrow H_2$ (Eq 2)

Overall reaction

$$Fe + 2H^+ \rightarrow Fe^{2+} + H_2 \quad \text{(Eq 3)}$$

As a second example, for the corrosion of iron in a solution containing dissolved oxygen, the following expressions are used:

Oxidation (anodic) $Fe \rightarrow Fe^{2+} + 2e^-$ (Eq 1)

Reduction (cathodic)

$$O_2 + 4H^+ + 4e^- \rightarrow 2H_2O \quad \text{(Eq 4)}$$

Overall reaction

$$2Fe + O_2 + 4H^+ \rightarrow 2Fe^{2+} + 2H_2O \quad \text{(Eq 5)}$$

The reaction for metal dissolution ($M \rightarrow M^{n+}$) driven by the cathodic reaction $O \rightarrow R$, is:

$$M + O \rightarrow M^{n+} + R \quad \text{(Eq 6)}$$

where M is a metal, O is oxygen or another oxidizing reagent, $n+$ is the multiple of the charge, and R is the reduced species or reduction. The corrosion process has been written as two separate reactions occurring at two distinct sites on the same surface (Fig. 1a). These two sites are known as the anode, or metal dissolution site, and the cathode, or the site of the accompanying reduction reaction.

As shown in Fig. 1(a), the corroding metal is equivalent to a short-circuited energy-producing cell in which the energy is dissipated during the consumption of cathodic reagent and the formation of corrosion products. To maintain a mass balance, the amount of cathodic reagent consumed must be equal, in chemical and electrochemical terms, to the amount of corrosion product formed. Because electrons are liberated by the anodic reaction and consumed by the cathodic reaction, corrosion can be expressed in terms of an electrochemical current. Expressing the mass balance requirement in electrochemical terms, it can be stated that the total current flowing into the cathodic reaction must be equal, and opposite in sign to, the current flowing out of the anodic reaction (Fig. 1b).

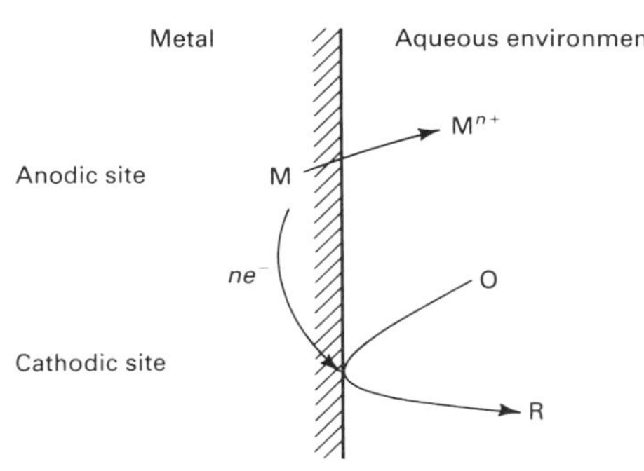

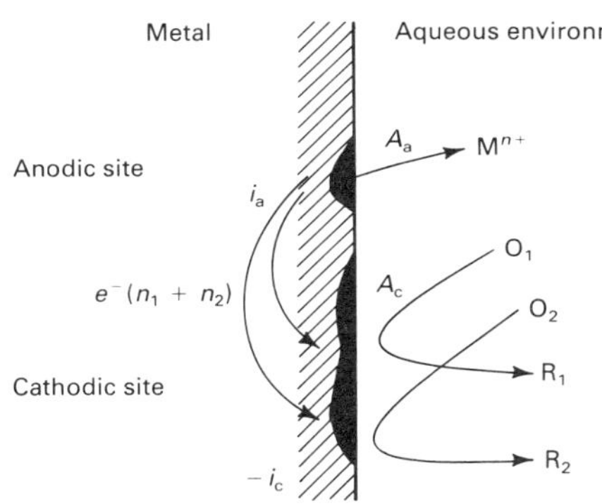

Fig. 1 Schematics of two distinct corrosion processes. (a) The corrosion process $M + O \rightarrow M^{n+} + R$ showing the separation of anodic and cathodic sites. (b) The corrosion process involving two cathodic reactions

If measurable, this current can be taken as a gage of the rate of the corrosion process and therefore the rate of metal wastage. The current, known as the corrosion current, i_{corr}, and the amount of metal corroded are related by Faraday's law:

$$i_{corr}t = \frac{nFw}{M} \quad \text{(Eq 7)}$$

where i_{corr} is expressed in amps; t is the time (in seconds) for which the current has flowed; nF is the number of coulombs (C) required to convert 1 mol of metal to corrosion product, where n is the number of electrons involved in the metal dissolution reaction ($n = 2$ for Eq 1), and F is the Faraday constant (96 480 C/mol); M is the molecular weight of the metal (in grams); and w is the mass of corroded metal (in grams).

Two additional observations can be made with regard to Fig. 1(b). First, several cathodic reactions may simultaneously support the metal corrosion; for example, in oxygenated acidic solutions, iron corrosion (Eq 1) could be simultaneously driven by the proton reduction (Eq 2) and the oxygen reduction (Eq 4). When complex alloys are involved, the metal corrosion process may also be the sum of more than one dissolution process. The corrosion current then equals the sum of the component partial currents:

$$i_{corr} = \Sigma i_a = -\Sigma i_c \quad \text{(Eq 8)}$$

Second, the area of the anodic and cathodic sites (A_a and A_c) may be very different (A_a is shown smaller than A_c in Fig. 1b). Therefore, although the anodic and cathodic currents must be equal, the respective current densities need not be:

$$i_a = -i_c \,;\, A_a \neq A_c$$

therefore:

$$\frac{i_a}{A_a} \neq \frac{i_c}{A_c} \quad \text{(Eq 9)}$$

The term i/A is a current density and will be designated I.

This inequality can have serious implications. For a smooth, single-component metal surface the anodic and cathodic sites will be separated, at any one instant, by only a few nanometers. The areas will shift with time so that the surface reacts evenly, thus undergoing general corrosion. However, such a situation often does not apply, and the presence of surface irregularities, alloy phases, grain boundaries, impurity inclusions, residual stresses, and high-resistance oxide films can often lead to the stabilization of discrete anodic and cathodic sites. Under these circumstances, metal dissolution can be confined to specific sites, and corrosion is no longer general but localized. The specific combination of a small anode and large cathode confines metal dissolution to a small number of localized areas, each dissolving with a large current density. Such a situation exists during such processes as pitting or cracking.

Aqueous corrosion is a complicated process that can occur in various forms and is affected by many chemical, electrochemical, and metallurgical variables, including:

- The composition and metallurgical properties of the metal or alloy

- The chemical (composition) and physical (temperature and conductivity) properties of the environment
- The presence or absence of surface films
- The properties of the surface films, such as resistivity, thickness, nature of defects, and coherence

The thermodynamic feasibility of a particular corrosion reaction is determined by the relative values of the equilibrium potentials, E_e, of the reactions involved. These potentials can be determined from the Nernst equation. The thermodynamics of a particular metal/aqueous system can be summarized in a potential-pH, or Pourbaix, diagram, as discussed in the section "Potential Versus pH (Pourbaix) Diagrams" of the article "Thermodynamics of Aqueous Corrosion" in this Volume. However, for this discussion, it is sufficient to state that if the corrosion reaction is to proceed such that metal M corrodes as M^{n+}, then:

$$(E_e)_{M/M^{n+}} < (E_e)_{O/R} \quad \text{(Eq 10)}$$

However, the most important questions for the corrosion engineer are, How fast does the corrosion reaction occur? Is it localized? Can it be prevented or at least slowed to an acceptable rate? To answer these questions and to determine a course of action, it is essential to have some knowledge of the steps involved in the overall corrosion process.

The overall process could be controlled by any one of several reactions, as shown in Fig. 2. Either the anodic (reaction area 1, Fig. 2) or cathodic (area 2) electron transfer reactions could be rate controlling. Alternatively, if these reactions are fast and the concentration of the cathodic reagent is low, then the rate of transport of the reagent O to the cathodic site (area 3) could be rate limiting. This situation is quite common for corrosion driven by dissolved oxygen that has limited solubility. If the metal dissolution reaction is reversible—that is, the reverse metal deposition reaction $M^{n+} + ne^- \rightarrow M$ can also occur—then the rate of transport of M^{n+} away from the anode (area 4) could also be the slow step.

The presence of corrosion films adds other complications. If the concentration of dissolved metal cations close to the electrode achieves a value at which oxides, hydroxides, or metal salts precipitate (area 5), then corrosion could become controlled by transport of M^{n+} (or O) through these porous precipitates (area 6). Alternatively, when coherent surface films form spontaneously on the metal surface by solid-state, as opposed to precipitation, reactions, then ionic transport of M^{n+}, or O^{2-}, to the film growth sites at the two interfaces (oxide/metal or oxide/solution) (area 7) will ensure very low corrosion rates. The presence of film defects in the form of pores and grain boundaries will affect the rates of these processes. Finally, it is possible, under certain circumstances, for the corrosion process to be controlled by the electronic conductivity of surface films (area 8) when the cathodic process occurs on the surface of the film. In light of these numerous possibilities for control of the corrosion process, the remainder of this article will discuss these individual processes and the laws that govern them.

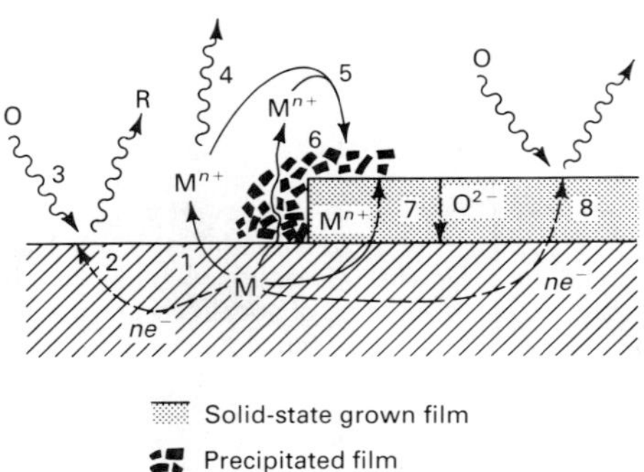

Fig. 2 Schematic of corrosion process showing various charge-transfer, film formation, and transport processes. See text for explanation of numbered reaction areas.

Activation Control

Activation control is the term used to describe control of the corrosion process shown in Fig. 2 by the electrochemical reactions given in Eq 1 and 2. The overall anodic reaction is the transfer of a metal atom from a site in the metal lattice to the aqueous solution as the cation M^{n+} or as some hydrolyzed or complexed metal cation species:

$$M_{\text{lattice}} \rightarrow M^{+}_{\text{surface}} \rightarrow M^{n+}_{\text{surface}} \rightarrow M^{n+}_{\text{solution}} \quad \text{(Eq 11)}$$

These steps are not necessarily separable experimentally. Similarly, the cathodic reaction—for example, oxygen reduction—consists of a number of steps:

$$O_2 + 2H^+ + 2e^- \rightarrow H_2O_2 \quad \text{(Eq 12)}$$

$$H_2O_2 + 2H^+ + 2e^- \rightarrow 2H_2O \quad \text{(Eq 13)}$$

The overall reactions are known as charge transfers. Either the anodic or cathodic charge-transfer reaction can control the overall corrosion rate. Both the anodic and cathodic reactions can be individually studied by using electrochemical methods in which the electrical potential applied to the electrode (or the current flowing through it) is controlled and the resulting current (or electrode potential) measured. Thus, the current-potential, or polarization, curves for both anodic and cathodic reactions can be determined. An example of an anodic polarization curve is shown in Fig. 3. This curve, for the anodic reaction, follows the Butler-Volmer equation:

$$i = i_o \left\{ \exp\left(\beta \frac{nF}{RT}\eta\right) - \exp\left(-(1-\beta)\frac{nF}{RT}\eta\right) \right\} \quad \text{(Eq 14)}$$

where R is the gas constant, T is the absolute temperature, and β is the symmetry coefficient taken to be close to 0.5. The term η is the overpotential, defined by:

$$\eta = E - E_e \quad \text{(Eq 15)}$$

and is a measure of how far the reaction is from equilibrium. At equilibrium ($E = E_e$, $\eta = 0$), no measurable current flows. However, the equilibrium is dynamic, with the rate of metal dissolution, i_a, equal to the rate of metal cation deposition, $-i_c$:

$$i_a = -i_c = i_o \quad \text{(Eq 16)}$$

where i_o is the exchange current.

If the potential is made more positive (anodic) than the equilibrium potential, then $i_a > |i_c|$ and metal dissolution proceeds. Similarly, for cathodic potentials, $i_a < |i_c|$ and metal cation deposition proceeds (Fig. 3). Over a short potential range, the two reactions oppose each other, but for sufficiently large overpotentials (η_a, anodic, and η_c, cathodic), one reaction occurs at a negligible rate, and the overpotential is then in the Tafel region, as indicated by point 1 in Fig. 3. The last

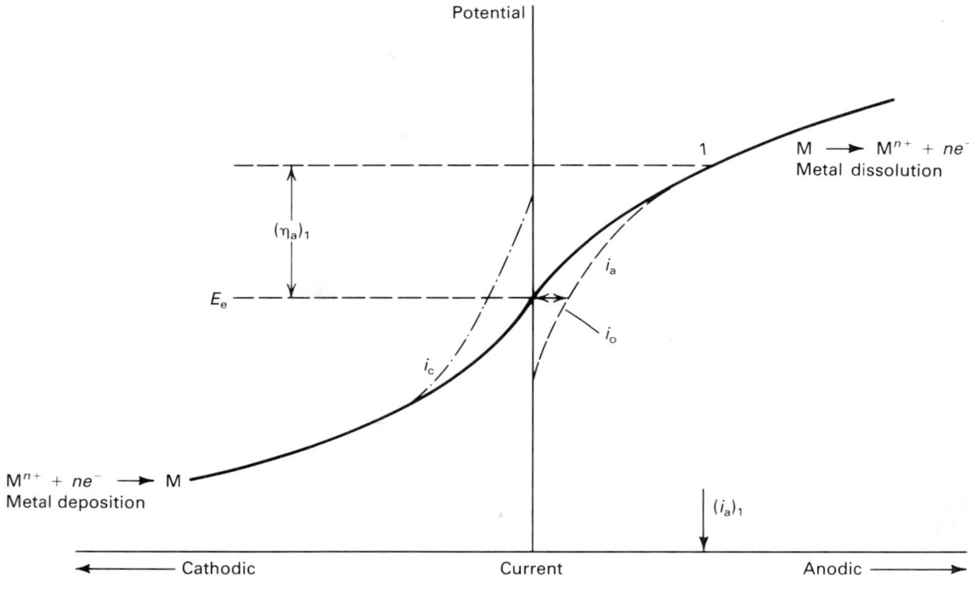

Fig. 3 Current-potential relationship for a metal dissolution ($M \rightarrow M^{n+}$)/deposition ($M^{n+} \rightarrow M$) process

term in Eq 14 can then be dropped, and the metal dissolution current density is given by:

$$i_a = i_o \exp\left(\beta \frac{nF}{RT} \eta\right) \qquad \text{(Eq 17)}$$

Taking logarithms and rearranging yields:

$$\eta_a = b_a \log\left(\frac{i_a}{i_o}\right) \qquad \text{(Eq 18)}$$

where b_a is the Tafel coefficient given by:

$$b_a = \frac{2.303RT}{\beta nF} \qquad \text{(Eq 19)}$$

and is obtained from the slope of a plot of η_a against log i_a. The intercept of this plot yields a value for i_o. Similarly, at cathodic overpotentials, a Tafel coefficient can be obtained for the metal cation deposition:

$$b_c = \frac{-2.303RT}{(1 - \beta)nF} \qquad \text{(Eq 20)}$$

A similar analysis can be performed for the cathodic process (O + $ne \rightarrow$ R), and Fig. 4 shows the two current-potential (polarization) curves.

If the two reactions are to couple together as a corrosion process, then the anodic current flowing because of metal dissolution must be counterbalanced by an equal cathodic current due to the reduction of O to R:

$$i_a = -i_c = i_{corr} \qquad \text{(Eq 21)}$$

where i_{corr} is the corrosion current. This condition can be achieved only at a single potential, the corrosion potential, E_{corr}, which must lie between the two equilibrium potentials, thus satisfying Eq 10:

$$(E_e)_a < E_{corr} < (E_e)_c \qquad \text{(Eq 22)}$$

such that the metal dissolution reaction is driven by an anodic activation overpotential:

$$\eta_a^A = E_{corr} - (E_e)_a \qquad \text{(Eq 23)}$$

and the cathodic reaction is driven by a cathodic activation overpotential:

$$\eta_c^A = E_{corr} - (E_e)_c \qquad \text{(Eq 24)}$$

The activation overpotential is a measure of how hard the anodic and cathodic reactions must be driven to achieve the corrosion current.

Two additional observations can be made with regard to Fig. 4. First the thermodynamic driving force for corrosion is equal to the difference in equilibrium potentials:

$$\Delta E_{therm} = (E_e)_c - (E_e)_a \qquad \text{(Eq 25)}$$

Generally, ΔE_{therm} is large, and the reverse reactions ($M^{n+} \rightarrow M$, $R \rightarrow O$) can be neglected. Consequently, E_{corr} is in the Tafel regions for both reactions (assuming no complications due to the presence of films).

Second, the two polarization curves are not necessarily symmetrical and are seldom identical (Fig. 4). The shape of a curve is determined by the exchange current and the Tafel coefficient. The latter is determined by n and β in Eq 19 and 20. As shown in Fig. 4, the metal dissolution/deposition reaction has a large i_o, and the anodic and cathodic branches are close to symmetrical, as expected for β close to 0.5. The consequence of a large i_o is that the current-potential curve is steep, and only small overpotentials are required to achieve large currents. By contrast, the current-potential relationship for the cathodic reaction is shallow due to a small i_o, and the anodic and cathodic branches are not symmetrical. Rather than attempt to interpret this lack of symmetry in terms of an ill-defined symmetry coefficient, it is sufficient for this discussion to know that it is taken care of in the values of the Tafel coefficients.

Because both reactions are occurring on different sites on the same surface (Fig. 1a), the corrosion current cannot be measured by coupling the material to a current-measuring device. The corrosion potential can be measured against a suitable reference electrode by using a voltmeter with an input impedance high enough to draw no current in the measuring circuit.

The actual value of E_{corr} cannot be predicted from the equilibrium potentials and therefore has no basic thermodynamic meaning. Figure 4 shows that its value is determined by the shape of the current-potential relationship for the two reactions and therefore by the kinetic parameters (i_o, β, n) for the two reactions. Because its value is determined by the properties of more than one reaction, the corrosion potential is often termed a mixed potential.

In the literature, diagrams such as Fig. 4 are often plotted in the form log i versus E. The algebraic sign of the cathodic current is neglected so that the anodic and cathodic currents can both be plotted in the same quadrant (Fig. 5). Such diagrams are generally called Evans diagrams. The two linear portions in the log $|i|$ versus E curves are the Tafel regions with slopes given by Eq 19 and 20. The exchange currents for the two reactions can be obtained by extrapolating the Tafel lines back to the respective equilibrium potentials (Fig. 5). Whether such diagrams are plotted linearly (i versus E) or in the logarithmic form is simply a matter of convenience. Sometimes, in the logarithmic plots, the nonlinearity

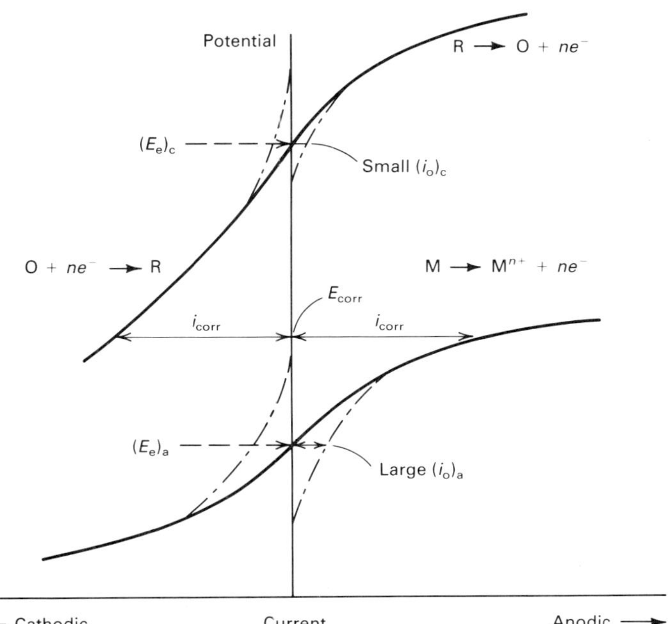

Fig. 4 Current-potential relationships for a metal dissolution/deposition and an accompanying redox reaction showing how the two reactions couple together at the corrosion potential, E_{corr}

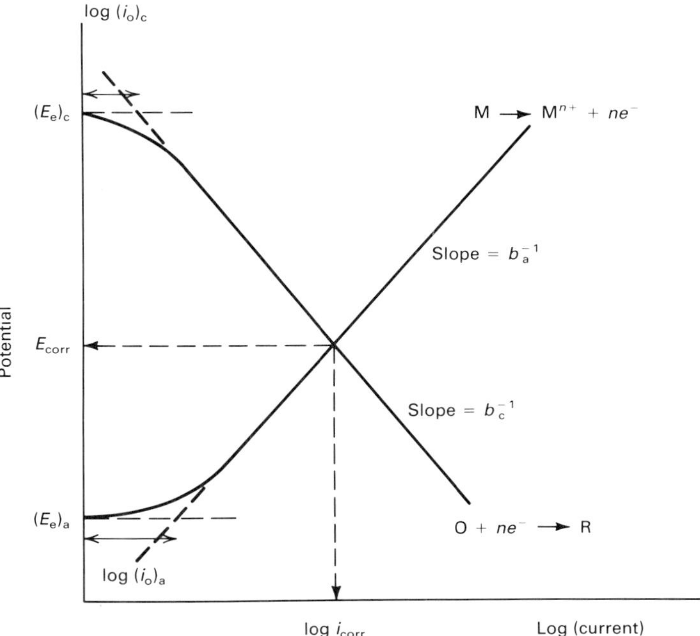

Fig. 5 Evans diagram for the corrosion process M + O → M^{n+} + R

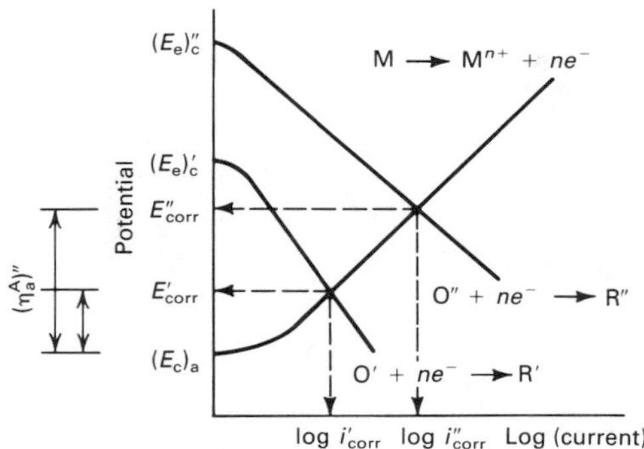

Fig. 6(a) Evans diagram for a metal dissolution coupled separately to two cathodic reactions with distinctly different equilibrium potentials, $(E_e)''_c$ and $(E_e)'_c$

close to the equilibrium potentials is ignored, and the curves are plotted as totally linear.

The intersection of the two polarization curves in the Evans diagrams gives a value for the corrosion current. This is true whatever the shape of the curves and irrespective of the rate-determining process. Such diagrams can be used to illustrate the impact of a variety of parameters on the corrosion process.

Thermodynamic Driving Force. Figure 6(a) shows the same dissolution process driven by two different cathodic reactions. Recalling the definition of ΔE_{therm} from Eq 25, the following can be written:

$$\Delta E'_{therm} < \Delta E''_{therm} \quad \text{(Eq 26)}$$

giving:

$$i'_{corr} < i''_{corr} \quad \text{(Eq 27)}$$

That is, the bigger the difference in equilibrium potentials, the larger the corrosion current. The anodic activation overpotential for the first reaction $[E'_{corr} - (E_e)_a]$ is less than that for the second reaction $[E''_{corr} - (E_e)_a]$, and therefore the corrosion current for the second reaction is larger than for the first reaction, as shown in Eq 27.

$$(\eta^A_a)'' > (\eta^A_a)' \quad \text{(Eq 28)}$$

Kinetics of the Charge Transfer Reactions. The value of ΔE_{therm} is not the only parameter controlling the corrosion rate. Figure 6(b) shows tne same situation as in Fig. 6(a) except the two cathodic reactions possess very different polarization characteristics. Despite the fact that $(E_e)''_c > (E_e)'_c$, the activation overpotential, $(\eta^A_a)''$, is less than $(\eta^A_a)'$; therefore, the corrosion couple with the largest thermodynamic driving force produces the lowest corrosion current. Figure 6(b) shows that this can be attributed to the differences in exchange current, i_o, and Tafel coefficient, b_c, for the two cathodic reactions. This situation often occurs for the corrosion of a metal in acid compared to its corrosion in dissolved oxygen. Even though the thermodynamic driving force is greater for corrosion in dissolved oxygen, corrosion often proceeds more quickly in acid. This is due to the slowness of the kinetics of oxygen reduction and can be appreciated by comparing the kinetic characteristics for the two processes on iron. Thus, $(I_o)_{H^+/H_2} = 10^{-3}$

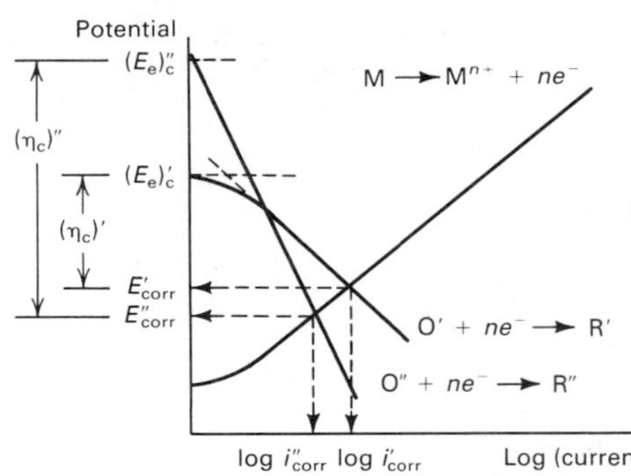

Fig. 6(b) Evans diagram for a metal dissolution coupled separately to two cathodic reactions, in which the impact of relative kinetics is greater than the thermodynamic driving force, ΔE_{therm}

to 10^{-2} A/m² and $(b_c)_{H^+/H^2} \sim 120$ mV/decade compared to $(I_o)_{O_2/H_2O} \sim 10^{-10}$ A/m² and $(b_c)_{O_2/H_2O} > 120$ mV/decade.

Rate Control by the Anodic or Cathodic Reaction. The overall rate of corrosion will be controlled by the kinetically slowest reaction, that is, the one with the smallest exchange current, i_o, and/or largest Tafel coefficient. The significance of this point and its importance in determining which reaction is rate controlling can be appreciated from Fig. 6(c), in which $(i_o)_a > (i_o)_c$ and $b_a < b_c$. This leads to a large difference in activation overpotentials with $\eta^A_a \ll \eta^A_c$. This means the cathodic reaction is strongly polarized and must be driven hard to achieve the corrosion current. However, the anodic reaction remains close to equilibrium, requiring only a small overpotential to achieve the corrosion current. Under these conditions, the corrosion potential lies close to the equilibrium potential for the kinetically fastest reaction. If the cathodic reaction was the faster, then $E_{corr} \rightarrow (E_e)_c$, and metal dissolution would be rate controlling. If the kinetics of the two reactions are close, the corrosion potential will be approximately equidistant between the two equilibrium potentials, and the corrosion reaction will be under mixed anodic/cathodic control, as shown in Fig. 5.

The corrosion of iron in dissolved oxygen can be used to illustrate this point. For the metal dissolution reaction, $(I_o)_{Fe/Fe^{2+}} \sim 10^{-5}$ to 10^{-4} A/m² and $(b_a)_{Fe/Fe^{2+}} \sim 50$ to 80 mV/decade, whereas for oxygen reduction, $(I_o)_{O_2/H_2O} \sim 10^{-10}$ A/m²

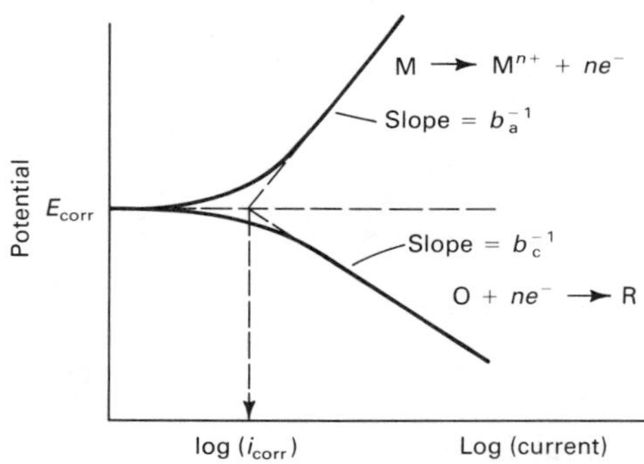

Fig. 7 Plot of the total current ($i_T = i_a + i_c$) versus potential showing the extrapolation of the Tafel regions to the corrosion potential, E_{corr}, to yield the corrosion current, i_{corr}

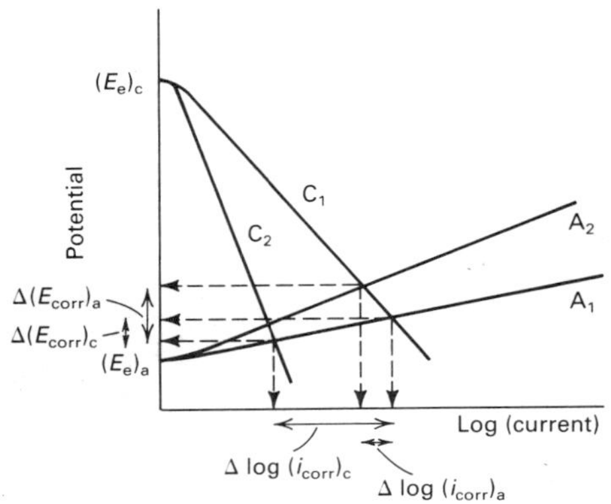

Fig. 6(c) Evans diagram showing the impact on the corrosion current, i_{corr}, and potential, E_{corr}, of varying the kinetics of a fast metal dissolution (A_1, A_2) or a slow cathodic process (C_1, C_2)

and $(b_c)_{O_2/H_2O} > 120$ mV/decade. Consequently, oxygen reduction should be rate controlling, and the corrosion potential would be expected to be close to the metal dissolution equilibrium potential.

Figure 6(c) shows the effects of changing the kinetics of the two reactions. Changes in the kinetics of the fast anodic reaction are reflected in variations in the value of E_{corr} but have little effect on i_{corr}; however, changes in the kinetics of the slow cathodic reaction have a large impact on the corrosion current but have little effect on the corrosion potential. Such effects can sometimes be used as diagnostic tests for ascertaining the rate-determining step. The maximum benefit in attempting to slow corrosion can be gained by attending to the rate-determining reaction. However, such measurements may not be unequivocal in the presence of corrosion films.

Measurement of Corrosion Rates

A common method of measuring corrosion rates is simply to expose a carefully weighed piece of the material to the corrosion environment for a known length of time, remove and reweigh it, and calculate the mass of metal lost. This is not always convenient in industrial applications because of the difficulty in placing, removing, and replacing metal coupons. However, it is possible to make use of the fact that corrosion is electrochemical in nature and to employ electrochemical methods to measure the corrosion rate.

When attempting an electrochemical measurement of the corrosion rate, one of the problems encountered is the desire to measure the current flowing at the corrosion potential. At this potential, no current will flow through an external measuring device (as discussed above). Consequently, any electrochemical attempt to measure i_{corr} will rely on current measurements at potentials other than the corrosion potential. An approximation or extrapolation is then made to estimate the current flowing internally at the corrosion potential.

The Tafel Method. As mentioned in the discussion of Fig. 4, the corrosion potential is generally in the Tafel region, in which the anodic and cathodic reactions are both proceeding under conditions appropriate for a Tafel analysis. Consequently, the polarization curves for both processes are determined by applying potentials well away from the corrosion potential, plotting the logarithm of the current against overpotential as for the Tafel analysis (Eq 18), and then extrapolating the currents in the two Tafel regions to the corrosion potential to obtain the corrosion current. The method is illustrated in Fig. 7, in which, as with the Evans diagram, both currents are plotted in the same quadrant. The polarization curve shown in Fig. 7 is in the form obtained experimentally; as a result, the current at E_{corr} passes through 0. In the Evans diagrams shown in Fig. 5 and 6, i_a and i_c are shown independently, although it is generally not possible to measure them experimentally. The current measured in the external circuit and plotted in Fig. 7 is always the sum $i_a + i_c$ (= 0 at E_{corr}).

A simplified application of this method can be used to estimate the corrosion current from a simple measurement of the corrosion potential, because the latter may be the only measurable parameter in an industrial system. In this case, values of the exchange current (i_o), Tafel coefficients (b_a,b_c), and equilibrium potential (E_e) for the metal dissolution reaction must be known from a previous experiment. Combining Eq 18 and 23 yields:

$$E_{corr} = (E_e)_a + b_a \log\left(\frac{i_{corr}}{i_o}\right) \qquad \text{(Eq 29)}$$

or

$$i_{corr} = i_o \exp\left\{2.303\,\frac{E_{corr} - (E_e)_a}{b_a}\right\} \qquad \text{(Eq 30)}$$

Complications arise when corrosion films are present or when corrosion is not uniform.

The linear polarization method is applicable when corrosion occurs under activation control. As opposed to the Tafel method, in which a large potential perturbation is applied to the system and seriously disturbs it, the linear polarization method uses only a small potential perturbation, $\pm\Delta E$ ($\lesssim$10 mV), for the freely corroding situation that occurs at E_{corr}. This small perturbation makes the method appropriate for *in situ* measurements. The current measured in the external circuit then equals the change in corrosion current, $\pm\Delta i$, caused by the small perturbation. Because both reactions proceed in their respective Tafel regions and in the vicinity of the corrosion potential, the currents are exponentially dependent on potential. For a small enough potential range ($\lesssim$20 mV), these exponentials can be linearized, giving an approximately linear current-potential relationship. The relationship between Δi and ΔE can then be obtained geometrically, as indicated in Fig. 8, which shows an expansion of the current-potential relationships around the corrosion potential. The terms s_a and s_c are the respective slopes of the anodic and cathodic curves at $E = E_{corr}$. Thus, the following can be written:

$$\Delta i = wy = wx + xy = \frac{s_a - |s_c|}{s_a|s_c|}\cdot \Delta E \qquad \text{(Eq 31)}$$

Rearranging and differentiating Eq 30 yields:

$$\frac{di_{corr}}{dE} = \frac{2.303}{b_a}\,i_{corr} = \frac{1}{s_a} \qquad \text{(Eq 32)}$$

for the anodic reaction. A similar process for the cathodic reaction gives:

$$\frac{2.303}{b_c}\,i_{corr} = \frac{1}{|s_c|} \qquad \text{(Eq 33)}$$

Substituting for s_a and s_c in Eq 31 yields:

$$i_{corr} = \frac{1}{2.303}\,\frac{b_a|b_c|}{b_a + |b_c|}\cdot\frac{\Delta i}{\Delta E} \qquad \text{(Eq 34)}$$

The quantity $\Delta E/\Delta i$ is termed the polarization resistance. Again, a knowledge of the Tafel coefficients is required before the method can be applied.

The corrosion current can be converted into a mass flux by the application of Faraday's law (Eq 7). If the density of the material is known, a penetration rate (distance/time) can be obtained. Corrosion rates are generally expressed in millimeters per year or mils per year.

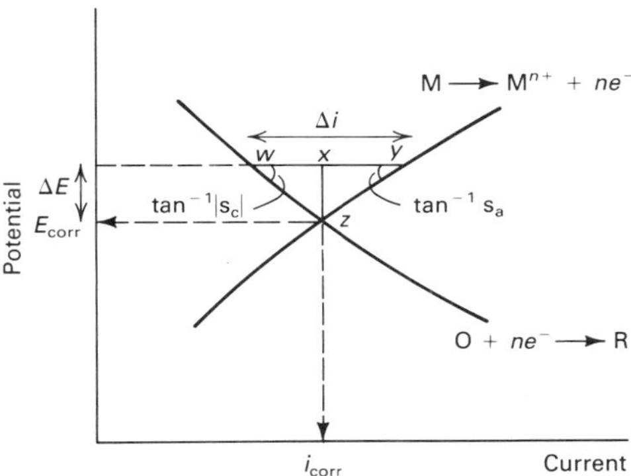

Fig. 8 Plot of the current-potential relationships, expanded around the corrosion potential, showing their linearization (for small values of ΔE) to obtain i_{corr} by using the linear polarization technique

Mass Transport Control

It has been assumed in this article that the corrosion rate is controlled by either the anodic or cathodic charge-transfer process (reaction area 1 or 2, Fig. 2). However, if the cathodic reagent at the corrosion site is in short supply, then mass transport of this reagent could become rate controlling (reaction area 3, Fig. 2). Under these conditions, the cathodic charge-transfer process is fast enough to reduce the concentration of the cathodic reagent at the corrosion site to a value less than that in the bulk solution. Because the rate of the cathodic reaction is proportional to the surface concentration of reagent, the reaction rate will be limited (polarized) by this drop in concentration. For a sufficiently fast charge transfer, the surface concentration will fall to zero, and the corrosion process will be totally controlled by mass transport.

Because the corrosion rate is now determined at least in part by the rate of transport (the flux) of reagent to the corrosion site, this flux needs to be calculated. This can be accomplished by using the Nernst diffusion layer treatment, a simplification of the Fick's diffusion law treatment. The model is illustrated in Fig. 9. Because the concentration of reagent O is lower at the surface than in the bulk of solution, O will be transported down the chemical gradient at a rate (the flux J) proportional to the gradient of the concentration-distance profile. This is a statement of Fick's first law, which applies under steady-state conditions, that is, surface concentration and concentration gradient constant with time:

$$J = -D\left(\frac{\partial C_O}{\partial x}\right) \qquad \text{(Eq 35)}$$

where D is the proportionality constant known as the diffusion coefficient (the negative sign accounts for the fact that the flux is down the gradient), and C_O is the reagent concentration at a point x. The solid line in Fig. 9 represents the concentration profile calculated from Fick's treatment.

A simpler analysis can be achieved by linearizing the profile according to the Nernst diffusion layer treatment, as represented by the dashed-dotted line in Fig. 9. The resistance to mass transport lies within this diffusion layer, and the linearization yields a demarcation line at a distance δ from the surface such that, for $x > \delta$, the bulk concentration is maintained by convective processes. By contrast, for $x \leq \delta$, reagent O is transported to the surface by diffusion only. This solution layer is called the diffusion layer, and its thickness is determined by the solution velocity.

Using this simplified treatment, Eq 35 can be written as:

$$J = \frac{-D(C_O^s - C_O^b)}{\delta} \qquad \text{(Eq 36)}$$

where C_O^s is the reagent concentration at the corroding surface ($x = 0$), and C_O^b is the concentration for $x \geq \delta$. For the steady state to be maintained, all the reagent transported down the gradient must react electrochemically, giving a current:

$$\frac{i_c}{nF} = \frac{-D_O(C_O^s - C_O^b)}{\delta} \qquad \text{(Eq 37)}$$

where the term nF takes care of the chemical to electrochemical conversion (see Eq 7, Faraday's

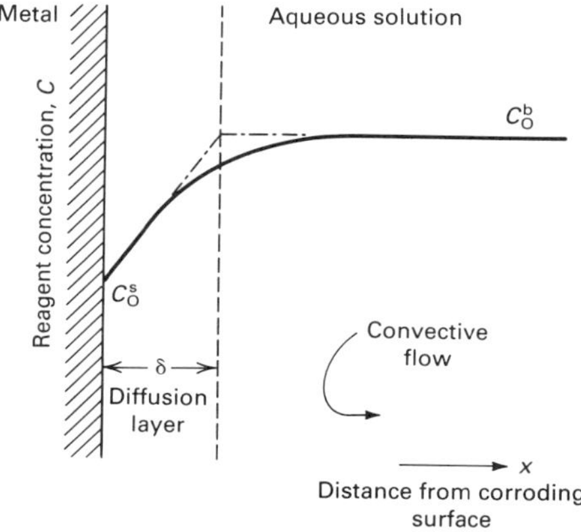

Fig. 9 Concentration-distance profile for the cathodic reagent O, depleted at the metal surface. The solid line shows Fick's treatment, and the dashed-dotted line indicates the approximation known as the Nernst diffusion layer treatment.

law). Under the limiting conditions $C_O^s \rightarrow O$, a limiting or maximum current is obtained:

$$(i_c)_{\text{lim}} = \frac{nFD_O C_O^b}{\delta} \qquad \text{(Eq 38)}$$

Because this is the maximum cathodic current that can flow, it represents the maximum achievable corrosion rate:

$$(i_{\text{corr}})_{\text{max}} = (i_c)_{\text{lim}} = \frac{nFD_O C_O^b}{\delta} \qquad \text{(Eq 39)}$$

When corrosion occurs at this limit, the corrosion rate can be increased or decreased only by varying the bulk concentration of reagent, C_O^b, or the diffusion layer thickness, δ. For nonlimiting conditions, the corrosion current will be given by Eq 37.

The effect of the concentration polarization can be seen by considering Fig. 10. For small shifts from the equilibrium potential (point 1), $C_O^s = C_O^b$, there is no limitation on the reagent supply. Charge transfer is completely rate controlling, and the overpotential is purely an activation overpotential:

$$\eta_T = \eta^A \qquad \text{(Eq 40)}$$

For larger shifts from the equilibrium potential, $C_O^s < C_O^b$ (point 2), and the current is correspondingly less than that expected on the basis of activation control; that is, the current follows the solid line as opposed to the dashed-dotted line. The current is both activation and concentration polarized, and the overpotential is the sum of an activation and a concentration overpotential:

$$\eta_T = \eta^A + \eta^C \qquad \text{(Eq 41)}$$

For a sufficiently large shift from equilibrium, the current becomes independent of potential, and the concentration overpotential becomes infinite (point 3). The corrosion rate is now at a maximum given by Eq 39.

The impact of various parameters on a corrosion process proceeding under mass transport or mixed activation-transport control can be assessed by the use of an Evans diagram, as shown in Fig. 11. Three situations are considered. For cathodic curve 1, corrosion occurs with the cathodic reaction totally mass transport controlled; that is, $C_O^s = 0$. If the solution is now stirred or made to flow, the thickness of the diffusion layer δ (Fig. 9) will decrease, and the corrosion current, given by Eq 39, will increase as shown (curve 2). The corrosion potential will shift to more positive values. This shift in E_{corr} is a consequence of the decrease in overpotential for the cathodic reaction due to the decrease in concentration overpotential:

$$\eta_T = E_{\text{corr}} - (E_e)_c = \eta^A + \eta^C \qquad \text{(Eq 42)}$$

because:

$$(\eta^C)_2 < (\eta^C)_1$$
$$(\eta_T)_2 < (\eta_T)_1 \qquad \text{(Eq 43)}$$

For more vigorous stirring, the concentration overpotential becomes zero because the flux of reagent O to the corroding surface is now fast enough to maintain the surface concentration equal to the bulk concentration. The reaction becomes activation controlled again (curve 3). Fluid velocity no longer affects corrosion rate.

Such changes in E_{corr} and i_{corr} with stirring or solution velocity can be used to indicate whether mass transport control is operative. If the anodic, as opposed to the cathodic, reaction was mass transport controlled, E_{corr} would shift to more cathodic (negative) values with increased stirring or flow rate.

Equation 38 indicates that for mass transport control by the cathodic reaction the rate is directly proportional to the concentration of cathodic reagent and is inversely proportional to the thickness of the diffusion layer, which is determined by the fluid velocity (assuming the solution properties do not change). Corrosion in dissolved oxygen often proceeds in this manner, because the concentration of oxygen in solution is limited. Using this situation to demonstrate how velocity affects corrosion in flowing environments, Eq 37 can be written as:

$$\frac{i_{\text{corr}}}{nF} = m_c\,(C_O^b - C_O^s) \qquad \text{(Eq 44)}$$

where m_c is a mass transport coefficient and would be given by D_O/δ if the Nernst diffusion layer treatment had been employed. As discussed above, to maintain the steady state, all the oxygen reaching the corroding surface is consumed, and the corrosion rate is given by:

$$\frac{i_{\text{corr}}}{nF} = k_c C_O^s \qquad \text{(Eq 45)}$$

where k_c is the potential-dependent rate constant for the electron transfer reaction. The relationship between k_c and i_o, η and b_a can be appreciated by comparing Eq 45 and 30. Eliminating C_O^s between Eq 44 and 45 yields:

$$i_{\text{corr}} = nF\, C_O^b \left\{\frac{m_c k_c}{(m_c + k_c)}\right\} \qquad \text{(Eq 46)}$$

where the constant k_c can be considered the activation control parameter, and m_c can be considered the mass transport control parameter. Whether or not activation kinetics or mass transport is rate determining is determined by the relative values of m_c and k_c. If $m_c \gg k_c$, the bracketed term in Eq 46 reduces to k_c, and the corrosion current is activation controlled. For $k_c \gg m_c$, the term reduces to m_c, and the corrosion current becomes mass transport controlled. For $m_c \sim k_c$, Eq 46 cannot be simplified, and corrosion would be under joint control.

If mass transport is a contributor to corrosion control, then a knowledge of the dependence of m_c on flow rate is required. This dependence is found experimentally, and its form varies, depending on the geometry of the system. In general, this dependence takes the form:

$$i_{\text{corr}} \propto f^n \qquad \text{(Eq 47)}$$

where f is the flow rate, and n is a constant that depends primarily on the geometry of the system. Confining attention to flow over a flat plate, n is 0.33 for laminar (smooth, Re < 2200) flow and ~0.7 for turbulent (Re > 2200) flow, where Re is the Reynold's number (Eq 48).

The variation of the diffusion layer thickness, δ, with flow conditions depends on flow rate as well as on solution properties, such as the kinematic viscosity (ν), the diffusion coefficient of the reagent (D), and the geometry of the system (L). These effects can be accounted for by introducing two dimensionless parameters, the Reynolds

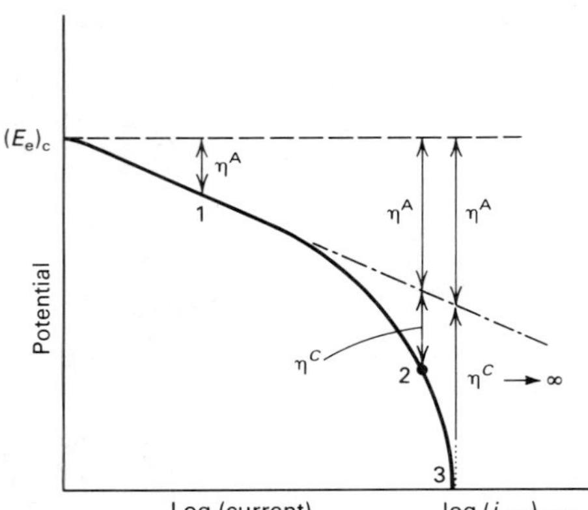

Fig. 10 Polarization curve for the cathodic process showing activation polarization (point 1), joint activation-concentration polarization (point 2), and transport-limited corrosion control (point 3)

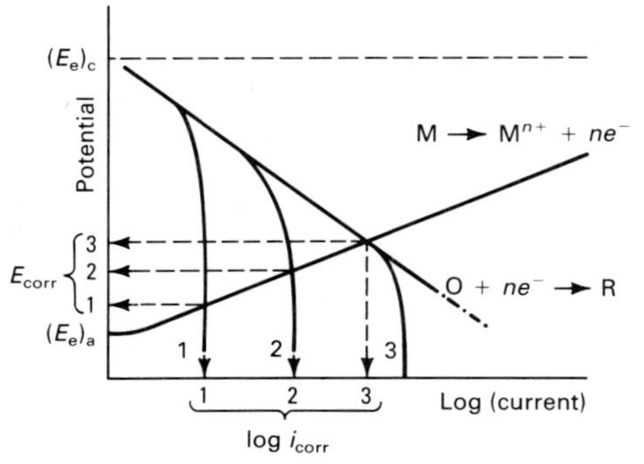

Fig. 11 Evans diagram for a corrosion process initially controlled by the transport of cathodic reagent to the corroding surface (line 1). Lines 2 and 3 show the effect of increasing the transport rate of reagent.

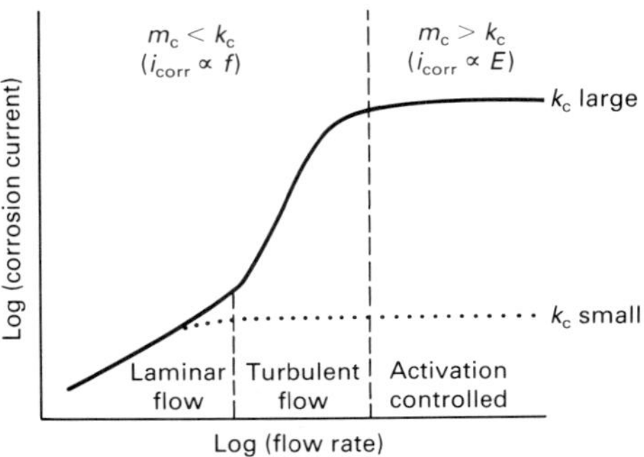

Fig. 12 Impact of flow rate on corrosion current showing the regions of laminar and turbulent flow and the switch from transport to activation control at high flow rates

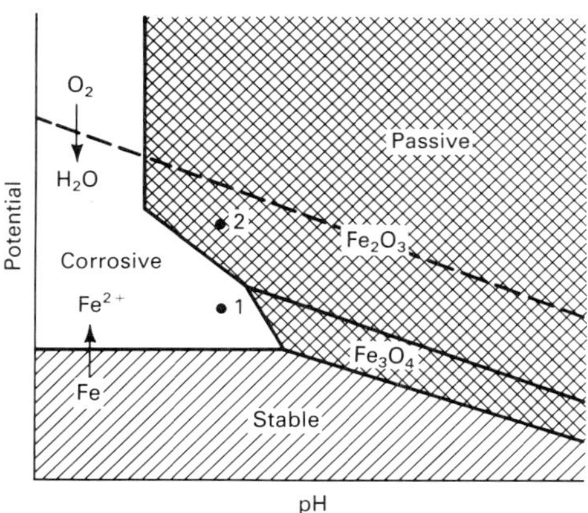

Fig. 13 Pourbaix diagram for the iron/water/dissolved oxygen system showing the effect of potential in moving the system from a corrosive (active) region (point 1) to a passive region (point 2)

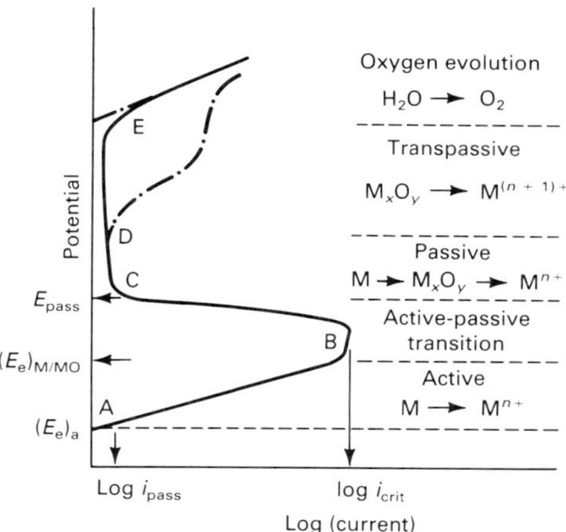

Fig. 14 Polarization curve for a metal/metal ion system that undergoes an active to passive transition. See text for details.

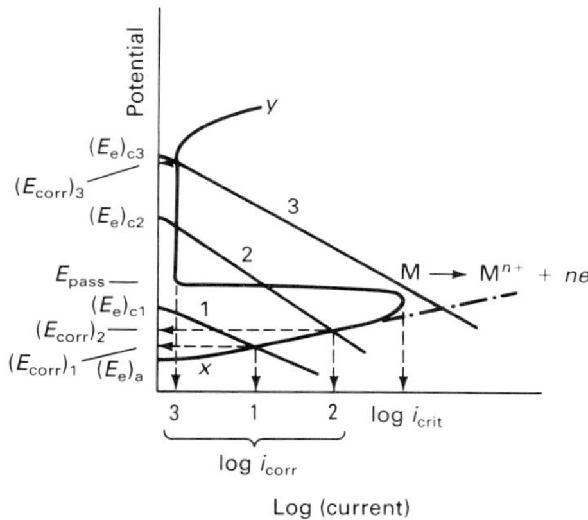

Fig. 15 Impact of various cathodic reactions on the corrosion current and potential for a metal capable of undergoing an active-passive transition

number, Re, and the Schmidt number, Sc, given by:

$$\text{Re} = \frac{fL}{\nu} \qquad \text{(Eq 48)}$$

$$\text{Sc} = \frac{\nu}{D} \qquad \text{(Eq 49)}$$

It can be shown that for flow over a smooth, flat, corroding surface:

$$(i_{\text{corr}})_{\max} = 0.62\ nFD_{\text{O}}C_{\text{O}}^{\text{b}}(\text{Re})^{0.5}(\text{Sc})^{0.33} \qquad \text{(Eq 50)}$$

showing that the corrosion rate is proportional to $f^{0.5}$. Laminar flow can be maintained only up to a certain Reynolds number (or flow rate if L and ν are constant), beyond which the flow becomes turbulent and the dependence on flow rate increases.

For still higher flow rates, the condition $m_c \gg k_c$ can be achieved, and the corrosion rate will become activation controlled and therefore independent of flow rate. This is equivalent to the situations discussed in Fig. 11, in which the corrosion rate (current) reached a constant value as the effect of concentration overpotential was removed.

These three regions are shown schematically in Fig. 12. The solid line shows the effect of flow rate when the anodic corrosion reaction is fast (k_c large) and a large flow rate is required to achieve activation control ($m_c \gg k_c$). The dotted line shows the behavior expected for a slow anodic reaction (k_c small) when only a low flow rate is required to achieve activation control.

Passivation

Thus far in the discussion on transport effects, only the transport of the cathodic reagent (area 3, Fig. 2) has been considered. The transport of metal dissolution product (area 4, Fig. 2) also affects the corrosion rate but in a different way. If the corrosion product is allowed to build up at the surface, supersaturation with regard to solid oxides and hydroxides can occur, leading to film formation reactions (area 5, Fig. 2).

The effects of film formation have been referred to above. With regard to the Evans diagrams shown in Fig. 5, 6, and 11, it can be seen that very substantial corrosion rates would be achieved if the kinetics of both the anodic and cathodic reactions were fast. Fortunately, in many cases, the metal dissolution rate decreases to low values once the potential is raised above a critical value. The metal is said to be passivated.

Passivation can occur when the corrosion potential exceeds (becomes more positive than) the potential corresponding to equilibrium between the metal and one of its oxides/hydroxides:

$$E_{\text{corr}} > (E_{\text{e}})_{\text{M/MO}} \qquad \text{(Eq 51)}$$

Inspection of the Pourbaix diagram for the particular metal/metal oxide/aqueous solution system shows that this condition moves the potential into the oxide stability region (Fig. 13). For point 1, $E_{\text{corr}} < (E_{\text{e}})_{\text{M/MO}}$ and corrosion of bare metal is expected, but for point 2, $E_{\text{corr}} > (E)_{\text{M/MO}}$, the metal should be oxide covered and passive. Under passive conditions, the corrosion rate will be dependent on the oxide film properties.

The current-potential, or polarization, curve for the anodic process is shown in Fig. 14 and can be divided into a number of regions. In region AB, the active region, metal dissolution occurs unimpeded by the presence of surface films. The current, i_a, should conform to the Tafel relationship (Eq 17), and its extrapolation back to $(E_{\text{e}})_{\text{a}}$ would yield a value of $(i_{\text{o}})_{\text{a}}$. At a potential B, shown in Fig. 14 to coincide with $(E_{\text{e}})_{\text{M/MO}}$, there is a departure from the Tafel relationship that becomes more pronounced as the potential increases, leading eventually to a decrease in current to a low value. The electrode is said to have undergone an active-passive transition and, by point C, has become passive. The potential at point B may or may not correspond to the potential $(E_{\text{e}})_{\text{M/MO}}$. Thermodynamics demands only that the condition given in Eq 51 be satisfied for passivation to occur. The maximum current achieved immediately before the transition is termed the critical passivating current density. This can be considered as the current density required to generate a sufficiently high surface concentration of metal cations such that the nucleation and growth of the surface film can proceed.

The potential at which the current falls to the passive value is called the passivation potential. It corresponds to the onset of full passivity and is sometimes called the Flade potential. In most cases, it has no thermodynamic significance. For gold, platinum, and silver, it is close to $(E_{\text{e}})_{\text{M/MO}}$, but for most other metals, the passivation potential is much more positive than this equilibrium value.

For $E > E_{\text{pass}}$, the metal is said to be in the passive region. In this region, the current is independent of potential, and metal dissolution occurs at a constant rate. Two possible explanations can be offered for this constancy. First, dissolution in the passive region occurs by the transport of ionic species through the film (reaction area 5, Fig. 2) under the influence of the electric field across the film. The increase in potential through the passive region is accompanied by a progressive thickening of the film such that the electric field within the oxide, and therefore the dissolution current, remain constant. Second, the current is controlled by the rate of dissolution of the film (a chemical, as opposed to an electrochemical, process) and is potential independent. The current is just sufficient to replace the dissolving film.

For potentials greater than point E, oxygen evolution can occur on the outside of the oxide film by the reaction:

$$4\text{OH}^- \rightarrow \text{O}_2 + 2\text{H}_2\text{O} + 4e^- \qquad \text{(Eq 52)}$$

For this last reaction to occur the film must be electronically conducting. This is possible because the passive films formed are commonly thin (nanometers) and possess semiconducting or even metallic properties.

The dashed-dotted line in Fig. 14, in the potential region D to E, corresponds to the phenomenon of transpassivity. In this region, the oxide film starts to dissolve oxidatively, generally as a hydrolyzed cation in a higher oxidation state. An example would be the further dissolution of the passive film on chromium, Cr_2O_3 with chromium in the +3 oxidation state, to chromate, CrO_4^{2-} with chromium in the +6 state.

The current in the passive region, then, is very dependent on the physical (conductivity, defect structure) and chemical (oxidation state) properties of the oxide. If the oxide were not present, then the current at potentials in the region C to E would be given by values obtained from the extrapolation of the active dissolution region, that is, line AB. These values would be extremely large. Any disruption of the passive film is a dangerous situation, and film breakdown at localized points leads to the initiation of such localized corrosion processes as pitting and cracking. These processes are characterized by very high local rates of metal dissolution and can lead to very rapid penetration of metal structures. Such processes will be discussed in the Section "Forms of Corrosion" in this Volume.

The following discussion will describe the properties of the cathodic reaction required to force the corrosion potential into the passive region, thus causing passivation and maintaining the corrosion current equal to the passive dissolution current. For passivation to occur, two conditions must be met:

- The equilibrium potential for the cathodic reaction must be greater than E_{pass}, the passivation potential
- The cathodic reaction must be capable of driving the anodic reaction to a current in excess of the critical passivation current, i_{crit}

Three possible situations are shown in Fig. 15. The dashed-dotted line shows the anodic polarization curve for the metal dissolution ($M \rightarrow M^{n+} + ne^-$), and lines 1, 2, and 3 show the cathodic polarization curves for three different cathodic processes ($O_n + ne^- \rightarrow R_n$).

Consider cathodic reaction 1 (Fig. 15), in which $(E_e)_{c1} < E_{pass}$. Because the corrosion potential must lie between $(E_e)_a$ and $(E_e)_{c1}$ for the two reactions to form a corrosion couple (Eq 22), the required condition for passivation, $E_{corr} > E_{pass}$, cannot be achieved. Therefore, the corrosion potential stays in the active region, and the metal will actively corrode.

For cathodic reaction 2 (Fig. 15), the condition $(E_e)_{c2} > E_{pass}$ is met, but the two polarization curves intersect at an anodic current less than i_{crit} (i_{crit} is the minimum current density required to supply a sufficient concentration of M^{n+} at the surface to initiate film growth by supersaturation with respect to the passivating oxide). Again, $E_{corr} < E_{pass}$, and the metal corrodes in the active region at a higher corrosion current than before.

For cathodic reaction 3 (Fig. 15), the conditions $(E_e)_{c3} > E_{pass}$ and $i > i_{crit}$ are both met. Therefore, $E_{corr} > E_{pass}$ and the metal passivates, with the corrosion current decreasing to a low value equal to the passive dissolution current.

Mild oxidizing agents ($\Delta E_{therm} = (E_e)_c - (E_e)_a$, small) will allow active corrosion, and strong oxidizing agents (ΔE_{therm} large) are required to force the metal or alloy into the passive region. As an example, steel corrosion in strong acid may proceed in the active region at a high rate, but in dissolved oxygen, the steel will passivate and corrode passively at an insignificant rate.

REFERENCES

1. L.S. Van Delinder, Ed., *Corrosion Basics—An Introduction*, National Association of Corrosion Engineers, 1984
2. L.L. Shrier, *Corrosion*, George Newnes Ltd., 1963
3. J.M. West, *Electrodeposition and Corrosion Processes*, 2nd ed., Van Nostrand Reinhold, 1970
4. J.M. West, *Basic Corrosion and Oxidation*, 2nd ed., Ellis Horwood, 1986
5. G. Wranglen, *An Introduction to Corrosion and Protection of Metals*, Institut für Metallskydd, 1972

Effects of Environmental Variables on Aqueous Corrosion

D.C. Silverman and R.B. Puyear, Monsanto Company

CORROSION involves the interaction (reaction) between a metal or alloy and its environment. Corrosion is affected by the properties of both the metal or alloy and the environment. In this discussion, only the environmental variables will be addressed, the more important of which include:

- pH (acidity)
- Oxidizing power (potential)
- Temperature (heat transfer)
- Velocity (fluid flow)
- Concentration (solution constituents)

The influence of biological organisms on these environmental variables is also an important consideration, as explained in the Appendix "Biological Effects" in this article. Additional information is available in the references cited in this article and in the Section "Specific Alloy Systems" in this Volume.

Before discussing the relationships, the expanded portion of the potential-pH diagram of iron at 25 °C (77 °F) shown in Fig. 1 should be considered. As discussed in the article "Thermodynamics of Aqueous Corrosion" (see the section "Potential Versus pH (Pourbaix) Diagrams") in this Volume, these diagrams are thermodynamic and show the most stable state of the metal in an aqueous solution. The dependence of iron corrosion on oxidizing power (emf), acidity (pH), temperature, and species concentration is illustrated in Fig. 1. For example, suppose the corrosion potential lies at −0.5 V (standard hydrogen electrode, SHE) at a pH of 8. The most stable state of iron is Fe^{2+}, indicating that iron dissolution is possible. If the pH is increased to 10 (the acidity is decreased), the most stable state becomes magnetite (Fe_3O_4), and most likely, iron corrosion would greatly decrease. If the pH is then decreased to about 8.5, the most stable state (Fe^{2+} or Fe_3O_4) becomes dependent on the concentration of the dissolved iron species. Thus, the corrosion rate may become dependent on the dissolved species. A change in temperature would change the entire diagram.

This simple example shows the dominating role that the environmental variables play in corrosion. Complex interrelationships can exist. The combined values of the variables pH, potential, concentration, and temperature not only affect corrosion but also affect the action of each variable. For example, with respect to Fig. 1, the effect of a pH change is dependent on the concentration of the dissolved species, and vice versa. Therefore, although the variables are discussed individually, the important point is to realize that the effect of one variable can be dependent on the magnitude of another. This point will be further discussed in this article.

Effect of pH (Acidity)

The concept of pH is complex. It is related to, but not synonymous with, hydrogen concentration or amount of acid. Before discussing how the magnitude of pH affects corrosion, some fundamentals are required. The pH is defined as the negative of the base ten logarithm of the hydrogen ion activity (Ref 2). This latter quantity is related to the concentration or molality through an activity coefficient. The term is expressed as

$$\text{pH} = -\log a_{\text{H}^+} = -\log \gamma_{\text{H}^+} m_{\text{H}^+} \quad \text{(Eq 1)}$$

where a_{H^+} is the hydrogen ion activity, γ_{H^+} is the hydrogen ion activity coefficient, and m_{H^+} is the molality (mol/1000 cm^3 of water). The value of the activity coefficient is a function of everything in the solution (ions, nonionized species, and so on).

The pH is usually measured with a pH meter, which is actually an electrometer. The voltage of a hydrogen ion specific electrode is measured relative to a reference electrode. This voltage is compared to the internally stored calibration obtained from a defined standard to yield the unknown pH. The actual hydrogen ion concentration (acidity level) can be calculated from this measured pH if the activity coefficient is known. Because the test solution usually has constituents that are far different from those of the buffer, the calculated hydrogen ion concentration is at best an estimate (Ref 3). Thus, the pH measured by a pH meter and the actual amount of acid as defined by the hydrogen ion concentration are related but not necessarily equal.

The importance of the hydrogen ion lies in its ability to interact with an alloy surface. Many alloys of commercial interest form an oxidized surface region, the outer most atomic layer of which often contains hydroxide-like species when water is present. Such a structure would tend to have a dependence on hydrogen ion concentration, possibly through a reaction that can be one step in corrosion (Ref 4):

$$H_2O \rightleftarrows OH_{\text{adsorbed}} + H^+ + e^- \quad \text{(Eq 2)}$$

Thus, under a number of conditions, the hydrogen ion concentration can influence corrosion through the equilibrium that exists among it, water, and the hydroxide ion formed on the alloy surface.

This interaction often results in a corrosion rate dependence on hydrogen ion concentration in the form of:

$$r = k\, C^n_{\text{H}^+} \quad \text{(Eq 3)}$$

where r is the corrosion rate, k is the rate constant, C_{H^+} is the hydrogen ion concentration, and n is an exponent. The value of n can be dependent on the hydrogen ion concentration. This type of dependence of the reaction rate on the hydrogen ion concentration is found in a number of systems, which are discussed below. This discussion is not meant to be all-encompassing, but is meant to provide a flavor for how this dependence is observed in practice.

Strongly Acid Conditions (pH < 5). Iron or carbon steel shows a complex dependence of the corrosion rate on pH. At low pH, the corrosion mechanism is dependent not only on the hydrogen ion concentration but also on the counterions present. Thus, all discussion must include the total constituency of the fluid. For example, the corrosion rate of iron in sulfuric acid (H_2SO_4) between a pH of less than 0 and about 4 tends to be limited by the diffusion of and saturation concentration of iron sulfate ($FeSO_4$) (Ref 5, 6). The metal dissolution rate is so high that the corrosion rate is equal to the mass transfer rate of iron from the saturated film of $FeSO_4$ at the metal surface. Because mass transfer rates are sensitive to fluid velocity, the corrosion rate is sensitive to fluid flow. This effect is well documented for concentrated H_2SO_4.

Corrosion of iron in hydrochloric acid (HCl) follows a different mechanism, and pH has a different effect on corrosion. The rate of corrosion is rapid at all acidic concentrations of pH $<$ 3. Unlike the sulfate ion in H_2SO_4, the chloride ion seems to participate in and accelerate the corrosion rate (Ref 7). The corrosion rate increases with hydrogen ion concentration (decreasing pH). These effects are reflected in Eq 3. This behavior indicates that in HCl hydrogen ion directly influences the reaction kinetics. The ion does not influence corrosion through mass transfer.

Corrosion of iron in phosphoric acid (H_3PO_4) solution follows a similar mechanism but with a subtle twist. Again, no passive film exists on the surface; however, the corrosion rate, at least between a pH of 0.75 and 4, seems to be independent of phosphate ion concentration at constant pH (Ref 8).

The important point is that the pH effect on corrosion of carbon steel at low pH is not simple.

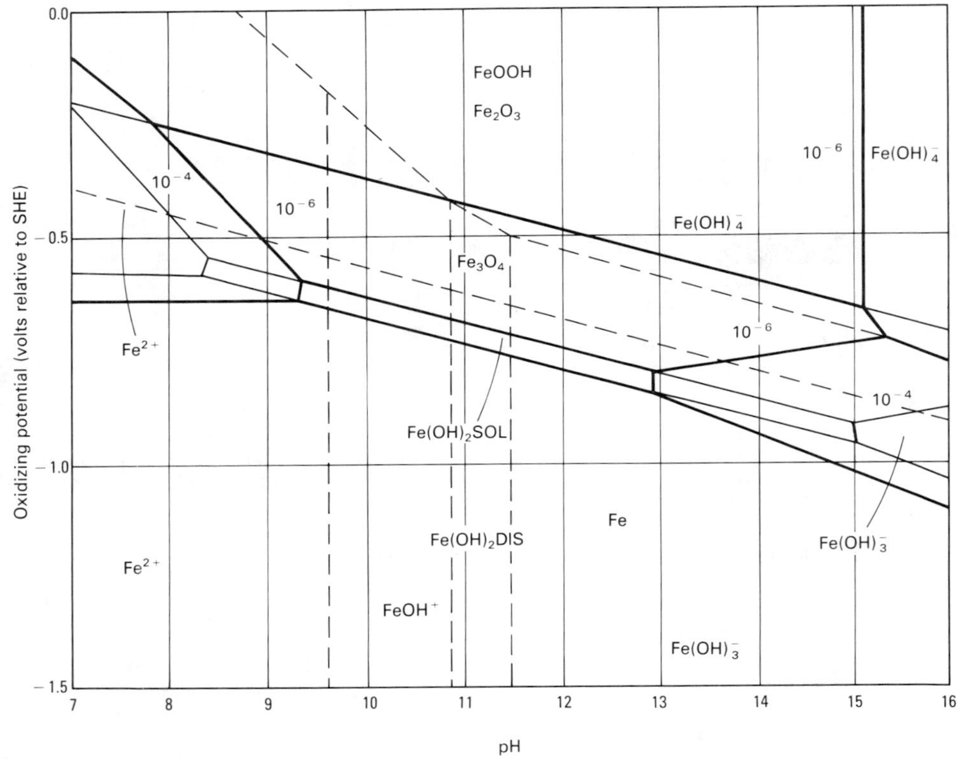

Fig. 1 Potential-pH (Pourbaix) diagram for iron at 25 °C (77 °F) in water. Ionic species are at activities of 10^{-6} and 10^{-4}. Source: Ref 1

Knowledge of how pH affects corrosion in one acid does not necessarily translate to knowledge in another acid. Very little information is available on the effect of acid mixtures on corrosion.

Ferritic iron-chromium alloys have been found to exhibit behavior in concentrated H_2SO_4 reminiscent of the behavior of carbon steel. A strong fluid velocity sensitivity has been noted in 1 *M* H_2SO_4 (5 to 10 wt%) (Ref 9) for those alloys with less than 12 wt% Cr and in the 68 to 93 wt% range (Ref 10) for E-Brite 26-1 (26 wt% Cr). The corrosion rate tends to be related to the rate of mass transfer of $FeSO_4$ from a saturated film on the surface. The one difference is that the presence of oxygen may impart a pseudopassivity that can be unstable. The major point is that the presence of chromium may provide little benefit in this environment. Both chromium content and H_2SO_4 concentration must be considered simultaneously, especially because an 18 wt% Cr ferritic alloy tends to be under activation control in 1 *M* H_2SO_4 (Ref 9).

The addition of nickel to create austenitic alloys alters this behavior in H_2SO_4 and eliminates, or at least diminishes, this velocity sensitivity, especially in the pH range of −0.5 to 3. At lower pH, the higher acid concentrations may produce a velocity sensitivity (Ref 11). Unfortunately, data are sparse on the effect of pH on the low corrosion rates expected for many of these alloys in this low pH range of −0.5 to 3. At still lower pH, the behavior is complex, and the particular literature on the alloy should be consulted. Impurities in the H_2SO_4 can significantly alter the corrosion resistance.

The behavior of austenitic alloys in HCl is far different from that in H_2SO_4, even at the same pH or hydrogen ion concentration. The change from sulfate to chloride anion tends to be detrimental. The presence of the chloride ion raises the possibility of localized attack, for example, crevice corrosion, pitting, and stress-corrosion cracking (SCC) (Ref 12). Once again, behavior with respect to pH is complex. The literature on the particular alloy should be consulted to determine the actual behavior as a function of pH in acidic solutions.

Non Group VIII base alloys show different types of pH dependencies at low pH. For example, in HCl, titanium is passive to a pH of about 0 or slightly lower. Then, a fairly abrupt change in mechanism occurs at still lower pH. There, titanium begins to corrode rather rapidly (Ref 13). The hypothesis is that the titanium valence changes from +4 to +3 and that Ti^{3+} is soluble (Ref 13, 14). The behavior in H_2SO_4 is somewhat different.

Other metals and alloys are affected by acidic pH in different ways. Unfortunately, mechanistic data are less plentiful than for iron-base alloys. A number of metals show a very strong dependence of corrosion on pH. With aluminum, the rate increases exponentially as pH decreases in the acidic region (Ref 15). Indeed, the corrosion rate tends to have a very sharp minimum at a pH of 7 to 9, with sharp corrosion rate increases with both increasing and decreasing pH (Ref 15, 16).

A similar effect of a sharp decrease in corrosion rate with increasing pH for pH < 4 has been noted for both zinc in HCl and lead in nitric acid (HNO_3) (Ref 17). Indeed, this type of behavior would be expected for any metal or alloy whose oxide is soluble in acids, such as zinc, aluminum, lead, tin, and copper.

Near-Neutral Conditions (5 < pH < 9). Corrosion behavior and alloy-environment interactions in the near-neutral pH region differ significantly from those under acidic conditions. In most cases, pH no longer plays a direct role in corrosion.

Iron (as carbon steel) has been one of the most extensively studied metals in this environment. Under acidic conditions, the oxide or hydroxide layers tend to dissolve. However, in the higher pH range, especially above a pH of about 5, these layers tend to remain on the surface. These layers have significant structure, which tends to be determined by the anions present in the solution (Ref 18). In addition, the corrosion kinetics become independent of pH, and hydrogen ion reduction is no longer an important reaction (Ref 19). The major reaction governing corrosion in most practical applications is the reduction of oxygen present in solution. Magnetite (Fe_3O_4) can be formed, which will tend to passivate iron (Ref 18). Thus, pH in this range no longer plays a major direct role in corrosion of iron, although the pH can still affect the solubility and equilibrium of other ions, such as sequestering agents. These other components can play a major role in corrosion in cooling water.

This characteristic of pH in the range of 5 < pH < 9 no longer playing a dominant role is found with other metals, such as zinc and lead (Ref 17). Aluminum shows a very sharp minimum in corrosion rate at about a pH of 7 to 9, with the minimum being somewhat dependent on the counter-ion in solution (Ref 15).

Alloys such as the austenitic iron-base and nickel-base alloys, ferritic alloys, and duplex alloys also tend to have general corrosion rates that are independent of pH in this range. Indeed, in pure water, these alloys would be passive. The presence of other constituents, such as chloride ions and oxygen, plays a much more dominant role, possibly changing the mechanism from uniform corrosion to localized attack.

Strongly Basic Conditions (pH > 9). Basic conditions offer yet another set of corrosion characteristics. In a number of cases, corrosion rate increases with pH (decreasing hydrogen ion concentration) or at least remains finite. In other cases, the increase in pH causes corrosion to occur when none was present at lower pH. These two types of behavior seem to encompass most metals and alloys, and representative examples will be described to demonstrate this behavior.

Iron corrosion persists even at high pH. This persistence is caused by soluble species ($Fe(OH)_3^-$ or, at elevated potentials, $Fe(OH)_4^-$) being the most thermodynamically stable corrosion products (Ref 1, 16). Even though a number of iron hydroxide species can be found that can create a porous barrier (Ref 20), corrosion still persists, although usually at a fairly low rate, until very high pH is reached (Ref 17).

At very high pH and especially at somewhat elevated temperatures, carbon steel can undergo SCC (Ref 21). Some environments that can cause SCC at high pH are sodium hydroxide (NaOH) at very high pH, carbonates and bicarbonates at moderately basic pH values, and possibly amines, although this point is controversial. Steel can also suffer SCC at lower pH, but this behavior is less prevalent. Examples of these environments are hydrogen fluoride (HF) vapors and hydrogen sulfide (H_2S). The mechanism in these cases may be one of hydrogen embrittlement (Ref 22).

A number of metals exhibit a sharp increase in corrosion rate with increasing pH. Among these are aluminum, zinc, and lead (Ref 15, 17). Aluminum corrosion increases very dramatically, changing by almost two orders of magnitude between a pH of 8 and 10. This increase is virtually independent of counter-ion and can be attributed to the formation of soluble aluminum hydroxide products (Ref 16).

Tantalum, which suffers virtually no corrosion under most acidic and neutral pH conditions, shows a significant increase in corrosion rate at high pH (Ref 23). The cause of this corrosion is believed to be a slow dissolution or flaking off of surface layers (Ref 23). This dissolution is probably caused by the formation of soluble tantalum hydroxide corrosion products (Ref 13).

Some metals, such as nickel and zirconium, are very resistant to corrosion at high pH. Possibly nickel and especially zirconium rely on the formation of insoluble oxides for their corrosion protection (Ref 24).

The austenitic and ferritic alloys tend to be immune to corrosion until very high pH is reached. One reason is that chromium, which is included in many of these alloys and which tends to accumulate on the surface, forms a passive oxide. This oxide, for example, chromium oxide (Cr_2O_3), is insoluble under these conditions. However, changes in temperature can affect corrosion at high pH.

Oxidizing Power (Potential)

Oxidizing power, or potential, relates to the ability to remove or add electrons from the metal so as to oxidize or reduce the surface. This variable is separated from the discussions on solution chemistry because such a potential can be applied by an external voltage source, by galvanic coupling of different metals, or by solution constituents. Practical applications include increasing passivity by altering the surface oxide (anodic protection) or preventing corrosion by supplying electrons to the metal that would normally be yielded by metal corrosion (cathodic protection). The anodic reaction rate is shifted or changed in the protected metal.

The alteration of the surface state to impart passivity is normally accomplished by anodic polarization of the metal or alloy surface to a potential noble to the corrosion potential. If an external voltage source is used to change the voltage, the technique is known as anodic protection (see the article "Anodic Protection" in this Volume). Among the practical examples of using externally applied anodic potentials to mitigate corrosion are mild steel and type 304 stainless steel in concentrated H_2SO_4, H_3PO_4, and NaOH (Ref 22).

The addition of constituents to the environment may alter the surface potential to create a passive film. In this case, the constituent reacts with the metal to form a tenacious metal-oxide compound that passivates the surface. There are several well-known examples of anodic polarization of the surface by changing the environment. For example, the addition of small amounts of ozone to water decreases the corrosion of carbon steel in water (Ref 25). The hypothesis is that the corrosion potential moves in a noble direction and the ozone reacts with the iron to create a more tenacious oxide. In another example, the addition of HNO_3 to H_2SO_4 has been shown to retard the corrosion of stainless steels. The hypothesis is that the potential is forced in the noble direction and the surface oxide layer becomes more protective.

Polarization of the surface potential in the active or cathodic direction can also be used to decrease corrosion. When the potential is lowered by means of an external voltage source, the technique is known as cathodic protection (Ref 22). Many practical examples exist, such as the protection of steel at coating defects in underground carbon steel pipelines (Ref 25) or the protection of ships hulls in seawater (see the articles "Cathodic Protection," "Marine Corrosion," and "Corrosion of Pipelines" in this Volume). The electrons are supplied from either an inert or active counterelectrode.

Direct electrical coupling of a metal to a more active metal is another example of using cathodic potentials to affect corrosion. Coupling zinc to steel to protect the steel is a major example. In this case, zinc corrosion liberates electrons to the steel, and the steel potential moves in an active direction (Ref 26).

Such cathodic polarization can be produced by constituents in the solution. Oxygen tends to polarize carbon steel in a noble direction and increase its corrosion. The addition of such species as sulfite (SO_3^{2-}) or hydrazine tends to cause a reaction with the oxygen and thus remove it (Ref 26). The effect of SO_3^{2-} tends to be to move the surface potential in the active direction. Such movement of potential may decrease the corrosion of iron (Ref 17). However, any change in potential may be dependent on other constituents, especially if they can interact with the inhibitor and the metal surface. Also, if the alloy is passive and this passivity is maintained by the oxygen, this addition could increase corrosion by moving the alloy into an active corrosion region (Ref 17).

Temperature and Heat Transfer

Temperature is a complex external variable. Temperature is analogous to potential. A potential difference creates a current flow, the objective of which is to eliminate the potential difference. In a similar manner, a temperature difference creates a heat flow, the objective of which is to eliminate the temperature difference. Both potential and temperature are measures of energy.

Temperature can affect corrosion in a number of ways. If the corrosion rate is governed completely by the elementary process of metal oxidation, the corrosion rate increases exponentially with an increase in temperature. This relationship is reflected in the Arrhenius expression:

$$r = A \exp\left(\frac{-E}{RT}\right) \qquad \text{(Eq 4)}$$

where r is the corrosion rate, A is a preexponential factor, E is an activation energy, R is the gas constant, and T is the absolute temperature.

The effect of temperature on corrosion rate is shown by solving Eq 4 at two temperatures and taking the ratio of the rates:

$$\ln\left(\frac{r_2}{r_1}\right) = \frac{-E}{R}\,\frac{\Delta T}{(T_2)(T_1)} \qquad \text{(Eq 5)}$$

where the subscripts 1 and 2 refer to the two temperatures and ΔT is the difference in temperature ($T_2 - T_1$). Equation 5 can be used to evaluate the effect of a temperature change on corrosion rate for this simple rate process. Examples of corrosion that follow this simple rate law are iron in HCl (Ref 27) and iron in sodium sulfate (Na_2SO_4) at a pH of about 2 (Ref 28). This situation is most common for corrosion under acidic conditions.

The temperature of the metal and the temperature of the solution often cannot be discussed separately from other variables. If a constituent in the solution that is important in corrosion has limited solubility, a temperature change can alter the concentration of that constituent. This alteration can have a profound effect on corrosion.

One classical example is the corrosion of iron in the presence of oxygen in systems both closed from the atmosphere and open to the atmosphere. The corrosion rate of iron in a system closed to the atmosphere has been shown to increase almost linearly with temperature from about 40 to 160 °C (105 to 320 °F). However, in the open system, the corrosion rate increases up to about 80 °C (175 °F) and then decreases (Ref 17). Oxygen mass transfer, which is proportional to the oxygen concentration in the liquid, controls the corrosion rate of steel in water. As temperature increases, oxygen solubility decreases so that the oxygen will tend to leave the liquid. In the closed system, the oxygen cannot escape from the vapor space above the liquid. As temperature increases, the water vapor pressure increases, which tends to maintain the oxygen concentration in the liquid. The corrosion rate (mass transfer rate) continues to increase with temperature because of temperature effects on viscosity, diffusivity, and so on. In the open systems, oxygen can escape from the immediate vicinity of the liquid. The vapor pressure remains constant. Above a certain temperature, the liquid-phase oxygen concentration in equilibrium with oxygen in the atmosphere has decreased to the extent that the corrosion (mass transfer) rate decreases.

Another point often overlooked is that the ionization constant of water increases with temperature. Pure water with pH of 7 at one temperature will have a lower pH at a higher temperature. Thus, an increase in temperature could affect corrosion by moving the pH from a neutral to an acidic value.

Fluid temperature changes can affect the polarity in galvanic corrosion. The corrosion potential of the anode might be more sensitive to temperature than that of the cathode. The anode potential can actually become noble with respect to that of the cathode (Ref 17). An example is the iron-zinc couple, the polarity of which can reverse as temperature increases. Iron will actually protect the zinc. The temperature of this reversal is as low as 60 °C (140 °F), but there is some dependence of temperature on constituents (Ref 26).

Solution temperature can also affect the onset of localized attack of passive alloys such as type 304 and 316 stainless steels. The solution usually contains a species, such as chloride ion, that aids in the initiation process (Ref 29). The time to initiation of crevice corrosion has been shown to be a function of temperature. There are indications that such initiation times do not always decrease with increasing temperature (Ref 29). In addition, a critical crevice temperature can be defined for many of these alloys (Ref 30). This critical temperature determines the temperature boundary at which crevice corrosion can initiate. Indeed, the critical crevice temperature has been shown to be a function of the chromium and molybdenum content of austenitic and ferritic alloys.

In practice, elevated or depressed temperatures are often created by heat transfer through a metal wall. Thus, the metal wall can be at a temperature different from that of the bulk fluid. There is a controversy over whether corrosion in the absence of heat transfer is identical to corrosion in the presence of heat transfer even if the metal temperatures are identical in the two situations (Ref 28).

The effect of a difference between wall and fluid temperatures on corrosion depends on the corrosion mechanism. If the corrosion rate is under activation control and follows Eq 4, the corrosion rate in the presence of heat transfer might be similar to that expected for corrosion at the same wall temperature in the absence of heat transfer. If the corrosion rate is controlled by the diffusion of a species, such as oxygen, to the surface then heat transfer may greatly change the corrosion rate. This effect has several possible causes (Ref 31, 32). First, a temperature difference between the wall and bulk solution can affect the solubility and diffusion coefficient of the diffusing species. Second, boiling near or on the wall can increase turbulence and possibly cause cavitation or increased diffusion (mass transfer). Third, heat transfer in the absence of fluid flow, as in stagnant tanks, can cause natural convection currents that can enhance mass transfer. Thus, if heat transfer is present, it must be considered an environmental variable.

Velocity/Fluid Flow Rate

Fluid flow rate, or fluid velocity, is also a complex variable (Ref 33). Its influence on corrosion is dependent on the alloy, fluid constituents, fluid physical properties, geometry, and corrosion mechanism. These relationships are best discussed in terms of specific examples. In a number of instances, the corrosion rate is determined by the rate of transfer of a species between the surface and the fluid. This situation arises when the corrosion reaction itself is very rapid and one of the corrosion reactants or products has low solubility in the bulk fluid. The corrosion rate becomes a function of the concentration gradient and is expressed by:

$$r = k\,(C_W - C_B) \qquad \text{(Eq 6)}$$

where r is the corrosion rate, k is a mass transfer coefficient, C_W is the concentration of the rate-limiting species at the metal wall, and C_B is the concentration of the rate-limiting species in the bulk fluid. The value of k can often be correlated with the dimensionless quantities Reynolds number (Re) and Schmidt number (Sc). The mass transfer coefficient is expressed in terms of the Sherwood number (Sh). These numbers are related to physical properties of the fluid and geometry by:

$$\text{Re} = \frac{vd}{\nu} \qquad \text{(Eq 7a)}$$

$$\text{Sc} = \frac{\nu}{D} \qquad \text{(Eq 7b)}$$

$$\text{Sh} = \frac{kd}{D} \qquad \text{(Eq 7c)}$$

where v is the fluid velocity, d is a characteristic length (for example, pipe diameter), ν is the kinematic viscosity (absolute viscosity divided by density), and D is the diffusion coefficient.

For many geometries, these quantities can be related by:

$$\text{Sh} = a\,\text{Re}^b\,\text{Sc}^c \qquad \text{(Eq 8)}$$

where a, b, and c are constants. Equations 6 to 8 indicate that the corrosion rate can be calculated if it depends on the mass transfer rate of a species from or to the bulk fluid. The only information required is the geometry, fluid velocity, and physical properties. There are a number of examples of corrosion that follow this behavior. The corrosion of carbon steel and E-Brite 26-1 in concentrated H_2SO_4 is governed by the rate of mass transfer of $FeSO_4$ from a saturated layer on the surface (Ref 5, 6, 10). Carbon steel corrosion in water in the near-neutral pH range is governed by the rate of mass transfer of dissolved oxygen from the bulk fluid to the surface (Ref 34). If a porous surface hydroxide layer forms, the mass transfer rate might become limited by diffusion through the porous film.

This effect of velocity has ramifications for localized attack, especially pitting and crevice corrosion. The presence of fluid flow can sometimes be beneficial in preventing or decreasing localized attack. For example, type 316 stainless steel has been shown to pit in quiescent seawater but not in moving seawater (Ref 35). When the seawater is moving, the mass transfer rate of oxygen is high enough to maintain a completely passive surface, but in the absence of flow, the mass transfer of oxygen is too slow and the surface cannot remain passive (Ref 36). This observation indicates that sometimes fluid velocity can be beneficial even if the corrosion rate involves the mass transfer of a reactant or product. The propensity for localized attack to occur can sometimes be decreased by maintaining sufficient fluid motion.

Under other circumstances, fluid flow can cause a type of erosion of a surface through the mechanical force of the fluid itself. This common process is called impingement. The process involves the removal of metal or alloy by the high wall shear stress created by the flowing fluid. Examples of such erosion occur either where fluid is forced to turn direction, for example, at pipe bends (Ref 22), or where high surface shear stresses can exist, for example, on ship hulls (Ref 35). Evidence exists that a critical wall shear stress can be defined for an alloy above which impingement causes erosion and below which such erosion is absent (Ref 37, 38). Thus, shear stress can be translated to a maximum velocity. This phenomenon has been demonstrated for copper-nickel alloys and aluminum alloys in salt water (Ref 39).

When solids are present in the liquid, they can cause wear or solid erosion corrosion (Ref 40). The wear is caused by the relative movement of the solids with respect to the surface. Again, such wear is more prevalent where fluid is forced to change direction or where high shear stresses occur. The particles must penetrate the laminar sublayer with enough force to remove the passive film on the alloy. Therefore, high shear stresses are often required for this type of erosion to occur. This problem can be significant in such systems as salt water carrying solids (for example, sand or coal) and carbon steel carrying air plus particulates.

Concentration

The concentration of constituents within the fluid often influences how the other variables manifest themselves. This discussion will focus on how the concentration of constituents works with other variables to influence corrosion behavior.

During the previous discussion, the point was made that pH plays a major role in corrosion. For iron, the corrosion rate is large at very low pH, is independent of pH in the neutral pH range, decreases with increasing pH, and finally increases again at very high pH. Additions of small amounts of other components can change this behavior.

For example, additions of chloride to H_2SO_4 increase the corrosion rate of iron. This increase is reported to be proportional to the chloride ion concentration raised to about the 0.5 power (Ref 41). A similar effect is reported for chloride ion in HCl (Ref 7). Thus, chloride ion accelerates the corrosion of iron in acidic solutions. However, bromide and iodide ions may inhibit corrosion (Ref 41), although this finding is controversial (Ref 17).

The dependence of the corrosion rate of iron on chloride ion concentration significantly decreases in neutral solutions when oxygen is present (Ref 17, 42). Oxygen accelerates the cathodic reaction far more than chloride can accelerate the anodic reaction. As salt concentration increases, the oxygen solubility decreases, masking the effect of chloride ion. The chloride ion effect is dependent on the cation, with the rate increasing in the order lithium chloride (LiCl), sodium chloride (NaCl), and potassium chloride (KCl) partially because of differences in oxygen solubility in the presence of these salts (Ref 17). These results illustrate that the effect of the concentration of one component on corrosion is often dependent on other environmental variables.

Small additions of certain inhibitors or passivators have a marked effect on corrosion. For example, as little as 0.0023 mol/L of sodium nitrite ($NaNO_2$) or sodium sulfite (Na_2SO_3) can decrease the pit initiation rate of aluminum. Little improvement is found at higher concentration for this system (Ref 43). This behavior is often found with many types of inhibitors. For example, small concentrations (10 ppm) of $NaNO_2$ (a passivator) can drastically inhibit the corrosion of iron, with little further decrease in corrosion found at higher concentrations (Ref 44). However, the critical concentration can depend on pH and on the presence of other constituents. Much higher concentrations may be required, depending on the other constituents. The actual concentration needed for a given system must be determined experimentally. Similarly, many organic inhibitors cause a drastic decrease in corrosion rate at very low concentrations, especially for iron in acidic solutions, with no benefit observed upon increasing the inhibitor concentration (Ref 45). All of these inhibitors tend to interact with the surface in one of three ways: a gettering of a finite amount of impurity in the solution (hydrazine), oxidation (passivation) of the surface (nitrite or chromate), or adsorption on the finite surface area to block corrosion (many organic inhibitors in acid).

However, although this type of behavior is common even for iron in neutral, aqueous environments, exceptions do exist. Sometimes, corrosion can increase with inhibitor concentration until a maximum is reached, followed by a rapid decrease with still further increases in concentration (Ref 44). An example is chromate ion. Chromate ion is normally considered to be an inhibitor. However, at very low concentrations and in the presence of strong activating ions such as chlorides in acidic media, chromate ion can actually accelerate corrosion until enough chromate is present. Also, synergistic action may be observed in which the efficacy of one inhibitor is dependent on the presence of another species, for example, oxygen or other oxidizing agents.

A question often asked is, What is the amount of chloride that is allowable before localized corrosion (crevice corrosion, pitting, or SCC) can occur in austenitic alloys? The answer is not straightforward. Work with boiling, saturated magnesium chloride suggests that 42 wt% Ni in the alloy prevents SCC (Ref 22). However, this rule of thumb does not answer the question. The maximum chloride concentration is dependent on the pH, other constituents, temperature, and other variables. Guidelines are available (Ref 29), and the articles in this Volume on the resistance of individual alloys to localized attack should be consulted.

Appendix: Biological Effects

Stephen C. Dexter
College of Marine Studies
University of Delaware

Biological organisms are present in virtually all natural aqueous environments. In seawater environments, such as tidal bays, estuaries, harbors, and coastal and open ocean seawaters, a great variety of organisms are present. Some of these are large enough to observe with the naked eye, while others are microscopic. In freshwater environments, both natural and industrial, the large organisms are missing, but there is still a great variety of microorganisms, such as bacteria and algae.

In all of these environments, the tendency is for organisms in the water to attach to and grow on the surface of structural materials, resulting in the formation of a biological film, or biofilm. The film itself can range from a microbiological slime film on freshwater heat transfer surfaces to a heavy encrustation of hard-shelled fouling organisms on structures in coastal seawater. There is a voluminous amount of literature on the formation of such films and their many adverse effects (Ref 46-48).

The biofilms that form on the surface of virtually all structural metals and alloys immersed in aqueous environments have the capability to influence the corrosion of those metals and alloys. This influence derives from the ability of the organisms to change the environmental variables discussed earlier in this article (pH, oxidizing power, temperature, velocity, and concentration). Thus, the value of a given parameter at the metal/water interface under the biofilm may be quite different from that in the bulk electrolyte away from the interface. The result can be the initiation of corrosion under conditions in which there would be none in the absence of the film, a change in the mode of corrosion (that is, from uniform to localized), or an increase or decrease in the corrosion rate. It is important to note, however, that the presence of a biofilm does not necessarily mean that there will always be a significant effect on corrosion. The purpose of this Appendix is to consider in general the characteristics of organisms that allow them to interact with corrosion processes and the general mechanisms by which organisms can influence the occurrence or rate of corrosion.

General Characteristics of Organisms

The organisms that are known to have an important impact on corrosion are mostly microorganisms such as bacteria, algae, and fungi (yeasts and molds). In this section, the general characteristics of the microorganisms that facilitate their influence on the electrochemistry of corrosion will be discussed (Ref 49-51). Information on the individual organisms can be found in the discussions of biological corrosion in the articles "General Corrosion" and "Localized Corrosion" in this Volume.

Physical Characteristics. Microorganisms range in length from 0.1 to over 5 μm (some filamentous forms can be several hundred micrometers long) and up to about 3 μm in width. Many of them are motile; that is, they can "swim" to a favorable, or away from an unfavorable, environment. Because of their small size, they can reproduce themselves in a short time. Under favorable conditions, it is common for bacterial numbers to double every 20 min or less. Thus, a single bacterium can produce a mass of over one million organisms in less than 7 h.

In addition to rapid reproduction, the bacteria as a group can survive wide ranges of temperature (−10 to > 100 °C, or 15 to 212 °F), pH (~0 to 10.5), dissolved oxygen concentration (0 to saturation), pressure (vacuum to > 31 MPa, or 4500 psi), and salinity (tolerances vary from the parts per billion range to about 30% salt). Despite these wide ranges of tolerance for the microorganisms as a whole, most individual species have much narrower ranges. Most bacteria that have been implicated in corrosion grow best at temperatures of 15 to 45 °C (60 to 115 °F) and a pH of 6 to 8. Oxygen requirements vary widely with species. Microbes may be obligate aerobes (require oxygen for growth), microaerophilic (require minute levels of oxygen for growth), facultative anaerobes (grow with or without oxygen), or obligate anaerobes (grow only in the complete absence of oxygen).

Some microbes can produce spores that are resistant to a variety of environmental extremes, such as drying, freezing, and boiling. Spores have been known to survive for hundreds of years under arctic conditions and then to germinate and grow when conditions become favorable. Many microbes can quickly adapt to a wide variety of compounds as food sources. This gives them high survivability under changing environmental conditions.

Metabolic Characteristics. Many of the microorganisms implicated in corrosion are able to have an influence on the electrochemical reactions involved by virtue of the products produced by their metabolism. A large percentage of them can form extracellular polymeric materials termed simply polymer, or slime. The slime helps glue the organisms to the surface, helps trap and concentrate nutrients for the microbes to use as food, and often shields the organisms from the toxic effect of biocides. The slime film can influence corrosion by trapping or complexing heavy-metal ions near the surface. It can also act as a diffusion barrier for chemical species migrating to or from a metal surface, thus changing the concentrations and pH at the interface where the corrosion takes place.

Some species of microbes can produce organic acids, such as formic and succinic, or mineral acids, such as H_2SO_4. These chemicals are corrosive to many metals. One series of bacteria is involved in metabolizing nitrogen compounds. As a group, they can reduce nitrates (NO_3^-) (often used as a corrosion inhibitor) to nitrogen (N_2) gas. Others can convert NO_3^- to nitrogen dioxide (NO_2), or vice versa, or they can break it down to form ammonia (NH_3). Still other series of bacteria are involved in the transformation of sulfur compounds (Fig. 2). They can oxidize sulfur or sulfides to sulfates (SO_4^{2-}) (or H_2SO_4), or they can reduce SO_4^{2-} to sulfides, often producing corrosive H_2S as an end product.

Organisms that have a fermentative type of metabolism produce carbon dioxide (CO_2) and hydrogen (H_2); others can utilize CO_2 and H_2 as sources of carbon and energy, respectively. Numerous species of bacteria and algae either produce or utilize oxygen. It is rare that a corrosion process would not depend on the concentration of at least one of these three dissolved gasses.

Finally, some bacteria are capable of being directly involved in the oxidation or reduction of metal ions, particularly iron and manganese. Such bacteria can shift the chemical equilibrium between Fe, Fe^{2+}, and Fe^{3+}, which will often influence the corrosion rate.

Community Structure. The ability of an organism to survive on a surface and to influence corrosion is often related to associations between that organism and those of other species. The bacteria implicated in corrosion may begin their lives on a metal surface as a scatter of individual cells, as shown in Fig. 3(a). As the biofilm matures, however, the organisms will usually be found in thick, semicontinuous films (Fig. 3b) or in colonies (Fig. 3c). It is in these latter two forms that there is the most potential for survival and growth of the organisms capable of influencing corrosion.

For example, the sulfate-reducing bacteria (SRB) are implicated in the corrosion of iron-base alloys in a variety of environments (Ref 52, 53). Most sulfate-reducing bacteria are obligate anaerobes, yet they are known to accelerate corrosion in aerated environments. This becomes possible when aerobic organisms form a film or colony and then, through their metabolism, create an anaerobic microenvironment with the organic acids and nutrients necessary for growth of the sulfate-reducing bacteria (Fig. 4). Thus, the organisms influencing corrosion can often flourish at the corrosion site by associating with other organisms in a microbial colony or consortium, even when the bulk environment is not conducive to their growth.

It should be noted that the dynamics of fluid flow past the metal surface can alter the form of the biofilm or can even prevent its formation. This can result in acceleration or deceleration of corrosion, depending on the role of the biofilm.

General Mechanisms of Influence

The presence of a biological film on a corroding metal surface does not introduce some new type of corrosion, but it influences the occurrence and/or the rate of known types of corrosion. These biological influences can be divided into three general categories:

- Production of differential aeration or chemical concentration cells
- Production of organic and inorganic acids as metabolic by-products

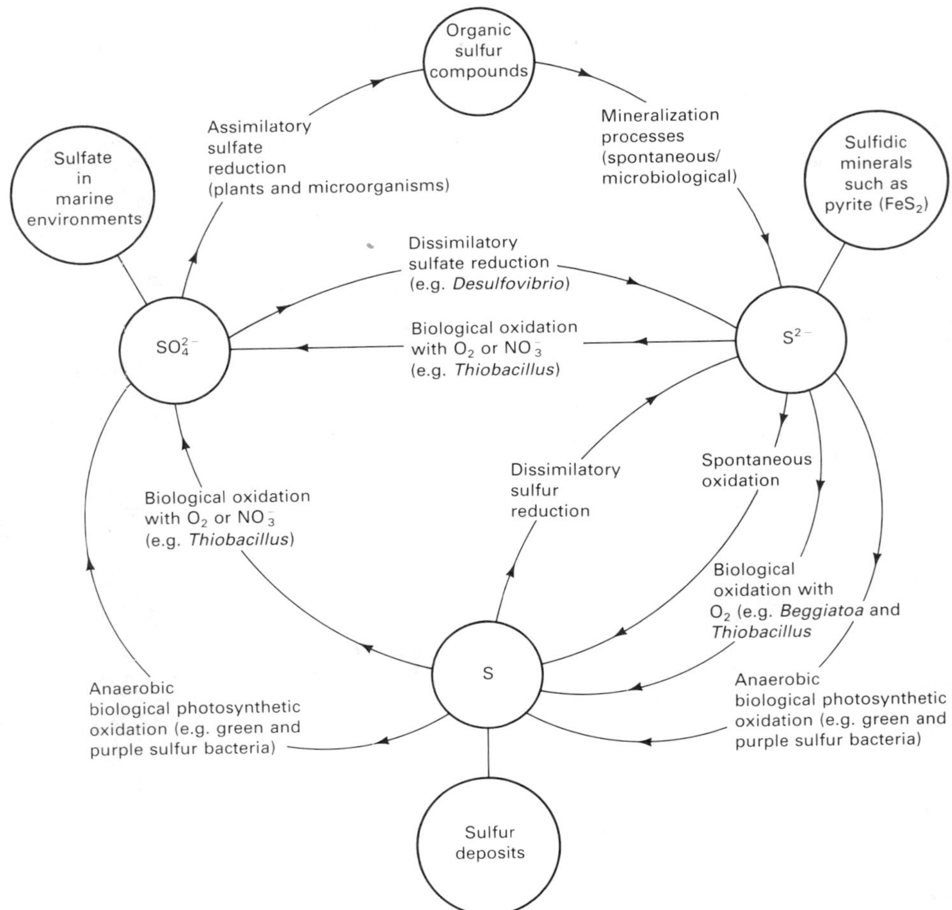

Fig. 2 The sulfur cycle showing the role of bacteria in oxidizing elemental sulfur to sulfate (SO_4^{2-}) and in reducing sulfate to sulfide (S^{2-}). Source: Ref 52

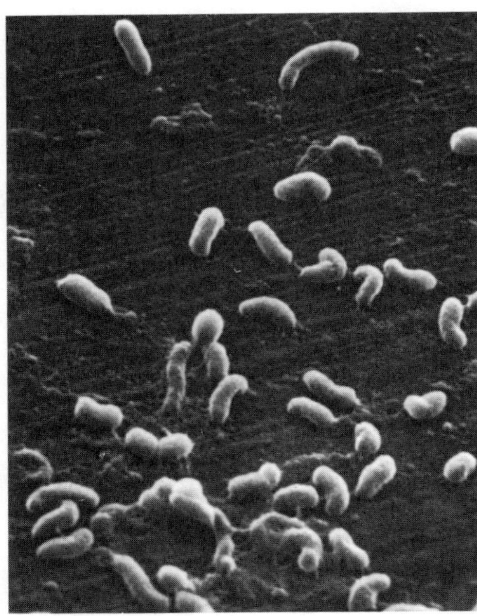

(a)

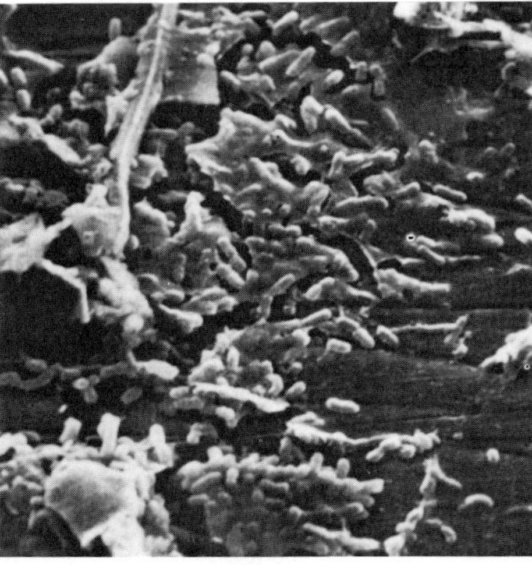

(b)

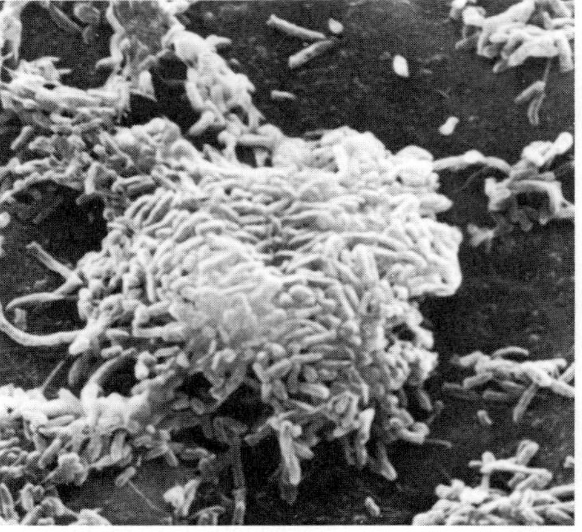

(c)

Fig. 3 Various forms of bacterial film that can influence corrosion. (a) Scatter of individual cells. 6050×. (b) Semicontinuous film of bacteria in slime. 3150×. (c) Bacterial cells in a colony. 2700×

- Production of sulfides under oxygen-free (anaerobic) conditions

Oxygen/Chemical Concentration Cells. Any biofilm that does not provide for complete, uniform coverage of the entire immersed surface of a metal or alloy has the potential to form concentration cells. In aerated environments, uncovered areas of the metal surface, in contact with oxygenated electrolyte, will be cathodic relative to those areas under the biofilm. Beneath the film or colony, oxygen is depleted as it is used by the organisms in their metabolism. Oxygen from the bulk electrolyte is unable to replenish those areas because of a combination of effects. First, oxygen migration through the film is slowed by the diffusion barrier effect, and second, oxygen that does penetrate the film is immediately utilized by the microbial metabolism. Formation of such a corrosion cell, as shown in Fig. 5, causes a pit to form at the anodic area under the bacterial colony.

As the pit grows, iron dissolves according to the anodic reaction:

$$Fe \rightarrow Fe^{2+} + 2e^-$$

The cathodic reaction is reduction of dissolved oxygen outside the pit to form OH^- according to:

$$O_2 + 2H_2O + 4e^- \rightarrow 4OH^-$$

The insoluble ferrous hydroxide corrosion product forms by the reaction:

$$3Fe^{2+} + 6OH^- \rightarrow 3Fe(OH)_2$$

Corrosion products mingle with bacterial film to form a corrosion tubercule, which itself may cause a problem with obstruction of fluid flow in piping systems.

In addition, if the above process takes place in the presence of bacteria capable of oxidizing ferrous ions to ferric ions, the corrosion rate will be accelerated because the ferrous ions are removed from solution as soon as they are produced. This depolarizes the anode and accelerates corrosion of iron under the deposit. The ferric ions form ferric hydroxide ($Fe(OH)_3$), which contributes to the rapid growth of the tubercule. This process has been responsible for corrosion and plugging of iron water pipes.

If chlorides are present in the system, the pH of the electrolyte trapped inside the tubercule may become very acid by an autocatalytic process similar to that described in the article "Localized Corrosion" in this Volume for crevice corrosion and pitting. Chloride ions from the environment combine with ferric ions produced by corrosion in the presence of the bacteria to form a highly corrosive, acidic ferric chloride solution inside the tubercule. This has been responsible for severe pitting of stainless steel

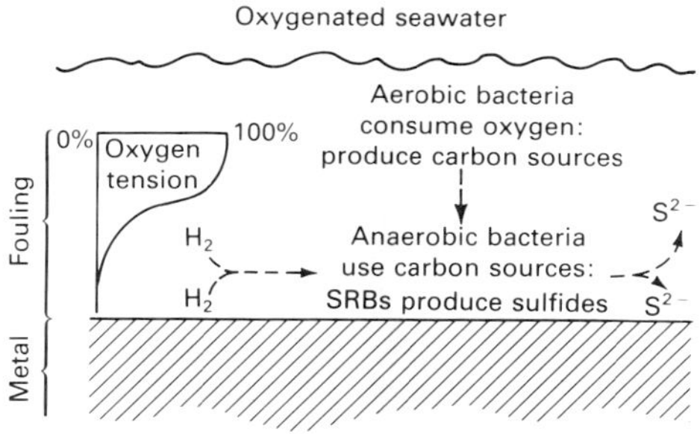

Fig. 4 Variations through the thickness of a bacterial film. Aerobic organisms near the outer surface of the film consume oxygen and create a suitable habitat for the sulfate-reducing bacteria at the metal surface. Source: Ref 52

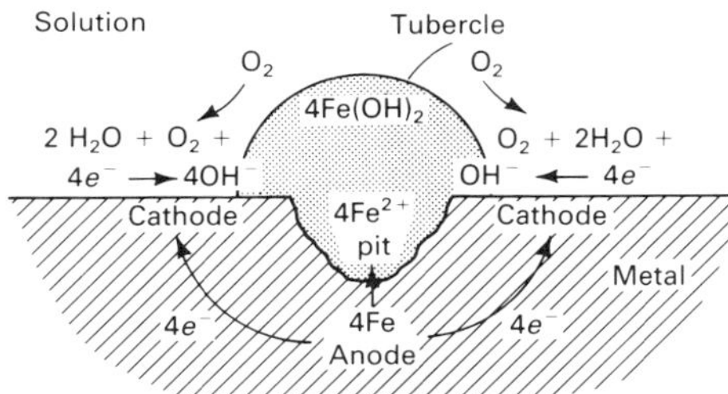

Fig. 5 Schematic of pit initiation and tubercule formation due to an oxygen concentration cell under a biological deposit. Source: Ref 53

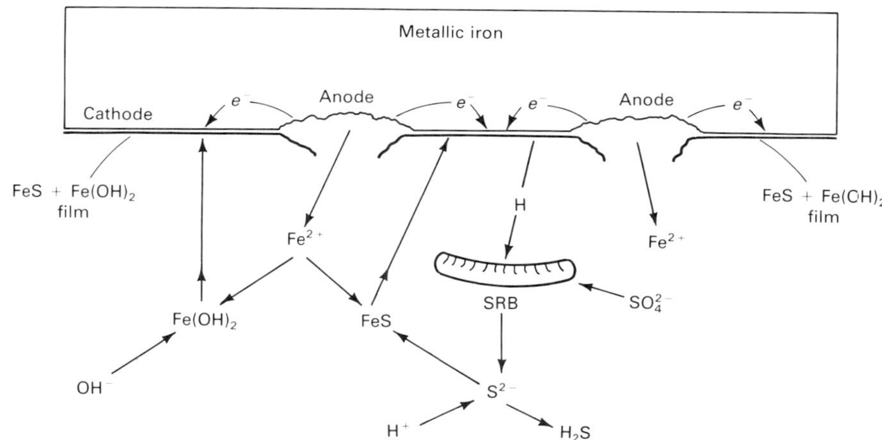

Fig. 6 Schematic of the anaerobic corrosion of iron and steel showing the action of sulfate-reducing bacteria in removing hydrogen from the surface to form FeS and H_2S

piping systems, as described in the section "Localized Biological Corrosion" in the article "Localized Corrosion" in this Volume.

Acid Production. The sulfur oxidizing bacteria can produce up to about 10% H_2SO_4. This mineral acid, with its accompanying low pH, is highly corrosive to many metals, ceramics, and concrete. Other species of bacteria produce organic acids that are similarly corrosive.

The acids produced by these organisms can also contribute to corrosion by aiding the breakdown of coatings systems. Alternatively, other organisms that have no direct influence on corrosion may be involved in the breakdown of coatings. The breakdown products are then sometimes usable as food by the acid-producing bacteria, ultimately leading to accelerated corrosion of the underlying metal.

Anaerobic Sulfide Production. The most thoroughly documented case in which microbes are known to cause corrosion is that of iron and steel under anaerobic conditions in the presence of sulfate-reducing bacteria. Based on electrochemistry, deaerated soils of near-neutral pH are not expected to be corrosive to iron and steel. However, if the soil contains sulfate-reducing bacteria and a source of sulfates, rapid corrosion has been found to occur.

The classical mechanism originally proposed for this corrosion involved the removal of atomic hydrogen from the metal surface by the bacteria using the enzyme hydrogenase (Ref 54). The removed hydrogen was then supposedly utilized by the bacteria in the reduction of sulfates to sulfides. The following set of equations was proposed to explain this mechanism:

$$4Fe \rightarrow 4Fe^{2+} + 8e^- \quad \text{(Eq 9)}$$
Anodic reaction

$$8H_2O \rightarrow 8H^+ + 8OH^- \quad \text{(Eq 10)}$$
Dissociation of water

$$8H^+ + 8e^- \rightarrow 8H \quad \text{(Eq 11)}$$
Cathodic reaction

$$SO_4^{2-} + 8H \xrightarrow{SRB} S^{2-} + 4H_2O \quad \text{(Eq 12)}$$
Cathodic depolarization

$$Fe^{2+} + S^{2-} \rightarrow FeS \quad \text{(Eq 13)}$$
Corrosion product

$$3Fe^{2+} + 6OH^- \rightarrow 3Fe(OH)_2 \quad \text{(Eq 14)}$$
Corrosion product

Without sulfate-reducing bacteria, the mechanism would stop after Eq 11, when the surface became covered by a monolayer of hydrogen. According to the theory, this hydrogen is stripped off by the bacteria, a process known as cathodic depolarization; this allows corrosion to continue.

It is now recognized that this original mechanism, although it undoubtedly plays an important role, does not represent the entire process (Fig. 6). It has been shown that the iron sulfide (FeS) film produced is protective if continuous but that it causes galvanic corrosion of the bare iron underneath if defective. Other corrosive substances, such as H_2S, can also be produced. The sulfate-reducing bacteria have been identified as contributors to the corrosion of stainless, copper, and aluminum alloys, but the details of the mechanism are still being debated (Ref 52, 53).

Additional information on the organisms involved in corrosion and the industries, environments, and alloy-electrolyte systems in which they have been active can be found in the articles "General Corrosion" and "Localized Corrosion" in this Volume. Information on detecting and characterizing biological corrosion in the laboratory and in the field can be found in the article "Evaluation of Microbiological Corrosion." Information on controlling biological corrosion can be found in the article "Control of Environmental Variables in Water Recirculating Systems."

REFERENCES

1. D.C. Silverman, Presence of Solid $Fe(OH)_2$ in EMF-pH Diagram for Iron, *Corrosion*, Vol 38 (No. 2), 1982, p 453
2. R.A. Robinson and R.H. Stokes, *Electrolyte Solutions*, 2nd ed., Academic Press, 1959
3. R.G. Bates, *Electrometric pH Determinations*, John Wiley & Sons, 1954
4. P. Lorbeer and W.J. Lorenz, The Kinetics of Iron Dissolution and Passivation in Solutions Containing Oxygen, *Electrochim. Acta*, Vol 25 (No. 4), 1980, p 375
5. B.T. Ellison and W.R. Schmeal, Corrosion of Steel in Concentrated Sulfuric Acid, *J. Electrochem. Soc.*, Vol 125 (No. 4), 1978, p 524
6. T.K. Ross, G.C. Wood, and I.J. Mahmud, The Anodic Behavior of Iron Carbon Alloys in Moving Acid Media, *J. Electrochem. Soc.*, Vol 113, 1966, p 334
7. R.J. Chin and K. Nobe, Electrodissolution Kinetics of Iron in Chloride Solutions, *J. Electrochem. Soc.*, Vol 119 (No. 11), 1972, p 1457
8. A.N. Katrevick, G.M. Florianovich, and Ya. M. Kolotyrkin, Clarification of the Kinetic Parameters of the Reaction of the Active Dissolution of Iron in Phosphate Solutions, *Zashch. Met.*, Vol 10 (No. 4), 1974, p 369
9. B. Alexandre, A. Caprani, J.C. Charbonnier, M. Keddam, and P.H. Morel, The Influence of Chromium on the Mass Transfer Limitation of the Anodic Dissolution of Ferritic Steels Fe-Cr in Molar Sulfuric Acid, *Corros. Sci.*, Vol 21 (No. 11), 1981, p 765
10. D.C. Silverman and M.E. Zerr, Application of Rotating Cylinder Electrode—E-Brite 26-1/Concentrated Sulfuric Acid, *Corrosion*, Vol 42 (No. 11), 1986, p 633
11. "The Corrosion Resistance of Nickel Containing Alloys in Sulfuric Acid and Related Compounds," CEB-1, International Nickel Company, Inc., 1983
12. "Resistance of Nickel and High Nickel Alloys to Corrosion by Hydrochloric Acid, Hydrogen Chloride, and Chlorine," CEB-3, International Nickel Company, Inc., 1974
13. D.C. Silverman, "Derivation and Application of EMF-pH Diagrams," Paper 66, presented at Corrosion/85, Boston, MA, National Association of Corrosion Engineers, 1985
14. A. Alon, M. Frenkel, and M. Schorr, Corrosion and Electrochemical Behavior of Titanium in Acidic, Oxidizing, Aqueous, and Alcoholic Systems, in *Proceedings of the Fourth*

International Congress on Metallic Corrosion, National Association of Corrosion Engineers, 1969, p 636
15. V. Vujicic and B. Lovrecek, A Study of the Influence of pH on the Corrosion Rate of Aluminum, *Surf. Technol.*, Vol 25, 1985, p 49
16. M. Pourbaix, *Atlas of Electrochemical Equilibria in Aqueous Solutions*, National Association of Corrosion Engineers, 1974
17. N.D. Tomashov, *Theory of Corrosion and Protection of Metals*, B.H. Tytell, I. Geld, and H.S. Preiser, Trans., Macmillan, 1966
18. M. Fischer, Possible Models of the Reaction Mechanism and Discussion of the Functions of Passivating Additions During the Chemical Passivation of Iron in Weakly Acidic, Neutral and Basic pH Ranges, *Werkst. Korros.*, Vol 29, 1978, p 188
19. B.J. Andrzejaczek, Influence of Oxygen on the Kinetics of Corrosion Processes on Iron in Water, *Korrosion*, Vol 15 (No. 5), 1984, p 239
20. R.N. Sylva, The Hydrolysis of Iron (III), *Rev. Pure Appl. Chem.*, Vol 22, 1972, p 115
21. W. Frank and L. Graf, Mechanism of Intergranular Corrosion and Intergranular Stress Corrosion Cracking in Mild Steels, *Res. Mech.*, Vol 13, 1985, p 251
22. M.G. Fontana and N.D. Green, *Corrosion Engineering*, 2nd ed., McGraw-Hill, 1978
23. D.R. Knittel, Zirconium: A Corrosion Resistant Material for Industrial Applications, *Chem. Eng.*, Vol 87 (No. 11), 1980, p 98
24. M. Schussler, *Corrosion Data Survey on Tantalum*, Fansteel, Inc., 1972
25. D.T. Merrill and J.A. Drago, "Evaluation of Ozone Treatment in Air Conditioning Cooling Towers," EPRI FP-1178, Electric Power Research Institute, Sept 1979
26. H.H. Uhlig, *Corrosion and Corrosion Control*, John Wiley & Sons, 1963
27. F.N. Speller, *Corrosion Causes and Prevention*, McGraw-Hill, 1951
28. Ya. M. Kolotyrkin, V.S. Pakhomov, A.G. Parshin, and A.V. Chekhovskii, Effect of Heat Transfer on Corrosion of Metals, *Khim. Neft. Mashinostr.*, Vol 16 (No. 12), 1980, p 20
29. J.W. Oldfield, T.S. Lee, and R.M. Kain, Avoiding Crevice Corrosion of Stainless Steels, in *Proceedings of Stainless Steels '84*, Institute of Metals, 1985, p 205
30. R.J. Brigham, The Initiation of Crevice Corrosion on Stainless Steels, *Mater. Perform.*, Vol 24 (No. 12), 1985, p 44
31. A.G. Parshin, V.S. Parkhomov, and Ya. M. Kolotyrkin, The Influence of Heat Transfer on the Kinetics of Cathode Processes With Limiting Diffusion Steps, *Corros. Sci.*, Vol 22 (No. 9), 1982, p 845
32. T.K. Ross, Corrosion and Heat Transfer—A Review, *Br. Corros. J.*, Vol 2, 1967, p 131
33. U. Lotz and E. Heitz, Flow Dependent Corrosion, *Werkst. Korros.*, Vol 34, 1983, p 451
34. B.K. Mahato, C.Y. Cha, and L.W. Shemilt, Unsteady State Mass Transfer Coefficients Controlling Steel Pipe Corrosion Under Isothermal Flow Conditions, *Corros. Sci.*, Vol 20, 1980, p 421
35. H.R. Copson, Effects of Velocity on Corrosion, *Corrosion*, Vol 16 (No. 2), 1960, p 86t
36. T.R. Beck and S.G. Chan, Experimental Observations and Analysis of Hydrodynamic Effects on Growth of Small Pits, *Corrosion*, Vol 37 (No. 11), 1981, p 665
37. P.A. Lush, S.P. Hutton, J.C. Rowlands, and B. Angell, The Relation Between Impingement Corrosion and Fluid Turbulence Intensity, in *Proceedings of the Sixth European Congress on Metallic Corrosion*, 1977, p 137
38. D.C. Silverman, Rotating Cylinder Electrode for Velocity Sensitivity Testing, *Corrosion*, Vol 40 (No. 5), 1984, p 220
39. B.C. Syrett, Erosion-Corrosion of Copper-Nickel Alloys in Sea Water and Other Environments—A Literature Review, *Corrosion*, Vol 32 (No. 6), 1976, p 242
40. B. Vyas, Erosion-Corrosion, in *Treatise on Materials Science and Technology*, Vol 16, Academic Press, 1979, p 357
41. S.S. Abd. El Rehim, M. Sh. Shalaby, S.M. Abd. El Halum, Effect of Some Anions on the Anodic Dissolution of Delta-S2 Steel in Sulfuric Acid, *Surf. Technol.*, Vol 24, 1985, p 241
42. E. McCafferty, Electrochemical Behavior of Iron Within Crevices in Nearly Neutral Chloride Solutions, *J. Electrochem. Soc.*, Vol 121 (No. 9), 1974, p 1007
43. H. Boehni, Pitting and Crevice Corrosion, in *Corrosion in Power Generating Equipment*, Proceedings of the Eighth International Brown Boveri Symposium, 1984, p 29
44. M. Cohen, Dissolution of Iron, in *Corrosion Chemistry*, G.R. Brubaker and P.B.P. Phipps, Ed., ACS Symposium Series, 89, American Chemical Society, 1979, p 126
45. R. Hausler, Corrosion Inhibition and Inhibitors, in *Corrosion Chemistry*, G.R. Brubaker and P.B.P. Phipps, Ed., ACS Symposium Series, 89, American Chemical Society, 1979, p 262
46. J.D. Costlow and R.C. Tipper, Ed., *Marine Biodeterioration: An Interdisciplinary Study*, Proceedings of the Symposium, Naval Institute Press, 1984
47. D.C. Marshall, *Interfaces in Microbial Ecology*, Harvard University Press, 1976
48. D.C. Savage and M. Fletcher, Ed., *Bacterial Adhesion*, Plenum Press, 1985
49. D.H. Pope, D. Duquette, P.C. Wayner, and A.H. Johannes, *Microbiologically Influenced Corrosion: A State-of-the-Art-Review*, Publication 13, Materials Technology Institute of the Chemical Process Industries, Inc., 1984
50. D.H. Pope, "A Study of Microbiologically Influenced Corrosion in Nuclear Power Plants and a Practical Guide for Countermeasures," EPRI NP-4582, Final Report, Electric Power Research Institute, 1986
51. J.D.A. Miller, Ed., *Microbial Aspects of Metallurgy*, Elsevier, 1970
52. *Microbial Corrosion*, Proceedings of the Conference, National Physical Laboratory, The Metals Society, 1983
53. S.C. Dexter, Ed., *Biologically Induced Corrosion*, Proceedings of the Conference, National Association of Corrosion Engineers, 1986
54. C.A.H. Von Wolzogen Kuhr and L.S. Van der Vlugt, *Water, Den Haag*, Vol 18, 1934, p 147-165

Effects of Metallurgical Variables on Aqueous Corrosion

D.W. Shoesmith, Fuel Waste Technology Branch, Atomic Energy of Canada Ltd.

THE STRUCTURE AND COMPOSITION of both metals and alloys are important in deciding their corrosion characteristics. Indeed, structure and composition are critical in many forms of localized corrosion. For a metal or alloy to corrode evenly, the anodic and cathodic sites must be interchangeable. This implies that every site on the surface is energetically equivalent and therefore equally susceptible to dissolution, but this is never the case.

This article will provide an introduction to the effects of crystal structure, alloying, heat treatments, and the resulting microstructures on corrosion properties. Detailed information on these metallurgical variables for a wide variety of ferrous and nonferrous metals and alloys can be found in the Section "Specific Alloy Systems" in this Volume. Reference should also be made to Volumes 9 and 10 of the 9th Edition of *Metals Handbook* for supplementary data on crystallographic/microstructural analysis and interpretation.

Metals and Metal Surfaces

Metals form as a series of irregular crystals. If these crystals or grains were perfect, the metal atoms would lie in regular close-packed planes. If this were true, the rate of metal dissolution would depend on which crystallographic planes were exposed to the corrosive environment. In addition to these perfect features, there are many sources of atomic disarray within the crystals that can lead to defects where they emerge at the surface. Some of the more significant crystal defects are described below. Additional information on defects can be found in the article "Crystal Structure of Metals" in Volume 9 of the 9th Edition of *Metals Handbook*.

Stacking Faults. The atoms in metals form close-packed layers that stack in various sequences. The most common crystal structures found in metals are the body-centered cubic (bcc), the face-centered cubic (fcc), and the hexagonal close-packed (hcp). The unit cells for these structures are shown schematically in Fig. 1 (a unit cell is a parallelepiped whose edges form the axes of a crystal; it is the smallest pattern of atomic arrangement). If the crystal structures are mixed, resulting in an error in the normal sequence of stacking of atomic layers, stacking faults are produced. These faults can extend for substantial distances through, and across, the crystal.

A slip plane is the lattice plane separating two regions of a crystal that have slipped relative to each other. Such permanent displacements occur under the influence of plastic deformation, as described in the article "Plastic Deformation Structures" in Volume 9 of the 9th Edition of *Metals Handbook*.

Dislocations. Slip regions can be caused by the movement of various small lattice dislocations, such as an additional layer of atoms or a stacking fault on one side of the defect. Dislocations are defects that exist in nearly all real crystals. An edge dislocation, which is the edge of an incomplete plane of atoms within a crystal, is shown in cross section in Fig. 2. In this illustration, the incomplete plane extends partway through the crystal from the top down, and the edge dislocation (indicated by the standard symbol $\perp$) is its lower edge.

If forces are applied (arrows, Fig. 3) to a crystal, such as the perfect crystal shown in Fig. 3(a), one part of the crystal will slip. The edge of the slipped region, shown as a dashed line in Fig. 3(b), is a dislocation. The portion of this line at the left near the front of the crystal and perpendicular to the arrows (Fig. 3b) is an edge dislocation, because the displacement involved is perpendicular to the dislocation.

The slip deformation in Fig. 3(b) has also formed another type of dislocation. The part of the slipped region near the right side, where the displacement is parallel to the dislocation, is termed a screw dislocation. In this part, the crystal no longer consists of parallel planes of atoms, but of a single plane in the form of a helical ramp (screw).

As the slipped region spread across the slip plane, the edge-type portion of the dislocation moved out of the crystal, leaving the screw-type portion still embedded (Fig. 3c). When all of the dislocation finally emerged from the crystal, the crystal was again perfect but with the upper part displaced one unit from the lower part (Fig. 3d).

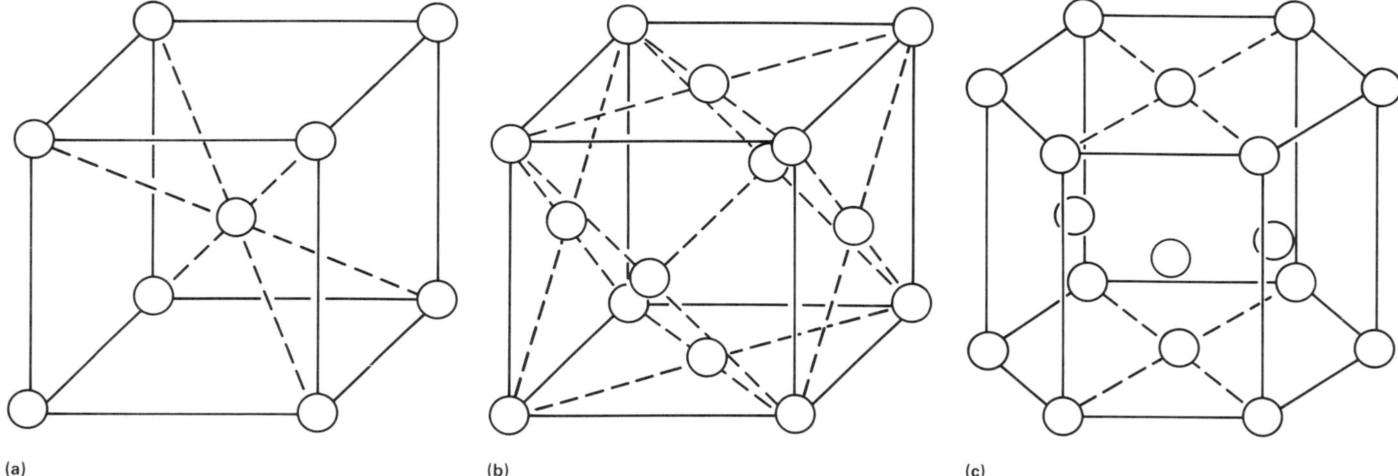

Fig. 1 Schematic of the unit cells for the most common crystal structures found in metals and alloys. (a) bcc. (b) fcc. (c) hcp

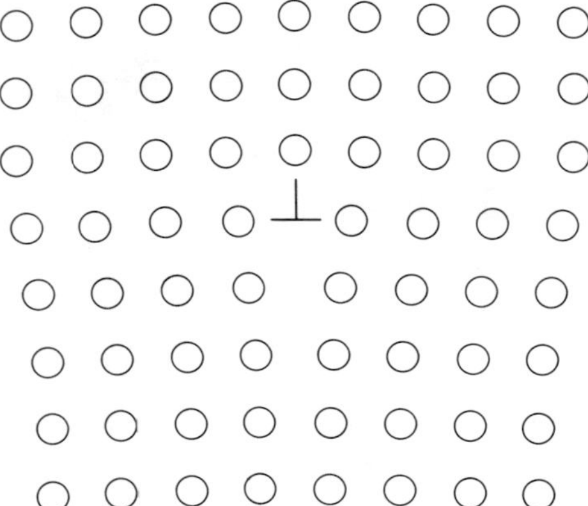

Fig. 2 Schematic of a section through an edge dislocation, which is perpendicular to the plane of the illustration and is indicated by the symbol ⊥

Therefore, Fig. 3 illustrates the mechanism of plastic flow by the slip process, which is actually produced by dislocation movement.

Point defects may be vacancies caused by the absence of one or more atoms in the crystal, impurity atoms of different sizes, and interstitial atoms (small atoms in spaces between the lattice atoms). Point defects can affect significant volumes of the crystal.

Grain Boundaries. The interface between grains is termed the grain boundary, and it is a region of major atomic disarray at which many faults and dislocations congregate. This disarray makes it energetically easier for impurities to concentrate at grain boundaries as opposed to the grain interior, where the atomic arrangement is more regular.

The areas at which these defects emerge on the surface constitute sites of high energy. These energetic sites possess increased chemical activity because each contains atoms with an incomplete number of nearest neighbors (Fig. 4). Atom A, lying within a relatively perfectly close-packed plane, is strongly coordinated on all but one side. Therefore, it has a lower chemical free energy compared to atom B in a step. When compared to atom C at a kink site, atom A has an even lower chemical free energy. Dissolution of kink sites, concentrated at dislocations or grain boundaries, will obviously be accompanied by a greater release of energy than dissolution of atoms from the planes.

Anodic metal dissolution sites are more likely to be found at dislocations and grain boundaries. In electrochemical terms:

$$(E_e)_{\text{kink}} < (E_e)_{\text{step}} < (E_e)_{\text{plane}} \quad \text{(Eq 1)}$$

and because these dislocations tend to concentrate at grain boundaries:

$$(E_e)_{\text{grain boun}} < (E_e)_{\text{grain}} \quad \text{(Eq 2)}$$

where E_e is the equilibrium corrosion potential, as described in the article "Kinetics of Aqueous Corrosion" in this Volume. Therefore, the thermodynamic driving force, ΔE_{therm}, for dissolution (the difference in cathodic and anodic equilibrium potentials):

$$\Delta E_{\text{therm}} = (E_e)_c - (E_e)_a \quad \text{(Eq 3)}$$

will be greater for grain-boundary sites than for the grains themselves. This does not mean that grain boundaries will always corrode preferentially, because the initial etching of the grain boundary will produce an increased surface area in this region. The corresponding additional interfacial energy will decrease the total free energy of the site. The chemical or mechanical activation of the grain boundaries determines whether they will be more active than the grains.

Alloys and Their Surfaces

Pure metals have a low mechanical strength and are rarely used in engineering applications. Stronger metallic materials, which are combinations of several elemental metals known as alloys, are most often used. Commonly used alloys have a good combination of mechanical, physical, fabrication, and corrosion qualities. The specific application determines which of these qualities is deemed most important for alloy selection.

Alloys can be single phase or polyphase, depending on the elements present and their mutual solubilities. For example, the addition of nickel to copper does not alter the fcc structure. The nickel occupies a lattice position within the copper host lattice, and the two metals are said to form a substitutional solid solution. By contrast, the alloying element can occupy an interstitial site in the host lattice and is said to form an interstitial solid solution. An example of such an alloy would be carbon steel, in which the small carbon atom is interstitially accommodated in the iron lattice.

It is often impossible to dissolve a large amount of one element in another. When this is attempted, two or more phases may form. The predominant phase is known as the primary phase, or matrix. The other, smaller phase is known as the secondary phase, or precipitate. Precipitates often contain the nonmetallic elements present in the alloy. If they are insoluble in the matrix, they concentrate, like impurities, at dislocations. This means they are often found at grain boundaries.

Phase Diagrams

Graphs of phase stability as a function of temperature and composition are called phase diagrams. They are based on the equilibrium conditions in the alloy. The stable phases at each temperature and composition are shown on the diagram. If two metallic components are involved, the graph is termed a binary phase diagram. Ternary and quaternary diagrams are necessary for more complex systems.

Figure 5 shows a portion of the binary phase diagram for the Fe-C system. Even for this system, the diagram is complex. As discussed below, this complexity is the key to the wide range of steel properties available.

Iron and Steels

As an example of the diversity of structures possible for a given material and its alloys, the structures and phases possible for iron, carbon steels, and stainless steels will be discussed. The primary purpose in this discussion will be to emphasize the principles, because these systems are covered in detail in the Section "Specific Alloy Systems" in this Volume (especially the articles "Corrosion of Cast Irons," "Corrosion of Carbon Steels," "Corrosion of Alloy Steels," and "Corrosion of Stainless Steels").

Iron. Depending on the temperature of formation, iron can exist in three different phase modifications, or allotropes: ferrite (α-iron), which is bcc; austenite (γ-iron), which is fcc; and δ-ferrite, which is bcc but of slightly different cell dimensions than normal ferrite. These allotropes are shown in Fig. 6.

Cast Irons and Carbon Steels. The most important alloying element of steel is carbon, whose solubility is different in the various phase modifications of iron. The small carbon atoms occupy interstitial positions between the iron

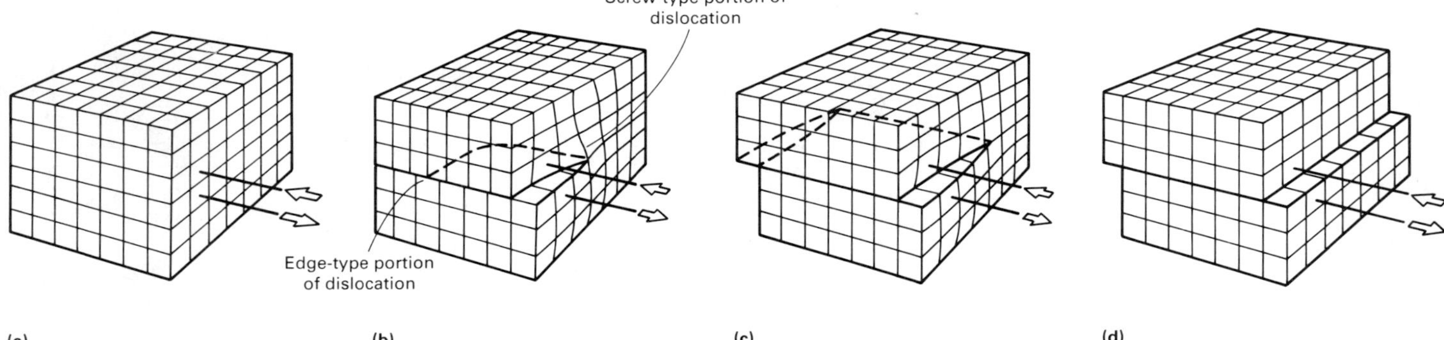

Fig. 3 Schematic representation of four stages of slip deformation by formation and movement of a dislocation (dashed line) through a crystal. (a) Crystal before displacement. (b) Crystal after some displacement. (c) Complete displacement across part of crystal. (d) Complete displacement across entire crystal

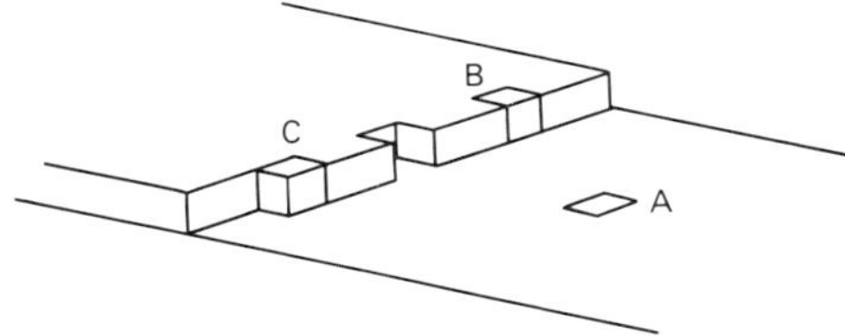

Fig. 4 Schematic of a dislocation emerging at a surface. A, plane atom; B, step atom; C, kink atom

atoms. Consequently, the less densely packed fcc lattice of austenite (γ-iron) can accommodate the carbon atom more readily than the bcc ferrite (α-iron). Therefore, austenite formation is promoted by alloying with carbon. Carbon is said to be an austenite-stabilizing element.

Carbon steels contain less than 2% C; cast irons contain more than 2% C. For steels, any composition can be heated until a homogeneous solid solution of austenite is obtained. This is apparent from the phase diagram shown in Fig. 5. Upon cooling, the four phases (ferrite, austenite, cementite, and martensite) can be formed. The relative proportions of these phases are determined by the carbon content, the rate of cooling, and any subsequent heat treatment.

Cementite is an iron carbide containing 6.67% C (by weight) with the composition Fe_3C. It forms as a mixture with ferrite when cooling slowly from the austenite region of the phase diagram (Fig. 5). The mixture, known as pearlite, forms separate grains, along with ferrite grains, in plain carbon steels. It possesses a lamellar structure with alternate bands of ferrite and cementite. Bainite, an austenite transformation product, is a lathlike aggregate of ferrite and cementite that forms under conditions intermediate to those that result in the formation of pearlite and martensite.

The way austenite is cooled determines the rate of segregation and the grain size of the ferrite and cementite phases. This provides the opportunity to produce a range of carbon steels, each having different mechanical properties. Very rapid cooling, or quenching, produces martensite. Under these conditions, the normal phase separation to produce ferrite and pearlite does not occur. A metastable forced solution of carbon in ferrite is obtained. The forcing of a martensite structure is known as hardening. Additional information on the microstructural constituents of carbon steels can be found in the article "Carbon and Alloy Steels" in Volume 9 of the 9th Edition of *Metals Handbook*.

For a carbon content greater than 2%, the phase diagram in Fig. 5 shows that heating will not bring the mixture into the single-phase region. In other words, a homogeneous solid solution cannot be achieved. Cast irons in this composition region are formed by casting from the molten state. They are used where hardness and corrosion resistance are required and where brittleness due to the cementite content poses no problem.

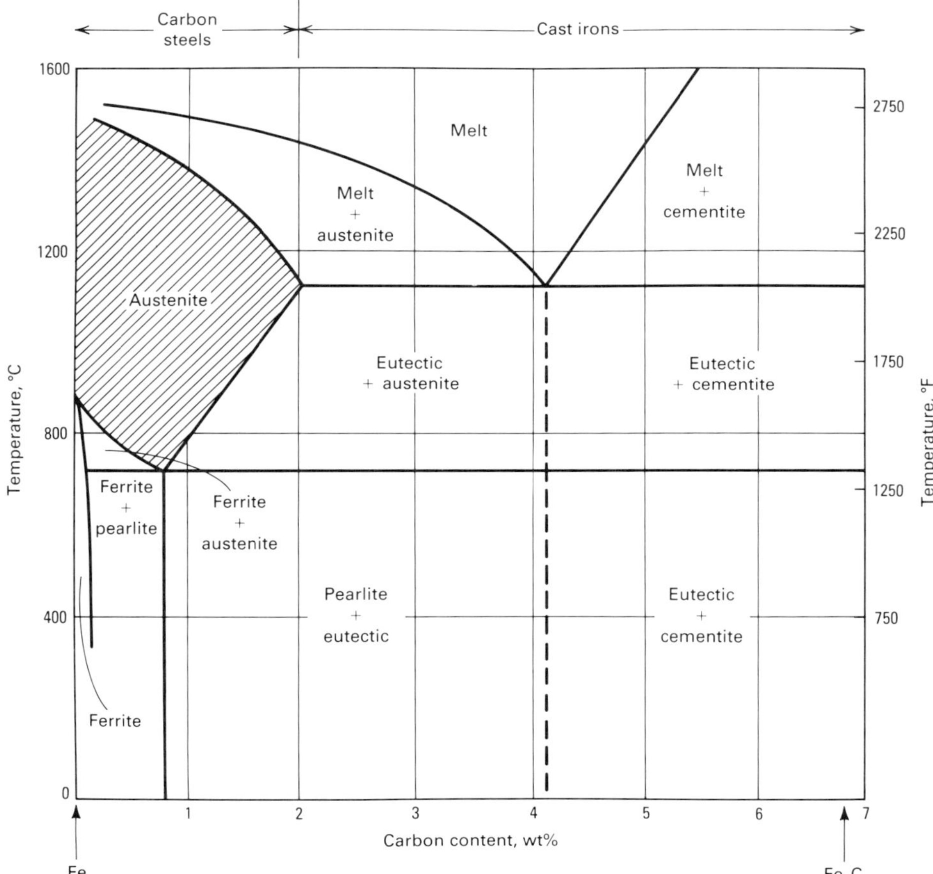

Fig. 5 Portion of the binary phase diagram for the Fe-C system

Stainless Steels. The three major phases in carbon steels—ferrite, austenite and martensite—are also formed in stainless steels. In addition, two other stainless steel categories, ferritic-austenitic (duplex) and precipitate-hardened, can be produced by specific heat treatments (Fig. 6). The microstructural characteristics of iron-chromium and iron-chromium-nickel alloys are discussed in the articles "Wrought Stainless Steels" and "Stainless Steel Casting Alloys" in Volume 9 of the 9th Edition of *Metals Handbook*.

The stability and the mechanical and physical properties of the various phases depend on the combination of alloying elements present. Alloying elements can be divided into two categories:

- *Austenite stabilizers*: carbon, nitrogen, nickel, and manganese
- *Ferrite stabilizers*: silicon, chromium, molybdenum, niobium, and titanium

The selection of a stainless steel for a particular engineering application depends on which mechanical or physical property is considered to be most important.

Effect of Alloying on Corrosion Resistance

One of the primary reasons for producing alloyed, or stainless, steels is to improve corrosion resistance. Alloying can affect corrosion resistance in many different ways.

Increased Nobility. Alloying can have a genuinely thermodynamic effect on corrosion resistance by increasing the nobility of the material. This is achieved by a decrease in ΔE_{therm}, which is expressed by Eq 3 and illustrated in the Evans diagram shown in Fig. 7. Thus:

$$[(E_e)_c - \{(E_e)_a\}_M] > [(E_e)_c - \{(E_e)_a\}_A] \quad \text{(Eq 4)}$$

or:

$$(\Delta E_{therm})_M > (\Delta E_{therm})_A \quad \text{(Eq 5)}$$

and:

$$(i_{corr})_M > (i_{corr})_A \quad \text{(Eq 6)}$$

where i_{corr} is the corrosion current, and the subscripts M and A indicate metal and alloy, respectively. The decrease in corrosion current is caused by an increase in the equilibrium potential for the anodic reaction. The equilibrium potential for the cathodic reaction may also change from metal to alloy, but this change is insignificant compared to the effect on the anodic reaction.

Noble systems of this intermetallic type are rare and generally produced only when the alloying element is a noble metal, such as gold, platinum, or palladium. For such alloys, the less noble constituent metal, for example, titanium in a Ti-2Pd alloy, often dissolves preferentially, leaving a protective film of noble metal on the alloying surface. Disruption of this surface film, which is often thin, will reinitiate corrosion.

Formation of a Protective Film. The addition of controlled amounts of selected alloying elements can often improve the stability and protectiveness of surface oxide films formed on the material surface. Thus, the addition of chromium to iron has a major effect on the corrosion

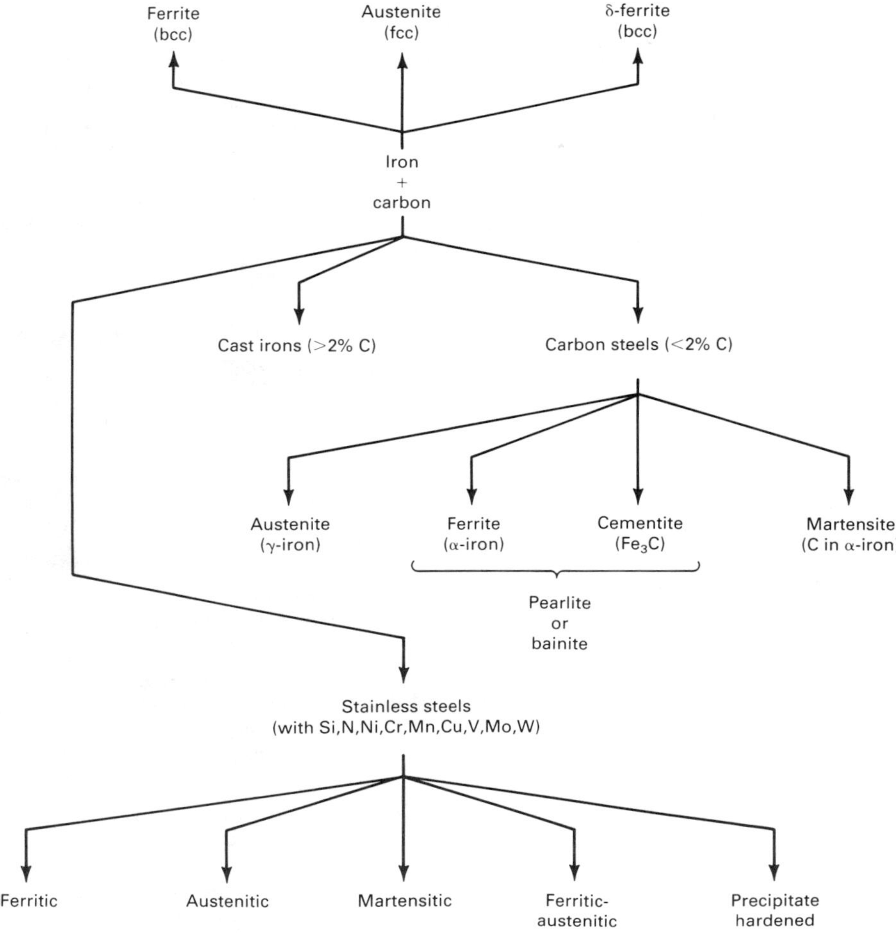

Fig. 6 Phases of iron, carbon steel, and stainless steel

resistance in acid. This can be appreciated by studying the anodic polarization curves for iron-chromium alloys shown in Fig. 8. The curves exhibit an active-passive transition, as discussed in the article "Kinetics of Aqueous Corrosion" in this Volume. The addition of chromium leads to decreases in the critical current for passivation (i_{crit}), the passivation potential (E_{pass}), and the passive current (i_{pass}), as indicated by the arrows in Fig. 8. Therefore, for the cathodic reaction shown, passivation is achievable only in case 3 for a steel containing more than approximately 12% Cr. The improved resistance to corrosion is due to an increase in chromium content of the iron-chromium oxide layer formed on the alloy surface.

The parameters i_{crit}, i_{pass}, and E_{pass} can be decreased even further by the addition of up to 8% Ni. This is the reason for the extensive use of 18-8 stainless steel (Fe-18Cr-8Ni).

Problems of Alloying. Alloying is not without its problems, one of which is illustrated in Fig. 8. As the chromium content is increased, the current due to transpassive dissolution, i_{trans}, at positive potentials also increases. Thus, under very aggressive oxidizing conditions, steels with high chromium contents become less resistant to corrosion because of oxidative dissolution of chromium (as Cr^{6+}, that is, $HCrO_4^-$) from the oxide (containing Cr^{3+}).

Other problems can also be introduced if the alloy is carelessly heat treated in the range 420 to 700 °C (790 to 1290 °F). A high carbon content in chromium steels can lead to the formation of chromium carbide. The carbides separate to the grain boundaries, leaving the steel deficient in chromium close to the grain boundary. This region is consequently less noble and is preferentially attacked. The steel is said to be sensitized to intergranular (intercrystalline) attack. One method of counteracting this process is to make further alloying additions of niobium or titanium, either of which will stabilize the carbon and prevent chromium carbide separation.

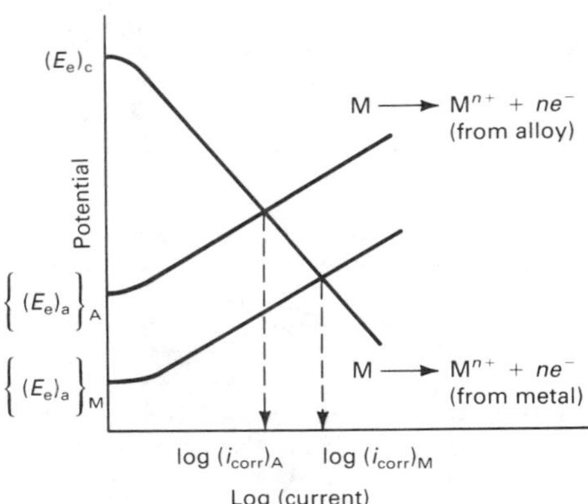

Fig. 7 Evans diagram for metal M in the pure form and incorporated into a more noble alloy

Impurities remaining after the fabrication process can also have a major effect on corrosion resistance. For example, sulfide inclusions in the form of conductive metal sulfides can act as local cathodes in steel and can promote corrosion. Sulfides catalyze the cathodic reactions, as shown by the Evans diagram in Fig. 9. The rate of the proton reduction reaction (determined by the exchange current, i_o, and the Tafel coefficient, b_c, as described in the article "Kinetics of Aqueous Corrosion" in this Volume) is increased at the sulfide inclusions, leading to a higher corrosion potential and increased corrosion current. The corrosion tends to occur locally, and pitting is observed near the inclusion.

The presence of precipitates with minor alloying elements and impurities can lead to problems, because phases with widely different electrochemical properties are then present. This can result in local variations in corrosion resistance. Also, the addition of alloying elements to improve the resistance to general, or uniform, corrosion may cause increased susceptibility to localized corrosion processes, such as pitting or intergranular corrosion.

Effect of Heat Treatment

Many of the mechanical properties of materials are improved by various heat treatments. Unfortunately, such properties as hardness and strength are often achieved at the expense of corrosion resistance. For example, the hardness and strength of martensitic steels are counterbalanced by a lower corrosion resistance than for the ferritic and austenitic steels. The very high strengths achieved for precipitation-hardened steels are due to the secondary precipitates formed during the solution heat treating and aging process. As discussed above, precipitates with electrochemical properties distinctly different from those of the matrix have a deleterious effect on corrosion.

Processes such as cold working, in which the material is plastically deformed into some desired shape, lead to the formation of elongated and highly deformed grains and a decrease in corro-

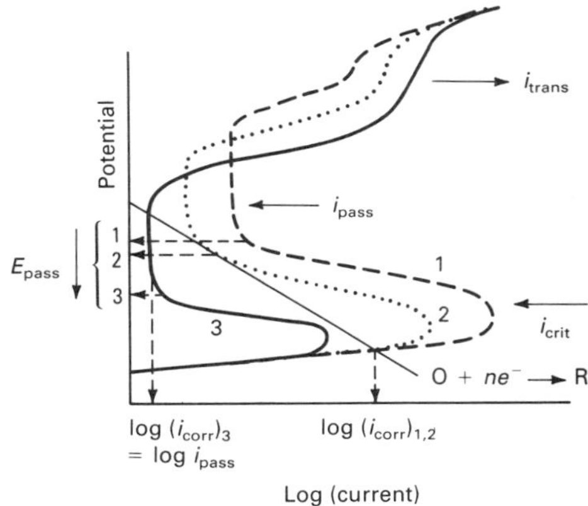

Fig. 8 Schematic anodic polarization diagrams for stainless steels containing various amounts of chromium. (1) 3% Cr; (2) 10% Cr; (3) 14% Cr. The polarization curve for the cathodic reaction $O + ne^- \rightarrow R$ is also shown. Arrows indicate the effect of chromium addition on i_{crit}, E_{pass}, i_{pass}, and i_{trans}.

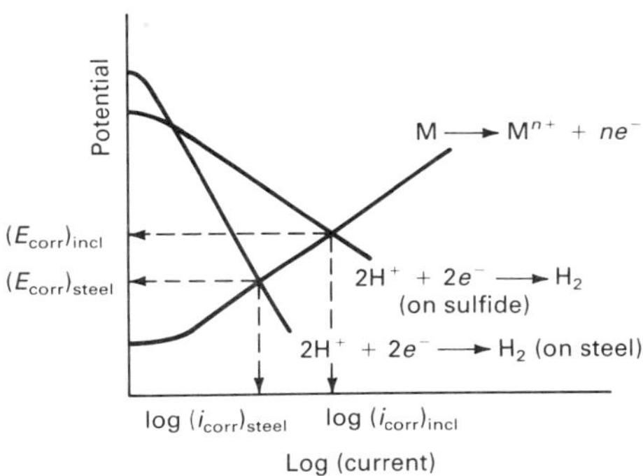

Fig. 9 Evans diagram for steel corrosion showing the increased corrosion rate when the cathodic reaction is catalyzed by a sulfide inclusion

sion resistance. Cold working can also introduce residual stresses that make the material susceptible to stress-corrosion cracking. An improvement in corrosion resistance can be achieved by subsequently annealing at a temperature at which grain recrystallization can occur. A partial anneal leads to stress relief without a major effect on the overall strength of the material.

From the corrosion viewpoint, welding is a particularly troublesome treatment. Because welding involves the local heating of a material, it can lead to phase transformations and the formation of secondary precipitates. It can also induce stress in and around the weld. Such changes can lead to significant local differences in electrochemical properties as well as the onset of such processes as intergranular corrosion. Therefore, the weld filler metal should be as close in electrochemical properties to the base metal as technically feasible, and the weld should be subsequently stress relieved. Detailed information on the corrosion problems associated with welded joints can be found in the article "Corrosion of Weldments" in this Volume.

REFERENCES

1. R.M. Brick, R.B. Gordon, and A. Phillips, *Structure and Properties of Alloys*, McGraw-Hill, 1965
2. L.S. Van Delinder, Ed., *Corrosion Basics—An Introduction*, National Association of Corrosion Engineers, 1984
3. L.L. Shrier, *Corrosion*, George Newnes Ltd., 1963
4. G. Wranglen, *An Introduction to Corrosion and Protection of Metals*, Institut für Metallskydd, 1972

Fundamentals of High-Temperature Corrosion in Molten Salts

John W. Koger, Martin Marietta Energy Systems, Inc.

MOLTEN SALTS, often called fused salts, can cause corrosion by the solution of constituents of the container material, selective attack, pitting, through electrochemical reactions, by mass transport due to thermal gradients, by reaction of constituents of the molten salt with the container material, by reaction of impurities in the molten salt with the container material, and by reaction of impurities in the molten salt with the alloy. Many hundreds of molten salt-metal corrosion studies have been documented, and predictions of corrosion are difficult if not impossible in engineering systems.

The most prevalent molten salts in use are nitrates and halides. Other molten salts that have been extensively studied but are not widely used include carbonates, sulfates, hydroxides, and oxides.

A somewhat general discussion of molten salt corrosion will be presented in this article, with emphasis on nitrates/nitrites and fluorides. Specific examples of results from experiments will be presented for some actual systems; these examples will indicate the scope of a program needed for a particular application.

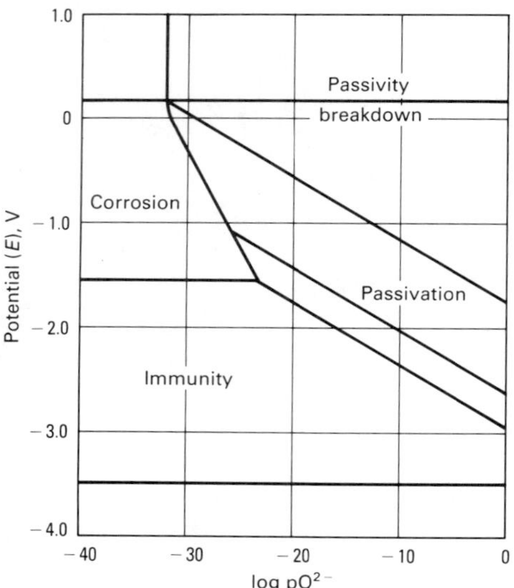

Fig. 1 Typical E versus pO^{2-} diagram for iron in a molten salt at an elevated temperature

Thermodynamics and Kinetics of Molten Salt Corrosion

The chemistry of molten salts can be as complicated as one wishes to make it, based on the definition of a molten salt and whether or not the media may be wholly ionic. For simplicity, most of the processes considered in this article involve electrode processes.

According to Inman and Lovering (Ref 1), except in rare cases in which hydrogen is a part of the molten salt or the melts are exposed to hydrogen atmosphere, the hydrogen ion plays a very small role. The oxygen ions are generally quite important in matters of corrosion. The function pO^{2-} (equivalent to pH in aqueous environments) defines the oxide ion activity. The higher the value of pO^{2-}, the more corrosion of metal will occur. Also, the concentration of oxide ions can influence the corrosive effects of certain nonoxygen-containing melts that have been subject to hydrolysis through contact with atmospheric moisture. In molten salt systems, corrosion is rarely inhibited because of the reactivity of the molten salts and the high temperatures. Molten salts often act as fluxes, thus removing oxide layers on container materials that generally might prove to be protective. Molten salts are generally good solvents for precipitates; therefore, passivation, because of deposits, generally does not occur.

One of the most familiar mechanisms of corrosion arises from ions of metals more noble than the container material, that is, the metal being corroded. In some cases, the more noble metal can be a constituent part of the molten salt, and in others, it can occur as an impurity in the system.

Another mechanism is best described by the example of silver in molten sodium chloride (NaCl). Thermodynamics would not predict a corrosion problem. However, the reaction occurs because sodium, as a result of the formation of silver chloride (AgCl), can dissolve in molten NaCl and distill out of the system. Thus, the reaction proceeds.

If a molten salt contains oxyanion constituents that can be reduced, oxide ions are released. Corrosion will occur on a metal in contact with the salt.

Lastly, oxygen itself can be reduced to oxide ions. However, uncombined oxygen is rarely found in molten salts because of limited solubility.

The potential, E, versus pO^{2-} diagram is often used as the equivalent to the E versus pH (Pourbaix) diagrams for aqueous corrosion (Ref 2). Both of these diagrams are used to establish the stability characteristics of a metal in the respective media. A typical E versus pO^{2-} dia-

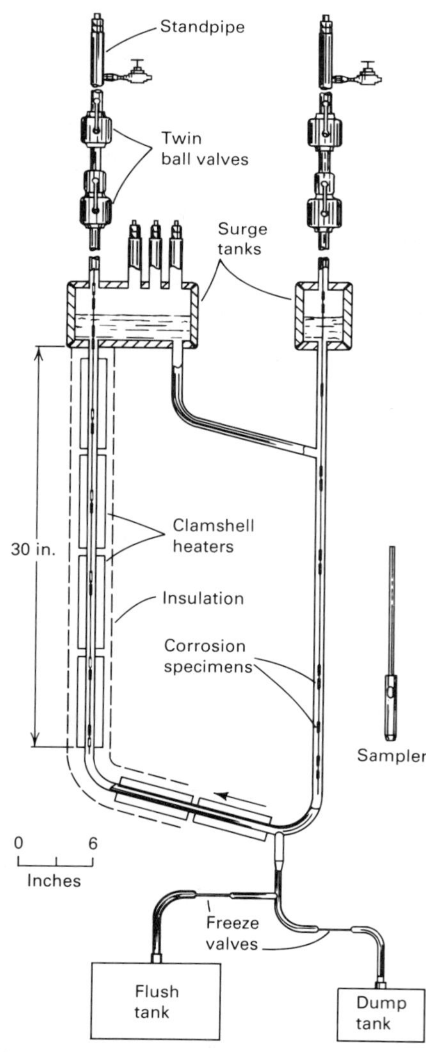

Fig. 2 Natural circulation loop and salt sampler

Table 1 Corrosion rates of iron-base alloys in eutectic molten salt mixtures

Salt mixture	Corrosion rate — Carbon steel µm/yr	Carbon steel mils/yr	Stainless steel µm/yr	Stainless steel mils/yr
$NaNO_3$-NaCl-Na_2SO_4 (86.3,8.4,5.3 mol%, respectively)	15	0.6	1	0.03
KNO_3-KCl (94,6 mol%, respectively)	23	0.9	7.5	0.3
LiCl-KCl (58,42 mol%, respectively)	63	2.5	20	0.8

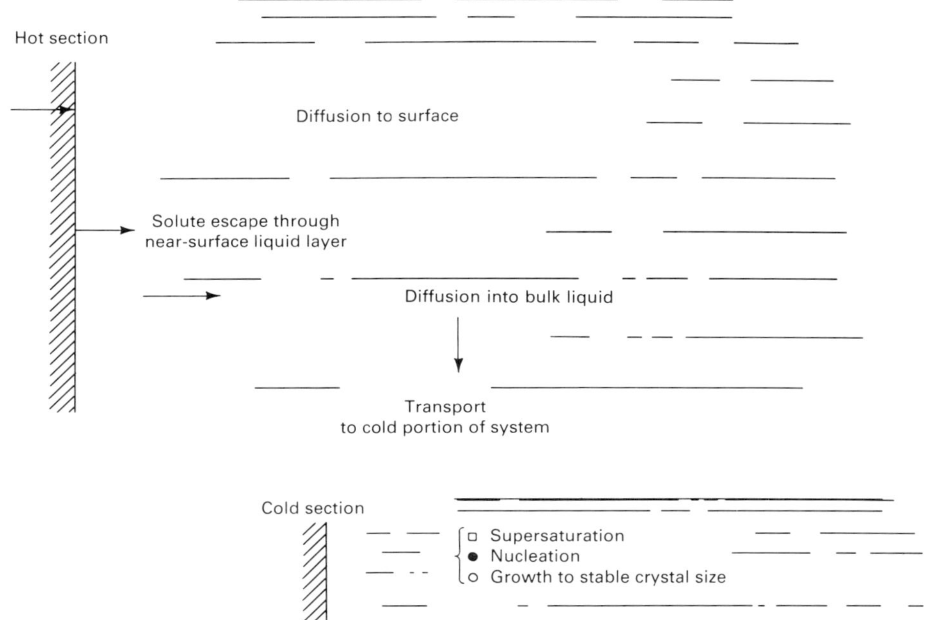

Fig. 3 Temperature-gradient mass transfer

gram for iron in a molten salt at an elevated temperature is shown in Fig. 1. Areas of corrosion, immunity, passivation, and passivity breakdown are evident. Additional information on Pourbaix diagrams used in aqueous corrosion studies is available in the article "Thermodynamics of Aqueous Corrosion" in this Volume.

Actually, the E versus pO^{2-} diagram is probably more useful than the Pourbaix diagram because of the absence of kinetic limitations at elevated temperatures. The following problems, however, do exist:

- Molten salt electrode reactions and the concomitant thermodynamic data are not readily available
- Products from the reactions are often lost by vaporization
- Diagrams based on pure component thermodynamic data are unrealistic because of departure from ideality
- Lack of passivity even where predictions would show passive behavior
- The stable existence of oxides other than the O^{2-} species

Test Methods

A number of kinetic and thermodynamic studies have been carried out in capsule-type containers. These studies can determine the nature of the corroding species and the corrosion products under static isothermal conditions and do provide some much-needed information. However, to provide the information needed for an actual flowing system, corrosion studies must be conducted in thermal convection loops or forced convection loops, which will include the effects of thermal gradients, flow, chemistry changes, and surface area effects. These loops can also include electrochemical probes and gas monitors. An example of the types of information gained from thermal convection loops during an intensive study of the corrosion of various alloys by molten salts will be given below. A thermal convection loop is shown in Fig. 2.

Purification

Molten salts, whether used for experimental purposes or in actual systems, must be kept free of contaminants. This task, which includes initial makeup, transfer, and operation, is specific for each type of molten salt. For example, even though the constituents of the molten fluoride salts used in the Oak Ridge Molten Salt Reactor Experiment were available in very pure grades, purification by a hydrogen/hydrogen fluoride (H_2/HF) gas purge for 20 h was necessary (Ref 3). For nitrates with a melting point of approximately 220 °C (430 °F), purging with argon flowing above and through the salt at 250 to 300 °C (480 to 570 °F) removes significant amounts of water vapor (Ref 4). Another purification method used for this same type of salt consisted of bubbling pure dry oxygen gas through the 350-°C (660-°F) melt for 2 h and then bubbling pure dry nitrogen for 30 min to remove the oxygen (Ref 5). All metals that contact the molten salt during purification must be carefully selected to avoid contamination from transfer tubes, thermocouple wells, the makeup vessel, and the container itself. This selection process may be an experiment in itself.

Nitrates/Nitrites

Nitrate mixtures have probably been studied and used more than any other molten salt group. This is perhaps because of the low operating temperatures possible (200 to 400 °C, or 390 to 750 °F). Steels of varying types are generally chosen for these systems.

As shown in Fig. 1, the E versus pO^{2-} diagram for iron indicates regions of corrosion, immunity, passivity, and passivity breakdown at temperatures of 240 to 400 °C (465 to 750 °F) (Ref 6). In general, the basicity of the melt prevents iron corrosion. Protection by passive films is less reliable, because oxide ion discharge may break down the passive film.

Electropolished iron spontaneously passivates in molten sodium nitrate-potassium nitrite in the temperature range of 230 to 310 °C (445 to 590 °F) at certain potentials (Ref 7). A magnetite (Fe_3O_4) film is formed, along with a reduction of nitrite or any trace of oxygen gas dissolved in the melt. At higher potentials, all reactions occur on the passivated iron. Above the passivation potentials, dissolution occurs with ferric ion soluble in the melt. At even higher potentials, nitrogen oxides are evolved, and nitrate ions dissolve in the nitrite melt. At higher currents, hematite (Fe_2O_3) is formed as a suspension, and NO_2 is detected. Carbon steel in molten sodium nitrate-potassium nitrate ($NaNO_3$-KNO_3) at temperatures ranging from 250 to 450 °C (480 to 840 °F) forms a passivating film consisting mainly of Fe_3O_4 (Ref 5).

Iron anodes in molten alkali nitrates and nitrites at temperatures ranging from 240 to 320 °C (465 to 610 °F) acquire a passive state in both melts. In nitrate melts, the protective Fe_3O_4 oxidizes to Fe_2O_3, and the gaseous products differ for each melt (Ref 8).

An interesting study was conducted on the corrosion characteristics of several eutectic molten salt mixtures on such materials as carbon steel, stainless steel, and Inconel in the temperature range of 250 to 400 °C (480 to 750 °F) in a nonflowing system (Ref 9). The salt mixtures and corrosion rates are given in Table 1. As expected, the corrosion rate was much higher for carbon steel than for stainless steel in the same mixture.

Low corrosion rates were found for both steels in mixtures containing large amounts of alkaline nitrate. The nitrate ions had a passivating effect.

Electrochemical studies showed high resistance to corrosion by Inconel. Again, the sulfate-containing mixture caused less corrosion because of passivating property of the nitrate as well as the preferential adsorption of sulfate ions.

Surface analysis by Auger electron spectroscopy indicated varying thicknesses of iron oxide layers and nickel and chromium layers. The Auger analysis showed that an annealed and air-cooled stainless steel specimen exposed to molten lithium chloride (LiCl)-potassium chloride (KCl) salt had corrosion to a depth five times greater than that of an unannealed stainless steel specimen. Chromium carbide precipitation developed during slow cooling and was responsible for the increased corrosion. The mechanism of corrosion of iron and steel by these molten eutectic salts can be described by the following reactions:

$$Fe \leftrightharpoons Fe^{2+} + 2e^- \quad \text{(Eq 1)}$$

$$LiCl + H_2O \leftrightharpoons LiOH + HCl \quad \text{(Eq 2)}$$

$$H^+ + e^- \leftrightharpoons \tfrac{1}{2}H_2,\ H_2O + 2e^- \leftrightharpoons O^{2-} + H_2 \quad \text{(Eq 3)}$$

$$\tfrac{1}{2}O_2 + 2e^- \leftrightharpoons O^{2-} \quad \text{(Eq 4)}$$

$$Fe^{3+} + e^- \leftrightharpoons Fe^{2+} \quad \text{(Eq 5)}$$

$$Fe^{2+} + O^{2-} \leftrightharpoons FeO \quad \text{(Eq 6)}$$

$$3FeO + O^{2-} \leftrightharpoons Fe_3O_4 + 2e^- \quad \text{(Eq 7)}$$

$$2Fe_3O_4 + O^{2-} \leftrightharpoons 3Fe_2O_3 + 2e^- \quad \text{(Eq 8)}$$

In an actual flowing operating system of KNO_3-NaO_2-NaO_3 (53, 40, and 7 mol%, respectively) at temperatures to 450 °C (840 °F), carbon or chromium-molybdenum steels have been used (Ref 10). For higher temperatures and longer times, nickel or austenitic stainless steels are used. Weld joints are still a problem in both cases.

Alloy 800 and types 304, 304L, and 316 stainless steels were exposed to thermally convective $NaNO_3$-KNO_3 salt (draw salt) under argon at 375 to 600 °C (705 to 1110 °F) for more than 4500 h (Ref 4). The exposure resulted in the growth of thin oxide films on all alloys and the dissolution of chromium by the salt. The weight change data for the alloys indicated that the metal in the oxide film constituted most of the metal loss, that the corrosion rate, in general, increased with temperature, and that, although the greatest metal loss corresponded to a penetration rate of 25 μm/yr (1 mil/yr), the rate was less than 13 μm/yr (0.5 mil/yr) in most cases. These latter rates are somewhat smaller than those reported for similar loops operated with the salt exposed to the atmosphere (Ref 11, 12), but are within a factor of two to five. Spalling had a significant effect on metal loss at intermediate temperatures in the type 304L stainless steel loop. Metallographic examinations showed no evidence of intergranular attack or of significant cold-leg deposits. Weight change data further confirmed the absence of thermal gradient mass transport processes in these draw salt systems.

Raising the maximum temperature of the type 316 stainless steel loop from 595 to 620 °C (1105 to 1150 °F) dramatically increased the corrosion rate (Ref 11, 12). Thus, 600 °C (1110 °F) may be the limiting temperature for use of such alloys in draw salt.

Table 2 Standard Gibbs free energies (ΔG^f) of formation for species in molten 2LiF·BeF_2

Temperature range: 733–1000 K

Material(a)	$-\Delta G^f$ (kcal/mol)	$-\Delta G^f$ (1000 K) (kcal/mol)
LiF(l)	141.8–16.6 × 10^{-3} T K	125.2
BeF_2(l)	243.9–30.0 × 10^{-3} T K	106.9
UF_3(d)	338.0–40.3 × 10^{-3} T K	99.3
UF_4(d)	445.9–57.9 × 10^{-3} T K	97.0
ThF_4(d)	491.2–62.4 × 10^{-3} T K	107.2
ZrF_4(d)	453.0–65.1 × 10^{-3} T K	97.0
NiF_2(d)	146.9–36.3 × 10^{-3} T K	55.3
FeF_2(d)	154.7–21.8 × 10^{-3} T K	66.5
CrF_2(d)	171.8–21.4 × 10^{-3} T K	75.2
MoF_6(g)	370.9–69.6 × 10^{-3} T K	50.2

(a) The standard state for LiF and BeF_2 is the molten 2LiF-BeF_2 liquid. That for MoF_6(g) is the gas at 1 atm. That for all species with d is that hypothetical solution with the solute at unit mole fraction and with the activity coefficient it would have at infinite dilution. Source: Ref 13

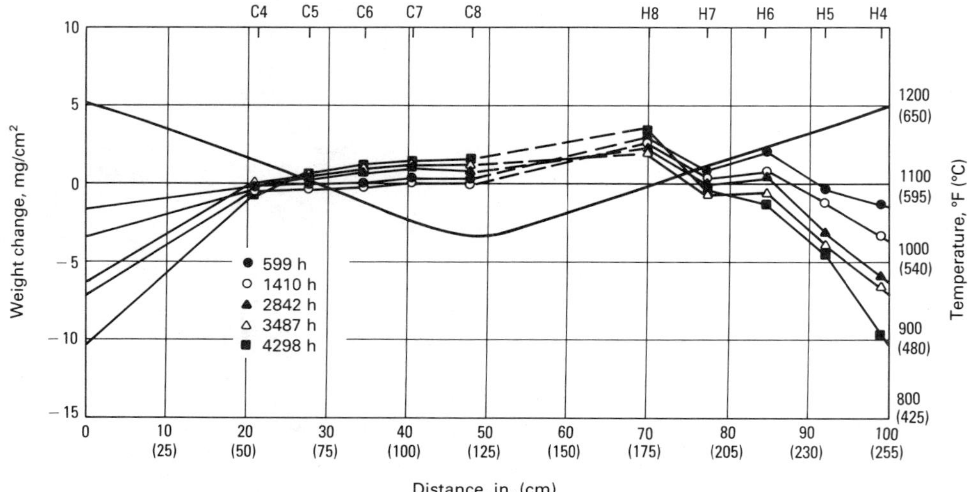

Fig. 4 Weight changes of type 316 stainless steel specimens exposed to LiF-BeF_2-ThF_4-UF_4 (68, 20, 11.7, and 0.3 mol%, respectively) as a function of position and temperature

Fluorides

Because of the Oak Ridge Molten Salt Reactor Experiment, a large amount of work was done on corrosion by molten fluoride salts (Ref 3). Because these molten salts were to be used as heat transfer media, temperature gradient mass transfer was very important. Very small amounts of corrosion can result in large deposits, given that the solubility of the corrosion product changes drastically in the temperature range in question. Many other variables can also cause this phenomenon. Thus, a corrosion rate in itself does not provide complete information about corrosion.

Because the products of oxidation of metals by fluoride melts are quite soluble in the corroding media, passivation is precluded, and the corrosion rate depends on other factors, including the thermodynamic driving force of the corrosion reactions. The design of a practicable system utilizing molten fluoride salts, therefore, demands the selection of salt constituents, such as lithium fluoride (LiF), beryllium fluoride (BeF_2), uranium tetrafluoride (UF_4), and thorium fluoride (ThF_4), that are not appreciably reduced by available structural metals and alloys whose components (iron, nickel, and chromium) can be in near thermodynamic equilibrium with the salt.

A continuing program of experimentation over many years has been devoted to defining the thermodynamic properties of many corrosion product species in molten LiF-BeF_2 solutions. Many of the data have been obtained by direct measurement of equilibrium pressures for such reactions as:

$$H_2(g) + FeF_2(d) \leftrightharpoons Fe(c) + 2HF(g) \quad \text{(Eq 9)}$$

and

$$2HF(g) + BeO(c) \leftrightharpoons BeF_2(l) + H_2O(g) \quad \text{(Eq 10)}$$

where g, c, and d represent gas, crystalline solid, and solute, respectively, using the molten fluoride (denoted l for liquid) as the reaction medium. All of these studies have been reviewed, and the combination of these data with those of other studies has yielded tabulated thermodynamic data for many species in molten LiF-BeF_2 (Table 2). From these data, one can assess the extent to which a uranium trifluoride (UF_3) bearing melt will disproportionate according to the reaction:

$$4UF_3(d) \leftrightharpoons 3UF_4(d) + U(d) \quad \text{(Eq 11)}$$

For the case in which the total uranium content of the salt is 0.9 mol%, as in the Oak Ridge Molten Salt Reactor Experiment, the activity of metallic uranium (referred to the pure metal) is

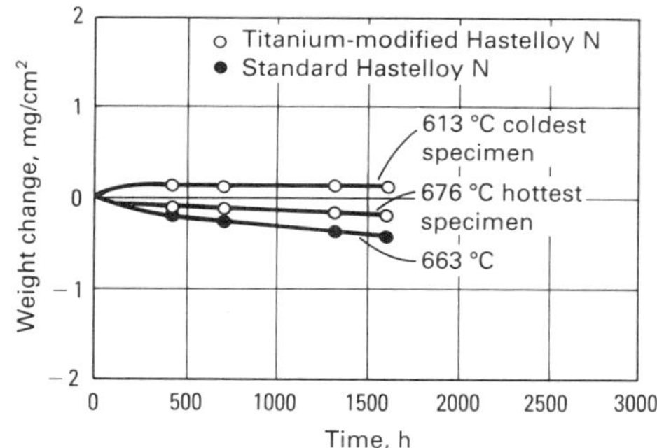

Fig. 5 Weight changes of Hastelloy N specimens versus time of operation in LiF-BeF_2-ThF_4 (73, 2, and 25 mol%, respectively)

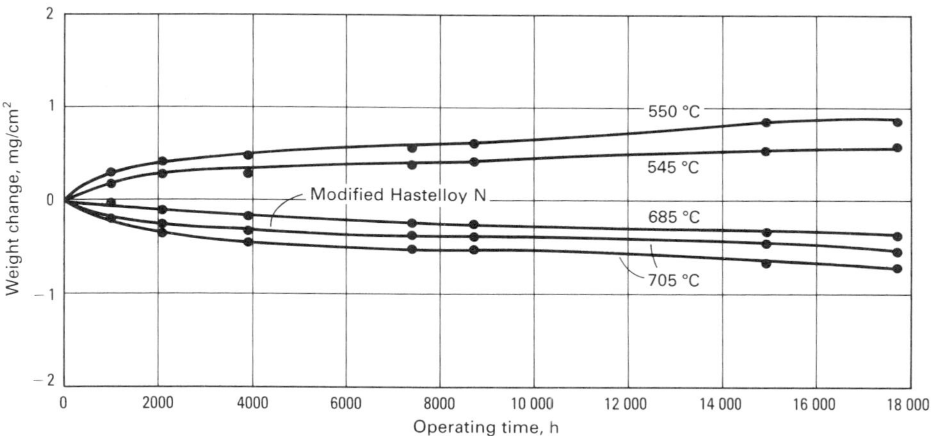

Fig. 6 Weight changes of Hastelloy N exposed to LiF-BeF_2-ThF_4-UF_4 (68, 20, 11.7, and 0.3 mol%, respectively) for various times

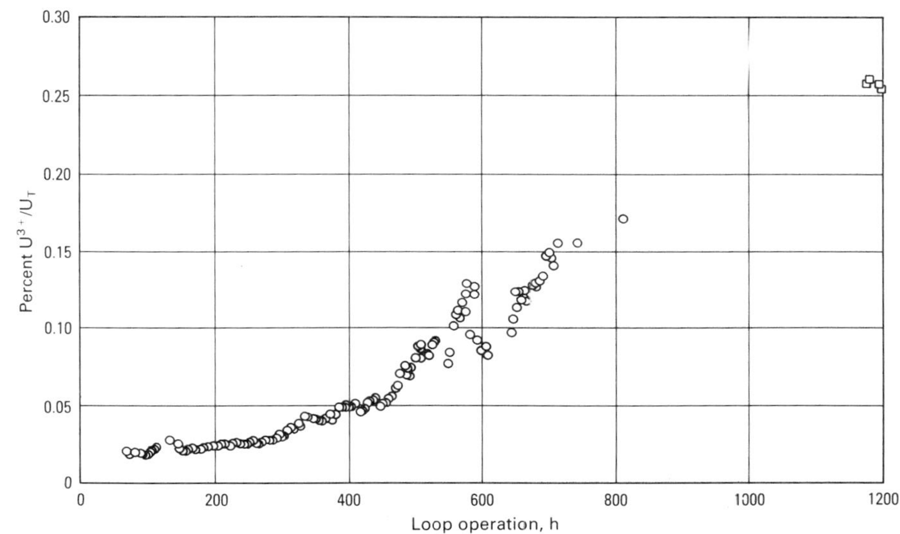

Fig. 7 Uranium (III) in fuel salt

near 10^{-15} with 1% of the UF_4 converted to UF_3 and is near 2×10^{-10} with 20% of the UF_4 so converted (Ref 14). Operation of the reactor with a small fraction (usually 2%) of the uranium present as UF_3 is advantageous insofar as corrosion and the consequences of fission are concerned. Such operation with some UF_3 present should result in the formation of an extremely dilute (and experimentally undetectable) alloy of uranium with the surface of the container metal. Operation with 50% of the uranium as UF_3 would lead to much more concentrated (and highly deleterious) alloying and to formation of uranium carbides. All evidence to date demonstrates that operation with relatively little UF_3 is completely satisfactory.

The data gathered to date reveal clearly that in reactions with structural metals, M:

$$2UF_4(d) + M(c) \leftrightharpoons 2UF_3(d) + MF_2(d) \quad \text{(Eq 12)}$$

chromium is much more readily attacked than iron, nickel, or molybdenum (Ref 14, 15).

Nickel-base alloys, more specifically Hastelloy N (Ni-6.5Mo-6.9Cr-4.5Fe) and its modifications, are considered the most promising for use in molten salts and have received the most attention. Stainless steels, having more chromium than Hastelloy N, are more susceptible to corrosion by fluoride melts, but can be considered for some applications.

Oxidation and selective attack may also result from impurities in the melt:

$$M + NiF_2 \leftrightharpoons MF_2 + Ni \quad \text{(Eq 13)}$$

$$M + 2HF \leftrightharpoons MF_2 + H_2 \quad \text{(Eq 14)}$$

or oxide films on the metal:

$$NiO + BeF_2 \leftrightharpoons NiF_2 + BeO \quad \text{(Eq 15)}$$

followed by reaction of nickel fluoride (NiF_2) with M.

The reactions given in Eq 13 to 15 will proceed essentially to completion at all temperatures. Accordingly, such reactions can lead (if the system is poorly cleaned) to a rapid initial corrosion rate. However, these reactions do not give a sustained corrosive attack.

On the other hand, the reaction involving UF_4 (Eq 12) may have an equilibrium constant that is strongly temperature dependent; therefore, when the salt is forced to circulate through a temperature gradient, a possible mechanism exists for mass transfer and continued attack. Equation 12 is of significance mainly in the case of alloys containing relatively large amounts of chromium.

If nickel, iron, and molybdenum are assumed to form regular or ideal solid solutions with chromium (as is approximately true) and if the circulation rate is very rapid, the corrosion process for alloys in fluoride salts can be simply described. At high flow rates, uniform concentrations of UF_3 and chromium fluoride (CrF_2) are maintained throughout the fluid circuit. Under these conditions, there exists some temperature (intermediate between the maximum and minimum temperatures of the circuit) at which the initial chromium concentration of the structural metal is at equilibrium with the fused salt. This temperature, T_{BP}, is called the balance point.

Table 3 Comparison of weight losses of alloys at approximately 663 °C (1225 °F) in similar flow fuel salts in a temperature gradient system

Alloy	Weight loss, mg/cm²		Average corrosion	
	2490 h	3730 h	µm/yr	mils/yr
Maraging steel	3.0	4.8	14	0.55
Type 304 stainless steel	6.5	10.0	28	1.1
Hastelloy N	0.4	0.6	1.5	0.06

Because the equilibrium constant for the chemical reaction with chromium increases with temperature, the chromium concentration in the alloy surface tends to decrease at temperatures higher than T_{BP} and tends to increase at temperatures lower than T_{BP}. At some point, the dissolution process will be controlled by the solid-state diffusion rate of chromium from the matrix to the surface of the alloy.

In some melts (NaF-LiF-KF-UF_4, for example), the equilibrium constant for Eq 12 with chromium changes sufficiently as a function of temperature to cause the formation of dendritic chromium crystals in the cold zone. For LiF-BeF_2-UF_4-type mixtures, the temperature dependence of the mass transfer reaction is small, and

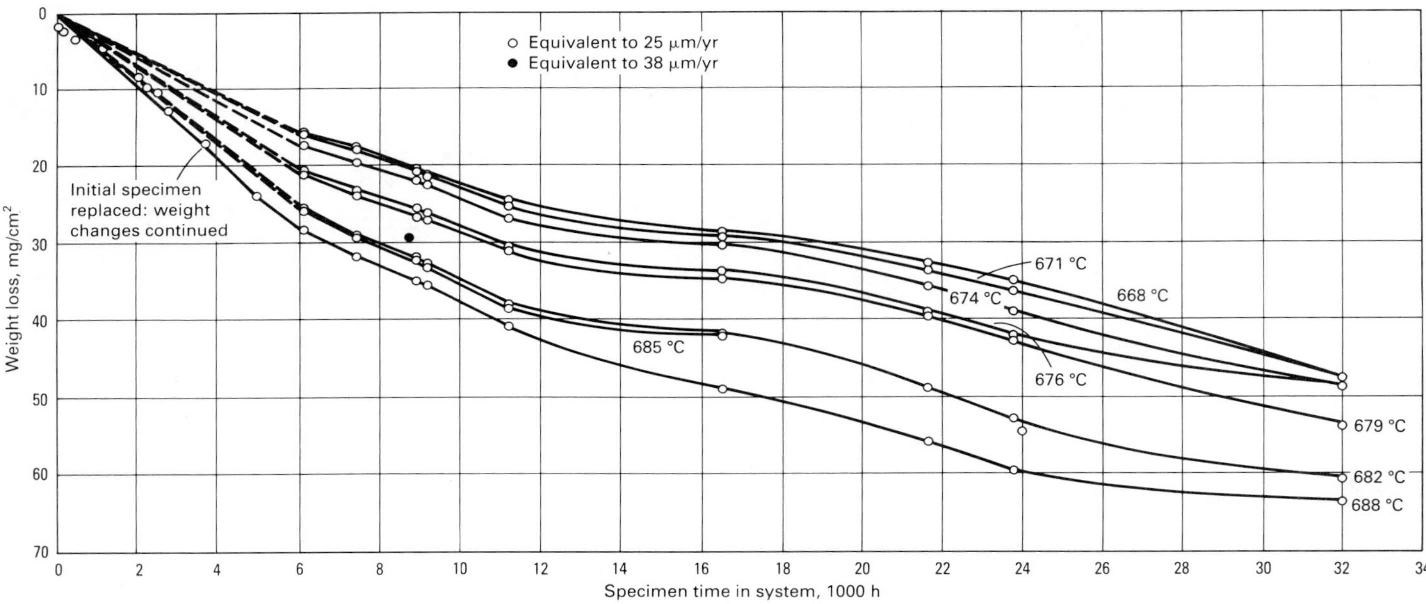

Fig. 8 Weight changes of type 304L stainless steel specimens exposed to $LiF-BeF_2-ZrF_4-ThF_4-UF_4$ (70, 23, 5, 1, and 1 mol%, respectively) for various times and temperatures

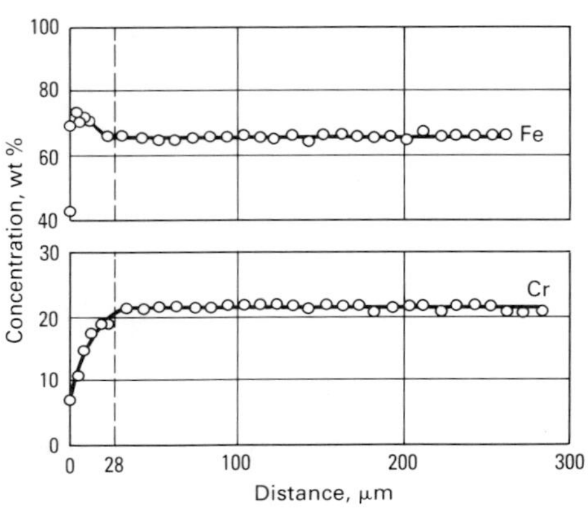

Fig. 9 Chromium and iron concentration gradient in a type 304L stainless steel specimen exposed to $LiF-BeF_2-ZrF_4-ThF_4-UF_4$ (70, 23, 5, 1, and 1 mol%, respectively) for 5700 h at 688 °C (1270 °F)

the equilibrium is satisfied at reactor temperature conditions without the formation of crystalline chromium. Thus, the rate of chromium removal from the salt stream by deposition at cold-fluid regions is controlled by the rate at which chromium diffuses into the cold-fluid wall; the chromium concentration gradient tends to be small, and the resulting corrosion is well within tolerable limits. A schematic of the temperature gradient mass transfer process is shown in Fig. 3.

Lithium fluoride-beryllium fluoride salts containing UF_4 or ThF_4 and tested in thermal convection loops showed temperature gradient mass transfer, as noted by weight losses in the hot leg and weight gains in the cold leg (Fig. 4). Hastelloy N was developed for use in molten fluorides and has proved to be quite compatible. The weight changes of corrosion specimens increased with temperature and time (Fig. 5 and 6). Electrochemical methods were used to determine the oxidation potential of molten fluoride salts in thermal convection loops. The values obtained correlated well with specimen weight change data (Fig. 7).

A type 304L stainless steel exposed to a fuel salt for 9.5 years in a type 304L stainless steel loop showed a maximum uniform corrosion rate of 22 μm/yr (0.86 mil/yr). Voids extended into the matrix for 250 μm (10 mils), and chromium depletion was found (Fig. 8 and 9).

The corrosion resistance of a maraging steel (Fe-12Ni-5Cr-3Mo) at 662 °C (1224 °F) was better than that of type 304L stainless steel, but was worse than that of a Hastelloy N under equivalent conditions. As shown in Table 3, the uniform corrosion rate was 14 μm (0.55 mil/yr). Voids were seen in the microstructure of the specimens after 5700 h, and electron microprobe analysis disclosed a definite depletion of chromium and iron.

Type 316 stainless steel exposed to a fuel salt in a type 316 stainless steel loop showed a maximum uniform corrosion rate of 25 μm/yr (1 mil/yr) for 4298 h. Mass transfer did occur in the system.

For selected nickel- and iron-base alloys, a direct correlation was found between corrosion resistance in molten fluoride salt and chromium and iron content of an alloy. The more chromium and iron in the alloy, the less the corrosion resistance.

Literature on Molten Salt Corrosion

A vast number of publications are noteworthy in connection with molten salt corrosion. Those mentioned below represent sources of information that are particularly helpful. Janz and Tompkins provide an extensive bibliography with over 400 references (Ref 16). Inman and Lovering give an excellent survey of the field with over 200 references (Ref 1). Allen and Janz discuss safety and health hazards (Ref 17). Gale and Lovering provide an overview for researchers considering working with molten salts (Ref 18).

Summary

In order to study the corrosion of molten salts or to determine what materials are compatible with a certain molten salt, the following questions must be answered. What is the purpose of the investigation? Is the researcher interested in basic studies, or is this work for information or work preliminary to assessment for a real system? For basic studies, capsule experiments or information from capsules is sufficient. Otherwise, flow systems or information from flow systems will be needed at some point to assess temperature gradient mass transfer. Salts to be used in either case need to be purified, and the same purity must be used in each experiment unless this factor is a variable. Analytical facilities must be used for the chemistry of the salt, including impurity content and surface analysis of the metals in question.

Vast amounts of useful information can be obtained from capsule and flow experiments. It is hoped that the preceding information on specific systems will provide an appreciation of the problems involved and the material that can be obtained from various experiments.

REFERENCES

1. D. Inman and D.G. Lovering, *Comprehensive Treatise of Electrochemistry*, Vol 7, Plenum Publishing, 1983
2. R. Littlewood, *J. Electrochem. Soc.*, Vol 109, 1962, p 525
3. J.W. Koger, Report ORNL-TM-4286, Oak Ridge National Laboratory, Dec 1972
4. P.F. Tortorelli and J.H. DeVan, Report ORNL-TM-8298, Oak Ridge National Laboratory, Dec 1982
5. A. Baraka, A.I. Abdel-Rohman, and A.A. El Hosary, *Br. Corros. J.*, Vol 11, 1976, p 44
6. S.L. Marchiano and A.J. Arvia, *Electrochim. Acta*, Vol 17, 1972, p 25
7. A.J. Arvia, J.J. Podesta, and R.C.V. Piatti, *Electrochim. Acta*, Vol 16, 1971, p 1797
8. A.J. Arvia, J.J. Podesta, and R.C.V. Piatti, *Electrochim. Acta*, Vol 17, 1972, p 33
9. H.V. Venkatasetty and D.J. Saathoff, *International Symposium on Molten Salts*, 1976, p 329
10. Yu. I. Sorokin and Kh. L. Tseitlin, *Khim. Prom.*, Vol 41, 1965, p 64

11. R.W. Bradshaw, "Corrosion of 304 Stainless Steel by Molten $NaNO_3$-KNO_3 in a Thermal Convection Loop," SAND-80-8856, Sandia National Laboratory, Dec 1980
12. R.W. Bradshaw, "Thermal Convection Loop Corrosion Tests of 316 Stainless Steel and IN800 in Molten Nitrate Salts," SAND-81-8210, Sandia National Laboratory, Feb 1982
13. C.F. Baes, Jr., "The Chemistry and Thermodynamics of Molten Salt Reactor Fuels," Paper presented at the AIME Nuclear Fuel Reprocessing Symposium, Ames, IA, American Institute of Mining, Metallurgical, and Petroleum Engineers, Aug 1969; see also *1969 Nuclear Metallurgy Symposium*, Vol 15, United States Atomic Energy Commission Division of Technical Information Extension
14. G. Long, "Reactor Chemical Division Annual Program Report," ORNL-3789, Oak Ridge National Laboratory, Jan 1965, p 65
15. J.W. Koger, "MSR Program Semiannual Progress Report," ORNL-4622, Oak Ridge National Laboratory, Aug 1970, p 170
16. G.J. Janz and R.P.T. Tompkins, *Corrosion*, Vol 35, 1979, p 485
17. C.B. Allen and G.T. Janz, *J. Hazard. Mater.*, Vol 4, 1980, p 145
18. R.J. Gale and D.G. Lovering, *Molten Salt Techniques*, Vol 2, Plenum Press, 1984, p 1

Fundamentals of High-Temperature Corrosion in Liquid Metals

P.F. Tortorelli, Oak Ridge National Laboratory

CONCERN ABOUT CORROSION of solids exposed to liquid-metal environments, that is, liquid-metal corrosion, dates from the earliest days of metals processing, when it became necessary to handle and contain molten metals. Corrosion considerations also arise when liquid metals are used in applications that exploit their chemical or physical properties. Liquid metals serve as high-temperature reducing agents in the production of metals (such as the use of molten magnesium to produce titanium) and because of their excellent heat transfer properties, they have been used or considered as coolants in a variety of power-producing systems. Examples of such applications include molten sodium for liquid-metal fast breeder reactors and central receiver solar stations as well as liquid lithium for fusion and space nuclear reactors. In addition, tritium breeding in deuterium-tritium fusion reactors necessitates the exposure of lithium atoms to fusion neutrons. Breeding fluids of lithium or lead-lithium are attractive for this purpose. Molten lead or bismuth can serve as neutron multipliers to raise the tritium breeding yield if other types of lithium-containing breeding materials are used. Liquid metals can also be used as two-phase working fluids in Rankine cycle power conversion devices (molten cesium or potassium) and in heat pipes (potassium, lithium, sodium, sodium-potassium). Because of their high thermal conductivities, sodium-potassium alloys, which can be any of a wide range of sodium-potassium combinations that are molten at or near room temperature, have also been used as static heat sinks in automotive and aircraft valves.

Whenever the handling of liquid metals is required, whether in specific uses as discussed above or as melts during processing, a compatible containment material must be selected. At low temperatures, liquid-metal corrosion is often insignificant, but in more demanding applications, corrosion considerations can be important in selecting the appropriate containment material and/or operating parameters. Thus, liquid-metal corrosion studies in support of heat pipe technology and aircraft, space, and fast breeder reactor programs date back many years and, more recently, are being conducted as part of the fusion energy technology program. In this article, the principal corrosion reactions and important parameters that control such processes will be briefly reviewed for materials (principally metals) exposed to liquid-metal environments. Only corrosion phenomena will be covered; liquid-metal cracking and environmental effects on mechanical properties are described in the article "Environmentally Induced Cracking" in this Volume (see in particular the sections "Liquid-Metal Embrittlement" and "Solid-Metal Embrittlement"). Furthermore, the discussion will be limited to corrosion under single-phase (liquid) conditions.

Corrosion Reactions in Liquid-Metal Environments

Liquid-metal corrosion can manifest itself in various ways. In the most general sense, the following categories can be used to classify relevant corrosion phenomena:

- Dissolution
- Impurity and interstitial reactions
- Alloying
- Compound reduction

Definitions and descriptions of these types of reactions are given below. However, it is important to note that this classification is somewhat arbitrary, and as will become clear during the following discussion, the individual categories are not necessarily independent of one another.

Dissolution

The simplest corrosion reaction that can occur in a liquid-metal environment is direct dissolution. Direct dissolution is the release of atoms of the containment material into the melt in the absence of any impurity effects. Such a reaction is a simple solution process and therefore is governed by the elemental solubilities in the liquid metal and the kinetics of the rate-controlling step of the dissolution reaction. The net rate, J, at which an elemental species enters solution, can be described as:

$$J = k(C - c) \quad \text{(Eq 1)}$$

where k is the solution rate constant for the rate-controlling step, C is the solubility of the particular element in the liquid metal, and c is the actual instantaneous concentration of this element in the melt. Under isothermal conditions, the rate of this dissolution reaction would decrease with time as c increases. After a period of time, the actual elemental concentration becomes equal to the solubility, and the dissolution rate is then 0. Therefore, in view of Eq 1, corrosion by the direct dissolution process can be minimized by selecting a containment material whose elements have low solubilities in the liquid metal of interest and/or by saturating the melt before actual exposure. However, if the dissolution kinetics are relatively slow, that is, for low values of k, corrosion may be acceptable for short-term exposures. The functional dependence and magnitude of the solution rate constant, k, depend on the rate-controlling step, which in the simplest cases can be a transport across the liquid-phase boundary layer, diffusion in the solid, or a reaction at the phase boundary. Measurements of weight changes as a function of time for a fixed $C - c$ (see discussion below) yield the kinetic information necessary for determination of the rate-controlling mechanism.

Corrosion resulting from dissolution in a nonisothermal liquid-metal system is more complicated than the isothermal case. Although Eq 1 can be used to describe the net rate at any particular temperature, the movement of liquid—for example, due to thermal gradients or forced circulation—tends to make c the same around the liquid-metal system. Therefore, at temperatures where the solubility (C) is greater than the bulk concentration (c), dissolution of an element into the liquid metal will occur, but at lower temperatures in the circuit where $C < c$, a particular element will tend to come out of solution and be deposited on the containment material (or it may remain as a suspended particulate). A schematic of such a mass transfer process is shown in Fig. 1. If net dissolution or deposition is measured by weight changes, a mass transfer profile such as the one shown in Fig. 2 can be established. Such mass transfer processes under nonisothermal conditions can be of prime importance when, in the absence of dissimilar-metal effects (see below), forced circulation (pumping) of liquid metals used as heat transfer media exacerbates the transport of materials from hotter to cooler parts of the liquid-metal circuit. Normally, the concentration in the bulk liquid, c, rapidly becomes constant with time such that, at given temperature, the concentration driving force ($C - c$, Eq 1) is then also constant. However, much more elaborate analyses based on Eq 1 are required to describe nonisothermal mass transfer precisely. Such

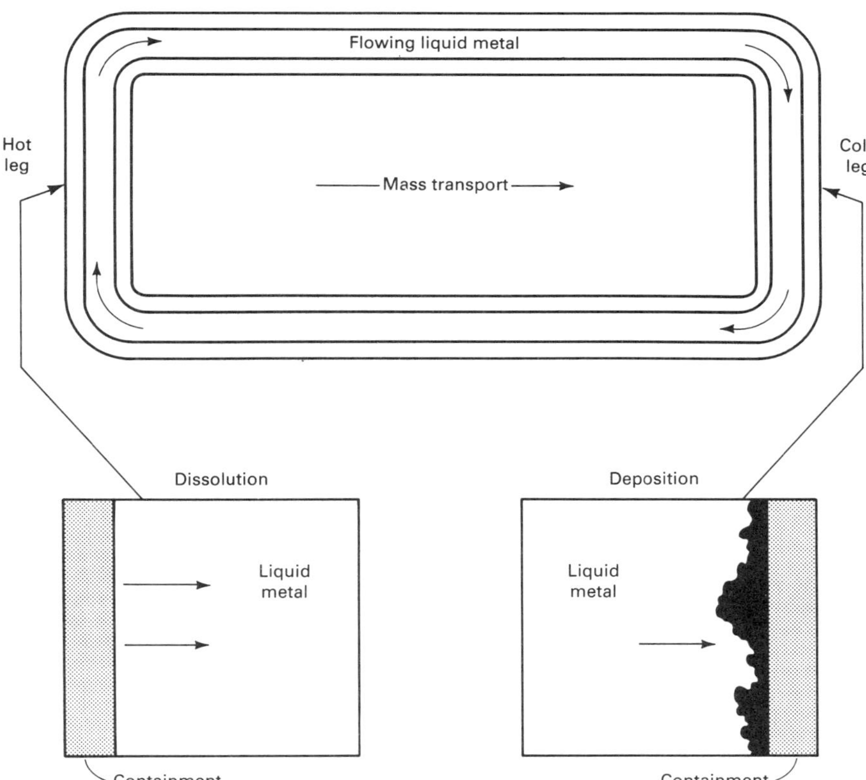

Fig. 1 Schematic of thermal gradient mass transfer in a liquid-metal circuit. Source: Ref 1

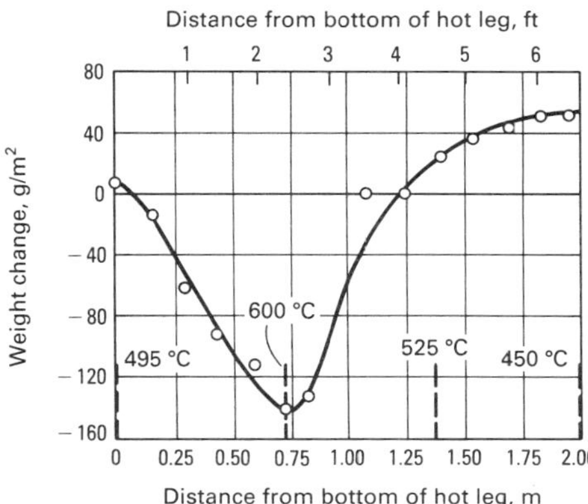

Fig. 2 Mass transfer as characterized by the weight changes of type 316 stainless steel coupons exposed around a nonisothermal liquid lithium type 316 stainless steel circuit for 9000 h. Source: Ref 2

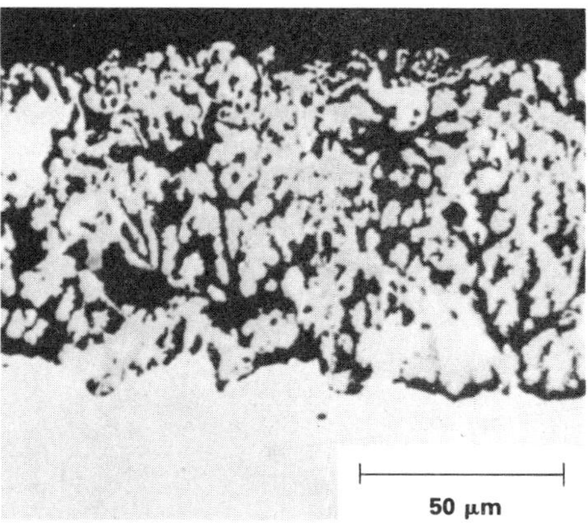

Fig. 3 Polished cross section of type 316 stainless steel exposed to thermally convective Pb-17at.%Li at 500 °C (930 °F) for 2472 h. Source: Ref 3

treatments must take into account the differences in k around the circuit as well as the possibility that the rate constant for dissolution (or deposition) may not vary monotonically with temperature because of changes in the rate-controlling step within the temperature range of dissolution (deposition). The presence of more than one elemental species in the containment material further complicates the analysis; the transfer of each element typically has to be handled with its own set of thermodynamic and kinetic parameters. Although a thermal gradient increases the amount of dissolution, plugging of coolant pipes by nonuniform deposition of dissolved species in cold zones often represents a more serious design problem than metal loss from dissolution (which sometimes may be handled by corrosion allowances). The most direct way to control deposition, however, is usually to minimize dissolution in the hot zone by use of more corrosion-resistant materials and/or inhibition techniques.

Mass transfer may even occur under isothermal conditions if an activity gradient exists in the system. Under the appropriate conditions, dissolution and deposition will act to equilibrate the activities of the various elements in contact with the liquid metal. Normally, such a process is chiefly limited to interstitial element transfer between dissimilar metals, but transport of substitutional elements can also occur. Elimination (or avoidance) of concentration (activity) gradients across a liquid-metal system is the obvious and, most often, the simplest solution to any problems arising from this type of mass transport process.

Under certain conditions, dissolution of metallic alloys by liquid metals can lead to irregular attack (Fig. 3). Although such localized corrosive attack can often be linked to impurity effects (see discussion below) and/or compositional inhomogeneities in the solid, this destabilization of a planar surface is due to preferential dissolution of one or more elements of an alloy exposed to a liquid metal. Indeed, the type of attack illustrated in Fig. 3 is thought to be caused by the preferential dissolution of nickel from an Fe-17Cr-11Ni (wt%) alloy (type 316 stainless steel). As such, this process resembles the dealloying phenomenon sometimes observed in aqueous environments (see the article "Metallurgically Influenced Corrosion" in this Volume for a description of dealloying corrosion). In contrast, an Fe-12Cr-1Mo steel, which did not undergo preferential dissolution of any of its elements, corroded uniformly when exposed under the same environmental conditions (Fig. 4)

Apart from possible effects on morphological development, the changes in surface composition due to preferential elemental dissolution from an alloy into a liquid metal are important in themselves. For example, in austenitic stainless steels exposed to sodium or lithium, the preferential dissolution of nickel causes a phase transformation to a ferritic structure in the surface region. In many cases, an equilibrium surface composition is achieved such that the net elemental fluxes into the liquid metal are in the same proportion as the starting concentrations of these elements in the alloy. Such a phenomenon has been rigorously treated and characterized for sodium-steel systems.

Impurity and Interstitial Reactions

For this discussion, impurity or interstitial reactions refer to the interaction of light elements present in the containment material (interstitials) or the liquid metal (impurities). Examples of such reactions include the decarburization of steel in lithium and the oxidation of steel in sodium or lead of high oxygen activity. In many cases, when the principal elements of the containment material have low solubilities in liquid metals (for example, refractory metals in sodium, lithium, and lead), reactions involving light elements such as oxygen, carbon, and nitrogen dominate the corrosion process. Impurity or interstitial reactions can be generally classified into two types: corrosion product formation and elemental transfer of such species.

Corrosion Product Formation. The general form of a corrosion product reaction is:

$$xL + yM + zI = L_xM_yI_z \quad \text{(Eq 2)}$$

where L is the chemical symbol for a liquid-metal atom, M is one species of the containment material, and I represents an interstitial or impurity atom in the solid or liquid (x, y, $z > 0$). The $L_xM_yI_z$ corrosion product that forms by such a reaction may be soluble or insoluble in the liquid metal. If it is soluble, the I species would cause greater dissolution weight losses and would result in an apparently higher solubility of M in L (Eq 1). This is a frequent cause of erroneous solubil-

Fig. 4 Polished cross section of Fe-12Cr-1MoVW steel exposed to thermally convective Pb-17at.%Li at 500 °C (930 °F) for 2000 h. Source: Ref 3

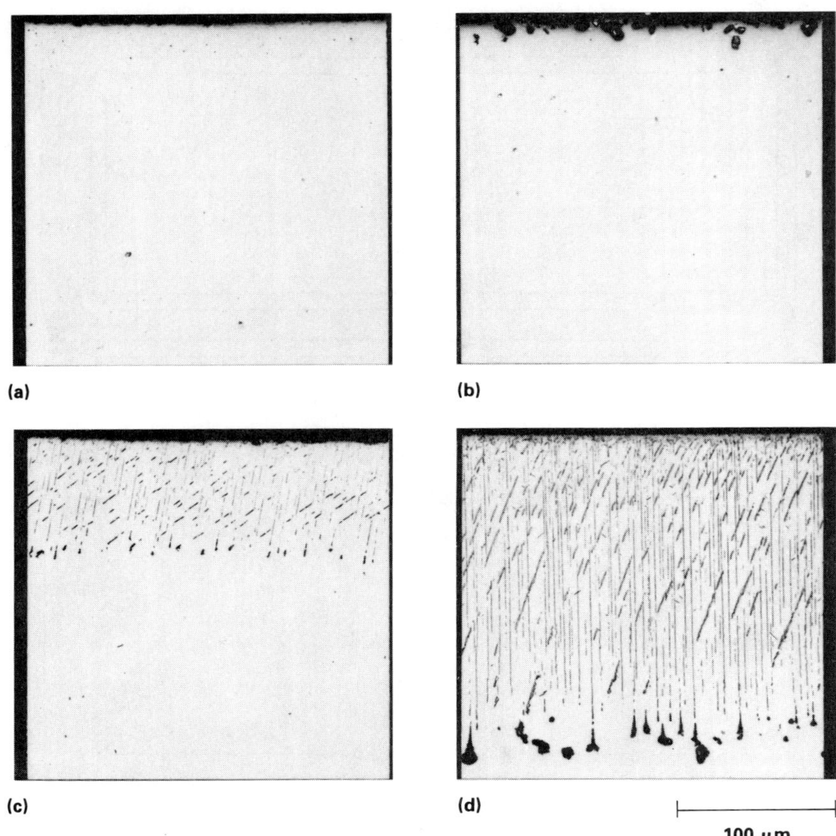

Fig. 5 Effect on initial oxygen concentration (150 to 1700 ppm) in niobium on the depth of attack by lithium. Polished and etched cross sections of niobium exposed to isothermal lithium at 816 °C (1500 °F) for 100 h. (a) 150 ppm. (b) 500 ppm. (c) 1000 ppm. (d) 1700 ppm. Etched with $HF-HNO_3-H_2SO_4-H_2O$. Source: Ref 4

ity measurements and is a good illustration of how dissolution and impurity reactions can be interrelated. Furthermore, if a soluble corrosion product forms at selected sites on the surface of the solid, localized attack will result. Under conditions in which a corrosion product is insoluble, a partial or complete surface layer will form. However, this does not necessarily mean that it can be observed. The product may be unstable outside the liquid metal environment or may dissolve in the cleaning agent used to remove the solidified residue of liquid metal from the exposed containment material.

A good example of the importance of impurity or interstitial reactions that form corrosion products can be found in the sodium-steel-oxygen system. It is thought that the reaction:

$$3Na_2O(l) + Fe(s) = (Na_2O)_2 \cdot FeO(s) + 2Na(l) \quad \text{(Eq 3)}$$

increases the apparent solubility of iron in sodium at higher oxygen activities, while the interaction of oxygen, sodium, and chromium can lead to the formation of surface corrosion products, for example:

$$2Na_2O(l) + Cr(s) = NaCrO_2(s) + 3Na(l) \quad \text{(Eq 4)}$$

This second type of reaction (Eq 4) is of primary importance in the corrosion of chromium-containing steels by liquid sodium. It can be controlled by reducing the oxygen concentration of the sodium to less than about 3 ppm and/or by modifying the composition of the alloy through reduction of the chromium concentration of the steel.

Such corrosion product reactions can also be observed in lithium-steel systems, in which nitrogen can increase the corrosiveness of the liquid-metal environment. In particular, the reaction:

$$5Li_3N(\text{in } l) + Cr(s) = Li_9CrN_5(s) + 6Li(l) \quad \text{(Eq 5)}$$

or an equivalent one with iron, can play an important role in corrosion by liquid lithium. The Li_9CrN_5 corrosion product tends to be localized at the grain boundaries of exposed steels. Such reaction products can probably also be formed when there is sufficient nitrogen in the solid; experimental observations have indicated that nitrogen can increase corrosion by lithium, whether it is in the liquid metal or in the steel.

Corrosion product formation is also important when certain refractory metals are exposed to molten lithium. Despite their low solubilities in lithium, niobium and tantalum can be severely attacked when exposed to lithium if the oxygen activities of these metals are not low. At temperatures below approximately 900 °C (1650 °F), the lithium reacts rapidly with the oxygen and niobium or tantalum (and their oxides and suboxides) to form a ternary oxide corrosion product. Such reactions result in localized penetration along grain boundaries and selected crystallographic planes. This form of corrosive attack can be eliminated, however, by minimizing the oxygen concentration of these refractory metals (Fig. 5) and by using alloying additions that form oxides that do not react with the lithium and that minimize the amount of uncombined oxygen in the material (1 to 2 at.% Zr in niobium and hafnium in tantalum).

A final example of a corrosion product reaction that can occur in a liquid-metal environment is the oxidation of a solid metal or alloy exposed to molten lead. In some cases, this reaction may actually be beneficial by providing a protective barrier against the highly aggressive lead. This barrier can act in a manner analogous to the behavior observed for the protective oxides formed in high-temperature oxidizing gases. However, this surface product will form and then heal only when the oxygen activity of the melt is maintained at a high level or when oxide formers, such as aluminum or silicon, have been added to the containment alloy to promote protection by the formation of alumina- or silica-containing surface products. Furthermore, reactions of additives to the melt with nitrogen in steel to form nitride surface films are thought to be the cause of reduced corrosion in lead and lead-bismuth systems.

Elemental Transfer of Impurities and Interstitials. The second general type of impurity or interstitial reaction is that of elemental transfer. In contrast to what is defined as corrosion product formation, elemental transfer manifests itself as a net transfer of interstitials or impurities to, from, or across a liquid metal. Although compounds may form or dissolve as a result of such transfer, the liquid-metal atoms do not participate in the formation of stable products by reaction with the containment material. For example, because lithium is such a strong thermodynamic sink for oxygen, exposure of oxygen-containing metals and alloys to this liquid often results in the transfer of oxygen to the melt. Indeed, for oxygen-contaminated niobium and tantalum, high-temperature lithium exposures result in the rapid movement of oxygen into the lithium.

The thermodynamic driving force for light element transfer between solid and liquid metals is

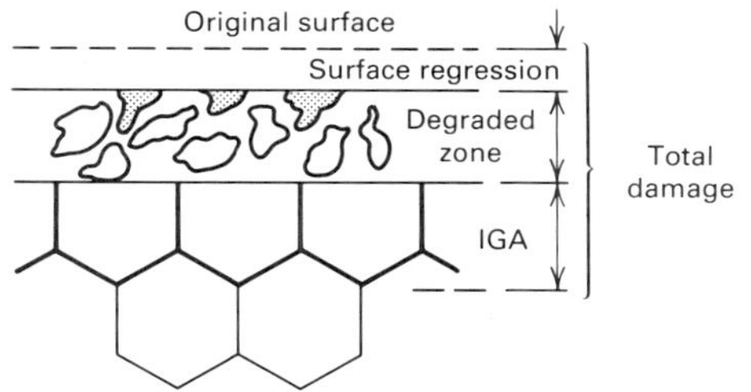

Fig. 6 Representative modes of surface damage in liquid-metal environments. IGA, intergranular attack. Source: Ref 5

normally expressed in terms of a distribution (or partitioning) coefficient. This distribution coefficient is the equilibrium ratio of the concentration of an element, such as oxygen, nitrogen, carbon, or hydrogen, in the solid metal or alloy to that in the liquid. Such coefficients can be calculated from knowledge or estimates of free energies of formation and activities based on equilibrium between a species in the solid and liquid. An example of this approach is its application to decarburization/carburization phenomena in a liquid-metal environment. Carbon transfer to or from the liquid metal can cause decarburization of iron-chromium-molybdenum steels, particularly lower-chromium steels, and carburization of refractory metals and higher-chromium alloys. There have been many studies of such reactions for sodium-steel systems. Although less work has been done in the area of lithium-steel carbon transfer, the same considerations apply. Specifically, the equilibrium partitioning of the carbon between the iron-chromium-molybdenum steel and the lithium can be described as:

$$\frac{C_C(s)}{C_C(Li)} = (a_{Cr})^{x/y}\left[\frac{C_C^\circ(s)}{C_C^\circ(Li)}\right]\exp\left(\frac{+\Delta F^\circ_{1/2Li_2C_2}}{RT}\right)\exp\left(\frac{-\Delta F^\circ_{(1/y)Cr_xC_y}}{RT}\right) \quad \text{(Eq 6)}$$

where $C_C(s)$, $C_C(Li)$ is the concentration of carbon in the steel and lithium, respectively; a_{Cr} is the chromium activity of the steel; $C_C^\circ(s)$, and $C_C^\circ(Li)$ represent the solubilities of carbon in the steel and lithium, respectively; ΔF° represents the free energies of formation of the indicated compounds; x,y is the stoichiometry of the chromium carbide; R is the gas constant; and T is the absolute temperature (in degrees Kelvin).

Equation 6 indicates that in order to decrease the tendency for decarburization of an alloy—that is, to increase the partitioning coefficient, $C_C(s)/C_C(Li)$—the chromium activity of the alloy must be increased or the free energy of formation of the matrix carbide(s) must be lowered (made more negative) by alloy manipulation or thermal treatment to form a more stable carbide dispersion. Experiments in lithium and sodium have shown that these factors have the desired effect. Tempering of iron-chromium-molybdenum steels to yield more stable starting carbides can significantly reduce decarburization by these two liquid metals. With very unstable microstructures, the steel can be severely corroded because of rapid lithium attack of the existing carbides. Furthermore, alloying additions, such as niobium, form very stable carbides and can dramatically reduce decarburization. In addition, as shown by Eq 6, increasing the chromium level of a steel effectively decreases the tendency for carbon loss. With higher-chromium steels, for example, austenitic stainless steels, carburization can then become a problem. If two dissimilar steels of significantly differing chromium activities and/or microstructures are exposed to the same liquid metal, the melt can act as a conduit for the relatively rapid redistribution of carbon between the two solids. Similar considerations would apply for any light element transfer across a liquid metal in contact with dissimilar materials; this can be further complicated by concentration (activity) gradient mass transfer of substitutional elements, as discussed above.

Alloying

Reactions between atoms of the liquid metal and those of the constituents of the containment material may lead to the formation of a stable product on the solid without the participation of impurity or interstitial elements:

$$xM + yL = M_xL_y \quad \text{(Eq 7)}$$

This is not a common form of liquid-metal corrosion particularly with the molten alkali metals, but it can lead to detrimental consequences if it is not understood or expected. Alloying reactions, however, can be used to inhibit corrosion by adding an element to the liquid metal to form a corrosion-resistant layer by reaction of this species with the contaminant material. An example is the addition of aluminum to a lithium melt contained by steel. A more dissolution-resistant aluminide surface layer forms, and corrosion is reduced.

Compound Reduction

Attack of ceramics exposed to liquid metals can occur because of reduction of the solid by the melt. In very aggressive situations, such as when most oxides are exposed to molten lithium, the effective result of such exposure is the loss of structural integrity by reduction-induced removal of the nonmetallic element from the solid. The tendency for reaction under such conditions can be qualitatively evaluated by consideration of the free energy of formation of the solid oxide relative to the oxygen/oxide stability in the liquid metal. Similar considerations apply to the evaluation of potential reactions between other nonmetallic compounds (nitrides, carbides, and so on) and liquid metals.

Table 1 Guidelines for materials selection and/or alloy development based on liquid-metal corrosion reactions

Corrosion reaction	Guidelines	Example
Direct dissolution	Lower activity of key elements.	Reduce nickel in lithium, lead, or sodium systems.
Corrosion product formation	Lower activity of reacting elements.	Reduce chromium and nitrogen in lithium systems.
	In case of protective oxide, add elements to promote formation.	Add aluminum or silicon to steel exposed to lead.
Elemental transfer	Increase (or add) elements to decrease transfer tendency.	Increase chromium content in steels exposed to sodium or lithium.
	Minimize element being transferred.	Reduce oxygen content in metals exposed to lithium.
Alloying	Avoid systems that form stable compounds.	Do not expose nickel to molten aluminum.
	Promote formation of corrosion-resistant layers by alloying.	Add aluminum to lithium to form surface aluminides.
Compound reduction	Eliminate solids that can be reduced by liquid metal.	Avoid bulk oxide-lithium couples.

Considerations in Materials Selection

The above types of corrosion reactions must be considered in materials selection for liquid-metal containment. In many cases, particularly at low temperature or with less aggressive liquids (such as molten steel), liquid-metal corrosion is not an important factor, and many materials, both metals and ceramics, would suffice. Under more severe conditions, however, an understanding of the various types of liquid-metal corrosion is necessary to select or develop a compatible containment material. For example, for applications in high-temperature molten lithium, most oxides would be unstable with respect to this liquid metal, low-chromium steels would decarburize, and alloys containing large amounts of nickel or manganese would suffer extensive preferential dissolution and irregular attack. Materials selection would then be limited to higher-chromium ferritic/martensitic steels or high-purity refractory metals and alloys.

A general summary of the types of the most common corrosion reactions and guidelines for materials selection and/or development is given in Table 1, which also includes typical examples for each category. Because two or more concurrent corrosion reactions are possible, and because consideration of all of the applicable materials consequences may lead to opposite strategies, materials selection for liquid-metal environments can become quite complex and may require optimization of several factors rather than minimization of any particular one. In addition, an assessment of the suitability of a given material for liquid-metal service must be based on the knowledge of its total corrosion response. As in many corrosive environments, a simple numerical rate is not an accurate measurement of the susceptibility of a material when reaction with the liquid metal results in more than one of the modes of attack shown in Fig. 6 and discussed above. Under such circumstances, a measurement reflecting total corrosion damage is much more appropriate for judging the ability of a material to resist corrosion by a particular liquid metal. Additional examples of liquid-metal corrosion can be found in the article "General Corrosion" (see the section "Liquid-Metal Dissolution") in this Volume.

ACKNOWLEDGMENT

Research sponsored by the Office of Fusion Energy, U.S. Department of Energy under Contract No. DE-AC05-84OR21400 with the Martin Marietta Energy Systems, Inc.

REFERENCES

1. J.E. Selle and D.L. Olson, in *Materials Considerations in Liquid Metal Systems in Power Generation*, National Association of Corrosion Engineers, 1978, p 15-22
2. P.F. Tortorelli and J.H. DeVan, *J. Nucl. Mater.*, Vol 85 and 86, 1979, p 289-293
3. P.F. Tortorelli and J.H. DeVan, *J. Nucl. Mater.*, Vol 141-143, 1986, p 592-598
4. J.R. DiStefano and E.E. Hoffman, Corrosion Mechanisms in Refractory Metal-Alkali Metal Systems, *At. Energy Rev.*, Vol 2, 1964, p 3-33
5. J.H. DeVan and C. Bagnall, in *Proceedings of the International Conference on Liquid Metal Engineering and Technology*, Vol 3, The British Nuclear Energy Society, 1985, p 65-72

SELECTED REFERENCES

- T.L. Anderson and G.R. Edwards, The Corrosion Susceptibility of 2¹/₄ Cr-1 Mo Steel in a Lithium-17.6 Wt Pct Lead Liquid, *J. Mater. Energy Syst.*, Vol 2, 1981, p 16-25
- R.C. Asher, D. Davis, and S.A. Beetham, Some Observations on the Compatibility of Structural Materials With Molten Lead, *Corros. Sci.*, Vol 17, 1977, p 545-547
- M.G. Barker, S.A. Frankham, and N.J. Moon, The Reactivity of Dissolved Carbon and Nitrogen in Liquid Lithium, in *Proceedings of the Third International Conference on Liquid Metal Engineering and Technology*, Vol 2, The British Nuclear Energy Society, 1984, p 77-83
- M.G. Barker, P. Hubberstey, A.T. Dadd, and S.A. Frankham, The Interaction of Chromium With Nitrogen Dissolved in Liquid Lithium, *J. Nucl. Mater.*, Vol 114, 1983, p 143-149
- N.M. Beskorovainyi, V.K. Ivanov, and M.T. Zuev, Behavior of Carbon in Systems of the Metal-Molten Lithium-Carbon Type, in *High-Purity Metals and Alloys*, V.S. Emel'yanov and A.I. Evstyukhin, Ed., Consultants Bureau, 1967, p 107-119
- N.M. Beskorovainyi and V.K. Ivanov, Mechanism Underlying the Corrosion of Carbon Steels in Lithium, in *High-Purity Metals and Alloys*, V.S. Emel'yanov and A.I. Evstyukhin, Ed., Consultants Bureau, 1967, p 120-129
- O.K. Chopra, K. Natesan, and T.F. Kassner, Carbon and Nitrogen Transfer in Fe-9Cr-Mo Ferritic Steels Exposed to a Sodium Environment, *J. Nucl. Mater.*, Vol 96, 1981, p 269-284
- O.K. Chopra and P.F. Tortorelli, Compatibility of Materials for Use in Liquid-Metal Blankets of Fusion Reactors, *J. Nucl. Mater.*, Vol 122 and 123, 1984, p 1201-1212
- L.F. Epstein, Static and Dynamic Corrosion and Mass Transfer in Liquid Metal Systems, in *Liquid Metals Technology—Part I*, Vol 53 (No. 20), Chemical Engineering Progress Symposium Series, American Institute of Chemical Engineers, 1957, p 67-81
- J.D. Harrison and C. Wagner, The Attack of Solid Alloys by Liquid Metals and Salt Melts, *Acta Metall.*, Vol 1, 1959, p 722-735
- E.E. Hoffman, "Corrosion of Materials by Lithium at Elevated Temperatures," ORNL-2674, Oak Ridge National Laboratory, March 1959
- A.R. Keeton and C. Bagnall, Factors That Affect Corrosion in Sodium, in *Proceedings of the Second International Conference on Liquid Metal Technology in Energy Production*, CONF-800401-P1, J.M. Dahlke, Ed., U.S. Department of Energy, 1980, p 7-18 to 7-25
- B.H. Kolster, The Influence of Sodium Conditions on the Rate for Dissolution and Metal/Oxygen Reaction of AISI 316 in Liquid Sodium, in *Proceedings of the Second International Conference on Liquid Metal Technology in Energy Production*, CONF-800401-P1, J.M. Dahlke, Ed., U.S. Department of Energy, 1980, p 7-53 to 7-61
- J. Konys and H.U. Borgstedt, The Product of the Reaction of Alumina With Lithium Metal, *J. Nuclear Mater.*, Vol 131, 1985, p 158-161
- K. Natesan, Influence of Nonmetallic Elements on the Compatibility of Structural Materials With Liquid Alkali Metals, *J. Nucl. Mater.*, Vol 115, 1983, p 251-262
- D.L. Olson, P.A. Steinmeyer, D.K. Matlock, and G.R. Edwards, Corrosion Phenomena in Molten Lithium, *Rev. Coatings Corros./Int. Quart. Rev.*, Vol IV, 1981, p 349-434
- A.J. Romano, C.J. Klamut, and D.H. Gurinsky, "The Investigation of Container Materials for Bi and Pb Alloys, Part I. Thermal Convection Loops," BNL-811, Brookhaven National Laboratory, July 1963
- E. Ruedl, V. Coen, T. Sasaki, and H. Kolbe, Intergranular Lithium Penetration of Low-Ni, Cr-Mn Austenitic Stainless Steels, *J. Nucl. Mater.*, Vol 110, 1982, p 28-36
- J. Sannier and G. Santarini, Etude de la Corrosion de Deux Aciers Ferritiques par le Plomb Liquide Circulant dans un Thermosiphon; Recherche d'un Modele, *J. Nucl. Mater.*, Vol 107, 1982, p 196-217
- S.A. Shields, C. Bagnall, and S.L. Schrock, Carbon Equilibrium Relationships for Austenitic Stainless Steel in a Sodium Environment, *Nucl. Technol.*, Vol 23, 1974, p 273-283
- S.A. Shields and C. Bagnall, Nitrogen Transfer in Austenitic Sodium Heat Transport Systems, in *Material Behavior and Physical Chemistry in Liquid Metal Systems*, H.U. Borgstedt, Ed., Plenum Press, 1982, p 493-501
- R.N. Singh, Compatibility of Ceramics With Liquid Na and Li, *J. Amer. Ceram. Soc.*, Vol 59, 1976, p 112-115
- D.L. Smith and K. Natesan, Influence of Nonmetallic Impurity Elements on the Compatibility of Liquid Lithium With Potential CTR Containment Materials, *Nucl. Technol.*, Vol 22, 1974, p 392-404
- A.W. Thorley, Corrosion and Mass Transfer Behaviour of Steel Materials in Liquid Sodium, in *Proceedings of the Third International Conference on Liquid Metal Engineering and Technology*, Vol 3, The British Nuclear Energy Society, 1985, p 31-41
- P.F. Tortorelli and J.H. DeVan, Mass Transfer Kinetics in Lithium-Stainless Steel Systems, in *Proceedings of the Third International Conference on Liquid Metal Engineering and Technology*, Vol 3, The British Nuclear Energy Society, 1985, p 81-88
- P.F. Tortorelli and J.H. DeVan, Effects of a Flowing Lithium Environment on the Surface Morphology and Composition of Austenitic Stainless Steel, *Microstruct. Sci.*, Vol 12, 1985, p 213-226
- J.R. Weeks and H.S. Isaacs, Corrosion and Deposition of Steels and Nickel-Base Alloys in Liquid Sodium, *Adv. Corros. Sci. Technol.*, Vol 3, 1973, p 1-66

Fundamentals of Corrosion in Gases

Samuel A. Bradford, Department of Mining, Metallurgical and Petroleum Engineering, University of Alberta

ENGINEERING METALS react chemically when exposed to air or to other more aggressive gases. Whether they survive or not depends on how fast they react. For a few metals, the reaction is so slow that they are virtually unattacked, but for others, the reaction can be disastrous. High-temperature service is especially damaging to most metals because of the exponential increase in reaction rate with temperature.

The most common reactant is oxygen in the air; therefore, all gas-metal reactions are usually referred to as oxidation, using the term in its broad chemical sense whether the reaction is with oxygen, water vapor, hydrogen sulfide (H_2S), or whatever the gas might be. Throughout this article, the process will be called oxidation, and the corrosion product will be termed oxide.

Corrosion in gases differs from aqueous corrosion in that electrochemical principles do not help greatly in understanding the mechanism of oxidation. For gaseous reactions, a fundamental knowledge of the diffusion processes involved is much more useful. The principles of high-temperature oxidation began to be understood only in the 1920s, whereas electrochemistry and aqueous corrosion principles were developed approximately 100 years earlier. The first journal devoted to corrosion in gases (*Oxidation of Metals*) began publication less than 20 years ago.

In this article, a short summary of thermodynamic concepts is followed by an explanation of the defect structure of solid oxides and the effect of these defects on conductivity and diffusivity. The commonly observed kinetics of oxidation will be described and related to the corrosion mechanisms. These mechanisms are shown schematically in Fig. 1. The gas first adsorbs on the metal surface as atomic oxygen. Oxide nucleates at favorable sites and most commonly grows laterally to form a complete thin film. As the layer thickens, it provides a protective scale barrier to shield the metal from the gas. For scale growth, electrons must move through the oxide to reach the oxygen atoms adsorbed on the surface, and oxygen ions, metal ions, or both must move through the oxide barrier. Oxygen may also diffuse into the metal.

Growth stresses in the scale may create cavities and microcracks in the scale, modifying the oxidation mechanism or even causing the oxide to fail to protect the metal from the gas. Improved oxidation resistance can be achieved by developing better alloys and by applying protective coatings. The basic principles of alloy oxidation, discussed in the section "Alloy Oxidation: The Doping Principle" in this article, are applicable to both alloy development and use of metallic coatings for protection against corrosive gases.

Fundamental Data. Essential to an understanding of the gaseous corrosion of a metal are the crystal structure and the molar volume of the metal on which the oxide builds, both of which may affect growth stresses in the oxide. For high-temperature service the melting point of the metal, which indicates the practical temperature limits, and the structural changes that take place during heating and cooling, which affect oxide adherence, must be known. These data are presented in Table 1 for pure metals. For the oxides, their structures, melting and boiling points, molar volume, and oxide/metal volume ratio (Pilling-Bedworth ratio) are shown in Table 2. The structure data were taken from many sources.

Thermodynamics of High-Temperature Corrosion in Gases

Free Energy of Reaction. The driving force for reaction of a metal with a gas is the Gibbs energy change, ΔG. For the usual conditions of constant temperature and pressure, ΔG is described by the Second Law of Thermodynamics as:

$$\Delta G = \Delta H - T\Delta S \qquad \text{(Eq 1)}$$

where ΔH is the enthalpy of reaction, T is the absolute temperature, and ΔS is the entropy change. No reaction will proceed spontaneously unless ΔG is negative. If $\Delta G = 0$, the system is at equilibrium, and if ΔG is positive, the reaction is thermodynamically unfavorable; that is, the reverse reaction will proceed spontaneously.

The driving force ΔG for a reaction such as aA + bB = cC + dD can be expressed in terms of the standard Gibbs energy change, ΔG° by:

$$\Delta G = \Delta G^\circ + RT \ln \frac{a_C^c \, a_D^d}{a_A^a \, a_B^b} \qquad \text{(Eq 2)}$$

where the chemical activity, a, of each reactant or product is raised to the power of its stoichiometric coefficient, and R is the gas constant. For example, in the oxidation of a metal by the reaction:

$$x\text{M} + \frac{y}{2}\text{O}_2 = \text{M}_x\text{O}_y$$

M is the reacting metal, M_xO_y is its oxide, and x and y are the moles of metal and oxygen, respectively, in 1 mol of the oxide.

The Gibbs energy change for the reaction is:

$$\Delta G = \Delta G^\circ + RT \ln \left[\frac{a_{\text{M}_x\text{O}_y}}{(a_\text{M})^x \cdot (a_{\text{O}_2})^{y/2}}\right] \qquad \text{(Eq 3)}$$

In most cases, the activities of the solids (metal and oxide) are invarient; that is, their activities = 1 for pure solids, and for the relatively high temperatures and moderate pressures encountered in oxidation reactions, a_{O_2} can be approximated by its pressure. Therefore, at equilibrium where $\Delta G = 0$:

$$\Delta G^\circ = -RT \ln \left[\frac{a_\text{prod}}{a_\text{react}}\right] \simeq +\frac{y}{2} RT \ln p_{O_2} \qquad \text{(Eq 4)}$$

where p_{O_2} is the partial pressure of oxygen.

In solid solutions, such as an alloy, the partial molar Gibbs energy of a substance is usually

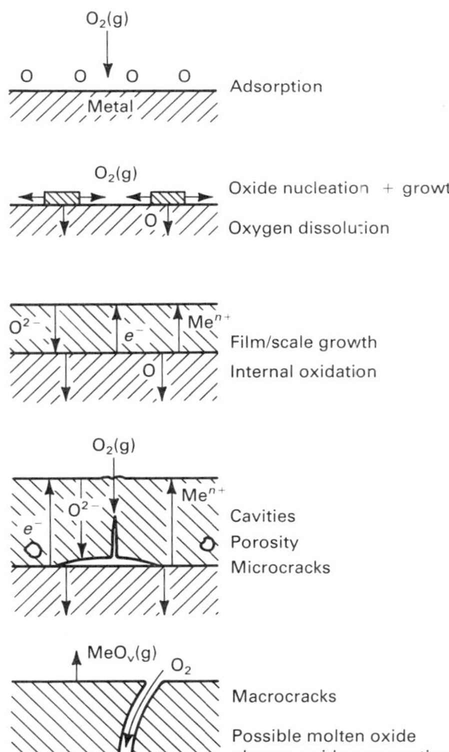

Fig. 1 Schematic illustration of the principal phenomena taking place during the reaction of metals with oxygen. Source: Ref 1

called its chemical potential μ. If 1 mol of pure A is dissolved in an amount of solution so large that the solution concentration remains virtually unchanged, the Gibbs energy change for the mole of A is:

$$\Delta\overline{G}_A = \mu_A - \mu_A^\circ = RT \ln a_A \qquad \text{(Eq 5)}$$

where μ_A° is the chemical potential of 1 mol of pure A, the chemical potential μ_A is the value in the solution, and a_A is the activity of A in the solution.

Metastable Oxides. Thermodynamically unstable oxides are often formed in corrosion by gases. The Gibbs energy of formation of the oxide, ΔG, is less negative than for a stable oxide, but in fact an unstable oxide can often exist indefinitely with no measurable transformation.

A common example is wustite (FeO), which is formed during the hot rolling of steel. Thermodynamically, it is unstable below 570 °C (1060 °F), but it remains the major component of mill scale at room temperature because the decomposition kinetics is extremely slow.

As another example, rapid kinetics can favor the formation of less stable oxide on an alloy. An alloy AB could oxidize to form oxides AO and BO, but if BO is more stable than AO, then any AO formed in contact with B should in theory convert to BO by the reaction:

$$B + AO \rightarrow BO + A$$

Nevertheless, if AO grows rapidly compared with BO and the conversion reaction is slow, AO can be the main oxide found on the alloy.

Thermodynamically unstable crystal structures of oxides are also sometimes found. A growing oxide film tends to try to align its crystal structure in some way with that of the substrate from which it is growing. This epitaxy can cause the formation of an unstable structure that fits the substrate best. For example, cubic aluminum oxide (Al_2O_3) may form on aluminum alloys instead of the stable rhombohedral Al_2O_3.

Free Energy-Temperature Diagrams. Metal oxides become less stable as temperature increases. The relative stabilities of oxides are usually shown on a Gibbs energy-temperature diagram, sometimes called an Ellingham diagram (Fig. 2), for common metals in equilibrium with their oxides. Similar diagrams are available for sulfides, nitrides, and other gas-metal reactions. In Fig. 2, the reaction plotted in every case is:

$$\frac{2x}{y} M + O_2 = \frac{2}{y} M_xO_y$$

That is, 1 mol of O_2 gas is always the reactant so that:

$$\Delta G^\circ = RT \ln p_{O_2} \qquad \text{(Eq 6)}$$

For example, the Gibbs energy of formation of Al_2O_3 at 1000 °C (1830 °F), as read from Fig. 2, is approximately −840 kJ (−200 kcal) for 2/3 mol of Al_2O_3.

The equilibrium partial pressure of O_2 is:

$$p_{O_2} = \exp\left(\frac{\Delta G^\circ}{RT}\right) \qquad \text{(Eq 7)}$$

and can also be read directly from Fig. 2 without calculation by use of the p_{O_2} scale along the bottom and right side of the diagram. A straight line drawn from the index point labeled O at the

Table 1 Structures and thermal properties of pure metals

Metal	Structure(a)	Transformation temperature °C	Transformation temperature °F	Volume change upon cooling(b), %	Melting point °C	Melting point °F	Molar volume(c) cm³	Molar volume(c) in.³
Aluminum	fcc	...	...	...	660.4	1220.7	10.00	0.610
Antimony	rhom	...	...	...	630.7	1167.3	18.18	1.109
Arsenic	rhom	...	...	...	Sublimation 615	1139	12.97	0.791
Barium	bcc	...	...	...	729	1344	39	2.380
Beryllium	(α) hcp	1250	2282	...	...	...	4.88	0.298
	(β) bcc	...	...	−2.2	1290	2354	4.99	0.304
Bismuth	rhom	...	...	...	271.4	520.5	21.31	1.300
Cadmium	hcp	...	...	...	321.1	610	13.01	0.793
Calcium	(α) fcc	448	838	...	...	...	25.9	1.581
	(β) bcc	...	...	−0.4	839	1542	...	...
Cerium	(γ) fcc	726	1339	...	...	...	20.70	1.263
	(δ) bcc	...	...	...	798	1468	...	...
Cesium	bcc	...	...	...	28.64	83.55	70.25	4.287
Chromium	bcc	...	...	...	1875	3407	7.23	0.441
Cobalt	(α) hcp	417	783	...	...	...	6.67	0.407
	(β) fcc	...	...	−0.3	1495	2723	6.70	0.408
Copper	fcc	...	...	...	1084.88	1984.78	7.12	0.434
Dysprosium	(α) hcp	1381	2518	...	...	...	19.00	1.159
	(β) bcc	...	...	−0.1	1412	2573	18.98	1.158
Erbium	hcp	...	...	...	1529	2784	18.45	1.126
Europium	bcc	...	...	...	822	1512	28.98	1.768
Gadolinium	(α) hcp	1235	2255	...	...	...	19.90	1.214
	(β) bcc	...	...	−1.3	1312	2394	20.16	1.230
Gallium	ortho	...	...	...	29.78	85.60	11.80	0.720
Germanium	diamond fcc	...	...	...	937.4	1719.3	13.63	0.832
Gold	fcc	...	...	...	1064.43	1947.97	10.20	0.622
Hafnium	(α) hcp	1742	3168	...	...	...	13.41	0.818
	(β) bcc	...	...	...	2231	4048	...	...
Holmium	hcp	...	...	...	1474	2685	18.75	1.144
Indium	tetr	...	...	...	156.63	313.93	15.76	0.962
Iridium	fcc	...	...	...	2447	4437	8.57	0.523
Iron	(α) bcc	912	1674	...	...	...	7.10	0.433
	(γ) fcc	1394	2541	1.0	...	...	7.26	0.443
	(δ) bcc	...	...	−0.52	1538	2800	7.54	0.460
Lanthanum	(α) hex	330	626	...	...	...	22.60	1.379
	(β) fcc	865	...	0.5	...	...	22.44	1.369
	(γ) bcc	...	...	−1.3	918	1684	23.27	1.420
Lead	fcc	...	...	...	327.4	621.3	18.35	1.119
Lithium	(β) bcc	−193	−315	...	180.7	357.3	12.99	0.793
Lutetium	hcp	...	...	...	1663	3025	17.78	1.085
Magnesium	hcp	...	...	...	650	1202	13.99	0.854
Manganese	(α) cubic	710	1310	...	...	...	7.35	0.449
	(β) cubic	1079	1974	−3.0	...	...	7.63	0.466
	(γ) tetr	...	...	−0.0	1244	2271	7.62	0.465
Mercury	rhom	...	...	...	−38.87	−37.97	14.81	0.904
Molybdenum	bcc	...	...	...	2610	4730	9.39	0.573
Neodymium	(α) hex	863	1585	...	...	...	20.58	1.256
	(β) bcc	...	...	−0.1	1021	1870	21.21	1.294
Nickel	fcc	...	...	...	1453	2647	6.59	0.402
Niobium	bcc	...	...	...	2648	4474	10.84	0.661
Osmium	hcp	...	...	...	~2700	~4890	8.42	0.514
Palladium	fcc	...	...	...	1552	2826	8.85	0.540
Platinum	fcc	...	...	...	1769	3216	9.10	0.555
Plutonium	α,β,γ	120,210,315	248,410,599	...	...	...	α 12.04	α 0.735
	δ,δ′,ε	452,480	846,896	...	640	1184	ε 14.48	ε 0.884
Potassium	bcc	...	...	...	63.2	145.8	45.72	2.790
Praseodymium	(α) hex	795	1463	...	...	...	20.80	1.269
	(β) bcc	...	...	−0.5	931	1708	21.22	1.295
Rhenium	hcp	...	...	...	3180	5756	8.85	0.540
Rhodium	fcc	...	...	...	1963	3565	8.29	0.506
Rubidium	bcc	...	...	...	38.89	102	55.79	3.405
Ruthenium	hcp	...	...	...	2310	4190	8.17	0.499
Samarium	(α) rhom	734	1353	...	...	...	20.00	1.220
	(β) hcp	922	1692	...	...	...	20.46	1.249
	(γ) bcc	...	...	...	1074	1965	20.32	1.240
Scandium	(α) hcp	1337	2439	...	...	...	15.04	0.918
	(β) bcc	...	...	...	1541	2806	...	...
Selenium	(γ) hex	209	408	...	217	423	16.42	1.002
Silicon	diamond fcc	...	...	...	1410	2570	12.05	0.735
Silver	fcc	...	...	...	961.9	1763.4	10.28	0.627
Sodium	(β) bcc	−237	−395	...	97.82	208.08	23.76	1.450
Strontium	(α) fcc	557	1035	...	...	...	34	2.075
	(β) bcc	...	...	...	768	1414	34.4	2.099
Tantalum	bcc	...	...	...	2996	5425	10.9	0.665

(continued)

(a) fcc, face-centered cubic; rhom, rhombohedral; bcc, body-centered cubic; hcp, hexagonal close-packed; ortho, orthorhombic; tetr, tetragonal; hex, hexagonal; bct, body-centered tetragonal. (b) Volume change upon cooling through crystallographic transformation. (c) Molar volume at 25 °C (77 °F) or at transition temperature for structures not stable at 25 °C (77 °F). Source: Ref 2

Table 1 (continued)

Metal	Structure(a)	Transformation temperature °C	Transformation temperature °F	Volume change upon cooling(b), %	Melting point °C	Melting point °F	Molar volume(c) cm^3	Molar volume(c) $in.^3$
Tellurium	hex	...	...	...	449.5	841.1	20.46	1.249
Terbium	(α) hcp	1289	2352	...	...	...	19.31	1.178
	(β) bcc	...	...	...	1356	2472.8	19.57	1.194
Thallium	(α) hcp	230	446	...	...	...	17.21	1.050
	(β) bcc	...	...	...	303	577	...	...
Thorium	(α) fcc	1345	2453	...	...	...	19.80	1.208
	(β) bcc	...	...	...	1755	3191	21.31	1.300
Thulium	hcp	...	...	...	1545	2813	18.12	1.106
Tin	(β) bct	13.2	55.8	27	231.9	449.4	16.56	1.011
Titanium	(α) hcp	882.5	1621	...	...	...	10.63	0.649
	(β) bcc	...	...	...	1668	3034	11.01	0.672
Tungsten	bcc	...	...	...	3410	6170	9.55	0.583
Uranium	(α) ortho	661	1222	...	...	...	12.50	0.763
	(β) complex tetr	769	1416	−1.0	...	...	13.00	0.793
	(γ) bcc	...	...	−0.6	1900	3452	8.34	0.509
Ytterbium	(β) fcc	7	45	0.1	819	1506	24.84	1.516
Yttrium	(α) hcp	1478	2692	...	...	...	19.89	1.214
	(β) bcc	...	...	...	1522	2772	20.76	1.267
Zinc	hcp	...	...	...	420	788	9.17	0.559
Zirconium	(α) hcp	862	1584	...	...	...	14.02	0.856
	(β) bcc	...	...	...	1852	3366	15.09	0.921

(a) fcc, face-centered cubic; rhom, rhombohedral; bcc, body-centered cubic; hcp, hexagonal close-packed; ortho, orthorhombic; tetr, tetragonal; hex, hexagonal; bct, body-centered tetragonal. (b) Volume change upon cooling through crystallographic transformation. (c) Molar volume at 25 °C (77 °F) or at transition temperature for structures not stable at 25 °C (77 °F). Source: Ref 2

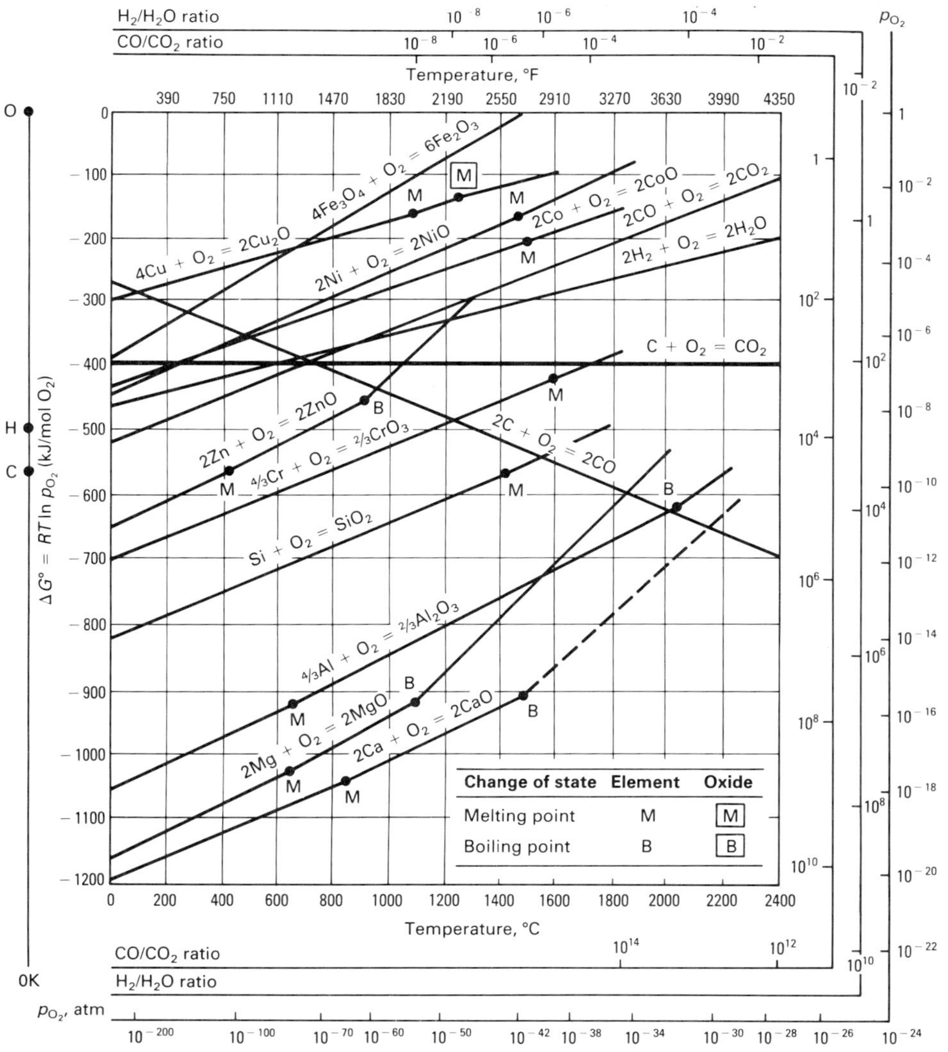

Fig. 2 Standard Gibbs energies of formation of selected oxides as a function of temperature. Source: Ref 4

upper left of the diagram, through the 1000-°C (1830-°F) point on the Al/Al_2O_3 line intersects the p_{O_2} scale at approximately 10^{-35} atm, which is the O_2 partial pressure in equilibrium with aluminum and Al_2O_3 at 1000 °C (1830 °F). This means that any O_2 pressure greater than 10^{-35} atm tends to oxidize more aluminum, while Al_2O_3 would tend to decompose to Al + O_2 only if the pressure could be reduced to below 10^{-35} atm. Obviously, Al_2O_3 is an extremely stable oxide.

The oxidation of a metal by water vapor can be determined in the same way. The reaction is:

$$xM + yH_2O = M_xO_y + yH_2$$

The equilibrium p_{H_2}/p_{H_2O} ratio for any oxide at any temperature can be found by constructing a line from the H index point on the left side of Fig. 2. For example, for the reaction:

$$2Al(l) + 3H_2O(g) = A_2O_3(s) + 3H_2(g)$$

at 1000 °C (1830 °F), the equilibrium H_2/H_2O ratio is 10^{10}. A ratio greater than this will tend to drive the reaction to the left, reducing Al_2O_3 to the metal. A ratio less than 10^{10} produces more oxide.

Similarly, the oxidation of metals by carbon dioxide (CO_2) is also shown on Fig. 2. For the reaction:

$$xM + yCO_2 = M_xO_y + yCO$$

the equilibrium carbon monoxide (CO)/CO_2 ratio is found from the index point marked C on the left side of the diagram. Oxidation of aluminum by CO_2 has an equilibrium CO/CO_2 ratio approximately 10^{10} at 1000 °C (1830 °F).

Isothermal Stability Diagrams. For situations that are more complicated than a single metal in a single oxidizing gas, it is common to fix the temperature at some practical value and plot the other variables of gas pressures or alloy composition against each other. This produces isothermal stability diagrams, or predominance area diagrams, which show the species that will be most stable in any set of circumstances.

One Metal and Two Gases. These diagrams, often called Kellogg diagrams, are constructed from the standard Gibbs energies of formation, $\Delta G°$, of all elements and compounds likely to be present in the system. For example, for the Ni-O-S system, the $\Delta G°$ values of nickel monoxide (NiO) (s), nickel monosulfide (NiS) (l), nickel sulfate ($NiSO_4$) (s), sulfur dioxide (SO_2) (g), sulfur trioxide (SO_3) (g), and S (l) are needed.

In Fig. 3, the boundary between the Ni (s) and NiO (s) regions represents the equilibrium Ni (s) + ½O_2 (g) = NiO (s); therefore, the diagram shows that at 1250 K any O_2 pressure above about 10^{-11} atm will tend to form NiO from metallic nickel if p_{S_2} is low. Similarly, S_2 gas pressure greater than about 10^{-7} atm will form NiS from nickel at low p_{O_2}. Also a mixed gas of 10^{-5} atm each of S_2 and O_2 should form nearly the equilibrium ratio of NiO (s) and $NiSO_4$ (s).

If the principal gases of interest were SO_2 and O_2, the same $\Delta G°$ data could be used to construct a diagram of log p_{O_2} versus p_{SO_2}, or as in Fig. 3, p_{SO_2} isobars can be added to the figure (the dotted lines). Thus, a mixed gas of 10^{-5} atm each of SO_2 and O_2 will form only NiO at 1250 K, with neither the sulfide nor sulfate being as stable.

When nickel metal is heated to 1250 K in the open air with sulfur-containing gases, p_{SO_2} + p_{S_2}

Table 2 Structures and thermal properties of selected oxides

Oxide	Structure	Melting point °C	Melting point °F	Boiling or decomposition, d. °C	Boiling or decomposition, d. °F	Molar volume(a) cm^3	Molar volume(a) $in.^3$	Volume ratio
α-Al_2O_3	$D5_1$ (corundum)	2015	3659	2980	5396	25.7	1.568	1.28
γ-Al_2O_3	(defect-spinel)	γ→α	...	...	...	26.1	1.593	1.31
BaO	*B*1 (NaCl)	1923	3493	~2000	~3632	26.8	1.635	0.69
BaO_2	Tetragonal (CaC_2)	450	842	d.800	d.1472	34.1	2.081	0.87
BeO	*B*4 (ZnS)	2530	4586	~3900	~7052	8.3	0.506	1.70
CaO	*B*1 (NaCl)	2580	4676	2850	5162	16.6	1.013	0.64
CaO_2	*C*11 (CaC_2)	...	...	d.275	d.527	24.7	1.507	0.95
CdO	*B*1 (NaCl)	~1400	~2552	d.900	d.1652	18.5	1.129	1.42
Ce_2O_3	$D5_2$ (La_2O_3)	1692	3078	...	...	47.8	2.917	1.15
CeO_2	*C*1 (CaF_2)	~2600	~4712	...	...	24.1	1.471	1.17
CoO	*B*1 (NaCl)	1935	3515	...	...	11.6	0.708	1.74
Co_2O_3	Hexagonal	...	...	d.895	d.1643	32.0	1.953	2.40
Co_3O_4	$H1_1$ (spinel)	→CoO		...	...	39.7	2.423	1.98
Cr_2O_3	$D5_1$ (αAl_2O_3)	2435	4415	4000	7232	29.2	1.782	2.02
Cs_2O	Hexagonal ($CdCl_2$)	...	...	d.400	d.752	66.3	4.046	0.47
Cs_2O_3	Cubic (Th_3P_4)	400	752	650	1202	70.1	4.278	0.50
CuO	*B*26 monoclinic	1326	2419	...	...	12.3	0.751	1.72
Cu_2O	*C*3 cubic	1235	2255	d.1800	d.3272	23.8	1.452	1.67
Dy_2O_3	Cubic (Tl_2O_3)	2340	4244	...	...	47.8	2.917	1.26
Er_2O_3	Cubic (Tl_2O_3)	...	...	...	...	44.3	2.703	1.20
FeO	*B*1 (NaCl)	1420	2588	...	...	12.6	0.769	1.78 on α-iron
α-Fe_2O_3	$D5_1$ (hematite)	1565	2849	...	...	30.5	1.861	2.15 on α-iron
	...	...	...	...	...	...	...	1.02 on Fe_3O_4
γ-Fe_2O_3	$D5_7$ cubic	1457	2655	...	...	31.5	1.922	2.22 on α-iron
Fe_3O_4	$H1_1$ (spinel)	...	...	d.1538	d.2800	44.7	2.728	2.10 on α-iron
	...	...	...	...	...	...	...	~1.2 on FeO
Ga_2O_3	Monoclinic	1900	3452	...		31.9	1.947	1.35
HfO_2	Cubic	2812	5094	~5400	~9752	21.7	1.324	1.62
HgO	Defect *B*10(SnO)	...	...	d.500	d.932	19.5	1.190	1.32
In_2O_3	$D5_3$(Sc_2O_3)	...	...	d.850	d.1562	38.7	2.362	1.23
IrO_2	*C*4(TiO_2)	...	...	d.1100	d.2012	19.1	1.166	2.23
K_2O	*C*1(CaF_2)	...	...	d.350	d.662	40.6	2.478	0.45
La_2O_3	$D5_2$ hexagonal	2315	4199	4200	7592	50.0	3.051	1.10
Li_2O	*C*1 (CaF_2)	~1700	~3092	1200	2192	14.8	0.903	0.57
MgO	*B*1 (NaCl)	2800	5072	3600	6512	11.3	0.690	0.80
MnO	*B*1 (NaCl)	...	...	...	...	13.0	0.793	1.77
MnO_2	*C*4 (TiO_2)	...	...	d.535	d.995	17.3	1.056	2.37
Mn_2O_3	$D5_3$ (Sc_2O_3)	...	...	d.1080	d.1976	35.1	2.142	2.40
α-Mn_3O_4	$H1_1$ (spinel)	1705	3101	...	...	47.1	2.874	2.14
MoO_3	Orthorhombic	795	1463	...	...	30.7	1.873	3.27
Na_2O	*C*1 (CaF_2)	Sublimation 1275	2327	...	...	27.3	1.666	0.57
Nb_2O_5	Monoclinic	1460	2660	...	...	59.5	3.631	2.74
Nd_2O_3	Hexagonal	~1900	~3452	...	...	46.5	2.838	1.13
NiO	*B*1 (NaCl)	1990	3614	...	...	11.2	0.683	1.70
OsO_2	*C*4 (TiO_2)	...	...	d.350	d.662	28.8	1.757	3.42
PbO	*B*10 tetragonal	888	1630	...	...	23.4	1.428	1.28
Pb_3O_4	Tetragonal	...	...	d.500	d.932	75.3	4.595	1.37
PdO	*B*17 tetragonal	870	1598	...	...	14.1	0.860	1.59
PtO	*B*17 (PdO)	...	...	d.550	d.1022	14.2	0.867	1.56
Rb_2O_3	(Th_3P_4)	489	912	...	...	62.0	3.783	0.56
ReO_2	Monoclinic	...	...	d.1000	d.1832	19.1	1.166	2.16
Rh_2O_3	$D5_1$ (α-Al_2O_3)	...	...	d.1100	d.2012	31.0	1.892	1.87
SiO	Cubic	~1700	~3092	1880	3416	20.7	1.263	1.72
SiO_2	β cristobalite *C*9	1713	3115	2230	4046	25.9	1.581	2.15
SnO	*B*10 (PbO)	...	...	d.1080	d.1976	20.9	1.275	1.26
SnO_2	*C*4 (TiO_2)	1127	2061	...	...	21.7	1.324	1.31
SrO	*B*1 (NaCl)	2430	4406	~3000	~5432	22.0	1.343	0.65
Ta_2O_5	Triclinic	1800	3272	...	...	53.9	3.289	2.47
TeO_2	*C*4 (TiO_2)	733	1351	1245	2273	28.1	1.715	1.38
ThO_2	*C*1 (CaF_2)	3050	5522	4400	7952	26.8	1.635	1.35
TiO	*B*1 (NaCl)	1750	3182	~3000	~5432	13.0	0.793	1.22
TiO_2	*C*4 (rutile)	1830	3326	~2700	~4892	18.8	1.147	1.76
Ti_2O_3	$D5_1$ (α-Al_2O_3)	...	...	d.2130	d.3866	31.3	1.910	1.47
Tl_2O_3	$D5_3$ (Sc_2O_3)	717	1323	d.875	d.1607	44.8	2.734	1.30
UO_2	*C*1 (CaF_2)	2500	4532	...	...	24.6	1.501	1.97
U_3O_8	Hexagonal	...	...	d.1300		101.5	6.194	2.71
VO_2	*C*4 (TiO_2)	1967	3573	...	...	19.1	1.166	2.29
V_2O_3	$D5_1$ (α-Al_2O_3)	1970	3578	...	...	30.8	1.879	1.85
V_2O_5	$D8_7$ orthorhombic	690	1274	d.1750	d.3182	54.2	3.307	3.25
WO_2	*C*4 (TiO_2)	~1550	~2822	~1430	~2606	17.8	1.086	1.87
β-WO_3	Orthorhombic	1473		...	...	32.4	1.977	3.39
W_2O_5	Triclinic	Sublimation, ~850	~1562	~1530	~2786	29.8	1.819	3.12
Y_2O_3	$D5_3$ (Sc_2O_3)	2410	4370	...	...	45.1	2.752	1.13
ZnO	*B*4 (wurtzite)	1975	3587	...	...	14.5	0.885	1.58
ZrO_2	*C*43 monoclinic	2715	4919	...	...	22.0	1.343	1.57

(a) Molar volume at 25 °C (77 °F) or at transition temperature for structures not stable at 25 °C (77 °C). Source: Ref 3

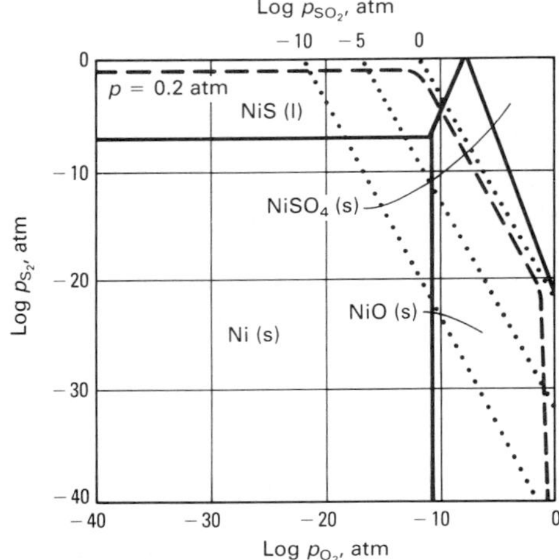

Fig. 3 The Ni-O-S system at 1250 K. Source: Ref 5

+ $p_{O_2} \simeq 0.2$ atm. The situation is shown by the dashed line in Fig. 3 labeled $p = 0.2$ atm.

An Alloy System and a Gas. Isothermal stability diagrams for oxidation of many important alloy systems have been worked out, such as that for the Fe-Cr-O system shown in Fig. 4. In this diagram, the mole fraction of chromium in the alloy is plotted against log p_{O_2} so that for any alloy composition the most stable oxide or mixture of oxides is shown at any gas pressure.

For an alloy system in gases containing more than one reactive component, the pressures of all but one of the gases must be fixed at reasonable values to be able to draw an isothermal stability diagram in two dimensions. Figure 5 shows an example of such a situation: the Fe-Zn system in equilibrium with sulfur and oxygen-containing gases with SO_2 pressure set at 1 atm and temperature set at 1164 K.

Limitations of Predominance Area Diagrams. Isothermal stability diagrams, like all predominance area diagrams, including Pourbaix potential-pH diagrams, must be read with an understanding of their rules:

- Each area on the diagram is labeled with the predominant phase that is stable under the specified conditions of pressure or temperature. Other phases may also be stable in that area, but in smaller amounts
- The boundary line separating two predominance areas shows the conditions of equilibrium between the two phases

Also, the limitations of the diagrams must be understood to be able to use them intelligently:

- The diagrams are for the equilibrium situation. Equilibrium may be reached quickly in high-temperature oxidation, but if the metal is then cooled, equilibrium is often not reestablished
- Microenvironments, such as gases in voids or cracks, can create situations that differ from the situations expected for the bulk reactant phases
- The diagrams often show only the major components, omitting impurities that are usually present in industrial situations and may be important
- The diagrams are based on thermodynamic data and do not show rates of reaction

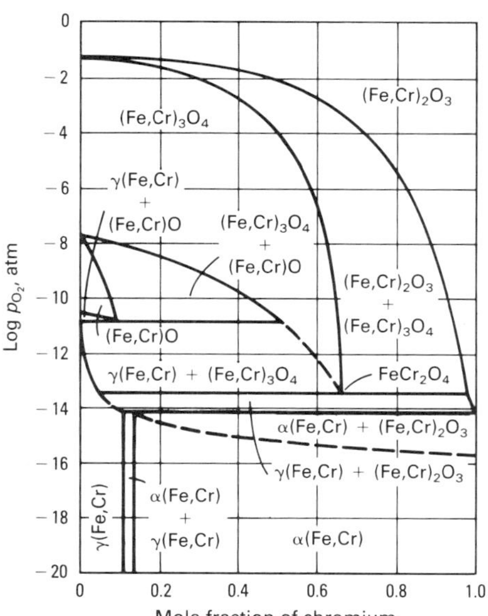

Fig. 4 Stability diagram for the Fe-Cr-O system at 1300 °C (2370 °F)

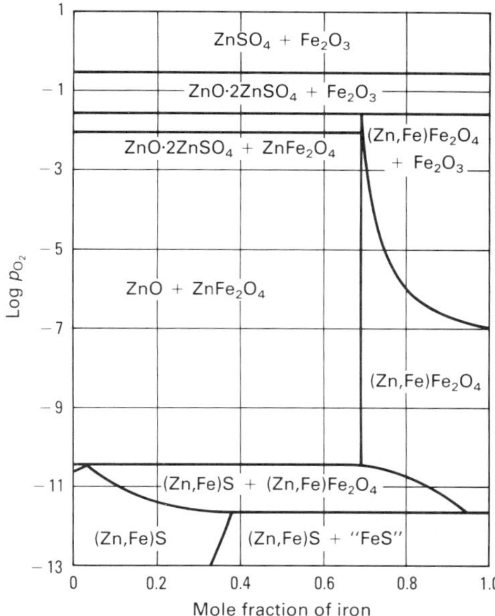

Fig. 5 The Fe-Zn-S-O system for p_{SO_2} = 1 atm at 1164 K. Source: Ref 6

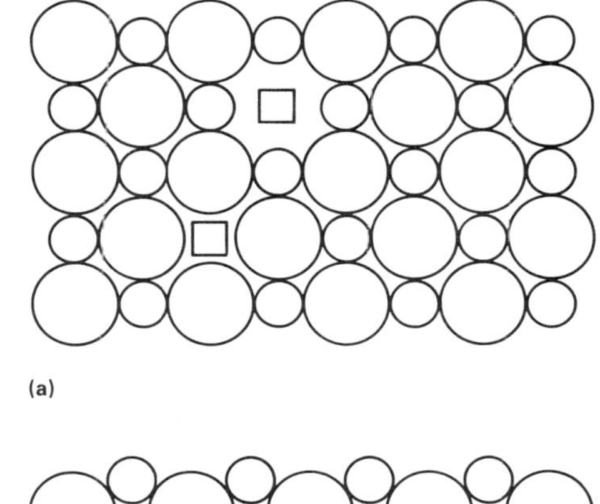

(a)

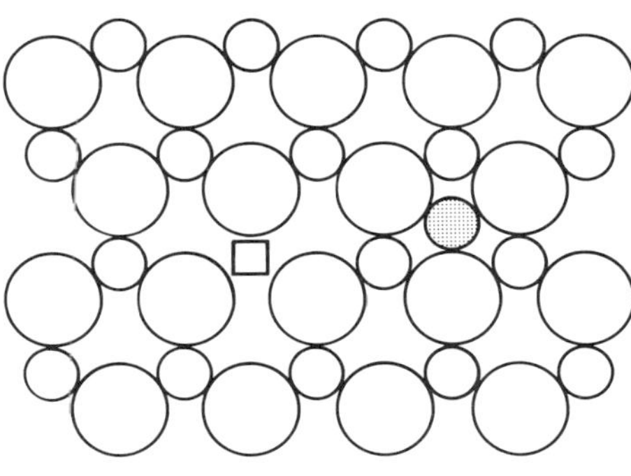

(b)

Fig. 6 Defects in ionic crystals. (a) Schottky defect. (b) Frenkel defect. Vacancies are indicated by open squares. Interstitial ion is shown as shaded circle.

Kinetics of Corrosion in Gases

Mechanisms of Oxidation. In 1923, N.B. Pilling and R.E. Bedworth classified oxidizable metals into two groups: those that formed protective oxide scales and those that did not (Ref 7). They suggested that unprotective scales formed if the volume of the oxide layer was less than the volume of metal reacted. For example, in the oxidation of aluminum:

$$2Al + \tfrac{3}{2}O_2 \rightarrow Al_2O_3$$

the Pilling-Bedworth ratio is:

$$\frac{\text{Volume of 1 mol of } Al_2O_3}{\text{Volume of 2 mol of Al}}$$

where the volumes can be calculated from molecular and atomic weights and the densities of the phases.

If the ratio is less than 1, as is the case for alkali and alkaline earth metals, the oxide scales are usually unprotective, with the scales being porous or cracked due to tensile stresses and providing no efficient barrier to penetration of gas to the metal surface. If the ratio is more than 1, the protective scale shields the metal from the gas so that oxidation can proceed only by solid-state diffusion, which is slow even at high temperatures. If the ratio is much over 2 and the scale is growing at the metal/oxide interface, the large compressive stresses that develop in the oxide as it grows thicker may eventually cause the scale to spall off, leaving the metal unprotected.

Exceptions to the Pilling-Bedworth theory are numerous, and it has been roundly criticized and rejected by many. Its main flaw is the assumption that metal oxides grow by diffusion of oxygen inward through the oxide layer to the metal. In fact, it is much more common for metal ions to diffuse outward through the oxide to the gas. Also, the possibility of plastic flow by the oxide or metal was not considered. Nevertheless, historically, Pilling and Bedworth made the first step in achieving understanding of the processes by which metals react with gases. And although there may be exceptions, the volume ratio, as a rough rule-of-thumb, is usually correct. The Pilling-Bedworth volume ratios for many common oxides are listed in Table 2.

Defect Structure of Ionic Oxides. Ionic compounds can have appreciable ionic conductivity due to Schottky defects and/or Frenkel defects. Schottky defects are combinations of cation vacancies and anion vacancies in the proper ratio necessary to maintain electrical neutrality. Figure 6(a) illustrates a Schottky defect in a stoichiometric ionic crystal. With Schottky defects, the ions must diffuse into the appropriate adjacent vacancies to allow mass transfer and ionic electrical conductivity.

Frenkel defects are also present in ionic crystals in such a way that electrical neutrality and stoichiometry are maintained (Fig. 6b). This type of defect is a combination of a cation vacancy and an interstitial cation. Metal cations are generally much smaller than the oxygen anions. Limited ionic electrical conductivity is possible in such crystals by diffusion of cations interstitially and by diffusion of cations into the cation vacancies.

Metallic oxides are seldom, if ever, stoichiometric and cannot grow by mere diffusion by Schottky and Frenkel defects. For oxidation to continue when a metal is protected by a layer of oxide, electrons must be able to migrate from the metal, through the oxide, to adsorbed oxygen at the oxide/gas interface. Nevertheless, Schottky and Frenkel defects may provide the mechanism for ionic diffusion necessary for oxide growth.

Defect Structure of Semiconductor Oxides. Oxides growing to provide protective scales are electronic semiconductors that also allow mass transport of ions through the scale layer. They may be conveniently categorized as *p*-type, *n*-type, and amphoteric semiconductors. Examples of the three types are listed in Table 3 (Ref 8).

The p-type metal-deficit oxides are nonstoichiometric with cation vacancies present. They will

Table 3 Classification of electrical conductors: oxides, sulfides, and nitrides

Type	Compounds
Metal-excess semiconductors (*n*-type)	BeO, MgO, CaO, SrO, BaO, BaS, ScN, CeO_2, ThO_2, UO_3, U_3O_8, TiO_2, TiS_2, (Ti_2S_3), TiN, ZrO_2, V_2O_5, (V_2S_3), VN, Nb_2O_5, Ta_2O_5, (Cr_2S_3), MoO_3, WO_3, WS_2, MnO_2, Fe_2O_3, $MgFe_2O_4$, $NiFe_2O_4$, $ZnFe_2O_4$, $ZnCo_2O_4$, ($CuFeS_2$), ZnO, CdO, CdS, HgS(red), Al_2O_3, $MgAl_2O_4$, $ZnAl_2O_4$, Tl_2O_3, (In_2O_3), SiO_2, SnO_2, PbO_2
Metal-deficit semiconductors (*p*-type)	UO_2, (VS), (CrS), Cr_2O_3 (<1250 °C, or 2280 °F), $MgCr_2O_4$, $FeCr_2O_4$, $CoCr_2O_4$, $ZnCr_2O_4$, (WO_2), MoS_2, MnO, Mn_3O_4, Mn_2O_3, ReS_2, FeO, FeS, NiO, NiS, CoO, (Co_3O_4), PdO, Cu_2O, Cu_2S, Ag_2O, $CoAl_2O_4$, $NiAl_2O_4$, (Tl_2O), Tl_2S, (GeO), SnS, (PbO), (Sb_2S_3), (Bi_2S_3)
Amphoteric conductors	TiO(a), Ti_2O_3(a), VO(a), Cr_2O_3 (>1250 °C, or 2280 °F), MoO_2, FeS_2, (OsS_2), (IrO_2), RuO_2, PbS

(a) Metallic conductors. Source: Ref 8

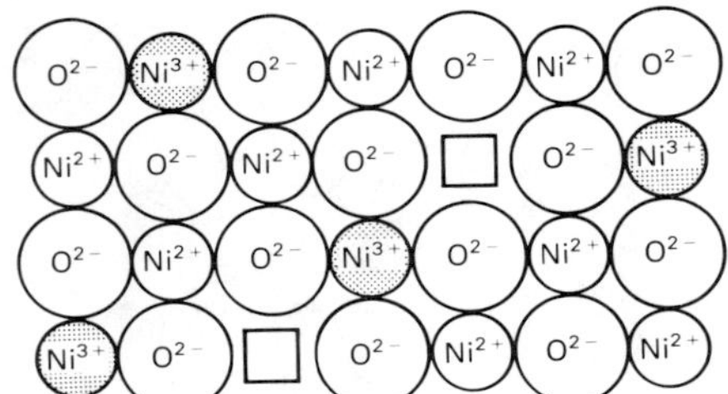

Fig. 7 Illustration of the ionic arrangement in *p*-type NiO scale. Cation vacancies are indicated as open squares. The N^{3+} cations are shaded.

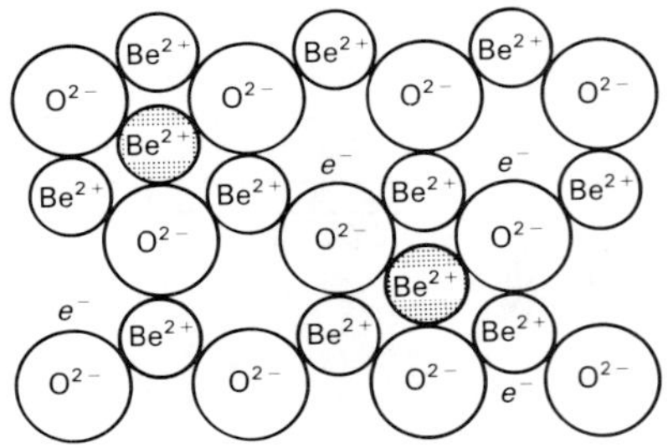

Fig. 8 Illustration of the ionic arrangement in *n*-type cation-excess BeO. Interstitial cations are shaded; free electrons are indicated as e^-.

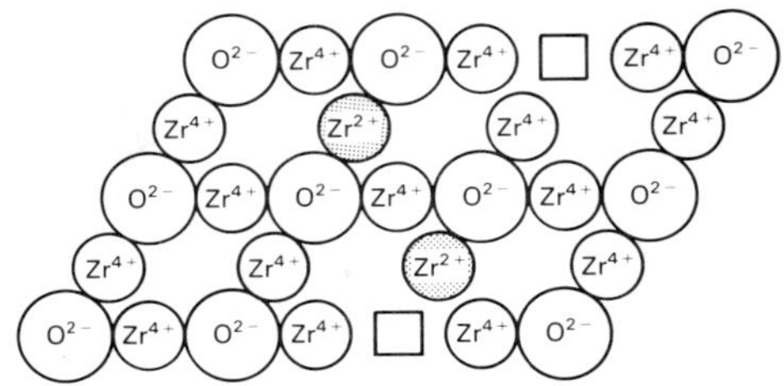

Fig. 9 Illustration of the ionic arrangement in *n*-type anion-deficient ZrO_2. Anion vacancies are indicated as open squares; Zr^{2+} ions are shaded.

also have some Schottky and Frenkel defects that add to the ionic conductivity. A typical example is NiO, a cation-deficient oxide that provides the additional electrons needed for ionic bonding and electrical neutrality by donating electrons from the 3*d* subshells of a fraction of the nickel ions. In this way, for every cation vacancy present in the oxide, two nickelic ions (Ni^{3+}) will be present (Fig. 7). Each Ni^{3+} has a low-energy positively charged electron hole that electrons from other nickelous ions (Ni^{2+}) can easily move into. The positive or *p*-type semiconductors carry most of their current by means of these positive holes.

Cations can diffuse through the scale from the Ni/NiO interface by cation vacancies, to the NiO/gas interface where they react with adsorbed oxygen. Electrons migrate from the metal surface, by electron holes, to the adsorbed oxygen atoms, which then become oxygen anions. In this way, while Ni^{2+} cations and electrons move outward through the scale toward the gas, cation vacancies and electron holes move inward toward the metal. Consequently, as the scale thickens, the cation vacancies tend to accumulate to form voids at the Ni/NiO interface.

The n-type semiconductor oxides have negatively-charged free electrons as the major charge carriers. They may be either cation excess or anion deficient. Beryllium oxide (BeO) typifies the cation-excess oxides because the beryllium ion (Be^{2+}) is small enough to move interstitially through the BeO scale. Its structure is shown in Fig. 8.

Oxygen in the gas adsorbs on the BeO surface and picks up free electrons from the BeO to become adsorbed O^{2-} ions, which then react with excess Be^{2+} ions that are diffusing interstitially from the beryllium metal. The free electrons coming from the metal surface as the beryllium ionizes travel rapidly through vacant high-energy levels. As with *p*-type oxides, the cation-excess *n*-type oxides grow at the oxide/gas interface as cations diffuse outward through the scale.

Another group of *n*-type semiconducting oxides is anion deficient, as exemplified by zirconium dioxide (ZrO_2). In this case, although most of the cations are contributing four electrons to the ionic bonding, a small fraction of the zirconium cations only contributes two electrons to become the zirconium ion Zr^{2+}. Therefore, to maintain electrical neutrality, an equal number of anion vacancies must be present in the oxide. This arrangement is shown in Fig. 9. The oxide grows at the metal/oxide interface by inward diffusion of O^{2-} through the anion vacancies in the oxide.

Amphoteric Oxides. A number of compounds can be nonstoichiometric with either a deficiency of cations or a deficiency of anions. An example is lead sulfide (PbS), which has a minimum in electrical conductivity at the stoichiometric composition. Thus, if the composition is $Pb_{<1}S$, it is *p*-type and if it is $PbS_{<1}$, it is *n*-type.

Similarly, intrinsic semiconductors, such as cupric oxide (CuO), have a few electron holes in their valence band and an equal number of free electrons in their nearly vacant conduction band. Current is carried both by migration of electron holes in the low-energy bonding levels and by free electrons in the higher-energy conduction levels.

Oxide Texture: Amorphous Oxides. In the very early stages of oxidation and especially at low temperatures, some oxides appear to grow with an amorphous structure. In general, these oxide glasses contain more oxygen than metal in their formulas so that oxygen triangles or tetrahedra form around each of the metal ions. The random network ring structures that result allow large anions or molecular oxygen to move through them more readily than the smaller cations do. Amorphous oxides tend to crystallize as they age. Examples are silicon dioxide (SiO_2), Al_2O_3, tantalum pentoxide (Ta_2O_5), and niobium pentoxide (Nb_2O_5).

In contrast, oxides with M_2O and MO formulas have structures in which the small cations can move readily. They are apparently always crystalline. Examples are NiO, cuprous oxide (Cu_2O), and zinc oxide (ZnO).

Oxide Texture: Epitaxy. As a crystalline oxide grows on a metal surface, it often aligns its crystal structure to be compatible with the structure of the metal substrate. This epitaxy finds the best fit, not a perfect fit, between the two crystal structures. For example, either (111) or (001) planes of Cu_2O grow parallel to the Cu (001) plane with the ⟨110⟩ directions of Cu_2O parallel to the ⟨110⟩ of copper (Ref 9).

Stress develops in an epitaxial oxide layer as it grows because of the slight misfit between the oxide and metal crystals. The stress is likely to produce dislocation arrays in the oxide that would be paths of easy diffusion for mass transport through the film. A mosaic structure may develop in the oxide because of the growth stresses. The mosaic structure consists of small crystallites with orientations very slightly tilted or twisted with respect to each other. The boundaries between the crystallites are dislocation arrays that serve as easy diffusion paths.

Stresses in epitaxial layers increase as the films grow thicker until at some point the bulk scale tends to become polycrystalline and epitaxy is gradually lost. Epitaxy may last up to about 50 nm in many cases, but it is seldom strong much over 100 nm.

Oxide Texture: Preferred Orientation. As oxidation produces thicker layers, the oxide grain size increases. Crystals that are favorably oriented for growth will grow at the expense of their neighboring grains until the oxide surface consists of a few large grains with similar orientation. The variation in growth rate of different oxide grains produces the roughening of the scaled surface that is commonly observed.

Linear Oxidation Reaction Rates. If the metal surface is not protected by a barrier of oxide, the oxidation rate usually remains constant with time, and one of the steps in the oxidation reaction is rate controlling rather than a transport process being rate controlling. This situation is to be expected if the Pilling-Bedworth ratio is less than 1, if the oxide is volatile or molten, if the scale spalls off or cracks due to internal stresses, or if a porous, unprotective oxide forms on the metals.

The linear oxidation rate is:

$$\frac{dx}{dt} = k_L \quad \text{(Eq 8)}$$

where x is the mass or thickness of oxide formed, t is the time of oxidation, and k_L is the linear rate constant. The rate constant is a function of the metal, the gas composition and pressure, and the temperature.

Integrated, the linear oxidation equation is:

$$x = k_L t \quad \text{(Eq 9)}$$

The oxidation never slows down; after long times at high temperatures, the metal will be completely destroyed. Figure 10 shows the relationship between oxide mass and time for linear oxidation.

Logarithmic and Inverse Logarithmic Reaction Rates. At low temperatures when only a thin film of oxide has formed (for example, under 100 nm), the oxidation is usually observed to follow either logarithmic or inverse logarithmic kinetics. Transport processes across the film are

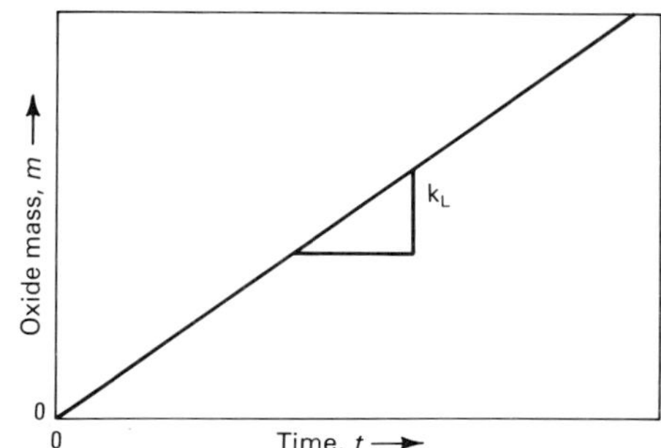

Fig. 10 Linear oxidation kinetics

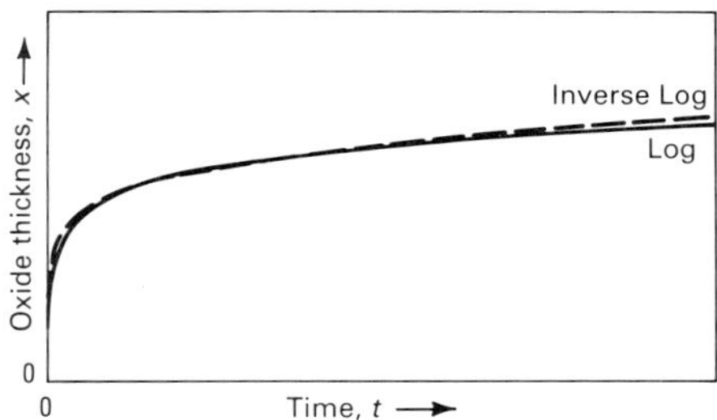

Fig. 11 Logarithmic and inverse logarithmic oxidation kinetics

rate controlling, with the driving force being electric fields across the film. The logarithmic equation is:

$$x = k_e \log (at + 1) \quad \text{(Eq 10)}$$

where k_e and a are constants.

The inverse logarithmic equation is:

$$\frac{1}{x} = b - k_i \log t \quad \text{(Eq 11)}$$

where b and k_i are constants. Under the difficult experimental conditions involved in making measurements in the thin film range, it is nearly impossible to distinguish between logarithmic and inverse logarithmic oxidation. Both equations have two constants that can be adjusted to fit the data quite well. Metals oxidizing with logarithmic or inverse log kinetics reach a limiting film thickness at which oxidation apparently stops. Figure 11 shows the curves for both logarithmic and inverse logarithmic kinetics.

Parabolic Kinetics. When the rate-controlling step in the oxidation process is the diffusion of ions through a compact barrier layer of oxide with the chemical potential gradient as the driving force, the parabolic rate law is usually observed. As the oxide grows thicker, the diffusion distance increases, and the oxidation rate slows down. The rate is inversely proportional to the oxide thickness, or:

$$\frac{dx}{dt} = \frac{k_p}{x} \quad \text{(Eq 12)}$$

Upon integration, the parabolic equation is obtained:

$$x^2 = \frac{k_p}{2} t \quad \text{(Eq 13)}$$

where k_p is the parabolic rate constant. Figure 12 shows the parabolic oxidation curve.

Other Reaction Rate Equations. A number of other kinetics equations have been fitted to the experimental data, but it is believed that they describe a combination of the mechanisms described above, rather than any new basic process. A cubic relationship:

$$x^3 = k't \quad \text{(Eq 14)}$$

has often been reported. It can be shown mathematically to be an intermediate stage between logarithmic and parabolic kinetics.

Initial Oxidation Processes: Adsorption and Nucleation. To begin oxidation, oxygen

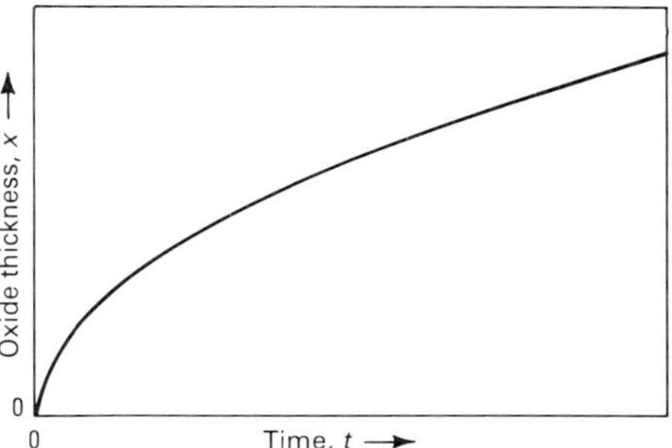

Fig. 12 Parabolic oxidation kinetics

gas is chemisorbed on the metal surface until a complete two-dimensional invisible oxide layer forms. Some atomic oxygen also dissolves into the metal at the same time. After the monolayer forms, discrete nuclei of three-dimensional oxide appear on the surface and begin expanding laterally at an ever-increasing rate. The nuclei may originate at structural defects, such as grain boundaries, impurity particles, and dislocations. The concentration of nuclei depends primarily on the crystal orientation of the metal, with more nuclei forming at high pressures and low temperatures.

These oxide islands grow outward rapidly by surface diffusion of adsorbed oxygen until a complete film three or four monolayers thick covers the metal. The oxidation rate then drops abruptly. If chemisorption were still the rate-controlling (slow) step in oxidation after the thin film is completed, a logarithmic rate law should be observed. A logarithmic rate law is found, but it is more likely the result of the strong electric field across the film that affects the oxidation.

Thin-Film Mechanisms. A large number of theories have been proposed to explain the oxidation mechanism at low temperatures or in the early stages of high-temperature oxidation where logarithmic kinetics is commonly observed. None of the theories is completely accepted yet, and perhaps none is completely correct, but they have common threads of agreement that indicate reasonably well what is happening. Some of the most important theories will be briefly described.

The Cabrera-Mott theory, probably the best established theory of thin film oxidation, applies to films up to about 10 nm thick (Ref 10). It proposes that electrons from the metal easily pass through the thin film by tunneling to reach adsorbed oxygen at the oxide/gas surface and form oxygen anions. A potential of approximately 1 V is set up between the external oxide surface and the metal. For a film 1 nm thick, the field strength would be 10^7 V/cm, powerful enough to pull cations from the metal and through the film. The rate-controlling step is the transfer of cations (or anions) into the oxide or the movement of the ions through the oxide. The electric field reduces this barrier. The structure of the oxide determines whether cations or anions migrate through the oxide. As the film grows thicker, the field strength decreases until it has so little effect on the ions that the rate-controlling mechanism changes.

N. Cabrera and N.F. Mott developed an inverse logarithmic kinetic equation to describe the mechanism. A logarithmic equation is more commonly observed, but it can be derived from the Cabrera-Mott mechanism if the activation energy for ionic migration is a function of film thickness. Such a situation would exist if the oxide film were initially amorphous and became more crystalline with aging, giving a constant field strength through the film instead of a constant voltage.

The Hauffe-Ilschner theory, a modification of Mott's original concept of a space charge developed across the oxide film, proposes that quantum-mechanical tunneling of electrons is the rate-controlling step (Ref 11). After the film thickness reaches about 10 nm, tunneling becomes increasingly difficult, and the observed reaction rate decreases greatly. For film thicknesses up to perhaps 20 nm, a logarithmic equation results. For films from 20 to 200 nm thick, the inverse logarithmic relationship holds. Potentials across thin films have been measured; a change in sign of the potential is interpreted as a change from electronic transport control to ionic transport control.

The Grimley-Trapnell Theory (Ref 12). T.B. Grimley and B.M.W. Trapnell used the Cabrera-Mott model, but assumed a constant electric field instead of a constant potential. They assumed that the adsorbed oxygen layer would always be complete, even at high temperatures and low pressure. The adsorbed oxygen would take electrons from cations in the oxide, not from the metal, so that a space charge would develop at the MO/O^-_{ads} interface and be independent of the oxide thickness. If the rate-controlling step is diffusion of cations through vacancies, logarithmic kinetics should be observed. If some other process is rate controlling, linear kinetics is most likely.

The Uhlig theory, developed by H.H. Uhlig and amplified by Fromhold, also predicts logarithmic kinetics at temperatures up to 600 K (Ref 13). The rate-controlling step is the thermal emission of electrons from the metal into the oxide (or electron holes from the adsorbed O^- to the oxide) under the combined effects of induced potential and applied field. The field is created by the diffusing ions. Because growth of the film depends on the electronic work function of the metal, the theory explains oxidation rate changes at crystal and magnetic transformations; most other theories do not.

Solid-State Diffusion. Diffusion processes in solids play a key role in the oxidation of metals. Mass transfer may be the result of diffusion of metal ions from the metal surface through the oxide layer to the adsorbed oxygen anions at the oxide/gas interface or the result of the diffusion of anions inward through the oxide to the metal. The diffusion of atomic oxygen into the metal from the oxide or the gas can also be involved. Within an alloy, the diffusion of oxidizable metal atoms toward the surface and the back diffusion of unreactive atoms from the metal surface inward to the unaltered alloy can occur.

Diffusion Mechanisms. Atoms or ions diffuse through solids by any of several mechanisms. The most common is vacancy diffusion. A metal crystal always contains large numbers of vacancies, while ionic oxides contain Schottky and Frenkel defects that also involve vacancies. An atom or ion sitting on a regular lattice site can diffuse by jumping to a vacant identical site nearby (Fig. 13a). For metal atoms, this is relatively easy because the jump distances are short. For ionic crystals, the jump distances are much longer because cation sites are surrounded by anion sites, and vice versa.

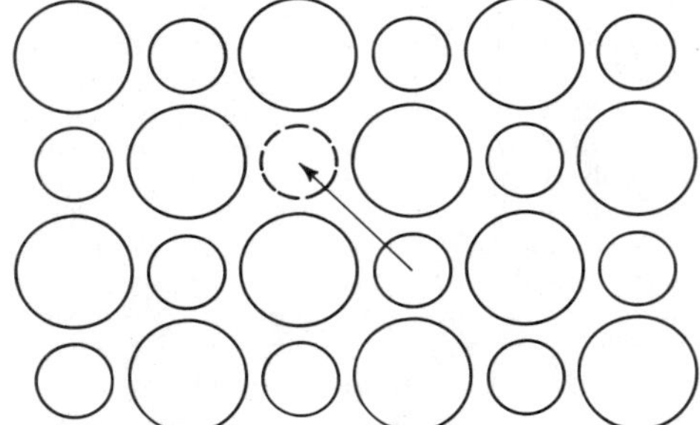
(a)

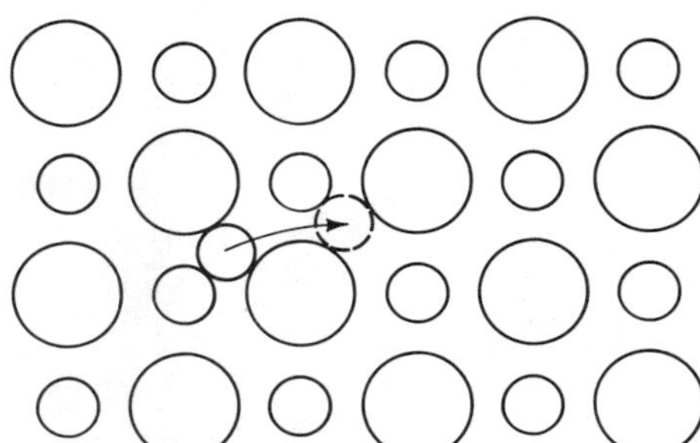
(b)

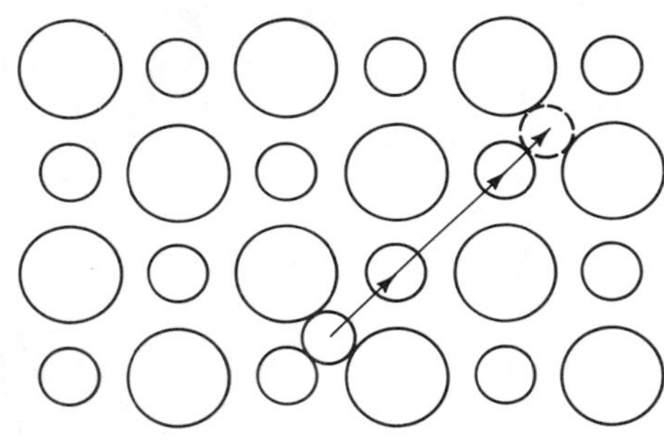
(c)

Fig. 13 Diffusion mechanisms. (a) Vacancy diffusion. (b) Interstitial diffusion. (c) Crowdion diffusion

Table 4 Selected diffusion data in metal oxides

Metal oxide	Temperature °C	Temperature °F	Frequency factor(a) cm^2/s	Activation energy for diffusion J/mol	Activation energy for diffusion BTU/mol
Copper in Cu_2O	800–1050	1470–1920	0.12	151.0	143
Nickel in NiO	740–1400	1365–2550	0.017	234	222
Oxygen in Fe_2O_3	1150–1250	2100–2280	10^{11}	610	578
Iron in Fe_3O_4	750–1000	1380–1830	5.2	230	218
Iron in $Fe_{0.92}O$	690–1010	1275–1850	0.014	126.4	120
Chromium in Cr_2O_3	1000–1350	1830–2460	4000	420	398
Oxygen in UO_2	450–600	840–1110	2.6×10^{-5}	124	118
Magnesium in MgO	1400–1600	2550–2910	0.25	330	313

(a) The frequency factor, D_o, is a function of the diffusing species and the diffusion medium. See Eq 19 and corresponding text. Source: Ref 6

Small interstitial atoms diffuse readily from one interstitial position to another. In ionic oxides, the cations may diffuse interstitially, but the anions are usually not small enough to do so. Interstitial diffusion is shown in Fig. 13(b).

In ionic crystals, an interstitial ion may crowd into a regular lattice site, displacing an ion, which is forced to move into an interstitial position or to the next lattice site. This "crowdion" effect may extend for several atomic spacings along a line or equivalent direction. Figure 13(c) shows "crowdion" diffusion.

Fick's Law. In 1855, A. Fick formulated his two laws of diffusion for the simplest sort of diffusion system: a binary system at constant temperature and pressure, with net movement of atoms in only one direction. This is the usual situation for diffusion through an oxide growing on a pure metal.

Fick's first law states that the rate of mass transfer is proportional to the concentration gradient. Mathematically:

$$J = -D\left(\frac{\partial c}{\partial x}\right) \quad \text{(Eq 15)}$$

where J is the flux or mass diffusing per second through a unit cross section in the concentration gradient ($\partial c/\partial x$), and D is the diffusion coefficient, or diffusivity in square centimeters per second, which is a function of the diffusing atoms, the structure through which they are diffusing, and the temperature. For diffusion of cations through a protective oxide, the entry of cations into the oxide at the metal/oxide interface will very nearly equal the flux of cations delivered to the oxygen at the oxide/gas interface.

For diffusion of oxygen into the metal, the concentration of oxygen changes with time inside the metal. Fick's second law describes this change as:

$$\frac{\partial c}{\partial t} = \frac{\partial}{\partial x}\left(D\frac{\partial c}{\partial x}\right) \quad \text{(Eq 16)}$$

Equation 16 must be solved for the particular geometry and boundary conditions involved (flat or round specimens, and so on). For oxygen atoms diffusing inward from a flat surface, with a constant diffusivity and a constant interfacial concentration, the solution to Eq 16 for the concentration of oxygen at any distance x from the metal surface is:

$$\frac{C_M - C_X}{C_M - C_O} = \text{erf}\left(\frac{x}{2\sqrt{Dt}}\right) \quad \text{(Eq 17)}$$

(assuming D is independent of composition) where C_M is the oxygen concentration at the metal/oxide interface, C_X is the concentration at time t and distance x from the surface, and C_O is the initial constant concentration at any distance x when $t = 0$. The error function, erf, is tabulated in many books on probability. Figure 14 shows the variation of C_X with x.

As oxidation proceeds, C_M and C_O remain constant; therefore, for any fixed value of C_X, Eq 17 reduces to:

$$X \propto \sqrt{t} \quad \text{(Eq 18)}$$

The Diffusion Coefficient. The diffusion coefficient D in rather stoichiometric compounds can usually be assumed to be proportional to the defect concentration. The diffusion coefficient can also vary with crystal orientation in noncubic crystals. For oxides that are epitaxial or have a

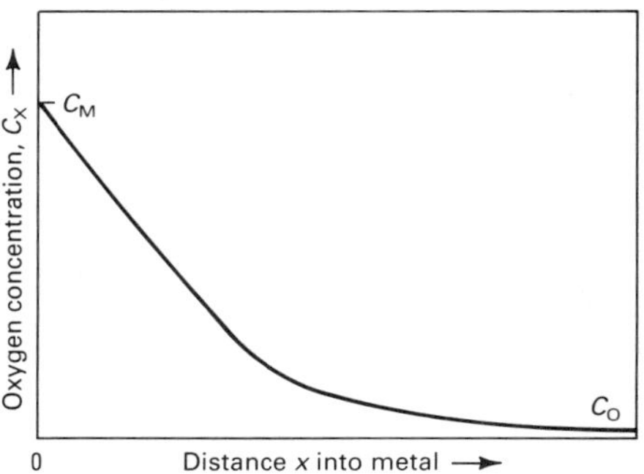

Fig. 14 Nonsteady-state diffusion. Fick's second law

preferred orientation, the diffusivity can be several times more or less than it would be for a random polycrystalline oxide. The diffusion coefficient D is independent of orientation in cubic crystals. Temperature has a major effect on the diffusion coefficient; D increases exponentially with temperature according to the Arrhenius equation:

$$D = D_o \exp\left(\frac{-Q}{RT}\right) \quad \text{(Eq 19)}$$

where D_o is a constant called the frequency factor that is a function of the diffusing species and the diffusion medium, Q is the activation energy for diffusion, R is the gas constant, and T is the absolute temperature. Activation energies for interstitial diffusion are much lower than those for vacancy diffusion.

The activation energy is a measure of the temperature dependence of a diffusion process. A high value of Q means that the diffusion proceeds much more rapidly at high temperatures, but very much slower at low temperatures. Typical values of D_o and Q for diffusion in oxides are listed in Table 4.

If the diffusivity D is plotted on a natural logarithm (base e) scale as a function of $1/T$, the slope of the resulting straight line is $-Q/R$. If the graph shows two intersecting lines, it indicates that one diffusion mechanism is operative at low temperatures, such as grain-boundary diffusion, and that another mechanism with a higher activation energy, such as volume diffusion, has gained control at high temperatures.

Effect of Impurities. All oxides contain some substitutional impurity cations from the alloy before oxidation or from the gas phase during oxidation. Although the solubility limit for foreign ions is low, they can have a great effect on diffusivity in the oxide, and consequently on oxidation rate.

In p-type oxides, such as NiO, any substitutional cation with a valence greater than the Ni^{2+} ion it replaces tends to increase the concentration of cation vacancies. Two aluminum ions (Al^{3+}) replacing two Ni^{2+} ions in the structure also supply two extra valence electrons to the oxygen so that an additional cation vacancy in the oxide structure will be necessary to maintain the balance of charge between anions and cations. Additional cation vacancies in the NiO increase the diffusivity of Ni^{2+} cations through the oxide. Conversely, substitutional cations with a valence lower than +2 should reduce diffusion in NiO, reducing the number of cation vacancies. Cations with a +2 valence should have little effect on diffusion if substituting for Ni^{2+} ions.

For n-type oxides, the effect is reversed. If Al^{3+} is substituted for some titanium ions (Ti^{4+}) in titanium dioxide (TiO_2), more anion vacancies will be present in the oxide. For every two Al^{3+} ions substituted, one additional O^{2-} vacancy must be present in the oxide to maintain the electronic charge balance. Diffusion of oxygen ions inward then increases because of the increase in anion vacancy concentration. Higher valent impurity ions would decrease oxygen diffusion in n-type oxides.

The impurity effect is especially important for diffusion at low temperatures at which the native defect concentration is low; therefore, the activation energy is associated with only the movement of ions. At high temperatures, the activation energy increases because it involves formation of defects as well as the motion of the ions.

Short-Circuit Diffusion. The activation energies for diffusion along line and surface defects in solids are much less than those for volume diffusion. Dislocations, grain boundaries, porosity networks, and external surfaces offer rapid diffusion paths at low temperatures at which volume diffusion has virtually stopped. In metals, diffusion along dislocations is more important than volume diffusion below about one-half of the absolute melting point. At high temperatures, of course, volume diffusion predominates in both metals and oxides.

Wagner Theory of Oxidation. C. Wagner derived the parabolic rate equation for scale growth on a metal in which diffusion of ions or electrons is rate controlling. Before the full derivation is presented, a simplified treatment will be given to emphasize the main features of diffusion-controlled oxidation.

It was assumed that a pure metal is oxidizing and that growth of a dense single-phase oxide occurs by a single diffusion mechanism, for example, the diffusion of cations outward through a p-type oxide. The rate of growth in oxide thickness, dx/dt, is then proportional to the flux of metal ions J_M described by Fick's first law (Eq 15):

$$\frac{dx}{dt} = k'J_M = -k'D_M \frac{(C_M - C_O)}{x} \quad \text{(Eq 20)}$$

or

$$\frac{dx}{dt} = \frac{\text{Constant}}{x} \quad \text{(Eq 21)}$$

Integrating with the limit that at time $t = 0$ the oxide thickness $x = 0$, the parabolic equation results:

$$x^2 = kt \quad \text{(Eq 22)}$$

The Wagner Derivation. The flow of ions in the x direction through a growing oxide layer depends on the concentration of ions and their drift velocity. The ion flux J_i is:

$$J_i = C_i u_i \text{ mol/cm}^2 \cdot \text{s} \quad \text{(Eq 23)}$$

where C_i is the molar concentration per cubic centimeter, and u_i is the average drift velocity. The drift velocity depends on the mobility of the ions and the force acting on them. The mobility B_i is the terminal ion velocity when it is acted upon by a unit force per ion.

A charged ion (valence z_i) moving through the oxide is acted upon by two forces: the chemical potential gradient, $d\mu_i/dx$, and an electric potential gradient, dE/dx. The flux is then:

$$J_i = \frac{-C_i B_i}{N}\left(\frac{d\mu_i}{dx} + z_i F \frac{dE}{dx}\right) \quad \text{(Eq 24)}$$

where N is Avogadro's number, and F is Faraday's constant.

Equation 24 can be used to calculate the oxidation rate in terms of more easily measured quantities, such as electrical conductivity or diffusivity of the ions. In terms of electrical conductivity:

$$\frac{C_i B_i}{N} = \frac{\sigma_i}{z_i^2 F^2} \quad \text{(Eq 25)}$$

where σ_i is the partial electrical conductivity due to the movement of the particular type of ion being considered. The total conductivity, σ, of the oxide would be the sum of the conductivity due to movement of cations (σ_c), the conductivity due to movement of anions (σ_a), and the conductivity due to movement of electrons or electron holes (σ_e):

$$\sigma = \sigma_c + \sigma_a + \sigma_e \quad \text{(Eq 26)}$$

In oxides with predominant electronic conduction: $\sigma_e \gg (\sigma_c + \sigma_a)$.

Substituting Eq 25 into Eq 24 yields:

$$J_i = \frac{-\sigma_i}{z_i^2 F^2}\left(\frac{d\mu_i}{dx} + z_i F \frac{dE}{dx}\right) \quad \text{(Eq 27)}$$

As the oxide forms, the flow of positive ions outward through the scale must be balanced by the flow of negative ions and electrons to maintain a charge balance, or:

$$z_c J_c - J_e = z_a J_a \quad \text{(Eq 28)}$$

where the subscripts c, a, and e refer to cations, anions, and electrons respectively. Wagner derived Eq 28 in a general form in which all these fluxes were considered (Ref 14). In most actual oxidation processes, the cation diffusion controls the scale growth, while anion diffusion is negligible; therefore, the diffusion equation will be derived for that situation, as demonstrated in Ref 4. It could be derived for anion diffusion in the same manner.

From Eq 27, the cation flux is:

$$J_c = \frac{-\sigma_c}{z_c^2 F^2}\left(\frac{d\mu_c}{dx} + z_c F \frac{dE}{dx}\right) \quad \text{(Eq 29)}$$

the anion flux is essentially zero, and the electron flux or electron hole flux is:

$$J_e = \frac{-\sigma_e}{F^2}\left(\frac{d\mu_e}{dx} - F\frac{dE}{dx}\right) \quad \text{(Eq 30)}$$

Combining Eq 28 to 30 to solve for the electric potential gradient yields:

$$\frac{dE}{dx} = \frac{-1}{F(\sigma_c + \sigma_e)}\left(\frac{\sigma_c d\mu_c}{z_c dx} - \sigma_e \frac{d\mu_e}{dx}\right) \quad \text{(Eq 31)}$$

Eliminating dE/dx from Eq 29 by substituting Eq 31 gives:

$$J_c = \frac{-\sigma_c \sigma_e}{z_c^2 F^2(\sigma_c + \sigma_e)}\left(\frac{d\mu_c}{dx} + z_c \frac{d\mu_e}{dx}\right) \quad \text{(Eq 32)}$$

Everywhere throughout the oxide, the equilibrium:

$$M = M_c^{z+} + z_c e^-$$

is established almost instantly; therefore, the chemical potential of the metal is:

$$\mu_M = \mu_c + z_c \mu_e \quad \text{(Eq 33)}$$

Thus, Eq 32 can be rewritten as:

$$J_c = \frac{-\sigma_c \sigma_e}{z_c^2 F^2(\sigma_c + \sigma_e)} \frac{d\mu_M}{dx} \quad \text{(Eq 34)}$$

which can be integrated over the entire scale thickness x from the inner, or metal/oxide, surface at which the metal chemical potential is μ_M^i to μ_M^o at the outer, or oxide/gas, surface. This gives a cation flux of:

$$J_c = \frac{-1}{z_c^2 F^2 x}\int_{\mu_M^i}^{\mu_M^o} \frac{\sigma_c \sigma_e}{(\sigma_c + \sigma_e)} d\mu_M \quad \text{(Eq 35)}$$

Because the oxide is an electronic semiconductor with $\sigma_c \ll \sigma_e \simeq 1$, Eq 35 can be rewritten as:

$$J_c = \frac{-1}{z_c^2 F^2 x}\int_{\mu_M^i}^{\mu_M^o} \sigma_c d\mu_M \quad \text{(Eq 36)}$$

From Eq 23, the flux is also:

$$J_c = C_c \frac{dx}{dt} \text{ mol/cm}^2 \cdot \text{s} \quad \text{(Eq 37)}$$

if the cation concentration C_c is expressed in moles per cubic centimeter. The rate-determining drift velocity of the cations, dx/dt is therefore also the growth rate of the scale. The growth rate is then:

$$\frac{dx}{dt} = \left[\frac{-1}{z_c^2 F^2 C_c}\int_{\mu_M^i}^{\mu_M^o} \sigma_c d\mu_M\right]\frac{1}{x} \quad \text{(Eq 38)}$$

where the bracketed term is the parabolic rate constant k_p, which can be mathematically evaluated. Agreement between calculated and experimental measurements of k_p has been found to be quite close for numerous oxidation systems.

If anion diffusion predominates when metal M reacts with nonmetal A so that $\sigma_a \gg \sigma_c$, the parabolic rate equation will be:

$$\frac{dx}{dt} = \left[\frac{1}{z_a^2 F^2 C_A}\int_{\mu_A^i}^{\mu_A^o} \sigma_A d\mu_A\right]\frac{1}{x} \quad \text{(Eq 39)}$$

Equation 23 can also be evaluated in terms of diffusivities D instead of conductivities by means of the Nernst-Einstein equation:

$$D_i = \frac{B_i RT}{N} \quad \text{(Eq 40)}$$

where B_i is the absolute mobility of the particle, R is the gas constant, and T is the absolute temperature. Comparison of Eq 40 with Eq 25 gives:

$$D_i = \frac{\sigma_i RT}{z_i^2 F^2 C_i} \quad \text{(Eq 41)}$$

so that Eq 38 can be rewritten as:

$$\frac{dx}{dt} = \left[\frac{-1}{RT}\int_{\mu_M^i}^{\mu_M^o} D_c d\mu_M\right]\frac{1}{x} \quad \text{(Eq 42)}$$

and Eq 39 becomes:

$$\frac{dx}{dt} = \left[\frac{1}{RT}\int_{\mu_A^i}^{\mu_A^o} D_a d\mu_A\right]\frac{1}{x} \quad \text{(Eq 43)}$$

The good agreement between parabolic rate constants calculated from conductivities or diffusivities and the rate constants measured in oxidation experiments indicates that Wagner's assumptions are generally valid. The principal assumptions about the scale are (Ref 4):

- The oxide scale is completely compact and adherent.
- Diffusion of ions through the scale is the rate-controlling process
- Thermodynamic equilibrium exists at both the metal/oxide and oxide/gas interfaces
- Thermodynamic equilibrium exists locally throughout the scale
- The oxide deviates only slightly from stoichiometry

Effects of Temperature and Pressure. The parabolic oxidation rate (Eq 12) increases exponentially with temperature, following the Arrhenius equation:

$$k_p = k_o \exp \frac{-Q}{RT} \quad \text{(Eq 44)}$$

where k_o is a constant that is a function of the oxide composition and the gas pressure. For cation-deficient or cation-excess oxides where $D_c \gg D_a$, the activation energy Q for oxide growth has been found to be the same as the activation energy for diffusion of cations in the oxide. For anion-deficient oxides, such as ZrO_2, where $D_a \gg D_c$, the activation energy for oxide growth is the same as that for anion diffusion, verifying that ionic diffusion is the rate-controlling process.

The activity of oxygen can be approximated by its partial pressure so that the chemical potential of oxygen becomes:

$$d\mu_o = \tfrac{1}{2} RT\, d \ln p_{O_2} \quad \text{(Eq 45)}$$

Equation 45 can be used to modify Eq 39 to:

$$\frac{dx}{dt} = \left[\frac{RT}{2z_a^2 F^2 C_A}\int_{p_{O2}^i}^{p_{O2}^o} \sigma_o\, d \ln p_{O_2}\right]\frac{1}{x} \quad \text{(Eq 46)}$$

where $p^o_{O_2}$ is the oxygen gas pressure, and $p^i_{O_2}$ is the dissociation pressure of the oxide in equilibrium with the metal. Equation 46 can be used if the variation in ionic conductivity with pressure is known.

For a cation-deficient p-type oxide, such as NiO, growth occurs at the oxide/gas interface where:

$$O_2 + 2Ni = 2NiO + 2V''_{Ni} + 4\,\dot{h}$$

The symbol V''_{Ni} stands for a doubly charged cation vacancy, and $\dot{h}$ represents an electron hole. Thermodynamic equilibrium is established at the interface so that at any time the equilibrium constant K is:

$$K = \frac{[V''_{Ni}]^2[\dot{h}]^4}{p_{O_2}} \quad \text{(Eq 47)}$$

where the brackets indicate the concentration (approximately the activity) of the species enclosed; thus, $[V''_{Ni}]$ is the concentration of cation vacancies. The solids nickel and NiO can be assumed not to change in concentration at the interface. Because there must be two electron holes for every cation vacancy, or $2\,[V''_{Ni}] = [\dot{h}]$, then:

$$p_{O_2} \propto [V''_{Ni}]^2\,[\dot{h}]^4 \propto [\dot{h}]^6 \quad \text{(Eq 48)}$$

The oxide conductivity σ is proportional to the hole concentration, so that:

$$\sigma \propto p^{1/6}_{O_2} \quad \text{(Eq 49)}$$

for a p-type oxide with divalent cations. In general, for any p-type oxide,

$$\sigma \propto p^{1/n}_{O_2} \quad \text{(Eq 50)}$$

where, theoretically, n is generally a number between 5 and 8 that depends on the metal valence.

Substituting into Eq 46, the oxidation rate is:

$$\frac{dx}{dt} \propto [(p^o_{O_2})^{1/n} - (p^i_{O_2})^{1/n}] \quad \text{(Eq 51)}$$

In most cases, the ambient oxygen pressure $p^o_{O_2}$ is much greater than the oxide dissociation pressure $p^i_{O_2}$ so that:

$$\frac{dx}{dt} \propto (p^o_{O_2})^{1/n} \quad \text{(Eq 52)}$$

for p-type oxides.

For n-type oxides, such as BeO, in which the oxide grows at the oxide/gas interface the reaction is:

$$O_2 + 2Be_i^{2+} + 4_e{}^- = 2BeO$$

where Be_i^{2+} is an interstitial cation, and e^- is a free electron. The equilibrium established will be:

$$K = \frac{1}{p_{O_2}[Be_i^{2+}]^2[e^-]^4} \quad \text{(Eq 53)}$$

For every Be^{2+} released into the oxide, two electrons are also freed so that $[e^-] = 2\,[Be_i^{2+}]$. The oxygen pressure is:

$$p_{O_2} \propto \frac{1}{[Be_i^{2+}]^2[e^-]^4} \propto \frac{1}{[e^-]^6} \propto \frac{1}{\sigma^6} \quad \text{(Eq 54)}$$

making the oxidation rate:

$$\frac{dx}{dt} \propto [(p^i_{O_2})^{-1/n} - (p^o_{O_2})^{-1/n}] \quad \text{(Eq 55)}$$

Because $p^i_{O_2} \ll p^o_{O_2}$, the oxidation rate should be practically independent of the ambient oxygen pressure for n-type oxides. In fact, this prediction has not been observed experimentally, because the oxides are invariably doped by electrically active impurity atoms.

Properties of Scales

Multiple Scale Layers. A pure metal that can oxidize to more than one valence will form a series of oxides, usually in separate layers. For example, iron has valences of +2 and +3 at high temperatures and forms wustite FeO, magnetite ($FeO \cdot Fe_2O_3$, commonly written as Fe_3O_4), and hematite (Fe_2O_3) scale layers. The layers will be arranged with the most metal-rich oxide next to the metal, progressively less metal in each succeeding layer, and finally the most oxygen-rich layer on the outside. Within each layer a concentration gradient exists with higher metal ion concentration closest to the metal.

If the oxygen partial pressure in the gas is so low that it is below the dissociation pressures of the outer oxygen-rich oxides, then only the thermodynamically stable, inner oxides will form. In general, the lowest valence, inner oxide will usually be p-type because of the ease with which electron holes and cation vacancies can form. The outermost oxide is often n-type because of its anion vacancies. A scale consisting of an inner layer with cations diffusing outward and an outer layer with anions diffusing inward will grow at the oxide/oxide interface.

Relative Thickness. When diffusion is rate controlling, the relative thickness of the layers is proportional to the relative diffusion rates if no porosity develops in the layers. For compact layers growing by a single diffusion mechanism, the ratio of thicknesses should be related to the ratio of the parabolic rate constants by:

$$\frac{x_1}{x_2} = \left(\frac{k_{p1}}{k_{p2}}\right)^{1/2} \quad \text{(Eq 56)}$$

where subscripts 1 and 2 refer to layers 1 and 2. The thickness ratio, consequently, is a constant and does not change with time. Because the ions diffusing through the various layers are likely to be different and because the crystal structures of the layers are certainly different, the thickness ratio is commonly found to be quite far from unity. One layer is usually much thicker than the others.

When diffusion controls the growth of each layer, the entire scale will appear to follow the parabolic oxidation equation with an effective parabolic rate constant k_p. However, this effective parabolic rate constant does not follow the Arrhenius equation (Eq 44) unless the thickness ratios remain far from unity throughout the temperature range, that is, unless one layer predominates at all temperatures. If an inner scale predominates, its growth is independent of oxygen pressure, but if the outermost scale is the major part of the scale, the rate constant will vary with oxygen pressure.

Paralinear Oxidation. With some metals, the oxidation begins as parabolic, but the protective scale gradually changes to a nonprotective outer layer. If the inner protective layer remains at a constant thickness, then the diffusion through the layer of constant thickness results in a linear rate of oxidation. The outer layer may become unprotective by sublimation, transformation to a

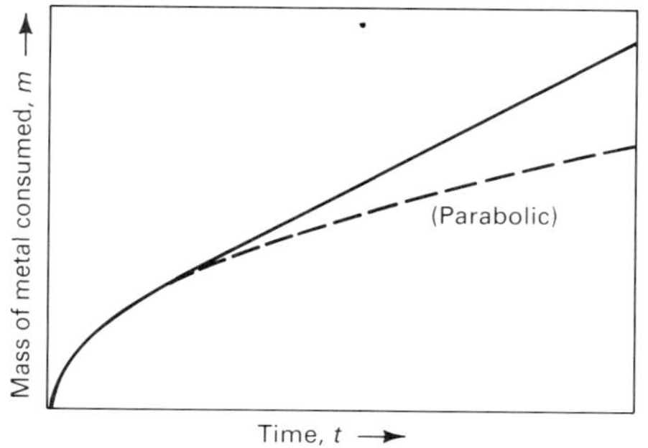

Fig. 15 Paralinear oxidation. Linear region is tangential to initial parabolic curve

porous layer, fracture, and so on. This type of oxidation behavior that is initially parabolic and gradually transforms to linear is termed paralinear oxidation (Fig. 15). The mass of metal m consumed at time t can be calculated from:

$$m = \frac{k_p}{k_l} \ln \frac{k_p}{k_p - k_l(m - k_l t)} \qquad \text{(Eq 57)}$$

where k_p is the parabolic rate constant for formation of the inner layer, and k_l is the linear rate constant for formation of the outer layer (Ref 15).

Oxide Evaporation. At very high temperatures, the evaporation of a protective oxide may limit its protective qualities or remove the oxide entirely, because evaporation rates increase exponentially with temperature. The platinum metals and refractory metals in particular tend to have volatile oxides. Suboxides and unusual valences are also often found at high temperatures; aluminum, for example, forms only one stable solid oxide, Al_2O_3, but it vaporizes as Al_2O and AlO. Evaporation may be much worse in gases containing water or halide vapor if volatile hydroxides (hydrated oxides) or oxyhalides form.

Theoretically, the evaporation rate should be directly proportional to the sublimation vapor pressure of the oxide if no evaporating molecules return to the surface, but at gas pressures above 10^{-3} to 10^{-4} atm, a gaseous stagnant boundary layer slows the escape of evaporated oxide molecules. The boundary layer becomes thinner at higher gas velocities, leading to higher evaporation losses.

As evaporation removes material from the oxide layer, diffusion through the oxide increases until the two rates finally become equal. The oxide thickness then remains constant, and the metal has oxidized paralinearly. If more than one oxide layer protects the metal, the higher-valent outermost oxide always has the higher vapor pressure and is the more volatile.

Stresses in Scales. As oxide scales grow, stresses develop that influence the protective properties of the scale. During growth, recrystallization may occur in either the metal or oxide to alter the stress situation radically. Changes in temperature usually have a great effect on the stress state for metals in engineering service. Stress relief, if it damages the scale, will result in partial or complete loss of oxidation protection.

Growth Stresses. The magnitude of growth stresses is not well defined by the Pilling-Bedworth ratio (discussed earlier in the section "Kinetics of Corrosion in Gases" in this article), indicating that other factors also play a major role. Crystalline oxides will attempt to grow on a metal substrate with an epitaxial relationship. However, because the fit of the oxide crystal on the metal crystal is never perfect, the stresses that develop tend to limit the epitaxy to about the first 50 nm of oxide. Therefore, the stresses in thick scales are not greatly affected by the original epitaxy.

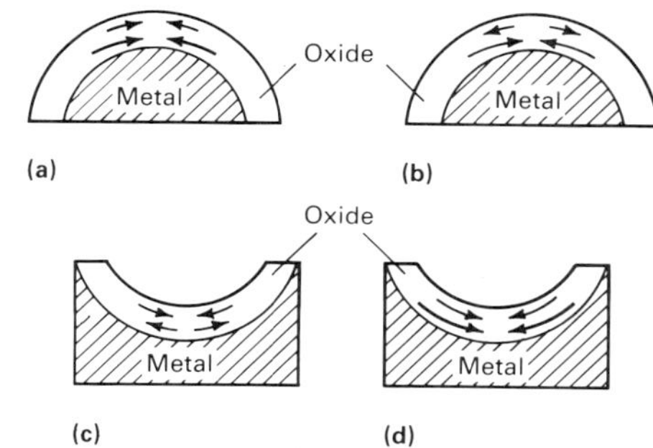

Fig. 16 Stress development on curved surfaces. (a) Oxide grows by cation diffusion on a convex surface. (b) Oxide grows by anion diffusion on a convex surface. (c) Oxide grows by cation diffusion on a concave surface. (d) Oxide grows by anion diffusion on a concave surface.

Polycrystalline oxides develop stresses along their grain boundaries because of more rapid growth of grains oriented in preferred directions. Short-circuit diffusion along oxide grain boundaries may lead to oxide formation at the boundaries that increases compressive stresses. Furthermore, any second phases or foreign inclusions in a metal may oxidize at a rate different from that of the parent metal and create high stresses within the oxide.

The compositional variation across a scale due to deviations from stoichiometry also create stresses within the oxide. Wustite is an important example, varying from $Fe_{0.95}O$ in equilibrium with the metal to as little as $Fe_{0.84}O$ in equilibrium with Fe_3O_4 at 1370 °C (2500 °F).

Composition will also tend to vary in a metal alloy in which one component is preferentially oxidized. If diffusion in the alloy is too slow to maintain a constant composition at the metal/oxide interface, stress develops in the metal. Similarly, any diffusion of oxygen into the metal from the oxide creates compressive stresses in the metal.

Surface geometry will contribute an additional effect to the growth stresses in an adherent oxide. Whether the contribution is compressive or tensile depends on whether the surface profile is convex or concave and whether the scale grows at the oxide/gas interface or at the metal/oxide interface. Figure 16 shows the four possibilities for changes in stress state in an oxide scale as it grows, assuming that the original growth stresses are compressive (Ref 16).

In Fig. 16(a), for a convex surface on which oxide grows at the oxide/gas interface by cation diffusion outward through the scale, the metal surface will gradually recede, increasing the compressive stresses at the metal/oxide interface as long as adhesion is maintained. Figure 16(b) shows oxidation of a convex surface on which the oxide grows at the metal/oxide interface by anion diffusion inward. The compressive stresses that develop at the metal/oxide interface are due only to the volume change of the reaction. Oxide that is pushed away from the growth area will gradually reduce its compressive stress until the outer surface may even be in tension.

For concave surfaces, Fig. 16(c) illustrates oxide growth at the oxide/gas interface by outward diffusion of cations. As the metal surface recedes, the compressive growth stresses are reduced and may eventually even become tensile if oxidation continues long enough. Figure (16(d) shows growth on a concave surface by anion diffusion inward for reaction at the metal/oxide interface. Very high compressive stresses develop during growth until they exceed the cohesive strength of the oxide.

Transformation Stresses. Preferential oxidation of one component in an alloy may alter the alloy composition to the point that a crystallographic phase transformation occurs. A change in temperature could also cause crystallographic transformation of either metal or oxide. The volume change accompanying a transformation creates severe stresses in both the metal and the oxide. Some oxides form initially in an amorphous structure and gradually crystallize as the film grows thicker. The tensile stress created by volume contraction may partially counteract the compressive growth stresses usually present.

Thermal Stresses. A common cause of failure of oxide protective scales is the stress created by cooling from the reaction temperature. The stress generated in the oxide is directly proportional to the difference in coefficients of linear expansion between the oxide and the metal. Examples of the coefficients are listed in Table 5 for a few important metal/oxide systems. In most cases, the thermal expansion of the oxide is less than that of the metal; therefore, compressive stress develops in the oxide during cooling. Multilayered scales will develop additional stresses at the oxide/oxide interface.

Stress Relief. Stresses develop in oxide scale during growth or temperature change. The oxide may develop porosity as it grows. If temperatures are high, the stress may reach the yield strength of either the metal or the oxide so that plastic deformation relieves the stress. A brittle oxide may crack. A strong oxide may remain intact until the internal stresses exceed the adhesive

Table 5 Coefficients of linear thermal expansion of metals and oxides

System	Oxide coefficient, m/m·K	Metal coefficient, m/m·K	Temperature range °C	Temperature range °F
Fe/FeO	12.2×10^{-6}	15.3×10^{-6}	100–900	212–1650
Fe/Fe_2O_3	14.9×10^{-6}	15.3×10^{-6}	20–900	70–1650
Ni/NiO	17.1×10^{-6}	17.6×10^{-6}	20–1000	70–1830
Co/CoO	15.0×10^{-6}	14.0×10^{-6}	25–350	75–660
Cr/Cr_2O_3	7.3×10^{-6}	9.5×10^{-6}	100–1000	212–1830
Cu/Cu_2O	4.3×10^{-6}	18.6×10^{-6}	20–750	70–1380
Cu/CuO	9.3×10^{-6}	18.6×10^{-6}	20–600	70–1110

Source: Ref 16

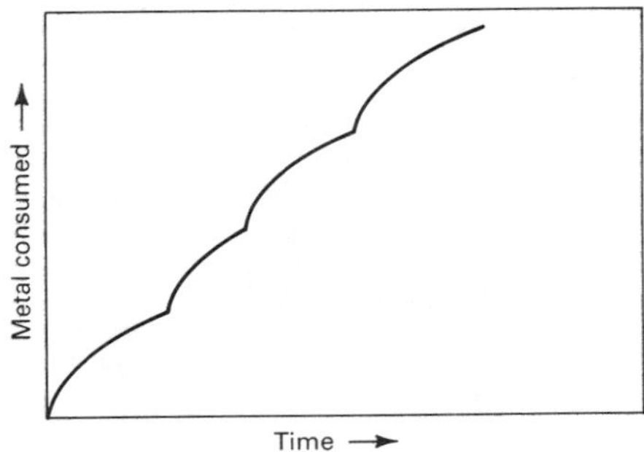

Fig. 17 Periodic cracking of scale

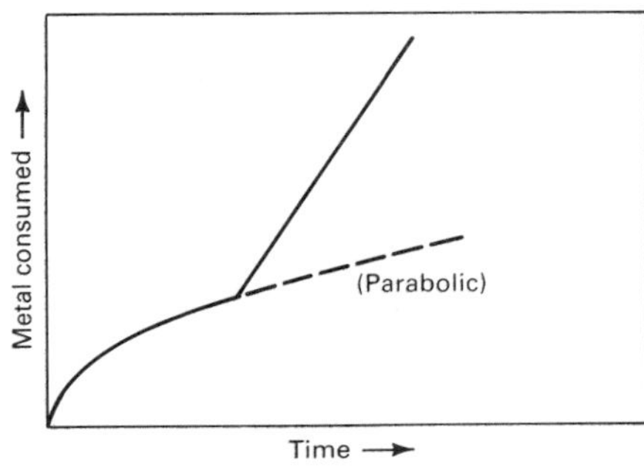

Fig. 18 Breakaway oxidation

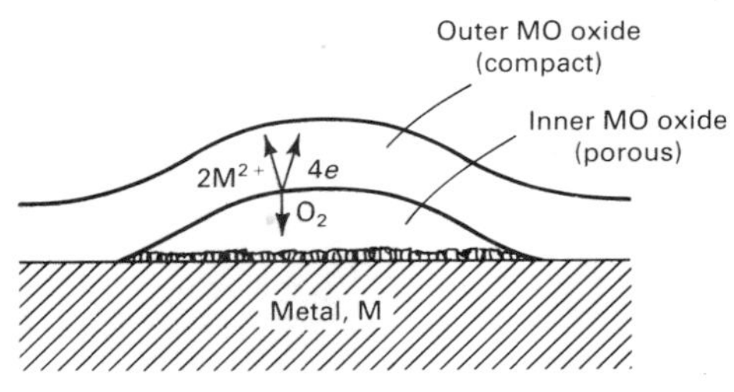

Fig. 19 Structure of double-layer single-phase scale

forces between metal and oxide so that the oxide pulls loose from the metal. The ways in which stresses can be relieved are discussed below.

Porosity. For oxides that grow by cation diffusion outward through cation vacancies (*p*-type oxides, such as NiO), the vacancies are created at the oxide/gas interface and diffuse inward through the scale as they exchange places with an equal number of outward-diffusing cations. The vacancies are annihilated within the oxide, at the metal/oxide interface, or within the metal, depending on the system.

In some oxides, the vacancies collect together within the oxide to form approximately spherical cavities. The preferred sites for cavity formation are along paths of rapid diffusion, such as grain boundaries and dislocation lines.

Vacancies that are annihilated at the metal/oxide interface may cause detachment of the scale because voids that form there reduce adhesion of oxide to metal. If detachment is only partial, the oxidation rate slows down because the cross-sectional area available for diffusion decreases.

Plastic Flow of Oxide. Dislocations form in the oxide as it grows epitaxially on the metal because of the crystallographic lattice mismatch between oxide and metal. As the growth continues, these glissile slip dislocations move out into the oxide by the process of glide. Once out in the oxide, they become sessile growth dislocations.

Although dislocations are present in the oxide, slip is not an important process in relieving growth stress (Ref 17). Plastic deformation of the oxide occurs only at high temperatures at which creep mechanisms become operative. The three important creep mechanisms in oxides are grain-boundary sliding, Herring-Nabarro creep, and climb. Grain-boundary sliding allows relative motion along the inherently weak boundaries. Herring-Nabarro creep allows grain elongation by diffusion of ions away from grain-boundary areas in compression over to boundaries in tension. Within the grains, dislocation climb is controlled by diffusion of the slower moving ions. The creep rate increases with amount of porosity in the oxide.

Cracking of Oxide. Tensile cracks readily relieve growth stresses in the oxide scale. As shown in Fig. 16(b) and (c), tensile stresses may eventually develop in oxide growing on curved surfaces and cause fracture. In the case of anion diffusion inward on convex surfaces, if oxygen diffuses on into the metal, the tensile stresses near the oxide/gas interface develop much more quickly.

Shear cracks can form in oxide having high compressive stresses near the metal surface, as shown in Fig. 16(a) and (d), if the scale cohesion is weak and adhesion to the metal is strong. If the metal/oxide interface is planar, a shear crack initiating at the interface can extend rapidly across the surface. However, if the interface is rough because oxidation has concentrated at the grain boundaries of the metal and keyed the oxide into the metal, rapid crack extension may be prevented and scale adherence may be improved.

Periodic cracking of a protective oxide results in the parabolic oxidation being interrupted by a sudden increase in rate when the gas can react directly with the bare metal surface. As oxide begins to cover the metal surface again, parabolic oxidation is resumed. A typical oxidation curve for this repeated process is shown in Fig. 17. The time periods between successive parabolic steps are sometimes fairly uniform, because a critical scale thickness is reached that causes cracks to be initiated. The overall oxidation of the metal becomes approximately a slow linear process.

Occasionally, a metal oxidizes parabolically until the scale cracks or spalls off, and from that time on, the oxidation is linear. The oxide, originally protective, completely loses its protective properties. The breakaway oxidation (Fig. 18) commonly occurs if many cracks form continuously and extend quickly through the oxide. It can also occur for alloys that have had one component selectively oxidized. When the protective scale spalls off, the metal surface is so depleted in the component that the same protective scale cannot re-form. Breakaway oxidation leaves bare metal continually exposed, unlike paralinear oxidation (Fig. 15), in which an inner protective scale always remains.

Decohesion and Double-Layer Formation. For scale growth by cation diffusion, a protective scale may eventually reach a thickness at which it can no longer deform plastically to conform to the receding metal surface. At this point, decohesion begins at some places along the metal/oxide interface. However, oxidation continues at the oxide/gas interface because cations continue to diffuse outward through the detached scale, driven by the chemical potential gradient across the scale.

At the inner scale surface, the cation concentration decreases, thus increasing the chemical potential of oxygen there. The increased chemical potential of oxygen has associated with it an increased pressure of O_2 gas that will form in the space between the scale and the metal. The O_2 then migrates to the metal surface, and because its pressure is greater than the dissociation pressure of oxide in equilibrium with metal, an inner layer of oxide begins to form on the metal.

The inner oxide layer forms a porous fine-grain structure, although it has essentially the same composition as the compact outer layer. Initially, it started forming as mounds from the nucleation sites. If the growth of these mounds were controlled by cation diffusion, the inner oxide would quickly thicken in the areas where it was thinnest and where diffusion distances were shortest. However, in this situation, the diffusion of O_2 from the outer layer to the inner layer is rate controlling so that the inner layer grows mainly at its high points and forms a porous scale (Ref 18). Figure 19 shows the mechanism of the double-layer scale.

Deformation of Metal. Foils and thin-wall tubes are often observed to deform during oxidation, thus relieving the oxide growth stresses. At high temperatures, slip and creep mechanisms can both be operative in metals while the temperature is still too low for plastic deformation of the oxide. The deformation processes are facilitated by accumulation of porosity in the metal, which is caused by cation diffusion outward from the metal and by selective oxidation of one component of an alloy.

Logarithmic Oxidation of Scales. A logarithmic rate law for thick oxide scales has been developed for situations in which cavities or precipitates hinder the diffusion so that only part of the oxide is available for diffusion (Ref 19). The equation developed for logarithmic thick scale growth should not be confused with thin-film logarithmic behavior, which has an entirely different mechanism.

Even where the whole oxide scale originally serves as a diffusion path for ions or vacancies, the area available for diffusion may reduce with time. This can occur where vacancies collect at the metal/oxide interface to cause partial detachment or cavity formation, or in the case of an A-B alloy, it can occur where particles of a very stable but slow-growing BO oxide form and restrict the growth of a much faster growing AO oxide layer. It has even been suggested that the mechanism might hold when growth stresses are so high that they cause the scale to crack in short cracks that run parallel to the metal surface.

The equation developed for these situations will approximate the standard logarithmic equation:

$$m = \mathrm{k} \log (at + 1) \qquad \text{(Eq 58)}$$

where m is the mass increase per unit area, and k and a are constants. The oxidation begins rapidly but then almost comes to a stop as the cross-sectional area available for diffusion decreases.

Catastrophic Oxidation. Although many oxidation failures can be described as catastrophies, the term catastrophic oxidation is reserved for the special situation in which a liquid phase is formed in the oxidation process. This can occur either when the metal is exposed to the vapors of a low-melting oxide or during oxidation of an alloy having a component that forms a low-melting oxide. The low-melting oxides that have commonly caused catastrophic oxidation are listed in Table 6.

The exact mechanism of catastrophic oxidation is disputed, but the evidence shows that it occurs at the metal/oxide interface. A liquid phase seems to be essential. The liquid usually forms at the oxide/gas surface and penetrates the scale along grain boundaries or pores to reach the metal. The penetration paths can also serve as paths for rapid diffusion of reacting ions. Once at the metal/oxide interface, the liquid spreads out by capillary action, destroying adherence of the solid scale.

Oxidation in the absence of any protective scale proceeds linearly or even at an ever increasing rate if the metal heats up from the exothermic oxidation reaction. The Wagner mechanism clearly does not apply to catastrophic oxidation.

Internal Oxidation. Internal oxidation is the term used to describe the formation of fine oxide precipitates found within an alloy. It is sometimes called subscale formation. Oxygen dissolves in the alloy at the metal/oxide interface or at the bare metal surface if the gas pressure is below the dissociation pressure of the metal oxides. The oxygen diffuses into the metal and forms the most stable oxide that it can. This is usually the oxide of the most reactive component of the alloy. Internal oxides can form only if the reactive element diffuses outward more slowly than the oxygen diffuses inward; otherwise, only surface scale would form.

Because diffusion of oxygen is usually the rate-controlling process in internal oxidation, parabolic behavior is observed. Wagner developed an equation that, with simplifications, is approximately (Ref 20):

$$x = \left(\frac{2N_o^s D_o t}{\nu N_B^i}\right)^{1/2} \quad \text{(Eq 59)}$$

where N_o^s is the mole fraction of oxygen in the alloy at its surface, N_B^i is the mole fraction of reactive metal B initially in the alloy, D_o is the diffusion coefficient of oxygen in the alloy, ν is the ratio of oxygen atoms to metal atoms in the oxide compound that forms, and x is the subscale thickness at time t. It is assumed that counterdiffusion of B atoms is negligible and that oxygen has a very low solubility limit in the alloy.

Because the diffusion coefficients of B atoms and oxygen both vary exponentially with temperature and because their activation energies are different, it is possible that internal oxidation will form in an alloy only in a certain temperature range. Because internal oxide precipitates reduce the cross section available for oxygen diffusion, only surface oxide forms above some critical solute concentration in the alloy.

Table 6 Low-melting oxides

Oxide	Melting point °C	°F
V_2O_5	674	1245
MoO_3	795	1463
MoO_2-MoO_3 eutectic	778	1432
Bi_2O_3	817	1503
PbO	885	1625
WO_3	1470	2678

In an A-B alloy, in which BO is a more stable oxide than AO, a mixed (A,B)O oxide may form as the internal oxide. This will occur if AO and BO have considerable mutual solubility so that the free energy of the system is lowered by precipitation of the mixed oxide. For example, the internal oxide that forms in unalloyed steels is (Fe,Mn)O (Ref 21).

Alloy Oxidation: The Doping Principle (Ref 22). For oxides that form according to the Wagner mechanism and contain wrong-valent impurity cations that are soluble in the oxide, the impurities alter the defect concentration of the scale. Consequently, the oxide growth rate may also be altered by the alloy impurities. Whether oxidation increases or decreases depends on the relative valences of the cations and on the type of oxide.

p-type Oxides. In a p-type semiconducting oxide (typified by NiO) the oxidation rate is controlled by cation diffusion through cation vacancies. If the number of cation vacancies can be decreased, the oxidation is slowed. The cation vacancies are present in the first place because a small fraction of the nickel cations have a higher-than-normal valence, contributing more than their share of valence electrons to the oxygen and thus allowing a small fraction of cations to be absent from the structure.

If a few cations with a higher valence are substituted for the regular cations in an oxide, the vacancy concentration is increased. For example, if a small amount of aluminum is alloyed with nickel and then oxidized to form a scale of NiO with a few substitutional Al^{3+} cations, the oxidation rate will be faster than for pure NiO because of the increased cation vacancy concentration. For every two Al^{3+} cations in the oxide, contributing one additional electron each, another cation vacancy V''_{Ni} must be present. For the same reason, adding lower-valent cations, such as lithium ions (Li^+), to the NiO will reduce the cation vacancy concentration. Substituting ions with the same valence as the rest of the cations in the oxide should have little effect.

n-type Oxides. The n-type semiconducting oxides behave exactly opposite to the p-type oxides. For the oxides that grow by anion diffusion through anion vacancies (typified by ZrO_2), the anion vacancies exist because some cations have a lower-than-normal valence, contributing fewer electrons to the oxygen than required by the structural arrangement. Consequently, anion vacancies are present to facilitate anion diffusion. If additional low-valent cations replace the regular higher-valent cations, the oxidation rate increases. Conversely, substituting higher-valent cations reduces the anion vacancy concentration and reduces the oxidation rate.

For n-type oxides that grow by interstitial diffusion of cations, as with BeO, substitution of higher-valent cations, for example, Al^{3+}, for a few of the Be^{2+} ions in the oxide structure leaves the oxide negatively charged with the excess electrons. The free electrons tend to prevent the ionization of the metal:

$$Be \rightarrow Be_i^{2+} + 2e^-$$

at the metal/oxide interface. The concentration of interstitial cations Be_i^{2+} is consequently reduced; therefore, the oxidation rate is reduced. However, if lower-valent cations are substituted in the n-type semiconducting oxide, for example, a small amount of Li^+ in BeO, the Li^+ supplies only one valence electron to the oxygen in place of the two electrons that would have been supplied by the replaced Be^{2+}. The oxygen must obtain more electrons from the surface of the metal, increasing the ionization and formation of interstitial Be_i^{2+} cations. Alloying with a metal that forms cations of the same valence as the rest of the cations in the oxide will have little effect.

Alloy Development. As it turns out, the doping principle has been useful in verifying the Wagner mechanism, but it is not very helpful in developing oxidation-resistant alloys. First, the concentration of foreign cations that can be put into solid solution in the oxide is severely restricted by low solubility limits. Second, the choices of foreign ions that could be used is extremely limited by their valence. For p-type oxides, the alloying element should have a lower valence than the metal being oxidized in order to slow corrosion, but p-type oxides are commonly formed by metals with +1 and +2 valences. The n-type oxides form on metals with high valences, but very few alloying elements have even higher valences to slow the oxidation.

For any stoichiometric oxides, nitrides, or sulfides that are ionic conductors instead of semiconducting electronic conductors, the doping principle can be applied in theory to reduce oxidation at room temperature. However, at high temperatures, perhaps all these protective scales carry current principally by free electrons (n-type) or electron holes (p-type).

Selective Oxidation. An alloy is selectively oxidized if one component, usually the most reactive one, is preferentially oxidized. The simplest case would be a binary alloy with a uniform scale composed entirely of the only oxide that one of the components can form. This situation will be described to illustrate the principles involved. The alloy is formed from metals A and B where B is more reactive than A is; that is, oxide BO is thermodynamically more stable than AO, and AO and BO are immiscible. Which scale, AO or BO, that will form on the A-B alloy depends on the relative nobility of A and B, their concentrations in the alloy, the oxygen pressure, and the temperature.

Alloying With a Noble Metal. An obvious example of selective oxidation would be scale formation on alloy A-B where A is so noble that AO is not thermodynamically stable at the environmental pressure and temperature. That is, the oxygen partial pressure in the gas is less than the equilibrium (dissociation) pressure of AO oxide. Then, only BO scale can form if it is stable.

Both Elements Reactive. For situations in which both A and B are reactive with oxygen at the temperature and gas pressure involved but A is somewhat more noble than B, the alloy composition determines which oxide forms. If the alloy contains some very low concentration of B, the activity of B could be so low that the free energy for formation of BO could actually be positive. That is, for the reaction:

$$B_{(alloy)} + \tfrac{1}{2}O_2 \rightarrow BO$$

the driving force is:

$$\Delta G \simeq \Delta G^\circ + RT \ln \frac{1}{N'_B \cdot p_{O_2}^{1/2}} \quad \text{(Eq 60)}$$

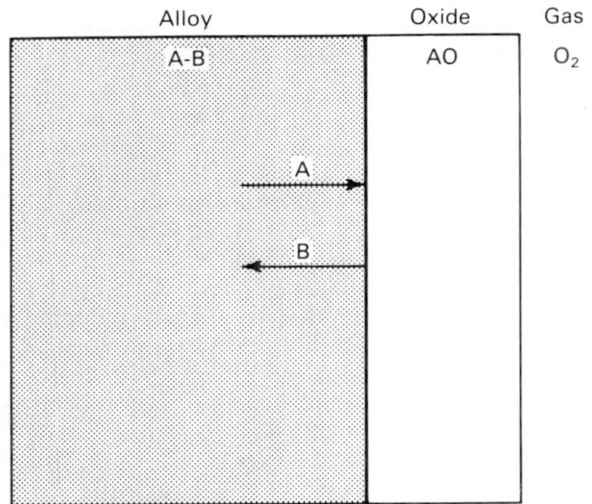

Fig. 20 Selective oxidation of alloy A-B, with oxide AO forming. Directions for diffusion of A and B are indicated.

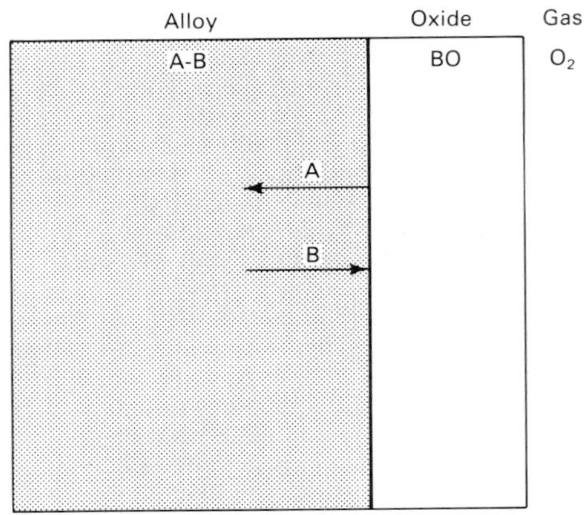

Fig. 21 Selective oxidation of alloy A-B for $N_B > N_B^*$

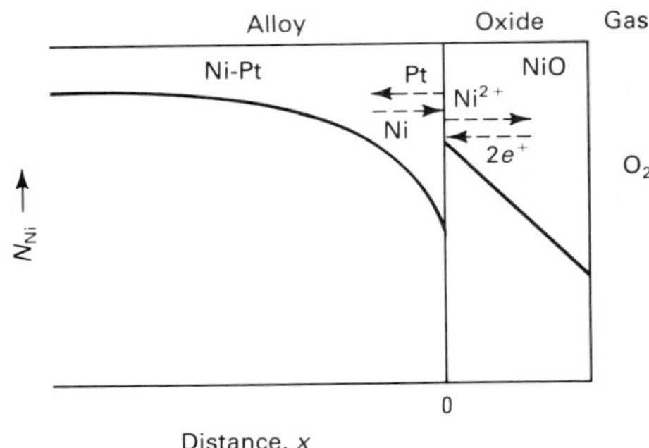

Fig. 22 Schematic diagram of concentration gradients of nickel in nickel-platinum alloy and NiO oxide. Electron holes are indicated as e^+.

where ΔG is the Gibbs energy change; $\Delta G°$ is the Gibbs energy change under standard conditions; the activity of the pure solid oxide is approximately 1 (invariant); the small mole fraction of B in the alloy is N'_B, which approximates its activity; and the pressure of $O_2(p_{O_2})$ closely approximates its fugacity. It can be seen that if N'_B is smaller than some critical value N_B^* described below, then the Gibbs energy for formation of BO could be positive even though $\Delta G°$ is negative. Because it is stable, AO will grow on the alloy, while BO cannot.

The situation for selective oxidation of A is illustrated in Fig. 20. Alloying element A becomes depleted as it reacts at the metal/oxide interface. This creates a concentration gradient of A, causing A to diffuse from the interior of the alloy toward the surface. At the same time, depleting A near the metal surface increases the concentration of B so that B should diffuse inward. If the diffusion rates of A and B in the alloy are similar to the rate-controlling diffusion through the oxide, then element A never becomes seriously depleted at the metal/oxide interface and AO continues to form. However, if diffusion through the alloy is much slower than through the oxide, the concentration of B will increase at the alloy surface until it reaches N_B^*, the critical concentration at which formation of BO is thermodynamically favorable. When that time comes, BO will form along with AO.

At a high concentration of B in an alloy, BO forms while A diffuses back into the alloy (Fig. 21). In this case, the mole fraction N_A is less than the minimum or critical concentration N_A^* at which AO oxide could be stable. However, if diffusion into the metal is slow, the concentration of A may eventually build up to N_A^*, and the AO oxide would then begin to form.

Reducing Oxidation Rates. It may at first seem that the oxidation rate of a reactive metal could be reduced by alloying it with a noble metal, but this is not always the case. If the concentration of the reactive element is far greater than the critical concentration needed to form an external scale, that is, if $N_B \gg N_B^*$, alloying with a noble metal will have very little effect on the oxidation rate.

For example, Wagner considered alloying nickel with a noble metal, such as platinum. For oxidation of pure nickel, the rate is proportional to the concentration gradient across the scale (Fick's first law); therefore, the rate constant k is given by:

$$k = \text{Constant}\ [(p^{o}_{O_2})^{1/n} - (p_d)^{1/n}] \qquad \text{(Eq 61)}$$

where $p^{o}_{O_2}$ is the oxygen pressure in the gas at the outer surface of the oxide, p_d is the dissociation pressure for NiO in equilibrium with the pure nickel, and n is a constant theoretically equal to 6 for NiO as indicated in Eq 49. Figure 22 shows the concentration gradients of nickel in the alloy and in the scale.

Oxidation of nickel-platinum alloy would have a rate constant of:

$$k' = \text{Constant} \cdot [(p^{o}_{O_2})^{1/n} - (p^{i}_{O_2})]^{1/n} \qquad \text{(Eq 62)}$$

where $p^{i}_{O_2}$ is the equilibrium oxygen pressure for NiO in equilibrium with the nickel-platinum alloy at the inner surface of the oxide. Comparing the oxidation rate of the alloy with the rate for pure nickel yields:

$$\frac{k'}{k} = \frac{(p^{o}_{O2})^{1/n} - (p^{i}_{O2})^{1/n}}{(p^{o}_{O2})^{1/n} - (p_d)^{1/n}} \qquad \text{(Eq 63)}$$

If activity of nickel in the alloy can be approximated by its concentration, Eq 63 can be written as:

$$\frac{k'}{k} = \frac{1 - (N^{*}_{Ni}/N^{i}_{Ni})^{2/n}}{1 - (N^{*}_{Ni})^{2/n}} \qquad \text{(Eq 64)}$$

where N^{*}_{Ni} is the critical concentration of nickel in an alloy in equilibrium with NiO, and N^{i}_{Ni} is the mole fraction of nickel in the alloy at the metal/oxide interface.

In considering Eq 64, it must be remembered that N^{*}_{Ni} is extremely small for a reactive metal, such as nickel. Therefore, alloying nickel with a small amount of platinum, but still keeping $N^{i}_{Ni} \gg N^{*}_{Ni}$, will have very little effect on the oxidation. Oxide growth will slow to a stop only when enough platinum is added to make N^{i}_{Ni} approach N^{*}_{Ni}. The oxidation rate is then no longer controlled by diffusion of nickel through the oxide but by interdiffusion of platinum and nickel in the alloy.

Composite External Scales. Wagner has shown that for an A-B alloy that is forming both AO and BO oxides, the mole fraction of B at the alloy surface must not exceed:

$$N^{i}_{B} \le (1 - N^{*}_{A}) = \frac{V}{z_B M_O}\left(\frac{\pi k_p}{D}\right)^{1/2} \qquad \text{(Eq 65)}$$

where N^{*}_{A} is the minimum concentration of A necessary to form AO, V is the molar volume of the alloy, z_B is the valence of B, M_O is the atomic weight of oxygen, k_p is the parabolic rate constant for growth of BO, and D is the diffusion coefficient of B in the alloy. Equation 65 does not take into account any complications, such as porosity and internal oxidation. If the concentration of B lies anywhere between N^{*}_{B} and $(1 - N^{*}_{A})$, thermodynamics predicts that both AO and BO form.

Competing Oxides. When oxides of both metals form, their relative positions and distribution depend on the thermodynamic properties of the oxides and the alloy, the diffusion processes, and reaction mechanisms. This section will discuss the two common situations in which both metals in a binary alloy oxidize to form two separate oxide phases.

The first situation involves immiscible oxides with the more stable oxide growing slowly. With both AO and BO stable but with rapid growth of AO and slow growth of BO, the more stable BO may nucleate first, but gradually becomes overwhelmed and surrounded by fast-growing AO. Figures 23(a) and (b) illustrate the situation. If diffusion in the alloy is rapid, the oxidation proceeds to form an AO scale with BO islands scattered through it. However, if diffusion in the alloy is slow, the metal becomes depleted of A near the metal/oxide interface, while the growth

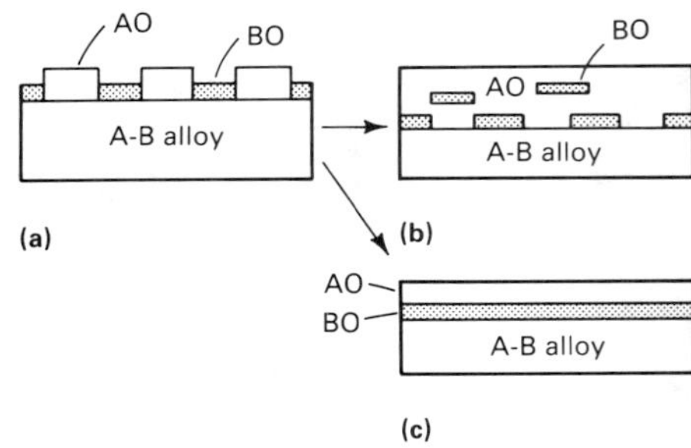

Fig. 23 Simultaneous growth of competing oxides. BO is more stable, but AO grows faster. (a) Early stage with nucleation of both oxides. (b) Later stage if diffusion in alloy is rapid. (c) Final stage if diffusion in alloy is slow

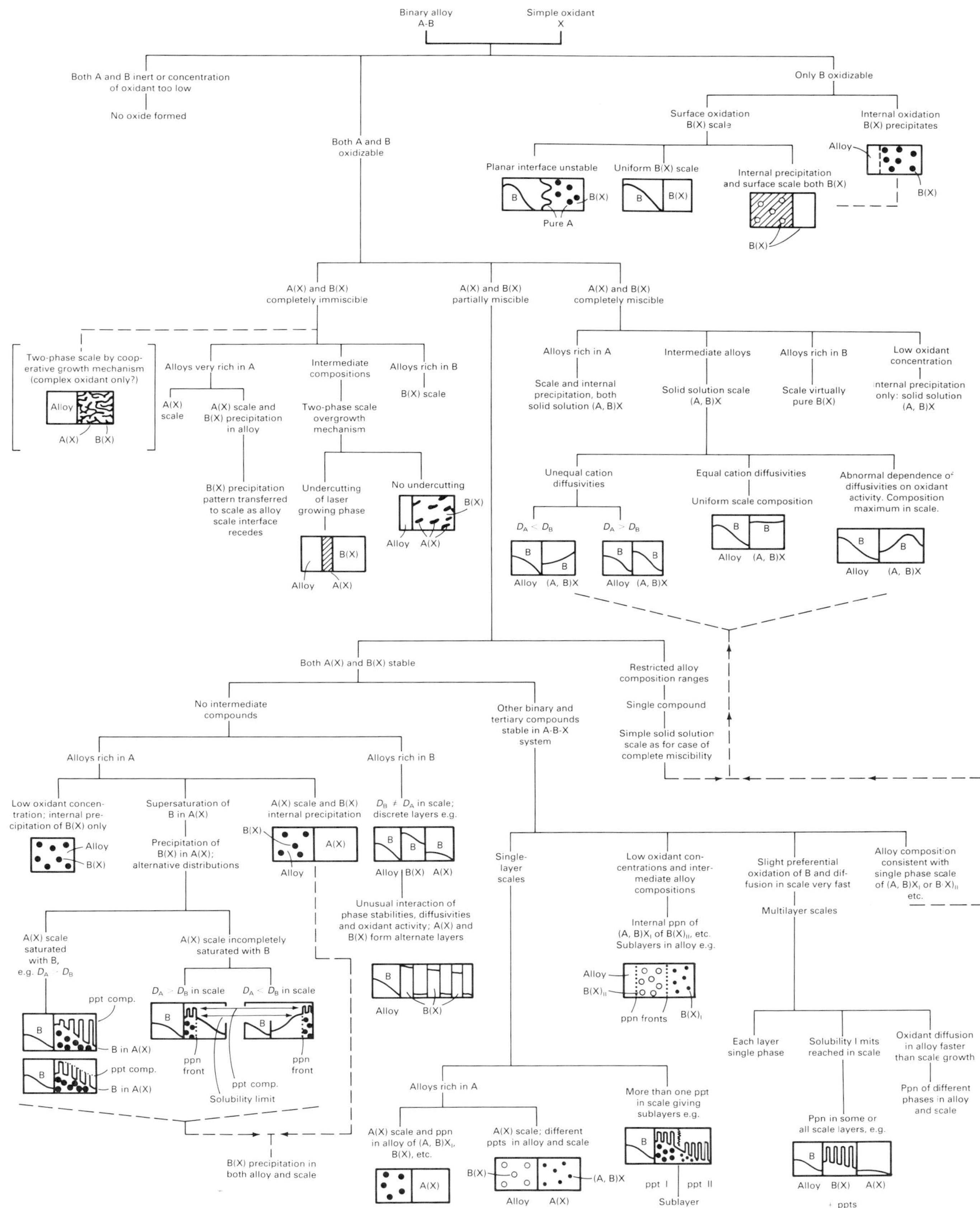

Fig. 24 Schematic showing the relationships between scale morphologies on binary alloys. ppn, precipitation, ppt, precipitate. Source: Ref 23

of BO continues until it forms a complete layer, undercutting the AO (Fig. 23c). Pockets of AO at the metal/oxide interface will gradually be eliminated by the displacement reaction:

$$AO + B_{(alloy)} \rightarrow BO + A_{(alloy)}$$

because BO is thermodynamically more stable than AO. This reaction continues even if the oxygen supply is cut off.

The second situation involves two oxides that are partially miscible. For alloys rich in A, an AO scale will form with some B ions dissolved substitutionally in the AO structure. If the solubility limit is exceeded when B ions continue to diffuse into the scale, BO precipitates as small islands throughout the AO layer. Even if the solubility limit is not reached, the more stable BO may nucleate within the AO scale and precipitate.

For alloys rich in B, a BO layer first forms. If B ions diffuse through the scale faster than A ions do, the concentration of A ultimately builds up in the scale close to the metal/oxide interface. An AO layer then forms underneath the BO. On the other hand, if A ions diffuse through the original BO scale more quickly than the B ions, the AO layer eventually forms on top of the BO layer. In addition, if A ions diffuse rapidly through BO and B ions diffuse rapidly through AO, alternate layers of BO/AO/BO/AO may even form.

Double Oxides. A great deal of research has been directed toward developing alloys that form slow-growing complex oxides. The silicates are particularly important because they can form glassy structures that severely limit diffusion of ions. Therefore, silicide coatings on metals have been very successful. In addition to forming a protective SiO_2 outer layer, much of the silicon diffuses into the underlying alloy, where it gradually oxidizes to ternary silicates.

Spinels often have extremely low diffusion rates. Spinels are double oxides of a metal with +2 valence and a metal with +3 valence, having the general formula $MO \cdot Me_2O_3$ and also having the crystal structure of the mineral spinel ($MgO \cdot Al_2O_3$). The iron oxide Fe_3O_4 has an inverse spinel structure. On iron-chromium alloys, the spinel phase can be either stoichiometric $FeO \cdot Cr_2O_3$ or the solid solution $Fe_{3-x}Cr_xO_4$. Although many ternary oxides tend to be brittle, much research has been devoted to minor alloy additions to improve the high-temperature mechanical properties of those ternary scales that are extremely protective.

Summary Outline of Alloy Oxidation. A schematic diagram has been constructed to show the morphologies of scale growth on binary alloys (Ref 23). It is shown in Fig. 24. The diagram illustrates those types of structures that are known to form, not all those that would be theoretically possible.

REFERENCES

1. P. Kofstad, Oxidation Mechanisms for Pure Metals in Single Oxidant Gases, *High Temperature Corrosion*, R.A. Rapp, Ed., National Association of Corrosion Engineers, 1983, p 123-138
2. Properties of Pure Metals, in *Properties and Selection: Nonferrous Alloys and Pure Metals*, Vol 2, 9th ed., *Metals Handbook*, American Society for Metals, 1979, p 714-831
3. R.C. Weast, Ed., Physical Constants of Inorganic Compounds, *Handbook of Chemistry and Physics*, 65th ed., The Chemical Rubber Company, 1984, p B68-B161
4. N. Birks and G.H. Meier, *Introduction to High Temperature Oxidation of Metals*, Edward Arnold, 1983
5. C.S. Giggins and F.S. Pettit, Corrosion of Metals and Alloys in Mixed Gas Environments at Elevated Temperatures, *Oxid. Met.*, Vol 14 (Nov. 5), 1980, p 363-413
6. T. Rosenqvist, Phase Equilibria in the Pyrometallurgy of Sulfide Ores, *Metall. Trans. B*, Vol 9B, 1978, p 337-351
7. N.B. Pilling and R.E. Bedworth, The Oxidation of Metals at High Temperatures, *J. Inst. Met.*, Vol 29, 1923, p 529-582
8. O. Kubashewski and B.E. Hopkins, *Oxidation of Metals and Alloys*, 2nd ed., Butterworths, 1962
9. K.R. Lawless and A.T. Gwathmey, The Structure of Oxide Films on Different Faces of a Single Crystal of Copper, *Acta Metall.*, Vol 4, 1956, p 153-163
10. N. Cabrera and N.F. Mott, Theory of Oxidation of Metals, *Rep. Prog. Phys.*, Vol 12, 1948-1949, p 163-184
11. K. Hauffe and B. Ilschner, Defective-Array States and Transport Processes in Ionic Crystals, *Z. Elektrochem.*, Vol 58, 1954, p 467-477
12. T.B. Grimley and B.M.W. Trapnell, The Gas/Oxide Interface and the Oxidation of Metals, *Proc. R. Soc. (London) A*, Vol A234, 1956, p 405-418
13. H.H. Uhlig, Initial Oxidation Rate of Metals and the Logarithmic Equation, *Acta Metall.*, Vol 4, 1956, p 541-554
14. C. Wagner, Contributions to the Theory of the Tarnishing Process, *Z. Phys. Chem.*, Vol B21, 1933, p 25-41
15. E.W. Haycock, Transitions from Parabolic to Linear Kinetics in Scaling of Metals, *J. Electrochem. Soc.*, Vol 106, 1959, p 771-775
16. P. Hancock and R.C. Hurst, The Mechanical Properties and Breakdown of Surface Films at Elevated Temperatures, in *Advances in Corrosion Science and Technology*, Vol 4, R.W. Staehle and M.G. Fontana, Ed., Plenum Press, 1974, p 1-84
17. D.L. Douglass, Exfoliation and the Mechanical Behavior of Scales, in *Oxidation of Metals and Alloys*, American Society for Metals, 1971, p 137-156
18. S. Mrowec and T. Werber, *Gas Corrosion of Metals*, National Center for Scientific, Technical and Economic Information, 1978
19. U.R. Evans, *The Corrosion and Oxidation of Metals*, Edward Arnold, 1960, p 836-837
20. C. Wagner, Types of Reaction in the Oxidation of Alloys, *Z. Elektrochem.*, Vol 63, 1959, p 772-782
21. S.A. Bradford, Formation and Composition of Internal Oxides in Dilute Iron Alloys, *Trans. AIME*, Vol 230, 1964, p 1400-1405
22. K. Hauffe, *Oxidation of Metals*, Plenum Press, 1965
23. B.D. Bastow, G.C. Wood, and D.P. Whittle, Morphologies of Uniform Adherent Scales on Binary Alloys, *Oxid. Met.*, Vol 16, 1981, p 1-28

Forms of Corrosion

Section Chairmen:
Bruce Craig, Metallurgical Consultants, Inc.
Steven L. Pohlman, Kennecott Corporation

Introduction

Bruce Craig, Metallurgical Consultants, Inc.
Steven L. Pohlman, Kennecott Corporation

OVER THE YEARS, corrosion scientists and engineers have recognized that corrosion manifests itself in forms that have certain similarities and therefore can be categorized into specific groups. However, many of these forms are not unique but involve mechanisms that have overlapping characteristics that may influence or control initiation or propagation of a specific type of corrosion.

The most familiar and often used categorization of corrosion is probably the eight forms presented by Fontana and Greene (Ref 1): uniform attack, crevice corrosion, pitting, intergranular corrosion, selective leaching, erosion corrosion, stress corrosion, and hydrogen damage. This classification of corrosion was based on visual characteristics of the morphology of attack. Fontana and Greene's introductory remarks in their chapter on forms of corrosion indicate that this classification is arbitrary and that many of the forms are interrelated, making exact distinction impossible. Other prominent corrosion authors such as Uhlig (Ref 2) and Evans (Ref 3) have avoided a classification format and have simply discussed the classical types of corrosion (for example, pitting and crevice corrosion) as they relate to specific metals and alloys.

Substantial advances in the field of corrosion science have begun to define the mechanisms of many forms of corrosion more clearly. However, rather than placing the mechanisms into distinct categories, the overlap between many of the forms has become greater. For example, there is evidence that hydrogen may dominate the crack initiation or crack propagation portion of fracture in some metal/solution systems where stress-corrosion cracking occurs. Additionally, in some metal systems where dealloying (selective leaching) occurs, this form of corrosion may be a precursor to stress-corrosion cracking.

In a similar vein, the magnitude of contribution of stress or corrosion to stress-corrosion cracking, hydrogen damage, or liquid metal embrittlement is not currently understood and can affect whether just pitting or crevice attack occurs or environmental cracking results. The transition from uniform corrosion to highly localized attack is not clearly understood, and there are conditions where a distinction cannot be drawn.

The forms of corrosion presented in this Section were categorized to represent the mechanisms of attack involved rather than to emphasize the visual characteristics. However, as with any classification system, these categories are not distinct or all-inclusive and do not necessarily represent the only mode of attack that may be observed. The forms of corrosion discussed in this Section are categorized in Table 1.

General Corrosion. In this Section, general corrosion is defined as corrosive attack dominated by uniform thinning. Although high-temperature attack in gaseous environments, liquid metals, and molten salts may manifest itself as various forms of corrosion, such as stress-corrosion cracking and dealloying, high-temperature attack has been incorporated into the article "General Corrosion" because it is often dominated by uniform thinning. However, subsequent articles in this Volume will discuss cases of highly localized high-temperature attack. Likewise, galvanic corrosion may appear to be highly localized, but if the anodic area were large and a highly conductive electrolyte existed, uniform thinning would occur.

Table 1 Forms of corrosion presented in this Section

General corrosion
Atmospheric corrosion
Galvanic corrosion
Stray-current corrosion
General biological corrosion
Molten salt corrosion
Corrosion in liquid metals
High-temperature corrosion
Oxidation
Sulfidation
Carburization
Other forms(a)
Localized corrosion
Filiform corrosion
Crevice corrosion
Pitting corrosion
Localized biological corrosion
Metallurgically influenced corrosion
Intergranular corrosion
Dealloying corrosion
Mechanically assisted degradation
Erosion corrosion
Fretting corrosion
Cavitation and water drop impingement
Corrosion fatigue
Environmentally induced cracking
Stress-corrosion cracking
Hydrogen damage
Liquid metal embrittlement
Solid metal induced embrittlement

(a) Other forms include hydrogen effects and hot corrosion.

Localized Corrosion. The forms of corrosion in the article "Localized Corrosion" need no explanation, even though other forms could be placed in this category. It should be noted, however, that localized biological corrosion often causes or accelerates pitting or crevice corrosion.

Metallurgically influenced corrosion was so classified as a result of the significant role that metallurgy plays in these forms of attack. It is well understood that metallurgy is important in all forms of corrosion, but this classification is meant to emphasize its role in these specific forms of attack.

Mechanically assisted degradation groups those forms of corrosion that contain a mechanical component, such as velocity, abrasion, and hydrodynamics, that has a significant effect on the corrosion behavior. Corrosion fatigue was included in this category because of the dynamic stress state; however, it could easily be categorized as a form of environmentally induced cracking.

The environmentally induced cracking article follows the current trend in the literature of combining those forms of cracking that are produced by corrosion in the presence of stress. As the introduction to this article explains, there are many differences as well as similarities among these forms of cracking. However, the distinction between each form is not always apparent; it is therefore easier to combine these different forms into one all-encompassing form.

As a result of the arbitrary nature of classifying forms of corrosion and the fact that corrosion science is a dynamic field with ever-changing theories and observations, the distinctions between many forms become highly personalized. Other articles in this Volume will contain variations of these forms of corrosion and are not always self-consistent. This does not imply a misunderstanding of the mechanisms of corrosion, but rather underscores the continuing evolution of corrosion theory.

REFERENCES

1. M.G. Fontana and N.D. Greene, *Corrosion Engineering*, McGraw Hill, 1967
2. H.H. Uhlig, *Corrosion and Corrosion Control*, J. Wiley & Sons, 1963
3. U.R. Evans, *The Corrosion and Oxidation of Metals*, Edward Arnold Ltd., 1960

General Corrosion

Chairman: Steven L. Pohlman, Kennecott Corporation

GENERAL CORROSION, as described in this article, refers to corrosion dominated by uniform thinning that proceeds without appreciable localized attack. Weathering steels and copper alloys are good examples of materials that typically exhibit general attack, while passive materials, such as stainless steels or nickel-chromium alloys, are generally subject to localized attack. Under specific conditions, however, each material may vary from its normal mode of corrosion. Examples describing the environmental conditions that promote uniform attack will be discussed throughout this article.

General corrosion has been divided into seven specific types of corrosion in this article. Atmospheric corrosion is probably the most common form of corrosion and may well be the most costly. Galvanic corrosion is an electrochemical form of corrosion that protects cathodic areas at the expense of anodic areas. Stray-current corrosion is similar to galvanic corrosion, but does not rely on electrochemically induced driving forces to cause rapid attack. Biological corrosion is a microbial-assisted form of attack that can manifest itself in a general or a localized form. Molten-salt corrosion and liquid-metal corrosion have become more of a concern as the demand for higher-temperature heat transfer fluids increases. High-temperature (gaseous) corrosion is an area of great concern, particularly for the industrial sector.

As noted in the Introduction to this Section, some of the categories of general corrosion described in this article also manifest themselves as other forms of corrosive attack, such as stress-corrosion cracking, dealloying, or pitting. However, because uniform thinning plays an important role in all of the categories described, each can and will be discussed under general corrosion.

Atmospheric Corrosion

Steven L. Pohlman
Kennecott Corporation

Atmospheric corrosion is defined as the corrosion or degradation of material exposed to the air and its pollutants rather than immersed in a liquid. This has been identified as one of the oldest forms of corrosion and has been reported to account for more failures in terms of cost and tonnage than any other single environment.

Dry corrosion. Many authors classify atmospheric corrosion under categories of dry, damp, and wet, thus emphasizing the different mechanisms of attack under increasing humidity or moisture. Supplementary information on atmospheric corrosion can be found in the article "Simulated Service Testing" (see the section "Corrosion Testing in the Atmosphere") in this Volume.

Types of Atmospheric Corrosion

In the absence of moisture, most metals corrode very slowly at ambient temperatures. Accelerated corrosion under dry conditions at elevated temperatures is covered in the section "High-Temperature Corrosion" in this article. Dry corrosion at ambient temperature occurs on metals that have a negative free energy of oxide formation and thus form a rapid thermodynamically stable film in the presence of oxygen. Typically, these films are desirable because they are defect free, nonporous, and self-healing and act as a protective barrier to further corrosive attack of the base metal. Metals such as stainless steels, titanium, and chromium develop this type of protective film. Porous and nonadhering films that form spontaneously on nonpassive metals as unalloyed steel are normally not desirable.

Tarnishing of copper and silver in dry air with traces of hydrogen sulfide (H_2S) is an example of a nondesirable film formation at ambient temperatures caused by lattice diffusion. For tarnishing to occur, sulfur impurities must be present. The sulfides increase the likelihood of defects in the oxide-lattice and thus destroy the protective nature of the natural film, which leads to a tarnished surface. Surface moisture is not necessary for tarnishing to occur, and in some cases, such as copper in the presence of trace amounts of H_2S, moisture can actually retard the process of tarnishing. In general, dry corrosion plays an insignificant part in atmospheric corrosion as a whole.

Damp corrosion requires moisture in the atmosphere and increases in aggressiveness with the moisture content. When the humidity exceeds a critical value, which is around 70% relative humidity, an invisible thin film of moisture will form on the surface of the metal, providing an electrolyte for current transfer. The critical value depends on surface conditions such as cleanliness, corrosion product buildup, or the presence of salts or other contaminants that are hygroscopic and can absorb water at lower relative humidities.

Wet corrosion occurs when water pockets or visible water layers are formed on the metal surfaces because of sea spray, rain, or drops of dew. Crevices or condensation traps also promote the pooling of water and lead to wet atmospheric corrosion even when the flat surfaces of a metal component appear to be dry (Fig. 1).

During wet corrosion, the solubility of corrosion product can affect the corrosion rate. Typically, when the corrosion product is soluble, the corrosion rate will increase. This occurs because the dissolved ions normally increase the conductivity of the electrolyte and thus decrease the internal resistance to current flow, which will lead to an increased corrosion rate. Under alternating wet and dry conditions, the formation of an insoluble corrosion product on the surface may increase the corrosion rate during the dry cycle by absorbing moisture and continually wetting the surface of the metal.

Fig. 1 Corroded steel formwork on the ceiling of a parking garage. The seams in this corrugated structure act as condensation traps and lead to wet atmospheric corrosion. Courtesy of R.H. Heidersbach, California Polytechnic State University

Fig. 2 Corroded weathering steel I-beam. Note how corrosion has thinned the bottom of the vertical web where corrosion products have fallen and formed a moist corrosive deposit. Courtesy of R.H. Heidersbach, California Polytechnic State University

Fig. 3 Corroded weathering steel gutter. Courtesy of R.H. Heidersbach, California Polytechnic State University

The rusting of iron and steel and the formation of patina on copper are examples of metals experiencing either damp or wet atmospheric corrosion. Figures 1 to 5 show examples of the damp/wet atmospheric corrosion of weathering steel components.

Atmospheric Contaminants

Wet atmospheric corrosion is often controlled by the level of contaminants found in the environment. For example, steel pillars 25 m (80 ft) from the seacoast will corrode 12 times faster than the same steel pillars 250 m (800 ft) further inland. The level of marine salts found at the two locations can explain the difference in the observed corrosion rates. More detailed information on marine atmospheres and their effect on the corrosivity of metals and alloys can be found in the article "Marine Corrosion" in this Volume.

Industrial atmospheres are more corrosive than rural atmospheres, primarily because of the sulfur compounds produced during the burning of fuels. Sulfur dioxide (SO_2) is selectively absorbed on metal surfaces, and under humid conditions the metal oxide surfaces catalyze the SO_2 to sulfur trioxide (SO_3) and promote the formation of sulfuric acid (H_2SO_4) according to the reaction $H_2O + SO_3 \rightarrow H_2SO_4$. An example of SO_2-induced corrosion of plain carbon steel is shown in Fig. 6.

Small additions of copper (0.1%) will increase the resistance of steel to a sulfur polluted environment by enhancing the formation of a tighter, more protective rust film (see the articles "Corrosion of Carbon Steels" and "Corrosion of Alloy Steels" in this Volume). Additions of nickel and chromium will accomplish the same end. Nickel and copper alloys form insoluble sulfates that help to protect the base metal and are therefore used extensively in industrial environments (see the articles "Corrosion of Copper and Copper Alloys" and "Corrosion of Nickel-Base Alloys" in this Volume). The remarkable longevity of ancient iron is probably due to a SO_2-free atmosphere rather than a high degree of resistance to general corrosive attack.

Fig. 4 Corroded weathering steel highway bridge girder, Courtesy of D. Manning, Ontario Ministry of Highways and Communications

Other major contaminants that promote atmospheric corrosion are nitrogen compounds, H_2S, and dust particles. Nitrogen compounds occur naturally during thunderstorms and are added to the environment by the use of ammonia (NH_3) base fertilizers. Hydrogen sulfide can be generated naturally by the decomposition of organic sulfur compounds or by sulfate-reducing bacteria (SRB) in polluted rivers. Detailed information on SRB and their effect on alloy corrosion behavior can be found in the articles "Effects of Environmental Variables on Aqueous Corrosion" (see the Appendix "Biological Effects") and "Localized Corrosion" (see the section "Localized Biological Corrosion").

Dust particles can be very detrimental to corrosion-resistant metals by adhering to the surface and absorbing water or H_2SO_4, and trapping the solution against the surface. Dust particles may also contain contaminants, such as chlorides, that can break down protective surface films and thus initiate corrosion.

Oxygen is not considered a contaminant, but is an essential element of the corrosion process. The normal cathodic reaction is the reduction of oxygen. In polluted areas with high concentrations of SO_2, the pH of the surface electrolyte may be low enough so that hydrogen reduction is the principal cathodic reaction. Once a suitable surface electrolyte has been formed by water vapor, oxygen will dissolve in the electrolyte solution and promote the cathodic reaction. Because the water layer on the surface of the metal is extremely thin, the diffusion of oxygen to the surface of the metal occurs very rapidly and does not slow the corrosion rate.

Carbon dioxide (CO_2) does not play a significant role in atmospheric corrosion, and in some cases, it will actually decrease corrosion attack.

Atmospheric Variables

Atmospheric variables such as temperature, climatic conditions, and relative humidity, as well as surface shape and surface conditions that affect the time of wetness, are important factors that influence the rate of corrosive attack. Additional information concerning variables is available in the article "Effects of Environmental Variables on Aqueous Corrosion" in this Volume.

Surface temperature is a critical variable. As the surface temperature increases, the corrosion

Fig. 5 Corroded regions on a painted highway bridge. Courtesy of R.H. Heidersbach, California Polytechnic State University

Fig. 6 Delamination of plain carbon steel due to SO_2 + ash deposit outside the boiler area of a coal-fired power plant. Courtesy of D.M. Berger, Gilbert/Commonwealth

rate will rise sharply to the point at which evaporation of the electrolyte takes place. At this temperature, the corrosion rate will decrease quickly.

Climatic Conditions. Metal surfaces located in areas where they become wet and retain moisture generally corrode more rapidly than surfaces exposed to rain. The rain has a tendency to wash the surface and remove particles of dust that can lead to differential aeration corrosion. Exceptions would be in areas that are subject to acid rain. Exposure of metals in different months of the year can have a pronounced effect on the corrosion rate. Winter exposure is usually the most severe because of increased combustion products in the air. The presence of SO_2 and other sulfur pollutants leads to an aggressive environment and the formation of a less protective corrosion product film on normally passive metals. One should be cautious regarding the month in which outdoor corrosion tests are performed. In locations in which sulfur-containing fuel is not burned during the winter months, the summer months may sometimes lead to higher rates of attack because of the increased surface temperatures.

Time of Wetness/Relative Humidity. Time of wetness is a critical variable with respect to the extent of corrosion experienced. The time of wetness determines the duration of the electrochemical process. The thickness and the chemical composition of the water film are both important.

The critical relative humidity is the humidity below which water will not form on a clean metal surface and thus electrochemical or wet corrosion will not occur. The actual relative humidity will change depending on the surface condition of the metal.

For iron, the critical relative humidity appears to be about 60%; at this level, rust slowly begins to form. At 75 to 80% relative humidity, there is a sharp increase in corrosion rate that is speculated to occur because of the capillary condensation of moisture within the rust corrosion product layer. At 90% relative humidity, there is another increase in the corrosion rate corresponding to the vapor pressure of ferrous sulfate. The critical relative humidity for copper, nickel, and zinc also appears to be between 50 and 70%, depending on surface conditions.

The nature of the corrosion product can greatly affect the time of wetness. If the corrosion product film is microporous in nature, capillary condensation can cause the condensation of moisture well below the critical relative humidity value. This occurs because of the differences in vapor pressure as measured over a curved surface as compared to a flat surface. For example, a 1.5-nm capillary will condense moisture at 50% relative humidity; a 36-nm capillary will condense moisture at 98% relative humidity. This phenomenon accounts for the formation of electrolyte in microcracks and in contact angles between dust particles and metal surfaces. The condensation of moisture on a metal surface can also be enhanced by the formation of a saturated solution, which will lower the equilibrium vapor pressure and allow condensation below 100% relative humidity.

Dew formation on metal surfaces can lead to accelerated corrosion because of the tendency of the dew to be acidic as a result of high SO_2 values near the ground. The dew can form on open or sheltered surfaces and leads to a corrosive attack of galvanized sheet called white rusting.

The thickness of the electrolyte layer is also an important factor in the corrosion process. Water begins to adhere to a polished metal surface at an estimated 55% relative humidity and will form a thin film, which will increase in thickness as the relative humidity increases. The thin water layers can support an electrochemical reaction, but polarization of the cathodic and anodic sites slows the process as the film thickness decreases and virtually stops at about 60%, the critical relative humidity value. The corrosion rate on a surface reaches a maximum when the water film thickness is above 150 μm. Therefore, not only is the time of wetness an important parameter but the thickness and conductivity of the surface electrolyte must also be known.

Atmospheric Corrosion of Specific Metal Systems

Irons and steel, zinc, copper, nickel, and aluminum are the metal systems of major economic importance when dealing with atmospheric corrosion. Table 1 provides an overview of atmospheric corrosion rates of various metals and alloys. Metals that are not particularly resistant to dilute H_2SO_4 such as copper, cadmium, nickel, and iron, show more rapid attack in industrial environments. Metals and alloys that are more resistant to H_2SO_4, such as lead, aluminum, and stainless steels, are less affected in the industrial environments. Copper forms a protective sulfate patina and is therefore more resistant than nickel. Copper also forms a basic copper chloride in seacoast environments. Nickel is very important in marine atmospheres, but is sensitive to the H_2SO_4 found in the industrial environments.

Low-alloy steels that resist atmospheric corrosion are called weathering steels (see the section "Weathering Steels" in the article "Corrosion of Carbon Steels" in this Volume). These alloyed steels form a protective rust film in alternating wet and dry environments. The weathering steels do not perform well under conditions of burial or total immersion. The atmospheric attack of wrought iron can sometimes progress along the internal planes formed during rolling and cause swelling of the material. For this reason, it is best not to cut across the grain boundaries and leave the face exposed to the environment. Stainless steels and aluminum alloys are normally very resistant to atmospheric conditions and will resist tarnishing in industrial, urban, and rural environments.

Lead, aluminum, and copper corrode initially, but form a protective film. In an urban atmosphere, nickel does not form a completely protective film and will experience a parabolic corrosion rate. Zinc attack appears to be linear after an initial period of decreasing corrosion rate. The corrosion rate of steel depends on the alloying elements typically attributed to the compact nature of the rust formed because of the alloying elements.

Copper, lead, and nickel form sulfates on the surface when attacked by dilute H_2SO_4. The lead forms a protective film, but the copper and nickel will slough off after a period of time. The protective carbonate film on zinc and cadmium is dissolved, and the metal is readily attacked. The oxide film on iron is formed by the hydrolysis of the ferrous sulfate. More detailed information on each of the metal systems mentioned above can be found in the Section "Specific Alloy Systems" in this Volume.

Prevention of Atmospheric Corrosion

Two approaches can be taken to prevent the onset of atmospheric corrosion. The first is a temporary fix that can be used during transport or storage. This consists of lowering the atmospheric humidity by using a desiccant, heating devices, or by treating the surface with a vapor phase or surface inhibitor. Permanent solutions to atmospheric corrosion can be accomplished by either changing the material or by applying a coating. Organic, inorganic, and metallic coatings have been effectively employed. The use of protective coatings and inhibitors to mitigate corrosion is discussed at length in the Section "Corrosion Protection Methods" in this Volume.

When using an alloy steel, the addition of small amounts of copper, phosphorus, nickel, and chromium are particularly effective in reducing atmo-

Table 1 Average atmospheric-corrosion rates of various metals for 10- and 20-year exposure times

Corrosion rates are given in mils/yr (1 mil/yr = 0.025 mm/yr). Values cited are one-half reduction of specimen thickness.

Metal	New York, NY (urban-industrial) 10 years	New York, NY (urban-industrial) 20 years	La Jolla, CA (marine) 10 years	La Jolla, CA (marine) 20 years	State College, PA (rural) 10 years	State College, PA (rural) 20 years
Aluminum	0.032	0.029	0.028	0.025	0.001	0.003
Copper	0.047	0.054	0.052	0.050	0.023	0.017
Lead	0.017	0.015	0.016	0.021	0.019	0.013
Tin	0.047	0.052	0.091	0.112	0.018	...
Nickel	0.128	0.144	0.004	0.006	0.006	0.009
65% Ni, 32% Cu, 2% Fe, 1% Mn (Monel)	0.053	0.062	0.007	0.006	0.005	0.007
Zinc (99.9%)	0.202	0.226	0.063	0.069	0.034	0.044
Zinc (99.0%)	0.193	0.218	0.069	0.068	0.042	0.043
0.2% C Steel(a) (0.02% P, 0.05% S, 0.05% Cu, 0.02% Ni, 0.02% Cr)	0.48	...	...	...	...	...
Low-alloy steel(a) (0.1% C, 0.2% P, 0.04% S, 0.03% Ni, 1.1% Cr, 0.4% Cu)	0.09	...	...	...	...	...

(a) Kearney, NJ (near New York City). Source: Ref 1

spheric corrosion. It has been reported that copper additives are more effective in temperate climates than in tropical marine regions. The combination of minor elements, such as the addition of chromium and nickel with copper and phosphorus, appears to be very effective for all locations. The effects of alloying additions in ferrous alloys (wrought carbon, alloy, and stainless steels, and cast irons and steels) are detailed in the first five articles of the Section "Specific Alloy Systems" in this Volume.

Galvanic Corrosion

Robert Baboian
Texas Instruments, Inc.
Steven L. Pohlman
Kennecott Corporation

Galvanic corrosion occurs when a metal or alloy is electrically coupled to another metal or conducting nonmetal in the same electrolyte. The three essential components are:

- Materials possessing different surface potential
- A common electrolyte
- A common electrical path

A mixed metal system in a common electrolyte that is electrically isolated will not experience galvanic corrosion, regardless of the proximity of the metals or their relative potential or size.

During galvanic coupling, corrosion of the less corrosion-resistant metal increases and the surface becomes anodic, while corrosion of the more corrosion-resistant metal decreases and the surface becomes anodic. The driving force for corrosion or current flow is the potential developed between the dissimilar metals. The extent of accelerated corrosion resulting from galvanic coupling is affected by the following factors:

- The potential difference between the metals or alloys
- The nature of the environment
- The polarization behavior of the metals or alloys
- The geometric relationship of the component metals or alloys

The differences in potential between dissimilar metals or alloys cause electron flow between them when they are electrically coupled in a conductive solution. The direction of flow, and therefore the galvanic behavior, depends on which metal or alloy is more active. Thus, the more active metal or alloy becomes anodic, and the more noble metal or alloy becomes cathodic in the couple. The driving force for galvanic corrosion is the difference in potential between the component metals or alloys.

Galvanic Series

A galvanic series of metals and alloys is useful for predicting galvanic relationships. Such a series is an arrangement of metals and alloys according to their potentials as measured in a specific electrolyte. The galvanic series allows one to determine which metal or alloy in a galvanic couple is more active. In some cases, the separation between the two metals or alloys in the galvanic series gives an indication of the probable magnitude of corrosive effect.

The potential of a metal or alloy is affected by environmental factors. Corrosion product films and other changes in surface composition can occur in some environments; therefore, no one value can be given for a particular metal or alloy. This requires a galvanic series to be measured in each environment of interest. Most commonly, however, the galvanic series has been constructed from measurements in seawater, as shown in Table 2. With certain exceptions, this series is broadly applicable in other natural waters and in uncontaminated atmospheres.

Because most engineering materials are alloys, the measurement of galvanic corrosion employing actual material is much more useful than predicting current flow from the electromotive force series. Therefore, tabulations such as Table 2 can be very useful.

Polarization

As stated above, electron flow occurs between metals or alloys in a galvanic couple. This current flow between the more active and more noble members causes shifts in potential due to polarization, because the potentials of the metals or alloys tend to approach each other.

The magnitude of the shift depends on the environment, as does the initial potential. If the more noble metal or alloy is more easily polarized, its potential is shifted more toward the more active metal or alloy potential. The shift in potential of the more active metal or alloy in the direction of the cathode is therefore minimized so that accelerated galvanic corrosion is not as great as would otherwise be expected. On the other hand, when the more noble metal or alloy is not readily polarized, the potential of the more active metal shifts further toward the cathode (that is, in the direction of anodic polarization) such that appreciable accelerated galvanic corrosion occurs. More detailed information on polarization can be found in the article "Kinetics of Aqueous Corrosion" in this Volume.

Area, Distance, and Geometric Effects

Factors such as area ratios, distance between electrically connected materials, and geometric shapes also affect galvanic-corrosion behavior.

Area effects in galvanic corrosion involve the ratio of the surface area of the more noble to the more active member(s). When the surface area of the more noble metal or alloy is large in comparison to the more active member, an unfavorable area ratio exists for the prevailing situation in which a couple is under cathodic control. The anodic current density on the more active metal or alloy is extremely large; therefore, the resulting polarization leads to more pronounced galvanic corrosion. The opposite area ratio—large active member surface, smaller noble member surface—produces only slightly accelerated galvanic effects because of the predominant polarization of the more noble material.

Effect of Distance. Dissimilar metals in a galvanic couple that are in close physical proximity usually suffer greater galvanic effects than those that are further apart. The distance effect is dependent on solution conductivity because the path of current flow is the primary consideration. Thus, if dissimilar pipes are butt welded with the electrolyte flowing through them, the most severe corrosion will occur adjacent to the weld on the anodic member.

Table 2 Galvanic series in seawater at 25 °C (77 °F)

Corroded end (anodic, or least noble)
Magnesium
Magnesium alloys
Zinc
Galvanized steel or galvanized wrought iron
Aluminum alloys
5052, 3004, 3003, 1100, 6053, in this order
Cadmium
Aluminum alloys
2117, 2017, 2024, in this order
Low-carbon steel
Wrought iron
Cast iron
Ni-Resist (high-nickel cast iron)
Type 410 stainless steel (active)
50-50 lead-tin solder
Type 304 stainless steel (active)
Type 316 stainless steel (active)
Lead
Tin
Copper alloy C28000 (Muntz metal, 60% Cu)
Copper alloy C67500 (manganese bronze A)
Copper alloys C46400, C46500, C46600, C46700 (naval brass)
Nickel 200 (active)
Inconel alloy 600 (active)
Hastelloy alloy B
Chlorimet 2
Copper alloy C27000 (yellow brass, 65% Cu)
Copper alloys C44300, C44400, C44500 (admiralty brass)
Copper alloys C60800, C61400 (aluminum bronze)
Copper alloy C23000 (red brass, 85% Cu)
Copper C11000 (ETP copper)
Copper alloys C65100, C65500 (silicon bronze)
Copper alloy C71500 (copper nickel, 30% Ni)
Copper alloy C92300, cast (leaded tin bronze G)
Copper alloy C92200, cast (leaded tin bronze M)
Nickel 200 (passive)
Inconel alloy 600 (passive)
Monel alloy 400
Type 410 stainless steel (passive)
Type 304 stainless steel (passive)
Type 316 stainless steel (passive)
Incoloy alloy 825
Inconel alloy 625
Hastelloy alloy C
Chlorimet 3
Silver
Titanium
Graphite
Gold
Platinum
Protected end (cathodic, or most noble)

Effect of Geometry. The geometry of the circuit also enters into the effect to the extent that current will not readily flow around corners. This is simply an extension of the principle described above, in which the current takes the path of least resistance.

Modes of Attack

Galvanic corrosion of the anodic member(s) of a couple may take the form of general or localized corrosion, depending on the configuration of the couple, the nature of the films induced, and the nature of the metals or alloys involved. Generally, there are five major categories.

Dissimilar Metals. The combination of dissimilar metals in engineering design by mechanical or other means is quite common—for example, in heating or cooling coils in vessels, heat exchangers, or machinery. Such combinations often lead to galvanic corrosion.

Nonmetallic Conductors. Less frequently recognized is the influence of nonmetallic conductors as cathodes in galvanic couples. Carbon brick in vessels is strongly cathodic to the common structural metals and alloys. Impervious graphite, especially in heat-exchanger applications, is cathodic to the less noble metals and alloys. Carbon-filled polymers can act as noble metals in a galvanic couple.

Another example is the behavior of conductive films, such as mill scale (magnetite, Fe_3O_4) or iron sulfides on steel, or of lead sulfate on lead. Such films can be cathodic to the base metal exposed at breaks or pores in the scale (Fig. 7) or even to such extraneous items as valves or pumps in a piping system.

Metallic Coatings. Two types of metallic coatings are used in engineering design: noble metal coatings and sacrificial metal coatings. Noble metal coatings are used as barrier coatings over a more reactive metal. Galvanic corrosion of the substrate can occur at pores, damage sites, and edges in the noble metal coating. Sacrificial metal coatings provide cathodic protection of the more noble base metal, as in the case of galvanized steel or Alclad aluminum.

Cathodic Protection. Magnesium, zinc, and aluminum galvanic (sacrificial) anodes are used in a wide range of cathodic protection applications. The galvanic couple of the more active metal and a more noble structure (usually steel, but sometimes aluminum, as in underground piping) provides galvanic (cathodic) protection, while accelerated corrosion of the sacrificial metal (anode) occurs. The article "Cathodic Protection" in this Volume contains information on the principles and applications of this method of corrosion prevention and the selection of anode materials.

Metal Ion Deposition. Ions of a more noble metal may be reduced on the surface of a more active metal—for example, copper on aluminum or steel, silver on copper. This process is also known as cementation, especially with regard to aluminum alloys. The resulting metallic deposit provides cathodic sites for further galvanic corrosion of the more active metal.

Predicting Galvanic Corrosion

The most common method of predicting galvanic corrosion is by immersion testing of the galvanic couple in the environment of interest. Although very time consuming, this is the most desirable method of investigating galvanic corrosion. Initially, screening tests are conducted to eliminate as many candidate materials as possible. These screening tests consist of one or more of the following three electrochemical techniques: potential measurements, current measurements, and polarization measurements. Additional information can be found in the article "Evaluation of Galvanic Corrosion" in this Volume.

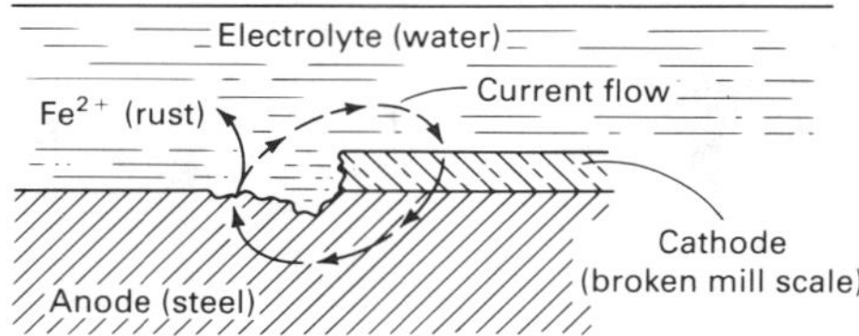

Fig. 7 Schematic showing how breaks in mill scale (Fe_3O_4) can lead to galvanic corrosion of steel

Potential measurements are made to construct a galvanic series of metals and alloys, as described above. As a first approximation, the galvanic series is a useful tool. However, it has serious shortcomings. Metals and alloys that form passive films will exhibit varying potentials with time and are therefore difficult to position in the series with certainty. Also, the galvanic series does not provide information on the polarization characteristics of the materials and so is not helpful in predicting the probable magnitude of galvanic effects.

Measurement of galvanic currents between coupled metals or alloys is based on the use of a zero-resistance milliammeter. Zero-resistance electrical continuity between the members of the galvanic couple is maintained electronically, while the resulting current is measured with the ammeter. Use of this technique should take into account certain limitations. First, when localized corrosion such as pitting or crevice corrosion is possible in the galvanic couple, long induction periods may be required before these effects are observed. Test periods must be of sufficient duration to take this effect into account. Also the measured galvanic current is not always a true measure of the actual corrosion current, because it is the algebraic sum of the currents due to anodic and cathodic reactions. When cathodic currents are appreciable at the mixed potential of the galvanic couple, the measured galvanic current will be significantly lower than the true current. Therefore, large differences between the true corrosion rate calculated by weight loss and that obtained by galvanic current measurements have been observed.

Polarization measurements on the members of a galvanic couple can provide precise information concerning their behavior. The polarization curves and the mixed potential for the galvanically coupled metals in a particular environment can be used to determine the magnitude of the galvanic-corrosion effects as well as the type of corrosion.

An important application in the use of polarization measurements in galvanic corrosion is the prediction of localized corrosion. Polarization techniques and critical potentials are used to measure the susceptibility to pitting and crevice corrosion of metals and alloys coupled in chloride solutions. In addition, this technique is valuable in predicting galvanic corrosion among three or more coupled metals or alloys.

Performance of Alloy Groupings

Light Metals. Magnesium occupies an extremely active position in most galvanic series and is therefore highly susceptible to galvanic corrosion as described in the article "Corrosion of Magnesium and Magnesium Alloys" in this Volume. It is widely used as a sacrificial anode in cathodic protection.

Aluminum and its alloys also occupy active positions in the galvanic series and are subject to failure by galvanic attack (Fig. 8 and 9). In chloride-bearing solutions, aluminum alloys are susceptible to galvanically induced localized corrosion, especially in dissimilar-metal crevices. In this type of environment, severe galvanic effects are observed when aluminum alloys are coupled with more noble metals and alloys. Cementation effects are also observed in the presence of dissolved heavy-metal ions such as copper, mercury, or lead. Some aluminum alloys are used for sacrificial anodes in seawater. An active, anodic alloy is used to clad aluminum, protecting it against pitting in some applications.

In the absence of chlorides or with low concentrations, as in potable water, aluminum and its alloys may be less active because of greater stability of the protective oxide film. Galvanic effects are not as severe under these conditions.

The galvanic-corrosion behavior of magnesium, aluminum, iron and steel, stainless steel, lead, tin, zinc, copper, nickel, cobalt, titanium, zirconium, tantalum, and noble metals are reviewed below. Detailed information on these metals and their alloys can be found in the Section "Specific Alloy Systems" in this Volume. Other articles of interest include "Cathodic Protection," which reviews metals and alloys used as sacrificial (galvanic) and impressed-current anodes; "Tarnish and Corrosion of Dental Alloys," which reviews corrosion-resistant materials used for inlays, implants, and other dental appliances; and "Corrosion of Metallic Implants and Prosthetic Devices," which reviews materials selection and corrosion behavior for orthopedic devices.

Iron and steel are fairly active materials and require protection against galvanic corrosion by the higher alloys. They are, however, more noble than aluminum and its alloys in chloride solutions. However, in low-chloride waters, a reversal of potential can occur that causes iron or steel to become more active than aluminum. A similar reversal can occur between iron and zinc in hot waters of a specific type of chemistry. Examples of galvanic corrosion of iron and steel are shown in Fig. 10 to 13.

Stainless Steels. Galvanic corrosion behavior of stainless steels is difficult to predict because of the influence of passivity. In the common galvanic series, a noble position is assumed by stainless steels in the passive state, while a more active position is assumed in the active state (Table 2). This dual position in galvanic series in chloride-bearing aqueous environments has been the cause of some serious design errors. More precise information on the galvanic behavior of stainless steels can be obtained by using polarization curves, critical potentials, and the mixed potential of the galvanic couple. In chloride-bearing environments, galvanically induced localized corrosion of many stainless steels occurs in couples with copper or nickel and their alloys and with other more noble materials. However, couples of stainless and copper alloys are often used with impunity in freshwater cooling systems. Iron and steel tend to protect stainless steel in aqueous environments when galvanically coupled. The passive behavior of stainless steels makes them easy to polarize; thus, galvanic effects on other metals or alloys tend to be minimized. However, galvanic corrosion of steel can be induced by stainless, particularly in aqueous environments and with adverse area ratios.

Lead, Tin, and Zinc. These three materials occupy similar positions in the galvanic series, although zinc is the most active. The oxide films formed on these materials can shift their potentials to more noble values. Thus, in some environments, they may occupy more noble positions than one might otherwise expect. For example, the tin coating in tin cans is anodic to steel under anaerobic conditions in the sealed container, but becomes cathodic when the can is opened and exposed to air. Zinc is an active metal. It is

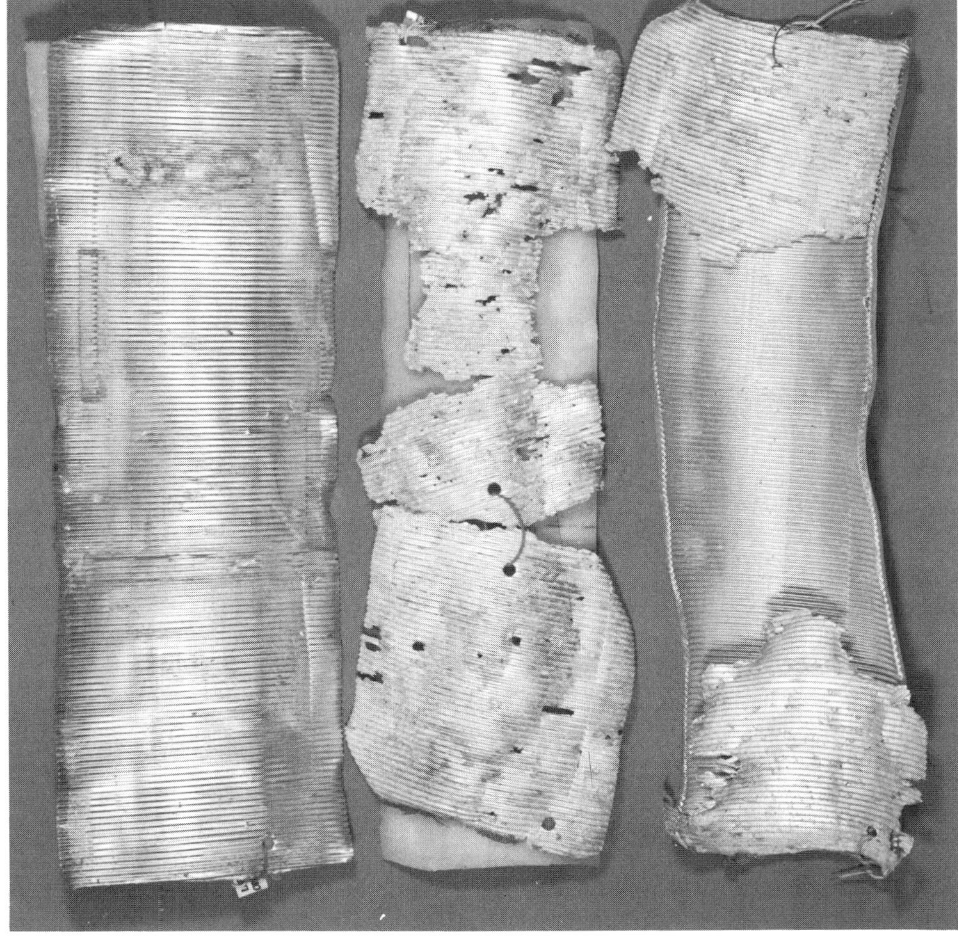

Fig. 8 Galvanic corrosion of aluminum shielding in buried telephone cable coupled to buried copper plates. Courtesy of R. Baboian, Texas Instruments, Inc.

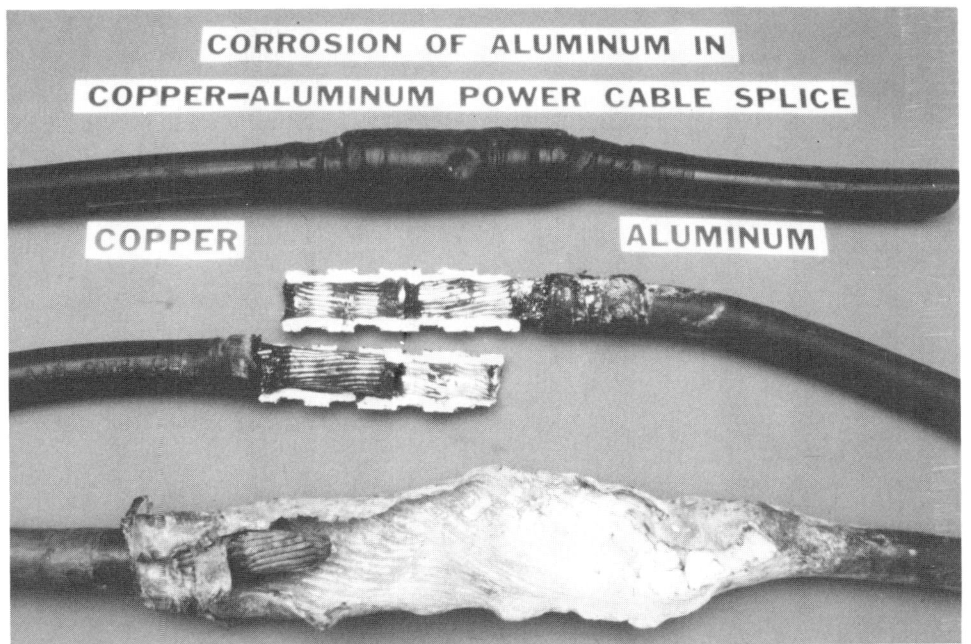

Fig. 9 Galvanic corrosion of aluminum in buried power cable splice (copper to aluminum). Courtesy of R. Baboian, Texas Instruments, Inc.

susceptible to galvanic corrosion and is widely used for galvanic anodes, in cathodic protection as a sacrificial coating (for example, galvanizing or electroplating) and as a pigment in certain types of coatings.

Copper Alloys. Copper and its alloys occupy an intermediate position in the galvanic series. They are not readily polarized in chloride-bearing aqueous solutions; therefore, they cause severe accelerated corrosion of more active metals, such as aluminum and its alloys and the ferrous metals. Somewhat similar to the nickel alloys, they lie between the active and passive positions for stainless steels (Table 2) and therefore induce localized corrosion of the active alloys.

Nickel Alloys. Nickel and its alloys are not readily polarized and will therefore cause accelerated corrosion of more active materials, such as aluminum and ferrous alloys. In chloride-bearing solutions, nickel is somewhat more noble than copper, and the cupronickels lie somewhere in between. Nickel and its alloys are similar to copper alloys in their effects on stainless steels. In some environments, the cast structure of a nickel weld may be anodic to the wrought parent metals.

The combination of a passive surface with the inherent resistance of nickel-chromium alloys, such as Inconel alloy 600 and Hastelloy alloy C-276, places them in more noble positions in the traditional galvanic series. In chloride-bearing solutions, Inconel alloy 600 is reported to occupy two positions because of existence of active and passive states in a manner similar to the stainless steels (Table 2). These alloys are readily polarized, and galvanic effects on other less noble metals and alloys therefore tend to be minimized.

Cobalt-base alloys, most of which are chromium bearing, are resistant to galvanic corrosion because of their noble position in the galvanic series. However, in environments in which their passive film is not stable, they occupy a more active position and can be adversely affected by more noble materials. The fact they they polarize readily tends to reduce their galvanic effects on less noble materials.

Reactive Metals. Titanium, zirconium, and tantalum are extremely noble because of their passive films. In general, these alloys are not susceptible to galvanic corrosion, and their ease of polarization tends to minimize adverse galvanic effects on other metals or alloys. Because of the ease with which they pick up hydrogen in the atomic state, they may themselves become embrittled in galvanic couples. Tantalum repair patches in glass-lined vessels have been destroyed by contact with cooling coils or agitators made of less noble alloys. Tantalum is susceptible to attack by alkalies, such as may form in the vicinity of a cathode in neutral solutions.

Noble metals. The term noble metal is applied to silver, gold, and platinum group metals. This designation in itself describes their position in the galvanic series and their corresponding resistance to galvanic corrosion. However, they do not polarize readily and can therefore have a marked effect in galvanic couples with other metals or alloys. This effect is observed with gold and silver coatings on copper, nickel, aluminum, and their alloys.

Methods of Control

Materials Selection. Combinations of metals or alloys widely separated in the relevant galvanic series should be avoided unless the more noble

Fig. 10 Rust staining of the Statue of Liberty torch due to galvanic corrosion of the iron armature in contact with the copper skin. Courtesy of R. Baboian, Texas Instruments, Inc.

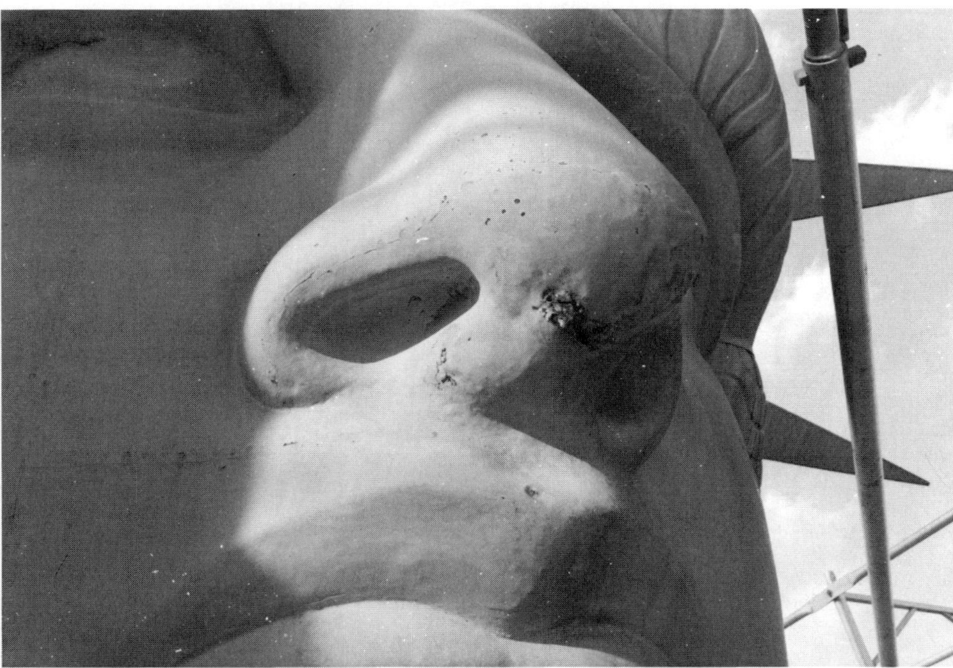

Fig. 11 Moisture that collected on the inside of the Statue of Liberty caused galvanic corrosion of the iron armature in contact with the copper skin. The copper skin on the nose was ruptured due to mechanical forces of the resulting corrosion products. Courtesy of R. Baboian, Texas Instruments, Inc.

material is easily polarized. Metallic coatings can be used to reduce the separation in the galvanic series, as described below. Additional information is provided in the article "Materials Selection" in this Volume.

Environmental Control. In particular cases, it is possible to reduce or eliminate galvanic-corrosion effects between widely dissimilar metals or alloys in a particular environment. The use of corrosion inhibitors is effective in some cases. Elimination of cathodic depolarizers (deaeration of water by thermomechanical means plus oxygen scavengers such as sodium sulfite or hydrazine) is very effective in some aqueous systems. Such methods are discussed in detail in the articles "Corrosion Inhibitors for Oil and Gas Production," "Corrosion Inhibitors for Crude Oil Refineries," and "Control of Environmental Variables in Water Recirculating Systems" in this Volume.

Barrier coatings of a metallic nature have already been discussed. Inert barrier coatings, organic or vitreous, can effectively isolate the metals from the environment. It is extremely dangerous to coat the anodic member of a couple because this may only reduce its area, with severely accelerated attack occurring at holidays in the otherwise protective coating. If inert barrier coatings are employed, both the anode and cathode must be protected—for example, the heads, tubesheets, and first 4 to 6 tube diameters on the tube side of a water-cooled heat exchanger.

Electrochemical techniques are comprised of three alternative methodologies: electrical isolation, use of transition materials, and cathodic protection.

Electrical Isolation. The joint between dissimilar metals can be isolated to break the electrical continuity. Use of nonmetallic inserts, washers, fittings, and coatings at the joint between the materials will provide sufficient electrical resistance to eliminate galvanic corrosion.

Transition Materials. In order to eliminate a dissimilar-metal junction, a transition piece can be introduced. The transition piece consists of the same metals or alloys as in the galvanic

Fig. 12 Galvanic corrosion of painted steel auto body panel in contact with stainless steel wheel opening molding. Courtesy of R. Baboian, Texas Instruments, Inc.

Fig. 13 Galvanic corrosion of steel pipe at brass fitting in humid marine atmosphere. Courtesy of R. Baboian, Texas Instruments, Inc.

couple bonded together in a laminar structure. The transition piece is inserted between the members of the couple such that the similar metals mate with one another. The dissimilar-metal junction then occurs at the bond interface, excluding the electrolyte.

Cathodic Protection. Sacrificial metals, such as magnesium or zinc, may be introduced into the galvanic assembly. The most active member will corrode while providing cathodic protection to the other members in the galvanic assembly (for example, zinc anodes in cast iron waterboxes of copper alloy water-cooled heat exchangers). Impressed-current systems can also provide the same effect. Both sacrificial anodes and impressed-current anodes are discussed in the article "Cathodic Protection" in this Volume.

Design. Unfavorable area ratios should be avoided. Metal combinations should be used in which the more active metal or alloy surface is relatively large. Rivets, bolts, and other fasteners should be of a more noble metal than the material to be fastened. Dissimilar-metal crevices, such as at threaded connections, are to be avoided. Crevices should be sealed, preferably by welding or brazing, although putties are sometimes used effectively. Replaceable sections of the more active member should be used at joints, or the corrosion allowance of this section should be increased, or both. Additional information is available in the article "Design Details to Minimize Corrosion" in this Volume.

Stray-Current Corrosion

Steven L. Pohlman
Kennecott Corporation

Stray-current corrosion, or stray-current electrolysis, is different from natural corrosion because it is caused by an externally induced electrical current and is basically independent of such environmental factors as oxygen concentration or pH. Environmental factors may enhance other corrosion mechanisms involved in the total corrosion process, but the stray-current corrosion portion of the mechanism is unaffected.

Stray currents are defined as those currents that follow paths other than their intended circuit. They leave their intended path because of poor electrical connections within the circuit or poor insulation around the intended conductive material. The escaped current then will pass through the soil, water, or any other suitable electrolyte to find a low-resistant path, such as a buried metal pipe or some other metal structure, and will flow to and from that structure, causing accelerated corrosion.

Sources of Stray Currents

The electric railways were the major source of stray direct current (dc), but since their demise, the problem has become less common. Other sources, such as cathodic protection systems, electrical welding machines, and grounded dc electric sources, create stray direct currents; therefore, stray-current corrosion does present problems occasionally. For example, stray-currents created by an electric welding machine on board ship with a grounded dc line located on shore will cause accelerated attack of the ship's hull as the stray currents generated at the welding electrodes pass out of the ship's hull through the water back to shore.

Stray currents cause accelerated corrosion to occur where they leave the metal structure and enter the surrounding electrolyte. At points where the current enters the structure, the site will become cathodic in nature because of changes in potential, while the area where the current leaves the metal will become anodic. These sites may be hundreds of yards apart. Houses in close proximity can experience dramatically different corrosion characteristics in their water lines. The pipes in one house may be protected, while those next door may be catastrophically failing.

Stray current flowing along a pipeline typically will not cause damage inside the pipe, because of the high conductivity of the metal compared to the fluid in the pipe. The damage occurring at the point where the current reenters the electrolyte will be localized and on the outside surface of the metal. In certain cases in which the pipe has insulated joints and the stray current enters the internal fluid, the corrosion will occur on the inside of the tube.

Damage caused by alternating current (ac) is less than that experienced by dc and decreases in severity as the frequency increases. Damage caused by alternating currents on active-passive metals, such as stainless steel and aluminum, is greater than damage to nonpassive metals, such as iron and zinc. The alternating reduction and oxidation of the surface layers caused by the ac may cause the passive layers to become porous and layered. A major source for ac stray currents is buried power lines.

Identifying Stray-Current Corrosion

Galvanic corrosion and stray-current corrosion are very similar in that they both show protected cathodic sites and preferentially corroded anodic sites. The major difference is that stray-current corrosion may vary over short periods of time, depending on the varying load of the power source, while galvanic corrosion proceeds at a constant rate because the electrochemical reaction is not dependent on an external current source.

Amphoteric metals such as aluminum and zinc can show signs of corrosion at the cathodic portion of the metal surface because of the build up of alkalies created by the cathodic electrochemical reactions. The extent of this type of corrosion is difficult to determine because it is dependent not only on the amount of current flow but also on the surrounding environment.

It is difficult to distinguish between ordinary corrosion and stray-current corrosion by visual inspection. In some cases, the attack is more localized, causing a concentration of pits that is not normally observed but under specific environmental conditions could occur even without the presence of stray currents. Some researchers have observed that stray-current corrosion tends to cause penetration along the grain boundaries—once again a phenomenon seen under many conditions. In the case of a gray cast iron, selective attack of the ferrite within the metal matrix has been observed. This type of attack can weaken the material and cause premature failure. The magnitude of stray currents is not easy to measure, but potential drops, potential differences, and the measurement of current flow along a buried structure are ways of determining the existence of stray currents.

Prevention of Stray-Current Corrosion

There are various ways to decrease the chances of stray-current corrosion from occurring. Basically, one should try to stop the leakage of the current from the intended circuit by maintaining good electrical connections and insulation. Therefore, a major factor in controlling stray-current corrosion is the testing for and controlling of the current before it enters the soil or surrounding electrolyte.

If current is escaping and nothing can be done to prevent it, the corrosion of the surrounding metal structures can be mitigated in several ways. One technique is called bonding, which consists of connecting the stray-current conductor with the source ground and thus eliminating the need for the current to leave the metal and enter the soil.

In addition, sacrificial anodes can also be placed in contact with the stray-current conductor to direct the corrosion to a preferred site. Also, dc power source can be placed on line to cause a flow of current in the opposite direction of the stray current and effectively protect the pipe by the impressed current.

Impeding the flow of the stray current along the metallic path by installing insulators is another way to protect an affected structure. Caution is required, however, because the current will flow around the insulators and cause corrosion damage at multiple sites if the source voltages are large. The insulator gaps must be numerous enough to make the current sufficiently small. Current densities passing through the circuit should be kept as low as possible.

Coating a pipe or structure is not an effective way to control the problem unless the coating is flawless. In fact, coatings that are cracked or contain pin holes will accelerate the attack at sites of imperfection.

General Biological Corrosion

Stephen C. Dexter
College of Marine Studies
University of Delaware

A number of metals, such as structural steel, copper alloys, magnesium alloys, and zinc, tend to corrode generally over the entire exposed surface in aqueous environments. This is particularly true in the absence of galvanic effects and crevices. The corrosion rates of these metals in aerated aqueous environments tend to be determined by the rate at which dissolved oxygen can be delivered to the metal surface. Thus, the corrosion rate will be affected by anything that changes the rate of oxygen transport. Biological organisms present in the environment have the potential to increase or decrease oxygen transport to the surface; consequently, they can either increase or decrease corrosion rates.

Although the focus of this section is on the influence of organisms on general corrosion rates, it should be noted that organisms are more likely to cause localized than general corrosion. More

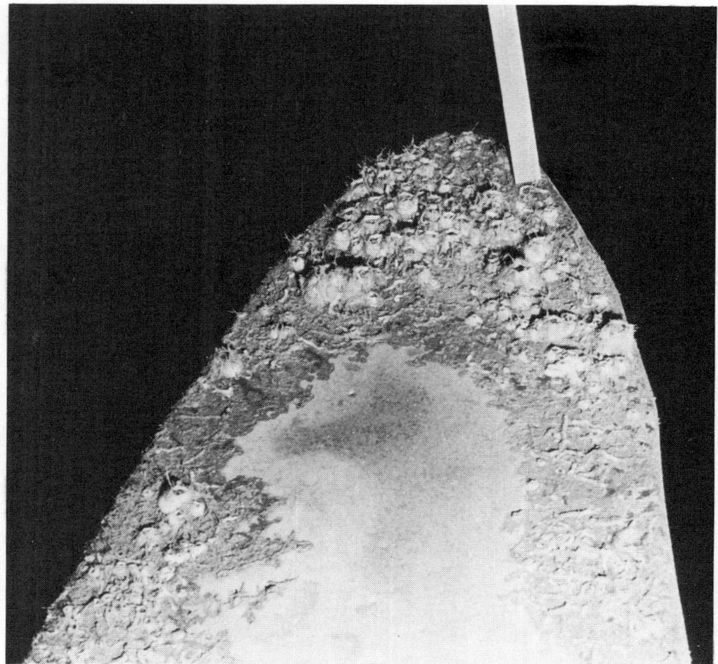

Fig. 14 Barnacles attached to the periphery of a high-strength steel rudder, which had originally been coated with an antifouling paint. During use, the paint around the edges had been removed by mechanical action, thus allowing the attachment of barnacles. Partial coverage of such macroorganisms can lead to localized corrosion. Complete coverage can sometimes provide a barrier film and limit corrosion. Courtesy of B. Little, Naval Ocean Research and Development Activity, Department of the Navy

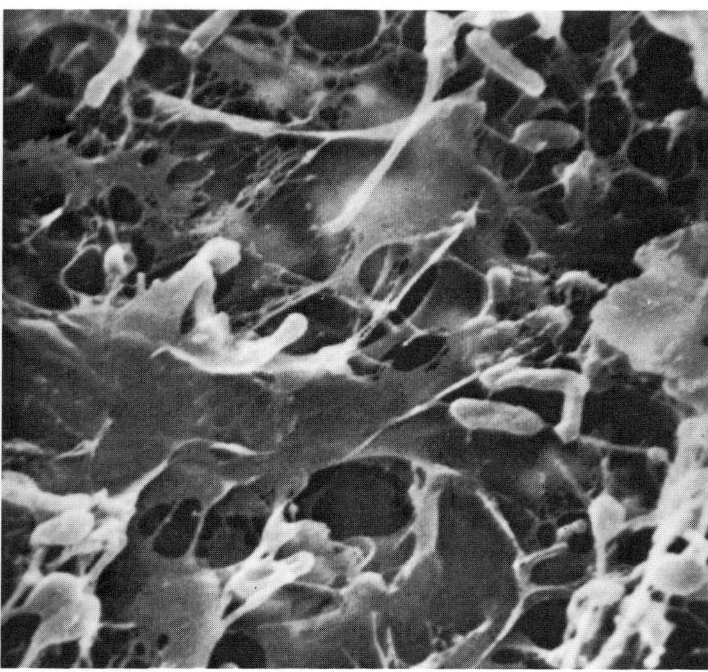

Fig. 15 Microbial film consisting of rod-shaped bacteria and slime. Courtesy of S.C. Dexter, University of Delaware

information on localized biological corrosion can be found in the article "Localized Corrosion" in this Volume.

Marine Environments

Effect of Macroorganisms. A heavy encrustation of biofouling organisms on structural steel immersed in seawater will often decrease the corrosion rate of the steel as long as the cover of organisms remains complete and relatively uniform. The heavy fouling layer acts as a barrier film in limiting the amount of dissolved oxygen reaching the metal surface. A layer of hard-shelled organisms, such as barnacles or mussels, on steel in the splash zone (just above the high tide level) also shields the metal from the damaging effect of wave action.

It should be emphasized that these beneficial effects on general corrosion occur under a complete fouling layer. If coverage is incomplete, as shown in Fig. 14, the fouling film is then more likely to cause the initiation of localized corrosion by creating oxygen concentration cells. The beneficial effects may also be lost if the fouling film leads to a high activity of sulfate-reducing bacteria at the metal surface. More information on this topic can be found in the Appendix "Biological Effects" to the article "Effects of Environmental Variables on Aqueous Corrosion" and in the section "Localized Biological Corrosion" in the article "Localized Corrosion" in this Volume. Additional information on both micro- and macrofouling in marine environments can also be found in the article "Marine Corrosion" (see the section "Seawater") and in Ref 2 to 4.

Effect of Microorganisms. A continuous film of bacteria, algae, and slime, as shown in Fig. 15, can have the same effect on oxygen transport and corrosion rates as that discussed above for the macroorganisms. It is rare, however, for a film of microorganisms in the marine environment to be continuous over large areas of exposed surface. Usually, microbial slime films are spotty. This is caused by a number of factors, including variability in the hydrodynamics of water flow over the surface and the natural tendency for the organisms to grow in discrete colonies rather than continuous films.

For these reasons, the influence of microbial films in marine environments is more likely to be a switch from general to localized corrosion than a change in the general corrosion rate. Microbial films are suspected of being capable of inducing pit initiation on stainless and copper alloys in marine environments. In addition, there are many references in the literature to the fact that natural seawater, with its full complement of organisms, is more corrosive than artificial seawater with fewer or different organisms. However, the effects are not yet well documented or understood. In general, microbial films do not seem to have the strong, well-documented effect on corrosion in marine environments that they do in freshwater and soil environments.

Freshwater Environments

As was the case in marine environments, microbial films will affect the general corrosion rate only if the film is uniformly distributed over the wetted surface. If microorganisms form in discrete deposits or colonies, the resulting corrosion is likely to be localized, rather than uniform. Again, the articles in this Volume that discuss biological corrosion should be consulted.

In some cases, a relatively uniform slime film does form. Examples are slime film formation on the piping of potable water handling systems and on the heat-transfer surfaces of low-temperature heat exchangers. In such cases, a small change in the general corrosion rate of the piping material is inconsequential, and the microbial slime film is of concern only if it leads to obstruction of flow, growth of organisms that would pose a health hazard, or localized corrosion.

Molten-Salt Corrosion

John W. Koger
Martin Marietta Energy Systems
Steven L. Pohlman
Kennecott Corporation

The corrosion of metal containers by molten, or fused, salts has been observed for an extended period of time, but over the last several decades, more effort has been directed toward understanding corrosion phenomena at higher temperatures. Annotated bibliographies of molten-salt corrosion for different media and metal combinations have been published and are cited in the selected references. Additional information on molten-salt corrosion can be found in the article "Fundamentals of High-Temperature Corrosion in Molten Salts" in this Volume.

Mechanisms of Molten-Salt Corrosion

Two general mechanisms of corrosion can exist in molten salts. One is the metal dissolution caused by the solubility of the metal in the melt. This dissolution is similar to that in molten metals, but is not common. The second and most common mechanism is the oxidation of the metal to ions that is similar to aqueous corrosion. For this reason, molten-salt corrosion has been identified as an intermediate form of corrosion between molten metal and aqueous corrosion.

General, or uniform, metal oxidation and dissolution is a common form of molten-salt corrosion, but is not the only form of corrosion seen. Selective leaching is very common at higher temperatures, as are pitting and crevice corrosion at lower temperatures. All the forms of corrosion observed in aqueous systems, including stress-assisted corrosion, galvanic corrosion, erosion-corrosion, and fretting corrosion, have been seen in fused salts.

Electrochemically, the molten salt/metal surface interface is very similar to the aqueous solution/metal surface interface. Many of the principles that apply to aqueous corrosion also apply to molten-salt corrosion, such as anodic reactions leading to metal dissolution and cathodic reduction of an oxidant.

The concept of acid and base behavior of the melt is very similar to its aqueous counterpart. The corrosion process is mainly electrochemical in nature because of the excellent ionic conductivity of most molten salts. Some investigators feel that dissolved water enhances the electrochemical corrosion nature of the molten salts.

Even though the corrosion mechanism is similar, there are major differences between molten-salt and aqueous corrosion. The differences arise mainly from the fact that molten salts are partially electronic conductors as well as ionic conductors. This fact allows for reduction reactions to take place in the melt as well as the metal/melt interface. This behavior also allows increase in frequency of cathodic reactions and can therefore lead to a substantial increase in corrosion rate over a similar electrochemically controlled aqueous system, especially if the corrosion media contain very few oxidants. Because of property differences between water and molten salt, the rate-controlling step in most molten-metal systems is ion diffusion into the bulk solution, not the charge transfer reaction that is typical of aqueous systems. Molten-salt systems operate at higher temperatures than aqueous systems, which leads to different forms of corrosion attack.

In aqueous systems in which specific elements are removed from alloys, such as the dezincification of zinc from brass, whether the nobler element is oxidized at some stage in the process and is subsequently plated out or whether the element agglomerates by a surface diffusion mechanism is not clear. In both cases, bulk diffusion is unlikely to play a major role. In high-temperature molten-salt systems, dissolution of the less noble element is the most probable mechanism and will even occur when the element is present at low concentration. Unlike aqueous systems, the rate of dissolution is therefore related to the bulk diffusion of the selectively leached element.

Because molten-salt corrosion reactions are reduction/oxidation controlled, the relative nobilities of the salt melts and the metals are important. The corrosion potential of the melt is often controlled by impurities in the melt or gas phase, which increases the cathodic reaction rate or changes the acidic or basic nature of the melt. The aggressiveness of the salt melts is typically governed by its redox equilibria.

Thermal gradients in the melt can cause dissolution of metal at hot spots and metal deposition at cooler spots. The result is very similar to aqueous galvanic corrosion, and like aqueous galvanic corrosion, a continuous electrical path is necessary between the hot and cold areas. Crevice corrosion has been observed, and wash-line attack caused by oxygen concentration corrosion is not uncommon at the metal/molten salt/air interface.

High-temperature corrosion in molten salts often exhibits selective attack and internal oxidation. Chromium depletion in iron-chromium-nickel alloy systems can occur by the formation of a chromium compound at the surface and by the subsequent removal of chromium from the matrix, leaving a depleted zone. Thus, the selectively removed species move out, while vacancies move inward and eventually form voids. The voids tend to form at grain boundaries in most chromium-containing metals, but in some high-nickel alloys, the voids form in the grains. Specific examples of the types of corrosion expected for the different metal-fused salt systems will follow.

Types of Molten Salts

Molten Fluorides. Fluoride melts are important because of their consideration for nuclear reactor cooling systems. Corrosion in many fluoride molten-salt melts is accelerated because protective surface films are not formed. In fact, the fluoride salts act as excellent fluxes and dissolve the various corrosion products.

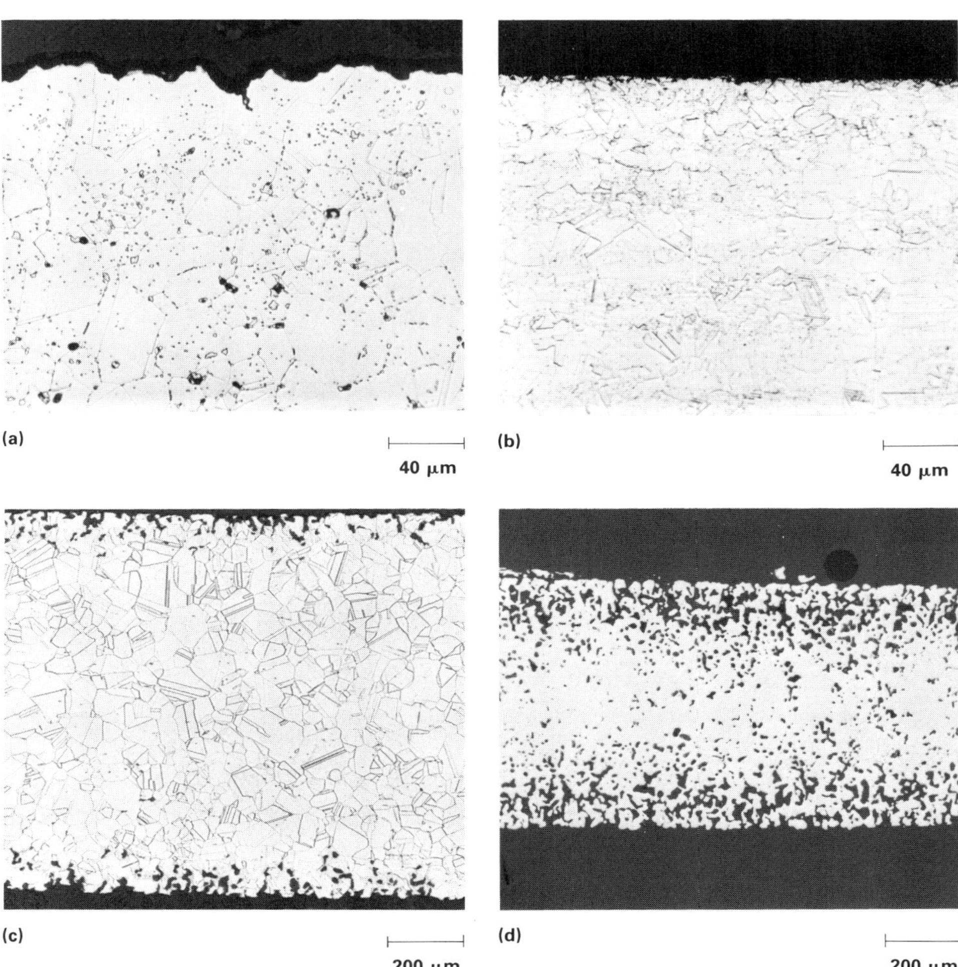

Fig. 16 Effect of molten-salt corrosion on nickel-base and stainless steel alloys. In all four examples, chromium depletion (dealloying) was the result of prolonged exposure. Accompanying chromium depletion was the formation of subsurface voids, which did not connect with the surface or with each other. As chromium was leached from the surface of the metal, a concentration (activity) gradient resulted and caused chromium atoms from the underlying region to diffuse toward the surface, leaving behind a zone enriched with vacancies. These vacancies would then agglomerate at suitable sites—primarily at grain boundaries and impurities. The vacancies became visible as voids, which tended to agglomerate and grow with increasing time and temperature. (a) Microstructure of Hastelloy alloy N exposed to $LiF\text{-}BeF_2\text{-}ThF_4\text{-}UF_4$ (68, 20, 11.7, 0.3 mol%, respectively) for 4741 h at 700 °C (1290 °F). (b) Microstructure of Hastelloy alloy N after 2000 h exposure to $LiF\text{-}BeF_2\text{-}ThF_4$ (73, 2, 25 mol%, respectively) at 676 °C (1249 °F). (c) Microstructure of type 304L stainless steel exposed to $LiF\text{-}BeF_2\text{-}ZrF_4\text{-}ThF_4\text{-}UF_4$ (70, 23, 5, 1, 1 mol%, respectively) for 5700 h at 688 °C (1270 °F). (d) Microstructure of type 304L stainless steel exposed to $LiF\text{-}BeF_2\text{-}ZrF_4\text{-}ThF_4\text{-}UF_4$ (70, 23, 5, 1, 1, mol%, respectively) for 5724 h at 685 °C (1265 °F). Courtesy of J.W. Koger, Martin Marietta Energy Systems

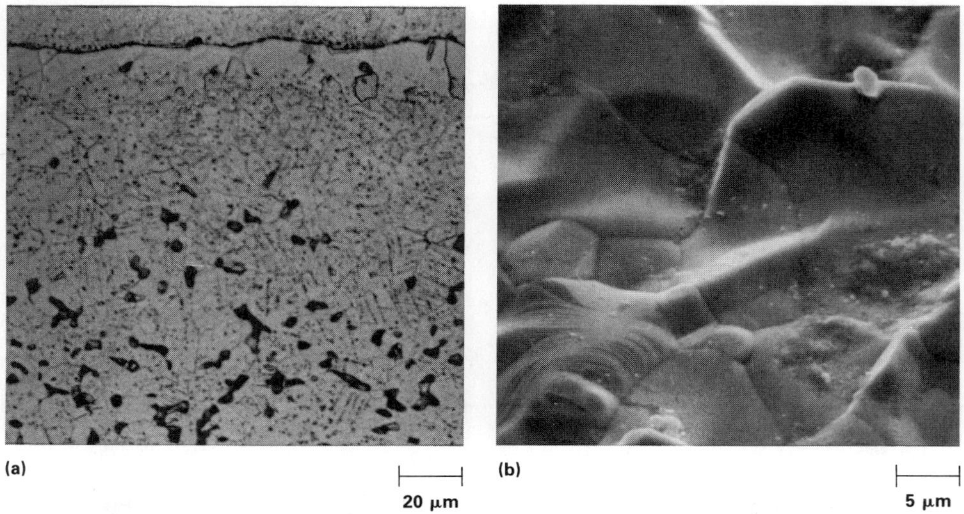

Fig. 17 Corrosion of type 316 stainless steel exposed to liquid sodium for 8000 h at 700 °C (1290 °F); hot leg of circulating system. Surface regression is uniform; a 10- to 15μm layer of ferrite (>95% Fe) has formed. Total damage depth: 27 μm. (a) Light micrograph of wall section showing σ phase (etched black) suppressed by composition changes to a depth of ~50 μm. (b) SEM micrograph of surface. Source: Ref 5

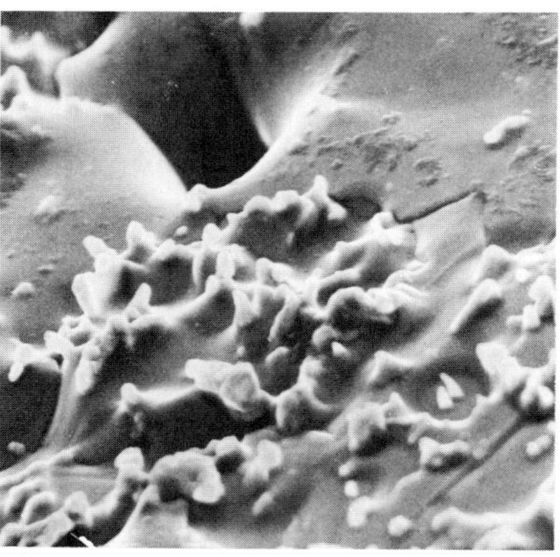

Fig. 19 Typical surface appearance of a stabilized stainless steel (X10CrNiMoTi 15 15) after a 5000-h exposure to flowing sodium at 700 °C (1290 °F). Cavities are formed at the grain corners; coral-like particles of a MoFe phase are on the grain surfaces. Courtesy of H.U. Borgstedt, Karlsruhe Nuclear Center

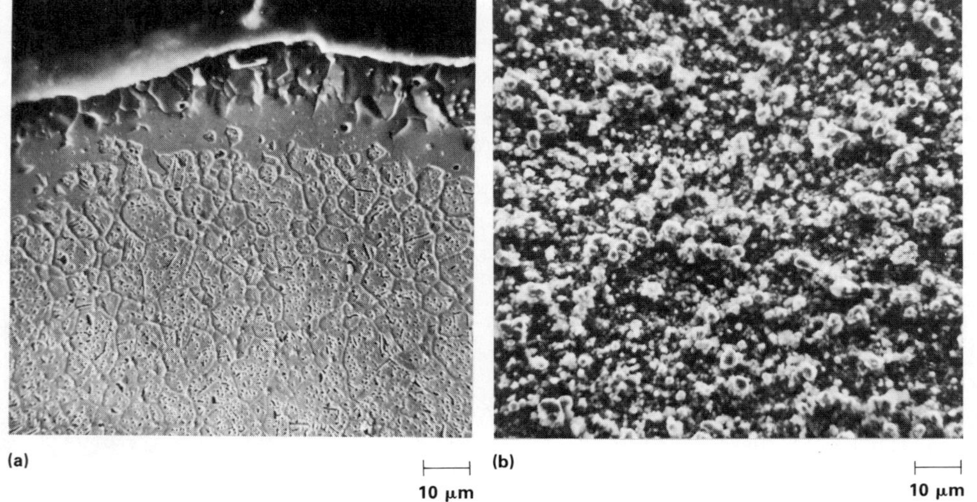

Fig. 18 Corrosion of X8CrNiMoVNb 16 13 stainless steel exposed to flowing sodium. (a) Ferrite formation in the surface layer and grain boundaries after 5000 h at 700 °C (1290 °F). SEM. Oxalic acid etch. (b) SEM of the deposition layer formed on the same steel after exposure to flowing sodium at 590 °C (1095 °F). Specimen taken downstream from the 700-°C (1290-°F) hot leg specimen shown in (a). The dark particles are rich in chromium and oxygen; some of the bright crystals are rich in calcium. Courtesy of H.U. Borgstedt, Karlsruhe Nuclear Center

Typically, nickel-base alloys show better corrosion resistance than iron-base alloys. Studies have also shown that most nickel- and iron-base alloys that contain varying amounts of chromium show void formation to varying depths. In nonisothermal flow systems, the metallic material shows void formation in the hot section and deposits in the cold section. In most alloys that contain chromium, depletion of the chromium accompanies the void formation. Analysis of the attacked metal and salt clearly shows that selective removal and outward diffusion of chromium result from oxidation by the fluoride mixture.

Inconel alloy 600 and certain stainless steels become magnetic after exposure to molten fluoride salt. This magnetism is caused by the selective removal of the chromium and the formation of a magnetic iron-nickel alloy covering the surface. The conditions of the melt need to be controlled to minimize the selective removal of the alloying elements. This control can be done by maintaining the melt in a reducing condition. Addition of beryllium metal to the melt is one way to slow the corrosion rate. Examples of microstructures of nickel-base and stainless steels as a result of exposure to molten salts are shown in Fig. 16.

Chloride Salts. Molten salts consisting of chlorides are important, but because they have limited use, they have been studied less than fluoride systems. In general, chloride salts attack steels very rapidly with preferential attack of the carbides. Aluminum coatings on steels are not effective, while the addition of nickel to steel is beneficial. Nickel-base alloys decrease in resistance with increased oxygen partial pressure. In the chloride salts, no protective oxide scale is formed on the nickel-base alloys. The attack of metal surfaces in pure sodium chloride has been observed at temperatures above 600 °C (1110 °F).

In most cases with iron-nickel-chromium alloys, the corrosion takes the form of intergranular attack. An increase of chromium in the alloy from 10 to 30% increases the corrosion rate by a factor of seven, while changes in nickel content have no effect. Thus, the intergranular attack is most likely selective with respect to chromium. The chromium removal begins at the grain boundary and continues with diffusion of the chromium from within the grain to the boundary layer, gradually enlarging the cavity in the metal. The gross corrosive attack is probably caused by the free chlorine, which is a highly oxidizing material, attacking the highly active structure-sensitive sites, such as dislocations and grain boundaries.

Selective attack of Inconel alloy 600 is observed in molten chloride melts. The resulting attacked region is a layer of porous spongelike material. The pores are not interconnected and are typically located at the grain boundaries.

Intergranular attack and rapid scale growth occur with zirconium alloys over 300 °C (570 °F). Platinum can be protected by the addition of oxide ions to the melt; this addition assists in the formation of a passive film.

Molten nitrates are commonly used for heat treatment baths; therefore, a great deal of material compatibility information exists. Plain carbon and low-alloy steels form protective iron oxide films that effectively protect the metal surface to approximately 500 °C (930 °F). Chromium additions increase the corrosion resistance of the steel, and hydroxide additions to the melt further increase the resistance of chromium containing steels. Aluminum and aluminum alloys should never be used to contain nitrate melts, because of the danger of explosion.

Molten Sulfates. High-temperature alloys containing chromium perform well in sulfate salts because they form a protective scale. If the chromium

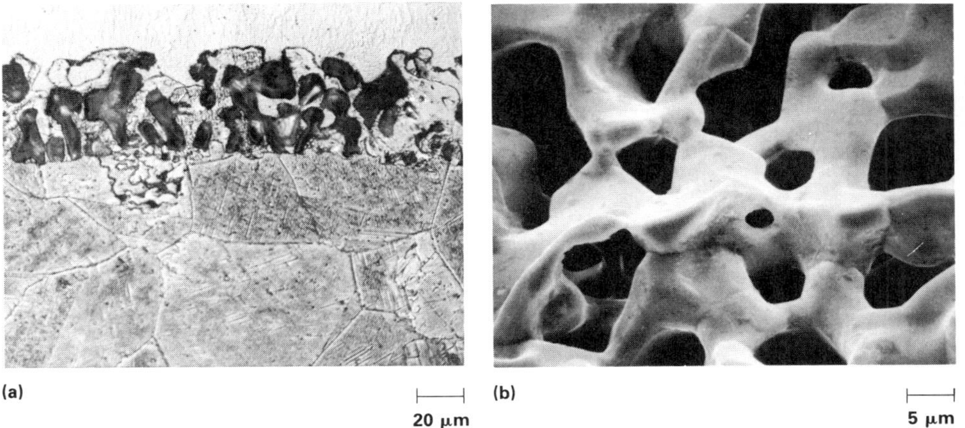

Fig. 20 Corrosion of Inconel alloy 706 exposed to liquid sodium for 8000 h at 700 °C (1290 °F); hot leg of circulating system. A porous surface layer has formed with a composition of ~95% Fe, 2% Cr, and <1% Ni. The majority of the weight loss encountered can be accounted for by this subsurface degradation. Total damage depth: 45 μm. (a) Light micrograph. (b) SEM of the surface of the porous layer. Source: Ref 5

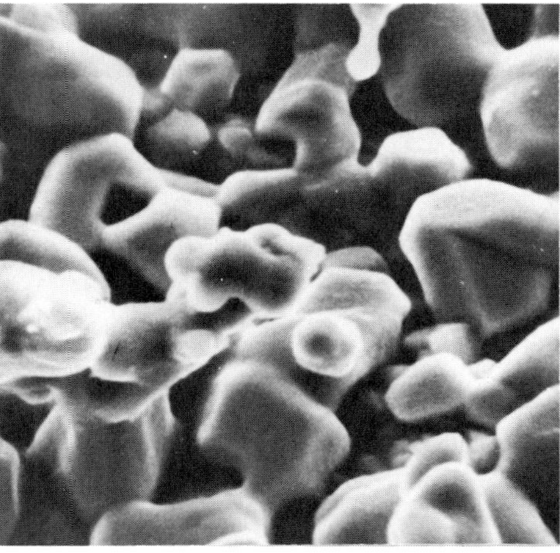

Fig. 22 Deposition of iron-rich crystals on Stellite 6 sheet after ~5000 h in flowing sodium at 600 °C (1110 °F). Courtesy of H.U. Borgstedt, Karlsruhe Nuclear Center

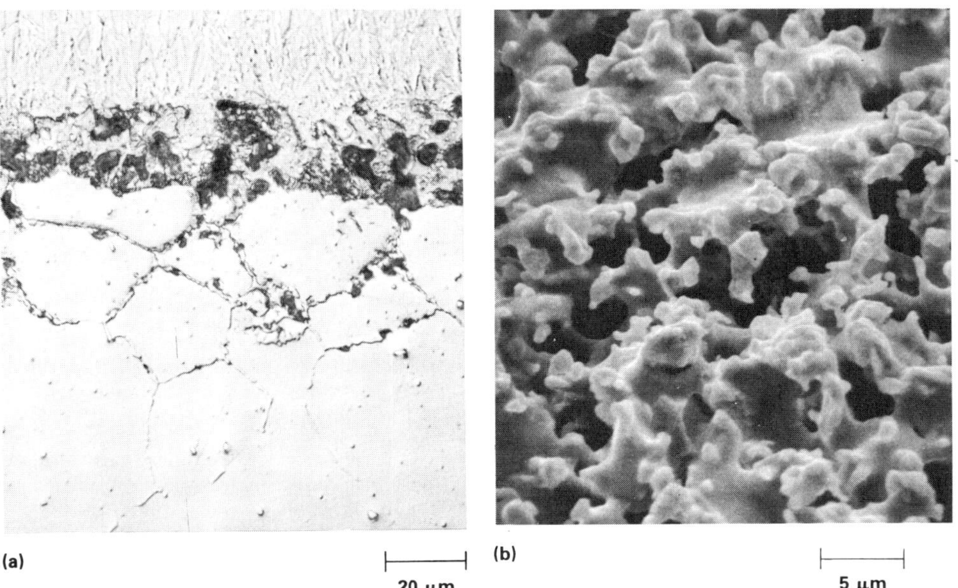

Fig. 21 Corrosion of Nimonic PE 16 exposed to the same conditions described for Fig. 20. A porous coral-like surface layer has formed with a composition similar to that of Inconel alloy 706, but with the addition of corrosion-resistant FeMo particles at the coral tips. Intergranular attack beneath this layer extends to a depth of 75 μm. Total damage depth: 135 μm. (a) Light micrograph. (b) SEM of the surface of the porous layer. Source: Ref 5

content is not sufficient, the alloy will suffer severe external corrosion and internal sulfidation. Non-film-forming metals, such as copper and silver, corrode very rapidly.

Hydroxide Melts. Stainless steels perform poorly in hydroxide melts because of selective oxidation of the chromium, which leads to pore formation in the metal. Nickel is more resistant than stainless steels or unalloyed steels.

The peroxide content of the melt controls the corrosion rate. An increase in water vapor and a decrease in oxygen pressure reduce the peroxide content and subsequently reduce the corrosion rate. An exception to this rule occurs with silver, which shows an increase in corrosion rate with increased water vapor. Most glass and silica are rapidly attacked by hydroxide melts, while alumina is more resistant.

Carbonate Melts. Austenitic stainless steels perform well in carbonate melts up to 500 °C (930 °F). If temperatures to 600 °C (1110 °F) are required, nickel-base alloys containing chromium are needed. For temperatures to 700 °C (1290 °F), high-chromium alloys containing at least 50% Cr are required. Above 700 °C (1290 °F), the passive films that form at lower temperatures will break down and preclude the use of metals. Aluminum coatings on steel structures perform well to 700 °C (1290 °F). For higher temperatures, alumina is required. Nickel does not provide adequate protection because of intergranular attack caused by the formation of nickel oxides.

Prevention of Molten-Salt Corrosion

There is a lack of reliable data for molten salts on structural material under industrial conditions, but several general rules should be observed. A material should be selected that will form a passive nonsoluble film in the melt if possible. Material selection is the key to successful containment. Minimizing the entry of oxidizing species such as oxygen and water into the melt is very important. The oxidizing power of the melt or its redox potential should be kept as low as possible, which can be accomplished by additives. The temperature of the bath should be kept as low as possible. The lower temperature not only decreases the diffusion rates of the ions into the melt but lowers the solubility of potentially passive surface films. Temperature gradients should be eliminated within the melts to decrease the selective dissolution and plating at hot and cold sites.

Corrosion in Liquid Metals

C. Bagnall
Westinghouse Electric Corporation
Advanced Energy Systems Division
W.F. Brehm
Westinghouse Hanford Company

This is a companion section to the article "Fundamentals of High-Temperature Corrosion in Liquid Metals" in this Volume. This section is intended to provide a pictorial account of the types of corrosion that may be found in a wide variety of liquid metal/containment/component combinations. The examples are not comprehensive, but will be informative and will provide a guide to appropriate literature sources for more specific information.

Liquid metals have long been considered for the improvement of efficiency in heat transfer

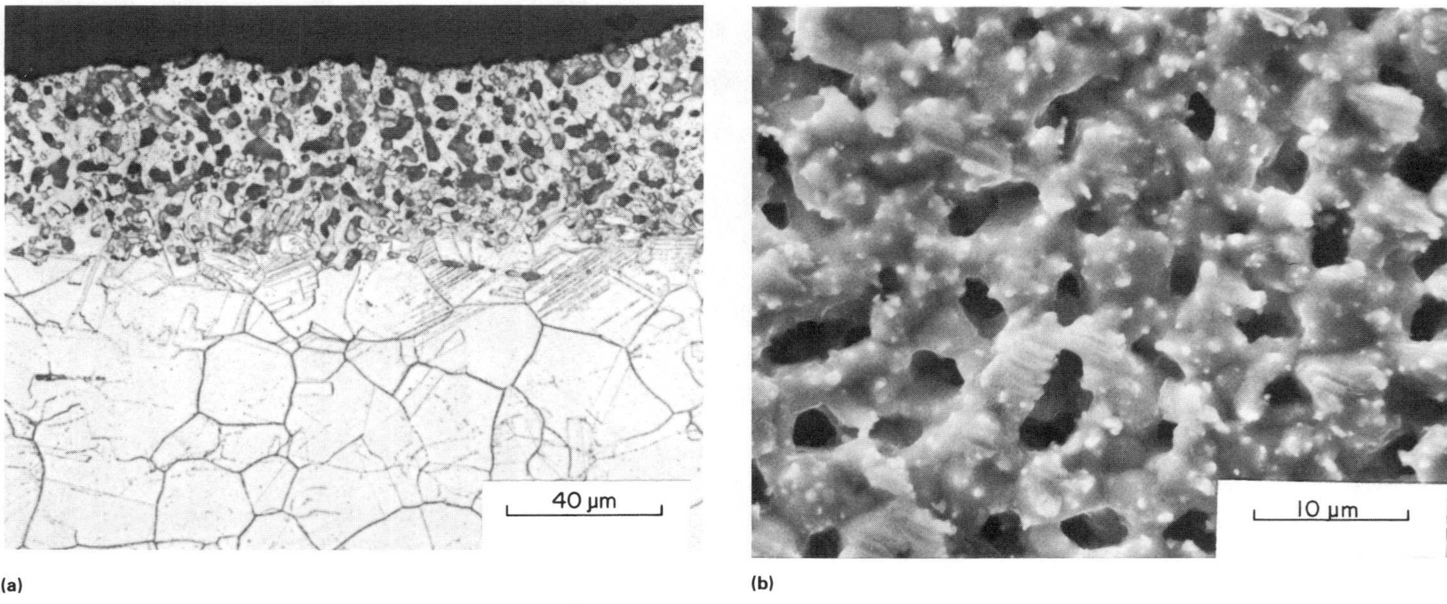

(a) (b)

Fig. 23 Corrosion of type 316 stainless steel exposed to thermally convective lithium for 7488 h at the maximum loop temperature of 600 °C (1110 °F). (a) Light micrograph of polished and etched cross section. (b) SEM showing the top view of the porous surface. Source: Ref 6

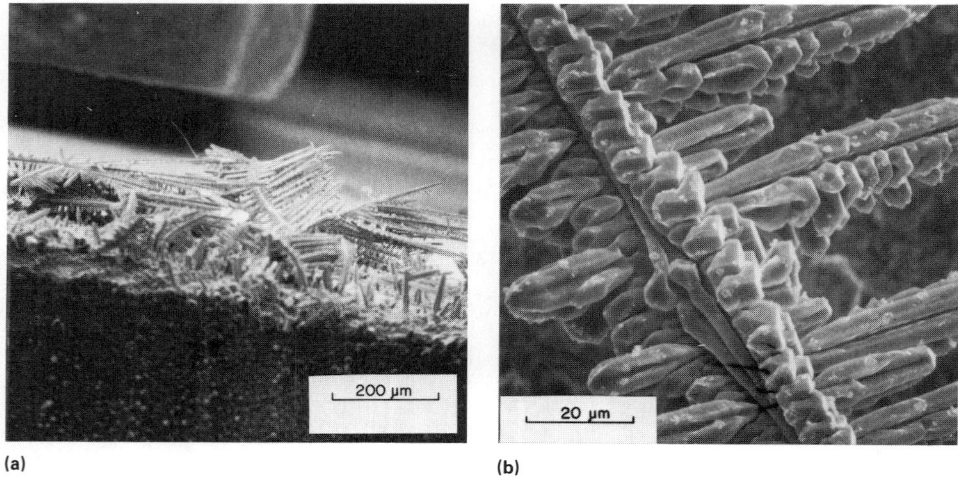

(a) (b)

Fig. 24 SEM micrographs of chromium mass transfer deposits found at the 460-°C (860-°F) position in the cold leg of a lithium/type 316 stainless steel thermal convection loop after 1700 h. Mass transfer deposits are often a more serious result of corrosion than wall thinning. (a) Cross section of specimen on which chromium was deposited. (b) Top view of surface. Source: Ref 7

systems. A recent example, from which many illustrations in this section are drawn (Fig. 17 to 22), is the work that has been performed around the world on the sodium-cooled fast breeder reactor. This concept led to large-scale research and development programs that have continued through more than two decades. The result is a vast wealth of knowledge related to sodium corrosion behavior in all its varied aspects. The body of information greatly overshadows the corrosion data compiled for other liquid metals. This type of reactor is now in commercial operation in France and the Soviet Union for power generation. Development reactors are operating in the United Kingdom, West Germany, Japan, India, and the United States.

Liquid lithium systems have been designed for two widely different areas: space nuclear power and fusion reactors. These two applications draw on unique properties of this liquid metal and have led to studies with a wide range of containment materials and operating conditions. Space power reactors require low mass; this in turn demands high-temperature operation. Lithium, with its low melting point/high boiling point and high specific heat, is an ideal candidate heat transfer medium. Refractory metal alloy containment is essential for these reactors, which may have design operating temperatures as high as 1500 °C (2730 °F).

Liquid lithium in fusion reactor concepts is selected because here the neutronics allow tritium fuel to be bred from the lithium; this is essential in order to make the economics of the reactor viable. Containment temperatures are below 700 °C (1290 °F); therefore, iron-base alloys can be used for construction.

The effects of liquid lithium on stainless steel, nickel, and niobium containment materials are shown in Fig. 23 to 32.

Liquid mercury, potassium, and cesium have also been used for space and terrestrial applications. In some cases, these have involved two-phase systems in which corrosion consideration became significantly altered. Lead, lead-bismuth, and lead-lithium alloys (Fig. 33) have received attention for topping cycle heat extraction systems, heat exchangers, reactor coolants, and, more recently, fusion reactor designs.

There are many other combinations of containment and liquid metals that have contributed to the knowledge of corrosion behavior; some have proved to be benign, while others have resulted in short-term catastrophic failures. In the discussion "Safety Considerations" in this section, some brief notes are given regarding safety precautions for handling liquid metals, operating circulating systems, dealing with fire and spillage, and cleaning contaminated components.

Forms of Liquid-Metal Corrosion

The forms in which liquid-metal corrosion are manifested can be divided into the following categories.

- Dissolution from a surface by (1) direct dissolution, (2) surface reaction, involving solid-metal atom(s), the liquid metal, and an impurity element present in the liquid metal, or (3) intergranular attack
- Impurity and interstitial reactions
- Alloying
- Compound reduction

All the variables present in the system play a part in the form and rate of corrosion that is established. There are ten key factors that have a major influence on the corrosion of metals and alloys by liquid-metal or liquid-vapor metal coolants. These are:

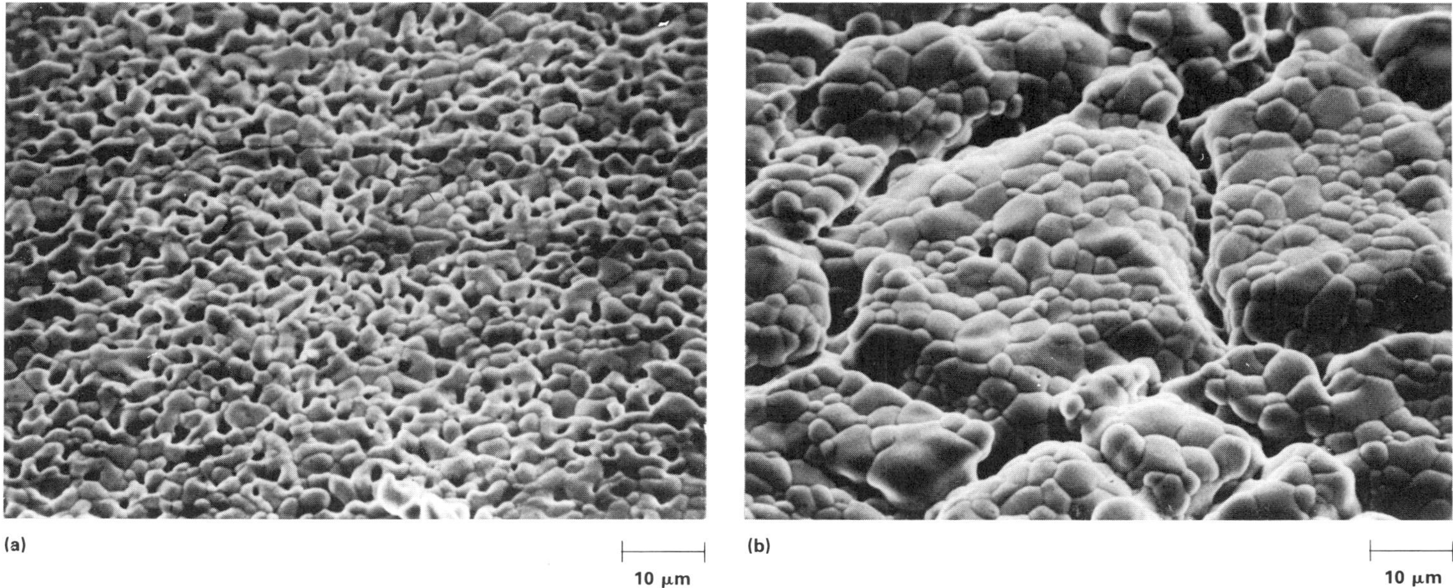

Fig. 25 Changes in surface morphology along the isothermal hot leg of a type 304 stainless steel pumped lithium system after 2000 h at 538 °C (1000 °F). Composition charges transform the exposed surface from austenite to ferrite, containing approximately 86% Fe, 11% Cr, and 1% Ni. (a) Inlet. (b) 7.7 m (25 ft) downstream. Source: Ref 8

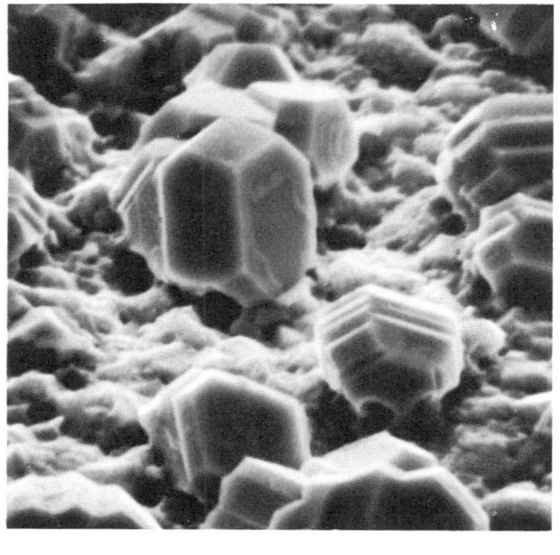

Fig. 26 Mass transfer deposits on X10CrNiMoTi 15 15 stainless steel after 1000-h exposure in static liquid lithium at 700 °C (1290 °F). Deposits are of the composition of the capsule steel (18Cr-8Ni). Courtesy of H.U. Borgstedt, Karlsruhe Nuclear Center

20 µm

Fig. 27 Corrosion of a capsule wall of 18 10 CrNiMoTi stainless steel by static lithium in the presence of zirconium foil. A porous ferritic surface layer has formed. Source: Ref 9

- Composition, impurity content, and stress condition of the metal or alloy
- Exposure temperature and temperature range
- Impurity content of the liquid metal
- Circulating or static inventory
- Heating/cooling conditions
- Single or two-phase coolant
- Liquid-metal velocity
- Presence/control of corrosion inhibition elements
- Exposure time
- Monometallic or multialloy system components.

These factors have a varied influence, depending on the combination of containment material and liquid metal or liquid-metal alloy. In most cases, the initial period of exposure (of the order of 100 to 1000 h, depending on temperature and liquid metal involved), is a time of rapid corrosion that eventually reaches a much slower steady-state condition as factors related to solubility and activity differences in the system approach a dynamic equilibrium. In some systems, this eventually leads to the development of a similar composition on all exposed corroding surfaces. High-nickel alloys and stainless steel exposed together in the high-temperature region of a sodium system will, for example, all move toward a composition that is more than 95% Fe.

Compatibility of a liquid metal and its containment varies widely, as is illustrated in Fig. 17 to 33. For a pure metal, surface attrition may proceed in an orderly, planar fashion, being controlled by either dissolution or a surface reaction. For a multicomponent alloy, selective loss of certain elements may lead to a phase transformation. For example, loss of nickel from austenitic stainless steel exposed to sodium may result in the formation of a ferritic surface layer (Fig. 17, 18a, 25, and 27). In high-nickel alloys, the planar nature of the corroding surface may be lost altogether, and a porous, spongelike layer may develop (Fig. 20 and 21). A more insidious situation can produce intergranular attack; liquid lithium, for example, will penetrate deep into refractory metals if precautions are not taken to ensure that the impurity element oxygen is in an oxide form more stable than Li_2O or LiO solutions, and is not left free in solid solution. Figure 31 illustrates intergranular attack in niobium.

Three factors—surface attrition, depth of depleted zone (for an alloy), and the presence of intergranular attack—should be evaluated collectively in any liquid-metal system. This evaluation will lead to an assessment of total damage, which may be presented either as a rate or as a cumulative allowance that must be made for the exposure of a given material over a given time. A large body of literature exists in which rate relationships for numerous liquid metal/containment combinations have been established. The more basic principles of liquid-metal corrosion are outlined in the article "Fundamentals of High-Temperature Corrosion in Liquid Metals" in this Volume and in the Selected References that follow this article.

One vitally important aspect of liquid-metal corrosion that is often overlooked is deposition. Corrosion itself is very often not a factor of major concern because surface recession rates in regions of maximum attack are often of the order of microns per year. The formation of compounds in the circulating liquid metal and the accumulation of deposits in localized regions where there is a

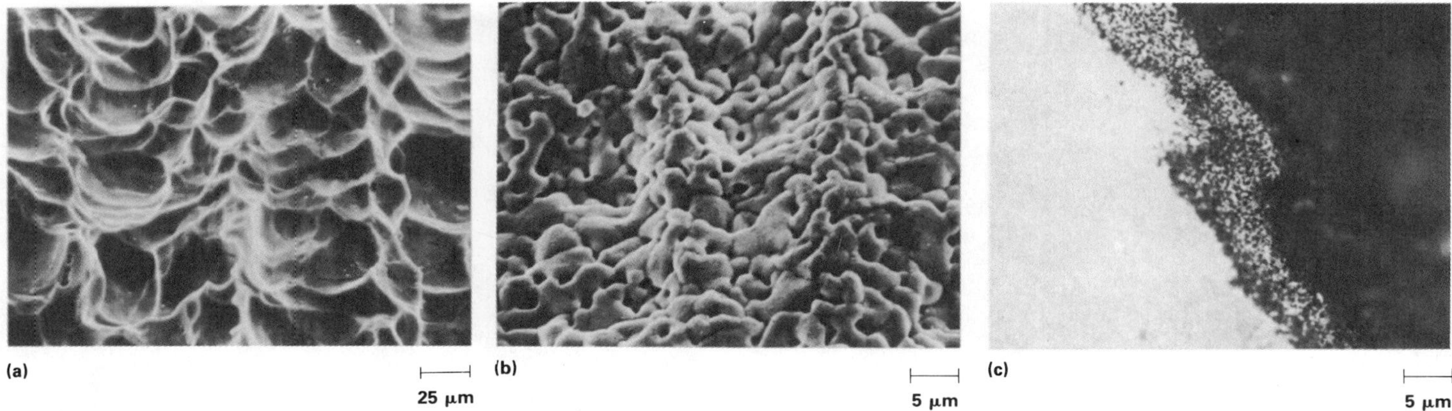

Fig. 28 Effects of flowing lithium on the inside surface of a type 316 stainless steel tubing. (a) Pickled surface before exposure. Composition: 65.8 Fe-18.0Cr-9.2Ni-3.3Mn-2.6Mo-0.9Si. (b) After exposure in flowing lithium (0.3 m/s, or 1 ft/s) for 1250 h at 490 °C (915 °F). Composition: 88.6Fe-7.5Cr-1.7Ni-0.6Mn. (c) After exposure to flowing lithium (1.3 m/s, or 4.3 ft/s) for 3400 h at 440 °C (825 °F). Taper section used to magnify damaged surface zone in metallographic mount. Courtesy of D.G. Bauer and W.E. Stewart, University of Wisconsin

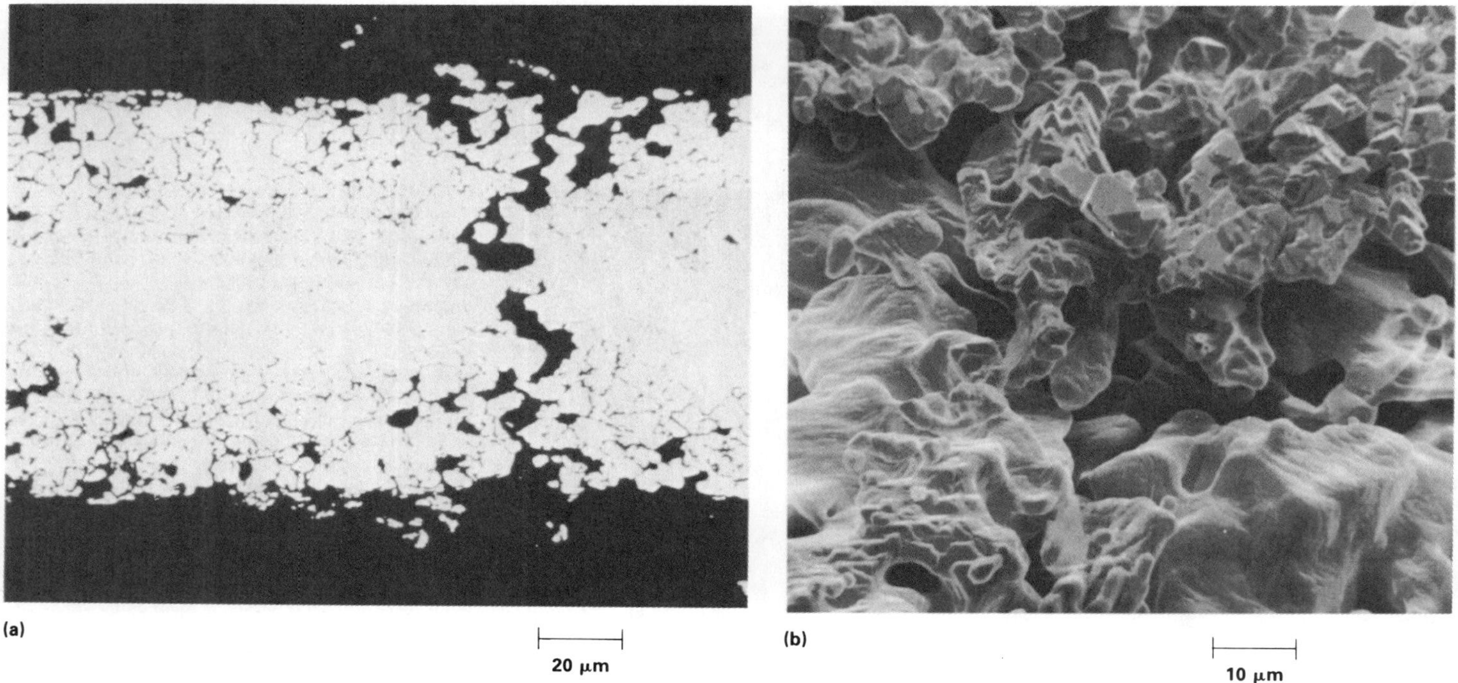

Fig. 29 Corrosion of nickel in static lithium after exposure for 300 h at 700 °C (1290 °F). (a) Light micrograph. (b) SEM micrograph. Source: Ref 10

drop in temperature, a change in flow rate or flow direction, or an induced change in surface roughness can, however, be very serious. If these deposits do not succeed in restricting flow channels completely, their nature is often such that they are only loosely adherent to deposition surfaces and may be dislodged by vibrations or thermal shock in the system, thus creating a major coolant flow restriction in a high-temperature region. Most deposits have a very low packing density; therefore, deposit growth can proceed at a rate that outstrips corrosion by several orders of magnitude. Examples of loosely adherent deposits are shown in Fig. 18(b), 22, 24, and 26. Figure 32 shows iron crystals that restricted flow in a pump channel of a lithium processing test loop.

Liquid-metal corrosion, as in other forms of corrosion, involves an appreciation for the source of corrosion in any system and an understanding of how potential sinks will operate on the corrosion burden, particularly if the liquid metal is not static but is circulated in a heat-transfer system either by pumping or by thermal convection.

Safety Considerations

The extensive work with the alkali and liquid metals has shown that such materials can be safely handled and used, provided certain precautions are heeded. The requirements for the safe use of liquid metals are in essence those of good industrial or laboratory practice, involving protection from contamination, chemical reactions, exposure to toxic or irritating substances, and protection from high temperatures. More specific information and details on safe operation and handling are available in Ref 15 and 16 and in the Materials Safety Data sheets issued by the Manufacturing Chemists' Association.

Chemical Reactivity. All the liquid metals react with oxygen and moisture to some degree; with the alkali metals, the reaction is vigorous enough to be potentially hazardous, particularly with potassium, rubidium, and cesium. Use of inert gas covers and the exclusion of moisture are the best defenses. Even with the nonalkali metals where the reactions with water are slow, water, such as that found on equipment that is not completely dry, must not be brought in contact with liquid metals because of the danger of a steam explosion that can scatter liquid metal over a considerable area, damaging equipment and inflicting severe burns on unprotected workers. It goes without saying that workers in the vicinity of liquid-metal systems should wear appropriate protective clothing (Ref 15).

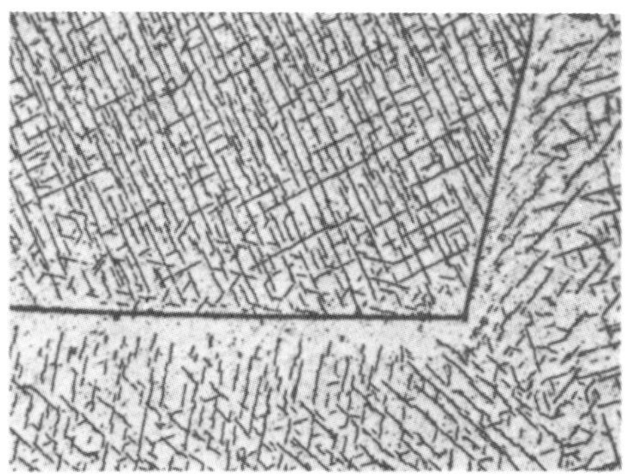

Fig. 30 Light micrograph of the polished and etched cross section of niobium containing 1500 wt ppm of oxygen showing the transcrystalline and grain boundary penetration that occurred after exposure to isothermal lithium for 100 h at 500 °C (930 °F). Source: Ref 11

50 μm

Fig. 31 Intergranular attack of unalloyed niobium exposed to lithium at 1000 °C (1830 °F) for 2 h. Light micrograph. Etched with 25% HF, 12.5 HNO_3, 12.5% H_2SO_4 in water. Source: Ref 12

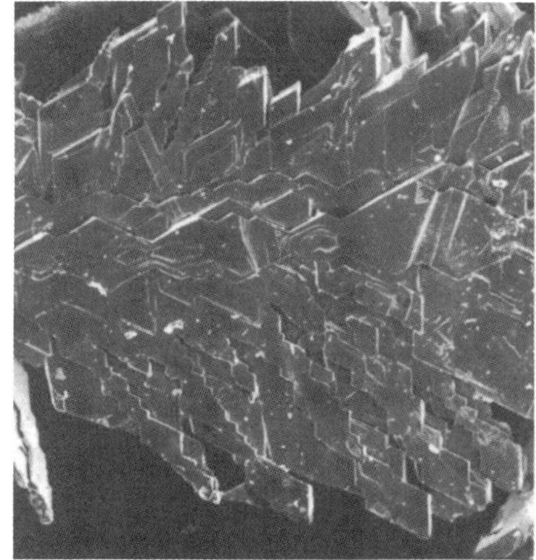

(a)

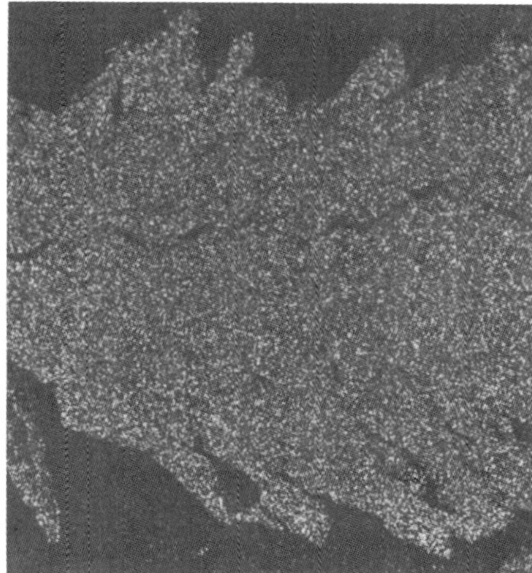

(b)

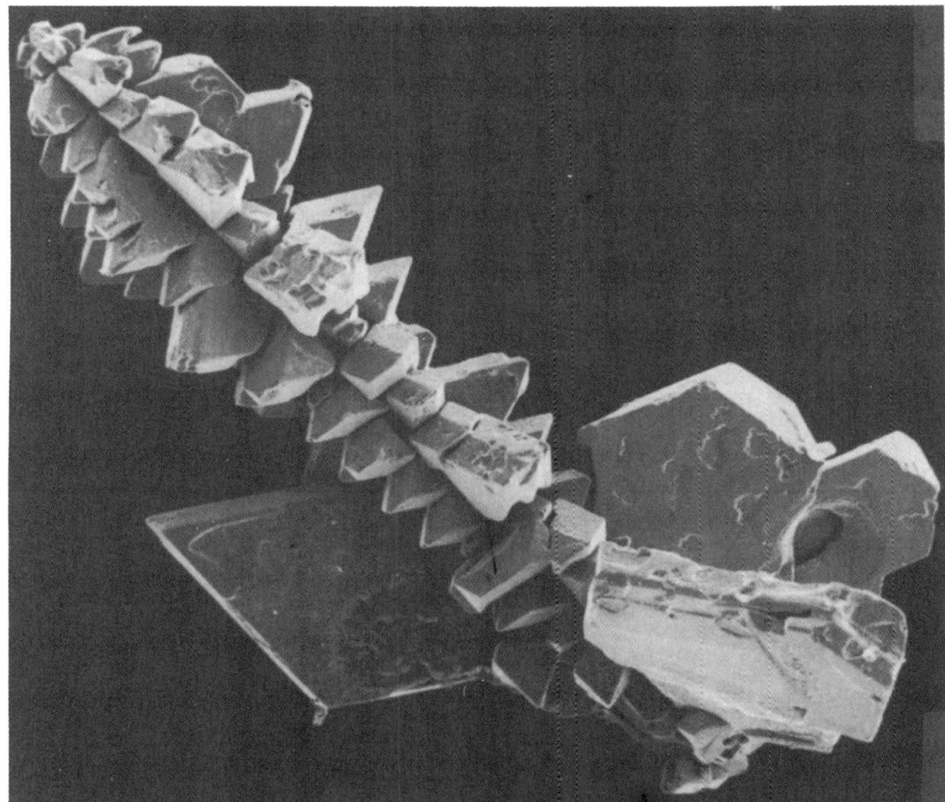

(c)

Fig. 32 Iron crystals found in a plugged region of a failed pump channel of a lithium processing test loop. Multifaceted platelike crystals are ~0.4 mm (0.015 in.) across. Composition: 86 to 93% Fe, 7 to 14% Ni, 0 to 1% Mn. (a) SEM. 70×. (b) Iron x-ray scan. 70×. (c) SEM. 90×. Source: Ref 13

Potassium, rubidium, and cesium form higher oxides than the monoxide when exposed to air; these compounds are powerful oxidizing agents, often shock sensitive, and definitely hazardous in the proximity of organic materials. They will form at room temperature in contact with the solid metals. The practical application of this information is the need for extreme caution when handling the metals or compounds of the heavier alkali metals, particularly when they have been stored for extended periods under less-than-ideal conditions, or when cleaning spills or fire residues. It is noted that the sodium-potassium eutectic alloy (NaK) will form potassium superoxide when exposed to air or oxygen. Because NaK is liquid at normal room temperature, small leaks at low temperature do not necessarily freeze and self-seal.

Circulating Systems. It is usually convenient to provide a closed circulating system, or loop, in which to perform liquid-metal corrosion experiments. Reference 16 describes a simple system; the principles involved are the same for even very complex specialized devices. The loop will provide the means for circulation of the liquid metal; it will contain devices for on-line purification (if needed), while the inert cover gas will protect against chemical contamination. Insulation and an enclosure will protect against high temperatures and the spread of reaction products in case of a containment breach. Such systems provide a safe, convenient environment for handling liquid metals. Detailed operating procedures, involving common-sense principles such as maintaining the inert cover gas at all times and melting frozen metals by directionally heating away from a free surface, must be worked out for each system. The tens of millions of hours of safe operation of such systems, ranging from 1-L capacity test rigs to 4000-MW (thermal) nuclear reactors, validate the concept. References 17 and 18 describe many such test systems and the experiments performed in them.

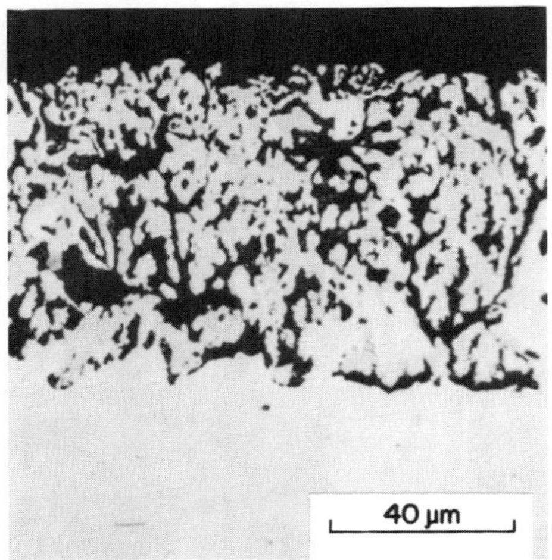

Fig. 33 Light micrograph of the polished cross section of a type 316 stainless steel exposed to thermally convective Pb-17at.% Li at 500 °C (930 °F) for 2472 h. Source: Ref 14

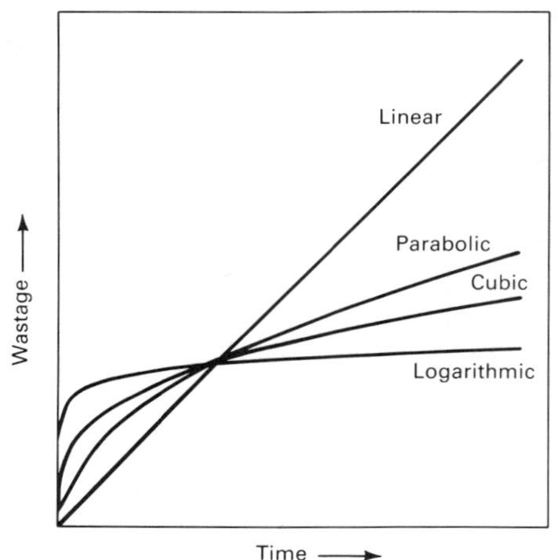

Fig. 34 Forms of kinetic curves that represent various thermal degradation processes

Recovery From Spills and Accidents. It must be remembered that spills of the nonalkali metals, even though the chemical reactivity hazard may not be great, must be handled with care because of the toxic nature of many of the metals and their vapors. Leaks and spills of the alkali metals, particularly when some of the leaked material has burned, present a special hazard because the spilled material often contains finely divided unreacted metal mixed with combustion products. Such mixtures can react vigorously with moist air, water, and alcohol. The products of the heavier alkali metals are the most reactive in this respect, but mixtures containing sodium and lithium are certainly not immune to violent reactions if carelessly handled. Cleanup of these residues must be approached with extreme care.

Removal of Residual Metals From Corrosion Specimens. Nonalkali metals can often be removed from corrosion specimens by draining, forming an amalgam or solution with an alkali metal, and then removing the mixture by a technique discussed below. Alkali metals can be removed from specimens by reaction with water or alcohols; the most vigorous reaction is with water, and the rate decreases as one progresses to heavier weight alcohols. The reaction becomes more vigorous with increasing atomic weight of the alkali metals. Cesium/water reactions are definitely explosive. Use of ethanol and methanol is generally safe for sodium reaction, but one must remember the flammability hazard with alcohol vapors. The glycol ethers, such as butyl cellosolve, can also be used; they react more slowly than water, present less of a fire hazard than ethanol or methanol, but have toxic liquid and vapors. The water, alcohol, and glycol ether reactions all generate hydrogen; adequate ventilation must be provided to prevent the buildup of the hydrogen and the attendant danger of an explosion.

Anhydrous liquid ammonia forms a true solution with the alkali metals and can be used to remove adherent alkali metals from corrosion specimens. Precautions against the hazards of liquid ammonia must be taken; the alkali metal in solution with ammonia is then usually reacted with water or alcohol before the ammonia mixture is discarded. The conditions must be maintained truly anhydrous in order to avoid hydrogen generation and contamination of the samples. Hydrogen can become implanted in refractory metal samples and embrittle them even at subzero temperatures. If the proper equipment is available, evaporation of the residual metal from the surface can be done with excellent results.

Removal of alkali materials from pipework, if hydrogen generation is not a problem, can be accomplished with alcohol or gycol ether reaction, optionally followed by water rinsing to wash away the reaction products. It must be remembered that these reagents react very slowly, if at all, with oxides of the alkali metals. Another successful method in use is reaction with water vapor/inert gas (argon or nitrogen) mixtures, or water spray in an inert carrier gas, followed by water rinsing. Evaporation has also been successfully used and could be considered where hydrogen generation is not permitted. Ammonia-base systems have also been used for refractory metal pipework where hydrogen generation was prohibited. References 16, 17, and 18 contain more detailed information.

Fire Protection and First Aid. Firefighting and medical treatment should, of course, be left to the professionals. There are, however, several factors to keep in mind. An alkali metal fire does not expand, as does, for example, a petroleum fire. It does produce vast quantities of caustic smoke that react with moisture in or on the body, and this produces severe burns. The smoke must be avoided unless respiratory protection and protective clothing are worn.

Lithium presents a special hazard because of the toxic nature of some of its compounds and because it reacts with nitrogen. The combustion products of a lithium fire contain nitride and acetylide, which react with water to form ammonia and acetylene, respectively.

The only way to extinguish liquid-metal fires reliably is to smother them; various drying agents, such as sodium carbonate, sodium chloride, powdered dry graphite, or carbon microspheres, are effective. The sodium compounds should not be used on a lithium fire because of the chemical displacement reaction producing sodium metal that takes place. Instead, the graphite products should be used for extinguishment. It should be noted that few substances float on lithium; most tend to sink. Reference 19, however, describes the use of a ternary eutectic BaCl-NaCl-KCl salt for effective lithium fire extinguishment.

Inert gas blanketing is also effective for extinguishing fires; argon is preferred to nitrogen because of its higher density. Very effective fire suppression can be achieved by using inert gas blanketing in combination with a catch pan having a perforated floor to let the escaping liquid metal run to a second argon-blanketed chamber underneath the first. Of course, the hazards of oxygen-deficient atmospheres must be recognized and controlled when using inert gas flooding as fire suppression. Reference 20 provides more information about fire suppression in the alkali metals.

Emergency first aid treatment should concentrate on removing the victim from danger (for example, out of the smoke generated by a fire) and then removing the metal or compounds from skin and eyes. Running water is the most effective treatment because the cooling and flushing effect overcomes any hazard from reaction of a small amount of metal on the skin or clothing. Safety showers and eyewash fountains should be readily available to workers involved with liquid metals; they should not be eliminated because of the presence of the alkali metals. More than one person has been spared serious injury because a safety shower was available to wash away a sodium spill.

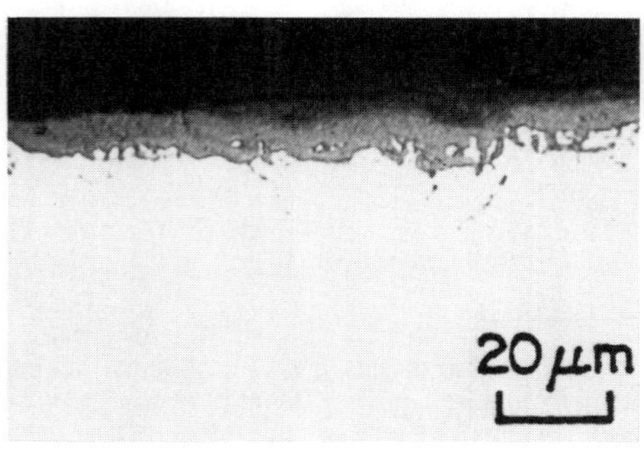

(a)

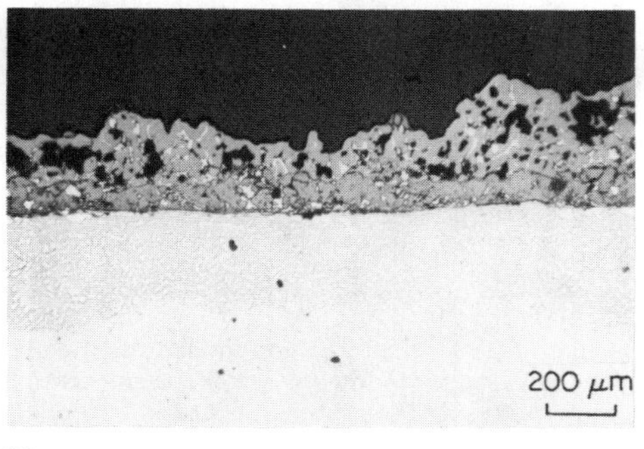

(b)

Fig. 35 Protective and nonprotective scales formed on Alloy 800. (a) Cr_2O_3-base protective oxide scale formed in sulfur-free oxidizing gas. (b) Sulfide-oxide scale formed in reducing conditions containing hydrogen sulfide. Courtesy of I.G. Wright, Battelle Columbus Division

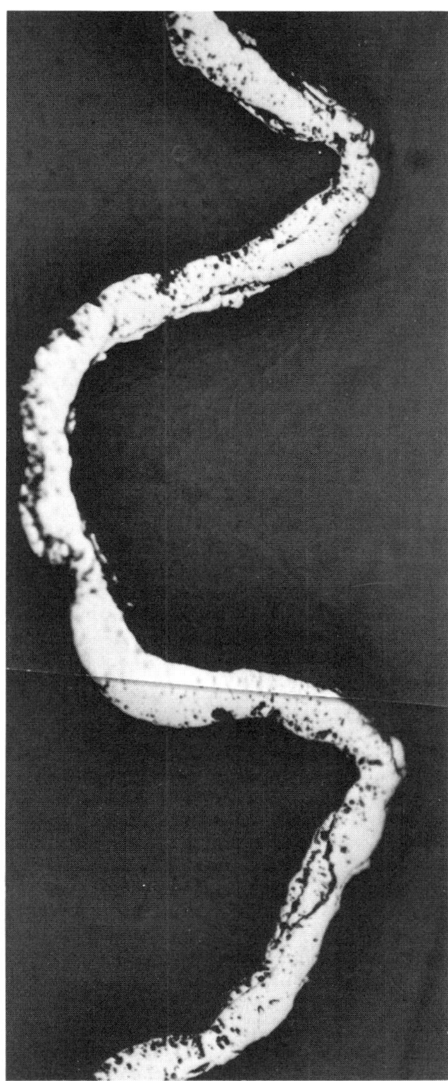

Fig. 36 Cr_2O_3 scale formed on pure chromium at 1100 °C (2012 °F). A Pilling-Bedworth ratio of 2.0 results in high compressive stress in the scale, which is relieved by buckling and spalling. Courtesy of I.G. Wright, Battelle Columbus Division

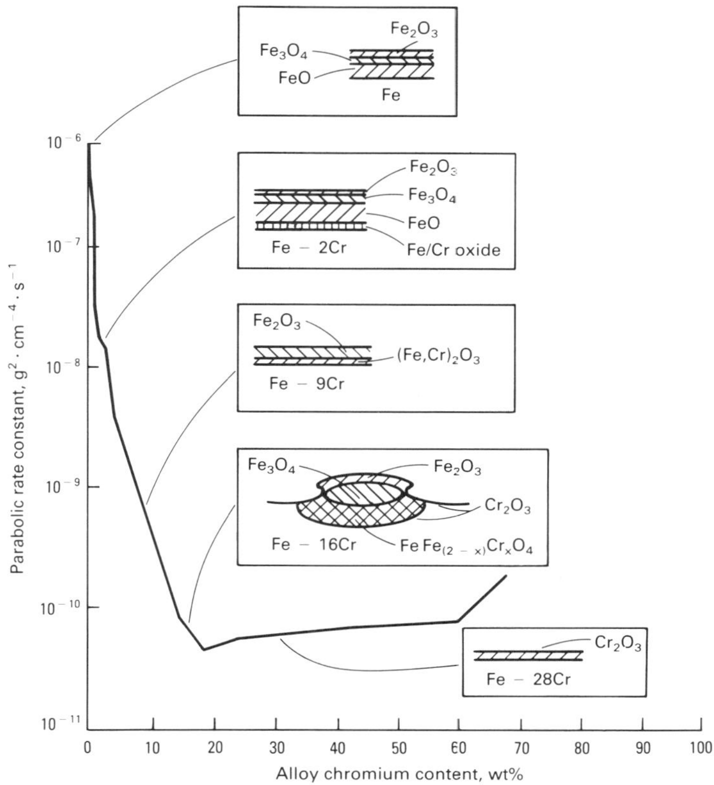

Fig. 37 Schematic of the variation with alloy chromium content of the oxidation rate and oxide scale structure (based on isothermal studies at 1000 °C, or 1832 °F, in 0.13 atm oxygen)

High-Temperature Corrosion

Ian G. Wright
Battelle Columbus Division

When metal is exposed to an oxidizing gas at elevated temperature, corrosion can occur by direct reaction with the gas, without the need for the presence of a liquid electrolyte. This type of corrosion is referred to as tarnishing, high-temperature oxidation, or scaling. The rate of attack increases substantially with temperature. The surface film typically thickens as a result of reaction at the scale/gas or metal/scale interface due to cation or anion transport through the scale, which behaves as a solid electrolyte. For continuous, nonporous scales, ionic transport through the scale is the rate-controlling process. The thermodynamic stability, the ionic defect structure, and certain morphological features of the scale formed are key factors in determining the resistance of an alloy to a specific environment.

Initial film growth is usually very rapid. If the scale is a nonporous solid and completely covers the metal surface, the reaction rate will decrease when the thickness reaches a few thousand angstroms as the transport of reactive species through the film becomes rate controlling. The subsequent corrosion rate depends on the details of this transport mechanism, which may be due to electrical potential or concentration gradients or to migration along preferential paths, and so may correspond to any of several rate laws, as shown in Fig. 34. Where a diffusion process is rate controlling, the kinetics usually follow a parabolic rate law, in which the rate progressively decreases with time. Figure 35(a) illustrates the compact, continuous protective scale of essentially chromium oxide (Cr_2O_3) formed on Alloy 800. If the scale is porous (or is formed as a vapor species) or does not completely cover the metal surface, a linear rate is usually experienced.

The latter circumstance can be assessed from the Pilling-Bedworth ratio, which is the ratio of the volumes of oxide produced to the metal consumed by oxidation; values of 1.0 or greater result in complete surface coverage by oxide and, usually, protective behavior. This is not a complete nor foolproof measure for assessing the likelihood of protective scaling behavior. At high temperatures, the growth of nominally protective oxides may be sufficiently rapid that the compressive stresses resulting from a Pilling-Bedworth ratio greater than 1 become sufficiently great that the scale (or alloy) deforms and possibly spalls as a relief mechanism; in some cases, the protection offered by such scales may be low at this point, as shown in Fig. 36.

The desired characteristics for a protective oxide scale include the following:

- High thermodynamic stability (highly negative Gibbs free energy of formation) so that it forms preferentially to other possible reaction products
- Low vapor pressure so that the oxide forms as a solid and does not evaporate into the atmosphere
- Pilling-Bedworth ratio greater than 1.0 so that the oxide completely covers the metal surface
- Low coefficient of diffusion of reactant species (metal cations, and corrodent anions) so that the scale has a slow growth rate
- High melting temperature
- Good adherence to the metal substrate, which usually involves a coefficient of thermal expansion close to that of the metal, and sufficient high-temperature plasticity to resist fracture from differential thermal expansion stresses

High-temperature scales are usually thought of as oxides, but may also be sulfides, possibly carbides, or mixtures of these species. Oxides and sulfides are nonstoichiometric compounds and semiconductors. There are essentially two types of semiconductors: *p*-type (or positive carrier)—which may have vacancies in its metal lattice, or an excess of anions contained interstitially—and *n*-type (or negative carrier)—which may have an excess of metal ions contained interstitially, or vacant anion lattice sites. For

Fig. 38 Multilayer oxide scale formed on Co-10Cr alloy at 1100 °C (2012 °F). Outer layer is CoO; inner (mottled gray) layer is CoO containing dissolved chromium and particles of Co-Cr spinel. The chromium level in this alloy is insufficient to form a fully protective Cr_2O_3 scale. Courtesy of I.G. Wright, Battelle Columbus Division

(a)

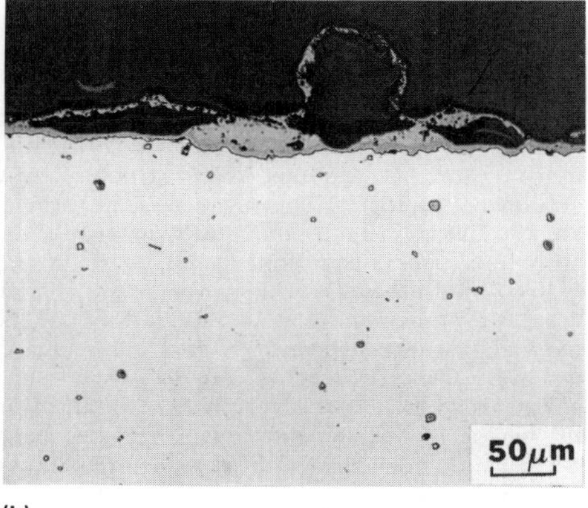

(b)

Fig. 39 Topography (a) and cross section (b) of oxide scale formed on Fe-18Cr alloy at 1100 °C (2012 °F). The bright areas on the alloy surface (a) are areas from which scale has spalled. The buckled scale and locally thickened areas (b) are iron-rich oxide. The thin scale layer adjacent to the alloy is Cr_2O_3, which controls the oxidation rate. Courtesy of I.G. Wright, Battelle Columbus Division

diffusion-controlled scaling, the rate of scale growth can be altered by modification of the concentration of the particular defects involved. For example, *p*-type oxides exhibit increased cationic transport rates (increased oxidation rates) at increased oxygen pressures, while transport in *n*-type oxides is essentially independent of oxygen pressure. Both types of oxide can be doped by the addition of specific ions to the oxide lattice. For *p*-type metal deficit oxides, for example, the addition of cations of higher valence than the native cations results in an increase in the number of cation vacancies and therefore an increase in the oxidation rate, while lower-valence cation additions have the opposite effect.

Sulfides typically exhibit an intrinsically greater rate of transport of anions and cations than the oxides of the same metal and so provide scales that are significantly less protective than oxides. Detailed information on the kinetics of high-temperature corrosion in gases and the thermodynamic stability of oxide/sulfide scales can be found in the article "Fundamentals of Corrosion in Gases" in this Volume.

High-Temperature Oxidation

Alloys intended for high-temperature applications are designed to have the capability of forming protective oxide scales. Alternatively, where the alloy has ultrahigh-temperature strength capabilities (which is usually synonymous with reduced levels of protective scale forming elements), it must be protected by a specially designed coating. Oxides that effectively meet the criteria for protective scales listed above and can be formed on practical alloys are limited to Cr_2O_3, alumina (Al_2O_3), and possibly silicon dioxide (SiO_2). In the pure state, Al_2O_3 exhibits the slowest transport rates for metal and oxygen ions and so should provide the best oxidation resistance.

Alloying requirements for the production of specific oxide scales have been translated into minimum levels of the scale-forming elements, or combinations of elements, depending on the base alloy composition and the intended service temperature. Figure 37 schematically represents the oxidation rate of iron-chromium alloys (1000 °C, or 1832 °F, in 0.13 atm oxygen) and depicts the types of oxide scale associated with various alloy types. Figure 38 illustrates the morphology of a semiprotective scale formed on a cobalt-chromium alloy. Alloys based on these minimum specifications will form the desired protective oxide upon initial exposure, but because of the accompanying deletion of the scale forming element, they will probably be unable to re-form the protective layer in the event of loss or failure of the initial scale.

A useful concept in assessing the potential high-temperature oxidation behavior of an alloy is that of the reservoir of scale-forming element contained by the alloy in excess of the minimum level (around 20 wt% for iron-chromium alloys at 1000 °C, or 1832 °F, according to Fig. 37). The more likely the service conditions are to cause repeated loss of the protective oxide scale, the greater the reservoir of scale-forming element required in the alloy for continued protection. Extreme cases of this concept result in chromizing or aluminizing to enrich the surface regions of the alloy or in the provision of an external coating rich in the scale-forming elements.

The breakdown of protective scales based on Cr_2O_3 or Al_2O_3 appears, in the majority of cases, to originate through mechanical means. The most common is spallation as a result of thermal cycling, or loss through impact or abrasion. Typical scale structures on an Fe-18Cr alloy after thermal cycling are shown in Fig. 39. Cases in which the scales have been destroyed chemically are usually related to reactions occurring beneath deposits, especially where these consist of molten species. An additional mode of degradation of protective Cr_2O_3 scale is through oxidation to the volatile chromium trioxide (CrO_3), which becomes prevalent above about 1010 °C (1850 °F) and is greatly accelerated by high gas flow rates.

Because these protective oxide scales will form wherever the alloy surface is exposed to the ambient environment, they will form at all surface discontinuities; therefore, the possibility exists that notches of oxide will form at occluded angles in the surface, which may eventually serve to initiate or propagate cracks under thermal cycling conditions. The ramifications of stress-assisted oxidation (and of oxidation assisting the applied stress) in the production of failure conditions are not very well understood, but constitute important considerations in practical failure analysis.

Sulfidation

When the sulfur activity (partial pressure, concentration) of the gaseous environment is sufficiently high, sulfide phases, instead of oxide phases, can be formed. The mechanisms of sulfide formation in gaseous environments and beneath molten-salt deposits have been determined in recent years. In the majority of environments encountered in practice by oxidation-resistant alloys, Al_2O_3 or Cr_2O_3 should form in preference to any sulfides, and destructive sulfidation attack occurs mainly at sites where the protective oxide has broken down. The role of sulfur, once it has entered the alloy, appears to be to tie up the chromium and aluminum as sulfides, effectively redistributing the protective scale-forming elements near the alloy surface and thus interfering with the process of formation or re-formation of the protective scale. If sufficient sulfur enters the alloy so that all immediately available chromium or aluminum is converted to sulfides, then the less stable sulfides of the base metal may form because of morphological and kinetic reasons. It is these base metal sulfides that are often responsible for the observed accelerated attack, because they grow much faster than the oxides or sulfides of chromium or aluminum; in addition, they have relatively low melting points, so that molten slag phases are often possible. Fig. 35 compares a protective (oxide) scale and a nonprotective (sulfide) scale formed on Alloy 800.

Sulfur can transport across continuous protective scales of Al_2O_3 and Cr_2O_3 under certain conditions, with the result that discrete sulfide precipitates can be observed immediately beneath the scales on alloys that are behaving in a protective manner. For reasons indicated above, as long as the amount of sulfur present as sulfides is small, there is little danger of accelerated attack. However, once sulfides have formed in the alloy, there is a tendency for the sulfide phases to be preferentially oxidized by the encroaching reaction front and for the sulfur to be displaced inward, forming new sulfides deeper in the alloy, often in grain boundaries or at the sites of other chromium- or aluminum-rich phases, such as carbides. In this way, finger-like protrusions of oxide/sulfide can be formed from the alloy surface inward, which may act to localize stress or otherwise reduce the load-bearing section. Such attack of an austenitic stainless steel experienced in a coal gasifier product gas is shown in Fig. 40. The sulfidation behavior of Alloy 800 at temperatures and oxygen and sulfur potentials representative of coal gasification processes is illustrated in Fig. 41 to 43.

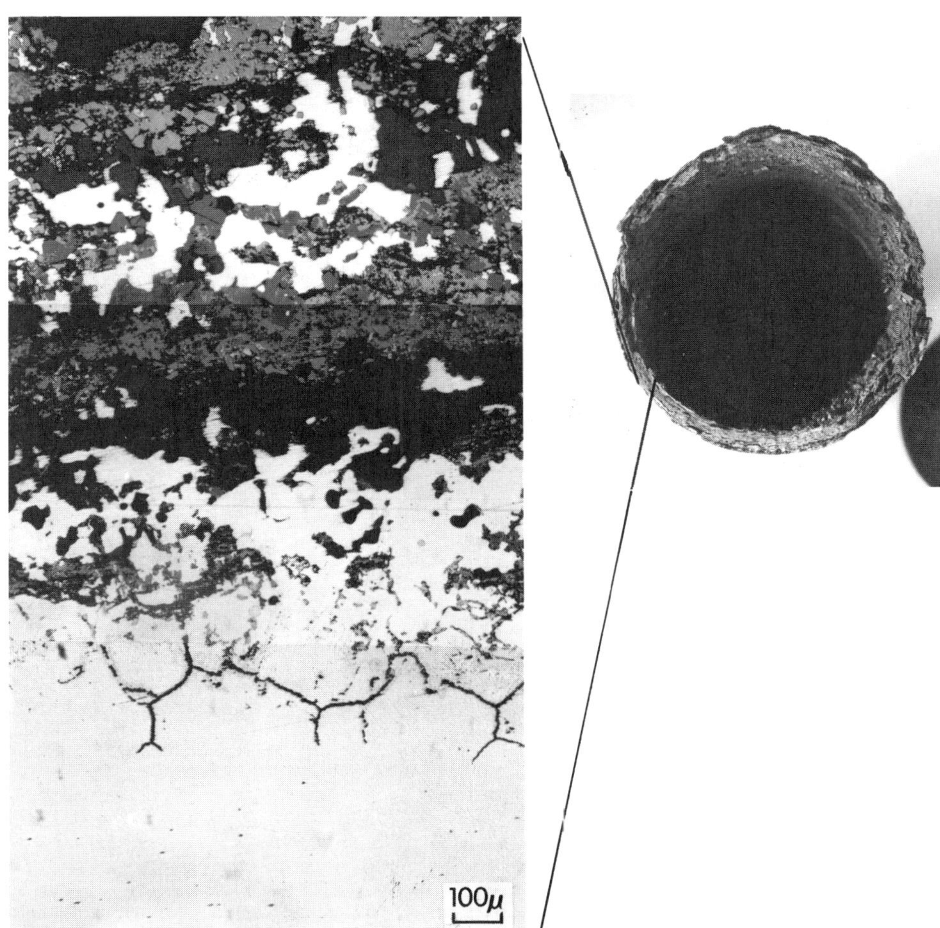

Fig. 40 Example of high-temperature sulfidation attack in a type 310 heat-exchanger tube after ~100 h at 705 °C (1300 °F) in coal gasifier product gas.

Carburization

As in the case of sulfide penetration, carburization of high-temperature alloys is thermodynamically unlikely except at very low oxygen partial pressures, because the protective oxides of chromium and aluminum are generally more likely to form than the carbides. However, carburization can occur kinetically in many carbon-containing environments. Carbon transport across continuous nonporous scales of Al_2O_3 or Cr_2O_3 is very slow, and alloy pretreatments likely to promote such scales, such as initially smooth surfaces or preoxidation, have generally been found to be effective in decreasing carburization attack. In practice, the scales formed on high-temperature alloys often consist of multiple layers of oxides resulting from localized bursts of oxide formation in areas where the original scale was broken or lost. The protection is derived from the innermost layer, which is usually richest in chromium or aluminum. Concentration of gaseous species such as carbon monoxide in the outer porous oxide layers appears to be one means by which sufficiently high carbon activities can be generated at the alloy surface for carburization to occur in otherwise oxidizing environments. The creation of localized microenvironments is also possible under deposits that create stagnant conditions not permeable by the ambient gas.

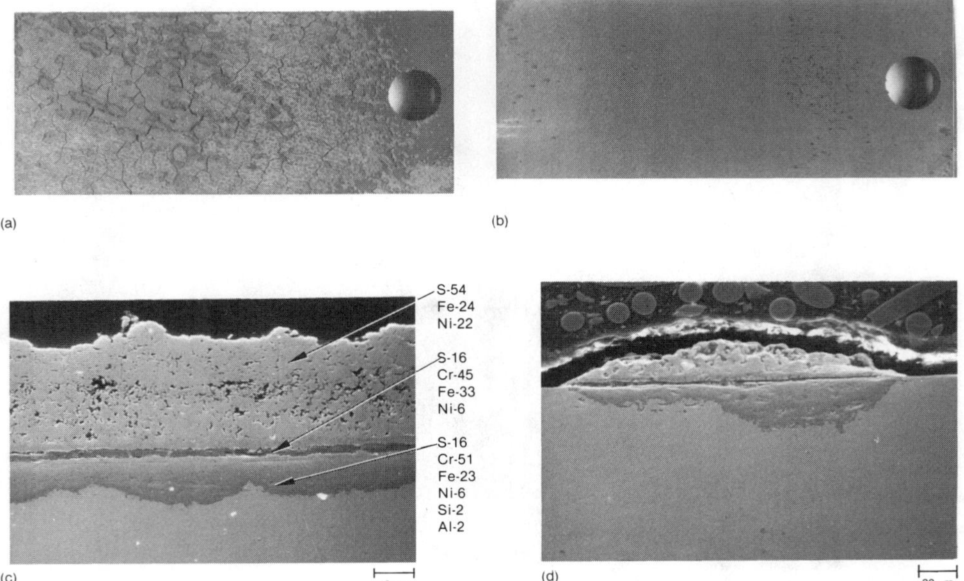

Fig. 41 Alloy 800 test coupons with a 0.254-mm (0.01-in.) diam grain size exposed to a coal gasifier environment for 100 h. (a) and (c) Tested at 650 °C (1200 °F) and oxygen and sulfur partial pressures of 3×10^{-24} atm and 1×10^{-8} atm, respectively. (b) and (d) Tested at 650 °C (1200 °F) and $pO_2 = 3 \times 10^{-24}$ atm and $pS_2 = 1 \times 10^{-9}$ atm. SEM micrographs show sulfide scale (c) and an external sulfide formation (d). (a) and (b) ~2×. Courtesy of G.R. Smolik and D.V. Miley, E.G. & G. Idaho, Inc.

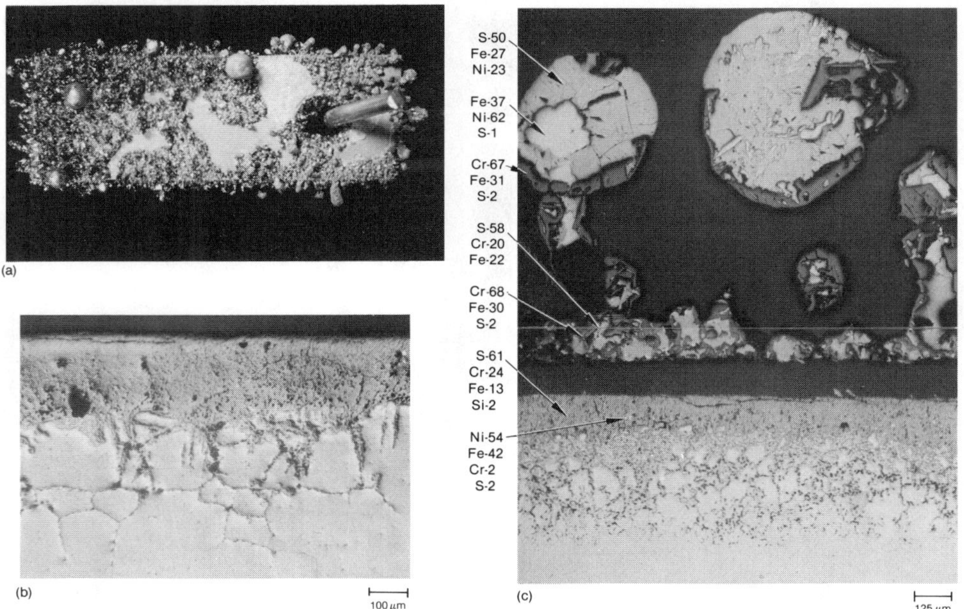

Fig. 42 Sulfidation attack of Alloy 800 test coupons exposed to a coal gasifier environment ($pO_2 = 3 \times 10^{-20}$ atm and $pS_2 = 1 \times 10^{-7}$ atm) at 870 °C (1600 °F) for 100 h. (a) and (b) Macrograph and micrograph, respectively, of a test coupon with a 0.254-mm (0.01-in.) diam grain size. (c) Micrograph showing external sulfides, sulfide scale, and intergranular sulfidation of a test coupon with a 0.022- to 0.032-mm (0.0008- to 0.0013-in.) diam grain size. (a) ~1.5×. Courtesy of G.R. Smolik and D.V. Miley, E.G. & G. Idaho, Inc.

Once inside the alloy, the detrimental effects of the carbon depend on the location, composition, and morphology of the carbides formed. Austenitic steels should carburize more readily than ferritic steels because of the high solubility of carbon in austenite. Iron-chromium alloys containing less than about 13% Cr contain various amounts of austenite, depending on temperature, and should be susceptible to carburization, while alloys with 13 to 20% Cr will form austenite as a result of absorption of small amounts of carbon. Iron-chromium alloys containing more than ~20% Cr can absorb considerable amounts of carbon before austenite forms, becoming principally $(CrFe)_{23}C_6$ and ferrite. An example of rapid high-temperature carburization attack of an austenitic stainless steel is shown in Fig. 44.

Minor alloying elements can exert an influence on the susceptibility to carburization of various alloys. In particular, silicon, niobium, tungsten, titanium, and the rare earths have been noted as promoting resistance to carburization. Experience with aluminum and manganese has been varied, while lead, molybdenum, cobalt, zirconium, and boron are considered detrimental.

Other Forms of High-Temperature Corrosion

Hydrogen Effects. In hydrogen at elevated temperatures and pressures, there is increasing availability of atomic hydrogen that can easily penetrate metal structures and react internally with reducible species. An example is the attack experienced by carbon steel, in which atomic hydrogen reacts with iron carbide to form methane, which then leads to fissuring of the steel. Alloy steels with stable carbides, such as chromium carbides, are less susceptible to this form of attack. For example, 2.25Cr-1Mo suffers some decarburization in high-temperature high-pressure hydrogen, but is less likely to fissure than carbon steel. The susceptibility of steels to attack by hydrogen can be judged from the Nelson Curves, which indicate the regions of temperature and pressure in which a variety of steels will suffer attack. Nelson curves and examples of high-temperature hydrogen attack of carbon and alloy steels are illustrated and discussed in the article "Corrosion in Petroleum Refining and Petrochemical Operations" in this Volume.

A further example of hydrogen attack is copper containing small amounts of cuprous oxide. This oxide reacts to form steam within the alloy, resulting in significant void formation.

Hot Corrosion. The term hot corrosion is generally used to describe a form of accelerated attack experienced by the hot gas path components of gas turbine engines. Two forms of hot corrosion can be distinguished; most of the corrosion encountered in turbines burning liquid fuels can be described as Type I hot corrosion, which occurs primarily in the metal temperature range of 850 to 950 °C (1550 to 1750 °F). This is a sulfidation-based attack on the hot gas path parts involving the formation of condensed salts, which are often molten at the turbine operating temperature. The major components of such salts are sodium sulfate (Na_2SO_4) and/or potassium sulfate (K_2SO_4), apparently formed in the combustion process from sulfur from the fuel and sodium from the fuel or the ingested air. Because potassium salts act very similarly to sodium salts, alkali specifications for fuel or air are usually taken to the sum total of sodium plus potassium. An example of the corrosion morphology typical of Type I hot corrosion is shown in Fig. 45.

Very small amounts of sulfur and sodium or potassium in the fuel and air can produce sufficient Na_2SO_4 in the turbine to cause extensive corrosion problems because of the concentrating effect of turbine pressure ratio. For example, a threshold level has been suggested for sodium in air of 0.008 ppm by weight below which hot corrosion will not occur. Type I hot corrosion, therefore, is possible even when premium fuels are used. Other fuel (or air) impurities, such as vanadium, phosphorus, lead, and chlorides, may combine with Na_2SO_4 to form mixed salts having

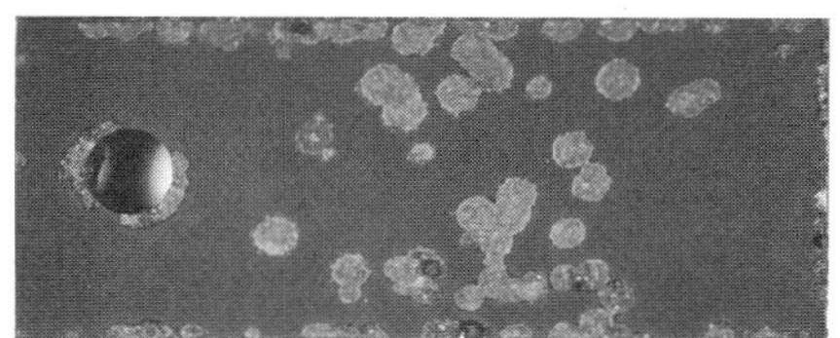

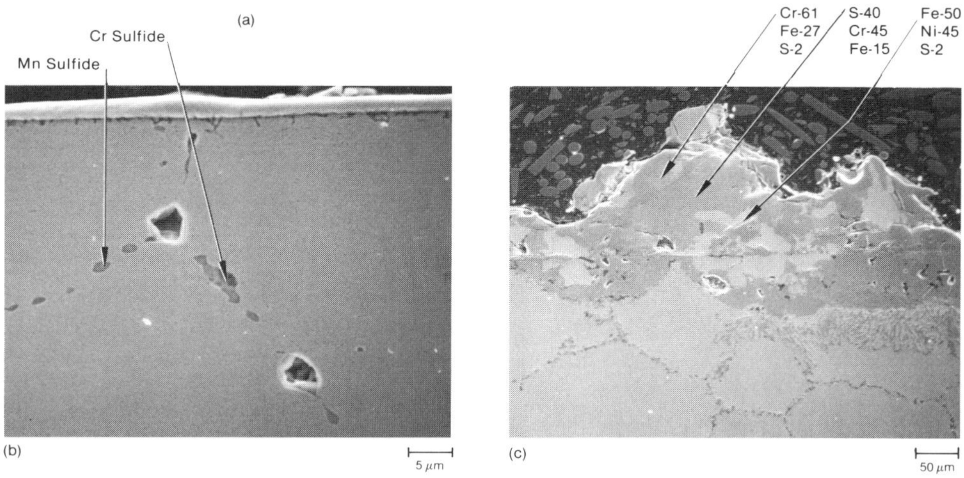

Fig. 43 Macrograph (a) of an Alloy 800 test coupon with a 0.254-mm (0.01-in.) diam grain size exposed to a coal gasifier environment ($pO_2 = 3 \times 10^{-19}$ atm and $pS_2 = 1 \times 10^{-7}$ atm) at 870 °C (1600 °F) for 100 h. ~1.5×. Micrographs (b) and (c) show cross sections through the Cr_2O_3 layer and disrupted oxide region having external sulfides. Courtesy of G.R. Smolik and D.V. Miley, E.G. & G. Idaho, Inc.

reduced melting temperature and thus broaden the range of conditions over which this form of attack can occur. Also, agents such as unburned carbon can promote deleterious interactions in the salt deposits.

Research over the past 15 years has led to greater definition of the relationships among temperature, pressure, salt concentration, and salt vapor-liquid equilibria so that the location and rate of salt deposition in an engine can be predicted. Additionally, it has been demonstrated that a high chromium content is required in an alloy for good resistance to Type I hot corrosion. The trend to lower chromium levels with increasing alloy strength has therefore rendered most superalloys inherently susceptible to this type of corrosion. The effects of other alloying additions, such as tungsten, molybdenum, and tantalum, have been documented, and their effects on rendering an alloy more or less susceptible to Type I hot corrosion are known and mostly understood.

Although various attempts have been made to develop figures of merit to compare superalloys, these have not been universally accepted. Nonetheless, the near standardization of such alloys as Alloy 738 and Alloy 939 for first-stage blades/buckets, and FSX-414 for first stage vanes/nozzles, implies that these are the accepted best compromises between high-temperature strength and hot corrosion resistance. It has also been possible to devise coatings with alloying levels adjusted to resist this form of hot corrosion. The use of such coatings is essential for the protection of most modern superalloys intended for duty as first-stage blades or buckets.

Type II, or low-temperature hot corrosion, occurs in the metal temperature range of 650 to 700 °C (1200 to 1400 °F), well below the melting temperature of Na_2SO_4, which is 884 °C (1623 °F). This form of corrosion produces characteristic pitting, which results from the formation of low-melting mixtures of essentially Na_2SO_4 and cobalt sulfate ($CoSO_4$), a corrosion product resulting from the reaction of the blade/bucket surface with sulfur trioxide (SO_3) in the combustion gas. The melting point of the Na_2SO_4-$CoSO_4$ eutectic is 540 °C (1004 °F). Unlike Type I hot corrosion, a partial pressure of SO_3 in the gas is critical for the reactions to occur. Knowledge of the SO_3 partial pressure-temperature relationships inside a turbine allows some prediction of where Type II hot corrosion can occur. Cobalt-free nickel-base alloys (and coatings) may be more resistant to Type II hot corrosion than cobalt-base alloys; it has also been observed that resistance to Type II hot corrosion increases with the chromium content of the alloy or coating.

REFERENCES

1. C.P. Larrabee, Corrosion Resistance of High Strength Low-Alloy Steels as Influenced by Composition and Environment, *Corrosion*, Vol 9 (No. 8), 1953, p 259-271

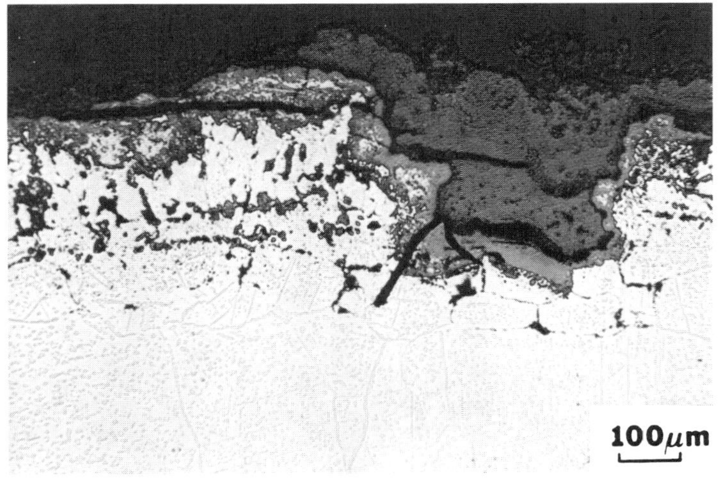

(a) (b)

Fig. 44 Example of high-temperature carburization attack pitting in type 310 reactor wall after ~4000-h exposure to coal gasification product gas. The pits were formed during operation under conditions of high carbon activity in the gas. (a) Overall view of pitting. (b) Section through a pit. Courtesy of I.G. Wright, Battelle Columbus Division

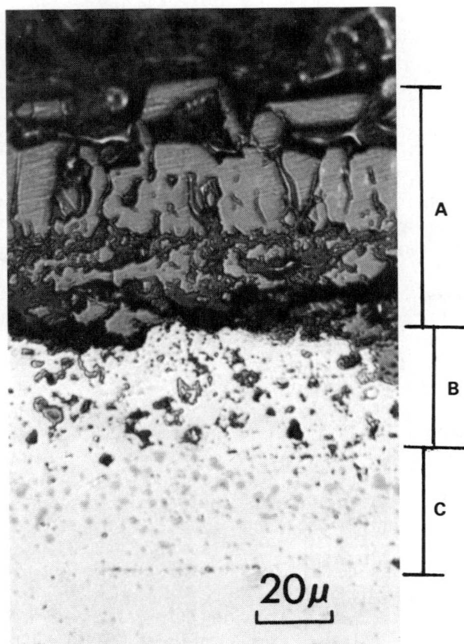

Fig. 45 Ni-20Cr-2ThO$_2$ after simulated Type I hot corrosion exposure (coated with Na$_2$SO$_4$ and oxidized in air at 1000 °C, or 1832 °F). A, nickel-rich scale; B, Cr$_2$O$_3$ subscale; C, chromium sulfides. Courtesy of I.G. Wright, Battelle Columbus Division

2. J.D. Costlow and R.C. Tipper, Ed., *Marine Biodeterioration: An Interdisciplinary Study*, Proceedings of the Symposium, U.S. Naval Institute Press, 1984
3. *Marine Fouling and Its Prevention*, Woods Hole Oceanographic Institution, U.S. Naval Institute Press, 1952
4. F.L. LaQue, *Marine Corrosion*, Wiley-Interscience, 1975
5. C. Bagnall and R.E. Witkowski, "Microstructural and Surface Characterization of Candidate LMFBR Fuel Cladding and Duct Alloys After Exposure to Flowing Sodium at 700 °C," WARD-NA-3045-53, U.S. Energy Research and Development Agency Technical Information Center, June 1978
6. P.F. Tortorelli and J.H. DeVan, Effects of a Flowing Lithium Environment on the Surface Morphology and Composition of Austenitic Stainless Steel, *Microstruct. Sci.*, Vol 12, 1985, p 213-226
7. P.F. Tortorelli and J.H. DeVan, Mass Transfer Deposits in Lithium-Type 316 Stainless Steel Thermal Convection Loops, in *Proceedings of the Second International Conference on Liquid Metal Technology in Energy Production*, CONF-800401, National Technical Information Service, 1980, p 13-56 to 13-62
8. C. Bagnall, A Study of Type 304 Stainless Steel Containment Tubing From a Lithium Test Loop, *J. Nucl. Mater.*, Vol 103 and 104, 1981, p 639-644
9. H.U. Borgstedt, *J. Nucl. Mater.*, Vol 103 and 104, 1981, p 693-698
10. H.U. Borgstedt, *Mater. Chem.*, Vol 5, 1980, p 95-108
11. J.R. DiStefano, "Corrosion of Refractory Metals by Lithium," M.S. thesis, University of Tennessee, 1963
12. W.F. Brehm, "Grain Boundary Penetration of Niobium by Lithium," Ph.D. thesis, Report HYO-3228-11, Cornell University, 1967
13. V.A. Maroni *et al.*, "Analysis of the October 5, 1979, Lithium Spill and Fire in the Lithium Processing Test Loop," ANL-81-25, Prepared for the U.S. Department of Energy under Contract W-31-109-Eng-38, Argonne National Laboratory, Dec 1981
14. P.F. Tortorelli and J.H. DeVan, Corrosion of Ferrous Alloys Exposed to Thermally Convective Pb-17 at.% Li, *J. Nucl. Mater.*, Vol 141-143, 1986, p 592-598
15. O.J. Foust, Ed., *Liquid Metals Handbook, Sodium and NaK Supplement*, U.S. Atomic Energy Commission, 1970
16. C.C. Addison, *The Chemistry of the Liquid Alkali Metals*, John Wiley & Sons, 1984
17. *Proceedings of the Second International Conference on Liquid Metal Technology in Energy Production*, CONF-800401, National Technical Information Service, 1980
18. *Proceedings of the Third International Conference on Liquid Metal Engineering and Technology*, Thomas Telford Ltd., 1985
19. D.W. Jeppson *et al.*, Lithium Literature Review: Lithium's Properties and Interactions, HEDL-TME-78-15, National Technical Information Service, 1978
20. L.D. Muhlestein, Liquid Metal Reactions Under Postulated Accident Conditions for Fission and Fusion Reactors, in *Proceedings of the Second International Conference on Liquid Metal Technology in Energy Production*, CONF-800401, National Technical Information Service, 1980

SELECTED REFERENCES

General References

- C.P. Dillon, Ed., *Forms of Corrosion Recognition and Prevention*, National Association of Corrosion Engineers, 1982
- M.G. Fontana and N.D. Greene, *Corrosion Engineering*, 2nd ed., McGraw-Hill, 1978
- H.H. Uhlig and R.W. Revie, *Corrosion and Corrosion Control*, 3rd ed., John Wiley & Sons, 1985

Atmospheric Corrosion

- W.H. Ailor, Ed., *Atmospheric Corrosion*, John Wiley & Sons, 1982
- S.W. Dean and E.C. Rhea, Ed., *Atmospheric Corrosion of Metals*, STP 767, American Society for Testing and Materials, 1982
- R.H. Heidersbach, "Corrosion Performance of Weathering Steel Structures," Paper presented at the Annual Meeting, Transportation Research Board, Jan 1987
- *Metal Corrosion in the Atmosphere*, STP 435, American Society for Testing and Materials, 1967
- I.L. Rozenfeld, *Atmospheric Corrosion of Metals*, E.C. Greco, Ed., B.H. Tytell, Trans., National Association of Corrosion Engineers, 1972

Galvanic and Stray-Current Corrosion

- R. Baboian, Ed., *Electrochemical Techniques for Corrosion*, National Association of Corrosion Engineers, 1977
- R. Baboian, W.D. France, L.C. Rowe, and J.F. Rynewicz, Ed., *Galvanic and Pitting Corrosion—Field and Laboratory Studies*, STP 576, American Society for Testing and Materials, 1974
- V. Chaker, Ed., Corrosion Effect of Stray Currents and the Techniques for Evaluating Corrosion of Rebars in Concrete, in *Corrosion of Rebars in Concrete*, STP 906, American Society for Testing and Materials, 1984

Molten-Salt Corrosion

- M.G. Fontana and N.D. Greene, *Corrosion Frequency*, McGraw-Hill, 1978
- G.J. Janz and R.P.T. Tomkins, *Corrosion*, Vol 35, 1979, p 485
- J.W. Koger, *Advances in Corrosion Science and Technology*, Vol 4, Plenum Press, 1974
- D.G. Lovering, Ed., *Molten Salt Technology*, Plenum Press, 1982
- A. Rahmel, Corrosion, in *Molten Salt Technology*, Plenum Press, 1982
- L.L. Sheir, Ed., *Corrosion*, Vol 1, Newnes-Butterworths, 1979, p 2.10

Corrosion in Liquid Metals

- C.C. Addison, *The Chemistry of the Liquid Alkali Metals*, John Wiley & Sons, 1984
- *Alkali Metal Coolants*, Symposium Proceedings, Vienna, International Atomic Energy Agency, 1967
- T.L. Anderson and G.R. Edwards, *J. Mater. Energy Syst.*, Vol 2, 1981, p 16-25
- C. Bagnall and D.C. Jacobs, "Relationships for Corrosion of Type 316 Stainless Steel in Sodium," WARD-NA-3045-23, National Technical Information Service, 1974
- W.E. Berry, *Corrosion in Nuclear Applications*, John Wiley & Sons, 1971
- H.U. Borgstedt, Ed., *Proceedings of the Conference on Material Behavior and Physical Chemistry in Liquid Metal Systems*, Plenum Press, 1982
- H.U. Borgstedt and C.M. Matthews, *Applied Chemistry of the Alkali Metals*, Plenum Press, 1986
- W.F. Brehm, "Grain Boundary Penetration of Niobium by Lithium," Ph.D. thesis, Report HYO-3228-11, Cornell University, 1967
- C.F. Cheng and W.E. Ruther, *Corrosion*, Vol 28 (No. 1), 1972, p 20-22
- M.H. Cooper, Ed., *Proceedings of the International Conference on Liquid Metal Technology in Energy Production*, CONF-760503, P1 and P2, National Technical Information Service, 1977
- J.M. Dahlke, Ed., *Proceedings of the Second International Conference on Liquid Metal Technology in Energy Production*, CONF-300401-P1 and P2, National Technical Information Service, 1981
- J.H. DeVan *et al.*, Compatibility of Refractory Alloys with Space Reactor System Coolants and Working Fluids, in CONF-830831D, National Technical Information Service, 1983
- J.F. DeStefano and E.E. Hoffman, *At. Energy Rev.*, Vol 2, 1964, p 3-33
- J.E. Draley and J.R. Weeks, Ed., *Corrosion by Liquid Metals*, Plenum Press, 1970
- R.L. Eichelberger and W.F. Brehm, "Effect of Sodium on Breeder Reactor Components," Paper 106, presented at Corrosion/78, National Association of Corrosion Engineers, 1978
- A.H. Fleitman and J.R. Weeks, Mercury as a Nuclear Coolant, *Nucl. Eng. Des.*, Vol 16 (No. 3), 1971
- J.D. Harrison and C. Wagner, *Acta Metall.*, Vol 1, 1959, p 722-735

- S.A. Jannson, Ed., *Chemical Aspects of Corrosion and Mass Transfer in Liquid Metals*, The Metallurgical Society, 1973
- C.J. Klamut, D.G. Schweitzer, J.G.Y. Chow, R.A. Meyer, O.F. Kammerer, J.R. Weeks, and D.H. Gurinsky, Material and Fuel Technology for an LMFR, in *Progress in Nuclear Engineering Series IV, V2-Technology Engineering and Safety*, Pergamon Press, 1960
- *Proceedings of the International Conference on Liquid Alkali Metals*, Proceedings of the British Nuclear Energy Society, Thomas Telford Ltd., 1973
- *Proceedings of the International Conference on Sodium Technology and Large Fast Reactor Design*, ANL-7520, Part I, National Technical Information Service, 1969
- *Proceedings of the Third International Conference on Liquid Metal Engineering and Technology*, Proceedings of the British Nuclear Energy Society, Thomas Telford Ltd., 1985
- M.C. Rowland *et al.*, "Sodium Mass Transfer XV. Behavior of Selected Steels Exposed in Flowing Sodium Test Loops," GEAP-4831, National Technical Information Service, 1965
- P.F. Tortorelli and J.H. DeVan, Corrosion of Fe-Cr-Mn Alloys in Thermally Convective Lithium and Corrosion of Ferrous Alloys Exposed to Thermally Convective Pd-17at.% Li, *J. Nucl. Mater.*, Vol 141-143, Proceedings of the Second International Conference on Fusion Reactor Materials (Chicago, April 1976), 1986, p 579-583, 592-598
- J.R. Weeks, *Nucl. Eng. Des.*, Vol 15, 1971, p 363-372
- J.R. Weeks and H.S. Isaacs, *Adv. Corros. Sci. Technol.*, Vol 3, 1973, p 1-65

High-Temperature Corrosion

- E.F. Bradley, Ed., *Source Book on Materials for Elevated Temperature Applications*, American Society for Metals, 1979
- B.R. Cooper and W.A. Ellingson, Ed., *The Science and Technology of Coal and Coal Utilization*, Plenum Press, 1984
- D.L. Douglass, Ed., *Oxidation of Metals and Alloys*, American Society for Metals, 1971
- U.R. Evans, *The Corrosion and Oxidation of Metals—First Supplementary Volume*, St. Martin's Press, 1968
- A.B. Hart and A.J.B. Cutler, Ed., *Deposition and Corrosion in Gas Turbines*, John Wiley & Sons, 1973
- U.L. Hill and H.L. Black, Ed., *The Properties and Performance of Materials in the Coal Gasification Environment*, Materials/Metalworking Technology Series, American Society for Metals, 1981
- D.R. Holmes and A. Rahmel, Ed., *Materials and Coatings to Resist High-Temperature Corrosion*, Applied Science, 1978
- *Hot Corrosion Problems Associated With Gas Turbines*, STP 421, American Society for Testing and Materials, 1967
- O. Kubaschewski and B.E. Hopkins, *Oxidations of Metals and Alloys*, 2nd ed., Academic Press, 1962
- D.B. Meadowcroft and M.I. Manning, Ed., *Corrosion-Resistant Materials for Coal Conversion Systems*, Applied Science, 1983
- S. Mrowec and T. Werber, *Gas Corrosion of Metals*, W. Bartoszewski, Trans. Foreign Science Publications, Department of the National Center for Scientific, Technical and Economic Information, available from National Technical Information Service, 1978
- R.A. Rapp, Ed., *High-Temperature Corrosion*, Publication 8, National Association of Corrosion Engineers, 1983
- M.F. Rothman, Ed., *High Temperature Corrosion in Energy Systems*, The Metallurgical Society, 1985
- H. Schmalzried, *Solid State Reactions*, A.D. Pelton, Trans., Academic Press, 1974
- I.G. Wright, Ed., *Corrosion in Fossil Fuel Systems*, Vol 83-5, Conference Proceedings, The Electrochemical Society, 1983

Localized Corrosion

Chairman: Stephen C. Dexter, College of Marine Studies, University of Delaware

THE FORMS OF CORROSION described in this article—filiform, crevice, pitting, and biological corrosion—all have the common feature that the corrosion damage produced is localized rather than spread uniformly over the exposed metal surface. This makes these forms of attack more difficult to deal with than those producing a generalized attack (see the article "General Corrosion" in this Volume). Instead of dealing with a slow, relatively uniform loss of metal thickness, the engineer is now faced with high rates of metal penetration at specific sites (or localized deterioration under a coating system in the case of filiform corrosion), while the remainder of the metal (or coating) goes largely unaffected. The attack can also be harder to detect because much of the damage may be subsurface, with only a small opening visible to the eye at the metal surface. Moreover, these forms of attack are economically important and dangerous because they can lead to premature failure of a structure by rapid penetration with little overall weight loss.

The purpose of this article is to provide the engineer with enough information to identify which form of corrosion is taking place on an existing structure or which form of corrosion is likely to occur on a new structure. Therefore, most of this article is devoted to illustrating the appearance of these localized forms of corrosion and to describing the classes of metals and alloys susceptible to them and the environmental conditions under which they occur. Some information is also given on the mechanisms of the attack and measures that can be used to prevent it. More information on these latter two topics can be found in the references given in this article as well as the Sections "Specific Alloy Systems" and "Fundamentals of Corrosion" in this Volume.

Filiform Corrosion

Christopher Hahin
Materials Protection Associates

Filiform corrosion occurs on metallic surfaces coated with a thin organic film that is typically 0.1 mm (4 mils) thick. The pattern of attack is characterized by the appearance of fine filaments emanating from one or more sources in semirandom directions. The source of initiation is usually a defect or mechanical scratch in the coating. The filaments are fine tunnels composed of corrosion products underneath the bulged and cracked coating. Filiforms are visible at an arm's length as small blemishes. Upon closer examination, they appear as fine striations shaped like tentacles or cobweblike traces (Fig. 1). A filiform has an active head, and a filamentous tail (Fig. 2). A close-up of the head/tail interface is shown in Fig. 3.

Filiform corrosion is often mistaken as having biological origins because of its wormlike appearance. Filiform corrosion routinely occurs on coated steel cans, aluminum foil laminated packaging, painted aluminum, and other lacquered metallic items placed in areas subjected to high humidity. Filiform corrosion has been observed on many organically coated metals and alloys, including steels, tin-plated steel, aluminum, and magnesium. The susceptibility of a metal to filiform corrosion can be determined by placing several coated and scratched panels in a salt fog chamber as described in ASTM D 2803 (Ref 1). If susceptible, filiform filaments will gradually grow out perpendicularly from the scratch. Many of these filaments will later orient themselves in the rolling direction of the panel.

Filiform attack occurs when the relative humidity is typically between 65 to 90% for most cases. The lowest reported relative humidity was 58% for nitrocellulose lacquer on steel, and even lower relative humidities were reported for aluminum in very corrosive environments (Ref 2). The average width of a filament varies between 0.05 to 3 mm (2 to 120 mils). Filament width depends on the coating, the ambient relative humidity, and the corrosive environment. Typical filament height is about 20 μm (0.8 mil). The filament growth rate can also vary widely, with rates observed as low as 0.01 mm/d (0.4 mil/d) and up to a maximum rate of 0.85 mm/d (35 mils/d). The depth of attack in the filiform tunnels can be as deep as 15 μm (0.6 mil). The fluid in the leading head of a filiform is typically acidic, with a pH from 1 to 4. In all cases, oxygen or air and water were needed to sustain filiform corrosion. This indicates that filiform corrosion is a specialized differential aeration cell (see the "Glossary of Terms" in this Volume for a definition of differential aeration cell).

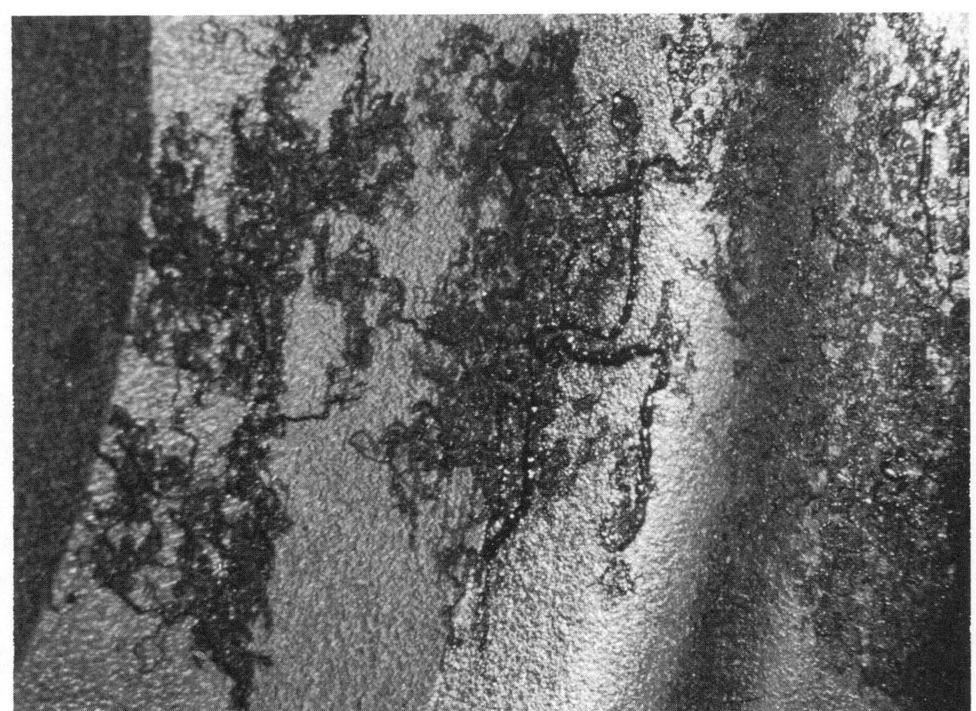

Fig. 1 A lacquered steel can lid exhibiting filiform corrosion showing both large and small filaments partially oriented in the rolling direction of the steel sheet. Without this 10× magnification by a light microscope, the filiforms look like fine striations or minute tentacles.

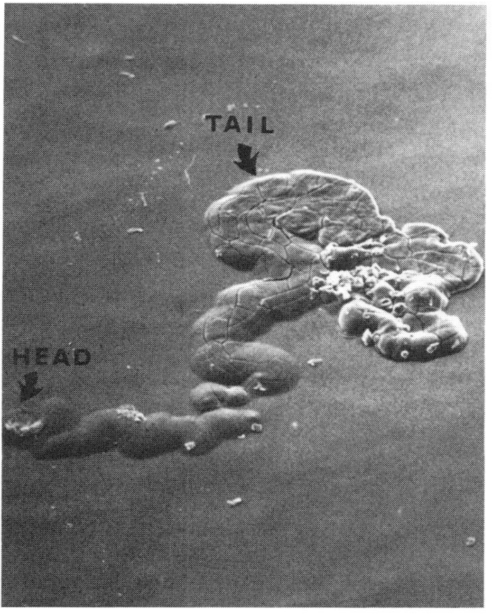

(a)

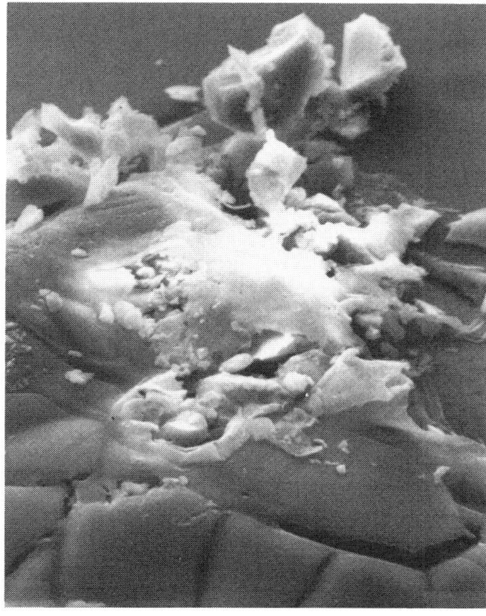

(b)

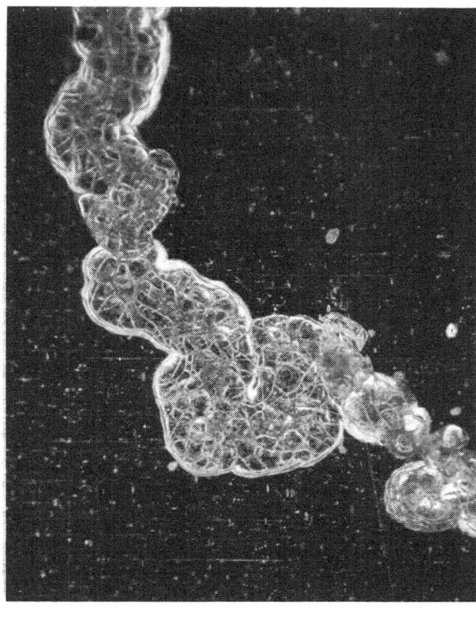

(c)

Fig. 2 Filiform corrosion on PVC-coated aluminum foil. (a) Advancing head and cracked tail section of a filiform cell. SEM. 80×. (b) The gelatinous corrosion products of aluminum oozing out of the porous end tail section of a filiform cell. SEM. 830×. (c) Tail region of a filiform cell. Tail appears iridescent due to internal reflection. Light microscopy. 60×

Filiform Corrosion of Coated Steels

Characteristics of Coated Steels. Steel and tin-plated steel are routinely coated to provide resistance to atmospheric corrosion, to isolate foodstuffs from their containers, and to provide an adherent surface for ink and paint. Transparent organic coatings can also be lightly tinted with various dyes to impart a brass, bronze, or bluish cast to improve the appearance of the steel. Tin plating on steel is very thin—of the order of 1.5 μm (0.06 mil) thick—and is a porous barrier (see the article "Corrosion of Tin and Tin Alloys" in this Volume). In some environments, tin is often cathodic to steel. In aggressive environments, tin plate provides little corrosion protection for the steel substrate. The corrosion rate of steel is fairly slow in alkaline environments, but increases markedly in acidic environments when the pH is less than 3. Steel is subject to pitting in concentrated chloride environments. Tin is amphoteric (it corrodes in alkaline and acidic media), but it is not rapidly attacked in certain deaerated acids. However, steel and tin are both readily corroded by aerated acids. Such conditions of low pH, aeration, and high chloride concentrations are often found to exist in the active heads of filiform corrosion cells in steel.

To prevent general rusting, many coatings are applied to steel to shield it from moist or humid environments. Filiform corrosion has been observed on steels coated with lacquers, varnish, polyurethane, boiled linseed oil, and various alkyd, urea, and epoxy paints. The coatings are applied by spraying, by direct contact with rubber rolls, by dipping, or by electrostatic methods. Cured and dried coatings are uneven in thickness, with numerous hills and valleys. Lacquers can contain entrapped solvents if they are insufficiently cured.

In general, organic coatings are flexible, and they tend to soften at higher temperatures. Coatings can separate from the steel substrate because of physical abrasion. The ease of separation depends on how well the surface was prepared before it was coated. Organic coatings can be semipermeable to water and can have numerous flaws caused by improper application of the coating, poor curing, or solvent entrapment. Coated articles may sustain some minor mechanical abrasions during their curing or storage. Coatings often crack when blistered by corrosion product expansion, gas evolution, or water retention. Coatings, therefore, have many potential defect sources at which filiform corrosion can begin. Heads of filiform cells will continue to remain active as long as the coating keeps expanding and cracking and if moisture and oxygen are available.

Conditions Leading to Filiform Attack. Filiform corrosion generally occurs in coated steels when the relative humidity is between 60 to 95% in a temperature range of 20 to 35 °C (70 to 95 °F). The surrounding atmosphere must contain air or oxygen. Attack usually begins at cuts, knicks, pores, or other disruptions in the coating. Carbon dioxide can also stimulate filiform corrosion by dissolving in water to create carbonic acid. If the condensing atmospheres contain concentrations of chlorides, sulfates, sulfides, or carbon dioxide, the likelihood of filiform corrosion is substantially increased because these constituents help to acidify the filiform head. The rate of advancing growth generally increases with greater condensed salt concentration and higher acidity in the head. Filiform corrosion has been observed on steels coated with various lacquers and other slower-drying resins. Table 1 lists filiform growth rates for steel at 23 to 25 °C (73 to 77 °F), the physical dimensions of filaments, and the active range of relative humidity for different coating systems.

General Appearance. Filiform cells on coated steels consist of an active head, which is blue, blue-green, or grayish in color, and a brownish, rusty filament-shaped tail. Filament growth tends to be more random when the steel surface has no burnished texture or pronounced rolling direction. Typical filament growth rates are about 0.2 to 0.4 mm/d (8 to 16 mils/d), with an average

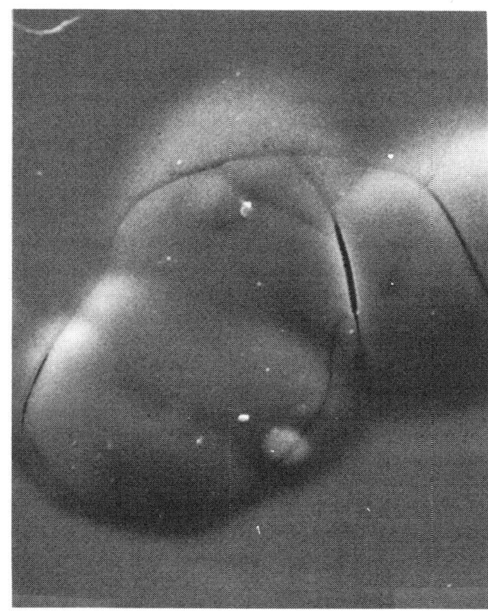

Fig. 3 Close-up of the advancing head shown in Fig. 2(a). Minute cracks can be seen at the head/tail interface of a filiform corrosion cell. These cracks are entry points for water and air to provide a source of hydroxyl ions and an electrolyte. Intermediate corrosion products are just beginning to form in the head, and they undergo further reaction to form an expanded tail. The tail region is a progressive reaction zone that ultimately forms spent corrosion products. Between the head and porous end, ions gradually react with water and oxygen and are slowly transported in the direction of the tail to form final corrosion products. SEM. 760×

filament width of 0.2 mm (8 mils). Filament heights vary with the type of coating and the corrosive environment. A moist environment of 80% relative humidity is typically needed at room temperature to initiate filiform attack. Warmer temperatures generally exacerbate the situation.

The solution in the advancing head of a filiform is usually acidic, with moisture and oxygen entering primarily through the long tail. The typical acidity in the filiform head is pH 1 to 4. Anions such as chloride, sulfide, or sulfate may enter the head from a corrosive atmosphere or by direct deposition on the surface and combine with the condensed water vapor that percolates through the cracked coating. Atmospheric corrosives may be supplemented by continual leaching of solvents or unreacted constituents of the coating that combine with water. The head is filled with ferrous ions and the tail with ferric hydroxides and hydrated iron oxides, depending on the longitudinal position in the tail and the age of the filaments. The advancing head of one filament is deflected or halted when it contacts a tail of another filiform, because a filiform tail is generally alkaline and filled with spent corrosion products. The head literally tunnels through the substrate, separating the coating from the steel and bulging it out by expansion of corrosion products or by hydrogen gas evolution if the head is very acidic (Ref 3). The constant forward motion of a filament will cease if the tail is not continually aerated and supplied with water vapor condensate.

Mechanism of Filiform Attack. Coated steels have numerous defects on their surfaces where air and water vapor condensate can penetrate through to the steel substrate. If the condensing atmosphere contains significant concentrations of salts, the condensate is more electrolytically conductive, and iron can more readily dissolve into solution. The activity of the head is initiated when its oxygen concentration or pH is considerably less than in the tail. Dissolved salts in the head solution can further decrease the solubility of oxygen. High salt concentrations can also lead to further acidification of the filiform head. In contrast, the tail is better aerated and receives a greater amount of freshly condensed water vapor. Voltage measurements between the head and tail regions normally indicate a potential difference of 0.1 to 0.2 V. The head advances as iron goes into solution; this undercuts the bond between the coating and the steel substrate.

The mechanism of coating expansion in the head region is not yet fully understood. Corrosion product expansion and the undercutting of the coating/steel interface in the head are major contributors to filament head bulging (Ref 4). Head and tail solution chemistry and temperature also affect the film strength of the coating and its bonding to the steel substrate. The head is a moving pool of acidic electrolyte. The trailing tail is a progressive zone of corrosion products and reactants that are gradually being converted to spent corrosion products. The end of the tail is where corrosion products have been completely reacted and fully expanded. The alkalinity of the tail region also stimulates coating cracking and debonding by softening and weakening the paint film.

If the filiform head solution is particularly acidic (pH 1 to 2), blistering can occur because of hydrogen evolution as a cathodic reaction:

$$2H^+ + 2e = H_2$$

The head or tail may bulge or blister considerably. If the bulged coating ruptures completely, a new initiation site is created, and this permits the formation of new filament heads.

The size of a filament is apparently governed by the original defect size and the flexibility of the film. The texture of the rolled surface and the nature of the coating also shape the geometry and the spread rate of filiform attack. The ferrous ions formed in the head are transported toward the tail and further oxidized in the first section of the tail by forming ferric hydroxide.

Water and oxygen can enter the cell at cracks in the coating in the tail region or at the head/tail interface. Water is either consumed by the head electrolyte or can be involved in cathodic reactions. Hydroxyl ions are provided by the oxygen reduction reaction, which takes place in the tail:

$$O_2 + 2H_2O + 4e = 4OH^-$$

There is a general transport of iron ions toward the tail region, where they combine with hydroxyl ions to form ferric hydroxides in the tail. Toward the end of the tail, ferric hydroxide decomposes to ferric oxide and water. Other iron hydrates may also form. Because the corrosion products of iron expand considerably, the coating bulges or splits. As the coating further cracks or splits, more oxygen and water can enter the tail, further stimulating the corrosion reactions. In some brittle coatings, the head may be very small, and cracks may form rapidly. A general diagram illustrating the mechanism of filiform corrosion in iron and steel is shown in Fig. 4(a).

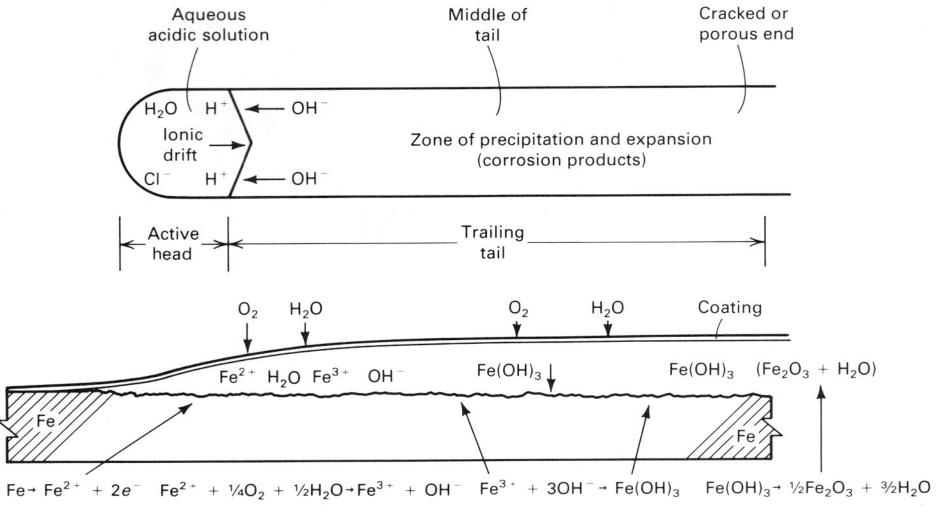

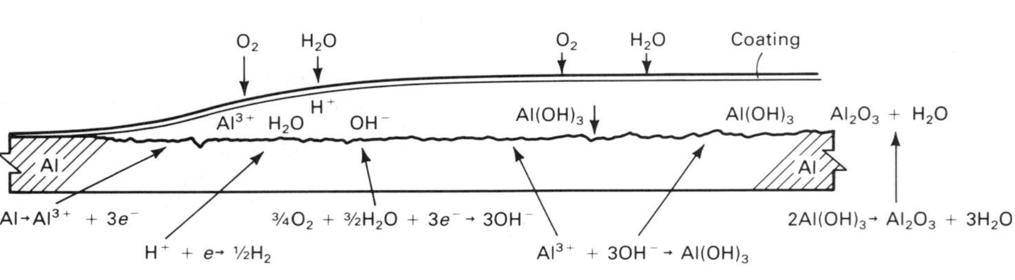

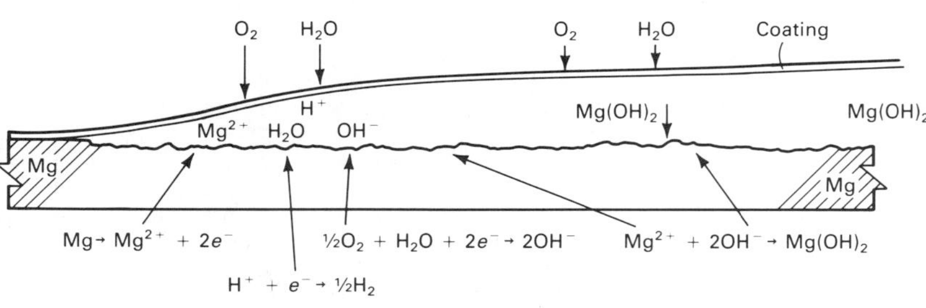

Fig. 4 Diagrams of the filiform corrosion cell in steel (a), aluminum (b), and magnesium (c). Corrosion products and predominant reactions are labeled. Filiform corrosion is a differential aeration cell driven by differences in oxygen concentration in the head versus the tail section. Potential differences between the head and tail are of the order of 0.1 to 0.2 V.

Water is required to form an electrolyte and to stimulate hydroxyl production. Water is necessary for the formation of new heads from a central defect source. The corrosivity of the head electrolyte is increased when salts are present on the surface or in the condensed water vapor. High salt concentrations in the head electrolyte decrease oxygen solubility and may further acidify the head solution. Several studies have shown that sealing the cracked tail halts filiform corrosion (Ref 4, 5, 6). By sealing the porous tail with epoxy, oxygen was prevented from entering. Similar effects were achieved by replacing oxygen with nitrogen gas. The driving force of the reaction—differential aeration—was stopped, and the filiform heads ceased to advance.

Filiform Corrosion of Coated Aluminum and Magnesium

Filiform corrosion is commonly observed on coated aluminum sheet, plate and foil, and magnesium sheet and plate. The appearance of filiform corrosion in aluminum and magnesium is similar to that of iron and steel, except that the corrosion products are gelatinous and milky in color. When dry, their filaments may take on an iridescent or clear appearance because of internal light reflection. Filament growth rates for aluminum are similar to those of coated steels. Magnesium has somewhat higher growth rates than aluminum. Filiform attack in both aluminum and magnesium is particularly severe in warm coastal and tropical regions that experience salt fall or in heavily polluted industrial areas.

Aluminum is susceptible to filiform corrosion in a relative humidity range of 75 to 95%, with temperatures between 20 to 40 °C (70 to 105 °F). Relative humidities as low as 30% in hydrochloric acid (HCl) vapors have been reported (Ref 6). Filiforms grow most rapidly at 85% relative humidity in aluminum. Typical filament growth rates average about 0.1 mm/d (4 mils/d). Filament width varies with increasing relative humidity—from 0.3 to 3 mm (12 to 140 mils). The depth of penetration in aluminum can be as deep as 15 μm (0.6 mil). Numerous coating systems used on aluminum are susceptible to filiform corrosion, including epoxy, polyurethane, alkyd, phenoxy, and vinyls. Condensates containing the chloride, bromide, sulfate, carbonate, and nitrate ions have stimulated filiform growth in coated aluminum alloys. Growth rates for filiform corrosion on aluminum and magnesium coated with lacquers and various slower drying resins are summarized in Table 1.

Mechanism of Filiform Attack. Filiform corrosion on aluminum and magnesium, as in iron and steel, is also a corrosion cell driven by differential aeration. The filiform cell consists of an active head and a tail that receives oxygen and condensed water vapor through cracks and splits in the applied coating. In aluminum, the head is filled with flowing flocs of opalescent alumina gel moving toward the tail. Gas bubbles may be present if the head is very acidic. In magnesium, the head appears blackish because of the etching of the magnesium, but the corrosive fluid is clear when the head is broken. Filiform tails in aluminum and magnesium filiforms are whitish in appearance. The corrosion products are hydroxides and oxides of aluminum and magnesium. Anodic reactions produce Al^{3+} or Mg^{2+} ions, which react to form insoluble precipitates with the hydroxyl ions produced in the oxygen reduction reaction occurring predominately in the tail.

The mechanism of initiation and activation in aluminum and magnesium are the same as for coated iron and steel, as shown in Fig. 4(b) and (c). The acidified head is a moving pool of electrolyte, but the tail is a region in which aluminum ion transport and gradual reaction with hydroxyl ions take place. The final corrosion products are partially hydrated and fully expanded in the porous tail. The head and middle sections of the tail are corresponding locations for the various initial reactant ions and the intermediate products of corroding aluminum in aqueous media.

However, in contrast to steel, aluminum and magnesium have shown a greater tendency to form blisters in acidic media, with hydrogen gas evolved in cathodic reactions in the head region. The corrosion products in the tail are either aluminum trihydroxide ($Al(OH)_3$), a whitish gelatinous precipitate, or magnesium hydroxide ($Mg(OH)_2$), a whitish precipitate.

Filiform Corrosion in Aircraft. Aircraft are routinely painted for corrosion protection, decreased drag resistance, and identification. Aircraft operating in warm, saline regions sustain considerable corrosion damage. In recent years, filiform corrosion was observed on 2024- and 7000-series aluminum alloys coated with polyurethane and other coatings (Ref 7, 8). Filiform corrosion increased in severity when chloride concentrations on the metal were high, particularly when the aircraft were frequently flying over ocean waters or based in coastal airfields and hangars. Prepainted surface treatment quality and the choice of primers were also influential. Two-coat polyurethane paint systems experienced far fewer incidences of filiform corrosion than single-coat systems did. Filiform corrosion rarely occurred when bare aluminum was chromic acid anodized or primed with chromate or chromate-phosphate conversion coatings. Rougher surfaces also experienced a greater severity of filiform attack. If left unchecked, filiform corrosion may lead to more serious structural damage caused by other forms of corrosion.

Filiform Corrosion in Packaging. Aluminum is widely used for cans and other types of packaging. Aluminum foil is routinely laminated to paperboard to form a moisture or vapor barrier. If the aluminum foil is consumed by filiform corrosion, the product may be contaminated, lost, or dried out because of breaks in the vapor barrier. Typical coatings on aluminum foil are nitrocellulose and polyvinylchloride (PVC), which provide a good intermediate layer for colorful printing inks.

Degradation of the foil-laminated paperboard may occur during its production or its subsequent storage in a moist or humid environment (Fig. 5). During the production of foil-laminated paperboard, moisture from the paperboard is released after heating in a continuous-curing oven. Heat curing dries the lacquer on the foil. Filiform corrosion can result as the heated laminate is cut into sheets and stacked on skids, while the board is still releasing stored moisture. As shown in Fig. 6, the hygroscopic paperboard is a good storage area for moisture. Packages later exposed to humidities above 75% in warm areas can also experience filiform attack. Coatings with water-reactive solvents, such as polyvinyl acetate, should not be used. Any solvents entrapped in the coating can weaken the coating, induce pores, or provide an acidic media for further filament propagation. Harsh curing environments can also result in the formation of flaws in the coating due to uneven shrinkage or rapid volatilization of the solvent. Rough handling can induce mechanical rips and tears.

Figure 7 shows typical flaws in PVC coating applied by a chromium-plated gravure. The tendency to follow flaws in the coated foil, such as hills and valleys or mechanical gouges in the

Table 1 Filiform corrosion growth rates on various coated metals

Coating	Initiating environment	Typical rate, mm/d	Typical rate, mils/d	Relative humidity, %	Filament width, mm	Filament width, mils
Steels						
Varnish	NaCl	0.33–0.53	13–21	65–85	0.1–0.3	4–12
	Acetic acid	0.5	20	86	0.15	6
Copol	NaCl	0.03	1.2	60–94	...	...
Lacquer	Acetic acid	0.85	33.5	...	0.1–0.5	4–20
Linseed oil	NaCl	0.04–0.08	1.6–3.1	...	0.05–0.1	2–4
Alkyds	NaCl	0.50	20	80	0.1–0.5	4–20
	Acetic acid	0.1	4	85	...	...
Alkyd urea	$FeCl_2$	0.26–0.43	10–17	80	0.25	10
Epoxy urea	NaCl/$FeCl_2$	0.01–0.46	0.4–18	80	0.25	10
Epoxy	Acetic acid	0.16	6.3	85	...	...
Acrylic	NaCl	0.19–0.86	7.5–34	80	0.25	10
	Acetic acid	0.1	4	85	...	...
Polyurethane	NaCl	0.16–0.50	6.3–20	90	0.1–0.3	4–12
	Acetic acid	0.09	3.5	85	...	...
Polyester	Acetic acid	0.08	3.1	85	...	...
Aluminum Alloys						
Alkyds	HCl vapor	0.1	4	85	0.5–1.0	20–40
Acrylic	HCl vapor	0.1	4	85	0.5–1.0	20–40
Polyurethane	HCl vapor	0.1	4	75–85	0.5–1.0	20–40
Polyester	HCl vapor	0.2	4	85	0.5–1.0	20–40
Epoxy	HCl vapor	0.09	3.5	85	0.5–1.0	20–40
Magnesium						
Alkyds	HCl vapor	0.2	8	75	...	...
Acrylic	HCl vapor	0.3	12	75	...	...
Polyurethane	HCl vapor	0.3	12	75	...	...
Polyester	HCl vapor	0.2	8	75	...	...
Epoxy	HCl vapor	0.3	12	75	...	...

Fig. 5 Penetration of the aluminum foil vapor barrier on laminated packaging. The interior of the package is back illuminated, showing the loss of aluminum foil to filiform attack. Light microscopy. 10×

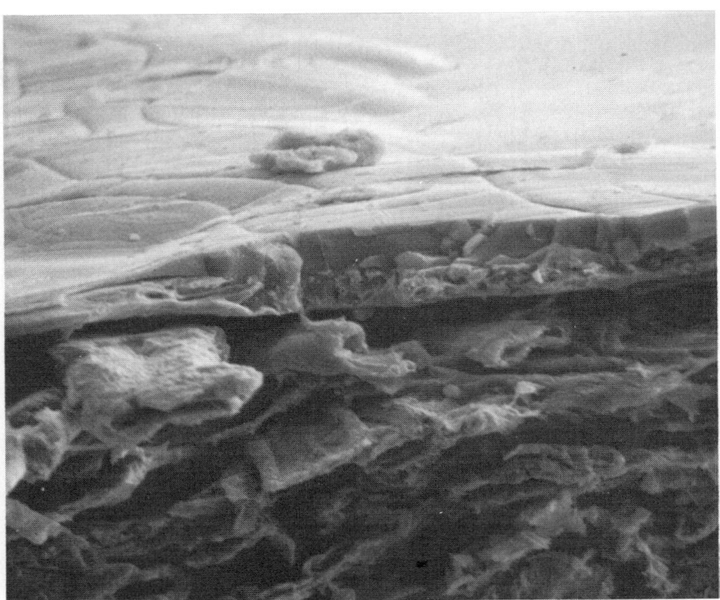

Fig. 6 Cross section of aluminum foil laminated on paperboard showing the expansion of the PVC coating by the corrosion products of filiform corrosion. Note the void spaces between the paperboard fibers that can entrap water. SEM. 650×

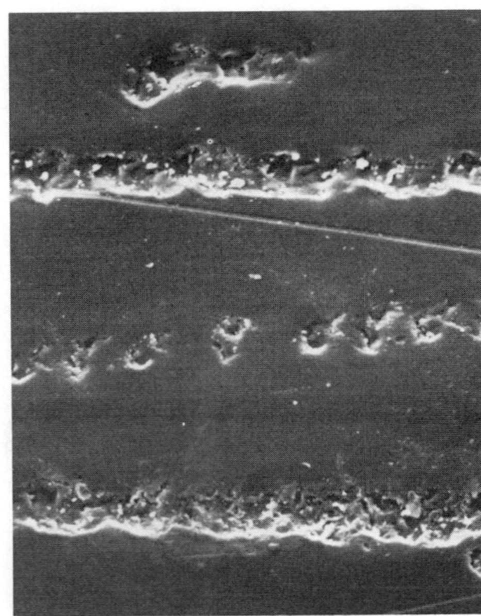

Fig. 7 Scratches in a nitrocellulose coating on aluminum induced by light abrasion. Hills and valleys in the foil are induced by a diamond-imprint gravure roll that applies the nitrocellulose as a lacquer. SEM. 200×

coating, is demonstrated in the filiforms on foil-coated paperboard observed by light microscopy (Fig. 8).

Prevention of Filiform Corrosion

Reduction in Relative Humidity Below 60%. Although the most direct method, reducing relative humidity is not practical if the metal is exposed to the elements or is in motion, as is the case for aircraft and automobiles. For articles in longer-term storage, controlled environments are beneficial—for example, the use of drying fans and humidistats or the addition of desiccants to plastic packaging or to small, confined areas. Structural designs can prevent moisture entrapment by better drainage or by attempting to exclude moisture entry. Reducing the ambient temperature can also be beneficial because it results in a decrease in the amount of moisture the air can hold.

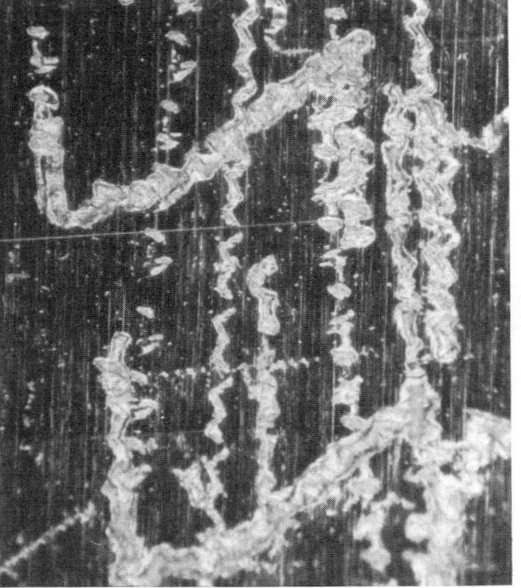

Fig. 8 Filiform in a nitrocellulose-coated aluminum foil laminated to paperboard showing a tendency to follow both the gravure imprint and the rolling direction of the sheet. Light microscopy. 75×

Use of Zinc and Zinc Primers on Steel. Filiform corrosion is reduced when the steel substrate is galvanized. Zinc chromate primers, chromic acid anodizing, and chromate or chromate-phosphate conversion coatings have provided some relief from filiform corrosion in aluminum alloys.

Multiple-Coat Paint Systems. Multiple coats on metal surfaces have fewer penetration and defect sites than single-coat paint systems. Multiple-coat systems resist penetration by mechanical abrasion and have fewer hills and valleys. Thicker coatings achieved by layer buildup have demonstrated substantially better resistance to filiform corrosion by decreasing oxygen and moisture penetration, decreased solvent entrapment, and fewer initiation sites. Smooth, well-prepared primed metal surfaces generally have better resistance than rougher surfaces.

Use of Less Active Metal Substrates. Steel, aluminum, and magnesium are all chemically active. The substitution of more resistant materials exposed to the initiating environment, such as copper, austenitic stainless steel, or titanium, may be necessary. Unless a coating is mandatory, a material substitution may eliminate the central part of the problem. Economics and structural considerations, however, will usually dictate whether a material substitution is a practical solution.

Crevice Corrosion

R.M. Kain
LaQue Center for
Corrosion Technology, Inc.

The presence of narrow openings or spaces (gaps) between metal-to-metal or nonmetal-to-metal components may give rise to localized corrosion at these sites. Similarly, unintentional crevices such as cracks, seams, and other metallurgical defects could serve as sites for corrosion initiation. Resistance to crevice corrosion can vary from one alloy-environment system to another. Passive alloys, particularly those in the

Fig. 9 Crevice corrosion at a metal-to-metal crevice site formed between components of type 304 stainless steel fastener in seawater

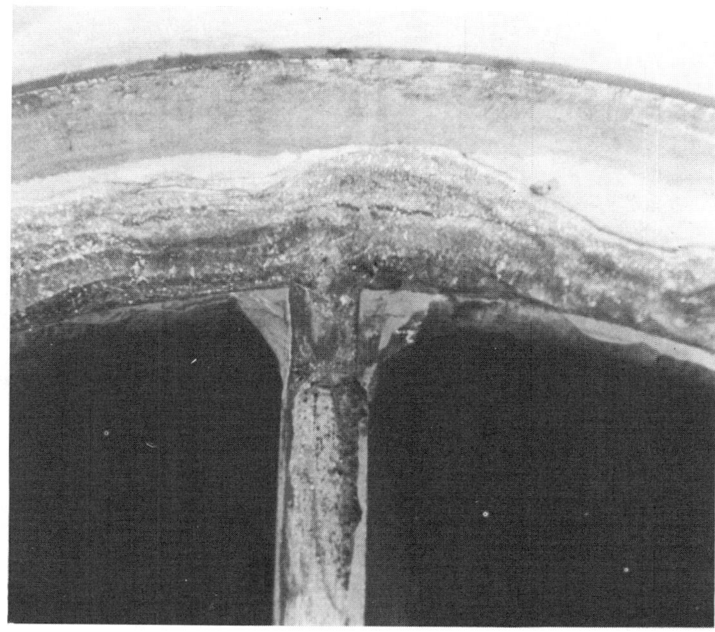

Fig. 10 Crevice corrosion at nonmetal gasket site on an alloy 825 seawater heat exchanger

(a)

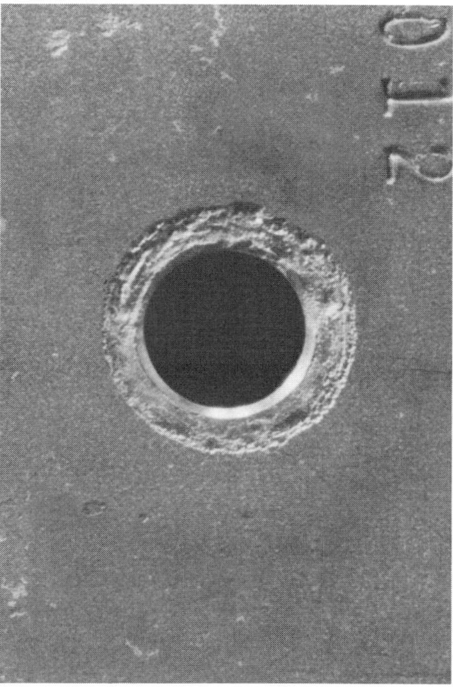

(b)

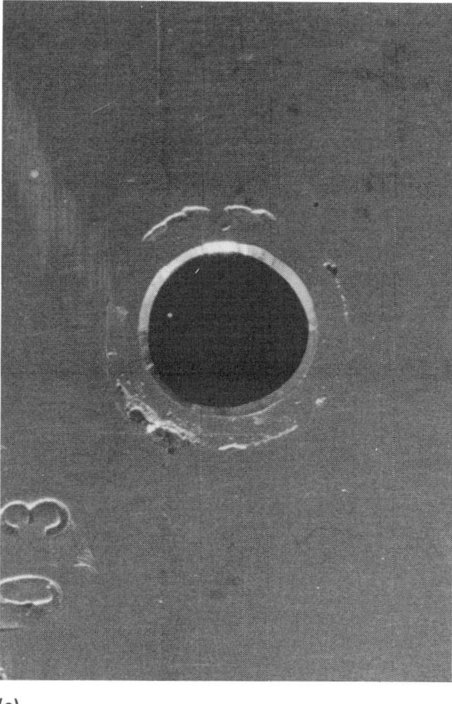

(c)

Fig. 11 Variation in stainless steel corrosion resistance in model SO_2 scrubber environments. (a) Type 304 in acid condensate. (b) Type 316 in acid condensate. (c) Type 304 in limestone slurry zone. Source: Ref 9

stainless steel group, are more prone to crevice corrosion than materials that exhibit more active behavior. Figure 9, for example, shows crevice corrosion of a type 304 stainless steel fastener removed from a seawater jetty after 8 years. Although the washer shows severe deterioration, the function of the fastener was not diminished. On the other hand, Fig. 10 shows crevice corrosion beneath the water box gasket of an alloy 825 (44Ni-22Cr-3Mo-2Cu) seawater heat exchanger that allowed sufficient leakage to warrant shutdown and replacement after only 6 months.

In cases in which the bulk environment is particularly aggressive, general corrosion may preclude localized corrosion at a crevice site. Figure 11 compares the behavior of type 304 and type 316 stainless steels exposed in different zones of a model sulfur dioxide (SO_2) scrubber. In the aggressive acid condensate zone, type 304 incurred severe general corrosion of the exposed surfaces, while the more resistant type 316 suffered attack beneath a polytetrafluoroethylene (PTFE) insulating spacer. In the higher pH environment of the limestone slurry zone, type 304 was resistant to

(a)

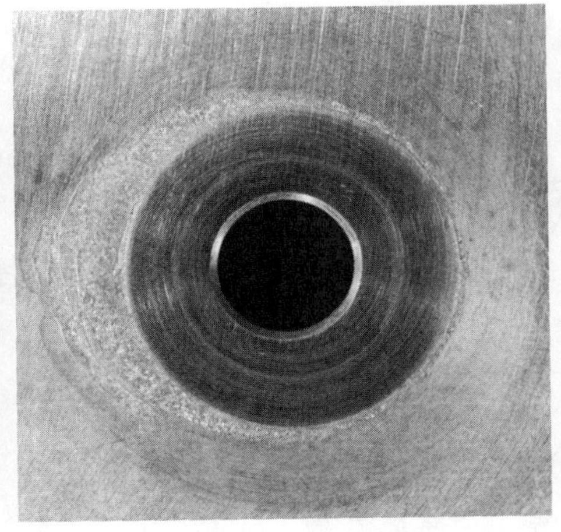

(b)

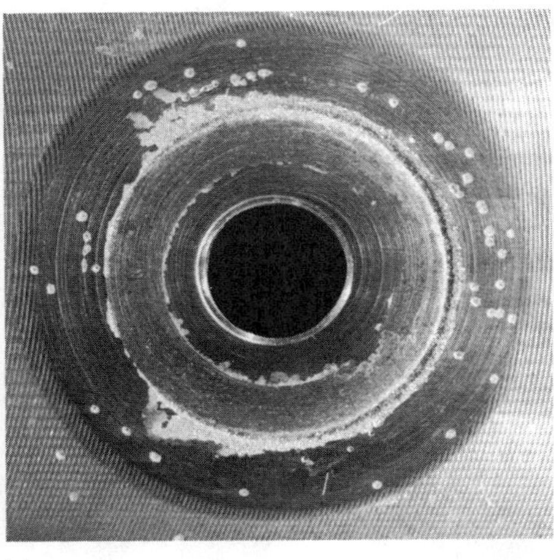

(c)

Fig. 12 Crevice-related corrosion for different alloys in natural seawater. (a) Alloy 904L (20Cr-25Ni-4.5Mo-1.5Cu) after 30 days. (b) 70Cu-30Ni after 180 days. (c) Alloy 400 (70Ni-30Cu) after 45 days

general corrosion, but was susceptible to crevice corrosion. Other alloy systems, such as aluminum and titanium, may also be susceptible to crevice corrosion (Ref 10). For aluminum, the occurrence of crevice corrosion would depend on the passivity of the particular alloy. In most cases, general corrosion would likely preclude crevice corrosion. Titanium alloys are typically quite resistant, but may be susceptible to crevice corrosion in elevated-temperature chloride-containing acidic environments.

In seawater, localized corrosion of copper and its alloys at crevices is different from that of stainless-type materials because the attack occurs outside of the crevice rather than within. In general, the degree of crevice-related attack increases as the resistance to general corrosion increases. Figure 12 compares the crevice corrosion behavior for several different materials exposed to ambient-temperature seawater for various periods. In each case, a nonmetallic washer created the crevice. The more classical form of crevice corrosion (that is, beneath the crevice former) is shown for type 904L stainless steel (20Cr-25Ni-4.5Mo-1.5Cu) after only 30 days of exposure (Fig. 12a). For 70Cu-30Ni, corrosion occurred just outside of the crevice mouth and was found to be quite shallow after 6 months (Fig. 12b). In contrast, crevice-related corrosion of alloy 400 (70Ni-30Cu) was more severe after only 45 days (Fig. 12c). In some cases, corrosion may occur within as well as outside of the crevice.

Mechanisms

Regardless of the material, a condition common to all types of crevice corrosion is the development of localized environments that may differ greatly from the bulk environment. In perhaps its simplest form, crevice corrosion may result from the establishment of oxygen differential cells. This can occur when oxygen within the crevice electrolyte is consumed, while the boldly exposed surface has ready access to oxygen and becomes cathodic relative to the crevice area. Crevice corrosion, for example, can be encountered by stainless-type alloys in some concentrations of sulfuric acid. Although the passivity of the exposed surfaces is maintained by dissolved oxygen in the acid, the presence of a crevice (for example, a gasket or O-ring seal) excludes oxygen and corrosion ensues in the active state.

Fig. 13 Crevice corrosion at tube/tubesheet interface after 3 months of exposure in a natural seawater test

Crevice corrosion in neutral chloride containing environments, such as natural waters and acid-chloride media, is more complex than the preceding example given for acid. It does, however, begin with the deoxygenation stage.

For stainless steels, numerous interrelated metallurgical, geometrical, and environmental factors affect both crevice corrosion initiation and propagation (Ref 11). A number of these factors are indicated in Table 2, and a more thorough discussion on the mechanism can be found in Ref 12. In short, however, the release of metal ions, particularly chromium, in the crevice produces an acidic condition as a result of a series of hydrolysis reactions. To effect charge neutrality with excess H^+ ions, Cl^- ions migrate and concentrate from the bulk environment. If the concentration of acid and chloride in the crevice solution becomes sufficiently aggressive to cause breakdown of the passive film, crevice corrosion initiation occurs. Although natural seawater, for example, is typically pH 8 and contains about 0.5 M Cl^-, crevice solutions may attain a pH of 1 or less and contain 5 to 6 M Cl^- (saturated) (Ref 12).

Table 2 Factors that can affect the crevice corrosion resistance of stainless steels

Geometrical
Type of crevice:
metal to metal
nonmetal to metal
Crevice gap (tightness)
Crevice depth
Exterior to interior surface area ratio
Environmental
Bulk solution:
O_2 content
pH
chloride level
temperature
agitation
Mass transport, migration
Diffusion and convection
Crevice solution: hydrolysis equilibria
Biological influences
Electrochemical reactions
Metal dissolution
O_2 reduction
H_2 evolution
Metallurgical
Alloy composition:
major elements
minor elements
impurities
Passive film characteristics

Crevice corrosion of the copper-containing alloys, as shown in Fig. 12, is frequently identified as metal ion concentration cell corrosion. A number of years ago, it was proposed that the concentration of metal ions in the crevice electrolyte rendered the crevice area cathodic to the area immediately outside the mouth of the crevice (Ref 13, 14). Corrosion outside of the crevice (anode) progressed because the bulk environment contained a much lower concentration of metal ions. In some cases, this has been supported by the observation of plated-out copper within the crevice. Other researchers have refuted this premise, suggesting that the mode of attack is merely a variation on the oxygen differential cell mechanism (Ref 15). In any event, the morphol-

(a)

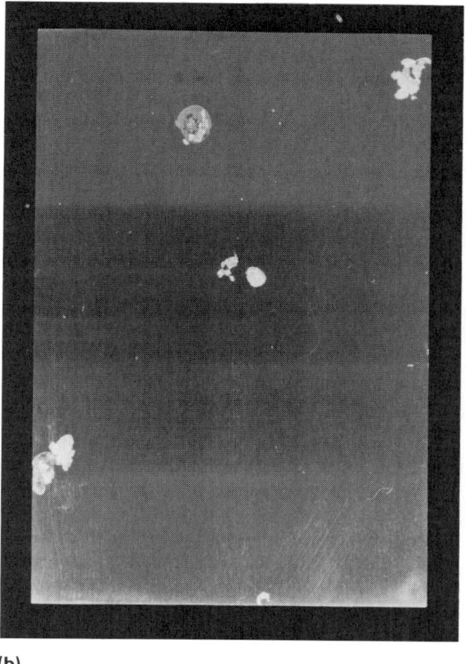

(b)

(c)

Fig. 14 Incidence of barnacle-related crevice corrosion for three nickel-base alloys after 1-year tidal zone exposures. Susceptible materials, alloy 600 (a) and alloy 690 (b), do not contain molybdenum. Resistant material, alloy 625 (c), contains about 9% Mo. Source: Ref 18

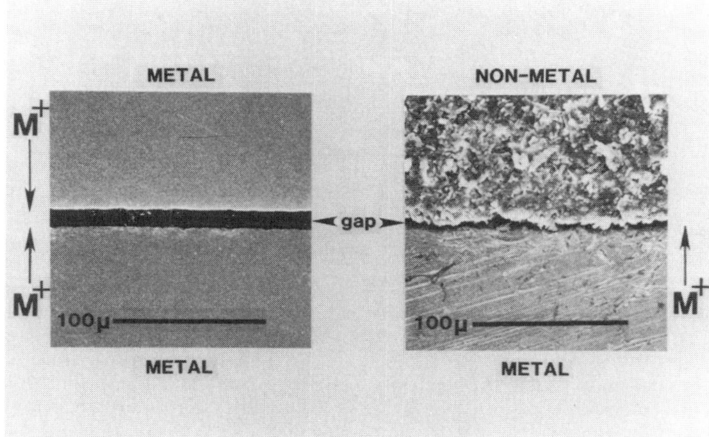

Fig. 15 Examples of crevice gaps attainable for metal-to-metal and nonmetal-to-metal crevices

Fig. 16 Example of shallow crevice corrosion incurred by type 316 stainless steel at lap joints after a 10-year exposure in aggressive marine atmosphere. Source: Ref 22

ogy of crevice-related attack for copper alloys is distinctly different from that for stainless steels and can be recognized accordingly.

Types of Crevices

Crevices fall into two categories: man-made and naturally occurring. The former may be unavoidable and may serve a particular design purpose, such as the fastener and gasket examples shown in Fig. 9 and 10. Other man-made crevices may result during fabrication or assembly. Some of these may be avoidable after consideration of the consequences of crevice corrosion. For example, Fig. 13 shows crevice corrosion at a stainless steel tube and tubesheet test assembly after only 3 months of exposure to seawater. In an actual service situation, this occurrence could be prevented by weld sealing the joints and/or by applying cathodic protection. Sealants, coatings, and greases can also promote localized corrosion by forming sites at the interface with some susceptible materials (Ref 16).

Naturally occurring crevices, such as those formed by debris, sand, and, in the case of marine applications, the attachment of barnacles or other biofouling organisms, may be a problem for some materials. It is well documented, for example, that type 304 and type 316 may incur crevice corrosion beneath barnacles (Ref 17). This may be a limiting factor in the use of these and other stainless alloys for condenser tube service if the surfaces are not kept free of fouling. Even higher-alloyed material may incur this attack. In a 1-year tidal zone test, nickel-base alloys 600 (15Cr-8Fe) and 690 (29Cr-8Fe), which have excellent resistance to chloride stress-corrosion cracking, incurred shallow crevice penetrations beneath barnacles (Ref 18). Figure 14 compares the resistance of these molybdenum-free alloys with the high molybdenum containing nickel-base alloy 625 (22Cr-4.5Fe-9Mo), which was totally resistant. Stainless steels with over 4 to 5% Mo may also be resistant to barnacle-related crevice corrosion.

Crevice Geometry

To gain a greater appreciation of crevice corrosion, one must recognize the importance of crevice geometry because it is frequently the

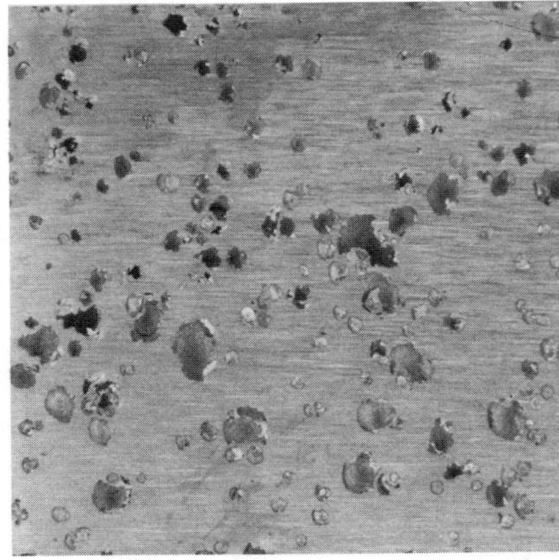

(a)

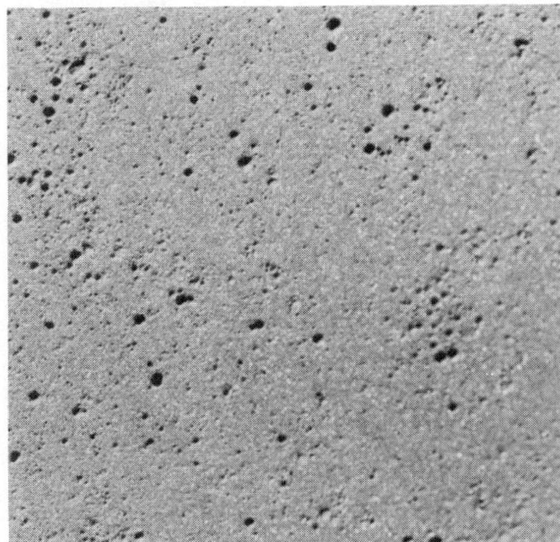

(b)

Fig. 17 Examples of deep pits (a) and shallow pits (b)

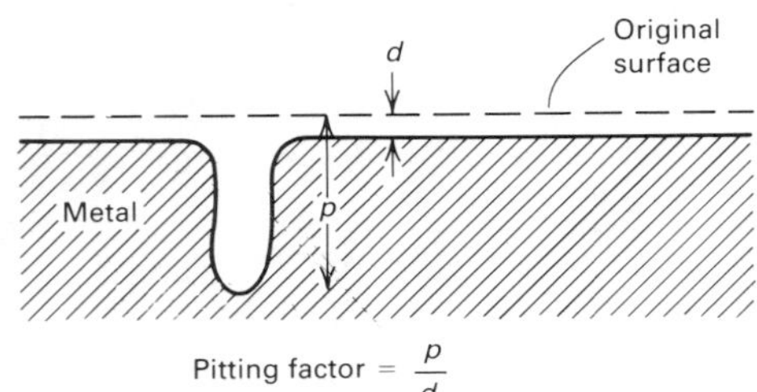

Fig. 18 Schematic illustrating the pitting factor *p/d*

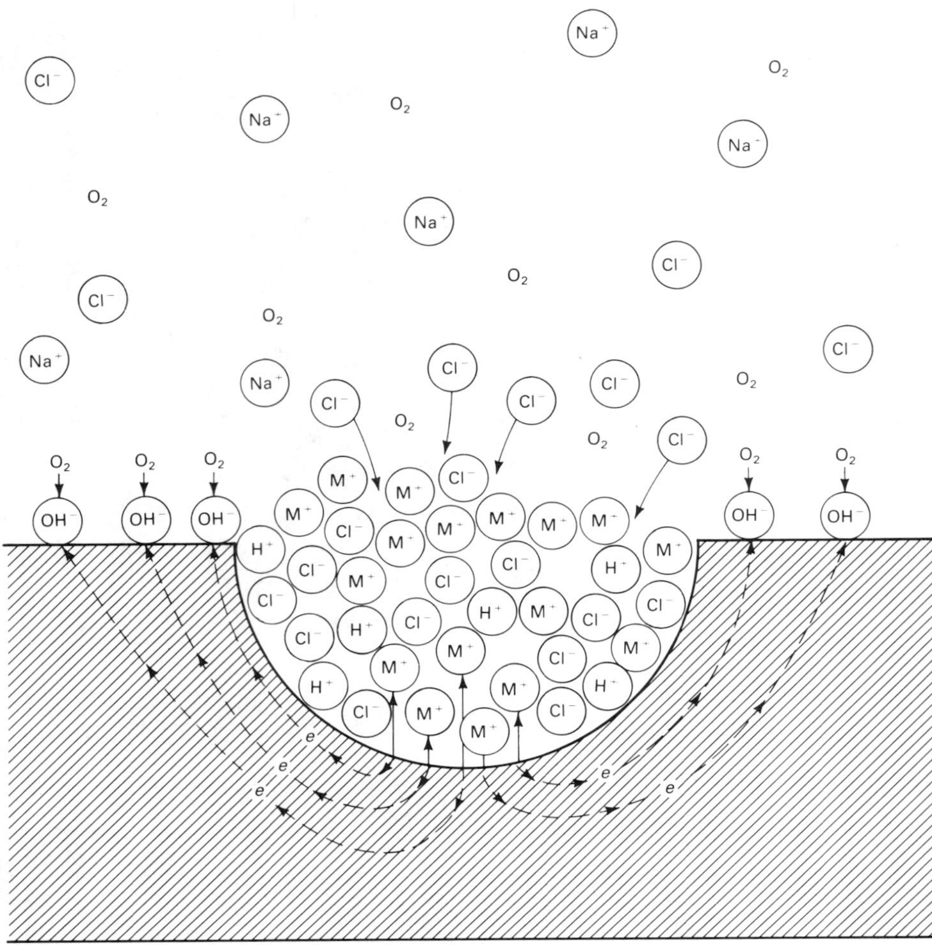

Fig. 19 Autocatalytic processes occurring in a corrosion pit. The metal, M, is being pitted by an aerated NaCl solution. Rapid dissolution occurs within the pit, while oxygen reduction takes place on the adjacent surfaces. A more detailed explanation of this self-sustaining process is given in Ref 34.

controlling factor governing resistance to corrosion in a particular situation. The occurrence or absence of crevice corrosion for a given alloy-environment system may or may not be indicative of performance under other conditions in which more severe crevices exist.

Crevices can be defined by the dimension of the gap (degree of tightness) and by the depth (distance from the mouth). For stainless steels in chloride-containing environments, tighter crevices reduce the volume of crevice electrolyte that must be deoxygenated and acidified and will generally cause more rapid initiation of attack. Typically, tighter crevices can be achieved between nonmetal and metal components than between two metal components. However, if gaps of equal dimension can be produced with metal-to-metal components, corrosion initiation can be more rapid because of metal ion production from both surfaces. This is illustrated in Fig. 15, which contains micrographs of the actual gap for crevices assembled under laboratory conditions. Dimensions of the order shown may or may not reflect those attainable in actual service. Variations in crevice geometry from one application to another can frequently account for variability in alloy performance.

Increasing the crevice tightness, depth, bulk environment chloride levels, and acidity increases the chances for crevice initiation (Ref 19, 20). In some cases, however, deep crevices may restrict propagation because of a voltage drop through the crevice solution (Ref 11). The use of higher-alloy stainless-type material offers the advantage of providing resistance over a broader range of conditions.

Other Environments

In addition to seawater, stainless steels may be susceptible to crevice corrosion in other chloride-containing natural waters. The occurrence of crevice corrosion is less likely as bulk chloride levels decrease. For a molybdenum-free grade of stainless steel, there may be no safe limit for chlorides. Laboratory tests have shown that crevice initiation can occur for type 304 in as little as 100 mg/L (ppm) chloride (Ref 21).

Crevice corrosion may also occur in severe marine atmospheres. In contrast to seawater immersion, however, the extent of attack is quite minimal. Figure 16 shows very shallow penetration found at lap joints of test rack fabricated of type 316 stainless steel after a 10-years exposure in the 25-m (80-ft) test site at Kure Beach, NC. In coastal architectural applications, rust staining from crevices may detract from the appearance of stainless steel if it is not periodically cleaned.

Prevention of Crevice Corrosion

Many of the factors noted in Table 2 must be considered if crevice corrosion is to be eliminated or minimized (Ref 23). Wherever possible, crevices should be eliminated at the design stage. When unavoidable, they should be kept as open and shallow as possible to allow continued entry of the bulk environment. Cleanliness is an important factor, particularly when conditions promote deposition on metal surfaces.

In critical areas, weld overlays with more corrosion-resistant alloys have proved to be effective measures in some cases (Ref 24). Similarly, the use of cathodic protection can be beneficial (Ref 25). The consequences of overprotection, however, should be recognized for those alloys that may be susceptible to hydriding and embrittlement (Ref 16).

Pitting Corrosion

A.I. Asphahani and W.L. Silence
Haynes International, Inc.

Pitting corrosion represents an important limitation to the safe and reliable use of many alloys in various industries. Pitting is a very serious type of corrosion damage because of the rapidity with which metallic sections might be perforated. The unanticipated occurrence of pitting and its unpredictable propagation rate make it difficult to take it into consideration in practical engineering designs.

Deterioration by pitting is one of the most dangerous and most common types of localized corrosion encountered in aqueous environments (Ref 26, 27). In the chemical-processing industries, localized corrosion is a major cause of repeated service failures (Ref 28, 29) and is estimated to account for at least 90% of metal damage by corrosion (Ref 30).

Pitting corrosion is defined as an extremely localized corrosive attack. Simply stated, pitting is the type of localized corrosion that produces pits, that is, sites of corrosive attack that are relatively small compared to the overall exposed surface. If appreciable attack is confined to a relatively small fixed area of metal acting as an anode, the resultant pits are described as deep (Fig. 17a). If the area of attack is relatively larger and not so deep, the pits are called shallow (Fig. 17b). Depth of pitting is sometimes expressed by the term pitting factor (Ref 31). This is the ratio of deepest metal penetration to average metal penetration as determined by the weight loss of the specimen (Fig. 18). A pitting factor of unity represents uniform, or general, corrosion.

Fig. 20 Two views of deep pits in a type 316 stainless steel centrifuge head due to exposure to $CaCl_2$ solution. Courtesy of N.W. Sachs, Allied Corporation

Mechanisms and Theories

It is stated that pits begin by the breakdown of passivity at favored nuclei on the metal surface (Ref 32). The breakdown is followed by the formation of an electrolytic cell. The anode of this cell is a minute area of active metal, and the cathode is a considerable area of passive metal. The large potential difference characteristic of this passive-active cell (for example, approximately 0.5 V for 300-series stainless steel) accounts for considerable flow of current with rapid corrosion at the small anode. The corrosion-resistant passive metal surrounding the anode and the activating property of the corrosion products within the pit account for the tendency of corrosion to penetrate the metal rather than spread along the surface.

Once pits are initiated, they may continue to grow by a self-sustaining, or autocatalytic, process; that is, the corrosion processes within a pit produce conditions that are both stimulating and necessary for the continuing activity of the pit (Ref 33). This process is illustrated schematically in Fig. 19. Pit growth is controlled by rate of depolarization at the cathode areas. In seawater, control is exercised by the amount and availability of dissolved oxygen. Ferric chloride, a more effective depolarizer than dissolved oxygen, cause greater numbers of pits and more rapid penetration in various industrial services.

The propagation of pits is thought to involve the dissolution of metal and the maintenance of a high degree of acidity at the bottom of the pit by the hydrolysis of the dissolved metal ions. The anodic metal dissolution reaction at the bottom of the pit ($M \rightarrow M^{n+} + ne$) is balanced by the cathodic reaction on the adjacent surface ($O_2 + 2H_2O + 4e \rightarrow 4OH^-$). The increased concentration of M^{n+} within the pit results in the migration of chloride ions (Cl^-) to maintain neutrality. The metal chloride formed, M^+Cl-, is hydrolyzed by water to the hydroxide and free acid ($M^+Cl^- + H_2O \rightarrow MOH + H^+Cl^-$). The generation of this acid lowers the pH values at the bottom of the pit (pH approximately 1.5 to 1.0), while the pH of the bulk solution remains neutral.

Most of the research focuses on understanding the factors that control pit initiation. Various theories have been proposed that attempt to explain the pitting phenomenon (Ref 35). These include kinetic theories, which explain the break-

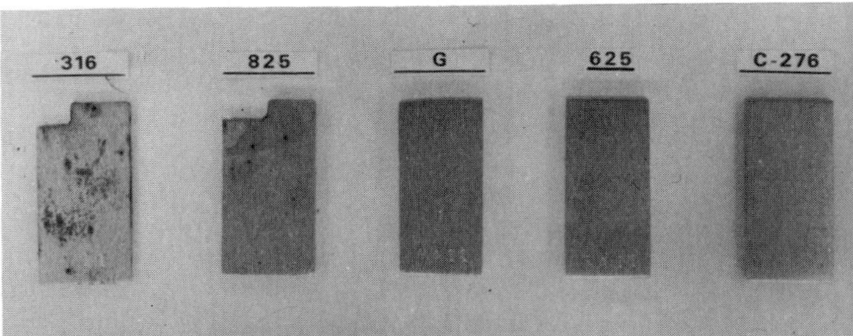

Fig. 21 Results of a 24-h test on five different alloys in 7 vol% H_2SO_4 + 3 vol% HCl + 1% $FeCl_3$ + 1% $CuCl_2$ solution at 25 °C (77 °F)

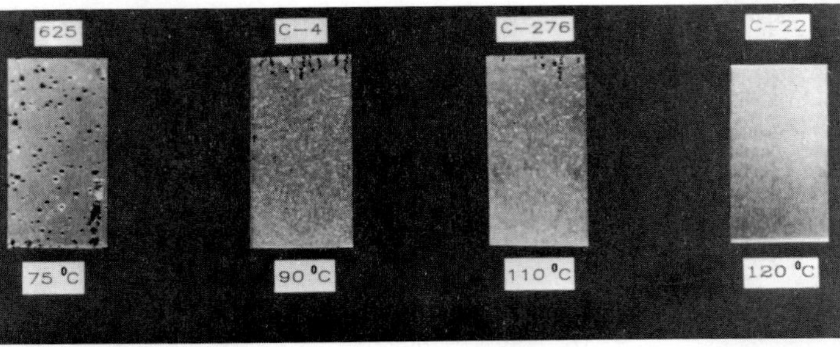

Fig. 23 Critical temperature above which pitting is observed for four different alloys. 24-h test in 7 vol% H_2SO_4 + 3 vol% HCl + 1% $FeCl_3$ + 1% $CuCl_2$ solution. Alloy C-4: Ni-16Cr-16Mo. Alloy C-22: Ni-22Cr-13Mo-3W. Remaining alloy compositions are given in text.

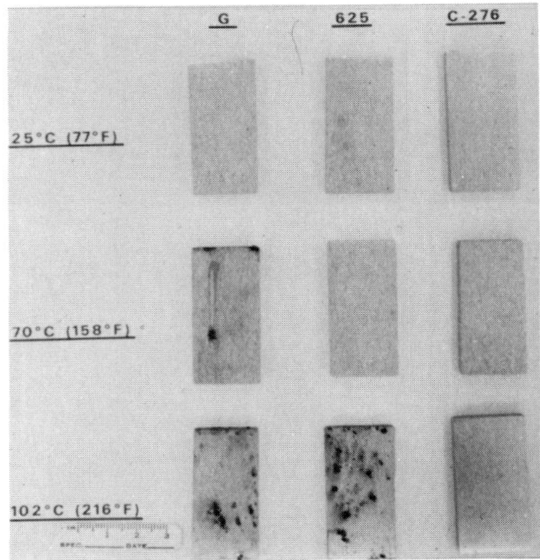

Fig. 22 Effect of temperature on pitting resistance of three different alloys. 24-h test in 7 vol% H_2SO_4 + 3 vol% HCl + 1% $FeCl_2$ + 1% $CuCl_2$ solution

down of passivity in terms of the competitive absorption between chloride ions and oxygen, and thermodynamic theories, which consider the pitting potential as that potential at which the chloride ion is in equilibrium with the oxide film.

Occurrence and Testing

Pitting corrosion occurs in most commonly used metals and alloys. Iron buried in the soil corrodes with the formation of shallow pits, but stainless steels immersed in seawater characteristically corrode with the formation of deep pits. Aluminum tends to pit in waters containing Cl^- (for example, at stagnant areas), and aluminum brasses are subject to pitting attack in polluted waters.

In environments containing appreciable concentrations of Cl^- or Br^-, most stainless materials (for example, iron-base, nickel-base, cobalt-base, and titanium alloys) tend to corrode at specific areas and tend to form deep pits. Figure 20 shows deep pits that formed in a type 316 stainless steel centrifuge head from a calcium chloride ($CaCl_2$) solution. Ions such as thiosulfate ($S_2O_3^{2-}$) may also induce pitting of stainless steels. In the absence of passivity, such as in deaerated alkali-metal chlorides and nonoxidizing metal chlorides (for example, stannous chloride, $SnCl_2$, or nickel chloride $NiCl_2$), pitting does not occur, although general, or uniform, corrosion may be appreciable in such environments.

The selection of materials for resistance to pitting corrosion in specific industrial services is better made when based on field testing and actual service performance. However, such empirical information is often generated under ill defined environmental conditions and is dependent on many uncontrollable factors that do not allow a precise comparison among the potential candidate alloys or metals. Accelerated laboratory tests might be useful in this respect. Such tests include the classic immersion tests (freely corroding conditions) and the advanced electrochemical techniques.

The immersion tests are very practical in providing decisive, direct comparison among various alloys, and these tests correlate very well with service experience. Figure 21 clearly distinguishes the improved pitting resistance of alloys G (Ni-20Fe-22Cr-6Mo), 625 (Ni-22Cr-9Mo), and C-276 (Ni-16Cr-16Mo-4W) from that of alloy 825 (Ni-30Fe-21Cr-3Mo) and type 316 stainless steel. Also, immersion tests allow emphasis on the effect of temperature on pitting resistance (Fig. 22) and allow establishment of the critical pitting temperature, that is, the temperature above which pitting occurs in a specific environment (Fig. 23). The higher the critical pitting temperature is (for example > 120 °C, or 250 °F, for alloy C-22), the more resistant the alloy is to pitting corrosion.

The electrochemical techniques used to investigate pitting corrosion include the simple monitoring of corrosion potential and its fluctuation with time due to pitting attack, potentiostatic tests, potentiodynamic tests (at relatively slow scan rates of a few millivolts per minute), rapid scan potentiodynamic techniques, potentiostatic induction time studies, potentiostatic scratch techniques, galvanostatic polarization techniques (or anodic pitting tests), constant potential techniques (or potentiostatic point-to-point methods), and cyclic polarization (or anodic hysteresis loop techniques). Reliable, comparative data can be generated with the electrochemical testing techniques to assess the effects of potential, temperature, pH, chloride ion concentration, heat treatment, and alloying elements (Ref 36). The potentiostatic polarization method is judged the most practical in generating data relevant to actual service performance (Ref 36, 37). More detailed information on testing and evaluation of pitting corrosion can be found in the articles "Laboratory Testing" and "Evaluation of Pitting Corrosion" in this Volume.

Prevention of Pitting Corrosion

Typical approaches to alleviating or minimizing pitting corrosion find their roots in the following major principles:

- Reduce the aggressivity of the environment—for example, chloride ions concentration, temperature, acidity, and oxidizing agents
- Upgrade the materials of construction—for example, addition of molybdenum/tungsten, overalloying of welds, and lining with high alloy
- Modify the design of the system—for example, avoid crevices, circulate/stir to eliminate stagnant solutions, and ensure proper drainage

In addition, based on the concept of critical potential to induce pitting corrosion, cathodic protection would be a good solution if design allows (Ref 38). Also, addition of extraneous ions (for example, OH^- or NO_3^-) would augment resistance to pit formation.

The applicability of any of the above mentioned remedial measures, among others, is dependent on the specific situation. A review and analysis of successful case histories of various remedial measures are recommended.

Localized Biological Corrosion

Stephen C. Dexter
College of Marine Studies
University of Delaware

Most of the documented cases in which biological organisms are the sole cause of, or an accelerating factor in, corrosion involve localized forms of attack. One reason for this is that organisms often do not form in a continuous film on the metal surface.

The large fouling organisms in marine environments settle as individuals, and it is often a period of months or even years before a complete cover is built up. A scatter of individual barnacles on a stainless steel surface will create oxygen concentration cells. The portion of the metal surface covered by the barnacle shell is shielded from dissolved oxygen in the water and thus becomes the anode through mechanisms described in the Section "Fundamentals of Corrosion" in this

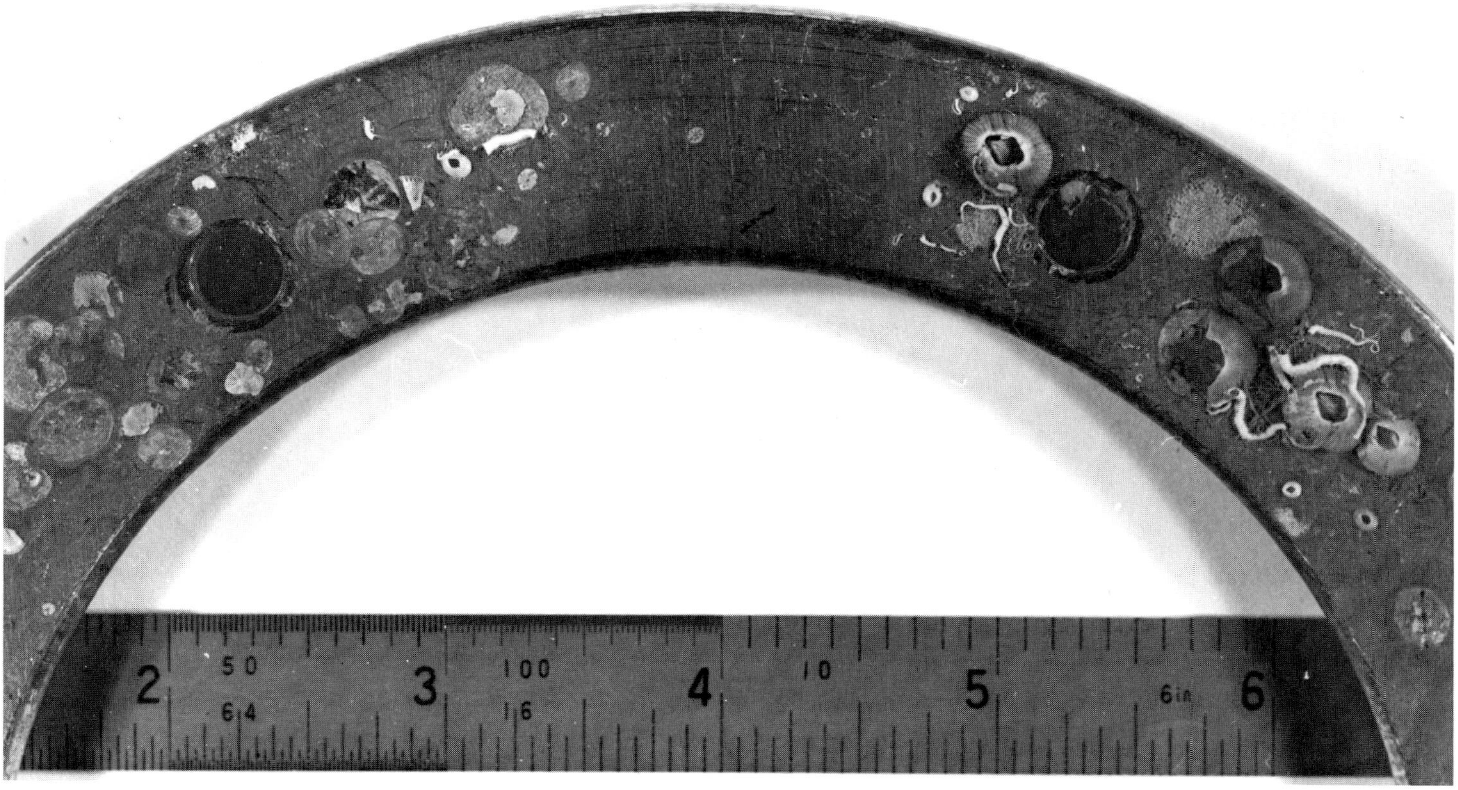

Fig. 24 Corrosion under a scatter of individual barnacles on a stainless steel surface

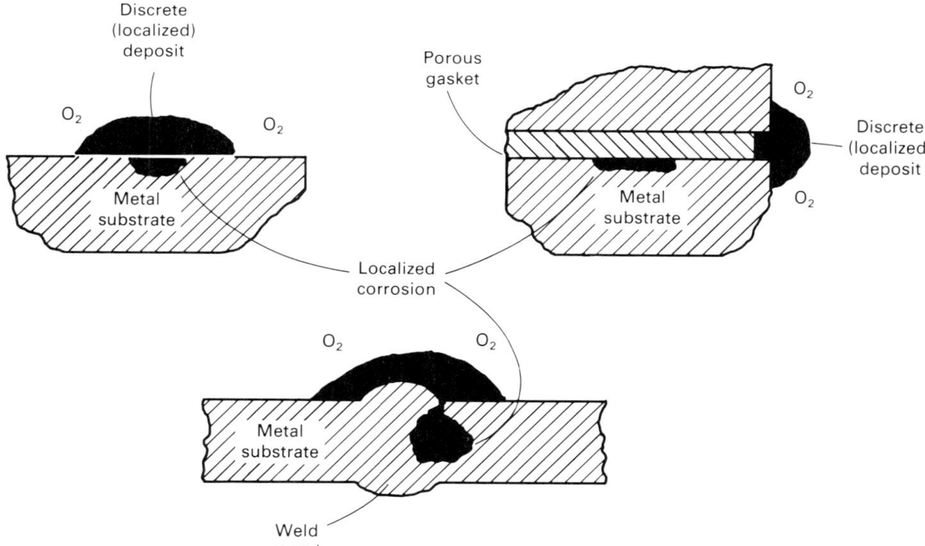

Fig. 25 The three most common forms of microbial corrosion. Source: Ref 39

Volume. The result is crevice corrosion under the base of the barnacles, as shown in Fig. 24. A similar effect can take place on aluminum and, to a lesser extent, on structural steel.

Microscopic organisms also tend to settle on metal surfaces in the form of discrete colonies or at least spotty, rather than continuous, films. Those few cases in which continuous films do form have been dealt with in the preceding article "General Corrosion" in this Volume. In the remainder of this section, the localized forms of attack influenced by microorganisms, as shown schematically in Fig. 25, will be discussed.

This section will catalog the industries most often reported as being affected by microbiological corrosion and the organisms usually implicated in the attack, and will illustrate the types of attack that have most commonly been documented by presenting generalized case histories for different classes of alloys. Cases that are still in the research stage or cases in which evidence for a primary role of the biofilm is not convincing will not be covered.

Further information on the mechanisms of localized biological corrosion can be found in the article "Effects of Environmental Variables on Aqueous Corrosion" (see the Appendix "Biological Effects") in this Volume. Additional information on the detection and control of biological corrosion can be found in the articles "Evaluation of Microbiological Corrosion" and "Control of Environmental Variables in Water Recirculating Systems," respectively, in this Volume.

Industries and Organisms Involved

The various industries that have been affected by microbiological corrosion problems are listed in Table 3. References to problems in the chemical-processing, nuclear power, oil field (both onshore and offshore), and underground pipeline industries are most common in the published literature (see also the articles on these industries in the Section "Specific Industries and Environments" in this Volume). This does not necessarily mean that these industries are more prone to problems of this type than other industries, but it does mean that these industries have been quick to recognize the biological causes of some of their corrosion problems, to take steps to solve these problems, and to make the results public.

(a) (b)

Fig. 26 Anaerobic biological corrosion of cast iron. (a) Cast iron pipe section exhibiting external pitting caused by bacteria. (b) Cast iron pipe showing penetration by bacteria-induced pitting corrosion. Source: Ref 42

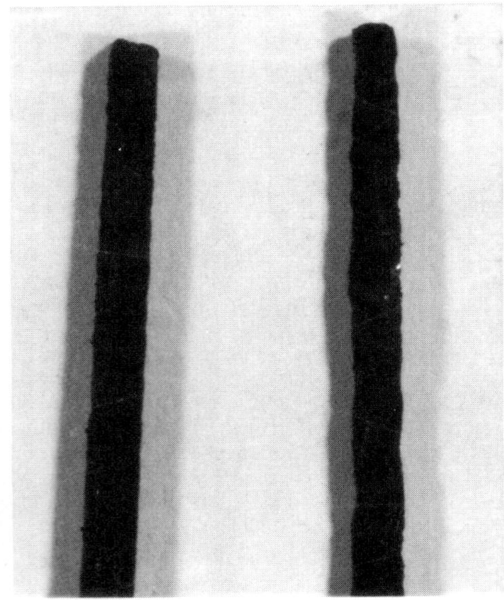

Fig. 27 Carbon steel wires from a prestressing tendon of a nuclear power plant showing the damage resulting from the formation of organic acids in the tendon due to the breakdown of grease by the bacteria present in the tendon. Source: Ref 46

The frequency of literature reports of microbiological corrosion problems in these industries also does not mean that corrosion is accelerated everywhere microorganisms are present in these industries. Many bacteria existing in natural and man-made environments do not cause or accelerate corrosion. Bacteria may exist at the corrosion site only because the electrochemistry of corrosion creates a favorable environment for their growth.

A select number of organisms, however, are repeatedly cited as causing corrosion in environments in which there would be none without them, or as accelerating corrosion, or as changing a relatively slow rate of general corrosion into one with rapid localized penetration of the metal. These organisms are listed in Table 4.

Biological Corrosion of Iron and Steel

A large number of case histories involving the microbial acceleration of the localized corrosion of iron and steel have been presented in the literature. These are far too numerous to list in this article. An excellent introduction to the literature can be gained by referring to Ref 40 to 46.

Anaerobic Corrosion. The corrosive action of the sulfate-reducing bacteria (SRB) from the genera *Desulfovibrio*, *Desulfotomaculum*, and *Desulfomonas* in anaerobic environments is well known. The morphology of attack is almost always localized and often looks much like the pitting shown for cast iron in Fig. 26. A general review of the mechanism and the fundamentals of biological corrosion in aqueous environments is given in the article "Effects of Environmental Variables on Aqueous Corrosion" (see the Appendix "Biological Effects") in this Volume.

In the early decades of this century, it was generally believed that the SRB could influence the corrosion of steel only in totally anaerobic environments such as deaerated soils and marine sediments. Recently, it has been recognized that anaerobic corrosion by SRB can take place in nominally aerated environments. In these latter cases, anaerobic microenvironments can exist under biodeposits of aerobic organisms, in crevices built into the structure, and at flaws in various types of coating systems. The most corrosive environments are often those in which alternate aerobic-anaerobic conditions exist because of the action of variable flow hydrodynamics or periodic mechanical action.

The anaerobic corrosion of iron and steel has been identified in such diverse environments as waterlogged soils of near-neutral pH; bottom muds of rivers, lakes, marshes, and estuaries (especially when these contain decaying organic material as a source of sulfates for the SRB); under marine fouling deposits and in various other offshore industrial environments (Ref 47); under nodules or tubercules in natural freshwaters and recirculating cooling waters; and under disbonded areas of pipeline coatings.

A considerable amount of work has been done to assess the relative aggressiveness of various soils and sediments. The work in soils prior to 1970 is reviewed in Ref 48. Factors such as the presence or absence of SRB, soil resistivity, and water content were considered. It was discovered that SRB activity correlated well with soil redox potential, E_h, on the normal hydrogen scale, a variable that is much easier to measure than numbers of SRB. Aggressive soils tended to have mean resistivities of less than 2000 Ω · cm and a mean redox potential more negative than 400 mV on the normal hydrogen scale corrected to pH 7. Soils that were borderline based on these two tests tended to be aggressive if their water content was over 20%. With regard to redox potential alone, soil corrosivity varied as follows (Ref 48):

Soil E_h	Corrosivity
<100 mV	Severe
100–200 mV	Moderate
200–400 mV	Slight
>400 mV	Noncorrosive

Other attempts to assess the risk of corrosion by SRB have been sporadic. One investigation attempted to assess the severity of the SRB hazard on the inside of submarine pipelines car-

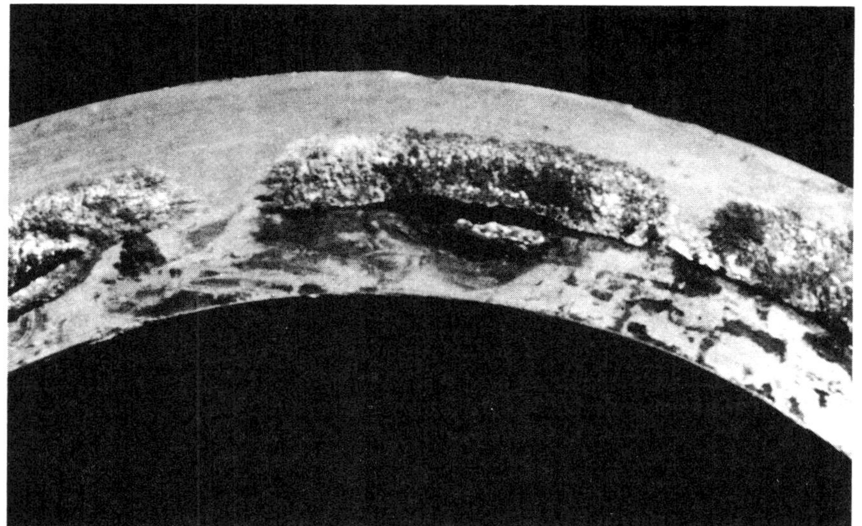

(a)

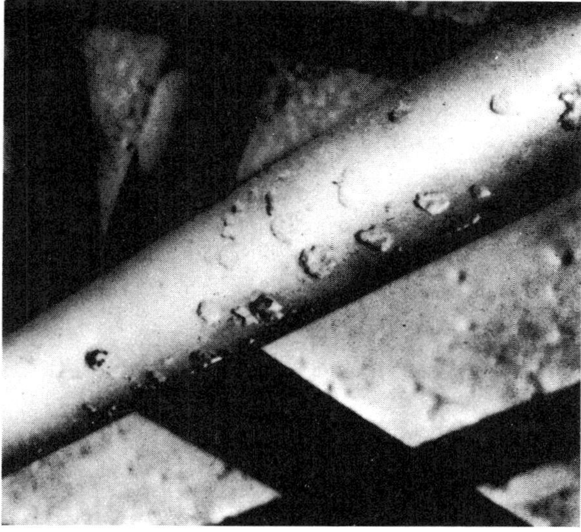

(b)

Fig. 28 Localized biological corrosion of austenitic stainless steel. (a) Crevice corrosion of type 304 stainless steel flange from a cooling water system. Staining shows evidence of adjacent biomounds. The corrosion attack reached a depth of 6 mm (1/4 in.). Courtesy of W.K. Link and R.E. Tatnall, E.I. Du Pont de Nemours & Co., Inc. (b) Pits on the underside of type 304 stainless steel piping used in a waste treatment tank (after sandblasting to remove biomounds). Courtesy of G. Kobrin and R.E. Tatnall, E.I. Du Pont de Nemours & Co., Inc.

rying North Sea crude oil by measuring both the numbers of SRB present in the oil and the activity (or vigor of growth) of the organisms (Ref 49). The risk was assessed as extreme if both the numbers of organisms and their activity were rated high, and the risk was considered to be minimal if both were rated low.

Efforts to solve the anaerobic iron and steel corrosion problem as outlined in Ref 48 include:

- Replacing the iron or steel with noncorrodible materials, such as fiberglass, PVC, polyethylene, and concrete
- Creating a nonaggressive environment around the steel by backfilling with gravel or clay-free sand to encourage good drainage (that is, oxygenating to suppress SRB), making the environment alkaline, or using biocides (in closed industrial systems)
- Using cathodic protection, although potentials of -0.95 V versus $Cu/CuSO_4$ (or even more negative) are often required; at these potentials, the risk of hydrogen cracking or blistering should be assessed
- Using various barrier coatings, some with corrosion inhibitors and/or biocides

Aerobic Corrosion. Corrosion of iron and steel under oxygenated conditions generally involves the formation of acidic metabolites. The aerobic sulfur-oxidizing bacteria *Thiobacillus* can create an environment of up to about 10% H_2SO_4, thus encouraging rapid corrosion. Other organisms produce organic acids with similar results. This corrosion can be localized or general, depending on the distribution of organisms and metabolic products. If all the bacterial activity is concentrated at a break or delamination in a coating material, the corrosion is likely to be highly localized. If, on the other hand, the metabolic products are spread over the surface, the corrosion may be general, as has been reported for carbon steel tendon wires used to prestress a concrete vessel in a nuclear power plant (Ref 46). In this case, the wires were coated with a hygroscopic grease prior to installation. A study to determine the cause of corrosion concluded that the wires, shown in Fig. 27, were corroded by formic and acetic acids excreted by bacteria in breaking down the grease.

Other cases of aerobic corrosion of iron and steel begin with the creation of oxygen concentration cells by deposits of slime-forming bacteria. Such corrosion is often accelerated by the iron-oxidizing bacteria in the formation of tubercules. This topic is addressed in the discussion "Tuberculation" in this section.

Biological Corrosion of Stainless Steel

There are two general sets of conditions under which localized biological corrosion of austenitic stainless steel occurs (Fig. 25). These will be illustrated by two generalized case histories. Typical examples of microbiologically induced localized corrosion of stainless steel are shown in Fig. 28.

Hydrotest or Outage Conditions. As originally reported in Ref 50, a new production facility required type 304L and 316L austenitic stainless steels for resistance to nitric and organic acids. All of the piping and flat-bottomed storage tanks were field erected and hydrostatically tested. The hydrotest water was plant well water containing 20 ppm chlorides and was sodium softened.

The pipelines were not drained after testing. The tanks were drained, but were then refilled for ballast because of a hurricane threat. Two to four months after hydrotesting, water was found dripping from butt welds in the nominally 3-mm (1/8-in.) wall piping. Internal inspection revealed numerous pits in and adjacent to welds under reddish-brown deposits in both piping and tanks. Upon cleaning off the deposit, the researchers found a large stained area with a pit opening. Metallographic sectioning showed a large subsurface cavity with only a small opening to the surface. Photographs of the weld corrosion deposits and the resulting pitting corrosion are shown in Fig. 26 to 35 in the article "Corrosion of Weldments" in this Volume (see the discussion "Microbiologically Induced Corrosion"). Pitted welds in a type 316L tank showed some evidence of preferential attack of the δ-ferrite stringers, as shown in Fig. 29. It is not yet known why such attack often concentrates at the weld line.

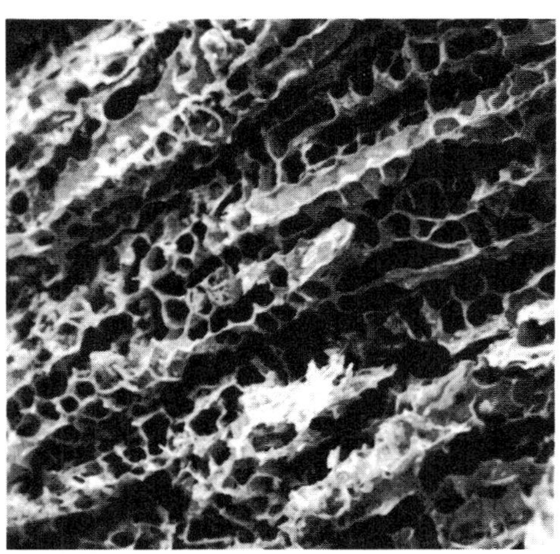

Fig. 29 SEM micrograph showing matrix remaining after preferential corrosion of the δ-ferrite phase in a type 316 stainless steel. 300×. Courtesy of J.G. Stoecker, Monsanto Company

Well water and deposits both showed high counts of the iron bacteria, *Gallionella*, and the iron/manganese bacteria, *Siderocapsa*. Deposits also contained thousands of parts per million of iron, manganese, and chlorides. Sulfate-reducing and sulfur-oxidizing bacteria were not present in either water or deposits. The proposed mechanism for the attack involves:

- Original colonization by the iron and manganese bacteria at the weld seams to create an oxygen concentration cell

Table 3 Industries known to be affected by microbiological corrosion

Industry	Problem areas
Chemical-processing industries	Stainless steel tanks, pipelines and flanged joints, particularly in welded areas after hydrotesting with natural river or well waters
Nuclear power generation	Carbon and stainless steel piping and tanks; copper-nickel, stainless, brass and aluminum bronze cooling water pipes and tubes, especially during construction, hydrotest, and outage periods
Onshore and offshore oil and gas industries	Mothballed and waterflood systems; oil and gas handling systems, particularly in those environments soured by sulfate reducing bacteria (SRB)-produced sulfides
Underground pipeline industry	Water-saturated clay-type soils of near-neutral pH with decaying organic matter and a source of SRB
Water treatment industry	Heat exchangers and piping
Sewage handling and treatment industry	Concrete and reinforced concrete structures
Highway maintenance industry	Culvert piping
Aviation industry	Aluminum integral wing tanks and fuel storage tanks
Metal working industry	Increased wear from breakdown of machining oils and emulsions

Table 4 Microorganisms most commonly implicated in biological corrosion

Genus or species	pH range	Temperature range °C	Oxygen requirement	Metals affected	Action
Bacteria					
Desulfovibrio					
Best known: *D. desulfuricans*	4–8	10–40	Anaerobic	Iron and steel, stainless steels, aluminum, zinc, copper alloys	Utilize hydrogen in reducing SO_4^{2-} to S^{2-} and H_2S; promote formation of sulfide films
Desulfotomaculum					
Best known: *D. nigrificans* (also known as *Clostridium*)	6–8	10–40 (some 45–75)	Anaerobic	Iron and steel, stainless steels	Reduce SO_4^{2-} to S^{2-} and H_2S, (spore formers)
Desulfomonas	...	10–40	Anaerobic	Iron and steel	Reduce SO_4^{2-} to S^{2-} and H_2S
Thiobacillus thiooxidans	0.5–8	10–40	Aerobic	Iron and steel, copper alloys, concrete	Oxidizes sulfur and sulfides to form H_2SO_4; damages protective coatings
Thiobacillus ferrooxidans	1–7	10–40	Aerobic	Iron and steel	Oxidizes ferrous (Fe^{2+}) to ferric (Fe^{3+})
Gallionella	7–10	20–40	Aerobic	Iron and steel, stainless steels	Oxidizes ferrous (and manganous) to ferric (and manganic); promotes tubercule formation
Sphaerotilus	7–10	20–40	Aerobic	Iron and steel, stainless steels	Oxidizes ferrous (and manganous) to ferric (and manganic); promotes tubercule formation
S. natans	...	...	...	Aluminum alloys	
Pseudomonas	4–9	20–40	Aerobic	Iron and steel, stainless steels	Some strains can reduce Fe^{3+} to Fe^{2+}
P. aeruginosa	4–8	20–40	Aerobic	Aluminum alloys	
Fungi					
Cladosporium resinae	3–7	10–45 (best at 30–35)	...	Aluminum alloys	Produces organic acids in metabolizing certain fuel constituents

- Dissolution of ferrous and manganous ions under the deposits
- Attraction of chloride ions as the most abundant anion to maintain charge neutrality
- Oxidation of the ferrous and manganous ions to ferric and manganic by the bacteria to form a highly corrosive acidic chloride solution in the developing pit

Many failures of this type have been reported in the chemical-processing industries in new equipment after hydrotesting but prior to commissioning in service. Similar failures have been reported in older equipment in both the chemical-processing and nuclear power industries when untreated well or river water was allowed to remain stagnant in the equipment during outage periods. Occasionally, the pitting will be accompanied by what appear to be chloride stress-corrosion cracks under the deposits (Ref 45, 46). Examples of transgranular cracks in a type 304 stainless steel tank are shown in Fig. 30.

Crevice or Gasket Conditions. A different set of conditions has lead to the localized corrosion of asbestos-gasketed flanged joints in a type 304 stainless steel piping system (Ref 52). Inspection of the system after about 3 years of service in river water revealed severe crevice corrosion in and near the flanged and gasketed joints. The corrosion sites were covered by voluminous tan-to-brown, slimy biodeposits, as shown in Fig. 31(a). Under the deposits were broad, open pits with bright, active surfaces (Fig. 31b). The surfaces under the gasket material and adjacent to the corroded areas were covered with black deposits, which emitted H_2S gas when treated with HCl.

The biodeposits were high in iron, silt- and slime-forming bacteria, and iron bacteria, but not chloride, manganese, and sulfur compounds. Sulfate-reducing bacteria were found only in the black deposits. These bacteria had survived continuous chlorination (0.5 to 1.0 ppm residual), caustic adjustment of pH to 6.5 to 7.5, and continuous additions of a polyacrylate dispersant and a nonoxidizing biocide (quaternary amine plus tris tributyl tin oxide) (Ref 52).

The suspected mechanism involves:

- Colonization by slime-forming bacteria at low-velocity sites near gasketed joints
- Trapping of suspended solids rich in iron by the growing biodeposit, thus creating an environment conducive to growth of the filamentous iron bacteria
- Rapid depletion of oxygen in the crevice area by a combination of biological and electrochemical mechanisms (Ref 53), creating an environment for the SRB
- Breakdown of passivity by a combination of oxygen depletion and SRB activity, causing localized corrosion

Standard approved methods for controlling the biological corrosion of stainless alloys are currently being developed. Some general guidelines for avoiding problems in hydrotesting, however, are given in Ref 50. These guidelines are summarized as follows. First, demineralized water or high-purity steam condensate is used for the test water. The equipment should be drained and dried as soon as possible after testing. Second, if a natural freshwater must be used, it should be filtered and chlorinated, and the equipment should be blown or mopped dry within 3 to 5 days after testing.

Biological Corrosion of Aluminum

Pitting corrosion of integral wing aluminum fuel tanks in aircraft that use kerosene-base fuels has been a problem since the 1950s (Ref 54). The fuel becomes contaminated with water by vapor condensation during variable-temperature flight conditions. Attack occurs under microbial deposits in the water phase and at the fuel/water interface. The organisms grow either in continuous mats or sludges, as shown in Fig. 32, or in volcanolike tubercules with gas bubbling from the center, as shown schematically in Fig. 33.

The organisms commonly held responsible are *Pseudomonas*, *Cladosporium*, and *Desulfovibrio*. These are often suspected of working together in causing the attack. *Cladosporium resinae* is usually the principal organism involved; it produces a variety of organic acids (pH 3 to 4 or lower) and metabolizes certain fuel constituents. These organisms may also act in concert with the slime-forming *Pseudomonads* to produce oxygen con-

(a)

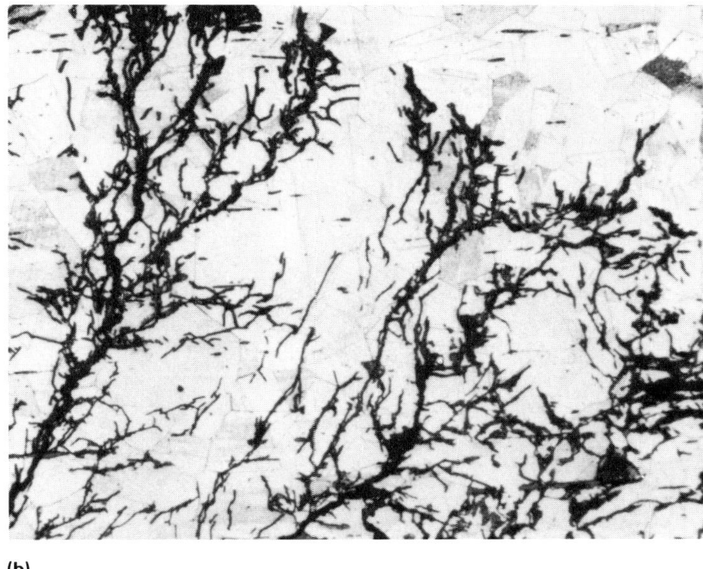

(b)

Fig. 30 Cracks emanating from pits in a type 304 stainless steel tank that was placed in hot demineralized water service with an operating temperature that fluctuated from 75 to 90 °C (165 to 195 °F). (a) Photomicrograph of a section through a typical biological deposit and pit in the wall of the tank. 25×. 10% oxalic acid etch. (b) Higher-magnification view of cracks. These branched transgranular cracks are typical of chloride stress-corrosion cracking of austenitic stainless steel. 250×. 10% oxalic acid etch. Source: Ref 51

Fig. 31(a) Single remaining biodeposit adjacent to resulting corrosion on a type 304 stainless steel flange. Numerous other similar deposits were dislodged in opening the joint. This flange was covered by a type 304 blind flange and was sealed with a bonded asbestos gasket. Source: Ref 52

centration cells under the deposit. Active SRB have sometimes been identified at the base of such deposits.

Control of this type of attack has usually focused on a combination of reducing the water content of fuel tanks; coating, inspecting, and cleaning fuel tank interiors; and using biocides and fuel additives. More information can be found in Ref 44 and 54.

Biological Corrosion of Copper Alloys

Far less is known about the influence of microorganisms on the corrosion of copper and copper alloys than was the case for iron and steel. The well-known toxicity of cuprous ions toward living organisms does not mean that the copper-base alloys are immune to biological effects in corrosion. It does mean, however, that only those organisms having a high tolerance for copper are likely to have a substantial effect. *Thiobacillus thiooxidans*, for example, can withstand copper concentrations as high as 2%. Most of the reported cases of microbial corrosion of copper alloys are caused by the production of such corrosive substances as CO_2, H_2S, NH_3, and organic or inorganic acids.

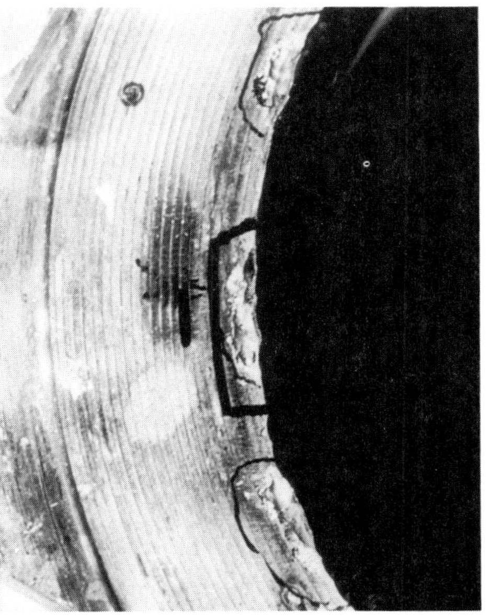

Fig. 31(b) Close-up of gouging-type corrosion under deposits shown in Fig. 31(a) after cleaning to remove black corrosion products. Source: Ref 52

Copper-nickel tubes from the fan coolers in a nuclear power plant were found to have pitting corrosion under bacterial deposits (Fig. 34). Slime-forming bacteria acting in concert with iron- and manganese-oxidizing bacteria were responsible for the deposits.

In another case, Monel heat-exchanger tubes were found to have severe pitting corrosion (Fig. 35) under discrete deposits rich in iron, copper, manganese, and silicon, with some nickel. Associated with the deposit were slime-forming bacteria, along with iron- and manganese-oxidizing bacteria. Several million SRB were found within each pit under the deposit. It was thought that the deposit-forming organisms created an environment conducive to growth of SRB, which then accelerated corrosion by the production of H_2S.

It is quite common to have bacterial slime films on the interior of copper alloy heat exchanger and condenser tubing. Usually, these films are a problem only with heat transfer as long as the organisms are living. When they die, however, organic decomposition produces sulfides, which are notoriously corrosive to copper alloys. Occasionally, NH_3-induced stress-corrosion cracking has been directly attributed to microbial NH_3 production.

Tuberculation

The formation of tubercules by biological organisms acting in conjunction with electrochemical corrosion occurs in many environments and on many alloys. An example of tuberculation in a steel economizer tube in sulfuric acid service is shown in Fig. 36. This example shows that it is possible for tubercules to form without the presence of any microorganisms; the phenomenon usually takes place in biologically active aqueous systems.

The process of tubercule formation is a complex one. A number of the reactions that can take place are illustrated for a ferrous alloy in Fig. 37. The volcanolike structure often starts with a deposit of slime-forming and iron-oxidizing bacteria at a point of low flow velocity. This creates an oxygen concentration cell, thus promoting dissolution of iron as Fe^{2+} under the deposit. As the Fe^{2+} ions move outward, they are oxidized to Fe^{3+}; this occurs electrochemically as they encounter higher oxygen concentrations and/or by the action of iron bacteria. The resulting corro-

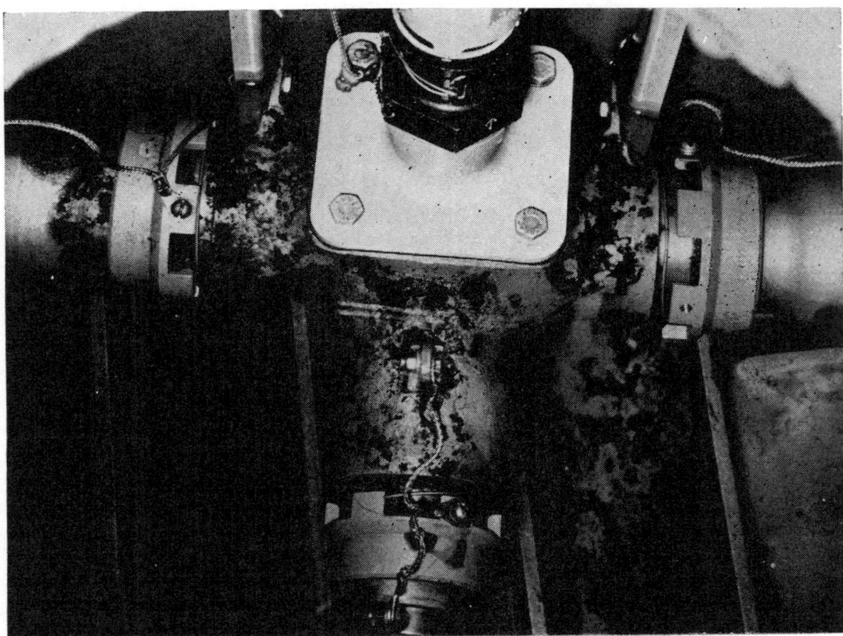

Fig. 32 Microbial growth in the integral fuel tanks of jet aircraft. Source: Ref 42

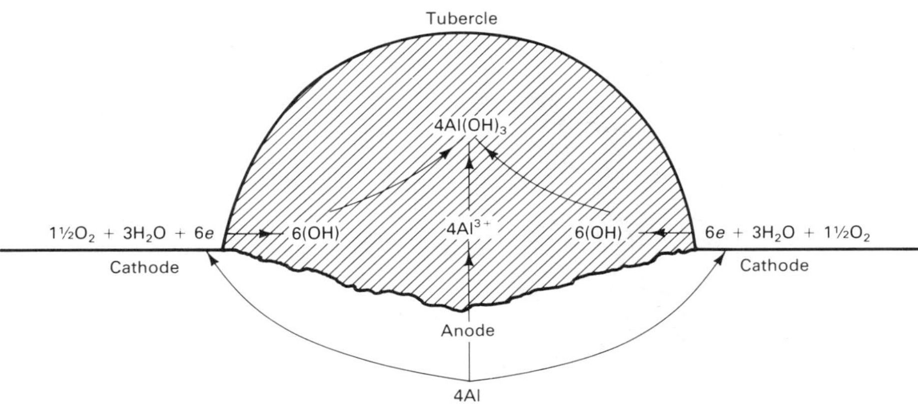

Fig. 33 Schematic of tubercule formed by bacteria on an aluminum alloy surface. Source: The Electrochemical Society

Fig. 34 Pitting corrosion in 90Cu-10Ni tubes from a fan cooler in a nuclear power plant. Pits are located under the small deposits associated with the deposition of iron and manganese by bacteria. Source: Ref 46

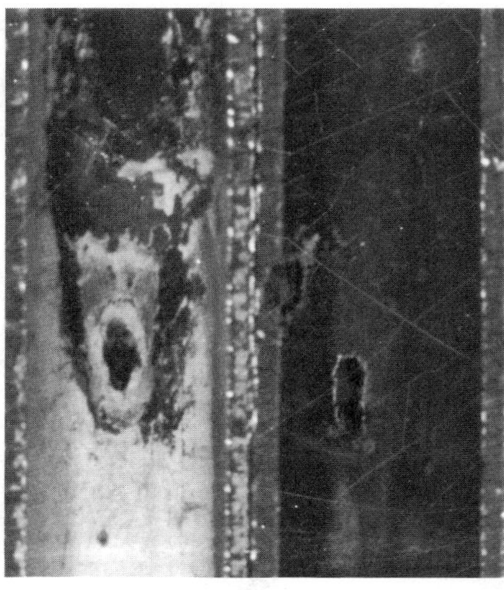

Fig. 35 Pitting corrosion in Monel tubes from a heat exchanger. Each pit was originally covered by a discrete deposit containing large numbers of SRB. Source: Ref 46

sion product, $Fe(OH)_3$, mingles with the biodeposit to form the wall of the growing tubercule. When bacteria are present, the tubercule structure is usually less brittle and less easily removed from the metal surface than when they are absent. The outside of the tubercule becomes cathodic, while the metal surface inside becomes highly anodic.

As the tubercule matures, some of the biomass may start to decompose, providing a source of sulfates for SRB to use in producing H_2S in the anaerobic interior solution. In some cases, the sulfur-oxidizing bacteria may assist in the formation of the sulfates. Depending on the ions available in the water, the tubercule structure may contain some $FeCO_3$ and, when SRB are present, some FeS. Finally, if there is a source of chlorides and if the iron-oxidizing bacteria *Gallionella* are present, a highly acidic, ferric chloride solution may form inside the tubercule.

Generally, not all of the above reactions will take place in any single environment. As the individual tubercules on a surface grow under the influence of any combination of reactions, they will eventually combine to form a mass that severely limits flow (or even closes it off altogether), leaving a severely pitted surface underneath.

REFERENCES

1. "Standard Test Method for Filiform Corrosion Resistance of Organic Coatings on Metal," D 2803, *Annual Book of ASTM Standards*, American Society for Testing and Materials
2. R. Preston and B. Sanyal, *J. Appl. Chem.*, Vol 6, 1956, p 26-44
3. W. Funke, *Prog. Org. Coatings*, Vol 9 (No. 1), April 1981, p 29-46
4. R. Ruggeri and T. Beck, *Corrosion*, Vol 39 (No. 11), Nov 1983, p 452-465
5. W. Slabaugh and E. Chan, *J. Paint Technol.*, Vol 38, 1966, p 417-420
6. W. Slabaugh, W. DeJager, S. Hoover, and L. Hutchinson, *J. Paint Technol.*, Vol 44 (No. 56), March 1972, p 76-83
7. W. Ryan, *Environment, Economics, Energy*, Vol 1, Society for the Advancement of Material and Process Engineering, May 1979, p 638-648
8. P. Bijlmer, *Adhesive Bonding of Aluminum Alloys*, Marcel Dekker, 1985, p 21-39
9. T.S. Lee and R.O. Lewis, *Mater. Perform.*, Vol 24 (No. 3), 1985, p 25
10. H.P. Godard, W.B. Jepson, M.R. Botwell, and R.L. Kane, Crevice Corrosion of Aluminum and Crevice Corrosion of Titanium, in *Corrosion of Light Metals*, John Wiley & Sons, 1967, p 45, 319
11. T.S. Lee, R.M. Kain, and J.W. Oldfield, "Factors Influencing the Crevice Corrosion Behavior of Stainless Steels," Paper 69, presented at Corrosion/83, Houston, TX, National Association of Corrosion Engineers, 1983
12. J.W. Oldfield and W.H. Sutton, *Br. Corros. J.*, Vol 13, 1978, p 13

Fig. 36 Steel hairpin bend tube used in the economizer of a sulfuric acid waste heat boiler. The tube exhibits tuberculation associated with oxygen attack. The bottom photograph shows the tubercules in greater detail. Source: Ref 55

13. U.R. Evans, Corrosion of Copper and Copper Alloys, in *The Corrosion of Metals*, 2nd ed., Edward Arnold, 1926
14. E.H. Wyche, L.R. Voight, and F.L. LaQue, *Trans. Electrochem. Soc.*, Vol 89, 1946, p 149
15. G.J. Schafer and P.K. Forster, *J. Electrochem. Soc.*, Vol 106, 1959, p 468
16. F.L. LaQue, Crevice Corrosion, in *Marine Corrosion, Causes and Prevention*, John Wiley & Sons, 1975, p 164-176
17. F.L. LaQue, Environmental Factors, in *Marine Corrosion, Causes and Prevention*, John Wiley & Sons, 1975, p 117
18. R.M. Kain, "Effect of Alloy Content on the Localized Corrosion Resistance of Several Nickel Base Alloys in Seawater," Paper 229, presented at Corrosion/86, Houston, TX, National Association of Corrosion Engineers, 1986
19. R.M. Kain, *Corrosion*, Vol 40 (No. 6), 1984, p 313
20. R.M. Kain, *Mater. Perform.*, Vol 33 (No. 2), 1984, p 24
21. R.M. Kain, A.H. Tuthill, and E.C. Hoxie, *J. Mater. Energy Syst.*, Vol 5 (No. 4), 1984, p 205
22. R.M. Kain, T.S. Lee, and J.R. Scully, Crevice Corrosion Resistance of Type 316 Stainless Steel in Marine Environments, in *Proceedings of the 9th International Congress on Metallic Corrosion*, Toronto, Canada, National Research Council of Canada, June 1984
23. J.W. Oldfield, R.M. Kain, and T.S. Lee, Avoiding Crevice Corrosion of Stainless Steels, in *Proceedings of Stainless Steel '84 Symposium*, Götenberg, Sweden, Chalmers University of Technology and Jernkontorte (Sweden) with The Metals Society (UK), Sept 1984
24. A.J. Sedriks, *Int. Met. Rev.*, Vol 27 (No. 6), 1972
25. T.S. Lee and A.H. Tuthill, *Mater. Perform.*, Vol 22 (No.1), 1983, p 48
26. Y.M. Kolotyrkin, *Corrosion*, Vol 19, 1963, p 261t
27. J.L. Crolet, J.M. Defranoux, L. Seraphin, and R. Tricot, *Mem. Sci. Rev. Met.*, Vol 71 (No. 12), 1974, p 797
28. H. Spahn, G.H. Wagoner, and U. Steinhoff, Paper A-2, Firminy Meeting, France, June 1973
29. M.G. Fontana, The 1977 Alpha Sigma M Lecture, *ASM News*, Vol 9 (No. 3), 1978, p 4
30. F.L. LaQue, *Localized Corrosion*, National Association of Corrosion Engineers, 1974, p i.47
31. H.H. Uhlig and R.W. Revie, *Corrosion and Corrosion Control*, 3rd ed., John Wiley & Sons, 1984, p 13-14
32. H.H. Uhlig, *Corrosion Handbook*, John Wiley & Sons, 1948, p 165
33. M.G. Fontana, *Corrosion Engineering*, McGraw-Hill, 1986, p 66
34. U.R. Evans, *Corrosion*, Vol 7 (No. 238), 1951
35. A.J. Sedriks, *Corrosion of Stainless Steels*, John Wiley & Sons, 1979, p 63
36. A.I. Asphahani, *Mater. Perform.*, Vol 19 (No. 8), 1980, p 9
37. P.E. Manning, *Corrosion*, Vol 39 (No. 3), 1983, p 98
38. H.H. Uhlig and R.W. Revie, *Corrosion and Corrosion Control*, 3rd ed., John Wiley & Sons, 1984, p 74
39. R.E. Tatnall, Experimental Methods in Biocorrosion, in *Biologically Induced Corrosion*, S.C. Dexter, Ed., Conference Proceedings, National Association of Corrosion Engineers, 1986, p 246-253
40. *Microbial Corrosion*, Conference Proceedings, National Physical Laboratory, The Metals Society, 1983
41. S.C. Dexter, Ed., *Biologically Induced Corrosion*, Conference Proceedings, National Association of Corrosion Engineers, 1986

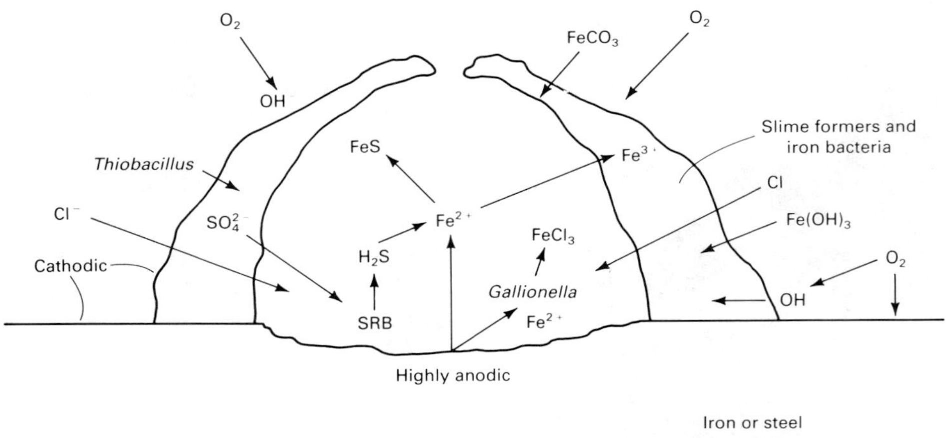

Fig. 37 Schematic diagram of electrochemical and microbial processes involved in tuberculation. Not all of these processes may be active in any given situation.

42. J.D.A. Miller, Ed., *Microbial Aspects of Metallurgy*, American Elsevier, 1970
43. H.A. Videla and R.C. Salvarezza, *Introduction to Microbiological Corrosion*, Biblioteca Mosaico, 1984 (in Spanish)
44. D.H. Pope, D. Duquette, P.C. Wayner, and A.H. Johannes, *Microbiologically Influenced Corrosion: A State-of-the-Art Review*, Publication 13, Materials Technology Institute of the Chemical Process Industries, Inc., 1984
45. D.H. Pope and J.G. Stoecker, Microbiologically Influenced Corrosion, in *Process Industries Corrosion—Theory and Practice*, B.J. Moniz and W.I. Pollock, Ed., National Association of Corrosion Engineers, 1986, p 227-242
46. D.H. Pope, "A Study of Microbiologically Influenced Corrosion in Nuclear Power Plants and a Practical Guide for Countermeasures," Final Report EPRI NP-4582, Electric Power Research Institute, 1986
47. P.F. Sanders and W.A. Hamilton, Biological and Corrosion Activities of Sulfate-Reducing Bacteria in Industrial Process Plant, in *Biologically Induced Corrosion*, S.C. Dexter, Ed., National Association of Corrosion Engineers, 1986, p 47-68
48. J.D.A. Miller and A.K. Tiller, Microbial Corrosion of Buried and Immersed Metal, in *Microbial Aspects of Metallurgy*, J.D.A. Miller, Ed., American Elsevier, 1970, p 61-106
49. R.A. King, J.D.A. Miller, and J.F.D. Stott, Subsea Pipelines: Internal and External Biological Corrosion, in *Biologically Induced Corrosion*, S.C. Dexter, Ed., National Association of Corrosion Engineers, 1986, p 268-274
50. G. Kobrin, Reflections on Microbiologically Induced Corrosion of Stainless Steels, in *Biologically Induced Corrosion*, S.C. Dexter, Ed., National Association of Corrosion Engineers, 1986, p 33
51. J.G. Stoecker and D.H. Pope, Study of Biological Corrosion in High Temperature Demineralized Water, *Mater. Perform.*, June 1986, p 51-56
52. R.E. Tatnall, Case Histories: Bacteria Induced Corrosion, *Mater. Perform.*, Vol 20 (No. 8), 1981, p 41
53. S.C. Dexter, K.E. Lucas, and G.Y. Gao, Role of Marine Bacteria in Crevice Corrosion Initiation, in *Biologically Induced Corrosion*, S.C. Dexter, Ed., National Association of Corrosion Engineers, 1986, p 144-153
54. J.J. Elpjick, Microbial Corrosion in Aircraft Fuel Systems, in *Microbial Aspects of Metallurgy*, J.D.A. Miller, Ed., American Elsevier, 1970, p 157-172
55. T.C. Breske, Sulfuric Acid Waste Heat Boiler Corrosion Control, *Mater. Perform.*, Sept 1979, p 9-16

Metallurgically Influenced Corrosion

Robert Steigerwald, Bechtel National, Inc.

THE PURPOSE of this article is to discuss corrosion as affected by metallurgical factors. These factors include alloy chemistry and heat treatment. Mechanical factors such as stress will not be covered. The metallurgical influences considered are the relative stability of the components of an alloy, metallic phases, metalloid phases such as carbides, and local variations in composition in a single phase. One example is given of the ways in which nonmetallic inclusions, such as oxides and sulfides, may influence corrosion.

Dealloying, selective leaching, and parting are terms used to describe that form of corrosion in which an element is selectively removed from an alloy. This phenomenon is discussed in a separate section of this article.

Stress corrosion and hydrogen embrittlement will not be discussed. However, they will be mentioned when they are influenced by one of the mechanisms under discussion.

The most common form of metallurgically influenced corrosion is intergranular corrosion. It occurs when corrosion is localized at grain boundaries. Often, this localized corrosion leads to the dislodgement of individual grains and a roughening, or sugaring, of the affected surface. It is typified by an apparent increase in the corrosion rate with time.

Mechanisms of Intergranular Corrosion

Intergranular corrosion takes place when the corrosion rate of the grain-boundary areas of an alloy exceeds that of the grain interiors. This difference in corrosion rate is generally the result of differences in composition between the grain boundary and the interior.

The differences in corrosion rate may be caused by a number of reactions. A phase may precipitate at a grain boundary and deplete the matrix of an element that affects its corrosion resistance. A grain-boundary phase may be more reactive than the matrix. Various solute atoms may segregate to the grain boundaries and locally accelerate corrosion. The metallurgical changes that lead to intergranular corrosion are not always observable in the microstructure; therefore, corrosion tests may sometimes be the most sensitive indication of metallurgical changes (see, for example, the article "Evaluation of Intergranular Corrosion" in this Volume).

Figure 1 illustrates the electrochemistry of intergranular corrosion. Polarization curves are shown for the grain-boundary and matrix areas. The system chosen is one that exhibits active-passive behavior, for example, a chromium-nickel-iron stainless steel in sulfuric acid (H_2SO_4). Several points must be noted. The difference in corrosion rate varies with potential. The rates are close or the same in the active and transpassive ranges and vary considerably with potential in the passive range.

Intergranular corrosion is usually not the result of active grain boundaries and a passive matrix. The corroding surface is at one potential. Differences in composition produce different corrosion rates at the same potential in the passive region. When more than one metallic phase is present in an alloy, its polarization behavior will be the volume average sum of the behavior of each phase (Fig. 2). Active-passive surfaces are possible in this case.

When an alloy is undergoing intergranular corrosion, its rate of weight loss will usually accelerate with time. As the grain-boundary area dissolves, the unaffected grains are undermined and fall out; this increases the weight loss. Typical weight loss versus time curves for an alloy undergoing intergranular corrosion are shown in Fig. 3.

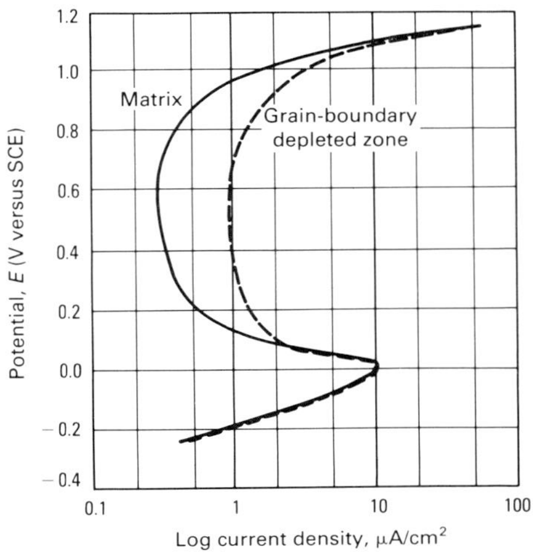

Fig. 1 Anodic polarization behavior of an active-passive alloy with grain-boundary depleted zones (schematic)

Intergranular Corrosion and Other Forms of Corrosion. Susceptibility to intergranular corrosion cannot be taken as a general indication of increased susceptibility to other forms of corrosion, such as pitting or general corrosion. The environments that cause intergranular corrosion for a particular alloy system are often very specific. Susceptibility to intergranular corrosion may mean susceptibility to intergranular stress-corrosion cracking, but some nickel-base alloys are actually more resistant to stress-corrosion cracking (SCC) when they are sensitized to intergranular corrosion (Ref 2).

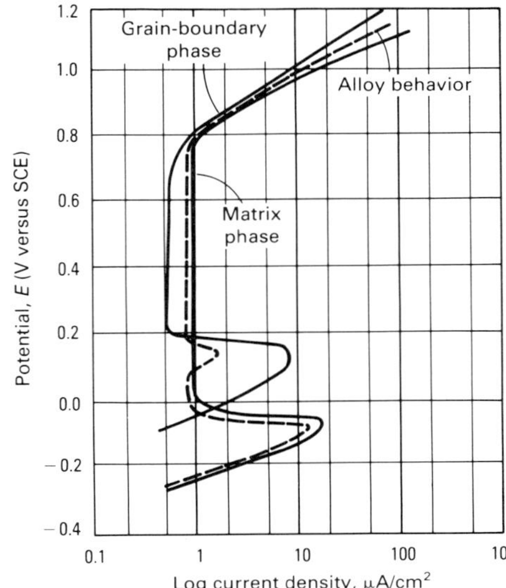

Fig. 2 Anodic polarization behavior of a two-phase active-passive alloy (schematic)

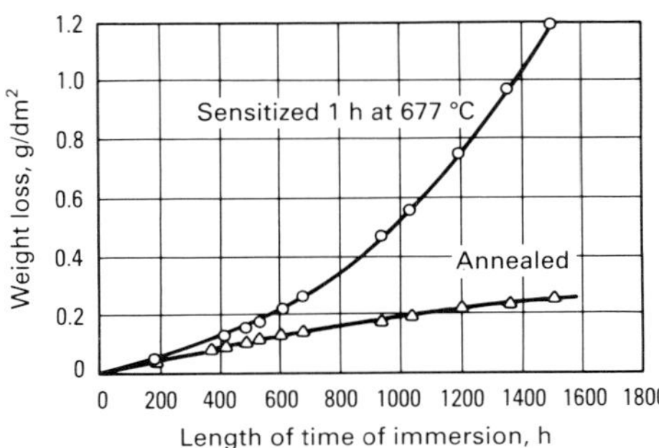

Fig. 3 Corrosion of type 304 steel in inhibited boiling 10% H_2SO_4. Inhibitor: 0.47 g Fe^{3+}/L of solution added as $Fe_2(SO_4)_3$. Source: Ref 1

Intergranular corrosion can occur in many alloy systems; comprehensive coverage of all such corrosion is beyond the scope of this article. Intergranular corrosion has been most widely investigated in stainless steels, and the behavior of these alloys will be the principal example of intergranular corrosion. Intergranular attack in nickel-base and aluminum alloys will also be discussed.

Metallurgical Effects on the Corrosion of Stainless Steels

As described in the article "Corrosion of Stainless Steels" in this Volume, metallurgical variables can influence the corrosion behavior of austenitic, ferritic, duplex, and martensitic stainless steels. The distribution of carbon is probably the most important variable influencing the susceptibility of these alloys to intergranular corrosion, but nitrogen and metallic phases are also important.

Austenitic Stainless Steels

Intergranular Corrosion. At temperatures above about 1035 °C (1900 °F), chromium carbides are completely dissolved in austenitic stainless steels. However, when these steels are slowly cooled from these high temperatures or reheated into the range of 425 to 815 °C (800 to 1500 °F), chromium carbides are precipitated at the grain boundaries. These carbides contain more chromium than the matrix does.

The precipitation of the carbides depletes the matrix of chromium adjacent to the grain boundary. The diffusion rate of chromium in austenite is slow at the precipitation temperatures; therefore, the depleted zone persists, and the alloy is sensitized to intergranular corrosion. This sensitization occurs because the depleted zones have higher corrosion rates than the matrix in many environments. Figure 4 illustrates how the chromium content influences the corrosion rate of iron-chromium alloys in boiling 50% H_2SO_4 containing ferric sulfate ($Fe_2(SO_4)_3$). In all cases, the alloys are in the passive state. The wide differences in the corrosion rate are the result of the differences in the chromium content.

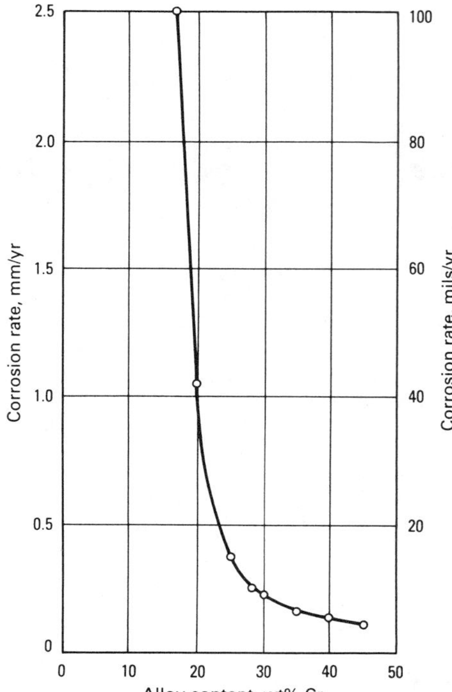

Fig. 4 The effect of chromium content on the corrosion behavior of iron-chromium alloys in boiling 50% H_2SO_4 with $Fe_2(SO_4)_3$. Source: Ref 3

If the austenitic stainless steels are cooled rapidly to below about 425 °C (800 °F), the carbides do not precipitate, and the steels are immune to intergranular corrosion. Reheating the alloys to 425 to 815 °C (800 to 1500 °F), as for stress relief, will cause carbide precipitation and sensitivity to intergranular corrosion. The maximum rate of carbide precipitation occurs at about 675 °C (1250 °F). Because this is a common temperature for the stress relief of carbon and low-alloy steels, care must be exercised in selecting stainless steels to be used in dissimilar-metal joints that are to be stress relieved.

Welding is the common cause of the sensitization of stainless steels to intergranular corrosion (see the article "Corrosion of Weldments" in this Volume). Although the cooling rates in the weld itself and the base metal immediately adjacent to it are sufficiently high to avoid carbide precipitation, the weld thermal cycle will bring part of the heat-affected zone (HAZ) into the precipitation range. Carbides will precipitate, and a zone somewhat removed from the weld will become susceptible to intergranular corrosion (Fig. 5). Welding does not always sensitize austenitic stainless steels. In thin sections, the thermal cycle may be such that no part of the HAZ is at sensitizing temperatures long enough to cause carbide precipitation. Once the precipitation has occurred, it can be removed by reheating the alloy to above 1035 °C (1895 °F) and cooling it rapidly.

Avoiding Intergranular Corrosion. Susceptibility to intergranular corrosion in austenitic stainless steels can be avoided by controlling their carbon contents or by adding elements whose carbides are more stable than those of chromium. For most austenitic stainless steels, restricting their carbon contents to 0.03% or less will prevent sensitization during welding and most heat treatment. This method is not effective for eliminating sensitization that would result from long-term service exposure at 425 to 815 °C (800 to 1500 °F).

Titanium and niobium form more stable carbides than chromium and are added to stainless steels to form these stable carbides, which remove carbon from solid solution and prevent precipitation of chromium carbides. The most common of these stabilized grades are types 321 and 347. Type 321 contains a minimum of 5 × C + N % titanium, and type 347 a minimum of 8 × C % niobium. Nitrogen must be considered when titanium is used as a stabilizer, not because the precipitation of chromium nitride is a problem in austenitic steels, but because titanium nitride is very stable. Titanium will combine with any available nitrogen; therefore, this reaction must be considered when determining the total amount of titanium required to combine with the carbon.

The stabilized grades are more resistant to sensitization by long-term exposure at 425 to 815 °C (800 to 1500 °F) than the low-carbon grades are, and the stabilized grades are the preferred materials when service involves exposure at these temperatures. For maximum resistance to intergranular corrosion, these grades are given a stabilizing heat treatment at about 900 °C (1650 °F). The purpose of the treatment is to remove carbon from solution at temperatures where titanium and niobium carbides are stable but chromium carbides are not. Such treatments prevent the formation of chromium carbide when the steel is exposed to lower temperatures.

Figure 6 illustrates how both carbon control and stabilization can eliminate intergranular corrosion in as-welded austenitic stainless steels. It also shows that the sensitized zone in these steels is somewhat removed from the weld metal.

Knife-Line Attack. Stabilized austenitic stainless steels may become susceptible to a localized form of intergranular corrosion known as knife-line attack or knife-line corrosion. During welding, the base metal immediately adjacent to the fusion line is heated to temperatures high enough to dissolve the stabilizing carbides, but the cooling rate is rapid enough to prevent carbide precipitation. Subsequent welding passes reheat this narrow area into the temperature range in which both the stabilizing carbide and the chromium carbide can precipitate. The precipitation of chromium carbide leaves the narrow band adjacent to the fusion line susceptible to intergranular corrosion.

Knife-line attack can be avoided by the proper choice of welding variables and by the use of stabilizing heat treatments. Additional information on knife-line attack can be found in the article "Corrosion of Weldments" in this Volume.

Testing for Intergranular Corrosion. Although testing for intergranular corrosion is discussed in the article "Evaluation of Intergranular Corrosion" in this Volume, a brief discussion is included here. The common methods of testing austenitic stainless steels for susceptibility to intergranular corrosion are described in ASTM A 262 (Ref 5). There are five acid immersion tests and one etching test. The oxalic acid etch test is used to screen samples to determine the need for further testing. Samples that have acceptable microstructures are considered to be insusceptible to intergranular corrosion and require no further testing. Samples with microstructures indicative of carbide precipitation must be subjected to one of the immersion tests.

Several electrochemical tests based on the polarization behavior of susceptible and insusceptible stainless steels have been proposed (Ref 6, 7). Although the tests have received considerable attention, none has yet been adopted in a national standard.

Intergranular Stress-Corrosion Cracking. Austenitic stainless steels that are susceptible to intergranular corrosion are also subject to intergranular SCC. The problem of the intergranular SCC of sensitized austenitic stainless steels in boiling high-purity water containing oxygen has received a great deal of study. This seemingly benign environment has led to cracking of sensitized stainless steels in many boiling water reactors, as described in the article "Corrosion in the Nuclear Power Industry" in this Volume.

Sensitized stainless alloys of all types crack very rapidly in the polythionic acid that forms during the shutdown of desulfurization units in petroleum refineries (Ref 8, 9). Because this service involves long-term exposure of sensitizing temperatures, the stabilized grades should be used. More detailed information can be found in the article "Corrosion in Petroleum Refining and Petrochemical Operations" in this Volume.

Effect of Ferrite and Martensite. Phases other than carbides can also influence the corrosion behavior of austenitic stainless steels. Ferrite, which is the result of an unbalanced composition, appears to reduce the pitting resistance of the steels. The presence of martensite may render the steels susceptible to hydrogen embrittlement under some conditions. The martensite can be produced by the deformation of unstable

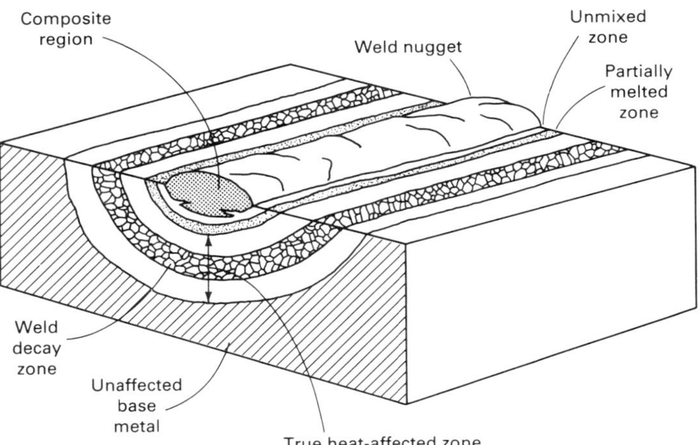

Fig. 5 Schematic diagram of components of weldment in austenitic stainless steel. Source: Ref 4

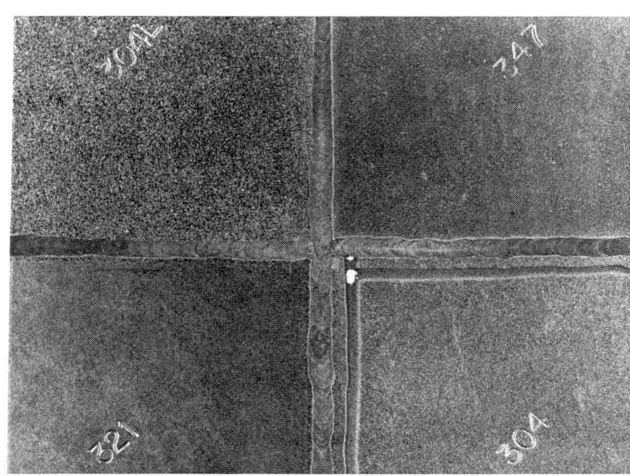

Fig. 6 Weld decay and methods for its prevention. The four different panels were joined by welding and then exposed to a hot solution of HNO_3/HF. Weld decay, such as that shown in the type 304 steel (bottom right), is prevented by reduction of the carbon content (type 304L, top left) or by stabilization with titanium (type 321, bottom left) or niobium (type 347, top right). Source Ref 1

austenite. Although this phenomenon can occur in a number of commercial stainless steels, it is most common in the lower-nickel steels such as type 301, in which the transformation is used to increase formability.

Effect of Sigma Phase. The effect of σ phase on the corrosion behavior of austenitic stainless steel has received considerable attention. This hard, brittle intermetallic phase precipitates in the same temperature range as chromium carbide and may produce susceptibility to intergranular corrosion in some environments.

Because it is hard and brittle, σ phase affects mechanical as well as corrosion properties. Although it is often associated with δ-ferrite, it can form directly from austenite.

The effects of σ phase on the corrosion behavior of austenitic stainless steels are most serious in highly oxidizing environments. The problem is of practical concern only if the phase is continuous. Although discrete particles of σ phase may be attacked directly, such corrosion does not seem to contribute significantly to the penetration of the steel.

The most important corrosion problem with σ phase in austenitic stainless steels occurs before it is microscopically resolvable (Ref 10). When the low-carbon molybdenum-containing austenitic stainless steels (such as type 316L and CF3M) or the stabilized grades (such as type 321 and type 347) are exposed at 675 °C (1245 °F), they may become susceptible to intergranular corrosion in nitric acid (HNO_3) and, in some cases, $Fe_2(SO_4)_3$-H_2SO_4. This susceptibility cannot be explained by carbide precipitation, and σ phase usually cannot be found in the optical microstructure. However, because some of the susceptible steels do exhibit continuous networks of σ phase, it has been assumed that this constituent is the cause of the intergranular corrosion. The hypothesis is that even when σ phase is not visible in the optical microstructure its effects are felt through some precursor or invisible phase. Invisible σ phase must be considered when testing for susceptibility to intergranular corrosion, but it seems to affect corrosion resistance only in very oxidizing environments, such as HNO_3.

Unsensitized austenitic stainless steels (that is, solution-annealed material containing no carbides or other deleterious phases) are subject to intergranular corrosion in very highly oxidizing environments, such as HNO_3 containing hexavalent chromium (Ref 11). None of the regularly controlled metallurgical variables influences this type of intergranular attack. Additional information on, and micrographs of, σ phase in austenitic stainless steels can be found in the article "Wrought Stainless Steels" in Volume 9 of the 9th Edition of *Metals Handbook*.

Ferritic Stainless Steels

Intergranular Corrosion. The mechanism for intergranular corrosion in ferritic stainless steels is largely accepted as being the same as that in austenitic stainless steels. Chromium compounds precipitate at grain boundaries, and this causes chromium depletion in the grains immediately adjacent to the boundaries (Ref 12, 13). This lowering of the chromium content leads to increased corrosion rates in the oxidizing solutions usually used to evaluate intergranular corrosion.

There are several differences between the sensitization of ferritic and austenitic stainless steels to intergranular corrosion. The first is that the solubility of nitrogen in austenite is great enough that chromium nitride precipitation is not a significant cause of intergranular corrosion in austenitic steels. It is, however, a significant cause in ferritic stainless steels. The second is the temperature at which it occurs. Sensitization in austenitic steels is produced by heating between 425 and 815 °C (800 and 1500 °F). In conventional ferritic alloys, sensitization is caused by heating above 925 °C (1700 °F). This difference is the result of the relative solubilities of carbon and nitrogen in ferrite and austenite.

Table 1 Results of ASTM A 763, Practice Z, on representative as-welded ferritic stainless steels

Welds were made using the gas tungsten arc welding technique with no filler metal added.

Alloy	Interstitial content, wt% C	N	Result
18Cr-2Mo	0.002	0.004	Pass
	0.010	0.004	Fail
	0.002	0.009	Fail
26Cr-1Mo	0.002	0.005	Pass
	0.004	0.010	Partial failure
	0.003	0.016	Fail
	0.013	0.006	Fail

Source: Ref 14

Table 2 Corrosion rates of 26% Cr ferritic stainless steels containing 0 to 3% Mo that were annealed for 15 min at 900 °C (1650 °F), water quenched, annealed for increasing times at 620 °C (1150 °F), then water quenched

Testing was performed according to recommendations in ASTM A 763, Practice X (ferric sulfate-sulfuric acid test).

	Corrosion rate, $mg/dm^2/d$						
		Annealing time at 620 °C (1150 °F)					
Alloy	900-°C (1650-°F) anneal	10 min	30 min	1 h	2 h	4 h	5 h
26-0	50	15 700	270	62	81	85	43
		15 600	264	50	67	85	43
26-1	43	5950(a)	8030(a)	990	50	40	53
	37	8220(a)	12 400(a)	890	50	37	50
26-2	78	15 600	940	138	80	74	270
	77	15 500	500	132	80	70	226
26-3	50	104	214	258	98	102	58
	50	95	160	96	93	97	58

(a) 56 h in test solution. Source: Ref 17

Because the sensitization temperatures are different for austenitic and ferritic steels, it is not surprising that the welding of susceptible steels produces different zones of intergranular corrosion. In austenitic steels, intergranular corrosion occurs at some distance from the weld, where the peak temperature reached during welding is approximately 675 °C (1250 °F). Because the sensitization of ferritic stainless steels occurs at higher temperatures, the fusion zone and the weld itself are the most likely areas for intergranular corrosion. Detailed information on sensitization and corrosion of ferritic stainless steels welds is available in the article "Corrosion of Weldments" in this Volume.

The mere presence of chromium carbides and nitrides in ferritic stainless steels does not ensure that they will be subject to intergranular corrosion. On the contrary, the usual annealing treatment for conventional ferritic stainless steels is one that precipitates the carbides and nitrides at temperatures (700 to 925 °C, or 1300 to 1700 °F) at which the chromium can diffuse back into the depleted zones. These same treatments would of course sensitize austenitic stainless steels because of the much slower rate of diffusion of chromium in austenite.

Avoiding Intergranular Corrosion. Clearly, the most straightforward method of preventing intergranular attack in ferritic stainless steels is to restrict their interstitial contents. The results shown in Table 1 give an indication of the levels of carbon and nitrogen required to avoid intergranular corrosion of iron-chromium-molybdenum alloys in boiling 16% H_2SO_4-copper-copper sulfate ($CuSO_4$) solutions. Evaluation was by bending. The samples that passed had no cracks.

For 18Cr-2Mo alloys to be immune to intergranular corrosion, it appears that the maximum level of carbon plus nitrogen is 60 to 80 ppm; for 26Cr-1Mo steels, this level rises to around 150 ppm. The notation of partial failure for the 26Cr-1Mo steel containing 0.004% C and 0.010% N indicates that only a few grain boundaries opened upon bending and that it probably represents the limiting composition. Using the 50% H_2SO_4-$Fe(SO_4)_3$ test, it was determined that the interstitial limits for the 29Cr-4Mo steel were 0.010% C (max) and 0.020% N (max), with the additional restriction that the combined total not exceed 250 ppm (Ref 13). As their alloy contents increase, the iron-chromium-molybdenum steels seem to grow more tolerant of interstitials with regard to intergranular corrosion.

The levels of carbon and nitrogen that are needed to keep 18Cr-Mo alloys free of intergranular corrosion are such that very low interstitial versions of 18% Cr alloys have received little commercial attention. The 26Cr-1Mo and 29Cr-4Mo steels have been made in considerable quantity with very low interstitials, for example, 20 ppm C and 100 ppm N.

The low-interstitial ferritic stainless steels respond to heat treatment in a manner somewhat similar to that of austenitic stainless steels. As the results of welding in Table 1 show, rapid cooling from high temperature will preserve resistance to intergranular corrosion. However, depending on alloy content and interstitial levels, these alloys may be sensitive to a cooling rate from temperatures above about 600 °C (1110 °F) (Ref 2, 15). Less pure iron-chromium-molybdenum alloys can also be affected by a cooling rate from around 800 °C (1470 °F), but at higher temperatures, it is impossible to quench them fast enough to avoid intergranular attack.

Table 3 Results of ASTM A 763, Practice Z, tests on as-welded ferritic stainless steels with titanium or niobium

Welds were made using gas tungsten arc welding with no filler metal added.

Alloy	(C + N), wt%	Ti, wt%	Nb, wt%	$\frac{\text{Ti or Nb}}{(C+N)}$, %	Result
18Cr-2Mo	0.022	0.16	...	7.3	Fail
	0.028	0.19	...	6.8	Fail
	0.027	0.23	...	8.5	Pass
	0.057	0.37	...	6.5	Pass
	0.079	0.47	...	5.9	Pass
18Cr-2Mo	0.067	...	0.32	4.8	Fail
	0.067	...	0.61	9.1	Pass
	0.030	...	0.19	6.3	Pass
26Cr-1Mo	0.026	0.17	...	6.5	Fail
	0.026	0.22	...	8.5	Fail
	0.026	0.26	...	10.0	Pass
26Cr-1Mo	0.026	...	0.17	6.5	Fail
	0.025	...	0.33	13.2	Pass

Source: Ref 14

Table 4 The effect of crystal structure on the corrosion behavior of an Fe-47Cr alloy

Solution	Corrosion rate, g/dm²/d: Ferrite	Corrosion rate, g/dm²/d: σ phase	Ratio(a)
Reducing			
10% HCl boiling	1461	7543	5.2
10% H_2SO_4 boiling	2939	7422	2.5
50% H_2SO_4 boiling	5088	5280	1.04
Oxidizing			
50% H_2SO_4 + $Fe_2(SO_4)_3$ boiling	0.0195	0.196	10
50% H_2SO_4 + $CuSO_4$ boiling	0.0170	0.415	24
65% HNO_3 boiling	0.0205	0.861	42
HNO_3 + HF at 65 °C (150 °F)	0.00	0.06	...
Pitting			
10% $FeCl_3{\cdot}6H_2O$ at room temperature	0.00	2.5	...

(a) Corrosion rate of σ phase ÷ corrosion rate of ferrite. Source: Ref 25

Isothermal heat treatments can also produce sensitivity to intergranular corrosion in low-interstitial ferritic stainless steels (Ref 16). For example, the effects of annealing at 620 °C (1150 °F) on the intergranular corrosion of 26% Cr alloys with 0 to 3% Mo were studied (Ref 17). The alloys contained 0.007 to 0.013% C and 0.020 to 0.024% N. As little as 10 min at temperature can lead to intergranular corrosion; however, continuing the treatment for 1 to 2 h can cure the damage (Table 2). Increasing the molybdenum content delays the onset of sensitization and makes it less severe. It does, however, delay recovery.

The very low levels of interstitials needed to ensure that ferritic stainless steels are immune to intergranular corrosion suggest that stabilizing elements might offer a means of preventing this type of corrosion without such restrictive limits on the carbon and nitrogen. Both titanium and niobium can be used, and each has its advantages (Ref 18). In general, weld ductility is somewhat better in the titanium-containing alloys, but the toughness of the niobium steels is better. As noted above, titanium-stabilized alloys are not recommended for service in HNO_3, but the niobium-containing steels can be used in this environment. Additional information on materials selection for HNO_3 environments is available in the section "Corrosion by Nitric Acid" in the article "Corrosion in the Chemical Processing Industry" in this Volume.

Table 3 shows the results of Cu-$CuSO_4$-16% H_2SO_4 tests on 26Cr-1Mo and 18Cr-2Mo steels with additions of either titanium or niobium. Inspection of the data suggests that the required amount of titanium cannot be described by a simple ratio as it is in austenitic steels. The amount of titanium or niobium required for ferritic stainless steels to be immune to intergranular corrosion in the $CuSO_4$-16% H_2SO_4 test has been investigated (Ref 19). It has been determined that for 26Cr-1Mo and 18Cr-2Mo alloys, the minimum stabilizer is given by:

$$Ti + Nb = 0.2 + 4\,(C + N) \qquad \text{(Eq 1)}$$

According to Ref 19, these limits are valid for combined carbon and nitrogen contents in the range of 0.02 to 0.05%. It should be emphasized that the limits set in Eq 1 are truly minima and are needed in the final product if intergranular attack is to be avoided.

This guideline is empirical and cannot be explained on the basis of stoichiometry. The alloys in the study (Ref 19) were fully deoxidized with aluminum before the stabilizing additions were made. Therefore, it is unlikely that the excess stabilizer is required because it reacts with oxygen.

The susceptibility of titanium-stabilized steels to intergranular attack in HNO_3 has been noted earlier. Because there is evidence that titanium carbide can be directly attacked by HNO_3, this mechanism is usually used to explain intergranular corrosion in titanium-containing steels. Another explanation that could be advanced about the intergranular attack of titanium-bearing steels under highly oxidizing conditions is an invisible σ phase such as that encountered in type 316L and discussed above.

Testing for Intergranular Corrosion. Standardized test methods for detecting the susceptibility of ferritic stainless steels to intergranular corrosion are described in ASTM A 763 (Ref 20). The methods are similar to those described in A 262 (Ref 5) for austenitic stainless in that there is an oxalic acid etch test and three acid immersion tests. The principal difference between the two standards is the introduction of microscopic examination of samples exposed to the boiling acid solutions. The presence or absence of grain dropping becomes the acceptance criterion for these samples.

Effects of Austenite and Martensite. The austenitic and martensitic phases are discussed together for ferritic stainless steels because they are interrelated; one can occur as the result of the other.

High-purity iron-chromium alloys are ferritic at all temperatures up to the melting point if they contain more than about 12% Cr. However, the γ loop in iron-chromium alloys can be greatly expanded by the addition of carbon and nitrogen. For example, it was found that the ferrite-austenite boundary was extended to 29% Cr in alloys that contained 0.05% C and 0.25% N (Ref 21).

Although the formation of austenite in ferritic stainless steels can be avoided by restricting their interstitial contents or by combining the interstitials with such elements as titanium or niobium, many of the ferritic stainless steels that are produced commercially will undergo partial transformation to austenite. Once the austenite is formed, the question is then what it will transform into. In one study, for example, the transformation products were dependent on the chromium content and the cooling rate (Ref 22). Slow cooling leads to the transformation of austenite into ferrite and carbides in all of the steels examined, but quenching can either produce martensite or retain the austenite.

In addition, the martensite start (M_s) temperature for a 17% Cr steel was measured at 176 °C (349 °F), and it was found that the transformation was 90% complete at 93 °C (199 °F) (Ref 22). The M_s for a 21% Cr steel was −160 °C (−255 °F), and martensite did not form in quenched 25% Cr alloys. Untempered martensite obviously reduces the toughness and ductility of ferritic stainless steels, and its presence is one cause of the poor ductility of welded type 430. In discussing this work (Ref 22), other researchers observed that welded type 430 (17% Cr) had poor ductility but that welded type 442 (21% Cr) had good ductility (Ref 23). These findings were attributed to the transformation of austenite to martensite in the lower-chromium steel but not in the 21% Cr steel. Both weldments were subject to intergranular corrosion, however.

The austenite retained in the higher-chromium steels is saturated with carbon, and when it is heated into the carbide precipitation region to, for example, 760 °C (1400 °F), it loses carbon and becomes unstable enough to transform to martensite upon cooling. This transformation product must then be tempered to restore ductility.

Another study found that martensite in type 430 corroded at a higher rate than the surrounding ferrite in boiling 50% H_2SO_4 + $Fe_2(SO_4)$ (Ref 13). This difference was attributed to the partitioning of chromium between ferrite and austenite at high temperatures. Because the austenite is lower in chromium, the martensite that forms from it would also be lower in chromium. The 50% H_2SO_4-$Fe_2(SO_4)$ test is quite sensitive to changes in chromium content in the 12 to 18% Cr range (Fig. 4). The test is less sensitive at higher chromium contents; therefore, no preferential attack was noted in austenite formed in type 446. This same austenite was preferentially attacked by boiling 5% H_2SO_4, presumably because of its higher interstitial content.

These corrosion experiments help to elucidate the effect of metallurgical factors on the corrosion behavior of ferritic stainless steels. However, these experiments describe situations rarely encountered in practice, because the mechanical properties of steels with such microstructures limit their usefulness.

Effect of Sigma and Related Phases. In contrast to the case of austenitic steels, the occurrence of σ phase in most commercial ferritic stainless steels can be predicted from the iron-chromium phase diagram. Fortunately, the kinetics of σ formation are very sluggish, and σ phase is not normally encountered in the processing of commercial ferritic stainless steels.

The formation of σ phase in the Fe-Cr system has been thoroughly researched, and the literature has been well summarized (Ref 24). The phase has the nominal composition of FeCr, but it can dissolve about 5% of either iron or chromium. It forms congruently from ferrite at 815 °C (1500 °F). The sluggishness of the reaction makes it difficult to define the low-temperature limits of the σ-phase field, but the ferrite/ferrite + σ phase boundary has been estimated at 9.5% Cr at 480 °C (895 °F) (Ref 24). Cold work accelerates the precipitation of σ phase.

There is relatively little information on how σ phase affects the corrosion behavior of ferritic stainless steels; however, continuous networks would be expected to be more troublesome than isolated colonies. Because σ phase contains more chromium than ferrite, its presence could also affect the corrosion behavior by either local or general depletion of the chromium content of the matrix.

One study investigated the corrosion behavior of an Fe-47Cr alloy that was heat treated so that it was either entirely ferrite or entirely σ phase (Ref 25). These data are shown in Table 4. The types of environments studied induced reducing (active), oxidizing (passive), and pitting corrosion conditions. The differences were greatest in the oxidizing and pitting environments. These results indicate that σ phase is more likely to corrode than ferrite in many instances and that no chromium depletion mechanism need be invoked to explain how σ phase can reduce the corrosion resistance.

In molybdenum-containing ferritic steels, χ phase, which is closely related to σ phase, can be found (Ref 26). It occurs in the temperature range of 550 to 950 °C (1020 to 1740 °F). It has the nominal composition Fe_2CrMo, but there are deviations from stoichiometry. In an investigation of the effect of heat treatment on the microstructure of 29Cr-4Mo alloys, both χ and σ phases were found in material held in the 700- to 925-°C (1290- to 1695-°F) range (Ref 27). Long-term aging of the 29Cr-4Mo steel did not render it susceptible to intergranular corrosion in the boiling 50% H_2SO_4 + $Fe_2(SO_4)_3$ solution.

This work also included 29Cr-4Mo-2Ni alloys, and χ and σ phases were seen to form much more quickly in these steels than in nickel-free materials. This observation is consistent with earlier results that nickel additions up to about 2% can accelerate the formation of σ phase in iron-chromium alloys (Ref 28). At higher levels, nickel decreases the rate of σ-phase precipitation. Sigma and χ reduce the ductility of the 29Cr-4Mo-2Ni alloys, but do not cause it to undergo intergranular corrosion. However, long-term aging at 815 °C (1500 °F) did render them susceptible to crevice corrosion in 10% hydrated ferric chloride ($FeCl_3 \cdot 6H_2O$) at 50 °C (120 °F). In this case, the ferrite was preferentially attacked—perhaps because it was depleted in chromium and molybdenum by precipitation of the second phase.

There is some evidence that the invisible χ or σ phase may affect the properties of stabilized 18Cr-2Mo ferritic stainless steels aged at approximately 620 °C (1150 °F). For example, it was shown that aging for even relatively short times could produce extensive intergranular corrosion in 18Cr-2Mo-Ti steels exposed to boiling 50% H_2SO_4 + $Fe_2(SO_4)_3$ (Ref 29). The steels were not subject to intergranular attack in 10% HNO_3 + 3% hydrofluoric acid (HF) or in boiling 16% H_2SO_4 + 6% $CuSO_4$ + Cu, and both of these solutions are known to produce intergranular attack in improperly stabilized ferritic stainless steels. Similar behavior has been noted in niobium-stabilized 18Cr-2Mo steels (Ref 30). In neither case was χ or σ phase clearly present at the grain boundaries.

Duplex Stainless Steels

Duplex stainless steels are those that are composed of a mixture of austenite and ferrite. The common cast stainless steels, such as CF-8 and CF-8M, are mostly austenite with some ferrite. These alloys are often considered to be simple analogs of wrought alloys with similar compositions; however, they do not always have the same response to heat treatment. The corrosion evaluation of these alloys deserves further study. Additional information on cast duplex stainless steels can be found in the article "Corrosion of Cast Steels" in this Volume.

Wrought duplex stainless steel may have either a ferrite matrix (type 329) or an austenitic matrix (U50). The most recent duplex alloys, such as 2205, are approximately 50:50 mixtures. The modern alloys are produced with low carbon contents, usually less than 0.03%, and intergranular corrosion resulting from carbide precipitation generally has not been a practical problem.

These alloys are usually high in chromium (25 to 27%) and molybdenum (2 to 4%). As a result, these alloys are prone to the formation of intermetallic phases such as σ and χ if they are not cooled rapidly through the 900- to 700-°C (1650- to 1290-°F) range (Ref 31). Although these intermetallic compounds do affect the corrosion resistance of the alloys, they have a more drastic effect on the mechanical properties. If a duplex alloy has satisfactory mechanical properties, it probably will not experience intergranular corrosion. In both wrought and cast alloys, it appears that the high rate of diffusion of chromium in the ferrite generally minimizes depleted zones and, therefore, intergranular corrosion.

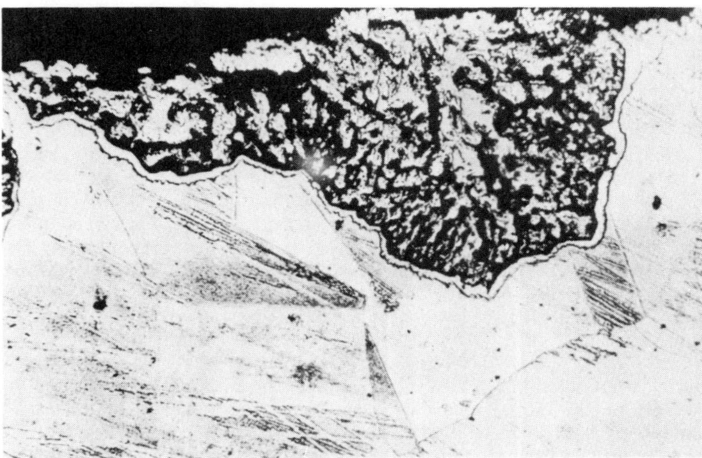

Fig. 7 Plug-type dezincification in an α-brass (70Cu-30Zn) exposed for 79 days in 1 *N* NaCl at room temperature. Note porous structure within the plug. The dark line surrounding the plug is an etching artifact. 160×. Source: Ref 61

Fig. 8 Uniform-layer dezincification in an admiralty brass 19-mm (3/4-in.) diam heat-exchanger tube. The top layer of the micrograph, which consists of porous, disintegrated particles of copper, was from the inner surface of the tube that was exposed to water at pH 8.0, 31 to 49 °C (87 to 120 °F), and 207 kPa (30 psi). Below the dezincified layer is the bright yellow, intact, admiralty brass outer tube wall. 35×. Courtesy of James J. Dillon. Permission granted by Nalco Chemical Company, 1987

(a)

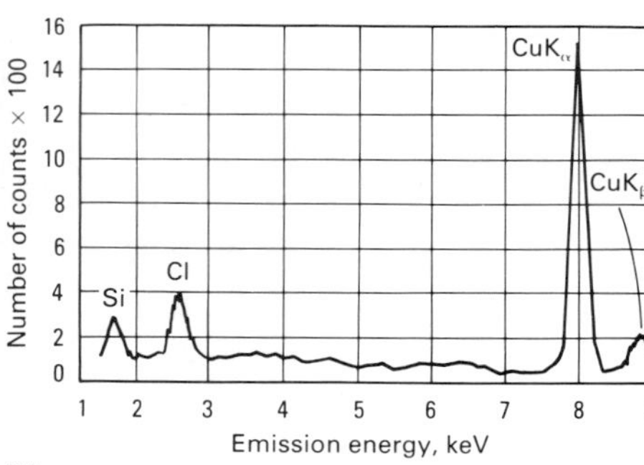

(b)

Fig. 9 Scanning electron micrograph (a) of copper deposit on the surface of 70Cu-30Zn brass specimen exposed for 10 days in 5 *N* HCl at 50 °C (120 °F). 780×. (b) Energy-dispersive x-ray spectra of the copper deposit. Source: Ref 61

Metallurgical Effects on the Corrosion of High-Nickel Alloys

The metallurgy of high-nickel corrosion-resistant alloys is more complicated than that of the iron-chromium-nickel and iron-chromium-nickel-molybdenum austenitic stainless steels. As the nickel content rises, carbon becomes less soluble in the matrix. In austenitic stainless steels, carbides usually precipitate as $M_{23}C_6$, where M is principally chromium. This form of carbide can also be found in sensitized high-nickel alloys, but M_3C_7 and M_6C carbides are also found. Molybdenum may be a constituent in these carbides.

Whether or not these carbides damage the corrosion resistance of these alloys depends on the temperature at which they precipitate. Continuous grain-boundary precipitates are likely to produce depleted zones that will lead to susceptibility to intergranular corrosion. Isolated carbides that persist from high temperatures will not affect corrosion behavior.

Various intermetallic phases, such as σ; Laves (or η), for example, Fe_2Mo; and $(Ni,Cr)_7Mo_6$, can also form in these alloys and affect their corrosion performance. These intermetallic phases are often the major cause of intergranular corrosion in nickel-base alloys.

The intergranular corrosion behavior of high-nickel alloys was reviewed in considerable detail (Ref 32, 33). Isothermal heat treatments produced susceptibility to intergranular corrosion in all the alloys examined in these studies.

It is clear that keeping the carbon as low as possible is beneficial to the resistance of these alloys to intergranular corrosion. Alloying additions of titanium and niobium are also beneficial, although they do not produce stabilized alloys in the same sense that stainless steels are stabilized.

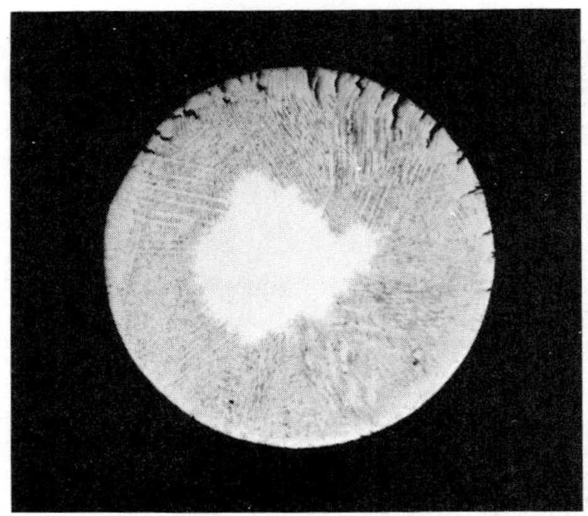

(a)

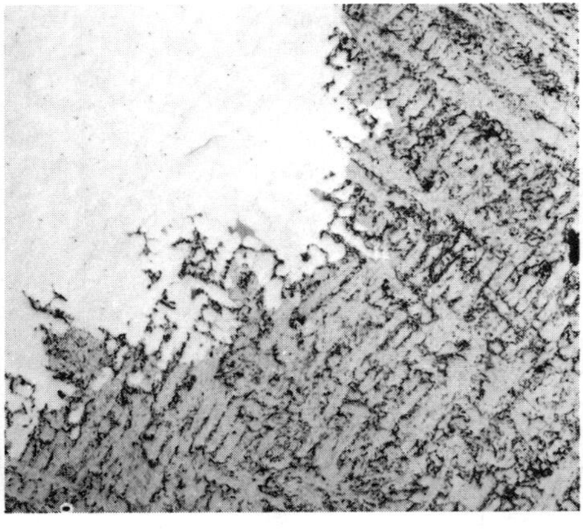

(b)

Fig. 10 Dezincification (a) of a silicon brass valve spindle. 4×. (b) Interface between sound metal (left) and dezincified region (right). 40×. Courtesy of P.J. Kenny, Ontario Research Foundation

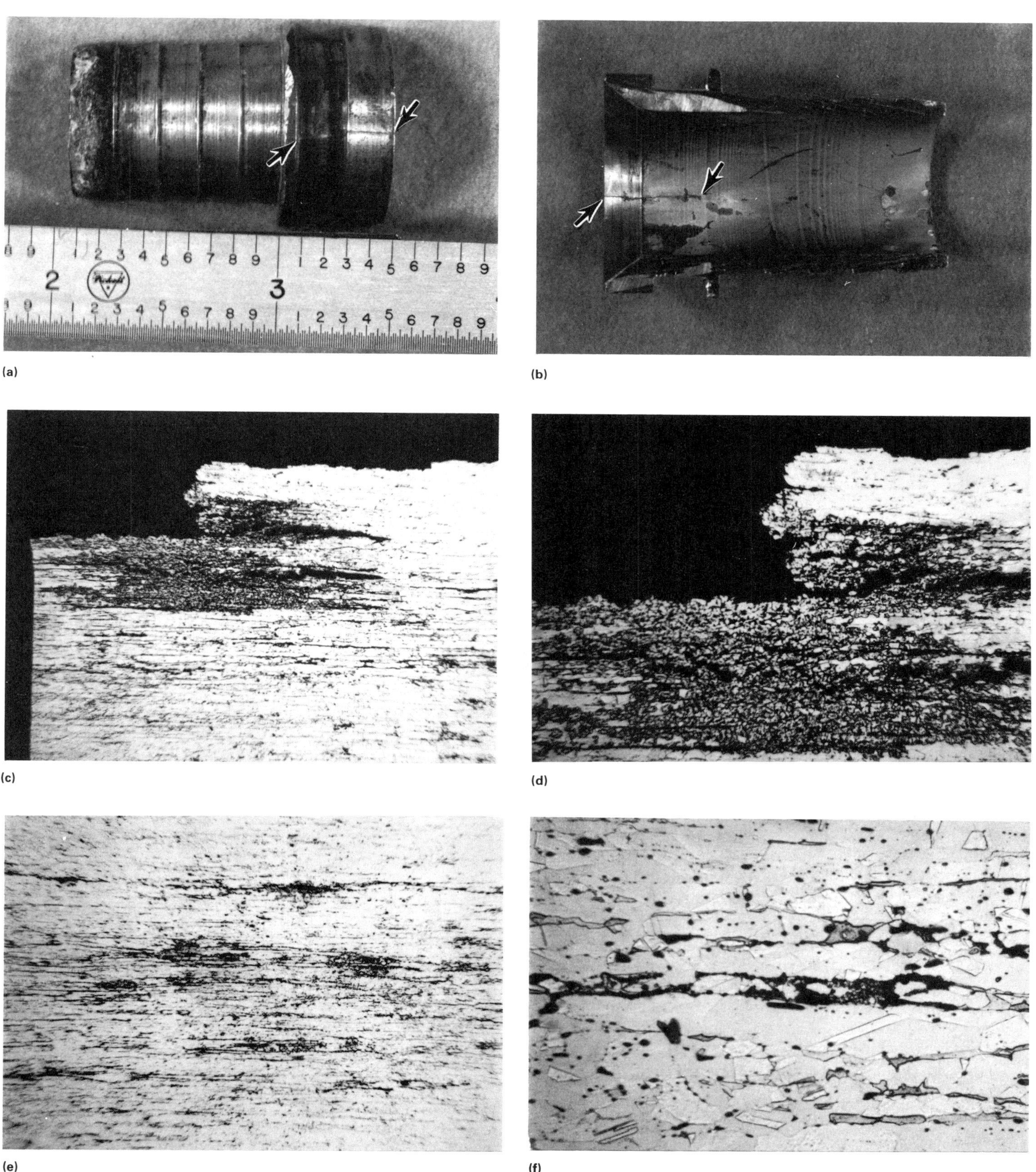

Fig. 11 Cracking due to dezincification of an α-β 60Cu-40Zn + Pb hose barb used to connect rubber hoses to aluminum flared tube fittings. (a) As-received barb with a crack (arrows) on the outside diameter. (b) Crack (arrows) on the inside diameter. (c) Area of plug-type dezincification. 35×. (d) Close-up of area in (c) showing the porous redeposited copper. 60×. (e) Small cracks prevalent throughout the fitting along stringers of β brass. 35×. (f) Detail of attack shown in (e). Initially, dezincification starts at β grains, then progresses along contiguous grain boundaries. 235×. Courtesy of W.W. Nash, Rockwell International

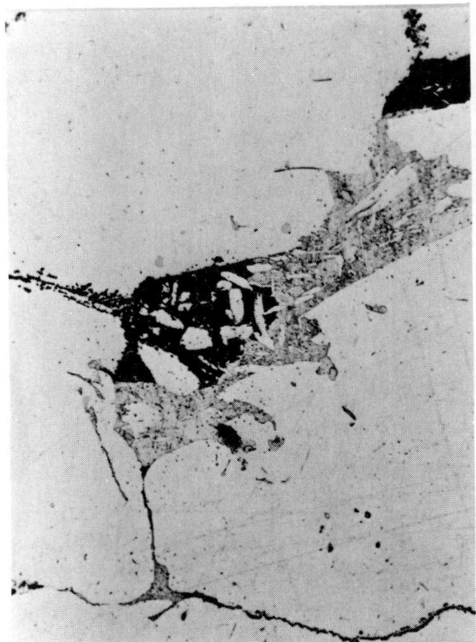

Fig. 12 Dealuminification of a cast aluminum bronze furnace electrode pressure ring exposed to recirculating cooling water (pH = 7.8 to 8.3, conductivity = 1000 to 1100 μS). The preferentially attacked γ phase left behind a residue of copper (darkened regions in eutectoid and along grain boundaries). The α particles within the eutectoid (light-gray areas) are unattacked. Etched with $FeCl_3$. 260×. Courtesy of Robert D. Port. Permission granted by Nalco Chemical Company, 1987

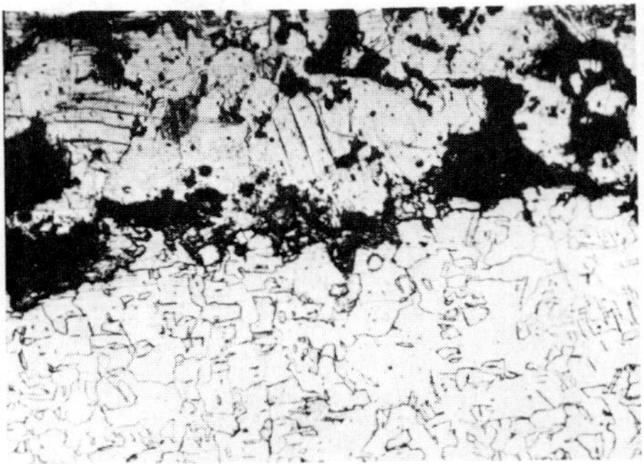

Fig. 13 Residual copper layer from a C71500 feedwater pressure tube that underwent denickelification. The tube was subjected to 205-°C (400-°F) steam on the external surface and boiling water on the internal surface (175 °C, or 350 °F, at pH 8.6 to 9.2). Courtesy of James J. Dillon. Permission granted by Nalco Chemical Company, 1987

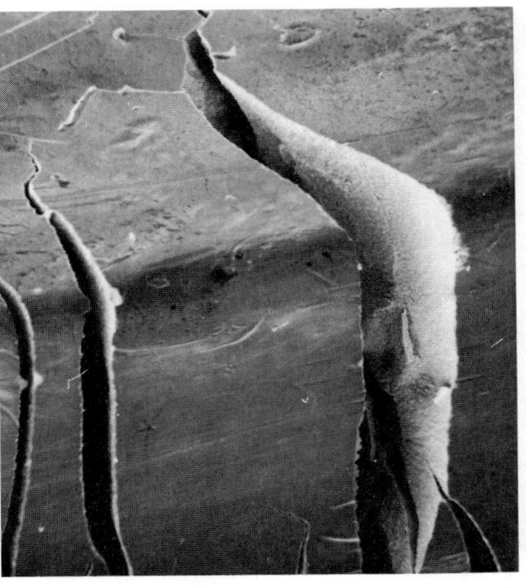

100 μm

Fig. 14 Cracks formed in gold sponge dealloyed layer of a Cu-25at.%Au single crystal upon bending in air following stress-free immersion in aqueous $FeCl_3$. Courtesy of B.D. Lichter, Vanderbilt University

In other words, no simple ratio of Ti,Nb to C can be given that will make a particular alloy immune to intergranular corrosion.

Fortunately, the problems produced by the isothermal heat treatment experiments are not usually manifested in welding. When welded by qualified procedures, most modern high-nickel alloys are resistant to intergranular corrosion. Additional information on the corrosion of nickel alloys and nickel alloy weldments can be found in the articles "Corrosion of Nickel-Base Alloys" and "Corrosion of Weldments," respectively, in this Volume.

Metallurgical Effects on the Corrosion of Aluminum Alloys

Intergranular corrosion in aluminum alloys can be the result of the direct attack of a precipitate that is less corrosion resistant (more active) than the matrix or the attack of a denuded zone adjacent to a noble phase.

Aluminum-magnesium alloys that contain more than 3% Mg (for example, 5083) may become susceptible to intergranular corrosion because of the preferential attack of Mg_2Al_8. In aluminum-magnesium-zinc alloys such as 7030, the compound $MgZn_2$ is attacked (Ref 34).

In aluminum-copper alloys such as 2024, $CuAl_2$ precipitates, which is more noble than the matrix. It appears to act as a cathode in accelerating the corrosion of a depleted zone adjacent to the grain boundary. A similar phenomenon seems to occur in aluminum-zinc-magnesium-copper alloys such as 7075 (Ref 34).

Although the aluminum alloys are more resistant to intergranular corrosion in the solution-treated condition, avoiding precipitates is not a practical means of avoiding intergranular corrosion in these systems. The precipitates are important for the strengthening of the alloys and are necessary for their performance. Whether or not the alloy will be subject to intergranular corrosion in a particular environment is an important part of the alloy selection process.

Exfoliation is a form of intergranular corrosion that may occur when aluminum alloys have their grains elongated in layers parallel to their surfaces. Intergranular corrosion can occur on the elongated grain boundaries. The corrosion product that forms has a greater volume than the volume of the parent metal. The increased volume forces the layers apart, and strips of metal exfoliate (delaminate). Additional information on the exfoliation and intergranular corrosion of aluminum alloys can be found in the articles "Corrosion of Aluminum and Aluminum Alloys" and "Evaluation of Exfoliation Corrosion" in this Volume.

Grooving Corrosion in Carbon Steel

When electric resistance welded (ERW) carbon steel pipe is exposed to aggressive waters, preferential corrosion, or grooving corrosion, of the weld is sometimes observed (Ref 35). This localized corrosion appears to be caused by the redistribution of sulfide inclusions along the weld line during the welding process.

It has been suggested that the welding process concentrates manganese sulfides in the weld and that the metal flow during the upset portion of the welding preferentially exposes the inclusions to the corrodent (Ref 35). The high temperatures in welding also break down the manganese sulfides, and this leads to the local enrichment of the matrix in sulfur or to the formation of iron sulfide.

Table 5 Combinations of alloys and environments subject to dealloying and elements preferentially removed

Alloy	Environment	Element removed
Brasses	Many waters, especially under stagnant conditions	Zinc (dezincification)
Gray iron	Soils, many waters	Iron (graphitic corrosion)
Aluminum bronzes	Hydrofluoric acid, acids containing chloride ions	Aluminum (dealuminification)
Silicon bronzes	High-temperature steam and acidic species	Silicon (desiliconification)
Tin bronzes	Hot brine or steam	Tin (destannification)
Copper nickels	High heat flux and low water velocity (in refinery condenser tubes)	Nickel (denickelification)
Copper-gold single crystals	Ferric chloride	Copper
Monels	Hydrofluoric and other acids	Copper in some acids, and nickel in others
Gold alloys with copper or silver	Sulfide solutions, human saliva	Copper, silver
High-nickel alloys	Molten salts	Chromium, iron,molybdenum, and tungsten
Medium- and high-carbon steels	Oxidizing atmospheres, hydrogen at high temperatures	Carbon (decarburization)
Iron-chromium alloys	High-temperature oxidizing atmospheres	Chromium, which forms a protective film
Nickel-molybdenum alloys	Oxygen at high temperature	Molybdenum

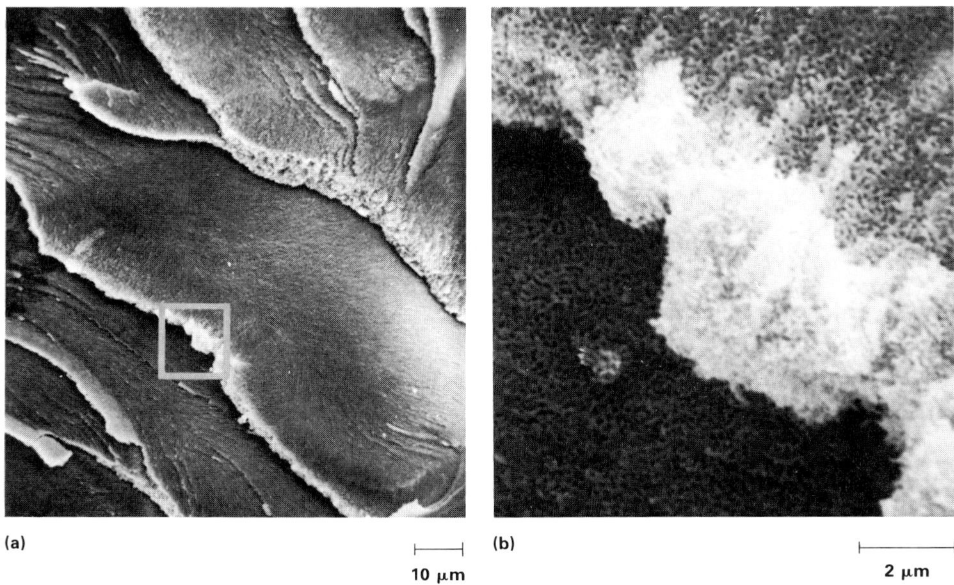

Fig. 15 Fracture surface of a Cu-25at.%Au single crystal upon bending in air following stress-free immersion in aqueous $FeCl_3$. (a) SEM micrograph of fracture surface and gold sponge. Note facet-step morphology. (b) High-resolution SEM micrograph of the boxed area shown in (a) showing facet-step structure in gold sponge. Note microporosity of surface. See also Fig. 16. Courtesy of B.D. Lichter, Vanderbilt University

Fig. 17 A 200-mm (8-in.) diam gray iron pipe that failed because of graphitic corrosion. The pipe was part of a subterranean fire control system. The external surface of the pipe was covered with soil; the internal surface was covered with water. Severe graphitic corrosion occurred along the bottom external surface where the pipe rested on the soil. The small-diameter piece in the foreground is a gray iron pump impeller on which the impeller vanes have disintegrated because of graphitic corrosion. See also Fig. 18. Courtesy of Robert D. Port. Permission granted by Nalco Chemical Company, 1987

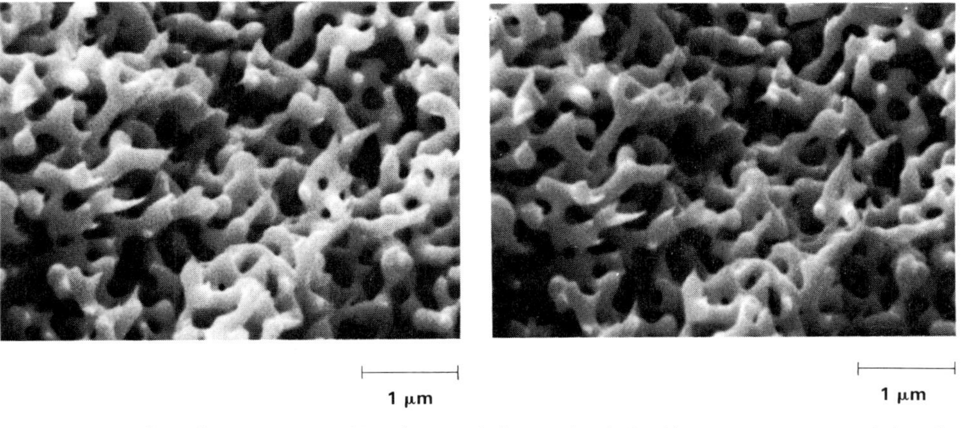

Fig. 16 High-resolution stereo pair SEM photograph showing detail of gold sponge structure revealed on the surface of the sample shown in Fig. 15. Average interpore spacing is ~200 nm. Evidence of ductile fracture can be seen on the surface (necking of the gold ligaments). Courtesy of B.D. Lichter, Vanderbilt University

Whether the redistribution of the sulfides makes the weld metal less corrosion resistant than the base metal or whether it makes the environment locally more corrosive is not clear. However, the weld metal is attacked at a greater rate than the base metal is. The grooves seem to form from the coalescence of pits that start near the sulfide inclusions.

In another study, the influence of sulfur content on the occurrence of grooving corrosion was investigated (Ref 36). It was concluded that sulfur content should not be a problem in steels containing less than 0.02% S. Postweld heat treatment also appears to influence susceptibility to grooving corrosion, although the data are somewhat mixed. Maximum susceptibility to grooving is produced by treatments at approximately 750 °C (1380 °F). Susceptibility decreases with higher-temperature heat treatments. Of the temperatures investigated, the highest, 1000 °C (1835 °F), was the most beneficial.

Dealloying Corrosion

Dealloying (also referred to as selective leaching or parting corrosion) is a corrosion process in which one constituent of an alloy is preferentially removed, leaving behind an altered residual structure (Ref 37). The phenomenon was first reported by Calvert and Johnson in 1866 on brass (copper-zinc) alloys (Ref 38). Since that time, dealloying has been reported in a number of copper-base alloy systems as well as in gray iron, noble metal alloys, medium- and high-carbon steels, iron-nickel-chromium alloys, and nickel-molybdenum alloys. Table 5 lists some of the alloy-environment combinations for which dealloying has been reported.

Mechanisms of Dealloying. Although numerous attempts have been made to clarify the mechanisms of dealloying, two theories are most prevalent. In the first, two metals in an alloy are dissolved, and one redeposits on the surface (Ref 39-52). In the second theory, one metal is selectively dissolved from an alloy, leaving a porous residue of the more noble species (Ref 53-60). Still others believe that both mechanisms take place (Ref 61-69).

Regardless of the mechanism, however, the metal in the affected area becomes porous and loses much of its strength, hardness, and ductility. Failure may be sudden and unexpected because dimensional changes do not always occur and the corrosion sometimes apears to be superficial, although the selective attack may have left only a small fraction of the orginal thickness of the part unaffected.

Dealloying in Aqueous Environments

Dezincification. As described in the article "Corrosion of Copper and Copper Alloys" in this Volume, copper-zinc alloys containing more than 15% Zn are susceptible to a dealloying process called dezincification. Dezincification is the most common form of dealloying. In the dezincification of brass, selective removal of zinc leaves a relatively porous and weak layer of copper and copper oxide (analyses of dezincified areas usually indicate 90 to 95% Cu, with the remainder being copper

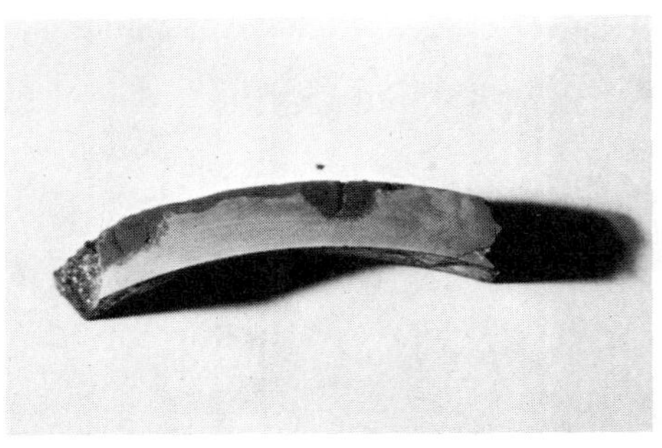
(a)

(b)

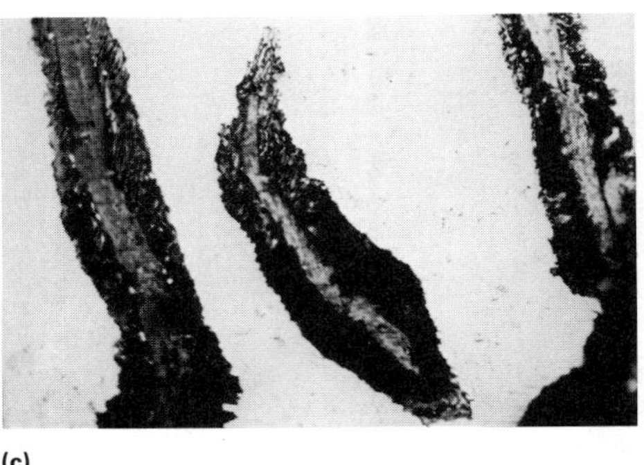
(c)

Fig. 18 External surface (a) of the gray iron pipe shown in Fig. 17 exhibiting severe graphitic corrosion. (b) Close-up of the graphitically corroded region shown in (a). (c) Micrograph of symmetrical envelopes of graphitically corroded cast iron surrounding flakes of graphite. Etched with nital. 530×. Courtesy of Harvey M. Herro. Permission granted by Nalco Chemical Company, 1987

oxide). Corrosion of a similar nature continues beneath the primary corrosion layer, resulting in gradual replacement of sound brass by weak, porous copper. Unless arrested, dealloying eventually penetrates the metal, weakening it structurally and allowing liquids or gases to leak through the porous mass in the remaining structure.

Dezincification may be either plug-type or uniform. The term plug-type dealloying refers to the dealloying that occurs in local areas; surrounding areas are usually unaffected or only slightly corroded. An example of plug-type dezincification in 70Cu-30Zn α-brass is shown in Fig. 7. In uniform-layer dealloying, the active component of the alloy is leached out over a broad area of the surface (Fig. 8). Dezincification is the usual form of corrosion for uninhibited brasses in prolonged contact with waters high in oxygen and carbon dioxide (CO_2). It is frequently encountered with quiescent or slowly moving solutions. Slightly acidic water, low in salt content and at room temperature, is likely to produce uniform attack, but neutral or alkaline water, high in salt content and above room temperature, often produces plug-type attack. Additional examples of dealloying corrosion are shown in Fig. 9 to 11.

Brasses with copper contents of 85% or more resist dezincification. Dezincification of brasses with two-phase structures is generally more severe, particularly if the second phase is continuous; it usually occurs in two stages: first in the high-zinc β phase, followed by the lower-zinc α phase.

Dezincification also appears to be one of the principal contributing factors in the SCC of copper-zinc and copper-zinc-tin alloys. The preferential dissolution or loss of zinc at the fracture interface during SCC results in the corrosion products having a higher concentration of zinc than the adjacent alloy. This dynamic loss of zinc near the crack aids in propagating the stress-corrosion fracture. Additional information on the role of dealloying in SCC behavior in copper-base alloys can be found in Ref 70 to 74 and in the article "Corrosion of Copper and Copper Alloys" in this Volume.

Tin tends to inhibit dealloying, especially in cast alloys. The addition of a small amount of phosphorus, arsenic, or antimony to admiralty metal (an all-α, 71Cu-28Zn-1Sn brass) inhibits dezincification. Inhibitors are not entirely effective in preventing dezincification of the α-β brass-

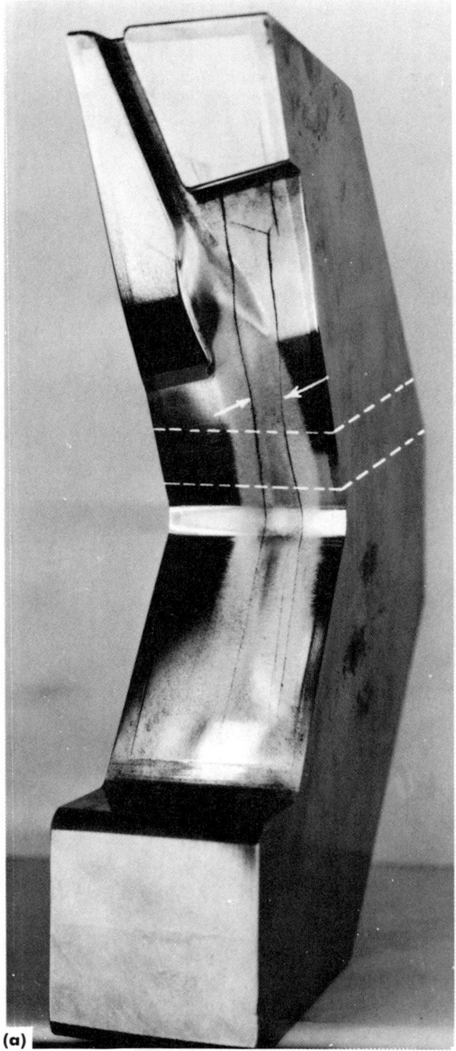
(a)

(b)

(c)

Fig. 19 Failed chromium-plated blanking die made from AISI A2 tool steel. (a) Cracking (arrows) that occurred shortly after the die was placed in service. (b) Cold etched (10% aqueous HNO_3 disk cut from the blanking die (outlined area) revealing a light-etching layer. Actual size. (c) Micrograph showing the decarburized layer that was unable to support the more brittle, hard chromium plating. Etched with 3% nital. 60×. Courtesy of G.F. Vander Voort, Carpenter Technology

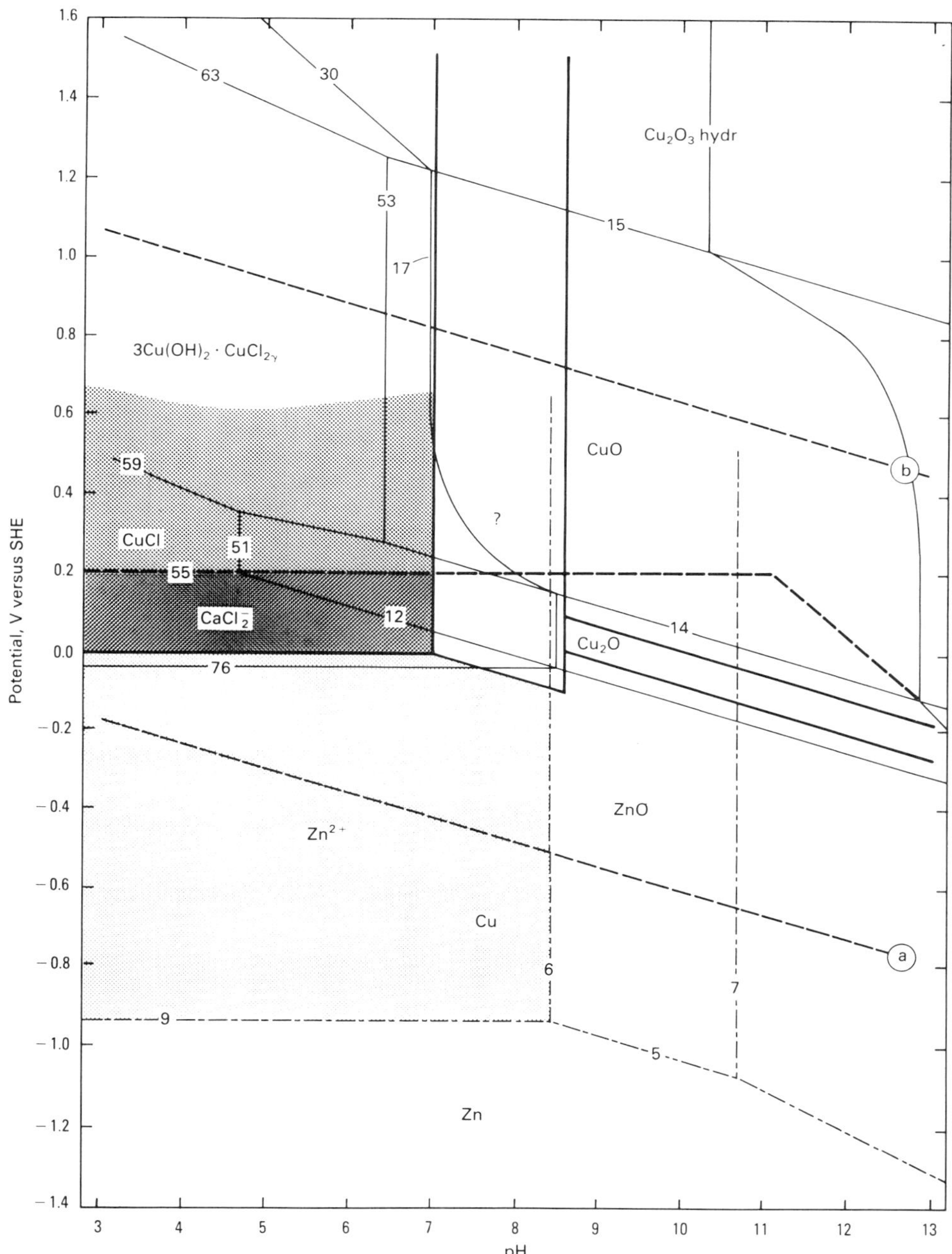

Fig. 20 Superimposed *E*-pH diagram of a 70Cu-30Zn alloy in 0.1 *M* NaCl. Lightly shaded area indicates the domain in which selective removal of zinc is expected in solutions free of copper ions. Intermediate shaded area indicates the domain in which both copper and zinc dissolve. Dark shaded area indicates the region in which copper is expected to deposit. Source: Ref 61

es, because they do not prevent dezincification of the β phase.

Where dezincification is a problem, red brass, commercial bronze, inhibited admiralty metal, and inhibited aluminum brass can be successfully used. Where selection of a low-zinc alloy is unacceptable, inhibited yellow brasses are generally preferred. Additional information on the effect of various alloying elements on dealloying corrosion and other forms of corrosion in copper-base materials can be found in the article "Corrosion of Copper and Copper Alloys" in this Volume.

Dealuminification. Dealloying occurs in some copper-aluminum alloys, particularly those containing more than 8% Al. It is especially severe in alloys with a continuous γ phase and usually occurs as plug-type dealloying. Figure 12 illustrates dealuminification of the γ-2 constituent in the β-eutectoid phase of an aluminum bronze (81Cu-11Al-4Fe-4Ni) casting. Heat treating to produce an α + β microstructure prevents dealuminification.

Denickelification. Dealloying of nickel in 70Cu-30Ni (C71500) is rare, but it has been observed at temperatures over 100 °C (212 °F), in low-flow conditions, and in high local heat flux. Figure 13 shows the residual copper layer of a 67.5Cu-31Ni-0.8Fe-0.7Mn alloy that underwent denickelification.

Destannification and Desiliconification. Dealloying of tin in cast tin bronzes has been observed as a rare occurrence in hot brine or steam. Silicon bronzes have been subject to desiliconification in isolated cases involving high-temperature steam plus acidic species (Ref 37).

Dealloying of Noble Metal Systems. In addition to the copper alloys discussed above, alloys such as copper-gold and silver-gold are subject to dealloying. Investigations into the failure mechanisms of Cu-25at.%Au single crystals have indicated that dealloying may be an important nucleation mechanism for transgranular SCC of this system (Ref 75). For example, cracks that formed on the tensile surface of a Cu-25at.%Au single-crystal sample after three-point bending in air and in the absence of a corroding medium are shown in Fig. 14. Prior to the bend test, the sample had been immersed in 2% ferric chloride ($FeCl_3$) for 10 days in the absence of applied stress. This resulted in the formation of an approximately 30-μm thick layer (gold-rich sponge). Most of the cracks in the sponge layer stopped at the boundary between the sponge and the unattacked alloy. However, some cracks propagated for distances up to 20 μm into the unattacked, normally ductile alloy.

The brittle fracture surface produced on an Cu-25at.%Au single-crystal sample afer three-point bending in air is shown in Fig. 15. The sample was produced by immersing a Cu-25at.%Au single crystal in 2% $FeCl_3$ for 30 days in the absence of applied stress. This resulted in a completely dealloyed (100% Au) sponge that was a single crystal having the same orientation as the origial Cu-25at.%Au alloy. The sponge sample failed in a brittle manner under an increasingly small applied load. However, it is likely that the actual failure occurred by ductile fracture of the gold ligaments constituting the sponge, as shown in Fig. 16.

In another study, microscopic examination of an electrochemically sulfidized polycrystalline Cu-13Au alloy (300-h anodic polarization at E_H = 0 mV in a sulfide-containing buffer solution of pH = 5 and pS^{2-} = 15) revealed the formation of a two-layer structure. This structure consisted of an outer layer of cuprous sulfide (Cu_2S) and an inner reaction layer that was comprised of porous gold-rich metal with Cu_2S inside its pores (Ref 76). It is believed that similar processes of preferential sulfidation are responsible for the tarnishing of silver-gold dental alloys in human saliva (Ref 77). Detailed information on the corrosion of noble metal alloys can be found in the articles "Corrosion of the Noble Metals" and "Tarnish and Corrosion of Dental Alloys" in this Volume.

Graphitic Corrosion. Perhaps the second most frequently observed type of dealloying is the graphitic corrosion of gray iron, which occurs in relatively mild aqueous environments and on buried pipe. The graphite in gray iron is cathodic to iron and remains behind as a porous mass when iron is leached out. Graphitic corrosion usually occurs at a low rate. The graphite mass is porous and very weak, and graphitic corrosion produces little or no change in metal thickness.

Graphitic corrosion does not occur in ductile iron or malleable iron, because no graphite network is present to hold together the residue. White iron has essentially no free carbon and is not subject to graphitic corrosion. Examples of graphitic corrosion are shown in Fig. 17 and 18. Three case histories of graphitic corrosion in gray iron components are presented in the article "Failures of Iron Castings" in Volume 11 of the 9th Edition of *Metals Handbook*.

Graphitic corrosion is often erroneously referred to as graphitization. Graphitization is a microstructural change that sometimes occurs in carbon or low-alloy steels that are subjected to moderately high temperatures (~455 to 595 °C, or 850 to 1100 °F) for extended periods of time (≥40 000 h). Graphitization results from the decomposition of pearlite into ferrite and carbon (graphite) and can embrittle steel parts, especially when the graphite particles form along a continuous zone through a load-carrying member.

Selective Dissolution in Molten Salts and Liquid Metals

Preferential removal of key alloying elements can also take place when multicomponent alloys are exposed to liquid-metal or molten-salt environments. For example, austenitic stainless steels exposed to liquid sodium lose nickel, and this results in the formation of a ferritic surface layer. More detailed explanations of corrosion phenomena in liquid-metal environments can be found in the articles "Fundamentals of High-Temperature Corrosion in Liquid Metals" and "General Corrosion" (see the section "Corrosion in Liquid Metals") in this Volume.

Selective dissolution in molten-salt environments differs from that in aqueous systems in that the rate of dissolution is related to the bulk diffusion of the selectively leached element. Bulk diffusion does not play a role in either of the two dealloying mechanisms discussed earlier in this article. In addition, selective dissolution in molten salts can occur even when the leached element is at low concentrations. The articles "Fundamentals of High-Temperature Corrosion in Molten Salts" and "General Corrosion" (see the section "Molten-Salt Corrosion") in this Volume discuss corrosion in molten fluorides, chloride salts, nitrates, sulfates, hydroxide melts, and carbonate melts in more detail.

Decarburization and Selective Oxidation

Preferential removal of alloying elements also occurs in high-temperature gaseous environments. Two examples are decarburizaton and selective oxidation of chromium.

Decarburization is a loss of carbon from the surface of a ferrous alloy as a result of heating in a medium that produces a carbon gradient. Unless special precautions are taken, the risk of losing carbon from the surface of steel is always present in any heating to high temperatures in an oxidizing atmosphere. A marked reduction in fatigue strength is noted in steels with decarburized surfaces.

Figure 19(a) shows a chromium-plated blanking die made from AISI A2 tool steel that cracked after limited service. Cold etching of a disk cut from the blanking die revealed a light-etching layer that is particularly prominent at the working face and along the adjacent sides (Fig. 19b). Microscopic examination (Fig. 19c) revealed that the surface at the working face was decarburized to a depth of about 0.05 mm (0.002 in.). The soft zone beneath the hard chromium plating permitted the plating to flex under the influence of the blanking stresses, and this caused cracking of the plating and surface region. Additional information on decarburization can be found in Volumes 11 and 12 of the 9th Edition of *Metals Handbook*.

Selective Oxidation of Chromium. Exposure of stainless steels to low-oxygen atmospheres at high temperatures (980 °C, or 1800 °F) has been shown to result in the selective oxidation of chromium (Ref 78). When there is competition for oxygen, the elements with higher free energies for their oxide formation (higher affinity for oxygen) are oxidized to a greater degree. In the case of stainless steels, this results in a more protective scale. More detailed information on oxide scale formation and high-temperature corrosion can be found in the articles "Fundamentals of Corrosion in Gases" and "General Corrosion" (see the section "High-Temperature Corrosion") in this Volume.

Evaluation of Dealloying Corrosion (Ref 61)

Reactions that lead to dealloying corrosion are relatively slow, and a lengthy exposure period is required to cause dealloying that is extensive enough to facilitate evaluation. Consequently, there is considerable interest in accelerated tests for evaluation of the dealloying tendencies of alloys. Many techniques have been implemented. For example, electrolyte compositions have been adjusted by using more concentrated solutions or solutions having variations in oxidizing power (Ref 67). Specific ions have been added to stimulate dealloying; for example, saturated cuprous chloride solutions have been used to accelerate the dezincification of copper-base alloys (Ref 79). Electrochemical stimulation has also been used. Unfortunately, the test methods employed can often be criticized as having biased the experimental result, and although specific techniques are now available that can cause dealloying to occur in the laboratory, there is still no firm basis for predicting the likelihood of dealloying in service based on these tests.

New techniques that may have predictive capacity in assessing the likelihood of dealloying involve the use of electrochemical hysteresis methods to generate experimental potential versus pH diagrams for alloys. Superposition of these experimental diagrams over the theoretical E-pH (Pourbaix diagrams) for the constituent metals of the alloy provides a basis for predicting the tendency for dealloying as a function of potential and pH. An example of such a superimposed diagram for an α-brass in 0.1 M sodium chloride (NaCl) is shown in Fig. 20. For initially copper-free chloride solutions, ranges of potential are indicated in which selective leaching of zinc predominates, in which alloy dissolution with replating is expected, and in which alloy dissolution without replating occurs. More detailed information on E-pH diagrams can be found in the article "Thermodynamics of Aqueous Corrosion" in this Volume.

REFERENCES

1. M.A. Streicher, in *Intergranular Corrosion of Stainless Alloys*, STP 656, R.F. Steigerwald, Ed., American Society for Testing and Materials, 1978, p 3-84
2. D. Van Rooyen, *Corrosion*, Vol 31, 1975, p 327-337
3. R.F. Steigerwald, *Metall. Trans.*, Vol 5, 1974, p 2265-2269
4. W.F. Savage, *Weld. Des. Eng.*, Dec 1969
5. "Standard Practices for Detecting Susceptibility to Intergranular Attack in Austenitic Stainless Steels," A 262, *Annual Book of ASTM Standards*, American Society for Testing and Materials
6. W.L. Clarke, R.L. Cowan, and W.L. Walker, in *Intergranular Corrosion of Stainless Alloys*, STP 656, R.F. Steigerwald, Ed., American Society for Testing and Materials, 1978, p 99-132
7. M. Akashi, T. Kawomoto, and F. Umemura, *Corros. Eng.*, Vol 29, 1980
8. A. Dravnieks and C.H. Samans, *Proc. API*, Vol 37 (No. 3), 1957, p 100
9. R.C. Scarberry, S.C. Pearman, and J.R. Crum, *Corrosion*, Vol 32, 1976, p 401-406
10. D. Warren, *Corrosion*, Vol 15, 1959, p 213t-220t
11. M.A. Streicher, *J. Electrochem. Soc.*, Vol 106, 1959, p 161-180
12. A.P. Bond, *Trans. Metall. Soc. AIME*, Vol 245, 1969, p 2127-2134
13. M.A. Streicher, *Corrosion*, Vol 29, 1973, p 337-360
14. R.F. Steigerwald, *Metalloved. Term. Obrab. Met.*, No. 7, 1973, p 16-20
15. R.J. Hodges, *Corrosion*, Vol 27, 1971, p 119-127
16. R.J. Hodges, *Corrosion*, Vol 27, 1971, p 164-167
17. C.R. Rarey and A.H. Aronson, *Corrosion*, Vol 28, 1972, p 255-258
18. A.P. Bond and E.A. Lizlovs, *J. Electrochem. Soc.*, Vol 116, 1969, p 1306-1311
19. H.J. Dundas and A.P. Bond, in *Intergranular Corrosion of Stainless Alloys*, STP 656, R.F. Steigerwald, Ed., American Society for Testing and Materials, 1978, p 154-178
20. "Standard Practices for Detecting Susceptibility to Intergranular Attack in Ferritic Stainless Steels," A 763, *Annual Book of ASTM Standards*, American Society for Testing and Materials
21. E.A. Baelecken, W.A. Fischer, and K. Lorenz, *Stahl Eisen*, Vol 81, 1961, p 768-778
22. A.E. Nehrenberg and P. Lillys, *Trans. ASM*, Vol 46, 1954, p 1176-1213
23. A.J. Lena, R.A. Lula, and G.C. Kiefer, *Trans. ASM*, Vol 46, 1954, p 1203-1205
24. D.C. Ludwigson and H.S. Link, in *Advances in the Technology of Stainless Steels and Related Alloys*, STP 369, American Society for Testing and Materials, 1965, p 249-310
25. R.F. Steigerwald and M.A. Streicher, Paper presented at the Annual Meeting, St. Louis, MO, National Association of Corrosion Engineers, 1965
26. J.G. McMullin, S.F. Reiter, and D.G. Ebeling, *Trans. ASM*, Vol 46, 1954, p 799-811
27. M.A. Streicher, *Corrosion*, Vol 30, 1974, p 115-125
28. A.J. Lena, *Met. Prog.*, Vol 66 (No. 1), 1954, p 86-90
29. E.A. Lizlovs and A.P. Bond, *J. Electrochem. Soc.*, Vol 122, 1975, p 589-593
30. H. Ogawa, Nippon Steel Corporation, unpublished research, 1974
31. H.D. Solomon and T.M. Devine, Paper 8201-089, presented at the ASM Metals Congress, St. Louis, MO, American Society for Metals, 1982
32. M.H. Brown, *Corrosion*, Vol 25, 1969, p 438
33. M.H. Brown and R.W. Kirchner, *Corrosion*,

Vol 29, 1973, p 470-474
34. H.P. Godard, W.E. Jepson, M.R. Bothwell, and R.L. Kane, *Corrosion of Light Metals*, John Wiley & Sons, 1967, p 70-73
35. C. Kato, Y. Otoguro, S. Kado, and Y. Hisamatsu, *Corros. Sci.*, Vol 18, 1978, p 61-74
36. C. Duran, E. Treiss, and G. Herbsleb, *Mater. Perform.*, Vol 25 (No. 9), 1986, p 41-48
37. R.H. Heidersbach, Jr., Dealloying, in *Forms of Corrosion—Recognition and Prevention*, C.P. Dillon, Ed., National Association of Corrosion Engineers, 1982, p 99-104
38. Calvert and J. Johnson, *J. Chem. Soc.*, Vol 19, 1866, p 436
39. W.D. Clark, *J. Inst. Met.*, Vol 73, 1947, p 263
40. M.G. Fontana, *Ind. Eng. Chem.*, Vol 39 (No. 5), 1947, p 87A
41. W. Lynes, *Proc. ASTM*, American Society for Testing and Materials, Vol 41, 1941, p 859
42. R.B. Abrams, *Trans. AES*, Vol 42, 1922, p 39
43. C.F. Nixon, *Trans. AES*, Vol 45, 1924, p 297
44. J.H. Hollomon and J. Wulff, *Trans. AIME*, Vol 147, 1942, p 297
45. D.B. Thompson, *Australas. Eng.*, Vol 48, Oct 1954
46. A.R. Zender and C.L. Bulow, *Heating, Piping, Air Cond.*, Vol 16, 1944, p 273
47. E.S. Dixon, *Bull. ASTM*, No. 102, 1940, p 21
48. F.H. Rhodes and J.T. Cary, *Ind. Eng. Chem.*, Vol 17, 1925, p 909
49. J.C. Scully, *The Fundamentals of Corrosion*, Pergamon Press, 1966
50. H.H. Uhlig, *The Corrosion Handbook*, John Wiley & Sons, 1948
51. K. Hashimoto, W. Ogawa, and S. Shimodaira, *J. Jpn. Inst. Met.*, Vol 4 (No. 1), 1963, p 42
52. R.M. Horton, *Corrosion*, Vol 26 (No. 7), 1971, p 160
53. F. Taylor and J.W. Wood, *Engineering*, Vol 149, 1940, p 58
54. L.E. Tabor, *J. Am. Water Works Assoc.*, Vol 48 (No. 3), 1963, p 239
55. S.S. Gastev, *Izvestiia, Akademii Nauk SSR Mettally*, Vol 3, 1965; referenced in *Corrosion Abstracts*, Vol 6, 1966, p 446
56. J.M. Bialosky, *Corros. Met. Prot.*, Vol 4, 1947, p 15
57. I.K. Marshakov, V.P. Bogdanov, and S.M. Aleikina, *Russ. J. Phys. Chem.*, Vol 38 (No. 7), 1964, p 960; Vol 38 (No. 8), 1964, p 104; Vol 39 (No. 6), 1965, p 804
58. W.H. Bassett, *Chem. Metall. Eng.*, Vol 27, 1922, p 340
59. K.D. Efrid, M.S. thesis, University of Florida, 1970
60. B.T. Rubin, Ph.D. thesis, University of Pennsylvania, 1969
61. E.D. Verink, Jr. and R.H. Heidersbach, Jr., in *Localized Corrosion—Cause of Metal Failure*, STP 516, American Society for Testing and Materials, 1972, p 303-322
62. S.C. Britton, *J. Inst. Met.*, Vol 67, 1971, p 119
63. L. Piatti and R. Grauer, *Werkst. Korros.*, Vol 14 (No. 7), 1963, p 551
64. G.T. Colegate, *Met. Ind.*, Vol 73 (No. 507), 1948, p 531
65. L. Kenworthy and W.G. O'Driscoll, *Corros. Technol.*, Vol 2, 1955, p 247
66. U.R. Evans, *The Corrosion and Oxidation of Metals*, E. Arnold, 1960
67. C.W. Stillwell and E.S. Turnipseed, *Ind. Eng. Chem.*, Vol 26, 1934, p 740
68. E.D. Verink, Jr. and P.A. Parrish, *Corrosion*, Vol 26, 1970, p 5
69. F.W. Fink, *Trans. Electrochem. Soc.*, Vol 75, 1939, p 441
70. A. Parthasarathi and N.W. Polan, *Metall. Trans. A*, Vol 13A, 1982, p 2027
71. H. Leideiser, Jr. and R. Kissinger, *Corrosion*, Vol 28, 1972, p 218
72. G.T. Burstein and R.C. Newmann, *Corrosion*, Vol 36, 1980, p 225
73. N.W. Polan, J.M. Popplewell, and M.J. Pryor, *J. Electrochem. Soc.*, Vol 126, 1979, p 1299
74. H.W. Pickering and P.J. Bryne, *J. Electrochem. Soc.*, Vol 116, 1969, p 1492
75. T.B. Cassagne, W.F. Flanagan, and B.D. Lichter, *Metall. Trans. A*, Vol 17A, April 1986, p 703-710
76. H.S. Kaiser, Alloy Dissolution, in *Corrosion Mechanisms*, F. Mansfield, Ed., Marcel Dekker, 1987, p 89-90
77. W. Popp, H. Kaiser, H. Kaesche, W. Brämer, and F. Sperner, in *Proceedings of the Eighth International Conference on Metallic Corrosion*, Dechema, Frankfurt/Main, 1981
78. M.G. Fontana and N.D. Greene, *Corrosion Engineering*, 2nd ed., McGraw-Hill, 1978, p 71
79. V.C. Lucey, *Br. Corros. J.*, Vol 1, 1965, p 7; Vol 2, 1965, p 53

Mechanically Assisted Degradation

William Glaeser and Ian G. Wright,
Battelle Columbus Laboratories

MECHANICALLY ASSISTED DEGRADATION of metals, in the context of this article, is defined as any type of degradation that involves both a corrosion mechanism and a wear or fatigue mechanism. This article will discuss five such forms of degradation: erosion, fretting, fretting fatigue, cavitation and water drop impingement, and corrosion fatigue. Only the mechanisms of these forms of degradation will be discussed. The ability of specific metals and alloys to withstand mechanically assisted degradation is treated in detail in many of the articles in the Section "Specific Alloy Systems" in this Volume. The analyses of failures involving these mechanisms, as well as means of failure prevention, are detailed in Volume 11 of the 9th Edition of *Metals Handbook*.

Erosion

Erosion can be defined as the removal of surface material by the action of numerous individual impacts of solid or liquid particles. Erosive wear should not be confused with abrasive or sliding wear, because the mechanisms of material removal, and therefore the materials selection criteria (though rudimentary), are different. In its mildest form, erosive wear is often manifested as a light polishing of the upstream surfaces of components penetrating the flowstream, or of bends or other stream-deflecting structures. This is illustrated in Fig. 1, which shows carbon steel heat transfer tubes in a fluidized-bed combustor. The tubes have been polished through the action of particles of sand impacting at a velocity of about 1.8 m/s (6 ft/s). The black appearance of these tubes is due to the oxide scale, which has been polished (that is, thinned) by the erosive action but not completely removed. In this case, metal wastage is probably a result of high-temperature corrosion assisted by erosion.

Where the erosive action is more severe, any scales or deposits are removed, and the polished, eroded surface is base metal. Such surfaces on low-alloy steels are often very susceptible to rusting; therefore, swabbing the tubes with a damp cloth can give a useful indication of the erosion zones and patterns. Figure 2 shows such rusting on the underside of a tube from a different fluidized-bed combustor. This rusting indicates that erosion occurred in a uniform zone around the bottom of the tube, up to the 9 o'clock position.

Erosion attack can be quite localized, as shown by the polished zones on the tube supports from a steam boiler (Fig. 3). Localized fly ash laden gas flow between the tube rows (supported on the lugs) eroded through the oxide scale and polished the metal surface beneath.

In more severe cases, erosion can result in very rapid attack that quickly leads to thinning and penetration. Conditions that give rise to such rapid erosion usually involve high velocities, large amounts of entrained solid or liquid particles, and abrupt changes in direction of the fluid. These conditions can lead to segregation and concentration of the eroding particles. Examples of such rapid erosion are shown in Fig. 4 to 6.

Erosion is typically thought to involve the action of a dilute dispersion of small solid (or liquid) particles entrained in a fluid jet, and it is from studies in this type of regime that most of the current understanding of erosion phenomena is derived. Strictly controlled laboratory studies of simplified systems or part processes in actual or simulated situations have provided a basis for the analysis of erosion damage produced under quite limiting circumstances: by the impacts of individual particles where particle motion is largely unconstrained in the approach to the target, during impact, and in rebounding from the target. There are many variables that can affect erosion. Important system variables are the relative velocities of the erodent and target; the angle of impact; the mass, size, size distribution, shape, hardness, and composition of the erodent particles; the number density of the particles in the conveying fluid or the frequency of impact; and the density and viscosity of the fluid.

The factors that affect the resistance of materials to erosion are difficult to define. Simple correlations of erosion rate have been attempted with target material hardness, melting temperature, elastic modulus, and so on, with some success for pure elements but less-than-universal application to steels and other alloys or ceramic materials. The relative importance of specific

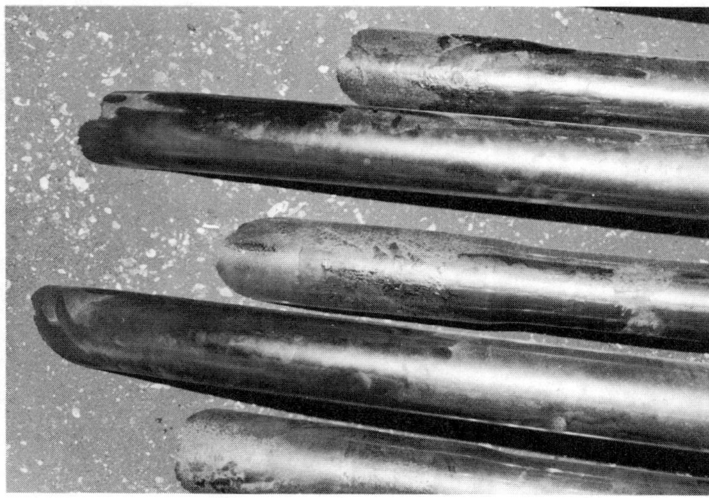

Fig. 1 Polishing of heat transfer tubes from erosion by sand in a fluidized-bed combustor

Fig. 2 Carbon steel heat transfer tube from a fluidized bed that was damaged by erosion and subsequent rusting

Fig. 3 Erosion of a tube support by ash-laden flue gas

materials properties will likely change, depending on the predominating mechanisms of erosion, which in turn depend on any of the large number of systems variables and on the way in which the target material absorbs the energy of impact. For discrete impacts at a shallow angle to the target surface, it appears that the primary mode of material loss is through the displacement of material in the path of the erodent and the eventual loss of the displaced material by subsequent cutting or by fracture, if the material is embrittled.

This form of erosion can be equated to micromachining. The simple, elegant model discussed in Ref 1 and 2 and its refinements, relating erosion essentially to material hardness, has been found to predict erosion rates of approximately the right magnitude, even if the ability of the model to rank different alloys is sometimes poor. This model can be simply stated as:

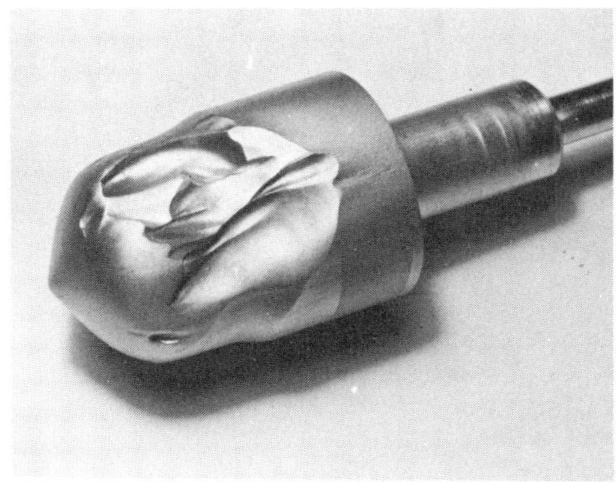

Fig. 4 Hardfaced stainless steel plug and seat, from a slurry flow control valve, that were eroded by high-velocity flow through the narrow orifice created during throttling

$$\frac{\text{Erosion loss}}{\text{Quantity of impacting erodent}} = C \cdot F(\theta)\,\frac{\rho v^2}{\text{HV}} \quad \text{(Eq 1)}$$

where C is a system constant, θ is the angle of impact, ρ is the density of the erodent, v is the erodent velocity, and HV is the hardness of the target.

Other refinements to this model, as well as approaches to sophisticated models, have been proposed and are fully discussed in other reviews (Ref 3-5). In the later refinements (Ref 2) of this model, the velocity dependence of erosion is increased from 2 to around 2.3. Nevertheless, the general relationship between erosive loss and erodent kinetic energy is not changed.

For impacts at larger angles to the target surface, fluid flow considerations indicate that in most practical cases the erodent particles will actually strike the surface at a variety of shallower angles, depending on the size and velocity of the particles and the drag exerted on them by the conveying fluid. For impacts normal to the target surface, therefore, it is likely that only a small fraction of the particles will actually impact at 90°. The forms of material loss will involve both surface displacement and cutting, as well as modes that result from normal impacts. For most metallic materials, normal impacts will result in indentation and local displacement of the surface by sharp particles, but less angular particles will cause deformation and/or eventual fatigue, depending on the rate at which the deformation can be accommodated. Most ceramic materials suffer cracking from normal impacts; the form and extent of the cracking depend on the intensity of the impact (particle size and velocity, and shape) and on the structure of the ceramic.

Erosion models that describe erosion by deformation and fracture are based on elastic-plastic indentation fracture mechanics. The volume of the target surface removed following an individual impact is usually taken to be the volume

Fig. 5 Eroded tube inserts from the inlet end of a fire-tube boiler. The inserts were eroded by particle-laden flue gas, which was forced to turn as it entered the boiler.

Fig. 6 Erosion of a rotary valve handling dust from a cyclone. The wear plates in the valve show some material loss, but the major damage is to the casing. Gaps between the casing and the valve allowed leakage of high-velocity air with entrained dust.

enclosed by the radial and lateral cracks resulting from the impact. The depth of the lateral cracks can be related to the hardness of the target, while the length of the radial crack is a function of both hardness and toughness. The exact form of the relationship among erosion loss, hardness, and fracture toughness depends on assumed details of the geometry of the damage and mode of material removal. The predicted velocity dependence of erosion is usually around 3. Further discussion of these erosion models can be found in Ref 6 and 7.

In practice, erosive conditions cover a wide spectrum. In situations where the concentration of erodent particles is so great that the independent action of individual particles is unlikely, the form of damage may be more like scouring from particles embedded in a massive backing than that from discrete particle impacts. Rules derived for abrasive wear (see the article "Wear Failures" in Volume 11 of the 9th Edition of *Metals Handbook*) may then be more appropriate. In a large number of systems of practical importance, the density and viscosity of the carrier fluid are significantly different from those associated with airborne dilute phase erosion; therefore, the motion of the erodent particles before, during, and after impact may be quite different from that considered in most erosion models. The actual mechanisms of erosion, as well as the relationships with, for example, nominal angle of impact, may not correspond with those where air is the carrier fluid. Overall, at the present state of development, analytical relationships between systems and materials variables, as well as material loss through erosion, can be used only as a guide to the relative importance of the variables encountered. Testing under conditions that accurately simulate the situation under consideration is probably the best method of forming a basis for decision making. More information on failures caused by erosion is available in the article "Liquid-Erosion Failures" in Volume 11 of the 9th Edition of *Metals Handbook*.

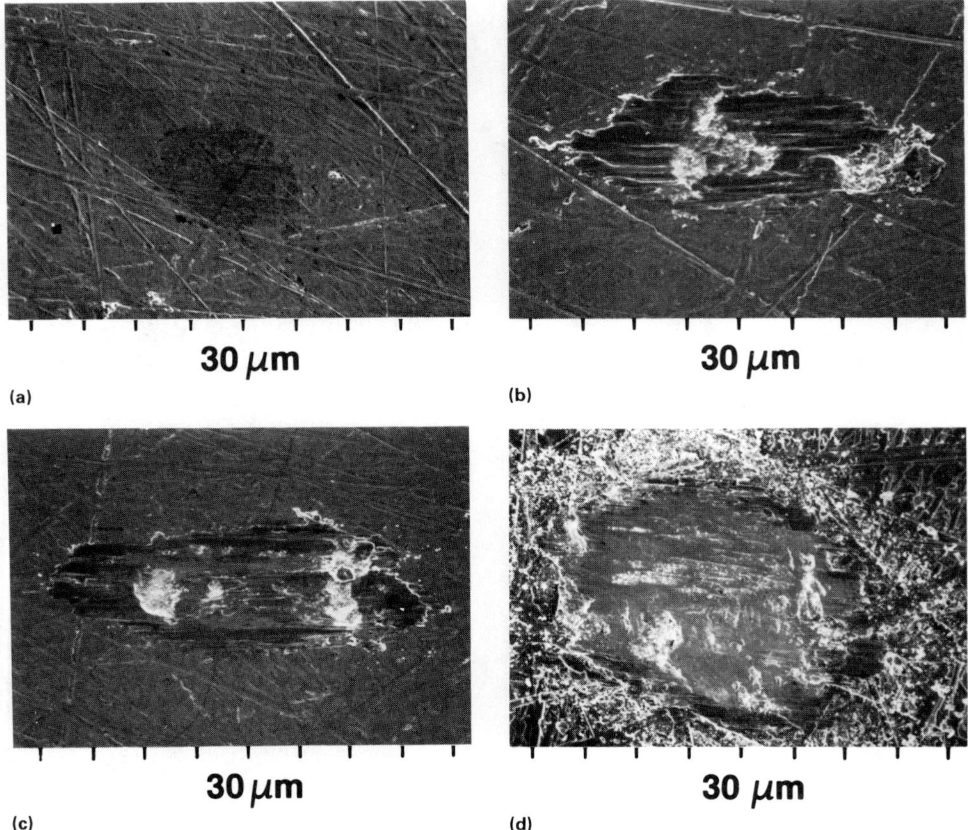

Fig. 7 Fretting of cobalt-gold plated copper flats in contact with solid gold in an electrical contact. (a) After 1000 cycles. (b) After 10^4 cycles. (c) After 10^5 cycles. (d) After 10^6 cycles. Source: Ref 8

Fretting Corrosion

Fretting corrosion is a combined wear and corrosion process in which material is removed from contacting surfaces when motion between the surfaces is restricted to very small amplitude oscillation (as low as 3 or 4 nm). Usually, the condition exists in machine components that are considered fixed and not expected to wear. Pressed-on wheels can often fret at the shaft/wheel hole interface.

Oxidation is the most common element in the fretting process. In oxidizing systems, fine metal particles removed by adhesive wear are oxidized and trapped between the fretting surfaces. The oxides act like an abrasive (such as lapping rouge) and increase the rate of material removal. This type of fretting in ferrous alloys is easily recognized by the red material oozing from between the contacting surfaces.

Fretting can also persist in contacts where no corrosion exists. For example, gold fretting against gold will produce a fine gold debris. Fretting occurs in the vacuum of outer space.

Fretting damage can be very serious and catastrophic. For example, in gas turbines, the blades are often anchored in dovetail slots in the hub. Vibrations in the blade system can cause fretting in the blade roots and loosening. The loose blades then rattle around and break out because of fatigue or impact wear. A piece of blade moving into the gas flow through the turbine can cause severe damage, destroying rows of blades and causing engine failure. Another fretting problem that has injury potential is the loosening of wheels or flywheels from shafts or axles. Railroad car wheels, for example, are shrink fitted onto their axles. If the wheel loosens from running vibrations and comes off during operation of the train, it can cause

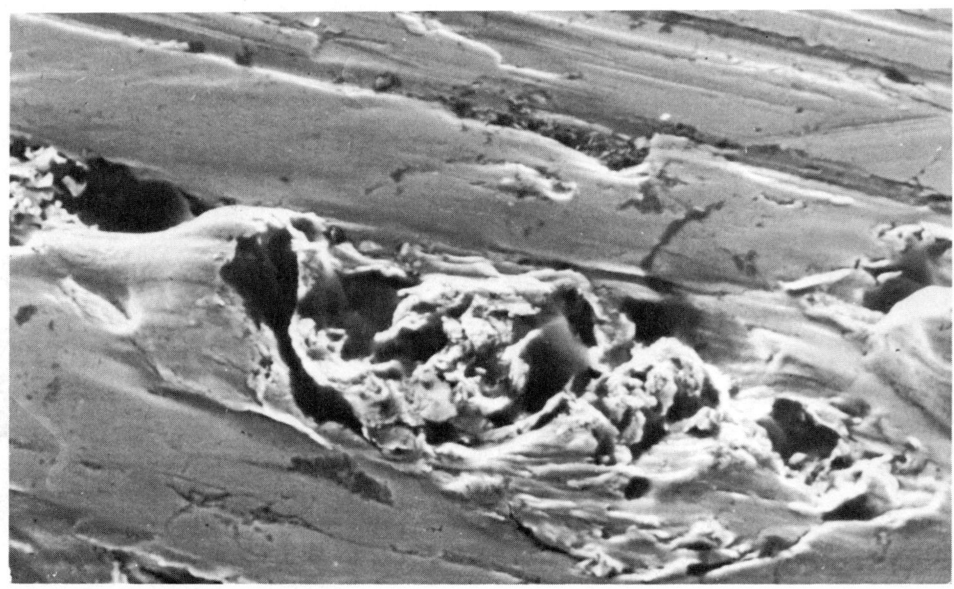

Fig. 8 Local cold welding on the surface of 0.2% C steel after 500 fretting cycles. Courtesy of R.B. Waterhouse, University of Nottingham

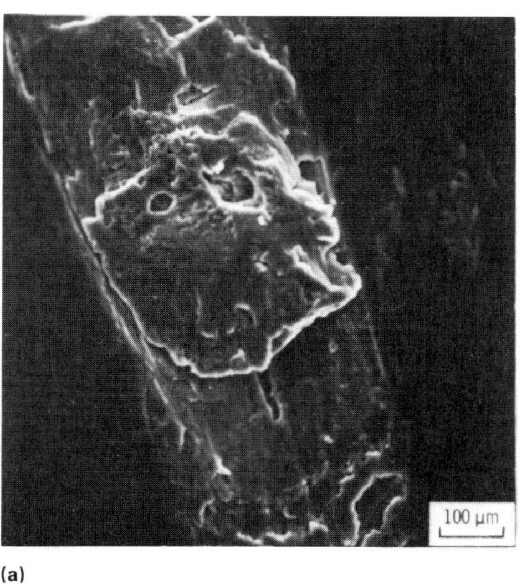

(a)

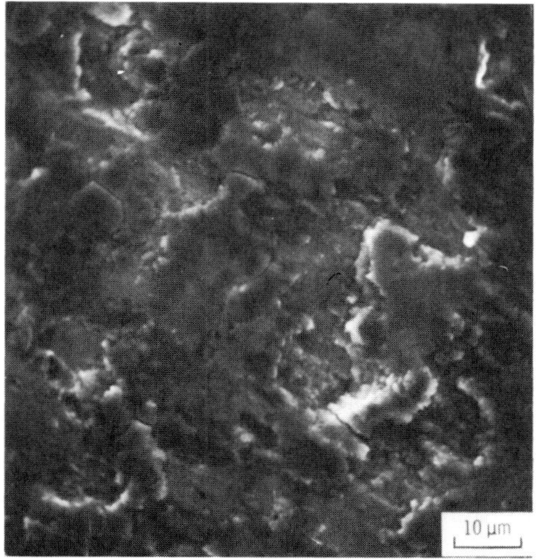

(b)

Fig. 9 Damage to Monel alloy after fretting in air at room temperature for (a) 1000 cycles and (b) 5 × 10⁴ cycles. Courtesy of R.C. Bill, NASA Lewis Research Center

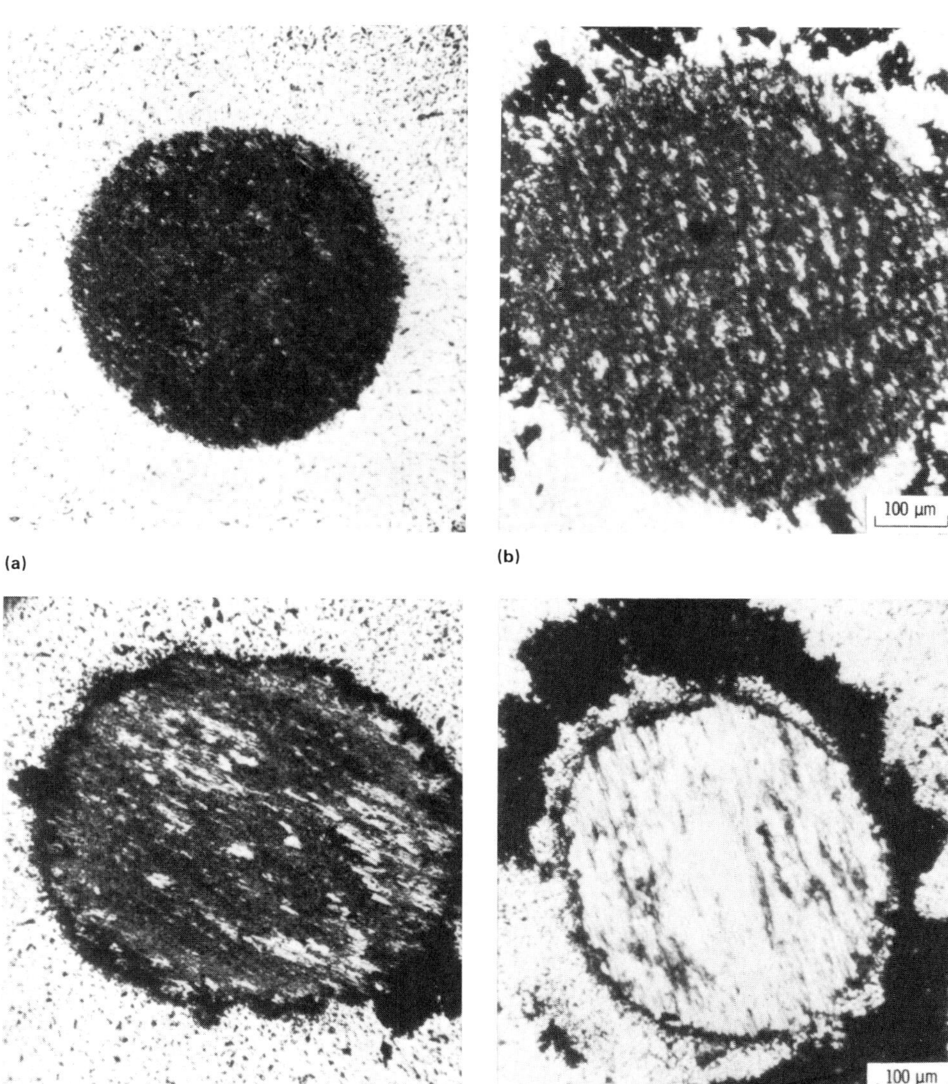

Fig. 10 Effect of relative humidity on fretting damage to high-purity iron tested in air. (a) Dry air. (b) 10% relative humidity. (c) 35% relative humidity. (d) Saturated air. All specimens shown after 3×10^5 cycles. See also Fig. 11. Courtesy of R.C. Bill, NASA Lewis Research Center

derailment and has the potential of becoming a loose rolling missile capable of penetrating nearby buildings.

Electrical connectors for low-current circuits are generally gold or gold-base alloys. Because of the high cost of gold, very thin electroplated gold coatings are used for contacts. Gold plate thicknesses as small as 0.25 µm (0.01 mil) are used. When electric contacts are subjected to vibrations of continual expansion and contraction from periodic thermal excursions, the small relative motion between the contacts produces fretting wear and the eventual removal of the gold plate. As the gold is removed, the substrate is subjected to atmospheric corrosion, and the contact resistance rises to intolerable levels. In addition, fretting debris becomes trapped between the surfaces and causes degraded conductivity. Fretting of electrical contacts is not anticipated in the design, and although it is common, it often comes as a surprise to the user. A typical progression of fretting damage induced between a solid gold rider on cobalt-gold plated copper is shown in Fig. 7.

Fretting corrosion has been a continuing problem in nuclear reactors. The condition is found on heat-exchanger tubes and on fuel elements. In both cases, long, flexible tubes are in contact with support surfaces and subjected to vibrations generated by fluid flow as the coolant flows around them. The supports for heat-exchanger tubes and fuel elements cannot be rigid, because the tubes must be able to expand or contract under thermal excursions without binding (this is also true for any tube heat-exchanger system). Tube impact fretting caused by flow-induced tube vibrations can reduce wall thickness, requiring eventual replacement. Extensive experimentation has revealed that the interaction of tube support clearance, excitation force, and type of tube motion controls wear rates (Ref 9). Other mechanical parts susceptible to fretting damage include couplings, riveted and pin joints, surgical implants, rolling contact bearings, and bridge bearings.

Factors Affecting Fretting

The following factors are known to influence the severity of fretting. If fretting conditions exist, fretting cannot be eliminated completely but can be reduced in severity.

Contact Load. As long as fretting amplitude is not reduced, fretting wear will increase linearly with increasing load.

Amplitude. There appears to be no measurable amplitude below which fretting does not occur. However if the contact conditions are such that deflection is only elastic, it is not likely that fretting damage will occur. Fretting wear loss increases with amplitude. The effect of amplitude can be linear, or there can be a threshold amplitude above which rapid increase in wear occurs (Ref 10). The transition is not well established and probably depends on the geometry of the contact.

Frequency. When the fretting is measured in volume of material removed per unit sliding dis-

(a)

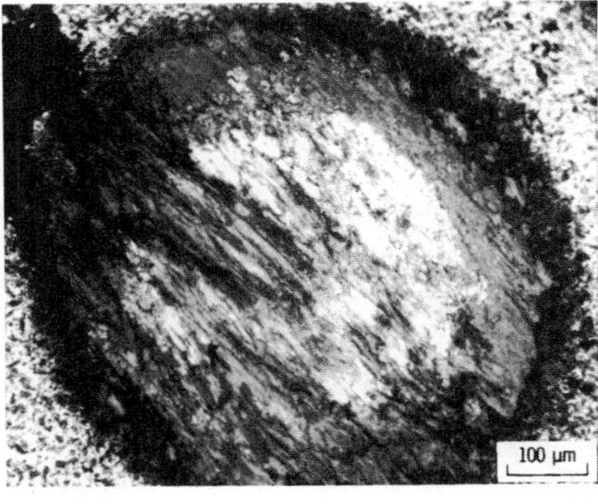

(b)

Fig. 11 Effect of relative humidity on fretting damage to high-purity nickel. Damage produced after 3 × 10^5 cycles in (a) dry air and (b) saturated air. See also Fig. 10. Courtesy of R.C. Bill, NASA Lewis Research Center

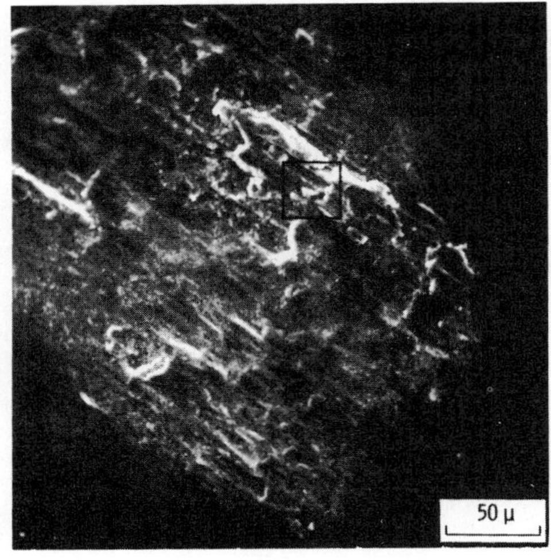

(a)

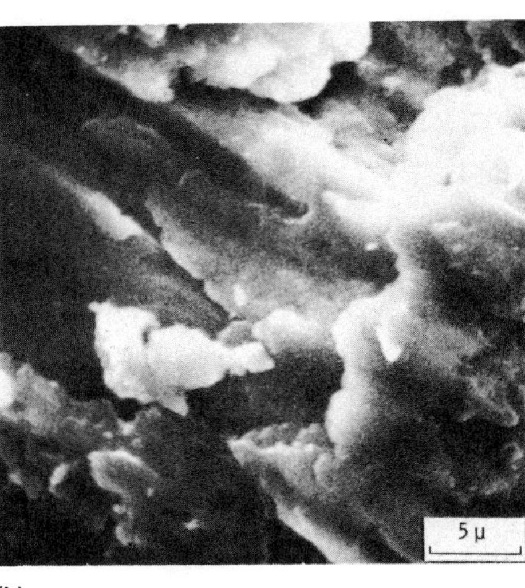

(b)

Fig. 12 SEM micrographs of titanium surfaces fretted at room temperature after 1000 cycles. (a) Overall view of damage. (b) Close-up of area in box in (a). Tested in air at fretting frequency of 55.8 Hz, amplitude of 70 µm, and normal load of 1.47 N (0.33 lbf). Compare with Fig. 13. Courtesy of R.C. Bill, NASA Lewis Research Center

tance, there does not appear to be a frequency effect.

Number of Cycles. An incubation period occurs during which fretting wear is negligible (Ref 11). After the incubation period, a steady-state wear rate is observed. Figure 8 shows local cold welding on a fretted steel surface early in the fretting process (500 cycles). As the fretting continues, a more general surface roughening occurs, as shown in Fig. 9.

Relative Humidity. For materials that rust in air, fretting wear is higher in dry air than in saturated air. The effect of humidity is shown in Fig. 10 and 11. In dry air, both the iron and nickel produce debris that remains in the contact region. The debris separates the surfaces and reduces the contacts. As humidity is increased, the debris becomes more mobile, escaping the contact and allowing metal-to-metal contact, as shown in Fig. 10 and 11. Diffraction analysis of the debris from the iron surfaces shows it to be α-Fe_2O_3 (Ref 12).

Temperature. The effect of elevated temperature on fretting depends on the oxidation characteristics of the metals. If increased temperature encourages the growth of a protective tough oxide layer that prevents metal-to-metal contact, the fretting rate will be lower. Low-carbon steel tends to show a sudden decrease in fretting wear rate at about 200 °C (390 °F) (Ref 13). For titanium alloys, an increase in wear rate has been observed between 350 and 500 °C (660 and 930 °F), and a rapid decrease in wear has been observed above 550 °C (1020 °F) (Ref 14, 15). Even the difference between room-temperature wear and wear at 650 °C (1200 °F) is significant (Fig. 12 and 13). Figure 12 shows the surface damage from room-temperature fretting, illustrating the inadequate protection afforded by the oxide. Figure 13 shows the effect of the tough, thick oxide layer formed at 650 °C (1200 °F), which protects the metal surface. It has been shown that superalloys behave similarly to titanium at elevated temperatures (Ref 16). A glassy layer of compacted oxide particles has been found on the fretted surface after high temperature. This layer reduces the fretting wear considerably. The formation of this glassy oxide layer has been linked to an increase of 130% in fretting fatigue strength of Inconel 718 (Ref 17).

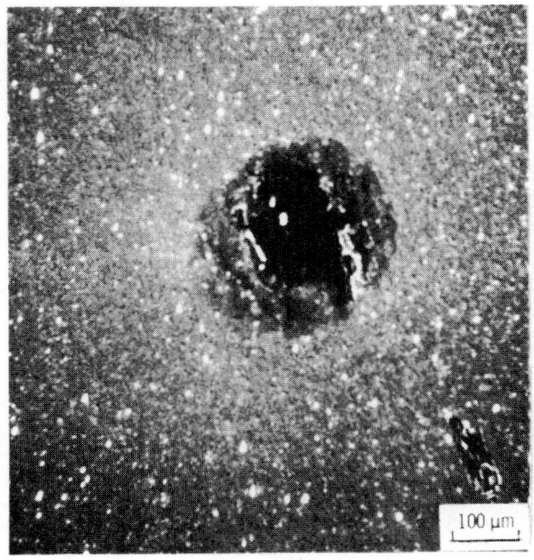

(a)

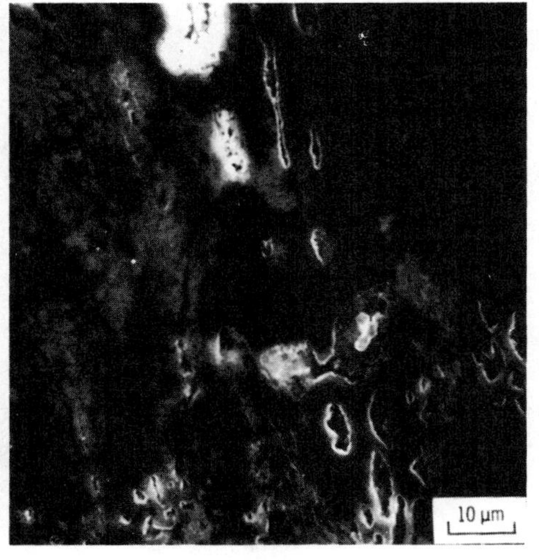

(b)

Fig. 13 Titanium surfaces fretted at 650 °C (1200 °F) after 10^5 cycles. The formation of a tough, thick oxide layer at this temperature decreased the amount of damage (compare with Fig. 12). (a) Overall view. (b) Close-up of central region. Courtesy of R.C. Bill, NASA Lewis Research Center

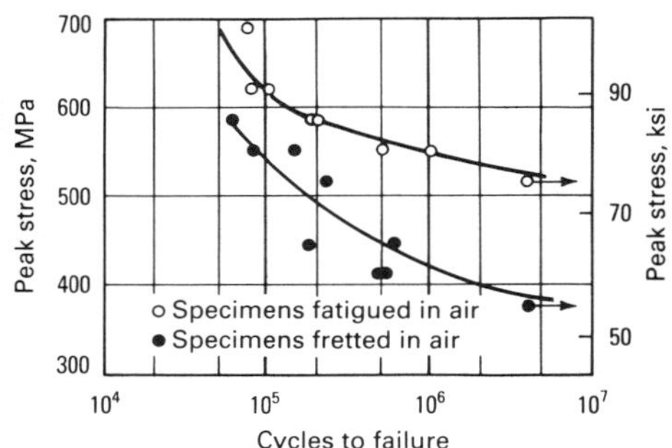

Fig. 14 Comparison of fatigue life for 4130 steel under fretting and nonfretting conditions. Specimens were water quenched from 900 °C (1650 °F), tempered 1 h at 450 °C (840 °F), and tested in tension-tension fatigue. Normal stress was 48.3 MPa (7 ksi); slip amplitude was 30 to 40 µm.

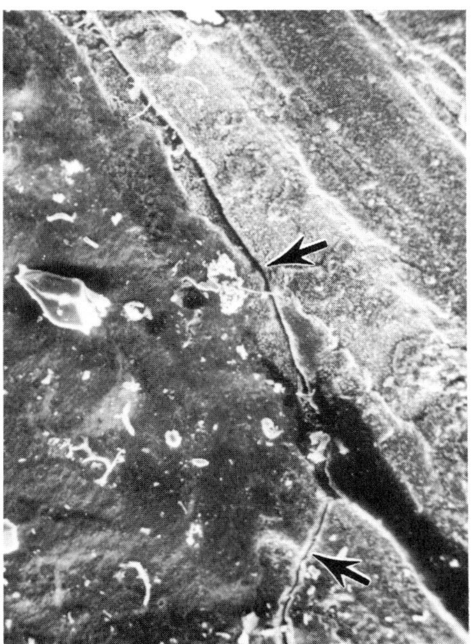

Fig. 15 Fretting scar on fatigued steel specimen showing location of fatigue crack (arrow). Source: Ref 19

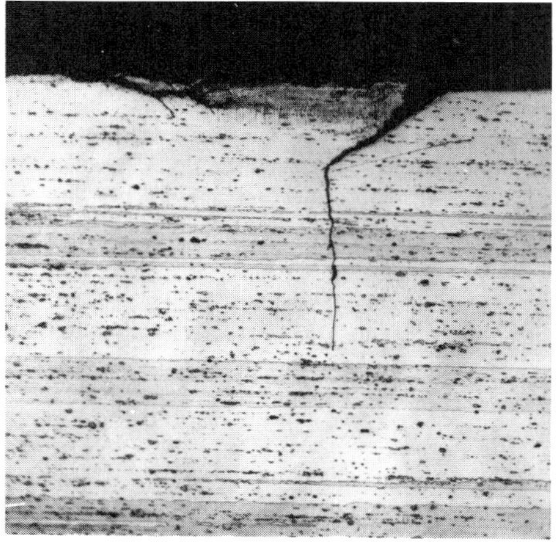

Fig. 16 Section through a bar of aged Al-4Cu alloy showing a crack initiated by fretting fatigue. Courtesy of R.B. Waterhouse, University of Nottingham

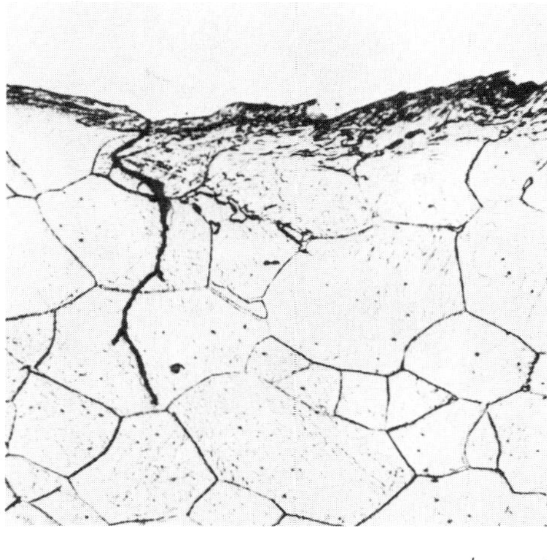

Fig. 17 Section showing fretting damage and fatigue crack initiation in 0.2% C steel. Courtesy of R.B. Waterhouse, University of Nottingham

Fretting Fatigue

Fretting decreases the fatigue life of parts operating under fatigue loading. Gas turbines experience this problem with clamped joints and shrink-fitted parts. These parts are subjected to high-frequency low-amplitude vibrations, often in combination with bending fatigue stress states (Ref 18). The combined action of fretting and reversing bending stress causes accelerated crack initiation and an increase in the rate of crack propagation. The result is a decrease in the fatigue strength of a given alloy. The effect on 4130 steel is shown in Fig. 14. In this case, the fatigue strength of the alloy is reduced under fretting, and the sensitivity to load is increased.

Fretting fatigue begins as a crack in the fretting scar zone. An example of a fretting fatigue crack is shown in Fig. 15. The crack is located at the boundary of the fretting scar. This is one attribute of fretting fatigue that identifies the origin of the fatigue. The fatigue crack, once initiated at the boundary of the fretted zone, propagates into the surface at an angle to the surface (Fig 16). Other examples of the crack habit are shown in Fig. 17 and 18. As the crack opens, fretting debris jams into it, adding to the propagation force. In a corrosive environment, the corroding medium can also penetrate the crack and add corrosion fatigue to the process. Once the crack propagation reaches a depth where it is no longer influenced by surface contact stress, it progresses as a fatigue crack normal to the surface. When the part breaks, the fretting fatigue fracture leaves a characteristic lip at the surface, as shown in Fig. 16.

Wire rope can fail by fretting fatigue. As the rope flexes, wire strands rub against each other and produce fretting at the wire-wire contacts. If the fretting occurs in a zone that is also subject to cyclic fatigue stresses, failure occurs. An example of wire rope fretting fatigue is shown in Fig. 19. Note the lip in the detail micrograph in Fig. 20.

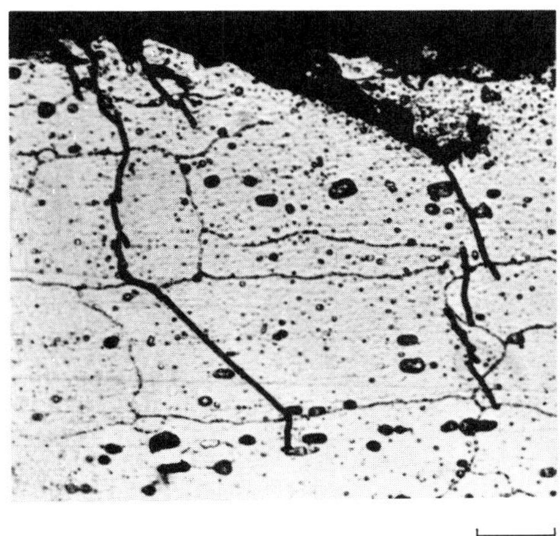

Fig. 18 Section showing fretting damage and fatigue cracks in Al-6Zn-3Mg alloy. Courtesy of R.B. Waterhouse, University of Nottingham

Fig. 19 Fretting fatigue failure of steel wire rope after seawater service. Wire diameter was 1.5 mm (0.06 in.). See also Fig. 20. Courtesy of R.B. Waterhouse, University of Nottingham

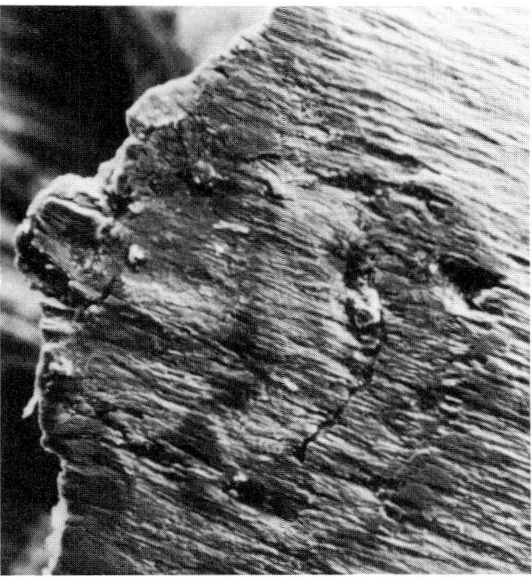

Fig. 20 Higher-magnification view of Fig. 19 showing fretting fatigue crack on the other side of the wear scar in Fig. 19. Courtesy of R.B. Waterhouse, University of Nottingham

Fig. 21 Cavitation damage to an ACI CN-7M cast pump impeller used to pump ammonium nitrate solution at 140 °C (280 °F). Courtesy of A.R. Wilfley and Sons, Inc., Pump Division

Cavitation Erosion and Water Drop Impingement

Cavitation erosion occurs on metal surfaces in contact with a liquid. Pressure differentials in the fluid generate gas or vapor bubbles in the fluid. When these bubbles encounter a high-pressure zone, they collapse and cause explosive shocks to the surface. These surface shocks cause localized deformation and pitting. Cavitation pits eventually link up and cause a general roughening of the surface and material removal. Cavitation is similar to particle erosion in its damage. However, surface features formed by cavitation are different from those formed by particle erosion. Cavitation produces rounded microcraters in the surface, while particle erosion produces imprints of the impacting particles. Crater formation moves surface material to the edges of the craters, and these extrusions eventually break off, causing loss of material from the surface. The sharp pressure pulses caused by the collapsing bubbles are highly localized and can remove weak or soft portions of microstructural phases (for example, ferrite from pearlite) (Ref 20).

Scattered shallow dimples or depressions are the earliest evidence of cavitation erosion in ductile surfaces. This represents an incubation period before the actual loss of material. Damage is seen first in the weaker elements of the microstructure.

Material removal is different for soft materials and hard materials. Soft materials suffer local plastic deformation and penetration. Babbitt, for example, will become pitted with steep-sided pits that group together in patterns reflecting the fluid dynamics.

Hard materials experience localized microcracking and chipping from the pressure pulses. Brittle materials such as glass and some plastics appear to develop surface cracking at minute surface flaws.

In materials that depend on passivating films for corrosion protection, cavitation will cause an apparent accelerated attack. In a corrosive medium, the cavitation will remove the protective film, and corrosion will weaken the material to the mechanical removal process. Figure 21 shows cavitation damage to an ACI CN-7M stainless steel impeller that was used to pump ammonium nitrate solution at 140 °C (280 °F).

Hardfacing materials and cemented carbides react differently to cavitation than to abrasion (Ref 21). The carbides are selectively eroded from the Stellite alloys, while cemented carbides suffer loss of the cobalt or nickel binder. Cavitation occurs in hydraulic equipment, fluid pump impellers, ship propellers, hydrodynamic bearings, fluid seals, inlets to heat-exchanger tubes, diesel engine wet cylinder liners, hydrofoils, liquid-metal power plants, and steam turbines.

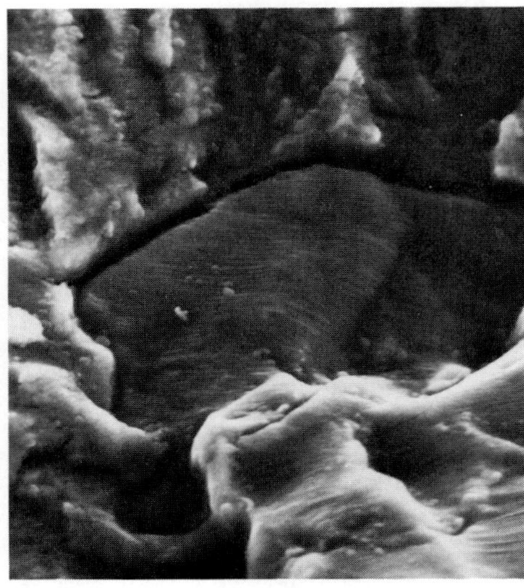

Fig. 22 Corrosion fatigue of Ti-6Al-4V alloy tested in ambient air. Intergranular cracking and fatigue striations are evident on the fracture surface; the grain appears to have separated from the rest of the microstructure. Source: Ref 22

Water drop impingement erosion is similar to cavitation in that it causes pitting of surfaces and may involve a cavitation mechanism. Two areas are most notable for water drop inpingement: steam turbines and helicopter rotor blades. In turbines, condensation of steam produces droplets that are carried into the rotor blades with consequent surface damage. Rain drop erosion on helicopter blades is the result of elastic compression waves produced by multiple impacts and their interaction; this generates tensile stresses just below the surface and causes cracking.

Water drop impingement damage can appear to be somewhat different from cavitation damage in ductile materials. The cavities in the surface show a directionality that is related to the angle of attack of the drops.

Corrosion Fatigue

Corrosion fatigue occurs in metals as a result of the combined action of a cyclic stress and a corrosive environment. Corrosion fatigue is dependent on the interactions among loading, environmental, and metallurgical factors. For a given material, the fatigue strength (or fatigue life at a given maximum stress value) generally decreases in the presence of an aggressive environment. The effect varies widely, depending primarily on the particular metal-environment combination. The environment may affect the probability of fatigue crack initiation, the fatigue crack growth rate, or both. Figure 22 shows an example of corrosion fatigue failure in Ti-6Al-4V alloy.

Corrosion Fatigue Crack Initiation. The influence of an aggressive environment on fatigue crack initiation of a material is illustrated in Fig. 23, which compares the smooth-specimen stress-life (*S-N*) curves obtained from inert and aggressive environments. Because as much as 95% of the structure life is spent on fatigue-crack initiation, *S-N* curve comparison will provide a good indication of the effect of environment on crack initiation. As shown in Fig. 23, an aggressive environment can promote crack initiation and can shorten the fatigue life of the structure.

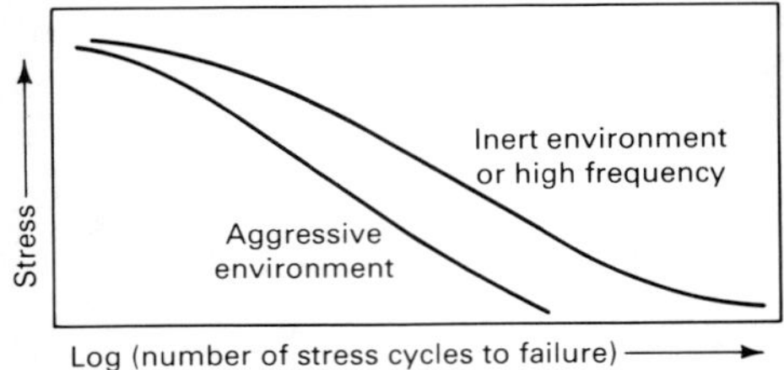

Fig. 23 Comparison of *S-N* curves for a material in an inert environment (top curve) and an aggressive environment (lower curve)

Corrosion fatigue cracks are always initiated at the surface unless there are near-surface defects that act as stress concentration sites and facilitate subsurface crack initiation. Surface features at origins of corrosion fatigue cracks vary with the alloy and with specific environmental conditions. In carbon steels, cracks often originate at hemispherical corrosion pits and often contain significant amounts of corrosion products. The cracks are often transgranular and may exhibit a slight amount of branching. Surface pitting is not a prerequisite for corrosion fatigue cracking of carbon steels, nor is the transgranular fracture path; corrosion fatigue cracks sometimes occur in the absence of pits and follow grain boundaries or prior-austenite grain boundaries.

In aluminum alloys exposed to aqueous chloride solutions, corrosion fatigue cracks frequently originate at sites of pitting or intergranular corrosion. Initial crack propagation is normal to the axis of principal stress. This is contrary to the behavior of fatigue cracks initiated in dry air, where initial growth follows crystallographic planes. Initial corrosion fatigue cracking normal to the axis of principal stress also occurs in aluminum alloys exposed to humid air, but pitting is not a requisite for crack initiation.

Corrosion fatigue cracks in copper and various copper alloys initiate and propagate intergranularly. Corrosive environments have little additional effect on the fatigue life of pure copper over that observed in air, although they change the fatigue crack path from transgranular to intergranular. Copper-zinc and copper-aluminum alloys, however, exhibit a marked reduction in fatigue resistance, particularly in aqueous chloride solutions. This type of failure is difficult to distinguish from stress-corrosion cracking (SCC) except that it may occur in environments that normally do not cause failures under static stress, such as sodium chloride or sodium sulfate solutions.

Environmental effects can usually be identified by the presence of corrosion damage or corrosion products on fracture surfaces or within growing cracks. Corrosion products, however, may not always be present. For example, corrosion fatigue cracking of high-strength steel exposed to a hydrogen-producing gas, such as water vapor, may be difficult to differentiate from some other forms of hydrogen damage. At sufficiently high frequencies, the fracture surface features produced by corrosion fatigue crack initiation and propagation do not

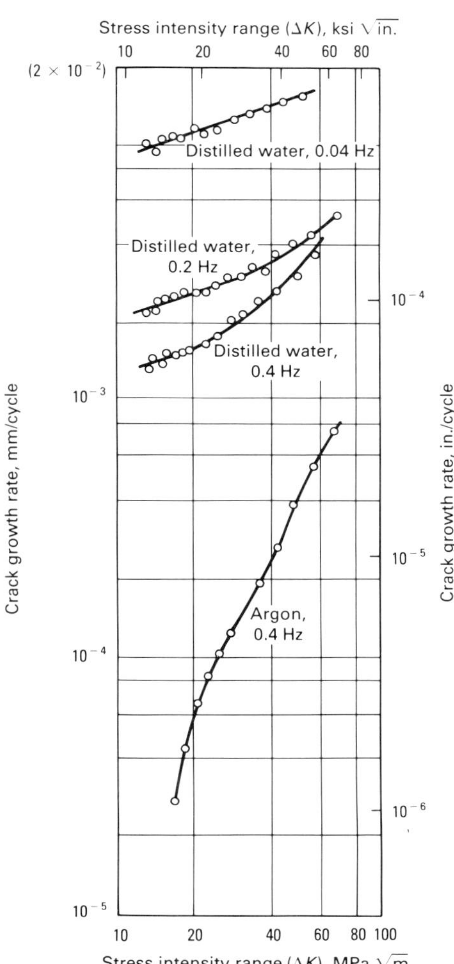

Fig. 24 Effect of stress intensity range and loading frequency on corrosion fatigue crack growth in ultrahigh-strength 4340 steel exposed to distilled water at 23 °C (73 °F)

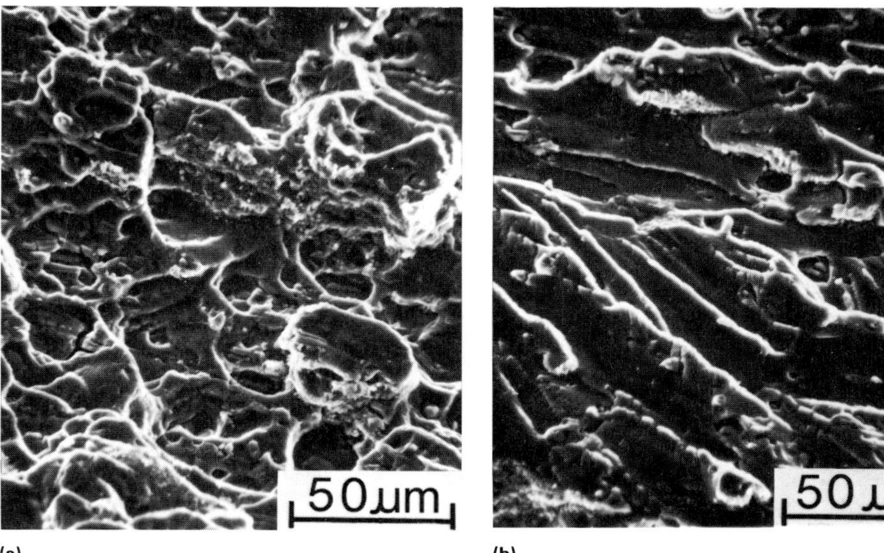

Fig. 25 Effect of water vapor on the fracture surface appearance of aluminum alloy 2219-T851 fatigue tested (a) in 1-atm dry argon and (b) in 0.2-torr water vapor. Testing conditions were the same except for frequency, which was 20 Hz in (a) and 5 Hz in (b). The magnifications, while too low to resolve fatigue striations clearly, indicate the general change in fracture morphology. Crack propagation was from left to right. Source: Ref 34

differ significantly from those produced by fatigue in nonaggressive environments.

Corrosion Fatigue Crack Propagation. Although corrosion fatigue phenomena are diverse, several variables are known to repeatedly influence crack growth rate:

- Stress intensity range
- Load frequency
- Stress ratio
- Aqueous environment electrode potential
- Environment
- Metallurgical variables

Effects of such variables as temperature, load history and waveform, stress state, and environment composition are unique to specific materials and environments (Ref 23-31).

Stress Intensity Range. For embrittling environments, crack growth rate generally increases with increasing stress intensity (ΔK); the precise dependence, however, varies markedly. Materials that are extremely environment sensitive, such as ultrahigh-strength steel in distilled water (Fig. 24), are characterized by high growth rates that depend on ΔK to a reduced power. Time-dependent corrosion fatigue crack growth occurs mainly above the threshold stress intensity for static load cracking and is modeled through linear superposition of SCC and inert environment fatigue rates (Ref 32, 33).

Frequency. Cyclic load frequency is the most important variable that influences corrosion fatigue for most material, environment, and stress intensity conditions. The rate of brittle cracking above that produced in vacuum generally decreases with increasing frequency. Frequencies exist above which corrosion fatigue is eliminated. The dominance of frequency is related directly to the time dependence of the mass transport and chemical reaction steps required for brittle cracking. Basically, insufficient time is available for chemical embrittlement at rapid loading rates; damage is purely mechanical, equivalent to crack growth in vacuum.

Stress Ratio. Rates of corrosion fatigue crack propagation generally are enhanced by increased stress ratio, R, which is the ratio of the minimum stress to the maximum stress. Stress ratio has only a slight influence on fatigue crack growth rates in a benign environment (see the article "Fatigue Crack Growth Data Analysis" in Volume 8 of the 9th Edition of *Metals Handbook*).

Electrode potential, like loading frequency, strongly influences corrosion fatigue crack propagation rates in aqueous environments. Controlled changes in the potential of a specimen can result in either the complete elimination or the dramatic enhancement of brittle fatigue cracking. The precise influence depends on the mechanism of the environmental effect and on the anodic or cathodic magnitude of the applied potential.

Environment. Increasing the chemical activity of the environment—for example, by lowering the pH of a solution, by increasing the concentration of the corrosive species, or by increasing the pressure of a gaseous environment—generally decreases the resistance of a material to corrosion fatigue. Decreasing the chemical activity of the environment improves resistance to corrosion fatigue.

In aluminum alloys and high-strength steels, for example, corrosion fatigue behavior is related to the relative humidity or partial pressure of water vapor in the air. Corrosion fatigue crack growth rates for these materials generally increase with increasing water vapor pressure until a saturation condition is reached. Figure 25 shows the appearance of the fracture surfaces of an aluminum alloy fatigue tested in argon and in air with water vapor present.

Temperature can have a significant effect on corrosion fatigue. The effect is complex and depends on temperature range and the particular material-environment combination in question, among other factors. The general tendency, however, is for fatigue crack growth rates to increase with increasing temperature.

Other factors, including the metallurgical condition of the material (such as composition and heat treatment) and the loading mode (such as uniaxial), affect corrosion fatigue crack propagation. These factors are discussed in the articles "Environmental Effects on Fatigue Crack Propagation" in Volume 8 of the 9th Edition of *Metals Handbook* and "Corrosion Fatigue Failures" in Volume 11 of the 9th Edition of *Metals Handbook*. The article "Evaluation of Corrosion Fa-

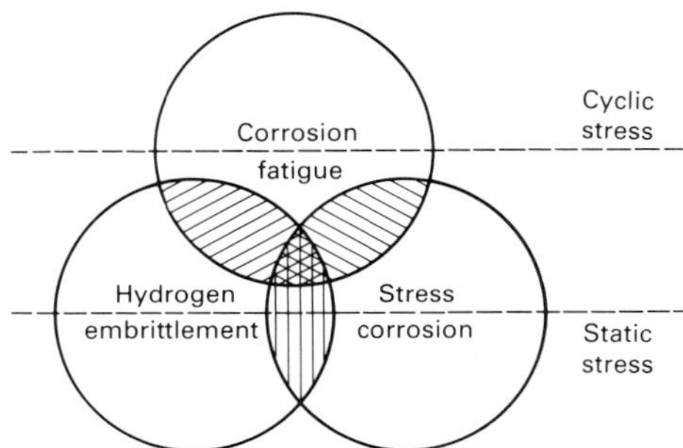

Fig. 26 Schematic showing the relationship among SCC, corrosion fatigue, and hydrogen embrittlement. Source: Ref 35

tigue'' in this Volume contains detailed information on testing methods and the use of fracture mechanics to predict corrosion fatigue behavior.

Relationship Between Corrosion Fatigue and SCC. The relationship between corrosion fatigue and two other environmental cracking mechanisms, SCC and hydrogen embrittlement, is shown in Fig. 26. Many investigations have attempted to link the mechanisms of corrosion fatigue and other environmental cracking processes (especially SCC); examples of these include Ref 35 to 37. There are, however, many unanswered questions about the mechanisms of these phenomena. Experimental evidence relating SCC and corrosion fatigue is presented in the article ''Evaluation of Corrosion Fatigue'' in this Volume.

REFERENCES

1. I. Finnie, *Wear*, Vol 3, 1960, p 87
2. I. Finnie and D.H. McFadden, *Wear*, Vol 48, 1978, p 181
3. C.M. Preece and N.H. Macmillan, *Ann. Rev. Mater. Sci.*, Vol 7, 1977, p 95
4. I.G. Wright, C.W. Price, and R.B. Herchenroeder, ''High-Temperature Erosion/Corrosion of Alloys,'' EPRI RP-557, Electric Power Research Institute, 1978
5. W.F. Adler, Ed., *Erosion: Prevention and Useful Applications*, STP 664, American Society for Testing and Materials, 1979
6. C.M. Preece, Ed., *Erosion*, Vol 16, *Treatise on Materials Science and Technology*, Academic Press, 1979
7. A.G. Evans, M.E. Gulden, and M. Rosenblatt, *Proc. R. Soc. (London)*, Vol A361, 1978, p 343
8. M. Antler and M.H. Drozdowicz, Fretting Corrosion of Gold-Plated Connector Contacts, *Wear*, Vol 74, 1981-1982, p 27-50
9. P.L. Ko, J.H. Tromp, and M.K. Weckworth, Heat Exchanger Tube Fretting Wear: Corrosion of Tube Motion and Wear, in *Materials Evaluations Under Fretting Conditions*, STP 780, American Society for Testing and Materials, 1981, p 86-105
10. R.B. Waterhouse, *Fretting Corrosion*, Pergamon Press, 1972, p 111-113
11. D.W. Hoeppner and G.L. Goss, *Wear*, Vol 27, 1974, p 61
12. R.C. Bill, Review of Factors That Influence Fretting Wear, in *Materials Evaluations Under Fretting Conditions*, STP 780, American Society for Testing and Materials, 1982, p 176
13. P.L. Hurricks, The Fretting Wear of Mild Steel From 200C to 500C, *Wear*, Vol 30, 1974, p 189
14. R.C. Bill, Review of Factors That Influence Fretting Wear, in *Materials Evaluation Under Fretting Conditions*, STP 780, American Society for Testing and Materials, 1982, 178-179
15. R.C. Bill, Technical Note TN D-7570, National Aeronautics and Space Administration, 1974
16. R.B. Waterhouse, The Fretting Wear of Nitrogen-Bearing Austenitic Stainless Steel at Temperatures to 600 C, *Trans. ASME*, Vol 08, p 359-363
17. D.E. Taylor, Fretting Fatigue in High Temperature Oxidizing Gases, in *Fretting Fatigue*, Applied Science, 1981, p 196-201
18. T.C. Lindley and K.J. Nix, The Role of Fretting in the Initiation and Early Growth of Fatigue Cracks in Turbo-Generator Materials, in *Multiaxial Fatigue*, STP 853, American Society for Testing and Materials, 1982, p 340-360
19. R.B. Waterhouse, Theories of Fretting Processes, in *Fretting and Fatigue*, Applied Science, 1981, p 207-111
20. E.H.R. Wade and C.M. Preece, Cavitation Erosion of Iron and Steel, *Metall. Trans. A*, Vol 9A, Sept 1978, p 1299-1310
21. C.J. Heathcock and A. Ball, Cavitation Erosion of Cobalt-Base Stellite Alloys, Cemented Carbides and Surface Treated Low Alloy Steels, *Wear*, Vol 74, 1981-1982, p 11-26
22. M. Yanishevsky and D.W. Hoeppner, ''Corrosion Fatigue Behavior of Ti-6Al-4V in Simulated Body Environments,'' Paper presented at the 16th Annual Technical Meeting of the International Metallographic Society, Calgary, Canada, July 1983
23. I.M. Bernstein and A.W. Thompson, Ed., *Hydrogen Effects in Metals*, The Metals Society of the American Institute of Mining, Metallurgical, and Petroleum Engineers, 1981
24. C.E. Jaske, J.H. Payer, and V.S. Balint, ''Corrosion Fatigue of Metals in Marine Environments,'' MCIC Report 81-42, Battelle Columbus Laboratories, 1981
25. S.W. Dean, E.N. Pugh, and G.M. Ugiansky Ed., *Environment-Sensitive Fracture: Evaluation and Comparison of Test Methods*, STP 821, American Society for Testing and Materials, 1984
26. Z.A. Foroulis, Ed., *Environment Sensitive Fracture of Engineering Materials*, The Metals Society of the American Institute of Mining, Metallurgical, and Petroleum Engineers, 1978
27. J. Hochmann, J. Slater, R.D. McCright, and R.W. Staehle, Ed., *Stress Corrosion Cracking and Hydrogen Embrittlement of Iron Base Alloys*, National Association of Corrosion Engineers, 1976
28. T.W. Crooker and B.N. Leis, Ed., *Corrosion Fatigue: Mechanics, Metallurgy, Electrochemistry and Engineering*, STP 801, American Society for Testing and Materials, 1984
29. O. Deveraux, A.J. McEvily, and R.W. Staehle, Ed., *Corrosion Fatigue: Chemistry, Mechanics and Microstructure*, National Association of Corrosion Engineers, 1973
30. H.L. Craig, Jr., T.W. Crooker, and D.W. Hoeppner, Ed., *Corrosion Fatigue Technology*, STP 642, American Society for Testing and Materials, 1978
31. M.H. Kamdar, Ed., *Embrittlement by Liquid and Solid Metals*, The Metallurgical Society, 1984
32. A.J. McEvily and R.P. Wei, Fracture Mechanics and Corrosion Fatigue, in *Corrosion Fatigue: Chemistry, Mechanics, and Microstructure*, O. Deveraux, A.J. McEvily, and R.W. Staehle, Ed., National Association of Corrosion Engineers, 1973, p 381-395
33. R.P. Wei and G. Shim, Fracture Mechanics and Corrosion Fatigue, in *Corrosion Fatigue: Mechanics, Metallurgy, Electrochemistry and Engineering*, STP 801, T.W. Crooker and B.N. Leis, Ed., American Society for Testing and Materials, 1984, p 5-25
34. R.P. Wei, P.S. Pao, R.G. Hart, T.W. Weir, and G.W. Simmons, *Metall. Trans. A*, Vol 11A, 1980, p 151
35. F.P. Ford, Current Understanding of the Mechanism of Stress Corrosion and Corrosion Fatigue, in *Environment-Sensitive Fracture: Evaluation and Comparison of Test Methods*, STP 821, S.W. Dean, E.N. Pugh, and G.M. Ugianski, Ed., American Society for Testing and Materials, 1984, p 32-51
36. R.P. Gangloff, Ed., *Embrittlement by the Localized Crack Environment*, Proceedings of the Symposium on Localized Crack Chemistry and Mechanics in Environment-Assisted Cracking, Philadelphia, PA, Oct 1983, The Metallurgical Society, 1984
37. D.A. Jones, A Unified Mechanism of Stress Corrosion and Corrosion Fatigue, *Metall. Trans. A*, Vol 16A, June 1985, p 1133-1141

Environmentally Induced Cracking

Chairman: Bruce Craig, Metallurgical Consultants, Inc.

THIS ARTICLE has been so titled to incorporate the forms of corrosion that produce cracking of metals as a result of exposure to their environment. This cracking may take the form of relatively slow, stable crack extension or, as is often the case, unpredictable catastrophic fracture.

The article is divided into four sections: stress-corrosion cracking, hydrogen damage (frequently referred to as hydrogen embrittlement), liquid-metal embrittlement, and solid metal induced embrittlement. In general, these different phenomena show many similarities, and it would at first seem appropriate to propose an all-encompassing mechanism to account for these behaviors. For example, all of these phenomena are generally dependent on yield strength and applied stress. As both of these factors increase, resistance to stress-corrosion cracking, hydrogen damage, liquid-metal embrittlement, and solid metal induced embrittlement decreases. However, as will be presented in the discussion of each of these phenomena, there are many differences encountered between the various forms of environmentally assisted cracking, and in fact, substantial differences are observed for behavior of metals and alloys within a specific form of cracking.

At this time, the understanding of each of these forms of cracking is largely phenomenological. No satisfactory theory exists for any of these mechanisms that totally explains all behavior observed either under laboratory or field conditions. Although many theories exist that are specific to the behavior of certain alloy systems or environments, none is universal enough to explain, for example, the diverse behavior of hydrogen damage for systems that develop hydrides versus alloys of iron that do not form hydrides. There are many of these contradictory factors that impede the development of an all-encompassing theory.

Additionally, there has been a continuing controversy over the last 20 years concerning the actual micromechanistic causes of stress-corrosion cracking, which some investigators consider to be related to hydrogen damage and not strictly an active-path corrosion phenomenon. Although certain convincing data exist for a role of hydrogen in stress-corrosion cracking of certain alloys, sufficient data are not available to generalize this concept.

As the understanding of these forms of cracking increases, the complexity of each also increases. For example, liquid-metal embrittlement has been a recognized form of cracking induced by liquid metals. However, it has recently been shown that certain metals can become embrittlers below their melting point and that embrittlement by these metals becomes progressively worse as the temperature approaches the melting point. This behavior, although similar to liquid-metal embrittlement, is sufficiently different to warrant reclassifying as solid metal induced embrittlement many of the environmentally assisted fractures that were previously termed liquid-metal embrittlement.

Because satisfactory mechanistic models have not been developed for any of these forms of environmental cracking, the prediction of environmentally assisted cracking is essentially nonexistent. However, prediction of these types of failures is most important, because measurable corrosion usually does not occur before or during crack initiation or propagation. When corrosion does occur, it is highly localized (that is, pitting, crevice attack) and may be difficult to detect.

The field of environmentally assisted cracking is one of the most active in corrosion research, and it is continually changing. Therefore, the reader is encouraged to consult the most current literature for more detail on specific theories and alloy behavior. Additional information can also be found in the articles "Evaluation of Stress-Corrosion Cracking" and "Evaluation of Hydrogen Embrittlement" in this Volume.

Stress-Corrosion Cracking

R.H. Jones,
Battelle Pacific Northwest Laboratories
R.E. Ricker, National Bureau of Standards

Stress-corrosion cracking (SCC) is a term used to describe service failures in engineering materials that occur by slow environmentally induced crack propagation. The observed crack propagation is the result of the combined and synergistic interaction of mechanical stress and corrosion reactions. This is a simple definition of a complex subject, and like most simplifications, it fails to identify the boundaries of the subject. As a result, before this problem can be discussed in detail, one must clearly define the type of loading involved, the types of materials involved, the types of environments that cause this type of crack propagation, and the nature of the interactions that result in this phenomenon. The term stress-corrosion cracking is frequently used to describe any type of environmentally induced or assisted crack propagation. However, in this discussion, the focus will be on the normal usage of the term as defined below.

One frequent misconception is that SCC is the result of stress concentration at corrosion-generated surface flaws (as quantified by the stress intensity factor, K); when a critical value of stress concentration, K_{crit}, is reached, mechanical fracture results. Although stress concentration does occur at such flaws, these do not exceed the critical value required to cause mechanical fracture of the material in an inert environment ($K_{SCC} < K_{crit}$). Precorrosion followed by loading in an inert environment will not result in any observable crack propagation, while simultaneous environmental exposure and application of stress will cause time-dependent subcritical crack propagation. The term synergy was used to describe this process because the combined simultaneous interaction of mechanical and chemical forces results in crack propagation where neither factor acting independently or alternately would result in the same effect. The exact nature of this interaction is the subject of numerous scientific investigations and will be covered in the discussion "Crack Propagation Mechanisms" in this section.

The stresses required to cause SCC are small, usually below the macroscopic yield stress, and are tensile in nature. The stresses can be externally applied, but residual stresses often cause SCC failures. However, compressive residual stresses can be used to prevent this phenomenon. Static loading is usually considered to be responsible for SCC, while environmentally induced crack propagation due to cyclic loading is defined as corrosion fatigue (see the section "Corrosion Fatigue" in the article "Mechanically Assisted Degradation" in this Volume). The boundary between these two classes of phenomena is vague, and corrosion fatigue is often considered to be a subset of SCC. However, because the environments that cause corrosion fatigue and SCC are not always the same, these two should be considered separate phenomena.

The term stress-corrosion cracking is usually used to describe failures in metallic alloys. However, other classes of materials also exhibit delayed failure by environmentally induced crack propagation. Ceramics exhibit environmentally induced crack propagation (Ref 1) and polymeric materials frequently exhibit craze cracking as a result of the interaction of applied stress and environmental reactions (Ref 2-5). Until recently, it was thought that pure metals were immune to SCC; however, it has been shown that pure metals are susceptible to SCC (Ref 6, 7). This discussion will focus on the SCC of metals and their alloys and will not be concerned with stress-environment interactions in ceramics and polymers.

Environments that cause SCC are usually aqueous and can be condensed layers of moisture or bulk solutions. Typically, SCC of an alloy is the result of the presence of a specific chemical

Table 1 Alloy-environment systems exhibiting SCC

Alloy	Environment
Carbon steel	Hot nitrate, hydroxide, and carbonate/bicarbonate solutions
High-strength steels	Aqueous electrolytes, particularly when containing H_2S
Austenitic stainless steels	Hot, concentrated chloride solutions; chloride-contaminated steam
High-nickel alloys	High-purity steam
α-brass	Ammoniacal solutions
Aluminum alloys	Aqueous Cl^-, Br^-, and I^- solutions
Titanium alloys	Aqueous Cl^-, Br^-, and I^- solutions; organic liquids; N_2O_4
Magnesium alloys	Aqueous Cl^- solutions
Zirconium alloys	Aqueous Cl^- solutions; organic liquids; I_2 at 350 °C (660 °F)

species in the environment. Thus, the SCC of copper alloys, traditionally referred to as season cracking, is virtually always due to the presence of ammonia in the environment, and chloride ions cause or exacerbate cracking in stainless steels and aluminum alloys. Also, an environment that causes SCC in one alloy may not cause SCC in another alloy. Changing the temperature, the degree of aeration, and/or the concentration of ionic species may change an innocuous environment into one that causes SCC failure. Also, an alloy may be immune in one heat treatment and susceptible in another. As a result, the list of all possible alloy-environment combinations that cause SCC is continually expanding, and the possibilities are virtually infinite. A partial listing of some of the more commonly observed alloy-environment combinations that result in SCC is given in Table 1.

In general, SCC is observed in alloy-environment combinations that result in the formation of a film on the metal surface. These films may be passivating layers, tarnish films, or dealloyed layers. In many cases, these films reduce the rate of general or uniform corrosion, making the alloy desirable for resistance to uniform corrosion in the environment. As a result, SCC is of greatest concern in the corrosion-resistant alloys exposed to aggressive aqueous environments. The exact role of the film in the SCC process is the subject of current research and will be covered in the discussion "Crack Propagation Mechanisms" in this section. Table 2 lists some alloy-environment combinations and the films that may form at the crack tip.

Table 2 Alloy-environment combinations and the resulting films that form at the crack tip

Metal or alloy	Environment	Initiating layer
α-brass, copper-aluminum	Ammonia	Dealloyed layer (Cu)
Gold-copper	$FeCl_3$	Dealloyed layer (Au)
	Acid sulphate	Dealloyed layer (Au)
Iron-chromium nickel	Chloride	Dealloyed layer (Ni)
	Hydroxide	Dealloyed layer or oxide
	High-temperature water	Dealloyed layer or oxide
α-brass	Nitrite	Oxide
Copper	Nitrite	Oxide
	Ammonia (cupric)	Porous dissolution zone
Ferritic steel	High-temperature water	Oxide
	Phosphate	Oxide (?)
	Anhydrated ammonia	Nitride
	$CO/CO_2/H_2O$	Carbide
	CS_2/H_2O	Carbide
Titanium alloys	Chloride	Hydride
Aluminum alloys, steels	Various media	Near-surface hydrogen

The Phenomenon of SCC

Stress-corrosion cracking is a delayed failure process. That is, cracks initiate and propagate at a slow rate (for example, 10^{-6} m/s) until the stresses in the remaining ligament of metal exceed the fracture strength. The sequence of events involved in the SCC process is usually divided into three stages:

- Crack initiation and stage 1 propagation
- Stage 2 or steady-state crack propagation
- Stage 3 crack propagation or final failure

Distinction among these stages is difficult because the transition occurs in a continuous manner and the division is therefore arbitrary. To enhance the understanding of this process, these steps will be discussed in the context of typical experimental techniques and results. Stress-corrosion cracking experiments can be classified into three different categories:

- Tests on statically loaded smooth samples
- Tests on statically loaded precracked samples
- Tests using slowly straining samples

More detailed information on these tests can be found in the article "Evaluation of Stress-Corrosion Cracking" in this Volume.

Tests on statically loaded smooth samples are usually conducted at various fixed stress levels, and the time to failure of the sample in the environment is measured. Figure 1 illustrates the typical results obtained from this type of test. In Fig. 1, the logarithm of the measured time to failure, t_f, is plotted against the applied stress, $\sigma_{applied}$, and the time to failure can be seen to increase rapidly with decreasing stress; a threshold stress, σ_{th}, is determined where the time to failure approaches infinity. The total time to failure at a given stress consists of the time required for the formation of a crack (the incubation or initiation time, t_{in}, and the time of crack propagation, t_{cp}. These experiments can be used to determine the maximum stress that can be applied to avoid SCC failure, to determine an inspection interval to confirm the absence of SCC crack propagation, or to evaluate the influence of metallurgical and environmental changes on SCC. However, the actual time for the formation or initiation of cracks is strongly dependent on a wide variety of parameters, such as surface finish and prior history. If a cracklike flaw or a crevice

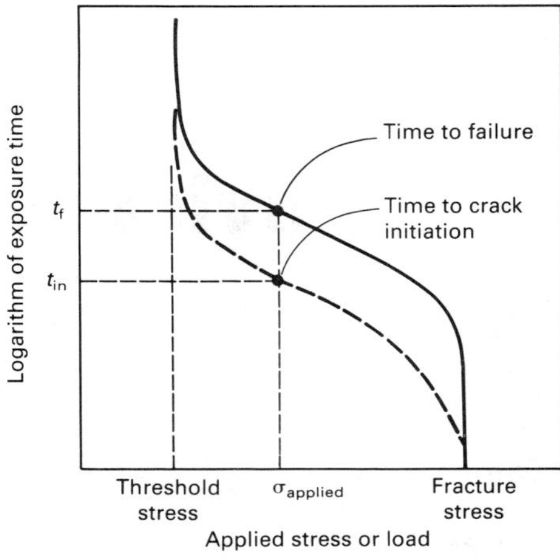

Fig. 1 Schematic of a typical time to failure as a function of initially applied stress for smooth sample stress-corrosion cracking tests

is present in the material, then the time to initiate a crack may be dramatically reduced (Ref 8).

Tests on statically loaded precracked samples are usually conducted with either a constant applied load or with a fixed crack opening displacement, and the actual rate or velocity of crack propagation, da/dt, is measured (Ref 9). The magnitude of the stress distribution at the crack tip (the mechanical driving force for crack propagation) is quantified by the stress intensity factor, K, for the specific crack and loading geometry. As a result, the crack propagation rate, da/dt, is plotted versus K as illustrated in Fig. 2. These tests can be configured such that K increases with crack length (constant applied load), decreases with increasing crack length (constant crack mouth opening displacement), or is approximately constant as the crack length changes (special tapered samples). Each type of test has its advantages and disadvantages. However, in service, most SCC failures occur under constant-load conditions such that the stress intensity increases as the crack propagates. As a result, it is usually assumed in SCC discussions that the stress intensity is increasing with increasing crack length.

Typically, three regions of crack propagation rate versus stress intensity level are found during crack propagation experiments. These are identified according to increasing stress intensity factor as stage 1, 2, or 3 crack propagation (Fig. 2). No crack propagation is observed below some threshold stress intensity level, K_{ISCC}. This threshold stress level is determined not only by the alloy but also by the environment and metallurgical condition of the alloy, and presumably, this level corresponds to the minimum required stress level for synergistic interaction with the environment. At low stress intensity levels (stage 1), the crack propagation rate increases rapidly with the stress intensity factor. At intermediate stress intensity levels (stage 2), the crack propagation rate approaches some constant velocity that is virtually independent of the mechanical driving force. This plateau velocity is characteristic of the alloy-environment combination and is the result of rate-limiting environmental processes such as mass transport of environmental species up the crack to the crack tip. In stage 3, the rate of crack propagation exceeds the plateau

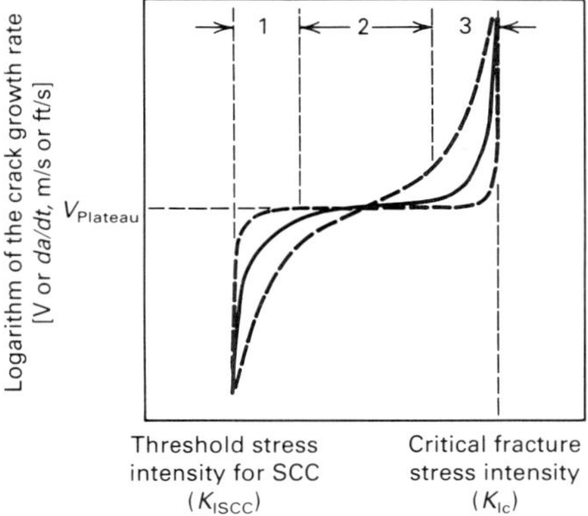

Fig. 2 Schematic diagram of typical crack propagation rate as a function of crack tip stress intensity behavior illustrating the regions of stage 1, 2, and 3 crack propagation as well as identifying the plateau velocity and the threshold stress intensity

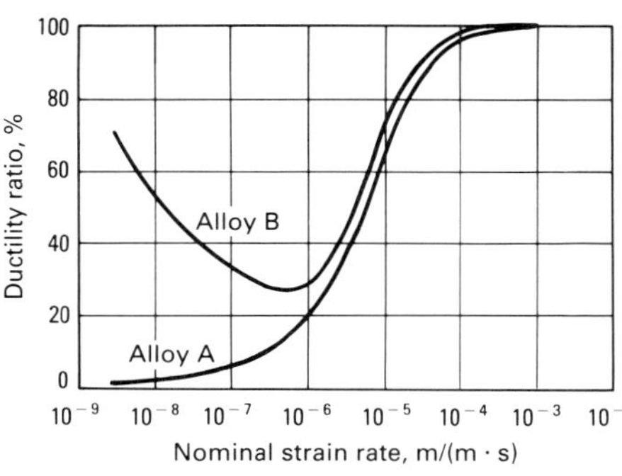

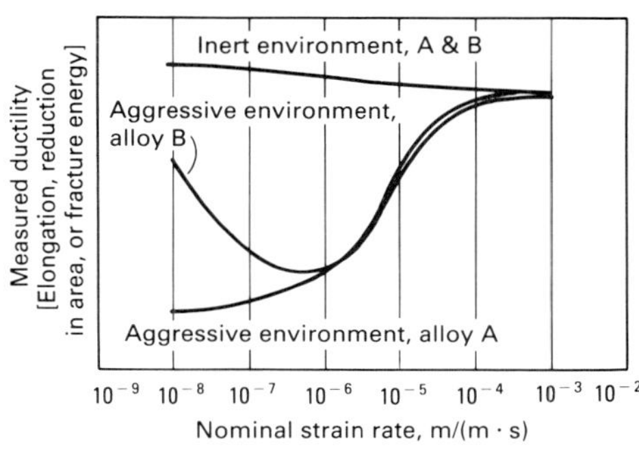

Fig. 3 Strain to failure plots resulting from slow strain rate testing. (a) Schematic of typical ductility versus strain rate behavior of two different types of alloys tested by the slow strain rate technique. (b) The ductility ratio is the ratio of a ductility measurement such as elongation, reduction in area, and fracture energy measured in the aggressive environment to that obtained in the inert reference environment.

velocity as the stress intensity level approaches the critical stress intensity level for mechanical fracture in an inert environment, K_{Ic} (Ref 10).

Slow Strain Rate Testing. Stress-corrosion tests can also be conducted by slowly increasing the load or strain on either precracked or smooth samples. Usually, a tensile machine pulls a smooth sample that is exposed to the corrosive environment at a low cross head speed (10^{-5} to 10^{-9} m/s). The strain to failure in the corrosive environment and the strain to failure in an inert environment can then be plotted against the strain rate, as shown in Fig. 3(a), or the ratio of these measurements can be plotted as shown in Fig. 3(b). The ratio of other tensile property measurements, such as reduction in area and ultimate tensile strength, may be plotted. These tests are called constant extension rate tests (CERT), slow strain rate tests, or straining electrode tests. These are excellent tests for comparing the relative susceptibility of alloys to cracking in an environment or for studying the influence of metallurgical variables on the susceptibility of an alloy. However, the application of this data to the prediction of actual in-service crack propagation rates is difficult and unreliable.

Overview of Mechanisms

Many different mechanisms have been proposed to explain the synergistic stress-corrosion interaction that occurs at the crack tip, and there may be more than one process that causes SCC. The proposed mechanisms can be classed into two basic categories:

- Anodic mechanisms
- Cathodic mechanisms

That is, during corrosion, both anodic and cathodic reactions must occur, and the phenomena that result in crack propagation may be associated with either class of reactions. The most obvious anodic mechanism is that of simple active dissolution and removal of material from the crack tip. The most obvious cathodic mechanism is hydrogen evolution, absorption, diffusion, and embrittlement. However, a specific mechanism must attempt to explain the actual crack propagation rates, the fractographic evidence, and the mechanism of formation or nucleation of cracks. Some of the more prominent of the proposed mechanisms will be covered in greater detail in the discussion "Crack Propagation Mechanisms" in this section, but they usually assume that breaking of the interatomic bonds of the crack tip occurs by one of the following mechanisms:

- Chemical solvation and dissolution
- Mechanical fracture (ductile or brittle)

Mechanical fracture includes normal fracture processes that are assumed to be stimulated or induced by one of the following interactions between the material and the environment:

- Adsorption of environmental species
- Surface reactions
- Reactions in the metal ahead of the crack tip
- Surface films

All of the proposed mechanisms contain one or more of these processes as an essential step in the SCC process. Specific mechanisms differ in the processes assumed to be responsible for crack propagation and the way that environmental reactions combine to result in the actual fracture process.

Controlling Parameters

The mechanisms that have been proposed for SCC require that certain processes or events occur in sequence for sustained crack propagation to be possible. These requirements explain the plateau region in which the rate of crack propagation is independent of the applied mechanical stress. That is, a sequence of chemical reactions and processes is required, and the rate-limiting step in this sequence of events determines the limiting rate or plateau velocity of crack propagation (until mechanical overload fracture starts contributing to the fracture process in stage 3). Figure 4 illustrates a crack tip in which crack propagation results from reactions in metal ahead of the propagating crack. Close examination of Fig. 4 will reveal that potential rate-determining steps include:

- Mass transport along the crack to the crack tip
- Reactions in the solution near the crack
- Surface adsorption at or near the crack tip
- Surface diffusion
- Surface reactions
- Absorption into the bulk
- Bulk diffusion to the plastic zone ahead of the advancing crack
- Chemical reactions in the bulk
- The rate of interatomic bond rupture

Changes in the environment that modify the rate-determining step will have a dramatic influence on the rate of crack propagation, while alterations to factors not involved in the rate-determining step or steps will have little influence, if any. However, significantly retarding the rate of any one of the required steps in the sequence could make that step the rate-determining step. In aqueous solutions, the rate of adsorption and surface reactions is usually very fast compared to the rate of mass transport along the crack to the crack tip. As a result, bulk transport into this region or reactions in this region are frequently believed to be responsible for determining the steady-state crack propagation rate or plateau velocity. In gaseous environments, surface reactions, surface diffusion, and adsorption may be rate limiting, as well as the rate of bulk transport to the crack tip (Ref 11, 12).

Several different environmental parameters are known to influence the rate of crack growth in aqueous solutions. These include, but are not limited to:

- Temperature
- Pressure
- Solute species
- Solute concentration and activity
- pH
- Electrochemical potential
- Solution viscosity
- Stirring or mixing

Altering any of these parameters may alter the rate of the rate-controlling steps, either accelerating or reducing the rate of crack propagation. Also, it may be possible to arrest or stimulate crack propagation by altering the rate of an environmental reaction. It is well known and generally accepted that the environment at occluded sites, such as a crack tip, can differ significantly from the bulk solution. If an alteration to the bulk environment allows the formation of a critical SCC environment at crack nuclei, then crack propagation will result. If the bulk environment cannot maintain this local crack tip environment, then crack propagation will stop. As a result, slight changes to the environment may

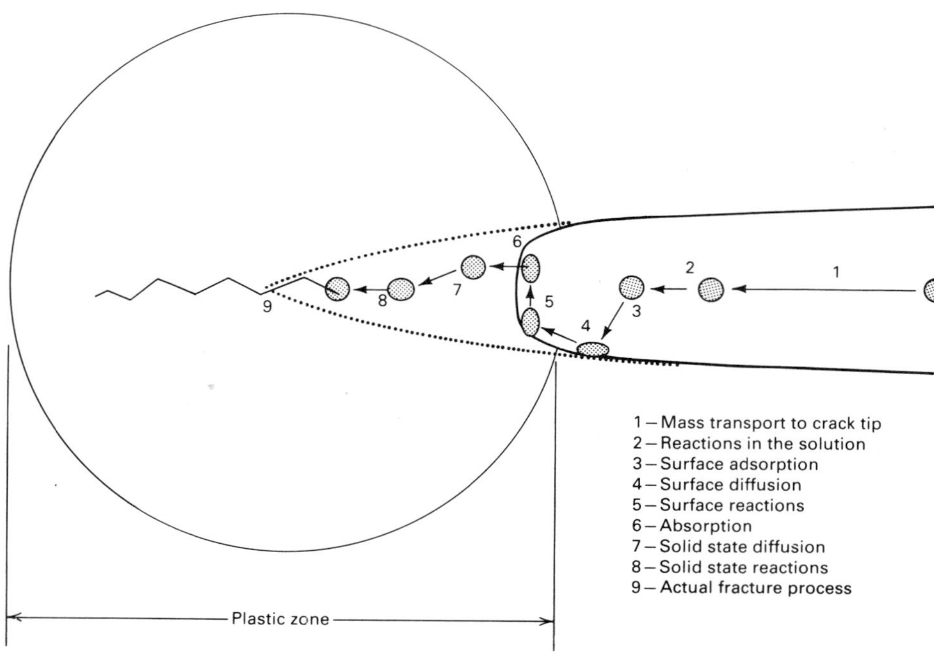

Fig. 4 Schematic of crack tip processes that may be the rate-determining step in environmentally assisted crack propagation

have a dramatic influence on crack propagation, while dramatic changes may have only a slight influence.

In addition to the environmental parameters listed above, stress-corrosion crack propagation rates are influenced by:

- The magnitude of the applied stress or the stress intensity factor
- The stress state, which includes (1) plane stress and (2) plane strain
- The loading mode at the crack tip
- Alloy composition, which includes (1) nominal composition, (2) exact composition (all constituents), and (3) impurity or tramp element composition
- Metallurgical condition, which includes (1) strength level, (2) second phases present in the matrix and at the grain boundaries, (3) composition of phases, (4) grain size, (5) grain-boundary segregation, and (6) residual stresses
- Crack geometry, which includes (1) length, width, and aspect ratio, and (2) crack opening and crack tip closure

Important Fracture Features

Stress-corrosion cracks can initiate and propagate with little outside evidence of corrosion and no warning as catastrophic failure approaches. The cracks frequently initiate at surface flaws that are either preexisting or formed during service by corrosion, wear, or other processes. The cracks then grow with little macroscopic evidence of mechanical deformation in metals and alloys that are normally quite ductile. Crack propagation can be either intergranular or transgranular; sometimes, both types are observed on the same fracture surface. Crack openings and the deformation associated with crack propagation may be so small that the cracks are virtually invisible except in special nondestructive examinations. As the stress intensity increases, the plastic deformation associated with crack propagation increases, and the crack opening increases. When the final fracture region is approached, plastic deformation can be appreciable because corrosion-resistant alloys are frequently quite ductile. The features of stress-corrosion fracture surfaces are covered in greater detail in the article "Modes of Fracture" and the "Atlas of Fractographs" in Volume 12 of the 9th Edition of *Metals Handbook*.

Phenomenology of Crack Initiation Processes

Crack Initiation at Surface Discontinuities. Stress-corrosion cracking frequently initiates at preexisting or corrosion-induced surface features. These features may include grooves, laps, or burrs resulting from fabrication processes. Examples of such features are shown in Fig. 5; these were produced during grinding in the preparation of a joint for welding. The feature shown in Fig. 5(a) is a lap, which subsequently recrystallized during welding and could now act as a crevice at which deleterious cations concentrate. The highly sensitized recrystallized material could also more readily become the site of crack initiation by intergranular corrosion. A

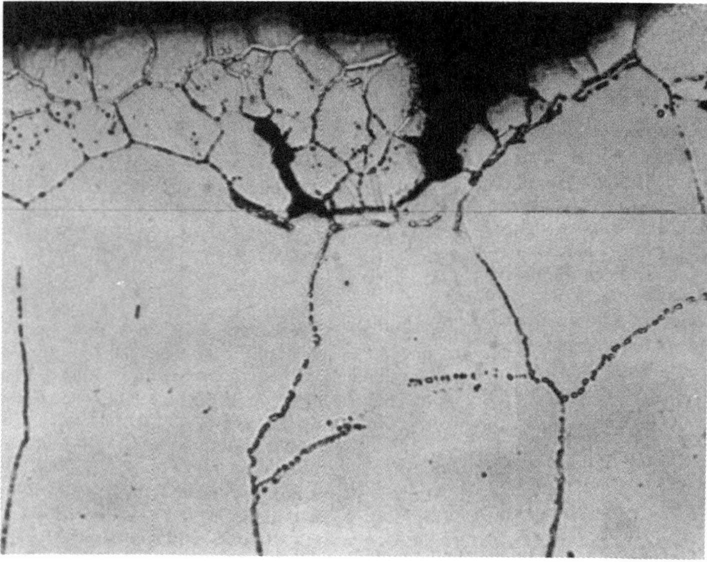
(a)

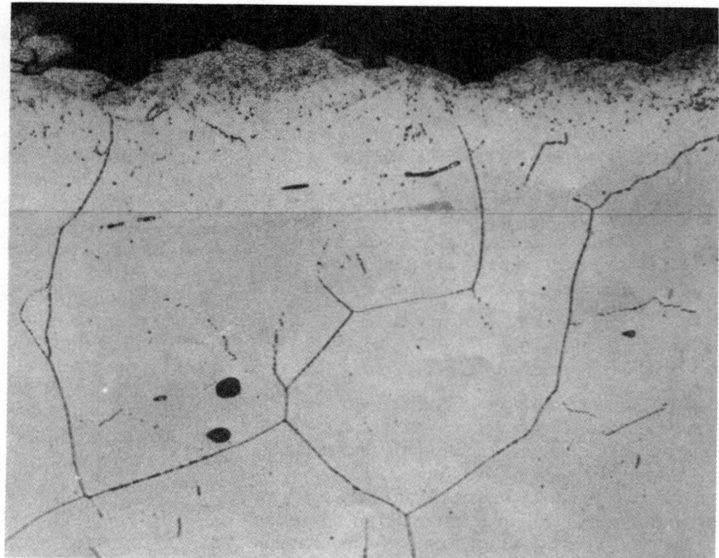
(b)

Fig. 5 Optical micrographs showing defects on the inner surface of type 304 stainless steel pipe near weld root (a) and near through crack (b). Both 1000×

cold-worked layer and surface burrs, shown in Fig. 5(b), can also assist crack initiation.

Crack Initiation at Corrosion Pits. Stress-corrosion cracks can also initiate at pits that form during exposure to the service environment (Fig. 6) or by prior cleaning operations, such as pickling of type 304 stainless steel before fabrication. Pits can form at inclusions that intersect the free surface or by a breakdown in the protective film. In electrochemical terms, pits form when the potential exceeds the pitting potential. It has been shown that the SCC potential and pitting potential were identical for steel in nitrite solutions (Ref 13).

The transition between pitting and cracking is dependent on the same parameters that control SCC, that is, the electrochemistry at the base of the pit, pit geometry, chemistry of the material, and stress or strain rate at the base of the pit. A detailed description of the relationship between these parameters and crack initiation has not been developed, because of the difficulty in measuring crack initiation. Methods for measuring short surface cracks are under development, but are limited to detecting cracks that are beyond the initiation stage. The pit geometry is important in determining the stress and strain rate at the base of the pit. Generally, the aspect ratio between the penetration and the lateral corrosion of a pit must be greater than about 10 before a pit acts as a crack initiation site. A penetration to lateral corrosion ratio of 1 corresponds to uniform corrosion, and a ratio of about 1000 is generally observed for a growing stress-corrosion crack. As in the case of a growing crack, the pit walls must exhibit some passive film forming capability in order for the corrosion ratio to exceed 1.

A change in the corrosive environment and potential within a pit may also be necessary for the pit to act as a crack initiator. Pits can act as occluded cells similar to cracks and crevices, although in general their volume is not as restricted. There are a number of examples in which stress-corrosion cracks initiated at the base of a pit by intergranular corrosion. In these circumstances, the grain-boundary chemistry and the pit chemistry were such that intergranular corrosion was favored. Crack propagation was also by intergranular SCC in these cases.

Although the local stresses and strain rates at the base of the pit play a role in SCC initiation, there are examples of preexisting pits that did not initiate stress-corrosion cracks. This observation has led to the conclusion that the electrochemistry of the pit is more important than the local stress or strain rate (Ref 13). A preexisting pit may not develop the same local electrochemistry as one grown during service because the development of a concentration cell depends on the presence of an actively corroding tip that establishes the anion and cation current flows. Similarly, an inability to reinitiate crack growth in samples in which active growth was occurring before the samples were removed from solution, rinsed, dried, and reinserted into solution also suggests that the local chemistry is very important.

Crack Initiation by Intergranular Corrosion or Slip Dissolution. Stress-corrosion crack initiation can also occur in the absence of pitting by intergranular corrosion or slip-dissolution processes. Intergranular corrosion-initiated SCC requires that the local grain-boundary chemistry differ from the bulk chemistry. This condition occurs in sensitized austenitic stainless steels or with the segregation of impurities such as phosphorus, sulfur, or silicon in a variety of materials. Slip-dissolution-initiated SCC results from local corrosion at emerging slip planes and occurs primarily in low stacking fault materials. The processes of crack initiation and propagation by the slip-dissolution process are in fact very similar.

Crack Initiation Mechanisms

Although the features causing SCC initiation, such as pits, fabrication defects, and intergranular corrosion, are easily observed and identified, there are few well-developed models of SCC initiation. This lack of models for crack initiation mechanisms is the result of several complicating factors. For example, the initiation of a crack is difficult to measure experimentally, even though it is not difficult to detect the location from which a growing crack has emanated. Furthermore, crack initiation has not been precisely defined, and it is difficult to determine at what point a pit is actually a crack and when intergranular corrosion becomes intergranular SCC. Also, the fracture mechanics concept of design assumes preexisting flaws in materials, although these may not be surface flaws that can become stress-corrosion cracks. It has been demonstrated that the corrosion fatigue threshold of 12% Cr and 2.0% NiCrMoV steels could be related to the minimum depth of surface pits, as shown in Fig. 7. Using a linear-elastic fracture mechanics approach and relating a pit to a half-elliptical surface crack, one researcher has shown that the critical pit dimension could be expressed by the following relationship (Ref 14):

$$a_0 = \frac{1}{\pi}\left(\frac{\Delta K_{th}}{F\Delta\sigma_0}\right)^2 \quad \text{(Eq 1)}$$

where ΔK_{th} is the corrosion fatigue threshold, F is a constant, and $\Delta\sigma_0$ is the alternating surface stress. A pit could be represented by a half-elliptical surface crack because it had intergranular corrosion at the base that caused the pit to have cracklike characteristics.

A model for stress-corrosion crack growth was developed from pits in brass tested in a nontarnishing ammoniacal solution (Ref 15). The

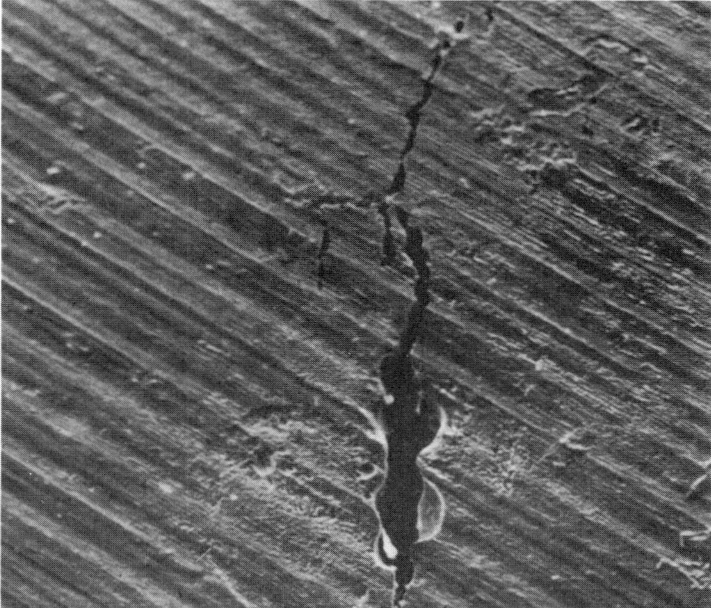

(a)

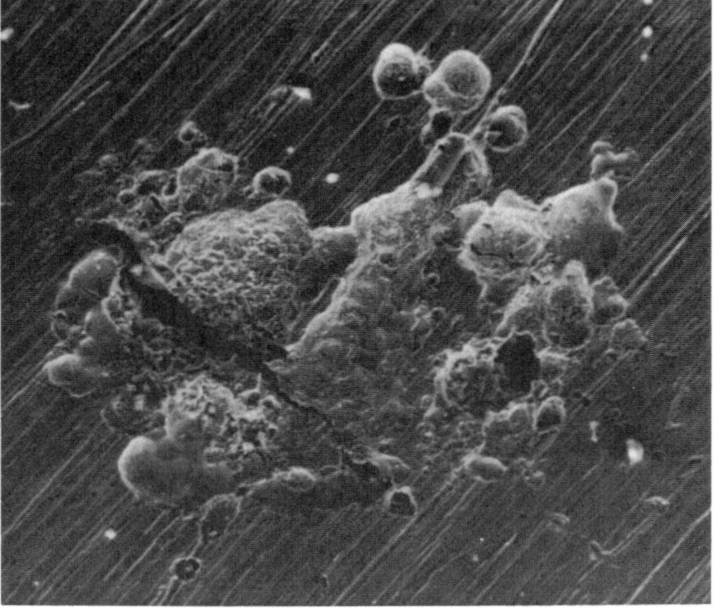

(b)

Fig. 6 Stress-corrosion crack initiating from a corrosion pit in a quenched-and-tempered high-strength turbine disk steel (3.39Ni-1.56Cr-0.63Mo-0.11V) test coupon exposed to oxygenated, demineralized water for 800 h under a bending stress of 90% of the yield stress. (a) 275×. (b) 370×. Courtesy of S.J. Lennon, ESCOM, and F.P.A. Robinson, University of the Witwatersrand

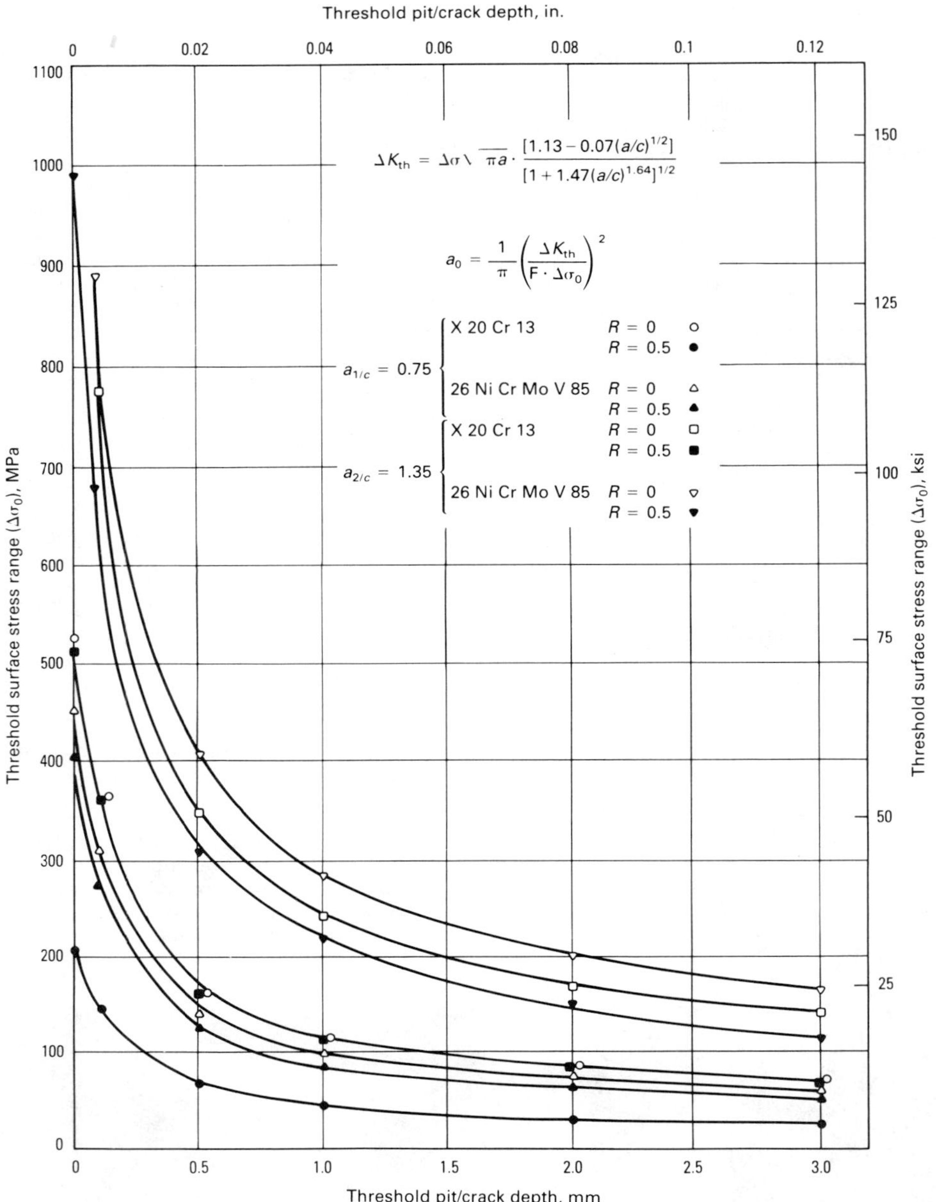

Fig. 7 Relationship between stress amplitude and minimum depth of surface defects for 12% Cr steel and a 2.0% NiCrMoV steel. Source: Ref 14

researchers evaluated the corrosion and stress aspects of a pit to develop a model for crack initiation. They assumed that the corrosion conditions in the base of the pit were essentially the same as those on a flat surface and that initiation required that a critical crack tip opening displacement be exceeded. Using linear-elastic fracture mechanics, the researchers developed the following relationship for the time to initiate a crack:

$$t_{inc} = \frac{(K_{ISCC})^2}{\pi B\,(\sigma^2 - \sigma_0^2)} \exp\left(\frac{-V_m}{V_0}\right) \quad \text{(Eq 2)}$$

where K_{ISCC} is the stress-corrosion threshold for crack growth for brass in ammoniacal solution, σ is the applied stress, σ_0 is the stress needed to close the crack, B is a constant, $-V_m$ is the electrochemical potential of the sample, and V_0 is the reversible potential. However, the assumption that the pit corrosion conditions are equal to the flat surface conditions limits the applicability of this model, because it is generally accepted that the electrochemistry in a pit is considerably different from that of a flat surface. Also, this model does not explicitly describe the transition from a pit to a crack, but treats a pit as a small crack in which the crack tip opening and crack depth are affected by corrosion.

The transition between intergranular corrosion and intergranular SCC was evaluated for nickel with segregated phosphorus and sulfur (Ref 16, 17). Because phosphorus and sulfur inhibit the formation of a passive film on nickel, it was expected that intergranular SCC would be similar

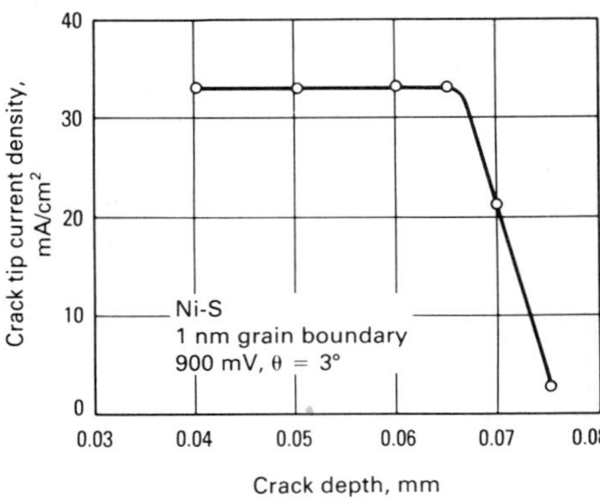

Fig. 8 Effect of crack depth on crack tip corrosion rates for nickel with actively corroding crack walls

in these two materials. It was observed, however, that only shallow intergranular corrosion to a depth of about 0.1 mm (0.004 in.) resulted in nickel in which sulfur was segregated to the grain boundary when tested at a strain rate of 10^{-6} in 1 *N* sulfuric acid (H_2SO_4). On the other hand, rapid intergranular SCC resulted in nickel in which phosphorus was segregated to the grain boundaries that was tested under similar conditions. Using crack tip chemistry modeling, it was shown that intergranular corrosion did not convert to a stress-corrosion crack in Ni+S because the sulfur remained on the crack walls, causing the electrolyte in the solution to become saturated and the crack tip corrosion rate to drop to zero. Starting from a flat surface, the corrosion current at the tip of the intergranular corrosion penetration was about 30 mA/cm^2 (195 mA/in^2) and decreased to 0 at a depth of about 0.08 mm (0.003 in.), as shown in Fig. 8. A similar corrosion depth was measured for samples held at 900 mV for extended periods or for straining electrode samples, while for Ni+P a crack growth rate of 10^{-4} mm/s was measured. These results indicate the importance of electrochemistry in the base of a pit or an intergranular penetration and show that the occurrence of intergranular corrosion does not *a priori* indicate that intergranular SCC will follow.

Phenomenology of Crack Propagation Processes

Crack initiation and propagation are related but different processes; however, by definition, if a crack initiates, it will propagate. As indicated in the discussion "Crack Initiation Mechanisms" in this section, cracks may initiate at preexisting surface flaws if the necessary electrochemical, mechanical, and metallurgical conditions are met, or corrosion processes may create a surface flaw by pitting or localized corrosion. The conditions that create a pit or localized corrosion at chemical inhomogeneities such as grain boundaries, inclusions, second phases, and interphase boundaries are not necessarily the same as those needed for sustained crack growth. For example, the electrochemical conditions near the surface of a material are similar to the bulk electrolyte conditions; electrolyte conditions in a propagating crack generally differ from the bulk electrolyte. If the conditions at the tip of the growing pit or localized corrosion penetration cannot achieve the proper pH, potential, or chemistry, crack propagation may not proceed. Also, if the pit aspect ratio is not greater than a critical value, then the local stresses and strains resulting from a sustained sharp crack are not attained and crack growth cannot occur. Lastly, for stress-corrosion crack growth resulting from cathodic hydrogen, a short crack may have to initiate by an anodic dissolution process, while propagation depends on hydrogen activity.

Although there are conditions under which local corrosion or pitting may not result in crack propagation, it is more frequent that they do lead to cracking. Three conditions frequently cited as requirements for sustained SCC are a susceptible material, a corrosive environment, and an adequate stress. The list of susceptible material conditions and corrosive environments that are known to cause SCC has been steadily expanding. For example, pure metals are known to crack when impurities are segregated to grain boundaries, nonsensitized stainless steels crack in high-purity water, and ferritic steels crack in environments other than nitrate and caustic solutions. Similarly, the details of the microstructure and microchemistry of a susceptible material, specific ions and chemistry, pH, potential, the corrosion rates of the local crack tip associated with a corrosive environment, and the crack tip stresses, strains, and strain rates associated with cracking are better known today than they were previously. Knowledge and understanding of the factors and mechanisms of stress corrosion are continuing to expand; however, a number of factors are well established. The phenomenological description of stress corrosion based on these known factors is presented in this section.

Environmental Factors

Environmental effects on SCC are frequently summarized simply by listing alloy-environment combinations in which SCC has been observed, such as that in Table 1. In recent years, the number of such combinations has increased. Added to this list is the observation of the transgranular SCC of copper, the intergranular SCC of pure metals such as iron and nickel, and the SCC of materials in high-purity water in the absence of specific anions. Stress-corrosion cracking of pure metals was thought to be impossible in early lists of material-environment combinations, but it is now realized that specific environment/metal reactions occur that allow crack advance. In the case of intergranular SCC of iron and nickel, this reaction occurs between the environment and impurities segregated to the grain boundaries (a condition that exists in many pure metals and engineering alloys).

This increase in known susceptible materials can be attributed to the use of new SCC tests, refined crack propagation monitoring equipment, improved electrochemical control equipment, and, perhaps most importantly, the increased research activity in this field over the last 20 to 30 years. Expectations for material performance have also increased with time as materials are used in more aggressive environments under more demanding loading conditions. Lists such as that given in Table 1 can be useful in materials selection for a design involving corrosive environments, because it can lead the materials engineer to seek more specific information on the materials and environments in question. However, such lists can also be misleading because service conditions may differ markedly from those in which the susceptibility listed in Table 1 was determined. Because SCC is dependent on bulk alloy chemistry, microstructure, microchemistry, loading parameters, and specific environmental factors such as oxidizing potential and pH, lists of the type given in Table 1 should be used only for a general overview of SCC.

A complete description of SCC must treat both the thermodynamic requirements and kinetic aspects of cracking. A knowledge of the thermodynamic conditions will help determine whether cracking is feasible; kinetic information describes the rate at which cracks propagate. There are thermodynamic requirements both for anodically assisted SCC (where crack propagation is dependent either directly or indirectly on the oxidation of metal atoms from the crack tip and their dissolution in the electrolyte) and for hydrogen-assisted crack growth resulting from the reduction of hydrogen ions at the crack tip.

Thermodynamics of SCC. The thermodynamic conditions for anodically assisted SCC are that dissolution or oxidation of the metal and its dissolution in the electrolyte must be thermodynamically possible and that a protective film, such as an oxide or salt, must be thermodynamically stable. The first condition becomes a requirement because, without oxidation, crack advance by dissolution would not result. That a process is controlled by anodic dissolution does not *a priori* indicate that the total crack extension is the sum of the total number of coulombs of charge exchanged at the crack tip. There are crack advance processes in which the crack advance is controlled by anodic dissolution but in which the total crack length is greater than can be accounted for by the total charge transfer. These processes are covered in the discussion "Crack Propagation Mechanisms" in this section. However, it is important to note that if the brittle crack advance process is initiated and controlled by anodic dissolution, the crack growth rate will be zero if the anodic current density is zero and will increase with increasing current density. This also holds true for crack advance resulting from brittle crack jump.

The thermodynamic requirement of simultaneous film formation and oxidation for stress-corrosion crack growth can be understood from the diagram shown in Fig. 9, in which the ratio of the corrosion currents from the walls relative to the crack tip is the critical parameter. This ratio must be substantially less than 1 for a crack to propagate; otherwise, the crack will blunt, or the crack tip solution will saturate. Crack initiation can also be controlled by this ratio, because a pit with a high wall corrosion rate will broaden as fast or faster than it will penetrate, resulting in general corrosion rather than crack growth. It is generally believed that the activity of the crack walls relative to the crack tip is a consequence of greater dynamic strain at the tip than along the walls.

A thermodynamic requirement of simultaneous film formation and oxidation of the underlying material led to the identification of critical potentials for the presence or absence of SCC. An example of these critical potentials is shown in Fig. 10 for a passive film forming material such as stainless steel. Zones 1 and 2 are those in which transgranular stress-corrosion crack growth is most likely to occur; intergranular stress-corrosion crack growth can occur over a wider range of potentials than these two zones. Transgranular SCC occurs in zone 1 because the material is in

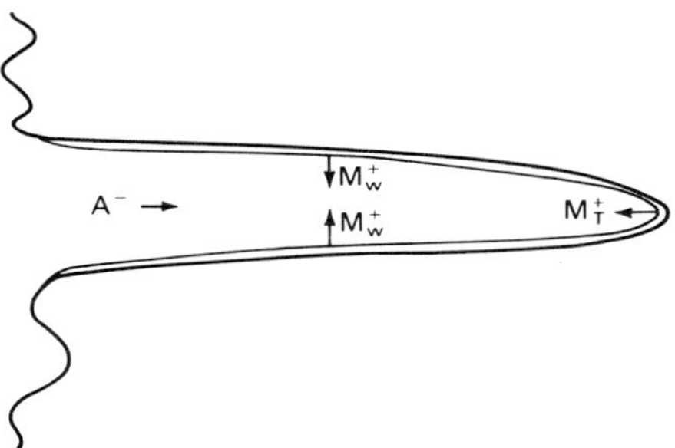

Fig. 9 Schematic of stress-corrosion crack showing important transport and corrosion reactions. A^- represents negatively charged anions migrating to the crack tip, M_W^+ represents metal ions entering the crack solution from the crack walls, and M_T^+ are metal ions entering the crack solution from the crack tip.

transition from active corrosion to passive film formation such that the simultaneous conditions for film formation on the crack walls and corrosion at the crack tip are met. A similar condition exists in zone 2, with the added factor that these potentials are at or above the pitting potential so that cracks can initiate by pitting.

Intergranular SCC occurs over a wider range of potentials than those shown for zones 1 and 2 because chemical inhomogeneities at the grain boundary produce a different electrochemical response relative to the bulk material. Therefore, passive crack walls and active crack tips can result over the potential range from zone 1 to zone 2.

Examples of the critical potentials for SCC are shown in Fig. 11 for several materials. Identification of critical potentials for SCC has led to the use of electrochemical methods for assessing stress-corrosion susceptibility. Zones 1 and 2 are identified by determining the electrochemical potential versus current curves, as shown in Fig. 10 and 11. The shapes of these curves determined at high and low sweep rates are also used to indicate potentials at which the

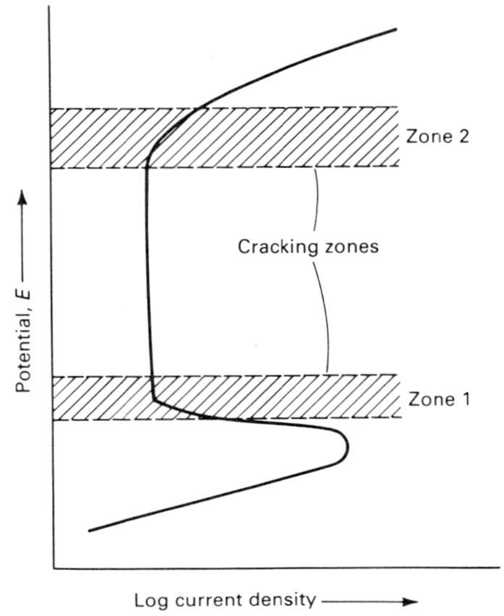

Fig. 10 Potentiokinetic polarization curve and electrode potential values at which stress-corrosion cracking appears

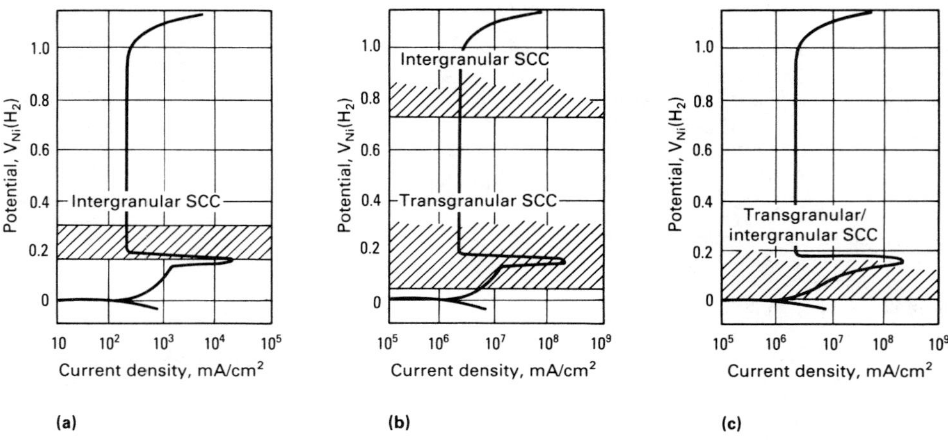

Fig. 11 Potentiokinetic polarization curve and electrode potential values at which intergranular and transgranular stress-corrosion cracking appears in a 10% NaOH solution at 288 °C (550 °F). (a) Alloy 600. (b) Alloy 800. (c) AISI type 304 stainless steel

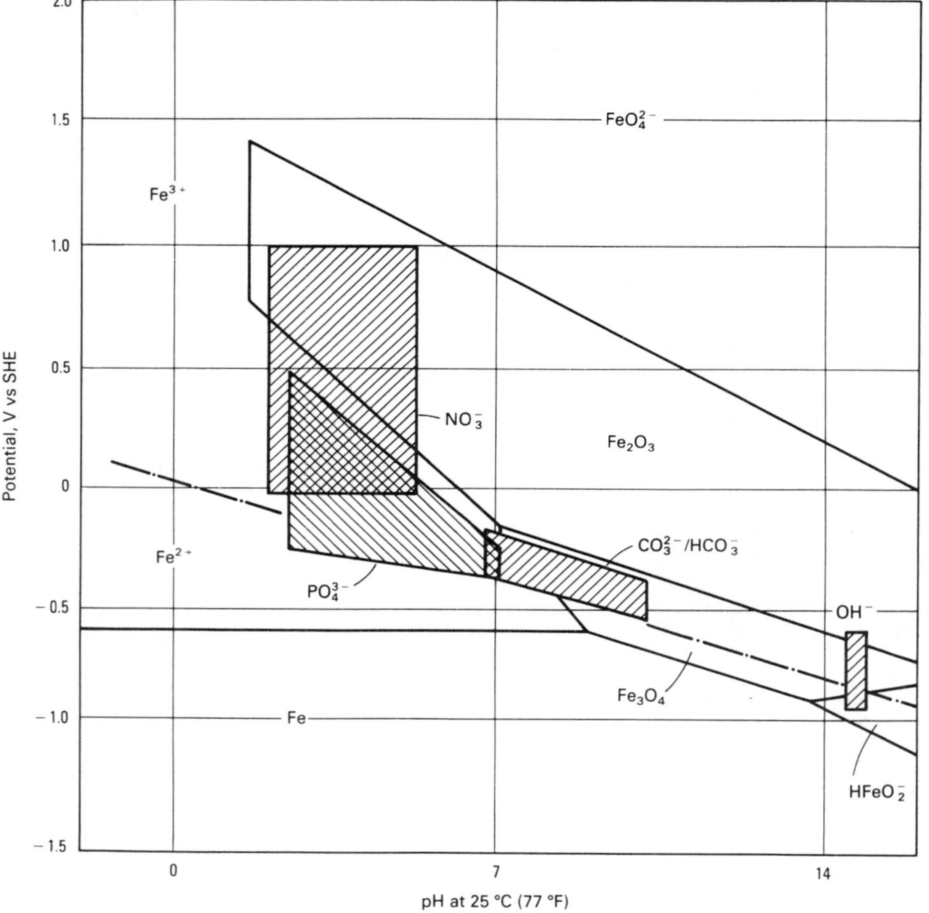

Fig. 12 Relationship between pH-potential conditions for severe cracking susceptibility of carbon steel in various environments and the stability regions for solid and dissolved species on the Pourbaix diagram. Note that severe susceptibility is encountered where a protective film (phosphate, carbonate, magnetite, and so on) is thermodynamically stable, but if ruptured, a soluble species (Fe^{2+}, $HFeO_2^-$) is metastable.

simultaneous conditions of film formation and metal oxidation occur.

Application of Potential-pH Diagrams. Critical potentials for SCC can also be related to potential-pH stability diagrams (Pourbaix diagrams), because these diagrams describe the conditions at which film formation and metal oxidation will occur. An example for carbon steel is given in Fig. 12, in which SCC is associated with potentials and pHs at which phosphate, carbonate, or magnetite films are thermodynamically stable while the species Fe^{2+} and $HFeO_2^-$ are metastable. A second example of the potential-pH regimes in which SCC occurs is given in Fig. 13 for a 70Cu-30Zn brass in a variety of solutions.

The effect of many environmental parameters, such as pH, oxygen concentration, and temperature, on the thermodynamic conditions for SCC can be related to their effect on the potential-pH diagrams (as shown in Fig. 12 and 13) or on the material potential relative to the various stability regions. Using the potential-pH diagram for iron in water at 25 °C (77 °F) shown in Fig. 14, the effects of changing the pH and oxygen concentration can be illustrated. A decrease in the pH from 9 to 6 at a potential of −0.2 V versus standard hydrogen electrode (SHE) shifts iron from a region of stability to one of active corrosion. Based on the thermodynamic criteria for simultaneous stability and active corrosion, the critical pH would be expected to be 7, with decreasing susceptibility at higher pHs because of increased film stability and at lower pHs because of decreased film stability and the increase of general corrosion. Changes in the oxygen concentration generally alter the electrode potential, with increasing oxygen concentration resulting in more oxidizing conditions. The effects of temperature on the potential-pH diagram must be determined for each temperature of interest as the regions of stability shift with temperature.

For materials in which SCC occurs by a hydrogen-induced subcritical crack growth mechanism, the thermodynamic requirement for crack growth is governed by the hydrogen reduction line shown by dashed line *a* in Fig. 14. Hydrogen reduction on iron in water at 25 °C (77 °F) occurs at potentials below this line, but not above it. Therefore, the range of potentials at which hydrogen is available to cause crack growth increases and becomes more oxidizing with decreasing pH.

Application of potential-pH diagrams for identifying specific conditions at which SCC will occur is limited by a number of factors, such as the availability of these diagrams for complex solutions and for the temperatures of interest and the substantial deviation of the chemistry of a crevice or crack from the bulk solution chemistry. Also, the electrode potential at the crack tip can differ from that of the free surface of the material. These differences arise from the need for diffusion of metal ions away from the actively corroding crack tip, migration of anions into the crack, reactions along the crack walls, convection of the electrolyte in the crack, reactions along the crack walls, and, in some cases, by gas bubble formation that causes potential drops. Efforts to measure the local crack tip chemistry and potentials are restricted by crack sizes, which can be substantially smaller than 1 μm. Therefore, mathematical modeling of transport within cracks and the resulting crack tip chemistry, reactions, and potentials have been actively pursued in recent years. The results of measurements in simulated crevices and modeling have clearly shown that the crack tip chemistry can differ from the bulk conditions, but the specifics of these differences are not well known and depend on the material and environment being considered.

For example, based on evaluations by one researcher, the pH within cracks and crevices of structural steels in marine environments ranged from 4 to 11, of stainless steel from 0 to 3, of aluminum from 3 to 4, and of titanium from 1 to 2

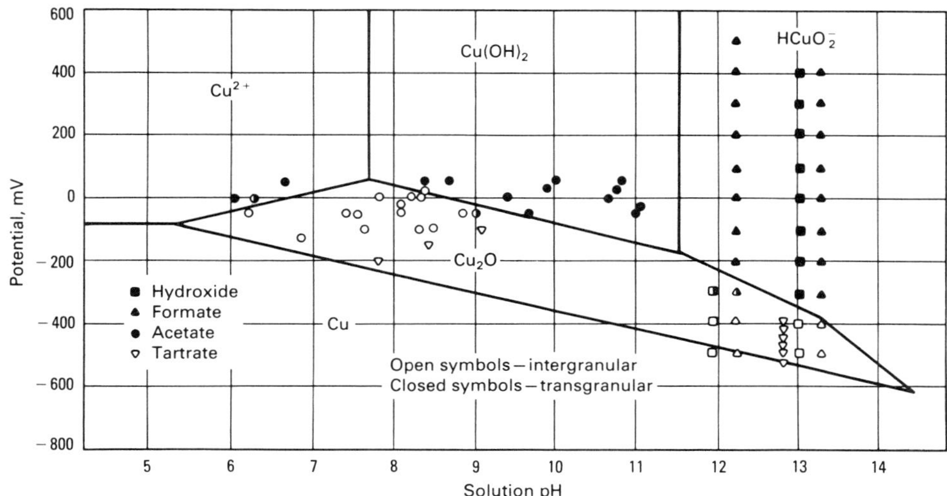

Fig. 13 Potential-pH diagram showing the domains of failure mode for 70Cu-30Ni brass in various solutions, together with the calculated positions of various boundaries relating to the domains of stability of different chemical species

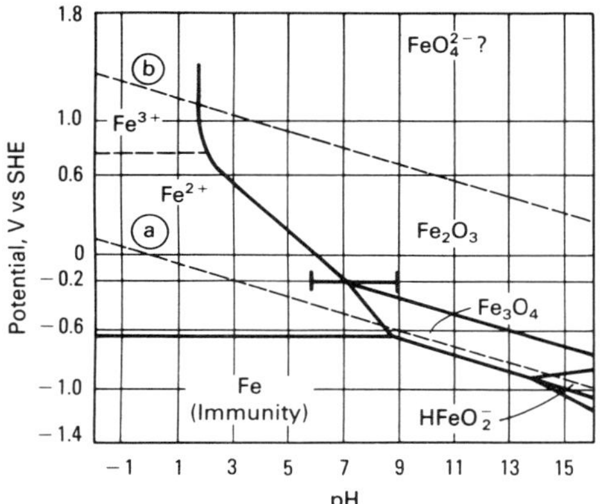

Fig. 14 Potential-pH diagram (Pourbaix) for iron in water at 25 °C (77 °F). A decrease in pH from 9 to 6 at potential of −0.2 V, which shifts iron from a region of stability to one of active corrosion, is indicated by the solid bar.

(Ref 18). Shifts of up to 300 mV in the crack tip potential in the anodic direction were observed for cathodically polarized steel, while other researchers found only a small shift in the crack tip potential relative to the external potential under steady-state conditions (Ref 16) or a 300-mV shift in the cathodic direction when the crevice walls were active, reducing to a small shift when the walls passivated (Ref 19). Therefore, the critical potentials at which cracking may occur can differ from those given by the bulk equilibrium conditions used to determine the potential-pH diagrams. Although it is possible to determine the equilibrium conditions based on the crack tip conditions, the current knowledge of crack tip conditions is fragmentary at best. Other factors that limit the usefulness of the equilibrium potential-pH diagrams are the lack of available diagrams for specific materials and environments and the variable temperatures, chemistries, and potentials common in many industrial applications.

Kinetics of SCC. A knowledge of the thermodynamic conditions at which SCC can occur is insufficient without a corresponding understanding of the kinetics of crack growth, because the life of a component may be adequate if the crack growth rate is sufficiently slow, even though SCC is thermodynamically possible. As in the thermodynamic conditions for SCC, environmental parameters such as potential, pH, oxygen concentration, temperature, and crack tip chemistry have a strong effect on the crack growth kinetics. The crack tip reactions and the rate-determining steps controlling the crack growth rate are specific to alloy-environment combinations; although a discussion of each system is beyond the scope of this section, some general observations will be made. Also, the crack growth rate depends on the crack advance process even though it is controlled by electrochemical reactions.

Detailed descriptions of various crack growth mechanisms are given in the discussion "Crack Propagation Mechanisms" in this section by categorizing these mechanisms as either anodic dissolution or mechanical fracture models. For the case of a crack growing by anodic dissolution alone, the total crack advance is a function of the total charge transfer at the crack tip, while the crack velocity is a function of the crack tip current density. For a crack growing with mechanical fracture, the total crack advance exceeds the total charge transfer at the crack tip, but the crack velocity may still be controlled by the crack tip current density. A limiting velocity can be described for a crack advancing under pure anodic dissolution by the following Faradaic relationship:

$$\frac{da}{dt} = \frac{i_a M}{z F_\rho} \qquad \text{(Eq 3)}$$

where i_a is the anodic current density of a bare surface, M is the atomic weight, z is the valence, F is Faraday's constant, and ρ is the material density. It has been shown (Ref 20) that this relationship (Eq 3) between the bare surface current density and the crack propagation rate is applicable to a wide variety of materials, as presented in Fig. 15. Equation 2 assumes that the crack tip is maintained in a bare condition, while the crack walls are relatively inactive. A bare crack tip and passive walls can result from the difference between the electrochemical conditions at the tip and other regions in the crack, a difference in the local chemistry of the material at the crack tip that causes the tip to be more active than the crack walls (such as at a sensitized or segregated grain boundary), or a crack tip strain rate that is sufficiently high to prevent the formation of a protective film.

A number of factors can reduce the crack velocity below that given by Eq 3 and in Fig. 15. The most widely examined crack growth retardation process is that resulting from the crack tip being covered by a film for some fraction of time. The process of crack growth in the presence of a film at the crack tip has been described by several mechanisms, such as slip dissolution and passive film rupture, covered in the discussion "Crack Propagation Mechanisms" in this section. In general, the crack growth rate depends on the rate at which the film is ruptured and reformed. The amount of corrosion that occurs between these two events has been used to describe the crack growth rate. This time period is determined by the crack tip strain rate, film fracture strain, repassivation rate of the surface, maximum corrosion rate while the tip is bare, and the corrosion rate decay with repassivation. Other factors that can reduce the crack growth rate below that given by Eq 3 are limits in the diffusion rate of species into and out of the crack tip, crack deflection away from the principal stress, and changes in the local material chemistry. Transport of species into and out of cracks is considered a major limitation to attaining the crack growth rate predicted by Eq 3. Factors such as the crack geometry or width, reactions or corrosion rate along the walls of the crack, diffusion rate of anions and cations, and metal salt solubility limits all contribute to transport-limited crack velocities.

A lack of clear knowledge about the specific conditions at the tip of a crack has limited the understanding of the role of specific species on crack growth rates. At this time, it is possible only to describe the effect of bulk electrolyte conditions on crack growth rate; the local conditions may vary significantly. However, knowledge of the local crack tip conditions is most important for understanding the mechanisms of cracking, although a knowledge of the external crack conditions is adequate for monitoring and controlling SCC. Using austenitic stainless steel as an example, the effects of electrochemical potential, oxygen content, and temperature are shown schematically in Fig. 16 through 18. Similar relationships exist for other anion concentrations, such as halides and sulfur species, but the data base is not as well developed as the cases shown. It is important to note that these are not independent effects, because the effects of oxygen concentration and temperature are probably related to the potential. The details of this relationship are complex and are beyond the scope of this section.

A mechanical fracture process in SCC can produce crack velocities exceeding those given by Eq 3 by some magnification factor, which can be as large as 100. The mechanism of brittle fracture induced by SCC, which is covered in the discussion "Crack Propagation Mechanisms" in this section, is thought to involve the formation of a corrosion product at the crack tip in which a

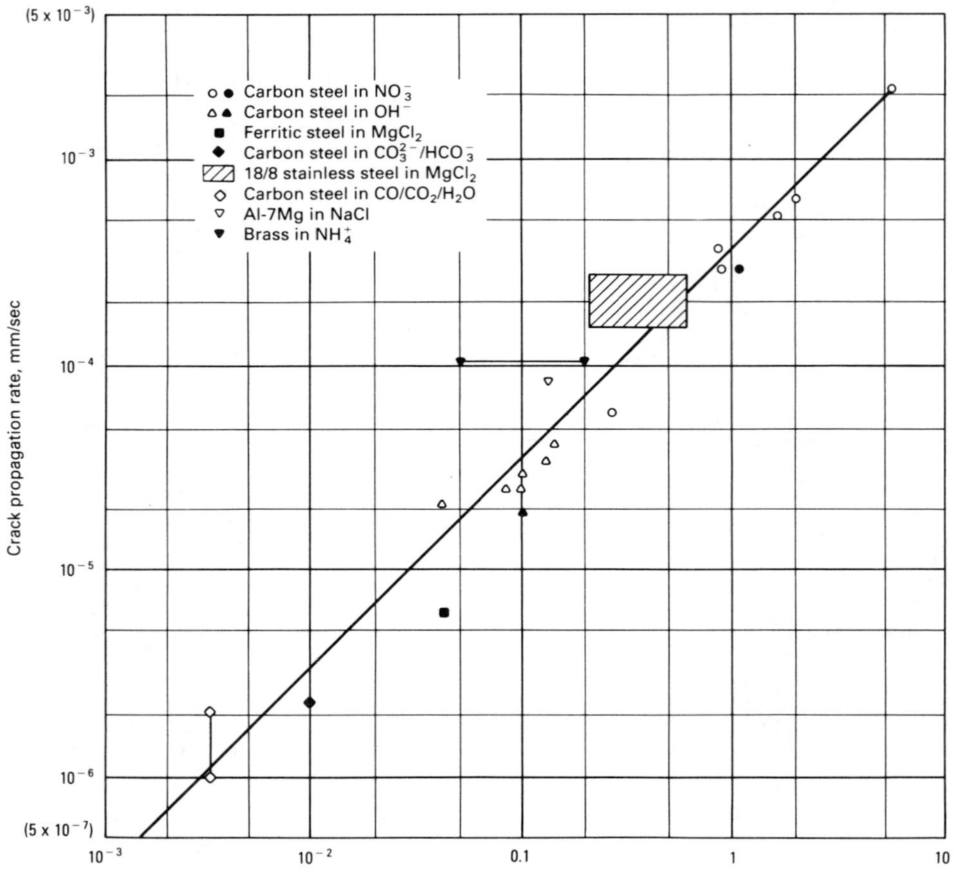

Fig. 15 Relationship between the average crack propagation rate and the oxidation (that is, dissolution and oxide growth) kinetics on a straining surface for several ductile alloy/aqueous environment systems

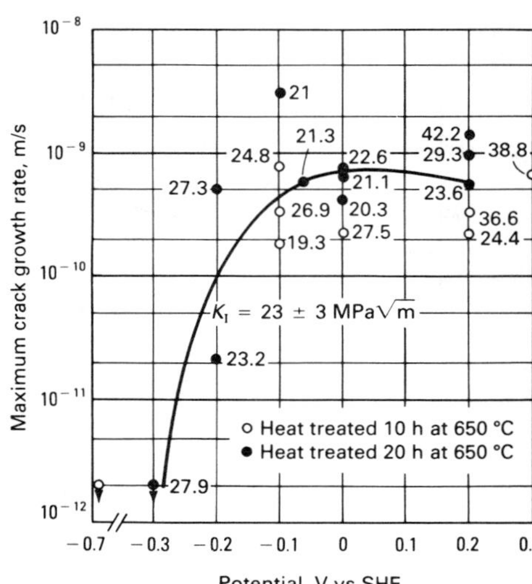

Fig. 16 Effect of potential on the maximum crack growth rate in sensitized type 304 stainless steel in 0.01 m Na_2SO_4 at 250 °C (480 °F). Numbers denote K_I values.

cleavage crack can initiate and propagate some depth into the ductile substrate. As mentioned earlier, this process would produce crack lengths that exceed those accounted for by the total charge transfer/metal oxidation process, but would be dependent on the crack tip corrosion rate because the formation rate of the corrosion product would depend on the corrosion rate.

Early observations and proposals of this process have been made (Ref 21-25). A current list of materials in which SCC is thought to occur through a film-induced cleavage process as suggested in Ref 25 is given in Table 2, which shows a wide range of materials and environments. An attempt has been made to identify the corrosion product or layer that initiates the cleavage cracking process, but this identification must be considered speculative. The electrochemical conditions controlling the brittle SCC process have not been carefully catalogued; however, for many systems, the main issue is whether cracking can be completely described by anodic dissolution or whether a mechanical fracture process is involved. The greatest uncertainty involves the possibility of a transition between anodic and mechanical SCC processes that is affected by electrochemistry, material chemistry, or mechanics.

The electrochemical factors described in the previous paragraphs were generally drawn from examples of intergranular SCC, but many of the effects would be similar for brittle transgranular SCC. For example, the brittle transgranular crack growth rate follows closely the dependence of the anodic current density on potential. Whether or not there are critical potentials at which transgranular SCC may switch from a brittle mechanical fracture mode to a purely anodic dissolution mode is unknown. However, for many systems, the critical potentials describing SCC are valid control parameters even though the process associated with these critical potentials may be uncertain. For an anodic dissolution controlled process, the critical potentials were associated with the need for an active tip and passive crack walls. These conditions are still necessary for a brittle SCC process because it is dependent on the anodic reaction to produce the corrosion reaction product. This product may be a film or a dealloyed layer, but the rate of formation and therefore the brittle crack growth rate are dependent on the crack tip current density.

Material Chemistry and Microstructure

The relationship between material chemistry and microstructure and SCC is equally as complex as the relationship between the environment and SCC. Bulk alloy composition can affect passive film stability and phase distribution (for example, chromium in stainless steel), minor alloying elements can cause local changes in passive film forming elements (for example, carbon in stainless steel causing sensitization), impurity elements can segregate to grain boundaries and cause local differences in the corrosion rate (for example, phosphorus in nickel or nickel-base alloys), and inclusions can cause local crack tip chemistry changes as the crack intersects them (for example, manganese sulfide in steel). Also, alloys can undergo dealloying, which is thought to be a primary method by which brittle SCC initiates. The following summary will be divided into intergranular and transgranular effects because some of the primary factors can be best described in this way; however, it is important to recognize that many of the material chemistry factors can affect both intergranular and transgranular cracking.

Intergranular Stress-Corrosion Cracking. Material chemistry and microstructure effects on intergranular SCC can generally be divided into the following two categories: grain-boundary precipitation and grain-boundary segregation. Grain-boundary precipitation effects include carbide precipitation in austenitic stainless steels and nickel-base alloys, which causes a depletion of chromium adjacent to the grain boundary and intermetallic precipitation in aluminum alloys, which are anodically active. Grain-boundary segregation of impurities such as phosphorus, sulfur, carbon, and silicon can produce a grain boundary that is up to 50% impurity within a 1- to 2-nm thick region. These impurities can alter the corrosion and mechanical properties of the grain boundary and can therefore cause cracking by anodic dissolution and perhaps mechanical fracture.

Grain-Boundary Precipitation in Stainless Steels. Chromium carbide precipitation in stainless steels occurs in the temperature range from 500 to 850 °C (930 to 1560 °F), with the rate of precipitation controlled by chromium diffusion. For intermediate times, such as occurs with heat treating and welding, chromium depletion occurs adjacent to the grain boundary during chromium carbide growth. This depletion can be described by the minimum chromium concentration adja-

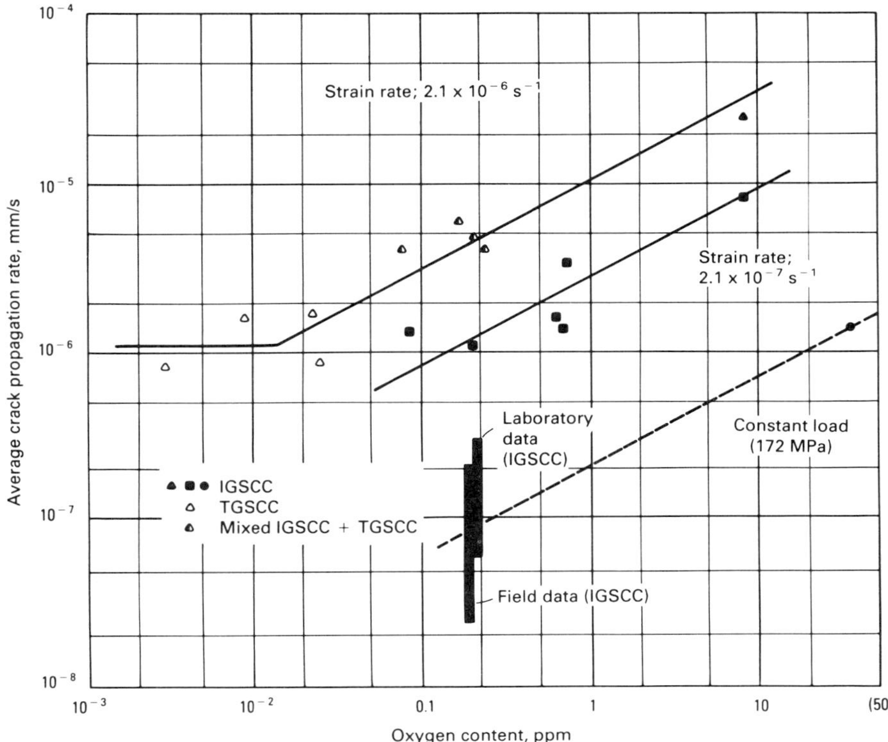

Fig. 17 Variation in the average crack propagation rate in sensitized type 304 stainless steel in water at 288 °C (550 °F) with oxygen content. Data are from both CERT tests, constant load, and field observations on boiling water reactor piping.

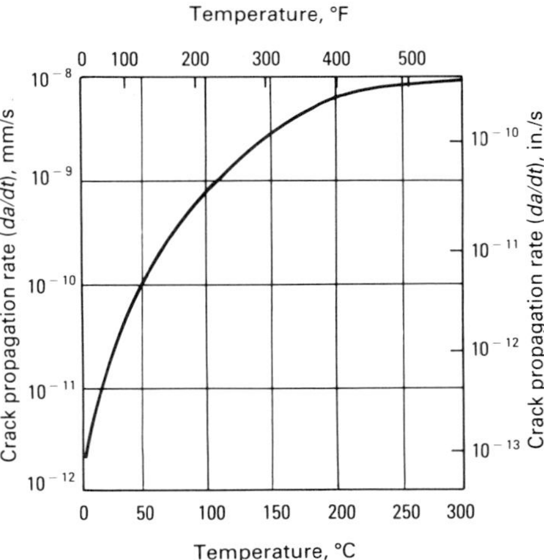

Fig. 18 Schematic of crack growth rate versus temperature for intergranular SCC of type 304 stainless steel

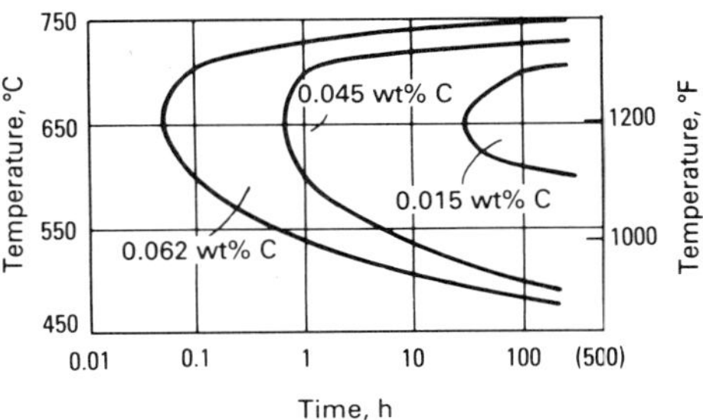

Fig. 19 Time/temperature/sensitization curves determined by EPR tests on type 304 stainless steel alloys of variable carbon contents

cent to the carbide and the width of the depleted zone. Minimum chromium concentrations of 8 to 10 at.% have been measured by analytical electron microscopy, while the width of the depleted zone has been measured to range from 10 nm to hundreds of nanometers. After times long enough for carbide growth to reach completion, the chromium profile is eliminated, and the chromium concentration returns to the bulk value.

The intergranular SCC of austenitic stainless steel is primarily dependent on the nature of the chromium-depleted zone, which is generally explained by the depletion of a passive film forming element along a continuous path through the material. The stress-corrosion susceptibility and crack growth rate of austenitic stainless steel can be described by the degree of sensitization (DOS) as measured by corrosion tests such as the Strauss or electrochemical potentiokinetic reactivation (EPR) tests. Quantitative comparisons between susceptibility as measured by the presence or absence of intergranular SCC in an SCC test or the crack growth rate and the DOS or chromium depletion parameters have been successful in cases in which sufficient data have been available, but these correlations are limited to specific alloys, environments, and stress conditions. Semiquantitative comparisons in which intergranular SCC is predicted for DOS values greater than some critical value are less costly to obtain and have been used more extensively.

An example of the time/temperature/DOS curves at several carbon concentrations is shown in Fig. 19. The curves represent the conditions necessary for the development of a constant degree of sensitization, with the strong effect of bulk carbon concentration on the time needed to develop a sensitized microstructure. Therefore, the most common method for reducing the possibility of developing a sensitized microstructure is to reduce the carbon concentration or to control the thermal history of the material. Given that the material is sensitized, control of the environment and stress conditions can be used to reduce the crack propagation rate.

Clearly, chromium and carbon are the two most important elements involved in the development of sensitization; however, other elements can play a role. Molybdenum has an effect similar to that of chromium—incorporation into the carbides and depletion adjacent to the grain boundaries—although it is generally present in smaller concentrations (for example, in type 316 stainless steel) than chromium. Nickel increases the activity of carbon in austenitic stainless steel; therefore, increases in nickel concentration can enhance sensitization for a given carbon concentration and thermal history. Manganese, silicon, and nitrogen have also been shown to affect sensitization, but the data that is available on these elements is comparatively sparse. Silicon has been shown to segregate to grain boundaries when present in sufficient quantities or by the nonequilibrium processes that occur when the material is irradiated with neutrons. The presence of silicon in the grain boundaries has been found to be deleterious in oxidizing environments.

Grain-Boundary Precipitation in Nickel Alloys. The cause of intergranular SCC in nickel-base alloys is more complex than in austenitic stainless steel. Chromium carbide precipitation and chromium depletion occur in nickel-base alloys, such as alloy 600, as they do in austenitic stainless steels, but a clear connection between the presence of this microstructure and intergranular SCC has not been made. Carbon solubility is considerably lower and chromium diffusion is faster in nickel-base alloys than in austenitic stainless steels; therefore, the nose in the carbide precipitation or sensitization curves shown in Fig. 19 is shifted to shorter times and higher temperatures. Significant heat-to-heat variations in the chromium depletion of alloy 600 have been observed for a given heat treatment. This variation is attributed to the sensitivity of alloy 600 to carbon concentration and to the variability in mill-anneal heat treatment conditions. Chromium depletion is thought to accelerate cracking in oxygenated water, but has not been identified as a controlling factor in deaerated water or caustic environments at 300 °C (570 °F). The rapid carbide precipitation kinetics in nickel-base alloys allows the development of a healed microstructure in which carbide growth is complete and the chromium profile is eliminated with relatively short heat treatment times.

Some significant differences between the intergranular SCC of nickel-base alloys and that of austenitic stainless steels include the positive effect of a semicontinuous distribution of carbides at the grain boundary in single-phase material and the galvanic couple between the γ' and γ phases in precipitation-hardened alloys. The beneficial effect of a semicontinuous distribution of carbides is thought to be related to dislocation generation from the carbides that results in a crack blunting or stress relaxation at the tip of a crack. The detrimental effect of the γ' phase is thought to result from a galvanic couple that polarizes the γ phase anodically relative to the γ' phase.

Grain-Boundary Precipitation in Aluminum Alloys. Aluminum alloys are also noted for the occurrence of intergranular SCC in aqueous environments, including sodium chloride (NaCl) solutions. The details of the intergranular SCC process in aluminum alloys are very complex and vary with alloy composition, but some features can be summarized. Grain-boundary precipitation has been identified as a contributing factor in the intergranular SCC in aluminum alloys; galvanic effects between the precipitates and matrix are considered to be important. In some cases, the precipitates are anodic, and in others, they are cathodic to the matrix. Peak-aged materials that give maximum hardnesses and microstructures in which slip bands are produced upon straining are considered to be vulnerable to SCC. Overaged structures are considered to be less susceptible. The mechanism of crack growth is thought to be a combination of local anodic dissolution and hydrogen embrittlement. Whether crack extension is by anodic dissolution or by hydrogen embrittlement remains an open question; however, in aqueous solutions, the primary source of hydrogen is anodic dissolution.

Grain-Boundary Segregation in Aluminum Alloys. Grain-boundary enrichment of magnesium without precipitation of a magnesium-rich phase has recently been identified as a factor in the intergranular SCC of aluminum alloys (Ref 26, 27). This enrichment is associated with increased corrosion activity at the grain boundary and the possible formation of magnesium hydride (MgH) or enhanced hydrogen entry along the grain boundaries. Crack advance is thought to occur by discontinuous crack jumps (possibly nucleated by hydrides) and crack arrest after about 150 to 350 nm of advance, followed by repetition of this process.

Grain-Boundary Segregation in Ferrous and Nickel Alloys. Grain-boundary enrichment of impurities can contribute to the intergranular SCC of iron-base alloys, austenitic stainless steels, and nickel-base alloys. The extent of their effect depends on the electrochemical potential, the presence of other stress-corrosion processes such as chromium depletion, and their concentration in the grain boundary. The enrichment of impurities to grain boundaries by equilibrium segregation can be described by the enrichment ratio, as shown in Fig. 20 for a variety of impurity-alloy combinations (Ref 28). The enrichment ratio defines the upper bound to the grain-boundary concentration that will be achieved under equilibrium conditions, which can be very long times at low temperatures. It can be seen from Fig. 20 that enrichment ratios as high as 10^5 are possible for some impurities and that ratios greater than 10^2 are not uncommon. For the bulk impurity concentrations present in most engineering alloys, grain-boundary impurity concentrations of 10 to 20 at.% are possible. Therefore, an intergranular stress-corrosion crack can propagate along a grain boundary that has a composition vastly different from that of the bulk alloy.

Ferritic steels exhibit intergranular SCC in hot nitrate, caustic, carbonate, and a variety of other environments. The presence of intergranular SCC is dependent on the electrochemical potential, because intergranular SCC is predominant at potentials in the active-passive transition. Early studies of this effect identified carbon segregation as the primary element of concern where the carbon atoms were said to provide suitable imperfection sites for adsorption of nitrate in an adsorption-induced crack growth mechanism (Ref 29). More recently, phosphorus segregation has been associated with the intergranular SCC of iron alloys in nitrate and caustic solutions (Ref

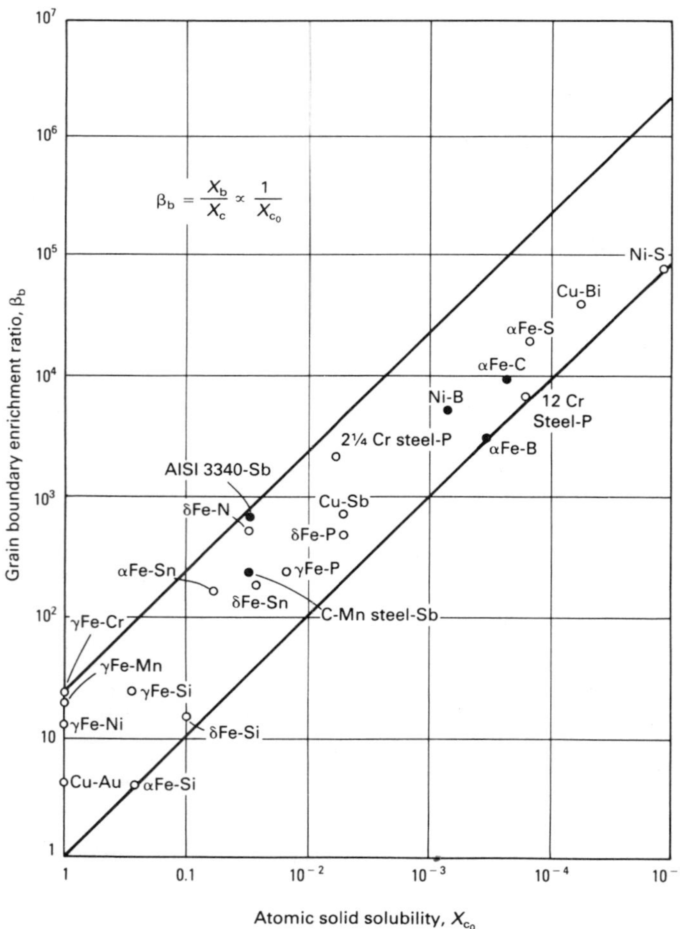

Fig. 20 Correlation between measured grain-boundary enrichment ratios, β, and the inverse of solid solubility, X_{c_0}. The symbols X_b and X_c represent the actual concentration of the species b and c in the bulk.

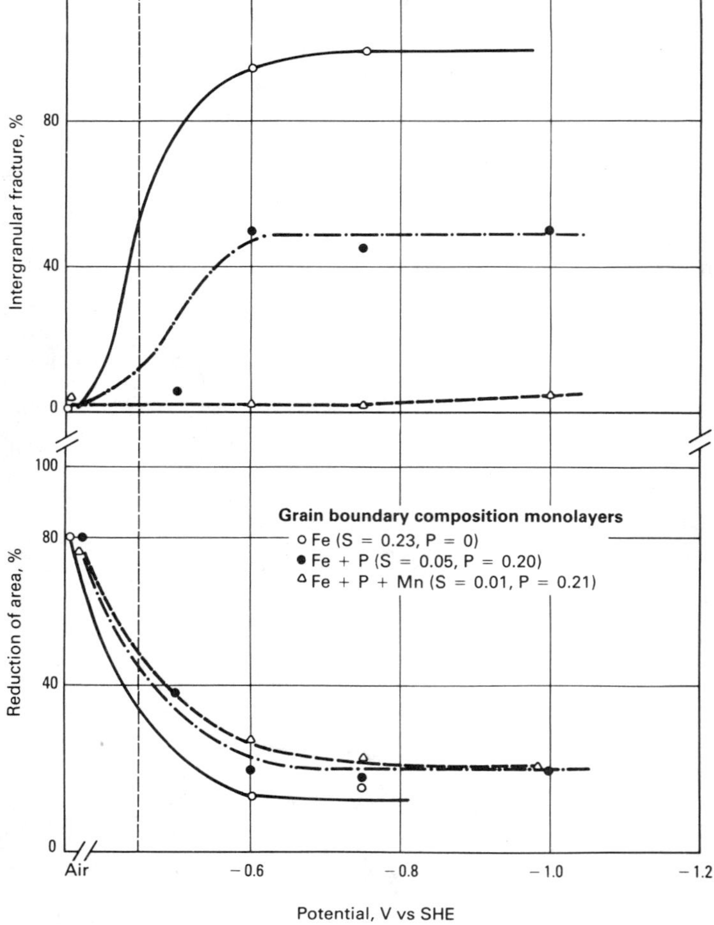

Fig. 21 Percent intergranular fracture, reduction of area, and strain to failure of iron, iron + phosphorus, and iron + phosphorus + manganese alloys tested at various cathodic potentials

30, 31); grain-boundary concentrations as low as 2 to 3 at.% have been sufficient to alter the passivity of iron in hot nitrate solutions. A complexity of SCC in ferritic steels is the susceptibility of such steels to intergranular SCC and to hydrogen-induced subcritical crack growth. The temperatures and electrochemical potentials at which intergranular SCC and cathodic hydrogen-induced subcritical crack growth occur are generally not the same. Stress corrosion tends to dominate at temperatures above about 50 °C (120 °F) and at potentials in the active-passive transition; hydrogen effects are predominant at temperatures below 50 °C (120 °F), at more cathodic potentials, and at lower pH values. Examples of the effect of sulfur and phosphorus on the tendency toward intergranular fracture are given in Fig. 21 and 22, from which it can be seen that sulfur was more effective than phosphorus in causing intergranular fracture of iron at cathodic potentials (Ref 32, 33). For the conditions of this test, a sulfur concentration of about 13 at.% was sufficient to change the fracture mode from transgranular to intergranular.

Grain-boundary impurity segregation of various impurities, including phosphorus, silicon, sulfur, and nitrogen, has been reported in austenitic stainless steels (Ref 34); however, no direct effect on intergranular SCC has been identified in high-temperature water. Phosphorus segregation has been shown to cause intergranular corrosion in highly oxidizing solutions, and impurity segregation of phosphorus and perhaps silicon has been suggested as a primary factor in irradiation-assisted SCC, which occurs in the oxidizing in-core environment of light water reactors (Ref 35). Phosphorus segregation can apparently contribute to the intergranular SCC of austenitic stainless steels in high-temperature water if the carbon concentration of the alloy is lower than 0.002%. At this low concentration, there is virtually no sensitization; therefore, the phosphorus segregation effect is observed.

Phosphorus is also the primary grain-boundary segregant in alloy 600 and has been shown to segregate by an equilibrium process (Ref 36). The grain-boundary composition versus temperature curves for alloy 600 and type 304 stainless steel are shown in Fig. 23, along with a calculated curve for equilibrium segregation in nickel (Ref 37). It can be seen that grain-boundary phosphorus concentrations to 15 at.% are possible. Grain-boundary segregation of sulfur, boron, nitrogen, and titanium has also been observed in alloy 600; however, a clear connection between the impurity segregation and intergranular SCC in nickel-base alloys has not been observed to date.

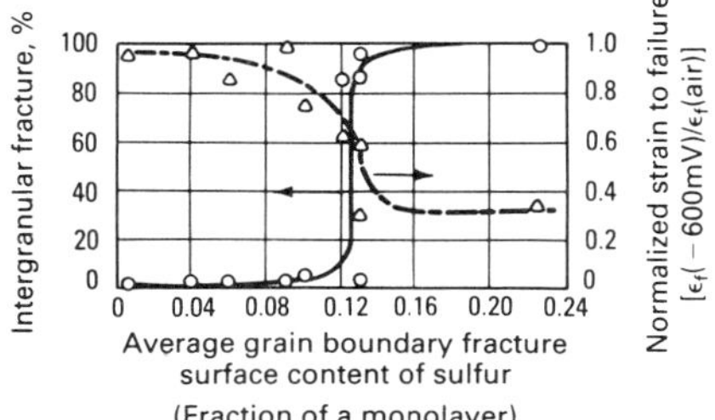

Fig. 22 Percent intergranular fracture and the normalized strain to failure plotted as a function of sulfur content at the grain boundary for straining electrode tests at a cathodic potential of −600 mV (SCE)

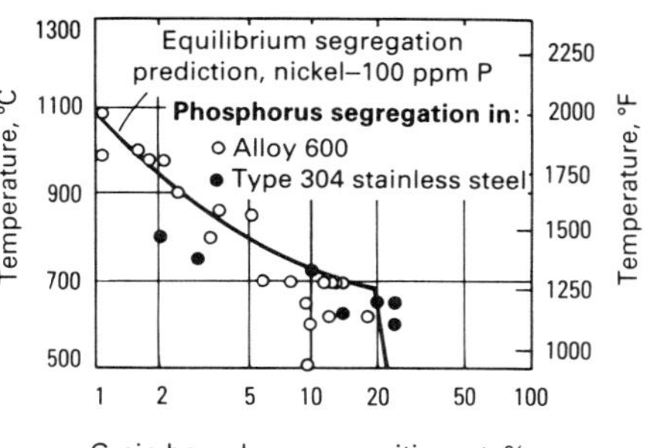

Fig. 23 Grain-boundary segregation measurements in Alloy 600 and type 304 stainless steel. Shown are Auger electron spectroscopy measurements of phosphorus segregation in the two alloys as compared to model prediction for phosphorus segregation in nickel.

In contrast to the high-temperature water results for alloy 600, a clear effect of phosphorus segregation on the intergranular SCC of nickel was observed at oxidizing potentials in acidic solutions at 25 °C (77 °F) (Ref 18), as shown in Fig. 24. Intergranular SCC was observed at anodic potentials ranging from the active-passive transition to transpassive potentials in nickel with phosphorus-enriched grain boundaries but not sulfur-enriched grain boundaries. The role of phosphorus was identified with degradation of the passive film formed on nickel in acidic solutions. This example is a clear case of active-path corrosion; however, the cracking was clearly stress dependent, as evidenced by a threshold stress intensity below which intergranular SCC did not occur and an alignment of cracks with the applied tensile stress. Because there were no grain-boundary carbides or chromium depletion in the nickel, these results illustrate the potential effect of impurities when other grain-boundary processes are absent. Therefore, it appears, as in the case of type 304 stainless steel cited earlier, that impurity segregation may induce intergranular SCC in the absence of carbides and chromium depletion.

Like ferritic materials, nickel-base alloys and austenitic stainless steels are susceptible to hydrogen-induced intergranular subcritical crack growth. Impurity segregation is thought to play a key role in this hydrogen-induced fracture, as the results for nickel shown in Fig. 25 illustrate. These results show the combined effect of sulfur segregation and cathodic hydrogen on the amount of intergranular fracture in nickel. For the conditions of these tests, a given amount of intergranular fracture resulted for different combinations of grain-boundary sulfur and hydrogen reduction rate. A decrease in the amount of sulfur could be compensated for by an increased hydrogen activity to give the same amount of intergranular fracture.

Transgranular Stress-Corrosion Cracking. Numerous metallurgical factors affect transgranular SCC, for example, crystal structure, anisotropy, grain size and shape, dislocation density and geometry, yield strength, composition, stacking fault energy, ordering, and phase composition. Some of these factors also affect intergranular SCC, as covered in the discussion "Intergranular Stress-Corrosion Cracking" in this section. Also, some of their effects on transgranular SCC are related to the corrosion

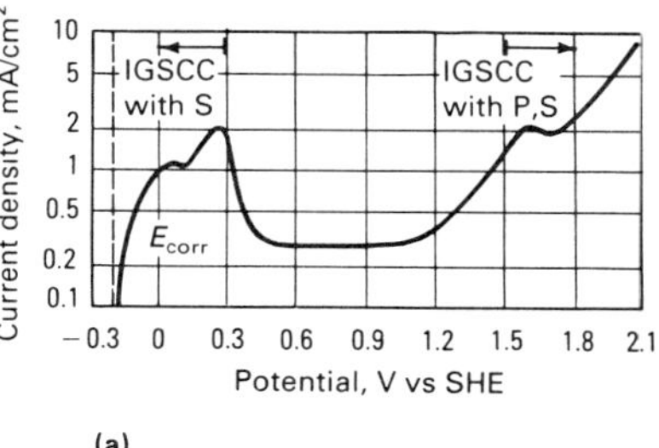

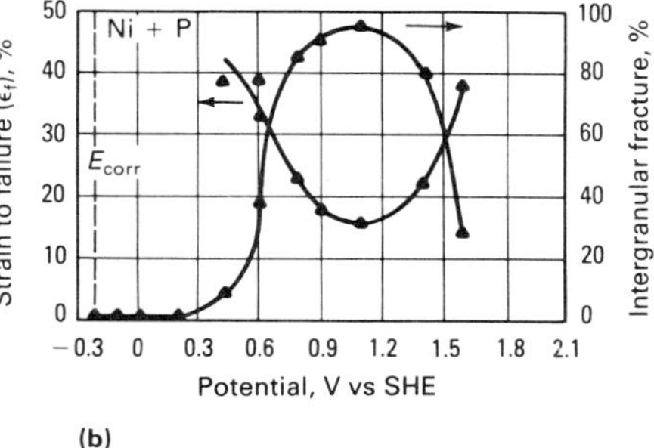

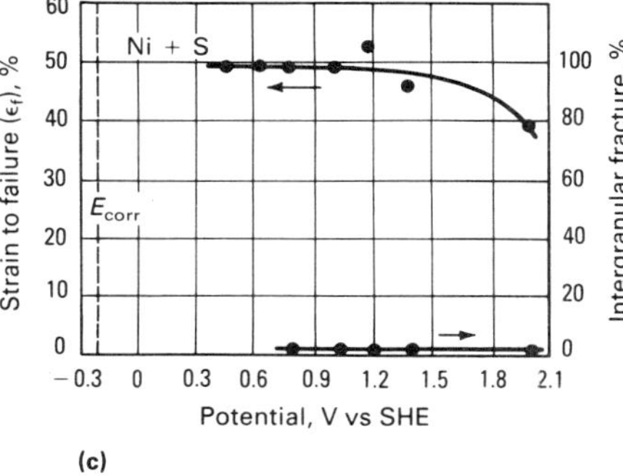

Fig. 24 Stress-corrosion cracking behavior of nickel with phosphorus and sulfur segregation. (a) Polarization curve for nickel in 1 *N* H_2SO_4 at 25 °C (77 °F). (b) Strain to failure and percent intergranular fracture for 26% phosphorus segregation at grain boundaries. (c) Strain to failure and percent intergranular fracture for 40% sulfur segregation at grain boundaries

behavior of the alloy, which can be understood from potential-pH diagrams or polarization curves. Because these effects are covered elsewhere in this Volume (see the Section "Specific Alloy Systems"), they will not be discussed here. Similarly, yield strength effects will be considered in the discussion "Mechanical Factors" in this section. Metallurgical factors in hydrogen-induced subcritical crack growth, such as crystal structure, hydride stability, and yield strength, are considered in the section "Hydrogen Damage" in this article. The effects of alloy composition on corrosion rate, hydrogen exchange current densities, and hydrogen absorption rates also influence the hydrogen-induced subcritical crack growth of materials in corrosive environments, but these are covered in the articles dealing with specific metals and alloys.

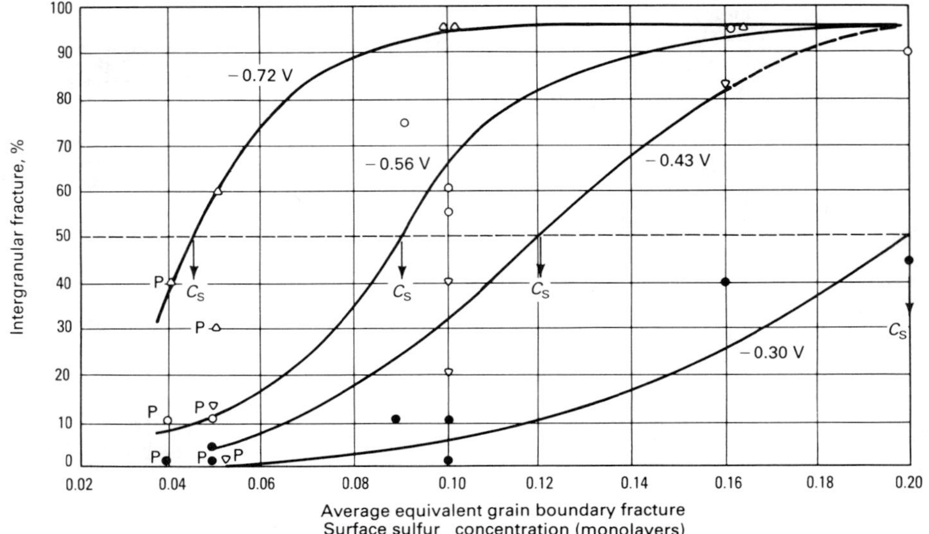

Fig. 25 Percent intergranular fracture and reduction of area versus grain-boundary composition of nickel for several cathodic test potentials. C_S is the critical sulfur concentration corresponding to 50% intergranular fracture. Points labeled P are equivalent sulfur concentrations for alloys with sulfur + phosphorus at the grain boundaries. Source: Ref 38

Alloying effects on slip planarity are a key metallurgical factor in transgranular SCC. Planar slip occurs in alloys with low stacking fault energy, alloys containing ordered phases, or alloys exhibiting short- or long-range ordering. The consequences of planar slip on transgranular SCC have been explained by the slip-dissolution model discussed in Ref 39. In this model, the passive film is ruptured by the emergence of a slip step. In high-chloride environments, evidence was presented that preferential corrosion occurred along the high dislocation density plane created by planar slip (Ref 40). A number of crack growth processes were suggested based on the planar slip localized corrosion process—for example, control of crack advance solely by anodic dissolution of the slip plane, brittle fracture of the corrosion product or tarnish film formed along the localized corrosion path, and the tunneling process by which corrosion along the slip plane branches out into tunnels accompanied by mechanical fracture of the remaining ligaments. The planar slip concept of transgranular SCC appears consistent with a number of observations, such as alloying and strain rate effects. Details of these mechanisms are presented in the discussion "Crack Propagation Mechanisms" in this section.

A significant difficulty associated with the slip-dissolution mechanism is the nature of the transgranular SCC fracture surfaces, which are generally not on the slip planes and have cleavage-like features. For example, for admiralty brass tested in aqueous ammonia (NH_3) and Al-5.5Zn-2.5Mg tested in aqueous NaCl, the primary fracture facets were (110) planes, while for stable austenitic stainless steel tested in aqueous magnesium chloride ($MgCl_2$) at 155 °C (310 °F), the primary facets were on (100) planes. There is also evidence that crack advance in brass occurs in a discontinuous manner (Ref 24). Several mechanisms for the development of a cleavage crack in ductile face-centered cubic (fcc) alloys have been presented, but a definite correlation has not emerged.

One concept that has received considerable attention is that proposed in Ref 25, in which a rapid crack advance that begins in a brittle film at the crack tip induces cleavage fracture of the ductile material ahead of the crack tip. The brittle films that are regarded as the initiating layers are given in Table 2. These include a dealloyed layer, oxide, nitride, carbide, or hydride. A dealloyed layer acting as a cleavage crack initiator is suggested for brass, copper-gold, and iron-chromium-nickel alloys. If this concept is supported by future research, the relative reactivity of the alloying elements will prove to be a key metallurgical factor in the transgranular SCC of materials. It has also been demonstrated that the dealloying rate of copper-aluminum and copper-zinc as a function of aluminum and zinc, respectively, corresponds to the crack growth rate in ammoniacal solution (Ref 25).

Mechanical Factors

There is more commonality in the mechanical aspects of stress corrosion between intergranular SCC and transgranular SCC and between various materials than there is in environmental and metallurgical aspects. As indicated earlier, many of the environmental or metallurgical factors are specific to a given material/environment combination such that a metallurgical factor may be significant in one environment, but not in another. Threshold stress intensities and stresses, the presence of a stress-independent crack growth regime, and a dependence of crack growth rate on strain rate are features common to a variety of environmentally induced crack growth processes and a variety of materials. Anodic and cathodic controlled processes show many of these common features.

The stress intensity dependence of environmentally induced subcritical crack growth is shown schematically in Fig. 26, along with a relationship for the linear-elastic mode I stress intensity. The applied stress intensity is a function of the uniform stress and the crack length; therefore, for a constant-stress test, the stress intensity increases with increasing crack extension. For low-strength materials that form a significant plastic zone ahead of the crack, nonlinear fracture mechanics relationships must be used to calculate K. The subcritical crack growth behavior of many material/environment combinations is characterized by a threshold and by stages 1, 2, and 3 (see the article "Fracture Mechanics" in Volume 8 of the 9th Edition of *Metals Handbook* for a detailed discussion of modes of crack formation).

In Fig. 26, the threshold is defined by the minimum detectable or reliably measured crack growth rate, while in some materials, stage 1 has a large K dependence such that a lower threshold does not result at lower crack velocities. There are a number of physical processes that can be associated with a threshold stress intensity, including a fracture strain for a passive film rupture mechanism, the critical resolved shear stress for the slip-dissolution mechanism, a fracture stress for a brittle film induced cleavage mechanism, or a critical crack tip opening for transport of species in the crack. The threshold stress intensity is generally associated with the development of a plastic zone at the crack tip; however, calculations of crack tip strains are not very accurate, and experimental correlations between plastic zone development and K_{ISCC} have not been made.

Stage 1 subcritical crack growth is marked by a rapid increase in crack velocity with a small increase in K. Dependencies of crack velocity on stress intensity to the fourth power have been reported for a number of material/environment combinations. Changes in crack velocity of two orders of magnitude with small changes in K are not uncommon for the stage 1 regime. Few explanations for the K dependence of stage 1 have been put forth. One explanation is that in stage 1, the crack tip plastic strain rate is increasing rapidly with K, and another explanation is that the transport of species into and out of the crack

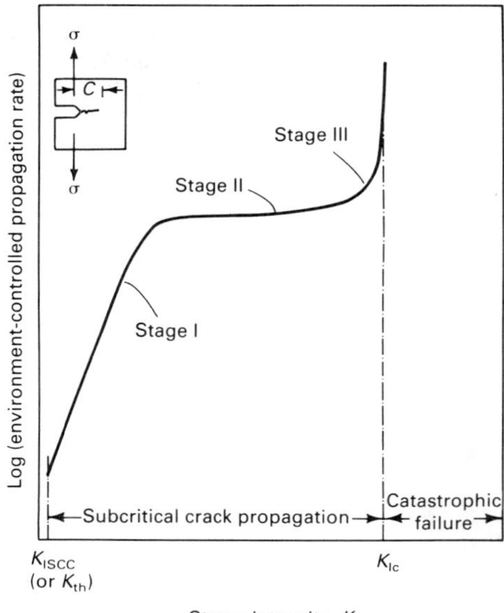

Fig. 26 Typical, subcritical crack propagation rate versus stress intensity relationship. Stress intensity, K, is defined as $K = A\sigma\sqrt{\pi C/B}$, where σ is the total tensile stress, C is the crack length, and A and B are geometrical constants.

increases rapidly with increasing crack volume in stage 1.

In metals tested in aqueous solutions, in transport of cations or anions in the crack electrolyte, or in passive film forming materials, the passive film rupture and repassivation rates are thought to be rate-limiting processes. Environmental parameters such as crack tip potential, corrosion rates, and pH also affect the stage 2 crack velocity so that in many systems the limiting velocity is a function of crack tip corrosion rate, strain rate, and transport rate. A steady-state crack transport calculation for intergranular SCC in nickel + phosphorus gave crack growth values within a factor of five of the measured stage 2 values (Ref 16). For transgranular SCC, there is increasing evidence that the crack velocity is a function of the corrosion rate, but that crack advance results from brittle crack jumps that exceed the depth of the corrosion reaction. In this case, the crack velocity may scale with the corrosion rate, but the crack length exceeds that determined from charge transfer.

Most stress-corrosion data have been obtained by using specimens loaded in a tensile mode in which the stress is perpendicular to the plane of the advancing crack; this condition is defined as mode I loading. In practice, a component may have a complex loading mode such that torsion and shear components are present. The effect of mixed loading modes on SCC is therefore an important consideration in component design; however, very few SCC data exist for other than mode I loading. Some examples of SCC in modes 1 and 3 for aluminum in 3.5% NaCl and type 304 stainless steel in boiling $MgCl_2$ are shown in Fig. 27, in which it can be seen that mode I loading gives the lowest threshold stress or stress intensity. Also shown in Fig. 27(a) is a schematic of the test specimen used to perform mixed-mode SCC tests. In both cases, the researchers concluded that the lower thresholds and higher crack velocities in mode I tests of these materials indicate a hydrogen embrittlement component to SCC because hydrogen effects are more prominent under tensile loading conditions (Ref 41, 42). However, anodic stress-corrosion processes are also aided by the crack opening that results from mode I loading.

Stress-corrosion data are frequently given as applied stress versus time to failure, as shown in Fig. 28. These data are obtained with notched or unnotched specimens statically loaded in tension or bending. The threshold stress is sensitive to the environment, alloy composition, and structure and is not strictly a material property. The threshold stress is related to the yield strength of the material, with thresholds generally being greater than one-half the yield strength, and in many cases, the threshold is a significant fraction (about 0.8) of the yield strength of the material. Therefore, most SCC tests performed with this technique are performed at values of about 80% of the yield strength or greater.

The data obtained with this method incorporate the stage 1, 2, and 3 regimes shown in Fig. 26 in addition to a crack initiation stage. Crack growth begins with the development of a pit or intergranular corrosion groove that raises the stress intensity to the K_{ISCC}. Because the stress is constant, the stress intensity increases with increasing crack length as given by the expression in Fig. 26 so that the crack progresses through stage 1 to stage 2 and ultimately to fracture in stage 3. Therefore, the failure time in a constant-

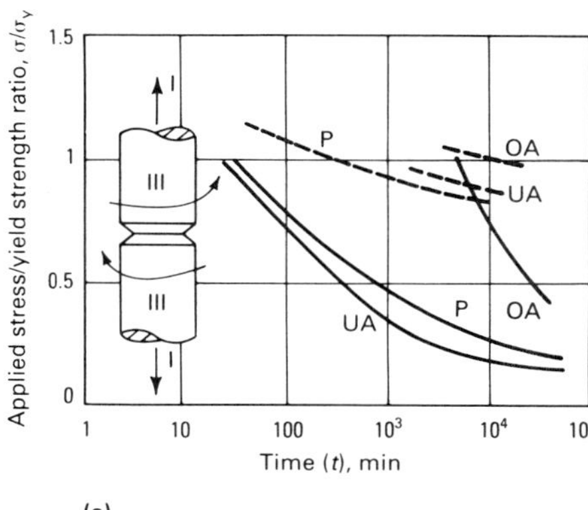

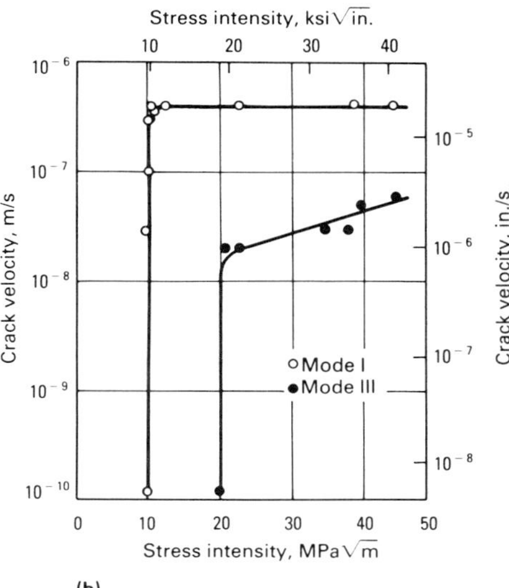

Fig. 27 Examples of SCC in mode 1 and mode 3. (a) Susceptibility of 7075 aluminum alloy to corrosion cracking in 3.5% NaCl solution tension. Mode 1 and torsion (solid lines). Mode 3 tests (dashed lines). P, peak of aging; OA, overaged; UA, underaged. (b) Crack velocity versus stress intensity curves for SCC of type 304 stainless steel in modes 1 and 3 in boiling $MgCl_2$

stress test is dependent on the initiation rate, the value of the K_{ISCC}, the crack growth rate in stage 2, and the fracture toughness K_{Ic}. These test data are useful because they relate to the sequence that may occur in practice, but it is difficult to relate them to physical processes and rate-controlling steps. Crack growth rates are frequently estimated from constant-load tests based on the crack depth divided by the total test time. Crack velocities obtained in this manner are minimum values because the time for initiation is included so that instantaneous velocities will always be greater.

The threshold stress is perhaps the least complicated value obtained from constant-load tests. For film-forming materials, the threshold can be related to the film formation and rupture rate. Below the threshold stress, the film formation rate is sufficiently high to maintain the surface in a filmed condition; at higher stresses, the film rupture rate exceeds the reformation rate such that a pit or an intergranular corrosion groove can grow. For transgranular SCC initiated by a brittle corrosion film, the threshold stress could be

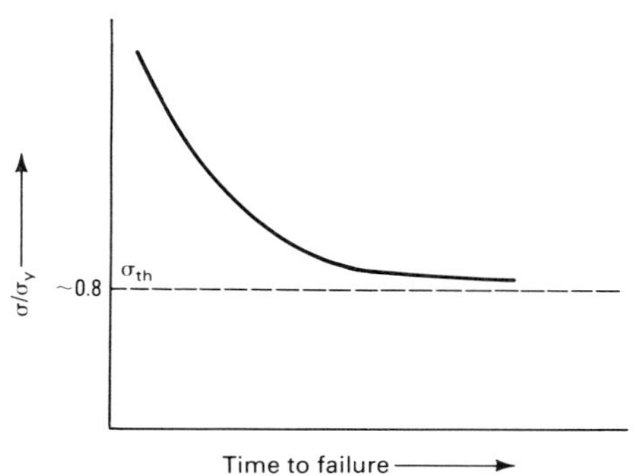

Fig. 28 Schematic diagram of time to failure versus applied stress, σ, normalized to the yield strength σ_y for stress corrosion

related to the fracture stress of the film; however, this type of relationship has not been established for the brittle film induced cleavage process.

Constant extension rate or slow strain rate tests have essentially replaced the use of the constant-stress test for stress-corrosion testing. This has occurred because of the variability in times of failure with a constant-stress test and the long test times needed to define the stress threshold. Also, the concept that the crack growth rate is a function of the strain rate, as shown in Fig. 29 for intergranular type 304 stainless steel, has led to the use of a constant extension rate test. The CERT specimens will fail within a given period of time, thus eliminating specimens that do not fail because they are loaded to a stress below the threshold. The crack growth rates given in Fig. 29 are minimum values obtained from unnotched specimens, and the strain rate is the average applied rate. Clearly, once a crack develops, the crack tip strain rate will differ from the average applied value. An example of the crack velocity as a function of the crack tip strain rate is given in Fig. 30 for intergranular SCC of a carbon-manganese steel tested in a carbonate solution. A threshold strain rate is evident in these results that is the same as a threshold stress intensity. A second threshold strain rate exists at high strain rates because mechanical fracture occurs before a crack can initiate and propagate to a measurable length. The two strain rate thresholds are illustrated in Fig. 31.

It is important to distinguish between static and dynamic effects. For example, a sample could be plastically strained 10% and the load dropped to below the threshold stress before being immersed in a corrosive environment and SCC will not occur. However, a sample can be slowly strained 10% in a corrosive environment with the development of numerous cracks and perhaps even complete failure. Therefore, it is the dynamic strain and creation of fresh surface or dynamic fracture of a brittle film that is crucial to SCC. Corrosion current measurements corroborate these observations; the current density of a prestrained sample is usually very similar to an unstrained sample, while the current density measured during dynamic strain is considerably greater than that in the unstrained sample.

Crack Propagation Mechanisms

As indicated in the previous discussion, it is unlikely that a single mechanism of SCC exists;

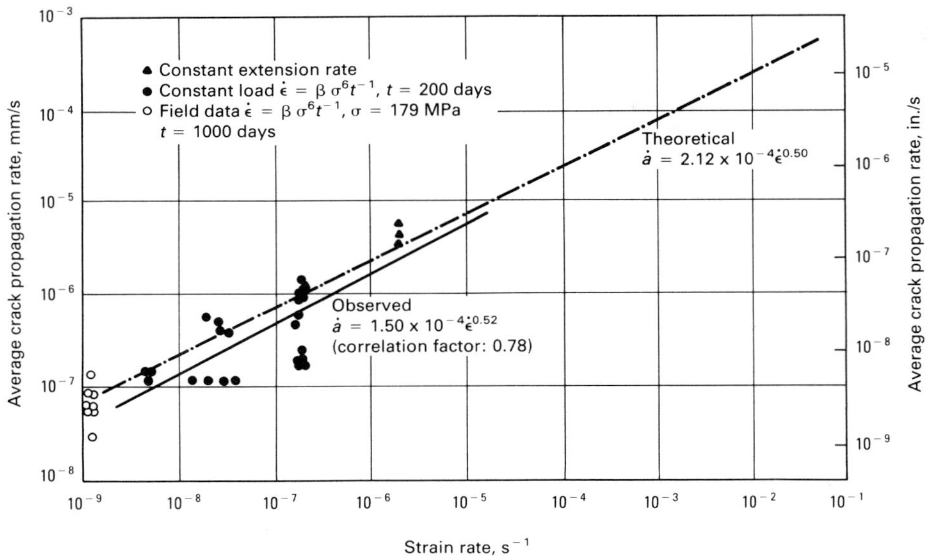

Fig. 29 Comparison between observed and theoretical crack propagation rate/strain rate ($\dot{a}/\dot{\epsilon}$) relationships for furnace-sensitized type 304 stainless steel in water/0.2 ppm oxygen at 288 °C (550 °F)

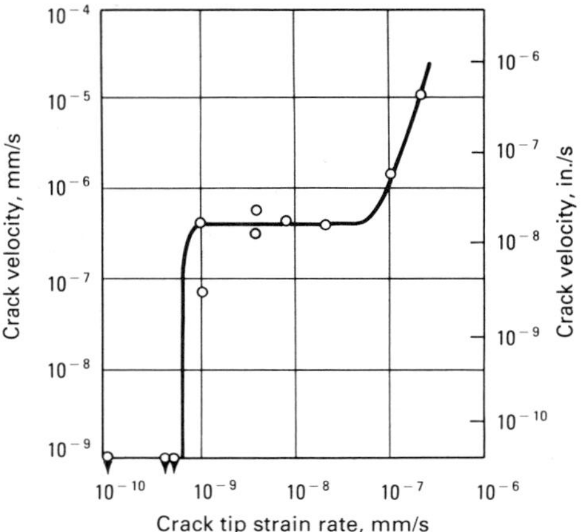

Fig. 30 Intergranular crack velocities for various applied crack tip strain rates in a carbon-manganese steel immersed in 1 *N* Na_2CO_3 + 1 *N* $NaHCO_3$ at −650 mV (SCE) and 75 °C (165 °F)

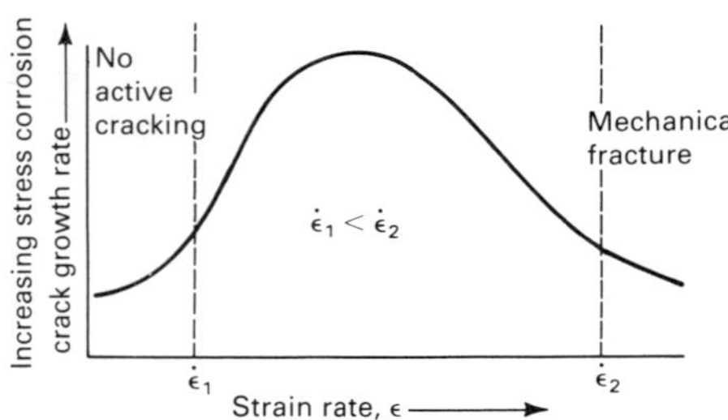

Fig. 31 Stress-corrosion crack growth as a function of the two strain rate thresholds, $\dot{\epsilon}_1$ and $\dot{\epsilon}_2$

instead, there are probably two or three different operative mechanisms. Many models have been proposed, but only the more prominent will be discussed. The proposed mechanisms for crack propagation fall into two basic classifications: those based on dissolution as the principal cause of crack propagation and those that involve mechanical fracture in the crack propagation process.

Dissolution Models

According to models of this type, the crack advances by preferential dissolution at the crack tip. A number of models have been proposed to account for this process. For example, preferential dissolution at the crack tip has been attributed to the formation of active paths in the material, stresses at the crack tip, and chemical-mechanical interactions (Ref 43-48). However, research has essentially eliminated all but one model from serious consideration, and the debate has recently centered on the details of this model.

Film Rupture. The film rupture model, also referred to as the slip-dissolution model, assumes that the stress acts to open the crack and rupture the protective surface film. Two investigators, working independently, first postulated that localized plastic deformation at the crack tip ruptures the passivating film, exposing bare metal at the crack tip (Ref 49, 50). The freshly exposed bare metal then dissolves rapidly, resulting in crack extension. Some investigators (Ref 51-53) assume that once propagation starts, the crack tip remains bare because the rate of film rupture at the crack tip is greater than the rate of repassivation (Fig. 32a). Others (Ref 54-57) assume that the crack tip repassivates completely and is periodically ruptured by the emergence of slip steps (Fig. 32b).

Considerable evidence has been found to support these mechanistic models, and intergranular corrosion may be considered the low stress limiting case of this mechanism (Ref 43-57). However, the observation of discontinuous cracking and crack arrest markings (discussed previously) is an indication that crack propagation can be, and frequently is, discontinuous. Also, transgranular SCC fracture surfaces are flat, crystallographically oriented, and match precisely on opposite sides of the fracture surface (indicating very little dissolution during crack advance). As a result, film rupture and dissolution are accepted as viable mechanisms of intergranular SCC in some systems, but are not generally accepted as mechanisms of transgranular SCC (Ref 58).

Mechanical Fracture Models

Mechanical fracture models originally assumed that stress concentration at the base of corrosion slots or pits increased to the point of ductile deformation and fracture (Ref 48). These early proposals assumed that the crack essentially propagated by dissolution and that the remaining ligaments then failed by mechanical fracture (ductile or brittle). A refinement of this approach has been proposed (Ref 40, 59); it is generally known as the corrosion tunnel model.

The Corrosion Tunnel Model. In this model, it is assumed that a fine array of small corrosion tunnels form at emerging slip steps. These tunnels grow in diameter and length until the stress in the remaining ligaments causes ductile deformation and fracture. The crack thus propagates by alternating tunnel growth and ductile fracture (Fig. 33). Cracks propagating by this mechanism should result in grooved fracture surfaces with evidence of microvoid coalescence on the peaks, as illustrated in Fig. 33(a). This is not consistent with the common fractographic features discussed previously. As a result, it was suggested (Ref 60) that the application of a tensile stress results in a change in the morphology of the corrosion damage from tunnels to thin flat shots, as shown in Fig. 33(b). The formation of slots was observed below the dealloyed sponge layer on the {110} type planes in austenitic stainless steels, and it was found that the formation of these slots required the presence of a tensile stress. The width of the corrosion slots was found to approach atomic dimensions, and as a result, close correspondence of matching fracture surfaces would be expected. It was concluded that transgranular SCC can be explained in terms of the formation and mechanical separation of corrosion slots (Ref 60).

Adsorption-Enhanced Plasticity. Studies of liquid-metal embrittlement (LME), hydrogen embrittlement, and SCC have led to the conclusion that similar fracture processes occur in each case (Ref 61-63). Because chemisorption is common to all three, this process was concluded to be responsible for the environmentally induced crack propagation (Ref 61). Based on fractographic studies, it was concluded that cleavage fracture is not an atomically brittle process, but occurs by alternate slip at the crack tip in conjunction with formation of very small voids ahead of the crack. It was also proposed that chemisorption of environmental species facilitated the nucleation of dislocations at the crack tip, promoting the shear processes responsible for brittle cleavagelike fracture (Ref 61-63). The origin of crack arrest markings by this mechanism is, however, uncertain. Nevertheless, this mechanism promises to explain many similarities among SCC, LME, and hydrogen embrittlement. Additional information on adsorption-enhanced plasticity can be found in the section "Liquid-Metal Embrittlement" in this article.

The Tarnish Rupture Model. This model was first proposed (Ref 64) to explain transgranular SCC, but was later modified by other researchers (Ref 65-67) to explain intergranular SCC. In the original model (Ref 64), a brittle surface film forms on the metal that fractures under the applied stress. Fracture of the film exposes bare metal, which rapidly reacts with the environment to re-form the surface film. The

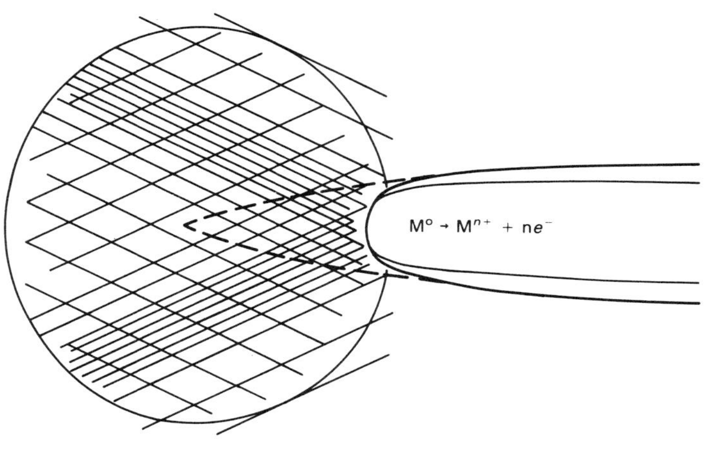

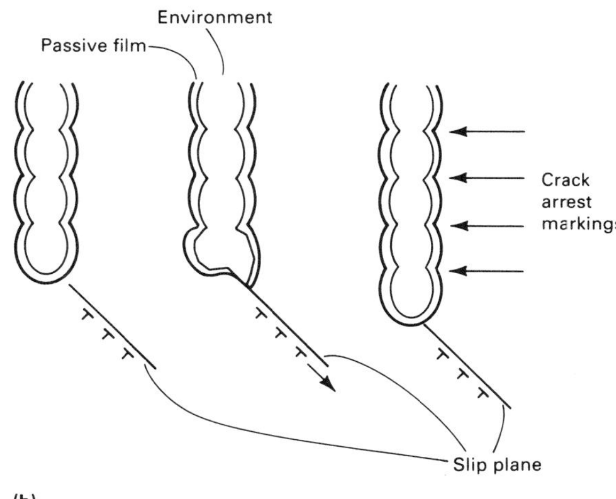

Fig. 32 Schematic representations of crack propagation by the film rupture model. (a) Ref 50. (b) Ref 56 and 57

crack propagates by alternating film growth and fracture, as shown in Fig. 34(a). This hypothesis was later modified to explain intergranular SCC based on the assumption that the oxide film penetrates along the grain boundary ahead of the crack tip, as shown in Fig. 34(b). Again, crack propagation consists of alternating periods of film growth and brittle film fracture. Film growth requires transport of species across the film and, as a result, the thickness of the film is limited in the absence of stress (Ref 67). This model predicts crack arrest markings on intergranular fracture surfaces and discontinuous acoustic emission during crack propagation that are not always observed during intergranular SCC. Also, it assumed penetration of the film into the grain boundary ahead of the crack tip, which may not be the case for all systems (Ref 68). At present, experimental results are insufficient to confirm or refute this model (Ref 66, 67).

The Film-Induced Cleavage Model. In 1959, the hypothesis was presented that dealloying and/or vacancy injection could induce brittle fracture (Ref 69). However, the exact nature of the interactions and how they could induce the observed crack propagation was not thoroughly evaluated. More recently, a model was developed based on the hypothesis that a surface film could induce cleavage fracture (Ref 70). This model assumes that:

- A thin surface film or layer forms on the surface
- A brittle crack initiates in this layer
- The brittle crack crosses the film/matrix interface with little loss in velocity
- Once in the ductile matrix, the brittle crack will continue to propagate
- This crack will eventually blunt and arrest, after which this process repeats itself

This model has the unique ability to explain the crack arrest markings, the cleavagelike facets on the fracture surface, and the discontinuous nature of crack propagation.

The hypothesis that a brittle crack will continue to propagate after it has entered the normally ductile matrix is a critical point. This allows a thin surface layer to induce brittle crack propagation over distances much greater than the film thickness. A critical examination of this hypothesis concluded that a brittle crack can propagate in a ductile matrix if the crack is sharp and is propagating at high velocities before entering the ductile matrix (Ref 71). A computer model was developed for this process, and it was concluded that a surface layer can initiate brittle fracture even if the layer is ductile (depending on lattice mismatch, and so on) (Ref 72). More research into surface films and brittle fracture is required before this model can be thoroughly evaluated.

Adsorption-Induced Brittle Fracture. This model, which was initially presented in Ref 46, is based on the hypothesis that adsorption of environmental species lowers the interatomic bond strength and the stress required for cleavage fracture. This model is frequently referred to as the stress-sorption model, and similar mechanisms have been proposed for hydrogen embrittlement and LME (Ref 73). This model predicts that cracks should propagate in a continuous manner at a rate determined by the arrival of the embrittling species at the crack tip. This model does not explain how the crack maintains an atomically sharp tip in a normally ductile material, because it does not include a provision for limiting deformation in the plastic zone. Also, the discontinuous nature of crack propagation is not explained by this model.

Hydrogen Embrittlement. Stress-corrosion cracking in some material/environment combinations can be a form of hydrogen-induced subcritical crack growth. Because the anodic reaction must have a corresponding cathodic reaction and because the reduction of hydrogen is frequently the cathodic reaction, hydrogen-induced subcritical crack growth can be the dominant stress-corrosion crack growth process in some materials. Many features of hydrogen-induced subcritical crack growth from cathodic hydrogen are very similar to those produced by gaseous or internal hydrogen. The mechanisms of hydrogen damage are discussed in the following section of this article. However, a few features of hydrogen embrittlement from cathodic hydrogen that are different from other forms of hydrogen embrittlement will be summarized.

Once hydrogen has been absorbed by a material, its effect, whether from a gaseous or cathodic source, is the same. This has been shown for a number of materials and a variety of properties. There are three primary differences between gaseous and cathodic hydrogen absorption processes, as follows. First, cathodic hydrogen adsorbs on the surface as atomic hydrogen (as reduced), while gaseous hydrogen adsorbs in the molecular form and must dissociate to form atomic hydrogen. Desorption of loosely bound molecular hydrogen is relatively easy, while the dissociation step can in some cases be the rate-determining step. Therefore, the desorption and absorption rates of gaseous and cathodic hydrogen may be substantially different for equal hydrogen activities. Second, the hydrogen activities produced by cathodic hydrogen can be quite large (thousands of psi) and are dependent on the anodic reaction rate, while gaseous hydrogen pressures are generally much lower. Lastly, the surface of the material at a crack tip may be substantially different under electrochemical corrosion conditions from that in the presence of gaseous hydrogen containing substantial quantities of other gases, such as O_2 and CO_2.

Hydrogen-induced crack growth as the dominant stress-corrosion mechanism has been suggested for ferritic steels, nickel-base alloys, titanium alloys, and aluminum alloys, although ferritic steels show the most evidence of this mechanism. The effects of such factors as yield strength, impurity segregation, and temperature on the crack growth behavior of ferritic materials in aqueous environments all follow the trends of gaseous hydrogen embrittlement. For example, the crack growth rate of a 3% Ni steel as a function of temperature in water is shown in Fig. 35(a), while the crack growth rate of 4340 steel in gaseous hydrogen is shown in Fig. 35(b). The maximum crack growth rate occurs at about 20 °C (70 °F), with similar decreases at both higher and lower temperatures. However, anodic stress-corrosion processes become active at temperatures above 100 °C (212 °F) for the steel tested in water. Similar trends have been shown for impurity segregation effects on hydrogen-induced crack growth of materials where cathodic and gaseous hydrogen produce essentially similar results. Some of these results have been presented in the discussion "Material Chemistry and Microstructure" in this section.

Specific mechanisms of cathodic hydrogen induced subcritical crack growth have not been

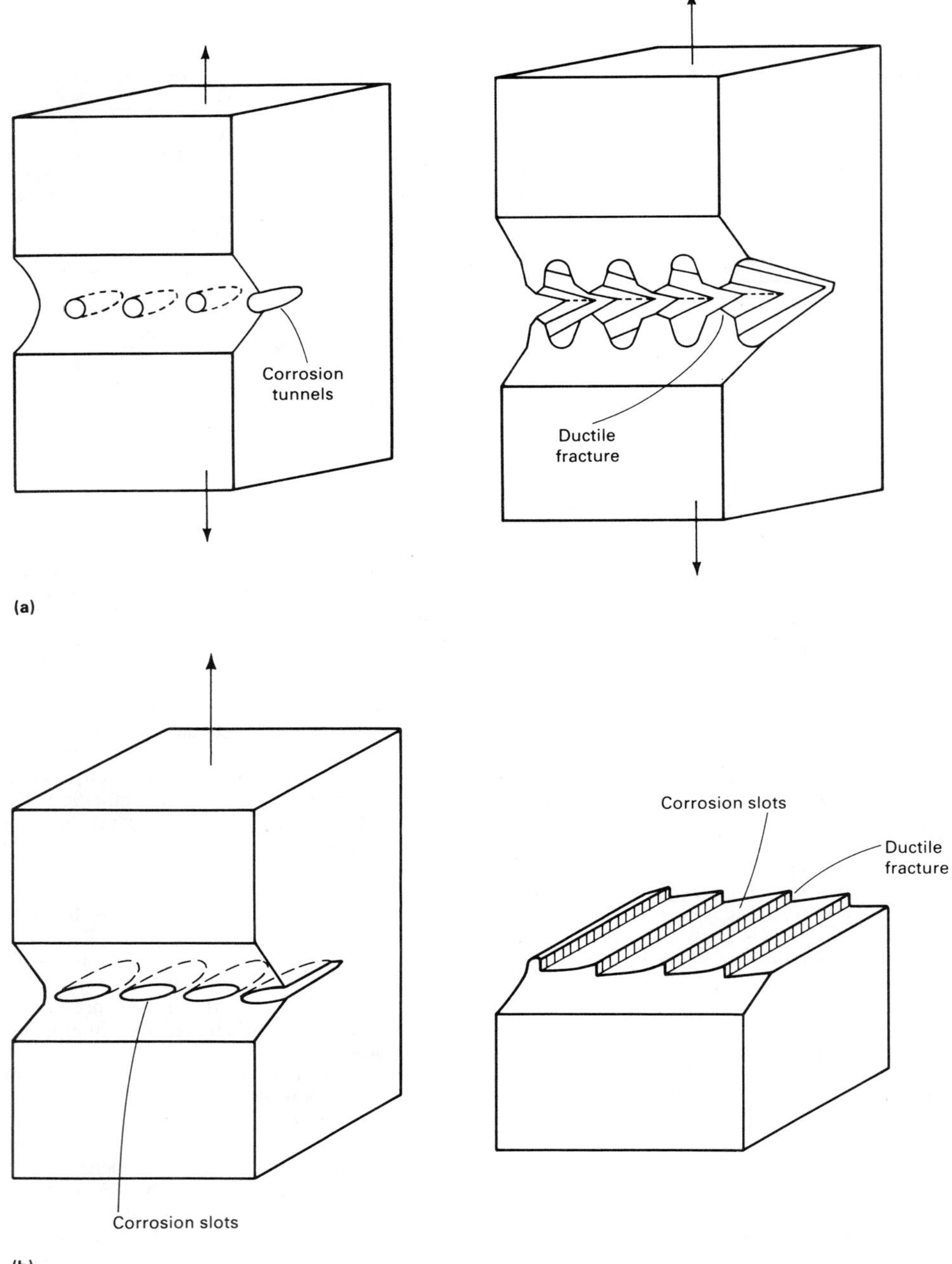

Fig. 33 Corrosion tunnel models. (a) Schematic of tunnel model showing the initiation of a crack by the formation of corrosion tunnels at slip steps and ductile deformation and fracture of the remaining ligaments. (b) Schematic diagram of the tunnel mechanism of SCC and flat slot formation as proposed in Ref 60

developed, because it has generally been sufficient merely to identify hydrogen as the cause for cracking. However, a mechanism was presented in which grain-boundary impurities act as hydrogen recombinant poisons and enhance the uptake of cathodic hydrogen (Ref 76). A schematic of this process is shown in Fig. 36, in which the tin and antimony that segregated to the grain boundaries of nickel enhance the hydrogen uptake kinetics. This mechanism does not propose a new mechanism by which hydrogen can cause cracking, but merely proposes a mechanism by which impurity segregation can enhance hydrogen uptake. There may be circumstances in which such a mechanism could tip the balance between an anodically driven process and a cathodically driven process or could accelerate cracking to a value that is measurable or of practical significance. However, a review of the combined effects of impurity segregation and hydrogen embrittlement concluded that grain-boundary impurities behave the same with cathodic and gaseous hydrogen in that they enhance crack growth by a combined grain-boundary embrittlement processes but not by enhanced hydrogen uptake (Ref 77). It should be noted, however, that a material with impurities such as sulfur, phosphorus, antimony, and tin segregated to their grain boundaries are more susceptible to all forms of hydrogen—cathodic, gaseous, or internal.

Summary

Stress-corrosion cracking is a phenomenon in which time-dependent crack growth occurs when the necessary electrochemical, mechanical, and metallurgical conditions exist. Corrosion fatigue is a related process in which the load is cyclic in corrosion fatigue rather than static as in stress corrosion. When hydrogen is generated as a product of the corrosion reaction, crack growth can occur by a hydrogen embrittlement process in much the same way as if hydrogen were in the gaseous form. A common feature of each of these processes is the subcritical crack growth in which cracks grow from existing flaws or initiation sites and grow to a size at which catastrophic failure occurs. Catastrophic failure occurs because the combination of crack length and applied stress increases the stress intensity to the fracture toughness of the material. A second common feature of stress corrosion, corrosion fatigue, and hydrogen-induced crack growth is that these mechanisms do not require the entire component to become embrittled; the effect is localized to the crack tip region.

This section was intended to familiarize the reader with the phenomenological and mechanistic aspects of stress corrosion in order to enhance the use of other discussions in this Handbook on stress-corrosion evaluation and occurrence in specific industries and environments. The phenomenological description of crack initiation and propagation describes well-established experimental evidence and observations of stress corrosion, while the discussions on mechanisms describe the physical process involved in crack initiation and propagation. The physical processes involved in crack growth have received more evaluation and are better understood than the processes responsible for crack initiation. There is some phenomenological understanding of stress-corrosion crack initiation, but the detailed mechanistic information is not well known.

Stress-corrosion cracking occurs when certain critical conditions are achieved. These conditions include electrochemical, mechanical, and metallurgical factors that must exist simultaneously. A change in any one of these three factors is adequate for eliminating SCC; therefore, a clear knowledge of these critical factors is important in system design. The important electrochemical parameters include oxidizing potential, pH, impurity concentration, and temperature. The important mechanical parameters include stress, stress intensity, and strain rate. The important metallurgical factors include localized microchemistry (such as the depletion of passive film forming elements or the enrichment of active corroding elements), bulk composition, deformation character, and yield strength.

There are a number of plausible mechanisms or physical processes that account for SCC. No single mechanism is adequate to describe stress corrosion in the variety of materials in which it has been observed. Stress-corrosion crack propagation mechanisms can be subdivided into dissolution and mechanical fracture models, with the mechanical fracture models further divided into ductile and brittle crack extension processes. Dissolution models include film-rupture and active-path processes, while ductile mechanical models include corrosion tunneling and adsorption-enhanced plasticity models. Brittle mechanical models include the tarnish rupture and film-induced cleavage models.

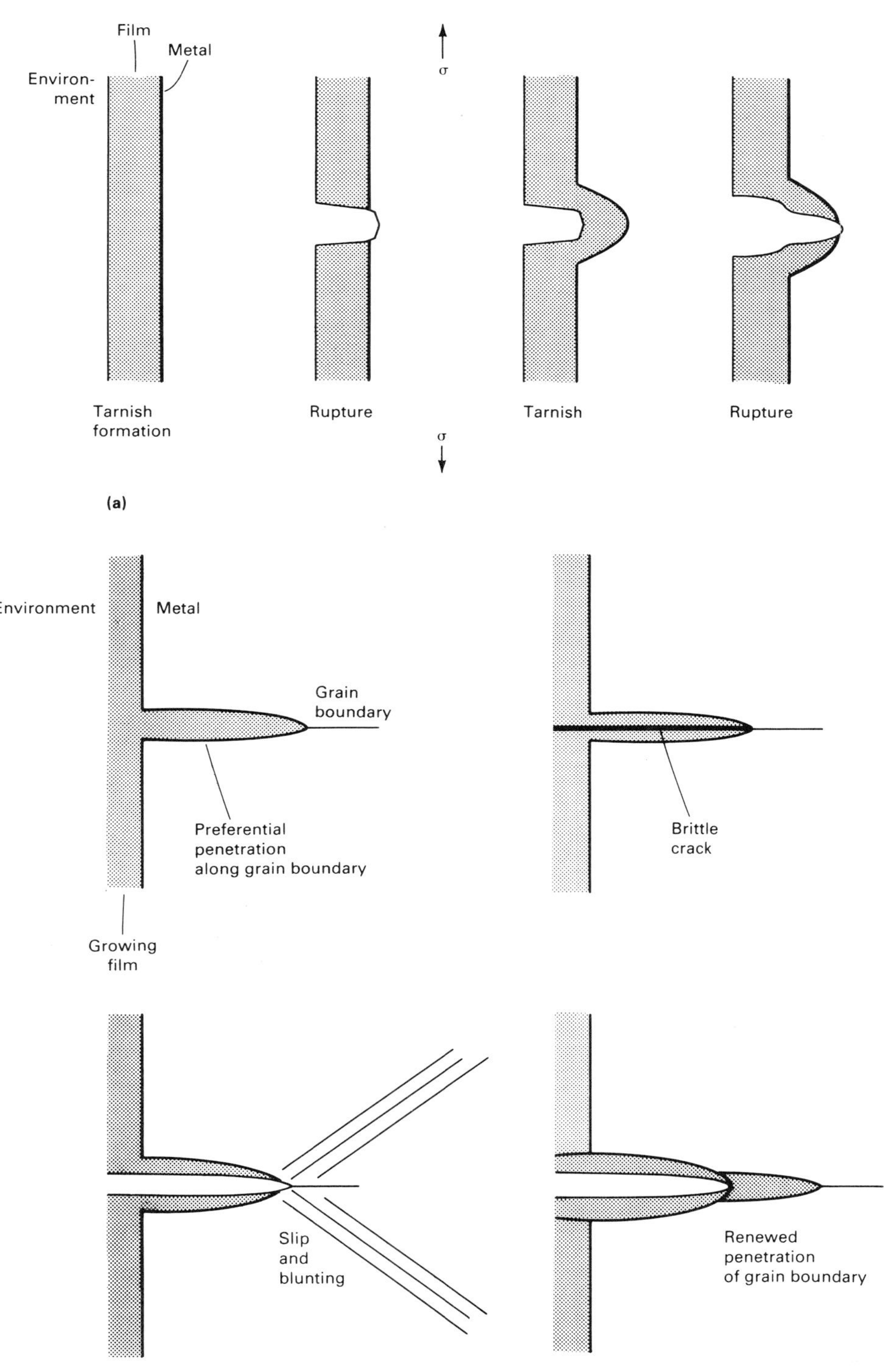

Fig. 34 Tarnish rupture models. (a) Schematic of the tarnish rupture model for SCC as proposed in Ref 64. (b) Modified tarnish rupture model of SCC for systems with intergranular oxide film penetration (Ref 66, 67)

Hydrogen Damage

Bruce Craig
Metallurgical Consultants, Inc.

Hydrogen damage is a form of environmentally assisted failure that results most often from the combined action of hydrogen and residual or applied tensile stress. Hydrogen damage to specific alloys or groups of alloys manifests itself in many ways, such as cracking, blistering, hydride formation, and loss in tensile ductility. For many years, these failures have been collectively termed hydrogen embrittlement; this term persists even though it is improperly used to describe a multitude of failure modes involving hydrogen, several of which do not demonstrate the classical features of embrittlement (that is, reduced load carrying capability or fracture below the yield strength). This section will classify the various forms of hydrogen damage, summarize the various theories that seek to explain hydrogen damage, and review hydrogen degradation in specific ferrous and nonferrous alloys. Information on the effect of hydrogen on fracture characteristics is available in the article "Modes of Fracture" in Volume 12 of the 9th Edition of *Metals Handbook*.

Classification of Hydrogen Processes

The specific types of hydrogen damage have been categorized in order the enhance the understanding of the factors that affect this behavior in alloys and to provide a basis for development and analysis of theories regarding different hydrogen damage mechanisms (Ref 78). Table 3 presents one of these classification schemes, describing the materials that are susceptible to the various forms of damage, the source of hydrogen, typical conditions for the occurrence of failure, and the initiation site. The mechanisms for each of these failure modes will also be described briefly. The first three classes are grouped together and designated hydrogen embrittlement because these are the failure modes that typically exemplify classical hydrogen embrittlement.

Hydrogen environment embrittlement occurs during the plastic deformation of alloys in contact with hydrogen-bearing gases or a corrosion reaction and is therefore strain rate dependent. The degradation of the mechanical properties of ferritic steels, nickel-base alloys, titanium alloys, and metastable austenitic stainless steels is greatest when the strain rate is low and the hydrogen pressure and purity are high.

Hydrogen stress cracking, often referred to as hydrogen-induced cracking or static fatigue, is characterized by the brittle fracture of a normally ductile alloy under sustained load in the presence of hydrogen. Most often, fracture occurs at sustained loads below the yield strength of the material. This cracking mechanism depends on the hydrogen fugacity, strength level of the material, heat treatment/microstructure, applied stress, and temperature. For many steels, a threshold stress exists below which hydrogen stress cracking does not occur. This threshold is a function of the strength level of the steel and the specific hydrogen-bearing environment. Therefore, threshold stress or stress intensity for hydrogen stress cracking is not considered a material property. Generally, the threshold stress decreases as the yield strength and tensile strength of an alloy increase. Hydrogen stress cracking is associated with absorption of hydrogen and a delayed time to failure (incubation time) during which hydrogen diffuses into regions of high triaxial stress. Hydrogen stress cracking may

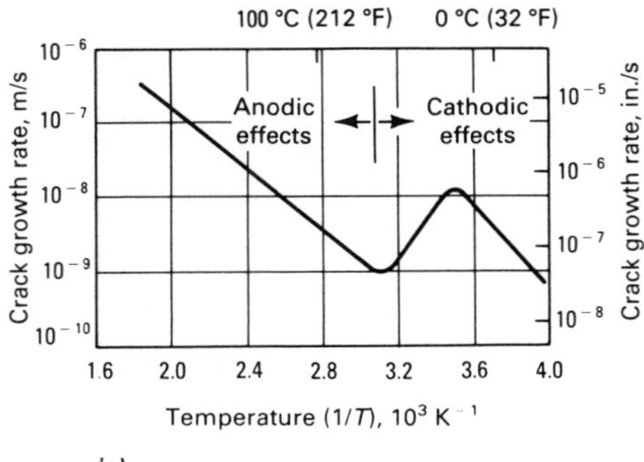

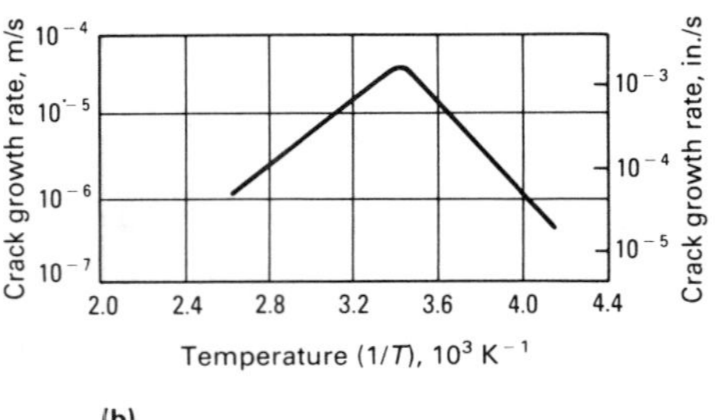

Fig. 35 Schematic of crack growth rate versus temperature for (a) 3% Ni steel in water (Ref 74) and (b) 4340 steel in gaseous hydrogen (Ref 75)

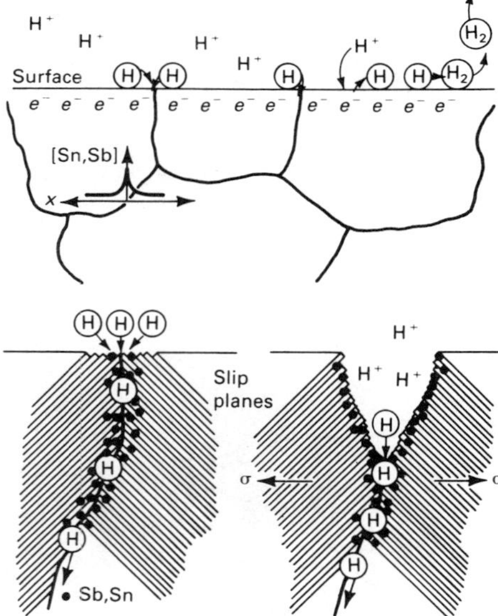

Fig. 36 Schematic showing effect of some impurities on mechanism by which intergranular embrittlement of nickel is presumed to occur at cathodic potentials

promote one mode of fracture in an alloy rather than another form normally observed in benign environments. Thus, all modes of cracking have been observed in most commercial alloy systems; however, hydrogen stress cracking usually produces sharp singular cracks in contrast to the extensive branching observed for SCC. The catastrophic cracking of steels in hydrogen sulfide (H_2S) environments referred to as sulfide stress cracking is a special case of hydrogen stress cracking.

Loss in tensile ductility was one of the earliest recognized forms of hydrogen damage. Significant decreases in elongation and reduction in area are observed for steels, stainless steels, nickel-base alloys, aluminum alloys, and titanium alloys exposed to hydrogen. This mode of failure is most often observed in lower-strength alloys, and the extent of loss in tensile ductility is a function of hydrogen content of the material. Loss in tensile ductility behavior is strain rate sensitive and becomes more pronounced as the strain rate decreases.

Hydrogen attack is a high-temperature form of hydrogen damage that occurs in carbon and low-alloy steels exposed to high-pressure hydrogen at high temperatures for extended time. Hydrogen enters the steel and reacts with carbon either in solution or as carbides to form methane gas; this may result in the formation of cracks and fissures or may simply decarburize the steel, resulting in a loss in strength of the alloy. This form of damage is temperature dependent, with a threshold temperature of approximately 200 °C (400 °F).

Blistering occurs predominantly in low-strength alloys when atomic hydrogen diffuses to internal defects, such as laminations or nonmetallic inclusions, and then precipitates as molecular hydrogen (H_2). The pressure of molecular hydrogen can attain such high values that localized plastic deformation of the alloy occurs, forming a blister that often ruptures. Blisters are frequently found in low-strength steels that have been exposed to aggressive corrosive environments (such as H_2S) or cleaned by pickling.

Shatter cracks, flakes, and fish eyes are features common to hydrogen damage in forgings, weldments, and castings. They are attributed to hydrogen pickup during melting operations when the melt has a higher solubility for hydrogen than the solid alloy. During cooling from the melt, hydrogen diffuses to and precipitates in voids and discontinuities, producing the features that result from the decreased solubility of hydrogen in the solid metal. In many aspects, these features are comparable to blistering, and this could be considered a special case of that class.

Microperforation by high-pressure hydrogen occurs at extremely high pressures of hydrogen near room temperature. Microperforation occurs predominately in steels. This form of hydrogen damage manifests itself as a network of small fissures that allow permeation of the alloy by gases and liquids.

Degradation in flow properties in hydrogen environments has been found at ambient temperatures for iron and steel and at elevated temperature for several alloy systems. The steady-state creep rate under constant load has been observed to increase in the presence of hydrogen for some nickel-base alloys.

Hydride formation produces embrittlement in magnesium, tantalum, niobium, vanadium, uranium, thorium, zirconium, titanium, and their alloys, as well as many other less common metals and alloys. The degradation of mechanical properties and the cracking of these metals and their alloys are attributable to the precipitation of metal hydride phases. Hydrogen pickup often results from welding, heat treating, charging from corrosion processes, or during melting of the alloy. Hydride formation is enhanced for some metal-hydrogen systems by the application of stress—the so-called stress-induced hydride formation. Alloy systems that form hydrides are generally ductile at high (>300 K) and low (<100 K) temperatures at which they fracture by ductile rupture. This temperature dependence is comparable to that observed for the hydrogen embrittlement of ferrous and nickel alloys. Some of these alloys are also susceptible to failure in hydrogen by mechanisms other than hydriding. Some evidence exists that nickel and aluminum alloys may also form a highly unstable hydride that could contribute to hydrogen damage of these alloys; however, this possibility has not been confirmed.

Theories for Hydrogen Damage

As may be appreciated from the numerous classes of hydrogen damage, there are many explanations or theories for these various forms of degradation. The preeminent theories for hydrogen damage are based on pressure, surface adsorption, decohesion, enhanced plastic flow, hydrogen attack, and hydride formation. Although many other theories have been presented, most are variations on these basic models.

The pressure theory of hydrogen damage, or more specifically, hydrogen embrittlement, is one of the oldest models for hydrogen damage (Ref 79). This theory attributes hydrogen embrittlement to the diffusion of atomic hydrogen into the metal and its eventual accumulation at voids or other internal surfaces in the alloy. As the concentration of hydrogen increases at these microstructural discontinuities, a high internal pressure is created that enhances void growth or initiates cracking. This model, although apparently reasonable for blistering and possibly appropriate for some aspects of loss in tensile ductility, does not explain many of the factors observed for classes of failure such as hydrogen stress cracking. However, it is a well-recognized phenomenon that charging hydrogen into steel or nickel alloys at high fugacity, either with high-pressure hydrogen gas or under extreme electrochemical charging, can create a significant density of voids and irreversible damage to the alloy consistent with a pressure-dependent model.

The surface adsorption theory suggests that hydrogen adsorbs on the free surfaces created adjacent to the crack tip, decreasing the surface free energy and thus the work of fracture (Ref 80). Reduction in the work of fracture would thus enhance crack propagation at stress levels below those typically experienced for a particular alloy in a benign environment. There are many arguments against this model. The principal criticism is that it greatly underestimates the work of fracture and does not account for the discontinuous crack growth that has been observed for hydrogen cracking.

Decohesion describes the effect of hydrogen on the cohesive force between atoms of the alloy matrix (Ref 81, 82). Sufficiently high hydrogen concentrations that accumulate ahead of a crack tip are assumed to lower the maximum cohesive force between metal atoms such that the local maximum tensile stress perpendicular to the plane of the crack then becomes equivalent to or greater than the lattice cohesive force and fracture results.

Table 3 Classifications of processes of hydrogen degradation of metals

	Hydrogen embrittlement: Hydrogen environment embrittlement	Hydrogen embrittlement: Hydrogen stress cracking	Hydrogen embrittlement: Loss in tensile ductility	Hydrogen attack	Blistering	Shatter cracks, flakes, fisheyes	Micro-perforation	Degradation in flow properties	Metal hydride formation
Typical materials	Steels, nickel-base alloys, metastable stainless steel, titanium alloys	Carbon and low-alloy steels	Steels, nickel-base alloys, Be-Cu bronze, aluminum alloys	Carbon and low-alloy steels	Steels, copper, aluminum	Steels (forgings and castings)	Steels (compressors)	Iron, steels, nickel-base alloys	V, Nb, Ta, Ti, Zr, U
Usual source of hydrogen (not exclusive)	Gaseous H_2	Thermal processing, electrolysis, corrosion	Gaseous hydrogen, internal hydrogen from electrochemical charging	Gaseous	Hydrogen sulfide corrosion, electrolytic charging, gaseous	Water vapor reacting with molten steel	Gaseous hydrogen	Gaseous or internal hydrogen	Internal hydrogen from melt; corrosion, electrolytic charging, welding
Typical conditions	10^{-6} to 10^8 N/m^2 (10^{-10} to 10^4 psi) gas pressure	0.1 to 10 ppm total hydrogen content	0.1 to 10 ppm total hydrogen content range of gas pressure exposure	Up to 10^8 N/m^2 (15 ksi) at 200–595 °C (400–1100 °F)	Hydrogen activity equivalent to 0.2 to 1 × 10^8 N/m^2 (3–15 ksi) at 0–150 °C (30–300 °F)	Precipitation of dissolved ingot cooling	2 to 8 × 10^8 N/m^2 (30–125 ksi) at 20–100 °C (70–200 °F)	1–10 ppm hydrogen content (iron at 20 °C, or 70 °F) up to 10^8 N/m^2 (15 ksi) gaseous hydrogen (various metals, T >0.5 melting point)	10^5 to 10^8 N/m^2 (15–15 000 psi) gas pressure hydrogen activity must exceed solubility limit near 20 °C (70 °F)
	Observed at −100 to 700 °C (−150 to 1290 °F); most severe near 20 °C (70 °F)	Observed at −100 to 100 °C (−150 to 212 °F); most severe near 20 °C (70 °F)	Observed at −100 to 700 °C (−150 to 1290 °F)	...	...	...	...	...	...
	Strain rate important; embrittlement more severe at low strain rate; generally more severe in notched or precracked specimens	Strain rate important; embrittlement more severe at low strain rate; always more severe in notched or precracked specimens	Occurs in absence of effect on yield stress; strain rate important	...	...	...	...	...	...
Failure initiation	Surface or internal initiation; incubation period not observed	Internal crack initiation	Surface and/or internal effect	Surface (decarburization); internal carbide interfaces (methane bubble formation)	Internal defect	Internal defect	Unknown	...	Internal defect
Mechanisms	Surface or subsurface processes	Internal diffusion to stress concentration	Surface or subsurface processes	Carbon diffusion (decarburization); hydrogen diffusion; nucleation and growth (bubble formation)	Hydrogen diffusion; nucleation and growth of bubble; steam formation	Hydrogen diffusion to voids	Unknown	Adsorption to dislocations; solid-solution effects	Hydride precipitation

Source: Ref 78

Enhanced plastic flow is associated with hydrogen dislocation interactions and is primarily based on fractographic observations (Ref 83). This approach proposes that atomic hydrogen enhances dislocation motion, generally screw dislocations, and the creation of dislocations at surfaces and/or crack tips, leading to softening of the material on a localized scale. Although this behavior has been observed in certain steels, hardening by hydrogen has also been found. Careful high-resolution electron microscopy of what appears to be brittle cleavage or intergranular fracture surfaces has revealed evidence of crack tip plasticity in support of this mechanism.

Hydride formation is the degradation of Group Vb metals (niobium, vanadium, and tantalum) and zirconium, titanium, and magnesium in hydrogen environments by the formation of a brittle metal hydride at the crack tip. When sufficient hydrogen is available in the alloy, a metal hydride precipitates. Cracking of the hydride occurs, followed by crack arrest in the more ductile matrix or continued crack growth between hydrides by ductile rupture. Because hydride formation is enhanced by the application of stress, the stress field ahead of the crack tip may induce precipitation of additional hydrides that cleave. Thus, in some alloys, brittle crack propagation occurs by repeated precipitation of hydrides ahead of the crack tip, cleavage of these hydrides, and precipitation of new hydrides and so on until fracture is complete (Ref 84).

Hydrogen attack is one of the better understood mechanisms of hydrogen damage, but is specific to a single class of damage by hydrogen at high temperature (Ref 85). Hydrogen attack may take two forms of damage: surface decarburization or internal decarburization. However, the mechanism is the same for both forms. At elevated temperatures, hydrogen diffuses into the steel or reacts at the surface with carbon in solid solution or that which has dissociated from carbides to form a hydrocarbon—typically methane. This chemical reaction is easily described thermodynamically, which sets this form of damage apart for the more complex forms of hydrogen damage. As expected, damage is dependent on temperature and hydrogen partial pressure. Surface decarburization occurs at temperatures above 540 °C (1000 °F) and internal decarburization at temperatures from 200 °C (400 °F) upward.

Hydrogen Trapping. Although numerous models exist, none adequately explains the behavior exhibited by alloys in different hydrogen-bearing systems. Until a universal theory is developed, one must rely on the phenomenological behavior between the more prominent alloy systems to understand hydrogen damage. One of the principal factors that determines the hydrogen damage susceptibility of ferrous alloys is a phenomenon referred to as trapping (Ref 86). Diffusion studies of iron and steels have shown an initial retardation in diffusion rate or lag time for hydrogen diffusion through these alloys before a steady-state diffusivity compatible with that expected theoretically is achieved. This lag time is generally considered to be related to the filling of traps by hydrogen. In fact, the apparent diffusivity of hydrogen in steels shows a precipitous decrease with increasing concentration of particles (traps), as shown in Fig. 37.

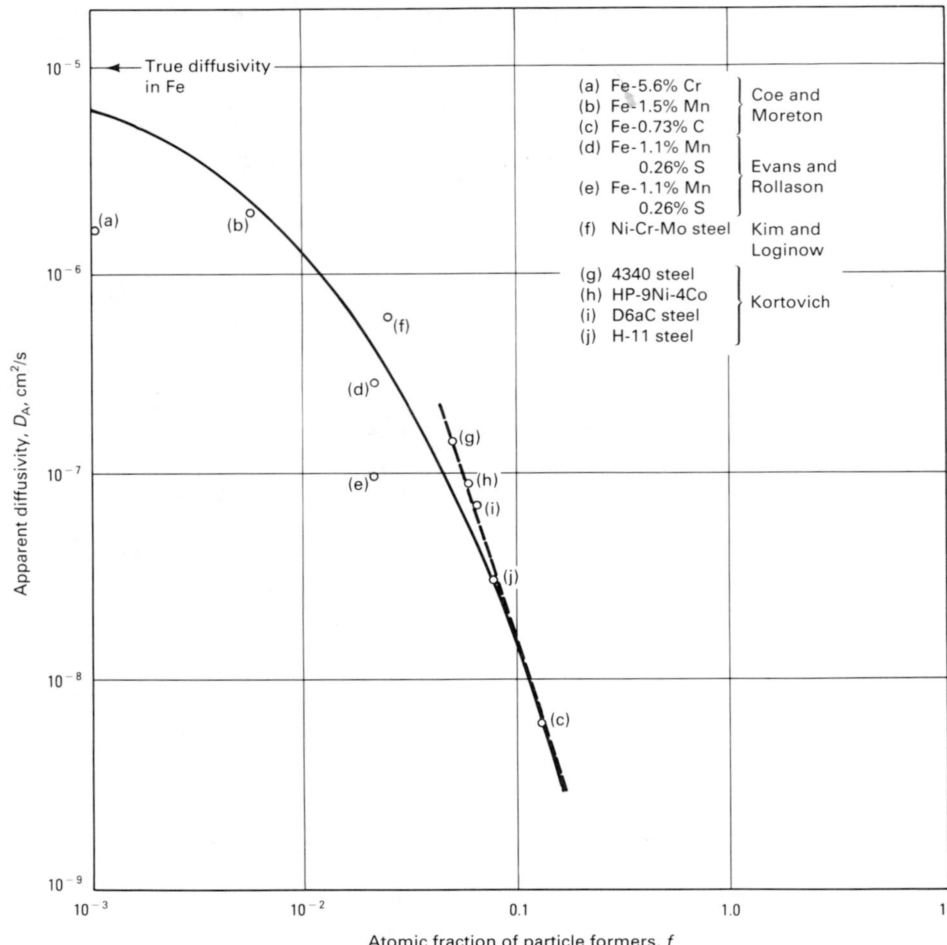

Fig. 37 Decrease in apparent hydrogen diffusivity with increased fraction of precipitate particles. Source: Ref 87

Hydrogen trapping may be considered the binding of hydrogen atoms to impurities, structural defects, or microstructural constituents in the alloy. Binding may be attributed to local electric fields and stress fields, temperature gradients, chemical potential gradients, or physical trapping. These hydrogen traps may be mobile (dislocations, stacking faults) or stationary (grain boundaries, carbide particles, individual solute atoms). They may also be reversible or irreversible traps. Short-duration trapping of hydrogen in which the occupancy time is limited is referred to as reversible. A long residency time for hydrogen characterized by a high binding energy is termed irreversible trapping. Table 4 presents a classification of hydrogen traps in steel. The concept and investigation of trapping have been developed primarily for steels; however, it may not be restricted to this system. Face-centered cubic alloys show a similar trapping behavior, although at a somewhat reduced efficiency for trapping compared to steels.

Hydrogen Damage in Iron-Base Alloys

Pure Irons. Hydrogen damage may occur in relatively pure irons, such as Ferrovac, Plastiron, and Armco Iron, producing either transgranular or intergranular fracture, depending on the presence of impurities and solutes (carbon, oxygen, and nitrogen) at the grain boundaries and the heat treatment. Hydrogen can also lower the yield strength and flow stress of high-purity iron around room temperature or lead to hardening. Although the yield strength is reduced for impure iron, the flow stress may be increased under conditions of low hydrogen fugacity.

Ferrous Alloys. Many factors affect the behavior of ferrous alloys in hydrogen-bearing environments. Hydrogen concentration, temperature, heat treatment/microstructure, stress level (applied and yield stress), solution composition, and environment are the primary factors involved in determining susceptibility to hydrogen embrittlement. Figure 38 shows the effect of hydrogen concentration on the time to failure for a high-strength steel. The longer the baking time, the lower the residual hydrogen in the steel matrix. In general, increasing the concentration of hydrogen in an alloy will reduce time to failure and the stress level at which failure will occur.

Hydrogen concentration in the alloy is a function of the fugacity or the approximate concentration of hydrogen at the surface exposed to the environment. Therefore, hydrogen embrittlement will be controlled by the hydrogen gas pressure or pH of the environment as well as constituents within the environment that may accelerate or inhibit the entry of hydrogen into the alloy. Elements such as sulfur, phosphorus, antimony, tin, and arsenic and their compounds have been found to inhibit the hydrogen recombination reaction in aqueous solutions, thus increasing the charging of atomic hydrogen into the alloy. In contrast, small amounts of oxygen in gaseous hydrogen environments have demonstrated an inhibitive effect on crack growth of high-strength steels subject to hydrogen cracking.

Figure 39 shows the dependence of both the threshold stress intensity and the crack growth rate of high-strength AISI 4130 steel on hydrogen

Table 4 Classification of hydrogen traps in steels according to size

Trap class	Example of trap: Elements at the left of iron	Example of trap: Elements with a negative ϵ^i_H(b)	Interaction energy(a), eV	Character if known	Influence diameter, D_i
	. . .	Ni	(0.083)	Most probably reversible	A few interatomic spacings
	Mn	Mn	(0.09)		
	Cr	Cr	(0.10)		
	V	V	(0.16)		
	. . .	Ce	(0.16)		
	. . .	Nb	(0.16)		
Point	Ti	Ti (vacancy)	0.27	Reversible	
	Sc	O	(0.71)		
	Ca	Ta	(0.98)	Getting more irreversible	
	K	Ia	(0.98)		
	. . .	Nd	(1.34)		
Linear	Dislocations		0.31 0.25 (average values)	Reversible Reversible	3 nm for an edge dislocation
	Intersection of three grain boundaries		. . .	Depends on coherency	. . .
Planar or bimensional	Particle/matrix interfaces TIC (incoherent) Fe_3C MnS		0.98 0.8–0.98	Irreversible, gets more reversible as the particle is more coherent	Diameter of the particle, or a little more as coherency increases
	Grain boundaries		0.27 Average value 0.55–0.61 (high angle)	Reversible Reversible or irreversible	Same as dislocation
	Twins		. . .	Reversible	A few interatomic spacings
	Internal surfaces (voids)		. . .	. . .	. . .
Volume	Voids		>0.22	. . .	Dimension of the defect
	Cracks		. . .	. . .	
	Particles		Depends on exothermicity of the dissolution of H by the particle	. . .	

(a) Values of interaction energies are either experimental or are calculated (when between parentheses) at room temperature. (b) ϵ^i_H is the interaction coefficient. A negative ϵ^i_H means hydrogen is attracted. Source: Ref 88

pressure. Increasing the hydrogen pressure reduces the threshold stress intensity for crack initiation and increases the crack growth rate for a specific stress intensity value.

Temperature also plays an important role in the hydrogen embrittlement of ferrous alloys. Embrittlement is most severe near room temperature (Ref 90) and becomes less severe or nonexistent at higher or lower temperatures (Fig. 40). At lower temperatures, the diffusivity of hydrogen is too sluggish to fill sufficient traps, but at high temperatures, hydrogen mobility is enhanced and trapping is diminished. As can be seen in Fig. 40, embrittlement is also strongly strain rate dependent. At high strain rates, fracture may proceed without the assistance of hydrogen because the mobility of hydrogen is not sufficient to maintain a hydrogen atmosphere around moving dislocations.

Figure 41 shows threshold stress intensity as a function of yield strength of AISI 4340 steel in aqueous and gaseous hydrogen. It can be seen that the threshold stress intensity for crack growth generally decreases with increasing yield strength, regardless of environment, and that very high-strength steels are not usable in hydrogen environments. Threshold stress intensities and crack growth rates are a function of the specific hydrogen environments, with H_2S being one of the most severe environments (Fig. 42). At lower yield strengths, the mechanism for hydrogen-assisted failure apparently changes, and blistering becomes the more common feature of failure. The threshold stress intensities for high-strength steels subjected to hydrogen environments are significantly less than those thresholds measured under benign conditions. These lower thresholds lead to subcritical crack growth when compared to critical values expected from fracture mechanics. Therefore, it is common to designate these thresholds as K_{ISCC} or K_{IH}.

In low-strength steels (700-MPa, or 100-ksi, yield or less), hydrogen damage occurs predominately by loss in tensile ductility or blistering. For loss in tensile ductility, hydrogen promotes the formation and/or growth of voids by enhancing the decohesion of the matrix at carbide particle and inclusion interfaces.

At higher hydrogen fugacities and often in the absence of stress, blistering or a form of cracking also associated with inclusions—referred to as stepwise, or blister cracking—can occur. Stepwise cracking has been observed frequently in low-strength steels subjected to H_2S-containing environments in the absence of stress (Fig. 43).

Fracture of low-strength steels in hydrogen environments may be characterized by ductile dimple-rupture, tearing, cleavage, quasi-cleavage, and, less frequently under certain conditions, intergranular cracking. The article "Modes of Fracture" in Volume 12 of the 9th Edition of *Metals Handbook* contains fractographs illustrating the effects of hydrogen on the fracture appearance of steels.

High-strength steels (>700-MPa, or 100-ksi, yield) are prone to fracture either in an intergranular fashion or by quasi-cleavage, depending on the stress intensity. These steels commonly display an incubation time before fracture initiates under sustained loading, usually in association with regions of high-stress triaxiality. Because triaxial stresses are created at notch roots or under plane strain, fracture initiates internally in the steel. Intergranular fracture is promoted by the presence of impurity elements at prior-austenite or ferrite grain boundaries. Elements such as phosphorus, sulfur, tin, antimony, and arsenic have been found to enhance the intergranular fracture of high-strength steels in hydrogen, and as expected, temper-embrittled steels are even more susceptible to hydrogen stress cracking than steels that are not embrittled.

Metallurgical structure can have a profound effect on the resistance of steels to hydrogen embrittlement. When compared at equivalent strength levels, a quenched-and-tempered fine-grain microstructure is more resistant to cracking than a normalized or bainitic steel. However, this is also dependent to large extent on the strength level at which this comparison is made. In general, the most resistant microstructure is a highly tempered martensitic structure with equiaxed ferrite grains and spheroidized carbides evenly distributed throughout the matrix. Because microstructure is dependent on heat treatment and composition, these factors are not easily separated and must be considered together. There is also a grain size effect that produces enhanced resistance to hydrogen with decreasing prior-austenite grain size (Fig. 44). However, if the grain size is significantly larger than the plastic zone size, the reverse may be true.

The role of alloying elements is quite complex and not easily distinguished from the effects of heat treatment, microstructure, and strength level. Depending on the microstructure and strength level, a specific alloying element may or may not contribute to the hydrogen embrittlement resistance of an alloy or may even increase susceptibility to cracking. The concentration of the alloying element is also a factor in the behavior of

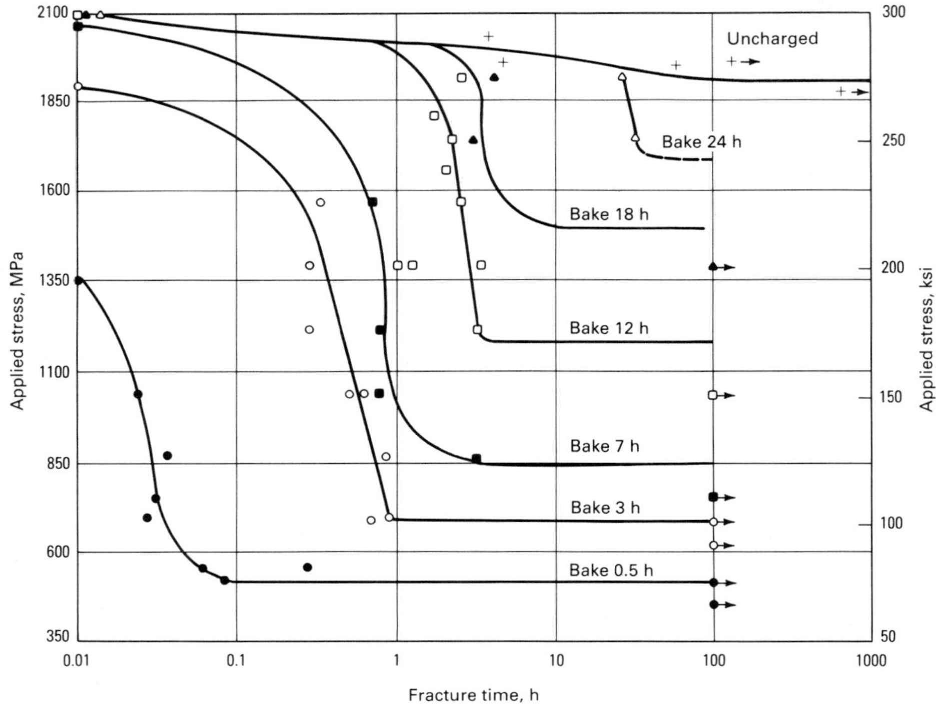

Fig. 38 Static fatigue curves for various hydrogen concentrations obtained by different baking times at 570 °C (300 °F). Sharp-notch high-strength steel specimens; normal notch strength: 2070 MPa (300 ksi). Source: Ref 81

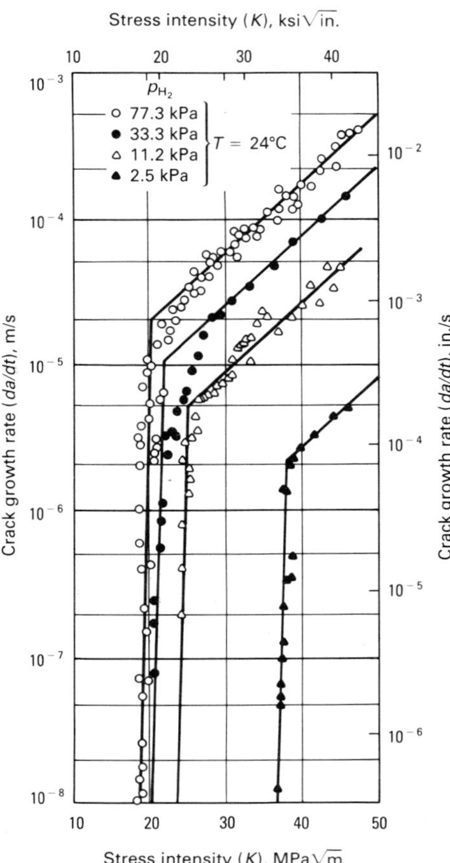

Fig. 39 The *K* dependence of *da/dt* at various hydrogen pressures at 24 °C (75 °F) for AISI 4340 steel. Source: Ref 89

alloys in hydrogen. Molybdenum, for example, is beneficial in reducing susceptibility to sulfide stress cracking, a form of hydrogen stress cracking, in AISI 4130 steels up to approximately 0.75 wt%. However, beyond this concentration, a separate Mo_2C phase precipitates in the alloy after tempering at above 500 °C (930 °F), signficantly reducing resistance to sulfide stress cracking.

In general, elements such as carbon, phosphorus, sulfur, manganese, and chromium impart greater susceptibility to hydrogen embrittlement in low-alloy steels. However, large increases in such elements as chromium, nickel, and molybdenum in order to produce stainless steels alter the crystal structure, microstructure, and subsequently the heat treatment requirements and therefore the hydrogen embrittlement behavior of this group of ferrous alloys.

The response of stainless steels to hydrogen-bearing environments is basically related to their strength level. Ferritic stainless steels have excellent resistance to hydrogen embrittlement because of their low strength and enhanced ductility. However, if the ferritic stainless steels are cold worked, they may become susceptible to cracking in hydrogen environments. Similarly, austenitic stainless steels are highly resistant to hydrogen cracking in the annealed or lightly cold-worked condition, but can become quite susceptible when heavily cold worked. This increased susceptibility to hydrogen cracking due to increasing yield strength from cold working is similar to the dependence of carbon and low-alloy steel on strength. Decreased resistance to hydrogen for highly cold-worked austenitic stainless steels is largely attributed to the deformation-induced formation of martensite. For those austenitic stainless steels having a very stable austenite phase and high yield strength (such as 21Cr-6Ni-9Mn) susceptibility is considered to be solely a function of yield strength similar to the body-centered cubic (bcc) low-alloy steel behavior.

Other factors that may affect the susceptibility of austenitic stainless steels to hydrogen damage are the possible formation of metastable hydride phase that would produce a hydride-based fracture path and the interaction of hydrogen with stacking faults to reduce stacking fault energy in the austenite, leading to planar slip and brittle fracture. The degree of participation of any of these factors has not been established.

Just as a similarity exists between austenitic stainless steels and low-alloy steels at the high-strength end of the spectrum, the lower-strength austenitics behave in the same manner as the low-alloy steels in hydrogen by a reduction in ductility. Figure 45 shows the loss in reduction in area for several austenitic stainless steels in high-pressure hydrogen. It is apparent that a wide variation in hydrogen damage exists between these various austenitic alloys. Type 304L is the most susceptible to loss in tensile ductility, and the stable austenitic alloys, such as 15Cr-25Ni, are almost unaffected. As observed in carbon and low-alloy steels, there is a temperature effect (~0 °C, or 32 °F) involved with the ductility loss in austenitic stainless steels, although it is somewhat lower than the room-temperature dependence observed for low-alloy steels.

It is apparent that there are many similarities in behavior between stainless steels and carbon and low-alloy steels in hydrogen. Similarly, the martensitic and precipitation-hardening stainless steels exhibit the same dependence on strength level and microstructure as observed in low-alloy steels. Martensitic and precipitation-hardening stainless steels are extremely susceptible to hydrogen embrittlement with increasing yield strength. Figure 46 compares several grades of precipitation-hardening stainless steel with type 410 martensitic stainless steel tested in an aqueous environment saturated with H_2S. The numbers adjacent to each data point represent the tempering or aging treatment. Generally, the same trend of decreasing time to failure with increasing yield strength is observed as for low-alloy steels. The poor performance of the type 410 martensitic stainless compared to precipitation-hardening stainless steels is typical of the behavior of most martensitic stainless steels, which also compare poorly with low-alloy steels at the same strength level.

Ultrahigh-strength (>1400 MPa, or 200 ksi) martensitic stainless steels, low-alloy steels, and maraging steels are extremely susceptible to cracking in hydrogen environments, including aqueous solutions containing NaCl. Although chlorides are the primary cause of SCC in many alloy systems, it is generally accepted that the mechanism of cracking in ultrahigh-strength steels is related to hydrogen embrittlement.

Figure 47 compares the crack growth rate of AISI 4340 in several environments against grade 250 maraging steel. The maraging steel has much better resistance to crack propagation in 3.5% NaCl than 4340 and displays a high threshold stress intensity for crack propagation. For maraging steels, there is evidence that peak-aged con-

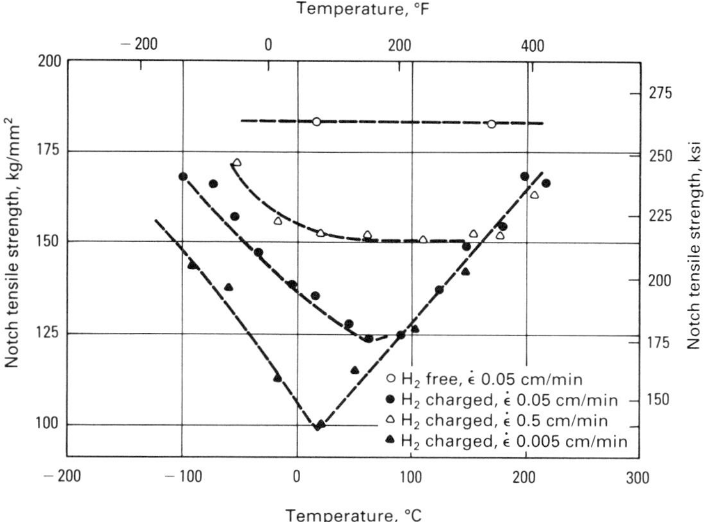

Fig. 40 Notch tensile strength of high-strength steel plotted against testing temperature for three strain rates (crosshead speeds). Source: Ref 90

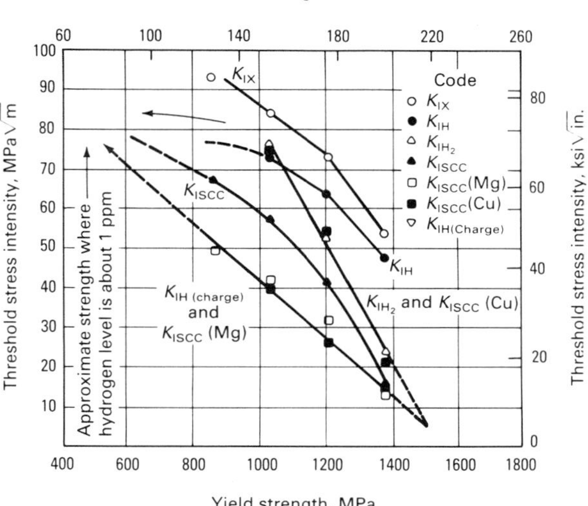

Fig. 41 Effect of yield strength on some threshold stress intensity parameters for crack growth in a commercial AISI 4340 plate. K_{IX} = threshold stress intensity as a function of the yield strength in air; K_{IH} is similar to K_{IX} but slowly loaded; K_{ISCC} (Mg) and K_{ISCC} (Cu) represent specimen coupled to magnesium and copper, respectively. Source: Ref 91

dition or slight overaging may improve resistance to hydrogen embrittlement, but underaging is detrimental to resistance.

Hydrogen attack is strictly a high-temperature form of hydrogen damage that primarily affects ferritic steels. Hydrogen attack is dependent on time, temperature, and hydrogen pressure. After a certain incubation time, a diminution of properties occurs with the onset of decarburization and cracking (Fig. 48). Resistance of steels to hydrogen attack is related to the stability of the carbides, and as such, the addition of carbide-stabilizing elements such as vanadium, titanium, niobium, and molybdenum enhance the resistance to this form of hydrogen damage.

Chromium-molybdenum steels have been found to be the most resistant to hydrogen attack for the cost involved; therefore, a great deal of industrial experience has been gained with these alloys. Much of this experience has been compiled and plotted to provide a guideline for steel selection as a function of temperature and hydrogen partial pressure and is presented as a series of curves often referred to as a Nelson diagram (Ref 96). As more experience is gathered on the use of these steels in hydrogen, this series of curves is periodically updated by the American Petroleum Institute.

Nickel Alloys

Nickel and its alloys are susceptible to hydrogen damage in both aqueous and gaseous hydrogen environments. The same factors that affect hydrogen embrittlement susceptibility in ferrous alloys are also prevalent in nickel alloys, although to a lesser degree. In general, fcc metals, because of their greater ease of slip and reduced solute diffusivities as compared to bcc materials, are less susceptible to hydrogen damage. As with ferrous alloys, hydrogen in nickel and its alloys may introduce intergranular, transgranular, or quasi-cleavage cracking, and although the macroscopic features appear to be brittle, on a microscopic scale there is a high degree of local plasticity, suggesting that hydrogen enhances flow at the crack tip. Figure 49 shows the reduction in ductility associated with hydrogen charging a 72Ni-28Fe alloy as a function of strain rate. Almost identical behavior was observed for unalloyed nickel.

Alloys based on the ternary Fe-Ni-Cr system (Incoloy) and Inconel alloys show reductions in ductility when charged with hydrogen, depending on the specific thermomechanical treatments performed on the alloy. Generally, these alloys are more resistant to hydrogen stress cracking in the cold-worked and unaged condition as compared to the cold-worked and aged condition. Age-hardenable alloys show the least resistance to hydrogen when aged to their peak or near-peak strength. Stabilization treatments, followed by aging, reduce the resistance of these alloys as compared to direct aging. Increased cold work also produces a loss in tensile ductility. Hydrogen stress cracking of these alloys may also occur, rather than a loss in tensile ductility, when they possess high yield strengths or are under high hydrogen fugacity. The nickel-copper alloys (Monels) have also been found to be susceptible to hydrogen embrittlement, with increasing strength obtained by cold working or aging.

Several nickel-base alloys are resistant to hydrogen-stress cracking when cold worked to yield strengths in excess of 1240 MPa (180 ksi). However, when these alloys (Hastelloy alloy C-276, Hastelloy alloy C-4, and Inconel alloy 625) are aged at low temperature, their resistance to hydrogen cracking is considerably diminished. This behavior is attributed to the segregation of phosphorus and sulfur to grain boundaries, which provide low-energy fracture paths (much the same as occurs in high-strength steels), or to an ordering reaction of the form Ni_2(Cr, Mo). Figure 50 presents data that relate the reduction in area loss to ordering for the ordering alloy Ni_2Cr charged with hydrogen.

Aluminum Alloys

Only recently has it been determined that hydrogen embrittles aluminum. For many years, all environmental cracking of aluminum and its alloys was represented as SCC; however, testing in specific hydrogen environments has revealed the susceptibility of aluminum to hydrogen damage. Hydrogen damage in aluminum alloys may take the form of intergranular or transgranular cracking or blistering. Blistering is most often associated with the melting or heat treatment of aluminum where reaction with water vapor produces hydrogen. Blistering due to hydrogen is frequently associated with grain-boundary precipitates or the formation of small voids. Blister formation in aluminum is different from that in ferrous alloys in that it is more common to form a multitude of near-surface voids that coalesce to produce a large blister.

In a manner similar to the mechanism in iron-base alloys, hydrogen diffuses into the aluminum lattice and collects at internal defects. This occurs most frequently during annealing or solution treating in air furnaces prior to age hardening.

Dry hydrogen gas is not detrimental to aluminum alloys; however, with the addition of water vapor, subcritical crack growth increases dramatically (Fig. 51). The threshold stress intensity for cracking of aluminum also decreases significantly in the presence of humid hydrogen gas at ambient temperature (Fig. 52).

Crack growth in aluminum in hydrogen is also a function of hydrogen permeability, as in the iron- and nickel-base alloys. Hydrogen permeation and the crack growth rate are a function of potential, increasing with more negative potentials, as expected for hydrogen embrittlement behavior. Similarly, the ductility of aluminum alloys in hydrogen is temperature dependent, displaying a minimum in reduction in area below 0 °C; this is similar to other fcc alloys (Ref 98).

Most of the work on hydrogen embrittlement of aluminum alloys has been on the 7000 series

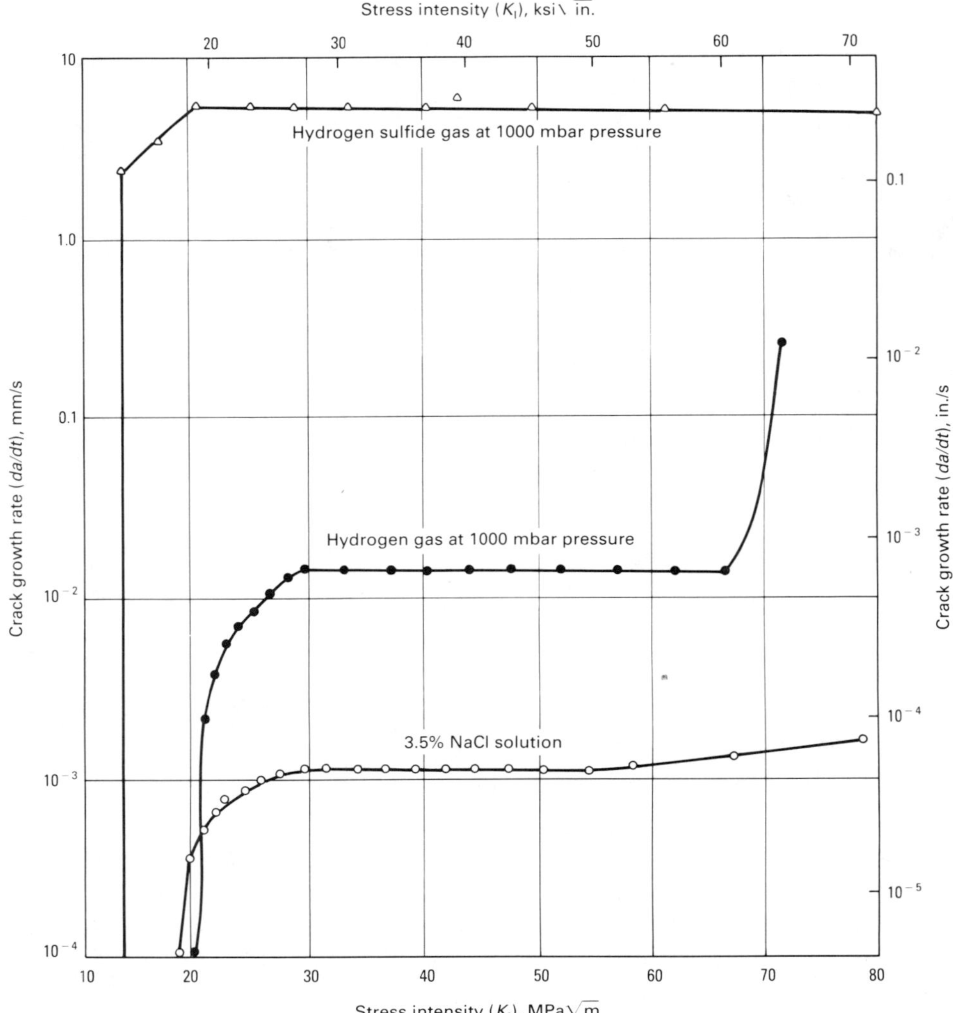

Fig. 42 Crack velocity as a function of stress intensity for a chromium-molybdenum-vanadium steel at 291 K (18 °C, or 64 °F). Source: Ref 92

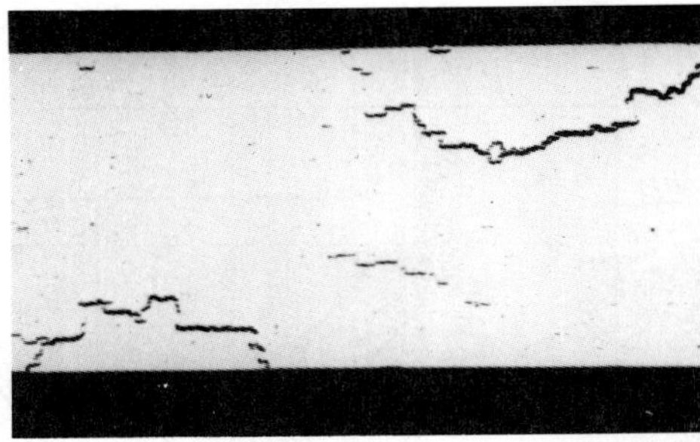

Fig. 43 Stepwise cracking of a low-strength pipeline steel exposed to H_2S. 6×

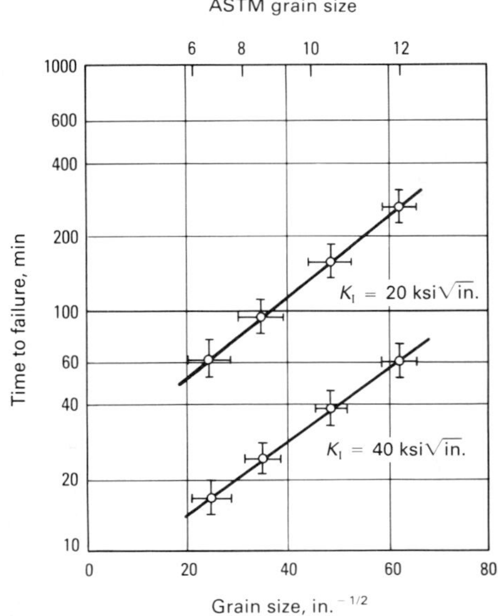

Fig. 44 Illustration of how a refinement in grain size improves resistance to hydrogen failure as measured by the time to failure of two strengths of AISI 4340 steels. Source: Ref 93

(Al-Zn-Mg); therefore, the full extent of hydrogen damage in aluminum alloys has not been determined or the mechanisms established. Some evidence for a metastable aluminum hydride has been found that would explain the brittle intergranular fracture of aluminum-zinc-magnesium alloys in water vapor. However, the instability of the hydride is such that it has been difficult to evaluate. Another explanation for intergranular fracture of these alloys is preferential decohesion of grain boundaries containing segregated magnesium. Overaging of these alloys increases their resistance to hydrogen embrittlement in much the same way as for highly tempered martensitic steels.

Copper Alloys

A common form of hydrogen damage in copper has been known for years as steam embrittlement and is observed only when copper contains oxygen. Hydrogen entering the metal reacts with oxygen either in solid solution or at oxide inclusions to form water. At temperatures above the critical temperature for water, steam forms, and the pressure generated is sufficient to produce microcavity formation and cracking. The equation for reaction with cuprous oxide particles is:

$$Cu_2O + 2H = 2Cu + H_2O(g)$$

The circumstances under which this form of hydrogen damage occurs are typically related to the annealing of copper in a hydrogen atmosphere. At lower temperatures, steam is not generated and therefore does not create the problem. Although the use of oxygen-free copper essentially eliminates susceptibility to steam embrittlement, the heavy cold working of oxygen-free copper can result in grain-boundary void formation when hydrogen is introduced.

Compared to the ferrous and nickel alloys, relatively little work has been performed specific to the hydrogen embrittlement of copper-base alloys. There is some indication that age-hardenable alloys such as beryllium copper are susceptible to embrittlement under severe charging conditions, producing a loss in tensile ductility. A systematic study of these alloys remains to be performed.

Titanium Alloys

Titanium and its alloys suffer hydrogen damage primarily by hydride-phase formation. Pure α-titanium is relatively unaffected by small concentrations (<200 ppm) of hydrogen; however, above this content, the impact toughness is impaired. The purity of the α-titanium is important to its behavior in hydrogen. Commercially pure titanium is much more sensitive to hydrogen than pure titanium is. The amount of hydrogen necessary to induce ductile-to-brittle transition behavior in commercially pure titanium is one-half the amount needed in pure titanium.

Loss in impact toughness also occurs in α/β-titanium and β-titanium alloys and is sometimes referred to as impact embrittlement. This hydrogen damage at high strain rates is the result of hydrides that precipitate after the high-temperature exposure of titanium to hydrogen.

Another mode of failure for titanium alloys in hydrogen predominates under slow strain rate loading. Figure 53 shows the strain rate behavior for an α/β alloy. As the strain rate increases, the effect of hydrogen is lost. As in steels, the reduction in area and the elongation of titanium alloys are diminished in hydrogen.

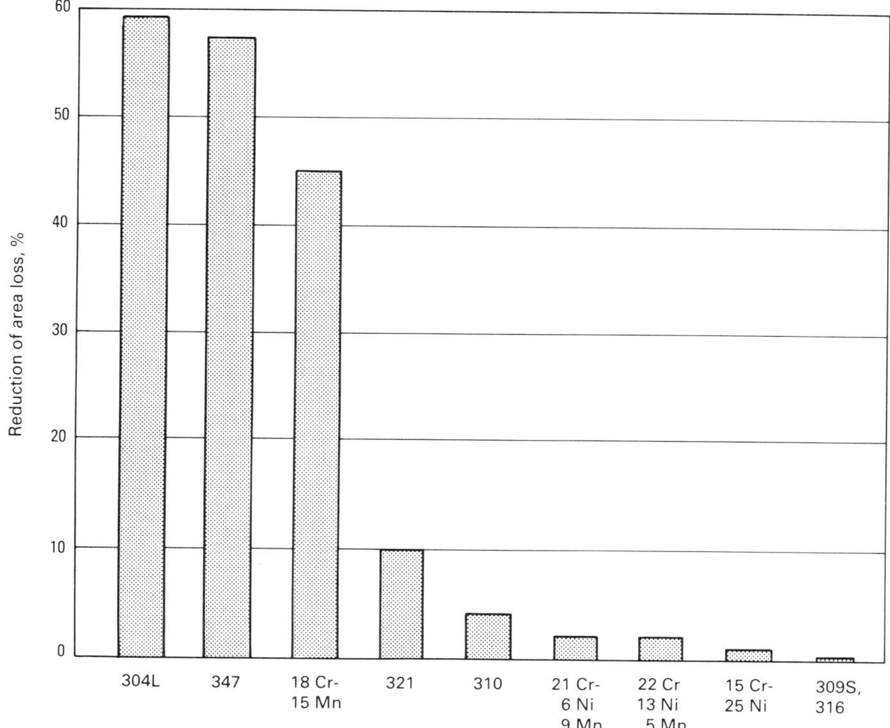

Fig. 45 Ductility loss for several austenitic stainless steels in high-pressure hydrogen. Source: Ref 94

Both types of failure for titanium alloys in hydrogen are attributed to hydride-phase precipitation. The low strain rate embrittlement is related to hydride formation caused by strain-enhanced precipitation, but embrittlement under impact is caused by hydride-phase formation after fabrication or heat treatment. Unlike many hydride-forming systems, titanium forms a stable hydride, but the kinetics of precipitation are slow compared to the Group Vb metals. Therefore, embrittlement is more prone to occur at low strain rates at which precipitation can proceed at a rate that is sufficient to provide a brittle crack path.

Because hydrogen solubility increases with temperature for these alloys, hydride embrittlement typically decreases as the temperature increases. Additionally, at higher temperatures, the hydride may become more ductile, reducing brittle crack initiation. As expected, the threshold stress intensity for crack propagation is also a function of the hydrogen content decreasing with increasing hydrogen.

The role of microstructure in the hydrogen damage of titanium is quite complex and is not fully understood. However, it has been determined that under slow strain rates the α/β alloys fail by intergranular separation along boundaries but that completely α alloys fracture by transgranular cleavage. Embrittlement is not as severe in α/β alloys with a continuous equiaxed α matrix as for those alloys with a continuous acicular β matrix (Ref 100). However, this behavior is a function of hydrogen pressure and may be reversed at lower pressures (Ref 101).

Zirconium Alloys

Zirconium and its alloys suffer hydrogen damage by hydride precipitation, especially in the presence of notches or at low temperatures in smooth specimens. Failure occurs by cracking of the brittle, precipitated hydride phase during straining. These fractured hydrides may then accelerate fracture by ductile microvoid formation and coalescence, primarily enhancing the latter. Similarly to titanium, zirconium alloys precipitate stable hydrides. Depending on the hydrogen content of the alloy, one of two stable or one metastable hydride may precipitate. The two stable zirconium hydrides are δ-hydride (fcc) and ϵ-hydride (face-centered tetragonal, fct). The metastable γ-hydride also has an fct structure (Ref 102).

As in other alloy systems, the tensile ductility generally diminishes with increasing hydrogen content (Fig. 54). Similarly, the hydrogen embrittlement of these alloys is dependent on stress state, becoming more pronounced as the stress state moves from uniaxial to biaxial to triaxial tension. For example, the local stress state ahead of a sharp notch can produce a region of biaxial or triaxial stress, thus increasing the susceptibility to hydrogen embrittlement.

The embrittlement by zirconium hydride formation is a strong function of temperature, because increasing the temperature will reduce the volume fraction of hydride due to the increased solid solubility of hydrogen with temperature. Ductile-to-brittle transition behavior has been observed in hydrided zirconium alloys, and the transition temperature has been found to be a function of hydrogen content, increasing with hydrogen concentration (Fig. 55).

Hydride distribution and morphology are other important factors in the extent of hydrogen damage in zirconium alloys. Because the brittle hydride phase provides an easy crack path for fracture, the distribution and morphology of this phase in relation to the sense of the applied tensile stress determine the degree of embrittlement. The shape and distribution of the hydrides are dependent on heat treatment prior to precipitation and cooling rate during precipitation. Rapid cooling produces a more uniform dispersion of hydrides, while slow cooling enhances the grain-boundary precipitation of the hydride platelets. Toughness is not as dramatically affected in the former case as in the latter.

Vanadium, Niobium, Tantalum, and Their Alloys

These metals and alloys are all embrittled by hydrogen—primarily by hydride-phase formation; however, unlike titanium and zirconium alloys, these systems do not form a stable hydride. The introduction of hydrogen into unalloyed vanadium, niobium, and tantalum increases the yield strength of these metals and creates ductile-to-brittle transition behavior (Fig. 56).

Many of these alloy systems display ductile-to-brittle transition behavior in the presence of hydrogen in much the same manner as that described for zirconium and titanium. However, these alloys have been found to exhibit grain-boundary cracking when low levels of hydrogen are present in solid solution below the terminal solid-solubility limit. In fact, all of the hydride-forming systems exhibit hydrogen damage from solute hydrogen at low concentrations of hydrogen at which hydrides are not expected to precipitate. Although fracture transition is correlated with hydride formation in vanadium, other systems, such as niobium, zirconium, titanium, and tantalum, display more classical hydrogen embrittlement without a corresponding hydride-phase precipitation (Ref 102, 105).

Once the hydrogen content exceeds terminal solid solubility, damage may proceed by hydride precipitate formation. Below this solubility limit, the relative resistance to cracking or loss in tensile ductility is largely a function of the hydrogen concentration and alloying elements (Fig. 57). However, the terminal solid solubility for a particular metal-hydrogen system that forms hydrides is an extremely strong function of stress so that reference to stress-free equilibrium phase diagrams for predicting hydride phase behavior may be inaccurate (Ref 107).

There are many other metals and alloys that form hydrides (for example, thorium, uranium, and beryllium), but far less is known about these systems. Therefore, a great deal of investigation into the metal/hydrogen interactions in many alloy systems remains to be done, especially considering the trend in new materials for advanced technologies.

Liquid-Metal Embrittlement

M.H. Kamdar
Benet Weapons Laboratory
U.S. Army Armament Research, Development, and Engineering Center

Liquid-metal embrittlement is the catastrophic brittle failure of a normally ductile metal when coated with a thin film of a liquid metal and subsequently stressed in tension. The fracture mode changes from a ductile to a brittle intergranular or brittle transgranular (cleavage) mode; however, there is no change in the yield and flow behavior of the solid metal. As shown in Fig. 58, embrittlement manifests itself as a reduction in fracture stress, strain, or both. Fracture can occur well below the yield stress of the solid. The stress needed to propagate a sharp crack or a flaw in liquid is significantly lower than that necessary to initiate a crack in the liquid-metal environment. In most cases, the initiation or the propagation of cracks appears to occur instantaneously, with the fracture propagating through the entire test specimen. The velocity of crack or fracture propagation has been estimated to be 10 to 100 cm/s (4 to 40 in./s).

Examination of LME fracture surfaces shows complete coverage by the liquid metal. Liquid metal is in intimate contact with the solid and is usually difficult to remove. The fracture mode becomes apparent when special techniques are used to remove the liquid. It should be emphasized that gross amounts of liquid are not necessary; even micrograms of liquid lead can cause LME in 75-mm (3-in.) thick steel tubes (Fig. 59). The fracture is usually brittle intergranular, with little indication of crack branching or striation to indicate slow crack propagation.

The presence of liquid at the moving crack tip appears necessary for fast fracture. However, in high-strength brittle metals, a crack initiated in the liquid may propagate in a brittle manner in the absence of liquid. In low-strength metals, a transition may occur from brittle-to-ductile failure under similar circumstances. Although far less frequent than brittle fracture, embrittlement can also occur by a ductile dimpled rupture mode in certain steels, copper alloys, and aluminum alloys. The embrittlement then manifests itself as the degradation of the mechanical properties in the solid metal. Fracture is not limited to polycrystalline metals and the presence of grain boundaries. Single crystals of zinc and cadmium are also embrittled by liquid metal and result in fracture by a brittle cleavage mode (Fig. 60). These are but a few of the manifestations of LME.

Embrittlement is not a corrosion, dissolution, or diffusion-controlled intergranular penetration process, but is considered to be a special case of brittle fracture that occurs in the absence of an inert environment and at low temperatures. Time- and temperature-dependent processes are not considered to be responsible for the occurrence of LME. Also, in most cases of LME, little or no penetration of liquid metal into the solid metal is observed. The embrittlement of the solid metal coated with liquid metal or immersed in the liquid does not depend on the time of exposure to the liquid metal before testing (Fig. 61) or on whether the liquid is pure or presaturated with the solid. Usually, one of the prerequisites is that the solid has little or no solubility in the liquid and forms no intermetallic compound to constitute an embrittlement couple. However, exceptions to this empirical rule have been noted.

An increase in test temperature decreases embrittlement, which leads to a brittle-to-ductile transition. Severe embrittlement occurs near the freezing temperature of the liquid. In fact, in iron-lead, iron-indium (Fig. 62), and many other metallic couples, embrittlement occurs below the melting temperature of the liquid. This new phenomenon is called solid metal induced embrittlement (SMIE) of metals and is described in the following section of this article.

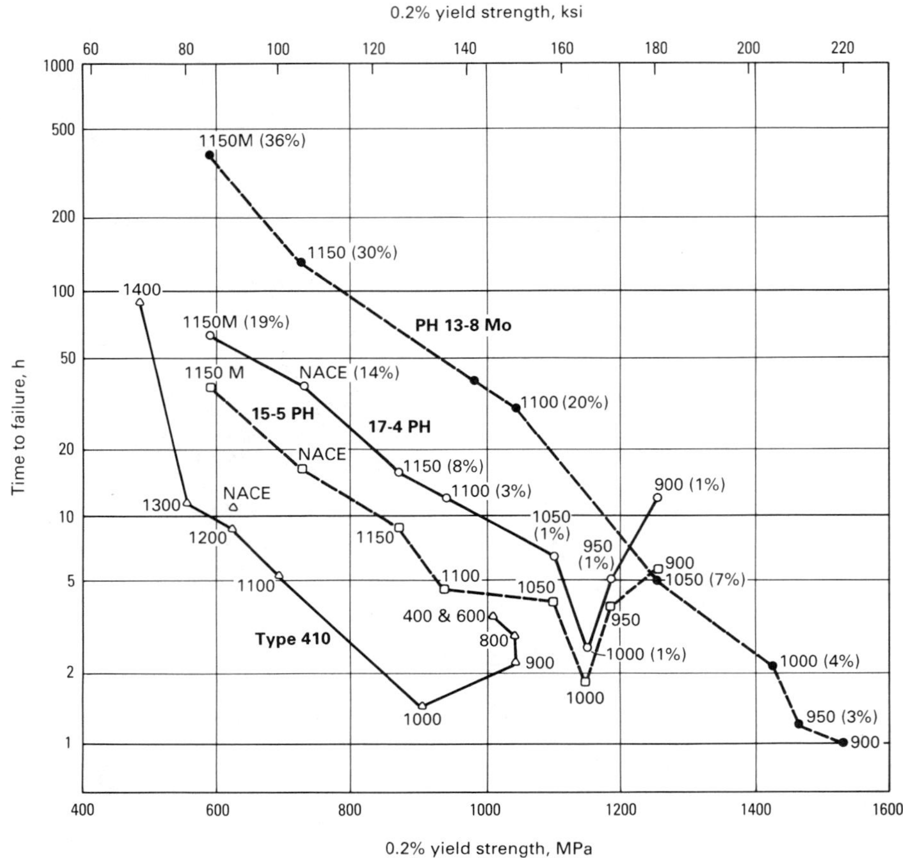

Fig. 46 Time to failure of various alloys as a function of yield strength when tested under 345 MPa (50 ksi) of applied stress in saturated H_2S. Numbers adjacent to data points represent tempering or aging treatment; parenthetical values indicate approximate amounts of austenite. Source: Ref 95

The occurrence of LME is not a laboratory curiosity. Such failures have been observed in metals and alloys in the following industrial applications and processes:

- Small amounts of alloying elements, such as lead and tellurium, added to steel to improve machinability may lead to embrittlement (Fig. 63). Internally leaded steels have cracked at lead inclusions. For example, leaded steel gears have cracked during induction-hardening heat treatments, and warm punching of leaded steel shafts has resulted in unexpected fracture during the forming operation
- Cadmium-plated titanium and steels are embrittled during high-temperature service by molten cadmium
- Indium, used as a high-vacuum seal in steel chambers, has caused cracking during bakeout operations
- Zircaloy tubes used in nuclear reactors have been cracked by both solid and liquid cadmium
- Although infrequent, LME also occurs in petrochemical plants and in the steel industry during heat treatment, hot rolling, brazing, soldering, and welding operations
- Embrittlement of steel occurs by electroplated or dipped cadmium, zinc, or tin—all of which provide corrosion resistance
- In liquid metal cooled reactors, liquid lithium can cause both corrosion and LME in metals and alloys

The distinctive features of embrittlment and of the resultant fracture surfaces are a significant loss in mechanical properties and usually a brittle fracture mode, although ductile fracture morphologies have been reported. The fracture surfaces are easily distinguished from those due to SCC, which are similar to fracture due to hydrogen or temper embrittlement. The embrittlement is severe, and the propagation of fracture is very fast in the case of LME as compared to that in SCC. Thus, the predominant occurrence of brittle fracture surfaces, fast or catastrophic fracture, significant loss in ductility and strength, and the presence of liquid at the tip of the propagating crack are some of the characteristics that may be used to distinguish LME from other environmentally induced failures. Detailed information on the mechanisms of LME and the susceptibility of specific metals and alloys to liquid metal induced failures can be found in the Selected References provided at the end of this article.

Mechanisms of Embrittlement

Several mechanisms have been proposed to explain LME, including stress-assisted dissolution of the solid at the crack tip and reduction in the surface energy of the solid by the liquid metal. It has been suggested that embrittlement is associated with liquid-metal adsorption induced localized reduction in the strength of the atomic bonds at the crack tip or at the surface of the solid metal at sites of stress concentrations.

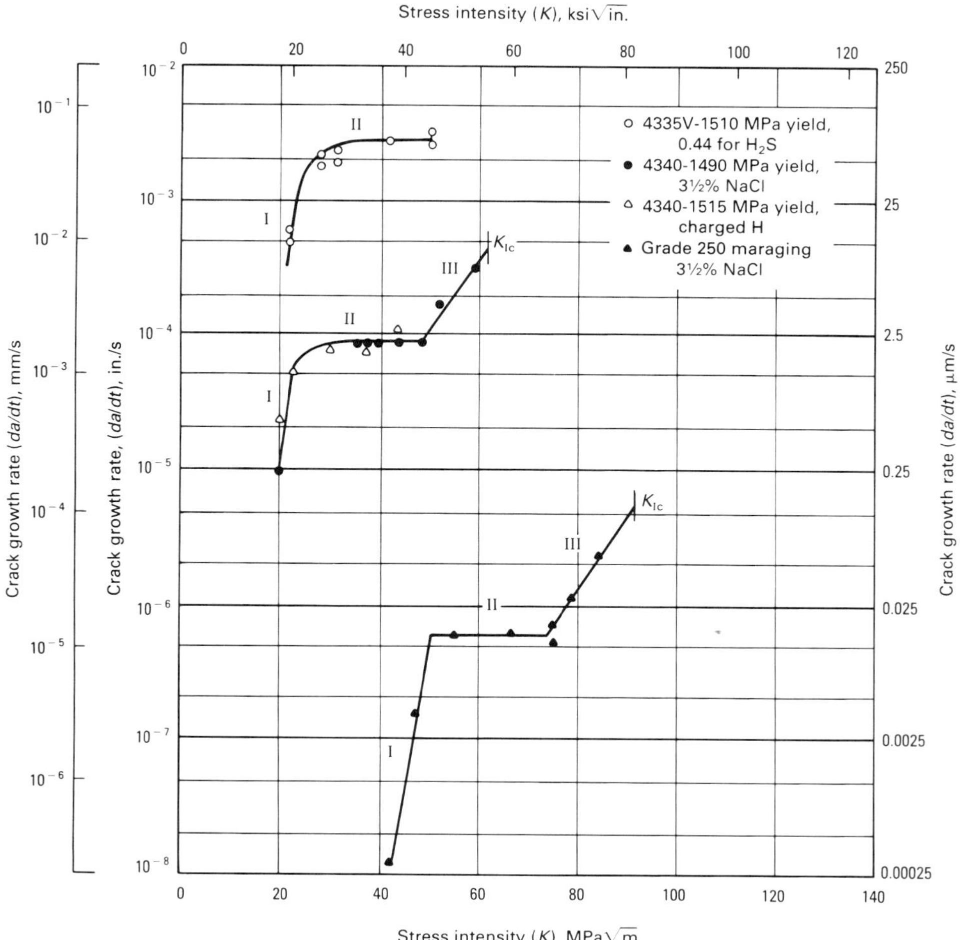

Fig. 47 Effect of material, environment, and stress intensity level on crack growth. Source: Ref 87

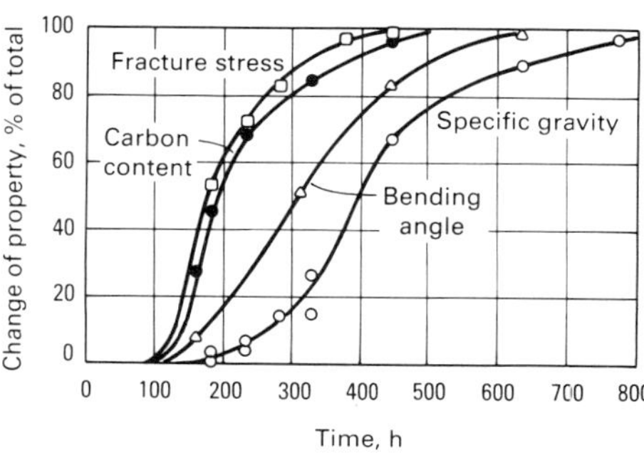

Fig. 48 Relative change of properties of an SAE 1020 steel as a function of time of exposure to hydrogen at 427 °C (800 °F) and 6.2 MPa (900 psi) partial pressure. Source: Ref 85

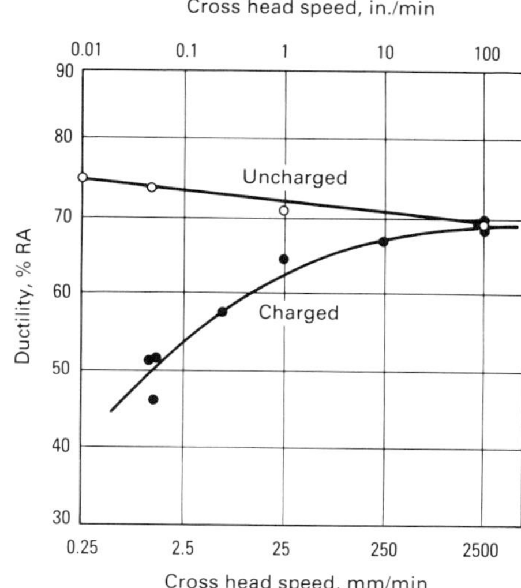

Fig. 49 Ductility at fracture as a function of strain rate in a hydrogen-charged and uncharged 72Ni-28Fe alloy. Source: Ref 81

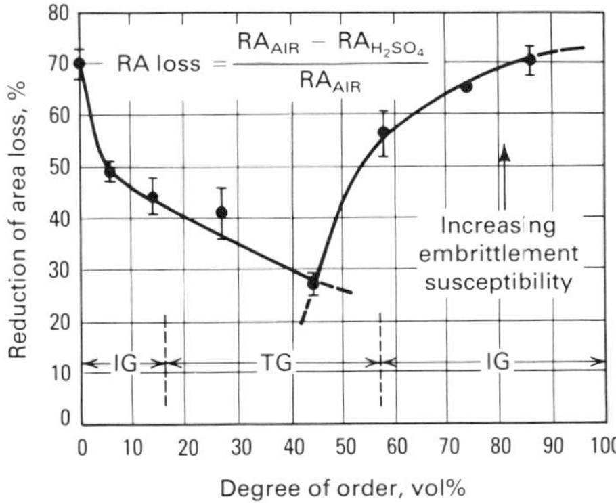

Fig. 50 Effect of the degree of order on the embrittlement susceptibility of Ni_2Cr. Regions of intergranular (IG) and ductile transgranular fracture (TG) are shown. Source: Ref 97

With this possibility in mind, consider the crack shown in Fig. 64. Crack propagation will occur by the breaking of A-A_0 bonds at the crack tip and, subsequently, the breaking of similar bonds at the propagating crack tip by the chemisorbed liquid-metal atom B (a vapor phase from a solid in a solid-solid metal embrittlement couple or an elemental gas, such as hydrogen, may also provide the embrittling atom B). Next, assume that liquid-metal atom B at the crack tip reduces the cohesive strength of A-A_0 bonds. The chemisorption process presumably occurs spontaneously or only after the A-A_0 bonds have been strained to some critical value. In any event, electronic rearrangement occurs because of adsorption, which weakens the bonds at the crack tip. When the applied stress is increased so that it exceeds the reduced breaking strength of A-A_0 bonds, then the crack propagates. The liquid-metal atom becomes stably chemisorbed on the freshly created surfaces. The surface diffusion of the liquid-metal atoms over the chemisorbed liquid-metal atoms feeds the advancing crack tip, thus propagating the crack at reduced stress and causing complete failure of the specimen.

In the above mechanism, it is assumed that reduction in the tensile cohesion is predominantly responsible for the occurrence of embrittlement, although both tensile as well as shear cohesion are reduced. Embrittlement can also occur by the adsorption-induced reductions in the shear strength of the atomic bonds at the crack tip. The reduced shear strength facilitates nucleation of dislocations or slip at low stresses at or near the crack tip. This localized increase in plasticity produces a plastic zone with sufficiently large strains such that a void is nucleated ahead of the crack tip at precipitates, inclusions, or at subboundaries in a single crystal. The void will grow, and crack growth will occur. The fracture occurs by the ductile rupture mode, with the appearance of dimples on the fracture surfaces. The localized increased plasticity results in an overall reduction in the strain at failure as compared to that in the absence of liquid metal, thus causing embrittlement. For embrittlement due to reduced shear cohesion, lower strain at failure or subcritical crack growth by linkage of voids is the measure of LME.

A schematic representation of the above process is given in Fig. 65, and supporting fractographic evidence is given for high-strength steel (Fig. 66a) and beryllium-copper alloys (Fig. 66b) broken in liquid-metal environments. Conversely, Fig. 66(c) to (e) and Fig. 67 provide strong support for the reduced tensile cohesion mechanism.

Most often, LME induces a brittle intergranular or cleavage fracture mode at reduced tensile stresses and thus supports the tensile decohesion mode. However, the dimpled ductile failure mechanism of LME should be considered an embrittling process in a failure analysis in which the reduction in the mechanical parameters at fracture is the measure of embrittlement. Al-

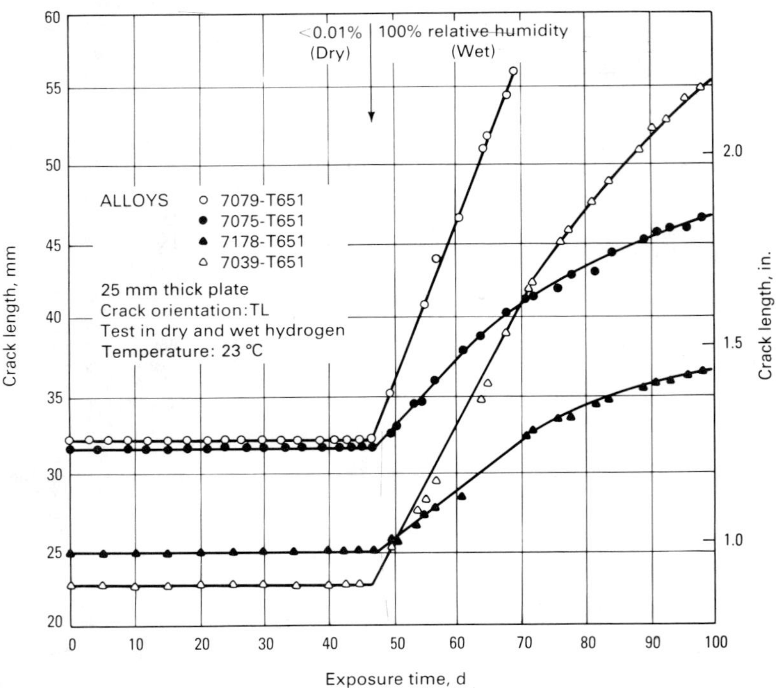

Fig. 51 Effect of humidity on subcritical crack growth of high-strength aluminum alloys in hydrogen gas. Source: Ref 98

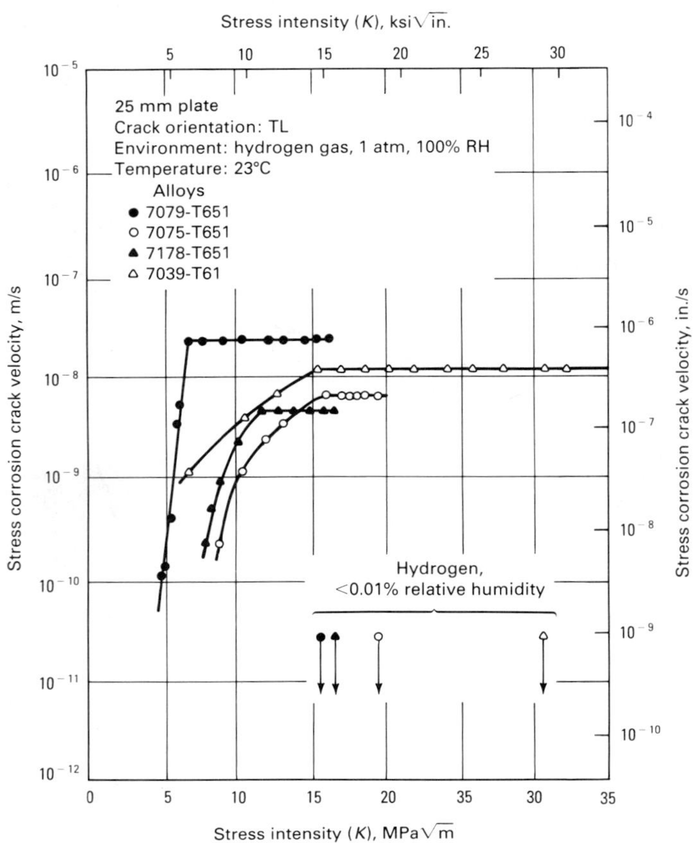

Fig. 52 Crack velocity of four high-strength aluminum alloys plotted as a function of crack-tip stress intensity in moist and dry hydrogen gas. Source: Ref 98

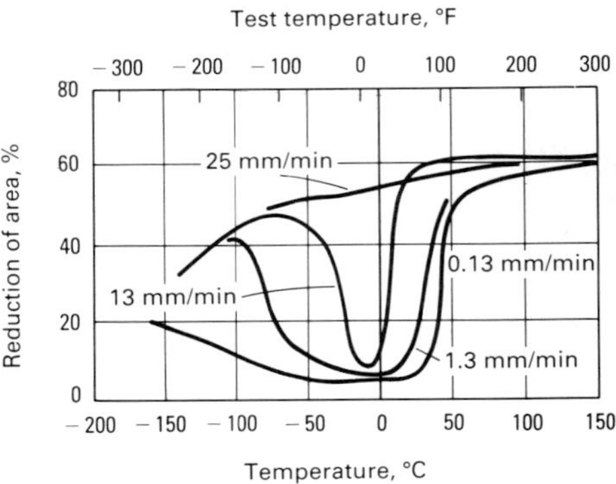

Fig. 53 Effects of hydrogen content (375 ppm), strain rate, and temperature on the tensile ductility of typical α/β titanium alloy unnotched tensile specimens. Source: Ref 99

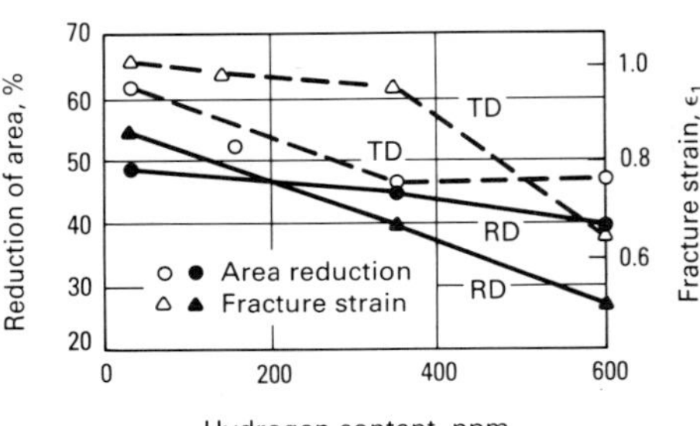

Fig. 54 Influence of hydrogen on the reduction of area at fracture and the true strain at fracture for Zircaloy-2. TD transverse direction; RD, rolling direction. Source: Ref 103

though LME can occur by either a brittle or a ductile fracture mode, both degrade the mechanical properties of the solid metal.

Role of Liquid in Crack Propagation

In the mechanism of embrittlement described above, it was implied that once a crack is nucleated subsequent crack propagation occurs mechanically with liquid absent at the crack tip or occurs by the continuous presence of the liquid-metal atoms at the propagating crack tip caused by the surface diffusion of liquid-metal atoms over chemisorbed liquid-metal atoms. The role of liquid in embrittlement should be investigated by measuring the crack growth rate as a function of temperature and stress intensity. The velocity of crack propagation, or crack growth rate, in SCC has been extensively investigated. However, such investigations in liquid metal have been reported only recently for brass (Fig. 68) and aluminum in liquid mercury. The velocity of crack growth in brass in liquid mercury was at least two orders of magnitude higher than that in an inert environment. The crack propagation activation energy for brass in liquid mercury is only 3 to 5 kcal/mol (Fig. 69). This corresponds to diffusion of liquid mercury over mercury adsorbed on brass. The very high velocities of crack propagation and very low activation energies are considered to be distinguishing features from similar investigations in SCC environments in the same metals. Such characteristics can be used to differentiate liquid

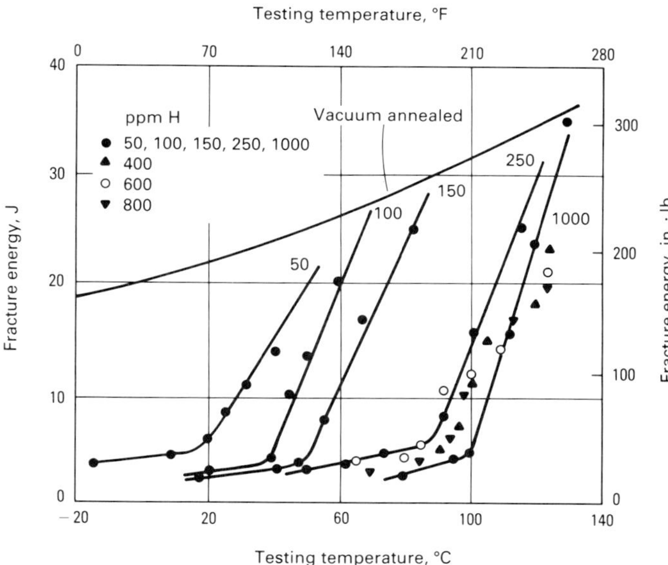

Fig. 55 Ductile-to-brittle transitions in hydrided zirconium. Source: Ref 104

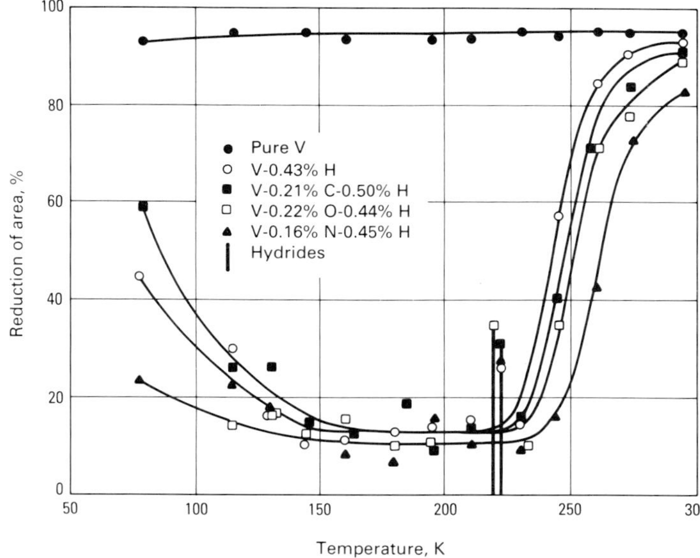

Fig. 56 Effect of hydrogen and combined carbon, nitrogen, or oxygen and hydrogen on the temperature dependence of ductility in vanadium. Source: Ref 105

metal induced crack propagation from stress-corrosion or hydrogen damage.

Occurrence of LME

Susceptibility to LME is unique to specific metals. For example, liquid gallium embrittles aluminum but not magnesium, and liquid mercury embrittles zinc but not cadmium. The equilibrium phase diagrams of most embrittlement couples show that they form simple binary systems with little or no solid solubility. Also, they form immiscible liquids in the liquid state, and they usually do not form intermetallic compounds. This is an empirical observation, and exceptions have been reported, although it is valid for many embrittlement couples.

The most critical and mandatory condition for LME is that the liquid should be in intimate contact with the surface of the solid to initiate embrittlement and should subsequently be present at the tip of the propagating crack to cause brittle failure. Even a thin oxide film several angstroms thick may prevent interaction between the solid and the liquid so that embrittlement is not observed. Thus, conditions that promote intimate contact and enhanced wetting of the liquid by the solid, such as freshly created surfaces by plastic deformation in a liquid environment, breaking of oxide films, or other such factors, will lead to LME of the solid.

The severity of embrittlement is not necessarily related to the above criteria and conditions, but it is related to the chemical nature of the embrittling species. For example, zinc is more severely embrittled by liquid gallium than by mercury. In addition, the severity of embrittlement is related to the properties of the solid and depends on such factors as strength, alloying elements, grain size, and strain rate. The severity or occurrence or nonoccurrence of embrittlement depends on the type of test and whether the solid contains stress concentrators, such as preexisting cracks or flaws, or is smooth and free of stress raisers. By increasing the possibilities of maintaining high stress concentrations, the susceptibility of the solid to LME is also increased. For example, smooth specimens of low-alloy low-strength steel are not embrittled by liquid lead, but the same steel containing a fatigue precrack is severely embrittled by liquid lead.

Certain other prerequisites must be fulfilled before fracture can initiate in a solid. For a ductile nonprecracked metal, these are an applied tensile stress, plastic deformation, and the presence of a stable obstacle to slip serving as a stress concentrator. This obstacle can be preexisting (for example, a grain boundary) or can be created during deformation (for example, a twin or kink band).

In addition, there should be an adequate supply of liquid metal to adsorb at the obstacle and subsequently at the propagating crack tip (only a few atomic monolayers of liquid-metal atoms are necessary for LME). Plastic deformation or yielding may mean localized deformation in few grains rather than general yielding of the solid metal. If a specimen contains a preexisting crack, then adsorption of the liquid at the crack tip and some tensile stress are necessary to propagate a crack. If the solid is notch brittle, LME will occur at reduced stress. However, a crack, once initiated, may propagate in a brittle manner in the absence of the liquid at the tip. Thus, it may not be necessary for the liquid to be present at the propagating crack tip. The occurrence of LME in an embrittlement couple requires good wetting of the surface of the solid by the liquid, the presence of sufficient stress concentration to initiate or propagate a crack in liquid, and possible concurrence with the empirical observations mentioned above.

Effects of Metallurgical, Mechanical, and Physical Factors

It has been shown that the prerequisites for liquid metal induced brittle fracture are the same as those for brittle fracture in an inert environment at low temperatures, that is, the need for plastic flow. A barrier for crack nucleation, sufficient tensile stress, and adsorption of liquid metal are necessary to initiate and/or propagate a crack. From the investigations reported for many embrittlement couples, including classic zinc-mercury couples, it has been concluded that adsorption-induced embrittlement can be regarded as a special case of brittle fracture that normally occurs in brittle materials at low temperatures in an inert environment (ductile fracture can also result by the adsorption-induced reductions in the shear strength of the atomic bonds at the crack tip). Thus, the effects of yield stress, grain size, strain rate, temperature, and so on, in a liquid-metal environment follow trends similar to those noted for metals tested in an inert environment.

Effects of Grain Size. The grain size dependence of fracture stress has been investigated for zinc-mercury, cadmium-gallium, brass and copper alloys in mercury, and low-carbon steel in lithium. In these and other instances, the fracture stress varied linearly with the reciprocal of the square root of grain size and followed the well-known Cottrell-Petch relationship of grain size dependence on fracture stress. For the zinc-mercury couple shown in Fig. 70, fracture is nucleation controlled in region I, whereas in region II it is propagation controlled.

Nucleation-controlled embrittlement means that once a crack is initiated it propagates to failure in the presence or absence of the liquid metal at the crack tip. In propagation-controlled failure, microcracks are formed at some low stress in a liquid-metal environment, but the microcracks propagate to failure only when a higher stress is reached. Thus, in propagation-controlled failure, unpropagated microcracks may be found beside the main fracture, and complete fracture of the specimen occurs only in the presence of liquid metal. The reverse is observed when fracture is initiation controlled.

Effects of Temperature and Strain Rate. Liquid-metal embrittlement occurs at the melting point of liquid metal (there are quite a few in-

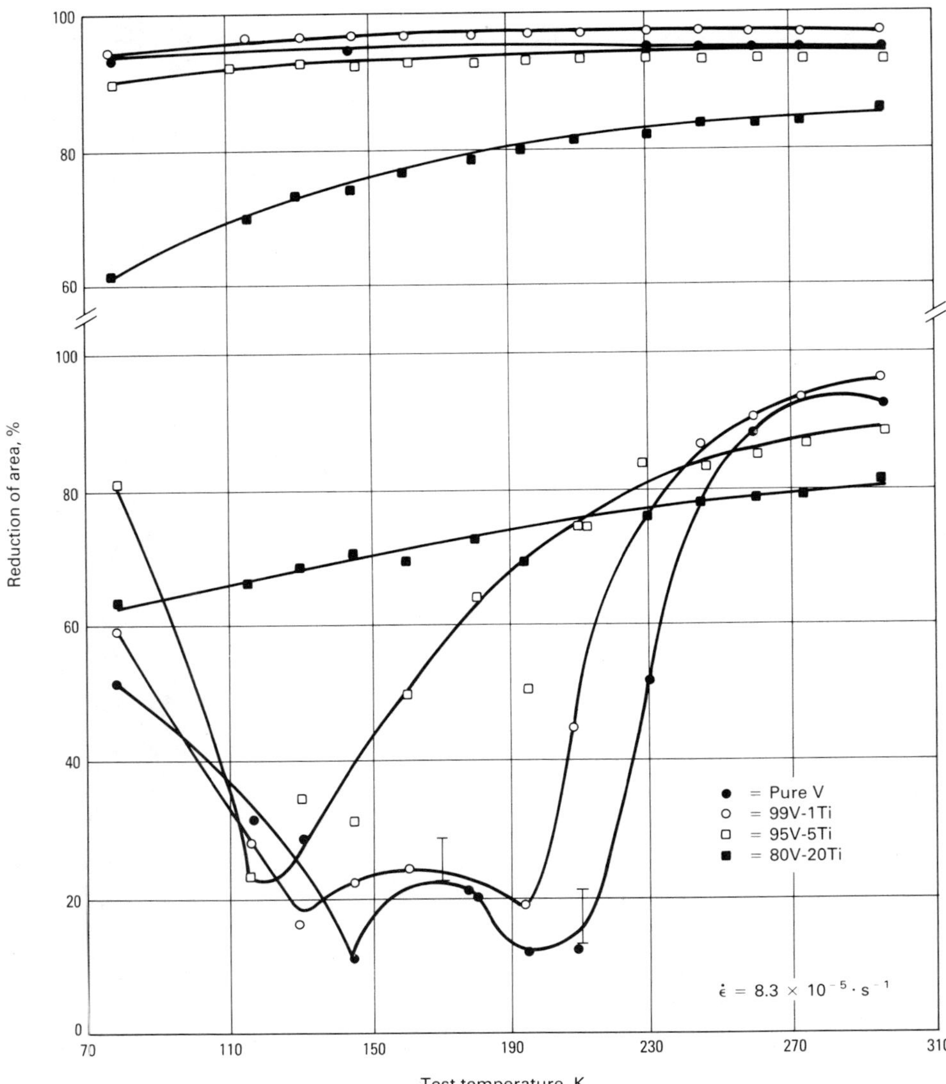

Fig. 57 Temperature dependence of reduction of area for selected hydrogen-charged (bottom) and uncharged (top) vanadium-titanium alloys. Source: Ref 106

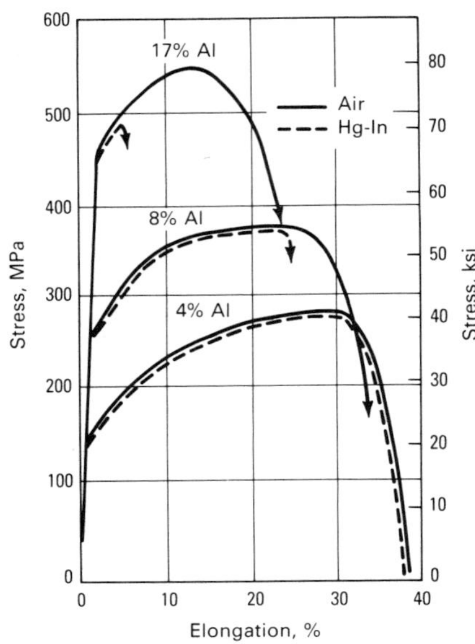

Fig. 58 Effects of environment on the yield stress and strain-hardening rate on various iron-aluminum alloys tested in air and mercury-indium solutions

Fig. 59 Liquid lead induced brittle fracture of a 75-mm (3-in.) section of a 4340 steel pressure vessel tube

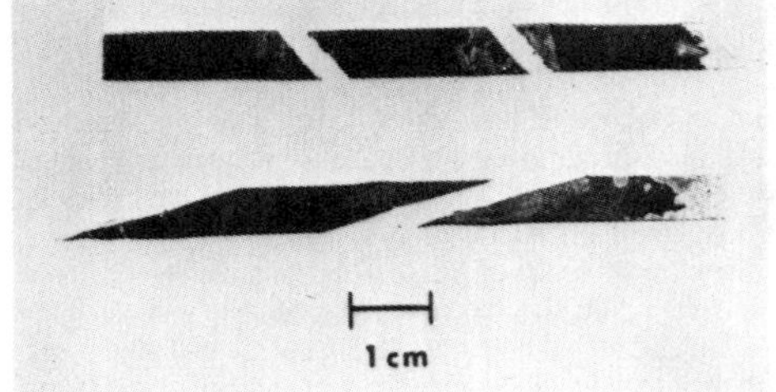

Fig. 60 Cleavage fracture of cadmium monocrystals at 25 °C (75 °F) following a coating with Hg-60In (at.%) solution

stances in which embrittlement occurs below the melting point of liquid by SMIE). In most cases, the susceptibility to embrittlement may remain essentially unchanged with temperature. At a sufficiently high test temperature, a brittle-to-ductile transition occurs when embrittlement ceases and ductility is restored to the solid. Such transitions do not occur at a sharply defined temperature and are not predictable on a theoretical basis or in terms of diffusion, dissolution, or other such embrittlement processes. It is generally accepted that a significant increase in ductility with temperature counteracts the inherent propensity for fracture in the embrittling liquid-metal environment.

In smooth steel specimens tested in liquid lead, brittle-to-ductile transition occurs at approximately 510 °C (950 °F), some 335 °C (600 °F) above the melting point of lead. In similar steel fracture mechanics type test specimens containing sharp notches tested in liquid lead, the transition occurs at approximately 650 °C (1200 °F), some 140 °C (250 °F) higher than that for smooth specimens. Transition temperature in this case depends on the presence or absence of a stress raiser in the specimen and on the type of test method used. Such brittle-to-ductile transitions have been reported for zinc-mercury, aluminum-mercury, brass-mercury, titanium-cadmium, steel-lead, and steel-lead-antimony solution embrittlement couples.

Transition temperature also varies with strain rate and grain size in a manner observed for metals tested in an inert environment. Thus, an increase in strain rate and a decrease in grain size increase the transition temperature. The effects of grain size on transition temperature have been reported for aluminum in Hg-3Sn solutions and brass-mercury embrittlement couples. The effects of strain rate on transition temperature have been reported for cadmium-gallium, zinc-mercury, aluminum-mercury, zinc-indium, brass-mercury, and titanium-cadmium embrittlement couples. Changes in the strain rate by orders of magnitude are usually required to change the transition temperature by 50 to 100 °C (90 to 180 °F). For example, in the titanium-cadmium couple, a change of approximately 60 °C (110 °F) occurs with a change of one to two orders of magnitude in the strain rate and corresponds to a comparable increase in yield stress with the strain rate. The effects of strain rate appear to be related to the increase in the yield stress. This

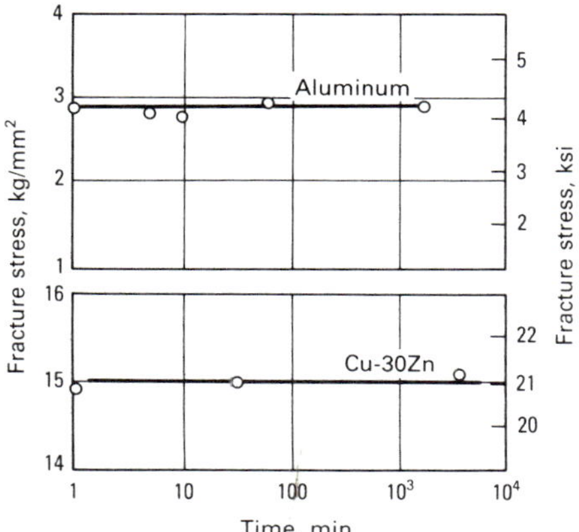

Fig. 61 Fracture stress of polycrystalline aluminum and Cu-30Zn brass as a function of exposure to liquid mercury before testing in this environment

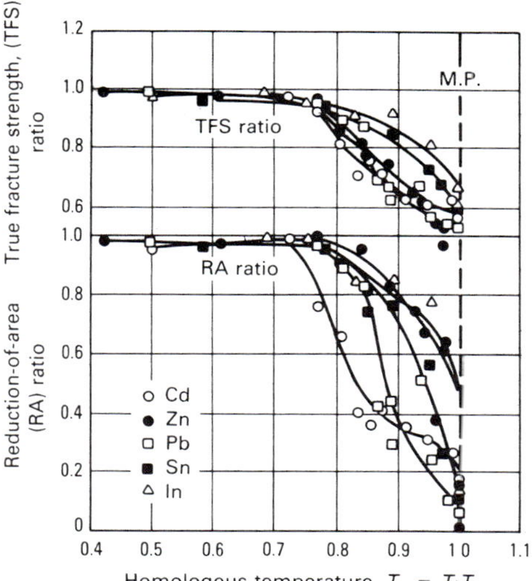

Fig. 62 Embrittlement of 4140 steel by various liquid metals below their melting point. Comparison of normalized true fracture strength (TFS) ratio and the reduction-of-area (RA) ratio as a function of homologous temperature T_H where T and T_M are the test and the melting temperature of the liquid metal, respectively. M.P., melting point

corresponds to an increase in the embrittlement susceptibility and a decrease in the brittle-to-ductile transition temperature.

Effect of Stress. Static fatigue or delayed-failure tests are conducted to study the effects of time-dependent processes, such as diffusion penetration of the grain boundary, as the cause of embrittlement. Also, such tests are performed because they simulate service conditions of constant stress in an embrittling environment as a function of time and assist in the failure analysis. Delayed failure has been reported for notch-insensitive aluminum-copper and beryllium-copper alloys in liquid mercury. Penetration of mercury did not occur, and embrittlement was independent of time to failure; but failure time increased with an increased stress level of the tests. In notch-sensitive zinc tested in mercury, the time to failure was very sensitive to the stress level, with complete failure occurring instantly at a stress slightly higher than a threshold stress. Penetration of mercury was not observed.

These tests indicate that time-dependent diffusion penetration or dissolution processes are not the cause of LME. In isolated instances, grain-boundary penetration by the liquid has been the cause of embrittlement. For example, unstressed aluminum or commercial aluminum alloys disintegrate into individual grains when contacted with liquid gallium. Liquid antimony embrittles steel, and embrittlement increases with temperature; thus, diffusional processes cause antimony to embrittle steel. Liquid copper and liquid lithium embrittle steel in a similar manner (Fig. 71). However, in these cases, both diffusional and classic LME have been observed. Such behavior can be differentiated by the temperature dependence, by the susceptibility to embrittlement, or by embrittlement occurring in solid-metal environments.

Inert Carriers and LME. In some cases, investigation of the embrittlement behavior of a potential solid-liquid metal couple is not possible. The reason is that the solid metal at temperatures just above the melting temperature of the liquid metal is either too ductile to initiate and propagate a brittle crack or the solid is excessively soluble in the liquid metal. It is possible, however, that a solid metal can be embrittled by a liquid metal of high melting point by dissolving it in an inert-carrier liquid metal of a lower melting point. Thus, the embrittling species is effectively in a liquid state at some temperature far below its melting point. In addition, it provides a means of investigating the variation in the degree of embrittlement induced in a solid metal as a function of the concentration and as a function of the chemical nature of the several active liquid-metal species dissolved separately in a common inert-carrier liquid metal.

For example, mercury does not embrittle cadmium, but indium may embrittle cadmium (melting point: 325 °C, or 617 °F). However, cadmium is quite ductile at 156 °C (313 °F), the melting temperature of indium. Mercury dissolves up to 70% In at room temperature. The indium in the mercury-indium solution at room temperature embrittles cadmium (Fig. 72). In fact, cadmium single crystals, which are ductile even at liquid helium temperature, can be embrittled by indium and cleaved in indium-mercury solutions (Fig. 60). Another example of embrittlement by inert carriers is that of Zircaloy nuclear fuel claddings used in liquid metal cooled nuclear reactors. In liquid cesium, Zircaloy claddings fail in a ductile manner; in cesium saturated with cadmium, Zircaloy is subject to brittle fracture.

The concept of inert carrier LME is useful in failure analysis where LME has occurred in a liquid solution—for example, solder rather than a pure liquid metal. The susceptibility of various species in the solder in causing LME can be investigated by incorporating each species in an inert liquid. This may suggest a means of preventing embrittlement by replacing the most potent embrittler with a nonembrittling species.

Fatigue in Liquid-Metal Environments

Most studies of LME have been concerned with the effects of tensile loading on fracture. Investigations of fatigue behavior are important because this test condition is more severe than other test conditions, including the tensile testing of metals. Thus, a solid tested in fatigue may become embrittled, but the same solid tested in tension may not exhibit embrittlement. This may be because at the high stress level corresponding to yield stress, which is a prerequisite for the occurrence of embrittlement in tough metals, the

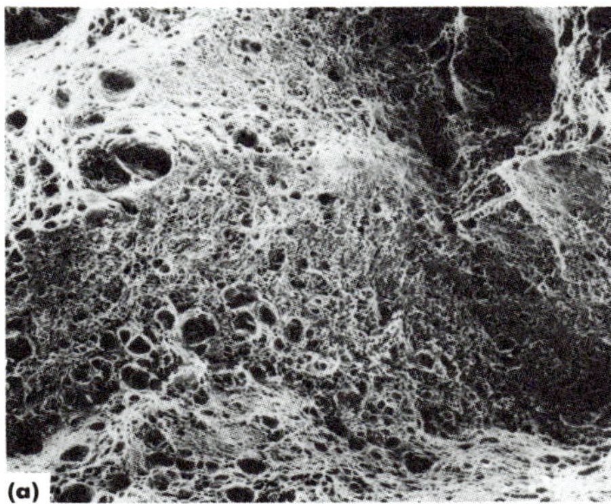

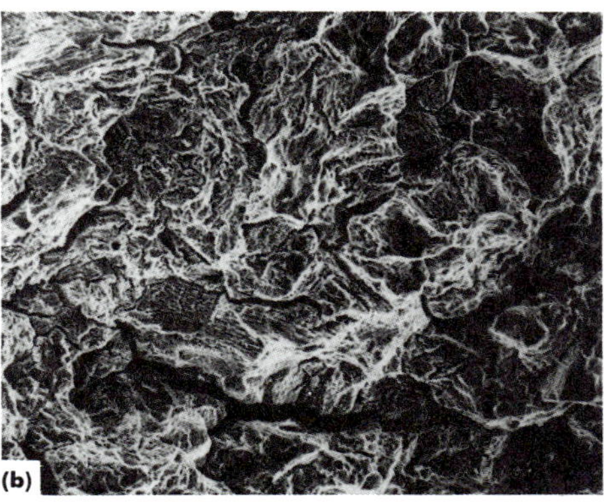

Fig. 63 Influence of lead on the fracture morphology of a 4340 steel. (a) Ductile failure after testing in argon at 370 °C (700 °F). (b) The same steel tested in liquid lead at 370 °C (700 °F) showing brittle intergranular fracture

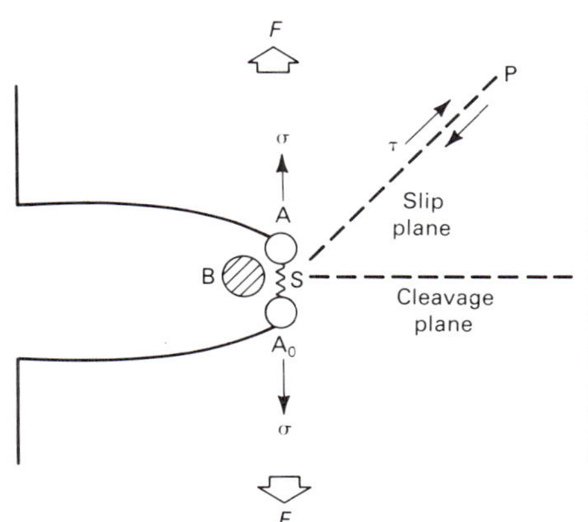

Fig. 64 Schematic illustrating displacement of atoms at the crack tip. The bond A-A_0 constitutes the crack tip, and B is the liquid-metal atom.

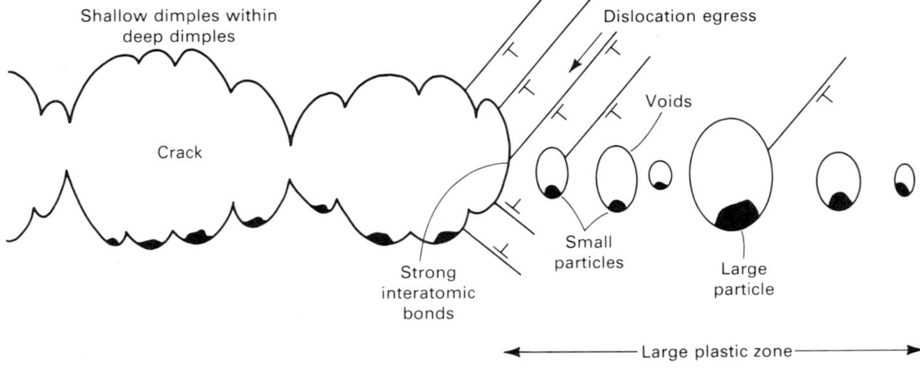

Fig. 65 Schematic illustrating the mechanisms of crack growth by microvoid coalescence. (a) Inert environment. (b) Embrittling liquid-metal environment

solid may be sufficiently ductile to prevent initiation of a crack in the liquid.

For example, smooth specimens of high-purity chromium-molybdenum low-alloy steel (yield stress: ~690 MPa, or 100 ksi) are not embrittled by liquid lead, but the same steel specimens containing a fatigue precrack are severely embrittled by liquid lead. The fatigue life in liquid lead is reduced to 25% of that in an inert argon environment. Crack propagation in a liquid-metal environment under fatigue and tensile loading can be significantly different. The stress intensity at failure of 7.7 MPa$\sqrt{\text{m}}$ (7 ksi$\sqrt{\text{in.}}$) for a 4340 steel specimen containing a fatigue precrack tested in cyclic fatigue in liquid lead was five times lower than that for the same specimen tested in tension in static fatigue and was twenty times lower than that in an inert argon environment. Furthermore, the 7.7-MPa$\sqrt{\text{m}}$ (7-ksi$\sqrt{\text{in.}}$) stress intensity was the same whether the specimen had a machined notch (0.13-mm, or 0.005-in., root radius) or had a fatigue precrack at the root of the notch; that is, embrittlement was independent of root radius. These results indicate that embrittlement is very severe and even blunt cracks can propagate to failure. Similar results have also been reported for the same steel tested in liquid mercury.

These results clearly indicate that fatigue testing of a notched or a fatigue precracked specimen provides the most severe test condition and therefore causes maximum susceptibility to embrittlement in a given liquid-metal environment. To determine whether a solid is susceptible to LME in a particular liquid, it is advisable to test the solid metal with a stress raiser or a notch in a tension-tension fatigue test.

Decrease or Elimination of LME Susceptibility

The reduction in the cohesion mechanism of embrittlement indicates that the electronic interactions, resulting in possible covalent bonding due to electron redistribution between the solid- and the liquid-metal atoms, reduces cohesion and thus induces embrittlement. Such interactions at the electronic level are the inherent properties of the interacting atoms and are therefore difficult or impossible to change in order to reduce the susceptibility to LME.

One possibility is to introduce impurity atoms in the grain boundary that have more affinity for sharing electrons with the liquid-metal atoms than for sharing electrons with the solid-metal atoms. For example, additions of phosphorus to Monel segregate to the grain boundaries and reduce the embrittlement of Monel by liquid mercury. Additions of lanthanides to internally leaded steels reduce lead embrittlement of steel. However, in general, the best alternative is to electroplate or clad the solid surface with a metal as a barrier between the embrittling solid-liquid metal couple, making sure that the barrier metal is not embrittled by the liquid metal.

Another possibility is that a ceramic or a covalent material coating on the solid surface will inhibit embrittlement. Apparently, only materials with metallic bonding are susceptible to LME. The severity of embrittlement could be reduced by decreasing the yield stress of the solid below the stress required to initiate a crack or by cladding with a high-purity metal of the alloy that is embrittled but that has a very low yield stress. Thus, Zircaloy, which is clad with a high-purity zirconium, becomes immune to embrittlement by cadmium. The obvious possibility is to replace embrittling liquid metal or solutions by nonembrittling metals or solutions.

Embrittlement of Nonferrous Metals and Alloys*

Zinc is embrittled by mercury, indium, gallium, and Pb-20Sn solder. Mercury decreases the fracture stress of pure zinc by 50% and dilute zinc alloys (0.2 at.% Cu or Ag) by five times that in an inert environment. The fracture propagation energy for a crack in zinc single crystals in mercury is 60% and in gallium is 40% of that in air. Zinc is embrittled more severely by gallium than by indium or mercury.

Aluminum. Mercury embrittles both pure and alloyed aluminum. The tensile stress is decreased by some 20%. Fatigue life of 7075 aluminum alloy is reduced in mercury, and brittle-to-ductile transition occurs at 200 °C (390 °F). Additions of gallium and cadmium to mercury increase the embrittlement of aluminum. Delayed failure by LME occurs in mercury. Dewetting of aluminum by mercury has been found to inhibit embrittlement. The possible cause of dewetting is the dissolution of aluminum by mercury and oxidation of fine aluminum particles by air and formation of aluminum oxide white flowers at the aluminum/mercury interface.

Aluminum alloys are embrittled by tin-zinc and lead-tin alloys. The embrittlement susceptibility is related to heat treatment and the strength level of the alloy. Gallium in contact with aluminum severely disintegrates unstressed aluminum alloys into individual grains. Therefore, grain-boundary penetration of gallium is sometimes used to separate grains and to study topographical features and orientations of grains in aluminum. There is some uncertainty about whether zinc embrittles aluminum. However, indium severely embrittles aluminum. Alkali metals, sodium, and lithium are known to embrittle aluminum. Aluminum alloys containing either lead, cadmium, or bismuth inclusions embrittle aluminum when impact tested near the melting point of these inclusions. The severity of embrittlement increases from lead to cadmium to bismuth.

Copper. Mercury embrittles copper, and the severity of embrittlement increases when copper is alloyed with aluminum and zinc. Antimony, cadmium, lead, and thallium are also reported to embrittle copper. The apparent absence of embrittlement of copper, if observed, should be attributed to the test conditions and metallurgical factors discussed in this article. Mercury embrittlement of brass is a classic example of LME. Embrittlement occurs in both tension and fatigue and varies with grain size and strain rate. Mercury embrittlement of brass and copper alloys has been extensively investigated. Lead and tin inclu-

*Adapted from M.G. Nicholas, A Survey of Literature on Liquid Metal Embrittlement of Metals and Alloys, in *Embrittlement by Liquid and Solid Metals*, M.H. Kamdar, Ed., The Metallurgical Society, 1984, p 27-50

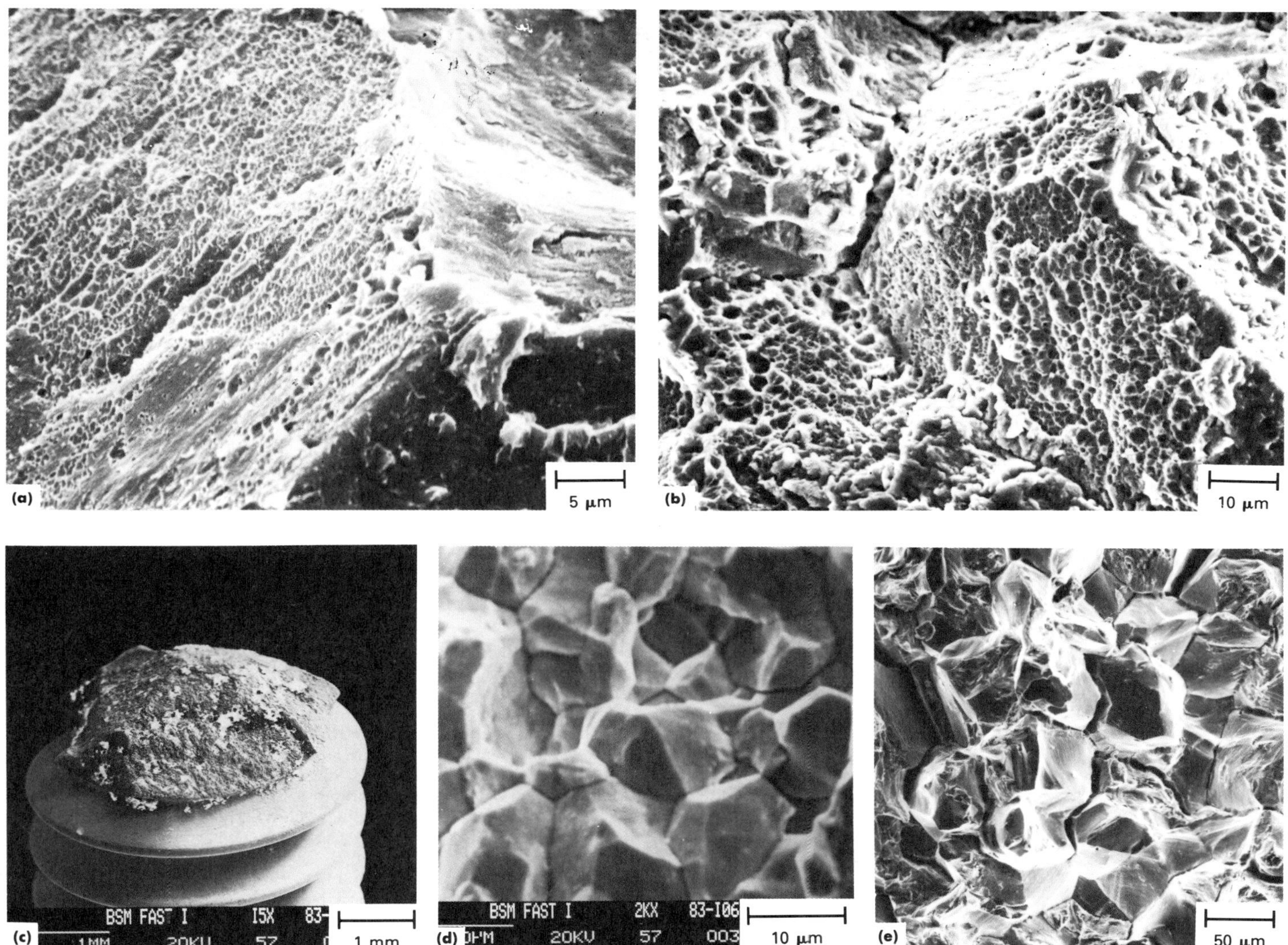

Fig. 66 Ductile and brittle fracture morphologies resulting from LME. (a) Fracture surface produced by subcritical cracking in D-6ac steel (tempered at 650 °C, or 1200 °F) in liquid mercury showing predominantly dimpled intercrystalline fracture along prior-austenite grain boundaries. (b) Fracture surface produced by rapid subcritical crack growth (~1 mm/s, or 0.4 in./s) in a Cu-1.9Be alloy in liquid mercury at 20 °C (70 °F) showing predominantly dimpled intercrystalline fracture. (c) Macrograph of a cadmium-plated fastener made from 1040 steel. Aerodynamic heating during descent of the solid fuel rocket engine resulted in brittle intergranular fracture. (d) Scanning electron micrograph of the failed machine screw shown in (c). (e) Fracture surface of a Monel specimen that failed in liquid mercury. The fracture is predominantly intergranular with some transgranular contribution.

sions in brass cause severe embrittlement when tested near the melting point of these inclusions. Lithium reduces the rupture stress and elongation at failure of solid copper. Sodium is reported to embrittle copper; however, cesium does not. The embrittling effects of bismuth for copper and their alloys are well documented.

Gallium embrittles copper at temperatures ranging from 25 to 240 °C (75 to 465 °F). Gallium also embrittles single crystals of copper. Indium embrittles copper at 156 to 250 °C (313 to 480 °F).

Other Nonferrous Materials. Tantalum and titanium alloys are embrittled by mercury and by Hg-3Zn solution. Refractory metals and alloys, specifically W-25Re, molybdenum, and Ta-10W, are susceptible to LME when in contact with molten Pu-1Ga. Cadmium and lead are not embrittled by mercury. However, indium dissolved in mercury embrittles both polycrystalline and single crystal cadium (Fig. 60). The embrittlement of titanium and its alloys by both liquid and solid cadmium is well recognized by the aircraft industry. Cadmium-plated fasteners of both titanium and steel are known to fail prematurely below the melting point of cadmium. Cadmium-plated steel bolts have good stress-corrosion resistance, but cadmium is known to crack steel. Therefore, such bolts are not recommended for use. Zinc is reported to embrittle magnesium and titanium alloys. Silver, gold, and their alloys are embrittled by both mercury and gallium. Nickel is severely embrittled by cadmium dissolved in cesium. The fracture mode is brittle intergranular with bright grain-boundary fracture.

Failure of Zircaloy tubes used as cladding material for nuclear fuel rods has been suspected to result from nuclear-interaction reaction products, such as iodine and cadmium carried by liquid cesium, which is used as a coolant in the reactor. Systematic investigation in the laboratory has shown that cadmium, both in the solid and the liquid state or as a carrier species dissolved in liquid cesium, causes severe liquid and solid metal induced embrittlement of zirconium and Zircaloy-2. Embrittlement of Zircaloy by calcium, strontium, zinc, cadmium, and iodine has also been reported.

Embrittlement of Ferrous Metals and Alloys

Embrittlement by Aluminum. Tensile and stress rupture tests have been conducted on steels in molten aluminum at 690 °C (1275 °F). For short-term tensile tests, a reduced breaking stress and reduction of area were found as compared to the values in air. In the stress rupture

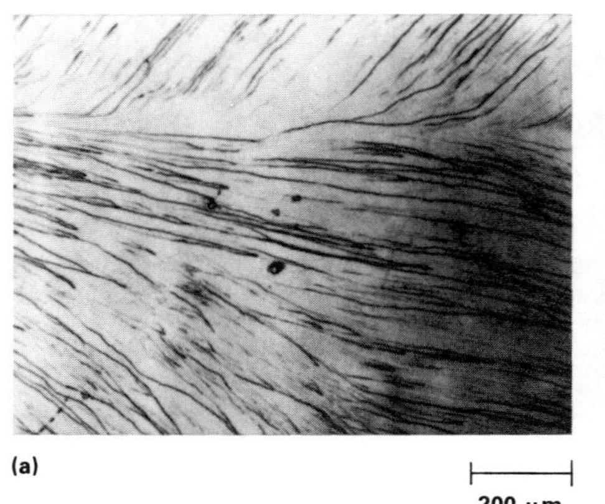

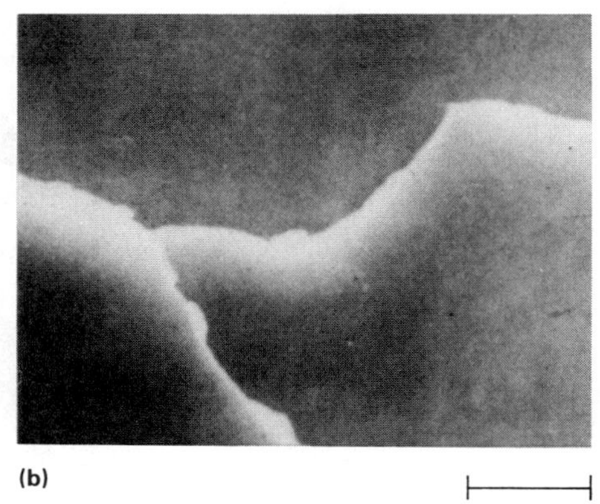

Fig. 67 Optical micrograph (a) and scanning electron micrograph (b) showing cleavage fracture surface produced by cracking zinc single crystal in liquid mercury

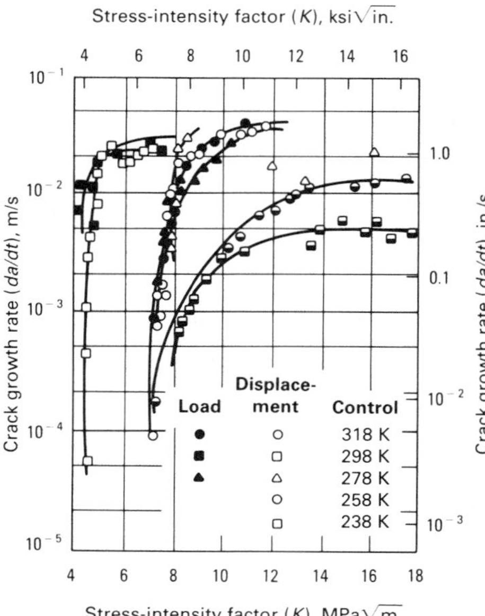

Fig. 68 Crack growth rate versus stress intensity factor for brass in liquid mercury at various temperatures under load and displacement control conditions

tests, the time to failure was dependent upon the applied stress.

Embrittlement by Antimony. AISI 4340 steel tested in fatigue in liquid Pb-35Sb at 540 °C (1000 °F) and in antimony at 675 °C (1250 °F) was very severely embrittled. The embrittlement in lead-antimony occurred 165 to 220 °C (300 to 400 °F) higher than that observed for high-purity lead. Small additions of antimony (~5 wt%) to lead had no effect on embrittlement when tested in fatigue, although 0.002 to 0.2% Sb additions have caused a significant increase in the embrittlement of smooth AISI 4140 steel specimens tested in tension. Embrittlement by antimony increases with temperature and is thought to occur by the grain-boundary diffusion of antimony in steel.

Embrittlement by Bismuth. Upon testing in liquid bismuth at 300 °C (570 °F), no embrittlement was noted in bend tests on a quenched-and-tempered steel. The stress rupture data obtained on the low-carbon steel showed that the time to failure and reduction of area increased with decreasing load, but no intercrystalline attack was noted.

Embrittlement by Cadmium. In several studies of the embrittlement of low-alloy AISI 4340 steel, embrittlement occurred at 260 to 322 °C (500 to 612 °F) but not at 204 °C (399 °F) for high-strength steel. Cracks were observed in samples loaded to 90% of their yield stress at 204 °C (399 °F). The threshold stress required for cracking decreases with an increase in temperature. Cracking at 204 °C (399 °F) was strongly dependent on the strength level of the steel, and embrittlement was not observed for strength levels less than 1241 MPa (180 ksi). Delayed failure occurs in cadmium-plated high-strength steels (AISI 4340, 4140, 4130, and an 18% Ni maraging steel). Failures were observed at 232 °C (450 °F), which is about 90 °C (160 °F) below the melting point of cadmium. Static failure limits of 10% and 60% of the room-temperature notch strength have been reported for electroplated and vacuum-deposited cadmium, respectively, at 300 °C (570 °F). A discontinuous crack propagation mode was observed, consisting of a series of crack propagation steps separated by periods of no apparent growth. This slow crack growth region was characterized by cracks along the prior-austenite grain boundaries. Once the cracks reached a critical size, a catastrophic failure occurred that was characterized by a transgranular ductile fracture.

Embrittlement of high-strength steels occurs when they are stress rupture tested in liquid and solid cadmium. Cadmium produces a progressive decrease in the reduction of area at fracture of AISI 4140 steel over the temperature range of 170 to 321 °C (338 to 610 °F). Cadmium was identified as a more potent solid-metal embrittler than lead, tin, zinc, or indium.

Embrittlement by Copper. The embrittlement of low-carbon steel by copper plate occurred at 900 °C (1650 °F) during a slow-bend test. The embrittling effects of the copper plate exceeded those encountered with brazing alloys. Similar observations have been made for plain carbon steels, silicon steels, and chromium steels at 1000 to 1200 °C (1800 to 2190 °F).

The surface cracking produced during the hot working of some steels at 1100 to 1300 °C (2010 to 2370 °F) also has several characteristics of LME. It is promoted by surface enrichment of copper and other elements during oxidation and subsequent penetration along the prior-austenite grain boundaries. Elements such as nickel, molybdenum, tin, and arsenic that affect the melting point of copper or its solubility in austenite also influence embrittlement. A ductility trough has also been noted, with no cracking produced at temperatures above 1200 °C (2190 °F). Steel plated with copper and pulse heated to the melting point of copper in milliseconds was embrittled by both liquid and solid copper. Hot tensile testing in a Gleeble testing machine at high strain rates produced severe cracking in AISI 4340 steel by both the liquid and solid copper.

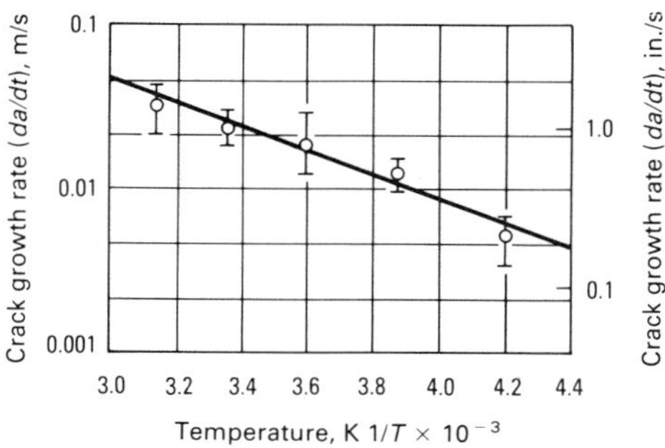

Fig. 69 Crack growth rate versus reciprocal of temperature for brass in liquid mercury. The activation energy, Q, is 4 cal/mol.

Embrittlement by Gallium. The alloy Fe-3Si is severely embrittled by gallium, as are the solid solutions of iron and AISI 4340 steel.

Embrittlement by Indium. Indium embrittles pure iron and carbon steels. Embrittlement depends on both the strength level and the microstructure. Pure iron was embrittled only at temperatures above 310 °C (590 °F), appreciably above the melting point of indium. However, other steels, such as AISI 4140 (ultimate tensile strength of 1379 MPa, or 200 ksi), were embrittled by both solid and liquid indium. Surface cracks were detected at temperatures below the melting temperature of indium. This was interpreted as a local manifestation of the underlying embrittlement mechanism, and it was assumed that the cracks must reach a critical size before the gross mechanical properties are affected.

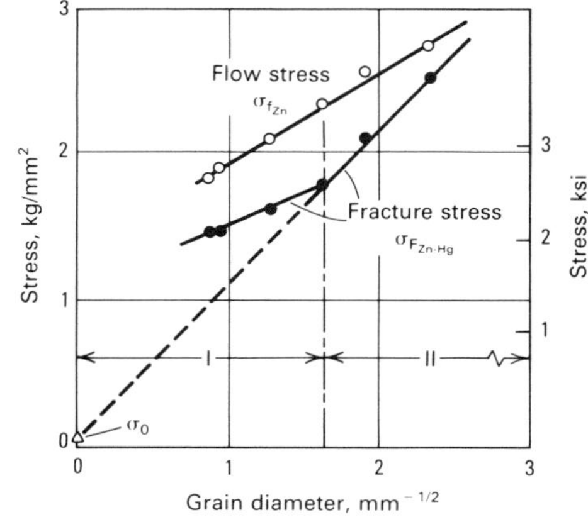

Fig. 70 Effect of grain size on LME. Variation of the flow stress of amalgamated zinc polycrystalline specimens, $\sigma_{f_{Zn}}$, and fracture stress of amalgamated zinc specimens, $\sigma_{F_{Zn\text{-}Hg}}$, as a function of grain size at 298 K

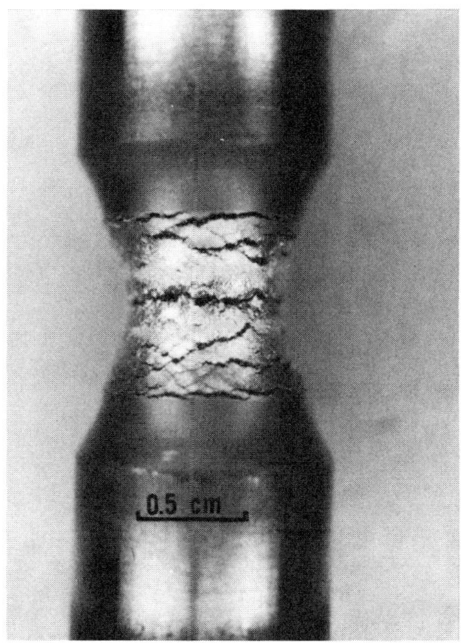

(a)

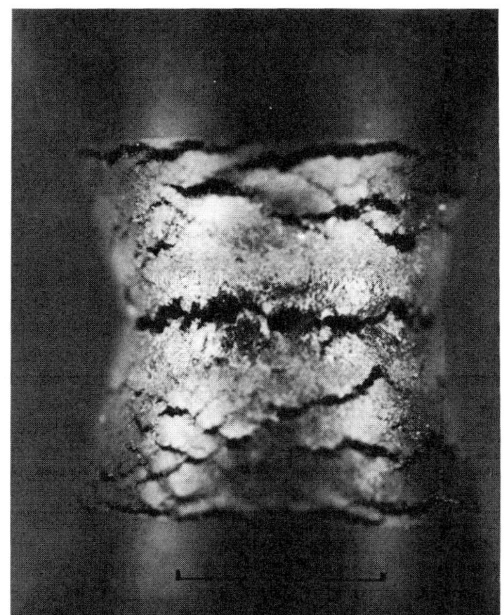

(b)

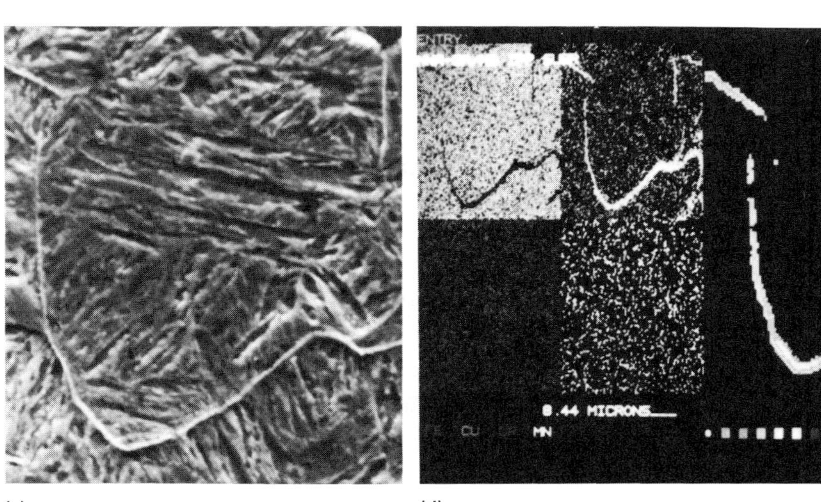

(c) (d)

Fig. 71 Copper-induced LME of 4340 steel. (a) and (b) Copper-plated specimen that was pulled at 1100 °C (2010 °F) in a Gleeble hot tensile machine showing liquid copper embrittlement of steel. (c) Scanning electron micrograph of the pulled specimen. 1600×. (d) Computer-processed x-ray map showing the presence of copper in prior-austenite grain boundaries.

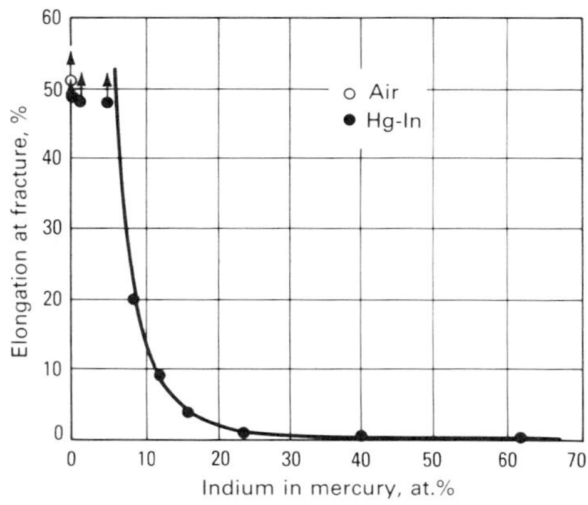

Fig. 72 Variation in ductility of polycrystalline cadmium as a function of indium content of mercury-indium surface coatings. Specimens were tested at 25 °C (75 °F) in air and in mercury-indium solution.

Embrittlement by Lead. The influence of lead on the embrittlement of steel has been extensively investigated and has been found to be sensitive to both composition and metallurgical effects. The studies fall into two major classifications:

- LME due to contact with an external source of liquid lead
- Internal LME in which the lead is present internally as inclusions or a minor second phase, as in leaded steels

Both external and internal lead-induced embrittlement exhibit similar characteristics, but for the purpose of simplicity, each will be treated individually.

LME Due to External Lead. AISI 4145 and 4140 steels exposed to pure lead exhibit classical LME, as shown by substantial decreases in both the reduction of area and the elongation at fracture. The fracture stress and the reduction of area decreased at temperatures considerably below the melting point of lead and varied continuously through the melting point, suggesting that the same embrittlement mechanism is operative for both solid- and liquid-metal environments.

Additions of zinc, antimony, tin, bismuth, and copper increase the embrittling potency of lead. Additions of up to 9% Sn, 2% Sb, or 0.5% Zn to lead increased the embrittlement of AISI 4145 steel. In some cases, the embrittlement and failure occurred before the UTS was reached. The extent of embrittlement increases with increasing impurity content. No correlation was observed between the degree of embrittlement and wettability. The lead-tin alloys readily wetted steels, but the more embrittling lead-antimony alloys did not.

LME Due to Internal Lead. Leaded steels are economically attractive because lead increases the machining speed and the lifetime of the cutting tools. The first systematic investigation of embrittlement of leaded steels was reported in 1968 (Mostovoy and Breyer). Embrittlement characteristics similar to those promoted by external lead were observed. The degradation in the ductility began at approximately 120 °C (215 °F) below the melting point of lead, with the embrittlement trough present from 230 to 454 °C (446 to 849 °F). This was followed by a reversion to ductile behavior at about 480 °C (895 °F).

The severity of the embrittlement and the brittle-to-ductile transition temperature T_R have been shown to be dependent upon the strength level of the steel, with the degree of embrittlement and T_R increasing with strength level. In these studies, intergranular fracture was produced, which was propagation controlled at low temperatures and nucleation controlled at high temperatures. The degree of embrittlement was critically dependent on the lead composition, and the influence of trace impurities completely masked any variations due to different carbon and alloy compositions of the steel. Lead embrittlement of a steel compressor disk was induced by bulk lead contents of 0.14 and 6.22 wt%. The lead is associated with the nonmetallic inclusions, and upon yielding, microcracks form at the weak inclusion/matrix interface, releasing a source of embrittling agent to the crack tip that aids subsequent propagation. An electron microprobe analysis of the nonmetallic inclusions identified the presence of zinc, antimony, tin, bismuth, and arsenic. With the exception of arsenic, all these trace impurities have been shown to have a significant effect on the external LME of steel.

The two most promising methods of suppressing LME are the control of sulfide composition and morphology and the cold working of the steel. The addition of rare-earth elements to the steel melt modifies the sulfide morphology and composition and can eliminate LME.

Embrittlement by Lithium. Exposure of AISI 4340 steel to lithium at 200 °C (390 °F) resulted in static fatigue, with the time to failure depending on the applied stress. A decreasing fracture stress and elongation to fracture were noted with increasing UTS of variously treated

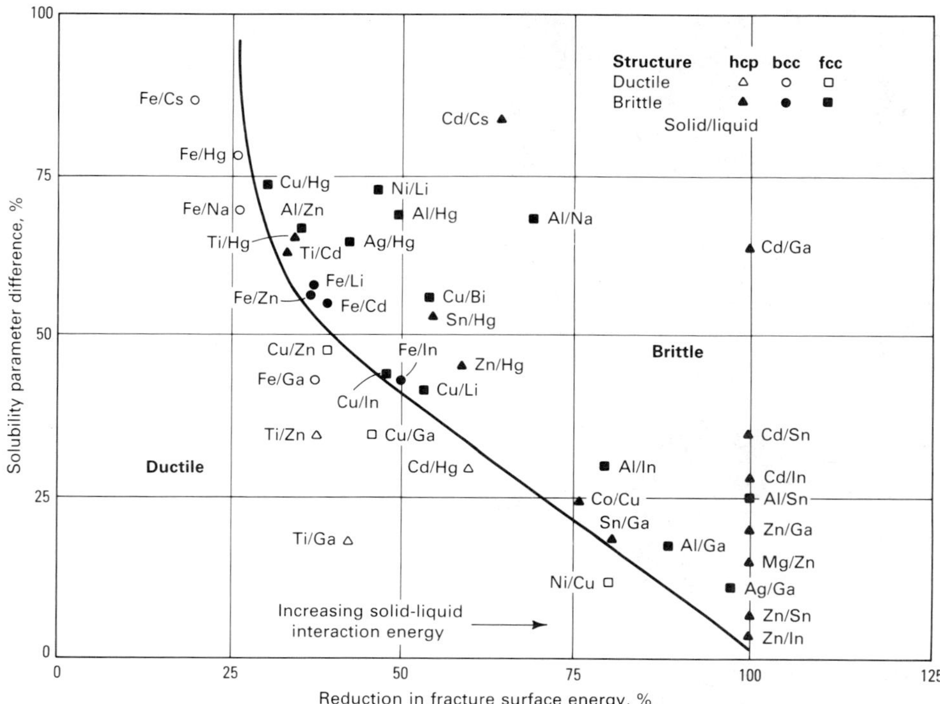

Fig. 73 Calculated reduction in the fracture surface energy relating to solubility parameter for many solid-liquid embrittlement couples. Note that the curve separates embrittlement couples from nonembrittled solid-liquid metal couples.

steels, and catastrophic failure occurred for those steels with tensile strengths exceeding 1034 MPa (150 ksi). The tensile ductility of low-carbon steel at 200 °C (390 °F) was drastically reduced in lithium, with intergranular failure after 2 to 3% elongation, but there was no effect on the yield stress or the initial work-hardening behavior. The fracture stress was shown to be a linear function of $d^{-1/2}$, where d is the average grain diameter, in accordance with the Petch relationship.

Embrittlement by Mercury. It has been shown that mercury embrittlement is crack nucleation controlled and can be induced in low-carbon steel samples by the introduction of local stress raisers. The fracture toughness of a notched 1Cr-0.2 Mo steel was significantly decreased upon testing in mercury. The effective surface energy required to propagate the crack was 12 to 16 times greater in air than in mercury. The fatigue life of 4340 steel in mercury is reduced by three orders of magnitude as compared to that in air.

The addition of solutes (cobalt, silicon, aluminum, and nickel) to iron, which reduced the propensity for cross-slip by decreasing the number of active slip systems and changed the slip mechanism from wavy to planar glide, increased the susceptibility to embrittlement. Iron alloys containing 2% Si, 4% Al, or 8% Ni and iron containing 20% V or iron containing 49% Co and 2% V have been shown to be embrittled by mercury in unnotched tensile tests. The degree of embrittlement behavior was extended to lower alloy contents by using notched samples. No difference in the embrittlement potency of mercury or a saturated solution of indium in mercury was noted.

Effect of Selenium. Selenium had no embrittling effect on the mechanical properties of a quenched-and-tempered steel (UTS: ~1460 MPa, or 212 ksi) bend tested at 250 °C (480 °F).

Embrittlement by Silver. Silver had no significant effect on the mechanical properties of a range of plain carbon steels, silicon steels, and chromium steels tested by bending at 1000 to 1200 °C (1830 to 2190 °F). However, a silver-base filler metal containing 45% Ag, 25% Cd, and 15% Sn has been reported to embrittle A-286 heat-resistant steel in static-load tests above and below 580 °C (1076 °F), the melting point of the alloy.

Effect of Sodium. Unnotched tensile properties of low-carbon steel remained the same when tested in air and in sodium at 150 and 250 °C (300 and 480 °F). Similarly, Armco iron, low-carbon steel, and type 316 steel were not embrittled by sodium at 150 to 1600 °C (300 to 2910 °F).

Embrittlement by Solders and Bearing Metals. A wide range of steels are susceptible to embrittlement and intercrystalline penetration by molten solders and bearing metals at temperatures under 450 °C (840 °F). Tensile tests have

Table 5 Summary of embrittlement couples

Solid		Liquid																				
		Hg		Cs	Ga		Na	In		Li	Sn		Bi		Tl	Cd	Pb		Zn	Te	Sb	Cu
		P	A	P	P	A	P	P	A	P	P	A	P	A	P	P	P	A	P	P	P	P
Sn	P	X	...	...	X	...	...	...	...	...	...	...	...	...	...	...	...	...	...	...	...	...
Bi	P	X	...	...	...	...	...	...	...	...	...	...	...	...	...	...	...	...	...	...	...	...
Cd	P	...	...	X	X	X	...	...	...	...	X	...	...	...	...	...	...	...	...	...	...	...
Zn	P	X	X	...	X	X	...	X	X	...	X	...	...	...	...	...	X	...	...	...	...	...
	LA	...	X	...	...	...	...	...	...	...	...	...	...	...	...	...	...	...	...	...	...	...
Mg	CA	...	...	...	...	...	X	...	...	...	...	...	...	...	...	...	...	...	X	...	...	...
Al	P	X	X	...	X	...	...	X	...	...	...	X	X	...	...	X	X	...	...	...	...	...
	CA	X	X	...	X	...	X	X	...	...	X	...	X	X	...	...	X	...	X	...	...	...
Ge	P	...	...	...	X	...	...	X	...	...	X	...	X	...	X	X	X	...	...	...	X	...
Ag	P	X	X	...	X	X	...	...	...	X	...	...	...	...	...	...	...	...	...	...	...	...
	LA	...	...	...	...	...	...	...	...	X	...	...	...	...	...	...	...	...	...	...	...	...
Cu	CP	X	X	...	...	...	X	...	...	X	...	...	X	X	...	...	X	...	...	...	...	...
	LA	X	...	...	X	...	...	...	...	...	...	...	...	...	...	...	...	...	...	...	...	...
	CA	X	X	...	X	...	...	X	X	X	X	(?)	X	...	...	...	X	...	...	...	...	...
Ni	P	X	...	...	...	...	...	...	...	X	...	...	...	...	...	...	X	...	...	...	...	...
	LA	...	...	...	...	...	...	...	...	X	...	...	...	...	...	...	...	...	...	...	...	...
	CA	...	...	...	...	...	...	...	...	...	X	...	...	...	...	...	X	...	...	...	...	...
Fe	P	...	...	...	...	...	...	...	...	...	...	...	...	...	...	...	...	...	...	...	...	X
	LA	X	X	...	X	...	...	X	X	X	...	...	...	...	...	...	...	...	...	...	...	...
	CA	...	...	...	...	...	...	X	...	X	X	...	...	...	...	X	X	X	X	X	X	X
Pd	P	...	...	...	...	...	...	...	...	X	...	...	...	...	...	...	...	...	...	...	...	...
	LA	...	...	...	...	...	...	...	...	X	...	...	...	...	...	...	...	...	...	...	...	...
Ti	CA	...	X	...	...	...	...	...	...	...	...	...	...	...	...	X	...	...	...	...	...	...

P, element (nominally pure); A, alloy; C, commercial; L, laboratory

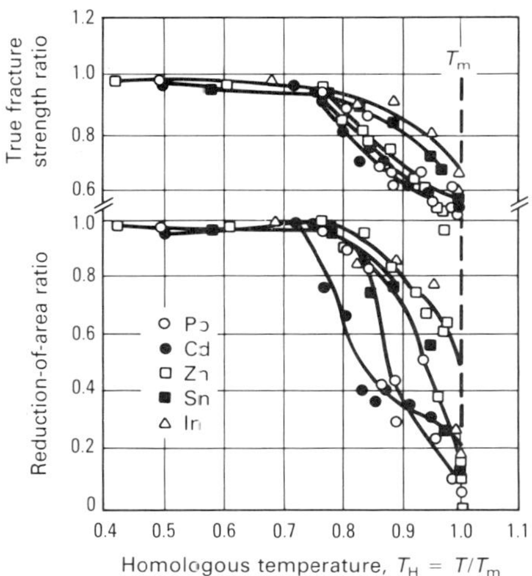

Fig. 74 Embrittlement of 4140 steel by various liquid metals below their melting point and their effects on normalized true fracture strength and reduction of area as a function of homologous temperature T_H. T and T_m are the test and melting temperatures of the liquid metal. Source: Ref 108

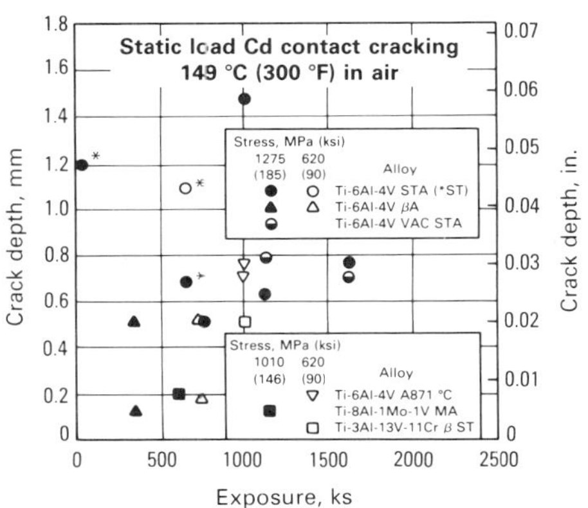

Fig. 75 Crack depth as a function of exposure time to solid cadmium environment for various titanium alloys at three stress levels. STA, solution treated and aged; ST, solution treated; VAC STA, vacuum solution treated and aged; βA, β-annealed; A, annealed; MA, mill annealed. Source: Ref 111

Fig. 76 Crack morphology for Ti-6Al-4V (solution treated and aged). (a) Typical specimen with multiple cracks in the indented area. (b) Fracture surface of (a) showing the depth of cadmium-induced cracking. (c) Cross section showing mixed intergranular cracking and α cleavage in cadmium-induced crack. Etched with Kroll's reagent. (d) TEM fractograph (two-stage replica) of similar area showing intergranular cracking and cleavage. Courtesy of D.A. Meyn

revealed embrittlement as a reduction in ductility. The embrittlement increased with grain size and strength level of the steel, except for temper-embrittled steels.

The tensile strength and ductility of carbon steel containing 0.13% C were decreased upon exposure to the molten solders and bearing alloys. The embrittlement was concomitant with a change to a brittle intergranular fracture mode and penetration along prior-austenite grain boundaries. Ductile failure was observed with samples tested in air or in the liquid metal at temperatures exceeding 450 °C (840 °F). No intercrystalline penetration of solder was noted in carbon steels containing 0.77% and 0.14% C at 950 °C (1740 °F).

It has been reported that solder embrittles steel more than Woods metal (a bismuth-base fusible alloy containing lead and tin), particularly if it contains 4% Zn. The bearing metals produced embrittlement similar to the solder containing 4% Zn alloy.

Embrittlement by Tellurium. Tellurium-associated embrittlement has been reported for carbon and alloy steels. Hot shortness occurs in AISI 12L14 + Te steel, with the most pronounced loss in ductility between 810 and 1150 °C (1490 and 2100 °F), embrittlement being most severe at 980 °C (1795 °F). The embrittlement has been shown to occur by the formation of a lead-telluride film at the grain boundary, which melts at 923 °C (1693 °F). The mechanical test data and the examination of fracture surfaces by Auger electron spectroscopy (AES) and scanning electron microscopy (SEM) indicated LME of steel by the lead-telluride compound.

Effect of Thallium. Thallium had no embrittling effects on the mechanical properties of a quenched-and-tempered steel (UTS: ~1460 MPa, or 212 ksi) tested in bending at 325 °C (615 °F).

Embrittlement by Tin. The embrittling effect of tin has been observed in a range of austenitic and nickel-chromium steels, the degree of embrittlement increasing with their strength level. Embrittlement depends on the presence of a tensile stress and is associated with intercrystalline penetration.

The embrittlement by solid tin occurs at approximately 120 °C (215 °F) below its melting point. The fracture surfaces exhibited an initial brittle zone perpendicular to the tensile axis that followed the prior-austenite grain boundaries. Layers of intermetallic compound present at the steel/tin interface did not impede the embrittlement process. Embrittlement has been observed in delayed-failure tests down to 218 °C (424 °F) (14 °C, or 25 °F, below the melting point of tin); however, in tensile tests, embrittlement by solid tin was effective at temperatures as low as 132 °C (270 °F), which is 100 °C (180 °F) below melting point.

AISI 3340 steel doped with 500 ppm of phosphorus, arsenic, and tin has been tested in the presence of tin while in the segregated (temper embrittled) and the unsegregated states. Temper-embrittled steels were found to be more susceptible to embrittlement than steel heat treated to a nontempered state.

A lower fatigue limit and lifetime at stresses below the fatigue limit for low-carbon steel and for 18-8 stainless steel have been noted when tested in tin at 300 °C (570 °F). The exposure time to the tin before testing had no influence on the fatigue life of the steel.

Table 6 Occurrence of SMIE in steels

Base metal	Embrittler (melting point)	Onset of embrittlement °C	Onset of embrittlement °F	Test type(b)	Specimen type(c)
1041	Pb (327 °C, or 621 °F)	288	550	ST	S
1041 leaded	Pb	204	399	ST	S
1095	In (156 °C, or 313 °F)	100	212	ST	S
3340	Sn (232 °C, or 450 °F)	204	399	ST	N
	Pb	316	601	ST	N
4130	Cd (321 °C, or 610 °F)	300	572	DF	N
4140	Cd	300	572	DF	N
	Pb	204	399	ST	S
	Pb-Bi (NA)(a)	Below solidus		ST	S
	Pb-Zn (NA)	Below solidus		ST	S
	Zn (419 °C, or 786 °F)	254	489	DF	N
	Sn	218	424	DF	N
	Cd	188	370	DF	N
	Pb	160	320	DF	N
	In	Room temperature		DF	N
	Pb-Sn-Bi (NA)	Below solidus		ST	S
	In	80	176	DF	S
	Sn	204	399	ST	S
	Sn-Bi (NA)	Below solidus		ST	S
	Sn-Sb (NA)	Below solidus		ST	S
	In	110	230	DF	S
	In	93	199	DF	S
	In-Sn (118 °C, or 244 °F)	93	199	DF	S
4145	Sn	204	399	ST	S
	In	121	250	ST	S
	Pb-4Sn (NA)	204	399	ST	S
	Pb-Sn (NA)	204	399	ST	S
	Pb-Sb (NA)	204	399	ST	S
	Pb	288	550	ST	S
4145 leaded	Pb	204	399	ST	S
4340	Cd	260	500	DF	N
	Cd	300	572	DF	N
	Cd	38	100	DF	S
	Zn	400	752	DF	N
4340M	Cd	38	100	DF	S
8620	Pb	288	550	ST	S
8620 leaded	Pb	204	399	ST	S
A-4	Pb	288	550	ST	S
A-4 leaded	Pb	204	399	ST	S
D6ac	Cd	149	300	DF	N

(a) NA, data not available. (b) ST, standard tensile test; DF, delayed-failure tensile test. (c) S, smooth specimen; N, notched specimen. Courtesy of Dr. A. Druschitz

Embrittlement of Austenitic Steels by Zinc. Two main types of interaction of zinc and austenitic stainless steel have been observed. Type I relates to the effects on unstressed material in which liquid-metal penetration/erosion is the major controlling factor, and Type II relates to stressed materials in which classic LME is observed.

Type I Embrittlement. Zinc slowly erodes unstressed 18-8 austenitic stainless steel at 419 to 570 °C (786 to 1058 °F) and penetrates the steel, with the formation of an intermetallic β nickel-zinc compound at 570 to 750 °C (1060 to 1380 °F). At higher temperatures, penetration along the grain boundaries occurs, with a subsequent diffusion of nickel into the zinc-rich zone. This results in a nickel-exposed zone adjacent to the grain boundaries, reducing the stability of the γ phase and causing it to transform to an α-ferrite; the associated volume change of the $\gamma \rightarrow \alpha$ transformation produces an internal stress that facilitates fracture along the grain boundaries. Similar behavior has been observed in an unstressed 316C stainless steel held 30 min at 750 °C (1380 °F), in which penetration occurred to a depth of 1 mm (0.4 in.), and in an unstressed type 321 steel held 2 h at 515 °C (960 °F), in which a penetration of 0.127 mm (0.005 in.) was observed.

Type II embrittlement occurs in stainless steel above 750 °C (1380 °F) and is characterized by an extremely fast rate of crack propagation that is several orders of magnitude greater than that of Type I, with cracks propagating perpendicular to the applied stress. In laboratory tests, an incubation period was observed before the propagation of Type II cracks, suggesting that they may be nucleated by Type I cracks formed during the initial contact with zinc.

At 800 °C (1470 °F), a stressed type 316C stainless steel failed catastrophically when coated with zinc. Cracking was produced at a stress of 57 MPa (8 ksi) at 830 °C (1525 °F) and 127 MPa (18 ksi) at 720 °C (1330 °F), but failure was not observed at a stress of 16 MPa (2 ksi) at 1050 °C (1920 °F).

Liquid-metal embrittlement may be produced by the welding of austenitic steels in the presence of zinc or zinc-base paints. Intercrystalline cracking has been observed in the heat-affected zone in areas heated from 800 to 1150 °C (1470 to 2100 °F), and electron microprobe analysis has been used to identify the grain-boundary enrichment of nickel and zinc, together with the formation of a low-melting nickel-zinc compound. The embrittlement of sheet samples of austenitic steel coated with zinc dust dye and zinchromate primer occurs at stresses of the order of 20 MPa (3 ksi).

Embrittlement of Ferritic Steels by Zinc. Embrittlement of ferritic steels and Armco iron by molten zinc has been reported in the temperature range of 400 to 620 °C (750 to 1150 °F). Long exposures and intercrystalline attack were needed to cause a reduction in the elongation to fracture; an iron-zinc intermetallic layer was formed that inhibited embrittlement until the layer was ruptured. High-alloy ferritic steels exhibit embrittlement by zinc at temperatures above 750 °C (1380 °F).

Delayed failure occurs in steel in contact with solid zinc at 400 °C (750 °F), which is 19 °C (34 °F) below the melting point of zinc. The slow crack growth region is characterized by intergranular mode of cracking.

Exposing AISI 4140 steel to solid zinc results in a decrease in the reduction of area and in fracture stress at 265 °C (510 °F), with no significant changes in the other mechanical properties. The tensile fracture initially propagates intergranularly, with the final failure occurring by shear. Liquid zinc embrittles 4140 steel at 431 °C (808 °F), and it has been shown that zinc is present at the crack tip. Figure 73 shows a compilation of liquid-metal embrittling and nonembrittling couples based on the theoretical calculations of the solubility parameter and the reduction in the fracture-surface energy. The embrittlement curve is separated by brittle and ductile fracture. A concise summary of embrittlement couples is provided in Table 5. Both pure and alloyed solids are listed.

Solid Metal Induced Embrittlement

M.H. Kamdar
Benet Weapons Laboratory
U.S. Army Armament Research, Development, and Engineering Center

Embrittlement occurs below the melting temperature of the solid in certain LME couples. The severity of embrittlement increases with temperature, with a sharp and significant increase in severity at the melting point, T_m, of the embrittler (Fig. 74). Above T_m, embrittlement has all the characteristics of LME. The occurrence of embrittlement below the T_m of the embrittling species is known as solid metal induced embrittlement of metals.

Although SMIE of metals has not been mentioned or recognized as an embrittlement phenomenon in industrial processes, many instances of loss in ductility, strength, and brittle fracture of metals and alloys have been reported for electroplated metals and coatings or inclusions of low-melting metals below their T_m (Ref 109). Delayed failure of cadmium-plated high-strength steel has been observed below the T_m of cadmium (Ref 110, 111). Accordingly, cadmium-plated steel bolts, despite their excellent resistance to corrosion, are not recommended for use above 230 °C (450 °F). Notched tensile specimens of various steels are embrittled by solid cadmium. Solid cadmium, silver, and gold embrittle titanium. Leaded steels are embrittled by solid lead, with considerable loss in ductility below the T_m of lead; this phenomenon accounts for numerous

Table 7 Occurrence of SMIE in nonferrous alloys

All test specimens were smooth type.

Base metal	Embrittler (melting point)	Onset of embrittlement °C	Onset of embrittlement °F	Test type(a)
Ti-6Al-4V	Cd (321 °C, or 610 °F)	38	100	DF
	Cd	149	300	BE
Ti-8Al-1Mo-1V	Cd	38	100	DF
	Cd	149	300	BE
Ti-3Al-14V-11Cr	Cd	149	300	BE
Ti-6Al-6V-2Sn	Cd	149	300	BE
	Ag (961 °C, or 1762 °F)	204–232	399–450	BE
	Au (1053 °C, or 1927 °F)	204–232	399–450	BE
Cu-Bi(b)	Hg (−39 °C, or −38 °F)	−84	−119	ST
Cu-Bi(c)	Hg	−87	−125	ST
Cu-3Sn(c)	Hg	−48	−54	ST
Cu-1Zn(c)	Hg	−46	−51	ST
Tin bronze	Pb (327 °Cor 621 °F)	200	392	IM
Zinc	Hg	−51	−60	ST
Inconel	In (156 °C, or 313 °F)	Room temperature		RE
Zircaloy-2	Cd	300	572	ST

(a) DF, delayed-failure tensile test; BE, bend test; ST, standard tensile test; IM, impact tensile test; RE, residual stress test. (b) Heat treated to uniformly distribute solutes. (c) Heat treated to segregate solutes to grain boundaries. Courtesy of Dr. A. Druschitz

Table 8 Delayed failure in SMIE and LME systems

Base metal	Liquid	Solid
Type A behavior: delayed failure observed		
4140 steel	Li	Cd
4340 steel	Cd	In
4140 steel	In	Cd
4140 steel		Pb
4140 steel		Sn
4140 steel		In
4140 steel		Zn
2024 Al	Hg	. . .
2424 Al	Hg-3Zn	. . .
7075 Al	Hg-3Zn	. . .
5083 Al	Hg-3Zn	. . .
Al-4Cu	Hg-3Zn	. . .
Cu-2Be	Hg	. . .
Cu-2Be	Hg	. . .
Type B behavior: delayed failure not observed		
Zn	Hg	. . .
Cd	Hg	. . .
Cd	Hg + In	. . .
Ag	Hg + In	. . .
Al	Hg	. . .

Courtesy of Dr. A. Druschitz

elevated-temperature failures of leaded steels, such as radial cracking of gear teeth during induction-hardening heat treatment, fracture of steel shafts during straightening at elevated temperature, and heat treatment failure of jet-engine compressor disks. Liquid and solid cadmium metal environments, as well as cadmium dissolved in inert nonembrittling coolant liquid, serve to embrittle the Zircaloy-2 nuclear fuel cladding tubes used in nuclear reactors (Ref 112). Inconel vacuum seals are cracked by solid indium. These reports of brittle failure clearly indicate the importance of SMIE in industrial processes.

Solid metal induced embrittlement was first recognized and investigated in the mid-1960s and early 1970s by studying the delayed failure of steels and titanium in a solid-cadmium environment (Ref 113–115). Results of these studies are shown in Fig. 75 and 76. A systematic investigation of SMIE of steel by a number of solid-metal embrittling species (Fig. 74) found that solid metal as an external environment can cause embrittlement and for steels represents a generalized phenomenon of embrittlement.

Solid metal induced embrittlement can also occur when the embrittling solid is an internal environment, that is, present in the solid as an inclusion (Ref 108). It has been clearly demonstrated that internally leaded steel is embrittled below the T_m of the lead inclusion of steel (Ref 108).

Brittle fracture in LME and SMIE is of significant scientific interest because the embrittling species are in the vicinity of or at the tip of the crack and are not transported by dislocations or by slip due to plastic deformation into the solid, as is hydrogen in the hydrogen embrittlement of steels. Also, embrittling species are less likely to be influenced by the effects of grain-boundary impurities, such as antimony, phosphorus, and tin, which cause significant effects on the severity of hydrogen and temper embrittlement of metals. Investigations of SMIE and LME, therefore, can be interpreted less ambiguously than similar effects in other environments, such as hydrogen and temper embrittlement of metals. Thus, solid-liquid environmental effects provide a unique opportunity to study embrittlement mechanisms in a simple and direct manner under controlled conditions.

It is conceivable that a common mechanism may underlie solid, liquid, and gas phase induced embrittlement. The interactions at the solid/environment interface and the transport of the embrittling species to the crack tip may characterize a specific embrittlement phenomenon. A study of SMIE and LME may provide insights into the mechanisms of hydrogen and temper embrittlement. It is apparent that the phenomenon of SMIE is of both industrial and scientific importance. A review of the investigations of the occurrence and mechanisms of SMIE follows.

Characteristics of SMIE

To date, SMIE has been observed only in those couples in which LME occurs, suggesting that LME is a prerequisite for the occurrence of SMIE. Solid metal induced embrittlement may occur in the absence of LME if a brittle crack cannot be initiated at the T_m of the embrittler. A recent compilation of SMIE couples is given in Tables 6 and 7, which show that all solid-metal embrittlers are also known to cause LME.

Solid metal induced embrittlement and liquid-metal embrittlement are strikingly similar phenomena. The prerequisites for SMIE are the same as those for LME: intimate contact between the solid and the embrittler, the presence of tensile stress, crack nucleation at the solid/embrittler interface from a barrier (such as a grain boundary), and the presence of embrittling species at the propagating crack tip. Also, metallurgical factors that increase brittleness in metals, such as grain size, strain rate, increases in yield strength, solute strengthening, and the presence of notches or stress raisers, all appear to increase embrittlement. The susceptibility to SMIE is stress and temperature sensitive and does not occur below a specific threshold value. Embrittlement by delayed failure is also observed for both LME and SMIE (Table 8).

Some differences also exist. Multiple cracks are formed in SMIE; in LME, a single crack usually propagates to failure. The fracture in SMIE is propagation controlled. However, crack propagation rates are at least two to three orders of magnitude slower than in LME. Brittle intergranular fracture changes to ductile shear because of the inability of the embrittler to keep up with the propagating crack tip. Incubation periods have been reported, indicating that the crack nucleation process may not be the same as in LME (Ref 116–118). Nucleation and propagation are two separate stages of fracture in SMIE.

These differences may arise because of the rate of reaction or interactions at the metal/embrittler interface and because the transport properties of solid- and liquid-metal embrittlers are of significantly different magnitudes. It has been suggested that reductions in the cohesive strength of atomic bonds at the tip are responsible for both SMIE and LME (Ref 109, 116, 117, 119, 120). However, transport of the embrittler is the rate-controlling factor in SMIE. Another possibility is that stress-assisted penetration of the embrittler in the grain boundaries initiates cracking, but surface self-diffusion of the embrittling species similar to that proposed for LME controls crack propagation.

Investigations of SMIE

The first investigations of delayed failure were reported in cadmium-, zinc-, and indium-plated tensile specimens of 4340, 4130, 4140, and 18% Ni maraging steel in the temperature range of 200 to 300 °C (390 to 570 °F), as shown in Fig. 77 and 78. The results indicated that 4340 was the most susceptible and that 18% Ni maraging steel was the least susceptible alloy to cadmium embrittlement. The activation energy for steel-cadmium embrittlement was 39 kcal/mol (Fig. 79), which corresponded to diffusion of cadmium in the grain boundary. A thin plated layer of nickel or copper has been reported to act as a barrier to the embrittler and to prevent SMIE of steel (Ref 110). Solid indium is reported to embrittle steels (Ref 117, 118), and an incubation period exists for crack nucleation.

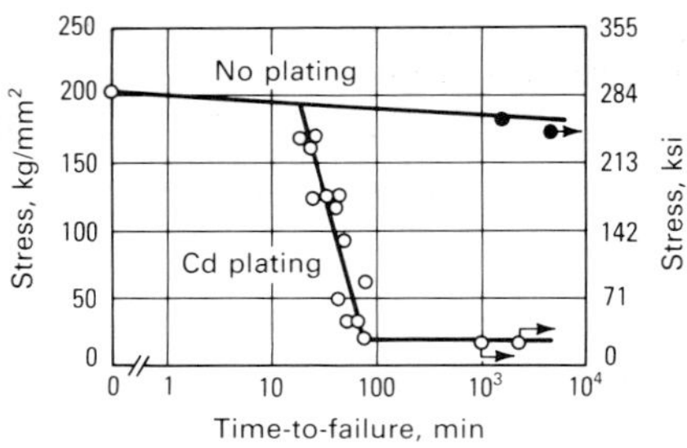

Fig. 77 Embrittlement behavior of cadmium-plated 4340 steel. Specimens were tested in delayed failure at 300 °C (570 °F) and unplated steel in air at 300 °C (570 °F). Source: Ref 110

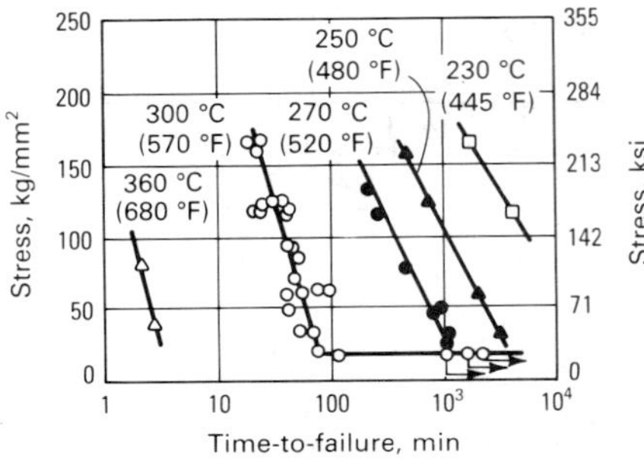

Fig. 78 Embrittlement behavior of cadmium-plated 4340 steel. Specimens were tested in delayed failure at temperatures ranging from 360 to 230 °C (680 to 445 °F). Source: Ref 110

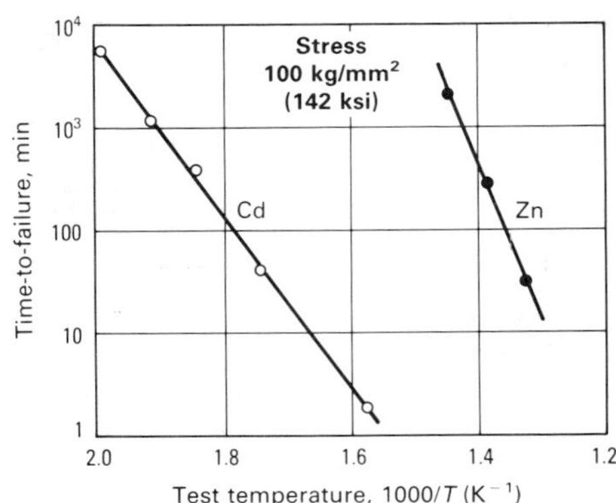

Fig. 79 Arrhenius plots of delayed failure in steel tested in cadmium and zinc. The activation energy Q is 39 kcal/mol for cadmium, 70 kcal/mol for zinc. Source: Ref 110

Embrittlement of different steels in lead was investigated as both an external and internal (leaded steel) environment; SMIE was conclusively demonstrated to be a reproducible effect (Ref 108, 109). It was also demonstrated that SMIE is an extension of LME (Ref 109, 117). Internally leaded high-strength steels are susceptible to severe embrittlement in the range of the T_m of lead, and this embrittlement is a manifestation of LME. However, it was found that the onset of embrittlement occurs some 95 °C (200 °F) below the T_m of lead (330 °C, or 625 °F) and is continuous up to the T_m, with no discontinuity or anomalies in the variation of the embrittlement with temperature (Ref 109). At the T_m of lead, a sharp increase occurs in the severity of embrittlement, and ductile-to-brittle transition occurs in the temperature range of 370 to 450 °C (700 to 840 °F).

The same behavior was noted for pure lead as an external environment soldered onto 4140 steel, and such embrittlement has also been observed for surface coatings of zinc, lead, cadmium, tin, and indium on steel (Ref 109). Embrittlement manifests itself as a reduction in tensile ductility over a range of temperatures extending from about three-quarters of the absolute T_m of the embrittler up to the T_m (Fig. 74). It has been shown that embrittlement is caused by the growth of stable subcritical intergranular cracks and that crack propagation is the controlling factor in embrittlement (Ref 109, 117). This indicated that transport of embrittlers to the crack tip was either by vapor phase or by surface or volume diffusion. In this study, the vapor pressures of the embrittling species at T_m varied widely and ranged from 20 to 6×10^{-36} Pa (1.5×10^{-1} to 4.5×10^{-38} torr) (Table 9). However, the crack propagation times for all embrittlers were similar. The estimated values of the diffusion coefficients ranged in the vicinity of 10^{-4} to 10^{-6} cm²/s (6.5×10^{-4} to 6.5×10^{-6} in.²/s). These values are comparable to surface or self-diffusion of embrittler over embrittler and suggest that diffusion, not vapor transport, is the rate-controlling process.

Table 9 Embrittler vapor pressures and calculated vapor-transport times at the embrittler-melting temperatures

Embrittler	Vapor pressure Pa	Vapor pressure torr	Time, s
Zn	20	1.5×10^{-1}	5×10^{-3}
Cd	10	7.5×10^{-2}	1×10^{-2}
Hg(a)	0.3	2.25×10^{-3}	3×10^{-1}
Sb	4×10^{-4}	3.0×10^{-6}	3×10^{2}
K	1×10^{-4}	7.5×10^{-7}	1×10^{3}
Na	2×10^{-5}	1.5×10^{-7}	5×10^{3}
Ti	4×10^{-6}	3.0×10^{-8}	3×10^{4}
Pb	5×10^{-7}	3.75×10^{-9}	2×10^{5}
Bi	2×10^{-8}	1.5×10^{-10}	5×10^{6}
Li	2×10^{-8}	1.5×10^{-10}	5×10^{6}
In	3×10^{-19}	2.25×10^{-21}	3×10^{17}
Sn	8×10^{-21}	6.0×10^{-23}	1×10^{19}
Ga	6×10^{-36}	4.5×10^{-38}	2×10^{35}

(a) At room temperature. Courtesy of P. Gordon

Delayed Failure and Mechanism of SMIE

If a metal in contact with the embrittling species is loaded to a stress that is lower than that for fracture and is tested at various temperatures, then either the environment-induced fracture initiates and propagates instantaneously, as in a zinc-mercury LME couple, or delayed failure or static fatigue is observed. Examples of these two types of fracture processes are given in Table 8. Investigations of delayed failure provide an opportunity to separate crack nucleation from crack propagation and, specifically, to evaluate the transport-related role of the embrittling species. Solid metal induced embrittlement is a propagation-controlled fracture process, and the time, temperature, and stress dependence of embrittlement presents an opportunity to study the kinetics of the cracking process.

The most critically investigated embrittlement system is 4140 steel embrittled by solid and liquid indium. Measurement of the decrease in electrical potential has been used to monitor crack initiation and propagation and to investigate the effects of temperature and stress level on delayed failure in 4140 steel by liquid and solid indium (Ref 116). The effects of temperature and stress level on the initiation time (incubation period for crack initiation) for both SMIE and LME are given in Fig. 80. The activation energy of the crack initiation process is approximately 37 kcal/mol and is essentially independent of the applied stress level. The activation energy represents the energy for stress-aided self-diffusion of both liquid and solid indium in the grain boundaries of the steel. Thus, crack initiation in both SMIE and LME occurs first by the adsorption of the embrittler at the site of crack initiation.

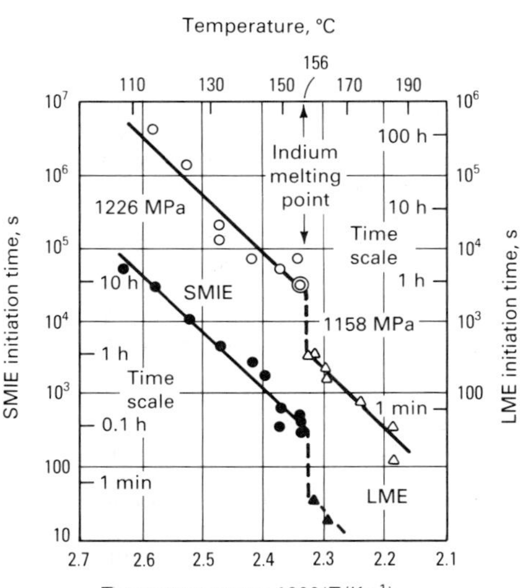

Fig. 80 Initiation time versus temperature in SMIE and LME of 4140 steel in indium at two stress levels. Source: Ref 117

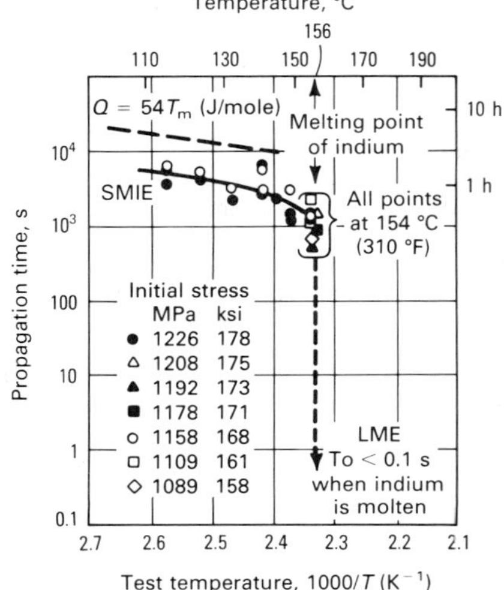

Fig. 81 Propagation time versus temperature for 4140 steel in indium at various temperatures and stress levels. Note the occurrence of SMIE and LME and little dependence on stress level of test. Source: Ref 117

However, because of the presence of an incubation period, the rate-controlling process is stress-aided diffusion penetration of the base-metal grain boundaries. Such penetration reduces the stress necessary to initiate a crack, causing embrittlement of the base metal. On this basis, it is difficult to explain embrittlement of single crystals, which do not contain grain boundaries. Thus, rather than stress-aided diffusion penetration along grain boundaries, the embrittling species may reduce cohesion of atoms at the grain boundaries in accord with the reduction-in-cohesion mechanism. This mechanism would also explain SMIE of single crystals. Crack propagation time as a function of temperature and stress level for both SMIE and LME of steel by indium is plotted in Fig. 81. The activation energy for crack propagation for this couple was 5.6 kcal/mol, which represents the energy for self-diffusion of indium over indium. Thus, propagation is controlled by the diffusion of indium over several multilayers of indium adsorbed on the crack surface or by the so-called waterfall mechanism of embrittlement. This mechanism is a variation of that proposed and discussed in detail as the reduction-in-cohesion mechanism for LME.

Adsorption of the embrittler at the solid surface is followed by the rate-controlling step of stress-aided diffusion of the embrittler in the grain boundary of the base metal. The embrittler reduces cohesion of the grain-boundary base metal atoms, thus initiating a crack at a lower stress than that observed for the alloy in the absence of the embrittling species, resulting in embrittlement. Crack propagation occurs by multilayer self-diffusion of the embrittling species. These processes are valid for both LME and SMIE in the steel-indium embrittlement couple. It is clear that SMIE is similar to LME except that SMIE is a slower, embrittler transport-limited process. In this regard, the study of SMIE is important in eliminating the possibility in LME that a crack, once nucleated, may propagate in a brittle manner in the absence of the embrittling species at the tip of the crack. Solid metal induced embrittlement is a recent phenomenon, and further research is needed concerning the effects of metallurgical, mechanical, and chemical parameters on embrittlement. However, it is clear that SMIE must be recognized as yet another phenomenon of environmentally induced embrittlement.

REFERENCES

1. T.A. Michalske and S.W. Freiman, *Nature*, Vol 295, 11 Feb, 1982, p 511
2. J.F. Fellers and B.F. Kee, *J. Appl. Polymer Sci.*, Vol 18, 1974, p 2355
3. E.H. Andrews, *Developments in Polymer Science*, Applied Science, 1979
4. H.T. Sumison and D.P. Williams, in *Fatigue of Composite Materials*, STP 569, American Society for Testing and Materials, 1975, p 226
5. C. Zeben, in *Analysis of the Test Methods for High Modulus Fibers and Composites*, STP 521, American Society for Testing and Materials, 1973, p 65
6. K. Sieradski and R.C. Newman, *Philos. Mag. A*, Vol 5 (No. 1), 1985, p 95
7. T. Cassaigne, E.N. Pugh, and J. Kruger, National Bureau of Standards and Johns Hopkins University, unpublished research, 1987
8. C.M. Chen, M.H. Froning, and E.D. Verink, in *Stress Corrosion—New Approaches*, STP 610, American Society for Testing and Materials, 1976, p 289
9. C.J. Beevers, Ed., *The Measurement of Crack Length and Shape During Fracture and Fatigue*, Chameleon Press, 1980
10. E.N. Pugh, U. Bertocci, and R.E. Ricker, National Bureau of Standards, unpublished research, 1987
11. D.P. Williams and H.G. Nelson, *Metall. Trans.*, Vol 1 (No. 1), 1970, p 63
12. R.E. Ricker and D.J. Duquette, The Role of Environment on Time Dependent Crack Growth, in *Micro and Macro Mechanics of Crack Growth*, K. Sadananda, B.B. Rath, and D.J. Michel, Ed., American Institute of Mining, Metallurgical, and Petroleum Engineers, 1982, p 29
13. R.M. Parkins, *Mater. Sci. Technol.*, Vol 1, 1985, p 480
14. L. Hagn, Lifetime Prediction for Parts in Corrosion Environments, in *Corrosion in Power Generating Equipment*, Plenum Press, 1983
15. O. Buck and R. Ranjan, Evaluation of a Crack-Tip-Opening Displacement Model Under Stress-Corrosion Conditions, in *Modeling Environmental Effects on Crack Growth Processes*, R.H. Jones and W.W. Gerberich, Ed., The Metallurgical Society, 1986, p 209
16. M.J. Danielson, C. Oster, and R.H. Jones, *J. Electrochem. Soc.*, in press
17. R.H. Jones, M.J. Danielson, and D.R. Baer, *J. Mater. Energy Syst.*, Vol 8, 1986, p 185
18. A. Turnbull, Progress in the Understanding of the Electrochemistry in Cracks, in *Embrittlement by the Local Crack Environment*, R.P. Gangloff, Ed., The Metallurgical Society, 1984, p 3
19. R.N. Parkins, Prevention of Environment Sensitive Fracture by Inhibition, in *Embrittlement by the Local Crack Environment*, R.P. Gangloff, Ed., The Metallurgical Society, 1984, p 385
20. R.N. Parkins, *Br. Corros. J.*, Vol 14, 1979, p 5
21. C. Edeleneau and A.J. Forty, *Philos. Mag.*, Vol 46, 1960, p 521
22. J.A. Beavers and E.N. Pugh, *Metall. Trans. A*, Vol 11A, 1980, p 809
23. M.T. Hahn and E.N. Pugh, *Corrosion*, Vol 36, 1980, p 380
24. E.N. Pugh, On the Propagation of Transgranular Stress Corrosion Cracks, in *Atomistics of Fracture*, R.M. Latanision and J.R. Pickens, Ed., Plenum Press, 1983, p 997
25. R.C. Newman and K. Sieradzki. "Film-Induced Cleavage During Stress-Corrosion Cracking of Ductile Metals and Alloys," NATO Advanced Research Workshop on Chemistry and Physics of Fracture, June 1986
26. N.J.H. Holroyd and G.M. Scamans, *Scr. Metall.*, Vol 19, 1985, p 915
27. E.C. Pow, W.W. Gerberich, and L E. Toth, *Scr. Metall.*, Vol 15, 1981, p 55
28. E.D. Hondros and M.P. Seah, *Int. Met. Rev.*, Dec 1977, p 262
29. L. Long and H. Uhlig, *J. Electrochem. Soc.*, Vol 112, 1965
30. J. Kuppa, H. Erhart, and H. Grabke, *Corros. Sci.*, Vol 21, 1981, p 227
31. N. Bandyopadhyay, R.C. Newman, and K. Sieradzdki, in *Proceedings of the Ninth International Congress on Metallic Corrosion* (Toronto, Canada), National Association of Corrosion Engineers, 1984
32. R.H. Jones, S.M. Bruemmer, M.T. Thomas, and D.R. Baer, Comparison of Segregated Phosphorus and Sulfur Effects on the Fracture Mode and Ductility of Iron Tested at Cathodic Potentials, *Scr. Metall.*, Vol 16, 1982, p 615
33. S.M. Bruemmer, R.H. Jones, M.T. Thomas, and D.R. Baer, Fracture Mode Transition of Iron in Hydrogen as a Function of Grain Boundary Sulfur, *Scr. Metall.*, Vol 14, 1980, p 137
34. A. Joshi and D.J. Stein, *Corrosion*, Vol 28 (No. 9), 1972, p 321
35. R.H. Jones, Some Radiation Damage-Stress Corrosion Synergisms in Austenitic Stainless Steels, in *Proceedings of the Second International Symposium on Environmental Degradation of Materials in Nuclear Power Systems-Water Reactors*, American Nuclear Society, Sept 1985, p 173
36. M. Guttman, P. Dumoulin, Nguyen Tan Tai, and P. Fontaine, *Corrosion*, Vol 37, 1981, p 416
37. S.M. Bruemmer, L.A. Charlot, and C.H. Henager, Jr., "Microstructural Effects on Microdeformation and Primary-Side Stress Corrosion Cracking of Alloy 600 Tubing," EPRI Final Report, Project 2163-4, Electric Power Research Institute, Aug 1986
38. R.H. Jones, S.M. Bruemmer, M.T. Thomas, and D.R. Baer, Influence of Sulfur, Phosphorus, and Antimony Segregation on the Intergranular Hydrogen Embrittlement of Nickel, *Metall. Trans. A*, Vol 14A, 1983, p 223
39. R.W. Staehle *et al.*, *Corrosion*, Vol 26 (No. 11), 1970, p 451
40. H.W. Pickering and P.R. Swann, *Corrosion*, Vol 19 (No. 3), 1963, p 373
41. A.W. Thompson and I.M. Bernstein, *Advances in Corrosion Sciences and Technology*, Vol 7, M.G. Fontana and R.W. Staehle, Ed., Plenum Press, 1980, p 53
42. R.M. Riecke, A. Athens, and I.O. Smith, *Mater. Sci. Technol.*, Vol 2, 1986, p 1066
43. E.H. Dix, *Trans. AIME*, Vol 137 (No 11), 1940
44. R.B. Mears, R.H. Brown, and E.H. Dix, *Symposium on Stress-Corrosion Cracking of Metals*, American Society for Testing and Materials and the American Institute of Mining, Metallurgical, and Petroleum Engineers, 1944, p 323
45. E.H. Dix, Jr., *Trans. ASM*, Vol 42, 1950, p 1057
46. H.H. Uhlig, *Physical Metallurgy of Stress Corrosion Fracture*, T.N. Rhodin, Ed., Interscience, 1959, p 1
47. L. Yang, G.T. Horne, and G.M. Pound, *Physical Metallurgy of Stress Corrosion Fracture*, T.N. Rhodin, Ed., Interscience, 1950, p 29
48. J.J. Harwood, *Stress Corrosion Cracking and Embrittlement*, W.D. Robertson, Ed., John Wiley & Sons, 1956, p 1
49. F.A. Champion, in *Symposium on Internal Stresses in Metals and Alloys*, Institute of Metals, 1948, p 468
50. H.L. Logan, *J. Res. Natl. Bur. Stand.*, Vol 48, 1952, p 99
51. E.W. Hart, *Surfaces and Interfaces II*,

Syracuse University Press, 1968, p 210
52. H.J. Engle, in *The Theory of Stress Corrosion Cracking in Alloys*, North Atlantic Treaty Organization, 1971
53. J.C. Scully, *Corros. Sci.*, Vol 15, 1975, p 207
54. D.A. Vermilyea, *J. Electrochem. Soc.*, Vol 119, 1972, p 405
55. D.A. Vermilyea, *Stress Corrosion Cracking and Hydrogen Embrittlement of Iron Based Alloys*, National Association of Corrosion Engineers, 1977, p 208
56. R.W. Staehle, in *The Theory of Stress Corrosion Cracking in Alloys*, North Atlantic Treaty Organization, 1971, p 223
57. R.W. Staehle, *Stress Corrosion Cracking and Hydrogen Embrittlement of Iron Based Alloys*, National Association of Corrosion Engineers, 1977, p 180
58. E.N. Pugh, *Corrosion*, Vol 41 (No. 9), 1985, p 517
59. P.R. Swann and J.D. Embury, *High Strength Materials*, John Wiley & Sons, 1965, p 327
60. J.M. Silcock and P.R. Swann, *Environment-Sensitive Fracture of Engineering Materials*, Z.A. Foroulis, Ed., The Metallurgical Society, 1979, p 133
61. S.P. Lynch, *Hydrogen Effects in Metals*, A.W. Thompson and I.M. Bernstein, Ed., The Metallurgical Society, 1981, p 863
62. S.P. Lynch, *Met. Sci.*, Vol 15 (No. 10), 1981, p 463
63. S.P. Lynch, *J. Mater. Sci.*, Vol 20, 1985, p 3329
64. A.J. Forty and P. Humble, *Philos. Mag.*, Vol 8, 1963, p 247
65. A.J. McEvily and P.A. Bond, *J. Electrochem. Soc.*, Vol 112, 1965, p 141
66. E.N. Pugh, in *Stress Corrosion Cracking and Hydrogen Embrittlement of Iron Based Alloys*, National Association of Corrosion Engineers, 1977, p 37
67. J.A. Beavers, I.C. Rosenberg, and E.N. Pugh, in *Proceedings of the 1972 Tri-Service Conference on Corrosion*, MCIC-73-19, Metals and Ceramics Information Center, 1972, p 57
68. T.R. Pinchback, S.P. Clough, and L.A. Heldt, *Metall. Trans. A*, Vol 7A, 1976, p 1241; *Metall. Trans. A*, Vol 6A, 1975, p 1479
69. A.J. Forty, *Physical Metallurgy of Stress Corrosion Fracture*, T.N. Rhodin, Ed., Interscience, 1959, p 99
70. K. Sieradzki and R.C. Newman, *Philos. Mag. A*, Vol 51 (No. 1), 1985, p 95
71. I.-H. Lin and R.M. Thomson, *J. Mater. Res.*, Vol 1 (No. 1), 1986
72. G.J. Dienes, K. Sieradzki, A. Paskin, and B. Massoumzadeh, *Surf. Sci.*, Vol 144, 1984, p 273
73. N.S. Stolloff, *Environment-Sensitive Fracture of Engineering Materials*, Z.A. Foroulis, Ed., The Metallurgical Society, 1979, p 486
74. M.O. Speidel and R.M. Magdowski, Stress Corrosion Cracking of Steam Turbine Steels—An Overview, in *Proceedings of the Second International Symposium on Environmental Degradation of Materials in Nuclear Power Systems-Water Reactors*, American Nuclear Society, 1986, p 267
75. D.P. Williams and H.G. Nelson, *Metall. Trans.*, Vol 1, 1970, p 63
76. R.M. Latanision and H. Opperhauser, *Metall. Trans.*, Vol 5, Scientific and Technical Book Service, 1974, p 483
77. R.H. Jones, A Review of Combined Impurity Segregation-Hydrogen Embrittlement Processes, in *Advances in the Mechanics and Physics of Surfaces*, R.M. Latanision and T.E. Fischer, Ed., Scientific and Technical Book Service, 1986
78. J.P. Hirth and H.H. Johnson, Hydrogen Problems in Energy Related Technology, *Corrosion*, Vol 32, 1976, p 3
79. C. Zapffe and C. Sims, Hydrogen Embrittlement, Internal Stress and Defects in Steel, *Trans. AIME*, Vol 145, 1941, p 225
80. N.J. Petch and P. Stables, Delayed Fracture of Metals Under Static Load, *Nature*, Vol 169, 1952, p 842
81. A.R. Troiano, The Role of Hydrogen and Other Interstitials in the Mechanical Behavior of Metals, *Trans. ASM*, Vol 52, 1960, p 54
82. R.A. Oriani, A Mechanistic Theory of Hydrogen Embrittlement of Steels, *Bunsen-Gesellschaft Phys Chemie*, Vol 76, 1972, p 848
83. C.D. Beachem, A New Model for Hydrogen Assisted Cracking (Hydrogen Embrittlement), *Metall. Trans.*, Vol 3, 1972, p 437
84. S. Gahr, M.L. Grossbech, and H.K. Birnbaum, *Acta Metall.*, Vol 25, 1977, p 125
85. F.H. Vitovec, Modeling of Hydrogen Attack of Steel in Relation to Material and Environmental Variables, in *Current Solutions to Hydrogen Problems in Steels*, C.G. Interrante and G.M. Pressouyre, Ed., American Society for Metals, 1982
86. G.M. Pressouyre and I.M. Bernstein, A Quantitative Analysis of Hydrogen Trapping, *Metall. Trans. A*, Vol 9A, 1978, p 1571
87. W.W. Gerberich, Effect of Hydrogen on High Strength and Martensitic Steels, in *Hydrogen in Metals*, I.M. Bernstein and A.W. Thompson, Ed., American Society for Metals, 1974, p 115
88. G.M. Pressouyre, A Classification of Hydrogen Traps in Steel, *Metall. Trans. A*, Vol 10A, 1979, p 1571
89. H.G. Nelson and D.P. Williams, Quantitative Observations of Hydrogen Induced Slow Crack Growth in a Low Alloy Steel, in *Stress Corrosion Cracking and Hydrogen Embrittlement of Iron Base Alloys*, National Association of Corrosion Engineers, 1977, p 390
90. B.A. Graville, R.G. Baker, and F. Watkinson, *Br. Weld. J.*, Vol 14, 1967, p 337
91. G. Sandoz, A Unified Theory for Some Effects of Hydrogen Source, Alloying Elements and Potential on Crack Growth in Martensitic AISI 4340 Steel, *Metall. Trans.*, Vol 3, 1972
92. P. McIntyre, The Relationship Between Stress Corrosion Cracking and Sub Critical Flaw Growth in Hydrogen and Hydrogen Sulphide Gases, in *Stress Corrosion Cracking and Hydrogen Embrittlement of Iron Base Alloys*, National Association of Corrosion Engineers, 1977, p 788
93. R.P.M Procter and H.W. Paxton, *Trans. ASM*, Vol 62, 1969, p 989
94. A.W. Thompson, "Hydrogen Embrittlement of Stainless Steels and Carbon Steels," Paper presented at the Midyear Meeting, Toronto, Canada, American Petroleum Institute, 1978
95. R.R. Gaugh, "Sulfide Stress Cracking of Precipitation Hardening Stainless Steels," Paper 109, presented at Corrosion/77, National Association of Corrosion Engineers, 1977
96. "Steels for Hydrogen Service at Elevated Temperatures and Pressures in Petroleum Refineries and Petrochemical Plants," Publication 941, 2nd ed., American Petroleum Institute, June 1977
97. B.J. Berkowitz, M. Kurkela, and R.M. Latanision, Effect of Ordering on the Hydrogen Permeation and Embrittlement of Ni_2Cr, in *Hydrogen Effects in Metals*, I.M. Bernstein and A.W. Thompson, Ed., American Society for Metals, 1981, p 411
98. M.O. Speidel, Hydrogen Embrittlement of Aluminum Alloys, in *Hydrogen in Metals*, I.M. Bernstein and A.W. Thompson, Ed., American Society for Metals, 1974
99. D.N. Williams, The Hydrogen Embrittlement of Titanium Alloys, *J. Inst. Met.*, Vol 91, 1963
100. H.G. Nelson, D.P. Williams, and J.E. Stein, Environmental Hydrogen Embrittlement of an α-β Titanium Alloy: Effect of Microstructure, in *Hydrogen Damage*, C.D. Beachem, Ed., American Society for Metals, 1977
101. H.G. Nelson, in *Hydrogen in Metals*, I.M. Bernstein and A.W. Thompson, Ed., American Society for Metals, 1974, p 445
102. D. Northwood and U. Kosasih, Hydrides and Delayed Hydrogen Cracking in Zirconium and Its Alloys, *Int. Met. Rev.*, Vol 28, 1983, p 92
103. F. Yunchang and D.A. Koss, The Influence of Multiaxial States of Stress on the Hydrogen Embrittlement of Zirconium Alloy Sheet, *Metall. Trans. A*, Vol 16A, 1985, p 675
104. D. Hardie, *J. Nucl. Mater.*, Vol 42, 1972, p 317
105. W.A. Spitzig, C.V. Owen, and T.E. Scott, The Effects of Interstitials and Hydrogen Interstitial Interactions on Low Temperature Hardening and Embrittlement in V, Nb, and Ta, *Metall. Trans. A*, Vol 17A, 1986, p 1179
106. C.V. Owen, T.J. Rowland, and O. Buck, Effects of Hydrogen on Some Mechanical Properties of Vanadium—Titanium Alloys, *Metall. Trans. A*, Vol 16A, 1985, p 59
107. H.K. Birnbaum, Hydrogen Related Second Phase Embrittlement of Solids, in *Hydrogen Embrittlement and Stress Corrosion Cracking*, American Society for Metals, 1984, p 153-177
108. J.C. Lynn, W.R. Warke, and P. Gordon, Solid Metal Induced Embrittlement of Steels, *Mater. Sci. Eng.*, Vol 18, 1975, p 51-62
109. S. Mostovoy and N.N. Breyer, Effect of Lead on the Mechanical Properties of 4145 Steel, *Trans. ASM*, Vol 61 (No. 2), 1968, p 219-232
110. Y. Asayama, Metal-Induced Embrittlement of Steels, in *Embrittlement by Liquid and Solid Metals*, M.H. Kamdar, Ed., American Institute of Mining, Metallurgical, and Petroleum Engineers, 1984
111. D.A. Meyn, Solid Cadmium Induced Cracking of Titanium Alloys, *Corrosion*, Vol 29, 1973, p 192-196
112. W.T. Grubb, Cadmium Metal Embrittlement of Zircaloy, *Nature*, Vol 265, 1977, p 36-37

113. Y. Iwata, Y. Asayama, and A. Sakamoto, Delayed Failure of Cadmium Plated Steels at Elevated Temperature, *J. Jpn. Inst. Met.*, Vol 31, 1967, p 73 (in Japanese)
114. D.N. Fager and W.F. Spurr, Solid Cadmium Embrittlement in Steel Alloys, *Corrosion*, Vol 27, 1971, p 72
115. D.N. Fager and W.F. Spurr, Solid Cadmium Embrittlement of Titanium Alloys, *Corrosion*, Vol 26, 1970, p 409
116. P. Gordon, Metal Induced Embrittlement of Metals—An Evaluation of Embrittler Transport Mechanisms, *Metall. Trans. A*, Vol 9, 1978, p 267-272
117. P. Gordon and H.H. An, The Mechanisms of Crack Initiation and Crack Propagation in Metal-Induced Embrittlement of Metals, *Metall. Trans. A*, Vol 13A, 1982, p 457-472
118. A. Druschitz and P. Gordon, Solid Metal Induced Embrittlement of Metals, in *Embrittlement by Liquid and Solid Metals*, M.H. Kamdar, Ed., American Institute of Mining, Metallurgical, and Petroleum Engineers, 1984
119. A.R.C. Westwood and M.H. Kamdar, Concerning Liquid Metal Embrittlement, Particularly of Zinc Monocrystals by Mercury, *Philos. Mag.*, Vol 8, 1963, p 787-804
120. N.S. Stoloff and T.L. Johnston, Crack Propagation in Liquid Metal Environments, *Acta Metall.*, Vol 11, 1963, p 251-256

SELECTED REFERENCES

Hydrogen Damage

- P. Azou, Ed., *Third International Congress on Hydrogen and Materials*, Pergamon Press, 1982
- C.D. Beachem, Ed., *Hydrogen Damage*, American Society for Metals, 1977
- I.M. Bernstein and A.W. Thompson, Ed., *Hydrogen in Metals*, American Society for Metals, 1972
- I.M. Bernstein and A.W. Thompson, Ed., *Hydrogen Effects in Metals*, American Institute of Mining, Metallurgical, and Petroleum Engineers, 1981
- R.W. Staehle *et al.*, Ed., *Stress Corrosion Cracking and Hydrogen Embrittlement of Iron Base Alloys*, National Association of Corrosion Engineers, 1977

Liquid-Metal Embrittlement

- M.H. Kamdar, *Prog. Mater. Sci.*, Vol 15, 1973
- M.H. Kamdar, Liquid Metal Embrittlement, in *Treatise on Materials Science and Technology*, Vol 25, C.L. Briant and S.K. Banerji, Ed., Academic Press, 1983, p 361-459
- M.H. Kamdar, Ed., *Embrittlement by Liquid and Solid Metals*, The Metallurgical Society, 1984
- V.I. Likhtman, P.A. Rebinder, and G.V. Karpeno, *Effects of Surface Active Medium on Deformation of Metals*, Her Majesty's Stationary Office, 1958
- V.I. Likhtman, E.D. Shchukin, and P.A. Rebinder, *Physico-Chemical Mechanics of Metals*, Academy of Sciences U.S.S.R., 1962
- S. Mostovoy and N.N. Breyer, The Effect of Lead on the Mechanical Properties of 4145 Steel, *ASM Quart.*, Vol 61 (No. 2), 1968
- W. Rostoker, J.M. McCaughney, and H. Markus, *Embrittlement by Liquid Metals*, Van Nostrand Reinhold, 1960
- N.S. Stoloff, Recent Developments in Liquid Metal Embrittlement in Environment Sensitive Fracture of Engineering Materials, in *Proceedings of the AIME Conference*, American Institute of Mining, Metallurgical, and Petroleum Engineers, 1980
- N.S. Stoloff, Metal Induced Embrittlement, in *Embrittlement by Liquid and Solid Metals*, M.H. Kamdar, Ed., The Metallurgical Society, 1984
- A.R.C. Westwood, C.M. Preece, and M.H. Kamdar, *Fracture*, Vol 13, H. Leibowitz, Ed., Academic Press, 1971

Corrosion Testing and Evaluation

Section Chairman:
Donald O. Sprowls, Consultant

Planning and Preparation of Corrosion Tests

Donald O. Sprowls, Consultant

CORROSION BEHAVIOR is a combined property of the metal and the environment to which it is exposed. Therefore, there is no universal corrosion test for all purposes. The factors associated with both the metal and the environment should be considered and controlled, when necessary, to establish appropriate exposure conditions during testing.

This article will discuss the basic corrosion-testing philosophy that impacts on all of the various types of testing described in this Section. These remarks have been drawn from several relatively recent books on the subject (Ref 1-4) and from practical experience in an industrial metallurgical research laboratory.

Properly conducted corrosion tests can mean the savings of millions of dollars. They are the means of avoiding the use of a metal under unsuitable conditions or of using a more expensive material than is required. Corrosion tests also help in the development of new alloys that perform more inexpensively, more efficiently, longer, or more safely than the alloys currently in use. Also, quality-control corrosion tests are a means of ensuring that alloys have the capabilities expected of them. Corrosion-testing programs can be simple ones that are completed in a few minutes or hours, or they can be complex, requiring the combined work of a number of investigators over a period of years.

Careful planning is essential to obtain meaningful test results. Preliminary time and effort spent considering metallurgical factors, environmental variations, statistical treatment, and the interpretation of accelerated test results are often the most useful part of the test procedures (Ref 5, 6).

Test Objectives

The first and most important part of test planning is to define clear objectives and significant criteria for interpreting the test results. The definition of the objective is important because it determines the kind of information that would be most pertinent to the goals of the test. Typical corrosion test objectives include:

- Determining the best material to fill a need (material selection)
- Predicting the probable service life of a product or structure
- Evaluating the new commercial alloys and processes
- Assisting with development of materials with improved resistance to corrosion
- Conducting lot-release and acceptance tests to determine whether material meets specifications (quality control)
- Evaluating environmental variations and controls (inhibitors)
- Determining the most economical means of reducing corrosion
- Studying corrosion mechanisms

In pursuit of most of the test objectives listed above, comparisons and predictions are involved for which statistical treatment of the test results is advantageous. Therefore, specific criteria, such as an acceptable significance level, and a plan of analysis should be determined in advance. Standardized test methods and practices are preferred when available.

In the field of research for alloy or process development and improvement, the investigator seeks to improve for tomorrow the material of today. He will usually begin with small specimens and laboratory tests and, as material variations show promise, will proceed through other types of testing, such as simulated service tests and testing of actual parts.

For some situations in which it is necessary to determine the best material for an application, special test conditions may have to be developed. One should have as thorough a knowledge as possible of all the service conditions that the product will be called on to withstand. Standardization of test methods has received the most attention, as might be expected, in problem areas that have become acute—for example, in the chemical-processing industries (Ref 7, 8).

After a material has been selected for an application, the fabricator and purchaser then need to know if the quality of incoming material is as specified. Quality-control corrosion tests ensure that the material meets specifications. Such tests must be sufficiently rapid to avoid undue delay of material shipments. Corrosion tests of this type do not require any particular correlation to end use. Alone, they do not necessarily reveal whether or not this alloy will make a better product. They simply ensure that the quality of the material shipped has corrosion resistance that is similar to the material on which the development tests were made.

In fundamental corrosion research, which focuses on determining how or why a particular form or mechanism of corrosion occurs, specialized techniques are often required; these may be quite sophisticated. Standardized test methods are usually not applicable. In recent years, however, considerable progress has been made in standardizing certain practices and techniques. A good example of this is electrochemical testing.

Procurement of Test Materials

In some cases, such as quality-control tests, the material to be tested is predetermined. In most other cases, there is considerable freedom of choice.

Metal Composition and Metallurgical Condition. If commercial alloys are to be tested, the best procedure is to obtain mill-fabricated material representative of current production. If possible, a fabrication history should be obtained listing the major fabricating steps, together with an accurate analysis of the metal composition. At the very least, the material should be certified as being within the composition limits specified for the alloy and meeting the strength or hardness guaranteed for the material. Metallographic examination may also be necessary to ensure that the material is in a proper metallurgical condition. Such basic information may prevent misleading conclusions as a result of off-composition or improper metallurgical treatment.

When nonstandard alloys or metallurgical treatments are to be evaluated, it may be economically impractical to have them fabricated with large production equipment. In such cases, small laboratory casts or heats are made, sometimes weighing only a few pounds. Although this is a valid procedure for initial tests, a promising alloy candidate should be evaluated several times by using specimens from quantities of material large enough to be representative of production before it is produced commercially. Several production lots must be evaluated because promising results from a single lot of material are often not reproduced on subsequent fabrication runs.

Metal Form (Mill Product). Another preliminary decision is the form of metal to be tested. Metals are available in two basic forms: cast and wrought (worked). These two forms should not be interchanged in testing. The various methods of casting (die, permanent mold, sand mold, and so on) and of working (drawing, extruding, forging, rolling, and so on) will affect grain structure and homogeneity, which in turn can affect corrosion resistance. The metal procured should be, as nearly as possible, the type that will be used in the intended application.

In certain types of corrosion tests, such as tests of compatibility with chemical solutions or evaluation of protective coatings, considerations of grain structure may not be critical. When this is

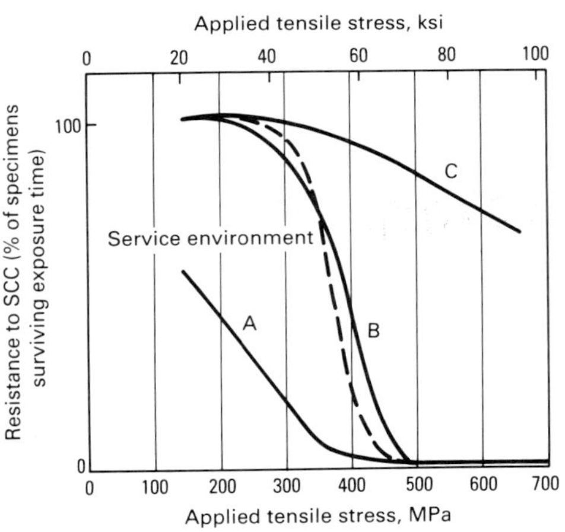

Fig. 1 Example illustrating the importance of correlating laboratory test data with field test data or service experience. A, B, and C represent laboratory SCC test media, but only medium B correlates well with the service environment. See text for explanation.

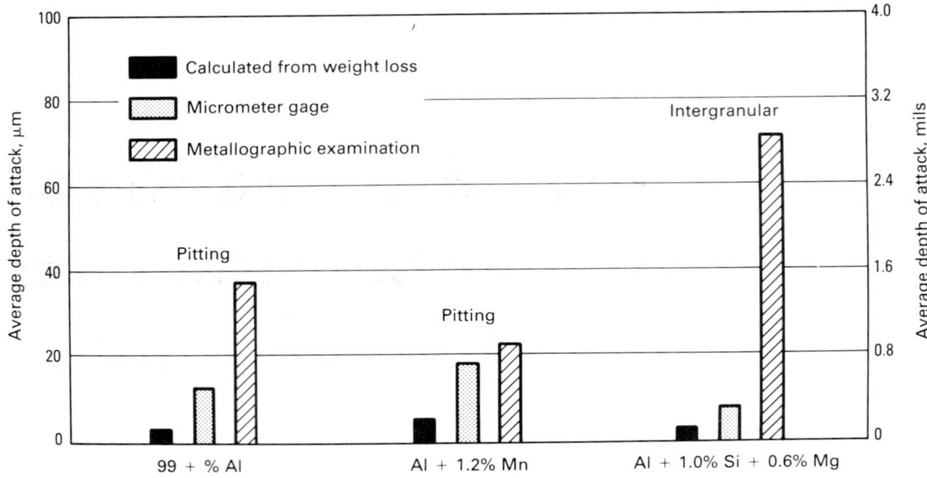

Fig. 2 Comparison of the wide variation in results obtained by different methods of measuring localized attack in aluminum alloys. The data shown are average results for each alloy exposed to seven atmospheric environments for a total of 10 years. Source: Ref 10

so, cast bar or rolled sheet is commonly used because these product forms are readily obtainable and are easy to fabricate into the desired test specimens.

However, if it is anticipated that the final use of a metal will be in a cast form, cast specimens should be evaluated. The same applies for wrought metal specimens. Data obtained from cast and wrought metal forms cannot be reliably interchanged. Similarly, when complex heat-treated alloys are being tested, one wrought form should not be substituted for another.

Preparation of Test Specimens

The sampling of test materials and the preparation of test specimens are extremely important variables in corrosion testing, because corrosion behavior can be significantly influenced by variations in metallurgical structure and the condition of the metal surface. These factors are especially pertinent when various forms of localized corrosion are under evaluation in complex alloys. The uniformity of the metal sample should be checked in advance as part of the plan for preparation of the test specimens. Problems resulting from this cause are less likely to be encountered with pure metals and homogeneous alloys.

The primary consideration should be the use of test specimens that are truly representative of the specified material, that is, alloy, metallurgical condition, and product form, rather than specimens that are the most expedient for the investigation. For example, it might seem advantageous to obtain flat, uniform disk specimens by machining slices from an extruded rod rather than by shearing or stamping test coupons from rolled sheet (and avoiding residual stresses in the sheared edges). However, the corrosion behavior of the end-grain surfaces of the machined disks could be different from that of the rolled surface of the sheet product.

Surface preparation by various machining or polishing methods and various chemical treatments can be a source of considerable variation in test results and must be controlled. In some cases, it may be desired to evaluate the metal in the mill-finished condition, although this approach will generally result in the greatest variability. As a general rule, corrosion-testing standards contain specific recommendations regarding appropriate surface treatments, depending on the metal alloy system. Judicious use of proper or standardized surface treatments will reduce variability in corrosion test results, as discussed in Ref 9.

Selection of Corrosive Media

The corrosion behavior of a given material depends on the environment and the conditions of exposure as well as the condition of the test material. The most reliable predictor of performance is service experience, followed closely by field testing, because both are based on the actual environment. When service history is lacking and time or budget constraints prohibit field testing, laboratory corrosion tests are used to predict or estimate corrosion performance. They are particularly useful for quality control, materials selection, and materials development.

Laboratory tests can be misleading if appropriate corrosive media relating back to the test objectives are not used. Part of the testing philosophy involves whether the test is intended to reproduce a certain environment accurately or whether it is more advisable to use a corrodent that represents a worst-case situation. In either case, the corrosion investigator must do everything possible to make the test reproducible by exercising explicit control over such environmental factors as concentration of reactants and contaminants, solution pH, temperature, aeration, velocity, impingement and bacteriological effects. Standard test methods should provide for such controls.

Accelerating Test Results. Almost invariably, it is desirable to have results from corrosion tests as quickly as possible, which necessitates speeding up the reactions to obtain quicker results. This is a legitimate approach if conditions are not changed to the point at which some corrosion mechanism other than that experienced in normal service becomes predominant. The important question is: How can conditions or reactions be accelerated, and how far can they be accelerated before a different reaction is induced? (Ref 5). Consider, for example, a test of the corrosion rate of a piece of steel in water. It is known that raising the temperature of the water increases the corrosion rate significantly. The tendency might be to accelerate the test by raising the temperature almost to the boiling point, but by doing so, the oxygen in the water—also a contributing factor to the corrosion rate—is driven out. The corrosion rate then decreases instead of continuing to rise with temperature, thus inducing a corrosion mechanism different from the one of interest.

Correlation With Field Experience. The soundness of an accelerated test medium can be shown by correlation with practical service experience or with results of appropriate field tests. However, some methods used to correlate test environments are inadequate and should be avoided.

For example, Fig. 1 shows a hypothetical example of correlations of three accelerated stress-corrosion cracking (SCC) test media (A, B, and C) with a service environment. The resistance to SCC of a given susceptible material is plotted as a function of an independent variable that influences the stress-corrosion performance. The curve labeled service environment represents the established performance of the material when stressed at various levels and exposed to the service environment. Medium A correlates well for specimens under high stress, but not for specimens under low stress; medium C gives the opposite performance. Only accelerated test medium B correlates well at all ranges of applied stress. An investigator should not assume that a good 1:1 correlation at a single level of the independent variable (138 MPa, or 20 ksi) will be similar at a different level (690 MPa, or 100 ksi). Medium A might be useful as a worst case, but medium C appears to be too mild for any case.

Assessment of Corrosion Damage

Measurement and evaluation of test results must be closely coordinated with the test objectives so that the measured results will make it possible to achieve the objectives with the great-

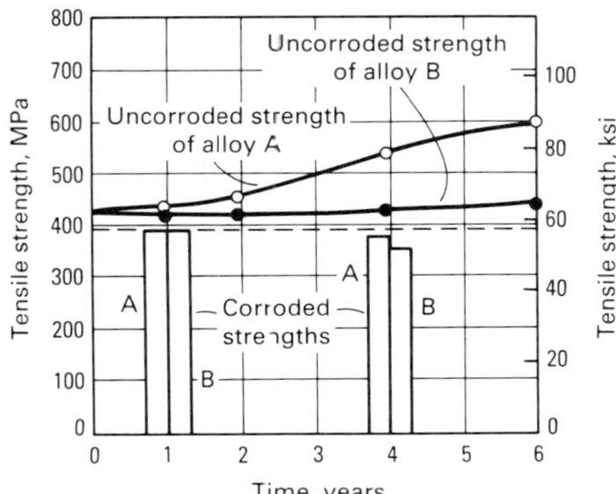

Fig. 3 Example showing the importance of using control specimens when determining strength reduction in alloys whose strength changes with time. Alloy A was appreciably strengthened by natural aging, but alloy B was not.

est degree of directness and clarity. The methods of assessment are limited only by the ingenuity of the investigator, who must try to find the best way to convey the practical meaning of the test results.

Appropriate methods of measuring degradation of the metal also depend on the form of corrosion observed; these methods are discussed in separate articles on the evaluation of specific corrosion behavior in this Volume. For example, the age-old method of measuring corrosion by determining weight loss and calculating average corrosion rates, which is adequate for metals that corrode uniformly, is not a realistic measure of localized forms of corrosion, such as pitting and intergranular attack. Figure 2 illustrates the wide variation in results obtained by measuring the localized corrosion of aluminum alloys in three different ways.

Controls. A vital consideration in the corrosion testing of metals is the use of suitable controls, both for the metal and for the test environment. This aspect of corrosion testing is a common source of error—even by experienced investigators (Ref 11).

With some materials, certain characteristics used for assessing corrosion damage will change with time, even in the absence of corrosion. In such cases, a set of control specimens must be provided for each period of corrosion testing. This set is then evaluated at the same time as the corroded specimens. Some alloys, for example, gradually change in strength over long periods of time at room temperature, and the strength of most metals is affected by time at elevated temperatures. Such a case is shown schematically in Fig. 3.

The bar graphs in Fig. 3 show the corroded strengths of two alloys after 1 and 4 years of exposure to a corrosive atmosphere. After 1 year of exposure, both alloys show the same strength and a similar reduction from the strength of uncorroded metal. After 4 years of exposure, the corroded specimens of alloy A are somewhat stronger than those of alloy B, but their reduction from the strength of uncorroded metal of the same age is considerably greater. If the corroded strength of alloy A had been compared with the strength of specimens tested at the beginning of the investigation, an erroneous conclusion would have been made regarding the effect of corrosion on this alloy.

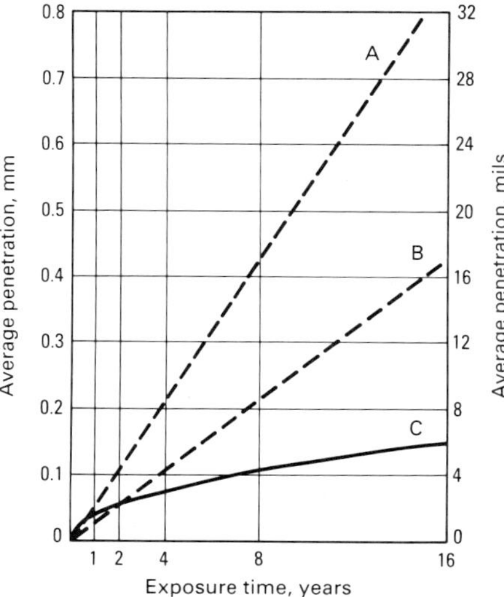

Fig. 4 Example of how short-term corrosion test results can be misleading. A, 6-month rate; B, 2-year rate; C, long-term corrosion of copper in seawater $\cong 1.5\ t^{0.5}$. Curves A and B were calculated from short-term test data; the long-term corrosion rate is more closely approximated by curve C, which is a function of the square root of exposure time t. Source: Ref 12

Specimens to be used for periodic controls should be protected from corrosion and stored at the same temperature at which the corrosion tests are conducted. In long-term field tests, in which temperature will fluctuate and cannot be controlled (such as tests in natural atmospheres), it may be desirable to store the controls on-site in dry, airtight containers. This ensures that both the corrosion specimens and the control specimens are exposed to the same time and temperature conditions. In tests at elevated temperatures, special precautions may be necessary to ensure that the control specimens themselves do not undergo high-temperature corrosion.

The other required control is a method of determining whether the corrosive medium maintains the intended degree of corrosivity. When an alloy is repeatedly tested by the same method, as in quality-control tests, the investigator may be sufficiently experienced to know whether the visual appearance and resultant data are of the order expected. This is often not the case with new alloys or products, because their corrosion performance is unpredictable. The usual procedure is to expose a well-established previously tested alloy as a control to determine whether the results on it are consistent with past experience. If they are, the new alloy is considered to have received a valid exposure. However, if the effects of corrosion on the well-known alloy are not typical, the investigator is alerted to the possibility of an error in test environment and the necessity for a rerun.

Reliability of Corrosion Test Results

Corrosion-testing methods (and the testers) are often criticized because usage of the test results has led to unfortunate decisions in the application of metals. Critics cite the difficulty in extracting from the results precise numbers that can be used in the design or prediction of service lives, questioning why corrosion tests cannot produce simple numerical results in a short period of time as mechanical tests do. They also assert that more reliable corrosion tests are needed; although this may be true, more careful and realistic interpretation of the test results is also necessary.

Corrosion rates should be used with caution. Corrosion engineers frequently take the results of relatively short-term one-period exposure tests, divide the weight loss (or pit depth) of the metal by the time of exposure, and from this determine a constant rate of corrosion for the particular metal-environment combination. Such early one-point average rates are valid only when the material corrodes linearly with time; however, this is the exception rather than the rule for metals in natural environments. The effect of film formation on the surface varies the corrosion rate with time in a large majority of cases. The creation of a film of corrosion product often shields the metal substrate from the corrosive environment, with a resulting decrease in corrosion rate. In this case, the projection of corrosion rates before a film has formed will be unduly pessimistic.

Figure 4 shows an example of such short-term evaluation errors for copper exposed by continuous immersion in tropical seawater. The actual corrosion rate (curve C) is a function of the square root of exposure time t. Decreasing-rate curves similar to that shown for copper in seawater have been observed for other metals in natural waters and in the atmosphere (Ref 10, 13). In other situations, such as low-carbon steels in aggressive marine atmospheres, the corrosion rate is nearly linear after formation of a protective film and, under some conditions, may even increase with time. To determine realistic corrosion rates, several replicate sets of specimens (at least three) should be exposed initially and tested after periods of increasing duration.

Specimen Replication. A certain amount of data scatter is inevitable in corrosion test results. The amount of scatter depends on the uniformity of the test material, the condition of the specimen surface, and the stability of the exposure conditions. All of these factors influence the precision of corrosion tests. Therefore, accurate testing procedures will provide some provision for double checking to reduce the likelihood of an incorrect conclusion based on a single nontypical result. This is usually accomplished by testing at least two replicate specimens. The number of replicates usually must be compromised among:

- The accuracy and precision necessary (statistical evaluation)
- The known reproducibility of the test
- The scope of the test program
- The cost of a test

Precision and Bias. At best, the precision of corrosion test results leaves something to be desired because of the relative complexity of corrosion systems. This is particularly true when evaluating forms of localized corrosion, which can often be related to heterogeneous microstructures in the test material. As progress is being made in the standardization of corrosion test methods, more attention is being directed toward providing potential users of the methods with information to help them assess in general terms its usefulness in proposed applications. The American Society for Testing and Materials (ASTM) recently made it mandatory that statements on the precision (reproducibility) and bias (systematic error) be included in every standard for a test method. If this information is not available to the corrosion investigator, he should include in his test program plans to obtain it. Assistance with this phase of corrosion testing is available in Ref 14.

The Unexpected. Because each corrosion experiment (excluding routine quality-control tests) probes the unknown, the experimenter must be alert for unexpected trends. During the test, frequent examinations should be made, and any unusual effects on the specimens or the test environment should be noted. Interim inspection will often permit corrections of problems with the specimens or the corrosive medium in the case of laboratory tests. A regular schedule of examinations is usually established at the beginning of the test, with the initial inspections being more frequent. Appropriate records of specimen condition and environment stability are often necessary to explain unexpected results.

REFERENCES

1. U.R. Evans, *The Corrosion and Oxidation of Metals: Scientific Principles and Practical Applications*, Edward Arnold, 1960
2. F.A. Champion, *Corrosion Testing Procedures*, 2nd ed., John Wiley & Sons, 1965
3. W.H. Ailor, *Handbook on Corrosion Testing and Evaluation*, John Wiley & Sons, 1971
4. G.S. Haynes and R. Baboian, *Laboratory Corrosion Tests and Standards*, STP 866, American Society for Testing and Materials, 1985
5. L.C. Wasson, Helpful Guidelines, Designing a Corrosion Experiment, *Mater. Prot.*, Vol 9 (No. 2), 1970, p 31-33
6. A. de S. Brasunas and C.B. Sonnino, Corrosion Data Reliability, in *Handbook on Corrosion Testing and Evaluation*, W.H. Ailor, Ed., John Wiley & Sons, 1971, p 101-112
7. "Laboratory Corrosion Testing of Metals for the Process Industries," TM-01-69, National Association of Corrosion Engineers
8. "Testing of Metals for Resistance to Sulfide Stress Cracking at Ambient Temperatures," TM-01-77, National Association of Corrosion Engineers
9. "Standard Practice for Preparing, Cleaning, and Evaluating Corrosion Test Specimens," G 1, *Annual Book of ASTM Standards*, Section 3, Vol 03.02, American Society for Testing and Materials
10. G. Sowinski and D.O. Sprowls, Weathering of Aluminum Alloys, in *Atmospheric Corrosion*, W.H. Ailor, Ed., John Wiley & Sons, 1982, p 297-328
11. B.W. Lifka and F.L. McGeary, Corrosion Testing, in *NACE Basic Corrosion Course*, National Association of Corrosion Engineers, 1970
12. C.R. Southwell, J.D. Bultman, and A.L. Alexander, Corrosion of Metals in Tropical Environments—Final Report of 16 Year Exposures, *Mater. Perform.*, Vol 15 (No. 7), 1976, p 9-25
13. R.A. Legault and V.P. Pearson, The Kinetics of the Atmospheric Corrosion of Aluminized Steel, *Corrosion*, Vol 34 (No. 16), 1978, p 344-349
14. "Standard Recommended Practice for Applying Statistics to Analysis of Corrosion Data," G 16, *Annual Book of ASTM Standards*, Section 3, Vol 03.02, American Society for Testing and Materials

In-Service Monitoring*

Sheldon W. Dean, Air Products and Chemicals, Inc.
Donald O. Sprowls, Consultant

IN-SERVICE MONITORING of industrial manufacturing operations is the type of service corrosion testing that presents the greatest challenge and for which there is a great need. In such operations, the expense of corrosion problems can be huge and the risks devastating (Ref 1). Corrosion monitoring has become an important aspect of the design and operation of modern industrial plants because it enables plant engineering and management personnel to be aware of the damage caused by corrosion and the rate of the deterioration. This article, therefore, will focus on methods of monitoring corrosion in industrial plants. The term monitoring, as used in this context, includes any technique for evaluating the progress or rate of corrosion.

Selecting a Corrosion-Monitoring Method

A large variety of techniques are available for corrosion monitoring in plant corrosion tests, and much has been written on the subject in recent years (Ref 2-11). The most widely used and simplest method of corrosion monitoring involves the exposure and evaluation of the corrosion in actual test coupons (specimens). The ASTM standard G 4 was designed to provide guidance for this type of testing (Ref 12). Additional detailed information on procedures from practical experience is given in Ref 10. Extensive overviews of the newer electrochemical methods of corrosion monitoring in industrial plants are given in Ref 7 and 8. However, in view of the growing number of methods available, the selection of corrosion-monitoring methods may be somewhat arbitrary.

In the selection of a corrosion-monitoring method, a variety of factors should be considered. First, the purpose of the test should be understood by everyone concerned with the corrosion-monitoring program. The cost and applicability of the methods under consideration should be known, and it is important to consider the reliability of the method selected. In many cases, it will be desirable to include more than one method in order to provide more confidence in the information generated.

Another question of importance is whether there is access to the process streams and equipment in question. If access is available, methods that involve probes or coupons become more feasible. Otherwise, nondestructive methods may be required. An important factor in the selection of monitoring methods is the response time required to obtain the desired information from the method. Coupon methods and techniques that require plant shutdown tend to be relatively slow in generating information. On the other hand, equipment that measures instantaneous corrosion rates can provide fast results. A final consideration is one of safety. In an operating plant, equipment failure can lead to a leak, which can result in loss of product, a hazard potential, and possible shutdown of the plant. It is important for the monitoring apparatus to minimize the possibility of such an incident.

Direct Testing of Coupons

Although they are not intended to replace laboratory tests completely, plant tests are specifically designed to monitor the life of existing equipment, to evaluate alternative materials of construction, and to determine the effects of process conditions that cannot be reproduced in the laboratory.

Advantages of Coupon Testing (Ref 10). Plant coupon testing provides several specific advantages over laboratory coupon testing. A large number of materials can be exposed simultaneously and can be ranked in actual process streams against a common set of process parameters. Testing can be used to monitor the corrosivity of process streams. The coupons can be designed for specific forms of corrosion, and the exposure time is usually unlimited.

Large Number of Coupons. Because many coupons can be exposed simultaneously, they can be tested in duplicate or triplicate (to measure scatter), and they can be fabricated to simulate such conditions as welding, residual stresses, or crevices. These permutations are then ranked; this gives the engineer increased confidence in selecting materials for new equipment, maintenance, or repair.

Actual process streams will reveal the synergistic effects of combinations of chemicals or contaminants. In addition, the possibility is remote that the corrosion of specific coupons will contaminate the process and affect the corrosion resistance of other coupons. However, one potential source of error that must be checked is contamination of the process stream by the corrosion of the existing equipment.

For example, in a hypothetical case, the equipment is fabricated from a nickel-base alloy, and the process is a reducing acid. Contamination of the process by nickel ions may result in an apparent improvement in the corrosion resistance of titanium test coupons. If the existing equipment is then replaced with titanium based on these false test results, it may corrode rapidly without the beneficial effect of the nickel ions.

Monitoring of Inhibitor Programs. Coupons are widely used to monitor inhibitor programs in, for example, water treatment or refinery overhead streams. With retractable coupon holders, the coupons can be extracted from the process without having to shut down in order to determine the corrosion rate.

Long Exposure Times. Some forms of localized corrosion—for example, crevice corrosion, pitting, and stress-corrosion cracking (SCC)—require time to initiate. To increase confidence in the test results, coupons can be exposed for as long as necessary; this will ensure that the initiation time has been exceeded.

Coupon design can be selected to test for specific forms of corrosion. Coupons can be designed to detect such phenomena as crevice corrosion, pitting, and dealloying corrosion. For example, some pulp mill bleach plant washer drums are electrochemically protected to mitigate crevice corrosion. Specifically designed crevice corrosion test coupons are used to monitor the effectiveness of the electrochemical protection program. These coupons are periodically removed from the equipment and examined for evidence of crevice corrosion.

Disadvantages of Coupon Testing (Ref 10). Coupon testing has four main limitations. First, coupon testing cannot be used to detect rapid changes in the corrosivity of a process. Second, localized corrosion cannot be guaranteed to initiate before the coupons are removed—even with extended test durations. Third, the calculated corrosion rate of the coupon cannot be translated directly into the corrosion rate of the equipment. Lastly, certain forms of corrosion cannot be detected with coupons.

Rapid Changes in Corrosivity. The calculated corrosion rate is an average over a specific period of time; therefore, field coupon testing cannot detect process upsets as they occur. For real-time monitoring, electrochemical methods, such as the polarization resistance technique, may be appropriate (see the section "Polarization Resistance Measurement" in this article).

Corrosion Rates of Coupons Versus Equipment. The corrosion rate of plant equipment seldom equals that calculated on a matching test coupon, because it is very difficult to duplicate the equipment with a coupon. Despite every effort to achieve equivalence, differences in mass

*Portions of this article were adapted with permission from S.W. Dean, Overview of Corrosion Monitoring in Modern Industrial Plants, in *Corrosion Monitoring in Industrial Plants Using Non-Destructive Testing and Electrochemical Methods*, STP 908, G.C. Moran and P. Labine, Ed., American Society for Testing and Materials, 1986, p 197-220 and coupon area/solution volume ratio are usually sufficient to render direct comparison meaningless. Useful correlations can be established by monitoring the corrosion rate of the equipment with ultrasonic thickness monitoring and by comparing this corrosion rate with the calculated rate for equivalent coupons.

Corrosion Forms Not Detected. The principal limitation in this area is the simulation of erosion-corrosion and heat transfer effects. Careful placement of the coupons in the process equipment can slightly offset these weaknesses.

Erosion-corrosion is related to process turbulence, and process turbulence is often a function of equipment design. Because coupons tend to shield one another from the effects of process turbulence, field coupon testing is not reliable as a method of simulating erosion-corrosion.

For heat transfer effects, specially designed coupons are required that simulate effects such as those found in heating elements or condenser tubes. Coupons range in design from thermowell-shaped devices to sample tubes in a test heat exchanger. Thermowell-shaped devices are heated or cooled on the inside and project into the process stream (Ref 12). Heat transfer tests can also be conducted in the laboratory. In this environment, the coupon forms part of the wall of the test vessel and can therefore be heated or cooled from one side. Because of the cost involved, heat transfer coupon tests are usually carried out on only one (or perhaps two) alloys that have been selected from a larger group.

Coupon Options (Ref 10). The design of the coupon is an important part of any plant corrosion-testing program. Proper selection of the coupon shape, surface finish, metallurgical condition, and geometry allows evaluation of specific forms of corrosion.

Uniform Corrosion Coupons. The most common coupon shape for the evaluation of uniform corrosion is rectangular. Circular shapes are also used. Rectangular coupons are the most common because most alloys are available in plate or sheet form. Other shapes are used when there are restrictions on available product forms or when a specific material condition is required. Coupon identification must be legible and permanent. The simplest method of identification is stamping and is preferred unless the test material is likely to be susceptible to SCC.

Coupon finish represents a significant contribution to the overall cost. The least expensive finish that is consistent with the test requirements should be selected. For example, an inexpensive surface finish is acceptable where carbon steel coupons are routinely used to monitor additions of inhibitor in water treatment programs. This may be achieved by punching or shearing, followed by glass bead blasting. On the other hand, when it is necessary to rank alloys in a process environment, the coupons must be finished with ground or machined parallel edges and sanded faces.

A wide variety of coupon finishes are available, such as:

- Mill finish
- Glass bead blasted
- Sandblasted
- Steel shotblasted
- Abrasive cloth or paper sanded
- Machined
- Electrolytically polished

The surface finish of the coupon should match the finish of the equipment. This match is very difficult to achieve for several reasons. For example, there is the difference in surface finish between different mill product forms and even between different heats of the same metal or alloy. In addition, there can be wide variation among mill scale and surface deposits from processing operations. The principal requirement of testing is to assess the corrosion resistance of the test alloy in the condition that would be used in actual equipment.

Abrasive cloth or paper sanding is the most common practice. Sanding removes such surface deposits as mill scale and such defects as scratches or pits. A 120-grit finish, which is standard, can be easily produced without specialized equipment. Metallurgical changes that are heat induced, such as the sensitization of stainless steels or nickel-base alloys, can be prevented by keeping the coupon cool through wet sanding with progressively finer abrasives.

Clean polishing belts should be used; this will avoid contamination of the coupon surface. For example, a belt that has been used to polish a brass coupon should not be used to polish an aluminum coupon.

The corrosion resistance of some alloys can be enhanced by a special surface condition. One example of this is the oxide films that are formed on titanium or zirconium. In this case, conditioning of the coupons would take place after mechanical surface preparation. However, some films that form during mill processing or chemical exposure actually reduce corrosion resistance. Magnetite and certain forms of iron sulfide on carbon steel are examples of these types of films. The coupons must be specially treated in such cases.

Coupons that are cut by punching or shearing will have cold-worked edges. The effects of cold work can extend back from the cut edge a distance equal to the material thickness. These effects must be removed by grinding or machining. Cold working can affect the corrosion rate significantly, and it may cause SCC in some materials. An important parameter in the coupon specification is the degree of edge preparation.

Galvanic Corrosion Coupons. Pairs of test coupons are electrically coupled to study the effects of galvanic corrosion. The relative areas exposed usually vary from 1:1 to 10:1 or greater. The area ratios should be reversed for complete assessment, that is, 10:1 and 1:10, although this may not be necessary when a specific effect—for example, simulation of the galvanic corrosion of dissimilar-metal fasteners in column trays—is being studied.

With metals that can become embrittled by hydrogen absorption—for example, titanium, zirconium, tantalum, and hardenable steels—the cathodic (protected) member of the galvanic couple may be subject to the greater damage. However, the typical mass loss measurements would not reveal such damage.

Crevice Corrosion Coupons. Equipment crevices, such as weld backing rings, tube-tubesheet joints, or flanged connections, are common sites for localized corrosion in the process environment. Many metals perform differently in crevices as opposed to unshielded areas. Behavior is dependent on several factors. These include how strongly oxygen reduction (cathode depolarization) controls the cathode reaction or how the crevice alters the bulk process chemistry by lowering the pH or concentrating aggressive species.

The most imaginative form of coupon corrosion testing is the simulation of crevice corrosion (Ref 13). The various techniques that can be used for crevice corrosion testing include rubber bands, spot-welded lap joints, and wire wrapped around threaded bolts. Each crevice test creates a particular crevice geometry between specific materials and has a particular anode/cathode area ratio. Thus, no crevice corrosion test is universally applicable (see the article "Evaluation of Crevice Corrosion" in this Volume).

The two most widely used crevice geometries in field coupon testing employ insulating spacers to separate and electrically insulate the coupons. Spacers are usually either flat washers or multiple-crevice washers. Either type of spacer can be made of materials ranging from hard ceramics to soft thermoplastic resins (Ref 14).

Reference 15 describes an electrochemical monitoring probe for detecting crevice corrosion. An installation consists of two electrodes—one held in an occluded area, and the other freely exposed to the process stream. These two electrodes are connected to a zero-resistance ammeter. When the condition of the system becomes favorable for crevice corrosion, the electrode in the occluded area begins to corrode, and the boldly exposed electrode serves as a cathode. The current flowing between these two electrodes is a measure of the extent of crevice corrosion. This device has been shown to be effective in laboratory tests for indicating the onset of crevice corrosion and for indicating when conditions become unfavorable for crevice corrosion. Prototype cells have been evaluated by several companies as part of a round robin testing program sponsored by the Materials Technology Institute of the Chemical Process Industries (Ref 16).

Stress-Corrosion Coupons. Typical sources of sustained tensile stress that cause SCC of equipment in service are the residual stresses resulting from forming and welding operations and the assembly stresses associated with interference-fitted parts. Therefore, the most suitable coupons for plant tests are the self-stressed bending and residual stress specimens described in the article "Evaluation of Stress-Corrosion Cracking" in this Volume. Convenient coupons are the cup impression, U-bend, C-ring, tuning fork, and welded panel. The method of stressing for all of these coupons results in decreasing load as cracks form and begin to propagate. Therefore, complete fracture is seldom observed, and careful examination is required to detect cracking.

Welded Coupons. Because welding is a principal method of fabricating equipment, welded coupons should be included in the corrosion test program when applicable. Aside from the effects of residual stresses, the primary concern is the behavior of the weld bead and the heat-affected zone (HAZ). In some alloys, the HAZ becomes sensitized to severe intergranular (sometimes called knife-line) attack, and in certain other alloys, the HAZ is anodic to the parent metal. Laboratory potential measurements can be conducted on the weldment to determine whether electrochemical effects are involved. If either the weld bead or the HAZ is anodic to the parent metal, the welded coupons should be made as large as possible so that the cathode/anode ratio will approach that in actual equipment. The concern in this case is the same as that discussed previously for galvanic couples. When possible,

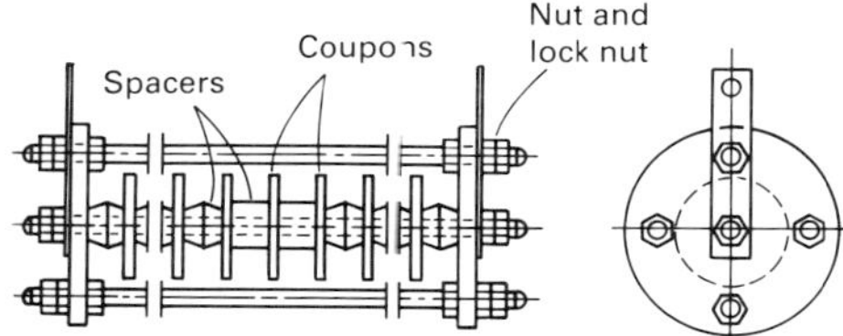

Fig. 1 Illustration of typical spool-type coupon rack. Source: Ref 12

it is more realistic to remove welded coupons from production-size weldments than to weld the small coupons.

Sensitized Metal. Sensitization is a metallurgical change that occurs when certain austenitic stainless steels, ferritic stainless steels, nickel-base alloys, and other alloys are heated under specific conditions. This results in the precipitation of carbides at grain boundaries, which reduces corrosion resistance. Any heat-inducing process (for example, stress relief or welding) may cause sensitization, which is time- and temperature-dependent. There is a specific temperature range over which each particular alloy will sensitize rapidly. Sensitization leads to intergranular corrosion in specific environments.

Welding is the most common cause of sensitization. However, welded coupons may not exhibit sensitization, because they may be given insufficient weld passes (compared with actual process equipment). As a result, they spend insufficient time in the sensitizing temperature range, and susceptibility to intergranular corrosion may not be detected.

An appropriate sensitizing heat treatment guarantees that any corrosion susceptibility induced by welding or heat treatment will be detected. The optimum temperature and time ranges for sensitization vary for different alloys. For example, 30 min at 650 °C (1200 °F) is usually sufficient to sensitize AISI type 316 stainless steel.

Some of the corrosion-resistant aluminum-magnesium (5000-series) alloys containing 3 to 6% Mg are also subject to sensitization when heated at temperatures in the range of 65 to 175 °C (150 to 350 °F). Reference 17 describes a test method for susceptibility to this phenomenon. The effective time and temperature conditions depend on the alloy content and metallurgical condition (Ref 18).

The advantage of using sensitized coupons is that, if corrosion occurs, they flag a potential problem. The user is thus obliged to consider how the equipment will be welded and heat treated and to determine whether high-temperature excursions will occur during operation.

Test Rack Design. It may be possible to expose corrosion coupons under stagnant or slowly flowing conditions by simply hanging them on an insulated wire or plastic cord, but this procedure is generally inadequate. Instead, specimen holders and test racks are usually used to support and insulate the coupons. These racks must hold coupons firmly in place to prevent mechanical damage and metal loss from causes other than corrosion. They must also electrically isolate the coupons from contact with each other and from the vessel itself to prevent unintentional electrochemical interactions. The basic types of racks will be discussed briefly below. They are described in detail in Ref 2, 5, 10, and 12.

Spool (Birdcage) Rack. A typical spool rack is useful in open vessels, such as reactors and tanks, for which access to the spool is readily achieved (Fig. 1). It has the advantage that a relatively large number of coupons can be accommodated.

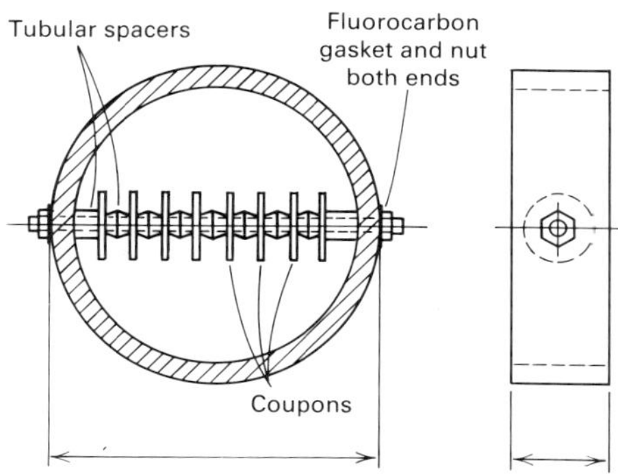

Fig. 2 Illustration of typical dutchman coupon rack. Source: Ref 12

The insert rack is convenient for use in pipes or other units that have flanged connections that allow easy access to the system. For larger-diameter pipes or nozzles, a dutchman-type rack can be used. Such a rack consists of a suitable spool-type piece with the coupons mounted crosswise on a bar (Fig. 2). In both cases, the equipment must be shut down during installation and removal of the racks.

A slip-in rack is designed to be inserted into and removed from equipment that is in operation. A retractable coupon holder makes this type of rack especially useful (Fig. 3). The slip-in rack requires a gate valve and a suitable-sized nozzle to serve as a retraction chamber. A rod-shaped coupon holder is contained in the retraction chamber, which is flanged to the gate valve. The other end of the retraction chamber contains a packing gland through which the coupon holder passes. Coupons are mounted on the rod in the extended position and are then drawn into the retraction chamber. The chamber is bolted to the gate valve, which is then opened to allow the coupons to be slid into the process stream. The sequence is reversed to remove the test coupons.

Coupon Cleaning and Evaluation. The test coupons should be cleaned as soon as possible after removal from the test. The procedures for cleaning and weighing, which depend on the test material and the extent of corrosion, are described in Ref 4, 10, and 19. These methods are overviewed in the article "Interpretation and Use of Corrosion Test Results" in this Volume.

Examination of the coupons after cleaning and weighing reveals the forms of corrosion that may be expected in equipment made of the coupon material. Because the coupons used for calculating corrosion rates are usually small, they should be examined very carefully for signs of localized corrosion effects that could invalidate the mass loss data. Coupons are examined with the unaided eye and then at increasing magnifications up to 30 to 50× with a binocular microscope. The scanning electron microscope is an extremely useful tool for detecting superficial localized effects.

In some cases, coupons must be bent and/or sectioned and metallographically examined to reveal certain types of corrosion damage. There are special localized corrosion effects that could not only jeopardize determination of realistic corrosion rates but also signal other serious types of behavior. These effects include edge attack, crevice corrosion, stress corrosion near stenciled identification numbers, pitting, selective corrosion (such as dealloying), knife-line attack, blistering, intergranular corrosion, embrittlement, and erosion-corrosion. In addition to the various corrosion effects, there may be inadvertent mechanical or galvanic corrosion damage. Techniques for evaluating specific forms of corrosion are described in various articles in this Section.

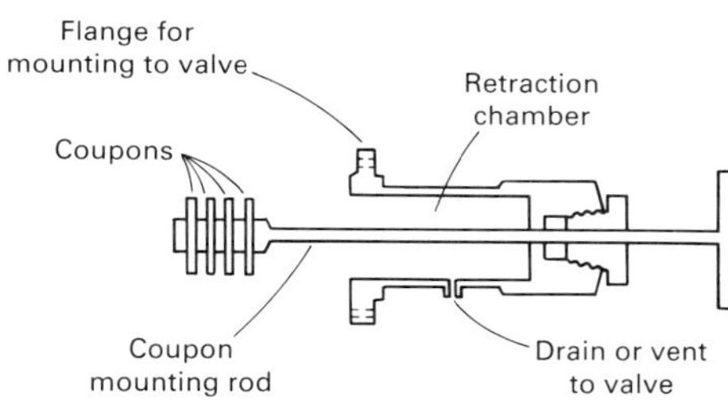

Fig. 3 Illustration of retractable coupon holder. Source: Ref 10

Electrical Resistance Probes

Electrical resistance probes (Fig. 4) are specially designed corrosion coupons. Their corrosion rate is calculated from measurement of electrical resistance rather than mass loss. These measurements are made by installing a wire or other device fabricated from the material in question in such a way that its electrical resistance can be conveniently measured. Corrosion reduces the cross section of the exposed element; therefore, its electrical resistance will increase with exposure time if corrosion is taking place. A temperature-compensating element should be incorporated in such a probe because the resistance of the probe is also influenced by the temperature.

Electrical resistance probes measure the remaining average metal thickness. To obtain the corrosion rate, a series of measurements is made over a period of time, and the results are plotted as a function of exposure time. The corrosion rate can be determined from the slope of the resulting plot.

There are several advantages to this approach. Because probes are relatively small, they can be installed easily. To determine the metal remaining, the probe can be wired directly to a control room location or to a portable resistance bridge at the probe location. For systems that are wired directly to control rooms, a computer system can be used to obtain the data and to reduce the results to corrosion rate values. Commercial electrical resistance equipment is widely available. Use of electrical resistance probes does not depend on the environment; there are two reasons for this. First, only the remaining metal is measured, and second, the conductivity of the corrosive environment is usually inconsequential. This is not true in, for example, molten salts or liquid metals; the high conductivity of the environment in such systems precludes the use of electrical resistance probes. Thus, no problems exist in measuring corrosion that occurs in liquid or vapor streams. Electrical resistance probes are used in various applications, such as monitoring the corrosivity of cooling water systems (Ref 20).

However, there are some disadvantages to this approach. Vessel and piping walls must be penetrated in order to install the probes; such penetration results in the potential for leaks. It is

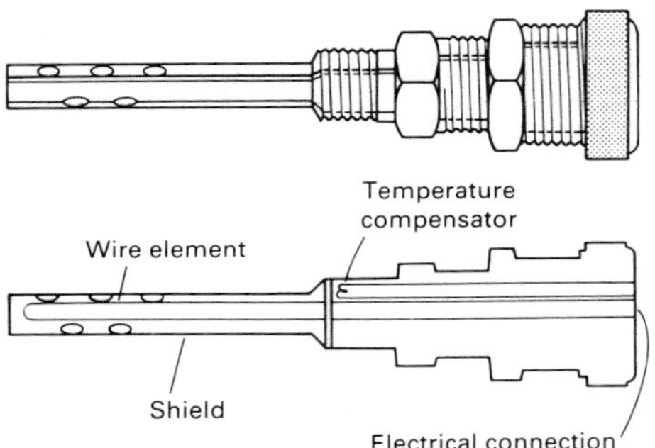

Fig. 4 Electrical resistance probe. Source: Ref 7

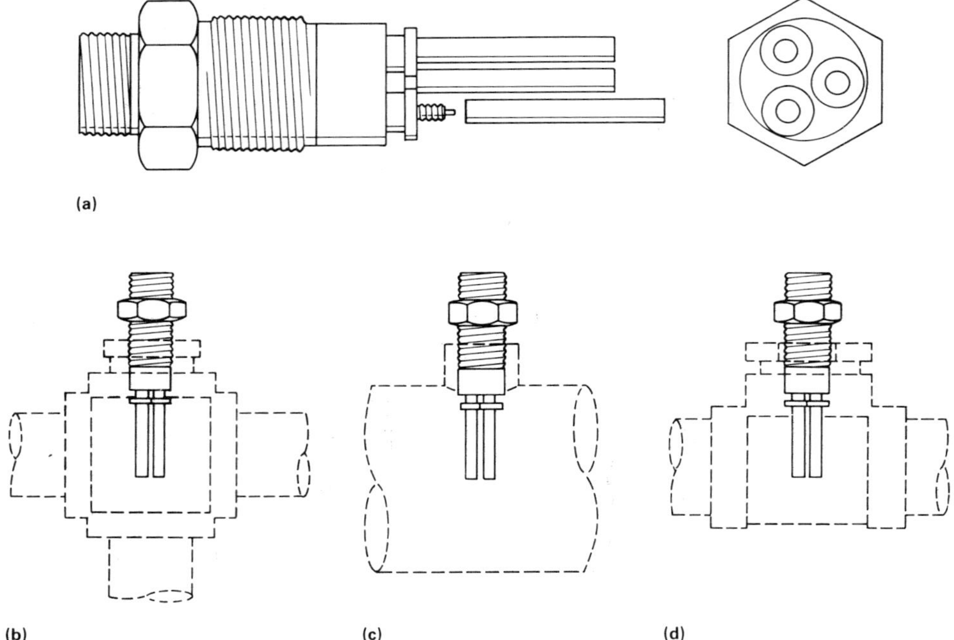

Fig. 5 Typical three-electrode polarization resistance probe (a) and installations of such probes. (b) Probe in a 38-mm (1½-in.) pipe fitting. (c) Probe in welded line. (d) Probe in 50-mm (2-in.) pipe tee. Source: Ref 7

expensive to direct wire the probe to a control room location, and such work must be carried out with care to avoid spurious signals and errors. On the other hand, it is time consuming and sometimes impossible to take measurements at the probe site with a portable bridge. The temperature compensation device reacts slowly, and it can be a source of error if the temperature varies when the measurement is taken. Corrosion rate measurements obtained in short periods of time can be inaccurate because the method measures only the remaining metal, not the rate of attack; this increases the signal-to-noise ratio in short exposures. This method provides no information on localized attack.

Ultrasonic Thickness Measurements

Ultrasonic thickness measurements can be used to monitor corrosion rates *in situ*. Ultrasonic thickness measurements involve placement of a transducer against the exterior of the vessel in question. The transducer produces an ultrasonic signal. This signal passes through the vessel wall, bounces off the interior surface, and returns to the transducer. The thickness is calculated by using the time that elapses between emission of the signal and its subsequent reception, along with the velocity of sound in the material. To obtain a corrosion rate, a series of measurements must be made over a time interval, and the metal loss per unit time must be determined.

One advantage of this method is that it is not necessary to penetrate the vessel to make the measurements. In addition, the results are obtained in terms of thickness. Small hand-held probes and reading devices are available for making these measurements and are relatively easy to use.

There are also several disadvantages. Ideally, a bare metal surface free of paint, thermal insulation, and corrosion products must be exposed. A coupling agent, such as grease, petroleum jelly, or oil, must be used so that the signal can pass from the probe into the metal and return. Problems may arise when vessel walls are at high or low temperatures. Serious problems may exist in equipment that has a metallurgically bonded internal lining, because it is not obvious from which surface the returning signal will originate.

Despite these drawbacks, the ultrasonic thickness approach is widely practiced where it is necessary to evaluate vessel life and suitability for further service. It must be kept in mind, however, that depending on the type of transducer used the ultrasonic thickness method can overestimate metal thicknesses when the remaining thickness is under approximately 1.3 mm (0.05 in.).

Polarization Resistance Measurement

Unlike the previously discussed methods, which provide information on remaining thickness, the technique of polarization resistance provides an estimate of the corrosion rate. This method is based on the Stearn-Geary equation (Ref 21). The theory behind the technique is that the corrosion rate of a probe is inversely proportional to its polarization resistance, that is, the slope of the potential-current response curve near the corrosion potential. It is necessary in a plant situation to use a probe that enters the vessel in the area where the corrosion rate is desired (Fig. 5). The electrodes of the probe are fabricated from the material in question. An electronic power supply polarizes the specimen about 10 mV from the corrosion potential. The resulting current is recorded as a measure of the corrosion rate. Polarization resistance yields an instantaneous estimate of corrosion rate.

Several commercially available probes and analyzing systems can be directly interfaced with remote computer data acquisition systems. Alarms can also be used to signal plant operators when high corrosion rates are experienced.

There are several limitations to this approach. The corroding environment must be an electrolyte with reasonably low resistivity. High-resistivity electrolytes produce erroneously low corrosion rates. The vessel wall must be penetrated, and this involves concerns regarding leaks, personnel safety, and other problems. The ability to use direct wiring from the probe location to a remote control room is desirable, but the installation of these wiring systems is costly. In addition, these systems do not provide information on localized corrosion, such as pitting and SCC. Also, the corrosion rate values are approximate at best, and the method is best suited for use during periods when substantial corrosion rate changes occur.

Measurement of Corrosion Potentials

The use of corrosion potential measurements for in-service corrosion monitoring is not as widespread as the use of polarization resistance. However, this approach can be valuable in some cases, particularly where an alloy could show both active and passive corrosion behavior in a given process stream. For example, stainless steels can provide excellent service as long as they remain passive. However, if an upset occurs that would introduce either chlorides or reducing agents into the process stream, stainless alloys may become active and may exhibit excessive corrosion rates. Corrosion potential measurements would indicate the development of active corrosion, and they may be coupled with polarization resistance measurements as additional confirmation of high corrosion rates (Ref 8).

The success of corrosion potential measurements depends on the long-term stable performance of a standard reference electrode. Such electrodes have been developed for continuous pH monitoring of process streams, and their application for measuring corrosion potentials is straightforward. However, the conditions of temperature, pressure, electrolyte composition, pH, and other possible variables can limit the applications of these electrodes for corrosion-monitoring service.

Alternating Current Impedance Measurements

Alternating current (ac) impedance-polarizing techniques are being increasingly used in electrochemical evaluation. In principle, the same probe elements used for polarization resistance measurements can be used for ac impedance studies. As shown in Fig. 6, the ac impedance approach permits the separation and independent analysis

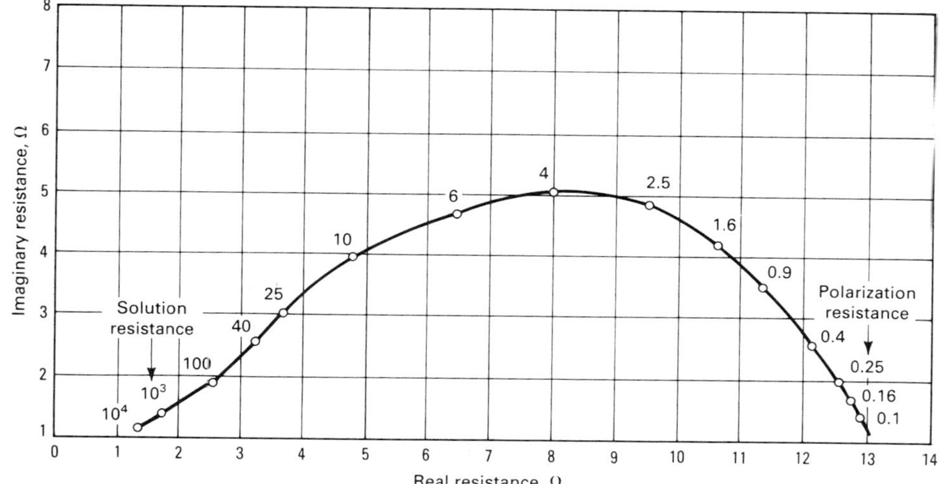

Fig. 6 Typical plot of ac impedance data in terms of real and imaginary resistance. The numbers on the curve indicate the frequency in hertz. Source: Ref 7

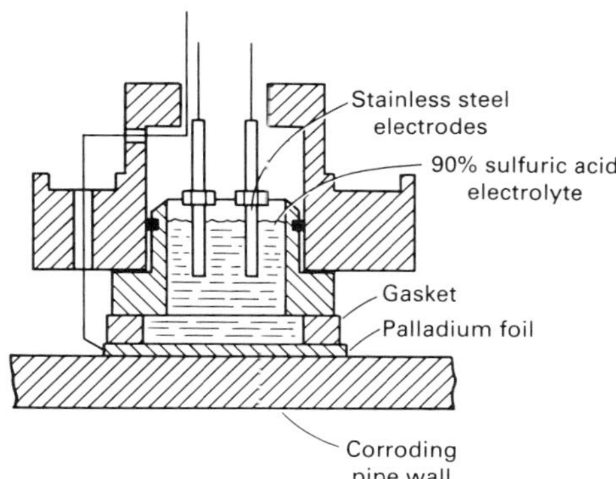

Fig. 7 Design of palladium foil hydrogen patch probe. Source: Ref 24

of the resistive and capacitive elements of the electrochemical corrosion reaction (Ref 22). The method is particularly useful for high-resistivity electrolytes, such as steam condensate. In these cases, polarization resistance measurements provide readings that are erroneously low, because the bulk resistance of the electrochemical cell adds to the measured resistance. This problem is avoided with the ac approach.

A sophisticated ac frequency generator and analyzer system is necessary to obtain results such as those shown in Fig. 6. Some investigators have suggested using only two frequencies to characterize ac impedance behavior (Ref 23). Although this will minimize the number of measurements, it does not reveal the details of the behavior. It is necessary to be familiar with the approach in order to interpret the results of ac impedance studies. The specimen designs used in commercial polarization resistance probes have not been optimized for ac analysis. For this reason, a full frequency response curve is desirable in order to verify that the behavior can be analyzed by conventional techniques, although it may take several hours to obtain a complete frequency response curve. Thus, more development is necessary before the ac impedance technique will achieve widespread acceptance for in-service corrosion monitoring. Another drawback is that the Stearn-Geary constant (Ref 21) or Tafel slopes must be known in order to convert ac impedance data into corrosion rate information.

Hydrogen Probe

The concept of the hydrogen probe is based on the fact that one of the cathodic reaction products in nonoxidizing acidic systems is hydrogen. The hydrogen atoms thus generated diffuse through the thickness of the vessel and are liberated at the exterior surface. A hydrogen probe—in one variation, the probe is a piece of palladium foil—is applied to the exterior surface. The probe serves as collector and catalyst for the subsequent oxidation of the hydrogen (Ref 24). The cell is attached to the palladium foil. The current employed in the oxidation reaction is supplied by a stainless steel working electrode (Fig. 7). The electrolyte is 90% sulfuric acid. The current that is required to oxidize the hydrogen is measured; it is directly proportional to the corrosion rate of the interior of the vessel.

Hydrogen probe analysis measures the corrosion rate, unlike ultrasonic thickness measurements and other techniques, which measure remaining wall thickness. However, hydrogen probe analysis is limited to systems in which the temperature is close to ambient and the diffusion rate of hydrogen is high. Gas pipeline service is the most common application. In this case, corrosion can occur when hydrogen sulfide, water, and sometimes carbon dioxide (CO_2) are present. This approach has another variation that consists of simply attaching a chamber to the exterior of the pipe and monitoring hydrogen liberation through increasing pressure (Ref 2).

One advantage of exterior hydrogen monitoring is that it does not require penetration of the pipe wall in order to obtain the corrosion rate. It is assumed in this method that all of the hydrogen liberated in the corrosion reaction diffuses through the steel vessel wall instead of being liberated as hydrogen gas at the surface, and this is true when hydrogen recombination poisons such as hydrogen sulfide (H_2S) are present. Commercial devices are available for corrosion monitoring by this technique, although it is questionable whether such devices could be positioned and allowed to operate unattended for extended periods of time. In addition, these units have all of the problems associated with any electrochemical measuring device, namely the need for complex electronic equipment and wiring and the need for operators and installers with a sensitivity to the requirements of such equipment. Also, this method is in practice limited to steel, which has a high hydrogen diffusivity and low solubility of hydrogen.

Exterior hydrogen monitoring does not supply a quantitative measurement of hydrogen damage. However, it does provide a means of measuring the relative severity of corrosion damage and of evaluating the effects of process changes on the severity of corrosion.

Analysis of Process Streams

Another useful in-service corrosion monitoring technique is analysis of the process streams for the presence of corrosion products. This straightforward approach usually does not require the installation of specialized equipment. For example, process streams from the bottom of the distillation column can be routinely sampled, and atomic absorption analysis techniques can be used to determine such heavy metals as iron, nickel, and chromium at very low levels. The concentration of such impurities is then directly proportional to the corrosion rate multiplied by the area of metal corroding if the only source of metal ions is corrosion and if the corrosion products are not precipitating. One problem is that the corroding area may not be known with certainty; therefore, the results are relative. However, they do help to determine whether conditions have improved.

Sentry Holes

In this monitoring technique, small sentry holes are drilled into the outside of a vessel or pipe at areas that are considered to be particularly susceptible to corrosion. The holes are drilled to the pressure design thickness. Thus, when corrosion has almost consumed the corrosion allowance, the appearance of a small leak indicates that action must be taken to prevent a major failure. Sentry holes may be threaded or may have nipples attached to facilitate plugging. Nondestructive testing is frequently performed in the area near the leak to determine the extent of the damage before repair or shutdown.

Side-Stream (Bypass) Loops

It is sometimes advantageous to operate a side-stream loop to determine the effect of chemical process changes without making these changes in the entire process system. This provides the advantage of an in-service corrosion test with real product streams, yet permits evaluation of the effects of additives, inhibitors, and other changes in the environment without affecting the main process stream.

In the simplest case, a side stream from a major piece of equipment is run through a small tank or a widened section of the line in order to permit changes in the product stream and insertion of corrosion-monitoring devices. In such a case, the effluent from the side stream is usually discarded. In a more complicated installation, the entire cycle of the operation may be duplicated on a small scale using a portion of the mainstream. Obviously, a more sophisticated side-stream apparatus can approach the complexity and the expense of a pilot plant.

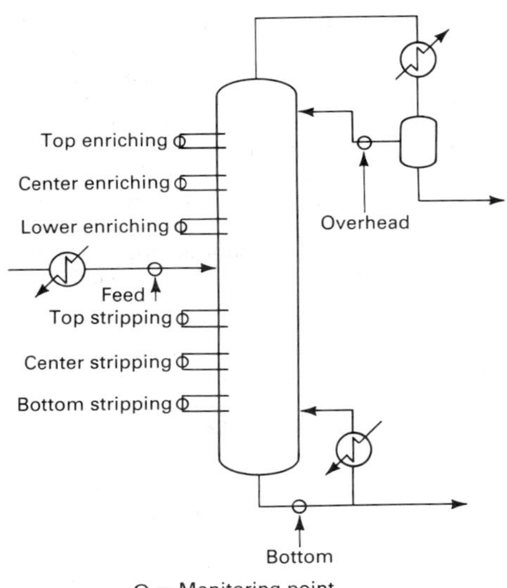

Fig. 8 Distillation column showing preferred locations of monitoring probes or other devices. Source: Ref 7

Tapped blind installed here in case of a leak
Block valve and disconnection point
Internal tubing run between chamber and tray level
Tray level *n*
Tray level *n*-5
Corrosion rate probe
Valve closed
Tapped blind installed here in case of leak
Sample valve
Sample bomb
To sample discharge point

Fig. 9 Use of a side-stream (bypass) loop to monitor corrosion in a distillation column. Source: Ref 7

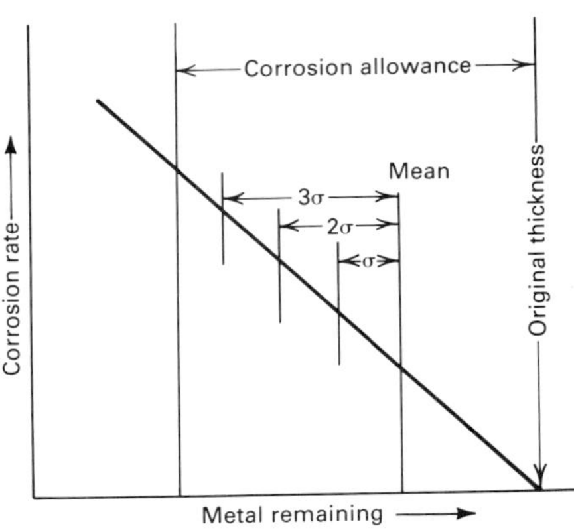

Fig. 10 Typical plot of metal thickness remaining versus calculated corrosion rate obtained from ultrasonic thickness data. Statistical analysis can be used to estimate the remaining corrosion allowance in terms of standard deviation σ. Source: Ref 7

Strategies in Corrosion Monitoring

Monitoring Locations. A principal part of any corrosion-monitoring program is deciding where to locate the corrosion-monitoring devices. Because corrosion will probably not occur uniformly throughout the plant, it is desirable in any corrosion-monitoring program to find sites at which the highest corrosion rates will be experienced.

The problems involved in developing corrosion-monitoring programs for a plant are illustrated in the example of a distillation column. The most logical points for corrosion monitoring in a distillation column are the feed point, the overhead product receiver, and the reboiler or bottoms product line. These points are the locations at which the highest and lowest temperatures are encountered, as well as the points at which the most and least volatile products are concentrated. However, these points are usually not the locations of the most severe corrosion.

The species causing the corrosion will often concentrate at an intermediate point in the column because of chemical changes within the column. Therefore, if there is a possibility of concentrating a corrosive species within the distillation column, several monitoring points would be required throughout the column for the corrosion-monitoring program to be comprehensive (Fig. 8). A monitoring program can be restricted to the most corrosive location within the column once this area has been identified (Ref 25).

Another problem with distillation columns is that the liquid on trays tends to be frothy. This creates difficulties for electrochemical methods. One solution to this problem is to install bypass loops that remove liquid from the column, pass it over the corrosion-monitoring probes, and reinject it at a lower point in the column (Ref 2). This practice avoids the problem of foam and froth and provides a more controlled flow rate over the corrosion-monitoring equipment. Use of a bypass loop also allows removal of liquid samples at times of high corrosion rates (Fig. 9).

The process stream can be sampled in locations other than distillation columns. A sample tap can be helpful in conjunction with polarization resistance devices. The alarm on polarization resistance monitoring equipment can be used to signal the need to remove samples. This is particularly desirable in the case of pilot plant operation, in which wide variations in processing conditions are encountered. It is also helpful in plants that produce different products in the same equipment.

High-velocity gas streams in pipes may cause problems with conventional monitoring systems. In this case, the presence of an aqueous phase is usually restricted to a thin layer on the surface of the pipe. A probe that protrudes into the pipe may miss the liquid layer that is present only close to the pipe wall. A flush-mounted surface probe can be used in such cases. This probe permits the measurement of polarization resistance in order to estimate the corrosion rate of the pipe wall (Ref 26).

Probe location is also critical in storage tanks containing nonaqueous liquids. The most corrosive location in these tanks may be at the liquid level if the liquid in the tank has a density exceeding that of water. In this case, the corrosion-monitoring probe should be mounted on a floating platform. In this way, the probe can detect the presence of a corrosive aqueous phase. However, when the liquid stored in the tank is less dense than the water, the corrosion-monitoring device should probably be positioned at the bottom of the tank.

Ultrasonic Thickness Data Analysis. Ultrasonic thickness measurements are often obtained during turnaround periods when the plant is not operating. A well-designed program will provide a series of thickness measurements at specified locations throughout the plant. Assuming that the original thickness of a vessel was constant, a diagram similar to that shown in Fig. 10 can be drawn. Figure 10 shows the inverse relationship between the average corrosion rate and the remaining thickness. It is then possible to estimate the remaining life of the vessel from the maximum corrosion rate and the remaining corrosion allowance. In cases in which corrosion rate varies randomly, the highest corrosion rate and the smallest remaining thickness can be estimated by using statistical analysis. This approach provides a rational method of estimating remaining life when faced with a quantity of thickness readings from an ultrasonic thickness survey.

Redundancy is also important in designing corrosion-monitoring programs. The use of at least two different types of corrosion-monitoring devices at any location is often desirable. For example, the use of an electrical resistance probe with a polarization resistance probe allows measurement of both instantaneous corrosion rates and an average corrosion rate. The data thus obtained can be correlated, and this is very helpful in identifying spurious or inaccurate readings.

In another approach, the polarization resistance probe is weighed before and after the test in order to correlate mass loss with the average corrosion rate that the probe suffered. There is reason to expect that the electrochemical value is in error if the average corrosion rate and the mass loss of the probe do not agree. Also, to obtain time average corrosion rate values, independent coupons can be installed together with polarization resistance probes.

A variety of corrosion-monitoring approaches must be used when designing pilot or demonstration plants. Coupon tests can be very helpful in selecting optimum materials for processes based on pilot plant experience. Polarization resistance monitoring is very useful for determining whether certain processing conditions cause corrosive situations to develop. Because the corrosion mechanisms are often not well understood and because the result of erroneous information can be serious overdesign or exposure to unexpected hazards, redundancy in the design of such monitoring systems is important in pilot plants.

Other Issues. A primary difficulty with corrosion-monitoring equipment is the need to install wires from the probes to the control room or to the instruments and data storage systems. Hard wire systems are usually more expensive than the probes and electronic instruments. In addition, wiring systems are often sources of problems due to breakage, moisture entry, and connection difficulties. One option is the use of devices to transform the data to coded CB radio signals. These small battery-operated units can be installed at probe locations, and they provide the desired information to the base station upon command. The base station would be located in the control room. Acceptance of this approach may increase as such systems become more available and more reliable.

The problem of leaks is another important issue in corrosion-monitoring systems. Leaks can cause hazards and can shut down operating plants. The corrosion-monitoring system must be installed such that leaks from the probe can be handled with minimal interruption. A device that penetrates the wall of the vessel may leak; therefore, it is essential to have contingency plans for dealing with leaks before the devices are installed.

A related problem concerns packing glands on pressure vessels that have removable devices. The pressure within a system exerts a force on any removable device. Therefore, it is important to prevent the device from being blown out, which would injure personnel attempting to remove the device or would expose such personnel to the fluid within the vessel. For this reason, restraining rods, chains, and so on, must be used.

Interpretation and Reporting

In-service monitoring, more than any other type of corrosion testing, requires the utmost skill in interpretation and reporting. Important economic decisions are often based on the test results. Although there are a number of standards that provide guidelines for certain procedures, there is no standard that is comprehensive. To plan an appropriate test program, the investigator must know or have good advice on both the chemistry and the mechanics of the processes involved; that is, the investigator must understand the entire corrosion system. There must be a strong emphasis on the strategy of the program and a searching analysis of the test results. It is important to prepare detailed records of what was done as part of the experiment, and it is also important to document any unplanned changes that occur in the process stream or the equipment during the investigation. Without a valid interpretation and effective (timely) reporting, the price of the work can be significantly greater than the cost of time and materials. However, the consequences of corrosion failures go beyond additional costs. Also involved are personal safety risks (and liability), hazard potentials, and product quality and pollution problems.

REFERENCES

1. J.H. Payer, W.K. Boyd, D.G. Dippold, and W.H. Fisher, NBS-Battelle Cost of Corrosion Study ($70 Billion!) Part I, Introduction, *Mater. Perform.*, Vol 19 (No. 5), May 1980, p 34
2. C.P. Dillon, A.S. Krisher, and H. Wissenberg, Plant Corrosion Tests, in *Handbook on Corrosion Testing and Evaluation*, W.H. Ailor, Ed., John Wiley & Sons, 1971, p 599-615
3. A.S. Couper, Methods in Oil-Refining Industries, in *Handbook on Corrosion Testing and Evaluation*, W.H. Ailor, Ed., John Wiley & Sons, 1971, p 617-624
4. "Laboratory Corrosion Testing of Metals for the Process Industries," NACE TM-01-69, National Association of Corrosion Engineers, 1976
5. J.N. Wanklyn, *Corrosion Monitoring in the Oil and Petrochemical and Process Industries*, J. Wanklyn, Ed., Oyez Scientific and Technical Service Ltd., 1982
6. *Corrosion Monitoring in Industrial Plants Using Nondestructive Testing and Electrochemical Methods*, STP 908, G.C. Moran and P. Labine, Ed., American Society for Testing and Materials, 1986
7. S.W. Dean, Overview of Corrosion Monitoring in Modern Industrial Plants, in *Corrosion Monitoring in Industrial Plants Using Nondestructive Testing and Electrochemical Methods*, STP 908, G.C. Moran and P. Labine, Ed., American Society for Testing and Materials, 1986, p 197-220
8. S.W. Dean, Electrochemical Methods for Corrosion Testing in the Process Industries, in *Electrochemical Corrosion Testing With Special Consideration of Practical Applications*, E. Heintz, J.C. Rowlands, and F. Mansfeld, Ed., Proceedings of International Workshop, Ferrara, Italy, DECHEMA, 1986, p 1-15
9. *Process Industries Corrosion—Theory and Practice*, National Association of Corrosion Engineers, 1986
10. B.J. Moniz, Field Coupon Corrosion Testing, in *Process Industries Corrosion—Theory and Practice*, National Association of Corrosion Engineers, 1986
11. S.J. Pikul, T.S. Lee, and D.B. Anderson, "Field Corrosion Tests for Materials Evaluation in Chemical Process Industries," Paper 119, presented at Corrosion/86, National Association of Corrosion Engineers, 1986
12. "Standard Method for Conducting Corrosion Coupon Tests in Plant Equipment," G 4, *Annual Book of ASTM Standards*, American Society for Testing and Materials
13. W.D. France, Jr., Crevice Corrosion of Metals, in *Localized Corrosion—Cause of Metal Failure*, STP 519, M. Henthorne, Ed., American Society for Testing and Materials, 1972, p 164-200
14. "Standard Guide for Crevice Corrosion Testing of Iron-Base and Nickel-Base Stainless Alloys in Seawater and Other Chloride Containing Aqueous Environments," G 78, *Annual Book of ASTM Standards*, American Society for Testing and Materials
15. R.B. Diegle, G.A. Breeze, and W.E. Berry, "Electrochemical Monitoring of Occluded Cell Corrosion," MTI Technical Report 3, MTI Project 2, Phase I, Materials Technology Institute of the Chemical Process Industries, March 1981
16. A.K. Agrawal and J.H. Payer, "A Sensor for Monitoring Crevice Corrosion—An Analysis and Evaluation," MTI Technical Report 8, MTI Project 2, Phase II, Materials Technology Institute of the Chemical Process Industries, March 1983
17. "Standard Practice for Determining the Susceptibility to Intergranular Corrosion of 5xxx Series Aluminum Alloys by Weight Loss After Exposure to Nitric Acid (NAWLT Test)," G 67, *Annual Book of ASTM Standards*, American Society for Testing and Materials
18. E.H. Dix, Jr., W.A. Anderson, and M.B. Shumaker, Influence of Service Temperature on the Resistance of Wrought Aluminum-Magnesium Alloys to Corrosion, *Corrosion*, Vol 15 (No. 2), 1959, p 55t-62t
19. "Standard Practice for Preparing, Cleaning, and Evaluating Corrosion Test Specimens," G 1, *Annual Book of ASTM Standards*, American Society for Testing and Materials
20. "Standard Test Methods for Corrosivity of Water in the Absence of Heat Transfer (Electrical Methods)," D 2776, *Annual Book of ASTM Standards*, American Society for Testing and Materials
21. M. Stern and R.M. Roth, *J. Electrochem. Soc.*, Vol 104, 1958, p 440t
22. S.W. Dean, in *Electrochemical Techniques for Corrosion*, R. Baboian, Ed., National Association of Corrosion Engineers, 1977, p 52-60
23. S. Haruyama and T. Tsuru, in *Electrochemical Corrosion Testing*, STP 727, F. Mansfeld and U. Bertocci, Ed., American Society for Testing and Materials, 1981, p 167-186
24. R. Martin and E.C. French, "Corrosion Monitoring in Sour Systems Using Electrochemical Hydrogen Potential Probes," Paper 6657, presented at the Sour Gas Symposium of the Society of Petroleum Engineers, Dallas, TX, American Institute of Mining, Metallurgical and Petroleum Engineers, 1977
25. C.G. Arnold and D.R. Hixson, *Mater. Perform.*, Vol 21 (No. 4), April 1982, p 21-24
26. E.C. French, *Oil Gas J.*, Vol 17, 1975, p 82-88

Simulated Service Testing

SIMULATED SERVICE TESTING is the most reliable predictor of corrosion behavior short of actual service experience. This includes exposures of either structural components or test specimens in outdoor environments that are representative of many general service situations. These so-called natural environments include exposures to the atmosphere, waters, and soil. Test materials are subjected to the cyclic effects of the weather, geographical influences, and bacteriological factors that cannot be realistically duplicated in the laboratory.

This type of testing is important for such objectives as materials selection, predicting the probable service life of a product or structure, evaluating new commercial alloys and processes, and calibrating laboratory corrosion tests. This type of information sought determines the selection of test specimens and the methods of assessing the corrosion effects.

Corrosion Testing in the Atmosphere

K.L. Money
LaQue Center for Corrosion Technology, Inc.

The corrosion or degradation of materials in the atmosphere occurs naturally. The rate of degree of degradation varies for different materials and is influenced by several environmental factors. Many of these factors are natural in origin, but some result from man-made sources. Among the latter sources, which are known to affect atmospheric degradation of materials, are the SO_x and NO_x compounds produced as fossil fuel combustion by-products. These species can react in the atmosphere and result in acid deposition. Although considerable public attention is focused on the effects of acid deposition on our ecosystem, the potential damage to materials may represent the largest economic impact of acid deposition.

Over the years, considerable attention has been devoted to the weathering of building materials. Observations of materials corrosion performance in the atmosphere have ranged from those made on existing structures to carefully designed experimental programs. These data represent an excellent base for understanding the behavior of given materials. However, much of the existing data suffer from two primary differences. First, many testing programs have been limited to a specific characterization of a material for a given application. This has resulted in variations in experimental methods such that the results of one testing program cannot be compared unambiguously to the results of others. Second, and most serious for present considerations, these programs have generally been conducted with an absence of comprehensive atmospheric monitoring. This has prevented the establishment of cause and effect relationships for some phenomena and has precluded the establishment of a baseline understanding of the atmospheric environmental characteristics for which a considerable materials degradation data base exists. To assess the impact of atmospheric variables on materials, it is necessary to monitor or have available relevant environmental parameters in association with controlled atmospheric exposures of materials of interest.

Types of Atmospheres

Before the materials or materials data can be evaluated, the atmospheric variables of the exposure site should be determined and periodically recorded during the test period. The variability of atmospheric-corrosion severity has been well documented (Ref 1-3). The severity of the environment is usually indicated by designating an environment as rural, urban, industrial, marine, or a combination of these. Table 1 lists the typical activity ranges of SO_x and Cl^- in these environments, which are discussed below.

A rural atmosphere is normally classified as one that does not contain chemical pollutants, but does contain organic and inorganic dusts. Its principal corrodent is moisture and, of course, oxygen and carbon dioxide. Arid or tropical atmospheres are special cases of the rural environment because of their extreme relative humidities and condensations (Ref 5). The rural atmosphere is generally the least corrosive.

An urban atmosphere is similar to the rural environment in that it is away from the industrial complexes. Materials exposed in these areas are subjected to the normal precipitation patterns and typical urban contaminants of SO_x and NO_x emitted by motor vehicles and home fuels.

An industrial atmosphere is typically identified with heavy industrial manufacturing facilities. These atmospheres can contain concentrations of sulfur dioxide, chlorides, phosphates, nitrates, or other specific industrial emissions. These emissions combine with precipitation or dew to form the liquid corrosive.

Table 1 Typical activity ranges of SO_x and chlorides measured in various atmospheres

These are average activity ranges measured over a 20–25 month period.

	mg $SO_x/dm^2/d$	mg $Cl^-/m^2/d$
Industrial	0.5–2	nil
Urban	0.5–4	nil
Rural (semi)	nil–2	nil
Marine	nil–0.5	25–150

Source: Ref 4

A marine atmosphere is laden with fine particles of sea salt carried by the winds and deposited on materials. The marine atmosphere is usually one of the more corrosive atmospheric environments. It has been shown that the amount of salt (chlorides) in the marine environment decreases with increasing distance from the ocean and is greatly influenced by wind direction and velocity (Ref 6). Marine atmospheres are discussed in detail in the article "Marine Corrosion" in this Volume.

Relative Corrosivity

Several programs have been conducted to assess the relative corrosivity at various locations over a finite period of time (Ref 4, 7-9). The use of calibration specimens to measure corrosivity can be traced to J.C. Hudson (Ref 8), who established the ranking of ten different environments based on the corrosion of wrought iron (Table 2). Table 3 summarizes additional site comparisons for the atmospheric-corrosion behavior of steel and zinc.

Atmospheric Factors

Methods have been developed for measuring many of the factors that affect atmospheric corrosion (Ref 4). The quantity and composition of atmospheric constituents and their variation with time have been determined. The factors that should be considered for measurement include, but are not limited to, those listed in Table 4.

Conducting Atmospheric-Corrosion Tests

The type of data or information required from the corrosion test often dictates the specifics of the test program. For example, the purpose of the program may be to evaluate the general corrosion behavior of the material or to determine its resistance to solar radiation, coatings discoloration, pitting, galvanic effects, loss of strength, or other changes in physical properties. The information one hopes to obtain from the exposure program must be determined in the planning stages of the work.

Materials To Be Exposed. Atmospheric-corrosion studies are usually carried out for a period of months and even years to ascertain the environmental effects on the degradation of the materials evaluated. Therefore, it is important to select standard or reference materials (control specimens or materials) that will be exposed alongside of the materials, alloys, or coating of interest. Control materials with a prior performance record in the exposure environment are extremely important for comparison purposes and for monitoring site corrosivity. For example, the International Organization for Standardization (ISO) has provided recommendations on low-carbon low-copper steel, commercial-purity

Table 2 Relative corrosivity of atmospheres at different locations

Location	Type of atmosphere	Average weight loss of iron specimens in 1 year, mg/cm^2	Relative corrosivity
Khartoum, Sudan	Dry island (arid)	0.08	1
Singapore	Tropical/marine	0.69	9
State College, PA	Rural	1.90	25
Panama Canal Zone	Tropical/marine	2.28	31
Kure Beach, NC (250-m, or 800-ft, lot)	Marine	2.93	38
Kearny, NJ	Industrial	3.92	52
Pittsburgh, PA	Industrial	4.88	65
Frodingham, UK	Industrial	7.50	100
Daytona Beach, FL	Marine	10.34	138
Kure Beach, NC (25-m, or 80-ft, lot)	Marine	35.68	475

Source: Ref 8

Table 3 Measured atmospheric-corrosion rates for steel and zinc

Site	Location	Type of atmosphere	Relative corrosivity Steel	Zinc
1	Normal Wells, Northwest Territory	Rural	0.02	0.2
2	Saskatoon, Saskatchewan	Rural	0.2	0.2
9	State College, PA(a)	Rural	1.0	1.0
17	Pittsburgh, PA (roof)	Industrial	1.8	1.5
18	London (Battersea), UK	Industrial	2.0	1.2
27	Bayonne, NJ	Industrial	3.4	3.1
28	Kure Beach, NC (250-m, or 800-ft, site)	Marine	3.6	1.9
31	London (Stratford), UK	Industrial	6.5	4.8
33	Point Reyes, CA	Marine	9.5	2.0
37	Kure Beach, NC (25-m, or 80-ft, site)	Marine	33.0	6.4

(a) The average weight losses on two 100- × 150-mm (4- × 6-in.) specimens after one year of exposure at the indicated site were used to calculate the relative corrosivity of the site. The losses in the rural atmosphere at State College, PA, were taken as unity and the relative corrosiveness at each of the other sites is given in this table as a fraction or a multiple of unity. Source: Ref 10

Table 4 Environmental parameters suggested for consideration of their influence on the atmospheric degradation of materials

Wet Deposition

pH
Conductivity
Cations: calcium (Ca^{2+}), magnesium (Mg^{2+}), sodium (Na^+), potassium (K^+), ammonium (NH_4^+), and hydrogen (H^+)
Anions: sulfates (SO_4^{2-}), nitrates (NO_3^-), and chlorides (Cl^-)

Dry Deposition

Sulfur dioxide (SO_2)
Nitrogen dioxide (NO_2), nitric acid (HNO_3)
Ammonia (NH_3)
Particulate matter, sulfates, nitrates

Meteorology

Wind speed
Wind direction
Relative humidity (dew point)
Temperature
Solar radiation
Rainfall volume and intensity

Others

Test specimen surface temperature
Time of wetness

Note: Time of wetness and the quantity of SO_2 and chloride are the most important variables in determining atmospheric corrosion. Such factors as hydrogen sulfide, nitrogen compounds, and other specific pollutants may be significant at specific sites if sources of these pollutants are located nearby.

Fig. 1 ISO wire helix specimen

aluminum, commercial-purity zinc, and commercial-purity copper as recommended control metals. The various reference materials selected by ISO for their exposure programs were chosen because they are representative engineering materials that are widely used and most frequently employed in external exposures. The materials were also selected, for comparison purposes, because they have known performance records in various environments and documented corrosion behaviors to different atmospheric constituents. Site corrosivity varies occasionally and underscores the need for simultaneous exposure of test and reference materials.

Fig. 2 Atmospheric-corrosion test rack

The number of duplicate or replicate specimens depends on the exposure period and the number of scheduled removals. For visual observations, two specimens for each environment are usually sufficient. Specimens that have been removed and cleaned should not be reexposed, because reexposure would in essence be starting the exposure period at time zero again.

Atmospheric Specimens. American Society for Testing and Materials (ASTM) and ISO guidelines for standard specimen designs for flat panels, stress corrosion, and other types of specimens such as the ISO wire helix specimen (Fig. 1) are outlined in Table 5. Guidelines from ASTM, the National Association of Corrosion Engineers (NACE), or ISO should be consulted for the proper cleaning procedures required before exposure and for cleaning and evaluation after exposure (these guidelines are summarized in Table 6).

A suitable means of identification should be used. Stamped code numbers are suitable for the more corrosion-resistant materials, but will be lost on the more corrodible materials. For these materials, drilled holes or side notches according to a number template can be used. A plastic tag affixed through a drilled hole with a nonmetallic tie has also been successfully used. In all cases, the method of identification of the specimen should not bias the corrosion evaluation. It is also good practice to map the specimens on the exposure rack; that is, one should draw, sketch, or list according to exposure the identity of each specimen in case the codes or tags are corroded away or lost.

Exposure Guidelines. Specimens are normally exposed on a test rack similar to that shown in Fig. 2. The rack is then attached to a frame or stand (ASTM G 50 or ISO DP8565). Specimen sizes are normally 100 × 150 mm (4 × 6 in.), but can be any size necessary to evaluate the behavior of the material properly. For example, large panels (Fig. 3) and stress-corrosion cracking (SCC) U-bend specimens (Fig. 4) are used in the sizes and configurations required to determine the performance of the material in the environment of interest.

For special situations, the exposure rack can be streamlined for small areas or for specialized exposures by using a corrosion spool rack similar to that shown in Fig. 5. In all cases, unless the objective of the exposure is to determine galvanic effects, the specimens should be isolated from

Table 5 Recommendations for specimen design for atmospheric-corrosion testing

Type	Reference	Typical size(a)
Flat panel	ASTM G 50 (Ref 12)	Ferrous: 100 × 150 mm (4 × 6 in.) Nonferrous: 100 × 200 mm (4 × 8 in.)
Stress corrosion		
U-bend	ASTM G 30 (Ref 13)	3 mm × 15 mm × 130 mm × 32 mm diam (0.12 in. × 0.6 in. × 5 in. × 1.25 in. diam)
Bent beam	ASTM G 39 (Ref 14)	3 point: 1 mm × 5 mm × 65 mm span (0.04 in. × 0.2 in. × 2.5 in. span); 2 point: 25 mm wide × 180 mm span (1 in. wide × 7 in. span)
C-ring	ASTM G 38 (Ref 15)	12 mm wide × 25 mm diam (0.5 in. wide × 1 in. diam)
Direct tension	ASTM G 49 (Ref 16)	Specimen size depends primarily on the dimensions of the product to be tested.
Welded	ASTM G 58 (Ref 17)	The thickness and size of the test specimen should represent the actual structural member.
Galvanic		
Disk	B-3 (Ref 18)	36 mm diam × 1.6 mm (1.4 in. diam × 0.06 in.), 33.5 mm diam × 1.6 mm (1.3 in. diam × 0.06 in.), 30 mm diam × 1.6 mm (1.2 in. diam × 0.06 in.), 25 mm diam × 1.6 mm (1 in. diam × 0.06 in.)
Plate	ISO/TC 156/WG3/N11 (Ref 19)	90 mm × 150 mm × 2 mm (3.5 in. × 6 in. × 0.08 in.), 70 mm × 25 mm × 2 mm (2.75 in. × 1 in. × 0.08 in.), 45 mm × 90 mm × 2 mm (1.8 in. × 3.5 in. × 0.08 in.), 25 mm × 70 mm × 2 mm (1 in. × 2.75 in. × 0.08 in.)
Open helix (metallic wire)	ISO/DP 9226 (Ref 20)	Wires with a diameter of 2-3 mm (0.08-0.12 in.) and 1000 mm (40 in.) long rolled into helix and attached to polyamide or metallic holder. See Fig. 1

(a) The appropriate ASTM or ISO document should be consulted before selecting a specimen size. Source: Ref 11

Table 6 Guideline for atmospheric exposures

Procedure or guideline	Society or organization reference: ASTM	NACE	ISO
Standard practice for conducting atmospheric tests on metals	G 50	. . .	DP8565(a)
Standard practice for recording data from atmospheric corrosion tests of metallic coated steel specimens	G 33	RP-02-81	. . .
Rating of electrodeposited panels subject to atmospheric exposure	B 537	. . .	. . .
Standard practices for preparing, cleaning, and evaluating corrosion test specimens	G 1	. . .	. . .
Standard definition of terms relating to corrosion and corrosion testing	G 15	. . .	. . .
Standard practice for applying statistics to analysis of corrosion data	G 16	. . .	. . .
Standard practice for examination and evaluation of pitting corrosion	G 46	. . .	. . .
Standard practice for making and using U-Bend corrosion test specimens	G 30	. . .	. . .
Corrosion of metals and alloys—determination of bi-metallic corrosion in outdoor exposure corrosion tests	. . .	. . .	ISO 7441-1984
Classification of corrosivity of atmospheres	. . .	. . .	DP9223
Corrosion of metals and alloys—guidelines of values for the categorization of corrosive atmospheres	. . .	. . .	DP9224
Corrosion of metals and alloys—aggressivity of atmospheres: methods of measurement of pollutant data(b)	G 91	. . .	DP9225
Corrosion of metals and alloys—corrosivity of atmospheres: methods of determination of corrosion rate of standard specimens for the evaluation of corrosivity	. . .	. . .	DP9226

(a) DP denotes draft proposed for an international standard currently in circulation within the responsible ISO committee. (b) This procedure applicable only to SO_x-containing atmospheres

Fig. 3 Large atmospheric-corrosion test panels

Fig. 4 U-bend specimens used for atmospheric-corrosion testing

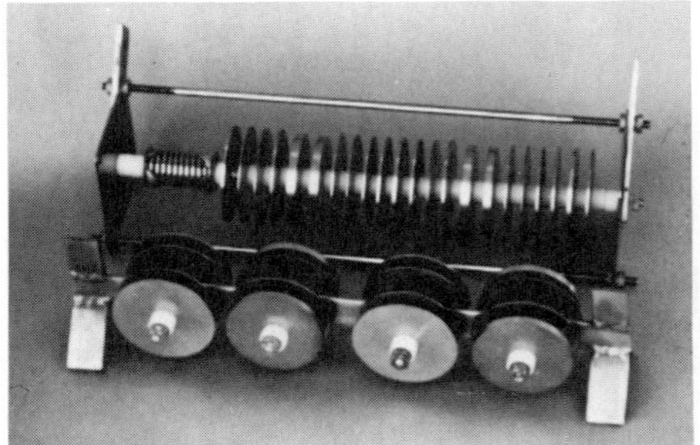

Fig. 5 Atmospheric-corrosion spool rack

each other and the exposure rack by some nonconductive material, such as the porcelain insulators visible in Fig. 2.

The normal convention used for specimen orientation is as follows: Specimens exposed in the northern hemisphere normally face south, and specimens exposed in the southern hemisphere face north. Also, it is common practice to expose specimens facing the most corrosive direction if the exposure site is close to a source of corrodent, such as seawater or power plant stacks. The ASTM standard G 50 recommends that specimens be exposed on an angle of 30° to the horizontal (this angle has been established to be 45° in Europe). If maximum exposure to sunlight is desired, the specimens should be exposed at an angle equal to the latitude of the site.

Evaluation of Results. After exposure to the atmosphere, a number of techniques are available evaluating the test panels and interpreting test results (Table 7). The most important step in atmospheric-corrosion testing is the documentation of results and observations for future reference and application. This reporting can take the form of internal company reports, technical papers, or presentations. In all cases, the author should attempt to outline the objectives of the work, the exposure details, and the conclusions of the exposure program.

Corrosion Testing in Water

Calvin H. Baloun
Chemical Engineering Department
Ohio University

As an environment in which corrosion occurs, water can be categorized as treated or natural. Regardless of the original intent for which a test has been developed, application can be extended to include any use. For example, ASTM G 78, which is described in the section "Methods of Analysis" in this article, is a guide for crevice corrosion testing of iron- and nickel-base alloys in seawater. The crevice assembly described could be installed on any alloy type in any environment if crevices are being studied. Only when quality control or acceptance stipulates a particular test method is there no room for departure. Some test methods involve the insertion of coupons or test panels into a conduit, with the environment at actual velocity and aeration. Other test methods involve the insertion of a test section into the flow line, especially if wall effects might be important. Still other tests require immersion of specimens in relatively open bodies of water.

Tests used as criteria for acceptance and/or quality control may require a very specific water chemistry and values of such factors as hardness, pH, aeration, temperature, alkalinity, and specific dissolved ion concentrations. It is usually valuable to monitor some of these factors in testing any water because compositions and contaminants change with time as sources (well, river, estuary, storm drains) and climatic conditions (wind direction, temperature, acid rain) vary. A variation in the mixture of sources might change test results significantly. When interpreting test results, it is important to be aware of the factors that can affect those results.

Unless galvanic action is being investigated (and it probably should be in most cases), specimens of differing alloy composition or thermomechanical treatment should be electrically isolated or insulated from each other and from the support framework. The factors that require consideration in water supplies are discussed in Ref 21 to 25. Many standardized and nonstandardized tests, methods, and practices are discussed in Ref 26 to 30.

Test Method Selection and Precautions

It is necessary to exercise sound judgment when selecting suitable testing procedures. For example, the results of static-immersion tests or alternate-immersion tests bear little relationship to pipelines flowing full within an operating plant. Test coupons inserted into an operating process line, or at least into a laboratory test loop using process fluids at operating temperatures and velocities, will produce results that agree much better with actual plant experience. Because corrosion is often accelerated at metal/liquid/vapor interfaces, these interfaces should be duplicated in the testing procedure if they are present in the plant or should be omitted from the testing if they are not an actual factor.

Test Specimens. Specimen geometry, ratio of environment solution volume to exposed specimen area, or other shape and area factors should be as similar as possible to true plant conditions. Exposed, cut specimen edges, which are often attacked more rapidly than other parts, should be examined and discounted if edges are not exposed in plant equipment. Similarly, stamped specimen identification characters may suffer increased rate of attack or may be sites of stress-assisted or accelerated corrosion.

Virtually every specimen size has been used in immersion testing. Popular laboratory test specimen sizes range from 1 cm^2 (0.15 $in.^2$) of exposed surface to 25 × 100 mm (1 × 4 in.), both of which are used frequently, to 100 × 150 mm (4 × 6 in.). In-plant and seawater tests sometimes involve specimens measuring 150 × 250 mm (6 × 10 in.) or larger. When weldments or projections are of concern, either the weldment or the projection should be incorporated into the specimen. Entire components are seldom tested in the laboratory, but may be installed in pilot plant or works scale operations (pumps, condensers, heat exchangers).

Effect of Water Variables. To date, no compilation of relative rankings of alloy types in various aqueous environments is available. However, for iron-base alloys, the corrosion rate is relatively uniform in most waters, whether they are of high or low purity. The rate seems to vary more with aeration, that is, from less than 0.05 mm/yr (<2 mils/yr) in deaerated water to more than 1.25 mm/yr (>50 mils/yr) in aerated water. In tap, brackish, or seawater, pitting tends to occur. In tap water and seawater, yellow brass tends to dezincify (especially when conditions are near stagnation), although corrosion rates are less than 0.05 mm/yr (<2 mils/yr) and more than 0.5 mm/yr (>20 mils/yr) for tap water and seawater, respectively. Detailed information on dezincification can be found in the articles "Metallurgically Influenced Corrosion" (see the section "Dealloying Corrosion") and "Corrosion of Copper and Copper Alloys" in this Volume. Information on the effects of various types of water (salt, potable, brackish, etc.) on alloy corrosion can be found in the Section "Specific Alloy Systems" in this Volume.

There are no definite rules concerning the effect of flow rates, temperatures, and/or pH on the corrosion of alloys in water systems, although in waters free of passivators (including dissolved oxygen) increasing flow rates tend to erode any protective corrosion films that may have formed. Corrosion rates of most alloys over the pH range of 4 to 10 show little effect of pH. Outside this range, that is, 0 to 4 and 10 to 14, the environment cannot technically be considered water, except for some highly unusual acid rain cases. Increasing temperatures sometimes precipitate protective salts, such as calcium carbonate, which decrease corrosion rates in normal-to-hard waters.

Very pure or very soft waters are often excellent solvents for metallic ions, such as copper. If these waters are used to prepare laboratory solutions for test purposes, dissolved copper from copper water lines can deposit or plate out on more active metal surfaces. In the case of aluminum alloys, this deposited copper can greatly accelerate corrosion immediately adjacent to the copper. The effect of dissolved copper ion on localized corrosion of aluminum is cited in ASTM Standards G 4, section 4.5 (Ref 31); G 52, section 6.1 (Ref 32); and G 71, section 4.2.1 (Ref 33). See also the article "Corrosion of Aluminum and Aluminum Alloys" in this Volume. Similar effects of lesser intensity have been observed on most active metals.

Temperature differentials between points in a flow system can produce accelerated attack due to differences in ionic activity. Although this attack usually occurs at the point of higher temperature, protective scales occasionally precipitate on the high-temperature surface, with attack occurring at the cooler sites. In any system involving differential temperature, the investigator should be aware of this possible behavior.

Another temperature effect in open, aqueous systems is the decreasing solubility of gases, particularly oxygen, with increasing temperature. This effect reduces the cathodic action, or more exactly that portion due to oxygen reduction, and thus decreases the amount of anodic reaction that occurs.

Similarly, the solubility of air and oxygen in saline solutions decreases with increasing con-

Table 7 Evaluation techniques for atmospheric-corrosion specimens

Technique	Value
Photographic documentation	Photographs of the specimens before and after cleaning give a permanent record of the performance of the material in the particular atmosphere.
Corrosion product analysis and surface deposits	Atmospheric-corrosion specimens usually have the corrosion product and airborne deposits on the surface at the time of removal. This adds a wealth of information on the observed behavior of the material.
Weight loss	For uniform corrosion, this is simple and can be converted to corrosion rate as $g/m^2/d$, mils per year, etc.
Pitting and localized corrosion	Yields information on the susceptibility of a material to localized attack. Pitting corrosion is often reported as average or maximum depth of attack and is usually measured with a dial depth gage or vernier microscope. Where possible, pitting data should be treated statistically with recognized methods covered in various standards. Weight loss data should not be used indiscriminately to calculate corrosion rates where the primary form of corrosion is localized.
Rust or rust stain	Data reveal the propensity of a material to rust and the degree of rust staining. Also, through cleaning procedures, it can be determined if the original appearance was retained.
Tensile test and other physical tests	Can often yield information on the atmospheric effect on the strength of materials, cracking behavior, etc.
Appearance	Effect of environment on appearance, color retention, etc.

centration of salt, but the solution conductivity increases with the dissolved salt concentration. The two effects combine in oxygen reduction cathodic systems to produce increasing corrosion rates up to about 3.5 wt% sodium chloride (NaCl) solutions and decreasing corrosion rates above that value. This particular effect is not observed in freshwater systems.

Natural waters, depending on the source, may contain variable concentrations of dissolved carbonates. These are principally calcium, magnesium, and/or sodium, with the solubility controlled by both the partial pressure of carbon dioxide and the solubility product of calcium carbonate as it varies with alkalinity and temperature. Deposition of a thin continuous film of carbonate can protect the underlying metal where a lack of the same would allow attack. On the other hand, a film that is too thick impedes heat transfer and can lead to reduced flow and ultimately to possible tube burnout. Carbonate scaling, measurement through the Langelier Index, and controlling treatments are discussed in Ref 24 and 34.

Methods of Analysis

As stated previously in this section, a number of references describe methods of analysis for testing waters and materials in waters. The ASTM and NACE standards that are directly or indirectly applicable include the following:

ASTM Standards

- Standard Test Methods for pH of Water (D 1293). These methods cover the determination of pH by electrometric measurement, using the glass electrode as the sensor. Two procedures are discussed. Method A covers the precise laboratory measurement of pH in water with the use of at least two of seven standard reference buffer solutions for equipment standardization. Method B concerns the routine or continuous measurement of pH in the laboratory and the measurement of pH under various continuous process conditions
- Standard Practice for Oxidation-Reduction Potential of Water (D 1498). This practice covers the apparatus and procedure for the electrometric measurement of oxidation-reduction potential in water
- Standard Test Methods for Particulate and Dissolved Matter, Solids, or Residue in Water (D 1888). These methods cover the determination of particulate (suspended) and dissolved matter in water containing more than 25 ppm of total matter and 25 ppm or less of total matter, respectively. The measured constituents are those that can be removed by filtration or are the residue upon evaporation to dryness of filtered or unfiltered samples
- Standard Test Methods for Corrosivity of Water in the Absence of Heat Transfer (Weight Loss Methods) (D 2688). These methods cover the determination of the corrosivity of water by evaluating pitting and by measuring the weight loss of metallic specimens. Additional information on pitting corrosion can be found in the articles "Localized Corrosion" and "Evaluation of Pitting Corrosion" in this Volume
- Standard Test Methods for Corrosivity of Water in the Absence of Heat Transfer (Electrical Methods) (D 2776). These methods cover the electrical resistance method and the polarization method for determining the corrosivity of water. The electrical resistance method (Method A) measures the corrosion rate of a metal sample, mounted on a probe, by periodically determining its change in electrical resistance as the cross-sectional area of the metal sample decreases because of corrosion. The observed measurement is the change in resistance ratio between a measuring element exposed to water and a reference element in proximity to, but protected from, the corrosive environment. The polarization resistance method (Method B) operates on the principle that when a prescribed voltage (usually 5 to 30 mV in water) is impressed across the interface boundary between a metal electrode surface and a conductive liquid, the resulting current flow will be directly proportional to the corrosion occurring on the metal electrode surface, if the current measurement includes both anodic and cathodic polarization characteristics
- Standard Test Method for Efficacy of Microbiocides Used in Cooling Systems (E 645). This method covers the effect of bactericides and fungicides in controlling microbial growth in cooling water by using water collected from operational cooling water systems. Additional information can be found in the articles "Control of Environmental Variables in Water Recirculating Systems" and "Evaluation of Microbiological Corrosion" in this Volume
- Standard Practice for Conducting Surface Seawater Exposure Tests on Metals and Alloys (G 52). This practice covers conditions for exposure of metals and alloys to surface seawater and presents the general procedures that should be followed in conducting seawater exposure tests so that meaningful comparisons can be made for different locations. Detailed information on seawater environments can be found in the article "Marine Corrosion" in this Volume
- Standard Guide for Crevice Corrosion Testing of Iron-Base and Nickel-Base Stainless Alloys in Seawater and Other Chloride Containing Aqueous Environments (G 78). This guide provides information for conducting crevice corrosion tests and identifies factors that may affect results and influence conclusions. These procedures can be used to identify the conditions that will most likely result in crevice corrosion and to provide a basis for assessing the relative resistance of various alloys to crevice corrosion under certain specified conditions. Additional information on crevice corrosion can be found in the articles "Localized Corrosion" and "Evaluation of Crevice/Concentration Cell Corrosion" in this Volume

NACE Standards

- Laboratory Methods for the Evaluation of Protective Coatings Used as Lining Materials in Immersion Service (TM-01-74). This standard gives guidelines to assist manufacturers and users of protective coatings in selecting materials by providing standard test methods for evaluating protective coatings used as linings for immersion service. Two test methods are discussed for the evaluation of protective coatings on any substrate, such as steel, copper, or aluminum, so that both the factors of chemical resistance and permeability can be considered
- Method of Conducting Controlled Velocity Laboratory Corrosion Tests (TM-02-70). This method covers laboratory corrosion tests that may be reproducibly conducted under velocity conditions. The method described is applicable only to the study of the effects of solution velocity on corrosion rate of a metal and is not to be used to determine the effect of a solution impinging on a metal surface
- Dynamic Corrosion Testing of Metals in High-Temperature Water (TM-02-74). This standard provides guidelines for corrosion testing of metals exposed to the high-temperature water used in high-pressure steam plants or water-cooled nuclear reactor plant systems. Standardized test methods for determining the effects of high-temperature water are also covered
- Preparation and Installation of Corrosion Coupons and Interpretation of Test Data in Oil Production Practice (RP-07-75). The use of uniform industry-proven methods to monitor corrosion in oil production systems are emphasized in this standard. Procedures for preparing, analyzing, and installing corrosion coupons are also outlined

Corrosion Testing in Soil

E. Escalante
Corrosion Group
National Bureau of Standards

An approach to evaluating the durability of a metal in soil will be described in this section. Specimen design, preparation, burial, and retrieval techniques will be discussed. The type of information sought during soil-induced corrosion evaluation controls the design configuration and the nature of the corrosion measurements. Consideration of these factors during the planning stage will help the corrosion engineer obtain a maximum amount of information with a minimum number of future problems in the program.

The corrosion of metals underground can be divided into two broad categories: corrosion in undisturbed soils and corrosion in disturbed soils. Corrosion in undisturbed soil is always low, regardless of soil conditions, and is limited only by the availability of the oxygen necessary for the cathodic reaction. Steel piles driven into soil fall under this category and therefore undergo limited corrosive attack.

Corrosion of metals in disturbed soils is strongly affected by soil conditions. Soil resistivity, pH, and soil composition play important roles in determining whether corrosion is a serious problem. Any metal buried by backfilling is in a disturbed soil and is subject to corrosion attack, depending on the characteristics of the soil. Most metals in soil are exposed to disturbed-soil conditions. Because it is the more common of the two, the disturbed-soil condition will be discussed in this section.

Soil Characteristics

Soil is a complex, dynamic environment that changes continuously, both chemically and physically, with the seasons of the year. Precipitation, plant growth, and animal life all have their effects on the environment, but perhaps a greater effect results from buildings, roads, electric power lines, landfills, farming, snow removal with salts, and so on. Characterizing a soil for its corrosivity is difficult at best. However, certain empirically derived facts are known to affect the corrosivity of soils.

Soil resistivity is recognized as an important parameter in underground corrosion (Ref 35).

The ions must migrate through the electrolyte in a soil to supply the metal surface with the electron donors or acceptors necessary for the corrosion reaction to proceed. The concentration of these salts in the electrolyte is important; soil resistivity is a measure of the concentration of the ions and how easily they move through the soil environment (Ref 36). A high soil resistivity suggests a low corrosion rate because of the low rate of ion diffusion.

The composition of these salts is also critical. Chlorides, for example, are known to enhance the breakdown of the protective oxide on some metals (Ref 37). Copper ions accelerate the attack on aluminum (Ref 34). Sulfates can provide oxidizers for bacterial corrosion (Ref 38). Other salts, such as carbonates, can affect the pH of the soil. Thus, the corrosivity of a soil can be strongly affected by the presence of certain chemical species, and a soil that is corrosive for one metal may not be corrosive for some other metal.

The acidity or the alkalinity of most soils is stable because of the buffering action of the soluble minerals available. However, soils across the country can differ in pH from 5 to 10. This pH range has a small effect on general and pitting corrosive attack, but has a large effect on the susceptibility to hydrogen embrittlement of ferrous alloys (Ref 39). Furthermore, some metals, such as zinc, copper and aluminum, are amphoteric and can undergo corrosive attack at high and low pH values; these metals may be susceptible to attack in acid and alkaline soils (Ref 34). These effects on the corrosion process make it desirable to measure the soil pH of a test site (Ref 40).

The ways in which soil resistivity, pH and soil composition interact to determine soil corrosivity is not well understood despite extensive amounts of data that have been collected over the years (Ref 35, 41, 42). Soil temperature has been found to have a strong effect on soil resistivity and an effect on oxygen solubility. Because these are opposing reactions competing to control the corrosion process, their cumulative effect on corrosion is minimized (Ref 39). Clearly, many factors affect soil corrosivity, and their interrelationship is complex.

Test Approach

Although soil characteristics can give some indication of the corrosivity of a soil environment, the optimum test is to bury some specimens of the metals in question in the soil of interest. The concept is simple, but there are some considerations that should be thought out. The type of data sought determines the design of the specimen. For example, provision must be made for electrochemical corrosion measurements to be carried out during the burial period. Periodic retrieval of specimens of an alloy requires that identical sets be buried, one set per retrieval. It is necessary to mark the specimens for future identification and to mark the location of the specimens at the site. The importance of maintaining permanent, clear, detailed records cannot be overemphasized. The longer the period of burial, the more important these records become.

Specimen Preparation

The design of the specimen is usually determined by the type of data needed. If the corrosion of a pipe is under consideration, then pipe material should be used. To avoid internal corrosion effects, the ends can be sealed by capping. Corrosion of a tank container suggests the use of sheet or plate material for testing. If welds are expected then welded material should be included. If dissimilar metals will be in contact, then galvanically coupled materials are required. Stressed U-bend or C-ring specimens, for example, are necessary for stress-corrosion or hydrogen embrittlement studies. Control specimens should also be included. These control specimens could be uncoated when the effect of coatings is being evaluated. Similarly, unstressed specimens should be included when stressed specimens are being studied.

Before cleaning and weighing, the specimen is marked for proper identification by stamping or notching the specimen. In addition to this identification on the specimen, plastic tags are attached to the specimen in case corrosion is sufficiently severe to destroy the identification markings. Because these tags are often lost during retrieval or transportation, they should not be a primary marking system. Where retrievals are planned on a periodic basis, duplicate specimens must be provided, and each must have its own unique identification mark.

Electrical contact to a specimen is required when electrochemical measurements are to be made during the burial period. This contact can be achieved with a 14-gage (~1.9-mm, or 0.075-in. thick) insulated conductor that extends above ground and is soldered or bolted to the specimen. Solder can be removed after retrieval and before reweighing by heating the joint and wiping away the solder, but this is possible only with high melting point metal specimens, such as ferrous alloys. To avoid galvanic attack of the specimen or the conductor, the joint should be coated. Similar electrical conductors to both electrodes of a galvanic couple allow galvanic currents to be measured.

Specimen Burial

A frequently used technique for burial is to dig a trench that is long enough to allow placement of the specimens about 300 mm (12 in.) apart along its length. If specimens are to be retrieved periodically, for example, 1, 2, or 4 years, then all specimens of a given retrieval increment should be grouped together. That is, all specimens to be recovered at the end of 1 year are grouped separately from those that are to be retrieved at the end of 2 years, and so on.

To facilitate retrieval of the specimens, it is useful to tie a small-diameter (3-mm, or 0.12-in.) nylon rope to every specimen in a retrieval group. Thus, when one specimen is found, digging can proceed by following the rope until all of the specimens are recovered. Wood posts can be used to mark the location of the ends of a group of specimens, and the rope can be tied to these posts. If the trench is in a straight line, it is a simple matter to dig up a post, find the rope, and dig between the posts. Where posts cannot be used, metal stakes positioned slightly below the surface of the ground can mark the location of the specimens. These metal markers can then be located with a metal detector. It is useful to have a clearly visible surface marker that identifies the site and an individual or company that can be contacted for information; this may also avoid future problems.

In every case, an accurate map of the area identifying the exact location of all specimens is mandatory. Although describing the details of burial and recovery may seem mundane, important data can be lost simply because a few precautions have not been taken.

Conductors attached to the specimens must be carefully and permanently marked for identification at the surface end of the conductor. These conductors can then be attached to contact points on the wood posts.

Corrosion Measurements

In general, engineers are interested in obtaining data on the corrosion rate of a metal in soil, and this requires knowledge about weight loss and length of exposure. Rate of pit growth, however, may be more important when a material is used as a container or a pipeline. For some stressed metals and alloys, SCC or hydrogen embrittlement may more likely be the mode of failure. Some or all of these factors may have to be considered in carrying out the exposure study.

Uniform Corrosion. Gravimetric mass loss is the reference by which all other weight loss measurements are judged when evaluating the general corrosive attack of a metal in soil. Gravimetric mass loss determinations involve weighing the specimen before the exposure period and then reweighing it after retrieval. When properly carried out, the loss in mass represents the material lost because of corrosion. This technique, however, generally requires long exposures (for example, months) so that the mass losses are measurable and significant. Specimens of 1000 g or less should be weighed to an accuracy of 0.5 mg. Sample size may be limited by the capacity of the balance available. An excellent source of information on preparing, cleaning, and evaluating specimens is ASTM G 1 (Ref 43).

If data are obtained only through periodic retrievals nothing can be learned about corrosion losses between retrieval times. One method of obtaining corrosion rate data in the intervening months is to use electrical measurements that take advantage of the electrochemical nature of underground corrosion.

In the polarization resistance technique, the potential of a metallic specimen is changed by 10 mV or less, and the amount of current applied to make this change is measured (Ref 44). Three electrodes are usually used: the working electrode (the metal of interest), the reference electrode (used to measure the potential of the working electrode), and the counter electrode (used to apply current to the working electrode). Figure 6 illustrates the configuration of the three electrodes in a typical measurement in soil. The effects of i_R, a measurement error that arises when measuring a potential in the presence of current flowing through a resistive medium, must be eliminated (Ref 45). Details of this polarization technique are described in the section "Electrochemical Methods of Corrosion Testing" of the article "Laboratory Testing" in this Volume.

Pitting corrosion, unlike uniform corrosion, cannot be evaluated by *in situ* electrical measurements. Therefore, the only option available is to measure the pit depth after retrieval of the specimens, as described in ASTM G 46 (Ref 46). Large, shallow pits can be easily measured with pit micrometers. A narrow, deep pit is more difficult to measure because a pit micrometer cannot be used. One approach is to insert a small-diameter wire of known length into the pit and measure the length of wire extending out of the pit. Although not very

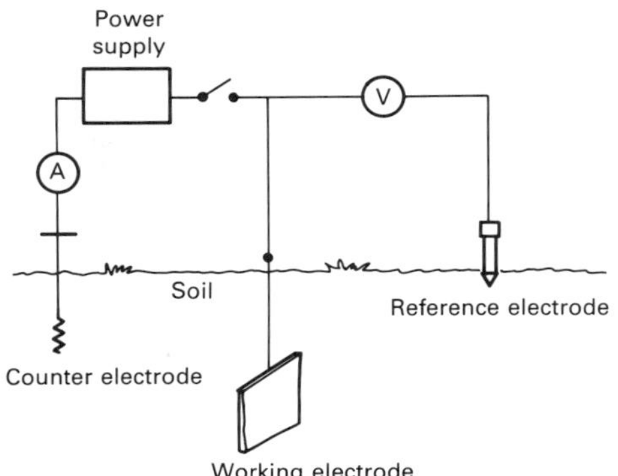

Fig. 6 Configuration of electrodes used in a polarization measurement of metals in soil

satisfactory, this method will reveal the minimum depth of a pit. A more accurate but time-consuming method is to cross section the specimen and use an optical means to measure the pit. A micrometer stage on a low-power microscope is useful for this type of measurement.

Whichever method is used, the objective is to measure the deepest pits. By measuring the maximum pit depth of duplicate specimens over a period of time, a rate of maximum pit growth can be obtained. The distribution of the pits (percentage of surface pitted) may also be of interest. Time to perforation data can be obtained by designing a specimen that is pressurized with air. This pressure is then monitored over a period of time until failure is indicated by a decrease (Ref 47). It is important to design the system so that the specimen being tested fails before the monitoring lines.

Stress-Corrosion Cracking and Hydrogen Embrittlement. Preparation of stressed specimens for underground performance studies is more time consuming because the specimen must be properly stressed before burial. Any fixtures or bolts attached to the specimen must be nonmetallic, electrically insulated from the specimen, or coated to avoid galvanic effects. Electrochemical measurements of the SCC specimens will not reveal information on the failure of the metal. However, if the hydrogen embrittlement specimen is cathodically charged with a sacrificial anode, it may be useful to monitor the current between the two electrodes and the electrical potential of the couple versus a reference electrode (Ref 48).

Time to failure data can be obtained by attaching a hollow tensile specimen to a stressing ring (Fig. 7) by monitoring the change in strain or diameter of the ring. Stress is applied to the tensile specimen by turning the stressing nuts so that the stressing ring is elastically deformed. The stressed tensile specimen is then imbedded in the soil, with the stressing ring extending above ground (Fig. 7). This and other special designs can provide failure data before retrieval. Retrieval of the specimen soon after failure will reduce subsequent destruction of the failure surface before it is obliterated by general corrosive attack. Examination of this surface may provide further information on the mode of failure.

Specimen Retrieval

If the specimens have been buried as described in the section "Specimen Burial" in this article, the retrieval process should proceed smoothly.

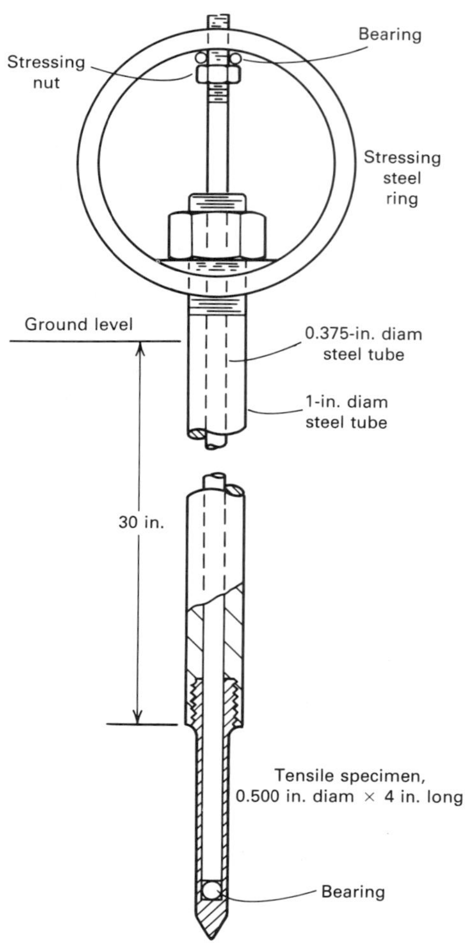

Fig. 7 Surface-monitored ring-stressed hollow specimen for SCC and hydrogen embrittlement studies of metals in soil. Source: Ref 49

Although the initial surface digging can be carried out with power equipment, the final few centimeters of soil must be removed by careful hand shoveling to reduce the possibility of specimen damage or loss of identification labels. A cursory examination of specimens, along with the maintenance of records and photographs of observations during this retrieval period, can be very useful—for example, Did the specimen fail during transit after retrieval? A final detailed examination of the specimens is then carried out in the laboratory. Depending on the situation, it may be desirable to clean and dry the specimen immediately after removal.

Summary

Corrosion of metals in undisturbed soil is very low and is independent of soil characteristics, but corrosion in disturbed soil is strongly influenced by the soil environment. Because of the difficulty in determining soil corrosivity through soil characterization or because data on specific alloys are needed, it is sometimes necessary to conduct an underground corrosion study. Information on the durability of metals in soil can be obtained by using physical and electrochemical measurements.

Gravimetric measurements of electrochemical polarization measurements will provide information on deterioration due to uniform corrosion. Similarly, pit depth measurements and surface monitoring of pressurized vessels provide data that can be used to determine perforation rates. The susceptibility of metals to hydrogen embrittlement and SCC can also be determined by physical measurements at the soil surface or by periodic retrieval of stressed specimens. Monitoring the potential and the galvanic current of cathodically charged stressed specimens reveals the susceptibility of an alloy to hydrogen embrittlement as a function of pH and galvanic current.

Specimen design depends on the type of data sought. It is essential to mark burial specimens properly, to map burial sites, and to maintain permanent records. This type of study consists of the following steps:

- Define the problem
- Identify the types of corrosion measurements that will supply the information needed
- Design the specimen to provide information on this need
- Choose a system for specimen identification
- Maintain good, permanent records at all stages of the program
- Bury the specimens with a connecting nylon cord for ease of retrieval
- Handle specimens with care during retrieval, and keep identification tags with the specimens
- Carry out final cleaning and evaluation in the laboratory with the utmost care

REFERENCES

1. "Notice of Proposed Standard for Sulfur Oxide, Particulate Matter, Carbon Monoxide, Photochemical Oxidants, Hydrocarbons, and Nitrogen Oxides," Environmental Protection Agency, 42CFR, Part 410, *Federal Register 36*, 1971, p 1502
2. F. Mansfield, "Regional Air Pollution Study Effects of Airborne Sulfur Pollutants on Materials," EPA 600/4-80-007, Environmental Protection Agency, Jan 1980
3. F.H. Haynie and J.B. Upham, Effects of Atmospheric Sulfur Dioxide on the Corrosion of Zinc, *Mater. Perform.*, Vol 9, 1970, p 35-40
4. H. Guttman and P.J. Sereda, Measurement of Atmospheric Factors Affecting the Corrosion of Metals, in *Metal Corrosion in the Atmosphere*, STP 345, American Society for Testing and Materials, 1968, p 355-360
5. K.G. Compton, Atmospheric Corrosion, in *NACE Basic Corrosion Course*, National Association of Corrosion Engineers, 1970, p 4-2
6. F.L. LaQue, Environment Factors in the Corrosion of Metals in Seawater and Sea Air, in *Marine Corrosion*, John Wiley & Sons, 1975, p 103
7. F.L. LaQue, *Proc. ASTM*, Vol 51, 1951, p 495
8. J.C. Hudson, *J. Br. Iron Steel Inst.*, Vol 148, 1943, p 161
9. S.K. Coburn *et al.*, Corrosiveness of Various Atmospheric Test Sites as Measured by Specimens of Steel and Zinc, in *Metal Corrosion in the Atmosphere*, STP 435, American Society for Testing and Materials, 1968, p 326
10. *ASTM Mater. Res. Stand.*, Dec 1961, p 977
11. S.W. Dean, Jr., Planning, Instrumentation and Evaluation of Atmospheric Corrosion Tests and a Review of ASTM Testing, in *Atmospheric Corrosion*, John Wiley & Sons, 1982, p 195-216

12. "Standard Recommended Practice for Conducting Atmospheric Corrosion Tests on Metals," G 50, *Annual Book of ASTM Standards*, American Society for Testing and Materials
13. "Standard Recommended Practice for Making and Using U-Bend Stress Corrosion Test Specimens," G 30, *Annual Book of ASTM Standards*, American Society for Testing and Materials
14. "Standard Practice for Preparation and Use of Bent Beam Stress Corrosion Test Specimens," G 39, *Annual Book of ASTM Standards*, American Society for Testing and Materials
15. "Standard Recommended Practice for Making and Using C-Ring Stress Corrosion Test Specimens," G 38, *Annual Book of ASTM Standards*, American Society for Testing and Materials
16. "Standard Recommended Practice for Preparation and Use of Direct Tension Stress Corrosion Test Specimens," G 49, *Annual Book of ASTM Standards*, American Society for Testing and Materials
17. "Standard Practice for Preparation of Stress Corrosion Test Specimens for Weldments," G 58, *Annual Book of ASTM Standards*, American Society for Testing and Materials
18. R. Baboian, Final Report on the ASTM Study: Atmospheric Galvanic Corrosion of Magnesium Coupled to Other Metals, in *Atmospheric Factors Affecting the Corrosion of Engineering Metals*, STP 646, S.K. Coburn, Ed., American Society for Testing and Materials, 1978, p 17-29
19. "Metals and Alloys—Test Methods for Contact Corrosion in Atmospheric Conditions," ISO/TC 156/WG 3/N11, Third Working Draft, Proposal of USSR to ISO Technical Committee 156 WG 3, Nov 1979
20. "Corrosion of Metals and Alloys—Corrosivity of Atmospheres—Methods of Determination of Corrosion Rates of Standard Specimens for the Evaluation of Corrosivity," ISO/TC 156 WG4/N131, German Proposal to ISO Technical Committee 156 WG3, 1979
21. F.N. Speller, in *Corrosion Handbook*, H. Uhlig, Ed., John Wiley & Sons, 1948, p 496
22. F.N. Speller, *Corrosion: Causes and Prevention*, McGraw-Hill, 1951, p 244
23. R.K. Swandby, *Corrosion Resistance of Metals and Alloys*, 2nd ed., F.L. LaQue and H.R. Copson, Ed., Reinhold, 1963, p 45
24. G. Butler and H.C.K. Ison, *Corrosion and Its Prevention in Waters*, Reinhold, 1966, p 18
25. *Prevention and Control of Water-Caused Problems in Building Potable Water Systems*, TPC 7, National Association of Corrosion Engineers, 1980
26. V.V. Kendall, in *Corrosion Handbook*, H. Uhlig, Ed., John Wiley & Sons, 1948, p 1058
27. F.L. LaQue, in *Corrosion Handbook*, H. Uhlig, Ed., John Wiley & Sons, 1948, p 1060
28. F.L. LaQue, in *Corrosion Resistance of Metals and Alloys*, 2nd ed., F.L. LaQue and H.R. Copson, Ed., Reinhold, 1963, p 107, 136
29. O.B.K. Fraser, in *Corrosion Handbook*, H. Uhlig, Ed., John Wiley & Sons, 1948, p 105
30. F.A. Champion, *Corrosion Testing Procedures*, 2nd ed., John Wiley & Sons, 1965
31. "Standard Practice for Conducting Plant Corrosion Tests," G 4, *Annual Book of ASTM Standards*, American Society for Testing and Materials
32. "Standard Practice for Conducting Surface Seawater Exposure Tests on Metals and Alloys," G 52, *Annual Book of ASTM Standards*, American Society for Testing and Materials
33. "Standard Practice for Conducting and Evaluating Galvanic Corrosion Tests in Electrolytes," G 71, *Annual Book of ASTM Standards*, American Society for Testing and Materials
34. H.H. Uhlig and R.W. Revie, *Corrosion and Corrosion Control*, 3rd ed., John Wiley & Sons, 1985, p 111-114, 415-420
35. M. Romanoff, "Underground Corrosion," NBS 579, NTIS PB 168 350, National Bureau of Standards, April 1957
36. "Standard Method for Field Measurement of Soil Resistivity Using the Wenner Four-Electrode Method," G 57, *Annual Book of ASTM Standards*, American Society for Testing and Materials
37. J.R. Galvele, Present State of Understanding of the Breakdown of Passivity and Repassivation, in *Passivity of Metals*, The Electrochemical Society, 1978, p 285-327
38. W.P. Iverson, An Overview of the Anaerobic Corrosion of Underground Metallic Structures: Evidence for a New Mechanism, in *Underground Corrosion*, American Society for Testing and Materials, 1981, p 33-52
39. E. Escalante, The Effect of Soil Resistivity and Soil Temperature on the Corrosion of Galvanically Coupled Metals in Soil, in *Galvanic Corrosion of Metals*, STP, American Society for Testing and Materials, to be published
40. "Standard Test Method for pH of Soil for Use in Corrosion Testing," G 5, *Annual Book of ASTM Standards*, American Society for Testing and Materials
41. W.J. Schwerdtfeger, Soil Resistivity as Related to Underground Corrosion and Cathodic Protection, *J. Res. Natl. Bur. Stand.*, Vol 69C (No. 1), Jan-March 1965
42. N.D. Tomashov, *Theory of Corrosion and Protection of Metals*, Macmillan, 1966, p 399-451
43. "Standard Practice for Preparing, Cleaning, and Evaluating Corrosion Test Specimens," G 1, *Annual Book of ASTM Standards*, American Society for Testing and Materials
44. D.A. Jones, Principles of Measurement and Prevention of Buried Metal Corrosion by Electrochemical Polarization, in *Underground Corrosion*, STP 741, American Society for Testing and Materials, 1981, p 123-132
45. M.A. Hubbe, Polarisation Resistance Corrosivity Test With a Correction for Resistivity, *Br. Corros. J.*, Vol 15 (No. 4), 1980, p 193-197
46. "Standard Recommended Practice for Examination and Evaluation of Pitting Corrosion," G 46, *Annual Book of ASTM Standards*, American Society for Testing and Materials
47. J.B. Vrable, Unbonded Polyethylene Film for Protection of Underground Structures, *Mater. Prot. Perform.*, Vol 11 (No. 3), March 1972, p 26-28
48. E. Escalante and W.F. Gerhold, The Galvanic Coupling of Some Stressed Stainless Steels to Dissimilar Metals Underground, in *Properties of Materials for Liquefied Natural Gas Tankage*, STP 579, American Society for Testing and Materials, 1976, p 88-93
49. J.B. Vrable, "Stress Corrosion Behavior of Line-pipe Steels in Underground Environments," Project 59.001-001(2), United States Steel Corporation, July 1970

Laboratory Testing

LABORATORY CORROSION TESTS are used to predict corrosion behavior when service history is lacking and time or budget constraints prohibit simulated service (field) testing. They can also be used as screening tests prior to simulated-service testing. Laboratory tests are particularly useful for quality control, materials selection, materials and environmental comparisons, and the study of corrosion mechanisms. These tests cover a spectrum ranging from simple immersion tests to various kinds of cabinet and autoclave controlled environments to sophisticated electrochemical tests. Most of these are accelerated tests by design; therefore, they must be used with caution. Because almost everything about these tests is based on assumptions, the testing conditions are arbitrary, and the problems of selecting the most pertinent testing method and a realistic interpretation are always present.

Standardized test methods are very useful for specifications and for routine tests used to compare experimental alloys and such products as inhibitors, coatings, and insulation materials. Many such tests are described in the Annual Book of ASTM Standards (Vol 03.02, *Metal Corrosion, Erosion, and Wear*) and in product specifications. However, it is often necessary for special tests to be set up that require creativity on the part of the investigator. This article will deal with basic procedures and approaches to laboratory corrosion testing but will not attempt to cover all of the published literature on specialized testing.

Electrochemical Methods of Corrosion Testing

John R. Scully
David W. Taylor Naval Ship Research and Development Center

Corrosion is an electrochemical process. Electrochemical processes require anodes and cathodes in electrical contact and an ionic conduction path through an electrolyte. The electrochemical process includes electron flow between the anodic and cathodic areas; the rate of this flow corresponds to the rates of the oxidation and reduction reactions that occur at the surfaces. Monitoring this electron flow provides the capability of assessing the kinetics of the corrosion process, not simply the thermodynamic tendencies for the process to occur spontaneously nor only the accumulated metal loss registered after the test. Measurements of this type have become known as electrochemical measurements of corrosion. Several textbooks cover many of the major aspects associated with the application of electrochemical methods in corrosion (Ref 1-8).

Electrochemical techniques are finding increased use in corrosion research and in engineering applications. Such methods are practical because the corrosion behavior of material-electrolyte combinations is a direct function of the mechanisms as well as the kinetics of the electrochemical reactions responsible. The availability of better electrochemical techniques and the demonstrated accuracy of such methods for investigating corrosion phenomena have been the subject of several recent technical symposia (Ref 9-14).

The earliest applications of electrochemical techniques were attractive because they offered a direct method of accelerating a corrosion process without changing the environment, that is, adding a strong oxidizer or increasing temperature as in other nonelectrochemical laboratory tests. However, recent advances have made the technique more of a nondestructive tool for the evaluation of corrosion phenomena and reaction rates, and this offers the possibility of *in situ* (field) as well as *ex situ* (laboratory) investigations. Thus, electrochemical techniques can be used to measure corrosion rates without removing the specimen from the environment or altering the sample itself. This capability offers distinct advantages over weight loss and visual observation testing procedures because the kinetics of the corrosion process can be quantitatively studied and because of the added convenience.

Most, if not all, of the typical forms of corrosion, including uniform, localized, galvanic, dealloying, stress corrosion, and hydrogen-induced failure, can be investigated by electrochemical techniques. Topics such as passivation, anodization, cathodic and anodic protection, and barrier and sacrificial coatings on metal substrates can also be investigated by using electrochemical techniques. Also available are electrochemical techniques that have been optimized for the study of hydrodynamic (mass transport controlled) corrosion processes (Ref 15) as well as those techniques designed to avoid dominance of the measured quantity by mass transport control, that is, transient electrochemical methods (Ref 16).

Although they have been limited to near ambient temperature aqueous corrosion studies in the past, electrochemical techniques are now being applied to high-temperature, high-pressure aqueous applications as well as to the study of corrosion processes in nonaqueous environments and aqueous environments of low conductivity. In the remainder of this section, specific examples will be given on the application of various electrochemical methods to the typical categories of corrosion attack mentioned above.

Fundamentals

Corrosion is an electrochemical process that is characterized by mass transport as well as the electrical and ionic charge transfer associated with a corroding solid or solids electrically in contact. The corroding system consists of anode and cathode sites electrically in contact and a surrounding electrolyte (Ref 17). According to mixed-potential theory, any net electrochemical reaction can be algebraically divided into two or more oxidation and reduction reactions, and there can be no net electrical charge accumulation. In a corroding system in the absence of applied potential, the oxidation of the metal and the reduction of some species in solution occur simultaneously at the metal/electrolyte interface (Ref 1, 3). Under these circumstances, the net measurable current is 0, however, a finite rate of corrosion is occurring at local anodic sites on the metal surface. Such a current-potential relationship is shown schematically in Fig. 1.

When the corrosion potential is located at a potential that is distinctly different from the reversible electrode potentials (redox potential, E_{redox}) of either the corroding metal or the species in solution that are cathodically reduced, the oxidation of the cathodic species or the reduction of any of the ions of the corroding metal that are in solution becomes negligible. The corrosion potential, E_{corr}, indicates the potential at which the oxidation rate equals the reduction rate and the net current measured externally is 0:

$$i_{app} = i_{ox} - i_{red} \qquad \text{(Eq 1)}$$

Because the oxidation current present at the corrosion potential is the quantity of interest in the corroding system, this parameter must be determined independently of the reaction rates of other adsorbed or dissolved reactants. Electrochemical polarization methods can be used to determine corrosion rates. More detailed information on the fundamentals associated with aqueous corrosion can be found in the articles "Thermodynamics of Aqueous Corrosion" and "Kinetics of Aqueous Corrosion" in this Volume.

Conducting an Electrochemical Polarization Experiment

The electrochemical polarization of metallic samples is accomplished with a power supply known as a potentiostat. An auxiliary electrode supplies the current to the working electrode (test specimen) in order to polarize it. The potential between the working electrode and a reference electrode is monitored or set at a fixed value, as in the potentiostatic test.

Figure 2(a) illustrates schematically a typical experimental arrangement. The system is designed so that only an extremely small current can pass between the reference electrode and the working electrode; the current needed to polarize

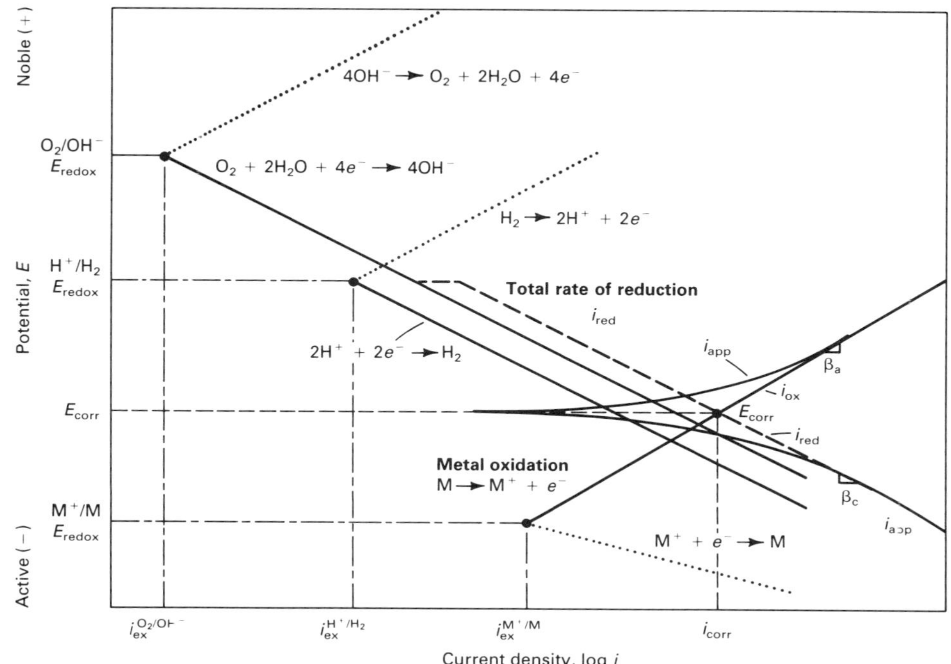

Fig. 1 Schematic electrochemical potential-current relationship for a corroding system consisting of two cathodic reactions and one anodic electrochemical reaction

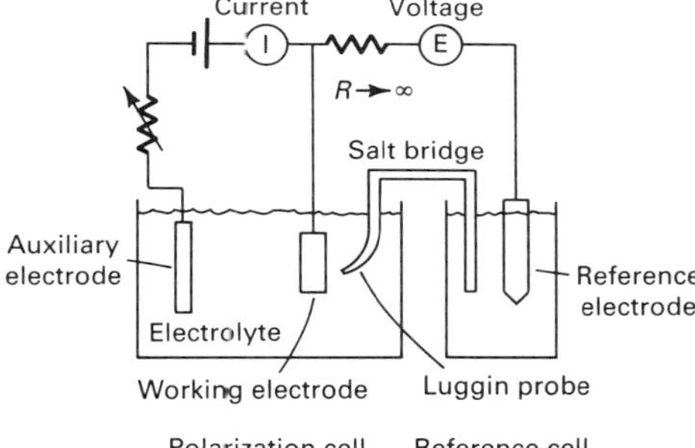

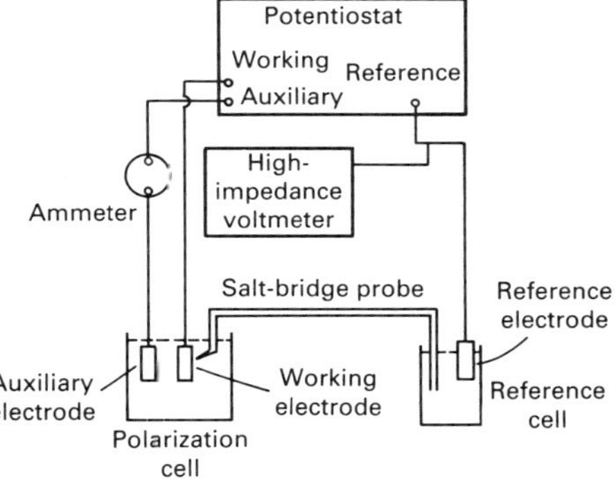

Fig. 2(a) Instrumentation setup for electrochemical polarization experiments

the working electrode is supplied from the auxiliary electrode (Ref 1). The potentiostat actually contains an operational amplifier or similar device in order to provide the feedback of the current necessary for maintaining the fixed potential between the working and reference electrode. Figure 2(b) illustrates a typical electrochemical cell. The potential can be applied in a ramp fashion, stepped as a function of time (sometimes called potentiostaircase), or held at a fixed level as in the potentiostatic method (Ref 18). Conversely, the current can be applied in a continuous galvanostatic or stepped galvanostaircase manner, and the change in potential monitored. Several American Society for Testing and Materials (ASTM) standards discuss methods for performing these experiments (Ref 19, 20).

Conversion of Electrochemical Current Data to Corrosion Rates

Measurement of I_{ox} at the potential of interest, where $I_{ox} = i_{ox} \cdot A$, over a known period of time leads to direct determination of the weight loss:

$$W = \frac{(I \cdot t \cdot M)}{nF} \qquad \text{(Eq 2)}$$

where W is the weight loss (grams), $I \cdot t$ equals the current multiplied by time or coulombs (C), n is the number of electrons involved in the electrochemical reaction, F is Faraday's constant (96 487 C/gram-equivalent), and M is the molecular weight of the electroactive species. This relation is known as Faraday's Law. Rearrangement of Eq 2 leads to a straightforward determination of the penetration rate (applicable only when I_{ox} is uniformly distributed over the entire wetted surface area or where the localized actively corroding area, A, is known), as follows:

$$\text{CR} = \frac{3.27 \times 10^{-3} \cdot i_{ox} \cdot \text{E.W.}}{d} \qquad \text{(Eq 3)}$$

where CR is the corrosion rate (in millimeters per year), where E.W. is the equivalent weight in grams, d is the metal or alloy density in grams per cubic centimeter, and i_{ox} is the corrosion current density in microamps per square centimeter.

For alloys, the equivalent weight E.W. should be calculated using the atomic fractions of each alloying element in an expression such as:

$$\text{Alloy E.W.} = \sum_{i=1}^{m} \frac{f_i M_i}{n_i} \qquad \text{(Eq 4)}$$

where f_i is the atomic fraction of the ith component of the alloy, M_i is the atomic weight of the ith component element, n_i is the electron loss required to oxidize the ith component element under the conditions of the corrosion process (n_i is usually equal to the stable valence of the element, or must be determined from either a Pourbaix (potential-pH) diagram or experimentally from an analysis of the corroding solution), and m is the number of component elements in the alloy.

This expression assumes that all the component elements oxidize when the alloy corrodes and that they are all oxidizing at essentially a uniform rate. In some situations, these assumptions are not valid; in these cases, the calculated corrosion rate will be in error. For example, if an alloy is composed of two or more phases and one phase preferentially oxidizes, the calculation must take this into consideration.

Electrochemical Methods For the Study of Uniform Corrosion

Polarization Methods. The following relationship observed between applied current and potential provides the basis for the electrochemical polarization technique (Ref 1, 2, 21):

$$i_{app} = i_{corr} \left[\exp \frac{2.3(E - E_{corr})}{\beta_a} - \exp \frac{-2.3(E - E_{corr})}{\beta_c} \right] + C\left(\frac{dE}{dt}\right) \qquad \text{(Eq 5)}$$

where i_{app} is the applied current density based on the electrochemical surface area; i_{corr} is the corrosion current density; E is the applied voltage; E_{corr} is the open-circuit, or freely corroding, potential; C is the interfacial capacitance associated with the electrochemical doublelayer (Ref 1, 2); β_a and β_c are the anodic and cathodic Tafel coefficients related to the slopes of the polarization curves in the anodic and cathodic regimes, respectively; and dE/dt is the time rate of change in applied potential, or voltage, scan rate. Ideally, the second term of the expression $C \cdot dE/dt$ approaches 0 at, for example, very low potential scan rates. Note that i_{app} becomes approximately equal to either i_{ox} or i_{red} at large overpotential. At very large anodic or cathodic overpotentials, Eq 5 can be rearranged in a more simplistic form (Ref 1-3, 5, 7):

$$\eta_a = \beta_a \log \left(\frac{i_{app}}{i_{corr}}\right) \qquad \text{(Eq 6)}$$

$$\eta_c = -\beta_c \log \left(\frac{i_{app}}{i_{corr}}\right) \qquad \text{(Eq 7)}$$

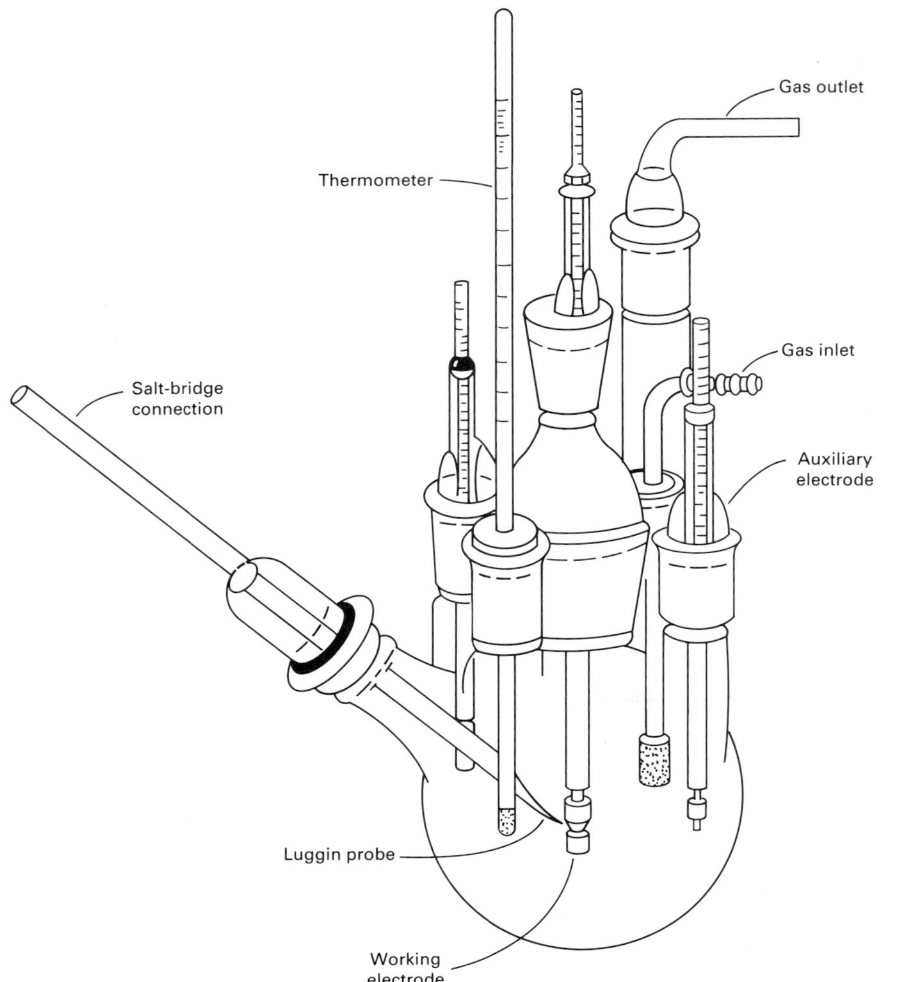

Fig. 2(b) Typical electrochemical polarization cell illustrating locations for working and auxiliary electrodes, and associated cell components

Tafel Extrapolation. By plotting potential versus log i_{app}, a linear relationship between current density and potential can be developed (Ref 1, 5, 7, 21). Extrapolation of the applied current from either the anodic or the cathodic Tafel region to the open-circuit potential or position of zero overpotential allows determination of i_{corr}. Figure 3 illustrates this method.

Complications of the Polarization Method Involving Solution Resistance. Polarization techniques can be complicated by several factors. One such factor arises from the resistivity of the solution, depending on the geometry, the location of the reference electrode, and the magnitude of applied current (Ref 2, 6). Placement of the reference electrode near the working electrode is facilitated through the use of a Luggin-Haber capillary. This arrangement is desired in order to minimize the solution resistance error, which can be estimated from the product of the applied current, the solution resistivity, and the distance from the Luggin probe to the specimen surface. The error contributes to the measured overpotential in the following manner:

$$\eta_{app} = \eta_{true} + \eta_{iR} \qquad \text{(Eq 8)}$$

where η_{iR} is equal to $i_{app} \cdot R_{soln}$ (the resistance of the solution). Thus, the true overpotential is overestimated at each applied current density, and the Tafel slope becomes inaccurate. The true scan rate in the potentiodynamic technique may also be altered. Several excellent reviews are available on the subject of the voltage error introduced from solution resistance (Ref 23-25).

Complications Involving Concentration Polarization Effects. The Tafel relationship established above is dependent on pure activation control, or charge transfer control. An additional consideration involves the concept of concentration polarization. In this case, the reaction rate is large enough that the electroactive species is depleted or concentrated at the reacting surface in order to maintain the reaction rate through diffusion. The reaction becomes diffusion controlled at the limiting current density. The deviation from activation control is described by (Ref 5):

$$\eta = \frac{2.3RT}{nF} \log\left(1 - \frac{i_{app}}{i_L}\right) \qquad \text{(Eq 9)}$$

where R is the gas constant, and T is the absolute temperature. Many corroding systems are frequently under what is termed mixed activation and mass transport control (Ref 5). This behavior can be described mathematically by the algebraic combination of Eq 5 and 9. To increase the reaction rate and thus minimize the contribution of Eq 9 to the overpotential, the diffusional boundary layer thickness is decreased by solution stirring or by the use of a rotating cylinder or disk electrode (Ref 15).

Other Issues Concerning Tafel Extrapolation. Experimentally, there are several problems with the Tafel extrapolation method. Large applied current densities may not be representative of the true corrosion situation occurring at open circuit. This is particularly true in the case of anodic polarization, in which the surface is changing because of corrosion and/or passivation of the metal. In the cathodic case, an excess of adsorbed hydrogen or a build up of hydroxyl ions in solution at the metal surface may result in alterations of surface chemistry not representative of the freely corroding case. In addition, concentration polarization may eliminate the Tafel region entirely.

Polarization Resistance Methods. For anodic and cathodic polarization within 10 mV of the corrosion potential, it is often observed that the applied current density is approximately linear with potential (Ref 1). This behavior is shown in Fig. 4 for the type 430 stainless steel in sulfuric acid (H_2SO_4). The slope of this plot is determined at a corrosion potential, as shown in Fig. 4 (Ref 26, 27). This slope has units of resistance area and is known as the polarization resistance. Stern and Geary simplified the fundamental equations describing charge transfer controlled reaction kinetics (Eq 5) for the case of small perturbations of the corrosion potential, that is, low polarization or overpotential (Ref 27, 28). This relationship has the form:

$$I_{corr} = \left(\frac{\Delta i_{app}}{2.3 \cdot A \cdot \Delta E}\right)\left(\frac{\beta_a \cdot \beta_c}{\beta_a + \beta_c}\right) \qquad \text{(Eq 10)}$$

$$i_{corr} = \left(\frac{1}{2.3 \cdot R_p}\right)\left(\frac{\beta_a \cdot \beta_c}{\beta_a + \beta_c}\right) \qquad \text{(Eq 11)}$$

where R_p is the polarization resistance, and β_a and β_c are the anodic and cathodic Tafel slopes, respectively. The value R_p is determined from the current density, not the current measurement; the surface area of the working electrode must be known. Knowledge of R_p and Tafel slope values permit direct determination of the corrosion rate at any instant in time (Ref 27-30). The ASTM standards G 59 (Ref 20) and D 2776 (Ref 31) describe procedures for conducting polarization resistance measurements. These standards do not address many of the complications that may arise.

Complications With Polarization Resistance Measurements. Numerous papers have addressed complications and possible alternative methods for use of the polarization resistance method (Ref 32-37). Three of the most obvious errors that are possible when using these methods involve invalidation of the results through oxidation of some other electroactive species besides the corroding metal in question, a change in the open-circuit or corrosion potential during the time taken to perform the measurement, and the application of a large applied potential resulting in inadvertent departure from linear current density versus potential behavior. Several tech-

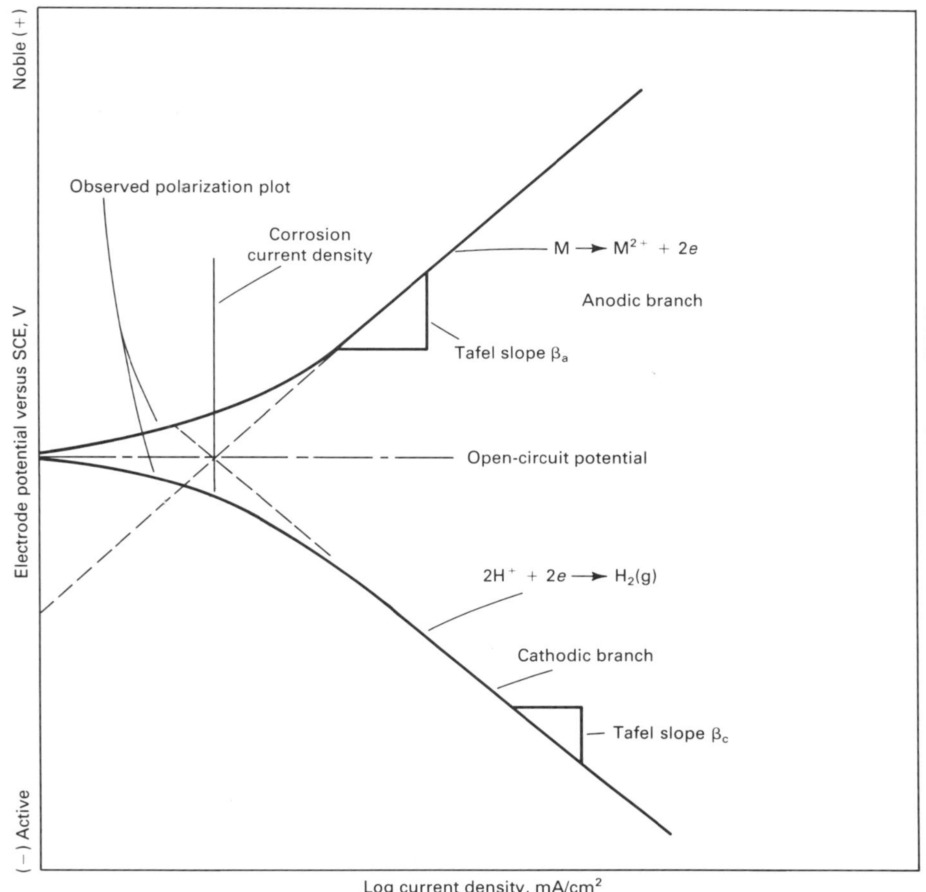

Fig. 3 Experimentally measured Tafel polarization plot. Source: Ref 22

niques actually take advantage of the departure from nonlinearity in order to calculate the anodic and cathodic Tafel slopes at a low overpotential and thus avoid some of the problems associated with large polarization (Ref 37).

Another source of error involves cases in which both the anodic and cathodic reactions are not under charge transfer control, as required for the derivation of Eq 10. Possible corrections to Eq 10 for cases in which pure activation control is not maintained, such as in the case of partial diffusion control or passivation (Ref 38). Other researchers have attempted to calibrate their results through the use of weight loss measurements (Ref 39). In fact, in one case, polarization resistance data for a number of alloy-electrolyte systems were compared to the observed corrosion currents (Ref 30). A linear correspondence was obtained over six orders of magnitude in corrosion rates.

Two other frequently encountered complications are the need to correct data for the electrolyte (solution) resistance R_s and the effects of capacitive hysteresis, C, occurring as a function of the voltage scan rate (Ref 2). Hysteresis in the current density-potential plot is observable when the time rate of change, dE/dt, of the applied potential is scanned in a cyclic manner under the conditions in which the sweep rate is large, the interfacial capacitance is large, or both. The effect can be minimized by performing the polarization resistance test by taking two or more dc current density-potential measurements instead of cycling the potential. The solution, or ohmic resistance, on the other hand, still complicates the measurement because the algebraic sum of the polarization resistance and the solution resistance is measured in a dc or pseudo-dc measurement.

Many treatments of this subject have used the electrical equivalent circuit model, which simulates the corroding metal/electrolyte interface (Ref 40, 41). The simplest form of such a model, the Randles circuit, is shown in Fig. 5. The three parameters discussed above (R_p, R_s, and C) are shown in the relationship to one another that best approximates a corroding electrochemical interface of a metal. The sum of the polarization resistance and the solution resistance is measured when a dc measurement is performed because the capacitive reactance approaches infinity. Thus, the true corrosion rate will be underestimated when R_s is appreciable. Conversely, any measurement at a cyclic applied potential that is scanned as a function of time causes the algebraic sum of the ohmic resistance and the resultant parallel impedance of the parallel resistive-capacitive network to be measured. If R_s is small compared to R_p, the capacitive error considered alone will usually result in an overestimation of the true corrosion rate. These complications can be overcome, as discussed below.

Electrochemical Impedance Methods. One approach to determining the corrosion rate of a metal without the complications discussed above involves the electrochemical impedance (sometimes known as ac impedance) method (Ref 40-43). In this technique, a small-amplitude sinusoidal potential perturbation is applied to the working electrode at a number of discrete frequencies. At each one of these frequencies, the resulting current waveform will exhibit a sinusoidal response that is out of phase with the applied potential signal by a certain amount. The electrochemical impedance is a frequency-dependent proportionality factor that acts as a transfer function by establishing a relationship between the excitation voltage signal and the current response of the system. As such, an electrochemical impedance is a fundamental characteristic of the electrochemical system it describes. A knowledge of the frequency dependence of impedance for a corroding system allows a determination of an appropriate equivalent electrical circuit describing that system. Figure 5 illustrates in the most simple case the equivalent electrical circuit model for an active corroding metal. The following expression describes the impedance for that system:

$$Z_{metal} = R_s + \frac{R_p}{(1 + \omega^2 \cdot R_p^2 \cdot C^2)} - \frac{j\omega\, C R_p^2}{(1 + \omega^2 \cdot R_p^2 \cdot C^2)} \quad \text{(Eq 12)}$$

where Z is the impedance magnitude, $\omega = 2\pi f$ is the frequency of the applied signal in radians per second, C is the interfacial capacitance, and $j = \sqrt{-1}$. Experimentally, it can be easily seen from Fig. 6 that at very low frequencies:

$$Z_{metal} = R_s + R_p \quad \text{(Eq 13)}$$

while at very large frequencies:

$$Z_{metal} = R_s \quad \text{(Eq 14)}$$

As evident from the above, a calculation of the polarization resistance is possible in media of quite high resistivity because R_p can be mathematically separated from R_s. In addition, capacitive hysteresis is eliminated. Other information determined from this technique includes the value of C, the capacitance. In the simple case of active corrosion, C may be found to scale linearly with the true electrochemical surface area from the relationship $C = c \cdot A$, where c is a specific capacitance per unit area, and A is the true electrochemical surface area. Determination of the corrosion rate also requires accurate values for the anodic and cathodic Tafel slopes β_a and β_c, as shown in Eq 10.

Electrochemical Methods for the Study of Galvanic Corrosion

Methods Based on Mixed-Potential Theory. Galvanic corrosion can be described in terms of mixed-potential theory, as schematically illustrated in Fig. 7. However, in the case of bimetal or multimetal galvanic attack in which two or more metals are electrically in contact with one another, there will be at least in theory a minimum of two cathodic and two anodic reactions. One of each of these reactions is occurring on each metal. In this case, the more noble of the two metals is cathodically polarized in the

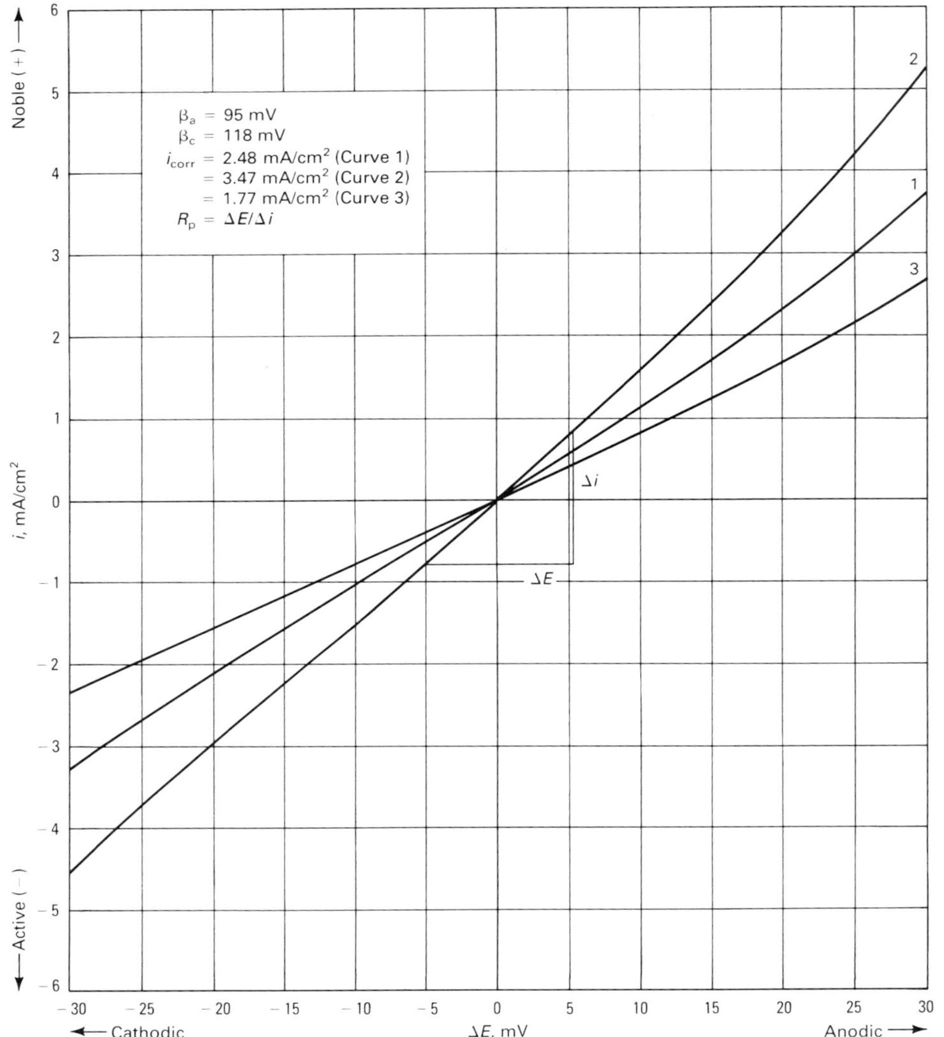

Fig. 4 Polarization curves for polarization resistance measurements based on the results from eight independent laboratories for type 430 stainless steel in 1 N H_2SO_4. Curve 1 is the mean result, with curves 2 and 3 showing ±2 standard deviations. Source: Ref 20

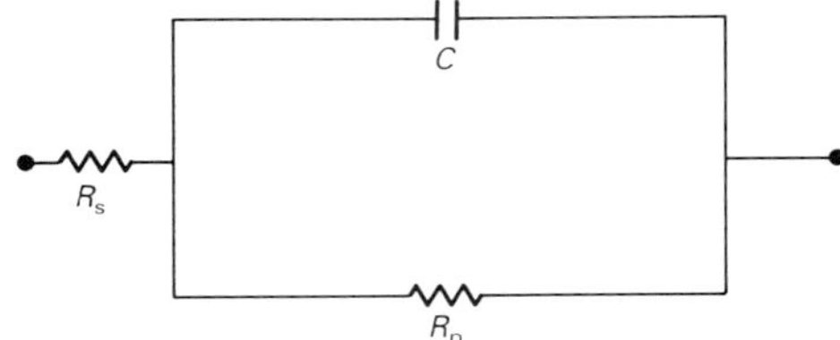

Fig. 5 Electrical equivalent circuit model simulating simple corroding metal/electrolyte interface. See also Figure 6

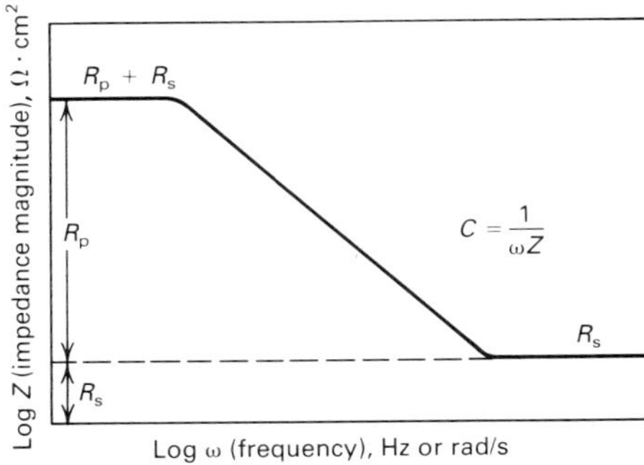

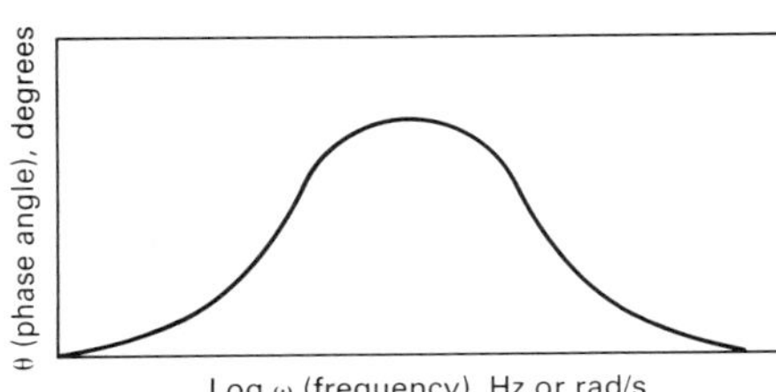

Fig. 6 Bode phase and magnitude plots demonstrating the frequency dependence of electrochemical impedance for the circuit model shown in Fig. 5

electronegative direction, and its anodic reaction rate will thus be suppressed. Conversely, the less noble, or more sacrificial, anodic material is anodically polarized, and the anodic reaction rate is accelerated. The mixed potential (the galvanic couple potential) of the galvanic couple and the resulting galvanic current can be uniquely determined from the sums of all of the anodic and cathodic currents for each material at each potential when the following condition is met:

$$\Sigma i_a \cdot A_a = \Sigma i_c \cdot A_c \quad \text{(Eq 15)}$$

where $\Sigma i_a \cdot A_a$ is the sum of the anodic currents (current density multiplied by area), and $\Sigma i_c \cdot A_c$ is the sum of the cathodic currents.

This relationship will be uniquely satisfied at one value of the applied potential and can be determined if potential versus anodic and cathodic current data are available for each material in the galvanic couple. Thus, this mixed potential will depend entirely on kinetic considerations rather than thermodynamic considerations. In simple bimetal cases, direct superposition of polarization data (corrected for wetted surface) will yield the same result (Ref 44). This technique is schematically illustrated in Fig. 7. However, because applied currents, not anodic or cathodic currents, are actually measured in the polarization experiment, the technique will introduce the least error when the cathodic reaction rate is negligible on the anode material and the anodic reaction rate is negligible on the cathode material at the galvanic couple potential (the mixed potential). Otherwise, a correction must be made. Obviously, when the open-circuit potentials of the anode and cathode materials are similar, this complication is more likely to arise, but in this case galvanic corrosion is less likely to be significant. In addition, special care must be taken in the procedures used to develop the polarization data (Ref 44), especially if time effects are to be taken into consideration when evaluating galvanic-corrosion behavior.

Direct Measurement of Galvanic-Corrosion Rates. A more straightforward procedure involves immersing the two dissimilar metals in an electrolyte and electrically connecting the materials together by using a zero-resistance ammeter to measure current (Ref 45-47). In this method, the galvanic current can be directly determined as a function of time. A reference electrode can be used in the usual manner to determine the galvanic couple potential.

Electrochemical Methods for the Study of Localized Corrosion

Pitting and crevice corrosion are usually associated with the breakdown of passivity (Ref 48). The anodic electrochemical behavior of a passive alloy in deaerated acid is schematically shown in Fig. 8(a). Figure 8(b) illustrates the results for type 430 stainless steel in H_2SO_4. The characteristics of this diagram will aid in determining the resistance of an alloy to localized attack.

Tests for evaluating the susceptibility of a material to pitting and crevice corrosion include potentiodynamic tests, galvanostatic tests, potentiostatic tests, scratch potentiostatic tests, triboellipsometric methods, pit-propagation rate curves, and electrochemical noise measurements (Ref 49). Several excellent review papers discuss these methods in more detail (Ref 48, 49, 50).

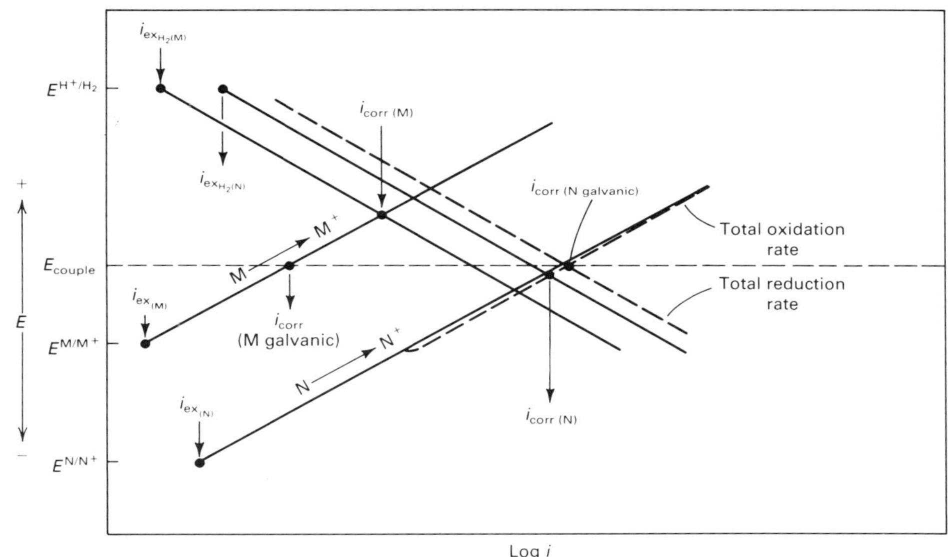

Fig. 7 Potential-current relationships for the case of a galvanic couple between two corroding metals. M, more noble metal; N, less noble metal. Source: Ref 3 and 5

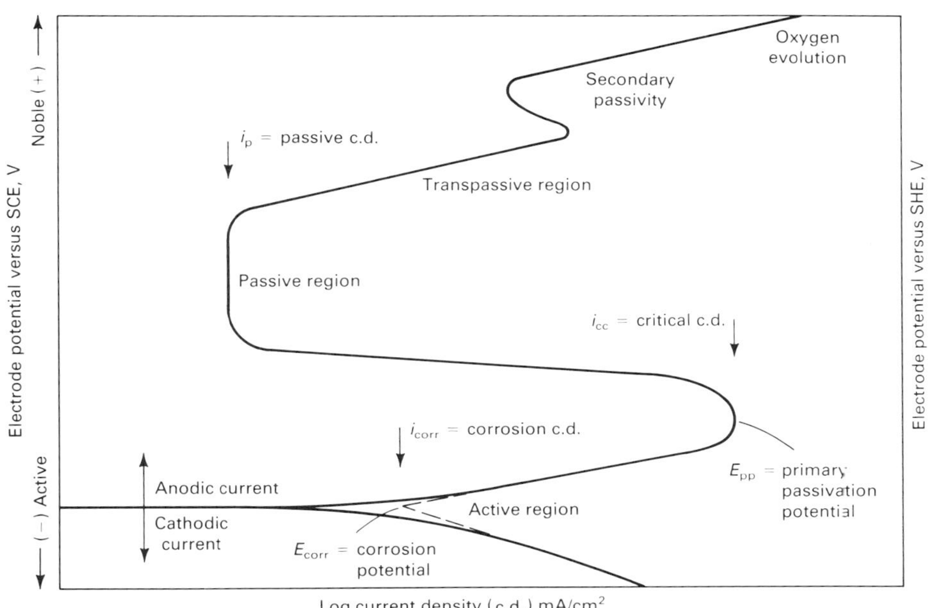

Fig. 8(a) Hypothetical anodic and cathodic polarization behavior for a material exhibiting passive anodic behavior. Source: Ref 22

Cyclic Potentiodynamic Polarization Methods. The ASTM standard G 61 describes a procedure for conducting cyclic potentiodynamic polarization measurements to determine relative susceptibility to localized corrosion (Ref 51). The method is designed for use with iron- or nickel-base alloys in chloride environments. In this test, a cyclic anodic polarization scan is performed at a fixed voltage scan rate.

Particular attention is focused on two features on the cyclic anodic polarization behavior diagram. These are the potential at which the anodic current increases considerably with applied potential or the breakdown potential. In general, the more noble this potential, obtained at a fixed scan rate in this test, the less susceptible the alloy to the initiation of localized attack. The second feature of great interest is the potential at which the hysteresis loop is completed upon reverse polarization scan. In general, once initiated, localized corrosion can propagate at some potential more electropositive than that at which the hysteresis loop is completed (when determined at a fixed scan rate). Therefore, the more electropositive the potential at which the hysteresis loop is completed, the less likely that localized corrosion will occur. This potential is known as the protection potential. Figure 9 illustrates cyclic polarization behavior for Hastelloy C-276 and type 304 stainless steel in 3.56% sodium chloride (NaCl) solution. Based on the above criteria for evaluation, it is evident that Hastelloy C-276 is relatively more resistant to localized corrosion in this environment.

Complications With Cyclic Potentiodynamic Polarization Methods. Although it is a reasonable method of screening variations in alloy composition and environments, the cyclic potentiodynamic polarization method has been found to have a number of shortcomings (Ref 50-54). One major problem concerns the effect of the potential scan rate. The values of both the protection potential and the breakdown potential are a strong function of the manner in which the tests are performed, particularly the potential scan rate employed. This problem is predominately related to the induction time required for pitting, the repassivation rates, and the complications arising from allowing too much pitting propagation to occur along with the accompanying chemistry changes before the reversal in the scan direction.

Potentiostatic and Galvanostatic Methods for Localized Corrosion. The shortcomings of the cyclic potentiodynamic polarization method have become the basis for several other electrochemical techniques. Four other methods are schematically illustrated in Fig. 10. Potentiostatic methods can be used to overcome the inherent problems involving scan rate. There are two general methods. The first is mainly associated with the study of initiation, and the second is mainly associated with the study of propagation and repassivation.

The first method (Fig. 10a) simply provides a means of determining the breakdown potential by polarizing individual samples at potentials above and below an approximate E_{bd} determined from the potentiodynamic method. Eventual initiation is indicated by a current increase. In the second method (Fig. 10b and c), initiation of pits is intentionally induced by applying a potential above the breakdown potential and then quickly shifting to a potential below that value. If the final applied potential is above the protection potential, propagation of the existing pits will continue and the current will increase. However, at E less than the protective potential, E_{prot}, the pits will eventually repassivate, and the current will subsequently decrease with time. The galvanostatic or galvanostaircase technique (Fig. 10d) measures potential versus time at a constant applied current. Measurements are often made until the time rate of change in potential approaches 0. The technique is under development in ASTM G 1 as a test method for application with aluminum alloys.

The Scratch-Repassivation Method for Localized Corrosion. One additional technique to be mentioned in the area of localized corrosion involves the scratch method (Ref 55). In this method, the alloy surface is scratched at a constant potential, and the current is measured as a function of time. The potential dependencies of the induction time and the repassivation time are determined by monitoring the current change over a range of different potentials. This is illustrated in Fig. 11. From this information, the critical pitting potential, thought to be located somewhere between the breakdown potential and the protection potential, can be determined. Other methods of studying localized corrosion are also available (Ref 49).

Environmental Cracking

There are numerous electrochemical techniques in the area of environmental cracking.

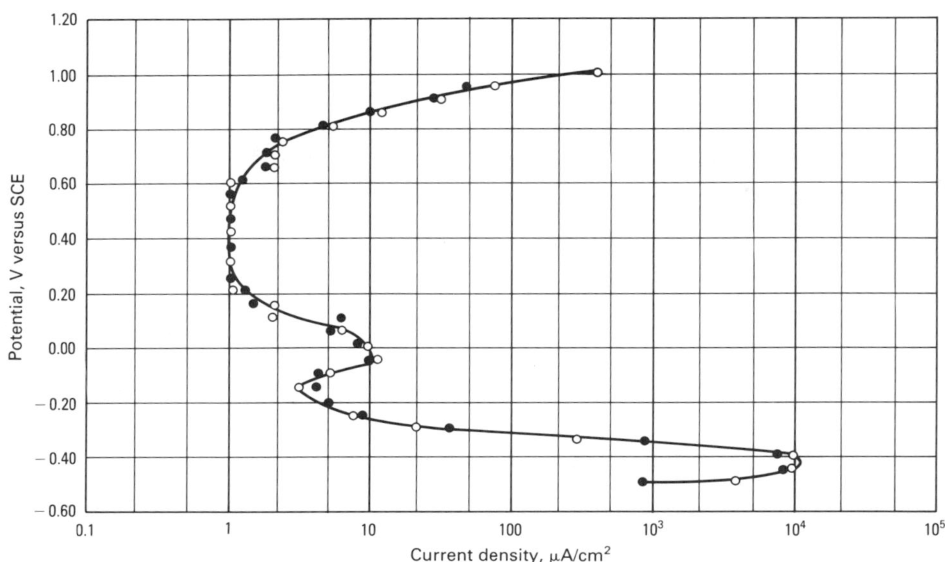

Fig. 8(b) Actual potentiostatic passive anodic polarization behavior for type 430 stainless steel in a 1 *N* H_2SO_4, H_2 purge, 30 °C (85 °F), 50 mV for 5 min (each specimen). Closed and open circles (data points) illustrate laboratory repeatability. Source: Ref 19

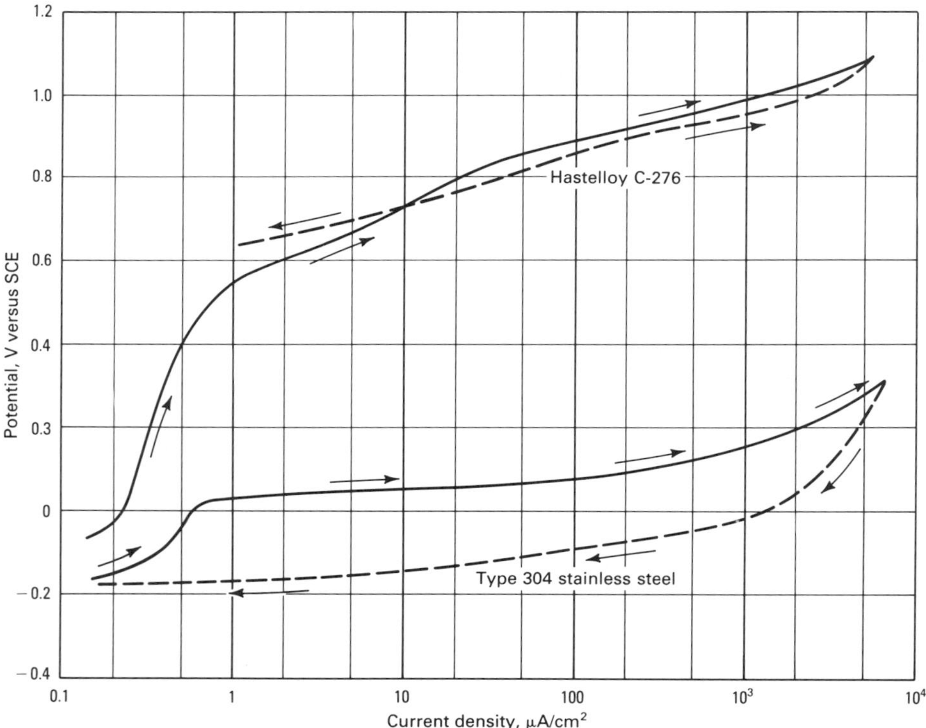

Fig. 9 Cyclic potentiodynamic polarization curves for Hastelloy C-276 and type 304 stainless steel in 3.56 wt% NaCl solution. Source: Ref 51

One of the techniques is for alloys that undergo passive film rupture and anodic dissolution, and the other is for alloys that are susceptible to hydrogen embrittlement. Both will be discussed below.

Stress Corrosion Cracking (SCC). In the area of electrochemical techniques for the study of SCC, the scratch-repassivation technique discussed above may be useful (Ref 56). This technique has been used to distinguish between the case where an alloy is susceptible to localized corrosion, such as pitting, and whether or not the alloy is susceptible to SCC. In general, once the passive film is mechanically disturbed at some potential, an electrochemical current can be measured that will decay back to a low level when repassivation has occurred. The total current will consist of the sum of the repassivation current and the current corresponding to dissolution of the exposed bare metal.

In this case, the electrochemical test is conducted in conjunction with the optical ellipsometric method to measure the rate of film growth. This combination offers a method of separating the two currents (dissolution-repassivation) from one another because it is a means of studying the kinetics of the film growth associated with repassivation. Based on the ratio of the charge involved in dissolution and the charge associated with repassivation, a criteria for stress corrosion is established:

$$R_{SCC} = 1 + \frac{Q_d}{Q_r} \quad \text{(Eq 16)}$$

where Q_d is the charge involved in dissolution, and Q_r is the charge associated with repassivation. For example, a low-carbon steel was found to be nonsusceptible to SCC in a solution in which R_{SCC} was 2.8. In addition, the steel was found to be susceptible when that ratio was 26, and it was nonsusceptible again in a solution in which R_{SCC} was 75 (Ref 56). In the latter case, widespread pitting occurred instead.

Hydrogen Embrittlement. The Devanathan-Stackurski method is used to study the parameters associated with the hydrogen embrittlement tendencies of metals and alloys (Ref 57). Essentially, a thin parallel-faced wafer of the metal of interest is positioned between two separate polarization cells such that there is no electrolyte path between the two cells. Adsorbed hydrogen is electrochemically generated through cathodic polarization and hydrogen evolution on one surface of the metal. The hydrogen that diffuses through the thin piece of metal is oxidized on the opposite surface by applying anodic polarization. Special care must be taken to avoid alloy dissolution on this surface. Under steady-state conditions, Fick's first law for diffusion is satisfied (Ref 57, 58). The hydrogen oxidation current is termed the permeation current. For a known specimen thickness and a measured permeation current density, the product of the diffusion coefficient and the hydrogen concentration in the metal at the charging surface can be determined:

$$J = \frac{nFDC_H}{L} \quad \text{(Eq 17)}$$

where J is the permeation current, D is the diffusion coefficient, C_H is the subsurface hydrogen concentration, and L is the specimen thickness. Special transient techniques have been devised that allow individual determination of D and C_H (Ref 58).

Evaluation of Alloy Sensitization. Measurement of the quantity of coulombs generated during the electrochemical polarization of a material from the passive range to the active corrosion potential can be used to detect the susceptibility to intergranular attack associated with the precipitation of chromium carbides and chromium nitrides at grain boundaries (Ref 59-62). A modification of this procedure, called the double-loop test (Ref 63, 64), involves polarizing the metal surface initially from the open-circuit potential in the active region to a potential in the passive range. Both variations of the method are illustrated in Fig. 12. This is followed by reverse polarization in the opposite direction back to the open-circuit potential. In this method, the degree of sensitization is determined by taking the ratio of the maximum current generated in the reactivation, or reverse, scan to that generated in the

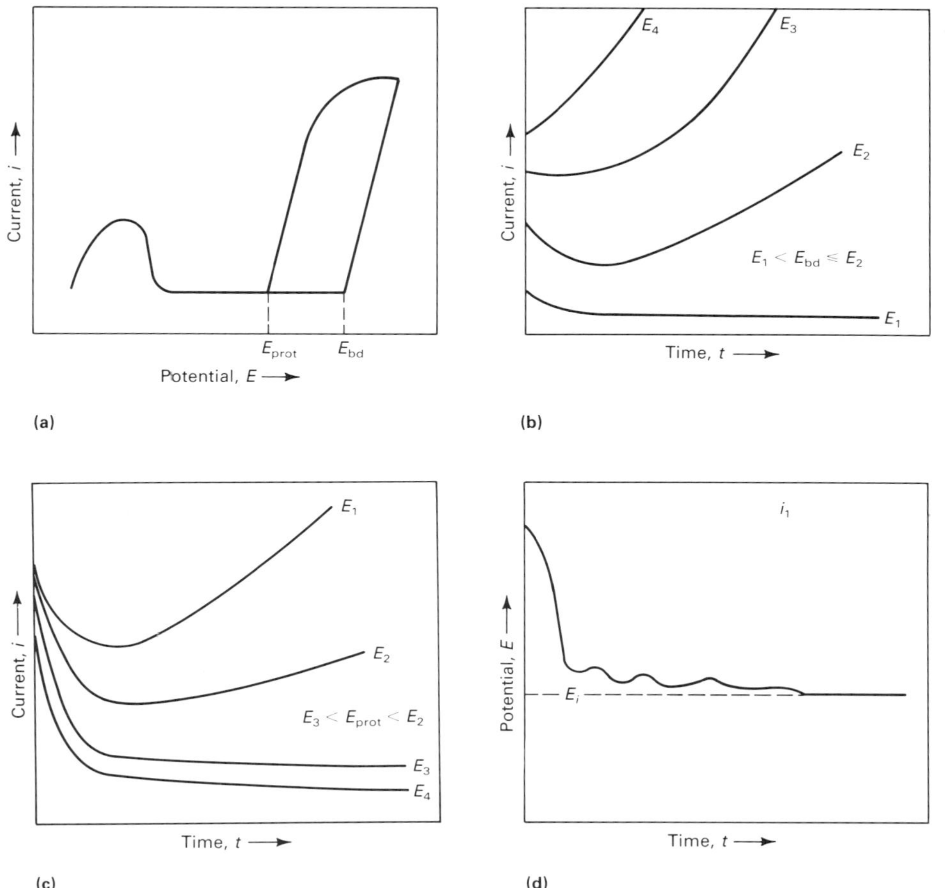

Fig. 10 Schematic representations for (a) a cyclic potentiodynamic polarization curve, (b) potentiostatic current-time curves with a passivated surface, (c) potentiostatic current-time curves for an activated surface, and (d) a galvanostatic potential-time curve. Source: Ref 48

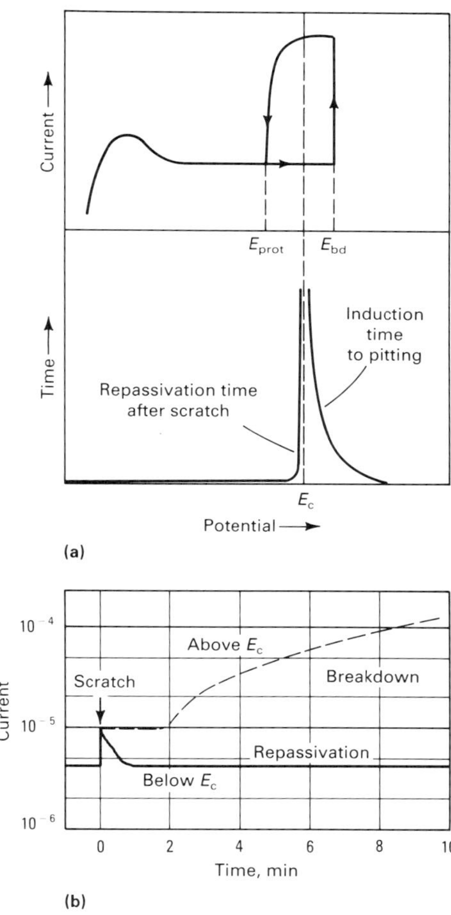

Fig. 11 Potential versus time plot of scratch test illustrating the critical potential, E_c, as it relates to the induction time (a) and the repassivation time (b). Source: Ref 48

initial anodic scan (I_r/I_a). The rationale for this procedure is contingent on the presence of current upon the reactivation scan that results mainly from incomplete passivation of the region adjacent to the grain boundaries because of chromium depletion. For nonsensitized material, the passive film remains essentially intact during the reverse scan, and the size of the reactivation polarization peak remains small. As a result, the charge Q is small. In one study, the charge was normalized by the grain-boundary area (GBA) because this is the area from which most of the current arises in the single reactivation scan (Ref 62):

$$P = \frac{Q}{\text{GBA (coulombs/cm}^2)} \quad \text{(Eq 18)}$$

$$\text{GBA} = A_s\,[5.0954 \times 10^{-3} \cdot \exp(0.34696 \cdot X)] \quad \text{(Eq 19)}$$

where P is the reaction charge density associated with the sensitized area, A_s is the specimen surface area, and X is the ASTM grain size. The same procedure was used for the ratio I_r/I_a (Ref 64). That is to say, the current peak, I_r, for the reactivation scan was normalized for the grain-boundary area as shown above, while the initial anodic current peak was normalized for the wetted surface area, A_s. The resulting expression is as follows:

$$\frac{i_r}{i_a} = \frac{I_r}{I_a}\,[5.0954 \times 10^{-3} \cdot \exp(0.34696 \cdot X)] \quad \text{(Eq 20)}$$

A number of investigators have correlated this electrochemically derived ratio with optical metallographic evaluations of the degree of material sensitization such as those outlined in ASTM A 262 (Ref 65). This has been accomplished for several different iron-nickel-chromium alloys (Ref 66-68). The technique is nondestructive.

Electrochemical Evaluation of Protective Coatings and Films

Numerous ac and dc electrochemical methods have been used to study the performance and the quality of protective coatings, including passive films on metallic substrates, and to evaluate the effectiveness of various surface pretreatments. Several are discussed below.

Anodized Aluminum Corrosion Test. One such method is the Ford anodized aluminum corrosion test (FACT) (Ref 69, 70). This test involves the cathodic polarization of the anodized aluminum surface by using a small cylindrical glass clamp-on cell and a special 5% NaCl solution containing cupric chloride ($CuCl_2$) acidified with acetic acid. A large voltage is applied across the cell by using a platinum auxiliary electrode. The alkaline conditions created by the cathodic polarization promote dissolution at small defects in the anodized aluminum. The coating resistance is decreased, more current begins to flow, and the voltage decreases. The cell voltage (auxiliary electrode to test specimen voltage) is monitored for 3 min, and the parameter cell voltage multiplied by time is recorded. The ASTM standard B 538 describes the method in greater detail (Ref 69).

A similar test, known as the cathodic breakdown test, involves cathodic polarization to −1.6 V (versus saturated calomel electrode, SCE) for a period of 3 min in acidified NaCl. Again, the test was designed for anodized aluminum alloys because the alkali created at the large applied currents will promote the formation of corroded spots at defects in the anodized film.

The electrolytic corrosion test was designed for electrodeposits of principally nickel and chromium on less noble metals, such as zinc or steel (Ref 71-73). Special solutions are used, and the metal is polarized to +0.3 V versus the standard

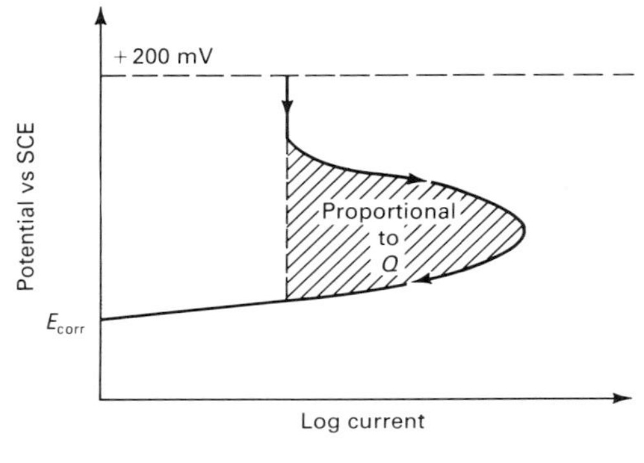

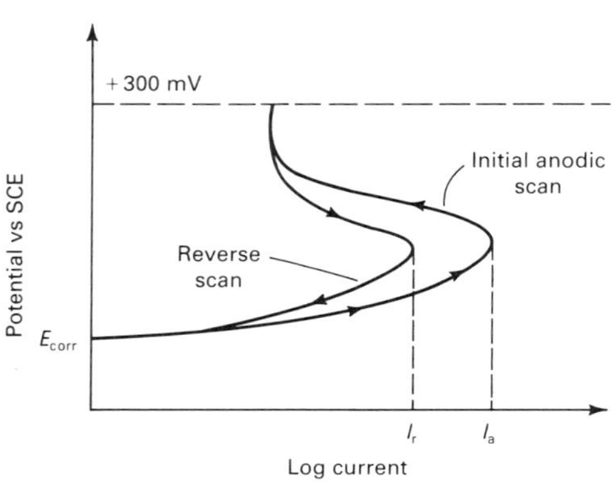

Fig. 12 Schematics of two procedures for anodic reactivation polarization testing. (a) Clarke *et al* method (Ref 59-62). (b) Akashi method (Ref 63, 64).

calomel electrode. The metal is taken through cycles of 1 min anodically polarized and 2 min unpolarized. An indicator solution is then used to detect the presence of pits that penetrate to the substrate. Each exposure cycle simulates 1 year of exposure under atmospheric-corrosion conditions. The ASTM standard B 627 describes the method in greater detail (Ref 74).

The paint adhesion on a scribed surface (PASS) test involves the cathodic polarization of a small portion of painted metal. The area exposed contains a scribed line that exposes a line of underlying bare metal. The sample is cathodically polarized for 15 min in 5% NaCl. At the end of this period, the amount of delaminated coating is determined from an adhesive tape pulling procedure.

The impedance test for anodized aluminum (ASTM B 457) is used to study the seal performance of anodized aluminum (Ref 75-77). In this sense, the test is similar to the FACT test except that this method uses a 1-V root mean square 1-kHz signal source from an impedance bridge to determine the sealed anodized aluminum impedance. The test area is again defined with a portable cell, and a platinum or stainless steel auxiliary electrode is typically used. The sample is immersed in 3.5% NaCl. The impedance is determined in ohms × 10^3. The ASTM standard B 457 covers this technique (Ref 75). In contrast to the methods discussed above, this test

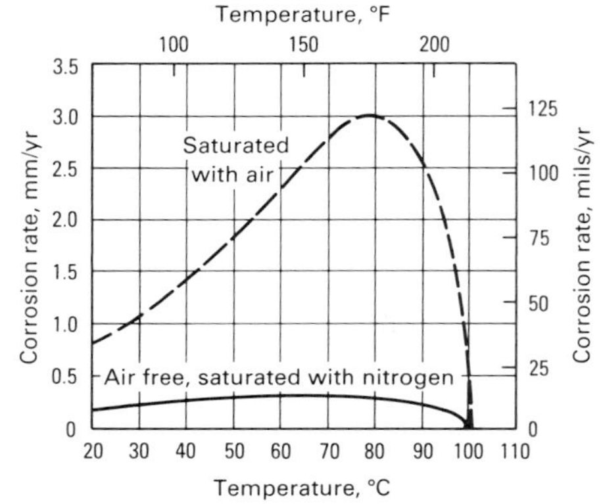

Fig. 13 Effect of temperature on the corrosion of Monel in 5 to 6% H_2SO_4. Solution velocity: 4.7 to 5 m/min (15.5 to 16.5 ft/min). Source: Ref 87 and 88

is essentially nondestructive and does not accelerate the corrosion process.

Electrochemical impedance spectroscopy, also known as the electrochemical impedance technique, offers an advanced method of evaluating the performance of metallic coatings (passive film forming or otherwise) and organic barrier coatings (Ref 78-83). The method does not accelerate the corrosion reaction and is nondestructive. The technique is quite sensitive to changes in the resistive-capacitive nature of coatings.

It is also possible to monitor the corrosion rate with this technique. In this respect, the electrochemical impedance technique offers several advantages over dc electrochemical techniques in that the polarization resistance related to the corrosion rate can be separated from the high dc resistance of the dielectric coating. This is not possible with the dc methods.

Because of the broad bandwidth in frequencies for the applied signal (usually from the mHz range to the kHz range), the electrochemical impedance technique also surpasses the capabilities of single-frequency impedance methods. The reason for this lies in the capability of the electrochemical impedance technique to discriminate between the resistive properties of the coating or passive film because of its weak ionic and or electronic conductance and the capacitive nature of the passive film or coating due to its dielectric constant, area, and thickness. A single-frequency technique depends exclusively on the *a priori* assumption that the frequency of measurement provides only resistive or capacitive (in phase or out of phase) information. What is more probable, however, is that the impedance determined at the single frequency is a combined resistive-capacitive response. Although impedance circuit models may become quite complex compared to the model shown in Fig. 5, frequency regimes in which information is primarily revealed on the coating capacitance or coating resistance can be separated from one another and analyzed independently by using a broad frequency bandwidth. If the coating or passive film capacitance can be determined and if the dielectric constant is known, then the film thickness can be estimated for a given exposed area:

$$C = \frac{\epsilon \cdot \epsilon^\circ A}{d} \qquad \text{(Eq 21)}$$

where A is the surface area, d is the dielectric thickness, and ε is the electric permittivity of free space (8.854×10^{-12} F/m). In addition to this correlation, the uptake of water in an organic coating can be monitored because the dielectric constant for water is over an order of magnitude greater than the dielectric constant for the dry organic coating. The quantity of water absorbed in the organic coating can also be estimated as follows (Ref 82):

$$\text{vol\% } H_2O = \frac{100 \cdot \log \left(\frac{C_t}{C_0}\right)}{\log (79)} \qquad \text{(Eq 22)}$$

where C_t is the coating capacitance at time t, and C_0 at time zero when the exposure begins.

The coating resistance can also be monitored as a function of exposure time. Large decreases signify permeation of ionic species through the coating or the presence of defects in the coating itself.

Limitations of Electrochemical Tests

Despite this impressive range of applications, electrochemical techniques are not always successful, and it is instructive at this point to identify some of the more general reasons for this situation. First and most obvious is the requirement that the corrosion process be electrochemical. Three more specific keys to successful application of these techniques include the implementation of relevant environment, relevant alloy and surface preparation, and relevant mechanical perturbations. Without these essential features, any electrochemical test designed to simulate, for example, a chemical plant process stream will be destined for failure. More subtle but equally important is the necessity that any acceleration of the corrosion process act only upon the prevailing mechanism of corrosion under normal nonaccelerated conditions. In summary, careful application of electrochemical methods used in conjunction with other test procedures can be a tremendous aid in solving corrosion problems.

Immersion Tests

Donald O. Sprowls
Consultant

Complete immersion of test specimens in a corrosive solution would appear to be a simple test indeed; however, in actuality, a number of test conditions must be controlled to ensure adequate reproducibility of test results. The actual conditions of the test will usually be determined by the nature of the problem at hand, the ingenuity of the investigator, and the budget of the test program. The following information is presented as a guide so that some of the pitfalls of such testing may be avoided.

The methods and procedures described are concerned with the basic factors that must be considered in sophisticated test procedures as well as in the simple tests that are discussed. Primary consensus standards for immersion corrosion testing of metals have been developed by

Table 1 Effect of temperature on corrosion of type 303 stainless steel by 65% HNO_3

Temperature °C	Temperature °F	Corrosion rate mm/yr	Corrosion rate mils/yr
Up to 120	250	<0.5	<20
125	260	2.5	100
135	280	5.0	200
160	320	12.5	500
165	330	25	1000
185	370	125	5000

Source: Ref 85

ASTM (Ref 84) and the National Association of Corrosion Engineers (NACE) (Ref 85), from which much of this information has been taken. Additional details can be found in the actual standards and in the other references cited in this section.

For proper planning of the test and interpretation of the test results, the specific influences of the following variables must be considered: solution composition, temperature, aeration, volume, velocity, and waterline effects; specimen surface preparation; method of immersion of specimens; duration of test; and method of cleaning specimens at conclusion of the exposure.

Total Immersion

Composition of Solution. Test solutions should be accurately prepared from chemicals that conform to the Specifications of the Committee on Analytical Reagents of the American Chemical Society* and from distilled water, except in those cases in which naturally occurring solutions or solutions taken directly from some plant process are used. The composition of the test solutions should be controlled to the fullest extent possible. They should also be described as completely and as accurately as possible when the results are reported.

Evaporation losses can be controlled by a constant level device or by frequent addition of appropriate solution to maintain the original volume within ±1%. The use of a reflux condenser ordinarily precludes the necessity of adding to the original kettle charge.

The composition of the test solution sometimes changes as a result of catalytic decomposition or by reaction with the test coupons. These changes should be determined if possible. Where required, the exhausted constituents should be added or a fresh solution provided during the course of the test. If problems are suspected, the composition of the test solutions should be checked by analysis at the end of the test to determine the extent of change in composition, such as might result from evaporation or depletion.

Corrosion products from the coupon may influence the corrosion rate of the metal or the corrosion rate of different metals exposed at the same time. For example, the accumulation of cupric ions in the testing of copper alloys in intermediate strengths of H_2SO_4 will accelerate the corrosion of copper alloys, as compared to the rates that would be found if the corrosion products were continually removed. Cupric ions may also exhibit a passivating effect on stainless steel coupons exposed at the same time. It is generally advisable to expose only alloys of the same general type in the testing apparatus.

Temperature of Solution. Perhaps the single most important factor in corrosion is the effect of temperature. Accordingly, the temperature of the surface of the specimen must be known because this is the corroding temperature. In many cases, corrosion increases rapidly with temperature. An example is the corrosion of type 303 stainless steel by nitric acid (HNO_3), as shown in Table 1. Corrosion increased from a low rate (<0.5 mm/yr or 20 mils/yr) to several thousand mils per year as a result of increasing the temperature 65 °C (120 °F). For some corrosion systems, there can be a double effect. For example, the corrosion rate of Monel alloy in air-saturated H_2SO_4 solutions increases exponentially with rising temperature to a maximum at 80 °C (175 °F) and then decreases with further increases in temperature because of the decrease in oxygen concentration (Fig. 13).

Laboratory tests are often conducted in controlled-temperature water or oil baths. The temperature of the corroding solution should be controlled within ±1 °C (±1.8 °F) and must be stated in the report of test results.

If no specific temperature, such as boiling point, is required or if a temperature range is to be investigated, the selected temperatures used in the test, as well as their respective durations, must be reported. Tests at ambient temperature should be conducted at the highest temperature anticipated. The variation in temperature should also be reported.

Accelerated corrosion tests are often conducted at temperatures above proposed operating temperatures in order to decrease the time of testing. This procedure is dangerous because the effect of temperature may be great, resulting in the elimination of more economical materials.

A common error involves the assumption that the environment temperature is the corroding temperature. This is particularly true in the case of materials for heating surfaces. The average temperature of the liquid in a tank may be 65 °C (150 °F), but the corroding or surface temperature of the steam heating coil will be considerably higher. Therefore, tests conducted at 65 °C (150 °F) may provide erroneous results. Surface temperatures can often be estimated by considering heat transfer coefficients.

Recent research has shown that dissolution of borosilicate glass vessels in distilled water at temperatures above 60 °C (140 °F) significantly lowers the corrosion rate of low-carbon steel. The method of aerating the solution is also important. A new test apparatus has been designed in which the above problems have been overcome and other parameters are capable of more precise control (Ref 89).

Aeration of Solution. Aeration or the presence of dissolved oxygen in a liquid environment can profoundly influence corrosion rates. Accordingly, this factor should be carefully considered in a corrosion test program. In general, some metals and alloys are more rapidly attacked in the presence of oxygen, but others may show better corrosion resistance. Although this discussion is primarily concerned with the oxidizing effects of aeration, these effects could also be produced by other oxidizing agents. Figure 14 shows a generalized picture of several materials in different environments.

Figure 13 also shows the effect of aeration and deaeration on Monel. An interesting feature is the rapid decrease in corrosion as the temperature approaches boiling. The rates at the boiling points are essentially the same under the two conditions of test. The reason for this behavior is that the solubility of oxygen in boiling water or dilute aqueous solutions is practically nil. This feature is of interest from the practical standpoint because a boiling test is more convenient to conduct than one at, for example, 80 °C (175 °F). The operator might think that he would accelerate the test by running it at boiling, but he may actually misinterpret the results by so doing.

Unless specified, the solution should not be aerated. Most tests related to process equipment should be conducted with the natural atmosphere inherent in the process, such as the vapors of the boiling liquid. If aeration is used, the specimens should not be located in the direct air stream from the sparger. Extraneous effects can be encountered if the air stream impinges on the specimens.

The simplest and most widely used aeration method consists of bubbling air through the solution. The solution is then assumed to be saturated with air. In most practical applications, air is involved; therefore, air is used in the test. If air is not bubbled through the solution, the aeration effect depends on the air present in the solution at

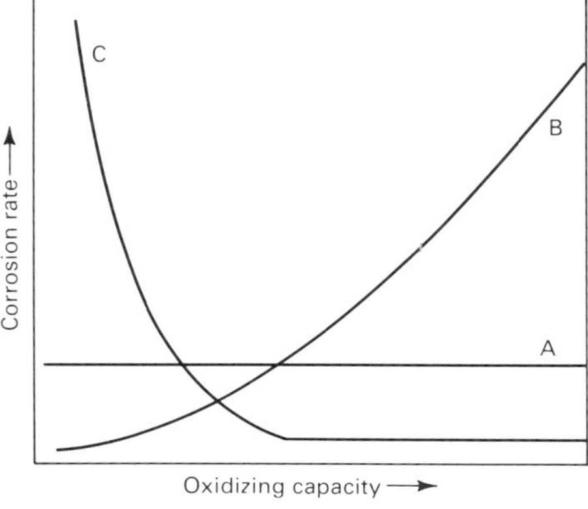

Curve A
1. Stainless steel in nitric acid
2. Aluminum in hydrogen peroxide
3. Hastelloy C in ferric chloride

Curve B
1. Monel in hydrochloric acid
2. Iron in water
3. Copper in sulfuric acid

Curve C
1. Stainless steel in sulfuric acid + copper sulfate
2. Stainless steel in sulfuric acid + nitric acid

Fig. 14 Effect of solution oxidizing capacity on corrosion rates of various metals. Source: Ref 86

*"Reagent Chemicals, American Chemical Society Specifications," American Chemical Society. For suggestions on the testing of reagents not listed by the American Chemical Society, see J. Rosin, Reagent Chemicals and Standards, Van Nostrand, and the "United States Pharmacopeia."

the start of the test and on its rate of removal or escape, if any. Air can also be absorbed from the atmosphere over the solution.

If oxygen saturation of the test solution is desired, this can best be achieved by sparging with oxygen. Appropriate precautions should be taken when oxygen-rich gases are used; many materials burn violently in pure oxygen. For other degrees of aeration, the solution should be sparged with synthetic mixtures of air or oxygen with an inert gas.

If complete exclusion of dissolved oxygen is necessary, specific techniques, such as prior heating of the solution and sparging with an inert gas (usually nitrogen), are required. A liquid atmospheric seal is often used on the test vessel to prevent air ingress.

Volume of Solution. The volume of the test solution should be large enough to avoid any appreciable change in its corrosivity during the test through exhaustion of corrosive constituents or by accumulation of corrosion products that may affect further corrosion. Two examples of a recommended minimum solution volume to specimen area ratio are 200 L/m^2 of specimen surface when detecting susceptibility to intergranular attack to austenitic stainless steel (Ref 90) and 100 L/m^2 of specimen surface when evaluating the exfoliation corrosion susceptibility of 5*xxx* aluminum alloys (Ref 91).

When the objective of the test is to determine the effect of a metal or alloy on the characteristics of the test solution (for example, to determine the effects of metals on dyes), it is desirable to reproduce the ratio of solution volume to exposed metal surface that exists in practice. The actual time of contact of the metal with the solution must also be taken into account. Any necessary distortion of the test conditions must be considered when interpreting the results.

Solution Velocity. The effect of velocity is not usually determined in normal laboratory tests, although specific tests have been designed for this purpose (Ref 87, 92). However, for the sake of reproducibility, some velocity control is desirable. Tests at the boiling point should be conducted with the least heat input possible, and boiling chips should be used to avoid excessive turbulence and bubble impingement. In tests conducted below the boiling point, thermal convection is generally the only source of liquid velocity. In test solutions with high viscosities, supplemental controlled stirring is recommended.

Method of Supporting Specimens. The supporting device and container should not be affected by or cause contamination of the test solution. The method of supporting specimens varies with the apparatus used for conducting the test, but should be designed to insulate the specimens from each other physically and electrically and to insulate the specimens from any metallic container or supporting device used within the apparatus.

The shape and form of the specimen support should ensure free contact of the specimen with the corroding solution, the liquid lines or the vapor phase (Fig. 15). If clad alloys are exposed, special procedures, such as coating of edges, will be required to ensure that only the cladding is exposed, unless the purpose is to test the ability of the cladding to protect cut edges. Common supports include glass or ceramic rods, glass saddles, glass hooks, fluorocarbon plastic strings, and various insulated or coated metallic supports.

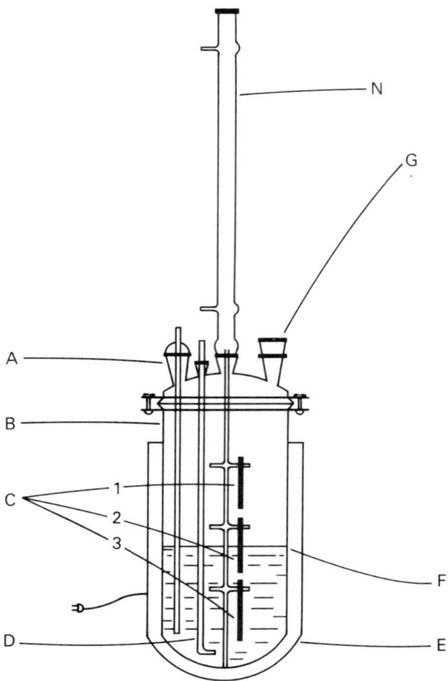

Fig. 15 Resin flask used to conduct simple immersion tests. A, thermowell; B, resin flask; C, specimens hung on supporting device (C-1, vapor phase; C-2, partial immersion; C-3, total immersion); D, air inlet; E, heating mantle; F, liquid interface; G, opening in flask for additional apparatus that may be required; H, reflux condenser. Source: Ref 84

In aqueous solutions under 1 atm of air or oxygen, it is desirable to position all specimens so that the oxygen supply is similar. This is especially important with corrosion systems in which the corrosion rate may be fairly high, such as for iron and steel in neutral salt solutions. In such tests, totally immersed specimens at a depth less than 20 mm (0.8 in.) had markedly higher corrosion rates, but with deeper immersion, the rate was lower and variations in depth were unimportant (Ref 93). More detailed discussion of the factors that can affect corrosion behavior in immersion tests can be found in Ref 87.

Partial Immersion

Laboratory tests under conditions of partial immersion are of practical importance because such conditions are commonly found in service. Tests conducted in three locations in a resin flask, as shown in Fig. 15, can be used to determine the relative susceptibility of a material to localized corrosion at the liquid line. Partial-immersion conditions provide a very suitable accelerated test for such metals as aluminum alloys and others that develop concentrated attack at the liquid line (or splash zone in certain equipment).

Intermittent Immersion

The term intermittent immersion refers to alternate immersion and emersion in a liquid corrodent. The conditions are of practical importance because they simulate, for example, the effects of the rise and fall of tidal waters and the movements of corrosive liquids in chemical plants. Also, they may provide a relatively rapid test for the effect of aqueous solutions because a thin film of the solution, frequently renewed and almost saturated with oxygen, can be maintained on the test specimen during most of the period of exposure, even when the shape of the specimen is complex. Reference 87 contains a detailed discussion of the rather considerable research that has been done on various alternate immersion cycles and drying conditions for tests of both ferrous and nonferrous alloys.

To simulate service conditions, the specimens should be dry before reimmersion because this has an important effect on the protective character of films on the metal. For maximum corrosion, this drying should be done slowly so that the film of the corrodent on the metal can be allowed the maximum time to act during immersion. In the interest of reproducibility, it is desirable to control the humidity and temperature of the atmosphere in order to ensure a constant rate of drying in successive cycles. When more thorough drying of the specimen is desired, forced-air circulation or radiant heating can be used to accelerate the drying. Artificial heating adds to the complexity of the test conditions, however, and may reduce the reproducibility of the test if it is not very carefully controlled.

Alternate Immersion in 3.5% NaCl. The conditions of this test that have gained the most widespread acceptance, particularly in the United States, are those described in ASTM G 44 (Ref 94). This practice utilizes a 1-h cycle that includes a 10-min period of immersion in an aqueous solution of 3.5% NaCl or in substitute ocean water, followed by a 50-min emersion period. This 1-h cycle is continued 24 h/day for the test duration required for the particular test material. Aluminum and steel alloys are typically exposed from 10 to 90 days or longer, depending on the resistance of the alloy to corrosion by saltwater. Although this alternate-immersion test is considered to be an accelerated test representative of certain natural conditions (particularly marine environments), it is not intended to relate to specialized chemical environments.

Air circulation is an important consideration because it affects the rate at which specimens dry and the loss of water by evaporation from the salt solution. The most important consideration consists of achieving moderate specimen drying conditions. Because various testing facilities use different immersion apparatuses and room sizes, individual experimentation is necessary to achieve adequate circulation. A mild circulation of air is recommended.

Specimen Drying. As with air circulation, no fixed procedure has been established and probably cannot be unless a standardized immersion apparatus and test chamber are adopted. The objective, however, is to ensure that all specimens dry slowly during the 50-min emersion period. Because they drain differently, specimens with different accrued corrosion films will dry at different rates. New specimens with little accumulated corrosion products become dry in about 15 min, but other specimens with an accumulation of corrosion product and salt should be allowed about 40 min to dry.

Apparatus. The usual methods of immersion involve:

- Placing specimens on a movable rack that is periodically lowered into a stationary tank containing the solution (Fig. 16)

- Placing specimens on a hexagonal Ferris wheel arrangement that rotates every 10 min through 60° and thus passes the specimens through a stationary tank of solution
- Placing specimens in a stationary tray open to the atmosphere and having the solution moved by air pressure, nonmetallic pump, or gravity drain from a reservoir to the tray

Rate of Immersion. The rate of immersion and removal of the specimens from the solution should be as rapid as possible without jarring them. For purposes of standardization, an arbitrary limit shall be adopted such that no more than 2 min elapse from the time the first portion of any specimen is covered or uncovered by solution.

Materials of construction that contact the salt solution shall be such that they are not affected by the corrodent to an extent that they can cause contamination of the solution and change its corrosiveness. Use of inert plastics or glass is recommended where feasible. Metallic materials of construction should be selected from alloys that are recommended for marine use and are of the same general type as the metals being tested. Preferably, all metal parts should be protected with a suitable corrosion-resistant coating.

Specimen holders should be designed to electrically insulate the specimens from each other and from any other bare metal. When this is not possible, as with certain stressing bolts or jigs, the bare metal contacting the specimen should be isolated from the corrodent by a suitable coating. Protective coatings should be of a type that will not leach inhibiting or accelerating ions or protective oils over the noncoated portions of the specimen. Coatings containing chromates should be avoided. They should not obstruct air flow over the specimens, which retards the drying rate.

Solution Conditions. The salt solution shall be prepared by dissolving 3.5 ± 0.1 parts by weight of NaCl in 96.5 parts of water. Reagent grade NaCl shall be used for conformance to the specifications of the Committee on Analytical Reagents of the American Chemical Society. Distilled or deionized water shall be used for compliance with the purity requirements of ASTM D 1193, Type IV, reagent water (Ref 95).

When tests are to be made in substitute ocean water, it shall be prepared without heavy metals in accordance with ASTM D 1141 (Ref 96). An advantage of the substitute ocean water in the case of stress-corrosion tests of certain high-strength aluminum alloys is that it causes less severe pitting than the 3.5% NaCl solution. It may be preferred in certain other cases because the test results do (or are thought to) relate better to real marine environments.

Temperature. A freshly prepared solution should be allowed to come within 3 °C (5 °F) of the specified air temperature before being used. Thereafter, no control is required on the solution temperature. Instead, the air temperature is controlled, and the solution is allowed to reach temperature equilibrium.

Minimum Volume. The volume of the test solution should be large enough to avoid any appreciable change in its corrosiveness through exhaustion of corrosive constituents or through the accumulation of corrosion products or other constituents that might significantly affect further corrosion. An arbitrary minimum ratio between the volume of the test solution and the surface area of the specimen(s) (including any uncoated accessories) of 320 L/m^2 (7.9 gal/ft^2) is recommended.

Fig. 16 Lift-type alternate-immersion apparatus. Specimens shown are in the emersion position, where they remain for 50 min. They are then lowered into the tanks of saltwater for 10 min to complete the 1-h cycle. Many shapes and sizes of specimens can be tested in this large equipment. Courtesy of the Aluminum Company of America

Air Conditions. The air temperature shall be maintained at 27 ± 1 °C (80 ± 2 °F). The percent relative humidity shall be controlled at 45 ± 6%.

Specimen Preparation and Duration of Tests

The type, size, and shape of specimens will vary with the purpose of the test, the nature of the test materials, and the apparatus used. Standard practices for preparing, cleaning, and evaluating corrosion test specimens are provided in ASTM G 1 (Ref 97). Proper selection of appropriate lengths of exposure is important for any corrosion test, and misleading results may be obtained if the time factor is not considered. An excellent technique for investigating changes in corrosion rate with length of exposure is described below (Ref 84-86, 98).

Planned interval tests involve the accumulated effects of corrosion at several times under a given set of conditions as well as the initial rate of corrosion of fresh metal, the more or less instantaneous corrosion rate of metal after long exposure, and the initial corrosion rate of fresh metal during the same period of time as the latter. The rates, or damage in unit time interval, are shown in the diagram in Table 2 as A_1, A_2, and B, respectively. Unit time interval can often be taken conveniently as 1 day in a planned interval test extended over a total period of several days. It would be desirable to have duplicate specimens for each interval. Further time extensions of the test should be made with similar added specimens and interval spacing if no changes occurred in the corrosion rate of the metal or the corrosiveness of the liquid during the first selection of test duration.

Comparison for corrosion damage A_1 for the unit time interval from 0 to 1, with corrosion damage B for the unit time interval from t to $t + 1$, shows the magnitude and direction of change in the corrosivity of the liquid that may have occurred during the total time of the test. Correspondingly, comparison of A_2 with B, where A_2 is the corrosion damage calculated by subtracting A_t from A_{t+1}, shows the magnitude and direction of change in corrodibility of the metal specimen during the test. These comparisons can be taken as criteria for the changes and are tabulated in Table 2. Also given in Table 2 are the criteria for all possible combinations of changes in the corrosivity of the liquid and corrodibility of the metal. The additional information obtained on occurrences in the course of the test justifies the additional effort involved. Table 3 lists the data obtained from a planned interval test of a carbon steel specimen in an aluminum chloride ($AlCl_3$)-antimony trichloride ($SbCl_3$) solution.

The causes for the changes in corrosion rate as a function of time are not given by the planned interval test criteria. The corrosivity of the liquid may decrease as a result of corrosion during the course of a test because of the reduction in concentration of the corrosive agent, the depletion of a corrosive contaminant, the formation of inhibiting products, or other metal-catalyzed changes in the liquid. The corrosivity of the liquid may increase because of the formation of autocatalytic products, the destruction of corrosion-inhibiting substances, or other catalyzed changes in the liquid. Changes in the corrosiveness of the liquid may also arise from changes in composition that would occur under the test conditions even in the absence of metal. To determine if the latter effect occurs, an identical test is run with-

Table 2 Planned interval test

Identical specimens are placed in the same corrosive liquid; imposed conditions of test are constant for the entire time ($t + 1$); A_1, A_t, A_{t+1}, B, represent the corrosion damage experienced by each test specimen; A_2 is a calculated value obtained by subtracting A_t from A_{t+1}.

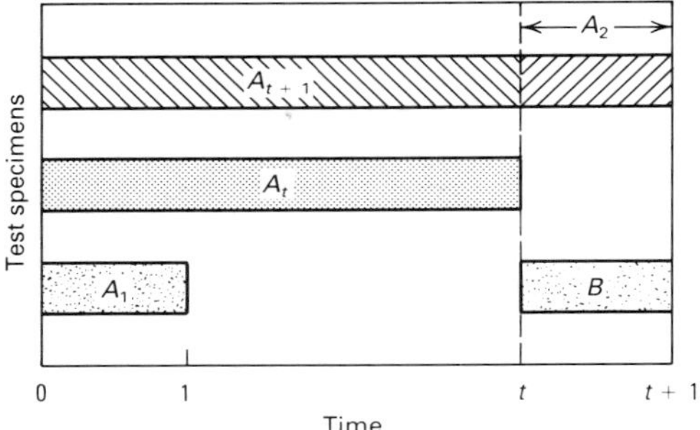

Corrosivity	Observed weight changes during corrosion testing	Criteria
Liquid		
corrosiveness	Unchanged	$A_1 = B$
	Decreased	$B < A_1$
	Increased	$A_1 < B$
Metal		
corrodibility	Unchanged	$A_2 = B$
	Decreased	$A_2 < B$
	Increased	$B < A_2$

Combinations of situations		
Liquid corrosiveness	Metal corrodibility	Criteria
Unchanged	Unchanged	$A_1 = A_2 = B$
Unchanged	Decreased	$A_2 < A_1 = B$
Unchanged	Increased	$A_1 = B < A_2$
Decreased	Unchanged	$A_2 = B < A_1$
Decreased	Decreased	$A_2 < B < A_1$
Decreased	Increased	$A_1 > B < A_2$
Increased	Unchanged	$A_1 < A_2 = B$
Increased	Decreased	$A_1 < B > A_2$
Increased	Increased	$A_1 < B < A_2$

Table 3 Planned interval corrosion test

Duplicate strips of low-carbon steel (19 × 75 mm, or ¾ × 3 in.) were immersed in 200 mL of 10% $AlCl_3$-90% $SbCl_3$ mixture through which dried hydrogen chloride gas was slowly bubbled at atmospheric pressure. Temperature: 90 °C (195 °F)

Value	Interval, days	Weight loss, mg	Penetration µm	Penetration mils	Apparent corrosion rate mm/yr	Apparent corrosion rate mils/yr
A_1	0–1	1080	42.9	1.69	15.7	620
A_t	0–3	1430	56.9	2.24	6.8	270
A_{t+1}	0–4	1460	58.2	2.29	5.3	210
B	3–4	70	2.8	0.11	1.02	40
A_2	calculated 3–4	30	1.3	0.05	0.46	18

$A_2 < B < A_1$
$0.05 < 0.11 < 1.69$

Conclusions: Liquid markedly decreased in corrosiveness during the test, and the formation of a partially protective scale on the steel was indicated; that is, metal became less corrodible.

out test strips for the total time, t. Test strips are then added, and the test is continued for unit time interval. Comparison with A_1 of corrosion damage from this test shows if the corrosive character of the liquid changes significantly in the absence of metal.

The corrodibility of the metal in a test may decrease as a function of time because of the formation of protective scale or the removal of a less resistant surface layer of metal. Metal corrodibility may increase because of the formation of corrosion-accelerating scale or the removal of a more resistant surface layer of metal. Indications of the causes of changes in corrosion rate can often be obtained from close observation of tests and corroded specimens as well as from special supplementary tests designed to reveal effects that may be involved.

Changes in liquid corrosiveness are not a factor in most plant tests that consist of once-through runs or where large ratios of solution volume to specimen area are involved. If the effect of corrosion on the mechanical properties of the metal or alloy is under consideration, a set of unexposed specimens is needed for comparison purposes.

Lengthy corrosion tests are generally not necessary to obtain accurate corrosion rates from materials that undergo severe corrosion. However, there are cases in which this assumption is not valid. For example, lead exposed to H_2SO_4 initially corrodes at an extremely high rate while building a protective film; the rates then decrease considerably so that further corrosion is negligible. The phenomenon of the formation of a protective film is observed with many corrosion-resistant materials. Therefore, short tests on such materials would indicate a high corrosion rate and would be completely misleading.

Short-term tests can also give misleading results on alloys that form passive films, such as stainless steels. With borderline conditions, a prolonged test may be needed to permit the breakdown of the passive film and subsequent more rapid attack. Consequently, tests conducted for long periods are considerably more realistic than those conducted for short durations. This statement must be qualified by stating that corrosion should not proceed to the point at which the original specimen size or the exposed area is drastically reduced or the metal is perforated.

If anticipated corrosion rates are moderate or low, the following equation gives the suggested tests duration:

$$\text{Hours} = \frac{2000}{\text{(corrosion rate in mils/yr)}}$$

For example, where the corrosion rate is 10 mils/yr (0.25 mm/yr), the test should run for at least 200 h. This method of estimating test duration is useful as an aid in deciding, after a test has been made, whether or not it is desirable to repeat the test for a longer period.

Reporting the Data

The importance of reporting all data as completely as possible cannot be overemphasized. Expansion of the testing program in the future or correlating the results with tests of other investigators will be possible only if all pertinent information is properly recorded. The following checklist is a recommended guide for reporting all important information and data:

- Corrosive media and concentration (and any changes during test)
- Volume of test solution
- Temperature (maximum, minimum, and average)
- Aeration (describe conditions or technique)
- Agitation (describe conditions or technique)
- Type of apparatus used for test
- Duration of each test
- Chemical composition or trade name of metals tested
- Form and metallurgical conditions of specimens
- Exact size, shape, and area of specimens
- Treatment used to prepare specimens for test
- Number of specimens of each material tested, and whether specimens were tested separately or which specimens were tested in the same container
- Method used to clean specimens after exposure and the extent of any error expected by this treatment
- Initial and final masses and actual mass losses for each specimen
- Evaluation of attack if other than general, such as crevice corrosion under support rod, pit depth and distribution, and results of microscopical examination or bend tests
- Corrosion rates for each specimen
- Minor occurrences or deviations from the proposed test program often can have significant effects and should be reported if known

Salt Spray Testing

Norman B. Tipton
The Singleton Corporation

Salt spray tests have been used for over 80 years as accelerated tests for determining the corrodibility of nonferrous and ferrous metals as well as the degree of protection afforded by both inorganic and organic coatings on a metallic base (Ref 99). This procedure has been extensively discussed since its inception because of the reproducibility variances and the questionable correlation of results as related to actual service performance. The primary objective is to provide an easily performed acceptable standard for comparing the performance of materials and coatings.

Many revisions to the salt spray test procedures and many improvements to the salt spray test cabinets have been made over the years through the joint efforts of the National Bureau of Standards, ASTM, equipment manufacturers, the automotive industry, and many governmental agencies. These revisions have eliminated many of the variables that have caused much of the criticism of this test procedure, with the result being a much more reliable and useful test. Even with the newly revised test procedures and modern designs of the salt spray test cabinets, there are still variables to be further investigated (Ref 99).

Applications of Salt Spray (Fog) Testing

The salt spray (fog) test has received its widest acceptance as a tool for evaluating the uniformity of thickness and degree of porosity of metallic and nonmetallic protective coatings, and it has served this purpose with a great deal of success. The test is useful for evaluating different lots of the same product, once a standard level of performance has been established, and it is especially helpful as a screening test for revealing a particularly inferior coating. In recent years, certain cyclic acidified salt spray (fog) tests have been implemented to test the resistance of aluminum alloys to exfoliation corrosion. The salt spray (fog) test is considered to be most useful as an accelerated laboratory corrosion test that simulates the effects of marine atmospheres on different metals, with or without protective coatings.

The most commonly used and accepted salt spray test methods in the United States are the various methods outlined in ASTM standards B 117 and G 85 (Ref 100, 101). Many of the governmental agencies and automotive companies have written their own standards and procedures, but in the interest of national standardization, these standards have been revised to conform with most of the details of ASTM. However, they still incorporate several statements relating to practices that experience has shown to be desirable or beneficial for achieving reliable, reproducible results and maximum correlation among laboratories.

Types of Salt Spray (Fog) Tests

The neutral salt spray (fog) test (ASTM B 117—Method 811.1 of Federal Test Method 151b) is perhaps the most commonly used salt spray test in existence for testing inorganic and organic coatings, especially where such tests are used for material or product specifications. The duration of this test can range from 8 up to 3000 h, depending on the product or type of coating. A 5% NaCl solution that does not contain more than 200 ppm total solids and with a pH range of 6.5 to 7.2 when atomized is used, and the temperature of the salt spray cabinet is controlled to maintain 35 + 1.1 or − 1.7 °C (95 + 2 or − 3 °F) within the exposure zone of the closed cabinet.

The Acetic Acid-Salt Spray (Fog) Test (ASTM G 85, Annex A1; Former Method B 287) is also used for testing inorganic and organic coatings, but is particularly applicable to the study or testing of decorative chromium plate (nickel-chromium or copper-nickel-chromium) plating and cadmium plating on steel or zinc die-castings and for the evaluation of the quality of a product.

This test can be as brief as 16 h, although it normally ranges from 144 to 240 h or more. As in the neutral salt spray test, a 5% NaCl solution is used, but the solution is adjusted to a pH range of 3.1 to 3.3 by the addition of acetic acid, and again, the temperature of the salt spray cabinet is controlled to maintain 35 + 1.1 or − 1.7 °C (95 + 2 or − 3 °F) within the exposure zone of the closed cabinet.

The Copper-Accelerated Acetic Acid-Salt Spray (Fog) Test or CASS test, which is covered in ASTM B 368 (Ref 102), is primarily used for the rapid testing of decorative copper-nickel-chromium or nickel-chromium plating on steel and zinc die-castings. It is also useful in the testing of anodized, chromated, or phosphated aluminum. The duration of this test ranges from 6 to 720 h. A 5% NaCl solution is used, with 1 g of copper II chloride ($CuCl_2 \cdot 2H_2O$) added to each 3.8 L of salt solution. The solution is then adjusted to a pH range of 3.1 to 3.3 by adding acetic acid. The temperature of the CASS cabinet is controlled to maintain 49 + 1.1 or − 1.7 °C (120 + 2 or − 3 °F) within the exposure zone of the closed cabinet.

Other Standard Tests. Many new salt spray test procedures have been developed in the past 20 years in order to achieve tests that are more closely aligned with a specific application. These modifications include a cyclic acidified salt spray (fog) test (ASTM G 85, Annex A2), an acidified synthetic seawater spray (fog) test (ASTM G 85, Annex A3; Former Method G 43), and a salt/sulfur dioxide (SO_2) spray (fog) test (ASTM G 85, Annex A4). The cyclic acidified salt spray (fog) test and the acidified synthetic sea water spray (fog) test are both primarily used for the production control of exfoliation-resistant heat treatments for various aluminum alloys (Ref 103). The salt/SO_2 spray (fog) test is mainly used to test for the exfoliation corrosion resistance of various aluminum alloys and a wide range of nonferrous and ferrous materials and coatings, both inorganic and organic, when exposed to an SO_2-laden salt spray (fog). As more of these cyclic-type tests are used in the near future, the development of the required sophisticated testing cabinets will be required.

Types of Salt Spray Cabinets and Their Construction

Salt spray cabinets are available from many manufacturers and range in size from extremely small bench-top cabinets to large walk-in types. The small bench-top models are not practical; they have been found to be difficult to control and should be avoided. The larger walk-in types have been developed to be controlled capably, but they are very expensive.

The most commonly used cabinet is the top-opening type (Fig 17), which can range in size from 0.25 to 4.5 m³ (9 to 160 ft³) and larger. The cabinet should be large enough to test the required number of parts adequately without overcrowding. Basic cabinets are made of plastic or, more commonly, of plastic-lined steel having no exposed metals or corrodible materials in the interior testing area. The cabinets consist of an air saturation tower with automatic level control,

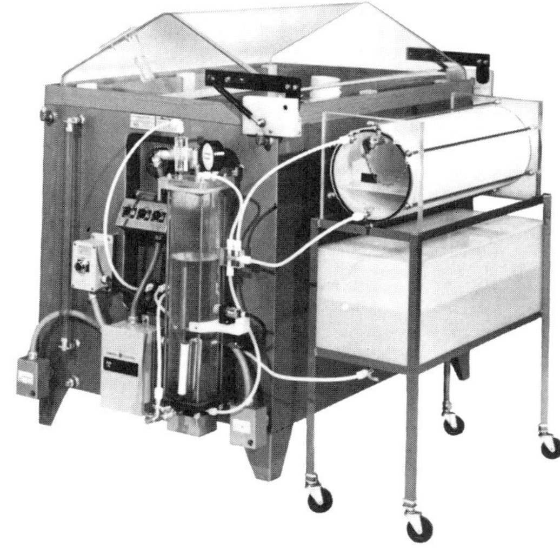

Fig. 17 Typical examples of top-opening salt spray cabinets with state-of-the-art features and pertinent accessories. Cabinets range in size from 0.25 to 4.5 m³ (9 to 160 ft³).

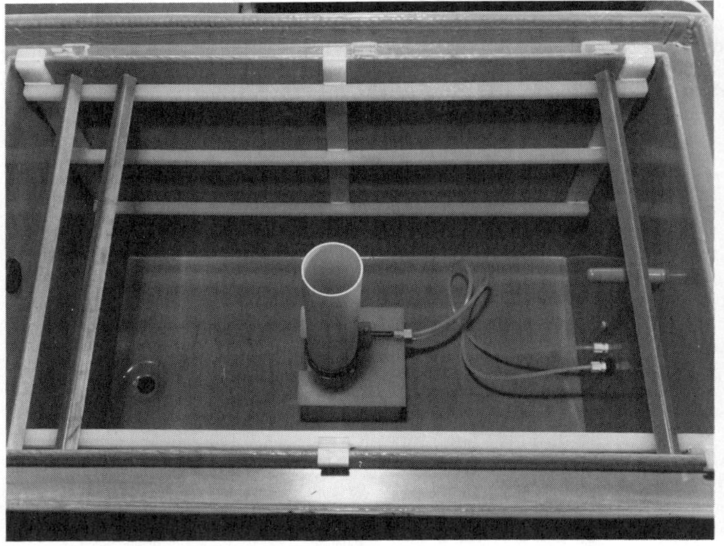

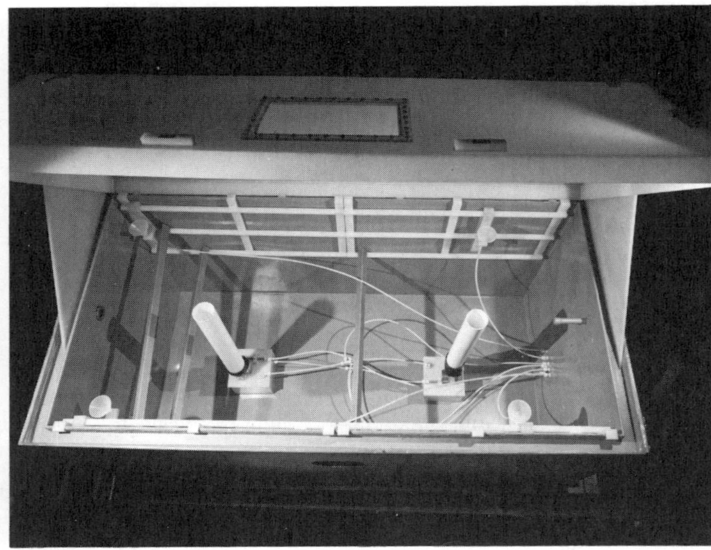

Fig. 18 Vertical-type dispersion towers are the most commonly used for ensuring an even distribution of a uniform free-falling salt mist (fog) over the test specimens. These typical dispersion towers are internally baffled and can be located in the most advantageous part of the cabinet. Single towers (left) are usually used in smaller cabinets up to 1.0 m^3, (36 ft^3), and multiple towers (right) are used in larger cabinets.

a salt solution reservoir with automatic level control, plastic atomizing nozzles that are suitably baffled or housed in a central fog generation tower with adequate internal baffling (Fig 18), specimen supports, and provisions for heating the cabinet and the air saturation tower along with suitable controls for maintaining temperatures.

Fig. 19 Chambers for testing materials under a variety of temperature and humidity conditions. Courtesy of the Aluminum Company of America

Miscellaneous Tests

Donald O. Sprowls
Consultant

There are a number of specialized corrosion tests that are very complex and can be given only brief mention in this article. These include tests in simulated atmospheres, tests in gases at elevated temperatures, aqueous corrosion tests at elevated temperatures, and tests conducted in liquid metals.

Simulated Atmospheres. Draft International Standard ISO 7384-1986 (E) describes general requirements for Corrosion Tests in artificial atmosphere (Ref 104). The corrosion processes are accelerated by intensifying such factors as temperature, relative humidity, condensation of the moisture, and corrosive agents (sulfur dioxide, chlorides, acids, ammonia, hydrogen sulfide, and so on). This standard applies to metals and alloys with and without permanent or temporary corrosion protection.

The ASTM designation G 87 is a standard practice for conducting moist SO_2 tests (Ref 105). Moist air that contains SO_2 quickly produces easily visible corrosion on many metals in a form resembling that which occurs in industrial environments. It is therefore a test environment that is well suited to the detection of pores or other sources of weakness in protective coatings as well as deficiencies in corrosion resistance associated with unsuitable alloy composition or treatments. Standard SO_2 chambers are available from several suppliers, but certain pertinent details are required before they will function according to this practice and provide consistent control for duplication of results.

Humidity-temperature chambers are commercially available for testing materials under a variety of conditions ranging in temperature from freezing to 65 °C (150 °F) and in relative humidity from 20 to 100% (Fig 19). Such tests are commonly used for evaluating various nonmetallic materials of construction that are used in contact with metals, such as insulations and adhesives. An example is the Owens Corning Fiberglas test method C-02A (Ref 106). Versions of this test method have been used in product specifications, such as ASTM C 665 for mineral fiber blanket thermal insulation for wood frame and light construction buildings (Ref 107).

The ASTM designation G 60 is a standard practice for conducting cyclic humidity tests (Ref 108). The procedure described is used to observe the behavior of steels under test conditions that retard the formation of a protective type of rust.

Tests in Gases at Elevated Temperatures. The deterioration of metals and alloys upon exposure to air or other gases at elevated temperatures is a specific type of corrosion that is commonly referred to as high-temperature oxidation. The metals may in fact form sulfides, nitrides, carbides, and oxides. This form of corrosion is a serious problem in various industries. Experimental test methods are required to study this phenomenon. Tests can elucidate kinetics, mechanisms, and chemistry; they can help develop more resistant alloys or qualify an alloy. The high-temperature oxidation of metals and the various oxidation test methods that have been used are extensively reviewed in Ref 109. Additional information is also available in the articles "Fundamentals of Corrosion in Gases" and "General Corrosion" (see the section "High-Temperature Oxidation/Sulfidation") in this Volume.

Aqueous Tests at Elevated Temperatures and Pressures. High-pressure equipment is necessary to conduct corrosion studies in high-temperature water and steam, and radioactive materials are often tested. Therefore, safety codes and practices should be strictly followed. Reference 110 contains a review of the procedures employed to evaluate the corrosion behavior of materials and concepts for nuclear reactor service. To date, two standards have been issued. The first is ASTM G 2, which addresses the testing of zirconium and zirconium alloys (Ref

111), and the second is NACE TM-01-71, which covers the autoclave corrosion testing of metals in high-temperature water (Ref 112). The NACE standard deals mainly with structural and pressure vessel materials, such as high-strength steel, stainless steel, and certain nickel-base alloys. Additional information on the evaluation of materials for nuclear reactor service can be found in the article "Corrosion in the Nuclear Power Industry" in this Volume.

Reference 113 describes testing in hot brine loops designed to provide quantitative information on corrosion rates that will be encountered during the desalination of seawater. The hot brine loop is essentially a device for circulating a heated 3.4% NaCl solution through tubular specimens or past flat specimens in order to determine the corrosion resistance of the alloy(s) under investigation. To accomplish this, the loop must possess certain characteristics. It must be chemically inert so as to prevent contamination of the brine by metal ions or by organic species. It must also be pressure tight, because it is usually operated above the atmospheric boiling point of water.

Liquid metals have large volumetric heat capacities, high heat transfer coefficients, and other properties that make them attractive as coolants for high-temperature nuclear reactors and in power generation systems that operate in conjunction with nuclear reactors. A comprehensive overview of the specialized test procedures required for testing the corrosiveness of liquid metals is given in Ref 114. The ASTM standard G 68 covers the liquid sodium corrosion testing of metals and alloys (Ref 115). Additional information can be found in the articles "Fundamentals of High-Temperature Corrosion in Liquid Metals" and "General Corrosion" (see the section "Corrosion in Liquid Metals") in this Volume.

REFERENCES

1. J.O'M. Bockris, *Modern Aspects of Electrochemistry*, Butterworths, 1954
2. E. Gileadi, E. Kirowa-Eisner, and J. Penciner, *Interfacial Electrochemistry An Experimental Approach*, Addison-Wesley, 1975
3. N.D. Tomashov, *Theory of Corrosion and Protection of Metals*, Macmillan, 1966
4. J.C. Scully, *The Fundamentals of Corrosion*, Pergamon Press, 1975
5. M.G. Fontana and N.D. Greene, *Corrosion Engineering*, McGraw-Hill, 1978
6. J. Newman, *Electrochemical Systems*, Prentice-Hall, 1973
7. H.H. Uhlig and R.W. Revie, *Corrosion and Corrosion Control*, John Wiley & Sons, 1985
8. J.O'M. Bockris, B.E. Conway, E. Yeager, and R.E. White, Ed., *Electrochemical Materials Science*, Vol 4, *Comprehensive Treatise of Electrochemistry*, Plenum Press, 1981
9. R. Baboian, Ed., *Electrochemical Techniques for Corrosion*, National Association of Corrosion Engineers, 1977
10. U. Bertocci and F. Mansfeld, Ed., *Electrochemical Corrosion Testing*, STP 727, American Society for Testing and Materials, 1979
11. G. Haynes and R. Baboian, Ed., *Laboratory Corrosion Tests and Standards*, STP 866, American Society for Testing and Materials, 1985
12. G.C. Moran and P. Labine, Ed., *Corrosion Monitoring in Industrial Plants Using Nondestructive Testing and Electrochemical Methods*, STP 908, American Society for Testing and Materials, 1986
13. R. Baboian, Ed., *Electrochemical Techniques for Corrosion Engineers*, National Association of Corrosion Engineers, 1986
14. R. Baboian, W.D. France, Jr., L.C. Rowe, and J.F. Rynewicz, Ed., *Galvanic and Pitting Corrosion—Field and Laboratory Studies*, STP 576, American Society for Testing and Materials, 1974
15. A.C. Riddiford, *Adv. Electrochem. Eng.*, Vol 4, 1966, p 47
16. D.D. MacDonald, *Transient Techniques in Electrochemistry*, Plenum Press, 1977
17. C. Wagner and W. Traud, *Z. Electrochem*, Vol 44, 1938, p 391
18. W.D. France, Jr., Controlled Potential Corrosion Tests, Their Application and Limitations, *Mater. Res. Stand.*, Vol 9 (No. 8), 1969, p 21
19. "Standard Practice for Standard Reference Method for Making Potentiostatic and Potentiodynamic Anodic Polarization Measurements," G 5, *Annual Book of ASTM Standards*, American Society for Testing and Materials
20. "Standard Practice for Conducting Potentiodynamic Polarization Resistance Measurements," G 59, *Annual Book of ASTM Standards*, American Society for Testing and Materials
21. Z. Tafel, *Physik. Chem.*, Vol 50, 1905, p 641
22. "Standard Practice for Conventions Applicable to Electrochemical Measurements in Corrosion Testing," G 3, *Annual Book of ASTM Standards*, American Society for Testing and Materials, 1985
23. D. Britz, *J. Electroanal. Chem.*, Vol 88, 1978, p 309
24. M. Hayes and J. Kuhn, *J. Power Sources*, Vol 2, 1977-1978, p 121
25. F. Mansfeld, *Corrosion*, Vol 38 (No. 10), 1982, p 556
26. M. Stern and R.M. Roth, *J. Electrochem. Soc.*, Vol 104, 1957, p 390
27. M. Stern and A.L. Geary, *J. Electrochem. Soc.*, Vol 105, 1958, p 638
28. M. Stern and A.L. Geary, *J. Electrochem. Soc.*, Vol 104, 1957, p 56
29. S. Evans and E.L. Koehler, *J. Electrochem. Soc.*, Vol 108, 1961, p 509
30. M. Stern and E.D. Weisert, Experimental Observations on the Relation Between Polarization Resistance and Corrosion Rate, in *ASTM Proceedings*, Vol 59, American Society for Testing and Materials, 1959, p 1280
31. "Test Methods for Corrosivity of Water in the Absence of Heat Transfer (Electrical Methods)," D 2776-79, *Annual Book of ASTM Standards*, American Society for Testing and Materials
32. F. Mansfeld and M. Kendig, *Corrosion*, Vol 37 (No. 9), 1981, p 556
33. R.L. Leroy, *Corrosion*, Vol 29, 1973, p 272
34. R. Bandy and D.A. Jones, *Corrosion*, Vol 32, 1976, p 126
35. M.J. Danielson, *Corrosion*, Vol 36 (No. 4), 1980, p 174
36. J.C. Reeve and G. Bech-Nielsen, *Corros. Sci.*, Vol 13, 1973, p 351
37. K.B. Oldham and F. Mansfeld, *Corros. Sci.*, Vol 13, 1973, p 813
38. I. Epelboin, C. Gabrielli, M. Keddam, and H. Takenouti, in *Electrochemical Corrosion Testing*, STP 727, F. Mansfeld and U. Bertocci, Ed., American Society for Testing and Materials, 1981, p 150
39. A.C. Makrides, *Corrosion*, Vol 29 (No. 9), 1973, p 148
40. D.D. MacDonald and M.C.H. McKubre, in *Electrochemical Corrosion Testing*, STP 727, F. Mansfeld and U. Bertocci, Ed., American Society for Testing and Materials, 1981, p 110
41. A.J. Bard and L.R. Faulkner, *Electrochemical Methods: Fundamentals and Applications*, John Wiley & Sons, 1980
42. F. Mansfeld, *Corrosion*, Vol 36 (No. 5), 1981, p 301
43. F. Mansfeld, M.W. Kendig, and S. Tsai, *Corrosion*, Vol 38, 1982, p 570
44. H. Hack and J.R. Scully, *Corrosion*, Vol 42 (No. 2), 1986, p 79
45. R. Baboian, in *Electrochemical Techniques for Corrosion*, R. Baboian, Ed., National Association of Corrosion Engineers, 1977, p 73
46. R. Baboian, in *Galvanic and Pitting Corrosion—Field and Laboratory Studies*, STP 576, R. Baboian, *et al.*, Ed., American Society for Testing and Materials, 1974, p 5
47. F. Mansfeld and J.V. Kenkel, in *Galvanic and Pitting Corrosion—Field and Laboratory Studies*, R. Baboian, *et al.*, Ed., STP 576, American Society for Testing and Materials, 1974, p 20
48. J. Kruger, in *Passivity and Its Breakdown on Iron and Iron Based Alloys*, R.W. Staehle and H. Okada, Ed., National Association of Corrosion Engineers, 1976
49. N. Sato and G. Okamoto, in *Electrochemical Materials Science*, Vol 4, *Comprehensive Treatise of Electrochemistry*, J.O'M. Bockris, B.E. Conway, E. Yeager, and R.E. White, Ed., Plenum Press, 1981, p 193
50. B.E. Wilde, *Corrosion*, Vol 28, 1972, p 283
51. "Standard Practice for Conducting Cyclic Potentiodynamic Measurements for Localized Corrosion," G 61, *Annual Book of ASTM Standards*, American Society for Testing and Materials
52. B.C. Syrett, *Corrosion*, Vol 33, 1977, p 221
53. N. Pessall and C. Liu, *Electrochim. Acta*, Vol 16, 1971, p 1987
54. B.E. Wilde and E. Williams, *J. Electrochem. Soc.*, Vol 118, 1971, p 1057
55. J.R. Ambrose and J. Kruger, *J. Electrochem. Soc.*, Vol 121, 1974, p 599
56. J.R. Ambrose and J. Kruger, in *Proceedings of the Fifth International Congress on Metallic Corrosion*, National Association of Corrosion Engineers, 1974, p 406
57. M.A. Devanathan and Z. Stackurski, *Proc. R. Soc. (London) A*, Vol 270A, 1962, p 90
58. J. McBreen, L. Nanis, and W. Beck, *J. Electrochem. Soc.*, Vol 113 (No. 11), 1966, p 1218
59. M. Prazak, *Corrosion*, Vol 19 (No. 3), 1963, p 75t
60. P. Novak, R. Stefec, and F. Franz, *Corrosion*, Vol 31 (No. 10), 1975, p 344
61. W.L. Clarke, V.M. Romero, and J.C. Danko, Paper (preprint 180), presented at Corrosion/77, National Association of Corrosion Engineers, 1977
62. W.L. Clarke, R.L. Cowan, and W.L. Walker, in *Intergranular Corrosion of Stainless*

Alloys, STP 656, R.F. Steigerwald, Ed., American Society for Testing and Materials, 1978, p 99

63. M. Akashi, T. Kawamoto, and F. Umemura, *Corros. Eng.*, Vol 29, 1980, p 163
64. A.P. Majidi and M.A. Streicher, *Corrosion*, Vol 40 (No. 11), 1984, p 584
65. "Standard Practices for Detecting Susceptibility to Intergranular Attack in Austenitic Stainless Steels," A 262, *Annual Book of ASTM Standards*, American Society for Testing and Materials
66. J.B. Lee, *Corrosion*, Vol 42 (No. 2), 1986, p 106
67. A. Roelandt and J. Vereecken, *Corrosion*, Vol 42 (No. 5), 1986, p 289
68. J.R. Scully and R. Kelly, *Corrosion*, Vol 42 (No. 9), 1986, p 537
69. "Standard Method of FACT (Ford Anodized Aluminum Corrosion Test) Testing," B 538, *Annual Book of ASTM Standards*, American Society for Testing and Materials
70. J. Stone, H.A. Tuttle, and H.N. Bogart, *Plating*, Vol 43, 1965, p 877
71. R.L. Saur and R.P. Basco, *Plating*, Vol 53, 1966, p 33
72. R.L. Saur and R.P. Basco, *Plating*, Vol 53, 1966, p 981
73. R.L. Saur and R.P. Basco, *Plating*, Vol 53, 1966, p 320
74. "Standard Method of Electrolytic Corrosion Testing (EC Test)," B 627, *Annual Book of ASTM Standards*, American Society for Testing and Materials
75. "Standard Method for Measurement of Impedance of Anodic Coatings on Aluminum," B 457, *Annual Book of ASTM Standards*, American Society for Testing and Materials
76. E.T. Englehart and G. Sowinski, Jr., *SAE J.*, Vol 72, 1974, p 51
77. E.T. Englehart and D.J. George, *Mater. Prot.*, Vol 3, 1964, p 25
78. J.D. Scantlebury, K.N. Ho, and D.A. Eden, in *Electrochemical Corrosion Testing*, STP 727, F. Mansfeld and U. Bertocci, Ed., American Society for Testing and Materials, 1981, p 187
79. S. Narian, N. Bonanos, and M.G. Hocking, *J. Oil Colour Chem. Assoc.*, Vol 66 (No. 2), 1983, p 48
80. T.A. Strivens and C.C. Taylor, *Mater. Chem.*, Vol 7, 1982, p 199
81. F. Mansfeld, M.W. Kendig, and S. Tsai, *Corrosion*, Vol 38 (No. 9), 1982, p 478
82. M. Kendig, F. Mansfeld, and S. Tsai, *Corros. Sci.*, Vol 23 (No. 4), 1983, p 317
83. R. Touhasaent and H. Liedheiser, *Corrosion*, Vol 28 (No. 12), 1982, p 435
84. "Recommended Practice for Laboratory Immersion Corrosion Testing of Metals," G 31, *Annual Book of ASTM Standards*, American Society for Testing and Materials
85. "Test Method—Laboratory Corrosion Testing of Metals for the Process Industries," NACE TM-01-69, National Association of Corrosion Engineers, 1976
86. M.G. Fontana, "Corrosion Testing," Lesson 8 in Home Study and Extension Courses, Metals Engineering Institute, American Society for Metals, 1969
87. F.A. Champion, *Corrosion Testing Procedures*, 2nd Ed., John Wiley & Sons, 1965
88. O.B.J. Fraser, D.E. Ackerman, and J.W. Sands, *Ind. Eng. Chem.*, Vol 19, 1927, p 332
89. P.E. Francis and A.D. Mercer, Corrosion of a Mild Steel in Distilled Water and Chloride Solutions: Development of a Test Method, in *Laboratory Corrosion Tests and Standards*, STP 866, G.S. Haynes and R. Baboian, Ed., American Society for Testing and Materials, 1985, p 184-196
90. "Standard Practices for Detecting Susceptibility to Intergranular Attack in Austenitic Stainless Steels," A 262, *Annual Book of ASTM Standards*, American Society for Testing and Materials
91. "Standard Test Method for Visual Assessment of Exfoliation Corrosion Susceptibility of 5xxx Series Aluminum Alloys (ASSET Test)" G 66, *Annual Book of ASTM Standards*, American Society for Testing and Materials
92. W.H. Ailor, Ed., *Handbook on Corrosion Testing and Evaluation*, John Wiley & Sons, 1971
93. G.D. Bengough, U.R. Evans, T.P. Hoar, and F. Wormwell, *Chem. Ind.*, Nov 1938
94. "Standard Recommended Practice for Alternate Immersion Stress Corrosion Testing in 3.5% Sodium Chloride Solution," G 44, *Annual Book of ASTM Standards*, American Society for Testing and Materials
95. "Standard Specification For Reagent Water," D 1193, *Annual Book of ASTM Standards*, American Society for Testing and Materials
96. "Standard Specification for Substitute Ocean Water," D 1141, *Annual Book of ASTM Standards*, American Society for Testing and Materials
97. "Standard Practice for Preparing, Cleaning, and Evaluating Corrosion Test Specimens," G 1, *Annual Book of ASTM Standards*, American Society for Testing and Materials
98. A. Wachter and R.S. Treseder, Corrosion Testing Evaluation of Metals for Process Equipment, *Chem. Eng. Prog.*, Vol 43, June 1947, p 315-326
99. W.D. McMaster, *A History of the Salt Spray Tests*, Publication GMR-497, General Motors Corporation, Sept 1965
100. "Standard Method of Salt Spray (Fog) Testing," B 117, *Annual Book of ASTM Standards*, American Society for Testing and Materials
101. "Standard Practice for Modified Salt Spray (Fog) Testing," G 85, *Annual Book of ASTM Standards*, American Society for Testing and Materials
102. "Standard Method for Copper-Accelerated Acetic Acid-Salt Spray (Fog) Testing (CASS Test)," B 368, *Annual Book of ASTM Standards*, American Society for Testing and Materials
103. S.J. Ketcham and I.S. Shaffer, Exfoliation Corrosion of Aluminum Alloys, in *Localized Corrosion—Cause of Metal Failure*, STP 516, American Society for Testing and Materials, 1972, p 3-16
104. "Corrosion Tests in Artificial Atmospheres," ISO 7384-1986 E, American National Standards Institute
105. "Standard Practice for Conducting Moist SO_2 Tests," ASTM G 87, *Annual Book of ASTM Standards*, American Society for Testing and Materials
106. S.V. Crume, A Corrosiveness Test for Fibrous Insulations, in *Laboratory Corrosion Tests and Standards*, STP 866, G.S. Haynes and R. Baboian, Ed., American Society for Testing and Materials, 1985
107. "Standard Specification for Mineral-Fiber Blanket Thermal Insulation for Light Frame Construction and Manufactured Housing," C 665, *Annual Book of ASTM Standards*, American Society for Testing and Materials
108. "Standard Practice for Conducting Cyclic Humidity Tests, G 60, *Annual Book of ASTM Standards*, American Society for Testing and Materials
109. L.R. Scharfstein and M. Henthorne, Testing at High Temperature in *Handbook on Corrosion Testing and Evaluation*, W.H. Ailor, Ed., John Wiley & Sons, 1971, p 291-366
110. W.E. Berry, Testing Nuclear Materials in Aqueous Environments, in *Handbook on Corrosion Testing and Evaluation*, W.H. Ailor, Ed., John Wiley & Sons, 1971, p 379-403
111. "Standard Practice for Aqueous Corrosion Testing of Samples of Zirconium and Zirconium Alloys," G 2, *Annual Book of ASTM Standards*, American Society for Testing and Materials
112. "Autoclave Corrosion Testing of Metals in High Temperature Water," NACE TM 01-71, National Association of Corrosion Engineers
113. R.J. Hart, Testing in Hot Brine Loops, in *Handbook on Corrosion Testing and Evaluation*, W.H. Ailor, Ed., John Wiley & Sons, 1971, p 367-378
114. R.L. Kluch and J.H. DeVan, Liquid Metal Test Procedures, in *Handbook on Corrosion Testing and Evaluation*, W.H. Ailor, Ed., John Wiley & Sons, 1971, p 405-433
115. "Standard Practice for Liquid Sodium Corrosion Testing of Metals and Alloys," G 68, *Annual Book of ASTM Standards*, American Society for Testing and Materials

Evaluation of Uniform Corrosion

Charles A. Natalie, Department of Metallurgical Engineering, Colorado School of Mines

TESTING FOR UNIFORM CORROSION can be broadly classified as laboratory long-term testing, laboratory accelerated testing, and service or field testing. Laboratory long-term testing usually consists of testing materials in simulated-service conditions on a relatively small scale. The advantage of this type of testing is that the tests can be controlled closely; thus, any unintentional disturbances that could occur during plant tests can be avoided. These tests usually attempt to simulate expected service conditions.

Accelerated laboratory tests, on the other hand, are short term in nature. They are designed to compare materials under severe conditions, and the test environment may not be directly related to the service environment. The environment and the test conditions must be carefully specified. The American Society for Testing and Materials (ASTM) has outlined standard methods for testing materials under accelerated conditions. These tests are usually for other modes of corrosion, such as stress-corrosion cracking, and are generally not applicable to uniform corrosion testing. Accelerated laboratory tests are described in other articles in this Section.

Service tests and plant tests involve placing test samples in actual service conditions for evaluation. The advantage of such testing is the use of the actual service environment. However, the problems involved in interrupting normal operations are also associated with this type of corrosion testing.

Uniform corrosion is one of the most common forms of corrosion; therefore, it must be designed for in many situations. The damage appears as the thickness of the metal decreases uniformly until failure occurs. Fortunately, uniform corrosion is usually easy to measure and predict; this facilitates proper design. The purposes of measuring uniform corrosion by experimental testing are:

- Evaluation and selection of materials for a specific environment or application
- Evaluation of metals and alloys to determine their effectiveness in new environments
- Routine testing to confirm the quality of materials
- Research and development

Experimental Measurements

In most cases, uniform corrosion rates are represented as a loss of metal thickness as a function of time. This value can be directly measured from experimental data, or as is often the case, it can be calculated from mass loss data. Mass loss is a measure of the difference between the original mass of the specimen and the mass when sampled after exposure. In measuring the mass after exposure it is important to remove any corrosion product adhering to the sample. This topic is discussed in more detail in the section "Sample Cleaning and Data Acquisition" in this article.

As mass loss is monitored, the reduction of thickness as a function of time can be calculated and monitored. Uniform corrosion rates are usually expressed as millimeters per year (mm/yr), mils per year (mils/yr), and/or inches per year (in./yr). A corrosion rate in mils per year can be calculated from weight loss data with the following expression:

$$\text{mils/yr} = \frac{534w}{dAt} \qquad \text{(Eq 1)}$$

where w is weight loss in milligrams, d is metal density in grams per cubic centimeter (g/cm^3), A is area of exposure in square inches (in.^{2}), and t is exposure time in hours. Conversion factors for some common uniform corrosion rate units are given in Table 1.

Other evaluations that are often included in testing for uniform corrosion include visual observations, ion concentration increase, hydrogen evolution, loss in tensile strength (according to ASTM G 50), and electrochemical testing. Initial observations in any corrosion testing should include visual observation of the corrosion activity because it is usually possible to observe the form of corrosion at low magnification. Visual evaluation can include reporting of the color or type of corrosion product as well as documentation with photographs.

In addition to the standard measurement of mass loss, the increase of metal ion in solution as the metal corrodes can be measured by analysis of samples of the corrodent removed periodically during the test (Ref 1). It may be possible to relate the increase in concentration to corrosion rate, depending on the structure of the test. In some cases, the amount of hydrogen generated in deaerated tests can be used to measure corrosion rates (Ref 2, 3).

Electrochemical tests can also be used to help evaluate corrosion rates. These methods are often successful in predicting uniform corrosion rates. Electrochemical methods include recording anodic and cathodic polarization curves, extrapolation of Tafel lines, and polarization break testing. Methods for conducting electrochemical tests include ASTM standards G 59, G 3, and G 5 (Ref 4). Reference 5 also explains some standard practices in this area, and electrochemical test methods are discussed in the article "Laboratory Testing" in this Volume.

Exposure Tests for Measuring Uniform Corrosion Rates

Sample Preparation. The first step in testing is the preparation of the samples. Clear and complete documentation of the test material is very important. This should include the chemical composition as well as the metallurgical treatment of the samples. Each sample should be clearly identified before testing.

For uniform corrosion testing, it is usually most convenient to use small test coupons. The size and shape of the test specimens are often specified in such a manner as to make them easy to handle and allow for ease of surface preparation. Specimens that are 6.4 mm (1/4 in.) thick and a few inches square are not uncommon. For some materials, it is advisable to protect specimen edges and identification marks in order to avoid extraneous localized corrosion.

Before exposure to the corrosive medium, the test samples should be cleaned and weighed. The surface can be cleaned with a mild abrasive paper. Ideally, the surface of the test specimen should be the same as that of the metal when in service. This is not always possible, because of the variations in surface condition produced during fabrication. However, surface preparation of the test specimens ensures a standard surface condition during testing.

Table 1 Relationships among some of the units commonly used for corrosion rates

d is metal density in grams per cubic centimeter.

Unit	Factor for conversion to mdd	g/m^2/d	μm/yr	mm/yr	mils/yr	in./yr
Milligrams per square decimeter per day (mdd)	1	0.1	36.5/d	0.0365/d	1.144/d	0.00144/d
Grams per square meter per day (g/m^2/d)	10	1	365/d	0.365/d	14.4/d	0.0144/d
Microns per year (μm/yr)	0.0274d	0.00274d	1	0.001	0.0394	0.0000394
Millimeters per year (mm/yr)	27.4d	2.74d	1000	1	39.4	0.0394
Mils per year (mils/yr)	0.696d	0.0696d	25.4	0.0254	1	0.001
Inches per year (in./yr)	696d	69.6d	25 400	25.4	1000	1

Source: Ref 1

After surface cleaning, the sample should be degreased with a solvent such as acetone, dried and weighed, and immediately put into the test medium. The standard procedures for treating test specimens of various metals are explained in detail in ASTM G 1 (Ref 4).

Testing Methods. The method and apparatus used to expose the test specimens to the test environment are very important and should meet the following guidelines:

- The corrosive medium should have easy access to samples
- The specimen should be well supported
- The sample should be electrically insulated from all other metals in the system
- The samples should be completely immersed unless other effects are being studied
- The samples should be readily accessible

An outline of standard procedures for designing these types of tests is provided in ASTM G 31 (Ref 4).

It is important to plan the experimental technique properly so that samples are removed and the weight is recorded over a proper duration of time. The corrosion rate may initially increase or decrease and may eventually remain constant with time. The intervals between samples should be selected so as to minimize misleading results. A general rule for selecting the total duration (in hours) of testing is as follows (Ref 6):

$$\text{Test duration} = \frac{2000}{\text{mils/yr}} \qquad \text{(Eq 2)}$$

where the mils/yr is the corrosion rate obtained in a laboratory test of short duration. The formula is based on the general rule that lower corrosion rates will require longer tests. For example, if the sample corrodes at a rate of 0.25 mm/yr (10 mils/yr), a corrosion test that is at least 200 h long should be designed (Ref 6).

In addition to the sequential removal of specimens with time, a planned interval testing schedule can be used to obtain more information on the initial corrosion rates and on accumulated effects (Ref 7). Planned interval testing involves exposing identical samples to the corrosion solution during different time intervals of the test. By evaluating the results, initial corrosion rates and changes in solution corrosivity can be determined. The planned interval testing program is an excellent example of good experimental planning.

Test Variables. Important variables that must be addressed during experimental planning include aeration, temperature, solution volume and flow, and degree of exposure. The presence of dissolved oxygen in the corrosion solution can have a dramatic effect on the rate of corrosion. If aeration is desired, the most common and simplest method consists of bubbling air or oxygen through the solution. If deaeration is required, purified nitrogen or argon can be bubbled through the solution in place of air. If a gas is not bubbled through the solution, the amount of dissolved oxygen will depend on the initial amount present and the rate of removal or adsorption from the atmosphere, which can depend on the configuration of the test apparatus.

Table 2 Some ASTM standard methods and practices for corrosion testing

Designation	Test method
G 85	Modified Salt Spray (Fog) Testing
G 4	Conducting Corrosion Coupon Test in Plant Equipment
B 117	Salt Spray (Fog) Testing
G 44	Alternate Immersion Stress-Corrosion Testing in 3.5% Sodium Chloride Solution
G 16	Applying Statistics to Analysis of Corrosion Data
G 61	Conducting Cyclic Potentiodynamic Polarization Measurements for Localized Corrosion
G 52	Conducting Surface Seawater Exposure Tests on Metals and Alloys
G 31	Laboratory Immersion Corrosion Testing of Metals
D 2776	Corrosivity of Water in Absence of Heat Transfer (Electrical Methods)
D 2688	Corrosivity of Water in Absence of Heat Transfer (Weight Loss Method)
G 2	Aqueous Corrosion Testing of Samples of Zirconium and Zirconium Alloys
G 50	Conducting Atmospheric Corrosion Test on Metals
G 60	Conducting Cyclic Humidity Tests

The influence of temperature can be very complicated in that it affects oxygen solubility, pH, corrosion product formation, and other factors. Therefore, the temperature of the test should be monitored throughout the duration of the experiments and should be controlled at the desired levels.

The volume of solution to be used during laboratory testing is often a problem for practical reasons. If the volume of the solution is too small, the concentration of metal ions may increase and influence the results as well as cause a depletion of the corrosive agent in the environment. At least 50 mL of test solution should be used per square centimeter of test area, and if the test solution is rapidly consumed, it should be periodically replenished (Ref 1).

The flow velocity of the solution also plays an important role in determining the corrosion rates. Corrosion rates may increase or decrease as the fluid velocity near the sample is increased. In any case, flow characteristics that represent the condition in service should be established during testing. This is often not a simple task, because the hydrodynamic conditions in service and their effect on the corrosion rate can be very complex. Rotating disks and cylinders are often used to study corrosion rates under controlled hydrodynamic conditions (Ref 8-11) (see also the article "Evaluation of Erosion and Cavitation" in this Volume).

The degree of exposure is also important in test design; total, partial, or intermittent exposure can change the mode and rate of corrosion. If the samples are immersed, they should be immersed at the same depth in the corrosive environment. Partially immersed samples can be used to study waterline corrosion. Table 2 lists ASTM standard tests for different types of exposures that are applicable for the evaluation of resistance to uniform corrosion.

Sample Cleaning and Data Acquisition. After a sample is removed from testing and before the final weight is recorded, the sample must be cleaned to remove any corrosion products. Also, the corrosion products must be removed without causing additional corrosion of the specimen. Because the weight change of the specimen is used to calculate the thickness loss of the metal, any corrosion product that is weighed will provide incorrect results. A common cleaning method consists of removing the corrosion product with a brush or rubber stopper under a stream of water. In many cases, it is necessary to follow this procedure with brief chemical or electrochemical cleaning. Detailed procedures for cleaning specimens after exposure are given in ASTM G 1 (Ref 4).

REFERENCES

1. G. Wranglén, *Corrosion and Protection of Metals*, Chapman & Hall, 1985, p 238
2. R.L. Martin and E.C. French, Corrosion Monitoring in Sour Systems Using Electrochemical Hydrogen Patch Probes, *J. Pet. Technol.*, Nov 1978, p 1566-1570
3. P.A. Schweitzer, *Corrosion and Corrosion Protection Handbook*, Marcel Dekker, 1983, p 483-484
4. "Metal Corrosion, Erosion, and Wear," Vol 03.02, *Annual Book of ASTM Standards*, American Society for Testing and Materials, 1986
5. F. Mansfeld and V. Bertocci, *Electrochemical Corrosion Testing*, STP 727, American Society for Testing and Materials, 1981
6. M. Fontana, *Corrosion Engineering*, 3rd ed., McGraw Hill, 1986, p 162
7. A. Wachter and R.S. Treseder, *Chem. Eng. Prog.*, Vol 43, 1947, p 315-326
8. A.J. Bard and L.R. Faulkner, *Electrochemical Methods: Fundamentals and Applications*, John Wiley & Sons, 1980, p 283-304
9. A.C. Riddiford, *Adv. Electrochem. Eng.*, Vol 4, 1966, p 47
10. H.W. Pickering and C. Wagner, *J. Electrochem. Soc.*, Vol 11, 1967, p 698
11. G.H. Feller, *Corros. Sci.*, Vol 8, 1968, p 259

Evaluation of Pitting Corrosion

Donald O. Sprowls, Consultant

PITTING is a form of localized corrosion that is often a concern in applications involving passivating metals and alloys in aggressive environments. Pitting can also occur in nonpassivating alloys with protective coatings or in certain heterogeneous corrosive media. It is a very damaging form of corrosion that is not readily evaluated by the methods used for uniform corrosion. Therefore, special accelerated tests have been devised for the evaluation of the relative resistance to pitting corrosion of passive alloys. The mechanisms of pitting corrosion are discussed in the section "Pitting" of the article "Localized Corrosion" in this Volume.

Test Methods

Method ASTM G 48 covers procedures for determining the pitting (and crevice) corrosion resistance of stainless steels and related alloys when exposed to an oxidizing chloride environment, namely 6% ferric chloride ($FeCl_3$) at 22 ± 2 or 50 ± 2 °C (70 ± 3.5 or 120 ± 3.5 °F) (Ref 1). Method A is a 72-h total-immersion test of small coupons that is designed to determine the relative pitting resistance of stainless steels and nickel-base chromium-bearing alloys. Method B is a crevice test under the same exposure conditions, and it can be used to determine both the pitting and crevice corrosion resistance of these alloys. These tests can be used for determining the effects of alloying additions, heat treatments, and surface finishes on pitting and crevice corrosion resistance.

Method ASTM F 746 covers the determination of the resistance to either pitting or crevice corrosion of passive metals and alloys from which surgical implants will be produced (Ref 2). This is a screening test that is used to rank surgical implant alloys in order of their resistance to localized corrosion in a dilute sodium chloride (NaCl) solution under the specific conditions of the test method. With this method, alloys are ranked in terms of the critical potential for pitting; the higher (more noble) this potential, the more resistant the alloy is to passive film breakdown and to localized corrosion. The method was intentionally designed to cause breakdown of at least one alloy (type 316L stainless steel) that is currently considered acceptable for surgical implant use. Those alloys that suffer pitting or crevice corrosion during the test do not necessarily suffer localized corrosion when placed within the human body as a surgical implant.

Practice ASTM G 61 describes a procedure for conducting cyclic potentiodynamic polarization measurements to determine susceptibility to localized corrosion (Ref 3). The procedure is preferably used for iron-, nickel-, and cobalt-base alloys in chloride environments. This standard practice uses both the critical potential for pitting and the protection potential in order to compare the pitting resistance of test materials. This and other electrochemical tests are described in more detail in the article "Laboratory Testing" in this Volume. In addition, a comparison of several different methods of determining the protection potential is presented in Ref 4.

Because pitting (or crevice) corrosion can perforate and destroy thin-wall industrial equipment, the protection potential, E_p, represents a limiting potential that should not be exceeded. Many authorities, however, have questioned the validity of such short-term test procedures. The most frequent criticism is that these tests may not adequately predict long-term corrosion behavior (Ref 5, 6). The lack of consistency among the protection potential results (Ref 4) and the lack of a single, most conservative test indicate a need for further research into the concept of a protection potential. At this point, no reason can be identified for the lack of consistency among the techniques reported. Therefore, conservative engineering practice would require that all three types of tests be performed and that the most conservative results be used (Ref 4).

Examination and Evaluation

ASTM G 46 provides assistance in the selection of procedures for the identification and examination of pits and in the evaluation of pitting corrosion to determine the extent of its effect (Ref 7). It is important to be able to determine the extent of pitting, either in a crevice application in which it is necessary to predict the remaining life of a metal structure or in laboratory test programs that are used to select the most pitting-resistant materials for service. The following is a summary of the procedures described in detail in Ref 7. Additional guidelines can be found in Ref 8.

Identification and Examination of Pits

Visual examination of the corroded surface can be performed with the unaided eye or a low-power microscope. The corroded surface is usually photographed, and the size, shape, and density of the pits are determined (Fig. 1 and 2).

Metallographic examination can be used to determine whether there is a correlation between pits and microstructure and whether the cavities are true pits or are the result of another mechanism, such as intergranular corrosion or dealloying.

Nondestructive Inspection. Reference 7 also includes procedures for the nondestructive evaluation of pitted specimens. These include radiographic, electromagnetic, ultrasonic, and dye-penetrant inspection. These methods can be used to locate pits and to provide some information on

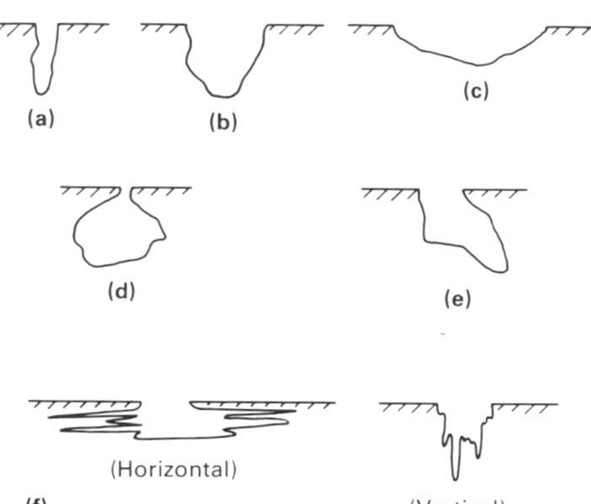

Fig. 1 Variations in the cross-sectional shape of pits. (a) Narrow and deep. (b) Elliptical. (c) Wide and shallow. (d) Subsurface. (e) Undercutting. (f) Shapes determined by microstuctural orientation. Source: Ref 7

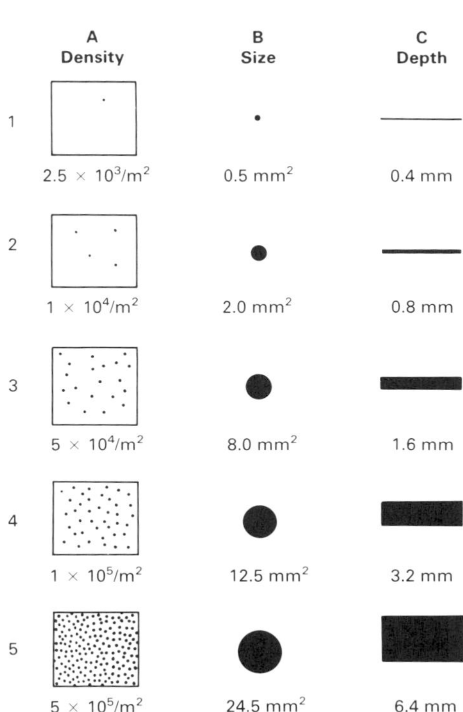

Fig. 2 Standard rating chart for pits. Source: Ref 7

their size, but they generally cannot detect small pits or differentiate between pits and other types of surface detects.

Determination of the Extent of Pitting

Mass loss is generally not a good indication of the extent of pitting unless uniform corrosion is slight and pitting is fairly severe. If there is significant uniform corrosion, the contribution of pitting to total mass loss is small. Mass loss should not be ignored in every case, however. For example, measurement of mass loss, along with visual comparison of pitted surfaces, may be sufficient to rank the relative resistances of alloys in laboratory tests.

Pit depth measurement is generally a better indicator of the extent of pitting than mass loss. Pit depth measurement can be accomplished by several methods, including metallographic examination, machining, use of a micrometer or depth gage, and the microscopical method. In the microscopical method, a metallurgical microscope is focused on the lip of the pit and then on the bottom of the pit. The difference between the initial and final readings on the fine-focusing knob of the microscope is the pit depth.

Evaluation of Pitting

Pitting can be described in several ways. Reference 7 includes procedures for the use of standard charts, metal penetration, statistical analysis, and loss in mechanical properties to quantify the severity of pitting damage. More than one of these methods can be used. In fact, it is often found that no one method is sufficient by itself.

Standard charts such as that shown in Fig. 2 can be used to rate pits in terms of density, size, and depth. Columns A and B in Fig. 2 are used to rate the density (that is, the number of pits per unit area on the specimen surface) and the average size of the pits, respectively. Column C rates the average depth of attack. An example of a rating using Fig. 2 might be A-3, B-2, C-3; this rating indicates a density of 5×10^4 pits/m^2, an average pit size of 2.0 mm^2, and an average pit depth of 1.6 mm.

Such charts facilitate communication among those who are familiar with the standard ratings and offer a simple means of storing data for comparison with other test results. However, it can be tedious and time consuming to measure all of the pits, and the time spent doing so is usually not justified, because maximum values (for example, pit depths) are often more significant than average values.

Metal Penetration. In this method, the deepest pits are measured and metal penetration is expressed in terms of the maximum pit depth, an average of the ten deepest pits, or both. Metal penetration is especially significant when the metal is intended for service as an enclosure for gas or liquid and when a hole could result in loss of fluid.

Metal penetration can also be expressed in terms of a pitting factor, which is the ratio of the deepest metal penetration to the average metal penetration (determined from mass loss):

$$\text{Pitting factor} = \frac{\text{Deepest metal penetration}}{\text{Average metal penetration}} \qquad \text{(Eq 1)}$$

A pitting factor of 1 represents uniform corrosion. The larger the number, the greater the depth of penetration. The pitting factor cannot be used when pitting or general corrosion is slight; values of 0 or infinity can be obtained in such situations.

The application of statistics to the analysis of corrosion data is covered in detail in ASTM G 16 (Ref 9). The subject is discussed briefly in Ref 7 to show that statistics can be used in the evaluation of pitting data.

The probability that pitting will initiate on a metal surface depends on a number of factors, including the pitting tendency of the metal, the corrosivity of the solution, the specimen area, and the duration of exposure. A pitting probability test can be used to determine the susceptibility of metals to pitting. However, this test will not provide information about the rate of propagation, and the results are applicable only to the conditions of exposure. Pitting probability P is expressed as a percentage after exposure of a number of specimens to a particular set of conditions (Ref 10, 11):

$$P = \frac{N_p}{N} \times 100 \qquad \text{(Eq 2)}$$

where N_p is the number of specimens that pit, and N is the total number of specimens.

The relationship between pit depth and area or time of exposure may vary with such factors as the environment and the metal exposed. Equations 3 and 4 are examples that have been found to apply under certain exposure conditions.

Equation 3 was found between maximum pit depth D and the area A of a pipeline exposed to soil (Ref 12-14):

$$D = \mathrm{b}A^{\mathrm{a}} \qquad \text{(Eq 3)}$$

where a and b > 0, and a and b are constants that were derived from the slope and the y-intercept of a straight line curve obtained when the logarithms of the mean pit depth for successively increasing areas on the pipe were plotted against the logarithms of the corresponding areas. The dependence on area is attributed to the increased chance for the deepest pit to be found when the size of the sample of pits is increased through an increased area of corroded surface.

The maximum pit depth D of aluminum exposed to various waters was found to vary as the cube root of time t, as shown in Eq 4 (Ref 10, 15):

$$D = \mathrm{K}t^{1/3} \qquad \text{(Eq 4)}$$

where K is a constant that is dependent on the composition of the water and the alloy. Equation 4 has been found to apply to several aluminum alloys exposed to different waters.

Extreme value probability statistics (Ref 16, 17) have been successfully applied to maximum pit depth data to estimate the maximum pit depth of a large area of material on the basis of the examination of a small portion of that area (Ref 8, 10, 15). The procedure consists of measuring maximum pit depths on several replicate specimens and then arranging the pit depth values in order of increasing rank. A plotting position for each order of ranking is obtained by substitution in the relation $M/(n + 1)$, where M is the order of ranking of the specimen in question, and n is the total number of specimens or values.

For example, the plotting position for the second value out of 10 would be $2/(10 + 1) = 0.1818$. These values are plotted on the ordinate of extreme value probability paper versus their respective maximum pit depths. A straight line indicates that extreme value statistics are applicable. Extrapolation of the straight line can be used to determine the probability that a specific pit depth will occur or the number of observations that must be made to find a particular pit depth.

Loss in Mechanical Properties. If pitting is the predominant form of corrosion and if the density of pitting is relatively high, the change in a mechanical property can be used to advantage for evaluation of the degree of pitting. The typical properties considered for this purpose are tensile strength, elongation, fatigue strength, impact resistance, and burst pressure (Ref 18, 19).

The precautions that must be taken in the application of these mechanical test procedures are covered in most standard methods. However, it must be stressed that the exposed and unexposed specimens should be as close to replicate as possible. Therefore, consideration should be given to such factors as edge effects, direction of rolling, and surface conditions.

Representative specimens of the metal are exposed to the same conditions except for the corrosive environment. The mechanical properties of the exposed and unexposed specimens are measured after the exposure, and the difference between the two results is attributed to corrosion damage.

Some of these methods are better suited to the evaluation of other forms of localized corrosion, such as intergranular or stress corrosion. Therefore, their limitations must be considered. The often erratic nature of pitting and the location of pits on the specimen can affect results. In some cases, the change in mechanical properties due to pitting may be too small to provide meaningful results. Perhaps one of the most difficult problems is to separate the effects due to pitting from those caused by some other form of corrosion.

REFERENCES

1. "Standard Test Methods for Pitting and Crevice Corrosion Resistance of Stainless Steels and Related Alloys by the Use of Ferric Chloride Solution," G 48, *Annual Book of ASTM Standards*, American Society for Testing and Materials
2. "Standard Test Method for Pitting or Crevice Corrosion of Metallic Surgical Implant Materials," F 746, *Annual Book of ASTM Standards*, American Society for Testing and Materials
3. "Standard Practice for Conducting Cyclic Potentiodynamic Polarization Measurements for Localized Corrosion," G 61, *Annual Book of ASTM Standards*, American Society for Testing and Materials
4. M. Hubbell, C. Price, and R. Heidersbach, Crevice and Pitting Corrosion Tests for Stainless Steels: A Comparison of Short-Term Tests With Longer Exposures, in *Laboratory Corrosion Tests and Standards*, STP 866, G.S. Haynes and R. Baboian, Ed., American Society for Testing and Materials, 1985, p 324-336
5. B.E. Wilde, Critical Appraisal of Some Popular Laboratory Tests for Predicting the Localized Corrosion Resistance of Stainless Alloys in Sea Water, *Corrosion*, Vol 28 (No. 8), Aug 1972, p 283
6. F.L. LaQue and H.H Uhlig, An Essay on Pitting, Crevice Corrosion and Related Potentials, *Mater. Perform.*, Vol 22 (No. 8), Aug 1983, p 34
7. "Standard Recommended Practice for Examination and Evaluation of Pitting Corro-

sion," G 46, *Annual Book of ASTM Standards*, American Society for Testing and Materials
8. F.A. Champion, *Corrosion Testing Procedures*, 2nd ed., John Wiley & Sons, 1985, p 197
9. "Standard Recommended Practice for Applying Statistics to Analysis of Corrosion Data," G 16, *Annual Book of ASTM Standards*, American Society for Testing and Materials
10. B.R. Pathak, Testing in Fresh Waters, *Handbook on Corrosion Testing and Evaluation*, W.H. Ailor, Ed., John Wiley & Sons, 1971, p 553
11. P.M. Aziz and H.P. Godard, Influence of Specimen Area on the Pitting Probability of Aluminum, *J. Electrochem. Soc.*, Vol 102, Oct 1955, p 577
12. G.N. Scott, Adjustment of Soil Corrosion Pit Depth Measurements for Size of Sample, in *Proceedings of the American Petroleum Institute*, Vol 14, Section IV, American Petroleum Institute, 1934, p 204
13. M. Romanoff, *Underground Corrosion*, National Bureau of Standards Circular 579, U.S. Government Printing Office, 1957, p 71
14. I.A. Denison, Soil Exposure Tests, in *The Corrosion Handbook*, H.H. Uhlig, Ed., John Wiley & Sons, 1948, p 1048
15. H.P. Godard, The Corrosion Behavior of Aluminum in Natural Waters, *Can. J. Chem. Eng.*, Vol 38, Oct 1960, p 1671
16. E.J. Gumbel, *Statistical Theory of Extreme Values and Some Practical Applications*, Applied Mathematics Series 33, U.S. Department of Commerce, 1954
17. P.M. Aziz, Application of the Statistical Theory of Extreme Values to the Analysis of Maximum Pit Depth Data for Aluminum, *Corrosion*, Vol 12, Oct 1956, p 495t
18. T.J. Summerson, M.J. Pryor, D.S. Keir, and R.J. Hogan, Pit Depth Measurements as a Means of Evaluating the Corrosion Resistance of Aluminum in Sea Water, in *Metals*, STP 196, American Society for Testing and Materials, 1957, p 157
19. R. Baboian, "Corrosion Resistant High-Strength Clad Metal System for Hydraulic Brake Line Tubing," SAE Preprint No. 740290, Society for Automotive Engineers, 1972

Evaluation of Galvanic Corrosion

Harvey P. Hack, David Taylor Naval Ship Research and Development Center

GALVANIC CORROSION, although listed as one of the forms of corrosion, should instead be considered as a type of corrosion mechanism, because any of the other forms of corrosion can be accelerated by galvanic effects. Therefore, any of the tests used for the more conventional forms of corrosion, such as uniform attack, pitting, or stress corrosion, can be used, with modifications, to determine galvanic-corrosion effects. The modifications can be as simple as connecting a second metal to the system or as complex as necessary to evaluate the appropriate parameters. A change in the method of data interpretation is often all that is needed to convert conventional test methods into galvanic-corrosion tests.

This article will discuss component, model, electrochemical, and specimen tests. Additional information on galvanic corrosion can be found in the article ''General Corrosion' in this Volume.

Component Testing

Component testing is an especially useful technique for galvanic corrosion prediction. The materials in a system are often selected primarily for reasons other than galvanic compatibility. In complex components, such as valves or pumps, many different materials can be used in a geometric configuration that is extremely difficult to model. In more complicated cases, even the most basic prediction, such as which materials will suffer increased corrosion due to galvanic effects, may not be possible from simple laboratory tests. Therefore, component testing becomes the best method for predicting material behavior in complex systems.

Conducting component tests for galvanic corrosion is similar to conducting component tests for any other type of corrosion. The same care must be taken to ensure that the materials, the operation of the component, and the environment are similar to those in service. However, one important difference with regard to galvanic corrosion is the relationship between the component being tested and the other elements of the system. For example, it would be a waste of effort to expose a complicated piece of machinery in order to look for galvanic corrosion when the whole device is cathodically protected as a result of being attached to a protected structure. Alternatively, incorrect results would be obtained for the exposure of an isolated bronze mixed-material valve when the ultimate use was in a piping system made of a more noble metal that could accelerate the corrosion of the entire valve galvanically. When outside interactions of this type are possible, the interacting materials must be made part of the corrosion system by exposing the appropriate surface area of those materials electrically connected to, and in the same electrolyte as, the component being tested.

The principal advantages of component testing are ease of interpretation of results and the lack of scaling or modeling uncertainties. The disadvantages include high cost and the need for extremely sensitive measures of corrosion damage to obtain results within reasonable time periods.

Modeling

Even when the galvanic behavior of panels of the materials of interest is known, the geometrical arrangement of these materials may make galvanic corrosion prediction difficult because of the effects of solution (electrolyte) resistance on the corrosion rates. An example of this is a heat-exchanger tube in a tubesheet. Assuming the tube to be anodic to the tubesheet, areas of the tube near the tubesheet will have low solution resistance to the cathode and will corrode rapidly, but areas away from the tubesheet will have a large solution resistance to the cathodic metal and will therefore corrode more slowly. In the reverse case, in which the tubesheet is anodic to the tube, the areas of the cathodic tube near the tubesheet will drive the galvanic corrosion of the tubesheet much more than distant areas will.

Computer Modeling. Geometrical effects can be modeled in computers by using such techniques as finite elements, boundary elements, and finite differences. The best computer models solve a version of the Laplace equation for the electrolyte surrounding the corroding materials and use the polarization behavior of the material in question as boundary conditions at the metal/electrolyte interface. The analysis is similar to the heat flow analysis, with potential analogous to temperature, current analogous to heat flux, and the polarization boundary condition analogous to a special nonlinear type of temperature-dependent convective flux.

The only data this type of model requires are the geometry, electrolyte conductivity, and polarization characteristics of the materials involved. The program generates potentials and current densities as a function of location, either of which can be related back to corrosion rate. The nonlinear boundary conditions make this type of computer modeling difficult to perform unless a large mainframe computer with sufficient computational capabilities is available. Computer modeling provides an excellent predictive tool for geometrical effects; however, it is still seen as less satisfying than physical scale model exposures.

Physical scale modeling must model the solution resistance effects and the relative effects of polarization resistance and solution resistance to obtain accurate geometrical predictive capability. When solution resistance is important, the best type of scale modeling is the scaled conductivity exposure. In this type of exposure, the model is reduced in size by some factor from the original. To maintain proper potential and current distribution scaling, the electrolyte conductivity must also be reduced by the same factor. Any resistive coatings, such as paints, must also have their conductivity scaled similarly. In the case of paints, this can be accomplished by applying a thinner layer, by the same scaling factor used for size, than the thickness used in practice.

For example, a one-tenth scale model of a heat exchanger designed to operate in seawater with a conductivity of 4 mho/cm should be placed in seawater diluted to a conductivity of 0.4 mho/cm. In this case, the observed potential and current distributions will be the same between the model and the full-scale heat exchanger. For physical scale modeling, measurements that can be taken include potential distribution by the use of a movable reference electrode, corrosion depth as a function of location, and, if the model design permits, current to different parts of the structure and mass loss of certain model components.

Although less expensive than full-scale component testing, physical scale modeling has many of the disadvantages of component testing. In addition, a great inaccuracy in conductivity scaling stems from the fact that the polarization resistance of the materials in the system of interest is often a function of solution conductivity. Thus, changing solution conductivity may influence polarization resistance sufficiently to make the results of the model inaccurate. There is no experimental way to avoid this shortcoming.

Laboratory Testing

Laboratory tests fall into two categories: electrochemical tests, in which the data are analyzed and reported in a way that assists galvanic-corrosion predictions, and specimen exposures, which may or may not be electrochemically monitored.

Electrochemical Tests

The use of electrochemical techniques to predict galvanic corrosion is summarized in Ref 1. The details that relate to testing techniques are discussed below.

Galvanic Series. When the only information needed is which of the materials in the system are possible candidates for galvanically accelerated corrosion and which will be unaffected or protected, the information from a galvanic series in the appropriate media is useful. Such a series is a list of freely corroding potentials of the materials of interest in the environment of interest arranged in order of potential (Fig. 1). The galvanic series is easy to use and is often all that is required to answer a simple galvanic-corrosion question. The material with the most negative, or anodic, corrosion potential has a tendency to suffer accelerated corrosion when electrically connected to a material with a more positive , or noble, potential. The disadvantages include:

- No information is available on the rate of corrosion
- Active-passive metals may display two, widely differing potentials
- Small changes in electrolyte can change the potentials significantly
- Potentials may be time dependent

Creating a galvanic series is a matter of measuring the corrosion potential of various materials of interest in the electrolyte of interest against a reference electrode half-cell, such as saturated calomel. This procedure is described in Ref 2. The details of such factors as meter resistance, reference cell selection, and measurement duration are also addressed in Ref 2. There is little difference from a normal reading of corrosion potential except for the measurement duration and the creation of a list ordered by potential.

To prepare a galvanic series that will be valid for the materials and environment of interest in service, all of the factors that affect the potential of those materials in that environment must be accounted for. These factors include material composition, heat treatment, surface preparation (mill scale, coatings surface finish, etc.), environmental composition (trace contaminants, dissolved gases, etc.), temperature, and flow rate. One important effect is exposure time, particularly on materials that form corrosion product layers. All of the precautions and warnings regarding the generation and use of a galvanic series are given in Ref 2.

Polarization Curves. More useful information on the rate of galvanic corrosion can be obtained by investigating the polarization behavior of the materials involved. This can be done by generating stepped potential or potentiodynamic polarization curves or by obtaining potentiostatic information on polarization behavior. The objective is to obtain a good indication of the amount of current required to hold each material at a given potential. Because all materials in the galvanic system must be at the same potential in systems with low solution resistivity, such as seawater, and because the sum of all currents flowing between the materials must equal 0 by Kirchoff's Law, the coupled potential of all materials and the galvanic currents flowing can be uniquely determined for the system. The corrosion rate can then be related to galvanic current by Faraday's Law if the resulting potential of the anodic materials is well away from their corrosion potential, or the corrosion rate can be found as a function of potential by independent measurement.

Potentiodynamic polarization curves are generated by connecting the specimen of interest to a scanning potentiostat. This device applies whatever current is necessary between the specimen and a counter electrode to maintain that specimen at a given potential versus a reference electrode half-cell placed near the specimen. The current required is plotted as a function of potential over a range that begins at the corrosion potential and proceeds in the direction (anodic or cathodic) required by that material. Such curves would be generated for each material of interest in the system. Additional information on the method for generating these curves is available in the article "Laboratory Testing" in this Volume and in Ref 3. The scan rate for potential must be chosen such that sufficient time is allowed for completion of electrical charging at the interface.

Potentiodynamic polarization is particularly effective for materials with time-independent polarization behavior. It is fast, relatively easy, and gives a reasonable, quantitative prediction of corrosion rates in many systems. However, potentiostatic techniques are preferred for time-dependent polarization. To establish polarization

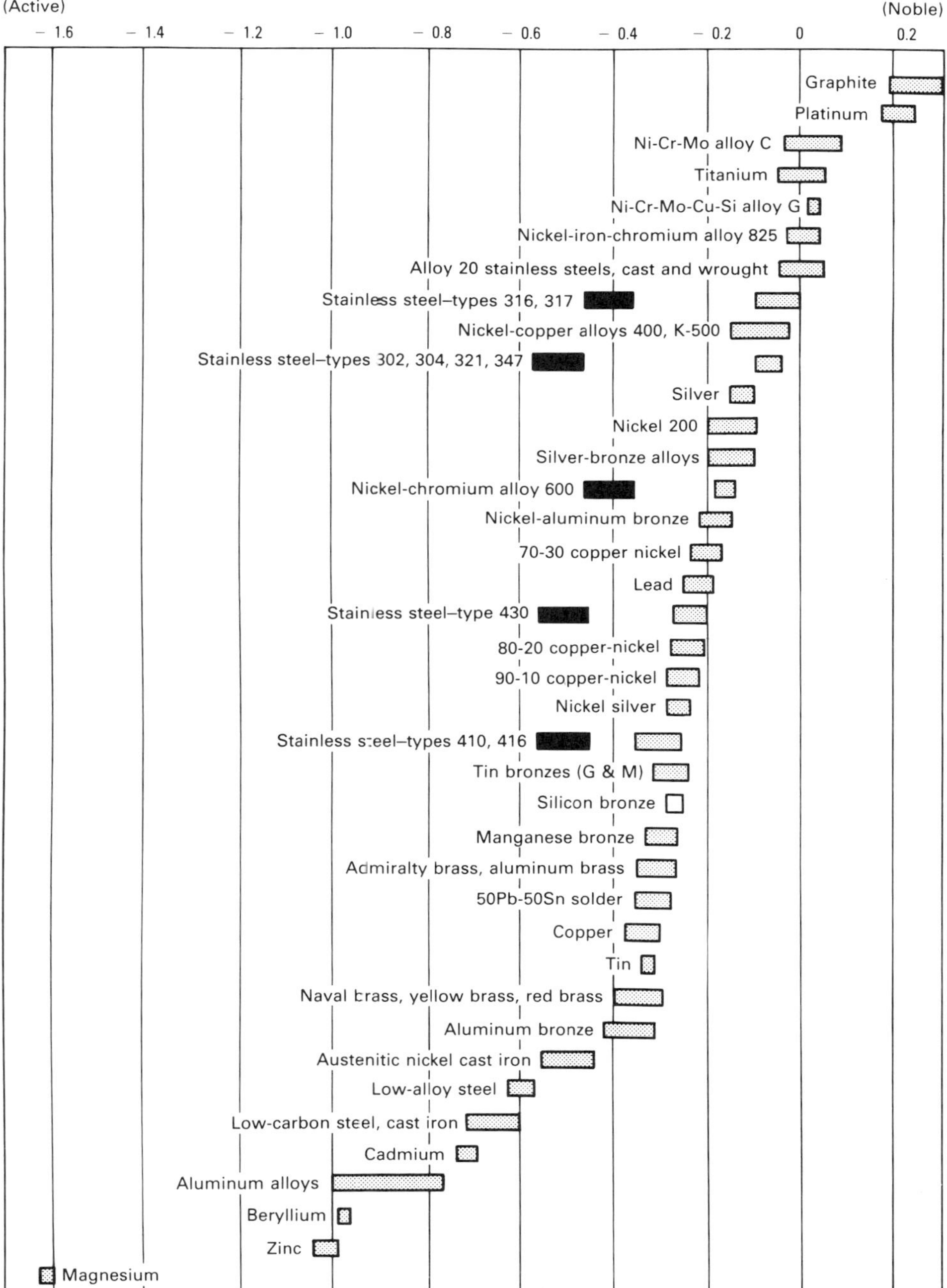

Fig. 1 Galvanic series for seawater. Dark boxes indicate active behavior of active-passive alloys.

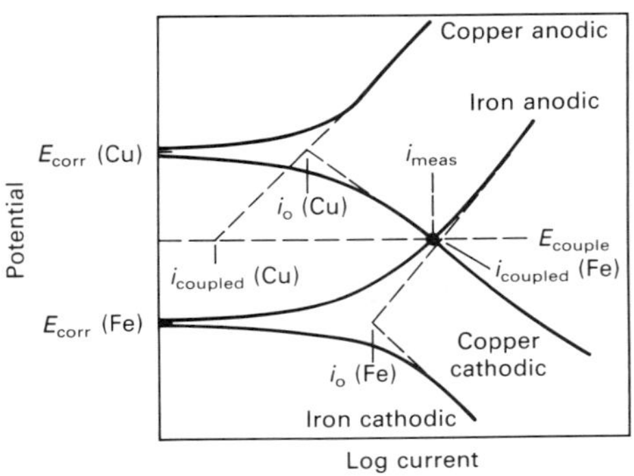

Fig. 2 Prediction of coupled potential and galvanic current from polarization diagrams. i, current; i_o, exchange current; E_{corr}, corrosion potential

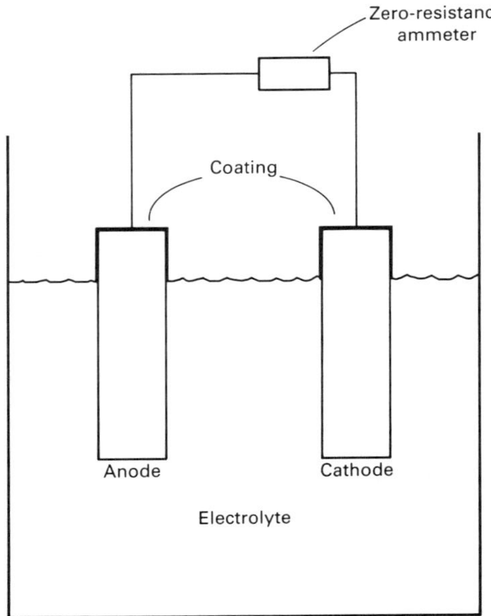

Fig. 3 Typical galvanic-corrosion immersion test setup using wire connections

characteristics for time-dependent polarization, a series of specimens is used, each held to one of a series of constant potentials with a potentiostat while the current required is monitored as a function of time. After the current has stabilized or after a pre-selected time period has elapsed, the current at each potential is recorded. Testing of each specimen results in the generation of one potential/current data pair, which gives a point on the polarization curve for that material. The data are then interpolated to trace out the full curve. This technique is very accurate for time-dependent polarization, but is expensive and time consuming. The individual specimens can be weighed before and after testing to determine corrosion rate as a function of potential, thus enabling the errors from using Faraday's Law to be easily corrected.

The process of predicting galvanic corrosion from polarization behavior can be illustrated by the example of a steel-copper system. Steel has the more negative corrosion potential and will therefore suffer increased corrosion upon coupling to copper, but the amount of this corrosion must be predicted from polarization curves. If the

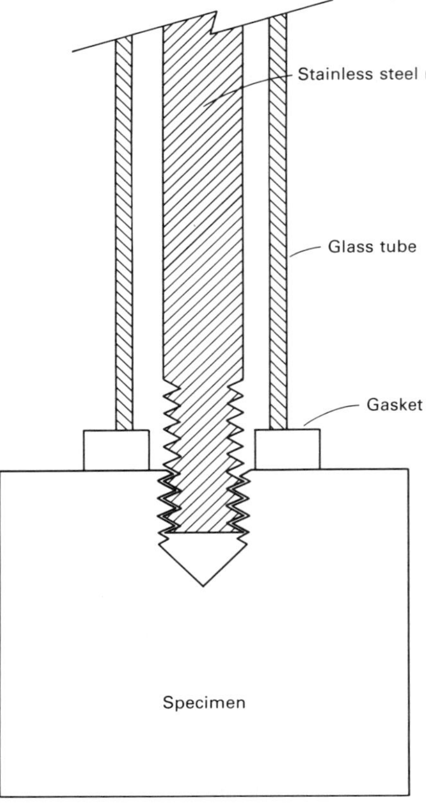

Fig. 4 Typical galvanic-corrosion test specimen using a threaded rod for mounting and electrical connection

polarization of each material is plotted as the absolute value of the log of current density versus potential and if the current density axis of each of these curves is multiplied by the wetted surface area of that material in the service application, then the result will be a plot of the total anodic current for steel and the total cathodic current for copper in this application as a function of potential (Fig. 2).

Furthermore, when the two metals are electrically connected, the anodic current to the steel must be supplied by the copper; that is, the algebraic sum of the anodic and cathodic currents must equal 0. If the polarization curves for the two materials, normalized for surface area as above, are plotted together, this current condition is satisfied where the two curves intersect. This point of intersection allows for the prediction of the coupled potential of the materials and the galvanic current flowing between them from the intersection point. This procedure works if there is no significant electrolyte resistance between the two metals; otherwise, this resistance must be taken into account in a complex manner that is beyond the scope of this article.

Specimen Exposures

Specimens for galvanic-corrosion testing include panels, wires, pieces of actual components, and other configurations of the materials of interest that are exposed in a process stream, a simulated service environment, or the actual environment. Specimens of the materials of interest are usually exposed in the same ratios of wetted or exposed area as in the service application. The

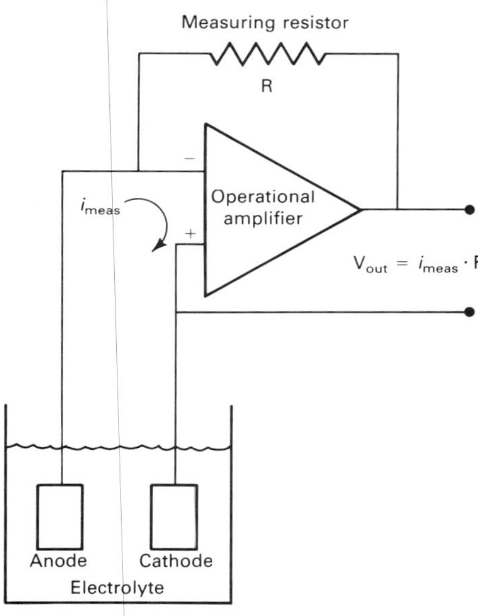

Fig. 5 Basic circuit for a zero-resistance ammeter

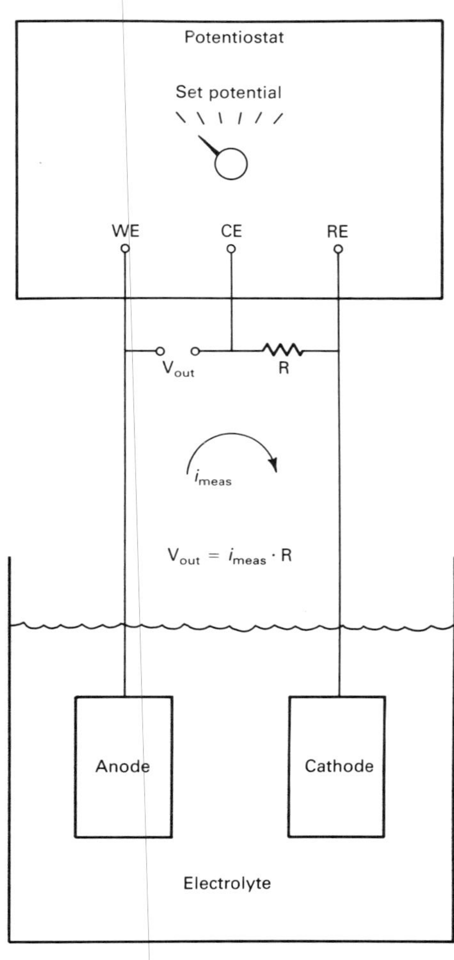

Fig. 6 Conversion of a potentiostat into a zero-resistance ammeter. WE, working electrode; CE, counter electrode; RE, reference electrode

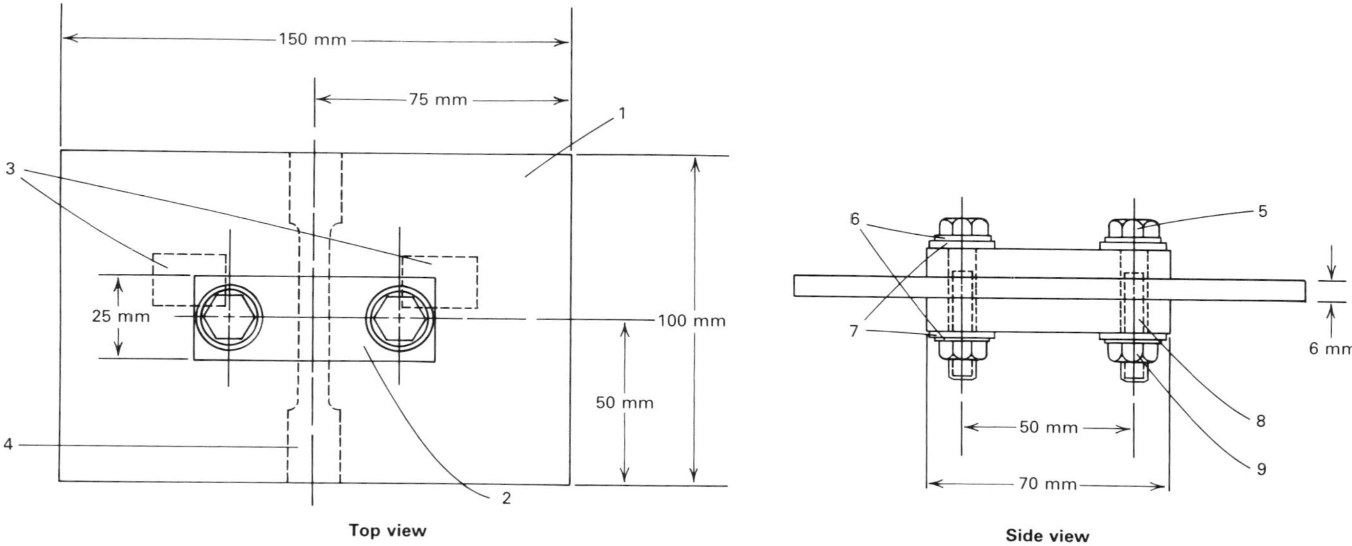

Fig. 7 Specimen configuration for the ISO test for atmospheric galvanic corrosion. 1, anodic plate, 1 piece; 2, cathodic plate, 2 pieces; 3, microsection, 2 pieces; 4, tensile test specimen; 5, bolt, 8 × 40 mm, 2 pieces; 6, washers, 1 mm thick, 16 mm diam, 4 pieces; 7, insulating washers, 1 to 3 mm thick, 18 to 20 mm diam, 4 pieces; 8, insulating sleeve, 2 pieces; 9, nut, 2 pieces. Dimensions given in millimeters

different materials are either placed in physical contact to provide electrical connection or are wired together such that the current between the materials can be monitored, usually as a function of time. Seldom can the effects of electrolyte resistance be included in this type of test, and the resistance is usually kept extremely low by appropriate relative placement of the materials.

Immersion. There are virtually no standardized tests for galvanic corrosion under immersion conditions, partly because the type of information needed, the extent of modeling of the service situation, and the type of system studied vary widely. This makes development of a standard test difficult. However, some general guidelines for galvanic-corrosion specimen testing in liquid electrolytes are given in Ref 4.

Immersion testing always involves an electrical connection between at least two dissimilar metals. This is usually accomplished with a wire, as in Fig. 3, although threaded mounting rods have also been used successfully for electrical connection, such as the assembly shown in Fig. 4. Soldered or brazed connections have the best electrical integrity.

The electrolyte must be excluded from the contact area by applying a sealant, such as silicone or epoxy, by keeping the joint area out of the electrolyte by partial immersion of the specimen, in which case a waterline area is created; or by use of a tube and gasket or O-ring seal in the case of a threaded mounting rod. Mounting the specimen in a specially formulated epoxy has been found to be effective in minimizing crevice corrosion while maintaining a dry electrical connection. However, selection of the best epoxy formulation is difficult. Care must be taken that the sealant or gasketing material is stable in the electrolyte being studied.

Almost any sealing procedure will create a potential area for crevice corrosion; thus, it is very difficult to study galvanic behavior independent of crevice corrosion behavior (see the article "Evaluation of Crevice/Concentration Cell Corrosion" in this Volume). Control specimens may be run with similar crevices and no electrical connection, but because the reproducibility of crevice corrosion behavior is not good, data scatter will be large. Under some circumstances, the galvanic effect of importance may be the acceleration of crevice corrosion attack.

The relative wetted surface areas of the materials being tested will have an important effect on the magnitude of the galvanic attack. The larger the cathode-to-anode area ratio is, the larger the attack will be; therefore, it would at first seem reasonable to accelerate the test by using a large ratio. This should not be done, because accelerating the attack may also change the mechanism of the attack, which would lead to erroneous conclusions. It is far more appropriate to use more accurate measurement techniques to determine the extent of the attack over a short period than to accelerate the test to obtain measurable attack quickly. If soldered or brazed connections are used for electrical connection, subsequent evaluation by weight loss is difficult; therefore, if weight loss is to be used to measure attack, threaded and sealed connections are preferred.

Measurement of the electrical current flowing between the metals can give a very sensitive indication of the extent of the galvanic attack and will allow the attack to be monitored over time. Coupled potential is another parameter that is useful to follow during the course of the exposure. The effect of exposure time on the rate of attack should be properly considered. Initially high rates of galvanic attack may decay to acceptable levels in a short period of time, or initially low attack rates may increase to unacceptable levels over time.

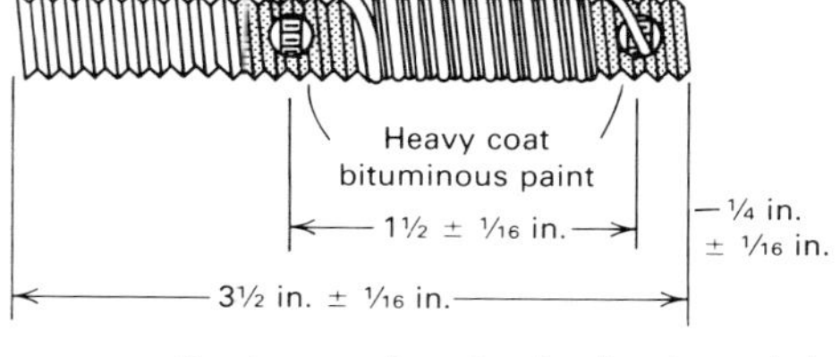

Cathodic element
Standard thread 3/8 in. – 16 U.S.
Anodic element 0.032-in. wire
Heavy coat bituminous paint
1½ ± 1/16 in.
¼ in. ± 1/16 in.
3½ in. ± 1/16 in.

Fig. 8 Specimen configuration for the wire-on-bolt test for atmospheric galvanic corrosion

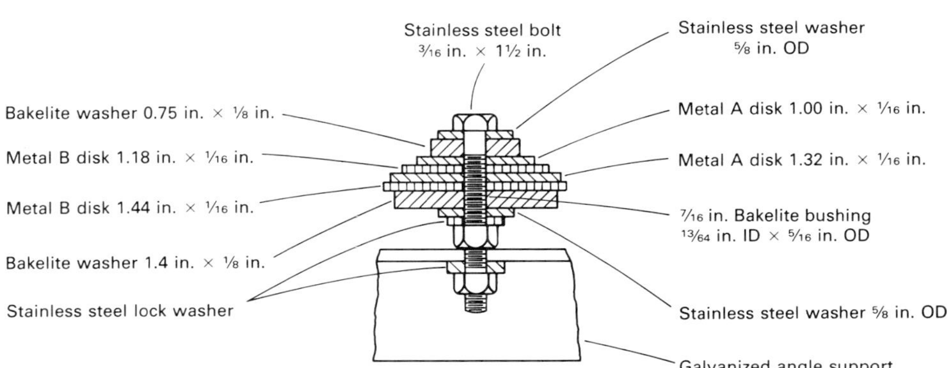

Fig. 9 Specimen configuration for the washer test for atmospheric galvanic corrosion

Current can be measured by inserting a resistor of 1 to 10 Ω in the current circuit and measuring the potential decrease across this resistor with a voltmeter having a resistance of at least 1000 Ω. The resistor should be sized such that the voltage across it does not exceed 5 mV; thus, the resistor will not significantly impede the current flow. Alternatively, a zero-resistance ammeter can be used instead of the resistor. This device is an operational amplifier connected to maintain 0 V across its input terminals (Fig. 5). A current-measuring resistor, placed in the feedback circuit, may be as large as the amplifier will allow, thus enabling currents in the nanoampere range to be easily measured. One simple way of creating a zero-resistance ammeter is by using a potentiostat with the counter electrode and reference electrode leads shorted together and set to a working electrode potential of 0 V (Fig. 6).

The importance of electrolyte flow in galvanic corrosion should not be overlooked in establishing the test procedure. A test apparatus should be used that reproduces the flow under service conditions. If this is not possible and flow must be scaled, the exact scaling method will depend on the assumed corrosion processes. Cathodic reactions, such as oxygen reduction, that are controlled by diffusion through a fluid boundary layer are likely to be properly scaled by reproducing the hydrodynamic boundary layer of the service application in the test. This should reproduce the diffusion boundary layer that controls the reaction.

Alternatively, the rate of reactions controlled by films such as anodic brightening of copper alloys, other passivation-type reactions, or control by calcareous deposit formation in seawater, may depend more on the shear stress at the surface required to strip off the film. In this case, surface shear stress may be a better hydrodynamic parameter to reproduce in the test.

Many different types of flow apparatus have been used, such as concentric tubes, in-line tubes, rotating cylinders, rotating ring-disks, rectangular flow channels with specimens mounted in the walls, and circulating foils. Each of these has its own hydrodynamic peculiarities, but one common area of concern is the leading edge of the specimen. It is difficult, even for specimens recessed in the walls of a flow channel, to avoid a step or gap that can create unexpected hydrodynamic conditions at the specimen surfaces downstream. Also, mounting to allow electrical connection must be considered, and crevice effects are essentially impossible to eliminate.

Atmospheric Tests. General testing guidelines become more complex when considering atmospheric or cabinet exposures. Testing in these environments differs markedly from immersion tests in a number of ways, most of which involve the insufficiency of electrolyte. Many of the variables that influence the behavior of specimens in the atmosphere are discussed in Ref 5.

The thinness of the electrolyte film and the normally low conductivity of the electrolyte combine to limit galvanic effects to within about 5 mm (0.2 in.) of the dissimilar-metal interface. Thus, area ratio effects become relatively unimportant. Sealing the electrical connections becomes relatively less important than in immersion testing if the connections are more than 5 mm (0.2 in.) from the area to be evaluated and if corrosion products will not interfere with the continuity of the connection. Periodic checks of electrical continuity in atmospheric galvanic-corrosion tests are recommended. Geometrical effects also become unimportant, except as they relate to the entrapment of moisture. However, specimen orientation effects must be considered. The behavior of the galvanic couples will depend on whether they are exposed on the top or the bottom of the panel, whether they are sheltered or not, or other considerations, such as the effect of specimen mass on condensation.

Because there are no standardized tests for galvanic corrosion immersed in electrolytes, it is somewhat surprising that several standard tests have emerged for atmospheric galvanic corrosion, even though less testing has been done in this area. One of these tests is an International Organization for Standardization (ISO) standard (Ref 6) and is also being developed by the American Society for Testing and Materials (ASTM). This test uses a 100- × 400-mm (4- × 16-in.) panel of the anodic material to which a 50- × 100-mm (2- × 4-in.) strip of the cathodic material is bolted (Fig. 7). After exposure, the anodic material can be evaluated for material degradation by weight loss and other corrosion measurements as well as by degradation of such mechanical properties as ultimate tensile strength.

This test is relatively easy to perform, but requires the availability of plate of the materials of interest and a prior knowledge of which material is anodic. Like any atmospheric galvanic-corrosion test, crevice effects cannot be adequately separated from galvanic effects in some cases; therefore, a coating is sometimes applied between the anode and cathode plates. The disadvantage of this test is the time required to obtain results; for systems with moderate corrosion rates, exposures of 1 to 5 years are not unusual.

Another commonly used atmospheric galvanic-corrosion test is the wire-on-bolt test, sometimes referred to as the CLIMAT test (Ref 7-9). In this test, a wire of the anodic material is wrapped around a threaded rod of the cathodic material (Fig. 8). Because corrosion can be rapid in this test, exposure duration should usually be limited. This makes the test ideal for measuring atmospheric corrosivity as well as material corrosion properties. Not all materials of interest are available in the required wire and threaded rod forms, and analysis is usually restricted to weight loss measurement and observation of pitting. When the required materials are available, this test is less expensive and easier to conduct than the ISO plate test.

A third atmospheric galvanic-corrosion test has been used extensively by ASTM, but has not been standardized. This test (Ref 10) involves the use of 25-mm (1-in.) diam washers of the materials of interest bolted together as shown in Fig. 9. The bolt that holds the washers together can also be used to secure the assembly in position. After exposure, the washers can be disassembled for weight loss determination. The materials needed for this test are not as large as those for the ISO plate test, but it can take as long and cannot provide mechanical properties data.

REFERENCES

1. R. Baboian, Electrochemical Techniques for Predicting Galvanic Corrosion, in *Galvanic and Pitting Corrosion—Field and Laboratory Studies*, STP 576, American Society for Testing and Materials, 1976, p 5-19
2. "Standard Guide for Development and Use of a Galvanic Series for Predicting Galvanic Corrosion Performance," G 82, *Annual Book of ASTM Standards*, American Society for Testing and Materials
3. "Standard Reference Test Method for Making Potentiostatic and Potentiodynamic Anodic Polarization Measurements," G 5, *Annual Book of ASTM Standards*, American Society for Testing and Materials
4. "Standard Guide for Conducting and Evaluating Galvanic Corrosion Tests in Electrolytes," G 71, *Annual Book of ASTM Standards*, American Society for Testing and Materials
5. "Standard Practice for Conducting Atmospheric Corrosion Tests on Metals," G 50, *Annual Book of ASTM Standards*, American Society for Testing and Materials
6. "Corrosion of Metals and Alloys—Determination of Bi-Metallic Corrosion in Outdoor Exposure Corrosion Tests," ISO 7441, International Standards Organization
7. K.G. Compton, A. Mendizza, and W.W. Bradley, Atmospheric Galvanic Couple Corrosion, *Corrosion*, Vol 11, 1955, p 383
8. H.P. Godard, Galvanic Corrosion Behavior of Aluminum in the Atmosphere, *Mater. Prot.*, Vol 2 (No. 6), 1963, p 38
9. D.P. Doyle and T.E. Wright, Rapid Methods for Determining Atmospheric Corrosivity and Corrosion Resistance, in *Atmospheric Corrosion*, W.H. Aylor, Ed., John Wiley & Sons, 1982, p 227
10. R. Baboian, Final Report on the ASTM Study: Atmospheric Galvanic Corrosion of Magnesium Coupled to Other Metals, in *Atmospheric Factors Affecting the Corrosion of Engineering Metals*, STP 646, S.K. Coburn, Ed., American Society for Testing and Materials, 1978, p 17-29

Evaluation of Intergranular Corrosion

Richard A. Corbett and Brian J. Saldanha, Corrosion Testing Laboratories, Inc.

IN THE ARTICLE "Localized Corrosion" in this Volume, intergranular corrosion is defined and the mechanisms are described. It is the purpose of this article to discuss when to evaluate for susceptibility to intergranular attack and how to determine which of the various evaluation tests are applicable. However, it may first be necessary to review the methodology of intergranular corrosion and its effect on the various alloy families.

Most alloys are susceptible to intergranular attack when exposed to specific environments. This is because grain boundaries are sites for precipitation and segregation, which makes them chemically and physically different from the grains themselves. Intergranular attack is defined as the selective dissolution of grain boundaries, or closely adjacent regions, without appreciable attack of the grains themselves. This is caused by potential differences between the grain-boundary region and any precipitates, intermetallic phases, or impurities that form at the grain boundaries. The actual mechanism differs with each alloy system.

Precipitates that form as a result of the exposure of metals at elevated temperatures (for example, during production, fabrication, and welding) often nucleate and grow preferentially at grain boundaries. If these precipitates are rich in alloying elements that are essential for corrosion resistance, the regions adjacent to the grain boundary are depleted of these elements. The metal is thus sensitized and is susceptible to intergranular attack in a corrosive environment. For example, in austenitic stainless steels such as AISI type 304, the cause of intergranular attack is the precipitation of chromium-rich carbides [$(Cr, Fe)_{23}C_6$] at grain boundaries. These chromium-rich precipitates are surrounded by metal that is depleted in chromium; therefore, they are more rapidly attacked at these zones than on undepleted metal surfaces.

Impurities that segregate at grain boundaries may promote galvanic action in a corrosive environment by serving as anodic or cathodic sites. Therefore, this would affect the rate of dissolution of the alloy matrix in the vicinity of the grain boundary. An example of this is found in aluminum alloys when they contain intermetallic compounds, such as Mg_5Al_8 and $CuAl_2$, at the grain boundaries. During exposures to chloride solutions, the galvanic couples formed between these precipitates and the alloy matrix can lead to severe intergranular attack. Susceptibility to intergranular attack depends on the corrosive solution and on the extent of intergranular precipitation, which is a function of alloy composition, fabrication, and heat treatment parameters.

Corrosion tests for evaluating the susceptibility of an alloy to intergranular attack are typically classified as either simulated-service or accelerated tests. The first laboratory tests for detecting intergranular attack were simulated-service exposures. These were first observed and used in 1926 when intergranular attack was detected in an austenitic stainless steel in a copper sulfate-sulfuric acid ($CuSO_4$-H_2SO_4) pickling tank (Ref 1). Another simulated-service test for alloys intended for service in nitric acid (HNO_3) plants is described in Ref 2. In this case, for accelerated results, iron-chromium alloys were tested in a boiling 65% HNO_3 solution.

Over the years, specific tests have been developed and standardized for evaluating the susceptibility of various alloys to intergranular attack. For example, tests for the low-alloy austenitic stainless steels have been standardized by the American Society for Testing and Materials (ASTM) in Standard A 262, with its various practices (A to E). Practice A is a screening test that uses an electrolytic oxalic acid etch combined with metallographic examination. The other practices involve exposing the material (possibly after a sensitizing treatment) to boiling solutions of 65% HNO_3, acidified ferric sulfate ($Fe_2(SO_4)_3$) solution, nitric-hydrofluoric acid (HNO_3-HF) solution, or acidified $CuSO_4$ solution, depending on the specific alloy and its application. Similar ASTM tests have been developed for other higher-alloyed stainless steels, ferritic stainless steels, high nickel-base alloys, and aluminum alloys (Table 1).

The Purpose of Testing

There is a perception in much of the industry that testing for susceptibility to intergranular attack is equivalent to evaluating the resistance of the alloy to general and localized corrosion. Although the tests used for evaluating susceptibility to intergranular attack are severe, they are not intended to duplicate conditions for the wide range of chemical exposures present in an industrial plant, even though some of these tests simulate service conditions.

Testing for susceptibility to intergranular attack, however, is useful for determining whether the correct material, in the proper metallurgical condition, has been supplied by a vendor. There are some problems associated with quality assurance programs for purchased materials. Such programs are sometimes based on faith in what is supplied by a vendor or production mill and what is certified in the documentation sent along with the material. However, such confidence may be misplaced. For example, there have been a number of accounts in which alloys have been substituted, resulting in premature failure. In one case, this occurred when Hastelloy B valves were substituted for the Hastelloy C-276 valves that were ordered to handle a hypochlorite solution. The Hastelloy B valves failed in about 3 months.

In addition, there are many examples in which the material supplied does not conform to its certified analysis. The problem of getting reliable certified analyses increases when documentation goes from a mill to an alloy supplier. In one case, for example, AISI type 304L stainless steel valves were ordered, but the vendor, having few orders for this alloy, substituted type 316L stainless steel valves and sent certifications that purposely omitted the molybdenum analysis. Normally, this would have been a good substitution for improved corrosion resistance at a bargain price, but these valves were destined for hot, concentrated HNO_3 service and failed prematurely.

These are just two examples of using a material that is incorrect or is not in the proper metallurgical condition; such problems, of course, are not limited to stainless steels. It should be realized that errors do occur and that for critical service the specified alloys must be in optimum metallurgical condition to resist intergranular attack and other forms of corrosion associated with precipitates at the grain boundaries.

Tests for Stainless Steels and Nickel-Base Alloys

The austenitic and ferritic stainless steels, as well as most nickel-base alloys, are generally supplied in a heat-treated condition such that they are free of carbide precipitates that are detrimental to corrosion resistance. However, these alloys are susceptible to sensitization from welding, improper heat treatment, and service in the sensitizing temperature range. The phenomenon of sensitization of these alloys is discussed further in the articles "Corrosion of Stainless Steels," "Corrosion of Weldments," and "Corrosion of Nickel-Base Alloys" in this Volume.

The theory and application of acceptance tests for detecting the susceptibility of stainless steels and nickel-base alloys to intergranular attack are extensively reviewed in Ref 3 and 4. It would be repetitive to review this work other than to discuss why and when it is necessary to evaluate the susceptibility of alloys to this form of attack

Table 1 Appropriate evaluation tests and acceptance criteria for wrought alloys

UNS number	Alloy name	Applicable tests (ASTM standards)	Sensitizing treatment	Exposure time, h	Criteria for passing, appearance or maximum allowable corrosion rate, mm/month (mils/month)
S43000	Type 430	Ferric sulfate (A 763-X)	None	24	1.14 (45)
S44600	Type 446	Ferric sulfate (A 763-X)	None	72	0.25 (10)
S44625	26-1	Ferric sulfate (A 763-X)	None	120	0.05 (2) and no significant grain dropping
S44626	26-1S	Cupric sulfate (A 763-Y)	None	120	No significant grain dropping
S44700	29-4	Ferric sulfate (A 763-X)	None	120	No significant grain dropping
S44800	29-4-2	Ferric sulfate (A 763-X)	None	120	No significant grain dropping
S30400	Type 304	Oxalic acid (A 262-A)	None	...	(a)
		Ferric sulfate (A 262-B)		120	0.1 (4)
S30403	Type 304L	Oxalic acid (A 262-A)	1 h at 675 °C (1250 °F)	...	(a)
		Nitric acid (A 262-C)		240	0.05 (2)
S30908	Type 309S	Nitric acid (A 262-C)	None	240	0.025 (1)
S31600	Type 316	Oxalic acid (A 262-A)	None	...	(a)
		Ferric sulfate (A 262-B)		120	0.1 (4)
S31603	Type 316L	Oxalic acid (A 262-A)	1 h at 675 °C (1250 °F)	...	(a)
		Ferric sulfate (A 262-B)		120	0.1 (4)
S31700	Type 317	Oxalic acid (A 262-A)	None	...	(a)
		Ferric sulfate (A 262-B)		120	0.1 (4)
S31703	Type 317L	Oxalic acid (A 262-A)	1 h at 675 °C (1250 °F)	...	(a)
		Ferric sulfate (A 262-B)		120	0.1 (4)
S32100	Type 321	Nitric acid (A 262-C)	1 h at 675 °C (1250 °F)	240	0.05 (2)
S34700	Type 347	Nitric acid (A 262-C)	1 h at 675 °C (1250 °F)	240	0.05 (2)
N08020	20Cb-3	Ferric sulfate (G 28-A)	1 h at 675 °C (1250 °F)	120	0.05 (2)
N08904	904L	Ferric sulfate (G 28-A)	None	120	0.05 (2)
N08825	Incoloy 825	Nitric acid (A 262-C)	1 h at 675 °C (1250 °F)	240	0.075 (3)
N06007	Hastelloy G	Ferric sulfate (G 28-A)	None	120	0.043 (1.7) sheet, plate, and bar; 0.05 (2) pipe and tubing
N06985	Hastelloy G-3	Ferric sulfate (G 28-A)	None	120	0.043 (1.7) sheet, plate, and bar; 0.05 (2) pipe and tubing
N06625	Inconel 625	Ferric sulfate (G 28-A)	None	120	0.075 (3)
N06690	Inconel 690	Nitric acid (A 262-C)	1 h at 540 °C (1000 °F)	240	0.025 (1)
N10276	Hastelloy C-276	Ferric sulfate (G 28-A)	None	24	1 (40)
N06455	Hastelloy C-4	Ferric sulfate (G 28-A)	None	24	0.43 (17)
N06110	Allcorr	Ferric sulfate (G 28-B)	None	24	0.64 (25)
N10001	Hastelloy B	20% Hydrochloric acid	None	24	0.075 (3) sheet, plate, and bar; 0.1 (4) pipe and tubing
N10665	Hastelloy B-2	20% Hydrochloric acid	None	24	0.05 (2) sheet, plate, and bar; 0.086 (3.4) pipe and tubing
A95005–A95657	Aluminum Association 5*xxx* alloys	Concentrated nitric acid (G 67)	None	24	(b)

(a) See A 262, practice A. (b) See G 67, section 4.1.

and to discuss acceptance criteria for the tests used.

Because sensitized alloys may inadvertently be used, acceptance tests are implemented as a quality control check to evaluate stainless steels and nickel-base alloys when:

- Different alloys, or regular carbon types of the specified alloy, are submitted for the low-carbon grades (for example, type 316 substituted for type 316L) and are involved in welding or heat treating
- An improper heat treatment during fabrication results in the formation of intermetallic phases
- The specified limits for carbon and/or nitrogen contents of an alloy are inadvertently exceeded

Some standard tests include acceptance criteria, but others do not (Ref 3). Some type of criterion is needed that can clearly separate material susceptible to intergranular attack from that resistant to attack. Table 1 lists evaluation tests and acceptance criteria for various stainless steels and nickel-base alloys that have been used by the DuPont Company, the U.S. Department of Energy, and others in the chemical-processing industry. Identifying such rates still leaves the buyer and seller free to agree on a rate that meets their particular needs.

Table 2 Media for testing susceptibility to intergranular corrosion

Alloy	Medium	Concentration, %	Temperature, °C (°F)
Magnesium alloys	Sodium chloride plus hydrochloric acid	...	Room
Copper alloys	Sodium chloride plus sulfuric or nitric acid	1 NaCl, 0.3 acid	40–50 (105-120)
Lead alloys	Acetic acid or hydrochloric acid	...	Room Room
Zinc alloys	Humid air	100% relative humidity	95 (205)

Source: Ref 13

Tests for Aluminum Alloys

The electrochemically active paths at the grain boundaries of aluminum alloy materials can be either the solid solution or closely spaced anodic second-phase particles. The identities of the specific active paths vary with the alloy composition and metallurgical condition of the product, as discussed in the article "Corrosion of Aluminum and Aluminum Alloys" in this Volume and in Ref 5 and 6. The most serious forms of such structure-dependent corrosion are stress-corrosion cracking (SCC) and exfoliation. Stress-corrosion cracking requires the presence of a sustained tensile stress, and exfoliation occurs only in wrought products with a directional grain structure. Not all materials that are susceptible to intergranular attack, however, are susceptible to SCC or exfoliation. Therefore, specific tests are required for the latter (see the article "Evaluation of Exfoliation Corrosion" in this Volume).

Strain-Hardened 5*xxx* Alloys. Alloys in this series that contain more than about 3% Mg are rendered susceptible to intergranular attack (sensitized) by certain manufacturing conditions or after being subjected to elevated temperatures up to about 175 °C (350 °F). This is the result of the continuous grain-boundary precipitation of the highly anodic Mg_2Al_3 phase, which corrodes preferentially in most corrosive environments.

The ASTM standard G 67 is a method that provides a quantitative measure of the susceptibility to intergranular attack of these alloys (Ref 7). This method consists of immersing test specimens in concentrated HNO_3 at 30 °C (85 °F) for 24 h and determining the mass loss per unit area as the measure of intergranular susceptibility. When this second phase is precipitated in a relatively continuous network along grain boundaries, the preferential attack of the network causes whole grains to drop out of the specimens. Such dropping out causes relatively large mass losses of the order of 25 to 75 mg/cm^2, although specimens of intergranular-resistant materials lose only about 1 to 15 mg/cm^2. Intermediate mass losses occur when the precipitate is randomly distributed. The parallel relationship between the susceptibility to intergranular attack and to SCC and exfoliation of these particular alloys makes ASTM G 67 a useful screening test for alloy and process development studies. A problem arises, however, in selecting a pass-or-fail value in relation to the performance of intermediate materials in environments other than HNO_3.

Heat-Treated High-Strength Alloys. Materials problems caused by SCC, exfoliation, or corrosion fatigue of the early 2*xxx* (aluminum-copper) alloys were identified with intergranular corrosion, and the blame came to be associated with improper heat treatment. In 1944, an accelerated test for detecting susceptibility to intergranular corrosion was incorporated into a U.S. Government specification for the heat treatment of aluminum alloys. This specification has been superseded by the current Military Specification MIL-H-6088F. Tests are required for periodic monitoring of 2*xxx* and 7*xxx* series alloys in all rivets and fastener components as well as sheet, bar, rod, wire, and shapes under 6.4 mm (0.25 in.) thick. Specimen preparation, test procedure, and evaluation criteria are detailed in Ref 8.

Other Tests for Aluminum Alloys. The volume of hydrogen evolved upon immersion of etched 2*xxx* series (aluminum-copper-magnesium) aluminum alloys in a solution containing 3% sodium chloride (NaCl) and 1% hydrochloric acid (HCl) for a stipulated time has been used as a quantitative measure of the severity of intergranular attack. A problem with this approach (which is quite valid) was that the correlation between the amount (or the rate) of hydrogen evolved is influenced by a number of factors, including alloy composition, temper, and grain size (Ref 9, 10).

Applied current or potential in neutral chloride solutions (for example, 0.1 *N* NaCl) provides another direct method of assessing the degree of susceptibility to intergranular attack when accompanied by a microscopic examination of metallographic sections (Ref 9, 11, 12). More sophisticated electrochemical approaches for studying systems involving active-path corrosion use potentiodynamic methods. Tests for SCC are discussed in the article "Evaluation of Stress-Corrosion Cracking" in this Volume.

Tests for Other Alloys

Although intergranular corrosion is present to some extent in alloys other than stainless and aluminum alloys, incidences of attack associated with this form of corrosion are few and are generally not of practical importance. Therefore, no attempts have been made to develop and standardize specific tests for detecting susceptibility to intergranular corrosion in these alloys. However, certain media have been commonly used for evaluating the susceptibility to intergranular corrosion of magnesium, copper, lead, and zinc alloys (Ref 13). These media are listed in Table 2. The presence or absence of attack in these tests is not necessarily a measure of the performance of the material in other corrosive environments.

Magnesium Alloys. There are rare instances of reported intergranular corrosion of magnesium alloys, as in the case of chromic acid contaminated with chlorides or sulfates.

The copper alloys that appear to be the most susceptible to intergranular corrosion are Muntz metal, admiralty metal, aluminum brasses, and silicon bronzes. Admiralty alloys have been observed to suffer intergranular corrosion upon exposure to saline cooling waters, although the incidence of attack is very low. The antimonial grades are reportedly superior to the arsenical grades in this respect. Similarly, arsenical and phosphorized grades of aluminum brass have been observed to suffer intergranular corrosion in seawater-type exposures.

Zinc die casting alloys have reportedly suffered intergranular corrosion in certain steam atmospheres. A laboratory test for simulating service failures, and particularly for alloy development work, has been in use for testing the susceptibility of zinc-base die castings to intergranular corrosion (Ref 14). The test consists of exposing samples to air at 95 °C (205 °F) and 100% relative humidity for 10 days under conditions permitting condensation of hot water on the metal. Susceptibility to intergranular corrosion is assessed by the effect on mechanical properties, such as impact strength. Experience has shown that castings with mechanical properties and dimensions that are not significantly altered by the 10-day exposure in this test will not suffer intergranular attack in atmospheric service.

REFERENCES

1. W.H. Hatfield, *J. Iron Steel Inst.*, Vol 127, 1933, p 380-383
2. W.R. Huey, *Trans. Am. Soc. Steel Treat.*, Vol 18, 1930, p 1126-1143
3. M.A. Streicher, in *Intergranular Corrosion of Stainless Alloys*, STP 656, American Society for Testing and Materials, 1978, p 3-84
4. M. Henthorne, in *Localized Corrosion—Cause of Metal Failure*, STP 516, American Society for Testing and Materials, 1972, p 66-119
5. T.J. Summerson and D.O. Sprowls, Corrosion Behavior of Aluminum Alloys, in *Aluminum Alloys: Their Physical and Mechanical Properties*, Vol III, E.A. Starke, Jr. and T.H. Sanders, Jr., Ed., Engineering Materials Advisory Services Ltd., 1986, p 1576-1662
6. B.W. Lifka and D.O. Sprowls, in *Localized Corrosion—Cause of Metal Failure*, STP 516, American Society for Testing and Materials, 1972, p 120-144
7. H.L. Craig, Jr., in *Localized Corrosion—Cause of Metal Failure*, STP 516, American Society for Testing and Materials, 1972, p 17-37
8. "Heat Treatment of Aluminum Alloys," Military Specification MIL-H-6088F, United States Government Printing Office
9. F.A. Champion, *Corrosion Testing Procedures*, 2nd ed., John Wiley & Sons, 1965, p 365, 366
10. G.J. Schafer, *J. Appl. Chem.*, Vol 10, 1960, p 138
11. S. Ketcham and W. Beck, *Corrosion*, Vol 16, 1960, p 37
12. M.K. Budd and F.F. Booth, *Corrosion*, Vol 18, 1962, p 197
13. F.A. Champion, *Corrosion Testing Procedures*, John Wiley & Sons, 1964
14. H.H. Uhlig, *Corrosion Handbook*, John Wiley & Sons, 1948

Evaluation of Exfoliation Corrosion

Donald O. Sprowls, Consultant

EXFOLIATION is a structure-dependent form of localized (usually) intergranular corrosion that is most familiar in certain alloys and tempers of aluminum. The mechanism of exfoliation is described in the article "Corrosion of Aluminum and Aluminum Alloys" in this Volume.

The occurrence of exfoliation in susceptible materials is influenced to a marked degree by environmental conditions. Figure 1 illustrates the broad range of behavior in different types of atmospheres. For example, forged truck wheels made of an aluminum-copper alloy (2024-T4) give corrosion-free service for many years in the warm climates of the southern and western United States, but they exfoliate severely in only 1 or 2 years in the northern states, where deicing salts are used on the highways during the winter months.

Accelerated laboratory tests do not precisely predict long-term corrosion behavior; however, answers are needed quickly in the development of new materials. For this reason, accelerated tests are used to screen candidate alloys before conducting atmospheric exposures or other field tests. They are also sometimes used for quality control tests. Several new laboratory tests for exfoliation corrosion have been standardized in recent years under the jurisdiction of American Society for Testing and Materials (ASTM) Committee G-1 on the Corrosion of Metals.

Spray Tests

Three cyclic acidified salt spray tests have been widely used in the aluminum and aircraft industries. These are covered by the procedures described in Annexes A2, A3, and A4 of ASTM G 85 (Ref 2). This standard does not prescribe the particular practice, test specimen, or exposure period to be used for a specific product, nor does it define the interpretation to be given to the test results. These considerations are prescribed by specifications covering the material or product being tested or by agreement between the purchaser and the seller.

Annex A2 describes a cyclic salt spray test that uses a 5% sodium chloride (NaCl) solution acidified to pH 3 with acetic acid in a spray chamber at a temperature of 49 °C (120 °F). This test is applicable for exfoliation testing of 2*xxx* (dry-bottom operation) and 7*xxx* (wet-bottom operation; that is, with approximately 25 mm, or 1 in., of water present in the bottom of the test chamber) aluminum alloys with a test duration of 1 to 2 weeks. Results with 7075 and 7178 alloys in various metallurgical conditions have been shown to correlate well with results obtained in a seacoast atmosphere (4-year exposure at Point Judith, RI) (Ref 3).

Annex A3 describes another cyclic-salt spray test that uses a 5% synthetic sea salt solution acidified to pH 3 with acetic acid in a spray chamber at a temperature of 49 °C (120 °F). The test is applicable to the production control of exfoliation-resistant tempers of the 2*xxx*, 5*xxx*, and 7*xxx* aluminum alloys (Ref 4, 5). Wet-bottom operating conditions are recommended with test durations of 1 to 2 weeks.

Annex A4 describes a salt-sulfur dioxide (SO_2) spray test that uses either 5% NaCl or 5% synthetic sea salt solution in a spray chamber at a temperature of 35 °C (95 °F). The spray may be either cyclic or constant; this, along with the type of salt solution and the test duration, is subject to agreement between the purchaser and the seller. The test is applicable for 2*xxx* and 7*xxx* aluminum alloys. Test duration is 2 to 4 weeks (Ref 1).

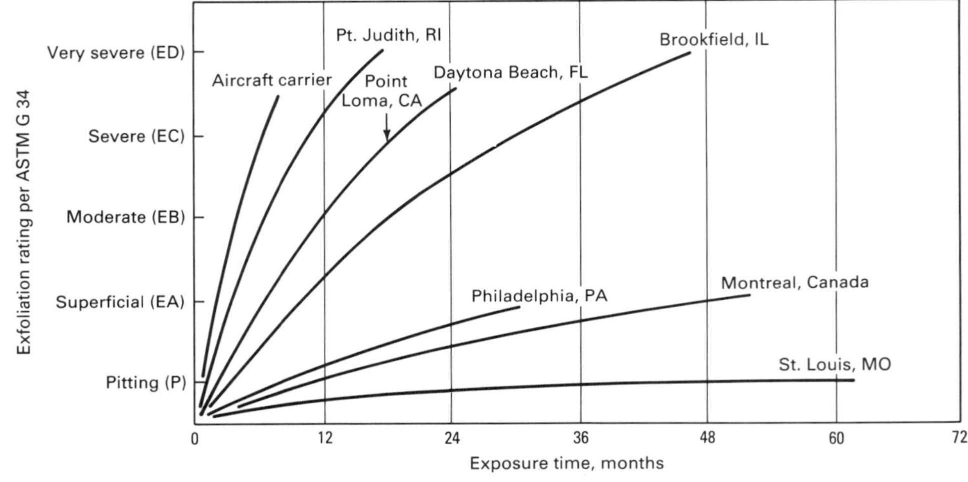

Fig. 1 Comparison of exfoliation of aluminum alloy 2124 (heat treated to be susceptible; EXCO ED rating) in various seacoast and industrial environments. Specimens were 13-mm (1/2-in.) plate. Source: Ref 1

Fig. 2 Examples of exfoliation rating EA (superficial): Specimens exhibit tiny blisters, thin slivers, flakes, or powder, with only slight separation of metal. Source: Ref 10

Immersion Tests

Total-immersion tests were developed to provide simpler, more easily controlled test methods. Chloride solutions did not cause exfoliation during reasonable periods of immersion; however, formulations of chloride-nitrate solutions were found that produced severe exfoliation of highly susceptible alloys of various types in only 1 or 2 days. Optimal test conditions differed for separate alloy families (Ref 6).

ASTM Standard G 66 describes a procedure for the continuous-immersion exfoliation testing of 5*xxx* alloys containing 2.0% or more magnesium (Ref 7). Specimens are immersed for 24 h at 65 °C (150 °F) in a solution containing ammonium chloride, ammonium nitrate, ammonium tartrate, and hydrogen peroxide. Susceptibility to exfoliation is determined by visual examination, using performance ratings established by reference to standard photographs. This method provides a reliable prediction of the exfoliation corrosion behavior of 5*xxx* alloys in marine environments (Ref 8). The test is also useful for alloy development studies and quality control of mill products such as sheet and plate (Ref 9).

ASTM Standard G 34 provides an accelerated exfoliation corrosion test for 2*xxx* and 7*xxx* aluminum alloys through the continuous immersion of test materials in an aqueous solution containing 4 *M* NaCl, 0.5 *M* potassium nitrate, and 0.1 *M* nitric acid at 25 °C (77 °F) (Ref 10). Susceptibility to exfoliation is determined by visual examination, using performance ratings established by reference to standard photographs. This method, also known as the EXCO test, is primarily used for research and development and quality control of such mill products as sheet and plate (Ref 9); however, it should not be construed as the optimal method for quality acceptance.

This method provides a useful prediction of the exfoliation behavior of these alloys in various types of outdoor service, especially in marine and industrial environments (Ref 3, 11). The test solution is very corrosive and represents the more severe types of environmental service (Fig. 1). However, it remains to be determined whether correlations can be established between EXCO test ratings and practical service conditions for a given alloy. Outdoor exposure tests are being conducted for this purpose. For example, it has been reported that samples of 7*xxx* (aluminum-zinc-magnesium-copper) alloys rated EA (superficial exfoliation) or P (pitting) in a 48-h EXCO test did not develop more than superficial exfoliation (EA rating) during 6- to 9-year exposures to seacoast atmospheres, while EC and ED (severe and very severe exfoliation, respectively) rated materials developed severe exfoliation within 1 to 7 years at the seacoast. (Specimens rated EA to ED are shown in Fig. 2 to 5) (Ref 11). It is anticipated that additional comparison will become available as the outdoor tests are extended.

However, performance differences have been noted, and this indicates that the EXCO test may be too severe for some of the more recently developed 2*xxx* and 7*xxx* alloys. Therefore, the testing program for evaluating the new alloy materials should consist of multiple tests with one of the less aggressive ASTM G 85 salt spray methods supplemented with outdoor tests. Also, caution must be exercised in setting limits for material procurement specifications based on accelerated tests (Ref 12).

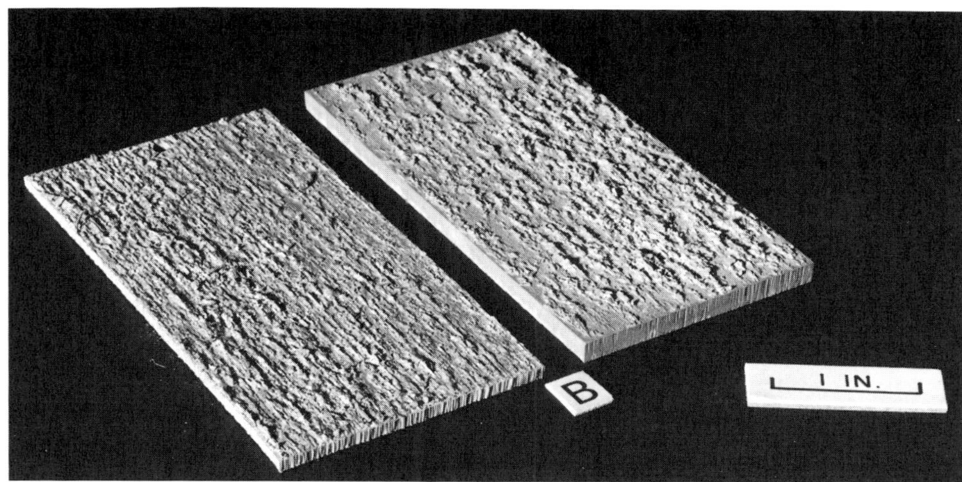

Fig. 3 Examples of exfoliation rating EB (moderate): Specimens show notable layering and penetration into the metal. Source: Ref 10

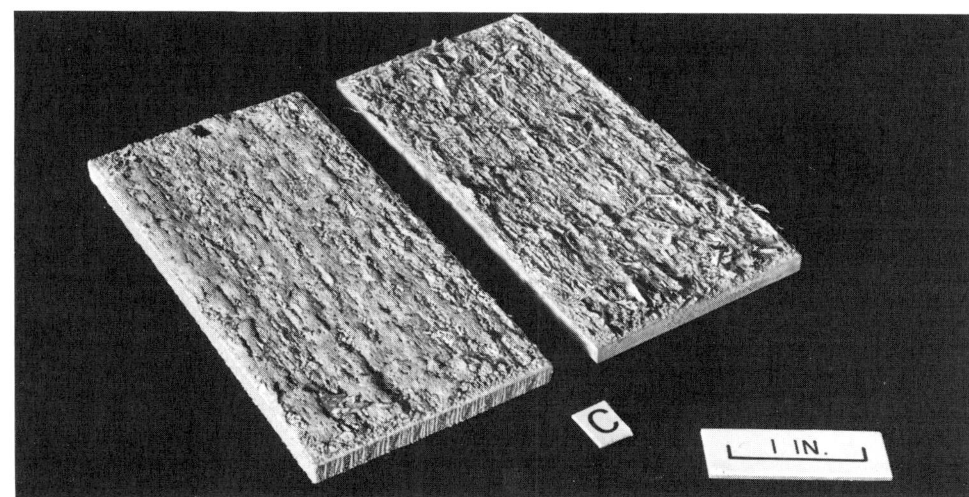

Fig. 4 Examples of exfoliation rating EC (severe): There is penetration to a considerable depth into the metal. Source: Ref 10

Fig. 5 Examples of exfoliation rating ED (very severe): Specimens appear similar to EC except for much greater penetration and loss of metal. Source: Ref 10

Visual Assessment of Exfoliation

One of the problems in evaluating the extent of damage due to exfoliation corrosion is the lack of a generally acceptable numerical measure of the corrosion. Therefore, the usual practice, as noted above for ASTM G 34 and G 66, is to assign visual ratings with reference to standard photographs, as shown in Fig. 2 to 5. The use of such ratings requires the inspector to make a judgment; because of this, the ratings are subject to variation among different inspectors.

Further, the lack of numerical measures of the corrosion damage hampers analysis of test results when a number of test materials must be compared. One approach is to assign numbers as substitutes for the letters. It is proposed for this purpose that a geometric scale (such as EA = 1, EB = 2, EC = 4, ED = 8) would be more consistent with the differences in damage illustrated by the standard photos than successive numbers would be (that is, 1, 2, 3, 4).

REFERENCES

1. S.J. Ketcham and E.J. Jankowsky, Developing an Accelerated Test: Problems and Pitfalls, in *Laboratory Corrosion Tests and Standards*, STP 866, G.S. Haynes and R. Baboian, Ed., American Society for Testing and Materials, 1985, p 14-23
2. "Standard Practice for Modified Salt Spray (Fog) Testing," G 85, *Annual Book of ASTM Standards*, American Society for Testing and Materials
3. B.W. Lifka and D.O. Sprowls, Relationship of Accelerated Test Methods for Exfoliation Resistance in 7*xxx* Series Aluminum Alloys with Exposure to a Seacoast Atmosphere, in *Corrosion in Natural Environments*, STP 558, American Society for Testing and Materials, 1974, p 306-333
4. H.B. Romans, An Accelerated Laboratory Test to Determine the Exfoliation Corrosion Resistance of Aluminum Alloys, *Mater. Res. Stand.*, Vol 9 (No. 11), 1969, p 31-34
5. S.J. Ketcham and P.W. Jeffrey, Exfoliation Corrosion Testing of 7178 and 7075 Aluminum Alloys, in *Localized Corrosion—Cause of Metal Failure*, STP 516, American Society for Testing and Materials, 1972, p 273-302
6. D.O. Sprowls, J.D. Walsh, and M.B. Shumaker, Simplified Exfoliation Testing of Aluminum Alloys, in *Localized Corrosion—Cause of Metal Failure*, STP 516, American Society for Testing and Materials, 1972, p 38-65
7. "Visual Assessment of Exfoliation Corrosion Susceptibility of 5*xxx*-Series Aluminum Alloys (ASSET Test)," G 66, *Annual Book of ASTM Standards*, American Society for Testing and Materials
8. T.J. Summerson, Interim Report, Aluminum Association Task Group on Exfoliation and Stress-Corrosion Cracking of Aluminum Alloys for Boat Stock, in *Proceedings of the Tri-Service Corrosion Military Equipment Conference*, Technical Report AFML-TR-75-42, Vol II, Air Force Materials Laboratory, 1975, p 193-221
9. "Standard Specification for Aluminum and Aluminum-Alloy Sheet and Plate," B 209, *Annual Book of ASTM Standards*, American Society for Testing and Materials
10. "Standard Test Method for Exfoliation Corrosion Susceptibility in 2xxx and 7xxx-Series Aluminum Alloys (EXCO Test)," G 34, *Annual Book of ASTM Standards*, American Society for Testing and Materials
11. D.O. Sprowls, T.J. Summerson, and F.E. Loftin, Exfoliation Corrosion Testing of 7075 and 7178 Aluminum Alloys—Interim Report on Atmospheric Exposure Tests (Report of ASTM G01.05.02 Interlaboratory Testing Program in Cooperation with the Aluminum Association), in *Corrosion in Natural Environments*, STP 558, American Society for Testing and Materials, 1974, p 99-113
12. B.W. Lifka, Corrosion Resistance of Aluminum Alloy Plate in Rural, Industrial, and Seacoast Atmospheres, *Mater. Prot.*, to be published

Evaluation of Stress-Corrosion Cracking

Donald O. Sprowls, Consultant

THERE ARE A NUMBER of corrosion-related causes of the premature fracture of structural components. The most common of these are compared in Fig. 1. Cracking due to corrosion fatigue occurs only under cyclic or fluctuating operating loads, while failure resulting from the other processes shown occurs under static or slowly rising loads. With certain alloy systems, hydrogen embrittlement (see the articles "Environmentally Induced Cracking" and "Evaluation of Hydrogen Embrittlement" in this Volume) may have a contributory role in each of these failure processes. Appropriate tests for the different failure modes are discussed in other articles in this Section.

This article will follow the broad outline listed below:

- General state of the art
- Static loading of smooth specimens
- Static loading of precracked specimens
- Dynamic loading: slow strain rate testing
- Selection of test environments
- Appropriate tests for various alloy systems
- Interpretation of test results

General State of the Art

Most stress-corrosion cracking (SCC) testing is performed either to determine the best material for a specific application or to compare the relative behaviors of material and environmental variations. Test conditions for the former should be representative of the most severe conditions anticipated in the intended service. For the latter, test conditions are usually chosen to produce various degrees of cracking in a reasonable time (Ref 1). The primary challenge in both cases is expressed well in the following statement, which was written a generation ago: "While it is relatively easy to determine if a product is susceptible to SCC, it is far more difficult to determine if it possesses a 'degree of susceptibility' which will restrict its general usefulness" (Ref 2).

Historically, service failures due to SCC have been identified with sustained tensile stress; thus, SCC testing has developed around the use of static loading. In some situations, it is advantageous to use an actual structural component for testing. However, this is usually not practical; more often, it is necessary to select smaller specimens that afford the required predictive capability. Before about 1965, only constant-load or constant-strain tests of smooth and notched test specimens of various configurations were used to assess SCC. More test methods are currently available than ever before.

During the 1960s, two accelerated testing techniques based on different mechanical approaches emerged. One technique tests and analyzes statically loaded, mechanically precracked test specimens by using linear elastic fracture mechanics concepts. The second technique consists of constant (slow) strain rate tests on smooth or precracked specimens. Laboratory testing with these techniques has frequently produced SCC, when the older, traditional tests have not.

Initiation and Propagation of SCC

The process of SCC is frequently discussed in terms of initiation (incubation and nucleation) and propagation, and illustrations similar to Fig. 2 can be found in the literature. However, an accepted model has not been established. There may be a gradual transition from localized corrosion to crack initiation and growth with no separation of stages, or there may be a repeated succession of short steps of initiation and growth. In any event, from an engineering standpoint, it is convenient to hypothesize the process in two generic stages: initiation and propagation. This terminology will be used throughout this article.

Two basic corrosion reactions, anodic and cathodic, dominate the SCC process in conjunction with mechanical stress. The chemical composi-

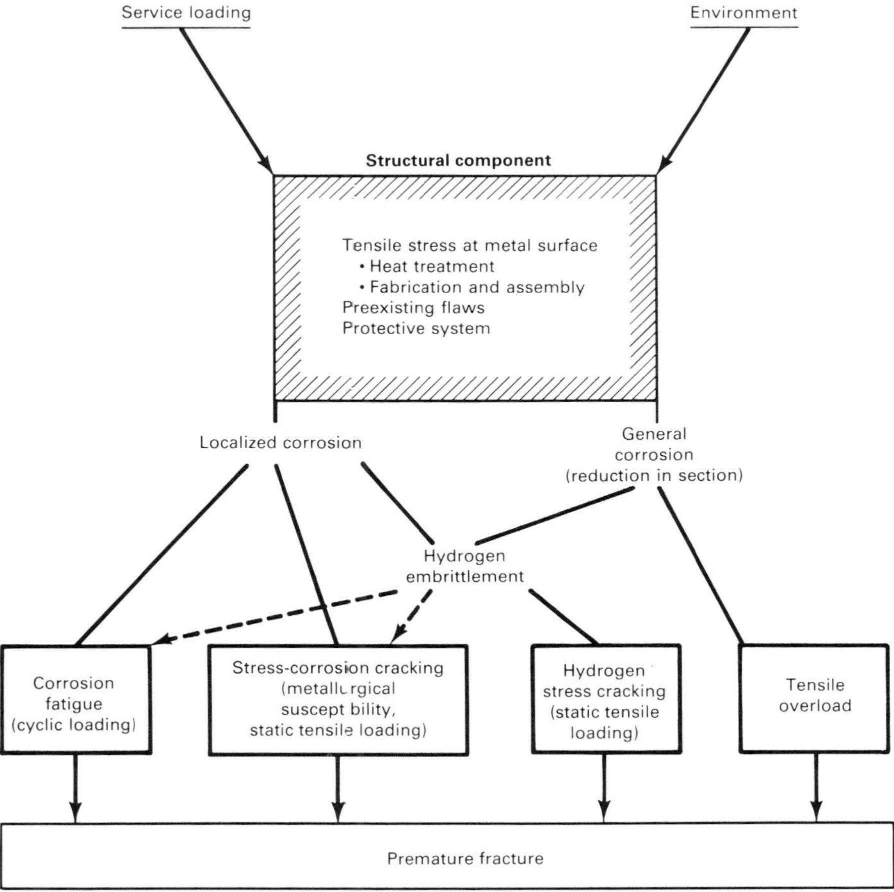

Fig. 1 Causes of premature fracture influenced by the corrosion of a structural component

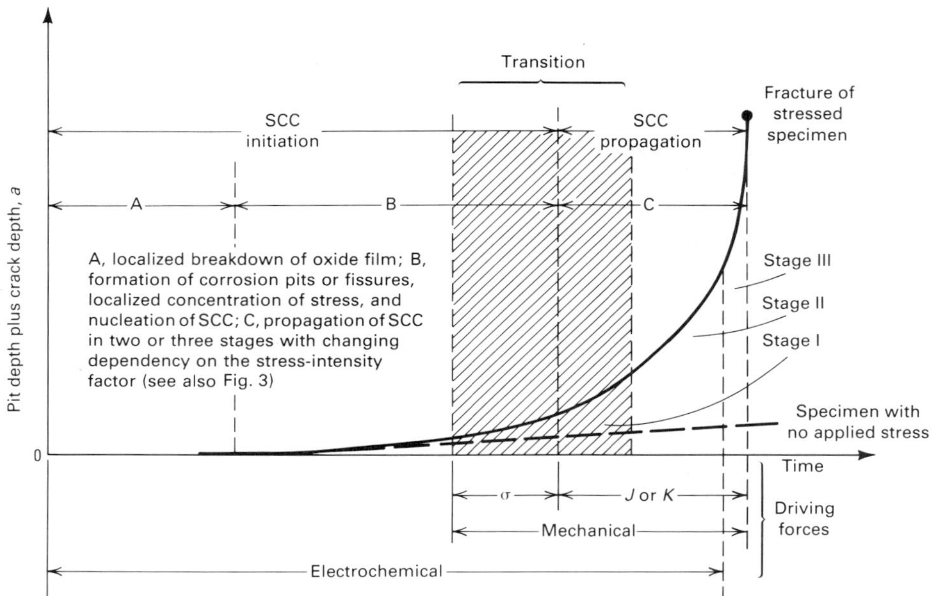

Fig. 2 The relative influences of electrochemical and mechanical factors in the corrosion and SCC damage of a susceptible material. The shaded area represents the transition of driving force from dominance by electrochemical factors to chiefly mechanical factors. Precise separation of initiation and propagation stages is experimentally difficult. Stimulation of cracking by atomic hydrogen may also become involved in this transition region.

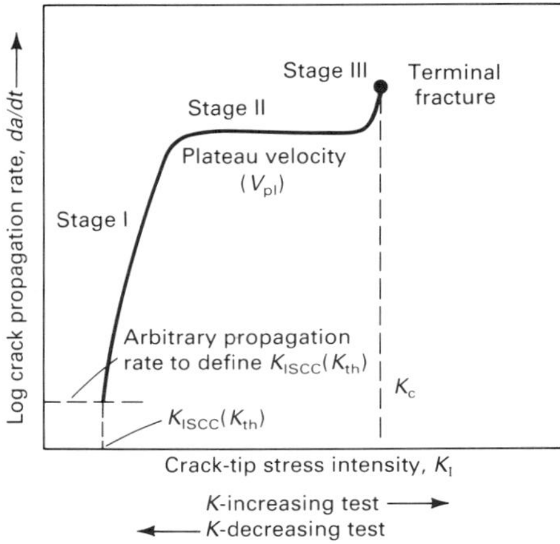

Fig. 3 Effect of stress intensity on the kinetics of SCC. Stages I and II may not always be straight lines but may be strongly curved, and one or the other may be absent in some systems. Stage III is of little interest and is generally absent in K-decreasing tests.

tion of the environment, including pH and the presence of hydrogen recombination poisons that affect the cathode reaction product, and the composition and metallurgical condition of the metal determine which of the two partial corrosion reactions is dominant. Anodic SCC (active path corrosion) involves the dissolution of metal during the initiation and propagation of cracks. Cathodic SCC (embrittlement by corrosion product hydrogen) involves the deposition of hydrogen at cathodic sites on the metal surface or on the walls of a fissure or crack and its subsequent absorption into the metal lattice. More information on the mechanisms of SCC is available in the article "Environmentally Induced Cracking" in this Volume.

Figure 2 also suggests the relative influences of the electrochemical and mechanical driving forces in the SCC process. Figure 2 indicates a change as SCC proceeds, with the role of stress being negligible at first and then becoming dominant as subcritical cracking advances. Environmental action must always be involved, although it may be dominant only at first. The preexistence of a mechanical flaw or crack in the stressed metal may of course alter the initiation stage. Application of the fracture mechanics based stress intensity factor (*J* for elastic-plastic fracture mechanics; *K* for linear elastic fracture mechanics) as a driving force for the propagation of SCC is illustrated schematically in Fig. 2 and 3 (Ref 3, 4).

Standardization of Tests

Standardization of SCC test methods in the United States was initiated in the 1960s by the American Society for Testing and Materials (ASTM), the National Association of Corrosion Engineers (NACE), and the federal government. Standard tests have also been developed in Europe (Ref 5), and uniform testing methods are currently under development on a broader basis through the International Organization for Standardization (ISO). Reference will be made throughout this article to available standards and to useful publications for details on the test methods.

There are several essential factors that must be given careful consideration in the design of all types of SCC tests:

- The composition of the test environment must remain constant throughout the test, unless changes are a part of the corrosion system of interest
- The materials used for SCC test fixtures must resist attack
- Stressing fixtures must remain dimensionally stable so as not to affect the stress placed on specimens during the test
- Galvanic action between the test specimens and ancillary equipment must be avoided; such action, if present, can either accelerate or retard SCC, depending on whether there is anodic or cathodic control

Static Loading of Smooth Specimens

Tests for predicting the stress-corrosion performance of an alloy in a particular service application should be conducted with a stress system similar to that anticipated in service. Table 1 lists the numerous sources of sustained tension that are known to have initiated SCC in service and the applicable methods of stressing. Most of the SCC service problems involve tensile stresses of unknown magnitude that are usually very high. The tests that incorporate a high total strain are usually the most realistic in terms of duplicating service.

The results are strongly influenced by the mechanical aspects of the tests, such as method of loading and specimen size. These mechanical aspects can have variable effects on the initiation and propagation lifetimes and can influence estimates of a threshold stress. Therefore, an apparent threshold stress for SCC is not a material property, and threshold estimates must be qualified with regard to the test conditions and the significance level.

Table 1 Stressing methods applicable to various sources of sustained tension in service

Source of sustained tension in service	Constant strain	Constant load
Residual stress		
Quenching after heat treatment	X	...
Forming	X	...
Welding	X	...
Misalignment (fit-up stresses)	X	...
Interference fasteners	X	...
Interference bushings		
Rigid	X	...
Flexible	...	X
Flareless fittings	X	...
Clamps	X	...
Hydraulic pressure	X	X
Dead weight	...	X
Faying surface corrosion	X	X

Note: The greatest hazard arises when residual, assembly, and operating stresses are additive.

Constant-Strain Versus Constant-Load Tests

Constant-strain (fixed-displacement) tests are widely used, primarily because a variety of simple and inexpensive stressing jigs can be devised. However, there is poor reproducibility of the exposure stress with some of these techniques. Therefore, sophisticated procedures have been developed to improve this facet of testing.

Constant-strain tests are sometimes called decreasing-load tests, because after the onset of SCC in small test specimens the gross section exposure stress decreases. This results from the opening of the crack (or cracks) under the high stress concentration at the crack tip (or tips) and causes some of the applied elastic strain to change to plastic strain, with an attendant reduction in

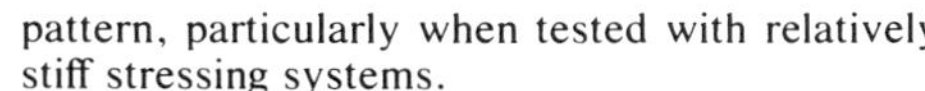

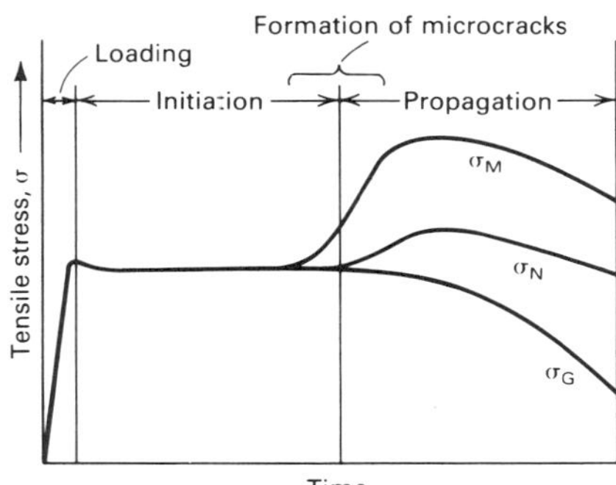

(a)

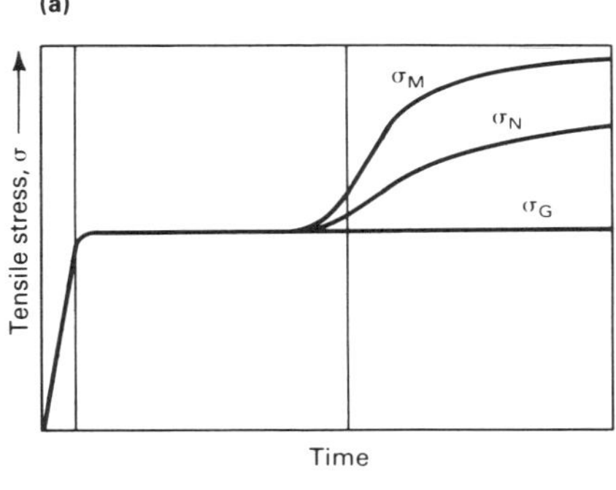

(b)

Fig. 4 Schematic comparison of changing stress during initiation and growth of isolated SCC in constant-strain and constant-load tests of a uniaxially loaded tension specimen. (a) Constant-strain test. (b) Constant-load test. σ_M is the maximum stress at crack tip, σ_N is the average stress in the net section, and σ_G is the applied stress to the gross section. Source: Ref 7

the initial load (Ref 6, 7). Such trends in changing stress during crack growth are shown in Fig. 4.

Comparison of the stress trends for a constant-strain test (Fig. 4a) with those for a constant-load test (Fig 4b) reveals that neither method of loading provides a constant-stress test after growth of microcracks has occurred. True constant-load (dead-load) tests result in increasing stress levels as cracking progresses, and are more likely to lead to earlier failure with complete fracture and lower estimates of a threshold stress than constant-strain tests. Figures 4(a) and (b) illustrate basic trends that may be applied to all types of test specimens, including precracked specimens. Specific curves, however, will differ depending on other test conditions.

The stiffness of the combined stressing frame/test specimen system can have a significant effect on materials evaluation if identical test procedures are not used (Ref 6). Many so-called constant-strain tests, particularly if a spring is included in the stressing system, are not actually constant-strain tests, because a significant amount of elastic strain energy may be contained in the stressing system. Depending on the "softness" of the spring or the elasticity of the stressing jig, the stiffness (compliance) of the stressing system can be varied greatly between zero stiffness (dead load) and infinite stiffness (true constant total strain). Figure 5 shows the typical change in net section stress with the onset of SCC in an intermediate-stiffness stressing frame.

The corrosion pattern on the test specimen, particularly the number and distribution of cracks, can impair the precision of results obtained by either constant-strain or constant-load tests. When isolated stress-corrosion cracks propagate in a specimen stressed by either method, the average tensile stress on the net section increases rapidly until the notch fracture strength is reached and the specimen breaks (Fig. 5a). Less penetration is required for fracture of specimens under dead load; this indicates that specimen life is shorter with lower-stiffness stressing frames. When microcracks initiate close to one another, their individual stress concentrations interact and are relaxed. Consequently, there may not be a sufficient stress concentration in the true constant-strain test to propagate further SCC, and the specimen will not break (Fig. 5b). Under a constant load, however, the growth of many cracks continues, and the specimen ultimately breaks.

With general cracking, crack propagation can be strongly influenced by frame stiffness. Therefore, SCC comparison of specimens tested at stress levels just above their thresholds is complicated by random variations in the cracking pattern, particularly when tested with relatively stiff stressing systems.

Although constant-load stressing appears to be advantageous for testing materials with relatively high resistance to SCC, difficulties arise when small-diameter specimens are utilized to avoid the use of massive loads or lever systems. In some test environments, highly stressed specimens may fail from general or pitting corrosion and an attendant increase in the effective stress. Such non-SCC failures complicate interpretation of test results, unless failure by SCC is confirmed by metallographic examination. Such extraneous failures are less likely to occur with specimens loaded under constant strain.

Therefore, small test specimens, which are generally preferred for laboratory screening tests and research studies, must be used with caution when estimates of serviceability are required. To determine serviceability, larger specimens should be used, and a stressing system should be selected that best duplicates the anticipated service conditions.

Bending Versus Uniaxial Tension

Historically, the most extensively used stressing systems have incorporated constant-deformation specimens stressed by bending. This method is versatile because of the variety of simple techniques that can be used to test most metal products in all types of corrosive environments. The state of stress in a bend specimen, however, is much more complex than in a tension specimen. Theoretically, tensile stress is uniform throughout the cross section in the tension specimen, except at corners in rectangular sections, but the tensile stress in bend specimens varies through the specimen thickness.

Tensile stress is at a maximum on the convex surface and decreases steeply to zero at the neutral axis. It then changes to a compressive stress, which reaches a maximum on the concave surface. Thus, only about 50% of the metal surface is under tension, and stress can vary from maximum to zero, depending on the stressing system. As SCC penetrates the metal, the stress gradient through the section thickness produces changes in stresses and strains that are different from those in a uniaxial tension specimen. This tendency yields significantly different SCC responses for the two types of stressing (Fig. 6).

Bending stress specimens experience other sources of variability in stress that are not present with direct tension stressing. Variations occur in the principal longitudinal stress across the width of the specimen as well as with the pres-

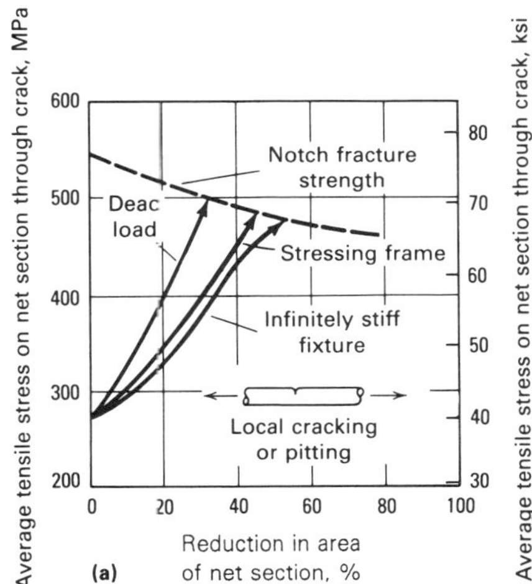

(a)

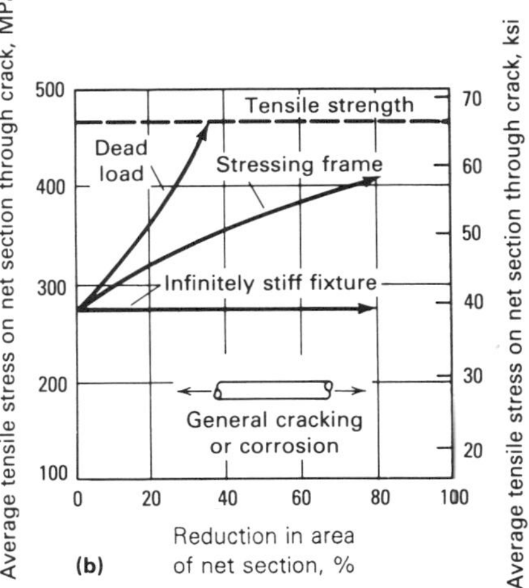

(b)

Fig. 5 Effect of loading method and extent of cracking or corrosion pattern on average net section stress in a uniaxially loaded tension specimen. Behavior is generally representative, but curves will vary with specific alloys and tempers. (a) Localized cracking. (b) General cracking. Source: Ref 8 (ASTM G 49)

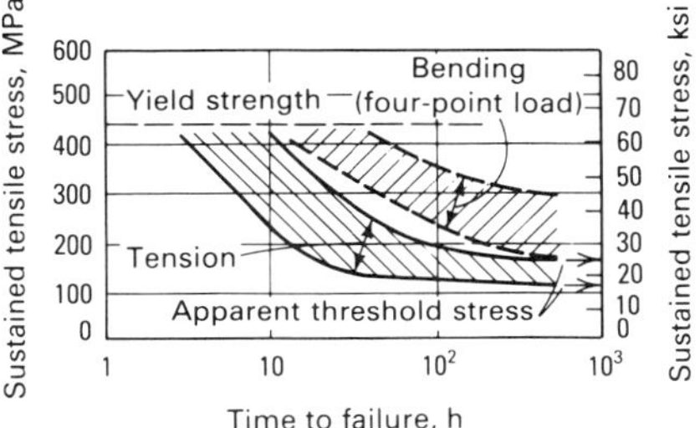

Fig. 6 Comparison of the SCC response with bending versus direct tension stressing under constant load for Al-5.3Zn-3.7Mg-0.3Mn-0.1Cr T6 temper alloy sheet. Tested to failure in 3% sodium chloride plus 0.1% hydrogen peroxide. Source: Ref 9

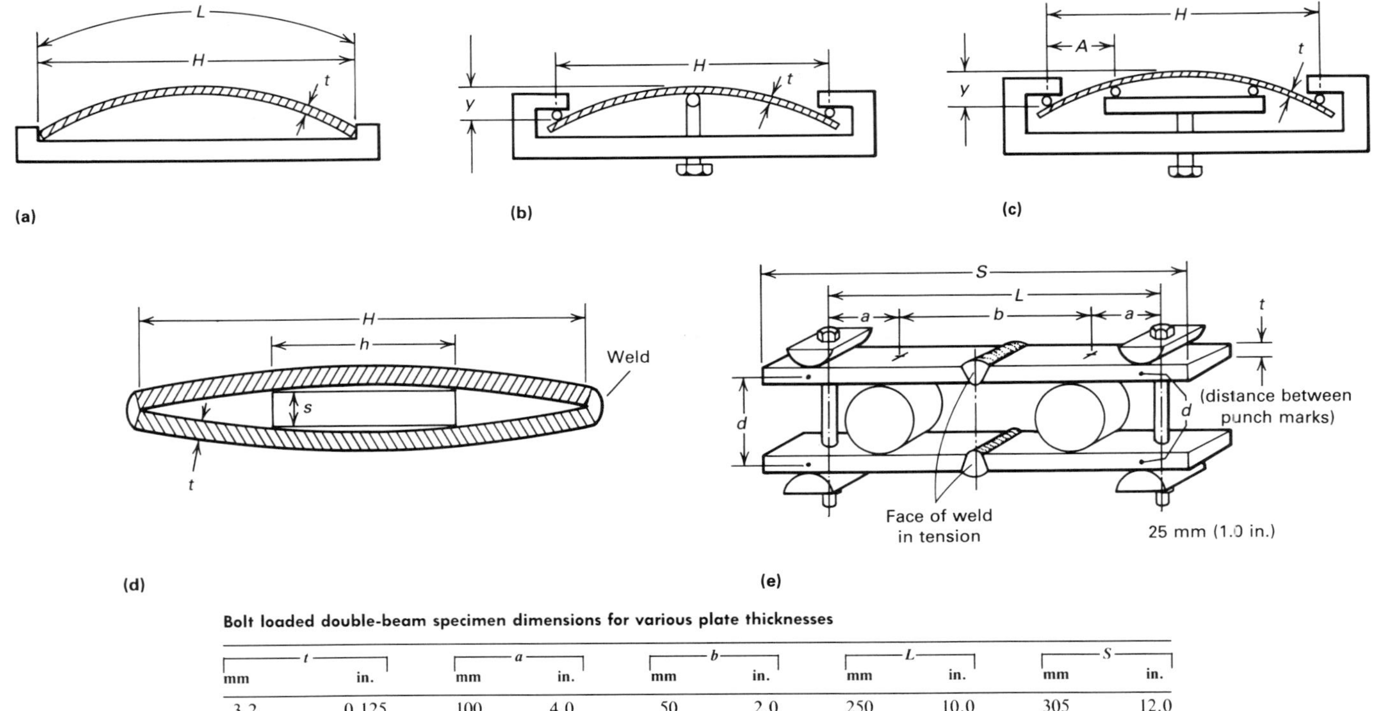

Bolt loaded double-beam specimen dimensions for various plate thicknesses

t, mm	t, in.	a, mm	a, in.	b, mm	b, in.	L, mm	L, in.	S, mm	S, in.
3.2	0.125	100	4.0	50	2.0	250	10.0	305	12.0
6.4	0.25	100	4.0	50	2.0	250	10.0	305	12.0
9.5	0.375	120	4.75	90	3.5	330	13.0	380	15.0
13	0.5	120	4.75	90	3.5	330	13.0	380	15.0
19	0.75	140	5.5	150	6.0	430	17.0	480	19.0
25	1.0	150	6.0	200	8.0	510	20.0	560	22.0
38	1.5	165	6.5	305	12.0	635	25.0	685	27.0

Fig. 7 Schematic specimen and holder configurations for bent-beam stressing. (a) Two-point loaded specimen. (b) Three-point loaded specimen. (c) Four-point loaded specimen. (d) Welded double-beam specimen. (e) Bolt-loaded double-beam specimen. Formula for stressing specimen (e): $\Delta d = 2fa/3Et(3L - 4a)$, where Δd is deflection (in inches), f is nominal stress (in pounds per square inch), and E is modulus of elasticity (in pounds per square inch). Source: Ref 10

ence of biaxial stresses, both of which are influenced by the design of the specimen. Therefore, just as in the case of constant-load stressing, optimal control of stress and more severe testing conditions are provided by uniaxial tension stressing.

Statically loaded, smooth test specimens for SCC tests can be divided into three general categories: elastic strain specimens, plastic strain specimens, and residual stress specimens. The commonly used specimen geometries for each of these categories are discussed below.

Elastic Strain Specimens

To control the surface tensile stress applied by deformation loading, strain is usually restricted to the elastic range for the test material. The magnitude of the applied stress can then be calculated from the measured strain and modulus of elasticity. In constant-load stressing, the load typically is measured directly, and the stress is calculated by using the appropriate formula for the specimen configuration and the method of loading. Load cells or calibrated springs may be useful for applying and monitoring possible changes in load during the test. The commonly used types of specimens for tests under elastic-range stress are described below.

Bent-beam specimens can be used to test a variety of product forms. The bent-beam configuration is primarily used for sheet, plate, or flat extruded sections, which conveniently provide flat specimens of rectangular cross section, but it is also used for cast materials, rod, pipe, or machined specimens of circular cross section. This method is applicable to specimens of any metal that are stressed to levels less than the elastic limit of the material; therefore, the applied stress can be calculated or measured accurately (ASTM G 39) (Ref 8).

Stress calculations by this method are not applicable to plastically stressed specimens. Bent-beam specimens are usually tested under constant-strain conditions, but constant-load conditions can also be used. In either case, local changes in the curvature of the specimen when cracking occurs result in changes in stress and strain during crack propagation. The "test stress" is taken as the highest surface tensile stress existing at the start of the test, that is, before the initiation of SCC.

Several configurations of bent-beam specimens and stressing systems are illustrated in Fig. 7 and are described in detail in ASTM G 39 (Ref 8). When specimens are tested at elevated temperatures, the possibility of stress relaxation should be investigated. More information on stress relaxation is available in the article "Creep, Stress-Rupture, and Stress-Relaxation Testing" in Volume 8 of the 9th Edition of *Metals Handbook*.

Two-point loaded specimens can be used for materials that do not deform plastically when bent to $(L - H)/H = 0.01$. The specimens should be approximately 25- × 250-mm (1- × 10-in.) flat strips cut to appropriate lengths to produce the desired stress after bending, as shown in Fig. 7(a). The maximum stress occurs at the midlength of the specimen and decreases to zero at specimen ends.

Three-point loaded specimens are flat strips that are typically 25 to 51 mm (1 to 2 in.) wide and 127 to 254 mm (5 to 10 in.) long. The thickness of a specimen is usually dictated by the mechanical properties of the material and the available product form. The specimen should be supported at the ends and bent by forcing a screw (equipped with a ball or knife-edge tip) against it at a point halfway between the end supports, as shown in Fig 7(b). In a three-point loaded specimen, the maximum stress occurs at the midlength of the specimen and decreases linearly to zero at the outer supports.

Two- and four-point loaded specimens are often preferred over the three-point loaded specimen, because crevice corrosion often occurs at the central support of the three-point loaded specimen. Because this corrosion site is very close to the point of highest tensile stress, it may cathodically protect the specimen and prevent possible crack formation, or it may cause hydrogen embrittlement. Furthermore, the pressure of the central support at the point of highest load introduces biaxial stresses at the area of contact and can introduce tensile stresses where compressive stresses are normally present.

Four-point loaded specimens are flat strips that are typically 25 to 51 mm (1 to 2 in.) wide and 127 to 254 mm (5 to 10 in.) long. The thickness of a specimen is usually dictated by the mechanical properties of the material and the available product form. The specimen is supported at the ends and is bent by forcing two inner supports against it, as shown in Fig. 7(c). The two inner supports are located symmetrically around the midpoint of the specimen.

In a four-point loaded specimen, the maximum stress occurs between the contact points of the inner supports; the stress is uniform in this area. From the inner supports, the stress decreases linearly toward zero at the outer supports. The four-point loaded specimen is preferred over the three-point and two-point loaded specimens, because it provides a large area of uniform stress.

Welded double-beam specimens consist of two flat strips 25 to 51 mm (1 to 2 in.) wide and 127 to 254 mm (5 to 10 in.) long. The strips are bent against each other over a centrally located spacer until both ends touch. The strips are held in position by welding the ends together, as shown in Fig. 7(d).

In a welded double-beam specimen, the maximum stress occurs between the contact points of the spacer; the stress is uniform in this area. From contact with the spacer, the stress decreases linearly toward zero at the ends of the specimen.

A bolt-loaded double-beam specimen is shown in Fig. 7(e), along with suggested specimen dimensions for various thicknesses of plate and the formula for stressing such specimens (Ref 10). The beam deflections required to develop the intended tensile stress are calculated with the formula and are then applied by bolting the ends of the beams together. The deflections are measured with a dial gage to within ±0.0127 mm (±0.0005 in.). Thus, the error in stress application—if the beams are of homogeneous material and if the cross sections are uniform—is within 2%. The precision of the deflection measurement is within 0.5%, and the error in determining the modulus of elasticity, E, is within 1%.

Constant-moment beam specimens are designed such that a constant moment exists from one end to the other when the specimen is bent in the manner shown in Fig. 8. This bending produces equal stress along the length of the specimen. The width-to-thickness ratio is less than 4 so that biaxial stresses are eliminated.

This type of specimen offers the advantage of a relatively large area of material under a uniform stress. Such specimens can be used when the dimensions of the specimen are too small for other bent-beam specimens—for example, when specimens are taken in the short-transverse direction in plate (see Fig. 18c). The elastic stress σ in the convex surface is calculated by using:

$$\sigma = \frac{4Ety}{h^2} \quad \text{(Eq 1)}$$

where h is the distance between inner edges of the supports, y is the maximum deflection between inner edges of the supports, t is the thickness of the specimen, and E is the modulus of elasticity.

C-Ring Specimens. As discussed in ASTM G 38, (Ref 8), the C-ring is a versatile, economical specimen for quantitatively determining the susceptibility to SCC of all types of alloys in a wide variety of product forms. It is particularly well suited for testing tubing and for making short-transverse tests on various product forms, as shown in Fig. 9. The sizes of C-rings can be varied over a wide range, but rings with outside diameters less than about 16 mm (5/8 in.) are not recommended because of increased difficulties in machining and decreased precision in stressing.

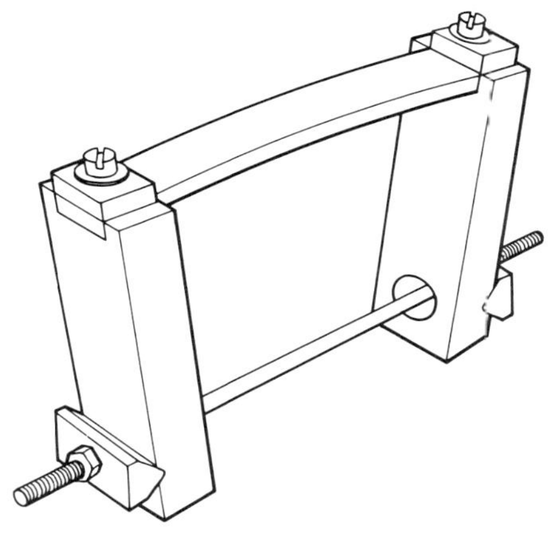

Fig. 8 Bent beam designed to produce pure bending. Source: Ref 11

The C-ring is typically a constant-strain specimen with tensile stress produced on the exterior of the ring by the tightening of a bolt centered on the diameter of the ring. However, an almost constant load can be developed by placing a calibrated spring on the loading bolt. C-rings can also be stressed in the reverse direction by spreading the ring and creating a tensile stress on the inside surface. These methods of stressing are illustrated in Fig. 10.

Circumferential stress is of principal interest in the C-ring specimen. This stress is not uniform (Ref 12), as discussed previously in the section "Elastic Strain Specimens" in this article. The stress varies around the circumference of the C-ring from zero at each bolt hole to a maximum at the middle of the arc opposite the stressing bolt. In a notched C-ring, a triaxial stress state is present adjacent to the root of the notch (Ref 13). For all notches, the circumferential stress at the root of the notch is greater than the nominal stress and can generally be expected to be in the plastic range.

Generally, the C-ring can be stressed with high precision. The most accurate stressing procedure consists of attaching circumferential and transverse electrical strain gages to the surface stressed in tension, followed by tightening the bolt until the strain measurements indicate the desired circumferential stress.

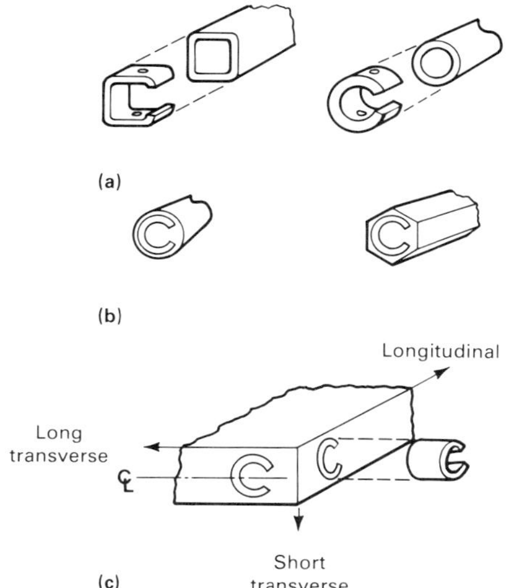

Fig. 9 Sampling procedure for testing various products with C-rings. (a) Tube. (b) Rod and bar. (c) Plate

The amount of compression required on the C-ring to produce elastic straining and the degree of elastic strain can be predicted theoretically. Therefore, C-rings can be stressed by calculating the deflection required to develop a desired elastic stress (ASTM G 38) (Ref 8). In notched specimens, a nominal stress is estimated using a ring outside diameter measured at the root of the notch and by taking into consideration the stress-concentration factor, K_t, for the specific notch.

O-ring specimens (Fig. 11) are used to develop a hoop stress in a particular part—for example, a cylindrical die forging in which a critical end-grain structure associated with the parting plane of the forging exists only at the surface of the forging. A relatively large surface area of metal is placed under a uniform tensile stress, and the O-ring stressing plug assembly simulates service conditions in structures containing interference-fit components. Stressed O-rings have also been used to evaluate protective treatments for the prevention of SCC (Ref 14).

An O-ring is stressed by pressing it onto an oversized plug that is machined to a predetermined diameter to develop the desired stress at

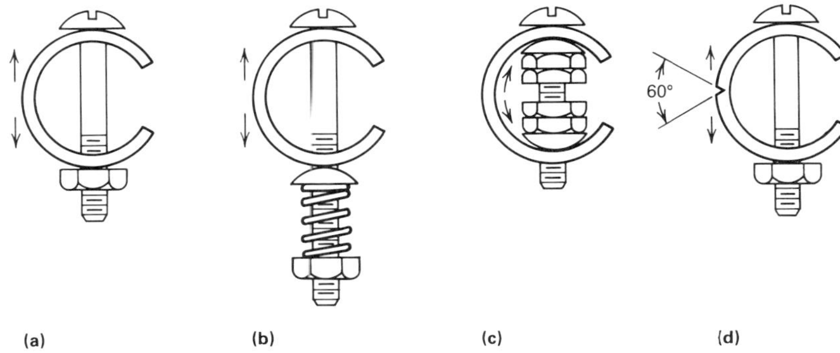

Fig. 10 Methods of stressing C-rings. (a) Constant strain. (b) Constant load. (c) Constant strain. (d) Notched C-ring; a similar notch could be used on the side of (a), (b), or (c).

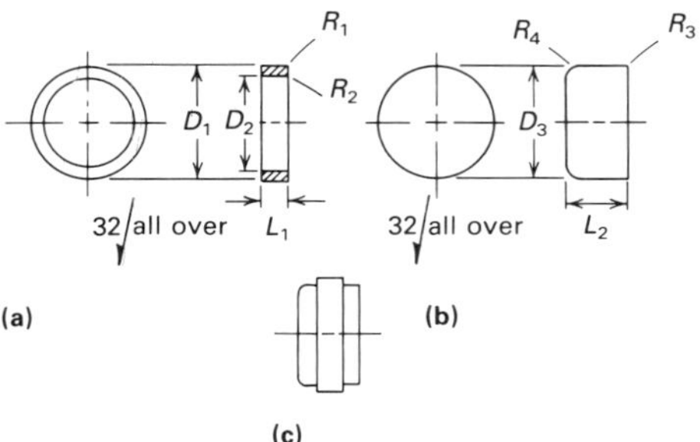

Fig. 11 O-ring SCC test specimen (a) and stressing plug (b). The O-ring is stressed by pressing it onto the plug, as shown in (c).

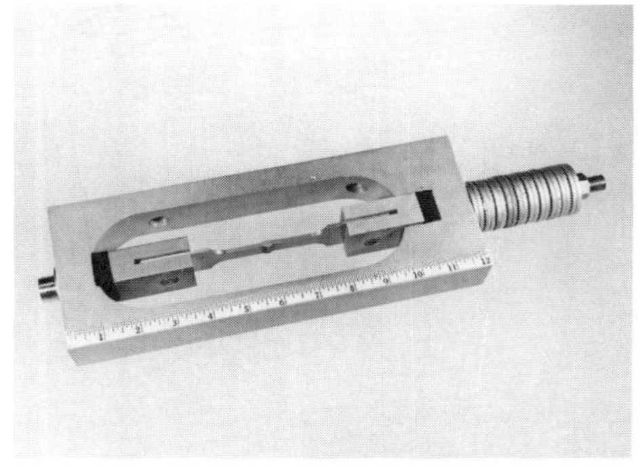

Fig. 12 Spring-loaded fixture used to stress 3.2-mm (0.125-in.) thick sheet tensile specimens in direct tension. Source: Ref 10

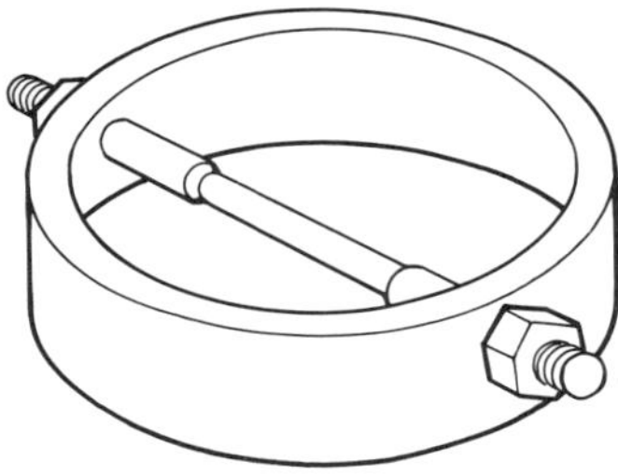

Fig. 13 Ring-stressed tension specimen for field testing. Source: Ref 1

the outside surface of the ring. The nominal dimensions of this specimen can be varied to suit the part being tested, but certain characteristics should be observed to achieve adequate control of the stresses. The ring width should not be more than four times the wall thickness in order to ensure maximum uniformity of the hoop stress from the centerline to the edges of the ring. The tensile stress varies through the thickness of the ring and is highest at the inside surface. Interference required for stressing an O-ring can be calculated by using:

$$I = \frac{F(\text{OD})^2}{E(\text{ID})} \qquad \text{(Eq 2)}$$

where I is the interference (on the diameter) between the O-ring and the plug, E is the modulus of elasticity, ID is the inside diameter, OD is the outside diameter, and F is the circumferential stress desired on the outside surface. Additional information regarding the design and stressing of O-ring specimens is given in Ref 15.

Tension Specimens. Specimens used to determine tensile properties in air are well suited and easily adapted to SCC, as discussed in ASTM G 49 (Ref 8). When uniaxially loaded in tension, the stress pattern is simple and uniform, and the magnitude of the applied stress can be accurately determined. Specimens can be quantitatively stressed by using equipment for application of either a constant load, a constant strain, or an increasing load or strain.

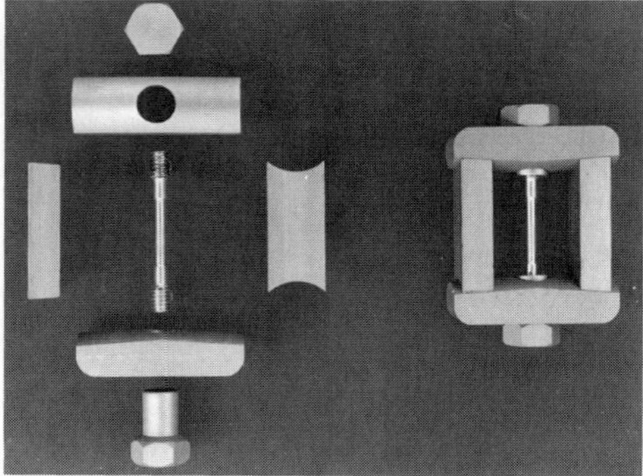

Fig. 14(a) Constant-strain SCC testing frame. Exploded view (left) showing the 3.2-mm (0.125-in.) diam tension specimen and various parts of the stressing frame. Final stressed assembly (right). Source: Ref 16

This type of test is one of the most versatile methods of SCC testing because of the flexibility permitted in the type and size of the test specimen, the stressing procedures, and the range of stress level. It allows the simultaneous exposure of unstressed specimens (no applied load) with stressed specimens and subsequent tension testing to distinguish between the effects of true SCC and mechanical overload.

A wide range of test specimen sizes can be used, depending primarily on the dimensions of the product to be tested. Stress-corrosion test results can be significantly influenced by the cross section of the test specimen. Although large specimens may be more representative of most structures, they often cannot be prepared from the available product forms being evaluated. They also present more difficulties in stressing and handling in laboratory testing.

Smaller cross-sectional specimens are widely used. They have a greater sensitivity to SCC initiation, usually yield test results rapidly, and permit greater convenience in testing. However, the smaller specimens are more difficult to machine, and test results are more likely to be influenced by extraneous stress concentrations resulting from nonaxial loading, corrosion pits, and so on. Therefore, use of specimens less than about 10 mm (0.4 in.) in gage length and 3 mm (0.12 in.) in diameter is not recommended, except when testing wire specimens.

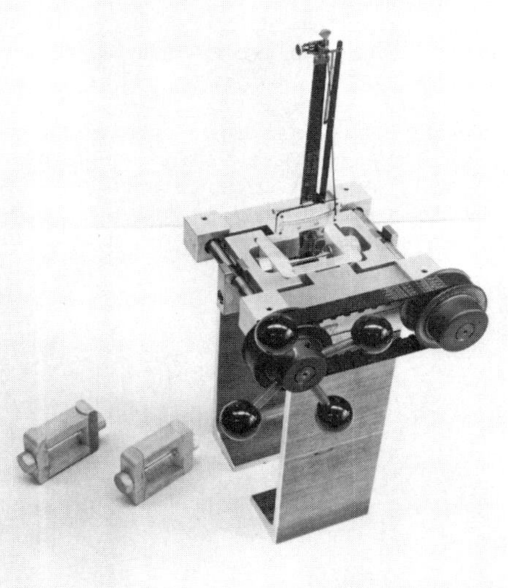

Fig. 14(b) Synchronous loading device used to stress specimens. The specimen is loaded to a prescribed strain value determined from a clip-on gage. The applied stress is given by the product of the strain and the material elastic modulus. A stressed assembly and one assembled finger-tight ready for stressing are shown.

Tension specimens containing machined notches can be used to study SCC and hydrogen embrittlement. The presence of a notch induces a triaxial stress state at the root of the notch, in which the actual stress will be greater by a concentration factor that is dependent on the notch geometry. The advantages of such specimens include the localization of cracking to the notch region and acceleration of failure. However, unless directly related to practical service conditions, the results may not be relevant.

Tension specimens can be subjected to a wide range of stress levels associated with either elastic or plastic strain. Because the stress system is intended to be essentially uniaxial (except in the case of notched specimens), great care must be exercised in the construction of stressing frames to prevent or minimize bending or torsional stresses.

The simplest method of providing a constant load consists of a dead weight hung on one end of the specimen. This method is particularly useful for wire specimens. For specimens of larger cross section, however, lever systems such as those used in creep-testing machines are more practical. The primary advantage of any dead-weight loading device is the constancy of the applied load.

A constant-load system can be modified by the use of a calibrated spring, such as that shown in Fig. 12. The proving ring, as used in the calibration of tension testing machines, has also been adapted to SCC testing to provide a simple, compact, easily operated device for applying axial load (Fig. 13). The load is applied by tightening a nut on one of the bolts and is determined by carefully measuring the change in ring diameter.

Constant-strain SCC tests are performed in low-compliance tension-testing machines. The specimen is loaded to the required stress level, and the moving beam is then locked in position. Other laboratory stressing frames have been used, generally for testing specimens of smaller cross section. Figure 14(a) shows an exploded view of such a stressing frame, and Fig. 14(b) illustrates a special loading device developed to ensure axial loading with minimal torsion and bending of the specimen.

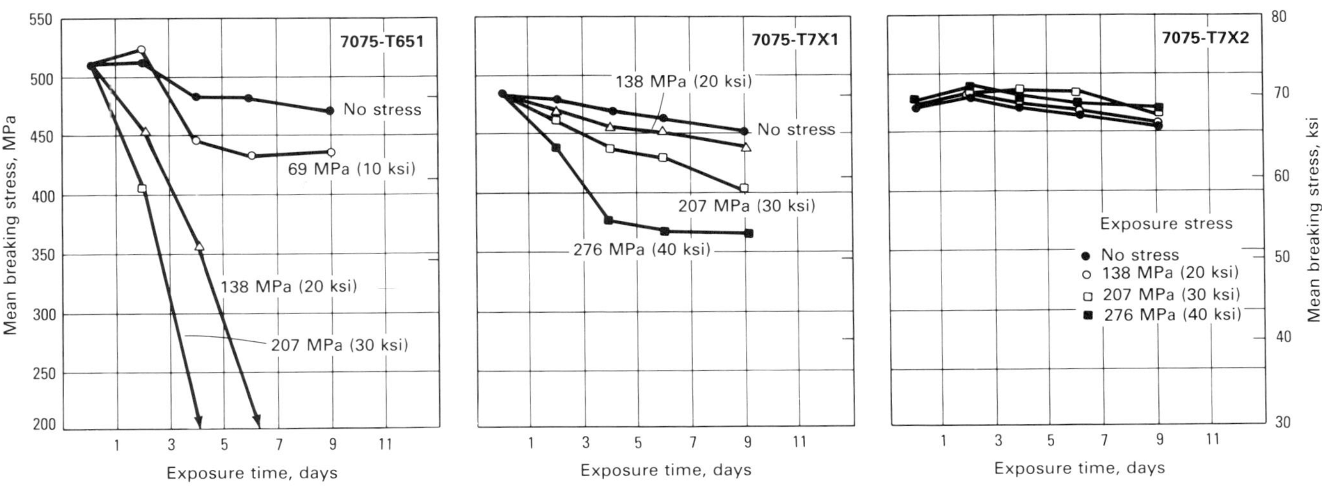

Fig. 15 Mean breaking stress versus exposure time for short-transverse 3.2-mm (0.125-in.) diam aluminum alloy 7075 tension specimens tested according to ASTM G 44 at various exposure stress levels. Each point represents an average of five specimens. Source: Ref 3

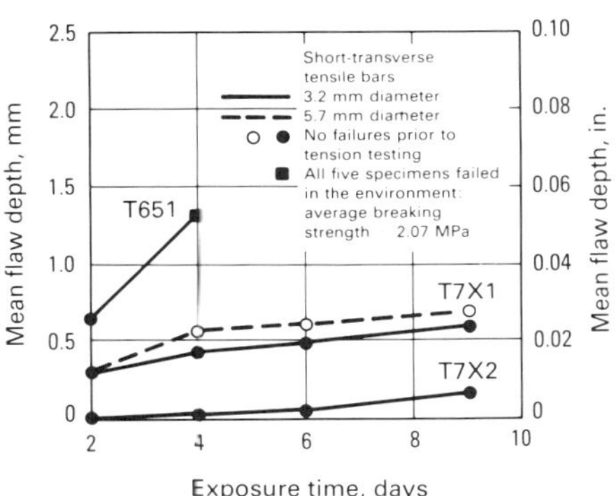

Fig. 16 Effect of temper on SCC performance of aluminum alloy 7075 subjected to alternate immersion in 3.5% NaCl solution at a stress of 207 MPa (30 ksi). Mean flaw depth was calculated from the average breaking strength of five specimens subjected to identical conditions. Source: Ref 17

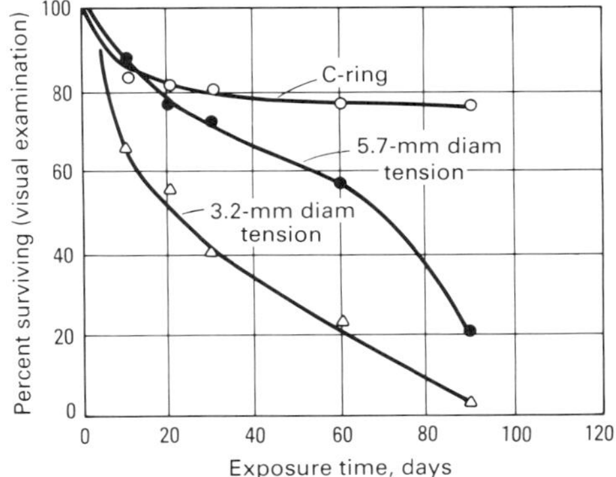

Fig. 17 Influence of specimen configuration on SCC test performance (alternate immersion in 3.5% sodium chloride per ASTM G 44). Aluminum alloy 7075-T7X51 specimens stressed 310 MPa (45 ksi); each point represents 60 to 90 specimens. Source: Ref 18

For stressing frames that do not contain any mechanism for the measurement of load, the stress level can be determined from measurement of the strain. However, only when the intended stress is below the elastic limit of the test material is the average linear stress (σ) proportional to the average linear strain (ϵ), $\sigma/\epsilon = E$, where E is the modulus of elasticity.

When tests are conducted at elevated temperatures with constant-strain loaded specimens, consideration should be given to the possibility of stress relaxation. When stress relaxation or creep occurs in the test specimen, some of the elastic strain is converted to plastic strain and the nominal applied test stress is reduced. This effect is particularly important when the coefficients of thermal expansion are different for the specimen and stressing frame. Frequently, nonmetallic (plastic) insulators are used between the specimen and stressing frame to avoid galvanic action. If such plastic insulators are part of the stress-bearing system, creep (even at room temperature) can significantly alter the applied load on the specimen.

Even though eccentricity in loading can be minimized to levels acceptable for tension-testing machines, tensile stress around the circumference of round specimens and at the corners of sheet-type specimens varies to some extent. Several factors may introduce bending moments on specimens, such as longitudinal curvature and misalignment of threads on threaded-end round specimens. These factors have a greater effect on specimens with smaller cross sections. Tests should be made on specimens with strain gages affixed to the specimen surface around the circumference in 90° or 120° intervals to verify strain and stress uniformity and to determine if machining practices and stressing jigs are of adequate tolerance and quality.

When SCC occurs, it generally results in complete fracture of the specimen, which is easy to detect. However, when testing relatively ductile materials at stress levels close to the threshold of susceptibility, fracture may not occur during the period of exposure. The presence of SCC in such cases must be determined by mechanical tests or by metallographic examination, as discussed previously.

To study trends in SCC susceptibility, such as in alloy development research, it is often necessary to detect small differences in susceptibility. For this purpose, it is advantageous to use replicate sets of specimens stressed at several levels, including zero applied stress. The sets are then removed for metallographic examination or tension tests after appropriate periods of exposure.

Figure 15 illustrates the use of this procedure with samples of 7075 aluminum alloy that have been given different thermal treatments to decrease susceptibility to SCC. Analysis of these breaking stress data by extreme value statistics enables calculation of survival probabilities and the estimation of a threshold stress, without depending on failures during exposure. By using an elastic-plastic fracture mechanics model, an effective flaw size is calculated from the mean breaking stress, the strength, and the fracture toughness of the test material. The effective flaw size corresponds to the weakest link in the specimen at the time of the tension test, and it therefore represents the maximum penetration of the SCC. An advantage to using flaw depth to examine SCC performance is that the effects of specimen size and alloy strength and toughness can be normalized. In contrast, the specimen lifetime and breaking strength are biased by those mechanical (non-SCC) factors.

Mean trends in the 207-MPa (30-ksi) exposure data for the three temper variants of aluminum alloy 7075 examined in Fig. 15 are shown in Fig. 16. These results clearly illustrate that the thermal treatments used to reduce the SCC susceptibility of the 7075-T651 decreased the SCC penetration (Ref 17). The equivalent performance of the 7075-T7X1 3.2- and 5.7-mm (0.125- and 0.225-in.) diam specimens is evident. In contrast, Fig. 17 shows the specimen biases in SCC ratings obtained by traditional pass-fail methods (Ref 18).

Tuning fork specimens are special-purpose specimens with numerous modifications (Fig. 18). In Europe, the metal is strained into the plastic range, and stresses and strains are usually not measured (Ref 19, 21). In the United States, however, these specimens have been used with measured strains in the elastic and plastic ranges. Specimens of the type shown in Fig. 18(b) are convenient when a small self-contained specimen is required that will afford some insight into the applied stresses. Such a specimen is particularly well suited for testing thin plate material in the longitudinal or long-transverse direction while keeping the original mill-finished surface intact.

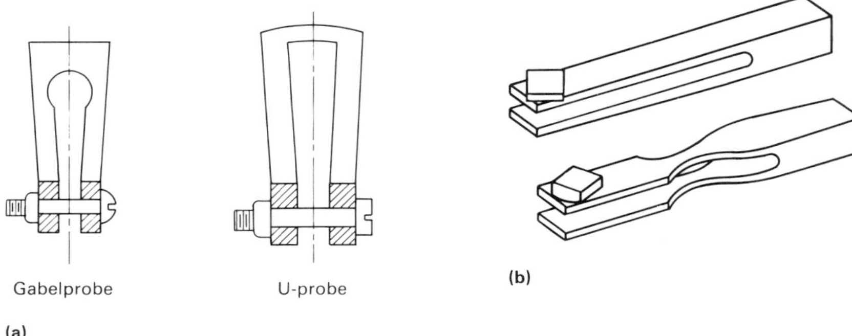

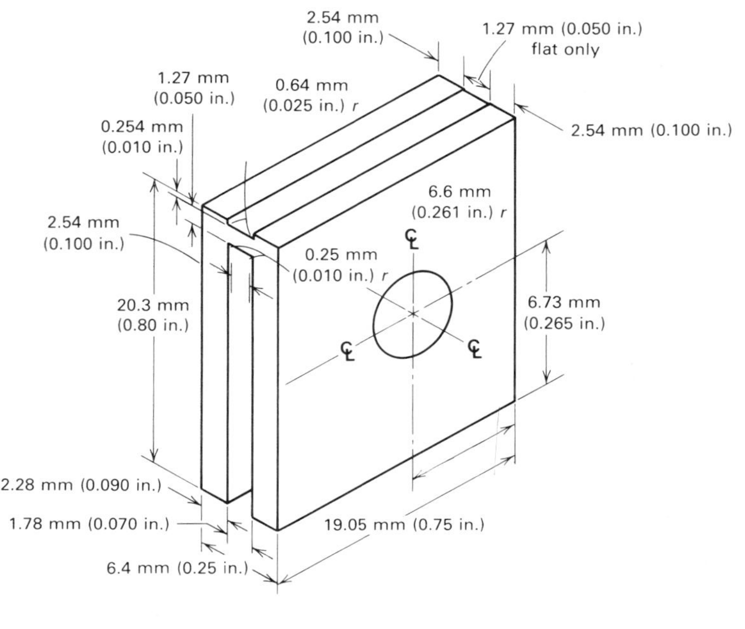

Fig. 18 Typical tuning fork SCC test specimens. (a) Source: Ref 19. (b) Source: Ref 1. (c) Source: Ref 20

Tuning fork specimens are stressed by closing the specimen tines and restraining them in the closed position with a bolt placed at the tine ends. The amount of closure is determined from Eq 3, which was derived from the data obtained with strain gages placed at the base of the tines on calibration specimens (Ref 1):

$$S = A\Delta t \qquad \text{(Eq 3)}$$

where S is the maximum tension stress in the outer fiber of either tine, A is the calibration constant, Δ is the total amount of closure at the tine ends, and t is the thickness of the tines.

The stress on tuning forks with straight tines is greatest in a small area at the base of the tines. In tuning forks with tapered tines, the maximum stress extends uniformly along the tapered section. Tuning forks must be given the same consideration with regard to biaxial stresses as other flexurally loaded specimens.

The miniature tuning fork shown in Fig. 18(c) was devised to conduct short-transverse tests on sections that are too thin for tensile specimens or C-rings to be obtained (Ref 20). As with other tuning fork specimens, the relationship between strain on the grooved surface and the deflection at the ends of the legs can be determined through the use of strain gages.

Plastic Strain Specimens

Many accelerated SCC tests are performed with plastically deformed specimens, because these specimens are simple and economical to manufacture and use. These specimens are convenient for multiple replication tests of self-stressed (fixed-deflection) specimens in all environments. Because they usually contain large amounts of elastic and plastic strain, they provide one of the most severe tests available for smooth SCC test specimens.

Generally, the stress conditions are not known precisely. However, the anticipated high level of stress can be obtained consistently only if the precautions described for each type of specimen are observed. Another consideration is that the cold work required to form the test specimen can change the metallurgical condition and the SCC behavior of certain alloys.

Tests of this type are primarily used as screening tests to detect large differences between the SCC resistance of one alloy in several environments, one alloy in several metallurgical conditions in a given environment, and different alloys in the same environment. These tests are sometimes claimed to be too severe and therefore unsuitable for many applications, but the stress conditions are nevertheless representative of the high locked-in fabrication and assembly stresses frequently responsible for SCC in service.

U-bend specimens are rectangular strips bent approximately 180° around a predetermined radius and maintained in this plastically (and elastically) deformed condition during the test. Standardized test methods for this type of specimen are described in ASTM G 30 (Ref 8). Bends slightly less than or greater than 180° are also used, but the term U-bend is generally applied to test specimens that are bent beyond their elastic limits. Figure 19 illustrates typical U-bend configurations showing several different methods of maintaining the applied stress.

U-bend specimens can be used for all materials sufficiently ductile to be formed into a U-configuration without cracking. A U-bend specimen is most easily made from strips of sheet, but specimens can be machined from plate, bar, wire, castings, and weldments. Of primary interest in U-bend specimens is circumferential stress, which is not uniform, as discussed previously in the section on "Bent-Beam Specimens" in this article. Stress distribution in the U-bend specimen is discussed in detail in Ref 22.

A good approximation of applied strain ϵ can be obtained by:

$$\epsilon = \frac{t}{2R} \text{ when } t < R \qquad \text{(Eq 4)}$$

where t is the specimen thickness, and R is the radius of curvature at the point of interest. Knowledge of the stress-strain curve is necessary to determine the stress. When a U-bend specimen is formed, the material in the outer fibers of the bend is strained into the plastic portion of the true stress/true strain curve, such as in section AB in Fig. 20(a). Several other stress-strain relationships that can exist in the outer fibers of a stressed U-bend test specimen are shown in Fig. 20(b) through (e). The actual relationship obtained depends on the method of stressing used.

Stressing is usually achieved by a one- or two-stage operation. Single-stage stressing is accomplished by bending the specimen into shape and maintaining it in that shape. The two types of stress conditions that can be obtained by single-stage stressing are defined by point X in Fig. 20(b) and (c). In Fig. 20(c), some elastic strain relaxation has occurred by allowing the U-bend legs to spring back slightly at the end of the stressing sequence.

Two-stage stressing involves forming the approximate U-shape and then allowing the elastic strain to relax completely before the second stage of applying the test stress. The applied test strain can be a percentage (from 0 to 100%) of the tensile elastic strain that occurred during preforming (Fig. 20d) or can involve additional plastic strain (Fig. 20e). The convex specimen surface is stressed in tension in the region 0NM (Fig. 20d), and the concave surface is in compression. In the region MP, the situation is reversed; that is, compression is on the convex surface, and tension is on the concave surface.

The slope MN of the curve shown in Fig. 20(d) is steep. Therefore, it is often difficult to apply reproducibly a constant percentage of the total

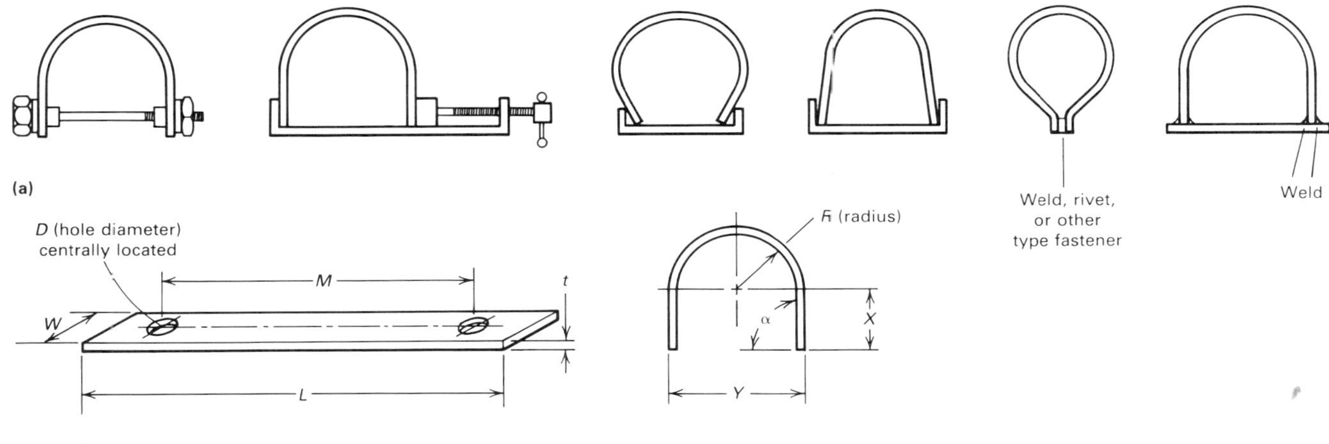

Alternative size	L		M		W		t		D		X		Y		R	
	mm	in.	mm	in.	mm	in.	mm	in.	mm	in.	mm	in.	mm	in.	mm	in.
A	80	3.2	50	2.0	20	0.8	2.5	0.098	10	0.4	32	1.26	14	0.55	5	0.2
B	100	4.0	90	3.5	9	0.35	3.0	0.12	7	0.28	25	0.98	38	1.50	16	0.6
C	120	4.7	90	3.5	20	0.8	1.5	0.06	8	0.31	35	1.4	35	1.4	16	0.6
D	130	5.1	100	4.0	15	0.6	3.0	0.12	6	0.24	45	1.77	32	1.26	13	0.51
E	150	5.9	140	5.5	15	0.6	0.8	0.03	3	0.12	61	2.40	20	0.8	9	0.35
F	310	12.2	250	9.8	25	0.98	13.0	0.51	13	0.51	105	4.13	90	3.5	32	1.26
G	510	20.1	460	18.1	25	0.98	6.5	0.26	13	0.51	136	5.35	165	6.5	76	3.0

Note: $\alpha = 1.57$ rad

Fig. 19 Typical U-bend SCC specimens. (a) Various methods of stressing U-bends. (b) Typical U-bend specimen dimensions

elastic prestrain, and the specimen surface may remain under compressive stress. Therefore, because they result in a more severe test (that is, higher applied stress), the stress conditions in Fig. 20(b) and (e) are recommended. Thus, the final applied strain prior to testing consists of plastic and elastic strain. To achieve the conditions illustrated in Fig. 20(b) and (e), springback of the U-bend legs after achieving the final plastic strain must be avoided. For materials with relatively low creep resistance, there will be some strain relaxation.

Residual Stress Specimens

Most industrial SCC problems are associated with residual stresses developed in the metal during such processes as heat treatment, fabrication, and welding. Therefore, residual stress specimens simulating anticipated service conditions are useful for assessing the SCC performance of some materials in particular structures and in specific environments.

Plastic Deformation Specimens. Residual stresses resulting from such fabricating operations as forming, straightening, and swaging that involve localized plastic deformation at room temperature can exceed the elastic limit of the material. Examples of specimens of this type that have been used are shown in Fig. 21 and 22. Other specimen types used include panels with sheared edges, punched holes, or stamped identification numbers and specimens that show evidence of other practical fabricating operations.

Weld Specimens. Residual stresses developed in and adjacent to welds are frequently a source of SCC in service. Longitudinal stresses in the vicinity of a single weld are unlikely to be as large as stresses developed in plastically deformed weldments, because stress in the weld metal is limited by the yield strength of the hot metal that shrinks as it cools. High stresses can be built up, however, when two or more weldments are joined into a more complex structure.

Test specimens containing residual welding stresses are shown in Fig. 23. In fillet welds, residual tensile stress transverse to the weld can be critical, as indicated in Fig. 23(a) for a situation in which the tension stress acts in the short-transverse direction in an aluminum-zinc-magnesium alloy plate.

Static Loading of Precracked (Fracture Mechanics) Specimens

The use of precracked (fracture mechanics) specimens is based on the concept that large structures with thick components are apt to contain cracklike defects. After a stress-corrosion crack begins to grow, or if the specimen is provided with a mechanical precrack, classical stress analysis is inadequate for determining the response of the material subjected to stress in the presence of a corrodent.

The mechanical driving force for cracks can be measured with linear elastic fracture mechanics theory in terms of the crack-tip stress intensity factor, K, which is expressed in terms of the remotely applied loads, crack depth, and test specimen geometry. At or above a certain level of K, SCC in a susceptible material will initiate and grow in certain environments, but below that level no measurable propagation is observed (Ref 23).

The apparent threshold stress intensity for the propagation of SCC (assuming that crack nuclei form in a manner that cannot be described by fracture mechanics, such as localized corrosion) is designated K_{ISCC} (or K_{th}). Therefore, in terms of linear elastic fracture mechanics theory, for a surface crack in a large plate remotely loaded in tension, the shallowest crack (of a shape that is long compared to its depth) that will propagate as a stress-corrosion crack is $a_{cr} = 0.2(K_{ISCC}/TYS)^2$, where TYS is tensile yield strength. Thus, a crack that is shallower than this critical value will not propagate under the given environmental conditions.

The value of a_{cr} incorporates the SCC resistance, K_{ISCC}, and the contribution of stress levels (of the order of the yield strength) to SCC due to residual or assembly stresses in thick component sections (Ref 4). Therefore, the application of fracture mechanics does not provide independent information about SCC; it simply provides a usable method for treating the stress factor in the presence of a crack.

When the rate of SCC propagation is determined and plotted as a function of K_I (the crack opening mode), the test results for a highly susceptible alloy will exhibit the general trend shown in Fig. 3. Actual curves vary depending on the SCC resistance and fracture toughness of the alloy.

Although precracking may shorten or modify the initiation period, it does not circumvent it. Therefore, this method of testing also requires arbitrary and sometimes long exposure periods.

Test Specimen Selection

Almost all standard plane-strain fracture toughness test specimens can be adapted to SCC testing. These standard configurations should be used to ensure valid fracture analyses. Comprehensive discussions on SCC testing with precracked specimens can be found in Ref 24-26. Precracked specimens are illustrated schematically in Fig. 24 where they are classified with respect to loading methods and the relationship with the stress intensity factor as stress-corrosion cracking propagates. Proportional dimensions and tolerances per ASTM E 399 (Ref 27) for the more commonly used specimens are given in Fig. 25. Minor modifications to accommodate different loading arrangements and to facilitate mechanical precracking can be made to these configurations without invalidating the plane-strain constraints on the specimens. Figure 26 illustrates alternative chevron-notch and face-groove designs.

Standards for SCC tests using precracked specimens have not yet been developed, although recommended test procedures have been published for certain uses (Ref 28). The best state-of-the-art stress intensity and compliance calibration relationships and guidelines for testing the specimens illustrated in Fig. 25 are discussed below. Standard names for these specimens and methods of loading per ASTM E 616, "Standard Terminology Relating to Fracture Testing," are used in Fig. 25 and in paragraph headings for the following discussion. Reference is also made to familiar names used in the literature that may appear elsewhere in this article.

Cantilever bend specimens (Fig. 25a), sometimes referred to as single-edge-notched cantilever bend specimens, have been used in constant-load tests (K-increasing) for characterizing high-strength steels and titanium alloys (Ref 28). Equations 5 to 7 are recommended (Ref 29, 30):

$$K = \frac{6M}{BW^2}\left(\frac{2W}{\pi a}\tan\frac{\pi a}{2W}\right)^{1/2} \times \left[\frac{0.923 + 0.199\left(1 - \sin\frac{\pi a}{2W}\right)^4}{\cos\frac{\pi a}{2W}}\right] \quad \text{(Eq 5)}$$

$$\frac{EBW\,(2V_0)}{M} = e^x \quad \text{(Eq 6)}$$

$$\frac{EBW\,V_{LL}}{M} = e^y \quad \text{(Eq 7)}$$

where e is the base of natural logarithm (2.718), $x = [0.1426 + 11.92(a/W) - 17.42(a/W)^2 + 15.84(a/W)^3 - 2.235(a/W)^4]$, $y = [6.188 + 12.98(a/W) - 41.19(a/W)^2 + 54.98(a/W)^3 - 22.28(a/W)^4]$. M is the applied bending moment, B is the specimen thickness (face grooves, when present, may be accounted for by replacing B with $\sqrt{BB_n}$, where B_n is the net thickness at the base of the face grooves; see Fig. 26b), W is the depth of the specimen, a is the depth of the notch plus crack, E is the modulus of elasticity, $2V_0$ is the total crack mouth opening displacement at the top face of the specimen, and V_{LL} is the total crack mouth opening displacement measured at the point of load application, which will vary depending on the load arm length.

Equation 5 is an expression for the stress intensity of a rectangular beam in pure bending and is valid over a wide range of a/W values. It applies to Mode I loading only, however, and the usual tests include a Mode II component from resulting shear stresses.

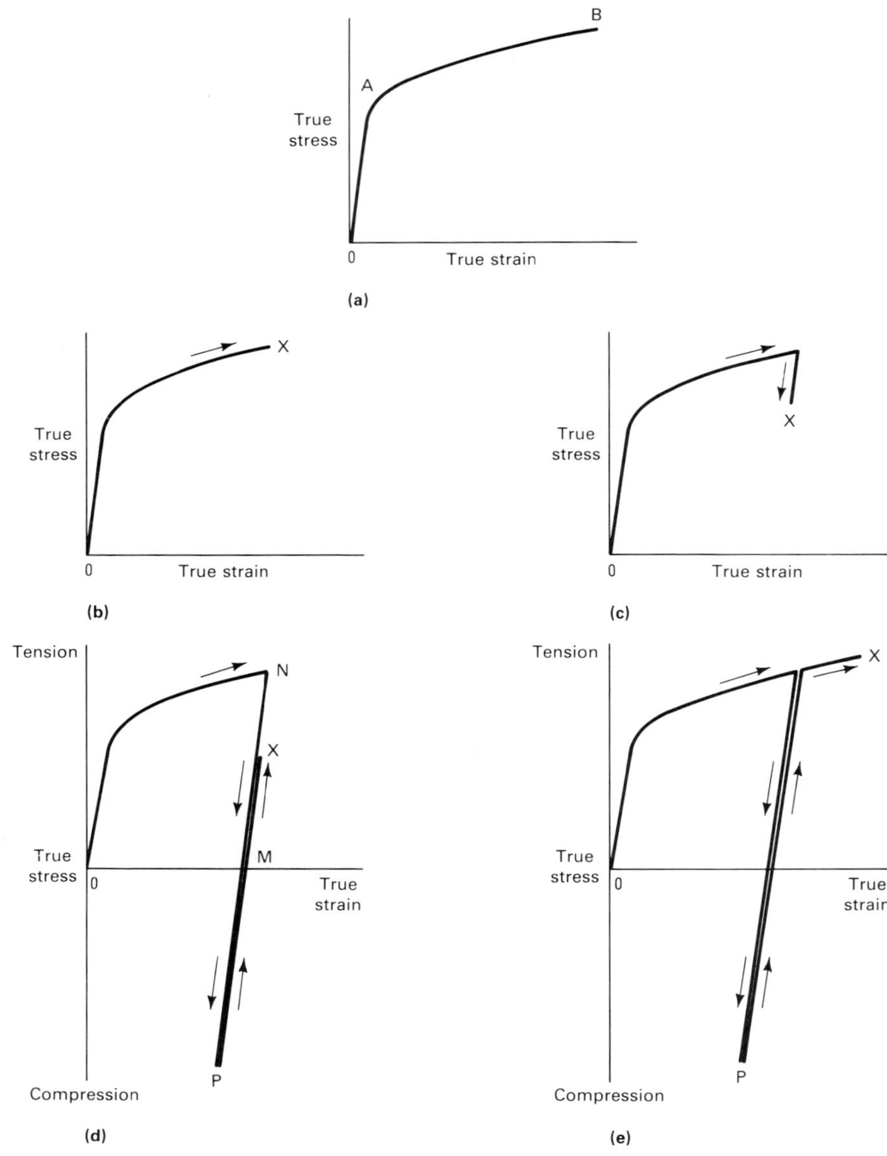

Fig. 20 True stress/true strain relationships for stressed U-bends. See text for discussion of (a) to (e).

Equations 6 and 7 were determined by fitting experimental compliance data for cantilever bend specimens with a polynomial equation expressing the natural log of the normalized compliance as a function of a/W. These experimental values are in excellent agreement with those determined from Eq 5 for pure bending, even though the stress state at the crack tip will differ for cantilever bending. It has been suggested that analyses using pure bending expressions related to compliance measurement are suitable for testing with the cantilever bend configuration (Ref 29).

Crack growth measurements can be made with clip gage readings in conjunction with the crack opening displacement calibrations given above or by any other method that can be verified within ±0.127 mm (±0.005 in.). Examples of various methods are given in Ref 29 and 31.

Modified compact specimens (K-decreasing or K-increasing), as shown in Fig. 25(b), are frequently referred to as 1T-WOL (wedge-opening loaded) or modified WOL specimens. Although most frequently used with constant-displacement (bolt) loading (Ref 28, 32), these specimens have also been used with constant load (Ref 3, 33, 34). The specimen configuration shown in Fig. 25(b) is similar to that adopted by the Navy (Ref 28), except that it does not incorporate face grooves.

Equations 8 to 11 can be used to calculate stress intensity levels and normalized crack opening displacements for fatigue precracking, for initiation of stress-corrosion testing, and for subsequent intervals during the test. These equations are based on boundary colocation values determined for this type of specimen configuration with face grooves and bolt loading (threaded bolt against a rigid loading tip) (Ref 29). The polynomial regression equation agrees with experimentally determined colocation values within 1% for $0.2 \leq a/W \leq 0.95$:

$$K_{Io} = \frac{P\left(2 + \frac{a_0}{W}\right)}{B\sqrt{W}\left(1 - \frac{a_0}{W}\right)^{3/2}} \times \left[1.308 + 5.278\left(\frac{a_0}{W}\right) - 19.67\left(\frac{a_0}{W}\right)^2 + 24.57\left(\frac{a_0}{W}\right)^3 - 10.27\left(\frac{a_0}{W}\right)^4\right] \quad \text{(Eq 8)}$$

$$2V_0 = \frac{P}{EB} \times e^x \quad \text{(Eq 9)}$$

where $x = [1.830 + 4.307(a/W) + 5.871\,(a_0/W)^2 - 17.53(a_0/W)^3 + 14.57(a_0/W)^4]$

$$2V_{LL} = 2V_0 \times \frac{e^y}{e^x} \quad \text{(Eq 10)}$$

where $y = [1.623 + 3.352(a_0/W) + 8.205(a_0/W)^2 - 19.59(a_0/W)^3 + 15.23(a_0/W)^4]$

$$K_{Ii} = \frac{E(2V_{LL})\left(2 + \frac{a_i}{W}\right)}{\sqrt{W}\left(1 - \frac{a_i}{W}\right)^{3/2}} \times \left\{\left[1.308 + 5.278\left(\frac{a_i}{W}\right) - 19.67\left(\frac{a_i}{W}\right)^2 + 24.57\left(\frac{a_i}{W}\right)^3 - 10.27\left(\frac{a_i}{W}\right)^4\right] \div e^z\right\} \quad \text{(Eq 11)}$$

where $z = [1.623 + 3.352(a_i/W) + 8.205(a_i/W)^2 - 19.59(a_i/W)^3 + 15.23(a_i/W)^4]$. In Eq 8 to 11, K_{Io} is the desired starting stress intensity, a_0 is the starting crack length, P is the load calculated to develop K_{Io} with measured a_0, W is the net width of the specimen measured from the load line, K_{Ii} is the stress intensity after time interval i, a_i is the crack length after time interval i, and $2V_{LL}$ is the total crack mouth opening displacement at the load line. All other quantities are as defined previously.

Double-beam specimens (K-decreasing or K-increasing), which are also referred to as double-cantilever beam specimens, are similar to modified compact specimens, but because of their greater width or length, they are well suited for studying SCC growth rates over a greater range of K_I values. The smaller height of these specimens (Fig. 25c) allows more versatility in performing short-transverse tests from moderate thicknesses of material. Like compact specimens, double-beam specimens are generally used with constant-displacement (bolt) loading for convenience, but they can also be used with constant load.

Bolt-loaded specimens used with a test procedure similar to that described in Ref 35 have been extensively employed for short-transverse tests of aluminum alloy products (Ref 34, 36, 37).

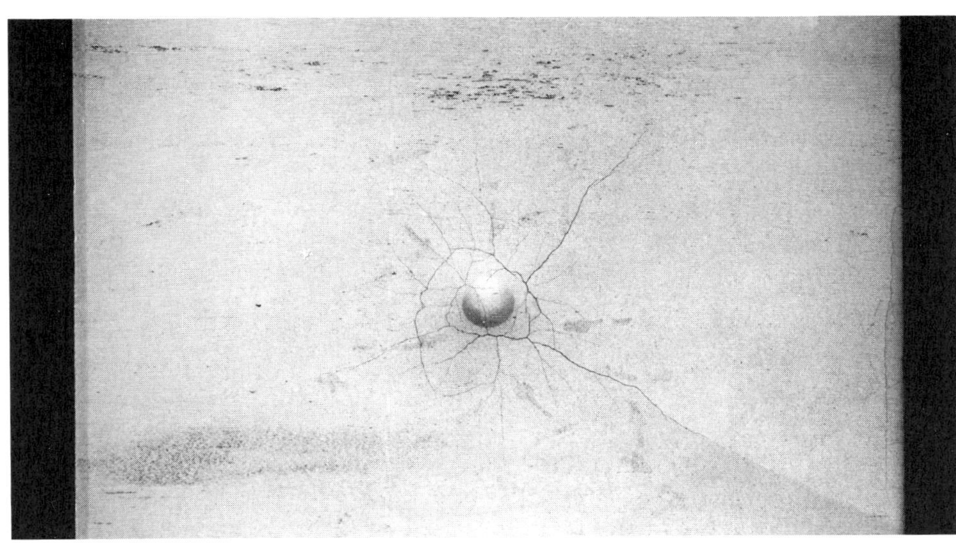

(a)

(b)

Fig. 21 SCC test specimens containing residual stresses from plastic deformation. (a) Cracked cup specimen (Ericksen impression). Source: Ref 1. (b) Joggled extrusion containing SCC in the plastically deformed region. Source: Ref 9

Equations 12 to 15 are recommended for general use with double-beam specimens:

$$K_{Io} = \frac{P\sqrt{12}}{B\sqrt{H}} \times \left[\frac{a_0}{H} + 0.6728 + 0.0377\left(\frac{H}{a_0}\right)^2\right] \quad \text{(Eq 12)}$$

$$2V_0 = \frac{4K_{Io}\sqrt{H}}{\sqrt{3}\,E}\left(\frac{a_0}{H} + 0.6728\right)^2 \times \left[1 + 1.5\left(\frac{C_0}{a_0}\right) - 1.15\left(\frac{C_0}{a_0}\right)^2\right] \quad \text{(Eq 13)}$$

$$2V_{LL} = \frac{2V_0}{1 + 1.5\left(\frac{C_0}{a_0}\right) - 1.15\left(\frac{C_0}{a_0}\right)^2} \quad \text{(Eq 14)}$$

$$K_i = \frac{\sqrt{3}\,E(2V_{LL})}{4\sqrt{H}\left(\frac{a_i}{H} + 0.6728\right)^2} \quad \text{(Eq 15)}$$

Equation 12 is an expression reported in Ref 38. The simplified Eq 15 provides more versatility with high accuracy for a wider range of specimen configurations and K values (crack growth) than equations previously published (Ref 35).

Two early K_I calibrations based on stress analysis (Ref 39) and compliance (Ref 40) are illustrated in Fig. 27 and are in excellent agreement. The shape of these curves can also be used as a design guide for preparing specimens. If the test must be completed in the shortest possible time, a_0 should be short to capitalize on the fact that the rate of decrease of K_I with crack extension is maximum for shallow cracks. However, if maximum accuracy is desired, a deeper crack (effective notch length M, Fig. 25c) should be chosen so that errors in crack length measurement do not cause significant errors in K_I.

Although in early work with aluminum alloys (Ref 35, 36) a relatively short effective notch length was used ($a_0/H \simeq 0.9$), deeper notches have been used recently ($a_0/H \simeq 1.2$ to 2.2), all with a $2H$ value of 25.4 mm (1.0 in.) (Ref 3, 34, 36,

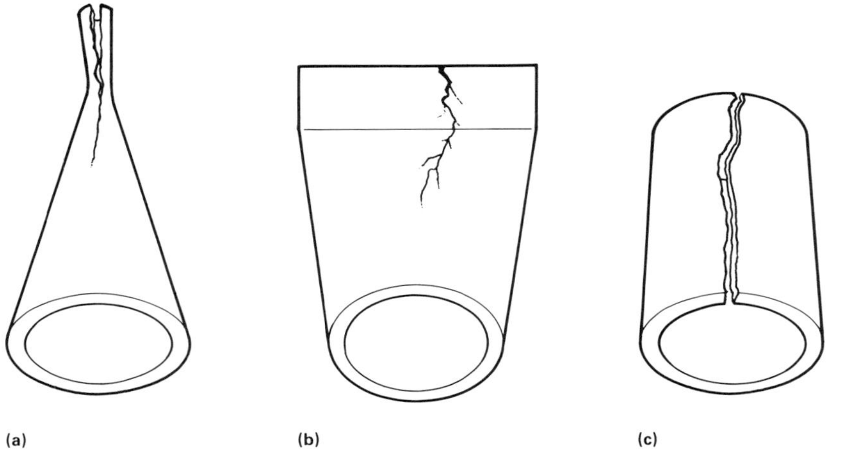

Fig. 22 SCC test specimens containing residual stresses from plastic deformation. Shown are 12.7-mm (0.5-in.) diam stainless steel tubular specimens after SCC testing. (a) and (b) Annealed tubing that was cold-formed before testing. (c) Cold worked tubing tested in the as-received condition. Source: Ref 1

37). The recommended starting a/H value shown in Fig. 25(c) is about 2 to 2.2, depending on the length of the precrack. Limited tests of a smaller beam height of $2H$ = 12.7 mm (0.5 in.) have shown little effect on the amount and rate of crack growth in aluminum alloy 7075 plate (Ref 41); however, additional study is needed in this area.

An alternative double-cantilever beam specimen has been developed for testing relatively thin sections (typically 6.4 mm, or 1/4 in., thick) of low-alloy steels (Ref 42). The specimen is stressed by forcing an appropriately dimensioned wedge into the slot. These specimens have been used to determine the effect of hardness of low-alloy steels on their resistance to SCC in environments containing hydrogen sulfide.

Constant K_I specimens are well suited for studying the mechanisms of SCC, because the stress intensity, K_I, is not dependent on crack depth and can be neglected in kinetic studies. Other attractive features are the relatively simple expressions for stress intensity and compliance and the apparent retention of plane-strain conditions in thin plate and sheet specimens. The cost of specimen preparation and instrumentation, however, prohibits its use for extensive SCC characterizations.

Reference 26 provides equations for the analysis of two types of constant K_I specimens: the tapered double-beam specimen and the double-torsion loaded single-edge cracked specimen. A recent evaluation of the double-torsion method (Ref 43) used Al-Zn-Mg alloy sheet 3.2 mm (0.125 in.) thick. By using the double-torsion specimen, *V-K* curves were produced for aluminum alloy 7075-T651 sheet with conventional two-stage growth and plateau velocities that were only slightly higher than those for conventional double-cantilever beam tests of plate.

Other precracked specimen configurations, such as those shown in Fig. 24, can be used for special testing conditions. Information on the preparation and use of these specimens and the related fracture mechanics equations are given in Ref 26 and 44 to 46.

Preparation of Precracked Specimens

When using precracked SCC test specimens, the investigator must consider the dimensional (size) requirements of the specimen, its crack configuration and orientation, and machining and precracking of the specimen. These considerations are discussed below. Additional guidelines and recommendations on specimen preparation in conjunction with fracture toughness testing are given in Ref 26, 27 and 44 to 46.

Dimensional Requirements. A basic requirement of all precracked specimen configurations is that the dimensions be sufficient to maintain predominantly triaxial stress (plane-strain) conditions, in which plastic deformation is limited to a very small region in the vicinity of the crack tip. Experience with fracture toughness testing has shown that for a valid K_{Ic} measurement neither the crack depth a nor the thickness B should be less than $2.5(K_{Ic}/YS)^2$, where YS is the yield strength of the material (Ref 28). Because of the uncertainty regarding a minimum thickness for which an invariant value of K_{ISCC} can be obtained, guidelines for designing fracture mechanics test specimens should be tentatively followed for SCC test specimens. The threshold stress intensity value should be substituted for K_{Ic} in the above expression as a test of its validity.

If specimens are to be used for determination of K_{ISCC}, the initial specimen size should be based on an estimate of the K_{ISCC} of the material. Overestimation of the K_{ISCC} value is recommended; therefore, a larger specimen should be used than may eventually be necessary. When determining stress-corrosion crack growth behavior as a function of stress intensity, specimen size should be based on the highest stress intensity at which crack growth rates are to be measured (substitute K_{Io} in the $2.5(K_{Ic}/YS)^2$ expression).

Notch Configuration and Orientation. For SCC testing, the depth of the initial crack-starter notch—that is, the machined slot with a fatigue or mechanical pop-in crack at its apex—can be as short as $0.2W$. Guidelines for the depth of the notch depend on the limits of accurate K_I calibration with respect to the range of a/W or a/H and

Fig. 23 SCC test specimen containing residual stresses from welding. (a) Sandwich specimen simulating rigid structure. Note SCC in edges of center plate. Source: Ref 10. (b) Cracked ring-welded specimen. Source: Ref 1

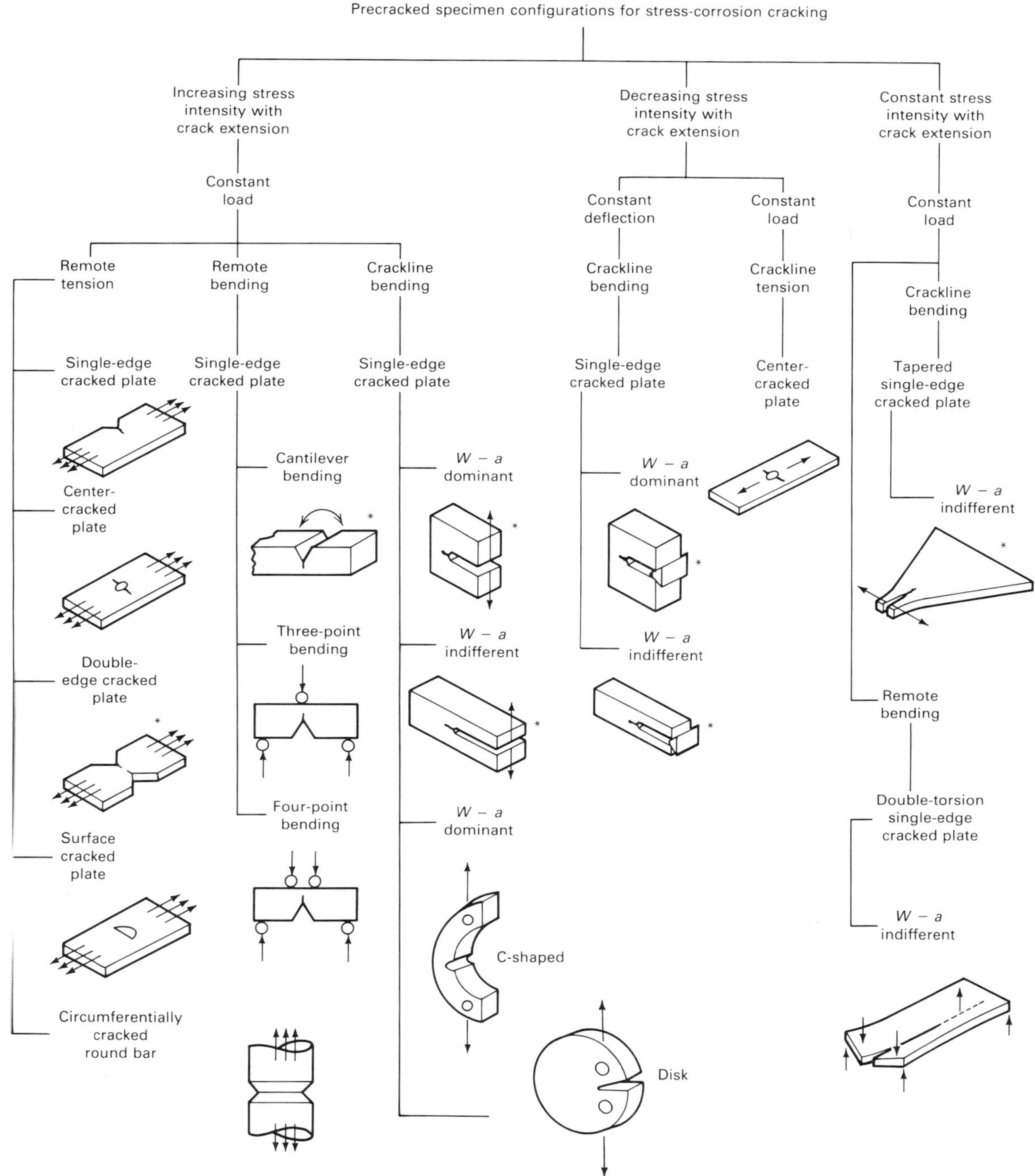

Fig. 24 Classification of precracked specimens for SCC testing. Asterisks denote commonly used configurations. Source: Ref 26

the considerations discussed previously for double-beam specimens.

Several designs of crack-starter notches are available for most plate specimens. The machined slot is used to simulate a crack, because it is impractical to produce plane cracks of sufficient size and accuracy in plate specimens. ASTM E 399 (Ref 27) recommends that the notch root radius should not be greater than 0.127 mm (0.005 in.), unless the chevron form is used, in which case it may be 0.25 mm (0.01 in.) or less (Fig. 26). This tolerance can be easily achieved with conventional milling and grinding equipment.

A significant factor in the SCC testing of thick sections of some metals, such as aluminum and titanium, is the direction of applied stress relative to the grain structure. A standardized plan for identifying the loading direction, the fracture plane, and the direction of crack propagation is shown in Fig. 28.

Machining. Specimens of the required orientation should be machined from products in the fully heat-treated and stress-relieved condition to avoid complications due to residual stresses in the finished specimens. Safeguards against the presence of residual stresses are especially important for precracked specimens because these specimens are usually bulky and contain notches that are machined deep into the metal. For specimens of material that cannot easily be completely machined in the fully heat-treated condition, the final thermal treatment can be given before the notching and finishing operations. However, fully machined specimens should be heat treated only when the heat treatment will not result in distortion, residual stress, quench cracking, or detrimental surface conditions.

Precracking. Fatigue precracking should be done in accordance with ASTM E 399 (Ref 27).

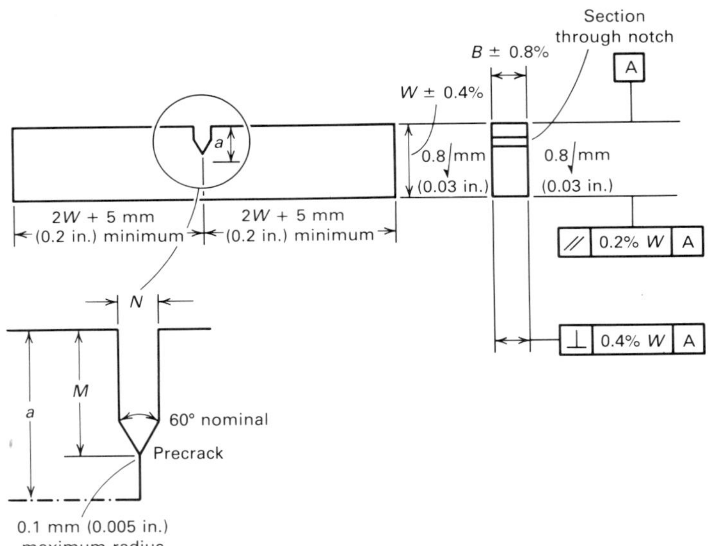

Fig. 25(a) Proportional dimensions and tolerances for cantilever bend test specimens. Width = W; thickness (B) = $0.5W$; half loading span (L) = $2W$; notch width (N) = $0.065W$ maximum if $W > 25$ mm (1.0 in.); $N = 1.5$ mm (0.06 in.) maximum if $W \leq 25$ mm (1.0 in.); effective notch length (M) = 0.25 to $0.45W$; effective crack depth (a) = 0.45 to $0.55W$

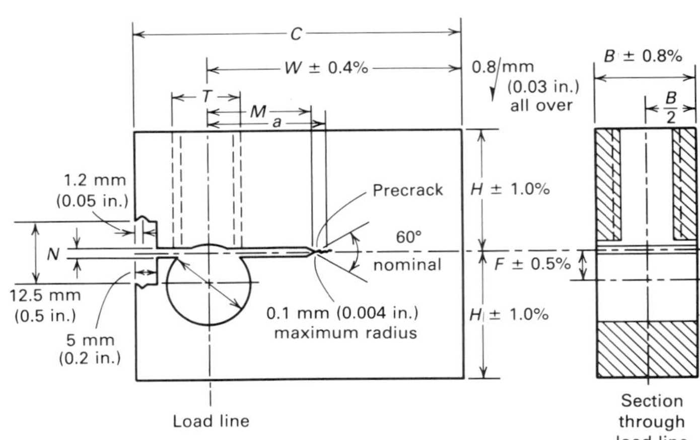

Fig. 25(b) Proportional dimensions and tolerances for modified compact specimens. Surfaces should be perpendicular and parallel as applicable to within $0.002H$ TIR. The bolt centerline should be perpendicular to the specimen centerline within 1°. Bolt of material similar to specimen where practical; fine threaded, square or Allen head. Thickness = B; net width (W) = $2.55B$; total width (C) = $3.20B$; half height (H) = $1.24B$; hole diameter (D) = $0.718B + 0.003B$; effective notch length (M) = $0.77B$; notch width (N) = $0.06B$; thread diameter (T) = $0.625B$

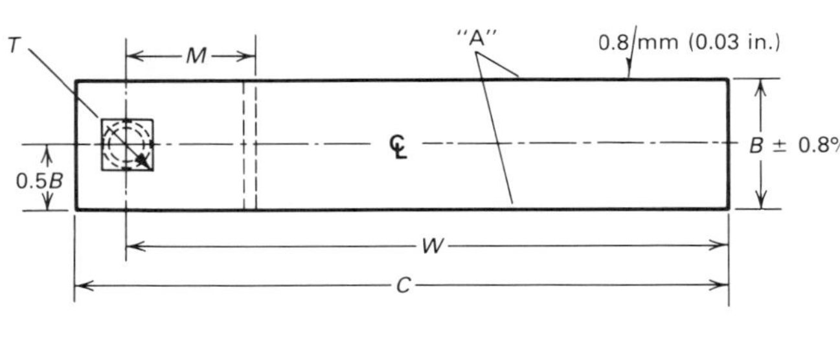

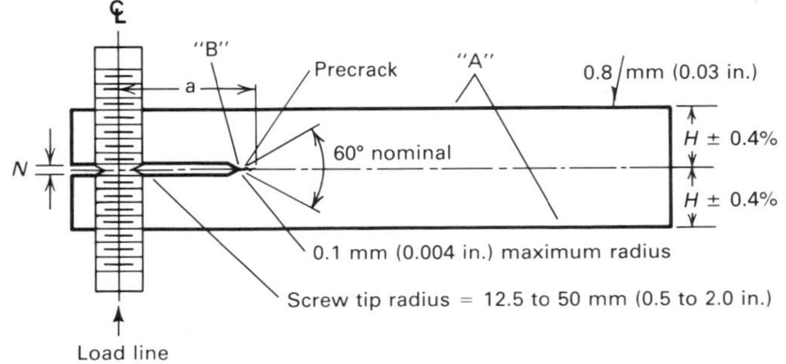

Fig. 25(c) Proportional dimensions and tolerances for double-beam specimens. "A" surfaces should be perpendicular and parallel as applicable to within $0.002H$ TIR. At each side, the point "B" should be equidistant from the top and bottom surfaces to within $0.001H$. The bolt centerline (load line) should be perpendicular to the specimen centerline to within 1°. Bolt of material similar to specimen where practical; fine threaded, square or Allen head. Half height = H; thickness (B) = $2H$; net width (W) = $10H$ minimum; total width (C) = $W + T$; thread diameter (T) = $0.75H$ minimum; notch width (N) = $0.14H$ maximum; effective notch length (M) = $2H$

The K level used for precracking each specimen should not exceed about two thirds of the intended starting K-value for the environmental exposure. This prevents fatigue damage or residual compressive stress at the crack tip, which may alter the SCC behavior, particularly when testing at a K level near the threshold stress intensity for the specimen.

Aluminum alloy specimens can also be precracked by pop-in methods (wedge-opening loaded to the point of tensile overload), but steel and titanium alloys are usually too strong and tough to pop in without breaking off one of the specimen arms. Chevron notches are usually used to facilitate starting such mechanical precracks, and face grooves are sometimes necessary to produce straight precracks in tougher alloys (Fig. 26). These modifications may also be necessary to control fatigue precracking of some materials.

When a specimen is mechanically precracked by pop in, the load should be maintained and should not be reduced for testing at a lower initial K-value. Reducing the load (crack mouth opening displacement) required for pop in will result in residual compressive stress at the crack tip, which could interfere with SCC initiation. When testing specimens at a relatively low fraction of K_{Ic}, fatigue precracking is recommended.

Testing Procedure

For all methods using precracked specimens, the primary objective is usually to determine K_{ISCC} or K_{th}, a threshold stress intensity for SCC for the alloy and environment combination. One procedure, similar to that used with smooth specimens, depends on the initiation of SCC at various levels of applied K_{Io} values. Both constant load (K-increasing) and constant-displacement (K-decreasing) tests can be used. The latter procedure, which is unique to precracked specimens, involves crack arrest. This technique requires a K-decreasing constant-displacement test. These methods are compared in Fig. 29, which illustrates the shift in the stress intensity factor as SCC growth occurs.

***K*-Increasing Versus K-Decreasing Tests.** In constant-load specimens (K-increasing tests), stress parameters can be quantified with confidence. Because crack growth results in an increasing crack opening, there is less likelihood that corrosion products will block the crack or wedge it open. Crack-length measurements can be made readily with several continuous-monitoring methods.

A wide selection of constant-load specimen geometries are available to suit the test material, experimental facilities, and test objective. Therefore, crack growth can be studied under either bend or tension loading conditions. Specimens can be used to determine K_{ISCC} by the initiation of a stress-corrosion crack from a preexisting fatigue crack using a series of specimens or to measure crack growth rates.

The principal disadvantages of constant-load specimens are the expense and bulk associated with the need for an external loading system. Bend specimens can be tested in relatively simple cantilever beam equipment, but specimens subjected to tension loading require constant-load creep-rupture equipment or similar testing machines. In this case, expense can be minimized by testing chains of specimens connected by loading links that are designed to prevent unloading upon failure of individual specimens. Because of the

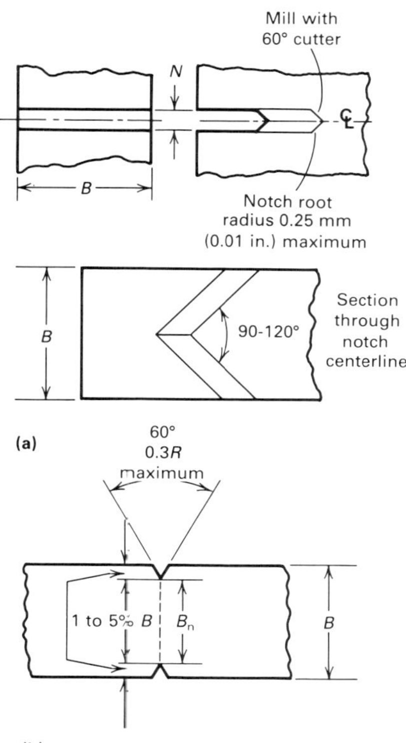

Fig. 26 Alternative chevron notch (a) and face grooves (b) for single-edge cracked specimens

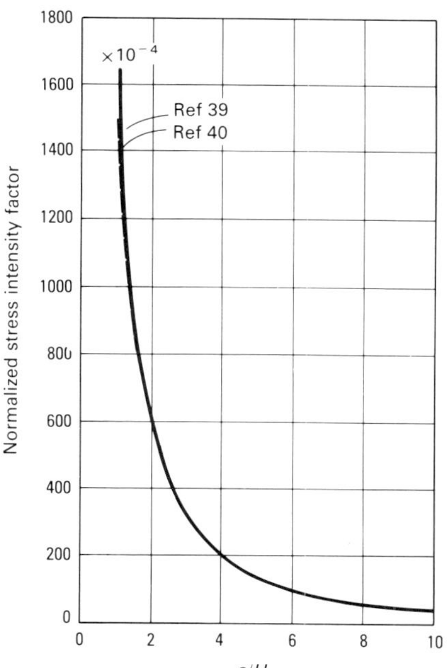

Fig. 27 Configuration and K_I calibration of a double-beam plate specimen. Normalized stress intensity K_I plotted against a/H ratio. ($W - a$) indifferent, crackline-loaded, single-edge cracked specimen. Source: Ref 26

size of these loading systems, it is difficult to test constant-load specimens under operating conditions, but they can be tested in environments obtained from operating systems.

Constant-displacement specimens (K-decreasing tests) are self-loaded; therefore, external stressing equipment is not required. Their compact dimensions also facilitate exposure to operating service environments. They can be used to determine K_{ISCC} by the initiation of stress-corrosion cracks from the fatigue precrack, in which case a series of specimens must be used to bracket the threshold value. This can also be achieved by the arrest of a propagating crack, because under constant-displacement testing conditions stress intensity decreases progressively as crack propagation occurs. In this case, a single specimen suffices in principle; in practice, the use of several replicate specimens is recommended to assess variability in test results.

Constant-displacement specimens are subject to several inherent disadvantages. Oxide formation or corrosion products can wedge the crack surfaces open, thus changing the applied displacement and load. Oxide formation or corrosion products can also block the crack mouth, thus preventing the entry of corrodent, and can impair the accuracy of crack length measurements by electrical resistance methods. Applied loads can be measured only indirectly by displacement changes or by other sophisticated instrumentation. Crack arrest must be defined by an arbitrary crack growth rate below which it is impractical to measure cracks accurately (commonly about 10^{-10} m/s, or 1.5×10^{-5} in./h).

Loading Arrangements and Crack Measurement. To monitor crack propagation rate as a function of decreasing stress intensity when testing constant-displacement loaded specimens, two of the three testing variables must be measured--crack depth (a_i) or load (P_i), and crack opening displacement at the load line (V_{LL}). Although crack initiation and growth can be detected from change in either load or crack length, load change is usually more sensitive to these conditions. Therefore, crack advance is easier to detect in specimens loaded in a testing machine, an elastic loading ring, or an instrumented bolt than in specimens loaded with a bolt or wedge. Figure 30 illustrates typical loading arrangements for which load changes can be automatically monitored (Ref 3, 33, 47).

Figure 31 illustrates an ultrasonic method of measuring crack length at the interior (midwidth and quarter widths) of a bolt-loaded double-beam specimen. This method provides a more accurate measure of crack depth than visual measurements made on the specimen surfaces. Various other techniques have been used, such as measurement of beam deflection for cantilever beam specimens (Ref 29) and changes in electrical resistance. Such arrangements, however, require calibration. It is feasible and desirable to obtain crack length measurements with a precision of at least ±0.127 mm (0.005 in.).

Exposure to Environment. When practical for laboratory accelerated testing, the test environment should be brought into contact with the specimen before it is stressed or immediately afterward; this enhances access of the corrodent to the crack tip to promote earlier initiation of SCC and to decrease variability in test results. Similarly, in certain cases, it may also be beneficial to introduce the corrodent even earlier, that is, during precracking. However, unless facilities are available to begin environmental exposure immediately after precracking, corrodent remaining at the crack tip may promote blunting due to corrosive attack. In addition, corrosion of the specimen surfaces in the small volume of the precrack or the advancing stress-corrosion crack will change the composition of the environment that is in contact with the crack tip and can significantly affect the test results. Therefore, hydrolysis reactions can drastically reduce the pH of the aqueous test environment (Ref 48) and can induce embrittlement of some steels by corrosion product hydrogen.

Selection of an appropriate test duration presents problems that vary with the testing system; this includes the alloy and metallurgical condition, the test environment, and the loading

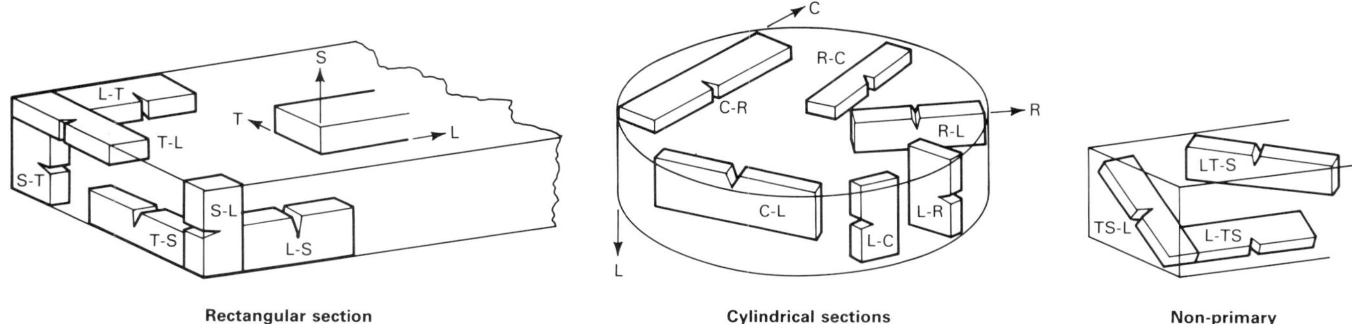

Fig. 28 Specimen orientation and fracture plane identification. L, length, longitudinal, principal direction of metal working (rolling, extrusion, axis of forging); T, width, long-transverse grain direction; S, thickness, short-transverse grain direction; C, chord of cylindrical cross section; R, radius of cylindrical cross section. First letter: normal to the fracture plane (loading direction); second letter: direction of crack propagation in fracture plane. Source: Ref 27

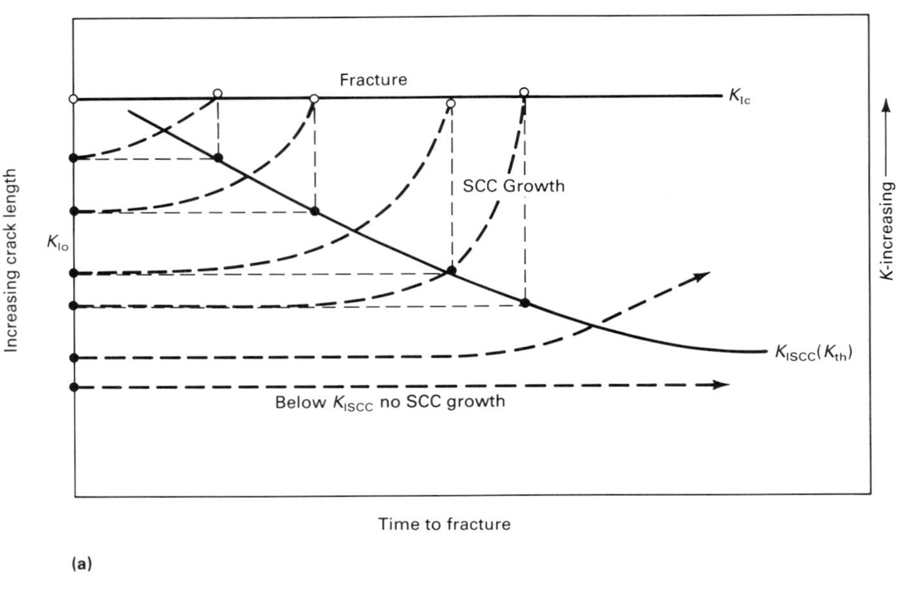

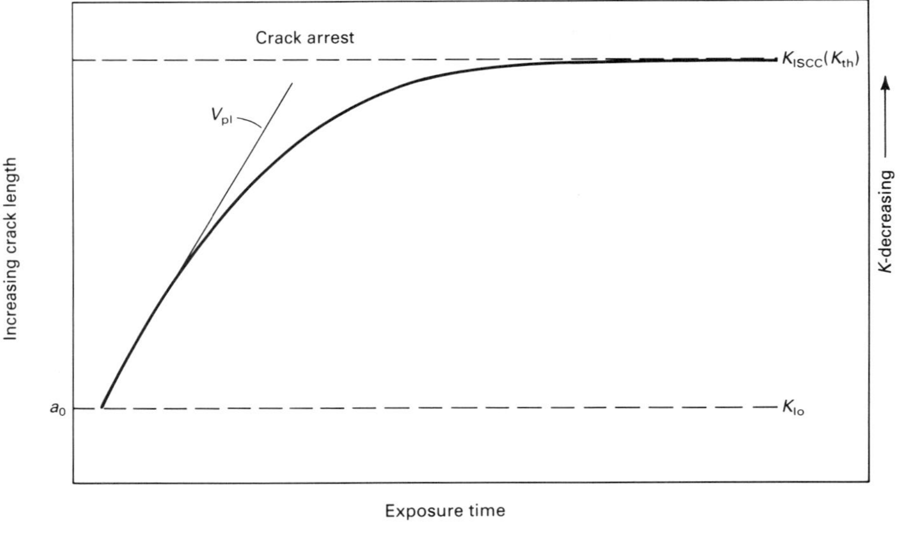

Fig. 29 Schematic comparison of determination of K_{ISCC} by crack initiation versus crack arrest. (a) Constant-load test. (b) Constant crack opening displacement test. a_0 = depth of precrack associated with the initial stress intensity K_{Io}; V_{pl} = plateau velocity.

method. Errors in interpretation of the test results can be caused by test durations that are either too short or too long. The optimum length of exposure can be best approached through recognition of meaningful crack propagation rates. What is considered meaningful depends on the available precision of measurement of crack lengths and an acceptably low rate for the criterion of a stress intensity threshold (Fig. 3). A problem also exists with the correlation of SCC crack growth rates in the laboratory test and in an anticipated service environment. The question leads ultimately to the intended application and a determination of what is a tolerable amount of SCC growth for a given length of time.

Calculation of Crack Growth Rates. There is no generally accepted procedure for calculating crack growth rate, da/dt, as a function of stress intensity from crack growth curves. Various approaches exist; the simplest is a graphical $\Delta a/\Delta t$ technique that may incorporate smoothing of the a versus t curve (Ref 35-37). Another widely used approach is smoothing of the crack growth curve by computer techniques for curve fitting the entire a versus t curve by a multiple-term polynomial function (Ref 29).

Other techniques include a secant method and an incremental polynomial method, in which derivatives of the smoothed crack growth curve are calculated at various points to determine instantaneous crack growth rates. Instantaneous growth rates are then plotted against the instantaneous stress intensities, K_{Ii}, at corresponding time intervals to obtain graphs similar to that shown in Fig. 3. Additional information on the secant and incremental methods, which are often used in fatigue studies, can be found in the article "Fatigue Crack Growth Data Analysis" in Volume 8 of the 9th Edition of *Metals Handbook*.

A limited study of the above four methods of treating crack growth data is presented for a high-strength aluminum alloy in Ref 3. All of the methods used to calculate crack growth rates produced the same general results, which were difficult to interpret because of large amounts of scatter resulting from the use of small crack growth increments. Moreover, the significance of such graphs is dubious when the corrosivity of the environment and the length of exposure can invalidate the estimate of K by causing gross corrosion product wedging effects and/or crack branching.

Reduction of crack length data becomes useless without prior subjective interpretation of crack length versus time curves. Allowances should be made for extraneous effects caused by erratic or apparent initiation of stress-corrosion crack growth, scatter in the measurement data due to excessive crack front curvature, multiple crack planes, crack-tip branching, and gross wedging caused by corrosion products.

A simple method of comparing materials by using crack growth curves is based on average growth rates taken from an exposure time of zero to an arbitrary time that is sufficient to achieve significant crack extension in the most SCC-susceptible materials being compared (Ref 41). This method not only rapidly identifies materials with relatively low resistance to SCC, but also provides numerical test results for highly resistant materials that may not develop a K_I versus da/dt curve with a definite plateau (see the section "Testing of Aluminum Alloys" in this article).

Dynamic Loading: Slow Strain Rate Testing

The most recently developed method for accelerating the SCC process in laboratory testing involves relatively slow strain rate tension testing of a specimen during exposure to appropriate environmental conditions. The application of slow dynamic strain exceeding the elastic limit assists in the SCC initiation. This accelerating technique is consistent with the various proposed general mechanisms of SCC, most of which involve plastic microstrain and film rupture.

Slow strain rate tests can be used to test a wide variety of product forms, including parts joined by welding. Tests can be conducted in tension, in bending, or with plain, notched, or precracked specimens. The principal advantage of slow strain rate testing is the rapidity with which the SCC susceptibility of a particular alloy and environment can be assessed.

Slow strain rate testing is not terminated after an arbitrary period of time. Testing always ends in specimen fracture, and the mode of fracture is then compared with the criteria of SCC susceptibility for the test material. In addition to its timesaving benefits, less scatter occurs in the test results. Comprehensive discussions on the slow strain rate testing technique can be found in Ref 49 to 52.

Critical Strain Rate. The most significant variable in slow strain rate testing is the magnitude of strain rate. If the strain rate is too high, ductile fracture will occur before the necessary corrosion reactions can take place. Therefore, relatively low strain rates must be used. However, at too low a strain rate, corrosion may be

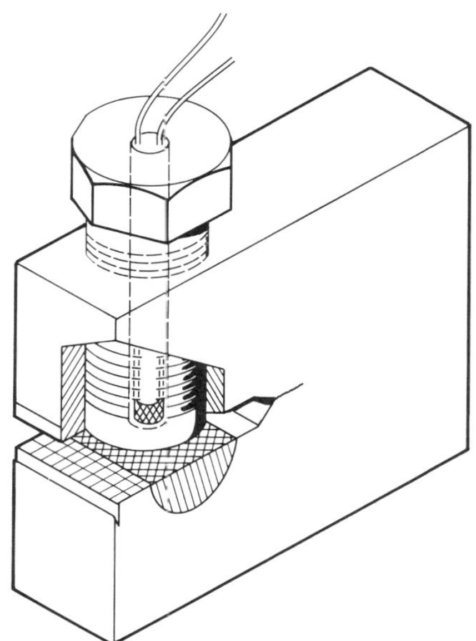

Fig. 30(a) Wedge-opening load specimen loaded with instrumented bolt. Source: Ref 47

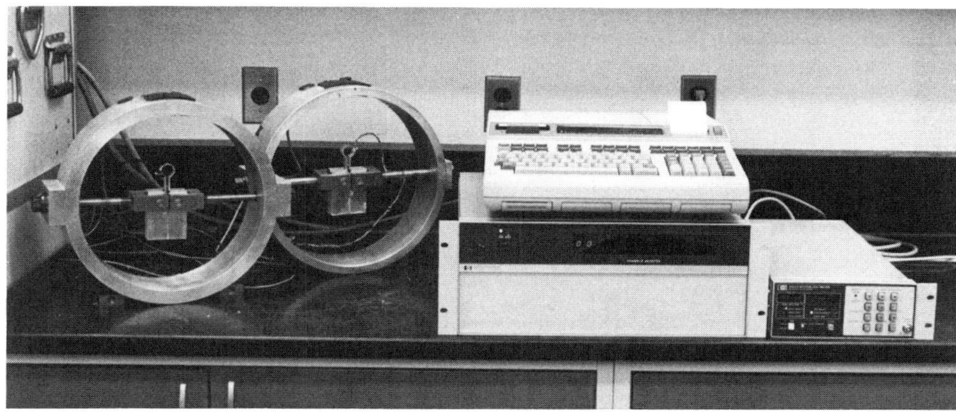

Fig. 30(b) Ring-loaded wedge-opening load specimen test setup. Box to the left of loading rings contains analog signal conditioning for load and displacement signals. The digital data acquisition system consists of a scanner connected to the analog load and displacement signals, a digital voltmeter, and a portable computer used to read and store data and to control the other instruments. Source: Ref 3

prevented because of repassivation or film repair so that the necessary reactions of bare metal cannot be sustained, and SCC may not occur. Although typical critical strain rates range from 10^{-5} to 10^{-7} s^{-1} depending on the alloy and environment system, the most severe strain rate must be determined in each case.

The repassivation reaction that is observed at very low strain rates and that prevents the formation of anodic SCC does not occur when cracking is the result of embrittlement by corrosion product hydrogen. This mechanistic difference can be used to distinguish between anodic SCC (active path corrosion) and cathodic SCC (hydrogen embrittlement) as illustrated in Fig. 32.

The fastest strain rate that will promote SCC in a given system depends on crack velocity. Generally, the lower the stress-corrosion cracking velocity, the slower the strain rate required. Applied strain rates known to have promoted SCC in metal/environment systems are listed in Table 2.

The most relevant strain rates for various aluminum alloys are illustrated in Fig. 33. These trends illustrate that slow strain rate tests should be performed in a strain rate regime that is appropriate for the given alloy and environment system.

Test Specimen Selection. Standard tension specimens (ASTM E 8) (Ref 54) are generally recommended for use with the specified conditions of gage lengths, radii, and so on, unless specialized studies are being conducted. For initially smooth specimens, the strain rate at the onset of the test is clearly defined; however, once cracks have initiated and grown, straining is likely to concentrate in the vicinity of the crack tip, and the effective strain rate is unknown. Rigorous solutions for determining the strain rate at crack tips or notches are not available, but effective strain rates are likely to be higher than for the same deflection rate applied to plain specimens.

Notched or precracked specimens can be used to restrict cracking to a given location—for example, when testing the heat-affected zone associated with a weld. Notched or precracked specimens can also be used to restrict load requirements where bending, as opposed to tensile loading, may offer an added benefit. The section thickness or diameter of such specimens is usually relatively small, so the testing duration is short.

Testing Equipment. Constant strain rate apparatus requirements include sufficient stiffness to resist significant deformation under the loads necessary to fracture the test specimens; a system to provide reproducible, constant strain rates over the range of 10^{-4} to 10^{-8} s^{-1}; and a cell to contain the test solution. Auxiliary equipment is used to control environmental conditions and to record test data. The testing equipment can also be instrumented to record load-elongation curves, which is convenient when testing at various strain rates. A typical constant strain rate unit is illustrated schematically in Fig. 34. Various types of

Fig. 31 Ultrasonic crack measurement system for double-beam specimens. Bolt-loaded specimen is mounted on translation stage at center. Ultrasonic transducer is located above specimen, and the oscilloscope at left indicates (left to right) the top of the specimen, the crack plane, and the bottom face reflection. Digital readouts of stage position and peak height for the crack front measurement used to make consistent positioning measurements are shown (right). This system has a crack growth resolution of approximately 0.127 mm (0.005 in.). Source: Ref 3

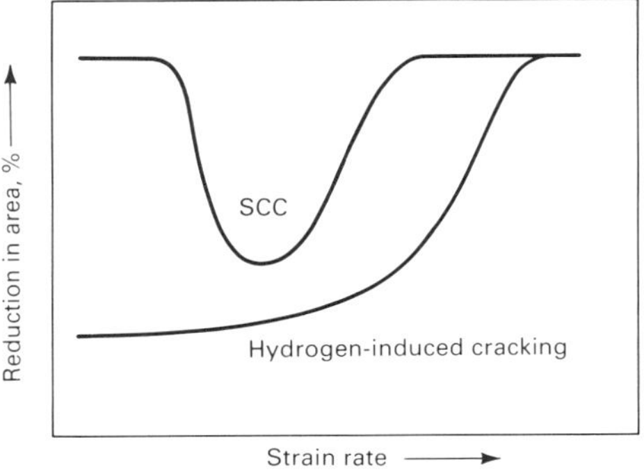

Fig. 32 Schematic showing the effect of strain rate on SCC and hydrogen-induced cracking. Source: Ref 53

Table 2 Critical strain rate regimes promoting SCC in various metal/environment systems

System	Applied strain rate, s^{-1}
Aluminum alloys in chloride solutions	10^{-4} and 10^{-7}
Copper alloys in ammoniacal and nitrite solutions	10^{-6}
Steels in carbonate, hydroxide, or nitrate solutions and liquefied ammonia	10^{-6}
Magnesium alloys in chromate/ chloride solutions	10^{-5}
Stainless steels in chloride solutions	10^{-6}
Stainless steels in high-temperature solutions	10^{-7}
Titanium alloys in chloride solutions	10^{-5}

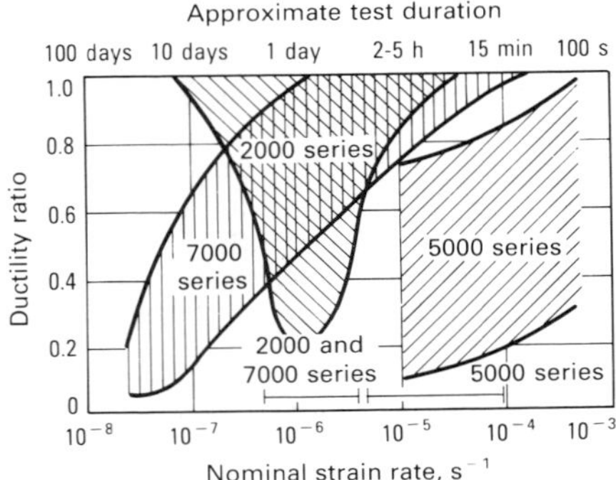

Fig. 33 Strain rate regimes for studying SCC of various aluminum alloys. Corrodent: 3% sodium chloride plus 0.3% hydrogen peroxide. Source: Ref 52

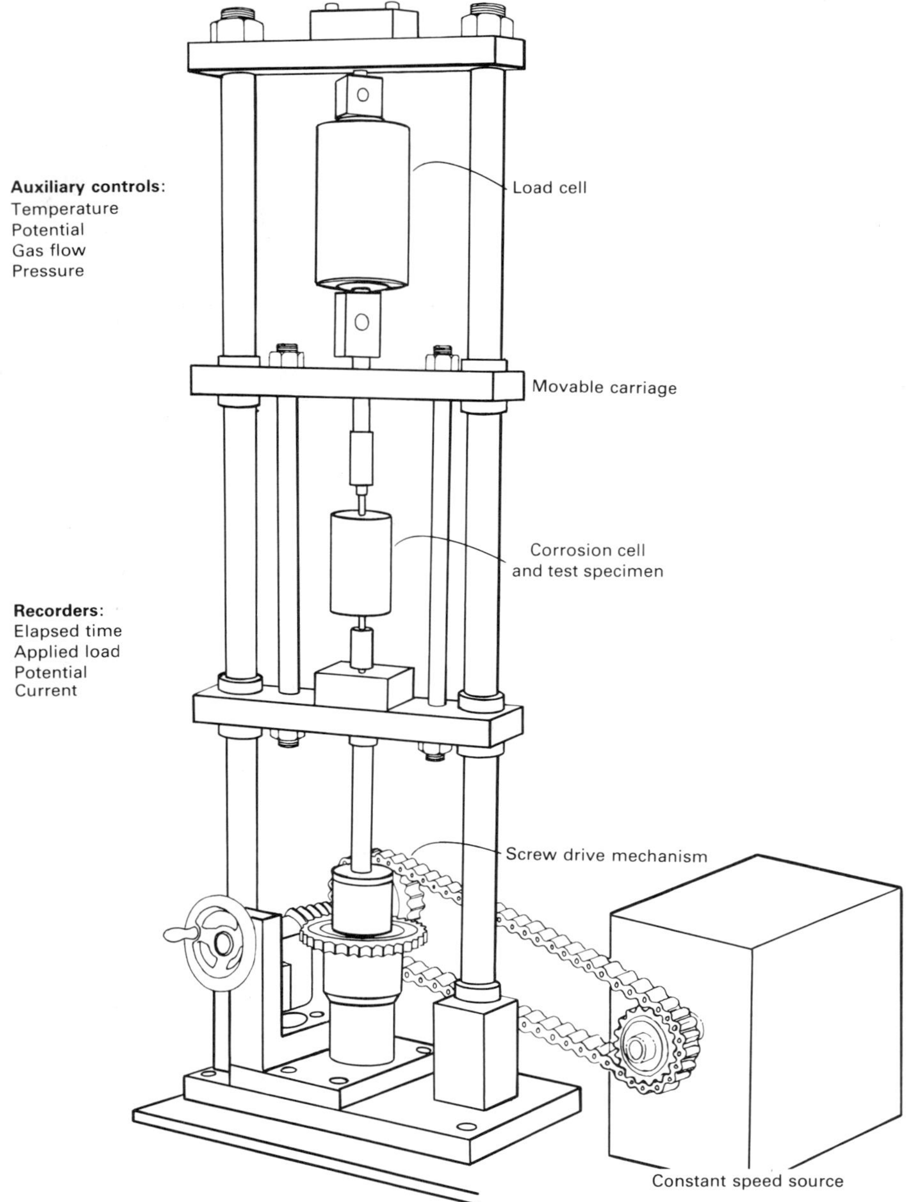

Fig. 34 Typical slow strain rate test apparatus. Source: Ref 51

corrosion cells may be required to control the test conditions for specific studies.

In addition to uniaxial tensile units, cantilever constant strain rate apparatus has also been used in which an extension arm attached to a cantilever beam specimen is lowered at a constant rate. This technique has been successfully used to study SCC of low-carbon steel in carbonate-bicarbonate environments to determine crack velocity, critical strain rates, and inhibitor effectiveness (Ref 55). Additional information on slow strain rate testing equipment and procedures is available in Ref 45 and 49.

Assessment of Results. Historically, the principal methods of SCC assessment derived from slow strain rate tension testing were based on time to failure, maximum gross section stress developed during the tension test, percent elongation, area bounded by the load-elongation curve, and reduction in area. Figure 35 depicts stress-elongation curves that illustrate how stress-corrosion cracks influence the elongation to fracture as well as the maximum load.

To eliminate non-SCC effects, parallel tests are conducted in an inert environment, and a ratio of the result obtained in the corrodent divided by the result obtained in the inert environment is commonly used as an index of SCC susceptibility. For example, in Fig. 33, higher SCC resistance is denoted by higher ductility ratios.

Figure 36 shows a stress-corroded specimen containing many secondary stress-corrosion cracks and reduced ductility at fracture. Some alloys experience rapid deterioration of mechanical properties on contact with certain corrosive environments; any additional effect of applied straining can best be assessed by comparison with the behavior of unstrained specimens. Therefore, it is essential that the cause of environmental degradation be verified as SCC.

Slow strain rate testing is very efficient in comparing environments in terms of their capability to produce SCC, for example, in steels having similar metallurgical characteristics. However, such comparisons are difficult and not very reliable when applied to groups of steels with different characteristics (Ref 53).

Slow strain rate testing as generally used does not provide data that can be used for design purposes. Recent work, however, has shown that average SCC velocities, threshold stresses, and threshold strain rates can be obtained with modified techniques combined with microscopy (Ref 50, 55, 56). For example, average SCC crack velocities can be determined from the depth of the largest crack measured on the fracture surfaces of specimens that have failed completely, or in longitudinal sections on the diameter of specimens that have not experienced total failure, divided by the time of testing. With this procedure, SCC is assumed to initiate at the start of the test, which is not always true.

With precracked specimens, other methods can be used to monitor crack growth and thus allow determination of crack velocities. The SCC behavior of a pipeline steel (Fig. 37) has been studied by using a precracked cantilever bend specimen in terms of threshold strain rate for crack growth and also in terms of crack growth rates analogous to the Stage II plateau velocity illustrated in Fig. 3. Material properties, such as

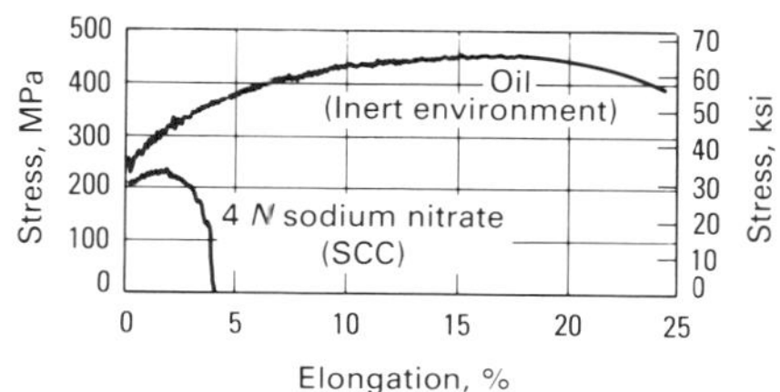

Fig. 35 Nominal stress versus elongation curves for carbon-manganese steel in slow-strain rate test in boiling 4 N sodium nitrate and in oil at the same temperature. Source: Ref 50

strength and toughness, that influence SCC performance when measured by tension testing are eliminated as factors; therefore, valid comparisons can be made of alloys with widely different structures and mechanical properties. Additional information on this method of assessment and the effects of strain rate can be found in Ref 57 to 59.

Selection of Test Environments

The primary environmental factors in SCC testing are the nature and concentration of anions and cations in aqueous solutions, electrochemical potential, solution pH, the partial pressure and nature of species in gaseous mixtures, and temperature. Separately or in combination, environmental variables can have a profound effect on the thermodynamics and kinetics of the electrochemical processes that control environmentally assisted fracture. Therefore, the choice of environmental conditions provides an important basis for developing accelerated SCC test methods.

The environmental requirements for SCC vary with different alloys. Although a mechanical precrack or a critical strain rate provides a worst case for SCC from a mechanical standpoint, there does not appear to be a generally applicable worst case from an environmental standpoint. However, because the presence of moisture and salt water is universal, the SCC characteristics of alloys in these environments—as well as in any special environment a given engineering structure may experience—are always of interest.

Figure 38 illustrates that electrochemical factors can override mechanical factors in determining SCC initiation sites. Three cantilever beam specimens of PH13-8Mo stainless steel were tested in salt water. Specimen A was tested at a high K level. With the participation of the chloride ions, the protective oxide film ruptured at the bottom of the precrack and initiated SCC, which was halted before the beam fractured completely. Specimen B was loaded at a lower K level. After 1300 h, a stress-corrosion crack initiated, but not in the precrack. Crack initiation occurred under the wall of the cell that surrounded the central portion of the specimen and contained the salt water.

Careful examination of this specimen and replicate specimens revealed small crevice corrosion pits under the wall that initiated SCC in an almost smooth surface. Even if these small pits had been as sharp as a fatigue crack, the K level would have been much lower than at the machined and fatigued notch. In the stagnant situation under the cell wall, the stainless steel reacted with the salt water to form hydrochloric acid and other corrosion products from the metal. Therefore, the low pH in a crevice, due to the hydrolysis of

Fig. 36 Photomacrographs of two carbon steel specimens after slow strain rate tests conducted at a strain rate of 2.5×10^{-6} s^{-1} and 80 °C (180 °F). The ductility ratio in this example was 0.74 (original diameter: 2.54 mm, or 0.100 in.). (left) Ductile fracture in oil. (right) SCC in carbonate solution

chromium corrosion product, overcame the mechanical disadvantage of the lack of a precrack. Specimen C was then tested to verify the effectiveness of electrochemical conditions in crack initiation. Saturated ferric chloride was selected to lower the pH to the range inside an active corrosion pit in the stainless steel; application of the solution to the unnotched beam resulted in the immediate initiation of many cracks in the smooth surface. Hydrochloric acid was found to be equally effective.

Stress-corrosion tests can be divided into two broad environmental classes: those conducted in actual service environments and those conducted under laboratory conditions.

Service Environments and Field Testing. The following examples illustrate the value, and in some cases the necessity, of exposure tests performed in actual service environments as an adjunct to laboratory evaluation. The standard 3.5% sodium chloride alternate immersion test data for aluminum alloys 2024 and 7075 proved useless in predicting the serviceability of these aluminum alloys for handling rocket propellant oxidizers such as nitrogen tetroxide and inhibited red fuming nitric acid (Ref 61). The alternate immersion test showed 2024-T351 and 7075-T651 to be susceptible to SCC at low short-transverse stresses, but 2024-T851 and 7075-T7351 were quite resistant even at high stresses. These data were supported by outdoor field tests in seacoast and industrial atmospheres.

However, in proof tests consisting of exposure to the actual service environment—inhibited red fuming nitric acid at 74 °C (165 °F)—SCC occurred in both tempers of 7075 alloy and did not occur in either temper of 2024 alloy (Fig. 39). There were no unexpected failures with the 2219-T87 and 6061-T651 materials, however.

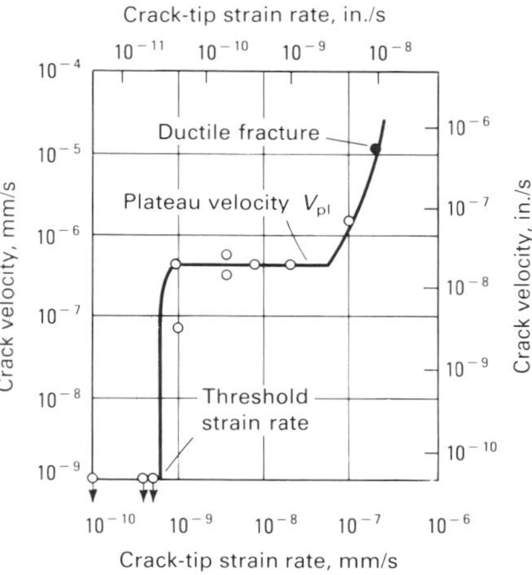

Fig. 37 Effects of beam deflection rate on stress-corrosion crack velocity in precracked cantilever bend specimens of a carbon-manganese steel. Tested in a carbonate-bicarbonate solution at 75 °C (165 °F) and at a potential of −650 mV versus SCE. Source: Ref 50

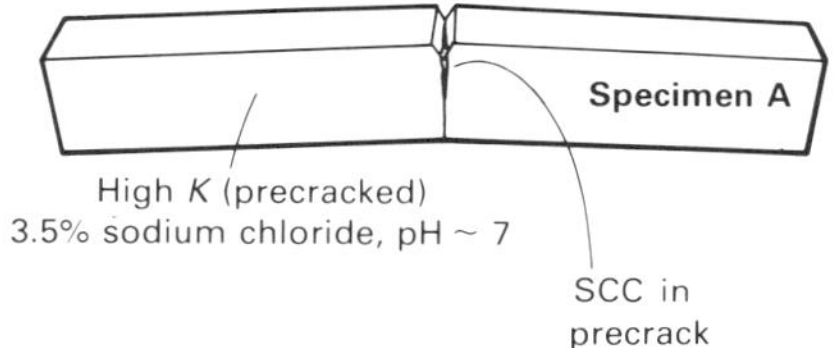

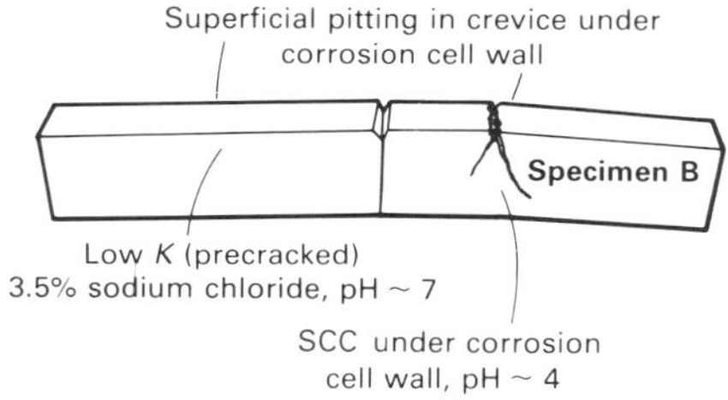

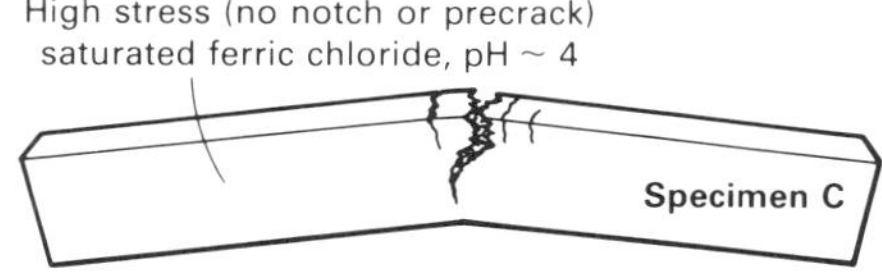

Fig. 38 Cantilever beam specimens of PH13-8Mo stainless steel after testing. Experiments demonstrate that electrochemical factors can override mechanical factors in determining initiation sites of SCC. See text for details. Source: Ref 60

Simulated-service tests should be conducted under conditions duplicating the service environment exactly, as illustrated by the following example with Ti-6Al-4V alloy pressure tanks for propellant-grade nitrogen tetroxide (<0.20 wt% moisture) (Ref 62). Preliminary laboratory SCC tests using specimens of Ti-6Al-4V demonstrated satisfactory compatibility with the nitrogen tetroxide and gave no indication of SCC.

In subsequent tests of actual pressurized tanks, however, SCC occurred rapidly, and the tanks failed. It was subsequently shown that the nitrogen tetroxide used in the tanks was of a higher-

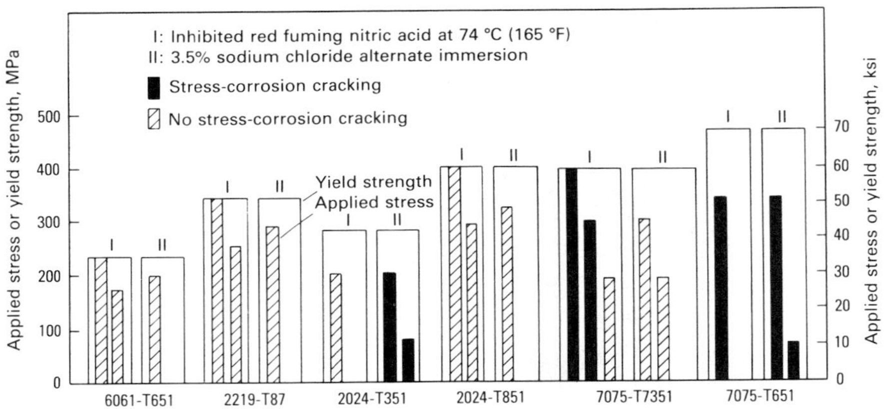

Fig. 39 SCC resistance of various aluminum alloys in inhibited red fuming nitric acid versus alternate immersion in 3.5% sodium chloride solution. Each bar graph represents an individual short-transvese C-ring test specimen machined from rolled plate and stressed at the indicated level. Source: Ref 61

purity grade than that used in the laboratory tests. When test specimens were exposed to this grade of nitrogen tetroxide, they also failed. The small quantities of water and nitric oxide (impurities) present in the nitrogen tetroxide used in previous studies were sufficient to inhibit SCC of Ti-6Al-4V in nitrogen tetroxide.

Field testing consists of placing a metal specimen in an environment in which conditions represent those anticipated in service. Typical examples include immersion in seawater, exposure to the atmosphere at marine or industrial sites, chemical plant streams, and so on. Field tests can be performed with test coupons or with actual or simulated structural components.

Laboratory Environments. Accelerated testing is conducted extensively in controlled laboratory environments for developmental studies in which it is not considered necessary that the test medium duplicate a particular service environment. Reliance should not be placed on such test results, however, unless they can be correlated with practical service experience or with the results of appropriate field tests.

Ideally, the results of a short exposure in a laboratory test will enable a reliable prediction of the SCC performance of an alloy over a long service period under particular environmental conditions. To fulfill this function, test conditions should be selected with regard to the anticipated service conditions. However, accelerated test media that yield reliable results for one alloy may not yield dependable results for another alloy, even though both alloys are of the same basic metal (Ref 9, 58). Therefore, caution must be exercised in the development and use of standardized test media. Recommended test media for various metals and alloys are discussed in later sections.

Mercury, other liquid-metal systems, and molten salts can also cause failure by cracking of metal under stress. In most cases of this type, oxidation-reduction reactions (corrosion) are not involved with the alloy, and the mechanisms of the cracking under such conditions are quite different from the mechanism of SCC.

Cracking caused by hydrogen embrittlement has been found to occur in titanium-base alloys and many high-strength ferrous alloys in various service environments. Hydrogen, under some conditions, enters the stressed metal without the presence of a corrosion reaction, and cracking ensues. Hydrogen can also enter stressed metal as a corrosion reaction product, and cracking may be stimulated or may ensue. Therefore, to characterize adequately the environmental cracking of such alloys under sustained tensile stress, testing should be performed both with and without conditions that could involve hydrogen damage. More information on testing for susceptibility to hydrogen embrittlement is available in the article "Evaluation of Hydrogen Embrittlement" in this Volume.

Length of Exposure. The question of appropriate test durations for the different material-environment combinations and the apparent need to establish them experimentally on a case-by-case basis has been a major stumbling block in the adoption of standard SCC test methods. The situation becomes acute when the results of an SCC test must be evaluated before all replicate specimens have cracked. In these cases, when a statistical probability evaluation must be performed, it should be recognized that time to failure is often normally distributed with respect to the logarithm of test time (see the section "Interpretation of Test Results" in this article). Exposure periods that are too short can result in a test that is trifling; when too long, they can introduce extraneous effects. An example is the excessive pitting of smooth test specimens that leads to failure by a different (and possibly unsuspected) mode of SCC, or by mechanical overload rather than by SCC (Ref 18). Problems in the case of precracked specimens can result from test periods being too short to develop a proper estimate of an apparent threshold stress intensity, especially in K-increasing tests; on the other hand, periods that are too long may result in corrosion product wedging or crack tip blunting, both of which can invalidate calculations of the effective stress intensity and SCC growth rates in K-decreasing tests (Ref 24, 36, 37). Therefore, it is necessary to scrutinize test results carefully so that extraneous effects can be excluded from the analysis of the test results.

Electrochemical Tests. Recognition of the importance of electrical potential as one of the controlling parameters in SCC has resulted in increasing use of tests with potentiostatic control or impressed electrical current. No standards exist for such electrochemical tests, although several methods are routinely used. These tests involve specific conditions that are applicable only to given alloy and environment systems under consideration. These types of tests offer greater rapidity and precision than free corrosion tests.

Sophisticated approaches for studying systems involving active path corrosion use potentiodynamic methods. Anodic polarization curves can provide reasonably accurate predictions of the critical potential ranges and kinetic factors controlling SCC in a given system. One procedure (Ref 63) involves first scanning a range of potentials in the anodic direction at a relatively high scan rate (~1000 mV/min) to determine regions with high current density, in which intense anodic activity is likely. This is followed by a relatively low scan rate (~20 mV/min), which indicates regions where relative inactivity is likely.

Scans should be started at a potential at which the surface is film free. The rapid scan minimizes film formation so that the observed currents relate to relatively film-free conditions. The slow scan allows time for filming to occur. Comparison of the two curves reveals any ranges of potential within which high anodic activity in the film-free region reduces to insignificant activity when the time requirements for film formation are met, thus indicating the critical potential range in which SCC is likely.

Figure 40 illustrates this type of experiment for a low-carbon steel in a carbonate-bicarbonate solution with the predicted domains of behavior. Establishing a precise range of critical potentials for SCC requires subjective determination of boundary conditions relating to the current densities.

A different approach with the use of anodic polarization curves has been used to characterize the SCC behavior of aluminum alloys (Ref 64). Corrosion of a susceptible microstructure in aqueous chloride solutions was exclusively intergranular only for a limited range of potentials between the first and second breakdown potentials, E_{BR}, determined from the anodic polarization curve, where E_{BR_1} approximates the critical pitting potential of the active corrosion path at the grain boundaries and E_{BR_2} approximates the critical pitting potential of the grain bodies. Figure 41 depicts a schematic anodic polarization curve for aluminum alloy 7075-T651 that illustrates the predicted domains of behavior.

The use of anodic polarization curves and related measurements is discussed in ASTM G 5 (Ref 8). Information obtained from the anodic polarization curve can be useful for screening

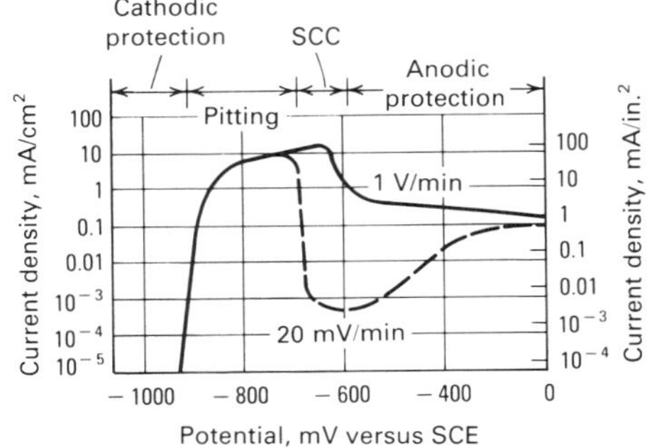

Fig. 40 Potentiodynamic polarization curves for carbon-manganese steel in 1 N sodium carbonate plus 1 N sodium bicarbonate at 90 °C (195 °F) showing the domains of behavior predicted from the curves. Source: Ref 63

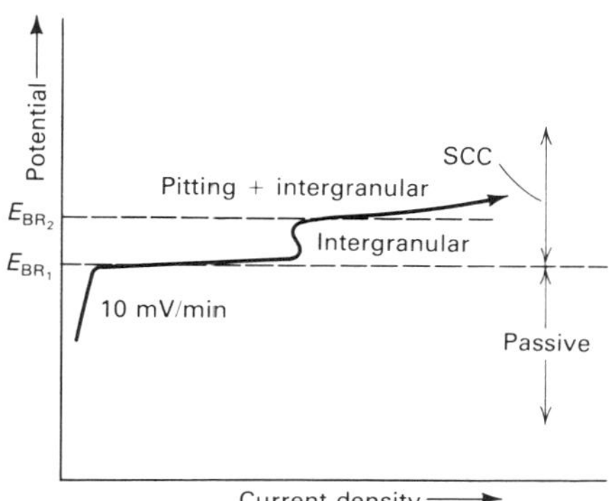

Fig. 41 Anodic polarization curves for aluminum alloy 7075-T651 in deaerated 3.5% sodium chloride solution showing the domains of behavior predicted from the curve. Source: Ref 65

environmental test media to reduce the number of actual SCC tests required. However, such electrochemical tests are concerned with testing a bulk environment in contact with an exposed surface, and it should not be assumed that such measurements represent the crack tip under conditions of crack growth. Therefore, even when an environment is identified as innocuous, a potent local environment may exist, such as in a pit when the pitting potential is exceeded or in a crevice or precrack.

Moreover, when laboratory SCC tests involve cracks or crevices, the potential at the crack tip may differ appreciably from that measured at the surface where the crack or crevice emerges. In laboratory tests, it must be considered whether or not environmental changes in composition or potential are involved and how they may relate to service conditions (Ref 58). Further discussion of electrochemical approaches to the evaluation of SCC is presented in the article "Laboratory Testing" in this Volume.

Testing of Aluminum Alloys

Stress-corrosion cracking of susceptible high-strength aluminum alloys can occur in moist air, seawater, and potable waters, and it varies with the alloy and temper and the magnitude of sustained tensile stress. Chloride solutions are generally favored for laboratory tests because sodium chloride is widely distributed in nature, and the test results are potentially relatable to stress-corrosion behavior in natural environments.

Testing With Smooth Specimens

Historically, stress corrosion testing of aluminum alloys has been done with smooth (notch-free) specimens, and most published SCC ratings of aluminum alloys are based on such tests.

Alternate Immersion in 3.5% Sodium Chloride. Exposure to 3.5% sodium chloride or to substitute ocean water (ASTM D 1141) by alternate immersion (ASTM G 44) (Ref 8) is a widely used procedure for testing smooth specimens of aluminum alloys. Aeration of the specimens, achieved by the alternate immersion, enhances the corrosion potential (Ref 66) and produces more rapid SCC of most aluminum alloys than continuous immersion. The ASTM G 44 standard practice consists of a 1-h cycle that includes a 10-min soak in the aqueous solution followed by a 50-min period out of solution in air at 27 °C (80 °F) and 45% relatively humidity, during which time the specimens are air dried. This 1-h cycle is repeated continuously for the total number of days recommended for the particular alloy being tested. Typically, aluminum alloys are exposed from 10 to 90 days, depending on the resistance of the alloy to corrosion by salt water. This test method is widely used for testing most types of aluminum alloys with all types of smooth specimens.

The alternate immersion test is primarily used for alloy development studies and for quality control of alloys with improved resistance to SCC (Ref 9, 67). This test method is specified in ASTM G 47 (Ref 8), which covers the method of sampling, type of specimen, specimen preparation, test environment, and method of exposure for determining the susceptibility to SCC of high-strength 2000-series alloys (1.8 to 7.0% Cu) and 7000-series alloys (0.4 to 2.8% Cu). Alternate immersion in 3.5% sodium chloride is also specified in ASTM G 64 (Ref 8).

The parallel SCC behavior of a variety of aluminum alloys in the ASTM G 44 test and in a severe seacoast atmosphere is shown in Table 3. Additional comparisons may be found in Ref 18 and 69. The relatively conservative estimate of the SCC behavior of an intermediate-resistance alloy in the ASTM G 44 test compared to that in various atmospheric environments is shown in Fig. 42, which also illustrates the wide range in behaviors in various atmospheric environments.

Although ASTM G 44 is a good general-purpose test for aluminum alloys, it is not equally discriminating of all alloys at near-threshold stress levels. This tendency has been reported for 7000-series (aluminum-zinc-magnesium-copper) alloys containing less than about 1% Cu (Ref 9, 16, 69). Also, the test is not representative of special chemical environments, such as inhibited red fuming nitric acid (Fig. 39).

Continuous Immersion in Boiling 6% Sodium Chloride. A 4-day exposure by continuous immersion in boiling 6% sodium chloride solution is widely used by U.S. aluminum producers for testing smooth specimens of 7000-series alloys containing no more than 0.25% Cu. Comparison of this test and a modified ASTM G 44 test to the industrial atmospheric exposure shown in Fig. 43 illustrates the advantage for the boiling salt test. An ASTM standard is in preparation at this writing.

This test is not effective for, and therefore not recommended for, the 2000-series (aluminum-copper) alloys, the 5000-series (aluminum-magnesium) alloys, or the 7000-series (aluminum-zinc-magnesium-copper) alloys containing more than about 1% Cu. Also, it is not recommended for testing precracked specimens because of the interference of wedges of corrosion products developed on the crack surfaces.

The impressed-current test for 5000-series alloys was developed for rapid evaluation of smooth coupon specimens of the 5000-series (aluminum-magnesium) alloys. The test solution is 3.5% sodium chloride, and the acceleration is provided by impressing on the test specimen a dc electrical current of 6.2×10^{-2} mA/mm^2 (40 mA/in.^{2}) of specimen surface. Good correlation with natural environment exposures is reported in Ref 70.

Other Testing Media. Although nitrates and sulfates, when dissolved in distilled water, tend to retard SCC, their presence in chloride environments can produce a synergistic stimulation of intergranular corrosion and SCC (Ref 65, 71, 72). Stress-corrosion cracking can also be accelerated for certain alloys by increasing acidity (lower pH), increasing temperature, adding oxidants, or electrochemically driving the SCC process by impressing an appropriate potential or electrical current density. These procedures, either singly or in combination, have been used to create various special-purpose tests:

- Continuous immersion test for 7000-series (aluminum-zinc-magnesium) alloys (Ref 73): aqueous solution containing 3% sodium chloride, 0.5% hydrogen peroxide (30%), 100 mL/L 1 *N*

Table 3 Comparison of the SCC behavior of various aluminum alloys in the ASTM G 44 test and after 5 years in a seacoast atmosphere

3.2-mm (1/8-in.) diam short-transverse tension specimens obtained from 64-mm (2.5-in.) thick hot-rolled plate; nine replicate specimens per test stress

Alloy and temper	Applied stress			Number of failures		Time to first and median failure, days			
						ASTM G 44(a)		Seacoast atmosphere(b)	
	MPa	ksi	% of yield strength	ASTM G 44(a)	Seacoast atmosphere(b)	First	Median	First	Median
2024–T351	145	21	50	9	9	7	7	37	37
	87	12.6	30	9	9	7	7	37	37
2024–T851	295	42.8	75	8	8	37	65	37	266
	197	28.6	50	0	2	...	...	643	...
5456–H116	156	22.6	75	0	0	...	...	...	...
6061–T651	254	36.8	90	0	0	...	...	...	...
7050–T7651	227	32.9	50	0	0	...	...	...	...
	136	19.7	30	0	0	...	...	...	...
	91	13.2	20	0	0	...	...	...	...
7050–T7451	217	31.5	50	0	0	...	...	...	...
	130	18.9	30	0	0	...	...	...	...
	87	12.6	20	0	0	...	...	...	...
7075–T651	221	32	50	...	9	...	...	7	7
	154	22.3	35	9	9	7	7	7	15
	88	12.8	20	9	9	7	67	7	37
7075–T7651	300	43.5	75	8	6	69	77	709	1491
	200	29	50	0	1	...	...	1069	...
	120	17.4	30	...	0	...	...	...	...
7075–T7351	335	48.6	90	6	2	67	80	1866	...
	273	39.6	75	0	0	...	...	...	...
	183	26.5	50	...	0	...	...	...	...

(a) Alternate immersion in 3.5% sodium chloride solution for 84 days; see Ref 8. (b) Point Judith, RI. Source: Ref 68

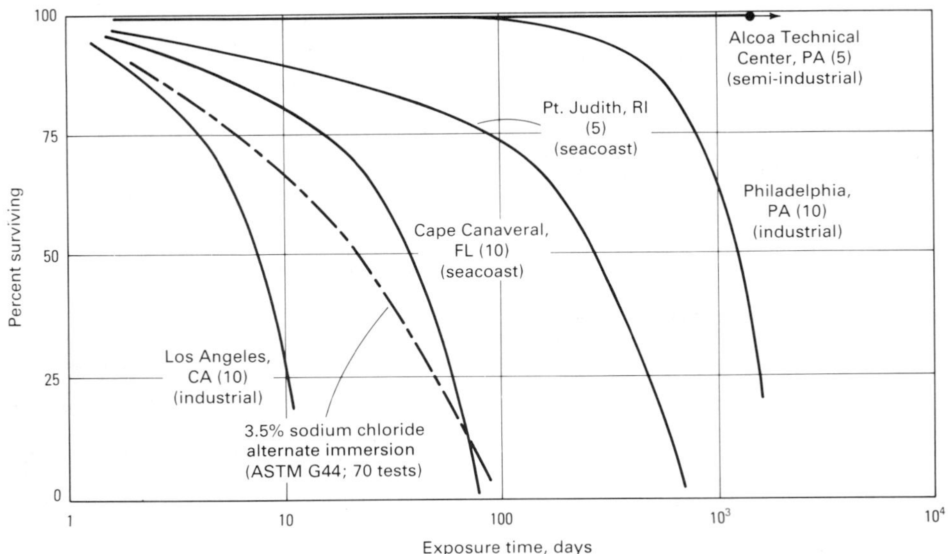

Fig. 42 Effect of variations in atmospheric environment on the probability and time to failure by SCC of a material with an intermediate susceptibility. Tests were made on short-transverse 3.2-mm (0.125-in.) diam tension specimens from 7075-T7651 type plate stressed 310 MPa (45 ksi). Parenthetical values indicate replication of tests. Source: Ref 3

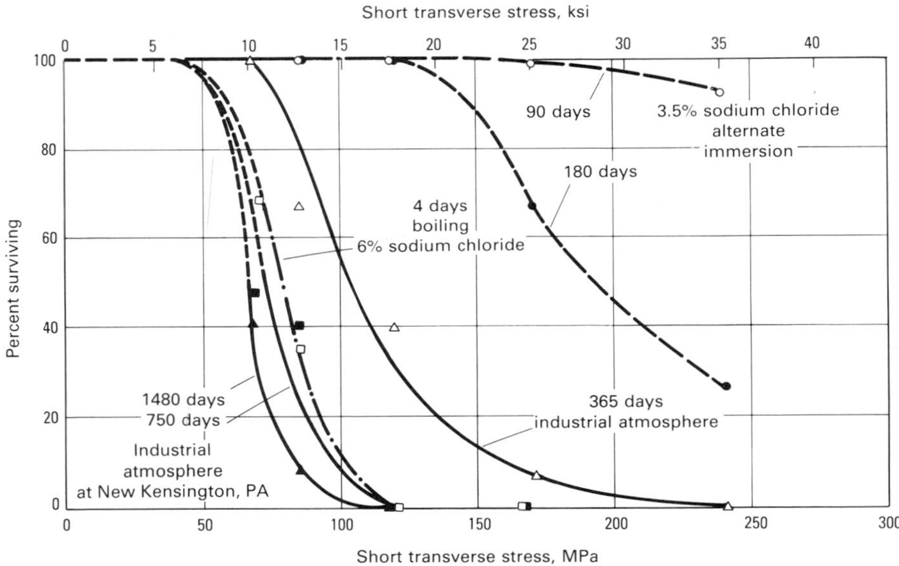

Fig. 43 Correlation of accelerated test media with service environment (industrial atmosphere). Combined data for five lots of rolled plate of aluminum alloy 7039-T64 (4.0Zn-2.8Mg-0.3Mn-0.2Cr). Tests in 3.5% sodium chloride were similar to ASTM G 44, except salt solution was made with commercial grade sodium chloride and New Kensington tap water. Source: Ref 9

sodium hydroxide, and 20 mL/L acetic acid (100%); pH 4.0

- Continuous immersion test for high-strength aircraft alloys (Ref 74): aqueous solution containing 2% sodium chloride plus 0.5% sodium chromate
- Impressed-current test for 7000-series (aluminum-zinc-magnesium) alloys (Ref 75): aqueous solution containing 2% sodium chloride plus 0.5% sodium chromate; pH 8.1, current density 4.65×10^{-4} mA/mm^2 (0.3 mA/in.^{2}); 30-day maximum exposure time
- Alternate immersion test using an aqueous solution containing 2.86% sodium chloride plus 0.52% magnesium chloride (total chloride equal to that in ocean water): proposed in Ref 76 as a less corrosive substitute for 3.5% sodium chloride solution for ASTM G 44
- Continuous immersion test for 2000-series (aluminum-copper) and 7000 series (aluminum-zinc-magnesium-copper) alloys (Ref 77): aqueous 1% sodium chloride plus 2% potassium dichromate at 60 °C (140 °F); 168-h maximum exposure time
- Continuous immersion test for 2000-series (aluminum-copper) and 7000-series (aluminum-zinc-magnesium-copper) alloys (Ref 36, 37): aqueous solution containing 0.6 *M* sodium chloride, 0.02 *M* sodium dichromate, 0.07 *M* sodium acetate, plus acetic acid to pH 4; used principally for testing precracked specimens

Testing With Precracked Specimens

Testing aluminum alloys with precracked specimens, especially the bolt-loaded double-beam type, has received widespread use in recent years, and the ranking of materials by this method is generally in good agreement with that established with smooth specimen tests. However, a number of problems in the interpretation of test results must be taken into account (Ref 34-37, 78-81). Subjective interpretations of the test results can be variable because there are as yet no standardized test procedures.

The bolt-loaded *K*-decreasing type of test is attractive because no complicated apparatus is required to perform the tests, and the results appear to be relatable to the control of SCC problems in engineering structures. Distinction among test materials or test environments is based on the amount and the rate of penetration by SCC, with the results being expressed in terms of crack depth, a threshold stress-intensity factor (K_{ISCC} or K_{th}), or plateau velocity (V_{pl}) (Fig. 3).

For example, the relative susceptibilities of various alloys can be determined from crack depth versus time curves after testing for exposure periods as short as 150 to 200 h (Ref 35). This is illustrated for an extreme range of susceptibilities in Fig. 44. Plateau velocities in this example are indicated by graphical estimates of the average slopes of the initial portions of the crack growth curves, beginning at the time when growth started and extending until the curves definitely started to bend over toward an arrest. An arrest would indicate K_{th}, but real arrests (zero crack growth) may not occur; therefore, it is customary to define K_{th} as the crack-tip stress intensity at which the crack growth rate has decreased to the limit of measuring capability. This is usually about 10^{-10} m/s (1 to 2×10^{-5} in./h), that is, where the growth is less than 0.2 mm (0.008 in.) within 30 days.

Plateau velocities can be readily determined for materials having a relatively low resistance to SCC, such as 7075-T651 and 7079-T651 alloy plates when stressed in the short-transverse direction. Such tests have been effectively used for the evaluation of corrosive environments and the study of SCC trends with the artificial aging of 7000-series (aluminum-zinc-magnesium-copper) alloys (Ref 78, 79). However, the use of plateau velocities for comparing materials with higher resistance to SCC becomes complicated when only small amounts of crack growth occur in normal exposure periods. In such cases, the initial penetration of SCC, even at near-critical stress intensities, may be delayed by an initiation (incubation) period and then may begin at small independent sites along an uneven mechanical crack front. The crack measurements are erratic, and the interpretation of the crack growth curves is subjective. Comparisons among relatively resistant materials are difficult.

Figure 45 shows a number of crack growth curves for several resistant materials. It is evident that the estimated plateau velocities are quite variable and do not correlate consistently with the total crack growth in a given exposure time. For these more SCC-resistant materials, average growth rates for the first 15 days of exposure appear to relate much better to the

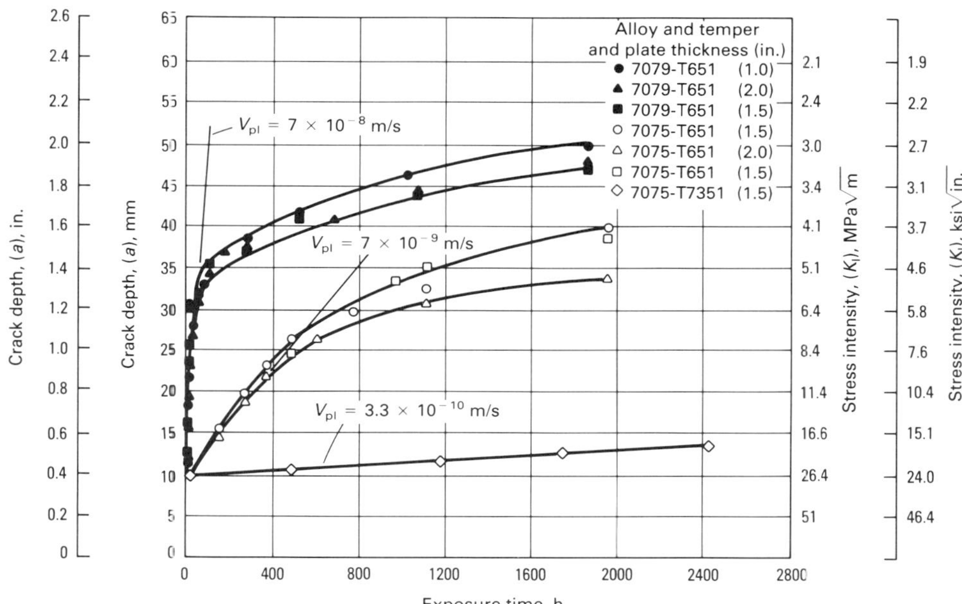

Fig. 44 Crack depth and stress intensity versus time curves for double-beam specimens of aluminum alloys 7075-T651, 7079-T651, and 7075-T7351 having nearly identical deflections and starting crack depths. Specimens with S-L orientation (see Fig. 28) measuring 25 × 25 × 127 mm (1 × 1 × 5 in.) bolt loaded to pop-in and wetted three times daily with 3.5% sodium chloride. Source: Ref 35

33, 34). Because crack growth results in increasing stress intensity and an increasing crack opening, corrosion product wedging is minimal, and each test usually has a definite end point (fracture). In these tests, fatigue precracked compact or modified compact specimens (Fig. 25b) are loaded to various initial stress intensities K_{Io} and exposed until fracture or until completion of a designated time period (Fig. 29). The designated cut-off period should be long enough for extended initiation times and yet not long enough to allow corrosion product wedging to exert a dominant influence.

The test results shown in Fig. 49 indicate that near-threshold values were reached within 1200 h, as judged by the flattening tendency of the curves. The slight downward slope of some of the curves after 1200 h may be the result of wedging by corrosion products, but this was not determined. The effect of such wedging would be to give lower estimates of the threshold stress intensity.

The testing of longitudinal (L-T, L-S in Fig. 28) and long-transverse (T-L, T-S in Fig. 28) specimens presents special problems with materials having typical directional grain structures. Stress-corrosion cracking growth is small and tends to be in the L-T plane, which is perpendicular to the plane of the precrack (Ref 36, 83). Such out-of-plane crack growth invalidates calculations of the plane-strain threshold stress intensity K_{ISCC}. On the other hand, the testing of materials having an equiaxed grain structure also presents problems with stress intensity calculations because of gross crack branching; this would be applicable to specimens of any orientation.

The most widely used corrodent for testing precracked specimens is 3.5% sodium chloride solution applied dropwise to the precrack two or (usually) three times daily (Ref 34-37). This intermittent wetting technique accelerates SCC growth but it also causes troublesome corrosion of the mechanical precrack. Less corrosive corrodents that have been used include substitute ocean water (ASTM D 1141) and an inhibited salt solution containing 0.06 *M* sodium chloride, 0.02 *M* sodium dichromate, 0.07 *M* sodium acetate, and acetic acid to pH 4 (Ref 36, 37, 81). Some investigators have tested 7000-series alloys in

actual amount of crack growth and to smooth specimen ratings according to ASTM G 64 (Table 4).

The performance of alloy 2 in Table 4 indicates another potential problem with tests performed at very high stress intensities; that is, with some very resistant materials, environmental crack growth will possibly be the result of mechanical fracture rather than by SCC. Therefore, it is necessary when testing SCC-resistant materials to verify that the crack growth is in fact SCC.

The determination of threshold stress intensities by the arrest method is frequently complicated by corrosion product wedging, which changes the stress state at the tip of the crack and invalidates the calculation of effective stress intensities from the crack lengths. With low-resistance alloys, such as 7075-T651, an arrest may never occur, because the crack is continually driven ahead by the advancing wedge of insoluble corrosion products (Ref 36, 79). An indication of this was shown by the initiation of SCC in precracked specimens exposed with no applied load for just a few months in a seacoast atmosphere (Ref 36). Experimental evidence of a threshold stress intensity will depend on the amount of corrosion occurring in a given alloy/environment system (Fig. 46).

With intermediate-resistance materials, the growth curves may develop prominent steps indicative of temporary arrests. Figure 47 shows some of the various curves that may be obtained, depending on the resistance to corrosion and SCC of the test material, the corrosivity of the test medium, the magnitude of the applied stress intensity, and the length of exposure. The significant portion of the curve is that which goes from the beginning of the test to the first appreciable cessation of the crack growth. It is assumed that if it were not for the intervention of the corrosion product wedging the curve would proceed to an arrest.

The threshold stress intensities determined by this method can be useful for ranking materials, but usually cannot be considered valid. Therefore, they cannot be used in design calculations based on fracture mechanics. Displays of complete *V-K* curves provide convenient comparisons of various materials, as shown in Fig. 48. Problems with the control of the testing procedure and of correlations with service conditions have impeded the standardization of this test method (Ref 24, 34).

Dead-weight loading, or a simulated dead-weight loading system used in conjunction with automatic data logging equipment (Fig. 30b), has proved to be a rigorous method for evaluating threshold stress intensities by SCC initiation (Ref

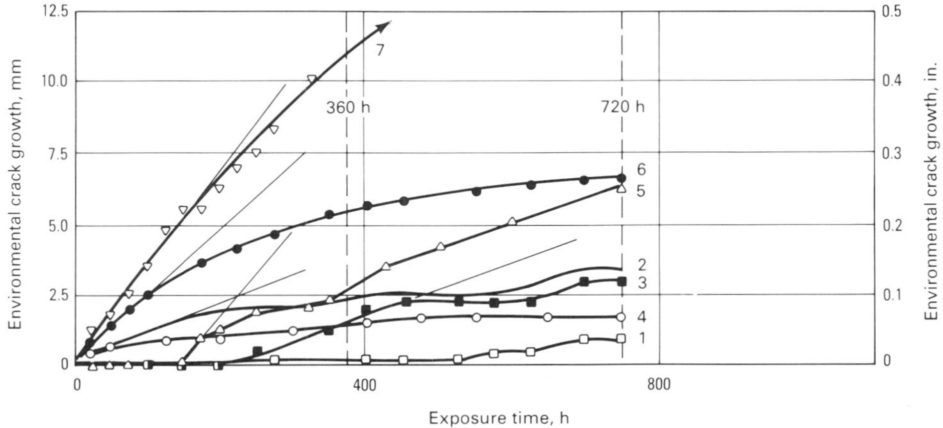

Fig. 45 Examples of SCC crack growth in various aluminum alloys with relatively high resistance to SCC. S-L (see Fig. 28) double-beam specimens bolt-loaded to pop-in and wetted with 3.5% sodium chloride three times daily; relative humidity 45%. The numbers 1 to 7 indicate different test materials listed in order of decreasing resistance to SCC (see Table 4), dashed lines indicate plateau velocities. The alloy 2 specimen failed by mechanical fracture rather than intergranular SCC. Source: Ref 36, 41

Table 4 Correlation of SCC plateau crack velocities with smooth specimen SCC ratings

Alloy	Smooth specimen rating(a)	Plateau crack velocity — First growth, m/s	First growth, in. × 10^{-5}/h	Average (0 to 15 days), m/s	Average (0 to 15 days), in. × 10^{-5}/h
1	A	6×10^{-10}	10	2×10^{-10}	3
2	A	7×10^{-9}	100(c)	1.8×10^{-9}	27(c)
3	B	2.1×10^{-9}	30	1.2×10^{-9}	19
4	B	4.2×10^{-9}	60	1.3×10^{-9}	20
5	B	7×10^{-9}	100	2.1×10^{-9}	30
6	C	6.3×10^{-9}	90	4.2×10^{-9}	60
7	D	1.1×10^{-8}	160	8.4×10^{-9}	120

(a) Short-transverse ratings per ASTM G 64 (Ref 8). (b) S-L (see Fig. 28) double-beam specimens bolt-loaded to pop-in and wetted three times daily with 3.5% sodium chloride. (c) Fractographic examination revealed mechanical fracture rather than the intergranular SCC verified in all other materials. Source: Ref 41

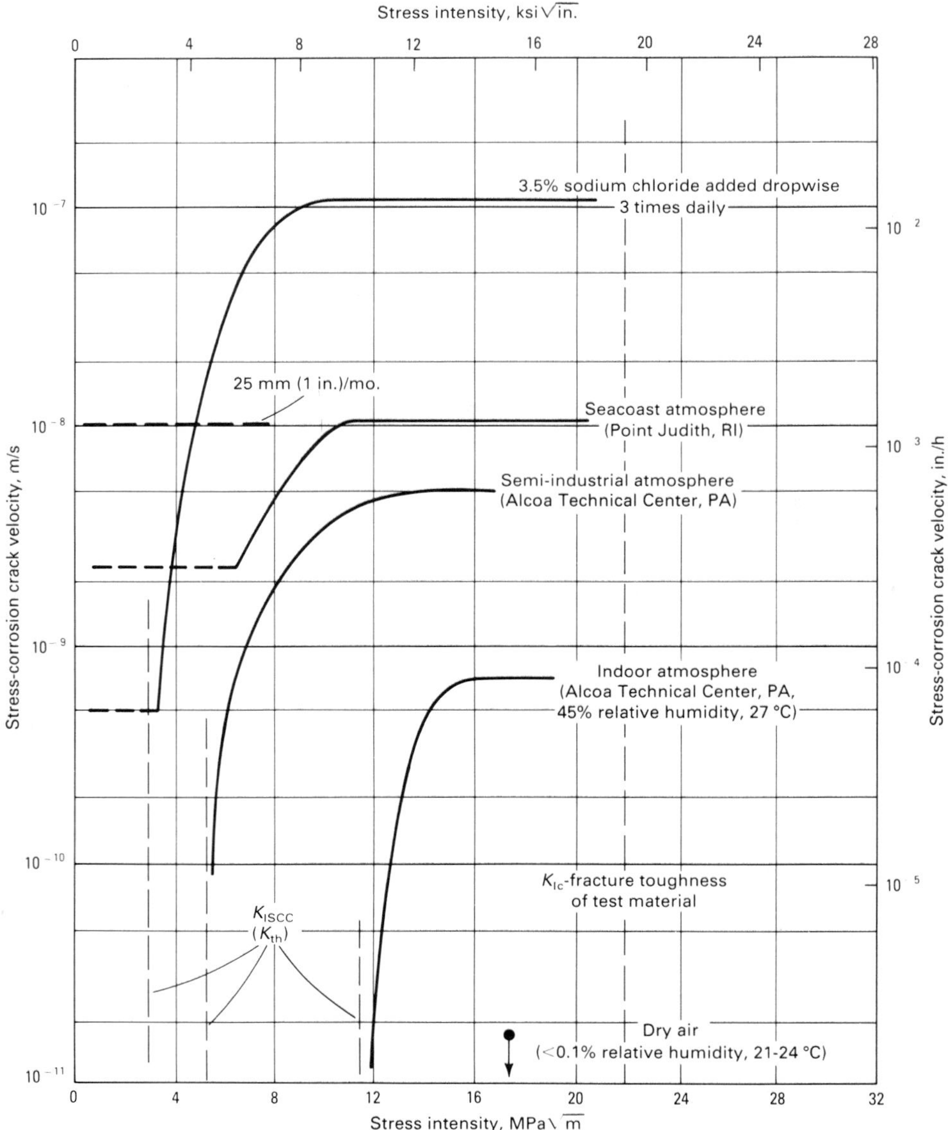

Fig. 46 Effect of corrosive environment on SCC velocity and threshold stress intensity for 7079-T651 plate (64 mm, or 2.5 in., thick) stressed in the short-transverse direction (S-L; see Fig. 28). Double-beam specimens bolt-loaded to pop-in. No SCC occurred during 3 years of exposure to dry air in a desiccator; however, the plateau velocity (horizontal part of each curve) and the apparent threshold stress intensity (K_{ISCC} or K_{th}) vary with the environment. Dashed portions of the curves represent the effect of corrosion product wedging. Source: Ref 41

distilled water (Ref 78) and in water vapor at 40 °C (104 °F) (Ref 84). Typical test durations that have been used range from 200 to 2500 h.

With low-resistance alloys, both of the first two corrodents listed in the preceding paragraph ranked alloys similarly and in agreement with exposure to a seacoast and an inland industrial atmosphere. Plateau velocities in the laboratory tests were about five to ten times faster than in the seacoast atmosphere and ten times faster than in the industrial atmosphere. In these K-decreasing laboratory tests, corrosion product wedging effects dominated after exposure periods of about 200 to 800 h. The length of exposure time before the intervention of corrosion product wedging varies with several factors, including the magnitude of K_{I0} and the inherent resistance to crevice corrosion of the test material in the corrosive environment (Ref 36, 41).

Slow Strain Rate Testing

Slow strain rate testing is not governed by any standards. Various aqueous solutions have been used in addition to 3.5% sodium chloride. Because the 3.5% sodium chloride solution did not appear aggressive enough for slow strain rate testing, more corrosive test mediums have been used, including oxidant additions to the sodium chloride solution or more acidic solutions, such as aluminum chloride (Ref 52, 85).

In a round-robin testing program using several aluminum alloy types and several corrodents, a solution containing 3% sodium chloride plus 0.3% hydrogen peroxide was considered the most promising candidate for possible standardization (Fig. 33). Additional study is needed to determine the optimum composition of these constituents. Another promising candidate was a solution of 2% sodium chloride plus 0.5% sodium chromate having a pH of 3.

Testing of Copper Alloys (Smooth Specimens)

Testing in Mattsson's Solution. According to ASTM G 37 (Ref 8), a stressed test specimen must be completely and continuously immersed in an aqueous solution containing 0.05 g-atom/L of Cu^{2+} and 1 g-mol/L of ammonium ion (NH_4^+) with a pH of 7.2. The copper is added as hydrated

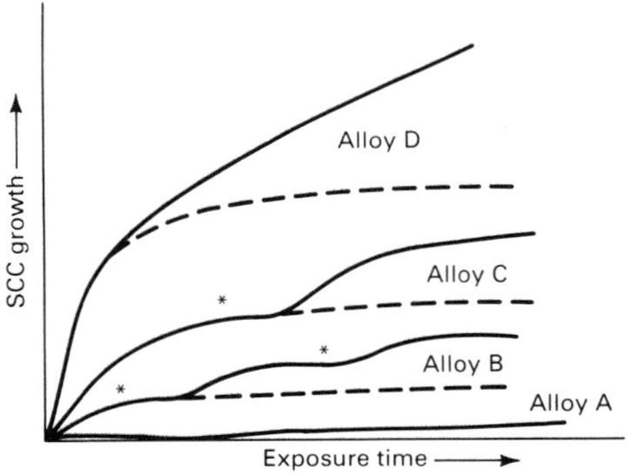

Fig. 47 Schematic of the variable effects of corrosion product wedging on SCC growth curves in a K-decreasing test. Solid lines: measured curve. Dashed lines: estimated true curve excluding the effect of corrosion product wedging. Asterisks indicate temporary crack arrests.

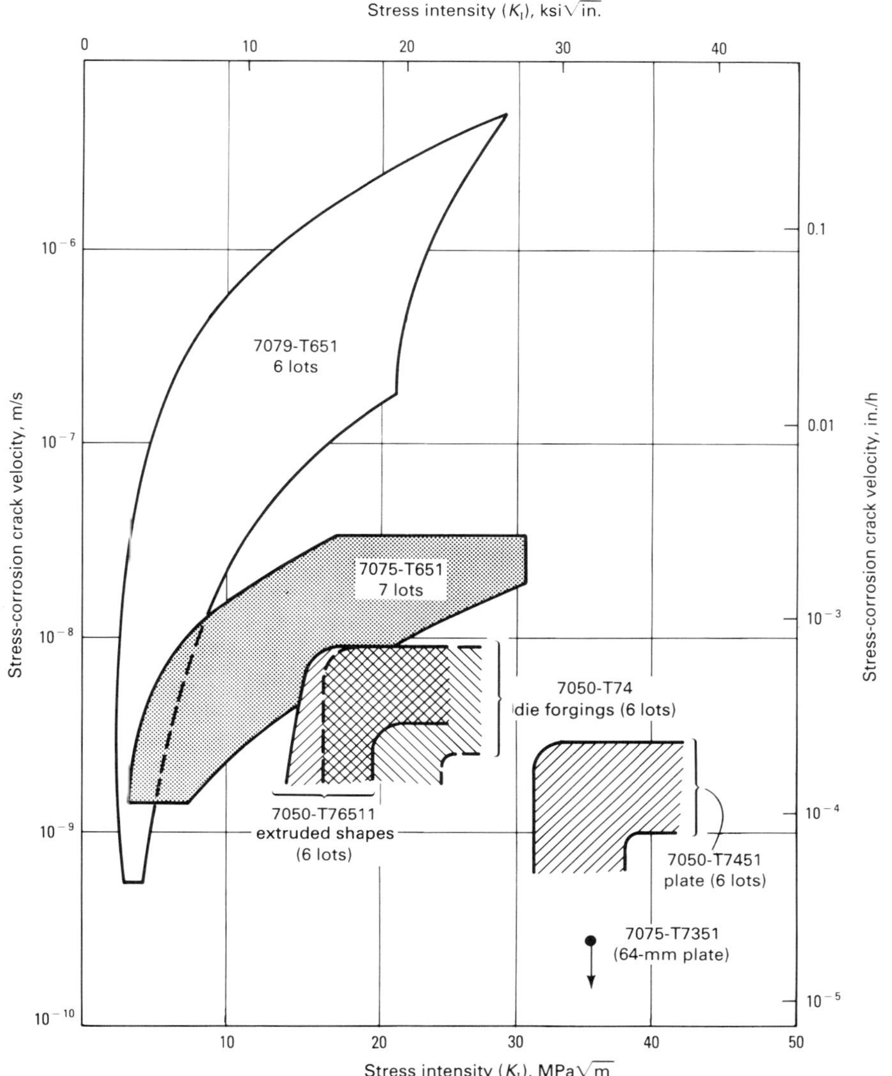

Fig. 48 SCC propagation rates for various aluminum alloy 7050 products. Double-beam specimens (S-L; see Fig. 28) bolt-loaded to pop-in and wetted three times daily with 3.5% NaCl. Plateau velocity averaged over 15 days. The right-hand end of the band for each product indicates the pop-in starting stress intensity (K_{Io}) for the tests of that material. Data for alloys 7075-T651 and 7079-T651 are from Ref 35. Source: Ref 82

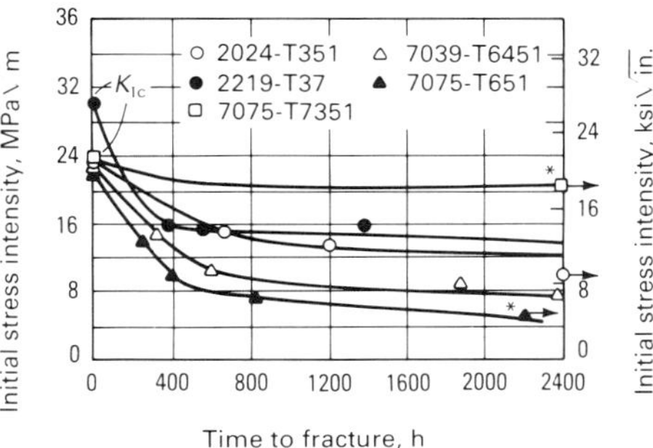

Fig. 49 Initial stress intensity versus time to fracture for S-L (see Fig. 28) compact specimens of various aluminum alloys exposed to an aqueous solution containing 0.06 *M* sodium chloride, 0.02 *M* sodium dichromate, 0.07 *M* sodium acetate, and acetic acid to pH 4. Asterisk indicates metallographic examination showed that SCC had started. Source: Ref 33

copper sulfate, and the NH_4^+ is added as a mixture of ammonium hydroxide and ammonium sulfate. The ratio of the latter two compounds is adjusted to achieve the desired pH.

Mattsson's pH 7.2 solution is recommended only for brasses (copper-zinc base alloys). This test environment may give erroneous results for other copper alloys and is not recommended. This is particularly true for alloys containing aluminum or nickel.

This test environment is believed to provide an accelerated ranking of the relative or absolute degree of susceptibility to SCC for different brasses. The test environment correlates well with the corresponding service ranking in environments that cause SCC, which may be due to the combined presence of traces of moisture and ammonia vapor. The extent to which the accelerated ranking correlates with the ranking obtained after long-term exposure to environments containing corrodents other than ammonia is not known. Such environments may be severe marine atmospheres (chloride), severe industrial atmospheres (predominantly sulfur dioxide), or superheated ammonia-free steam.

It is currently not possible to specify a time to failure in Mattsson's pH 7.2 solution that corresponds to a distinction between acceptable and unacceptable SCC behavior in brass alloys. Such correlations must be determined on an individual basis.

Mattsson's pH 7.2 solution may also cause some stress-independent general and intergranular corrosion of brasses. Therefore, SCC failure may possibly be confused with mechanical failure induced by corrosion-reduced net cross section. This is most likely with small cross-sectional specimens, high applied stress levels, long exposure times, and SCC-resistant alloys. Careful metallographic examination is recommended for accurate determination of the cause of failure. Alternatively, unstressed control specimens can be exposed to corrosive environments in order to determine the extent to which stress-independent corrosion degrades mechanical properties.

Other Testing Media. The most widely used SCC agent for copper and copper alloys is ammonia (NH_3) (Ref 86). The NH_4^+ ion does not appear to cause cracking in a stable salt, such as ammonium sulfate. Cracking will occur in a salt that dissociates (such as ammonium carbonate) to form ammonia.

The $Cu(NH_3)_x^{2+}$ ion (x is usually 4 to 5) is thought to be necessary to induce SCC in copper metals (Ref 87). Amine groups also cause cracking, or are easily converted to ammonia. Amines and sulfamic acid also cause cracking. Dry ammonia does not cause SCC of brass, as demonstrated by the successful use of brass valves and gages on tanks of anhydrous ammonia.

Stress-corrosion cracking of copper metals in ammonia will not occur in the absence of oxygen or an oxidizing agent. Carbon dioxide is also a requisite (Ref 88). Therefore, air rather than pure oxygen is necessary, and as a practical matter, moisture is essential. When other factors are favorable, a very small amount of NH_3 is sufficient to cause cracking. The controlling factor may therefore be moisture, because cracking may appear to be caused by the presence of a condensed moisture film.

Other than ammonia, the most effective agents for causing cracking are the fumes from nitric acid or moist nitrogen dioxide. Sulfur dioxide will also crack brass; but both maximum and minimum concentration limits exist, and the reaction is slow (Ref 86). Alloy development studies have been conducted with a moist ammoniacal test atmosphere containing 80% air, 16% NH_3, and 4% water vapor at 35 °C (95 °F). However, none of these corrodents has received the attention that ammonia has garnered (Ref 87).

Historically, immersion of a copper alloy product in a mercurous nitrate solution has been used to test for residual stresses (Ref 89, 90). Because these residual stresses are possible sources of failure by SCC in other environments, some have regarded this test as a stress-corrosion test. However, it is only an indirect method of identifying SCC tendencies and does not correlate to the presence of SCC as well as test methods based on specific attack by ammonia (Ref 86). It does indicate, however, that mercury and other low-melting liquid metals can cause embrittlement and failure due to cracking.

Testing of Carbon and Low-Alloy Steels

Generally, steels with lower strengths are susceptible to SCC only upon exposure to a small number of specific environments, such as the hot caustic solutions encountered in steam boilers, hot nitrate solutions, anhydrous ammonia, and hot carbonate-bicarbonate solutions (Ref 91, 92).

Boiler Water Embrittlement Detector Testing. Caustic cracking failures frequently originate in welded structures in the vicinity of faying surfaces, where small leaks cause soluble salts to accumulate in high local concentrations of caustic soda and silica. As a general rule, crevices or splash areas on hot metal surfaces where the concentration of dissolved soluble salts can occur are likely sites for SCC. This type of intergranular cracking failure has been produced with concentrations of sodium hydroxide as low as 5%, but a concentration of 15 to 30% is usually required at 200 to 250 °C (390 to 480 °F) to produce this phenomenon. The apparatus and procedures used to determine the embrittling or nonembrittling characteristics of the water in an operating boiler are detailed in ASTM D 807 (Ref 8).

Other Testing Media. Caustic cracking occurs in digester vessels used in the chemical-processing industries, and laboratory studies have been conducted using sodium hydroxide concentrations of about 30 to 35% (Ref 93). Tests in boiling nitrate solutions have frequently been used to study the effects of composition and metallurgical variables (Ref 92). In studies of low-carbon steel in boiling nitrate solutions having different cations, solutions containing the more acidic cations in greater concentrations were found to be the most potent. This tendency is illustrated by the apparent threshold stresses for failure of a 0.05% C steel in nitrate solutions with a range of concentrations, as shown in Table 5.

Cracking can be accelerated by the addition of small amounts of acid or oxidizing agents, such as potassium permanganate, manganese sulfate, sodium nitrite, and potassium dichromate, but hydroxides and other salts, particularly those forming insoluble iron products, such as sodium carbonate or sodium hydrogen phosphate, retard or prevent failure. Sodium nitrite is also a known inhibitor if the nitrite concentration is equal to that of the nitrate ion. A standard test environment has not been established, and conditions should be tailored to individual testing requirements.

The ranking of a given series of alloys may vary with exposure conditions (Ref 58). Consequently, selection of a particular alloy for use in an environment that varies from that used in laboratory ranking tests may result in unexpected service failure. This tendency is illustrated by the effects of alloying additions in ferritic steels on cracking in two different environments (Fig. 50). Figure 50(a) illustrates that each of the alloying additions is beneficial in the carbonate-bicarbonate solutions, with molybdenum having the greatest effect. However, the molybdenum addition has an adverse effect in the 35% sodium hydroxide solution, although the beneficial effects of nickel and chromium additions remain the same (Fig. 50b). Although nickel additions are beneficial in the above example, a similar addition of nickel to a carbon-manganese steel produced susceptibility to SCC in boiling magnesium chloride; this did not occur in the steel without the addition of nickel (Ref 95).

The use of laboratory testing media that duplicate service conditions is equally important when accelerated tests are used for quality control through the acceptance or rejection of production lots of a particular alloy. Reference 96 discusses tests of prestressing steels intended for use as concrete reinforcing bars (rebars) in which an ammonium thiocyanate solution was used to discriminate between heats of steel.

Use of the carbonate-bicarbonate solutions for testing pipeline steels by the slow strain rate method revealed that the susceptibility to SCC was dependent on the electrochemical potential of the specimen surface in the test environment, as shown in Fig. 50(a). A critical range in which SCC occurred was established. The critical range varies with the test environment and alloy composition. Several tests at various carbonate-bicarbonate concentrations, temperatures, pH levels, and corrosion potentials indicated that test conditions using an impressed potential of −650 mV versus the saturated calomel electrode (SCE) and a temperature of 75 °C (165 °F) were optimal (Fig. 37).

Testing of High-Strength Steels (Ref 4, 97)

For steels with yield strengths greater than about 690 MPa (100 ksi)—such as low-alloy and alloy steels, hot-work die steels, maraging steels, and martensitic and precipitation-hardenable stainless steels—the environments that cause SCC are not specific. In many alloy systems, the phenomena of SCC and hydrogen embrittlement cracking are indistinguishable (Fig. 1). This is particularly the case in environments that contain sulfides or other promoters of hydrogen entry.

Environments of major concern are natural waters—for example, rainwater, seawater, and atmosphere moisture. Any of these environments may become contaminated, which significantly increases the likelihood of SCC. Contamination

(a)

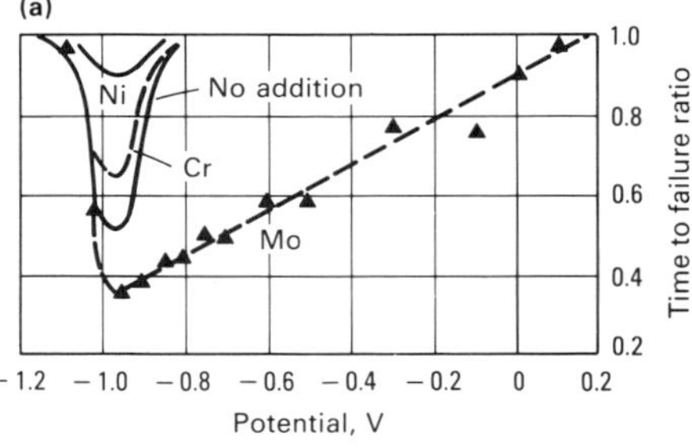

(b)

Fig. 50 Effect of various alloying elements on the SCC behavior of a low-alloy ferritic steel in two different corrosive environments. Behavior indicated by time to failure ratios in a slow strain rate test. (a) Immersed in 1 *N* sodium carbonate plus 1 *N* sodium bicarbonate at 75 °C (165 °F). (b) Immersed in boiling 35% sodium hydroxide. Source: Ref 94

with hydrogen sulfide is particularly serious; consequently, the presence of hydrogen sulfide in high concentrations in salt water associated with certain deep oil wells (termed sour wells; see the article "Corrosion in Petroleum Production Operations" in this Volume) places an upper limit of approximately 620 MPa (90 ksi) on the yield strength that can be tolerated in stressed steel in such environments without cracking.

Sulfide Stress Cracking. Determination of sulfide stress cracking is covered in NACE TM-01-77 (Ref 98). Stressed specimens are immersed in acidified 5% sodium chloride solution saturated with hydrogen sulfide at ambient pressure and temperature. The solution is acidified with the addition of 0.5% acetic acid, yielding an initial pH of approximately 3. Applied stress at convenient increments of the yield strength is used to obtain cracking data that are plotted as shown in Fig. 51. A 30-day test period is considered sufficient to reveal failure of susceptible material in most cases.

The purpose of this test standard is to facilitate conformity in testing. Evaluation of data requires individual judgment on several points based on the specific requirements of the end use. Consequently, the test should not be used as a single criterion for evaluating an alloy for use in environments containing hydrogen sulfide or other hydrogen charging elements. Attention should be paid to other factors that may affect SCC, such as pH, temperature, hydrogen sulfide concentration, corrosion potential, and stress level, when determining the suitability of a metal for use.

Table 5 Apparent threshold stress values for 0.05% C steel in nitrate solutions of varying concentrations

	Apparent threshold stress values at a solution concentration of:							
	8 *N*		4 *N*		2.5 *N*		1 *N*	
Nitrate solution	MPa	ksi	MPa	ksi	MPa	ksi	MPa	ksi
Ammonium nitrate	14	2	21	3	48	7	83	12
Calcium nitrate	34	5	48	7	83	12	159	23
Lithium nitrate	34	5	55	8	131	19 at 2 *N*	159	23
Potassium nitrate	41	6	62	9	97	14	165	24
Sodium nitrate	55	8	131	19	152	22	179	26

Source: Ref 92

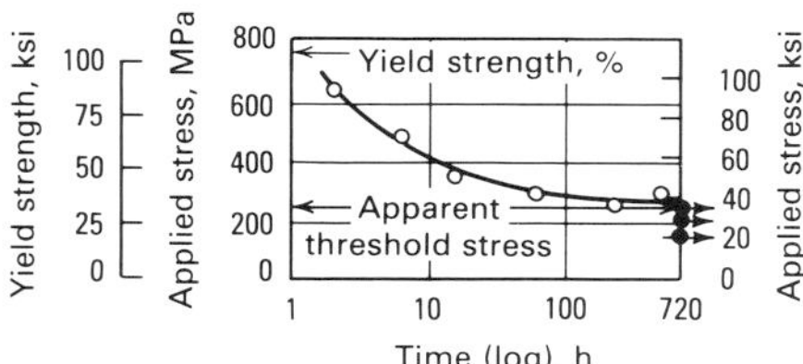

Fig. 51 Method of plotting results of sulfide stress cracking tests. Open symbols indicate failure; closed symbols indicate runouts. Source: Ref 98

Table 6 Influence of cutoff time on apparent K_{ISCC} using the SCC initiation method

Exposure time, h	Apparent K_{ISCC} MPa $\sqrt{m}$	ksi $\sqrt{in.}$
100	187	170
1000	127	110
10 000	28	25

Note: The initiation method was used on a constant-load cantilever bend specimen (K-increasing) of alloy steel with a yield strength of 1240 MPa (180 ksi). Test environment was synthetic seawater at room temperature. Source: Ref 24

Table 7 Comparison of K_{ISCC} values determined by initiation and arrest methods

Steel alloy	K_{ISCC}, MPa $\sqrt{m}$ (ksi $\sqrt{in.}$) Initiation	Arrest
10Ni, normal purity	24 (22)	26 (24)
10Ni, high purity	59 (54)	57 (52)
18Ni, normal purity	22–33 (20–30)	28 (25)
18Ni, high purity	<33 (<30)	<33 (<30)

Note: Based on a crack growth rate of 2.5×10^{-4} mm/h (10^{-5} in./h). Modified compact specimens: constant load for initiation and wedge-loaded with a bolt for arrest. Test environment: salt water at room temperature. Source: Ref 24

The NACE test method recommends the use of smooth, small-diameter tension specimens stressed with constant-load or sustained-load devices (Ref 98). However, different types of beam and fracture mechanics specimens may be included in the testing standard in the future.

Another test method, known as the Shell Bent Beam Test, has been used for over 25 years in the petrochemical industry to rank various materials for use in sour environments (Ref 99). However, acceptance has not been sufficient to generate the interest for standardization.

Testing in sodium chloride solution constitutes a worst-case determination for high-strength steels; as such, it is generally considered unrealistically aggressive for the useful ranking of steels in service environments that do not contain hydrogen sulfide or other conditions favoring entry of hydrogen. Tests are usually performed in water containing about 3.5% sodium chloride, artificial seawater, natural seawater (rarely), or a marine atmosphere (Ref 4), unless specific environmental conditions are under study. ASTM G 44 (Ref 8) is used where applicable.

In salt water and freshwater, a true threshold K_{ISCC} exists for high-strength steels that is useful for characterizing resistance to SCC. Ideally, K_{ISCC} defines the combination of applied stress and defect size below which SCC will not occur under static loading conditions in a given alloy and environment system. However, the reported value of K_{ISCC} for a given system often reflects the initial K_I level and the exposure time associated with the testing. Table 6 illustrates the risk of overestimating K_{ISCC} by terminating the exposure test too soon when using the SCC initiation method (Ref 23, 24). A similar risk exists in tests conducted with the arrest method. Table 7 shows that K_{ISCC} values determined by the initiation and arrest methods may be the same when testing times are sufficiently long and when compatible criteria are used for establishing the threshold (Ref 24).

Figure 52 illustrates a method used to compare various high-strength steels (Ref 4, 100). Data were obtained in salt water or seawater, and K_{ISCC} values are plotted versus yield strength. Envelopes are used to enclose all known valid data for the various steels. The crosshatched envelopes or individual data points represent the featured steels, which allows comparison with characteristics of the other steels. The straight lines in Fig. 52 illustrate how K_{ISCC} values relate to the maximum depth of long surface flaws that can be tolerated without stress-corrosion crack growth.

Electrochemical Polarization. Although the mechanism of cracking in hydrogen sulfide environments is predominantly one of hydrogen embrittlement, the mechanism of environmentally induced failures in environments not containing sulfides or other promoters of hydrogen entry is not clearly agreed upon (Ref 97). Time to failure in a sodium chloride solution depends on the corrosion potential (Ref 4, 101), which determines whether failure results from active path corrosion or hydrogen embrittlement cracking. Electrochemical studies have shown that embrittlement of high-strength steels by corrosion product hydrogen occurs when, for a given environment, the electrochemical potential of the metal is equal to or more anodic than the reversible hydrogen potential, that is, for thermodynamic conditions that favor the deposition of hydrogen on the surface of the steel.

Figure 53 compares the various types of cracking behavior that can be expected from electrochemical polarization (Ref 102). All of the curves except curve G were obtained experimentally. Curve A represents the case in which only hydrogen embrittlement is obtained; curve B shows only active path corrosion. Both processes are shown in curves C and D.

When both anodic and cathodic polarization shorten the cracking time, as in curve E, it is not

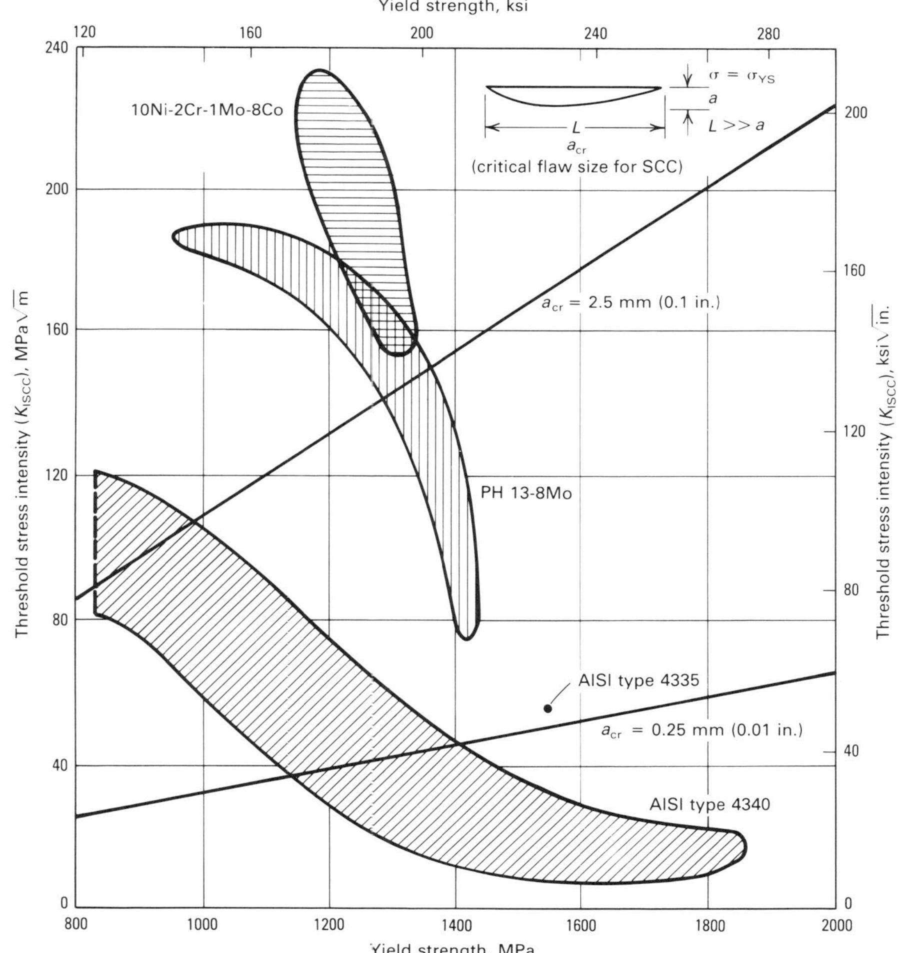

Fig. 52 Comparison of SCC behavior of several high-strength steels based on threshold stress intensity (K_{ISCC}) values in salt water. Source: Ref 100

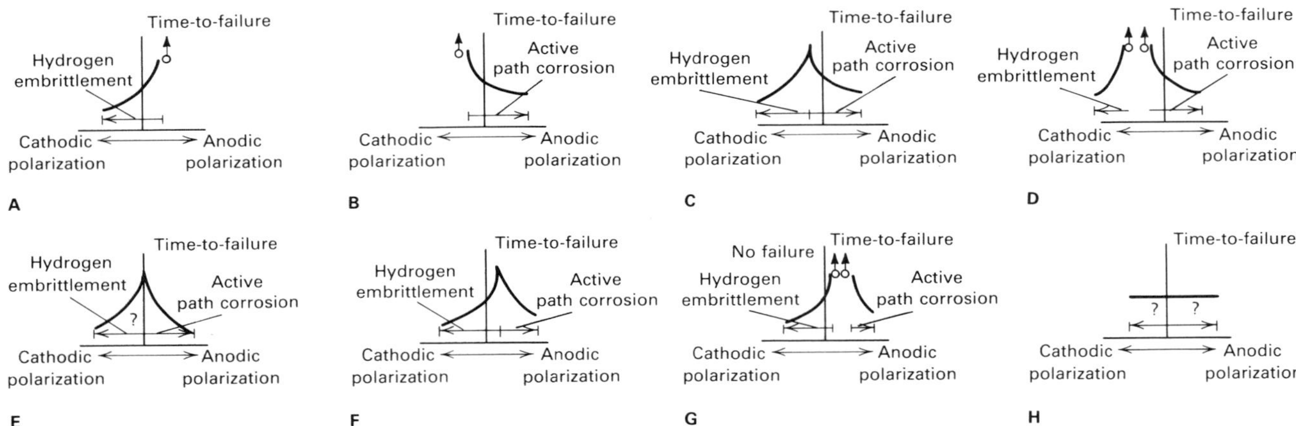

Fig. 53 Use of electrochemical polarization to distinguish between SCC and hydrogen embrittlement mechanisms in a high-strength steel immersed in sodium chloride solution. See text for explanation of curves A through H. Source: Ref 102

possible to determine which mechanism prevails without applied current. Curves F and G can be expected in acid solutions when the corrosion potential is anodic to the reversible hydrogen potential.

In curve H, neither anodic nor cathodic polarization has any effect on cracking time. Therefore, it is possible that a hydrogen embrittlement mechanism is involved. However, the mechanism by which hydrogen enters the steel is not electrochemical. To perform realistic accelerated tests, the end use of the material and the environmental conditions involved should be considered so that the test procedure involves the appropriate cracking mechanism. It should be noted that hydrogen embrittlement cracking can also occur as a result of galvanic action between the test specimen and components of the stressing system. In all SCC testing, therefore, all electrical contact between the specimen and ancillary fixtures must be avoided, except when galvanic effects are desired.

Testing of Nonheat-Treatable Stainless Steels

The environments causing SCC that are encountered in the chemical industry are specific and are limited primarily to chloride and caustic solutions at elevated temperatures and sulfide environments at ambient temperatures. In seawater at or near room temperature, austenitic (iron-chromium-nickel) and ferritic (iron-chromium) steels do not experience SCC. Fully ferritic stainless steels are highly resistant to SCC in chloride and caustic environments that cause austenitic stainless steels to crack. However, laboratory studies have shown small additions of nickel or copper to ferritic steels may render them susceptible to SCC in severe environments (Ref 4).

Testing in Boiling Magnesium Chloride Solution. ASTM G 36 (Ref 8) is applicable to wrought, cast, and welded austenitic stainless steels and related nickel-base alloys. This method determines the effects of composition, heat treatment, surface finish, microstructure, and stress on the susceptibility of these materials to chloride SCC. Although this test can be performed with various concentrations of magnesium chloride, ASTM G 36 specifies a test solution maintained at a constant boiling temperature of 155.0 ± 1.0 °C (311.0 ± 1.8 °F), that is, approximately 45% magnesium chloride. Also described is a test apparatus capable of maintaining solution concentration and temperature within the recommended limits for extended periods of time. Typical exposure times are up to 1000 h. However, historically, most of the SCC data on austenitic stainless steels were obtained by using a boiling 42% $MgCl_2$ solution (boiling point: 154 °C, or 309 °F). For this reason, much current testing is still done at the lower concentration.

Most chloride cracking testing has been carried out in accelerated test media such as boiling magnesium chloride (Ref 4, 103, 104). All austenitic stainless steels are susceptible to chloride cracking as shown in Fig. 54. It is noteworthy, however, that the higher-nickel types 310 and 314 were appreciably more resistant than the others (Fig. 55). Although this solution causes rapid cracking, it does not necessarily simulate the cracking observed in field applications.

Other ions in addition to chloride can cause cracking. Of all halogen ions, chlorides cause the most cases of SCC in austenitic stainless steels. Known cases of fluoride and bromide SCC are few, and iodide is not known to produce SCC. In addition, cations, such as Li^+, Ca^{2+}, Zn^{2+}, NH_4^+, Ni^{2+}, and Na^+, affect test results to varying degrees (Ref 107). Although chloride SCC occurs primarily at temperatures above about 90 °C (190 °F), acidified chloride solutions can produce SCC at low temperatures (Ref 107-109). Therefore, in diagnosing service failures, it is necessary to establish which ions (and other environmental and stress conditions as well) have caused the failure. In this manner, an appropriate test procedure can be designed for the evaluation of alternative materials.

Reference 110 discusses laboratory reproduction of an environment that caused SCC at the top of a distillation tower in a crude oil refinery. The service environment consisted of a very dilute hydrochloric acid solution (36 ppm chlo-

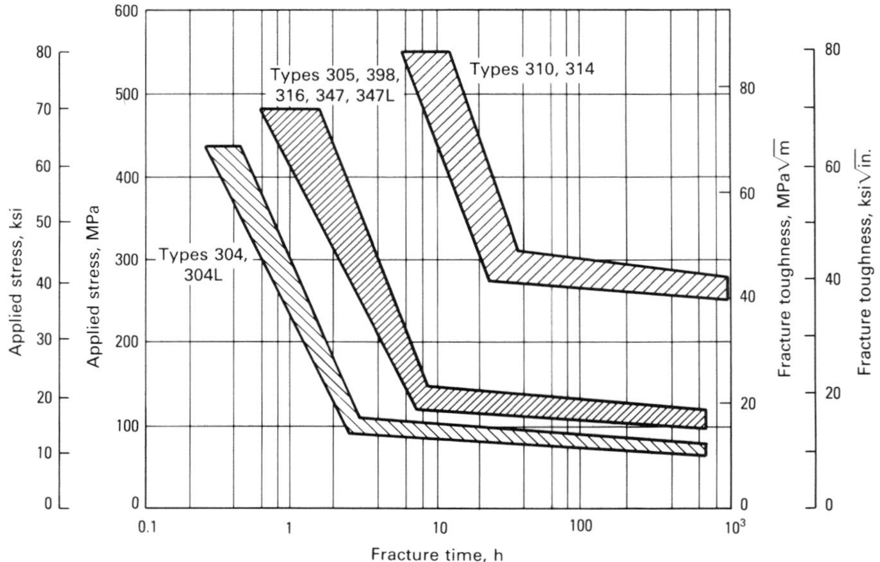

Fig. 54 Relative SCC behavior of austenitic stainless steels in boiling magnesium chloride. Source: Ref 105

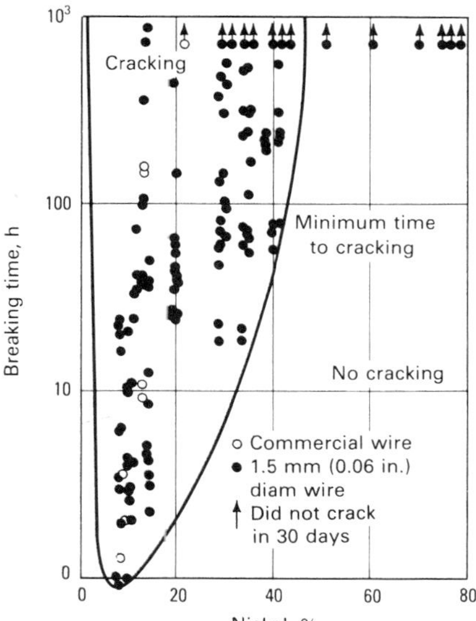

Fig. 55 Effect of nickel additions to a 17 to 24% Cr steel on resistance to SCC in boiling 42% magnesium chloride. 1.5-mm (0.06-in.) diam wire specimens dead-weight loaded to 228 or 310 MPa (33 or 45 ksi). Source: Ref 106

ride) with a pH of 3 saturated with hydrogen sulfide gas at 80 °C (175 °F). In this test environment, austenitic stainless steels, such as type 304 or 316 failed, but the ferritic types 430 and type 434 did not.

Testing in Polythionic Acids. Petrochemical refinery equipment is subject to polythionic acid cracking, which may occur after shutdown. Polythionic acid forms by the decomposition of sulfides on metal walls in the presence of oxygen and water. ASTM G 35 (Ref 8) describes procedures for preparing and conducting exposures to polythionic acids ($H_2S_nO_6$, where n is usually 2 to 5) at room temperature to determine the relative susceptibility of sensitized stainless steels or related materials (high nickel-chromium-iron alloys) to intergranular SCC.

This test method can be used to evaluate stainless steels or other materials in the as-received condition or after high-temperature service (480 to 815 °C, or 900 to 1500 °F) for prolonged periods of time. Wrought products, castings, and weldments of stainless steels or other related materials used in environments containing sulfur or sulfides can also be evaluated. Other materials that are capable of being sensitized can also be tested.

A variety of smooth SCC test specimens, surface finishes, and methods of applying stress can be used. Stressed specimens are immersed in the polythionic acid solution, which can be prepared by passing a slow current of hydrogen sulfide gas for 1 h through a fritted glass tube into a flask containing chilled (0 °C, or 32 °F) 6% sulfurous acid, after which the liquid is kept in a stoppered flask for 48 h at room temperature. Solutions can also be prepared by passing a slow current of sulfur dioxide gas through a fritted glass bubbler submerged in a container of distilled water at room temperature. This is continued until the solution becomes saturated. The hydrogen sulfide gas is then slowly bubbled into the sulfurous acid solution.

Prior to use, the polythionic acid solution should be filtered to remove elemental sulfur and then tested for acid content. This can be done by analytical tests or by using a control test specimen of sensitized type 302 stainless steel. The control should fail by cracking in less than 1 h.

The wick test can be used to evaluate the chloride cracking characteristics of thermal insulation for applications in the chemical process industry. ASTM C 692 (Ref 8) covers the methodology and apparatus used to conduct this procedure. When a dilute aqueous solution is transmitted to a metal surface by capillary action through an absorbent fibrous material, the process is called wicking. Cracking occurs at much lower temperatures when alternate wetting and drying is used than when the specimens are kept wet continuously.

Other Testing Media. Hot concentrated caustic solutions are another type of environment encountered in chemical industries that causes SCC of stainless steels. However, the conditions leading to caustic cracking are more restrictive than those leading to chloride cracking, and caustic environments have not received the attention that chlorides have. There is little difference in the susceptibilities among types 304, 304L, 316, 316L, 347, and USS 18-18-2 austenitic steels. All of these alloys crack rapidly in solutions of 10 to 50% sodium hydroxide at 150 to 370 °C (300 to 700 °F) (Ref 4, 104, 111).

Certain strong acid solutions containing chlorides, such as 5 N sulfuric acid plus 0.5 N sodium chloride, 3 N perchloric acid plus 0.5 N sodium chloride, and 0.5 N to 1.0 N hydrochloric acid, are capable of causing SCC in austenitic stainless steels at room temperature (Ref 4). Cracking in these environments is similar to the type of cracking that occurs in hot chloride environments.

Electrochemical Polarization. Stress-corrosion cracking in austenitic and ferritic stainless steels can be delayed or prevented by the application of cathodic current; however, if ferritic steels are overprotected by relatively large cathodic current, they are apt to blister or crack due to the hydrogen discharged by the cathodic protection action. Anodic polarization significantly accelerates the initiation of SCC, but appears to have a smaller accelerating effect on crack propagation (Ref 112).

Testing of Magnesium Alloys

There is no standard accelerated test environment recommended for assessing the susceptibility of magnesium-base alloys to SCC. Exposure of stressed specimens to the atmosphere has generally been used to determine the SCC susceptibility of specific products.

The chloride-containing solutions typically used in accelerated tests for aluminum alloys are unsatisfactory for SCC tests of magnesium alloys because of excessive general corrosion. In one investigation, a chromate-inhibited chloride solution (35 g/L sodium chloride plus 20 g/L potassium chromate; pH 8) was found to be suitable for testing magnesium alloys (Ref 113). Good correlation was observed between the SCC behavior of magnesium-aluminum-zinc alloys exposed by total immersion in this solution and the behavior of the same alloys exposed to an industrial atmosphere. Cracking of highly stressed susceptible alloys occurs within a few hours, but exposures can be continued up to 1000 h without incurring excessive pitting. Laboratory tests also have been conducted using potassium hydrogen fluoride and a dilute solution of sodium chloride plus sodium bicarbonate as the test medium (Ref 114).

Testing of Nickel Alloys

Nickel-base alloys are highly resistant to the chloride SCC that affects stainless steels. Iron-chromium-nickel alloys with nickel contents greater than 50% are immune to cracking in boiling 42% magnesium chloride (Fig. 55). However, SCC of nickel and high-nickel alloys has been experienced in high-temperature caustic soda and caustic potash solutions and in molten caustic.

Cracking of some nickel-base alloys has also occurred under special conditions in fluosilicic acid, hydrofluoric acid, mercuric salt solutions, and high-temperature water and steam that are contaminated with trace amounts of oxygen, lead, fluorides, or chlorides (Ref 103, 106, 115). Sensitized alloys are susceptible to SCC in sulfur compounds such as sodium sulfite, sodium thiosulfate, and polythionic acids.

The standard test environments that are most frequently used for high-nickel alloys are the same as those employed for stainless steels. In a study of sulfur-induced SCC of sensitized Inconel alloy 600 steam generator tubing in water contaminated by air and sodium thiosulfate at temperatures from 22 to 95 °C (72 to 203 °F), a solution of 0.1 M sodium tetrathionate with a pH of 3.5 to 4.0 at 22 °C (72 °F) appeared to be an excellent test medium for sensitization in nickel alloys and stainless steels. Slow strain rate testing was also found to be more effective than tests with statically loaded U-bend specimens (Ref 116).

Slow strain rate testing was also effective for evaluating several nickel- and cobalt-base alloys in hot chloride and hot caustic solutions. The average length of secondary stress-corrosion cracks, as determined by metallographic examination, appeared to be a more appropriate parameter for quantifying the severity of SCC behavior than loss in ductility or loss in fracture strength parameters; this is illustrated in Table 8 for Hastelloy alloy C-276. However, when using slow strain rate testing methods, care must be taken not to confuse stress-assisted localized corrosion with SCC (Ref 117).

Testing of Titanium Alloys

Although titanium alloys are not susceptible to SCC in either boiling 42% magnesium chloride or boiling 10% sodium hydroxide solutions, which are commonly used to study SCC in stainless steels, the susceptibility of titanium and its alloys to SCC has been demonstrated in several environments. This information is given in Table 9.

Testing in a Hot Salt Environment. The hot salt test consists of exposing a stressed salt-coated test specimen to an elevated temperature for various predetermined lengths of time. The exposure periods are determined by the alloy, stress level, temperature, and selected damage criterion (that is, embrittlement, cracking, or rupture, or a combination of these phenomena). Exposures are typically carried out in laboratory ovens or furnaces equipped with loading equip-

Table 8 Results of slow strain rate tests on Hastelloy alloy C-276

Alloy condition	Strain rate, s^{-1}	Environment	Reduction in area, %	Ultimate tensile strength MPa	Ultimate tensile strength ksi	Time to failure, h	Average length of secondary stress-corrosion cracks
Mill annealed	3.4×10^{-6}	Air	71	745	108	60	0
	3.4×10^{-6}	50% sodium hydroxide, 147 °C (297 °F)	61	593	86	52	13×10^{-5} m (5 mils)
50% cold swaged	3.4×10^{-6}	Air	49	1524	221	17	0
	3.4×10^{-6}	50% sodium hydroxide, 147 °C (297 °F)	51	1503	218	18	$<1 \times 10^{-5}$ m (<0.1 mil); no obvious SCC
	9×10^{-7}	Air	53	1558	226	29	0
	9×10^{-7}	50% sodium hydroxide, 147 °C (297 °F)	47	1524	221	30	2.5×10 m (1 mil)
	5.3×10^{-7}	Air	51	1593	231	51	0
	5.3×10^{-7}	50% sodium hydroxide, 147 °C (297 °F)	47	1565	227	60	4.8×10^{-5} m (1.9 mils)

Source: Ref 116

Table 9 Environments and temperatures conducive to SCC of titanium alloys

Environment	Temperature, °C (°F)
Hot dry chloride salts	260–480 °C (500–900 °F)
Seawater, distilled water, and aqueous solutions	Ambient
Nitric acid, red fuming	Ambient
Nitrogen tetroxide	Ambient to 75 °C (165 °F)
Methanol, ethanol	Ambient
Chlorine	Elevated
Hydrogen chloride	Elevated
Hydrochloric acid, 10%	Ambient to 40 °C (105 °F)
Trichloroethylene	Elevated
Trichlorofluoroethane	Elevated
Chlorinated diphenyl	Elevated

ment for stressing specimens. Environmental conditions, the degree of control required, and the means for obtaining control are described in ASTM G 41 (Ref 8).

This test method can be used to test all metals if service conditions warrant. The test limits maximum operating temperatures and stress levels, or it categorizes different alloys according to their susceptibility if hot salt damage has been found to accelerate failure by creep, fatigue, or rupture. Although limited evidence relates this phenomenon to actual service failures, cracking under stress in a hot salt environment is a potential design-controlling factor.

The hot salt test should not be construed as being related to the SCC of materials in other environments. It should be used only in an environment that may be encountered in service.

Hot salt testing can be used for alloy screening to determine the relative susceptibility of metals to embrittlement and cracking and to determine the time-temperature-stress threshold levels for the onset of embrittlement and cracking. However, certain types of specimens are more suitable for each of these types of characterizations. Precracked specimens are unsuitable for testing of titanium alloys, because cracking reinitiates at salt/metal/air interfaces and results in many small cracks that extend independently. Therefore, smooth specimens are recommended.

Testing in Water and Aqueous Solutions. Water, seawater, and almost any neutral aqueous solution (except atmospheric water vapor) can cause SCC in many titanium alloys in the presence of preexisting cracklike flaws, although susceptibility in these environments cannot be detected by smooth specimens. Therefore, fracture mechanics type characterizations are necessary. For titanium alloys, the extremely rapid growth of stress-corrosion cracks in salt water and the dependency on specimen geometry preclude the possibility of using crack growth rate data for design purposes.

Therefore, ranking of materials must be based on K_{ISCC} values, and a true threshold stress intensity for SCC apparently does exist (Ref 118). Titanium alloys do not exhibit stage I type crack growth kinetics (Fig. 3) in neutral aqueous solutions. Tests have been performed for sufficient periods of time to allow detection of crack growth rates of 10^{-9} m/s (1.4×10^{-4} in./h), but SCC has not been observed. The slowest crack velocity that has been detected is 10^{-8} m/s (1.4×10^{-3} in./h). Therefore, in neutral aqueous solutions, a threshold K_{ISCC} exists at which SCC will not propagate (Ref 4, 118). The above rates, however, are not as slow as those observed in high-susceptibility aluminum alloys (Fig. 46). Tests are commonly performed in water containing about 3.5% sodium chloride, artificial seawater, or natural seawater unless specific environments are being tested.

Electrochemical Polarization. The halide ions (chloride, bromide, and iodide) are SCC agents unique for titanium alloys in aqueous solutions at room temperature. The crack initiation load and velocity are controlled by the applied potential, as illustrated for the crack initiation load in Fig. 56. At potentials more negative than about −700 to −1400 mV, depending on the solution, specimens were cathodically protected. Sodium fluoride solution and solutions of the other anions that do not produce SCC (hydroxide, sulfide, sulfate, nitrite, nitrate, perchlorate, cyanide, and thiocyanate) yielded results at all potentials in the same scatterband as the air values.

At potentials more positive than the above values, susceptibility in varying degrees occurred in the chloride, bromide, and iodide solutions. The width of the critical potential range and the potential for maximum susceptibility varies with the anion. A region of anodic protection occurred in the chloride and bromide solutions, but not in the iodide solution.

Crack propagation can be halted by switching the potential to either the anodic or cathodic protection zone. The corrosion potential of titanium alloys in 3.5% sodium chloride and seawater—about −800 mV versus SCE—is similar (slightly more negative) to the potential at which SCC susceptibility reaches a maximum (Ref 118).

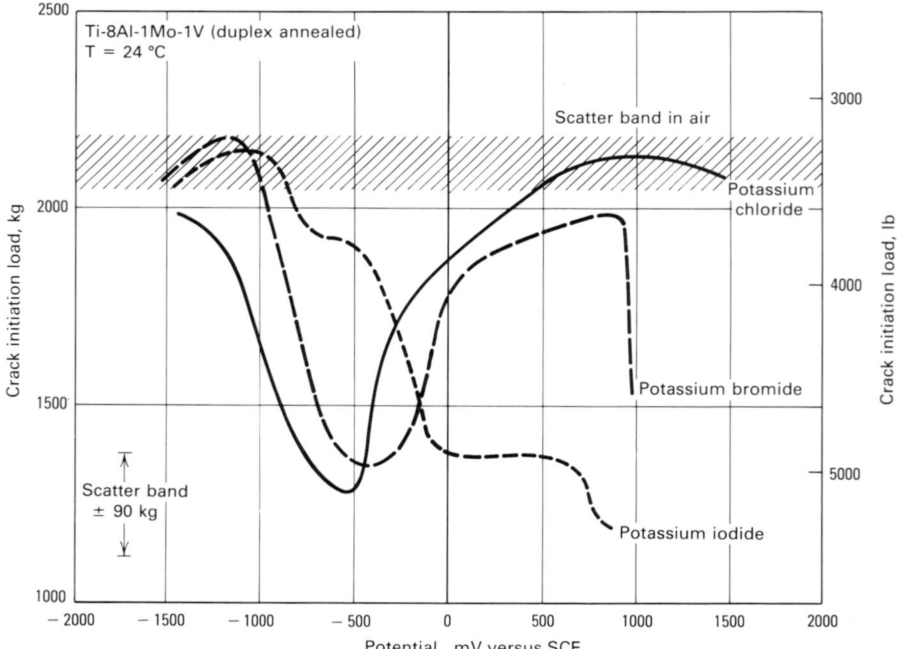

Fig. 56 Variation of crack initiation load with potential in 0.6 *M* halide solutions for Ti-8Al-1Mo-1V. Specimen: single-edge cracked sheet that was tension loaded by constant displacement. Source: Ref 118, 119

Testing in Organic Fluids. A wide variety of organic fluids can cause SCC in some titanium alloys under specific test conditions (Table 9). Most of these fluids attack the passive surface film that is characteristic of titanium alloy products. Consequently, precracked specimens do not have to be used to accelerate the SCC initiation. A standard environment does not exist; test conditions must be selected with appropriate consideration given to the type of environmental service required.

Sustained-Load Cracking in Inert Environments. High-strength titanium alloys for use in highly stressed components for military aircraft and other similar applications may be susceptible to sustained-load cracking in inert environments (including dry air). Sustained-load cracking is similar to SCC except that it is much slower and occurs in the total absence of a reactive environment. Sustained-load cracking is caused by, or is greatly aggravated by, hydrogen dissolved in the titanium during processing. Vacuum annealing can reduce the hydrogen level to less than 10 ppm, at which concentration the tendency toward sustained-load cracking is greatly reduced (Ref 4, 120).

Figure 57 illustrates an example of sustained-load cracking in mill-annealed plate of Ti-8Al-1Mo-1V containing 48 ppm hydrogen. As shown in Fig. 57, the threshold stress intensity factor for sustained-load cracking in dry air is designated K_{IH} because it is attributed to hydrogen in the metal. When the hydrogen concentration was reduced to 2 ppm by vacuum annealing, the K_{IH} value was increased to equal the inherent plane-strain fracture toughness, K_{Ic}. However, the K_{Iscc} value was not affected (Ref 121). Therefore, in addition to the practical importance of sustained-load cracking, its potential contribution to cracking should be taken into account when evaluating environmental effects, particularly in mechanistic studies.

Special Considerations for Testing of Weldments

ASTM G 58 (Ref 8) covers test specimens in which stresses are developed by the welding process only (that is, residual stress, Fig. 23), an externally applied load in addition to the stresses due to welding (Fig. 7e), and an externally applied load only, with residual welding stresses removed by annealing.

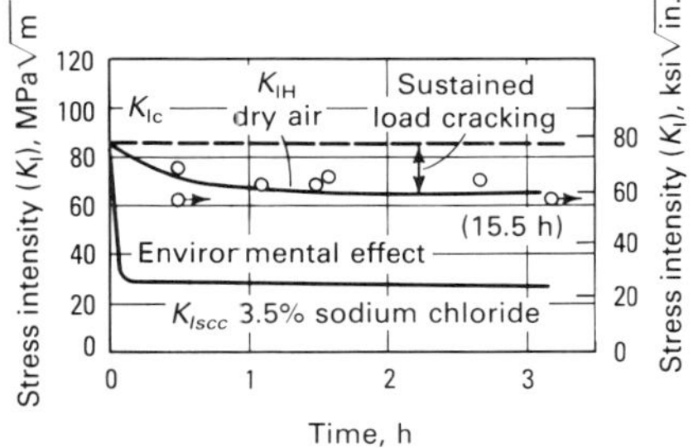

Fig. 57 Effect of sustained-load cracking compared to SCC in Ti-8Al-1Mo-1V mill-annealed sheet. Hydrogen concentration, 48 ppm; yield strength, 850 MPa (123 ksi); cantilever bend specimen (T-S); B = 6.35 mm (0.25 in.). See Fig. 28 for an explanation of specimen orientation and fracture plane identification. Source: Ref 121

The National Materials Advisory Board Committee on Environmentally Assisted Cracking Test Methods for High-Strength Weldments recently published the following guidelines on SCC testing of weldments (Ref 24). Fracture mechanics of cracked bodies was found to be a valid and useful approach for designing against environmentally assisted cracking, although several limitations and difficulties must be taken into consideration. For static loading, K_{Iscc} and da/dt versus K_I are useful parameters. They are specified to a material, temperature, and metal/environment system and are functions of local chemical composition, microstructure, and so on.

Superimposed minor load fluctuations and infrequent changes in load can alter environmental cracking response. This effect, which cannot be predicted from K_{Iscc} and da/dt values, may be significant and detrimental. Reexamination of static loading as a design premise may be required. Existing test methodology or environmentally assisted cracking tendency is applicable to the evaluation of weldments. As in other structural components, residual stress must be treated in a quantitative and realistic manner.

The National Materials Advisory Board report supports current design emphasis based on the presumption of preexisting cracklike flaws in the structure and covers testing with precracked (fracture mechanics) specimens only. It contains a critical assessment of the problems associated with environmentally assisted cracking in high-strength alloys and of state-of-the-art design and test methodology.

Surface Preparation of Smooth Specimens

The pronounced effect of surface conditions on the time required to initiate SCC in test specimens is well known (Ref 5). Unless the as-fabricated surface is being studied, the final surface preparation generally preferred is a mechanical process followed by degreasing. However, chemical etches or electrochemical polishes can be used to remove heat-treating films or thin layers of surface metal that may have become distorted during machining.

Care should be exercised to select an etchant that will not selectively attack constituents or phases in the metal and that will not deposit undesirable residues on the surface. Etching or pickling should not be used with alloys that are susceptible to hydrogen embrittlement.

Precautions should be taken when machining specimens to avoid overheating, plastic deformation, or the development of residual stress in the metal surface. Machining should be performed in stages so that the final cut leaves the principal surface with a clean finish of 0.7 μm (30 μin.) rms or smoother. The required machining sequences, types of tools, and feed rate depend on the alloy and metallurgical condition of the testpiece. Lapping, mechanical polishing, and similar operations that produce flow of the metal should be avoided.

Interpretation of Test Results

This is the most fallible part of SCC testing and evaluation; it includes the analysis that leads to the conclusions and recommendations. Stress-corrosion test data are at best imprecise and test dependent, and they must be qualified with the testing conditions. It is important to verify the mode of environmental cracking (Fig. 1) and then to review the data to exclude all extraneous results, as discussed previously with the individual test methods. Following are some comments on the nature of the test dependency of the most commonly used criteria of SCC behavior.

Criteria of SCC Behavior

Specimen Life (Time to Failure). Stress-corrosion testing frequently involves determining the lives of specimens under specific test conditions. This includes the initiation (or incubation) of a stress-corrosion crack and its propagation to the point of fracture (Fig. 2). Such a determination is easily accomplished when only a single crack forms and the specimen fractures within the chosen test period. However, it often happens that SCC occurs but the specimen does not fracture. This is especially likely when testing relatively low-strength materials by constant strain loading (Fig. 5b) and when testing at applied stress or stress intensity levels only slightly above the threshold (Fig. 29a). Cracks may initiate at multiple sites in constant-strain loaded smooth specimens with relatively low applied stress, and a difficult problem arises in deciding when to consider a specimen failed if it does not fracture visibly.

It is often found that the majority of specimens in a set of replicates in a test fail rapidly; this leaves a few specimens that fail at much longer times or do not fail at all before the test is discontinued. Such behavior presents difficulties, both theoretical and practical, in deciding when to terminate a test, choosing a satisfactory representative value, and comparing such values.

The arithmetic mean specimen life is widely used for smooth specimens because it can be manipulated algebraically and can be used in many standard statistical tests of significance. It should be remembered, however, that extremely large or extremely small values may cause the mean to be atypical of the true distribution. Moreover, in using the arithmetic mean, it is assumed that the population is normally or very nearly normally distributed. The median, on the other hand, has the advantages that it is influenced less by extreme values, requires no assumption about the population distribution, and can be obtained much faster than arithmetic mean values because only about half the number of replicates exposed need to be tested to failure. The median is used in a German specification (Ref 19) that also provides for the use of the geometric mean if the replication is small.

References 1 and 122 contain examples for highly susceptible steel and aluminum alloys, respectively; these examples demonstrate the normal distributions for the logarithms of the specimen lives. With such distributions, a geometric mean would be the best representative value of the specimen life. It has also been shown that a Weibull distribution can be appropriate for the non-normally distributed test data for a relatively resistant aluminum alloy (Ref 123). Thus, it should not be assumed that any one distribution is applicable for all testing situations.

Comparisons of alloys with differing strength and fracture toughness based on time to failure can be completely misleading. For example, SCC growth curves are illustrated schematically in Fig. 58 for alloys with different fracture toughnesses. Curves A, B, and C represent materials with decreasing toughness, with curve C showing fast fracture initiated by corrosion pits or fissures with no SCC.

The behavior of a material that does not develop localized pitting or intergranular attack is represented by a line coincident with the abscissa, designated D in Fig. 58. The time-to-failure ranking above the graph indicates D as best and A, B, and C as poorest. Actually, the SCC responses of C and D were not measured, and the true SCC ranking of A and B (indicated by depth of SCC at the time of fracture) exhibits a trend opposite to that inferred from the time to failure data above.

Further difficulties may arise, because the total time to fracture is also influenced by non-SCC factors, such as specimen type and size, method of loading, initial stress level, and initiation behavior of the alloy. Consequently, the SCC ranking of materials may vary among investigators using different testing techniques. Nevertheless, comparisons of specimen lives derived from smooth specimens can be useful in certain mechanistic studies and in tests for comparing environmental variations if the mechanical aspects of the investigation are held constant.

Threshold Stress (Stress-Time Curve). More information about the resistance to SCC of a material can be obtained when testing smooth specimens by using a range of applied stresses. Such data are usually presented graphically with the applied gross section stress plotted against specimen life (Fig. 54). The primary interest is generally in the long-life portion of the curve to obtain an estimate of the threshold stress.

A common method of estimating threshold stress involves the experimental determination of the lowest stress at which cracking occurs in at least one specimen and the highest stress at which cracking does not occur in several specimens (for example, three or more, depending on variability). An average of the lowest failure and highest no-failure stresses is usually taken as the threshold stress. Such determinations of critical stresses are carried out at specified test times that are known through experience or preliminary tests to be sufficient to produce SCC in the alloy-environment system of interest. Statistical methods are available for determining threshold stresses more precisely (for example, the Probit method or staircase method) and are commonly used for the determination of fatigue limits. However, the additional testing involved can be quite extensive.

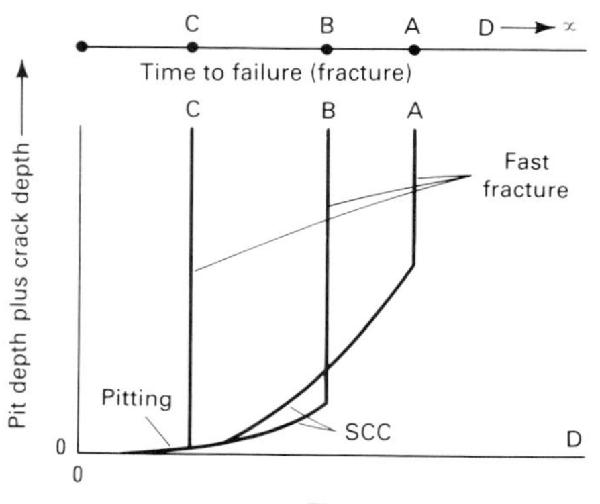

Fig. 58 Various processes in SCC as influenced by the fracture toughness of the metal. Kinetics for pitting (or, in material D, nonpitting), SCC (materials A and B only), and fast fracture. Line at top illustrates how time to failure data can be misleading. Source: Ref 124

Apparent threshold stresses determined in laboratory tests of coupon specimens are useful for ranking the SCC susceptibility of various materials, but, because such data are dependent on test conditions, they are not realistic for the purpose of engineering design (Ref 4). Also, when using such data as an aid in selecting the material for a specific structure, caution should be exercised in trying to relate the laboratory test conditions to the anticipated service conditions. Not only the environmental condition but also the geometry and size of the test specimens and the method of stressing should be compared (Fig. 6, 17). Further, threshold stresses obtained with statically loaded smooth specimens are likely to be nonconservative; it has been shown in tests on carbon-manganese steel that the threshold stress obtained by static loads is reduced by applied constant slow strain rates and that it can be reduced even further with cyclic loading (Ref 49). With appropriate frequency, load change, and temperature, average creep rates can be sustained over extended periods, but with static loading, the creep rate may fall below the level needed to promote SCC. In general, experimentally determined threshold stresses for materials with only limited susceptibility to SCC are more sensitive to variations in testing conditions than in the case of highly susceptible materials.

Percent Survival (Curve). This method of analyzing stress corrosion test results is especially useful when some of the specimens in a group survive the duration of the test. Examples of various comparisons by this technique are shown in Fig. 17, 42, and 43. Although the curves can be drawn on regular coordinate graph paper, the percent survival values will often lie along a straight line when plotted on normal probability paper, as illustrated in Fig. 59. The linearity of this plot indicates that the statistical distribution of the test results is logarithmic-normal. The vertical positions of the lines indicate the cracking ability of the environment, which can be represented by median cracking times. The slopes of the lines correspond to the variance, which can be used to calculate confidence limits.

Threshold Stress Intensity (K_{ISCC}, K_{th}). Linear elastic fracture mechanics is well established as a basis for materials characterization, including environmental cracking (Ref 24, 126, 127). In practice, it is most practical to define K_{ISCC} as the K_I level associated with some generally acceptable and definably low rate of crack growth that is commensurate with the design service life. When K_{ISCC} values are reported, the criterion for their assessment and the exposure time in the environment must accompany the threshold values. A rational approach to the development of useful data for design is to establish an operational definition of K_{ISCC} that is appropriate for the structure under consideration.

Such characterization requires that linear elastic fracture mechanics and plane-strain conditions be satisfied. However, for certain low-strength steels and aluminum alloys, existing data show that SCC can occur under conditions that deviate substantially from plane-strain, and that SCC is by no means limited to or is most severe under plane-strain loading conditions (Ref 24, 36, 128, 129). In these cases, the application of linear elastic fracture mechanics is no longer valid, and the parameter K_{ISCC} is no longer meaningful.

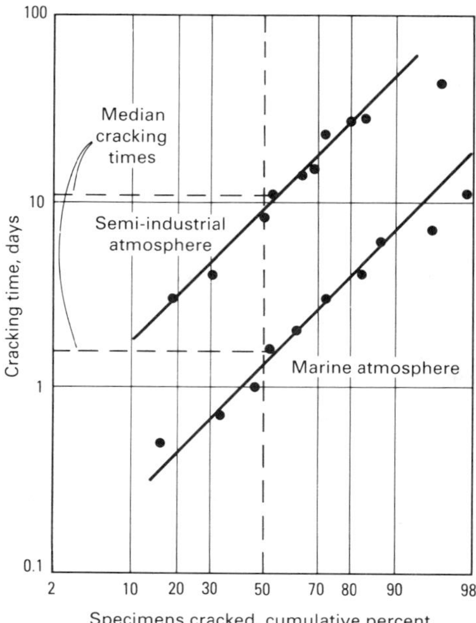

Fig. 59 Distribution of SCC test results for a stainless steel. Source: Ref 1, 125

Similarly, when testing materials with a high resistance to SCC, loading to high percentages of K_{Ic} may cause a relaxation of stress due to creep. In this case also, the apparent K_{ISCC} values are meaningless. Constant-load tests, therefore, are preferred for lower strength materials (Ref 130).

The symbol K_{th} has been used to identify threshold stress intensity factors developed under test conditions that do not satisfy all the requirements for plane-strain stress. Design calculations using such values should not be employed unless it is clear that the laboratory tests exhibit the same stress state as that for the intended application. Nevertheless, properly determined K_{th} values can be useful for ranking materials.

In principle, experimentally determined K_{ISCC} values should be the same whether they are determined by the initiation or the arrest test method (Fig. 29). In both tests, there are dimensional requirements for ensuring that the test results are independent of geometrical effects (see the section "Preparation of Precracked Specimens" in this article). However, precautions must be exercised during testing to avoid the potential problems involved with the environmental exposure (incubation, corrosion product wedging, crack branching, crack tip blunting, and so on). Comparisons of K_{ISCC} values determined for selected steels by both methods, along with examples of overestimated values resulting from insufficient length of exposure, are shown in Tables 6 and 7. It is advisable, when practicable, to use a test that matches the type of loading encountered in the anticipated service.

From the parameter K_{ISCC}, a value of a_{cr} can be calculated using the relation $a_{cr} = 0.2\ (K_{ISCC}/\mathrm{TYS})^2$ (see the section "Static Loading of Precracked (Fracture Mechanics) Specimens" in this article). This is the shallowest crack (surface length is long compared to its depth) that will propagate as a stress-corrosion crack at a yield strength level of gross stress under the given

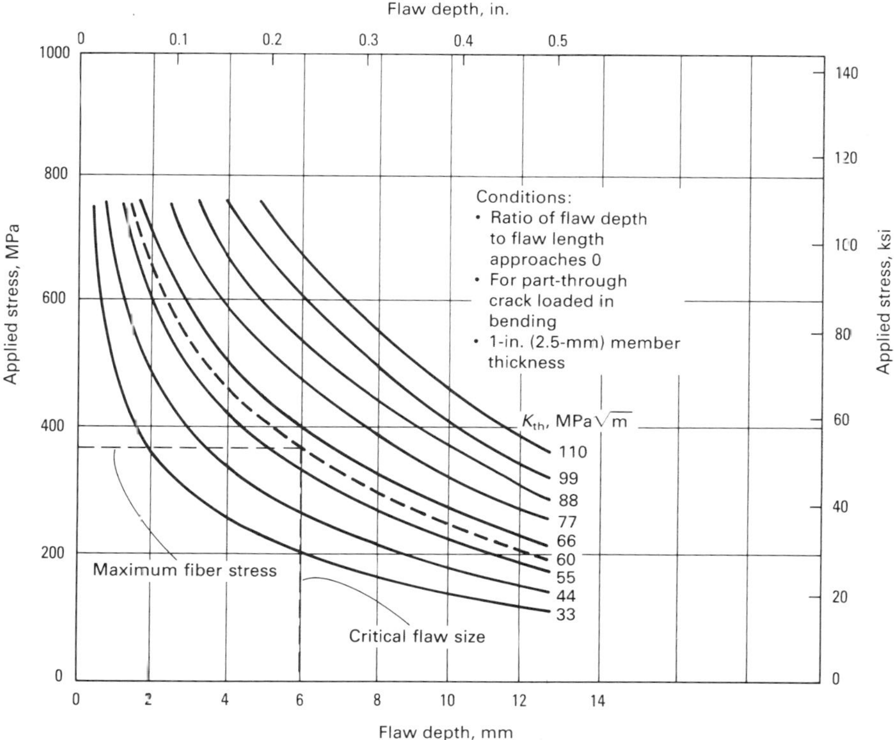

Fig. 60 Relationship of applied stress and flaw depth to crack propagation in hydrogen gas. Dashed lines show an example of the use of such a chart for a steel with K_{th} of 60.5 MPa$\sqrt{m}$ (55 ksi$\sqrt{in.}$) at an operating stress of 359 MPa (52 ksi). Source: Ref 131

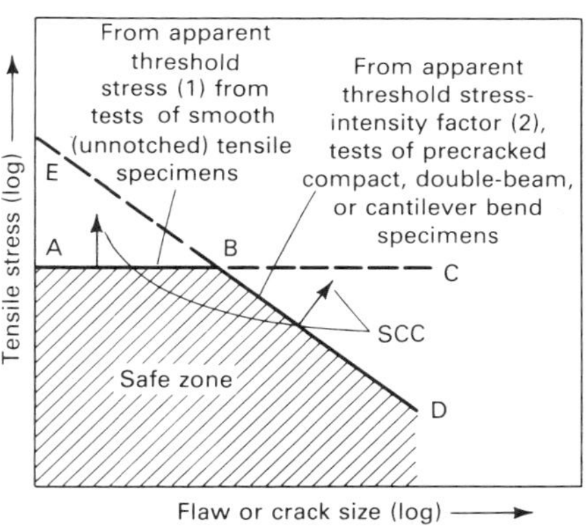

Fig. 61 Concept for combining SCC thresholds obtained on smooth and linear elastic fracture mechanics specimens to yield a conservative assessment of materials. (1) Minimum stress at which small tensile specimens fail by SCC when stressed in environment of interest. (2) Minimum stress intensity at which significant stress-corrosion crack growth occurs in environment of interest. Source: Ref 136

environmental conditions. This can be a very useful parameter for comparing materials, especially when the measured a_{cr} values can be related to the capability of the flaw inspection system used for a given engineering structure (Fig. 52). Straight lines representing assumed values of a_{cr} in Fig. 52 illustrate how K_{ISCC} values for the various steel alloys relate to the maximum depth of long surface flaws that can be tolerated without growth of SCC.

Such a plot can be used as follows. If the inspection system to be used can detect all long surface flaws deeper than 0.25 mm (0.01 in.), then the materials engineer would select an alloy with a K_{ISCC} above the 0.25-mm (0.01-in.) line. Conversely, if the K_{ISCC} and tensile yield strength of a material are known, the equation can be used to estimate the maximum tolerable flaw size. Substitution of an anticipated design stress in terms of percentage of tensile yield strength in the formula for a_{cr} will generate a new series of a_{cr} lines of lower slope.

Alternatively, the following method, which is specific for a given loading method, can be used. Inasmuch as the flaw depth and applied stress are uniquely related for a specific loading situation, a family of curves for constant K_{th} values can be developed within these parameters. Figure 60 shows such curves for long, shallow flaws; the curves were generated according to the following equations (Ref 132-134):

$$K_I = M_B\,\sigma\sqrt{\frac{\pi a}{Q}} \quad \text{(Eq 16)}$$

$$M_B = \frac{Y}{6}\sqrt{\frac{t}{\pi a}} \quad \text{(Eq 17)}$$

$$Y = \sqrt{139\left(\frac{a}{t}\right) - 221\left(\frac{a}{t}\right)^2 + 783\left(\frac{a}{t}\right)^3} \quad \text{(Eq 18)}$$

where K_I is stress intensity, σ is applied stress, a is flaw depth, Q is a shape parameter, and t is thickness. When K_{th} is substituted into Eq 16 and a flaw shape is assumed (Q = 0.8), this parameter is related to the flaw size and the applied stress. In this representation, SCC growth will not occur below the curve for the appropriate K_{th}. For example, for a steel with K_{th} of 60.5 MPa$\sqrt{m}$ (55 ksi$\sqrt{in.}$), a very deep flaw (6 mm, or 0.25 in.) would be required to cause crack propagation in hydrogen gas of a steel component stressed to 360 MPa (52 ksi) in bending. Such representations are useful for relating test data and K_{th} in design and in the development of crack inspection requirements, as well as for ranking alloys.

The requirements for K_{ISCC} tests, as well as for other fracture mechanics tests, include very explicit criteria regarding the minimum crack length for ensuring that the test results can be analyzed properly using existing linear fracture mechanics concepts. Therefore, a typical K_{ISCC} test uses a relatively large starting crack of the order of 25 mm (1 in.) long in a 25-mm (1-in.) thick specimen. In many types of service, however, initial defects of this size are rare. For example, damage-tolerant design criteria for military aircraft specify a flaw size of the order of 1.27 mm (0.05 in.) as the initial worst-case damage assumption upon introduction of a new part into service (Ref 126). Experimental work on high-strength steels exposed to hydrogen sulfide gas indicates that for a given combination of materials and applied stress there may indeed be a defect size below which the direct applicability of linear elastic fracture mechanics is questionable (Ref 135). Because of their susceptibility to SCC, however, high-strength steels should not be contemplated for service in the presence of hydrogen sulfide.

For example, Fig. 61 represents a concept of combining SCC thresholds based on smooth specimen and linear elastic fracture mechanics tests of aluminum alloy plate to give a conservative estimate of materials for design. As shown in Fig. 61, the threshold stress intensity analysis breaks down in the small flaw region (ABE) when the smooth specimen threshold stress is exceeded. Therefore, the definition of a safe zone requires results from both types of tests; the exclusive use of either one of the test methods can yield nonconservative conclusions. It is anticipated that application and further development of elastic-plastic fracture mechanics theory will lead to improved estimates of critical stress/flaw size combinations for the onset of SCC and tensile fracture, as proposed in Fig. 62.

SCC Velocity (*V*-K_I Curves). The issues of crack initiation and crack growth must be addressed for proper consideration of SCC response (Fig. 2). The use of mechanically precracked specimens provides a convenient approach for kinetic measurements where the crack growth rate can be determined as a function of the crack-tip stress intensity factor (Fig. 3). Current emphasis, however, is on the identification of a steady-state response (plateau velocity) for the ranking of materials (Fig. 46). A disadvantage of this approach is that in some testing situations a plateau velocity is not observed (Fig. 63). However, for design purposes, the phenomena of incubation and the non-steady state response (Stage I, Fig. 3) must also be taken into account. However, these latter requirements are not well understood.

An approximate approach to this problem of involving both the time of incubation (initiation) and the maximum crack growth rate of SCC at high stress intensities (plateau velocity) was proposed earlier in this article for use with testing of aluminum alloys with constant crack opening displacement tests (Fig. 45 and Table 4). This involves a simple average rate taken from time

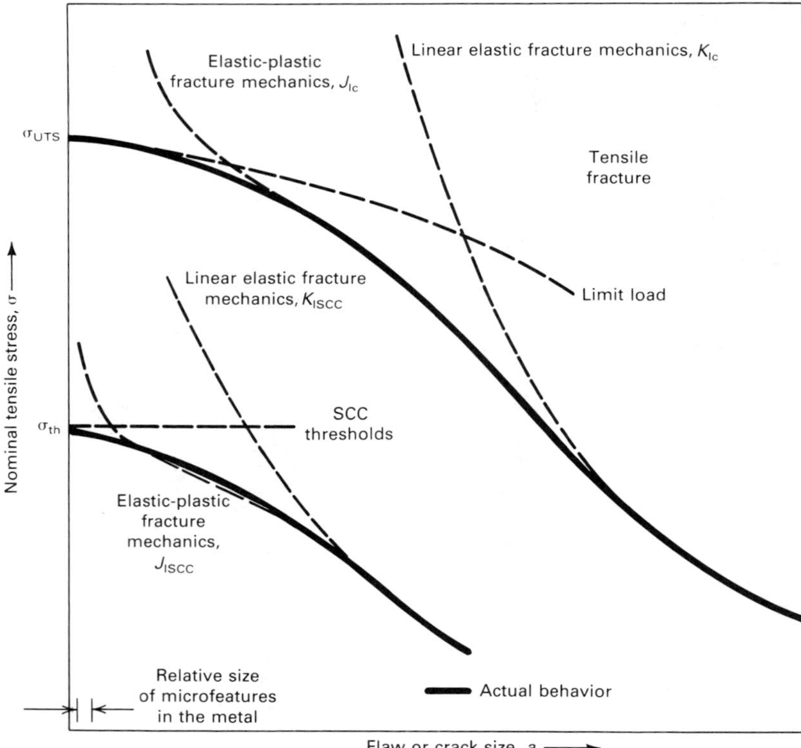

Fig. 62 Proposed linear elastic and elastic-plastic models for describing critical combinations of stress and flaw size at SCC thresholds and at the onset of rapid tensile fracture. Source: Ref 3

zero to a judiciously chosen test duration. Useful ranking of materials can be displayed on V-K_I diagrams such as that shown in Fig. 48.

Another approach to the average velocity concept is available through an elastic-plastic fracture mechanics interpretation of breaking load test data obtained from stress corrosion tests of smooth tension specimens (Ref 17). The stress-corrosion crack growth rates estimated from the slopes of the curves in Fig. 16 from 0 to 4 or 6 days are shown in Table 10 to agree reasonably well with the plateau (K-independent) growth rates determined from conventional fracture mechanics tests using bolt-loaded double-beam specimens. It is also noteworthy that a distinction between the intermediate resistance T7X1 temper and the nearly immune T7X2 temper was made more readily using either the smooth specimen test or the average crack growth over 42 days with the double-beam specimen. Fractographic examination confirmed that this was a correct distinction.

The implication here, which needs further investigation, is that from a single test method it is possible to compare materials (1) on the basis of their probabilities of initiating and propagating SCC flaws to an arbitrary depth or (2) by their respective crack growth rates, both being meaningful engineering descriptors of SCC damage. An additional advantage to this approach is that the effects of specimen size and alloy strength and toughness can be normalized (Fig. 16).

Ductility Ratio (DR versus Strain Rate Curves). Various ratio criteria, such as reduction of area, elongation, fracture stress, fracture energy, and time to failure, have been found to be useful in environmental studies with the slow strain rate test method. However, such criteria have limited use in comparing various materials because of their dependence on the strength and toughness of the alloy and the specificity of the critical strain rate and the environmental species (Fig. 33 and Table 2). Recent work has shown that average SCC growth rates, threshold stresses, and threshold strain rates can be obtained with modified techniques combined with microscopy. (Ref 50, 55, 56).

Other Criteria. Several other more specialized criteria can be found in the literature, such as:

- The Jones Stress Corrosion Index (Ref 70, 137)
- Critical Strain (Ref 99)
- Mean Critical Stress (Ref 138)

Precision of SCC Data

Variability in the measured values for SCC behavior arises from three primary sources: uncertainties associated with the measurement methods, variation in the test materials, and variation in the test environment. Suitable investigations must minimize the contributions from the first source and allow for quantitative assessment of the latter.

In the production of sophisticated high-strength and high-toughness alloys, close metallurgical control of the fabricating and thermal treatments is necessary to ensure that the required mechanical properties satisfy specifications. It is equally important from the standpoint of SCC that the metallurgical condition of the alloy be properly controlled. Just as there is a range of applicable mechanical properties for a given alloy and metallurgical condition (temper), a range in the SCC behavior can be expected from one heat (or lot) to another. Also, there can be an appreciable variation in the behavior of different-sized mill products of the same material. An example of the variations in the SCC behavior of different lots and various mill products is shown in Fig. 48 for aluminum alloy 7050. When selecting a material for a particular structural component SCC tests should be made on the specific products and sizes that will be required. Costly mistakes have been made in the past in the evaluation of prototype structures by testing parts fabricated from different products for economic expediency.

It is an unfortunate circumstance that the precision of SCC test results is generally lower for materials with an intermediate resistance to SCC than for materials with a very low or a very high resistance. This presents a special challenge in the comparison of competitive materials with improved resistance to SCC. Unfortunately, there is a scarcity of test data for determining what portion of the scatter-bands shown in Fig. 48 is due to material variations and how much is due to the precision of the test measurements.

Variability can occur even in carefully controlled investigations. Reference 139 reports results of measurements of K_{ISCC} for a 4340 steel made in numerous laboratories by different test methods.

In another investigation, the precision in the measurement of plateau velocities was determined for a number of replicate bolt-loaded double-beam specimens of short-transverse orientation from a sample of 25-mm (1-in.) thick 7075-T651 aluminum alloy plate (Ref 140). Tension pop-in specimens were exposed by continuous immersion in aqueous solutions of 1 M sodium chloride and 1 M sodium perchlorate. For six tests in the chloride solution, the mean plateau velocity was 1.4×10^{-8} m/s (1×10^{-3} in./h) with a standard deviation of 0.2×10^{-8} m/s (0.1×10^{-3} in./h), and the range was 1 to 2×10^{-8} m/s (7×10^{-4} to 1.4×10^{-3} in./h) (±35%). For nine tests in the perchlorate solution, the mean plateau velocity was 8×10^{-9} m/s (6×10^{-4} in./h) with a standard deviation of 0.2×10^{-9} m/s (1.4×10^{-4} in./h), and the range was 5×10^{-9} to 1×10^{-8} m/s (4 to 8×10^{-4} in./h) (±31%).

Normalizing SCC Data

It is often necessary to normalize test results with respect to one of the mechanical properties of critical interest for a given engineering structure in order to place the SCC response in proper perspective. This is commonly done by expressing the exposure stress (or stress intensity) and the apparent threshold values in terms of percent yield strength (or percent critical stress intensity) rather than in absolute units. This process can sometimes result in a different ranking order of alloys, and controversy can arise over which ranking is more pertinent. The use of normalized data is usually most appropriate when the utmost resistance to SCC is required; that is, the material should be resistant even when stressed to nearly 100% of the normalizing property. For alloy development screening tests, consideration should be given to evaluating materials in both ways.

Predicting Service Life

Life predictions are difficult and should be made with caution because there are no workable mathematical models. The field is plagued with confusion created to a large extent by:

- The complex, multifaceted nature of the phenomenon, which involves metallurgy, mechanics, chemistry, and time

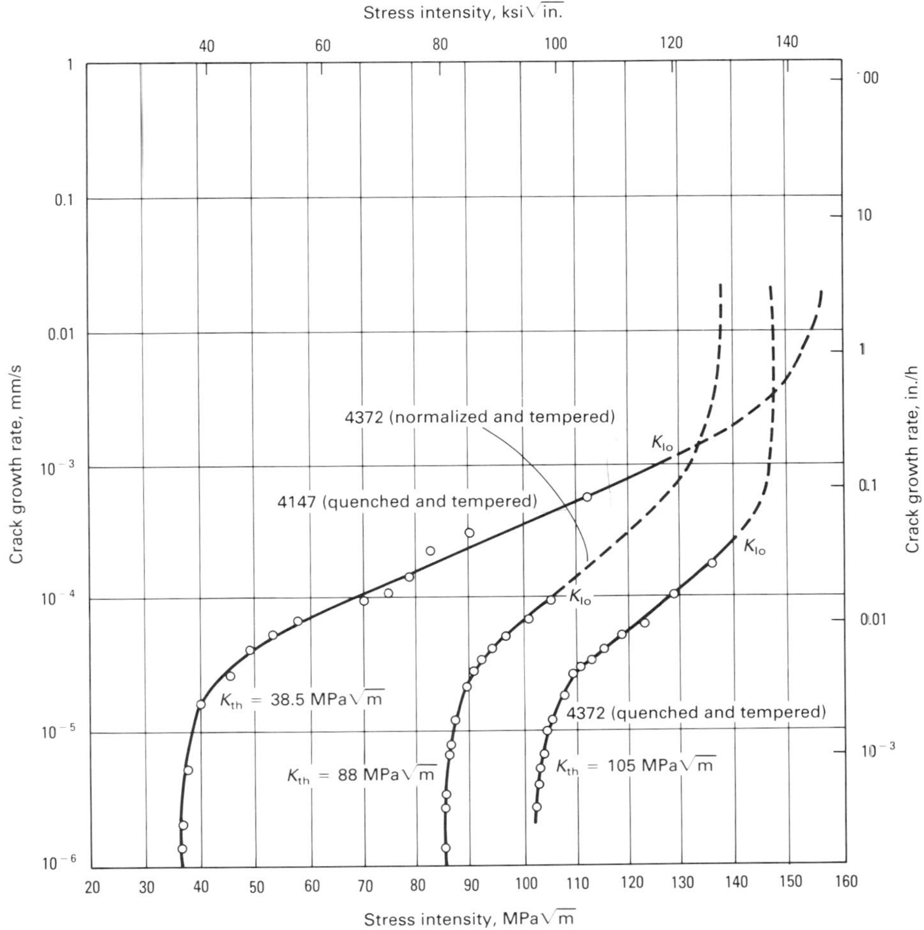

Fig. 63 Crack growth kinetics of three steels in hydrogen at 21 MPa (3000 psi). Source: Ref 131

Table 10 Stress-corrosion crack growth rates in aluminum alloy 7075 obtained by different test methods

	Stress-corrosion crack growth rate					
	7075–T651		7075–T7X1		7075–T7X2	
Test method	m/s	in. × 10^{-5}/h	m/s	in. × 10^{-5}/h	m/s	in. × 10^{-5}/h
Breaking load test using smooth tensile bar stressed 207 MPa (30 ksi); 4 or 6 days	>3.8 × 10^{-9}	>54(a)	7 × 10^{-10}	10(b)	1.3 × 10^{-10}	2(b)
Bolt-loaded double-beam, pop-in stress; plateau velocity obtained from V-K_I curves	3.1 × 10^{-9}	44	7.5 × 10^{-10}	11	(c)	
Bolt-loaded double-beam, pop-in stress; average growth 0–15 days	3.4 × 10^{-9}	48	4 × 10^{-10}	6	4 × 10^{-10}	6
Bolt-loaded double beam, pop-in stress; average growth 0–42 days	2.8 × 10^{-9}	40	4 × 10^{-10}	6	2 × 10^{-10}	3

(a) Over 4 days. (b) Over 6 days. (c) No plausible estimate could be made, because of the slight crack growth. Source: Ref 3

- The large number of variables known to affect SCC behavior
- Relatively poor correlation between laboratory test results and service experience
- Extensive data scatter
- Difficulty in assessing precisely the service conditions that the part must withstand

Designing to avoid SCC has traditionally taken the approach of preventing the initiation of SCC (safe-life concept). There is considerable service experience to justify this; one factor is that the materials that have given the most service problems are capable of developing relatively high SCC propagation rates (of the order of 25 mm, or 1 in., per month as shown in Fig. 46). Growth rates of this order, combined with relatively low thresholds of stress or stress intensity, make the fail-safe concept rather impractical. However, with the availability of advanced materials with higher thresholds for initiation of SCC and lower propagation rates, the damage-tolerant concept of design may become more practical for SCC and corrosion fatigue.

Selection of Test Method

The SCC test method selected should not be so severe that it rejects a material that is adequate for a particular application; on the other hand, the test should not be so mild that it passes materials that will fail in service. For tests of new and unfamiliar materials or environments, it is expedient to perform more than one type of test. Although standardized tests, which can be specified for screening tests in alloy or process development or quality control, are an essential link between research and engineering, there is a need for much freer choice of test conditions for research studies of SCC mechanisms.

The appropriate SCC test method is the one that is best adapted to test the material product form to be evaluated and the one that will yield the type of test results that best address the test objective(s). The newer methods that use fracture mechanics type specimens and loading by means of a constant slow strain rate are more severe, when applicable, than the older techniques that use smooth (defect-free) specimens. However, the results of all tests require interpretation. The application of linear elastic fracture mechanics has opened the way to the correlation of the mechanical aspects of SCC test methods. It is anticipated that further development of the science of elastic-plastic fracture mechanics will enhance the application of test data to service needs.

REFERENCES

1. A.W. Loginow, Stress Corrosion Testing of Alloys, *Mater. Prot.*, Vol 5 (No. 5), 1966, p 33-39
2. E.H. Dix, Jr., Aluminum-Zinc-Magnesium Alloys, Their Development and Commercial Production, *Trans. ASM*, Vol 42, 1950, p 1057-1127
3. D.O. Sprowls *et al.*, "A Study of Environmental Characterization of Conventional and Advanced Aluminum Alloys for Selection and Design: Phase II—The Breaking Load Test Method," Contract NASI-16424, NASA Contractor Report 172387, Aug 1984
4. B.F. Brown, *Stress Corrosion Cracking Control Measures*, National Bureau of Standards Monograph 156, U.S. Department of Commerce, June 1977
5. R.N. Parkins *et al.*, Report Prepared for the European Federation of Corrosion Working Party on Stress Corrosion Test Methods, *Br. Corros. J.*, Vol 7, July 1982, p 154-167
6. R.N. Parkins, Stress Corrosion Test Methods—Physical Aspects, in *The Theory of Stress Corrosion Cracking in Alloys*, J.C. Scully, Ed., NATO Scientific Affairs Division, 1971, p 449-468
7. G. Vogt, Comparative Survey of Type of Loading and Specimen Shape for Stress Corrosion Tests, *Werkst. Korros.*, Vol 29, 1978, p 721-725
8. *Metal Corrosion, Erosion, and Wear*, Vol 03.02, Section 3, *Annual Book of ASTM Standards*, American Society for Testing and Materials
9. H.L. Craig, Jr., D.O. Sprowls, and D.E. Piper, Stress-Corrosion Cracking, in *Hand-*

book on Corrosion Testing and Evaluation, W.H. Ailor, Ed., John Wiley & Sons, New York, 1971, p 231-290

10. M.B. Shumaker *et al.*, Evaluation of Various Techniques for Stress Corrosion Testing Welded Aluminum Alloys, in *Stress Corrosion Testing*, STP 425, American Society for Testing and Materials, 1967, p 317-341
11. R.A. Davis, Stress Corrosion Cracking Investigation of Two Low Alloy, High Strength Steels, *Corrosion*, Vol 19 (No. 2), 1963, p 45t-55t
12. S.O. Fernandez and G.F. Tisinai, Stress Analysis of Un-notched C-rings Used for Stress Cracking Studies, *J. Eng. Ind.*, Feb 1968, p 147-152
13. F.S. Williams, W. Beck, and E.J. Jankowsky, A Notched Ring Specimen for Hydrogen Embrittlement Studies, *Proceedings ASTM*, Vol 60, American Society for Testing and Materials, 1960, p 1192
14. D.O. Sprowls *et al.*, "Investigation of the Stress Corrosion Cracking of High Strength Aluminum Alloys," Final Technical Report for U.S. Government NASA Contract NAS-8-5340, Control No. 1-4-50-001167-01(lf), CPB-02-1215-64, 1967
15. Report of Task Group 1, of ASTM Subcommittee B-3/X, Stress Corrosion Testing Methods, in *Stress Corrosion Testing*, STP 425, 1967, p 3-20; *Proceedings ASTM*, Vol 65, American Society for Testing and Materials, 1965, p 182-197
16. B.W. Lifka and D.O. Sprowls, Stress Corrosion Testing of 7079-T6 Aluminum Alloy in Various Environments, in *Stress Corrosion Testing*, STP 425, American Society for Testing and Materials, 1967, p 342-362
17. R.J. Bucci *et al.*, The Breaking Load Method: A New Approach for Assessing Resistance to Growth of Early Stage Stress Corrosion Cracks, in *Corrosion Cracking*, V.S. Goel, Ed., Proceedings of International Conference and Exposition on Fatigue, Corrosion Cracking, Fracture Mechanics, and Failure Analysis, American Society for Metals, 1986, p 267-277
18. D.O. Sprowls *et al.*, Evaluation of a Proposed Standard Method of Testing for Susceptibility to SCC of High Strength 7XXX Series Aluminum Alloy Products, in *Stress-Corrosion—New Approaches*, STP 610, H.L. Craig, Jr., Ed., American Society for Testing and Materials, 1976, p 3-31
19. "Testing of Light Metals, Stress Corrosion Test," German Standard DIN 50908, 1964
20. F.H. Haynie *et al.*, "A Fundamental Investigation of the Nature of Stress Corrosion Cracking in Aluminum Alloys," Technical Report AFML 66-267, USAF Contract No. AF 33(615)-1710 Air Force Materials Laboratory, June 1966
21. P. Brenner, Realistic Stress Corrosion Testing, *Metallurgy*, Vol 23 (No. 9), 1969, p 879-886
22. H. Nathorst, Stress Corrosion Cracking in Stainless Steels, Part II—An Investigation of the Suitability of the U-Bend Specimen, *Weld. Res. Counc. Bull.*, Series No. 6, Oct 1950
23. R.P. Wei, S.R. Novak, and D.P. Williams, Some Important Considerations in the Development of Stress Corrosion Cracking Test Methods, *Mater. Res. Stand.*, Vol 12, 1972, p 25
24. "Characterization of Environmentally Assisted Cracking for Design—State of the Art," National Materials Advisory Board Report No. NMAB-386, National Academy of Sciences, 1982
25. B.F. Brown, The Application of Fracture Mechanics to Stress Corrosion Cracking, *Met. Rev.*, Vol 13, 1968, p 171-183
26. H.R. Smith and D.E. Piper, Stress Corrosion Testing with Precracked Specimens, in *Stress Corrosion Cracking in High Strength Steels and in Titanium and Aluminum Alloys*, B.F. Brown, Ed., Naval Research Laboratory, 1972, p 17-78
27. "Standard Method Test for Plane-Strain Fracture Toughness of Metallic Materials," E 399, *Annual Book of ASTM Standards*, Vol 03.01, American Society for Testing and Materials
28. J.A. Hauser, II, R.W. Judy, Jr., and T.W. Crooker, "Draft Standard Method of Test for Plane-Strain Stress-Corrosion-Cracking Resistance of Metallic Materials in Marine Environments," NRL Memorandum Report 5295, Naval Research Laboratory, March 1984
29. W.B. Lisagor, Influence of Precracked Specimen Configuration and Starting Stress Intensity on the Stress Corrosion Cracking of 4340 Steel, in *Environment-Sensitive Fracture: Evaluation and Comparison of Test Methods*, STP 821, S.W. Dean, E.N. Pugh, and G.M. Ugiansky, Ed., American Society for Testing and Materials, 1984, p 80-97
30. H. Tada, P. Paris, and G. Irwin, *The Stress Analysis of Cracks Handbook*, Del Research Corporation, 1973
31. J.A. Joyce, D.F. Hasson, and C.R. Crowe, Computer Data Acquisition Monitoring of the Stress Corrosion Cracking of Deplated Uranium Cantilever Beam Specimens, *J. Test. Eval.*, Vol 8 (No. 6), 1980, p 293-300
32. S.R. Novak and S.T. Rolfe, Modified WOL Specimen for K_{ISCC} Environmental Testing, *J. Met.*, Vol 4 (No. 3), 1969, p 701-728
33. J.G. Kaufman, J.W. Coursen, and D.O. Sprowls, An Automated Method for Evaluating Resistance to Stress-Corrosion Cracking With Ring-Loaded Precracked Specimens, in *Stress Corrosion—New Approaches*, STP 610, H.L. Craig, Jr., Ed., American Society for Testing and Materials, 1976, p 94-107
34. C. Micheletti and M. Buratti, New Testing Methods for the Evaluation of the Stress-Corrosion Behavior of High-Strength Aluminum Alloys by the Use of Precracked Specimens, in *Symposium Proceedings, Aluminum Alloys in the Aircraft Industry*, Turin, Italy, Oct 1976, Technicopy Ltd., 1978, p 149-159
35. M.V. Hyatt, Use of Precracked Specimens in Stress Corrosion Testing of High Strength Aluminum Alloys, *Corrosion*, Vol 26 (No. 11), 1970, p 487-503
36. D.O. Sprowls *et al.*, "Evaluation of Stress Corrosion Cracking Susceptibility Using Fracture Mechanics Techniques," Contract NAS 8-21487, Contractor Report NASA CR-124469, May 1973
37. R.C. Dorward and K.R. Hasse, "Flaw Growth of 7075, 7475, 7050, and 7049 Aluminum Plate in Stress Corrosion Environments," Final Technical Report for U.S. Government Contract NAS 8-30890, Oct 1976; *Corrosion*, Vol 34 (No. 11), 1978, p 386-395
38. W.B. Fichter, The Stress Intensity Factor for the Double Cantilever Beam, *Int. J. Fract.*, Vol 22, 1983, p 133-143
39. J.E. Srawley and B. Gross, Stress Intensity Factors for Crackline-Loaded Edge-Crack Specimens, *Mater. Res. Stand.*, Vol 7 (No. 4), 1967, p 155-162
40. S. Mostovoy, R.B. Crosley, and E.J. Ripling, Use of Crackline Loaded Specimens for Measuring Plane Strain Fracture Toughness, *J. Mat.*, Vol 2 (No. 3), 1967, p 661-681
41. D.O. Sprowls and J.D. Walsh, Evaluating Stress-Corrosion Crack Propagation Rates in High Strength Aluminum Alloys with Bolt Loaded Precracked Double Cantilever Beam Specimens, in *Stress Corrosion—New Approaches*, STP 610, H.L. Craig, Jr., Ed., American Society for Testing and Materials, 1976, p 143-156
42. R.B. Heady, Evaluation of Sulfide Corrosion Cracking Resistance in Low Alloy Steels, *Corrosion*, Vol 33 (No. 3), 1977, p 98-107
43. T.L. Bond, R.A. Yeske, and E.N. Pugh, Studies of Stress Corrosion Crack Growth in Al-Zn-Mg Alloys by the Double Torsion Method, in *Environment-Sensitive Fracture: Evaluation and Comparison of Test Methods*, STP 821, S.W. Dean, E.N. Pugh, and G.M. Ugiansky, Ed., American Society for Testing and Materials, 1984, p 128-149
44. J.C. Lewis and G. Sines, Ed., *Fracture Mechanics: Fourteenth Symposium, Volume II: Testing and Application*, STP 791, American Society for Testing and Materials, 1983
45. S.W. Dean, E.N. Pugh, and G.M. Ugiansky, Ed., *Environment-Sensitive Fracture: Evaluation and Comparison of Test Methods*, STP 821, American Society for Testing and Materials, 1984
46. J.H. Underwood, S.W. Freiman, and F.I. Baratta, Ed., *Chevron-Notched Specimens: Testing and Stress Analysis*, STP 855, American Society for Testing and Materials, 1984
47. W.B. Gilbreath and M.J. Adamson, Aqueous Stress Corrosion Cracking of High Toughness D6AC Steel, in *Stress-Corrosion—New Approaches*, STP 610, H.L. Craig, Jr., Ed., American Society for Testing and Materials, 1976, p 176-187
48. B.F. Brown, Concept of Occluded Corrosion Cells, *Corrosion*, Vol 26 (No. 8), 1970, p 249
49. G.M. Ugiansky and J.H. Payer, Ed., *Stress Corrosion Cracking—The Slow Strain-Rate Technique*, STP 665, American Society for Testing and Materials, 1979
50. R.N. Parkins, Development of Strain-Rate Testing and Its Implications, in *Stress-Corrosion Cracking—The Slow Strain-Rate Technique*, STP 665, G.M. Ugiansky and J.H. Payer, Ed., American Society for Testing and Materials, 1979, p 5-25
51. J.H. Payer, W.E. Berry, and W.K. Boyd, Constant Strain Rate Technique for Assessing Stress-Corrosion Susceptibility, in *Stress-Corrosion—New Approaches*, STP 610, H.L. Craig, Jr., Ed., American Society for Testing and Materials, 1976, p 82-93
52. N.J.H. Holroyd and G.M. Scamans, Slow-

Strain-Rate Stress Corrosion Testing of Aluminum Alloys, in *Environment-Sensitive Fracture: Evaluation and Comparison of Test Methods*, STP 821, S.W. Dean, E.N. Pugh, and G.M. Ugiansky, Ed., American Society for Testing and Materials, 1984, p 202-241

53. C.D. Kim and B.E. Wilde, A Review of Constant Strain-Rate Stress Corrosion Cracking Test, in *Stress Corrosion Cracking—The Slow Strain Rate Technique*, ASTM STP 665, G.M. Ugianski and J.H. Payer, Ed., American Society for Testing and Materials, p 97-112, 1979
54. "Standard Methods of Tension Testing of Metallic Materials," E 8, *Annual Book of ASTM Standards*, Vol. 03.01, American Society for Testing and Materials
55. R.N. Parkins, Fifth Symposium on Line Pipe Research, Catalog No. L30174, U1-40, American Gas Association
56. W.R. Wearmouth, G.P. Dean, and R.N. Parkins, Role of Stress in the Stress Corrosion Cracking of a Mg-Al Alloy, *Corrosion*, Vol 29 (No. 6), 1973, p 251-258
57. R.N. Parkins and Y. Suzuki, Environment Sensitive Cracking of a Nickel-Aluminum Bronze Under Monotonic and Cyclic Loading Conditions, *Corros. Sci.*, Vol 23 (No. 6), 1983, p 577-599
58. R.N. Parkins, A Critical Evaluation of Current Environment-Sensitive Fracture Test Methods, in *Environment-Sensitive Fracture: Evaluation and Comparison of Test Methods*, STP 821, S.W. Dean, E.N. Pugh, and G.M. Ugiansky, Ed., American Society for Testing and Materials, 1984, p 5-31
59. J. Yu, N.J.H. Holroyd, and R.N. Parkins, Application of Slow-Strain-Rate Tests to Defining the Stress for Stress Corrosion Crack Initiation in 70/30 Brass, in *Environment-Sensitive Fracture: Evaluation and Comparison of Tests Methods*, STP 821, S.W. Dean, E.N. Pugh, and G.M. Ugiansky, Ed., American Society for Testing and Materials, 1984, p 288-309
60. B.F. Brown, The Contributions of Physical Metallurgy and of Fracture Mechanics to Containing the Problem of Stress Corrosion Cracking, *Philos. Trans. R. Soc.* (London) A, Vol 282 (No. 1307), 1976, p 235-245
61. D.O. Sprowls and R.H. Brown, Stress Corrosion Mechanisms for Aluminum Alloys, in *Fundamental Aspects of Stress-Corrosion Cracking*, R.W. Staehle, A.J. Forty, and D. van Rooyen, Ed., National Association of Corrosion Engineers, 1969, p 466-512
62. J.D. Jackson, W.K. Boyd, and R.W. Staehle, "Corrosion of Titanium," DMIC Memorandum 218, Defense Metals Information Center, Battelle Memorial Institute, Sept 1966
63. R.N. Parkins, Predictive Approaches to Stress Corrosion Cracking Failure, *Corros. Sci.*, Vol 20, 1980, p 147-166
64. J.R. Galvele and S.M. DeMichelli, Mechanism of Intergranular Corrosion of Al-Cu Alloys, *Corros. Sci.*, Vol 10, 1970, p 795-807
65. S. Maitra and G.C. English, Mechanism of Localized Corrosion of 7075 Alloy Plate, *Metall. Trans. A*, Vol 12, March 1981, p 535-541
66. H. Bohni and H.H. Uhlig, Environmental Factors Affecting the Critical Pitting Potential of Aluminum, *J. Electrochem. Soc.*, Vol 116, Part II, 1969, p 906-910
67. R.H. Brown, D.O. Sprowls, and M.B. Shumaker, The Resistance of Wrought High Strength Aluminum Alloys to Stress Corrosion Cracking, in *Stress-Corrosion Cracking of Metals—A State of the Art*, STP 518, American Society for Testing and Materials, 1972, p 87-118
68. B.W. Lifka, Corrosion Resistance of Aluminum Alloy Plate in Rural, Industrial, and Seacoast Atmospheres, *Mater. Prot.*, to be published
69. D.O. Sprowls and R.H. Brown, What Every Engineer Should Know About the Stress Corrosion of Aluminum, *Met. Prog.*, Vol 81 (No. 4), April 1962, p 79-85; Vol 81 (No. 5), May 1962, p 77-83
70. F.F. Booth and H.P. Godard, An Anodic Stress-Corrosion Test for Aluminum-Magnesium Alloys, in *First International Congress on Metallic Corrosion*, Butterworths, 1962, p 703-712
71. A.H. Le, B.F. Brown, and R.T. Foley, The Chemical Nature of Aluminum Corrosion, IV: Some Anion Effects on SCC of AA7075-T651, *Corrosion*, Vol 36 (No. 12), Dec 1980, p 673-679
72. D.O. Sprowls, J.D. Walsh, and M.B. Shumaker, Simplified Exfoliation Testing of Aluminum Alloys, in *Localized Corrosion—Cause of Metal Failure*, STP 516, American Society for Testing and Materials, 1972, p 38-65
73. W. Pistulka and G. Lang, Accelerated Stress Corrosion Test Methods for Al-Zn-Mg Type Alloys, *Aluminum*, Vol 53 (No. 6), 1977, p 366-371
74. "Stress-Corrosion Cracking Testing of Aluminum Alloys for Aircraft Parts," German Aircraft Standard LN 65666, July 1974 (in German)
75. P.W. Jeffrey, T.E. Wright, and H.P. Godard, An Accelerated Laboratory Stress Corrosion Test for Al-Zn-Mg Alloys, in *Proceedings of the Fourth International Congress on Corrosion*, National Association of Corrosion Engineers, 1969, p 133-139
76. T.S. Humphries and J.E. Coston, "An Improved Stress Corrosion Test Medium for Aluminum Alloys," NASA Technical Memorandum NASA TM-82452, George C. Marshall Space Flight Center, Nov 1981
77. W.J. Helfrich, "Development of a Rapid Stress Corrosion Test for Aluminum Alloys," Final Summary Report, Contract No. NAS 8-20285, George C. Marshall Space Flight Center, May 1968
78. M.V. Hyatt and M.O. Speidel, High Strength Aluminum Alloys, in *Stress Corrosion Cracking in High Strength Steels and in Titanium and in Aluminum Alloys*, B.F. Brown, Ed., Naval Research Laboratory, 1972, p 148-244
79. M.O. Speidel, Stress Corrosion Cracking of Aluminum Alloys, *Metall. Trans. A*, Vol 6, April 1975, p 631-651
80. L. Schra and J. Faber, "Influence of Environments on Constant Displacement Stress-Corrosion Crack Growth in High Strength Aluminum Alloys," NLR TR 81138 U, National Aerospace Laboratory NLR, 1981
81. J.R. Pickens, Techniques for Assessing the Corrosion Properties of Aluminum Powder Metallurgy Alloys, in *Rapidly Solidified Powder Aluminum Alloys*, STP 890, M.E. Fine and E.A. Starke, Jr., Ed., American Society for Testing and Materials, 1986, p 381-409
82. R.E. Davies, G.E. Nordmark, and J.D. Walsh, "Design Mechanical Properties, Fracture Toughness, Fatigue Properties Exfoliation and Stress Corrosion Resistance of 7050 Sheet, Plate, Extrusions, Hand Forgings and Die Forgings," Final Report, Naval Air Systems, Contract N00019-72-C-0512, July 1975
83. R.C. Dorward and K.R. Hasse, Long-Transverse Stress-Corrosion Cracking Behavior of Aluminum Alloy AA7075, *Br. Corros. J.*, Vol 13 (No. 1), 1978, p 23-27
84. G.M. Scamans, Discontinuous Propagation of Stress Corrosion Cracks in Al-Zn-Mg Alloys, *Scr. Metall.*, Vol 13, 1979, p 245-250
85. S. Maitra, Determination of SCC Resistance of Al-Cu-Mg Alloys by Slow Strain Rate and Alternate Immersion Testing, *Corrosion*, Vol 37 (No. 2), 1981, p 98-103
86. D.H. Thompson, Stress Corrosion Cracking of Copper Metals, in *Stress Corrosion Cracking of Metals—A State of the Art*, STP 518, American Society for Testing and Materials, 1972, p 39-57
87. E.N. Pugh, J.V. Craig, and A.J. Sedricks, The Stress-Corrosion Cracking of Copper, Silver, and Gold Alloys, in *Proceedings of Conference on Fundamental Aspects of Stress-Corrosion Cracking*, National Association of Corrosion Engineers, 1969, p 118-158
88. G. Edmunds, E.A. Anderson, and R.K. Waring, Ammonia and Mercury Stress-Corrosion Cracking Tests for Brass, in *Symposium on Stress-Corrosion Cracking of Metals*, American Society for Testing and Materials and the American Institute of Mining, Metallurgical, and Petroleum Engineers, 1944, p 7-18
89. "Standard Method of Mercurous Nitrate Test for Copper and Copper Alloys," B 154, *Annual Book of ASTM Standards*, American Society for Testing and Materials, 1984, p 352-354
90. "Tubing of Copper and Copper Alloys for Condensers and Heat Exchangers," German Standard DIN-1785 (available from ANSI), 1983
91. H.L. Logan, *The Stress Corrosion of Metals*, John Wiley & Sons, 1966
92. R.N. Parkins, Stress Corrosion Cracking of Low Carbon Steels, in *Proceedings of Conference on Fundamental Aspects of Stress Corrosion Cracking*, National Association of Corrosion Engineers, 1969, p 361-373
93. M.J. Humphries and R.N. Parkins, Stress Corrosion Cracking of Mild Steels in Sodium Hydroxide Solutions Containing Various Additional Substances, *Corros. Sci.*, Vol 7, 1967, p 747-761
94. R.N. Parkins, P.W. Slattery, and B.S. Poulson, The Effects of Alloying Additions to Ferritic Steels Upon Stress-Corrosion Cracking Resistance, *Corrosion*, Vol 37 (No. 11), 1981, p 650-664
95. B.S. Poulson and R.N. Parkins, Effect of Nickel Additions Upon the Stress Corrosion of Ferritic Steels in a Chloride Environment, *Corrosion*, Vol 29, Nov 1973, p 414-422
96. "Stress Corrosion Cracking Resistance Test

for Prestressing Tendons," Technical Report 5, *Proceedings of the Federation Internationale de la Preconstrainte*, Eighth Congress, Cement and Concrete Association, 1978

97. E.H. Phelps, A Review of the Stress Corrosion Behavior of Steels with High Yield Strength, in *Proceedings of Conference on Fundamental Aspects of Stress Corrosion Cracking*, National Association of Corrosion Engineers, 1969, p 398-410
98. "Test Method for Testing of Metals for Resistance to Sulfide Stress Cracking at Ambient Temperatures," TM-01-77, National Association of Corrosion Engineers, 1977, p 77-84
99. J.P. Fraser, G.G. Eldredge, and R.S. Treseder, Laboratory and Field Methods for Quantitative Study of Sulfide Corrosion Cracking, *Corrosion*, Vol 14, 1958, p 517t-523t
100. G. Sandoz, High Strength Steels, in *Stress Corrosion Cracking in High Strength Steels and in Titanium and in Aluminum Alloys*, B.F. Brown, Ed., Naval Research Laboratory, Washington, DC, 1972, p 79-145
101. H.J. Bhatt and E.H. Phelps, The Effect of Electrochemical Polarization on the Stress Corrosion Behavior of Steels with High Yield Strength, in *Proceedings of the Third International Congress on Metallic Corrosion*, May 16-26, 1966
102. H.J. Bhatt and E.H. Phelps, Effect of Solution pH on the Mechanism of Stress Corrosion Cracking of a Martensitic Stainless Steel, *Corrosion*, Vol 17, 1961, p 430t-434t
103. R.M. Latanision and R.W. Staehle, Stress Corrosion Cracking of Iron-Nickel-Chromium Alloys, in *Proceedings of Conference on Fundamental Aspects of Stress Corrosion Cracking*, National Association of Corrosion Engineers, 1969, p 214-307
104. S.W. Dean, Review of Recent Studies on the Mechanism of Stress Corrosion Cracking in Austenitic Stainless Steels, in *Stress Corrosion—New Approaches*, STP 610, H.L. Craig, Jr., Ed., American Society for Testing and Materials, 1976, p 308-337
105. E. Denhard, Effect of Composition and Heat Treatment on the Stress Corrosion Cracking of Austenitic Stainless Steels, *Corrosion*, Vol 16 (No. 7), 1960, p 131-141
106. H.R. Copson, Effect of Composition on Stress Corrosion Cracking of Some Alloys Containing Nickel, in *Physical Metallurgy of Stress Corrosion Fracture*, T.N. Rhodin, Ed., Interscience, New York, 1959, p 247-272
107. A.W. Loginow, J.F. Bates, and W.L. Mathay, New Alloy Resists Chloride Stress Corrosion Cracking, *Mater. Prot. Perform.*, Vol 11 (No. 5), 1972, p 35-40
108. N.A. Nielsen, Observations and Thoughts on Stress Corrosion Mechanism, *J. Met.*, Vol 5, 1970, p 794-829
109. J.D. Harston and J.C. Scully, Stress Corrosion of Type 304 Steel in H_2SO_4/NaCl Environments at Room Temperature, *Corrosion*, Vol 25 (No. 12), 1969, p 493-501
110. S. Takemura, M. Onoyama, and T. Ooka, Stress Corrosion Cracking of Stainless Steels in Hydrogen Sulfide Solutions, *Corrosion*, Vol 16 (No. 7), 1960, p 338-348
111. J.L. Wilson, F.W. Pement, and R.G. Aspden, Effect of Alloy Structure, Hydroxide Concentration, and Temperature on the Caustic Stress-Corrosion Cracking of Austenitic Stainless Steel, *Corrosion*, Vol 30 (No. 4), 1974, p 139-149
112. H. Kohl, A Contribution to the Examination of Stress-Corrosion Cracking of Austenitic Stainless Steel in Magnesium Chloride Solution, *Corrosion*, Vol 23 (No. 12), 1967, p 39-49
113. G.F. Sager, R.H. Brown, and R.B. Mears, Tests for Determining Susceptibility to Stress Corrosion Cracking, in *Symposium on Stress Corrosion Cracking of Metals*, American Society for Testing and Materials and the American Institute of Mechanical Engineers, 1945, p 255-272
114. H.B. Romans, Stress Corrosion Test Environments and Test Durations, in *Stress Corrosion Testing*, STP 425, American Society for Testing and Materials, 1967, p 182-208
115. W.K. Boyd and W.E. Berry, Stress Corrosion Cracking of Nickel and Nickel Alloys, in *Stress Corrosion Cracking of Metals—A State of the Art*, STP 518, American Society for Testing and Materials, 1972, p 58-78
116. R.C. Newman, R. Roberge, and R. Bandy, Evaluation of SCC Test Methods for Inconel 600 in Low-Temperature Aqueous Solutions, in *Environment-Sensitive Fracture: Evaluation and Comparison of Test Methods*, STP 821, S.W. Dean, E.N. Pugh, and G.M. Ugiansky, Ed., American Society for Testing and Materials, 1984, p 310-322
117. A.I. Asphahani, Slow Strain-Rate Technique and Its Applications to the Environmental Stress Cracking of Nickel-Base and Cobalt-Base Alloys, in *Stress Corrosion Cracking—The Slow Strain Rate Technique*, STP 665, G.M. Ugiansky and J.H. Payer, Ed., American Society for Testing and Materials, 1979, p 279-293
118. M.J. Blackburn, W.H. Smyrl, and J.A. Sweeny, Titanium Alloys, in *Stress Corrosion Cracking in High Strength Steels and in Titanium and in Aluminum Alloys*, B.F. Brown, Ed., Naval Research Laboratory, 1972, p 246-363
119. T.R. Beck, Electrochemical Aspects of Titanium Stress Corrosion, in *Proceedings of Conference on Fundamental Aspects of Stress Corrosion Cracking*, National Association of Corrosion Engineers, 1969, p 605-619
120. D.A. Meyn, Effect of Hydrogen on Fracture and Inert Environment Sustained Load Cracking Resistance of Alpha-Beta Titanium Alloys, *Metall. Trans.*, Vol 5 (No. 11), 1974, p 2405-2414
121. G. Sandoz, Subcritical Crack Propagation in Ti-8Al-1Mo-1V Alloy in Organic Environments, Salt Water and Inert Environments, in *Proceedings of Conference on Fundamental Aspects of Stress Corrosion Cracking*, National Association of Corrosion Engineers, 1969, p 684-690
122. F.F. Booth and G.E. Tucker, Statistical Distribution of Endurance in Electrochemical Stress-Corrosion Tests, *Corrosion*, Vol 21 (No. 5), 1965, p 173-177
123. J.H. Harshbarger, A.I. Kemppinen, and B.W. Strum, Statistical Treatment of Nonnormally Distributed Stress-Corrosion Data, in *Handbook on Corrosion Testing and Evaluation*, W.H. Ailor, Ed., John Wiley & Sons, 1971, p 87-100
124. B.F. Brown, Chapter 1, in *Stress Corrosion Cracking in High Strength Steels and in Titanium and in Aluminum Alloys*, B.F. Brown, Ed., Naval Research Laboratory, 1972, p 13
125. B.F. Brown, "Stress Corrosion Cracking and Related Phenomena in High Strength Steels," NRL Report 6041, Naval Research Laboratory, Nov 1963
126. "Military Specification for Airplane Damage Tolerance Requirements," MIL-A-83444 U.S. Air Force, 2 July 1974
127. *Damage Tolerant Handbook, A Compilation of Fracture and Crack-Growth Data for High-Strength Alloys*, Vol 1, MCIC-HB-01R, Battelle Columbus Laboratories, Dec 1983
128. S.R. Novak, Effect of Prior Uniform Plastic Strain on the K_{ISCC} of High-Strength Steels in Sea Water, *J. Eng. Fract. Mech.*, Vol 5, 1973, p 727-763
129. W.G. Clark, Jr., Some Problems in the Application of Fracture Mechanics, in *Fracture Mechanics—13th Conference*, STP 743, R. Roberts, Ed., American Society for Testing and Materials, 1981, p 269
130. M.O. Speidel, Fracture Mechanics and Stress Corrosion, *Blech, Rohre, Profile*, Vol 25 (No. 1), 1978, p 14-18 (in German)
131. A.W. Loginow and E.H. Phelps, Steels for Seamless Hydrogen Pressure Vessels, *Corrosion*, Vol 31 (No. 11), 1975, p 404-412
132. C.F. Tiffany, J.N. Masters, and F. Pall, Some Fracture Considerations in the Design of Spacecraft Pressure Vessels, in *Proceedings of the ASM Metals Congress*, American Society for Metals, 1966
133. G.M. Sinclair and S.T. Rolfe, Analytical Procedure for Relating Subcritical Crack Growth to Inspection Requirements, in *Proceedings of the ASME Metals Engineering Division Conference*, American Society of Mechanical Engineers, 1969
134. F.W. Smith, "Stress-Intensity Factors for a Semi-Elliptical Surface Flaw," Structural Development Research Memorandum 17, The Boeing Company, 1966
135. W.G. Clark, Jr., Applicability of the K_{ISCC} Concept to Very Small Defects, in *Cracks and Fracture (Ninth Conference)*, STP 601, American Society for Testing and Materials, 1976, p 138-153
136. J.G. Kaufman, Stress Corrosion—Traditional Versus Fractured Mechanicians, *Corrosion*, Vol 35 (No. 4), April 1979, p i
137. E.L. Jones, Stress Corrosion of Aluminum Magnesium Alloys II—Method for Expressing Stress Corrosion Susceptibility on a Comparative Basis, *J. Appl. Chem.*, Jan 1954, p 7-10
138. J.T. Staley, Evaluating New Aluminum Alloy Forging Alloys, *Met. Eng. Quart.*, Vol 14 (No. 4), 1974, p 50-55
139. R.P. Wei and S.R. Novak, Interlaboratory Evaluation of K_{ISCC} Measurement Procedures for Steels, in *Environment-Sensitive Fracture: Evaluation and Comparison of Test Methods*, STP 821, S.W. Dean, E.N. Pugh, and G.M. Ugiansky, Ed., American Society for Testing and Materials, 1984, p 75-79
140. A.H. Le and R.T. Foley, Stress Corrosion Cracking of AA7075-T651 in Various Electrolytes—Statistical Treatment of Data Obtained Using DCB Precracked Specimens, *Corrosion*, Vol 39 (No. 10), 1983, p 379-383

Evaluation of Hydrogen Embrittlement

Louis Raymond, L. Raymond & Associates

HYDROGEN EMBRITTLEMENT of metals is an old, frequently encountered, and often misunderstood phenomenon. It is most commonly thought to occur by subcritical crack growth, often producing time-delayed fractures in production parts with no externally applied stress. Many problems still exist, starting with a basic definition of hydrogen embrittlement, in addition to identifying its source, controlling its effects, and preventing its occurrence.

This wide range of problems makes the evaluation of hydrogen embrittlement a multifaceted technical activity. Additional information on the mechanism of hydrogen embrittlement and susceptibility of a variety of ferrous and nonferrous alloys to hydrogen damage can be found in the section "Hydrogen Damage" in the article "Environmentally Induced Cracking" in this Volume.

Research investigations on the phenomenon range from studies of crack nucleation and growth, including such parameters as incubation time, crack growth rates, and threshold stress intensities, to studies on the relative susceptibility of materials to hydrogen embrittlement.

Toward a Definition (Ref 1)

Much confusion exists in the published literature over the definition of hydrogen embrittlement. Metal processing, chemical, and petrochemical industries have experienced various types of hydrogen problems for many years. The aerospace industry has experienced new and unexpected hydrogen embrittlement problems, principally in dealing with high-strength steels. There are many sources of hydrogen, several types of embrittlement, and various theories for explaining the observed effects.

Hydrogen embrittlement is often classified into three types:

- Internal reversible hydrogen embrittlement
- Hydrogen environment embrittlement
- Hydrogen reaction embrittlement

If specimens have been precharged with hydrogen from any source or in any manner and embrittlement is observed during mechanical testing, then embrittlement is caused by either internal reversible embrittlement or by hydrogen reaction embrittlement. If hydrides or other new phases containing hydrogen form during testing in gaseous hydrogen, then embrittlement is attributed to hydrogen reaction embrittlement. For all embrittlement determined during mechanical testing in gaseous hydrogen other than internal reversible and hydrogen reaction embrittlement, hydrogen environment embrittlement is assumed to be responsible.

Internal reversible hydrogen embrittlement has also been termed slow strain rate embrittlement and delayed failure. This is the classical type of hydrogen embrittlement that has been studied quite extensively. Widespread attention has been focused on the problem resulting from electroplating, particularly of cadmium in high-strength steel components. Other sources of hydrogen are processing treatments, such as melting and pickling. More recently, the embrittling effects of many stress-corrosion processes have been attributed to corrosion-produced hydrogen. Hydrogen that is absorbed from any source is diffusible within the metal lattice. To be fully reversible, embrittlement must occur without the hydrogen undergoing any type of chemical reaction after it has been absorbed within the lattice.

Internal reversible hydrogen embrittlement can occur after a very small average concentration of hydrogen has been absorbed from the environment. However, local concentrations of hydrogen are substantially greater than average bulk values. For steels, embrittlement is usually most severe at room temperature during either delayed failure or slow strain rate tension testing. This time-dependent nature (incubation period) of embrittlement suggests that diffusion of hydrogen within the lattice controls this type of embrittlement. Cracks initiate internally, usually below the root of a notch at the region of maximum triaxiality. Embrittlement in steel is reversible (ductility can be restored) by relieving the applied stress and aging at room temperature, provided microscopic cracks have not yet initiated. Internal reversible hydrogen embrittlement has also been observed in a wide variety of other materials, including nickel-base alloys and austenitic stainless steels, provided they are severely charged with hydrogen.

Hydrogen environment embrittlement was recognized as a serious problem in the mid-1960s when the National Aeronautics and Space Administration (NASA) and its contractors experienced failure of ground-based hydrogen storage tanks. These tanks were rated for hydrogen at pressures of 35 to 70 MPa (5 to 10 ksi). Consequently, the failures were attributed to high-pressure hydrogen embrittlement. Because of these failures and the anticipated use of hydrogen in advanced rocket and gas turbine engines and auxiliary power units, NASA has initiated both in-house and contractural research. The contractural effort generally has been to define the relative susceptibility of structural alloys to hydrogen environment embrittlement. A substantial amount of research has concerned the mechanism of the embrittlement process. There is marked disagreement as to whether hydrogen environment embrittlement is a form of internal reversible hydrogen embrittlement or is truly a distinct type of embrittlement.

Hydrogen Reaction Embrittlement. Although the sources of hydrogen may be any of those mentioned previously, this type of embrittlement is quite distinct from hydrogen environment embrittlement. Once hydrogen is absorbed, it may react near the surface of diffuse substantial distances before it reacts. Hydrogen can react with itself, with the matrix, or with a foreign element in the matrix. The chemical reactions that comprise this type of embrittlement or attack are well known and are encountered frequently. The new phases formed by these reactions are usually quite stable and embrittlement is not reversible during room temperature aging treatments.

Atomic hydrogen (H) can react with the matrix or with an alloying element to form a hydride (MH_x). Hydride phase formation can be either spontaneous or strain induced. Atomic hydrogen can react with itself to form molecular hydrogen (H_2). This problem is frequently encountered after steel processing and welding; it has been termed flaking or "fisheyes." Atomic hydrogen can also react with a foreign element in the matrix to form a gas. A principal example is the reaction with carbon in low-alloy steels to form methane (CH_4) bubbles. Another example is the reaction of atomic hydrogen with oxygen in copper to form steam (H_2O) resulting in blistering and a porous metal component.

Although hydrogen reaction embrittlement is not a major topic in this article, its definition is included for the sake of completeness and in the hope of establishing a single definition for each of the various hydrogen embrittlement phenomena to avoid problems with semantics.

Further confusion results from the relation of stress-corrosion cracking (SCC) to hydrogen embrittlement, because the crack growth mechanism is often found to be the same. On the surface, the active corrosion process produces the hydrogen that is the cause of the failure. In SCC, the pits or crevices (polarized anodically) are initiation sites, and therefore, although the growth mechanisms are the same, the method of prevention based on initiation can be different. Detailed information on the mechanisms of SCC and the methods of testing and/or evaluating susceptibility to SCC can be found in the articles "Environmentally Induced Cracking" (see the section "Stress-Corrosion Cracking") and "Evaluation of Stress-Corrosion Cracking" in this Volume.

Sources of Hydrogen

In contrast to most forms of corrosion, hydrogen embrittlement generally occurs in service when the part is being protected from corrosion or when corrosion on the part is absent, that is, when a high-strength steel is cathodically protected. The corrosion is usually taking place elsewhere (at the anode), but atomic hydrogen is being generated at the surface of the steel (cathode) by the dissociation of water. Failures that occur after a period of time are related to the diffusion of atomic hydrogen into the steel. The fracture path is usually intergranular in steels; however, not all intergranular failures in steels are due to hydrogen embrittlement. In addition, not all hydrogen embrittlement failures in other metals and alloys are intergranular. The effects of hydrogen on the fracture appearance of metals are reviewed and illustrated in the article "Modes of Fracture" in Volume 12 of the 9th Edition of *Metals Handbook*.

Time-delayed embrittlement failures are caused by the residual atomic hydrogen in the steel from the making or melting process and the atomic hydrogen introduced into the steel during processing and subsequent manufacture (for example, plating, machining with hydrocarbon-base oils, pickling, and welding). The atomic, diffusible, or nascent hydrogen (H) is the cause of the problem, not the total hydrogen that includes molecular hydrogen (H_2). Also necessary is applied stress or residual stress from welding or heat treating. Hydrogen embrittlement is generally found in high-strength steels, but hydrogen stress cracking (HSC) has been reported for other materials, such as refractory metals, superalloys, and even austenitic steels, when tested under high-pressure hydrogen gas.

Source of Stress

Cracking studies of high-strength steels in aqueous chloride solutions at a potential exceeding the open-circuit potential of about −0.6 V versus a saturated calomel electrode (SCE) are, in reality, hydrogen embrittlement tests, because the steel is at a cathodic potential; that is, hydrogen is produced at the surface of the steel while under stress. The worst case, or the most severe condition that produces cracking is generated when atomic hydrogen is produced while the part (or test coupon) is under stress. This is the reason hardware is processed (plated) with no stress (except unavoidable manufacturing-induced residual stress) and then stress relieved or baked before the external stress is applied. If residual stress exists during plating, the part could break or at least could have microcracks that subsequently lead to an early service failure.

Hydrogen relief or baking treatments can be effective in removing the atomic hydrogen from the steel; however, this does not ensure that hydrogen is removed from the part. The hydrogen can reside as molecular hydrogen at the interface between the steel part and a plating or coating. Sulfur, in the form of manganese sulfide inclusions commonly found in steel, can act as a poison to dissociate the molecular hydrogen to atomic hydrogen. At the point of high stress, atomic hydrogen will then diffuse back into the steel part and eventually lead to a time-delayed hydrogen embrittlement failure.

Because there are three sources of hydrogen—the manufacture of steel (melting), processing and manufacturing, and in-service (environment)—and because there are three sources of stress—applied, residual from heat treatment, and residual from welding or plastic deformation—nine possible combinations exist. Therefore, it is not always easy to identify the specific cause of a hydrogen embrittlement service failure. A comprehensive evaluation for the prevention of hydrogen embrittlement service failures should include all nine possibilities. An evaluation of the possible controls that avoid in-service hydrogen embrittlement failures must focus on either eliminating the sources of hydrogen or operating at a sufficiently low stress (below the threshold) to prevent cracking.

Current hydrogen embrittlement prevention and control procedures are primarily directed at the plating process and maintenance chemicals. These procedures are covered in ASTM F 519 (Ref 2). Over 30 other military and federal specifications include hydrogen embrittlement relief treatments. Hardware, such as springs or structural fasteners, is tested directly by sustained or step-load stress tests to evaluate the effectiveness of the hydrogen embrittlement relief treatments.

Testing

Tests for hydrogen embrittlement are performed to determine the effect of hydrogen damage in combination with residual or applied stresses. In the past decade, conventional testing methods have been modified to incorporate fracture mechanics, and the various types of hydrogen damage have been further classified in terms of crack nucleation, crack growth rates, and threshold stress intensity measurements.

This section will discuss the current methods of hydrogen embrittlement testing and will focus on accelerated small-specimen testing methods for failure analysis and production control of hydrogen embrittlement. Additional information on hydrogen damage in metals and on test methods for hydrogen embrittlement can be found in Ref 1 to 9 and in the article "Environmentally Induced Cracking" in this Volume.

Standardized Tests. Currently, the only standards for hydrogen embrittlement testing are ASTM F 519 (Ref 2) and F 326 (Ref 3). These standards are based on (1) not putting hydrogen into the steel by keeping the hydrogen in the plating bath at acceptably low levels (ASTM F 326) and (2) using mechanical tests to ensure that the amount of residual hydrogen after baking is under acceptably low levels (ASTM F 519).

ASTM F 326. This standard method covers an electronic hydrogen detection instrument procedure for the measurement of plating permeability to hydrogen, a variable that is related to hydrogen absorbed by steel during plating and to the hydrogen permeability of the plate during post plate baking. A specific application of this method involves controlling cadmium-plating processes in which the plate porosity relative to hydrogen is critical, such as with cadmium plating of high-strength steel.

This method uses a metal-shelled vacuum probe as an ion gage. A section of the probe shell is cadmium plated at the lowest current density encountered during the electroplating process. During subsequent baking, the probe ion current that is proportional to hydrogen pressure is recorded as a function of time. The slope of this curve has an empirical relationship to failure data, such as those discussed in ASTM F 519.

ASTM F 519. This method covers the evaluation of the hydrogen-generating potential of fluids (aircraft maintenance chemicals) and the hydrogen embrittlement control of electroplating processes. Test specimens are installed into the plating bath during the plating of hardware to monitor indirectly the amount of hydrogen in the plating bath. The acceptable level of hydrogen is determined by a go/no-go situation established by the failure of a sustained loaded, stressed specimen that has been baked at 191 °C ± 14 °C (375 °F ± 25 °F) for a minimum of 23 h. The procedures and requirements are specified for the following five types of AISI 4340 steel test specimens:

- Type 1a: notched round bars, stressed in tension, under constant load
- Type 1b: notched round bars loaded in tension with stressed O-rings
- Type 1c: notched round bars loaded in bending with loading bars
- Type 1d: notched C-rings loaded in bending with loading bolt
- Type 2a: unnotched ring specimens loaded in bending with displacement bars

For platings, no stress is applied until the parts have been baked; baking is specified to occur within 1 h after plating. For maintenance chemicals and cleaners, stress is applied before the test specimens are exposed to the environment. The latter condition is obviously much more severe and discriminates against much lower levels of hydrogen, but is more representative of their end use.

The cantilever beam test is a constant-load test in which a V-notched specimen is inserted along a portion of the beam and enclosed by an environmental chamber (Fig. 1). A crack at the root of the V-notch is initiated and extended by fatigue before testing. Notch-root thickness is prescribed by the American Society for Testing and Materials (ASTM), although the requirement is often excessive for high-toughness steels. The specimen is subjected to a constant load over a predetermined time period. As the crack grows, the stress intensity increases. Time to failure is plotted versus applied stress intensity. The lower limit of the resultant curve is a threshold stress intensity for hydrogen embrittlement, K_{IHE}, as shown in Fig. 2.

The K_{IHE} results of a cantilever beam test depend on how much time elapses before the test is terminated. Recommended test periods for establishing the true stress intensity threshold range from 200 h, which is typical for hydrogen embrittlement testing, to as long as 5000 h (Ref 11). Another limitation of this testing method is that it can be expensive in terms of materials and machining. As many as 12 specimens, placed under different loads in separate test machines, are needed for each test in order to obtain valid K_{IHE} values.

The wedge-opening load test applies a constant wedge- or crack-opening displacement; as the crack extends, stress intensity decreases until crack arrest occurs (Fig. 3). The initial load is assumed to be slightly above K_{IHE}. The specimen is maintained under these conditions for about 5000 h to establish the threshold. The crack

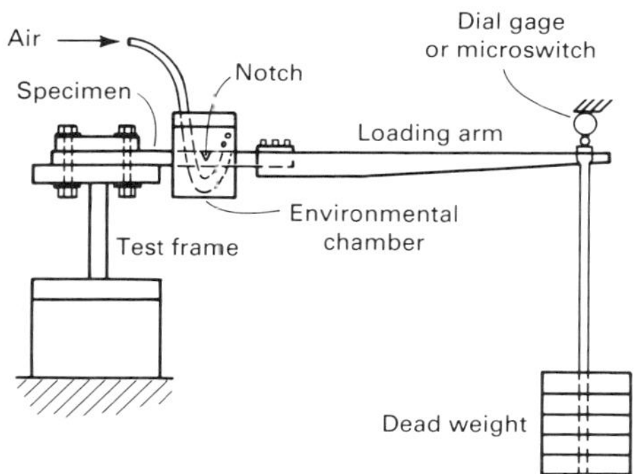

Fig. 1 Fatigue-cracked cantilever beam test specimen and fixtures. Source: Ref 10

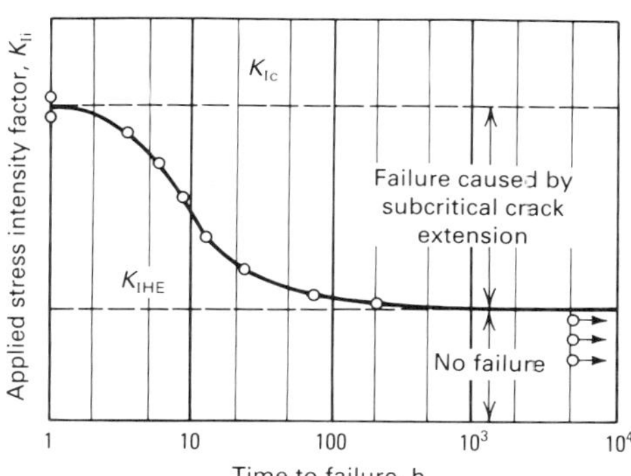

Fig. 2 Procedure to obtain K_{IHE} with precracked cantilever beam test specimen. Source: Ref 10

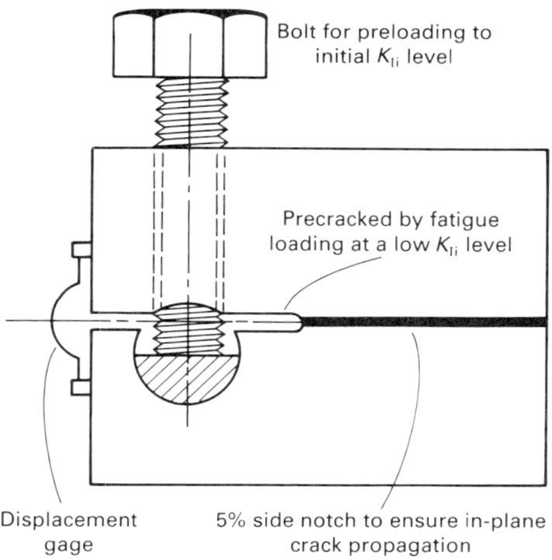

Fig. 3 Schematic showing basic principle of modified wedge-opening load test specimen

grows to a point after which further growth is not measured (K_{IHE}). However, it is difficult to determine precisely when the no-growth criterion is met. Crack tip opening displacement should also be monitored. Corrosion reactions, accompanied by expansion in volume, may occur at the crack tip. This changes the opening displacement and increases the load, thus altering the desired testing conditions.

As subcritical crack extension occurs, stress intensity increases in the cantilever beam test and decreases in the wedge-opening load test (Fig. 4). Generally, the threshold stress intensity measured with the wedge-opening load test is lower than that measured with the cantilever beam test (Ref 12). The advantage of the wedge-opening load test is that only one specimen is required to measure K_{IHE}.

In long-term tests, it is essential to ensure that the concentration and the composition of the environmental solution do not change over time. For example, evaporation may cause the solution level to drop below the crack line of the specimen; this would render the test invalid.

Figure 5 shows the results of wedge-opening load and cantilever beam tests on 25.4-mm (1-in.) thick iron-nickel-cobalt alloy steel specimens. The data imply that the wedge-opening load K_{IHE} crack arrest result is the lower limit for the cantilever beam K_{IHE} threshold.

The open circles in Fig. 5 represent "no fracture" at various exposure times for an overaged Fe-10Ni-8Co alloy in a cantilever beam test. The open squares indicate failure at increased stress intensities, following step loading at various exposure times at the lower stress intensity. The results suggest that increasing the load during the test produces more aggressive hydrogen embrittlement conditions than a constant load.

This may be due to the possible formation of an oxide film on the surface. When the load is increased, the oxide film is broken, exposing fresh metal, and more hydrogen is produced at the crack tip. For this reason, the test should use a rising load, because the constant-load cantilever beam test does not provide worst-case (fresh metal exposed) loading conditions.

The data also suggest that the cantilever beam test can generate an artificially high K_{IHE} threshold, depending on the time limit selected. If the test had been terminated at 200 h rather than at 5000 h, the reported K_{IHE} values from the cantilever beam test (Fig. 5) would have been four times higher than those measured after the longer time period. Similarly, if insufficient time is allowed in the wedge-opening load test, the incubation period may not be exceeded, and no crack growth will result.

Both testing methods require costly and time-consuming steps and result in a parameter, K_{IHE}, whose design significance is questionable. However, the parameter does provide a relative ranking of susceptibility to hydrogen embrittlement or, more generally, SCC.

The contoured double-cantilever beam test is used to measure crack growth rate at a constant stress intensity factor. This test simplifies the calculation of stress intensity by using a contoured specimen so that stress intensity is proportional to the applied load and is independent of the crack length. Under a constant load, stress intensity also remains constant with crack extension. For the test geometry shown in Fig. 6, the stress intensity factor equals 20 times the load ($K = 20P$).

Data on hydrogen embrittlement can be obtained with subthickness specimens (Ref 14), even in excess of the ASTM requirement of $K^2_{max} < 0.4\ B/YS^2$ (where B is the thickness and YS is the yield strength of the specimen), by using side grooves. Side grooves provide additional constraint on the material being tested. They also enable the maintenance of a plane-strain condition in a thin specimen by enhancing stress triaxiality. This method has been extensively used to study the effect of heat treatment (hardness) and environment on the hydrogen stress cracking of AISI 4340 steels (Fig. 7).

The contoured double-cantilever beam test has also been used to study the stress history effect that produces an incubation time before hydrogen stress cracking. Figure 8 shows that incubation time is dependent on the type of steel. A decrease in the stress intensity factor from 44 to 22 MPa$\sqrt{m}$ (40 to 20 ksi$\sqrt{in.}$) may change the

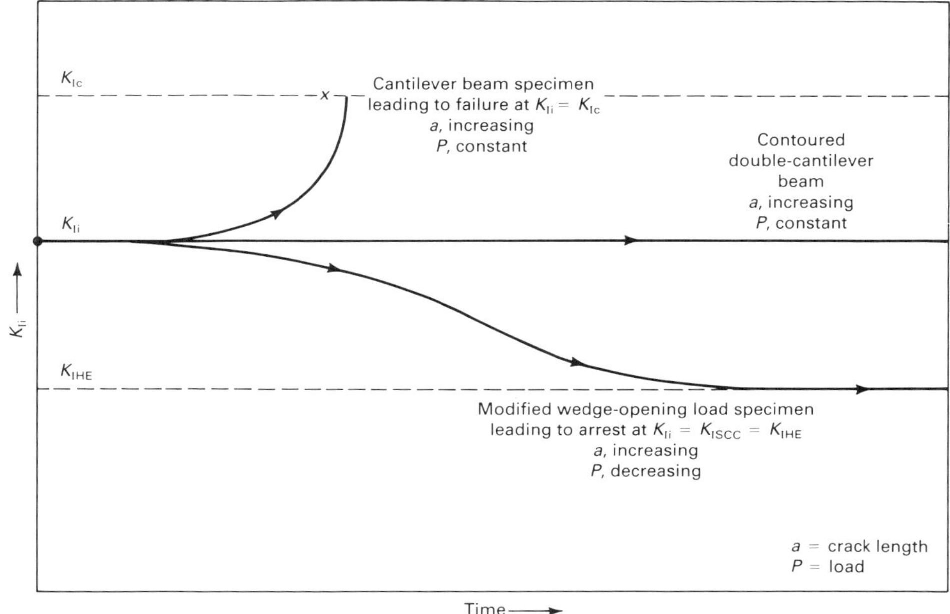

Fig. 4 Influence of time, crack extension, and load on stress-intensity behavior of modified wedge-opening load, cantilever beam, and contoured double-cantilever beam test specimens. Source: Ref 10

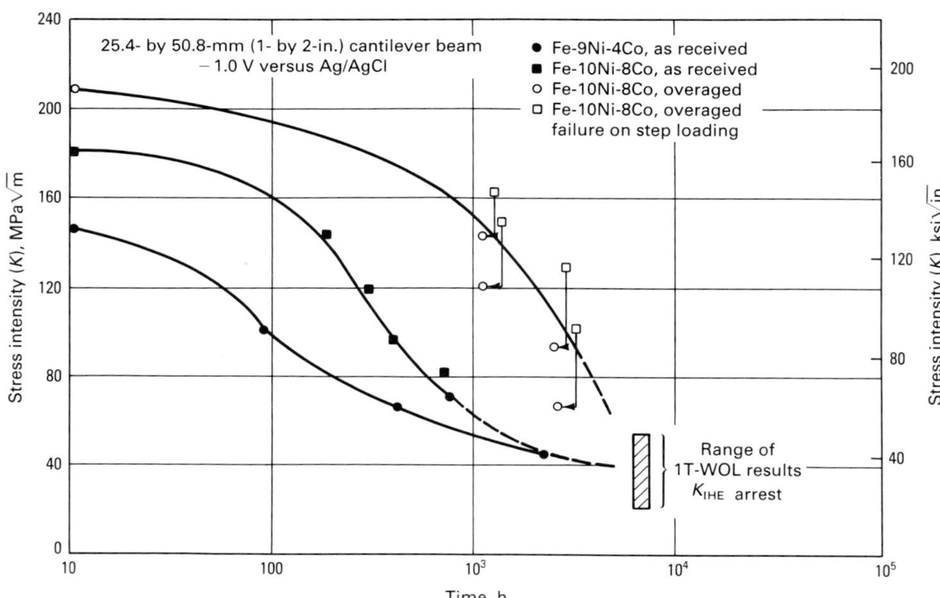

Fig. 5 Comparison of single-edge-notched cantilever beam and wedge-opening load test results for hydrogen embrittlement cracking of iron-nickel-cobalt steels. Source: Ref 13

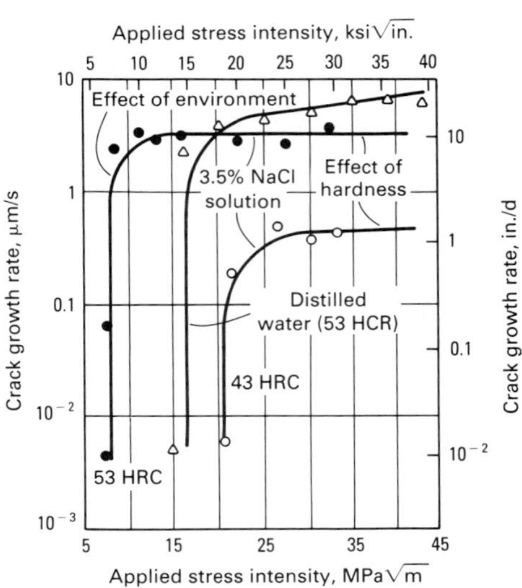

Fig. 7 Hydrogen embrittlement crack growth rate as a function of applied stress intensity for two different hardnesses and environments for an AISI 4340 steel contoured double-cantilever beam test specimen

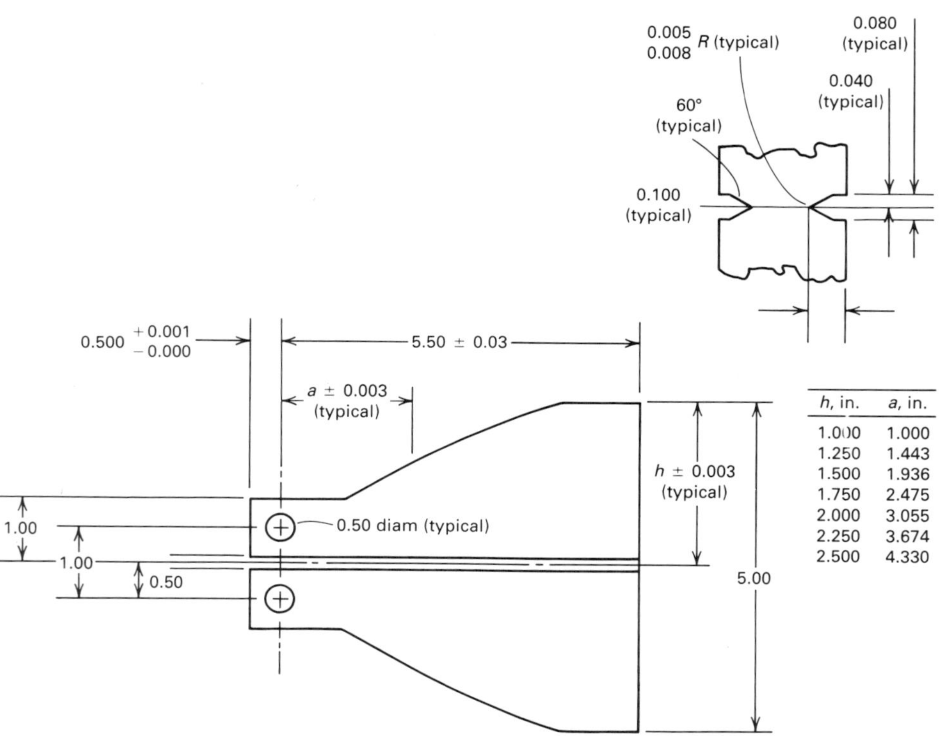

h, in.	a, in.
1.000	1.000
1.250	1.443
1.500	1.936
1.750	2.475
2.000	3.055
2.250	3.674
2.500	4.330

Fig. 6 Dimensions and configuration for double-cantilever beam test specimen. Specimen contoured to $3a^2/h^3 + 1/h = C$, where C is a constant. All values given in inches (1.0 in. = 25.4 mm)

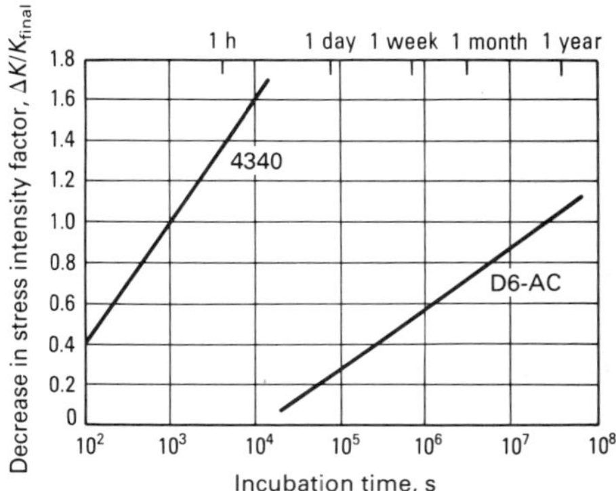

Fig. 8 Incubation time for crack growth in AISI type 4340 and type D6-AC steel contoured double-cantilever beam test specimens as a function of decrease in stress intensity. Source: Ref 15

incubation time from less than 1 h for AISI 4340 steel to about 1 year for type D6-AC steel.

Three-Point and Four-Point Bend Tests. The contoured double-cantilever beam test uses a constant load to maintain a constant stress intensity factor with crack extension. The same effect can be produced by using a three- or four-point bend test under displacement control. These tests use heavily side-grooved Charpy V-notch specimens (Fig. 9). Because crack-opening displacement is constant as the crack extends, the load decreases; therefore, there is a slight initial increase in stress intensity to a maximum value that drops slightly as the ratio of crack depth to specimen width exceeds 0.5. Typically, stress intensity is constant within a small range. Figure 10 compares the change in stress intensity factor with crack extension as a function of load control to that of displacement control for a three-point bend specimen.

The rising step-load test provides a stress intensity that is different at each load but that remains constant with crack extension as each load level is sustained. Crack initiation is signaled by a drop in load (Fig. 11). The rising step-load test was developed as an accelerated low-cost test for measuring the resistance of steels (particularly weldments) to hydrogen embrittlement (Ref 13, 17). The threshold obtained by this method could be slightly higher if the test duration of each load is too short, but the test duration can be extended near the initiation loads in duplicate tests in order to obtain a more accurate measurement.

To index susceptibility to hydrogen-assisted cracking, the test should last no longer than 24 h, and the hydrogen source should reflect the most aggressive environment. In one experiment, a 3.5% sodium chloride solution was selected to simulate seawater, and a cathodic potential of −1.2 V versus SCE was used to generate hydrogen in order to reproduce the extreme conditions

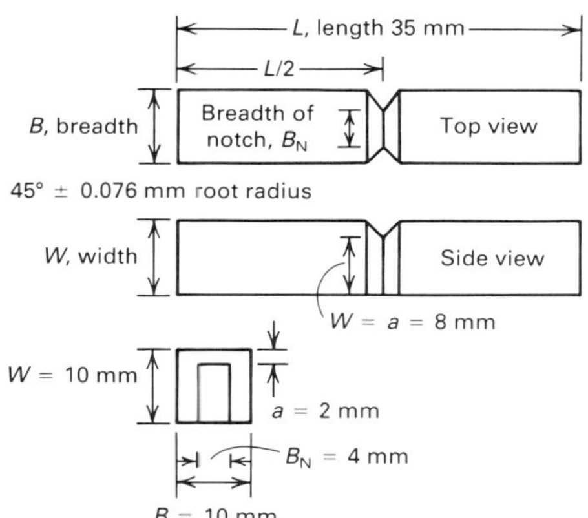

Fig. 9 Side-grooved Charpy V-notch test specimen used for three- and four-point bend tests

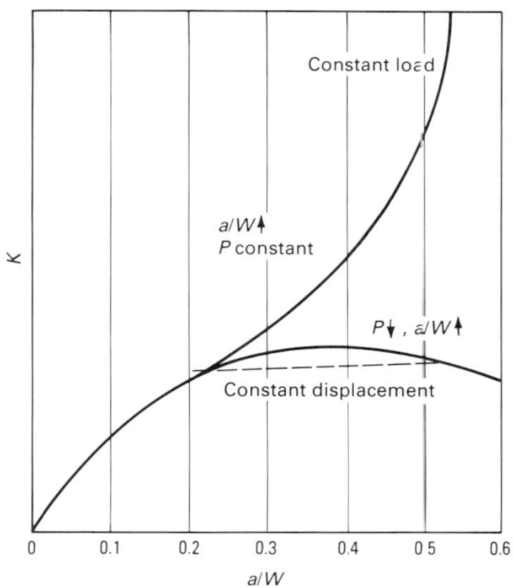

Fig. 10 Change in stress intensity factor with crack extension as a function of load control and displacement control for a three-point bend specimen

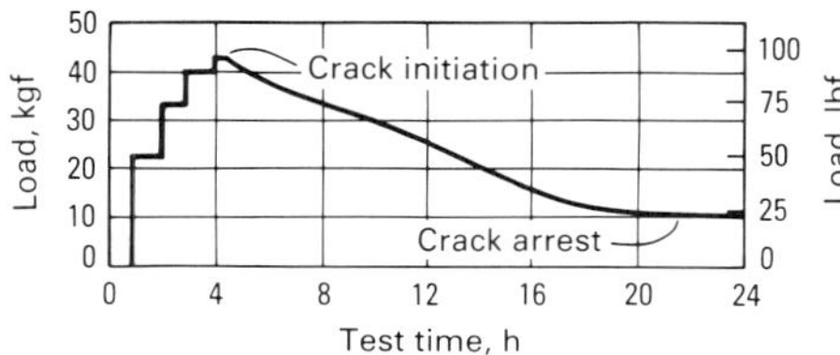

Fig. 11 Typical load-time record for four-point rising step-load test. Source: Ref 16

of sacrificial anodic protection generally found on a ship hull.

A Charpy specimen was chosen because such specimens are small and easy to machine and handle. In this test, however, the specimen was modified. Instead of using a fatigue precrack, the notch-root radius was machined to less than 7.6 μm (3 mils). This was done to lower the cost and to give less ambiguous environmental conditions at the crack tip. Also, hydrogen cracks nucleate below the surface.

The specimen was deeply side grooved, a common practice used in crack growth rate tests to prevent the crack from branching. Side grooves are also used in crack-opening displacement or *J*-integral testing in order to cause load displacement curves to increase monotonically to fracture by inducing a highly triaxial stress field at the crack tip. Because a Charpy specimen is small, deep side grooves produce a triaxial stress field at the notch, and thus promote crack initiation. The extent of the side grooving is such that the remaining ligament is only 40% of the original thickness. The modified Charpy specimen dimensions are shown in Fig. 9.

The specimen was loaded by means of beams and an instrumented bolt (Fig. 12). Four-point bending under constant displacement control and stress intensity produced crack growth. Once cracking initiated at the notch ($a/W = 0.2$, where a is crack length and W is width of the specimen), arrest did not occur until the crack was nearly through the specimen. The load was manually increased at 1-h intervals. An environmental chamber encompassed the specimen and included a potentiostat to produce hydrogen while under stress.

The rising step-load test was used to evaluate high-strength HY ship steels and weldments in an environment simulating seawater under conditions of cathodic protection commonly employed to protect ship hulls (Ref 17). Samples from the heat-affected zone (HAZ) and other locations in the weld metal were tested. Interlayer gas tungsten arc heating was evaluated as a means of providing a refined, homogeneous, tempered microstructure with improved resistance to HSC. As a baseline, HY-130 and HY-180 steels were compared.

Figure 13 plots rising step-load test results for HY-130 and HY-180 base metals, in addition to combinations of modified HY steel compositions and programmed-cooling-rate thermal cycles for the base metal and the weld wire. The vertical axis is a plot of a parameter derived from the specimen strength ratio in ASTM E 399—that is, $6 P_{max}/B(W - a)^2$YS, where P_{max} is the maximum load that the specimen is able to sustain, B is the specimen thickness, W is the specimen width, a is the crack length, and YS is the yield strength in tension. For the data shown in Fig. 13, P_{max} was replaced by the crack initiation load. The horizontal axis is a ratio of K_{IHE}/YS, which was measured in a separate test program with cantilever beam and wedge-opening load specimens.

The resistance to hydrogen embrittlement of the two base metals and six locations in HY-130 weldments was ranked by using this testing method. Test results showed that HY-180 is more susceptible to HSC than HY-130 and that the resistance to hydrogen embrittlement of specimens taken from the HAZ and the fusion line is consistently higher than that of weld metal specimens. The resistance of the weld metal is affected by grain structure; interlayer gas tungsten arc reheating homogenized the weld structure, but it did not temper the weld metal. Specimens from the gas tungsten arc reheated weldment consistently exhibited higher hardness and lower resistance to hydrogen embrittlement than similar specimens from the standard HY-130 weld metal (Ref 12, 17).

The disk-pressure testing method measures the susceptibility to hydrogen embrittlement of metallic materials under a high-pressure gaseous environment (Ref 18). The test is used for the selection and quality control of materials, protective coatings, surface finishes, and other processing variables.

In this test, a thin disk of the metallic material to be tested is placed as a membrane in a test cell and subjected to helium pressure until it bursts. Because helium is inert, the fracture is caused by mechanical overload; no secondary physical or chemical action is involved. An identical disk is placed in the same test cell and subjected to hydrogen pressure until it bursts. Metallic materials that are susceptible to environmental hydrogen embrittlement fracture under a pressure that is lower than the helium-burst pressure; materials that are not susceptible fracture under the same pressure for both hydrogen and helium. The

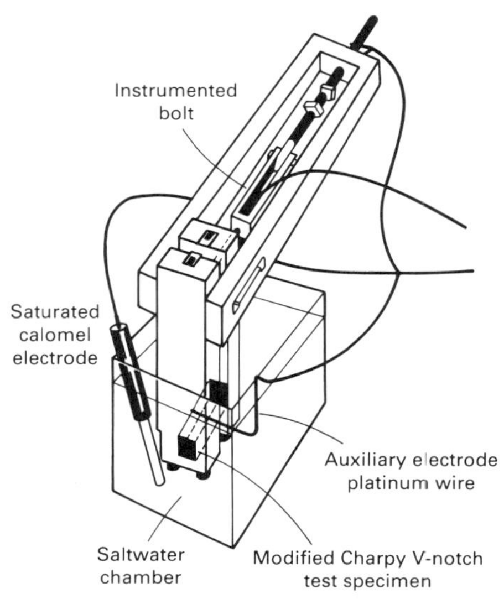

Fig. 12 Loading frame used for rising step-load test

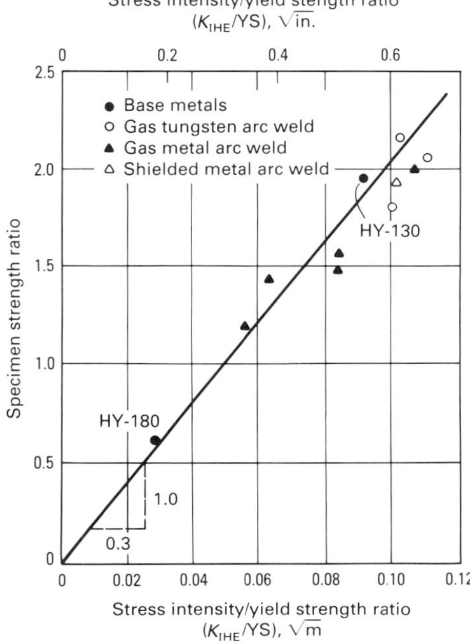

Fig. 13 Analytical correlation of strength ratio with threshold stress intensity data

ratio, S_{H_2}, between the helium-burst pressure, P_{He}, and the hydrogen-burst pressure, P_{H_2}, indicates the susceptibility of the material to environmental hydrogen embrittlement:

$$S_{H_2} = \frac{P_{He}}{P_{H_2}}$$

If S_{H_2} is equal to or less than 1, the material is not susceptible to environmental hydrogen embrittlement. When S_{H_2} is greater than 2, the material is considered to be highly susceptible. At values between 1 and 2, the material is moderately susceptible, with failure expected after long exposure to hydrogen; therefore, the material must be protected against exposure.

A compilation of test results is shown in Fig. 14. The alloys having little or no sensitivity are 7075-T6 aluminum; Haynes 188 (cobalt base); beryllium copper (copper base); types 304, 316, and 310 austenitic stainless steel; type 430 ferritic steel; and age-hardened austenitic A286 steel. Titanium-base alloy Ti-6Al-4V exhibits moderate sensitivity. Alloys with high sensitivity are Haynes 25 (cobalt base) and iron-base alloys, including medium- and high-strength steels. Conventional testing methods, such as the cantilever beam, wedge-opening load, and contoured double-cantilever beam tests, have also been adapted for testing in high-pressure gaseous hydrogen environments.

Slow strain rate tensile tests can be used to evaluate many product forms, including plate, rod, wire, sheet, and tubing, as well as welded parts. Smooth, notched, or precracked specimens can be used. The principal advantage of this standardized test is that the susceptibility to hydrogen stress cracking (HSC) for a particular metal-environment combination can be rapidly assessed (Ref 19).

A variety of specimen shapes and sizes can be used; the most common is a smooth bar tensile coupon, as described in ASTM E 8 (Ref 20). The specimen is exposed to the environment and is stressed under displacement control. For stainless steel in chloride solution, the strain rate is 10^{-6}/s. One or more of the following parameters are applied to the tensile test at the same initial strain rate:

- Time to failure
- Ductility, as assessed, for example, by reduction of area or elongation to fracture
- Maximum load achieved
- Area bounded by a nominal stress-elongation curve or a true stress/true strain curve

Additional information on slow strain rate tensile tests can be found in the article "Evaluation of Stress-Corrosion Cracking" in this Volume.

Potentiostatic Slow Strain Rate Tensile Testing. Studies have been conducted on the use of dissociated water under potentiostatic conditions that produce hydrogen on the surface of the tensile test specimen while under slow strain rate displacement control. Results suggest that hydrogen is the most significant parameter in stress cracking under the conditions of hydrogen sulfide SCC found in oil fields (Ref 21).

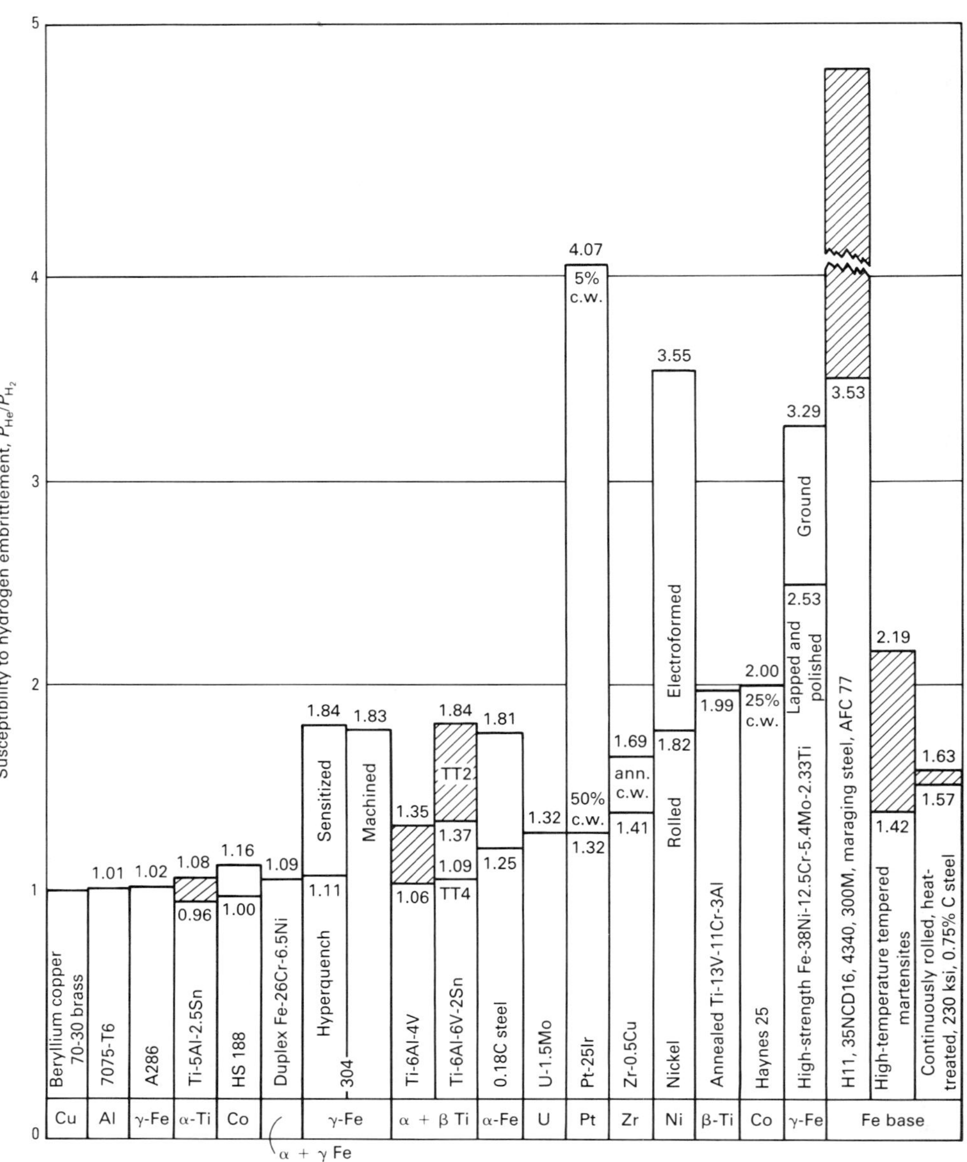

Fig. 14 Relative hydrogen susceptibility of various metals and alloys tested at a rate of 65 bars/min at room temperature. c.w., cold worked; ann., annealed. Source: Ref 18

Interpretation of Test Results

The phenomenon of hydrogen embrittlement or SCC is very complex. The test results depend on environmental conditions (potential, pH, oxygenation level, temperature, environment), material and melting procedure (air, vacuum, electroslag remelting, vacuum arc remelting, and so on); test specimen geometry (notched tensile or bend specimens, cantilever beam specimen, wedge-opening specimen), and specimen preparation (crush grinding, fatigue precracking). Therefore, evaluation of the test results, on a quantitative scale, is impossible. The purpose of the test is to address a very specific set of conditions.

The existing standard, ASTM F 519, is based on the assumption that air-melted AISI 4340 at 53 ±1 HRC is a worst-case condition. If no fracture occurs in this specimen after 200 h at 75% of the notched strength, then other 4340 steels processed by alternate melting processes (even steels with higher hardness levels because of vacuum melting) will not have residual levels of atomic hydrogen that would produce sustained load failure in the environment of concern. Data to support this claim are shown in Fig. 15, which shows the extreme susceptibility of air melted 4340 steel as compared to electroslag remelted or vacuum arc remelted 4340 steel. This test is not intended to be a relative susceptibility test, nor is it intended to be used with any material other than air-melted 4340 according to MIL S-5000E.

This quality control method suffers from the fact that manufacturing-induced stress in the part might not be the same as in the representative test coupon. This is why stress relief before plating is advised. If the coupon is stress free but the parts have a high residual stress, the coupons will pass the 200-h sustained load test, but the parts, which might have cracked during plating, may inadvertently be put into service. Conversely, if care is not taken in the manufacture of the coupons and if high residual stresses are intro-

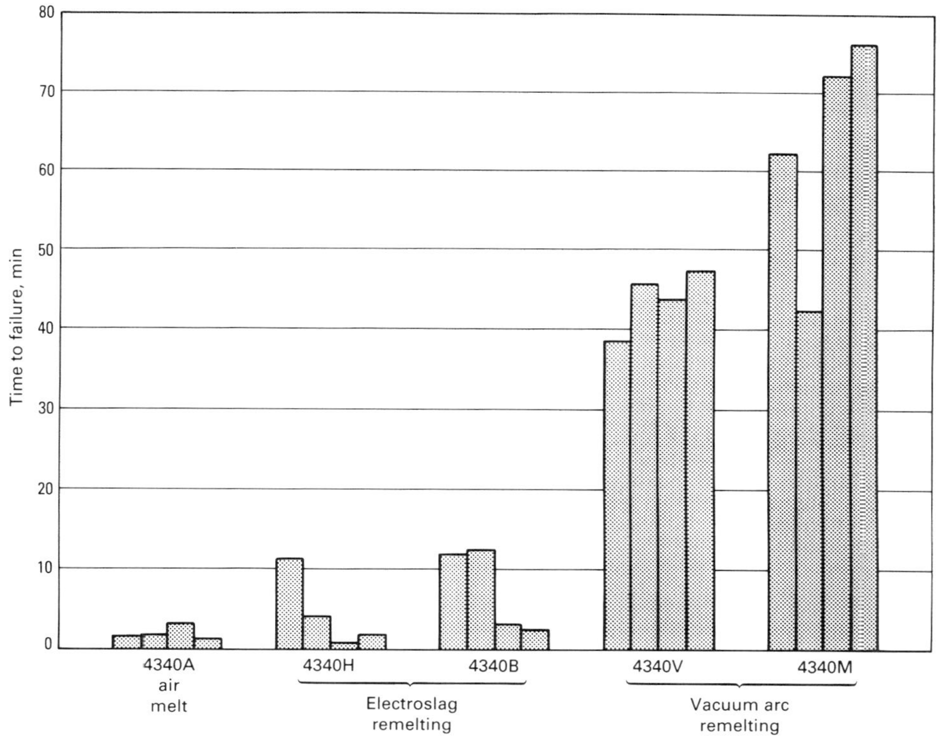

Fig. 15 Time to failure for AISI 4340 high-strength steels with an ultimate tensile strength range of 1790-2070 MPa (260-300 ksi) using ASTM F 519, Type 1c bend specimens. Source: Ref 22

duced, the coupons will fail the 200-h sustained load test, but the parts will be stress or crack free. Based on the latter situation, all of these acceptable parts would be discarded. Obviously, before such action is taken, the parts themselves are generally tested.

To prevent any misrepresentation of test sampling with separately manufactured test coupons, the hardware itself is often tested directly, such as in the case of fasteners or springs, and the use of representative test coupons is completely avoided. The problem with testing actual hardware is twofold. First, the test is destructive and therefore can be costly, depending on the type of hardware being evaluated, and second, the method of applying the stress and the magnitude of the applied stress are not always easily obtained. It can often be very dangerous to reuse parts that have been proof tested by the application of a stress above the anticipated service stresses. Revising these parts can produce conditions that adversely affect structural integrity by consuming a large portion of the life of the part instead of ensuring its longevity.

Prevention and Control

The potential for hydrogen embrittlement must be evaluated before any prevention and control procedures are implemented (Ref 23). The sources of hydrogen must be examined with regard to three possibilities: the manufacture of the steel, the manufacture and processing of the steel part, and the environmental service conditions of the steel part. A source of hydrogen and a sufficient stress must exist in order for hydrogen embrittlement to occur. If either condition is eliminated, hydrogen embrittlement will not occur.

From a control viewpoint, the attempt is to eliminate the hydrogen or to ensure that no hydrogen is introduced during the manufacture and processing of the steel. It is then anticipated that no problem will exist in service. This approach focuses on the hydrogen embrittlement process that has classically been referred to as internal hydrogen embrittlement. This possibility is introduced during the manufacture and processing of the steel. Once the part is in service, any potential hydrogen embrittlement failure can result from hydrogen produced from a corrosion reaction or from the steel operating in high-pressure hydrogen. This is referred to as environmental hydrogen embrittlement.

Hydrogen embrittlement failures are prevented by controlling the amount of hydrogen introduced during the manufacture and processing of the hardware. Consideration must be given to the amount of hydrogen introduced during the melting of the steel, the manufacture of the part, and the in-service environment, which should also include cleaners and paint strippers. The necessary controls are covered in ASTM F 519. The major difficulty appears to involve identifying the manufacturing-induced residual stresses that result from heat treatment and, especially, welding (if it is part of the manufacturing process). Much more attention must be given to evaluating the potential for hydrogen embrittlement failures in the presence of residual stresses. The source of the residual stress must be identified in each piece of production hardware, or the assumption of high residual stresses in the manufactured part, especially in welded structures, must be included in the design analysis.

The usefulness of the threshold stress intensity K_{IHE} parameter has yet to be established. Its main function appears to be its use in designing against the service-induced or application-induced possibilities of hydrogen embrittlement failures. This can be done only by controlling the applied stress and geometrical factors, such as root radius, in order to maintain the effective applied stress intensity at a level that is below threshold for subcritical crack growth in a given environment. This approach is a relatively new frontier for technological advancement in that the method of measuring K_{ISCC} has not yet been standardized and its significance in design has not yet been completely understood.

REFERENCES

1. L. Raymond, Ed., *Hydrogen Embrittlement Testing*, STP 543, American Society for Testing and Materials, 1972
2. "Standard Method for Mechanical Hydrogen Embrittlement Testing of Plating Processes and Aircraft Maintenance Chemicals," F 519, *Annual Book of ASTM Standards*, American Society for Testing and Materials
3. "Standard Method for Electronic Hydrogen Embrittlement Test for Cadmium-Electroplating Processes," F 326, *Annual Book of ASTM Standards*, American Society for Testing and Materials
4. I.M. Bernstein and A.W. Thompson, Ed., *Hydrogen in Metals*, American Society for Metals, 1973
5. C.D. Beachem, Ed., *Hydrogen Damage*, American Society for Metals, 1977
6. C.G. Interrante and G.M. Pressouyre, Ed., *Current Solutions to Hydrogen Problems in Steels*, American Society for Metals, 1983
7. R. Gibala and F. Hehemann, Ed., *Hydrogen Embrittlement and Stress Corrosion Cracking*, American Society for Metals, 1984
8. L. Raymond, Ed., *Hydrogen Embrittlement: Prevention and Control*, STP 962, American Society for Testing and Materials, 1987
9. A.R. Troiano, The Role of Hydrogen and Other Interstitials in the Mechanical Behavior of Metals, *Trans. ASM*, Vol 52, 1960, p 54-80
10. S.T. Rolfe and I.M. Barsom, *Fracture and Fatigue Control in Structures*, Prentice-Hall, 1977
11. R.P. Wei, S.R. Novak, and D.R. Williams, Some Important Considerations in the Development of Stress Corrosion Cracking Test Methods, *Mater. Res. Stand.*, Vol 12 (No. 9), Sept 1972, p 25
12. C.A. Zanis, Subcritical Cracking in High Strength Steel Weldments—A Materials Approach, *SAMPE Quart.*, Jan 1978, p 8-12
13. C.A. Zanis, P.W. Holsberg, and E.C. Dunn, Jr., Seawater Subcritical Cracking of HY-Steel Weldments, *Weld. J. Res. Suppl.*, Vol 59, Dec 1980
14. "Rapid Inexpensive Tests for Determining Fracture Toughness," NMAB-328, National Academy of Sciences, 1976
15. D.L. Dull and L. Raymond, Stress History Effect on Incubation Time for Stress Corrosion Crack Growth in AISI 4340 Steel, *Metall. Trans.*, Vol 3, Nov 1972, p 2943-2947
16. D.L. Dull and L. Raymond, Electrochemical Techniques, in *Hydrogen Embrittlement and Testing*, STP 543, American Society for Testing and Materials, 1974, p 20-33
17. P.J. Fast, C.S. Susskind, and L. Raymond, "Charpy V-Notched Specimens for Indexing Stress Corrosion Cracking in HY Ship Steels

and Weld Metals," Final report under contract F04701-78-C-0079, The Aerospace Corporation, Aug 1979

18. J.P. Fiddle, R. Bernardi, R. Broudeur, C. Roux, and M. Rapin, Disk Pressure Testing of Hydrogen Environment Embrittlement, in *Hydrogen Embrittlement Testing*, STP 543, American Society for Testing and Materials, 1974, p 221-253
19. "Recommendations for Conducting Slow Strain Rate Stress Corrosion Tests," Document ISO/TC 156/WG 2, International Standards Organization, British Standards Institute
20. "Standard Methods of Tension Testing of Metallic Materials," E 8, *Annual Book of ASTM Standards*, American Society for Testing and Materials
21. D.L. Dull and L. Raymond, Surface Cracking of Inconel 718 During Cathodic Charging, *Metall. Trans.*, Vol 4, June 1973, p 1635-1638
22. L. Raymond and C. Beneker, "Evaluation of the Relative Hydrogen Embrittlement Susceptibility of ESR 4340 and Its Heat Treat Distortion Properties," Final Report AMMRC TR 82-49, Material Technology Laboratory, Sept 1982
23. L. Raymond, Hydrogen Embrittlement Control, *ASTM Standard. News*, Vol 13 (No. 12), Dec 1985, p 46

Evaluation of Corrosion Fatigue

Donald O. Sprowls, Consultant

A FREQUENT CAUSE of the premature fracture of structural components (see Fig. 1 in the article "Evaluation of Stress-Corrosion Cracking" in this Volume) is corrosion fatigue cracking (CFC), which is not limited to certain metallurgical conditions of the metal or to critical environmental species, as are other forms of environmentally assisted cracking. Classical CFC and stress-corrosion cracking (SCC) failures (which may be clearly distinguished by fractographic features) are separated by a spectrum of behaviors dictated by a large number of conditions related to the metallurgical condition of the material, the method of stressing, the electrochemistry of the metal/environment system, and the interaction kinetics.

Relationship Between CFC and SCC

Figure 1 illustrates a conceptual interrelationship of corrosion fatigue, stress corrosion, and hydrogen embrittlement used in Ref 1 in a discussion of the current understanding of the mechanisms of stress corrosion and corrosion fatigue. The most serious practical situations involving ductile alloy/environment systems are in the crosshatched regions, which indicate the combination of any two failure mechanisms, and especially in the center, in which all three phenomena interact. Although much research in this area has been published in recent years (Ref 1-11), many unanswered questions remain. One researcher has concluded that the danger of cracking in practical situations and the applicability of various testing techniques can be assessed from a mechanistic base as well as from the existing foundation of empirical experience (Ref 1). Some of the experimental evidence relating CFC to SCC will be cited in later sections of this article that discuss specific aspects of corrosion fatigue testing.

Prediction of Corrosion Fatigue Life

According to a recent assessment of standards for CFC and SCC, the methodology for the interpretation and use of standard test method data for environmental cracking is not well developed. Moreover, to the extent that the two phenomena can interact synergistically under some conditions, existing approaches to standard test development may produce nonconservative data in some cases (Ref 12). Similarly, according to another researcher, the research and testing community is still unable to deliver a scientific basis for a reliable estimate of fatigue life for all conceivable load and environmental combinations,

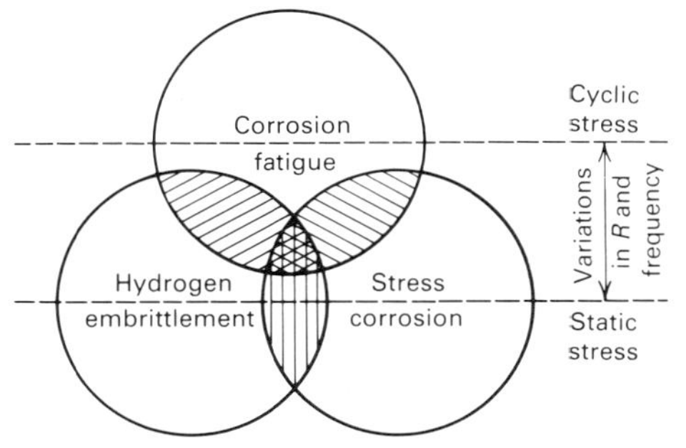

Fig. 1 Venn diagram illustrating the interrelationship among stress corrosion, corrosion fatigue, and hydrogen embrittlement. Source: Ref 1

despite the estimated billions of dollars spent on fatigue (Ref 13).

Therefore, the designer may rely on experience with similar components in service rather than a laboratory evaluation of mechanical test specimens. Laboratory tests, however, are essential for understanding fatigue behavior, and current studies with fracture mechanics test specimens are beginning to provide satisfactory design criteria.

Types of Corrosion Fatigue Tests

Laboratory corrosion fatigue tests can be classified as either cycles to failure (crack initiation) or crack propagation tests. In cycles to failure testing, specimens or parts are subjected to the number of stress cycles required for CFC to initiate and subsequently grow large enough to produce failure. Such data are usually obtained by testing smooth or notched specimens. With this type of testing, however, it is difficult to distinguish between CFC initiation life and CFC propagation life.

In crack propagation testing, fracture mechanics methods are used to determine the crack growth rates of preexisting cracks under cyclic loading. Preexisting cracks or sharp defects in a material reduce or may eliminate the crack initiation portion of the fatigue life of the component. Both types of testing are important; however, it appears that crack initiation is of more significance in the failure process of relatively thin sections, while crack growth appears to dominate thick-section component endurance (Ref 14).

The usual approach to corrosion fatigue testing is to perform a state-of-the-art fatigue test in the presence of the environment of interest. This article will emphasize important factors that should be recognized and controlled in corrosion fatigue tests and will provide references to published standard methods and recommended practices. Information on the planning and design of fatigue tests and the statistical analysis of test results is available in the article "Fatigue Data Analysis" in Volume 8 of the 9th Edition of *Metals Handbook*.

Standards and Recommended Practices for Fatigue Testing

Only a few standardized procedures for fatigue testing are available because many of the testing machines are custom built. However, several types of fatigue-testing machines, testing methods, and test specimens are accepted as standard. In the United States, the American Society for Testing and Materials (ASTM) issues voluntary standards and recommended practices. Those related to fatigue testing are:

- ASTM E 206, "Standard Definitions of Terms Relating to Fatigue Testing and the Statistical Analysis of Fatigue Data"
- ASTM E 466, "Standard Practice for Conducting Constant Amplitude Axial Fatigue Tests of Metallic Materials"
- ASTM E 467, "Standard Practice for Verification of Constant Amplitude Dynamic Loads in an Axial Load Fatigue Testing Machine"
- ASTM E 468, "Standard Practice for Presentation of Constant Amplitude Fatigue Test Results for Metallic Materials"
- ASTM E 513, "Standard Definitions of Terms Relating to Constant Amplitude Low-Cycle Fatigue Testing
- ASTM E 606, "Standard Practices for Constant Amplitude Low-Cycle Fatigue Testing
- ASTM E 647, "Standard Test Method for Constant-Load-Amplitude Fatigue Crack Growth Rates above 10^{-8} m/Cycle"
- ASTM E 739, "Standard Practice for Statistical Analysis of Linear or Linearized Stress-Life (S-N) and Strain-Life (ϵ-N) Fatigue Data"
- ASTM E 742, "Standard Definitions of Terms Relating to Fluid Aqueous and Chemical Environmentally Affected Fatigue Testing"
- ASTM E 912, "Definitions of Terms Relating to Fatigue Loading"

Cycles to Failure Tests

Cycles to failure (crack initiation) tests are procedures in which a specimen or part is subjected to cyclic loading to failure. A large portion of the total number of cycles in these tests is spent initiating the crack. Although cycles to failure tests conducted on small specimens do not precisely establish the fatigue life of a large part, such tests do provide data on the intrinsic fatigue crack initiation behavior of a metal or alloy. As a result, such data can be used to develop criteria to prevent fatigue failures in engineering design. Examples of the use of small-specimen fatigue test data can be found in the basis of the fatigue design codes for boilers and pressure vessels; complex welded, riveted, or bolted structures; and automotive and aerospace components.

Fatigue-Testing Regimes

The magnitude of the nominal stress on a cyclically loaded component is often measured by the amount of overstress, that is, the amount by which the nominal stress exceeds the fatigue limit or the long-life fatigue strength of the material used in the component. The number of load cycles that a component under low overstress can endure is high; therefore, the term high-cycle fatigue is often applied. Because high-cycle fatigue tests require uninterrupted operation for long periods of time, test machines that are simple and reliable are used. Either constant-load amplitude or constant deflection tests can be conducted.

As the magnitude of the nominal stress increases, the initiation of multiple cracks is more likely. Also, spacing between fatigue striations, which indicate the progressive growth of the crack front, is increased, and the region of final fast fracture is increased in size.

Low-cycle fatigue is the regime characterized by high overstress. The commonly accepted dividing line between high-cycle and low-cycle fatigue is considered to be between 10^4 and 10^5 cycles. In practice, this distinction is made by determining whether the dominant component of the strain imposed during cyclic loading is elastic (high cycle) or plastic (low cycle), which in turn depends on the properties of the metal as well as the magnitude of the nominal stress.

Different test techniques may be required for the control and monitoring of low-cycle fatigue tests. Typically, strain-controlled tests (constant strain amplitude) are used.

In routine low- and high-cycle fatigue crack initiation testing, complete fracture of a small specimen is generally the failure criterion. Approximately 30 to 40% of the low-cycle fatigue life and about 80 to 90% of the high-cycle fatigue life measured by cycles to failure involve nucleation of the dominant fatigue crack that eventually causes failure.

Loading Parameters

Most laboratory fatigue testing is done either with axial loading or in bending, thus producing only tensile and compressive stresses. The stress is usually cycled between a maximum and a minimum tensile stress or between a maximum tensile stress and a maximum compressive stress. The latter is considered a negative tensile stress, is given an algebraic minus sign, and is therefore known as the minimum stress.

Applied stresses are described by three parameters. The mean stress, S_m, is the algebraic average of the maximum and minimum stresses in 1 cycle, $S_m = (S_{max} + S_{min})/2$. In the completely reversed cycle test, the mean stress is zero. The range of stress, S_r, is the algebraic difference between the maximum and minimum stresses in 1 cycle, $S_r = S_{max} - S_{min}$. The stress amplitude, S_a, is one-half the range of stress, $S_a = S_r/2 = (S_{max} - S_{min})/2$.

During a fatigue test, the stress cycle is usually maintained constant so that the applied stress conditions can be written $S_m \pm S_a$, where S_m is the static or mean stress, and S_a is the stress amplitude, which is equal to half the stress range. The nomenclature used to describe test parameters involved in cyclic stress testing is shown in Fig. 2.

The stress ratio is the algebraic ratio of two specified stress values in a stress cycle. Two commonly used stress ratios are the ratio A of the stress amplitude to the mean stress ($A = S_a/S_m$) and the ratio R of the minimum stress to the maximum stress ($R = S_{min}/S_{max}$).

If the stresses are fully reversed, the stress ratio R becomes -1; if the stresses are partially reversed, R becomes a negative number less than 1. If the stress is cycled between a maximum stress and no load, R becomes 0. If the stress is cycled between two tensile stresses, R becomes a positive number less than 1. A stress ratio R of 1 indicates no variation in stress, making the test a sustained-load SCC test rather than a corrosion-fatigue test.

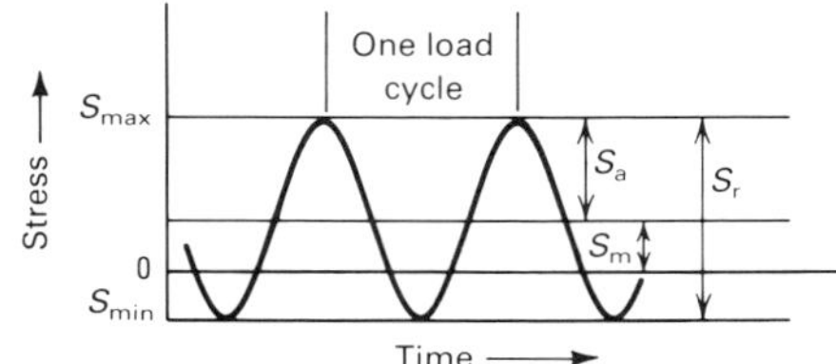

Fig. 2 Nomenclature used to describe test parameters involved in cyclic stress testing

Presentation of Fatigue Data

High-cycle fatigue data are presented graphically as stress, S, versus cycles to failure, N, in S-N diagrams or S-N curves.

***S-N* Curves.** The results of fatigue cycles to failure tests are usually plotted as maximum stress, minimum stress, or stress amplitude versus number of cycles, N, to failure, using a logarithmic scale for the number of cycles. Stress is plotted on either a linear or a logarithmic scale. The resulting plot of the data is termed an S-N curve. Three typical S-N curves are shown in Fig. 3.

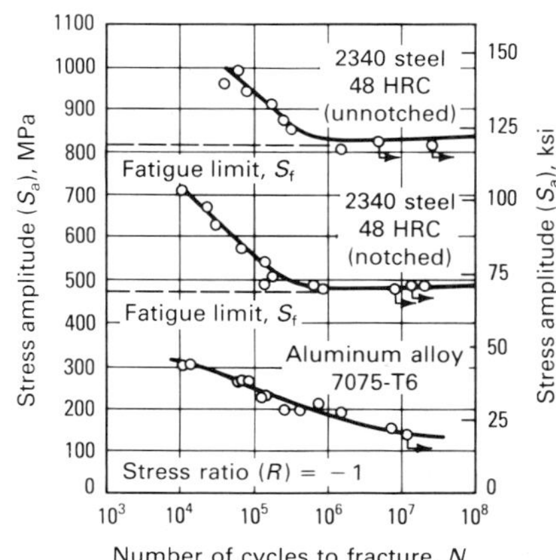

Fig. 3 Typical S-N curves for constant amplitude and sinusoidal loading

Fatigue Limit and Fatigue Strength. The horizontal portion of an S-N curve represents the maximum stress that the metal can withstand for an infinitely large number of cycles with 50% probability of failure. This maximum stress is known as the fatigue (endurance) limit, S_f. Most nonferrous metals do not exhibit a fatigue limit. Instead, their S-N curves continue to drop at a slow rate at high numbers of cycles, as shown by the curve for aluminum alloy 7075-T6 in Fig. 3.

For these types of metals, fatigue strength, rather than fatigue limit, is reported. Fatigue strength is the stress to which the metal can be subjected for a specified number of cycles. Because there is no standard number of cycles, each table of fatigue strengths must specify the number of cycles for which the strengths are reported. The fatigue strength of nonferrous metals at 10^8 or 5×10^8 cycles is sometimes erroneously called the fatigue limit.

Low-Cycle Fatigue. For the low-cycle fatigue region ($N < 10^4$ cycles), tests are conducted with controlled cycles of elastic plus plastic (total) strain range, rather than with controlled load or stress cycles. Under controlled-strain testing, fatigue life behavior is represented by a log-log plot of the total strain range, $\Delta\epsilon_t$, versus the number of cycles to failure (Fig 4).

The total strain range may be separated into elastic and plastic components. For many metals and alloys, the elastic strain range, $\Delta\epsilon_e$, is equal to the stress range divided by the modulus of elasticity. The plastic strain range, $\Delta\epsilon_p$, is the difference between the total strain range and the elastic strain range.

When lines of curves are presented with fatigue data, the equation and the method of the fit should be indicated. Any presentation of corrosion fatigue data should include the following pertinent information, when applicable, regarding the material and the test:

- Material identification (product form)
- Tensile strength
- Orientation of the specimen
- Surface condition
- Notch description (stress concentration factor)
- Type of fatigue test (mode of loading)
- Controlled test parameters
- Stress ratio
- Stress cycle frequency
- Stress wave shape
- Test temperature and environment

Fatigue Test Specimens

A typical fatigue test specimen has three areas: the test section and the two grip ends. The grip

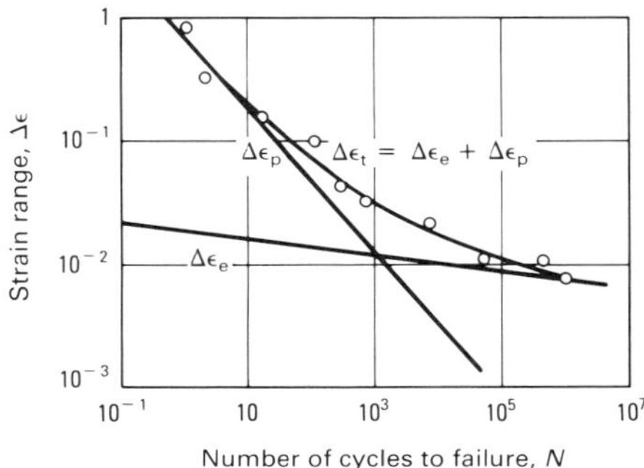

Fig. 4 Typical plot of strain range versus cycles-to-failure for low-cycle fatigue.

ends are designed to transfer load from the test machine grips to the test section and may be identical, particularly for axial fatigue tests. The transition from the grip ends to the test area is designed with large, smoothly blended radii to eliminate any stress concentrations in the transition.

The design and type of specimen used depend on the fatigue-testing machine used and the objective of the fatigue study. The test section in the specimen is reduced in cross section to prevent failure in the grip ends and should be proportioned to use the middle to upper ranges of the load capacity of the fatigue machine, that is, to avoid very low load amplitudes where sensitivity and response of the system are decreased. Several types of fatigue test specimens are illustrated in Fig. 5.

Cylindrical Specimens. Three types of specimens with circular cross sections are commonly used:

- Specimens with a continuous radius between the grip ends with the minimum diameter at the center (Fig. 5c and e); referred to as hourglass specimens
- Specimens with tangentially blending fillets between the test section and the grip ends (Fig. 5a)
- Specimens for use in cantilever beam loading with tapered diameters proportioned to produce nominally constant stress along the test section (Fig. 5b)

The design of the grip ends depends on the machine design and the gripping devices used. Round specimens for axial fatigue machines may be threaded, buttonhead, or constant-diameter types for clamping in V-wedge pressure grips.

For rotating-beam machines, short, tapered grip ends with internal threads are used, and the specimen is pulled into the grip by a draw bar. A long constant-diameter grip end shank is used on machines with lathe-type collet chucks as grips. Torsional fatigue specimens are generally cylindrical; however, there is usually a flat or keyway in the grip ends to transmit torque from the machine into the specimen (Fig. 5a).

Flat Sheet and Plate Specimens. Generally, flat specimens for either axial or bending fatigue tests are reduced in width in the test section, but may have thickness reductions. The most commonly used types include:

- Specimens with tangentially blending fillets between the test section and the grip ends; used in both axial and bending fatigue tests
- Specimens with a continuous radius between the grip ends; also used in both axial and bending fatigue tests
- Specimens for use in cantilever reverse bending tests with tapered widths (Fig. 5d).

Flat specimens are generally clamped in flat wedge-type grips, or they can be held with a stiff bolted clamp/joint friction grip for reversed axial loading. Pin loading can be used when compression loads are not encountered. When pin loading is used, the holes drilled in the grip end must be designed to avoid shear or bearing failures at the holes, tensile failure between the holes at maximum load, and fatigue cracking at the holes in the grip end. In the axial fatigue testing of flat sheet specimens, the test length must be designed to prevent premature buckling of the specimen.

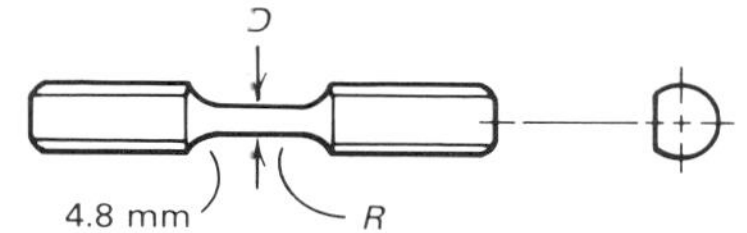

D, selected on basis of ultimate strength of material
R, 12.7 mm

(a)

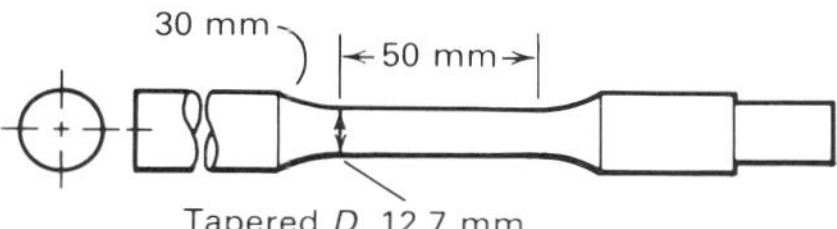

(b)

90 mm
19 mm
D
R
12 mm

D, 5 to 10 mm selected on basis of ultimate strength of material
R, 90 to 250 mm

(c)

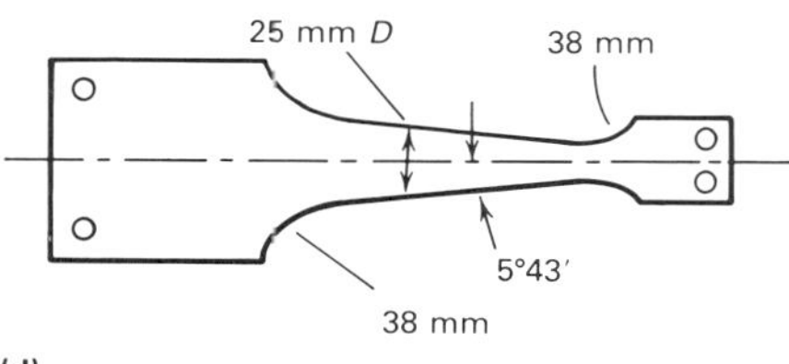

(d)

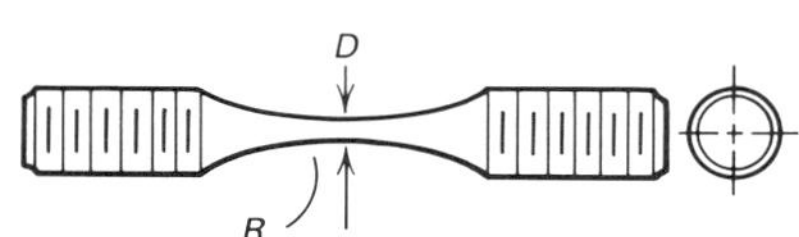

D, selected on basis of ultimate strength of material
R, 75 to 250 mm

(e)

Fig. 5 Typical fatigue test specimens. (a) Torsional specimen. (b) Rotating cantilever beam specimen. (c) Rotating beam specimen. (d) Plate specimen for cantilever reverse bending. (e) Axial loading specimen

Effect of Test Specimen Size (Ref 15)

It is difficult to predict the fatigue performance of large machine members directly from the results of laboratory tests on small specimens. In most cases, a size effect exists; that is, the fatigue strength of large members is lower than that of small specimens. Precise determination of this phenomenon is difficult. It is extremely difficult to prepare geometrically similar specimens of increasing diameter that have the same metallurgical structure and residual stress distribution throughout the cross section. The problems in fatigue testing of large specimens are considerable, and few fatigue machines can accommodate specimens with a wide range of cross sections.

Changing the size of a fatigue specimen usually results in variations of two factors. First, increasing the diameter increases the volume or surface area of the specimen. The change in amount of surface is significant because fatigue failures usually initiate at the surface. Second, for plain or notched specimens loaded in bending or torsion, an increase in diameter usually decreases the stress gradient across the diameter and increases the volume of material that is highly stressed.

Experimental data on the size effect in fatigue typically show that the fatigue limit decreases with increasing specimen diameter. The data for steel shafts tested in reversed bending given in Table 1 show that the fatigue limit can be appreciably reduced in large section sizes.

No size effect was found for smooth fatigue specimens of plain carbon steel with diameters ranging from 5 to 35 mm (0.2 to 1.4 in.) when tested in axial tension-compression loading (Ref 17). However, when a notch was introduced into the specimen, so that a stress gradient was produced, a definite size effect was observed. Therefore, it can be concluded that a size effect in fatigue is primarily due to the existence of a stress gradient.

The fact that large specimens with shallow stress gradients have lower fatigue limits supports the concept that a critical value of stress must be exceeded over a given finite depth of material for failure to occur. This appears to be a more realistic criterion of size effect than the ratio of the change in surface area to the change in specimen diameter. The importance of stress gradients in size effect explains why correlation between laboratory results and service failure is often poor. Actual failures in large parts are usually directly related to stress concentrations, either intentional or accidental, and it is usually impossible to duplicate the same stress concentration and stress gradient in a small laboratory specimen.

Effect of Stress Concentration

Fatigue strength is significantly reduced by the introduction of a stress raiser, such as a notch, hole, or corrosion pit. Because actual machine elements invariably contain such stress raisers as fillets, keyways, screw threads, press fits, and holes, fatigue cracks in structural parts usually initiate at such geometrical irregularities.

An optimal means of minimizing fatigue failure is the reduction of avoidable stress raisers through careful design and the prevention of accidental stress raisers by careful machining and fabrication. Stress concentration can also arise from surface roughness and metallurgical stress raisers, such as porosity, inclusions, local overheating in grinding, and decarburization. Some of these conditions will be discussed in the section "Surface Effects and Fatigue" in this article.

The effect of stress raisers on fatigue is generally studied by testing specimens containing a notch, usually a V-notch or a U-notch. The presence of a notch in a specimen under uniaxial load introduces three effects:

- There is an increase or concentration of stress at the root of the notch

Table 1 Effect of specimen size on the fatigue limit of normalized plain carbon steel in reversed bending

Specimen diameter		Fatigue limit	
mm	in.	MPa	ksi
7.6	0.30	248	36
38	1.50	200	29
152	6.00	144	21

Source: Ref 16

- A stress gradient is set up from the root of the notch toward the center of the specimen
- A triaxial state of stress is produced at the notch root

The ratio of the maximum stress in the region of the notch (or other stress concentration) to the corresponding nominal stress is the stress concentration factor, K_t. In some situations, values of K_t can be calculated by using the theory of elasticity or can be measured by using photoelastic plastic models. Values for K_t are reported in Ref 18 to 21.

The effect of notches on fatigue strength is determined by comparing the *S-N* curves of notched and unnotched specimens (Fig. 3). The data for notched specimens are usually plotted in terms of nominal stress based on the net cross section of the specimen. The effectiveness of the notch in decreasing the fatigue limit is expressed by the fatigue notch factor, K_f. This factor is the ratio of the fatigue limit of unnotched specimens to the fatigue limit of notched specimens.

For materials that do not exhibit a fatigue limit, the fatigue notch factor is based on the fatigue strength at a specified number of cycles. Values of K_f have been found to vary with the severity of the notch, the type of notch, the material, the type of loading, and the stress level. The published values of K_f are subject to considerable variability and should be carefully examined for their limitations and restrictions. However, two general trends are usually observed for test conditions of completely reversed loading. First, K_f is usually less than K_t, and second, the ratio of K_f/K_t decreases as K_t increases.

The notch sensitivity of a material in fatigue is expressed by a notch sensitivity factor, q:

$$q = \frac{K_f - 1}{K_t - 1} \qquad \text{(Eq 1)}$$

A material that experiences no reduction in fatigue due to a notch ($K_f = 1$) has a factor of $q = 0$, but a material in which the notch has its full theoretical effect ($K_f = K_t$) has a factor of $q = 1$. However, q is not a true material constant, because it varies with the severity and type of notch, the size of the specimen, and the type of loading. As shown in Fig. 6, notch sensitivity increases with tensile strength. Therefore, it is sometimes possible to decrease fatigue performance by increasing the hardness or tensile strength of a material.

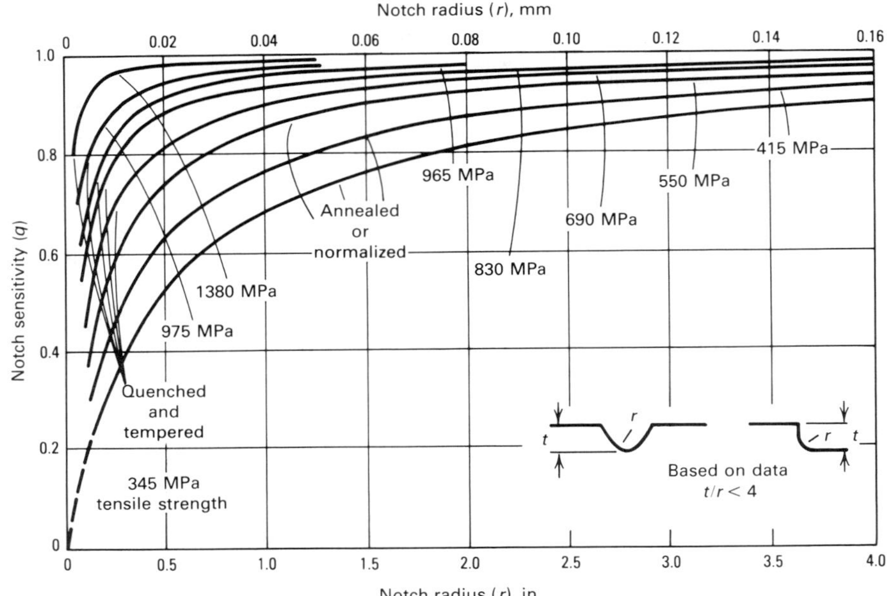

Fig. 6 Variation of notch sensitivity index with notch radius for steels tested in bending or axial fatigue loading. Source: Ref 22

Surface Effects and Fatigue

Fatigue properties are generally very sensitive to surface conditions. Except in special cases involving internal defects or case hardening, all fatigue cracks initiate at the surface. Factors that affect the surface of a corrosion-fatigue specimen can be divided into four categories:

- Surface roughness or stress raisers at the surface
- Changes in the properties of the surface metal
- Changes in the residual stress condition of the surface
- Corrosion

Surface Roughness. In general, fatigue life increases as the magnitude of surface roughness decreases. Decreasing surface roughness minimizes local stress raisers. Therefore, special attention must be given to the surface preparation of fatigue test specimens. Typically, a metallographic finish, free of machining grooves and grinding scratches is required. Figure 7 illustrates the effects that various surface conditions have on the fatigue properties of steel.

Changes in Surface Properties. Because fatigue failure is dependent on surface condition, any phenomenon that changes the fatigue strength of the surface material will greatly alter fatigue properties. Decarburization of the surface of heat-treated steel is particularly detrimental to fatigue performance. Similarly, the fatigue strength of aluminum alloy sheet is reduced when a soft aluminum coating (cladding) is applied to the stronger age-hardenable aluminum alloy sheet.

Marked improvements in fatigue properties can result from the formation of harder and stronger surfaces on steel parts by carburizing and nitriding (Ref 23). However, because favorable compressive residual stresses are produced in the surface by these processes, the higher fatigue properties are not exclusively due to the formation of higher-strength material on the surface. The effectiveness of carburizing and nitriding in improving fatigue performance is greater when a high stress gradient exists (as in bending or torsion) than in an axial fatigue test.

The greatest increase in fatigue performance occurs in notched fatigue specimens that are nitrided. The amount of strengthening depends on the diameter of the part and the depth of surface hardening. Improvements in fatigue properties similar to those caused by carburizing and nitriding may also be produced by flame hardening and induction hardening. In surface-hardened parts, failure initiates at the interface between the hard case and the softer core, rather than at the surface.

Electroplating of the surface generally decreases the fatigue limit of steel because the platings inherently contain cracks. The particular plating conditions used to produce an electroplated surface can have a significant effect on fatigue properties because large changes in the residual stress, adhesion,

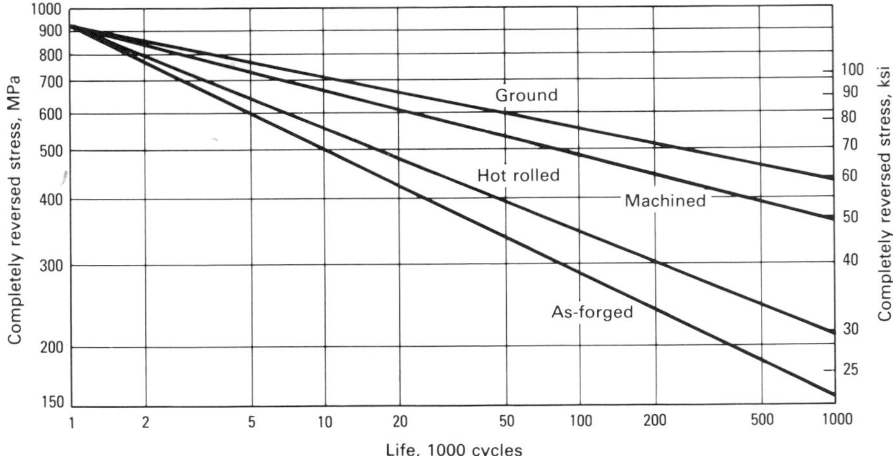

Fig. 7 Effect of surface conditions on the fatigue properties of steel (302 to 321 HB)

porosity, and hardness of the plate can be produced. Anodized coatings on aluminum alloys can be similarly detrimental to fatigue strength.

Surface Residual Stresses. Processing, such as grinding, polishing, and machining, that work hardens or increases residual stress on the surface can influence fatigue strength, although there is no generalization that predicts the extent of improved fatigue strength that can be derived from work hardening and residual stress. Compressive residual surface stresses generally increase fatigue strength, but tensile residual surface stresses do not. There may be a gradual decrease in residual stress if the cyclic stresses cause some plastic deformation. Compressive residual surface stress provides greater improvement in the fatigue strength of harder materials (such as alloy spring steel). In softer materials (such as low-carbon steel), work hardening effectively improves fatigue strength.

In a notched high-strength steel, the beneficial effect of prestretching and the detrimental effect of precompression are much greater than in a plain carbon steel because of the type of residual stress present at the notch. A compressive residual stress introduced during quenching from a tempering temperature will increase the fatigue strength, particularly in notched specimens.

In general, residual stresses are introduced by the poor fit of structural parts, a change in the specific volume of a metal accompanying phase changes, a change in shape following plastic deformation, or thermal stresses resulting from rapid temperature changes, such as occur in quenching. The influence of residual stress on fatigue strength is similar to that of an externally applied static stress. A static compressive surface stress increases the fatigue strength, and static tensile surface stress reduces it.

Specimen Preparation. Because both fatigue crack initiation and corrosion are typically surface dependent, proper machining and surface preparation of test specimens is critical. Unless care is taken, scatter caused by variable surface conditions will mask the scatter that is typical for the material being studied.

One of the primary aims of corrosion-fatigue testing is the comparison of materials. Therefore, uniform preparation procedures must be established. Machining operations must not alter the surface structure of the metal; therefore, heat generation, heavy cutting, and severe grinding are prohibited. For as-machined surfaces, the final machining should be transverse to the plane of stress. Transition fillets must be blended into the test area without steps or undercutting. Surface polishing using metallographic techniques is preferred for smooth specimens, in which machining marks are removed by a sequence of grinding steps.

The final polishing is not a buffing operation, but a cutting operation that uses lapping compounds or aluminum oxide powder in a liquid medium to remove grinding scratches. For flat sheet or plate specimens, edges should be slightly rounded and ground to eliminate nicks, dents, cuts, and sharp edges, which cause early crack initiation.

Environmental Effects and Fatigue

In corrosion fatigue, the magnitude of the cyclic stress and the number of times it is applied are not the only critical loading parameters. Time-dependent environmental effects are also of primary importance. When failure occurs by corrosion fatigue, stress cycle frequency, stress wave shape, stress amplitude, and stress ratio all affect the cracking processes, depending on the resistance to SCC of the material.

Environmental Effects on Fatigue Strength. For any given material, the fatigue strength, or fatigue life at a given value of maximum stress, generally decreases in the presence of an aggressive environment. This effect varies widely, depending primarily on the characteristics of the material/environment combination. The environment affects the probability of fatigue crack initiation, crack growth rate, or both. For many materials, the stress range required to cause fatigue failure diminishes progressively with increasing time and with the number of cycles of applied stress.

A corrosive environment can clearly reduce the crack initiation time by any form of localized attack to form a stress concentration. Corrosion attack would clearly be favored at new unfilmed metal surfaces formed at a persistent slip band. A clear example of the importance of stress concentration due to corrosion is illustrated in Fig. 8, which shows fatigue endurance results for plain and notched specimens of a 13% Cr ferritic steel in distilled water and salt water. These environments are seen to cause a massive reduction in the endurance limit of the smooth specimens but only a relatively modest effect on the notched specimens. The importance of surface finish and the adverse effect of corrosion, even when only allowed to occur before a fatigue test, is of course well known and included in standard texts on fatigue design (Ref 25).

Effect of Frequency and Stress Wave Form. In nonaggressive environments, cyclic frequency generally has little effect on fatigue behavior. On the other hand, in aggressive environments, fatigue strength is strongly dependent on frequency. Corrosion fatigue strength (endurance limit at a prescribed number of cycles) will generally decrease as the cyclic frequency is decreased. This effect is most important at frequencies less than 10 Hz.

The frequency dependence of corrosion fatigue is thought to result from the fact that the interaction of a material with its environment is essentially a rate-controlled process. Low frequencies, especially at low strain amplitudes or when there is substantial elapsed time between changes in stress levels, allow time for interaction between material and environment; high frequencies do not, particularly when high strain amplitude is also involved.

The effect of stress frequency on low-cycle corrosion fatigue is also very dependent on susceptibility to SCC. For example, at low intermediate frequencies, material sensitive to SCC fractured by the propagation of fatigue cracks, but at lower frequencies, damage due to SCC during the period of maximum strain became greater and the fatigue strength at extremely low frequencies was markedly decreased. For materials that are insensitive to SCC, failure was due to the propagation of CFC in the entire range of stress frequencies tested (0.02 to 17.3 cycles/min, or cpm) (Ref 26).

Historically, CFC is not considered to be a problem at very high frequencies. However, a new ultrasonic frequency corrosion fatigue testing system has been developed for examining the corrosion fatigue strength of engineering materials in the very high cycle regime (10^6 to 10^{10} cycles) typical of many design lifetimes. Central to this resonant fatigue-testing system is a high-efficiency large-displacement, piezoelectric transducer that operates at a frequency of 20 kHz; this enables the accumulation of large numbers of cycles within a reasonable amount of time. The apparatus, environmental controls, and testing method are described in Ref 27.

Experimental 20-kHz and low-frequency corrosion fatigue results obtained with this method for a number of engineering materials, including a 12% Cr stainless steel, show similar reductions in fatigue limit at a specified number of cycles, despite the difference in frequency. This result holds for a variety of aqueous environments with and without electrochemical controls imposed during testing. The data illustrate the capabilities of the ultrasonic corrosion fatigue technique and are analyzed to assess its advantages and limitations.

In some cases, corrosion fatigue strength is strongly affected by the stress wave form as well as by the stress frequency. The period of zero stress, the period during which the stress changes, and the period of application of maximum stress in a single cycle each have an effect on the specimen life. The damaging effect of the period during which the stress is changing is greatest, while the protective effect of the stress-free time decreases the corrosion fatigue damage.

Effect of Stress Amplitude. In general, a low amplitude of cyclic stress favors relatively long fatigue life, permitting greater opportunity for involvement of the environment in the fatigue process. However, environmental interaction may be insignificant unless the strain rate is in a critical range for SCC in certain alloy/environment systems.

Stress amplitude must be considered together with mean stress and frequency. Low stress levels may allow more time for environmental interaction, but if the frequency is high, the crack tip may not be exposed to the environment for a time sufficient for the corrosion processes to do significant damage.

Crack Propagation Tests

In large structural components, the existence of a crack does not necessarily imply imminent failure of the part. Significant structural life may remain in the cyclic growth of the crack to a size at which a critical failure occurs. The objective of corrosion fatigue crack propagation testing is to determine the rates at which subcritical cracks grow under cyclic loading before reaching a size that is critical for fracture under specified environmental conditions.

The article "Environmental Effects on Fatigue Crack Propagation" in Volume 8 of the 9th Edition of *Metals Handbook* contains information on the effects of environmental variables. Detailed information on specimen types, crack-measuring techniques, fatigue crack growth data analysis, and fracture mechanics is also available in Volume 8. The following discussion contains many excerpts from these excellent articles.

Fracture Mechanics Approach to Corrosion Fatigue

Fracture mechanics provides the basis for many modern fatigue crack growth studies. The growth or extension of a fatigue crack under cyclic loading is principally controlled by maximum load and stress ratio. However, as in crack initiation, there are a number of additional factors that may exert a strong influence, especially with the presence of an aggressive environment. Most CFC growth rate investigations attempt to follow

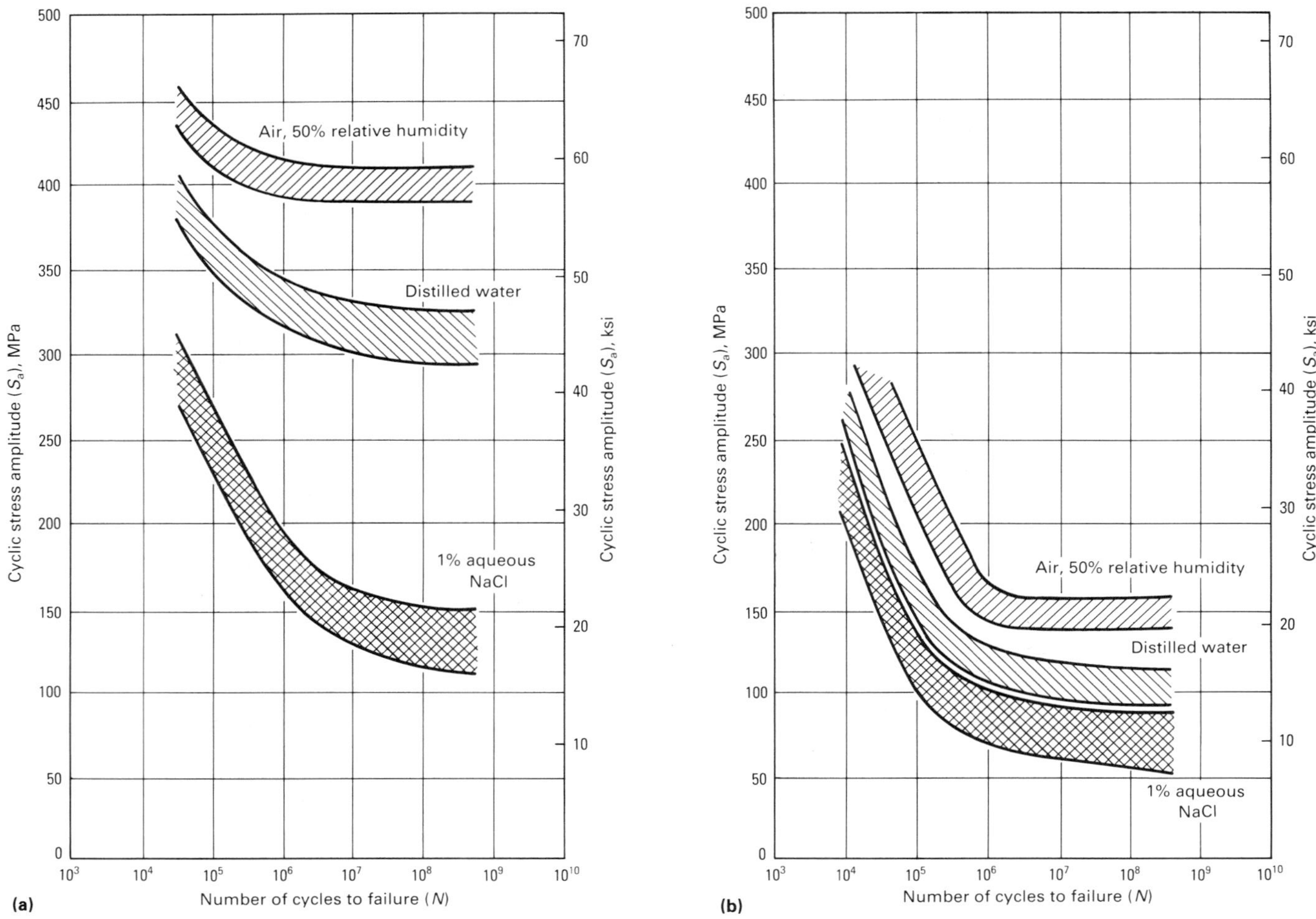

Fig. 8 Corrosion fatigue endurance data for specimens of 13% Cr steel. Rotating bending tested (mean load zero) at a frequency of 50 Hz and temperature of 23 °C (73 °F). (a) Smooth specimens. (b) Notched specimens. Source: Ref 24

the general provisions of standard test method ASTM E 647.

In this constant-load-amplitude method, crack length is measured visually or by an equivalent method as a function of elapsed cycles, and these data are subjected to numerical analysis to establish the rate of crack growth. Crack growth rates are then expressed as a function of crack tip stress intensity range ΔK, which is calculated from expressions based on linear-elastic stress analysis.

Background information on the rationale for employing linear-elastic fracture mechanics for this purpose is given in Ref 28. Expressing the crack growth rate da/dN as a function of ΔK provides results that are independent of specimen geometry, and this enables the exchange and comparison of data obtained from a variety of specimen configurations and loading conditions. Moreover, this feature enables da/dN versus ΔK data to be used in the design and evaluation of engineering structures. It is important in the generation of CFC growth data, however, that there be judicious selection, monitoring, and control of mechanical, chemical, and electrochemical test variables in order to ensure that the data are truly applicable to the intended use.

Results of fatigue crack growth rate tests for many metallic structural materials have shown that complete da/dN versus ΔK curves have three distinct regions of behavior (Fig. 9). In an inert (or benign) environment, the rate of crack growth depends strongly on K at K levels approaching K_c or K_{Ic} at the high end (Region III) and at levels approaching an apparent threshold, ΔK_{th}, at the lower end (Region I), with an intermediate Region II that depends on some power of K or ΔK of the order of 2 to 10 (Ref 28). This is described by the power-law relationship:

$$\frac{da}{dN} = C(\Delta K)^n \quad \text{(Eq 2)}$$

where C and n are constants for a given material and stress ratio. In an aggressive environment, the CFC growth curve can be quite different from the pure fatigue curve, depending on the sensitivity of the material to the given environment and the occurrence of various static stress fracture mechanisms. The environmental effects are quite strong above some threshold for SCC (K_{ISCC}) and may be negligible below this level. In addition, certain loading factors, such as cyclic frequency, stress ratio, and stress wave form, can have marked effects on the crack growth curves in aggressive environments. Therefore, a variety of curves can be expected from the broad range of material-environment-loading systems that will produce various CFC behaviors.

Figure 10 shows an example of a high-strength aluminum alloy with a high resistance to SCC. This alloy had a CFC growth rate ranging up to one order of magnitude higher in 3.5% sodium chloride (NaCl) solution than that in dry air. These data also illustrate the similar behavior when tested with either a K-increasing (remote load) or a K-decreasing (wedge force) loading method.

Corrosion fatigue growth rate versus stress intensity data have been extensively produced (Ref 2-8, 31, 32) and can be used for mitigating corrosion fatigue from several perspectives. These laboratory crack growth rate data are scalable quantitatively for predicting component performance. The da/dN versus ΔK curve for a given material and environment is integrated in conjunction with the stress intensity solution for a component in order to predict life (Ref 28). A specific example reported in Ref 33 is illustrated in Fig. 11. The predicted 85-year life of a welded pipe is based on week-long laboratory measurements of da/dN versus ΔK for steel in an oil environment.

Alloy development for optimized corrosion fatigue resistance, defined in terms of da/dN versus ΔK for different metallurgical conditions, is directly related to component life goals based on fracture mechanics. Similarly, the effects of mechanical and chemical variables on corrosion

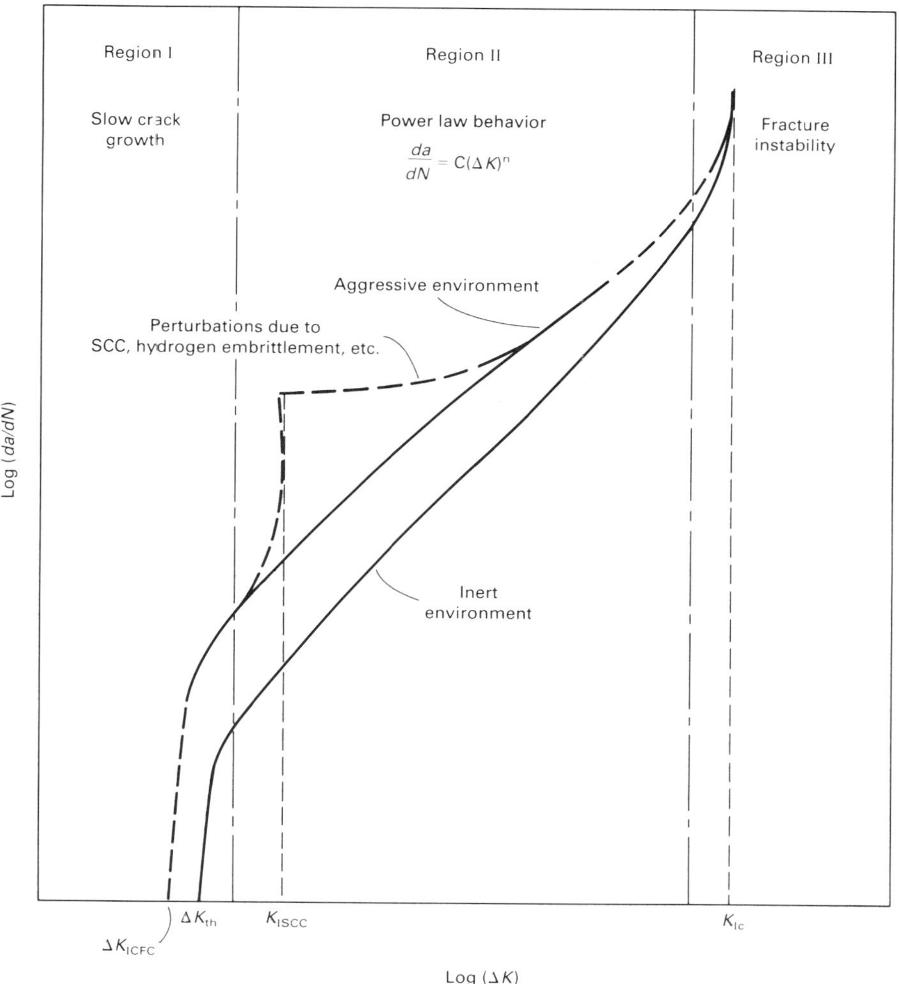

Fig. 9 Corrosion fatigue crack propagation rate as a function of the cyclic crack tip stress intensity factor. Source: Ref 29

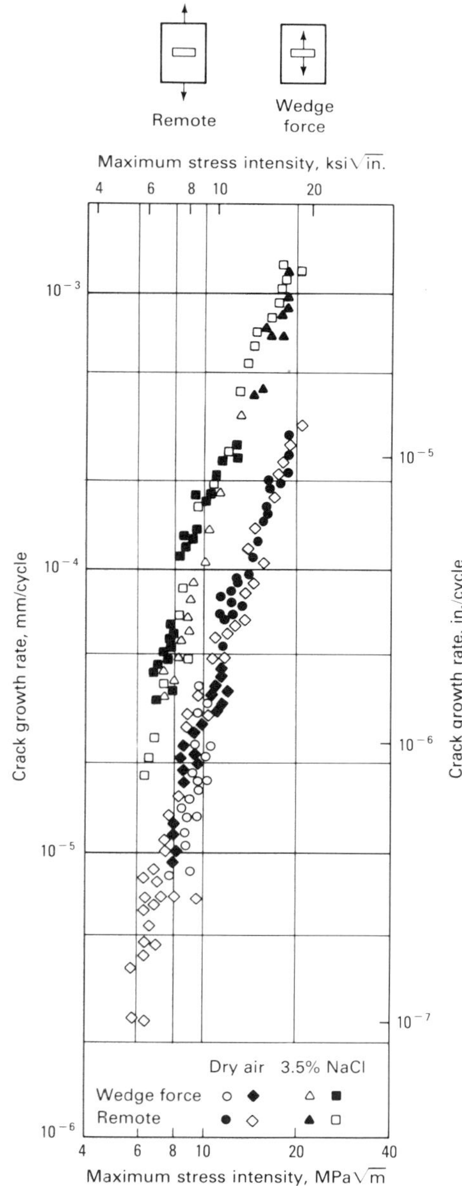

Fig. 10 Crack tip stress intensity control of fatigue crack propagation in 7075-T6 aluminum alloy sheet—long-transverse loading. Remote and wedge force methods of loading specimens in aqueous 3.5% sodium chloride environment and benign dry air environment. Source: Ref 30

fatigue and component performance are defined through measured changes in growth rate at different applied stress intensity levels.

Crack growth rate data are important to fundamental studies of corrosion fatigue mechanisms. The fracture mechanics approach isolates crack propagation from initiation in terms of a precise near-tip mechanical driving force, ΔK. Crack growth rates are related directly to the kinetics of mass transport and the chemical reaction that constitute embrittlement. As shown in Fig. 12, prediction of the effect of loading frequency on crack growth rate in salt water (normalized to vacuum) identifies important rate-limiting crack tip electrochemical reactions. Modeling and measurements in Fig. 12 provide a sound basis for extrapolating short-term laboratory data in order to predict long-term component cracking.

The fracture mechanics approach to corrosion fatigue may be invalidated by several factors. Discussion in this article is limited to crack growth in conjunction with net section elastic loading and small-scale plasticity confined to the crack tip. Studies of crack growth under large-scale cyclic plasticity have not been extended to consider environmental effects (Ref 35, 36). Also, stress intensity provides a mechanical description of similitude, which may not completely describe crack growth due to interacting chemical and mechanical driving forces.*

Investigations of this phenomenon are currently underway. Results indicate that simple K-based approaches can be compromised for (1) small corrosion fatigue cracks below 5 mm (0.2 in.), (2) cases in which varying crack shape and load transients alter crack chemistry and embrittlement, and (3) situations in which surface roughness or corrosion products impede crack displacement (Ref 37-39). References 35 to 39 should be consulted if an application includes plasticity, small crack, load transient, or crack closure effects.

Variables Influencing Corrosion Fatigue

Although corrosion fatigue phenomena are diverse and specific to the environment, several variables are known to influence crack growth rate. The following factors must be considered in any study of corrosion fatigue:

- Stress intensity range
- Load frequency
- Stress ratio
- Aqueous environment electrode potential
- Environment composition
- Alloy composition, microstructure, and yield strength

Moreover, the effects of such variables as temperature, load history and waveform, stress state, and environment composition may be unique to specific materials and environments.

Stress Intensity Range. Corrosion fatigue crack growth rates generally increase with increasing stress intensity; however, the precise

*Fracture mechanics is based on the concept of similitude, in which cracks of different geometries grow at equal rates when subjected to equal near-tip driving forces, usually ΔK.

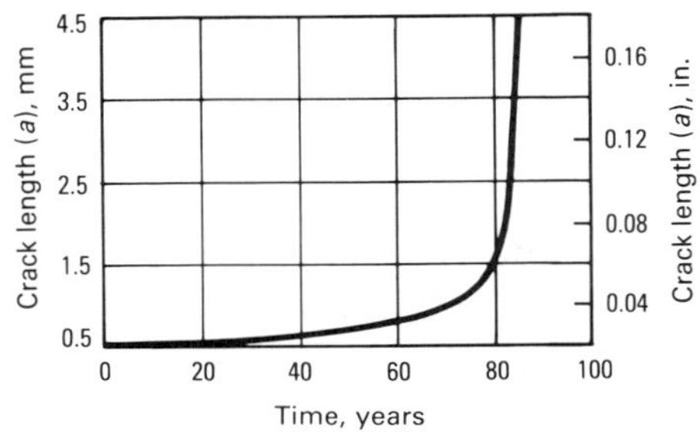

Fig. 11 Predicted fatigue crack extension from a weld toe crack in an API 5LX52 carbon steel pipeline carrying hydrogen sulfide contaminated oil. Temperature: 23 °C (73 °F). Source: Ref 33

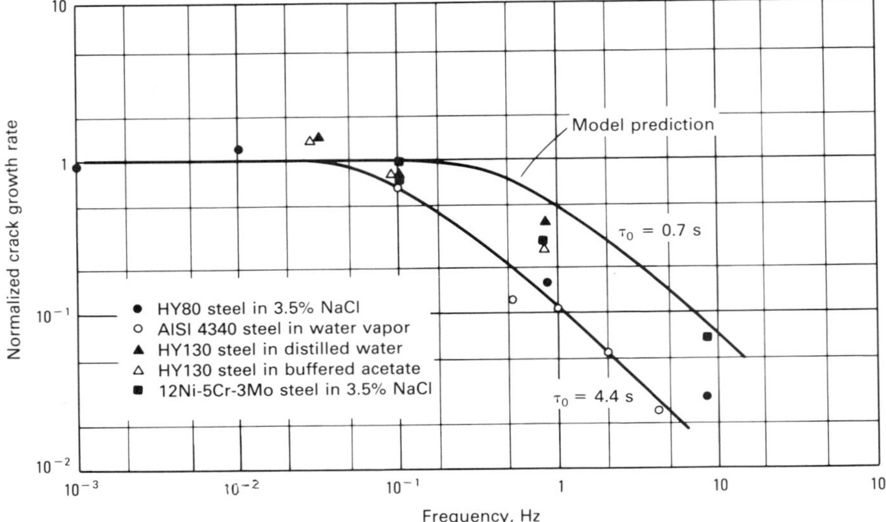

Fig. 12 Modeled effect of loading frequency on corrosion fatigue crack growth in alloy steels in an aqueous chloride solution. The determination of the normalized crack growth rate and the time constants, τ_0, from the model can be found in Ref 34.

dependence varies markedly. It is incorrect to assume that the three regimes (near threshold, power law, and fast fracture; see Fig. 9) of fatigue cracking observed for benign environments simply shift to higher crack speeds at all ΔK levels. Although such behavior may actually occur, as illustrated in Fig. 10 for the aluminum/dry air system, the data provided in Fig. 13 and 14 are more typical of complex stress intensity dependencies for aggressive environments.

Materials that are extremely environment sensitive, such as ultrahigh-strength steel in distilled water (Fig. 13), are characterized by high crack growth rates that are influenced less by ΔK. Time-dependent corrosion fatigue crack growth occurs predominantly above the threshold stress intensity for static load cracking and is modeled through linear superposition of SCC and inert environment fatigue rates (Ref 28, 34). Time-dependent crack growth is the part of the crack extension that is attributable to the constant or rising-load portion of the cyclic load cycle, when SCC and hydrogen embrittlement mechanisms may be operative. Therefore, such factors as stress wave form and stress hold-time are particularly significant under such circumstances.

Cycle-dependent corrosion fatigue crack propagation often occurs below the threshold for time-dependent stress corrosion (Ref 41). Typical ΔK dependencies for this mode of cracking are complex, as illustrated in Fig. 14 for Ti-6Al-4V exposed to aqueous sodium chloride. The mechanistic implications of the various stress intensity dependencies are detailed in Ref 28, 41, and 42. The influences of such variables as frequency, stress ratio, and metallurgical factors depend on the proportions of time- and cycle-dependent corrosion fatigue.

Frequency. Cyclic load frequency is the most important variable that influences corrosion fatigue for most material, environment, and stress intensity conditions. The rate of environmental cracking above that produced in vacuum generally increases with decreasing frequency. Frequencies exist above which corrosion fatigue is diminished. The dominance of frequency is directly related to the time dependence of the mass transport and chemical reaction steps required for environmental cracking. Basically, insufficient time is available for chemical embrittlement at rapid loading rates; damage is only mechanical and is equivalent to crack growth in vacuum. It is impossible to predict the frequency range at which corrosion fatigue is severe, because of the numerous chemical processes. It is also difficult to extrapolate short-term (high-frequency) laboratory crack growth rate data in order to predict long-term component performance.

The literature provides qualitative guidance in terms of frequency effects on corrosion fatigue. Crack growth in environmentally sensitive materials stressed above the stress-corrosion threshold proceeds at rapidly increased rates with decreasing frequency, as illustrated in Fig. 13. The frequency effect is predicted through the integration of static load data throughout each fatigue cycle (Ref 34). For such systems, the chemical contribution to fatigue cracking is suppressed above 0.5 to 5 Hz, with higher critical frequencies associated with more aggressive environments and sensitive microstructures.

Frequency effects on cycle-dependent cracking are complex and unpredictable. The data given in Fig. 12 and 14 provide typical examples. For steels in salt water (Fig. 12), corrosion fatigue is suppressed at about 10 Hz. Crack speed increases with decreasing frequency and reaches a plateau between 0.5 and 0.1 Hz. In contrast, corrosion fatigue in Ti-6Al-4V occurs at loading frequencies at least as high as 10 Hz (Fig. 14).

Stress Ratio. The rates of corrosion fatigue crack propagation generally are increased by higher stress ratios. The deleterious effect of increased stress ratio is illustrated in Fig. 15 for carbon steel stressed cyclically in pressurized nuclear reactor water at 288 °C (550 °F). The extent of the corrosion fatigue effect relative to dry air increased from about a factor of four at low R (0.11 to 0.24) to as much as 20- to 30-fold at high R (0.61 to 0.71) (Ref 43). Corrosion fatigue in this system probably proceeds by repeated passive film formation and rupture. Increased stress ratio at constant ΔK results in increased crack tip strain and strain rate, enhanced film rupture, and therefore increased corrosion fatigue crack propagation (Ref 44).

Electrode Potential. Like loading frequency, electrode potential strongly influences rates of corrosion fatigue crack propagation for alloys in aqueous environments. Controlled changes in the potential of a specimen can result in either the complete elimination or the dramatic increase of fatigue cracking. The precise influence depends on the mechanism of the environmental effect and on the anodic or cathodic magnitude of the applied potential.

For ferrous alloys that crack by hydrogen embrittlement when stressed in aqueous solutions, corrosion fatigue is increased by high cathodic polarization from the free corrosion potential. Specific data are given in Fig. 16 for three steels stressed at constant ΔK in aqueous chloride (Ref 35, 45, 46). For each steel, the crack growth rate in salt water is about three times faster than that reported for air at the free corrosion potential and is enhanced by a factor of five at a very high cathodic potential. For the low-strength BS4360:50D grade, intermediate potentials appear to be mildly beneficial. Corrosion fatigue parallels electrode potential induced changes in the amount of cathodically evolved hydrogen in the crack tip. Precise modeling of this reaction sequence is complex and only partially completed (Ref 47).

Electrode potential control can suppress corrosion fatigue for alloys that crack through anodic dissolution/film rupture or anion adsorption mechanisms (Ref 48). For example, corrosion fatigue of austenitic stainless steel in chloride solutions at elevated temperatures is probably suppressed by polarization active to the critical potential for SCC.

Another example is illustrated in Fig. 17. Aluminum alloy 7079-T651 is degraded by corrosion fatigue in several aqueous halide solutions at the free corrosion potential. Compared to dry argon, cracking in potassium iodide is enhanced by more than an order of magnitude. The CFC growth rate was further increased by anodic polarization above about −0.6 V versus standard hydrogen electrode, but is suppressed by cathodic polarization.

Electrode potential should be monitored and, if appropriate, maintained constant during corrosion fatigue experimentation. Apparent effects of variables, such as the dissolved oxygen content of the solution, flow rate, ion concentration, and alloy composition on corrosion fatigue, can often be traced to changing electrode potential.

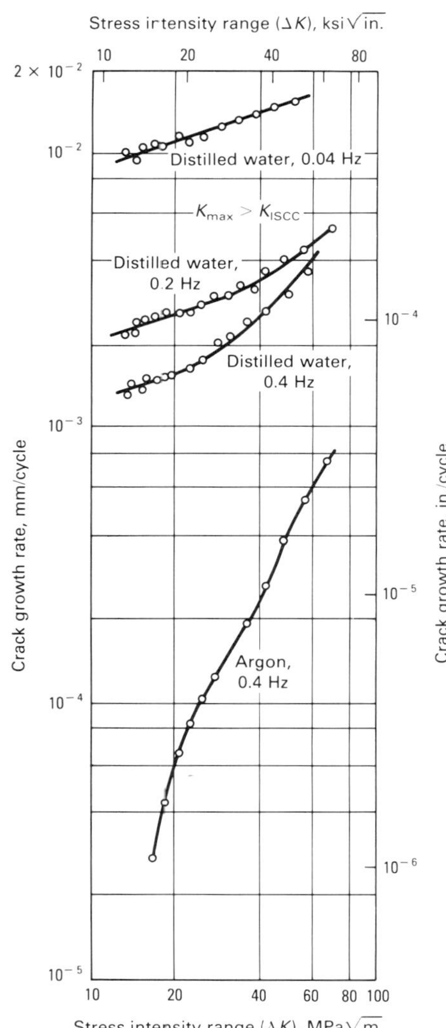

Fig. 13 Effect of stress intensity range and loading frequency on corrosion fatigue crack growth in ultrahigh-strength 4340 steel exposed to distilled water at 23 °C (73 °F). Source: Ref 40

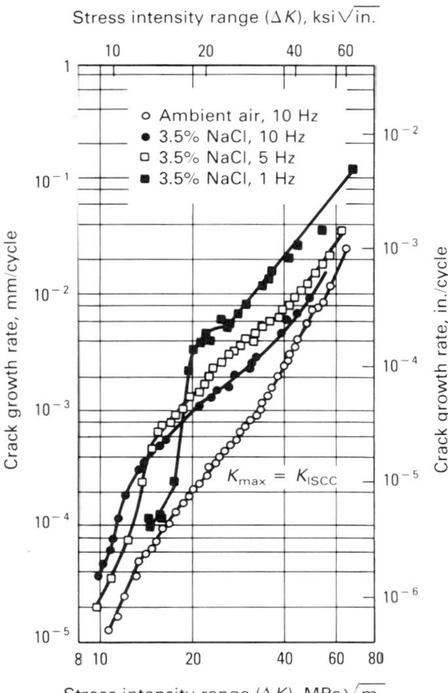

Fig. 14 Effects of stress intensity and frecuency on corrosion fatigue in Ti-6Al-4V in aqueous sodium chloride at 23 °C (73 °F). Stress ratio: R = 0.1. Source: Ref 41

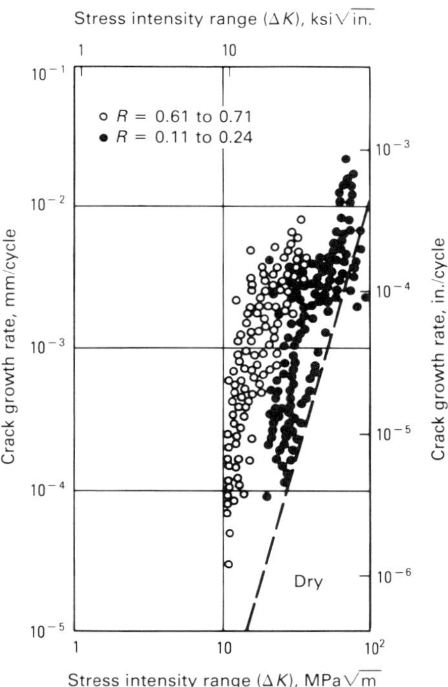

Fig. 15 Effect of stress ratio on corrosion fatigue crack propagation in ASTM A533 B and A508 carbon steels exposed to pressurized high-purity water at 288 °C (550 °F). Frequency: 0.017 Hz. Average behavior in air is represented by the dashed line labeled "Dry." Source: Ref 43

Crack Closure Effects. Premature crack surface contact during unloading, or crack closure, can greatly reduce rates of fatigue crack propagation. The true (or effective) crack tip driving force is reduced below the applied ΔK because of the reduced crack tip displacement range. Closure phenomena are produced by a variety of mechanisms and are particularly relevant to fatigue crack propagation in the near-threshold regime, after large load excursions, or for embrittling corrosive mechanisms, as reviewed in Ref 37, 50, and 51.

Two mechanisms of crack closure are relevant to corrosion fatigue. Rough, intergranular crack surfaces, typical of environmental embrittlement, promote crack closure because uniaxially loaded cracks open in a complex three-dimensional mode, thus allowing for surface interactions and load transfer. Roughness-induced closure is most relevant to corrosion fatigue at low ΔK and at stress ratio levels at which absolute crack opening displacements (0.5 to 3 μm) are less than fractured grain heights (5 to 50 μm).

Alternatively, crack closing is impeded by dense corrosion products within the pulsating fatigue crack. For mildly oxidizing environments, such as moist air, this closure mechanism is relevant at low stress intensity levels and contributes to the formation of a threshold, as described in Ref 51.

For corrosive bulk environments, cracking at high ΔK values may be retarded below growth rates observed for air or vacuum due to corrosion product formation within the crack. The engineering significance of beneficial crack closure influences depends on the stability of the corrosion product during complex tension-compression loading and fluid conditions. The beneficial effects of environmentally induced crack closure are reported in the literature for several material environment systems.

An example of crack closure is presented in Fig. 18 for ASTM A471 steel exposed to either moist air or steam at 100 °C (212 °F). Crack growth is reduced for the latter environment because of crack surface oxidation and enhanced closure. It should be noted that oxide-induced closure influences cracking even at a relatively high stress ratio of 0.35. The data presented in Fig. 18 were obtained at a high loading frequency of 100 Hz. Chemically enhanced crack growth is essentially precluded.

Crack growth at lower frequencies could be deleteriously influenced by metal embrittlement and beneficially influenced by increased oxide formation and closure. The complexity of predicting environmentally enhanced cracking is clear, as emphasized by data given in Fig. 19 for a 2.25Cr-1Mo steel (ASTM A542, class 3) stressed in dry hydrogen at 23 °C (73 °F). At high ΔK levels and low loading frequencies, cracking was accelerated in hydrogen as compared to moist air; this is due to classical hydrogen embrittlement. For low stress intensities, cracking was also increased by hydrogen exposure; however, the effect is not due to hydrogen embrittlement because of the rapid loading frequencies. Oxides form on crack surfaces through a fretting mechanism during cycling in air, and crack growth rates are reduced by oxide-induced closure. However, oxide formation is precluded for cycling in pure hydrogen, crack closure is absent, and growth rates are increased relative to those in moist air. Equally rapid rates of fatigue cracking have been reported for low ΔK cycling in hydrogen and in helium. Equal rates of cracking are observed for hydrogen, helium, and moist air at high R levels where closure is absent (Ref 50, 51).

CFC Growth Characterization

Experimental characterizations of corrosion fatigue crack propagation are complicated by the numerous variables that influence the failure process. Both the mechanics of loading and the composition of the environment must be controlled. Standard methods of measurement and fracture mechanics analysis of fatigue crack propagation in benign environments can be found in ASTM Standard E 647.

Four problem areas are relevant to corrosion fatigue experimentation. First, the environment must be contained about the cracked specimen without affecting loading, crack monitoring, or specimen-environment composition. Parameters such as environmental purity, composition, temperature, and electrode potential must be monitored and controlled frequently.

Second, the deleterious effect of low cyclic frequency dictates that crack growth rates be measured at low frequencies (often <0.2 Hz); this leads to long test times—often several days

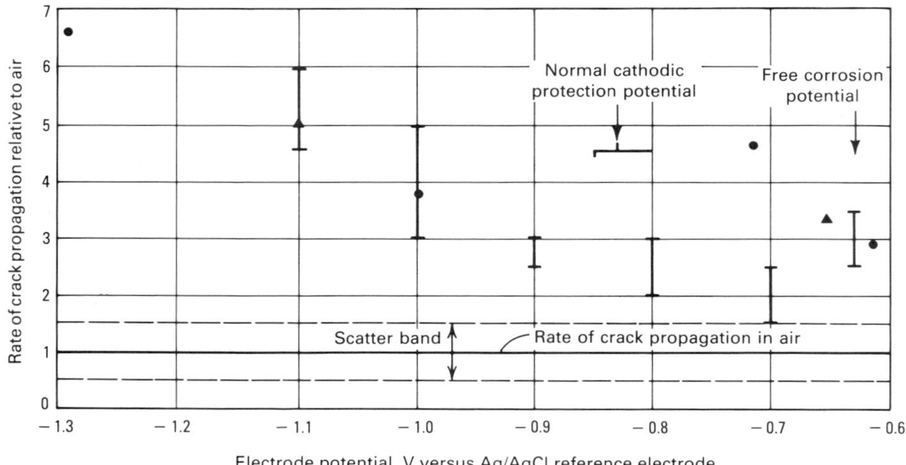

Material tested	Symbol	Yield strength MPa	Yield strength ksi	Environment	Ref
BS4360:50D	I	450	65	Seawater	44
HY80	●	600	85	3% NaCl	45
AISI 4130	▲	750	110	3% NaCl	46

Fig. 16 Effect of applied electrode potential on corrosion fatigue crack propagation in several steels exposed to seawater or 3% sodium chloride at constant ΔK between 20 and 40 MPa$\sqrt{m}$ (18 and 36 ksi$\sqrt{in.}$). Temperature: 23 °C (73 °F); frequency: 0.1 Hz; stress ratio: $R = 0.1$

to weeks. Load-control and crack-monitoring electronics and environment composition must be stable throughout long-term testing.

Third, crack length must also be measured for calculations of stress intensity and crack growth rate. Optical methods are often precluded by the environment and test chamber. Indirect methods, which are based on specimen compliance (Ref 53) or electrical potential difference (Ref 54), have been successfully applied to the monitoring of crack growth in a wide variety of hostile environments. Experimental and analytical requirements, however, are complex for indirect crack monitoring.

Nonvisual techniques will often have greater accuracy and less variability in crack length measurement than optical observations. In addition, the nonvisual techniques are generally more consistent and reliable, especially where long periods of unattended testing are anticipated. Beyond that, nonvisual techniques are absolutely essential where the test temperature or the environment prohibits direct visual observations. The nonvisual method is more precise than visual techniques, and it allows for automation of the test beyond the mere acquisition of the crack growth rate data (Ref 55, 56).

Finally, specimen geometry and size requirements for ΔK-based crack propagation data, which are scalable to components through similitude, have not been established completely for subcritical crack growth. In-plane yielding must be limited to the crack tip by ensuring that net section stress is below yield and that the maximum plastic zone size, defined as approximately $0.2\ (K_{max}/\sigma_{YS})^2$, is much less (for example, 10- to 50-fold) than the uncracked ligament. Specimen thickness, as it influences the degree of plane-strain constraint, and crack size, as it influences the chemical driving force, may affect corrosion fatigue crack speeds. Currently, such effects are unpredictable; specimen thickness and crack geometry must be treated as variables.

CFC Growth in Specific Environments

Corrosion fatigue studies in aqueous solutions and gaseous environments at ambient temperatures present fewer problems experimentally than many of the other environments considered in this article. Nevertheless, simulation of practical conditions in relation to a specific engineering application requires careful consideration if the data are to be relevant. Similarly in an academic investigation, comparison of the results of different pieces of work is possible only if the mechanical, metallurgical, and chemical variables are properly characterized and controlled. It is often the case, however, that the most frequent problem in determining the validity of corrosion fatigue data lies with the control and monitoring of the bulk environment chemistry and the monitoring and recording of the electrochemical potential in aqueous solutions.

Corrosion fatigue crack growth data are required for practical engineering applications in many diverse and specialized environments. An excellent overview of the following eight specific types of environments is given in the article "Environmental Effects on Fatigue Crack Propagation" in Volume 8 of the 9th Edition of *Metals Handbook* and will not be repeated in this article:

- Vacuum and gaseous environments at ambient temperatures
- Vacuum and oxidizing gases at elevated temperatures
- Aqueous solutions at ambient temperatures
- Acidified chloride environments at ambient and elevated temperatures
- High-temperature pure water under aerated conditions
- High-temperature pure water under deaerated conditions
- Liquid-metal environments
- Steam or boiling water with contaminants

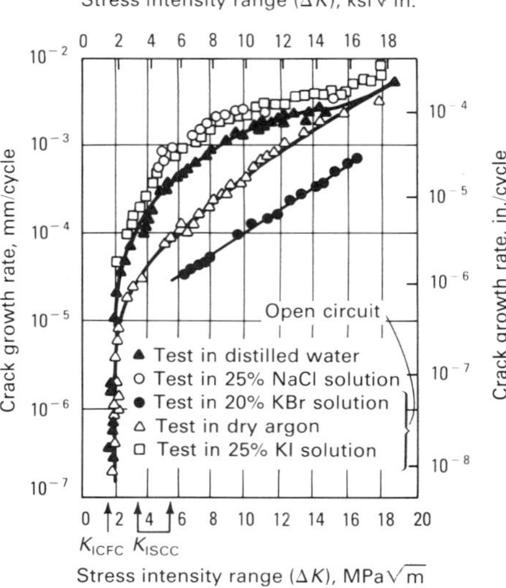

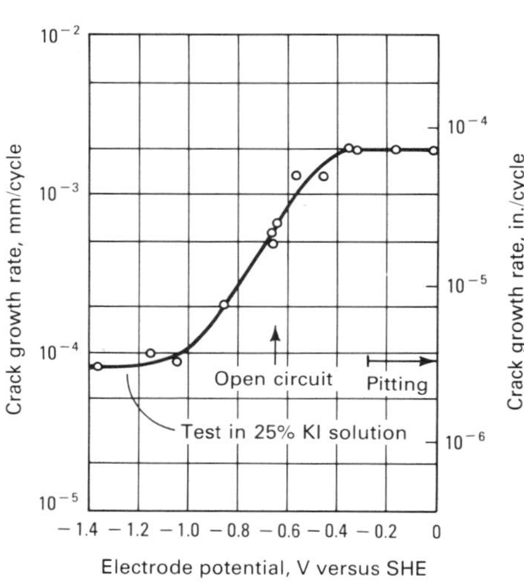

Fig. 17 Corrosion fatigue behavior of aluminum alloy 7079-T651 plate (S-L orientation). Temperature: 23 °C (73 °F); frequency: 4 cycles/s; stress ratio: $R = 0$. (a) Effect of stress intensity range on crack growth rate. K_{ICFC} and a range of K_{ISCC} are indicated at the bottom. (b) Effect of electrode potential at ΔK of 6.7 MPa$\sqrt{m}$ (6 ksi$\sqrt{in.}$) in 25% potassium iodide solution. Source: Ref 49

REFERENCES

1. F.P. Ford, Current Understanding of the Mechanism of Stress Corrosion and Corrosion Fatigue, in *Environment-Sensitive Fracture*, STP 821, S.W. Dean, E.N. Pugh, and G.M. Ugianski, Ed., American Society for Testing and Materials, 1984, p 32-51
2. O. Devereux, A.J. McEvily, and R.W. Staehle, Ed., *Corrosion Fatigue—Chemistry, Mechanics, and Microstructure*, Proceedings of NACE Conference, Storrs, CT, June 1971, National Association of Corrosion Engineers, 1972
3. H.L. Craig, Jr, T.W. Crooker, and D.W. Hoeppner, Ed., *Corrosion-Fatigue Technol-*

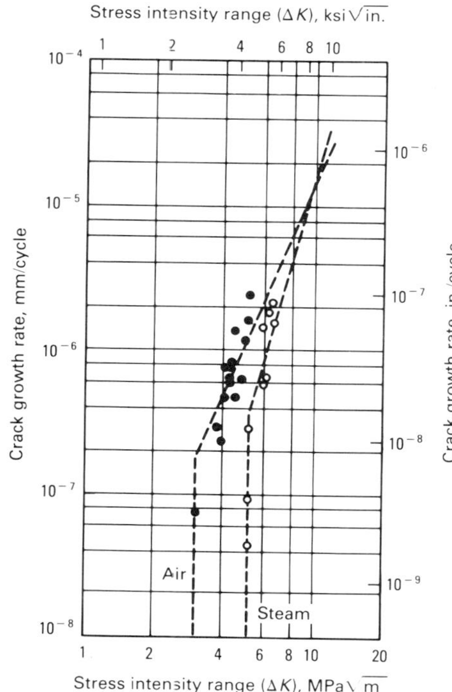

Fig. 18 Corrosion fatigue crack propagation in ASTM A471 steel exposed to moist air and steam. Temperature: 100 °C (212 °F); frequency: 100 Hz; stress ratio: R = 0.35. Source: Ref 52

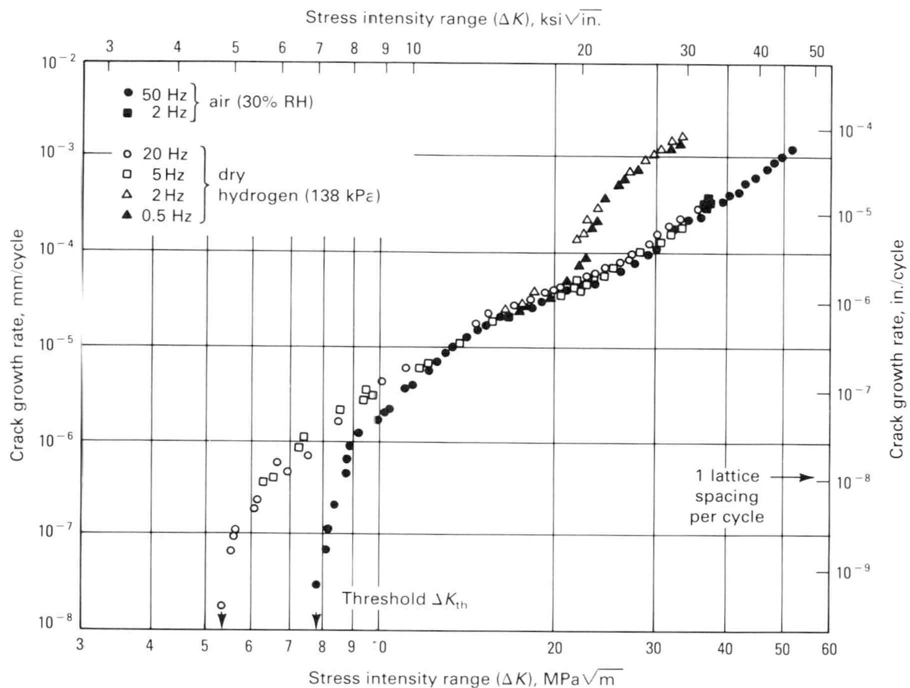

Fig. 19 Corrosion fatigue in 2.25Cr-1Mo pressure vessel steel in dry hydrogen at 23 °C (73 °F) due to hydrogen embrittlement at high ΔK and to reduced oxide-induced closure at low ΔK. Stress ratio: R = 0.05. Source: Ref 50

ogy, STP 642, American Society for Testing and Materials, 1978

4. R.W. Staehle and M.O. Speidel, Ed., *Stress Corrosion and Corrosion Fatigue Handbook*, ARPA Contract M00014-75-C-0703, Advanced Research Project Agency of the Department of Defense
5. P.R. Swann, F.P. Ford, and A.R.C. Westwood, Ed., *Mechanisms of Environment Sensitive Cracking of Materials*, Proceedings of Metals Society Conference, University of Surrey, U.K., April 1977, The Metals Society, 1979
6. R.N. Parkins and Ya. M. Kolotyrkin, Ed., *Corrosion Fatigue*, Proceedings of the First USSR-UK Seminar on Corrosion Fatigue of Metals, Lvov, USSR, May 1980, The Metals Society, 1983
7. T.W. Crooker and B.N. Leis, Ed., *Corrosion Fatigue: Mechanics, Metallurgy, Electrochemistry and Engineering*, STP 801, American Society for Testing and Materials, 1983
8. S.W. Dean, E.N. Pugh, and G.M. Ugiansky, Ed., *Environment-Sensitive Fracture*, STP 821, American Society for Testing and Materials, 1984
9. R.N. Parkins, "A Critical Evaluation of Current Environment-Sensitive Fracture Test Methods," in *Environment-Sensitive Fracture*, STP 821, S.W. Dean, E.N. Pugh, and G.M. Ugiansky, Ed., American Society for Testing and Materials, 1984, p 5-31
10. R.P. Gangloff, Ed., *Embrittlement by the Localized Crack Environment*, Proceedings of the TMS-AIME/MDS-ASM Symposium on Localized Crack Chemistry and Mechanics in Environment-Assisted Cracking, Philadelphia, PA, Oct 1983, The Metallurgical Society, 1984
11. D.A. Jones, A Unified Mechanism of Stress Corrosion and Corrosion Fatigue Cracking, *Metall. Trans. A*, Vol 16A, June 1985, p 1133-1141
12. T.W. Crooker, Environmental Cracking in Structural Alloys: A Look at Standards for Corrosion Fatigue and Stress-Corrosion Cracking, *ASTM Stand. News*, Vol 13 (No. 11), Nov 1985, p 54-58
13. J.T. Fong, Fatigue Research: Needs and Opportunities, *ASTM Stand. News*, Vol 13 (No. 11), Nov 1985, p 59-63
14. H.J. Westwood and W.K. Lee, Corrosion-Fatigue Cracking in Fossil-Fueled Boilers, in *Corrosion Cracking*, Proceedings of the International Conference on Fatigue, Corrosion Cracking, Fracture Mechanics and Failure Analysis, Salt Lake City, UT, Dec 1985, American Society for Metals, 1986, p 23-34
15. G.E. Dieter, *Mechanical Metallurgy*, McGraw-Hill, 1976, p 426-427
16. O.J. Horger, Fatigue Characteristics of Large Sections, in *Fatigue*, American Society for Metals, 1953
17. C.E. Phillips and R.B. Heywood, *Proc. Inst. Mech. Eng. (London)*, Vol 165, 1951, p 113-124
18. R.E. Peterson, *Stress Concentration Design Factors*, John Wiley & Sons, 1974
19. H. Neuber, "Theory of Notch Stresses Principles for Exact Calculation of Strength With References to Structural Form and Material," Springer, 1958, AEC-Tr-4547; available through NTIS, U.S. Dept. of Commerce
20. R.J. Roark, *Formulas for Stress and Strain*, 4th ed., McGraw-Hill, 1965
21. T. Topper, R. Wetzel, and J. Morrow, Neuber's Rule Applied to Fatigue of Notched Specimens, *J. Mater.*, Vol 4 (No. 1), 1969
22. G. Sines and J.R. Weisman, Ed., *Metal Fatigue*, McGraw-Hill, 1959
23. J.B. Bidwell *et al.*, *Fatigue Durability of Carburized Steel*, American Society for Metals, 1957
24. M.O. Speidel, Corrosion Fatigue in Fe-Ni-Cr Alloys, in *Proceedings of the International Conference on Stress Corrosion and Hydrogen Embrittlement of Iron-Base Alloys*, Unieux-Firminy, France, June 1973, National Association of Corrosion Engineers, 1977, p 1071-1094
25. "Metallic Materials and Elements for Aerospace Vehicle Structures," MIL-HDBK-5B, Vol I, U.S. Department of Defense, Sept 1971, p 2-29
26. K. Endo and K. Komai, Effects of Stress Wave Form and Cycle Frequency on Low Cycle Corrosion Fatigue, in *Corrosion Fatigue: Chemistry, Mechanics, and Microstructure*, O. Devereux, A.J. McEvily, and R.W. Staehle, Ed., Conference held at the University of Connecticut, June 1971, National Association of Corrosion Engineers, 1972, p 437-450
27. L.D. Roth and L.E. Willertz, Application of Ultrasonic Fatigue Testing Techniques to the Evaluation of the Corrosion-Fatigue Strength of Materials, in *Environment-Sensitive Fracture*, STP 821, S.W. Dean, E.N. Pugh, and G.M. Ugianski, Ed., American Society for Testing and Materials, 1984, p 497-512
28. A.J. McEvily and R.P. Wei, Fracture Mechanics and Corrosion Fatigue, in *Corrosion Fatigue: Chemistry, Mechanics and Microstructure*, O. Devereux, A.J. McEvily, and R.W. Staehle, Ed., National Association of Corrosion Engineers, 1973, p 381-395
29. P.M. Scott, Chemistry Effects in Corrosion Fatigue, in *Corrosion Fatigue: Mechanics,*

Metallurgy, Electrochemistry and Engineering, STP 801, T.W. Crooker and B.N. Leis, Ed., American Society for Testing and Materials, 1983, p 319-350

30. J.A. Feeney, J.C. McMillan, and R.P. Wei, Environmental Fatigue Crack Propagation of Aluminum Alloys at Low Stress Intensity Levels, *Metall. Trans.*, Vol 1, 1970, p 1741-1757
31. *Corrosion Cracking*, Proceedings of International Conference on Fatigue, Corrosion Cracking, Fracture Mechanics and Failure Analysis, Salt Lake City, UT, Dec 1985, American Society for Metals, 1986
32. Z.A. Foroulis, Ed., *Environment-Sensitive Fracture of Engineering Materials*, The Metals Society and the American Institute of Mining, Metallurgical, and Petroleum Engineers, 1978
33. O. Vosikovsky and R.J. Cooke, An Analysis of Crack Extension by Corrosion Fatigue in a Crude Oil Pipeline, *Int. J. Pressure Vessel Piping*, Vol 6, 1978, p 113-129
34. R.P. Wei and G. Shim, Fracture Mechanics and Corrosion Fatigue, in *Corrosion Fatigue: Mechanics, Metallurgy, Electrochemistry and Engineering*, STP 801, T.W. Crooker and B.N. Leis, Ed., American Society for Testing and Materials, 1984, p 5-25
35. M.H. El Haddad, T.H. Topper, and B. Mukherjee, Review of New Developments in Crack Propagation Studies, *J. Test Eval.*, Vol 9, 1981, p 65-81
36. N.E. Dowling, Crack Growth During Low Cycle Fatigue of Smooth Axial Specimens, in *Cyclic Stress-Strain and Plastic Deformation Aspects of Fatigue Crack Growth*, STP 637, American Society for Testing and Materials, 1977, p 97-121
37. R.P. Gangloff and R.O. Ritchie, Environmental Effects Novel to the Propagation of Short Fatigue Cracks, in *Fundamentals of Deformation and Fracture*, K.J. Miller, Ed., Cambridge University Press, 1987
38. S.J. Hudak, Jr. and R.P. Wei, Consideration of Nonsteady State Crack Growth in Material Evaluation and Design, *Int. J. Pressure Vessel Piping*, Vol 9, 1981, p 63-74
39. R.P. Gangloff, Crack Size Effects on the Chemical Driving Force for Aqueous Corrosion Fatigue, *Metall. Trans. A*, 1987
40. C.S. Kortovich, Corrosion Fatigue of 4340 and D6AC Steels Below K_{ISCC}, in *Proceedings of the 1974 Triservice Conference on Corrosion of Military Equipment*, AFML-TR-75-42, Air Force Materials Laboratory, Wright-Patterson Air Force Base, 1975
41. D.B. Dawson and R.M. Pelloux, Corrosion Fatigue Crack Growth of Titanium Alloys in Aqueous Environments, *Metall. Trans. A*, Vol 5A, 1974, p 723-731
42. J.M. Barsom, Corrosion Fatigue Crack Propagation Below K_{ISCC}, *Eng. Fract. Mech.*, Vol 3, 1971, p 15-25
43. B. Tompkins and P.M. Scott, Environment Sensitive Fracture: Design Considerations, *Met. Tech.*, Vol 9, 1982, p 240-248
44. P.M. Scott, Effects of Environment on Crack Propagation, in *Developments in Fracture Mechanics—II*, G.G. Shell, Ed., Applied Science, 1979, p 221-257
45. J.P. Gallagher, "Corrosion Fatigue Crack Growth Rate Behavior Above and Below K_{ISCC}," Report NRL-7064, Naval Research Laboratory, 1970
46. R.P. Gangloff, Exxon Research and Engineering Company, unpublished research, 1984
47. A. Turnbull, Progress in the Understanding of the Electrochemistry in Cracks, in *Embrittlement by the Localized Crack Environment*, R.P. Gangloff, Ed., The Metals Society and the American Institute of Mining, Metallurgical, and Petroleum Engineers, 1984, p 3-31
48. R.N. Parkins, Prevention of Environment Sensitive Cracking by Inhibition, in *Embrittlement by the Localized Crack Environment*, R.P. Gangloff, Ed., The Metals Society and the American Institute of Mining, Metallurgical, and Petroleum Engineers, 1984, p 385-404
49. M.O. Speidel, M.J. Blackburn, T.R. Beck, and J.A. Feeney, Corrosion Fatigue and Stress Corrosion Crack Growth in High Strength Aluminum Alloys, Magnesium Alloys and Titanium Alloys Exposed to Aqueous Solutions, in *Corrosion Fatigue: Chemistry, Mechanics and Microstructure*, O. Devereux, A.J. McEvily, and R.W. Staehle, Ed., National Association of Corrosion Engineers, 1973, p 324-345
50. R.O. Ritchie, Application of Fracture Mechanics to Fatigue, Corrosion Fatigue and Hydrogen Embrittlement, in *Analytical and Experimental Fracture Mechanics*, G.C. Sih, Ed., Sithoff and Noorohoff, 1981, p 81-108
51. S. Suresh and R.O. Ritchie, The Propagation of Short Fatigue Cracks, *Int. Met. Rev.*, 1987
52. L.K.L. Tu and B.B. Seth, Threshold Corrosion Fatigue Crack Growth in Steels, *J. Test. Eval.*, Vol 6, 1978, p 66-74
53. S.J. Hudak, Jr. and R.J. Bucci, Ed., *Fatigue Crack Growth Measurement and Data Analysis*, STP 738, American Society for Testing and Materials, 1981
54. R.P. Gangloff, Electrical Potential Monitoring of the Formation and Growth of Small Fatigue Cracks in Embrittling Environments, in *Advances in Crack Length Measurement*, C.J. Beevers, Ed., Engineering Materials Advisory Services, 1982, p 175-230
55. J.K. Donald, Fatigue Crack Propagation and the Role of Automated Testing, *ASTM Stand. News*, Vol 13 (No. 11), Nov 1985, p 50-53
56. C.J. Beevers, Ed., *Advances in Crack Length Assessment*, Engineering Materials Advisory Service, 1982

Evaluation of Crevice Corrosion

R.M. Kain, LaQue Center for Corrosion Technology, Inc.

CREVICE CORROSION is a form of localized corrosion that primarily affects passive-type alloys. Stainless steels, particularly those with little or no molybdenum, are especially prone to crevice corrosion. In neutral and acidic chloride-containing environments, higher-alloyed stainless-type materials and related nickel-base alloys may also be susceptible in some cases. Although other alloy systems, such as aluminum, copper, and titanium, may also be susceptible to crevice-related corrosion, testing and test development have primarily focused on assessing the behavior of the stainless-type alloys. Much of this attention has been associated with the identification and development of more resistant alloys for marine applications and in certain process industries, such as pulp and paper. Over the years, diverse crevice corrosion testing methodologies have evolved and continue to find significant utility in simple immersion studies, electrochemical testing, and mathematical modeling.

Early experience with stainless steel in seawater, for example, revealed that localized corrosion occurred beneath marine barnacles and other deposits. Comparative immersion tests of available stainless alloys also produced crevice corrosion beneath insulating washers on test racks. Simple testing and service experience both revealed considerable variability in stainless steel performance. Recognizing the limitations of the 300-series stainless steels, alloy producers saw potential markets for new, more corrosion-resistant marine alloys. Several high-molybdenum austenitic (some with nitrogen enhancement), duplex, and superferritic alloys were subsequently developed. Prior to these developments, a number of alloy 20-type stainless alloys (20% Cr) that gained acceptance in the chemical-processing industry were also considered for marine service. This was particularly true in the case of cast stainless pump and valve materials. The merits of high-molybdenum nickel-base alloys, such as alloy 625 and alloy C-276, were recognized early, but use was generally limited to high-performance applications.

Few alloy development laboratories had access to natural seawater. Those researchers who used seacoast test facilities were often left to the mercies of seemingly long-term natural exposure in order to obtain data. Both situations prompted the use and development of laboratory screening tests. Some used simple salt solutions or recipes simulating ocean water. It was recognized for some time that the environment within the crevice was different from that of the bulk environment. Oxygen differential cells could be established between cathode surfaces exposed to oxygenated seawater and anodic crevice areas. Others recognized that hydrolysis reactions within crevices could produce changes in pH and chloride concentration in the crevice environments. As such, investigators sought the use of ferric chloride ($FeCl_3$) solutions to simulate crevice conditions. Results could now be obtained within a matter of days rather than months or years of natural exposure testing. The development of economical and reliable electrochemical test equipment also provided another tool for investigators.

Aspects of Crevice Corrosion Testing

This article will discuss a number of the more frequently used crevice corrosion testing procedures. It is recognized, however, that other techniques exist and that their merits should be examined.

Because crevices are real space dimensions between mating materials, use of the term artificial crevice test is discouraged in favor of descriptors that identify the crevice former as naturally occurring (for example, a barnacle) or man-made (for example, a washer). The use of the latter as either a metal or nonmetal device can recreate the geometry of some fabricated crevices, but not all. Similarly, a man-made crevice former device may or may not reproduce the conditions generated by attachment and growth of a barnacle.

Another aspect of crevice corrosion test selection is based on the two accepted phases of crevice corrosion: initiation and propagation. Initiation refers to the breakdown of passivity and the actual occurrence of crevice corrosion as determined by some finite degree of propagation or penetration. Under some conditions, breakdown may be followed by repassivation without any outward evidence of attack. For a given alloy-environment combination, the occurrence of crevice corrosion is highly dependent on the crevice geometry. Tighter crevices are more conducive to promoting breakdown, but this factor may affect the propagation behavior of materials in different ways (Ref 1). Environmental factors may affect the rate of crevice corrosion propagation. Although natural and synthetic seawater, for example, will both cause AISI type 316 and other stainless alloys to initiate, different rates of propagation are likely to occur (Ref 2). The overall size of the specimen in relation to the crevice area can also affect propagation behavior. Regardless of the purpose of crevice corrosion testing, the investigator should be aware that these and many other factors could affect test results. A number of metallurgical, environmental, and geometrical factors that affect crevice corrosion are discussed in the section "Crevice Corrosion" of the article "Localized Corrosion" in this Volume.

Guidelines for Crevice Corrosion Testing

ASTM Specification G 78 provides guidance in the conduct of crevice corrosion tests for stainless steels and related nickel-base alloys in seawater and other chloride-containing environments (Ref 3). This guide does not promote any particular test technique, but advises the reader to be aware that crevice corrosion test results can be affected by a number of interrelated factors. In particular, attention is advised in the area of crevice geometry and specimen preparation. Importantly, Ref 3 points out that the occurrence or absence of crevice corrosion in a given test is no assurance that it will or will not occur under other conditions. This is especially true if the test conditions do not consider factors that may affect initiation and/or propagation behavior. Laboratory screening tests play a valuable role in the ultimate use of stainless materials in corrosive environments. Laboratory immersion tests and electrochemical studies are important tools for investigating the effects of alloy content and environmental variables. However, test results should be used judiciously so as not to limit materials that may not meet artificially contrived conditions selected to accelerate results.

Immersion Tests

Depending on accessibility, laboratory immersion tests may be able to utilize natural or actual process environments. This provides a convenient opportunity to study, for example, the effects of temperature, crevice geometry, and such factors as surface finish. These tests can be conducted at marine corrosion test centers and at laboratories associated with chemical-processing facilities.

Whether preparing samples for field or laboratory exposures, the researcher must be sure to provide flat-parallel surfaces for attachment of the crevice formers. Two such formers, one to each side of a test panel, are generally attached with a corrosion-resistant fastener. A consistent level of tightness should be used to minimize variability in crevice gap. The mechanical properties of some nonmetallic materials used to form crevices may relax with time. In some cases, this may be sufficient to render the crevice ineffective. Some materials are more deformable than others (for example, polymer versus ceramic) and may allow for greater tightness (smaller crevice gap). Crevice formers may vary in size and may take the shape of continuous annular washers and O-rings or of some type of serrated or multiple-crevice assembly.

To gain a greater appreciation of alloy performance, the researcher should test several different types and sizes of crevice-forming materials whenever possible. Similarly, if different product forms (plate, pipe, casting, and so on) are to be used in service, differences in thermomechanical processing and finishing may influence corrosion resistance. Heavily pickled materials, for example, may not have the same surface composition and roughness as an as-rolled product. For specific applications, the metallurgical condition of the material should be representative of that in service. If alloying effects are to be investigated, it may be advisable to test with a common surface finish. Conversely, if competitive alloys are to be compared, both as-produced and ground surfaces should be tested.

Specific Crevice Corrosion Tests

Spool Specimen Test Racks. A considerable amount of crevice corrosion data have been obtained from test rack exposures of candidate alloys in various process streams and other industrial environments. Simple tetrafluoroethylene (TFE) fluorocarbon sleeves and washers, intended primarily as electrical insulating spacers between candidate alloy specimens, have proved to be effective crevice formers. Test results for metal coupons and spool disk specimens have identified the occurrence of crevice corrosion and the extent of penetration incurred during the exposure period. Although these and other tests lack definition with regard to actual times to initiation and rates of propagation, they are nonetheless quite effective in screening materials for further consideration.

One area in which test rack exposures have been extensively used is the pulp and paper industry. For example, Ref 4 discusses results from bleach plant exposure in which crevice corrosion susceptibility was identified for 24 of the 26 alloys tested. However, the crevices found in bleach plant washers may be less severe than those on the test specimens. This is supported by the observation that some materials that exhibited less than 0.4 to 0.46 mm (15 to 18 mils) of crevice penetration in a 3-month test have actually performed satisfactorily in service for up to 20 years. More information on the use of test rack exposure tests is available in the article "In-Service Monitoring" in this Volume.

Ferric Chloride Tests. The ferric chloride test described in ASTM G 48 involves exposure to a highly oxidizing acid chloride environment (Ref 5). Crevices are created at sites of contact with TFE-fluorocarbon blocks secured by rubber bands. A test is generally conducted at 22 or 50 ± 2 °C (72 or 120 ± 3.5 °F) for 72 h.

The pH of the standard 6% $FeCl_3$ solution is about 1.2 and is less than the pH of the critical crevice solution, which can cause a breakdown of passivity for type 304 and type 316 stainless steels. Critical crevice solution compositions for these alloys with respect to pH and chloride concentration have been identified at levels of pH 2.0 at 1.5 M Cl^- and pH 1.7 at 3.5 M Cl^-, respectively (Ref 6). Use of $FeCl_3$ eliminates the time factor for development of the aggressive crevice solution from the neutral pH of, for example, a chloride-containing natural water. The crevice prevents oxygen from maintaining the passive film. The combination of pH and high Cl^- will lower the breakdown or critical potential for pitting, while the presence of the ferric ions (Fe^{3+}) maintains the potential of the alloy in this critical region.

More resistant alloys do not exhibit a breakdown potential in the domain associated with Fe^{3+} reduction at ambient temperature. In some cases, the Fe^{3+} serves as a chemical potentiostat and holds the corrosion potential well within the passive region in the absence of oxygen. Increasing the $FeCl_3$ solution temperature to 50 °C (120 °F), for example, promotes breakdown and allows for ranking of more resistant alloys. Figure 1 shows schematically the polarization characteristics for stainless steels exposed to $FeCl_3$.

Materials Technology Institute Tests. The Materials Technology Institute of the Chemical Process Industry (MTI) has identified five corrosion tests for iron- and nickel-base corrosion resistant alloys (Ref 7). The results of these tests are intended to provide a reasonable assessment of the probable utility of new alloys in a variety of corrosive environments. Two of the methods address resistance to crevice corrosion. Method MTI-2 proposes alloy ranking on the merits of increased critical crevice temperature, and method MTI-4 fosters a concept of critical chloride concentration. Both methods are concerned with initiation resistance; neither method provides guidance in the area of crevice corrosion propagation.

Method MTI-2, originating from ASTM G 48, also involves the use of a 6% $FeCl_3$ solution for determining the relative resistance of alloys to crevice corrosion in oxidizing chloride environments. In the MTI procedure, crevices are formed by the application of two serrated TFE washers, each having 12 plateaus or contact sites in which corrosion may initiate. The washers, one on each side, are initially secured at a specified torque of 0.28 N · m (2.5 in. · lb). The critical crevice temperature is determined by increasing the $FeCl_3$ temperature in 2.5-°C (4.5-°F) increments within the range of 0 to 100 °C (32 to 212 °F) and by noting the closest temperature that produces attack (less than 0.025 mm, or 1 mil) in depth within a 24-h test period. Specimens are disassembled and inspected after each increment. The number of sites incurring this degree of attack is also reported, along with the minimum temperature for initiation. Results for materials with a standard surface finish (wet 80-grit or dry 120-grit) can be compared with recommended controls of type 316 stainless steel and alloy C-276.

Table 1 provides critical crevice temperature data for a number of nickel alloys. As described in Ref 8, some differences in test procedure were considered in regard to $FeCl_3$ makeup and specimen surface conditioning. In contrast to the specified MTI procedure, the modified procedure involved fresh surface grinding between exposure at each temperature increment. Reportedly, this minimized scatter and lowered the critical crevice temperature. Differences in critical crevice temperature as a function of procedure increased with increased alloy content. This is shown in Fig. 2 for the four alloys common to all three tests. Considering the diverse composition with respect to molybdenum content, it is reasonable to expect a similar ranking in $FeCl_3$ regardless of procedure.

The value of this ranking with regard to service expectation remains limited. Some agreement has been cited among critical crevice temperature and specific seawater test data (Ref 9, 10), but this agreement does not take into account the effect of

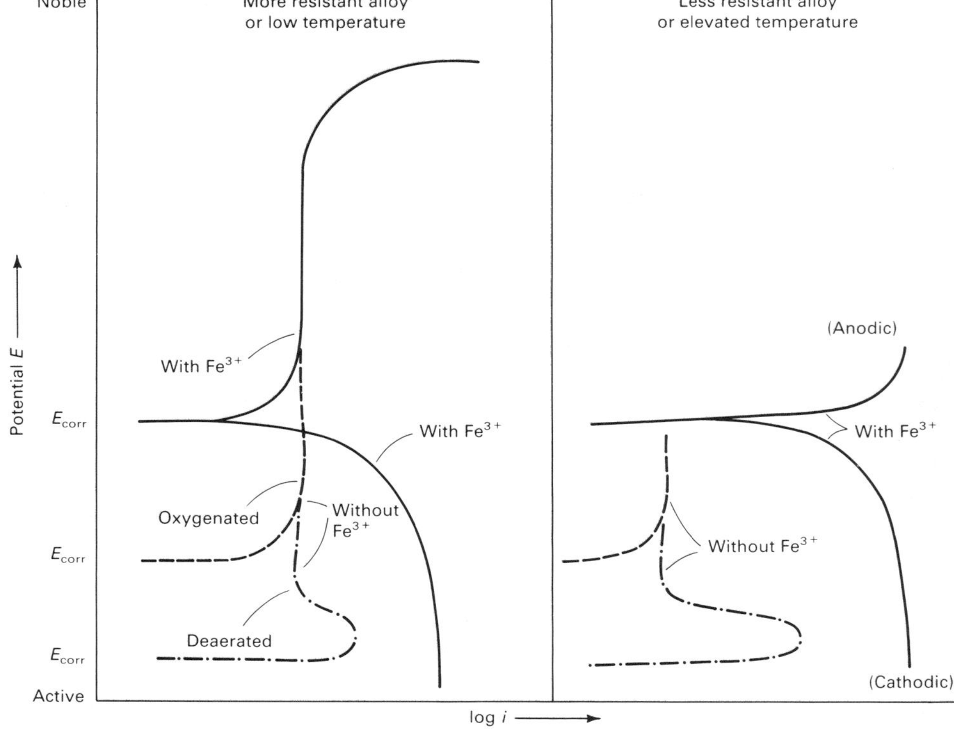

Fig. 1 Schematic showing the polarization characteristics of stainless-type materials in an acid chloride solution with and without Fe^{3+}

Table 1 Critical crevice temperature data for various alloys evaluated for 24-h periods in 6% $FeCl_3$ (10% $FeCl_3 \cdot 6H_2O$) using MTI and modified MTI procedures

Data are compared to literature critical crevice temperature values obtained in testing in 10% $FeCl_3$ (not hexahydrate).

Alloy	Critical crevice temperature					
	MTI		Modified MTI		Literature	
	°C	°F	°C	°F	°C	°F
825	0.0, 0.0	32, 32	2.5, 2.5	36.5, 36.5	−3	26.5
904L	2.5, 5.0	36.5, 40	2.5, 2.5	36.5, 36.5	...	...
317LM	2.5, 2.5	36.5, 36.5	2.5, 2.5	36.5, 36.5	...	...
HX	7.5, 7.5	45.5, 45.5	5.0, 5.0	40, 40	...	...
G	23.0, 25.0	73.5, 77	20.0, 20.0	70, 70	15–21	60–70
G-3	25.0, 25.0	77, 77	20.0, 20.0	70, 70	...	...
C-4	37.5, 37.5	99.5, 99.5	20.0, 20.0	70, 70	...	...
625	35.0, 40.0	95, 105	30.0, 30.0	85, 85	45–50	115–120
C-22	70.0, 70.0	160, 160	52.5, 55.0	126.5, 130	...	...
ALLCOR	52.5, 52.5	126.5, 126.5	40.0, 42.5	105, 110	...	...
C-276 (heat 1)	60.0, 65.0	140, 150	45.0, 42.5	115, 110	52–58	125.5–136.5
C-276 (heat 2)	62.5, 65.0	145, 150	45.0, 45.0	115, 115	...	...
C-276 (heat 3)	62.5, 65.0	145, 150	47.5, 47.5	117.5, 117.5	...	...
C-276 (heat 4)	55.0, 60.0	130, 140	47.5, 47.5	117.5, 117.5	...	...
C-276 (heat 5)	...	...	50.0, 50.0	120, 120	...	...

Source: Ref 8

crevice geometry and the interrelationship with the natural environment. Although increasing $FeCl_3$ temperature (for example, from 25 to 50 °C, or 77 to 120 °F) is certainly detrimental, the same is not necessarily true in seawater when bulk oxygen levels decline with increasing temperature and when biological factors may be altered (Ref 11).

Method MTI-4 uses increases in neutral bulk Cl^- concentration at eight levels ranging from 0.1 to 3.0% NaCl to establish a minimum (critical) Cl^- concentration in order to produce crevice corrosion at room temperature (20 to 24 °C, or 68 to 75 °F). The MTI-4 method is consistent with MTI-2 in the area of specimen requirements, crevice assembly procedure, and evaluation technique. The recommended test period, however, is 1000 h, and suggested controls are type 304 and 316 stainless steel.

Both methods are acceptable screening tools and are well suited for alloy development purposes. Method MTI-2, for example, has been frequently used to demonstrate the beneficial effects of chromium and molybdenum in promoting resistance to crevice corrosion initiation. However, extension of either method beyond its intended scope may be misleading. In the case of MTI-4, for example, differences in crevice geometry from that of the serrated washer could significantly alter resistance to a given bulk Cl^- concentration. Tighter and/or larger crevices could promote initiation at lower Cl^- levels (Ref 12). Use of the serrated washer identified earlier, for example, has produced crevice initiation at Cl^- levels as low as 100 to 300 mg/L in a matter of a few days when assembled at an initial torque of 8.5 N · m (75 in. · lb) (Ref 13). As noted in the section "Crevice Corrosion" of the article "Localized Corrosion" in this Volume, initiation at metal-to-metal sites can sometimes promote more rapid initiation and at lower Cl^- levels in comparison to nonmetal-to-metal crevices of comparable geometry.

Multiple-Crevice Assembly Testing

Background. Since its inception in the mid-1970s, the multiple-crevice assembly test has been one of the most popular and most controversial procedures available for evaluating the crevice corrosion resistance of stainless steels and related alloys. Although frequently cited in the literature, its present status is that of an acknowledged test method in the Standard Guide covered by ASTM G 78. Mutliple-crevice assembly devices generally consist of two serrated washers that provide a number of plateau-contact sites when fastened to a sheet or plate specimen. The acetal resin washer multiple-crevice assembly shown in Fig. 3 contains 20 plateaus, thus producing a total of 40 sites on each specimen. Typically, triplicate sheet or plate specimens are tested. The design shown is one of several in use that evolved from the multiple-crevice assembly originally developed by D.B. Anderson (Ref 14). Other multiple-crevice assemblies may have fewer plateaus of somewhat different size. Procedures generally call for attachment with an insulated, corrosion-resistant fastener and tightening in a reproducible manner with a calibrated torque wrench. The actual level of tightness may vary as a function of initial torque, crevice-forming material (that is, washer), and any relaxation. The fastener can also be used to attach the specimen to a suitable support and thus prevent the creation of any other undesirable crevice sites.

Multiple-crevice assemblies were developed as a rapid and economical screening tool for establishing resistance to crevice corrosion in natural seawater. Testing was specifically intended to be severe enough to produce some measure of alloy behavior before other factors, such as fouling and seasonal variation in temperature, could come into play. It was also desirable that the multiple-crevice assembly demonstrate recognized differences in alloy capabilities, such as those between type 304 and type 316 stainless steels, as well as the exceptional degree of performance expected of highly corrosion-resistant alloys. Unlike other crevice corrosion tests, the multiple-crevice assembly test relied solely on naturally occurring processes and required neither outside electrochemical nor chemical stimulation.

Because crevice corrosion appeared to be random in its occurrence, the creation of a number of identical crevice sites on a given set of specimens would provide the basis for statistical analysis. Therefore, the multiple-crevice design and the use of the probability concept gained appeal.

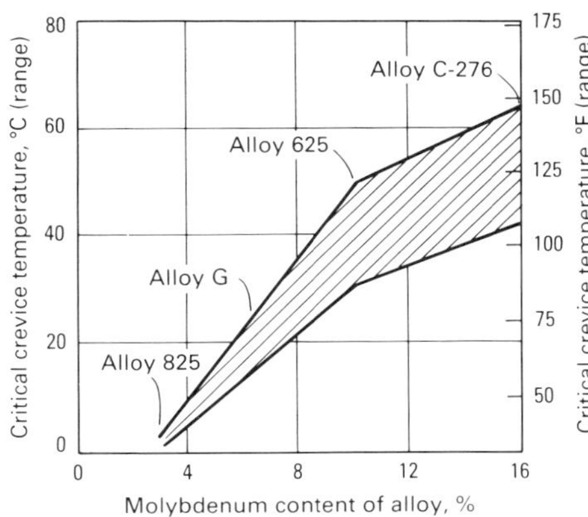

Fig. 2 Molybdenum content versus critical crevice temperature range for several nickel-base alloys. The range of critical crevice temperatures was obtained by varying test procedure.

Early concerns with geometrical factors primarily addressed the size of the specimen in terms of boldly exposed area (cathode) to shielded or crevice area (anode). Subsequent research, especially through mathematical modeling (Ref 15), showed that the occurrence of crevice corrosion was dependent on a number of interrelated factors but the probability concept suggested only a material (that is, alloy) property. Because such crevice geometry factors as crevice tightness could be overriding considerations, proponents and users of the multiple-crevice assembly generally abandoned the probability concept in describing the occurrence of crevice corrosion.

Examples of Use. The multiple-crevice assembly remains popular because it enables the investigator to report results in terms of initiation and propagation. For example, resistance to initiation can be expressed as the number of sides or sites or the percentage thereof exhibiting attack in a given test period. Because attack, when it occurs, is located at specific sites, penetration measurements can be made at each site and can be reported as a maximum depth. This can be accomplished by using an appropriate dial depth gage or microscope. Results have been reported as a range of these maximum values or their average value. Table 2, for example, lists data describing the effect of alloy molybdenum con-

Fig. 3 Serrated washers used for multiple-crevice assembly testing

Table 2 Multiple-crevice assembly test results showing the beneficial effects of molybdenum in nickel-base alloys containing 30% Cr

Results of 30-day test in filtered, natural seawater at 30 °C (85 °F)

Alloy	Alloy molybdenum content, %	Initiation resistance: Time observed, h — Earliest	Average	Average number of sites visibly corroded	Penetration resistance: Maximum depth — mm	mils	Average depth — mm	mils
A	0	52	76	10–11	1.75	69	0.8	31.5
B	0.5	52	76	11	1.65	65	0.48	19
C	0.9	100	115	9–10	1.2	47	0.4	16
D	2.6	100	115	8–9	0.8	31.5	0.2	8
E	4.9	ND(a)	ND(a)	1	0.05	2	0.05	2

(a) ND, not detected during 30-day test. Source: Ref 16

tent on both initiation and propagation resistance for a series of controlled chemistry heats of a nickel-base alloy containing 30% Cr.

Other methods of data presentation have been used to describe crevice corrosion initiation and propagation behavior. Figures 4 to 7 show multiple-crevice assembly results plotted in bar graph form as mean values plus and minus one standard deviation of the percentage of sites initiated and maximum depth of attack. Figures 4 and 5 provide a ranking of alloys based on their resistance to crevice corrosion when tested in their respective mill conditions and with a common surface finish (120-grit SiC, wet ground). The second examples (Fig. 6 and 7) are used to illustrate the influence of the initial crevice assembly torque level. Observations on different materials responding to differences in crevice tightness help to explain further the apparent variability in multiple-crevice assembly tests and in other types of crevice tests.

Furthermore, it has been postulated that the multiple-crevice assembly can produce a range of tight crevice gaps (Ref 17). This may vary depending on the actual degree of tightness at the time of exposure. As mentioned earlier, relaxation may be a factor in this regard. It has been shown that a multiple-crevice assembly made of acetal resin experiences more relaxation than, for example, assemblies of PTFE or ceramic (Ref 18). Despite this shortcoming, however, the acetal washers still provided the most severe conditions in comparative tests. Multiple-crevice assemblies of TFE are now widely used in variations of the ASTM G 48 and MTI $FeCl_3$ test procedures, replacing the certainly less controllable rubber band technique.

Variability. Because of the criticality of crevice geometry, some degree of variability should be expected in data not only from different sources but also from a single location or a single investigator. In this regard, several round robin, comparative test studies have been conducted over the years. One such program involved exposure of several grades of stainless steel to pulp-bleaching environments. Testing was performed in accordance with a procedure developed by the National Association of Corrosion Engineers T-5H-6 Task Group ("Test Methods for Measuring Crevice Corrosion Rates"). Multiple exposures involved the use of a Rulon device similar to that shown in Fig. 3, but with somewhat larger plateaus. Participants used specimens of prescribed size and assembly torque of prescribed levels. Because the initiated sites were few and nonreproducible, it was concluded that the method was not useful in warm (up to 82 °C, or 180 °F), low-pH (1 to 5), low-chloride (50 to 2000 ppm) environments. However, this actually fits well with the premise that a range of tight crevice gaps exists with each assembly. Mathematical modeling has shown an interrelationship between bulk chloride level and crevice gap. For the relatively small crevice depth of a plateau site and the low-level chlorides present in the bleach plant, only the tightest of crevice sites should be expected to initiate.

The above procedure evolved from that used in the American Society for Testing and Materials (ASTM) G1.09 Task Group round robin testing of multiple-crevice assemblies. In these tests, triplicate specimens with wet ground (120-grit SiC) surfaces were exposed for 30 days with multiple-crevice assemblies tightened to an initial torque of 8.5 N · m (75 in. · lb). Of the eight participants with access to natural seawater, seven were located on the U.S. eastern coast at latitudes

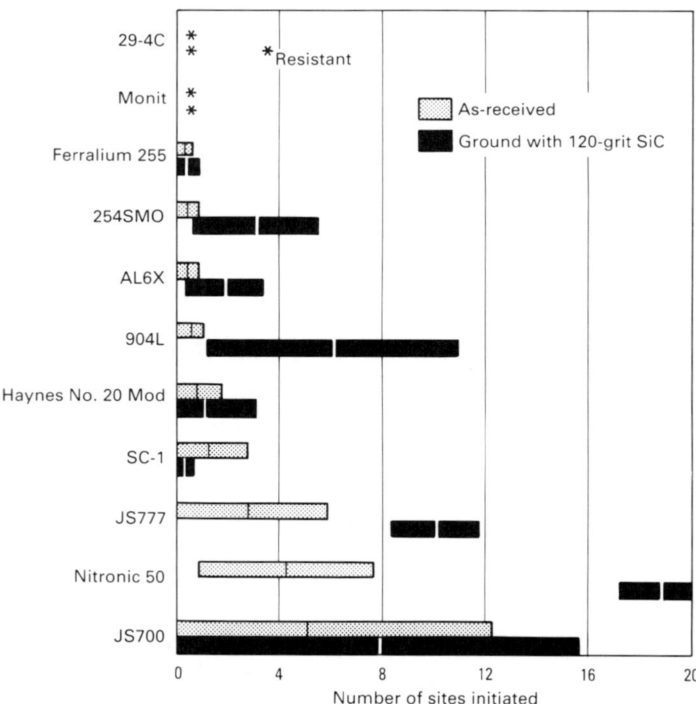

Fig. 4 Summary of number of initiated crevice sites beneath multiple-crevice assembly washers for a series of alloys tested in natural seawater. Bar graphs represent the mean values (plus and minus one standard deviation) for replicate 30-day tests of both as-received and ground (120-grit SiC) specimens. Assembly torque: 8.5 N · m (75 in. · lb)

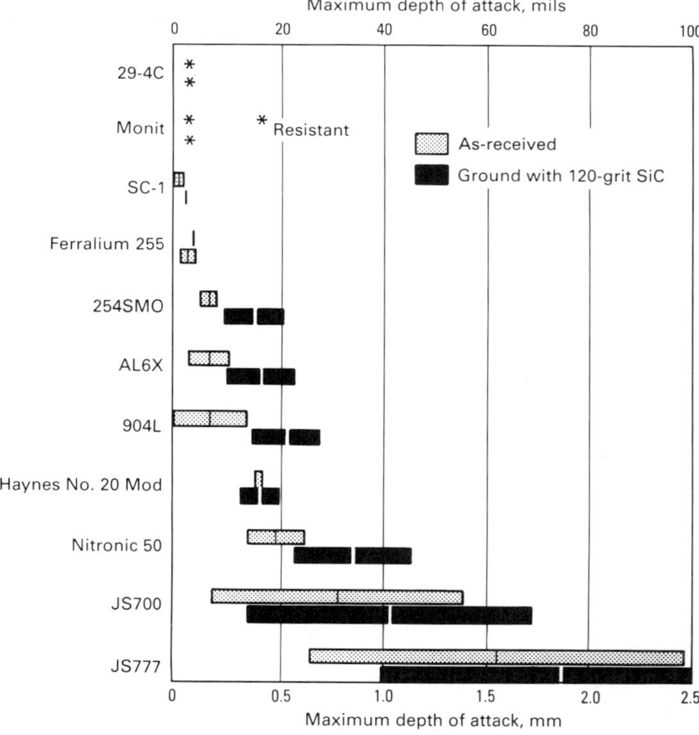

Fig. 5 Summary of maximum depth of attack beneath multiple-crevice assembly washers for a series of alloys tested in natural seawater. Bar graphs represent the mean values (plus and minus one standard deviation) for 30-day replicate tests on both as-received and ground (120-grit SiC) specimens. Assembly torque: 8.5 N · m (75 in. · lb)

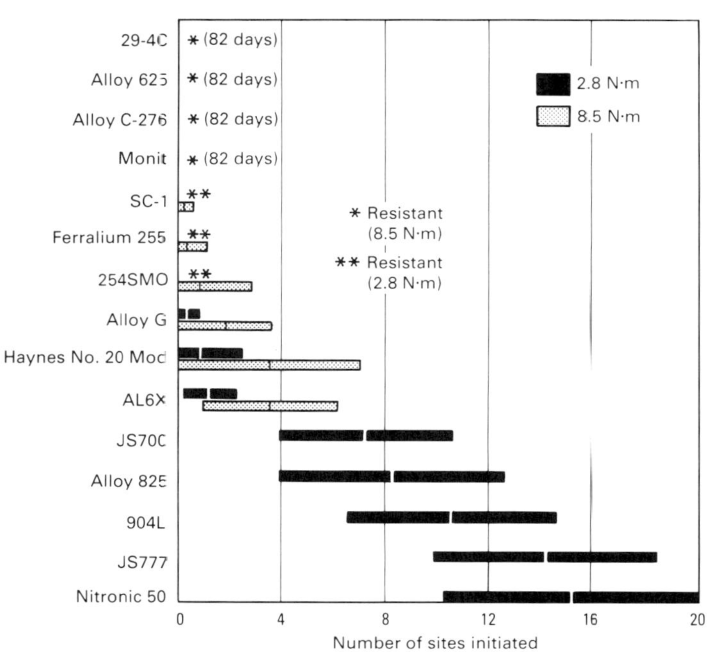

Fig. 6 Summary of number of initiated crevice sites beneath multiple-crevice assembly washers for a series of alloys tested in natural seawater. Bar graphs represent mean values (plus and minus one standard deviation) for replicate 60-day tests using an assembly torque of either 2.8 N · m (25 in. · lb) or 8.5 N · m (75 in. · lb). All specimen surfaces were ground with 120-grit SiC.

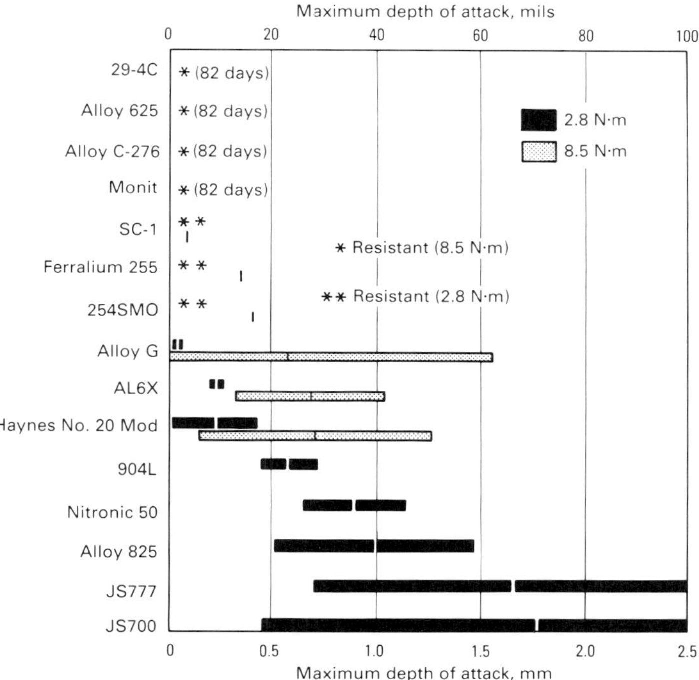

Fig. 7 Summary of maximum depth of attack beneath multiple-crevice assembly washers for a series of alloys tested in natural seawater. Bar graphs represent the mean values (plus and minus one standard deviation) for 60-day replicate tests conducted using an assembly torque of either 2.8 N · m (25 in. · lb) or 8.5 N · m (75 in. · lb). All specimen surfaces were ground with 120-grit SiC.

ranging from Key West, FL, to New England. The eighth test location was situated on the coast of California. All except two sites provided some degree of filtration, and three were unable to provide temperature control. As expected, some differences in seawater chemistry (for example, chloride and dissolved oxygen) and ambient temperatures were reported.

Although the purpose of these tests was to identify the reproducibility of data for different classes of materials, an alloy ranking was also identified. None of the participants reported any incidence of crevice corrosion for nickel alloy C-276 (~15% Mo). On the other hand, at least one specimen of alloy G (~6% Mo) initiated attack at all eight test locations. Overall, the actual number of sites attacked as well as the range and maximum depth of penetration were both found to be quite minimal. For type 304 and type 316 stainless steels, however, crevice corrosion occurred on all specimens except for one of the 2% Mo containing type 316. From this general point of view, identification of alloy susceptibility was quite reproducible. However, in terms of the actual number of sites that initiated, much variability was identified, particularly on a location-to-location basis or a participant-to-participant basis.

Figure 8 shows a plot of the percentage of crevice sites that incurred visible attack after 30 days. Each data point represents a single specimen with 40 crevice sites. Test locations are identified on the basis of ascending order of initiation for type 304. Although results showed considerable variation from some locations to others, repeatability at each site was generally good to excellent in most cases. Three of the locations (1, 2, and 7) ranked type 316 as more resistant than type 304, as is typically assumed. Two other locations (3 and 5) showed mixed behavior, but still favored type 316; the opposite is shown by results from locations 4 and 6. As can be seen, results from location 8 actually showed greater initiation resistance for type 304.

Figure 9 shows the range of penetration incurred by both alloys at all eight test locations. With the exception of location 7, which consistently reported low incidences of attack, the range of penetration measured elsewhere was more or less comparable, considering the variation of two orders of magnitude in site-to-site propagation. On the basis of maximum depth of penetration seven of eight participants reported greater resistance for type 316.

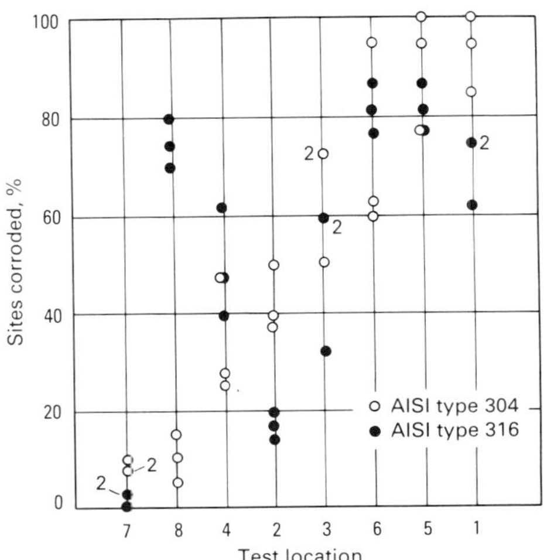

Fig. 8 Crevice corrosion initiation results from 30-day multiple-crevice assembly tests on triplicate specimens in natural seawater at various test sites (ASTM round robin). Source: Ref 3

Because of the variability in test results from this round robin, a second set of experiments with five participants was conducted at one of the test locations. Actual assembly and conduct of the test were performed by individuals with different degrees of experience as well as others with no prior experience in multiple-crevice assembly testing. All of the tests were conducted at the same time in seawater controlled at 25 ± 2 °C (77 ± 3.5 °F). The results of these tests are given in Fig. 10. Overall, the degree of variability is consistent with that observed from data provided from different test locations. Overall, penetrations measured in the latter test were comparable to those illustrated in Fig. 9.

At least one other series of comparative tests has been performed (Ref 19). In this case, 13 participants used multiple-crevice assemblies in round robin testing of type 304 in 3.5% NaCl at 30 °C (85 °F). Crevice assembly design and initial torque were the same as the above ASTM G1.09 procedure. Again, considerable variability was evident from one participant to another. Based on a percentage of total sites initiated out of 120 available, a range of 20 to 86% was reported. Although some exceptions were noted, a number of individual participants reported good reproducibility among replicate specimens. In contrast to natural seawater, penetrations in 3.5% NaCl tests did not exceed 0.23 mm (9 mils). Average penetrations from triplicate specimens varied by only a few hundredths of a millimeter.

The degree of variability described above is likely to be encountered in any type of crevice

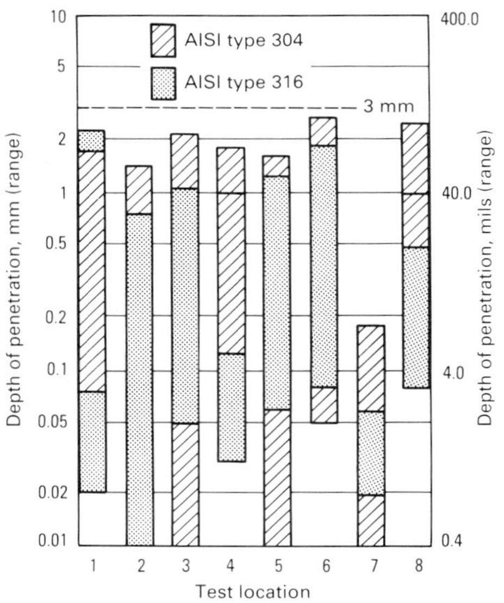

Fig. 9 Crevice corrosion propagation results from 30-day multiple-crevice assembly tests on triplicate specimens in natural seawater at various test sites (ASTM round robin)

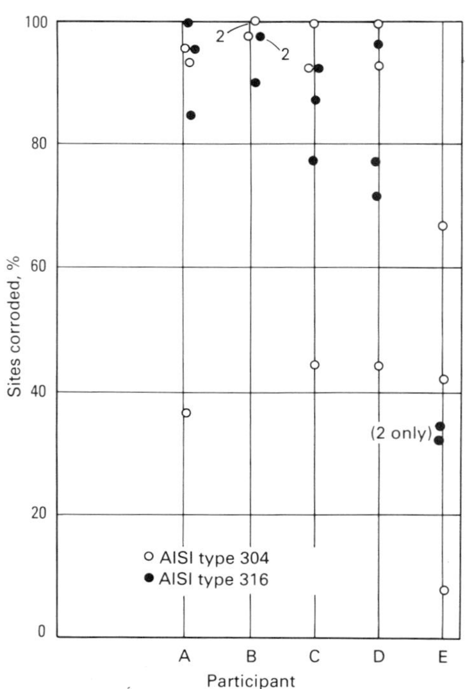

Fig. 10 Crevice corrosion initiation results from 30-day multiple-crevice assembly tests on triplicate specimens in natural seawater at site 5 in ASTM round robin

corrosion test. When conducted in a conscientious manner, multiple-crevice assembly can generate meaningful results toward the ranking of alloys and can identify the effect of other exposure variables. The multiple-crevice assembly test cannot and should not be used to predict alloy performance under other conditions.

Other Tests. Simple immersion tests of materials of interest can be performed with a variety of crevice formers. In reality, this may be a more acceptable way of characterizing alloy performance than conducting numerous experiments targeted at a single parameter, such as critical crevice temperature. Specification ASTM G 78 shows a number of different types of crevice formers that have been applied to sheet and plate materials by different investigators. Other types may be more suited to other product forms, such as pipe and tubing. For example, the investigator may use different O-ring materials of different diameter and thickness to produce a range of geometries. Similarly, sleeves of nonmetallic tubing or nylon compression fittings could be used in an evaluation of tubular materials. Such devices may, for example, replicate the depth and tightness of a tube-tubesheet assembly but not the electrochemical reactions of dissimilar metals. Other assemblies, for example, may replicate the geometry of an O-ring or gasket seal, but may not consider the full dynamic impact of a piping system. Use of any of these devices can be applied to materials other than stainless steel. The investigator should become familiar with the localized corrosion tendency of each family of alloys. Copper-base alloys, for example, typically exhibit attack adjacent to the mouth of the effective crevice area.

Electrochemical Tests

ASTM Standard Methods. The *Annual Book of ASTM Standards* identifies two practices for investigating localized corrosion. Standard Practice G 61 is recommended primarily for iron-, nickel-, or cobalt-base alloys (Ref 20). Crevices are formed on a 16-mm (0.63-in.) diam disk electrode by a TFE-fluorocarbon gasket/mounting assembly. The electrode is made the anodic member of a polarization cell containing a deaerated 3.5% NaCl solution. (This procedure has been used for other environments of interest.) After a 1-h period of free corrosion, the crevice electrode potential is increased in the noble direction at a scan rate of 0.6 V/h. The current measured and the potential charge are both continuously plotted (or data collected by computer for subsequent analysis). Upon reaching a current of 5 mA, the scan direction is reversed and continued to its potential of origin. Susceptibility to crevice corrosion is identified by the occurrence of hysteresis during the reverse scan. Relative alloy resistance can be established by comparing the forward and reverse scan potential-current domains with alloy C-276 and type 304 stainless steel standards. Specification ASTM G 61 provides standard polarization plots (forward and reverse scan) for comparison and equipment checkout. As with other crevice corrosion tests, some degree of variability is expected. In this test, several factors may affect results, most notably the actual time between specimen preparation and exposure as well as the degree of tightness used in assembling the electrode mount.

Nonstandard variations, such as use of different electrode assemblies and/or polarization scan rates, may significantly affect the measured response (Ref 21). Some correlation has been made between cyclic polarization tests according to ASTM G 61 and actual immersion tests in seawater (Ref 22). For the most part, cyclic polarization tests are able to differentiate between highly alloyed, resistant materials (for example, the high-molybdenum nickel-base alloys 625 and C276) and lower-alloy materials (for example, 300-series stainless steel). Many alloys with intermediate compositions may be less easily correlatable because they exhibit resistance to a broader range of environments (that is, Cl^- concentration and crevice geometry) than the 300-series stainless steels but less than that for nickel-base high-molybdenum alloys.

Another ASTM practice, F 746, describes procedures for testing the pitting and crevice corrosion of metallic surgical implant materials (Ref 23). Testing is again limited to making relative rankings of performance. The procedure calls for the use of a cylindrical electrode with crevices formed by a mounting compression gasket and an intentional crevice-forming collar. Corrosion is induced by a polarization step to +0.8 V versus the saturated calomel electrode (SCE). Provisions are established for current monitoring and subsequent potential-time increments. The test is designed to produce corrosion for a control electrode of type 316L stainless steel. Alloys can be compared relative to breakdown, propagation, or repassivation tendencies.

Other Potentiostatic and Potentiodynamic Polarization Tests. A potentiostatic test procedure identified with the Santron CTD 400 potentiostat is discussed in Ref 24. This technique, like the MTI procedure, identifies alloy crevice corrosion resistance according to an established critical crevice temperature. Tests have been conducted in neutral NaCl solution and synthetic seawater under constant applied potentials, for example, +600 mV versus SCE. In this automated test, the equipment is programmed to increase the solution temperature in 5-°C (9-°F) increments if a specific critical current level is not reached in a given period, for example, 15 to 20 min.

Test results obtained with the above method using 12-serration multiple-crevice assemblies are summarized in Table 3. Also given in Table 3 is a critical crevice temperature ranking determined in 72-h $FeCl_3$ tests according to ASTM G 48 (rubber band test) and another ranking based on the total number of sites initiated in 30-day natural seawater multiple crevice assembly (20 serrations) tests. The three procedures provided the same order of merit for only 3 of the 12 materials, numbers 1, 2, and 6. In several of the cases, at least two procedures provided the same ranking for a given alloy. In the Santron test, the noble potential is intended to mimic the redox potential of $FeCl_3$. In natural seawater, such potentials would never be achieved without chemical stimulation. Nickel-base chromium-molybdenum alloys, for example, may reach potentials of only +350 mV versus SCE in ambient temperature seawater (Ref 26).

Potentiodynamic polarization tests for crevice-free electrodes in a series of increasingly aggressive simulated crevice solutions have been used to rank alloys according to a criterion associated with anodic peak current density (Ref 27). Such information has been used in mathematical modeling to identify the localized environment that can cause breakdown of passivity and what conditions lead to its development (Ref 15). In addition, as shown in Fig. 11, a plot of the above log current versus pH has been used to characterize propagation resistance for several cast alloys (Ref 6). Alloy propagation behavior can be ranked according to the slope of the log *i*/pH plot.

Remote Crevice Assemblies. An electrochemical procedure requiring no stimulation other than that provided by the bulk environment-alloy reaction has been described in the literature.

Table 3 Initiation of crevice corrosion in immersion tests in seawater, $FeCl_3$, and synthetic seawater

All specimens were ground with 120-grit SiC.

Alloy	Filtered seawater(a) Number of sites	Rank	$FeCl_3$(b) Failure temperature °C	°F	Rank	Synthetic seawater Failure temperature °C	°F	Rank
29-4C	0	1	55	131	1	90.0	195	1
Monit	0	2	47	117	2	67.5	155	2
SC-1	1	3	45	113	4	60.0	140	5
Ferralium 255	2	4	37	99	5	60.0	140	4
Haynes No. 20 Mod	6	5	28	82	8	47.5	120	7
AL6X	11	6	37	99	6	57.5	135	6
254SMO	18	7	46	115	3	62.5	145	3
904L	36	8	22	72	10	42.5	110	9
JS700	47	9	31	88	7	45.0	115	8
JS777	60	10	14	57	12	30.0	85	12
AISI type 329	73	11	25	77	9	40.0	105	11
Nitronic 50	112	12	15	59	11	40.0	105	10

(a) 30-day test at 30 °C (85 °F). (b) 72-h test in 10% $FeCl_3 \cdot 6H_2O$. (c) Santron test; 20-min measuring time. Source: Ref 25

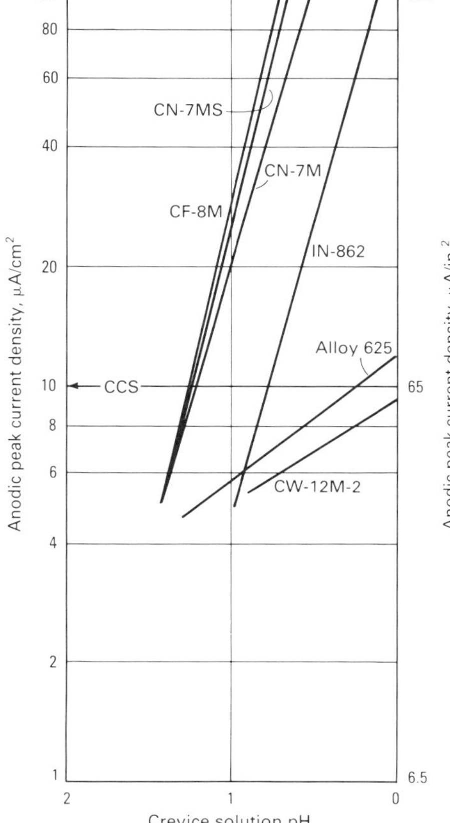

Fig. 11 Plot of anodic peak current density versus simulated crevice solution pH used to determine the composition (pH) of the critical crevice solution (CCS) according to the 10-μA/cm² (64.5-μA/in.²) criterion. Source: Ref 6

These techniques have been identified as either remote crevice or remote cathode tests (Ref 28, 29). Remote crevice assembly tests involve the physical separation but electrical connection of a small anode or crevice member and a larger member (cathode). Both are exposed in the bulk environment. Current between the two members can be monitored through a zero-resistance ammeter. The technique is quite capable of accurately identifying the time to initiation and subsequent propagation. Examples of its utility have been demonstrated in both natural seawater and other chloride-containing environments (Ref 2).

Unlike other techniques, this procedure has the capability of separating the initiation and propagation phases of crevice corrosion. This capability is summarized by the plot of corrosion current normalized for initiation time in Fig. 12. Results show very good reproducibility in both the trend of increasing current once initiation has occurred as well as the magnitude of current. The total charge (coulombs) reflects the total amount of propagation, which is directly proportional to mass loss due to crevice corrosion.

With larger crevice areas, such as the 15- × 15-mm (0.6- × 0.6-in.) electrodes used in these remote crevice assembly tests, crevice propagation may progress across the face of the electrode as well as penetrate it. A range of depths of attack is likely to be encountered within an apparent crevice area just as a range is encountered beneath the sites of a multiple-crevice assembly device.

Other Electrochemical Techniques. Other specialized tests are under investigation. One involves the use of compartmentalized cells with anode and cathode members exposed to their respective environments or simulations thereof (Ref 2, 6). Such tests have been used to describe the effects of changes in solution chemistry and surface area ratios. In addition, a vibrating electrode technique has been used to map variations in current density above a creviced stainless steel specimen of known crevice geometry (Ref 30). Necessarily, such tests are more expensive and perhaps more conducive to mechanistic studies rather than mass alloy characterizations. A number of electrochemical techniques that may be considered for crevice corrosion testing are reviewed in Ref 31 and 32. More information on electrochemical testing is available in the article "Laboratory Testing" in this Volume.

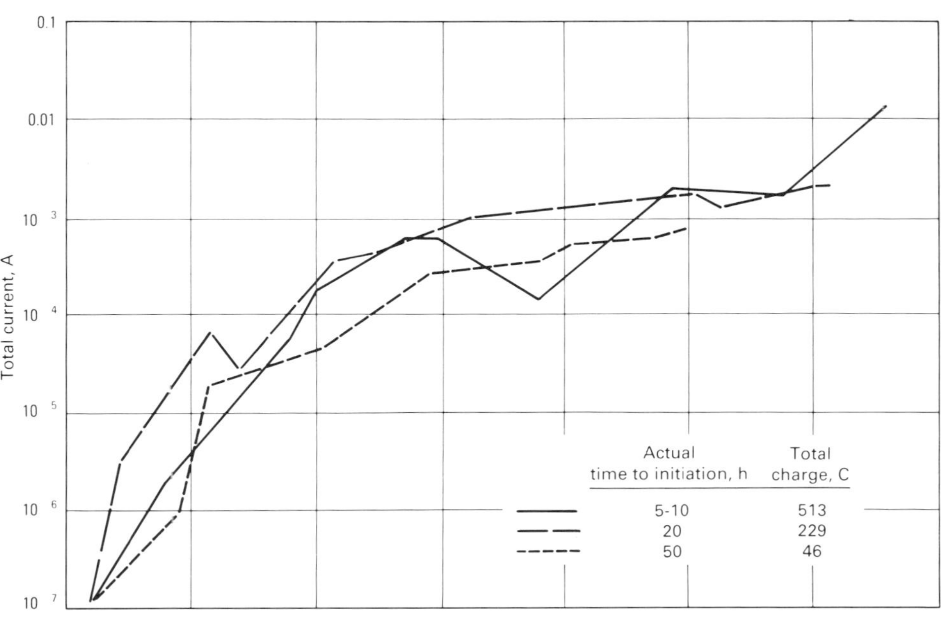

Fig. 12 Comparison of crevice corrosion propagation currents for type 316 stainless steel remote crevice assemblies after normalizing initiation times. Source: Ref 2

REFERENCES

1. T.S. Lee *et al.*, Mathematical Modelling of Crevice Corrosion of Stainless Steels, in *Corrosion and Corrosion Protection Proceedings*, Vol 81-8, The Electrochemical Society, 1981, p 213-224
2. R.M. Kain and T.S. Lee, Recent Developments in Test Methods for Investigating Crevice Corrosion, in *Laboratory Corrosion Tests and Standards*, STP 866, American Society for Testing and Materials, 1985, p 299-323
3. "Standard Guide for Crevice Corrosion Testing of Iron-Base and Nickel-Base Stainless

Alloys in Seawater and Other Chloride Containing Aqueous Environments," G 78, *Annual Book of ASTM Standards*, American Society for Testing and Materials

4. A.H. Tuthill, Resistance of Highly Alloyed Materials and Titanium to Localized Corrosion in Bleach Plant Environments, *Mater. Perform.*, Vol 24 (No. 9), 1985, p 43-49
5. "Standard Method for Pitting and Crevice Corrosion Resistance of Stainless Steels and Related Alloys by the Use of Ferric Chloride Solution," G 48 (reapproved 1980), *Annual Book of ASTM Standards*, American Society for Testing and Materials
6. R.M. Kain *et al.*, "Use of Electrochemical Techniques for the Study of Crevice Corrosion in Natural Seawater," Paper 60, presented at Corrosion/85, Houston, TX, National Association of Corrosion Engineers, 1985
7. R.S. Treseder and E.A. Kachik, MTI Corrosion Tests for Iron and Nickel-Base Corrosion Resistant Alloys, in *Laboratory Corrosion Tests and Standards*, STP 866, American Society for Testing and Materials, 1985, p 373-399
8. E.L. Hibner, Modification of Critical Crevice Temperature Test Procedures for Nickel Alloys in a Ferric Chloride Environment, *Mater. Perform.*, Vol 26 (No. 3), 1987, p 37-40
9. A.P. Bond and H.J. Dundas, Resistance of Stainless Steels to Crevice Corrosion in Seawater, *Mater. Perform.*, Vol 23 (No. 7), 1984, p 39-43
10. A. Garner, Crevice Corrosion of Stainless Steel in Seawater: Correlation of Field Data with Laboratory Ferric Chloride Tests, *Corrosion*, Vol 37 (No. 3), 1981, p 178-184
11. T.S. Lee *et al.*, The Effect of Environmental Variables on Crevice Corrosion of Stainless Steels in Seawater, *Mater. Perform.*, Vol 23 (No. 7), 1984, p 9-15
12. R.M. Kain, Crevice Corrosion Behavior of Stainless Steel in Seawater and Related Environments, *Corrosion*, Vol 40 (No. 6), 1984, p 313-321
13. R.M. Kain, A.H. Tuthill, and E.C. Hoxie, The Resistance of Type 304 and Type 316 Stainless Steel to Crevice Corrosion Natural Waters, *J. Mater. Energy Syst.*, Vol 5 (No. 4), 1984, p 205-211
14. D.B. Anderson, Statistical Aspects of Crevice Corrosion in Seawater, in *Galvanic and Pitting Corrosion—Field and Laboratory Studies*, STP 576, American Society for Testing and Materials, 1976, p 261
15. J.W. Oldfield and W.H. Sutton, New Technique for Predicting the Performance of Stainless Steels in Seawater and Other Chloride Containing Environments, *Br. Corros. J.*, Vol 15 (No. 1), 1980, p 31-34
16. R.M. Kain, "Effect of Alloy Content on the Localized Corrosion Resistance of Several Nickel Base Alloys in Seawater," Paper 229, presented at Corrosion/86, Houston, TX, National Association of Corrosion Engineers, 1986
17. R.M. Kain, "Crevice Corrosion and Metal Ion Concentration Cell Corrosion Resistance of Candidate Materials for OTEC Heat Exchangers," ANL/OTEC-BCM-022, Argonne National Laboratory and the U.S. Department of Energy, May 1981
18. G.O. Davis and M.A. Streicher, "Initiation of Chloride Crevice Corrosion on Stainless Alloys," Paper 205, presented at Corrosion/85, Houston, TX, National Association of Corrosion Engineers, 1985
19. Japan Corrosion Society, private communication, 1986
20. "Standard Practice for Conducting Cyclic Potentiodynamic Polarization Measurements for Localized Corrosion," G 61, *Annual Book of ASTM Standards*, American Society for Testing and Materials
21. R.M. Kain, "Localized Corrosion Behavior in Natural Seawater: A Comparison of Electrochemical and Crevice Testing of Stainless Steel," Paper 70, presented at Corrosion/80, Houston, TX, National Association of Corrosion Engineers, 1980
22. B.E. Wilde, A Critical Appraisal of Some Popular Laboratory Electrochemical Tests for Predicting the Localized Corrosion Resistance of Stainless Alloys in Sea Water, *Corrosion*, Vol 28 (No. 8), 1972, p 283
23. "Standard Test Method for Pitting or Crevice Corrosion of Metallic Surgical Implant Materials," F 746, *Annual Book of ASTM Standards*, American Society for Testing and Materials
24. S. Bernhardsson, Paper 85, presented at Corrosion/80, Houston, TX, National Association of Corrosion Engineers, 1980
25. N.S. Nagaswami and M.A. Streicher, "Accelerated Laboratory Tests for Crevice Corrosion of Stainless Alloys," Paper 7, presented at Corrosion/83, Houston, TX, National Association of Corrosion Engineers, 1983
26. J.M. Kroughman and F.P. Ijsseling, Crevice Corrosion of Stainless Steels and Nickel Alloys in Seawater, in *Proceedings of 5th International Congress on Marine Corrosion and Fouling*, Barcelona, Spain, Editorial Garsi Londres, 17, Madrid-28, España, May 1980, p 214
27. J.W. Oldfield and W.H. Sutton, Crevice Corrosion of Stainless Steels, II. Experimental Studies, *Br. Corros. J.*, Vol 13 (No. 3), 1978, p 104
28. T.S. Lee, A Method of Quantifying the Initiation and Propagation Stages of Crevice Corrosion, in *Electrochemical Corrosion Testing*, STP 727, American Society for Testing and Materials, 1981, p 43-68
29. R.M. Kain, Electrochemical Measurement of the Crevice Corrosion Propagation Resistance of Stainless Steels: Effect of Environmental Variables, *Mater. Perform.*, Vol 23 (No. 2), 1984, p 24
30. H.S. Issacs, "Application of the Vibration Probe to Localized Current Measurements," Paper 55, presented at Corrosion/85, Houston, TX, National Association of Corrosion Engineers, 1985
31. J. Postlewaite, *Can. Metall. Quart.*, Vol 22 (No. 1), 1983, p 133
32. F.P. Ijsseling, Electrochemical Methods in Crevice Corrosion Testing, *Br. Corros. J.*, Vol 15 (No. 2), 1980, p 51

Evaluation of Erosion and Cavitation

EROSION AND CAVITATION are mechanically assisted forms of material degradation. Erosion, in the context of this article, is defined as the progressive loss of material from a solid surface due to mechanical interaction between that surface and a fluid; therefore, any equipment that is exposed to moving fluids may be susceptible to erosion. Cavitation is caused by the formation and collapse of vapor bubbles in a liquid near a metal surface. More information on the mechanisms of these forms of attack is available in Ref 1 and in the article "Mechanically Assisted Degradation" in this Volume.

Laboratory tests used to evaluate erosion and cavitation damage in metals include the following (Ref 2):

- High-velocity flow tests, including venturi tubes, rotating disks, and ducts containing specimens in throat sections
- High-frequency vibratory tests using either magnetostriction devices or piezoelectric devices
- Impinging jet tests using either stationary or rotating specimens exposed to high-speed jet or droplet impact

These tests are generally designed to provide high erosion intensities on small specimens in relatively short times. These methods may not closely simulate service conditions, but they are useful for ranking candidate materials.

Standard Test Methods

There are two American Society for Testing and Materials (ASTM) standards for the evaluation of erosion and cavitation. These are designations G 32 (Ref 3), which is a vibratory test method using either a magnetostrictive or piezoelectric device that vibrates at a frequency of 20 kHz, and G 73 (Ref 4), which is much broader in scope than G 32. Both of these standards include definitions of terms related to cavitation and erosion as well as information on specimen preparation, test conditions and procedures, and data interpretation. In both cases, the test apparatus is calibrated, and comparisons of the relative resistances of materials are made by the use of reference materials.

ASTM Standard G 32 is based on the generation and collapse of cavitation bubbles on a specimen surface vibrating at high frequency (Ref 3). It is used to evaluate the relative resistances of different materials to cavitation erosion.

Test Specimen. Reference 3 specifies a test specimen measuring 15.9 mm ± 0.05 mm (0.625 ± 0.002 in.) in diameter and not less than 3.2 mm (0.125 in.) in thickness. More details on specimen requirements are provided in Ref 3.

The test apparatus (Fig. 1) produces axial oscillation of the test specimen, which is partially immersed in the test liquid. This is done by either a magnetostrictive or piezoelectric transducer driven by an electronic oscillator or amplifier. The transducer vibrates at a frequency of 20 kHz. The apparatus must include some method of measuring the displacement amplitude of the transducer as well as a means of maintaining the specified test temperature.

Test Conditions. The standard test liquid is distilled or other reagent water meeting the specifications of ASTM D 1193 ("Standard Specification for Reagent Water") maintained at a temperature of 22 ± 1 °C (72 ± 2 °F). Air is maintained over the test liquid at a pressure of 96 ± 12 kPa (28.4 ± 3.5 in. of mercury). A peak-to-peak displacement amplitude of 0.05 mm (0.002 in.) ± 5% is specified for the duration of the test.

Test conditions can be varied by using different test liquids, temperatures, and pressures. Any such variations must be noted when results are reported. Liquids other than reagent water that have been used include petroleum derivatives, glycerin, and liquid metals. Water has been used at temperatures other than 22 °C (72 °F), and these tests have shown that the erosion rate peaks strongly at a temperature about midway between the freezing point and the boiling point of the test liquid. Therefore, the erosion rate of a material in water at normal atmospheric pressure peaks at about 50 °C (122 °F) and falls off on either side of this value. The decrease in erosion rate is most dramatic as temperature rises above 50 °C (122 °F) (see the section "Effects of Test Parameters" in this article).

Test Procedure. The specimen must be cleaned and accurately weighed before testing, then immersed in the test liquid to a depth between 3.2 mm (0.125 in.) and 12.7 mm (0.5 in.). The peak-to-peak displacement amplitude should be monitored constantly. The test should be interrupted periodically to determine the mass loss of the specimen, which should be carefully cleaned and dried before reweighing. The time between measurements should be such that a plot of cumulative mass loss versus exposure time can be established. The time between measurements, therefore, depends on the specimen material; softer materials (for example, aluminum alloys) can be examined every 15 min, while the interval between measurements for a harder material, such as Stellite alloy 6B, may range from 8 to 10 h. Testing should be continued at least until the erosion rate reaches a maximum and begins to diminish. When several materials are being compared, all should be tested until they reach comparable mean depths of erosion.

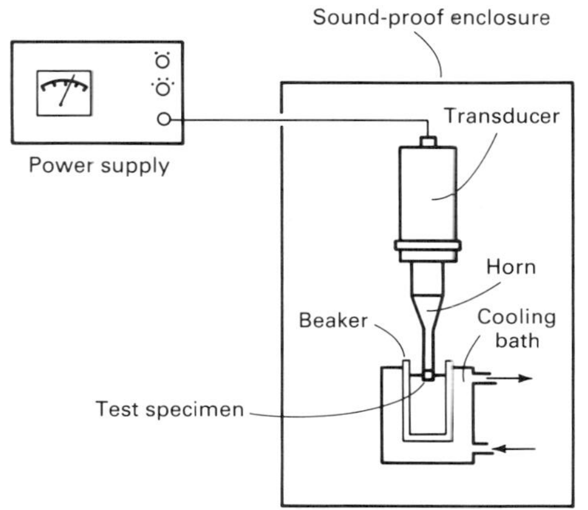

Fig. 1 Schematic of a typical vibratory erosion/cavitation test apparatus. Source: Ref 3

Data Acquisition and Reporting. The mean depth of erosion of the specimen should be calculated based on the full area of the test surface of the specimen. The report should include the following information:

- Alloy type, metallurgical condition, composition, and mechanical properties (including hardness)
- Specimen preparation, preferably including initial surface roughness measurement
- The number of specimens tested
- Tabulation of cumulative mass losses and cumulative exposure time for each specimen
- Plots of cumulative mean depth versus cumulative exposure time for each specimen

To facilitate comparisons between different materials, results are often expressed in terms of a single number based on the erosion resistance of the test specimen relative to that of a standard reference material. These types of ratings are discussed more fully in Ref 4.

The report should also include information on any deviations from the standard test procedure (for example, test liquid, temperature, or pressure, as discussed earlier) as well as any unusual occurrences or observations. In addition, each test should include at least one of the standard reference materials to facilitate calculation of the normalized erosion resistance of the test materials. Reference materials include aluminum alloy 1100-0, commercially pure annealed nickel (for example, Nickel 270), and type 316 stainless steel with a hardness of 150 to 175 HV. As previously mentioned, these materials are also used to calibrate the testing apparatus.

ASTM Standard G 73 provides guidelines for liquid drop impingement testing that, in addition to evaluating the resistance of metals, can be used to evaluate the degradation of optical properties of window materials and the destruction of coatings by the impinging liquid (Ref 4). The standard does not outline a single test method or apparatus, but instead gives guidelines for setting up tests and specifies test and analysis procedures and the reporting of data.

Test Specimens. Metallic test specimens can be chosen to present a curved (airfoil or cylindrical) or flat surface to the impinging liquid. They can be machined from a solid bar or can be cut from sheet. If specimens are machined, care should be taken to avoid work hardening of the surface. Surface finish should range from 0.4 to 2.6 μm (16 to 63 μin.) rms (root mean square). If another surface finish is employed, it should be noted in the test report.

Test Apparatus. ASTM Standard G 73 mainly addresses erosion test devices of the rotating disk type. In these tests, the specimen or specimens are attached to a rotating disk or arm, and their circular path passes through one or more liquid jets or sprays, resulting in discrete impacts between the specimen and the droplets or the cylindrical surface of the jets. Peripheral velocities (and therefore impact velocities) ranging from about 50 m/s (165 ft/s) to as high as 1000 m/s (3280 ft/s) may be obtained, depending on the particular apparatus used. Figure 2 illustrates two examples of rotating disk test apparatuses.

Droplet or jet diameters also vary from about 0.1 to 5 mm (0.004 to 0.2 in.). Droplets can be generated by spray nozzles, vibrating hollow needles, or rotating disks with water fed onto their surface. The typical droplet or jet diameter and the volume of liquid impacting the specimen per unit time should be determined within 10%. Jet diameter is usually assumed to be the same as the nozzle diameter. Other types of liquid impact erosion test devices are also briefly described in Ref 4.

Test procedures for various materials vary to some extent. The standard gives procedures for structural materials and coatings, including metals, structural plastics, composites, and metals with ceramic or metallic coatings; elastomeric coatings; window materials; and transparent thin-film coatings on window materials. Details on these procedures are outlined in the standard.

Calculation of Erosion Resistance. Because it is currently not possible to define "absolute" erosion resistance, most investigators use comparative evaluations of different materials to quantify relative erosion resistance. ASTM Standard G 73 details several approaches for calculating and presenting relative or normalized measures of erosion resistance for the various classes of materials considered. These include evaluation based on time to failure, total material loss, erosion rate-time patterns, incubation period and maximum erosion rate, and terminal erosion rate. Results for different materials can also be normalized by direct comparison to one of the designated reference materials (for metals, aluminum alloys 1100-0 or 6061-T6; annealed commercially pure nickel, such as Nickel 270; and type 316 stainless steel with a hardness between 155 and 170 HV) or by indirect comparison to a standardized reference scale. Details on these procedures, as well as on the calculation of an apparatus severity factor, are given in ASTM G 73.

Other Tests

Numerous other techniques are available for evaluating cavitation and erosion in the laboratory. Most of these are variations on the methods already described, employing rotating disks, vibrating devices, or venturi tubes to achieve the required fluid velocities (Ref 5). Another device used to produce cavitation and erosion is the liquid gun, which projects short, discrete slugs of liquid out of a nozzle onto the specimen (Ref 6). Details of other nonstandard test procedures are available in Ref 7 to 9.

Effects of Test Parameters (Ref 5)

Test conditions—including the temperature, pressure, and flow velocity of the test liquid and the frequency, amplitude, and distance between the specimen and the vibrating surface in a vibrating fluid—all influence the incubation and intensity of cavitation. The effects of varying these test parameters are described below.

Variations in velocity and pressure should not be considered separately, because the intensity of erosion cavitation is a function of both parameters. Generally, erosion rates increase with velocity. A common technique used to evaluate the effect of velocity is to determine the number and size of pits produced in a soft mate-

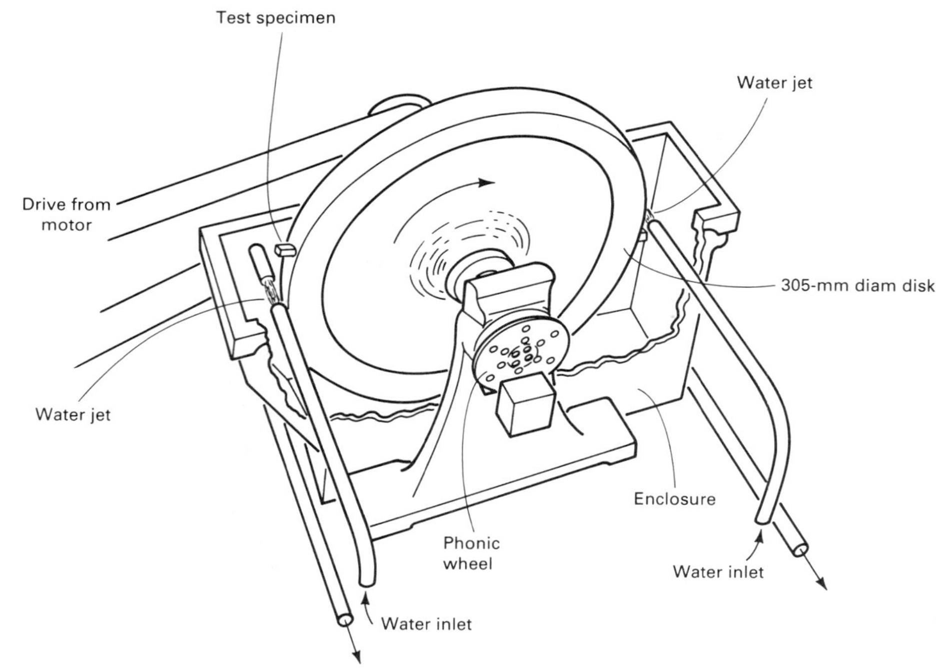

Fig. 2 Examples of rotating disk and rotating arm erosion/cavitation test apparatuses. (a) Small, relatively low-speed rotating disk and jet apparatus. (b) Large, high-speed rotating arm spray apparatus. Source: Ref 4

rial, such as aluminum, per unit time as velocity is varied (Ref 10). Although the number of pits observed increases over a wide range with velocity, the variation in pit size is relatively insignificant.

The influence of pressure has been investigated for both flow and vibratory cavitation (Ref 11, 12). In the former investigation, velocity was held constant and pressure was increased. Cavitation damage increased to a maximum and then decreased to zero at the pressure corresponding to cavitation inception (Ref 11). In vibratory testing, the erosion rate increased steadily with increasing pressure up to 4 atm, the maximum pressure obtainable in the device used (Ref 12). The authors noted, however, that at sufficient pressure the cavitation rate would in fact drop to zero.

Temperature. As mentioned earlier in this article, erosion rate increases with temperature to a maximum, then decreases near the boiling point of the test liquid. The decrease in erosion rate near the boiling point is attributed to the increase in vapor pressure of the test liquid (Ref 5). The effect of temperature on the erosion rate of plain carbon steel in three aqueous environments is shown in Fig. 3.

Amplitude and frequency in vibratory devices influence the degree of damage. In tests at constant frequency, erosion rate increased slightly with increasing amplitude (Ref 13). The incubation period for damage decreased with increasing amplitude.

Most vibratory devices operate at the the resonant frequency of a piezoelectric crystal. Therefore, frequency cannot be easily varied over a significant range. In one investigation, three different devices with frequencies of 10, 20, and 30 kHz were used to determine the effect of frequency on erosion rate (Ref 13). The stress produced by the cavitation increased as frequency decreased, but the effect was not as significant as that caused by changes in amplitude.

Distance Between Specimen and Vibrating Surface. The amount of erosion damage to a specimen rises to a maximum as the distance between the specimen and the vibrating surface decreases. At a frequency of 20 kHz, the position that results in maximum erosion rates is approximately 0.5 mm (0.02 in.) from the vibrating surface (Ref 5). The position of the maximum is independent of amplitude, but no data are available on the influence of frequency on this position.

The influence of fluid properties has been studied by several investigators. Unfortunately, most of them made little attempt to identify those properties of the fluid that affect erosion rates. One investigator, however, considered fluid viscosity, vapor pressure, surface tension, and specific gravity for water, gasoline, and the sodium-potassium eutectic (Ref 14). The temperature dependence of erosion rates was attributed to the effects of velocity and surface tension at low temperatures and to the increase in vapor pressure at high temperatures. By comparing the temperature dependence of properties of the different fluids, he was able to predict the variation in erosion damage in the three fluids. The resulting equation, however, cannot be used to predict erosion rates in other fluids.

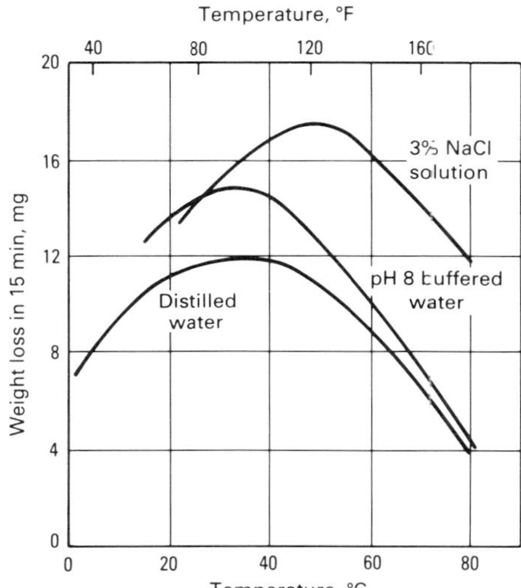

Fig. 3 Influence of temperature on the erosion rate of plain carbon steel in a vibratory cavitation device. Source: Ref 5

Data Correlations

Correlations Among Test Methods (Ref 2). Investigations have indicated that there are considerable differences in the erosion intensities produced by different test techniques (that is, high-velocity flow, vibratory, and impingement) and test parameters (vibratory frequency; specimen size and shape). There is, however, general agreement on the ranking of test materials. When the erosion resistance of a material is expressed as a normalized erosion resistance (NER), which is the ratio of the erosion rate of the test material to that of a standard material, results from different test methods and different laboratories can be directly compared and are reasonably consistent (Ref 15). Calculation of NER values and statistical analysis of erosion rate data are discussed in detail in Ref 4, which also reports on the results of interlaboratory testing using a liquid impingement testing method at three specified velocities.

Correlations Between Laboratory Results and Service (Ref 2). The erosion rates of materials that can be anticipated in service cannot be quantified at this time; however, methods have been developed to identify service erosion intensities under conditions of controlled operation. These include aluminum panels attached to hydraulic turbine vanes (Ref 16), a strain gage technique for measuring the relative intensities of laboratory and service devices (Ref 17), and the use of a radioactive paint to correlate the erosion intensity of an operating turbine with an erosion rate index (Ref 18). This latter investigation led to the establishment of relative resistance scales, in which laboratory exposures of 1 to 2 h in a vibratory apparatus can be compared with 16- to 100-h exposures in a venturi apparatus and with months of operation in full-scale turbines.

REFERENCES

1. M.G. Fontana, *Corrosion Engineering*, 3rd ed., McGraw-Hill, 1986
2. J.Z. Lichtman, D.H. Kallas, and A. Rufolo, Evaluating Erosion (Cavitation) Damage, in *Handbook of Corrosion Testing and Evaluation*, W.H. Ailor, Ed., John Wiley & Sons, 1971, p 453-472
3. "Standard Method of Vibratory Cavitation Erosion Test," G 32, *Annual Book of ASTM Standards*, American Society for Testing and Materials
4. "Standard Practice for Liquid Impingement Erosion Testing," G 73, *Annual Book of ASTM Standards*, American Society for Testing and Materials
5. C.M. Preece, Ed., *Erosion*, Vol 16, *Treatise on Materials Science and Engineering*, Academic Press, 1979
6. M.C. Rochester and J.H. Brunton, Influence of Physical Properties of the Liquid on the Erosion of Solids, in *Erosion, Wear, and Interfaces With Corrosion*, STP 567, American Society for Testing and Materials, 1974, p 128-147
7. M. Matsumura, Y. Oka, S. Okumoto, and H. Furuya, Jet-in-Slit Test for Studying Erosion-Corrosion, in *Laboratory Corrosion Tests and Standards*, STP 866, G.S. Haynes and R. Baboian, Ed., American Society for Testing and Materials, 1985, p 358-372
8. J.M. Hobbs, Report 69, National Engineering Laboratory, 1962
9. J.W. Holt and G.M. Wood, Ed., *Cavitation Research Facilities and Techniques*, American Society of Mechanical Engineers, 1964
10. R.T. Knapp, J.W. Daly, and F.G. Hammitt, *Cavitation*, McGraw-Hill, 1970
11. M.J. Robinson and F.G. Hammitt, *Trans. ASME D*, Vol 89, 1967, p 161
12. F.G. Hammitt and D.O. Rogers, *J. Mech. Eng. Sci.*, Vol 12, 1970, p 432
13. B. Vyas and C.M. Preece, *J. Appl. Phys.*, Vol 47, 1976, p 5133
14. W.C. Leith, *Proc. ASTM*, Vol 65, 1965, p 789
15. F.J. Heymann, "Toward Quantitative Prediction of Erosion Damage," Report E-1463, Westinghouse Electric Corporation, Development Engineering Department, June 1969
16. R.T. Knapp, "Accelerated Field Tests of Cavitation Intensities," Paper 56-A-57, American Society of Mechanical Engineers, 1956
17. R. Canavelis, The Investigation of Cavitation Damage by Means of Resistance Gages, in *Cavitation Research Facilities and Techniques*, J.W. Holt and G.M. Wood, Ed., American Society of Mechanical Engineers, 1964
18. S.L. Kerr and K. Rosenberg, An Index of Cavitation Erosion by Means of Radioisotopes, *Trans. ASME*, Vol 80 (No. 6), Aug 1958, p 1308-1314

Evaluation of Microbiological Corrosion

John G. Stoecker II, Monsanto Company

THE MANY CASE HISTORIES reported in the literature point out that microbiological corrosion can occur in any aqueous environment. An aqueous environment is defined as any environment that contains even a trace amount of water. For example, the small amount of water in aviation fuel qualifies that environment as aqueous. Therefore, microbiological corrosion is a potential problem. Because microorganisms are present in virtually all aqueous systems, they have the potential to influence corrosion in any aqueous system anywhere in the world.

An evaluation of the possibility for microbiological corrosion must be considered a part of good engineering practice when planning a new facility or process. This consideration must include water used for such processes as hydrostatic testing and water batching even though the final process environment may not contain water even in trace amounts. The water used by fabricators to hydrostatically test new equipment, along with the mud and soil in the construction area, must also be considered sources of microbes and associated corrosion problems. Microbiological corrosion should be avoided from all of these various sources, because the damage initiated by the microbes may continue as crevice or under-deposit attack after the organisms are killed in a nonaqueous process. There are currently no standard tests for determining whether or not the microbes present at a given site will be involved in initiating or accelerating corrosion. Nevertheless, several factors should be considered in identifying the potential for biological corrosion, as discussed below.

Use of Historical Data

The history of biological corrosion problems with the intended water source at the proposed site can provide a great deal of information regarding the potential for problems. An investigation at neighboring facilities can also yield valuable information, particularly when planning a new installation on an undeveloped site. Because many cases of microbiological corrosion have been misidentified as crevice or under-deposit corrosion, the investigation should center on the history of these problems. The investigation should include problems associated with microbes in the water and in the ground.

During the investigation, it is helpful to have several case histories with photographs of microbiologically influenced corrosion on hand. Some information of this type is available in this Volume in the articles "General Corrosion" (see the section "General Biological Corrosion") and "Localized Corrosion" (see the section "Localized Biological Corrosion") and in the Appendix to the article "Effects of Environmental Variables on Aqueous Corrosion." These articles also contain information on the mechanisms of microbiological corrosion and the environmental conditions under which it occurs. The section "Localized Biological Corrosion" mentioned above presents a table listing the industries known to be affected by microbiological corrosion and another table showing the microorganisms involved. Considerably more information is available in Ref 1 and in the references listed in the articles mentioned above.

The investigation of a water source can include many users if the source is a large municipal system. The list of users to be investigated can be reduced to a manageable number by selecting those that use the water in a manner similar to the intended use. Similar, in this context, means at the same temperature and pH, because the various microbes are generally active and troublesome only between specific temperatures and ranges of pH. Microbes seem to be the most troublesome at temperatures equivalent to the temperature of the human body and at pH 7, but there are cases in which they have created problems at extremes of both temperature and pH.

It should be noted that a filter-treated and/or chlorinated water source does not eliminate the potential for microbiologically influenced corrosion. Chlorination is not totally effective in eliminating all microbes from water, and it has been shown that several species use chlorides for metabolic purposes after the free chlorine has been reacted. In addition, microbiological activity is a persistent problem in ultrapure deionized water in which extremely fine filters are used; this proves that filtration will not remove all microbes or eliminate the problems they may cause. Additional information on the control or elimination of microbes in waters can be found in the article "Control of Environmental Variables in Water Recirculating Systems" in this Volume.

Testing for Microbiological Activity

The water source and soil at the proposed site should be tested for microbiological activity. These tests should be conducted under the direction of a microbiologist who has experience in microbiologically influenced corrosion. Field test kits are currently available that employ an antibody tagging technique to identify and facilitate the determination of the population of troublesome microbes, as described in Ref 2 and 3.

Chemical Analysis

Chemical analysis of the water source and the soil at the site is helpful in determining whether the critical nutrients and energy sources necessary to support microbiological activity are present. For example, the presence or absence in the environment of sulfur and its compounds to support the growth of sulfate-reducing or sulfur-oxidizing bacteria can be determined by chemical analysis. Chemical analysis should be used with caution, however, because the availability of nutrients is not a totally controlling factor in microbiological activity. Microbes are adaptable and can use both organic and inorganic sources for energy (Ref 4). Moreover, growth of a harmless bacterial species can sometimes result in production of a nutrient, not present in the ambient environment, that will allow the damaging species to fluorish at a later time. In addition, microbiological activity is a major problem on filters in ultrapure deionized water, in which nutrients would seem to be rather scarce.

Corrosion Testing

The preceding discussion has dealt with determining the potential for microbiological corrosion problems at a new and undeveloped site where utility services are not available. At a developed site where energy sources such as electricity and steam are available, corrosion testing can be a valuable tool for assessing the potential for future problems or analyzing a present corrosion problem to determine if biological activity is involved (Ref 5, 6). Utilities are usually required to design tests so that the environmental conditions of the intended process can be reproduced as closely as possible.

There are no standard tests designed specifically for investigating susceptibility to microbiological corrosion. The presence of microorganisms in an environment, however, does not introduce some new type of corrosion, but rather influences the types of corrosion that are already known. Therefore, if the types of corrosion that a given material of construction is susceptible to are known, the standard test methods for those types (specified elsewhere in this Volume) can

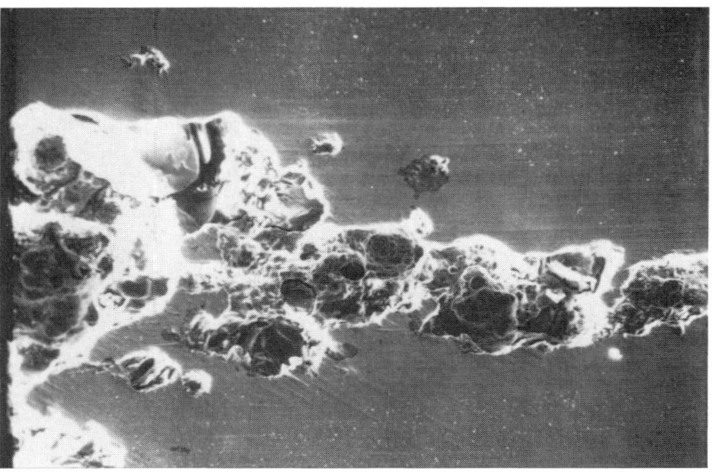

(a)

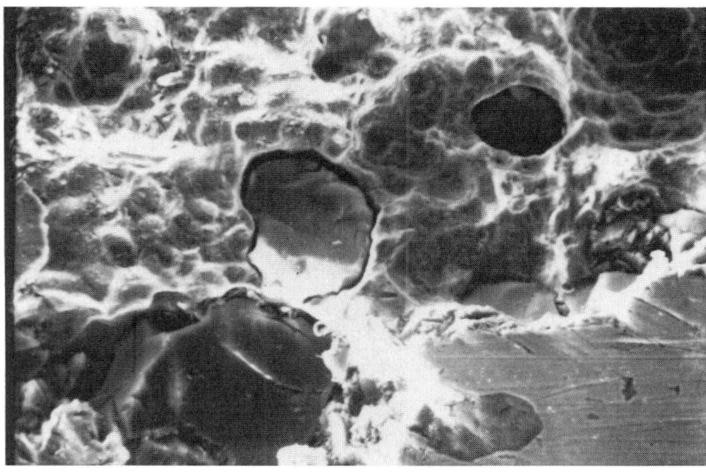

(b)

Fig. 1 Intergranular corrosion in a type 304 stainless steel associated with a typical pit and microbiological deposit. (a) View at 100×. (b) Higher-magnification view at 400×. Source: Ref 7

often be helpful—if supplemented with the proper biological techniques for the organisms involved.

As with most corrosion testing, the environmental conditions in the test vessels must match those in the proposed process in order to obtain valid results. Coupon and test probe procedures (described elsewhere in this Volume) should not be used to determine and/or confirm the potential for microbiological corrosion without the benefits of microbiological analysis. Laboratory testing is not recommended, because a population of mixed species is extremely difficult to maintain in this type of setting.

The corrosion tests will yield some very unusual results if the damage was microbiologically influenced. The damage will almost always be localized in the form of a pit (Fig. 1), whether under a deposit, tubercule, or scale; in a crevice; or nearly surrounded by a stain on the metal surface. In addition, there is usually a peculiar color associated with the damage.

For example, the deposits, tubercules, and stains associated with the damage on carbon and stainless steel will usually be reddish brown. The pit in a copper alloy under scale will usually be vivid green. Black spots of iron sulfide will usually be present in association with the damage where sulfate-reducing bacteria are involved. A few drops of hydrochloric acid on these black spots will produce the pungent odor of hydrogen sulfide as a test for sulfur in this case.

Chemical analysis of the residue associated with the damage will usually show a depletion of iron and nickel and an increase in the concentration of chromium and copper as compared to the normal levels in the base material. Usual levels of chlorine, sulfur, manganese and other tramp elements may also be present in the residue.

Risk Analysis

A risk analysis should be performed based on the information collected according to the preceding outline. The information gathered from the case histories of the water and neighboring sites, along with the corrosion test results, should receive the heaviest weighting in this analysis. Much of the work that has been done to assess the risk of microbiological corrosion in soils and in North Sea oil pipeline environments is summarized in the article "Localized Corrosion" in this Volume.

Actual experiences are better suited to uncovering problems caused by dormant spores that have become active than are field microbiological and chemical tests. For example, spores that have been dormant for centuries have been found in the permafrost of Antarctica. These spores need only the proper environment to become active, as in the case with oil facilities in the North Sea and Arctic Ocean.

With this in mind, a safe, conservative approach during the risk analysis is a good policy. In this context, a safe, conservative approach involves the pessimistic view that microbiological corrosion will be a problem when definite facts and data are not available. Actual methods for reducing the potential for microbiological corrosion are covered in the articles "Localized Corrosion" and "Control of Environmental Variables in Water Recirculating Systems" in this Volume.

Summary

Several factors that should be considered in identifying the potential for biological corrosion have been presented. It should be noted that this is a relatively new and unexplored corrosion phenomenon for industry. Because it is a new field of study, the knowledge concerning it is expanding very rapidly; this makes the numbered references in this article extremely important as sources of detailed, current information.

REFERENCES

1. J.G. Stoecker and D.H. Pope, Microbiologically Influenced Corrosion, in *Process Industries Corrosion—The Theory and Practice*, National Association of Corrosion Engineers, 1986, p 227-241
2. D.H. Pope, Discussion of Methods for the Detection of Microorganisms Involved in Microbiologically Influenced Corrosion, in *Biologically Induced Corrosion—Proceedings of the International Conference*, National Association of Corrosion Engineers, 1986, p 275
3. D.H. Pope, "The Work of MTI-TAC Committee 36 on Microbiologically Influenced Corrosion Test Kits," Materials Technology Institute of the Chemical Process Industries, to be published
4. J.G. Stoecker, Guide for the Investigation of Microbiologically Induced Corrosion, *Mater. Perform.*, Vol 23 (No. 8), Aug 1984, p 48
5. R.E. Tatnall, Fundamentals of Bacteria Induced Corrosion, *Mater. Perform.*, Vol 20 (No. 9), Sept 1981, p 32
6. R.E. Tatnall, Experimental Methods in Biocorrosion, in *Biologically Induced Corrosion—Proceedings of the International Conference*, National Association of Corrosion Engineers, 1986, p 246
7. J.G. Stoecker and D.H. Pope, Study of Biological Corrosion in High Temperature Demineralized Water, *Mater. Perform.*, Vol 25 (No. 6), June 1986, p 51-56

Interpretation and Use of Corrosion Test Results

Donald O. Sprowls, Consultant

IMPORTANT ASPECTS of any corrosion investigation are the credibility of the test results and user acceptance of the test methodology. When commercially competitive materials are evaluated or when corrosion tests are used for material acceptance, the testing techniques come under close scrutiny and quick criticism in the marketplace. Public safety and producer liability are potent considerations in decisions involving material selection and the determination of safe environments.

After the appropriate test (or tests) is completed, the data must be analyzed and presented in a manner that facilitates proper decision making. To be useful, an expression of the test results must have engineering significance, must be mechanistically sound, and must be comprehensible (Ref 1). Finally, there is the question of the amount of confidence that can be placed in the decision.

Credibility of Test Results

The question of the reliability of corrosion test results is discussed in the first article in this Section, "Planning and Preparation of Corrosion Testing." The following 13 articles are devoted to methods of generating reliable data. This article will consider the cost effectiveness of corrosion testing. One of the conclusions reached in Ref 2 is that perhaps the greatest problem is the misapplication of reliable laboratory corrosion data. An accelerated test can reliably indicate only how a given material will behave under the actual conditions of testing. Confidence should not be placed in accelerated corrosion test data when they are applied to other conditions, nor should a one-to-one correlation be expected. A significant degree of confidence is justified only after an accumulation of sufficient experience in correlating test data with actual performance behavior under known conditions (Ref 2).

Data that can be directly used in design or for control of operations will be obtained only in certain situations in which a corrosion test is required to provide an answer to a specific corrosion problem in a specific environment. Most often, the results must be interpreted in terms of comparisons or rankings. A criticism of such interpretations is that they are subjective; that is, they depend too much on the judgment of the investigator. Therefore, it is advisable when planning a corrosion-testing program to select techniques carefully in order to provide ranking parameters that are subject to minimal influence from testing conditions and that have engineering significance. This is not possible in all cases. It is particularly difficult in tests for stress-corrosion cracking (SCC) and corrosion fatigue cracking because there are so many confusing variables in these tests.

Interpretation of corrosion test results is also complicated by the time factor and by the consequences in the event of a corrosion failure. In addition, there are more difficult questions than ever before due to the increased use of toxic materials—questions such as:

- Can service life be predicted, and with what level of confidence?
- Will serious corrosion occur? In what form? Is there an incubation period?
- What is the corrosion rate, and will it change with time?
- What are the consequences of corrosion failure in terms of public safety and inconvenience, loss of life and property damage, and liability and cost?
- How can an acceptable degree of risk be defined in the selection of new, more corrosion-resistant materials?
- Is supplemental corrosion protection needed? If so, what type is best, and how long will it be effective?

Therefore, credible interpretation in many cases of material and environmental evaluation is ultimately a shared responsibility of the corrosion engineer, the material engineer/designer, and possibly a statistician. It should be recognized in the planning stage that the corrosion engineer needs input from the equipment designer to help identify the corrosion system correctly. A corrosion system consists of one or more metals and all parts of the environment that influence corrosion during the expected lifetime of a structure.

User Acceptance

Reference 1 contains a discussion of the user acceptance problem, with emphasis on the rank-ordering concept for interpreting material and environmental evaluation tests. A more critical approach to rank ordering is one way of improving user acceptance of laboratory test results. A preferred ranking parameter would be a dominant factor in the corrosion test and would be recognizable to the user/engineer as a type of data that would be useful.

Development of an appropriate corrosion-testing plan requires considerable research and engineering effort. Research aspects involve defining the limits of the corrosion system, identifying the dominant corrosion factors, and developing/selecting an appropriate corrosion test procedure. The engineering aspect involves collecting service experience data and correlating it with laboratory corrosion test data to permit establishment of acceptability criteria. An example of how this was done in the case of sulfide stress cracking is shown in Table 1. In this example, critical stress, S_c, for 50% probability of failure in a standard laboratory test was the rank-ordering parameter for materials considered for use in sour gas environments. Accumulated field experience with various materials was used to set an S_c acceptability limit. The validity of this limit was confirmed by showing that specimens cut from actual field service failures had S_c values below the acceptability criterion.

The lack of an acceptability limit for the rank-ordering parameter can delay engineering acceptance of a proposed standard corrosion test. A cooperative industry effort is usually required to obtain the necessary field experience information.

Statistical Planning and Analysis

Many investigators have traditionally avoided the use of statistical techniques because the added reliability did not seem to offset the effort and time required to become familiar with the methodology or to perform the calculations. Some individuals within larger organizations have access to company statisticians and to central computers, but because of lack of communication,

Table 1 Correlation of service sulfide stress cracking experience with typical laboratory test data that allowed establishment of the acceptability criterion of $S_c > 10$

Type of steel	Field experience failure frequency	Laboratory tests S_c (typical)(a)
API grade J-55	nil	11–17
API grade C-75	nil	10–17
API grade N-80	Moderate	5–14
API grades P-105, P-110	High	3–10

(a) S_c is defined as the "critical stress" (ksi × 0.1) calculated from the deflection of a beam specimen (three-point loading), which corresponds to a 50% probability of failure in a standardized hydrogen sulfide solution as calculated from results obtained at varying beam deflections. Source: Ref 3 these services have often not been used to advantage.

For example, the corrosion engineer performs the experiment and then turns over the data to a statistician for analysis. The statistician either does the best job that can be done with the available data or informs the engineer that the experiment was not properly designed. In the latter case, the engineer either decides to learn about statistical design or no longer seeks the services of a statistician.

Fortunately, personal computers have replaced the engineers' slide rule. The software that is available includes comprehensive statistical analysis programs. From these, methods can be selected to suit particular needs. Therefore, data can now be statistically analyzed almost as fast as the numbers can be punched on a keyboard. The engineer need only become familiar with statistical methods of experimental design and data analysis in order to take advantage of these expanded capabilities. With these techniques, the engineer can maximize both the amount of information obtained and the reliability of the results for any set amount of expended effort.

Reference 4 contains an introduction to the techniques of experiment design and data analysis. The specific application of statistics to corrosion problems often involves simplification and the use of a limited number of methods. Specification ASTM G 16 provides a set of sample procedures that is currently used in the statistical analysis of corrosion experiments (Ref 5). The examples included are intended to provide a method for planning corrosion experiments, analyzing data obtained, and establishing the degree of confidence that can be placed in the results of specific experimental or field applications data. Alternative methods or improved approaches are constantly being developed that may provide more complete analysis and understanding of specific experiments. More detailed information is given in Ref 6 and in the Section "Statistics and Data Analysis" in Volume 8 of the 9th Edition of *Metals Handbook*.

Computers and Data Utilization

With the advent of digital circuitry and the widespread use of microcomputers in the laboratory, new and extremely powerful techniques are now available for data acquisition, numerical processing, data management, and data communication. These techniques can provide higher resolution and can be less labor intensive, faster, and more interactive than conventional data recording (Ref 7, 8).

Corrosion will continue to be an important subject because it involves processes that cause serious deterioration of engineering structures. However, in the words of a British corrosion worker, "Existing technical data are not being properly used. The problem centers on how to translate the complexities of the corrosion processes into tabular form and simple equations that can be handled by artificial intelligence techniques. Since computer-aided systems have been developed for machine designs and chemical process flow sheets, it surely follows that this will be a major effort in the next few years. It will be necessary to look away from the laboratory bench and re-examine what the present needs of the user engineer community are" (Ref 9).

REFERENCES

1. R.S. Treseder, An Engineering View of Laboratory Corrosion Tests, in *Laboratory Corrosion Tests and Standards*, STP 866, G.S. Haynes and R. Baboian, Ed., American Society for Testing and Materials, 1985, p 5-13
2. A. deS. Brasunas and C.B. Sonnino, Corrosion Data Reliability, in *Handbook on Corrosion Testing and Evaluation*, W.H. Ailor, Ed., John Wiley & Sons, 1971, p 112
3. R.S. Treseder and T.M. Swanson, *Corrosion*, Vol 24 (No. 2), Feb 1968, p 31
4. F.H. Haynie, Statistical Planning and Analysis, in *Handbook on Corrosion Testing and Evaluation*, W.H. Ailor, Ed., John Wiley & Sons, 1971, p 31-38
5. "Standard Practice for Applying Statistics to Analysis of Corrosion Data," G 16, *Annual Book of ASTM Standards*, American Society for Testing and Materials
6. C.E. Makepeace, Statistical Design of Experiments, in *Handbook on Corrosion Testing and Evaluation*, W.H. Ailor, Ed., John Wiley & Sons, 1971, p 59-85
7. D.G. Tipton, Microcomputer Data Acquisition for Corrosion Research, in *Laboratory Corrosion Tests and Standards*, STP 866, G.S. Haynes and R. Baboian, Ed., American Society for Testing and Materials, 1985, p 24-35
8. M.W. Kendig, U. Berocci, and J.E. Strutt, Ed., *Proceedings of the Symposium on Computer Aided Acquisition and Analysis of Corrosion Data*, Vol 85-3, The Electrochemical Society, 1985
9. C. Edeleanu, Corrosion Information: The Engineer's Need, *Br. Corros. J.*, Vol 20 (No. 3), 1985, p 101-103

Designing To Minimize Corrosion

Section Chairman:
James Q. Lackey, E.I. Du Pont de Nemours & Company, Inc.

Materials Selection*

Gregory Kobrin, E.I. Du Pont de Nemours & Company, Inc.

FACTORS THAT INFLUENCE the service life of equipment, for example, a chemical process heat exchanger or tank, a bridge, or an automobile chassis, from a corrosion standpoint include the following (Ref 1):

- Design
- Materials of construction
- Specification
- Fabrication and quality control
- Operation
- Maintenance
- Environmental conditions

In weighing the relative importance of these factors, design and materials are of major and equal importance in achieving the desired performance and life. The designer and materials engineer must work closely together to ensure that premature failure will not occur because of design defects or improper material selection. It is also inefficient and costly to use these elements of design and material selection in a compensatory manner.

This article will discuss the step-by-step process by which materials are selected in order to avoid or minimize corrosion and will include information on materials that are resistant to the various forms of corrosion. Subsequent articles in this Section will discuss design details that can be applied to minimize corrosion, the application of fracture mechanics to equipment design, corrosion of weldments, and the economics of corrosion.

The Materials Selection Process

Review of Operating Conditions. The first step in the materials selection process is a thorough review of the corrosive environment and equipment operating conditions. This review requires input from knowledgeable process engineers. Precise definition of the chemical environment, including the presence of trace compounds, is vital. For example, the nickel-molybdenum alloy Hastelloy B-2 (UNS N10665) is highly resistant to hydrochloric acid (HCl) up to the atmospheric boiling point. However, the presence of small quantities of oxidizing metal ions, such as ferric ion (Fe^{3+}), will result in severe corrosion. Other operating conditions that require definition, especially for equipment used in the chemical-processing industry, include temperatures, pressures, flow rates, liquid versus gaseous phases, aqueous versus anhydrous phases, continuous versus intermittent operation, media used for cooling or heating, external versus internal environment, and product purity.

Abnormal or upset conditions are often overlooked during the selection process. For example, plain carbon steel may be the optimum choice for vessels and piping that must contain noncorrosive hydrocarbon gases, such as ethylene, under pressure at normal temperatures. However, the cooling effect that occurs during venting to the atmosphere, for whatever reason, may lower the temperature of vessels, piping, and relief valves to below the ductile-to-brittle transition point and result in a catastrophic brittle fracture. Thus, the selection of special steels, qualified by impact testing at the lowest expected temperature, would be appropriate.

Review of Design. Next, the type and design of the equipment and its various components should be considered, along with size, complexity, and criticality in service. Selecting a material for a simple storage tank generally does not require the same attention and effort as choosing the material of construction for a highly sophisticated chemical process reactor. This is especially true when considering critical, unique pieces of equipment in large, single-train, continuous process plants in which a failure would shut down the entire operation. In this case, great effort is expended to select the optimum material for safe low-maintenance service.

The materials used to join the components into an assembly will require as much attention as the component materials themselves. Many bolted agitator assemblies in reactors, as well as riveted wheels in centrifugal compressors, have failed catastrophically because the bolts or rivets did not have adequate strength or corrosion resistance.

When welding is the joining method, the materials engineer is challenged to ensure that the welds are as corrosion resistant as the base metals. Generally, the weld metal must equal the base metal in chemical composition and must be virtually free of surface defects, such as porosity, slag inclusions, incomplete penetration or lack of fusion, for long maintenance-free service. The challenge is even greater when dissimilar-metal welds are required. Improper selection may allow local attack due to weld metal dilution or may allow hydrogen-assisted cracking due to hard heat-affected zones (HAZs). More information on preventing corrosion of welds is available in the article "Corrosion of Weldments" in this Volume.

Selection of Candidate Materials. Once the chemical environment, operating conditions, and type and design of the equipment have been defined, consideration of materials of construction is in order. Occasionally, the selection is based on reliable, pertinent past experience and, as such, is well defined. More often, however, selection is anything but straightforward for a number of reasons, such as complex chemical environments and stringent code requirements.

The list of materials to choose from is large and continues to increase. Ferrous and nonferrous metals and alloys, thermoplastics, reinforced thermosetting plastics (RTP), nonmetallic linings, glass, carbon and graphite, and catalyzed resin coatings are among the various materials available. Many materials will be immediately excluded because of service conditions, that is, pressures too high for RTP, temperatures too high for nonmetallic linings and coatings such as rubber or epoxy resins, environment too aggressive for carbon steel, and so on. Remaining choices may still be great in number.

It is always desirable to minimize the list of materials; this allows in-depth evaluation. In other cases, the initial list may be exceptionally small because of limited knowledge about the operating conditions or the complex chemical environment. A search of data sources should follow in either case.

Literature Survey. One might begin with a literature survey, using applicable sources such as *Corrosion Abstracts* (Ref 2), along with the *Corrosion Data Surveys* (Ref 3) published by the National Association of Corrosion Engineers (NACE). This technical society, in conjunction with the National Bureau of Standards, has initiated an on-going program to computerize reference data on the performance of materials in corrosive environments.

Other data sources include a wide variety of handbooks, conference proceedings, and literature compilations published by the American Society for Testing and Materials (ASTM), ASM INTERNATIONAL, and NACE. In addition, expert systems, which are computer programs containing methods and information developed by experts, are the latest tools available for materials selection (Ref 4-7). Such systems are designed to solve problems, make predictions, suggest possible treatments, and offer materials and corrosion advice with a degree of accuracy equaling that of their human counterparts.

Experience and data generated in-house often serve as the most reliable bases for materials selection. Ideally, this is coupled with outside experience, when available, from materials vendors and equipment fabricators to complete this

*The section "Hydrogen Damage" was written by George E. Kerns, E.I. Du Pont de Nemours & Company, Inc.; the section "Erosion-Corrosion" was written by J.Q. Lackey, E.I. Du Pont de Nemours & Company, Inc.

initial screening process. Contacts with clients referred by vendors should not be overlooked for added experience.

At this point, the list of candidate materials should be narrowed to a reasonable number for in-depth evaluation. Final selections should not be based solely on the above data sources, because in most cases the data provided are insufficient for the complete characterization of an environment or a set of conditions.

Evaluation of Materials. The in-depth evaluation of each candidate material should begin with a thorough understanding of the product forms available, along with the ease of fabrication by standard methods. For example, it would be wasteful of time and money to evaluate an Fe-14.5Si alloy for anything but a cast component such as a pump casing or valve body. The alloy is unavailable in any other form. Because of its poor weldability, this alloy should also be ruled out for applications involving welding.

Corrosion testing in representative environments is generally the next step. The extent of the investigation (and determination of test conditions) depends on such factors as (Ref 8):

- Degree of uncertainty after available information has been considered
- The consequences of making a less-than-optimum selection
- The time available for evaluation

Laboratory testing of candidate materials is common and in some cases may be the only means available for final determination. Wherever possible, the actual process fluids should be used. Otherwise, mixtures simulating the actual environment must be selected. There is considerable risk in using the latter, because undefined constituents can have a significant effect on the performance of a particular material.

Depending on the application, weighed and measured coupons of candidate materials are exposed to the corrosive fluids under a variety of conditions ranging from simple static immersion at a controlled temperature to complex testing under combined heat transfer and velocity conditions. Guidance for conducting laboratory corrosion tests is available in Ref 9 and 10. After exposure for a specified length of time (generally a minimum of 1 week), the coupons are, in the case of metals and alloys, cleaned and reweighed, and a corrosion rate is calculated based on weight loss and exposed surface area. The rate is commonly expressed in millimeters of penetration per year or inches or mils (1 mil = 0.001 in.) of penetration per year.

In addition, coupons are examined under a microscope for evidence of local attack, such as pitting, crevice corrosion, and exfoliation. Special coupons, such as galvanic couples, welded, and stressed coupons, are often exposed to determine if other forms of corrosion may occur on certain metals and alloys. These coupons may require metallographic examination for evidence of dealloying (parting), stress-corrosion cracking (SCC), intergranular corrosion, and other corrosion phenomena.

Nometallic materials, such as thermoplastics, coatings, reinforced thermosetting resins, elastomers, and ceramics, are also evaluated in laboratory tests, but the criteria are different from those used for metals. First, exposure time must generally be longer (often a minimum of 1 to 3 months) before significant changes occur. Exposure times of 6 months to 1 year are common. Also, corrosion rate calculations based on weight loss and surface area are not applicable in most cases. Of more importance are changes in weight, volume, hardness, strength, and appearance before and after exposure. Corrosion testing and evaluation are covered in detail in the Section "Corrosion Testing and Evaluation" in this Volume.

If possible, candidate materials should be tested under conditions more like the final application rather than in laboratory glassware, that is, in a semi-works or pilot operation or in full-scale equipment. Generally, the results are more reliable because test coupons are integrated into the process and are exposed to the same conditions as the actual equipment. Because of nonuniform conditions (flows, composition) within process equipment, coupon locations should be carefully selected (see the article "In-Service Monitoring" in this Volume).

Reliability is further enhanced when it is possible to test full-size components fabricated from candidate materials (Ref 11). Examples include:

- Flanged sections of selected alloys and/or nonmetallics installed in a pipeline
- Experimental alloy impellers in pumps for corrosion and cavitation studies
- Tubing installed in a full-size operating or miniature test heat exchanger to evaluate materials with optimum resistance to corrosion under heat transfer conditions
- Paddles of candidate materials bolted to a reactor agitator for erosion-corrosion studies

The primary disadvantages of this method of testing are the cost of fabrication, installation, removal, and evaluation; the downtime resulting from equipment being taken out of service and dismantled for evaluation; and the fact that a test component could fail prematurely and cause a unit shutdown and/or equipment damage.

Specifications. At this point, all candidate materials have been thoroughly evaluated (along with the economics, to be discussed later), and the materials of construction have been selected for the particular application. Clear and concise specifications must now be prepared to ensure that the material is obtained as ordered and that it meets all the requirements of the application. Perhaps the best known and most widely used specifications are the standards of ASTM. Thousands of specifications for virtually all metal and nonmetal materials of construction are covered in 15 sections encompassing 65 volumes.

Similar standards in countries other than the United States include DIN (Germany), BS (Great Britain), AFNOR (France), UNI (Italy), NBN (Belgium), and JIS (Japan). Other materials specifications that are well known but are more limited in application are those of the Society of Automotive Engineers (SAE) and its Aerospace Materials Specifications (AMS), the American Welding Society (AWS), the American Petroleum Institute (API), and the American National Standards Institute (ANSI).

Fabrication requirements must also be spelled out in detail to avoid mistakes that could shorten the life of the equipment and to satisfy the requirements of state and federal regulatory agencies and insurance companies. The American Society for Mechanical Engineers (ASME) code governs the fabrication of equipment for the chemical, power, and nuclear industries, and the API code governs the fabrication of equipment for the refining industry. Piping for these industries is generally fabricated per applicable ANSI codes. In these codes, allowable stresses for design calculations have been determined for virtually all metals and alloys that might be selected for corrosive (and noncorrosive) service. Where welding is the primary joining method, welding procedures and welders must be qualified before fabrication begins. Testing and quality assurance requirements, such as radiography, hydrostatic testing, and ultrasonic inspection, are also covered in the codes and are specified where applicable to ensure compliance.

The fabricator is generally required to provide detailed drawings that list dimensions, tolerances, all pertinent materials specifications, fabrication and welding details, and testing and quality assurance requirements for review. Prefabrication meetings are held for final review of all drawings and details so that customer and vendor are in agreement. Thus, problems or errors that could lead to costly delays in fabrication or failures in service can be detected early and corrected.

Money spent on inspection and monitoring during equipment fabrication/erection to ensure compliance with specifications is repaid by trouble-free startup and operation of the fabricated assembly. In some cases, every component of an assembly must be tested to avoid excessive corrosion and/or premature failure. For example, an additional quality check of a vessel fabricated from AISI type 316L stainless steel for hot acetic acid service might be to test every plate, flange, nozzle, weld, and so on, for the presence of molybdenum by using a chemical spot test method (Ref 12). The absence of molybdenum, which might indicate the mistaken use of a different stainless steel, such as type 304L, would result in accelerated corrosion in this service. Another example is the testing of every component (including weld metal) of a heat exchanger fabricated from chromium-molybdenum steels for hot high-pressure hydrogen service to avoid the possibility of catastrophic failure by hydrogen attack. The use of portable x-ray fluorescence analyzers for this type of quality assurance testing of critical service components has become quite popular in recent years (Ref 13).

Follow-Up Monitoring. Once built, installed, and commissioned in service, the equipment, piping, reactor, heat exchanger, and so on, should be monitored by the materials engineer to confirm the selection of materials of construction and all other requirements for the intended application. Frequent shutdowns for thorough inspections (visually and with the aid of applicable nondestructive examination methods) and periodic evaluation of corrosion coupons exposed at key locations in the equipment represent both the ideal and most difficult monitoring techniques to achieve. In actual practice, equipment is generally kept on-stream continuously for long periods of time between shutdowns, so on-stream monitoring techniques must also be used.

In the petroleum industry, the internal corrosion in oil and gas production operations is often monitored with hydrogen probes (Ref 14). These instruments measure hydrogen created by corrosion reactions. A portion of the hydrogen penetrates the vessel or pipeline wall, and the rest of the hydrogen is dissolved in the process fluid or released as gas bubbles. Hydrogen probes measure hydrogen permeation and provide information on the rate of corrosion.

Other on-stream corrosion-monitoring techniques that are used in petroleum and chemical industries include

- The electrical resistance and linear polarization methods (Ref 15). The former determines corrosion trends with time, and the latter determines an instantaneous corrosion rate (see the article "In-Service Monitoring" in this Volume)
- Ultrasonic thickness measurement. This is a useful monitoring tool, especially when baseline readings are taken at selected locations before the equipment is placed in service. The inspection locations can be changed if erosion or corrosion areas are localized

With these methods, the materials engineer is able to determine the adequacy of the materials selection and to predict the remaining life so that replacements and/or repairs can be scheduled well in advance of failure. The corrosion test methods discussed above can also be used to evaluate alternate materials that might be more cost effective at the time of replacement of the vessel or a component.

Selecting Materials to Avoid or Minimize Corrosion

General Corrosion. Of the many forms of corrosion, general, or uniform, corrosion is the easiest to evaluate and monitor. Materials selection is usually straightforward. If a material shows only general attack, a low corrosion rate, and negligible contamination of the process fluid and if all other factors, such as cost, availability, and ease of fabrication, are favorable, then that is the material of choice. An acceptable corrosion rate for a relatively low-cost material such as plain carbon steel is about 0.25 mm/yr (10 mils/yr) or less. At this rate and with proper design and adequate corrosion allowance, a carbon steel vessel will provide many years of low-maintenance service.

For more costly materials, such as the austenitic (300-series) stainless steels and the copper- and nickel-base alloys, a maximum corrosion rate of 0.1 mm/yr (4 mils/yr) is generally acceptable. However, a word of caution is in order. One should never assume, without proper evaluation, that the higher the alloy, the better the corrosion resistance in a given environment. A good example is seawater, which corrodes plain carbon steel fairly uniformly at a rate of 0.1 to 0.2 mm/yr (4 to 8 mils/yr) but severely pits certain austenitic stainless steels.

At times, nonmetallic coatings and linings ranging in thickness from a few tenths to several millimeters are applied to prolong the life of low-cost alloys such as plain carbon steels in environments that cause general corrosion. The thin-film coatings that are widely used include baked phenolics, catalyzed cross-linked epoxy-phenolics, and catalyzed coal tar-epoxy resins (see the article "Organic Coatings and Linings" in this Volume). It is advisable not to use thin-film coatings in services where the base metal corrosion rate exceeds 0.5 mm/yr (20 mils/yr), because corrosion is often accelerated at holidays (for example, pinholes) in the coating. Thick-film linings include glass, fiber- or flake-reinforced furan, polyester and epoxy resins, hot-applied coal tar enamels, and various elastomers such as natural rubber.

A special case for materials selection under general corrosion conditions is that of contamination of the process fluid by even trace amounts of corrosion products. In this case, product purity, rather than corrosion rate, is the prime consideration. One example is storage of 93% sulfuric acid (H_2SO_4) in plain carbon steel at ambient temperature. The general corrosion rate is 0.25 mm/yr (10 mils/yr) or less, but traces of iron impart a color that is objectionable in many applications. Therefore, thin-film baked phenolic coatings are used on carbon steel to minimize or eliminate iron contamination. In the same way, thin-film epoxy coated carbon steel or solid or clad austenitic stainless steels are used to maintain the purity of adipic acid for various food and synthetic fiber applications. More information on this form of corrosion is available in the article "General Corrosion" in this Volume.

Localized Corrosion. Although general corrosion is relatively easy to evaluate and monitor, localized corrosion in such forms as pitting, crevice corrosion, and weld metal attack is at the opposite end of the scale, and materials selection is difficult. Localized corrosion is insidious and often results in failure or even total destruction of equipment without warning.

All metals and alloy systems are susceptible to most forms of localized corrosion by specific environments. For example, carbon or alloy steel pipelines will pit in aggressive soils because of local concentrations of corrosive compounds, differential aeration cells, corrosive bacteria, stray dc currents, or other conditions, and these pipelines generally require a combination of nonmetallic coatings and cathodic protection for long life. Also, holidays in mill scale left on plain carbon steels are sites for pitting because the mill scale is cathodic to the steel surface exposed at the holiday. For this reason, it is advisable to remove all mill scale by sand- or gritblasting before exposing plain carbon steels to corrosive environments.

Pitting. Aqueous solutions of chlorides, particularly oxidizing acid salts such as ferric and cupric chlorides, will cause pitting of a number of ferrous and nonferrous metals and alloys under a variety of conditions. The ferritic (400-series) and austenitic stainless steels are very susceptible to chloride pitting (as well as to crevice corrosion and SCC, which are discussed later in this section). Molybdenum as an alloying element is beneficial, so molybdenum-containing stainless steels, such as types 316 and 317, are more resistant than the nonmolybdenum alloys. However, most chloride environments require higher alloys containing greater amounts of chromium and molybdenum, such as Hastelloy alloy G-3 (UNS N06985), Inconel alloy 625 (UNS N06625), and Hastelloy alloy C-22 (UNS N06022), for optimum performance. Exceptions are titanium and its alloys, which show exceptional resistance to aqueous chloride environments (including the oxidizing acid chlorides), and copper, copper-nickel, and nickel-copper alloys, which are widely used in marine applications.

Other noteworthy combinations of metals and corrosive fluids to avoid when selecting materials because of pitting tendencies include:

- Aluminum and aluminum alloys in electrolytes containing ions of such heavy metals as lead, copper, iron, and mercury
- Plain carbon and low-alloy steels in waters containing dissolved oxygen or in waters and soils infected with sulfate-reducing bacteria
- Austenitic stainless steel weldments exposed to stagnant natural waters, particularly U.S. Gulf Coast well waters, which are infected with iron and/or manganese bacteria (Ref 16)

An unusual form of localized corrosion known as end-grain attack has occurred in chemical-processing plants with such specific metal/fluid combinations as austenitic stainless steels in hot nitric acid (HNO_3) and organic acids, and plain carbon steels in decanoic acid.

When certain forms of these metals (for example, plates, threaded rods, or pipe nozzle ends) are cut normal to the rolling direction, the ends of nonmetallic inclusions at the cut edges are attacked by the process fluid; this results in small-diameter but deep pits. The solution is to seal the cut edge with a layer of weld metal that is equal to the base metal in composition.

Crevice corrosion can occur not only at metal/metal crevices, such as weld backing rings, but also at metal/nonmetallic crevices, such as asbestos-gasketed pipe flanges or under deposits. In some fabricated assemblies, it is possible and cost effective to avoid crevices by careful design. For example, crevice corrosion occurred behind a weld backing strip at the closing seam in a type 304L stainless steel reactor handling hot HNO_3. A small amount of corrosion by stagnant acid in the crevice created hexavalent chromium ions (Cr^{6+}), which caused accelerated attack; other exposed surfaces in the vessel were unaffected. The closing seam could have been welded from both sides or from one side with a consumable insert ring, which would have avoided the problem. Similar attack has occurred in stainless shell and tube heat exchangers at the rolled tube-to-tubesheet joints and has been solved by seal welding the joints with appropriate weld filler metal and process.

However, in many cases, crevices are either too costly or impossible to design out of a system, so careful selection of materials is the answer. Titanium is susceptible to crevice corrosion in hot seawater and other hot aqueous chloride environments. Therefore, for a flanged and gasketed piping system in these fluids, commercially pure titanium grade 55 (UNS R50550) may be acceptable for piping, but flanges will require the more crevice attack resistant grade 7 (UNS R52400), which contains 0.15% (nominal) Pd, or grade 12 (UNS R53400), which contains small amounts of molybdenum and nickel. This is more cost effective than selecting the more expensive alloys for the piping as well. Another approach that has been successfully used in these fluids is installation of nickel-impregnated gaskets with grade 55 titanium flanges.

The austenitic stainless steels are susceptible to crevice corrosion in media other than HNO_3 solutions. For example, type 304L stainless steel exhibits borderline passivity in hot acetic acid solutions, particularly in crevices. Accordingly, the materials engineer will specify the more crevice corrosion resistant type 316L stainless steel where crevices cannot be avoided, such as piping and vessel flanges, or for the entire fabricated assembly because the cost differential between materials in this case may be negligible.

Preferential attack of weld metal is another form of localized corrosion that can be avoided by judicious selection of weld filler metal or weld process. Weld metal corrosion in high nickel-chromium-molybdenum Hastelloy alloy C-276 (UNS N10276) in a hot oxidizing H_2SO_4 process stream containing chlorides was eliminated by repair welding with Hastelloy alloy C-22 (UNS N06022). In applications in which type 304L stainless steel is selected for use in hot HNO_3 solutions, welds exposed to the process are frequently made by an inert gas process, such as gas tungsten arc or gas metal arc, rather than a flux-utilizing process, such as shielded metal arc. This prevents weld corrosion by eliminating the minute particles of trapped slag that are the sites for initiation of local corrosion. Many other examples of preferential weld metal attack and their solutions are discussed in the article "Corrosion of Weldments" in this Volume.

Nonmetallic materials of construction are widely used where temperatures, pressures, and stresses are not limiting and in such media as aqueous chloride solutions, which cause localized corrosion of metals and alloys. Examples in which lower-cost nonmetallic constructions are selected over expensive high alloys include the following:

- Rubber-lined steel for water treatment ion exchange resin beds, which must be periodically regenerated with salt brine or dilute mineral acids or caustic solutions
- Glass-lined steel for reaction vessels in chlorinated hydrocarbon service
- Acid-proof brick and membrane lined steel for higher temperature, and solid RTP polyester and vinyl-ester construction for lower temperature, flue gas and chlorine neutralization scrubbers

It should be apparent that an in-depth evaluation of candidate materials for environments that can cause localized corrosion is imperative in order to select the optimum material of construction. In particular, corrosion test coupons should reflect the final fabricated component—that is, include crevices and weldments where applicable—and should be examined critically under the microscope for evidence of local attack. In cases in which the more common 300- and 400-series stainless steels fall short, the newer ferritics, such as 26Cr-1Mo (UNS S44627) and 27Cr-3Mo-2Ni (UNS S44660), and the duplex ferritic-austenitic alloys, such as 26Cr-1.5Ni-4.5Mo (UNS S32900) and 26Cr-5Ni-2Cu-3.3Mo (UNS S32550), should be evaluated as potentially lower-cost alternatives to higher alloys. Finally, proven nonmetallic materials (for example, the RTPs), used either as linings for lower-cost metals (such as plain carbon steel) or for solid construction, should not be overlooked. More information on the mechanisms of pitting, crevice corrosion, and other forms of localized attack is available in the article "Localized Corrosion" in this Volume; the articles "Corrosion of Weldments" and "Metallurgically Influenced Corrosion" contain information on the preferential corrosion of weld metal.

Galvanic corrosion is quite possibly the only form of corrosion that can be beneficial as well as harmful. The materials engineer will frequently select galvanic corrosion—that is, cathodic protection using sacrificial metal anodes or coatings of magnesium, zinc, or aluminum—to stifle existing, or prevent new, corrosion of structures fabricated primarily from plain carbon or low-alloy steels—for example, bridges, underground and underwater pipelines, auto frames, off-shore drilling rigs, and well casings. All too often, however, galvanic corrosion caused by contact between dissimilar metals in the same environment is harmful. Examples are:

- Unprotected underground plain carbon steel pipelines connected to above-ground tanks and other structures that are electrically grounded with buried copper rods or cables
- Stainless steel shafts in "canned" pumps rotating in carbon or graphite bushings in a strong electrolyte
- Copper-nickel or stainless steel heat exchanger tubes rolled in plain carbon steel tubesheets exposed to river water for cooling
- Aluminum thermostat housings on cast iron auto engine blocks in contact with glycol-water mixtures

Newer designs that require several different metals for various reasons such as cost and physical, mechanical, and/or electrical properties present a challenge to the materials engineer with regard to selection for the avoidance of galvanic corrosion. Knowledge of the galvanic series of metals based on the electrochemical potential between a metal and a reference electrode in a given environment is essential. A practical galvanic series for metals and alloys in seawater is given in Table 1.

A word of caution: Metals behave differently in different environments; that is, the relative positions of metals and alloys in the galvanic series can vary significantly from one environment to another. In fact, variations within the same environment can occur with changes in such factors as temperature, solution concentration, degree of agitation or aeration, and metal surface condition. Thus, galvanic series that are based on seawater or other standard electrolytes are worthwhile for initial materials selection for multiple metal/alloy systems in a given environment. However, additional tests should be carried out in the stated environment by using the anodic polarization measurements described in ASTM G 5 (Ref 10). Suitable metal combinations can be determined by examining and superimposing polarization curves of candidate metals and by estimating the mixed potential values (Ref 18).

Table 1 Practical galvanic series of metals and alloys in seawater

Least noble; most anodic; most susceptible to corrosion
Magnesium and its alloys
Zinc
Aluminum and its alloys
Cadmium
Plain carbon and low-alloy steels
Gray and ductile cast irons
Nickel cast irons
Type 410 stainless steel (active)
50Pb-50Sn solder
Type 304 and 316 stainless steels (active)
Lead
Tin
Muntz metal, manganese bronze, naval brass
Nickel (active)
Inconel 600 (active)
Yellow and red brasses, aluminum and silicon bronzes
Copper and copper-nickel alloys
Nickel (passive)
Inconel 600 (passive)
Monel 400
Titanium
Type 304 and 316 stainless steels (passive)
Silver, gold, platinum
Most noble; most cathodic; least susceptible to corrosion

Source: Ref 17 (condensed and modified from the original to include the precious metals)

In conclusion, the selection of dissimilar metals that are far apart on any galvanic series should be avoided unless provisions are made for sacrificial corrosion of one for another, as in cathodic protection. Otherwise, dissimilar-metal applications in corrosive environments should be approached with extreme caution and should be thoroughly investigated before making the final selections. The section "Galvanic Corrosion" of the article "General Corrosion" in this Volume contains more information on galvanic attack.

Intergranular corrosion is a selective form of corrosion that proceeds along individual grain boundaries, with the majority of the grain being unaffected. Intergranular corrosion can affect certain alloys that are highly resistant to general and localized attack; noteworthy are several of the 300- and 400-series stainless steels and austenitic higher-nickel alloys. These alloys are made susceptible to intergranular corrosion by sensitization—that is, the precipitation of chromium carbides and/or nitrides at grain boundaries during exposure to temperatures from 450 °C (840 °F) to 870 °C (1600 °F), with the maximum effect occurring near 675 °C (1250 °F). Exposure to such temperatures can occur during processing at the mill, welding and other fabrication operations, or by plant service conditions. The resulting depletion in chromium adjacent to the chromium-rich carbides/nitrides provides a selective path for intergranular corrosion by specific media, such as hot oxidizing (nitric, chromic) and hot organic (acetic, formic) acids.

Susceptible stainless steels are those that have normal carbon contents (generally >0.04%) and do not contain carbide-stabilizing elements (titanium and niobium). Examples are AISI types 302, 304, 309, 310, 316, 317, 430, and 446. Susceptible higher-nickel alloys include Inconel alloys 600 (UNS N06600) and 601 (UNS N06601), Incoloy alloy 800 (UNS N08800) despite the presence of titanium, Incoloy 800H (UNS N08810), Nickel 200 (UNS N02200), and Hastelloy alloys B (UNS N10001) and C (UNS N10002). Intergranular corrosion in these alloys is avoided by one or a combination of the following:

- Keep the alloy in the solution heat-treated condition at all times
- Limit interstitial elements, primarily carbon and nitrogen, to the lowest practical levels
- Add carbide-stabilizing elements, such as titanium, niobium, and tantalum, along with a stabilizing heat treatment where necessary

In general, these alloys are purchased from the mill in the solution heat-treated condition, a requirement of most specifications. With regard to the austenitic stainless steels, solution heat treating consists of heating to a minimum temperature of 1040 °C (1905 °F) to dissolve all the carbides, followed by rapid cooling in water or air to prevent sensitization.

Preserving the solution-treated condition is difficult, except for certain applications in which reheating is not a requirement of fabrication. One example is a pump shaft machined from forged or rolled and solution heat-treated bar stock. However, most applications for these alloys, such as piping components and pressure vessels, require hot rolling (for example, of plate) or hot bending (for example, of pipe) and welding; these practices, as discussed earlier, can cause sensitization. Solution heat treating after fabrication is generally impractical, because of the possibility of irreparable damage due to distortion or excessive scaling or because of the inability to cool rapidly enough through the critical sensitization temperature range.

Therefore, in most applications involving exposure to environments that cause intergranular corrosion, the low-carbon/nitrogen or stabilized alloy grades are specified. The new ferritic stainless steel UNS S44800 with 0.025% C (maximum) plus nitrogen is replacing the higher-carbon 446 grade in applications in which resistance to intergranular corrosion is a requirement. Types 304L, 316L, and 317L, with carbon and nitrogen contents limited to 0.03% (maximum) and 0.10% (maximum), respectively, are used instead of their higher-carbon counterparts. The newer nickel-chromium alloy 904L (UNS N08904), with 0.02% maximum carbon, and the high-nickel Hastelloy alloys C-22 (UNS N06022) and B-2 (UNS N10665), which have maximum carbon contents of 0.015% and 0.01%, respectively, are now being used almost exclusively in the as-welded condition without intergranular corrosion problems.

These low carbon levels are readily and economically achieved with the advent of the argon oxygen decarburization (AOD) refining process used by most alloy producers. By limiting the interstitial element content, sensitization is limited or avoided entirely during subsequent welding and other reheating operations. However, designers should be aware of the fact that lowering the carbon/nitrogen content also lowers the maximum allowable design stresses, as noted in appropriate sections of applicable fabrication codes, such as ASME Section 8, Division 1 for unfired pressure vessels and ANSI B31.3 for process piping.

Commercially pure Nickel 200 is a special case. With a maximum carbon content of 0.15%, Nickel 200 will precipitate elemental carbon or graphite in the grain boundaries when heated in the range of 315 to 760 °C (600 to 1400 °F). This results in embrittlement and susceptibility to intergranular corrosion in certain environments, such as high-temperature caustic. Where embrittlement and intergranular corrosion must be avoided, Nickel 201 (UNS N02201) with a maximum carbon content of 0.02% is specified.

Titanium as a carbide-stabilizing element is used in several ferritic and austenitic stainless steels, including types 409, 439, 316Ti, and 321, as well as the higher-nickel Incoloy alloy 825 (UNS N08825), at a minimum concentration of about five times the carbon plus nitrogen content. In the same way, niobium, generally with tantalum, is used in types 309Cb, 310Cb, and 347 austenitic stainless steels at a minimum combined concentration of about ten times the carbon content.

The higher-nickel alloys 20Cb-3, Inconel 625, and Hastelloy G contain even higher concentrations of niobium—up to a maximum of about 4% in the case of alloy 625. In general, when stabilized alloys are heated in the sensitizing temperature range, chromium depletion at the grain boundaries does not occur, because the stabilizing elements have a greater affinity for carbon than does chromium.

Under certain conditions, however, stabilized alloys will sensitize, especially during multipass welding or cross welding. They are also susceptible to a highly localized form of intergranular corrosion known as knife-line attack, which occurs in base metal at the weld fusion line. In some cases, these alloys are given stabilizing heat treatments after solution heat treatment for maximum resistance to intergranular corrosion in the as-welded condition. For example, type 321 stainless steel is stabilize annealed at 900 °C (1650 °F) for 2 h, and alloys 825 and 20Cb-3 at 940 °C (1725 °F) for 1 h, before fabrication to avoid sensitization and knife-line attack. So treated, type 321 may still be susceptible because titanium has a tendency to form an oxide during welding; therefore, its role as a carbide stabilizer may be diminished. For this reason, type 321 is always welded with a niobium-stabilized weld filler metal, such as type 347 stainless.

Some specialty alloys have low interstitial element content plus the addition of stabilizing elements for resistance to intergranular corrosion. These alloys include the higher-nickel Hastelloy alloy G-3 (UNS N06985), which contains 0.015% C (maximum) and niobium plus tantalum up to 0.5%, and the newer ferritic stainless steels (Ref 19) listed below:

- UNS S44627: 26% Cr, 1% Mo, 0.01% C (max), 0.015% N (max), 0.2% Nb (max)
- UNS S44635: 25% Cr, 4% Ni, 4% Mo, 0.025% C (max), 0.035% N (max), 0.80% Ti+Nb (max)
- UNS S44660: 26% Cr, 2% Ni, 3% Mo, 0.03% C (max), 0.04% N (max), 1.0% Ti+Nb (max)
- UNS S44735: 29% Cr, 0.5% Ni, 4% Mo, 0.03% C (max), 0.045% N (max), 1.0% Ti + Nb (max)

The new ferritic stainless steels were developed primarily for heat-exchanger tubing applications for use in place of the higher-carbon unstabilized type 446 stainless steel. These ferritic stainless steels have many useful properties.

As discussed previously, specific corrosion environments cause intergranular corrosion in specific metal and alloy systems. A wealth of information, both published (Ref 20) and unpublished, has been developed on corrodents that cause intergranular corrosion in sensitized austenitic stainless steels; a partial listing appears in Table 2. The low-carbon/nitrogen or stabilized grades are specified for applications in which the austenitics have satisfactory general and localized corrosion resistance in these environments and in which sensitization by such operations as welding and hot forming will undoubtedly occur.

Evaluation tests for intergranular corrosion are conducted to determine if purchased materials have the correct chemical composition and are in the properly heat-treated condition to resist intergranular corrosion in service (Ref 21). Evaluation testing is imperative for a number of reasons:

- The stainless steels and nickel-base alloys to which these tests are applied are relatively expensive
- These materials are frequently specified for critical applications in the petrochemical, process, and power industries
- The principal cost associated with a corrosion failure is generally that of production loss, not replacement
- For maximum cost effectiveness, these materials should be used in their best possible metallurgical and corrosion-resistant conditions

Most of the evaluation tests are described in detail in ASTM standards. Appropriate evaluation tests and acceptance criteria for wrought alloys are listed in Table 1 in the article "Evaluation of Intergranular Corrosion" in this Volume.

Table 2 Partial listing of environments known to cause intergranular corrosion in sensitized austenitic stainless steels

Environment	Concentration, %	Temperature, °C (°F)
Nitric acid	1	Boiling
	10	Boiling
	30	Boiling
	65	60 (140) to boiling
	90	Room to boiling
	98	Room to boiling
Lactic acid	50–85	Boiling
Sulfuric acid	30	Room
	95	Room
Acetic acid	99.5	Boiling
Formic acid	90	Boiling
	10 (plus Fe^{3+})	Boiling
Chromic acid	10	Boiling
Oxalic acid	10 (plus Fe^{3+})	Boiling
Phosphoric acid	60–85	Boiling
Hydrofluoric acid	2 (plus Fe^{3+})	77 (170)
Ferric chloride	5	Boiling
	25	Room
Acetic acid/anhydride mixture	Unknown	100–110 (212–230)
Maleic anhydride	Unknown	60 (140)
Cornstarch slurry, pH 1.5	Unknown	49 (120)
Seawater	...	Room
Sugar liquor, pH 7	66-67	75 (167)
Phthalic anhydride (crude)	Unknown	232 (450)

Intergranular corrosion is rare in nonsensitized ferritic and austenitic stainless steels and nickel-base alloys, but one environment known to be an exception is boiling HNO_3 containing an oxidizing ion such as dichromate (Ref 22), vanadate, and/or cupric. Intergranular corrosion has also occurred in low-carbon, stabilized and/or properly solution heat-treated alloys cast in resin sand molds (Ref 23). Carbon pickup on the surface of the castings from metal-resin reactions has resulted in severe intergranular corrosion in certain environments. Susceptibility goes undetected in the evaluation tests mentioned above because test samples obtained from castings generally have the carbon-rich layers removed. This problem is avoided by casting these alloys in ceramic noncarbonaceous molds.

Other metals, such as magnesium, aluminum, lead, zinc, copper, and certain alloys, are susceptible to intergranular corrosion under very specific conditions. Very few case histories are reported in the literature. An unusual form of intergranular corrosion known as exfoliation, which occurs in aluminum-copper alloys, is discussed in the section "Other Types of Corrosion" in this article. Intergranular SCC is discussed in the following section. More information on intergranular corrosion is available in the article "Metallurgically Influenced Corrosion" in this Volume.

Stress-corrosion cracking is a type of environmental cracking caused by the simultaneous action of a corrodent and sustained tensile stress. The following discussion deals primarily with anodic SCC. Anodic SCC is believed to be a delayed cracking phenomenon that occurs in normally ductile materials under the stress resulting from accelerated electrochemical corrosion at anodic sites of the material as well as at the crack tip. Other types of environmental cracking, such as hydrogen stress cracking and liquid-metal embrittlement, are discussed later in this article (see also the article "Environmentally Induced Cracking" in this Volume).

The National Association of Corrosion Engineers, the Materials Technology Institute of the Chemical Process Industries, and others have published tables of corrodents known to cause SCC of various metal alloy systems (Ref 24-26). Table 3 lists these data in condensed form and covers the SCC environments of major importance to the materials engineer. This table, as well as those published in the literature, should be used only as a guide for screening candidate materials for further in-depth investigation, testing, and evaluation.

Stress-corrosion cracking is not a certainty in the listed environments under all conditions. Metals and alloys that are indicated as being susceptible can give good service under specific conditions. For example, referring to Table 3:

- Anhydrous ammonia will cause SCC in carbon steels, but rarely at temperatures below 0 °C (32 °F) and only when such impurities as air or oxygen are present; addition of a minimum of 0.2% H_2O will inhibit SCC
- Aqueous fluorides and hydrofluoric acid (HF) primarily affect Monel alloy 400 (UNS N04400) in the nickel alloys system; others are resistant
- Steam is known to cause SCC only in aluminum bronzes and silicon bronzes in the copper alloys system
- Polythionic acid only cracks sensitized austenitic stainless steels and nickel alloys; SCC is avoided by solution annealing heat treatments or selection of stabilized or low-carbon alloys

Stress-corrosion cracking is often sudden and unpredictable, occurring after as little as a few hours exposure or after months or even years of satisfactory service. Cracking occurs frequently in the absence of other forms of corrosion, such as general attack or crevice attack. Virtually all alloy systems are susceptible to SCC by a specific corrodent under a specific set of conditions, that is, concentration, temperature, stress level, and so on. Only the ferritic stainless steels as a class are resistant to many of the environments that cause SCC in other alloy systems, but they are susceptible to other forms of corrosion by some of these environments.

The combination of aqueous chlorides and austenitic stainless steels is probably the most important from the standpoints of occurrence, economics, and investigation. Although the mechanism and boundary conditions for chloride SCC are still not fully defined, it is reasonably safe to state that chloride SCC of austenitic stainless steels:

- Seldom occurs at metal temperatures below 60 °C (140 °F) and above 200 °C (390 °F)
- Requires an aqueous environment containing dissolved air or oxygen or other oxidizing agent
- Occurs at very low tensile stress levels such that stress-relieving heat treatments are seldom effective as a preventive measure
- Affects all the austenitic stainless steels about equally with regard to susceptibility, time to failure, and so on
- Is characterized by transgranular branchlike cracking as seen under a metallurgical microscope

So many materials and environments are sources of chlorides that they are hard to avoid (Ref 27). Significant costs for repairs and replacements, as well as lost utility, have occurred through the years in the petrochemical industry as a result of chloride SCC:

- In water-cooled heat exchangers from chlorides in the cooling water
- Under thermal insulation allowed to deteriorate and become soaked with water that leached chlorides from the insulation
- Under chloride-bearing plastics, elastomers, and adhesives on tapes

In the case of shell and tube heat exchangers, in which chloride-bearing waters are used for cooling, a number of preventive measures are available. In vertical units with water on the shell side, cracking occurs most often at the external surfaces of the tubes under the top tubesheet (Ref 28). This is a dead space (air pocket) at which chlorides are allowed to concentrate by alternate wetting and drying of tubing surfaces. Adding vents to the top tubesheet sometimes alleviates this problem by eliminating the dead space and allowing complete water flooding of all tubing surfaces (Ref 29).

However, in many cases, this approach results in only a nominal increase in time to failure or no benefit whatsoever, because of other inherent deficiencies, such as low water flows or throttling water flow to control process temperatures. Thus, a material more resistant to chloride SCC is required, such as the austenitic higher-nickel Incoloy alloys 800 (UNS N08800) and 825 (UNS N08825); Inconel 600 (UNS N06600); ferritic stainless steels such as type 430 (UNS S43000), 26Cr-1Mo (UNS S44627), and SC-1 (UNS S44660); duplex stainless steels such as Ferralium 255 (UNS S32550) and AISI type 329 (UNS S32900); or titanium. In this case, selection depends primarily on economics, that is, the least expensive material that will resist process-side corrosion as well as water-side SCC.

Alloys are said to be either resistant or immune to chloride SCC, depending on how they perform in accelerated laboratory tests. In general, an alloy is immune if it passes the boiling 42% magnesium chloride test conducted according to

Table 3 Some environment-alloy combinations known to result in SCC

	Alloy system								
					Stainless steels				
Environment	Aluminum alloys	Carbon steels	Copper alloys	Nickel alloys	austenitic	duplex	martensitic	Titanium alloys	Zirconium alloys
Amines, aqueous	...	●	●	...	...	...	...	...	...
Ammonia, anhydrous	...	●	...	...	...	...	...	...	...
Ammonia, aqueous	...	...	●	...	...	...	...	...	...
Bromine	...	...	...	...	...	...	...	...	●
Carbonates, aqueous	...	●	...	...	...	...	...	...	...
Carbon monoxide, carbon dioxide, water mixture	...	●	...	...	...	...	...	...	...
Chlorides, aqueous	●	...	...	●	●	●	...	...	●
Chlorides, concentrated, boiling	...	...	...	●	●	●	...	...	...
Chlorides, dry, hot	...	...	...	●	...	...	...	●	...
Chlorinated solvents	...	...	...	...	...	...	...	●	●
Cyanides, aqueous, acidified	...	●	...	...	...	...	...	...	...
Fluorides, aqueous	...	...	...	●	...	...	...	...	...
Hydrochloric acid	...	...	...	...	...	...	...	●	...
Hydrofluoric acid	...	...	...	●	...	...	...	...	...
Hydroxides, aqueous	...	●	...	...	●	●	●	...	...
Hydroxides, concentrated, hot	...	...	...	●	●	●	●	...	...
Methanol plus halides	...	...	...	...	...	...	...	●	●
Nitrates, aqueous	...	●	●	...	...	...	●	...	...
Nitric acid, concentrated	...	...	...	...	...	...	...	...	●
Nitric acid, fuming	...	...	...	...	...	...	...	●	...
Nitrites, aqueous	...	...	●	...	...	...	...	...	...
Nitrogen tetroxide	...	...	...	...	...	...	...	●	...
Polythionic acids	...	...	...	●	●	...	...	...	...
Steam	...	...	●	...	...	...	...	...	...
Sulfides plus chlorides, aqueous	...	...	...	...	●	●	●	...	...
Sulfurous acid	...	...	...	...	●	...	...	...	...
Water, high-purity, hot	●	...	...	●	...	...	...	...	...

ASTM G 36. Examples are Inconel alloy 600, 26Cr-1Mo, and grade 3 commercially pure titanium (UNS R50550). Industry has recognized the severity of this test and has devised other accelerated laboratory tests, such as boiling 25% sodium chloride and the Wick Test (ASTM C 692), that are more representative of actual conditions in the field (Ref 30). Thus, alloys that fail the G 36 test but pass the C 692 test, such as Incoloy 800, alloy 2205 (UNS S31803), and 20 Mo6 (UNS N08026), typically provide many years of service as tubes in water-cooled heat exchangers. (The austenitic 300-series stainless steels, as a class, fail all of the accelerated laboratory chloride SCC tests mentioned above.)

Unfortunately, all of the immune and resistant alloys are more expensive than most of the austenitic stainless steels. A more cost-effective application of some of these alloys involves a procedure known as safe ending, in which short lengths, (for example, 0.3 to 0.6 m, or 12 to 24 in.) are butt welded to the austenitic stainless steel tubes. The SCC-resistant ends are positioned in the exchanger at the point of greatest exposure to SCC conditions, that is, under the top tubesheet. Safe-ended tubes have extended the life of austenitic stainless steel tubing several-fold, with only a nominal increase in cost. Of course, the dissimilar metals must be weld compatible; this eliminates the use of titanium for safe ending.

Another cost-effective answer to chloride SCC in heat exchangers is bimetallic tubing, in which the austenitic stainless steel required for process-side corrosion resistance is clad with a water-side SCC- and corrosion-resistant material such as 90Cu-10Ni (UNS C70600) (Ref 31). Still another answer is cathodic protection with a sacrificial metal coating such as lead containing 2% Sn and 2% Sb applied by hot dipping or flame spraying (Ref 32).

Chloride SCC under insulation can be prevented by keeping it dry, but this is easier said than done in many cases. It is particularly difficult in humid, high annual rainfall climates, such as the Gulf Coast area of the United States, and where insulated equipment must be washed down periodically or is exposed to fire control deluge systems that are periodically activated (Ref 33). Under these conditions, three preventive measures are applied, as follows.

The first preventive measure is the addition of sodium metasilicate as an SCC inhibitor to the insulation at a minimum concentration of ten times the chloride content. The inhibitor is activated when the insulation becomes wet and is effective only when it wets the stainless steel surface. For maximum protection, metasilicate is painted on the vessel or piping before the installation of inhibited insulation. However, SCC has occurred after many years of service under inhibited insulation that was allowed to become so wet that the water-soluble inhibitor was leached out to a point below the minimum concentration required for prevention of SCC.

Second is protective coating of the vessel or piping before insulation. Catalyzed high build epoxy paints are effective to about 100 °C (212 °F), catalyzed coal tar-epoxy enamels to about 150 °C (300 °F), and silicone-base coatings to about 200 °C (390 °F). These coatings are even more effective if the stainless steel surfaces are heavily sandblasted before coating. Sandblasting peens the surface to a depth of 0.01 to 0.1 mm (0.4 to 4 mils), and this results in a layer under compressive stresses that counteract the tensile stresses required for SCC (peening will be discussed later in this article).

The third preventive measure is cathodic protection of the vessel or piping with aluminum foil under insulation (Ref 34). This method is claimed to provide both a physical barrier to chloride migration to stainless steel surfaces as well as cathodic protection when the insulation becomes wet, and it is effective at vessel temperatures between 60 and 500 °C (140 to 930 °F). However, the foil is attacked by the alkaline (generally) leachates from the insulation and must be renewed periodically.

Other important environments that cause SCC of stainless steels are hydroxide (caustic) solutions, sulfurous acid, and polythionic acids. Caustic SCC of austenitic stainless steels can be both transgranular and intergranular and is a function of solution concentration and temperature. It seldom occurs at temperatures below 120 °C (250 °F). At higher temperatures, the newer ferritic stainless steels, nickel, and high-nickel alloys provide outstanding service; 26Cr-1Mo stainless steel has found widespread application in heat-exchanger tube bundles serving caustic evaporators at 170 to 200 °C (340 to 390 °F). Nickel 200 and 201 as well as Inconel alloy 600 are resistant to 300 °C (570 °F) in caustic concentrations to 70%.

Polythionic acid and sulfurous acid will cause SCC in sensitized nonstabilized austenitic stainless steels and nickel-base alloys. Cracking is always intergranular and requires relatively low tensile stresses for initiation and propagation. As-welded, normal carbon grades, such as types 304 and 316 and Incoloy alloy 800, are particularly susceptible to SCC in weld HAZs. Low-carbon (<0.03% C) and stabilized grades, such as types 321 and 347, are resistant, especially after receiving a stabilizing heat treatment. The normal carbon grades in the solution heat-treated condition are also resistant. Susceptibility to polythionic acid SCC can be determined by laboratory corrosion testing according to ASTM G 35.

Polythionic acid and sulfurous acid SCC are major considerations in the petroleum-refining industry, especially in desulfurizer, hydrocracker, and reformer processes (Ref 35, 36). These acids form in process units during shutdowns when equipment and piping containing sulfide deposits and scales are opened and exposed to air and moisture. Preventive measures include flushing with alkaline solutions to neutralize sulfides before shutdown and purging with dry nitrogen during shutdown according to recommended practices established by NACE (Ref 37).

Cast austenitic stainless steels such as Alloy Casting Institute (ACI) types CF-8 (UNS J92600) and CF-8M (UNS J92900) are inherently more resistant (but not immune) to chloride SCC than their wrought counterparts (types 304 and 316, respectively) for several reasons. Castings generally have lower residual stresses than wrought alloys after solution-annealing heat treatments, and service-applied stresses are often lower because of heavy section thicknesses. However, the principal reason for improved SCC resistance is the presence of varying amounts of free ferrite, resulting in a duplex austenitic-ferritic microstructure. Free ferrite is primarily attributed to additions of 1 to 2% Si to improve fluidity during pouring and to resist hot cracking during cooling of the casting.

During work conducted some years ago on the development of the duplex cast ACI alloy CD-4MCu (UNS J93370), researchers noted improved resistance to chloride SCC in cast alloys containing high ferrite; this confirmed observations in the field by materials engineers (Ref 38). A significant improvement in SCC resistance occurred wth ferrite contents in the range 13 to 20 vol% and greater. This led to the development of alloys with controlled ferrite content by balancing chemical composition. Ferrite formation is promoted by chromium and by elements that act like chromium, such as silicon, niobium, and molybdenum. Nickel and elements that act like nickel (carbon, manganese, nitrogen) retard ferrite formation. When specified, ferrite is typically controlled at the foundry by adjusting chromium toward the upper end of the specification range, and nickel toward the lower end, along with the silicon addition.

Resistance to SCC in these alloys is believed to be a result of the keying action of ferrite particles (Ref 39). This action blocks direct propagation of SCC through the austenitic matrix. Along with this benefit, however, are improvements in strength, weldability, and resistance to general corrosion, particularly in hot concentrated nitric, acetic, phosphoric, and sulfuric acids and mixed nitric-hydrofluoric acid (Ref 40). The American Society for Testing and Materials has recognized alloys produced with controlled ferrite content in specification A 351. However, users are cautioned to limit applications to a maximum temperature of 425 °C (800 °F) because of the thermal instability of these grades.

A number of measures for preventing SCC of austenitic stainless steels have already been discussed. Two additional measures worthy of consideration are stress-relieving heat treatments and shot peening. The typical stress-relieving temperature for plain carbon steels—595 °C (1100 °F)—is only slightly effective for the austenitics, which require slow cooling from about 900 °C (1650 °F) for effective relief of residual stresses from such operations as welding. However, such treatments may not prevent, but only prolong, the time to failure; chloride SCC of austenitic stainless steels can occur at very low stress levels. Also, these elevated temperatures may cause unwanted distortion of complex and/or highly stressed structures and will sensitize susceptible alloys. This sensitization will result in intergranular corrosion or intergranular SCC.

Shot peening is the controlled bombardment of a metal surface with round, hard steel shot for the purpose of introducing compressive stresses in surface layers. These compressive stresses counteract the tensile stresses required for SCC. The depth of the resultant cold-worked layers is generally in the range 0.1 to 0.5 mm (4 to 20 mils); 100% coverage is required. In laboratory tests in boiling 42% magnesium chloride, shot-peened type 304 stainless steel U-bend samples showed no SCC after more than 1000 h of exposure, but nonpeened control samples all cracked in approximately 1 h (Ref 41). Successful applications in the chemical-processing industries include a type 316 stainless steel centrifuge exposed to an organic chloride process stream (Ref 42) and numerous storage tanks exposed to a variety of SCC conditions (Ref 43).

The exposure of shot-peened surfaces to excessive temperatures or to environments that cause excessive general or pitting corrosion should be avoided. Temperatures above 565 °C (1050 °F) will relieve the beneficial compressive stresses and reduce the overall benefits of shot peening. Once the relatively thin layers of residual compressive stress are penetrated by general corrosion or pitting, SCC can occur.

Plain carbon steels are susceptible to SCC by several corrodents of economic importance, including aqueous solutions of amines, carbonates, acidified cyanides, hydroxides, nitrates, and anhydrous ammonia. Susceptible steels in common use throughout the petrochemical-processing industry include ASTM A106 grade B for piping, A285 grade C for tanks, and A 515 grade 70 for pressure vessels. Cracking is both intergranular and transgranular; the former occurs in hot hydroxides and nitrates, and the latter in warm acidified cyanide solutions.

The temperature and concentration limits for the SCC susceptibility of carbon steels in caustic soda (sodium hydroxide) are fairly well defined; they have been derived from field experience and reproduced in chart form (Fig. 1). These limits for the other aqueous SCC environments mentioned above are not nearly as well defined. For example, SCC was not considered to be a problem at temperatures below 88 °C (190 °F) in aqueous monoethanolamine and diethanolamine solutions used for scrubbing carbon dioxide and hydrogen sulfide out of natural gas and hydrogen-rich synthesis gas streams. However, recent experience in petroleum refineries has resulted in general agreement that SCC preventive measures, such as thermal stress relief, must be applied to plain carbon steels in aqueous amines at all temperatures and concentrations.

With regard to anhydrous ammonia, the SCC of plain carbon and low-alloy steels occurs at ambient temperatures when air or oxygen is present as a contaminant at concentrations of only a few parts per million (Ref 45). Most of the adverse experience has been with the nonstress-relieved higher-strength quenched-and-tempered steels, such as the grades covered in ASTM A 517. Cracking seldom occurs at temperatures below 0 °C (32 °F) and is virtually nonexistent at or below −33 °C (−28 °F), a common temperature for storage at atmospheric pressure. Water is an effective SCC inhibitor at a minimum concentration of 0.2% by weight.

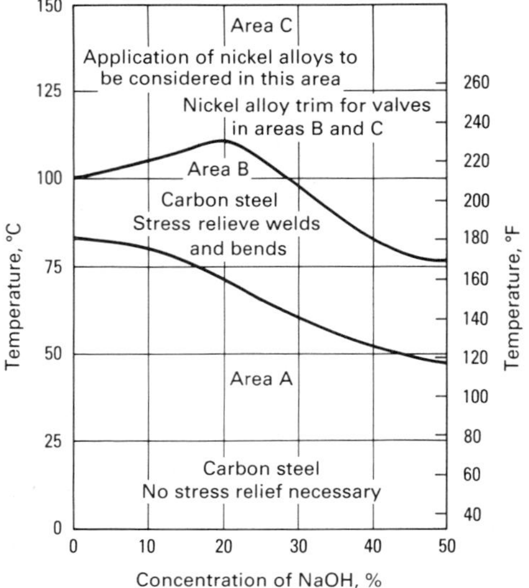

Fig. 1 Temperature and concentration limits for SCC susceptibility of carbon steels in caustic soda. Source: Ref 44

Thermal stress relief is perhaps the single most effective measure for preventing, or at least greatly prolonging, the time to SCC of plain carbon and low-alloy steels in all of the environments mentioned above. However, temperature is a more important factor than time at temperature. There is general agreement that stress-relieving temperatures below 595 °C (1100 °F), regardless of the hold time, are not effective for preventing SCC. In fact, the recommended minimum temperature for some environments, such as aqueous acidified cyanides, is 650 °C (1200 °F). As with any thermal treatment, special precautions must be observed, such as adequate support to avoid distortion and the removal of dirt, grease, and other foreign material that might react with the structure at elevated temperatures.

The above comments on thermal stress relief apply to new carbon steel piping and vessels. Stress-corrosion cracking has been found in such vessels as horizontal pressure storage tanks ("bullets") that were stress relieved after SCC was found in the original as-welded vessels and were repaired by grinding and welding (Ref 45). Undoubtedly, the stress-relieving temperatures propagated tiny cracks that were not detected during repairs.

Just as aqueous chloride is a potent SCC environment for austenitic stainless steels, aqueous ammonia causes extensive cracking in copper-base alloys. Virtually all copper alloys, as well as pure copper itself, can be made to crack in this environment. The most susceptible alloys are the brasses containing more than 15% Zn, such as admiralty brass (UNS C44300, 28% Zn), yellow brass (UNS C27000, 34% Zn), and Muntz metal (UNS C28000, 40% Zn), followed by red brass (UNS C23000, 15% Zn) and several bronzes. The least susceptible alloys are 90Cu-10Ni (UNS C70600), 70Cu-30Ni (UNS C71500), and unalloyed copper.

The conditions that are conducive to SCC by aqueous ammonia are water, ammonia, air or oxygen, and tensile stress in the metal. Cracking is always intergranular, requires only trace quantities of ammonia in many cases, and occurs at ambient temperature. For example, U-bends in admiralty brass condenser tubes cracked in warm water from a cooling tower that contained 15 to 25 ppm ammonia because of close proximity to an ammonia plant. Other sources of ammonia include the decomposition of amines and the microbiological breakdown of organic matter.

Thermal stress relief is generally not one of the better preventive measures, because ammoniacal SCC occurs at relatively low stress levels. In fairly mild ammoniacal environments, such as the cooling tower water system mentioned above, the copper-nickel alloys, particularly 90Cu-10Ni, give good service. Small quantities of hydrogen sulfide have been found to inhibit SCC of brasses in petroleum refinery process streams, probably by reducing the dissolved oxygen content (Ref 46).

Another potent SCC agent for copper alloys is steam, but susceptibility is limited primarily to the higher-strength bronzes alloyed with aluminum (for example, UNS C61400) or silicon (for example, UNS C65500). In this regard, one noteworthy design to avoid is dimpled jacket construction. Severe SCC occurred at plug welds in a silicon bronze steam jacket on a similar alloy vessel used for heating a waste H_2SO_4 stream.

Stress-corrosion cracking of aluminum and its alloys in the chemical-processing industries is rare, because the lower-strength alloys in predominant use are resistant. These include the commercially pure alloy 1100, manganese alloy 3003, and magnesium alloys 5052 and 5083. Susceptible alloys are the high-strength grades of economic importance in the aircraft and aerospace industries, such as the copper alloys 2014 (UNS A92014) and 2024 (UNS A92024) and the copper-zinc-magnesium alloys 7075 (UNS A97075) and 7178 (UNS A97178). These alloys, heat treated to maximum strength levels, are susceptible to SCC in humid air containing traces of chlorides (Ref 47).

Titanium (UNS R50550 and other unalloyed grades) has found increasing application in the chemical-processing industries, primarily as a replacement for austenitic stainless steels, because of outstanding resistance to corrosion by hot, concentrated oxidizing acids, such as HNO_3, and virtual immunity to pitting and SCC by hot aqueous chloride solutions. However, titanium and, to a greater extent, its higher-strength alloys (for example, UNS R56400) are susceptible to SCC in a variety of environments, including anhydrous alcohols (methanol, ethanol) containing traces of halides, anhydrous red fuming HNO_3, anhydrous hot chlorides, some important industrial organic solvents (carbon tetrachloride, trichloroethane, and so on), anhydrous nitrogen tetroxide, and HCl.

In general, SCC preventive measures, such as thermal stress relief, cathodic protection, or inhibition, have not been used to any great extent with titanium and its alloys, although water additions are reported to inhibit cracking in anhydrous red fuming HNO_3 and anhydrous nitrogen tetroxide (Ref 48). Rather, SCC is prevented by avoiding contact with specific cracking agents.

Zirconium (UNS R60702) and, to a greater extent, its alloys (for example, UNS R60705) exhibit SCC behavior similar to that of titanium and its alloys in anhydrous alcohols containing traces of halides and in chlorinated organic solvents. Stress-corrosion cracking also occurs in aqueous ferric and cupric chloride solutions and hot concentrated HNO_3, especially when sustained tensile stresses are high. Because threshold stresses for SCC are high in comparison to yield strengths, thermal stress relief is a practical preventive measure. Holding temperatures between 650 and 850 °C (1200 and 1560 °F) are reported to be effective (Ref 49).

As a class, nickel-base alloys are susceptible to SCC by a wide variety of corrodents. However, in most cases, the corrodents are specific to a few, but not all, of the alloys in this class. For example, sensitized alloys 800 and 600 crack in thiosulfate solutions and polythionic acids, but the stabilized alloys Incoloy 825 and Inconel 625 are resistant. The latter alloys are particularly applicable in petrochemical process streams containing chlorides along with polythionic acids. In the nuclear power industry, Inconel 600 and weld metal alloys 82 (UNS N06082) and 182 (UNS W86182) crack intergranularly in crevices in high-purity water containing oxygen at elevated temperatures and pressures (Ref 50). In the absence of crevices, these alloys are resistant. Inconel alloy 690 (UNS N06690) is resistant under all conditions.

Nickel 200 and 201, Monel 400, Inconel 600 and 690, and Incoloy 800 are susceptible to SCC in caustic solutions over a wide range of concentrations and at temperatures above 290 °C (550 °F). Cracking is predominantly intergranular. The presence of oxygen and chlorides, which are common contaminants, tends to accelerate cracking, especially in Inconel alloy 600. Fairly high stress levels are generally required. Failures have occurred in steam generator tubes serving nuclear power plants from caustic concentrated on tube surfaces in crevices or as a result of a hot wall effect and/or poor water circulation. This experience prompted considerable research (Ref 51-57). In general, Nickel 200 and 201 and the nickel-chromium-iron-Inconel alloys 690 and 600 appear to be the most resistant under the various conditions tested, although they are not immune to caustic SCC.

Other SCC environments of importance with regard to nickel-base alloys, specifically Monel 400, are acidic fluoride solutions and HF. Monel 400 is susceptible to SCC in cold-worked or as-welded conditions, but is resistant in annealed or thermally stress-relieved conditions (Ref 58). Free air or oxygen accelerates corrosion rates and SCC tendency. However, Monel 400 has excellent general corrosion resistance over a broad range of temperatures and concentrations, so its use in HF alkylation units and other process streams containing HF or fluorides is widespread (Ref 59). Thermal stress relief at a minimum temperature of 540 °C (1000 °F), followed by slow cooling, will prevent SCC or will greatly prolong time to failure.

Corrosion-resistant polymers, such as thermoplastic and thermosetting resins, that are of interest to the materials engineer are susceptible to two forms of cracking in specific environments (Ref 60):

- Environmental stress cracking, which occurs when the polymer is stressed (residual or applied) and exposed to an organic solvent or aqueous solution of a wetting agent
- Craze cracking, which consists of a multitude of fine cracks that develop in contact with organic liquids or vapors, with or without the presence of mechanical stresses

For example, high-density polyethylene may crack in the presence of benzene or ethylene dichloride, and polystyrene may crack in the presence of aliphatic hydrocarbons (Ref 61). The mechanisms for these phenomena are not completely understood. Cracking may initiate at minute particles of low molecular weight polymers in the higher molecular weight polymer matrix. Thus, when evaluating polymers for corrosive service, test specimens should be stressed before exposure and should be carefully examined after exposure for evidence of environmental stress cracking and craze cracking. For guidance, the Non-Metals Section of Ref 3 uses the footnote "May Stress Crack" to indicate potential problems with a specific polymer in a specific environment. Additional information on SCC is available in the article "Environmentally Induced Cracking" in this Volume.

Hydrogen Damage. The entry of hydrogen into metals and alloys can result in several forms of damage:

- Loss of ductility and/or fracture strength
- Internal damage due to defect formation
- Sustained propagation of defects at stresses well below those required for mechanical fracture
- Macroscopic damage due to entrapment at mechanical interfaces

Although atomic, rather than molecular, hydrogen is the detrimental species within metals, it may be absorbed from a molecular hydrogen gas atmosphere. Hydrogen itself may be introduced during several stages of equipment or component manufacture (before any period of service). Heat treating of the metal (or alloy) in a hydrogen-containing furnace atmosphere, such as cracked ammonia, can result in absorption. Acid pickling, plating, or welding operations can each introduce hydrogen into the lattice of the metal.

Subsequent chemical service involving aqueous corrosion or high-temperature, hydrogen-containing environments may also introduce hydrogen. It is important to realize that many ductile metals and alloys, such as copper or austenitic stainless steels, will show a definite loss in ductility when exposed to environments that strongly promote the entry of nascent hydrogen (Ref 62).

Because iron-base alloys are principal materials of construction, these alloys have been the focus of most of the studies relating to hydrogen effects. In addition, ferritic (body-centered cubic, or bcc) steels have a particular sensitivity to hydrogen. For these reasons, this discussion will focus on hydrogen effects on steel. Such hydrogen effects have been thoroughly described in a recent review of hydrogen damage (Ref 63).

In ferrous alloys, embrittlement by hydrogen is generally restricted to those alloys having a hardness of 22 HRC or greater. However, other forms of hydrogen damage, such as hydrogen attack or hydrogen blistering, are associated with unhardened low-alloy or carbon steels.

As the result of hydrogen absorption, embrittlement of a metal or alloy may influence subsequent mechanical behavior without producing immediate and resolvable damage within the metal structure. In this respect, it is an insidious and somewhat reversible process, unlike hydrogen attack or blistering.

Hydrogen embrittlement may occur as a result of acid pickling, electroplating, and aqueous corrosion, which are electrochemical processes involving the discharge of hydrogen ions. The resulting nascent hydrogen is a chemisorbed species on the metal surface and, if not evolved as a molecular product, can enter the metal. This entry (or charging) of hydrogen has been described in detail (Ref 64). For the case of pure iron, introduction of hydrogen into the metal at high rates can produce irreversible effects (Ref 64). The solubility of hydrogen in the lattice is known to increase with applied tensile stress (Ref 65). Its solubility in metals obeys Sievert's law, with the concentration being directly proportional to the square root of the pressure (or fugacity). Within the lattice, hydrogen has been shown to interact with dislocations in iron (Ref 66-69) and can thus affect subsequent plastic deformation.

In the case of pure titanium, as well as in alloys of titanium or zirconium, the entry of hydrogen can result in the formation of a hydride phase, thus reducing the ductility (Ref 62). A surface hydride phase has also been produced by hydrogen charging of nickel (Ref 70). An iron-chromium-nickel hydride has been proposed for austenitic stainless steel (Ref 71). Hydrogen charging of austenitic stainless steels is known to produce a martensite phase responsible for microcracking (Ref 72). Similar microcracks have been reported for cold-worked Inconel 600 (Ref 73). Studies of hydrogen effects on aluminum alloys have not been conclusive (Ref 74).

It is known for the case of low-carbon steel that hydrogen absorption is enhanced by plastic deformation (Ref 75). This occurs up to 76% cold reduction, as shown in Fig. 2. However, the hydrogen permeation rate is increased by reduction up to only 15% and is decreased by further plastic deformation (Ref 63). In the presence of an applied stress, the deleterious effect of hydrogen on delayed failure is increased by the triaxiality of the stress state, as shown in Fig. 3.

For heat-treated steels, sensitivity to hydrogen is a strong function of strength level. Figure 4 shows substantial embrittlement in a quenched-and-tempered steel with a tensile strength of 1860 MPa (270 ksi). The deleterious hydrogen content of 0.5 cm^3/100 g corresponds to approximately 1 hydrogen atom per 40 000 metal atoms (Ref 77). Other studies have shown substantial embrittlement of high-strength steels at hydrogen contents of 2 to 3 ppm (Ref 78, 79).

The effects of temperature and strain rate on the embrittlement of carbon steel have been carefully studied (Ref 80). For a spheroidized AISI 1020 steel charged with hydrogen, the maximum embrittling effect appears to occur in the temperature range of −31 to 24 °C (−25 to 75 °F). Embrittlement is also enhanced by lower strain rates (5×10^{-2} in./in./min) and is insignificant at higher strain rates of the order of 2×10^4 in./in./min. Figure 5 shows such an effect of strain rate. A similar effect of strain rate has been shown for nickel and an iron-nickel alloy charged with hydrogen (Ref 81).

In aqueous systems, the entry of hydrogen is promoted by cathodic poisons that inhibit the recombination of adsorbed, discharged nascent hydrogen on the surface of the corroding metal. These poisons include cyanide and ionic species of sulfur, arsenic, selenium, bismuth, tellurium, phosphorus, iodine, and antimony.

Embrittlement by hydrogen-containing gas species may manifest itself in the stress-assisted propagation of preexisting surface defects

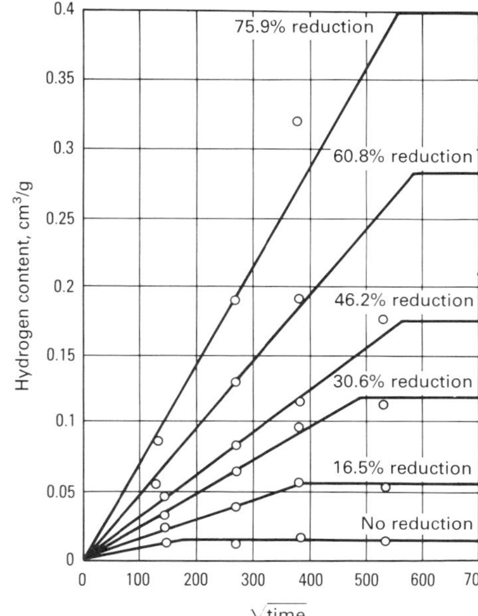

Fig. 2 Effect of cold reduction on rate of hydrogen absorption in carbon steel. Test performed on wire stock in 1 *N* H_2SO_4 at 35 °C (95 °F); original rod diameter was 13 mm (1/2 in.). Source: Ref 75

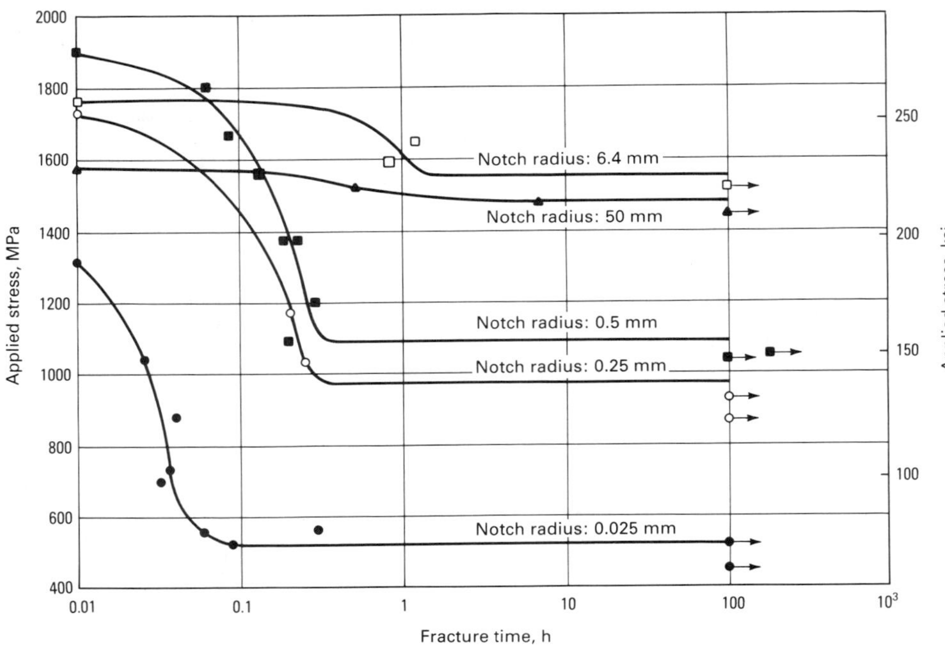

Fig. 3 Static fatigue curves for specimens of different notch sharpness. All specimens were baked for 30 min at 150 °C (300 °F). Source: Ref 76

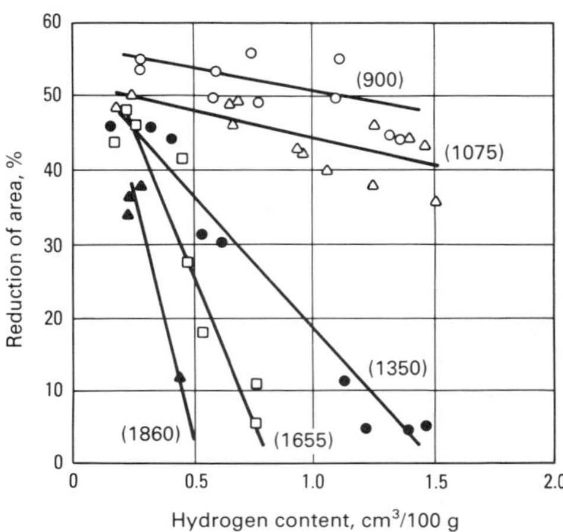

Fig. 4 Ductility (measured as percent reduction of area) versus hydrogen content for quenched-and-tempered steel at various strength levels. Ultimate tensile strength in megapascals is indicated in parentheses beside the curves. Source: Ref 77

(cracks). In the case of high-strength steels, such a phenomenon has been observed in H_2, H_2S, HCl, and HBr gas environments (Ref 82-85). A hydrogen sulfide (versus H_2) environment will result in a much higher crack velocity for the same steel, applied mechanical stress, and gas pressure (Ref 84).

One key to removing detrimental atomic hydrogen derives from its mobility at higher temperatures. A bake-out cycle, involving temperatures of 175 to 205 °C (350 to 400 °F), allows the diffusion and escape of hydrogen from the metal or alloy. If the hydrogen charging conditions were not severe enough to cause internal damage, the bake-out cycle (described in Federal Specification QQC-320) will restore full ductility.

In aqueous systems, the entry of hydrogen can be promoted by galvanic coupling to a more active metal. Therefore, cadmium or zinc platings containing defects or holidays can promote the hydrogen charging and embrittlement of susceptible metals or alloys. Such platings, or hot-dip galvanizing, should be avoided in the case of low-alloy high-strength steels. As another example, titanium has been found to embrittle by hydriding when it is coupled with carbon steel in acidic environments or when it is exposed to moist environments after the application of a zinc chromate paint primer.

Susceptibility to hydrogen embrittlement is strongly influenced by the strength level of the metal or alloy. Wet hydrogen sulfide environments are considered to be among the most aggressive in promoting hydrogen entry. In such environments, common metals and alloys are qualified according to strength level and/or heat treatment in terms of their resistance to hydrogen-induced failure. These qualifications are summarized in NACE standards MR-01-75, MR-01-76, and RP-04-75. In general, iron-base alloys with a ferritic or martensitic structure are restricted to a maximum hardness of 22 HRC. Most other alloys are restricted to a maximum hardness of 35 HRC. There are exceptions in both cases. The procedure for materials testing in a wet hydrogen sulfide environment are discussed in NACE TM-01-77.

Face-centered cubic (fcc) metals or alloys generally have more resistance to hydrogen embrittlement than those having a bcc lattice structure. However, cold work can increase the susceptibility of either structure. As a more direct effect of hydrogen entry, formation of a hydride phase can be expected to reduce the ductility of any metal or alloy.

In the pickling of steel, the level of hydrogen absorption is strongly affected by both the bath temperature and the nature of the acid (Fig. 6). Cathodic poisons have been ranked according to their effectiveness in increasing the permeation rate of hydrogen through low-carbon steel, as follows (Ref 87):

$$As > Se > Te > S > Bi$$

The role of sulfur as a poison is particularly important because sulfur is commonly encountered and because the chemical form of the sulfur greatly influences its effectiveness as a hydrogen entry promoter. The susceptibility to embrittlement by hydrogen can be demonstrated by the relative resistance to cracking in such environments as wet hydrogen sulfide. In such tests, microstructure has a definite effect on suscepti-

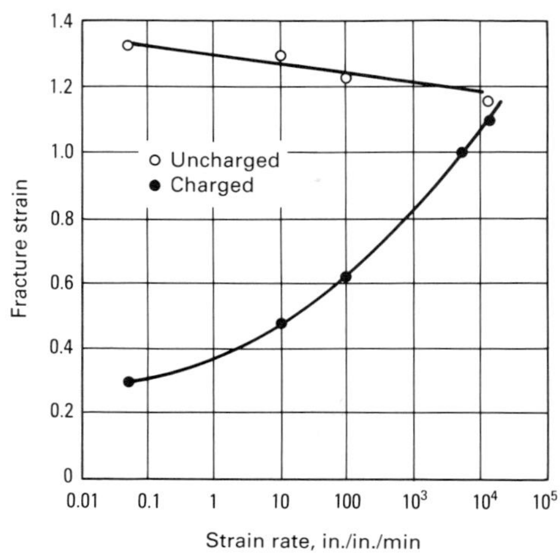

Fig. 5 Fracture strain as a function of strain rate in hydrogen-charged and uncharged 1020 steel at room temperature. Source: Ref 80

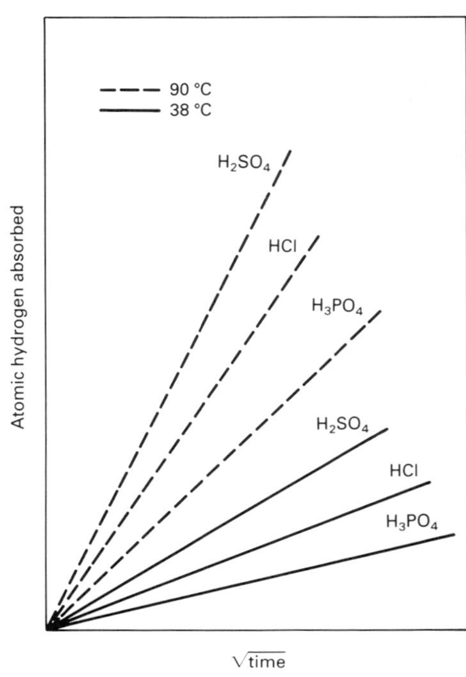

Fig. 6 Effect of anion and temperature on hydrogen absorption in a low-carbon steel. All acid concentrations were 2 *N*. Source: Ref 86

bility. In steels, untempered martensite is the most susceptible phase. Lamellar carbide structures are less desirable than those with spheroidized structures. Quenched-and-tempered microstructures are more resistant than those that have been normalized and tempered (Ref 88). For the same strength level in low-alloy steel, it has been shown that a bainitic structure is more resistant to hydrogen-assisted cracking than a quenched-and-tempered martensitic structure (Ref 89).

Embrittlement by gaseous hydrogen environments at ambient temperature has been effectively inhibited by the addition of 0.4 to 0.7 vol% oxygen (Ref 82, 90, 91). However, similar additions to a hydrogen sulfide gas environment did not halt the growth of cracks (Ref 91, 92).

Because of the higher hydrogen solubility in the high-temperature fcc structure of iron (versus the low-temperature bcc structure), cooling of steel in hydrogen atmospheres from temperatures of the order of 1100 °C (2010 °F) can result in internal damage. Exceeding the solubility limit for hydrogen will result in the embrittlement of hydrogen-sensitive microstructures, such as martensite, formed by rapid cooling of some ferritic alloys. The internal precipitation of hydrogen is believed to be responsible for the generation of fissures, delaminations, or other defects. Such defects have been termed flakes, shatter-cracks, fisheyes, or snowflakes. The defects are generally associated with hydrogen precipitation at voids, laminations, or inclusion/matrix interfaces already present in the steel. A reduced cooling rate, which allows hydrogen to be slowly released from the steel, is a general solution to the problem. Slower cooling will also inhibit the formation of hydrogen-sensitive microstructures.

Underbead cracking is an embrittlement phenomenon that is associated with absorption of hydrogen by molten metal during the welding process. Sources of hydrogen include moisture or organic contaminants on the surface of the prepared joint, moisture in low-hydrogen coated electrodes (such as E7018), moisture in flux-cored wire (such as M16), or a high-humidity environment. Upon rapid cooling of the weld, entrapped hydrogen can produce internal fissuring or other damage, as described earlier. In addition, the weld HAZ may contain the martensite phase in quench-hardenable alloys. The HAZ is then embrittled by high levels of entrapped hydrogen. Several steels have exhibited susceptibility to such embrittlement—for example, carbon steels containing 0.25 to 0.35 wt% C, low-alloy steels (such as AISI 4140 or 4340), and martensitic or precipitation-hardening stainless steels. Solutions to the hydrogen damage problems associated with welding include the use of dry welding electrodes, proper cleaning and degreasing procedures for prepared weld joints, the use of an appropriate preheat before welding, and an adequate postweld heat treatment.

Welding electrodes should be kept dry by using a heated rod box. The electrodes should be removed only as needed. If they are moistened or exposed in the ambient atmosphere for prolonged periods, low-hydrogen coated electrodes must be heated at 370 to 425 °C (700 to 800 °F) to remove moisture (Ref 63). Recommended preheat temperatures for steels, as a function of steel composition, section thickness, and electrode type, have been published (Ref 93). Welding procedures for the avoidance of hydrogen cracking in carbon-manganese steels have also been published (Ref 94). Appropriate postweld heat treatments for steels can range from a hydrogen bake-out at 175 °C (350 °F) to a martensite tempering treatment at temperatures as high as 705 °C (1300 °F) (Ref 63).

Hydrogen attack is a damage mechanism that is associated with unhardened carbon and low-alloy steels exposed to hydrogen-containing environments at temperatures above 220 °C (430 °F) (Ref 63). Exposure to the environment is known to result in a direct chemical reaction with the carbon in the steel. The reaction occurs between absorbed hydrogen and the iron carbide phase, resulting in the formation of methane:

$$2H_2 + Fe_3C \rightarrow CH_4 + 3Fe$$

Unlike nascent hydrogen, the resulting methane gas does not dissolve in the iron lattice. Internal gas pressures develop, leading to the formation of voids, blisters, or cracks. The generated defects lower the strength and ductility of the steel. Because the carbide phase is a reactant in the mechanism, its absence in the vicinity of generated defects serves as direct evidence of the mechanism itself. The recommended service conditions (temperature, hydrogen pressure) for carbon and low-alloy steels are shown by the respective Nelson curves in Fig. 7. Chromium and molybdenum are beneficial alloying elements. This is most likely the result of their high affinity for carbon as well as the stability of their carbides. Hydrogen attack does not occur in austenitic stainless steels (Ref 63). In carbon or low-alloy steels, the extent of hydrogen attack is a function of exposure time.

Hydrogen blistering is a mechanism that involves hydrogen damage of unhardened steels near ambient temperature. It is known that the entry of atomic hydrogen into steel can result in its collection, as the molecular species, at internal defects or interfaces. If the entry kinetics are substantial (promoted by an acidic environment, high corrosion rates, and cathodic poisons), the resulting internal pressure will cause internal separation (fissuring or blistering) of the steel. Such damage typically occurs at large, elongated inclusions and results in delaminations known as hydrogen blisters. Field experience indicates that fully killed steels are more susceptible than semikilled steels (Ref 95), but the nature and size of the original inclusions appear to be the key factors with regard to susceptibility. Rimmed steels or free-machining grades with high levels of sulfur or selenium would most likely show a high susceptibility to blistering. Stepwise cracking at the ends of blisters indicates an effect of elongated inclusions in the delamination process (Ref 63, 95). Similar stepwise cracking occurs in the hydrogen-induced failure of low-alloy pipeline steels (Ref 96). Both stepwise cracking and blistering appear to be limited to environments in which acidic corrosion occurs and in which cathodic poisons, such as sulfide, are present to promote hydrogen entry.

Solutions to the blistering problem include the use of low-sulfur calcium-treated argon-blown steels. Hot-rolled or annealed (as opposed to cold-rolled) steel is preferred (Ref 63). Silicon-killed steels are preferable to aluminum-killed steels. Also, treatment with synthetic slag or the addition of rare-earth metals can favor the formation of less detrimental globular sulfides (Ref 97). Ultrasonic inspection of the steel (according to ASTM E 114 and A 578) should be done before fabrication to detect laminations and other discontinuities that will promote blister formation. Equipment inspection and blister-venting procedures require unusual care (Ref 63). In services in which blistering can be expected, external support pads should not be continuously welded to the vessel itself. This will prevent hydrogen entrapment at the interface.

The permeation of hydrogen through ferritic steels can produce physical separation at mechanical joints. For example, bimetallic tubes, with a carbon steel inner liner, exhibited collapse of the liner due to its exposure to HF. Acid corrosion of the inside surface allowed nascent hydrogen to permeate the steel. Molecular hydro-

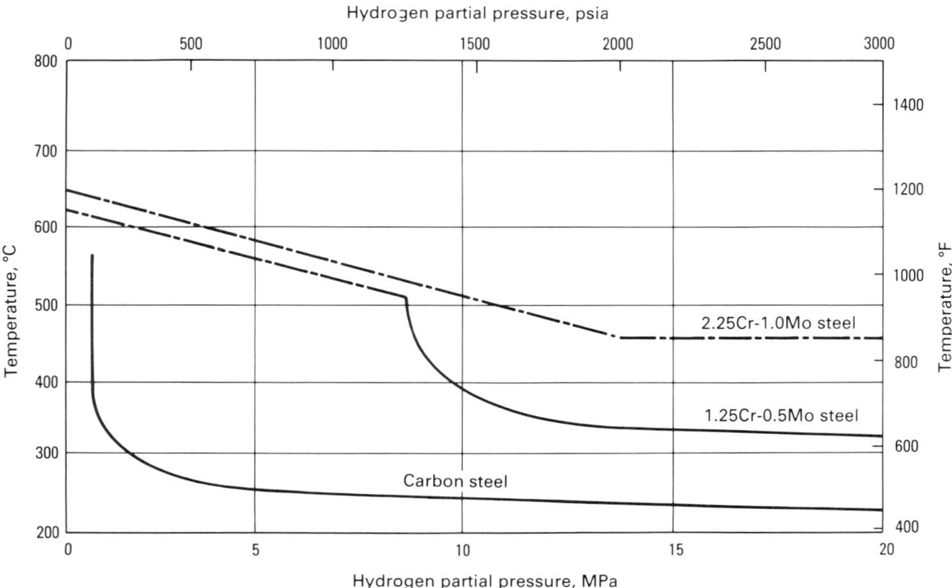

Fig. 7 Operating limits for three steels in hydrogen service to avoid hydrogen attack. Dashed lines show limits for decarburization, not hydrogen attack. Source: Ref 63

gen gas was formed, and trapped, at the interface with the outer tube section (brass). The accumulation of pressure was found to collapse the inner steel liners (Ref 63).

In high-temperature H_2/H_2S service, weld overlaid 2.25Cr-1Mo steel was found to disbond at the weld interface (Ref 98). In this case, a weld overlay of type 309 stainless steel, followed by type 347 stainless steel, was applied. Hydrogen-induced cracking was found to occur in the transition zone below the weld metal after approximately 3½ years of service. The disbonding was found to be more severe with higher cooling rates after hydrogen absorption. Out-gassing treatments during the cool down were found to prevent disbonding (Ref 99).

Figure 8 shows an example of hydrogen-assisted SCC failure of four AISI 4137 steel bolts having a hardness of 42 HRC. Although the normal service temperature (400 °C, or 750 °F) was too high for hydrogen embrittlement, the bolts were also subjected to extended shutdown periods at ambient temperatures. The corrosive environment contained trace hydrogen chloride and acetic acid vapors as well as calcium chloride if leaks occurred. The exact service life was unknown. The bolt surfaces showed extensive corrosion deposits. Cracks had initiated at both the thread roots and the fillet under the bolt head. Figure 8(b) shows a longitudinal section through the failed end of one bolt. Multiple, branched cracking was present, typical of hydrogen-assisted SCC in hardened steels. Chlorides were detected within the cracks and on the fracture surface. The failed bolts were replaced with 17-4 PH stainless steel bolts (Condition H1150M) having a hardness of 22 HRC (Ref 63).

As an example of hydrogen attack, a section of plain carbon steel (0.22% C and 0.31% Si) had been mistakenly included as a part of a type 304 stainless steel hot-gas bypass line used to handle hydrogen-rich gas at 34 MPa (5000 psi) and 320 °C (610 °F). After 15 months of service, the steel pipe section ruptured, causing a serious fire. Figure 9 shows a section of the 44-mm (1¾-in.) OD pipe near the fracture. The pipe had been weakened by hydrogen attack through all but 0.8 mm (1/32 in.) of the 8-mm (5/16-in.) thick wall. As a result of the hydrogen attack and the internal methane formation, the microstructural damage consisted of holes or voids near the outer surface as well as interconnected grain-boundary fissures in a radial alignment near the inner surface (Fig. 9b). The radially aligned voids preceded both the circumferential crack and pipe rupture (Ref 63).

Hydrogen blistering is illustrated in Fig. 10, which shows a cross section of a 152-mm (6-in.) diam blister that formed in the wall of a steel sphere. The sphere had been used to store anhydrous HF for 13.5 years at ambient temperatures. The source of nascent hydrogen gas was the cathodic hydrogen generated by the corrosion reaction between the acid and the steel. The corrosion rate was less than 0.05 mm/yr (2 mils/yr). Figure 10(b) shows the propagation of the blister, with the stepwise cracking (arrow) at its edge caused by the buildup of hydrogen pressure within the blister itself (Ref 63). More information on hydrogen attack is available in the article "Environmentally Induced Cracking" in this Volume.

(a)

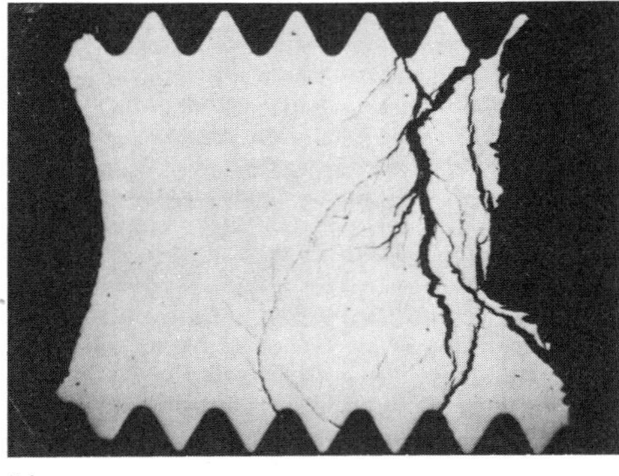

(b)

Fig. 8 4137 steel bolts (hardness: 42 HRC) that failed by hydrogen-assisted SCC caused by acidic chlorides from a leaking polymer solution. (a) Overall view of failed bolts. (b) Longitudinal section through one of the failed bolts in (a) showing multiple, branched hydrogen-assisted stress-corrosion cracks initiating from the thread roots. Source: Ref 63

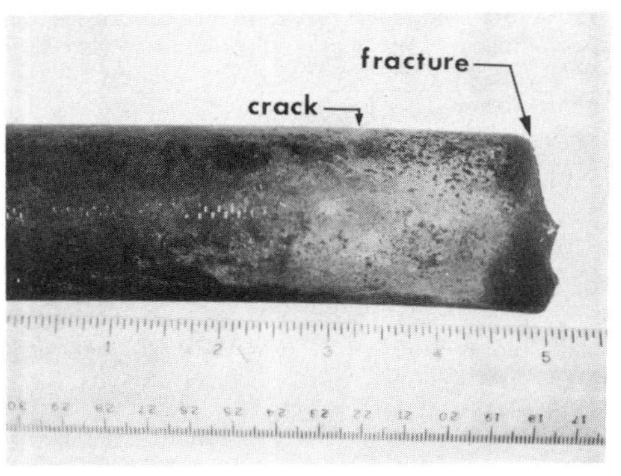

(a)

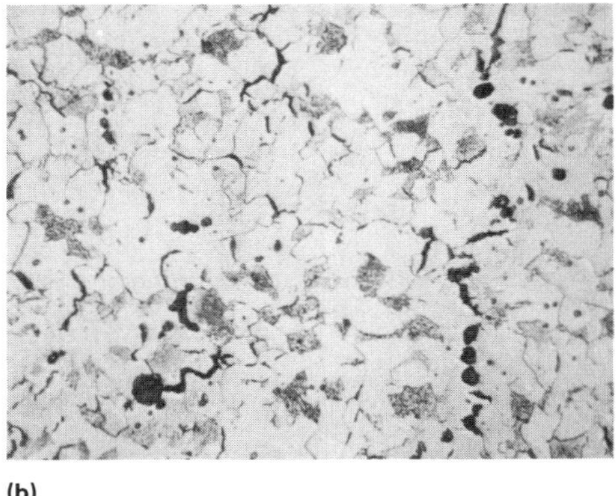

(b)

Fig. 9 Section of ASTM A106 carbon steel pipe with wall severely damaged by hydrogen attack. The pipe failed after 15 months of service in hydrogen-rich gas at 34.5 MPa (5000 psig) and 320 °C (610 °F). (a) Overall view of failed pipe section. (b) Microstructure of hydrogen-attacked pipe near the midwall. Hydrogen attack produced grain-boundary fissures that are radially aligned. Source: Ref 63

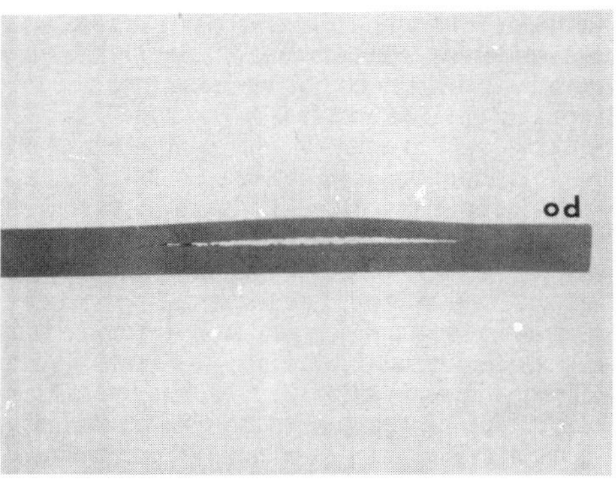

(a)

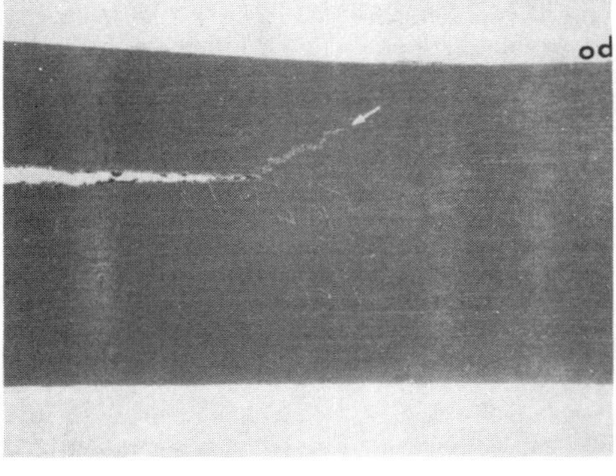

(b)

Fig. 10 Hydrogen blister in 19-mm (¾-in.) steel plate from a spherical tank used to store anhydrous HF for 13.5 years. (a) Cross section of 152-mm (6-in.) diam blister. (b) Stepwise cracking (arrow) at edge of hydrogen blister shown in (a). Source: Ref 63

Erosion-corrosion is a frequently misinterpreted type of metal deterioration that results from the combined action of erosion and corrosion. This section will be limited to a discussion of three types—liquid erosion-corrosion, cavitation, and fretting. Abrasive wear, which is erosion without corrosion, will also be discussed for comparison purposes.

Liquid erosion-corrosion is the accelerated wastage of a metal or material attributed to the flow of a liquid (Ref 100-102). Liquid erosion-corrosion damage is characterized by grooves, waves, gullies, rounded holes, and/or horseshoe-shaped grooves. Analysis of these marks can help determine the direction of flow. Most metals are susceptible to liquid erosion-corrosion under specific conditions. Carbon steels, for example, can be severely damaged by steam containing entrained water droplets. By contrast, the 300-series stainless steels at about the same hardness and strength level are very resistant to flowing wet steam. Virtually anything that is exposed to a moving liquid is susceptible to liquid erosion-corrosion. Examples include piping systems, particularly at bends, elbows, or wherever there is a change in flow direction or increase in turbulence; pumps; valves, especially flow control and pressure let-down valves; centrifuges; tubular heat exchangers; impellers; and turbine blades.

Surface films that form on some metals and alloys are very important in their ability to enhance resistance to liquid erosion-corrosion. Titanium is a reactive metal, but is resistant to liquid erosion-corrosion in many environments because of its very stable titanium dioxide surface film. The 300-series stainless steels, as mentioned above, are also resistant because of their stable passive surface films.

Both carbon steel and lead have relatively good resistance to certain concentrations of H_2SO_4 under low-to-moderate flow conditions. Both depend on a metal sulfate corrosion product film for resistance; however, both will fail fairly rapidly after removal of the sulfate film, even at low velocities.

Another example is the carbon steel and some low-alloy steels used to handle petroleum refinery fluids that contain hydrogen sulfide. At low velocities or under stagnant conditions, these materials are normally satisfactory because of formation of a tenacious protective iron sulfide film. However, with increased velocity, the film is eroded away, and very rapid attack occurs.

Velocity often increases attack, but it may also decrease attack, depending on the material of construction and the corrosive environment. For example, increasing the velocity causes accelerated attack of carbon steel in steam condensate by increasing the supply of dissolved oxygen and/or carbon dioxide to the steel surface. In cooling water, however, increased velocity often reduces the attack of carbon steel by improving the effectiveness of inhibitors and by reducing deposits and pitting in stagnant areas.

Many 300-series stainless steels are subject to pitting and crevice corrosion in seawater. However, they may exhibit good resistance if the seawater is kept flowing at a minimum critical velocity. This prevents the formation of deposits and retards general corrosion, pitting, and crevice attack. Table 4 shows the effects different seawater velocities have on the liquid erosion-corrosion of various metals.

Cavitation is a form of erosion-corrosion that is caused by the formation and collapse of vapor bubbles in a liquid against a metal surface (Ref 102-106). Cavitation occurs in hydraulic turbines, on pump impellers, on ship propellers, and on many surfaces in contact with high-velocity liquids subject to changes in pressure. Cavitation can occur if the pressure on a liquid is reduced sufficiently to cause boiling even at room temperature. Boiling produces bubbles that collapse on the high-pressure cycle. Rapidly collapsing vapor bubbles produce shock waves that have developed pressures as high as 414 MPa (60 ksi).

The appearance of cavitation (Fig. 11) is similar to pitting except that surfaces in the pits are usually much rougher. Most investigators believe that cavitation damage is caused by a combination of corrosion and mechanical effects. Apparently, the collapsing vapor bubbles mechanically destroy the protective surface films. Thus, fresh surfaces are exposed to corrosion and the reestablishment of protective films, which is followed by more cavitation, and so on. Damage occurs when the cycle is allowed to repeat over and over again, for example, when a pump is operated "dead headed" against a closed valve. Table 5 shows the relative cavitation resistance of a variety of metals and alloys in water and seawater.

Table 4 Corrosion of metals and alloys in seawater as a function of velocity

Material	Typical corrosion rate, mg/dm²/d 0.3 m/s (1 ft/s)(a)	1.2 m/s (4 ft/s)(b)	8.2 m/s (27 ft/s)(c)
Carbon steel	34	72	254
Cast iron	45	...	270
Silicon bronze	1	2	343
Admiralty brass	2	20	170
Hydraulic bronze	4	1	339
G bronze	7	2	280
10% aluminum bronze	5	...	236
Aluminum brass	2	...	105
90Cu-10Ni (0.8% Fe)	5	...	99
70Cu-30Ni (0.5% Fe)	<1	<1	39
Monel 400	<1	<1	4
Type 316 stainless steel	1	0	<1
Hastelloy C	<1	...	3
Titanium	0	...	0

(a) Immersed in tidal current. (b) Immersed in seawater flume. (c) Attached to immersed rotating disk. Source: International Nickel Company

Fretting is erosion-corrosion that occurs at the contact area between two metals under load and subject to slight relative movement by vibration or some other force (Ref 108-110). Damage begins with local adhesion between mating surfaces and progresses when adhered particles are ripped from a surface and react with air or other corrosive environment. Affected surfaces show pits or grooves with surrounding corrosion products. On ferrous metals, corrosion product is usually a very fine, reddish iron oxide; on aluminum, it is usually black.

Fretting is detrimental not only because of the destruction of metallic surfaces but also because of a severe effect on the fatigue life. It has been shown that fretting can reduce the endurance limit of a metal by 50 to 70% (Ref 109).

The relative motion necessary to produce fretting is very small. Displacements as small as 10^{-8} cm have produced fretting. Fretting generally does not occur on contacting surfaces in continuous motion, such as ball or sleeve bearings.

Fig. 11 Internal surface of carbon steel pipe section damaged by cavitation

Fretting can be minimized or eliminated in many cases by one or more of the following:

- Increasing the hardness of contacting surfaces. This may mean increasing the hardness of both or just one of the components. Surface-hardening treatments such as shot peening, nitriding, chrome plating, and carburizing are beneficial
- Increasing the friction between the mating members by roughening or by plating (lead, copper, nickel, silver, gold)
- Applying phosphate coatings to exclude air or applying anaerobic sealants or adhesives to increase the tightness of the fit
- Increasing the fit interference, which reduces slippage by increasing the force on mating components
- Switching to materials with more fretting resistance, as shown in Table 6

Abrasive wear is damage that results from the action of hard particles on a surface under the influence of a force that is oblique to the surface (Ref 112-117). This is not, strictly speaking, a form of erosion-corrosion, but will be briefly discussed for comparison with the forms of erosion-corrosion mentioned above.

Three common forms of abrasive wear are erosion abrasion, grinding abrasion, and gouging abrasion. Erosion abrasion usually involves low velocities and weak support of the abrasive material. Examples are wear on a plowshare in sandy soil and polishing of a metal surface with an abrasive held in a soft cloth. Thus, the energy of the abrasive is quite low, and impact is absent.

Grinding abrasion is the fragmentation of the abrasive, usually between two strong surfaces. Examples are a lapping operation in a machine shop and ball/rod mill grinding. Thus, impact is low to moderate, but the gross stress may be quite high, at least on a microscopic scale.

Gouging abrasion is recognized by the prominent grooves or gouges that are present on the wearing surfaces. Examples are abrasive disk grinding, machine tool cutting, and wear of power shovel bucket teeth. Heavy impact is generally associated with this type of abrasion, along with gross stress.

To alleviate these forms of abrasion, a careful study of the type of abrasion and an understanding of the service conditions are required. The material selection should be based on the known properties of materials versus service requirements. More information on metal wear and corrosion is available in the article "Mechanically Assisted Degradation" in this Volume; the article "Wear Failures" in Volume 11 of the 9th Edition of *Metals Handbook* contains information on abrasion and wear.

Other Forms of Corrosion. Selective leaching, also known as dealloying or parting corrosion, occurs when one element is preferentially removed from an alloy, leaving an often porous residue of an element that is more resistant to the environment. It is a problem of commercial significance in copper alloy systems (Ref 118), primarily copper-zinc and copper-aluminum and, to a lesser extent, copper-nickel. The terms dezincification, dealuminization, and denickelification describe the selective leaching of zinc, aluminum, and nickel, respectively, from the alloys. In these cases, a porous residue of copper remains, either as a fairly uniform layer or in plugs. The latter is more damaging in that the effect is similar to pitting corrosion.

Table 5 Resistance of metals and alloys to cavitation damage in vibratory testing at 25 °C (75 °F)

Metal or alloy	Product form	Composition, %	Weight loss for last 60 min of exposure, mg/h Freshwater	Seawater
Ferrous				
Iron	Cast	Fe-3.1C-2.3Si-0.75Mn-0.12S-0.07P	50.1	80.9
Iron	Cast	Fe-3.4C-1.3Si-0.75Mn-0.25P-0.08S	69.8	115.3
Iron	Cast	Fe-3.4C-2.3Si-0.59Mn	89.7	100.2
Iron	Cast	Fe-3C-6Cu-4Cr-14.4Ni-1.9Si	41.6	51.4
Iron	Cast	Fe-3.3C-1.3Si-0.4Mo-0.51Mn	54.1	63.9
Iron	Cast	Fe-3C-6Cu-13.5Ni-2Cr-1.5Si-1Mn-0.1S-0.04P	85.3	95.3
Steel	Rolled	Fe-0.35C-0.45P-0.67Mn	34.2	39.6
Steel	Rolled	Fe-0.27C-0.4S-0.45P-0.48Mn	68.3	77.8
Steel	Rolled	Fe-0.2C-0.03S-0.02P-0.5Mn	78.2	82.4
Steel	Cast	Fe-0.37C-0.31Si-0.04S-0.04P-1.1Mn	44.8	53.6
Steel	Cast	Fe-0.26C-0.32Si-0.04S-0.04P-0.6Mn	72.9	80.9
Steel	Rolled	Fe-0.34C-1.18Ni-0.6Cr-0.52Mn-0.2Si-0.03S-0.02P	20.0	22.0
Steel	...	Fe-0.19C-2.2Ni-0.6Mn-0.02S-0.02P	61.3	64.0
Stainless steel	Rolled	Fe-17.2Cr-0.08C-0.57Si-0.47Mn-0.34Ni-0.02S-0.03P	11.8	10.8
Stainless steel	Rolled	Fe-12.2Cr-0.09C-0.38Si-0.43Mn-0.32Ni-0.02S-0.02P	20.6	23.0
Stainless steel	Cast	Fe-18Cr-10Ni-0.15C-0.5Si-0.5Mn	13.5	13.4
Stainless steel	Rolled	Fe-18.4Cr-8.7Ni-0.07C-0.37Si-0.48Mn-0.14S-0.19P	16.1	15.3
Nonferrous				
Bronze	Rolled	Cu-39Zn-1Sn	69.5	65.2
Brass	Rolled	Cu-40Zn	77.8	68.7
Brass	Rolled	Cu-15Zn	115.2	101.3
Brass	Rolled	Cu-10Zn	134.9	122.8
Bronze	Cast	Cu-10Al(a)	15.3	14.5
Bronze	Cast	Cu-11Sn-1.5Si	54.6	62.4
Bronze	Cast	Cu-10Sn-2Pb	60.4	48.5
Bronze	Cast	Cu-3.5Si(a)	42.6	40.4
Bronze	Cast	Cu-5Si-1Mn	52.4	54.5
Bronze	Forged	Cu-25Zn(a)	19.2	19.9
Bronze	Cast	Cu-40Zn-1Fe	53.0	55.4
Bronze	Cast	Cu-10Sn-2Zn	65.8	57.4
Nickel alloy	Cast	Ni-32.5Cu-4Si-2Fe	20.0	21.4
Nickel alloy	Drawn	Ni-29Cu-1Mn-1Fe	53.3	53.2
Copper-nickel	Rolled	Cu-30Ni	86.2	87.6

(a) 1.0% maximum present, but not determined analytically. Source: Ref 107

Selective leaching in copper alloy systems occurs primarily in certain waters, especially under deposits in stagnant areas in heat exchangers. Alloy additions of arsenic, antimony, or phosphorus are effective in inhibiting this attack, but only in copper-zinc alloys. Thus, arsenical or antimonial admiralty brass (UNS C44300 and C44400, respectively) is specified, for example, where this alloy is required for water service.

Graphitic corrosion of cast iron is another commercially important form of selective leaching. In this case, the iron matrix corrodes, leaving behind a porous graphite mass that can be carved with a pocket knife. Cast iron underground municipal watermains (Ref 119) and fire watermains at petrochemical plant sites are affected by graphitic corrosion from both the soil and water sides. Internal cement linings and external protective coatings, with cathodic protection in severely corrosive soils, are relatively low-cost solutions to watermain corrosion problems. The section "Dealloying Corrosion" of the article "Metallurgically Influenced Corrosion" in this Volume contains more information on the phenomenon of dealloying.

Exfoliation is a form of localized corrosion that primarily affects aluminum alloys. Corrosion proceeds laterally from initiation sites on the surface and generally proceeds intergranularly along planes parallel to the surface. The corrosion products that form in the grain boundaries force metal away from the underlying base material, resulting in a layered or flakelike appearance. Extruded products from the 2000-series copper-magnesium alloys, the 7000-series zinc-copper-magnesium alloys, and, to a lesser extent, the 5000-series alloys are particularly susceptible to exfoliation in both marine and industrial environments. Also, at least one case affecting 6000-series magnesium-silicon alloys in freshwater service has been reported (Ref 120).

This attack is generally associated with the alloy fabrication method and temper, impurities in the alloy matrix, and the distribution of intermetallic compounds at the surface and in grain boundaries. Aluminum alloys 1100, 3003, and 5052 are resistant. Standard test methods for determining susceptibility to exfoliation corrosion in aluminum alloys are covered in ASTM standards G 34 and G 66.

Liquid-metal embrittlement (LME), also known as liquid metal assisted cracking, is not considered to be a corrosion phenomenon, except in cases involving aqueous mercury compounds (Ref 121). However, LME is discussed here because it is a problem frequently encountered by materials engineers. Liquid-metal embrittlement is the penetration, usually along grain boundaries, of metals and alloys by such metals as mercury, which are liquid at room temperature, and metals that have relatively low melting points, such as bismuth, tin, lead, cadmium, zinc, aluminum, and copper. Stress, temperature, and time are the factors that facilitate and accelerate LME. Virtually all metal and alloy systems are subject to LME by one or more of these metals at or above their melting points.

Zinc is a prime offender because of widespread use throughout industry in the form of corrosion-resistant coatings applied to carbon steels by hot-dip galvanizing, electroplating, tumbling, and spray painting. Plain carbon steels are embrittled by zinc at temperatures above 370 °C (700 °F) for long periods of time, especially when the steel is heavily stressed or cold worked.

Austenitic stainless steels and nickel-base alloys will also crack in the presence of molten zinc. These alloys usually crack instantly when welded to galvanized steel, a fairly common occurrence in the chemical-processing industry. In addition, austenitic alloy failures have occurred:

- In high-temperature bolting fastened with galvanized steel nuts
- During welding or heat treating of components contaminated by grinding with zinc-loaded grinding wheels, contact with zinc-coated structurals or slings, or exposure to zinc paint overspray
- During process industry plant fires involving piping and vessels (thin-wall expansion joint bellows are especially susceptible) sprayed with molten zinc from coated steel structures

Thus, it is imperative that all traces of zinc be removed from coated steel members before welding to austenitic alloys and before intimate contact with these alloys at temperatures above 370 °C (700 °F). Also, austenitic stainless steels and nickel-base alloys should be handled with non-coated steel hoist chains, cables, and structurals; they should be dressed and cleaned with new grinding wheels and stainless steel brushes, and they should be marked with materials (paints,

Table 6 Relative fretting resistance of various material combinations

Combination	Fretting resistance
Aluminum on cast iron	Poor
Aluminum on stainless steel	Poor
Bakelite on cast iron	Poor
Cast iron on cast iron, with shellac coating	Poor
Cast iron on chromium plating	Poor
Cast iron on tin plating	Poor
Chromium plating on chromium plating	Poor
Hard tool steel on stainless steel	Poor
Laminated plastic on cast iron	Poor
Magnesium on cast iron	Poor
Brass on cast iron	Average
Cast iron on amalgamated copper plate	Average
Cast iron on cast iron	Average
Cast iron on cast iron, rough surface	Average
Cast iron on copper plating	Average
Cast iron on silver plating	Average
Copper on cast iron	Average
Magnesium on copper plating	Average
Zinc on cast iron	Average
Zirconium on zirconium	Average
Cast iron on cast iron with coating of rubber cement	Good
Cast iron on cast iron with Molykote lubricant	Good
Cast iron on cast iron with phosphate conversion coating	Good
Cast iron on cast iron with rubber gasket	Good
Cast iron on cast iron with tungsten sulfide coating	Good
Cast iron on stainless steel with Molykote lubricant	Good
Cold-rolled steel on cold-rolled steel	Good
Hard tool steel on tool steel	Good
Laminated plastic on gold plating	Good

Source: Ref 111

crayons, and so on) free from zinc and other low-melting metals.

Cadmium is probably second to zinc in importance as an agent of liquid-metal embrittlement, because of its application as a corrosion-resistant coating to a variety of hardware, particularly fasteners. Failures by cadmium LME of bolting operating at temperatures above 300 °C (570 °F) and fabricated from such high-strength alloy steels as AISI 4140 and 4340 and austenitic stainless steels are fairly common. In fact, some high-strength steels and high-strength titanium alloys are embrittled by cadmium at temperatures below its melting point by mechanisms not yet understood. The solution to LME by cadmium is similar to that of zinc, that is, avoidance of contact with, and contamination of, susceptible metal and alloy systems at temperatures above the 321 °C (610 °F) melting point of cadmium (and at all temperatures at which high-strength steels and titanium alloys are involved).

Metal systems that are embrittled by contact with mercury include copper and its alloys, aluminum and its alloys, Nickel 200 (at elevated temperatures) and Monel alloy 400, and titanium and zirconium and their alloys. Cracking is intergranular except in zirconium alloys; in these alloys, cracking is transgranular. Mercury LME of aluminum and copper alloys was more common years ago in the petrochemical industry when mercury-filled manometers and thermometers were extensively used. Failures or upsets would release mercury into process or service (steam, cooling water, and so on) streams, causing widespread cracking of piping, heat exchanger tube bundles, and other equipment. Under these conditions, even pure aluminum and pure copper are susceptible. With regard to the titanium system, the commercially pure grades used in the chemical-processing industry are less sensitive than the alloys. In addition, LME in aqueous solutions of mercurous salts, such as mercurous nitrate, is possible because the mercurous ion can be reduced to its elemental form at local cathodic sites.

Although not a metal, sulfur will penetrate the grain boundaries of nickel and nickel alloys at elevated temperatures in much the same way as in the low-melting metals mentioned above. Sulfur forms a very aggressive nickel-nickel sulfide eutectic alloy that melts at about 625 °C (1157 °F). Sources other than elemental sulfur include organic compounds (greases, oils, cutting fluids) and sulfates. Thus, contamination from these sources before welding, hot forming, annealing, and other heating operations must be avoided (see the section "Liquid-Metal Embrittlement" of the article "Environmentally Induced Cracking" in this Volume).

Economics

Cost-Effective Materials Selection. The two extremes for selecting materials on an economic basis without consideration of other factors are (Ref 122):

- *Minimum cost:* Selection of the least expensive material, followed by scheduled periodic replacements or correction of problems as they arise
- *Minimum corrosion:* Selection of the most corrosion-resistant material regardless of installed cost or life of the equipment

Cost-effective selection generally falls somewhere between these extremes and includes consideration of other factors, such as availability and safety. For example, critical components in a large single-train chemical-processing plant should be fabricated from materials that tend toward the minimum corrosion extreme because failure could shut down the entire operation. However, component materials for a multitrain or batch operation, especially one that processes a relatively short-lived product, might tend toward the minimum cost extreme, even to the point of purchasing used equipment at a fraction of the cost of new fabrication. Thus, different strategies are appropriate for different situations. Additional information is available in the article "Corrosion Economic Calculations" in this Volume.

REFERENCES

1. R.J. Landrum, Designing for Corrosion Resistance, *Chem. Eng.*, 24 Feb 1969, p 118-124; 24 March 1969, p 172-180
2. *Corrosion Abstracts*, National Association of Corrosion Engineers
3. *Corrosion Data Survey—Metals Section*, 6th ed., 1985; *Corrosion Data Survey—Nonmetals Section*, National Association of Corrosion Engineers, 1975
4. C. Westcott *et al.*, "The Development and Application of Integrated Expert Systems and Data Bases for Corrosion Consultancy," Paper 54, presented at Corrosion/86, Houston, TX, National Association of Corrosion Engineers, March 1986
5. E.H. Schmauch *et al.*, "Expert Systems for Personal Computers," Paper 55, presented at Corrosion/86, Houston, TX, National Association of Corrosion Engineers, March 1986
6. S.E. Marschand *et al.*, "Expert Systems Developed by Corrosion Specialists," Paper 56, presented at Corrosion/86, Houston, TX, National Association of Corrosion Engineers, March 1986
7. W.F. Bogaerts *et al.*, "Artificial Intelligence, Expert Systems and Computer-Aided Engineering In Corrosion Control," Paper 58, presented at Corrosion/86, Houston, TX, National Association of Corrosion Engineers, March 1986
8. R.B. Puyear, Material Selection Criteria for Chemical Processing Equipment, *Met. Prog.*, Feb 1978, p 40-46
9. "Laboratory Corrosion Testing of Metals for the Process Industries," NACE TM-01-69 (1976 Revision); "Method of Conducting Controlled Velocity Laboratory Corrosion Tests," NACE TM-02-70, National Association of Corrosion Engineers
10. *Metal Corrosion, Erosion, and Wear*, Vol 03.02, Section 3, *Annual Book of ASTM Standards*, American Society for Testing and Materials, 1986
11. G. Kobrin, Evaluate Equipment Condition by Field Inspection and Tests, *Hydrocarbon Process.*, Jan 1970, p 115-120
12. B.J. Moniz, Field Identification of Metals, in *Process Industries Corrosion—The Theory and Practice*, National Association of Corrosion Engineers, 1986, p 839
13. B.J. Moniz, Field Identification of Metals, in *Process Industries Corrosion—The Theory and Practice*, National Association of Corrosion Engineers, 1986, p 842
14. Monitoring Internal Corrosion in Oil and Gas Production Operations With Hydrogen Probes, NACE Publication 1C 184, *Mater. Perform.*, June 1984, p 49-56
15. J.C. Bovankovich, On-Line Corrosion Monitoring, *Mater. Prot. Perform.*, June 1973, p 20-23
16. G. Kobrin, Reflections on Microbiologically Induced Corrosion of Stainless Steels, in *Biologically Induced Corrosion*, NACE-8, S.C. Dexter, Ed., National Association of Corrosion Engineers, 1986, p 33-46
17. *Guide to Engineered Materials*, Vol 1, ASM INTERNATIONAL, June 1986, p 79
18. R. Baboian, New Methods for Controlling Galvanic Corrosion, *Mach. Des.*, 11 Oct 1979, p 78-85
19. R.M. Davison, H.E. Deverell, and J.D. Redmond, Ferritic and Duplex Stainless Steels, in *Process Industries Corrosion—The Theory and Practice*, National Association of Corrosion Engineers, 1986, p 427-443
20. "Intergranular Corrosion of Chromium-Nickel Stainless Steels—Final Report," Bulletin 138, Welding Research Council, 1969
21. M.A. Streicher, Tests for Detecting Susceptibility to Intergranular Corrosion, in *Process Industries Corrosion—The Theory and Practice*, National Association of Corrosion Engineers, 1986, p 123-159
22. J.S. Armijo, Intergranular Corrosion of Nonsensitized Austenitic Stainless Steels, *Corrosion*, Jan 1968, p 24-30

23. W.H. Herrnstein, J.W. Cangi, and M.G. Fontana, Effect of Carbon Pickup on the Serviceability of Stainless Steel Alloy Castings, *Mater. Perform.*, Oct 1975, p 21-27
24. *Corrosion Data Survey—Metals Section*, 5th ed., National Association of Corrosion Engineers, 1974, p 268-269
25. D.R. McIntyre and C.P. Dillon, *Guidelines for Preventing Stress Corrosion Cracking in the Chemical Process Industries*, Publication 15, Materials Technology Institute of the Chemical Process Industries, 1985, p 8-14
26. *The Role of Stainless Steels in Petroleum Refining*, American Iron and Steel Institute, 1977, p 41
27. D.R. McIntyre and C.P. Dillon, *Guidelines for Preventing Stress Corrosion Cracking in the Chemical Process Industries*, Publication 15, Materials Technology Institute of the Chemical Process Industries, 1985, p 21-22
28. D.R. McIntyre and C.P. Dillon, *Guidelines for Preventing Stress Corrosion Cracking in the Chemical Process Industries*, Publication 15, Materials Technology Institute of the Chemical Process Industries, 1985, p 208-209
29. D.R. McIntyre and C.P. Dillon, *Guidelines for Preventing Stress Corrosion Cracking in the Chemical Process Industries*, Publication 15, Materials Technology Institute of the Chemical Process Industries, 1985, p 216-217
30. R.M. Davison, H.E. Deverell, and J.D. Redmond, Ferritic and Duplex Stainless Steels, in *Process Industries Corrosion—The Theory and Practice*, National Association of Corrosion Engineers, 1986, p 434-435
31. D.R. McIntyre and C.P. Dillon, *Guidelines for Preventing Stress Corrosion Cracking in the Chemical Process Industries*, Publication 15, Materials Technology Institute of the Chemical Process Industries, 1985, p 150-151
32. D.R. McIntyre and C.P. Dillon, *Guidelines for Preventing Stress Corrosion Cracking in the Chemical Process Industries*, Publication 15, Materials Technology Institute of the Chemical Process Industries, 1985, p 179
33. D.R. McIntyre, Factors Affecting the Stress Corrosion Cracking of Austenitic Stainless Steels Under Thermal Insulation, in *Corrosion of Metals Under Thermal Insulation*, STP 880, American Society for Testing and Materials, 1985, p 29-33
34. J.A. Richardson and T. Fitzsimmons, Use of Aluminum Foil for Prevention of Stress Corrosion Cracking of Austenitic Stainless Steel Under Thermal Insulation, in *Corrosion of Metals Under Thermal Insulation*, STP 880, American Society for Testing and Materials, 1985, p 188-198
35. D.R. McIntyre and C.P. Dillon, *Guidelines for Preventing Stress Corrosion Cracking in the Chemical Process Industries*, Publication 15, Materials Technology Institute of the Chemical Process Industries, 1985, p 69
36. *The Role of Stainless Steels in Petroleum Refining*, American Iron and Steel Institute, 1977, p 42-44
37. "Protection of Austenitic Stainless Steels in Refineries Against Stress Corrosion Cracking by Use of Neutralizing Solutions During Shutdown," NACE RP-01-70, (1970 Revision), National Association of Corrosion Engineers
38. M.G. Fontana, F.H. Beck, and J.W. Flowers, Cast Chromium Nickel Stainless Steels for Superior Resistance to Stress Corrosion, *Met. Prog.*, Dec 1961
39. J.W. Flowers, F.H. Beck, and M.G. Fontana, Corrosion and Age Hardening Studies of Some Cast Stainless Alloys Containing Ferrite, *Corrosion*, May 1963, p 194t-195t
40. J.W. Flowers, F.H. Beck, and M.G. Fontana, Corrosion and Age Hardening Studies of Some Cast Stainless Alloys Containing Ferrite, *Corrosion*, May 1963, p 195t-196t
41. W.H. Friske, *Shot Peening to Prevent the Corrosion of Austenitic Stainless Steels*, AI-75-52, Rockwell International, 1975
42. D.R. McIntyre and C.P. Dillon, *Guidelines for Preventing Stress Corrosion Cracking in the Chemical Process Industries*, Publication 15, Materials Technology Institute of the Chemical Process Industries, 1985, p 164
43. Internal Report, Accession No. 15925, E.I. Du Pont de Nemours & Company, Inc., p 6, 7
44. *Corrosion Data Survey—Metals Section*, 6th ed., National Association of Corrosion Engineers, 1985, p 176
45. O.L. Towers, SCC in Welded Ammonia Vessels, *Met. Constr.*, Aug 1984, p 479-485
46. D.R. McIntyre and C.P. Dillon, *Guidelines for Preventing Stress Corrosion Cracking in the Chemical Process Industries*, Publication 15, Materials Technology Institute of the Chemical Process Industries, 1985, p 53
47. J.D. Jackson and W.K. Boyd, "Stress-Corrosion Cracking of Aluminum Alloys," DMIC Memorandum 202, Battelle Memorial Institute, 1965, p 2, 3
48. J.R. Myers, H.B. Bomberger, and F.H. Froes, Corrosion Behavior and Use of Titanium and Its Alloys, *J. Met.*, Oct 1984, p 52, 53
49. D.R. McIntyre and C.P. Dillon, *Guidelines for Preventing Stress Corrosion Cracking in the Chemical Process Industries*, Publication 15, Materials Technology Institute of the Chemical Process Industries, 1985, p 88
50. R.A. Page, Stress Corrosion Cracking of Alloys 600 and 690 and Nos. 82 and 182 Weld Metals in High Temperature Water, *Corrosion*, Vol 39 (No. 10), Oct 1983, p 409-421
51. A.R. McIlree and H.T. Michels, Stress Corrosion Behavior of Fe-Cr-Ni and Other Alloys in High Temperature Caustic Solutions, *Corrosion*, Vol 33 (No. 2), Feb 1977, p 60-67
52. Ph. Berge *et al.*, Caustic Stress Corrosion of Fe-Cr-Ni Austenitic Alloys, *Corrosion*, Vol 33 (No. 12), Dec 1977, p 425-435
53. R.S. Pathania, Caustic Cracking of Steam Generator Tube Materials, *Corrosion*, Vol 34 (No. 5), May 1978, p 149-156
54. R.S. Pathania and J.A. Chitty, Stress Corrosion Cracking of Steam Generator Tube Materials in Sodium Hydroxide Solutions, *Corrosion*, Vol 34 (No. 11), Nov 1978, p 369-378
55. K.H. Lee *et al.*, Effect of Heat Treatment Applied Potential on the Caustic Stress Corrosion Cracking of Inconel 600, *Corrosion*, Vol 41 (No. 9), Sept 1985, p 540-553
56. R.S. Pathania and R.D. Cleland, The Effect of Thermal Treatments on Stress Corrosion Cracking of Alloy-800 in a Caustic Environment, *Corrosion*, Vol 41 (No. 10), Oct 1985, p 575-581
57. J.R. Crum, Stress Corrosion Cracking Testing of Inconel Alloys 600 and 690 Under High Temperature Caustic Conditions, *Corrosion*, Vol 42 (No. 6), June 1986, p 368-372
58. Summary of Questionnaire Replies on Corrosion in HF Alkylation Units, *Corrosion*, Vol 15, May 1959, p 237t-240t
59. Materials for Receiving, Handling and Storing Hydrofluoric Acid, NACE Publication 5A171 (1983 Revision), *Mater. Perform.*, Nov 1983, p 9-12
60. F.W. Billmeyer, *Textbook of Polymer Science*, 2nd ed., John Wiley & Sons, 1971, p 133
61. R.B. Seymour, *Plastics vs Corrosives*, John Wiley & Sons, 1982, p 46
62. D.R. McIntyre, Environmental Cracking, in *Process Industries Corrosion*, National Association of Corrosion Engineers, 1986, p 21-30
63. D. Warren, Hydrogen Effects on Steel, in *Process Industries Corrosion*, National Association of Corrosion Engineers, 1986, p 31-43
64. J. McBreen *et al.*, The Electrochemical Introduction of Hydrogen Into Metals, in *Fundamental Aspects of Stress Corrosion Cracking*, National Association of Corrosion Engineers, 1967, p 51-63
65. W. Beck *et al.*, Hydrogen Permeation in Metals as a Function of Stress, Temperature, and Dissolved Hydrogen Concentration, in *Hydrogen Damage*, American Society for Metals, 1977, p 191-206
66. R. Gibala, "Internal Friction in Hydrogen-Charged Iron," Case Institute of Technology, 1967
67. R. Gibala, *AIME Abstract Bull. (Inst. of Metals Div.)*, Vol 1, 1966, p 36
68. R. Gibala, Hydrogen-Dislocation Interaction in Iron, *Trans. Met. Soc. AIME*, Vol 239, 1967, p 1574
69. R. Gibala, "On the Mechanism of the Köster Relaxation Peak," Case Institute of Technology, Department of Metallurgy, 1967
70. A. Szummer, *Bull. Acad. Polon Ser. Sci. Chem.*, Vol 12, 1964, p 651
71. D.A. Vaughan *et al.*, *Corrosion*, Vol 19, 1963, p 315t
72. M.L. Holzworth *et al.*, *Corrosion*, Vol 24, 1968, p 110-124
73. N.A. Nielsen, Observations and Thoughts on Stress Corrosion Mechanisms, in *Hydrogen Damage*, American Society for Metals, 1977, p 219-254
74. M.O. Spiedel, Hydrogen Embrittlement of Aluminum Alloys?, in *Hydrogen Damage*, American Society for Metals, 1977, p 329-351
75. L.S. Darken *et al.*, Behavior of Hydrogen in Steel During and After Immersion in Acid, in *Hydrogen Damage*, American Society for Metals, 1977, p 60-75
76. H.H. Johnson *et al.*, Hydrogen Crack Initiation and Delayed Failure in Steel, *Trans. AIME*, Vol 212, 1958, p 526-536
77. K. Farrell *et al.*, Hydrogen Embrittlement

of an Ultra-High-Tensile Steel, *J. Iron Steel Inst.*, Dec 1964, p 1002-1011
78. C.D. Kim *et al.*, Techniques for Investigating Hydrogen-Induced Cracking of Steels With High Yield Strength, *Corrosion*, Vol 24, 1968, p 313-318
79. K. Farrell, Cathodic Hydrogen Absorption and Severe Embrittlement in a High Strength Steel, *Corrosion*, Vol 26, 1970, p 105-110
80. W.M. Baldwin, Jr. *et al.*, Hydrogen Embrittlement of Steels, *Trans. AIME (J. Met.)*, Vol 200, 1954, p 298-303
81. P. Blanchard *et al.*, "La Fragilisation des Metaus par L'hydrogen. Influence de la Structure Cristallographic et Electronique," Paper presented before the Societe Francaise de Metallurgie, Oct 1959
82. G.G. Hancock *et al.*, Hydrogen Cracking and Subcritical Crack Growth in a High Strength Steel, *Trans. AIME*, Vol 236, 1966, p 447
83. D.P. Williams *et al.*, Embrittlement of 4130 Steel by Low-Pressure Gaseous Hydrogen, *Metall. Trans.*, Vol 1, 1970, p 63
84. G.E. Kerns *et al.*, Slow Crack Growth of a High Strength Steel in Chlorine and Hydrogen Sulfide Gas Environments, *Scr. Metall.*, Vol 6, 1972, p 631-634
85. G.E. Kerns *et al.*, Slow Crack Growth of a High Strength Steel in Chlorine and Hydrogen Halide Gas Environments, *Scr. Metall.*, Vol 6, 1972, p 1189-1194
86. R.M. Hudson, *Corrosion*, Vol 20, 1969, p 24t
87. T.P. Radhakrishnan *et al.*, *Electrochim. Acta*, Vol 11, 1966, p 1007
88. G.E. Kerns, Effects of Sulfur and Sulfur Compounds on Aqueous Corrosion, in *Process Industries Corrosion*, National Association of Corrosion Engineers, 1986, p 353-365
89. M.T. Wang *et al.*, Effect of Heat Treatment and Stress Intensity Parameters on Crack Velocity and Fractography of AISI 4340 Steel, in *Hydrogen in Metals*, International Conference in Paris, 1972, p 342
90. V. Sawicki, Ph.D. dissertation, Cornell University, 1972
91. G.E. Kerns, Ph.D. dissertation, Ohio State University, 1973
92. P. McIntyre *et al.*, "Accelerated Test Technique for the Determination of K_{ISCC} in Steels," Corporate Laboratories Report MG/31/72, British Steel Corporation, 1972
93. Bulletin 191, Welding Research Council, March 1978
94. N. Bailey, Welding Carbon Manganese Steels, *Met. Constr.*, Vol 2, 1970, p 442-446
95. R.L. Schuyler, III, Hydrogen Blistering of Steel in Anhydrous Hydrofluoric Acid, *Mater. Perform.*, Vol 18 (No 8), 1979, p 9-16
96. G. Herbsleb *et al.*, Occurrence and Prevention of Hydrogen Induced Stepwise Cracking and Stress Corrosion Cracking of Low Alloy Pipeline Steels, *Corrosion*, Vol 37 (No. 5), 1981, p 247-255
97. S.A. Golovanenko *et al.*, Effect of Alloying Elements and Structure on the Resistance of Structural Steels to Hydrogen Embrittlement, in *H_2S Corrosion in Oil and Gas Production—A Compilation of Classic Papers*, National Association of Corrosion Engineers, 1981, p 198
98. J. Watanabe *et al.*, "Hydrogen-Induced Disbonding of Stainless Weld Overlay Found in Hydrodesulfurizing Reactor," Paper presented at ASME Conference on Performance of Pressure Vessels with Clad and Overlaid Stainless Steel Linings, Denver, CO, American Society of Mechanical Engineers, 1981
99. J. Watanabe *et al.*, "Hydrogen Induced Disbonding of Stainless Steel Overlay Weld," Paper presented at the Pressure Vessel Research Committee Meeting, New York, NY, 1980
100. M.G. Fontana and N.D. Greene, *Corrosion Engineering*, McGraw-Hill, 1967, p 72-91
101. F.J. Heymann, Erosion by Liquids—The Mysterious Murderer of Metals, *Mach. Des.*, Dec 1970
102. I.M. Hutchings, *The Erosion of Materials by Liquid Flow*, Publication 25, Materials Technology Institute of the Chemical Process Industries, 1986
103. P. Eisenbery *et al.*, How to Protect Materials Against Cavitation Damage, *Mater. Des. Eng.*, March 1967
104. T.E. Backstrom, "A Suggested Metallurgical Parameter in Alloy Selection for Cavitation Resistance," Report CHE 72, Department of the Interior, Dec 1972
105. R.W. Hinton, "Cavitation Damage of Alloys—Relationship to Microstructure," Paper presented at NACE Conference, National Association of Corrosion Engineers, March 1963
106. R.E.A. Arndt, "Cavitation and Erosion: An Overview," Paper presented at NACE Conference, National Association of Corrosion Engineers, March 1977
107. *Trans. ASME*, Vol 59, 1937
108. R.B. Waterhouse and M. Allery, The Effect of Non-Metallic Coatings on the Fretting Corrosion of Mild Steel, *Wear*, Vol 8, 1965, p 421-447
109. R.B. Waterhouse *et al.*, The Effect of Electrodeposited Metals on the Fatigue Behavior of Mild Steel Under Conditions of Fretting Corrosion, *Wear*, Vol 5, 1962, p 235-244
110. Fretting and Fretting Corrosion, *Lubrication*, Vol 52 (No. 4), 1966
111. J.R. McDowell, in *Symposium on Fretting Corrosion*, STP 144, American Society for Testing and Materials, 1952
112. H.S. Avery, *Abrasive Wear—The Nature of the Abrasive*, Publication RCR CR340, Abex Corporation
113. H.S. Avery, "Hard Facing Alloys," Paper presented at the ASM Wear Conference, Boston, MA, American Society for Metals, 1969
114. T.E. Norman, New Austenitic Alloy for Ultra-Abrasive Applications, *J. Eng. Mining*, April 1965
115. K.J. Blensali and W.L. Silence, Metallurgical Factors Affecting Wear Resistance, *Met. Prog.*, Nov 1977
116. G.L. Sheldon, Effect of Surface Hardness and Other Materials Properties on Erosion Wear of Metals, *J. Eng. Mater. Test.*, April 1977
117. W.A. Stauffer, Wear of Metals by Sand Erosion, *Met. Prog.*, Jan 1956
118. R. Heidersbach, Clarification of the Mechanism of the Dealloying Phenomenon, *Corrosion*, Feb 1968, p 38-44
119. R.A. Gummow, The Corrosion of Municipal Iron Watermains, *Mater. Perform.*, March 1984, p 39-42
120. J. Zahavi and J. Yahalom, Exfoliation Corrosion of Al Mg Si Alloys in Water, *J. Electrochem. Soc.*, Vol 129 (No. 6), June 1982, p 1181-1185
121. D.R. McIntyre and C.P. Dillon, *Guidelines for Preventing Stress Corrosion Cracking in the Chemical Process Industries*, Publication 15, Materials Technology Institute of the Chemical Process Industries, 1985, p 88-96
122. R.B. Puyear, Material Selection Criteria for Chemical Processing Equipment, *Met. Prog.*, Feb 1978, p 42

SELECTED REFERENCES

- S.K. Coburn, Ed., *Corrosion Source Book*, American Society for Metals, 1984
- M.G. Fontana, *Corrosion Engineering*, 3rd ed., McGraw-Hill, 1986
- *Guide to Engineered Materials*, Vol 1, ASM INTERNATIONAL, June 1986, p 73-87
- I.M. Hutchings, *The Erosion of Materials by Liquid Flow*, Publication 25, Materials Technology Institute of the Chemical Process Industries, 1986
- D.R. McIntyre and C.P. Dillon, *Guidelines for Preventing Stress Corrosion Cracking in the Chemical Process Industries*, Publication 15, Materials Technology Institute of the Chemical Process Industries, 1985
- B.J. Moniz and W.I. Pollock, Ed., *Process Industries Corrosion—The Theory and Practice*, National Association of Corrosion Engineers, 1986
- G.A. Nelson, Criteria for Selecting Metals Used in Chemical Plants, *Met. Prog.*, May 1960, p 80-88, 134, 166-174
- P.A. Schweitzer, Ed., *Corrosion and Corrosion Protection Handbook*, Marcel Dekker, 1983
- H.H. Uhlig and R.W. Revie, *Corrosion and Corrosion Control*, 3rd ed., John Wiley & Sons, 1985
- F.L. Whitney, Jr., Factors in the Selection of Corrosion Resistant Materials, *Met. Prog.*, June 1957, p 90-95

Design Details to Minimize Corrosion

Peter Elliott, Cortest Engineering Services Inc.

GOOD ENGINEERING DESIGN is fundamental to reliable and effective material use, given adequate data and the availability of a suitable material in a suitable form. Designing to minimize corrosion can be reliable only if it is part of an overall design philosophy.

Design can never be absolute. There will often be a tendency for compromise based on cost and the availability of materials and resources. For example, one designer may recognize a potential corrosion hazard and may plan for early replacement or more regular maintenance. Such design options may be relatively inexpensive. Designers may have little experience with regard to corrosion and the subtleties of material fabrication and assembly. In these cases, incorrect decisions can be extremely costly.

A designer cannot be a corrosion engineer; therefore, it is necessary to convey a basic knowledge of corrosion to the designer. This information must relate to the total design requirements. In some cases, the process conditions may not be known with sufficient exactness to permit a technical design decision.

The basic principles of corrosion control by design to be discussed in this article are based on practical situations. Texts are available on the subject (Ref 1-3), but several of them are confined to simplistic geometric shapes that may not always be readily applicable to the design of process plants or operating equipment. It is important to be aware of design details that can lead to early deterioration. These fine details of design, often compounded by human error, account for many significant failures. Therefore, corrosion engineering design must be an integral part of the total design, which in turn includes aspects ranging from appraisal of the design concept to inspection and quality control in installation and operation. In addition, it cannot be overemphasized that poor design may render corrosion-resistant materials susceptible to premature corrosion.

Figure 1 illustrates the frequency of failures attributed to design errors. The results, based on the responses of personnel in the chemical-processing industries, support the importance of being aware of the circumstances that relate a material to its specific environment.

Why Failures Occur

In the context of design, there are several areas that relate to material/component failure:

- *Overload* suggests a weakness in plant control instrumentation or operation
- *Abnormal conditions* can result from a lack of process control or variations in raw material
- *Poor fabrication* may relate to inadequate instructions or inspection, for example, excessive cold work, overmachining, and excessive torque loading
- *Poor handling*. Scratches or machine marks can result from poor detailing or instruction
- *Assembly*, if incorrect (for example, welds and fastening) can seriously affect stress, flow, and compatibility
- *Storage and transportation* can be significant for many ultimate applications. Some structures or components may not be accessible for remedial work even if a corrosion risk is recognized. Proper design should accommodate these aspects of material handling.

Design Function

Good engineering design attempts to provide a product that satisfies the user and is profitable to the manufacturer and the distributor. The objective is a maximum return on capital expended in the shortest return period.

Optimum designs are a direct function of the limits of an organization, the available resources, equipment, manpower, and the experience and expertise of the designer. It is necessary to be aware of specifications and working codes of practice and to recognize that quality control and assurance procedures are basic to reliability and therefore to efficiency and safety.

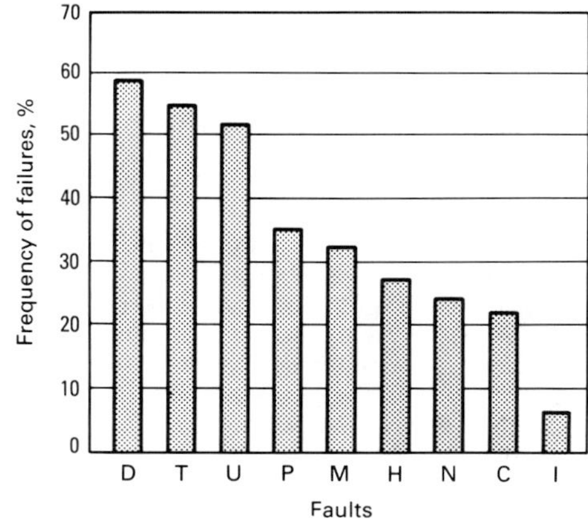

Fig. 1 Results of a survey of causes of failure in the chemical-processing industry showing that plant design faults (D) were the most frequently cited cause of failure. Other causes: T, incorrect application of protective treatment; U, unforeseen operating conditions; P, poor process control; M, materials faults; H, human errors; N, lack of awareness of corrosion risk; C, contamination of product; I, instrument failure. Source: Ref 4

Corrosion control by design is not necessarily effective for the variety of factors described above. Corrosion data must be interpreted carefully and must be relevant to plant or process conditions. The designer is in the most effective and logical position to provide ideal and realistic options. Versed in technical, political, and economic strategies, the designer can achieve most of the total requirements of design. In reality, there must be a systematic approach involving direct input from multidisciplinary sources. The communication channels include not only the design stage and operation but also failure analysis and monitoring of maintenance procedures.

Design and Materials Selection

Materials are selected to perform a basic function or to provide a functional requirement (see article "Materials Selection" in this Volume). Therefore, in many cases, a corrosion-resistant alloy or metal may not satisfy such primary requirements as strength, reflectivity, wear resistance, and dimensional stability. Tungsten, for

Table 1 Functional requirements of materials

Characteristic	Functional requirements
Strength under actual service conditions	Tensile strength; strength in torsion, compression, shear, etc.
Toughness	Ability to cope with an overload, impact, etc.
Dimensional stability	Practical assembly; stiffness of structure; effect of sustained loading, etc.
Wear and abrasion	Lubrication or low friction; loss of surface products (passive films) of benefit in corrosion control
Physical properties	Design may require certain properties peculiar to the material (thermal, electrical, acoustic, optical, magnetic, etc.)
Machinability	Related to mechanical properties; notches, work hardening, scratches, etc.

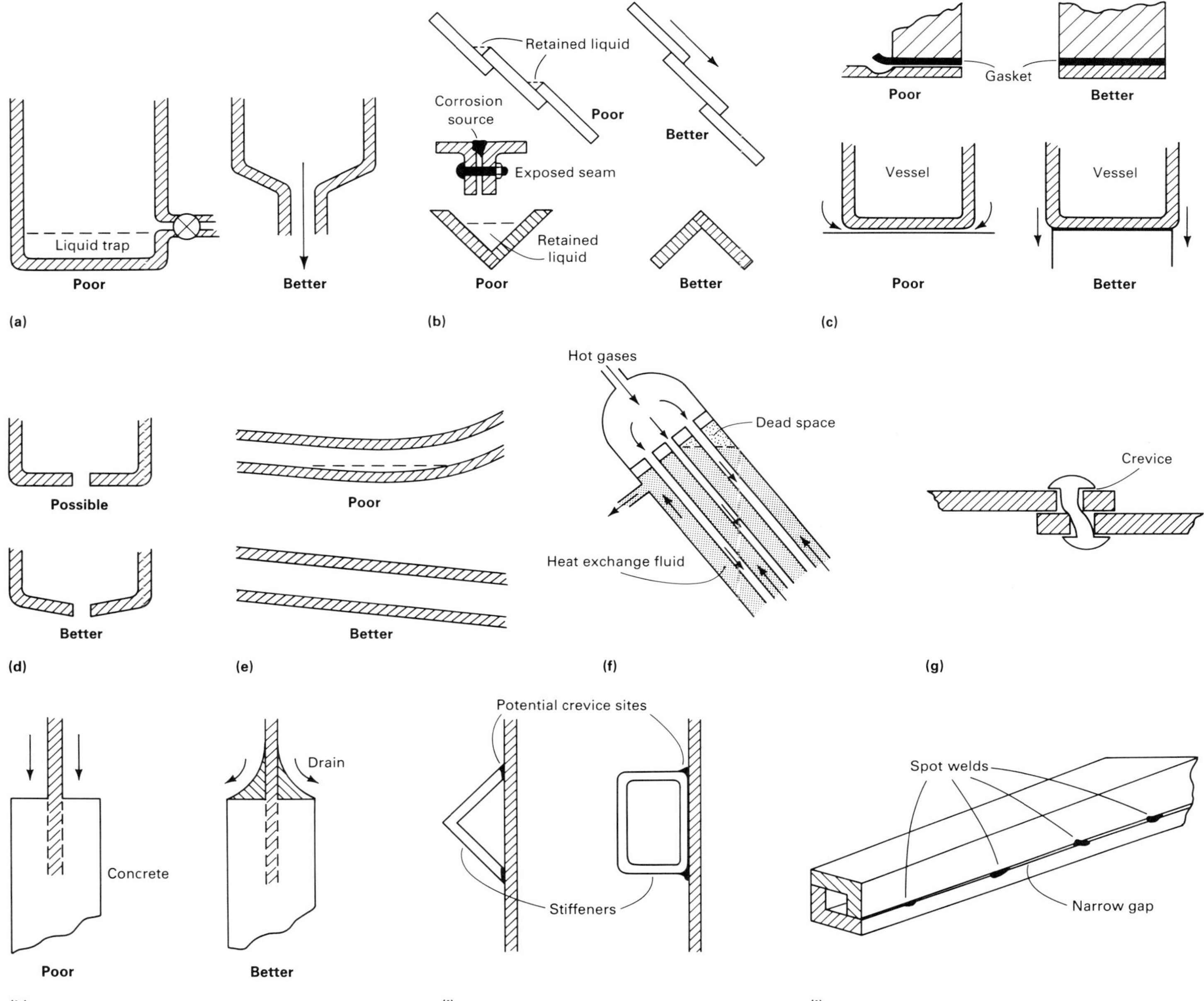

Fig. 2 Examples of how design and assembly can affect localized corrosion by creating crevices and traps where corrosive liquids can accumulate. (a) Storage containers or vessels should allow complete drainage; otherwise, corrosive species can concentrate in bottom of vessel, and debris may accumulate if the vessel is open to the atmosphere. (b) Structural members should be designed to avoid retention of liquids; L-shaped sections should be used with the open side down, and exposed seams should be avoided. (c) Incorrect trimming or poor design of seals and gaskets can create crevice sites. (d) Drain valves should be designed with sloping bottoms to avoid pitting of the base of the valve. (e) Nonhorizontal tubing can leave pools of liquid at shutdown. (f) to (j) Examples of poor assembly that can lead to premature corrosion problems. (f) Nonvertical assembly of heat exchanger permits a dead space that may result in overheating if very hot gases are involved. (g) Nonaligned assembly distorts the fastener, which creates a crevice and may result in a loose fitting that can contribute to vibration, fretting, and wear. (h) Structural supports should allow good drainage; use of a slope at the bottom of the member allows liquid to run off, rather than impinging directly on the concrete support. (i) Continuous welding is necessary for horizontal stiffeners to prevent the formation of traps and crevices. (j) Square sections formed from two L-shaped members require continuous welding to seal out the external environment.

example, may be a strong, high melting point metal, but its impact properties make it delicate for handling. Furthermore, any temperature excursion would render the material nonresistant to oxidation, because the products that form above about 750 °C (1380 °F) are loose and powdery.

Corrosion data sources tend to overemphasize chemical properties and make less reference to those parameters that affect material suitability for a particular service condition. Some of these are given in Table 1.

Design Details

The following sections will demonstrate the aspects of design detail that may accelerate corrosion (Ref 5).

Shape. Geometrical form is basic to design. The objective is to minimize or avoid situations that worsen corrosion. These situations can range from stagnation (for example, retained fluids and/or solids) to sustained fluid flow (for example, erosion/cavitation in components moving in or contacted by fluids as well as splashing or droplet impingement).

Common examples of stagnation include nondraining structures, dead ends, badly located components, and poor assembly or maintenance practices (Fig. 2). The general problems include localized corrosion associated with differential aeration (oxygen concentration cells), crevice corrosion, and deposit corrosion.

Fluid movement need not be excessive to damage a material. Much depends on the nature of

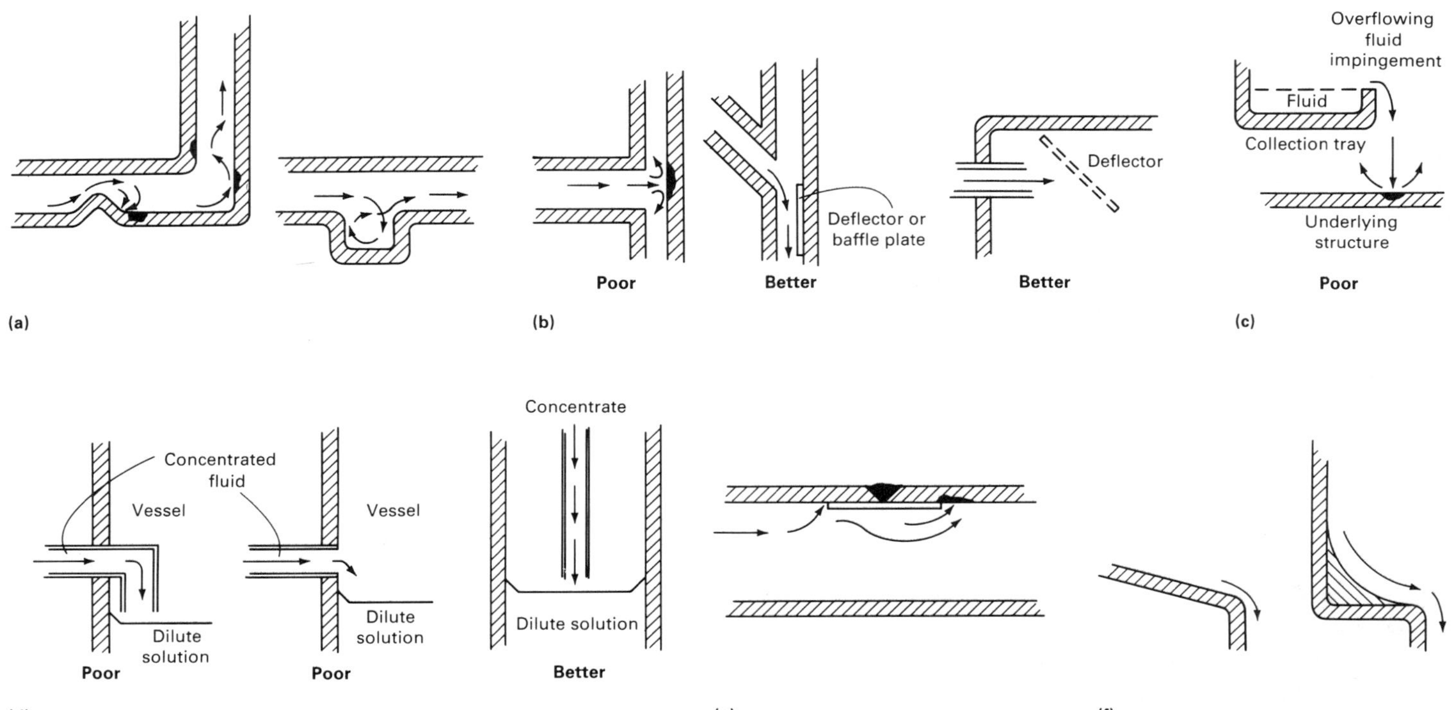

Fig. 3 Effect of design features on flow. (a) Disturbances to flow can create turbulence and cause impingement damage. (b) Direct impingement should be avoided; deflectors or baffle plates can be beneficial. (c) Impingement from fluid overflowing from a collection tray can be avoided by relocating the structure, by increasing the depth of the tray, or by using a deflector. (d) Splashing of concentrated fluid on container walls should be avoided. (e) Weld backing plates or rings can create local turbulence and crevices. (f) Slopes or modified profiles should be provided to permit flow and minimize fluid retention.

the fluid and the hardness of the material. A geometric shape may create a sustained delivery of fluid or may locally disturb a laminar stream and lead to turbulence. Replaceable baffle plates or deflectors are beneficial where circumstances permit their use. They effectively resolve the problem of impingement damage to the structurally significant component.

Careful fabrication with inspection will alleviate the effects of such factors as poor profiles (welding, bolting), rubbing surfaces (wear, fretting), and galvanic effects due to incompatible assembly of components. Figure 3 shows typical situations in which geometric detail is significant to flow.

Accessibility may become important in cases in which shape could have been better considered at the design stage. Common engineering structural steelwork requires regular preventive maintenance, and access may often be restricted. Figure 4 shows situations in which surface cleaning and/or painting is difficult or impossible. Condensation in critical areas may also contribute to corrosion. Typical structures susceptible to this phenomenon include chimneys or exhausts to high-temperature plants, boilers, or incinerators.

Site location relative to winds and airborne particulates can lead to adverse deterioration of structures. The significance of pollution and the synergistic effects on corrosion are illustrated in Fig. 5, which demonstrates poor layout relative to geographic topography. Design geometries that leave structures exposed to the elements should be carefully reviewed because atmospheric corrosion is significantly affected by temperature, relative humidity, rainfall, and pollutants. Also related to this aspect are the date and location of on-site fabrication, assembly, and painting. Codes of practice must be adapted to the location and the season.

Table 2 Galvanic corrosion sources and design considerations

Source	Design considerations
Metallurgical sources (both within the metal and for relative contact between dissimilar metals)	Difference in potential of dissimilar materials; distance apart; relative areas of anode and cathode; geometry (fluid retention); mechanical factors (for example, cold work, plastic deformation, sensitization)
Environmental sources	Conductivity and resistivity of fluid; changes in temperature; velocity and direction of fluid flow; aeration; ambient environment (seasonal changes, etc.)
Miscellaneous sources	Stray currents; conductive paths; composites (for example, concrete rebars)

Insulation represents another area for potential corrosion attack, although the form and requirements for insulating media differ considerably. Moisture-absorbing tendencies will vary, as will the extent of crevicing from compaction and shrinkage or chloride buildup for certain materials. Wet-dry cycling can lead to concentration effects that may result in stress-corrosion cracking (SCC) of certain stainless steels or pitting on other materials, such as aluminum, that contact the insulation barriers. Figure 6 shows some typical examples in which design and installation procedures could have been improved.

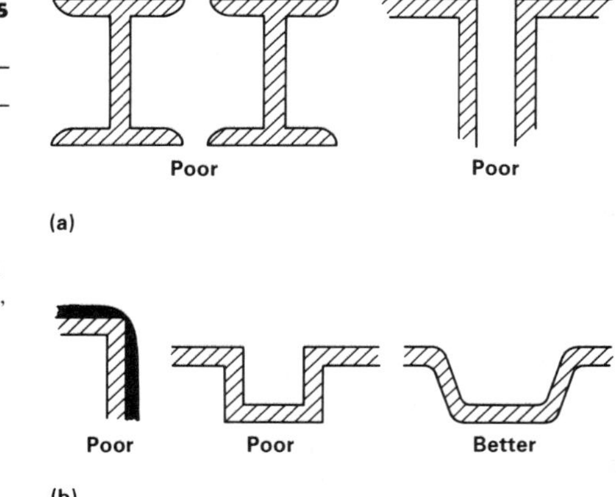

Fig. 4 Effects of design on cleaning or painting of equipment. (a) Poor access in certain structures makes surface preparation and painting difficult; access to the types of areas shown should be maintained at a minimum of 45 mm ($1^3/_4$ in.) or one-third of the height of the structure. (b) Sharp corners and profiles should be avoided if the structure is to be painted or coated.

Compatibility. In plant environments, it is often necessary to use different materials in close proximity. Direct contact of dissimilar metals introduces the possibility of galvanic corrosion, and small anode areas should be avoided wher-

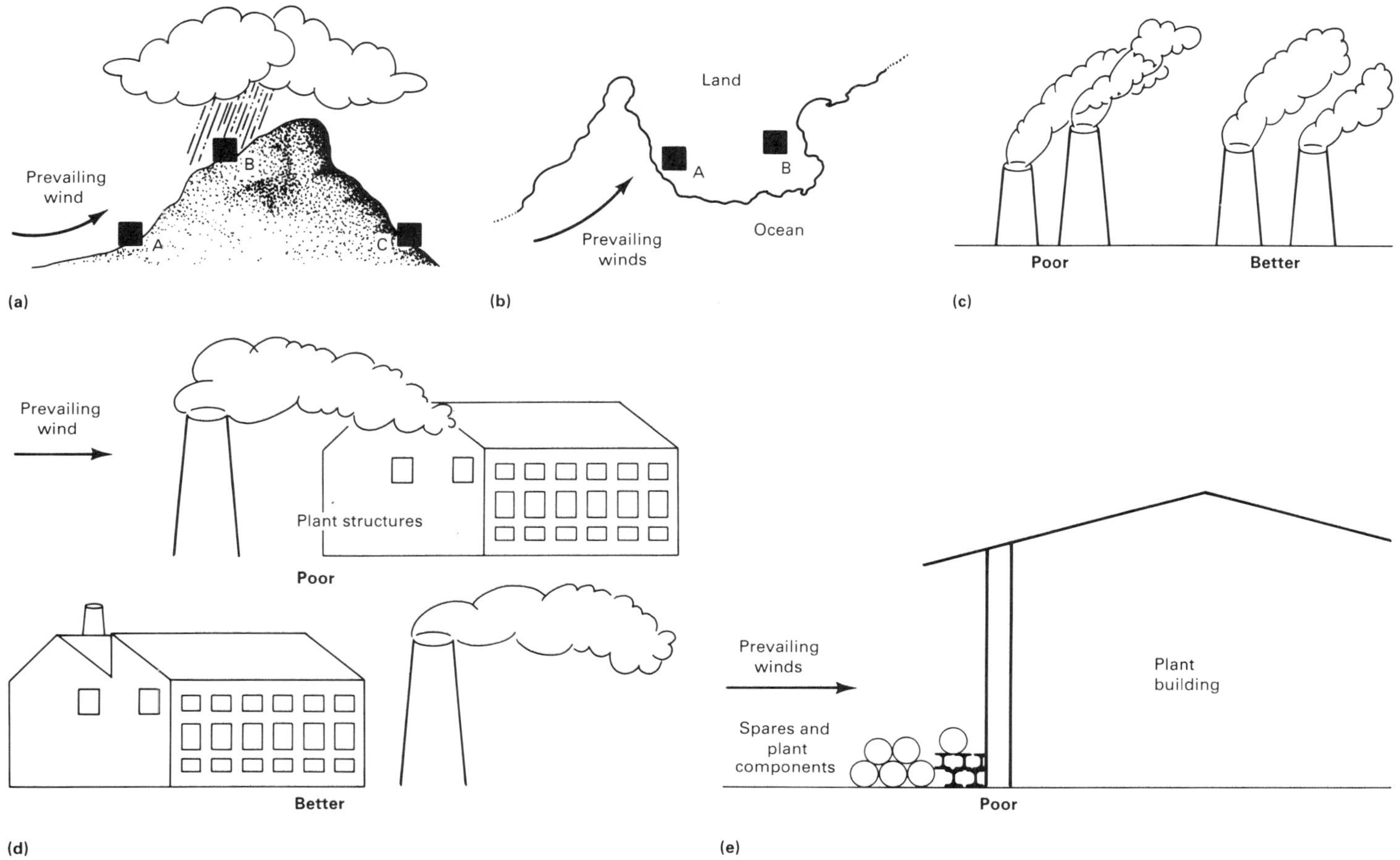

Fig. 5 Site location as a design consideration. (a) Topographic features must be considered in choosing a site. Location C would be the preferred site. (b) In marine atmospheres, prevailing winds should be taken into account; location B is the preferred site. (c) Local industry can affect corrosion of chimney stacks and similar structures. "Lick-over" of gases, which relates to stack height, location, and prevailing winds, should be avoided. (d) Plant structures should be located upwind from stacks. (e) Spares and components should be stored away from the prevailing wind.

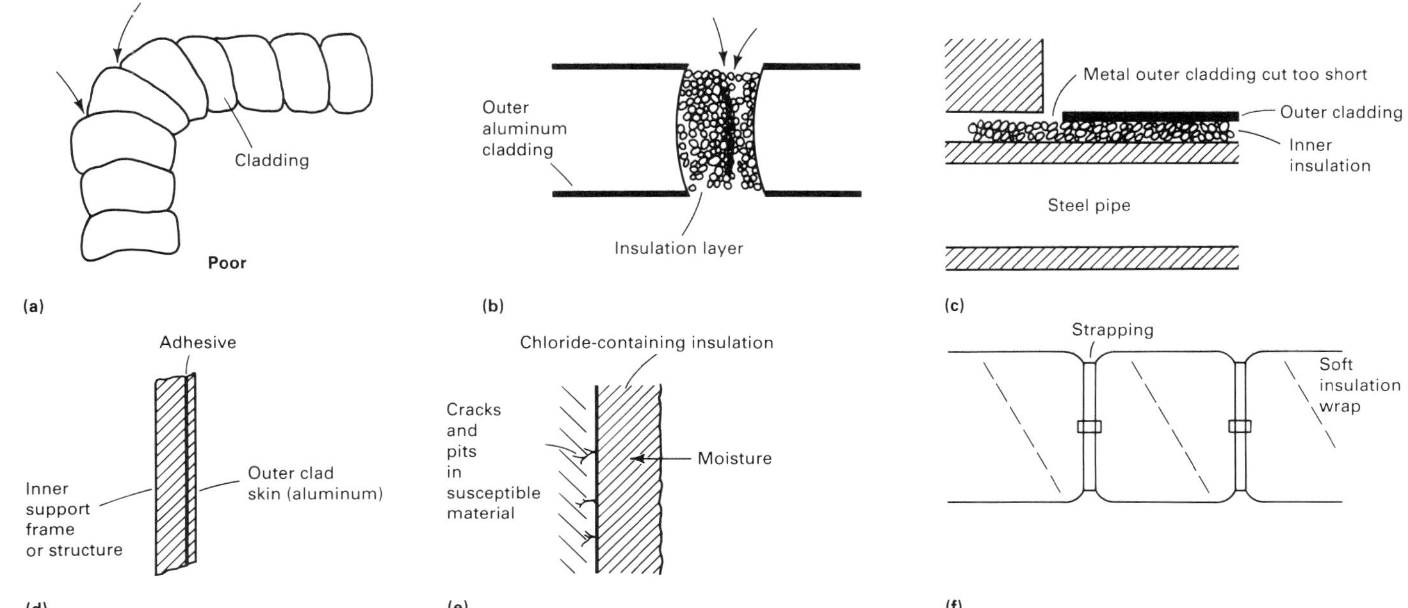

Fig. 6 Corrosion problems associated with improper use of insulation and lagging. (a) Incorrect overlap in lobster-back cladding does not allow fluid runoff. (b) Poor installation left a gap in the insulation that allows easy access to the elements. (c) Outer metal cladding was cut too short, leaving a gap with the inner insulation exposed. (d) Poor or noncontinuous contact of adhesives can lead to a crevice or capillary entry of fluid; also, adhesives may not have sealing properties. (e) Insufficient insulation may allow water to enter; chloride in some insulation can result in pitting or SCC of susceptible materials. (f) Overtightened strapping can damage the insulation layer.

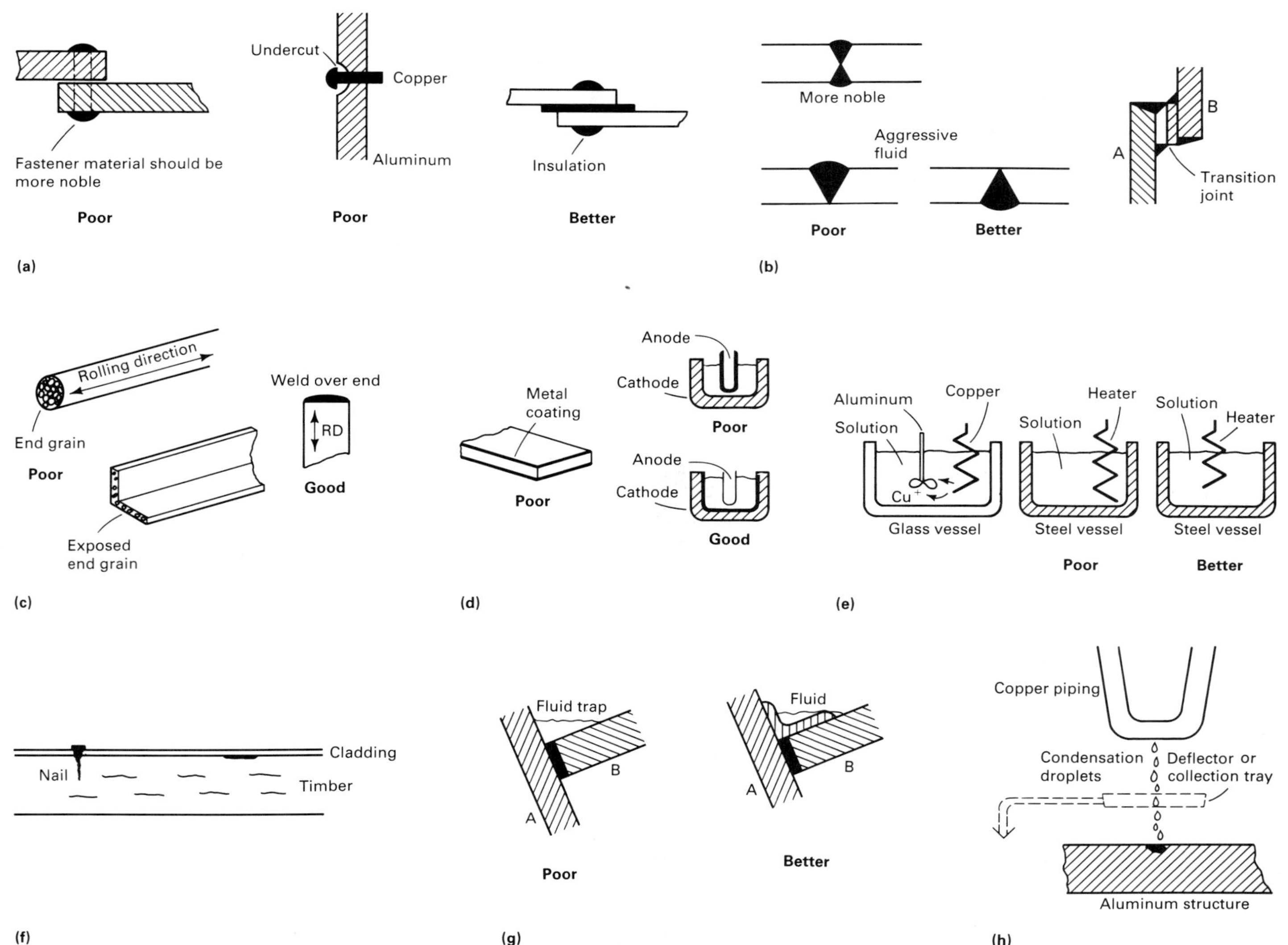

Fig. 7 Design details that can affect galvanic corrosion. (a) Fasteners should be more noble than the components being fastened; undercuts should be avoided, and insulating washers should be used. (b) Weld filler metals should also be more noble than base metals. Transition joints can be used when a galvanic couple is anticipated at the design stage, and weld beads should be properly oriented to minimize galvanic effects. (c) Local damage can result from cuts across heavily worked areas. End grains should not be left exposed. (d) Galvanic corrosion is possible if a coated component is cut. When necessary, the cathodic component of a couple should be coated. (e) Ion transfer through a fluid can result in galvanic attack of less noble metals. In the example shown at left, Cu^+ ions from the copper heater coil could deposit on the aluminum stirrer. A nonmetallic stirrer may be necessary. At right, the distance from a metal container to a heater coil should be increased to minimize ion transfer. (f) Wood with copper preservatives can be corrosive to certain nails, especially those with nobility different from that of copper. Aluminum cladding may also be at risk. (g) Contact of two metals through a fluid trap can be avoided by using an extended seal, mastic, or a coating. (h) Condensation droplets from copper piping can impinge on an underlying aluminum structure; such contact can be avoided by the use of collection trays or deflectors.

ever this is apparent (see the section "Galvanic Corrosion" of the article "General Corrosion" in this Volume). Components that perhaps were designed in isolation may end up in direct contact in the plant. In such cases, the ideals of a total design concept become especially apparent in hindsight.

Galvanic corrosion resulting from metallurgical sources is well documented. Problems such as weld decay and sensitization can generally be avoided by material selection or suitable fabrication techniques. Less obvious instances include end-grain attack and stray-current effects, which can render designs ineffective.

Designers, when aware of compatibility effects, need to exercise their ingenuity to minimize the conditions that most favor galvanic corrosion (Ref 1). Table 2 provides some relevant parameters in this context.

The most common design details relating to galvanic corrosion include jointed assemblies (Fig. 7). Where dissimilar metals are to be used, some consideration should be given to compatible materials known to have similar potentials (Ref 6) (for more information see the article "Materials Selection" and the Section "Specific Alloy Systems" in this Volume). Care should be exercised in that galvanic series are limited and refer to specific environments. The confusion of terminology can also be problematic; such terms as mild steel, stainless steel, Hastelloy, and Inconel are too vague and do not provide sufficient assurance about material performance in a corrosive environment.

Where noncompatible materials are to be joined, it is necessary to use a more noble metal in a joint (Fig. 7). Effective insulation can be useful if it does not introduce crevice corrosion possibilities. Some difficulties arise in the use of adhesives, which may not be sealants.

The relative surface areas of anodic and cathodic surfaces should not be underestimated, because instances of corrosion failure may result from a combination of galvanic and crevice attack. Corrosion in a small anodic zone can be several hundred times greater than that in similar bimetallic components of similar area. Anodic components may on occasion be overdesigned (thicker) to allow for the anticipated corrosion loss. In other cases, easy replacement is a cost-effective option, given an awareness by the designer of such information.

Where metallic coatings are used, there is always a risk of galvanic corrosion, especially along the cut edges. Rounded profiles and effective sealants or coatings can be beneficial. Transition joints can be introduced when different

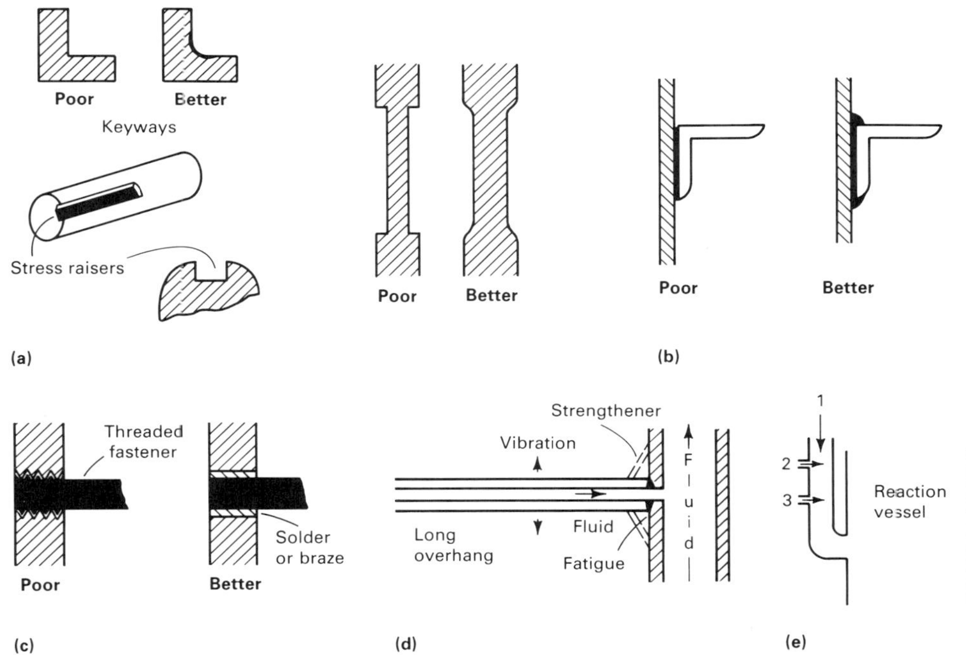

Fig. 8 Design details that can minimize local stress concentrations. (a) Corners should be given a generous radius. (b) Welds should be continuous to minimize sharp contours. (c) Sharp profiles can be avoided by using alternative fastening systems. (d) Too long of an overhang without support can lead to fatigue at the junction. Flexible hose may help alleviate this situation. (e) Side-supply pipework may be too rigid to sustain thermal shock from air under pressure (1), steam (2), and cold water in sequence (3).

metals will be in close proximity. These and other situations are illustrated in Fig. 7. Another aspect is the coating of the cathodic material for corrosion control. Ineffective painting of an anode in an assembly can significantly reduce the desired service lifetime because local defects will effectively multiply the risk of anodic sites and localized corrosion.

Mechanical Factors. Environments that promote metal dissolution can be considered more damaging if stresses are involved (see the section "Stress-Corrosion Cracking" of the article "Environmentally Induced Cracking" in this Volume). In such circumstances, materials may fail catastrophically and unexpectedly. Safety and health may be significantly affected.

Figure 8 shows cases in which design detail is used to minimize stress. Perfection is rarely attained in general practice, and some compromise on materials limitations, both chemical and mechanical, is necessary. The difficulty is that mechanical fault can contribute to corrosion and that corrosion (as a corrosive environment) can initiate or cause mechanical failure. Quality control and assurance can eliminate the former.

Designs that introduce local stress concentrations directly or as a consequence of fabrication should be carefully considered. Of particular importance are stress levels for the selected material; the influence of tensile, compressive, or shear stressing; alternating stresses; vibration or shock loading; service temperatures (thermal stressing); fatigue; and wear (fretting, friction). Profiles and shapes contribute to stress-related corrosion if material selection dictates the use of materials susceptible to failure by SCC or corrosion fatigue.

Materials selection is especially important wherever critical components are used. Also important is the need for correct procedures at all stages of operation, including fabrication, transport, startup, shutdown, and normal operation. Less obvious cases of failure can arise from vibration transfer, poor surface finish, nonuniform application of surface coatings, or the application of coatings to poorly prepared surfaces.

Surfaces. Corrosion is a surface phenomenon, and the effects of poorly prepared surfaces, rough textures, and complex shapes and profiles can be expected to be deleterious. Figure 4 shows some examples in which design specification could have considerably reduced the onset of corrosive damage. Design limitations include surfaces exposed to deposits, retained soluble salts (because of poor access for preparation before painting), nondraining assemblies, poorly handled components (distortion, scratches, dents), and poorly located components (relative position to adjacent equipment, and so on).

Painting and surface-coating technology have advanced in recent years and have provided sophisticated products that require careful mixing and application. Maintenance procedures frequently require field application; in such cases, control is not anticipated. This is significant, for example, in the offshore locations of the oil and gas industry. Inspection codes and procedures are necessary, and total design should incorporate these wherever possible. In critical areas, design for on-line monitoring and inspection will also be important. The human factor in such procedures is often overlooked. The need for better techniques, standardization, and mechanization or full automation has been stated, and adequate training and motivation are of primary importance. Reporting and assessment may sometimes be inaccurate (Ref 7).

REFERENCES

1. V.R. Pludek, *Design and Corrosion Control*, Macmillan, 1977
2. R.N. Parkins and K.A. Chandler, *Corrosion Control in Engineering Design*, Department of Industry, Her Majesty's Stationery Office, 1978
3. L.D. Perrigo and G.A. Jensen, Fundamentals of Corrosion Control Design, *North. Eng.*, Vol 13, 1982, p 16-34
4. P. Elliott, Corrosion Survey, supplement to The Chemical Engineer, Sept. 1973
5. P. Elliott and J.S. Llewyn-Leach, *Corrosion Control Checklist for Design Offices*, Department of Industry, Her Majesty's Stationery Office, 1981
6. C.J. Smithells, Ed., Corrosion Control, in *Metals Reference Book*, Butterworths, 1977
7. J. Jelinek and B. Studman, Inspection Offshore, *Gas Eng. Mgmt.*, Nov-Dec, 1983, p 395-404

Corrosion of Weldments

The ASM Committee on Corrosion of Weldments*
Chairman: Kenneth F. Krysiak, Hercules Inc.

IT IS NOT UNUSUAL to find that, although the wrought form of a metal or alloy is resistant to corrosion in a particular environment, the welded counterpart is not. Further, welds can be made with the addition of filler metal or can be made autogenously (without filler metal). However, there are also many instances in which the weld exhibits corrosion resistance superior to that of the unwelded base metal. There also are times when the weld behaves in an erratic manner, displaying both resistance and susceptibility to corrosive attack. Corrosion failures of welds occur in spite of the fact that the proper base metal and filler metal have been selected, industry codes and standards have been followed, and welds have been deposited that possess full weld penetration and have proper shape and contour.

It is sometimes difficult to determine why welds corrode; however, one or more of the following factors often are implicated:

- Weldment design
- Fabrication technique
- Welding practice
- Welding sequence
- Moisture contamination
- Organic or inorganic chemical species
- Oxide film and scale
- Weld slag and spatter
- Incomplete weld penetration or fusion
- Porosity
- Cracks (crevices)
- High residual stresses
- Improper choice of filler metal
- Final surface finish

Metallurgical Factors

The cycle of heating and cooling that occurs during the welding process affects the microstructure and surface composition of welds and adjacent base metal. Consequently, the corrosion resistance of autogenous welds and welds made with matching filler metal may be inferior to that of properly annealed base metal because of:

- Microsegregation
- Precipitation of secondary phases
- Formation of unmixed zones
- Recrystallization and grain growth in the weld heat-affected zone (HAZ)
- Volatilization of alloying elements from the molten weld pool
- Contamination of the solidifying weld pool

Corrosion resistance can usually be maintained in the welded condition by balancing alloy compositions to inhibit certain precipitation reactions, by shielding molten and hot metal surfaces from reactive gases in the weld environment, by removing chromium-enriched oxides and chromium-depleted base metal from thermally discolored (heat tinted) surfaces, and by choosing the proper welding parameters (Ref 1).

Weld Solidification. During the welding process, a number of important changes occur that can significantly affect the corrosion behavior of the weldment. Heat input and welder technique obviously play important roles. The way in which the weld solidifies is equally important to understanding how weldments may behave in corrosive environments (Ref 2).

A metallographic study has shown that welds solidify into various regions, as illustrated in Fig. 1. The composite region, or weld nugget, consists of essentially filler metal that has been diluted with material melted from the surrounding base metal. Next to the composite region is the unmixed zone, a zone of base metal that melted and solidified during welding without experiencing mechanical mixing with the filler metal. The weld interface is the surface bounding the region within which complete melting was experienced during welding, and it is evidenced by the presence of a cast structure. Beyond the weld interface is the partially melted zone, which is a region of the base metal within which the proportion melted ranges from 0 to 100%. Lastly, the true HAZ is that portion of the base metal within which microstructural change has occurred in the absence of melting. Although the various regions of a weldment shown in Fig. 1 are for a single-pass weld, similar solidification patterns and compositional differences can be expected to occur in underlying weld beads during multipass applications.

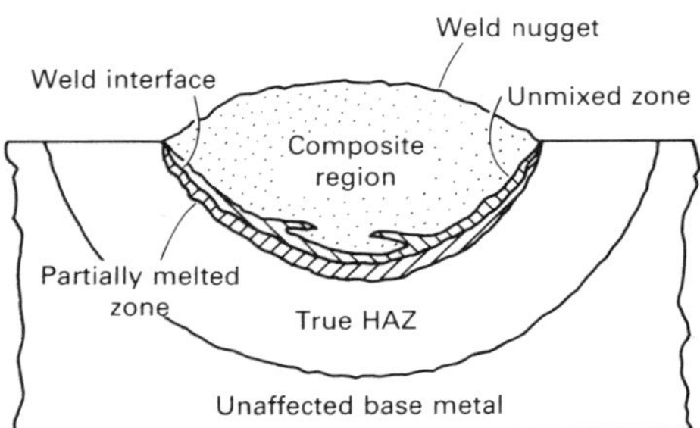

Fig. 1 Schematic of a weld cross section. Source: Ref 3

Corrosion of Aluminum Alloy Weldments

Variations in microstructure across the weld and HAZ of aluminum weldments are known to produce susceptibility to corrosion in certain environments (Ref 4). These differences can be measured electrochemically and are an indication of the type of corrosion behavior that might be expected. Although some aluminum alloys can be autogenously welded, the use of a filler metal is preferred to avoid cracking during welding and to optimize corrosion resistance.

The variations in corrosion potential (equilibrium potential) across three welds are shown in Fig. 2 for alloys 5456, 2219, and 7039. These differences can lead to localized corrosion, as demonstrated by the corrosion of the HAZ of an as-welded structure of alloy 7005 shown in Fig. 3. In general, the welding procedure that puts the least amount of heat into the metal has the least influence on microstructure and the least chance of reducing the corrosion behavior of aluminum weldments.

Tables are available in Ref 4 that summarize filler alloy selection recommended for welding various combinations of base metal alloys to obtain maximum properties, including corrosion resistance. Care must be taken not to extrapolate the corrosion performance ratings indiscriminately. Corrosion behavior ratings generally pertain only to the particular environment tested, usually rated in continuous or alternate immersion in fresh or salt water. For example, the highest corrosion rating (A) is listed for use of filler alloy 4043 to join 3003 alloy to 6061 alloy. In strong (99%) nitric acid (HNO_3) service, however, a weldment made with 4043 filler alloy would experience more rapid attack than a weldment made using 5556 filler metal. With certain alloys, particularly those of the heat-treatable 7*xxx* series, thermal treatment after welding is sometimes

*Andrew Garner, Pulp and Paper Research Institute of Canada; John Grocki, Haynes International, Inc.; J.R. Kearns, Allegheny Ludlum Steel Division of Allegheny Ludlum Corporation; Hermann Kernberger, Vereinigte Edelstahlwerke (VEW) AG; Gregory Kobrin, E.I. Du Pont de Nemours & Company, Inc.; Curtis W. Kovach, Trent Tube Division, Crucible Materials Corporation; Bernard W. Lifka, Alloy Technology Division, Alcoa Laboratories; Christer Martenson, Sandvik Steel Company; Peter Norberg, AB Sandvik Steel; Charles Pokross, Fansteel, Inc.; Mortimer Schussler (retired), Fansteel, Inc.; N. Sridhar, Haynes International, Inc.; Robert E. Tatnall, E.I. Du Pont de Nemours & Company, Inc.; National Association of Corrosion Engineers

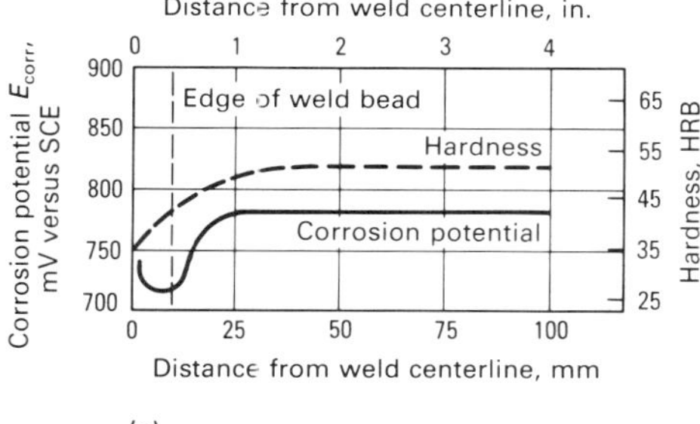

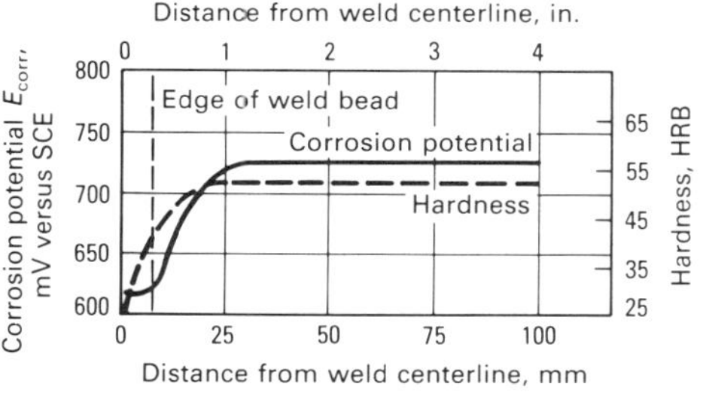

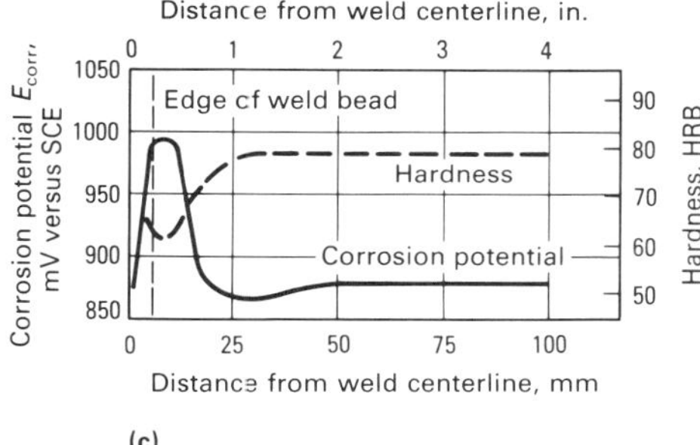

Fig. 2 Effect of the heat of welding on microstructure, hardness, and corrosion potential of welded assemblies of three aluminum alloys. The differences in corrosion potential between the HAZ and the base metal can lead to selective corrosion. (a) Alloy 5456-H321 base metal with alloy 5556 filler; 3-pass metal inert gas weld. (b) Alloy 2219-T87 base metal with alloy 2319 filler; 2-pass tungsten inert gas weld. (c) Alloy 7039-T651 base metal with alloy 5183 filler; 2-pass tungsten inert gas weld. Source: Ref 4

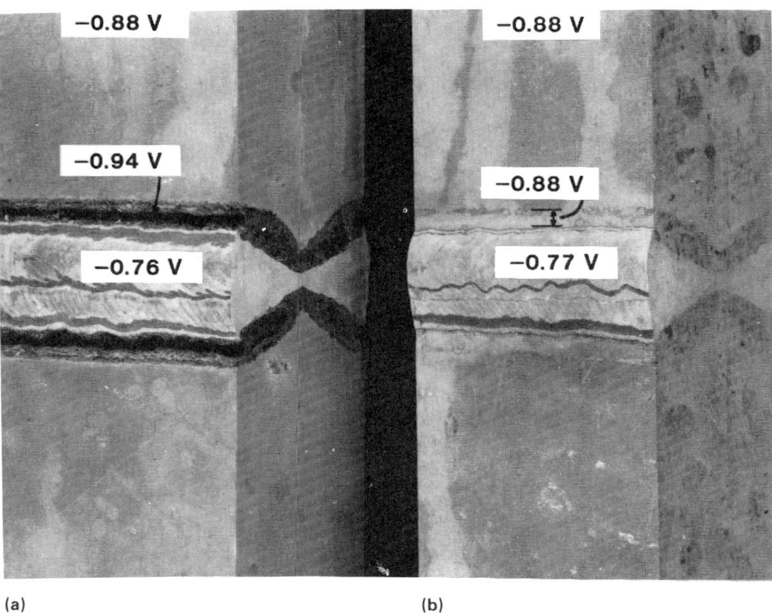

Fig. 3 Welded assemblies of aluminum alloy 7005 with alloy 5356 filler metal after a 1-year exposure to seawater. (a) As-welded assembly shows severe localized corrosion in the HAZ. (b) Specimen showing the beneficial effects of postweld aging. Corrosion potentials of different areas of the weldments are shown where they were measured. Electrochemical measurements performed in 53 g/L NaCl plus 3 g/L H_2O_2 versus a 0.1 *N* calomel reference electrode and recalculated to a saturated calomel electrode (SCE). Source: Ref 4

used to obtain maximum corrosion resistance (Fig. 3) (Ref 5-7).

As with many other alloy systems, attention must be given to the threat of crevice corrosion under certain conditions. Strong (99%) HNO_3 is particularly aggressive toward weldments that are not made with full weld penetration. Although all of the welds shown in Fig. 4 appear to be in excellent condition when viewed from the outside surface, the first two welds (Fig. 4a and b), viewed from the inside, are severely corroded. The weld made using standard gas tungsten arc (GTA) welding practices with full weld penetration (Fig. 4c) is in good condition after 2 years of continuous service.

Researchers have shown that aluminum alloys, both welded and unwelded, have good resistance to uninhibited HNO_3 (both red and white) up to 50 °C (120 °F). Above this temperature, most aluminum alloys exhibit knife-line attack (a very thin region of corrosion) adjacent to the welds. Above 50 °C (120 °F), the depth of knife-line attack increases markedly with temperature. One exception was found in the case of a fusion-welded 1060 alloy in which no knife-line attack was observed even at temperatures as high as 70 °C (160 °F). In inhibited fuming HNO_3 containing at least 0.1% hydrofluoric acid (HF), no knife-line attack was observed for any commercial aluminum alloy or weldment even at 70 °C (160 °F).

More information on the welding of aluminum and aluminum alloys is available in the articles "Arc Welding of Aluminum Alloys" and "Resistance Welding of Aluminum Alloys" in Volume 6 of the 9th Edition of *Metals Handbook*. The corrosion of these materials is discussed in detail in the article "Corrosion of Aluminum and Aluminum Alloys" in this Volume.

Corrosion of Tantalum and Tantalum Alloy Weldments

Examination of equipment fabricated from tantalum that has been used in a wide variety of service conditions and environments generally shows that the weld, HAZ, and base metal display equal resistance to corrosion. This same resistance has also been demonstrated in laboratory corrosion tests conducted in a number of different acids and other environments. However, in applications for tantalum-lined equipment, contamination of the tantalum with iron from underlying backing material, usually carbon steel, can severely impair the corrosion resistance of tantalum. About the only known reagents that rapidly attack tantalum are fluorine; HF and acidic solutions containing fluoride; fuming sulfuric acid (H_2SO_4) (oleum), which contains free sulfur trioxide (SO_3); and alkaline solutions.

An exception to the generalization that base metal and weldments in tantalum show the same corrosion resistance under aggressive media is discussed in the following example. Because tantalum is a reactive metal, the pickup of interstitial elements, such as oxygen, nitrogen, hydrogen, and carbon, during welding can have a damaging effect on a refractory metal such as tantalum.

Preferential Pitting of a Tantalum Alloy Weldment in H_2SO_4 Service. A 76-mm (3-in.) diameter tantalum alloy tee removed from the bottom of an H_2SO_4 absorber that visually showed areas of severe etching attack was examined. The absorber had operated over a period of several months, during which time about 11 400 kg (25 000 lb) of H_2SO_4 was handled. The absorber was operated at 60 °C (140 °F) with nominally 98% H_2SO_4. There was a possibility that some of the H_2SO_4 fed into the process stream may have been essentially anhydrous or even in the oleum range. Oleum is known to attack tantalum very rapidly at temperatures only slightly higher than 60 °C (140 °F). In addition, the H_2SO_4 effluent was found to contain up to 5 ppm of fluoride.

Investigation. The materials in both the flange and the corrugated portion of the tee were verified by x-ray fluorescence to be Tantaloy "63" (Ta-2.5W-0.15Nb). Corrosive attack was visible to the unaided eye on the radius of the flange, on the first corrugation of the part and in two bands—one on each side of the GTA weld about 13 mm (0.5 in.) from the weld centerline and running the full length of the piece. Other areas of the part, such as the lip weld joining the corrugated tube section to the flange, other areas of the base metal away from the weld, and even the weld metal itself (including the adjacent HAZ), appeared on a cursory visual basis to be free from significant attack.

Stereo microexamination showed that the corrosive attack took the form of pitting. The areas of most severe attack that were observed were parallel to the longitudinal weld, circumferentially around the first corrugation of the part, and on

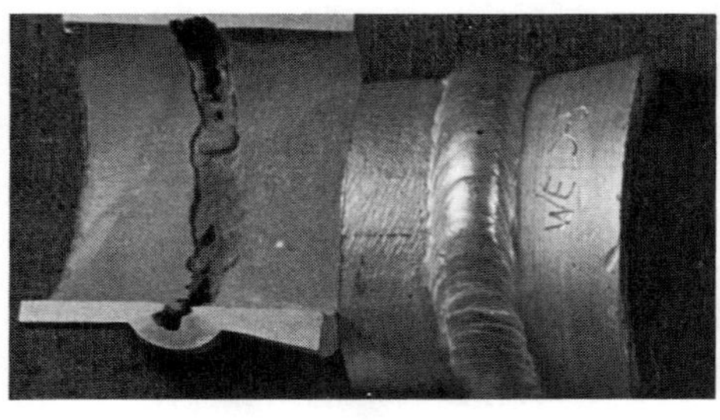
(a)

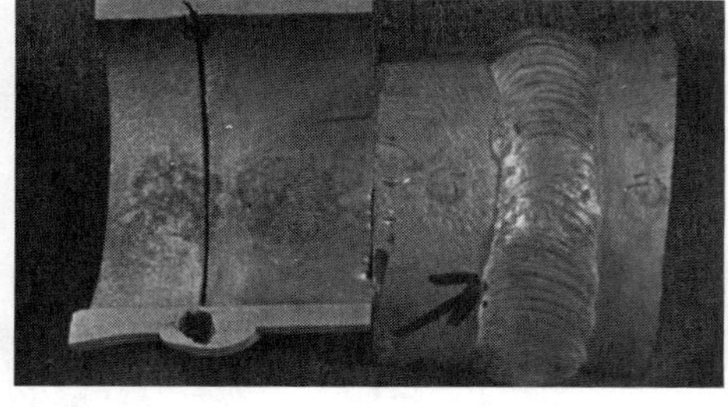
(b)

(c)

Fig. 4 Corrosion of three aluminum weldments in HNO_3 service. (a) and (b) GTA and oxyacetylene welds, respectively, showing crevice corrosion on the inside surface. (c) Standard GTA weld with full penetration is resistant to crevice corrosion.

the radius of the flange. Corrosion was characterized by a large number of closely occurring pits. What appeared to be markings resembling lines or scratches during an examination with the unaided eye were actually found at magnifications up to 30× to be rows of corrosion pits aligned in the longitudinal direction and parallel to the weld. However, fewer somewhat shallower pits were found generally over the entire part.

In some locations, the pitting was extensive in the weld metal, in the HAZ, and even in the base metal 180° away from the weld. The extent of pitting appeared to be most severe on the inside diameter at sites that had been abraded by the tool used in forming the corrugations and in some areas containing scratches. There was no noticeable corrosion anywhere on the outside of the part.

Metallographic examination showed classical corrosion pits that were nearly spherical in shape. The maximum pit depth observed was 0.06 mm (2.5 mils). Pits did not appear to be typical of erosion or impingement-type attack, because the pits did not show the typical undercutting or undermining. Pitting did not follow the grain boundaries. No cracks were found propagating from the base of the pits in any of the samples; therefore, there was no evidence of corrosion fatigue or stress-corrosion cracking (SCC).

A transverse section taken from another area of severe attack 13 mm (0.5 in.) from the weld metal was bent in the transverse direction with the inside diameter of the sample in tension. The sample was fully ductile; it was cold flattened 180° on itself with a sharp bend radius.

Conclusions. The corrosion on the tee was pitting that occurred generally over the entire part. Pitting was more severely concentrated parallel to the weld metal at a distance of 13 mm (0.5 in.) from the centerline of the weld (which was well outside the HAZ of the weld), on the radius of the flange, and on the first corrugation at the inlet. There was no evidence that the attack was due to cavitation erosion, corrosion fatigue, or SCC. The attack did not reduce the ductility of the material. All areas of the part were fully ductile, as evidenced by the soft, ductile nature of the part during sawing and cutting and by bend tests in the area of most severe attack.

The specific corrosion agents responsible were believed to be H_2SO_4 in the oleum concentration range in the presence of some fluoride ion (F^-). Corrosion tests and years of industrial experience with equipment indicate that in the absence of F^- (or free fluorine) and in H_2SO_4 concentrations of 98% and below such pitting attack does not occur on tantalum or Tantaloy "63" metal at 60 °C (140 °F).

Related Laboratory Experiments. Some laboratory experiments were performed on weldments of tantalum and Tantaloy "63" that may relate to and suggest why preferential corrosion attack was observed parallel to the longitudinal weld of the tee in Example 1 at locations considerably beyond the weld HAZ.

Gas tungsten arc butt welds were made in specimens of Tantaloy "63" and tantalum in a copper welding fixture operated in open air with an argon-flooded torch and trailing shield and with an argon flood on the backside of the root of the weld. Both materials showed surface oxidation parallel to the weld at about the location of the hold-down clamps of the welding fixture. The presence of this oxide film or heat-tint was revealed by electrolytically anodizing the samples in a dilute phosphoric acid (H_3PO_4) solution at 325 V. This film has been arbitrarily designated as a heat-tint oxide. The zone of oxidation could have occurred either by air leaking past the hold-down clamps and reacting with the hot tantalum at this location or by oxygen that may have been present in the welding atmosphere at this location. Heat-tint oxides were not noted on the weld metal or the HAZ. This is perhaps because these regions of the weld reached a much higher temperature—sufficiently high to volatilize oxygen from the tantalum weld metal and HAZ as tantalum suboxide (TaO).

The heat-tint oxide layer was removed from the welded specimens by swabbing with a strong pickling solution of HNO_3, HF, and H_2SO_4 before the samples were anodized. The heat-tint oxide layer, as well as the clean base metal, was found to be unaffected by aqua regia (3 parts concentrated HCl and 1 part concentrated HNO_3 by volume) after 6 h of exposure at 80 °C (175 °F).

During pickling in the HNO_3-HF-H_2SO_4 mixture, it was noted that the heat-tint oxide was attacked more rapidly than the weld metal, the HAZ, or the base metal outside of the thermally oxidized region. Such heat-tint oxide was often barely visible on the as-welded samples, appearing only as a slight yellowish tinge on the surface in many cases. However, minimal immersion (a few seconds) in the acid pickle revealed the heat-tinted area as whitish or grayish, hazy or smoky bands parallel to the weld. After the oxide was completely removed by pickling, either by swab pickling or by completely immersing the sample in an HF-containing medium, all parts of the sample appeared to corrode at approximately the same rate.

Therefore, the pitting observed at certain regions parallel to the welds was believed to be associated with the initial removal of the heat-tint oxide. Once the heat-tint oxide was removed, this selective attack no longer occurred, and corrosion was uniform.

Oxygen Tolerance of Tantalum Weldments. Tantalum reacts with oxygen, nitrogen, and hydrogen at elevated temperatures. The absorption of these interstitial elements, often called a gettering reaction, produces a sharp reduction in ductility and can cause embrittlement. This impairment in ductility (and also in notch toughness, as manifested by an increase in ductile-to-brittle transition temperature) can be considered a form of corrosion. The other Group Va refractory metals (niobium and vanadium) and the Group IVa reactive metals (titanium, zirconium, and hafnium) can also suffer similar attack.

An investigation was conducted to determine the approximate tolerances of tantalum and Tantaloy "63" weldments for oxygen contamination that may be permitted during fabrication or subsequent service. Weldments of the materials were doped with various amounts of oxygen added either by anodizing or by oxidation in air. This was followed by vacuum annealing treatments to diffuse the oxygen through the sample cross section. The oxygen concentration was monitored principally by hardness tests. Hardness is generally believed to be a better indicator of the extent of interstitial contamination than chemical analysis, which is subject to scatter and inaccuracy because of sampling difficulty. Bend tests (at room and liquid argon temperatures) and room-temperature Olsen cup formability tests were conducted to determine the hardness levels at which the materials embrittled.

The results showed that weldments of both materials remain ductile when hardened by interstitial contamination by oxygen up to a Rockwell 30T hardness in the low 80s. Above this hardness, embrittlement may be expected. The hardness level at which embrittlement occurs is substantially above the typical maximum allowable hardness of 65 HR30T specified for Tantaloy "63" or the 50 HR30T for tantalum flat mill products. Thus, if the extent of interstitial contamination by oxygen (and/or nitrogen) is controlled so that these maximum allowable hardness limits are not exceeded, embrittlement of weldments should not occur.

On the basis of chemical composition, the maximum oxygen tolerance for tantalum weldments appears to be about 400 to 550 ppm; for Tantaloy "63" weldments, it is about 350 to 500 ppm. Although commercially pure tantalum exhibits a somewhat higher tolerance for oxygen (and total interstitial contamination) than Tantaloy "63", the latter material appears to have somewhat better resistance to oxidation; this tends to offset the advantage tantalum has of

Fig. 5 Corroded type 316 stainless steel pipe from a black liquor evaporator. Two forms of attack are evident: preferential attack of the weld metal ferrite, suffered during HCl acid cleaning, and less severe attack in the sensitized HAZ (center). Source: Ref 8

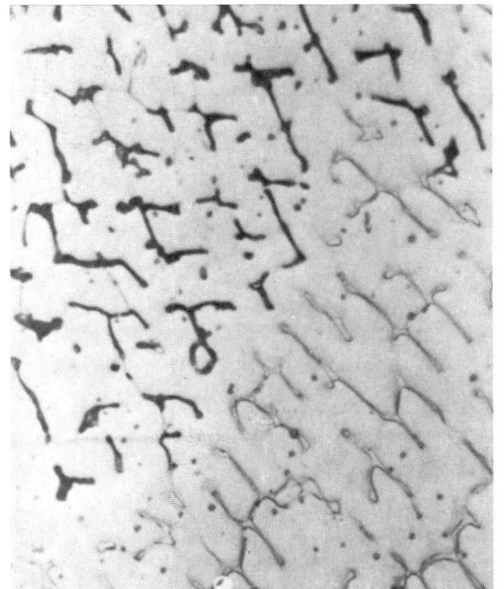

Fig. 6 Preferential corrosion of the vermicular ferrite phase in austenitic stainless steel weld metal. Discrete ferrite pools that are intact can be seen in the lower right; black areas in upper left are voids where ferrite has been attacked. Electrolytically etched with 10% ammonium persulfate. 500×. Source: Ref 9

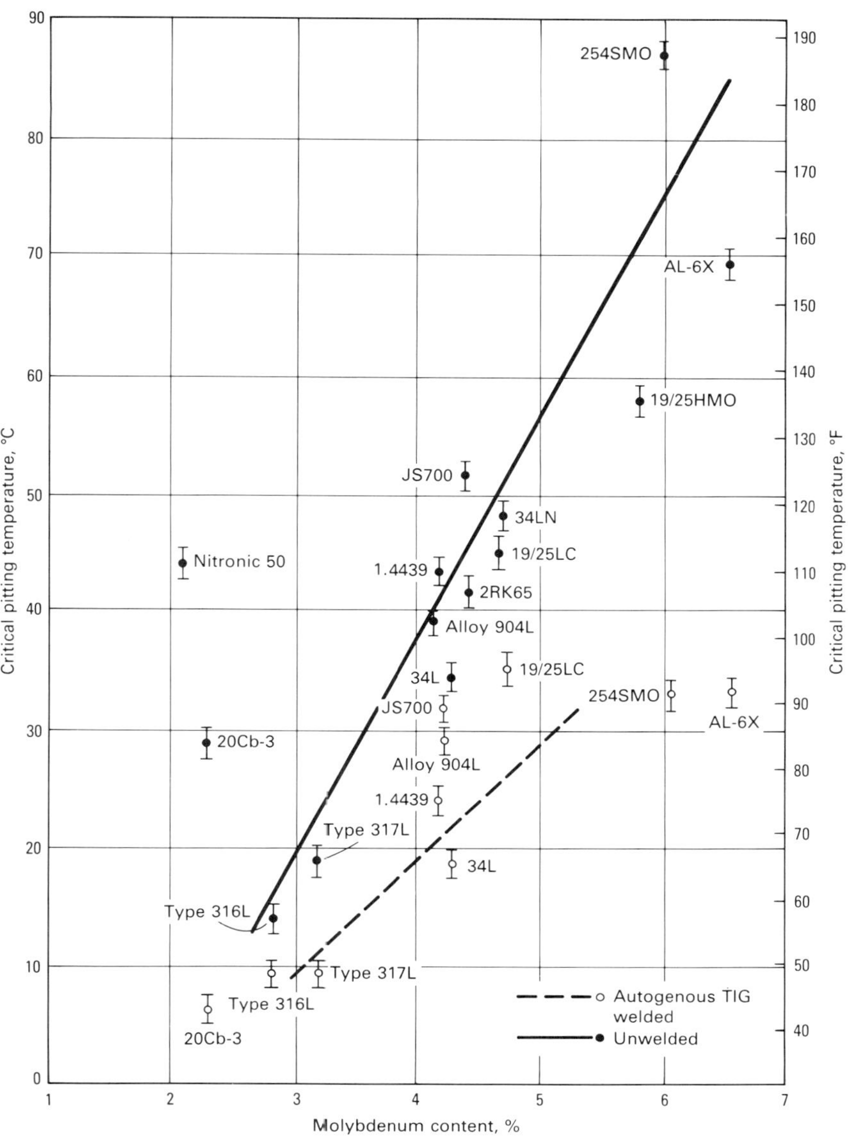

Fig. 7 Critical pitting temperature versus molybdenum content for commercial austenitic stainless steels tested in 10% $FeCl_3$. Resistance to pitting, as measured by the critical pitting temperature, increases with molybdenum content and decreases after autogenous tungsten inert gas welding. Source: Ref 8

a higher allowable oxygen pickup before embrittlement occurs. It should be further emphasized that the results are based on the assumption that oxygen was believed to be distributed relatively uniformly throughout the cross section in all parts of the weldment. A locally high concentration, such as a high surface contamination of oxygen or nitrogen, could result in a severe loss in ductility and could possibly even produce embrittlement. Therefore, all handling, cleaning, and fabrication practices on tantalum and its alloys should avoid producing such surface contamination as well as gross contamination. The article "Corrosion of Tantalum" in this Volume gives more detailed information on the corrosion of tantalum and tantalum alloys.

Corrosion of Austenitic Stainless Steel Weldments

The corrosion problems commonly associated with welding of austenitic stainless steels are related to precipitation effects and chemical segregation. These problems can be eliminated or minimized through control of base metal metallurgy, control of the welding practice, and selection of the proper filler metal.

Preferential Attack Associated With Weld Metal Precipitates. In austenitic stainless steels, the principal weld metal precipitates are δ-ferrite, σ phase, and $M_{23}C_6$ carbides. Small amounts of M_6C carbide may also be present. Sigma phase is often used to describe a range of chromium- and molybdenum-rich precipitates, including χ and laves (η) phases. These phases may precipitate directly from weld metal, but are most readily formed from weld metal δ-ferrite in molybdenum-containing austenitic stainless steels.

The δ-ferrite transforms into brittle intermetallic phases, such as σ and χ at temperatures ranging from 500 to 850 °C (930 to 1560 °F) for σ and 650 to 950 °C (1200 to 1740 °F) for χ. The precipitation rate for σ and χ phases increases with the chromium and molybdenum contents. Continuous intergranular networks of σ phase reduce the toughness, ductility, and corrosion resistance of austenitic stainless steels.

It is extremely difficult to discriminate between fine particles of σ and χ phases by using conventional optical metallographic techniques; hence the designation σ/χ phase. The use of more sophisticated analytical techniques to identify either phase conclusively is usually not justified when assessing corrosion properties, because the precipitation of either phase depletes the surrounding matrix of crucial alloying elements. Grain-boundary regions that are depleted in chromium and/or molybdenum are likely sites for attack in oxidizing and chlorine-bearing solutions. The damage caused by preferential corrosion of alloy-depleted regions ranges from the loss of entire grains (grain dropping) to shallow pitting at localized sites, depending on the distribution and morphology of the intermetallic precipitate particles at grain boundaries.

Because these precipitates are usually chromium- and molybdenum-rich, they are generally more corrosion resistant than the surrounding austenite. However, there are some exceptions to this rule.

Preferential attack associated with δ-ferrite and σ can be a problem when a weldment is being used close to the limit of corrosion resistance in environments represented by three types of acidic media:

- Mildly reducing (for example, HCl)
- Borderline active-passive (for example, H_2SO_4)
- Highly oxidizing (for example, HNO_3)

Acid cleaning of AISI type 304 and 316 stainless steel black liquor evaporators in the pulp and paper industry with poorly inhibited HCl can lead to weld metal δ-ferrite attack (Fig. 5 and 6). Attack is avoided by adequate inhibition (short cleaning times with sufficient inhibitor at low enough temperature) and by specifying full-finished welded tubing (in which the δ-ferrite networks within the weld metal structure are altered by cold work and a recrystallizing anneal). The latter condition can easily be verified with laboratory HCl testing, and such a test can be specified when ordering welded tubular products.

Sulfuric acid attack of σ phase or of chromium- and molybdenum-depleted regions next to σ-phase precipitates is commonly reported, although it is difficult to predict because the strong influence of tramp oxidizing agents, such as ferric (Fe^{3+}) or cupric (Cu^{2+}) ions can inhibit preferential attack. Type 316L weld filler metal has been formulated with higher chromium and lower molybdenum to minimize σ-phase formation, and filler metals for alloys such as 904L are balanced to avoid δ-ferrite precipitation and thus minimize σ phase.

Highly oxidizing environments such as those found in bleach plants could conceivably attack δ-ferrite networks and σ phase. However, this mode of attack is not often a cause of failure, probably because free-corrosion potentials are generally lower (less oxidizing) than that required to initiate attack. Preferential attack of δ-ferrite in type 316L weld metal is most often reported after prolonged HNO_3 exposure, as in nuclear fuel reprocessing or urea production. For these applications, a low corrosion rate in the Huey test (ASTM A 262, practice C) is specified (Ref 10).

Pitting Corrosion. Under moderately oxidizing conditions, such as a bleach plant, weld metal austenite may suffer preferential pitting in alloy-depleted regions. This attack is independent of any weld metal precipitation and is a consequence of microsegregation or coring in weld metal dendrites. Preferential pitting is more likely in autogenous (no filler) GTA welds (Fig. 7), in 4 to 6% Mo alloys (Table 1), when the recommended filler metal has the same composition as the base metal (Fig. 8), and when higher heat input

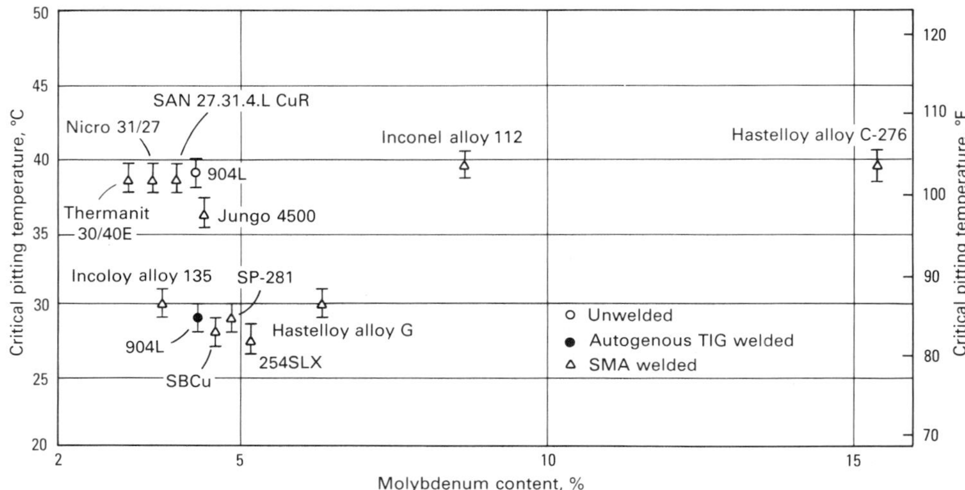

Fig. 8 Effects of various welding techniques and filler metals on the critical pitting temperature of alloy 904L. Data for an unwelded specimen are included for comparison. Source: Ref 8

welding leaves a coarse microstructure with surface-lying dendrites (Fig. 9). Such a microstructure is avoided by use of a suitably alloyed filler metal (Fig. 8).

Filler metals with pitting resistance close to or better than that of corresponding base metals include:

Base metal	Filler metals
Type 316L	316L, 317L, 309MoL
Type 317	317L, 309MoL
Alloy 904	Sandvik 27.31.4.LCuR Thermanit 30/40 E, Nicro 31/27 Fox CN 20 25 M, IN-112, Avesta P12, Hastelloy alloy C-276
Avesta 254 SMO	Avesta P12, IN-112, Hastelloy alloy C-276

Even when suitable fillers are used, preferential pitting attack can still occur in an unmixed zone of weld metal. High heat input welding can leave bands of melted base metal close to the fusion line. The effect of these bands on corrosion resistance can be minimized by welding techniques that bury unmixed zones beneath the surface of the weldment.

When the wrong filler metal is used, pitting corrosion can readily occur in some environments. In the example shown in Fig. 10, the type 316L base metal was welded with a lower-alloy filler metal (type 308L). Tap water was the major environmental constituent contributing to crust formation on the weld joint. The type 316L base metal on either side of the joint was not affected.

Crevice Corrosion. Defects such as residual welding flux and microfissures create weld metal crevices that are easily corroded, particularly in chloride-containing environments. Some flux formulations on coated shielded metal arc or stick electrodes produce easily detached slags, and others give slags that are difficult to remove completely even after gritblasting. Slags from rutile (titania-base) coatings are easily detached and give good bead shape. In contrast, slags from the basic-coated electrodes for out-of-position welding can be difficult to remove; small particles of slag may remain on the surface, providing an easy initiation site for crevice attack (Fig. 11).

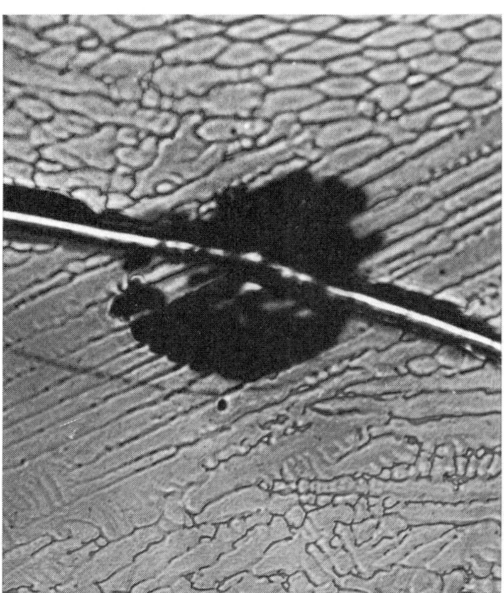

Fig. 9 A scratch-initiated pit formed in type 317L weld metal at 190 mV versus SCE in 0.6 *N* NaCl (pH 3) at 50 °C (120 °F). Pitting occurred at a grain with primary dendrites lying parallel to the surface rather than in grains with dendrites oriented at an angle to the surface.

Microfissures or their larger counterparts, hot cracks, also provide easy initiation sites for crevice attack, which will drastically reduce the corrosion resistance of a weldment in the bleach plant. Microfissures are caused by thermal contraction stresses during weld solidification and are a problem that plagues austenitic stainless steel fabrications. These weld metal cracks are more likely to form when phosphorus and sulfur levels are higher (that is, more than 0.015% P and 0.015% S), with high heat input welding, and in austenitic weld metal in which the δ-ferrite content is low (<3%).

Small-scale microfissures are often invisible to the naked eye, and their existence can readily explain the unexpectedly poor pitting performance of one of a group of weldments made with

Table 1 Amounts of principal alloying elements in stainless steels tested for pitting resistance

Test results are shown in Fig. 7 and 8.

Alloy	Composition, % Cr	Ni	Mo	N
Base metals				
Type 316L	16	13	2.8	...
Type 317L	18	14	3.2	...
34L	17	15	4.3	...
34LN	18	14	4.7	...
1.4439	18	14	4.3	0.13
Nitronic 50	21	14	2.2	0.20
20Cb-3	20	33	2.4	...
Alloy 904L	20	25	4.2	...
2RK65	20	25	4.5	...
JS700	21	25	4.5	...
19/25LC	20	25	4.8	...
AL-6X	20	24	6.6	...
254SMO	20	18	6.1	0.20
19/25HMO	21	25	5.9	0.15
Filler metals				
Type 316L	19	12	2.3	...
Type 317L	19	13	3.8	...
309MoL	23	14	2.5	...
Batox Cu	19	24	4.6	...
254SLX	20	24	5.0	...
SP-281	20	25	4.6	...
Jungo 4500	20	26	4.4	...
Nicro 31/27	28	30	3.5	...
Thermanit 30/40E	28	35	3.4	...
SAN 27.31.4.LCuR	27	31	3.5	...
Incoloy alloy 135	27	31	3.5	...
Hastelloy alloy G	22	38	3.7	...
P 12	21	61	8.6	...
Inconel alloy 112	21	61	8.7	...
Hastelloy alloy C-276	15	58	15.4	...

filler metals of apparently similar general composition. The microfissure provides a crevice, which is easily corroded because stainless alloys are more susceptible to crevice corrosion than to pitting. However, microfissure-crevice corrosion is often mistakenly interpreted as self-initiated pitting (Fig. 12 and 13).

Crevice corrosion sites can also occur at the beginning or end of weld passes, between weld passes, or under weld spatter areas. Weld spatter is most troublesome when it is loose or poorly adherent. A good example of this type of crevice condition is the type 304 stainless system shown in Fig. 14.

Microfissure corrosion in austenitic stainless steel weldments containing 4 to 6% Mo is best avoided with the nickel-base Inconel 625, Inconel 112, or Avesta P12 filler metals, which are very resistant to crevice attack. Some stainless electrodes are suitable for welding 4% Mo steels, but they should be selected with low phosphorus and sulfur to avoid microfissure problems.

Hot tap water is not thought to be particularly aggressive; however, Fig. 15 shows what can happen to a weld that contains a lack-of-fusion defect in the presence of chlorides. In this case, the base metal is type 304 stainless steel, and the weld metal type 308.

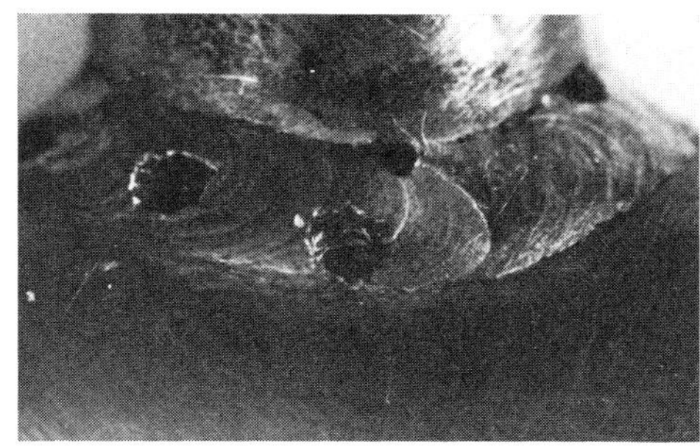

Fig. 10 Pitting of underalloyed (relative to base metal) type 308L weld metal. The type 316L stainless steel base metal is unaffected. About 2.5×

Carbide Precipitation in the HAZ. The best known weld-related corrosion problem in stainless steels is weld decay (sensitization) caused by carbide precipitation in the weld HAZ. Sensitization occurs in a zone subject to a critical thermal cycle in which chromium-rich carbides precipitate and in which chromium diffusion is much slower than that of carbon. The carbides are precipitated on grain boundaries that are consequently flanked by a thin chromium-depleted layer. This sensitized microstructure is much less corrosion resistant, because the chromium-depleted layer and the precipitate can be subject to preferential attack (Fig. 16). In North America, sensitization is avoided by the use of low-carbon grades such as type 316L (0.03% C max) in place of sensitization-susceptible type 316 (0.08% C max). In Europe, it is more common to use 0.05% C (max) steels, which are still reasonably resistant to sensitization, particularly if they contain molybdenum and nitrogen; these elements appear to raise the tolerable level of carbon and/or heat input. However, low-carbon stainless steels carry a small cost premium; therefore, they are not universally specified.

Thiosulfate ($S_2O_3^{2-}$) pitting corrosion will readily occur in sensitized HAZs of type 304 weldments in paper machine white-water service (Fig. 17). This form of attack can be controlled by limiting sources of $S_2O_3^{2-}$ contamination, the principal one of which is the brightening agent sodi-

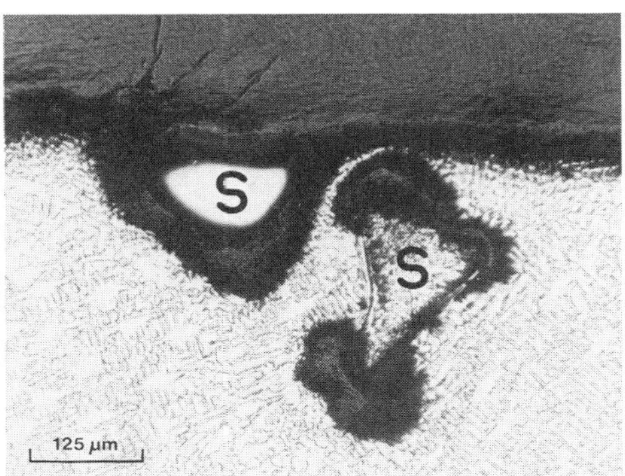

Fig. 11 Crevice corrosion under residual slag (S) in IN-135 weld metal after bleach plant exposure. Etched with glyceregia. Source: Ref 8

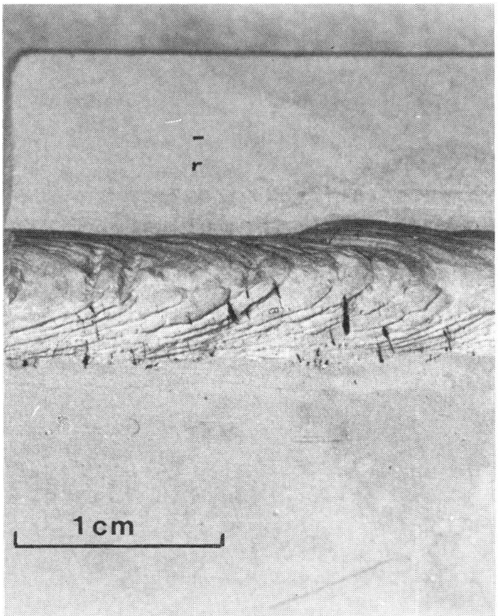

Fig. 12 Microfissure corrosion of IN-135 weld metal on an alloy 904L test coupon after bleach plant exposure. See also Fig. 13. Source: Ref 8

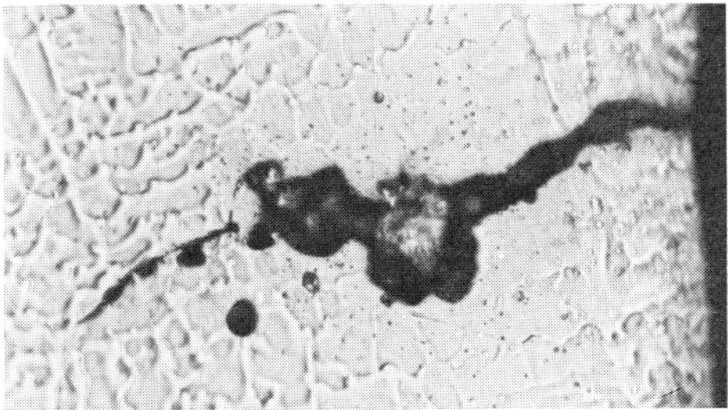

35 μm

Fig. 13 Section from the bleach plant test coupon in Fig. 12 showing crevice corrosion that has almost obliterated evidence of a microfissure. This form of attack is often mistakenly interpreted as self-initiated pitting; more often, crevice corrosion originates at a microfissure. Etched with glyceregia. Source: Ref 8

Fig. 14 Cross section of a weldment showing crevice corrosion under weld spatter. Oxides (light gray) have formed on the spatter and in the crevice between spatter and base metal.

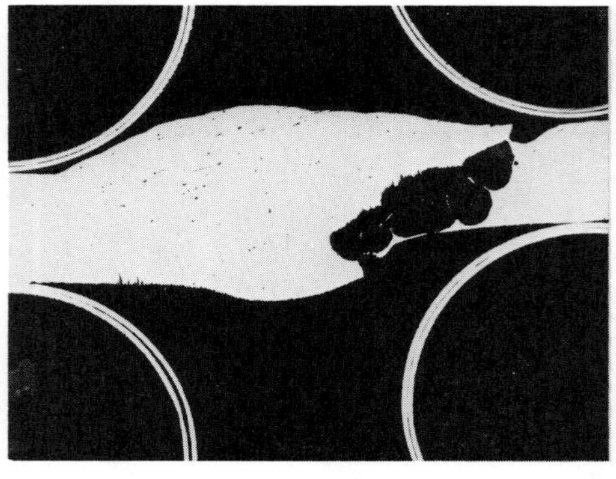

(a)

(b)

Fig. 15 Unetched (a) and etched (b) cross sections of a type 304 stainless steel weldment showing chloride pitting attack along a crevice created by a lack-of-fusion defect. Service environment: hot tap water

um hyposulfite ($Na_2S_2O_4$). However, nonsensitized type 304 will also be attacked, and type 316L is the preferred grade of stainless steel that should be specified for paper machine service.

At higher solution temperatures, sensitized type 304 and type 316 are particularly susceptible to SCC—whether caused by chlorides, sulfur compounds, or caustic. For example, type 304 or 316 kraft black liquor evaporators and white liquor tubing are subject to SCC in sensitized HAZs. In many cases cracking occurs after HCl acid cleaning. Although the initial crack path may be intergranular, subsequent propagation can have the characteristic branched appearance of transgranular chloride SCC. Intergranular SCC caused by sulfur compounds can also occur during the acid cleaning of sensitized stainless steels in kraft liquor systems.

Sigma Precipitation in HAZs. When the higher molybdenum alloys such as 904L, AL-6XN, and 254SMO were first developed, one of the anticipated corrosion problems was attack of single-phase precipitates in weld HAZs. This form of attack has subsequently proved to be either superficial or nonexistent in most applications, probably because the compositions of the alloys have been skillfully formulated to minimize σ phase-related hot-rolling problems.

More recently, nitrogen has been added to molydenum-bearing austenitic stainless steels to retard the precipitation of chromium- and molybdenum-rich intermetallic compounds (σ or χ phases). The incubation time for intermetallic precipitation reactions in iron-chromium-nickel-molybdenum stainless alloys is significantly increased by raising the alloy nitrogen content. This has allowed the commercial production of thick plate sections that can be fabricated by multipass welding operations. In addition to suppressing the formation of deleterious phases, nitrogen, in cooperation with chromium and molybdenum, has a beneficial effect on localized corrosion resistance in oxidizing acid-chloride solutions.

Corrosion Associated With Postweld Cleaning. Postweld cleaning is often specified to remove the heat-tinted metal formed during welding. Recent work has shown that cleaning by stainless steel wire brushing can lower the corrosion resistance of a stainless steel weldment (Fig. 18). This is a particular problem in applications in which the base metal has marginal corrosion resistance. The effect may be caused by inadequate heat-tint removal, by the use of lower-alloy stainless steel brushes such as type 410 or 304, or by the redeposition of abraded metal or oxides.

Any cleaning method may be impaired by contamination or by lack of control. Results of a study in bleach plants suggest that pickling and glass bead blasting can be more effective than stainless wire brushing and that brushing is more difficult to perform effectively in this case.

Corrosion Associated With Weld Backing Rings. Backing rings are sometimes used when welding pipe. In corrosion applications, it is important that the backing ring insert be consumed during the welding process to avoid a crevice. In the example shown in Fig. 19, the wrong type of backing ring was used, which left a crevice after welding. The sample was taken from a leaking brine cooling coil used in the production of nitroglycerin. The cooling coils contained calcium chloride ($CaCl_2$) brine inhibited with chromates. Coils were made by butt welding sections of seamless type 304L stainless steel tubing. This failure was unusual because several forms of corrosion had been observed.

A metallographic examination of a small trepanned sample revealed the following:

- *Microstructure:* The base metal and weld metal microstructures appeared satisfactory
- *Pitting:* Irregular corrosion pits were seen on the inside tube surface at crevices formed by the tube and the backing ring adjacent to the tube butt weld. The deepest pits extended 0.1 to 0.2 mm (4 to 8 mils) into the 1.65-mm (0.065-in.) thick tube wall
- *Cracking:* There were numerous brittle, branching transgranular cracks originating on the inside surface at the crevice under the backing ring
- *Preferential Weld Corrosion:* Extensive preferential corrosion of the ferrite phase (vermicular morphology) in the tube weld had occurred and penetrated almost completely through the tube wall. Corrosion originated on the outside surface of the tube

It was concluded that the preferential weld corrosion from the process side was the most probable cause of the actual leak in the nitrator coil. The preferential corrosion of ferrite in nitrating mixtures of HNO_3 and H_2SO_4 is well known. Whether this corrosion causes a serious problem depends on the amount of ferrite present in the weld. If the amount of ferrite is small and the particles are not interconnected, the overall corrosion rate is not much higher than that of a completely austenitic material. If the particles are interconnected, as in this case, there is a path for fairly rapid corrosion through the weld, causing failure to occur.

To minimize this problem, two possible solutions were considered. The first was to weld the coils with a filler metal that produces a fully austenitic deposit, and the second was to solution anneal at 1065 °C (1950 °F) after welding to dissolve most of the ferrite. It would also help to select stainless steel base metal by compositions (for example, high nickel, low chromium content) to minimize the production of ferrite during welding. Welding with a fully austenitic filler metal was considered to be the best approach.

Cracking on the brine side was caused by chloride SCC. Most probably, the cracking did not happen during operation at 15 °C (60 °F) or lower. It is thought that the cracking most likely occurred while the coil was being decontaminated at 205 to 260 °C (400 to 500 °F) in preparation for weld repairing of the leak. Brine that was trapped in the crevice between the tube wall and the backing ring was boiled to dryness. Under these conditions, SCC would occur in a short time.

There probably were stress cracks behind all of the backing rings. The future life of this coil was questionable. A new coil was recommended.

The pitting corrosion caused by the brine was not considered to be as serious as it first appeared. If this were the only corrosion (and the sample was representative of the coil), the coil would not have failed for considerable length of time. The decontamination process, which evaporated the trapped brine, produced some of the observed corrosion and made the pitting appear worse than it was before decontamination.

Chromates are anodic inhibitors and therefore can also greatly increase the corrosion (usually

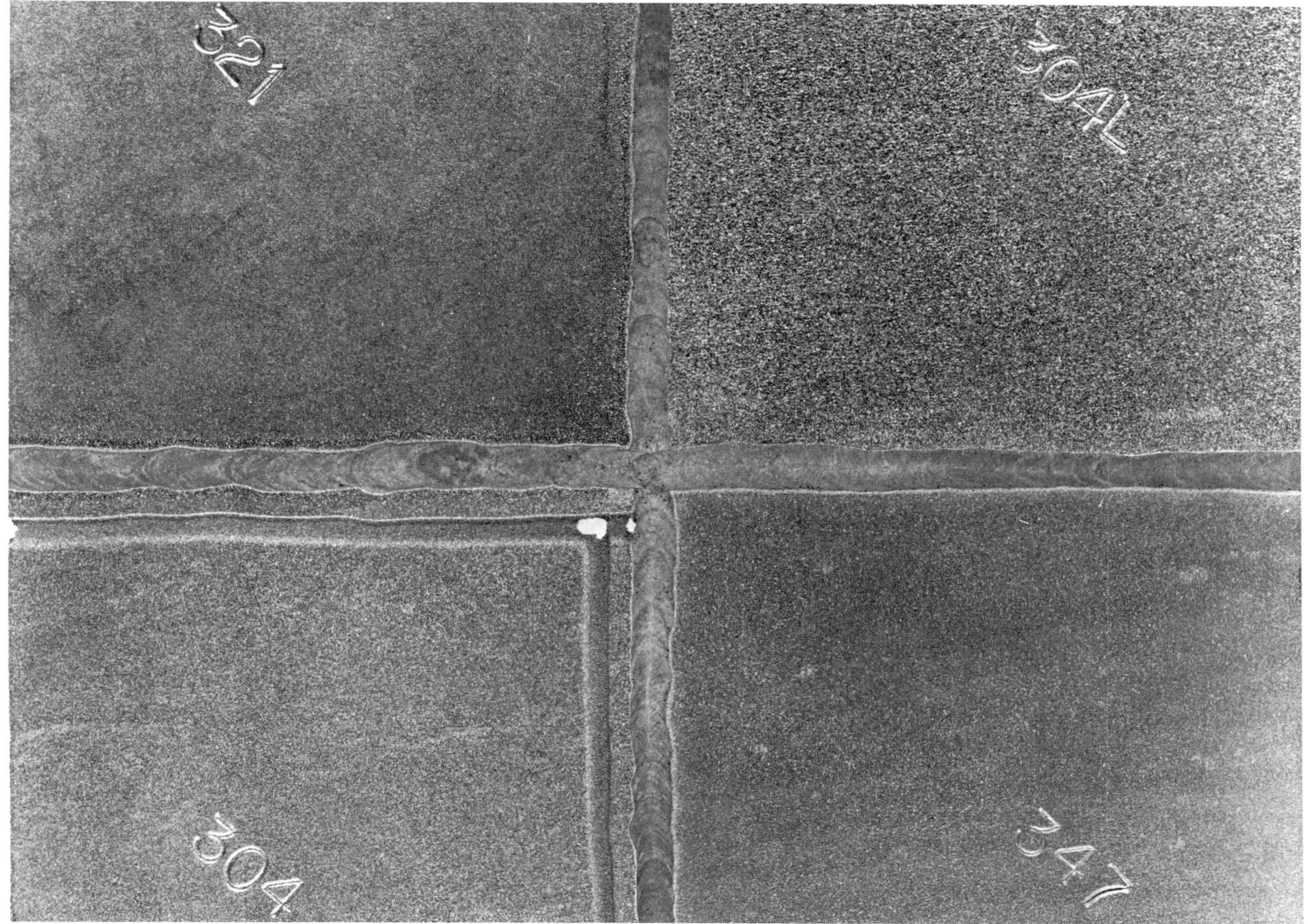

Fig. 16 Weld decay (sensitization) in austenitic stainless steel and methods for its prevention. Panels of four different AISI 300-series stainless steels were joined by welding and exposed to hot HNO_3 + HF solution. The weld decay evident in the type 304 panel was prevented in the other panels by reduction in carbon content (type 304L) or by addition of carbide-stabilizing elements (titanium in type 321, and niobium in type 347). Source: Ref 2

by pitting) in the system if insufficient quantities are used. This may also have occurred in the crevices in the nitrator coil butt welds, regardless of the bulk solution concentration. The best solution to this problem was to eliminate the crevices, that is, not to use backing rings.

Effects of GTA Weld Shielding Gas Composition. The chromium in a stainless steel has a strong chemical affinity for oxygen and carbon. Weld pools formed by electric arc processes must be shielded from the atmosphere to prevent slag formation and oxidation (Fig. 20), to maintain a stable arc, and to reduce contamination of the molten metal by the weld environment. Argon or argon plus helium gas mixtures are commonly used in GTA welding processes to create a barrier between the solidifying weld and the atmosphere. In other cases, nitrogen is commonly used as a backing gas to protect the backside of the root pass.

The composition of a shielding gas can be modified to improve the microstructure and properties of GTA welds in austenitic stainless steels. More specifically, the use of argon mixed with small volumes of nitrogen (10 vol% N_2 or less) in a GTA welding process enhances the corrosion resistance of iron-chromium-nickel-molybdenum-nitrogen stainless alloys in oxidizing acid chloride solutions (Fig. 21). In certain nonoxidizing solutions, argon-nitrogen shielding gas reduces the δ-ferrite content of weld metal, and influences weld metal solidification behavior.

The nitrogen content of weld metal increases with the partial pressure of nitrogen in the GTA weld shielding gas. The increase in weld metal nitrogen content is greater when nitrogen is mixed with an oxidizing gas, such as CO_2, than with either a reducing (hydrogen) or a neutral (argon) gas. Porosity and concavity are observed in austenitic stainless steel weld metals when more than 10 vol% nitrogen is added to an argon shielding gas. Although solid-solution additions of nitrogen are not detrimental to the SCC resistance of unwelded molybdenum-containing austenitic stainless steels, an increased weld metal nitrogen content tends to increase susceptibility to SCC.

Effects of Heat-Tint Oxides on the Corrosion Resistance of Austenitic Stainless Steels. Under certain laboratory conditions, a mechanically stable chromium-enriched oxide layer can be formed on a stainless steel surface that enhances corrosion resistance. In contrast, the conditions created by arc-welding operations produce a scale composed of elements that have been selectively oxidized from the base metal. The region near the surface of an oxidized stainless steel is depleted in one or more of the elements that have reacted with the surrounding atmosphere to form the scale. The rate of oxidation for a stainless steel, and consequently the degree of depletion in the base metal, is independent of the alloy composition and is controlled by diffusion through the oxide. The oxidized, or heat-tinted, surface of a welded stainless steel consists of a heterogeneous oxide composed primarily of iron and chromium above a chromium-depleted layer of base metal. The properties of such a surface depend on:

- The time and temperature of the thermal exposure
- The composition of the atmosphere in contact with the hot metal surface
- The chemical composition of the base metal beneath the heat-tint oxide
- The physical condition of the surface (contamination, roughness, thermomechanical history) prior to heat tinting

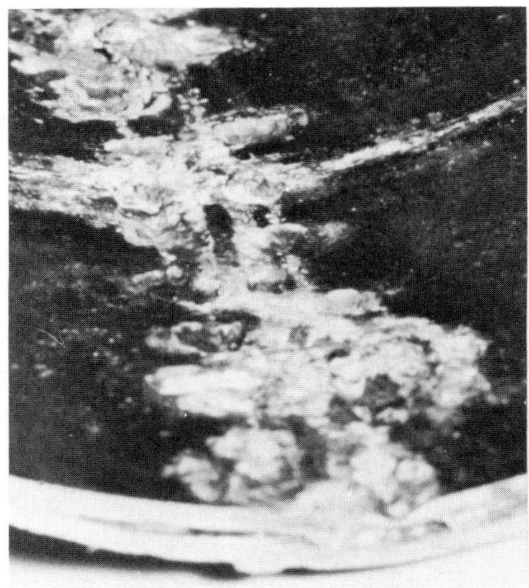

Fig. 17 Thiosulfate pitting in the HAZ of a type 304 stainless steel welded pipe after paper machine white-water service. 2×. Source: Ref 8

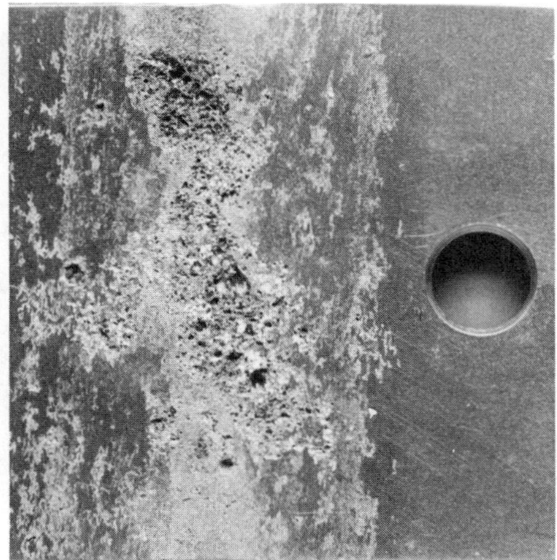

Fig. 18 Pitting corrosion associated with stainless steel wire brush cleaning on the back of a type 316L stainless steel test coupon after bleach plant exposure. Source: Ref 8

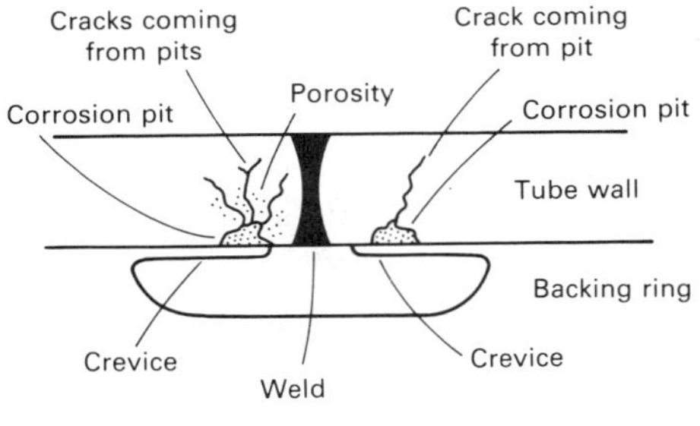

Fig. 19 Schematic of a stainless steel nitrator cooling coil weld joint. Failure was caused by improper design of the backing ring, which was not consumed during welding and left a crevice. Source: Ref 11

- The adherence of the heat-tint oxide to the base metal

The defects, internal stresses, and composition of the heat-tint oxide make it a poor barrier to any corrosive media that might initiate localized corrosion in the chromium-depleted layer of base metal.

The severity of localized corrosion at heat-tinted regions exposed to oxidizing chloride solutions is directly related to the temperature of the hot metal surface during welding. A heat-tint oxide on an austenitic stainless steel exposed in air first becomes obvious at approximately 400 °C (750 °F). As the surface temperature is increased, differently colored oxides develop that appear to be superimposed upon the oxides formed at lower temperatures (Table 2). Dark blue heat-tint oxides are the most susceptible to localized corrosion. It should be noted that gas-shielded surfaces do not form the same distinctly colored oxides as surfaces exposed to air during welding, but gas-shielded surfaces can also be susceptible to preferential corrosion.

Whether a weld heat tint should be removed prior to service depends on the corrosion behavior of the given alloy when exposed to the particular environment in question. Preferential corrosion at heat-tinted regions is most likely to occur on an alloy that performs near the limit of its corrosion resistance in service, but certain solutions do not affect heat-tinted regions. Even when heat-tinted regions are suspected of being susceptible to accelerated corrosion in a particular environment, the following factors should be considered:

- The rate at which pits, once initiated in the chromium-depleted surface layer, will propagate through sound base metal
- The hazards associated with the penetration, due to localized corrosion, of a process unit
- The cost and effectiveness of an operation intended to repair a heat-tinted stainless steel surface

The corrosion resistance of heat-tinted regions can be restored in three stages. First, the heat-tint oxide and chromium-depleted layer are removed by grinding or wire brushing. Second, the abraded surface is cleaned with an acid solution or a pickling paste (a mixture of HNO_3 and HF suspended in an inert paste or gel) to remove any surface contamination and to promote the reformation of a passive film. Third, after a sufficient contact time, the acid cleaning solution or pickling paste is thoroughly rinsed with water, preferably demineralized or with a low chloride ion (Cl^-) content.

Grinding or wire brushing may not be sufficient to repair a heat-tinted region. Such abrading operations may only smear the heat-tint oxide and embed the residual scale into the surface, expose the chromium-depleted layer beneath the heat-tint oxide, and contaminate the surface with ferrous particles that were picked up by the grinding wheel or wire brush. A stainless steel surface should never be abraded with a wheel or brush that has been used on a carbon or low-alloy steel; wire brushes with bristles that are not made of a stainless steel of similar composition should also be avoided. Conversely, attempting to repair a heat-tinted region with only a pickling paste or acid solution may stain or even corrode the base metal if the solution is overly aggressive or is allowed to contact the surface for an extended time. If the acid is too weak, a chromium-depleted scale residue may remain on a surface. Even if the chromium-depleted layer were completely removed by a grinding operation, mechanically

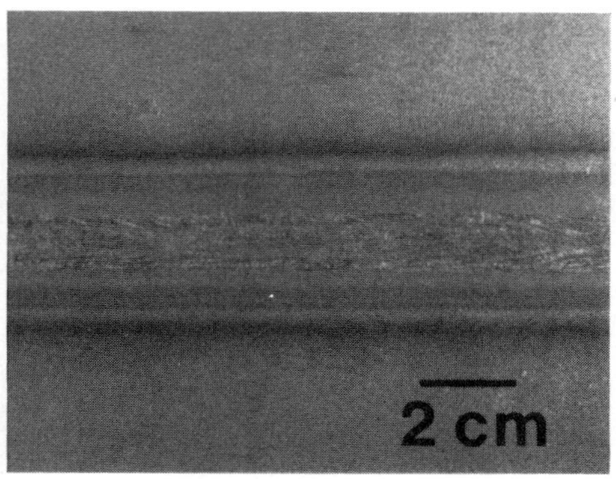

(a)

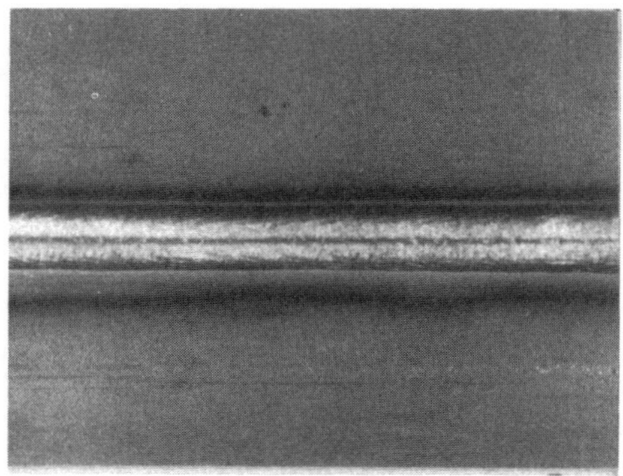

(b)

Fig. 20 Examples of properly shielded (a) and poorly shielded (b) autogenous GTA welds in type 304 stainless steel strip. Source: Ref 11

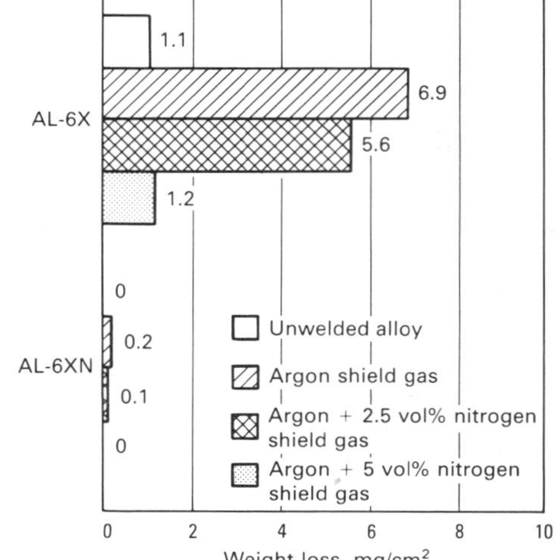

Fig. 21 Effect of GTA weld shielding gas composition on the corrosion resistance of two austenitic stainless steels. Welded strip samples were tested according to Ref 17; test temperature was 35 °C (95 °F). Source: Ref 11

Table 2 Welding conditions and corrosion resistance of heat-tinted UNS S31726 stainless steel plate

Welding conditions(a)				Corrosion test results(b)		
Heat input				Maximum pit depth,		
kJ/mm	kJ/in.	Welding current, A	Centerline heat-tint color	mm	mils	Number of pits on heat-tinted surface
0.3	7.525	50	None	0.1	4	2
0.59	15.050	100	Straw	0.7	28	10
0.89	22.576	150	Rose	0.8	31	50
1.19	30.101	200	Blue	0.7	28	>70
1.48	37.626	250	White	0.9	35	>70

(a) Single-pass autogenous bead-on-plate GTA welds were made to heat tint the root surface of 6.4-mm (1/4-in.) thick plate samples. (b) Duplicate coupons, each with one 25- × 51-mm (1- × 2-in.) heat-tinted surface, were exposed to 10% $FeCl_3$ solutions at 50 °C (120 °F). The weld face and edges of each coupon were covered with a protective coating. Source: Ref 12

ground surfaces generally have inferior corrosion resistance compared to properly acid-pickled surfaces.

Unmixed Zones. All methods of welding stainless steel with a filler metal produce a weld fusion boundary consisting of base metal that has been melted but not mechanically mixed with filler metal and a partially melted zone in the base metal. The weld fusion boundary lies between a weld composite consisting of filler metal diluted by base metal and an HAZ in the base metal (Fig. 1). The width of the unmixed zone depends on the local thermal conditions along the weld fusion line. For a GTA welding process, the zone is most narrow at the weld face and is broadest near the middle of the weld thickness.

An unmixed zone has the composition of base metal but the microstructure of an autogenous weld. The microsegregation and precipitation phenomena characteristic of autogenous weldments decrease the corrosion resistance of an unmixed zone relative to the parent metal. Unmixed zones bordering welds made from overalloyed filler metals can be preferentially attacked when exposed on the weldment surface (Fig. 22). The potential for preferential attack of unmixed zones can be reduced by minimizing the heat input to the weld and/or by flowing molten filler metal over the surface of the unmixed zone to form a barrier to the service environment. Care must be taken in this latter operation to avoid cold laps and lack-of-fusion defects. In both cases, preferential attack is avoided as long as the surface of the unmixed zone lies beneath the exposed surface of the weldment.

Chloride SCC. Welds in the 300-series austenitic stainless steels, with the exception of types 310 and 310Mo, contain a small amount of δ-ferrite (usually less than 10%) to prevent hot cracking during weld solidification. In hot, aqueous chloride environments, these duplex weldments generally show a marked resistance to cracking, while their counterparts crack readily (Fig. 23). The generally accepted explanation for this behavior is that the ferrite phase is resistant to chloride SCC and impedes crack propagation through the austenite phase. Electrochemical effects may also play a part; however, under sufficient tensile stress, temperature, and chloride concentration, these duplex weldments will readily crack. An example is shown in Fig. 24.

Caustic Embrittlement (Caustic SCC). Susceptibility of austenitic stainless steels to this form of corrosion usually becomes a problem when the caustic concentration exceeds approximately 25% and temperatures are above 100 °C (212 °F). Because welding is involved in most fabrications, the weld joint becomes the focus of attention because of potential stress raiser effects and because of high residual shrinkage stresses. Cracking occurs most often in the weld HAZ.

In one case, a type 316L reactor vessel failed repeatedly by caustic SCC in which the process fluids contained 50% sodium hydroxide (NaOH) at 105 °C (220 °F). Failure was restricted to the weld HAZ adjacent to bracket attachment welds used to hold a steam coil. The stresses caused by the thermal expansion of the Nickel 200 steam coil at 1034 kPa (150 psig) aggravated the problem. Figure 25 shows the cracks in the weld HAZ to be branching and intergranular. Because it was not practical to reduce the operating temperature below the threshold temperature at which caustic SCC occurs, it was recommended that the vessel be weld overlayed with nickel or that the existing vessel be scrapped and a replacement fabricated from Nickel 200.

Microbiologically Induced Corrosion (MIC). Microbiological corrosion in the process industries is most often found in three areas: cooling water systems, aqueous waste treatment, and groundwater left in new equipment or piping systems after testing. Nearly all confirmed cases of MIC have been accompanied by characteristic deposits. These are usually discrete mounds. Deposit color can also be an indication of the types of microorganisms that are active in the system. For example, iron bacteria deposits on stainless steel, such as those produced by *Gallionella*, are often reddish.

Investigators have shown that in almost all cases the environment causing the damage was a natural, essentially untreated water containing one or more culprit species of microbiological organisms. In the case of austenitic stainless steel weldments, corrosion generated by bacteria takes a distinctive form, that is, subsurface cavities with only minute pinhole penetration at the surface. The following case history involving well water in Texas illustrates these characteristics (Ref 13).

New production facilities at one plant site required austenitic stainless steels, primarily types 304L and 316L, for resistance to HNO_3 and organic acids and for maintaining product purity. The piping was shop fabricated, field erected, and then hydrostatically tested. All of the large (>190 000 L, or 25 000 gal) flat-bottom storage tanks were field erected and hydrostatically tested. During the early stages of construction, sodium-softened plant well water (also used for drinking) containing 200 ppm of chlorides was used for testing.

No attempts were made to drain the pipelines after testing. Tanks were drained, but then refilled to a depth of approximately 0.5 to 1 m (2 to

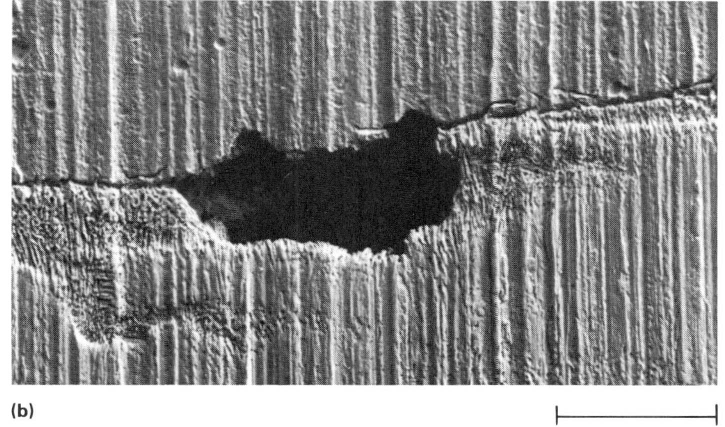

Fig. 22 Optical (a) and scanning electron (b) micrographs of pitting in the unmixed zone of iron-chromium-nickel-molybdenum stainless steel plates that were GTA welded with an overalloyed filler metal. The unmixed zones were preferentially attacked in an oxidizing acid chloride solution at elevated temperature.

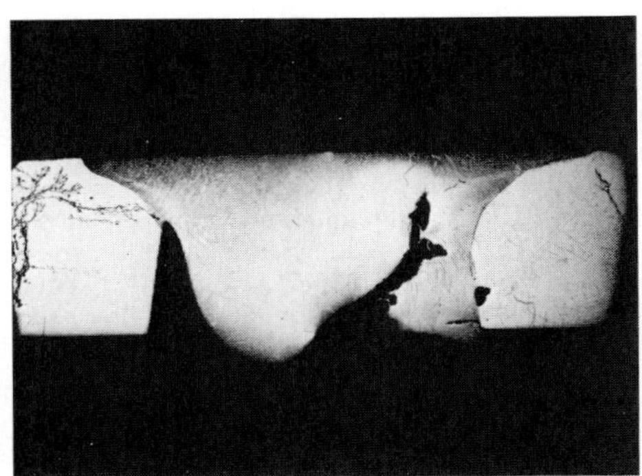

Fig. 23 Selective attack of a type 317L stainless steel weldment and chloride SCC of the adjacent 317L base metal. The environment was a bleaching solution (7 g/L Cl_2) at 70 °C (160 °F).

3 ft) for ballast because of a hurricane threat. The water was left in the tanks to evaporate.

The problem became evident when water was found dripping from butt welds in type 304L and 316L piping (nominal wall thickness: 3.2 mm, or ⅛ in.) approximately 1 and 4 months, respectively, after the hydrotest. Internal inspection showed pits in and adjacent to welds under reddish-brown deposits. Tank manways were uncovered, and similar conditions were found. As shown in Fig. 26, moundlike deposits were strung out along weld seams in the tank bottoms.

Figure 27 shows a closeup view of a typical deposit still wet with test water. The brilliantly colored deposit was slimy and gelatinous in appearance and to the touch, and it measured 76 to 102 mm (3 to 4 in.) in width. At one point during the investigation, a similar deposit on a weld that was covered with about 150 mm (6 in.) of water was thoroughly dispersed by hand. Twenty-four hours later, the deposit had returned in somewhat diminished form to exactly the same location.

Figure 28 shows a nearly dry deposit. After wiping the deposit clean, a dark ring-shaped stain outlining the deposit over the weld was noted (Fig. 29). There was, however, no evidence of pitting or other corrosion, even after light sanding with emery. Finally, probing with an icepick revealed a large deep pit at the edge of the weld, as shown in Fig. 30. Figure 31, a radiograph of this weld seam, shows the large pit that nearly consumed the entire width of the weld bead, as well as several smaller pits. A cross section through a large pit in a 9.5-mm (⅜-in.) thick type 304L tank bottom is shown in Fig. 32.

The characteristics of this mode of corrosion were a tiny mouth at the surface and a thin shell of metal covering a bottle-shaped pit that had consumed both weld and base metal. There was no evidence of intergranular or interdendritic attack of base or weld metal. However, pitted welds in a type 316L tank showed preferential attack of the δ-ferrite stringers (Fig. 33).

This type 316L tank was left full of hydrotest water for 1 month before draining. The bottom showed severe pitting under the typical reddish-brown deposits along welds. In addition, vertical rust-colored streaks (Fig. 34) were found above and below the sidewall horizontal welds, with deep pits at the edges of the welds associated with each streak (Fig. 35).

Analyses of the well water and the deposits showed high counts of iron bacteria (*Gallionella*) and iron/manganese bacteria (*Siderocapsa*). Both sulfate-reducing and sulfur-oxidizing bacteria were absent. The deposits also contained large amounts (thousands of parts per million) of iron, manganese, and chlorides.

As indicated, nearly all biodeposits and pits were found at the edges of, or very close to, weld seams. It is possible that the bacteria in stagnant well water were attracted by an electrochemical phenomenon or surface imperfections (oxide or slag inclusions, porosity, ripples, and so on) typically associated with welds. A sequence of events for the corrosion mechanism in this case might be the following:

- Attraction and colonization of iron and iron/manganese bacteria at welds
- Microbiological concentration of iron and manganese compounds, primarily chlorides, be-

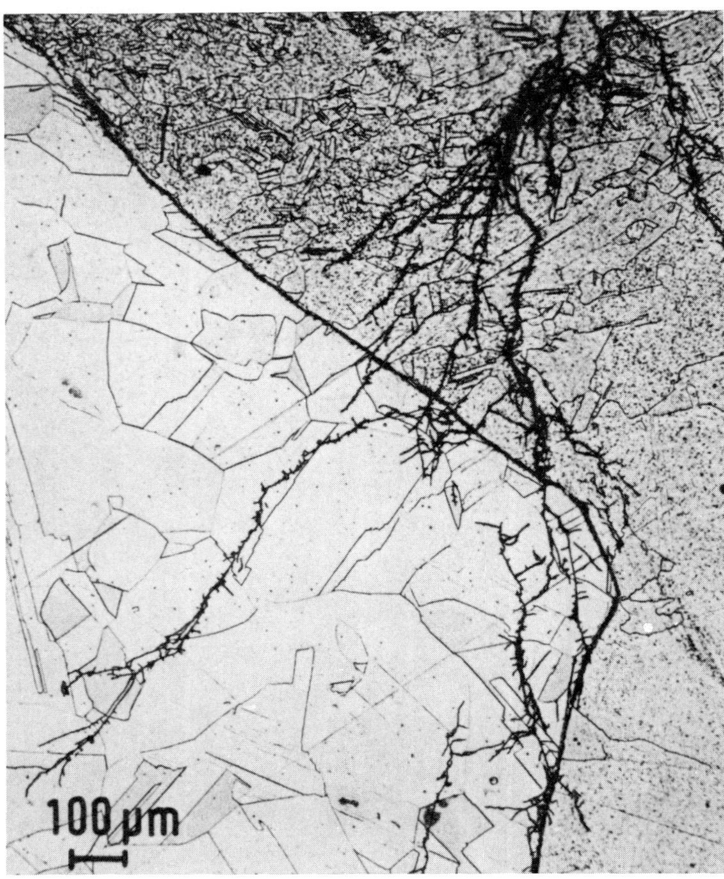

Fig. 24 Chloride SCC of type 304 stainless steel base metal and type 308 weld metal in an aqueous chloride environment at 95 °C (200 °F). Cracks are branching and transgranular.

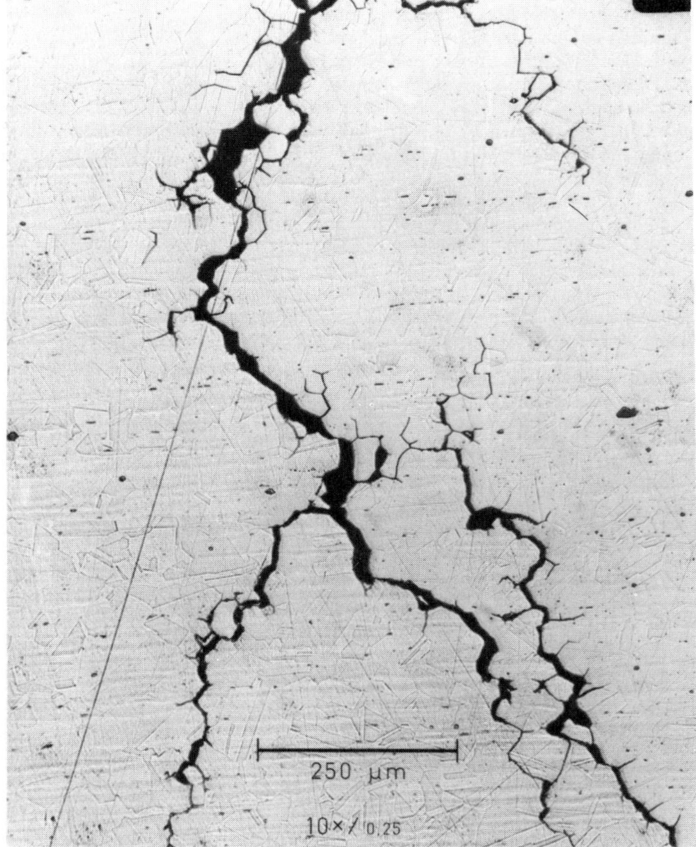

Fig. 25 Caustic SCC in the HAZ of a type 316L stainless steel NaOH reactor vessel. Cracks are branching and intergranular.

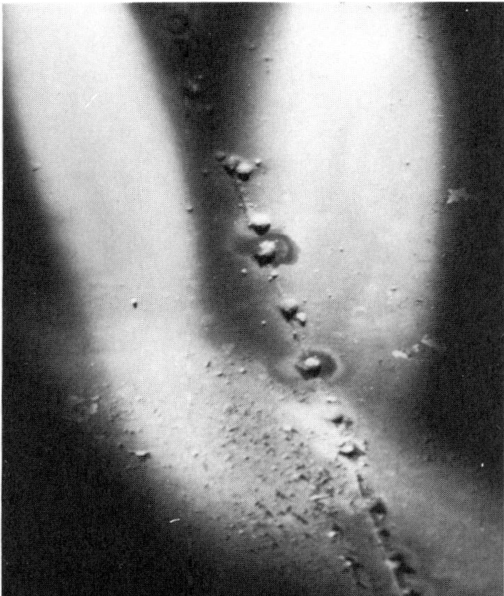

Fig. 26 Moundlike microbiological deposits along a weld seam in the bottom of a type 304L stainless steel tank after several months of exposure to well water at ambient temperature. Source: Ref 13

cause Cl^- was the predominant anion in the well water

- Microbiological oxidation to the corresponding ferric and manganic chlorides, which either singly or in combination are severe pitting corrodents of austenitic stainless steel
- Penetration of the protective oxide films on the stainless steel surfaces that were already weakened by oxygen depletion under the biodeposits

All affected piping was replaced before the new facilities were placed in service. The tanks were repaired by sandblasting to uncover all pits, grinding out each pit to sound metal, and then welding with the appropriate stainless steel filler metal. Piping and tanks have been in corrosive service for about 19 years to date with very few leaks, indicating that the inspection, replacement, and repair program was effective.

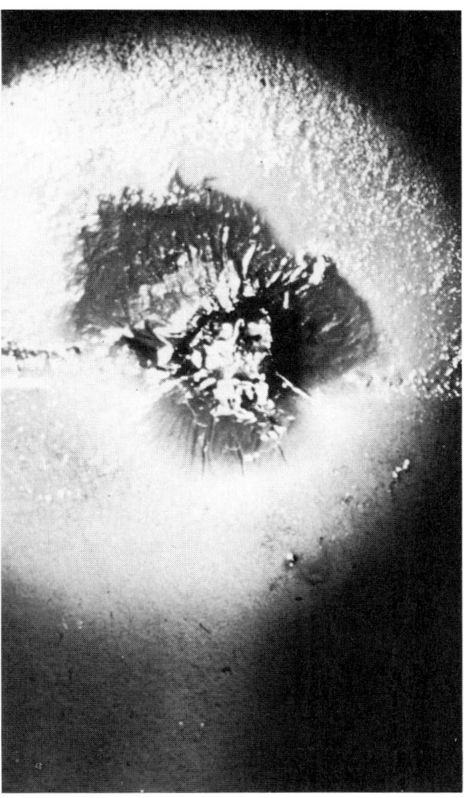

Fig. 27 Closeup of a wet deposit as shown in Fig. 26. Source: Ref 13

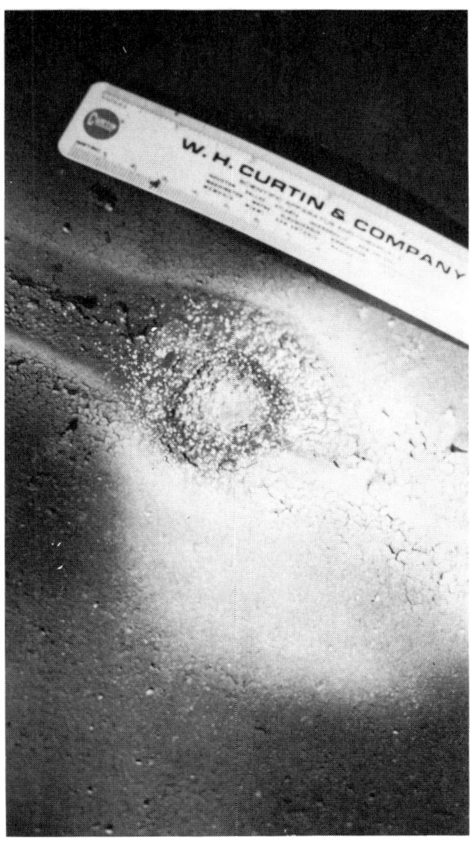

Fig. 28 Closeup of a dry deposit. See also Fig. 26 and 27. Source: Ref 13

Fig. 29 Ring-shaped stain left around a weld after removal of the type of deposit shown in Fig. 26 to 28. Source: Ref 13

Fig. 30 Large pit (center) at the edge of the weld shown in Fig. 29. The pit was revealed by probing with an icepick. Source: Ref 13

Corrosion of Ferritic Stainless Steel Weldments

Conventional 400-series ferritic stainless steels such as AISI types 430, 434, and 446 are susceptible to intergranular corrosion and to embrittlement in the as-welded condition. Corrosion in the weld area generally encompasses both the weld metal and weld HAZ. Early attempts to avoid some of these problems involved the use of austenitic stainless steel filler metals; however, failure by corrosion of the HAZ usually occurred even when exposure was to rather mild media for relatively short periods of time.

Figure 36 shows an example of a saturator tank used to manufacture carbonated water at room temperature that failed by leakage through the weld HAZ of the base metal after being in service for only 2 months. This vessel, fabricated by welding with a type 308 stainless steel welding electrode, was placed in service in the as-welded condition. Figure 37 shows a photomicrograph of the weld/base metal interface at the outside surface of the vessel; corrosion initiated at the inside surface. Postweld annealing—at 785 °C (1450 °F) for 4 h in the case of type 430 stainless steel—restores weld area ductility and resistance to corrosion equal to that of the unwelded base metal.

To overcome some of these earlier difficulties and to improve weldability, several of the standard grade ferritic stainless steels have been modified. For example, type 405, containing nominally 11% Cr, is made with lower carbon and a small aluminum addition of 0.20% to restrict the

Fig. 31 Radiograph of a pitted weld seam in a type 304L stainless steel tank bottom. Source: Ref 13

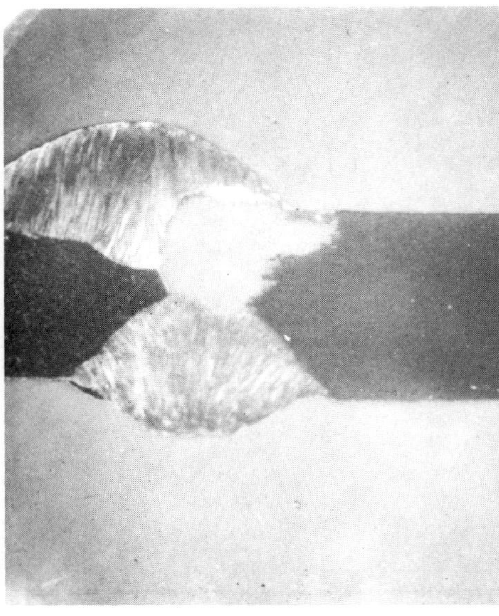

Fig. 32 Cross section through a pitted weld seam from a type 304L tank showing a typical subsurface cavity. Source: Ref 13

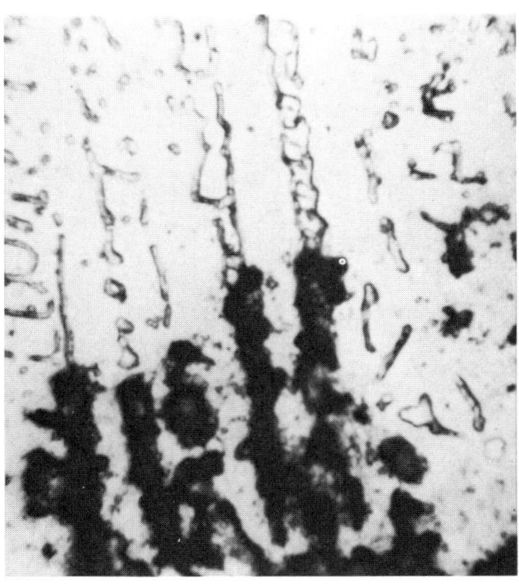

Fig. 33 Micrograph showing preferential attack of δ-ferrite stringers in type 316 stainless steel weld metal. 250×. Source: Ref 13

formation of austenite at high temperature so that hardening is reduced during welding. For maximum ductility and corrosion resistance, however, postweld annealing is necessary. Recommendations for welding include either a 430- or a 309-type filler metal, the latter being used where increased weld ductility is desired.

A New Generation of Ferritic Stainless Steels. In the late 1960s and early 1970s, researchers recognized that the high chromium-molybdenum-iron ferritic stainless steels possessed a desirable combination of good mechanical properties and resistance to general corrosion, pitting, and SCC. These properties made them attractive alternatives to the austenitic stainless steels commonly plagued by chloride SCC.

It was reasoned that by controlling the interstitial element (carbon, oxygen, and nitrogen) content of these new ferritic alloys, either by ultrahigh purity or by stabilization, the formation of martensite (as well as the need for preheat and postweld heat treatment) could be eliminated, with the result that the welds would be corrosion resistant, tough, and ductile in the as-welded condition. To achieve these results, electron beam vacuum refining, vacuum and argon-oxygen decarburization, and vacuum induction melting processes were used. From this beginning, two basic ferritic alloy systems evolved:

- *Ultrahigh purity:* the (C + N) interstitial content is less than 150 ppm (Ref 15)
- *Intermediate purity:* the (C + N) interstitial content exceeds 150 ppm (Ref 15)

Although not usually mentioned in the alloy chemistry specifications, oxygen and hydrogen are also harmful, and these levels must be carefully restricted. Table 3 lists the compositions of some ultrahigh purity, intermediate purity, and standard-grade ferritic stainless steels.

The unique as-welded properties of the new ferritic stainless steels have been made possible by obtaining very low levels of impurities, including carbon, nitrogen, hydrogen, and oxygen, in the case of the alloys described as ultralow interstitials and by obtaining a careful balance of niobium and/or titanium to match the carbon content in the case of the alloys with intermediate levels of interstitials. For these reasons, every precaution must be taken, and welding procedures that optimize gas shielding and cleanliness

Fig. 34 Rust-colored streaks transverse to horizontal weld seams in the sidewall of a type 316L stainless steel tank. Source: Ref 13

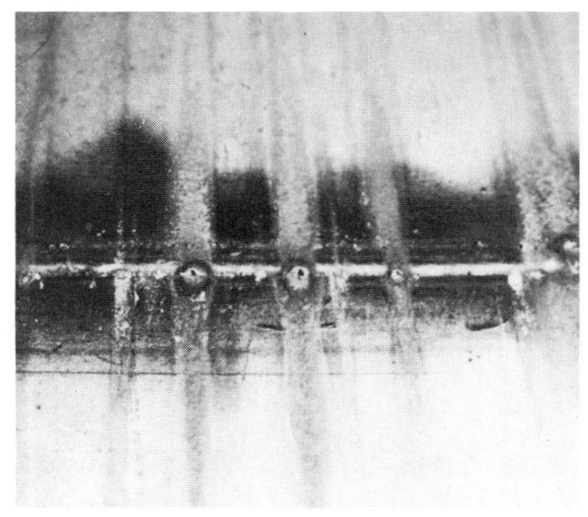

Fig. 35 Closeup of the rust-colored streaks shown in Fig. 34. Source: Ref 13

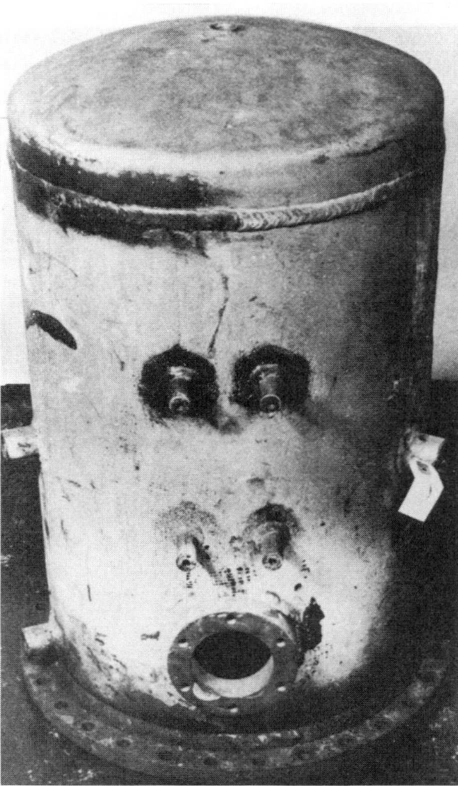

Fig. 36 As-welded type 430 stainless steel saturator tank used in the manufacture of carbonated water that failed after 2 months of service. The tank was shielded metal arc welded using type 308 stainless steel filler metal. Source: Ref 14

Table 3 Typical compositions of some ferritic stainless steels

Alloy	Composition, % C(max)	Cr	Fe	Mo	N	Ni	Other
Standard grades (AISI 400 series)							
Type 405	0.08	13	bal	...	...	...	0.2Al
Type 430	0.12	17	bal	...	...	...	...
Type 430Ti	0.10	17	bal	...	...	...	Ti 6×C min
Type 434	0.12	17	bal	0.75–1.25	...	...	...
Type 446	0.20	25	bal	...	...	...	...
Intermediate purity grades							
26-1Ti	0.02	26	bal	1	0.025	0.25	0.5Ti
AISI type 444	0.02	18	bal	2	0.02	0.4	0.5Ti
SEA-CURE	0.02	27.5	bal	3.4	0.025	1.7	0.5Ti
Monit	0.025	25	bal	4	0.025	4	0.4Ti
Ultrahigh purity grades							
E-Brite 26-1	0.002	26	bal	1	0.01	0.1	0.1Nb
AL 29-4-2	0.005	29	bal	4	0.01	2	...
SHOMAC 26-4	0.003	26	bal	4	0.005	...	...
SHOMAC 30-2	0.003	30	bal	2	0.007	0.18	...
YUS 190L	0.004	19	bal	2	0.0085	...	0 15Nb

must be selected to avoid pickup of carbon, nitrogen, hydrogen and oxygen.

To achieve maximum corrosion resistance, as well as maximum toughness and ductility, the GTA welding process with a matching filler metal is usually specified; however, dissimilar high-alloy weld metals have also been successfully used. In this case, the choice of dissimilar filler metal must ensure the integrity of the ferritic metal system. Regardless of which of the new generation of ferritic stainless steels is to be welded, the following precautions are considered essential.

First, the joint groove and adjacent surfaces must be thoroughly degreased with a solvent, such as acetone, that does not leave a residue. This will prevent pickup of impurities, especially carbon, before welding. The filler metal must also be handled carefully to prevent it from picking up impurities. Solvent cleaning is also recommended. *Caution:* Under certain conditions, when using solvents, a fire hazard or health hazard may exist.

Second, a welding torch with a large nozzle inside diameter, such as 19 mm (3/4 in.), and a gas lens (inert gas calming screen) is necessary. Pure, welding grade argon with a flow rate of 28 L/min (60 ft^3/h) is required for this size nozzle. In addition, the use of a trailing gas shield is beneficial, especially when welding heavy-gage materials. Use of these devices will drastically limit the pickup of nitrogen and oxygen during welding. Back gas shielding with argon is also essential. *Caution:* Procedures for welding austenitic stainless steels often recommend the use of nitrogen backing gas. Nitrogen must not be used when welding ferritic stainless steels. Standard GTA welding procedures used to weld stainless steels are inadequate and therefore must be avoided.

Third, overheating and embrittlement by excessive grain growth in the weld and HAZ should be avoided by minimizing heat input. In multipass welds, overheating and embrittlement should be

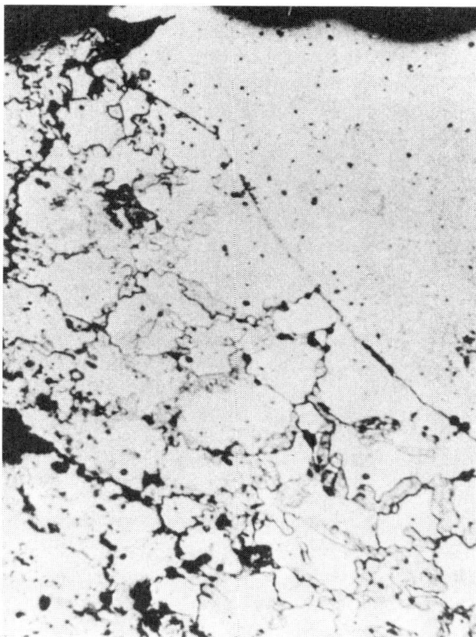

Fig. 37 Micrograph of the outside surface of the saturator tank in Fig. 36 showing intergranular corrosion at the fusion line. Source: Ref 14

Fig. 38 Top view of a longitudinal weld in 6.4-mm (1/4-in.) E-Brite ferritic stainless steel plate showing intergranular corrosion. The weld was made with matching filler metal. About 4×

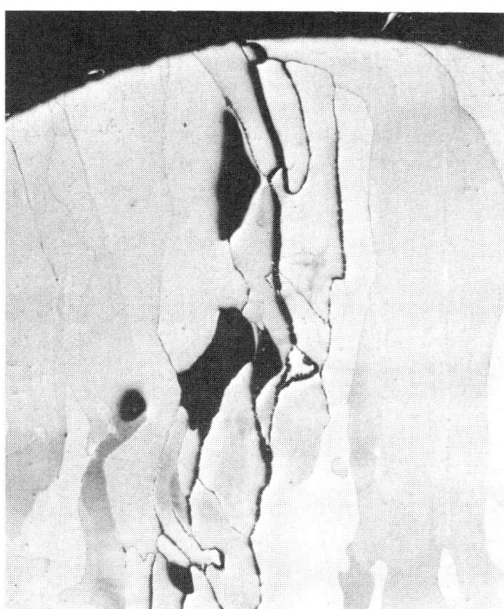

Fig. 39 Intergranular corrosion of a contaminated E-Brite ferritic stainless steel weld. Electrolytically etched with 10% oxalic acid. 200×

Fig. 40 Intergranular corrosion of the inside surface HAZ of E-Brite stainless steel adjacent to the weld fusion line. Electrolytically etched with 10% oxalic acid. 100×

avoided by keeping the interpass temperature below 95 °C (200 °F).

Lastly, to avoid embrittlement further, preheating (except to remove moisture) or postweld heat treating should not be performed. Postweld heat treatment is used only with the conventional ferritic stainless alloys. The following example illustrates the results of not following proper procedures.

Leaking Welds in a Ferritic Stainless Steel Wastewater Vaporizer. A nozzle in a wastewater vaporizer began leaking after approximately 3 years of service with acetic and formic acid wastewaters at 105 °C (225 °F) and 414 kPa (60 psig).

Investigation. The shell of the vessel was weld fabricated in 1972 from 6.4-mm (1/4-in.) E-Brite stainless steel plate. The shell measured 1.5 m (58 in.) in diameter and 8.5 m (28 ft) in length. Nondestructive examination included 100% radiography, dye-penetrant inspection, and hydrostatic testing of all E-Brite welds.

An internal inspection of the vessel revealed that portions of the circumferential and longitudinal seam welds, in addition to the leaking nozzle weld, displayed intergranular corrosion. At the point of leakage, there was a small intergranular crack. Figure 38 shows a typical example of a corroded weld. A transverse cross section through this weld will characteristically display intergranular corrosion with grains dropping out (Fig. 39). It was also noted that the HAZ next to the weld fusion line also experienced intergranular corrosion a couple of grains deep as a result of sensitization (Fig. 40).

The evidence indicated weldment contamination; therefore, effort was directed at finding the levels of carbon, nitrogen, and oxygen in the various components present before and after welding. The averaged results were as follows:

E-Brite
Base plate
C = 6 ppm
N = 108 ppm (C+N = 114 ppm)
O = 57 ppm

Corroded longitudinal weld
C = 133 ppm
N = 328 ppm (C+N = 461 ppm)
O = 262 ppm

Corroded circumferential weld
C = 34 ppm
N = 169 ppm (C+N = 203 ppm)
O = 225 ppm

E-Brite
Weld wire
C = 3 ppm
N = 53 ppm (C+N = 56 ppm)
O = 55 ppm

Sound longitudinal weld
C = 10 ppm
N = 124 ppm (C+N = 134 ppm)
O = 188 ppm

Sound circumferential weld
C = 20 ppm
N = 106 ppm (C+N = 126 ppm)
O = 85 ppm

These results confirmed suspicions that failure was due to excessive amounts of nitrogen, carbon, and oxygen. To characterize the condition of the vessel further, Charpy V-notch im- tests were run on the unaffected base metal, the HAZ, and the uncorroded (sound) weld metal. These tests showed the following ductile-to-brittle transition temperatures:

Specimen	Ductile-to-brittle transition temperature, °C	°F
Base metal	40 ± 3	105 ± 5
HAZ	85 ± 3	180 ± 5
Weld	5 ± 3	40 ± 5

Comparison of the interstitial levels of the corroded welds, sound welds, base metal, and filler wire suggested that insufficient joint preparation (carbon pickup) and faulty gas shielding were probably the main contributing factors that caused this weld corrosion failure. Discussions with the vendor uncovered two discrepancies. First, the welder was using a large, 19-mm (3/4-in.) inside diameter ceramic nozzle with a gas lens, but was flowing only 19 L/min (40 ft^3/h) of argon; this was the flow rate previously used with a 13-mm (1/2-in.) inside diameter gas lens nozzle. Second, a manifold system was used to distribute pure argon welding gas from a large liquid argon tank to various satellite welding stations in the welding shop. The exact cause for the carbon pickup was not determined.

Conclusions. Failure of the nozzle weld was the result of intergranular corrosion caused by the pickup of interstitial elements and subsequent precipitation of chromium carbides and nitrides. Carbon pickup was believed to have been caused by inadequate joint cleaning prior to welding. The increase in the weld nitrogen level was a direct result of inadequate argon gas shielding of the molten weld puddle. Two areas of inadequate shielding were identified:

- Improper gas flow rate for a 19-mm (3/4-in.) diam gas lens nozzle
- Contamination of the manifold gas system

In order to preserve the structural integrity and corrosion performance of the new generation of ferritic stainless steels, it is important to avoid the pickup of the interstitial elements carbon, nitrogen, oxygen, and hydrogen. In this particular case, the vendor used a flow rate intended for a smaller welding torch nozzle. The metal supplier recommended a flow rate of 23 to 28 L/min (50 to 60 ft^3/min) of argon for a 19-mm (3/4-in.) gas lens nozzle. The gas lens collect body is an important and necessary part of the torch used to weld these alloys. Failure to use a gas lens will result in a flow condition that is turbulent enough to aspirate air into the gas stream, thus contaminating the weld and destroying its mechanical and corrosion properties.

The manifold gas system also contributed to this failure. When this system is first used, it is necessary to purge the contents of the manifold of any air to avoid oxidation and contamination. When that is done, the system functions satisfactorily; however, when it is shut down overnight or for repairs, air infiltrates back in, and a source of contamination is reestablished. Manifold systems are never fully purged, and leaks are common.

The contaminated welds were removed, and the vessels were rewelded and put back into service. Some rework involved the use of covered electrodes of dissimilar composition. No problems have been reported to date.

Recommendations. First, to ensure proper joint cleaning, solvent washing and wiping with a clean lint-free cloth should be performed immediately before welding. The filler wire should be wiped with a clean cloth just prior to welding. Also, a word of caution: Solvents are generally flammable and can be toxic. Ventilation should be adequate. Cleaning should continue until cloths are free of any residues.

Second, when GTA welding, a 19-mm (3/4-in.) diameter ceramic nozzle with gas lens collect body is recommended. An argon gas flow rate of 28 L/min (60 ft^3/min) is optimum. Smaller nozzles are not recommended. Argon back gas shielding is mandatory at a slight positive pressure to avoid disrupting the flow of the welding torch.

Third, the tip of the filler wire should be kept within the torch shielding gas envelope to avoid contamination and pickup of nitrogen and oxygen (they embrittle the weld). If the tip becomes contaminated, welding should be stopped, the contaminated weld area ground out, and the tip of the filler wire that has been oxidized should be snipped off before proceeding with welding.

Fourth, a manifold gas system should not be used to supply shielding and backing gas. Individual argon gas cylinders have been found to provide optimum performance. A weld button spot test should be performed to confirm the integrity of the argon cylinder and all hose connections. In this test, the weld button sample should be absolutely bright and shiny. Any cloudiness is an indication of contamination. It is necessary to check for leaks or to replace the cylinder.

Fifth, it is important to remember that corrosion resistance is not the only criterion when evaluating these new ferritic stainless steels. Welds must also be tough and ductile, and these factors must be considered when fabricating welds.

Lastly, dissimilar weld filler metals can be successfully used. To avoid premature failure, the dissimilar combination should be corrosion tested to ensure suitability for the intended service.

Corrosion of Duplex Stainless Steel Weldments

In the wrought condition, duplex stainless steels have microstructures consisting of a fairly even balance of austenite and ferrite. The new generation of duplex alloys are now being produced with low carbon and a nitrogen addition. These alloys are useful because of their good resistance to chloride SCC, pitting corrosion, and intergranular corrosion in the as-welded condition. Nominal compositions of some duplex stainless steels are given in Table 4.

The distribution of austenite and ferrite in the weld and HAZ is known to affect the corrosion properties and the mechanical properties of duplex stainless steels. To achieve this balance in properties, it is essential that both base metal and weld metal be of the proper composition. For example, without nickel enrichment in the filler rod, welds can be produced with ferrite levels in excess of 80%. Such microstructures have very poor ductility and inferior corrosion resistance. For this reason, autogenous welding (without the addition of filler metal) is not recommended unless postweld solution annealing is performed, which is not always practical. To achieve a balanced weld microstructure, a low carbon content and the addition of nitrogen (with Alloy 2205 at least 0.12% N) should be specified for the base metal. Low carbon helps to minimize the effects of sensitization, and the nitrogen slows the precipitation kinetics associated with the segregation of chromium and molybdenum during the weld-

Table 4 Compositions of various duplex stainless steels

UNS No.	Typical alloy	Composition, %(a) C	Cr	Cu	Fe	Mn	Mo	Ni	N	Si	Others
S31500	SAF 3RE60	0.03 max	18.5	...	bal	1.6	2.7	4.9	0.07	1.7	...
S32404	Uranus 50	0.04 max	21.5	1.5	bal	2.0 max	2.5	7.0	0.1	1.0 max	...
S31803	Alloy 2205	0.03 max	22	...	bal	2.0 max	3.0	5.5	0.15	1.0 max	...
S32304	SAF 2304	0.03 max	23	...	bal	2.5 max	0.5	4.0	0.1	1.0 max	...
S32900	Type 329 SS	0.2 max	25.5	...	bal	1.0 max	1.5	3.75	...	0.75 max	...
S31100	IN-744	0.05 max	26	...	bal	1.0 max	...	6.5	...	0.6 max	...
S31200	44LN	0.03 max	25	...	bal	2.0 max	3.0	6.5	0.17	1.0 max	...
S32950	7Mo-Plus	0.03 max	27.5	...	bal	2.0 max	1.8	4.4	0.25	0.6 max	...
S31260	DP-3	0.3 max	25	0.5	bal	1.0 max	3.0	6.5	0.2	0.75 max	0.3W
S32250	Ferralium alloy 255	0.04 max	25.5	1.7	bal	1.5 max	3.0	5.5	0.17	1.0 max	...

(a) Nominal unless otherwise indicated.

Table 5 Chemical compositions of alloy 2205 specimens tested and filler metals used in Ref 16

Specimen size and configuration	Element, % C	Si	Mn	P	S	Cr	Ni	Mo	Cu	N
Parent metals										
48.1-mm (1.89-in.) OD, 3.8-mm (0.149-in.) wall tube	0.015	0.37	1.54	0.024	0.003	21.84	5.63	2.95	0.09	0.15
88.9-mm (3.5-in.) OD, 3.6-mm (0.142-in.) wall tube	0.017	0.28	1.51	0.025	0.003	21.90	5.17	2.97	0.09	0.15
110-mm (4.3-in.) OD, 8-mm (0.31-in.) wall tube	0.027	0.34	1.57	0.027	0.003	21.96	5.62	2.98	0.09	0.13
213-mm (8.4-in.) OD, 18-mm (0.7-in.) wall tube	0.017	0.28	1.50	0.026	0.003	21.85	5.77	2.98	0.10	0.15
20-mm (3/4-in) plate	0.019	0.39	1.80	0.032	0.003	22.62	5.81	2.84	...	0.13
Filler metals										
1.2 mm (0.047 in.) diam wire; 1.6 mm (0.063 in.) diam rod; 3.2 mm (0.125 in.) diam wire	0.011	0.48	1.61	0.016	0.003	22.50	8.00	2.95	0.07	0.13
3.25 mm (0.127 in.) diam covered electrode	0.020	1.01	0.82	0.024	0.011	23.1	10.4	3.06	...	0.13
4.0 mm (0.16 in.) diam covered electrode	0.016	0.94	0.78	0.015	0.011	23.0	10.5	3.13	...	0.11

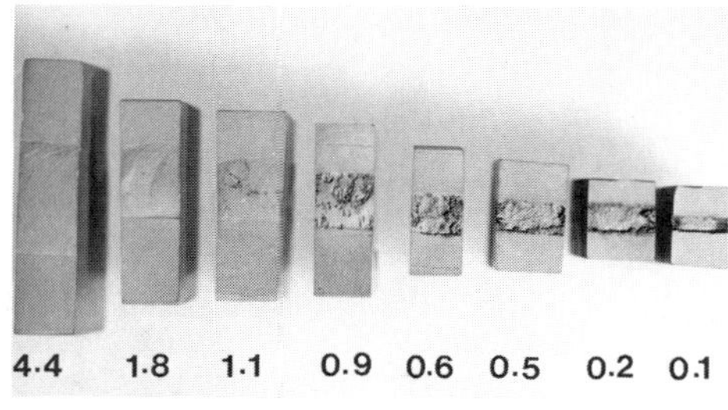

Fig. 41 Effect of welding heat input on the corrosion resistance of autogenous GTA welds in Ferralium alloy 255 in 10% $FeCl_3$ at 10 °C (40 °F). The base metal was 25.4 mm (1 in.) thick. Source: Ref 18

ing operation. Nitrogen also enhances the reformation of austenite in the HAZ and weld metal during cooling.

These duplex alloys have been used in Europe for many years; therefore, guidelines relating to austenite-ferrite phase distribution are available. It has been shown that to ensure resistance to chloride SCC welds should contain at least 25% ferrite. To maintain a good phase balance for corrosion resistance and mechanical properties (especially ductility and notch toughness) comparable to the base metal, the average ferrite content of the weld should not exceed 60%. This means using welding techniques that minimize weld dilution, especially in the root pass. Conditions that encourage mixing of the lower-nickel base metal with the weld metal reduce the overall nickel content. Weld metal with a lower nickel content will have a higher ferrite content, with reduced mechanical and corrosion properties.

Once duplex base metal and welding consumables have been selected, it is then necessary to select joint designs and weld parameters that will produce welding heat inputs and cooling rates so as to produce a favorable balance of austenite and ferrite in the weld and HAZ.

Researchers have shown that the high-ferrite microstructures that develop during welding in lean (low-nickel) base metal and weld metal compositions can be altered by adjusting welding heat input and cooling rate. In these cases, a higher heat input that produces a slower cooling rate can be used to advantage by allowing more time for ferrite to transform to austenite. There are, however, some practical aspects to consider before applying higher heat inputs indiscriminately. For example, as heat input is increased, base metal dilution increases. As the amount of lower-nickel base metal in the weld increases, the overall nickel content of the deposit decreases; this increases the potential for more ferrite, with a resultant loss in impact toughness, ductility, and corrosion resistance. This would be another case for using an enriched filler metal containing more nickel than the base metal. Grain growth and the formation of embrittling phases are two other negative effects of high heat inputs. When there is uncertainty regarding the effect that welding conditions will have on corrosion performance and mechanical properties, a corrosion test is advisable.

The influence of different welding conditions on various material properties of Alloy 2205 has been studied (Ref 16). Chemical compositions of test materials are given in Table 5, and the results of the investigation are detailed in the following sections.

Intergranular Corrosion. Despite the use of very high arc energies (0.5 to 6 kJ/mm, or 13 to 152 kJ/in.) in combination with multipass welding, the Strauss test (ASTM A 262, practice E) (Ref 10) failed to uncover any signs of sensitization after bending through 180°. The results of Huey tests (ASTM A 262, practice C) on submerged-arc welds showed that the corrosion rate increased slightly with arc energy in the studied range of 0.5 to 6.0 kJ/mm (13 to 152 kJ/in.). For comparison, the corrosion rate for parent metal typically varies between 0.15 and 1.0 mm/yr (6 and 40 mils/yr), depending on surface finish and heat treatment cycle.

Similar results were obtained in Huey tests of specimens from bead-on-tube welds produced by GTA welding. In this case, the corrosion rate had a tendency to increase slightly with arc energy up to 3 kJ/mm (76 kJ/in.).

Pitting tests were conducted in 10% ferric chloride ($FeCl_3$) at 25 and 30 °C (75 and 85 °F) in accordance with ASTM G 48 (Ref 17). Results of tests on submerged-arc test welds did not indicate any significant change in pitting resistance when the arc energy was increased from 1.5 to 6 kJ/mm (38 to 152 kJ/in.). Pitting occurred along the boundary between two adjacent weld beads. Attack was caused by slag entrapment in the weld; therefore, removal of slag is important.

Gas tungsten arc weld test specimens (arc energies from 0.5 to 3 kJ/mm, or 13 to 76 kJ/in.) showed a marked improvement in pitting resistance with increasing arc energy. In order for duplicate specimens to pass the $FeCl_3$ test at 30 °C (85 °F), 3 kJ/mm (76 kJ/in.) of arc energy was required. At 25 °C (75 °F), at least 2 kJ/mm (51 kJ/in.) was required to achieve immunity. Welds made autogenously (no nickel enrichment) were somewhat inferior, but improvements were achieved by using higher arc energies.

For comparison with a different alloy, Fig. 41 shows the effect of heat input on the corrosion resistance of Ferralium alloy 255 welds made

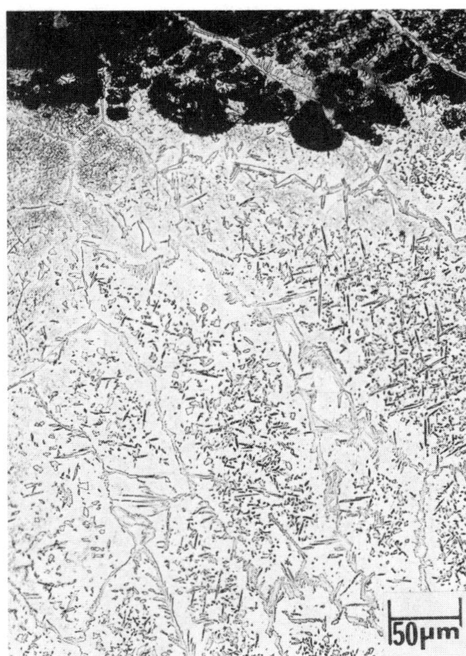

Fig. 42 Preferential corrosion of the ferrite phase in the weld metal of Ferralium alloy 255 GTA welds in 10% $FeCl_3$ at room temperature. Base metal was 3.2 mm (1/8 in.) thick.

autogenously and tested in $FeCl_3$ at 15 °C (60 °F). Preferential corrosion of the ferrite phase is shown in Fig. 42. In a different test, Ferralium alloy 255 was welded autogenously and tested in a neutral chloride solution according to ASTM D 1141 (Ref 19) at 60 to 100 °C (140 to 212 °F). In this case, preferential attack of the austenite phase was observed. An example is shown in Fig. 43. Similar results would be expected for alloy 2205.

A study of the alloy 2205 weld microstructures (Ref 16) revealed why high arc energies were found to be beneficial to pitting resistance. Many investigations have indicated that the presence of chromium nitrides in the ferrite phase lowers the resistance to pitting of the weld metal and the HAZ in duplex stainless steels. In this study, both weld metal and HAZ produced by low arc energies contained an appreciable amount of chromium nitride (Cr_2N) (Fig. 44). The nitride precipitation vanished when an arc energy of 3 kJ/mm (76 kJ/in.) was used (Fig. 45).

The results of $FeCl_3$ tests on submerged-arc welds showed that all top weld surfaces passed the test at 30 °C (85 °F) without pitting attack, irrespective of arc energy in the range of 2 to 6 kJ/mm (51 to 152 kJ/in.). Surprisingly, the weld metal on the root side, which was the first to be deposited, did not pass the same test temperature.

The deteriorating effect of high arc energies on the pitting resistance of the weld metal on the root side was unexpected. Potentiostatic tests carried out in 3% NaCl at 400 mV versus SCE confirmed these findings. Microexamination of the entire joint disclosed the presence of extremely fine austenite precipitates, particularly in the second weld bead (Fig. 46) but also in the first or root side bead. The higher the arc energy, the more austenite of this kind was present in the first two weld beads. Thus, nitrides give rise to negative effects on the pitting resistance, as do fine austenite precipitates that were presumably re-formed at as low a temperature as approximately 800 °C (1470 °F).

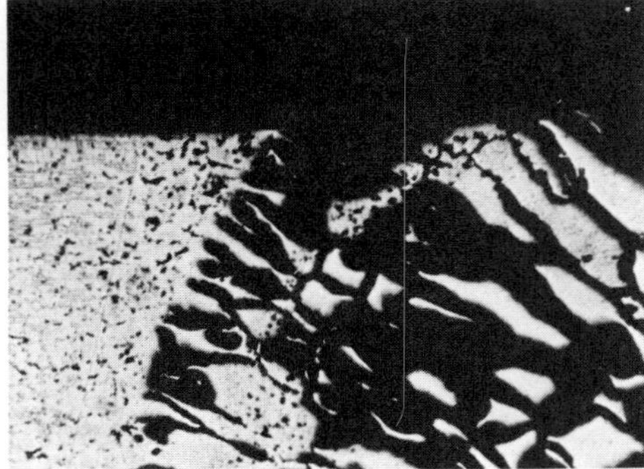

Fig. 43 Preferential attack of the continuous austenite phase in an autogenous GTA weld in Ferralium alloy 255. Crevice corrosion test was performed in synthetic seawater according to ASTM D 1141 (Ref 19) at 100 °C (212 °F). Etched with 50% HNO_3. 100×

Therefore, the resistance of alloy 2205 to pitting corrosion is dependent on several factors. First, Cr_2N precipitation in the coarse ferrite grains upon rapid cooling from temperatures above about 1200 °C (2190 °F) causes the most severe impairment to pitting resistance. This statement is supported by a great number of $FeCl_3$ tests as well as by potentiostatic pitting tests. Generally, it seems difficult to avoid Cr_2N precipitation in welded joints completely, particularly in the HAZ, the structure of which can be controlled only by the weld thermal cycle. From this point of view, it appears advisable to employ as high an arc energy as practical in each weld pass. In this way, the cooling rate will be slower (but not slow enough to encounter 475 °C (885 °F) embrittlement), and the re-formation of austenite will clearly dominate over the precipitation of Cr_2N.

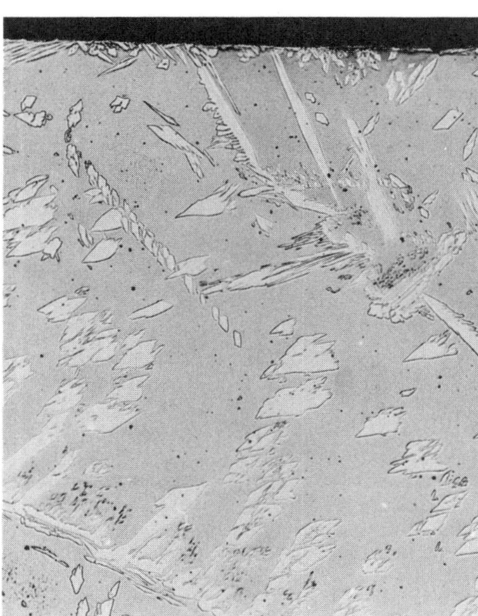

Fig. 45 Microstructure of bead-on-tube weld made by autogenous GTA welding with an arc energy of 3 kJ/mm (76 kJ/in.). Virtually no chromium nitrides are present, which results in adequate pitting resistance. 200×. Source: Ref 16

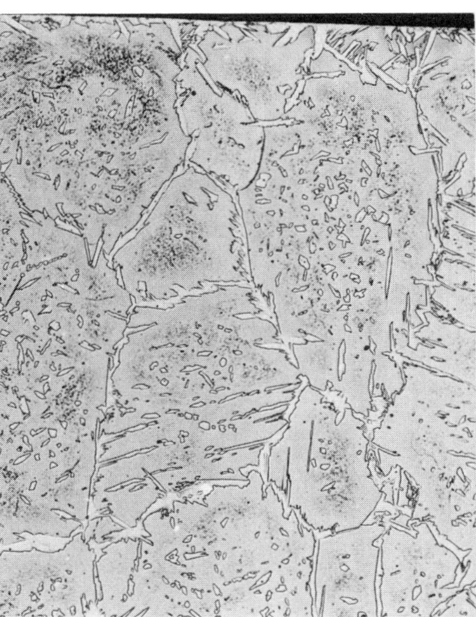

Fig. 44 Microstructure of bead-on-tube weld made by autogenous GTA welding with an arc energy of 0.5 kJ/mm (13 kJ/in.). Note the abundance of chromium nitrides in the ferrite phase. See also Fig. 45. 200×. Source: Ref 16

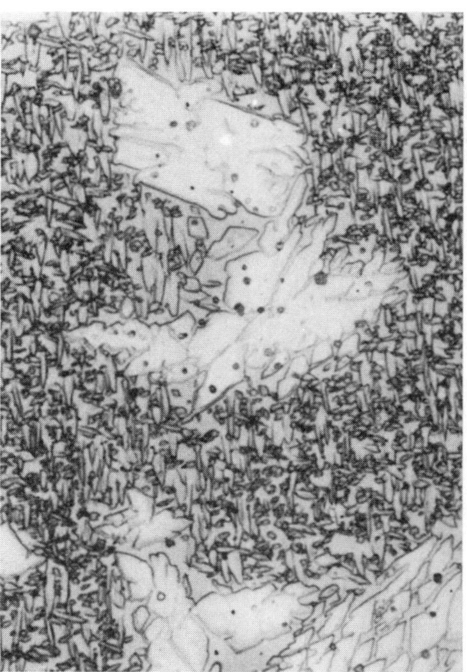

Fig. 46 Microstructure of the second weld bead of a submerged-arc weld joint in 20-mm (3/4-in.) duplex stainless steel plate. The extremely fine austenite precipitate was formed as a result of reheating from the subsequent weld pass, which used an arc energy of 6 kJ/mm (152 kJ/in.). 1000×. Source: Ref 16

In addition, if there were no restriction on maximum interpass temperature, the heat produced by previous weld passes could be used to decrease the cooling rate further in the critical temperature range above about 1000 °C (1830 °F). Preliminary tests with preheated workpieces have shown the significance of temperature in suppressing Cr_2N precipitation. Currently, the maximum recommended interpass temperature for alloy 2205 is 150 °C (300 °F). This temperature limit does not appear to be critical, and it is suggested that this limit could be increased to 300 °C (570 °F). The maximum recommended interpass temperature for Ferralium alloy 255 is 200 °C (390 °F). Excessive grain growth as a result of too much heat input must also be considered to avoid loss of ductility and impact toughness.

Second, the fine austenite precipitates found in the reheated ferrite when high arc energies and multipass welding were combined are commonly referred to as γ_2 in the literature. The harmful influence of γ_2 on the pitting resistance has been noted with isothermally aged specimens, but as far as is known, it has never been observed in connection with welding. It is felt, however, that γ_2 is less detrimental to pitting than Cr_2N. Moreover, γ_2 formation is believed to be beneficial to mechanical properties, such as impact strength and ductility.

A third factor that lowers pitting resistance is oxide scale. Where possible, all surface oxides should be removed by mechanical means or, preferably, by pickling. Root surfaces (in pipe), however, are generally inaccessible, and pitting resistance must rely on the protection from the backing gas during GTA welding. It is therefore advisable to follow the current recommendation for stainless steels, which is a maximum of 25 ppm oxygen in the root backing gas.

Stress-Corrosion Cracking. The SCC resistance of alloy 2205 in aerated, concentrated chloride solutions is very good. The effect of welding on the SCC resistance is negligible from a practical point of view. The threshold stress for various welds, as well as for unwelded parent metal in the $CaCl_2$ test, is as high as 90% of the tensile strength at the testing temperature. This is far above all conceivable design limits.

Also, in environments containing both hydrogen sulfide (H_2S) and chlorides, the resistance of welds is almost as high as for the parent metal. In this type of environment, however, it is important to avoid too high a ferrite content in weld metal and HAZ. For normal welding of joints, the resulting ferrite contents should not cause any problems. For weld repair situations, however, care should be taken so that extremely high ferrite contents (>75%) are avoided. To preserve the high degree of resistance to SCC, the ferrite content should not be less than 25% (Ref 20).

Another reason to avoid coarse weld microstructures (generated by excessive welding heat) is the resultant nonuniform plastic flow, which can locally increase stresses and induce preferential corrosion and cracking effects.

Use of High-Alloy Filler Metals. In critical pitting or crevice corrosion applications, the pitting resistance of the weld metal can be enhanced by the use of high nickel-chromium-molybdenum alloy filler metals. The corrosion resistance of such weldments in Ferralium alloy 255 is shown in Table 6. For the same weld technique, it can be seen that using high-alloy fillers does improve corrosion resistance. If high-alloy fillers are used, the weld metal will have better corrosion resistance than the HAZ and the fusion line. Therefore, again, proper selection of welding technique can improve the corrosion resistance of the weldments.

Table 6 Corrosion resistance of Ferralium alloy 255 weldments using various nickel-base alloy fillers and weld techniques

3.2-mm (0.125-in.) plates tested in 10% $FeCl_3$ for 120 h

Filler metal	Critical pitting temperature: Gas tungsten arc, °C	°F	Gas metal arc, °C	°F	Submerged arc, °C	°F
Hastelloy alloy G-3	30–35	85–95(a)	30	85(a)	30–35	85–95(b)
IN-112	30	85(a)	...	...	35–40	95–105(b)
Hastelloy alloy C-276	...	...	...	...	25–30	75–85(a)
Hastelloy alloy C-22	30	85(a)	...	...	35–40	95–105(a)

(a) HAZ. (b) HAZ plus weld metal

Corrosion of Nickel and High-Nickel Alloy Weldments

The corrosion resistance of weldments is related to the microstructural and microchemical changes resulting from thermal cycling. The effects of welding on the corrosion resistance of nickel-base alloys are similar to the effects on the corrosion resistance of austenitic stainless steels. For example, sensitization due to carbide precipitation in the HAZ is a potential problem in both classes of alloys. However, in the case of nickel-base alloys, the high content of such alloying elements as chromium, molybdenum, tungsten, and niobium can result in the precipitation of other intermetallic phases, such as μ, σ, and η.

Therefore, this section is concerned with the characteristics of the various nickel-base alloys and the evolution of these alloys. The corrosion resistance of weldments is dictated not only by the HAZ but also by the weld metal itself. The effect of elemental segregation on weld metal corrosion must also be examined. The nickel-base alloys discussed in this section are the solid-solution alloys.

The nickel-molybdenum alloys, represented by Hastelloy alloys B and B-2, have been primarily used for their resistance to corrosion in nonoxidizing environments such as HCl. Hastelloy alloy B has been used since about 1929 and has suffered from one significant limitation: weld decay. The welded structure has shown high susceptibility to knife-line attack adjacent to the weld-metal and to HAZ attack at some distance from the weld. The former has been attributed to the precipitation of molybdenum carbide (Mo_2C); the latter, to the formation of M_6C-type carbides. This necessitated postweld annealing, a serious shortcoming when large structures are involved. The knife-line attack on an alloy B weldment is shown in Fig. 47. Many approaches to this problem were attempted, including the addition of carbide-stabilizing elements, such as vanadium, titanium, zirconium, and tantalum, as well as the lowering of carbon.

The addition of 1% V to an alloy B-type composition was first patented in 1959. The resultant commercial alloys—Corronel 220 and Hastelloy alloy B-282—were found to be superior to alloy B in resisting knife-line attack but were not immune to it. In fact, it was demonstrated that the addition of 2% V decreased the corrosion resistance of the base metal in HCl solutions. During this time, improvements in melting techniques led to the development of a low-carbon low-iron version of alloy B called alloy B-2. This alloy did not exhibit any propensity to knife-line attack (Fig. 48).

Segregation of molybdenum in weld metal can be detrimental to corrosion resistance in some environments. In the case of boiling HCl solutions, the weld metal does not corrode preferentially. However, in H_2SO_4 + HCl and H_2SO_4 + H_3PO_4 acid mixtures, preferential corrosion of as-welded alloy B-2 has been observed (Fig. 49). No knife-line or HAZ attack was noted in these tests. During solidification, the initial solid is poorer in molybdenum and therefore can corrode preferentially. This is shown in Fig. 49 for an autogenous GTA weld in alloy B-2. In such cases, postweld annealing at 1120 °C (2050 °F) will be beneficial.

The nickel-chromium-molybdenum alloys represented by the Hastelloy C family of alloys and by Inconel 625 have also undergone evolution because of the need to improve the corrosion resistance of weldments. Hastelloy alloy C (UNS N10002) containing nominally 16% Cr, 16% Mo, 4% W, 0.04% C, and 0.5% Si had been in use for some time but had required the use of postweld annealing to prevent preferential weld and HAZ attack. Many investigations were carried out on the nature of precipitates formed in alloy C, and two main types of precipitates were identified. The first is a Ni_7Mo_6 intermetallic phase called μ, and the second consists of carbides of the Mo_6C type. Other carbides of the $M_{23}C_6$ and M_2C were also reported. Another type, an ordered Ni_2Cr-type precipitate, occurs mainly at lower temperatures and after a long aging time; it is not of great concern from a welding viewpoint.

Both the intermetallic phases and the carbides are rich in molybdenum, tungsten, and chromium and therefore create adjacent areas of alloy depletion that can be selectively attacked. Carbide precipitation can be retarded considerably by lowering carbon and silicon; this is the principle behind Hastelloy alloy C-276. The time-temperature behaviors of alloys C and C-276 are compared in Fig. 50, which shows much slower precipitation kinetics in alloy C-276. Therefore, the evolution of alloy C-276 from alloy C enabled the use of this alloy system in the as-welded condition. However, because only carbon and silicon were controlled in C-276, there remained the problem of intermetallic μ-phase precipitation, which occurred at longer times of aging. Alloy C-4 was developed with lower iron, cobalt, and tungsten levels to prevent precipitation of μ phases.

The effect of aging on sensitization of alloys C, C-276, and C-4 is shown in Fig. 51. For alloy C, sensitization occurs in two temperature ranges (700 to 800 °C, or 1290 to 1470 °F, and 900 to 1100 °C, or 1650 to 2010 °F) corresponding to carbide

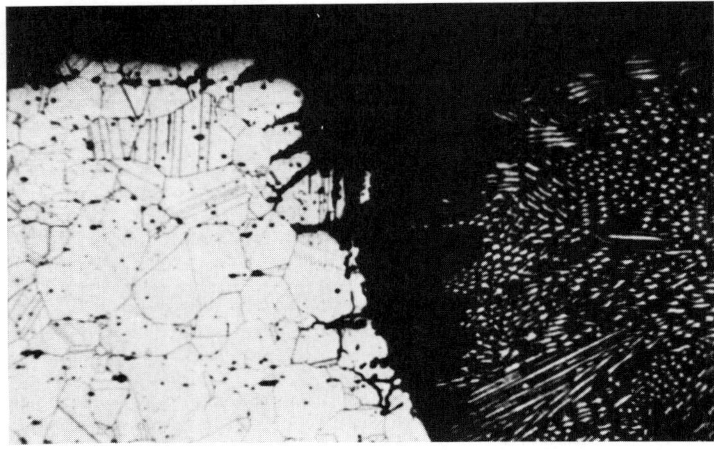

Fig. 47 Cross section of a Hastelloy alloy B weldment corroded after 16 days of exposure in boiling 20% HCl. 80×. Source: Ref 21

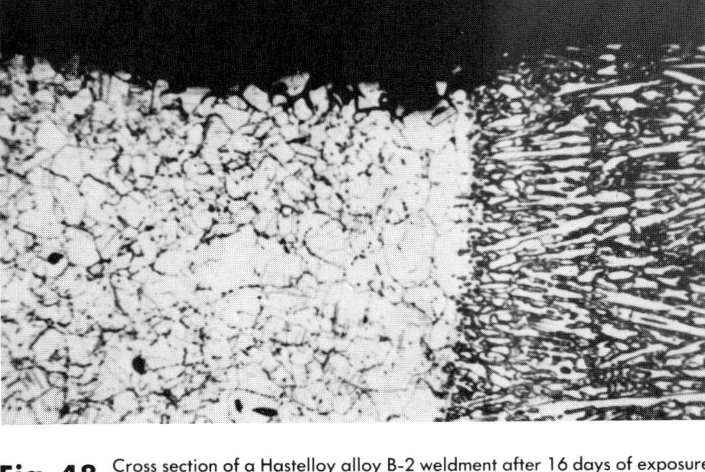

Fig. 48 Cross section of a Hastelloy alloy B-2 weldment after 16 days of exposure to boiling 20% HCl. 80×. Source: Ref 21

and μ-phase precipitation, respectively. For alloy C-276, sensitization occurs essentially in the higher temperature region because of μ-phase precipitation. Also, the μ-phase precipitation kinetics in alloy C-276 are slow enough not to cause sensitization problems in many high heat input weldments; however, precipitation can occur in the HAZ of alloy C-276 welds (Fig. 52). Because C-4 has lower tungsten than C-276, it has lower pitting and crevice corrosion resistance, for which tungsten is beneficial. Therefore, an alternate solution to alloy C-4 was needed in which both corrosion resistance and thermal stability are preserved. Hastelloy alloy C-22 has demonstrated improved corrosion resistance and thermal stability.

Because of the low carbon content of alloy C-22, the precipitation kinetics of carbides were slowed. Because alloy C-22 has lower molybdenum and tungsten levels than alloy C-276, μ-phase precipitation was also retarded.

From a weld HAZ point of view, this difference is reflected in lower grain-boundary precipitation even in a high heat input weld (Fig. 53). The HAZ microstructure of alloy C-4 was similar to this. This difference in the sensitization of the HAZ is also illustrated in Fig. 54, which shows that the HAZ and weld metal of alloy C-276 are attacked to a considerable extent in an oxidizing mixture of H_2SO_4, ferric sulfate ($Fe_2(SO_4)_3$), and other chemicals.

Corrosion of Carbon Steel Weldments

The corrosion behavior of carbon steel weldments is dependent on a number of factors. Consideration must be given to the compositional effects of the base metal and welding consumable and to the different welding processes used. Because carbon steels undergo metallurgical transformations across the weld and HAZ, microstructures and morphologies become important. A wide range of microstructures can be developed based on cooling rates, and these microstructures are dependent on energy input, preheat, metal thickness (heat sink effects), weld bead size, and reheating effects due to multipass welding. As a result of their different chemical compositions and weld inclusions (oxides and sulfides), weld metal microstructures are usually significantly different from those of the HAZ and base metal. Similarly, corrosion behavior can also vary.

In addition, hardness levels will be lowest for high heat inputs, such as those produced by submerged-arc weldments, and will be highest for low-energy weldments (with faster cooling rates) made by the shielded metal arc processes. Depending on the welding conditions, weld metal microstructures generally tend to be fine grained with basic flux and somewhat coarser with acid or rutile (TiO_2) flux compositions.

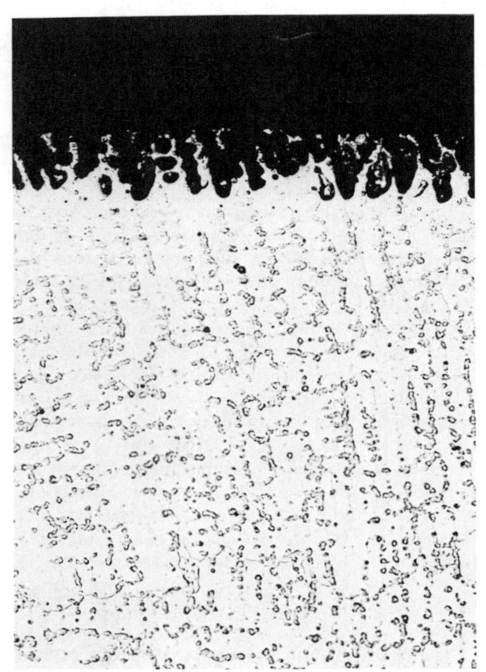

Fig. 49 Preferential corrosion of autogenous GTA weld in Hastelloy alloy B-2 exposed to boiling 60% H_2SO_4 + 8% HCl

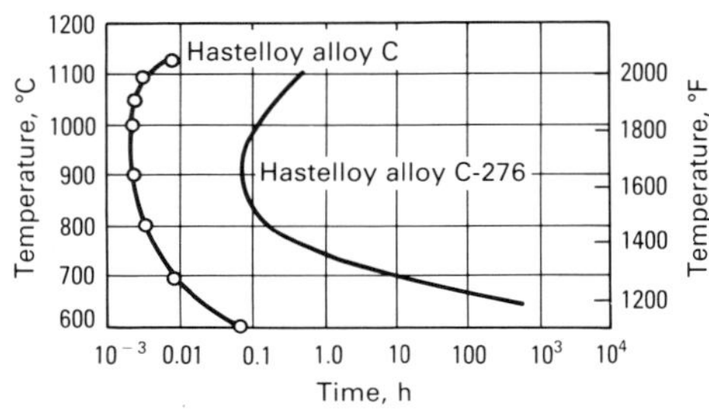

Fig. 50 Time-temperature transformation curves for Hastelloy alloys C and C-276. Intermetallics and carbide phases precipitate in the regions to the right of the curves. Source: Ref 22

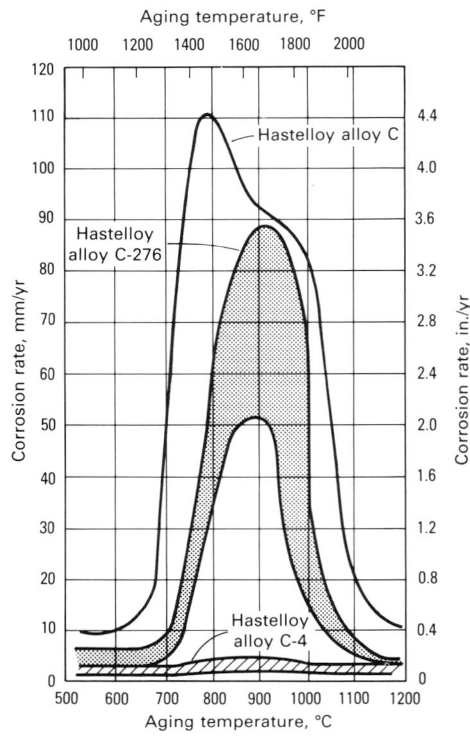

Fig. 51 Effect of 1-h aging treatment on corrosion resistance of three Hastelloy alloys in 50% H_2SO_4 + 42 g/L $Fe_2(SO_4)_3$. Source: Ref 23

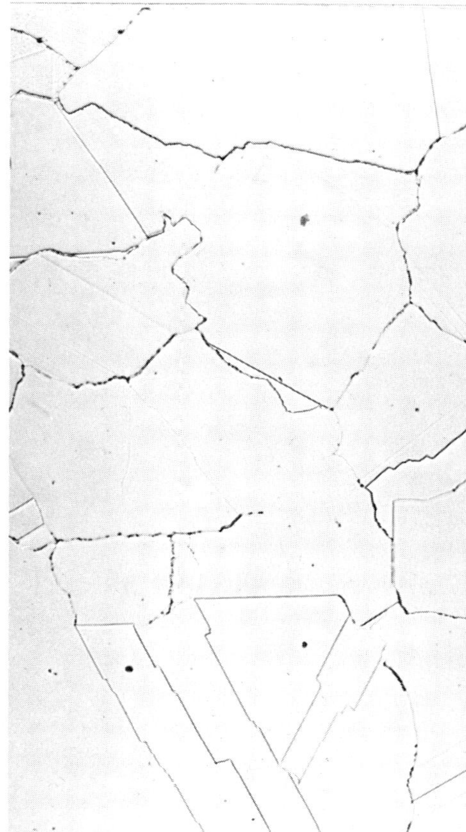

Fig. 52 Typical microstructure of the HAZ of a multipass submerged-arc weld in Hastelloy alloy C-276. Source: Ref 24

Fig. 53 Typical microstructure of the HAZ of a multipass submerged-arc weld in Hastelloy alloy C-22. Matching filler metal was used. Source: Ref 24

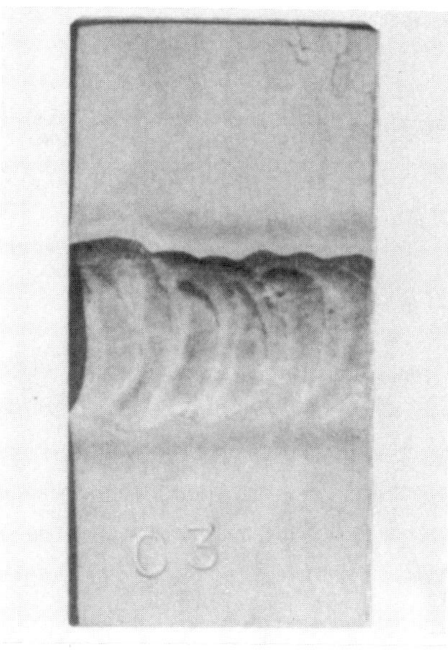

(a) (b)

Fig. 54 Corrosion of the weld metal and the HAZ in Hastelloy alloys C-22 (a) and C-276 (b) in an aerated mixture of 6 vol% H_2SO_4 + 3.9% $Fe_2(SO_4)_3$ + other chemicals at 150 °C (300 °F). Source: Ref 24

During welding, the base metal, HAZ, and underlying weld passes experience stresses due to thermal expansion and contraction. Upon solidification, rather high levels of residual stress remain as a result of weld shrinkage. Stress concentration effects as a result of geometrical discontinuities, such as weld reinforcement and lack of full weld penetration (dangerous because of the likelihood of crevice corrosion and the possibility of fatigue cracking), are also important because of the possibility of SCC. Achieving full weld penetration, minimizing excessive weld reinforcement through control of the welding process or technique, and grinding (a costly method) can be effective in minimizing these geometric effects. A stress-relieving heat treatment is effective in reducing internal weld shrinkage stress and metal hardness to safe levels in most cases.

Preferential HAZ Corrosion. An example of preferential weld corrosion in the HAZ of a carbon steel weldment is shown in Fig. 55. This phenomenon has been observed in a wide range of aqueous environments, the common link being that the environments are fairly high in conductivity, while attack has usually, but not invariably, occurred at pH values below about 7 to 8.

The reasons for localized weldment attack have not been fully defined. There is clearly a microstructural dependence, and studies on HAZs show corrosion to be appreciably more severe when the material composition and welding are such that hardened structures are formed. It has been known for many years that hardened steel may corrode more rapidly in acid conditions than fully tempered material, apparently because local microcathodes on the metal surface stimulate the cathodic hydrogen evolution reaction. On this basis, water treatments ensuring alkaline conditions should be less likely to induce HAZ corrosion, but even at pHs near 8, hydrogen ion (H^+) reduction can account for about 20% of the total corrosion current; pH values substantially above this level would be needed to suppress the effect completely.

Fig. 55 Preferential corrosion in the HAZ of a carbon steel weldment after service in an aqueous environment. 5×. Source: Ref 25

Preferential Weld Corrosion. It is probable that similar microstructural considerations also apply to the preferential corrosion of weld metal, but in this case, the situation is further complicated by the presence of deoxidation products, their type and number depending largely on the flux system employed. Consumable type plays a major role in determining weld metal corrosion rate, and the highest rates of metal loss are normally associated with shielded metal arc electrodes using a basic coating. In seawater, for example, the corrosion rate for a weld made using a basic-coated consumable may be three times as high as for weld metal from a rutile-coated consumable. Fewer data are available for submerged-arc weld metals, but it would appear that they are intermediate between basic and rutile shielded metal arc electrodes and that a corrosion rate above that of the base steel can be expected.

Galvanic-corrosion effects have also been observed and have caused unexpected failure of piping tankage and pressure vessels where the welds are anodic to the base metal (Ref 26). The following examples illustrate the point.

In one case, premature weld failures were experienced in a 102-mm (4-in.) ASTM A53 pipe

Table 7 Compositions of carbon steel base metals and some filler metals subject to galvanic corrosion

See Tables 8 and 9 for corrosion rates of galvanic couples.

Metal	Composition, %: C	Mn	Si	Cr	Ni	Fe	Others
Base metals							
ASTM A53, grade B	0.30	1.20	...	...	...	bal	...
ASTM A285, grade C	0.22	0.90	...	...	...	bal	...
Filler metals							
E6010	—No specific chemical limits—						
E6013	—No specific chemical limits—						
E7010-Al	0.12	0.60	0.40	...	...	bal	0.4–0.65Mo
E7010-G	...	1.00(a)	0.80(a)	0.30(a)	0.50(a)	bal	0.2Mo, 0.1V
E7016	...	1.25(b)	0.90	0.20(b)	0.30(b)	bal	0.3Mo(b), 0.08V(b)
E7018	...	1.60(c)	0.75	0.20(c)	0.30(c)	bal	0.3Mo(c), 0.08V(c)
E8018-C2	0.12	1.20	0.80	...	2.0–2.75	bal	...
ENiCrFe-2 (Inco Weld A)	0.10	1.0–3.5	1.0	13.0–17.0	bal	12.0	1–3.5Mo, 0.5Cu, 0.5–3(Nb + Ta)
Incoloy welding electrode 135	0.08	1.25–2.50	0.75	26.5–30.5	35.0–40.0	bal	2.75–4.5Mo, 1–2.5Cu

(a) The weld deposit must contain only the minimum of one of these elements. (b) The total of these elements shall not exceed 1.50%. (c) The total of these elements shall not exceed 1.75%. Source: Ref 26

that was used to transfer a mixture of chlorinated hydrocarbons and water. During construction, the pipeline was fabricated with E7010-A1 welding electrodes (see Table 7 for the composition limits for all materials discussed in these examples). Initial weld failures and subsequent tests showed the following welding electrodes to be anodic to the A53 grade B base metal: E7010-A1, E6010, E6013, E7010-G, and E8018-C2. Two nickel-base electrodes—Inco-Weld A (AWS A5.11, class ENiCrFe-2) and Incoloy welding electrode 135—were tested; they were found to be cathodic to the base metal and to prevent rapid weld corrosion. The corrosion rates of these various galvanic couples are listed in Table 8.

Another example is the failure of low-carbon steel welds in seawater service at 25 °C (75 °F). Fabrications involving ASTM A285, grade C, plate welded with E6013 electrodes usually start to fail in the weld after 6 to 18 months in seawater service at this temperature. Welds made with E7010 electrodes do not fail. Tests were conducted in seawater at 50 °C (120 °F) using A285, grade C, plate welded with E6010, E7010-A1, and E7010-G. It was determined that E7010-A1 was the best electrode to use in seawater and that E6010 and E7010-G were not acceptable (although they were much better than E6013), because they were both anodic to the base metal. A zero resistance ammeter was used to determine whether the electrodes were anodic or cathodic in behavior.

In another case, welds made from E7010-A1 electrodes to join ASTM A285, grade C, base metal were found to be anodic to the base metal when exposed to raw brine, an alkaline-chloride (pH > 14) stream, and raw river water at 50 °C (120 °F). When E7010-G was exposed to the same environment, it was anodic to the base metal in raw brine and raw river water and was cathodic to ASTM A285, grade C, in the alkaline-chloride stream. When the base metal was changed to ASTM A53, grade B, and A106, grade B, it was found that E7010-A1 weld metal was cathodic to both when exposed to raw brine at 50 °C (120 °F).

Finally, routine inspection of a column in which a mixture of hydrocarbons was water washed at 90 °C (195 °F) revealed that E7016 welds used in the original fabrication were corroding more rapidly than the ASTM A285, grade C, base metal. Corroded welds were ground to sound metal, and E7010-A1 was used to replace the metal that was removed. About 3 years later, during another routine inspection, it was discovered that the E7010-A1 welds were being selectively attacked. Tests were conducted that showed E7010-A1 and E7016 weld metals to be anodic to A285, grade C, while E7018 and E8018-C2 would be cathodic. Corrosion rates of these various galvanic couples are given in Table 9.

These examples demonstrate the necessity for testing each galvanic couple in the environment

Table 8 Corrosion rates of galvanic couples of ASTM A53, grade B, base metal and various filler metals in a mixture of chlorinated hydrocarbons and water

The areas of the base metal and the deposited weld metal were equal.

Galvanic couple	Corrosion rate: mm/yr	mils/yr
Base metal	0.4	15
E6010	0.9	35
Base metal	0.18	7
E6013	0.9	35
Base metal	1.3	50
E7010-A1	4.3	169
Base metal	1.7	68
E7010-G	2.8	112
Base metal	0.36	14
E8018-C2	1.7	66
Base metal	0.48	19
Inco Weld A	0.013	0.5
Base metal	0.36	14
Incoloy welding electrode 135	<0.0025	<0.1

Source: Ref 26

Table 9 Corrosion rates of galvanic couples of ASTM A285, grade C, base metal and various filler metals at 90 °C (195 °F) in water used to wash a hydrocarbon stream

Galvanic couple	Corrosion rate: mm/yr	mils/yr
Base metal	0.69	27
E7010-A1	0.81	32
Base metal	0.46	18
E7016	0.84	33
Base metal	1.3	50
E7018	1.2	48
Base metal	2.2	85
E8018-C2	1.04	41

Source: Ref 26

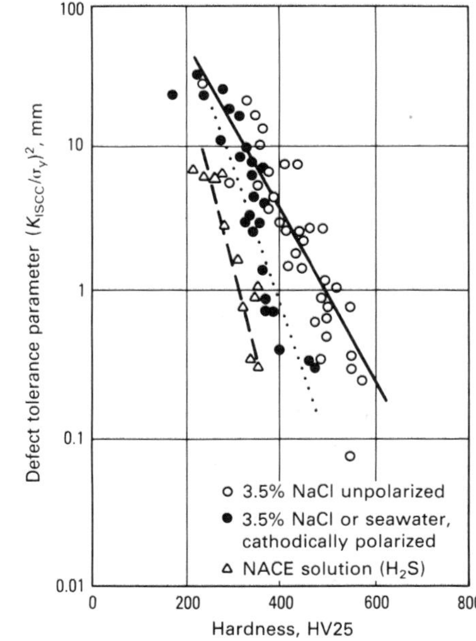

Fig. 56 SCC defect tolerance parameter versus hardness for carbon steel weldments in three environments. Data are derived from published tests on precracked specimens of various types of carbon steel base metals, HAZs, and weld metals. SCC defect tolerance parameter is dependent on crack length; details are available in Ref 25. Source: Ref 25

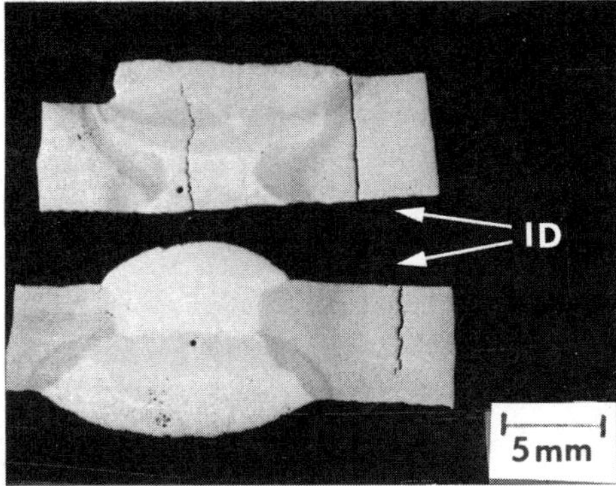

Fig. 57 Cross sections of pipe-to-elbow welds showing stress-corrosion cracks originating from the inside surface of the weld metal and the base metal. Source: Ref 27

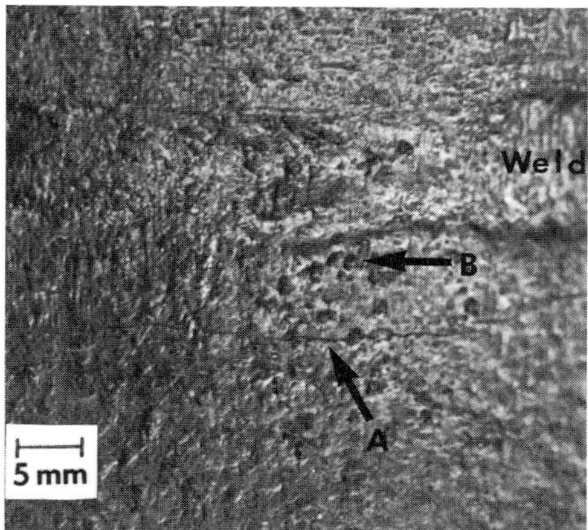

Fig. 58 Photograph of inside surface of a pipe showing 38-mm (1.5-in.) stress-corrosion crack (A) next to and parallel to a circumferential weld. Also shown are shallow corrosion pits (B). Source: Ref 27

for which it is intended. Higher-alloy filler metals can sometimes be used to advantage to prevent rapid preferential weld corrosion.

Stress-Corrosion Cracking. There is no doubt that residual welding stresses can cause SCC in environments in which such failure represents a hazard. This is the case for failure by both active path and hydrogen embrittlement mechanisms, and in the latter case, failure may be especially likely at low heat input welds because of the enhanced susceptibility of the hardened structures inevitably formed. Most SCC studies of welds in carbon and carbon-manganese steels have evaluated resistance to hydrogen-induced SCC, especially under sour (H_2S) conditions prevalent to the oil and gas industry. Although full definition of the effect of specific microstructural types has not been obtained, an overriding influence of hardness is evident (Fig. 56). The situation regarding active path cracking is less clear, but there are few, if any, cases in which SCC resistance increases at higher strength levels. On this basis, it is probable that soft, transformed microstructures around welds are preferable.

Carbon and low-alloy steels are also known to fail by SCC when exposed to solutions containing nitrates (NO_3^-). Refrigeration systems using a 30% magnesium nitrate ($Mg(NO_3)_2$) brine solution, for example, are commonly contained in carbon steel. In this case, pH adjustment is important, as is temperature. Failures in the HAZ due to SCC have been reported when brine temperatures have exceeded 30 °C (90 °F) during shutdown periods. To avoid these failures, carbon steel is being replaced with type 304L stainless. Others have stress relieved welded carbon steel systems and have operated successfully, although elevated-temperature excursions are discouraged.

More recently, it has been shown that cracking can occur under certain conditions in carbon dioxide (CO_2) containing environments, sometimes with spectacular and catastrophic results. Processes in the oil, gas, and chemical industries require removal of CO_2 from process streams by a variety of absorbants. In most cases, process equipment is fabricated from plain carbon steel.

SCC in Oil Refineries. Monoethanolamine (MEA) is an absorbant used to remove acid gases containing H_2S and CO_2 in oil refining operations. Recent failures in several refineries have shown that cracks can be parallel or normal to welds, depending on the orientation of principal tensile stresses. Cracking has been reported to be both transgranular and intergranular.

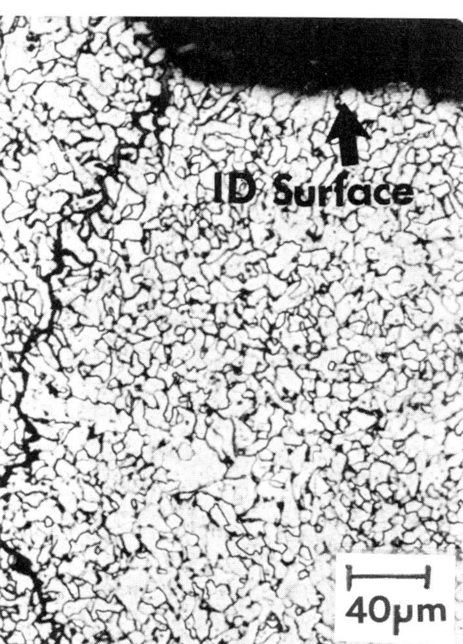

Fig. 59 Micrograph showing tight intergranular SCC originating at the inside surface of a pipe. Source: Ref 27

Before 1978, postweld stress relief of carbon steel weldments in MEA systems was performed only when the metal temperature of the equipment was expected to exceed 65 °C (150 °F) and the acid gas contained more than 80% CO_2 or when temperatures were expected to exceed 95 °C (200 °F) in any acid gas concentration.

Currently, any equipment containing MEA at any temperature and at any acid gas concentration is being postweld stress relieved. This is the result of surveys conducted by several refineries to define the extent of the SCC problem in this environment. These inspection programs showed that leaks were widespread and were found in vessels that ranged in age from 2 to 25 years. However, there were no reports of cracking in vessels that had been postweld stress relieved. In addition, it was found that all concentrations of MEA were involved and that MEA solutions were usually at relatively low temperatures (below 55 °C, or 130 °F). Equipment found to suffer from cracking included tanks, absorbers, carbon treater drums, skimming drums, and piping. The following example of a metallurgical investigation conducted by one oil refinery illustrates the problem of SCC of carbon steel in amine service (Ref 27).

Leaking Carbon Steel Weldments in a Sulfur Recovery Unit. In December 1983, two leaks were discovered at a sulfur recovery unit. More specifically, the leaks were at pipe-to-elbow welds in a 152-mm (6-in.) diam line operating in lean amine service at 50 °C (120 °F) and 2.9 MPa (425 psig). Thickness measurements indicated negligible loss of metal in the affected areas, and the leaks were clamped. In March 1984, 15 additional leaks were discovered, again at pipe-to-elbow welds of lean amine lines leading to two major refining units. The piping had been in service for about 8 years.

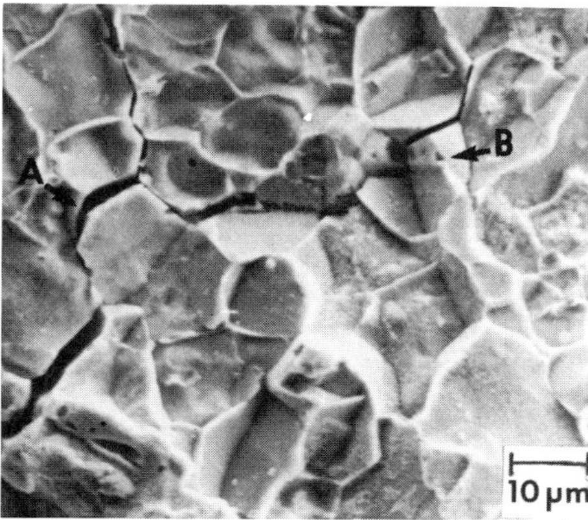

Fig. 60 SEM micrograph showing intergranular SCC (A) and initiation sites for pitting (B) on the inside surface of a pipe. Source: Ref 27

Investigation. Metallurgical examination of several of the welds revealed that leaking occurred at what appeared to be stress-corrosion cracks originating from the inside surface. Cracks were present in weld metal and base metal approximately 5 mm (0.2 in.) away from the weld, and they passed through the HAZ, as shown in Fig. 57. In other cases, stress-corrosion cracks also originated in the HAZ. The cracks typically ran parallel to the weld (Fig. 58).

Brinell hardness values, obtained by conversion of Knoop microhardness readings, were 133 to 160 (pipe base metal), 160 to 230 (weld metal), 182 to 227 (HAZs), and 117 to 198 (elbow base metal). The pipe base metal had an equiaxed fine-grain microstructure typical of low-carbon steel, and the elbow base metal had a nonequiaxed microstructure typical of hot-finished fittings. Carbon contents ranged from 0.25 to 0.30% by weight. Cracking was intergranular, as shown in Fig. 59 and 60.

The refinery operators immediately embarked on a program of visual inspection of all amine lines. As of June 1985, a total of 35 leaks in lean amine piping had been discovered. All leaks were at cracks in or around pipe-to-elbow welds, except for two leaks at welds that connected a tee and reducer, respectively. Piping size ranged from 76 to 305 mm (3 to 12 in.). Service temperature ranged from 40 to 60 °C (100 to 140 °F), with most leaks having occurred in lines carrying lean amine at 55 °C (130 °F). Pressures ranged from atmospheric to 2.9 MPa (425 psig), with most leaks having occurred between 2.8 and 2.9 MPa (400 and 425 psig). All piping had been in service for about 8 years, except two leaks at piping welds that had been in service for only 4 years.

As had been generally accepted industry practice, the specifications called for stress relieving or postweld heat treatment of piping and vessels in amine service at temperatures above 95 °C (200 °F). Therefore it was highly unlikely that any of the leaking welds had received postweld heat treatment. Further metallurgical examination of leaking welds from various lines conclusively confirmed that the leaking originated at stress-corrosion cracks. No leaks were found in rich amine piping. The characteristics of the mode of

Table 10 Chemistry limits on deaerator feedwater

Control parameter	Limit
Total hardness	<0.5 ppm as $CaCO_3$
Phenolphthalein alkalinity	Trace (max)
Methyl orange alkalinity	14–18 ppm as $CaCO_3$
Chloride	7.6–8.8 ppm
Total dissolved solids	70–125 ppm

Source: Ref 28

Table 11 Chemical analyses of steels and weld deposit

Sample	Analysis, %									
	C	Mn	Si	P	S	Ni	Cr	Mo	Al	Fe
Plate 1	0.25	0.88	<0.05	0.029	0.036	<0.05	<0.05	<0.03	<0.01	bal
Plate 2	0.21	0.83	<0.05	0.03	0.024	<0.05	<0.05	<0.03	<0.01	bal
Weld deposit	0.14	0.53	0.14	0.035	0.031	<0.05	<0.05	<0.03	<0.01	bal

Source: Ref 28

fracture suggested that the failure mechanism was a form of caustic SCC.

It is interesting to note that other researchers also have metallographically examined numerous samples of similar cracks; their results can be summarized as follows:

- Cracks were essentially intergranular and were filled with gray oxide scale
- Hardness of welds and HAZ's was less than 200 HB
- Cause of fracture was believed to be a form of caustic SCC
- Cracking occurs whether or not MEA solutions contain corrosion inhibitors

Preventive Measures. As a result of this particular investigation and others, all welds in equipment in MEA service are being inspected. Wet fluorescent magnetic-particle inspection after sandblasting to remove oxides and scale appears to be the most effective technique. Shear-wave ultrasonic (SWU) inspection has also been used for piping, but it does not always distinguish SCC and other defect indications, such as shrinkage cracks, slag inclusions, lack of fusion, or fatigue cracks. Nevertheless, SWU is considered helpful, because these other types of defects also can pose a threat to the structural integrity of the system in question. Inspection frequency is dependent on the critical nature of the particular equipment in question, and most important, all welds in these systems are now being postweld stress relieved.

Corrosion of Welds in Carbon Steel Deaerator Tanks. Deaerator tanks, the vessels that control free oxygen and other dissolved gases to acceptable levels in boiler feedwater, are subject to a great deal of corrosion and cracking. Several years ago, there were numerous incidences of deaerator tank failures that resulted in injury to personnel and property damage losses. Since that time, organizations such as the National Board of Boiler and Pressure Vessel Inspectors and the Technical Association of the Pulp and Paper Industry have issued warnings to plant operators, and these warnings have resulted in the formation of inspection programs for evaluating the integrity of deaerator tanks. As a result, many operators have discovered serious cracking problems. The following example illustrates the problem (Ref 28).

Weld Cracking in Oil Refinery Deaerator Vessels. Two deaerator vessels with associated boiler feedwater storage tanks operated in similar service at a refinery. The vertical deaerator vessels were constructed of carbon steel (shell and dished heads), with trays, spray nozzles, and other internal components fabricated of type 410 stainless steel. Boiler feedwater was treated by sand filtration using pressure filters, followed by ion-exchange water softening. Hardness was controlled at less than 0.5 ppm calcium carbonate ($CaCO_3$). A strong cationic primary coagulant (amine) was used to aid the filtering of colloidal material. Treated water was blended with condensate containing 5 ppm of a filming amine corrosion inhibitor. Final chemistry of the feedwater was controlled to the limits given in Table 10. Oxygen scavenging was ensured by the addition of catalyzed sodium bisulfite ($NaHSO_3$) to the storage tanks. Treated water entered the top of the tray section of the deaerators through five or six spray nozzles and was stored in the horizontal tanks below the deaerators.

Inspection Results. Deaerator vessel and storage tank A were inspected. All tray sections were removed from the deaerator. With the exception of the top head to shell weld in the deaerator, all internal welds were ground smooth and magnetic particle inspected. No cracks were found. Corrosion damage was limited to minor pitting of the bottom head in the deaerator vessel.

Inspection of deaerator vessel B revealed cracking at one weld. Tray sections were removed from the deaerator vessel, and shell welds were gritblasted. Except for the top head to shell weld in the deaerator, all internal welds in both B units were then ground smooth and magnetic particle inspected. Three transverse cracks were found at the bottom circumferential weld in the deaerator vessel. These were removed by grinding to a depth of 1.5 mm (0.06 in.).

Inspection of storage tank B revealed numerous cracks transverse to welds. With the shell constructed from three rings of plate, the longitudinal ring welds were located just below the water level. These longitudinal welds exhibited no detectable cracking. One circumferential crack was found above the working water level in the vessel. The remaining cracks were located at circumferential welds below the working water level. Numerous cracks transverse to circumferential welds were detected, but only one longitudinal crack was detected. All cracks were removed by grinding to a depth of 2 mm (0.08 in.).

Unlike deaerator vessel A, it was noted that none of the spray nozzles in deaerator vessel B was operational at the time of inspection. In addition, two valves had fallen to the bottom of the deaerator vessel. The bottom section of trays in deaerator vessel B had fallen to the bottom of the storage vessel. Corrosion damage in deaerator vessel B was limited to underdeposit pitting attack at circumferential welds in the bottom.

Metallurgical Analysis. A section was cut from a circumferential weld region in storage tank B. As shown in Fig. 61, the cracking was predominantly transverse to the weld. Chemical analysis was performed on samples cut from weld metal and base metal; the results are given in Table 11. The results show that the steel plate was not aluminum- or silicon-killed, but was most likely a rimmed grade. Cross sections were cut perpendicular to both transverse and longitudinal cracks and were examined metallographically.

As shown in Fig. 62, metallographic examination of the base metal structures revealed ferrite and lamellar pearlite phases with a nearly equiaxed grain structure. The approximate grain size was ASTM 6 to 7. Figure 63 shows a longitudinal crack in a weld HAZ, with associated grain refinement. Cracking initiated from the bottom of a pit. The oxide associated with the major crack was extensive and contained numerous secondary cracks. Analysis of the oxide deposit within the crack by wavelength-dispersive spectroscopy revealed slightly less oxygen than an Fe_2O_3 standard. Therefore, it was assumed that the oxide deposit was a mixture of Fe_3O_4 and Fe_2O_3.

Figure 64 shows a crack extending into base metal, transverse to the weld, with secondary cracking to the periphery of the oxidized region. It was clear that the oxide exhibited extensive internal cracking. Figure 64 also shows the entrainment of lamellar pearlite phase (dark) within the oxide corrosion product. In addition, the crack tips are blunt.

Discussion. The cracks described in this example are very similar to those found in many other investigations, despite a variety of deaerator ves-

(a)

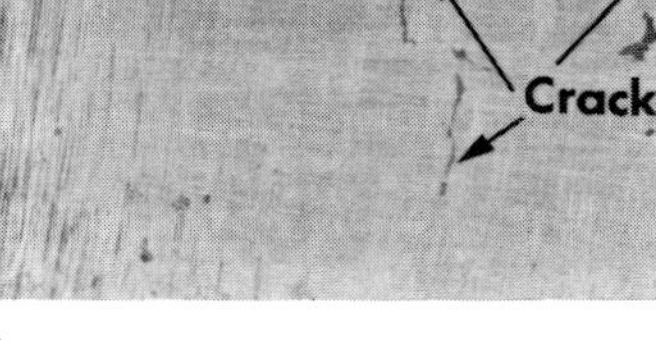

(b)

Fig. 61 Transverse and longitudinal cracks on as-ground weld areas on the inside surface of storage vessel B. (a) Transverse and longitudinal cracks. (b) Transverse cracks. Source: Ref 28 sel designs and operating conditions. Cracks typically display the following characteristics:

- Cracks occur most often in welds and HAZs, but can also occur in the base metal
- Cracks are generally transverse to the weld HAZ, and occur both parallel and perpendicular to the hoop stress direction
- The worst cracks appear to be located in circumferential and head-to-shell welds in horizontal vessel designs
- Cracks are concentrated at, but not solely located within, the working water level in the vessel
- Cracks are perpendicular to the vessel plate surface
- Cracks are predominantly transgranular with minor amounts of branching
- Cracks are filled with iron oxide. Cracking of the oxide corrosion product is followed by progressive corrosion. The ferrite phase is selectively attacked, with retention of the pearlite phase within the oxide corrosion product
- Cracks initiate from corrosion pits. Weld defects, however, can also become active sites for crack initiation
- Crack tips are blunt

Conclusions. These findings suggest that the failure mechanism is a combination of low-cycle corrosion fatigue and stress-induced corrosion. Extensive oxide formation relative to the depth of cracking is a key feature. The formation of oxide was associated with corrosion attack of the ferrite phase. The lamellar pearlite phase remained relatively intact and was contained within the oxide product. The oxide itself exhibited numerous cracks, allowing aqueous corrosion of fresh metal to occur at the oxide/metal interface. Mechanical or thermal stresses are most likely responsible for this network of cracks within the oxide product. The mechanism appears to be stress-assisted localized corrosion. Sharp, tight cracks were not found in fresh metal beyond the periphery of the oxide corrosion product. It therefore appears reasonable that cracking could have occurred subsequent to corrosion and within the brittle oxide.

Cracking at welds and HAZs suggests that residual weld shrinkage stresses play a major role. Welds in deaerator vessels typically have not been postweld stress relieved. It is not unusual to find residual welding stresses of yield strength magnitude. This problem can be aggravated by vessel design (high localized bending stresses around saddle supports that fluctuate with water level and are accelerated by operational upsets).

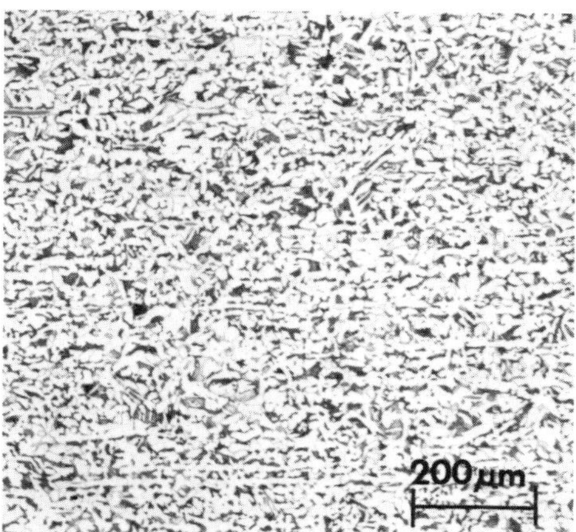

Fig. 62 Micrograph of the typical base metal microstructure of storage vessel B. Etching with nital revealed ferrite (light) and lamellar pearlite (dark). Source: Ref 28

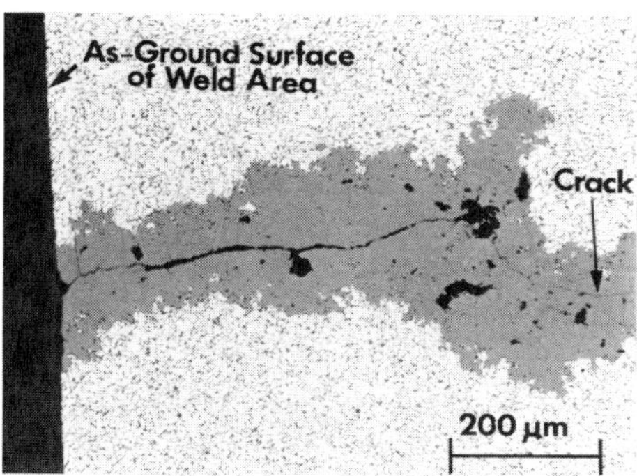

Fig. 63 Micrograph of a longitudinal crack in the HAZ of a weld from storage vessel B. Etched with nital. Source: Ref 28

No fault was found with the steel plate chemical composition or with welding consumables. There was no evidence of embrittlement or caustic SCC (that is, no branched intergranular cracks).

Recommendations. All welds in deaerator vessels should be postweld stress relieved. Operational upsets should be avoided, and water chemistry must be maintained with acceptable limits. This is especially true with regard to water oxygen levels, which should be kept low to minimize pitting corrosion.

REFERENCES

1. F.C. Brautigam, Welding Practices That Minimize Corrosion, *Chem. Eng.*, 17 Jan 1977, p 145-147
2. M.A. Streicher, Theory and Application of Evaluation Tests For Detecting Susceptibility to Intergranular Attack in Stainless Steels and Related Alloys—Problems and Opportunities, in *Intergranular Corrosion of Stainless Alloys*, STP 656, American Society for Testing and Materials, 1978, p 70
3. W.F. Savage, New Insight Into Weld Cracking and a New Way of Looking at Welds, *Weld. Des. Eng.*, Dec 1969
4. *Aluminum: Properties and Physical Metallurgy*, J.E. Hatch, Ed., American Society for Metals, 1984, p 283
5. *Welding Aluminum*, American Welding Society—The Aluminum Association, 1972
6. J.G. Young, BWRA Experience in the Welding of Aluminum-Zinc-Magnesium Alloys, *Weld. Res. Suppl.*, Oct 1968
7. "Alcoa Aluminum Alloy 7005," Alcoa Green Letter, Aluminum Company of America, Sept 1974
8. A. Garner, How Stainless Steel Welds Corrode, *Met. Prog.*, Vol 127 (No 5), April 1985, p 31
9. K.F. Krysiak, "Cause and Prevention of Unusual Failures of Materials," Paper 19, presented at Corrosion/83, Anaheim, CA, National Association of Corrosion Engineers, April 1983
10. "Standard Practices for Detecting Susceptibility to Intergranular Attack in Austenitic Stainless Steels," A 262, *Annual Book of ASTM Standards*, American Society for Testing and Materials
11. J.R. Kearns and H.E. Deverell, "The Use of Nitrogen to Improve Fe-Cr-Ni-Mo Alloys for

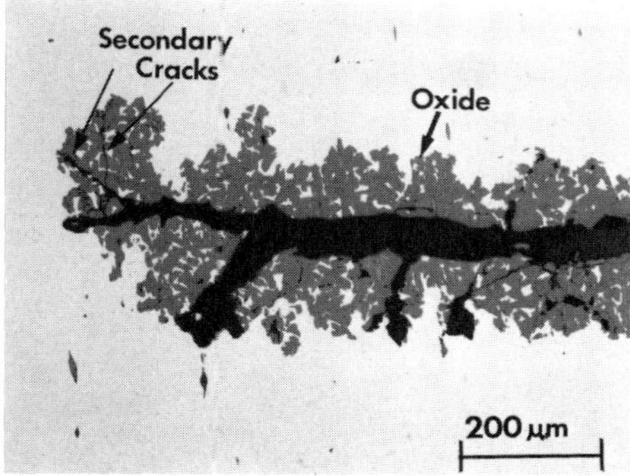

(a)

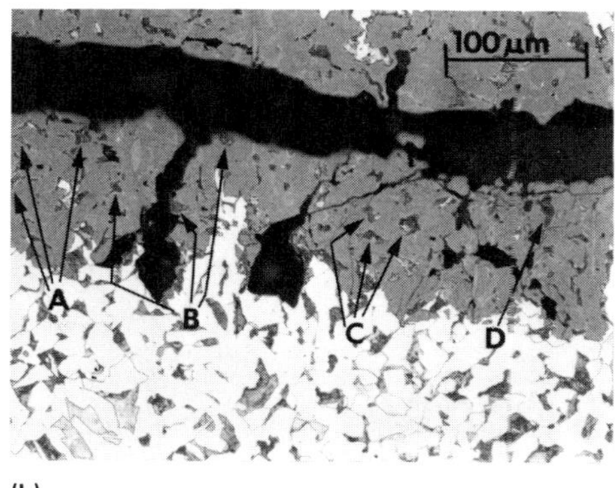

(b)

(c)

Fig. 64 Micrographs of a transverse crack in storage vessel B. (a) Crack extending into base metal. As-polished. (b) Lamellar pearlite phase (dark) entrained in the oxide corrosion product. (c) Microcracks and entrained pearlite phase in the oxide corrosion product. (b) and (c) Etched with nital. Source: Ref 28 the Chemical Process Industries," Paper 188, presented at Corrosion/86, Houston, TX, National Association of Corrosion Engineers, March 1986

12. J.R. Kearns, "The Corrosion of Heat Tinted Austenitic Stainless Alloys," Paper 50, presented at Corrosion/85, Boston, MA, National Association of Corrosion Engineers, March 1985
13. G. Kobrin, Corrosion by Microbiological Organisms in Natural Waters, *Mater. Perform.*, Vol 15 (No. 7), 1976
14. R.H. Espy, "How Corrosion and Welding Conditions Affect Corrosion Resistance of Weldments in Type 430 Stainless Steel," Paper 22, presented at Corrosion/68, Houston, TX, National Association of Corrosion Engineers, 1968
15. K.F. Krysiak, Welding Behavior of Ferritic Stainless Steels—An Overview, *Weld. J.*, Vol 65 (No. 4), April 1986, p 37-41
16. B. Lundquist, P. Norberg, and K. Olsson, "Influence of Different Welding Conditions on Mechanical Properties and Corrosion Resistance of Sandvik SAF 2205 (UNS S31803)," Paper 10, presented at the Duplex Stainless Steels '86 Conference, the Hague, Netherlands, Oct 1986
17. "Standard Test Methods for Pitting and Crevice Corrosion Resistance of Stainless Steels and Related Alloys by the Use of Ferric Chloride Solution," G 48, *Annual Book of ASTM Standards*, American Society for Testing and Materials
18. N. Sridhar, L.H. Flasche, and J. Kolts, Effect of Welding Parameters on Localized Corrosion of a Duplex Stainless Steel, *Mater. Perform.*, Dec 1984, p 52-55
19. "Standard Specification for Substitute Ocean Water," D 1141, *Annual Book of ASTM Standards*, American Society for Testing and Materials
20. E. Perteneder, J. Tosch, and G. Rabensteiner, "New Welding Filler Metals for the Welding of Girth Welds on Pipelines of Corrosion-Resistant Cr-Ni-Mo-N-Duplex Steels," Paper presented at the International Conference on Welding in Energy Related Projects, Toronto, Canada, Welding Institute of Canada, Sept 1983
21. F.G. Hodge and R.W. Kirchner, Paper 60, presented at Corrosion/75, Toronto, Canada, National Association of Corrosion Engineers, April 1975
22. R.B. Leonard, *Corrosion*, Vol 25 (No. 5), 1969, p 222-228
23. F.G. Hodge and R.W. Kirchner, Paper presented at the Fifth European Congress on Corrosion, Paris, Sept 1973
24. P.E. Manning and J.D. Schobel, Paper presented at ACHEMA '85, Frankfurt, West Germany, 1985; See also *Werkst. Korros.*, March 1986
25. T.G. Gooch and P.H.M. Hart, "Review of Welding Practices For Carbon Steel Deaerator Vessels," Paper 303, presented at Corrosion/86, Houston, TX, National Association of Corrosion Engineers, March 1986
26. C.G. Arnold, "Galvanic Corrosion Measurement of Weldments," Paper 71, presented at Corrosion/80, Chicago, IL, National Association of Corrosion Engineers, March 1980
27. J. Gutzeit and J.M. Johnson, "Stress-Corrosion Cracking of Carbon Steel Welds in Amine Service," Paper 206, presented at Corrosion/86, Houston, TX, National Association of Corrosion Engineers, March 1986
28. G.E. Kerns, "Deaerator Cracking—A Case History," Paper 310, presented at Corrosion/86, Houston, TX, National Association of Corrosion Engineers, March 1986

Corrosion Economic Calculations

Ellis D. Verink, Jr., Department of Materials Science and Engineering, University of Florida

ENGINEERING ECONOMY is a discipline that can be used to assist engineers and engineering managers in measuring the economic impact of their decisions on the financial goals of a business. Basically, engineering economy is concerned with money as a resource and as the price of other resources. Business success requires the prudent and efficient use of all resources, including money. The principles of engineering economy permit direct comparisons of potential alternatives in monetary terms. In this way, they encourage efficient use of resources.

Corrosion is basically an economic problem. Thus, the corrosion behavior of materials is an important consideration in the economic evaluation of any project. It is not always wise to select the material with the lowest initial cost, because the initial cost is not necessarily the last cost. Overall costs include maintenance, downtime, time value of money, tax aspects, and obsolescence.

This article will discuss the principles and terminology of engineering economy and their application to a number of generic corrosion-related problems. Several of these problems appeared in NACE Standard RP-02-72 (Ref 1).

Money and Time

Consider the effects of time and earning power on \$20. If \$20 is placed in the bank and earns interest at a rate of 5% per year, it grows to \$21 when the interest, \$1, is paid at the end of the year. Thus, \$20 today at 5% is equivalent to \$21 a year from now. Stated another way, in order to have \$21 one year from now, only \$20 need be deposited today if the interest rate is 5%. The \$20 is termed the discounted present value of the \$21 needed 1 year from now. The initial deposit as well as the earned interest left in the account have earning power because the interest is compounded, which means that it is computed on both the principal and the accrued interest.

Suppose the \$20 were used to pay four equal, annual installments of \$5 each. Without interest, the \$20 would be exhausted after making the last payment. If the \$20 is deposited in the bank at 5% interest, it will be worth \$21 at the end of the first year. Paying out \$5 would leave \$16 to be held at interest the second year. The \$16 invested at 5% will earn \$0.80 in 1 year. Subtracting \$5 from \$16.80 leaves \$11.80 to be held at 5% interest for the third year. A sum of \$11.80 will earn \$0.59 by the end of the third year at 5%. Another annual payment of \$5 leaves \$7.39 to earn interest during the fourth year. Adding the \$0.37 interest earned during the fourth year and subtracting the final \$5 annual payment leaves a balance of accrued interest of \$2.76.

The earning power of money permits another strategy. If the initial deposit were reduced to \$17.73 at 5% interest, \$5 can be paid out each year for 4 years, leaving nothing. This example illustrates the distinction between the terms equivalent and equal. Twenty dollars is equal to four payments of \$5 each. It would also be equivalent to four \$5 annual payments (only) if the interest rate were 0. The \$17.73 is not equal to the sum of four payments of \$5 each. However, when \$17.73 is invested at 5%, it is equivalent to four annual payments of \$5.

The term equivalent implies that the concept of the time value of money is applied at some specific interest rate. Therefore, for an amount of money to have a precise meaning, it must be fixed both in time and amount. Mathematical formulas and tables are available to translate an amount of money at any particular time into an equivalent amount at another date.

Many kinds of translations are possible. For example, a single amount of money can be translated into an equivalent amount at either a later or an earlier date. This is accomplished by calculating the present worth (PW) or the future worth (FW) as of the present date. Single amounts of money can be translated into equivalent annuities, A, involving a series of uniform amounts occurring each year. Conversely, annuities can be translated into equivalent single amounts at an earlier or later date. The present worth of an annuity, P/A, is the single amount of money equivalent to a future annuity. The single amount equivalent to a past annuity is referred to as the future worth of an annuity, F/A.

It also is possible to calculate the amount of money that would be equivalent to a nonuniform series of cash flows. Two types of nonlinear series that find application are an arithmetic progression, in which the series changes by a constant amount, and a geometric progression, in which the series changes by a constant rate. The arithmetic progression is considered to be representative of variable costs, such as maintenance costs, which may increase as equipment ages. The geometric progression is used to represent the effects of inflation or deflation.

Notation and Terminology

American National Standards Institute (ANSI) standard Z94.5 (Ref 2), a compilation of the symbology and terminology of the field, offers to practitioners the improved communication benefits of standardization. With the development and publication of this standard and its adoption by the Institute of Industrial Engineers and the Engineering Economy Division of the American Society of Engineering Education, it is expected that future books and articles requiring symbols common to engineering economy will use these, because they represent the consensus choice of the prominent authors and educators in this field. This should eliminate one of the significant deterrents to the use of these methods in the past.

Table 1 lists the definitions and symbols used for parameters. Table 2 lists functional forms of compound interest factors. Table 3 is an explanatory supplement to Table 2, and it represents diagrams, algebraic forms, and uses for compound interest factors.

The basic form of the notation used for all time value factors consists of a ratio of two letters representing two amounts of money—for example, P/A or F/P—plus an interest rate, i%, and a number of periods, n. These are customarily written as (P/A, i%, n) or (F/P, i%, n). The present worth, P, of a known annuity, A, can also be expressed as:

$$P = A\left(\frac{P}{A}\right) \quad \text{(Eq 1)}$$

Table 1 Suggested standard definitions and symbols used for parameters

Definition of parameter	Symbol
Effective interest rate per interest period (discount rate)	i
Nominal interest rate per year	r
Number of compounding periods	n
Number of compounding periods per year	m
Present sum of money. The letter P implies present (or equivalent present value)	P
Future sum of money. The letter F implies future (or equivalent future value)	F
End-of-period cash flows (or equivalent end-of-period values) in a uniform series continuing for a specified number of periods. The letter A implies annual or annuity	A
Uniform period-by-period increase or decrease in cash flows (or equivalent values); the arithmetic gradient	G
Amount of money (or equivalent value) flowing continuously and uniformly during a given period	$\bar{P}$ or $\bar{F}$
Amount of money (or equivalent value) flowing continuously and uniformly during each and every period continuing for a specific number of periods	$\bar{A}$

Source: Ref 2

Table 2 Functional forms of compound interest factors

See Table 3 for further explanation.

Factor number	Factor name	Functional format
Group I: all cash flows discrete: end-of-period compounding		
1	Compound amount factor (single payment)	$(F/P, i\%, n)$
2	Present worth factor (single payment)	$(P/F, i\%, n)$
3	Sinking fund factor	$(A/F, i\%, n)$
4	Capital recovery factor	$(A/P, i\%, n)$
5	Compound amount factor (uniform series)	$(F/A, i\%, n)$
6	Present worth factor (uniform series)	$(P/A, i\%, n)$
7	Arithmetic gradient conversion factor (to uniform series)	$(A/G, i\%, n)$
8	Arithmetic gradient conversion factor (to present value)	$(P/G, i\%, n)$
Group II: all cash flows discrete: continuous compounding		
9	Continuous compounding compound amount factor (single payment)	$(F/P, r\%, n)$
10	Continuous compounding present worth factor (single payment)	$(P/F, r\%, n)$
11	Continuous compounding sinking fund factor	$(A/F, r\%, n)$
12	Continuous compounding capital recovery factor	$(A/P, r\%, n)$
13	Continuous compounding compound amount factor (uniform series)	$(F/A, r\%, n)$
14	Continuous compounding present worth factor (uniform series)	$(P/A, r\%, n)$
Group III: continuous, uniform cash flows: continuous compounding (payments during one period only)		
15	Continuous compounding present worth factor (single, continuous payment)	$(P/\bar{F}, i\%, n)$
16	Continuous compounding compound amount factor (single, continuous payment)	$(F/\bar{P}, i\%, n)$
Group IV: continuous, uniform cash flows: continuous compounding (payments during a continuous series of periods)		
17	Continuous compounding sinking fund factor (continuous, uniform payments)	$(\bar{A}/F, i\%, n)$
18	Continuous compounding capital recovery factor (continuous, uniform payments)	$(\bar{A}/P, i\%, n)$
19	Continuous compounding compound amount factor (continuous, uniform payments)	$(F/\bar{A}, i\%, n)$
20	Continuous compounding present worth factor (continuous, uniform payments)	$(P/\bar{A}, i\%, n)$

Source: Ref 2

Or the future worth, F, of a known present amount P, as:

$$F = P\left(\frac{F}{P}\right) \quad \text{(Eq 2)}$$

Other forms are also common, such as PW( ) or FW( ), which are called operators because they represent some computational operation, such as the present worth or future worth of whatever is inside the parentheses.

It should be evident that (F/P) is the reciprocal of (P/F) and that (F/A) and (A/F) are also reciprocals. This observation is useful because it means that only three time factors need to be tabulated in order to conduct six operations. Another algebraic relationship shows that if two time value factors are multiplied, the product is a third time value factor. For example:

$$\left(\frac{A}{F}\right)\left(\frac{P}{A}\right) = \left(\frac{P}{F}\right) \quad \text{(Eq 3)}$$

Methods of Economic Analysis

Economic analysis methods that concern the entire service life are sometimes called life-cycle-cost methods. Those methods that lead to single measure numbers include:

- Internal rate of return (IROR)
- Discounted payback (DPB)
- Present worth (PW), also referred to as net present value (NPV)
- Present worth of future revenue requirements (PWRR)
- Benefit-cost ratios (BCR)

All five methods use the concept of present worth. Although each method has certain advantages, the individual methods vary considerably with regard to their application and complexity.

The IROR method compares the initial capital investment with the present worth of a series of net revenues or savings over the anticipated service life. Expenses include operation, maintenance, taxes, insurance, and overheads, but do not include return on (or of) the invested capital. From an economic standpoint, IROR consists essentially of the interest cost on borrowed capital plus any existing (positive or negative) profit margin. The disadvantage of this measure of economy is that it ignores benefits extending beyond the assumed life of the equipment and thus may omit a substantial part of the actual service life. This may lead to unnecessarily pessimistic measures of long-range economy.

The PWRR method is particularly applicable to regulated public utilities, for which the rates of return are more or less fixed by regulation. It is also applicable where it has already been determined by IROR analysis that a project is economically viable; the engineer must determine which is the most economic alternative under circumstances in which several alternatives produce the same revenue, with some creating less expense (requirement for revenue) than others and consequently a greater profit margin (or lower losses). The principal objection to the PWRR method is that it is inadequate where alternatives are competing for a limited amount of capital, because it does not identify the alternative that produces the greatest return on invested capital.

The DPB and BCR methods are somewhat more complicated than the PWRR method. The BCR method is similar to the IROR method in that both require that alternatives be examined not only for economic measures compared to a "do nothing" scenario but also for incremental measures associated with incremental capital investments.

The PW method, also referred to as the net present value method, is considered the easiest and most direct of the five methods and has the broadest application to engineering economy problems. Many industries refer to this method as the discounted cash flow method of analysis. This method is often used to test the results of other methods of analysis. Therefore, because this method is often preferred, it will be given primary attention in this article.

Annual Versus Continuous Compounding

For completeness, Tables 1 to 3 contain formulas involving annual and continuous compounding. Because actual cash flow (both inward and outward) is continuous, continuous compounding might seem to be the more accurate assumption for engineering economy studies. In reality, although the overall cash flows tend to be continuous, the cash flow data are seldom precise enough in economy studies to take full advantage of continuous compounding. It is also true that the normal purpose of the economy study is to analyze some specific event that will occur at a specific time; therefore, a procedure that is readily associated with a specific time seems appropriate. For these reasons, and because it is conceptually easier to understand and apply, annual compounding will be used in this article.

The present worth method is a form of discounted cash flow in which cash flow data (which must include dates of receipts and disbursements) are discounted to present worth. Before applying these methods, a management decision must be made regarding the desired life (usually expressed as a number of years, n) and the minimum acceptable rate of return on invested capital for a project (expressed in terms of the effective interest rate, i, or the nominal rate or return, r). Rates of return (before taxes) vary among industries, ranging from 10 to 15% where obsolescence is not high to 25 to 40% (or perhaps higher) for more dynamic industries. It would be convenient if the minimum acceptable rate of return were known in advance of making engineering economy analyses; however, it is not always easy to learn what rate of return is considered to be acceptable. Under such circumstances, it is helpful to prepare a series of economic alternatives in which the rate of return is varied in order to provide management with options.

Applying the Present Worth Method. To illustrate the features of this method, it will be determined whether an expenditure of \$15 000 should be made to reduce labor and maintenance from \$8200 to \$5100 per year. Money is worth 10%, and the life of the project is 10 years. The effects of taxes and depreciation have been omitted for simplicity.

Table 4 shows the projected pattern of cash flow for Plan A (the present method, which consists of paying \$8200 per year for labor and maintenance) and for Plan B (the proposed method, which consists of an initial expenditure of \$15 000 plus \$5100 per year for labor and maintenance over the life of the project). A minus sign

Table 3 Diagrams, algebraic forms, and uses for compound interest factors

Factor number	Factor name	Algebraic form	Use when:
Group I: all cash flows discrete: end-of-period compounding			
Cash flow diagram for factors 1–6 (and 9–14):			
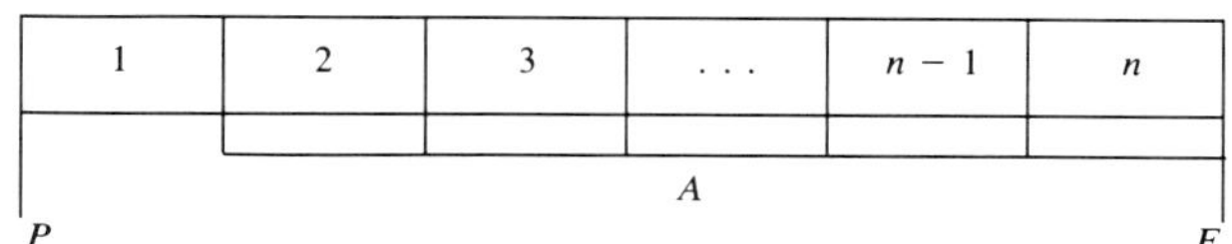			
1.	Compound amount factor (single payment)	$(1+i)^n$	Given P, to find F
2.	Present worth factor (single payment)	$(1+i)^{-n}$	Given F, to find P
3.	Sinking fund factor	$\frac{i}{(1+i)^n-1}$	Given F, to find A
4.	Capital recovery factor	$\frac{i(1+i)^n}{(1+i)^n-1}$	Given P, to find A
5.	Compound amount factor (uniform series)	$\frac{(1+i)^n-1}{i}$	Given A, to find F
6.	Present worth factor (uniform series)	$\frac{(1+i)^n-1}{i(1+i)^n}$	Given A, to find P
Cash flow diagram for factors 7 and 8:			
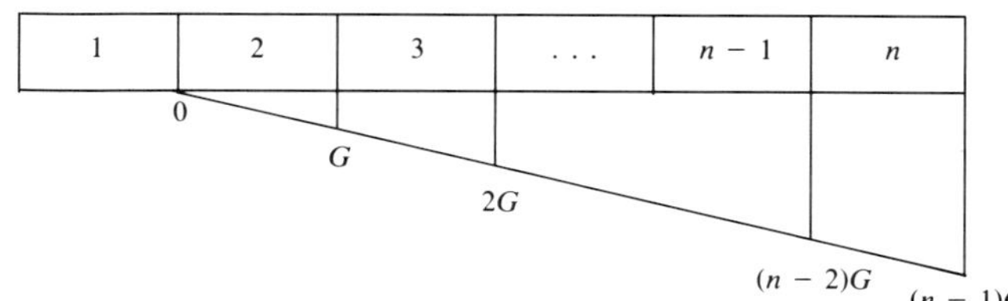			
7.	Arithmetic gradient conversion factor (to uniform series)	$\left[\frac{1}{i}-\frac{n}{(1+i)^n-1}\right]$	Given G, to find A
8.	Arithmetic gradient conversion factor (to present value)	$\left[\frac{1}{i}-\frac{n}{(1+i)^n-1}\right]\left[\frac{(1+i)^n-1}{i(1+i)^n}\right]$	Given G, to find P
Group II: all cash flows discrete: continuous compounding			
9.	Continuous compounding compound amount factor (single payment)	e^{rn}	Given P, to find F
10.	Continuous compounding present worth factor (single payment)	e^{-rn}	Given F, to find P
11.	Continuous compounding sinking fund factor	$\frac{e^r-1}{e^{rn}-1}$	Given F, to find A
12.	Continuous compounding capital recovery factor	$\frac{e^{rn}(e^r-1)}{e^{rn}-1}$	Given P, to find A
13.	Continuous compounding compound amount factor (uniform series)	$\frac{e^{rn}-1}{e^r-1}$	Given A, to find F
14.	Continuous compounding present worth factor (uniform series)	$\frac{e^{rn}-1}{e^{rn}(e^r-1)}$	Given A, to find P

(continued)

means that money leaves the bank; a plus sign means that the bank balance increases in size. The net cash flow for each year appears in the right hand column under the heading "B − A."

Selection of Plan B will result in a net positive cash flow, that is, the net amount of money in the bank is increased over the life of the project when Plan B is selected. Before concluding that Plan B should be implemented, it is necessary to reduce these cash flows to a common basis for comparison and to determine whether the objective of a 10% rate of return has been achieved. The present worth (or present value) of these cash flows provides such a basis. The rate of return for Plan B is calculated by iteration, using interest tables and interpolating between values.

First Iteration. Assume a rate of return of 10% and refer to Table 5:

$$PW = -15\,000 + 3100\left(\frac{P}{A}, 10\%, 10 \text{ years}\right)$$
$$= -15\,000 + 3100(6.145)$$
$$= +\$4047.95 \quad \text{(Eq 4)}$$

The rate of return for which the discounted cash flow is equal to 0 (that is, the first term on the right in Eq 4 is balanced by the second term) is the actual rate of return. From the first iteration, it is already evident that Plan B returns in excess of 10% because the net cash flow is positive. Thus, Plan B is preferred. The numerical value of the actual rate of return can be determined by additional iterations. Such an exercise will reveal that the rate of return in this case is 16.1%.

In this example, Plans A and B could represent alternative materials of construction having different corrosion rates, with the annual dollar difference being related to the consequences of corrosion on maintenance costs. This presupposes the availability of corrosion data that can be used to estimate expected life, maintenance costs, and so on. Other examples will illustrate how to account for the effects of salvage value, taxes, and depreciation.

Depreciation

Depreciation has been defined as the lessening in value of an asset with the passage of time. All physical assets (except land) depreciate with time. There are several types of depreciation. Two of the more common types are physical depreciation and functional depreciation. Accidents can also cause loss of value, but this cause is often accommodated in other ways, such as insurance and reserves, and therefore will not be considered in this article.

Physical depreciation includes such phenomena as deterioration resulting from corrosion, rotting of wood, bacterial action, chemical decomposition, and wear and tear, which can reduce the ability of an asset to render its intended service. Functional depreciation does not result from the inability of an asset to serve its intended purpose but rather from the fact that some other asset is available that can perform the desired function more economically. Thus, obsolescence, inadequacy, and/or the inability to meet the demands placed on the asset lead to functional depreciation. Technological advances produce improvements that often result in the obsolescence of existing assets.

The manner in which depreciation can be accounted for is largely a tax question. The tax laws specify the procedures that are permissible. Because tax laws change occasionally, procedures that are attractive under a given set of circumstances may become unattractive (or even forbidden) under other circumstances. Some of the more common methods of depreciation include:

- Straight line
- Declining balance
- Declining balance switching to straight line
- Sum of the year's digits
- Accelerated cost recovery system

The straight-line depreciation method assumes that the value of an asset declines at a constant rate. If the original cost P of the asset

Table 3 (continued)

Factor number	Factor name	Algebraic form	Use when:
Group III: continuous, uniform cash flows: continuous compounding (payments during one period only)			
Cash flow diagram for factors 15 and 16:			
15.	Continuous compounding present worth factor (single continuous payment)	$\frac{e^r - 1}{re^{rn}} = \frac{i(1+i)^{-n}}{\ln(1+i)}$	Given F, to find P
16.	Continuous compounding compound amount factor (single continuous payment)	$\frac{e^{rn}(e^r - 1)}{re^r} = \frac{i(1+i)^{n-1}}{\ln(1+i)}$	Given P, to find F
Group IV: continuous, uniform cash flows (payments during a continuous series of periods)			
Cash flow diagram for factors 17–20:			
17.	Continuous compounding sinking fund factor (continuous, uniform payments)	$\frac{r}{e^{rn} - 1} = \frac{\ln(1+i)}{(1+i)^n - 1}$	Given F, to find A
18.	Continuous compounding capital recovery factor (continuous, uniform payments)	$\frac{re^{rn}}{e^{rn} - 1} = \frac{(1+i)^n \ln(1+i)}{(1+i)^n - 1}$	Given P, to find A
19.	Continuous compounding compound amount factor (continuous, uniform payments)	$\frac{e^{rn} - 1}{r} = \frac{(1+i)^n - 1}{\ln(1+i)}$	Given A, to find F
20.	Continuous compounding present worth factor (continuous, uniform payments)	$\frac{e^{rn} - 1}{re^{rn}} = \frac{(1+i)^n - 1}{(1+i)^n \ln(1+i)}$	Given A, to find P

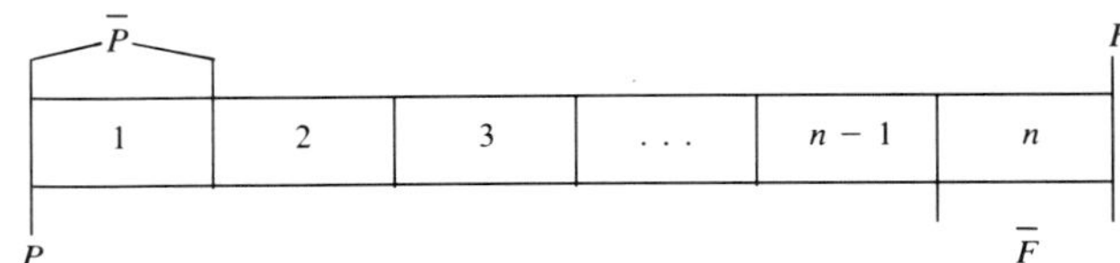

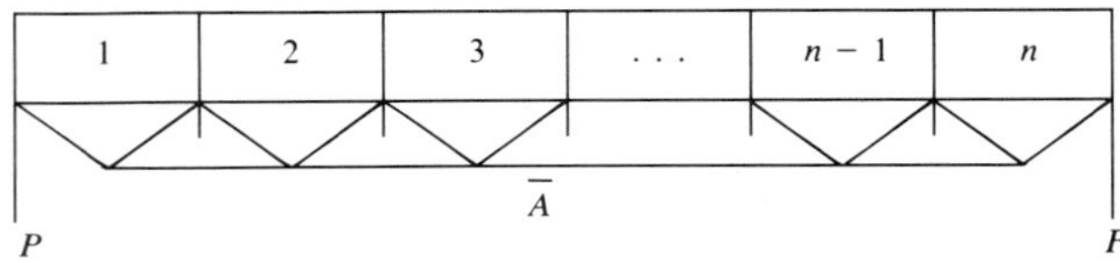

was \$6000 and the asset had a salvage value S of \$1000 after a service life of 5 years, the annual depreciation would be (\$6000 − 1000) /5 — \$1000 per year. Annual depreciation D can be expressed as:

$$D = \frac{(P - S)}{n} \quad \text{(Eq 5)}$$

The book value of the asset would decrease at the end of each year by the amount of the annual depreciation until, at the end of the fifth year, the book value would be the same as the salvage value.

The declining-balance method assumes that the asset depreciates more rapidly during the early years of its service life than in the later years. A certain percent depreciation is applied each year to the remaining book value of the asset. Under these circumstances, the size of the depreciation declines with each successive year until the asset is fully depreciated. When the declining-balance method is used, the maximum rate that has been permissible for tax purposes is double the straight-line rate. This accounts for the term double-declining balance. In addition, before 1981, businesses were allowed to depreciate an asset by using declining-balance depreciation for the early years, followed by straight-line depreciation when the allowable depreciation (using declining balance) fell below the amount permissible under straight-line depreciation.

The sum of the year's digits method assumes that the value of an asset decreases at a decreasing rate. Assume that an asset has a 5-year life. The sum of the year's digits equals 1 + 2 + 3 + 4 + 5 = 15. For the \$6000 asset mentioned above, which has a salvage value after 5 years of \$1000, depreciation for the first year would be (\$6000 − 1000) (5/15) = \$1666.67. Depreciation for the second year would be (\$6000 − 1000) (4/15) = \$1333.33, and so on, until the asset is fully depreciated.

Most of the property acquired before 1981 is still being depreciated by one of the methods mentioned above. Each of these (pre-1981) methods requires the taxpayer to estimate a useful life based on experience or on an Internal Revenue Service (IRS) Guideline. The guidelines are presented as ranges of values in the class life asset depreciation range (ADR) system. The midpoint of the range is referred to as the ADR life.

Other Methods. For property placed in service after 1981, the accelerated cost recovery system (ACRS), which is a part of the Economy Recovery Act of 1981, prescribed a different method for recovering the cost of depreciable property. The accelerated cost recovery system provides that eligible depreciable assets be assigned to one of only four classes of property based on expected useful life:

- *Three-year property* includes automobiles, light-duty trucks, research and experimental equipment, and other equipment having an ADR life of 4 years or less
- *Five-year property* includes all personal property as well as most production equipment and public utility property with an ADR life between 5 and 18 years
- *Ten-year property* includes public utility property with an ADR life between 18 and 25 years as well as real property with an ADR life less than or equal to 12½ years
- *Fifteen-year property* includes depreciable real property with an ADR life greater than 12½ years as well as public utility property with an ADR life greater than 25 years

Table 6 lists depreciation percentages for the ACRS classes. After property has been assigned to the proper ACRS class, the depreciation rate is prescribed by the IRS and is based on the declining-balance method and a depreciation rate of (150%/n). The method switches to straight-line depreciation when the declining-balance depreciation falls below the straight-line depreciation. For 15-year property, the rate for the first year is one-half that of the second year on the assumption that the property will be used for only one-half of the first year (Table 6).

The percentages listed in Table 6 are applied to the original cost of the asset with no consideration for future salvage value. If there is an actual salvage value at the time the asset is retired, the tax liability is determined after taking into account the actual salvage amount and the (then) book value. Table 7 gives an example of ACRS depreciation.

The only other option under ACRS is straight-line depreciation using the recovery periods shown in Table 8. The taxpayer may select one of the three optional recovery periods appropriate for the applicable ACRS property class. Salvage values are ignored.

Table 4 Tabulation of cash flow

Period, years	Plan A (the present method)	Plan B (the proposed method)	B − A, dollars
0.........	...	−15 000	−15 000
1.........	−8200	−5100	+3100
2.........	−8200	−5100	+3100
3.........	−8200	−5100	+3100
4.........	−8200	−5100	+3100
5.........	−8200	−5100	+3100
6.........	−8200	−5100	+3100
7.........	−8200	−5100	+3100
8.........	−8200	−5100	+3100
9.........	−8200	−5100	+3100
10.........	−8200	−5100	+3100
Totals	**−\$82 000**	**−\$66 000**	**+\$16 000**

Generalized Equations

It is possible to prepare generalized equations that will simplify the solution of a large percentage of engineering economy problems. These

Table 5 Annual compounding at 10% effective interest rate, *i*, per year

	Single payment: Compound-amount factor	Single payment: Present worth factor	Equal payment series: Compound-amount factor	Equal payment series: Sinking-fund factor	Equal payment series: Present worth factor	Equal payment series: Capital recovery factor	Uniform gradient-series factor
n	To find *F* given *P* *F*/*P*, *i*, *n*	To find *P* given *F* *P*/*F*, *i*, *n*	To find *F* given *A* *F*/*A*, *i*, *n*	To find *A* given *F* *A*/*F*, *i*, *n*	To find *P* given *A* *P*/*A*, *i*, *n*	To find *A* given *P* *A*/*P*, *i*, *n*	To find *A* given *G* *A*/*G*, *i*, *n*
1	1.100	0.9091	1.000	1.0000	0.9091	1.1000	0.0000
2	1.210	0.8265	2.100	0.4762	1.7355	0.5762	0.4762
3	1.331	0.7513	3.310	0.3021	2.4869	0.4021	0.9366
4	1.464	0.6830	4.641	0.2155	3.1699	0.3155	1.3812
5	1.611	0.6209	6.105	0.1638	3.7908	0.2638	1.8101
6	1.772	0.5645	7.716	0.1296	4.3553	0.2296	2.2236
7	1.949	0.5132	9.487	0.1054	4.8684	0.2054	2.6216
8	2.144	0.4665	11.436	0.0875	5.3349	0.1875	3.0045
9	2.358	0.4241	13.579	0.0737	5.7590	0.1737	3.3724
10	2.594	0.3856	15.937	0.0628	6.1446	0.1628	3.7255
11	2.853	0.3505	18.531	0.0540	6.4951	0.1540	4.0641
12	3.138	0.3186	21.384	0.0468	6.8137	0.1468	4.3884
13	3.452	0.2897	24.523	0.0408	7.1034	0.1408	4.6988
14	3.798	0.2633	27.975	0.0358	7.3667	0.1358	4.9955
15	4.177	0.2394	31.772	0.0315	7.6061	0.1315	5.2789
16	4.595	0.2176	35.950	0.0278	7.8237	0.1278	5.5493
17	5.054	0.1979	40.545	0.0247	8.0216	0.1247	5.8071
18	5.560	0.1799	45.599	0.0219	8.2014	0.1219	6.0526
19	6.116	0.1635	51.159	0.0196	8.3649	0.1196	6.2861
20	6.728	0.1487	57.275	0.0175	8.5136	0.1175	6.5081
21	7.400	0.1351	64.003	0.0156	8.6487	0.1156	6.7189
22	8.140	0.1229	71.403	0.0140	8.7716	0.1140	6.9189
23	8.954	0.1117	79.543	0.0126	8.8832	0.1126	7.1085
24	9.850	0.1015	88.497	0.0113	8.9848	0.1113	7.2881
25	10.835	0.0923	98.347	0.0102	9.0771	0.1102	7.4580
26	11.918	0.0839	109.182	0.0092	9.1610	0.1092	7.6187
27	13.110	0.0763	121.100	0.0083	9.2372	0.1083	7.7704
28	14.421	0.0694	134.210	0.0075	9.3066	0.1075	7.9137
29	15.863	0.0630	148.631	0.0067	9.3696	0.1067	8.0489
30	17.449	0.0573	164.494	0.0061	9.4269	0.1061	8.1762
31	19.194	0.0521	181.943	0.0055	9.4790	0.1055	8.2962
32	21.114	0.0474	201.138	0.0050	9.5264	0.1050	8.4091
33	23.225	0.0431	222.252	0.0045	9.5694	0.1045	8.5152
34	25.548	0.0392	245.477	0.0041	9.6086	0.1041	8.6149
35	28.102	0.0356	271.024	0.0037	9.6442	0.1037	8.7086
40	45.259	0.0221	442.593	0.0023	9.7791	0.1023	9.0962
45	72.890	0.0137	718.905	0.0014	9.8628	0.1014	9.3741
50	117.391	0.0085	1163.909	0.0009	9.9148	0.1009	9.5704
55	189.059	0.0053	1880.591	0.0005	9.9471	0.1005	9.7075
60	304.482	0.0033	3034.816	0.0003	9.9672	0.1003	9.8023
65	490.371	0.0020	4893.707	0.0002	9.9796	0.1002	9.8672
70	789.747	0.0013	7887.470	0.0001	9.9873	0.1001	9.9113
75	1271.895	0.0008	12708.954	0.0001	9.9921	0.1001	9.9410
80	2048.400	0.0005	20474.002	0.0001	9.9951	0.1001	9.9609
85	3298.969	0.0003	32979.690	0.0000	9.9970	0.1000	9.9742
90	5313.023	0.0002	53120.226	0.0000	9.9981	0.1000	9.9831
95	8556.676	0.0001	85556.760	0.0000	9.9988	0.1000	9.9889
100	13780.612	0.0001	137796.123	0.0000	9.9993	0.1000	9.9928

Table 6 Depreciation percentages for ACRS classes

Recovery year is	Classes of property: 3-year	5-year	10-year	15-year public utility
1	25	15	8	5
2	38	22	14	10
3	37	21	12	9
4	...	21	10	8
5	...	21	10	7
6	...	...	10	7
7	...	...	9	6
8	...	...	9	6
9	...	...	9	6
10	...	...	9	6
11	...	...	...	6
12	...	...	...	6
13	...	...	...	6
14	...	...	...	6
15	...	...	...	6

Table 7 An example of ACRS depreciation

End of year	Depreciation charges during year	Book value at end of year, dollars
0	...	6000
1	(0.15)($6000) = $ 900	5100
2	(0.22)($6000) = 1320	3780
3	(0.21)($6000) = 1260	2520
4	(0.21)($6000) = 1260	1260
5	(0.21)($6000) = 1260	0

Table 8 Straight-line recovery periods for ACRS

Class of property	Optional recovery periods, years
3-year	3, 5, or 12
5-year	5, 12, or 25
10-year	10, 25, or 35
15-year	15, 35, or 45

equations take into account the influence of taxes, depreciation, operating expenses, and salvage value in the calculation of present worth and annual cost. Using these equations, a problem can be solved merely by entering data into the equations with the assistance of the compound interest tables and solving for the unknown value. The tax rate t is expressed as a decimal. Operating expense is represented by X. The letter S represents salvage value. The other symbols are explained in Tables 1 to 3.

For straight-line depreciation:

$$\text{PW} = -P + \left\{\frac{t(P - S)}{n}\right\}\left(\frac{P}{A}, i\%, n\right) - (1 - t)(X)\left(\frac{P}{A}, i\%, n\right) + S\left(\frac{P}{F}, i\%, n\right) \quad \text{(Eq 6)}$$

where P, the first term, is the cost at time zero of the initial investment. The value at time zero is a definition of present worth (at time zero). Therefore, it is not necessary to translate this amount to another date, because the evaluation will be made as of time zero. The investment involves flow of money out of the bank; therefore, the sign of this term is minus.

The second term in Eq 6 concerns depreciation. The depreciation method will influence the mathematical formulation of this term. The portion enclosed in braces expresses the annual amount of tax credit permitted by this method of depreciation—in this case, straight-line depreciation. The portion in parentheses translates these equal annual amounts back to time zero by converting them to present worth. The ACRS depreciation methods allow specific straight-line depreciation options. When one of these options is chosen, the formulation of this term remains the same as shown above, with the exception that salvage is ignored. If the straight-line option is not chosen, this term in the generalized formula becomes a series of terms, each having the tax rate multiplied by the appropriate depreciation percentage of the initial cost (in accordance with Table 6) multiplied by the appropriate single-payment present worth factor, P/F, for the applicable year and interest rate. The sum of these terms will represent the effect of depreciation on present worth.

The third term in Eq 6 actually consists of two terms. One is $[(X)\,(P/A, i\%, n)]$, which represents the cost of items properly chargeable as expense, such as the cost of maintenance or insurance and the cost of inhibitors. Because this term involves expenditure of money from the bank, it carries a minus sign. The term $[t(X)\,(P/A, i\%, n)]$ accounts for the tax credit for this business expense. Because it represents a savings, the sign is plus.

The fourth term translates the anticipated (future) value of salvage to present value. This is a one-time event rather than a uniform series and therefore involves the single-payment present worth factor. As noted above, the ACRS depreciation method ignores the salvage value of an asset; therefore, when using the ACRS method, the fourth term is 0.

Present worth can be converted to equivalent annual cost A by:

$$A = (\text{PW})\left(\frac{A}{P}, i\%, n\right) \quad \text{(Eq 7)}$$

Alternative materials with the same life can be compared from an engineering economy standpoint merely by comparing the magnitude of their present values. However, if the candidate materials have different life expectancies, it is necessary to convert present worth to equivalent annual cost A to compare the materials from an engineering economy standpoint. The material with the lowest annual cost is the material of choice.

Examples and Applications

Example 1. A new heat exchanger is required in conjunction with a rearrangement of existing facilities. Because of corrosion, the expected life of a carbon steel exchanger is 5 years. The installed cost is $9500. An alternative to the carbon steel heat exchanger is a unit fabricated of AISI type 316 stainless steel with an installed cost of $26 500 and an estimated life of 15 years to be written off in 11 years. The minimum acceptable rate of return is 10%, the tax rate is 48%, and the depreciation method is straight line. It is necessary to determine which unit would be more economical based on annual costs.

Because the lives of the two heat exchangers are unequal, the economic choice cannot be based merely on the discounted cash flow over a single life of each alternative. Instead, comparison must be made on the basis of equivalent uniform annual costs, commonly referred to as annual cost, as mentioned above. In this example, data are given for only the first two terms of Eq 6. The third term, which involves maintenance expense, and the fourth term involving salvage value are both assumed to be 0.

The first step is to compute the discounted cash flow over one life span for each alternative. This involves calculation of the present worth for each material:

$$\text{PW}_{\text{steel}} = -\$9500 + \left\{0.48\,\frac{(9500 - 0)}{5}\right\}(3.791)$$
$$= -\$6043 \quad \text{(Eq 8)}$$

$$\text{PW}_{316} = -\$26\,500 + \left\{0.48\frac{(26\,500 - 0)}{11}\right\}(6.495)$$
$$= -\$18\,989 \quad \text{(Eq 9)}$$

To compare the two alternatives, both discounted cash flows (the present worths) must be converted to annual costs:

$$A_{\text{steel}} = -\$6043(0.2638)$$
$$= -\$1594 \quad \text{(Eq 10)}$$

$$A_{316} = -\$18\,989(0.15396)$$
$$= -\$2924 \quad \text{(Eq 11)}$$

Therefore, the carbon steel heat exchanger, which has the lower annual cost, is the more economical alternative under these conditions.

Example 2. In this case, the carbon steel heat exchanger in Example 1 will require $3000 in yearly maintenance (painting, use of inhibitors, cathodic protection, and so on). It will be determined whether carbon steel would remain the preferred alternative under these conditions.

Maintenance costs are treated as expense items with a 1-year life (end-of-year costs). This example involves the third term in Eq 6. In Example 1, the annual cost without the yearly maintenance cost was calculated; therefore, in this example, the annual cost including maintenance ($3000 per year) can be obtained simply by subtracting the after-tax maintenance costs from the result given in Example 1:

$$A_{\text{steel}} = -\$1594 - (1 - 0.48)(\$3000)$$
$$= -\$3154 \quad \text{(Eq 12)}$$

Because this is an after-tax cash flow, the quantity [(1 − 0.48)($3000)] is the after-tax maintenance cost; 0.48 is the tax rate expressed as a decimal. The $3000 is the present value of the maintenance for the first year; therefore, no further discounting is necessary in this case. Comparing this result with the annual cost of AISI type 316 stainless steel in Example 1, it is evident that if $3000 is required each year to keep the carbon steel unit operative, the stainless steel unit would be more economical.

Example 3. Given the conditions described in Example 1, this example will determine the service life at which the carbon steel heat exchanger is economically equivalent to the type 316 stainless steel unit if it is not certain that a 5-year life will be attained for unprotected carbon steel.

The carbon steel heat exchanger will be economically equivalent to the type 316 stainless steel heat exchanger when their annual costs are equal. Therefore, the problem can be solved by determining how many years of life for steel will be equivalent to an annual cost of $2924. Trial and error methods are useful for such problems. Trying n = 3 years gives:

$$\text{PW}_{\text{steel}} = -\$9500 + 0.48\left\{\frac{(9500 - 0)}{3}\right\}(2.487)$$
$$= -\$5719 \quad \text{(Eq 13)}$$

$$A_{\text{steel}} = -\$5719(0.40211)$$
$$= -\$2300 \quad \text{(Eq 14)}$$

Trying n — 2 years yields:

$$\text{PW}_{\text{steel}} = -\$9500 + 0.48\left\{\frac{(9500 - 0)}{2}\right\}(1.736)$$
$$= -\$5542 \quad \text{(Eq 15)}$$

$$A_{\text{steel}} = -\$5542(0.57619)$$
$$= -\$3193 \quad \text{(Eq 16)}$$

Thus, a carbon steel heat exchanger must last more than 2 years, but will be economically favored in less than 3 years under the conditions given.

Example 4. Under the conditions described in Example 3, it will be determined how much product loss X could be tolerated after 2 of the 5 years of anticipated life, for example, from roll leaks or a few tube failures, before the selection of type 316 stainless steel could have been justified. Equating:

$$A_{316} = A_{\text{steel}} + A_{\text{product loss}}$$
$$-\$2924 = -\$1594 + [(1 - 0.48)(X)(0.8264)][0.2638] \quad \text{(Eq 17)}$$

Solving for X:

$$-\$1330 = 0.1134(X)$$
$$X = -\$11\,728 \quad \text{(Eq 18)}$$

The term (1 − 0.48) will be recognized from the third term of Eq 6 as $(1 - t)$ where t is the tax rate. The quantity 0.8264 is the single-payment present worth factor, (P/F, 10%, 2 years) and translates the product loss X to its present worth at time zero. The quantity [0.2638] is the uniform series capital recovery factor, (A/P, 10%, 5 years) and translates the present worth of the product loss to annual cost so that it can be added to the annual cost of steel for comparison with the annual cost of AISI type 316 stainless steel. If production losses exceed $11 728 in year two (no losses in other years), the AISI type 316 heat exchanger could be justified.

REFERENCES

1. "Recommended Practice for Direct Calculation of Economic Appraisals of Corrosion Control Measures," RP-02-72, National Association of Corrosion Engineers
2. "Engineering Economy," Z94.5, American National Standards Institute/Institute of Industrial Engineers

SELECTED REFERENCES

- *AT&T Engineering Economy*, 3rd ed., McGraw-Hill, 1977
- E.L. Grant, W.G. Ireson, and R.S. Leavenworth, *Principles of Engineering Economy*, 7th ed., John Wiley & Sons, 1982
- F.C. Jelen, *Cost and Optimization Engineering*, McGraw-Hill, 1983
- "Recommended Practice for Direct Calculation of Economic Appraisals of Corrosion Control Measures," RP-02-72, National Association of Corrosion Engineers
- G.J. Thuesen and W.J. Fabrycky, *Engineering Economy*, 6th ed., Prentice-Hall, 1984
- E.D. Verink, Economic Appraisals of Corrosion Control Measures, *J. Educ. Mod. Mater. Sci. Eng.*, Vol 3 (No. 2), 1981, p 239

Corrosion Protection Methods

Section Chairman:
Herbert E. Townsend, Bethlehem Steel Corporation
Section Co-chairmen:
Thomas W. Cape, Chemfil Corporation
Kenneth B. Tator, KTA-Tator, Inc.

Fundamentals of Corrosion Protection in Aqueous Solutions

Henry Leidheiser, Jr., Lehigh University

THE IMPORTANT METALS in structures—aluminum, copper, iron, zinc, and so on—are inherently unstable. Their ionic solutions and their carbonates, hydroxides, oxides, sulfides, sulfates, and many other salts are more stable than the free metal under the many conditions to which these metals are exposed. The free energy change for the conversion of these metals to a compound is a large negative value, and the thermodynamic driving force to convert the elemental metal to an oxide or salt is great. Thus, to retain a required physical property of the metal, such as strength, thermal conductivity, magnetic nature, or electrical conductivity, it is essential to protect the metal from the environments to which it is expected to be exposed. This article will describe the various methods that can be used to protect a metallic system against corrosion. These methods can be divided into the following: thermodynamic protection, kinetic protection, barrier protection, structural design, environmental control, and metallurgical design.

The objective of most protective systems is to reduce the rate of corrosion to a value that is tolerable or that will allow the material to attain its normal or desired lifetime. Only in a limited number of cases must corrosion protection be designed to eliminate corrosion completely.

Protection on a Thermodynamic Basis

This type of protection is based on the requirement that the metal have a high positive value for the free energy change for conversion of the metal to the corrosion products that can form in the environment to which the metal is exposed. For example, the free energy change for the conversion of gold to its oxide is 163 000 J/mol (39 000 cal/mol), and for the conversion of silver to its ion at standard state in an aqueous phase, the free energy change is 77 000 J/mol (18 430 cal/mol). These two metals could *a priori* be expected to resist corrosion when exposed to oxygenated pure water, because the tendency in these systems is for the metal to be more stable than its aquated ion or its oxide. As a general rule, a metal M whose thermodynamic potential in contact with its ions is a high positive value for the reaction $M - ne^- \rightarrow M^{n+}$ is stable in aqueous solutions at room temperature. No blanket statement can be made without full knowledge of the composition of the aqueous phase because some ions and other complexing agents may form complexes that are more thermodynamically stable than the free metal.

Another form of thermodynamic protection is obtained when an external potential is applied such that the metal is stable with respect to the metal ion concentration in the aqueous phase with which it is in contact. For example, if nickel immersed in an aqueous solution of pH 4 is polarized externally to a potential of −0.5 V, elemental nickel is stable with respect to its ions, and nickel will be protected from corrosion. However, this potential is more negative than the potential for the generation of hydrogen, and hydrogen formation may occur. These facts are illustrated in the Pourbaix (potential versus pH) diagram for nickel shown in Fig. 1. The dashed line labeled A represents the phase boundary for the equilibrium, $H_2 \rightarrow 2H^+ + 2e^-$, with hydrogen at unit fugacity being stable below this line and the hydrogen ion (H^+) at unit activity stable above this line. More detailed information about Fig. 1 can be found in Ref 1. This type of diagram is very useful for appraising the relative thermodynamic stabilities of the metal with respect to its ions and its hydrous and anhydrous oxides. Stability diagrams for metals in contact with water are given in Ref 1.

Protection on a Kinetic Basis

The corrosion rate of an actively corroding metal is determined by the intersection of the kinetic curves that characterize the anodic and cathodic halves of the corrosion reaction. If the rates of either of these reactions can be changed such that the intersection point is at a lower current density, the corrosion rate is reduced.

A widely used method for taking advantage of the corrosion kinetics is known as cathodic protection. In some cases, the applied potential is not sufficient to polarize the metal into its thermodynamic stability region, such as in the case of nickel described previously. In the kinetics case, a small potential is applied between an inert electrode and the metal to be protected. The potential of the metal is made more negative, and the rate of the anodic reaction is reduced. Figure 2 shows a diagram illustrating the phenomenon of cathodic protection. If a metal exhibiting the anodic and cathodic polarization curves shown in Fig. 2 is polarized to the potential denoted by A—B, the cathodic current density is that shown as A. The extrapolated anodic polarization curve allows calculation of the corrosion rate B. This type of diagram enables calculation of the corrosion rate under any specific applied cathodic potential. More detailed descriptions of the principles and practice of cathodic protection are given in the article "Cathodic Protection" in this Volume.

As stated above, the rate of the overall corrosion reaction is limited by the rate at which the slower of the anodic or cathodic reactions occurs. One of the two reactions is rate controlling. A knowledge of the rate-controlling reaction can often be applied to reducing the corrosion rate. An example is the use of inhibitors that selectively reduce the cathodic hydrogen evolution reaction in the acid pickling of steel. The pickling solution effectively attacks the oxide, with little attack on the base steel, because the anodic corrosion reaction is limited by the low rate of the cathodic hydrogen evolution reaction.

A second example is the corrosion of zinc in neutral chloride solution. As shown in Fig. 3, the anodic reaction occurs at the metal/oxide interface, and the cathodic oxygen reduction reaction occurs at the oxide/solution interface. The electrons formed by the anodic reaction must pass through the oxide to be available for the cathodic reaction on the oxide surface. A reduction in the electron conductivity of the oxide causes a reduction in the rate of electron transfer and a consequent reduction in the corrosion rate. The conductivity of the oxide can be reduced by doping the oxide with cobalt by using a zinc-cobalt alloy or by introducing cobalt into the oxide on the surface by immersion in a solution of cobalt ions (Ref 2).

Barrier Protection

The concept of barrier protection is very simple: The metal is protected from the environment by coating it with a barrier that resists penetration by aggressive environmental constituents. Five types of barrier coatings will be discussed: anodic oxides, ceramic and inorganic coatings, inhibitors, organic coatings, and phosphate and other conversion coatings.

Anodic oxides are widely used to protect aluminum from corrosion. The metal and many of its alloys are anodized in such acids as boric ($B_2O_3 \cdot 3H_2O$), oxalic, phosphoric (H_3PO_4), and sulfuric (H_2SO_4) under conditions in which an oxide is formed on the surface. The oxides formed in $B_2O_3 \cdot 3H_2O$ are relatively thin and nonporous in nature, but they are not suitable for protecting aluminum against corrosion in aggressive environments. The oxides formed in H_3PO_4 or H_2SO_4 are many microns thick and are very porous. The porosity is reduced in a second step in which the oxide is sealed in steam, boiling water, or

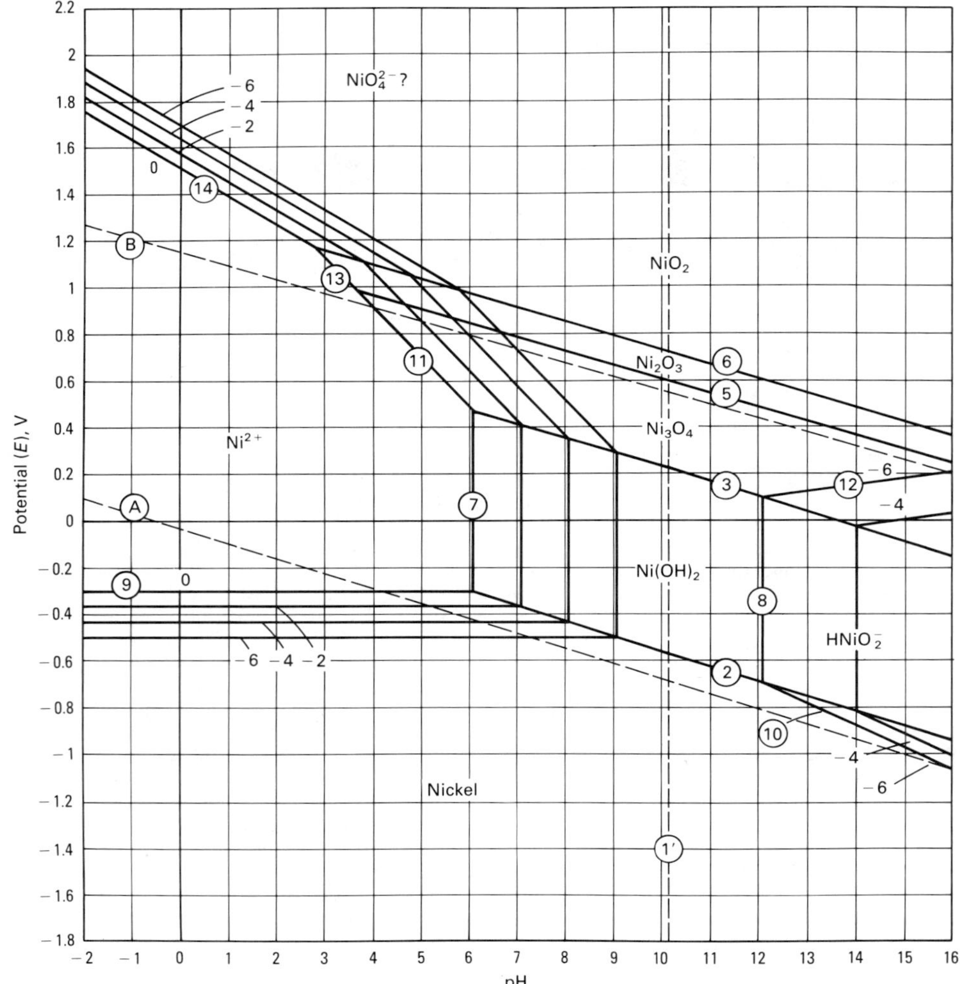

Fig. 1 The Pourbaix diagram for the nickel-water system at 25 °C (75 °F). See Ref 1 for details. Source: Ref 1

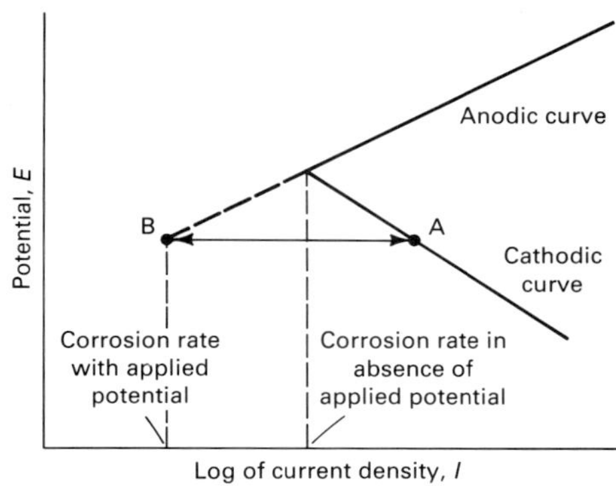

Fig. 2 Diagram showing how the corrosion rate can be calculated when a cathodic potential is applied. The situation shown is schematic in nature, and other factors must be considered when cathodic protection is applied in practice.

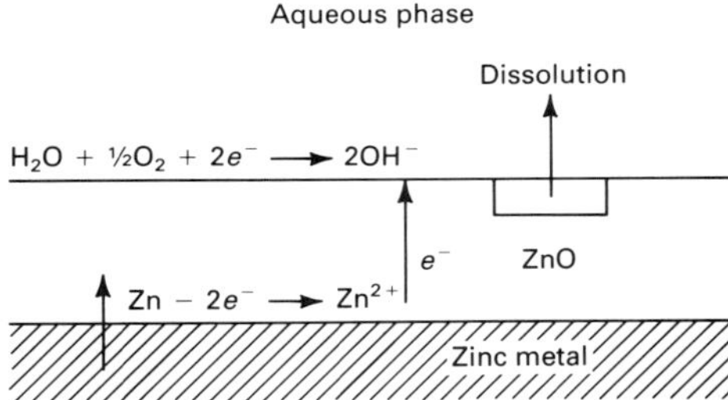

Fig. 3 The mechanism by which zinc corrodes in a neutral chloride solution. Source: Ref 2

aqueous solutions containing nickel acetate ($Ni(C_2H_3O_2)_2$) or other salts.

The anodizing of aluminum is widely utilized to protect aluminum components on automobiles, buildings, storm doors, and windows (see the article "Aluminum Anodizing" in this Volume). Other metals are protected to a limited extent by anodization before the component is placed in service. Stainless steels and other iron alloys that form passive oxides in aggressive environments are also protected from corrosion by applying an anodic potential while the part is in service. This potential is sufficient to maintain a barrier oxide on the surface. This method, known as anodic protection, is discussed in detail in the article "Anodic Protection" in this Volume.

Ceramic coatings based on special cements are used to protect steel from corrosion. They serve as a barrier and also maintain an alkaline environment at the steel/ceramic interface in which the corrosion rate is low. Inorganic coatings, such as silica, are used to protect stainless steel from tarnishing, and silicon nitride coatings are used to protect some types of components in electronic devices. Many other ceramic and inorganic coatings are used industrially to protect vital components against corrosion. More information on the use of ceramic coatings for corrosion protection is available in the articles "Porcelain Enamels" and "Chemical-Setting Ceramic Linings" in this Volume.

Corrosion Inhibitors. In many cases, the role of inhibitors is to form a surface coating one or several molecular layers thick that serves as a barrier. Many effective organic inhibitors have a reactive group attached to a hydrocarbon. The reactive group interacts with the metal surface, and the hydrocarbon portion of the molecule is in contact with the environment. Cinnamates (salts or esters of cinnamic acid) are used to protect steel surfaces against corrosion in neutral or slightly alkaline media by this mechanism.

The behavior of the corrosion potential can be used to determine which of the half reactions is most affected by the inhibitor. Figure 4 illustrates three cases. In Fig. 4(a), the corrosion inhibitor affects the anodic and cathodic curves equally. The corrosion rate is reduced to the value I_{inhib}, but the corrosion potential is unaffected. In Fig. 4(b), the cathodic reaction is affected to a greater extent, and the corrosion potential moves to a more negative value. In Fig. 4(c), the anodic reaction is affected to a greater extent, and the corrosion potential moves to a more positive value. The change in corrosion potential in the presence of an inhibitor can thus be used to estimate the kinetics of the inhibition. Generally, additional information is needed to draw firm conclusions. More information on the use of inhibitors is available in the articles "Corrosion Inhibitors for Oil and Gas Production," "Corrosion Inhibitors for Crude Oil Refineries," and "Control of Environmental Variables in Water Recirculating Systems" in this Volume.

Organic coatings are the most widely used barrier coating for protecting aluminum, steel, and zinc against atmospheric corrosion. The main function of the coating is to serve as a barrier to water, oxygen, and ions and thus prevent the cathodic reaction $H_2O + {}^1\!/_2O_2 + 2e^- \rightarrow 2OH^-$ from occurring beneath the coating. The protective barrier properties may be lost by the collection of water at the interface in the form of a thin layer or as blisters. Once water is present as an aqueous phase at the interface, the electrochemical corrosion reaction is possible by charge transfer laterally between neighboring anodic and cathodic areas or by charge transfer through the coating. Therefore, it is important to have coatings that are good barriers for water. In practice, a second line of defense is used to provide protection if the barrier properties of the organic coating are inadequate. A corrosion inhibitor, such as lead chromate ($PbCrO_4$), is incorporated into the primer. As a water phase develops, some of the chromate is solubilized and is available for inactivating the portions of the surface that are in contact with the aqueous phase. For additional information, see the article "Organic Coatings and Linings" in this Volume.

Conversion Coatings. Organic coatings are effective only if they protect the entire metal

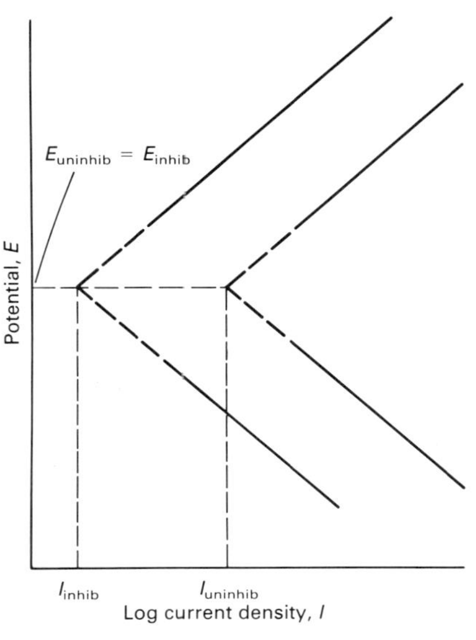

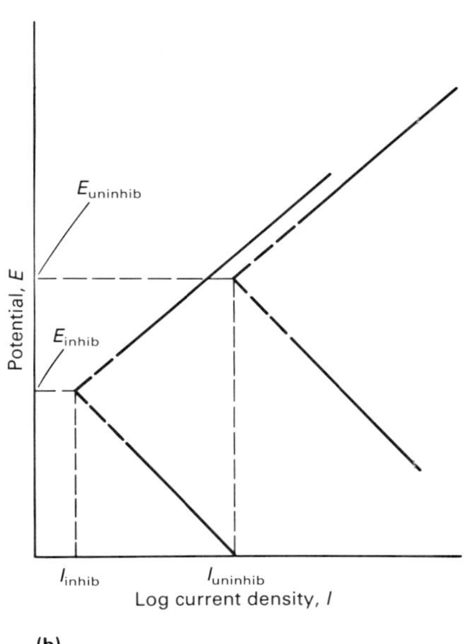

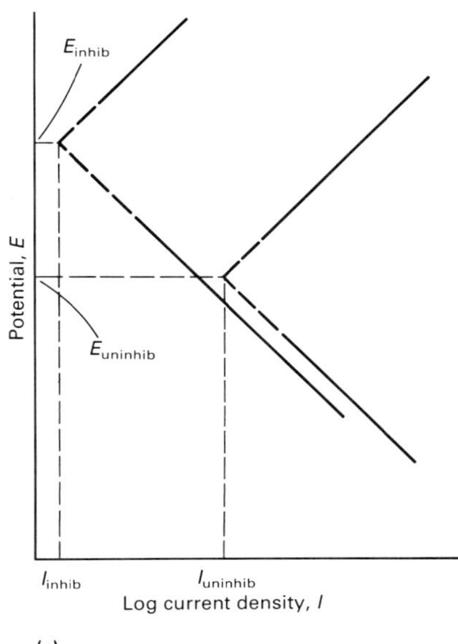

Fig. 4 Examples of three extreme cases in which corrosion inhibitors affect corrosion kinetics. (a) The inhibitor affects the anodic and cathodic curves to an equal extent. The corrosion rate is reduced, but the corrosion potential remains the same. (b) The corrosion inhibitor affects the cathodic curve extensively, and the potential in the presence of the inhibitor becomes more negative. (c) The corrosion inhibitor affects the anodic curve extensively, and the potential in the presence of the inhibitor becomes more negative.

surface. Because many uses for organic coatings are on automobiles and appliances that may be damaged during service, the damaged area must not lead to total loss of the protective property near the damaged region. Certain conversion coatings provide resistance to lateral loss of corrosion protection. Phosphate coatings, chromate coatings, and mixed-oxide coatings are applied before the organic coating. These coatings, intermediate between the metal and the organic coating, are known as conversion coatings because they convert some of the base metal to a coating in which ions of the base metal are a component. Additional information on conversion coatings is available in the articles "Phosphate Conversion Coatings" and "Chromate Conversion Coatings" in this Volume.

Protection by Structural Design

Inappropriate design is often the source of severe corrosion. Crevices provide locations where an occluded corrosion cell may develop. Liquid is retained in a crevice long after the atmosphere has a low moisture level; the chemical nature of the ions formed by the corrosion process is acidic, and the corrosion rate is accelerated. The corrosion product occupies a larger volume than the metal from which it is formed, and the expansion forces are often sufficient to break a clamping bolt or cause serious deformation.

Designs that have recesses where rainwater can collect result in longer periods of time during which the metal can support an electrochemical reaction and consequently a higher corrosion rate. Erosive conditions, when combined with a corrosive medium, can lead to rapid corrosion rates. Drain water containing a sediment should not be allowed to drip or run over metal.

Structures should be designed, installed, and used such that the length of time they are wet with water is minimized. The corrosion reaction will only occur when liquid water is present on the metal. Liquid water can occur at relative humidities below 100% when deliquescent salts are present on the surface; therefore, it is useful to remove contaminants at regular intervals. Erosive agents dispersed in flowing liquids should also be avoided. An excellent summary of designing to minimize corrosion is given in Ref 3, and more information on this subject is available in the article "Design Details to Minimize Corrosion" in this Volume.

Protection by Environmental Control

This method of protection is limited to closed systems in which changes in the composition of the corrosion medium can be tolerated. The method depends on removing a constituent of the corrosion reaction from the medium or, in some cases, adding a component to the medium. Constituents that can be removed from the corrosion medium include oxygen; living materials, such as bacteria; erosive components in flowing systems; solid matter that may provide crevicelike situations at the bottom of pipes or tanks; and specific ions, such as chloride (Cl^-), that accelerate corrosion processes. Additives that may reduce corrosion in closed systems include agents that control pH, oxidizing agents that destroy organic matter, surface-active agents that maintain sediment in suspension, inhibitors, agents that reduce dissolved oxygen concentration, and materials that remove scale from the surface. Environmental control is a complicated subject, and specialists in cooling water treatment, boiler water treatment, and waste treatment should be consulted.

Protection by Metallurgical Design

The most obvious aspect of metallurgical design is the use of the proper alloy for the environment against which protection is sought. Economic and lifetime considerations are important factors in the choice of the alloy, and each situation must be considered individually (see the article "Materials Selection" in this Volume). Metallurgical considerations are also necessary once the alloy has been selected. Variables that influence the corrosion rate or sensitivity to fatigue or corrosion cracking include grain size, preferred orientation, presence of inclusions, method of fabrication, surface preparation method, and heat treatment.

REFERENCES

1. M. Pourbaix, *Atlas of Electrochemical Equilibria in Aqueous Solutions*, Pergamon Press, 1966
2. H. Leidheiser, Jr. and I. Suzuki, *J. Electrochem. Soc.*, Vol 128, 1981, p 242-249
3. V.R. Pludek, *Design and Corrosion Control*, John Wiley & Sons, 1977

Cleaning for Surface Conversion

William D. Krippes, J.M.E. Chemicals

CLEANING in preparation for applying a conversion coating is of the utmost importance. It enhances the coating weight and quality of the conversion coating. Cleaning the metal surface before contact with the conversion coating bath will provide the maximum corrosion protection and paint adhesion. Improper cleaning will inhibit the reaction of the conversion coating and the metal surface. Valuable reaction time is lost if the conversion coating must first penetrate contaminant soil. More information on the application and corrosion resistance of conversion coatings is available in the articles "Phosphate Conversion Coatings" and "Chromate Conversion Coatings" in this Volume.

Cleaners consist of different components that work together to produce the desired result. Time, temperature, equipment, and concentrations are variables that must be considered and adjusted in order to maximize the cleaning and to minimize the attack of the cleaner on the metal surface. To prepare the surface for a conversion coating, the cleaner should:

- Remove all contaminants from the metal surface and prevent their redeposition
- Prohibit or minimize the formation of cleaner-produced metal compounds on the metal surface that are detrimental to the conversion coating
- Rinse free with clean, hot water

Contaminants and Some Methods of Removal

Any substance that inhibits or interferes with the application of the conversion coating, the adhesion and reactivity of the conversion coating, and/or the subsequent adhesion of the organic coating is a contaminant that must be removed. Some contaminants are easier to remove than others. Most nonadhering contaminants can be flushed off with the cleaner, but contaminants that are part of the metal surface must first be put into solution by the cleaner. Some contaminants are too difficult for the cleaner alone to remove; therefore, they are mechanically removed with abrasive brushes (see the article "Power Brush Cleaning and Finishing" in Volume 5 of the 9th Edition of *Metals Handbook*). In some cases, an acid pickle may be required to remove such surface contaminants as aluminum and carbon smut, metal fines, surface rust, and scale.

A nonuniform reaction of the conversion coating with the metal surface will cause adhesion failure of the conversion coating and the subsequent organic coatings. Different contaminants prevent reaction of the conversion coating to the metal surface in different ways. Residual organic soil prevents contact of the aqueous conversion coating with the metal surface. Inorganic soil undermines the integrity of the conversion coating. Metal oxides that remain on the surface will be sites of future galvanic action under the subsequent organic coating. This can result in premature failure of the finished product because of underfilm corrosion.

Oils. Several different types of oils may be deposited on the surface of the metal. Stamping oils, rolling oils, machining oils, corrosion protection oils, and combinations of these are some of the types that may be encountered. The quantity of oil and the ease of removal depend on the length of time and the conditions of storage, the original use of the oil, and the type of metal to be protected. Lubricating oils are easier to remove than rust-preventive oils. Lubricating oils usually have emulsifiers to facilitate cleaning, but rust-preventive oils have additives that enhance their water-repellent qualities.

Because oil is insoluble in water, it interferes with the proper deposition of aqueous conversion coatings. To remove oils from the metal surface, several different surfactants (surface-active agents), such as detergents or wetting agents, can be added to the cleaner. The surfactants reduce the surface tension of the water and permit the cleaning solution to contact the metal surface. The concentration of surfactants in a cleaner has significant effect on the time required to penetrate the soil. Inadequate quantities will slow the cleaning process.

Cleaning time is decreased as the concentration of surfactants is increased, until a point is reached at which the addition of more surfactant does not reduce the cleaning time. This point may vary with changes in temperature, soil conditions, and type of metal being cleaned. The optimum surfactant concentration is determined by the percentage that will penetrate the soil in a required time at a minimum cost (Fig. 1).

Detergents, in addition to being surfactants, may emulsify the oils. Most of the strong synthetic detergents can be categorized as anionic, cationic, or nonionic. Anionic detergents are surfactants in which the active portion is a negative ion (anion). This type of detergent usually has the best detergency, along with high-foam characteristics. Cationic detergents are surfactants in which the positive ion (cation) is the active portion. These are usually good emulsifiers. Nonionic detergents do not ionize. They are valued for their low-foaming characteristics and are the most commonly used type of detergent.

Carbon is often found on products that are rolled from hot ingots or are continuously cast. One source of carbon is when rolling oil is subjected to the excessive temperatures encountered during the reduction process. Carbon residue is very difficult to remove because of its nonreactive nature. Carbon that is left on the metal surface can cause voids or bridges within the conversion coating that appear later as poor adhesion. One method for testing the amount of carbon on the metal surface is the Ford LECO Swab Test (Ford specification ESB-M2PX).

If the amount of carbon on the surface is not too high, it can be removed by mechanical means, such as brushing. The type of brushes used will vary with the type of metal to be cleaned. Generally, if the surface carbon is determined to exceed 6.5 mg/m^2 (0.6 mg/ft^2) as measured by the Ford test, additional acid pickling must be done to clean the surface adequately. For a chemical cleaner to remove carbon, additives that have a greater attraction for carbon than the attractive force between carbon and the metal surface must be incorporated into the cleaner.

Metal Fines and Scale. Certain rolling practices often leave metal dust or small pieces of metal adhering to the metal surface. These are called metal fines. Scale is the oxidized surface of steel, produced during heating for working the steel, that has been rolled into the surface of the metal. These contaminants shield the metal surface from the conversion coating, but when under stress, they may later separate from the metal surface, leaving it unprotected. A cleaner can lower the surface tension to wash off the metal fines, or it can dissolve them. Mechanical means are also used. The method of removal is determined by the quality of metal fines as well as the type of metal being cleaned.

Scale, if it cannot be dissolved, must also be mechanically removed. A preliminary cleaning in an acid-pickling solution is often necessary. More information on pickling ferrous metals is avail-

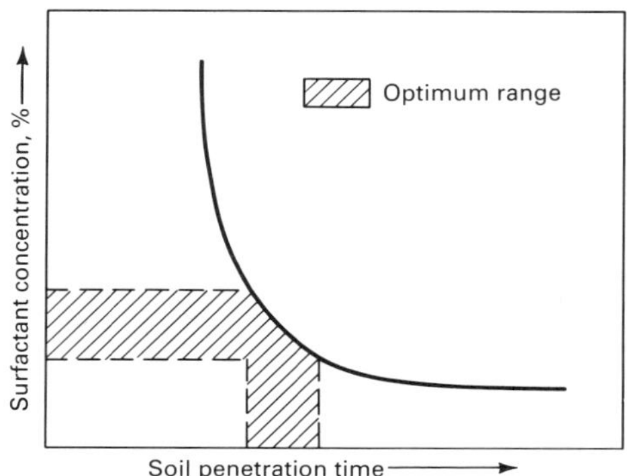

Fig. 1 Relationship between concentration of surfactants in a cleaner and cleaning time

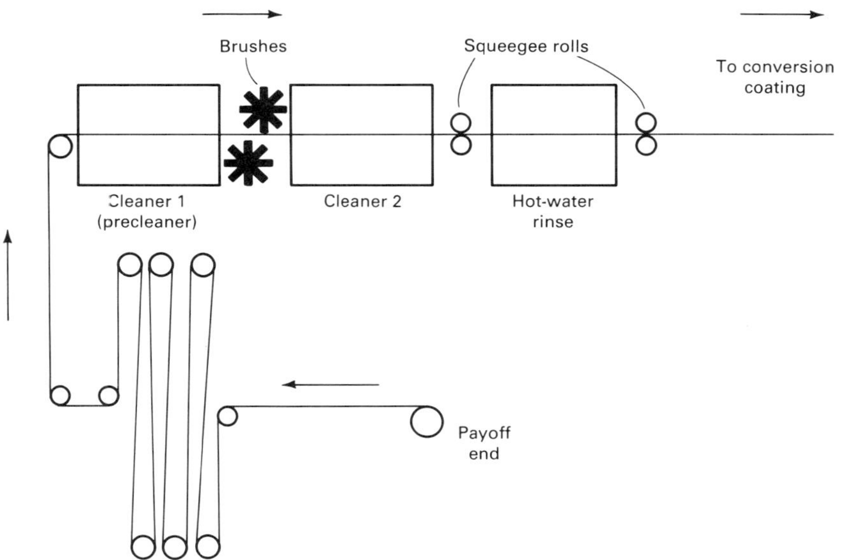

Fig. 2 Schematic of the cleaning section of a coil-coating line

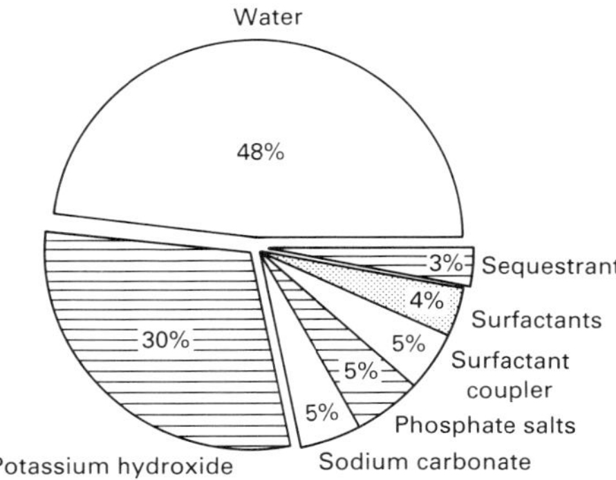

Fig. 3 Composition of a typical liquid alkaline cleaner

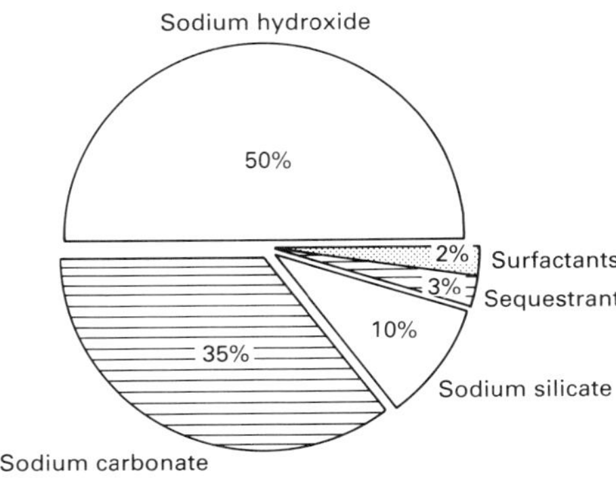

Fig. 4 Composition of a typical powder alkaline cleaner

able in the article "Pickling of Iron and Steel" in Volume 5 of the 9th Edition of *Metals Handbook*.

Passivation coatings are applied to electroplated zinc, hot-dip zinc, and zinc-aluminum alloy coatings applied on steel substrates. Passivation treatments are applied at the time of production to prevent white rust during transportation and storage. Examples of some types of passivation coatings are straight chromic acid (H_2CrO_4), sodium dichromate-sodium silicate ($Na_2Cr_2O_7$-Na_2SiO_3), and blends of various additives in conjunction with $Na_2Cr_2O_7$ and/or H_2CrO_4). Some types of conversion coatings can be properly applied over some types of passivation coatings, but others cannot. Passivation coatings containing resins or silicate will not permit any conversion coating over the passivation coating. If the conversion coating cannot be applied to the particular passivation coating, the passivation coating must be removed.

Removal of the passivation coating from the metal surface can be extremely difficult, especially if the passivation coating has aged and completely reacted. Cleaner solutions have difficulty in removing reacted passivation coatings. Mechanical means must be relied on; therefore, uniformity of cleaning is very difficult to achieve with passivated substrates.

Metal oxides are the products of corrosion or high-temperature oxidation of the metal surface. Metal oxides are formed through the natural reaction of the metal surface with oxygen, usually in the presence of moisture. They form a loose, spongy matrix that is not conducive to good adhesion of organic coatings on the metal surface. Metal oxides will not react with the conversion coating to form a corrosion-inhibiting barrier. Left on the surface within a conversion coating, they will accelerate further corrosion of the remaining substrate. For these reasons, metal oxides must be removed before the conversion coating is applied.

Selection of the cleaner is based on the metal to be cleaned and the type of equipment available. Metal substrates such as zinc, aluminum, steel, or combinations of these metals vary in solubility in the type of cleaner used. For example, aluminum oxide is usually only slightly soluble in acid cleaners, but it is very soluble in alkaline cleaners. Zinc oxide is soluble in both acid and alkaline cleaners. Iron oxide is soluble in acid cleaners.

The primary constituents of an alkaline cleaner may be sodium hydroxide (NaOH) or potassium hydroxide (KOH). An acid cleaner can be based on an organic phosphate or an acid, such as phosphoric acid (H_3PO_4) or acetic acid. These constituents are usually referred to as the backbone of the cleaner. Care should be taken to ensure that the backbone does not react with nonoxidized metal on the surface. When this is permitted to occur, metal compounds are deposited on the surface that interfere with the formation of the conversion coating and its adhesion. Aluminum hydroxide ($Al(OH)_3$) (aluminum smut), zinc hydroxide ($Zn(OH)_2$), and zinc phosphate ($Zn_3(PO_4)_2$) are some examples of these metal compounds. The formation of these metal compounds is prevented by the use of buffers that inhibit the reaction of the cleaner with the metal surface. If this is not effective, additional processing steps, such as smut removal, must be taken.

A buffer consists of an acid or an alkali in the form of a salt, such as borate, phosphate, or silicate. A buffering agent enables the use of a higher concentration of the cleaner. If silicates are used as the buffer, adequate rinsing should be provided because of their poor rinsing qualities. More information on alkaline and acid cleaners is available in the articles "Alkaline Cleaning" and "Acid Cleaning of Iron and Steel," respectively, in Volume 5 of the 9th Edition of *Metals Handbook*.

Equipment

Equipment will vary depending on the type and shape of metal to be processed. Cleaning is usually accomplished by dip or spray, followed by a hot-water rinse. Mechanical assistance, such as the use of brushes and/or the use of spray pressure and impingement of the cleaning solution, is important. Pressure, nozzle size, and pattern are all considered when designing equipment for various cleaning applications. Preformed parts are usually suspended from a conveyor; coiled metal is continuously unwound through the cleaning stages and then rewound at the end.

The speed at which the metal is moved through the stages will vary, depending on its shape as well as the amount and type of soil on the surface. Generally, the cleaning time allotted is twice the amount of time required to apply the conversion coating. Rinsing time is one-fourth to one-half the cleaning time. Temperatures range from room temperature to 75 °C (170 °F), depending on soil conditions and length of cleaning time.

Cleaning times for preformed parts are usually 1 to 2 min. Because preformed parts cannot be squeegeed between stages, a weaker cleaner is used to minimize contamination of the rinse water. Also, it is difficult to reach all sections of a preformed part. For these reasons, preformed parts require more time in the cleaner bath for soil removal than coiled metal does.

The time required to clean coiled metal is 6 to 12 s, which is substantially less than that for preformed parts. Metal in a coil can receive a uniform spray pattern, which enhances cleaning. The use of squeegees between stages permits higher concentrations of cleaners with minimum contamination of the rinse. Figure 2 shows the equipment used to clean coiled sheet metal.

To maximize cleaning efficiency and to minimize contamination of the treatment tanks, various arrangements of cleaner and rinse tanks are used. Some possible combinations are:

- Preclean, clean, hot rinse
- Clean, clean, hot rinse
- Clean, brush, clean, hot rinse
- Clean, hot rinse

The use of two cleaning tanks or a preclean and cleaning tank serves several purposes (Fig. 2). First, it maintains a relatively clean and uncontaminated cleaner immediately before the water rinse. This reduces the chance of redepositing soil on the surface of the metal before rinsing. Second, two cleaner tanks allow the use of different types of cleaners. The first tank is usually a cleaner that will lift and float easily removed soil to the surface of the tank to be skimmed off later. The second cleaner is one that will emulsify oil and finalize the cleaning process.

Equipment maintenance is very important. Sequestrants (agents that help to keep contaminants in solution) and water softeners are incorporated into cleaners to aid in maintenance. Their purpose is to prevent or reduce the precipitation of hard water salts, metals salts, and flocculents out of the cleaner bath. Spray nozzles must not be allowed to clog, or heat exchangers to cake over with these precipitates. Sequestrants and water softeners prevent a precipitate from forming before contaminants can be removed by a controlled throwaway. A controlled throwaway is when a percentage of the old bath is removed and new bath is added. The old bath is usually skimmed off the top in conjunction with the removal of any contaminants that have floated to the surface.

Cleaner Formulation

Liquid Cleaners. When formulating a cleaner, the decision must be made regarding whether the cleaner will be a liquid or a powder. Liquid cleaners can be more easily incorporated into automatic controls. They can also contain a greater percentage of surfactants. However, in liquid cleaners, the surfactants are not always soluble in the cleaner concentrate. This causes the surfactants to split off and form a layer on the top of the cleaner. Couplers are added to hold the surfactants in solution in the concentrated cleaner. The selection of the coupler is based on the ratio of surfactants to coupler; some couplers also contribute other benefits to the cleaner, such as detergency or wetting capabilities, in addition to coupling.

Figure 3 illustrates the composition of a typical liquid cleaner. The surfactant percentage usually consists of two or more different surfactants. Use of more than one surfactant provides a wider range of cleaning capability.

Powder cleaners are usually less expensive than liquid cleaners and can be shipped more economically, but they have a tendency to harden during storage. Therefore, care should be taken to avoid absorption of moisture from the air during blending and storage. It may be necessary to incorporate an anticake agent into the powder product. The percentage of liquid surfactants should also be limited. Some surfactants are available in a powder form and are more easily incorporated into a powder product. Figure 4 illustrates the composition of a typical powder cleaner. The sodium carbonate is usually referred to as a filler; other than some benefit to waste treatment, it contributes very little to cleaning. The specific percentage used in the example may vary, depending on the desired strength of the cleaner concentrate.

High-Temperature Versus Low-Temperature Cleaners. Low-temperature cleaners have been developed to conserve energy and to lower costs. Typically, the only differences between a high-temperature cleaner (60 to 75 °C, or 140 to 170 °F) and a low-temperature cleaner (40 to 55 °C, or 100 to 130 °F) are the types and quantities of surfactants. Surfactant selection for low-temperature cleaners is primarily based on the low-temperature foaming characteristics and cleaning capabilities of the surfactant. Low-temperature cleaners are usually run at higher concentrations and are limited in the variety of metals and types of soil that can be cleaned.

When the cleaning job is difficult, high temperatures are a requirement. Heat mobilizes the soil so that it may be removed faster and more easily.

Control of the Cleaner Bath

Control of the cleaner bath is usually achieved by titrating for free alkalinity and total alkalinity (in alkaline cleaners) or by titrating for free acid and total acid (in acid cleaners). For alkaline cleaners, the free alkalinity reading indicates the concentration of the cleaner solution. Phenolphthalein is often used as the indicator for this reading. The total alkalinity reading is then taken, usually with methyl-orange as the indicator. The difference between the readings indicates the concentration of dissolved metals in the cleaner bath. The acceptable level of dissolved metals for a particular cleaner will be determined by experimentation. Typically, when the difference between the readings equals or exceeds the free alkalinity reading, it is time to dump the bath.

Automatic control of the cleaner bath is usually accomplished by measuring the conductivity of the bath. Conductivity increases with the concentration of the cleaner solution. If a liquid cleaner is used, the controller may be connected to a pump, which will activate below a certain conductivity to replenish the bath and shut off above that conductivity. Powdered cleaners usually require a small off-line tank in which a slurry is made by adding water to the powder cleaner. Constant agitation is necessary to maintain the slurry. In this form, the controller may activate a slurry pump for additions to the bath.

Testing for Surface Cleanliness

The evaluation of whether the material is clean and ready for the conversion coating is sometimes very difficult. The speed of the line and the shape of the equipment does not allow easy examination of the material. To determine if the bath is capable of cleaning, it is advisable to obtain a small sample of the material and to hand dip it. The sample is cleaned and rinsed according to the time and sequence of the line. The surface of the sample is then wet with fresh cold water, and the wet film is observed for water break (discontinuities in the water film). If the sample is not free of water break, then it is not clean.

The other test consists of wiping the cleaned and rinsed surface with a wet, white cloth. If smut and soils remain on the surface, the cloth will darken from this residue.

Waste Treatment of Old Cleaner Bath

In the waste treatment of cleaner baths, one of the first tasks to be accomplished is the removal of oils from the bath. The bath is given time to cool; this allows the oil to rise to the surface where it can be skimmed. If an emulsion has been formed, acids may be used to lower the pH and to aid in splitting off the oil.

The bath is then treated so that the metals and other soil in the bath precipitate out. The precipitate is drawn off the bottom through a filter. The bath is given time to settle before filtration. Flocculents and adjustments to the pH are used to enhance the formation of this precipitate. It may be necessary to lower the levels of sequestrants in the cleaner in order to aid in the precipitation of dissolved metals in the cleaner bath. These additions and adjustments also improve the quality of the filtration cake. The filter cake must be firm, dry, and easily removed from the filter. Carbonate salts are used for this purpose.

To reduce the amount of waste to be treated and to lower the cost of cleaning, it is sometimes advisable to reclaim the usable portion of the old bath. This can be accomplished by selecting a cleaner that does not form an emulsion with oils. The equipment necessary is an off-line holding tank with a cone bottom that can decant the top layer of the bath. By decanting the top oily layer and removing the bottom precipitate, the remainder of the bath may be reused with only a small addition of fresh cleaner.

Phosphate Conversion Coatings

Thomas W. Cape, Chemfil Corporation

PHOSPHATE CONVERSION COATINGS are applied to various metal substrates to enhance corrosion resistance, increase paint adhesion, or both. This article will discuss the types, uses, and theory of phosphate coatings; the tests for bath control and coating performance; and the equipment used in processing. More detailed information is available in the Selected References given at the end of this article.

Much information on phosphating bath formulations and processes is also contained in patents. Although these can be sources for much information, it must be remembered that the use of such information is limited by the patent.

The primary use of phosphate coatings is to enhance corrosion resistance, but there are other applications. These include the use of phosphate conversion coatings for their electrical insulating properties (for example, in electric motor laminations) and for lubricity (for example, to increase the formability of sheet metals). However, such applications are beyond the scope of this article.

Types of Phosphate Coatings

Phosphating processes are categorized in a number of different ways: by the types of metal ions present, by gross variations in coating weight, by the type of accelerators used, by the number of stages used, or by the end use of the phosphate coating. In this article, the basic categorization of phosphate coatings will be as follows:

- Iron phosphates—lightweight, amorphous phosphate coatings that do not contain significant amounts of divalent metal ions from solution. Coating weights range from 0.16 to 0.80 g/m^2 (15 to 75 mg/ft^2)
- Zinc phosphates—medium-weight, crystalline phosphate coatings that contain divalent metal ions from the solution and/or the metal surface. Coating weights range from 1.4 to 4.0 g/m^2 (130 to 370 mg/ft^2)
- Heavy phosphates (manganese phosphates)—heavy coatings that contain divalent metal ions from solution and from the metal surface. Coating weights range from 7.5 to 30 g/m^2 (700 to 2800 mg/ft^2)

There are two major categories for the uses of phosphate coatings: bare corrosion protection and painted corrosion protection. Bare phosphate means simply that the phosphate coating is not painted.

Because of the crystalline nature of zinc phosphates and manganese phosphates, these coatings can hold oils and waxes so well that the corrosion resistance of the metal is increased much more than is expected based on either the phosphate or the oil separately. Similarly, the phosphate coating holds paint physically. The fact that painted applications of phosphate coatings are categorized separately from bare phosphate is a reflection of the great importance of painted applications, rather than any fundamental difference in the mechanism of corrosion inhibition.

Processing Sequence

The basic processing sequence for applying a phosphate conversion coating is cleaning, phosphating, and sealing. Depending on the type of phosphate coating being applied, some of these basic operations may be eliminated. Typical processing sequences for the three basic types of phosphate coatings are described below.

Iron Phosphate. A standard iron phosphate coating is applied by cleaning and rinsing the metal, coating the metal in the phosphate bath, rinsing again, and sealing for enhanced corrosion protection and improved paint adhesion. The sealing step is sometimes omitted to reduce costs.

Zinc Phosphate. Metals that are to be given a zinc phosphate coating are cleaned (twice, if possible), rinsed, and conditioned in some cases before the phosphate coating is applied. After the coating is applied, the metal is again rinsed; it is then sealed and rinsed again (the rinse after sealing is omitted in some situations).

In heavy phosphate coating, the metal surface is cleaned, rinsed, and pickled, if necessary, to remove rust. The processing sequence is then the same as that for zinc phosphate coatings, with conditioning and the rinse after sealing being optional for some systems.

General Phosphate Theory

Coating Formation. The basic process involved in the formation of any phosphate coating is the precipitation of a divalent metal and phosphate ions (PO_4^{3-}) on a metal surface. Phosphate salts, particularly divalent metal salts, are soluble in acid solutions and insoluble in neutral or basic solutions. The phosphate baths are acidic enough to keep the ions in solution. When metal is exposed to the solution, the acid attacks the metal surface. Two changes occur in the solution directly adjacent to the metal surface. First, the acid is neutralized and the pH rises. Second, the concentration of metal ions increases. Typical metals phosphated, such as steel or galvanized steel, will increase the concentration of iron or zinc ions, respectively, in this boundary layer between the metal surface and the bulk of solution.

The basic chemical reactions involved in the phosphating process are:

$$2H^+ + M \rightarrow H_2 + M^{2+} \quad \text{(Eq 1)}$$

$$2H_2PO_4^- + 3M^{2+} \rightarrow M_3(PO_4)_2 + 4H^+ \quad \text{(Eq 2)}$$

where M^{2+} is one of several possible metal ions. The consumption of acid (H^+) causes Eq 2 to proceed to precipitation. The production of metal ions (M^{2+}) pushes Eq 2 to precipitation of $M_3(PO_4)_2$. Equation 2 is applicable to many crystalline phosphate coatings, but other phosphates that might precipitate are shown in Eq 3:

$$5M^{2+} + 4H_2PO_4^- \rightarrow M_5H_2(PO_4)_4 + 6H^+ \quad \text{(Eq 3)}$$

The composition of the phosphate coating depends on the composition of the metal surface and on the bath exposed to the metal. Common phosphate compositions are given in Table 1. For example, a zinc phosphate bath that is treating steel will contain hopeite ($Zn_3(PO_4)_2{\cdot}4H_2O$) and phosphophyllite ($Zn_2Fe(PO_4)_2{\cdot}4H_2O$). The relative proportion of phosphophyllite in the phosphate coating is called the P-ratio:

$$\text{P-ratio} = \frac{\%\ \text{phosphophyllite}}{\%\ \text{phosphophyllite} + \%\ \text{hopeite}} \quad \text{(Eq 4)}$$

The P-ratio will be affected by the substrate. Iron in the metal being phosphated is necessary for phosphophyllite formation. Similarly, a phosphate bath that does not have zinc will not form phosphophyllite on steel. Table 2 gives the general composition of the phosphate coating as a function of substrate and divalent metal in the bath. In addition to these obvious effects, more subtle changes in bath chemistry affect the amounts of phosphophyllite. Some of these factors, as well as factors that affect composition in other phosphate baths, will be discussed later in this article.

Table 1 Phosphate compounds found in conversion coatings

Compound name	Chemical formula
Vivanite	$Fe_3(PO_4)_2{\cdot}8H_2O$
Iron-hureaulite(a)	$Fe_5H_2(PO_4)_4{\cdot}4H_2O$
Strengite	$FePO_4$
Hopeite	$Zn_3(PO_4)_2{\cdot}4H_2O$
Phosphophyllite(a)	$Zn_2Fe(PO_4)_2{\cdot}4H_2O$
Scholzite	$Zn_2Ca(PO_4)_2{\cdot}4H_2O$
Phosphonicollite(b)	$Zn_2Ni(PO_4)_2{\cdot}4H_2O$
Phosphomangallite(a)	$Zn_2Mn(PO_4)_2{\cdot}4H_2O$
Manganese-hureaulite(a)	$Mn_5H_2(PO_4)_4{\cdot}4H_2O$
	$AlPO_4$

(a) Phosphophyllite is $(Fe,Mn)Zn_2(PO_4)_2{\cdot}4H_2O$; Hureaulite is $(Mn,Fe_5)H_2(PO_4)_4{\cdot}4H_2O$. The names used in this article were chosen to avoid confusion as to the actual composition. (b) No mineralogical name known

Table 2 Phosphate coatings formed on various substrates

Type of coating (Divalent metal ion)	Metal substrate: Iron/steel	Zinc	Aluminum
Iron phosphate (none)	$Fe_3(PO_4)_2{\cdot}8H_2O$(a), iron oxides	$Zn_3(PO_4)_2{\cdot}4H_2O$	$AlPO_4$
Zinc phosphate (Zn^{2+})	$Zn_3(PO_4)_2{\cdot}4H_2O$, $Zn_2Fe(PO_4)_2{\cdot}4H_2O$	$Zn_3(PO_4)_2{\cdot}4H_2O$	$Zn_3(PO_4)_2{\cdot}4H_2O$
Zinc phosphate (Zn^{2+}, Ni^{2+})	$Zn_3(PO_4)_2{\cdot}4H_2O$, $Zn_2Fe(PO_4)_2{\cdot}4H_2O$, $Zn_2Ni(PO_4)_2{\cdot}4H_2O$	$Zn_3(PO_4)_2{\cdot}4H_2O$, $Zn_2Ni(PO_4)_2{\cdot}4H_2O$	$Zn_3(PO_4)_2{\cdot}4H_2O$, $Zn_2Ni(PO_4)_2{\cdot}4H_2O$
Zinc phosphate (Zn^{2+}, Ca^{2+})	$Zn_3(PO_4)_2{\cdot}4H_2O$, $Zn,Fe(PO_4)_2{\cdot}4H_2O$, $Zn_2Ca(PO_4)_2{\cdot}4H_2O$	$Zn_3(PO_4)_2{\cdot}4H_2O$, $Zn_2Ca(PO_4)_2{\cdot}4H_2O$	$Zn_3(PO_4)_2{\cdot}4H_2O$, $Zn_2Ca(PO_4)_2{\cdot}4H_2O$
Zinc phosphate (Zn^{2+}, Mn^{2+})	$Zn_3(PO_4)_2{\cdot}4H_2O$, $Zn_2Fe(PO_4)_2{\cdot}4H_2O$, $Zn_2Mn(PO_4)_2{\cdot}4H_2O$(b)	$Zn_3(PO_4)_2{\cdot}4H_2O$, $Zn_2Mn(PO_4)_2{\cdot}4H_2O$(b)	$Zn_3(PO_4)_2{\cdot}4H_2O$, $Zn_2Mn(PO_4)_2{\cdot}4H_2O$(b)
Manganese phosphate (Mn^{2+})	$Mn_5H_2(PO_4)_4{\cdot}H_2O$, $Fe_5H_2(PO_4)_4{\cdot}4H_2O$	$Zn_3(PO_4)_2{\cdot}4H_2O$, $Zn_2Mn(PO_4)_2{\cdot}4H_2O$(b)	$Mn_5H_2(PO_4)_4{\cdot}H_2O$
Iron phosphate (Fe^{2+})	$Fe_5H_2(PO_4)_2{\cdot}4H_2O$, $FePO_4$	$Zn_3(PO_4)_2{\cdot}4H_2O$, $Zn_2Fe(PO_4)_2{\cdot}4H_2O$, $Fe_5H_2(PO_4)_2{\cdot}4H_2O$	...

(a) Because of the amorphous nature of this coating, x-ray diffraction cannot provide a complete description of the coating. Vivanite has been observed. (b) The form of manganese in the phosphate coating has not been positively identified. In addition to phosphomangallite, manganese-hureaulite ($Mn_5H_2(PO_4)_2{\cdot}H_2O$) could be the manganese-containing species.

Table 3 Accelerators used for phosphate coatings

Accelerator	Advantages	Disadvantages
Nitrite	No harmful by-products; removes iron from bath; very active	Unstable in bath—constant additions must be made
Chlorate	Stable in bath; removes iron from bath; very active	Harmful by-products
Nitrate	No harmful by-products; stable in bath	Low activity; does not remove iron
Hydrogen peroxide	No harmful by-products; very active; removes iron from bath	Unstable in bath—constant additions must be made; high sludging; bath conditions critical
Nitroguanidine	Very active	Not available as liquid concentrate; does not remove iron; cannot be shipped pure
Nitrobenzene sulfonic acid	Very active	Does not remove iron; harmful by-products

Accelerators. Modern phosphating uses additives called accelerators to hasten the phosphating process. These accelerators often help remove iron from the bath. The acceleration is accomplished by replacing H^+ in Eq 1 with a better oxidizing agent. The most common accelerators are nitrite (NO_2^-) and chlorate (ClO_3^-). Other common accelerators are nitrate (NO_3^-), peroxide (H_2O_2), and organic nitro compounds (such as nitrobenzene sulfonic acid and nitroguanidine). The advantages and disadvantages of each individual accelerator are listed in Table 3. Most baths that treat steel will build up iron in the bath as a harmful contaminant unless treated. Chlorate, peroxide, and nitrite will readily oxidize iron so that it will precipitate as $FePO_4$. Many phosphates use combinations of two or more accelerators. Common combinations are nitrite/nitrate, nitrite/chlorate/nitrate, and chlorate/nitrobenzene sulfonic acid. Some accelerators form by-products that over time will build up in a bath and could be detrimental to performance.

If it is economically feasible to change baths on a regular basis, then this problem is minimized. Chlorate, for example, is a good accelerator for an iron phosphate bath because the bath is inexpensive enough to replace often as the chloride builds up. On the other hand, chlorate is a poor accelerator for a high-nickel zinc phosphate bath, because the bath is too expensive to replace often.

There are three methods of acceleration beyond the chemical methods described above. First, higher agitation produces a coating more quickly. If an immersion phosphate process is converted to a spray process, the processing time is decreased. Similarly, an immersion process can be accelerated by increasing agitation. Second, the temperature can be raised to increase the rate of coating formation. Most phosphating processes have an optimum operating range for temperature, but the upper end may be more applicable than the lower end. Third, electrical currents are known to accelerate some processes. Currently, this process is not in widespread use, because chemical acceleration is more desirable.

Sludge and Scale. Phosphating is a precipitation process, and as such, the precipitation sometimes occurs in undesirable ways that are classified as sludge and scale. Sludge is the precipitation that occurs in the bath but does not go onto the workpiece. One source of this sludge is the oxidation of iron or manganese to form highly insoluble phosphates. In addition, the precipitation for coating formation may produce some phosphate that does not deposit on the workpiece. Scale is the phosphate coating formed on processing equipment. A major cause of scale is excessive heating of the bath. Because most divalent metal phosphates are less soluble at higher temperatures, the localized heating that occurs with raising or maintaining the bath temperature will cause scale to form on heating elements.

Testing

There are three basic types of tests for phosphating: phosphate bath testing, phosphate coating analysis, and paintability testing.

Phosphate Bath Testing

Acid Type and Concentration. The primary methods of bath control are acid-base titrations that measure the amount and type of acid present. Free acid is the amount (in milliliters) of 0.1 *N* sodium hydroxide (NaOH) needed to neutralize the strong acidity from 10 mL of the bath. Total acid is the amount (in milliliters) of 0.1 *N* NaOH needed to neutralize all the acid in 10 mL of solution. Consumed acid is the amount (in milliliters) of 0.1 *N* sulfuric acid (H_2SO_4) needed to restore weak acid to 10 mL of bath so that any more acid added will be strong acid. Results are stated in points—for example, 2 mL = 2 points.

Nitrite Concentration. Another important bath-testing parameter is NO_2^- concentration. Because NO_2^- decomposes in solution, it is critical to be able to measure the concentration. Two general methods can be used:

- Permanganate titration: This reaction has some interferences; therefore, it is important to follow the directions of the supplier exactly
- Gas generation: There are two chemicals used for this reaction: urea and sulfamic acid. A standard volume of bath is exposed to an excess of sulfamic acid or urea, and the volume of resultant gas is measured

In deference to tradition, the results are stated in points.

Other Bath Parameters. In addition to the basic control procedures, a variety of ion-specific tests can be run. First, fluoride (F^-) is determined most efficiently by the use of an ion-sensitive electrode. This would be of particular interest in a zinc or heavy phosphate bath that is processing a significant portion of aluminum, because one of the important uses of F^- is to keep the concentration of aluminum down by the formation of AlF_6^{3-} ions. Second, the zinc concentration can be checked by an EDTA (a strong metal complex) titration. Because other divalent metal ions (Ni^{2+}, Mn^{2+}, Ca^{2+}) can interfere with the titration, the supplier should be consulted for an exact procedure. Third, a test for ferrous ions (Fe^{2+}) may be desirable as a monitor of Fe^{2+} level or as a check for the presence of Fe^{2+}. Typically, Fe^{2+} can be titrated with the permanganate solution used for checking NO_3^- or with ammonium dichromate, such as that often used for standardizing the ferrous ammonium sulfate solution in the titration for chromate final rinses. If only a check for Fe^{2+} ion is desired, the phosphate supplier should have papers that will indicate the presence of Fe^{2+}.

The purpose of all of the above tests is to help control the replenishment of the phosphate bath. Typically, the replenishment is accomplished by metering and checking free acid, total acid, or the concentration of the accelerator. Bath monitoring and control can be accomplished by two automated feedback systems. One is based on automatic titration. The results of these titrations automatically control bath replenishment.

Although this is the proper automatic method, it suffers from mechanical complexity and therefore reduced reliability. The second method, conductivity measurement, overcomes this problem. Conductivity can be related to the concentration of the basic constituents for a given bath and thus can serve as a test for replenishment. Conductivity measurements are simple to run on a phosphate bath; therefore, they are more widely used than titrometric-type controllers. Unfortunately, because conductivity does not directly measure the relevant parameters, some systems require adjustment of replenishment levels as the phosphate bath ages and require recalibration when a new bath is built up.

Phosphate Coating Tests

Coating Weight. The most common method of evaluating phosphate coatings is determination of the coating weight. This is usually accomplished by the gravimetric method, in which test panels are weighed before and after stripping to determine coating weight. Gravimetric determination of coating weight involves the following steps:

- The area to be stripped (MA) is measured
- The panel is baked at 95 °C (200 °F) for 5 min
- The panel is accurately weighed to obtain initial weight (IW)
- All areas that are not to be stripped are masked using electroplater's tape
- For steels, the panel is stripped in 5% chromic acid (H_2CrO_4), typically at 70 °C (160 °F) for 15 min
- For galvanized steels, the panel is stripped with ammonium hydroxide (NH_4OH) inhibited with 1 g/L ammonium dichromate ($(NH_4)_2Cr_2O_7$) at room temperature for 15 min
- The panel is rinsed, and any tape that was previously applied is removed
- The panel is baked again at 95 °C (200 °F) for 5 min
- The panel is reweighed to obtain the final weight (FW)
- Coating weight is calculated using the equation (IW − FW)/MA

The panel must be clean of all oils; gloves should always be worn for handling test panels.

Two other methods are commonly used for the determination of phosphate coating weights. First, it is common to determine zinc phosphate coating weights, especially on steel, by specular reflectance infrared absorption. This method is especially effective for quick determination of relative coating weights. Unfortunately, the range of coating weights that can be measured is limited to that of commonly used zinc phosphates (not iron phosphates or heavy phosphates). In addition, there is a dependency on the type of metal substrate used—for example, steel, galvanized steel, or aluminum. A more general method is x-ray analysis for phosphorus. This can be done by using a portable analyzer or the x-ray analyzer on a scanning electron microscope. Both the infrared and x-ray techniques must be carefully calibrated.

The use of coating weight as a quality control technique is limited by the fact that within fairly broad ranges coating weight alone is a poor indicator of quality. It is more useful as a test of the consistency of the bath. Thus, if there are wide variations in coating weight, quality is suspect. Coating weights are more important for iron phosphate than for zinc phosphate or heavy phosphate simply because other quality control tests are less useful.

Coating Morphology. Of primary consideration in the quality of zinc and heavy phosphate coatings is the structure of the coating. This includes two factors: the shape and the size of the phosphate coating. Unfortunately, there are few general rules that are applicable to all systems. The interaction of primer and phosphate dictates optimum performance, but once a good crystal size and crystal shape are determined, holding them constant will hold good performance. This of course assumes that the other properties do not deteriorate.

Typically, coating morphology is checked using optical microscopy. Zinc phosphate coatings can usually be observed adequately at 400×. Heavy phosphate can be seen at lower magnifications. Iron phosphate generally cannot be observed by optical microscopy. Scanning electron microscopy can be used for all phosphates, but it is generally used only when a problem has been identified.

Completeness of Coating. Complete coverage of the metal surface is the most important test parameter for phosphate coatings. Typically, this can be checked by one of three methods:

- Visual observation of a matte finish. A shiny area would typically be an indication of unphosphated areas. In general, iron phosphates will have a shinier surface than zinc phosphate, which is shinier than heavy phosphate. Also finer, more uniform crystal size will appear shinier
- Microscope observation. Often, uncoated areas are readily observable during a check of coating morphology
- The Ferro test. A solution of sodium chloride (NaCl), potassium ferricyanide ($K_3Fe(CN)_6$), and a surfactant on filter paper is placed on a phosphated steel panel. The blue color indicates uncoated spots

The composition of a phosphate coating can affect performance and therefore is often monitored. Generally, only heavy or zinc phosphates are analyzed, not iron phosphates. One composition test is the determination of phosphophyllite (Table 1) content in a phosphate coating. There are three methods for this determination:

- Direct analysis of the iron content of the solution obtained by determination of coating weight
- X-ray crystallographic determination of the relative proportions of hopeite and phosphophyllite
- X-ray analysis for zinc and phosphorus. This is based on the difference in ratio of zinc to phosphorus between hopeite and phosphophyllite.

The first and third methods have the advantage of also allowing analysis for other elements that may be present, such as calcium, magnesium, or nickel. These other elements are of greater importance for galvanized steel.

Phosphate coating adhesion is a measure of how well the phosphate coating stays intact on the metal surface. Two common techniques are the tape pull test and the thumbnail test. The tape pull consists of determining how much, if any, of the phosphate coating can be removed by a standard tape, such as Scotch No. 600. A good coating will not lose any phosphate to the tape. The thumbnail test consists of scratching the phosphate coating and observing the mark left. The more obvious the mark, the worse the adhesion of the phosphate coating. It should be noted, however, that all phosphate coatings leave some mark; this is often used as a test to determine if a piece has been phosphated.

Corrosion Resistance. These tests measure the ability of the phosphate coating to resist corrosion. Typical tests would expose the phosphated panels to a corrosive environment and measure the amount of rust after a set period of time. Common corrosive environments include salt fog or condensing humidity. The phosphate coating may be oiled or waxed if its final use will be oiled or waxed. Corrosion tests of unpainted phosphate coatings are generally not used if the final application is painted.

Paintability Tests

These tests can generally be divided into three categories: accelerated corrosion testing, exposure corrosion testing, and adhesion testing. It is beyond the scope of this article to detail all of the commonly used test procedures, but a general outline can be given.

Paint Weakening. Many paint systems are so strong that in order to measure phosphate failure in a timely manner the paint system must be weakened. Common methods of weakening the system include scribing with a carbide tool, bending, drawing, undercuring the paint, and applying a thinner-than-standard paint film. The type of paint weakening used or whether it is used at all depends on the system.

Painted Corrosion Resistance. The commonly used accelerated tests can be divided into three categories: continuous testing, cyclic testing, and actual-exposure testing.

Continuous tests, such as the salt spray (ASTM B 117), water soak (ASTM D 870), or condensing humidity tests, have one basic corrosion condition. A variation on this procedure is that some tests have an initiation procedure prior to the test.

Cyclic tests use a variety of conditions cycled on a set schedule. Typically, these tests use saltwater exposure, controlled humidity or water exposure, and a drying period. Often, one or more of the conditions are at elevated temperatures. Among the conditions less commonly used are cold exposure and gravelometer damage. Cycles typically last from 6 h to a week, and the entire test lasts from 3 to 12 weeks.

Atmospheric exposure tests often use outdoor exposure under conditions of actual use or of most severe use. Examples of these tests include scribing automotive test panels and exposing them to an industrial atmosphere with salt solution twice a week (an accelerated test) and exposing painted building panels to typical outdoor exposure (not an accelerated test).

Iron Phosphating

The general characteristics of iron phosphate coatings are given in Table 4. There are three types of iron phosphating that will be discussed: iron phosphates that do not require precleaning (cleaner/coaters), iron phosphates that require precleaning (standard iron phosphates), and organic phosphates. Before describing these differ-

Table 4 Characteristics of phosphate coatings

Characteristic	Type of coating: Iron phosphate	Zinc phosphate	Heavy phosphate
Coating weight	0.16–0.80 g/m² (0.0005–0.0026 oz/ft²)	1.4–4.0 g/m² (0.0045–0.013 oz/ft²)	7.5–30 g/m² (0.025–0.1 oz/ft²)
Types	Cleaner/coater Standard Organic phosphate	Standard Nickel-modified Low-zinc Calcium-modified Manganese-modified	Manganese phosphate Zinc phosphate Ferrous phosphate
Common accelerators	Nitrite/nitrate Chlorate Molybdate	Nitrite/nitrate Chlorate Nitrobenzene sulfonic acid	None Chlorate Nitrate Nitroguanidine
Operating temperatures	Room-70 °C (160 °F)	Room-70 °C (160 °F)	60-100 °C (140–212 °F)
Free acid, points	−2.0 to 2.0	0.5–3.0	3.6–9.0
Total acid, points	5–10	10–25	20–40+
Prephosphate conditioners	None	Titanium phosphate None	Manganese phosphate Titanium phosphate None
Primary use	Paint base for low-corrosion environments	Paint base for high-corrosion environments	Unpainted applications
Limitations	Low painted corrosion resistance; low unpainted corrosion resistance	Poor unpainted corrosion resistance	Expensive, long processing times
Materials needed for tanks	Low-carbon steel	Low-carbon steel, stainless steel, or plastic-lined steel	Stainless steel or low-carbon steel
Application method	Spray and immersion	Spray and immersion	Immersion only

ent systems, it is appropriate to discuss some aspects common to all systems:

- Iron phosphates historically have been used primarily on steel, but some systems can be used on galvanized steel or aluminum
- The standard accelerators used are NO_3^-, NO_2^-, and ClO_3^-, although molybdate (MoO_4^{2-}) is also used
- Final rinses are considered optional, although the improvement seen with final rinses is most dramatic with iron phosphates

Cleaner/Coaters. These iron phosphates combine the functions of cleaning and phosphating into one stage. These have the advantage of requiring less floor space in the plant because of the elimination of the cleaner and rinse stages. In addition, maintenance is proportionally easier. Iron phosphates are generally used because they are less expensive than zinc or heavy phosphates; therefore, the cost savings of these cleaners/coaters greatly enhances their popularity. The disadvantages of cleaner/coaters include lower phosphate quality and the fact that two cleaner/coater stages are sometimes required so that the workpiece is cleaned better.

Standard iron phosphates are generally five-stage systems (see the section "Processing Sequence" in this article). These systems usually produce the best phosphate coating quality among the three types of iron phosphates. In addition, the phosphate bath lasts longer, and the workpiece can be cleaned better. On the other hand, there is additional work in maintaining both a cleaner and phosphate stage, and more floor space is required for the extra stage.

The organic phosphating process involves an iron phosphate process in a chlorinated solvent. The workpiece is vapor degreased by the refluxing solvent prior to entry into the bath. In addition, it is rinsed by vapor degreasing after phosphating. Organic phosphating has the advantage of being a one-stage phosphating process. The primary disadvantages are lower phosphate quality and exposure to chlorinated organic vapors. Typically, the system is run without a final rinse.

Applications of Iron Phosphates. Iron phosphates are usually used when the primary consideration is good paint adhesion and low cost. Applications that are not exposed to corrosive environments or that require forming of parts after painting are suitable for iron phosphates. Parts formed after painting can be, and often are, pretreated with a zinc phosphate, chromate, or mixed oxide coating when galvanized steel or aluminum is used as the metal substrate. If low cost was the primary consideration, a steel substrate would have been used.

Zinc Phosphate

Zinc phosphates (Table 4) are based on the formation of crystalline tertiary phosphates with the general formula $Zn_2M(PO_4)_2{\cdot}4H_2O$, where M can be zinc, iron, nickel, calcium, or manganese. The inclusion of nickel, calcium, or manganese comes from the addition of these elements to the phosphating baths, and the inclusion of iron comes from inclusion of the iron in steel into the phosphate coating. These additions delineate some of the different types of zinc phosphates. In addition, several of these modifications can be made to the same bath. One example would be a nickel-modified low-zinc zinc phosphate.

Standard Zinc Phosphate. This type of phosphate has no divalent metal ions other than zinc. In general, lower performance and lower costs are seen with these baths as opposed to the other zinc phosphate baths. Extensive cleaning and preconditioning are often not used to save floor space. Typically, these standard zinc phosphates are a compromise between low-cost iron phosphates and the higher performance of other types of zinc phosphates. Further, they are often restricted to use on steel.

Nickel-Modified Zinc Phosphate. Nickel in the zinc phosphate bath produces inclusion of small quantities of nickel in the phosphate coating. This causes large improvements in the painted performance of the phosphate coating, especially on galvanized products. Most zinc phosphates contain some nickel, which gives the liquid concentrates their familiar green color. Traditionally, relatively low levels of nickel have been used (0.15 to 0.5 g/L), primarily because of the high cost of nickel. Higher levels of nickel (0.6 to 4.0 g/L) are being introduced because of the greater performance of these phosphate baths.

Low-Zinc Zinc Phosphate. Traditional phosphate baths have zinc concentrations in the range of 1.0 to 3.0 g/L. Low-zinc baths have zinc concentrations of 0.5 to 1.0 g/L. These baths were developed to optimize the formation of phosphophyllite on steel. It was subsequently found that these baths also use nickel more efficiently. Thus, low-zinc baths were found to improve performance on all substrates through greater formation of phosphophyllite or phosphonicollite. The only significant disadvantage with these baths is that the zinc concentration must be controlled carefully.

Calcium-Modified Zinc Phosphate. The addition of calcium to zinc phosphate baths produces many of the performance advantages of nickel in the phosphate baths without the associated higher costs. Calcium-modified zinc phosphate coatings have found application in the appliance industry.

Manganese-Modified Zinc Phosphate. Manganese-modified zinc phosphates have recently been introduced, particularly in automotive phosphating. One advantage obtained has been lower operating temperatures. Currently, there is debate as to the degree of improvement in performance afforded by these phosphate systems.

Prephosphate Conditioners. Titanium phosphates are known to have substantial benefits to the phosphate coatings when applied prior to phosphating. Although a number of other chemicals, such as oxalic acid or finely divided zinc phosphate salts, have a similar effect, virtually all prephosphate conditioners used in zinc phosphate systems are based on titanium phosphate technology. Currently, many zinc phosphate systems are totally dependent on the conditioning stage to deposit a beneficial zinc phosphate coating. The conditioner can be applied either as a portion of the final cleaning stage, or preferably, as a separate stage between cleaning and phosphating.

The theory as to why these titanium salts help the formation of a phosphate coating is not well developed. What is known is that the use of these colloidal suspensions produces phosphate coatings that have smaller crystal size and a more complete coating. It is believed that this is accomplished by the formation of many more active sites for initiation of phosphate coating formation. Because there are more sites for initiation, the crystals do not get a chance to grow as large, and the probability that a given area will not be coated decreases.

Applications of Zinc Phosphates. Zinc phosphate coatings can be used as a paint base or as a structure to hold oils or waxes. These two applications cannot be implemented at the same time. If an oil or wax has been used, paint will not adhere to the metal. In protection from corrosive environments, paint usually shows much better performance than waxes and oils. There are applications, however, in which paint is not preferred over oils and waxes.

Zinc phosphates are used when good paint adhesion and good painted corrosion resistance are desired. Companies usually find it most cost effective to put their effort and money into applying the most effective zinc phosphate possible. In most cases, the cost of the paint system is much greater than the cost of the phosphate system. Zinc phosphates usually cannot be used for coil-coating applications in which the painted sheet must be formed. Zinc phosphates can be effectively applied to coiled steels only if the end use does not require drawing of the steel for the final product. In addition, zinc phosphates are not used in applications in which maximum corrosion resistance is needed and paint cannot be used.

Heavy Phosphates

Heavy phosphates (Table 4) are divided into three categories: zinc phosphates, manganese phosphates, and ferrous phosphates. These phosphates form a coarse crystalline coating. Typically, the heavy phosphates are much more expensive than zinc phosphates in chemical, heating, and floor space costs. Their advantage is in increased unpainted corrosion resistance.

Manganese phosphates are used when maximum corrosion resistance is needed for unpainted applications. If adequate rinsing can be obtained, an accelerated process is used. When accelerators are used, NO_3^- is usually the one of choice, with small amounts of nickel being beneficial.

Manganese phosphates build up Fe^{2+} ions with use, which can lead to lower performance. The iron level in the bath is typically controlled with H_2O_2. The phosphate supplier should be contacted regarding the nature of this addition.

Manganese phosphates are sensitive to the effects of cleaning. Use of an acid or strong alkali cleaner produces coarser coatings. This can be overcome to a large extent by conditioning the surface immediately before phosphating with a suspension of finely divided manganese phosphate. This is especially important if the workpiece is rusted and requires pickling prior to phosphating.

Zinc Phosphate. The same basic chemicals used to make a zinc phosphate coating for a paint base can be used to make a heavy phosphate. The concentration, temperature, and time are all increased to accomplish this. Despite these increases, a zinc phosphate can be applied faster and at a lower temperature than a manganese phosphate. As with zinc phosphates used for a paint base, a titanium phosphate or (less likely) an oxalate conditioning stage is very beneficial.

Ferrous phosphates in many ways combine the worst attributes of the other heavy phosphates. Their corrosion resistance is not as good as manganese phosphates. Neither can they operate at the lower temperatures or shorter times of zinc phosphates. However, ferrous phosphates operate at a higher acidity, which means that rust can be more effectively removed during the phosphating process, eliminating the need for pickling.

Final Rinses

To achieve maximum corrosion resistance for any of these phosphates, the coatings should be sealed before use. The process of phosphating and either painting or applying an oil or wax provides the best corrosion resistance if the pores in the phosphate coating where corrosion is most likely to be initiated are treated to make those areas more corrosion resistant. Historically, the sealer of choice is chromate (hexavalent chromium, CrO_4^{2-}). Regardless of which final rinse is used, the optimum pH is in the range of 3.8 to 4.8. Recently, trivalent chromium (Cr^{3+}) and nonchromium final rinses have become increasingly popular for environmental reasons.

Chromate Rinses. There are two basic types of chromate sealing rinses. One type contains only CrO_4^{2-}, and the other is a mixture of CrO_4^{2-} and Cr^{3+}.

Hexavalent chromium (chromate) sealing rinses are the oldest type of currently used sealing rinses. They are used without rinsing just prior to painting and can be used for any type of phosphate, but poorer quality coatings benefit the most.

Hexavalent/trivalent chromium rinses are used when the workpiece is rinsed again after this stage. The corrosion resistance of CrO_4^{2-} final rinses is greatly diminished upon rinsing, probably because of the good solubility of CrO_4^{2-}. Addition of Cr^{3+} eliminates this effect because an insoluble chromate compound is formed.

Both of these baths are controlled by monitoring CrO_4^{2-} concentration. The CrO_4^{2-} is titrated with a standardized ferrous sulfate ($FeSO_4$) solution.

Trivalent Chromium Rinses. The concern over the high cost of environmentally safe final rinses have led to development of chromate-free rinses. One effective group is those rinses that contain Cr^{3+}. This group has produced very effective final rinses, some of which are as good as chromate final rinses.

Nonchromium Final Rinses. Because even Cr^{3+} must be treated for disposal, much work has been devoted to developing chromium-free final rinses. There are two basic types of nonchromium rinses that are being developed: environmentally safe metal salts and polymers or oligamers. At this point, it should be noted that in order for a final rinse to be proved effective it must be better than if it were not used. To show that on a given test it is equivalent to a chromate final rinse is insufficient, because most accelerated tests cannot differentiate between effective and ineffective final rinses.

Methods of Application

The two basic methods of applying phosphate conversion coatings are spraying and immersion. In general, spray coating is less expensive, but immersion coating provides better quality.

Spray phosphating involves pumping the bath through a series of risers that have spray nozzles aligned to coat all areas of the workpiece. For complex parts, this sometimes means that internal portions will receive little or no coating. Spray coating results in rapid coating formation with lower bath concentrations than those needed for immersion coating. Spraying also gives better cleaning of external surfaces of parts because of impingement of the coating bath, has lower equipment costs, requires less floor space, and uses smaller bath volumes than immersion coating (this reduces recharging costs).

Immersion coating gives more uniform coatings than spraying. It is slower and requires much larger bath sizes than spraying, but these disadvantages can be partially offset by the use of higher bath concentrations. Immersion baths require more floor space and higher equipment costs than spraying, and because of their greater volume, they are more expensive to recharge and to dispose of. The advantages of immersion coating include better cleaning and coating on internal surfaces of the workpiece, better coating uniformity, better drainage between stages with less drag-through of contaminants from one stage of the process to another, less maintenance and repair, lower bath heating costs, and generally better coating performance than sprayed coatings. Also, the large baths required for immersion coating are less sensitive to such changes as depletion of chemicals or bath contamination, and temperatures are more stable in the larger baths. Nickel can be included in the coating more efficiently in immersion baths than in spray baths, which improves the corrosion performance of zinc and manganese phosphate coatings on galvanized steels. Immersion phosphation produces changes in the phosphate coating composition that results in better painted performance with some paint systems.

The selection of an application method should be based on the particular requirements of the workpiece. However, the following general guidelines may be helpful in choosing a coating process:

- Heavy phosphate coatings are usually applied by immersion because of the prohibitive heat loss involved in spray operations
- Iron phosphate coatings are usually applied by spraying because the emphasis is on low cost. The exception is organic phosphates, which are typically applied by one-stage immersion
- The more complex the configuration of the workpiece, the more advantageous immersion becomes. Thus, spraying is often used for strip applications, but complex parts are more likely to be immersion phosphated

Equipment Considerations

Materials. Alkali cleaners, conditioners, and final rinses may use low-carbon steel for the bulk of the tank. Phosphate systems may require stainless steel, depending on the presence of F^- or NO_3^-, acidity, operating temperature, and the desired lifetime of the tank. Type 316 stainless steel is preferred. Nozzles must be stainless steel. Plastic or plastic-lined steel can be substituted for stainless steel in some circumstances, but the use of plastics is limited to lower-temperature applications.

Heating and Circulation. The optimum method of heating depends on the specifics of the operation, but the following points should be kept in mind:

- Because phosphates precipitate at higher temperature, any source of heat should be easy to clean
- Spot heating should be avoided
- The heat source should not be on the bottom of the tank, because the sludge settling will insulate it

The common methods of heating are steam or hot-water coils, electric heating, gas heating, water jackets, and external heat exchangers. For immersion phosphating, the last method would also be a useful source of agitation. The solution must be pumped in spray phosphating, but solution pumping is also necessary to obtain optimum results in some immersion systems. Barrel phosphating can provide considerable agitation around the part by rotation of the barrel; additional agitation is usually not required. Conveyor applications may or may not need additional agitation.

Sludge and Scale. Because phosphating is a precipitation process and is not 100% efficient, some precipitation occurs in the bath. Precipitation on the equipment is called scale, and precipitation that is suspended in the bath is called sludge. The sludge usually settles to the bottom; it can then be pumped to a filtration unit. In some cases, most of the solution is pumped to a temporary storage tank, and the sludge-rich solution from the bottom is removed. Scale is more difficult to remove. The most expedient method is to use an inhibited acid to dissolve it.

SELECTED REFERENCES

- T. Biestek and J. Weber, *Electrolytic and Chemical Conversion Coatings: A Concise Survey of Their Production Properties and Testing*, Trans. by A. Kozlowski, Port Callis Press, Ltd, 1976
- D.B. Freeman, *Phosphating and Metal Pretreatment: A Guide to Modern Processes*, Industrial Press, Inc., 1986
- J.J. Khain, *Theory and Practice of Metal Phosphating*, Finishing Publications, Ltd, 1978 (available from Metal Finishing Systems, San Diego)
- G. Lorain, *Phosphating of Metals*, Finishing Publications, Ltd, 1974 (available from Metal Finishing Systems, San Diego)
- N. Rausch, *Die Phosphatieraung von Metallen*, Eugen G. Leuze Verlag, 1974
- K. Woods and S. Spring, *Met. Finish.*, Vol 76, June 1978, p 17-22
- K. Woods and S. Spring, *Met. Finish.*, Vol 77, Feb 1979, p 24-28
- K. Woods and S. Spring, *Met. Finish.*, Vol 77, March 1979, p 56-60
- K. Woods and S. Spring, *Met. Finish.*, Vol 78, Sept 1980, p 41-46

Chromate Conversion Coatings

Karl A. Korinek, Parker Chemical Company

CHROMATE CONVERSION COATINGS are formed by a chemical or an electrochemical treatment of metals or metallic coatings in solutions containing hexavalent chromium (Cr^{6+}) and, usually, other components. The process results in the formation of an amorphous protective coating composed of the substrate, complex chromium compounds, and other components of the processing bath.

Chromate conversion coatings are applied primarily to enhance bare or painted corrosion resistance, to improve the adhesion of paint or other organic finishes, and to provide the metallic surface with a decorative finish. Chromating processes are widely used to finish aluminum, zinc, steel, magnesium, cadmium, copper, tin, nickel, silver, and other substrates. Chromate conversion coatings are most frequently applied by immersion or spraying, but other methods of application, such as brushing, roll coating, dip and squeegee, electrostatic spraying, or anodic deposition, are used in special cases.

Technology of Chromating Processes

Chromate coatings are applied by contacting the processed surfaces with a sequence of processing solutions. The processing baths are arranged in a series of tanks, and the surfaces to be processed are transferred through the sequence of stages by using manual, semiautomatic, or automatic control. The chromate coatings are usually applied to metal parts or to a continuous metal strip running at speeds to 5 m/s (1000 ft/min).

The basic processing sequence consists of the following six steps: cleaning, rinsing, conversion coating, rinsing, posttreatment rinsing or decorative color rinsing, and drying. In many applications, this sequence is expanded to accommodate pickling, deoxidizing, dyeing, brightening, and other rinsing stages, or the sequence can be shortened when cleaning or posttreatment rinsing is not necessary. Some typical processing sequences are given in Tables 1 to 3. The key processing steps are discussed in the following sections.

Cleaning. Surface preparation before chromating is very important for the quality and performance of the coatings. Alkaline cleaners are used to remove organic soils and metallic impurities from the surface. Where oils and greases are very heavy, a precleaning step may be necessary before alkaline cleaning. In some cases, an acid deoxidizer, acid pickle, or mechanical brushing is needed to reduce the oxide layer substantially and to activate the surface. Freshly prepared surfaces, such as electroplated or hot-dip coated metals, can be chromated directly or after rinsing. Occasionally, the freshly electroplated surfaces are treated with an acidic or alkaline neutralizing rinse before chromating.

Rinsing. The alkali-cleaned metal must be thoroughly rinsed, preferably in hot water. The rinse water should be overflowed at a rate that will keep it free of scum and contamination. Maintaining the rinse in good condition not only improves the quality of the processed parts but also prevents contamination of subsequent stages. This is very important because the replenishing chemicals for the chromating bath are formulated to maintain the operating baths in the recommended ranges under normal rinse drag-in conditions. Excessive drag-in could lead to acid or alkali contamination, which, for the processes controlled by pH or conductivity, could result in the addition of too much or too little replenishing chemical. In the long term, this could lead to seemingly unexplainable problems with coating quality.

Chromating. The composition of the chromating bath depends on the metal to be treated, the type of application, and the specific requirement for the final product. The chromating bath is usually built up from one or more chemical concentrates (makeup chemicals) and is maintained by the addition of one or more concentrates that constitute the replenishing chemicals. The replenishing chemicals serve to maintain the concentrations of the key bath components and the bath pH in the desirable range and to bind any reaction products in a form that does not interfere with the coating reaction or the quality of the produced chromate coating. Several different replenishers may be required to maintain a constant bath composition during the processing of different substrates.

A great variety of solution compositions, many of them proprietary, are available for chromating. Selection of the proper process usually begins with the coating performance definition for the metal substrate to be treated, and it depends on the type of processing equipment available. For a given coating, maximum performance is achieved when the process is maintained within the specification of the supplier and when the coating weights are within the recommended range. The most important parameters for the control of chromating processes are time, temperature, chromate concentration, pH, reaction products, and accelerator concentration.

To maintain optimum quality in chromate coatings, it is important to produce coatings that have the desired composition and are within the desired coating weight range. The optimum processing parameters are usually established for each production facility. Once these processing conditions are established, it is good operating practice not to deviate substantially without conducting an appropriate evaluation, which should confirm that the proposed process modification does not affect final product quality. Control of the specified parameters within the recommended ranges will normally ensure that the coating composition is also maintained. The amount of chromate coating applied to the surface can be estimated from the coating appearance or can be accurately measured by using methods for determining coating weight (see the section "Properties of Chromate Coatings" in this article).

The coating weight increases with time and temperature. It usually also increases with chromate concentration, but a substantial increase above the recommended concentration range could lead to a different coating composition or could change other bath components and result in lower coating weights. The pH of the chromating solution must be in the range in which the substrate metal is dissolved, but also must allow for deposition of the coating. In zinc chromating processes, the zinc dissolution rate reaches an appreciable level below pH 4. Below pH 1, coating deposition is slow, but it increases with pH to a specific pH value, after which a further increase in pH will gradually decrease the rate of coating formation. Few chromating processes for such metals as magnesium or silver can be practiced in the neutral or alkaline pH range.

For the chromating processes that use accelerators (see the section "Control and Testing" in this article), it is important to maintain their level in order to ensure a good coating reaction. This is accomplished by controlled addition of an appropriate replenishing chemical.

A special group of chromate coatings, mainly used in striplined applications, is applied by no-rinse or dry-in-place processes. Unlike conventional chromating baths, which must be autodrained or occasionally dumped when reaction product accumulation prevents coating formation, these processes result in full use of the chemical. The coatings are formed by an application of a uniform wet chromate film of controlled thickness by using roll coating, dip and squeegee, or other methods and must be dried completely prior to being exposed to a physical contact. The applied solutions are more concentrated than those used in conventional processes. The chromium coating weight depends only on the wet film thickness and the concentration of chromium. These processes do not require rinsing or posttreatment, but drying and in some cases a degree of curing are necessary. Their primary advantage lies in the elimination of chromium-containing process wastewaters, which makes them very attractive from an environmental standpoint.

Table 1 Typical process sequences for chromating of aluminum, zinc, and their alloys

Step	Decorative finish	Step	Painting
1	Vapor degreasing: solvent (optional for heavy oils)	1	Vapor degreasing: solvent (optional for heavy oils)
2	Cleaning: alkaline (single or double stage)	2	Cleaning: alkaline (single or double stage)
3	Rinsing: hot water	3	Rinsing: hot water
4	Deoxidizing or desmutting: inorganic acids (optional)	4	Conversion coating: chromate or chromium phosphate
5	Rinsing: water	5	Rinsing: cold or warm water
6	Conversion coating: chromate	6	Posttreatment: chromate or deionized water (optional)
7	Rinsing: cold or warm water	7	Drying: warm air
8	Dyeing: dye solution (optional)	8	Painting: single or double coat
9	Rinsing: cold or warm water		
10	Drying: warm air		

Table 2 Typical process sequences for chromating of continuous galvanized strip

Step	Electrogalvanized strip	Step	Hot-dip galvanized strip
1	Electrogalvanizing	1	Hot-dip galvanizing
2	Rinsing: warm water, multiple stages, neutralizing (optional)	2	Surface conditioning: heat treatments
3	Conversion coating: chromate	3	Conversion coating: chromate
4	Rinsing: cold or warm water	4	Drying: warm air
5	Posttreatment: chromate (optional)	5	Oiling (optional)
6	Drying: warm air		
7	Oiling or painting (optional)		

Table 3 Typical process sequences for no-rinse treatments

Step	Process
1	Cleaning: alkaline (single or double stage)
2	Rinsing: hot water
3	No-rinse chromating
4	Drying or curing
5	Painting: single or double coat

Rinsing. With the exception of no-rinse chromating treatments and some chromate posttreatments that are dried without rinsing, good rinsing after chromating is essential. The chromated surfaces should be rinsed with cold water immediately after treatment. The rinse should be continuously overflowed at a rate that will keep it free of excessive drag-in contamination. Double cold rinsing is frequently recommended. In spray operations, fresh water can be introduced through the last riser to induce tank overflow. Static rinse tanks and acidic or hot rinses must be avoided.

Posttreatment. After rinsing, chromate conversion coatings are often treated with a posttreatment solution. The chromium-containing posttreatments are applied to increase the corrosion resistance of chromate coatings substantially in bare and painted applications. These posttreatments also provide an improved substrate for adhesion of paint or other organic finishes. For optimum performance, the control ranges recommended by the supplier should be maintained. Chromate-base posttreatments are also used in phosphating (see the article "Phosphate Conversion Coatings" in this Volume).

Coloring. The appearance of chromate coatings can be modified for decorative purposes by coloring with dyestuffs or pigments. Dyeing is usually carried out, after chromating and rinsing, by immersion in a dilute solution of a dyestuff. The dyestuffs are either physically absorbed into the coating or can chemically react with the film. Some colored chromate films are applied directly by a no-rinse process in which the applied chromate coatings contain a dye or a pigment. After the posttreatment or coloring step, some processes may require water rinsing, while others produce better results without rinsing.

Drying is the final processing step in the formation of chromate coatings. The coatings produced by conventional processes require only enough heat to dry the surface. Drying should be carried out in a stream of clean, warm air. For thick chromate films, it is important to avoid high drying temperatures, which can cause the formation of brittle, cracked coatings that provide reduced corrosion protection. A maximum processing temperature of 60 °C (140 °F) is often recommended for chromate coatings on aluminum. No-rinse processes often require curing and must be processed through an oven to reach the specified level of cure.

Methods of Application and Equipment

In chromating, the processed surfaces are contacted with a sequence of processing solutions. The choice of equipment depends on the nature of the articles being processed. Most chromating processes can be applied by spray or immersion.

A typical spray line consists of a completely enclosed tunnel in which various processing solutions are applied to the processed surfaces in separate stages. These stages are equipped with reservoir tanks, pumps, spray risers, and nozzles. The sprayed solutions are collected and returned to the reservoir tanks, which are usually located under the spraying section. Various types of monorail conveyors are used to move the processed parts through the tunnel. Figure 1 shows a schematic of a section of a spray chromating line. To minimize drag-in, drainage zones are provided between the individual stages. Misting is sometimes necessary to keep the surfaces wet during processing. The primary advantage of spray processing is improved cleaning uniformity due to the spray impingement. Spray line processing is applicable only to parts that can be suspended on hooks, jigs, or other structures or to the processing of a continuous strip. Figure 2 shows a schematic of a continuous coil-coating line.

Several types of nozzles can be used on the spray chromating line. The cleaner and cleaner-rinse stages use high-impingement nozzles having flat spray patterns with uniform distribution and tapered edges. This provides uniformity of spray distribution as patterns overlap in multiple-nozzle installations. In coil-coating installations, flat spray nozzles are usually positioned throughout all stages of the line. In monorail installations, the first line of sprays in the chromating stage will normally use wide-angle flat-flooding nozzles, followed by hollow-cone whirl-type nozzles in the rest of the section. The rinses after chromating generally use flooding-type spray nozzles to prevent damage to or removal of the chromate coating. If there is a possibility of drying between the spray zones, wetting nozzles should be provided. Smaller whirl-type nozzles are often used for this purpose.

Processing by immersion is more common, and it is preferred for its equipment simplicity, easier access of the chromating solution to recessed areas, simple handling of small parts, and ease of maintenance. Electrochemical conversion coatings must be applied by immersion. The parts, mounted on jigs or hooks, are transferred from one processing tank to another by a monorail conveyor. The conveyors have different degrees of automation, ranging from a simple manual to a fully automated programmed transfer control. Small parts are treated in baskets or barrels. The barrels are rotated during immersion. Rotation is important for the development of good, complete coatings. Rotation rates of 0.2 to 1.0 rpm provide for good mixing and do not damage the coatings.

In the processing of a continuous strip, the immersion treatment is accomplished in specially designed reaction cells in which the strip is submersed in a recirculating treatment bath. The treatment solution is stored in a reservoir tank; it is then usually pumped to both ends of the reaction cell. An adjustable weir maintains the required solution level. The recirculating pump is sized to change the solution in the cell approximately once per minute.

The processing tank, piping, and reservoir tank used for chromating solutions should be constructed of AISI type 316 stainless steel or low-carbon steel lined with type I (normal impact) polyvinyl chloride (PVC). The heat exchanger, pump, and nozzles should be of type 316 stainless steel. Chemical-metering pumps for adding concentrates should have type I PVC machined reagent heads and fittings with diaphragms faced using polytetrafluoroethylene (PTFE) or a similar substance. Polyethylene tubing is recommended for the concentrate supply lines to the main tank.

Properties of Chromate Coatings

The composition of chromate coatings is not a well-defined subject. The coatings are three-dimensional films, and their compositions depend largely on the metal substrate, the chemical com-

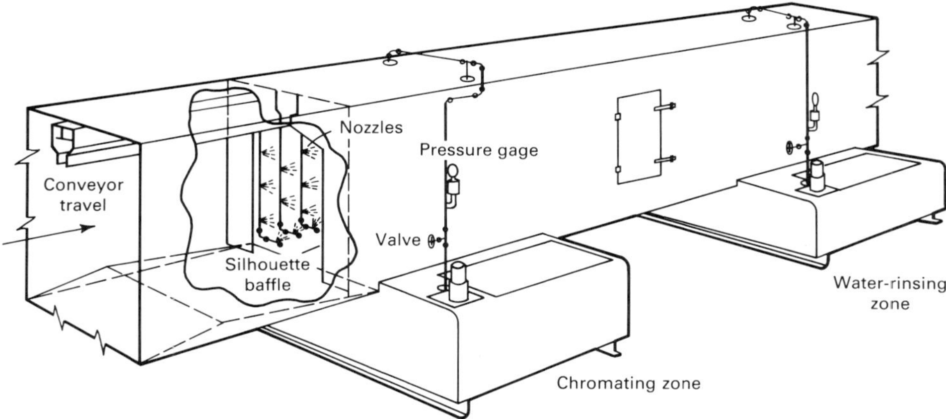

Fig. 1 Schematic layout of a spray chromating section. Courtesy of Parker Chemical Company

position of the bath, and other process parameters. During formation of these films, the chemical composition and the surfaces involved are changed, which results in continuously changing elemental composition throughout the coating. The changing composition profiles are more apparent in thinner coatings (10 to 20 nm), in which the metal/coating and coating/solution interfaces form a substantial part of the coating. For thicker films (≥100 nm), the bulk of the coating can have a relatively constant composition.

The main constituents of chromate coatings are trivalent chromium (Cr^{3+}) and Cr^{6+}, the basis metal, various oxides, water, and other components, such as phosphate, sulfate, and fluoride. New instrumentation for surface analysis has enhanced the ability to characterize the compositions of chromate coatings. The use of x-ray photoelectron spectroscopy (XPS) and Auger electron spectroscopy (AES), combined with depth profiling using argon etching and other techniques, has improved the understanding of the composition of three-dimensional structures.

Determination of the composition is complicated by the changes that occur during aging. Many freshly formed coatings contain a significant amount of Cr^{6+}, are hydrophilic, and can be dissolved in nitric acid (HNO_3). After several hours, the coatings are only partially soluble in HNO_3, and after 24 h or heating to 50 °C (120 °F), they become hydrophobic and insoluble in HNO_3. The aging factor is probably not very significant in applications in which the properties of bare coatings are important. When the coatings are used as a paint base, the paint is often applied to the freshly prepared surfaces, which differ in some ways from surfaces that are usually characterized by surface analysis.

To obtain the most accurate information about the composition of the coatings, it is necessary to conduct a surface analysis, preferably by several complementary techniques, for the given substrate, chromating bath, and the actual process. It can be shown that the actual composition of the coating may differ with the degree of oxidation of the substrate, the condition of the chromating bath, and other processing features, such as the type of cleaner used, processing temperatures, pressures in the spray operations, and the method of drying.

A recent study of the composition of chromate passivation films on zinc, aluminum, and aluminum-zinc alloy coated sheet steel clearly illustrated the substrate effects (Ref 1). The XPS analysis of aluminum-zinc alloy coated steel surfaces before and after chromating is shown in Fig. 3 and 4. After chromating, a passivation film is formed that comprises chromium in different oxidation states. A thin layer of Cr^{6+} is present at the outer surface of the film. Following its removal by sputtering, a layer consisting mostly of trivalent oxides was observed on all three substrates. The presence of metallic chromium was detected on aluminum and on aluminum-zinc alloy coated sheet. The aluminum and aluminum-zinc chromate coatings differ from the chromate coating on zinc by the presence of aluminum oxide in the intermediate layer and metallic chromium in the inner layer. These two differences may account for the greater bare corrosion resistance of passivated aluminum-zinc coatings as compared to zinc coatings.

Another recent study of a chromate coating used as a paint base demonstrated that the surface of aluminum-zinc alloy coated steel changes significantly during each step of the pretreatment process (Ref 2). During the alkaline cleaning step, aluminum is removed, forming a zinc-rich oxide film. In the chromating step, the oxide film is removed, and a chromate coating is deposited primarily on the zinc-rich areas of the surface. The chromate posttreatment, the final step of the pretreatment process, deposits chromate coating over the entire surface, substantially enhancing the chromate coating thickness in aluminum-rich areas. The chromate coating resulting from this process forms an excellent paint base, providing improved paint adhesion and increased corrosion resistance.

The above examples illustrate that the compositions of chromate coatings depend on all the steps of the chromating process. For the many chromating processes, bath compositions, and substrates, the best information on the composition of the coatings can be obtained only from surface analysis of a broad range of samples prepared under conditions corresponding to the actual processing conditions. This is not a simple task, because it is not always easy to simulate production conditions in the laboratory. In addition, even the most advanced analysis techniques can generate artifacts that complicate data interpretation.

Only a small number of coatings and chromating processes have been characterized by surface analysis techniques. Most of the studies reached conclusions about the chemical composition of the coating from some method of chemical identification. The following chemical formulas are most commonly ascribed to the compounds in chromate coatings:

- Substrate oxides and hydroxides: M_xO_y, $M_x(OH)_y$
- Substrate chromates: M_xCrO_4, $M_xCr_2O_7$
- Chromium oxides and hydroxides: Cr_2O_3, Cr_xO_y, CrOOH, $Cr(OH)_3$, $Cr_2O_3 \cdot xH_2O$
- Chromium chromates: $Cr(OH)CrO_4$, $Cr_x(CrO_4)_y$
- Metallic chromium
- Substrate with other anions: M_xX_y, for example, $AlPO_4$
- Chromium with other anions: Cr_xX_y, for example, $CrPO_4$

Coating weight is one of the most important parameters for the monitoring of the desirable qualities of chromate coatings. It describes the weight of the chromate coating per unit surface area or the weight of chromium in the chromate coating per unit area. It is typically given in milligrams per square foot, milligrams per square decimeter, or grams per square meter. Coating weights are usually measured by gravimetry, atomic absorption spectrometry (AAS), or x-ray fluorescence.

The gravimetric procedure is used to determine chromate coating weights on aluminum. It uses an HNO_3 solution or a sodium nitrite ($NaNO_2$) molten salt bath as the stripping reagent. The coating weights are calculated from the weight loss of the measured sample. The HNO_3 stripping

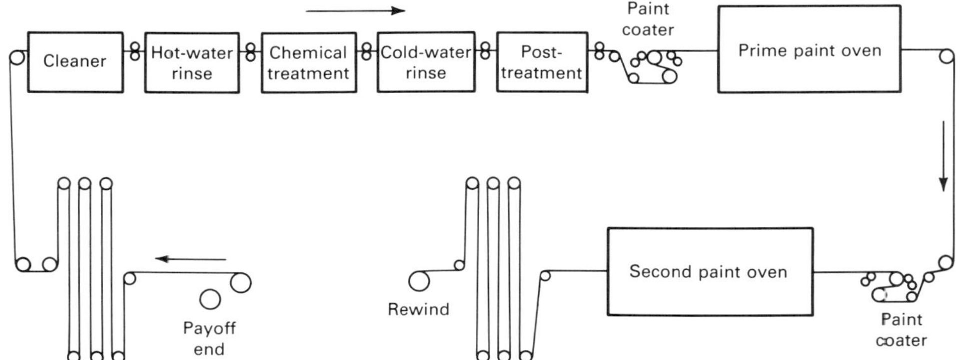

Fig. 2 Schematic of a continuous coil-coating line

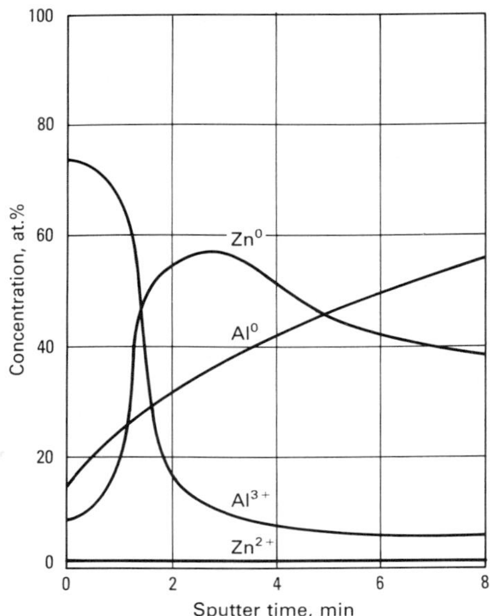

Fig. 3 XPS analysis of aluminum-zinc coated sheet steel before chromating. See Fig. 4 for analysis after chromating. Courtesy of The Electrochemical Society

can be used only for freshly prepared coatings that have not been oven dried.

The AAS method can use different stripping solutions that dissolve the coating and partially dissolve the substrate. A solution of HNO_3 and hydrofluoric acid (HF) can be used to determine chromium coating weights on aluminum, while diluted hydrochloric acid (HCl) is usually selected as the stripping solution for chromate coatings on zinc and zinc-aluminum alloys. The coating weight is measured in milligrams of chromium per square foot. To obtain the total coating weight, the percentage of chromium in the coating must be known.

X-ray Fluorescence. A more recent method for measuring coating weight uses x-ray fluorescence. This procedure can be used to determine chromium in many types of coatings. The measurement of coating weight requires a calibration procedure to convert the chromium counts obtained to the chromium coating weight. As in the AAS method, the percentage of chromium in the coating must be known to obtain the total coating weight.

The thickness of chromate conversion coatings can range from 10 to 1000 nm. For a given combination of substrate and conversion coating, the color of the coating can be used as an indicator of coating thickness. With the exception of very thin films, chromate coatings exhibit a characteristic crazed structure after drying that can be observed by high-magnification electron microscopy (Fig. 5). The thinner films are often used as a paint base or when a colorless appearance is required. Thicker coatings are used for bar corrosion protection and decorative applications.

The color of chromate coatings depends considerably on the substrate and the type of chromating process used. On zinc, very thin coatings are transparent; as thickness increases, the color becomes blue with an iridescent pattern, followed by green, slightly yellow, yellow, brown, bronze, olive, and black. To improve their decorative finish, freshly formed chromate coatings can be easily colored with pigments or dyes to create various color finishes with a range of yellow, gold, red, green, blue, purple, violet, grey, and black. After drying, the films lose their absorbency, become hydrophobic, and cannot be dyed.

Corrosion Protection. Chromate conversion coatings provide excellent bare or painted corrosion protection to the metal. The level of protection depends on the substrate metal, the type of chromate coating used, and the chromium coating weight. In unpainted applications, corrosion protection for the different conversion coatings generally increases with coating weight, and the upper limit of the coating weight is determined by the process limitations or by the color requirement. Table 4 lists typical corrosion data measured by the ASTM B 117 method (Ref 3).

In painted applications, the conversion coating must improve corrosion resistance and provide for good paint adhesion. The upper limit of the coating weight for the painted surfaces is normally defined by the onset of weaker paint adhesion or of corrosion problems related to paint delamination.

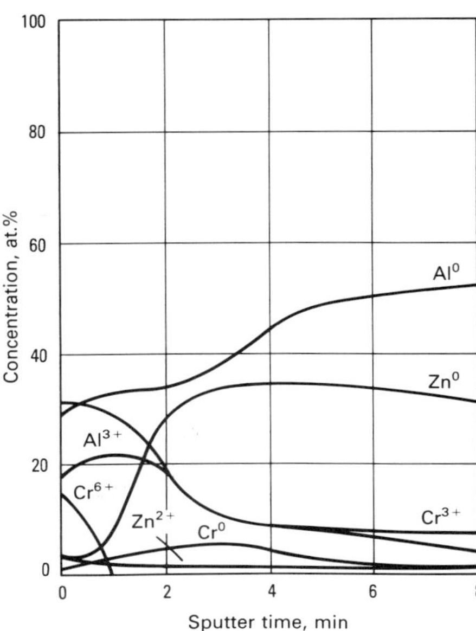

Fig. 4 XPS analysis of aluminum-zinc coated sheet steel after chromating. See Fig. 3 for analysis before chromating. Courtesy of The Electrochemical Society

Opinions differ widely regarding the mechanism of corrosion protection provided by the chromate coatings. The most widely advanced concepts suggest that the chromate coatings provide a barrier insulation from the environment and inhibit the cathodic corrosion reactions.

Electrical Resistance. Chromate films are characterized by a relatively low electrical resistance. The reported values range from about 10 to several hundred microOhms per square centimeter. The resistance of thin films is sufficiently small to allow for the use of chromated metals in electric or electronic applications. In corrosive environments, the electrical resistance of thick films is lower than that of an uncoated surface. Thin chromate films do not interfere with resistance welding, but machine settings may need to be adjusted as film thickness increases.

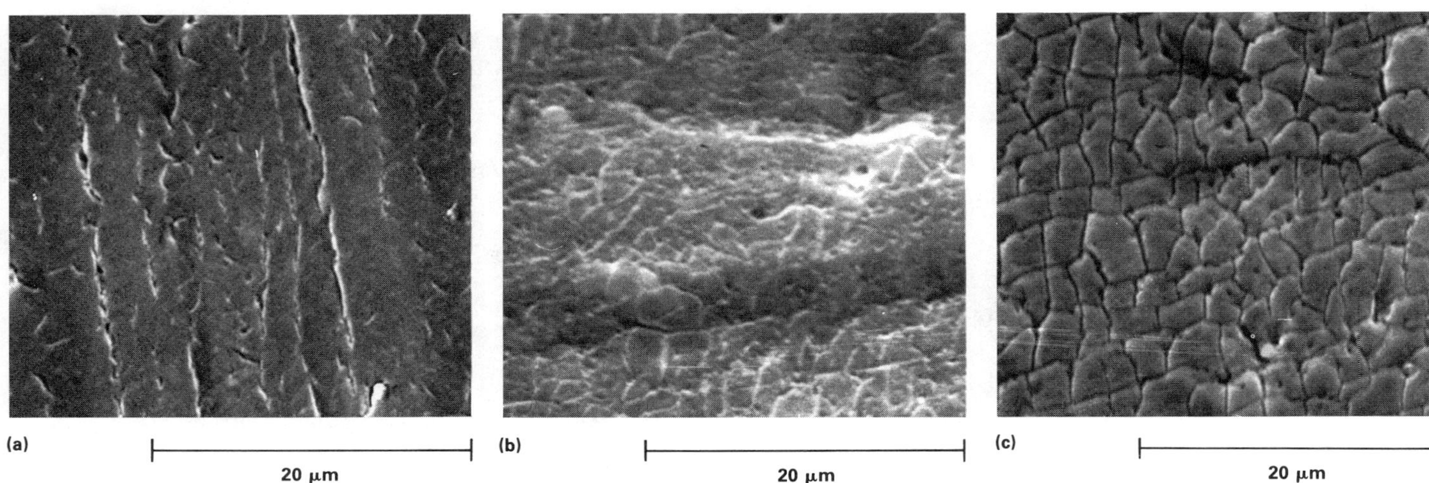

Fig. 5 Scanning electron micrographs of chromate coatings. (a) Chromium phosphate: 645 mg/m² (60 mg/ft²). (b) Chromate/ferricyanide: 377 mg/m² (35 mg/ft²). (c) Chromate/ferricyanide: 1345 mg/m² (125 mg/ft²)

Table 4 Typical salt spray data for chromate coatings on zinc and aluminum

Substrate	Type of chromate coating	Time to corrosion stain, h
Electroplated zinc	Untreated	<4
	Clear	24–48
	Iridescent	100–200
	Olive drab	100–400
	Electrolytic	1000
Hot-dip zinc	Untreated	<4
	Clear passivate	24–100
Aluminum alloy 3003.......	Untreated	<24
	Clear	60–120
	Yellow-brown	250–800

Hardness and Abrasion Resistance. The hardness of chromate coatings depends strongly on the temperature during chromating and drying. Freshly made wet films are very soft and can be easily damaged by abrasion. After drying, the films develop good hardness, which allows for safe handling. However, even the dry films are susceptible to severe scratching or abrasion.

Control and Testing

As the metal is processed through the bath, its composition undergoes changes resulting from the coating reactions, bath drag-out, and mechanical losses. Maintenance of bath composition requires analytical monitoring and replenishing with chemical concentrates and water. The buildup of reaction products often slows the coating reaction and therefore must be controlled by a continuous discharge of a small volume of the working bath or by periodic dumpings of the bath.

Ideally, the chromating bath is operated under steady-state conditions. The reaction products are continuously removed by autodraining, and the replenisher is added to replace consumed or discharged chemicals. The main parameters used in monitoring and controlling the chromating bath are Cr^{6+} concentration, free acid level, and total acid level. Monitored less frequently are pH and concentration of accelerators, free fluoride, and Cr^{3+} concentration.

Hexavalent Chromium Concentration. Chromic acid (H_2CrO_4) and its salts are present in solution in several ionic forms. Above pH 6, H_2CrO_4 forms chromate ion (CrO_4^{2-}); between pH 2 and 6, $HCrO_4^-$ and dichromate ion ($Cr_2O_7^{2-}$) are in equilibrium; below pH 1, the main species is H_2CrO_4. The equilibria and equilibrium constants, K, are:

Table 5 Test performance of chromate/molybdate coatings

The paint system consisted of an epoxy primer and a polyester top coat.

Type of test	Ref	Ratings(a): Hot-dip galvanized steel	Ratings(a): Aluminum alloy 3003
Adhesion			
Impact (ASTM D 2794)	15	10	10
T-bend (ASTM D 3794)	14	0T,1T(b)	0T,1T
Accelerated corrosion			
Salt spray (ASTM B 117)	3	0–1(c)	N
Humidity (ASTM D 2247)	12	10	10

(a) 10, best; 0, worst; N, perfect. (b) Designation of bend configuration; see Ref 14. (c) Number given is the creepage from the scribed mark in 1/16 of an inch.

$$H_2CrO_4 \rightleftharpoons HCrO_4^- + H^+ \quad K = 4.1 \qquad \text{(Eq 1)}$$

$$HCrO_4^- \rightleftharpoons CrO_4^{2-} + H^+ \quad K = 1.3 \times 10^{-8} \qquad \text{(Eq 2)}$$

$$H_2O + Cr_2O_7^{2-} \rightleftharpoons 2HCrO_4^- \quad K = 3 \times 10^{-2} \qquad \text{(Eq 3)}$$

$$2CrO_4^{2-} + 2H^+ \rightleftharpoons Cr_2O_7^{2-} + H_2O \quad K = 4.2 \times 10^{14} \qquad \text{(Eq 4)}$$

A common method for measuring the Cr^{6+} concentration is by a redox titration using, for example, a 0.1 N ferrous sulfate ($FeSO_4$) solution. The method is simple, accurate, and provides a rapid means of monitoring Cr^{6+}. A good redox indicator is a 0.1% solution of 1,10 phenanthroline ferrous complex (ferroin). According to:

$$CrO_3 + 3Fe^{2+} + 6H^+ \rightarrow Cr^{3+} + 3Fe^{3+} + 3H_2O \qquad \text{(Eq 5)}$$

each millimeter, or point, of 0.1 N $FeSO_4$ corresponds to 3.33 mg of CrO_3 or 1.73 mg of Cr^{6+}.

Free and Total Acid. The concept of free acid and total acid monitoring is the most common approach to bath control. The values correspond to the number of milliliters of a standard alkali solution (usually 0.1 N NaOH) required to titrate a standard bath sample (normally 10 mL) to the end point at a pH of 4.5 (bromcresol green indicator end point) for the free acid, and to the end point at a pH of 9.5 (phenolphthalein indicator end point) for the total acid. Chromic acid, expressed as anhydride, will titrate:

$$2CrO_3 + 2NaOH \rightarrow Na_2Cr_2O_7 + H_2O \quad \text{Free acid} \qquad \text{(Eq 6)}$$

$$CrO_3 + NaOH \rightarrow NaHCrO_4 \qquad \text{(Eq 7)}$$

$$CrO_3 + 2NaOH \rightarrow Na_2CrO_4 + H_2O \quad \text{Total acid} \qquad \text{(Eq 8)}$$

According to Eq 6 to 8, each milliliter of 0.1 N NaOH solution corresponds to 10 mg CrO_3 (5.2 mg Cr^{6+}) in the free acid titration and to 5 mg CrO_3 (2.6 mg Cr^{6+}) in the total acid titration.

Other acids and salts may also contribute to the free or total acid titration values. For example, phosphoric acid (H_3PO_4) titrates according to:

$$H_3PO_4 + NaOH \rightarrow NaH_2PO_4 + H_2O \quad \text{Free acid} \qquad \text{(Eq 9)}$$

$$H_3PO_4 + 2NaOH \rightarrow Na_2HPO_4 + H_2O \quad \text{Total acid} \qquad \text{(Eq 10)}$$

From Eq 9 and 10, it can be seen that each milliliter of 0.1 N NaOH corresponds to 9.5 mg of phosphate ion (PO_4^{3-}) in the free acid titration and to 4.75 mg PO_4^{3-} in the total acid titration.

The free acidity points are measures of bath acidity, which is important in the attack on the metal. For optimum performance, the bath must be operated in the recommended range for the free acid.

When the bath is operated at a constant level of Cr^{6+} by the addition of replenishing chemicals, reaction products build up, and the total acid value increases. To maintain uniform coating reaction, it is often desirable to keep the total acid value below the limit established for the process by discarding part of the solution.

Other processing parameters that must be monitored include the concentrations of accelerators (when used), free fluoride, Cr^{3+}, and the bath pH.

Accelerators. In some processes, the level of accelerator must be monitored, usually by spectrophotometric procedures. Solution standards that correspond to the low and high limits are prepared, and a color comparison is made for the working bath. This method is used, for example, in the ferricyanide ($Fe(CN)_6^{3-}$) accelerated processes for aluminum.

Fluoride. In the processing of aluminum and its alloys, it is often recommended to monitor free

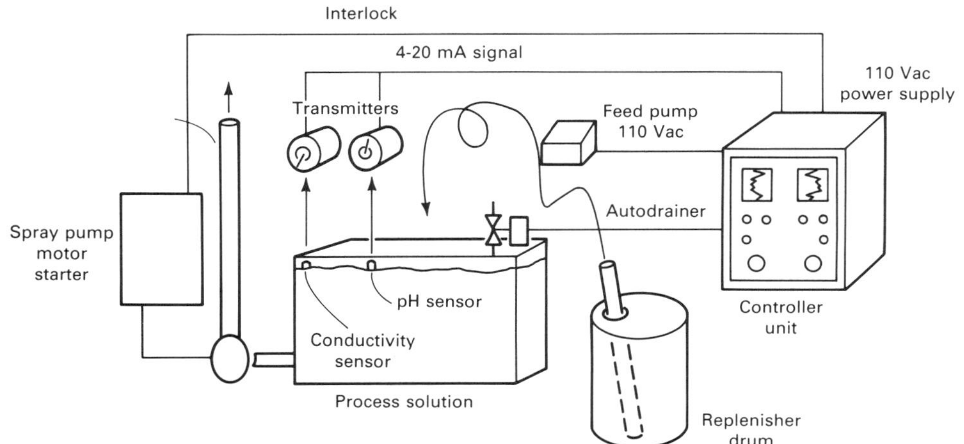

Fig. 6 Schematic diagram of an automatic process control system. Courtesy of Parker Chemical Company

Table 6 Standard recommended practices and specifications for chromate conversion coatings

Substrate	Designation	Title
Aluminum	ASTM B 449	Chromate Treatments on Aluminum
	ASTM D 1730	Preparation of Aluminum and Aluminum Alloy Surfaces for Painting
	AMS 2473	Chemical Treatment for Aluminum Base Alloys—General Purpose Coating
	AMS 2474	Chemical Treatment for Aluminum Base Alloys—Low Electrical Resistance Coating
	MIL-C-5541	Chemical Films and Chemical Film Material for Aluminum and Aluminum Alloys
	MIL-C-6858	Welding Resistance: Aluminum, Magnesium, etc: Spot and Seam
	MIL-C-81706	Chemical Conversion Materials for Coating Aluminum and Aluminum Alloys
Zinc	ASTM B 201	Testing Chromate Coatings on Zinc and Cadmium Surfaces
	ASTM D 2092	Preparation of Zinc-Coated Steel Surfaces for Painting
	AMS 2402	Zinc Plating
	MIL-A-81801	Anodic Coatings for Zinc and Zinc Alloys
	MIL-C-17711	Chromate Coatings for Zinc Alloy Castings and Hot-Dip Galvanized Surface
	MIL-T-12879	Chemical Treatments, Prepaint and Corrosion Inhibitive, for Zinc Surfaces
Magnesium	AMS 2475	Protective Treatments, Magnesium Base Alloys
	MIL-M-3171	Magnesium Alloy, Process for Pretreatment and Prevention of Corrosion on
	MIL-W-6858	Welding Resistance: Aluminum, Magnesium, etc: Spot and Seam
Cadmium	ASTM B 201	Testing Chromate Coatings on Zinc and Cadmium Surfaces

fluoride in the bath. This can be accomplished by using a fluoride ion (F^-) selective electrode and a suitable pH meter.

pH is another useful function for monitoring bath acidity/alkalinity. It should be measured at constant temperature or on a pH meter having temperature compensation.

Trivalent Chromium. In some processes, the level of Cr^{3+} may be important for coating quality. Spectrophotometric methods with low and high standards can be used for such monitoring.

Automatic Control. A well-established practice for monitoring the pretreatment section, particularly in smaller plants, consists of making one or two titrations per day for each bath. The bath control then relies on operator adjustments or on a timer-controlled pump that will deliver the replenisher according to adjustable settings.

With increasing demand for more continuous process control, automatic process monitoring and control are rapidly being recognized as superior to the manual methods. All of the above-mentioned bath-monitoring parameters can be adapted for continuous bath control. The instrumentation for automatic titration and spectrophotometric control is available, but it has not been widely accepted by the industry. Most of the automatic process control systems rely on conductivity and pH sensing. For most compositions, conductivity is related to bath concentration. It depends on temperature, but temperature compensation can be incorporated into the monitoring system. The low maintenance requirements and good reliability of electrodeless conductivity probes contributed to the broad acceptance of this control method. Continuous pH monitoring is also used for some chromating baths, but the electrodes require more frequent checks. Figure 6 shows a schematic of a chromating-stage controlling system.

The conductivity and pH controls can also be applied to other stages of the pretreatment section, and the information from the individual stages can be linked to a central process control unit. The unit can be equipped to provide complete process control, documentation reports, and communication with other plant computer control systems.

Chromate Coatings on Specific Metals

Aluminum. Because of its good corrosion resistance and attractive appearance, aluminum has found significant applications in the building, packaging, transportation, defense, and appliance industries. In most of these applications, aluminum is painted with decorative or functional organic finishes. The natural oxide, which is always present on aluminum surfaces, does not provide adequate corrosion protection or paint adhesion. Paint adhesion and under-paint corrosion resistance can be substantially increased by chromate conversion coatings. The four types in use are alkaline oxide, chromium phosphate, chromate, and no-rinse.

Alkaline Oxide. The alkaline chromating process was the first chemical treatment for aluminum, and it is still used for some appliances and for military equipment (Ref 4). The coatings are applied by immersion in 20- to 30-point alkali chromate/carbonate baths of pH 10 to 11 for up to 20 min at temperatures approaching 95 °C (200 °F). Typical coating weights are between 100 and 500 mg/ft^2, with colors ranging from light to brownish green.

Chromium phosphate coatings were first introduced in 1945 (Ref 5). The coatings are used as a paint base for architectural extrusions in doors, windows, and other exterior applications. Because the coatings do not contain Cr^{6+}, they are widely used for aluminum can end stock and rigid aluminum food containers made from prepainted coil sheet.

The coatings are applied by spray or immersion from processing baths that contain H_2CrO_4, H_3PO_4, and F^-, that have 6 to 30 chromium concentration points, and that usually have a pH less than 2. The coating weights range from 5 to 500 mg/ft^2, and the colors range from colorless to emerald green. Paint base coatings are applied for 5 to 60 s at 25 to 50 °C (80 to 120 °F), depending on the coating weight required. Decorative coatings require dwell times of 1 to 3 min and temperatures of 40 to 60 °C (100 to 140 °F).

The can stock coatings are usually applied in the 5- to 15-mg/ft^2 range, are colorless, and provide excellent lacquer adhesion. In architectural applications, coating weights from 15 to 100 mg/ft^2 form an excellent base for paint. The higher coating weights, up to 500 mg/ft^2, have good bare corrosion resistance and are also suitable for decorative applications. Recent work on the composition of chromium phosphate films has shown that they consist primarily of hydrated chromium phosphate ($CrPO_4$), Cr_2O_3, and aluminum oxides (Ref 6).

Chromate coatings were first introduced in the early 1950s and are now widely accepted by the aluminum-finishing industry for such applications as domestic appliances, small parts, aircraft and electronic equipment, and continuous coil coating of architectural aluminum (Ref 7). The films provide excellent paint adhesion and superior painted and unpainted corrosion resistance. The low contact resistance of bare films is useful in spot welding.

The processing baths contain H_2CrO_4, HF, other mineral acids, and accelerators; they are typically run between 6 and 30 points. The original accelerator was $Fe(CN)_6^{3-}$. Other accelerators, such as molybdate (MoO_4^{2-}) (Ref 8), have recently become more accepted because they eliminate the problem of $Fe(CN)_6^{3-}$ waste treatment and disposal.

Coating weights range from 15 to 200 mg/ft^2, with colors ranging from iridescent yellow to brown. For most paint base applications, coating weights are from 15 to 30 mg/ft^2. The coatings from the $Fe(CN)_6^{3-}$-accelerated process were characterized by XPS and reported to consist of microcrystallites of hydrated chromium oxides covered with an adsorbed monolayer of the accelerator (Ref 9). The MoO_4^{2-}-accelerated coatings have similar compositions, with the accelerator uniformly distributed through the film. Paint base coatings are applied within 5 to 60 s at 25 to 60 °C (80 to 140 °F). Longer times may be required for bare corrosion coatings applied by immersion.

No-rinse processes are finding increasing use in the coil coating of aluminum. In terms of corrosion protection and adhesion, these processes can often provide quality equaling that of conventional processes. An interesting comparison of the options for conversion coatings for aluminum is given in Ref 10.

The applied compositions contain Cr^{6+} and Cr^{3+} as well as other ingredients, such as F^- or PO_4^{3-}. Some formulations include organic compounds. For most paint base applications, the coating weights range from 5 to 25 mg/ft^2. Because the process does not include rinsing after the treatment, the coating weights are directly proportional to the thickness of the applied wet film and the solids content of the coating solution.

Zinc and Galvanized Steels. Although zinc and zinc-coated steels provide better atmospheric-corrosion resistance than bare cold-rolled steel, the natural resistance of zinc surfaces to atmospheric corrosion remains low. Most articles made of, or coated with, zinc by electroplating or hot-dipping are further protected. Chromating is widely used to provide corrosion protection and a decorative finish in bare applications; it can also be used as a pretreatment before painting. The chromate coatings are formed on zinc surfaces from acid solutions containing Cr^{6+}, usually other mineral acids, and accelerators. Chromate coatings on zinc can be categorized as clear, iridescent, and colored.

The largest area for the application of clear coatings is the passivation of hot-dip galvanized steel. Most unpainted zinc and zinc-aluminum coated steel sheet products used for roofs and walls of industrial and farm buildings are chromated on hot-dip galvanizing lines. These chemical treatments are applied in the coating weight range of 1 to 2 mg Cr/ft^2, with the upper limit usually defined by the color acceptable to the coated sheet users. On the strip line, coatings are applied by immersing or spraying the freshly prepared galvanized strip in chromating solutions for 1 to 3 s immediately before a set of squeegee rolls, which remove the excess solution from the strip. Broad ranges of chromium concentration (10 to 80 points) and temperatures of 40 to 70 °C (100 to 160 °F) are used to achieve the desired coating weights. The corrosion protection provided by the chromate coatings increases with their color. Typical salt spray corrosion data are given in Table 4.

Similar bath compositions are used in monorail applications, and the coatings find use in domestic appliances, automotive parts, military hardware, and other areas. The chromate coatings on zinc are also especially suitable for dyeing.

Chromate coatings can be deposited electrolytically from baths containing CrO_4^{2-}, PO_4^{3-}, F^-, and other anions. The resulting coatings, which are grey to black in appearance, have superior corrosion resistance and hardness when compared to the conventional conversion coatings (Table 4).

The use of chromates as a paint base is well established. Chromate treatments are often the best choice when the production schedule requires the coating of zinc and aluminum (also known as mixed production). The use of MoO_4^{2-}-, vanadate-, and tungstate-accelerated chromating baths is also well established. These baths can be used over a concentration range of 5 to 30 points at temperatures between 40 and 70 °C (100 and 160 °F), and with contact times from 5 to 60 s. Table 5 compares the corrosion and adhesion test results of chromate/molybdate coatings on hot-dip galvanized steel and aluminum. No-rinse chromate processes with compositions similar to those used for aluminum are finding increasing acceptance because of the environmental advantages.

Steel. Both chemical and electrochemical methods for chromating steel have been used. The chemical method requires high temperatures and prolonged contact times. No-rinse chromate treatments are used on continuous strip lines for paint base applications. They are applied using chemical compositions and wet film thicknesses to give coating weights in the 15- to 35-mg/ft^2 range and provide excellent paint base properties. Electrolytically produced chromium/chromium oxide coatings, better known as tin-free steel, are extensively used in the metal-packaging industry.

Magnesium is more susceptible to oxidation than aluminum, and it is frequently protected against corrosion by chromating during storage. Before chromating, the natural oxides should be removed by mechanical cleaning or acid pickling in dilute HNO_3 or HF. A mixture of HNO_3 and sodium dichromate ($Na_2Cr_2O_7$), used at ambient temperatures and contact times of 30 to 120 s, will etch the surface and produce greyish chromate coatings.

Electrochemical coatings can be deposited from solutions of H_3PO_4, $Na_2Cr_2O_7$, and ammonium bifluoride (NH_4HF_2). The coating characteristics can be modified by varying the time or current density to form thin paint base coatings or heavy films, which provide maximum bare corrosion protection.

Cadmium chromating processes use the same or similar compositions and treatment conditions as for zinc.

Testing Methods and Standards for Coatings

Testing methods for chromate coatings have been developed to assess the quality of the bare coatings and, in the case of painted applications, the performance of the entire coating system. The tests are used to evaluate several characteristics of the chromate coatings that are important to performance, as follows.

Appearance. The coatings are visually examined for uniformity, color, scratches, and visible defects. The apparent nonuniformity of iridescent coatings is usually not significant.

Coating weight is the most important quantitative parameter measurable for the coating. Corrosion protection and paint adhesion vary, depending on coating weight. The recommended ranges should be maintained.

Abrasion Resistance. The stability of the coatings can be tested by rubbing the surface with a finger wrapped in a soft white tissue. Removal of a large portion of the coating may be significant for adhesion of the coating or a paint or for dusting or buildup on dies.

Corrosion. Many different accelerated corrosion tests are used. For bare coatings, the most frequently conducted tests are the neutral salt spray test (Ref 3), the different water immersion tests, the humidity test (Ref 12), and the stack tests. For painted surfaces, the neutral salt spray test and the humidity test are the most common, but acetic acid salt spray (Ref 13) is often preferred for aluminum.

Adhesion. Standard adhesion tests for painted chromate coatings include T-bend (Ref 14), impact (Ref 15), cross hatch adhesion, various forming tests, and adhesion tests after exposure to salt spray or boiling water.

Outdoor exposure is the ultimate test for bare and painted surfaces. Outdoor exposure tests are conducted at selected sites representing rural, coastal, and industrial environments. These tests provide the best information about the performance of any coating system.

Many of the testing methods are established as national and international standards, while others are a part of well-established quality control practices developed by metal producers, finishers, and end users. Some of the standard recommended practices and specifications for chromate conversion coatings are listed in Table 6.

Safety and Waste Treatment

When handling any chemicals, it is a good working practice to consult the Material Safety Data Sheets (MSDS), which contain useful information on the chemicals and the possible hazards associated with handling and storage. The MSDS will also list all hazardous ingredients, special precautionary information, and first aid procedures in case of an accidental exposure.

Chromating baths and the concentrates used for their buildup are usually acidic and can irritate the skin and eyes upon contact. Some chromium compounds have also been identified as possible carcinogens. Both the National Toxicology Program and the International Agency for Research on Cancer report that there is sufficient evidence for the carcinogenity of chromium and certain chromium compounds in animals and humans. Recent studies have shown that aqueous solutions of Cr^{6+} administered directly into the lung have caused an increased incidence of lung cancer in animals. Although the potential for inhalation exposure with the chemicals in the concentrated form is very low, heating and spray processing may increase the possibility of inhalation.

The disposal of spent solutions and rinse waters requires waste treatment. Hexavalent chromium must be reduced to Cr^{3+} before neutralization and precipitation. Sodium pyrosulfite ($Na_2S_2O_5$) is usually used as the reducing agent in smaller operations, while for larger plants, sulfur dioxide (SO_2) is preferred for economic reasons. Wastewater treatment sludges from chromating operations are considered hazardous waste unless classified otherwise.

REFERENCES

1. H.E. Townsend and R.G. Hart, *J. Electrochem. Soc.*, Vol 131, 1984, p 1345
2. J.A. Kramer and L. Salvati, Jr., in *Proceedings of NCCA Technical Meeting*, National Coil Coaters Association, Oct 1982, p 23
3. "Standard Method of Salt Spray (Fog) Testing," B 117, *Annual Book of ASTM Standards*, American Society for Testing and Materials
4. British Patent 226,776
5. U.S. Patent 2,438,877
6. J.A. Treverton and N.C. Davies, *Mater. Technol. (London)*, Vol 4, 1977, p 480
7. U.S. Patent 2,796,370
8. U.S. Patent 3,404,043
9. J.A. Treverton, in *Production and Use of Coil-Coated Strip*, The Metals Society, 1981, p 64
10. J.I. Maurer, in *Proceedings of the Aluminum Finishing Seminar*, Vol 2, The Aluminum Association, 1982, p 453
11. U.S. Patent 3,404,044
12. "Standard Method for Testing Coated Metal Specimens at 100% Relative Humidity," D 2247, *Annual Book of ASTM Standards*, American Society for Testing and Materials
13. "Standard Method of Acetic Acid-Salt Spray (Fog) Testing," B 287, *Annual Book of ASTM Standards*, American Society for Testing and Materials
14. "Standard Practice for Testing Coil Coatings," D 3794, *Annual Book of ASTM Standards*, American Society for Testing and Materials
15. "Standard Test Method for Resistance of Organic Coatings to the Effects of Rapid Deformation (Impact)," D 2794, *Annual Book of ASTM Standards*, American Society for Testing and Materials

Aluminum Anodizing

Jeff Pernick, International Hardcoat, Inc.

ALUMINUM ANODIZING is an electrochemical method of converting aluminum into aluminum oxide (Al_2O_3) at the surface of the item being coated. It is accomplished by making the workpiece the anode while suspended in a suitable electrolytic cell. Although several metals can be anodized, including aluminum, titanium, and magnesium, only aluminum anodizing has found widespread use in industry.

Because a wide variety of coating properties can be produced through variations in the process, anodizing is used in almost every industry in which aluminum can be used. The broadest classification of types of anodize is according to the acid electrolyte used. Various acids have been used to produce anodic coatings, but the most common ones in current use are sulfuric (H_2SO_4) and chromic (CrO_3) acids. Although CrO_3 anodizing is standardized, there are two main types of H_2SO_4 anodizing. The first is a room-temperature H_2SO_4 process termed conventional anodizing, and the second is a low-temperature H_2SO_4 process termed hardcoat anodizing. In addition to CrO_3, conventional, and hardcoat anodizing, a process known as sealing can be used to enhance certain characteristics. A number of standard tests are used in the industry to test the quality and characteristics of anodic coatings.

The three common types of anodize described above are usually controlled and described through the use of military specification MIL-A-8625 (Table 1). It has become standard in the industry to describe anodic coatings with the type and class nomenclature outlined in this specification.

The articles "Corrosion of Magnesium and Magnesium Alloys" and "Corrosion of Aluminum and Aluminum Alloys" in this Volume contain information on the corrosion resistance of anodized magnesium alloys and aluminum alloys. More information on the anodizing process for aluminum is available in the article "Cleaning and Finishing of Aluminum and Aluminum Alloys" in Volume 5 of the 9th Edition of *Metals Handbook*.

Chromic Anodize

Chromic anodize (type I; see Table 1) is formed by immersing the workpiece in an aqueous solution of CrO_3. Current is then applied, with the workpiece being positively charged. Typical operating parameters for the CrO_3 anodizing process are:

- *Electrolyte concentration*: 50 to 100 g/L CrO_3
- *Temperature*: 37 ± 5 °C (100 ± 9 °F)
- *Time in bath*: 40 to 60 min
- *Voltage*: Increase from 0 to 40 V in 10 min; hold at 40 V for balance of time
- *Current density*: 0.15 to 0.30 A/dm^2 (1.4 to 4.3 A/ft^2)

Chromic Anodized Coatings. The CrO_3 anodizing process produces a coating that is nominally 2 μm (0.08 mils) thick. It is relatively soft and susceptible to damage through abrasion or handling. The color of the class 1 coating ranges from clear to gray, depending on whether the coating is sealed and on the alloy coated. The coating can be dyed to produce a class 2 coating; however, this is not generally done, because the coating is thin and does not retain the dye color well. About two-thirds of the coating thickness penetrates the base metal; one-third of the coating builds above the original base metal dimension. Thus, for a coating thickness of 2 μm (0.08 mils) per side, the dimensional change of the workpiece would be 0.7 μm (0.028 mils) per side.

Although the industry has adopted the penetration/buildup terminology, the terms are somewhat misleading. Actually, when the aluminum is converted to Al_2O_3 it takes up more space—approximately 133% of the space previously occupied by the aluminum converted. The penetration/buildup terms are used only as a convenience in predicting dimensional change in a coated article. The corrosion resistance of this coating is very good. The coating will pass in excess of 336 h in 5% salt (NaCl) spray per ASTM B 117.

Advantages. Although CrO_3 anodizing is the least used of the three types of anodize, it has several advantages that make its use desirable. First, because CrO_3 is much less aggressive toward aluminum than H_2SO_4, it should be used whenever part design is such that rinsing is difficult. Difficult rinsing designs would include welded assemblies, riveted assemblies, and porous castings. Second, a typical CrO_3 anodize buildup is 0.7 μm (0.028 mils) per side with good repeatability. Therefore, it is a very good coating to use when it is necessary to coat a precise dimension to size. Third, because CrO_3 anodize produces the least reduction in fatigue strength of the three coatings, it should be used where fatigue strength is a critical factor. Fourth, the color of CrO_3 anodize will change with different alloy compositions and heat-treat conditions; this makes it useful as a test of the homogeneity of structural components. Lastly, when properly applied, CrO_3 anodize can be used as a mask for subsequent hardcoat anodize operations.

Suitable Alloys. Most alloys can be successfully coated by the CrO_3 process. Exceptions are high-silicon die-cast alloys and high-copper alloys. The rule for suitability is that any alloy containing more than 5% Cu, 7% Si, or total alloying elements of 7.5% should not be coated by this process.

Relative Costs. Chromic anodize costs more than H_2SO_4 but less than hardcoat anodize.

Sulfuric Anodize

Sulfuric anodize, or type II anodize, is formed by immersing the item in an aqueous solution of H_2SO_4. Current is then applied, and the workpiece is positively charged. Typical operating parameters for the H_2SO_4 anodizing process are:

- *Electrolyte concentration*: 15% H_2SO_4
- *Temperature*: 21 ± 1 °C (70 ± 2 °F)
- *Time in bath*: 30 to 60 min
- *Voltage*: 15 to 22 V, depending on the alloy
- *Current density*: 1 to 2 A/dm^2 (9.3 to 18.6 A/ft^2)

Sulfuric Anodized Coatings. This process produces a coating that is normally 8 μm (0.31 mils) in minimum thickness. Although harder than type I coatings, H_2SO_4 anodize may still be damaged by moderate handling or abrasion. The color of the class 1 coating is yellow-green because of the preferred sealing method of immersion in sodium dichromate ($Na_2Cr_2O_7$). Clear coatings can also be produced by sealing in hot water. Clear coatings should be specified by the notation "class 1, clear." This coating can also be dyed to produce a class 2 coating. This type of anodize produces the most pleasing colors of the three anodizing methods. Dyed H_2SO_4 anodize coatings have deep colors with good repeatability. Like CrO_3 anodize, H_2SO_4 anodize coatings penetrate the base metal for two-thirds of their thickness and build above the original base metal dimension for one-third the total thickness. As with all types of anodize, the corrosion resistance of H_2SO_4 anodize is very good; it has an ASTM B 117 salt spray resistance of at least 336 h.

Table 1 Classification of anodize according to MIL-A-8625

Type	Class	Description	Dye	Seal	Thickness μm	Thickness mils
I	1	CrO_3 anodize	No	Yes	1.3–2.5	0.05–0.1
I	2	CrO_3 anodize, dyed	Yes	Yes	1.3–2.5	0.05–0.1
II	1	H_2SO_4 anodize	No	Yes	7.5–15	0.3–0.6
II	2	H_2SO_4 anodize, dyed	Yes	Yes	7.5–15	0.3–0.6
III	1	Hardcoat anodize	No	No	46–56	1.8–2.2
III	2	Hardcoat anodize, dyed	Yes	Yes	46–56	1.8–2.2

Advantages. Sulfuric anodize is the most widely used type of anodize and has many desirable benefits. First, because it has a fairly hard surface, it can be used in situations that require light-to-moderate wear resistance. Applications include lubricated sliding assemblies and items subject to handling wear, such as front panels. Second, because it is the most aesthetically pleasing type of anodize, it should be used where final appearance is important. It can be dyed almost any color and produces deep, rich shades that make the item appear to be made of a material bearing a color throughout, rather than an applied coating. Lastly, because corrosion resistance is good, it should be used whenever corrosion resistance is needed and the specialized benefits of the other two anodize types are not required.

Suitable Alloys. With the exception of high-silicon die-cast alloys, all alloys can be successfully coated with H_2SO_4 anodize. Clarity and depth of color of the anodize increase with the purity of the alloy. Therefore, alloys should be chosen for maximum purity consistent with the physical requirements needed in the item.

Relative Cost. Sulfuric anodize is the least costly and most widely available type of anodize.

Hardcoat Anodize

Hardcoat anodize, or type III anodize, is formed by immersing the item in an aqueous solution of H_2SO_4. Current is then applied, with the workpiece being the anode. The operating parameters for a generic hardcoat anodize process are:

- *Electrolyte concentration*: 22 to 24% H_2SO_4
- *Temperature*: 0 ± 1 °C (32 ± 2 °F)
- *Time in bath*: 20 to 120 min
- *Voltage*: constantly increased to maintain current density at 2.5 to 4.0 A/dm^2 (23.2 to 37 A/ft^2)

Hardcoat Anodize Coatings. This process produces a coating that is normally 50 μm (2 mils) thick, although other thicknesses can be specified. The coating is extremely hard. It is described as file hard (equal to about 60 to 70 HRC). The color of the class 1 coating ranges from gray to bronze to almost black, depending on the alloy coated, the coating thickness, and the electrolyte temperature. The coating can be dyed to produce a class 2 coating. Because thick coatings are naturally very dark, only colors darker than natural are possible. Generally, this limits the dying of hardcoat to black in common processes. If a more extensive color choice is required, there are several proprietary hardcoat processes available to accomplish this.

Hardcoat penetrates the base metal for one-half of its thickness and builds above the original base metal dimension for one-half of its thickness. Thus, for a thickness of 50 μm (2 mils) per side, the dimensional change of the workpiece would be 25 μm (1 mil) per side. Commercially available coating thickness tolerances are the greater of ±5 μm or ±10% of the total targeted thickness. The corrosion resistance of the unsealed class 1 coating is very good and comparable to the other types of anodize. When the hardcoat anodize is sealed, as in a class 2 coating, it becomes the most corrosion-resistant type of anodize.

Advantages. Hardcoat anodize, because of its variety of desirable properties, has found widespread use in manufactured products. First, because of its extreme hardness, it is used in situations in which wear resistance is required. Applications include valve/piston assemblies, drive belt pulleys, tool holders and fixtures, and many other items requiring wear resistance.

Second, because of its excellent resistance to corrosion, hardcoat is used on aluminum components in harsh environments. These include outside exposure in salt air, marine components, automobile wash equipment, components for the aircraft and aerospace industries, and food preparation machines.

Third, because hardcoat is an excellent electrical resistor, it can be used to insulate heat sinks for direct mounting of electrical or electronic equipment. Also, it is used in welding fixtures where some areas may need to be insulated from work.

Fourth, because hardcoat is a naturally porous substance, it is used in many areas in which the bonding or impregnation of other materials to aluminum is needed. This coating bonds very well with paints and adhesives. Also, it can be impregnated with teflon (polytetrafluoroethylene, or PTFE) and many dry film lubricants to impart lubricating properties to the coating.

Lastly, because of its desirable properties and also because it produces a buildup of coating, it is widely accepted as a salvage coating to restore worn or improperly machined parts to usable dimensions. Coating thicknesses in excess of 250 μm (10 mils) per side are possible on some alloys with certain proprietary hardcoat processes.

Suitable Alloys. Although almost all alloys can be coated, the 6000-series aluminum alloys produce the best hardcoat properties. As with the other anodize types, high-silicon die castings produce the lowest-quality coatings. Also, because the hardcoat process is sensitive to copper, alloys in the 2000 series should be avoided if possible. Alloys containing copper can be hardcoated, but only a relatively few commercial sources have the ability to coat these alloys with reliability.

Relative Costs. Hardcoat anodize is the most expensive type of anodize. It is generally twice the cost of H_2SO_4 anodize and 50% more than CrO_3 anodize.

Sealing of Anodized Coatings

Because all of the anodic processes produce porous Al_2O_3 coatings, it is often desirable to seal the coating to close these pores and to eliminate the path between the aluminum and the environment. Sealing involves immersing the coating in hot water; this hydrates the Al_2O_3 and causes the coating to swell in order to close the pores. Conventional sealing is generally done at a minimum temperature of 95 °C (200 °F) for not less than 15 min. There are also several proprietary nickel-base sealing agents available that are said to produce sealing at low temperature through catalytic action. Chromic and sulfuric anodizes are almost always sealed. However, because sealing softens the coating somewhat, hardcoat anodize is usually not sealed unless criteria other than hardness have the maximum importance in the finished coating.

Testing of Anodized Coatings

There are six commonly used tests to determine the quality of anodized coatings. These are visual, corrosion resistance, wear resistance, adhesion, thickness, and coating weight. Only a brief overview will be given here; extensive instructions are available in specification MIL-A-8625.

Visual inspection often indicates the overall quality of a coating. The anodic film should be uniform in appearance and free from breaks, scratches, and powdery areas.

Corrosion resistance is most often tested by salt spray. A coated panel is suspended in a salt fog for a period of time (typically 336 h) and then examined for pits and corrosion.

Wear resistance is tested through an abrasive cycle. A test panel is weighed, abraded for a number of cycles, and weighed again to determine the coating weight lost through the abrasion.

Adhesion is tested by bending a coated panel around a mandrel and checking for delamination.

Thickness is commonly checked by using one of three methods. The first is by metallographic examination of the coating under a calibrated microscope. Second, thickness can be determined by measuring a dimension of the coated part, stripping the coating, and measuring again to determine dimensional change. Third, coatings can be measured using eddy-current instruments.

Coating weight is an indication of the density of the coating in relation to its thickness. Coating weight is determined by weighing a coated panel of known area, stripping the coating, reweighing the panel, and dividing the weight loss by the panel area for the indication of weight loss per unit area.

Table 2 lists standard test methods for anodize. More information on the testing of anodize is available in the section "Electrochemical Methods of Corrosion Testing" of the article "Laboratory Testing" in this Volume.

Stripping of Anodized Coatings

Stripping a part that has been anodized always results in some loss of dimensions as compared to the original sizes of the part. This is because the aluminum that was consumed to form the coating is removed since it has now become part of the coating. Thus, while a type II coating 7.5 μm (0.3 mils) thick would result in a 2.5-μm (0.1-mil) increase from original dimensions, stripping would decrease this by at least 7.5 μm (0.3 mils), depending on the precision of the operation.

There are three main methods of stripping, with varying degrees of controllability. Controllability is defined as the ability to remove only the anodize and not damage the aluminum base met-

Table 2 ASTM standard test methods for anodized coatings

Method	Standard
Coating thickness	
Eddy current	B 244
Metallographic	B 487
Light section microscope	B 681
Coating weight	B 137
Sealing	
Dye stain	B 136
Acid dissolution	B 680
Impedance/admittance	B 457
Voltage breakdown	B 110
Corrosion resistance	
Salt spray	B 117
Copper-accelerated, acetic acid salt-spray	B 368
Ford anodized aluminum corrosion test	B 538

al. The least controllable method is by immersion in warm sodium hydroxide (NaOH). This is known as caustic etching. In addition to removing anodize, this process also dissolves aluminum at a fast rate. A more controllable method is by immersion in a H_2SO_4-CrO_3 solution. These solutions are generally classified as deoxidizers. This process will also dissolve aluminum, but at a much slower rate than etching. The most controllable method is by immersing the part to be stripped in a CrO_3-H_3PO_4 solution at a minimum temperature of 95 °C (200 °F). This solution will dissolve only the coating and will not harm the aluminum.

SELECTED REFERENCES

- "Anodic Coatings for Aluminum and Aluminum Alloys," MIL-A-8625, Naval Publication and Forms Department, June 1985
- S. Wernick and R. Pinner, *Surface Treatment of Aluminum*, Robert Draper Ltd., 1972

Organic Coatings and Linings

Kenneth B. Tator, KTA-Tator, Inc.

THE TOTAL national yearly cost of metallic corrosion in the United States was estimated at $167 billion in 1985. This total includes replacement costs and lost production costs as well as the decrease in lifetime and the expected replacement value of a given component subject to damage by corrosion. The cost of corrosion also includes the means by which the effects of corrosion are mitigated, such as the use of cathodic protection, inhibitors, alternative materials of construction, overdesign, and protective coatings.

This latter category, the use of protective coatings, is in itself a very substantial category. In 1975, in a more detailed survey, the National Association of Corrosion Engineers (NACE) estimated the cost of direct expenditures by NACE members to combat corrosion at $9.67 billion. Of this amount, $2.9 billion was spent for protective coatings and services. Coating application accounted for $805 million; purchase of coatings for atmospheric service, $531 million; marine coatings, $502 million; external pipeline coatings, $310 million; and miscellaneous expenses, $750 million.

Therefore, the use of protective coatings and linings and the costs associated with such use are considerable. More metal surfaces are protected by organic coatings and linings than by all other methods combined. In addition to protecting against corrosion, coatings often beautify and provide an aesthetic appeal. Safety colors are used to mark pipes, to indicate their contents, and to provide warnings of hazardous or dangerous work areas.

The Effect of Legislation on the Coatings Industry

Legislation concerning worker health and environmental protection has had a marked impact on the coatings industry. For example, although the practice of blast cleaning a steel surface and applying an alkyd, vinyl, or epoxy coating system remains one of the best methods of corrosion protection for many steel surfaces, the alkyds, epoxies, and vinyls themselves have changed as a result of legislation restricting the release of volatile organic compounds (solvents) and the use of toxic pigments

Surface preparation techniques have also changed drastically; silica sand as a blast-cleaning abrasive has been banned in virtually all Western countries except the United States and Canada (and there is a strong movement to ban it in these countries as well). The containment and safe disposal of spent blast-cleaning abrasives are required in many localities to prevent environmental damage caused by the leaching into water supplies of lead, chromate, and other toxic paint pigments removed during blast cleaning. Therefore, although a paint layer over a properly cleaned surface still acts as a barrier against a corrosive environment in most cases, the components that constitute this barrier have changed considerably within the past 10 years, and the means by which the surface is properly cleaned are rapidly changing at the present time.

This article will present an overview of the different types of coating and lining materials available, along with information on the various means of surface preparation and the equipment and techniques of coating application. However, it must be recognized that all facets of coating technology are strongly influenced by legislative decree; the alkyds, vinyls, and epoxies currently in use are being reformulated, and within a few years, although the names will be the same, the compositions of these coatings (and their potential protective capabilities) may be considerably different. Furthermore, surface preparation and application techniques in current use are also under considerable legislative pressure and are expected to change considerably in the near future.

Coating Industry Response to Legislative Pressure

The coating industry has been very responsive to enacted legislation and to proposed legislation limiting the use of potentially harmful or toxic raw materials or surface preparation and/or application techniques. As a result of this positive industry reaction, the legislators as a whole have been pragmatic in their laws. For example, it is widely recognized that lead, when ingested by the body, results in anemia, damage to the central nervous system, disruption of the reproductive system, and retardation in child development. Therefore, in 1973, Congress, under the impetus of the Food and Drug Administration, limited the lead content in all paints used for consumer use to 0.06% solids by weight.

This law effectively banned the use of lead pigments in all paints used around the home. However, because lead-containing paint pigments, such as red lead, lead chromate, and lead suboxide, are perhaps the best pigments in use for corrosion protection, they were not outlawed for paints used for corrosion protection purposes in industry. In fact, even today, no legislation restricts the use of such pigments for industrial and marine corrosion protection purposes.

However, recently enacted legislation regarding worker and environmental protection is drastically curtailing the use of lead pigments as well as other toxic pigments, such as those in the chromate, mercury, and tin families. Although such pigments can be legally used in a paint, the substantial costs due to worker protection and the high costs associated with containment and disposal of paints containing such pigments are effectively precluding their use in most new coating formulations.

Similarly, restrictions regarding volatile organic compounds in coatings have been gradually enacted over the last 10 years in order to allow formulators to develop comparable paint materials with complying solvents. Although solvent restrictions were originally enacted against what were believed to be photochemically reactive solvents (solvents that reacted under the influence of ultraviolet light and degraded to form smog and to deplete the ozone layer), it was later found that virtually all volatilizing solvents photochemically degraded and were therefore environmentally detrimental. This realization, together with the potential for further legislative restriction, has motivated coating formulators to develop water-base paints or high-solids (low-solvent) coating materials.

Volatile organic content (VOC) legislation has been mandated in California for many architectural coatings. This rule prohibits the manufacture, sale, or use of designated industrial maintenance primers and topcoats if the VOC content exceeds 420 g/L (3.5 lb/gal.). Most coating formulators were prepared for this ruling, and the quality of water-base and high-solids coating formulations is fast approaching, and in some cases exceeding, the performance of conventionally formulated solvent-containing coating systems.

Conversely, there are many industrial facilities and highway bridges that are coated with old lead-containing paints. Environmental Protection Agency (EPA) legislation requires the disposal of removed paints (and spent abrasive) in hazardous waste disposal sites if the leachate after acid digestion (pH 5) contains more then 5 ppm of lead or chromate and 2 ppm of mercury. The costs of such a disposal, not including collection costs, are estimated by many painting contractors to be from six to ten times as much as disposal costs in a normal sanitary landfill. The cost of containing the spent abrasive and paint, as opposed to letting the spent abrasive fall to the ground during blast cleaning, may in itself double or triple the cost of a paint job.

To date, enactment and enforcement of legislation have been very spotty, and many paint contractors or plant painting forces have not had to strictly comply with the law. However, this situation is expected to change in the very near future, and it is felt that the high costs associated with the repainting of many existing bridges and structures will lead to the development of new surface preparation concepts and enclosure techniques. The surface preparation equipment required for large-scale coating removal projects will become much more capital intensive, and it is possible that those coating contractors who tend to disregard specification requirements regarding the extent of surface preparation or coating thickness, and so on, will find another means of evasion, namely the violation of environmental legislation regarding containment and disposal of toxic paint residues.

Despite the pending health and environmental legislative influences and the rapidly occurring changes in the coatings industry, corrosion continues, and painting for corrosion protection must be done. Therefore, the following sections will discuss the more commonly used coating and lining materials; their characteristics, advantages, and disadvantages; surface preparation equipment and techniques; application and inspection methods; and equipment.

Coating and Lining Materials

Paints or linings that act as a protective film to isolate the substrate from the environment exist in a number of different forms. Sheet linings, commonly of the vinyl or vinylidene chloride family, are one such type of coating that can be either adhered to the surface to be protected or suspended as a bag within a tank, for example, to provide protection. Hot-applied organisols, or plastisols, again usually of the vinyl family, can also be applied to a surface, typically by dipping or flow coating, to provide a protective film.

Powder coatings are being increasingly used to protect concrete-reinforcing rod, as pipeline coatings, and as coating materials in the original equipment manufacturing markets. Fine powders produced from high molecular weight resins of the thermoplastic vinyl and fluorinated hydrocarbon families or from thermoset resins of the epoxy and polyester families are applied to the surface to be protected by either electrostatic spray or fluidized-bed deposition. The metal being protected is usually preheated at the time of application, and after application it is reheated to an elevated temperature (generally from 150 to 315 °C, or 300 to 600 °F). The specific time/temperature baking schedule depends on the metal temperature at the time of application and the type of powder being applied.

Alternatively, some coating systems are characterized by the application method used. For example, for coil-coated metal sheet (commonly steel or galvanized steel), very specialized high-speed roller application equipment is used to coat the sheet steel as it is unwound from a coil. The paint used in the coil-coating process can be of virtually any generic type, although alkyds, polyesters, epoxies, and zinc-rich epoxy coatings are the most prevalent.

Certain lining materials, such as hand lay-up fiberglass-reinforced plastics, are also used to protect steel from corrosion. Such coating systems usually consist of an epoxy primer applied to a blast-cleaned steel surface, followed by one or more polyester gel coats, with one or more layers of a fiberglass veil or woven roving mat laid within the gel coats as reinforcement. The system is then sealed with a layer of the polyester gel coat (a semiclear, 100% solids resin coat). Similarly, rubber linings are used to protect against corrosion. There are various types of rubbers, but they can generally be categorized as prevulcanized or postvulcanized (vulcanized after application). Similarly, rubbers can be formulated with different hardnesses and chemical resistances. Commonly, a rubber lining is a composite of two or three different types of rubbers adhered to each other and to the surface.

The coatings and linings discussed above are mentioned here in summary only; a more thorough description is beyond the scope of this article. Powder coatings and coil coatings are discussed in more detail in Volume 5 of the 9th Edition of *Metals Handbook*.

This article will be limited to the liquid-applied (usually by brush, roller, or spray) coating and lining materials that are commonly used for corrosion protection in atmospheric or immersion service. The rate of base metal corrosion where such coatings or linings are used should not exceed approximately 1.3 mm/yr (50 mils/yr). For corrosion rates above this, both in atmospheric and immersion service, or where catastrophic failure is of concern, liquid-applied coatings probably should not be used, and corrosion-protective measures should include the use of more corrosion-resistant alloys, sheet or rubber coatings and linings, fiberglass lay-ups, and so on.

The coating systems discussed in this article will be categorized by the generic type of binder or resin and will be grouped according to the curing or hardening mechanism inherent in that generic type. Although the resin or organic binder of the coating material is most influential in determining the resistances and properties of the paint, it is also true that the type and amount of pigments, solvents, and additives such as rheological aids will dramatically influence the application properties and protective capability of the applied film. Furthermore, hybridized systems can be formulated that are crosses between the categories. For example, an acrylic monomer or prepolymer can be incorporated with virtually any other generic type of resin to produce a product with properties that are a compromise between the acrylic and the original polymer. In many cases, this is advantageous, as in the mixing of vinyls and acrylics or heat-curing alkyds and acrylics. In other cases, such as with an epoxy, acrylic modification may be a detriment. Table 1 lists the advantages and limitations of the principal coating resins.

Auto-Oxidative Cross-Linked Resins

These coating types dry and ultimately cross link by reaction with oxygen from the atmosphere. All such coatings in this class contain drying oils, which consist mainly of polyunsaturated fatty acids and undergo film formation by oxidative drying. The drying reaction is accelerated by the presence of metallic salts such as cobalt, manganese, and lead napthenates or octoates. The drying reaction results in a reduction of unsaturated groups, commonly ethylenic carbon double bonds. The auto-oxidation reaction occurs at a relatively fast rate shortly after application of the wet paint, and it continues throughout the life of the coating, although at a much slower rate with time.

The auto-oxidative reaction is a decomposition reaction that results in the formation of hydroperoxides. Although auto-oxidative cross linking is necessary to attain the ultimate resistant properties of the film (such properties are generally attained within weeks after application), with time, the hydroperoxides formed as a result of the auto-oxidative cross linking will decompose into organic acid constituents, leading to chemical deterioration of the film. Similarly, the auto-oxidative reaction results in the hardening and ultimate embrittlement of the film, causing reduced distensibility and flexibility with the passage of time. This brittleness and accompanying hydroperoxide decomposition result in an aging deterioration of the film and the ultimate loss of protective capability. The rate at which the coating will lose its protective capability is determined by temperature, thickness, solar exposure, and the types of pigmentation (for example, zinc oxide and lead-containing pigments are particularly useful in absorbing hydroperoxides and slowing the rate of film deterioration).

Auto-oxidatively cross-linked films, in every case, contain a drying oil; this is usually a vegetable oil, such as tung, oiticica (similar to tung; obtained from the seeds of a South American tree), castor, linseed, safflower, soybean, and tall oil, or a fish oil (menhaden). The drying oil can be mixed directly with the pigment (such as red or white lead, with the addition of the metallic dryer) to form the complete coating. More commonly, however, the drying oil is combined with a resin, such as an alkyd, epoxy ester, or polyurethane, by heating or cooking. The resin adds toughness and chemical resistance to the oil, increasing its chemical and moisture resistance.

The amount of oil combined with the resin will influence the protective capability of the applied film. Long oil modifications result in less chemical resistance and longer drying times, but increased penetrability and a greater ability to protect over a poorly cleaned surface. A lesser oil modification results in a short oil that must be applied over a relatively clean substrate and dries quickly. Short oils are commonly used in fast-drying coatings that fully cure after baking for a few minutes at about 95 °C (200 °F). Short oil coatings have good moisture and chemical resistance but are relatively hard and brittle. The medium oil modifications are a compromise between the long and short oil modifications. Most commonly, alkyds, epoxy esters, and urethane resins are used in combination with drying oils to form auto-oxidatively cross-linked coatings.

Alkyd resins are derived as the reaction product of polyhydric alcohols and polybasic acids. This definition also includes polyester resins, of which alkyds are a specific type. The basic difference between an alkyd and a polyester is that the alkyd uses a polybasic acid derived from semidrying or drying oils such that the resin so formed can undergo auto-oxidation at ambient temperatures through reaction with atmospheric oxygen at the unsaturated groups present in the fatty acid molecules. Polyesters, on the other hand, do not undergo the auto-oxidative drying mechanism, because the dibasic acid used, commonly a glycerol, has very few free unsaturated groups available for auto-oxidative cross linking. The curing reaction of polyester resins is a free radical initiated cross linking, and polyesters are therefore described as chemically cross-linked coating systems.

The properties of alkyd coatings are derived predominantly from the properties of the drying oil used in the manufacture of the resin. Drying time, hardness, color, and moisture sensitivity all depend

Table 1 Advantages and limitations of principal coating resins

Resin type	Advantages	Limitations	Comments
Alkyds	Good resistance to atmospheric weathering and moderate chemical fumes; not resistant to chemical splash and spillage. Long oil alkyds have good penetration but are slow drying; short oil alkyds are fast drying. Temperature resistant to 105 °C (225 °F)	Not chemically resistant; not suitable for application over alkaline surfaces, such as fresh concrete or for water immersion	Long oil alkyds make excellent primers for rusted and pitted steel and wooden surfaces. Corrosion resistance is adequate for mild chemical fumes that predominate in many industrial areas. Used as interior and exterior industrial and marine finishes
Epoxy esters	Good weather resistance; chemical resistance better than alkyds and usually sufficient to resist normal atmospheric corrosive attack	Generally the least resistant epoxy resin. Not resistant to strong chemical fumes, splash, or spillage. Temperature resistance: 105 °C (225 °F) in dry atmospheres. Not suitable for immersion service	A high-quality oil-base coating with good compatibility with most other coating types. Easy to apply. Used widely for atmospheric resistance in chemical environments on structural steel, tank exteriors, etc.
Vinyls	Insoluble in oils, greases, aliphatic hydrocarbons, and alcohols. Resistant to water and salt solutions. Not attacked at room temperature by inorganic acids and alkalis. Fire resistant; good abrasion resistance	Strong polar solvents redissolve the vinyl. Initial adhesion poor. Relatively low thickness (0.04 to 0.05 mm, or 1.5 to 2 mils) per cost. Some types will not adhere to bare steel without primer. Pinholes in dried film are more prevalent than in other coating types.	Tough and flexible, low toxicity, tasteless, colorless, fire resistant. Used in potable water tanks and sanitary equipment; widely used industrial coating. May not comply with VOC regulations
Chlorinated rubbers	Low moisture permeability and excellent resistance to water. Resistant to strong acids, alkalies, bleaches, soaps and detergernts, mineral oils, mold, and mildew. Good abrasion resistance	Redissolved in strong solvents. Degraded by heat (95 °C, or 200 °F, dry; 60 °C, or 140 °F, wet) and ultraviolet light, but can be stabilized to improve these properties. May be difficult to spray, especially in hot weather	Fire resistant, odorless, tasteless, and nontoxic. Quick drying and excellent adhesion to concrete and steel. Used in concrete and masonry paints, swimming pool coatings, industrial coatings, marine finishes
Coal tar pitch	Excellent water resistance (greater than all other types of coatings); good resistance to acids, alkalies, and mineral, animal, and vegetable oils	Unless cross linked with another resin, is thermoplastic and will flow at temperatures of 40 °C (100 °F) or less. Hardens and embrittles in cold weather. Black color only, will alligator and crack upon prolonged sunlight exposure, although still protective	Used as moisture-resistant coatings in immersion and underground service. Widely used as pipeline exterior and interior coatings below grade. Pitch emulsions used as pavement sealers. Relatively inexpensive
Polyamide-cured epoxies	Superior to amine-cured epoxies for water resistance. Excellent adhesion, gloss, hardness impact, and abrasion resistance. More flexible and tougher than amine-cured epoxies. Temperature resistance: 105 °C (225 °F) dry; 65 °C (150 °F) wet	Cross linking does not occur below 5 °C (40 °F). Maximum resistances generally require 7-day cure at 20 °C (70 °F). Slightly lower chemical resistance than amine-cured epoxies	Easier to apply and topcoat, more flexible, and better moisture resistance than amine-cured epoxies. Excellent adhesion over steel and concrete. A widely used industrial and marine maintenance coating. Some formulations can be applied to wet or underwater surfaces.
Coal tar epoxies	Excellent resistance to saltwater and freshwater immersion. Very good acid and alkali resistance. Solvent resistance is good, although immersion in strong solvents may leach the coal tar.	Embrittles upon exposure to cold or ultraviolet light. Cold weather abrasion resistance is poor. Should be topcoated within 48 h to avoid intercoat adhesion problems. Will not cure below 10 °C (50 °F). Black or dark colors only. Temperature resistance: 105 °C (225 °F) dry; 65 °C (150 °F) wet	Good water resistance. Thicknesses to 0.25 mm (10 mils) per coat. Can be applied to bare steel or concrete without a primer. Low cost per unit coverage
Polyurethanes (aromatic or aliphatic)	Aliphatic urethanes are noted for their chemically excellent gloss, color, and ultraviolet light resistance. Properties vary widely, depending on the polyol coreactant. Generally, chemical and moisture resistances are similar to those of polyamide-cured epoxies, and abrasion resistance is usually excellent.	Because of the versatility of the isocyanate reaction, wide diversity exists in specific coating properties. Exposure to the isocyanate should be minimized to avoid sensitivity that may result in an asthmaticlike breathing condition upon continued exposure. Carbon dioxide is released upon exposure to humidity, which may result in gasing or bubbling of the coating in humid conditions. Aromatic urethanes may darken or yellow upon exposure to ultraviolet radiation.	Aliphatic urethanes are widely used as glossy light-fast topcoats on many exterior structures in corrosive environments. They are relatively expensive, but extremely durable. The isocyanate can be combined with other generic materials to enhance chemical, moisture, low-temperature, and abrasion resistance.

(continued)

to a great extent on the drying oil used and its type and degree of unsaturation (available cross-linking sites). For example, the use of pentaerythritol as the polyhydric alcohol leads to faster drying, greater hardness, better gloss and gloss retention, and better water resistance than alkyds based on glycerols of equal fatty acid content. Similarly, the drying oil also affects resin properties. Soybean oil has been shown to give good drying rates and good color retention; therefore, the standard alkyds are usually soybean alkyds. Linseed oils, on the other hand, generally dry faster but darken upon light exposure. The chemical structure of a linseed oil alkyd is shown in Fig. 1. Castor and coconut oils have good color-retentive properties and are used as plasticizing resins because of their nonoxidizing character.

Phthalic anhydride and ter- and isophthalic acids are the most common polybasic acids used in alkyd preparation. The main advantage of ter- and isophthalic acids compared with phthalic anhydrides is that they provide polymers with a higher molecular weight and higher viscosity. Furthermore, they exhibit somewhat faster-drying, more flexible, tougher films, and greater thermal stability than orthophthalic alkyds.

As a class, alkyd coating systems are still the workhorse of the corrosion protection industry, accounting for perhaps as much as two-thirds of all paints sold for corrosion protection. Although they have limited chemical and moisture resistance and relatively poor alkaline resistance, their low cost, ease of mixing and application, and excellent ability to penetrate and adhere to relatively poorly prepared, rough, dirty, or chalked surfaces make them the coating system of choice on steel exposed to nonchemical atmospheric service. Thus, alkyds are widely used for structural steel exposed in chemical plants away from the chemical processes. They are also used in such applications as water tank exteriors, highway bridges, boat and ship superstructures above

Table 1 (continued)

Resin type	Advantages	Limitations	Comments
Asphalt pitch	Good water resistance and ultraviolet stability. Will not crack or degrade in sunlight. Nontoxic and suitable for exposure to food products. Resistant to mineral salts and alkalies to 30% concentration	Black color only. Poor resistance to hydrocarbon solvents, oils, fats, and some organic solvents. Do not have the moisture resistance of coal tars. Can embrittle after prolonged exposure to dry environments or temperatures above 150 °C (300 °F), and can soften and flow at temperatures as low as 40 °C (100 °F)	Often used as relatively inexpensive coating in atmospheric service, where coal tars cannot be used. Relatively inexpensive. Most common use is as a pavement sealer or roof coating.
Water emulsion latex	Resistant to water, mild chemical fumes, and weathering. Good alkali resistance. Latexes are compatible with most generic coating types, either as an undercoat or topcoat.	Must be stored above freezing. Does not penetrate chalky surfaces. Exterior weather and chemical resistance not as good as solvent or oil-base coatings. Not suitable for immersion service	Ease of application and cleanup. No toxic solvents. Good concrete and masonry sealers because breathing film allows passage of water vapor. Used as interior and exterior coatings
Acrylics	Excellent light and ultraviolet stability, gloss, and color retention. Good chemical resistance and excellent atmospheric weathering resistance. Resistant to chemical fumes and occasional mild chemical splash and spillage. Minimal chalking, little if any darkening upon prolonged exposure to ultraviolet light	Thermoplastic and water emulsion acrylics not suitable for any immersion service or any substantial acid or alkaline chemical exposure. Most acrylic coatings are used as topcoats in atmospheric service. Acrylic emulsions have limitations described under "Water emulsion latex."	Used predominantly where light stability, gloss, and color retention are of primary importance. With cross linking, greater chemical resistance can be achieved. Cross-linked acrylics are the most common automotive finish. Emulsion acrylics are often used as primers on concrete block and masonry surfaces.
Amine-cured epoxies	Excellent resistance to alkalies, most organic and inorganic acids, water, and aqueous salt solutions. Solvent resistance and resistance to oxidizing agents are good as long as not continually wetted. Amine adducts have slightly less chemical and moisture resistance.	Harder and less flexible than other epoxies and intolerant of moisture during application. Coating will chalk on exposure to ultraviolet light. Strong solvents may lift coatings. Temperature resistance: 105 °C (225 °F) wet; 90 °C (190 °F) dry. Will not cure below 5 °C (40 °F); should be topcoated within 72 h to avoid intercoat delamination. Maximum properties require curing time of about 7 days.	Good chemical and weather resistance. Best chemical resistance of epoxy family. Excellent adhesion to steel and concrete. Widely used in maintenance coatings and tank linings
Phenolics	Greatest solvent resistance of all organic coatings described. Excellent resistance to aliphatic and aromatic hydrocarbons, alcohols, esters, ethers, ketones, and chlorinated solvents. Wet temperature resistance to 95 °C (200 °F). Odorless, tasteless, and nontoxic; suitable for food use	Must be baked at a metal temperature ranging from 175 to 230 °C (350 to 450 °F). Coating must be applied in a thin film (approximately 0.025 mm, or 1 mil) and partially baked between coats. Multiple thin coats are necessary to allow water from the condensation reaction to be removed. Cured coating is difficult to patch due to extreme solvent resistance. Poor resistance to alkalies and strong oxidants	A brown color results upon baking, which can be used to indicate the degree of cross linking. Widely used as tank lining for alcohol storage and fermentation and other food products. Used for hot water immersion service. Can be modified with epoxies and other resins to enhance water, chemical, and heat resistance
Organic zinc-rich	Galvanic protection afforded by the zinc content, with chemical and moisture resistance similar to that of the organic binder. Should be topcoated in chemical environments with a pH outside the range 5–10. More tolerant of surface preparation and topcoating than inorganic zinc-rich coatings	Generally have lower service performance than inorganic zinc-rich coatings, but ease of application and surface preparation tolerance make them increasingly popular	Widely used in Europe and the Far East, while inorganic zinc-rich coatings are most common in North America. Organic binder can be closely tailored to topcoats (for example, epoxy topcoats over epoxy-zinc-rich coatings) for a more compatible system. Organic zinc-rich coatings are often used to repair galvanized or inorganic zinc-rich coatings.
Inorganic zinc-rich	Provides excellent long-term protection against pitting in neutral and near-neutral atmospheric, and some immersion, services. Abrasion resistance is excellent, and dry heat resistance exceeds 370 °C (700 °F). Water-base inorganic silicates are available for confined spaces and VOC compliance.	Inorganic nature necessitates thorough blast-cleaning surface preparation and results in difficulty when topcoating with organic topcoats. Zinc dust is reactive outside the pH range of 5–10, and topcoating is necessary in chemical fume environments. Somewhat difficult to apply; may mudcrack at thicknesses in excess of 0.13 mm (5 mils)	Ethyl silicate zinc-rich coatings require atmospheric moisture to cure and are the most common type. Widely used as a primer on bridges, offshore structures, and steel in the building and chemical-processing industries. Used as a weldable preconstruction primer in the automotive and shipbuilding industries. Use eliminates pitting corrosion.

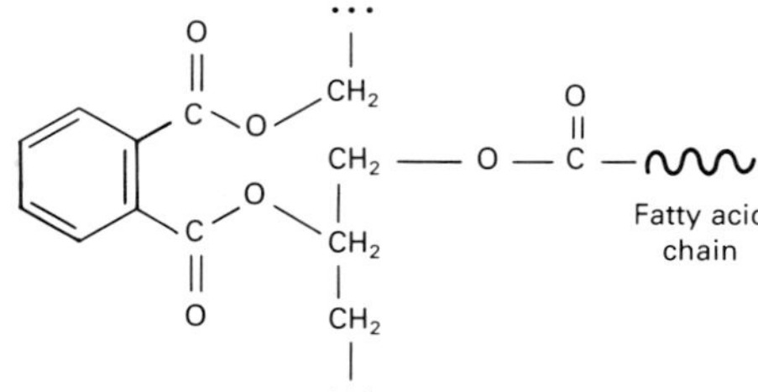

Fig. 1 Molecular structure of a linseed oil alkyd. The molecule is obtained through the reaction of a linseed oil monoglyceride and phthalic anhydride.

the waterline, and in baking formulations as coatings for containers, appliances, and machinery and equipment housings.

Alkyd Modification. Alkyds are amenable to modification with a variety of other different resins to improve specific properties, such as drying time, color retention, and moisture and chemical resistance. For example, modification with a chlorinated rubber resin results in improved toughness and adhesion as well as an increase in acid, alkali, and water resistance. Such modified paints are used on concrete floors and as highway-marking paints.

Modification with a phenolic resin improves gloss retention and water and alkali resistance. Phenolic alkyd resins have performed satisfactorily in water immersion—a service in which nonphenolic-modified alkyd resins are not suitable.

Alkyd resins with vinyl modification are commonly formulated as universal primers (primers that, after application, can be topcoated with most generic-type intermediate and topcoats). The alkyd constituent improves adhesion, film build, solvent resistance, and thermal resistance, and the vinyl modification enhances recoatability, chemical, and moisture resistance.

Silicone modification of alkyd resins is perhaps the most widely promoted modification for corrosion-protective coatings. A silicone intermediate is added to the alkyd resin in quantities up to 30% to provide polymers with greatly improved durability, gloss retention, and heat resistance.

Moisture resistance is greatly improved by the silicone modification, and such paints are extensively used as marine and maintenance paints.

Unmodified alkyds of medium-to-long oil length are familiar to homeowners as the oil-base paints used extensively around the home. Their ability to penetrate and adhere to wood and masonry surfaces, their availability in wide color ranges and glosses ranging from flat to high gloss, and their ease of application and cleanup make them ideal for a do-it-yourself homeowner.

Highly thixotropic alkyds have recently been promoted for both home and industrial use. Thixotropic paints remain in a semisolid or jellylike state when at rest, but become thinner and spread easily with the shearing action caused by brushing. As soon as the brushing action is stopped, the paints again recover their jellylike characteristics. Such paints reduce pigment settling, reduce tendencies to run or sag, and make possible the application of thicker coats that have a longer wet-edge time due to the use of slower-evaporating solvents, which otherwise cannot be used because they may cause sagging. Dripping from the brush is minimized, and the thixotropy prevents absorption on porous surfaces. Thixotropic paints are usually not recommended for spray application. The thixotropy is introduced by modification with a polyamide resin, and the jellylike behavior of the combined resin has been attributed to hydrogen bonding between the carbonyl group of the alkyd and the amide group of the polyamide coreactant.

Alkyds can be modified with polyisocyanates to improve drying rates and chemical and abrasion resistance. Such resins, commonly called uralkyds, are available in a variety of colors and generally exhibit excellent gloss and color retention. Similarly, epoxy resins are combined with the alkyd to obtain coating formulations having improved adhesion to metal, better gloss and color retention and much improved water and chemical resistance.

Epoxy esters differ from epoxy-modified alkyds in that the epoxy is a direct esterification product of an epoxy resin and a fatty acid, such as a vegetable oil or rosin. Like alkyds, epoxy ester resins are usually prepared by reacting the drying oil with the epoxy resin at a temperature of approximately 220 to 290 °C (425 to 550 °F) in the presence of esterification catalysts. The same drying oils used to prepare alkyd resins are also used to prepare epoxy esters. As with alkyds, the oil length can be categorized as long, medium, and short, and the properties accruing to the resin derive predominantly from the amount and type of drying oil modification to the resin. As a general rule, compared with alkyds, epoxy resins have better adhesion and moisture and chemical resistance, although they are slightly more expensive.

Thermoplastic Resins

Thermoplastic resins are characterized, as their name implies, by a softening at elevated temperatures. Although the resins within thermoplastic paint systems are commonly co- or ter-polymers, the molecular structure is not cross linked into a rigid molecule, as are the chemically cross-linked or the auto-oxidized coating systems. Rather, the resinous binder is dissolved in a suitable solvent, mixed with the pigment and other constituents comprising the formulated paint, and packaged. Upon application, the solvent volatilizes, and the resinous binder, along with the pigment, is deposited onto the surface being coated as a solid film. The molecular entanglement of the relatively linear (although perhaps coiled or tangled) molecule segments provides a layering effect that, with pigmentation and sufficient thickness, provides the characteristic moisture and chemical resistance of the resin system used.

The most useful thermoplastic resins for corrosion protection are the vinyls, chlorinated rubbers, thermoplastic acrylics, and bituminous resins (coal tar and asphalt). Each resin type has its own characteristic qualities, specific resistances, and susceptibilities, and each will be discussed below in greater detail. However, in all cases, because no chemical change takes place to the resin upon application and drying, it stands to reason that the applied coating will redissolve if it is exposed to the same solvents that it was originally dissolved in upon formulation. This is precisely what happens, and such coatings have no resistance to the solvents in which they were originally dissolved or in solvents of equal or greater solvency for that particular resin. This, however, is not usually a disadvantage; although solvent resistance may be poor, such coatings are rarely, if ever, exposed in a solvent-laden atmosphere. Instead, they are called on to resist weathering, acid or alkali exposures, and moisture. As a rule, all have good resistance in these applications, and the vinyls, chlorinated rubbers, and coal tars have excellent moisture resistance as well.

Conversely, the solvent susceptibility of these thermoplastic coatings is often an advantage. When maintenance repainting is necessary, the same system can be reapplied, and the solvents in the new topcoat will partially redissolve the existing old coat, resulting in intermolecular entanglement of the new coating with the old and giving excellent adhesion. Thus, solvent solutions of these thermoplastic resins have excellent recoatability with the same system, even after extended periods of time. This cannot be said for either the auto-oxidative cross-linking coatings or the chemically cured coatings.

While the predominant characteristics of auto-oxidative cross-linking resins derive from the type and amount of drying oil incorporated into the coating, the characteristics of the thermoplastic resins derive chiefly from their molecular structure. Thus, apart from the observations described above for thermoplastic resins, a further description of the characteristics of these resins must be individually categorized.

Vinyls. Vinyl coatings are usually applied as a solvent-deposited solution consisting of either a polyvinyl chloride-polyvinyl acetate (PVC-PVA) copolymer in the ratio of approximately 86% PVC to 13 to 14% PVA or as a high molecular weight emulsion of the same copolymer in what is commonly called a latex or water-base paint.

The vinyl solution coatings are generally very low-solids high-solvent materials that, although quite widespread in use at present, will be increasingly influenced by restrictive VOC legislation. Most thermoplastic vinyl solution coatings contain solvents (generally oxygenated solvents such as ketones and glycol ethers) ranging from 75 to 90% of the total volume of the formulated paint. As a result, the VOC content far exceeds the widely used 420-g/L (3.5-lb/gal.) maximum, and where such legislation is enforced, solution vinyl coatings cannot be used.

Most vinyl resins consist of the copolymer of PVC and PVA. Polyvinyl chloride is a very inert material with high symmetry, an extremely stable carbon-to-carbon backbone, and stable carbon-to-hydrogen and carbon-to-chlorine pendant sidegroups. As a consequence, it is very inert and cannot be dissolved even by some of the stronger solvents used in coating manufacture. Polyvinyl chloride is usually heat extruded into sheets or formed by heat. Without modification by PVA or other copolymers, PVC is unsuitable for use in paint manufacturing. Polyvinyl acetate, on the other hand, while retaining the stable carbon-to-carbon backbone, does have an ester sidechain that is susceptible to solvent attack. Consequently, PVA is readily dissolved by the stronger solvents. Although PVA coatings are used, they have only limited moisture and chemical resistance and are suitable only for interior decorative coatings without further modification.

However, a copolymer of PVC and PVA in a ratio of approximately 86% PVC to approximately 13% PVA, combined with approximately 1% or less maleic acid, is the basis for most solvent-deposited vinyl coatings used for corrosion protection. Figure 2 shows an example of a PVC-PVA vinyl. Because the vinyl resin is somewhat sensitive to ultraviolet light, it is essential to use an ultraviolet scattering pigment such as rutile titanium dioxide in order to minimize chalking and ultraviolet degradation of the coating. Other pigments can be used in conjunction with titanium dioxide, but they act more as filling and reinforcing pigments. Because of the acidic nature of the vinyl resin, zinc oxide and zinc dust pigments must be used with care; they have a tendency to gel or curdle the paint while in solution.

Vinyl solution coatings are noted for their outstanding toughness and water resistance. The U.S. Army Corps of Engineers has used a five-coat vinyl system for years on steel exposed both in fresh- and saltwater immersion and in atmospheric service on locks and dam gates under their jurisdiction throughout the United States. Although the five-coat system is somewhat expensive to apply, it is extremely durable when properly applied. Total repainting intervals of over 20 years are commonplace with the Army Corps of Engineers, although maintenance repainting and touch up of deteriorated or damaged paint may be necessary at more frequent intervals. The Army Corps of Engineers formulations are presented below:

Ingredient	Percent by weight
Formula V-766e	
Vinyl resin, type 3 ((VYHH)(a)	5.6
Vinyl resin, type 4 (VMCH)(a)	11.6
Titanium dioxide	13.0
Diisodecyl phthalate	2.9
Methyl isobutyl ketone	32.0
Toluene	34.7
Phosphoric acid	0.2
Formula V-766e(AP)	
Vinyl resin, type 3 (VYHH)(a)	5.4
Vinyl resin, type 4 (VMCH)(a)	11.1
Titanium dioxide	12.5
Diisodecyl phthalate	2.9
Toluene	11.2
Nitropropane solvent	48.0
Methyl ethyl ketone	8.7
Phosphoric acid	0.2

(a) VYHH and VMCH are proprietary designations for these vinyl resins. Approximate compositions: VYHH, 86% polyvinyl chloride, 14% polyvinyl acetate; VMCH, 86% polyvinyl chloride, 13% polyvinyl acetate, 1% maleic acid. Source: U.S. Army Corps of Engineers specification for vinyl formula V-766e

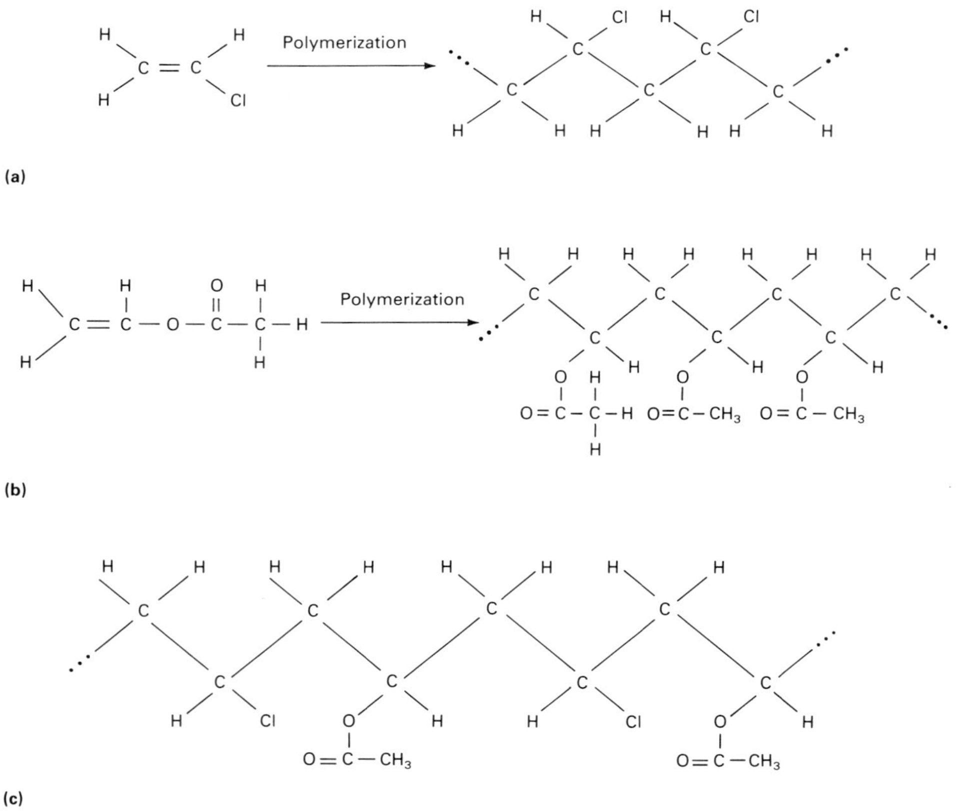

(a)

(b)

(c)

Fig. 2 Reaction of PVC and PVA polymers to form a PVC-PVA copolymer. (a) Polymerization of PVC. (b) Polymerization of PVA. (c) PVC-PVA copolymer

Vinyl paints are used as complete systems (vinyl primer, intermediate, and topcoat) or as topcoats over zinc-rich primers for the protection of highway bridges throughout the United States. The relative inertness of the vinyl topcoat, coupled with good adhesion (when formulated with maleic acid), protects the relatively reactive zinc in an organic or inorganic zinc-rich primer from reaction with environmental moisture, acid rain, and other atmospheric contaminants. The zinc-rich primer provides galvanic protection to the underlying steel (see the section "Zinc-Rich Coatings" in this article).

Chlorinated rubber coatings are also considered to be solvent-deposited coatings in that there is no chemical change that occurs during application, and the coating film forms on the surface after application as the solvent evaporates. Chlorinated rubber coating systems are much more widely used in Europe than in the United States, although systems based on chlorinated rubbers have excellent resistance to acids, alkalies, oxidizing agents and have very low water vapor transmission rates (approximately one-tenth that of an alkyd resin coating). In addition, they are generally nonflammable (due to their chlorine content) and nontoxic. Chlorinated rubber paints are used principally as chemical- and corrosion-resistant coatings, marine coatings, building masonry and swimming pool paints, highway marking paints, fire-retardant and mold-resistant paints, and primers for hot-applied coal tar or asphaltic enamels (to provide good adhesion to steel).

Chlorinated rubber resins are manufactured by chlorinating natural rubber with approximately 65 to 68% chlorine. The resulting resin must be plasticized and stabilized against light, heat, and gellation in order to be used as a binder in protective coatings. Heat stability is obtained by incorporating low molecular weight epoxy compounds or epoxidized oils into the resin. Lightfast plasticizers such as dialkyl phthalates and acrylic resins are used. For severe chemical service, chlorinated paraffins or chlorinated diphenyls should be used because of their high resistance to saponification.

The amount of plasticizer is very important in these coatings because it affects the properties of the film. If the coating is underplasticized, it will be harder and more brittle and will exhibit poor adhesion. If the resin is overplasticized, the coating will be softer and thermoplastic and will exhibit a tackiness that may result in a higher dirt retention. Overplasticization will also increase water transmission rates or water permeability through the film, diminishing the corrosion protection. The chemical structure of a chlorinated rubber resin is shown in Fig. 3.

The presence of ferrous or ferric ions may cause gelation of the chlorinated rubber coating. As a consequence, care must be taken during the manufacture to prevent excessive contact with iron or steel. Certain reactive pigments, metal chlorides, or ketone solvents that will aggravate gelling should not be used, and the cans in which the paint is packaged should be coated to minimize corrosion and to prevent possible contamination by ferrous or ferric ions.

Chlorinated rubber resins are sold as an off-white powder and are commonly pigmented with most of the usual pigments used in chemical-resistant coatings. Rutile titanium dioxide is a widely used pigment for white and tinted paints. Metallic pigments, particularly aluminum and copper flake powders, as described in Volume 7 of the 9th Edition of *Metals Handbook*, may cause gelation. Also, zinc oxide should not be used as a pigment in major amounts, because it will degrade chlorinated rubber at elevated temperatures. Zinc-rich paints based on chlorinated rubber resins are used successfully. The formulations typically contain 75% or more of zinc dust and are plasticized with a small amount of chlorinated paraffin.

Chlorinated rubber coatings are soluble in most solvents, including aromatic hydrocarbons, chlorinated hydrocarbons, esters, ketones, and some alcohol ethers. Care must be taken in the solvent blend used to form a chlorinated rubber coating; a solution that dries too rapidly may result in cob-webbing or a stringy application when sprayed or in pulling of the brush or roller when applied by brushing or rolling.

Coatings with thicknesses in excess of 0.13 mm (5 mils) should be applied in three or more coats to provide optimum protection. After drying, properly plasticized chlorinated rubber coatings will develop excellent adhesion to both steel and concrete substrates.

Because of their extremely low water vapor transmission rates, as well as their flexibility, extensibility, and ability to bridge or seal over pores or cracks, chlorinated rubber paints are excellent coatings for use in concrete swimming pools. In the United States, this is a major market for chlorinated rubber coatings. These same attributes, however, are also valuable when coating steel and other substrates (such as wood); accordingly, such coatings have found extensive use worldwide in protecting steel from corrosion in various chemical atmospheres and in salt- and freshwater immersion.

Acrylic coatings can be formulated as thermoplastic solvent-deposited coatings, cross-linked thermoset coatings, and water-base emulsion coatings. Acrylic resins that are used for protective coatings consist of polymers and copolymers of the esters of methacrylic and acrylic acid. The resulting resins, with suitable pigmentation, provide excellent film-forming coatings characterized by excellent light fastness, gloss, and ultraviolet stability. Chemical resistance to weathering environments is generally excellent, as is resistance to moisture. Most acrylic coatings are not suitable for immersion service or strong chemical environments, because of attack of the ester group present on both the polymethyl methacrylate and ethyl acrylate resins. This ester group, however, is present only as a pendant sidechain; accordingly, the carbon-to-carbon backbone chain of the acrylic resins does provide very good chemical and moisture resistance, which enables these resins to be used in most chemically moderate atmospheric services.

Coatings can be formulated with acrylate, methacrylate, or higher homopolymers. Blends of both resin types are also commonly used for corrosion protection. Generally speaking, those coatings formulated from the acrylates or from copolymers with a preponderance of the acrylate ester are softer and more flexible than the corresponding methacrylates. The ultraviolet resistance of the acrylate is somewhat lower than that of the methacrylate because of the presence of a tertiary hydrogen attached directly to a carbon

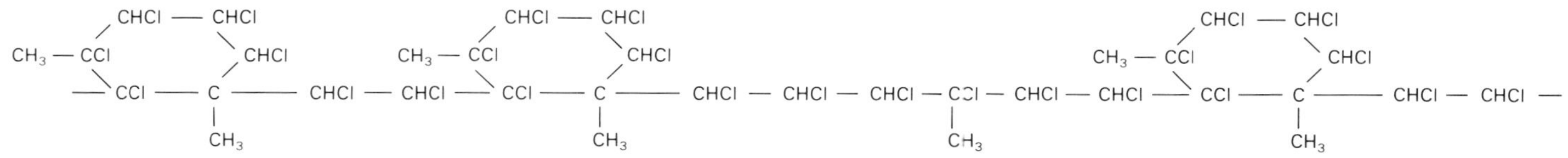

```
      CHCl — CHCl                  CHCl — CHCl                                                   CHCl — CHCl
     /          \                 /          \                                                  /          \
CH3 — CCl        CHCl       CH3 — CCl        CHCl                                          CH3 — CCl        CHCl
     \          /                 \          /                                                  \          /
    — CCl —— C —— CHCl — CHCl —— CCl —— C —— CHCl — CHCl — CHCl — CCl — CHCl — CHCl —— CCl —— C —— CHCl — CHCl —
             |                              |                             |                              |
            CH3                            CH3                           CH3                            CH3
```

Fig. 3 Molecular structure of a chlorinated rubber resin. Source: Ref 1

comprising the molecular backbone. This active hydrogen is vulnerable to photo-oxidative and thermo-oxidative attack. Consequently, acrylate esters have a greater susceptibility to ultraviolet light and oxygen, leading to yellowing or darkening in comparison with the methacrylates. The methacrylates, although providing a harder and more brittle film with less adhesion, are more chemical and moisture resistant and have a greater heat tolerance.

With both resin types, greater flexibility, toughness, and resistance to abrasion can be obtained by increasing the molecular weight. Hardness decreases, although flexibility and elongation increase in both resin families. However, with too great an increase in molecular weight, solution viscosity becomes excessive, and solvent release in a formulated paint becomes excessively slow.

Acrylic resins are characteristically soluble in moderately hydrogen-bonded solvents, such as ketones, esters, aromatic hydrocarbons, and certain chlorinated hydrocarbons. Excellent color retention, gloss, and ultraviolet stability, coupled with very good moisture and chemical resistance, make acrylic coatings quite popular in protecting steel in atmospheric service.

Thermoplastic Acrylics. Like all thermoplastic coatings, thermoplastic acrylics are essentially solvent-deposited resins that do not undergo any chemical change after, or at the time of, application. Upon application, the solvent carrier volatilizes, and the coating system sets, hardens, and attains its final resistant properties as the solvent evaporates. Heating or baking is often used to hasten solvent evaporation.

Probably the major use of thermoplastic acrylics is in automobile finishing and refinishing. Advantages include the ease of tinting and coloration, excellent dispersion of metallic reflective pigments, high gloss and ultraviolet stability, and thermal reflow. Thermal reflow is important because, after the original factory coating is applied, minor imperfections, sanding marks, and other irregularities can be overcome by reheating the coating under heat lamps or in an oven. The surface of the coating will partially melt and reflow to form the smooth, glossy appearance one expects with an automotive finish.

Another major market is aerosol can paint, which is widely sold throughout the world. These paints are thermoplastic acrylics, and they dry and harden through solvent evaporation.

Solvent-deposited acrylics are also used for maintenance finishes, but more commonly, the acrylic monomer or copolymer is added to an alkyd. This results in an acrylic-modified alkyd with the excellent penetration properties, adhesion, and ease of application of the alkyd, combined with the improved ultraviolet resistance, gloss, and color retention of the acrylic.

Thermosetting Acrylics. Thermosetting, or cross-linking, acrylics can be obtained by combining the methacrylate or acrylate esters with reactive nonacrylic polymers. The most common cross-linking agents used are the epoxy resins, the urea or melamine-formaldehyde resins, and the vinyl resins. Generally, the nonacrylic polymer is the predominant structure present in the coating film, and the acrylic modification is done only to enhance the color, gloss, and ultraviolet stability of the nonacrylic constituent. The resulting thermoset acrylic finishes are generally harder, tougher, and more resistant to heat and solvents than thermoplastic acrylics. On the other hand, thermoset acrylic finishes are also generally less resistant to ultraviolet light and usually require heating or baking after application to attain the desired physical properties.

Acrylic finishes based on urea or melamine-formaldehyde resins cross linked with acrylic monomers are widely used for coating such home appliances as refrigerators, washing machines, dryers, dishwashers, freezers, and air conditioners. The coatings have excellent adhesion and detergent resistance, and they have short curing times at relatively low curing temperatures. The major advantages of such coatings are the high gloss, range of colors available, and lack of yellowing or darkening upon exposure to sunlight.

Water-Base Acrylic Coatings. A major coating market for acrylics has been the acrylic emulsion paints widely used by the homeowner for indoor and exterior painting purposes. Compared to oil-base coatings, the other major homeowner-type paint, the water-base acrylics have a number of advantages. They have a much faster drying time, and can be recoated within 1 h. They have better alkali resistance, excellent adhesion, come in a variety of colors and maintain their color and gloss in the presence of sunlight and ultraviolet light, and have excellent long-term flexibility and toughness. The major advantage, of course, is the ease of cleanup with water-base paints.

The disadvantages are that the oil-base paints have greater penetrability, particularly on wood and porous surfaces, and have better initial adhesion. On smooth surfaces, water-base paints may peel or loose adhesion, while an oil-base paint may wet sufficiently to maintain good adhesion. However, acrylic paints have better resistance to blistering because of their breathing ability and have much better resistance to chalking and yellowing. Water-base coatings may provide better adhesion to damp wood and masonry substrates than oil-base paints.

The chemistry of water emulsion coatings is extremely complex. Generally, high molecular weight acrylic polymers are polymerized in a water emulsion. The desired pigments are dispersed in the emulsion, and the required additives are added, followed by tinting and color pigments. Titanium dioxide is the most widely used white-hiding pigment, although most other water-insoluble alkali-resistant organic and inorganic pigments can also be used. Wetting agents are added to ensure complete pigment wetting in the water vehicle, and antifoaming additives are used to prevent foaming during manufacture and at the time of application. Thickening additives (for proper application viscosity) are used to prevent pigment settling as well as running and sagging after application.

Surfactants are added to enhance pigment dispersion, and it is these additives that provide the water sensitivity that most water-base coatings have after application. Briefly stated, most surfactants have a hydrophobic (water-hating) and a hydrophilic (water-loving) end, which orients around the emulsified resin and the pigment particles. The hydrophobic end is adjacent to the resin particle surface, while the hydrophilic end is attracted to the water carrier. After application of the paint, many of these surfactants remain entrapped within the film cross section, and the hydrophilic end of the surfactant attracts water from the environment, which tends to soften and swell the coating. In atmospheric service, there is insufficient water—or insufficient wet time—to cause a problem. However, in prolonged water immersion, water may be attracted by the surfactant, penetrate the film, and cause softening and swelling of the coating with a loss of adhesion. Consequently, water-base coatings, as currently formulated, are not suitable for prolonged water immersion.

For most air dry (nonbaking) applications, a coalescing solvent is added. This very slow-evaporating organic solvent, which is miscible with water, is added to the water carrier. As the water evaporates, the concentration of the solvent becomes higher and higher in the water/solvent mixture. Ultimately, in the drying film, the solvent reaches a sufficient concentration that it can partially solvate the resin particles, causing them to soften and meld together upon further drying. The coalescing solvent helps convert the emulsified particles into a relatively smooth, homogeneous film. However, residual solvent may be entrapped within the film, and it too can lead to water sensitivity.

For wood or masonry surfaces where water vapor transmission may be a problem, latex paints may be suitable because they are said to breathe. This breathing occurs as a result of irregularities during the coalescence part of the film forming; the particles may not align themselves properly relative to other emulsified particles. A series of pores or defects in the coalesced film will result. Such discontinuities are extremely small, and the surface tension of water in liquid form will not enable penetration through the coating to the underlying substrate. However, water vapor may readily pass through such discontinuities, resulting in its breathing ability.

Water-base paints develop poor film properties when applied during cold, damp weather. In order for proper weather and chemical resistances to develop in a water emulsion film, water evaporation and proper drying must occur. If water evaporation is retarded and if the drying sequence is upset by becoming either too slow (or in some cases too rapid), proper coalescence and film formation do not occur. The result can be a soft, poorly adherent film that, when drying does finally occur, will show cracking; this film may also be brittle and adhere poorly. Too fast a water evaporation rate on a hot surface can lead to a powdery, poorly adherent paint film or a film with many voids, pinholes, or cross-sectional porosities.

Acrylic paints for interior use are similar to those for exterior use, except that a higher pigment volume can be used, and the pigments that are not as light-fast as those for exterior use can be used. The major competition for acrylic interior paints comes from lower-cost PVA or butadiene-styrene base latex coatings.

Photopolymerization of acrylic homo- and copolymers is for the painting industry an emerging—and very interesting—technology. The wavelength ranges of ultraviolet light (about 200 to 400 nm) and visible light (approximately 400 to 800 nm) can initiate cross linking in a susceptible binder system. Acrylic esters show a very rapid response to photoinitiation as compared to all other commonly used coating resins. Within the acrylic family, the acrylate is at least ten times more reactive to ultraviolet or visible light radiation than the methacrylate. All types of resins used in conventional paint systems (such as alkyds, epoxies, polyesters, and urethanes) can be modified with acrylate esters, and the combination can be combined with a suitable photoinitiator to aid in the radical polymerization mechanism.

High-solids, low molecular weight, light-cured coatings can be applied by brush, roller, and spray. After setting up and exposure to an intense source of ultraviolet or visible light, they cross link through a photoinitiated free radical polymerization, which leads to thorough cross linking, hardening, and almost immediate attainment of moisture and chemical resistance. The potential of this technology for exterior painting is evident; the sun itself might act as the cross-linking agent. However, such technology with regard to corrosion-protective coatings is far from perfected. Current problems include the extremely high cost of formulated paints, the requirement for intense ultraviolet or light energy in order to initiate curing, and the curing problems with pigmented coatings or coatings more than a few mils thick.

Therefore, although photochemically cured coatings for corrosion protection have not been developed, the technology is viable, and some radiation-curable coatings have been developed. Commercial applications include clear coatings, thin films of pigmented printing inks, coatings pigmented primarily with inert pigments (for example, paste fillers for particleboard), and paper coatings. Light-cured acrylic laminates and overlays, applied as an adhesive or paste on tooth enamel, have been developed for dental practice.

Bituminous Coatings. The bitumens used in the coatings industry are coal tar and asphalt. Although these materials are distinctly different physically and chemically, in appearance they are essentially identical black thermoplastic tar materials. As coatings, the bitumens can be applied as hot melts, solvent cutbacks, or emulsions.

Hot melt application involves heating the bitumen to a temperature of 175 to 245 °C (350 to 475 °F) such that its viscosity at that temperature is very low—almost waterlike. Hot melt asphalts and coal tar materials are usually applied with mops or swabs to the surface being coated, although brush, roller, or spray application is sometimes used. Flow coating of the interiors of pipes and small vessels is also done. In this method, the hot melt is flowed into the pipe or vessel, and the item is rotated or turned to cover all surfaces before the coating cools and hardens. The interiors and exteriors of pipeline sections are often flow coated at pipeline-coating facilities (see the article "Corrosion of Pipelines" in this Volume).

Asphaltic or coal tar bitumens dissolved in a suitable solvent (normally aliphatic and aromatic hydrocarbons) are termed solvent cutbacks. Dissolving the bitumen in a solvent lowers its viscosity sufficiently that the cutback can be applied by spray, brush, or roller, as appropriate. After application, the solvent volatilizes and the bitumen resolidifies into a film. The coating thicknesses of solvent cutbacks are generally considerably less than those achieved by hot melt application, but the convenience of not having to heat the bitumen at the job site immediately before application is a major advantage.

Water emulsions are prepared by suspending minute particles of the bitumen in water using emulsifying agents and by combining with inert fillers, such as pulverized talcs (hydrated magnesium and aluminum silicates), coal dust, powder silica, mica, and limestone dust. After application, the water evaporates. Coalescence occurs in the fashion typical of any emulsion coating, and a protective film is formed.

In general, hot melt application provides the best moisture and chemical resistance, followed in order by the solvent cutback and water emulsions. The use of asphaltic and coal tar bitumens as roof coatings, highway or pavement sealers, and underground coatings or waterproofing compounds is quite common. Asphaltic materials have much better atmospheric weathering resistance and are less deteriorated by ultraviolet light than coal tar coatings are. On the other hand, coal tar coatings have much better resistance to moisture, acids, and alkalies than the asphaltics. The coal tars and asphalts are generally incompatible as a result of their chemical makeup and usually should not be used as mixtures or applied one type over another.

Coal tar enamels or pitches are derived from the coking of coal. When coal is heated in the absence of air to a temperature of approximately 1095 °C (2000 °F), it decomposes partially into a gas and coke. The gas is subsequently condensed, and coal tar is formed. Lighter fractions are removed from this tar by subsequent heating and gas extraction until the desired coal tar fraction for use in coatings is obtained. The compounds in coal tar range from low-boiling, low molecular weight benzene to complex, high molecular weight, aromatic hydrocarbons.

Essentially, there are three types of coal tar enamels:

- A straight pitch base that is relatively hard and brittle but has the greatest moisture and chemical resistance
- A plasticized enamel that is more flexible and has greater cold weather application tolerance, but the least chemical and moisture resistance
- An intermediate grade (containing about 50% of each of the first two) that is also intermediate in its properties

For best adhesion, preheating of the object being coated is preferred; if this is not feasible, a primer can be used to promote adhesion of the coal tar. Many of the primers are solvent cutbacks of coal tar pitch combined with a chlorinated rubber resin, plasticizers, highly volatile solvents, and some color pigment. The primer is of a lower viscosity and can more readily wet the steel surface. In addition, it dries faster and leaves a firmly bonded layer on the steel. Subsequently applied coal tar adheres more readily to steel that is primed in this manner.

Coal tar coatings (whether applied as hot melts, cutbacks, or emulsions) are subject to ultraviolet light degradation and resultant cracking. The cracking results from the volatilization or evaporation of lower molecular weight constituents from the surface and attendant shrinking. The surface of the coating contracts and cracks, while the underlying thicknesses remain plastic. Such "alligator" cracking (Fig. 4) is not detrimental, although it does diminish the thickness of the coating within the crack fissures. Because of this phenomenon, coal tar coatings are used in underground or immersion service, in which they are not subject to ultraviolet light. As roof coatings, they must be shielded by a subsequent coat of non-coal tar material or by gravel or pebbles applied for this purpose.

Asphalts. Natural asphalt, or gilsonite, is a hard, brittle resin that is mined predominantly in Utah, Colorado, and Trinidad. Increasingly, asphalts obtained as residues from the distillation of crude petroleum are replacing the natural asphalts because of better uniformity, fewer impurities, and lower transportation costs. Asphaltic compounds are comprised of complex polymeric aliphatic hydrocarbons, and they have good water and chemical resistance; like coal tars, they are sensitive to oils and solvents. Asphaltic coatings have much greater ultraviolet resistance than coal tars do and are therefore used for above-grade applications in which coal tars will not be suitable.

Cross-Linked Thermosetting Coatings

Chemically cured coatings, in the context of this article, refer to coatings that harden or cure and attain their final resistant properties by virtue of a chemical reaction either with a copolymer or with moisture. Examples of coatings that chemically cross link by copolymerization include epoxies, unsaturated polyesters, urethanes in which the isocyanate is reacted with a polyol, high-temperature curing silicones, and phenolic linings. Chemically cured coatings that react with water include moisture-cured polyurethanes and most of the inorganic zinc-rich coatings. Although drying-oil-base coatings (see the section "Auto-Oxidative Cross-Linked Resins" in this article) are also chemically cross linked upon

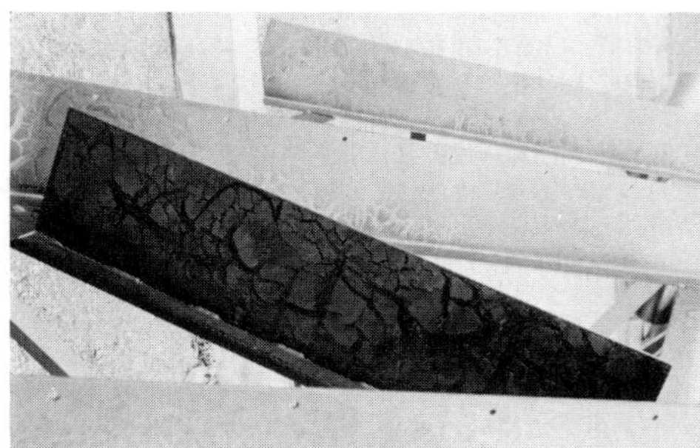

Fig. 4 "Alligator" cracking of a coal tar coating caused by volatilization of low molecular weight constituents from the surface and by coating shrinkage

their reaction with oxygen from the air, they are excluded from this category because the autooxidative cross-linking mechanism is a sufficiently distinct category of chemical cross linking to merit a separate discussion.

The fundamental principle of chemical cross linking is to formulate a coating using a resin or binder of low molecular weight. When mixed and applied, reaction with moisture from the atmosphere or with a coreactant takes place, resulting in cross linking and a much larger three-dimensional molecular structure. In fact, it is possible to begin the application of a cross-linked coating system to the interior of a large tank or vessel, and upon completion of the application many hours later, the film may cure into one large macromolecule that may cover many thousands of square feet of surface area. Such a coating or lining, if suitably formulated, will provide a tough, flexible, and highly chemical-resistant corrosion protection barrier over the underlying substrate. Such a large macromolecule with the attendant corrosion-resistant properties would be impossible to obtain in a prereacted binder system.

Coatings and linings based on chemically converted binders can be formulated to have excellent resistance to acids, alkalies, and moisture and can be formulated to resist abrasion, ultraviolet degradation, and thermal degradation. As a general rule, toughness and flexibility of the coating increase as the molecular length increases. Similarly, chemical and moisture resistance increase as the cross-linking density increases within the macromolecule. However, as with any coating, constituent chemical groups within any molecular structure may result in characteristic strengths or weaknesses within the molecular chain. Furthermore, the arrangement and composition of pendant sidegroups may further alter the properties of any given molecular structure. Also, unreacted moieties (reactive sites) within the molecule may lead to decreased chemical, moisture, or thermal resistance.

The rate at which the molecule cross links is dependent not only on the reacting moieties, but also on the cross-linking mechanism (for example, addition, condensation, or free radical initiated cross linking). Most importantly, external factors such as temperature and (with moisture reactions) atmospheric humidity affect the rate and extent of cross linking. Thus, chemically converted coatings must set after application through solvent or water volatilization and then harden and attain their final cured properties by the cross-linking reaction, which is temperature and/or moisture dependent. Too fast a rate of reaction may lead to an overcured, hard, impervious coating that cannot be recoated or topcoated with a properly adherent subsequent coat.

This is always a problem in maintenance repainting, in which a renewal coat is applied to the original cross-linked coating system after an extended period of time. In extreme cases, however, the recoat interval within which no recoat problems may be anticipated could be as short as a few hours for certain coatings or under certain ambient conditions. Amine-cured coal tar epoxy coatings are a good example, and the recoat window may be less than 16 h, after which intercoat adhesion may become a problem.

The curing and hardening of some chemically cross-linked systems, such as most polyesters and vinyl-esters, is so rapid that plural component spray application systems are required. Such systems mix the coreactants at the time of spray application within the spray gun or outside the spray gun nozzle. The chemical cross-linking reaction occurs so rapidly that by the time the coating has deposited on the surface the coating is beginning to set up. Although such coatings are hard to the touch within minutes, solvent volatilization and additional cross linking should be allowed to continue for days before to exposure in a severe chemical environment. As a general rule, curing of most chemically cross-linked systems should proceed for approximately 7 days at 25 °C (75 °F) before the coating system is exposed to severely corrosive conditions, immersion, or acid or alkali splash or spillage. The more common chemically cross-linked binders or resins used for coatings and linings will be discussed below.

Epoxies. Chemically cross-linked epoxies usually come in two packages; the first usually consists of the epoxy resin, pigments, and some solvent, and the second is the copolymer curing agent. The two packages are mixed immediately before application and, upon curing, develop the large macromolecule structure. The properties of the epoxy coating derive both from the type and molecular weight of the epoxy resin and from the copolymer curing agent that is used to cross link with it.

Epoxy Resins. For industrial maintenance coatings, the most common epoxy resin is the glycidyl ether type, particularly that derived from bisphenol-A and epichlorohydrin. The molecular structure of this type of epoxy resin is shown in Fig. 5.

Cycloaliphatic epoxies have been developed that offer improvements in light stability and ultraviolet degradation, but do not exhibit the adhesion, chemical resistance, and flexibility of the resins derived from epichlorohydrin and bisphenol-A. Epoxy cresol novolacs have also been introduced, and these are said to provide greater high-temperature resistance and chemical resistance at the expense of brittleness and a lack of flexibility.

Copolymer Curing Agents. Although epoxies are generally thought of as catalyzed, the cross-linking reaction is in fact a copolymerization. The reaction occurs primarily through the epoxy ring endgroups or the mid-chain hydroxyls of the epoxy resin. The curing agent is usually an amine or polyamide, with cross linking derived through the active hydrogens attached to the amine nitrogen. Although a variety of cross-linking agents can be used (including mercaptans, polybasic acids and anhydrides, and phenol-formaldehyde and phenol resins), the most common are the polyamines, the amine adducts, and the polyamides.

The polyamines (for example, diethylenetriamine, hydroxyethyl diethylene triamine, bishydroxydiethylene triamine) are relatively small molecules with a low molecular weight compared to the epoxy resin. As a consequence, when reacted, they lead to tight cross linking and high chemical and moisture resistance. However, during the cross-linking reaction, any unreacted amine may be squeezed out of the cross-linked film to the surface, developing the so-called amine blush, a hazy white coloration on the coating surface. The blush is not detrimental in atmospheric service and can be allowed to remain on the surface. However, for immersion service, it is a good practice to wipe it from the surface (it is water soluble) before application of a subsequent epoxy coat. To minimize the formation of the amine blush, many formulators will specify a 15- to 30-min induction period after mixing prior to application. This allows the reaction to begin and initial cross linking to occur before the paint is applied. Some of the small amine molecules will partially cross link with epoxy resin molecules, reducing the tendency toward squeezing out of the smaller molecules and minimizing amine blush.

Recognizing this problem, many coating manufacturers supply the amine as a prereacted amine adduct. In this case, the epoxy pigment and solvents are packaged as before in one container, but an excess of the amine is prereacted with some of the epoxy resin to increase its molecular size. The prereacted amine adduct is then packaged in a separate container, sometimes with additional pigment and solvent. Chemical cross linking of the resultant applied film is not considered to be quite as tight as that provided by a nonprereacted amine, and the chemical resistance is said to be somewhat lower. However, application is much easier and not nearly as dependent on thorough mixing and induction time of the reactants.

Polyamide curing agents are the condensation products of a dimerized fatty acid with a polyamine. Terminal amine functionality allows cross linking to occur as with a straight amine, although the polyamide molecule is much larger. The cross-linked film has improved flexibility, improved gloss and flow, excellent water resistance, and good chemical resistance. Polyamide-cured coatings, however, have somewhat less solvent and alkali resistance than amine and amine adduct cured epoxies do.

Specially formulated polyamide-cured epoxies have the ability to displace water from the substrate surface. Such materials can even be applied and cured underwater to form corrosion-resistant coatings.

One other type of curing agent, ketimine, is worthy of note. The ketimine curing agent allows the application of a high molecular weight, high-solids, or solventless epoxy resin with standard spray equipment. These more viscous epoxies

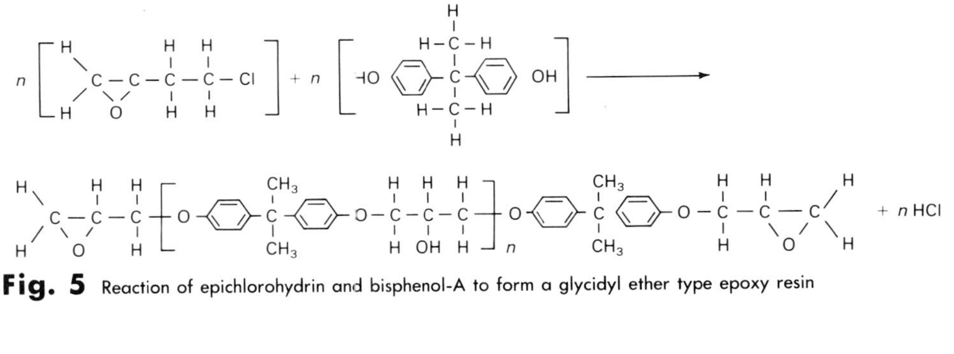

Fig. 5 Reaction of epichlorohydrin and bisphenol-A to form a glycidyl ether type epoxy resin

can be mixed with the low molecular weight ketimine curing agent and spray applied to the surface. No reaction will occur until the ketimine decomposes in the presence of atmospheric moisture to form a polyamine and a ketone. The ketone volatilizes from the film, and the amine reacts as described above to cross link the epoxy resin after application. These systems result in a tightly cross-linked epoxy film, but require relative humidity in excess of 50% and good ventilation to expose the ketimine to moisture and to remove the residual ketone and other solvents from the paint film. The ketimine curing reaction is shown in Fig. 6.

Coal tar epoxies are a combination of a coal tar combined with an epoxy resin packaged separately in one container. The curing agent is an amine, amine adduct, or polyamide (as described above) packaged in a separate container. The cross-linking reaction is the same as that described above. The coal tar acts as a filler within the cross-linked epoxy matrix, and the resulting film has the good toughness, adhesion, ultraviolet resistance, and thermal stability of the epoxies, combined with the extremely high moisture resistance afforded by coal tar. As might be expected, the amine-cured coal tar epoxies generally have greater chemical and moisture resistance, but are more brittle and harder to apply and topcoat than the amine adduct- and polyamide-cured coal tar epoxies. On the other hand, the polyamide coal tar epoxies are more flexible, easier to topcoat, and more tolerant of application variables.

Epoxy-Phenolic Coatings. Phenolic cross-linked epoxy coatings are extremely resistant to acids, alkalies, and solvents, and they are widely used as linings for tank cars, drums, cans, process tanks, and equipment. High molecular weight epoxies with a high proportion of hydroxyl functional groups available for reaction are used. During manufacture, the epoxy is packaged with a phenolic resin (commonly a phenolformaldehyde resin), pigment, and an acid catalyst. After mixing, the coating is applied by spraying, and it must be baked at temperatures from 150 to 205 °C (300 to 400 °F) in order for cross linking to occur. Multiple-coat systems require brief intermediate baking between coats; longer baking times are commonly used after application of the final coat.

The resulting coatings have the excellent acid and solvent resistance associated with the phenolics as well as the flexibility, adhesion, and toughness of epoxies. Epoxy phenolics have the good alkali resistance of the epoxy constituent, but straight phenolic linings have very poor alkali resistance.

The cross linking that occurs is between the hydroxyl groups along the epoxy chain and the methylol groups present in the phenolic resin. A possible secondary reaction is the reaction between the terminal epoxide groups of the epoxy resin and the hydroxyl groups of the phenolic resin.

$$(R_1)(R_2)C{=}N{-}R{-}N{=}C(R_1)(R_2) + 2H_2O \rightarrow (H)_2N{-}R{-}N(H)_2 + 2R_1{-}\overset{O}{\overset{\|}{C}}{-}R_2$$

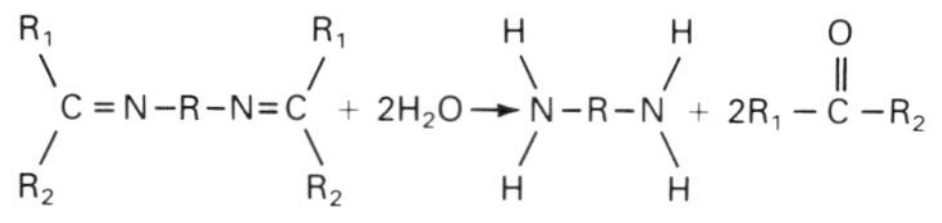

Fig. 6 Curing reaction of ketimine-cured epoxies. Ketimine decomposes in the presence of atmospheric moisture to form a polyamine plus a ketone. R, R_1, and R_2 are alkyl groups. Source: Ref 2

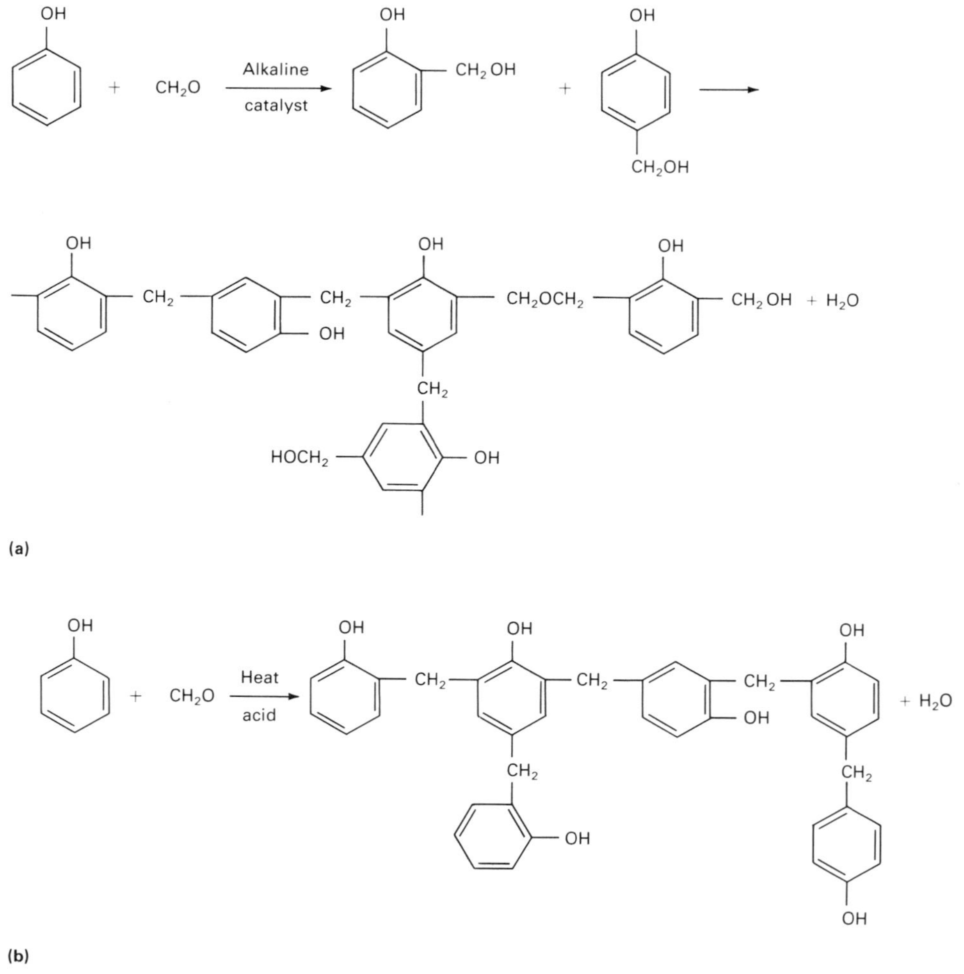

Phenolic Linings. Phenolic resins are formed by the reaction of phenol with formaldehyde to form phenolformaldehyde. The methylol group can react on the two ortho positions and the paraposition on the benzene ring to form phenolformaldehydes with functionalities (possible reaction sites) up to 3. Upon heating, a condensation reaction occurs between methylol groups on adjacent molecules that cross link the film and liberate water. The reaction of phenol and formaldehyde to form a cross-linked film is shown in Fig. 7.

The phenolformaldehyde resins are usually dissolved in alcohol and applied by spray, dip, or roller. The fact that water is liberated during the cross-linking reaction is important. The coatings must be applied in multiple-coat systems; each coat is approximately 0.025 mm (1 mil) thick and must be baked for a few minutes at 120 to 205 °C (250 to 400 °F) between coats. This partially cross links the coating, but more importantly, it volatilizes the water formed during the cross-linking reaction. Subsequent coats are applied and baked until, after the final application, a postbake at higher temperatures and longer durations is performed. It is easy to determine when the coating has been baked sufficiently to ensure proper curing and cross linking. The phenolic resin darkens upon heat exposure, and a thoroughly cured coating is a relatively uniform, light chocolate brown color.

Because they are odorless, tasteless, and nontoxic after full curing, phenolics are often applied as linings for vessels and tanks used for the processing and storage of such food products as beer, wine, and sugar syrups. They also have excellent resistance to boiling water and hot aqueous solutions of mild acids. Thus, they often find application as linings for steel subject to high-temperature freshwater and salt water as well as acid. A major weakness, however, is the lack of alkali resistance, and unless phenolic coatings are modified with epoxies, they should not be used for alkaline service.

Urethane coatings have chemical- and moisture-resistant properties similar to those of the epoxies, but they can also be formulated in a variety of light-stable colors that maintain their gloss and "wet look" upon prolonged exposure. The chemistry involved in urethane coating curing is not unduly complex and typically consists essentially of the reaction of an isocyanate-containing (—N=C=O) material with a polyhydroxylated (—OH containing) coreactant (Fig. 8). Cross linking occurs because of the high reactivity and affinity of the isocyanate group for the active hydrogen of the polyol hydroxyl or any active hydrogen atom attached to a nitrogen or

(a)

(b)

Fig. 7 Reaction of phenol and formaldehyde to form a phenolic resin. (a) Reaction using an alkaline catalyst. (b) Reaction using an acid catalyst. Source: Ref 3

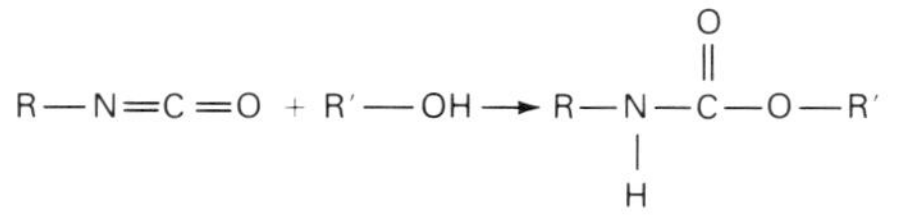

$R—N{=}C{=}O + R'—OH \rightarrow R—NH—C(=O)—O—R'$

Fig. 8 Reaction of an isocyanate and a polyol to form a urethane. R and R′ are different aromatic or aliphatic groups. Source: Ref 4

Fig. 9 Molecular structures of TDI, an aromatic isocyanate used in urethane coatings. (a) 2,4 isomer. (b) 2,6 isomer. Source: Ref 4

Fig. 11 Molecular structure of HDI, an aliphatic isocyanate used in urethane coatings. Source: Ref 4

oxygen atom. The rate of this cross-linking reaction depends on a number of factors, such as the type and configuration of both the isocyanate and polyol materials and the temperature. However, the reaction is such that with most formulations cross linking can occur at temperatures as low as −18 °C (0 °F) or less.

A number of different types of isocyanate materials are used in urethane coatings. The molecular weight and structure as well as isocyanate functionality (the number of isocyanate groups available for reaction) can affect the final properties of the coating. Similarly, different polyol structures, such as acrylics, epoxies, polyesters, polyethers, and vinyls, when reacted with a given isocyanate to form a cross-linked coating, will result in variations of physical and chemical properties. Described below are some of the general properties of the isocyanate and polyol constituents that are coreacted to form urethane coatings.

The isocyanates are commonly categorized as either aromatic (containing the benzene ring) or aliphatic (straight-chain or cyclical) hydrocarbons. The first isocyanate used in urethane coatings was an aromatic isocyanate, toluene diisocyanate (TDI) (Fig. 9). When reacted, TDI resulted in a suitable chemical-resistant urethane coating, but it was generally undesirable because of its low viscosity and toxic irritating vapors. To form a safer, more suitable coreactant, TDI was further reacted to provide a higher molecular weight polyisocyanate that contained less than 1% of the original TDI monomer (Fig. 10).

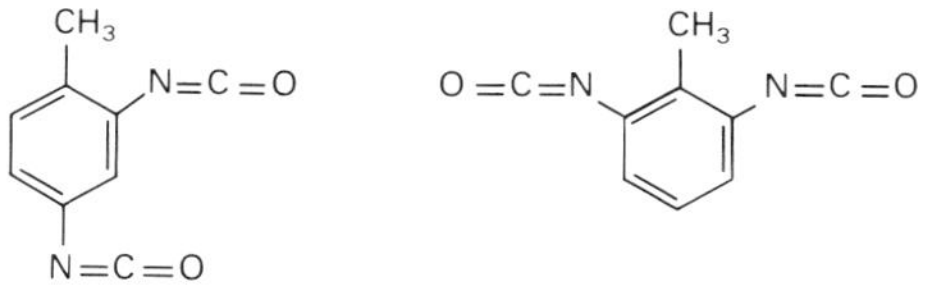

These higher molecular weight aromatic isocyanates, when reacted with a polyol (usually a polyester, polyether, or epoxy), formed useful chemical-resistant coatings that could be cured at low temperatures. However, the urethanes so formed tend to chalk, yellow, and darken somewhat upon exposure to sunlight.

In the early 1970s, aliphatic isocyanates were commercially developed. Although they were considerably more expensive than the aromatic isocyanates, they allowed the formulation of nonyellowing, light-stable, high-gloss finish coats. The appearance of urethane coatings formulated with aliphatic isocyanates was unmatched by the epoxies, acrylics, or any other coating material then (and now) available in the industrial marketplace. In spite of the high cost of the aliphatic polyurethanes, the coatings were increasingly specified in high-visibility applications, such as city water tanks, railroad tank and hopper car exteriors, and other areas where visibility or aesthetic appeal were paramount.

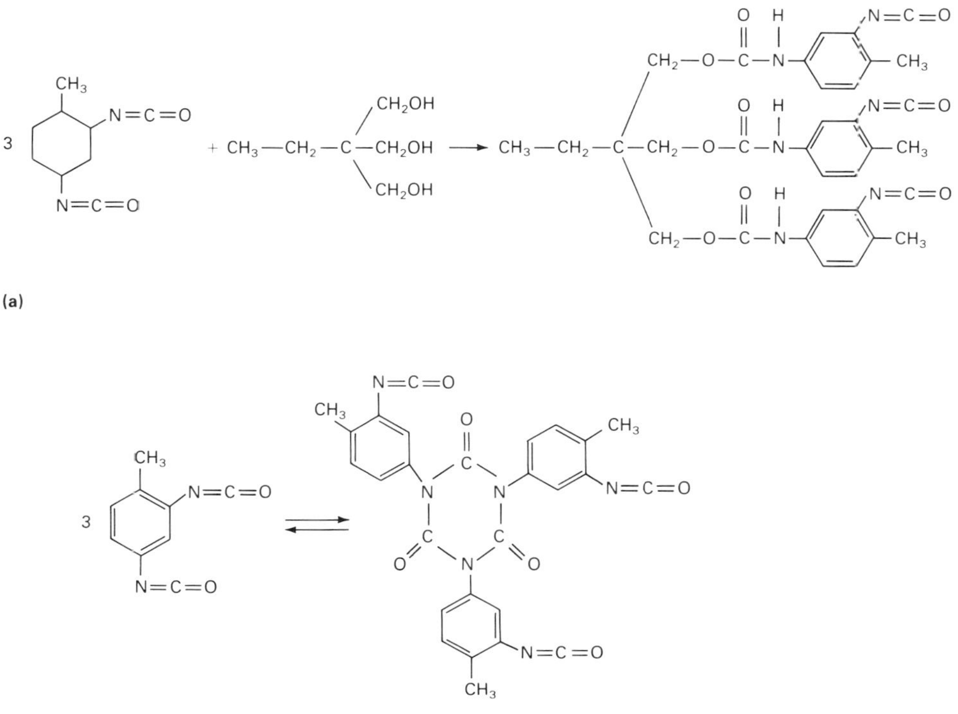

An important aliphatic isocyanate is hexamethylene diisocyanate (HDI) (Fig. 11). In its monomeric form, HDI is an irritant. Like TDI, however, HDI can be reacted (commonly with water) to form a larger molecule.

Various combinations of aromatic and aliphatic isocyanates can be produced, including specialty isocyanates, to make high-solids or solvent-free coatings, elastomeric coatings, or coatings designed for baking or curing at elevated temperatures. Specialty isocyanates can be either aromatic or aliphatic; if aromatic, they will discolor. Both straight-chain and cyclical aliphatic isocyanates, although nonyellowing, are much slower to react and cross link than aromatic isocyanates.

Polyols. As stated previously, polyols coreact with isocyanates in order to form the polyurethane film. Accordingly, the polyol is packaged separately from the isocyanate, and the polyol package always contains the pigment, most of the solvent, and any additives used for flow, thixotropy, antisetting properties, and so on. The major properties of the cross-linked urethane film, such as chemical resistance, toughness, lightfastness, and flexibility, derive principally from the polyol constituent. Long, repeating carbon-to-carbon backbone chains with little branching (for example, vinyl) result in tough, flexible resins. Aromatic groups add rigidity to the chain, as does cross linking. The presence of numerous ester, ether, and urethane groups reduces chemical and moisture resistance.

For corrosion-protective coatings, the polyol coreactant is usually one of the following: acrylic, polyester, polyether, epoxy, vinyl, or alkyd. Asphalts and coal tars can be added with any of the polyols. Because they are essentially nonreactive, however, it is felt the bitumen acts predominantly as a filler in the urethane matrix, providing the cured coating with the attendant properties of good chemical resistance, high moisture resistance, and high film build at a relatively low cost.

Acrylic urethanes are perhaps the most widely used urethanes for corrosion protection in atmospheric service. These coatings, when properly formulated, have excellent weatherability, gloss, and color retention and have good chemical and moisture resistance. They can be readily tinted and pigmented to provide a variety of deep and pastel colors at a lower cost per gallon than the next most popular class, the polyester urethanes.

As with conventional acrylic coatings, the acrylate polymer family is softer and more flexible than the corresponding methacrylates, although an increase in the molecular weight of the monomers, such as methyl acrylate to butyl acrylate, leads to softer, more flexible polymers for both families. Acrylate and methacrylate blends are also used as reactive polyols.

Acrylic urethanes are not used for immersion service, and for the most part, they do not have the chemical resistance of the polyester urethanes. However, they do have excellent weathering properties when an aliphatic isocyanate is used.

Fig. 10 Modifications of TDI to form higher molecular weight polyisocyanates. (a) Polymerization of TDI with an alcohol to produce a TDI alcohol adduct. (b) Condensation of TDI monomer to produce an isocyanate ring. Source: Ref 4

Polyesters result from the condensation reaction of a dibasic acid and a polyhydric alcohol. To form a hydroxylated polyester, an excess of the hydroxyl functionality from the alcohol is allowed over that of the acid. The result is a polyester molecule with a mid-chain and end-chain hydroxyl functionality.

Because of their high isocyanate demand when coreacted, polyesters result in relatively hard chemical-resistant polymer films. Impact resistance is not as great as with the acrylic urethanes, but as a rule, chemical and moisture resistance are better. However, because ester linkages occur in the molecular backbone, alkaline attack and water sensitivity may occur at locations where the ester group is exposed, that is, at areas of low cross-link density or where less branching of adjacent molecules occurs. Accordingly, polyester urethanes, like acrylic urethanes, are not normally used for immersion service.

Polyethers are often the polyol of choice for use as elastomeric coatings, coal tar urethanes, and other urethane coatings that are sheltered or protected from light. The polyether prepolymer is considerably less expensive than acrylic or polyester polyols; polyethers, however, are sensitive to ultraviolet-induced oxidative degradation. Furthermore, the ether linkages (—C—O—C—) are water sensitive, and in polyethers, they are repeated throughout the polymer chain without separation by long water-insensitive hydrocarbon chains or aromatic groups that can minimize water sensitivity.

Epoxies. The reaction of an epoxy with an isocyanate would seem to be a natural one, because of the mid-chain hydroxyl functionality of the epoxy molecule. However, there are very few epoxy urethanes on the market, and those being promoted are used predominantly as primers or intermediate coats. This is due to:

- The tendency of the epoxy and the epoxy urethane to chalk
- The lower moisture resistance resulting from the urethane link, which gives a less chemical- and moisture-resistant polymer than a straight epoxy coating
- The greater expense of a urethane epoxy compared to a conventional amine or polyamide cross-linked epoxy

As a result, most epoxy urethanes are formulated with the less expensive aromatic isocyanate and are promoted primarily as low-temperature curing epoxies or fast-curing, chemically resistant urethanes for interior use.

Vinyl urethanes use a long-chain, linear, hydroxyl-bearing PVC/PVA resin reacted with a polyfunctional isocyanate prepolymer. Urethane coatings using vinyl polyols combine the abrasion resistance of the urethane with the toughness, flexibility, and chemical resistance of the vinyl. Such urethane coatings are promoted for use where flexibility and abrasion resistance are important; vinyl urethane coatings, however, are subject to some chalking or fading upon exterior exposure and do not have the color, gloss, weatherability, or solvent resistance of the acrylics and polyesters. Because the vinyl is thermoplastic and is attacked and softened by solvents, recoatability after extended times is not a problem and is a major advantage of vinyl urethane systems.

Alkyds are commonly reacted with isocyanates to form a so-called uralkyd or urethane oil coating. The isocyanate reaction decreases the drying time of the coating and provides enhanced resistance to chemicals, moisture, weathering, and abrasion. Uralkyd (urethane oil) coatings harden by autooxidation of their unsaturated oleoresinous sidechains. The result is an upgraded coating system with enhanced performance over the non-urethane system. Because of the isocyanate reaction, however, the cost of such a system is somewhat higher than those systems that are not urethane modified. A uralkyd is prepared by replacing some of the dibasic acid with an isocyanate to form a urethane cross-linked polymer. The resulting urethane alkyd resin still has the oleoresineous sidechains, and although it is cross linked and upgraded by the isocyanate, it is still sensitive to water and chemicals through the ester and urethane links.

Moisture-Cured Urethanes. Isocyanates can also react with the hydroxyl group in water to form a class of coatings known as moisture-cured urethanes. Single-package moisture-cured urethanes use an isocyanate prepolymer that, when applied, reacts with the humidity in the air to form a hard, tough, resinous film. Because of their rapid rate of reaction, aromatic isocyanates are used almost exclusively in moisture-cured urethanes. The pigments must be essentially nonreactive with the isocyanate; although it is possible to use a number of pigments, aluminum leaf is most common. The reaction of a single-package urethane with moisture is illustrated in Fig. 12. It should be noted that the isocyanate/water reaction produces gaseous carbon dioxide. In fact, some carbon dioxide is evolved in the curing of all urethane coatings, because moisture (humidity) is present in the atmosphere during curing.

In areas where the relative humidity is expected to be low or where assurance of a more complete cure is necessary, a small amount of a tertiary amine catalyst, such as dimethylethanolamine, can be added. This catalyst is packaged separately from the isocyanate and added just before use. Although it comes as a two-package system, this type of catalyzed urethane can be considered to be an extension of a single-package moisture-cured urethane.

Zinc-Rich Coatings

Zinc-rich coatings, often called zinc-rich primers, are a unique class of cross-linked coatings that provide galvanic protection to a ferrous substrate. As the name implies, the binder is highly loaded with a metallic zinc dust pigment. After the coating is applied to a thoroughly cleaned substrate, the binder holds the metallic zinc particles in contact with the steel and with each other. Thus, metal-to-metal contact of two dissimilar metals is made, resulting in a galvanic cell. In this couple, zinc becomes the anode and sacrifices itself to protect the underlying cathodic steel (the articles "Hot Dip Coatings" and "Corrosion of Zinc" in this Volume contain more information on corrosion of galvanized steels).

The major advantage of corrosion protection using zinc-rich coatings is that pitting corrosion and subfilm corrosion are eliminated, even at voids, pinholes, scratches, and abrasions in the coating system. This cannot be said of any other type of protective coating, and it is this protective capability that makes zinc-rich coatings unique and widely used.

This advantage, however, comes with certain disadvantages. The underlying steel substrate must be cleaned of all rust, old paint, and other contaminants that may interfere with metal-to-metal contact. Thus, the degree of surface preparation must be quite thorough; blast cleaning should produce a Commercial Blast Cleaning minimum and, for immersion service, a White or Near-White surface. The Steel Structures Painting Council (SSPC), NACE, and other organizations issue standards for the surface preparation of metals for organic coatings; these are discussed in more detail in the section "Surface Preparation and Coating Application" in this article. When power tool cleaning, cleaning to bright metal is necessary. Zinc-rich coatings can be applied to pickled steel, although, again, the surface must be perfectly clean.

Because of the high reactivity of the zinc-dust pigment, zinc-rich coatings are not suitable in condensing or hydrolyzing environments outside the pH range of approximately 5 to 10. Acids and alkalies will attack the zinc-dust pigment; even if topcoats are used, chemicals may penetrate through pinholes, scratches, voids, or discontinuities within the topcoat and cause aggressive attack of the zinc-rich primer.

In addition, because of the high zinc loading, the zinc-dust pigment should be suspended in the spray pot during spray application. Thus, the use of an agitator or stirrer in the spray pot is mandatory. Similarly, the weight of the paint, as a result of its high pigment loading (often 2.2 to 3 kg/L, or 18 to 25 lb/gal.), prohibits long hose lengths from the spray pot to the spray gun in order to minimize settling in the lines. If the spray gun is elevated too high above the spray pot, air pressure to the spray pot may not be sufficient to push the heavy zinc-rich coating material through the spray line to the gun.

When spraying on warm days, care must be taken that the primer does not begin to set up in the spray lines. This is a particular problem with most inorganic zincs, especially if a spray application is interrupted for more than just a few minutes without first draining the fluid (paint) hose. With no movement in a relatively warm

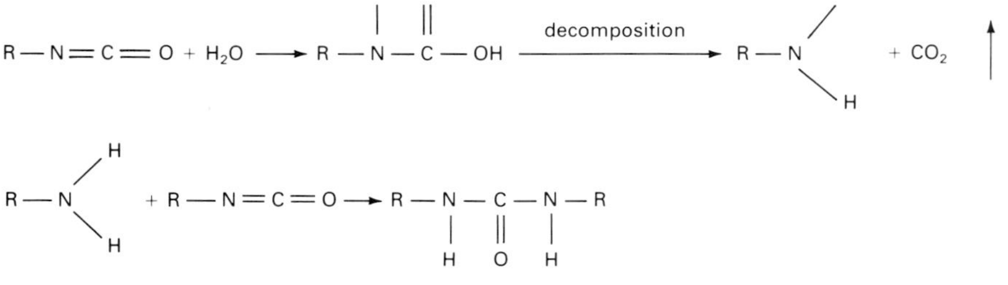

$$\mathrm{R{-}N{=}C{=}O} + \mathrm{H_2O} \longrightarrow \mathrm{R{-}\underset{H}{\overset{}{N}}{-}\overset{O}{\overset{\|}{C}}{-}OH} \xrightarrow{\text{decomposition}} \mathrm{R{-}NH_2} + \mathrm{CO_2}\uparrow$$

$$\mathrm{R{-}NH_2} + \mathrm{R{-}N{=}C{=}O} \longrightarrow \mathrm{R{-}\underset{H}{N}{-}\underset{O}{\underset{\|}{C}}{-}\underset{H}{N}{-}R}$$

Fig. 12 Curing reaction of a single-package moisture-cured urethane. The isocyanate reacts with water to form unstable carbamic acid, which decomposes into an amine plus carbon dioxide gas. The amine then reacts with isocyanate to form a urea derivative. R indicates an aromatic or aliphatic group. Source: Ref 4 paint hose, plugs of paint may begin drying and hardening such that when spraying operations begin the paint does not readily move through the fluid line to the gun. Blockages occur, the gun works only intermittently, and, in the worst cases, hardening of the paint may occur to such an extent that the entire line must be discarded.

This rapid-drying characteristic, which is particularly evident with the inorganic zincs, may also result in a greater amount of dry spray than usually occurs with conventional nonzinc-rich coatings. Thus, particular care should be taken during application to ensure that the zinc-rich primer is deposited in a smooth, wet application. Because of high zinc loading over the pigment volume concentration of the binder, the zinc-rich coatings are more porous than any other generic type of paint. Because of this porosity, they are harder to topcoat. Topcoats applied over both organic and inorganic zinc-rich primers will often have voids or bubbles in their cross-sectional thickness and bubbles or pinholes extending down through their surface to the underlying zinc-rich coating. This is because of entrapped air, solvents, and moisture within the zinc-rich primer before or during topcoating. The entrapped liquids or gases must pass through the topcoat in order to escape into the atmosphere. As the topcoat dries, escaping gases are held as bubbles or blisters, or the path of such an escaping gas hardens into a pinhole.

Insufficient curing of the zinc-rich binder before application of the topcoat can sometimes lead to poor adhesion or cohesion within the zinc and can ultimately lead to peeling or disbonding of the topcoat. However, despite these disadvantages, the advantage that accrues by the elimination of subfilm and pitting corrosion makes zinc-rich coatings, either topcoated or untopcoated, one of the most widely used corrosion prevention coatings for painting steel in industrial, marine, and other severe service environments.

Zinc-rich coatings used as primers with topcoats or as complete one-coat systems are widely used for the corrosion protection of steel substrates. Organic zinc-rich systems are used in the automotive industry to protect car body rocker panels, wheel wells, and other corrosion-prone interior steel parts (see the article "Corrosion in the Automotive Industry" in this Volume). Inorganic zinc-rich coatings are also used extensively in the marine industry for the protection of exterior hulls, cargo and ballast tanks, offshore platforms, and other areas of high marine corrosion (see the articles "Marine Corrosion" and "Corrosion in Petroleum Production Operations" in this Volume). Inorganic zinc-rich coatings are used in these same environments and in applications requiring high abrasion and high heat resistance (to 370 °C, or 700 °F) when untopcoated. Both organic and inorganic zinc-rich coatings are extensively used for steel protection on bridges and highway structures, chemical and petrochemical plants, sewage and water treatment, and any area where freshwater or saltwater corrosion, mild fumes, and high humidity and resultant corrosion is a problem. The articles "Corrosion in the Chemical Processing Industry," "Corrosion in Petroleum Refining and Petrochemical Operations," and "Corrosion in Structures" in this Volume contain more information on the use of coatings in those applications.

Zinc-rich coatings can be subcategorized according to the type of binder material used. Zinc-rich coatings with organic binders are similar in many ways to the coating systems previously discussed, except that sufficient zinc-dust pigment is added to provide galvanic protection. The inorganic zinc-rich coatings, on the other hand, use an entirely different binder chemistry and are therefore quite unlike the organic zinc-rich coatings. As a general rule, organic zinc-rich coatings are more widely used in Europe and Asia than inorganic zinc-rich coatings; in the United States and Canada, the inorganic zinc-rich coatings are more widely used.

Organic zinc-rich coatings are most commonly formulated from epoxy polyamide, urethane, vinyl, and chlorinated rubber binders. Alkyds have been used for some air dry formulations, but they are most commonly used with baking formulations, notably in the automotive industry.

To manufacture a successful organic zinc-rich coating, one does not simply substitute the zinc-dust pigment for the normal inhibitive and filler pigments commonly used. Instead, skillful formulation is required to achieve galvanic protective capability while also maintaining suitable storage, application, and physical properties of the applied film. When such formulation is achieved, the result is a protective coating having the physical attributes determined by the organic binder, along with the galvanic capability and chemical susceptibilities inherent in the use of the zinc-dust pigment.

For example, vinyl and chlorinated rubber zinc-rich primers are easier to topcoat than epoxy and urethane zinc-rich primers for the same reasons that vinyl and chlorinated rubber primers are easier to topcoat than overcatalyzed or excessively hardened epoxies and urethanes. Similarly, the heat resistance of the thermoset epoxies and urethanes is greater than that of the thermoplastic vinyls and chlorinated rubbers. However, in all cases, the untopcoated zinc-dust pigment is attacked by acids and alkalies, and even though the binder is resistant to such chemicals, organic zinc-rich primers should not be used in aggressive environments.

The drying, hardening, and ultimate curing of the organic zinc-rich coating is determined by the type of binder used. The hardening and curing mechanisms of the various binders have been described previously and are the same for zinc-rich coatings. Organic zinc-rich primers are more tolerant of deficient surface preparation because they wet more readily and seal poorly prepared surfaces where residues of rust or old paint may remain. Similarly, topcoating with the same generic type of topcoat (for example, a vinyl zinc-rich coating topcoated with a vinyl topcoat) is more readily accomplished, because organic zinc-rich coatings of all types generally have a less porous surface and are more like conventional organic coatings than the inorganic zinc-rich coatings.

Inorganic Zinc-Rich Coatings. The SSPC has categorized inorganic zinc-rich coatings as being of three major groups:

- Postcured water-base alkali metal silicates
- Self-cured water-base alkali metal silicates
- Self-cured solvent-base alkyl silicates

Although the binder, in all cases, is an inorganic silicate (essentially the same material as glass or sand), the curing of the binder is different in all three cases.

Postcured water-base inorganic silicates come as three-package systems; the zinc dust, silicate binder, and curing solution are packaged separately. If elevated baking temperatures can be achieved, the curing solution may not be necessary.

The earliest postcured zinc-rich coatings were based on a sodium silicate vehicle, and the most popular commercial products available have essentially the same chemistry. There are other inorganic zinc-rich coatings based on lithium or potassium silicates that will self-cure, but curing can be accelerated by the use of a curing solution, if desired. The following steps summarize the curing of postcured water-base inorganic zinc-rich coatings.

The first step is primarily a dehydration process in which the water, which is the basic solvent for the silicate, evaporates from the clean steel surface, leaving the zinc and the alkali silicate on the surface as a coating. At this point, when the coating is free of water, it is very hard and metallic. Scraping with a coin merely polishes the zinc in the coating.

Second, once the initial drying process has taken place, an acid phosphate curing solution is sprayed or brushed over the zinc-rich coating in order to wet out the surface thoroughly. The initial reaction is for the acid to gradually neutralize the sodium in the sodium silicate solution and to create a mildly acidic surface condition.

Third, during initial curing, the pigment constituents may react with the silica matrix. This, together with further neutralization of the silicate alkalinity, insolubilizes the film.

The addition of the acidic curing solution on the surface of the zinc coating ionizes the zinc in the coating as well as the lead oxide pigment mixed with it. As this takes place, there is a rapid reaction between the active silicate groups with the lead and zinc, and this reaction insolubilizes the silica matrix.

The acid phosphate also reacts with the zinc and lead to form very insoluble zinc and lead phosphates within the silicate matrix. At this point, the coating has become insoluble to water and resistant to exposure to weather.

Lastly, after initial curing, further hydrolysis and improvement of film properties continue over time. Following the application and reaction of the curing agent, the remaining soluble salts on the surface are removed with clean water or by weathering. The coating at this point is dense and relatively nonporous, insoluble in water, and resistant to marine and mild chemical atmospheres.

Postcured coatings are in the author's opinion the best performing of the inorganic zinc-rich materials. The reason for this is most likely the relatively complete, insolubilizing hard cure that is attained after application of the curing solution. By applying the curing solution and thoroughly curing the coating, the maximum protective properties of the coating are attained shortly after application. In contrast, self-cure coatings rely on ambient atmospheric conditions in order to cure completely. Recent research indicates that many self-cured coatings do not achieve their ultimate resistance and protective capability until long after their original application.

Self-Curing Water-Base Alkali Silicates. The most common of these silicate binders is based on potassium and lithium silicates or combinations of these. In addition, lithium hydroxide-colloidal silica and quaternary ammonium silicate binders are also included in this category.

The reaction chemistry of the water-base self-cure inorganic zinc-rich silicates is similar to that of the postcured sodium silicates, with the exception that the curing solution is not applied. The metal silicate vehicle, before field addition of the zinc-dust pigment, has a higher ratio of silica to metal oxides and a lower alkalinity than the postcured vehicles. Therefore, after water evaporation and insolubilization resulting from the silica reaction with the pigment, sufficient neutralization and cure are attained by further reaction with acid formed by the atmospheric hydrolysis of carbon dioxide (reaction of CO_2 and atmospheric moisture to form a weak carbonic acid).

Self-curing alkali silicate zinc-rich coatings become hard within minutes and are considered generally resistant to precipitation within 30 min after application. If a more rapid insolubilization or cure is desired, heat or an acidic curing solution can be applied. The reaction chemistry would then be essentially identical to that of the postcured sodium silicate matrix. When final curing is ultimately attained, most water-base zinc-rich coatings, whether post- or self-cured, experience a color change—often from a reddish gray or light gray to a darker bluish gray.

Self-Cured Solvent-Base Alkyl Silicates. The binders for this class of coatings are essentially modified alkyl silicates consisting of partially hydrolyzed members of the series methyl through hexyl or glycol ether silicates. Of these, the ethyl silicate type is most commonly used.

The reaction chemistry of ethyl silicates can be best demonstrated by the hydrolysis of tetraethyl orthosilicate in the presence of an acid or alkali catalyst. The reaction is carried to partial completion with a degree of hydrolysis ranging from 70 to 95%. The reaction is irreversible, because of the volatilization of the alcohol that results from the ester exchange; however, should premature gelling of the silicate occur, alcohol may be added to redissolve the gel. During the condensation phase of the reaction, the partially polymerized silicate combines with atmospheric moisture to eliminate alcohol, which vaporizes.

After complete hydrolysis, the cross-linked network forms a matrix to hold the pigment particles. In addition, it is conceivable that the zinc and other metallic pigment additives (and perhaps the underlying ferrous surface) may react with the silicate in a manner similar to that described for postcured sodium silicate.

Single-package inorganic zinc-rich primers based on the ethyl silicate chemistry described above have been developed. Instead of packaging the zinc dust separately from the ethyl silicate binder, the coating manufacturer adds the zinc dust in the manufacturing plant, along with additives that keep the zinc dust in suspension. Careful quality control of raw materials, particularly zinc-dust pigment and additives, is exercised to ensure they are dry and moisture free. Also, the ethyl silicate vehicle is somewhat less prehydrolyzed and is usually catalyzed with an acid catalyst such that upon application and curing the reaction with moisture occurs at a faster rate than with two-package ethyl silicate zinc-rich primers.

The properties of the single-package ethyl silicate zinc-rich primers are similar to those of the dual-package primers (where the zinc dust is packaged separately from the liquid silicate binder); but in general, single-package inorganic zinc-rich coatings do not cure as hard or as fast as dual-package products, and they do not provide quite as long a corrosion protection, although corrosion protection is still excellent. On the other hand, they are much easier to mix and apply, and these application advantages have resulted in their increased use.

Comparisons of Inorganic and Organic Zinc-Rich Coatings. In general, the inorganic zinc-rich coatings provide somewhat longer galvanic protection than the organics do, but they are less tolerant of deficient surface preparation, are somewhat harder to mix and apply, generally have a greater tendency toward mudcracking at excessive thickness, and are more difficult to topcoat. The organic zinc-rich coatings have limited temperature resistances that are determined by their organic binder. In contrast, inorganic zinc-rich coatings will resist dry heat to over 370 °C (700 °F), and when topcoated with a silicone or ceramic coating, they can be used at temperatures of 650 °C (1200 °F) or higher. The abrasion and impact resistance of the inorganics are unsurpassed.

All water-base zinc-rich coatings, whether post- or self-cured, require somewhat greater surface cleanliness in order to achieve suitable adhesion and environmental protection. Water-base zinc-rich coatings have a greater tendency to mudcrack at excessive thicknesses than solvent-base inorganics do, and there is also an increased tendency to form zinc surface reaction products, particularly with sodium and lithium silicate vehicles. A major advantage of all water-base zinc-rich coatings is that they do not have a flash point and can be used in tanks, ship holds, and other confined or poorly ventilated areas without danger. However, such coatings do require good ventilation or dehumidification to cure properly. Solvent-base inorganic zinc-rich coatings that are modified to have a flash point higher than their normal 20- to 30-°C (70- to 90-°F) tag open cup often form a softer, less resistant film.

Coating System Selection

The selection of the proper corrosion-resistant coating system depends on a number of factors. This section will highlight the more important considerations.

Environmental Resistance. The coating system should be resistant to the chemical, temperature, and moisture conditions expected to be encountered in service.

Appearance. In high-visibility areas (for example, water tanks, railroad cars, and appliances), color, gloss, and a pleasing appearance may be very important.

Safety. Toxic pigments and solvents may prohibit the use of some coatings; future removal and disposal problems should be considered. Coatings with volatile explosive solvents may be dangerous in enclosed, poorly ventilated spaces.

Surface Preparation. Blast cleaning may be prohibited in some refineries or near electrical and hydraulic machinery.

Skill of the Labor Force. Certain coating systems require more application expertise than others. Generally, inorganic zincs, vinyl-esters, polyesters, vinyls, and chlorinated rubbers are less application tolerant than some of the other coating systems.

Substrate to be Coated. Coating systems for aluminum, lead, copper, and so on, may be different from those for ferrous metals.

Available Equipment. The more resistant coating systems generally require spray application over a blast-cleaned surface. For some polyesters and vinyl-esters, plural component spray equipment may be required.

Design Life. If a structure is intended to have a long service life, the use of a more resistant coating system may be justified than if a shorter service life is intended.

Cost. Generally speaking, the applied cost of a more resistant coating system is greater than the applied cost of a less resistant system.

Accessibility for Future Repair. If future repair will be difficult or expensive, perhaps a longer lasting coating system should be specified.

Consequences of Coating Failure. If a coating failure will be disastrous, such as in tanks holding highly corrosive materials or in nuclear power plants where peeling or disbonding coating may clog sumps and screens, a more resistant coating system should be selected.

Each of these factors will, to varying degrees, directly influence the choice of the coating system selected to protect a metal in a given environment. Once the choice is made, it is usually necessary to prepare appropriate specifications detailing the surfaces to be painted, and the surfaces to be protected from paint. In addition, the surface preparation and coating application details should be delineated. At a minimum, the thickness of each coat and total coating thickness should be specified, and the surface profile of a blast-cleaned surface, the interval between coats, and other factors (such as testing for holidays and pinholes and for adhesion) should also be specified. Although the preparation of a proper specification is beyond the scope of this article, a good specification is not only an important contractual document but is also the means of conveying the owner's (or specifier's) intentions and instructions regarding the painting to be accomplished. Thus, the potential importance of a good specification cannot be overestimated.

Surface Preparation

The selection of the coating material to be used to protect a given metal from an environment usually determines the surface preparation that will be required. For example, zinc-rich coatings almost always require blast cleaning, while many alkyd and oil-base coating systems can be applied over rust, mill scale, or a poorly cleaned surface. The SSPC has prepared a series of surface preparation standards that are widely used throughout North America. The National Association of Corrosion Engineers has a set of encapsulated panels showing the four degrees of blast cleaning.

Similar standards based on photographs developed by the Swedish Standards Institute are widely used in Europe and Asia. These so-called Swedish standards are photographs of four starting grades of steel that are wire brush cleaned or blast cleaned to various degrees of cleanliness. Table 2 presents an abstract of the SSPC standards with a cross reference to the appropriate NACE and Swedish standards.

Although there are other national or corporate standards, the SSPC and Swedish standards are perhaps the most widely used. However, these standards (and most others) specify only the visual degree of cleaning and do not define the surface roughness or profile obtained by cleaning or by the removal of invisible contaminants.

In addition to cleanliness, the surface must be roughened to provide for a mechanical bond of the paint to the substrate. This roughening is

Table 2 Uses and applicable standards for various surface preparation techniques

Technique	Applicable standards	Uses
Solvent cleaning	SSPC-SP1	Used to remove oil, grease, dirt, soil, drawing compounds, and various other contaminants. Does not remove rust or mill scale. No visual standards are available.
Hand tool cleaning	SSPC-SP2	Used to remove loose rust, mill scale, and any other loose contaminants. Standard does not require the removal of intact rust or mill scale. Visual standards: SSPC-VIS I—BSt3, CSt3, and DSt3(a)
Power tool cleaning	SSPC-SP3	Same as hand tool cleaning. Visual standards: SSPC-VIS I—BSt3, CSt3, and DSt3(a)
White-metal blast cleaning	SSPC-SP5; NACE 1	Used when a totally cleaned surface is required; blast-cleaned surface must have a uniform, gray-white metallic color and must be free of all oil, grease, dirt, mill scale, rust, corrosion products, oxides, old paint, stains, streaks, or any other foreign matter. Visual standards: SSPC-VIS I—ASa3, BSa3, CSa3, and DSa3(a); NACE 1
Commercial blast cleaning	SSPC-SP6; NACE 3	Used to remove all contaminants from surface, except the standard allows slight streaks or discolorations caused by rust stain, mill scale oxides, or slight, tight residues of rust or old paint or coatings. If the surface is pitted, slight residues of rust or old paint may remain in the bottoms of pits. The slight discolorations allowed must be limited to one-third of every square inch. Visual standards: SSPC-VIS I—BSa2, CSa2, DSa2(a); NACE 3
Brush-off blast cleaning	SSPC-SP7; NACE 4	Used to remove completely all oil, grease, dirt, rust scale, loose mill scale, and loose paint or coatings. Tight mill scale and tightly adherent rust and paint or coatings may remain as long as the entire surface has been exposed to the abrasive blasting. Visual standards: SSPC-VIS I—BSa1, CSa1, DSa1(a); NACE 4
Pickling	SSPC-SP8	Used for complete removal of all mill scale, rust, and rust scale by chemical reaction, electrolysis, or both. No available visual standards
Near-white blast cleaning	SSPC-SP10; NACE 2	Used to remove all oil, grease, dirt, mill scale, rust, corrosion products, oxides, paint, or any other foreign matter. Very light shadows, very slight streaks, and discolorations caused by rust stain, mill scale oxides, or slight, tight paint or coating residues are permitted to remain, but only in 5% of every square inch. Visual standards: SSPC-VIS I—ASa2-1/2, BSa2-1/2, CSa2-1/2, and DSa2-1/2(a); NACE 2

(a) The Swedish standards (SSPC-VIS I) are the most commonly used visual standards for evaluating the cleanliness of a prepared surface. The use of these standards requires a determination of the extent of rust on the uncleaned steel; this is graded from A to D.

termed anchor pattern, profile, or tooth, and it is essentially a pattern of peaks and valleys on the steel surface. This pattern is obtained by abrasive blasting, and it must be carefully controlled according to the coating system being applied. If the peaks are too high, they will project above the coating and cause pinpoint rusting. A rule of thumb for surface profile is that it should be approximately one-third of the required coating system thickness up to a coating thickness of 0.3 mm (12 mils).

The roughness or surface profile imparted to the metal during the course of cleaning preparatory to painting influences adhesion. As a rule, greater roughness results in greater adhesion, although excessive roughness may result in high spots or peaks that are not adequately covered by the thickness of the applied coating. Such peaks may be initiation sites for pinpoint rusting and corrosion. As a general rule, most mineral abrasive blast-cleaning materials impart a profile or roughness ranging from 13 to 100 μm (0.5 to 4 mils), depending on particle size and impact velocity. Metallic grits may have deeper profiles, approaching 178 μm (7 mils) or more, but metallic shot and grit, when sized properly, will usually impart profiles within the 50- to 115-μm (2- to 4.5-mil) range. Such surface roughnesses are suitable for most of the coating systems discussed in this article, although some of the more highly stressed systems, such as the vinyl-esters and polyesters, may require a deeper profile to ensure adequate adhesion.

The removal of invisible contaminants is more difficult to assess. It should be obvious that chemical contamination after surface preparation and before coating may result in blistering, poor adhesion, or other defects in the applied coating system. It is less evident that such chemical contamination may be inherent in a metal after it has been exposed and corroded in a chemical environment. Even thorough blast cleaning to white metal may not remove chemical species (notably chloride ions) that may have penetrated into the grain boundaries or microstructure of the metal. If such penetrating anions can react to form a water-soluble salt, such as ferrous chloride, then an aggressive underfilm corrosion may occur if water permeates the coating.

As might be expected, the presence of invisible water-soluble contaminants is more detrimental in immersion service than in dry atmospheric service. Methods for removing such contaminants are usually based on water blasting or water flushing in order to dissolve and flush the water-soluble contaminants from the surface. Other approaches are the use of chelating primers, which react with and complex the soluble salts and render them inactive and noncorrosive; and the application of a porous coating, such as an inorganic zinc-rich primer, that will allow passage of water-soluble corrosion products.

A major detriment to removal is, of course, the inability to adequately detect the presence of contaminants. Although indicator kits and chemically treated papers have been developed for this purpose, more work is needed in this area. Furthermore, very little work has been done to establish the tolerance of various coatings, surfaces, or exposure environments to the presence of differing amounts of contaminants.

Cleanliness is essential in the preparation of a surface that will receive protective coatings. Paint applied over rust, dirt, or oil will bond poorly (or not at all) to the substrate, and early paint failure will usually result. A clean surface is free of such contaminants as rust, flash rust, dirt and dust, salts, oil and grease, old paint, and mill scale.

Other areas of the surface also must be addressed before the actual cleaning. Coatings will generally not cover weld spatter properly. Weld spatter may also become dislodged in service, exposing the unprotected substrate. Sharp edges cause paint to draw thin because of surface tension effects and should be eliminated by grinding. Inside corners provide a collection site for excess paint and/or abrasive and should be filled. Crevices and pits should be filled with weld metal, if necessary, and smoothed. Some pits, because of their sharp, angular edges, will require extensive welding and grinding before surface preparation. Other pits, because they consist of smooth, even depressions, need only be cleaned for coating.

In addition, for maintenance painting, thick layers of stiff, old paint should be removed, and glossy painted surfaces must be roughened. Nonferrous surfaces also frequently require some type of pretreatment. In the case of aluminum or galvanized steel, degreasing and the application of a wash primer are often employed to provide proper adhesion.

Methods of Surface Preparation

A variety of surface preparation methods are available, including:

- Solvent or chemical washing
- Steam cleaning
- Hand tool cleaning
- Power tool cleaning
- Water blasting
- Abrasive blast cleaning

Each of these is discussed below.

Solvent cleaning (covered in SSPC SP1) removes oil and grease and is usually used in conjunction with or before the mechanical methods of preparation. Because of rapid contamination, fresh solvent must be used continuously, and the rags or cloths must be turned and/or replaced frequently. If this is not done, the grease and oil will be spread across the surface rather than removed. The final wash should be made with clean solvent. The organic solvents used should not contact the eyes or skin and should not be used near sparks or open flames. The disadvantages of solvent cleaning are that it will have no effect on rust or mill scale, it is very slow and labor intensive, and the materials are rapidly used up and need constant replacement. In addition, many of the solvents used are highly flammable and toxic. Natural or mechanical exhaust ventilation for diluting the concentration of vapors in an enclosed working space should be employed during the entire work period.

Steam cleaning uses high-temperature high-velocity wet steam to remove heavy soil, oil, or grease. It does not remove rust or mill scale, nor does it etch or roughen the surface. Commercial detergents can be added to the steam to improve cleaning. On steel, steam cleaning is sometimes followed by spot blast cleaning or wire brushing on areas exhibiting coating failure or rusting.

Hand tool cleaning (SSPC SP2) is used only for removing loosely adherent paint, rust, or mill scale. Because this method is slow, it is primarily used for spot cleaning. Hand tools include scrapers, wire brushes, abrasive pads, chisels, knives, and chipping hammers. The disadvantages of using these tools, in addition to the lack of speed, are that they will not remove tightly adherent contaminants (for example, dirt trapped in crevices, or oil or grease) and that they may raise burrs or dent the surface, actually causing some damage to the surface. In addition, these tools will generally not provide a surface profile.

Power tool cleaning (SSPC SP3) is faster than hand tool cleaning, it removes loose paint, rust, and scale, and it is good for the preparation of welded surfaces. Power tool cleaning is also good for removing old paint that has been lifted by rust. It provides a duplication of hand tools in power-driven equipment (electric or pneumatic), such as sanders, wire brushes, grinders, clippers, needle guns, and rotary descalers. Generally, power tool cleaning is suitable only for small areas because it is relatively slow. Power tools do not leave as much residue or produce as much dust as abrasive blasting, and they are frequently used where blasting dust could damage sensitive surroundings. However, they may polish the surface if they are used at too high a speed or kept in one spot for too long.

Water blast cleaning (sometimes called hydroblasting) may be high or low pressure, hot or cold, and with or without a detergent, depending on the type of cleaning desired. This process removes loose, flaky rust, paint, and mill scale. It has gained wide acceptance where abrasive blast cleaning, dust, and contamination present a hazard to either personnel or machinery. Low-pressure washing (<14 MPa, or 2000 psi) is effective in removing dirt, mildew, and chalk from the coating surface, and it is generally safe on concrete or masonry surfaces. Low-pressure water blast units use the same components as high-pressure water blast equipment. Low-pressure cleaning is referred to as power washing, and it should be recognized as such. For cleaning steel, water pressures as high as 69 MPa (10 000 psi) or more and volumes from 30 to 38 L/min (8 to 10 gal./min) are used.

Water alone does not etch a metal surface or remove tight paint, rust, or mill scale. Therefore, abrasives can be injected into the water stream to remove tightly adhering materials, to hasten the cleaning, or to roughen the surface profile. Any type of abrasive that is commonly used with abrasive blast cleaning can be used in water blast cleaning. Because the abrasive is normally not dried, screened, and recycled, less expensive abrasives are commonly selected in this application. The abrasive is injected into the system after water is pressurized by means of a suction head to prevent pump damage. Injecting abrasives into the water eliminates the dust that normally accompanies dry use of such abrasives.

Flash rusting is a concern when using these methods. Flash rust is a light rust layer that forms on steel after cleaning, particularly after water or wet abrasive blasting, or in humid marine environments. To avoid flash rusting, rust inhibitors such as sodium and/or potassium dichromate or phosphate are often used during or after water blasting. These inhibitors may retard rusting for up to 7 days. The disadvantage of using these corrosion inhibitors is that the solutions, upon drying, leave salts that can produce adhesion problems for protective coatings. The primary consideration, therefore, should be to determine if the protective coating is compatible with the inhibitor that was used.

Abrasive blast cleaning is the preferred surface preparation method for paints and coatings that require an anchor pattern and a high degree of cleanliness. Blast cleaning is the only method that can completely remove intact mill scale and give an even roughness with a controlled anchor pattern. Abrasive blasting is the propelling, or shooting, of sand or other types of small, hard particles at a surface. In nozzle blasting, the force that propels the abrasive is compressed air. The abrasive strikes the surface at speeds of 320 to 640 km/h (200 to 400 mph), breaking and loosening the rust or scale.

The cleaning principle is the same in centrifugal wheel blasting. In this case, the spinning of large paddle wheels creates the force that throws the abrasive at the surface. This method is usually confined to shop use with units approximately the size of a large trailer, although smaller one-man units have been developed for preparing floors or ship hulls. In contrast with open blast cleaning, the abrasives used are recyclable steel grit or shot.

Vacuum blasting is another option for dust-free blast cleaning. With this technique, the blast nozzle is surrounded by a brush, and a vacuum is created within the annular space. The spent abrasive and removed paint, rust, and debris are contained by the brush, removed by the vacuum, and transported to a separator. The debris and fine particles are removed, and the abrasive is returned for reuse. This method is very slow, because the blast pattern is approximately 25 mm (1 in.) in diameter. However, for small areas or where dust cannot be tolerated, it may be the best alternative.

The grades of cleaning are identified as Brush-Off (SSPC SP7), Commercial (SSPC SP6), Near-White (SSPC SP10), and White (SSPC SP5). The specific definitions are provided later in this section. Abrasive blasting is ideally suited for high-production work because it rapidly cleans large areas. In spite of its efficiency and good results, some problems still occur, such as the accumulation of used abrasives in low-lying areas and the attendant difficulty in removing them. Airborne dust and abrasive can interfere with machinery and the work of persons nearby and can contaminate adjacent locations. Airborne dust can be a hostile environment for the blast operator, and an air-fed hood must be worn while operating the air blast cleaning equipment. In addition, open nozzle blast abrasives are expensive and are usually not recycled in the field.

Open nozzle blasting equipment has six basic components:

- Air compressor
- Air hose
- Blasting machine, with a safety valve
- Blast hose
- Nozzle
- Moisture and oil separator

The function of each part is described below.

The air compressor is the source of energy for the blasting job. The constant supply of a high-volume high-pressure air stream hour after hour is the most critical part of blast operations.

Work is done in direct proportion to the volume and pressure of air at the nozzle. The larger the compressor, the larger the nozzle it can operate. The larger the nozzle at the proper pressure, the faster the job can be completed.

The air hose connects the compressor to the blast pot. For efficient blasting, the air hose should have as large an inside diameter as practical in order to reduce friction and avoid air pressure loss (a minimum 32-mm, or 1¼-in., ID air hose is usually recommended). A 15% production loss can result from a 69-kPa (10-psi) drop in pressure. The air hose should also be as short as practical to reduce leakage and should contain as few couplings as possible.

The blasting machine or sandpot is a pressurized container that holds the abrasive. A valve at the bottom measures and controls the amount of abrasive fed into the blast hose. Efficient abrasive blast operations require the right size pot for the job, which enables very few refill stops, and the releasing of an even flow of abrasive with the air stream. A deadman control valve is a required safety feature for operation of the blast pot.

To start the machine, the operator presses down on the deadman valve with his hand; this allows the air inlet valve to open and the air outlet valve to close, at which time blasting begins. Should he want to stop the operation, the operator releases his hold on the deadman valve, thus closing the air inlet valve, shutting off any incoming air from the compressor to the machine, and instantly opening the bleed-off or depressurizing valve.

Consider the possibility of an operator who faints, collapses, or falls off the scaffold. He may be within the range of the nozzle, which in effect is an automatic shot gun. Many injuries have been caused by such incidents. With remote control deadman valves, the machine will shut off the moment the grip on the nozzle is relaxed.

The blast hose connects the blast pot and nozzle and carries both air and abrasive. It is treated to prevent electrical shock by the installation of a grounding line. Sturdily constructed, multiple-ply hose with a minimum inside diameter of 32 mm (1¼ in.) is required in most work. A short length of lighter, more flexible hose with a 25- or 32-mm (1- or 1¼-in.) inside diameter is sometimes joined in at the end. These sections, called whips, are easier to handle and are effective for work in areas with many angles, pipes, and stiffeners. The use of whips will, however, effectively reduce the pressure at the nozzle and cancel the advantage gained by a large-diameter blast hose; therefore, their use should be discouraged except where necessary.

The nozzle is a major element in the blasting operation. Nozzle sizes are identified by the diameter of the orifice. Diameters are measured in sixteenths of an inch. A 4.8-mm (3/16-in.) diam orifice is designated No. 3; a 7.9-mm (5/16-in.) diam nozzle, No. 5; and so forth. Nozzles are commonly available up to 12.7 mm (½ in.) (No. 8) in diameter. The rate of surface cleaning is a function of the volume of air pushed through the nozzle at high pressure. For example, if a cleaning rate of 9 m^2/h (100 ft^2/h) can be attained with a 6.4-mm (1/4-in.) nozzle, a rate of 37 m^2/h (400 ft^2/h) can be attained with a 12.7-mm (1/2-in.) nozzle. However, the nozzle can be too large for the air volume. This will result in insufficient air pressure, and production will drop. Therefore, nozzle size should be as large as permitted by the air supply to maintain air pressure of 620 to 690 kPa (90 to 100 psi) at the nozzle.

Nozzles should use the venturi design, have as long a nozzle length as practical, use a tungsten carbide or other wear-resistant lining, and have as large an orifice size as possible with a given compressor. For difficult-to-reach places, nozzles also come in shapes such that they can propel abrasives around corners, in a 360° arc, or even backwards.

Moisture and Oil Separators. The large consumption of compressed air in an abrasive blasting operation introduces the problem of contamination by moisture (especially in high-humidity areas) and oil mists from compressors. This is especially true of portable compressors. To combat this, an adequately sized moisture and oil separator should be installed at the blast machine (at the most distant point from the compressor). Separators are usually of the cyclone type and employ expansion chambers and small micron filters to eliminate up to 95% of moisture and contaminants. They require solvent cleaning to remove oil and regular replacement of filters.

Abrasives. The proper abrasive is the one that provides the necessary cleanliness and profile. The characteristics of such an abrasive related to performance include size, hardness, shape, and material.

Size. A large abrasive particle will cut deeper than a smaller particle of the same composition and shape and will therefore provide a deeper surface profile. However, the greatest cleaning rate is achieved with the smaller abrasive because of a larger number of impacts per unit area.

Hardness. Hard abrasives generally cut deeper and faster than softer or brittle abrasives. A hard but brittle abrasive will shatter upon impact; this reduces its cleaning power accordingly.

Shape. The shape and size of abrasive grains determine the type of surface profile achieved by blast cleaning. Steel shot, because it is round, peens the surface to give a wavy profile. Therefore, shot is particularly effective at removing brittle deposits such as mill scale. Grit, on the other hand, is angular, and when blast cleaning produces a jagged finish that is generally preferred for coating adhesion. A wide variety of surface patterns are available from different grits. Use of sand and slag abrasives that are semiangular results in a pattern somewhat between that of shot and grit.

Materials. The abrasive should have a neutral pH, that is, between 6 and 8. It should not be washed with seawater or other contaminated water. Four types of abrasive materials are commonly used on a nonrecycled basis, as follows.

The first type is the natural oxides. Silica sand is the most widely used natural oxide because it is readily available, low in cost, and effective. The hazards of sand have recently been elucidated by the EPA and by the Occupational Safety and Health Administration, and the use of silica sand abrasives has been restricted in many areas. Another natural silica material has received attention in recent years. It consists of a blend of coarse and fine staurolite sands mined from certain mineral deposits in Florida. It is characterized as efficient cutting, and it has less dusting and a lower breakdown rate than silica sand.

The second type is the metallic abrasives. Steel shot and grit abrasives are efficient, hard, and dust free, but care must be taken in their storage to prevent rusting. Although their initial cost is higher, they may be recycled many times to make them cost effective. A mixture of both shot and grit abrasives in blasting may combine the advantages of both.

The third type is the slag abrasives. Copper and nickel abrasives are a by-product of the ore-smelting industry. Coal slags result from the burning of coal. These abrasives are fast cutting, but they have a high breakdown rate and are generally not recycled. Slag abrasives are most likely to exhibit an acidic pH.

The fourth type is the synthetic abrasives. Aluminum oxide and silicon carbide are nonmetallic abrasives that have cleaning properties similar to the metallics without the problem of rusting. They are very hard, fast cutting, and low dusting, but they are quite costly and must be recycled for economical use.

Coating Application

In a most basic sense, coating application can be described as getting the paint from the can to the surface being coated. There are a number of ways of accomplishing this, and often the type of paint material determines the selection of the most appropriate method. For example, zinc-rich coatings should always be sprayed (except for very small areas), and agitation must be used when spraying to keep the heavy pigment particles from settling. Vinyls and chlorinated rubbers are not usually formulated for brush or roller application, and they too should be sprayed. On the other hand, alkyds and oil-base coatings are often applied by brush or roller, but can be readily sprayed as well. For most corrosion-protective coating systems, airless or conventional spraying is preferred; spraying deposits a more uniform film with greater thickness control and a better appearance. The rate of application is much faster with spraying, particularly airless spray, than with brush and roller application. The more common application techniques and equipment are described below.

Brushing

Brushing is an effective, relatively simple method of paint application, particularly with primers, because of the ability to work the paint into pores and surface irregularities. Because brushing is slow, it is used primarily for smaller jobs, surfaces with complex configurations (edges, corners, cuts, and so on), or where overspray may constitute a serious problem. The other disadvantage of brushing is that it does not produce a very uniform film thickness. Natural brush bristles of good quality are preferred, but synthetic bristles resistant to strong paint solvents can also be used. Bristles that are naturally or artificially flagged, that is, with split tips, are preferred because they hold more paint. Nylon and polyester bristles are more water resistant than natural fibers and are preferred for latex paints.

Paint Mitts. A paint mitt can hold more paint than a brush and is faster, but it produces a very nonuniform thickness. Therefore, paint mitts are generally used only for hard-to-reach areas or oddly shaped objects, such as small pipes and railings.

Rolling

Rollers are best used on large, flat areas that do not require the surface smoothness or uniformity achievable by spraying. Rollers are also used in interior areas where overspray presents a cleaning and masking problem. However, the brush is preferred over a roller for applying primers because of the difficulties in penetrating pores, cracks, and other surface irregularities. In rolling, air mixes with the paint and leaves points for moisture to penetrate the cured film. Rolling is generally more suitable for topcoating a primer that has been applied by some other method.

The nap (fiber length) of a roller normally varies from 6.4 to 19 mm (1/4 to 3/4 in.). A longer nap holds more paint, but does not give a smooth finish; therefore, it is used on rough surfaces. Rollers with extra-long naps (32 mm, or 1 1/4 in.) are used for chain-link fencing. Handle extensions up to 3 m (10 ft) in length or longer allow a painter to reach otherwise inaccessible areas, but the net effect is usually a reduction in the quality of the application. High-performance coatings for immersion service are seldom applied by roller because of nonuniform thickness, wicking caused by roller residue, and the rapid drying properties of some coatings, such as vinyls.

Spray Painting

Spray painting of coatings results in a smoother, more uniform surface than that obtained by brushing or rolling, because these application methods tend to leave brush or stipple marks and irregular thicknesses. The most common methods of spray painting are conventional and airless.

The conventional spray method of spraying relies on air for paint atomization. Jets of compressed air introduced into the stream of paint at the nozzle break the stream into tiny droplets that are carried to the surface by the current of air.

Because large amounts of air are mixed with the paint during conventional spray application, paint losses from bounce back or overspray can be high. Such losses have been estimated to be as much as 30 to 40%, depending on the configuration of the substrate. However, the ability to vary fluid and air independently gives conventional spray the versatility to provide a wide choice of pattern shapes and coating wetness by infinitely varying the atomization at the gun. Spray gun triggering is more easily controlled for precise spraying of irregular shapes, corners, and pipes than with airless spray. Conventional spraying provides a finer degree of atomization and a higher quality surface finish than airless spray.

However, some of the disadvantages of conventional air spray application are:

- It is slower than airless application
- More overspray results than with other methods
- It is hard to coat corners, crevices, and so on, because of blowback

Nonetheless, conventional spray is still one of the most frequently used application methods. The equipment required for conventional spray application of coatings is discussed below.

Air compressors can be of various types, with the size usually dependent on the amount of air required to operate the spray gun. Hoses must be properly sized to deliver the proper amount of air volume and pressure to the gun. It takes approximately 276 to 414 kPa (40 to 60 psi) and 0.25 m^3/

min (8.5 ft^3/min) to operate most production spray guns with medium-viscosity paint.

An oil and moisture separator should be set up between the air compressor and the paint pot to prevent moisture and oil from getting into the paint and ruining the quality of the finish. This component of the conventional spray set-up should be considered as a standard item that cannot be omitted, because of potential paint contamination and equipment damage.

Pressure Feed Tanks (Paint Pots). The amount of fluid material delivered to the spray gun is regulated by the pressure set by the fluid pressure regulator of the feed tank pressure pot. The pressure pot should contain 19 to 38 L (5 to 10 gal.) of paint for most jobs. Pressures of 55 to 83 kPa (8 to 12 psi) at the gun are commonly used; excessive fluid pressure can cause the fluid stream to exit the fluid nozzle at too high a velocity for the air jets in the nozzle to atomize the fluid stream properly.

Connecting Hoses. Two hoses are used in conventional air spray painting: the air hose and the fluid hose. Air hoses supply air from the compressor to the paint pot and should be at least 12.7 mm (1/2 in.) in inside diameter. The air hose from the paint pot to the spray gun should be 6.4 to 7.9 mm (1/4 to 5/16 in.) in inside diameter and as short as possible.

The fluid hose usually has a solvent-resistant liner. It should have a 7.9- to 9.5-mm (5/16- to 3/8-in.) inside diameter for medium-viscosity materials and should be as short as possible. Hoses with inside diameters up to 12.7 mm (1/2 in.) are commonly used. Excessive hose length allows the solids to settle in the line before reaching the spray gun. This leads to clogging and the application of a nonhomogeneous film.

The spray gun is used to atomize the paint by mixing it with air under pressure in order to apply the coating material to the surface. Various spray patterns can be obtained by changing the volume of both air and fluid. The selection of the fluid nozzle and the needle size is another way to regulate the amount of paint exiting the fluid nozzle. If it is found that excessive amounts of paint are flowing through the fluid nozzle at low pressures (55 to 83 kPa, or 8 to 12 psi), the fluid-adjusting knob can be used to reduce the flow, or a smaller fluid nozzle/needle combination may be needed. Paint manufacturers normally recommend at least one set of sizes known to work for their product.

Airless spray provides a rapid film build, greater paint flow into surface irregularities, and rapid coverage. In airless spraying, paint is forced through a very small nozzle opening at very high pressures to break the existing paint stream into tiny particles (similar to water exiting a garden hose). Because of the high fluid pressure of airless spray, paint can be applied more rapidly and at greater film builds than in conventional spraying. The high-pressure paint stream generated by airless spray penetrates cavities and corners with little surface rebound. If airless spray is used, proper spraying technique is essential for quality paint films. Poor technique can result in variations in paint thickness, holidays, and a host of other film defects in addition to wasted time and materials.

Some of the advantages of airless spray as compared to conventional spray include:

- Higher film builds are possible (heavier coatings)
- Less fog or rebound
- Easier to use by the operator, because there is only one hose
- Higher viscosity paints can be applied
- Easier clean up

The disadvantages of the airless spray process include:

- Relies on dangerous high pressure
- Fan pattern is not adjustable
- Many more working parts that can cause difficulty
- Higher initial cost than other forms of spraying
- Requires special care to avoid excessive build-up of paint that causes solvent entrapment, pinholes, runs, sags, and wrinkles

The equipment and materials for airless spraying are discussed below.

Power Source. An electric motor can be used instead of an air compressor to drive the fluid pump. The electric airless is a self-contained spray outfit mounted on wheels and designed to use commercial 115-V electrical power.

An air compressor, usually with a 1-hp, 690-kPa (100-psi) electric motor or a 1 1/2-hp gasoline engine, is required to operate one gun with a 0.46-mm (0.018-in.) tip. For two or more guns or heavy materials, a greater capacity is needed.

Air Hose and Siphon Hose. A 12.7-mm (1/2-in.) ID hose is generally used to deliver air from the compressor to the pump. The most common hose length is 15 m (50 ft). However, as hose length and pump size increase, the inside diameter of the hose must be increased as well.

A siphon hose is required, and it must have an inside diameter of 12.7 to 19 mm (1/2 to 3/4 in.) to provide adequate fluid delivery. Also, the hose material must be resistant to the fluid solvent and other materials. A paint filter should be used to prevent dirt from clogging the fluid hose and tip. This hose is often eliminated, and the pump is immersed directly into the paint.

High-Pressure Fluid Pump. The fluid pump is the most important part of the hydraulic airless system; it multiplies the air input pressure to deliver material at pressures up to 31 MPa (4500 psi). Pumps are offered with delivery ranges from 0.8 L/min (28 oz/min) (one gun) up to 11.4 L/min (3 gal./min) (three to four guns). The pump used for many airless outfits will convert 0.45 kg (1 lb) of air input into 13.5 kg (30 lb) of fluid pressure. This is referred to as a 30:1 ratio. Pump ratios as high as 45:1 are common.

Some airless spray units use an electric-hydraulic power pack. These units can supply two guns with airless spray tips of up to 0.53 mm (0.021 in.) and long spray lines, or one gun for high-volume delivery with an airless tip of up to a 0.79-mm (0.031-in.) orifice. These electric units are wheel mounted and use a long-stroke slow-speed pump. The unit is equipped with an explosion-proof motor and can deliver as much as 3.8 L/min (1 gal./min) at pressures up to 21 MPa (3000 psi).

A paint filter should be attached to the pump at the fluid hose connection. The filter prevents dirt and particles from clogging the gun and being sprayed onto the work surface.

The fluid hose is made to withstand high operating pressures. It is made of vinyl-covered, reinforced nylon braid and can handle about 2040 kg (4500 lb) of pressure. The hose is designed to resist the solvent action of the various coatings being sprayed. A wire is often molded into the hose assembly to prevent any possible static electrical charge. The hose inside diameter should be at least 6.4 mm (1/4 in.), and the length should not be longer than necessary to do the work (ideally, 15 to 30 m, or 50 to 100 ft). Fluid hose diameters up to 12.7 mm (1/2 in.) are commonly used.

The airless spray gun is designed for use with high fluid pressures. It is similar in appearance to the conventional gun, but has a single hose for the fluid. The hose may be attached to the front of the gun or to the handle. Atomization occurs when fluid is forced through the small orifice in the spray tip at high pressures.

A variety of spray tips are available to suit the type of material being sprayed and the size of spray pattern desired. The orifice, or hole, in the tip and the fan angle control the pattern size and fluid flow. There are no controls on the gun itself. Tip orifices vary to fit the viscosity of the paint being applied; fan angles range from 10 to 95°.

Quality Assurance

To obtain the desired protection of a metal substrate from a coating system, it is important not only to choose the proper coating materials but also to ensure that they are properly applied. On most jobs, such assurance is provided by a reputable paint force, or contractor, but on many other jobs, surveillance by plant personnel or thorough inspection by independent third-party organizations is done. In all cases, the rudiments of quality are the same:

- Proper masking and protection of surfaces not to be blast cleaned or painted
- Removal of rust and contaminants and suitable roughening of the surface
- Application of the specified coating system to the proper thickness
- Observation of application parameters, such as minimum and maximum temperatures, interval between coats, and induction times
- Verification of proper hardening or cure of the coating
- Testing to ensure that defects such as pinholes, skips, or holidays are minimized or avoided
- Tests for adhesion, color, gloss, and other parameters that may affect appearance or protective capability of the coating

Coating inspection requires training, experience, and familiarity with the instrumentation the inspector must use. Inspector training courses are available, and certification of inspectors is available from some training organizations. For example, NACE and the National Institute of Accreditation and Certification both certify coating inspectors. A brief overview of some of the more common coating inspections and inspection equipment is presented below.

Inspection Sequence

Inspection often begins with a pre-job conference at which the ground rules are set, but when the work begins, the inspector is responsible for witnessing, verifying, inspecting, and documenting the work at various inspection points. The inspector should be aware of the following inspection points and should witness or conduct the operation if feasible:

- Pre-surface preparation inspection
- Measurement of ambient conditions
- Evaluation of compressed air cleanliness and surface preparation equipment

- Determination of surface preparation, cleanliness, and profile
- Witnessing coating mixing and thinning
- Inspection of application equipment
- Inspection of coating application
- Determination of wet-film thickness
- Determination of dry-film thickness
- Evaluating cleanliness between coats
- Pinhole and holiday testing
- Evaluation of adhesion and cure

Pre-Surface Preparation Inspection. Prior to the start of surface preparation and other coating activities, it is necessary to inspect in order to determine if the work is ready to be prepared and painted. Heavy deposits of grease, oil, dust, dirt, and other contaminants must be removed. Removal of this miscellaneous debris before blast cleaning ensures that these materials are not redeposited onto the freshly blast-cleaned surfaces.

In addition, the specifications may require that weld spatter be ground or otherwise removed and that sharp edges be rounded (customarily to a 3.2-mm, or 1/8-in., radius). Unusual pitting in the steel substrate should be examined; the pits will either be accepted as they are or will be ground or filled. Areas that do not need to be cleaned or coated should be masked to protect them from the cleaning and coating operations.

Measurement of Ambient Conditions. Before final blast cleaning and before any paint is applied, the inspector or painter should check the ambient or surrounding weather conditions. This is especially important in the early morning when the weather is changing or during seasonal changes when condensation is common. The presence of a thin film of condensed moisture on the surface may be visually imperceptible, but a quick check of the dew point and the surface temperature will determine if such a condition exists.

Dew point is calculated by using temperature readings from a psychrometer and the appropriate psychrometric tables. A psychrometer is an instrument used for determining relative humidity and dew point. The temperature of the surface to be coated should be at least 3 °C (5 °F) higher than the dew point.

Evaluation of Compressed Air Cleanliness and Surface Preparation Equipment. The air supply should be tested for cleanliness at frequent intervals during the day. This is done by holding a blotter or cheesecloth in the air stream for a minimum of 2 min. The blotter or cheesecloth should be held at a distance of 0.5 to 0.6 m (18 to 24 in.) from the source of the air stream being tested. The test material should be examined immediately after the test is complete for evidence of contamination by oil or moisture. Examination should be made visually as well as by touching and smelling the test material. If there is no discoloration on the blotter, the air quality is excellent, while moisture and oil on the blotter indicate unsatisfactory air.

Determination of Surface Preparation, Cleanliness, and Profile. Surface preparation is believed to be the most important factor affecting the performance of the coating. Any compromise in surface preparation usually reduces the performance of the applied material as well. The primary goals of surface preparation are, first, to provide a satisfactory, clean surface free of detrimental contaminants and, second, to provide an anchor pattern and to increase the steel surface area to improve bonding of the applied coating or lining to the steel surface.

Surface cleanliness should be visually inspected to verify that it is in conformance with the specified cleanliness standard. Surface profile can be determined by using any of several instruments, including surface profile comparators, which give a visual and tactile comparison to a known profile; replica tape with a spring micrometer, in which a tape is pressed into the valleys of a profile and the profile height measured with a micrometer; and dial surface profile gages, where the base of a depth micrometer rests on the profile peaks while a needle penetrates into the valleys.

Witnessing Coating Mixing and Thinning. Before a coating can be applied to a substrate, it must be suitably mixed to ensure that the material is homogeneous. During storage, most coatings tend to show some separation of the lighter and heavier components. Heavier pigments settle to the bottom of the can and sometimes appear as a hard cake, while other materials settle as a soft sludge. Some binders and some paints curdle or gel when stored for too long a period. Stirring may reconstitute the paint to a homogeneous liquid; if not, then the paint must be discarded.

Inspection of Application Equipment and Techniques. A lack of cleanliness is the predominant failure associated with spray guns and spray pumps. Paint chips or residues of previous coating materials can lodge in the pump, lines, or guns and cause clogging. Cleanliness of mixing equipment, lines, spray pots, guns, tips, and other application equipment is important and necessary.

During application, the safety and respiration equipment of the operator should be verified to be in working order, and the spray technique should be reviewed. For example, is the spray gun the proper distance from the work? Is there pattern overlap, uniform speed, and triggering before and after each stroke? Is the gun perpendicular to the surface?

Wet-film thickness readings are used as an aid to the painter in determining how much material to apply to the surface in order to achieve the specified dry film. Wet-film thickness gages are of value only if one knows how heavy a wet film to apply. The wet-film thickness/dry-film thickness ratio is based on knowing the percent solids by volume of the specific coating material:

$$\text{Wet-film thickness} = \frac{\text{Desired dry-film thickness}}{\text{\% solids by volume}}$$

If the paint is thinned at the time of use:

$$\text{Wet-film thickness} = \frac{\text{Desired dry-film thickness}}{\left(\dfrac{\text{\% solids by volume}}{100\% + \text{\% thinner added}}\right)}$$

Dry-film thickness measurements are made after complete drying of the coating. They are used to determine if the desired coating thickness is obtained. Dry-film thickness gages for steel surfaces are almost always of the magnetic type. If properly calibrated and operated, they have an error of only approximately 5 to 15% at normal temperatures (−18 to 100 °C, or 0 to 212 °F). Dry-film thickness instruments fall into four basic categories, as follows.

Magnetic pull-off instruments rely on the attraction of a magnet to an underlying steel surface. The stronger the attraction, the greater the force necessary to pull off the magnet. This pulling force can be calibrated to indicate coating thickness.

Magnetic fixed probe instruments measure the strength of the magnetic flux in a steel substrate exerted by a permanent magnet. Thickness can be measured by nulling the wheatstone bridge circuitry at various known thicknesses.

In eddy current inspection, a fluctuating electrical current induces residual eddy currents in an underlying metallic (aluminum, copper, stainless steel, and so on) substrate. The strength of the induced currents is a function of the distance of the induction probe from the substrate, that is, separation by paint thickness. The strength of the induced eddy current is measured and converted into paint thickness.

Destructive Inspection. The most common instrument for destructive coating thickness measurement cuts a small "V" into the paint at a known angle. By measuring the width of the slope of the "V" incision, a measurement of coating thickness can be obtained. The advantage is that each coat of a multiple-coat system can be measured and that thickness measurement is independent of substrate (steel, concrete, wood, and so on).

Evaluating Cleanliness Between Coats. When a multiple-coat system is to be applied, the surface between coats should be examined to ensure that all grease, oil, dirt, bird droppings, pebbles, sand from abrasives, dry spray, and the like have been removed before the application of the next layer of coating.

Pinhole and Holiday Testing. Many coating systems, particularly those for immersion or splash zone service, require additional testing with holiday detectors after the coating has dried. Pinhole and holiday detectors are of two basic types: low-voltage wet sponges and dc high voltage. The basic function of a holiday tester is to detect minute pinholes, holidays, skips, or misses in coating systems, all of which can cause premature coating failure.

The low-voltage (30 and 67.5 V are common) wet-sponge holiday detectors are used for locating discontinuities in nonconductive coatings applied to metal bases. The low-voltage detector is suitable for use on coatings up to 0.5 mm (20 mils) in thickness. The basic unit consists of the detector, a ground cable, and a sponge electrode. The ground cable is firmly attached to the bare substrate, and the sponge electrode is saturated with tap water. The electrode is moved across the entire surface at a speed of approximately 4.5 to 6 m/min (15 to 20 ft/min); the water permits a small current to flow through the pinholes down to the substrate. Once the current reaches the substrate, the circuit is completed to the detector unit, and an audible signal indicates that a pinhole or discontinuity has been found.

High-voltage holiday detectors basically function on the same operating principle as the low-voltage instruments described above except that a sponge is not used. The instrument consists of a testing unit that can produce various voltage outputs, a ground cable, and an electrode made of conductive neoprene, brass, or steel. High-voltage units are available up to 20 000 V and more. The rule of thumb for high-voltage detectors is to apply approximately 100 V per mil of dry-film thickness. High-voltage detectors are used for nonconductive coatings applied to conductive substrates. The ground wire should be firmly attached to a section of the bare substrate and the electrode passed over the entire surface. A spark will jump from the electrode through the air gap down to the substrate at pinholes, holidays, and so on, simultaneously activating a signaling device in the unit.

Adhesion Testing. Occasionally, there is a need to test the adhesion of the coatings after application in the field. Different types of adhesion-testing methods are used, ranging from the simple penknife to the more elaborate tensile-testing units. The use of a penknife generally requires a subjective evaluation of the coating adhesion based on some previous experience. Generally, one cuts through the coating and probes at it with the knife blade, trying to lift it from the surface to ascertain whether or not the adhesion is adequate.

A modified version of this type of testing is the cross-cut test. The cross-cut test consists of cutting an X, or a number of small squares or diamonds, through the coating down to the substrate. Tape is then rubbed vigorously onto the scribed marks and removed briskly. The crosshatch pattern is evaluated according to the percentage of squares delaminated or remaining intact. A cross-cut guide can be used to cut a precision crosshatch pattern.

Another technique involves adhering a small dolly to the coated surface, scoring around the dolly periphery, and using a tensile-pulling device to pull it from the surface. The tensile pressure required to disbond, as well as the nature of the disbondment, is used to characterize adhesion.

Tests for Cure. The rate of cure of organic coating materials is affected by surface temperature, ambient conditions, coating formulation, and film thickness. Laboratory testing of paint chips is the only true means for verifying cure. Field techniques include the following.

Solvent Rub. On thermosetting cross-linked coatings, a clean cloth saturated with a strong solvent such as methyl ethyl ketone (MEK) or methyl isobutyl ketone (MIBK) can be used to rub the surface of the coating. If the material is properly mixed and cured, no color will be transferred to the cloth. If the coating is improperly mixed or cured, it may partially redissolve, and the color will be seen on the cloth. Of course, the solvent rub test cannot be used on certain solvent-sensitive coatings, such as alkyds, vinyls, and chlorinated rubbers.

Sandpaper Test. When a properly cured film is abraded with fine sandpaper, a fine powdery residue is observed. Coating material that remains on the face of the sandpaper may indicate an uncured, slightly tacky coating.

Hardness Tests. A penetrating needle hardness tester or pencil hardness tester can be used to check film cure. The use of such instruments entails exerting a light perpendicular pressure on an instrument, which holds a hardened steel indenter ground to microscopic accuracy. The penetration of the spring-loaded indenter, which indicates hardness, is read directly from a scale dial. Pencil hardness is established by the lead hardness that will just scratch the paint surface.

REFERENCES

1. C.R. Martens, Ed., *Technology of Paints, Varnishes, and Lacquers*, Reinhold, 1968
2. K.B. Tator, Can Failures Still Occur When the Correct Coating is Selected and Applied Properly, in *Corrosion Control by Organic Coatings*, H. Leidheiser, Ed., National Association of Corrosion Engineers, 1981
3. S.H. Richardson and K.V. McCullough, Phenolic Resins, in *Technology of Paints, Varnishes, and Lacquers*, C.R. Martens, Ed., Reinhold, 1968
4. K.B. Tator, Urethane Coatings: Influence of Chemical Composition Upon Performance Properties, *J. Protec. Coatings Linings*, Vol 2 (No. 2), Feb 1985

SELECTED REFERENCES

- Chapters 3 to 5, in *Coating Inspector Training Course Manual*, KTA-Tator, Inc., 1985
- "Federation Series on Coatings Technology," Federation of Societies of Coating Technology
- C.R. Martens, *Technology of Paints, Varnishes, and Lacquers*, Reinhold, 1968
- C.G. Munger, *Corrosion Prevention by Protective Coatings*, National Association of Corrosion Engineers, 1984
- M.K. Snyder and D. Bendersky, "Removal of Lead-Base Bridge Paints," NCHRP Report 265, Transportation Research Board, 1983
- *Surface Cleaning, Finishing, and Coating*, Vol 5, 9th ed., *Metals Handbook*, American Society for Metals, 1982
- P. Swaraj, *Surface Coatings*, John Wiley & Sons, 1985
- K.B. Tator, Urethane Coatings: Influence of Chemical Composition Upon Performance Properties, *J. Protec. Coatings Linings*, Vol 2 (No. 2), Feb 1985

Electroplated Coatings

Joseph Mazia, Mazia Tech-Com Services, Inc.
David S. Lashmore, National Bureau of Standards

ELECTROPLATED COATINGS can be very effective in reducing the annual $167 billion (as of 1985) cost of corrosion. Even when metallic coatings are applied to a substrate to provide decorative effects or to enhance material properties, such as wear or electrical contact resistance, the corrosion performance of the coating-substrate system must be carefully considered to avoid future degradation. Metallic coatings can be very useful in providing the optimum properties for a part not obtainable if it were fabricated from a single material.

The overall properties of coated parts are determined by the properties of the coating itself and by defects in the coating. The understanding of how defects in a coating affect the corrosion performance is critical in designing an effective coating system.

Electroplating is the electrodeposition of an adherent metallic coating upon an electrode for the purpose of securing a surface with properties or dimensions different from those of the basis metal (Ref 1). Electrodeposits are applied to metal substrates for decoration, protection, corrosion resistance, chemical inertness, wear resistance, buildup of substrate dimensions, electrical properties, magnetic properties, solderability, reflectance, and reduction of friction (Ref 2). Electrodeposited coatings are sometimes applied for two or more reasons. For example, decorative and protective plating are combined in such applications as steel automobile bumpers, trim items made of zinc die castings or steel, trim items on large and small appliances, costume jewelry, electronic circuits, piano strings, construction and architectural hardware, plumbing fittings, electrical contacts, plastic items, and magnetic memory devices (Ref 2).

This ability to be decorative as well as protective to substrates distinguishes electrodeposited coatings from metallic coatings applied by hot dipping or spraying. Metallic luster, achieved by bright plating or by polishing after plating, lends a distinctive appearance not provided by other types of coatings. However, corrosion resistance is the primary reason for the use of electrodeposited coatings. Electrodeposited coatings are also applied to protect substrate metals. Examples of this application include:

- Tin, as well as chromium, plating of steel strip for food packaging and other container uses
- Electrogalvanizing, or zinc plating, of steel strip, sheet, stampings, forgings, wire, and screw machine products
- Zinc and cadmium plating of fasteners and other hardware items
- Chromium plating of gun bores

Electrodeposits are used to enhance properties other than corrosion protection. For example, many fasteners are coated with a bright cadmium or zinc deposit supplemented with a clear or colored chromate conversion film. This results in coatings that are both protective and decorative. In addition, chromium is applied to the interior of gun barrels to provide lubricity and erosion resistance as well as corrosion protection.

This article will discuss the various factors that affect the corrosion performance of electroplated coatings, the effects of environment and the deposition process on substrate coatings, and the electrochemical techniques capable of predicting the corrosion performance of a plated part. Because the practical aspects of electroplating are covered in such publications as Volume 5 of the 9th Edition of *Metals Handbook* and Ref 3 to 6, this article will be primarily concerned with the corrosion resistance of electrodeposited coatings. An exception will be the discussion of mill-applied finishes, about which the open literature is sparse and limited in scope. Special attention will be given to the design of coating systems for optimum protection of the substrate. An ideal electroplated coating system will be defined. Finally, attention will be given to controlled weathering tests and accelerated tests used to predict and determine the relative durability of the coating.

Idealized Coating

Coatings are either anodic or cathodic to the substrate. Only a very rough idea of which situation applies can be obtained from the electromotive force (emf) series (Table 1). Coatings that are anodic to a substrate in one environment may be cathodic to the same substrate in a second environment. The environment itself may change from the surface of the coating to one that exists within any defects, such as scratches, cracks, or corrosion pits. These defects are particularly important if the coating is cathodic to the substrate; in this case, if the substrate becomes exposed, rapid corrosion of the substrate may occur as it attempts to protect the coating. If the coating is anodic to the substrate, defects are somewhat less important, although they still must be considered. Thus, if the substrate becomes exposed, it will be protected by the corrosion of the coating. However, if the corrosion should proceed to such an extent that a large area of the substrate is exposed, the coating will lose its effectiveness, because the electrical resistance of the substrate limits the area that the coating can protect.

The nature of any corrosion products formed also must be taken into account. For example, in industrial environments, zinc is more protective of steel than cadmium is because the zinc sulfate ($ZnSO_4$) corrosion product formed has a lower solubility than cadmium sulfate ($CdSO_4$); $ZnSO_4$ thus provides a blocking action to reduce subsequent corrosion. This situation is reversed in marine environments, in which the corrosion products are more likely to be carbonates and chlorides. Zinc produces more soluble corrosion products than cadmium does in this environment. The most desirable situation is to have the coating anodic to the material beneath it and at the same time to have a very low corrosion rate.

This general rule has been exploited to a high degree of sophistication by the automotive industry. Chromium-plated exterior steel parts have a coating system consisting of several layers of metals, each of which is anodic to the layer beneath it. In fact, there are accurate lifetime predictions based on the electrochemical potential differences and the thicknesses of the layers. Also, multilayer corrosion standards are available from the National Bureau of Standards that ensure that the coatings meet specifications. Even the chromium layer has been designed to be microcracked or microporous to spread the corrosion reactions over a large area. Corrosion is then distributed over a large area, and localized pitting is minimized. This will be discussed in more detail in the section "Applications for Electrodeposited Coatings" in this article.

Table 1 Electromotive force series of metals and alloys

Solution concentration: 1 *N*; temperature: 25 °C (75 °F)

Metal	Ion	Standard potential (E_0), V
Calcium	Ca^{2+}	−2.87
Sodium	Na^+	−2.71
Magnesium	Mg^{2+}	−2.37
Aluminum	Al^{3+}	−1.66
Zinc	Zn^{2+}	−0.763
Chromium	Cr^{3+}	−0.71
	Cr^{6+}	−0.56
Iron	Fe^{2+}	−0.44
Cadmium	Cd^{2+}	−0.40
Nickel	Ni^{2+}	−0.25
Tin	Sn^{2+}	−0.136
Lead	Pb^{2+}	−0.126
Hydrogen	H^+	0
Copper	Cu^{2+}	0.337
	Cu^+	0.521
Silver	Ag^+	0.799
Platinum	Pt^{4+}	0.90
Gold	Au^+	1.68

Source: Ref 2

Substrates

The emf series is a table that lists in order the standard electrode potentials of specified electrochemical reactions (Ref 1). The potentials are measured against a standard reference electrode when the metal is immersed in a solution of its own ions at unit activity (Table 1).

Similarly, the galvanic series is a list of metals and alloys arranged according to their relative potentials in a given environment (Ref 1). Figure 1 shows a galvanic series of commercially used metals and alloys in seawater. The positions of metals and alloys in a galvanic series—and the corrosion behavior of metals and alloys—depend on the specific environment to which the materials are exposed. Therefore, there are as many galvanic series as there are aqueous environments. Generally, the relative positions, of metals and alloys in both the emf and galvanic series are the same. An exception is the position of cadmium with respect to iron and its alloys. In the emf series, cadmium is cathodic to iron, but in the galvanic series (at least in seawater) cadmium is anodic to iron. Thus, if only the emf series were used to predict the behavior of a ferrous metal system, cadmium would not be chosen as a sacrificial protective coating, yet this is the principal use for cadmium plating on steel.

The importance of knowing the position of substrates in the emf and a specific galvanic series is frequently overlooked. The sole consideration is often the position of the protective coating in these series. It is prudent, however, to select a coating as close as possible in potential to the substrate in moist environments, because few coatings are completely free of pores, cracks and other defects. Even if a coating does start out defect free, scratching, normal wear, or gouging can change this condition. Accordingly, in a corrosive environment, plated components present at least two (not one) metals to the hostile galvanic-corrosion forces (see the section "Electrochemical Predictions of Corrosion Performance" in this article).

Cleanliness of the Substrate

Regardless of the composition and metallurgical condition of a substrate, its surface must be free from soils and oxides before being plated. The ASM International Committee on Selection of Cleaning Processes classifies soil into six groups: pigmented drawing compounds, unpigmented oil and grease, chips and cutting fluids, polishing and buffing compounds, rust and scale, and miscellaneous surface contaminants, such as lapping compounds and residue from magnetic-particle inspection. Cleaning methods are discussed in Ref 2 to 6. Although these methods are beyond the scope of this article, the point is made that unclean substrates often defeat the purpose of electrodeposited or other plated coatings.

Freedom from water break after the activating rinse and before immersion of a part in the plating tank is an essential prerequisite for satisfactory plating. Water break is defined as the appearance of a discontinuous film of water on a surface, signifying nonuniform wetting and usually associated with surface contamination (Ref 1). A false reading of freedom from water break often occurs if the inspection is made after the rinse following a surfactant-laden alkaline cleaner. This misleading conclusion is nullified by inspection after acid dipping and rinsing.

Knowledge of the Nature of Substrates

Before designing or applying a protective coating to a metal, the plater must know the nature and essential properties of the substrate. Generally, a substrate is known to be ferrous or nonferrous, but these generic terms are inadequate. For example, there are thousands of ferrous metals and alloys, ranging from cast iron (high carbon) to ingot iron (very low carbon) and the entire range of carbon and alloy steels. Also, the microstructure, hardness, yield and tensile strengths, and internal stress of the substrate have major influences on the protective quality and durability of plated coatings. High-strength, high-carbon, high-hardness steels are susceptible to hydrogen embrittlement when subjected to alkalies, acids, and plating processes. Aluminum and zinc alloys and, to a lesser extent, chromium are amphoteric (soluble in both acidic and alkaline solutions). Accordingly, these metals must be treated with inhibited solutions to prevent chemical attack. Methods of determining the nature and condition of the substrates are described in the following sections.

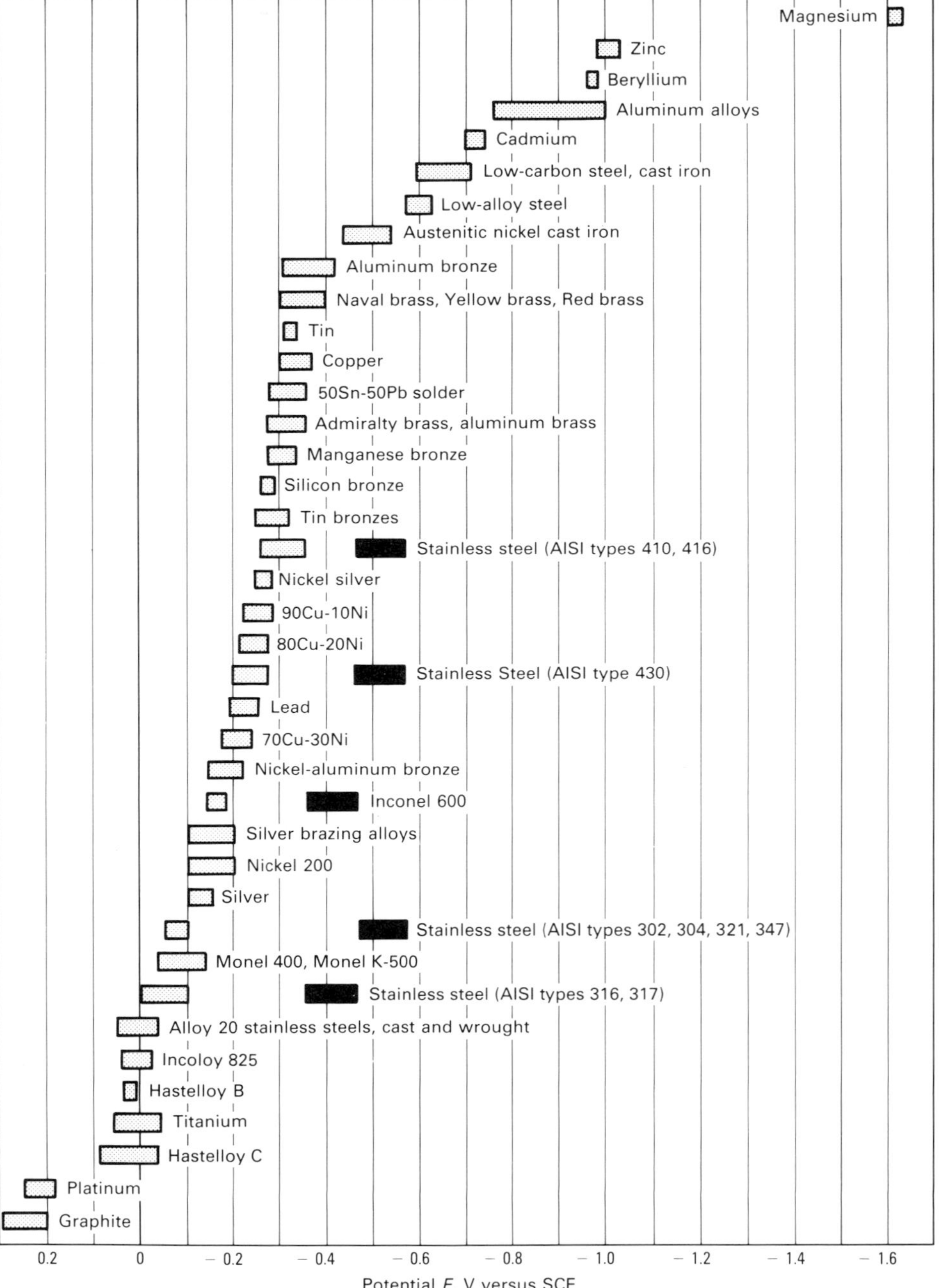

Fig. 1 Galvanic series of metals and alloys in seawater. Alloys are listed in order of the potential they exhibit in flowing seawater; those indicated by the black rectangle were tested in low-velocity or poorly aerated water and at shielded areas may become active and exhibit a potential near −0.5 V. Source: Ref 7

Information on the Substrate. A design engineer, draftsman, engineer, or project manager in an organization with in-house plating facilities should have detailed information on the analysis, microstructure, and mechanical properties of the substrate. In a plating job shop, the customer should have someone informed on the subject. The plater is advised to ask these potential sources for as much information about the substrate as can be obtained.

Engineering Drawings and Associated Lists. Study of engineering drawings, parts lists, bills of material, and pertinent manuals will usually clear up questions concerning the substrate submitted for plating. The bill of material, for example, may simply specify 1095 steel. This informs the plater that he is dealing with a high-carbon (0.95% C) steel that can be heat treated to a high degree of hardness (for example, 50 to 60 HRC). This alerts the plater to be concerned with hydrogen embrittlement. As another example, if a drawing or associated list specifies Aluminum Alloy 6061T, the plater can deduce that the substrate is heat treatable by precipitation hardening to various tempers (degrees of hardness). Such an alloy is subject to developing a smut or to leaving a heavy oxide layer after being alkaline etched or cleaned. An experienced plater will recognize this and will know to treat the cleaned and etched part in a desmutter/deoxidizer in the treating sequence before anodizing or chemically converting the surface.

Chemical Spot Tests. A simple chemical spot reaction test is documented in Ref 6 and 7. This method is based on the use of a magnet and five chemicals. Three of the chemicals are different concentrations of nitric acid (HNO_3), another is concentrated hydrochloric acid (HCl), and the fifth is a 10% solution of potassium ferricyanide ($K_3Fe(CN)_6$). Although it tests for only one element at a time, this simple test distinguishes among a large number of commonly used steels and nickel alloys.

Electrolytic Spot Test Method. This is a commercially available method, complete with chemicals, a power source, and instructions. The operation is divided into four steps. First, a test paper moistened with a conductive solution is placed in contact with the unknown metal. Second, using a dc power source with the unknown metal connected to the anode terminal (+) and a metal probe to the cathode terminal (−), a small amount of the unknown metal is dissolved into the test paper. Dissolution is brought about by electrolytic action when the dc current is switched on. Third, a few drops of a specific color developer are added to the wet spot on the paper. The developer reacts chemically with a characteristic component of the dissolved metal. Fourth, almost immediately, a color appears on the paper if a specific suspected alloying element is present in the unknown. This method is semiquantitative because the relative color intensity is proportional to the amount of alloying element or major metallic constituent. By comparing the results of the unknown with those of standard alloy samples, the nature of the unknown substrate can be adequately identified.

Electronic Alloy Identification. This method is quite rapid and depends on the Peltier Effect (also the basis of the thermocouple). If a current of electricity is sent through a circuit containing two dissimilar metals, heat is emitted at one junction and absorbed at the other junction. A characteristic voltage (potential) is generated at the thermocouple junction that causes an electrical current to flow. This voltage varies with the metals that constitute the couple. The electronic alloy identification meter operates on this principle. It measures the voltage generated by the action of a file of known composition scratched along the surface of an unknown alloy. The file acts as the positive meter input probe, with a clip-on providing negative reference. In a commercially available apparatus, the thermocouple voltage is shown on a battery-powered digital display. This voltage is characteristic of specific alloys or classes of alloys. Again, the instrument can be calibrated against known alloy specimens that are sold as standards.

Physical and Mechanical Properties of Substrates

The susceptibility to corrosion of a plated substrate depends on physical properties and mechanical condition. For example, surface profile (roughness, presence of scratches, pits, lapped metal, and so on) can profoundly affect the durability of plated ware. Some of the harmful effects can be avoided by mechanical or electropolishing, polishing and buffing, shot peening, or glass beading. Internal stresses can also be harmful. These may dictate heating for stress relief, or shot peening or glass beading to convert surface tensile stresses to compression. The physical properties of a substrate, particularly at the surface, are often as important to the metal finisher as its composition. Porosity, hardness, melting point, coefficient of thermal expansion, and permeability to hydrogen (nascent, or atomic, and gas) are vital to many operations. Porosity of substrates, particularly in castings and sintered powder metal compacts, causes many problems because of the difficulty of rinsing processing solutions. Capillary action tends to hold solutions with forces that defy normal rinsing efforts. This results in spotting out of corrosive liquids, which appear as stains and corrosion products on the plated surface. It also causes contamination of processing baths.

Some surface irregularities of substrates can be compensated for by the leveling properties of specially formulated plating solutions. Leveling action is the ability of a plating solution to produce a surface smoother than that of the substrate. The property of the plating solutions that produces leveling is also referred to as microthrowing power.

Another property of substrates that can cause difficulties is susceptibility to hydrogen absorption. High-carbon and, particularly, hardened steels and titanium are susceptible to embrittlement when processed by methods that subject them to atomic or gaseous hydrogen. Recognizing this weakness of such substrates, the designer or specifier of plating sequences must select alkaline cleaning and oxide removal (pickling agents, activators) procedures that lessen or avoid hydrogen evolution. Mechanical cleaning and mechanical plating (also known as peen plating) are examples of the methods used to avoid hydrogen embrittlement. Because total avoidance of hydrogen evolution is often not possible, a step or steps must be added to the plating sequence to remove hydrogen. This usually consists or baking plated parts at low temperature (205 °C, or 400 °F) to release absorbed hydrogen. Time at temperature is also an important variable in ensuring the release of hydrogen and restoration of the mechanical properties of the plated substrate.

Substrate metals that have been heavily worked by such operations as deep drawing, swaging, polishing and buffing, grinding, machining, forging, and die drawing often come to the plating shop with a damaged layer on the surface. This heavily worked layer is termed a Beilby Layer in honor of Sir George T. Beilby, who stated that this layer was amorphous or vitreous rather than crystalline. He compared this weak, somewhat brittle layer to substances that resemble the glass-like form assumed by the silicates when they are solidified from the molten state (Ref 8). X-ray analyses in later years proved that severe cold working causes fragmentation of crystals into particles of different orientations but that the surface metal is not vitreous or amorphous. However, this smeared, shattered layer is inherently low in ductility and fatigue strength and is therefore a weak foundation for plated coatings. This is particularly true of metal objects that require a surface of substantial corrosion resistance and high mechanical properties as well as metal objects that will be further formed or distorted in service. To avoid the weaknesses introduced by such layers, abrasive cleaning, machining, or electrolytic polishing are required to remove them before plating.

Stresses in Substrates

Substrates with locked-in stresses reduce the corrosion resistance of electroplated coatings relative to the same coating on an annealed substrate. Stressed metal is anodic to annealed or lesser stressed metals and is therefore more prone to corrosion in unfavorable service conditions. Substrates with flaws on the surface, stress-raising or stress-localizing scratches, or abrupt changes in section are less resistant to corrosion than smooth, streamlined parts. The coefficients of thermal expansion of the coating and substrate should also be matched as closely as possible to minimize stresses caused by differential expansion during temperature cycling.

Stress-Corrosion Cracking of Substrates

Stress-corrosion cracking (SCC) is a cracking process that requires the simultaneous action of a corrodent and sustained tensile stress. Certain alloys are susceptible to SCC when subjected to certain corrosive environments. Examples are copper alloys, such as Cu-30Zn (cartridge brass), and nickel-chromium alloys. Ammoniacal materials, such as aqueous ammonia (NH_4) and amines, produce such failures in stressed copper alloys. Halides, such as chlorides and bromides, attack certain nickel-chromium alloys with this type of corrosion.

The basic characteristics of SCC are summarized in Ref 9. There is no available evidence that plated coatings on substrates susceptible to SCC will prevent its occurrence. Stress relief of susceptible alloys is often adequate for avoiding SCC. For example, brass cartridge cases, which are made by a deep-drawing process (leaving high residual stress), are stress relieved in a temperature range that produces recovery without recrystallization. Stress relieving at a higher temperature would produce recrystallization and an unacceptable reduction in mechanical properties. Avoidance of the use of SCC-susceptible alloys in known hostile environments is another approach to avoiding this mode of failure, but this is not always feasible. For example, for the brass cartridge case cited above, amine stabilizers in the smokeless powder propellant obviated control of the environment; therefore, stress relief was required. However, use of SCC-susceptible metals in corrosion cracking environments should be avoided whenever possible.

Design for Plating

Much can be done at the design stage to improve the plateability of metal stampings, castings, forgings, and extruded shapes. In doing so, the corrosion protection afforded by the plating can be enhanced. Corrosion resistance of a plated object depends on coating thickness and distribution, and these factors depend largely on the design of the parts. The ASTM standard B 507 (Ref 10) covers practices for the design of articles to be electroplated on racks with nickel, but its significance goes beyond nickel plating. It also applies in most respects to all rack-plated coatings. The general principles of design are discussed in Ref 10, which states that the electrodeposition of one metal onto another follows a pattern that can be illustrated by a sketch showing current distribution and the distribution of the resultant electrodeposit (Fig. 2).

Distribution of the deposit will vary slightly as a function of bath type, additives, current density, and temperature. The coating is preferentially deposited at protuberances, sharp corners, edges, fins, and ribs on the component being plated; to avoid coating buildup, these areas should be rounded with a minimum radius of 0.4 mm (0.015 in.). A 0.8-mm (0.03-in.) radius is preferable. Conversely, the base of any cavity, recess, or depression in the component being plated will be deficient in plate thickness; these features should be designed to avoid sharp fillets. In general, such features should be designed so that the ratio of width to depth of the depression is held to a minimum of 3, and the minimum radius at the edge and base of an indentation should be one-fourth of the depth of the indentation. When sharp angles are required for functional purposes, the electroplater cannot be expected to meet minimum coating thickness requirements in these areas and corrosion resistance may be inadequate.

When designs cannot be changed, electroplaters resort to one or more of the following methods to overcome the aforementioned difficulties: provision of inside anodes, use of thieves or robbers (nonconductive shields that slow or prevent deposition) to lower the cathodic current density at protuberances, bipolar electrodes, changing electroplating baths to ones with greater covering power, and switching from electroplating to autocatalytic (electroless) plating. This last method is widely used on parts that cannot be properly plated by electrodeposition. The limitation is that the most prevalent coating applied autocatalytically—electroless nickel—deposits an alloy of nickel and phosphorus (phosphorus content varies with the solution used and the deposition condition), not unalloyed nickel. There is also an economic disadvantage in switching to these processes. However, in most cases in which electroless processes have been adopted, they have been specified for reasons other than merely to compensate for part designs that cannot be readily electroplated with uniform coatings.

An article should not be designed for plating, but rather to perform its intended function (Ref 11). If the part is successfully designed, provision will have been made for corrosion resistance,

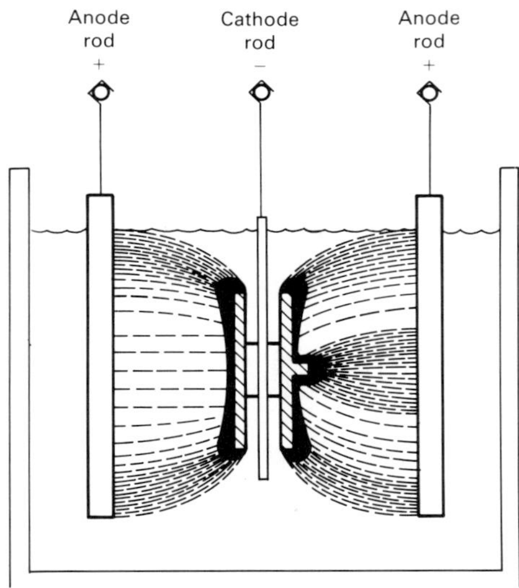

(a)

120°

90°

60°

(b)

Fig. 2 Current distribution (a) and effect of substrate geometry (b) on thickness of electroplated nickel deposits. For the 120° substrate in (b), the ratio of average coating thickness to minimum coating thickness is 1.9; for the 90° substrate, 2.7; and for the 60° substrate, 3.3. Source: Ref 10

hardness, lubricity, decorative appeal, or whichever required property the finish process is to supply. Function, fit, form, and aesthetics—the major dictates of design criteria—often take precedence over the practicality of corrosion prevention by plated coatings; this need not be the case. Design manuals must be instructional with regard to the effects of geometry on function, and designers should be taught these principles.

Additional information on design for electroplating is available in Ref 12. Table 2 lists proper and improper designs for various features of electroplated components. Plating of relatively small parts in rotating containers is termed barrel plating. This differentiates it from rack plating, in which individual parts are held in a rack or fixture during plating. Fastening hardware, such as nuts, bolts, washers, screws, and studs, are barrel plated, as are numerous critical electrical and electronic components. Many of these parts are as important from the corrosion resistance standpoint as rack-plated parts are—perhaps more so because such parts as fasteners and connectors are used in, and therefore coupled to, other materials. This raises the problem of galvanic corrosion.

Bulk parts in a plating barrel are subject to a greater variation in cathodic current density than parts plated on racks. Cathode contact is made by danglers, chains, buttons, or studs near the bottom of the load, while the plating occurs only at the surface of the load. This produces a large statistical variation in thickness of the electrodeposits on barrel-plated parts; this is not the case for rack-plated parts. Corrosion of a plated part is often controlled by the minimum thickness of protective coating; therefore, if parts are to be barrel plated, special design practices should be observed, including (Ref 12):

- Avoid blind holes, recesses, and joint crevices that can retain tumbling compounds and metal debris from previous operations
- Avoid intricate surface patterns that will be blurred by barrel finishing
- Design parts to withstand the multiple impacts of barrel rotation
- Design small, flat parts with ridges or dimples to prevent nesting
- Design for good entry and drainage of solutions during rotation by using simple shapes
- Design surfaces to be plated so that they undergo proper mechanical preparation and cleaning and to ensure that they receive their share of metal deposit. They should be convex, if possible, rather than recessed

In a larger sense, the design of parts for plating parallels good design practice for corrosion resistance. Avoidance of sharp edges, burrs, or other areas of high free energy, avoidance of abrupt changes in section, recesses, and other traps for moisture, and avoidance of scratches, notches and other anodic-site producers are common design criteria for parts in general as well as for plated parts. More information on designing to minimize corrosion is available in the article "Design Details to Minimize Corrosion" in this Volume.

Advantages and Disadvantages of Electroplated Coatings

Alternatives to electroplating for the protection of metal surfaces against corrosion include organic finishing, hot-dip coating, metal spraying, chemical vapor deposition (CVD) and physical vapor deposition (PVD), vacuum deposition by sputtering and ion implantation, and ceramic coating (see the articles "Organic Coatings and Linings," "Hot Dip Coatings," "Thermal Spray Coatings," "CVD/PVD Coatings," "Surface Modification," and "Chemical-Setting Ceramic Linings," respectively, in this Volume). The decision to use one or more of these methods must consider at least the following factors:

- Applicability to the substrate in question (for example, zinc plating is applicable as a protective coating to steel, but not cast iron)
- Purposes of the finish (Is color a requirement? Will the object to be finished fit into available plating equipment?)
- Cost of suitable electroplating versus alternative methods of coating the part (What finish will be functionally adequate and applicable at the least cost?)

Table 2 Influence of substrate design features on electroplateability

Feature	Poor design	Influence on electroplateability	Better design
Convex surfaces		Ideal shape. Easy to plate uniformly, especially where edges are rounded	
Flat surfaces		Use a 0.4-mm (0.015-in.) crown to minimize undulations caused by uneven buffing.	
Sharply angled edges		Undesirable. Reduced coating thickness at center areas requires increased plating time to obtain a minimum thickness of durable electroplate. All edges should be rounded. Edges that will contact painted surfaces should have a minimum radius of 0.8 mm (0.03 in.).	
Flanges	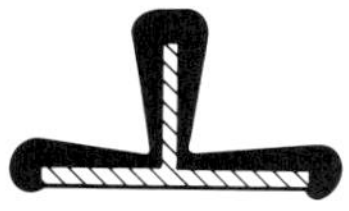	Large flanges with sharp inside angles should be avoided to minimize plating costs. Use a generous radius on inside angles and taper the abutment.	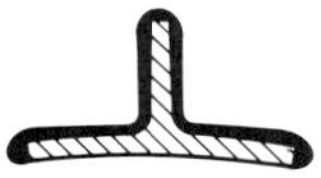
Slots		Narrow, closely spaced slots and holes reduce electroplateability and cannot be properly plated unless corners are rounded.	
Blind holes	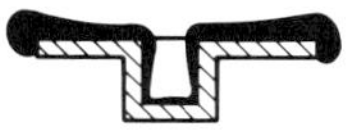	Must usually be exempted from minimum thickness requirements. Where necessary, limit depth to 50% of width. Avoid diameters of less than 6 mm (0.24 in.).	
Sharply angled indentations		Increase plating time and costs for a specified minimum thickness, and reduce the durability of the plated part.	
Flat-bottom grooves		Inside and outside angles should be rounded generously.	
V-shaped grooves		Deep V-shaped grooves cannot be satisfactorily plated and should be avoided. Shallow, rounded grooves are better.	
Fins	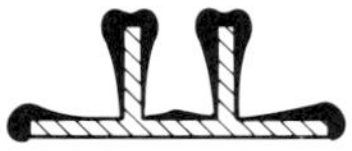	Increase plating time and costs for a specified minimum thickness and reduce the durability of the plated part.	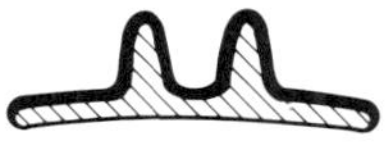
Ribs		Narrow ribs with sharp angles usually reduce electroplateability; wide ribs with rounded edges pose no problem. Taper each rib from its center to both sides and round off edges. Increase spacing if possible.	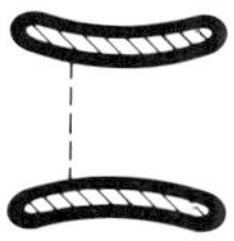
Concave recesses		Electroplateability depends on dimensions.	
Deep scoops		Increase plating time and costs for a specified minimum thickness.	
Spearlike juts		Buildup on jut will rob corners from their share of electroplate. Crown the base and round off all corners.	
Rings	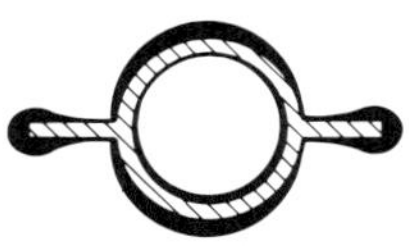	Electroplateability depends on dimensions. Round off corners and crown from center line, sloping towards both sides.	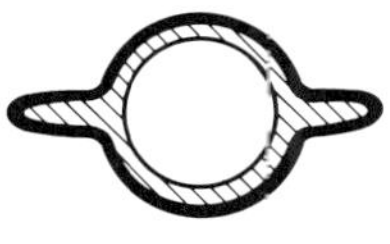

Source: Ref 12

- Anticipated service conditions and length of service desired under these conditions (SC5, extended very severe; SC4, very severe; SC3, severe; SC2, moderate; SC1, mild) (Ref 13)

Advantages of Electroplating

Variety of Coatings. The principal metals that are widely plated are cadmium, zinc and its alloys, chromium, copper, gold, nickel, tin, and silver. Other metals that are plated by in-house shops and some job platers are cobalt, indium, iron, lead, palladium, platinum, rhenium, and rhodium. Plateable alloys add versatility to the catalog of plated coatings. The commonly plated alloys are brass, bronze, copper-cadmium, tin-lead (solder), tin-zinc, gold alloyed with many metals, and nickel alloys with iron and cobalt. Ternary alloys, such as karat gold alloys and copper-tin-zinc, are also available. Nickel-phosphorus and nickel-boron are readily plated by the autocatalytic (electroless) process. Nickel-phosphorus can also be electrodeposited. Coatings of many types can be codeposited with occluded particles by electrodeposition or electroless deposition to modify friction and wear properties. Particles used for these applications range from such codeposited or occluded abrasives as diamond, silicon carbide (SiC), and such lubricious materials as polytetrafluoroethylene (PTFE).

Versatility of Application. Plated coatings can be applied to racked parts, in barrels (both by immersion in tanks and manually), semiautomatically, or on microprocessor-controlled fully automatic machines. Local, selective application of a coating, if necessary, can be accomplished by such methods as brush plating (also known as electrochemical metallizing and tampon plating). Similarly, precious metals, such as gold and palladium, have been conserved in electronics applications by selective plating on continuous strips. This is referred to as reel-to-reel plating. The adaptability of plating methods to the job at hand is limited only by the imagination and skill of the plater.

Availability of Applicators. Plating job shops and in-house applicators are plentiful in the United States and throughout the world. The technology is popularized and nurtured by such technical societies and trade associations as the American Electroplaters and Surface Finishers Society, the National Association of Metal Finishers, the Metal Finishing Suppliers' Association, and the Society of Automotive Engineers Airline Plating Forum.

The electrical conductivity of electrodeposited and other metallic coatings gives them an advantage over most organic and ceramic-base coatings in applications that require this characteristic. Applications in which electrical conductivity is necessary include the welding, electrical bonding, or grounding of objects as well as electrical and electronic circuitry.

Galvanic Protection of Substrates. Zinc and cadmium deposits protect steel substrates by sacrificial action in many wet environments. In aircraft and other aerospace applications, cadmium plating is used on steel components in contact with aluminum alloys because cadmium is closer in potential to aluminum than steel is (Fig. 1). On coil-coating lines for prepainted steel sheet and strip, electrogalvanizing stages are provided in the wet section to protect the substrate at points of failure of the organic overcoating. Electrogalvanized steel, often further treated with chromate conversion coatings, is used for various purposes because of the sacrificial protective value of the zinc coating. Duplex or triplex coatings of nickel, often with a final thin coating of chromium, are another example of a more complex galvanic protective system (see the section "Nickel-Chromium Coatings" in this article).

Weldability and Solderability. Metal deposits are amenable to welding and soldering—an advantage over most organic coatings and porcelain enamel. Certain coatings are more weldable and solderable than others, which leads to the application of special types of plating to facilitate a specific type of joining or fabricating method.

Colored Anodic Coatings on Aluminum. Anodic oxidation of aluminum provides a protective coating that can be colored. The coating is integral to the substrate and can be colored by dyeing, integral color anodizing, or electrolytic coloring (Ref 6, 14). These coatings are highly resistant to abrasion and corrosion and are widely used on architectural forms. More information on these types of coatings is available in the article "Aluminum Anodizing" in this Volume.

Control of Thickness. Unlike hot-dip coating, electrodeposition affords a means of predicting and controlling the amount of metal deposited. This is possible through the validity and knowledge of Faraday's laws, which are as follows (Ref 2):

- The quantity of any element discharged at an electrode is proportional to the quantity of electricity that is passed
- The quantities of different ions of elements discharged by a given amount of electricity are proportional to the electrochemical equivalents of those elements

The quantity of electricity is measured by the coulomb (C; equivalent to one A · s), and the electrochemical equivalent of an element is its atomic weight divided by its valence. It is known that the 96 480 C, or 1 Faraday will deposit 1 gram equivalent weight of any element at 100% electrode efficiency. Electrode efficiencies are a function of plating electrolytes and conditions. They can be determined by coulometric measurement and are published for most plating processes. Because most plants control plating by measuring ampere hours, the amount in grams of metal plated, M_d, is calculated by the relationship:

$$M_d = \frac{A \cdot h}{26.8} EE_M \, CE_M$$

where EE_M is the electrochemical equivalent of metal M (atomic weight of M divided by valence of M), and CE_M is the cathode efficiency of the plating bath and conditions.

Ductility and Formability. Substrate metals coated by electrodeposition can be formed more readily than some metals coated by hot dipping or certain organic coatings. They do not result in intermetallic alloy layers of the type common to ordinary hot-dip galvanizing, for example. Diffusion of some electrodeposits into substrates, and vice versa, occurs after plating as a function of temperature, time, and the diffusion tendencies of the metals involved. These diffused layers are normally thin and require protracted periods of time to develop, usually after the component has been formed to a desired shape.

Electroless Plating. The coating of the interiors of tubes or other recessed areas uniformly and with controlled thickness is the special province of autocatalytic, or electroless, plating.

Mechanical Plating. Mechanical, or peen, plating is a method of coating substrates, principally steel, with malleable metals, such as zinc, cadmium, tin, or their alloys. In mechanical plating, hard, small, spherical objects, such as glass shot, are tumbled against the parts to be plated in the presence of finely divided metal powder, such as zinc dust, and appropriate chemicals. Special advantages of this method over electroplating are the relative absence of problems with hydrogen embrittlement, the simplicity of operation with small parts in a single rotating container, and the ability to deposit desirable alloys that cannot be electroplated. As an example of the latter, tin and cadmium alloys have been found to provide excellent protection of steel against corrosion in marine environments.

Disadvantages of Electroplating

Color Limitations. The colors of electrodeposited coatings are limited to copper, silver metallic, brass, bronze, and gold and its alloys; black, iridescent, or olive drab chromate conversion coatings; and patinas formed on copper and its alloys by chemical and electrochemical treatments.

Inhospitable Substrates. Some substrates, such as cast iron, and certain substrate conditions, such as porosity, require special plating procedures. Unimpregnated powder-metal compacts and porous castings that can be plated are subject to spotting out, or bleeding, because of retention in pores of cleaning and plating alkalies, acids, and salts.

Limitations of Design. Parts are designed to function in the most economical manner. This is not always compatible with plating. Where the geometry of a part is such that cathodic current distribution will be unfavorable to uniform coatings, it is sometimes wise to seek an alternative protective coating. The criteria here are that the coating must achieve the required minimum thickness on significant surfaces, must meet all functional and dimensional requirements, and must accomplish these objectives at a reasonable cost.

Size Limitations. Large metal structures are often beyond the capabilities of electrodeposition. The limitation lies in the capacity of tanks to contain the electrolyte. Exceptions are certain architectural structures and statuary, which may be brush plated. Such coatings are almost always on copper or bronze and consist of nickel topped with gold electrodeposits. Steel structures, such as bridges, dams, and building elements, are more advantageously hot-dip galvanized and painted, or painted without zinc but with specially designed organic coating systems.

Effects of Deposition Parameters

The process used to apply the coating may have an important effect on the properties of the entire coating system. One of the more important considerations is the effect of hydrogen formation during electroplating. Hydrogen may embrittle the substrate. This is particularly important when cadmium, chromium, or zinc is used as a coating over high-strength steel. Atomic hydrogen can combine to form high-pressure hydrogen gas at defects or regions of poor adhesion, causing blisters in the coating. Appropriate standards have been developed to cover relief from hydrogen embrittlement in high-strength steels (Ref 15). Hydrogen can also be introduced into the coating system during one of the cleaning steps conducted before electroplating.

Mechanisms of Corrosion Prevention of Plated Coatings

Coatings applied by electrodeposition or electroless plating protect substrate metals in three ways. The first is cathodic protection of the substrate by sacrificial corrosion of the coating—for example, zinc and cadmium coatings on steel. Second is barrier action, that is, interposing a more corrosion-resistant coating between the environment and a less corrosion-resistant substrate. Examples of this are copper-nickel-chromium and nickel-chromium coating systems over steel and zinc alloy automotive parts. Third is environmental modification or control in combination with a nonimpervious barrier coating. An example of this type of protection is the electrolytic tinplate used in food packaging.

Anodic Coatings. Zinc and cadmium deposits will protect steel at pores and other discontinuities by cathodic protection. Observation of the emf series (Table 1) shows that zinc is electronegative to steel in potential when these metals are immersed in solutions of their own ions. Under the same conditions, cadmium is more noble than iron, but in actual service under several environmental influences, cadmium is anodic to iron and is therefore sacrificially protective. Figure 3 shows the galvanic protection offered by a zinc coating to a steel substrate. This mechanism of galvanic protection gave rise to the

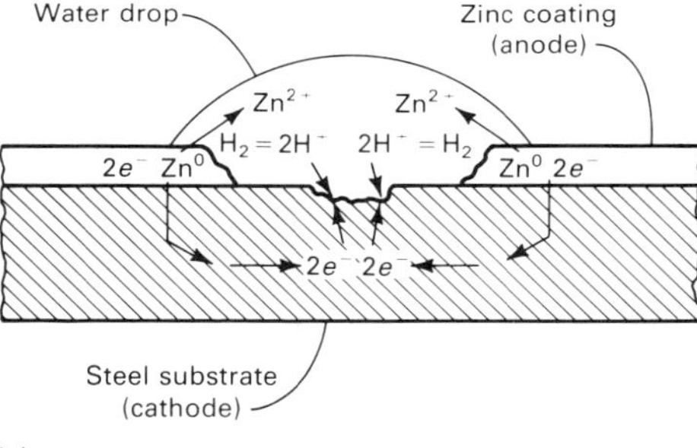

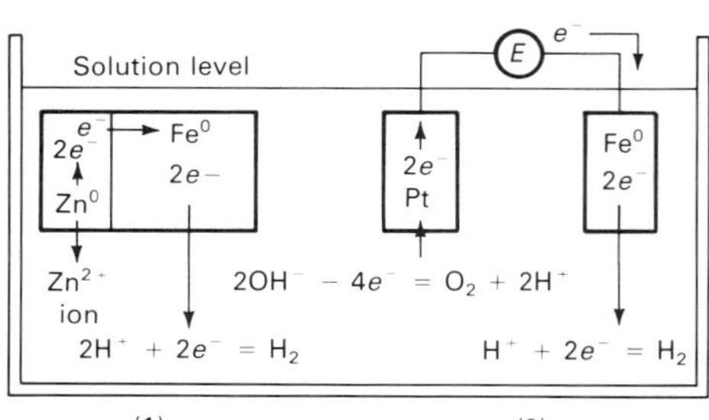

Fig. 3 Principles and mechanism of galvanic protection of a substrate by a coating. (a) Galvanic protection of a steel substrate at a void in a zinc coating. Corrosion of the substrate is light and occurs at some distance from the zinc. (b) Galvanic protection of a steel member (1) by a zinc member and by an impressed electric current flowing from an external dc power source (E) using an insoluble platinum electrode (2). Source: Ref 2

Table 3 Applicability of thickness measurement methods to coatings

Substrate	Coating/Method(a) Copper	Nickel	Chromium	Electroless nickel	Zinc	Cadmium	Gold	Palladium	Rhodium	Silver	Tin	Lead	Tin-lead alloys	Nonmetals	Vitreous and porcelain enamels
Magnetic steel (including corrosion-resistant steel)	C, M	C, M(b)	C, M	C(c), M(b)	C, M	B, C, M	B, M	B, M	B, M	B, C, M	B, C, M	B, C, M	B(d), C(d), M	B, M	M
Nonmagnetic stainless steels	C, E(e)	C, M(b)	C	C(c)	C	B, C	B	B	B	B, C, E(e)	B, C	B, C	B(d), C(d)	B, E	E
Copper and alloys	C, only on brass and Cu-Be	C, M(b)	C	C(c)	C	B, C	B	B	B	B, C	B, C	B, C	B(d), C(d)	B, E	E
Zinc and alloys	C	M(b)	...	...	...	B	B	B	B	B	B	B	B(d)	B, E	...
Aluminum and alloys	B, C	B, C, M(b)	B, C	B, C(c), E(b)(c)	B, C	B, C	B	B	B	B, C	B, C	B, C	B(d), C(d)	E	E
Magnesium and alloys	B	B, M(b)	B	B	B	B	B	B	B	B	B	B	B(d)	E	...
Nickel	C	...	C	...	C	B, C	B	B	B	B, C	B, C	B, C	B(d), C(d)	B, E	...
Silver	B	B, M(b)	B	B	B	...	B	...	...	...	...	B, C	B(d)	B, E	E
Glass sealing nickel-cobalt-iron alloys (UNS K94610)	M	C, M(b)	M	C(c), M(b)	M	B, M	B, M	B, M	B, M	B, M	B, M	B, C, M	B(d), C(d), M	B, M	...
Nonmetals	B, C, E(e)	B, C, M(b)	B, C	B, C(c)	B, C	B, C	B	B	B	B, C	B, C	B, C	B(d), C(d)	...	...
Titanium	B	B, M(b)	...	B, E(b)(c)	B	B	B	B	B	B	B	B	B(d)	B, E	...

(a) B, β backscatter; C, coulometric; E, eddy current; M, magnetic. (b) Method is sensitive to permeability variations of the coating. (c) Method is sensitive to variations in phosphorus content of the coating. (d) Method is sensitive to alloy composition. (e) Method is sensitive to conductivity variations of the coating. Source: Ref 17

term galvanized steel, which refers only to zinc-coated steel. This old, but commonly used term is modified to electrogalvanizing when referring to electrodeposited zinc in order to differentiate it from hot-dip galvanizing.

Cathodic Coatings. With reference to Table 1, it is apparent that the metals commonly plated, other than zinc (and cadmium, despite the inversion mentioned above), are electropositive to iron. This is also evidenced by Fig. 1. Such coatings on steel are therefore expected to act strictly as barriers to prevent corrosion of the steel substrate. For this to be successful, the coatings must be pore free and flaw free. It is very difficult to obtain impervious coatings of more noble metals on steel or zinc alloys. Substantial metal thicknesses, usually 25 μm (1 mil) or more, are required. With some of these metallic deposits, particularly the noble metals, thinner coatings applied by pulse plating are more impervious than conventionally plated metals, but this is the exception.

Generally, electroplated or hot-dipped coatings that are completely free of pores and other discontinuities are not commercially feasible (Ref 2). Pits eventually form at coating flaws, and the coating is penetrated. The substrate exposed at the bottom of the resulting pit corrodes rapidly (Fig. 4). A crater forms in the substrate, and because of the large area ratio between the more noble coating and the anodic crater, the crater becomes anodic at high current density. Electrons flow from the substrate to the coating as the steel dissolves. Hydrogen ions (H^+) in the moisture accept the electrons and, with dissolved oxygen, form water at the noble metal surface near the void. Use of an intermediate coating that is less noble than a surface coating but more noble than the basis metal may result in the mode of corrosion shown in Fig. 5. This would be typical of a costume jewelry item with the substrate being brass, the intermediate coating nickel, and the topcoat tarnish-resistant gold. It is also exemplified by nickel-chromium coating systems.

Exceptions. The durability of a plated coating system cannot always be accurately predicted by conventional analysis of the potentials shown in the emf or galvanic series. Much depends on environmental exposure and the oxidized films formed in pores and discontinuities of plated coatings.

For example, several corrosion specialists predicted that copper-plated zinc one-cent pieces would be unsuitable coinage because of rapid corrosion. The coating is barrel-plated copper on a zinc-rich alloy, with the coating thickness varying from 4 to 8 μm (0.15 to 0.3 mil). In this thickness range, the coating is far from impervious, or pore free. In the 1970s, before the minting of these coins, pictures were published showing exfoliated, voluminous zinc corrosion products on sample pennies after exposure to the ASTM B 117 neutral salt spray test (Ref 16). The predicted result did not occur after more than 10 years of service. This indicates that the conditions to which these coins have been subjected are more forgiving than a salt spray test.

Another example of an apparent exception is the case of lead-plated steel, which is quite durable despite the fact that lead is more noble than iron. This is the case because a lead corrosion product develops in the pores (most likely lead sulfate $PbSO_4$ and arrests the progress of corrosion.

Importance of Thickness of Plated Coatings. In general, the thicker a metallic coating is, the less porous it is and the better it serves as a barrier against corrosion. Therefore, the measurement of thickness is a most important feature of testing electrodeposited and electroless coatings. Table 3 indicates the applicability of coating thickness measuring methods. American Society for Testing and Materials (ASTM) and International Standards Organization (ISO) standards covering thickness measurement methods are given in Ref 17.

Porosity is another coating property that is very sensitive to deposition parameters and to substrate preparation. For very thin gold coatings (less than about 1 mm, or 0.04 mil), substrate texture controls coating porosity. At greater

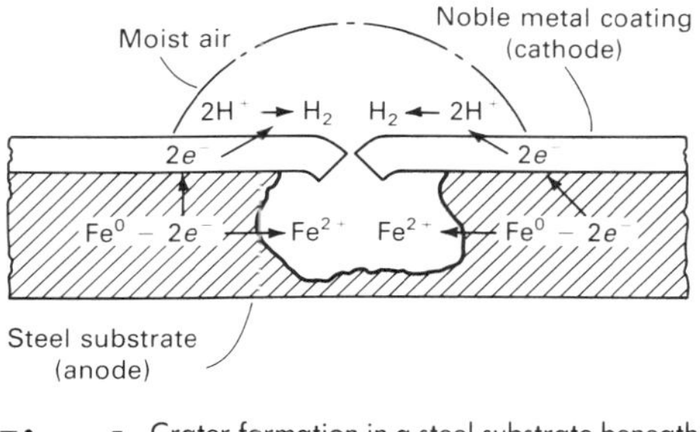

Fig. 4 Crater formation in a steel substrate beneath a void in a noble metal coating, for example, passive chromium or copper. Corrosion proceeds under the noble metal, the edges of which collapse into the corrosion pit. Source: Ref 2

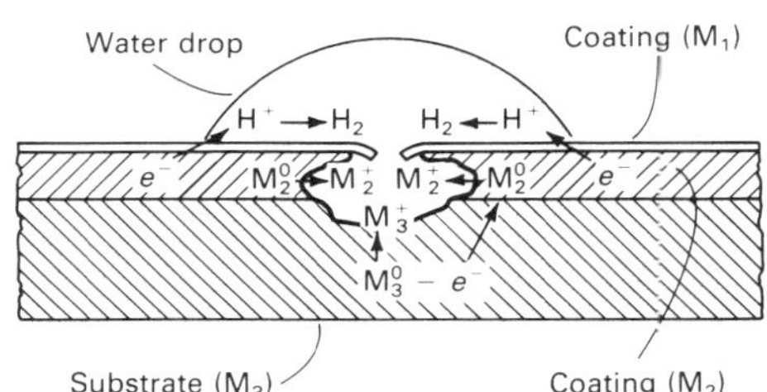

Fig. 5 Corrosion pit formation in a substrate beneath a void in a duplex noble metal coating. The top coating layer (M_1) is cathodic to the coating underlayer (M_2), which is in turn cathodic to the substrate (M_3). As in Fig. 4, the coating tends to collapse into the pit. Source: Ref 2

Table 4 Suggested thicknesses for electrodeposited tin coatings on copper alloys (50% Cu min)

Purpose of coating	Minimum local thickness μm	mils
Contact with food or water where a complete cover of tin must be maintained against corrosion and abrasion	30	1.2
Protection in atmosphere and in less aggressive immersion conditions	15	0.6
To provide solderability and protection in mild atmospheric conditions	5(a)	0.2(a)
Coatings flow brightened by fusion (solderability and protection in mild atmospheres)	2.5–8(a)	0.1–0.3(a)

(a) On brass, an undercoat of 2.5-μm (0.1-mil) thick, copper, nickel, or bronze is required. Source: Ref 28

Table 5 Suggested thicknesses for electrodeposited tin coatings on ferrous components

Purpose of coating	Minimum local thickness μm	mils
Contact with food or water where a complete cover of tin must be maintained against corrosion and abrasion	30	1.2
Protection in atmosphere	20	0.8
Protection in moderate atmospheric conditions with only occasional condensation of moisture	10	0.4
To provide solderability and protection in mild atmospheres	5	0.2
Coatings flow brightened by fusion (solderability and protection in mild atmospheres)	2.5–8	0.1–0.3

Source: Ref 28

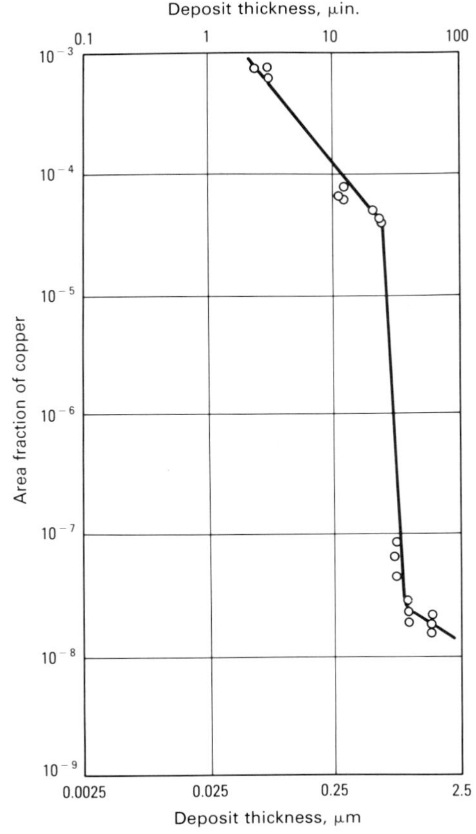

Fig. 6 Porosity versus deposit thickness for electrodeposited, unbrightened gold on a copper substrate. Compare with Fig. 7.

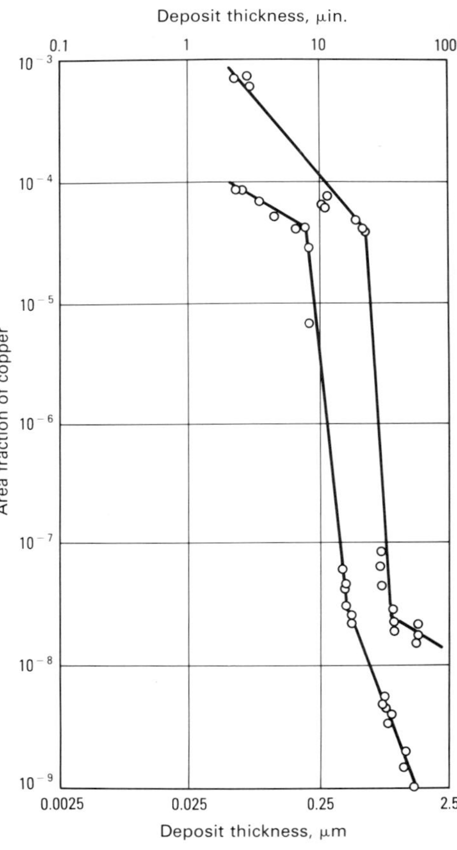

Fig. 7 Porosity versus deposit thickness for pulse-plated gold on a copper substrate. The curve for an unbrightened gold deposit on a copper substrate (top) is shown for comparison (see also Fig. 6).

thicknesses, porosity is controlled by the coating properties. The variation in porosity with coating thickness is illustrated in Fig. 6, which shows three distinct phases: substrate dominated, transition, and coating dominated. Porosity can also be controlled to a large extent through deposition parameters, and this effect is attributed to a reduction in deposit grain size. The effects of porosity with thickness for pulsed-plated gold are shown in Fig. 7; again, the same three phases are apparent, although they are displaced downward.

Electrodeposited coatings applied to thicknesses of less than about 25 μm (1 mil) are porous. Porosity decreases with thickness, but it is present even at greater thicknesses. One of the reasons for this is the codeposition of hydrogen at cathodes, owing to the fact that most plating processes operate at less than 100% cathode efficiency. With sacrificially protective coatings, such as zinc and cadmium on steel, porosity is of little significance. Within limits, substrate steel exposed in the pores will be cathodic to the zinc or cadmium coating. Thus, the coatings will corrode under adverse conditions and protect the exposed cathodic steel, at least until the coating has been consumed by conversion to corrosion products. With the exception of chromium, coatings that are cathodic to the substrate depend on freedom from porosity for protection of the less noble substrate. Accordingly, such coatings must be tested for porosity as well as for thickness. Examples of ASTM standards for porosity testing are given in Ref 16 and 18 to 21.

Adhesion is very important in the electrodeposition of a metallic coating, as implied in the definition of electroplating given earlier in this article. Therefore, it is logical for many end product specifications for plated components to contain adhesion tests for verification of this requirement. The ASTM standard B 571 (Ref 22) covers the following adhesion tests for metallic coatings: the burnishing test, the chisel-knife test, the draw test, the file test, the grind-saw test, and the heat-quench test.

Corrosion. Testing for corrosion resistance is an important part of evaluating electroplated coatings and allied finishing techniques. Standard methods of corrosion testing are given in Ref 16 and 23 to 27. Accelerated tests of corrosion resistance of electroplated coatings can often lead to misleading conclusions. Claims that a certain number of hours in a particular test is equivalent to so many months or years of actual service should be viewed with suspicion.

Applications for Electrodeposited Coatings

Nickel is widely used as a corrosion-resistant coating and as an undercoat for subsequent coatings. It is not resistant to HNO_3 or to environments containing chloride ions (Cl^-). However, it corrodes slowly in environments that do not contain Cl^- ions. Nickel is widely used by the automotive industry as an underlayer for microcracked chromium to protect steel. As previously discussed, nickel is usually plated as part of a multilayer coating system in which the electrode potential of each layer is different from those of other layers. The outer layer in such systems is often 130 mV (versus the silver-silver chloride, Hg-$HgCl_2$, electrode) more anodic than the layer beneath it. Nickel is also used over copper as an underlayer for gold to prevent rapid diffusion of the gold into the copper substrate. Nickel and its reaction products are toxic. Nickel with organic additives to produce a very bright finish is widely used, although these coatings are much more brittle than the semibright nickel.

Nickel-phosphorus can be electrolessly or electrolytically deposited as a metallic coating. In general, its corrosion performance exceeds that of nickel in all environments. If the phosphorus content is above about 10 wt%, the coating is amorphous and therefore lacks grain boundaries or other crystalline defects at which corrosion can be initiated. Electroless nickel has some advantage over electrolytic nickel in that its thickness and phosphorus content are uniform over the substrate and independent of substrate geometry. Electroless nickel is not resistant to hot caustic solutions. Heat treating nickel-phosphorus coatings above 350 to 400 °C (660 to 750 °F)

Table 6 Recommended minimum thicknesses and typical applications for zinc and cadmium coatings electrodeposited on iron and steel

Service conditions	Coating thickness µm	mils	Chromate finish	Time to white corrosion in salt spray, h	Typical applications
Electrodeposited zinc					
Mild (indoor atmosphere; minimum wear and abrasion)	5	0.2	None	...	Screws, nuts and bolts, buttons, wire goods, fasteners
			Clear	12–24	
			Iridescent	24–72	
			Olive drab	72–100	
Moderate (mostly dry indoor atmosphere; occasional condensation, wear, and abrasion)	8	0.3	None	...	Tools, zipper pulls, shelves, machine parts
			Clear	12–24	
			Iridescent	24–72	
			Olive drab	72–100	
Severe (exposure to condensation; infrequent wetting by rain; and cleaners)	13	0.5	None	...	Tubular furniture, window screens, window fittings, builders' hardware, military hardware, appliance parts, bicycle parts
			Clear	12–24	
			Iridescent	24–72	
			Olive drab	72–100	
Very severe (exposure to bold atmospheric conditions; frequent exposure to moisture, cleaners, and saline solutions; likely damage by abrasion or wear)	25	1	None	...	Plumbing fixtures, pole line hardware
Electrodeposited cadmium					
Mild (see above)	5	0.2	None	...	Springs, lock washers, fasteners, tools, electronic and electrical components
			Clear	12–24	
			Iridescent	24–72	
			Olive drab	72–100	
Moderate (see above)	8	0.3	None	...	Television and radio chassis, threaded parts, screws, bolts, radio parts, instruments
			Clear	12–24	
			Iridescent	24–72	
			Olive drab	72–100	
Severe (see above)	13	0.5	None	...	Appliance parts, military hardware, electronic parts for tropical service
			Clear	12–24	
			Iridescent	24–72	
			Olive drab	72–100	
Very severe (see above)(b)	25	1	None	...	...
			Clear	24	
			Iridescent	24–72	
			Olive drab	72–100	

(a) Thickness specified is after chromate conversion coating, if used. (b) There are some applications for cadmium coatings in this environment; however, these are normally satisfied by hot-dipped or sprayed coatings. Source: Ref 12

results in recrystallization and loss of some corrosion resistance even though other properties, such as wear or hardness, may be enhanced.

Cadmium and zinc were compared earlier in this article. Cadmium is generally preferred for the protection of steel in marine environments and zinc is preferred in industrial environments. Cadmium tends to embrittle high-strength steel somewhat less than zinc does (being process related, new deposition processes may change this). The coefficient of friction of cadmium is less than that of zinc; therefore, cadmium is preferred for fastening hardware and connectors that have to be taken on and off repeatedly. Cadmium is much more toxic than zinc, and applications in which its corrosion products may get into the environment should be avoided. The corrosion performance of both zinc and cadmium is greatly enhanced by chromate conversion coatings (see the article "Chromate Conversion Coatings" in this Volume).

Chromium is very resistant to atmospheric corrosion, but is soluble in HCl or in alkaline (caustic) solutions. Its resistance to corrosion is thought to be due to the formation of an amorphous chromium oxide that acts as a passive film to protect the metal. Chromium is deliberately deposited for decorative and wear applications as a microcracked coating over nickel so that corrosion currents are uniformly distributed over a large area. Chromium may be deposited as a noncracked coating both from hexavalent chromium (Cr^{6+}) and from trivalent chromium (Cr^{3+}) electrolytes. More information on corrosion of chromium plating is available in the article "Corrosion of Electroplated Hard Chromium" in this Volume.

Tin is widely used as a barrier and as an anodic coating for steel and copper. Because it is readily solderable and its oxide is conducting, it finds application in electrical conductors and contacts. Its corrosion products are not toxic; therefore, tinned containers are widely used in the food industry. Recommended thicknesses for application over copper components are given in Table 4 and over ferrous components in Table 5. More information on corrosion of tin and tinplate is available in the article "Corrosion of Tin and Tin Alloys" in this Volume.

Lead is stable in most atmospheres and in both chromic (H_2CrO_4) and sulfuric acid (H_2SO_4) electrolytes; however, it is affected by chlorides. Lead electrodeposits form very inert passive films that tend to reduce the effects of galvanic corrosion.

Copper is not particularly corrosion resistant in the atmosphere, and it tarnishes rapidly. When used alone, it should be protected by an appropriate corrosion inhibitor, such as benzotriazole. On the other hand, copper is often present in a coating system because bright copper is leveling, and its small grain size tends to reduce the porosity and thus improve the corrosion resistance of subsequent coatings.

Gold is often used decoratively and for electrical contact applications. Gold that is plated directly over copper will increase the corrosion rate of the copper through the unavoidable porosity of the gold deposit. Gold will also rapidly diffuse into copper. Therefore, electroplated nickel or cobalt barrier coatings are normally plated underneath the gold coating. For wear applications, alloys with nickel or cobalt are preferred; however, pure gold has a lower contact resistance. Pulse plating will reduce porosity and increase the hardness of electrodeposited gold.

Selection of Coatings

Tables 6 and 7 represent current practice in guiding designers and platers to minimum thickness of the two principal classes of protective and decorative/protective coatings used in product finishing; zinc and cadmium coatings and bright nickel-chromium. Specification guides for these coatings are given in Ref 1 and are also contained in Federal and Military Specifications (Ref 6, 29). Frequently cited specifications in these documents are:

Specification	Title
ASTM A 165	Standard Specification for Electrodeposited Coatings of Cadmium on Steel
MIL-QQ-P-416	Plating, Cadmium Electrodeposited
ASTM B 633, MIL-QQ-P-416	Standard Specification for Electrodeposited Coatings of Zinc on Iron and Steel
ASTM B 456	Standard Specification for Electrodeposited Coatings of Copper plus Nickel plus Chromium and Nickel plus Chromium
MIL-QQ-C-320	Chromium Plating (Electrodeposited)
MIL-A-8625	Anodic Coatings for Aluminum and Aluminum Alloys

Table 7 Recommended minimum thicknesses and typical applications for electrodeposited bright nickel-chromium coatings

Service conditions	Classification(a)		Minimum thickness				Typical applications
			Nickel		Chromium		
	Nickel	Chromium	µm	mils	µm	mils	
Steel, iron, or zinc die-cast substrates							
Mild (normally warm, dry, indoor atmospheres; minimum wear or abrasion)	b	r	10	0.4	0.1	0.004	Household appliances, interior auto hardware, hair dryers, fans, inexpensive cooking utensils, coat and luggage racks, standing ashtrays, interior trash receptacles, inexpensive light fixtures
	p	r	10	0.4	0.1	0.004	
	d	r	10	0.4	0.1	0.004	
	b	mc	10	0.4	0.8	0.03	
	p	mc	10	0.4	0.8	0.03	
	d	mc	10	0.4	0.8	0.03	
	b	mp	10	0.4	0.3	0.012	
	p	mp	10	0.4	0.3	0.012	
	d	mp	10	0.4	0.3	0.012	
Moderate (indoor exposure where condensation may occur, as in kitchens or bathrooms)	b	r	20	0.8	0.3	0.012	Steel and iron: stove tops, oven liners, home, office, and school furniture; bar stools, golf club shafts. Zinc alloys: bathroom accessories, cabinet hardware
	p	r	20	0.8	0.3	0.012	
	d	r	20	0.8	0.3	0.012	
	b	mc	15	0.6	0.8	0.03	
	p	mc	15	0.6	0.8	0.03	
	d	mc	15	0.6	0.8	0.03	
	b	mp	15	0.6	0.3	0.012	
	p	mp	15	0.6	0.3	0.012	
	d	mp	15	0.6	0.3	0.012	
Severe (occasional or frequent wetting by rain or dew; possible exposure to cleaners and saline solutions)	d	r	30	1.2	0.3	0.012	Patio, porch, and lawn furniture; bicycles; scooters; wagons; hospital furniture; fixtures; cabinets
	d	mc	20	0.8	0.8	0.03	
	d	mp	25	1.0	0.3	0.012	
	p	r	40	1.6	0.3	0.012	
	p	mc	30	1.2	0.8	0.03	
	p	mp	30	1.2	0.3	0.012	
Very severe (damage from wear or abrasion likely in addition to corrosive media)	d	r	40	1.6	0.3	0.012	Auto bumpers, grilles, hubcaps, and lower body trim; light housings
	d	mc	30	1.2	0.8	0.03	
	d	mp	30	1.2	0.3	0.012	
Copper and copper alloy substrates							
Mild (see above)	b	r	5	0.2	0.13	0.005	Household appliances, oven doors and liners, interior auto hardware, trim for major appliances, receptacles, light fixtures
	p	r	5	0.2	0.13	0.005	
	d	r	5	0.2	0.13	0.005	
	b	mc	5	0.2	0.8	0.03	
	p	mc	5	0.2	0.8	0.03	
	d	mc	5	0.2	0.8	0.03	
	b	mp	5	0.2	0.25	0.01	
	p	mp	5	0.2	0.25	0.01	
	d	mp	5	0.2	0.25	0.01	
Moderate (see above)	b	r	15	0.6	0.25	0.01	Plumbing fixtures, bathroom accessories, hinges, light fixtures, flashlights, and spotlights
	p	r	15	0.6	0.25	0.01	
	d	r	15	0.6	0.25	0.01	
	b	mc	10	0.4	0.8	0.03	
	p	mc	10	0.4	0.8	0.03	
	d	mc	10	0.4	0.8	0.03	
	b	mp	10	0.4	0.25	0.01	
	p	mp	10	0.4	0.25	0.01	
	d	mp	10	0.4	0.25	0.01	
Severe (see above)	d	r	25	1.0	0.25	0.01	Patio, porch, and lawn furniture; light fixtures; bicycle parts; hospital and laboratory fixtures
	d	mc	20	0.8	0.8	0.03	
	d	mp	20	0.8	0.3	0.012	
	p	r	25	1.0	0.3	0.012	
	p	mc	20	0.8	0.8	0.03	
	p	mp	20	0.8	0.25	0.01	
	d	r	30	1.2	0.25	0.01	
	d	mc	25	1.0	0.8	0.03	
	d	mp	25	1.0	0.25	0.01	
Very severe (see above)	d	r	30	1.2	0.25	0.01	Boat fittings, auto trim, hubcaps, lower body trim
	d	mc	25	1.0	0.8	0.03	
	d	mp	25	1.0	0.25	0.01	

(a) Nickel classifications: b, fully bright; p, dull or semibright; d, double layer or triple layer nickel coating, with the bottom layer containing less than 0.005% S and the top layer more than 0.04% S. If there are three layers, the middle layer should contain more sulfur than the top layer. Chromium classifications: r, regular (conventional) chromium; mc, microcracked chromium having more than 750 cracks/in.; mp, microporous chromium containing a minimum of 64 500 pores/in.^{2} that are visible to the unaided eye. Source: Ref 12

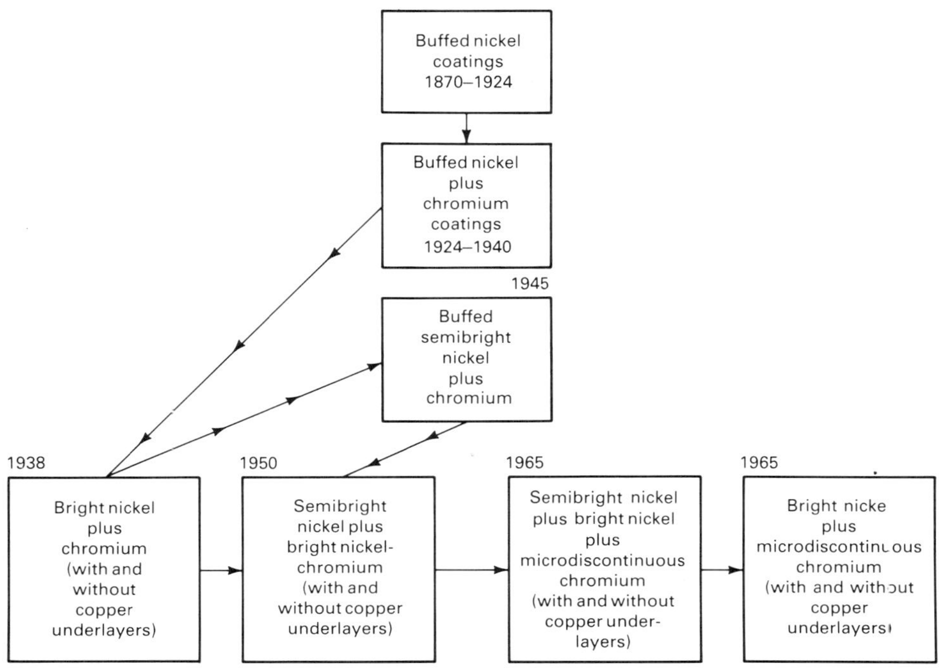

Fig. 8 History of the development of nickel and nickel-chromium coatings showing time of introduction and periods of use. Source: Ref 30

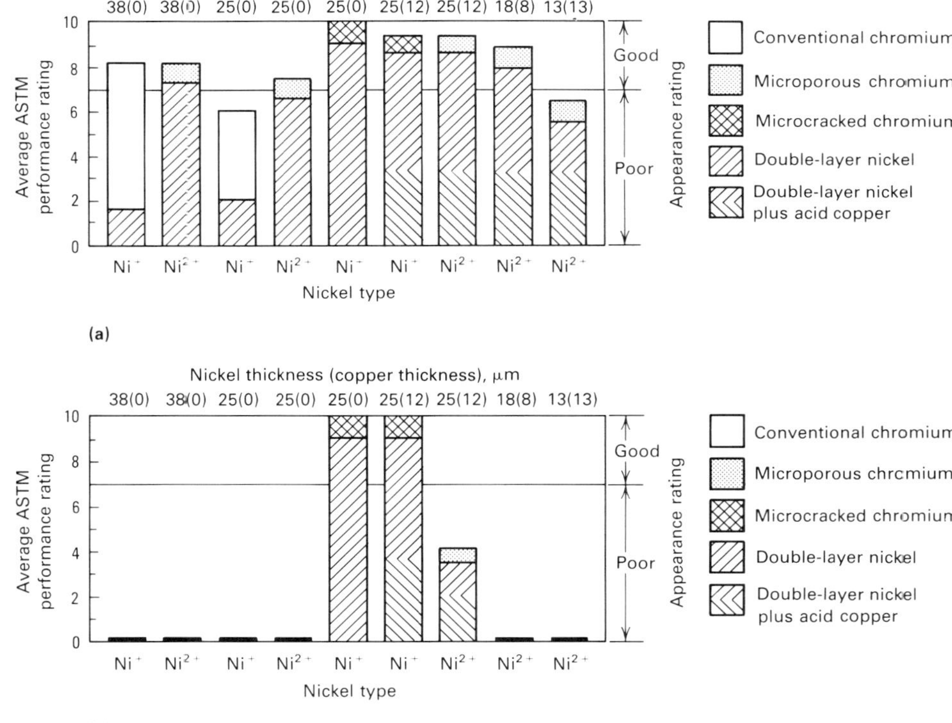

Fig. 9 Performance in a marine atmosphere of various types of nickel-chromium and copper-nickel-chromium coatings on flat (a) and contoured (b) steel panels. ASTM performance rating: 10, best; 0, worst. Test duration: 36 months. Source: Ref 30

International Standards Organization standards often govern international transactions involving the buying and selling of plated parts. These standards are available from the American National Standards Institute or from the International Standards Organization.

Nickel-Chromium Coatings. Because of their importance in the automotive industry, nickel-chromium (with or without a copper underlayer) coatings have been subjected to many changes and exhaustive testing over a long span of years. Results of extensive field experience and testing of variations dating from the days of buffed nickel coatings (1870 to 1924) are reported in Ref 30. Figure 8 shows a history of the development and use of nickel-chromium coatings. Significant developments in this history were the determination that the basic corrosion resistance of these systems is controlled by the thickness and composition of the nickel layer, the invention of the leveling copper and nickel plating processes, the development of semibright and bright nickel plating, the emergence of crack-free and microcracked chromium, and the evolution of duplex and triplex (multilayer) nickel-chromium systems.

Reference 30 also discusses the application of these coatings on steel, zinc, plastics, and aluminum. In these systems, the chromium layer is very thin—of the order of only 0.25 μm (0.01 mils). However, the corrosion resistance of chromium plate depends not so much on its thickness as on its physical state. Crack-free, or nonporous, chromium is excellent, but does not remain crack-free in service. A small number of cracks are detrimental, but the presence of many fine microcracks may be beneficial (Ref 2). Microcracked chromium deposits cause the galvanic-corrosion action to be spread over a very wide area and therefore do not suffer from localized corrosion. Microcracked chromium, in combination with duplex or triplex nickel, is the most durable of these coatings. The results of marine exposure for 36 months on variations of these coatings on steel are shown in Fig. 9. Figure 10 shows results of a similar program for nickel-chromium and copper-nickel-chromium coatings on zinc.

Plating of Mill Products

A trend toward applying tin and zinc coatings by electrodeposition to sheet and strip in tandem with continuous pickling and cold-rolling lines developed just before and during World War II. Much earlier, processes were developed in Germany and the United States to electrogalvanize wire. Advantages claimed over hot-dip galvanizing were control of thickness, the fact that the substrate is not affected by heat, and that the ductility and formability of the substrate is not limited by the iron-zinc intermetallic compound layer formed on hot-dip galvanized steel. Another advantage claimed for electrogalvanizing is that the thickness can be controlled, whereas in hot-dip galvanizing, the edges, especially of strip steel, tend to destroy wipes. Therefore, special care is required to secure a uniform smooth coating (Ref 31).

Electrolytic tinning was commercially introduced in 1927 (Ref 31). Before that time, all of the tinplate on the market was manufactured by hot-dip processes. Again, because of the amenability of electrodeposition to close thickness control and the development of highly efficient pretreating and plating processes, electrotinning has captured practically all of the food pack business. Various electrolytic processes have been used for producing tinplate, including the acid processes—sulfate, fluoborate, phenolsulfonic, and halide (fluoride-chloride)—and the alkaline sodium stannate bath.

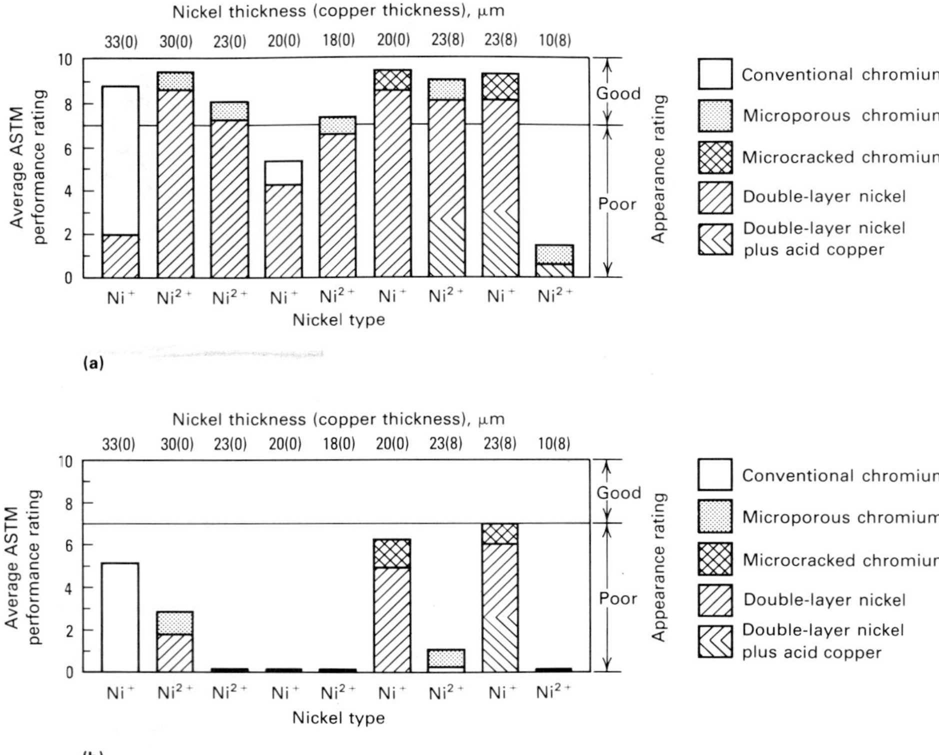

Fig. 10 Performance in a marine atmosphere of various types of nickel-chromium and copper-nickel-chromium coatings on flat (a) and contoured (b) zinc panels. ASTM performance rating: 10, best; 0, worst. Test duration: 36 months. Source: Ref 30

In addition to electrogalvanizing and electrotinning, continuous plating of cold-rolled sheet and strip is performed by integrated steel mills as well as specialty continuous strip producers and coil coaters. The specialty houses continuously plate nickel on steel for the battery (mostly alkaline dry cell) industry. They also plate alloy coatings such as nickel-zinc for protective/decorative uses and copper- and copper alloy-plated steel for postfabrication of decorative items. Copper-nickel-chromium on nonferrous metals and wide-strip steel is also available. Reference 32 lists typical examples of decorative/protective applications of continuously plated copper-nickel-chromium plated steel strip. Among these are various components for household appliances, dispensers, handles, lamps, cover plates for switches and outlet boxes, and furniture trim. Functional applications for copper-plated strip include flashlight shells and alligator clips for jumper cables. Nickel-plated strip is used in battery cans and caps and for paint brush ferrules. There are few companies in the continuous strip plating field because of the high initial investment required and the specialized knowlege needed to operate this process successfully.

Coil coaters, the name given to manufacturers of prepainted sheet and strip (mostly steel and aluminum substrates), apply thin zinc coatings plus chemical conversion coatings before applying organic coatings. Products of this segment of the industry are used in postforming operations.

Excellent control of plating conditions and the absence of complex substrate geometries provide a high degree of uniformity in coatings on preplated strip. This is reflected in predictable corrosion resistance because uniform thickness leads to uniform corrosion resistance.

Chromium plating has also penetrated the tin mill product field in such applications as food can ends. Thin coatings of tin and chromium act as a base for can varnishes and lacquers and for pigmented outer coatings. On the interior of cans, tin or chromium coatings are claimed to prevent filiform corrosion under the clear organic coatings—an insidious type of underfilm corrosion characteristic of clear and sparsely pigmented organic coatings. Information on equipment and processing is available in Ref 4, 31, and 33; Ref 33 also discusses corrosion of tinplate.

Immersion Plating

Immersion, or chemical, deposition has a long history of limited industrial use—for example, tin immersion deposits on mill wire products (called liquor finishing), immersion nickel plating of steel before porcelain enameling and many types of ceramic coatings, and the immersion copper plating of wire to be used in inert-gas shielded arc welding. In the last application mentioned, pure copper and copper-tin (gold colored) coatings have both been used as aids to wet-drawing of the wire to finished size and to furnish a highly conductive surface in the welding gun. Both tin and copper have been used in wire mills for the wire used for such products as paper clips and staples. According to the corrosion theory, these coatings, which are much more noble than the substrate, would be expected to hasten corrosion of the substrate. However, this is not the case, probably because the porosity is so widely distributed that whatever current is available for corrosion is so spread out as to be ineffective. This is similar to the effect of microdiscontinuous chromium (MDC).

Electrochemical Predictions of Corrosion Performance

Accelerated tests used for the prediction of corrosion performance for a particular application of a coating system have a certain risk associated with them. It is very difficult to design a test environment to simulate the actual environment, but there is often no alternative. A number of accelerated electrochemical tests are given in Ref 34 to 37.

Sample Preparation. The preparation of the sample to be tested requires a great deal of care for a number of reasons. On an actual part, the geometry is rarely designed so that the current density is uniform over the entire part. Because such coating properties as grain size and porosity are critically affected by local current density, the ability of a coating to protect the part would be expected to vary from one point on the part to another. When conducting accelerated tests on flat samples (usually required by the tests listed in Ref 34-37), the current density used to apply the coating(s) must be typical of that used on the real part. Equally important, the surface finish of the specimen substrate should be comparable to that of the actual substrate. The above tests usually require that the surface finish be produced by wet grinding on 600-grit SiC paper. Electrodeposited coatings are sometimes used without additional surface finishing. Grinding or certainly electropolishing will affect test results.

Test Selection. Very high localized currents can exist during these tests and may result in rapid removal of the coating. The desired information is usually the corrosion current i_{corr}, which can be correlated with actual weight loss. The polarization resistance measurement (Ref 34) is probably the safest of these tests for this determination. The suitability of the test should be determined by correlating actual performance with test results whenever possible. Crevice corrosion under the gaskets used may affect results and should be considered.

REFERENCES

1. "Standard Definitions of Terms Relating to Electroplating," B 374, *Annual Book of ASTM Standards*, American Society for Testing and Materials
2. C.L. Faust, "Electroplating," Course 22, Metals Engineering Institute, American Society for Metals, 1983
3. F.A. Lowenheim, Ed., *Modern Electroplating*, 3rd ed., John Wiley & Sons, 1974
4. L.J. Durney, Ed., *Electroplating Engineering Handbook*, 4th ed., Van Nostrand Reinhold, 1984
5. F.A. Lowenheim, *Electroplating*, McGraw-Hill, 1978
6. M. Murphy, Ed., *Metal Finishing Guidebook & Directory*, Metals and Plastics Publications, 1986
7. R.S. Treseder, Ed., *The NACE Corrosion Engineer's Reference Book*, National Association of Corrosion Engineers, 1980
8. G.T. Beilby, The Hard and Soft States in Metals, *J. Inst. of Met.*, No. 2, 1912, p 149-185

9. R.G. Baker, "Corrosion of Electrodeposited Coatings," Slide Lecture 46, American Electroplaters and Surface Finishers Society, 1981
10. "Standard Practice for Design of Articles to be Electroplated on Racks with Nickel," B507, *Annual Book of ASTM Standards*, American Society for Testing and Materials
11. J. Hyner, Design for Plating, in *Electroplating Engineering Handbook*, L.J. Durney, Ed., Van Nostrand Reinhold, 1984, p 50
12. *Quality Metal Finishing Guide*, Metal Finishing Suppliers' Association
13. "Standard Specification for Electrodeposited Coatings of Copper Plus Nickel Plus Chromium and Nickel Plus Chromium," B 456, *Annual Book of ASTM Standards*, American Society for Testing and Materials
14. S. Wernick and R. Pinner, *The Surface Treatment and Finishing of Aluminium and Its Alloys*, Robert Draper, 1972
15. "Method for Mechanical Hydrogen Embrittlement Testing of Plating Processes and Aircraft Maintenance Chemicals," F 519, *Annual Book of ASTM Standards*, American Society for Testing and Materials
16. "Standard Method of Salt Spray (Fog) Testing," B 117, *Annual Book of ASTM Standards*, American Society for Testing and Materials
17. "Thickness of Metallic and Inorganic Coatings," B 659, *Annual Book of ASTM Standards*, American Society for Testing and Materials
18. "Standard Specification for Electroplated Engineering Nickel Coatings," B 689, *Annual Book of ASTM Standards*, American Society for Testing and Materials
19. "Standard Specification for Electrodeposited Coatings of Tin-Nickel Alloy," B 605, *Annual Book of ASTM Standards*, American Society for Testing and Materials
20. "Standard Test Methods for Porosity in Gold Coatings on Metal Substrates," B 583, *Annual Book of ASTM Standards*, American Society for Testing and Materials
21. "Standard Specification for Electrodeposited Coatings of Tin," B 545, *Annual Book of ASTM Standards*, American Society for Testing and Materials
22. "Standard Test Methods for Adhesion of Metallic Coatings," B 571, *Annual Book of ASTM Standards*, American Society for Testing and Materials
23. "Standard Recommended Practice for Rating of Electroplated Panels Subjected to Atmospheric Exposure," B 537, *Annual Book of ASTM Standards*, American Society for Testing and Materials
24. "Standard Method of Testing FACT (Ford Anodized Aluminum Corrosion Test)," B 538, *Annual Book of ASTM Standards*, American Society for Testing and Materials
25. "Standard Methods for Corrosion Testing of Decorative Chromium Electroplating by the Corrodkote Procedure," B 380, *Annual Book of ASTM Standards*, American Society for Testing and Materials
26. "Standard Method for Copper-Accelerated Acetic Acid-Salt Spray (Fog) Testing (CASS Test)," B 368, *Annual Book of ASTM Standards*, American Society for Testing and Materials
27. "Standard Method of Acetic Acid-Salt Spray (Fog) Testing," B 287, *Annual Book of ASTM Standards*, American Society for Testing and Materials
28. L.L. Shreir, Ed., *Corrosion Control*, Vol 2, *Corrosion*, Newnes-Butterworths, 1976
29. *Department of Defense Index of Specifications and Standards*, DODISS, United States Navy, Naval Publications and Forms Center
30. G.A. DiBari, Decorative Electrodeposited Nickel-Chromium Coatings, *Met. Finish.*, Vol 75 (No. 6 and 7), June-July, 1977
31. H.E. McGannon, Ed., *The Making, Shaping and Treating of Steel*, United States Steel Company, 1964
32. H. Wettwer, Continuous Plating of Copper, Nickel and Chromium on Wide Steel Strip, *Plat. Surf. Finish.*, Vol 61 (No. 5), May 1974
33. C.H. Mathewson, *Zinc*, ACS Monograph Series, Reinhold, 1959
34. "Standard Practice for Conducting Potentiodynamic Polarization Measurements," G 59, *Annual Book of ASTM Standards*, American Society for Testing and Materials
35. "Standard Practice for Conducting Cyclic Potentiodynamic Polarization Measurements for Localized Corrosion," G 61, *Annual Book of ASTM Standards*, American Society for Testing and Materials
36. "Practice for Standard Reference Method for Making Potentiostatic and Potentiodynamic Anodic Polarization Measurements," G 5, *Annual Book of ASTM Standards*, American Society for Testing and Materials
37. "Standard Recommended Practice for Conventions Applicable to Electrochemical Measurements in Corrosion Testing," G 3, *Annual Book of ASTM Standards*, American Society for Testing and Materials

Hot Dip Coatings

HOT DIP COATING is a process in which a protective coating is applied to a metal by immersing it in a molten bath of the coating metal. Although hot dip coatings can be used to protect a number of metals, this article will consider only those used to protect steel.

Hot dip coatings have a number of advantages, including the ability to coat recessed or difficult areas (such as corners and edges) with a standard minimum coating thickness, resistance to mechanical damage (because the coating is metallurgically bonded to the steel), and good resistance to corrosion in a number of environments. However, the process has two limiting factors. First, the coating must melt at a reasonably low temperature, and second, the steel base metal must not undergo undesirable property changes during the coating process.

Hot dip coatings may be applied by continuous or batch processes. Materials such as steel sheet and wire may be hot dipped by continuously passing the steel through the molten metal bath. Continuous processing is highly automated and mechanized and is often associated with steel mill operations. The section "Continuous Hot Dip Coating" in this article discusses the corrosion resistance of zinc, aluminum, and zinc-aluminum alloy hot dip coatings on steel. Materials that are batch processed are generally fabricated before hot dipping, and the process may be performed manually or semiautomatically. Articles that can be hot dipped by the batch process range in size from large steel structural members to such small items as fasteners. The section "Hot Dip Galvanizing by the Batch Process" in this article provides details on the use of hot dip zinc coatings to protect steel. Other articles in this Volume that contain information on the use of hot dip coatings for corrosion prevention include "Corrosion of Zinc," "Corrosion in the Automotive Industry," and "Corrosion in Structures."

Continuous Hot Dip Coatings

A.J. Stavros
Union Carbide Corporation

Steel sheet and wire are coated by a continuous hot dip process; that is, they enter the coating bath in an unending strip. In theory, all continuous hot dip processes are similar in that the steel sheet or wire is subjected to successive cleaning, coating, and postcoating steps.

Typical cleaning steps may include alkaline cleaning or acid pickling (both of which may be electrolytic), oxidation (usually gaseous for sheet, but often in molten lead for wire), and reduction (gaseous). If a gaseous reduction is the final cleaning step, the steel must enter the molten coating bath directly without being exposed to air. When the final cleaning step is an acid pickle (this is usually the case for wire), the steel is then immersed in a liquid flux, which dissolves any remaining oxides, before entering the molten bath. Similarly, gaseous reduction can also be considered a flux treatment. All-gaseous cleaning is used on about 60% of the sheet-coating lines. The remaining 40% is approximately half liquid cleaning/flux and half liquid cleaning/gaseous cleaning. Most wire-coating lines use the liquid/flux technique.

The clean steel is then immersed in the molten coating bath long enough to allow the coating metal to wet and react with the steel surface. As the coated sheet or wire emerges from the molten bath, it pulls coating metal up from the surface, which can then be smoothed or wiped to the desired thickness by a variety of methods. Most sheet-coating lines use a gas-wiping technique in which a jet of steam, air, or gas (such as nitrogen) is directed against the emerging sheet. Wire galvanizers either contact the wire with solid devices lined with asbestos or similar materials to produce thin coatings or use floating mounds of charcoal for heavier coatings, although other techniques are also used.

The coated steel can be given any number of subsequent mechanical, thermal, or chemical posttreatments designed to impart specific properties. Typically, a coated sheet might be oiled or coated with a chromate solution to inhibit staining or superficial corrosion during storage and transit (see the article "Chromate Conversion Coatings" in this Volume). Waxing would serve the same purpose on wire and would facilitate handling during subsequent processing.

Zinc-Coated (Galvanized) Steel

Galvanized steel has been in use for over 100 years. Continuous galvanizing of sheet steel was introduced into the United States in 1936. In 1985, an estimated 5.9×10^6 Mg (6.5×10^6 tons) of galvanized sheet steel were produced. Because of its long history of use, galvanized steel has been exposed to, and studied in, a wide range of corrosive environments. An excellent summary of the behavior of galvanized steel is provided in Ref 1.

Microstructure of Galvanized Coatings

A hot dip zinc coating consists of a series of layers. Starting at the steel surface, each layer is an iron-zinc alloy with successively lower iron content until the outer layer of pure zinc is reached. In continuous galvanizing, however, it is typical to add about 0.1 to 0.2% Al, which suppresses the formation of the alloy layers so that the coating is mostly pure zinc (Fig. 1). Other alloying elements, in addition to aluminum, can influence appearance, mechanical properties, and durability. Additional information is available in Ref 2. Zinc coatings are sometimes given an additional in-line heat treatment to produce a fully alloyed structure; this is often desirable if paint adhesion is important.

Protection by Galvanized Coatings

Zinc coatings protect steel from corrosion by two methods. First, in many environments, zinc corrodes much more slowly than steel; thus, the zinc coating essentially forms a barrier between the steel and the corrosive environment. Second, zinc protects steel electrochemically. When zinc is coupled to steel, the steel is polarized to such a potential that it becomes the cathode of the steel-zinc couple and is immune to further corrosion for the life of the zinc. In practice, this means that steel exposed at a coating defect or at a cut edge will not rust until the nearby zinc is consumed.

Atmospheric Corrosion Resistance

The atmospheric corrosion resistance of galvanized steel has been monitored for many years in numerous environments. Zinc corrosion has been found to be approximately linear with time. Examples of such behavior are shown in Fig. 2, in which the years to initial rust are plotted against the average coating weight for galvanized sheet in three different atmospheres. Similar behavior is observed on wire (Fig. 3).

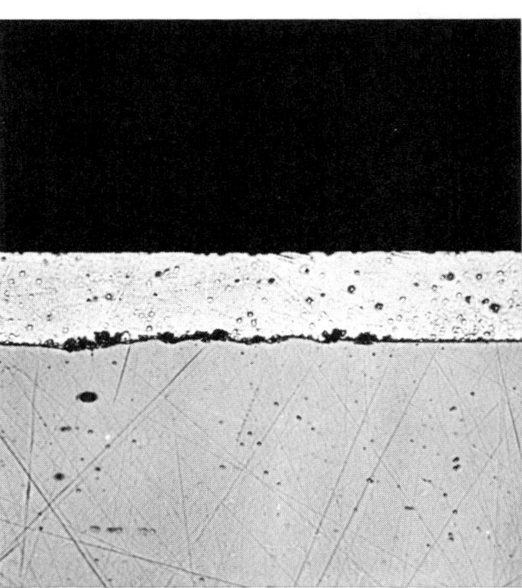

Fig. 1 Microstructure of continuously galvanized steel. In continuous hot dip galvanizing, the formation of various iron-zinc alloy layers is suppressed by the addition of 0.1 to 0.2% Al.

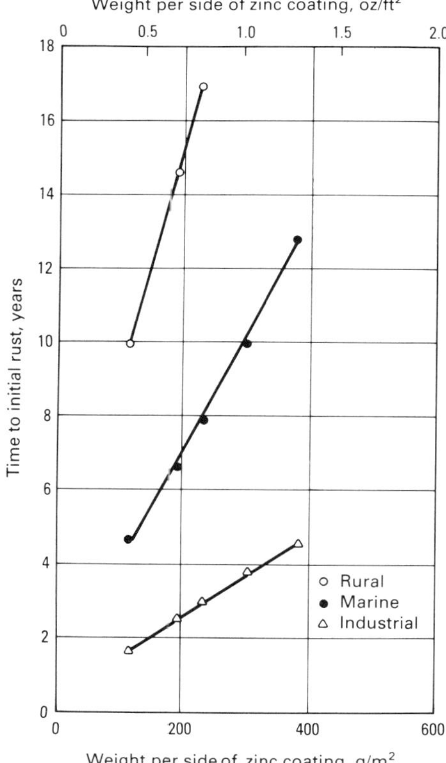

Fig. 2 Corrosion of galvanized steel in rural (State College, PA), marine (Sandy Hook, NJ), and industrial (Altoona, PA) atmospheres. Source: Ref 3

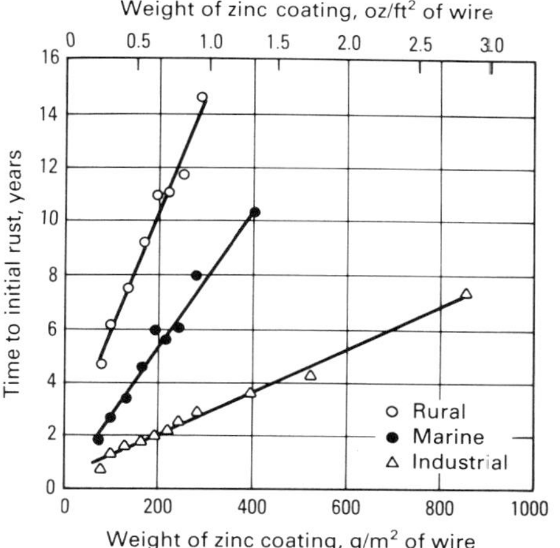

Fig. 3 Corrosion of galvanized wire in rural (State College, PA), marine (Sandy Hook, NJ), and industrial (Pittsburgh, PA) atmospheres. Source: Ref 4

Effects of Alloy Additions. There is good evidence that small amounts of alloying elements can affect the corrosion resistance of galvanized coatings. The mechanisms for these effects are complex and not fully understood, but current literature suggests that alloying elements segregate to grain boundaries, where they can accelerate or retard intergranular attack (Ref 5). Considering the Zn/0.1-0.2Al alloy typically used for continuous hot dip galvanizing, magnesium and copper tend to reduce intergranular corrosion, but bismuth, cadmium, lead, and tin increase attack. There is also some evidence that chromium, nickel, and titanium are beneficial.

Effect of Heat Treatment. Heating (galvannealing) a galvanized coating to produce an all iron-zinc alloy coating appears to be beneficial for corrosion resistance. Various investigators have reported increased atmospheric corrosion resistance of zinc coatings heat treated to give 10 to 15% Fe in the coating, but have not agreed on whether the effect was due to greater coating thickness resulting from zinc-iron diffusion or an inherent increase in corrosion resistance (Ref 6, 7).

Corrosion Mechanism. In dry air, zinc is oxidized to zinc oxide (ZnO). In the presence of moisture, ZnO reacts further to form zinc hydroxide ($Zn(OH)_2$). Neither of these corrosion products is very protective to the underlying zinc. However, $Zn(OH)_2$ can react further with carbon dioxide (CO_2) from the atmosphere to produce a basic zinc carbonate ($ZnCO_3$), which is protective.

In recent work, corrosion products identified on zinc coatings in industrial and marine atmospheres were mostly hydrozincite ($Zn_5(CO_3)_2(OH)_6$) (Ref 8). Zinc oxide was also present, as was a chloride in the marine environment. Surprisingly, zinc sulfate ($ZnSO_4$) was also found, even in the marine environment, indicating a definite role for sulfur dioxide (SO_2) in corrosion. Other investigations obtained a good correlation between the amount of SO_2 reacting with zinc and weight loss (Ref 9).

Aqueous Corrosion Resistance

Aqueous corrosion of galvanized coatings is largely determined by the impurities in the water. Water is usually classified into two major types: natural water and seawater.

The term natural water is used to describe groundwater surface water, and rainwater. Groundwater, by definition, flows from subsurface sources—springs and wells, for example. It often contains dissolved minerals, such as iron and manganese, salts of calcium and magnesium, and dissolved CO_2. Surface water refers to creeks and rivers (freshwater) and to runoff occurring after storms. It usually carries suspended organic and inorganic matter and many dissolved minerals picked up from soils. Even rainwater is not pure because the water droplets are often formed around dust and other particles suspended in the air. These droplets can absorb additional particles and dissolve atmospheric gases, such as oxygen, nitrogen, and CO_2, while falling to earth.

Seawater is of course characterized by its high sodium chloride (NaCl) content. It also contains many other salts.

Aqueous Corrosion Mechanism. Zinc corrodes in distilled water by the reaction:

$$Zn + 2H_2O \rightarrow Zn(OH)_2 + H_2$$

If oxygen is present, it combines with the evolved hydrogen to form water. This depolarizes and accelerates the corrosion reaction so that the corrosion rate in an oxygen-saturated solution depends on the diffusion of oxygen.

Corrosion in Natural Waters. Natural waters contain many impurities, and the reaction of zinc with these impurities largely controls the corrosion rate. The lowest corrosion rates are observed in hard waters in which calcium and magnesium salts can act as cathodic inhibitors. Their nearly insoluble hydroxides precipitate on cathodic regions, and this retards oxygen reduction. Also, the CO_2 in the water reacts with $Zn(OH)_2$ to form $ZnCO_3$. All of these corrosion products mix to form a very protective scale on the zinc surface. A typical corrosion rate in hard water is 0.25 mg/dm²/d, but in a soft water with no protective scale, zinc may corrode ten times as fast.

Carbon dioxide generally has an accelerating effect on corrosion in both hard and soft waters (much more so in soft water). Table 1 illustrates the result of increasing the CO_2 content on the corrosion rate of galvanized steel. Also shown are the results when the zinc overlay is removed and the underlying alloy layers are exposed directly. Although the corrosion rates of the alloy layers are not much different from the corrosion rate of pure zinc, the layers did not provide the same degree of galvanic protection to exposed steel.

Both pH and temperature also affect the corrosion of galvanized coatings in natural waters. Both effects are thought to result from changes in the protective scales that occur as these variables change. Generally, corrosion will be lowest in the pH range of 6 to 12.5 and will increase with temperature to a maximum near 70 °C (160 °F). The potential of the zinc electrode changes with increasing temperature; therefore, it may no longer provide galvanic protection to steel at higher temperatures.

Corrugated Steel Pipe. Perhaps the most extensive effort to quantify the effect of natural waters on the corrosion of galvanized steel was carried out to predict the durability of corrugated steel drainage pipe (Ref 11). Based on a survey of thousands of culverts in California, the time to perforation of 16-gage corrugated steel pipe was related to water pH and resistivity. An abbreviated diagram of the results is shown in Fig. 4, which also gives correction factors for steel thickness. Although the time to perforation includes the steel thickness as well as the galvanized thickness, the variation in corrosion rate with pH and resistivity is mostly due to zinc. A note of caution: The fact that this method seems to work in the geographic area from which the data were derived does not guarantee similar results in other areas. There are numerous instances in the literature in which the predictions from this diagram were not borne out by service records.

Corrosion in Seawater. Limited data have been generated on the resistance of galvanized coatings to seawater. It was quickly evident that zinc coatings are inadequate by themselves for long-term protection. The high chloride content of seawater suggests a much higher corrosion rate than freshwater, but this higher rate is not

Table 1 Corrosion of galvanized steel in water with varying CO_2 contents

CO_2 content, ppm	Corrosion rate, mg/dm²/d: Zinc overlay	Zn-4.8Fe layer	Zn-7.6Fe layer
0	1.0	...	1.1
5	2.3	2.6	2.6
18	3.3	7.0	5.4
37	4.2	11.3	9.1

Source: Ref 10

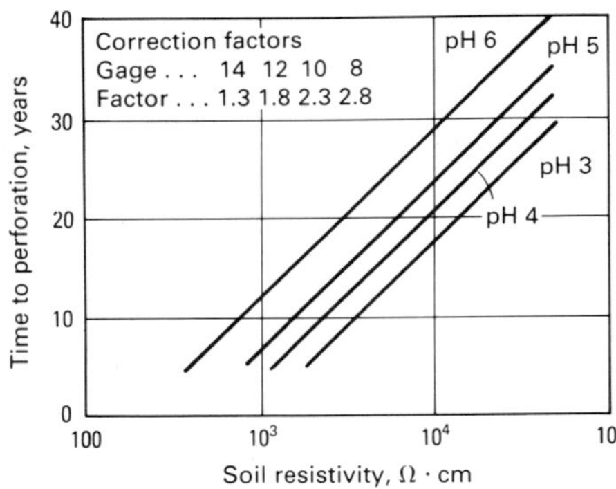

Fig. 4 Corrosion of 16-gage galvanized, corrugated steel pipe in soils of varying pH and resistivity. Correction factors for different steel thicknesses are also given. Source: Ref 11

obtained. Magnesium and calcium salts often inhibit corrosion sufficiently through scale formation to make seawater less corrosive than soft freshwater. Typical corrosion rates are 5 to 10 mg/dm²/d.

Corrosion in Soils

Soil corrosion is complex and not well understood. The physical and chemical characteristics of soils are important in determining corrosiveness. Coarse, sandy soils tend to permit easy access for air and water, but clay or silt show poor aeration and drainage. Considering the soluble chemicals that may be present, soil corrosivity can range from equivalent to atmospheric or aqueous media to virtually inert. Similarly, if conditions favor the formation of insoluble, protective films on the galvanized surface, a longer service life can be obtained.

The National Bureau of Standards studies initiated in the 1920s remain the primary source for underground corrosion data (Ref 12). A simplified summary for galvanized coatings is given in Table 2.

Aluminum-Coated (Aluminized) Steel

The desirability of combining the properties of aluminum and steel through coatings has been recognized since before the turn of the century. Because of the higher temperature and aggressiveness of molten aluminum and its tenacious oxide, hot dip aluminum coating of steel was much more difficult to put into practice than hot dip galvanizing. Although aluminizing remains a more demanding process than galvanizing, aluminum-coated steel is readily available throughout the world, with an estimated sheet production capacity of over 363 000 mg/yr (400 000 tons/yr) in the United States alone. Major uses include roofing panels, automotive exhaust components, and chain link fence.

Table 2 Corrosion of galvanized steel in various soils

Coating weight: 300 g/m² (1 oz/ft²)

Soil description	Soil corrosivity	Coating life, years
Poor drainage, low resistivity	High	1–2
Poor drainage, high salt content	Moderate	4–6
Good drainage, high resistivity, low salt content	Low	18–25

Source: Ref 13

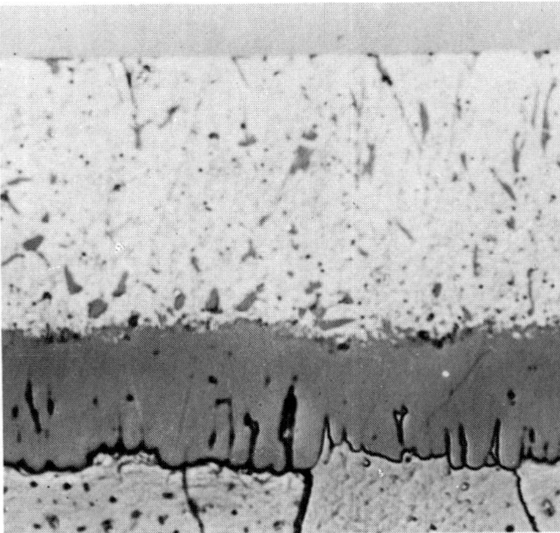

Fig. 5 Microstructure of type 2 aluminum coating on steel. This coating forms a layer of essentially pure aluminum (top) with scattered gray particles of aluminum-iron; the light gray center layer is aluminum-iron, and the bottom layer is the base steel. 1000×

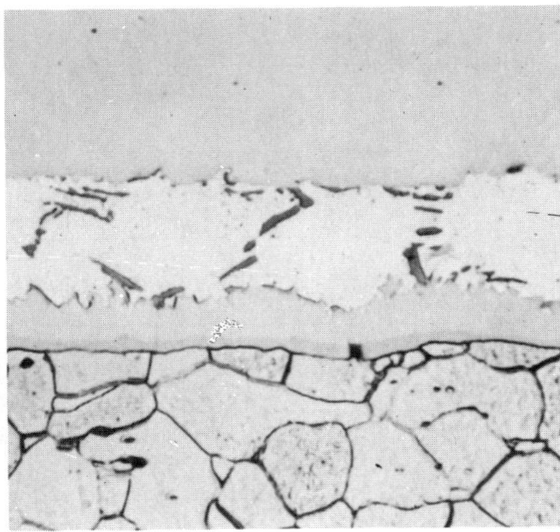

Fig. 6 Microstructure of type 1 aluminum coating on steel. From top: a nickel filler, aluminum-silicon alloy, aluminum-silicon-iron alloy, and steel base metal. 1000×

Microstructure of Aluminum Coatings

Two types of aluminum coatings are commercially significant. Type 2 uses commercially pure aluminum for the coating, and Type 1 uses an aluminum alloy containing 5 to 11% Si. The microstructure of the Type 2 coating shows a layer of aluminum, often with scattered iron-aluminum intermetallic particles, bonded to the steel substrate by an iron-aluminum intermetallic layer (Fig. 5). This intermetallic layer forms a distinctive serrated boundary with the steel and is generally identified as Fe_2Al_5, although some investigations have found additional iron-aluminum compounds (Ref 14).

When silicon is added to form a Type 1 coating, a different microstructure results. The intermetallic layer becomes narrower and smoother (Fig. 6), resulting in increased coating ductility relative to the Type 2 coating. With increasing silicon additions, the coating bath temperature can also be lowered, and the growth of the intermetallic layer is further inhibited.

Effect of Alloy Additions. Beryllium, copper, and some other elements have also been found to retard the growth of the intermetallic layer. Additions of these and other elements to the steel itself can also retard the growth of the alloy layer. The way in which these elements reduce the growth is not understood, although silicon appears to act after being incorporated into the intermetallic layer itself (Ref 15).

Protection by Aluminum Coatings

Aluminum coatings protect steel from attack by forming a very resistant barrier between the corrosive atmosphere and the steel. The aluminum oxide that forms on the aluminum surface is highly resistant to a wide range of environments.

In some aggressive environments containing chlorides, in which the aluminum oxide barrier is breached, the aluminum, like zinc, can also electrochemically protect steel from rusting. Because of this limited ability to protect steel galvanically, aluminum coatings are usually not used at very low thicknesses when the likelihood of coating defects is increased.

Atmospheric Corrosion Resistance

Numerous tests have shown that Type 1 and Type 2 aluminum coatings provide excellent atmospheric corrosion protection for steel. The major difference between the two coatings is the darkening and pinpoint rust formation that occur very early on Type 1 coatings. Weight loss measurements show that overall corrosion losses are nearly identical for both coatings (Ref 16). However, because of this appearance factor, the major use for Type 1 coatings is for high-temperature resistance, especially in automotive exhaust components. Typical atmospheric corrosion data for Type 2 aluminum-coated steel are given in Table 3. Galvanized steel data are included for comparison purposes. On the basis of this and similar investigations, a Type 2 aluminum coating will provide substantially longer life than a galvanized coating of equivalent thickness.

In 1961, the American Society for Testing and Materials (ASTM) exposed aluminum-coated wire (all hot dip coated wire is made with a Type 1 coating) and galvanized wire at marine, industrial, and rural atmosphere test sites. Results from unfabricated wire show that (Ref 17):

- Aluminum coatings are rust free 1.3 to 2.2 times longer than zinc coatings of the same thickness (43 μm) in industrial and marine atmospheres
- Zinc coatings are rust free longer at the rural test site, an unexpected result that is possibly explained by the nonuniformity of the aluminum coating. Fabricated wire products made from precoated wire showed virtually no difference between aluminum and zinc coatings

Corrosion Mechanism. Because contact with oxygen results in the formation of a protective

Table 3 Coating thickness losses for galvanized steel and type 2 aluminized steel in atmospheric exposure

Years exposed	Coating thickness loss: G90(a) µm	mils	Type 2 µm	mils	G90 µm	mils	Type 2 µm	mils
1	2.6	0.1	0.5	0.02	7.0	0.28	1.4	0.04
2	5.2	0.2	0.7	0.028	8.6	0.34	2.4	0.09
4	9.3	0.37	1.2	0.047	12.7	0.5	3.8	0.15
6	14.5	0.57	3.1	0.12	16.2	0.64	4.5	0.18
10	24.4	0.96	2.9	0.11	23.5	0.93	6.0	0.24
15	...	...	5.3	0.21	26.0	1.02	6.7	0.26

(a) G90 galvanized steel has a coating weight of 0.90 oz/ft^2 (270 g/m^2). Source: Ref 16

aluminum oxide layer, other components of the atmosphere must be responsible for the corrosion of the aluminum coating. Careful analysis of the corrosion products formed on aluminum coatings in industrial and marine atmospheres has shown that the predominant product is an amorphous, hydrated aluminum sulfate (Ref 8). This suggests that SO_2 is an important factor in atmospheric corrosion. This sulfate compound also appears to be more protective than the sulfates found on zinc coatings, which is consistent with observed corrosion behavior.

Aqueous Corrosion Resistance

Aluminum-coated steel has not generally been used in situations requiring aqueous corrosion resistance. There are no counterparts to galvanized water tanks, galvanized pipe, or galvanized pails. However, since 1979 the Type 2 aluminum coating has been used in the manufacture of corrugated steel pipe.

Corrosion in Natural Waters. Numerical data on the corrosion resistance of aluminum coatings are few because of the lack of use. Behavior is usually inferred from aluminum alloy data, but this could be inappropriate because of differences in structure and composition. Nonetheless, studies on aluminum alloys have shown that the air-formed aluminum oxide is destroyed after immersion (Ref 18). Corrosion resistance then depends on the oxide being re-formed from dissolved oxygen quicker than the aluminum is attacked by other ions, such as chlorides (Cl^-), nitrates (NO_3^-), or sulfates (SO_4^{2-}). With the protective aluminum oxide layer in place, the normal mode of failure is pitting, not general dissolution.

In general, soft waters are the least aggressive toward aluminum. The oxide is regarded as stable from pH 5 to 9. However, specific ions can change this range of stability. There is also some evidence that pitting may increase as the pH varies from neutral, but increasing the flow rate, especially at the more extreme pH values, can alleviate pitting. Small concentrations of copper and some other heavy metals can also lead to accelerated pitting. Pitting and corrosion have been observed in less than 3 years on aluminum-coated roofing panels exposed to stagnant water (Ref 19). On the other hand, when under a constant flow of relatively soft neutral pH water, corrugated steel pipe made from Type 2 aluminum-coated steel gives excellent service (Ref 20).

Corrosion in Seawater. Unlike aluminum alloys, aluminum-coated steel is not used in seawater. Corrosion rates are too high to provide economical use except as part of a more complex protection system. Typical thickness loss after 1 year of immersion in seawater has been reported to be 198 µm (7.8 mils) for Type 1 and 38 µm (1.5 mils) for Type 2 aluminum-coated sheets (Ref 16).

Corrosion in Soils

As with aqueous environments, aluminum-coated steel does not have a history of use in soils. Behavior would be expected to depend on pH, resistivity, and especially the chemistry of the soil. The mode of failure should be pitting. To date, no detrimental soil experience has been reported for the corrugated steel pipe use discussed previously.

Aluminum-Zinc Alloy Coated Steel

As discussed earlier, galvanized coatings provide an acceptable degree of resistance to atmospheric, aqueous, and soil corrosion. Because of galvanic protection, the rusting and staining of steel exposed at cut edges or coating defects are prevented until the nearby zinc is consumed. The major inadequacies of zinc coatings are the limited protection in the more severe environments and any significant high-temperature protection. Aluminum coatings have addressed these two factors. However, because they cannot provide cathodic protection to exposed steel in most environments, early rusting occurs at coating defects and cut edges. Even though this rusting seldom progresses, it precludes use where appearance is important. Aluminum coatings are also subject to crevice corrosion in marine environments.

Many attempts have been made to improve the corrosion resistance of both galvanized and aluminum coatings through alloying. Although combinations of these two elements with each other were known to provide an attractive degree of corrosion resistance, hot dip coatings did not become feasible until the discovery that silicon inhibits the rapid alloying reaction with steel (Ref 21). The 55Al-Zn (by weight, 55Al-43.4Zn-1.6Si) coating was first available commercially in 1972. This composition was selected from a systematic investigation of aluminum-zinc alloys, with up to 70% Al providing the best combination of galvanic protection and low corrosion rate. Current use varies from metal roofing, for which it is the major coated steel used, to automotive components, appliances, and, most recently, corrugated steel pipe. Production capacity of sheet in the United States in over 726 000 Mg (800 000 tons). The 55Al-Zn coating is now the fastest-growing hot dip coating worldwide in both production capacity and use.

Microstructure of 55Al-Zn Coating

The 55Al-Zn coating has a two-phase structure of cored aluminum-rich dendrites and a zinc-rich interdendritic constituent. This overlay is bonded to the steel substrate by a thin intermetallic layer whose composition is 48% Al, 24% Fe, 14% Zn, and 11% Si. X-ray diffraction suggests a structure similar to $Al_{13}Fe_4$. In addition, silicon particles are often found in the interdendritic region. By volume, the coating is approximately 80% Al + Si and 20% Zn. The effect of cooling rate during solidification is manifested in the spacing between the dendritic arms. Faster cooling (used commercially) results in finer spacing, which improves corrosion resistance.

Protection by Aluminum-Zinc Alloy Coatings

The 55Al-Zn coating provides both barrier and galvanic protection. Because the zinc-rich constituent is intimately distributed throughout the coating, it will be in contact with exposed steel at any break in the coating and at cut edges. Although less galvanic protection is available than with pure galvanized coatings, the alloy coating lasts longer because the overall corrosion rate is controlled by the aluminum-rich phase, which corrodes much more slowly than zinc.

Atmospheric Corrosion Resistance

Samples of 55Al-Zn-coated steel have been tested in atmospheric exposure for over 20 years. Figure 7 shows thickness loss with time for the first 13 years of exposure in four different atmospheres. Compared to galvanized panels exposed at the same time, the 55Al-Zn coating provides two to six times the corrosion resistance (based on equal coating thicknesses). Although these results were based on pilot line samples, subsequent testing of commercial 55Al-Zn sheet steel for 10 years shows a slightly greater advantage over galvanized steel (Ref 23).

Corrosion Mechanism. The zinc-rich constituent of the coating has been observed to corrode preferentially. As these regions are removed, their space is taken by corrosion products that become mechanically locked into the interdendritic spaces. These corrosion products are mostly amorphous aluminum or hydrated aluminum-zinc sulfates—similar to the corrosion products found on the coating surface of aluminum and 55Al-Zn coatings. These sulfates are adherent and may help explain the improved durability of the aluminum-zinc coating. Support for this mechanism is also obtained from aqueous corrosion studies in which the corrosion potential is observed to change that of a galvanized coating upon immersion to a value approaching that of aluminum after subsequent corrosion (Ref 24).

Aqueous Corrosion Resistance

The 55Al-Zn coating is finding increased use in applications demanding resistance to aqueous corrosion, especially where wet/dry cycles are obtained.

Corrosion in Natural Waters. As with other coatings, the corrosion of 55Al-Zn coating would be expected to vary with the specific properties of the water. It is not known how water hardness will affect corrosion, but in distilled water (very soft) and distilled water containing 85 mg/L of Cl^- ion, the 55Al-Zn coating is much more resistant than a galvanized coating (Table 4). In similar tests, 55Al-Zn and galvanized panels were immersed for 90 days in distilled water containing 45 ppm of SO_4^{2-} and 10 ppm of Cl^- at pH values from 3 to 11 (Ref 26). The pH was maintained

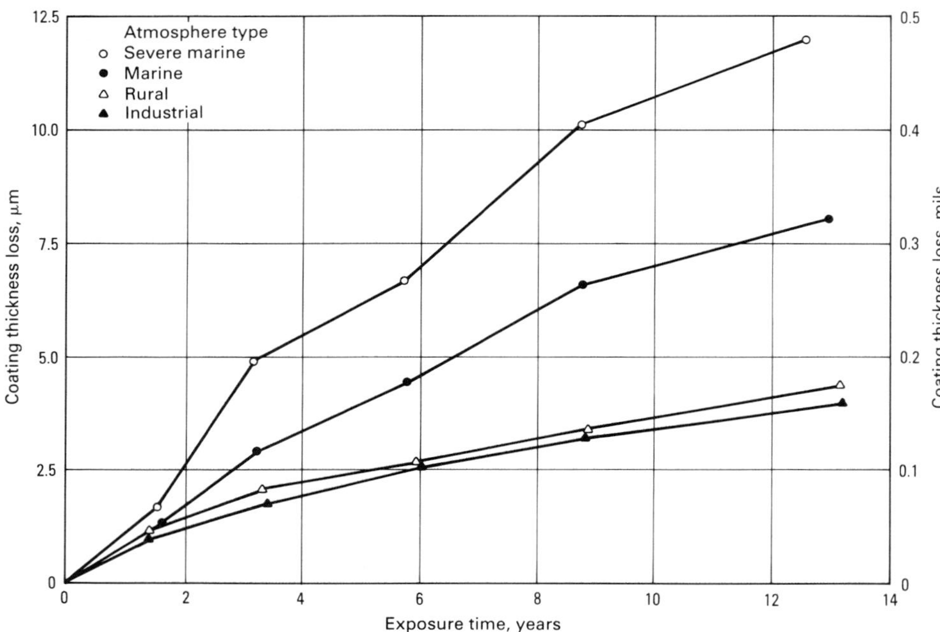

Fig. 7 Coating thickness loss of 55Al-Zn-coated steel in four atmospheres. Source: Ref 22

Table 4 Average coating thickness loss of galvanized and 55Al-Zn-coated steel after 56 days of immersion

Material	Thickness loss: Distilled water, μm	Distilled water, mils	85 mg/L NaCl, μm	85 mg/L NaCl, mils
Galvanized	1.06	0.042	1.26	0.049
55Al-Zn	0.015	0.0005	0.133	0.005

Source: Ref 25

Table 5 Coating weight losses of galvanized and 55Al-Zn-coated steels after a 90-day immersion in water of various pHs

pH	Coating weight loss, %: Galvanized	55Al-Zn
3	32	31
5	45	2
7	41	4
9	40	8
11	33	26

Source: Ref 26

through sulfuric acid (H_2SO_4) or sodium hydroxide (NaOH) additions. As Table 5 demonstrates, the 55Al-Zn retains more coating than the galvanized at all pH values, especially within the 5 to 9 range most characteristic of natural waters.

The longest service history of exposure to natural water for the 55Al-Zn coating is obtained from corrugated steel pipe installed between October, 1973, and October, 1974 (Ref 27). Water chemistry, pH, and resistivity varied widely from site to site and sometimes changed considerably with time. Erosion and abrasion caused additional wear factors.

Overall, 55Al-Zn applied at a coating weight of 180 g/m^2 provides greater durability than a 600-g/m^2 galvanized coating. Continued monitoring of these sites suggests an average of 10 years of additional life for the 55Al-Zn coating in the pipe inverts, the point of severest corrosion and wear.

Corrosion in Seawater. No results of seawater exposures have been published. It is unlikely that the 55Al-Zn coating would provide service that differs substantially from that of a galvanized or aluminum coating. Supplemental protection would be required for long-term use in seawater.

Corrosion in Soils

Again, there are few performance data for the 55Al-Zn coating in soil. The corrugated steel pipe exposures described previously provide the longest history of soil exposure, but because most culverts fail from the inside, exterior soil corrosion was not monitored closely in these tests. Some data are available from laboratory tests in which 16-gage coated steel panels were buried in test soils and monitored for coating loss (Ref 27). Figure 8 shows coating loss and the soil characteristics. These data suggest that the 55Al-Zn coating should provide corrosion resistance in soil similar to that of a galvanized coating.

Hot Dip Galvanizing by the Batch Process

J.W. Gambrell
American Hot Dip Galvanizers Association

The metallic coating produced by the hot dip galvanizing process is the result of a metallurgical reaction called diffusion. Diffusion occurs when the steel is immersed in the molten zinc. In this reaction, a series of intermetallic iron-zinc phases is formed as zinc diffuses inward and iron diffuses outward. The finished product consists of four layers on the steel; the outer layer is free zinc, and the inner three layers are separate intermetallic layers that are metallurgically bonded to each other and the steel. Thus, a galvanized coating has extremely high bond strength relative to other coatings, which rely on mechanical or weak electrical forces for bonding.

For the diffusion reaction to occur, the zinc must wet the steel surface. The presence on the steel surface of foreign materials, such as rust, scale, paints, or lubricants, can prevent the zinc from wetting and reacting with the steel surface; therefore, all contaminants must be removed before galvanizing.

The batch hot dip galvanizing process consists of two basic steps: surface preparation and immersion in the bath of molten zinc. Within each of these broad steps is a series of operations essential to the production of a quality galvanized coating.

Surface Preparation

Surface preparation essentially involves cleaning the steel. Dirt, oils, lubricants, greases, and other impurities, as well as normal oxidation products, such as mill scale or rust, may be deposited on the steel during fabrication, transportation, or storage. These impurities must be removed for the galvanizing reaction to occur.

Alkaline Degreasing. If required by the condition of the material, the first step in the process is treatment in an alkaline solution. Oils, greases, and other saponifiable compounds are removed in this step. Most of the various proprietary compounds available for this purpose consist of a mixture of basic sodium salts.

Acid Pickling. The steel article is usually rinsed after alkaline degreasing and before pickling to avoid neutralization and weakening of the pickle acid. The pickling process removes surface oxides and mill scale. The pickle acid is an aqueous solution of a mineral acid, usually hydrochloric acid (HCl) or sulfuric acid (H_2SO_4); acid concentrations and pickling temperatures vary from galvanizer to galvanizer, but are within a narrow range. An inhibitor is sometimes used to avoid attack of the base metal.

Abrasive Cleaning. As an alternative to alkaline degreasing and pickling, abrasive gritblasting may be used, followed by a flash pickle to remove any surface oxides that may develop between cleaning and further processing. Abrasive cleaning may not be practical for the galvanizer who handles a variety of sizes and shapes of material. Abrasive cleaning provides rapid and complete cleaning and is particularly effective in cleaning iron castings because casting sands are easily removed and surface uniformity is promoted. The method is also useful for cleaning fabricated products composed of different ferrous materials that may pickle at different rates or for removing weld slags or other impurities that are difficult to remove chemically. Abrasive cleaning may result in increased coating thicknesses because of changes in the steel surface profile (increased effective surface area and roughness).

Fluxing is required to dissolve any oxide films formed on the steel after pickling but before galvanizing and to ensure that a clean metal

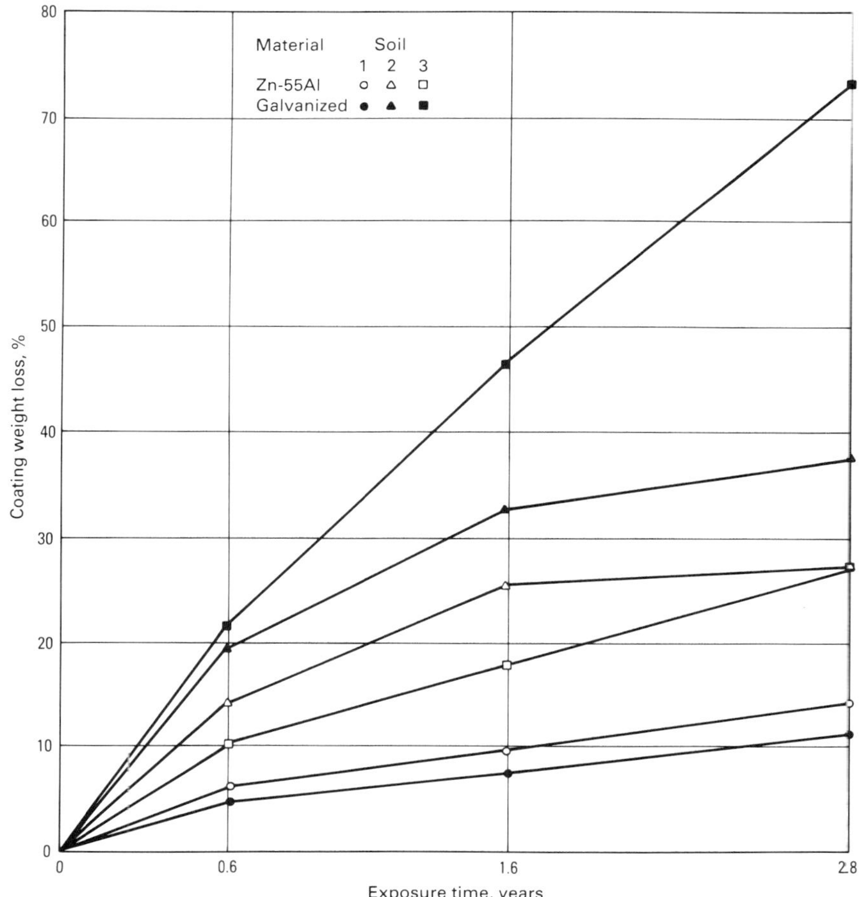

Soil number	Description	pH	Resistivity, Ω · cm
1	Native shale, clay; wet to dry	4	76 000
2	Native shale with chloride and sulfate salts; wet and dry	5	35 000
3	Native shale, clay, and bentonite with chloride and sulfate salts; wet	6	1700

Fig. 8 Corrosion of galvanized steel and 55Al-Zn-coated steel in three soils. Soil characteristics are a so given. Source: Ref 27

surface contacts the molten zinc. The fluxing procedures used include the wet process and the dry process. In the wet process, steel is passed through a layer of molten flux floating on the surface of the molten zinc. The dry process (prefluxing), involves immersion of the steel in an aqueous flux solution and drying before galvanizing. In both cases, a flux composed of zinc ammonium chlor de ($ZnCl_2 \cdot 2NH_4Cl$) is used. Because both processes have advantages, selection of a method is based on individual choice and the types of material to be processed.

Galvanizing

The molten zinc metal is contained in a kettle made of firebox-quality steel, although ceramic-lined kettles are increasing in use, particularly in Europe. The zinc is maintained at a temperature of 445 to 460 °C (830 to 860 °F). The kettle should be large enough to accommodate the material expected to be handled and should have sufficient heat content to prevent deep thermal cycling, with a resulting decrease in the kettle life.

The clean iron or steel is immersed (dipped) directly into the molten zinc or is passed through a molten flux blanket into the zinc. The material is immediately wetted, and as it reaches the temperature of the molten zinc, the diffusion reaction begins, resulting in the formation of a series of iron-zinc alloys. For most steels, the reaction is rapid at first, but it subsequently slows. The thickness of the coating does not increase substantially with longer immersion times.

The coating produced consists of a series of zinc-iron alloys, with increasing zinc content occurring toward the coating surface. When the material is withdrawn from the kettle, a thin layer of free zinc remains on the zinc-iron alloys, resulting in the characteristic bright, shiny, galvanized coating finish when allowed to cool quickly in air. The thickness of the free zinc layer varies with the speed of withdrawal from the molten bath.

Most galvanized articles are air cooled after coating, although centrifuged parts and smaller items are often water quenched. Water quenching halts the alloying reaction by diminishing the heat retained by the workpiece; this prevents the continued reaction that could convert the free zinc layer to iron-zinc alloy.

Coating Weight and Thickness

Galvanizing specifications in the United States require the coating to be measured in ounces per square foot. For wire, sheet, or small parts and fasteners, the procedure involves stripping the coating, determining the weight loss of the item, then calculating the coating weight. For large iron and steel structural materials, this may be impossible or impractical. The coating weight can be determined by measuring the coating thickness with a calibrated magnetic thickness gage. The thickness-to-weight conversion is 1 oz of zinc/ft^2 (305 g/m^2 of zinc) equals 1.7 mils (43 μm) of thickness.

Factors Affecting Coating Thickness and Structure

For most steels, the coating thickness and structure are relatively insensitive to variations in the galvanizing process. Variations in steel chemistry; zinc bath temperature; the physical condition of the steel surface (grain size, stresses, microstructure, roughness); zinc chemistry; the amount of wiping, shaking, or centrifuging; and the rate of cooling after galvanizing can effect the thickness, metallurgical characteristics, and the appearance of the coating.

Steel Chemistry. Although almost any steel can be galvanized, steel chemistry can have a marked effect on the thickness, structure, and appearance of the galvanized coating. Silicon, phosphorus, carbon, and manganese may be present in the steel and can influence the zinc-iron reaction mechanism, depending on their concentration.

The most influential steel constituent is silicon, which is added to steel as ferro-silicon to remove oxygen from the molten steel before casting. Figure 9 shows the effect of increasing silicon content in the steel versus time at two different galvanizing temperatures. In high-silicon (reactive) steels, the increase in coating thickness results from the accelerated growth of the zinc-iron alloy layers. This growth is due to the formation of loosely packed small grains or long-stem crystals in the outermost alloy layer of the coating, allowing zinc from the bath to penetrate to the steel surface, with continuation of the reaction. In general, steels with the following maximum impurity levels are best suited to galvanizing: 0.05% Si, 0.05% P, 0.25% C, and 1.3% Mn.

Zinc Bath Temperature. For normal steels, an increase in the zinc bath temperature, if it does not exceed the normal operating range, results in reduced immersion time and improved drainage of free zinc from the coating. The coating thickness for silicon-containing steels is very temperature dependent (Fig. 9).

Steel Surface Condition. The condition of the steel surface with respect to roughness and microstructure can affect the thickness of the galvanized coating. Roughening of the surface increases the profile and thus the effective surface area of the material, resulting in increased coating thickness. Fine-grain microstructures can result in lower coating weights (Ref 29). Coating thickness specifications generally provide for lower coating thicknesses for thinner steel sections.

Zinc Chemistry. Trace elements in the molten zinc, either present as impurities or intentionally added to the zinc to achieve certain beneficial results, can affect the appearance of the galvanized coating. Prime Western zinc, the grade normally used for batch process galvanizing, can contain up to 1.4% Pb, 0.2% Cd, 0.05% Fe, and must contain a minimum of 98% Zn.

Although the lead content of Prime Western zinc can be as high as 1.4%, a typical galvanizing bath will contain 1% or less. Lead has no effect on the viscosity of molten zinc, but it does affect the surface tension; this results in better drainage of zinc from the galvanized product, an important consideration when galvanizing complex shapes or wire fabrics.

Aluminum is usually not found in the grades of zinc normally used for galvanizing, but is occasionally added to improve the finish of the work. Aluminum at a concentration of 0.005% in zinc will result in coatings with improved brightness and uniformity. It also retards the surface oxidation of the molten zinc, resulting in a smoother flow of zinc over the work and improved drainage characteristics.

The typical galvanizing zinc bath may contain other trace elements that were originally present in the zinc or that resulted from the chemistry of the steel processed. These trace elements, including cadmium and copper, are usually present at too low a concentration to affect the coating structure or appearance.

Degree of Wiping, Shaking, or Centrifuging. These techniques are used for the efficient removal of excess zinc after galvanizing and result in a coating with a very uniform surface appearance. These techniques are particularly important when galvanizing threaded materials.

Rate of Cooling. Silicon-containing steels that are allowed to cool slowly after galvanizing will continue to react until the temperature decreases to below about 300 °C (570 °F). This may result in the iron-zinc alloy layers extending to the coating surface and an almost total absence of the free zinc layer. The result is a gray coating instead of the normal bright finish. The service life of the coating is not adversely affected and may in fact be greater with these coatings because of their increased thicknesses.

If the material is not water quenched after galvanizing, then close stacking of material while still hot should be avoided because of heat retention, which could allow the zinc-iron alloying reaction to continue. Material should be well spaced with good air circulation for cooling efficiency.

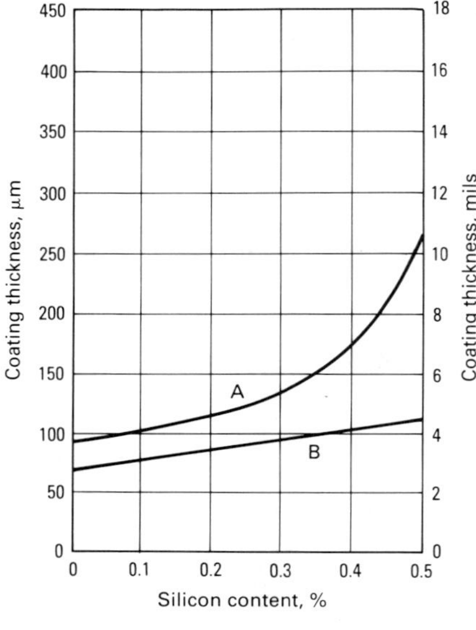

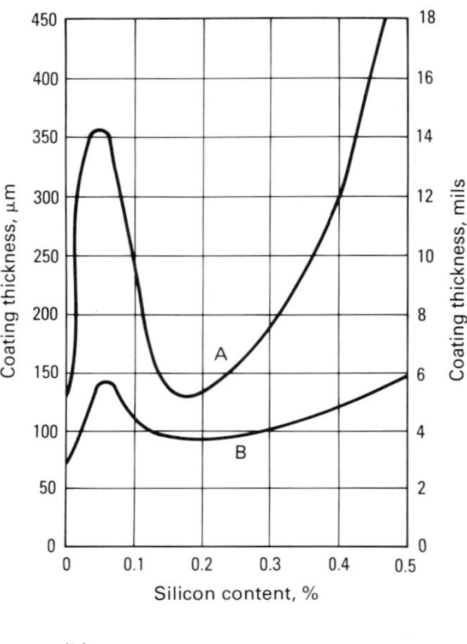

Fig. 9 Effect of silicon content of steel, immersion time, and galvanizing temperature on the thickness of hot dip galvanized coatings. Curves A and B are for 9-min and 3-min immersions, respectively. (a) Galvanizing temperature: 430 °C (805 °F). (b) Galvanizing temperature: 460 °C (860 °F). Source: Ref 28

Table 6 Properties of alloy layers of hot dip galvanized steels

Layer	Alloy	Iron, %	Melting point °C	Melting point °F	Crystal structure	Diamond pyramid microhardness	Alloy characteristics
Eta (η)	Zinc	0.03	419	787	Hexagonal	70–72	Soft, ductile
Zeta (ζ)	$FeZn_{13}$	5.7–6.3	530	986	Monoclinic	175–185	Hard, brittle
Delta (δ)	$FeZn_7$	7.0–11.0	530–670	986–1238	Hexagonal	240–300	Ductile
Gamma (γ)	Fe_3Zn_{10}	20.0–27.0	670–780	1238–1436	Cubic	...	Thin, hard, brittle
Steel base metal	Iron	...	1510	2750	Cubic	150–175	...

Mechanical Properties of the Coating and Steel Substrate

Hardness and Abrasion Resistance. The layers that compose the galvanized coating, being discrete zinc-iron alloys, vary in hardness. The free zinc layer (η) is relatively soft, but the alloy layers are very hard, harder even than ordinary structural steels. Typical values for the microhardness and other properties of the various alloy layers are given in Table 6.

The alloy layers are from four to six times more resistant to abrasion than pure zinc. Galvanized coatings exhibit better abrasion resistance compared to paints with the same coating thickness and can be effectively used where excessive abrasive wear is expected, for example, floor gratings, stairs, conveyors, and storage bins.

Adhesion and Impact Resistance. Unlike other coatings, which are mechanically or chemically bonded to the steel, the galvanized coating is metallurgically bonded to and integral with the steel, making conventional measures of bond strength inappropriate for this coating. As a result of the metallurgical bond, the galvanized coating is very adherent.

The structure of the galvanized coating, particularly the relative thicknesses of the δ and ζ layers (Table 2), is primarily influenced by the steel chemistry and, to a lesser extent, by the galvanizing temperature and the duration of immersion. Coating structure has the greatest effect on the impact resistance of the coating. A high relative proportion of ζ phase in the iron-zinc alloy may result in localized flaking if the coating is subjected to heavy impact or excessive twisting or bending. Semikilled steels with silicon contents of 0.05 to 0.12% are the most susceptible to coating brittleness and less adherent coatings.

Steel Strength and Ductility. The effects of hot dip galvanizing on the mechanical properties of steels have been thoroughly researched. This research has shown that galvanizing causes no significant alteration in the bend, tensile, and impact properties of the commonly galvanized steels produced throughout the world (Ref 30). Galvanizing reduces slightly the impact toughness of cold-rolled steels, but this reduction is generally not sufficient to affect the applicability of the steel.

Weld Stresses. Welded structures have higher strength in the galvanized condition than in the uncoated condition. Hot dip galvanizing reduces weld stresses by 50 to 60% (Ref 30).

Fatigue Strength. For most steels, little reduction in fatigue strength occurs as a result of galvanizing. Fatigue tests comparing hot dip galvanized steels with uncoated steels usually compare an uncoated steel with mill scale and the same steel after galvanizing. These tests do not consider that an uncoated steel exposed to an outside environment begins to rust immediately, with the resulting formation of pits that may be five to seven times deeper than the general corrosion and a rapid decrease in fatigue strength. The reduction in fatigue strength due to corrosion attack is much more significant than that due to galvanizing.

Embrittlement. It is relatively rare for a steel to be in the embrittled condition after hot dip galvanizing. The amount of cold working (plastic straining) of steels susceptible to strain aging is the single most important factor where strain-age embrittlement is of concern. The temperature of the galvanizing process accelerates the aging process; generally, susceptible materials fail almost immediately after removal from the bath. Research has shown that a minimum cold bend radius of three times the section thickness will preclude sufficient cold working and prevent the embrittled condition from occurring (Ref 29, 30).

If possible, steel fabrications should use steels with a low susceptibility to strain-age embrittlement. When cold working is necessary, the guidelines and limitations contained in ASTM A 143 (Ref 31) should be observed.

Hydrogen embrittlement does not occur when ordinary carbon or low-alloy steels are galvanized. Hardened steels may become embrittled if the hydrogen picked up during pickling is not expelled during immersion in the molten zinc bath. Precautions should be taken with steels having an ultimate tensile strength exceeding 1034 MPa (150 ksi). Steels of medium-to-high strength that have been severely cold worked may have local areas that also approach this threshold limit for susceptibility to hydrogen embrittlement.

Fabrication Details for Galvanizing

The design and fabrication of materials to be hot dip galvanized involve a few basic rules that are easy to apply and are necessary to the production of quality galvanized articles meeting intended functional requirements. Foremost among these rules is the interrelationship among the design engineer, fabricator, and galvanizer. Although this interrelationship becomes increasingly important with the complexity of the configuration of the material to be galvanized, most problems relating to design can be eliminated if it becomes an integral part of the design and fabrication program.

Steel Selection. As previously discussed, steel chemistry can affect the nature and extent of the zinc-iron alloys formed during galvanizing. Some steels that are commonly used for galvanized products are ASTM A36 (with $<0.05\%$ Si specified) for structural shapes, A120 for pipe, A569 for sheet, and A615 for reinforcing bar.

Combining Different Materials. Optimum galvanizing quality is seldom obtained when ferrous materials that have different chemistries, different surface conditions, or are constructed by different fabrication methods are combined. Different pickling and immersion times are required for different materials, and combinations may result in over- or undertreatment of the combined parts of the fabricated article. Different material types for which combinations should be avoided include excessively rusted items, material with machined or pitted surfaces, cast iron and steel, malleable iron, hot-rolled steel, cold-rolled steel, and reactive steels.

Castings. The cleaning of castings requires special attention. Burned-on sand from the casting process is difficult to remove by conventional HCl or H_2SO_4 pickling methods, and the use of hydrofluoric acid (HF), although effective, can be extremely dangerous to personnel. Castings should be shot- or gritblasted before delivery to the galvanizer.

Castings that consist of two or more different metals may pickle and galvanize differently. The designer and fabricator should be aware of this; abrasive cleaning may moderate these effects and provide for a more uniform coating appearance. Alternatively, the materials may be galvanized separately and assembled after galvanizing.

Surface contaminants resulting from fabrication should be removed before delivery of the material to the galvanizer, unless special arrangements have been made. Typical contaminants include welding slag and paints or other labeling compounds. Welding slag results from the use of coated welding rod. Weld slags are not removed by pickling, and if the use of uncoated rod is not possible, slags should be removed by mechanical methods. Paints or other labeling compounds, often used to identify the work, are sometimes difficult or impossible to remove by alkaline cleaning methods. Permanent markings for identification, if needed, should be punched or embossed in the steel, applied to the part as weld metal, or punched into 12-gage or thicker steel tags and wired securely to the part with heavy-gage steel wire.

Size of Material. The size of material to be galvanized is limited by the size of the galvanizing kettle and the capacity of the lifting equipment. If the material is too long or too deep to be completely immersed in the molten zinc in one dip, it may still be possible to galvanize by immersing one-half of the article in the kettle, withdrawing, then immersing the other half. By using this procedure, known as double end dipping, materials up to 75% longer than the kettle may be successfully galvanized, depending on the depth of the item and other kettle dimensions. Although double end dipping is an accepted galvanizing practice, special attention to galvanizing details and proper inspection are essential to its use on tubular items.

Threaded and Movable Parts. For threaded parts, such as nut and bolt assemblies, it is advisable to retap the nut after galvanizing to ensure a proper fit. For movable parts, such as shackles or shafts, a radial clearance of 1.6 mm ($^1/_{16}$ in.) is usually sufficient to allow free movement after galvanizing.

Drainage. When designing for hot dip galvanizing, consideration must be given by the designer and fabricator to providing a way for cleaning solutions and zinc to flow into, through, and out of the item so that all surfaces can be properly galvanized.

Venting of hollow objects is necessary for proper treatment of exterior and interior surfaces. More important, venting is essential for avoiding the extreme explosion hazard that can result from heating a closed or improperly vented hollow object to the galvanizing temperature. Discussions of proper venting techniques become very complex when intricate configurations are under consideration. Venting practices for various configurations, as well as additional fabrication techniques, are discussed in Ref 32 to 34.

Overlapping or contacting surfaces. When designing parts that are to be hot dip galvanized, it is best to avoid narrow gaps between plates, overlapping surfaces, and back-to-back angles and channels. When overlapping or contacting surfaces cannot be avoided, the edges should be completely sealed by welding; otherwise, pickling can enter and result in conditions that would permit the generation of high pressures and steam, delayed weeping, or improperly cleaned and galvanized areas.

Welding for Hot Dip Galvanizing. When materials are welded before hot dip galvanizing, the cleanliness of the weld area and the composition of the weld metal can affect the quality and appearance of the galvanized coating. Specific welding information can be obtained from the American Welding Society (Ref 35) or welding suppliers. Several welding techniques have been found to be satisfactory for materials to be galvanized:

- An uncoated electrode should be used to prevent flux deposits
- If a coated electrode is used, all flux residues must be removed, because they are chemically inert to pickling acids
- A welding process, such as metal inert gas, tungsten inert gas, or CO_2 shielded arc, should be used because virtually no slag is produced
- In the case of heavy weldments, a submerged arc method is recommended
- If none of the above is available, a coated rod specifically designed for self-slagging, as recommended by welding equipment distributors, should be selected
- A welding rod that will provide a deposited weld composition as close as possible to the parent steel should be chosen. This will prevent differential reaction rates during cleaning
- High-silicon welding rods may cause excessively high coating weights or darkened coatings

Galvanized Coatings in Corrosion Service

Protection Mechanism. Galvanized coatings protect steel in corrosion service in two ways: barrier protection and cathodic protection. Barrier protection is provided by the galvanized coating and is further enhanced by the formation of a thin, tightly adherent layer of zinc corrosion products on the coating surface. Upon initial weathering of a freshly galvanized surface, ZnO is formed, and it is converted to $ZnOH_2$ in the presence of moisture. Further reaction with CO_2 in the air results in the formation of basic $ZnCO_3$, which is relatively insoluble and impedes further corrosion. The gray patina normally associated with weathered galvanized coatings is the result of this thin layer of basic $ZnCO_3$.

Cathodic protection is provided to the steel by the fact that zinc is anodic to steel in most environments. Minor discontinuities or small areas of exposed steel resulting from drilled holes or cut edges are protected from corrosion by the sacrificial protection afforded by zinc. The corrosion products that result from this action provide further protection.

Atmospheric Exposure. Zinc, steel, and hot dip galvanized coatings have been the subject of long-term atmospheric studies conducted throughout the world (Ref 36, 37). From these studies, the behavior of these materials in a specific atmospheric environment can be reasonably estimated. An exact determination of corrosion behavior is complicated by several factors: the frequency and duration of exposure to moisture (rain, sleet, snow, and dew), the type and concentration of corrosive pollutants, the prevailing wind direction and velocity, and exposure to sea spray or windborne abrasives. All atmospheres contain some type of corrosive agent, and the concentration of these agents, as well as the frequency and duration of moisture contact, determines the corrosion rate of galvanized coatings.

Figure 10 shows the results of outdoor atmospheric-exposure tests designed to measure the protective life of galvanized coatings in various atmospheric environments. The sites were selected as representative of various broad environmental classifications: heavy industrial, moderate industrial (urban), suburban, rural, and marine. Within these broad classifications, the following factors are most significant in influencing the rate of corrosion of the galvanized coating.

Industrial and Urban Environments. Corrosive conditions are most pronounced in areas with highly developed industrial complexes that release sulfurous gases and corrosive fumes and mists to the atmosphere. These corrodents react with the normally impervious basic $ZnCO_3$ film to produce $ZnSO_4$ and other soluble zinc salts that, in the presence of moisture, are washed from the surface. This exposes fresh zinc to the atmospheres, and another corrosion cycle begins.

Rural and Suburban. The corrosion rate of zinc in these areas is relatively slow compared to that in industrial settings. Once the original weathering occurs, there is little in the atmosphere to convert the basic zinc salts to water-soluble compounds.

Marine Atmospheres. The corrosion rate of zinc and galvanized steel in marine atmospheres is influenced by several factors. Zinc forms a soluble corrosion product, zinc chloride ($ZnCl_2$), in marine atmospheres; therefore, the corrosion rate is influenced by salt spray, sea breezes, topography, and proximity to the coastline. For example, one investigation found that the corrosion rate for zinc exposed 24.4 m (80 ft) from the ocean was three times that for zinc exposed 244 m (800 ft) from the ocean (Ref 37).

The corrosion product formed in a given type of atmosphere (industrial, marine, and so on) determines the corrosion rate of zinc in that atmosphere. Results from exposures in a variety of atmospheres show that zinc is 20 to 30 times more resistant to corrosion than steel is (Table 7).

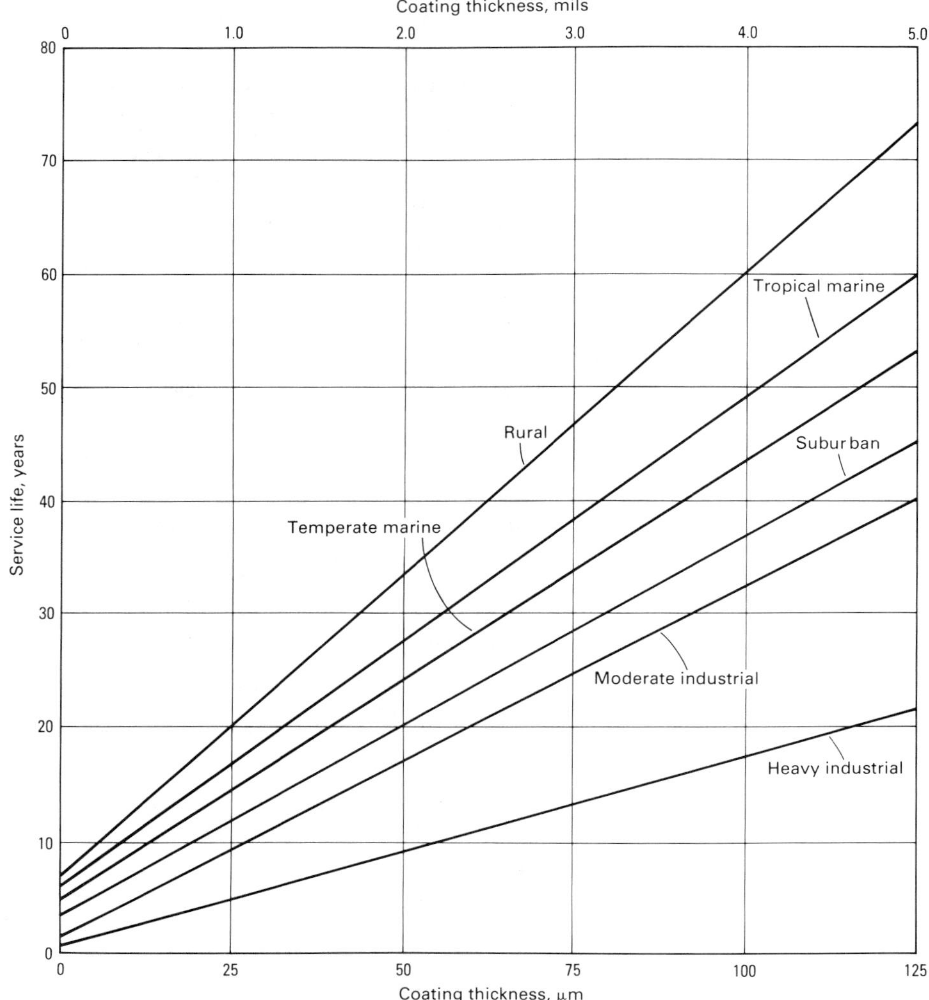

Fig. 10 Service life versus coating thickness for hot dip galvanized steel in various atmospheres. Source: Ref 36

Seawater and Salt Spray Performance. Table 8 gives the approximate corrosion rates of zinc in various waters, and Fig. 11 illustrates the expected service life of galvanized coatings in areas exposed to salt spray influences. Sea salts are mainly NaCl, with small amounts of calcium, magnesium, and manganese salts. Typical pH is about 8. Compared to other metals and alloys, galvanized coatings provide considerably more protection to steel than many other metals and alloys, even though the anticipated coating life is shorter in seawater and salt spray exposures than a number of others.

Freshwater Performance. The corrosion protection mechanism of zinc in freshwater is similar to that in atmospheric exposures. The corrosion rate depends on the ability of the coating to develop a protective layer of adherent basic zinc salts. This layer denies access to the coating by oxygen and slows the rate of attack. The ability of the water to form scale depends on a number of variables, such as the pH of the water, hardness, total alkalinity, and total dissolved solids. Table 9 demonstrates the effects of various water chemistries on the relative corrosion rates of zinc.

Water Temperature. The corrosion rate of zinc in water, and therefore that of the galvanized coating, increases with temperature to between 65 and 70 °C (150 and 160 °F), at which point the rate begins to decrease (Fig. 12).

Water pH. Zinc is an amphoteric metal with the capacity to passivate by means of protective layers. The corrosion rate of zinc decreases with increasing pH and reaches a minimum at 12.0 to 12.5. Most waters are in the pH range of 6 to 8. The scale-forming ability of the water and the concentration of dissolved ions in the water are more important influences on the corrosion rate than pH in this practical exposure range.

Performance in Soils. The corrosion rate and performance of galvanized steels in soils are a function of the type of soil in which the steel is located. Soils can vary considerably in composition and can contain bound and unbound salts, organic compounds, products of weathering, bacteria and other microorganisms, dissolved gases (such as hydrogen, oxygen, and methane), acids, and alkalies. Soils vary in permeability, depending on the soil structure. Although the concentration of oxygen is lower in soils than in air, the CO_2 concentration is higher. Variation among soils is high, and corrosion conditions are complicated.

In general, soils in coarse, open textures are often aerated, and the performance of galvanized steel would be expected to be similar to that in air. In soils with fine textures and high water-holding capacities, such as clay and silt-bearing soils, corrosion rates are likely to be higher.

Soil resistivity is recognized as a reliable method of predicting the corrosivity of soils. High-resistivity (poor conducting) soil would be less corrosive than low-resistivity (good conducting) soil. Dry soils are poor conductors and are the least corrosive to zinc.

Performance in Contact With Other Metals. The galvanic behavior of dissimilar metals in electrical contact is a complex corrosion problem, partly because of the large number of metals used in various applications. Copper and brass in contact with zinc or a galvanized coating in a moist environment can cause rapid galvanic corrosion of the zinc. Installations that use zinc and copper or copper alloys should provide for electrical insulation between the two, and because zinc may also be attacked by dissolved copper, copper materials should be located downstream of the zinc in nonrecirculating water systems. With aluminum or stainless steel, the performance of hot dip galvanized steel is good if the

Table 7 Weight losses of steel and zinc in various locations
Results are from 2-year atmospheric exposures.

Location	Weight loss, g: Zinc	Steel	Steel/zinc loss ratio
Norman Wells, N.W.T., Canada	0.07	0.73	10.4
Phoenix, AZ	0.13	2.23	17.2
Saskatoon, Sask., Canada	0.13	2.77	21.3
Esquimalt, Vancouver Is., Canada	0.21	6.50	31.0
Fort Amidor Pier, Panama C.Z.	0.28	7.10	25.4
Ottawa, Ontario, Canada	0.49	9.60	19.6
Miraflores, Panama C.Z.	0.50	20.90	41.8
Cape Kennedy, 0.8 km (0.5 miles) from ocean	0.50	42.0	84.0
State College, PA	0.51	11.17	21.9
Morenci, MI	0.53	7.03	13.3
Middletown, OH	0.54	14	25.9
Potter County, PA	0.55	10	18.2
Bethlehem, PA	0.57	18.30	32.1
Detroit, MI	0.58	7.03	12.1
Point Reyes, CA	0.67	244.0	364.2
Trail, B.C., Canada	0.70	16.90	24.1
Durham, NH	0.70	13.30	19.0
Halifax, NS (York Redoubt)	0.70	12.97	18.5
South Bend, PA	0.78	16.20	20.8
East Chicago, IN	0.79	41.10	52.0
Brazos River, TX	0.81	45.40	56.0
Monroeville, PA	0.84	23.80	28.3
Daytona Beach, FL	0.88	144.0	163.6
Kure Beach, NC (244-m, or 800-ft), site	0.89	71.0	79.8
Columbus, OH	0.95	16.00	16.8
Montreal, Quebec, Canada	1.05	11.44	10.9
Pittsburgh, PA	1.14	14.90	13.1
Waterbury, CN	1.12	11.00	9.8
Limon Bay, Panama C.Z.	1.17	30.30	25.9
Cleveland, OH	1.21	19.0	15.7
Newark, NJ	1.63	24.7	15.2
Cape Kennedy, 55 m (180 ft) from ocean			
Ground level	1.83	215.0	117.5
9 m (30 ft) elevation	1.77	80.2	45.3
18 m (60 ft) elevation	1.94	64.0	33.0
Bayonne, NJ	2.11	37.70	17.9
Kure Beach, NC (24.4-m (80-ft) site)	2.80	260.0	92.9
Halifax, NS (Federal Building)	3.27	55.30	16.9
Galeta Point, Panama C.Z.	6.80	336.0	49.4

environment is dry. In humid environments, isolation may be required.

Performance of Fully Alloyed Coatings. Coatings that consist entirely of iron-zinc alloy layers (without the free zinc η layer) result from the galvanizing of reactive steels. These coatings are characterized by a gray matte or mottled appearance. Fully alloyed coatings may exhibit premature rust-colored staining due to iron in the coating, but this should not be considered failure of the coating, because base metal rusting has not occurred. Fully alloyed coatings with thicknesses similar to normal coatings will give similar corrosion protection service.

High-Temperature Performance. When galvanized steel is continuously exposed to temperatures above about 200 °C (390 °F), the free zinc η layer may peel (Ref 41). The remaining alloy layers will not be affected and will continue to provide useful service life.

Joining of Galvanized Structural Members

Bolting is the most widely used method of joining galvanized steel. A number of standard galvanized fasteners are available for use with galvanized steel structural members (Table 10).

Galvanized coatings in bearing-type connections, which develop shear resistance by allowing the bolts to bear on the plates, are not detrimental to performance and have a long history of use—for example, in electrical utility transmission line towers. In friction-type connections, all loads in the plane of the joint are transferred by the friction developed between the connected surfaces. The load that can be transmitted is determined from the clamping force applied to the bolts and the coefficients of friction of the faying surfaces. Clean, galvanized mating surfaces have a coefficient of friction that is slightly lower than that of as-rolled steel. The coefficient of friction of galvanized surfaces can be made equal to that of uncoated steel surfaces by wire brushing or light grit blasting (Ref 40, 41). Neither treatment should be severe enough to produce breaks or discontinuities in the galvanized coating. The fatigue behavior of galvanized steel connections equals that of uncoated steel connections, regardless of whether the galvanized surfaces were wire brushed or gritblasted after galvanizing (Ref 40, 41, 42).

Welding. Zinc-coated steel can be satisfactorily welded by all common welding methods, but attention must be given to the possible generation of zinc fumes. Adequate ventilation, operator respiration units, or a fume-extracting welding unit should be used to avoid potential harmful effects. Extensive tensile, bend, radiographic, and fatigue tests show that the properties of sound metal inert gas or metallic arc welds on galvanized steel are equivalent to those on uncoated steel (Ref 35).

Table 8 Corrosion of zinc in various waters

Water type	Approximate material loss: μm/yr	mils/yr
Seawater		
Global oceans, average	15–25	0.6–1.0
North Sea	12	0.5
Baltic Sea and Gulf of Bothnia	10	0.4
Freshwater		
Hard	2.5–5	0.1–0.2
Soft River Water	20	0.8
Soft Tap Water	5–10	0.2–0.4
Distilled Water	50–200	2.0–8.0

Source: Ref 38

Penetration of molten zinc into the weld metal is the primary factor in cracking of galvanized steel weldments. The crack begins at the root of the weld and may or may not extend to the surface. Recommendations to avoid fillet weld cracking include (Ref 35):

- Treat the base metal to reduce the amount of available zinc by, for example, beveling the standing plate in a tee joint at an angle of 15 to 45°
- Remove zinc from both faying surfaces by burning with an oxygen fuel gas torch or by shotblasting
- Provide a parallel gap of 1.6 mm (1/16 in.) between the weld elements
- Choose consumables that will give a low silicon content weld, for example, manganese silicate flux and 2% low manganese/low silicon electrodes for submerged arc welding. For CO_2 welding, low-silicon filler wire gives freedom from zinc penetration cracking, but causes a small amount of porosity. For shielded arc welds, use a low-silicon electrode and a rutile covering

Weld porosity can occur because of the volatilization of zinc during welding. Porosity can be reduced by making adequate provision for the escape of gases evolved during welding. Normal porosity does not reduce the static tensile strength below that specified for satisfactory low-carbon steel weld metal, but it will affect the fatigue strength of a fillet weld.

Weld Damage Repair. Galvanized materials damaged by field welding can be touched in with an organic or inorganic zinc-rich paint. Organic paint does not require a high degree of surface preparation and dries quickly. The paint must have a zinc dust content of 95% and should be applied in several coats to a thickness of three times the galvanized thickness to provide equivalent protection.

Low-melting zinc-cadmium or zinc-tin-lead alloy rods can also be used for repair. The rod is heated to about 330 °C (625 °F) with an oxyacetylene torch, and the melted alloy is rubbed and spread over the damaged area. This gives equal protection, but is more expensive than paint touch up. The damaged area can also be zinc metallized by thermal spraying, but the equipment requirements may prevent convenient field use. Additional information on repair techniques is given in Ref 43.

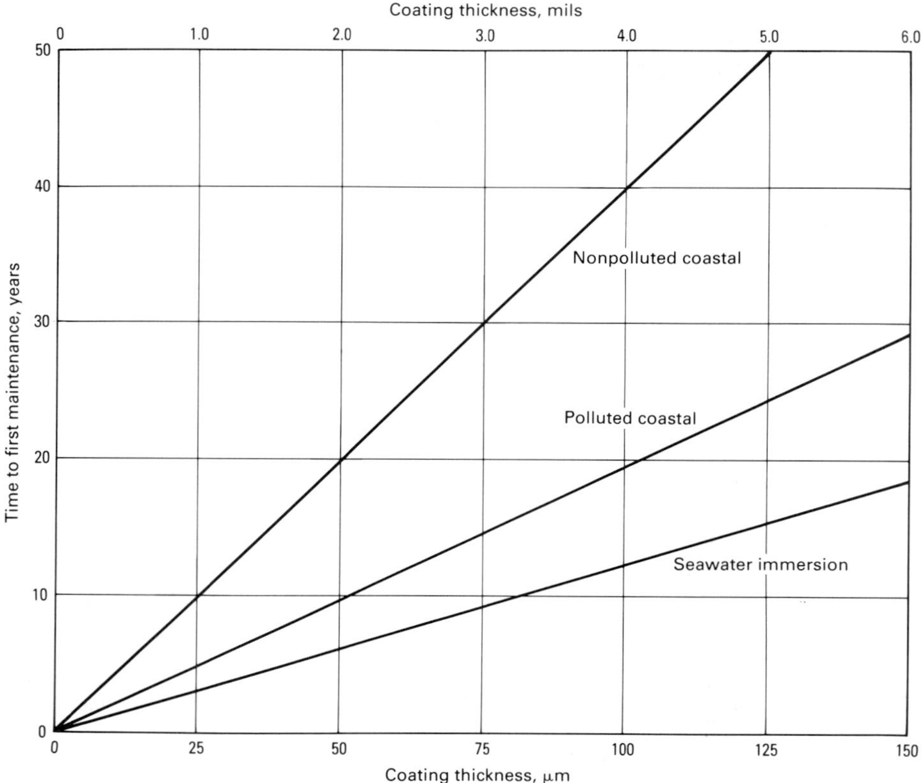

Fig. 11 Time to first maintenance versus coating thickness for hot dip galvanized coatings in seawater immersion and sea spray exposures. Source: Ref 39

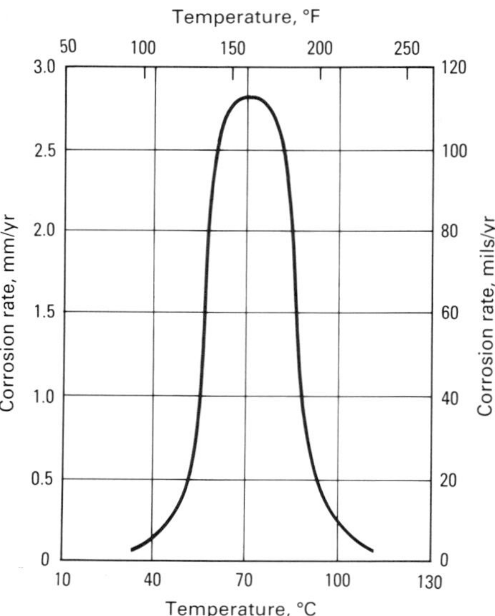

Fig. 12 Influence of water temperature on the corrosion rate of zinc in distilled, aerated water. Source: Ref 40

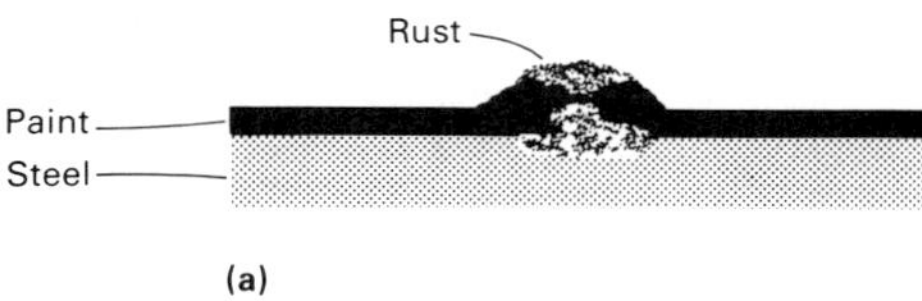

(a)

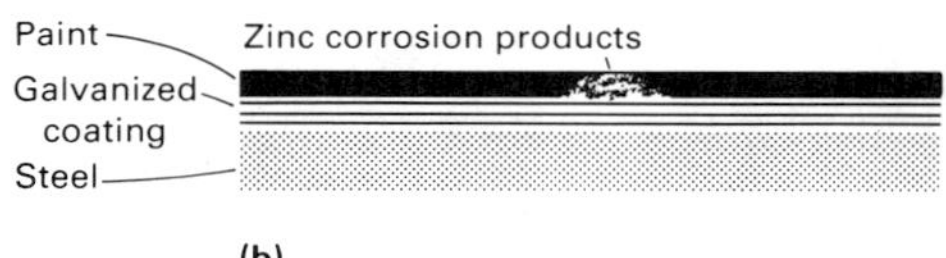

(b)

Fig. 13 Illustration of the mechanism of corrosion for painted steel (a) and painted galvanized steel (b). (a) A void in the paint results in rusting of the steel, which undercuts the paint coating and results in further coating degradation. (b) A void in the coating of a painted galvanized steel is sealed with zinc corrosion products; this avoids the undercutting seen in (a) and prevents further deterioration of the painted coating.

Painting Galvanized Steel

Galvanized coatings, when used without further treatment, offer the most economical corrosion protection for steel in many environments. The galvanized coating makes an excellent base on which to develop a paint system. Painting of galvanized steel is desirable for aesthetics, as camouflage, as warning or identification markings, to prevent bimetallic corrosion, or when the anticipated environment is particularly severe (see the article "Organic Coatings and Linings" in this Volume for supplementary information on painting for corrosion protection).

In corrosive atmospheres, a duplex system of galvanized steel top coated with paint has several advantages that make it an excellent system for corrosion prevention:

- The life of the galvanized coating is extended by the paint coating
- The sacrificial and barrier properties of the zinc coating are used if a break occurs in the paint film
- Undercutting of damaged paint coatings, a major cause of failure of paints on steel, does not occur with a zinc substrate (Fig. 13)
- Surface preparation of a weathered zinc surface for maintenance painting is easier than that for rusted steel

Synergistic Effects of Galvanized and Painted Systems. The galvanized coating prevents rusting of steel by acting as a barrier against the environment and by sacrificially corroding to provide cathodic protection. Painting the galvanized coating extends the service life of the underlying zinc because the barrier property of the paint delays the reaction of zinc with the environment. If a crack or other void occurs in the paint and exposes the galvanized coating, the zinc corrosion products formed tend to fill and seal the void; this delays further reaction.

When painted steel is exposed to the environment, rust forms at the steel/paint interface. Because rust occupies a volume several times that of the steel, the expansion resulting from rusting leads to rupture of the steel/paint bond. Further, rust is porous; it accumulates moisture and other reactants, and this increases the rate of attack on the steel. The result is undercutting, flaking, and blistering of the paint film, leading to failure of the paint coating (Fig. 13). Zinc corrosion products occupy a volume only slightly greater (20 to 25%) than zinc; this reduces the expansive forces and conditions that lead to paint failure.

A coating system consisting of painted galvanized steel provides a protective service life up to 1.5× that predicted by adding the expected lifetimes of the paint and the galvanized coating in a severe atmosphere (Ref 44). This is demonstrated in Table 11. The synergistic improvement is even greater for mild environments (Ref 45, 46).

Paint adhesion is the primary concern in painting galvanized steel. The surface of the zinc is nonporous and does not allow mechanical adhesion of the paint. Surface contaminants, such as oils, waxes, or postgalvanizing treatments, also effect adhesion. A fresh zinc surface is reactive to certain paint ingredients, such as fatty acids; this can produce zinc soaps and disrupt the zinc-paint bond.

Galvanized coatings can be successfully painted immediately after galvanizing or after extended weathering. The deliberate use of weathering is not recommended, because weathering may not be uniform, the time required is long (6 to 12 months), hygroscopic impurities can form that may be difficult to remove, and there is exposure to atmospheric pollutants.

Chemical etchants, such as acids or copper sulfate, should not be used. The action of these chemicals is difficult to control, surface preparation may be nonuniform, and the galvanized coating could be damaged if allowed to remain in extended contact with the chemicals. Long-term adhesion will suffer with this type of treatment, although initial adhesion may be obtained.

Mechanical roughening of the zinc surface through the use of a light blast can provide a good surface for painting. However, careful control of the blast pressure and flow rate must be exercised

Table 9 Corrosion of zinc in different types of water

Water type	Attacking substances	Passivating substances	Properties of corrosion products: Solubility	Adhesion	Relative corrosion rate
Hard water	Oxygen, CO_2	Calcium, magnesium	Very low	Very good	Very low
Seawater	Oxygen, CO_2, Cl^-	Calcium, magnesium	Low	Very good	Moderate
Soft with free air supply	Oxygen, CO_2	...	High	Good	High
Soft or distilled with poor air supply	Oxygen	...	Very high	Very poor	Very high

Source: Ref 38

Table 10 Some standard galvanized fasteners

Description	ASTM specification	Grade or type
Carbon steel bolts	A 307	Grade A, B
High-strength bolts	A 325	Type 1
Transmission tower bolts	A 394	Type 0, 1
Quenched-and-tempered alloy steel bolts	A 354	Grade BC

to avoid excessive removal of the galvanized coating.

Initial adhesion of the paint can be achieved through the use of a pretreatment primer to provide an adequate base for further coating. Long-term adhesion is obtained by the selection of top coat that is compatible with the primer and galvanized steel.

Pretreatment Methods. As with all painting operations, the surface to be painted must be free of contaminants that could affect the adhesion of the paint or the appearance of the painted article. Non-oily substances can be removed by brushing or scrubbing, then rinsing with water. Oily materials should be removed with a solvent, such as naphtha, turpentine, or mineral spirits, followed by a final wiping with clean cloths and clean solvent, to avoid spreading oil films on the surface.

Two-component wash primers meeting Steel Structures Painting Council (SSPC) specification SSPC-Paint 27 are ideal pretreatments. The primer is applied by spraying in thin coats according to the recommendations of the manufacturer. The freshly prepared primer should be applied to a dry film thickness of 7.6 to 13 μm (0.3 to 0.5 mils).

Finish Coating. The primed material should be finish coated as soon as possible after priming treatment. Wash primers are moisture sensitive, and the vehicle may gel under high-humidity conditions and lose adhesion. The primer usually dries in 15 to 30 min, and it is dry enough to recoat after 30 to 60 min. Almost all paints will adhere to the wash primer.

Direct Application Systems. A more convenient alternative to natural weathering or pretreatment of the galvanized surface is the use of a paint system directly compatible with the surface. Alternatives are discussed below.

Zinc Dust-Zinc Oxide Paints. Federal specification TT-P-641G describes three types of zinc dust-zinc oxide paints. All three contain the same pigmentation of 4 parts zinc dust to 1 part zinc oxide; the only variation is in the paint base material. Types 1 and 2 (linseed oil and alkyd resin bases, respectively) are recommended for general use, and type 3 (phenolic resin base) is especially formulated for severe moisture exposure or underwater service. All are useful as primers for adherence and are satisfactory as finish coats. If color is required, the top coat can be pigmented. Other compatible top coats may be used, but those with very strong solvent systems should be avoided, particularly if used before proper aging of the base film.

Portland Cement in Oil Paints. These paints are compatible with either fresh or weathered galvanized coatings. Although they tend to be brittle, adherence is excellent. They are not as versatile as the zinc dust-zinc oxide paints and are usually limited to applications in which a high gloss is not required and an oil base is suitable. They are available in a wide choice of colors.

Other Direct Application Systems. Newer paint systems that have been successfully used on direct application to galvanized steel include epoxy resin based paints, chlorinated rubber based paints, vinyl copolymer based paints, coal tar epoxy paints, and acrylic latex emulsion paints.

Economics of Hot Dip Galvanizing

Corrosion control results in a negative cash flow for the owner of any facility. In selecting a system for corrosion control, the economic consequences of the selection should be determined. Initial cost should not be the determining factor in selecting a system. Instead, the desired service life of the project should be reasonably estimated, and the life cycle cost of several systems should be evaluated to determine the most economical system for the particular project.

A number of models have been developed for the economic evaluation of corrosion protection systems (Ref 47-49). The projected structure life, inflation and discount (cost of capital) rates, number of years in the maintenance cycle, estimated costs for future maintenance, and original system costs must be determined based on discounted cash flow techniques. In the private sector, tax and investment incentives must also be considered. Table 12 illustrates a simple analysis that does not take tax or investment incentives into account. Although the initial galvanizing cost is set at 25% above the paint cost, this is often not the case; in reality, the cost of galvanized steel is frequently lower than that of painted steel. More information on the use of engineering economy is available in the article "Corrosion Economic Calculations" in this Volume.

Selected Applications of Hot Dip Galvanized Steel

Hot dip galvanized coatings are found in a wide variety of applications requiring long-term maintenance-free corrosion protection. The examples in this section will demonstrate the scope of galvanized steel use.

Bridges. The first all hot dip galvanized bridge in the United States was erected in 1966 at Stearns Bayou, Ottawa County, MI. The bridge is 128 m (420 ft) long and has a 9.1-m (30-ft) wide roadway with a 1.5-m (5-ft) walkway on each side. All the components of this bridge were hot dip galvanized. To avoid possible effects from road salting, telescoping splash plates in the joints were installed to divert deck drainage away from the beams.

When inspected in 1986, the average coating thickness on stringers and diaphragms was 112 μm (4.4 mils), which is enough to last another 60 to 100 years. The bearing pads, the most deteriorated component of the bridge, showed a coating thickness of 90 μm (3.5 mils). The article "Corrosion in Structures" contains detailed information on the use of hot dip galvanized steel structural members.

Pulp and paper mills contain a number of areas that, in the absence of corrosion protection, would rapidly deteriorate because of exposure to various chemicals. A paper mill in the northwest United States used galvanized structural steel pipe supports, ladders, cages, and miscellaneous steel items in the stock tank, black liquor, lime kiln, and paper machine areas. Galvanized steel was used during the original construction 19 years ago and for subsequent expansions. All of the galvanized steel is located outside in a humid environment subject to salt spray. When inspected in 1985, coating thicknesses ranged from 90 to 160 μm (3.5 to 6.5 mils). More information on materials of construction for pulp and paper mills is available in the article "Corrosion in the Pulp and Paper Industry" in this Volume.

Recreation. The Gettysburg Observation Tower overlooks the historic Gettysburg battleground in Gettysburg, PA. The structure was erected in 1974 with galvanized pipes and rolled-steel sections. An inspection in 1984 revealed coating thicknesses of 100 to 150 μm (4 to 6 mils); this is above the specification for newly galvanized material. Service life can be expected to be 100 years or more, assuming the environment does not change appreciably.

Utility Industry. A galvanized substation located near Knoxville, TN, and owned by the Tennessee Valley Authority was constructed in 1936 to handle 100 800 kW of electricity produced by a nearby dam. Substation structures are of a bolted lattice construction. The minimum coating thicknesses measured in a recent inspection were 70 μm (2.75 mils) on lattice angles and 50 μm (2.5 mils) on bolt heads. Based on estimations of the original coating weight, this substation will last another 30 years before maintenance coating will be required.

Other Applications. Hot dip galvanized steel coated by the batch process is used in oil refineries and petrochemical industries and for miscellaneous highway uses, such as guard rail, light and sign standards, and fencing. The coating has found extensive use in water and wastewater treatment plants, both in atmospheric and immersion service. Such applications of electrical utility transmission towers, microwave transmission stations, pole line hardware, cooling towers and cooling tube bundles, nuts, bolts, and various fasteners all involve extensive use of galvanizing.

Galvanized reinforcement bars, ties, and lintels for concrete and masonry reinforcement and support provide long-term corrosion protection in

Table 11 Synergistic protective effect of galvanized steel/paint systems in atmospheric exposure

Type of atmosphere	Galvanized steel Thickness μm	mils	Service life(a), years	Paint Thickness μm	mils	Service life(a), years	Galvanized plus paint Thickness μm	mils	Service life(a), years
Heavy industrial	50	2	10	100	4	3	150	6	19
	75	3	14	150	6	5	225	9	29
	100	4	19	100	4	3	200	8	33
	100	4	19	150	6	5	250	10	36
Metropolitan (urban)	50	2	19	100	4	4	150	6	34
	75	3	29	150	6	6	225	9	52
	100	4	39	100	4	4	200	8	64
	100	4	39	150	6	6	250	10	67
Marine	50	2	20	100	4	4	150	6	36
	100	4	40	100	4	4	200	8	66
	100	4	40	150	6	6	250	10	69

(a) Service life is defined as time to about 5% red rust. Source: Ref 45

Table 12 Discounted cash flow analysis of galvanized versus painted steel

The time value of money is an important consideration when analyzing the economics of different coatings. This example assumes an inflation rate of 4%, a discount rate of 10%, a repaint cycle of 10 years, and an expected service life of 50 years. No tax or investment considerations are made. A repaint cost of 75% of the original cost is assumed, and the galvanizing cost is 25% greater than the paint cost.

Year	Galvanizing Original cost	NPW(a)	Paint Original cost	NPW(a)
0	$1.25	$1.25	$1.00	$1.00
10	...	...	...	$0.43
20	...	...	...	$0.24
30	...	...	...	$0.14
40	...	...	...	$0.08
Total lifetime cost	...	$1.25	...	$1.89

(a) NPW, net present worth of inflated future maintenance costs

vital areas that are not normally visible. The use of galvanized materials in concrete reinforcement and masonry applications is ideal, because the normal pH of these materials before setting is about 12 to 12.5, which corresponds to the pH range in which corrosion of zinc is at a minimum.

Galvanizing Specifications

Galvanized coatings produced by the hot dip (batch) process are covered by numerous specifications. In addition to those already discussed in the text, the following are pertinent specifications under the authority of ASTM:

- A 123 "Standard Specification for Zinc (Hot-Galvanized) Coatings on Products Fabricated from Rolled, Pressed, and Forged Steel Shapes, Plates, Bars, and Strip"
- A 153 "Standard Specification for Zinc Coating (Hot Dip) on Iron and Steel Hardware"
- A 384 "Standard Recommended Practice for Safeguarding Against Warpage and Distortion During Hot-Dip Galvanizing of Steel Assemblies"
- E 376 "Standard Recommended Practice for Measuring Coating Thickness by Magnetic-Field or Eddy-Current (Electromagnetic) Test Methods"

REFERENCES

1. G.J. Slunder and W.K. Boyd, *Zinc: Its Corrosion Resistance*, 2nd ed., International Lead Zinc Research Organization, 1983
2. R.J. Krepski, *The Influence of Bath Alloy Additions in Hot Dip Galvanizing—A Review*, St. Joe Minerals Corporation, 1980
3. Report of Subcommittee XIV on Inspection of Black and Galvanized Sheets, Committee A-5, in *ASTM Proceedings*, American Society for Testing and Materials, 1952, p 113
4. F.M. Reinhart, in *Twenty-Year Atmospheric Corrosion Investigation of Zinc-Coated and Uncoated Wire and Wire Products*, STP 290, American Society for Testing and Materials, 1961
5. L.P. Devillers and P. Niessen, The Mechanism of Intergranular Corrosion of Dilute Zinc-Aluminum Alloys in Hot Water, *Corros. Sci.*, Vol 16, 1976, p 243-252
6. S.E. Hadden, Effect of Annealing on the Resistance of Galvanized Steel to Atmospheric Corrosion, *J. Iron Steel Inst.*, Vol 171, 1952, p 121-127
7. H.S. Campbell *et al.*, Effect of Heat Treatment on the Protective Properties of Zinc Coatings on Steel, *J. Iron Steel Inst.*, Vol 203, 1965, p 248-251
8. J.J. Friel, Atmospheric Corrosion Products on Al, Zn, and Al-Zn Metallic Coatings, *Corrosion*, Vol 42, 1986, p 422-426
9. G. Schikorr, *Corrosion Behavior of Zinc*, Vol 1, English ed., American Zinc Institute and Zinc Development Association, 1965, p 72
10. L. Kenworthy and M.D. Smith, Corrosion of Galvanized Coatings and Zinc by Waters Containing Free Carbon Dioxide, *J. Iron Steel Inst.*, Vol 70, 1944, p 463-489
11. "Method for Estimating the Service Life of Metal Culverts," Test Method 643-B, California Department of Public Works, 1963
12. M. Romanoff, "Underground Corrosion," NBS 579,227, National Bureau of Standards, 1957
13. R.M. Burns and W.W. Bradley, *Protective Coatings for Metals*, 3rd ed., Reinhold, 1967, p 165
14. S.G. Denner *et al.*, Hot Dip Aluminizing of Steel Strip, *Iron Steel Int.*, June 1975, p 241-252
15. G. Eggeler *et al.*, On the Influence of Silicon on the Growth of the Alloy Layer During Hot Dip Aluminizing, *J. Mater. Sci.*, Vol 21, 1986, p 3348-3350
16. H.F. Graff, Aluminized Steel, in *Encyclopedia of Materials Science and Engineering*, Pergamon Press, 1986, p 138-141
17. V.I. Kelley, in *Atmospheric Corrosion Investigation of Aluminum-Coated, Zinc-Coated, and Copper Bearing Steel Wire and Wire Products (A 12 Year Report)*, STP 585, American Society for Testing and Materials, 1975
18. H.P. Godard *et al.*, *The Corrosion of Light Metals*, John Wiley & Sons, 1967, p 11
19. L. Allegra *et al.*, Resistance of Galvanized, Aluminum-Coated, and 55% Al-Zn Coated Sheet Steel to Atmospheric Corrosion Involving Standing Water, in *Atmospheric Corrosion*, W.H. Ailor, Ed., John Wiley & Sons, 1982, p 595-606
20. G.E. Morris and L. Bednar, Comprehensive Evaluation of Aluminized Steel Type 2 Pipe Field Performance, in *Transportation Research Record 1001*, National Research Council, Transportation Research Board, 1984, p 49-60
21. J.C. Zoccola *et al.*, Atmospheric Corrosion Behavior of Aluminum-Zinc Alloy Coated Steel, in *Atmospheric Factors Affecting the Corrosion of Engineering Metals*, STP 646, American Society for Testing and Materials, 1978, p 165-184
22. H.E. Townsend and J.C. Zoccola, Atmospheric Corrosion Resistance of 55% Al-Zn Coated Sheet Steel: 13-Year Test Results, *Mater. Perform.*, Vol 18, 1979, p 13-20
23. H.E. Townsend and A.R. Borzillo, Twenty-Year Atmospheric Corrosion Tests of Hot Dip Coated Sheet Steel, *Mater. Perform.*, to be published
24. J.H. Payer, Electrochemical Methods for Coatings Study and Evaluation, in *Electrochemical Techniques for Corrosion*, R. Baboian, Ed., National Association of Corrosion Engineers, 1977
25. J.B. Horton *et al.*, Corrosion Characteristics of Zinc, Aluminum, and Al-Zn Alloy Coatings on Steel, in *Proceedings of the Sixth International Congress on Metallic Corrosion* (Sydney, Australia), 1975
26. S.A. Kriner, unpublished research, 1985
27. A.J. Stavros, Galvalume Corrugated Steel Pipe: A Performance Summary, in *Transportation Research Record 1001*, National Research Council, Transportation Research Board, 1984, p 69-76
28. D. Horstmann, Reaction Between Liquid Zinc and Silicon-Free and Silicon-Containing Steels, in *Proceedings of the Seminar on Galvanizing of Silicon-Containing Steels*, International Lead-Zinc Research Organization, 1975, p 94
29. R.W. Sandelin, "Effects of Microstructure on the Galvanizing Characteristics of Steel," Paper presented at the Annual Meeting, American Hot Dip Galvanizers Association, Sept 1964

30. *Galvanizing Characteristics of Structural Steels and Their Weldments*, International Lead Zinc Research Organization, 1975
31. "Standard Recommended Practice for Safeguarding Against Embrittlement of Hot-Dip Galvanized Structural Steel Products and Procedure for Detecting Embrittlement, A 143, *Annual Book of ASTM Standards*, American Society for Testing and Materials
32. *The Design of Products To Be Hot Dip Galvanized After Fabrication*, American Hot Dip Galvanizers Association, 1985
33. *Recommended Details of Galvanized Structures*, American Hot Dip Galvanizers Association, 1983
34. "Standard Recommended Practice for Providing High-Quality Zinc Coatings (Hot-Dip)," A 385, *Annual Book of ASTM Standards*, American Society for Testing and Materials
35. *Welding Zinc Coated Steels*, American Welding Society, 1973
36. R.M. Burns and W.W. Bradley, *Protective Coatings for Metals*, 2nd ed., Reinhold, 1955, p 128
37. S.K. Coburn, C.P. Larrabee, H.H. Lawson, and G.B. Ellis, Corrosiveness at Various Atmospheric Test Sites as Measured by Specimens of Steel and Zinc, in *Metal Corrosion in the Atmosphere*, STP 435, American Society for Testing and Materials, 1968, p 371-372
38. C.J. Sunder and W.K. Boyd, *Zinc: Its Corrosion Resistance*, 2nd ed., International Lead Zinc Research Organization, 1983, p 113-150
39. "Code of Practice for Protective Coating Iron and Steel Structures Against Corrosion," BS 493, British Standards Institution, 1977, p 23
40. P.C. Birkemoe, W.D. Crouch, and W.H. Munse, "Design Criteria for Joining Galvanized Structurals," Annual Report, ZM-96, International Lead Zinc Research Organization, April 1969
41. R.A. Sanderson, "Fatigue Behaviour of High Strength Bolted Galvanized Joints," M.A. thesis, University of Toronto, 1968
42. D.J.L. Kennedy, High Strength Bolted Galvanized Joints, *J. Struct. Div.*, Vol 98 (No. St12), Dec 1972, p 2723-2738
43. "Standard Practice for Repair of Damaged Hot-Dip Galvanized Coatings," A 780, *Annual Book of ASTM Standards*, American Society for Testing and Materials
44. J.F.H. van Eijnsbergen, Twenty Years of Duplex Systems, *Metallwissenschaft Technik*, Vol 29 (No. 6), June 1975
45. J.F.H. van Eijnsbergen, Supplement (to Twenty Years of Duplex Systems), *Thermisch Verzinken*, Vol 8, 1979
46. D.S. Carr, "Performance of Painted Galvanized Steel," Paper presented at the Semi-Annual Meeting, Houston, TX, American Hot Dip Galvanizers Association, Sept 1982
47. B.R. Appleman, Economics of Corrosion Protection by Coatings, *J. Protec. Coatings Linings*, Vol 2 (No. 3), March 1985
48. T.J. Kinstler, "Probability Functions in Corrosion Economics—or a Corrosion Engineer Goes to Monte Carlo," Paper presented at Corrosion/82, National Association of Corrosion Engineers, March 1982
49. "Recommended Practice: Direct Calculation of Economic Appraisals of Corrosion Control Measures," NACE RP-02-72, National Association of Corrosion Engineers

Porcelain Enamels

PORCELAIN ENAMELS are glass coatings that are applied primarily to fabricated sheet steel, cast iron, or aluminum parts to improve appearance and to protect the metal surface. Porcelain enamels are distinguished from other ceramic coatings by their predominantly vitreous nature and the types of applications for which they are used, and they are distinguished from paint by their inorganic composition and the fusion of the coating matrix to the substrate metal. Porcelain enamels of all compositions are matured at 425 °C (800 °F) or above. Because they offer only barrier protection to the metal substrate, porcelain enamel coatings must be free from defects and coating discontinuities to provide optimum protection.

The most common applications of porcelain enamels include major appliances, water heater tanks, sanitary ware, and cookware. Porcelain enamels are also used in a wide variety of applications ranging from chemical-processing vessels, heat exchangers, agricultural storage tanks, piping and pump components, and barbeque grills to architectural panels, signing, specially executed murals, and microcircuitry components. Table 1 lists some additional applications for porcelain enamels. Normally, porcelain enamels are selected for products or components where there is a need for one or more of the special service requirements that porcelain enamel can provide, such as chemical resistance, corrosion protection, weather resistance, specific mechanical or electrical properties, appearance or color needs, cleanability, or thermal shock capability.

Types of Porcelain Enamels

Porcelain enamels for sheet steel and cast iron are classified as either ground-coat or cover-coat enamels. Ground-coat enamels contain oxides that promote adherence of the enamel to the metal substrate. Cover-coat enamels are applied over ground coats to improve the appearance and properties of the coating. Cover coats can also be applied directly to properly prepared decarburized steel substrates. The color of ground coats is limited to various shades of blue, black, brown, and gray. Cover coats, which may be clear, semiopaque, or opaque, can be pigmented to take on a great variety of colors. Colors can also be smelted into the basic coating material.

For aluminum, neither ground coats nor adherence-promoting oxides are required. Single-coat systems are used for most applications. When two coats are desired, the first coat can be of any color. Porcelain enamels for aluminum are usually transparent and can be pigmented and opacified inorganically to produce the desired appearance.

The basic material of the porcelain enamel coating is called frit. Frit is a special glass of small friable particles produced by quenching a molten glassy mixture. Because porcelain enamels are usually designed for specific applications, the compositions of the frits from which they are made vary widely.

Table 1 Some applications for porcelain enamels

Industrial products
- Chemical reactors
- Commercial heat exchangers
- Food-processing vessels
- Induction heating coils
- Ion gun parts
- Jet engine components
- Microcircuitry boards
- Mufflers
- Transformer cases

General products
- Camping equipment
- Cooking and serving utensils
- Grills
- Hospital ware
- Kitchen cabinets
- Lighting reflectors
- Meat scales
- Silos
- Tabletops
- Telephone booths
- Venetian blinds

Household appliances
- Air conditioners
- Broilers
- Dishwashers
- Freezers
- Home laundry equipment
- Ranges, gas and electric
- Refrigerators
- Space heaters
- Water heaters

Architectural
- Awnings
- Baseboards
- Chalkboards
- Exterior building panels
- Floor, roof, and wall tiles
- Tunnel panels
- Wall murals

Plumbing fixtures
- Bathtubs and lavatories
- Kitchen sinks
- Laundry tubs
- Tub enclosures

Signs
- Advertising signs
- Highway and traffic signs
- House numbers
- Street signs

Enamel Frits for Sheet Steel. Alkali borosilicates are often used as ground coats on sheet steel. Their compositions vary with the intended application of the enameled product. For example, acid resistance is obtained by the addition of titanium dioxide (TiO_2) and a large increase in silicon dioxide (SiO_2) with a corresponding decrease in boron trioxide (BO_3). The resistance of the enamel to alkalies or water can be improved by adding zirconium oxide, usually as zircon, to the frit and by maintaining a high content of silicon dioxide. Table 2 lists compositions of frits for a regular porcelain enamel and of alkali-, acid-, and water-resistant ground-coat enamels for sheet steel.

Weather resistance is usually a function of acid resistance. Porcelain enamels for outdoor use are made from various types of frits that produce the resistance and color desired. Resistance to thermal shock and high temperature is obtained by controlling the expansion of the glass coating.

Cover coats for sheet steel are applied over ground coats or directly to properly prepared decarburized steel. Electrostatic dry-powder cover coats can be applied over an electrostatic dry-powder ground coat, and the entire system can be matured in a single firing.

Cover-coat enamels made from titania-opacified frits are generally quite acid resistant; even in amounts too small to impart any opacity, titania imparts acid resistance. For alkali resistance, zirconium oxide is a desirable constituent.

Enamel Frits for Cast Iron. Compositions of frits for enamels for cast iron vary depending on whether the frit is applied by the dry process (the article to be coated is heated above the firing temperature, and the enamel is applied to the hot metal as a dry powder) or the wet process (an enamel slip is applied to the metal at ambient temperatures, dried, and fired). Dry-process enamels are commonly used for large cast iron fixtures because of their brilliance and ability to

Table 2 Melted-oxide compositions of frits for ground-coat enamels for sheet steel

	Composition, wt%			
Constituent	Regular blue-black enamel	Alkali-resistant enamel	Acid-resistant enamel	Water-resistant enamel
SiO_2	33.74	36.34	56.44	48.00
B_2O_3	20.16	19.41	14.90	12.82
Na_2O	16.74	14.99	16.59	18.48
K_2O	0.90	1.47	0.51	...
Li_2O	...	0.89	0.72	1.14
CaO	8.48	4.08	3.06	2.90
BaO	9.24	8.59	...	...
ZnO	...	2.29	...	...
Al_2O_3	4.11	3.69	0.27	...
ZrO_2	...	2.29	...	8.52
TiO_2	...	...	3.10	3.46
CuO	...	...	0.39	...
MnO_2	1.43	1.49	1.12	0.52
NiO	1.25	1.14	0.03	1.21
Co_3O_4	0.59	1.00	1.24	0.81
P_2O_5	1.04	...	...	0.20
F_2	2.32	2.33	1.63	1.94

cover small surface irregularities. Acid resistance is imparted to these enamels by reducing the alumina content, by increasing silica, and by adding up to about 8% titanium dioxide. Dry-process enamels are seldom used in applications requiring resistance to severe thermal shock.

Ground coats are usually necessary to fill surface voids in castings. Ground coats for wet-process enamels are often mixtures of frit, enamel reclaim, and refractory raw material used at very low application weight. Ground coats for dry-process enamels are applied by the wet process and are fused to thin, viscous coatings that protect the casting surface from excessive oxidations while it is heated to enameling temperature.

Enamel frits for aluminum are usually based on lead silicate and on cadmium silicate, but they can also be based on phosphate or barium. The high-lead enamels for aluminum have a high gloss, good acid and weather resistance, and good mechanical properties. The phosphate enamels generally are not alkali resistant or water resistant, but may have good acid resistance. They melt at relatively low temperatures and are useful in many applications. The barium enamels are not as low melting as the lead or phosphate glasses, but they do have good chemical durability.

Surface Preparation

The adhesion and appearance of porcelain enamel depend on closely controlled cleaning and roughening of the metal surface. Complete removal of oil, sand, drawing compounds, weld oxide, and other surface contaminants is required. Steel can be prepared by chemical or mechanical procedures. Special steels for porcelain enameling are discussed in the article "Selection of Steel Sheet for Porcelain Enameling" in Volume 1 of the 9th Edition of *Metals Handbook*.

Preparation of Steel

Chemical treatment usually involves the use of mechanized equipment in production operations. The parts to be enameled are dipped in or sprayed with a series of solutions.

When special low-carbon decarburized steel is direct-cover-coat enameled, it must be etched to remove a minimum of 22 g/m^2 (2 g/ft^2) of metal surface, and a nickel deposit of 0.9 to 1.3 g/m^2 (0.08 to 0.12 g/ft^2) of surface is required. Ferric sulfate ($Fe_2(SO_4)_3$) etching solution is often used to remove the required amount of metal surface; this results in a better bond than that obtained after etching with sulfuric acid (H_2SO_4).

A modification of the $Fe_2(SO_4)_3$ system—called oxy-acid—is also used. Oxy-acid etchant solution is a mixture of H_2SO_4, $Fe_2(SO_4)_3$, and ferrous sulfate ($FeSO_4$). The reactions involved are the same as with the $Fe_2(SO_4)_3$ system. The advantage of this system is that all of the reactions take place in one etching tank; two tanks are required for the $Fe_2(SO_4)_3$ method.

Metal preparation for "no nickel/no pickle" enameling requires at least the same amount of cleaning as conventional metal preparation does for conventional enamel. There is, however, no acid etching or nickel deposition required with the "no nickel/no pickle" system. One advantage of using this system is the reduction in wastewater treatment problems.

Mechanical preparation consists of abrasive blasting with steel shot or steel grit. Grit- or shotblasting is used on parts designed without pockets or crevices when configuration and thickness permit blasting without distortion. The flat areas of parts made with sheet steel thinner than 1.5 mm (0.06 in.) distort excessively when cleaned by this method.

Abrasive blasting is especially useful for preparing hot-rolled steel and parts that are to be enameled on one side only. The process is also used for preparing large parts and when enamels with poor bonding characteristics are used. Before blasting, oil and drawing compounds are removed by alkaline cleaning or by heating at 425 to 455 °C (800 to 850 °F) to burn off the organic contaminants.

Preparation of Cast Iron and Aluminum

Cast iron is prepared by blasting to remove adhering mold sand and the thin surface layer of chilled iron. Because the surface contains more combined carbon than is present in the remainder of the casting, the surface layer must be removed to prevent excessive evolution of gas during firing of the enamel.

Quartz sand is commonly used for abrasive cleaning of cast iron; however, steel shot, steel grit, and chilled cast iron grit propelled centrifugally from rotating wheels are generally used for cleaning sanitary ware. Zircon sand and fused alumina grit are used for special purposes.

After blasting, the casting is inspected for cracks, sand holes, slag holes, blow holes, fins, and washes. Cracks and larger holes are repaired by welding. Fins and washes are removed by grinding, and the repaired casting is blasted a second time before enameling.

Aluminum. The preparation of heat-treatable aluminum alloys for porcelain enameling involves the removal of soil and surface oxide and the application of a chromate conversion coating. Final drying removes all surface moisture; drying must be accomplished without contaminating the cleaned surface of the aluminum. Parts made of nonheat-treatable aluminum alloys require only the removal of soil, which can be done by alkaline cleaning or vapor degreasing.

The Porcelain-Enameling Process

Several basic methods are used to apply porcelain enamels to the base metal. These include dipping, flow coating, electrodeposition, manual spray, electrostatic spray, and dry-powder spray. The best method of application for a particular part is determined by quantity and quality requirements, the type of material being applied, units produced per hour, capital investment, labor cost, and, ultimately, part cost.

Application techniques can be manual or mechanized. Mechanization is used for high-volume part requirements of the same or similar shape. Hand application is necessary if a variety of parts must go through the same process system.

Regardless of the method of application, good porcelain-enameling techniques must be used to ensure uniform coverage in the areas requiring porcelain enamel protection. Excessive thickness, beads, or pooling of the porcelain enamel reduce product quality, and the product is more prone to chipping. Areas at which the coating is too thin do not receive the full protective and decorative capabilities of the porcelain enamel.

Dipping is widely used as a method of applying porcelain enamel, particularly when both sides of the parts require coverage. This method can be used for both ground-coat and cover-coat applications. Dipping is performed by immersing the part in the prepared porcelain enamel slip (a suspension of finely divided ceramic material in liquid), then withdrawing it and allowing the excess material to drain from the part. It is sometimes necessary to rotate, tilt, spin, or shake complex shapes to ensure uniform coverage. In areas where excessive porcelain enamel slip is retained on the part after draining, the excess should be removed by siphoning or wiping before the part is dried. Dip porcelain enamels are normally applied at thicknesses of 50 to 100 μm (2 to 4 mils). The AISI 300-series stainless steels are the preferred materials for dipping equipment.

In flow coating, the porcelain enamel slip is flowed onto the surface of the part. The process is applicable to high-volume continuous operations for parts requiring the same porcelain enamel. In automatic flow coating, the parts are placed on hangers at the correct angle for draining and are carried by conveyor through the flow coating chamber. To ensure complete coverage, the porcelain enamel slip is pumped at high volume and low pressure though nozzles that are directed at various areas of the part.

Another version of automatic flow coating involves the use of a constant-head tank to supply slip at a constant velocity to headers and nozzles that flood parts with slip as they are conveyed though the flow-coat chamber. The advantage of this system is that the flow to the nozzles is constant and not subject to variations present in pumped systems.

Automated flow coating is favored over hand dipping because it offers higher rates of production, improved coating quality, and reduced cost of the applied film. Control of the porcelain enamel slip and proper operation of the machine are important functions of flow coating. It is common practice to check the specific gravity and pickup of the porcelain enamel slip three to four times each hour. All parts of the flow-coating machine that contact the porcelain enamel slip should be constructed of 300-series stainless steel.

Spraying of the porcelain enamel slip is done primarily for one-side coverage. It is also used to reinforce enamel bisque (the dried, unfired coating) and to repair enameled surfaces. Spraying is ideal for parts that are too large for hand or mechanical manipulation, particularly where service and appearance requirements do not permit drain lines, beading, or buildup of the porcelain enamel. Spraying is the method that is most commonly used to apply porcelain enamels to aluminum; thicknesses of 65 to 90 μm (2.5 to 3.5 mils) are desirable.

Wet electrostatic spraying of porcelain enamel is used to reduce losses in material by charging the porcelain enamel slip during atomization to a potential of 100 000 to 120 000 V. The electrostatically charged droplets are attracted to the grounded parts being sprayed. A well-operated electrostatic unit can deposit up to 85% of the sprayed material on the part as compared to 30 to 50% in conventional spraying operations.

Electrostatic powder spraying is another method that can be used when many parts are being produced that require the application of the same porcelain enamel. The parts must be of a configuration that can be properly and evenly coated by this process. When these conditions are met, this is a very efficient method of applying porcelain enamel. Up to 99% of the material is used, with little or no direct labor required for the application operation. Smooth-running conveyors are required with this method of application to prevent loss of powder prior to firing the parts.

Powder is delivered to the spray guns from a feeder unit where it is diffused by clean compressed air into a fluidlike state. The fluidized powder is then siphoned by the movement of high-velocity air flowing through a venturi and is propelled through powder feed tubes to the spray gun. The powder feeder provides a steady, controlled flow of powder to the guns. Independent control of powder and air volume ensures the proper ratios to provide the desired thickness coverage on the product.

The powder is propelled toward the workpiece in the form of a diffused cloud. A high-voltage low-amperage power unit supplies current to the charging electrode, and this causes the powder to seek out and attach itself to the grounded workpiece.

The recovery equipment booth serves to collect and return the powder that is not held on the workpiece. The powder moves through a closed-loop system with the use of filters and final filters; thus, none of the airborne powder escapes into the environment.

Electrodeposition is another process that can be used to apply enamel to steel. The process uses a series of tanks in which the parts are submerged, and the enamel is deposited electrophoretically. This process is basically limited to direct-on enameling, but can be considered for two-coat/one-fire applications. The main advantages of this system are the very uniform appearance and the exceptionally thin enamel layers.

Drying and Firing. Parts coated with porcelain enamels are dried (if the enamel was applied by a wet process) and fired. Drying permits the application of an additional coating of enamel, if required, and allows the parts to be handled more easily. Parts are either air dried or dried with radiant or convection dryers. Convection drying consists of gradual heating of the parts to 120 °C (250 °F); cycle times range from 2 to 5 min. The coating is still wet during the initial stages of drying, so drying must be accomplished in an atmosphere free of dirt, scale, or dust. The parts are then fired in either continuous or batch-type furnaces at temperatures of about 425 °C (800 °F) for steel parts and 525 to 550 °C (980 to 1020 °F) for aluminum. More information on the porcelain-enameling process is available in the article "Porcelain Enameling" in Volume 5 of the 9th Edition of *Metals Handbook*.

Process Variables

The thickness of the applied layer of porcelain enamel, the firing time, and the firing temperature markedly affect the properties of the coating. Increasing the thickness of the coating increases the resistance to burn-off and produces truer colors; however, thin coatings have the greatest flexibility.

Coating Thickness. The optimum thickness of porcelain enamel depends on the substrate metal and the service requirements of the part. On aluminum, porcelain enamel is applied to produce a fired enamel thickness ranging from about 65 to 125 μm (2.5 to 5 mils). A tolerance of ±13 μm (±0.5 mil) is required in order to maintain uniform opacity for a white enamel coating 115 μm (4.5 mils) thick.

On sheet steel, a ground coat about 50 to 100 μm (2 to 4 mils) thick is used to promote adhesion. To cover the ground coat, a very opaque white or pastel cover coat about 100 to 150 μm (4 to 6 mils) thick is required. Thus, a two-coat system on these products has a thickness ranging from 150 to 255 μm (6 to 10 mils).

Coatings for cast iron products are much thicker than those for sheet steel or aluminum. Dry-process coatings on cast iron products such as sanitary ware range from 1020 to 1780 μm (40 to 70 mils) in thickness. Coatings applied by the wet process are thinner than dry-process coatings; wet-process coatings range in thickness from 255 to 635 μm (10 to 25 mils).

Coating thicknesses for hot-water tanks normally range from 150 to 230 μm (6 to 9 mils), with 150 μm (6 mils) a generally accepted minimum thickness. Heat-exchanger surfaces, depending on end use, are sometimes double coated for added durability.

The thickness of the porcelain enamel on a large part of simple configuration can be closely controlled when application is by a mechanical spraying system that is adapted to the part. For example, mechanically applied porcelain enamel on curved silo panels measuring 2 × 3 m (6 × 9 ft) can be maintained within ±13 μm (±0.5 mil); however, when application is by hand spraying, the variation in enamel thickness is ±50 μm (±2 mils).

An enamel thickness of 65 to 180 μm (2.5 to 7 mils) is desirable for aluminum architectural panels. When white or light-colored enamel is used, however, the enamel thickness ranges above 75 μm (3 mils) in order to produce acceptable opacity. Two coats with a total thickness of about 125 μm (5 mils) result in more uniform opacity than one coat that is 125 μm (5 mils) thick. Additional details are available in Ref 1.

Firing Time and Temperature. Firing of porcelain enamel involves the flow and consolidation of a viscous liquid and the escape of gases through the coating during its formation. Within limits, time and temperature are varied in a compensating manner. For example, similar properties and appearances develop when liners for household refrigerators are fired at 805 °C (1480 °F) for 2½ min or at 790 °C (1450 °F) for 4 min. In all cases, there is a minimum practical temperature for the attainment of complete fusion, acceptable adherence, and desired appearance. Most ground-coat enamels for high-production steel parts exhibit acceptable properties over a firing range of 55 °C (100 °F) at an optimum firing time. However, control within 11 °C (20 °F) is ordinarily maintained to produce uniform appearance and allow interchangeability of parts. As the combined effects of firing time and temperature increase, resulting in more thorough firing, up to a maximum, the following conditions occur:

- Colors shift dramatically, particularly reds and yellows. In general, white and colors shift toward yellow. Usually, furnace temperature is changed to achieve minor adjustments in color matching
- Gloss of the enamel coating increases
- Chemical resistance of the enamel coating increases
- Gas bubbles are eliminated
- Enamel coating becomes more dense and brittle and less resistant to chipping
- Maximum adherence is attained in the optimum portion of the firing range

Color Matching and Control. In color matching, coloring oxides are usually used. Although preblended oxides for a specific color are available, they are more difficult to adjust. In most cases, two or three oxides are sufficient to match any specific color. Usually, the proper color intensity is obtained first; adjustment is then made for the desired color shade. Cadmium-sulfoselenide pigments, for red and yellow, are generally used with cadmium-stabilized clear frits.

Color stability can be adversely affected by improper mill additions. However, a color with only fair stability can be improved by the proper mill additions, and minor color adjustments, particularly white, are possible.

Sometimes, gum must be used to control bisque strength; sodium nitrite and urea are used to control tearing. Because these additions have a marked effect on some colors, they should be used in initial color matching.

Finer grinding reduces the intensity of the color. It is imperative that the fineness of the milled color be controlled within specific limits. Milling is usually stopped before completion to permit sample firing of the enamel and comparison with a color standard. Adjustments to the mill can then be made. Color can be controlled to some extent by variations in the fineness of the grinding. The thickness of the fired enamel coating affects many colors. In general, thick coatings produce lighter colors, and thin coatings result in darker colors.

The set and specific gravity of a colored enamel slip are important to the finished results. Mottling or color separation is possible if the colored enamel is applied too wet or too dry. Color corrections of electrostatic dry powder cannot be made by the enameler.

Flatness and Distortion. Sag and distortion of sheet steel parts result from low metal strength at the firing temperature, thermal stresses due to nonuniform heating and cooling, and transformation to austenite. Changes in design of the parts and firing practice alleviate the first two causes, and the use of extra-low carbon content or of special stabilized steels minimizes transformation to austenite.

Ground-coat enamels have a limited effect on the distortion of parts because their coefficients of thermal expansion approach that of steel. When a ground coat is applied to both sides of the metal, there is a counterbalancing of expansion and contraction stresses.

The effect of cover-coat enamels on the configuration and flatness of porcelain-enameled parts can be pronounced as a result of low coefficients of expansion and one-side application. The likelihood of distortion is greatly increased when multiple or thicker coats of cover-coat enamels are necessary on one surface. Sometimes, cover coats must be applied to the back side of parts to equalize the stresses.

Adjustments in the firing cycle can sometimes help to minimize distortion. A cycle with relatively slow heating and cooling rates is preferable to rapid heating and cooling.

Variations in the method of supporting the work during firing can often change the sagging characteristics to an appreciable degree. Furnace supports and fixtures can be designed to distribute the load and equalize heating and cooling rates. The design and fabrication of sheet steel parts for porcelain enameling are discussed in Ref 2.

Process Control

Proper workability during the application of wet-process porcelain enamels depends on:

- Control of the porcelain enamel slip, particularly with respect to stability of suspension
- Weight of enamel slip deposited and retained per unit area
- Specific gravity, consistency, and particle size of the enamel slip
- Stability upon aging at ambient temperature

Stability of suspension, or the ability of the various mill additions to keep the milled frit in suspension, is determined by both slip measurements and visual observation of any separation that occurs. Stability of the suspension is a function of many factors, but is usually controlled by the quantity of colloid, in the form of clay or bentonite, and electrolytes used to deflocculate the clay. Enamel slips for aluminum have a shorter shelf life than those for sheet steel.

Pickup weight of enamel and enamel retained per unit area are measured by draining the enamel on a flat or cylindrical shape of known weight and area and actually weighing the pickup of enamel in wet or dry form. This is a very useful test, particularly for dipping enamels, and it closely simulates actual production operations. During the test, the operator can observe any tendency toward sliding, excessively long or short drain time, and variations in setting time. The pickup of an enamel is a function of specific gravity, colloid content, total salts content and type, and consistency. These are controlled by varying water content, addition of salts, and fineness of grind.

The specific gravity of enamel slips is measured either by weighing a known volume in comparison with the weight of an equivalent volume of water or by the use of a hydrometer. Control of specific gravity is almost entirely a function of the ratio of water to solids. To ensure uniformity, testing for specific gravity is required for the preparation of all porcelain enamel slips.

The consistency of a porcelain enamel slip for spraying is commonly determined by the slump test. In this test, a fixed volume of the porcelain enamel slip is allowed to flow out suddenly in a circular pattern on a calibrated plate, and the diameter of the resulting pool is measured immediately. This is a simple and useful test for porcelain enamels that will be sprayed because it indicates uniformity of slip conditions between various millings. Other tests for consistency involve the use of viscosimeters of various types, including those that use the rotational, flow, and falling piston methods.

For porcelain enamel slips applied by dipping, a measure of drain time is a useful test. Drain time is the total elapsed interval between the time a standard size sample plate is removed from a container of well-stirred porcelain enamel slip and the time at which the draining motion of the slurry on the sample has stopped.

The particle size of the frit for porcelain enamel slips is commonly determined by standard screen analysis. Reproducible measurements are easily obtained when a standardized shaking device is used. The particle size of the frit is important to the suspension characteristics of the porcelain enamel slip, and slight solubility of the frits shows a major change with variation in the size of the particles.

Stability toward aging of porcelain enamel slips is measured by exposing a tested sample of the enamel to whatever temperatures are expected in normal service. Exposure is for many hours and days, and retests of the critical properties are made at intervals during testing. Aging usually has an effect on the stability of the suspension, pickup, setting time, and the consistency of the porcelain enamel slip. Aging causes bubbly glass and poor surface quality of the fired enamel. Leaching of soluble elements such as sodium or boron from the frit is a cause of aging. This problem is more frequently encountered with less water-resistant frits. The effect is greater at higher temperatures.

Corrosion Resistance of Porcelain Enamels

Porcelain enamels possess excellent resistance to corrosion in a variety of environments. Enamels are formulated to have resistance to specific environments, along with ease of processing and minimum cost. Table 3 lists corrosion applications for porcelain enamels.

Atmospheric Exposure. Porcelain enamels have excellent resistance in atmospheric exposure, including corrosive industrial atmospheres, gases, smoke, salt spray, and seacoast exposures. The weather resistance of porcelain enamels is usually measured by the degree to which the coating maintains its original color and gloss. Enamels formulated for acid resistance usually have better atmospheric-corrosion resistance than other types. These enamels have shown no appreciable change in appearance after 15 years of exposure.

Waters. All porcelain enamels are completely resistant to water at room temperature. Resistance decreases at higher temperatures. Special porcelain enamels have been developed for hot-water storage tanks that can withstand continuous exposure to hot water for periods of 10 to 20 years. Natural waters, because of their varying compositions, have varying effects on porcelain enamels. Aerated water with a low dissolved solids content, for example, has been found to be more corrosive than hard water. Freezing and thawing cycles can cause some porcelain enamels to spall or disintegrate; however, properly formulated and applied enamels can withstand thousands of freezing and thawing cycles without failure. Porcelain enamels can withstand salt spray tests (ASTM B 117) for days and even weeks without evidence of corrosion (Ref 3), and they can provide excellent service in intermittent or continuous exposure to seawater.

Soils. Porcelain enamels formulated to withstand both acid and alkaline attack provide good service in soils.

Acids. Resistance to acids varies widely, depending on composition and the application process used. The degree of attack by acids appears to depend less on the type of acid (with the exception of hydrofluoric) than on pH. Special formulations can provide excellent protection against aqueous solutions of most acids except hydrofluoric. The highest degree of acid resistance is obtained by sacrificing resistance to other media, such as alkalies. Figure 1 shows weight loss of a porcelain enamel in boiling mineral acids and in boiling water.

Alkalies. Most porcelain enamels are unaffected by alkalies at room temperature. Special formulations provide alkali resistance to solutions with a maximum pH of 12 at temperatures to 100 °C (212 °F).

Organics. Porcelain enamels are completely resistant to attack by common organic solvents, dyes, greases, and oils. Enamels are not dissolved by these materials and do not absorb them. Acid-resistant porcelain enamels are re-

Table 3 Applications in which porcelain enamels are used for resistance to corrosive environments

Application	Corrosive environment: Temperature °C	°F	pH	Corrosive medium
Bathtubs	To 49	To 120	5–9	Water, cleansers
Chemical ware	To 100	To 212	12	Alkaline solutions
	To 100	To 212	1–2	All acids except hydrofluoric
	175–230	350–450	1–2	Concentrated sulfuric acid, nitric acid, and hydrochloric acid
Home laundry equipment	To 71	To 160	11	Water; detergents; bleach
Range exteriors	21–66	70–150	2–10	Food acids; cleaners
Range oven liners, conventional	66–315	150–600	2–10	Food acids; cleaners
Range burner grates	66–590	150–1100	2–10	Food acids; cleaners
Refrigerators	−18 to 66	0–150	2–10	Food acids; cleaners
Kitchen sinks	To 71	To 160	2–10	Food acids; water; cleansers
Water heaters	To 71	To 160	5–8	Water
Industrial heat exchangers	Depends on application(a)			

(a) Applications include coal- and oil-fired boilers and black liquor evaporators; corrosive media would include ash from coal-fired boilers, corrosive condensates, and exhaust from black liquor evaporators.

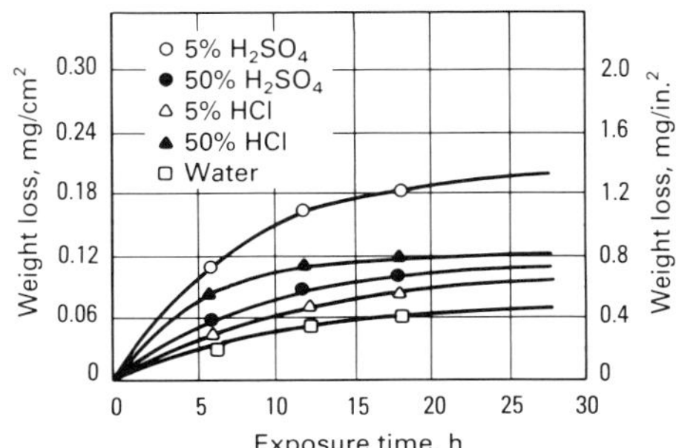

Fig. 1 Corrosion of a porcelain enamel in boiling water and boiling mineral acids. Source: Ref 4

Table 4 Maximum service temperatures for porcelain enamels

Service temperature °C	°F	Limiting conditions
425	800	Usual limit for enamels maturing at about 815 °C (1500 °F)
540	1000	Maximum for enamels maturing at about 815 °C (1500 °F), without reboil
760	1400	Operating limit for special high-temperature enamels
1095	2000	Refractory enamels useful for short periods for protection of stainless steels and special alloys

Source: Ref 5

quired for organic material that hydrolyzes upon contact with moisture to form acid solutions.

High Temperatures. Porcelain enamels greatly reduce high-temperature oxidation of the base metal. This ability is largely due to the fact that the enamels themselves are fully oxidized and do not suffer further oxidation at elevated temperatures. They also form an effective barrier to the diffusion of oxygen into the metal. Protection depends on the temperature at which the enamel begins to soften and become more fluid. This temperature is normally about 205 °C (400 °F) below the firing temperature, but specially formulated enamels can provide oxidation protection to metals at temperatures to 1095 °C (2000 °F). Maximum service temperatures for porcelain enamels are shown in Table 4.

The resistance of an enamel to thermal shock varies inversely with its thickness. Thermal shock failure occurs when the temperature gradient perpendicular to the surface is large enough to cause excessive differential shrinkage and tensile stress. Thermal shock resistance also depends on the design and section thickness of the coated part. Flexing of the metal due to localized thermal gradients parallel to the surface can produce bending and tensile stresses in the coating; therefore, an increase in the strength or rigidity of the part increases the resistance of the coating to thermal shock. Most porcelain enamels applied at conventional thickness can withstand abrupt temperature changes of 110 to 165 °C (200 to 300 °F).

Evaluation of Porcelain Enameled Surfaces

Specifications and quality control for porcelain enamel coatings require the evaluation of a range of properties, depending on the intended service of the porcelain-enameled product. Although material and process variables can be brought into approximate control by using small test panels, process control is maintained by the evaluation of finished parts, even though the mechanical and chemical tests entailed are destructive.

Standard test procedures are available for most porcelain enamels. Specific test methods for various properties are listed in Table 5. Some of these properties and tests are discussed below (refer to Table 5 for the title of the test, specification, or standard).

Table 5 Test methods, specifications, and standards for porcelain enamels

Designation(a)	Title of test, specification, or standard
Adherence	
ASTM C 313	Adherence of Porcelain Enamel and Ceramic Coatings to Sheet Metal
ASTM C 703	Spalling Resistance of Porcelain Enameled Aluminum
PEI T-29	Adherence of Porcelain Enamel Copper Coats Direct To Steel
Thickness	
ASTM C 664	Thickness of Diffusion Coating
ASTM D 1186	Dry Film Thickness of Non-Magnetic Organic Coatings Applied on a Magnetic Base, Measurement of
ASTM E 376	Coating Thickness by Magnetic Field or Eddy-Current (Electromagnetic) Test Methods, Rec. Practice for Measuring
Color and gloss	
ASTM C 346	45-deg Specular Gloss of Ceramic Materials
ASTM C 540	Image Gloss of Porcelain Enamel Surfaces
ASTM E 97	Reflectance Factor of Opaque Specimens by Broad-Band Filter Reflectometry, 45 deg, 0 deg Directional
ASTM C 347	Reflectivity and Coefficient of Scatter of White Porcelain Enamels
ASTM D 2244	Color Differences of Opaque Materials, Instrumental Evaluation of
ASTM D 1535	Color by the Munsell System, Specifying
ASTM C 538	Color Retention of Red, Orange and Yellow Porcelain Enamels
Chemical resistance and weather characteristics	
ASTM C 282	Acid Resistance of Porcelain Enamels (Citric Acid Spot Test)
ASTM C 614	Alkali Resistance of Porcelain Enamels
ASTM C 756	Cleanability of Surface Finishes
ASTM C 283	Boiling Acid, Resistance of Porcelain Enameled Utensils to
ASTM D 1567	Detergent Cleaners for Evaluation of Corrosive Effects on Certain Porcelain Enamels, Testing
ASTM C 872	Lead and Cadmium Releases from Porcelain Enamel Surfaces
Chipping resistance	
ASTM C 409	Torsion Resistance of Laboratory Specimens of Porcelain Enameled Iron and Steel
Abrasion resistance	
ASTM C 448	Abrasion Resistance of Porcelain Enamels
Thermal shock	
ASTM C 385	Thermal Shock Resistance of Porcelain Enameled Utensils
Tests related to preparation of coatings and substrates	
ASTM C 374	Fusion Flow of Porcelain Enamel Frits (Flow-Button Methods)
ASTM C 539	Linear Thermal Expansion of Porcelain Enamel and Glaze Frits and Ceramic Whiteware Materials by the Interferometric Method
ASTM C 285	Sieve Analysis of Wet-Milled and Dry-Milled Porcelain Enamel
ASTM C 839	Compressive Stress of Porcelain Enamels by Loaded-Beam Method
ASTM C 715	Nickel on Steel for Porcelain Enameling by Photometric Analysis
ASTM C 810	Nickel on Steel for Porcelain Enameling by X-Ray Emission Spectrometry
ASTM C 632	Reboiling Tendency of Sheet Steel for Porcelain Enameling
ASTM C 694	Weight Loss of Sheet Steel During Immersion in Sulfuric Acid Solution
ASTM C 774	Yield Strength of Enameling Steels After Straining and Firing
ASTM C 660	Production and Preparation of Gray Iron Coatings for Porcelain Enameling
Tests related to continuity of coating	
ASTM C 536	Continuity of Coatings in Glassed Steel Equipment by Electrical Testing
ASTM C 743	Continuity of Porcelain Enamel Coatings
Specifications	
PEI S-100	Specification for Architectural Porcelain Enamel on Steel for Exterior Use
PEI ALS-105	Recommended Specifications for Architectural Porcelain Enamel on Aluminum for Exterior Use
PEI ALS-106	Recommended Specifications for Porcelain Enamel Finishes on Aluminum Cookware
WH-196-J	Federal Specification—heater, water, electric and gas fired residential. (This covers resistance of porcelain enamels to hot water under "Solubility of Glass Lining.")
Plumbing fixtures standards	
ANSI A 112.19.4	Porcelain Enameled Formed Steel Plumbing Fixtures
ANSI A 112.19.1M	Enameled Cast Iron Plumbing Fixtures

(a) ASTM, American Society for Testing and Materials; PEI, Porcelain Enamel Institute; ANSI, American National Standards Institute

Adherence refers to the degree of attachment of enamel to the metal substrate. Although none of the adherence tests in common use results in the quantified force per unit area required to detach the enamel by tensile force normal to the interface, various tests aimed at evaluating adherence are regularly used in the industry.

The standard adherence tests for porcelain enamel on steel are ASTM C 313 and PEI Bulletin T-29. The ASTM C 313 test is applicable only to steel substrates between 0.4 mm (0.016 in.) and 2 mm (0.082 in.) in thickness. The PEI T-29 test applies to direct-on cover coats on substrates with a thickness range of 0.7 mm (0.028 in.) to 1.3 mm (0.050 in.).

Both adherence tests for porcelain enamel on steel include deforming the metal and measuring the area from which the porcelain enamel is removed. The indicator of adherence is the adherence index, which is the ratio of the porcelain enamel remaining in the deformed area to that in the same measured area before deformation.

Enamels for cast iron pose a special problem because of the relatively greater thickness and rigidity of the metal substrate and the brittleness of the iron. Simple, unstandardized impact tests are used in these cases.

Resistance to spalling, a defect characterized by the separation of porcelain enamel from the base metal without apparent external cause, is the indicator used to measure the adherence of porcelain enamel on aluminum. Spalling can result from the use of improper alloys or enamel formulations, incorrect pretreatment of the base metal, or faulty application and firing procedures.

Two methods for determining resistance to spalling are outlined in ASTM C 703. Method A, which uses a 5% solution of ammonium chloride, requires a 96-h immersion of the test specimen. Method B uses a 1% solution of antimony trichloride and requires a 20-h immersion of the test specimen.

The spall test is a pass/fail test. Failure is determined by either of two criteria. The first is the existence of spall areas of specified dimensions at specimen edges. The second criterion is the existence on the specimen interior (away from the edges) of spots exceeding specified dimensions or a spot level exceeding a specified density, usually spots/m^2 (spots/ft^2).

Thickness. The specifications for architectural porcelain ename (PEI S-100 for steel and PEI ALS-105 for aluminum), the specification for porcelain-enameled aluminum cookware (PEI ALS-106), and the standards for porcelain-enameled formed steel plumbing fixtures (ANSI A112.19.4) and porcelain-enameled cast iron plumbing fixtures (ANSI A112.19.1M) all require a specific thickness for the porcelain enamel coating.

The procedure used for measuring the porcelain enamel thickness depends on the type of base metal used. For porcelain-on-steel products, enamel thickness is measured through procedures outlined in ASTM D 1186. For porcelain-on-aluminum products, coating thickness is measured according to test procedures specified in ASTM E 376. In some cases, primarily laboratory investigations, measurement of porcelain enamel thickness is accomplished by using the procedures specified in ASTM C 664.

Color and Gloss. Porcelain enamel finishes are produced in literally hundreds of colors and many textures. This capability provides the manufacturers of appliances, cookware, outdoor grills, architectural panels, signs, decorative products, and many other applications with unusual design versatility as well as desirable performance properties.

The common method of specifying color is based on the capacity of the observed article to reflect light of different wavelengths (different colors). A physical standard, such as a plaque of porcelain-enameled steel, is provided as the color to be matched within stated limits. The difference in color between the control standard and the test specimen can be measured instrumentally by using the procedures specified in ASTM D 2244.

Gloss is particularly desirable in such products as appliances. However, high-gloss enamels capable of reflecting distinct images are not recommended for architectural porcelain enamel for exterior use.

The gloss of porcelain enamel can be measured by following procedures from two standards: ASTM C 346 and ASTM C 540. The test chosen for evaluating gloss depends on the purpose for which the porcelain enamel is used.

Acid Resistance. There are two standard tests for determining the acid resistance of porcelain enamels: ASTM C 282 (generally referred to as the citric acid spot test) and ASTM C 283 (generally known as the boiling acid test). The citric acid spot test is used as a testing and grading system for such porcelain enamel applications as appliances, plumbing fixtures, and architectural products. Evaluation is based on visual examination following wet or dry rubbing and blurring-highlight tests. The boiling acid test is designed primarily for cookware applications, and the results are expressed in weight loss per unit area.

Alkali Resistance. Home laundry equipment, dishwashers, and other porcelain enamel applications in which the surface is normally exposed to an alkaline environment at elevated temperatures require an alkali-resistant coating. The standard test for alkali resistance is ASTM C 614. This test covers the measurement of the resistance of a porcelain enamel to a hot solution of tetrasodium pyrophosphate. Alkali resistance is expressed in terms of weight loss for the area exposed to the test solution.

Weather Resistance. The long-term weatherability of porcelain enamel is of primary interest to specifiers and manufacturers of architectural porcelain enamel products intended for exterior use. The actual weathering performance of the material has been documented in a series of on-site exposure tests conducted by the National Bureau of Standards in cooperation with the Porcelain Enamel Institute.

The weathering of porcelain enamel is evaluated in terms of the changes in gloss and color that occur during outdoor exposure. Changes in gloss and color are measured by using the procedures outlined earlier. Weathering tests of up to 30 years show that porcelain enamels have considerable inherent gloss and color stability; rates of change are primarily influenced by enamel compositions, choice of colors (reds and yellows are most susceptible to color change), and the severity of the exposure site (seacoast and corrosive industrial environments have been found to have the most aggressive effect).

Spalling of enamels during exposure to weather occurs if improper materials and processing procedures have been used. The spalling resistance of weathered porcelain enamel on aluminum, for example, is best ascertained by using ASTM C 703.

In exposure tests, good correlation has been observed between acid resistance, as determined by the citric acid spot test, and the color retention of steel enamels. However, use of the acid spot test has shown an even stronger reliability in predicting color change in porcelain-on-steel. The correlation between acid resistance and color retention in porcelain-on-aluminum is somewhat less definitive but still observable. The best indicator for predicting the weatherability of red, orange, and yellow porcelain enamels is ASTM C 538. Because architectural porcelain enamel for exterior use is subjected to weathering for long periods of time, the specifications for porcelain enamel on aluminum (PEI ALS-105) and porcelain enamel on steel (PEI S-100) require compliance with specific levels of acid resistance, which are considered indicators of good weatherability.

Chipping Resistance. Relatively thick layers of porcelain enamel cannot be subjected to severe bending or other substrate deformation without fracture. However, coatings of 125 μm (5 mils) or less that are well bonded to a relatively thin metal substrate—for example, 26 gage 0.4546 mm (0.0179 in.) or less—can withstand the bending of the substrate to radii of curvature within its elastic limit and the bending to return to the original shape with little or no apparent damage. Chipping of typical porcelain enamel on sheet iron occurs at about the strain required for permanent deformation of the base metal.

Abrasion Resistance. The test for determining the resistance of porcelain enamel to various types of abrasion is ASTM C 448. The test consists of three parts. The first determines the resistance to surface abrasion of porcelain enamels for which the unabraded 45° specular gloss is more than 30 gloss units. Here the specular gloss of the specimens is measured before and after a specified abrasive treatment of the surface; the percentage of the original specular gloss that is retained is the surface abrasion index.

The second part of this test determines the resistance to surface abrasion of porcelain enamels for which the unabraded 45° specular gloss is 30 gloss units or less. The weight loss by a specific abrasive treatment—modified by an adjustment factor for each abrasive tester, lot of abrasive, and lot of calibrated plate glass standards used—results in an adjusted weight loss value. This is recorded as the index of resistance to surface abrasion.

The final portion of the test measures the resistance of porcelain enamels to subsurface abrasion. In this case, the scope of the linear portion of the abrasion time-weight loss curve is determined and then multiplied by an adjustment factor associated with each abrasion tester, lot of abrasive, and lot of calibrated plate glass standards used. The adjusted scope is taken as an index of the resistance to subsurface abrasion.

Thermal Shock Resistance. The standard used for evaluating the durability of porcelain-enameled utensils subjected to thermal shock is ASTM C 385. In this test, cooking utensils are subjected to a series of dry heating and quenching cycles until the utensil fails by removal of the enamel from the utensil.

Continuity of Coatings. Ensuring the continuity of the coating after manufacture is important in porcelain enamel or so-called glassed steel applications in which a prime purpose of the coating is to protect the substrate against corrosion.

There are two principal test methods for determining either discontinuity of coverage or potential discontinuity through too-thin coverage. For glassed steel equipment, the prescribed test procedure is ASTM C 536. For conventional porcelain-enameled products, such as appliances, plumbing fixtures, and architectural panels, the test most commonly used is ASTM C 743. Both tests essentially involve the use of electrical probes of relatively high voltage to discern either discontinuities in the coating or insufficient coverage for coating integrity in service use.

Resistance to Hot Water. Federal specification WH-196-J specifies a solubility test for determining the resistance of porcelain enamels to hot water. In this test, 89- × 89-mm (3.5- × 3.5-in.)

sections cut from the outer wall of a water heater are tested at a rolling boil for 8 cycles of 18 h each in a special apparatus containing 400 mg of reagent grade sodium bicarbonate dissolved in 1 L of water. The resistance of the porcelain enamel is specified in terms of weight loss in milligrams per square inch exposed area. The weight loss allowed by WH-196-J is 2.3 mg/cm^2 (15 mg/in.^{2}).

ACKNOWLEDGMENT

Thanks to John C. Oliver of the Porcelain Enamel Institute, Jeffrey F. Wright of Ferro Corporation, and Cullen L. Hackler and Vernon C. Jett of Mobay Corporation for their assistance in obtaining data for this article.

REFERENCES

1. "Manual of Dipping and Flow Coating for Porcelain Enameling," Bulletin P-302, Porcelain Enamel Institute
2. "Design and Fabrication of Sheet Steel Parts for Porcelain Enameling," Bulletin P-306, Porcelain Enamel Institute
3. "Standard Method of Salt Spray (Fog) Testing," B 117, *Annual Book of ASTM Standards*, American Society for Testing and Materials
4. C.L. Hackler and M. Dinulescu, Porcelain Enameled Flat Plate Heat Exchangers—Engineering and Application, in *Industrial Heat Exchangers*, Proceedings of the 1985 Symposium on Industrial Heat Exchanger Technology, A.J. Hayes *et al.*, Ed., American Society for Metals, 1985
5. "Development of Porcelain Enamel Coatings," Bulletin E-6, Porcelain Enamel Institute

SELECTED REFERENCES

- "Alloy, Design and Fabrication Considerations for Porcelain Enameling Aluminum," Bulletin P-402, Porcelain Enamel Institute
- "Design and Fabrication of Sheet Steel Parts for Porcelain Enameling," Bulletin P-306, Porcelain Enamel Institute
- Glass; Ceramic Whitewares; Porcelain Enamels, Vol 15.02, *Annual Book of ASTM Standards*, American Society for Testing and Materials
- *Glass Linings and Vitreous Enamels*, Publication 6H160, National Association of Corrosion Engineers, 1960
- F.W. Nelson, Engineering Properties of Inorganic Coatings, *Mater. Res. Stand.*, Vol 7 (No. 7), July 1967, p 309-311
- "1939 Exposure Test of Porcelain Enamels on Steel, 30-Year Inspection," Building Science Series 38, National Bureau of Standards
- "Preparation of Sheet Steel for Porcelain Enameling," Bulletin P-307, Porcelain Enamel Institute
- "Pretreatment of Alloys for Porcelain Enameling Aluminum," Bulletin P-403, Porcelain Enamel Institute
- "Properties of Porcelain Enamel: High Temperature Properties," Data Bulletin PEI 504, Porcelain Enamel Institute
- "Properties of Porcelain Enamel: Resistance to Corrosion," Data Bulletin PEI 503, Porcelain Enamel Institute
- "Quality Control Procedures for Porcelain Enameling Aluminum," Bulletin P-405, Porcelain Enamel Institute
- "Weathering of Porcelain Enamels on Aluminum, 12-Year Inspection, Exposure period 1964-1976," Department of Ceramic Engineering, University of Illinois at Urbana-Champaign, 1977
- "Weather Resistance of Porcelain Enamels, 15-Year Inspection of the 1956 Exposure Test," Building Science Series 50, National Bureau of Standards

Chemical-Setting Ceramic Linings

Gregory D. Maloney, Saureisen Cements Company

INORGANIC CHEMICAL-SETTING CERAMIC LININGS have become one of the most widely used construction materials in designing protective linings for industrial installations in which high temperatures, aggressive corrosive media, and complicated substrate geometry exist, such as floors, trenches, sumps, reaction vessels, tanks, scrubbers, ducts, chimneys, and other air pollution control equipment. They are used in various industries, including power, steel- and metalworking, chemical, pulp and paper, refinery, waste treatment, and mining.

Inorganic monolithic linings have proved themselves in these industries because of their chemical resistance to both high and low concentrations of strong acids and solvents, thermal insulation that protects the substrates from extremely high temperatures, temperature resistance to 870 °C (1600 °F), good compressive and flexural strength for environments in which stress and strain are factors, and abrasion resistance. Monolithic linings can be applied by cast or gunite (shotcreting) methods over old and new steel or concrete as well as brick and mortar masonry. This article will discuss the function of monolithic linings, the use of these materials, the types of applications in which these materials can be successfully used, and the limitations of these linings.

History of Chemical-Setting Silicates

The progress of silicate cement development has changed over the years with growing technology to meet industry needs. The first silicate cements were composed of a sodium silicate (Na_2SiO_3) liquid and a combination of fillers, such as silica flour, clay, silica aggregates, and barytes, and were formulated as chemical-resistant mortars for use in ambient or high-temperature acid lining construction. This type of silicate cement was very slow in setting and had to be exposed to the open air or heat cured, which created construction delays. Another problem was that these cements were not water resistant. These problems were resolved by the use of acid washing, which helped set the cements and make them water resistant.

In the early 1920s, Na_2SiO_3 mortars were introduced that used an acid catalyst to insolubilize the silica gel; this produced a mortar that cured faster and was water resistant. The physical properties of the silicate mortar were unchanged; that is, the acid, temperature, and solvent resistance were maintained. Because the acid-catalyzed mortar was chemically activated upon mixing, it could set within 24 to 48 h. This was a distinct advantage for the construction industry because it allowed brick to be laid continuously without concern over the mortar being squeezed out of the joints or the brick sliding out of line. Typical setting agents used are ethyl acetate ($C_4H_8O_2$), zinc oxide (ZnO), sodium fluorosilicate (Na_2SiF_6), glyceryl diacetate, formamide (CH_3ON), metallic polyphosphates, and other amides or amines.

It was later determined that when Na_2SiO_3 mortars were exposed to sulfuric acid (H_2SO_4) the reaction product was sodium sulfate (Na_2SO_4), which is a salt that expands and grows through hydration. This sodium salt can pick up as much as 10 mol of water of crystallization with a resultant size increase of 150%, thus creating internal stresses in the structure in which it was used. As a result, monolithic linings and brick surfaces would sometimes crack or spall.

As time passed, industry required silicate technologies that would meet changing applications and structural designs. By the 1950s, potassium silicate (K_2SiO_3) cements were introduced using methods of insolubilizing the silicate. The K_2SiO_3 cements possess enhanced corrosion resistance, particularly in H_2SO_4 environments. Potassium silicate cements react with H_2SO_4 to form potassium sulfate (K_2SO_4), which is not a growth salt. This eliminates the problem of internal stresses when the mortar is exposed to H_2SO_4 solutions.

The introduction of K_2SiO_3 cements was a positive step in producing a material that was more suited for application technology and corrosive environments. The K_2SiO_3 cements offer improved physical properties, which are particularly beneficial in monolithic applications. Furthermore, K_2SiO_3 cements provide better workability with less tackiness and longer working times than Na_2SiO_3 cements. Additional advantages include greater resistance to strong acid solutions and sulfation, more refractoriness, and no efflorescence as with Na_2SiO_3 cements. By this time, Na_2SiO_3 and K_2SiO_3 materials were not only being used as mortars but were also being introduced in new areas of monolithic application as both castable and gunite grades.

Within the last decade, the modified silicates have been developed. Modified silicates are manufactured by the addition of a powder form of Na_2SiO_3 in conjunction with a proprietary ingredient added directly to the powder fillers. This process eliminates the need to have both powder and liquid on the job site; water is the only required addition. Although these products have been commercially available, high cost and potential for sulfation have limited their use. The modified silicates do not have the chemical properties of the other silicates. The only advantage appears to be their simplicity of mixing.

Advantages and Disadvantages

The Na_2SiO_3 and K_2SiO_3 cements were developed because a material was needed that was acid resistant in dilute-to-strong concentrations and that offered higher temperature resistance than chemical-resistant organic materials. The development of these silicate-base acidproof cements resulted in many advantageous characteristics, such as a 100% K_2SiO_3-bonded system that had resistance to most solvents and acids over a pH range of 0 to 7; that was water and vapor resistant without special treatment; and that could withstand all concentrations of H_2SO_4, nitric acid (HNO_3), hydrochloric acid (HCl), and phosphoric acid (H_3PO_4). In addition, the Na_2SiO_3- and K_2SiO_3-bonded cements can be applied over damp acid-attacked concrete or brick surfaces as long as these substrates have acid pH surfaces, and they cure chemically within 36 h, thus decreasing construction delays. These cements can be applied by gunite or cast methods. Detailed information on guniting (shotcreting) can be found in American Concrete Institute publication ACI 547R on refractory concrete.

Certain disadvantages were encountered during the development of the acid-resistant silicate cements. It was common knowledge that the silicate cements were not resistant to alkalies, hydrofluoric acid (HF), and fluoride salts. As previously mentioned, the Na_2SiO_3 cements formed a growth salt when exposed to H_2SO_4 that put undue internal stresses on the structure of the material.

Inorganic monolithic linings also have a certain amount of permeability compared to organic surfacing materials. Over time, acid can penetrate the lining and eventually reach the surface of the substrate. This problem is now being combated by using a dual-lining system, which includes a chemically resistant elastomeric membrane applied to the surface of the substrate.

Current Technologies

Monolithic and Membrane Dual Linings. Organic linings or coatings have been used in stacks and ducts, but they will usually deteriorate if temperatures exceed 260 °C (500 °F) in high-sulfur gases. There are many types of lining materials for stacks and ducts; however, because of unpredicted environmental conditions within a stack or duct, it may be difficult to select the best candidate material. Stacks that are designed with heat recovery units and/or scrubbers, combined with one or more flues inside the shell that can control the temperature, should be considered for dual-lining applications.

In recent years, the trend toward using the superior technology of dual linings has emerged and is being recommended where corrosion problems occur throughout industry. Figure 1 illustrates the design of a typical membrane/monolithic system in the chemical industry.

There are numerous reasons for this new construction process. Condensation occurs not only on the face of the acid-resistant lining but can also penetrate and condense on the substrate to be protected. Although acidproof monolithic linings offer the proper chemical resistance, they are inherently inelastic, or brittle. In time, monolithic linings may tend to crack and absorb acids or acid gas condensate; therefore, it is advantageous to have a backup membrane. In many applications, the coefficient of thermal expansion of the monolithic lining may not match that of the substrate. Therefore, a flexible membrane will help accommodate stresses resulting from these differences in thermal expansion as well as other mechanically induced stresses.

For many applications, the monolithic lining should not be bonded to the substrate, and the membrane acts as a bond barrier. The concept of placing an impervious membrane between the substrate and inorganic linings has become a recommended practice. Some corrosion-resistant gunite manufacturers have installation specifications for use with membranes.

Membrane choices include asphaltics, resins, and synthetic elastomers. The membrane selected should resist the maximum acid concentrations and temperature expected. Organic linings can fail by disbonding, swelling, abrasion, and blistering from high temperature. Protection of an organic membrane with an inorganic barrier will minimize thermal exposure and mechanical abrasion to the membrane as well as the corrosive media that reach the membrane. An organic membrane compensates for some of the shortcomings of inorganic linings, such as cracking or spalling due to mechanical stresses from shrinkage, induced stresses, insufficient thermal expansion allowances, the anchoring system, vibration, and thermal or other stresses. To provide an effective barrier, inorganic linings are applied much thicker than typical organic coatings. They often contain fillers to act as reinforcement and to decrease shrinkage. Fillers are also incorporated to provide wear resistance for abrasive conditions.

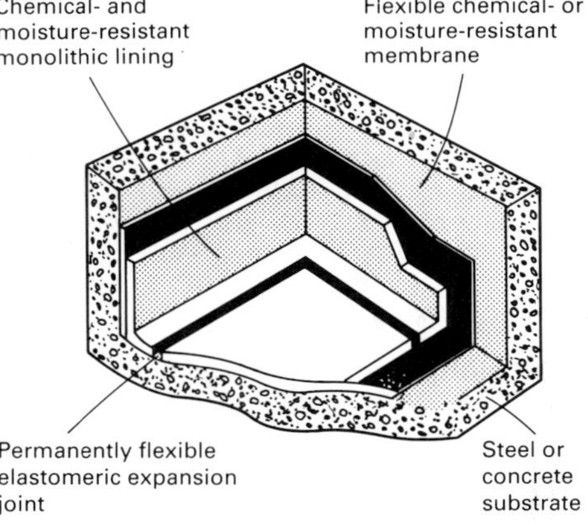

Fig. 1 Schematic of a chemical-resistant dual-lining system that provides double protection to the substrate in the form of a flexible membrane and a rigid surface layer. The flexible corrosion-resistant membrane is applied in direct contact with steel or concrete substrates. It is then covered by the monolithic cement lining, which provides protection over a broad pH range as well as against high temperatures.

The installation procedure for a dual-lining system is basically the same as if each component were being applied separately. These procedures are established for both organic coatings and inorganic monolithic linings. The condition of the working area and preparation of the surfaces follow standard practices used throughout the industry. Anchors are usually installed on the steel or concrete substrates before the organic membrane is applied.

Lightweight Insulating Materials. Industry has recently seen the development and application of improved lower-density, lightweight, insulating corrosion-resistant lining materials. The evolution of these products has followed the need to accommodate load limitations of structural supports and improved thermal protection of membrane coatings where substrate temperatures can range from 40 to 150 °C (100 to 300 °F) and even higher if process limits or gas cleanup systems experience frequent excursion or bypass conditions. Membranes require adequate thermal protection to retain physical and chemical properties throughout their service life. Because the membrane will also coat the studs, the inorganic coating must be thick enough to protect the membrane on the highest point of the studs. Therefore the inorganic lining material must not only protect the membrane from physical abuse but must also protect it from thermal deterioration.

When membranes are not used, condensation of chemicals from hot process streams should be designed to occur within a lining and not at the substrate. However, when membranes are used, the condensation can occur at the membrane, thus possibly reducing the required lining thickness. Therefore, inorganic silicate lining materials with lower thermal conductivity and density have been developed to meet these needs.

Application Procedures. Through the use of current technology, such as the dual-lining system and the availability of improved application techniques, corrosion protection practices have been greatly enhanced. Observations of successes and failures in various applications have led to the development of the following parameters, which must be addressed to specify the proper material. In any application, the primary consideration is the exposure environment. This includes the types and concentrations of chemicals, physical abuse, and temperatures. In this evaluation, it is important to remember that both the silicate monolithic lining and the membrane must resist the corrosive environment. The next consideration is the thickness of the monolithic lining. In high-temperature environments, the monolithic lining must be able to decrease the temperature through its thickness to a level that will not burn out the membrane applied to the substrate or the anchoring system. Once the materials that can resist the corrosive environment have been selected, the physical structure must be considered.

Steel. When a steel substrate is being used, sources of variation in electrical potential on the metal surfaces must be considered. These include mill scale, metal impurities, localized strains on parts of the metal, junctures of dissimilar metals, movement of electrolytes, weld joints, concentration cells, and externally imposed currents. Some or all of these differences in potential can occur on the surface and must be considered in the recommendation of a protective coating. The protective coating must provide the following properties to protect the steel:

- Physical protection to surface
- Prevention of concentration cells
- Dielectric properties
- Lower surface temperatures
- Prevention of electrolyte flow
- Deprivation of free oxygen from the surface
- Stress relief

Stress relief is an important factor in the design of a system. If a system does not compensate for relief in areas of high stresses, cracking will occur that can severely damage the entire system. Stresses on the system of major concern could result from vibration, unsupported surfaces, changes in planes, and welds.

The following five steps are application procedures that should be followed to ensure proper installation of a dual lining system. First, all areas of high stress that will result in movement of the steel structure, such as oil canning or vibration, must be externally supported. Second, the anchoring system should then be applied to the substrate. Anchors, such as V-type or longhorn studs, should be installed on the proper centerlines in a diamond or staggered pattern with the tines randomly oriented. Third, the steel surface must be stripped of all oil and grease by chemical cleaning. The metal should then be sandblasted to a Steel Structures Painting Council SP 10 Near-White Blast with a nominal 64-μm (2.5-mil) profile. Welds should be ground to a smooth, rounded radius with no sharp edges before blasting. Fourth, the appropriate membrane should be applied, ensuring that all of the studs are completely coated and that the system is free of pin holes.

The final step is to apply the monolithic lining. The two available methods of application should be considered. A castable material will shrink more than the gunitable material; therefore, expansion joints must be planned into the system in areas exceeding 6 × 6 m (20 × 20 ft) or where changes in planes will cause stresses to occur. A gunitable material will shrink less and therefore does not require expansion joints. Figure 2 illustrates the gunite application of K_2SiO_3 cement over steel ductwork in a fossil fuel power plant.

Concrete. In applications over concrete, as with steel, the first considerations are the corrosive environment and the thickness of the material. Once the materials have been selected, the physical structure must be considered. When the substrate is concrete, factors different from those associated with steel must be considered. Inorganic monolithics can be applied over old or new concrete surfaces. When concrete substrates are being used, the following questions should be answered:

- Is the concrete old acid-attacked concrete or new concrete?
- Is the concrete capable of supporting loads?
- Is the structure above grade or below grade?
- Does the concrete need to be waterproofed from the outside?
- Does the concrete structure have existing cracks?
- Is the concrete surface a high or low pH?

Fig. 2 Gunite application of 50 mm (2 in.) of K_2SiO_3 cement over steel ductwork in a coal-fired power plant

A combination of many of these items may apply and should be considered in the recommendation of the proper system. The following are the criteria for a concrete surface before a system may be applied. When working with old acid-attacked concrete, all loose and deteriorated concrete should be removed. The surface should be firm, hard, and at a minimum pH of 5, and all surfaces should be brought back to grade and the slopes reestablished. When working with new concrete, the concrete should be firm and sound. All structural cracks should be repaired, and it should be cured for a minimum of 7 days. Slopes should be a minimum 3.2 mm (1/8 in.) to a maximum 6.4 mm (1/4 in.) per linear foot and should have attained a minimum compressive strength of 21 MPa (3000 psi).

The following five steps are application procedures that should be followed to ensure proper installation of a dual-lining system. First, the concrete should meet the standards outlined above. Second, the concrete should be cleaned of all oil, grease, and form release compounds by chemical cleaning, waterblasting, or scarifying. Third, the anchoring system should be placed at the specified centerline distance in a diamond-shape pattern with a random orientation of anchor tines. The anchoring system should consist of V-type or longhorn studs. Fourth, the membrane should be applied to the recommended thickness, ensuring that all studs are completely coated and that the coating is free of pinholes.

The final step is to apply the monolithic. The castable and the gunitable methods should both be considered. As stated above, when applying by the gunite method there is less shrinkage; therefore, there is no need for expansion joints in the monolithic. However, the lining will require control joints in the normal manner. Expansion joints are recommended with cast applications and should be installed at the perimeter of floor areas, equipment pads, and floor trenches; around floor drains, column bases, and protrusions; at changes in planes; and over expansion joints in concrete slabs not to exceed 6 m (20 ft) on centerlines.

Brick. A monolithic lining can be applied over acid-attacked masonry construction. The first step is the surface preparation of the brick. Flyash and other contaminants should be removed by sandblasting or waterblasting. All attacked or unsound mortar should be removed from the joints. Mortar joints should be cleaned to a depth of at least 13 mm (1/2 in.) to provide support for the monolithic lining. If the joints cannot be cleaned to this depth, it will be necessary to set longhorn or V-type anchors in the joints. The preferred method of application in this case is to gunite the material. Guniting allows the material to be applied under pressure, thus enabling the material to be packed into the 13-mm (1/2-in.) open joints and allowing the system to be supported by the studs or keying into the joints.

Corrosion Resistance

Monolithic linings are among the fastest growing products in the corrosion protection field because of their resistance to a wider range of corrosive media than most other construction materials. The following industrial applications illustrate the corrosion resistance and some uses of monolithic linings.

Wastewater treatment systems in the southern United States have experienced corrosion problems in areas of high temperatures, high humidity, and slow drainage. Consequently, millions of dollars have been spent in replacing the corroded concrete infrastructures in these wastewater treatment systems in wet wells, grit chambers, sewer lines, and aeration basins. This corrosion went unnoticed throughout the years because the corrosion mechanism was not understood. Most of the corrosion was not caused by discharged chemicals but was microbiologically induced. The concrete infrastructures are seldom completely filled and the combination of high temperatures and high humidity creates a perfect breeding ground for bacteria, fungi, algae, and gaseous products from decomposing sewage.

Carbon dioxide (CO_2), combined with hydrogen sulfide (H_2S) from the slightly warm and fermenting sewage, reacts with the damp concrete surfaces to form carbonates and calcium sulfate ($CaSO_4$). As this process proceeds, the pH becomes more acidic. Bacterial action results in the production of elemental sulfur from various sulfate reactions, with subsequent oxidation of elemental sulfur to H_2SO_4, which the bacteria secretes.

The H_2SO_4 then directly attacks the underlying portland cement concrete substrate and causes destruction of the infrastructure. There are documented cases of microbiologically induced corrosion having proceeded to the point at which structures collapsed in as few as 6 years or required major rehabilitation in as few as 4 years.

A proven method of rehabilitating deteriorated concrete wastewater structures involves applying a urethane-base membrane to the properly prepared substrate, followed by the pneumatic application of a 100% K_2SiO_3-bonded acid-resistant concrete. The urethane-base membrane is a two-component material with a high solids content and very low permeability. To ensure pinhole-free coatings, the membrane is applied in two or more coats to a wet film thickness ranging from 1.6 to 3.3 mm (63 to 130 mils). These membranes are applied by airless spray with a spray tip pressure of approximately 28 MPa (4000 psi). The high pressure further helps to ensure pinhole-free application of the membrane. The substrate must be dry enough to accommodate the urethane membrane without reacting with the polyisocyanate catalysts usually used.

Therefore, it is sometimes desirable to apply a moisture-tolerant primer, such as a suitable formulated epoxy, to the damp concrete or to apply a fresh layer of concrete first. Following the proper application and cure of the urethane membrane, a veneer of a 100% K_2SiO_3 acidproof cement is applied over the membrane and anchors.

The chemical industry can be the most difficult of all industries in which to design a protective lining. The conditions within one plant can be staggering; there may be acids, alkalies, and solvents, along with extremely high temperatures and physical abuse. Potassium silicate base cements satisfy most of these conditions while maintaining economy. The combination of a wide range of chemical resistance, high physical strength, and refractoriness makes these cements leading candidates for many chemical plant applications.

For example, at a major Midwest chemical plant, the action of HCl by-products and ambient moisture on floors, digester tank supports, trenches, and sumps in $NiCl_2$ and $NiSO_4$ processing units produced very corrosive conditions. Restoration of the existing structures was accomplished with a K_2SiO_3 cement and an appropriate membrane.

In the $NiCl_2$ unit, the existing concrete floor was chipped out, and new concrete was poured; the steel beam supports for the digester tanks were also encapsulated with regular concrete. Both floor and digester foundations then received a trowel-applied asphaltic membrane. The membrane provided a flexible acid-, water-, and alkali-resistant barrier between the concrete substrate and the monolithic surfacing. The trowel-applied membrane system is applied at room temperature and offers the chemical resistance of hot-applied membranes for those areas where a hot membrane may not be practical or advisable.

The next step was to apply a K_2SiO_3 monolithic lining by casting into forms. The lining was 38 mm (1 1/2 in.) thick over foot traffic areas near the digesters; 38 to 50 mm (1 1/2 to 2 in.) thick around the encapsulated digester foundation, creating a collar; and 50 mm (2 in.) thick in forklift truck aisles. Finally, at the interface of the digester tank wall and the collar, an elastomeric urethane-base compound was used as a permanently flexible expansion joint.

In the $NiSO_4$-processing unit, the problem areas included trenches running among process equipment in two different buildings. In addition, the various collection sumps into which these trenches drained required restoration. As was done in the $NiCl_2$ area, the deteriorated concrete was reconstructed with new concrete. The asphaltic membrane was trowel applied. A K_2SiO_4 monolithic lining was cast over the membrane to 25 mm (1 in.) in the trenches, 25 to 38 mm (1 to 1 1/2 in.) in the sumps, and 38 mm (1 1/2 in.) on the floors, and the elastomeric urethane expansion joint compound was applied.

The service conditions for the two processing areas included a pH of 1 to 2 due to concentrations of HCl up to 32%, water, temperatures to 95 °C (200 °F), and vibrations from imposed loads as great as 3630 kg (8000 lbs) in the truck aisles. These applications were done by the casting method, but they could have easily been done by the gunite method.

CVD/PVD Coatings

Hugh O. Pierson, Ultramet

VAPOR-DEPOSITED COATINGS modify the surface properties of materials. They are widely used and have been particularly successful in improving such mechanical properties as wear, friction, and hardness in cutting applications. Their use as corrosion-resistant coatings is also becoming widespread.

Deposition Processes

The vapor deposition processes fall into two major categories—physical vapor deposition (PVD) and chemical vapor deposition (CVD)—which in turn comprise various subcategories. Although these divisions often appear confusing and arbitrary, each has its own well-defined advantages and disadvantages, whether technical or economic. This is also true of the materials used in vapor deposition. Therefore, the need for a thorough understanding of both processes and materials is essential in order to select the optimum combination.

The major deposition processes that will be reviewed in this article are sputtering, evaporation, ion plating, and CVD. These processes all have one characteristic in common: The deposition species are transferred and deposited in the form of individual atoms or molecules. In this respect, they are fundamentally different from particulate or liquid deposition processes, such as thermal spraying or electroplating (see the articles "Thermal Spray Coatings," and "Electroplated Coatings" in this Volume). A primary benefit of vapor deposition is that it is essentially nondamaging to the environment, because there are no solvents or electrochemicals to dispose of. At a time when environmental-protection regulations are continually being tightened, this crucial factor favors vapor deposition over other processes that may be more harmful to the environment.

Sputtering is the principal PVD process. It involves the transport of a material from a source (target) to a substrate by means of the bombardment of the target by gas ions that have been accelerated by a high voltage. Atoms from the target are ejected by momentum transfer between the incident ions and the target. These ejected particles move across the vacuum chamber to be deposited on the substrate.

Figure 1 shows a schematic of the sputtering process. In its simplest form, the process occurs in an inert (noble) gas at low pressure (0.13 to 13 Pa, or 1×10^{-3} to 100×10^{-3} torr); in most cases, the gas is argon. Argon has higher mass than other noble gases, such as neon or helium, and is easier to ionize. Higher mass gives a higher sputtering yield, especially if the mass of the bombarding particle is of the same order of magnitude or is greater than that of the target atom. Other gases, such as oxygen or nitrogen, can also be used, but they may react chemically with the target. The sputtering process begins when an electric discharge is produced and the argon becomes ionized. The low-pressure electric discharge is known as glow discharge, and the ionized gas is termed plasma.

The argon ions hit the solid target, which is the source of the coating material (and not to be confused with the substrate, which is the item to be coated). The target is negatively biased and therefore attracts the positively charged argon ions, which are accelerated in the glow discharge. This attraction of the ions to the target (also known as bombardment) causes the target to sputter, which means that material is dislodged from the target surface because of momentum energy exchange. The higher the energy of the bombarding ions, the higher the rate of material dislodgement.

Sputtering developed very rapidly in the 1970s in the semiconductor industry, in which the technique is essential for mass production. With the advent of the technique of magnetron, or magnetron-enhanced sputtering, two original primary disadvantages—nonuniform coverage and low deposition rate—have been largely eliminated. Sputtering is making rapid inroads in such corrosion applications as high-chromium alloy coatings on turbine blades and many other new applications that require high-quality coatings with good adhesion.

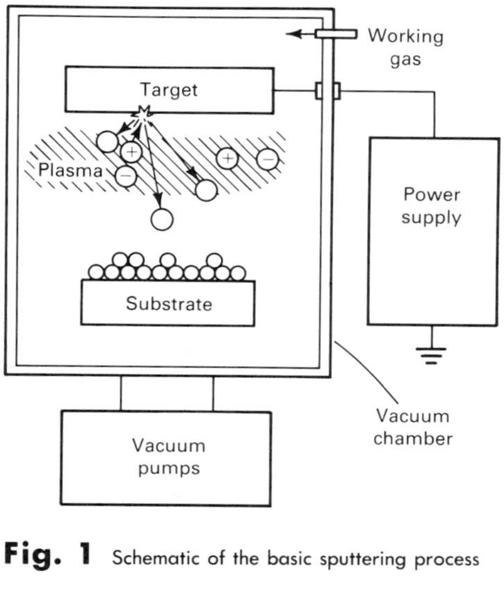

Fig. 1 Schematic of the basic sputtering process

For the deposition of thin metal films (in the 1000-Å range), sputtering is the best technique. Deposition is feasible in a controllable manner for both compound and elemental targets (the source of the coating material). Adhesion is good and can be further improved by sputter cleaning the substrate or by bias sputtering. Large equipment is readily available, and sputtering has become a highly automated process. The quality, structure, and uniformity of the deposit are excellent. The disadvantages of sputtering include its thickness limitation and high cost. It is also a line-of-sight process that requires special fixtures and shaped targets to coat uniformly; furthermore, the coating of compound curves, recesses, and holes is difficult.

Evaporation was the first PVD process used on an industrial basis for the aluminum metallization of plastics and glass for decorative purposes. Originally the most widely used process, evaporation has now been overtaken by sputtering. The metallization process is relatively simple and well adapted to mass production.

The basic evaporation process involves the transfer of material to form a coating by physical means alone, essentially vaporization. Practically all metals can be evaporated, making evaporation a universal process with regard to metals. Like sputtering, it is a line-of-sight process; therefore, to achieve efficiency of coverage and uniformity, it is necessary to use multiple evaporation sources and to rotate or move the substrate uniformly to expose all areas.

Unlike other vapor deposition processes, evaporation is a low-energy process, with particle energy averaging 0.2 to 0.3 eV as compared to 10 to 40 eV for sputtering, 200 eV or more for ion plating, and 100 keV for ion implantation. The term low energy means that adhesion of the particle to the substrate may be marginal. In addition, the temperature of the substrate is not increased by the deposition, which means there is little or no diffusion. Because no strong physical or chemical action takes place at the interface (as opposed to ion plating or CVD, for example), particular care must be taken to remove all impurities in the chamber and to have a perfectly clean substrate. The basic process (with an electron beam source) is illustrated in Fig. 2.

Very large evaporation equipment, which was developed for the optical industry, is commercially available. Coatings of several microns in thickness can be rapidly deposited. The disadvantages of this method are nonuniform coverage (limited to line of sight) and relatively low adhesion unless glow discharge cleaning is used (glow discharge cleaning removes surface atoms from the substrate). Evaporation is very successful in applications in which adhesion and good struc-

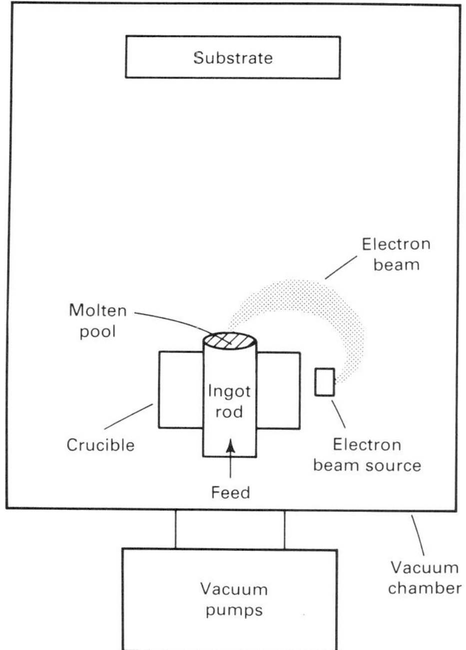

Fig. 2 Schematic of the basic evaporation process

ture are not critical, such as decorative and optical uses. Aluminum is the principal evaporation material; chromium and stainless steel are also widely used. Because improvements in technique have improved quality, evaporation is not used for critical corrosion applications.

Ion plating is actually a hybrid concept based on the evaporation (or sputtering) mechanism coupled with a glow discharge (Fig. 3). There are three possible deposition systems: resistance evaporation, electron beam evaporation, and sputtering. All three systems generate a plasma by a glow discharge, thus imparting a large increase in the energy of the deposition species.

It is generally accepted that plasma plays the same part as high temperature in improving the adhesion and increasing the reactivity of the coating; an ion passing through a potential of 100 eV acquires an equivalent temperature of 10^6 K. The simplest forms of ion plating are ion nitriding and ion carburizing, in which nitrogen or carbon ions are obtained from the ionization of nitrogen or a hydrocarbon. These react with a steel substrate to form nitrides or carbides. If the deposition species is ionized in a reactive gas, a compound is formed with some of the gas atoms, and deposition of the compound occurs. Ion plating is widely used and is gradually replacing pack cementation or diffusion.

Because ion plating is a glow discharge process in which a plasma is formed, it is necessary to have at least a threshold gas pressure to maintain the plasma. This is different from standard nonionized evaporation, in which the minimum pressure obtainable is preferred. The net effect of this higher gas pressure is to increase the number of collisions with gas molecules and therefore produce more scattering of the deposition species, which in turn diminishes the importance of the line-of-sight principle and improves coverage.

The deposited species in ion plating have higher energy than those of evaporation or sputtering

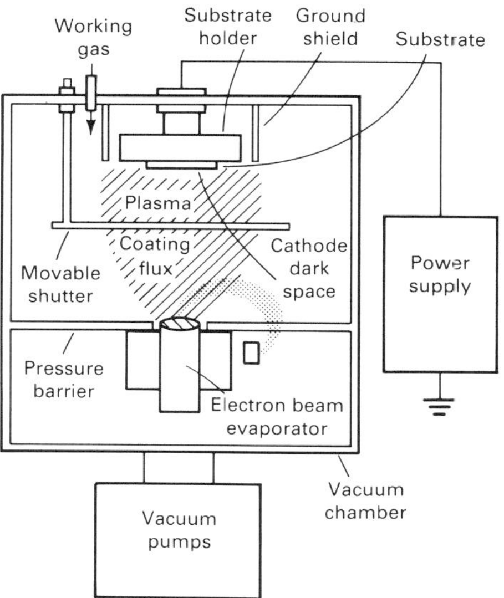

Fig. 3 Schematic of the basic ion plating process

processes; this results in improved adhesion as well as a deposit with improved structure and fewer imperfections. In addition, if a sputtering system is used, it is easy to sputter the surface of the substrate to clean it before deposition.

On the negative side, the equipment is more complicated and therefore more expensive than either evaporation or sputtering equipment, and the process is more exacting. The level of ionization can be difficult to control, which may result in uneven deposition. Although the process has not yet reached the total reliability necessary for automated operation and total acceptance, its advantages are such that it is rapidly gaining in popularity.

Chemical vapor deposition is a well-established deposition process that is extensively used in the semiconductor and cutting tool industries. It has the unique ability to deposit thick, dense, high-quality films (up to 6.4 mm, or $^1/_4$ in. in thickness or more) at a cost that is generally lower than that of PVD processes. The high temperature (about 1000 °C, or 1830 °F) necessary for CVD promotes good adhesion but also considerably restricts the type of substrate that can be coated to ceramics, refractory metals, and special alloys.

In most cases, steel will require heat treatment in a vacuum or protected atmosphere after CVD coating and will often change dimensions sufficiently to require postdeposition machining if tolerances are close. Another constraint resulting from the high temperature of deposition is the stress in the deposit due to the difference in thermal expansion between substrate and deposit. These stresses may be sufficient to cause cracking, spalling, and loss of adhesion. The new technique of plasma CVD can be used at much lower temperatures in, for example, semiconductor applications, but the resulting films tend to incorporate impurities.

The materials that can be easily deposited by CVD are more limited than with PVD. Chemical vapor deposition is particularly useful for depositing compounds and refractory metals and lends itself very well to corrosion applications.

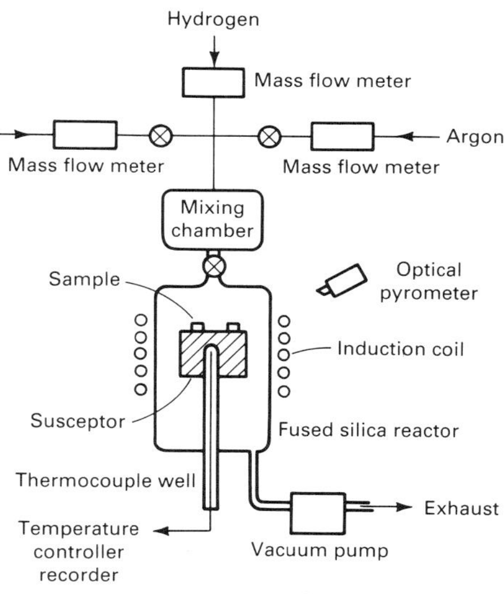

Fig. 4 Schematic of a laboratory CVD reactor

Figure 4 shows a schematic of CVD laboratory equipment. The reactive gases are metered in the mixing chamber. They then pass into the deposition chamber and contact the heated part to be coated (sample); at this point, they react and deposition occurs. By-products are removed through the vacuum system.

Chemical vapor deposition reactions can occur over the full range of pressure. Low pressure increases boundary-layer diffusion and uniformity, often at the expense of deposition efficiency. Each application and each reaction must be analyzed to determine the optimum conditions for deposition.

Coating Materials and Applications

The list of materials that can be vapor deposited is extensive and covers almost any coating requirement; the most important materials are given in Table 1. Aluminum and aluminum alloys are among the most widely used deposition materials and are gradually replacing cadmium in many corrosion applications. Sputtered chromium and stainless steel are also making great inroads in corrosion applications. A promising material is titanium nitride. It is hard, very stable, and refractory, and it has good lubricating characteristics and a pleasing gold appearance. It is widely used in wear applications, such as cutting tools, and in decorative applications. It is usually applied by sputtering or ion plating, and it holds potential for corrosion protection.

The properties of these materials vary widely and reflect the complexity of selecting a corrosion-resistant coating for a given substrate or a given environment and the need to conduct a thorough analysis of the problem. This is especially true when cost is the primary factor. For example, gold is an ideal material for many corrosion applications, but would of course be cost prohibitive in most cases. Before selecting a material and process for a corrosion resistance application, the following considerations should be reviewed:

- Availability and cost of the coating material

Table 1 Some corrosion-resistant vapor-deposited materials

Material	Deposition process
Aluminum	Evaporation
Chromium	Sputtering
Nickel	CVD, sputtering
Rhenium	CVD
Tantalum	Sputtering, CVD
Titanium	Evaporation, sputtering
Tungsten	CVD, sputtering
Boron	CVD
Carbon	CVD, ion plating
Titanium diboride	Evaporation, CVD
Silicon carbide	Sputtering, CVD
Titanium carbide	Sputtering, evaporation, CVD
Tungsten carbide	Sputtering, CVD
Boron nitride	CVD
Silicon nitride	Sputtering, ion plating
Titanium nitride	CVD, sputtering, ion plating
Aluminum oxide	Sputtering, ion plating
Molybdenum silicide	Sputtering, ion plating
Nickel-chromium alloys	Sputtering
Gold alloys	Evaporation
Borazon	CVD
MCrAlY	Evaporation, sputtering
Bronze	Evaporation, sputtering

- Cost, size, complexity, and availability of the deposition equipment
- Resistance to the chemical environment of the coating material
- Compatibility with the substrate, especially thermal expansion mismatch
- In special applications, such as aerospace, high-temperature resistance, radiation resistance, and high-vacuum stability

Chemical vapor deposited and PVD coatings have the advantage that the film (if thick enough) is essentially pore free and fully dense. Penetration by moisture and gases to the substrate is greatly reduced, if not eliminated. Despite this, the role played by CVD/PVD coatings in corrosion protection remains small. Some current applications are discussed below.

Coating for Graphite. Graphite is perhaps the best material for applications requiring high strength and low density at very high temperatures. It is used as a hot-pressed material or as a carbon-carbon composite, which is a combination of graphite fibers and a carbonized resin matrix. Graphite is used exclusively in such applications as rocket nozzles or advanced turbine components. However, its disadvantage is a lack of oxidation resistance above 500 °C (930 °F). By using CVD to coat the graphite with silicon carbide, it is possible to provide oxidation protection up to 1650 °C (3000 °F). Silicon carbide oxidizes to form a silicon oxide surface that will prevent further oxidation. Advanced studies to further increase the oxidation temperature of graphite are in progress using such oxides as hafnia.

Aluminum Coating of Steel. The coatings of steel (and other metals) with aluminum by ion plating is an industrial process with excellent potential for large-scale production. It replaces hot dipping of aluminum on steel, which has been in use for many years. During immersion in hot dipping, a brittle intermetallic compound layer of iron-aluminum is formed; this limits the formability of the steel. In addition, the coating contains iron impurities, which may be a severe shortcoming in highly corrosive environments (Ref 1).

Electroplated cadmium is another commonly used system for the corrosion protection of steel and other metals. However, it can cause hydrogen embrittlement of high-strength steel and titanium. In addition, cadmium has a low melting point (320 °C, or 608 °F) and presents environmental problems because of its toxicity. A process has recently been developed for applying aluminum on metals by ion vapor deposition, which has none of the restrictions of hot dipping or cadmium electroplating. This process falls under military specification MIL-C 83488.

Vapor deposition of aluminum does not produce hydrogen; therefore, it does not cause embrittlement. Vapor-deposited aluminum protects against stress-corrosion cracking (SCC) and has a temperature limit of 495 °C (925 °F). It is less expensive than most other barrier coatings. Applications for vapor-deposited aluminum in the aerospace industry include fuel and pneumatic line fittings (in place of anodizing), electrical connectors to replace cadmium, powder metallurgy parts, fasteners, electrical black boxes, corrosion and SCC protection of depleted uranium and titanium alloys, and electromagnetic interference compatibility (Ref 2).

Alloy Coatings for Gas Turbines. Gas turbines can be divided into three general categories: aircraft turbines using highly refined fuels with operating temperatures between 900 and 1150 °C (1650 and 2100 °F) (new designs require even higher temperatures); marine turbines using moderately refined fuels, which operate at 700 to 870 °C (1290 to 1600 °F) and are exposed to salt corrosion; and industrial turbines operating at 760 to 925 °C (1400 to 1700 °F) and using moderately refined fuels, often with high sulfur contents. The primary material problems in these applications are oxidation and high-temperature corrosion. These problems are becoming more acute with the trend toward higher operating temperatures for better performance and improved fuel economy.

Until recently, thermal spraying of refractory alloys to protect exposed parts constituted the bulk of the market. An estimated $25 million worth of refractory powder was used in 1985 for gas turbines alone (Ref 3). In the last few years, PVD coating by electron beam evaporation and sputtering have gained wide acceptance. These coatings consist of alloys of chromium, aluminum, and yttrium with cobalt, nickel, or iron and are known generically as MCrAlY. They are successfully used on aircraft turbines. Recent developments have shown an improvement factor of four over previously used diffusion coatings for protection against sulfidation corrosion (Ref 4). Large production units are available. For example, 100-μm (4-mil) thick films, usually NiCoCrAlY or CoCrAlY, are deposited by electron beam evaporation.

Marine gas turbines are subject to severe sulfidation, oxidation, and hot corrosion, especially in the presence of sodium sulfate (Na_2SO_4). Electron beam evaporated CoCrAlY coatings have considerably increased the service lives of first-stage blades. Further improvements have been achieved by using high-chromium low-yttrium alloys.

Industrial turbines normally do not require corrosion protection unless high-sulfur fuels are used, for example, as in Saudi Arabia. In such cases the hot corrosion problem can be solved by using thin-film CoCrAlY coatings and a very large (200-kW) electron beam coater capable of handling industrial turbine parts weighing up to 40 kg (90 lbs).

REFERENCES

1. F. Dunbar, "Aluminum Coated Steels: Past, Present and Future," Paper 8201-037, presented at the ASM Metals Congress, St. Louis, MO, American Society for Metals, Oct 1982
2. D. Muehlberger and J. Reilly, "Improved Equipment, Productivity Increase, Applications for Ion Vapor Deposition of Aluminum," Technical Report 83-0691, Society of Automotive Engineers, 1985
3. "Wear Resistant Coatings Database," Gorham International Inc.
4. L. Bianchi, Electron Beam PVD Corrosion Resistant Coatings for Extended Life of Gas Turbine Parts, *Ind. Heat.*, June 1980

Thermal Spray Coatings

Robert H. Unger, TAFA Inc.

THERMAL SPRAY COATINGS provide effective long-term corrosion protection of iron and steel over a wide range of corrosive environments. Corrosion protection can be obtained for up to 20 years without maintenance and up to 40 years with minimal maintenance. Such long-term protection has been documented for both atmospheric and immersion service in numerous industrial and marine applications.

Thermal spraying is a particularly versatile technique because it can be used both in-shop or on-site. This article will discuss the various thermal spray processes, the types of coatings produced, and the protection these coatings can provide in various environments. Some of the advantages of thermal spraying are the ability to apply significantly thicker coatings than galvanizing, excellent paintability, no heat distortion from processing, no curing requirements, and the capability of applying coatings in the field.

Fundamentals of the Process

Thermal spraying comprises a group of processes in which finely divided molten metallic or nonmetallic material is sprayed onto a prepared substrate to form a coating. The sprayed material is originally in the form of wire or powder. As the coating materials are fed through the spray unit, they are heated to a molten or plastic state and propelled by a stream of compressed gas onto the substrate. As the particles strike the surface, they flatten and form thin platelets that conform and adhere to the irregularities of the prepared surface and to each other. They cool and accumulate, particle by particle, into a lamellar, castlike structure. The spray gun generates the necessary heat for melting through combustion of gases, an electric arc, or a plasma. Figure 1 illustrates a general thermal spraying process.

The deposited structures of thermal spray coatings differ from those of the same material in the wrought form because of the incremental nature of the coating buildup and because the coating composition is often affected by reaction with the process gases and the surrounding atmosphere while the materials are in the molten state. For example, where air or oxygen is used as the process gas, oxides of the material applied may be formed and become part of the coating. The as-applied structures of all thermal spray coatings are similar in their lamellar nature; the variations in structure depend on the particular thermal spray process used, the processing parameters and techniques employed, and the material applied.

The bond between the sprayed coating and the substrate is generally mechanical. Proper surface preparation of the substrate before spraying is often the most critical influence on the bond strength of the coating. Surface preparation techniques will be discussed in the section "Surface Preparation" in this article.

Methods of Deposition

The most common methods of depositing thermal spray coatings can be classified as wire flame, powder flame, electric arc, and plasma.

Wire flame spraying, the oldest spray process, uses the heat from a chemical reaction as the melting source. Common fuel gases are acetylene, methyl-acetylene-propadiene (MAPP), propane, propylene, and natural gas, each combined with oxygen. Acetylene is the most widely used because of the higher temperatures it produces. Any material available in the form of wire and capable of being melted below 2480 °C (4500 °F) can be flame sprayed. Figure 2 shows a schematic of the wire flame spraying process.

The gas flame is used only to melt the material. Spraying is accomplished by surrounding this flame with a coaxial stream of compressed gas—for example, air—to atomize the molten material and to propel it onto the workpiece.

Wire flame spraying is widely used in the on-site spraying of zinc, aluminum, and zinc/aluminum alloys because of its low capital investment, portability, and ease of operation. In general, wire flame sprayed coatings exhibit lower bond strengths, higher porosity, a narrower working-temperature range, and higher heat transmittal to the substrate than plasma or electric arc sprayed coatings.

The powder flame spray process also uses combustion gases as the heat source, but with a powder feedstock (Fig. 3). This offers the advantage of being able to spray nonmetallic materials that are not available as wire. In its simplest form, the gun requires no compressed air supply. The powder feedstock is stored in a hopper mounted on top of the gun, and a small amount of oxygen from the gas supply is diverted to carry the powder by aspiration into the oxygen-fuel gas flame, in which it is melted and carried by the flame onto the workpiece. The equipment is lighter and more compact than other forms of thermal spray equipment, but because of the lower particle velocities obtained, the coatings generally have lower bond strength, higher porosity, and a lower overall cohesive (interparticle) strength than the coatings produced by other thermal spray processes.

Electric arc spraying uses metal feedstock in wire form without the need for combustion gases. Figure 4 shows a schematic of the electric arc spraying processes. Two feedstock wires serve as electrodes, which are initially insulated from each other and advance to meet at a point in an atomizing gas stream. A potential is applied across the wires so that an arc is formed at the intersection tips where the wires melt. A stream of gas, usually compressed air, flows across the arc zone and strips off the liquid metal to form molten droplets that are propelled onto the substrate.

Because of the high temperatures in the arc zone, the coatings have excellent adhesion and high cohesive strength. The superheating of the particles leads to metallurgical interactions after impact of the particles on the metal substrate. These localized interactions can often lead to

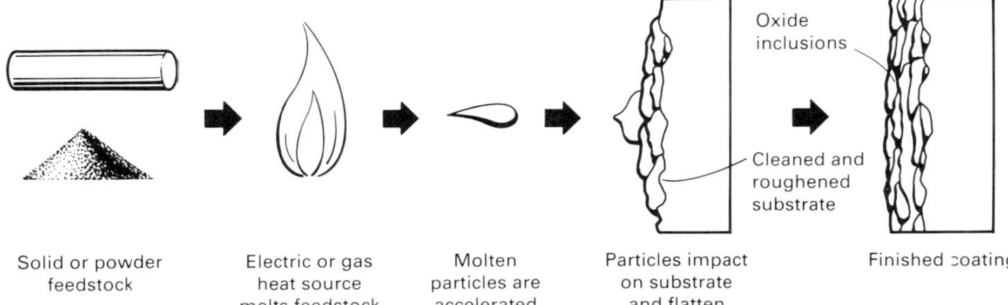

Fig. 1 Schematic of the general thermal spray process

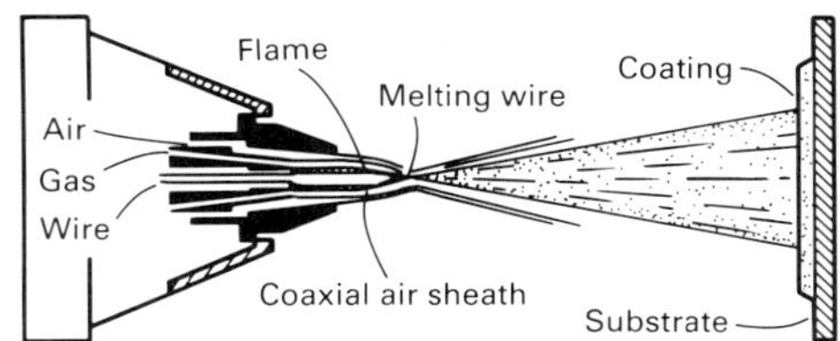

Fig. 2 Schematic of the wire flame spray process

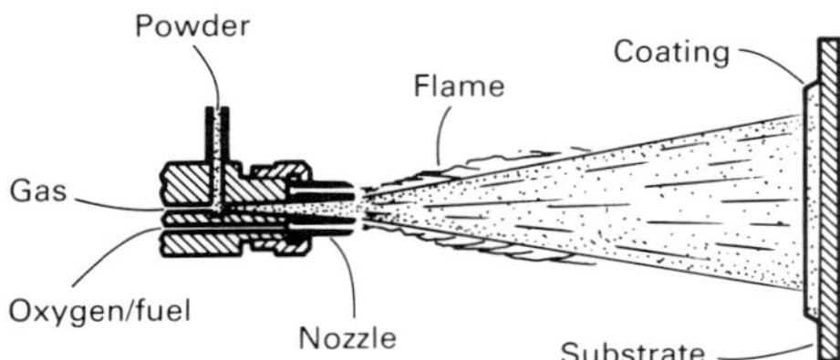

Fig. 3 Schematic of the powder flame spray process

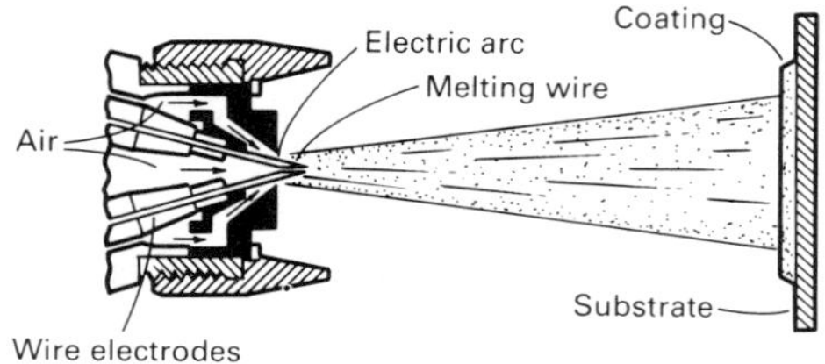

Fig. 4 Schematic of the electric arc spray process

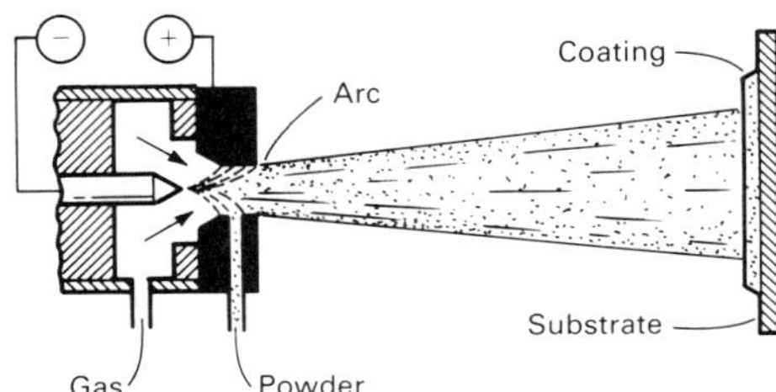

Fig. 5 Schematic of the plasma spray process

small weld spots; this increases bond and cohesive strengths considerably compared to flame spraying. Substrate heating is also significantly lower than in other processes because of the absence of the heat from a flame.

The electric arc spray system, like the flame spray system, is light and portable and is commonly used for on-site anticorrosion spraying. A portable generator can be used to supply electricity for remote jobs.

The plasma spray process uses coating materials in the form of powders, which are melted with a plasma heat source (Fig. 5). Plasma provides controllable temperatures that range well above the melting point of any known substance. To generate the plasma, inert gas is passed through an electric arc contained within the gun; the arc, in turn, heats the gas. The hot gas issuing from the gun is called plasma and resembles a bright flame. Powder feedstock is introduced into the plasma stream, becomes molten, and is accelerated and carried to the workpiece by the free plasma stream.

The plasma system produces excellent coatings, but the equipment is complex and expensive to operate. For most anticorrosion spraying, particularly on-site work, the other spray processes usually provide adequate coatings, and plasma is not generally used. Plasma spraying is often used where coatings are needed to provide corrosion protection against chemical or high-temperature corrosion and where the work can be done in-shop.

Surface Preparation

Thermal spray coatings rely primarily on mechanical bonding to the substrate surface. Because bonding relies on this mechanical interlocking of spray particles with the substrate surface, preparation of the surface is critical to a successful coating. Preparation of the substrate before spraying is virtually the same for all thermal spray processes.

Cleaning and Degreasing. Chemical cleaning is used on parts that are contaminated or impregnated with material that cannot be removed by other methods. Surfaces can be chemically cleaned by vapor degreasing, steam, hot detergent washing, or application of an industrial cleaning solvent. Degreasing is usually the most economical and safest way to remove lubricants and oils.

Grit blasting is the most frequently used surface preparation procedure. The roughening cleans the surface of contamination that may inhibit bonding and creates an irregular profile of minute surface irregularities. This enhances the adhesion of the coating. Before application of the thermal spray coating, a steel substrate should be blasted to a white-metal condition (according to Steel Structures Painting Council specification SSPC-SP5-82) with an appropriate surface profile (75 μm, or 3 mils). Sharp grit should be used, because the substrate must be cleaned and given an anchor tooth profile for a proper mechanical bond. Coarse-grit (18 to 24 mesh) materials are also used, such as aluminum oxide, chilled iron grit, coal slag, copper slag, and silica sand.

Rough threading is often used in conjunction with grit blasting for cylindrical surfaces. It consists essentially of machining a thread form on the area to be sprayed. This type of preparation provides excellent adhesion of the coating to the substrate.

Bond coats are often used when an extremely high bond strength is desired or when mechanical deformation of the substrate is not possible or practical. This involves spraying materials that are exothermic and self-bonding. Nickel-aluminum alloys, aluminum bronze, and molybdenum are the most commonly used bond materials. A thin coating (normally about 50 to 125 μm, or 2 to 5 mils) of bond material is generally applied before the corrosion-resistant overcoat. Care should be taken to ensure that the bond material is compatible with the overcoat and that a galvanic cell will not be created, which may be detrimental to the performance of the coating system.

Preheating of the substrate is often beneficial. Preheating to 65 to 95 °C (150 to 200 °F) will eliminate surface condensation and will reduce shrink differentials between the coating and the substrate. It is generally employed only with the flame spray processes, which use their flame as the heat source for warming the substrate.

Coating Selection for Atmospheric and Immersion Environments

Zinc and aluminum and their alloys are the metals most widely used for thermal spray anticorrosion coatings. They are extensively used for the corrosion protection of iron and steel in a wide range of environments and have been shown to provide very long-term protection (over 20 years) in both marine and industrial locations. The British Standards Institute Code of Practice (Ref 1) specifies that only these electrochemically active metals will give protection of greater than 20 years for every type of environment considered and that, in areas of immersed seawater or in seawater splash zones, only sealed sprayed zinc or aluminum will give such protection. Corrosion tests of thermal spray coated steel conducted by the American Welding Society confirm the remarkable effectiveness of sprayed zinc and aluminum coatings over long periods of time in various hostile environments (Ref 2).

Under most conditions, zinc and aluminum are more corrosion resistant than steel; therefore, they form an effective barrier film between the steel substrate and the atmosphere. More important, because they are anodic to steel, they also act galvanically to protect the ferrous substrate from corrosion in electrolytic solutions, such as saline environments. In effect the metallized coating serves as a sacrificial anode. This prevents corrosion of the steel substrate even where the coating coverage is incomplete or suffers mechanical damage.

Zinc and aluminum coatings both provide excellent protection in a wide variety of marine and industrial environments. In general, aluminum corrodes less rapidly than zinc in highly acidic conditions, but zinc performs better than aluminum in alkaline conditions. Aluminum is chosen for the protection of steel in chemical plants or in other applications in which the temperature is likely to exceed 120 °C (250 °F). Sprayed zinc coatings are used in coal mines because of the possibility of impact spark generation between aluminum and rusty steel (Ref 3). Zinc is normally the preferred metal for the protection of steel in fresh cold waters; aluminum is used in aqueous solutions above 65 °C (150 °F).

Thermal spray zinc and aluminum coatings are progressively supplanting both painting and hot-dip galvanizing for reasons of effectiveness and economics. Thermal spray coatings have a predictable life, require a single application, protect damaged areas cathodically, have good abrasion resistance, and do not require drying.

The size or shape of the structure to be protected is not a limiting factor. Components can be metal sprayed in-plant or on-site, and automation is possible for long runs of identical work. The thickness of the metal coating can be controlled according to the degree of protection required. The life of the metal coating is proportional to the coating weight per unit area. For a very long service life or in highly corrosive conditions, it is possible to increase the coating thickness to enhance corrosion protection.

Sprayed zinc or aluminum coatings can be applied in thicknesses ranging from 50 to 500 μm (2 to 20 mils) (Ref 4). Furthermore, parts that are not to be coated can be masked during spraying, and the coating thicknesses can be varied in different areas of the same structure. Areas of coating discontinuity or insufficient thickness can easily be rectified by additional spraying.

Both aluminum and zinc coatings have good adhesion to grit-blasted steel. Spraying does not cause excessive heating of the substrate; therefore, there is no distortion or effect on the mechanical properties of the steel. All grades of steel can be sprayed.

A limited amount of porosity (usually $<15\%$) is an inherent feature of thermal spray coatings. In the case of active metal coatings, such as zinc or aluminum, porosity is not a deficiency. Pores will not result in substrate attack, because of the protection afforded by galvanic action. In fact,

Table 1 Thermal spray coatings for atmospheric and immersion service

Exposure environment	Thickness of coating systems for 20-year expected life: Unsealed zinc, μm	mils	Sealed zinc, μm	mils	Unsealed aluminum, μm	mils	Sealed aluminum, μm	mils
Atmospheric								
Inland (Nonpolluted)	150	6	150	6	150	6	100	4
Inland (Polluted)	150	6	150	6	150	6	150	6
Coastal (Nonpolluted)	250	10	150	6	150	6	150	6
Coastal (Polluted)	350	14	250	10	250	10	150	6
Immersion								
Seawater splash zone	. . .		250	10	. . .		150	6
Seawater immersion	. . .		250	10	. . .		150	6

Source: Ref 1

the natural surface porosity provides an excellent base for sealers or topcoats where these are desired for decorative purposes or to extend the life of the coating.

Over the years, zinc and aluminum thermal sprayed coatings have been used in a broad variety of corrosion protection programs. In Europe, where metal spraying for corrosion protection has been used far more widely than in the United States, there are numerous case histories of successfully sprayed structures, some as long as 40 years ago, that have been kept in good condition with minimal maintenance (Ref 5).

The selection of a zinc or aluminum coating system depends on the service environment and the desired life. Table 1 lists such coatings for atmospheric and immersion service, but is intended to serve only as a guide in choosing a coating system.

Zinc-Aluminum Alloys. Corrosion protection can also be provided by the use of zinc-aluminum alloys, such as Zn-15Al. This alloy does not have the performance record of pure zinc or aluminum, but studies in Europe and Japan and, more recently, in the United States suggest that such alloys combine the advantages of zinc and aluminum and may outperform both in certain environments (Ref 6). The Zn-15Al alloy is now available in either wire or powder form. Installed costs are comparable to either pure zinc or aluminum.

Other Coating Materials. Austenitic stainless steels, aluminum bronze, nickel-base alloys, and MCrAlY, among other materials, are also thermal sprayed to combat corrosion in certain applications. Selection of a specific alloy depends on the particular environment. When using these alloys, it must be understood that these coatings will not galvanically protect the underlying steel. This can be a particular problem due to the porosity of thermal spray coatings. Great care must be taken to ensure that these coatings are properly sealed in order to prevent penetration of the corrosive medium to the underlying steel and subsequent corrosion at the coating/substrate interface.

Coating Selection for High-Temperature Environments

Thermal spray coatings are extensively used by industry to protect steel components and structures from heat oxidation at surface temperatures to 1095 °C (2000 °F). By ensuring long-term protection, thermal spray coatings show real economic advantages during the service lives of such items. Coatings are particularly effective in protecting low-alloy and carbon steels. Table 2 lists thermal spray coatings for use in high-temperature applications.

Aluminum has been widely used to protect such steels in a number of applications involving high surface temperatures. Nickel-chromium alloys and some of the MCrAlY alloys have also provided protection in severe environments.

Coatings for Temperatures to 550 °C (1020 °F). Sprayed aluminum coatings provide excellent oxidation resistance and cathodic protection against corrosion by industrial atmospheres, hot gases, and condensation products at temperatures to 550 °C (1020 °F). Sealing of the sprayed aluminum coatings, although not essential, may be desired for decorative purposes or to maintain a clean appearance. A high-temperature aluminum-pigmented silicone paint is generally recommended.

Nickel-chromium alloys provide good protection over this temperature range in the presence of sulfurous gases. To achieve adequate protection, nickel-chromium alloys should have a chromium content of at least 40% (Ref 8). The addition of a small amount of titanium is also advantageous for increasing corrosion resistance and strengthening the coating-to-substrate bond (Ref 9).

Coatings for Temperatures From 550 to 900 °C (1020 to 1650 °F). For service temperatures in this range, sprayed aluminum or aluminum alloy coatings 150 to 200 μm (6 to 8 mils) thick have proven to be effective. At these temperatures, sprayed aluminum diffuses into the steel surface. These aluminized coatings are nonporous, consisting essentially of an aluminum-iron alloy containing about 15% Al. During prolonged service at 550 to 900 °C (1020 to 1650 °F), the aluminum-iron diffusion process continues, resulting in the depletion of aluminum at the steel surface; therefore, oxidation resistance is progressively impaired.

The aluminum-sprayed steel is often heat treated before being put into service in order to bring about the required diffusion. If a high-bonding thermal spray process, such as electric arc spraying is used, the heat-treating step is not always necessary. Electric arc sprayed steel can be put directly into service and can be heat treated during operation.

Coatings for Temperatures From 900 to 1000 °C (1650 to 1830 °F). Nickel-chromium alloy coatings 375 μm (15 mils) thick can be used for protection at temperatures in this range. In the absence of sulfurous gases, the alloys generally range in composition from Ni-15Cr-25Fe to Ni-20Cr. For environments containing sulfurous gases, a nickel-chromium alloy, followed by a supplementary protective coating of 100 μm (4 mils) of sprayed aluminum, has shown to be an effective system. During heat treatment or service, the aluminum diffuses into the nickel-chromium alloy, forming an aluminum-rich layer that is highly resistant to sulfurous gas attack (Ref 7).

Work with superalloys has shown that FeCrAlY alloys, when thermally sprayed in an inert atmosphere, offer a stable system with good ductility and oxidation resistance at these high temperatures (Ref 10). Specifically, inert electric arc sprayed FeCrAlY is applied to tungsten alloy fibers to produce a very strong corrosion-resistant fiber-reinforced superalloy for use in turbine blades at very high temperatures.

Sealing of Thermal Sprayed Coatings

Sealers or topcoats are commonly used to extend the life of thermal spray coatings and for decorative purposes. They are generally applied in a minimum of two coats; the first coat is thinned to allow proper penetration into the coating pores. Common sealers include vinyls, epoxies, phenolics, and polyurethanes. Such materials are discussed in detail in the article "Organic Coatings and Linings" in this Volume.

Sealers are not always necessary for zinc or aluminum coatings, because of the galvanic protection afforded. However, they are commonly used in most environments to seal the pores of the coating and to extend surface life. A sealed coating also generally gives a more pleasing appearance.

Sealers are essential for most immersion or chemical corrosion applications. In particular, materials other than zinc or aluminum, which do not galvanically protect the steel, must be sealed to prevent coating failure due to the porosity of the coating.

Table 2 Thermal spray coatings for elevated-temperature service

Service temperature	Coating metal or alloy	Coating thickness, μm	mils
Up to 550 °C (1020 °F)	Aluminum	175	7
Up to 550 °C (1020 °F) in the presence of sulfurous gases	Ni-43Cr-2Ti	375	15
550-900 °C (1020-1650 °F)	Aluminum or aluminum-iron	175	7
900-1000 °C (1652-1830 °F)	Nickel-chromium or MCrAlY	375	15
900-1000 °C (1650-1830 °F) in the presence of sulfurous gases	Nickel-chromium, followed by aluminum	375 100	15 4

Source: Ref 7

Quality Control and Inspection

Inspection and quality control procedures for thermal sprayed coatings are similar to those used for other types of coatings.

Quality of Surface Preparation. The abrasives used for surface preparation should be inspected to ensure that they meet specifications for grit size, type, and cleanliness. After the blasting operation, the surface should be visually inspected to ensure that the substrate is clean. A profilometer can be used to inspect for adequate surface roughness. Standard coupons or photographs may also be used as visual standards to ensure proper surface preparation. These are readily available from the SSPC, or they can be prepared in advance by the applicator.

Coating Quality. Generally, certification of the coating material can be readily obtained from the material supplier. This certifies that the material meets proper specifications for such qualities as chemical composition and physical characteristics. The spray equipment should be operated and maintained according to the standards of the manufacturer.

The installed coating can be visually inspected for surface finish and defects. Coating thickness can be verified by destructive testing, which consists of removing the coating with a knife or chisel. More commonly, the thickness of zinc or aluminum coatings is verified by using a magnetic dry-film thickness gage.

Typical Applications

Thermal spray coatings, primarily zinc and aluminum, have been successfully used to combat corrosion in a wide range of applications. Steel structures and components that have been zinc or aluminum sprayed include TV towers, antennae, radar, bridges, light poles, girders, ski lifts, and countless other similar structures. In addition, thermal spray coatings, primarily aluminum, offer years of protection in marine applications, such as buoys and pylons. Aluminum spraying has been used in offshore oil rigs for well head assemblies, flare stacks, walkways, and other structural steel components.

Shipboard components, both above and below deck, commonly use aluminum spraying. The United States Navy, in particular, uses aluminum spraying extensively to combat corrosion (Ref 11). There are countless approved applications for sprayed metal coatings aboard Navy ships (Ref 12).

Numerous immersion applications have also employed zinc or aluminum spraying, for example, dams and sluice gates. The interiors of potable water storage tanks are also sprayed with zinc to provide corrosion protection of the steel without the threat of contaminating the water with a solvent that may be present in an epoxy coating system.

Aluminum has also been used to control chemical corrosion in such applications as storage tanks for fuels or other liquids. The interiors of railroad hopper cars are often sprayed to protect them against sulfuric acid (H_2SO_4) corrosion when hauling coal.

Stainless steels, Hastelloy alloys, and other alloys can also be effective against chemical corrosion for storage vessels, rolls, pumps, and other structures. The coatings must be properly sealed when using these alloys.

Common applications involving high-temperature corrosion include coating exhaust stacks, chimneys, flues, rotary kilns and dryers, catalytic crackers, and furnace parts. These generally involve use of an aluminum or nickel-chromium alloy coating such as those previously described.

One specific high-temperature corrosion application that has met with great success in recent years is the coating of boilers in paper mills, power plants, and chemical plants. Water-wall tubes in these applications suffer severe corrosion because of the high sulfur content of the burning fuel, the high operating pressures, and abrasion (Ref 13). A 375-μm (15-mil) coating of Ni-43Cr-2Ti alloy offers extremely good protection against this very severe corrosion at temperatures up to 550 °C (1020 °F) (Ref 8).

REFERENCES

1. "Code of Practice for Protective Coating of Iron and Steel Structures Against Corrosion," BS 5493:1977, British Standards Institute
2. "Corrosion Tests of Flame-Sprayed Coated Steel: 19-Year Report," C2.14-74, American Welding Society, 1974
3. J.C. Bailey, Aluminum-Coated Steel: Sparking and the Fires Hazard, *Met. Mater.*, Oct 1978, p 26-27
4. "Selection of Sprayed Zinc and Aluminum Coatings for the Protection of Iron and Steel Against Corrosion," Technical Bulletin 2.3.1.5, TAFA, Inc., 1983
5. J.C. Bailey, "U.K. Experience in Protecting Large Structures by Metal Spraying," Paper presented at the Eighth International Thermal Spraying Conference, Miami Beach, FL, American Welding Society, 1976
6. M. Leclercq and R. Bensimon, "New Zinc-Based Alloy for Metallizing," Paper presented at the Eighth International Metal Spray Conference, Miami Beach, FL, American Welding Society, 1976
7. "Sprayed Metal Coatings to Combat High Temperature Corrosion," Information Sheet 3, Association of Metal Sprayers
8. R.H. Unger and F.B. Easterly, "A New Coating for Corrosion Protection in Boilers," Paper presented at the TAPPI Engineering Conference, Seattle, WA, Technical Association of the Pulp and Paper Industry, 1986
9. E.P. Sadowski and P.C. Shah, "Nickel-Chromium Filler Metal," U.S. Patent 4,025,314, 1977
10. D. Petrasek, D. McDanels, L. Westfall, and J. Stephens, Fiber-Reinforced Superalloys, *Mater. Prog.*, Aug 1986, p 27
11. "Metal Sprayed Coating Systems for Corrosion Protection Aboard Naval Ships," DOD-STD-2138(SH), United States Department of Defense, Nov 1981
12. A.R. Parks, "Metal Sprayed Coating Systems on Board U.S. Navy Ships," Paper presented at the Second National Conference on Thermal Spray, Long Beach, CA, American Society for Metals, 1984
13. A. Plumbley and W. Roczniak, "Recovery Unit Waterwall Protection," Paper presented at the British Columbia Black Liquor Recovery Boiler Advisory Committee Meeting, Vancouver, BC, 1974

Anodic Protection

Carl E. Locke, University of Kansas

ANODIC PROTECTION is the most recently developed of all the various corrosion control methods available. This method was first used in the field in the late 1950s. Anodic protection did not become commercially successful until the early 1970s, and it is currently used on a smaller scale than other corrosion control techniques. This article will provide a brief history of the technique, will discuss anodic protection use, and will compare anodic and cathodic protection. A more complete description of all aspects of anodic protection is available in Ref 1.

History

The scientific principles of passivity on which anodic protection is based can be traced to experiments by Faraday and Schobein in the 19th century (Ref 1). Work by C. Edeleanu published in 1954 (Ref 2) was responsible for further investigations and developments at the Continental Oil Company. This later work resulted in several installations of anodic protection to control the corrosion of chemical plant equipment (Ref 3-6). Simultaneous investigations at the Pulp and Paper Institute of Canada led to installation of anodic protection to protect pulp and paper digesters (Ref 7-9).

Commercialization of anodic protection was begun by Continental Oil Company through licensing agreements, and only a few companies are currently marketing anodic protection. The most active of these companies are Canadian-based corporations that sell, engineer, and install anodic protection systems in the United States.

Anodic Protection Use

Anodic protection has been most extensively applied to protect equipment used to store and handle sulfuric acid (H_2SO_4) (Ref 10). Sales of anodically protected heat exchangers used to cool H_2SO_4-manufacturing plants have represented one of the more successful ventures in the past few years. These heat exchangers are sold complete with the anodic protection systems installed and have a commercial advantage in that less costly materials can be used.

Protection of steel H_2SO_4 (>78% concentration) storage vessels is perhaps the most common application of anodic protection. Companies in North America and Europe sell systems primarily for these storage vessels and the heat exchangers. There is little activity directed toward developing applications to protect metals from corrosion by other chemicals. However, in the USSR, a substantial amount of effort is devoted to installing anodic protection in a wide variety of systems. Several recent applications are discussed in Ref 11.

Anodic protection is used to a lesser degree than the other corrosion control techniques, particularly cathodic protection (see the article "Cathodic Protection" in this Volume). This is because of the limitations on metal-chemical systems for which anodic protection will reduce corrosion. In addition, it is possible to accelerate corrosion of the equipment if proper controls are not implemented. The understanding of anodic protection is also not as widespread as for other techniques. Anodic protection does have a place in the corrosion control area, for which it is the most feasible and economical technique, and use of this technique is expected to increase slowly in coming years.

Comparison of Anodic and Cathodic Protection

Cathodic protection has been used since the middle of the 19th century and has gained widespread acceptance. Anodic protection is sometimes confused with cathodic protection, but the two techniques are fundamentally different. Basically, the difference concerns which electrode is protected; the cathode is protected in cathodic protection, and the anode is protected in anodic protection. However, there are other, more significant theoretical differences. The background and theory of both methods are discussed in Ref 12.

Cathodic protection is based on using electrochemistry to slow or stifle the corrosion reaction. Currents generated by galvanic metallic couples or external dc current supplies flow in such a direction as to shift the cathode potential to a value at which the corrosion reaction will not occur. Other descriptions of cathodic protection state that the currents result in the metal to be protected becoming a cathode over the entire metal surface and shifting the anodic reaction to an external electrode. In principle, any corrosion system can be cathodically protected as long as there is a continuous ionic path between the external anode and the metal to be protected. The only limitations are ones of economics and practicality due to current and voltage requirements.

Anodic protection, like cathodic protection, is an electrochemical method of controlling corrosion, but is based on a different electrochemical principle. Anodic protection is based on the phenomenon of passivity. A limited number of metals in a limited number of chemicals have the property of passivity. The electrochemical nature of achieving passivity requires that the potential of the metal be controlled. This is not necessary for most of the applications of cathodic protection. Current requirements for anodic protection can be much lower than those required for cathodic protection, but this is not always the case for all systems.

Background and Theory

Anodic protection can be used to control the corrosion of metals in chemicals that exhibit very interesting behavior when subjected to anodic polarization. This behavior can be studied with an experimental setup, as illustrated in Fig. 1. When the potential of the working electrode relative to the reference electrode is controlled and shifted in the more anodic (positive) direction, the current required to cause that shift will vary. If the current required for the shift has the general type of behavior with respect to potential shown in Fig. 2, the metal is termed active-passive and can be anodically protected. Few systems exhibit this type of behavior. The metals and solutions that have been found to have active-passive properties and that can be anodically protected include:

Solutions	Metals
Sulfuric acid	Steels
Phosphoric acid	Stainless steels
Nitric acid	Nickel
Nitrate solutions	Nickel alloys
Aqueous ammonia	Chromium
Organic acids	
Caustic solutions	

The corrosion rate of an active-passive metal can be significantly reduced by shifting the potential of the metal so that it is at a value in the passive range shown in Fig. 2. The current required to shift the potential in the anodic direction from the corrosion potential E_{corr} is several orders of magnitude greater than the current necessary to maintain the potential at a passive value. The current will peak at the potential value shown as E_p (Fig. 2). The current required to achieve passivity (protection) is typically a few hundred milliamps per square foot wetted area. The current necessary to maintain passivity is usually 10.76 $\mu A/m^2$ (1 $\mu A/ft^2$) or less.

This interesting anodic polarization behavior results from the formation of a surface layer on the metal that is relatively insoluble in the chemical environment. This passive film is not yet completely understood.

Passivity is the phenomenon responsible for the high corrosion resistance of stainless steels. It can be variously defined, but a reasonable definition is that passivity is the loss of chemical reactivity experienced by some metals in special environments. These metals and alloys, when passive, behave as noble metals. This noblelike behavior is attributed to the presence of a thin oxide film on the metal surface. The mechanism of formation, the composition, and the structure of the passive film have been extensively studied, but are not well understood.

Passivity can be achieved by alloying and by chemical means in addition to the electrochemical method described in this section. Anodic protection can be used to form the passive film on metals in chemical systems that would normally be corrosive; at other times, anodic protection can be used to maintain the passivity of the metal so that process upsets or changes do not force the metal to become active and corrode.

Equipment Required for Anodic Protection

Figure 3 shows a schematic of an anodic protection system for a storage vessel. Each of these components has specific requirements that will be discussed below. In addition, the various items used for each component of the system will be briefly described.

Electrodes

The cathode should be a permanent-type electrode that is not dissolved by the solution or the currents impressed between the vessel wall and electrode. The cathodes used in most of the first applications of anodic protection were made of platinum-clad brass (Ref 1, 2, 3-6). These electrodes were excellent electrochemically, but they were costly, and the area contacting the solution was limited by this cost. Because the overall circuit resistance between cathode and vessel wall is proportional to the electrode surface area, it is advantageous to use large surface area electrodes. Therefore, many other, less costly metals have been used for cathodes. Metals that have been used for cathodes in anodic protection systems, as well as the chemical environments in which they were used, include:

Metals	Environment
Platinum on brass	Various
Steel	Kraft digester liquid
Illium G	H_2SO_4 (78–105%)
Silicon cast iron	H_2SO_4 (89–105%)
Copper	Hydroxylamine sulfate
Stainless steel	Liquid fertilizers (nitrate solutions)
Nickel-plated steel	Chemical nickel plating solutions
Hastelloy C	Liquid fertilizers (nitrate solutions), H_2SO_4, Kraft digester liquid

The electrode size is chosen to conform to the geometry of the vessel and to provide as large a surface area as possible. The location of the cathode is not a critical factor in simple geometries, such as storage vessels, but in heat exchangers, it is necessary to extend the electrode around the surface to be protected. Multiple cathodes can be used in parallel to distribute the current and to decrease circuit resistance.

Reference electrodes must be used in anodic protection systems because the potential of the vessel wall as the anode must be controlled. The reference electrode must have an electrochemical potential that is constant with respect to time and that is minimally affected by changes in temperature and solution composition. Several reference electrodes have been used for anodic protection, including:

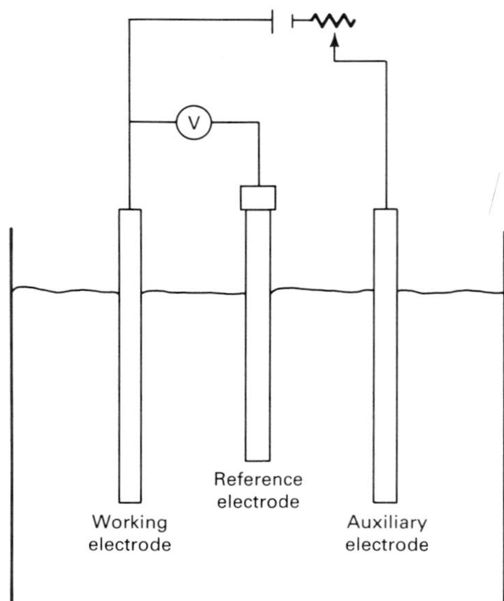

Fig. 1 Schematic of experimental apparatus used for anodic polarization studies. Current flow between the working electrode and the auxiliary electrode forces a shift in potential between the working electrode and the reference electrode. V, voltmeter. Source: Ref 15

Electrode	Solution
Calomel	H_2SO_4
Ag-AgCl	H_2SO_4, Kraft solutions, fertilizer solutions
Mo-MoO_3	Sodium carbonate solutions
Bismuth	NH_4OH
Type 316 stainless steel	Fertilizer solutions, oleum
Hg-$HgSO_4$	H_2SO_4, hydroxylamine sulfate
Pt-PtO	H_2SO_4

The reference electrode has been a source of many problems in anodic protection installations because of its more fragile nature with respect to the cathode.

Potential Control and Power Supply

Potential Control. As mentioned above, the potential of the vessel wall with respect to the reference electrode must be controlled in anodic protection installations. The potential control circuitry has two functions. First, the potential must be measured and compared to the desired preset value. Second, a control signal must then be sent to the power supply to force the dc current between the cathode and vessel wall. In early systems, this control function was done in an on-off method because of the high costs of electronic circuitry. The more sophisticated and extremely low-cost circuitry currently available has resulted in all systems having a continuous proportional-type control. The amount of current forced through the circuitry is that required to maintain the potential at the preset control point.

The dc power supplies have the identical design and requirements as the rectifiers for cathodic protection with one exception. Because of the nature of the active-passive behavior of the vessel, the currents required to maintain the potential of the vessel wall in the passive range can become very small. Some designs of dc power supplies must be specially modified to reduce the minimum amount of current put out of the power supply.

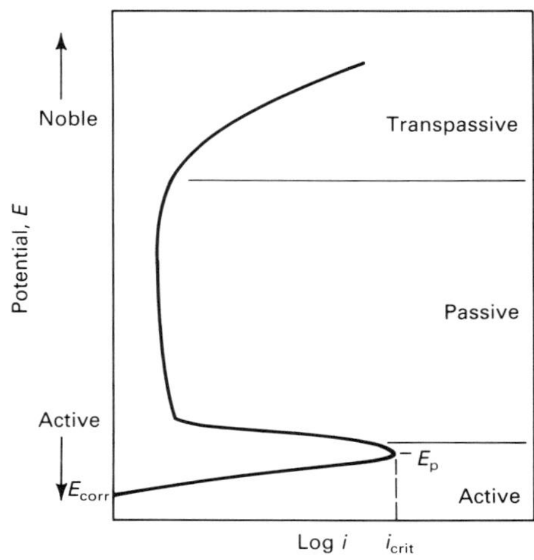

Fig. 2 Schematic anodic polarization curve. Metal-environment systems that have this type of anodic polarization behavior are termed active-passive and can be anodically protected. Source: Ref 15

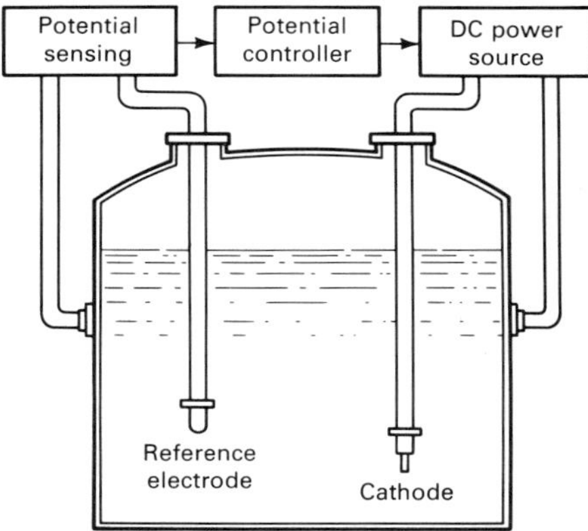

Fig. 3 Schematic of an anodic protection system. One or more cathodes, a reference electrode, a potential sensing and controlling circuit, and a dc power supply are required for each anodic protection system. The vessel wall is made the anode of the circuit by current forced between the cathode and the tank wall. The currents are controlled so that the potential of the wall with respect to the reference electrode is shifted and maintained in the passive region (see Fig. 2). Source: Ref 1

The packaging of these electronic components occasionally involves special requirements because most of the installations are made in chemical plants. Explosion-proof enclosures are sometimes required, and chemically resistant enclosures are necessary in other installations.

Design Concerns

Design of an anodic protection system requires knowledge of the electrochemical parameters of the metal-solution system, the geometry of the equipment to be protected, any special operational conditions, and the special requirements of the environment around the system. The electrochemical parameters of concern are the potential at which the vessel must be maintained for corrosion protection, the currents required to establish passivity, the currents required to maintain passivity, and the solution resistivity. The electrode potential can be determined directly from the polarization curve. The currents can be estimated from the polarization data, but they are time dependent, and the variation of the currents with respect to time must be estimated. Empirical data available from field installations are the best source of this information (Ref 1).

Solution resistivity is important in the determination of the overall circuit resistance. The power requirements for the dc power supply should be as low as possible to reduce operating costs. The solution resistivity will usually be sufficiently low that the circuit resistance will be controlled by the cathode surface area in contact with the solution.

Applications

Anodic protection has been used for storage vessels, process reactors, heat exchangers, and transportation vessels containing various corrosive solutions. Sulfuric acid is used in the largest number of systems in operation throughout the world. Anodic protection has been successful in preserving product quality through reduction of metal pickup by the solution, in extending the useful life of the vessels, and in allowing the use of a lower-cost alloy. Examples of these applications will be given below.

Storage Vessels. Storage of H_2SO_4 in strengths of 93% and above in low-carbon steel vessels has met with some success in terms of vessel life. Anodic protection has been successful in reducing the amount of iron picked up during storage. This enhances the economic worth of the product in that pure acid is more valuable than contaminated acid. Field studies have shown that the iron content of H_2SO_4 in concentrations of 93% and above increases at rates of 5 to 20 ppm per day of storage, depending on acid concentration, vessel size, acid residence time, and storage temperature (Ref 1). Several anodic protection systems have been successful in reducing the rates of iron pickup to 1 ppm per day or less. The level of purity of the acid has been sufficient to meet market demands for low iron content acid.

However, tanks constructed since the 1970s have been made with steel containing low amounts of copper. Steel with low copper content will corrode readily in these acids (Ref 13). Anodic protection of the newer vessels has been successful in lowering the corrosion rate by a factor of four to five, and vessel life has been greatly extended.

Heat Exchangers. A large market has developed for anodically protected heat exchangers as replacements for cast iron coolers. Shell and tube, spiral, and plate-type exchangers have been sold complete with anodic protection as an integral part of the equipment (Ref 1, 10, 11). Sulfuric acid of 96 to 98% concentration at temperatures up to 110 °C (230 °F) has been handled in AISI type 316 stainless steel exchangers by the use of anodic protection. Corrosion rates have been reduced from unprotected rates of more than 5 mm/yr (200 mils/yr) to less than 0.025 mm/yr (1 mil/yr), and cost savings have been substantial because of extended equipment life and the higher-purity acid that was produced by using these protected heat exchangers. Several other corrosive systems have also been handled in anodically protected heat exchangers (Ref 1).

Transportation Vessels. Sulfuric acid will continue to pick up iron during transportation in trucks, railroad cars, and barges. Anodic protection has been applied to these vessels to maintain the purity of the acid and to extend storage time. These applications are described in Ref 1, 11, and 12.

Galvanic Cathodes. The potentials of metals have been maintained in the passive region by using the potential differences between metals to achieve anodic protection by galvanic methods. Unlike cathodic protection, nothing is sacrificed, but the protection is achieved. A titanium pipe has been used to protect a steel vessel containing ammonium hydroxide (NH_4OH) solution and platinum has been used to protect a stainless steel tank containing H_2SO_4 (Ref 14, 15). A combination of galvanic and impressed current protection that uses graphite cathodes for the anodic protection system is described in Ref 16. The graphite supplies sufficient current to maintain protection, but additional current can be supplied through the graphite electrodes to reestablish protection, if needed.

Economics

Economic justification of anodic protection is based on the same factors as in any comparison of corrosion control methods. Protection of H_2SO_4 storage vessels can be accomplished with baked phenolic linings or anodic protection. Because an anodic protection system has some basic requirements (electrodes, potential controller, and power supply) independent of tank size, anodic protection is more economically feasible for larger tanks. A detailed comparison of the economics of lining and anodically protecting steel storage vessels can be found in Ref 1. This comparison indicates that anodic protection is 42.5% less expensive per ton of acid storage capacity than a baked phenolic lining for a vessel with a capacity of about 18 000 Mg (20 000 tons). Costs of the lining are 74.3% less expensive per ton of acid storage capacity than the anodic protection cost of a 907-Mg (1000-ton) capacity vessel.

Anodic protection for other equipment when a less costly alloy can be used has been compared in terms of economics to the absence of protection. The anodically protected heat exchangers used in H_2SO_4-manufacturing plants have been shown to reduce maintenance costs by almost 95% (Ref 17). These reduced maintenance costs, combined with the improved quality of the acid produced in the plant, have provided an attractive return on the investment required for the anodically protected exchangers.

Many other applications of anodic protection have also been found to be economically feasible, by using standard methods of engineering economics comparisons (see the article "Corrosion Economic Calculations" in this Volume). As experience is gained and confidence in this new method increases, there should be several other applications for which anodic protection will be shown to be the best economic alternative.

REFERENCES

1. O.L. Riggs, Jr. and C.E. Locke, *Anodic Protection: Theory and Practice in the Prevention of Corrosion*, Plenum Press, 1981
2. C. Edeleanu, *Metallurgia*, Vol 50, 1954, p 113-116
3. J.D. Sudbury, O.L. Riggs, and D.A. Shock, *Corrosion*, Vol 16, 1960, 47t-54t
4. D.A. Shock, O.L. Riggs, and J.D. Sudbury, *Corrosion*, Vol 16, 1960, p 55t-58t
5. O.L. Riggs, M. Hutchison, and N.L. Conger, *Corrosion*, Vol 16, 1960, p 58t-62t
6. C.E. Locke, M. Hutchison, and N.L. Conger, *Chem. Eng. Prog.*, Vol 56 (No. 11), 1960, p 50-55
7. W.A. Mueller, *Can. J. Chem.*, Vol 38, 1960, p 576-587
8. T.R.B. Watson, *Tappi*, Vol 44, 1961, p 208-210
9. T.R.B. Watson, *Mater. Prot.*, Vol 3 (No. 6), 1965, p 54
10. C.E. Locke, Status of Anodic Protection: Twenty-Five Years Old, in *Proceedings of the International Congress on Metallic Corrosion*, Vol 1, National Research Council of Canada, 1984, p 316-319
11. V. Kuzub and V. Novitskiy, Anodic Protection and Corrosion Control of Industrial Equipment, in *Proceedings of the International Congress on Metallic Corrosion*, Vol 1, National Research Council of Canada, 1984, p 307-310
12. C.E. Locke, Corrosion: Cathodic and Anodic Protection, in *Encyclopedia of Chemical Processing and Design*, Vol 12, Marcel Dekker, 1981, p 13-59
13. D. Fyfe, R. Vanderland, and J. Rodda, *Chem. Eng. Prog.*, Vol 73 (No. 3), 1977, p 65
14. W.A. Szymanski, *Mater. Perform.*, Vol 16 (No. 11), 1977, p 16
15. G. Bianchi, A. Barosi, and S. Trasatti, *Electrochim. Acta*, Vol 10, 1965, p 83
16. V.G. Moisa and V.S. Kuzub, Anodic Protection With Additional Protectors, *Zashch. Met.*, Vol 16 (No. 1), 1980, p 83
17. F.W.S. Jones, *Anti-Corros. Methods Mater.*, 12 Dec 1976

Cathodic Protection

Robert H. Heidersbach, Metallurgical Engineering Department, California Polytechnic State University

CATHODIC PROTECTION is an electrochemical means of corrosion control in which the oxidation reaction in a galvanic cell is concentrated at the anode and suppresses corrosion of the cathode in the same cell. Figure 1 shows a simple cathodic protection system. The steel pipeline is cathodically protected by its connection to a sacrificial magnesium anode buried in the same soil electrolyte.

Cathodic protection is different from anodic protection (see the article "Anodic Protection" in this Volume). In cathodic protection, the object to be protected is the cathode, but in anodic protection, the object to be protected is the anode.

Anodic protection can be used on only a limited number of alloys in certain restricted environments, but cathodic protection can, in principle, be applied to any metal. In practice, cathodic protection is primarily used on carbon steel. The effectiveness of cathodic protection allows carbon steel, which has little natural corrosion resistance, to be used in such corrosive environments as seawater, acid soils, salt-laden concrete, and many other corrosive environments. Properly designed and maintained cathodic protection systems can prevent corrosion indefinitely in these environments.

Cathodic protection was first suggested by Sir Humphrey Davy in the 1820s as a means of controlling corrosion on British naval ships (Ref 1). It became common in the 1930s on the Gulf Coast of the United States, where it was used to control the corrosion of pipelines carrying high-pressure hydrocarbons (natural gas and petroleum products). Much of the terminology of cathodic protection still relates to corrosion control of onshore buried steel pipelines.

Virtually all modern pipelines are coated with an organic protective coating that is supplemented by cathodic protection systems sized to prevent corrosion at holidays in the protective coating (see the article "Corrosion of Pipelines" in this Volume). This combination of protective coating and cathodic protection is used on virtually all immersed or buried carbon steel structures, with the exception of offshore petroleum production platforms and reinforced concrete structures.

Offshore platforms are usually uncoated but cathodically protected (see the article "Corrosion in Petroleum Production Operations" in this Volume). Cathodic protection causes changes in the chemistry of seawater near the protected structure, and this causes the precipitation of a natural coating on the structure that reduces the need for cathodic protection current. Concrete structures normally rely on the protectiveness of the concrete cover to prevent the corrosion of embedded steel (see the article "Corrosion in Structures" in this Volume). When corrosion of embedded steel occurs because of a loss of this protectiveness, cathodic protection is sometimes used to extend the life of the already deteriorated structure.

Fundamentals of Cathodic Protection

Table 1 shows the theoretical electrochemical potentials obtained by pure metals in 1 N solutions of their own ions. This abbreviated electromotive series is described in most chemistry or corrosion textbooks (Ref 2) and in the article "Thermodynamics of Aqueous Corrosion" in this Volume.

Figure 2(a) shows two of these metals—iron and zinc—separately immersed in a weak mineral acid. The chemical reactions that occur in Fig. 2(a) are:

$$Fe \rightarrow Fe^{2+} + 2e^- \quad \text{Oxidation reaction} \qquad \text{(Eq 1a)}$$

$$2H^+ + 2e^- \rightarrow H_2 \quad \text{Reduction reaction} \qquad \text{(Eq 1b)}$$

$$2H^+ + Fe \rightarrow Fe^{2+} + H_2 \quad \text{Net reaction} \qquad \text{(Eq 1c)}$$

$$Zn \rightarrow Zn^{2+} + 2e^- \quad \text{Oxidation reaction} \qquad \text{(Eq 2a)}$$

$$2H^+ + 2e^- \rightarrow H_2 \quad \text{Reduction reaction} \qquad \text{(Eq 2b)}$$

$$2H^+ + Zn \rightarrow Zn^{2+} + H_2 \quad \text{Net reaction} \qquad \text{(Eq 2c)}$$

Both metals corrode, and both corrosion (oxidation) reactions are balanced by an equal reduction reaction, which in both cases involves the liberation of hydrogen gas from the acid environments. The two corrosion reactions are independent of each other and are determined by the corrosivity of hydrochloric acid on the two metals in question.

If the two metals were immersed in the same acid and electrically connected, (Fig. 2b), the reactions for zinc would then become:

$$Zn \rightarrow Zn^{2+} + 2e^- \quad \text{Oxidation} \qquad \text{(Eq 3a)}$$

$$2H^+ + 2e^- \rightarrow H_2 \quad \text{Reduction} \qquad \text{(Eq 3b)}$$

Almost all of the oxidation reaction (corrosion of zinc) has been concentrated at the zinc electrode (anode) in Fig. 2(b), and almost all of the reduction reaction (hydrogen liberation) has been concentrated at the iron electrode (cathode). The oxidation of the zinc anode in Fig. 2(b) is much faster than that in Fig. 2(a). At the same time, most of the corrosion of iron in Fig. 2(a) has stopped in Fig. 2(b). As shown schematically, the zinc anode in Fig. 2(b) has been used to cathodically protect the iron cathode in Fig. 2(b).

Of course, some corrosion of the iron may still occur; whether or not this happens depends on

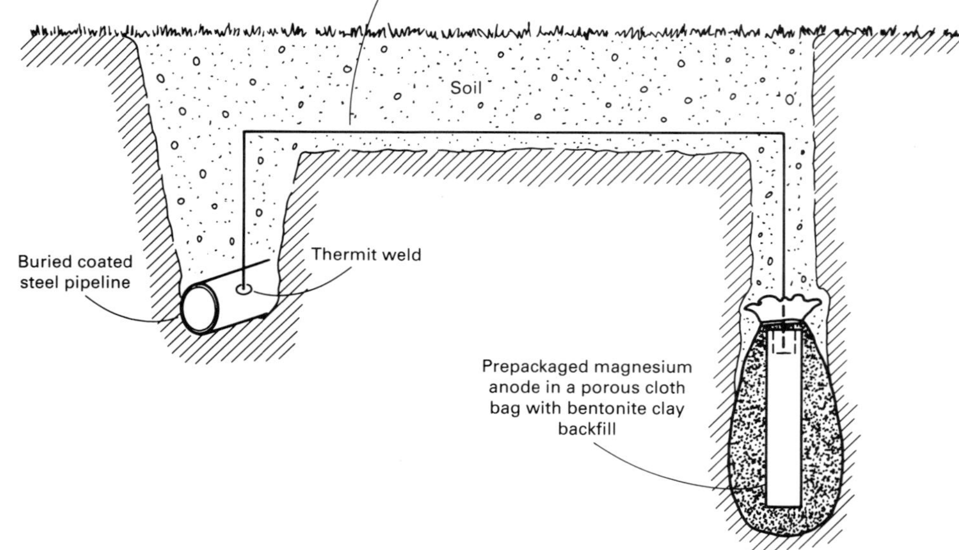

Fig. 1 Cathodic protection of buried pipeline using a buried magnesium anode

Table 1 Standard electromotive force series for selected metals

Metal-metal ion equilibrium (unit activity)	Potential at 25 °C (77 °F), V
Noble or cathodic (protected)	
Ag/Ag^{+}	+0.80
Cu/Cu^{2+}	+0.34
H_2/H^{+}	(reference) 0
Fe/Fe^{2+}	−0.44
Zn/Zn^{2+}	−0.76
Al/Al^{3+}	−1.66
Mg/Mg^{2+}	−2.36
More active or anodic	

Table 2 Comparison between sacrificial anode and impressed-current cathodic protection systems

Sacrificial anode system	Impressed-current system
Simple	Complex
Low/no maintenance	Requires maintenance
Works best in conductive electrolytes	Can work in low-conductivity electrolytes
Lower installation costs for smaller installations	Remote anodes possible
Higher capital investment for large systems	Low capital investment for large systems
	Can cause the following problems: Stray current corrosion; Hydrogen embrittlement; Coating debonding; Cathodic corrosion of aluminum

the relative sizes of the zinc and iron electrodes. Some reduction of hydrogen may still occur on the zinc anode. The anode is the electrode at which a net oxidation reaction occurs, whereas cathodes are electrodes at which net reduction reactions occur. With proper design, the oxidation rate on the cathode will be suppressed to the point at which it becomes negligible. If this takes place, cathodic protection has been achieved.

All cathodic protection systems require an anode, a cathode, an electric circuit between the anode and cathode, and an electrolyte. Thus, cathodic protection will not work on structures exposed to air environments. The air is a poor electrolyte, and it prevents current from flowing from the anode to the cathode.

Types of Cathodic Protection

There are two types of cathodic protection: sacrificial anode, or passive, systems and impressed-current, or active, systems. Both types are widely used.

Sacrificial anode systems are simpler. They require only a material anodic to the protected steel in the environment of interest. A simple sacrificial anode cathodic protection system used to control corrosion on a buried pipeline is shown in Fig. 1.

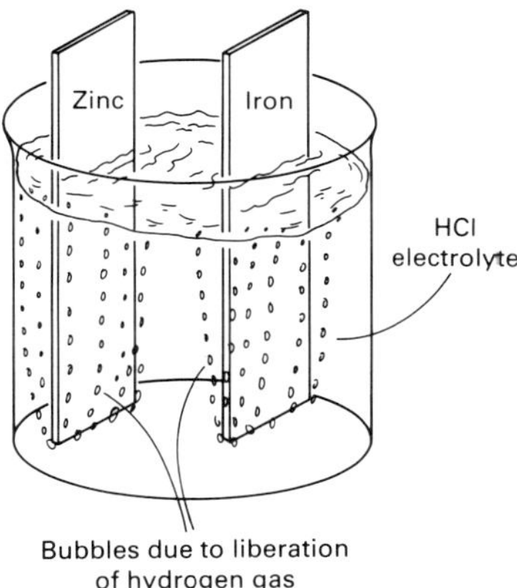

Fig. 2(a) Corrosion of zinc and iron in hydrochloric acid

Figure 3 shows an impressed-current system used to protect a pipeline. The buried anodes and the pipeline are both connected to an electrical rectifier, which supplies direct current to the buried electrodes (anodes and protected cathode) of the system. Unlike sacrificial anodes, impressed-current anodes need not be naturally anodic to steel, and in fact, they seldom are. Most impressed-current anodes are made from nonconsumable electrode materials that are naturally cathodic to steel. If these electrodes were wired directly to a structure, they would act as cathodes and would cause accelerated corrosion of the structure they are intended to protect. The direct current source reverses the natural polarity and allows the materials to act like anodes. Instead of corrosion of the anodes, some other oxidation reaction, that is, oxygen or chlorine evolution, occurs at the anodes, and the anodes are not consumed.

Impressed-current systems are more complex than sacrificial anode systems. The capital expenses necessary to supply direct current to the system are higher than for a simple connection between an anode and a cathode.

The voltage differences between anode and cathode are limited in sacrificial anode systems to approximately 1 V or even less, depending on the anode material and the specific environment. Impressed-current systems can use larger voltage differences. The larger voltages available with impressed-currents allow remote anode locations, which produce more efficient current distribution patterns along the protected cathode. These larger voltages are also useful in low-conductivity environments, such as freshwater and concrete, in which sacrificial anodes would have insufficient throwing power.

Large voltages can have disadvantages. It is possible to overprotect high-strength steels and cause hydrogen embrittlement. Coating debonding is also possible (Fig. 4). Debonding occurs when moisture penetrates a coating and hydrogen is generated at the metal surface beneath the coating. The gas can accumulate until pressure causes blisters or cracks in the protective coating. Once this coating damage occurs, the demands for protective current increase and may exceed the capabilities of the system.

Aluminum is especially vulnerable to overprotection. All cathodic reactions cause the immediate environment to become somewhat more basic (less acidic) than they are in the absence of cathodic

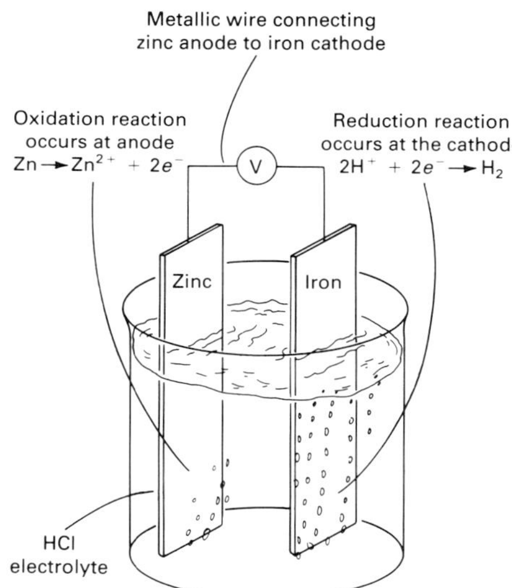

Fig. 2(b) Cathodic protection of iron by zinc in hydrochloric acid

protection. Unlike steel, aluminum is an amphoteric metal with increased corrosion susceptibility in acids and bases. If the environment around an aluminum structure becomes too basic, it will corrode at an accelerated rate. Thus, the cathodic protection of aluminum can cause cathodic corrosion if too much current is supplied to the cathode (see also the article "Corrosion of Aluminum and Aluminum Alloys" in this Volume).

Table 2 lists some of the important differences between impressed-current and sacrificial anode cathodic protection systems. Selection of a cathodic protection system usually depends on tradeoffs among the advantages of each type of system. For example, most offshore petroleum production platforms use sacrificial anodes because of their simplicity and reliability, even though the capital costs would be lower with impressed-current systems.

Criteria for Cathodic Protection

A number of criteria have been used for determining whether or not a structure is cathodically protected from corrosion. This is an area of controversy within the industry, and it is likely that existing standards will change over the next few years. Criteria also differ for buried utilities, offshore structures, and concrete structures.

The original National Association of Corrosion Engineers specification for buried utility pipelines proposed the following criteria for determining when a steel or cast iron structure is cathodically protected (Ref 3):

- A voltage of −0.85 V relative to a copper/saturated copper sulfate electrode
- A negative (cathodic) voltage shift of at least 300 mV caused by the application of cathodic protection current
- A minimum negative (cathodic) voltage shift of 100 mV determined by interrupting the current and measuring the voltage decay
- A voltage at least as negative (cathodic) as that originally established at the Tafel segment of the *E*-log *I* curve (Fig. 5)

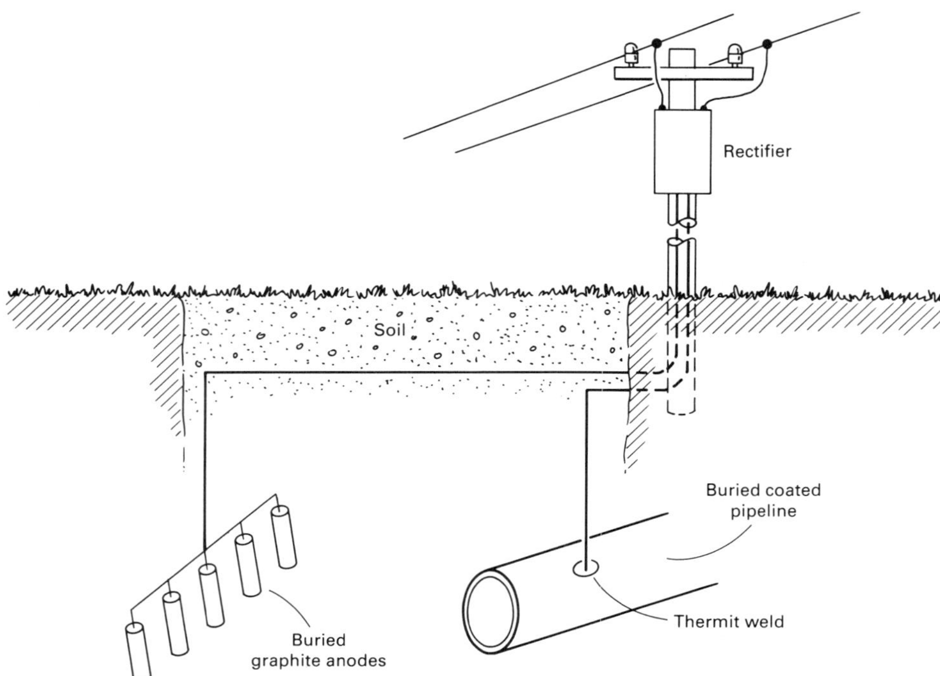

Fig. 3 Impressed-current cathodic protection of a buried pipeline using graphite anodes

- A net protective current from the electrolyte into the surface

All of the above criteria are currently in use, but the most common and most widely applicable criterion is the −0.85 V versus a copper/copper sulfate reference cell. Most structures can be inspected to determine if they are protected relative to this standard. The only equipment necessary is a reference cell (Fig. 6) and a wire lead that can be connected to the structure in question. The other criteria require recordkeeping, the ability to interrupt current (impossible for most sacrificial anode designs), and more sophisticated survey equipment.

Copper sulfate electrodes can become contaminated by seawater, and it is common to use −0.805 V versus silver/silver chloride as the protection potential for marine structures (Ref 4). Both criteria are the same; only the reference electrode material is different. These criteria are based on early research by the National Bureau of Standards (Ref 3).

The reader is cautioned that the sign convention used in this article agrees with that used by the U.S. cathodic protection industry. Much of the scientific and international literature uses the opposite sign convention for positive and negative electrical terminals. Regardless of the sign convention chosen, the important point is that electrons should flow into the protected structure from the external circuit.

Fig. 4 Debonded organic coating near a high-silicon cast iron button anode

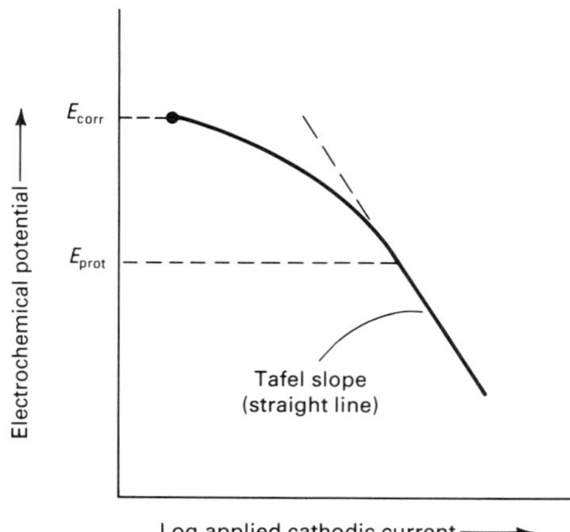

Fig. 5 Tafel slope criterion for determining cathodic protection. See the article "Kinetics of Aqueous Corrosion" in this Volume for detailed explanations of these diagrams.

Anode Materials

Different requirements for sacrificial anodes and impressed-current anodes lead to the use of widely different materials for these applications.

Sacrificial anodes must be anodic to steel in the environment of interest. They must also corrode reliably (avoid passivation). Impressed-current anodes can be cathodic to steel, but they must have low consumption rates when connected to a cathodic protection power source.

Sacrificial Anodes

Commercial sacrificial anodes include magnesium, zinc, and aluminum alloys. The energy characteristics of these alloys are given in Table 3.

Magnesium anodes are the only sacrificial anodes that are routinely specified for use in buried soil applications. Most magnesium anodes in the United States are supplied with a prepackaged bentonite clay backfill in a permeable cloth sack (Fig. 1). This backfill ensures that the anode will have a conductive environment and will corrode reliably. The additional material is less expensive than the soil resistivity surveys that would be needed to determine if the backfill is necessary.

Some magnesium anodes have been used offshore in recent years in an attempt to polarize the structures to a protected potential faster than would occur if zinc or aluminum alloy anodes were used. Magnesium tends to corrode quite readily in salt water, and most designers avoid the use of magnesium for permanent long-term marine cathodic protection applications.

Figure 7 shows a magnesium anode used to control corrosion on a glass-lined domestic water

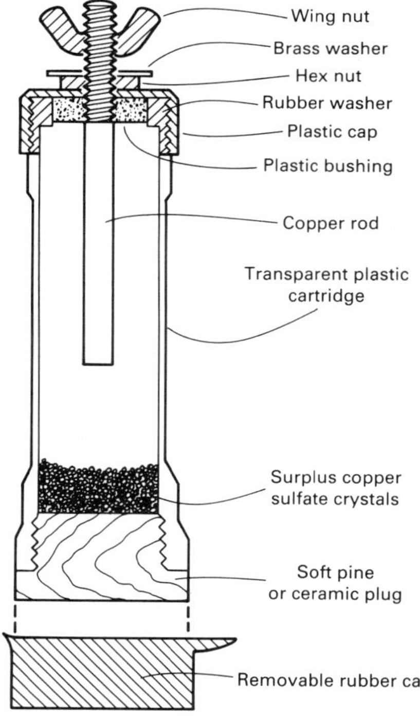

Fig. 6 Copper-saturated copper sulfate reference electrode

heater. Similar cathodic protection arrangements are found on most domestic water heaters in the United States.

Zinc is used for cathodic protection in freshwater and marine water. Zinc is especially well suited to cathodic protection on ships that move between salt water and harbors in brackish rivers or estuaries. Figure 8 shows zinc anodes on the underside of a small fishing boat. Aluminum anodes would passivate in the harbors and might not work when they return to sea. Zinc anodes are also used to protect ballast tanks, heat exchangers, and many mechanical components on ships, coastal power plants, and similar structures.

The weight of zinc is an advantage for marine pipelines. Bracelet anodes (Fig. 9) are attached at pipe joints to provide ballast and to prevent corrosion in the water-mud environment, in which aluminum might passivate.

Aluminum is used on offshore structures where its light weight provides significant advantages. Welded-on aluminum anodes for an offshore platform are shown in Fig. 10. Aluminum does not passivate in salt water if certain alloying additions, such as tin, antimony, and mercury, are present. Toxicity questions prevent the use of aluminum alloys with mercury additions in U.S. waters.

Impressed-Current Anodes

Impressed-current anodes must be corrosion resistant and otherwise durable in the environment in which they are used. Consumption rates for lead alloy, platinum, graphite, and high-silicon cast iron impressed-current anode materials are given in Table 4.

High-silicon cast iron (Fe-0.95C-0.75Mn-14.5Si-4.5Cr) is used for onshore cathodic protection applications and in other locations where abrasion resistance and other mechanical damage considerations are important. High-silicon cast iron anodes are available in solid rods, tubular form, and various cast shapes in a wide variety of sizes to meet the demands of specific applications. For example, solid rods range from 30 mm in diameter x 230 mm in length ($1^1/_8 \times 9$ in.) at 0.45 kg (1 lb) to 115 mm in diameter × 1525 mm in length (4.5 × 60 in.) at 100 kg (220 lb). The smaller rods are used for protecting underground freshwater storage tanks, and the larger rods are used in more aggressive seawater or ground bed applications. Figure 4 shows a high-silicon cast iron button anode on a navigational lock on the Tennessee River. This anode must withstand collisions with small vessels, trees, rocks, and so on. The coating debonding shown in Fig. 4 is due to improper circuit design and cannot be attributed to the use of high-silicon cast iron anodes. Detailed information on designing impressed-current cathodic protection systems with high-silicon iron anodes is available in Ref 5.

Graphite anodes are extensively used for onshore pipeline cathodic protection applications in which they can be buried in multiple-anode ground beds (Fig. 3). Graphite anodes, which have very low electrode-to-environment resistances, are normally available in 75-mm diam × 1525-mm long (3- × 60-in.) and 100-mm diam × 2000-mm (4- × 80-in.) rods. Because of the brittle nature of the material, graphite must be stored and handled carefully.

Polymeric anodes are used to mitigate the corrosion of reinforcing steel in salt-contaminated concrete (Fig. 11). The system consists of a mesh of wirelike anodes, which are made of a conductive polymer electrode material coated onto copper conductors. The conductive polymer not only serves an an active anode material but also shields the conductors from chemical attack. These mesh anodes are designed and optimally spaced to provide long-term uniform protection at low current densities. The anode mesh is placed on the surface of a reinforced concrete structure, covered with an overlay of Portland cement or polymer-modified concrete, and then connected to a low-voltage dc power source. The properties of polymeric anodes used for construction applications are given in Table 5.

Precious metals are used for impressed-current anodes because they are highly efficient electrodes and can handle much higher currents than anodes fabricated from other materials. Precious metal anodes are actually platinized titanium or tantalum anodes; the platinum is either clad to or electroplated on the substrate. The small precious metal anode shown in Fig. 12 performs the same function as materials weighing several times more.

Lead alloy anodes, containing 2% Ag, or 1% Ag and 6% Sb, are used for cathodic protection systems in seawater. Lead alloy anodes should not be buried in the sea bottom or used in freshwater applications.

Ceramic anodes are the newest materials available for cathodic protection anodes. These anodes are supplied either as oxide coatings on transition metal substrates or as bulk ceramics. Their use as cathodic protection anodes for protecting reinforcing steel in concrete is new but the oxide coated transition metal anodes have been used since the late 1960s in the chlor-alkali industry as anodes for chlorine production and seawater electrolysis, and cathodic protection of water tanks and buried steel structures, among other industrial processes.

The oxide/metal composite anodes for chlorine environments consist of a mixed ruthenium dioxide (RuO_2) and titanium oxide (TiO_2) coating sintered onto a pure titanium (grade 1) substrate (Ref 6). Such materials, which are produced in sheet, mesh, and wire form, are based on patented dimensionally stable anode (DSA) technology (Ref 6). These anodes are so named because they remain unchanged with regard to their shape, geometry, and dimensions during their entire operating life. Figure 13 shows examples of a mixed oxide/titanium mesh anode, based on DSA technology, that is used to prevent corrosion of reinforcing steel in concrete.

The basic chemical constituents of the bulk ceramic anodes are suboxides of titanium, with Ti_4O_7 and Ti_5O_9 being the principal components. Such conductive ceramics are resistant to both oxidation and reduction in strong acid and basic environments. Some characteristic properties of conductive ceramics are given in Table 6.

Table 3 Energy characteristics of sacrificial anode alloys

Alloy	Energy capability		Consumption rate	
	A·h/kg	A·h/lb	kg/A·yr	lb/A·yr
Aluminum-zinc-mercury	2750–2840	1250–1290	3.2–3.0	7.0–6.8
Aluminum-zinc-indium	1670–2400	760–1090	5.2–3.6	11.5–8.0
Aluminum-zinc-tin	920–2600	420–1180	9.4–3.4	20.8–7.4
Zinc	810	370	10.7	23.7
Magnesium	1100	500	7.9	17.5

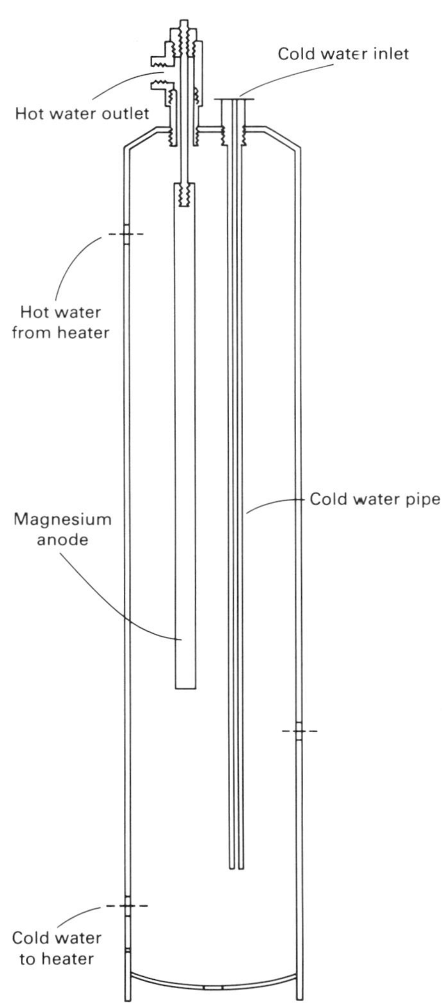

Fig. 7 Magnesium anode used to cathodically protect glass-lined steel water heater

Fig. 8 Zinc anodes on the underside of a fishing boat

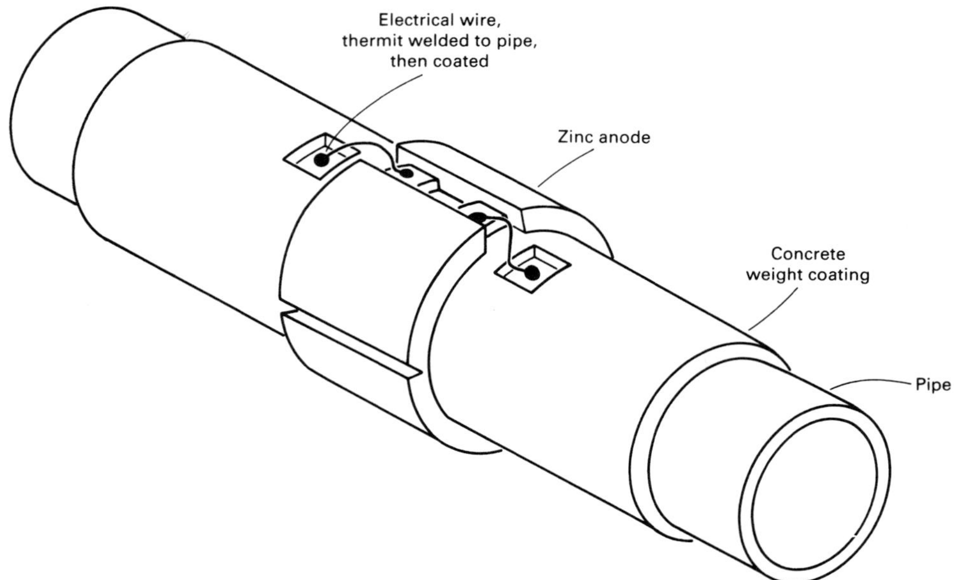

Fig. 9 Zinc bracelet anode at a joint in an offshore pipeline

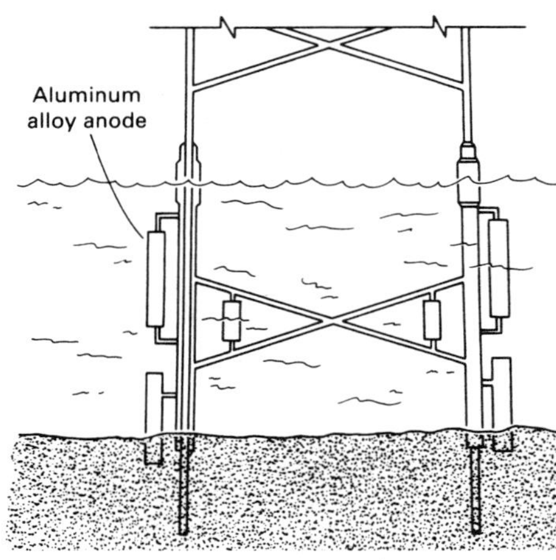

Fig. 10 Welded-on aluminum sacrificial anodes used to protect an offshore platform

Power Sources

Impressed-current cathodic protection requires external power sources. The most common source of electricity is a local power utility. This will normally involve the dc rectifier arrangement shown in Fig. 3. Remote locations can use solar cells (Fig. 14), thermoelectric current sources, special fuel-driven electric generators, or even windmills.

Design

Most cathodic protection design is conducted by consulting companies that specialize in corrosion control. Design procedures vary among organizations, but the following general guidelines are frequently followed for onshore buried structures (Ref 7):

- Decide whether impressed-current or sacrificial anodes will be used
- Decide on the design life of the system
- Determine or assume the condition of the coating. From this, the current density for cathodic protection can be estimated
- Calculate the maximum electric current required
- Determine the number and type of anodes required and their respective spacing
- Calculate the ground bed resistance. Figure 15 shows the effect of soil resistivity and pH on the corrosion of zinc sacrificial anodes
- Calculate the lead wire size
- Calculate the required dc voltage
- Determine the rectifier size
- Locate the ground bed

The advent of computers has changed some cathodic protection design, test, and inspection procedures (Ref 8), but most cathodic protection is still designed and tested in the manner described in U.S. government publications from the 1960s and 1970s (Ref 7, 9-11). Additional information on computer-aided cathodic protection can be found in the article "Marine Corrosion" in this Volume (see the section "Cathodic Protection").

Case Histories

The following examples provide specific details that supplement the general guidelines listed above. These examples have been selected to familiarize the design engineer with the steps to be followed in selecting a specific corrosion control method. An additional case history of an offshore platform cathodic protection design procedure can be found in the article "Marine Corrosion" in this Volume. Other examples are available in Ref 7 and 9 to 14.

Resistance Calculations*

The resistance of impressed-current anodes buried in soil can be lowered by surrounding them with carbonaceous backfill material, such as coke breeze (crushed coke) or flake graphite. This is particularly beneficial in high-resistance soils. If soil resistance is ten or more times the backfill resistivity—500 to 1000 Ω · cm—the voltage drop of anode current passing through the backfill may become negligible with respect to the voltage drop through the soil. Thus, the resistance of a backfilled anode can be considered to be lower than that of a nonbackfilled anode because the backfilled anode is effectively longer and of greater diameter.

Increasing the diameter and/or the length of a cylindrical anode will decrease its anode-to-electrolyte resistance. However, changes in length have a greater effect than changes in diameter.

Many engineers first measure structure-to-anode resistance by actually impressing current into installed anodes. In this way, a rectifier can be sized to fit the circuit resistance without relying on questionable test and empirical data. Testing installed anodes for the purpose of sizing rectifiers has the disadvantage of delaying completion of an installation. However, it means that sufficient cathodic protection current will be supplied in an efficient manner.

*Adapted with permission by John W. McKinney, Jr., Durichlor 51 Anode Company, from W.T. Bryan, Ed., *Designing Impressed Current Cathodic Protection Systems With Durco Anodes*, The Duriron Company, 1970, p 56-64

Single Anode. The following formulas are relations developed for a single cylindrical anode (Ref 5):

$$R_v = \frac{0.0052\rho}{L}\left(2.3 \log 10 \frac{8L}{d} - 1\right)$$

$$R_h = \frac{0.0052\rho}{L}\left(2.3 \log 10 \frac{4L^2 + 4L\sqrt{s^2 + L^2}}{ds} + \frac{s}{L} - \frac{\sqrt{s^2 + L^2}}{L} - 1\right)$$

where R_v is the electrolyte-to-anode resistance, single vertical anode to remote reference (in ohms), R_h is the electrolyte-to-anode resistance, single horizontal anode to remote reference (in ohms), ρ is the electrolyte resistivity (in ohm centimeters), L is the length of anode (in feet), d is the diameter of anode (in feet), and s equals twice the depth of the anode (in feet).

For a single vertical anode, the following simplified expression can be used:

$$R_v = \frac{\rho}{L} K$$

where R_v, ρ, and L are as given above, and K is the shape function representing the anode length to anode diameter ratio, which can be obtained from Table 7.

One-Row Vertical Anode Group. The total anode-to-electrolyte resistance for a group of vertical anodes, connected in parallel and equally spaced in one row, is expressed as follows:

$$R_n = \frac{1}{n} R_v + \frac{\rho P}{S}$$

where R_n is the total anode-to-electrolyte resistance for a group of vertical anodes equally

Fig. 11 Two views of polymer mesh anodes used to protect reinforcing steel in bridge decks, parking garages, and other large structural surfaces. Courtesy of Raychem Corporation, Cathodic Protection Division

Fig. 12 Precious metal impressed-current anode on offshore platform prior to launching in the North Sea. Courtesy of W.H. Thomason, Conoco, Inc.

Table 4 Consumption rates of impressed-current anode materials

Material	Typical anode current density, A/m^2	Typical anode current density, A/ft^2	Consumption rate per A·yr
Pb-6Sb-1Ag	160–220	15–20	0.045–0.09 kg (0.1–0.2 lb)
Pt (plated on substrate)	540–1080	50–100	0.006 g
Pt (wire or clad)	1080–5400	100–500	0.01 g
Graphite	10.8–40	1–4	0.225–0.45 kg (0.5–1.0 lb)
Fe-14Si-4Cr	10.8–40	1–4	0.225–0.45 kg (0.5–1.0 lb)

Table 5 Properties of polymeric mesh anodes used for construction applications

Properties	Typical data
Recommended maximum design current output in soil	52-mA/m (16-mA/ft) length(a)
Recommended maximum design current output in water	10-mA/m (3-mA/ft) length
Maximum pressure rating (hydrostatic)	7 MPa (1000 psi)
Maximum temperature rating	65 °C (150 °F)
Minimum installation temperature	−18 °C (0 °F)
Chemical resistance per ASTM D 543 for 7 days at ambient conditions with weight gain less than 1%	
3% NaCl	Pass
3% Na_2SO_4	Pass
10% NaOH	Pass
ASTM oil #1	Not recommended
Crude oil	Not recommended

(a) Average current output for anode in coke breeze backfill when maximum current density is 82 mA/m (25 mA/ft). ASTM D 543, "Standard Test Method for Resistance of Plastics to Chemical Reagents," can be found in Vol 08.01 of the *Annual Book of ASTM Standards*.

spaced and in one row (in ohms) (a remote cathode is assumed), n is the number of anodes, $\rho\rho$ is the soil resistivity, measured with pin spacing equal to S (in ohm-centimeters), R_v is as given above, P is the paralleling factor obtained from Table 8, and S is the spacing between adjacent anodes (in feet).

More Than One Row Vertical Anode Group. An anode group composed of two or more rows of vertical anodes, separated by a distance substantially larger than that between the anodes within a single row, has a total resistance approximately equal to the total parallel resistance of all the rows. The usual formula for paralleling resistances:

$$\left(\frac{1}{R} = \frac{1}{R_1} + \frac{1}{R_2} + \frac{1}{R_3} + \ldots\right)$$

is used.

Vertical and Horizontal Anode Groups—Simplified. If vertical anode dimensions are assumed to be 8 to 12 in. in diameter and 10 ft in length and if the horizontal anode dimensions are assumed to be 1 ft^2 in cross section, 10 ft in length, and 6 ft below the surface of the electrolyte (these assumptions are useful when designing anode system with high-silicon cast iron anodes in soils in which effective backfill has been installed around the anodes), then the following empirical formulas can be used:

$$R_v = \frac{\rho F}{537} \qquad R_h = \frac{\rho F}{483}$$

where R_v is the electrolyte-to-anode resistance, any number of vertical anodes to remote reference (in ohms), R_h is the electrolyte-to-anode resistance, any number of horizontal anodes to remote reference (in ohms), ρ is the electrolyte resistivity (in ohm-centimeters), and F is the adjusting factor for groups of anodes (from Table 9). Note: where only one anode is used, $F = 1.0$.

Approximate anode-to-electrolyte resistance values for single anodes are quickly and easily obtained from the following:

- $R_v = 0.002\rho$, for a vertically installed 60-in. anode in a 10-ft column of backfill, 1 ft in diameter
- $R_v = 0.005\rho$, for a vertically installed 2- × 60-in. anode, without backfill
- $R_v = 0.006\rho$, for a vertically installed 1½- × 60-in. anode, without backfill
- R = anode-to-electrolyte resistance for a single anode (in ohms)
- ρ = resistivity of electrolyte (in ohm centimeters)

Example 1: Cathodic Protection System for Steel Storage Tanks. An impressed-current cathodic protection system is being designed to

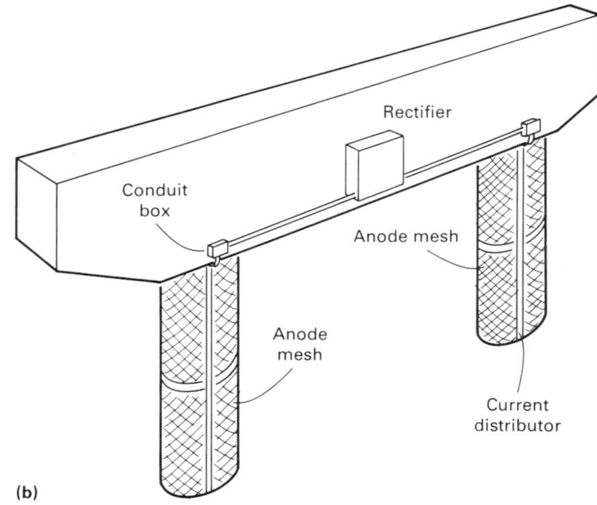

Fig. 13 Use of mixed oxide/titanium anode mesh for cathodic protection. (a) Sidewalk and barrier wall installation. (b) Installation of anode mesh on a bridge substructure. Courtesy of ELTECH Systems Corporation

Fig. 14 Solar cells used to provide electricity for the cathodic protection of a buried pipeline

stop corrosion on the bottoms of several large-diameter steel storage tanks that are in contact with the soil. In order to distribute current to all surfaces, seventeen $1\frac{1}{2}$- × 60-in. high-silicon cast iron anodes will be installed vertically in coke breeze backfill. These anodes will be spaced 20 ft apart; the backfill is approximately 8 ft in length and 10 in. in diameter. The average soil resistance is 4500 Ω · cm at anode depth. What will be the total anode to soil resistance after installation?

Solution. Because anodes are to be installed around the periphery of tanks, single-row configuration is assumed, and Table 7 must be consulted for the shape function K, where:

$$\frac{L}{D} = \frac{8}{0.833} = 9.604$$

$$K = 0.0175$$

$$R_v = \frac{\rho}{L} K = \frac{4500}{8} \times 0.0175 = 9.844\ \Omega$$

$$R_n = \frac{1}{n} R_v + \frac{\rho\rho P}{S} = \frac{1}{17} 9.844 + \frac{4500 \times 0.00150}{20}$$

$R_n = 0.579 + 0.338 = 0.917\ \Omega =$ total resistance between all anodes and soil

Example 2: Cathodic Protection of a Pipeline. An impressed-current cathodic protection unit is to be connected to a pipeline. The right-of way is limited, and space permits only three $1\frac{1}{2}$- × 60-in. anodes, which must be installed horizontally because of a rock shelf. These anodes are to be backfilled with coke breeze (effective size is 10 ft in length × 1 ft^2 in cross section) and laid side by side parallel to each other, 6 ft below grade and 10 ft apart. The cathodic protection unit must supply 5.0 A to protect the pipeline adequately. If cable, connection, and pipe-to-soil resistance are neglected, what dc voltage must the rectifier power supply be capable of producing? Soil resistivity is 12 000 Ω · cm at a depth of 6 ft.

Solution:

$$R_h = \frac{\rho F}{483} = \frac{12\,000 \times 0.460}{483} = 11.42\ \Omega$$

From Ohm's law: $E = IR = 5.0 \times 11.42 = 57.10$.

Therefore, the rectifier must produce 58 V.

Cathodic Protection of Steel Pile Structures*

Example 3: Use of Zinc Sacrificial Anodes to Protect a Pipe Pile Structure. A pier structure 20 ft wide extending 150 ft from the shore consists of sixteen 16-in. diam steel pipe piles, 40 ft long, that support the pier deck. The piles will be driven so that 14 ft will be in the soil zone, 18 ft in the submerged zone, 3 ft in the tidal zone, and 5 ft in the atmospheric/splash zone (Fig. 16).

The pier is located in a calm saltwater bay. The water is not badly polluted and has an average temperature of 62 °F. The water samples taken at three different locations show an average resistivity of 25 Ω·cm. A decision has been made to coat the piles with 16-mil dry-film thickness of coal tar epoxy and to apply cathodic protection in the form of a sacrificial anode system.

Calculation of Steel Areas Exposed to the Specified Zones:

Area in atmospheric/splash zone $= \pi D L n = \pi\left(\frac{16}{12}\right)(5)(16) = 335$ ft^2

Area in tidal zone $= \pi D L n = \pi\left(\frac{16}{12}\right)(3)(16) = 201$ ft^2

Area in submerged zone $= D L n = \pi\left(\frac{16}{12}\right)(18)(16) = 1206$ ft^2

Area in soil zone $= \pi D L n = \pi\left(\frac{16}{12}\right)(14)(16) = 938$ ft^2

where D is the diameter of the pile (in feet), L is the length of pile in the specified zone (in feet), and n equals the number of piles.

Current Requirement Calculations (Table 10):

Tidal Zone:
The total area of steel in this zone is 201 ft^2.
Assume 10% coating damage during installation:
Area of bare steel = 201 × 0.10 = 20 ft^2
Area of coated steel = 201 × 0.90 = 181 ft^2
Current required to polarize the bare steel = 20 ft^2 × 35 mA/ft^2 = 0.7 A

*Adapted with permission from T.D. Dismuke, S.K. Coburn, and C.M. Hirsh, Ed., *Handbook of Corrosion Protection for Steel Pile Structures in Marine Environments*, American Iron and Steel Institute, 1981, p 130-145

Table 6 Properties of bulk ceramic anode materials

Electrical resistivity

250–1000 × 10^{-6} Ω·cm. Resistivity decreases as density increases.

Thermal expansion

5 × 10^{-6}/°C

Thermal conductivity

10 to 20 W/m·K

Modulus of rupture

100 MPa (14.5 ksi) and higher. Some grades are 24–34 MPa (3.5–5.0 ksi) for special applications.

Melting point

1400 °C (2552 °F)

Density

2.5–4.0 g/cm^3

Thermal stability

Stable to 350 °C (660 °F) in air
Stable to 1350 °C (2460 °F) for nonoxidizing atmosphere

Overpotentials

Evolution	Overpotential at 20 °C (68 °F), mV	Current density mA/cm^2	Electrolyte
Oxygen	1712	10	1 M H_2SO_4
Hydrogen	784	10	
Oxygen	1574	10	1 M NaOH
Hydrogen	524	10	

Table 7 The shape function, *K*, obtained by dividing the anode length by the anode diameter in any units

L/d	K
5	0.0140
6	0.0150
7	0.0158
8	0.0165
9	0.0171
10	0.0177
12	0.0186
14	0.0194
16	0.0201
18	0.0207
20	0.0213
25	0.0224
30	0.0234
35	0.0242
40	0.0249
45	0.0255
50	0.0261
55	0.0266
60	0.0270

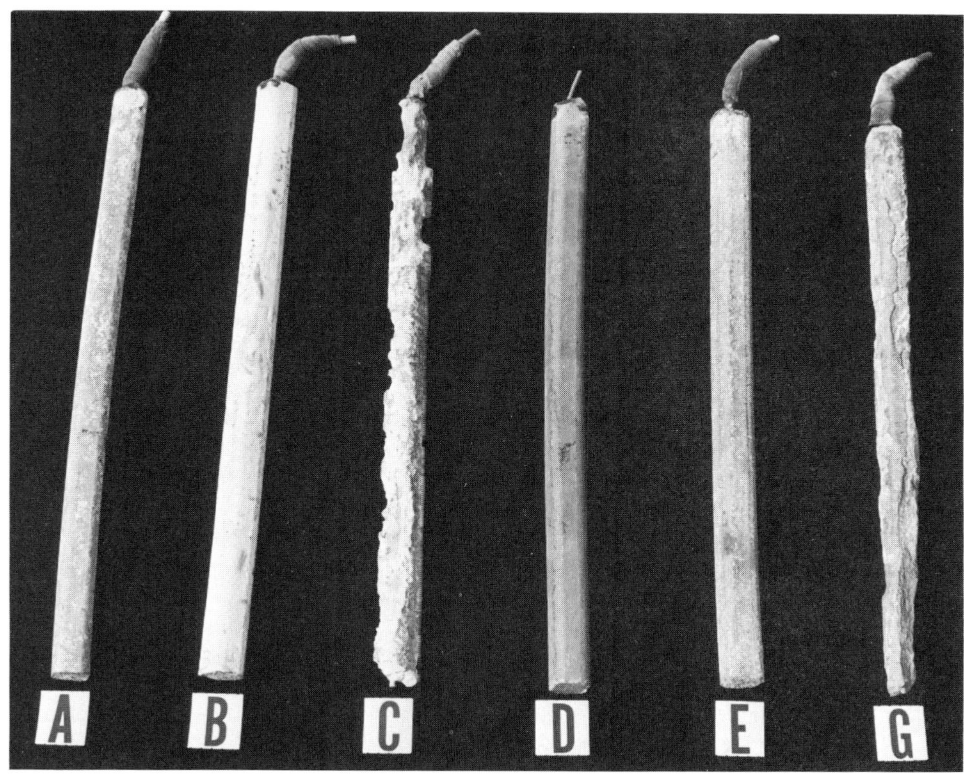

Site	Soil	Location	Internal drainage	Range of resistivity, Ω · cm	pH
A	Sagemoor sandy loam	Toppenish, WA	Good	400	8.8
B	Hagerstown loam	Loch Raven, MD	Good	12 600–37 300	5.3
C	Clay	Cape May, NJ	Poor	400–1150	4.3
D	Lakewood sand	Wildwood, NJ	Good	13 800–57 500	5.7
E	Coastal sand	Wildwood, NJ	Poor	1320–49 500	7.1
G	Tidal marsh	Patuxent, MD	Poor	400–15 500	6.0

Fig. 15 Corrosion of zinc anodes that were galvanically connected to type 304 stainless steels for 4 years at six different underground test sites. Courtesy of E. Escalante, National Bureau of Standards

Current required to polarize the coated steel
= 181 ft^2 × 5 mA/ft^2 = 0.9 A
Total current required to maintain protection of the steel during high tide
= 0.7 + 0.9 = 1.6 A

Submerged Zone:
Area of steel in submerged zone = 1206 ft^2
Assume 15% coating damage during installation:
Area of bare steel = 1206 × 0.15 = 181 ft^2
Area of coated steel = 1206 × 0.85 = 1025 ft^2
Current required to polarize the bare steel
= 181 ft^2 × 35 mA/ft^2 (corrosion rate in the area known to be high)
= 6.3 A

Current required to polarize the coated steel
= 1025 ft^2 × 5 mA/ft^2
= 5.1 A
Total current required to polarize the steel in the submerged zone
= 6.3 + 5.1 = 11.4 A
Current required to maintain protection of the bare steel
= 181 ft^2 × 10 mA/ft^2
= 1.8 A
Current required to maintain protection of the coated steel
= 1025 ft^2 × 1.5 mA/ft^2
= 1.5 A
Total current required to maintain protection of the steel in the submerged zone
= 1.8 + 1.5 = 3.3 A

Soil Zone:
The total area of steel in this zone is 938 ft^2.
Assume 50% coating intact after pile-driving operations:
Area of bare steel = 938 × 0.50 = 469 ft^2
Area of coated steel = 938 − 469 = 469 ft^2
Current required to polarize the bare steel
= 469 ft^2 × 5 mA/ft^2
= 2.3 A

Current required to polarize the coated steel
= 469 ft^2 × 1.0 mA/ft^2
= 0.5 A
Total current required to polarize the steel in the soil zone
= 2.3 + 0.5 = 2.8 A
Current required to maintain protection of the bare steel
= 469 ft^2 × 1.5 mA/ft^2
= 0.7 A
Current required to maintain protection of the coated steel
= 469 ft^2 × 0.5 mA/ft^2
= 0.2 A
Total current required to maintain protection of the steel in the soil zone
= 0.7 + 0.2 = 0.9 A

Total Current Requirements:
Total current required to polarize the entire structure:
I_{pol} = 1.6 + 11.4 + 2.8
= 15.8 A
Total current required to maintain protection of the entire structure:
I_{prot} = 1.6 + 3.3 + 0.9
= 5.8 A

Table 8 Paralleling factor used for determining anode resistance

n	P
2	0.00261
3	0.00289
4	0.00283
5	0.00268
6	0.00252
7	0.00237
8	0.00224
9	0.00212
10	0.00201
12	0.00182
14	0.00168
16	0.00155
18	0.00145
20	0.00135
22	0.00128
24	0.00121
26	0.00114
28	0.00109
30	0.00104

Table 9 Adjusting factors, *F*, for parallel anodes

Number of anodes in parallel	Adjusting factors—anode spacing, ft: 5	10	15	20	25
2	0.652	0.576	0.551	0.538	0.530
3	0.586	0.460	0.418	0.397	0.384
4	0.520	0.385	0.340	0.318	0.304
5	0.466	0.333	0.289	0.267	0.253
6	0.423	0.295	0.252	0.231	0.218
7	0.387	0.265	0.224	0.204	0.192
8	0.361	0.243	0.204	0.184	0.172
9	0.332	0.222	0.185	0.166	0.155
10	0.311	0.205	0.170	0.153	0.142
11	0.292	0.192	0.158	0.141	0.131
12	0.276	0.180	0.143	0.132	0.122
13	0.262	0.169	0.139	0.123	0.114
14	0.249	0.160	0.131	0.116	0.107
15	0.238	0.152	0.124	0.109	0.101
16	0.226	0.144	0.117	0.103	0.095
17	0.218	0.138	0.112	0.099	0.091
18	0.209	0.132	0.107	0.094	0.086
19	0.202	0.127	0.102	0.090	0.082
20	0.194	0.122	0.098	0.086	0.079
22	0.182	0.114	0.091	0.079	0.073
24	0.171	0.106	0.085	0.074	0.067
26	0.161	0.100	0.079	0.069	0.063
28	0.152	0.094	0.075	0.065	0.059
30	0.145	0.089	0.070	0.061	0.056

The Total Current Required to Maintain Protection as the Coating Deteriorates. Calculate the expected current requirement after 20 years of service. Assume coating deterioration at the rate of 2% per year:

Intact coating in tidal zone after 20 years

= 90% initial − 40% deterioration = 50%

Current required to maintain protection of the bare steel in the tidal zone after 20 years

$= 201\ \text{ft}^2 \times 0.50 \times 35\ \text{mA/ft}^2$
$= 3.5\ \text{A}$

Current required to maintain protection of the coated steel in the tidal zone after 20 years

$= 201\ \text{ft}^2 \times 0.50 \times 5\ \text{mA/ft}^2$
$= 0.5\ \text{A}$

Intact coating in submerged zone after 20 years

= 85% initial − 40% deterioration = 45%

Current required to maintain protection of the bare steel in the submerged zone after 20 years

$= 1206\ \text{ft}^2 \times 0.55 \times 10\ \text{mA/ft}^2$
$= 6.6\ \text{A}$

Current required to maintain protection of the coated steel in the submerged zone after 20 years

$= 1206\ \text{ft}^2 \times 0.45 \times 1.5\ \text{mA/ft}^2$
$= 0.8\ \text{A}$

Total current required to maintain protection of the entire structure in the tidal, submerged, and soil zones after 20 years

$= 3.5 + 0.5 + 6.6 + 0.8 + 0.9$
$= 12.3\ \text{A}$

Assume no coating deterioration in the soil zone.

Anode Design Calculations:

Average total current required to maintain protection of the entire structure for 20-year life

$$= \frac{\text{Initial current (after polarization)} + \text{final current}}{2}$$

$$= \frac{5.8 + 12.3}{2} = 9.1\ \text{A}$$

Total weight of high-purity zinc anode material required for 20-year life expectancy:

$$W = \frac{\text{CR} \times \text{L} \times \text{ATC}}{\text{E} \times \text{U}}$$

where W is the weight (in pounds), CR is the consumption rate (in lb/A·yr), L is the life (in years), ATC is the average total current (in amps), E is the efficiency factor, and U is the utilization factor:

$$W = \frac{23.585 \times 20 \times 9.1}{0.90 \times 0.85}$$
$$= 5611\ \text{lb}$$

Based on the structure description and the amount of zinc required, it can be determined that for this structure eight strings of two 390-lb (10- × 10- × 15-in.) zinc anodes (6240 lb) suspended at the locations shown in Fig. 16 should afford adequate protection for the 20-year life expectancy. The other factor to consider in a galvanic anode system is the amount of current that can be discharged from the anodes.

Resistance of the Anode System. The 390-lb anode measures 10 × 10 × 15 in. The outside perimeter of the anode is 40 in. The diameter of a cylindrical object whose circumference is 40 in. is 12.73 in., or 1.06 ft. The anode spacing is 25 ft.

Using the formula from the section "Resistance Calculations" for the resistance of a single anode, it can be calculated that the resistance of one 390-lb anode is:

$$R_v = \frac{0.0052\rho}{L}\left[2.3 \log \frac{8L}{d} - 1\right]$$

$$= \frac{(0.0052)(25)}{(15/12)}\left[2.3 \log \frac{(8)(15/12)}{1.06} - 1\right]$$

$$= 0.129\ \Omega$$

The resistance of the anode system is:

$$R_s = \frac{0.0052\rho}{n\ L}\left[2.3 \log \frac{8L}{d} - 1 + \frac{2L}{S} 2.3 \log 0.656n\right]$$

where R_v is the resistance of one vertical anode (in ohms), R_s is the resistance of the anode system (in ohms), n is the number of anodes, L is the length of anode (in feet), ρ is the water resistivity (in ohm-centimeters), d is the diameter of anode (in feet), S is the spacing between anodes (in feet):

$$R_s = \frac{(0.0052)(25)}{(8)(2)(15/12)}\left[2.3 \log \frac{(8)(2)(15/12)}{1.06} - 1 + \frac{(2)(2)(15/12)}{25} 2.3 \log 0.656(8)\right]$$

$$= 0.0147\ \text{ohm}$$

Anode system output:

$$I = \frac{E}{R}$$

where E is the driving potential, and R is the resistance of the anode system to the electrolyte.

Driving voltage of galvanic zinc anode to polarized coated steel is of the order of 0.10 V:

Driving potential = 1.1 V − 1.00 V = 0.10 V

$$I \text{ available after polarization} = \frac{0.10}{0.0147} = 6.8\ \text{A}$$

The 6.8-A output after polarization is greater than the 5.8 A required for the total current requirements given above. Therefore, the current requirement is satisfied.

To verify that life expectancy is met:

Average total current output for 20-year life

$$= \frac{6.8 + 12.3}{2} = 9.6\ \text{A}$$

Total weight of anode material required:

$$W = \frac{23.585 \times 20 \times 9.6}{0.90 \times 0.85}$$
$$= 5919\ \text{lb}$$

The total weight of the anode material used is 8 × 2 × 390 = 6240 lb. Therefore, the life expectancy is satisfied.

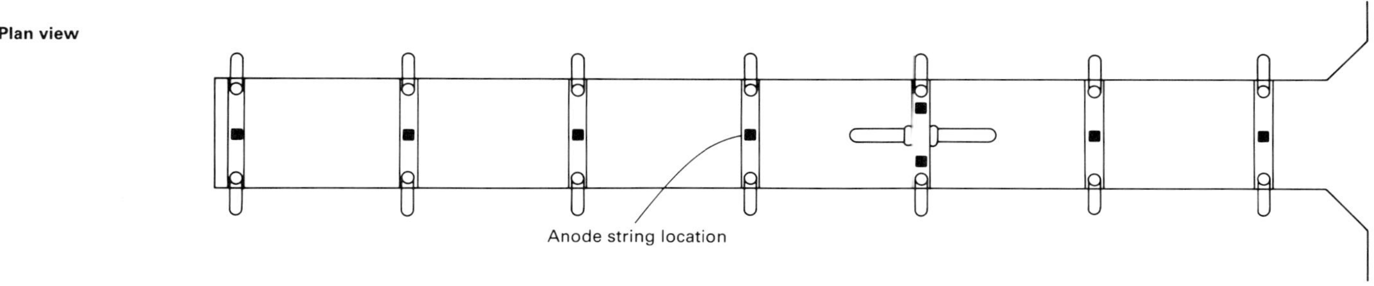

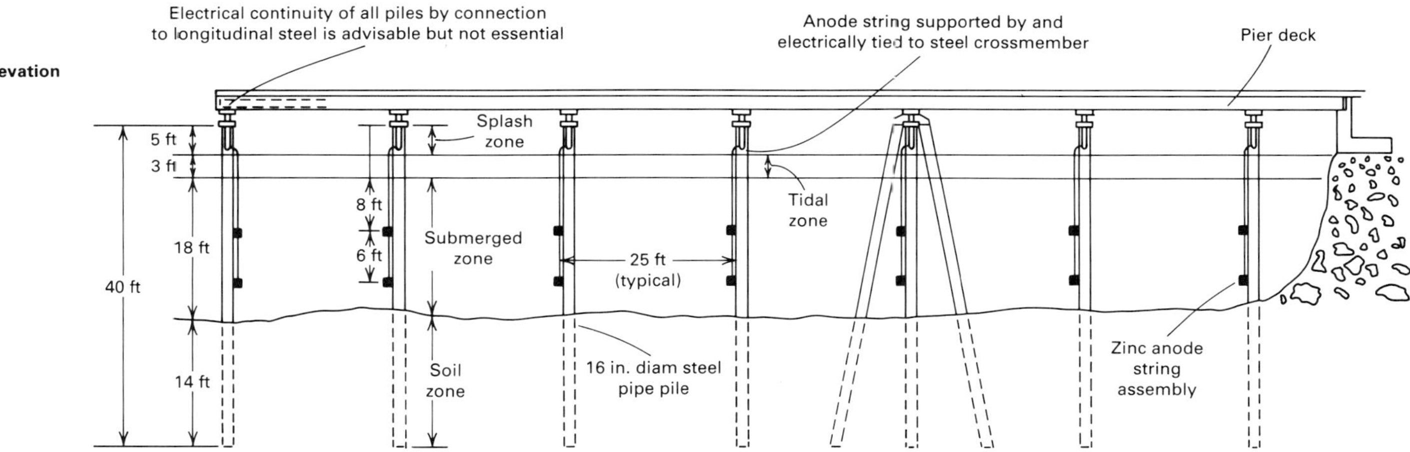

Fig. 16 Protection of a steel pipe pile pier with a protective coating and zinc sacrificial anodes

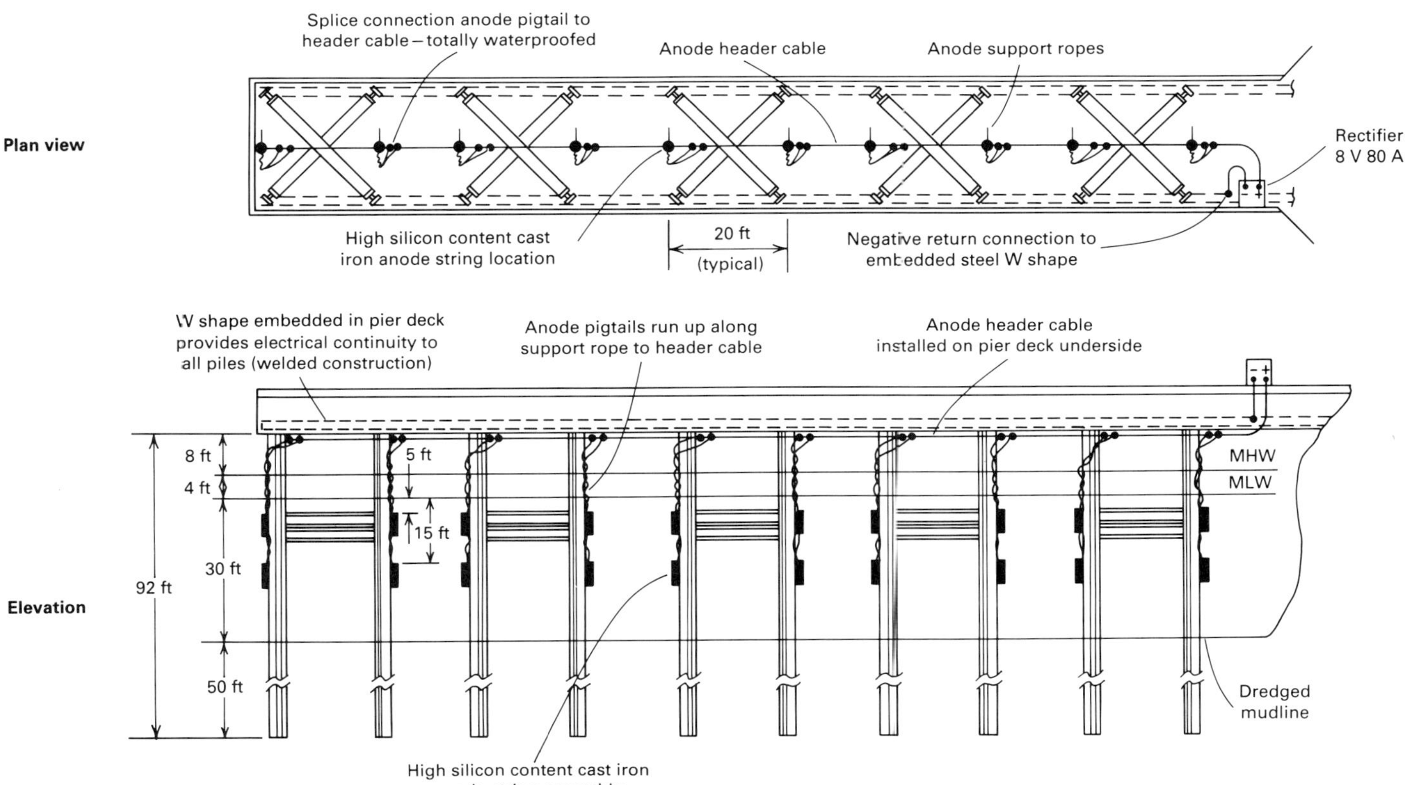

Fig. 17 Protection of H-pile steel pier structure using a protective coating and high-silicon cast iron impressed-current anodes

Table 10 Current densities required for coated and uncoated steel in moving and stagnant seawater and in soil areas

Environment	Current density, mA/ft² — Bare steel: To polarize	Bare steel: After polarization	Coated steel: To polarize	Coated steel: After polarization
Moving seawater	30–35	7–10	3–5	1.0–1.5
Stagnant seawater	15–25	4–7	1–3	0.5–1.0
Soil zone	4–5	1–1.5	0.5–1.0	0.1–0.5

Example 4: Use of Impressed-Current Anodes to Protect an H-Pile Structure. A 30-ft wide pier structure that extends 200 ft from shore is supported by 20-H-piles that are 92 ft long. The H-piles are connected in four-pile clusters by W shapes. The piles are to be driven so that 50 ft will be below the dredged mud line, 30 ft in the submerged zone, 4 ft in the tidal zone, and 8 ft in the atmospheric/splash zone (Fig. 17). The pier is located in a saltwater bay in a highly commercial and industrial area. There are commercial, industrial, chemical, and sewage pollutants. The water samples taken at three different locations show an average resistivity of 30 Ω·cm.

A decision has been made to coat the H-piles and beams with a vinyl coating of 10-mil dry-film thickness and to apply cathodic protection in the form of an impressed-current system. The cathodic protection system should be designed and a cost estimate of the coating and cathodic protection system should be made for a 20-year life expectancy.

Calculation of Steel Areas Exposed to the Specified Corrosion Zones. The 10 cross braces and the 20 vertical H-piles have a surface area of 8 ft² per linear foot:

Area in atmospheric/splash zone $= L\,n\,A = (8)(20)(8) = 1280\ \text{ft}^2$

Area in tidal zone $= L\,n\,A = (4)(20)(8) = 640\ \text{ft}^2$

Area in submerged zone: Piles $= L\,n\,A = (30)(20)(8) = 4800\ \text{ft}^2$

Cross braces $= L\,n\,A = (30)(10)(8) = 2400\ \text{ft}^2$

Total $= 4800 + 2400 = 7200\ \text{ft}^2$

Area in soil zone $= L\,n\,A = (50)(20)(8) = 8000\ \text{ft}^2$

where L is the length of pile or cross brace in each zone, n is the number of piles or cross braces, and A is the surface area of 1 linear foot of H-pile or cross brace.

Current Requirement Calculations. As shown in Table 10, a greater amount of current is initially required to polarize the structure and bring it to protective levels than is required to maintain these levels.

Tidal Zone:

The total area of steel in the tidal zone is 640 ft². Coating damage of 5% is assumed during installation:

Area of bare steel = 640 × 0.05 = 32 ft²
Area of coated steel = 640 × 0.95 = 608 ft²

Current required to polarize the bare steel
= 32 ft² × 35 mA/ft²
= 1.1 A

Current required to polarize the coated steel
= 608 ft² × 5 mA/ft²
= 3.0 A

Total current required to maintain protection of the steel during high tide
= 1.1 + 3.0 = 4.1 A

Submerged Zone:

The total area of steel in the submerged zone is 7200 ft². If one assumes that after installation, the coating is 95% intact, there will be 6840 ft² of coated steel and 360 ft² of bare steel. The current required to polarize this steel is as follows:

Bare steel current requirement:
= 360 ft²× 35 mA/ft²
= 12.6 A

Coated steel current requirement
= 6840 ft² × 5 mA/ft²
= 34.2 A

Total current required to polarize the steel in the submerged zone
= 12.6 + 34.2 = 46.8 A

The current required to maintain protection is as follows:

Bare steel current requirement
= 360 ft² × 10 mA/ft²
= 3.6 A

Coated steel current requirement
= 6840 ft² × 1.5 mA/ft²
= 10.3 A

Total current required to maintain protection of the steel in the submerged zone
= 3.6 + 10.3 = 13.9 A

Soil Zone:

The total area of all piles in the soil zone is 8000 ft². Assuming that the coating is 40% intact, the area of coated steel will be 3200 ft², and the area of bare steel will be 4800 ft². The current required to polarize this steel is as follows:

Bare steel current requirement
= 4800 ft² × 5 mA/ft²
= 24.0 A

Coated steel current requirement
= 3200 ft²× 1.0 mA/ft²
= 3.2 A

Total current required to polarize the steel in the soil zone
= 24.0 + 3.2 = 27.2 A

The current required to maintain protection is as follows:

Bare steel current requirement
= 4800 ft² × 1.5 mA/ft²
= 7.2 A

Coated steel current requirement
= 3200 ft² × 0.5 mA/ft²
= 1.6 A

Total current required to maintain protection of the steel in the soil zone
= 7.2 + 1.6 = 8.8 A

Total Current Requirements. Total current required to polarize the entire structure:

$I_{pol} = 4.1 + 46.8 + 27.2$
$= 78.1$ A

Total current required to maintain protection of the entire structure:

$I_{prot} = 4.1 + 13.9 + 8.8$
$= 26.8$ A

It must be remembered that the above current requirement calculations are for the initial period following construction of the pier. The coating can be expected to deteriorate in the post-construction years. The amount of current required to maintain cathodic protection of the structure will then increase unless an extensive maintenance program of coating repair is followed for the submerged portions.

Assuming 2% per year additional coating deterioration in the tidal zone, 1% per year in the submerged zone, and 0.5% per year in the soil zone, the final current requirements at the end of the 20-year life expectancy would be:

Tidal Zone:

Intact coating in the tidal zone after 20 years
= 95% initial − 20 × 2% deterioration = 55%

Bare steel current requirement = 640 ft² × 0.45 × 35 mA/ft²
= 10.1 A

Coated steel current requirement = 640 ft² × 0.55 × 5 mA/ft²
= 1.8 A

Total current required to polarize the steel in the tidal zone
= 10.1 + 1.8 = 11.9 A

Submerged Zone:

Intact coating in the submerged zone after 20 years
= 95% initial − 20 × 1% deterioration = 75%

Bare steel current requirement = 7200 ft² × 0.25 × 10 mA/ft²
= 18.0 A

Coated steel current requirement= 7200 ft² × 0.75 × 1.5 mA/ft²
= 8.1 A

Total current required to maintain protection in the submerged zone
= 18.0 + 8.1 = 26.1 A

Soil Zone:

Intact coating in the soil zone after 20 years
= 40% initial − 20 × 0.5% deterioration = 30%

Bare steel current requirement = 8000 ft² × 0.70 × 1.5 mA/ft²
= 8.4 A

Coated steel current requirement = 8000 ft² × 0.30 × 0.5 mA/ft²
= 1.2 A

Total current required to maintain protection in the soil zone
= 8.4 + 1.2 = 9.6 A

Total current required to maintain protection of the entire structure at the end of the 20-year period
= 11.9 + 26.1 + 9.6
= 47.6 A

Anode Design Calculations. The anode material is high-silicon content cast iron. Consumption rate is 0.75 lb/A·yr:

Average total current required for 20-year life

$$= \frac{\text{Initial current} + \text{Final current}}{2}$$

$$= \frac{26.8 + 47.6}{2}$$

= 37.2 A

Total weight of anode material required for 20-year life expectancy:

$$W = \frac{\text{CR} \times \text{L} \times \text{ATC}}{\text{U}}$$

where CR, L, ATC, and U are as given in Example 3. In impressed-current systems, the efficiency factor, E, is 100% (compared to 90% for sacrificial anode systems):

$$= \frac{0.75 \text{ lb/A·yr} \times 20 \text{ yr} \times 37.2 \text{ A}}{0.50}$$

$$= 1116 \text{ lb}$$

The total number of anodes will be determined by taking into consideration the water depth, number and spacing of piles, amount of current available from each anode for a uniform consumption, and current distribution to the structure.

For this structure, twenty 2-in. diam × 60-in. long anodes with enlarged heads have been chosen; the anodes are to be hung in ten strings of two anodes each. The top of the upper anode is to be 5 ft below the mean low water level (MLW), and the top of the lower anode will be 15 ft below MLW. See Fig. 17 for anode locations.

Resistance of Anodes (Ground bed resistance). The resistance of one vertical anode is:

$$R_v = \frac{0.0052\rho}{L}\left[2.3 \log \frac{8L}{d} - 1\right]$$

where R_v is the anode-to-electrolyte resistance (in ohms), ρ is the electrolyte resistivity (in ohm-centimeters), L is the length of the anode (in feet), and d is the diameter of the anode (in feet):

$$R_v = \frac{(0.0052)(30)}{(60/12)}\left[2\ 3 \log \frac{(8)(60/12)}{(2/12)} - 1\right]$$

$$R_v = 0.140\ \Omega$$

The calculations for the resistance of several anodes in parallel are made by using the following formula:

$$R_s = \frac{0.0052\rho}{nL}\left[2.3 \log \frac{8L}{d} - 1 + \frac{2L}{S} 2.3 \log 0.656n\right]$$

where R_s is the resistance of vertical anodes in parallel to electrolyte (in ohms), and ρ, n, L, d, and S are as given previously.

Where two anodes are installed vertically in the same string, assume length, L, as equal to length of both anodes combined. From Fig. 19, the spacing is 20 ft:

$$R_s = \frac{(0.0052)(30)}{(10)(120/12)}\left[2.3 \log \frac{(8)(120/12)}{(2/12)} - 1 + \frac{(2)(120/12)}{20} 2.3 \log 0.656(10)\right]$$

$$R_s = 0.011\ \Omega$$

DC Circuit Cable Resistance. The circuit of a rectifier system consists of one or more anode header cables that run from the rectifier positive terminal to the anodes. Current from the anodes flows through the electrolyte and is collected on the structure. A negative return cable then returns this current to the rectifier negative terminal. The cable size must be large enough to carry the current flowing through it and large enough to avoid an excessive voltage drop.

For this installation, a No. 4 American Wire Gage copper wire has sufficient current-carrying capacity. The voltage drop in this wire with the rectifier at the shore end of the pier will be as follows:

Voltage drop = Resistance of wire (Ω) × Current (A)

where:

$$\text{Cable resistance} = \left[\frac{\text{Header cable length}}{2} + \text{Negative return length}\right] \times \text{Wire resistance/ft}$$

The header cable length is divided by 2 because the total current is not flowing through the entire length of cable. The current is being dropped off in equal amounts at equal distances. The wire unit resistance is 0.000254 Ω/ft:

$$\text{Circuit cable resistance} = \left[\frac{200 \text{ ft}}{2} + 20 \text{ ft}\right] \times 0.0002542\ \Omega/\text{ft}$$

$$= 0.0305\ \Omega$$

$$\text{Voltage drop} = 0.0305\ \Omega \times 78.1 \text{ A} = 2.38 \text{ V}$$

where 78.1 A is the greatest current flow expected (polarization) from the total current requirements given previously.

Total dc Circuit Resistance. The total dc circuit resistance is the total of the following:

- Anodes-to-electrolyte resistance
- Circuit cable resistance
- Structure-to-electrolyte resistance (negligible)

Anode resistance + Cable resistance
= 0.011 + 0.0305 = 0.0415 Ω

Rectifier dc outputs:

Voltage required = Maximum current required × Total circuit resistance
= 78.1 A × 0.0415 Ω
= 3.24 V

Add 2 V for anode back voltage = 5.24 V
Allow 20% for stack aging = 1.05 V

Rectifier voltage required = 6.29 V
Rectifier output required = 6.29 V and 78.1 A
Closest available rectifier = 8 V dc, 80 A dc

This rating is for a single phase, 120-V air-cooled selenium rectifier.

Rectifier Efficiency:

$$\text{Percent efficiency} = \frac{\text{dc output}}{\text{ac input}} \times 100$$

REFERENCES

1. H. Davy, *Philos. Trans. R. Soc. (London)*, Vol 114, 1824-1825, p 151-158, 242-246, 328-346
2. M.G. Fontana, *Corrosion Engineering*, McGraw-Hill, 1986, p 497-499
3. "Control of External Corrosion on Underground or Submerged Metallic Piping Systems," NACE RP-01-69, National Association of Corrosion Engineers
4. "Corrosion Control on Steel, Fixed Offshore Platforms Associated With Petroleum Production," NACE RP-01-76, National Association of Corrosion Engineers
5. W.T. Bryan, Ed., *Designing Impressed Current Cathodic Protection Systems With Durco Anodes*, 2nd ed., The Duriron Company, Inc., 1970
6. L.M. Ernes, "Characteristics and Advantages of the Dimensionally Stable Anode (DSA[R])," ELTECH Systems Corporation, March 1987
7. *Electrical Design—Corrosion Control*, Technical Manual TM5-811-4, U.S. Army, 1962
8. R. Heidersbach, J. Fu, and R. Erbar, Ed., *Computers in Corrosion Control*, National Association of Corrosion Engineers, 1986
9. L. West and T. Lewicki, *Corrosion Control—General*, Vol I, *Civil Engineering Corrosion Control*, NTIS AD/A-004 082, Air Force Civil Engineering Center, TR 74-6, Jan 1975
10. L. West and T. Lewicki, *Cathodic Protection Testing Methods and Instruments*, Vol II, *Civil Engineering Corrosion Control*, NTIS AD/A-004 083, Air Force Civil Engineering Center, TR 74-6, Jan 1975
11. L. West and T. Lewicki, *Cathodic Protection Design*, Vol III, *Civil Engineering Corrosion Control*, NTIS AD/A-006 400, Air Force Civil Engineering Center, TR 74-6, Jan 1975
12. M.E. Parker, *Pipeline Corrosion in Cathodic Protection*, Gulf Publishing, 1962
13. A.W. Peabody, *Control of Pipeline Corrosion*, National Association of Corrosion Engineers, 1967
14. J. Morgan, *Cathodic Protection*, Macmillan, 1960

Corrosion Inhibitors for Oil and Gas Production

Paul J. Stone, Chevron U.S.A.

CORROSION INHIBITORS used in the oil field can be grouped into several common types or mechanistic classes: passivating, vapor phase, cathodic, anodic, film forming, neutralizing, and reactive. These inhibitors are designed for use with carbon steel only. Inorganic inhibitors, such as sodium arsenite (Na_2HAsO_3) and ferrocyanide, have been used to inhibit carbon dioxide (CO_2) corrosion in oil wells, but the treatment frequency and effectiveness have not been satisfactory. This has led to the development of many organic chemical formulations that could almost be reduced to a single type of organic molecule: film-forming amines and their salts. These organic corrosion inhibitors can be classified as cathodic, anodic, or cathodic-anodic.

In the mid-1940s, long-chain polar compounds were shown to have inhibitive properties. This discovery dramatically altered the inhibitor practices on primary production oil wells and gas wells. It permitted operation of wells that, because of the corrosivity and volume of water produced along with the hydrocarbons, would not have been used due to economics. Perhaps entire reservoirs would have been abandoned because of the high cost of corrosion. Inhibitors also allowed the injection and production of high volumes of corrosive water resulting from the secondary-recovery concept of waterflooding.

The consensus is that organic compounds inhibit corrosion by adsorbing at the metal/solution interface. Three possible types of adsorption are associated with organic inhibitors: π-bond orbital adsorption, electrostatic adsorption, and chemisorption. A more simplistic view of the mechanism of corrosion inhibitors can be described as controlled precipitation of the inhibitor from its environment (water and hydrocarbons) onto metal surfaces.

During the past 25 years, the primary improvements in inhibitor technology have been the refinement of formulations and the development of improved methods of applying inhibitors. The methods of evaluating the performance during their use have also advanced considerably.

Inhibitor Formulations

A wide variety of inhibitor formulations are available. However, most of these inhibitors are produced from only a few basic types of starting molecules. Fatty acids and some form of basic nitrogen-containing precursor are the principal active-ingredient sources. Historically, the first proprietary organic inhibitors were fatty imidazolines made from by-product fatty acids and polyethylene amines. The reaction is a condensation reaction that produces the following structure:

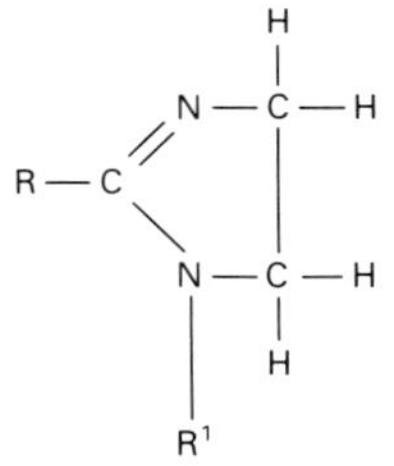

```
               H
               |
         N  —  C — H
       //      |
R — C          |
       \       |
         N  —  C — H
         |     |
         |     H
         |
         R¹
```

in which R is a chain of carbon molecules (average: C_{18}), and R^1 is a hydrogen atom or C_2H_4N group.

The molecules produced from these products were dissolved in hydrocarbon solvent or water-alcohol-base solvent, depending on whether they were further reacted. Typical further reactions were salted with acetic acid (CH_3COOH) or quaternized with a short-chain alkyl chloride. Highly corrosive wells, such as those found in the Talco Field (Texas), required daily batch treatments with these early inhibitors.

The inhibitors currently in use are generally more complex mixtures or reaction products and have been formulated to meet the demands of a very competitive industry. Some of these demands on inhibitor formulation are discussed below.

Pour Point. Because inhibitors are usually stored and used outdoors, they must remain liquid at low temperatures. A pour point of −30 °C (−20 °F) is usually required. Some areas of the world may have an even lower pour point requirement (−40 to −45 °C, or −40 to −50 °F). The required pour point often dictates the activity and solvent system of a particular inhibitor.

Solubility of the inhibitors is dictated by the intended use. By their very nature, inhibitors cannot be truly soluble in either hydrocarbons or water; degrees of dispersibility are more descriptive.

Performance. The end user of corrosion inhibitors will often specify a laboratory test that the inhibitor must pass before a field trial or purchase is considered. The wheel test is commonly used in the oil- and gas-producing industry. Therefore, many inhibitors are formulated to pass the wheel test (see the section "Laboratory Testing of Corrosion Inhibitors" in this article).

Emulsion Tendencies. The application of the inhibitor must not cause secondary problems. Batch treatments have often caused emulsions of the hydrocarbons and water that, relative to normal operations, are extremely difficult to break. In some cases, the emulsions resulting from batch treatments were so viscous that the surface separation equipment was literally stopped up by the emulsion formed by the presence of high inhibitor concentrations. Therefore, inhibitors are specifically formulated to be nonemulsifying. Alternatively, emulsion-breaker chemicals are added to formulations in small amounts to prevent emulsions.

Table 1 lists the cationic molecular structures found in many commercial inhibitors manufactured in the past 25 years. These amines or cationic molecules are often neutralized with an organic acid or quaternized to achieve a final basic product. In approximately 70% of the inhibitor formulations, the choice of the acid or anionic molecule is critical to the performance of the final product. Also, a mixture of acids is sometimes used to obtain a desired property. In addition, Table 2 lists the anionic molecular structures found in many commercial inhibitors manufactured in the past 25 years, and the compositions of fatty acids used in inhibitor production are given in Table 3.

Varying Characteristics of Oil and Gas Wells

The varying characteristics and number of organic inhibitors are explained by the varying characteristics of oil and gas wells. In this article, oil and gas wells will be characterized as oil wells, gas wells, or water injection systems.

Oil wells, which produce liquid hydrocarbons and water, can be divided into two basic types: flowing wells, which have a natural flow of hydrocarbons and water to the surface, and artificially lifted wells, which require some form of pump. The latter type includes rod-pumped wells, which use a positive-displacement downhole pump connected to the surface by rods; gas-lifted wells, in which gas is injected at some point down the hole to lighten the fluid and cause flow; electrical submerged-pump wells, which use multistage centrifugal pumps placed near the bottom of the well; and hydraulically pumped wells, which use positive-displacement pumps driven hydraulically by oil or water from a surface motor and pump.

Gas Wells. Most gas wells flow naturally. These include dry gas wells, which produce mainly gaseous hydrocarbons, and gas-condensate wells, which produce significant quantities of liquid hydrocarbons, liquid petroleum gas (LPG), and methane at high pressures from high-temperature reservoirs.

Water injection systems are used for disposal, waterflood, or tertiary recovery.

Each of the above categories can be divided into a sweet or sour designation, which is a function of the absence or presence of hydrogen sulfide (H_2S). Corrosion is usually defined as sour when any phase of the produced fluids contains measurable H_2S. Sweet corrosion is caused by the presence of CO_2, formic acid (HCOOH), CH_3COOH, or other short-chain acids in the produced gas and/or fluids. Because corrosivity is usually related to the ratio of produced water to hydrocarbons, the volume and composition of the produced water and oil will influence the application method and performance of the inhibitor.

Influence of Well Depth and Completion Method

The industry is continuously seeking new oil and gas reserves. New finds are generally deeper and/or more inaccessible. As the depth increases in the newer wells, the bottom hole temperature increases. Bottom hole temperatures of 205 °C (400 °F) have been encountered in several deep reservoirs. High labor and equipment costs in offshore production and remote areas have caused the risk of equipment malfunction due to inhibitor failure (incorrect method of application or poor inhibitor performance) to become increasingly important. This has forced the industry to screen inhibitors carefully and to select application methods that are highly reliable. However, in the case of batch treatment, application must not be too frequent.

Factors Influencing Corrosivity of Produced Fluids

Characteristics of Fluids. Produced fluids vary from noncorrosive gases to extremely corrosive brines. The amount of water in the produced fluids may vary from 1 to 99%. Methods of monitoring corrosivity are discussed in the articles "Corrosion in Petroleum Production Operations" and "In-Service Monitoring" in this Volume.

Temperature, Pressure, and Velocity Effects. The effect of temperature increase on corrosion rate is similar to its effect on other chemical reactions. That is, the rate of reaction doubles with each 10-°C (18-°F) increase in temperature. The effect of increased pressure is to raise the solubility of acid gases, thus increasing the corrosivity of the water. Bottom hole pressures exceeding 138 MPa (20 000 psi) are occasionally encountered.

Cyclic Loading, Stress, and Wear. Corrosion inhibitors can be specially formulated to control corrosion that is accelerated by cyclic loading, stress, and wear. Methods of evaluating inhibitors for inhibition of corrosion accelerated by stress and fatigue are discussed in Ref 2 to 5. Because most oil wells are not perfectly vertical holes, wear of tubing by rods becomes a problem in wells that are highly deviated or crooked. Rod guides and inhibitors are sometimes effective in controlling wear and corrosion.

Erosion and Abrasion. Many wells in geologically young formations produce fine sand along with the fluids. This fine sand may remove inhibitor films, depending on the velocity of the

Table 1 Cationic molecular structures in commercial oil field inhibitors

The letter R denotes fatty acids derived from such oils as soya, coconut, tallow, and tall oil. Sources of these fatty acids are listed in Table 3.

Imidazolines

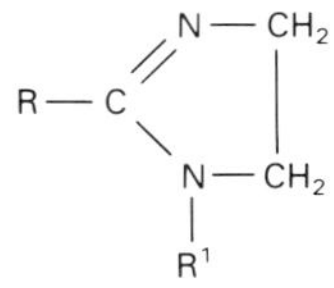

```
        N — CH2
      //      |
R — C         |
      \       |
        N — CH2
        |
        R1
```

Primary amine

$R-NH_2$

Diamines

$R-NH-C_3H_6-NH_2$

Amido-amines

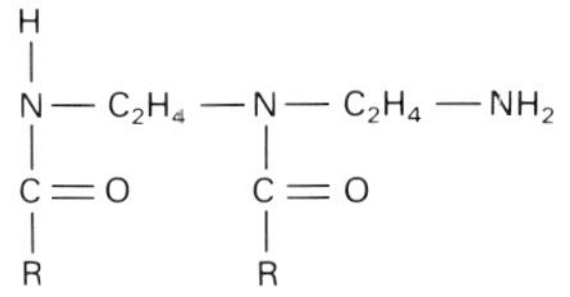

```
H
|
N — C2H4 — N — C2H4 — NH2
|          |
C = O      C = O
|          |
R          R
```

Dimerized amido-amines

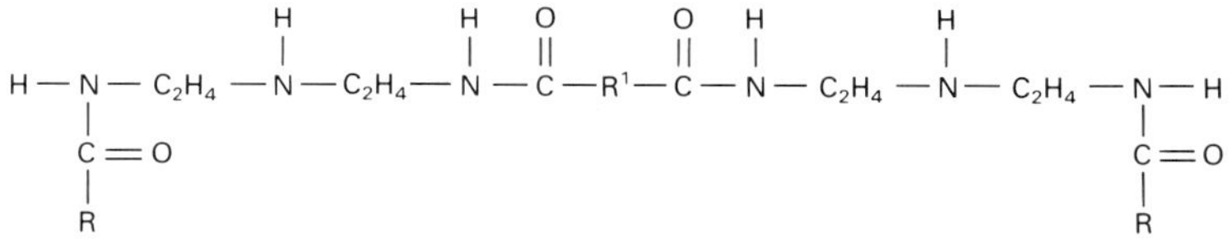

```
                   H           H   O        O   H                  H
                   |           |   ||       ||  |                  |
H — N — C2H4 — N — C2H4 — N — C — R1 — C — N — C2H4 — N — C2H4 — N — H
    |                                                             |
    C = O                                                         C = O
    |                                                             |
    R                                                             R
```

Dimer acid amido-amines

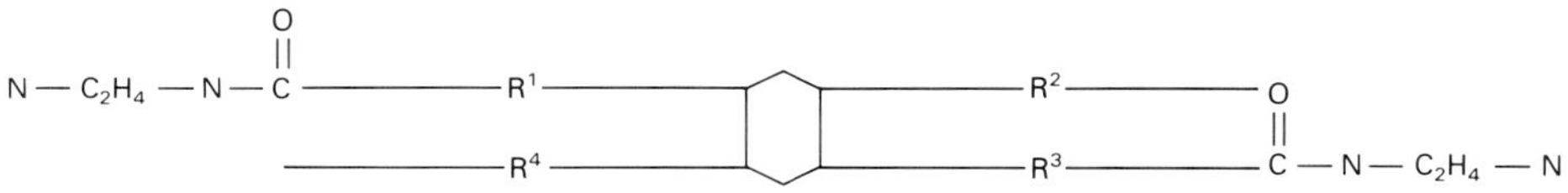

```
              O
              ||
N — C2H4 — N — C ———— R1 ———— [ring] ———— R2 ———— O
                                                  ||
             ———— R4 ———— [ring] ———— R3 ———— C — N — C2H4 — N
```

Oxyethylated primary amines

$R-N-[(C_2H_4O)_n-H]_2$

Alkyl pyridine

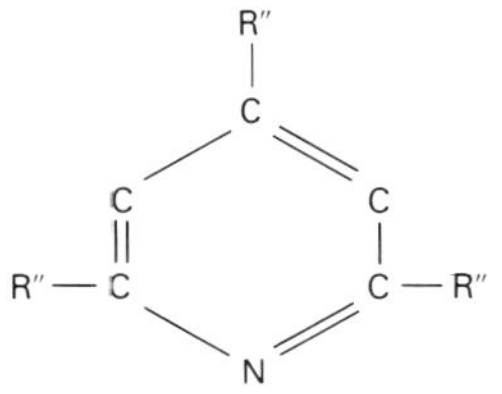

```
          R″
          |
          C
        /   \\
      C       C
      ||      |
R″ — C        C — R″
        \   //
          N
```

R″ may be CH_3 to C_3H_7 or higher

Quaternized amines

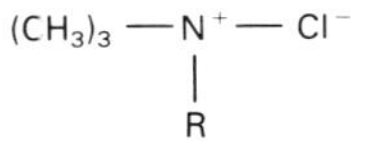

```
(CH3)3 — N+ — Cl−
         |
         R
```

Table 2 Anionic molecular structures used in commercial oil field inhibitors

The letter R denotes fatty acids derived from such oils as soya, coconut, tallow, and tall oil. Sources of these fatty acids are listed in Table 3.

Dimer-trimer acids (C_{18} average)

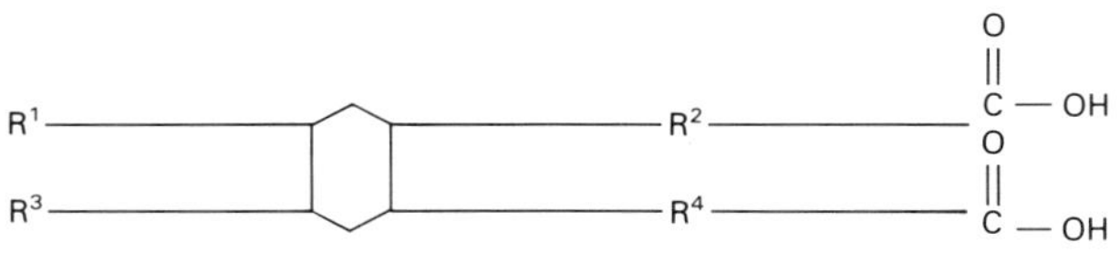

```
                                              O
                                              ||
R¹ ----------[ring]---------- R² ------------ C — OH
                                              O
                                              ||
R³ ----------[ring]---------- R⁴ ------------ C — OH
```

Fatty acids

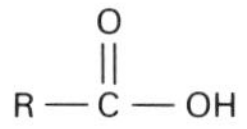

```
       O
       ||
R — C — OH
```

Naphthenic acids

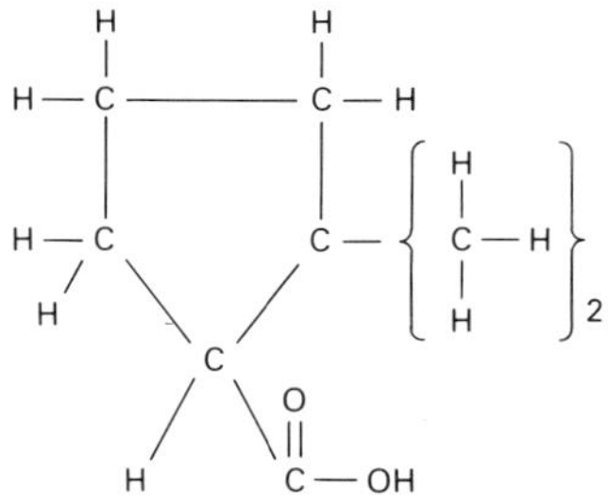

```
      H             H
      |             |
H — C ———————— C — H
      |             |        H
      |             |        |
H — C             C — { C — H }
     /               \         |      2
    H        C         H
            /  \   O
           H    \  ||
                  C — OH
```

Dodecyl benzene sulfonic acid

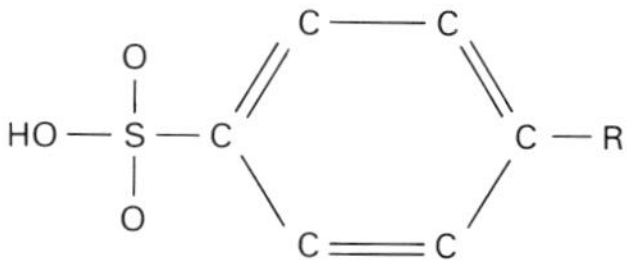

```
            C ——— C
   O       //       \\
   |      /           \
HO — S — C             C — R
   |      \           /
   O       \         /
            C ═══ C
```

R = C_{12} branched or straight chained

Petroleum oxidates

Petroleum-derived oxidates (from petrolatum, paraffin wax, lubricating oil, or fuel distillate) are complex mixtures composed primarily of organic acids and esters, but also contain alcohols, ketones, hydroxy and keto acids, estolides, lactides, and lactones.

Phosphate esters of ethoxylated alcohol

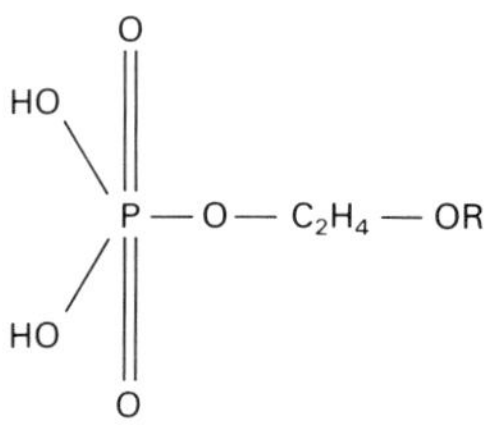

```
         O
HO       ||
   \     ||
     P — O — C2H4 — OR
   /     ||
HO       ||
         O
```

R = C_{10} to C_{16} (average C_{12})

Other organic acids

1. Acetic
2. Hydroxyacetic
3. Benzoic

fluids. Also, this fine sand may be a primary cause of equipment failure in artificially lifted wells. Sand control, a process of permanently preventing movement of fine sand particles from the formation, is often necessary for good corrosion inhibition.

Water Volume. Early studies by the National Association of Corrosion Engineers (NACE) and the National Gas Producers Association (NGPA) resulted in the designation of two types of corrosion: water independent and water dependent. Water-independent corrosion occurs in wells producing 0.1 to 1% H_2O. Severe corrosion begins at initial water production (Ref 6). Water-dependent corrosion develops in wells as increasing percentages of water are produced. Corrosion in this type of well may begin after several years of production. Pressure is related to corrosivity in gas wells as follows (Ref 7):

- A partial pressure of CO_2 below 48 kPa (7 psi) is considered noncorrosive
- A partial pressure of CO_2 between 48 and 207 kPa (7 and 30 psi) may indicate corrosion
- A partial pressure of CO_2 above 207 kPa (30 psi) usually indicates severe corrosion. Although this general rule is widely accepted, many exceptions have been noted

A 1953 survey of corrosive oil wells showed that no oil wells were corrosive when the water production was under 40%. Water contents from 0 to 40% usually represent well-emulsified water. Because oil is the continuous phase, the production equipment is considered to be oil-wet (Ref 8-12). However, recent oil production in the North Sea has shown that water contents of approximately 4% may be highly corrosive (Ref 13). As waterfloods have matured, many fields have produced sour water from the outset, although many have changed from sweet to sour during the initial water injection phase. (A waterflood is the injection of water into a field to displace hydrocarbon from the field.) Sour fluids in these mature waterfloods are generally regarded as being highly corrosive, but this notion is not borne out by the facts. There are simply more sour waterfloods that are handling large amounts of water.

Methods of Inhibitor Application

Continuous treatment of producing wells and waterflood injection systems is accomplished by a chemical-proportioning pump that operates constantly. Continuous treatment of artificially lifted wells is carried out by continuously injecting a small quantity of inhibitor into the annulus with a portion of the produced fluid. This technique is known as slip-stream flushing. After an initial treatment of several gallons of inhibitor, inhibitor is continuously injected in order to maintain an inhibitor concentration of 25 to 100 ppm in the produced fluid. The mechanical setup of continuous treatment is simple in artificially lifted wells because the annulus is open to the tubing. Gas-lifted wells prove to be an exception in that the dry gas may cause the solvent to be stripped from the inhibitor, resulting in gunking of the inhibitor and plugging of the system. A treating system of inhibiting gas-lifted wells is described in Ref 14. Continuous treatment of

Table 3 Sources of fatty acids used in corrosion inhibitors

Data obtained from literature sources and from Procter & Gamble analyses. Crude tall oil, distilled tall oil, and distilled tall oil fatty acids are the most common sources for oleophilic groups of inhibitor molecules. Tall oil is a by-product of pulping wood chips (usually pine wood) by the kraft process. Although tall oil is removed from the wood as a sodium salt, 10 to 15% inert unsaponifiable material is present. Crude tall oil is one of the cheapest raw materials known to the industry, but due to the variability of composition from tree to tree, experience and care are required in its use.

Fatty acid	Number of double bonds	Number of carbon atoms	Source, %			
			Coconut oil	Soya oil	Tallow	Tall oil fatty acids(a)
Caprylic	0	8	7	...	...	...
Capric	0	10	6	...	...	...
Lauric	0	12	50	0.5	...	...
Myristic	0	14	18	0.5	3	...
Palmitic	0	16	8.5	12	24	...
Stearic	0	18	3	4	20	4
Oleic	1	18	6	25	43	52
Linoleic	2	18	1	52	4	40
Linolenic	3	18	0.5	6	0.5	..
Others(b)	NA	NA	...	...	4.5	4

(a) Tall oil acids consist of 60 to 70% fatty acids and 30 to 40% rosin acids. (b) NA, not applicable

flowing oil and gas wells requires certain types of completion. The following completion techniques are engineered for continuous treating:

- *Dual completion* (either concentric or parallel). In this completion technique, two separate strings of tubing are run in the same hole and may or may not be the same size
- *Capillary or small-bore treating string* involves using a string of continuous, small-diameter tubing strapped to the outside of the well tubing or pipe. This type of completion provides excellent inhibitor injection control with superior results (Ref 15-18). Disadvantages are the cost and the mechanical difficulties
- *Side pocket mandrel valve.* The annulus is filled with liquid inhibitor solution, and continuous injection on the surface is used to apply pressure to the annulus so that the injection valve opens. Mechanical difficulties with the valve and the stability of the inhibitor solution are disadvantages
- *Packerless completion.* No downhole valves are involved, but inhibitor solution stability must be considered (Ref 19, 20). At high bottom hole temperatures, continuous treatment is generally selected when production rates are such that the expected treatment life of batch treatments is economically prohibitive
- *A low-cost continuous method* for marginal gas wells involves perforating the tubing above the packer, filling the annulus with an inhibitor solution, and then continuously pumping inhibitor solution into the annulus at the surface (Ref 21)

Intermittent Treatment. Batch treatment of flowing gas or oils wells may be effective at low fluid production volumes. Treatment is accomplished by pumping a 10% inhibitor into hydrocarbons or water, then shutting in the well long enough for the inhibitor solution to fall to the bottom of the well. Because the liquid fluid level is unknown, the depth to which the inhibitor solution will fall is unknown; consequently, the results of the application are difficult to predict. Regular treatment intervals from 3 to 6 months can be obtained under optimum conditions in low-volume wells. A more positive approach to batch treating is full tubing displacement. This is accomplished by pumping a solution into the well and displacing the solution with water or hydrocarbon to the bottom of the well. Another approach would be to displace the inhibitor solution with an inert gas, such as nitrogen; nitrogen is less likely than water or hydrocarbon to stop the flow of the well.

Squeeze Treating. Successful squeeze treatments were first reported in 1955 in the Placido Field of Victoria County, TX (Ref 22, 23). Since that time, the method of squeezing inhibitor into the formation has been defined and refined.

Squeeze treating is applicable to all oil and gas wells with sufficient porous and permeable producing zones. It will result in essentially continuous treatment because it has been found that some inhibitors will adsorb rapidly onto sand and will desorb slowly into produced fluids from the producing sand. The inhibitors used for this technique should be very film persistent. The most widely used technique is to mix one to two 208-L (55-gal) drums with 10 to 20 barrels of diesel oil or kerosene. The inhibitor solution is displaced to the bottom of the well, and an overflush of 200 to 500 barrels or more of oil or brine is then used to push the inhibitor into the formation. This technique often leads to continuous feedback lasting anywhere from 3 months to 2 years. The advantages of this method are that it will successfully treat wells with high fluid levels, the frequency of treatment is drastically reduced, and it is more reliable than some batch methods. Also, the entire length of the tubing is treated. However, squeeze treating is not recommended for wells with known mechanical uncertainties.

The major concern with squeeze treating is the possibility of plugging the formation by emulsion blocks or wettability reversal in the producing formation. The possible plugging effect of inhibitor squeezes on cores were studied in the laboratory, and it was reported that various sandstone formations differed in adsorption characteristics and that limestones were markedly different from sandstones (Ref 24, 25). This work was expanded upon in a field study, which is discussed in Ref 26. Tracer studies on formation squeezes were also conducted (Ref 27).

In a unique variation of the formation squeeze, the corrosion inhibitor/hydrocarbon diluent is atomized with an inert gas, such as nitrogen, and displaced down the tubing and into the formation with the same inert gas (Ref 28). Longer squeeze life and better inhibition have been claimed. The method is advantageous in wells with low bottom hole pressures because the low density of the nitrogen eases the resumption of production in such wells.

Batch Treatment of Pumping Wells. Corrosion problems in rod-pumped wells are directly related to the production volume and water content of producing wells. Severe corrosion problems can be identified by excessive rod, pump, and tubing failures. Embrittlement and pitting result in reduced rod life and fatigue failures, which are greatly accelerated by the presence of H_2S. Hydrogen sulfide and high water volumes are common to most waterfloods.

The most common way of treating rod-pumped wells consists of batch treating with a pump truck, equipped to flush the inhibitor down the annulus with produced fluids or water, followed by further flushing with the same fluids. In the 1950s and 1960s, batch treatment was accomplished without treating trucks by the use of wellhead inhibitor lubricator pots and manual flushing with produced fluids.

The methodology of batch treating involves the development of guidelines for each field. The required guidelines are volume of inhibitor per treatment, volume of flush, and treatment frequency. General rules are based on barrels of fluid produced per day (BFPD). General rules for frequency are (Ref 29, 30):

Volume of total fluid, BFPD	Frequency
0–50	Monthly
50–150	Every 2 weeks
150–350	Weekly
350–800	Twice weekly

Another set of recommended guidelines for treating frequency is given in Ref 31.

Successful batch treating of rod-pumped wells also involves the following factors:

- Circulation of high-volume wells is often necessary. In high gas-oil ratio (GOR) wells, it may be necessary to isolate the annulus for approximately 2 h after treatment to allow the inhibitor slug to fall
- Calculation of inhibitor volume based on the frequency of treatment and fluid production is an accepted procedure; inhibitor concentration is generally maintained at 25 to 35 ppm
- Determination of the correct minimum overflush
- Selection of the proper inhibitor

Automatic computer-controlled chemical injectors are available for batch treating. The cost advantages and disadvantages of computer-controlled chemical injectors are discussed in Ref 31. Corrosion fatigue in rod-pumped wells is as important as weight loss corrosion. A corrosion fatigue testing apparatus is described in Ref 32 that can be used to field test corrosion inhibitors specially formulated to prevent corrosion fatigue.

Treatment With High-Density Corrosion Inhibitors. Most liquid inhibitors have a density of 910 to 980 g/L (7.6 to 8.2 lb/gal). High-density liquid corrosion inhibitors are formulations that have been coupled with weighting agents. Their high density (1200 g/L, or 10 lb/gal) and their immiscibility with hydrocarbons and water enable them to fall through static columns of hydrocarbons. Continuous treating can be accomplished by using frequent small-volume batch treatments. Some applications for these unique inhibitors are high fluid level pumping wells and/or high- and low-pressure gas wells.

The weighting agent in high-density inhibitors is often zinc chloride ($ZnCl_2$), which may precipitate as zinc sulfide (ZnS) or zinc oxide (ZnO) when applied to some wells. The precipitation problem may become apparent in the form of plugging of downhole equipment and/or failure of the inhibitor to reach a desired point. Data on the fall rates of high-density and regular inhibitors are given in Ref 33 and 34. Other forms of weighted inhibitors, such as solid sticks, are available. More detailed information on liquid weighted inhibitors is available in Ref 35.

Miscellaneous Treatments. Slow-dissolving solids, sticks, pellets, microencapsulated droplets, and other configurations are used as vehicles for corrosion inhibitors. In addition, downhole dump bailers and concentric kill strings are used for both batch and continuous treatment.

Corrosion Problems and Inhibition in Waterfloods

Corrosion inhibition in waterfloods involves protecting surface-gathering lines and tanks used to recycle produced water, water-treating equipment, and the surface lines and downhole tubing of injection wells. The primary causes of corrosion in waterfloods are oxygen contamination and the acidity of the water. Due to the large amount of water handled, the cost of corrosion failures in repair and downtime can be substantial.

Oxygen Corrosion. Oxygen may be present in the supply water when the water source is surface or shallow ground water. Seawater is commonly used in near-shore or offshore fields and usually contains approximately 8 ppm dissolved oxygen. Another oxygen source is small amounts of contamination into closed water-handling systems through nongas-blanketed holding tanks, open vents, or thief hatches of water holding tanks, as well as around the shaft of the suction side of centrifugal transfer pumps (Ref 36). Crude oil or diesel oil blankets on the water surface in tanks do not stop oxygen from contaminating the water. Oxygen corrosion can be controlled by removing the oxygen or by using inhibitors.

Oxygen removal by mechanical means applies mainly to high oxygen containing waters, such as seawater. Oxygen is removed by the following mechanical processes:

- Countercurrent gas-stripping towers represent the simplest and most economical method if large amounts of natural gas are available at low cost
- Vacuum tower removal is applicable where no gas is available (Ref 37)
- The water can be gas lifted from its source—for example, lifting seawater from a depth of 30 m (100 ft) by the same technology used to lift fluids in gas-lift oil wells

Usually, none of the above methods will reduce the oxygen to the required point; therefore, chemical oxygen scavengers must be used to remove the last traces.

Oxygen scavenging refers to removal of the oxygen by chemical reaction. Oxygen scavengers can reduce the oxygen content of the water to less than 10 ppb, a level that is considered insignificant from the corrosivity standpoint. The most commonly used oxygen scavengers are:

- Sulfite ion (SO_3^{2-}), which may come from sulfur dioxide (SO_2) gas generated on the site, from solutions of sodium sulfite (Na_2SO_3) or sodium bisulfite ($NaHSO_3$), or from a solution of ammonium bisulfite (NH_4HSO_3). Ammonium bisulfite is the most commonly used because of its stability in storage and its ability to exist as a highly concentrated liquid solution. Solutions containing 60 to 70% NH_4HSO_3 are commercially available
- Hydrazine, which is a practical oxygen-scavenging chemical only at elevated temperatures, such as in boilers or steam generators
- Sodium hydrosulfite ($NaHSO_2$) is recommended for scavenging oxygen, with much less polymer degradation than SO_3^{2-}, in polymer flood applications. However, $NaHSO_2$ is very unstable in solution and requires storage of a solid and solution preparation daily

The reaction of the SO_3^{2-} oxygen scavengers listed above involves a free radical mechanism that necessitates an initiating step. Very small additions (<1 ppm) of transition metal ions, such as cobalt, will catalyze the reaction. The variables affecting the rate and completeness of the SO_3^{2-} reaction are important in designing a waterflood. Sodium sulfite used in a 1- to 2-ppm stoichiometric excess plus 0.1 to 0.2 ppm cobalt chloride ($CoCl_2$) at ambient temperature often completely removes oxygen in an acceptable time period (usually a few minutes). When the rate or degree of completeness of the reaction is not satisfactory, an investigation into possible interferences with the reaction may be necessary. The most common problem is deactivation of the catalyst. All transition metals are not equally effective, and pH affects the various possible metal ions differently (Ref 38). In waters containing sulfide, the catalyst can be precipitated as an insoluble solid (Ref 39), or it can be complexed with a chelating agent, such as ethylenediamine tetraacetic acid (EDTA). In both cases, the catalyst is rendered inactive.

Water used for cyclic steam injection and steam floods must be given greater attention with regard to oxygen corrosion. Corrosion and other water-treating problems in steam injection systems are discussed in Ref 40.

When removal of oxygen is not practical because of small volumes of water to be treated or other factors, two organic inhibitors are available. The first, a zinc amino methyl phosphonate, reportedly achieves excellent oxygen corrosion control in fresh to slightly brackish supply waters (Ref 41-43). The second, an organic sulfophosphate, exhibits inhibition of produced water handling systems processing sour brine contaminated with oxygen (Ref 44). The inhibitor is dispersible only in water and is said to inhibit better when H_2S and hydrocarbons are present in the water in addition to oxygen.

Corrosion by Acid Gases. The corrosive acids found in waterflood systems are usually encountered because produced water is recycled and mixed with the supply water. In a few fields, the produced water has been kept separate and theoretically should always be kept separate from the usually less problematic supply water.

Experience has shown that the presence of H_2S in injection water is much more troublesome with regard to corrosivity and handling problems than waters containing only CO_2 or short-chain acids. This is the case primarily because iron sulfide (FeS), the product of H_2S corrosion, is very insoluble. The FeS corrosion product may deposit on downstream equipment, may plug injection wells, and may cause difficulties in oil-water separation. An inhibitor has not yet been discovered to prevent the precipitation of FeS.

The corrosion inhibitors designed for acidic corrosion are usually film-forming amine salts. Water-soluble or dispersible modifications of these materials are used in waterflood applications. Experience has shown that the system must be free of oxygen contamination for these materials to be effective. Therefore, the above materials are sometimes used in conjunction with oxygen scavengers.

Film-forming ability and stability are determined by the interplay of a number of factors involving both the oleophilic hydrocarbon chain and the polar portion of the molecule. Changes in the oleophilic portion of the molecule that promote water solubility tend to decrease filming ability. The conflict between solubility and film stability is one of the basic obstacles in formulating inhibitors for waterfloods. Water compatibility or solubility is important if the water is to be injected into a low-permeability reservoir.

The chemistry and ideas used in inhibitor molecules for producing wells have been extended or modified for use in waterfloods. Sufficient dispersibility (added to the inhibitor so that the inhibitor may be effectively transported from the injection point to the farthest injection well) must be balanced with the film-forming ability of the inhibitor. Aside from the solubility or dispersibility requirement, the same molecules can be used for waterfloods. The required concentration of inhibitor, however, is drastically increased by the absence of the hydrocarbon phase. For example, addition of a hydrocarbon phase to sour brine was found to lower the inhibitor requirement for equal inhibition from 25 to 5 ppm (Ref 45). Many other investigations and field experience show that the presence of hydrocarbons augments film formation and persistence.

Alternatives to oxygen removal and/or organic corrosion inhibitors are the use of internally coated steel pipe with either plastic or cement lining, corrosion-resistant alloys, fiberglass-reinforced plastic, or other nonmetals in place of carbon steel. Finally, inorganic inhibitors used in aerated water—for example, cooling towers and radiators—are usually not effective and are too expensive to be applied to waters on a once-through basis, as would be the case in a waterflood.

Bacteria-Induced Corrosion

Sulfate-reducing bacteria (SRB) (genus *Desulfovibrio*) reduce sulfate (SO_4^{2-}) to H_2S, often converting a noncorrosive source water into an aggressively corrosive water. Changes in the biological environment, such as temperature, velocity, pressure, shielding debris (in the bottom of tanks), deposits (organic and inorganic), and nutrients, cause bacteria to grow. The change that appears to cause the greatest increase in growth consists of transporting the water to a surface holding tank having many static areas and letting the water warm up or cool to 40 ± 6 °C (100 ± 10 °F). Sulfate-reducing bacteria also produce tubercles or biomass in conjunction with other bacteria. The SRB have in many cases produced enough H_2S to make a sweet surface system sour and have been thought to be the cause of producing a reservoir change from sweet to sour. The injection of seawater into offshore fields has probably been the worst offender in introducing SRB to production equipment. Changes in temperature can increase SRB growth from zero to high levels. Additional information on this subject is available in the section "Biological Corrosion" of the article "Localized Corrosion" in this Volume.

Solutions to bacteria corrosion problems include avoiding static or dead areas in initial surface system design, keeping the system clean, and using a bactericide. Cleaning the physical surface system consists of yearly removal of settled solids in surface tankage and separation vessels. Tanks left uncleaned may contain 1 to 2 m (4 to 6 ft) of solids and sludge after 5 to 10 years of operation. Surface water transmission lines are kept clean by regular pigging, which is the most effective method of controlling bacteria corrosion. The surface piping system must be designed for pig launchers and traps.

Bactericides are chemicals that kill or control microorganisms. Many of the chemicals are surface-active cationic materials and must be used with care relative to compatibility with anionic chemicals, such as scale inhibitors. The most common surface-active bactericides are dimethyl coco amine quaternized with methyl chloride; coco diamine acetate, benzoate, or adipate; 3-alkoxy*-2hydroxy-*n*-propyl trimethylammonium chloride (where * denotes a linear primary alcohol, C_{12}-C_{15}) (Ref 46); and dimethyl coco amine quaternized with 2,2-dichloro-diethyl ether. The most common nonsurface-active bactericides are formaldehyde, glutaraldehyde, acrolein, chlorine dioxide (ClO_2), chlorine, sodium hypochlorite (NaClO), isothiazolone, dibromonitrile proionamide, and sodium dimethyldithiocarbamate. The mechanisms and examples of biological corrosion are discussed in Ref 47 to 49.

A method of enumerating bacteria is discussed in Ref 50. Other useful procedures include a quick field method for enumerating total bacteria and studying the effect of bactericides (Ref 51), a method of studying biofilms and the effect of bactericides on them (Ref 52), and a rating system for evaluating bacterial problems in waterfloods (Ref 53).

Laboratory Testing of Corrosion Inhibitors

Many testing methods designed to simulate field conditions have been reported in the literature. However, only a few of these methods survive, and none has become the standard test.

Static Test. After several years of use, a static test was formalized based on work described in Ref 54. In the test, coupons are exposed for about 1 week in fluids with and without inhibitors and are evaluated on a weight loss basis (Ref 55). Another static inhibitor screening test consists of short-term static exposure in field fluids, followed by immersion in a copper ion solution to determine filming ability (Ref 56).

Wheel Test. One of the first written reports on the wheel test dates from 1963 (Ref 57). After 25 years of use and many series of multi-laboratory comparative inhibitor tests, no consensus has been reached on a standardized version of the wheel test. Results of round-robin testing are available in Ref 58 and 59. A wheel test procedure (not a standard) is described in Ref 60.

The wheel test is a dynamic test performed by placing synthetic or field fluids in a 7-oz beverage bottle containing a metal coupon. The bottle and its contents are purged with CO_2 or H_2S, and the bottle is capped. The bottles are then agitated for approximately 2 h by securing them to the circumference of a wheel and rotating the wheel. After the first agitation period, the coupons are transferred to a rinse bottle containing fluids with inhibitor and are agitated again. The coupons are then transferred to another bottle containing only corrosive fluids (no inhibitor) and are agitated for a longer period of time, usually 24 h. At the end of this time, the metal coupon is removed and cleaned, and weight loss is measured. The maximum temperature at which this test can be safely conducted is 80 °C (180 °F). High-temperature versions of the wheel test are discussed in Ref 61. Temperatures of 150 to 205 °C (300 to 400 °F) and pressures of several thousand pounds per square inch are achieved by the use of high-alloy pressure bonds. Agitation of the bombs is usually accomplished in the same manner as in the low-temperature wheel test. The results of one series of high-temperature wheel tests indicate a maximum corrosion rate at 105 °C (225 °F) with an unexpected decrease at 165 °C (325 °F) (Ref 62).

Monitoring Results of Inhibition in the Field

The following methods are used to monitor corrosion rates and inhibitor effectiveness:

- Coupons (NACE standard method)
- Spools, pump joints, and pony rods
- Iron counts
- Copper ion displacement
- Radioactive tracer methods
- Caliper survey
- Copra correlation (Ref 63), a quantitative assessment of deep, hot gas well corrosion
- Electrochemical methods, such as resistance measurements, PAIR (polarization admittance instantaneous rate measurements), or potentiodynamic polarization measurements (Ref 64)

Quality Control of Inhibitors

The materials used in most corrosion inhibitors are of necessity by-products of many chemical industries. Although many inhibitor formulators and manufacturers adhere to rigid specifications and procedures it is impossible to correlate product-manufacturing specifications with field performance. One report on quality control procedures used by inhibitor manufacturers showed that only 19% of the manufacturers checked performance and only 10% made compositional analyses (Ref 65).

Computerization of Inhibitor Treating Programs

Problem areas and problem wells have existed since the beginning of corrosion inhibition. Wells or batteries (geographical groups of wells) that do not respond satisfactorily to current chemical corrosion programs are defined as problem wells. Computers can effectively handle large volumes of data and, coupled with their sorting capability, can identify problem wells, changes in conditions, and other factors that influence repair costs. Computer programs can also be used as an accounting tool to monitor corrosion costs and other chemical expenses (Ref 66).

REFERENCES

1. C.C. Nathan, *Corrosion Inhibitors*, National Association of Corrosion Engineers, 1973
2. J.F. Chittum, Corrosion Fatigue Cracking of Oilwell Sucker Rods, *Mater. Perform.*, Vol 7 (No. 12), Dec 1968, p 37-38
3. J.F. Bates, Sulfide Cracking of High Strength Steels in Sour Crude Oils, *Mater. Perform.*, Vol 8 (No. 1), Jan 1969, p 33-39
4. C.M. Hudgins, A Review of Sulfide Corrosion Problems in the Petroleum Industry, *Mater. Perform.*, Vol 8 (No. 1), Jan 1969, p 41-47
5. C.C. Patton, Petroleum Production—Stringent Corrosion Control Procedures Key to Extended Fatigue Life, *Mater. Prot. Perform.*, Vol 11 (No. 1), June 1972, p 17-18
6. J.I. Bregman, *Corrosion Inhibitors*, MacMillan, 1963
7. *Condensate Well Corrosion*, Natural Gasoline Association of America, 1953
8. G.L. Farrar, Combatting Corrosion in Oil and Gas Wells, *Oil Gas J.*, Vol 51 (No. 49), April 1953, p 106-109, 111, 113
9. H.L. Bilhartz, Sweet Oil Well Corrosion, *World Oil*, Vol 134, April 1952, p 208-216
10. H.L. Bilhartz, How to Predict and Control Sweet Oil Well Corrosion, *Oil Gas J.*, Vol 50 (No. 50), April 1952, p 116-118, 151, 153
11. H.L. Bilhartz, *Sweet Oil Well Corrosion—API Drilling and Production Practice*, American Petroleum Institute, 1952, p 54
12. H.L. Bilhartz, High Pressure Sweet Oil Well Corrosion, *Corrosion*, Vol 7 (No. 8), Aug 1951, p 256-264
13. C.J. Houghton and R.V. Westermark, Downhole Corrosion Mitigation in Ekofis K (North Sea) Field, *Mater. Perform.*, Vol 22 (No. 1), Jan 1983, p 16
14. D.H. Mutti, J.E. Atwood, C.R. LaFayette, and A.O. Landrum, "Corrosion Control of Gas-Lift Well Tubulars by Continuous Inhibitor Injection into the Gas-Lift Gas Stream," Paper 5612, presented at the 50th Annual Meeting, Dallas, TX, Society of Petroleum Engineers, 1975
15. R.B. Todd, J.H. Cannon, H.J. EnDean, and K. Belanus, Corrosion Protection by Downhole Continuous Inhibitor via External Capillary, *Mater. Perform.*, Vol 20 (No. 2), 1981, p 32
16. J.B. Bradburn and R.B. Todd, Continuous Injection Method Controls Downhole Corrosion, Pt. 1, *Petrol. Eng. Int.*, Vol 53, July 1981, p 44-46
17. J.B. Bradburn and R.B. Todd, Continuous Injection Method Controls Downhole Corrosion, Pt. 2, *Petrol. Eng. Int.*, Vol 53, Aug 1981, p 54
18. L.M. Cenegy and C.L. Chin, A Test for Corrosion Inhibitors to be Used in Oilfield Capillary Injection Systems, *Mater. Perform.*, Vol 22 (No. 4), April 1983, p 15-19
19. T.W. Hamby, Jr., "Development of High Pressure Sour Gas Technology," Paper 8309, presented at the 54th Conference, Las Vegas, NV, Society of Petroleum Engineers, 1979
20. M.C. Place, Jr., "Corrosion Control—Deep Sour Gas Production," Paper 8310, presented at the 54th Conference, Las Vegas, NV, Society of Petroleum Engineers, 1979
21. R.H. Hausler and S.G. Weeks, Low Cost, Low Volume, Continuous Inhibition of Gas

Production Tubulars, *Mater. Perform.*, Vol 25 (No. 6), 1986, p 27-37
22. R.H. Poetker and J.D. Stone, Squeezing Inhibitor Into Formation, *Petrol. Eng.*, Vol 28, May 1956, p B29-B34
23. R.H. Poetker and J.D. Stone, Inhibition Improved 17 Percent While Cost Dropped 50 Percent, *Oil Gas J.*, Vol 54, July 1956
24. J.K. Kerver and F.A. Morgan III, Corrosion Inhibitor Squeeze Technique—Laboratory Study of Formation Permeability Damage, *Mater. Perform.*, Vol 2 (No. 4), April 1963, p 10-22
25. J.K. Kerver and F.A. Morgan III, Corrosion Inhibitor Squeeze Technique—Laboratory Adsorption-Desorption Studies, *Mater. Perform.*, Vol 4 (No. 4), July 1965, p 69-79
26. J.K. Kerver and H.R. Hanson, Corrosion Inhibitor Squeeze Technique-Field Evaluation of Engineered Squeezes, *J. Petrol. Technol.*, Vol 17, Jan 1965, p 50-57
27. P.J. Raifsnider, C.L. Guinn, C.L. Barr, and D.L. Lilly, "Radioactive Tracer Studies on Squeeze Inhibition of Oil Wells," Paper presented at the 19th annual conference, New York, NY, National Association of Corrosion Engineers, March 1963
28. G.L. Nunn and B.E. Hamilton, Well Treatment With Inert Gas-Inert Gas Squeeze for Corrosion Control, *Mater. Perform.*, Vol 6 (No. 5), May 1967, p 37-40
29. S. Evans and C.R. Doran, "Batch Treatment of Sucker Rod Pumped Wells," Paper presented at the Southwestern Petroleum Short Course, Department of Petroleum Engineering, Texas Tech University, April 1983
30. S. Evans and C.R. Doran, Batch Treatment Controls Corrosion in Pumping Wells, *World Oil*, Vol 198, Feb 1984, p. 55-57
31. W.J. Frank, Here's How to Deal With Corrosion Problems in Rod-Pumped Wells, *Oil Gas J.*, Vol 74, May 1976, p 63-72
32. L.A. Phillips and J.R. Cowden, "Corrosion Fatigue Test Selects Effective Corrosion Inhibitors," Paper 266, presented at Corrosion/80, Chicago, IL, National Association of Corrosion Engineers, March 1980
33. G.B. Farquhar, M.J. Michnick, and R.R. Annand, Tracer Experiments During Batch Treatment of Gas Wells With Corrosion Inhibitors, *Mater. Prot. Perform.*, Vol 10 (No. 8), Aug 1971, p 41-45
34. C.C. Patton, D.A. Deemer, and H.M. Hillard, Jr., Field Study of Fall Rate-Oilwell Liquid Inhibitor Effectiveness, *Mater. Perform.*, Vol 9 (No. 2), Feb 1970, p 37-41
35. C.O. Bundrant, High Density Corrosion Inhibitors Simplify Oil Well Treatments, *Mater. Perform.*, Vol 8 (No. 9), Sept 1969, p 53-55
36. L.C. Case, Oxygen Traces May Be Corroding Your Waterflood Piping, *Oil Gas J.*, Vol 62, Jan 1964
37. B.L. Carlberg, "Vacuum Deaeration—A New Unit Operation for Waterflood Treatment Plants," Paper 6096, presented at the Fall Meeting, Society of Petroleum Engineers, 1976
38. E.S. Snavely and F.E. Blount, Rates of Reaction of Dissolved Oxygen With Scavengers in Sweet and Sour Brines, *Corrosion*, Vol 25 (No. 10), 1969, p 397
39. C.C. Templeton, S.S. Rushing, and J.C. Rogers, Solubility Factors Accompanying Oxygen Scavenging With Sulfite in Oil Field Brines, *Mater. Perform.*, Vol 2 (No. 8), Aug 1963, p 42
40. N.G. Haseltine and C.M. Beeson, Steam Injection Systems and Their Corrosion Problems, *Mater. Perform.*, Vol 4 (No. 10), 1965, p 57
41. G.B. Hatch, U.S. Patent 3,532,639, granted 6 Oct 1970, Inhibiting (Oxygen) Corrosion With Zinc Salt-Methanol Phosphonic Acid Derivative Combinations
42. P.H. Ralston, U.S. Patent 3,393,150, granted 16 July 1968, Amine Phosphonate Scale Inhibitor
43. G.B. Hatch, A. Park, and P.H. Ralston, U.S. Patent 3,483,133, granted 9 Dec 1969, Method of Inhibiting Corrosion With Aminomethylenephosphonic Acid Compounds
44. R.L. Martin, R.R. Annand, D. Wilson, and W.E. Abrahamson, Inhibitor Control of Oxygen Corrosion: Application to a Sour Gas Gathering System, *Mater. Prot. Perform.*, Vol 10 (No. 12), Dec 1971, p 33
45. L. Riggs, Jr. and F.J. Radd, Physical and Chemical Study of an Organic Inhibitor for Hydrogen Sulfide Attack, *Corrosion*, Vol 19, Jan 1963, p 1t
46. J.R. Stanford, Inhibitor for Oxygen-Free Flood Waters, U.S. Patent No. 3,424,681, Jan 1969
47. R.M. Jordan and L.T. Shearer, "Aqualin Biocide in Injection Waters," Paper 280, presented at the Research Meeting, Tulsa, OK, Society of Petroleum Engineers, 1962
48. A.E. Baumgartner, Microbiological Corrosion—What Causes It and How It Can Be Controlled, *J. Petrol. Technol.*, Vol 14 (No. 10), Oct 1962, p 1074
49. J.M. Sharpley, "Elementary Petroleum Microbiology," Gulf Publishing Co., Houston, TX, 1966
50. "Recommended Practice for Biological Analysis of Subsurface Injection Waters," API RP38, American Petroleum Institute
51. E.S. Littmann, "Oilfield Bactericide Parameters as Measured by ATP Analysis," Paper SPE 5312, presented at the International Symposium on Oilfield Chemistry, Dallas, TX, Society of Petroleum Engineers, Jan 1975
52. I. Ruseka, J. Robbins, J.W. Costerton, and E.S. Lasken, Biocide Testing Against Corrosion—Causing Oil-Field Bacteria Helps, *Oil Gas J.*, Vol 80 (No. 10), March 1982, p 253
53. C.C. Wright, Rating Water Quality and Corrosion Control in Water Floods, *Oil Gas J.*, Vol 61, May 1963
54. A Proposed Standardized Laboratory Procedure for Screening Inhibitors for Use in Sour Oil and Gas Wells, T-1K NACE Publication 55-2, *Corrosion*, Vol 11 (No. 3), March 1955, p 143t
55. A Proposed Laboratory Screening Test for Materials to be Used as Inhibitors in Sour Oil and Gas Wells, T-1K NACE Publication 60-2, *Corrosion*, Vol 16 (No. 2), Feb 1960, p 63t-64t
56. W.B. Hughes, A Copper Ion Displacement Test for Screening Corrosion Inhibitors, *J. Petrol. Technol.*, Vol X, Jan 1958, p 54-56
57. E.D. Junkin and D.R. Fincher, Oil Field Corrosion Inhibitors Evaluated by Film Persistency Test, *Mater. Prot.*, Vol 2, Aug 1963, p 18-23
58. C.C. Nathan, Correlations of Oil-Soluble, Water Dispersible Corrosion Inhibitors in Oil Field Fluids, *Corrosion*, Vol 18 (No. 8), Aug 1962, p 282T-285T
59. D.R. Fincher *et al.*, Cooperative Evaluation of Inhibitor Film Persistency Test, T-1D-2 NACE Publication 1D166, *Mater. Perform.*, Vol 5 (No. 10), Oct 1966, p 69
60. B.F. Davis, Wheel Test Method Used for Evaluation of Film Persistent Inhibitors for Oilfield Applications, T-1D-8 NACE Publication 1D182, Item 54238, *Mater. Perform.*, Vol 21 (No. 12), Dec 1982, p 45-47
61. J.L. Magnon, "Laboratory Testing of Corrosion Inhibitors," Paper 56, presented at Corrosion/73, Anaheim, CA, National Association of Corrosion Engineers, March 1973
62. J.D. Garber, R. Perkins, and H. Su, High-Temperature Wheel Test Simulates CO_2 Corrosion, *Oil Gas J.*, Vol 84, April 1986, p 62-64
63. L.K. Gatzke and R.H. Husler, The Copra Correlation—A Quantitative Assessment of Deep, Hot Gas Well Corrosion and Its Control, Paper 48, presented at Corrosion/83, Anaheim, CA, National Association of Corrosion Engineers, April 1983
64. R.L. Martin, Potentio Dynamic Polarization Studies in the Field, *Mater. Perform.*, Vol 15, March 1979, p 3
65. Survey of Quality Control Procedures Used in the Manufacture of Oil Field Inhibitors, T-1D-5 NACE Publication 1D267, *Mater. Perform.*, Vol 6 (No. 6), June 1965, p 82-84
66. J. Fu, R. Erbar, and R. Heidersbach, Ed., *Computers in Corrosion Control*, National Association of Corrosion Engineers, 1986

Corrosion Inhibitors for Crude Oil Refineries

Glenn L. Scattergood, Nalco Chemical Company

THE REFINING of crude oil results in a variety of corrosive conditions. Some of these conditions, such as initial condensate corrosion, are well defined, but others, such as SO_x corrosion, are difficult to define precisely. Refinery corrosion is generally caused by a strong acid attacking the equipment surface. The most common strong acid is hydrogen chloride, although other strong acids have recently been identified. Given the general title of SO_x, these consist of sulfurous (H_2SO_3), sulfuric (H_2SO_4) thiosulfurous, and other polythionic acids.

The costs of corrosion and the savings gained through the use of appropriate corrosion inhibitors are considerable in the refinery, where conditions tend to be severe because of high flow rates, temperatures, and pressures. Four major categories of corrosion inhibitors have been identified based on the mechanism of operation: neutralizers, filming inhibitors (or barrier layer formers), scavengers, and miscellaneous types (Ref 1). The largest applications in the refinery are for neutralizers and filming inhibitors. These will be covered in detail. Scavengers and miscellaneous types will also be defined.

Inhibitor Mechanisms

Neutralizers lessen the corrosivity of the environment by decreasing hydrogen ion (H^+) concentration, which reduces the concentration of the corrosive reactant. Neutralizers function by controlling the corrosion caused by acidic materials, such as hydrogen chloride, carbon dioxide (CO_2), sulfur dioxide (SO_2), carboxylic acids, and related compounds. These materials are found in small quantities in many process streams. However, because of such separation processes as distillation, one or more of these acidic species can concentrate in specific areas and cause severe corrosion. The area most susceptible to corrosion in the refinery is the heat exchanger, where the first drops of water condense (the initial condensate). An effective neutralizer will exhibit the same distillation/condensation properties as the acid it is designed to control.

Filming inhibitors function by strong adsorption, or chemisorption, and decrease attack by forming a barrier on the metal surface that resists penetration by acid solutions. A filming inhibitor must possess a hydrocarbon portion attached to a strongly polar group. The molecules are oriented on the metal surface, with the polar group adsorbed onto the metal surface and the hydrocarbon part extending away from the surface. The hydrocarbon end will attract the molecules of the process stream to provide an additional barrier to a potentially corrosive aqueous solution.

Scavengers. Perhaps the most widely used scavenger system is employed in boilers to remove oxygen from the feedwater. Techniques such as steam stripping can be used to remove most of the dissolved oxygen from water; however, such methods become increasingly costly when the last traces of oxygen must be removed from the boiler feedwater. In these cases, chemical techniques for oxygen removal become more attractive. Hydrazine and sodium sulfite are the two most widely used scavengers in boiler systems (Ref 1).

Miscellaneous inhibitors include such materials as scale inhibitors, which minimize deposition of scale on the metal surface, and biocides, which kill living organisms that can foul equipment.

Inhibitor Composition

Neutralizing inhibitors. The primary function of a neutralizing inhibitor is to reduce the concentration of H^+ ion in the environment. A variety of neutralizers are used in many applications in the refinery. The list includes ammonia (NH_3), sodium hydroxide (NaOH), and various proprietary alkylamines and polyamines. The physical characteristics of each neutralizer determine its application. A strong alkali, such as NaOH, is an excellent neutralizer when injected into the desalted crude, but it cannot be used in overhead heat exchangers. Ammonia is an inexpensive overhead neutralizer, but it has no solubility in the initial condensate.

Filming Inhibitors. Most of the inhibitors used are of the film-forming type. Instead of reacting with or removing an active corrodent species, filming inhibitors function by creating a barrier between the metal and the environment. They consist of one or more polar groups based on nitrogen, sulfur, or oxygen that are attached to the metal surface by chemisorption or electrostatic forces (Ref 2).

Filming amine chemistry in the refinery includes amides, diamides, and imidazoline salts. Each type is known to be effective in selected environments. The amino group is the important functional and salt-forming species. For readily handled commercial products, the amide intermediate is reacted with an imidazoline salt to enhance solubility in carrier solvents and to decrease gelling or phase separation. However, both groups are effective inhibitors.

Application of Inhibitors

Neutralizers. The primary cause of corrosion in an atmospheric distillation tower overhead system is the hydrochloric acid (HCl) produced by the hydrolysis in the furnace of water-soluble magnesium chloride and calcium chloride salts found in the crude oil and escaping the desalter. Corrosion occurs when water in the system condenses, such as in the heat exchangers and connected piping. The hydrogen chloride gas dissolves in the initial condensate, forming hot HCl. To be an effective neutralizer, the inhibitor must be present when and where the acid is formed. Only a neutralizer that exhibits the same condensation profile as HCl can inhibit HCl corrosion. A neutralizer that is too volatile, such as NH_3, has no solubility in the initial condensate. A nonvolatile neutralizer, such as the commercially available polyamines, will have already condensed and taken the path through the exchangers of the condensed hydrocarbon.

To apply a neutralizer correctly to an overhead system, the point of initial condensation must be defined by a thorough study of the partial pressure relationship between the stripping steam and the hydrocarbons. The neutralizer must be added at a point in the system such that it will completely vaporize, allowing the neutralizer to condense in accordance with its condensation profile. An initial condensate neutralizer will not allow a low-pH environment to exist inside the heat exchangers and connected piping.

Serious consideration must be given to other areas in the atmospheric distillation system that may experience water condensation. Because of localized cooling, liquid water may exist in tower-top pumps or on the top tray, where a cool reflux is returned to the tower. In these areas, small amounts of water may become concentrated with HCl, H_2SO_3, or H_2SO_4. Where these conditions exist, the problem must be addressed by use of an effective inhibitor.

A filming inhibitor should be applied to an overhead system with an understanding of the physical dynamics of the system. The filming inhibitor is capable of complete inhibition as it forms a protective barrier on equipment surfaces. Unlike neutralizers, the filmer need not contact the corrosive species to render it inactive. Therefore, filmers are much more cost effective than neutralizers.

Filming inhibitors are not volatile in an overhead system. They will follow the path of condensed hydrocarbon through the series of heat exchangers. An oil-soluble filmer has the opportunity to contact all of the equipment parts that are washed over by condensed hydrocarbon, and it is a very effective inhibitor once contact has been made. Portions of the heat exchanger are not contacted by condensed hydrocarbon and cannot be protected by filming inhibitors. The most common approach to overhead corrosion control in a refinery is to use a filmer in conjunction with a neutralizer.

The distillation tower may benefit from applying a filming inhibitor to the reflux injection, thus providing tray protection from the top of the tower to the bottom. Again, the filmer is only present where liquid hydrocarbon is available to carry it. The areas of a tower that contact only vapor, such as the underside of trays and the dome of the tower, cannot benefit from a filming inhibitor.

High-temperature corrosion (>205 °C, or 400 °F) caused by a concentration of naphthenic acids is successfully inhibited by the use of filming inhibitors that are stable at high temperatures. Naphthenic acids are cyclic carboxylic acids that are present in some crude oils and are common in lower fractions of the atmospheric and vacuum towers, such as the gas oils. Neutralization values can become quite high as the acids are distilled to their boiling points, and gas oil equipment may experience severe corrosion. This corrosion is lessened by injecting a filming inhibitor designed for naphthenic acid corrosion into the tower draws.

Monitoring Corrosion

Corrosion monitoring in the refinery is intended to provide quantitative information that will enable the service life of the process equipment to be predicted. A corrosion test program that is soundly designed, conducted, and evaluated will provide this information, but a poorly conceived program can generate information that is useless or misleading. Seven types of tests are listed below in order of decreasing reliability; each test encompasses fewer variables, and is therefore less reliable, than the one above it (Ref 3):

- Full-scale equipment
- Small side-stream equipment in an operating unit
- Equipment in a pilot unit
- Plant coupon tests
- Pilot plant coupon tests
- Laboratory tests using plant process samples
- Laboratory tests using synthetic solutions

Full-scale equipment is the only source of completely reliable information. However, it is a very slow and expensive way of determining whether a fractionating tower fabricated of a given alloy will be serviceable. Thorough, quantitative inspection data are needed to estimate corrosion rates properly on existing equipment.

Small side-stream equipment is advantageous in that inspections for corrosion evaluation can be scheduled independently of production and maintenance schedules. Small side-stream equipment is available for almost any unit in the refinery. Coupled with a continuously monitoring computer, corrosion control effectiveness is available around the clock.

For equipment in a pilot unit, the feedstock and recycle streams may differ from those in a plant unit, and operation is usually less stable. An additional limitation is that a pilot plant must usually be operated under varying conditions in order to develop process data. It is seldom held at one condition long enough to develop good corrosion information.

Plant coupons are not reliable, because of the placement of the coupon. The coupon may not be in the area of most severe corrosion, or it may receive a coating of corrosion inhibitor that is not representative of the process equipment.

REFERENCES

1. S.W. Dean, R. Derby, and G.T. von dem Bussche, "Inhibitor Types," Paper 253, presented at Corrosion/81, National Association of Corrosion Engineers, 1981
2. Z.A. Foroulis, Corrosion and Corrosion Inhibition in the Petroleum Industry, *Phys. Rev. B*, Vol 21 (No. 11), 1980, p 5432-5437
3. B.J. Moniz and W.I. Pollock, Ed., *Process Industries Corrosion*, National Association of Corrosion Engineers, 1985

Control of Environmental Variables in Water-Recirculating Systems

Bennett P. Boffardi, Calgon Corporation

THE CORROSION PROCESS that occurs in industrial systems is insidious and all-embracing. This process is often difficult to discern until extensive deterioration has occurred. Corrosion robs industry of millions of dollars annually through loss or contamination of products, replacement cost and overdesign of equipment, reductions in efficiency, high maintenance expense, and waste of valuable resources. Corrosion also restrains technological progress and jeopardizes human safety.

Water, the most commonly used cooling fluid, removes unwanted heat from heat transfer surfaces. However, water is corrosive to most alloys and contributes to the buildup of insulating deposits through water-formed and water-borne foulants.

The Cooling System

An understanding of the relationship between cooling water and the buildup of deposits and corrosion of heat transfer surfaces requires an awareness of cooling system characteristics. There are basically three types of cooling systems: once-through, open recirculating, and closed recirculating systems (Fig. 1).

In once-through systems, the cooling water passes through the heat transfer equipment only once before it is discharged. Large volumes of water are used with only a small temperature increase across the exchanger. The mineral content of the water remains essentially the same. Water is usually drawn from such sources as rivers, lakes, and wells.

Open recirculating systems continuously reuse the water that passes through the heat transfer equipment. Circulated water can be drawn from spray ponds or cooling tower basins. Evaporative cooling to the atmosphere dispels the unwanted heat transferred to the cooling water. The water then returns to the source and recirculates. In cooling tower operations, the water cascades over and down the tower, where evaporative heat transfer takes place. Makeup water is added to replace the evaporative losses. Additional makeup water replaces the intentionally discharged water (blowdown) to maintain an acceptable level of dissolved minerals and suspended solids in the cooling water.

Open recirculating cooling systems are oxygen saturated and may contain a high level of dissolved solids. Inlet temperatures to the heat transfer equipment are usually higher than those for once-through cooling systems. Also, there may be a larger temperature differential across the exchanger. These factors can significantly affect the buildup of deposits and deterioration of the heat transfer equipment.

Closed recirculating systems have little water loss and continuously recirculate the same water. The heat absorbed from the heat transfer equipment is dissipated to another heat sink, which is cooled by other methods. Because there are no evaporative losses, the makeup water is minimal, and the mineral content remains essentially constant. However, corrosion by-products can easily accumulate and foul heat transfer equipment because there are no methods of removing them from the system. Operating temperatures of the closed recirculating systems range from cold (as in chilled-water systems) to hot (for engine cooling jackets).

Corrosion Processes in Water-Recirculating Systems

The corrosion processes and the fouling of heat transfer equipment must be understood if long-term reliability is to be achieved. Corrosion can be defined in a very practical sense as the deterioration of metal caused by the reaction with its surrounding environment. Because water is one of the most common heat transfer fluids, it is not surprising that most of the problems associated with corrosion and deposits are water related. However, dissolved gases (such as oxygen, carbon dioxide, CO_2, ammonia, and chlorine), dissolved salts (such as calcium, magnesium, chloride, sulfate, and bicarbonate), and suspended solids make water something other than pure H_2O.

For the corrosion reactions to take place between water and the metal surface, a potential difference must exist between different areas on the surface. This causes the passage of electrical current through the metal from the area of high potential to low potential. Thus, corrosion of metals in contact with water is electrochemical in nature. The basic reactions occur at the region of lower potential, which is the anode. The dissolution reaction causes metal ions to form and go into solution. The anodic oxidation reaction can be generally represented by:

$$M \rightarrow M^{n+} + ne^-$$

where M represents the metal that has been oxidized to its ionic form having a valence of $n+$ and the release of n electrons. For the more common heat transfer materials, the individual reactions are:

$$Fe \rightarrow Fe^{2+} + 2e^-$$

$$Cu \rightarrow Cu^+ + e^-$$

$$Al \rightarrow Al^{3+} + 3e^-$$

The liberated electrons that migrate through the metal to areas of higher potential are used in the reduction of other ions or oxygen in the water. These reactions occur at the cathodic site on the metal surface:

$$O_2 + 4H^+ + 4e^- \rightarrow 2H_2O$$

(Reduction of oxygen in acid solution)

$$O_2 + 2H_2O + 4e^- \rightarrow 4OH^-$$

(Reduction of oxygen in neutral or alkaline solution)

$$2H^+ + 2e^- \rightarrow H_2$$

(Hydrogen evolution)

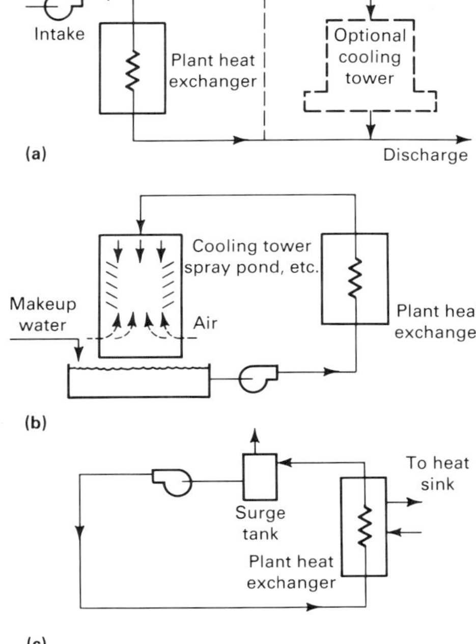

Fig. 1 Schematics of three types of cooling systems. (a) Once-through system. (b) Open recirculating system. (c) Closed recirculating system

$$Cu^{2+} + 2e^- \rightarrow Cu$$
(Metal deposition)

$$Fe^{3+} + e^- \rightarrow Fe^{2+}$$
(Reduction of metal ions)

Interaction between the products of the anodic and cathodic reactions can occur, forming solid corrosion products on the metal surface. For example, ferrous ions (Fe^{2+}) coming from the corrosion of metallic iron will react with the hydroxyl ions (OH^-) produced from the reduction of dissolved oxygen:

$$Fe^{2+} + 2OH^- \rightarrow Fe(OH)_2$$

Ferrous hydroxide ($Fe(OH)_2$) is further oxidized to form ferric hydroxide ($Fe(OH)_3$), which is unstable and subsequently transformed to hydrated ferric oxide (Fe_2O_3)—common red rust:

$$4Fe(OH)_2 + O_2 + 2H_2O \rightarrow 4Fe(OH)_3$$

$$2Fe(OH)_3 \rightarrow Fe_2O_3 + 3H_2O$$

A buildup of rust occurs at the anodic sites, forming mounds known as tubercles. Under these mounds, localized corrosion continues to accelerate (Fig. 2) (Ref 1). Not all corrosion products are detrimental. The protective oxide films on copper (cuprous oxide, Cu_2O) and aluminum (aluminum oxide, Al_2O_3) are the result of corrosion:

$$2Cu^+ + 2OH^- \rightarrow Cu_2O + H_2O$$

$$Al^{3+} + 3OH^- \rightarrow Al(OH)_3$$

$$2Al(OH)_3 \rightarrow Al_2O_3{\cdot}3H_2O$$

These films are self-limiting, inhibiting the corrosion process once they are fully developed.

The corrosion process between water and metal surfaces can take many forms (Ref 2). The more common forms that have been observed in heat transfer equipment have the characteristics discussed in the following sections (additional information is available in the Section "Forms of Corrosion" in this Volume).

General corrosion, or uniform attack, occurs when the anodic areas on the metal surface keep shifting to different sites. This continual shifting results in relatively uniform metal removal. Because this type of corrosion can often be predicted, material loss can be taken into account.

Galvanic corrosion occurs when dissimilar materials are in contact in a conducting fluid (water). Accelerated corrosion occurs with the least resistant alloy, while the more resistant alloy is protected. The resistance of alloys can generally be described by the galvanic series, which ranks materials according to their chemical reactivity in seawater. The metals closer to the active end of the series will behave as the anode and will corrode, but those closer to the noble end will behave as the cathode and will be protected. Table 1 lists an abridged galvanic series. The intensity of attack is related to the relative surface areas of the metals in electrical contact. Large cathodic areas coupled to small anodic areas will aggravate galvanic corrosion and cause severe dissolution of the more active metal. The reverse situation—large anodic areas coupled to small cathodic areas—produces very little galvanic current.

Galvanic corrosion obviously must be considered in the design of heat transfer equipment. Alloys close to one another in the galvanic series should be used. Less obvious are the damaging effects that can occur when a dissolved noble metal is transported through the water and is capable of depositing on an active metal. For example, copper ions can plate out onto steel heat transfer tube surfaces, setting up local corrosion cells.

Erosion-corrosion is normally restricted to copper-base alloys. It occurs in areas where turbulence intensity at the metal surface is high enough to cause mechanical or electrochemical disruption of the protective film. Corrosion occurs at these sites and forms horseshoe-crescent-shaped indentations facing upstream of the water flow. The process is usually accelerated when abrasive solid particles, such as sand, are entrained in the water. Because turbulence increases with velocity, areas having higher water velocities are prone to attack. For example, turbulence intensity is much higher at tube inlets than it is several feet down the tube, resulting in the phenomenon of inlet-end erosion-corrosion.

Crevice corrosion is an electrochemical attack that is due to differences in the corrosive environment between a shielded area and its surroundings. Attack usually occurs in areas having a small volume of stagnant solution, such as at tube sheet supports, under deposits or tubercles, and at threaded joints. Corrosion is usually initiated because the oxygen concentration within the crevice is lower than that of the surrounding area. The outside area is higher in oxygen concentration and becomes the predominant cathodic region. Anodic dissolution occurs at the stagnant area. Once attack is underway, the area in the crevice or under a deposit becomes increasingly more aggressive because of pH depression and an increase in electrolyte concentration.

Pitting corrosion is one of the most insidious forms of attack. It takes place at small discrete areas where overall metal loss is negligible. The pit develops at a localized anodic site on the surface and continues to grow because of a large cathodic area surrounding the anode. High concentrations of metal chlorides often develop within the pit and hydrolyze to produce an acidic pH environment. This solution remains stagnant,

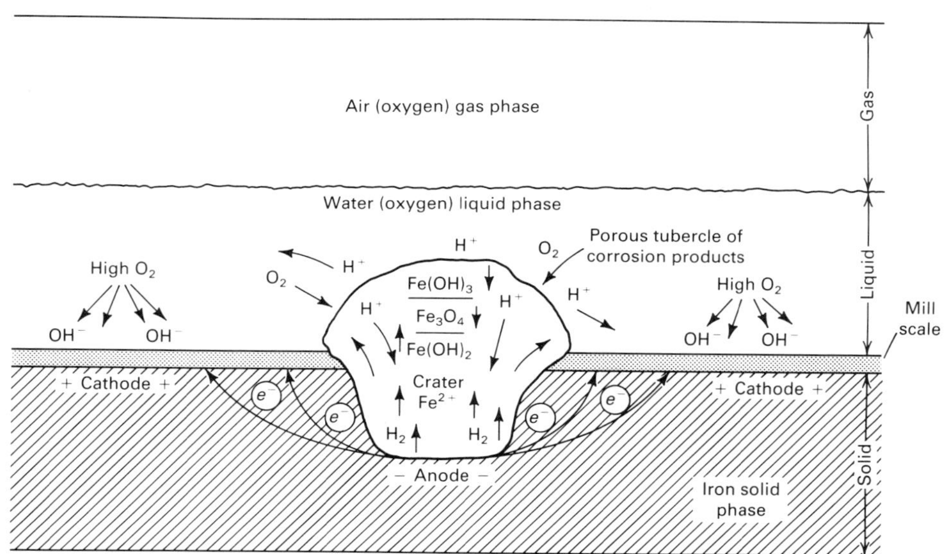

Fig. 2 Schematic of localized corrosion under a rust tubercle. Source: Ref 1

Table 1 Galvanic series of some commercial metals and alloys in seawater

Noble or cathodic
Platinum
Gold
Graphite
Titanium
Silver
Chlorimet 3 (62Ni-18Cr-18Mo)
Hastelloy C (62Ni-17Cr-15Mo)
18-8Mo stainless steel (passive)
18-8 stainless steel (passive)
Chromium stainless steel 11–30% Cr (passive)
Inconel (passive) (80Ni-13Cr-7Fe)
Nickel (passive)
Silver solder
Monel (70Ni-30Cu)
Cupronickels (60-90Cu, 40-10Ni)
Bronzes (Cu-Sn)
Copper
Brasses (Cu-Zn)
Chlorimet 2 (66Ni-32Mo-1Fe)
Hastelloy B (60Ni-30Mo-6Fe-1Mn)
Inconel (active)
Nickel (active)
Tin
Lead
Lead-tin solders
18-8Mo stainless steel (active)
18-8 stainless steel (active)
Ni-Resist (high-nickel cast iron)
Chromium stainless steel, 13% Cr (active)
Cast iron
Steel or iron
Aluminum alloy 2024
Cadmium
Aluminum alloy 1100
Zinc
Magnesium and magnesium alloys
Active or anodic

Source: Ref 2

having a high salt content and low oxygen concentration. The reactions within the pit become self-sustaining (autocatalytic) with very little tendency for them to be suppressed, ultimately causing penetration through the base metal. Pitting corrosion has also been associated with both crevice and galvanic corrosion. Metal deposition (copper ions plated on a steel surface) can also create sites for pitting attack.

Influence of Water Quality

The corrosivity of water is significantly influenced by concentrations of dissolved species, including gases, pH, temperature, suspended matter, and bacteria. The effects of these entities on corrosion are assumed to be independent; although this is not totally correct—interactions do exist—it is easier to visualize their contribution when they are considered separately.

Dissolved Gases

Dissolved oxygen is a major factor contributing to the natural corrosion of steel. It has been shown that oxygen is part of the overall electrochemical reactions occurring at the interface between the aqueous phase and the metal surface. Therefore, it is not surprising that steel corrosion is proportional to the oxygen content in the water. Oxygen solubility in water will vary with temperature, pressure, and electrolyte concentration. Increasing the temperature decreases oxygen solubility, but increasing the partial pressure of oxygen increases its solubility. Concentrated electrolytes decrease oxygen solubility (molar salt concentrations exist in pit cavities and crevices). For open recirculating cooling water systems, the concentration of dissolved oxygen is approximately 6 mg/L. The presence of oxygen is necessary for the formation of protective oxides on copper (Cu_2O), aluminum (Al_2O_3) and steel (γ-Fe_2O_3). In the absence of dissolved oxygen, the corrosion of steel is greatly reduced. A uniform protective film of magnetite (Fe_3O_4) is formed according to the following reaction:

$$3Fe + 4H_2O \rightarrow Fe_3O_4 + 4H_2$$

Overall, oxygen can have either a negative or a positive impact on the corrosion of steel. In aerated systems, uneven distribution of corrosion products can form on the metal surface, giving rise to localized corrosion. Under deaerated conditions, the Fe_3O_4 film isolates the base metal from the water, drastically reducing corrosion.

Carbon dioxide is more soluble than oxygen in pure water (1.3 g/L at 30 °C, or 85 °F) and will convert to carbonic acid (H_2CO_3), producing a solution having a pH of less than 6 where acid attack can predominate:

$$CO_2 + H_2O \rightleftarrows H_2CO_3 \rightleftarrows H^+ + HCO_3^-$$

Adjusting the pH of the water upward redistributes the ratio of dissolved carbonic species. Carbonic acid will dissociate to form bicarbonate ions (HCO_3^-) and subsequently carbonate ions (CO_3^{2-}). The ratio of the various components can be calculated from the pH of the system. In open recirculating cooling water systems, the pH is usually controlled within the 7 to 8.5 range. Within this range, only the CO_2/HCO_3^- ratio is important (the CO_3^{2-} concentration is negligible). Calcium ions (Ca^{2+}) in the water will react with bicarbonate species to produce calcium carbonate ($CaCO_3$). This salt has a low solubility and will precipitate onto heat transfer surfaces (see the section "Calcium Carbonate" in this article). The effect of CO_2 is most important in boiler condensate systems (see the articles "Corrosion in the Nuclear Power Industry" and "Corrosion in Fossil Fuel Power Plants" in this Volume). Here, H_2CO_3 reacts with the steel to form ferrous bicarbonate ($Fe(HCO_3)_2$), which is a highly soluble salt. Rapid general thinning of steel can occur. The addition of soluble amine inhibitors neutralizes the H_2CO_3 to suppress the corrosive attack. Copper alloys are also susceptible to increased attack in the presence of CO_2.

Chlorine is not a natural constituent of cooling waters, but is added for biological control. When dissolved in water, chlorine will convert to hypochlorous acid (HClO) and hydrochloric acid (HCl), which will suppress the pH:

$$Cl_2 + H_2O \rightarrow HClO + HCl$$

Acid attack on steel is a concern when the pH falls below 7. Above this pH, the deleterious effects of chlorine are reduced. Adequate biological control can be achieved if the pH is maintained at approximately 7.5. However, chlorine will accelerate the corrosion of copper alloys, even at alkaline pHs. It is one of the most aggressive species to copper alloys in cooling waters; it can induce localized attack and degrade the protective Cu_2O film.

Ammonia. Another gas that can affect heat transfer equipment is ammonia (NH_3). Although its effect on ferrous alloys is minimal, it has a drastic impact on copper-rich brasses and alloys. These alloys can experience both rapid general thinning and stress-corrosion cracking (SCC). Ammonia is originally formed from the thermal degradation of various nitrogen-containing compounds added to a boiler to reduce ferrous corrosion. It forms a soluble complex with copper that can autocatalytically deteriorate the alloy in a short period of time.

Temperature

As previously stated, corrosion is an electrochemical phenomenon. It is not surprising that an increase in temperature will cause an increase in corrosion rates. Temperature plays a dual role with respect to oxygen corrosion. Increasing the temperature will reduce oxygen solubility. In open systems, in which oxygen can be released from the system, corrosion will increase up to a maximum at 80 °C (175 °F) where the oxygen solubility is 3 mg/L. Beyond this temperature, the reduced oxygen content limits the oxygen reduction reaction, preventing occurrence of the iron dissolution process. Thus, the corrosion rate of carbon steel decreases, and at boiling water conditions, the temperature effect is similar to room temperature with a high oxygen content. For closed systems, in which oxygen cannot escape, corrosion continues to increase linearly with temperature. The other physicochemical properties affected by temperature are the diffusion of oxygen to the metal surface, the viscosity of water, and solution conductivity. Increasing the temperature will increase the rate of oxygen diffusion to the metal surface, thus increasing corrosion rate because more oxygen is available for the cathodic reduction process. The viscosity will decrease with increasing temperature, which will aid oxygen diffusion. Ionic mobility will also increase with temperature, increasing the overall conductivity of water.

An unusual temperature effect can occur with copper alloys (Ref 3). Temperature differences of at least 65 °C (115 °F) between the ends of copper conduits will cause the cold end to be cathodic to the hot end. Copper ions will migrate to the cold end and dissolve (corrode) at the hot end. At the cathode, copper ions will plate out, but at the anode, the surface will become rough and will pit. This effect is known as thermogalvanic attack.

One other effect of temperature should be noted, although it is not related to corrosion. An increase in temperature will decrease the solubility of many sparingly soluble inorganic salts. The solubility of $CaCO_3$ and calcium sulfate ($CaSO_4$) will decrease with an increase in temperature, precipitating and forming a thick barrier deposit at the hottest areas.

Suspended Solids

Suspended matter, such as clays, silt, and corrosion products, is always present in open recirculating cooling water systems. Particulates scrubbed from the air add to the suspended solids loading. These materials are usually soft and nonabrasive. They are capable of depositing in low-flow areas, forming a physical barrier, and preventing oxygen from reaching the metal/solution interface. This buildup will contribute to the formation of differential aeration cells (crevice corrosion) and will promote localized attack.

Effect of pH

The normal pH range for an open recirculating cooling water system is 6.5 to 9. Closed systems operate at a pH of 8.5 to 9; in boilers, the pH of the water is often 11. These pH values are the bulk pH of the water, but the actual pH at the metal surface can be different, depending on prevailing surface reactions. Oxygen reduction will produce OH^- ions, raising the pH, but underdeposit corrosion products can depress the pH.

When the bulk water pH is moderately acidic (pH 5), uniform corrosion is the predominant form of attack, which increases with decreasing pH. In mineral acid (pH 4 or below), the protective oxide film dissolves, exposing bare metal surfaces. Corrosion is further accelerated when dissolved oxygen is reduced at the metal surface at low pH. Both hydrogen evolution and oxygen reduction become the prevalent cathodic reactions. As the pH increases above 4, iron oxides precipitate from solution to form deposits. Uniform corrosion gradually decreases, but underdeposit attack begins because of the formation of Fe_2O_3 adhering to the surface. These deposits impede the diffusion of oxygen to the metal surface. As the pH increases, the nature of the iron oxide deposits changes from loosely adherent at pH 6 to hard and tenacious at pHs above 8. Although the corrosion of steel in aerated waters decreases within the normal operating pH range of 6.5 to 9, the rates are sufficiently high that chemical treatments must be added to these systems to bring the rates within a manageable level of less than 0.13 mm/yr (5 mils/yr).

Copper alloys are not as sensitive to pH as carbon steel is. Acid pHs will accelerate general corrosion. As the pH increases, uniform corrosion decreases significantly. However, general thinning is not as severe a problem as the formation of cupric ions (Cu^{2+}), which can cathodically deposit on steel and create active sites for pitting attack.

Aside from the effect of pH on corrosion, increasing the pH of waters having moderate levels of calcium and alkaline values can result in the precipitation of $CaCO_3$. The deposition of this alkaline scale can impede the diffusion of oxygen to the metal surface in addition to forming a heat transfer barrier.

Dissolved Salts

Dissolved constituents in water can have a variety of effects, both individually and through their interactions. The effects include increased corrosion in addition to scale and deposit formation. Increasing the dissolved solids content of the waters increases its conductivity. Galvanic effects due to the coupling of dissimilar metals are extended in waters having a higher salt content compared to waters of low conductivity.

Hardness ions (calcium and magnesium) and HCO_3^- ions are inhibitive and will suppress corrosion, but chloride (Cl^-) and sulfate (SO_4^{2-}) ions are antagonistic and will increase the rate of some forms of corrosion. The aggressiveness of water can be reduced by increasing the concentration of hardness ions. Most waters used for cooling have pHs in the vicinity of neutral or above. This environment is very conducive to the formation of a protective film of $CaCO_3$ even when the water is below the saturation level for this salt. The actual mechanism that occurs at the water/metal surface interface is more involved than simple deposition of deposit. It is quite probable that $CaCO_3$ formation is due to electrochemical changes at the metal surface. The pH at the surface is usually higher than that of the bulk water because of an increase in OH^- ion concentration, a reaction product of oxygen reduction.

The overall aggressiveness of water is related to its hardness and alkalinity. Soft waters which are low in calcium, are more corrosive than hard waters. Laboratory studies using three different waters of various compositions have shown that increased calcium concentration reduces general corrosion (Ref 4), as illustrated in Fig. 3.

The HCO_3^- ion is the predominate alkaline species in natural waters within the 6.5 to 9 pH range. Alone, it is a mild inhibitive species. Its reaction with Ca^{2+} ions is quite obvious, producing bulk precipitation of $CaCO_3$ once saturation is reached. Localized corrosion cells normally develop under the deposits. Increasing the bicarbonate concentration beyond that required for $CaCO_3$ saturation can be aggressive to steel due to the formation of a nonprotective iron carbonate ($FeCO_3$) film.

Copper alloys are also attacked by high HCO_3^- concentrations. Greenish-blue nodules that consist of basic copper salts usually form at active pit sites.

Chloride and SO_4^{2-} ions are known to have a deleterious effect on steel. Much of the antagonistic nature of Cl^- ions is due to their ability to adsorb on the metal surface and interfere with the formation of passive films. Pitting is the most common form of attack. The small exposed area where Cl^- ions have adsorbed are anodic to the large cathodic passive oxide surface. High current densities are generated at the Cl^- site. Once corrosion begins, hydrolysis of the metal ions from the anodic reaction causes a decrease in pH, which discourages film repair and accelerates attack. The level of Cl^- ions needed to initiate attack can be as low as a few milligrams per liter

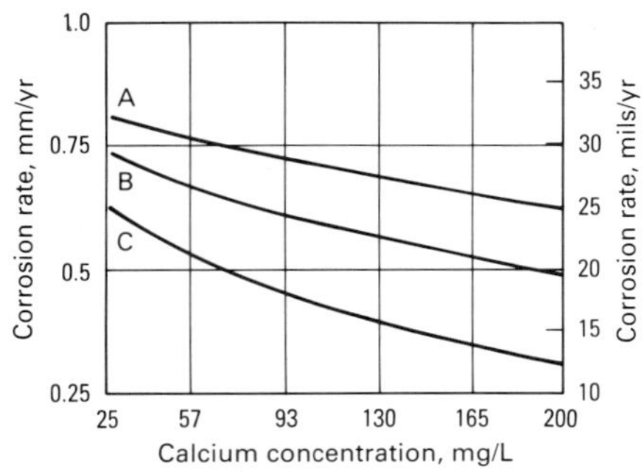

Fig. 3 Corrosion rate versus calcium concentration in three different waters. A, HCO_3^- = 38 mg/L, Cl^- = 78 mg/L, SO_4^{2-} = 78 mg/L; B, HCO_3^- = 100 mg/L, Cl^- = 224 mg/L, SO_4^{2-} = 224 mg/L; C, HCO_3^- = 266 mg/L, Cl^- = 644 mg/L, SO_4^{2-} = 644 mg/L

for some stainless steels in high-purity waters. The more susceptible an alloy is to general attack, the less effect there is from Cl^- ions. Carbon steel will corrode in chloride-containing waters primarily from uniform corrosion rather than localized attack.

Copper alloys are subject to degradation by Cl^- ions that modify the Cu_2O structure. The small, negatively charged Cl^- ion can migrate through the Cu_2O film to areas of high positive charge density (Cu_2O is a *p*-type semiconductor). Substitution of a monovalent charged chloride species for a divalent charged oxygen species can occur. To maintain electroneutrality, cuprous ions (Cu^+) are ejected (dissolved) into the aqueous phase. The loss of Cu^+ ions is, by definition, corrosion.

Sulfate ions are somewhat more elusive in their effect on corrosion. These ions do not appear to have the film-piercing properties of Cl^- ions. In fact, there is evidence that SO_4^{2-} ions may inhibit corrosion of some stainless steels (Ref 5).

Other ions found in cooling waters that have an effect on corrosion are manganese, sulfide (S^{2-}), phosphate (PO_4^{3-}), and nitrate (NO_3^-). Sodium (Na^+) and potassium (K^+) ions are considered neutral species and have no discernible effect. Manganese is found in both well and surface waters and can cathodically deposit on stainless steels and copper-base alloys. Attack usually occurs under these deposits. Sulfide ions from putrefaction of organic matter, sulfate-reducing bacteria, or pollution will attack copper-base alloys and steels.

The Cu_2O film becomes nonprotective through a substitution of S^{2-} ions for the oxygen ion, accelerating corrosion. Attack occurs on steel through the formation of ferrous sulfide (FeS). Deep pits will develop in both alloys. Phosphate ions can act as an accelerator or as an inhibitor for steel corrosion, depending on its concentration. At low concentrations, PO_4^{3-} ions will cause pits to develop on the surface of the metal. Higher concentrations of 15 to 20 mg/L reverse this role, and the ion contributes to the stabilization of γ-Fe_2O_3. Nitrate ions can be reduced on steel surfaces, producing nitrites, NH_3, and OH^- ions. Nitrite ions (NO_2^-) are inhibitory to copper and steel alloys, but NH_3 will attack copper.

Scale Deposition

Water-formed deposits, commonly referred to as scale, can be defined as a crystalline growth of an adherent layer (barrier) of insoluble salt or oxide on a heat exchanger surface. The rate of formation is a complicated function of many variables, including temperature, concentration of scale-forming species, pH, water quality, and hydrodynamic conditions. The normal solubilities of scales increase with temperature, but few, for example, $CaCO_3$ and $CaSO_4$, have the reverse trend. Unfortunately, these scales are commonly found in cooling water systems.

Calcium carbonate is perhaps the most commonly found scale in cooling water systems. Calcium and bicarbonate alkalinity are both needed to form this extremely tenacious scale (alkalinity is the concentration of HCO_3^-, CO_3^{2-}, and OH^- ions present in the water). An increase in heat and/or pH will cause the HCO_3^- ions to decompose to CO_2 and $CaCO_3$:

$$Ca(HCO_3)_2 \rightarrow CaCO_3 \downarrow + CO_2 \uparrow + H_2O$$

The greatest concentration of $CaCO_3$ will occur at the hottest areas along the heat transfer surfaces.

Many methods have been proposed to predict the formation of $CaCO_3$. However, they are all based on the thermodynamic equilibria of H_2CO_3 and are alkalinity corrected for temperature and dissolved solids (ionic strength). The more commonly used equations or indices are the Ryznar Stability Index (RSI) (Ref 6) and the Langelier Saturation Index (LSI) (Ref 7). The LSI is defined as:

$$LSI = pH - pH_s \qquad \text{(Eq 1)}$$

where pH is the actual measured value in the water, and pH_s is the pH of saturation calculated from the expression:

$$pH_s = (pK_2' - pK_{sp}') + pCa + pAlk \qquad \text{(Eq 2)}$$

where K_2' is the apparent second dissociation constant of H_2CO_3, K_{sp}' is the apparent solubility product of $CaCO_3$, pCa is $-\log_{10}(Ca^{2+})$ in moles per liter, and pAlk = $-\log_{10}$[total alkalinity] in equivalents per liter. The complexity of Eq 1 and 2 has been reduced to nomographs (Ref 8). The RSI is an empirical expression:

$$RSI = 2pH_s - pH \qquad \text{(Eq 3)}$$

where pH and pH_s have the same meaning as previously described. Interpretations of LSI and RSI values are listed in Table 2.

These indices indicate only the tendency for $CaCO_3$ to deposit, not the rate or capacity for deposition. Also, these values do not take into

Table 2 Prediction of water characteristics by LSI and RSI

Index: LSI	Index: RSI	Tendency of water
2.0	<4	Heavy scale forming, nonaggressive
0.5	5–6	Slightly scale forming and mildly aggressive
0	6–6.5	Balanced or at $CaCO_3$ saturation
−0.5	6.5–7	Nonscaling and slightly aggressive
−2.0	>8	Undersaturated, very aggressive

account the tendency for $CaCO_3$ to supersaturate, its rate of formation, or whether the water contains any inhibitor to prevent deposition.

Calcium carbonate formation can be controlled by adding acids or specific chemicals tailored to inhibit its formation or modify the crystal lattice. Sulfuric acid (H_2SO_4), which is inexpensive, is most often used. Other acids, such as HCl, citric, or sulfamic acid (NH_2SO_3H), are also suitable. Acid addition produces salts that are more soluble than $CaCO_3$. These salts can reach saturation and must be controlled to prevent precipitation on heat transfer surfaces:

$$Ca(HCO_3)_2 + H_2SO_4 \rightarrow CaSO_4 \downarrow + 2CO_2 + 2H_2$$

Calcium carbonate precipitation can also be inhibited by chemical treatment. A cost-effective group of materials is the polymeric inorganic phosphates. This class of compounds includes the salts of pyrophosphate, tripolyphosphate, and hexametaphosphate. Approximately 1 mg/L of active material can stabilize 44 times the $CaCO_3$ equilibrium saturation. This concentration is well below the stoichiometric values required for complexing calcium hardness and is termed threshold treatment. At these concentrations, the polyphosphates inhibit the crystallization of $CaCO_3$ crystallites by suppressing both nucleation and crystal growth. The polyphosphates are partly adsorbed on the surface of the growing crystals and partly included in incipient crystal nuclei. A disadvantage of the polyphosphates is their ability to hydrolyze or revert to PO_4^{3-} ions, which have no scale-inhibiting properties (see the section "Calcium Phosphate" in this article).

Other phosphate compounds found to be effective belong to the class known as phosphonates. Two of the more common compounds are AMP (nitrilo tris (methylene phosphonic acid)) and HEDP (1-hydroxyethylidine-1,1-diphosphonic acid). Threshold treatment inhibitor concentrations range from 0.25 to 0.5 mg/L, although higher concentrations are often used. The compounds AMP and HEDP have greater hydrolytic stability than the polyphosphates. It should be noted that AMP will degrade in the presence of Cl^- to PO_4^{3-} ions. The addition of zinc ions (Zn^{2+}) significantly stabilizes the AMP molecule while still maintaining its control over $CaCO_3$ scale. Although HEDP is less affected by Cl^- ion than AMP, at least 50% of the HEDP is degraded to PO_4^{3-} at residual chlorine concentrations of 0.2 to 0.5 mg/L (Ref 9).

Organic polymers have also been found to be effective $CaCO_3$ inhibitors. These include polycarboxylates, such as polyacrylates, polymethacrylates, polymaleates, and their copolymers. Treatment levels are higher than the phosphorus-bearing materials, usually in the range of 2 to 4 mg/L. The molecular weights of these polymers should range from 1000 to 10 000 amu for effective scale control.

The polymers adsorb onto the $CaCO_3$ crystal structure, limiting the growth of $CaCO_3$ and ultimately limiting scale formation. These polymers are more frequently considered dispersants. They retard $CaCO_3$ scale by maintaining small particles of distorted crystalline material in suspension.

Bifunctional compounds are a recent development for $CaCO_3$ control. Phosphinocarboxylic acids contain both organic phosphorus and carboxylic acid groups (Ref 10). Simple molecules combining both the phosphono and carboxylic groups have also emerged and are very effective $CaCO_3$ inhibitors (Ref 9). However, these bifunctional compounds may be more expensive than the traditional low molecular weight polymers, polyphosphates, or phosphonates.

Calcium sulfate can exist in various forms in cooling water systems, the most common form being gypsum ($CaSO_4 \cdot 2H_2O$). The hemihydrate and anhydrous forms are much less common. Their solubilities as a function of temperature are shown in Fig. 4. Because $CaSO_4 \cdot 2H_2O$ is more soluble than $CaCO_3$ by at least a factor of 50, it will precipitate only after the latter scale has been formed, within the normal pH range of 7 to 9. This phenomenon provides the basis for H_2SO_4 addition to control $CaCO_3$ in recirculating cooling water systems. The normal upper limit for calcium and sulfate concentration in the absence of an inhibitor is expressed by:

$$[Ca^{2+}] \cdot [SO_4^{2-}] = 500\ 000 \quad \text{(Eq 4)}$$

where the bracketed values are the ionic concentrations expressed in milligrams per liter.

Calcium sulfate scale can be most effectively controlled with polyacrylates (Ref 12), their copolymers, and phosphinocarboxylates (Ref 10). As with $CaCO_3$, excellent calcium scale control is achieved with polymers having low molecular weights in the range of 1000 to 10 000 amu. Concentrations of 1 to 2 mg/L can increase $CaSO_4$ solubility by a factor of 20. This is similar to maintaining the $[Ca] \cdot [SO_4] > 10^7$ compared to 500 000 in the absence of chemical treatment. These polycarboxylates are stable over a wide range of pH and temperature. The basic mode of inhibition is through a combination of threshold crystal distortion and dispersancy mechanism that prevents or delays precipitation.

Another chemical that controls $CaSO_4$ precipitation efficiently is the phosphonate AMP. In fact, the family of aminophosphonates has been found to be extremely effective. The level of AMP needed is approximately 0.5 mg/L.

Calcium phosphate ($Ca_3(PO_4)_2$) scale has become more common in recirculating cooling water systems. Increasing pH, calcium concentration, and phosphate addition from chemical treatment have provided the potential for this deposit to form on heat transfer surfaces. Other water sources have also contributed to the increased level of phosphate. Makeup waters obtained from agricultural runoff and partly treated sewage can have high levels of PO_4^{3-} ions (at least 10 mg/L).

The solubility of $Ca_3(PO_4)_2$ decreases with increasing pH. It is essentially unaffected over normal temperature ranges (about 25 to 75 °C, or 75 to 165 °F). These deposits are usually amorphous and eventually transform to a more crystalline hydroxyapatite—$Ca_5(PO_4)OH$. Because of the low solubility of $Ca_3(PO_4)_2$ (about 10^{-30}), deposits can easily form in waters containing 5 mg/L of PO_4^{3-} ions and 300 mg/L of Ca^{2+} ions at pH 7 to 7.5. The scale-forming tendency of $Ca_3(PO_4)_2$ is a complex function of pH, calcium hardness, PO_4^{3-} concentration, ionic strength, and temperature. Rule-of-thumb relationships between these variables do not exist. Also, in the absence of any phosphate deposit, the PO_4^{3-} ions can contribute to the corrosion inhibition of carbon steel.

Calcium phosphate is an extremely difficult scale to inhibit. Intensive research has brought

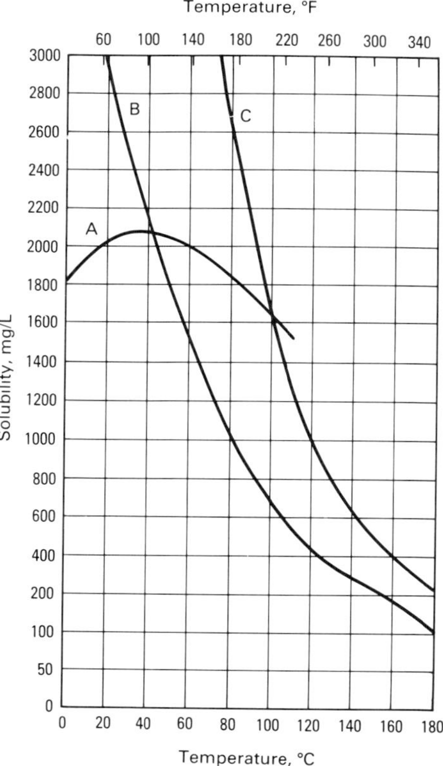

Fig. 4 Solubility of three forms of $CaSO_4$ versus water temperature. A, $CaSO_4 \cdot 2H_2O$; B, anhydrous $CaSO_4$; C, hemihydrate ($CaSO_4 \cdot \frac{1}{2}H_2O$). Source: Ref 11

forth new chemistry to control this tenacious deposit. These new inhibitors include acrylic acid-hydroxypropylacrylate copolymer (Ref 13), styrene sulfonic acid-maleic acid copolymer, acrylic acid-sulfonic acid copolymer (Ref 14), and phosphino/acrylic acid-organo-sulfonic acid copolymer. They function by markedly modifying the morphology and size of $Ca_3(PO_4)_2$ deposit while also acting as a dispersant to prevent adherence of $Ca_3(PO_4)_2$ to heat transfer surfaces. Normal concentrations of these copolymers are 10 to 15 mg/L under conditions of moderate calcium hardness (about 500 mg/L as $CaCO_3$) and 10 mg/L of PO_4^{3-} ions at pH 8 to 8.5. Increasing any of these parameters will require an increase in polymer concentration.

Silicate Scales. Calcium silicate ($CaSiO_3$) and magnesium silicate ($MgSiO_3$) scales tend to develop under more alkaline cooling water conditions, in which the pH is approximately 8.5 or greater. These scales are very tenacious, dense, and difficult to remove from heat transfer surfaces. Although the solubility of silica (SiO_2) increases with pH, the solubility of the alkaline silicates decreases with increasing pH.

An upper limit for SiO_2 concentration is 150 mg/L in the recirculating water, although other factors affect this value. Magnesium silicate can precipitate on heat transfer surfaces with magnesium concentrations as low as 50 mg/L and 150 mg/L SiO_2. A rule-of-thumb pseudosolubility product of $[Mg] \cdot [SiO_2] \leq 8400$ has been developed (bracketed values are in milligrams per liter). The addition of chemical treatment as a preventive measure is essentially nonexistent.

The most effective method of control is to keep the SiO_2 concentration in the recirculating cooling water below the 150-mg/L limit.

Fouling Deposits. Water-borne deposits, commonly known as foulants, are loose, porous, insoluble materials suspended in water. They include such diverse substances as particulate matter scrubbed from the air; migrated corrosion products; silt, clays, and sand suspended in makeup water; organic contaminants (oils); biological matter; floc carryover from clarifiers; and such extraneous materials as leaves, twigs, and wood fibers from cooling towers. Fouling interferes with the flow of cooling water as compared to the reduction in heat transfer caused by barrier scales. However, fouling can reduce heat transfer efficiency because of plugging of the exchanger. Thus, adequate water flow through the tubes is essential.

High flow rates (1.5 to 2.5 m/s, or 5 to 8 ft/s) can sweep away ordinary deposits, but low flow rates (less than 0.6 m/s, or 2 ft/s) cause the suspended foulants to drop out and deposit. Regions of low velocity include shell-side coolers, compressor jackets, water boxes, and cooling tower basins.

Mechanical methods can be used to reduce fouling. In once-through cooling systems, coarse filters (bar screens, trash racks, traveling screens) are used to remove large debris. Fine filtration is not practical. This is not the case with open recirculating systems. When the makeup water contains an appreciable concentration of suspended matter, it is advantageous to use sidestream filtration. In general, passing a few percent of the recirculating water through the sidestream filter will reduce the suspended solids loading 80 to 90%.

Another effective mechanical method used to control foulants is to pass scrapers, brushes, or balls through the heat exchanger tubes, wiping them free of deposits. This technique is most frequently used in power utility surface condensers or other critical exchangers.

Chemical treatment has also been found to be an effective control method. Synthetic polymers are the chemicals more commonly used to disperse foulants. Polyacrylates, polymaleates, partially hydrolyzed polyacrylamides, and their copolymers constitute the majority of dispersant chemicals. The most important aspect of these materials is their molecular weights. An effective dispersant has a molecular weight of 1000 to 10 000 amu—the same range as for scale control. Concentrations of a few milligrams per liter are required in open recirculating systems, but lower levels are normally used in once-through systems.

Natural dispersants, such as tannins, lignin sulfonate, and carboxymethyl cellulose, are occasionally used, but they are not as effective as the synthetic polymers. The natural materials require higher concentrations to produce good dispersion of foulants, may not be cost effective, and may contribute to foaming. The synthetic polymers are not as easily degraded by biological organisms as the natural polymers are.

Increasing the molecular weight of the synthetic polymers into the million range changes the characteristics of the treatment from dispersion to flocculation. These high molecular weight polymers consist of polyacrylamides, polyamines, or polyacrylates and their various copolymers. Fouling is controlled by agglomeration of the suspended solids into nonadherent larger particles. To settle deposits, the polymers are usually added to concentrations of 0.2 to 0.5 mg/L to clarifiers or thickeners. Similar concentrations are used in open recirculating systems. However, the deposit will accumulate in low-flow or stagnant areas in the recirculating systems. Normally, heat transfer will not be seriously reduced, because of the low bulk density of these deposits. Buildup of sludge in the cooling tower basins will necessitate periodic cleaning.

Biological Fouling. The presence and growth of living matter is commonly referred to as biofouling. Biofouling can interfere with the flow of water through heat exchangers and other conduits. This inhibits heat transfer and contributes to corrosion and general deterioration of the entire cooling system.

Recirculating cooling water systems are ideal incubators for promoting the growth and proliferation of microorganisms. Water saturated with oxygen, exposed to sunlight, maintained at a temperature of 30 to 60 °C (85 to 140 °F), and having a pH of 6 to 9 ensure abundant nutrients and appropriate environment for life-sustaining growth. Although biofouling interferes with heat transfer, it does so in a manner quite different from that of inert deposits.

The buildup of a biofilm is initiated with the adsorption of organic material on the metal surface from the bulk water. Transport of microbial particles to the surface is due to the turbulent flow. The microorganisms can attach to the surface, with growth occurring through the assimilation of nutrients. Eventually, some of the biofilm is sheared away and reentrained in the flowing water to repeat the process of biofilm development. These steps are shown schematically in Fig. 5.

The biofilm can cause losses in heat transfer because of its insulating properties. Kinetic energy is absorbed from the flowing water through the ripple surface. Increased pumping energy is required to overcome the frictional resistance of the biofilm. Even though the film is 95 to 98% H_2O, the pressure decrease produced by a 500-μm thick film is greater than would be expected.

Three major classes of microorganisms are associated with recirculated cooling water systems: algae, fungi, and bacteria. Algae can range from very simple single-cell (unicellular) plants to multicellular species. The latter include diverse forms and shapes, including slimy masses composed of several cells or long strands (filaments) of algae. All algae contain colored pigments, the most important of which is chlorophyll. Algae usually flourish on wetted surfaces exposed to oxygen and sunlight, such as cooling tower lumber, mist eliminators, screens, and distribution trays.

Their impact on metal corrosion can be severe. The large slime mass contributes to crevice corrosion and pitting. Massive growth can inhibit proper water distribution by plugging screens, restricting flow, and interfering with pump suction.

Fungi are similar to algae, but do not contain chlorophyll. The major forms of interest are yeast and molds. They require moisture and air but not sunlight and exist on nutrients found in water or substances to which they are attached, for example, bacteria and algae. Mold fungi are filamentous in form, but most yeasts are unicellular. Fungi consume wood components, causing serious surface deterioration and internal decay (wood rot).

Bacteria are unicellular microscopic plantlike organisms that are similar to algae but lack chlorophyll. They exist in three basic forms: rod-shaped (bacillus), spherical (coccus), and spiral (spirillus). The nutrients required for growth must be in solution, which necessitates a certain amount of moisture for their environment. Some bacteria require other carbon forms in addition to CO_2 in order to produce carbohydrates for food. Therefore, water or wet environments high in organic content are suitable for the proliferation of bacteria slime. Heat exchangers are ideal incubators for this biomass, which can significantly reduce heat transfer efficiency and aggravate underdeposit corrosion attack.

The presence of oxygen is not required by all species of bacteria. Aerobic bacteria require free oxygen for growth, but anaerobic bacteria grow in the absence of oxygen.

Acidic waters produced by some forms of bacteria will directly attack metal surfaces. One species that is quite common is sulfate-reducing bacteria. These organisms are anaerobic and convert dissolved sulfur compounds, that is, SO_4^{2-}, to hydrogen sulfide (H_2S). Carbon steel, stainless steel, and copper-base alloys can be severely corroded by H_2S. *Desulfovibrio desulfuricans* is the most prevailing sulfate-reducing species. It

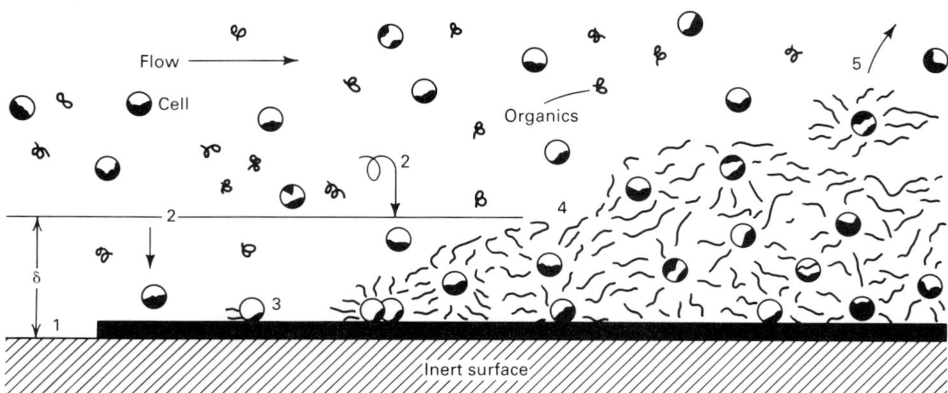

Fig. 5 Steps in biofilm formation. Formation is initiated when small organic molecules become attached to an inert surface (1) and microbiological cells are adsorbed onto the resulting layer (2). The cells send out hairlike exopolymers to feed on organic matter (3), adding to the coating (4). Flowing water detaches some of the formation (5), producing an equilibrium layer δ. Source: Ref 15

exists mainly under deposits devoid of oxygen. The form of corrosive attack on carbon steel by this bacteria is quite distinctive; it is recognizable by the smooth disk-shaped concentric rings formed on the metal surface. Corrosion rates as high as 2.5 mm/yr (100 mils/yr) can occur under optimum growth conditions. The following general reactions occur:

$$10H^+ + SO_4^{2-} + 4Fe \rightarrow 4Fe^{2+} + H_2S + 4H_2O$$

$$H_2S + Fe^{2+} \rightarrow FeS + 2H^+$$

The formation of black iron sulfide deposits in conjunction with a rotten egg odor are also characteristic of attack by sulfate-reducing bacteria.

The aerobic sulfur bacteria *Thiobacillus* can oxidize sulfur, sulfides, or sulfates to H_2SO_4. Localized pH depression as low as 1 can occur, causing severe general thinning of steels where these organisms contact the metal. *Thiobacillus* and *Desulfovibrio* bacteria can both exist simultaneously in close proximity. The anaerobic sulfate-reducers can survive beneath the aerobic sulfur bacteria deposits.

Another organism common to cooling water systems is the nitrifying bacteria, which can oxidize NH_3 to nitrate. This is accompanied by a decrease in pH and occurs according to the reaction:

$$NH_3 + 2O_2 \rightarrow HNO_3 + H_2O$$

Rapid general thinning can occur with copper-base alloys and steels. Nitrite-base steel corrosion inhibitors can be rendered ineffective because of their oxidation to nitrate by this species of bacteria.

Several other groups of bacteria exist in cooling water environments. Iron-depositing bacteria can oxidize water-soluble Fe^{2+} ions to insoluble Fe_2O_3, which will subsequently deposit on the inside of piping, reduce flow, and aggravate crevice corrosion. Slime-forming bacteria form dense, sticky biomasses that impede water flow and contribute to fouling by sustaining the growth of other organisms.

The most practical and efficient method of controlling microbiological activity in cooling waters is through the use of biocides (Ref 16). These chemicals have the ability to kill the organisms or inhibit their growth and reproductive cycles. Biocides perform their function in various ways. Some biocides alter the permeability of the microbe cell wall, thus interfering with their vital life processes. Others damage the cell by interfering with the normal flow of nutrients and discharge of waste (Table 3).

Biocides can be either oxidizing or nonoxidizing toxicants. Chlorine is the most prevalent industrial oxidizing biocide. It rapidly hydrolyzes in water to form HClO and HCl:

$$Cl_2 + H_2O \rightarrow HClO + HCl$$

The HClO is the active species and dissociates as a function of pH:

$$HClO \rightarrow H^+ + ClO^-$$

At pH 7.5, equal concentrations of HClO and hypochlorous ions (ClO^-) exist. Above this pH, ClO^- predominates, with essentially total ionization at pH 9.5. Therefore, chlorine becomes less effective in the more alkaline environments. Generally, the pH range of 6.5 to 7.5 is considered practical when chlorine is used as a biocide. Lower pHs will accelerate corrosion. Continuous treatment levels of 0.1 to 0.2 mg/L are common, but intermittent treatment requires levels of 0.5 to 1.0 mg/L. Chlorine is both an excellent algicide and bactericide. However, *Desulfovibrio* can develop a strong resistance, requiring an increase in chlorine concentration or a change to an alternate biocide. Other sources of chlorine are salts of HClO, such as sodium hypochlorite (NaClO) and calcium hypochlorite ($Ca(ClO)_2$). Both of these salts function in much the same way as chlorine gas.

Another oxidizing biocide is chlorine dioxide (ClO_2). This gas does not form HClO in water, but exists solely as ClO_2 in solution. It is used more extensively in the pulp and paper industry for bleaching than as a biocide. However, ClO_2 is extensively used in cooling waters contaminated with NH_3 or phenols due to the low demand for reaction with these species.

Brominated compounds that form hypobromous acid (HBrO) are very effective over a broader pH range than HClO. At pH 7.5, 50% of the HClO is present, but over 90% of the HBrO is present. Increasing the pH to 8.7 reduces the HClO concentration to 10%, but 50% of the HBrO remains.

Nonoxidizing biocides can be more effective than oxidizing biocides because of their overall control of algae, fungi, and bacteria. They also have greater persistence, with many of them being pH independent.

An organo-bromine broad spectrum nonoxidizing toxicant is 2,2-dibromo-3-nitrilopropionamide (DBNPA). This molecule is an extremely potent bactericide and is only slightly effective as an algicide. It has little fungicide activity. The toxicity of DBNPA decreases with an increase in alkaline pH.

Organo-sulfur compounds include a wide variety of different biocides of which methylene bisthiocyanate (MBT) is the most common. Their mode of activity is through the inhibition of cell growth by preventing the transfer of energy or life-sustaining chemical reactions from occurring within the cell. Methylene bisthiocyanate is effective in controlling algae, fungi, and bacteria, most notably *Desulfovibrio*. A shortcoming of methylene bisthiocyanate is its pH sensitivity and rapid hydrolysis in the alkaline pH range. Therefore, methylene bisthiocyanate is not recommended for use in cooling water systems having a pH above 8.

Under broader alkaline conditions, sulfur-base biocides, such as bistrichloromethylsulfone and tetrahydro-3,5-dimethyl-2H-1,3,5-thiadiazine-2-thione, are more appropriate. The former is active in the pH range of 6.5 to 8, and the latter in more alkaline cooling water systems.

Isothiazolinone is a relatively new sulfur-containing biocide. It is very effective in controlling algae and bacteria. Isothiazolinone can be used over a broad pH range with no decrease in activity.

Quaternary ammonium salts are generally most effective against algae and bacteria in alkaline pH range. The quaternary compounds cause cell death by reducing the permeability of the cell wall, preventing the typical intake of nutrients necessary to sustain life. Because of their surface-active nature, these compounds are easily rendered ineffective in systems heavily fouled with dirt, oil, and debris.

Organic tin compounds, such as bistributyl-tin oxide are very effective against algae and fungi. They function best in the alkaline pH range and provide synergistic biocidal activity when combined with quaternary ammonium salts.

Glutaraldehyde (1,5-pentanedial) is an effective broad-spectrum biocide capable of controlling

Table 3 Microbiocides used in cooling water systems

Microbiocide	Effectiveness(a) Bacteria	Fungi	Algae	Comments
Chlorine	E	S	E	Oxidizing; reacts with $-NH_2$ groups; effective at neutral pH; loses effectiveness at high pH. Use concentration: 0.1–0.2 mg/L continuous free residual; 0.5–1.0 mg/L intermittent free residual
Chlorine dioxide	E	G	G	Oxidizing; pH insensitive; can be used in presence of $-NH_2$ groups. Use concentration: 0.1–1 mg/L intermittent free residual
Bromine	E	S	E	Oxidizing; substitute for Cl_2 and ClO_2; effective over broad pH range. Use concentration: 0.05–0.1 mg/L continuous free residual; 0.2 to 0.4 mg/L intermittent free residual
Organo-bromide (DBNPA)	E	NA	S	Nonoxidizing; pH range 6–8.5. Use concentration: 0.5–24 mg/L intermittent feed
Methylene bisthiocyanate	E	S	S	Nonoxidizing; hydrolyzes above pH 8. Use concentration: 1.5–8 mg/L intermittent feed
Isothiazolinone	E	G	E	Nonoxidizing; pH insensitive; deactivated by HS^- and $-NH_2$ groups. Use concentration: 0.9–13 mg/L intermittent feed
Quaternary ammonium salts	E	G	E	Nonoxidizing; tendency to foam; surface active; ineffective in highly oil- or organic-fouled systems. Use concentration: 8–35 mg/L intermittent feed
Organo-tin/quaternary ammonium salts	E	G	E	Nonoxidizing; tendency to foam; functions best in alkaline pH. Use concentration: 7–50 mg/L
Glutaraldehyde	E	E	E	Nonoxidizing; deactivated by $-NH_2$ groups; effective over broad pH range. Use concentration: 10–75 mg/L intermittent feed

(a) E, excellent; G, good; S, slight; NA, not applicable

slime-forming and sulfate-reducing bacteria, fungi, and algae (Ref 17). It functions over broad pH and temperature ranges and is compatible with chlorine. Glutaraldehyde is deactivated in systems containing NH_3 and other primary amines, that is $-NH_2$ groups.

Corrosion Control

Aqueous corrosion has been shown to consist of electrochemical processes (see the section "Corrosion Processes in Water Recirculating Systems" in this article). The detrimental effects of these processes in cooling water environments can be significantly reduced by various methods, such as designing systems with more corrosion-resistant materials, applying protective coatings (paints, epoxy), and using sacrificial anodes (cathodic protection) and chemical treatments.

For chemical treatment or corrosion inhibitor programs to be effective, they must protect all exposed metal from corrosive attack, must be effective at low concentration, must not cause deposits on the metal surface, must remain effective under broad range of pH, temperature, water quality, and heat flux, must prevent scale formation and disperse deposits, and must have minimal toxicological effect when discharged. Economics should also be considered when assessing the merits of a chemical treatment program. In once-through cooling systems, protection must be achieved with a few milligrams per liter, or the cost may be prohibitive. For these types of systems, use of more corrosion-resistant alloys is often a more economical approach to corrosion control. With respect to open recirculating cooling systems, chemical addition to the makeup water can be in the range of 10 to 50 mg/L (active). Cost is not as prohibitive, because of the recycling of the water system, which also includes the inhibitor. Closed recirculating cooling water systems require high active treatment levels (in the range of several thousand milligrams per liter). These systems have very small water losses.

Inhibitors can be broadly classified according to the rate process being controlled. Anodic inhibitors suppress anodic reactions; that is, the rate of metal ions being transferred into the aqueous environment is reduced. Cathodic inhibitors impede the cathodic reaction—for example, oxygen reduction reaction. Mixed inhibitors hinder both reactions.

Inorganic inhibitors usually affect the anodic process. Rapid suppression of general corrosion occurs. However, when these inhibitors are below a critical concentration, they can stimulate localized pitting attack and therefore must be used with caution.

Cathodic inhibitors reduce corrosion primarily by interfering with oxygen reduction reaction. This family of inhibitors decreases general corrosion and does not stimulate pitting attack. When the anodic and cathodic reactions are both affected by the chemical treatment, a mixed mode of inhibition is in effect.

The performance of an inhibitor can be assessed by its effectiveness in reducing the corrosion rate. This is generally expressed as a rate of thinning or penetration in microns per year or mils per year.

The current approach to cooling water treatment is toward the more alkaline or high-pH conditions. This was not always the case. Historically, acid treatment was widely used to prevent scale deposition on heat transfer surfaces. The system pH was slightly acidic (pH 6.5 to 7.0), and corrosion was controlled by using highly effective anodic inhibitors, the most effective being chromates.

Anodic Inhibitors

Chromates are highly effective in protecting ferrous and many nonferrous alloys against corrosion. The protective film is formed on ferrous metals through the oxidation of iron to form a mixed oxide ($Cr_2O_3 + Fe_2O_3$) (Ref 18).

There is a critical chromate concentration that is necessary for protection, because this inhibitor is sensitive to aggressive ions, such as Cl^- and SO_4^{2-}. Below this minimum, severe pitting attack can occur even though uniform corrosion is low. Above the minimum chromate concentration, local attack is arrested, producing excellent overall protection. Normal control concentrations of chromate for proper steel protection require a high initial dosage of 500 to 1000 mg/L to meet the Cl^- and SO_4^{2-} demands, followed by a reduction in concentration to a maintenance level of 200 to 250 mg/L. There is no discernible change in corrosion protection within normal bulk water temperatures (38 to 65 °C, or 100 to 150 °F) and pH ranges (6 to 11).

When chromates are used at low concentrations of approximately 15 to 25 mg/L, their mode of inhibition is cathodic (Ref 19). Steel protection is provided in conjunction with the dissolved oxygen in the water to produce the protective oxide layer of γ-Fe_2O_3. The reduction product of chromate—Cr_2O_3—does not become part of the protective oxide. The pH range remains the same as for high concentration—pH 6 to 11. Low levels of chromate are insensitive to electrolyte concentration.

Nitrites. Sodium nitrite ($NaNO_2$) is an anodic inhibitor and requires a critical concentration for the protection of steel. Nitrites are extensively used as inhibitors in closed recirculating systems. The level of inhibitor depends on the aggressive species in solution. Sulfate ions interfere with nitrite protection to a greater extent than Cl^- ions. Treatment level should be at least the concentration of $NaNO_2$ required to produce a weight ratio of 1 with aggressive ion concentrations ($[NaNO_2]/[NaCl + Na_2SO_4] = 1$). This concentration is normally in the range of 500 to 750 mg/L at pH above 7.5. Nitrite may contribute to pitting of carbon steel if the concentration falls below this critical level.

Nitrites are easily oxidized to nitrates in open recirculating cooling water systems and therefore are not suitable. However, when formulated with borax, nitrites are excellent corrosion inhibitors for closed systems. The borax buffers the water to a pH above 8.5. Dosage levels for the borax-nitrite systems are 1500 to 2000 mg/L, with a nitrite concentration near 800 mg/L. It should be noted that NO_2^- ions are nutrients for some species of bacteria, which can oxidize the NO_2^- ions and render the treatment ineffective, producing slime deposits and a low pH. Biocides (isothiazolinone) are often used in conjunction with borax-nitrite programs for optimum protection.

Molybdates have been available for corrosion protection for over 50 years, but have never been widely used. Molybdates are classified as anodic inhibitors. In waters having moderate Cl^- concentration (200 mg/L), the level of sodium molybdate (Na_2MoO_4) needed for protection is at least 1000 mg/L. In dilute electrolytes, the concentration can be reduced to 150 mg/L. Normal treatment practices consist of combinations of molybdates with other inhibitors to produce synergistic treatments. This reduces the high level of molybdate ions (MoO_4^{2-}) needed when used alone. Combinations with zinc salts; phosphonates (AMP and HEDP); inorganic phosphates; nitrites; and carboxylates, such as long-chain acrylates and azoles (benzo- or tolyltriazoles), have been shown to be effective in controlling corrosion of multimetal systems (Ref 20).

Sodium molybdate is a nonoxidizing inhibitor and requires a suitable oxidizing agent to augment the inhibitor and to impart a protective film. In aerated systems, the most abundant oxidizer is oxygen. Any of the above combinations are applicable except nitrites (see the section "Nitrites" in this article). A wide range of ratios have been used in specific formulations. A typical treatment consists of MoO_4, HEDP, zinc, and benzotriazole in a 3:3:1:1 weight ratio (Ref 20).

The use of molybdates in closed systems requires an oxidizing salt, such as $NaNO_2$. The optimum composition for the Na_2MoO_4:$NaNO_2$ system consists of a 60:40 weight ratio of the two salts (Ref 20).

Molybdate treatments have minimal pH dependency and can be used over a pH range of 5.5 to 8.5. They are sensitive to electrolyte concentration and are adversely affected by aggressive ions, such as Cl^- and SO_4^{2-}, when used alone. Temperature dependency is minimal. Molybdate treatments are known to inhibit both pitting and differential aeration attack. The precipitation of calcium molybdate ($Ca(MoO_4)$) is a concern in waters with moderate-to-high calcium hardness.

The mechanism by which molybdates inhibit the corrosion of ferrous metals is uncertain and complex. Simplistically, when iron corrodes, MoO_4^{2-} ions, in conjunction with other anions, adsorb to form a nonprotective complex with Fe^{2+} ions. Because of dissolved oxygen or other oxidizers in the water, some of the Fe^{2+} ions are oxidized to the ferric (Fe^{3+}) state, and the ferrous-molybdate is transformed to ferric-molybdate, which is both insoluble and protective in neutral and alkaline waters.

Phosphates. Sodium phosphate (Na_3PO_4) is an anodic inhibitor and is effective in the presence of oxygen. Its protective properties toward steel are a function of pH. The monosubstituted phosphate (MH_2PO_4, where M is a metal) is the least protective, but the trisubstituted phosphate (M_3PO_4) is the most protective. Use of phosphate as a corrosion inhibitor should be relegated to more alkaline environments (pH $>$ 8).

In open recirculating waters, phosphate treatment levels should be 15 to 20 mg/L. At lower concentrations of a few milligrams per liter, PO_4^{3-} ions will cause pitting attack. The oxygen content of the system is primarily responsible for the ability of the phosphate to inhibit steel (Ref 21). The dissolved oxygen produces a defective thin film of γ-Fe_2O_3. The PO_4^{3-} ions fill in the voids and accelerate film growth. These plugs prevent any further diffusion of Fe^{2+} ions from the metal surface. The primary reactant responsible for steel inhibition is oxygen, which forms the thin oxide film.

Inhibition by PO_4^{3-} ions is sensitive to electrolyte concentration; Cl^- ions can promote pitting attack. Because the protective oxide film contains voids and other inclusions, Cl^- ions are easily adsorbed to soluble complexes. Hydrolysis of these complexes produces acid domains, which leads to localized acidic attack. Film breakdown is a function of the aggressive ion concentration, but film repair will depend on PO_4^{3-} level and oxygen concentration.

Overall, inhibition by PO_4^{3-} ions is sensitive to water quality and pH. There is minimal temperature sensitivity. One word of caution regarding PO_4^{3-} ions as steel corrosion inhibitors: With waters of high calcium hardness, the potential for deposit formation increases with hardness, phosphate level, pH, and temperature. These deposits will stimulate crevice (underdeposit) corrosion. The need for an inhibitor to prevent calcium deposits becomes exceedingly more important as cooling waters increase in cycles of concentration and pH (see the section "Scale Deposition" in this article).

Cathodic Inhibitors

Cathodic inhibitors suppress the corrosion rate by reducing the effectiveness of the cathodic process. They do not cause intense local attack and are generally considered to be safe inhibitors. Cathodic inhibitors are not as effective as the anodic inhibitors.

Precipitating inhibitors fall under the domain of a cathodic inhibitor. They produce insoluble films on the cathode under conditions of locally high pH and isolate the cathode from the solution. Calcium bicarbonate ($Ca(HCO_3)_2$) will react with the alkaline medium at the cathode to form $CaCO_3$. The increase in alkalinity is due to the oxygen reduction process, generating OH^- ions:

$$Ca(HCO_3)_2 + OH^- \rightarrow CaCO_3 + HCO_3^- + H_2O$$

At the appropriate pH, $CaCO_3$ will precipitate to form a hard, smooth deposit that prevents oxygen from diffusing onto the metal surface. The tendency of a water to form a $CaCO_3$ deposit is given by the LSI (see the section "Scale Deposition" in this article).

Zinc ions are used to achieve a general reduction in corrosion by precipitating as zinc hydroxide ($Zn(OH)_2$) at the cathode due to locally elevated pH. Zinc is usually combined with chromates, phosphonates, or polyphosphates. The durability of $Zn(OH)_2$ is tenuous at best. Although not normally used alone in cooling systems, zinc is synergistic when combined with other inhibitors as part of a multicomponent treatment program.

Zinc causes rapid development of a protective film over the metal surface. Above pH 7.5, the solubility of zinc is rapidly reduced to a few tenths of 1 mg/L. Its solubility can be significantly enhanced when combined with phosphonates or with some of the polymers used for $Ca_3(PO_4)_2$ control.

Polyphosphates are widely used cathodic inhibitors. They have been in use for more than 40 years and are among the most economical of all inhibitor treatments. Sodium salts of the polyphosphates are normally used for corrosion control. They exist as linear polymers having the general structure shown in Fig. 6. The lower members of the series having x equal to or less than 2 are crystalline in nature, that is, $x = 0$, orthophosphate; $x = 1$, pyrophosphate; $x = 2$, tripolyphosphate. Higher members of the series are glassy and have no definitive structure.

Varying the ratio of n Na_2O: m P_2O_5 will produce species of different chain lengths and physicochemical properties, even though the inhibitive action of these polyphosphates is not significantly altered. The pH of the environment directly affects the protective properties of the phosphates. Orthophosphate inhibits corrosion in a more alkaline environment than pyrophosphate does. The crystalline phosphates protect steel at higher pHs than the glassy phosphates, which function better near neutral pH. Normal concentrations in recirculating waters are 15 to 20 mg/L. The pH should be maintained in the range of 6.5 to 7.5 if steel and copper alloys are both part of the system metallurgy.

The polyphosphates are relatively insensitive to electrolyte concentrations, but do require an increase in dosage level with an increase in water corrosivity. They are effective inhibitors in controlling galvanic attack between two dissimilar metals (Ref 22), but do not prevent deposition of cathodic species on the more active metal, that is, copper deposition on steel.

Divalent metal ions, and Ca^{2+} in particular, are needed with the polyphosphates for effective inhibition of steel. The ratio of Ca^{2+} ion concentration to polyphosphate concentration should be at least 0.2, and preferably 0.5. The protective film develops through the formation of a positively charged colloidal complex that migrates to the cathode, forming an amorphous protective barrier. The cationic complex is accounted for by assuming that the calcium intercalates the polyphosphate chains. The colloidal particles are the result of numerous cationic complexes loosely knit one to another through the Ca^{2+} ions. The polyphosphates will revert to PO_4^{3-} ions, which is a potential scale in the presence of Ca^{2+} ions and alkaline pH (see the section "Scale Deposition" in this article).

The phosphonates differ from the polyphosphates by a direct bond formation between the phosphorus and carbon atoms, rather than an oxygen-phosphorus bond in the polyphosphate. This linkage creates greater hydrolytic stability, which reduces the problem of reversion common to the polyphosphates. Two classes of materials are extensively used: AMP and HEDP. Their structures are shown in Fig. 7 and 8. When used alone, both compounds are marginally good steel inhibitors, especially at high pH (pH 8.0). They protect ferrous metals through a mixed mode of inhibition. Incipient local attack occurs but is rapidly arrested. Normal treatment levels are in the range of 15 to 20 mg/L.

The hydrolytic stability of the phosphonates eliminates the problem of $Ca_3(PO_4)_2$ deposition in moderately to highly alkaline waters. However, AMP will degrade in the presence of chlorine, producing PO_4^{3-} ions, and therefore should not be fed concurrently. The phosphonates form very stable complexes with a variety of metal ions and thus accelerate their corrosion. Increased attack occurs on nonferrous alloys, especially the copper-base alloys. The addition of zinc significantly suppresses this antagonistic attack. Calcium phosphonate deposits will form in moderate-hardness high-alkaline waters, that is, 400 to 500 mg/L calcium hardness at pH 8 to 8.5. These deposits can be controlled with some of the new polymers available for $Ca_3(PO_4)_2$ control.

The phosphonates are sensitive to overall water quality. The protection afforded to carbon steel decreases with increasing aggressive ion concentration. Sensitivity to temperature is also exhibited. The major difference between AMP and HEDP is that the latter resists oxidation to a greater extent and can therefore be used in the presence of chlorine.

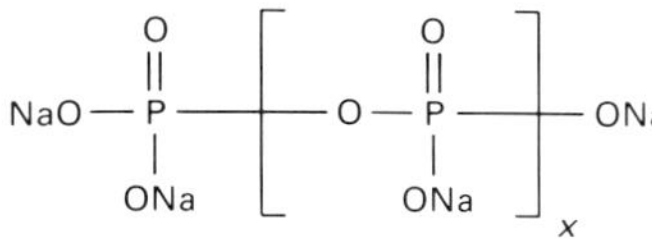

Fig. 6 General structure of sodium polyphosphate. $x = 0$, sodium phosphate; $x = 1$, sodium pyrophosphate; $x = 2$, sodium tripolyphosphate; $x = 12$ to 14, sodium polyphosphate

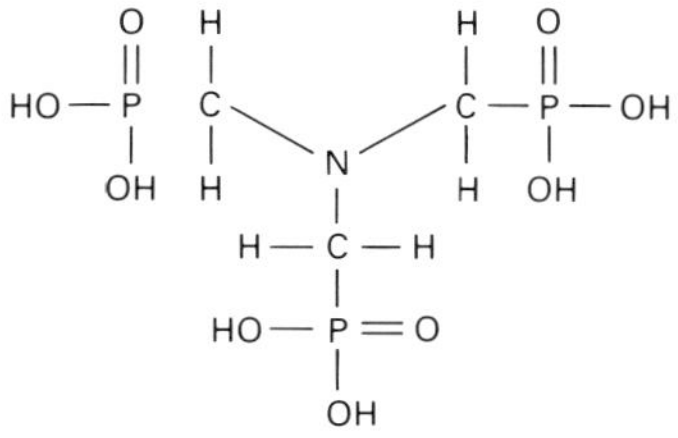

Fig. 7 Structure of AMP

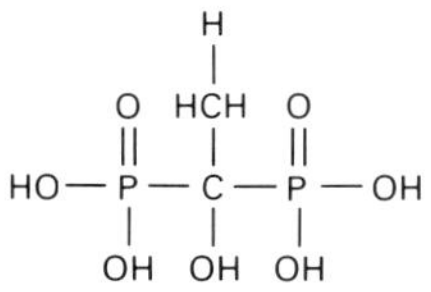

Fig. 8 Structure of HEDP

Multicomponent Systems. Cooling water formulations containing mixtures of inhibitors usually offer increased protection to ferrous metals. Such mixtures are synergistic in their action. Many combinations have been developed to achieve enhanced protection under a wide variety of plant operating conditions. Heavy-metal formulations containing zinc and/or chromate have been widely used in industry, where environmentally acceptable, particularly for severe corrosion problems. Nonheavy-metal treatment programs (no zinc or chromium) are receiving increased attention because of government discharge restraints. Complex blends of PO_4^{3-}, polyphosphate, phosphonate, copper corrosion inhibitor (see the section "Copper Inhibitors" in this article), and $Ca_3(PO_4)_2$ dispersants are probably the most widely used multicomponent inhibitors.

Zinc chromate ($ZnCrO_4$) is one of the most effective multicomponent treatments. It inhibits carbon steel corrosion by stifling the oxygen reduction reaction and is therefore classified as a cathodic inhibitor. Both components of this formulation—low-level zinc and low-level chromate—are also cathodic inhibitors. As little as 5% of either component will show significant synergistic action to ferrous metals, although at least 20% of each ingredient appears to be optimum (Ref 23). Both zinc and chromate exist as individual species in solution without the formation of a specific compound or intermediate.

The effect of zinc addition (5 to 10 mg/L as Zn^{2+} ion) to low-level chromate (up to 20 mg/L of CrO_4^{2-}) is synergistic and significantly reduces local attack. This formulation has broad applicability to multimetal systems. It is capable of protecting copper-base metals, aluminum alloys, and galvanized steel. The protection afforded by

$ZnCrO_4$ to these multimetal systems is the reduction of general attack and the prevention of galvanic attack due to the reduction of noble metal ions on more active base metal.

Normal concentrations are approximately 10 mg/L of each constituent. The treatment program is insensitive to normal operating temperatures and overall water aggressiveness. Therefore, $ZnCrO_4$ has broad application to a multitude of water compositions and levels of aggressiveness in addition to a wide range of operating temperatures. The pH of the system can range from 5.5 to 7.5, with a pH of 7 being most acceptable. Even after major upsets due to pH excursions, the protective film is rapidly reestablished. However, there are some disadvantages to this treatment. Zinc chromate has no threshold-inhibitive effect on hardness scales. It lacks dispersion action to keep the metal surfaces free of deposits and debris, and pH excursion above 7.5 will cause precipitation of zinc to form insoluble basic zinc salts.

Zinc Chromate/Phosphonate. The disadvantages of $ZnCrO_4$ formulation are circumvented by the addition of such a phosphonate as AMP. Approximately 5 mg/L of the phosphonate supplements the $ZnCrO_4$ program to provide threshold treatment to prevent $CaCO_3$ and $CaSO_4$ scales. It can extend the pH range up to 9 because of the increased stability the phosphonate gives Zn^{2+} ions. The phosphonate adds detergent power to the system, flushes away suspended solids, and helps keep the metal surface clean.

Zinc chromate is not a nutrient for biological growth; this simplifies biological control. The Zn^{2+} ions, which are stabilized at high pH, also improve the resistances to phosphonate degradation. All of the advantages of $ZnCrO_4$ are maintained: rapid film formation, insensitivity to electrolyte concentration, and insensitivity to temperature.

Zinc Polyphosphate. The addition of zinc to the polyphosphates does not appreciably change the general nature of the polyphosphates. This system retains its insensitivity to electrolyte concentration, its threshold inhibition of $CaCO_3$ and $CaSO_4$, its ability to protect ferrous and nonferrous metals, and its detergent properties. Zinc polyphosphate inhibitors also enable multivalent metal ions to form positively charged colloidal complexes.

Zinc increases the rate at which the protective film is formed on the metal surfaces (Ref 24). This rapid protective film formation improves the general corrosion protection to the system. It is also synergistic in combination with polyphosphates. The amount of treatment needed using a zinc/polyphosphate inhibitor combination is less than that required for the polyphosphates alone.

The mode of inhibition is cathodic, similar to the polyphosphates in calcium-containing waters. The zinc accelerates film formation, restraining attack until a thin, tenacious, durable film is developed. The deposition of zinc phosphate ($Zn_3(PO_4)_2$) does not seem to be involved in the inhibition process.

Approximately 10 to 20% Zn is usually incorporated into the polyphosphate for synergism. Beyond this level, little improvement is observed. Maintenance concentrations are usually 10 mg/L as polyphosphate. Good practice requires a dosage of two to three times the maintenance level as a pretreatment to the system for a short period of time, that is, less than 1 week. A pH range of 6.8 to 7.2 is required for good control. This limitation in pH is necessary to prevent excessive attack on copper-base alloys (the lower the pH, the greater the attack on copper). Sensitivity to bulk water temperatures is minimal.

Zinc Phosphonates. The combination of zinc with phosphonates provides significantly improved protection compared to phosphonates alone. The addition of zinc makes this formulation synergistic in its protection to carbon steel.

Good corrosion control is attained by using 20 to 80% Zn, with 30 to 60% being optimum (Ref 25). Overall treatment levels are 8 to 10 mg/L phosphonate with 2 to 3 mg/L Zn^{2+} ions. The need for zinc becomes paramount in the presence of copper-base alloys. When used alone, the phosphonates are aggressive to copper and form a strong copper-phosphonate complex. Zinc negates this antagonistic effect by forming a stronger and more stable complex, significantly reducing attack on the alloys. The effectiveness of the zinc-phosphonate combination is due to increased cathodic protection. The zinc effectively counteracts the anodic character of the phosphonates through the formation of a phosphonate-zinc complex, which is cathodic compared to the phosphonates alone.

Due to this complex, broad pH application becomes possible. The zinc phosphonate treatment can be used over a pH range of 6.5 to 9. Zinc is held in solution at these more alkaline pH levels, and protection actually improves with an increase in pH.

Little sensitivity is shown with increased salt or electrolyte concentration, and temperature effects are minimal. Thus, the zinc-phosphonate systems can be used over a wide range of water quality, at high bulk water temperatures (70 to 75 °C, or 160 to 170 °F), and with pH levels up to 9.0.

Degradation of phosphonates to PO_4^{3-} ions by chlorine is significantly reduced with a phosphonate/zinc system, and the latter can be used in chlorine environments because of diminished chlorine demand. The zinc stabilizes the complex, preventing the deterioration of the carbon-phosphorus bond in the oxidizing environment.

Nonheavy-Metal Systems. Many combinations of building blocks are currently being used in nonheavy-metal formulations. Some of the systems in use include combinations of the phosphonates, AMP/HEDP; polyphosphate/phosphonate mixtures, polyphosphate/HEDP; polyphosphate or phosphonate/orthophosphate; and polyphosphate or HEDP/PO_4^{3-}. These systems are synergistic in protecting carbon steel. They combine the many advantages of the individual components, including:

- Complex formation with calcium and numerous polyvalent metal cations required to form the protective film over the metal surface. The polyphosphates still form the colloidal species for cathodic protection, and the phosphonate serves as a cathodic polarizer
- Threshold inhibition of slightly soluble inorganic salts, such as $CaCO_3$ and $CaSO_4$
- Detergent and dispersive action on surface deposits

Most nonheavy-metal treatments for carbon steel function best in more alkaline or elevated-pH systems. Waters are less corrosive but more prone to form scale or deposit. Therefore, good scale control is absolutely necessary. These treatments are generally not as effective as heavy-metal formulations. However, because of the less corrosive nature of the recirculating cooling system, good overall protection can be achieved. More care and attention are required to operate these nonheavy-metal treatments because they are less tolerant of system upsets and function under more contained ranges of water corrosivity, pH, and temperature.

The AMP/HEDP nonheavy-metal inhibitor pair is classified as a cathodic inhibitor. A critical AMP/HEDP ratio of 1.5:1 is needed for optimum carbon steel protection (Ref 26). Performance improves with pH and should be used in systems having a pH of at least 7.5. The combined phosphonate program is not as sensitive to temperature as the individual components, nor is it adversely affected by high levels of Cl^- and SO_4^{2-} ions.

Normal treatment levels of at least 15 mg/L total phosphonate can be used to achieve good protection (<75 μm/yr, or 3 mils/yr). Adequate pretreatment of at least two to three times normal concentration for 1 week is required before reducing the concentration to a maintenance level. With good pH control, normal treatment levels can be reduced and still achieve the desired corrosion protection. Concern over attack on copper-base alloys still exists such that the addition of a specific copper inhibitor is highly recommended.

Polyphosphate/HEDP. The second nonheavy-metal system—polyphosphate/HEDP—is also cathodic in nature while being synergistic in protecting carbon steel. Approximately 40 to 80% by weight of the polyphosphate is needed for good protection of carbon steel. Corrosion control of approximately 50 μm/yr (2 mils/yr) can be achieved with 15 mg/L total phosphate content. The sensitivity of polyphosphate/HEDP to pH is minimal over the range of pH 6 to 8. Temperature sensitivity is minor. Good corrosion control is achieved at bulk water temperatures exceeding 60 °C (140 °F). As with the AMP/HEDP system, sensitivity to water aggressiveness is not a major concern. The antagonistic effects of Cl^- and SO_4^{2-} ions are subdued by this treatment program.

Treatment levels are approximately 15 to 20 mg/L, which can be reduced if the system is pretreated at two to three times maintenance concentration for at least 1 week. Attack on copper-base alloys is reduced compared to the AMP/HEDP system; however, as with the combined phosphonate system, there is a need for a specific copper inhibitor. Concern over reversion still exists, because the polyphosphates are subject to hydrolysis. The PO_4^{3-} ions generated will improve the overall corrosion protection, assuming $Ca_3(PO_4)_2$ precipitation is controlled.

The polyphosphate/PO_4 combination functions in a mixed mode. The cathodic portion comes from either the polyphosphate or phosphonate (HEDP), while the anodic portion derives from the PO_4^{3-} ion. The addition of PO_4^{3-} improves protection to carbon steel through a synergistic interaction with either polyphosphate (Ref 14) or phosphonate. The level of polyphosphate or phosphonate is not critical and can range from 20 to 80%. The corrosion rate of carbon steel can be easily controlled to 25 μm/yr (1 mil/yr) with 15 to 18 mg/L total phosphate.

This system is applicable over a broad pH range of 6.0 to 8.5. There is little sensitivity to pH. The lack of sensitivity to pH refers only to corrosion control, and the inhibitors remain active in solution. The formation of a $Ca_3(PO_4)_2$ precipitate is prevented by the addition of an appropriate additive that inhibits its growth and adherence to the metal surface.

Sensitivity to temperature is also minimal. The inhibitor pair has applicability to very hot systems having bulk water temperatures as high as 70 °C (160 °F). Reversion of the polyphosphate molecules to PO_4^{3-} will be accelerated at the higher temperature, but will also enhance overall protection due to the inhibitive activity of its product (PO_4^{3-} ion).

The system is moderately sensitive to aggressive ions, specifically Cl^-, that can promote pitting attack. Because PO_4^{3-} is an anodic inhibitor, there is a critical concentration that must be maintained that is a function of the electrolyte. Treatments containing polyphosphate will revert to PO_4^{3-}, and a sufficient amount of a $Ca_3(PO_4)_2$ inhibitor should be present to suppress its formation.

Attack on copper-base alloys is not a major concern when polyphosphates are used. However, the phosphonate can be aggressive to copper-base alloys. The addition of a specific copper inhibitor is recommended.

Copper Inhibitors

Most of the steel inhibitors that have been discussed exert some degree of control over corrosion of copper-base alloys. However, system upsets, such as pH excursions and process leaks, can dissolve copper into the cooling water, where it can interfere with steel protection. Concentrations of at least 0.1 mg/L can deposit on steel conduits, accelerating localized attack. Ancillary inhibitors are available that are very effective in controlling corrosion of copper-base alloys and preventing galvanic deposition or dissolved copper onto ferrous metals.

Three specific inhibitors have been extensively used: mercaptobenzothiazole (MBT), benzotriazole, and tolyltriazole. Mercaptobenzothiazole is an extremely effective inhibitor for copper-base alloys. Its inhibitive properties are attributed to formation of an adherent protective film on the metal oxide surface. The inhibitor reacts with the metal surface to form a chemisorbed barrier. The initial corrosion product of Cu^+ ions reacts with the MBT molecule to form a 3-dimensional complex of Cu(I)MBT. Rate of film growth is rapid, reaching a self-limiting thickness within a short period of time.

Mercaptobenzothiazole is susceptible to oxidizing agents, such as air, chlorine, and ultraviolet light from sunlight, that can degrade the molecule into a disulfide having no inhibitory properties. This degradation occurs only to the molecule in solution, not the complex on the metal surface. Because of the sensitivity of MBT toward chlorine, the inhibitor should be used only if no chlorine residual is present. Thus, MBT should not be added to open recirculating cooling water systems, in which continuous chlorination is often practiced.

Cupric ions can be effectively inhibited from cathodically depositing onto more active metals, such as steel and aluminum, by reacting with MBT in a 1:2 molar ratio. In the absence of complicating factors, such as copper concentrations exceeding 0.1 mg/L in the cooling water or a chlorine residual, the concentration of MBT required for good copper inhibition is approximately 4 mg/L. A feeding procedure based on 3 mg/L, followed by 1 mg/L addition 12 h later, has been very successful.

Benzotriazole and tolyltriazole function similarly in controlling copper corrosion. Structurally, tolyltriazole differs from benzotriazole in that the former has a methyl group attached to the benzene ring. Both materials behave similarly in cooling waters, reacting with the metal surface to produce a three-dimensional chemisorbed layer.

The azoles are classified as cathodic inhibitors because they adsorb at cathodic sites and interfere with the oxygen reduction reaction. These molecules are more resistant to oxidation than MBT. In the presence of chlorine, 1-chlorobenzotriazole (or 1-chlorotolyltriazole) is probably formed, having minimal inhibitory properties. Upon dissipation of residual chlorine, the chlorocompound reverts back to the active azole. Chlorination does not affect the integrity of the complex on the metal surface.

The azoles can deactivate dissolved copper in cooling water systems and prevent its deposition onto steel or other active metals. A 2:1 molar ratio of inhibitor to copper ions is required, similar to the MBT reaction.

The concentration of azoles necessary for copper inhibition is approximately 1 to 2 mg/L in the absence of any residual chlorine. The optimum practice is to pretreat the system with two to three times normal concentration for 1 to 2 days, then control at a maintenance dosage. Under chlorination conditions, the azole should be fed after the dissipation of residual chlorine. If continuous chlorination is practiced and cannot be changed to intermittent practice, the level of treatment should be increased to 4 to 5 mg/L. In the presence of continuous chlorination, overall copper corrosion protection will be less than with intermittent chlorination.

REFERENCES

1. M.F. Obrecht and M. Pourbaix, *J. Am. Water Works Assoc.*, Vol 59, 1967, p 977
2. M.G. Fontana and D.N. Green, *Corrosion Engineering*, 2nd ed., McGraw-Hill, 1978, p 28-113
3. V.G. Ereneta, *Corros. Sci.*, Vol 19, 1979, p 507
4. B.P. Boffardi and G.W. Schweitzer, Water Quality, Corrosion Control and Monitoring, in *The Ninth International Congress on Metallic Corrosion*, National Research Council of Canada, June 1984
5. G.T. Burnstein and D.M. Davis, *Corros. Sci.*, Vol 20, 1980, p 1143
6. J.W. Ryznar, *J. Am. Water Works Assoc.*, Vol 36 (No. 4), 1944, p 472
7. W.F. Langelier, *J. Am. Water Works Assoc.*, Vol 28, 1936, p 1500
8. M.L. Riehl, *Water Supply and Treatment*, 9th ed., National Lime Association, 1976
9. R.H. Ashcraft, "Scale Inhibition Under Harsh Conditions by 2-Phosphonobutane-1,2,4 tricarboxylic Acid (PBTC)," Paper 123, presented at Corrosion/85, Boston, MA, National Association of Corrosion Engineers, March 1985
10. M.J. Smith and P. Miles, U.S. Patent 4,046,707
11. E.P. Partridge and A.H. White, *J. Am. Chem. Soc.*, Vol 51, 1929, p 360
12. P.H. Ralston, *Mater. Perform.*, Vol 11 (No. 6), 1972, p 39
13. I.T. Godlewski, J.J. Schuck, and B.L. Libutti, U.S. Patent 4,029,577
14. B.P. Boffardi and G.W. Schweitzer, "Advances in the Chemistry of Alkaline Cooling Water Treatment," Paper 132, presented at Corrosion/85, Boston, MA, National Association of Corrosion Engineers, March 1985
15. J.F. Gary *et al.*, Ed., *Condenser Biofouling Control*, Ann Arbor Science Publications, 1980, p 49-76
16. J.W. McCoy, *Microbiology of Cooling Water*, Chemical Publishing, 1980
17. R.G. Eagar, A.B. Theis, M.H. Turakhia, and W.G. Characklis, "Glularaldehyde: Impact on Corrosion Causing Biofilms," Paper 125, presented at Corrosion/86, Houston, TX, National Association of Corrosion Engineers, March 1986
18. M. Cohen and A.F. Beck, *J. Electrochem. Soc.*, Vol 62, 1958, p 969
19. G.B. Hatch, "Influence of Inhibitors on Differential Aeration Attack on Steel II—Dichromate and Orthophosphate," Paper presented at Corrosion/64, National Association of Corrosion Engineers, March 1964
20. M.S. Vakasovich and J.P.G. Farr, *Mater. Perform.*, Vol 25 (No. 5), 1986, p 9
21. M. Cohen and M.J. Pryor, *J. Electrochem. Soc.*, Vol 98, 1951, p 263
22. B.G. Hatch, *Ind. Eng. Chem.*, Vol 44 (No. 8), 1952, p 1780
23. G.B. Hatch, *Mater. Prot.*, Vol 4 (No. 7), 1965, p 52
24. G.B. Hatch and P.H. Ralston, *Mater. Prot.*, Vol 3 (No. 8), 1964, p 35
25. G.B. Hatch and P.H. Ralston, *Mater. Prot. Perform.*, Vol 11 (No. 1), 1972, p 39
26. B.P. Boffardi, U.S. Patent 4,206,075

Surface Modification

R.D. Granata, Lehigh University
P.G. Moore, Naval Research Laboratory

SURFACE MODIFICATION is the alteration of surface composition or structure by the use of energy or particle beams. Elements may be added to influence the surface characteristics of the substrate by the formation of alloys, metastable alloys or phases, or amorphous layers. Surface-modified layers are distinguished from conversion or coating layers by their greater similarity to metallurgical alloying versus chemically reacted, adhered, or physically bonded layers. However, surface structures are produced that differ significantly from those obtained by conventional metallurgical processes. This latter characteristic further distinguishes surface modification from other conventional processes, such as amalgamation or thermal diffusion. Two different modification methods will be discussed. The first, ion implantation, is the introduction of ionized species (usually elements, for example, Ti^+) into the substrate using kilovolt to megavolt ion accelerating potentials. The second method, laser processing, is high-power laser melting with or without mixing of materials precoated on the substrate, followed by rapid melt quenching.

The advantages of the surface modification approach to promoting corrosion resistance include (Ref 1):

- Alteration of the surface without sacrifice of bulk properties
- Conservation of scarce, critical, or expensive alloying elements
- Production of novel surface alloys (unattainable by conventional metallurgical techniques) with superior properties
- Avoidance of coating adhesion problems

The disadvantages of the surface modification techniques include:

- Processing materials having deep or hidden contours
- Cost of processing equipment
- Substrate sensitivity to high energy input (for example, thermal instability)
- The thinness (susceptibility to damage) of the coatings

Ion implantation and laser processing will be discussed separately in the following sections.

Ion Implantation

Surface modification by ion implantation is a technique that was derived from the semiconductor industry. Specimens for corrosion research were initially prepared by using high-energy research instrumentation or commercial semiconductor implanters. Equipment for surface modification of metal parts by batch processing has been built, and production capabilities of the equipment are being evaluated. Ion implantation is reaching a stage of development in which specific advantages are being recognized and exploited for enhancing the corrosion and wear resistance of critical metal parts (Ref 2). For wear resistance, ion implantation has been used for the hardening and friction reduction of metal surfaces (Ref 3). Related techniques are also used in conjunction with ion implantation to increase the ratio of material introduced into the substrate per unit area, to provide appropriate mixtures of materials, or to overcome other difficulties involved in surface modification by ion implantation alone. These techniques include:

- *Ion beam sputtering*: An ion beam of argon or xenon directed at a target sputters material from the target to a substrate; the sputtered material arrives at the substrate with enough energy to promote good adhesion of the coating to substrate
- *Ion beam mixing*: Deposited layers (electroplating, sputtering) tens or hundreds of nanometers thick are mixed and bonded to the substrate by an argon or xenon ion beam
- *Plasma ion deposition*: Ion beams are used to create coatings having special phases, especially ion beam formed carbon coatings in the diamond phase or ion beam formed boron nitride coatings
- *Ion beam assisted deposition*: Ion beams are combined with physical vapor deposition

Materials Processing and Product Characteristics. The ion implantation technique consists of introducing the target (substrate material) into the implanter vacuum chamber, vaporizing and ionizing the implant species, and electrostatically accelerating the ionized species into the target. Figure 1 shows schematics of research and semiconductor production ion implantation devices. A production-type implanter for materials surface modification that has recently come on-line is shown in Fig. 2.

The implant species may be most any element that can be vaporized and ionized in a vacuum chamber. The material used to generate the implant species is usually in the form of a chloride salt of a metal, a gaseous compound, or the pure element. Solid materials are heated to provide an adequate vapor pressure of the implant species. The vaporized material is ionized and accelerated toward the target. Magnetic fields separate various mass/charge ratios and enable the selection of specific implant species. A selected species impacts the target material and is implanted within, producing a surface-modified substrate. The depth to which the species is implanted depends on the accelerating voltage and the atomic numbers of the implant versus target species (Ref 4). Commonly used acceleration voltages are 20 to 200 keV. The typical average implantation depth (range) is 10 to 100 nm. The distribution of implanted species versus range is roughly Gaussian.

Figure 3 shows a diagram illustrating calculated concentration profiles of cobalt, aluminum, and boron implanted in iron at 50 keV to a dose of 10^{17} ion/cm^2 (6.45 × 10^{17} ion/in.^{2}). Implantation causes removal (sputtering) of the surface layers, but the thickness of the sputtered layer is generally less than the implantation thickness. Sputtering rates of elements differ and depend on such factors as substrate composition, chemical bonding, metallurgical structure, beam energy, and angle of the incident beam. In some cases, sputtering rate becomes equal to the implantation rate; an equilibrium is achieved (saturation), and no net addition of ions to the substrate occurs. Maximum concentrations of implanted species may be as high as 50 at.%.

The advantages of the ion implantation surface modification process include (Ref 1):

- No sacrifice of bulk properties
- Solid-solubility limit can be exceeded
- Alloy preparation is independent of diffusion constants
- No coating adhesion problems, because there is no interface
- No significant change in sample dimensions
- Depth concentration distribution is controllable
- Composition can be changed without affecting grain sizes
- Precise location of implanted area(s)
- Clean vacuum process
- Allows screening of the effects of changes in alloy composition

The implantation process can be viewed as a means of changing the material properties of an item only at the surface. An item can be fabricated from a material having desirable bulk properties (machinable, low-cost) and can be modified to provide the requisite surface properties (corrosion resistance, wearability). However, the disadvantages of the process must be kept in mind:

- The ion beam cannot process deep or hidden contours (±30° incidence)
- Processing equipment may be too costly
- The substrate may be excessively heated by high energy input; for example, the typical energy flux of 2 W/cm^2 (13 W/in.^{2}) can heat a substrate 500 °C (930 °F)

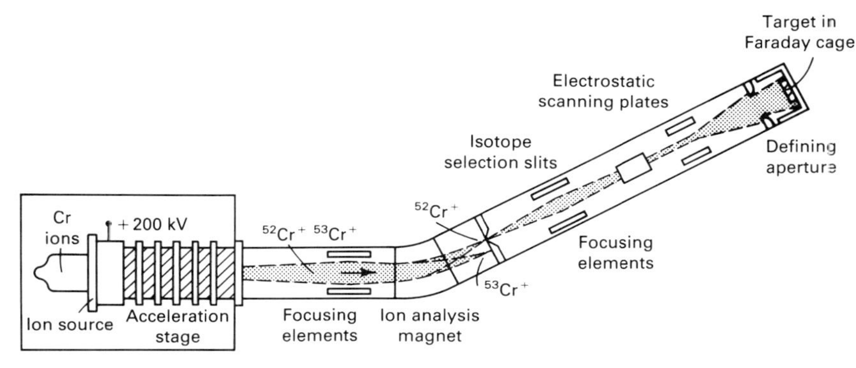

(a)

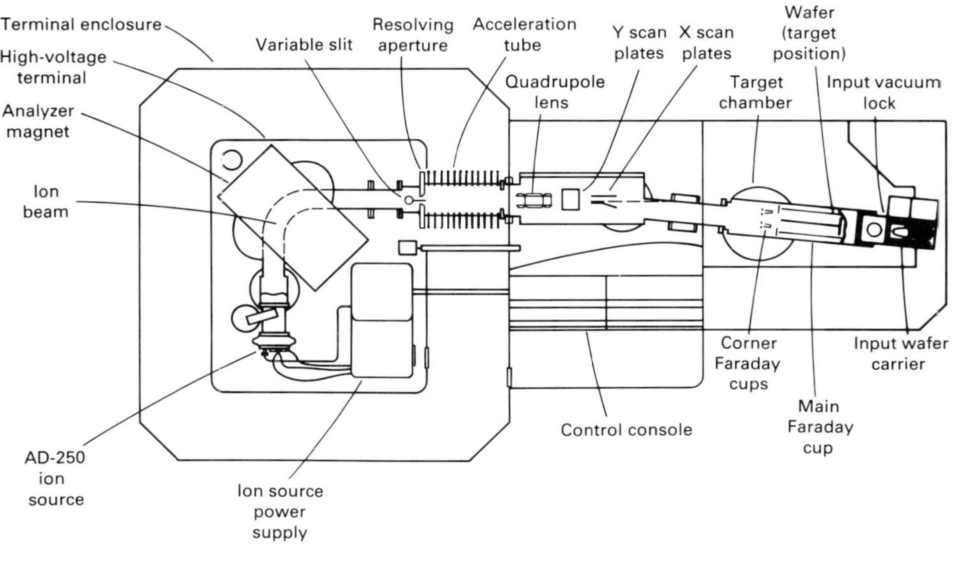

(b)

Fig. 1 Schematics of research-type ion implantation system (a) and a production-type semiconductor implanter (b)

Fig. 2 Commercial ion implanter for materials modification. Courtesy of Spire Corporation

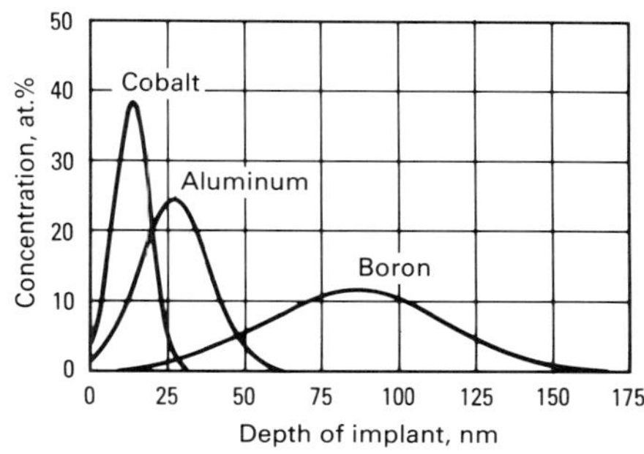

Fig. 3 Concentration profiles for cobalt, aluminum, and boron implants in iron approximated using methods described in Ref 4. Energy: 50 keV. Dose: 10^{17} ion/cm^2 (6.45×10^{17} ion/in.^{2})

- Inclusions in substrate material may bridge the depth of the surface-modified region, producing sites with undesirable properties
- Treatment depth may be inadequate for some applications, resulting in galvanic effects with localized corrosion or in implant layer removal by abrasion or erosion

Problems associated with thermal effects and sample handling have been discussed and technical solutions proposed (Ref 4-6). For example, nitrogen-implanted surfaces must be kept below 200 °C (390 °F), the temperature at which outward diffusion of nitrogen occurs. Many sample types have been implanted while keeping their surface temperatures below 100 °C (212 °F) by using heat sinks and by modifying processing conditions.

Dry Corrosion Properties. Dry corrosion or thermal oxidation in the absence of a liquid phase is primarily governed by the transport processes in the oxide and the ability of the oxide to resist spalling and breakdown or abrasion and erosion. The ion implantation process has provided a means for studying and modifying the surface layers active in dry corrosion. Interesting results have been obtained with implanted materials, as in the following five examples.

First, in one study, the effects of the implanted material were found to persist to a much greater depth than the thickness of the implanted layer (Ref 7). Oxide layer growths of more than 100 times the implant depth showed the persistence of the implanted element effects. An Ni-20Cr alloy developed a complete, protective Cr_2O_3 scale more rapidly when implanted with cerium, yttrium, calcium, or aluminum, but implanted zirconium or chromium had no effect. Scale adherence was increased by ion implantation with cerium, yttrium, calcium, or zirconium. For yttrium-implanted Ni-20Cr, the weight change was approximately 25% that of unimplanted material after 24 h at 1000 °C (1830 °F) in 101-kPa (1-atm) oxygen.

Second, a stainless steel implanted with 10^{17} ion/cm^2 (6.45×10^{17} ion/in.^{2}) of yttrium gave no observable spalling of the oxide after over 6500 h versus unimplanted material, which spalled more than 2 mg/cm^2 (Ref 8). Third, implantation of aluminum into an Fe-Cr-1.4Al alloy increased the surface aluminum concentration to 10% and promoted the formation of a thick, protective, pure Al_2O_3 scale (Ref 9).

Fourth, high doses of silicon ion implanted into low silicon containing substrates did not provide the desired corrosion resistance associated with their conventional high-silicon counterpart materials. However, ion beam mixing of deposited silicon films with argon at 500 °C (930 °F) produced silicon-rich films on iron or titanium that are thicker than those obtained by ion implantation alone (Ref 10). These thick silicon layers should provide enhanced corrosion resistance. Attempts are being made to produce highly oxidation-resistant $MoSi_2$ by ion beam mixing of molybdenum and silicon layers (Ref 11). Lastly, zirconium was implanted with such elements as iron, cobalt, or nickel to help relieve internal stresses within the oxide film that result in the break-away oxidation associated with protective oxide fracture (Ref 12).

Wet Corrosion Properties. There are two approaches to using ion implantation for increasing the corrosion resistance of substrates; the first is to enhance passivation characteristics, and the second is to create novel materials. In the first approach, implantation creates a surface composition based on conventional metallurgical experience that reacts to form a passive layer. In the second approach, implantation creates a surface composition and a structure that are difficult, impossible, or impractical to achieve by conventional metallurgy.

The two main problems in applying ion implantation methods for creating corrosion-resistant surfaces are producing a surface layer that has a corrosion rate low enough to prevent penetration to the base material during the designed lifetime of the item and producing a surface that is self-repairing. The first problem is approached by introducing elements that enhance corrosion resistance, as in conventional metallurgy—for example, chromium (Ref 13) or molybdenum (Ref 14) in ferrous metals. Another approach is to create combinations of elements in the surface layers that cannot be produced by conventional metallurgical methods—for example, tantalum in iron (Ref 15) or phosphorus in iron-chromium alloys (Ref 16). Tantalum forms intermetallic compounds with iron, but conventional alloys with sufficiently high tantalum content are not readily produced. Other examples are chemically resistant amorphous layers produced, for example, on iron-chromium implanted with phosphorus, carbon, silicon, or boron. These materials passivate spontaneously in hot 25% hydrochloric acid (HCl) (Ref 16).

The second problem is approached by the introduction of elements that are not soluble in or removed by the environment, for example, palladium in titanium. Titanium passivation is strongly promoted by palladium and can be alloyed with it to provide lasting self-passivation properties in acidic media. However, palladium is required only at the interface as a catalyst for the hydrogen reduction reaction and is not readily soluble. Thus, it may be more cost-effective to modify the surface by implantation than to use palladium in the bulk alloy. The palladium-assisted passivation of titanium is a persistent effect with research specimens (Ref 17).

Substantial efforts have been made in the search for effective applications of ion implantation for enhancing corrosion resistance (Ref 18-24). The greatest difficulty lies in the thinness of the implanted layer; corrosion-resistant layers must not be entirely consumed or damaged. Many research efforts have attempted to mimic conventional alloy compositions. However, the corrosion rates of conventional alloy systems are usually too high relative to the thinness of the implanted layer. The most successful approaches have involved detailed knowledge of a specific material and environmental corrosion conditions or have created unconventional materials. The palladium-implanted titanium, tantalum-implanted iron, and phosphorus or boron implanted in iron-chromium systems described above are good examples. Another example of adapting ion implantation to a specific application is its use for interfacial reactions. Implantation of aluminum or titanium in 1010 steel significantly decreased the cathodic delamination rates of organic coatings from the implanted substrates (Ref 25).

Wear resistance enhancement by ion implantation is included in the discussion of surface-modified corrosion-resistant surfaces because erosion, cavitation, and fretting corrosion are important forms of corrosion. Ion implantation has been shown to have a major effect on wear resistance properties. The combined effects of ion implantation on corrosion and wear resistance are particularly well suited in applications in which bearings or sliding surfaces are subjected to intermittent use or experience long periods of storage.

Wear resistance is a property that has shown significant improvement for some substrates after surface modification by ion implantation. Activity in this area has been increasing at research laboratories, and specialized equipment has been developed for commercialization. Some of the work on the development of wear-resistant ion-implanted materials is focusing on quantification and lifetime predictions.

Adhesion and abrasion are the most specific wear mechanisms, with corrosive and surface fracture related mechanisms making contributions in certain cases. In adhesive wear, surfaces under sufficient local pressure deform plastically and weld at the contact points. When the surfaces separate, the contact point welds break, resulting in the formation of debris and in surface roughening. The oxidized debris causes most of the subsequent wear. Abrasive wear is simply the cutting of one surface by another edged or pointed surface. Abrasive wear decreases as the hardness of the cut surface approaches that of the cutting surface. Abrasion is dramatically lower when both material hardnesses are equal. Adhesive and abrasive wear as well as tribomechanical properties for ion-implanted materials are discussed in Ref 26 and 27, respectively.

There are many examples of ion implantation studies for increasing the wear resistance of materials. Many studies have used nitrogen-implanted steel because of the readily available high-current nitrogen beams and because direct comparison to thermal nitriding processes is possible. Increases of the order of two to three times in the microhardness of steels due to nitrogen implantation have been reported (Ref 28). However, large reductions in wear rate often result from relatively small increases in microhardness. Extrusion dies for TiO_2-pigmented plastics fabricated from low-carbon steel often wear rapidly, yet nitrogen implantation increases the die lifetime by up to ten times. This modification suggests that the microhardness of the steel has been raised above that of TiO_2 (Ref 26). Extended tool life has been demonstrated for plastic molding machines with estimated ion implantation costs as low as 10% of the ordinary part costs (Ref 3). Nitrogen implantation of a titanium alloy (Ti-6Al-4V) has been found to decrease the wear rate by more than two orders of magnitude (Ref 29). Data suggest that wear rates have been reduced by almost three orders of magnitude in a titanium alloy designed for hip joint prostheses (Ref 26).

Wear reduction can be attributed to factors other than microhardness. The residual carbon films introduced during implantation of other elements may decrease the occurrence of adhesive wear by providing lower friction coefficients. Implantation of titanium or tantalum usually results in the coimplantation or inward diffusion of carbonaceous matter present in the vacuum system. In the case of titanium-implanted steel, the wear reduction has been attributed to the formation of an amorphous iron-titanium-carbon film. Other systems are being studied—for example, tin-implanted iron, yttrium, and nitrogen in chromium steels and nitrogen implantation in cobalt-cemented carbide. It has been suggested that deep-penetrating interstitially active elements, such as nitrogen, be used to ion beam mix thin coatings of oxide-modifying materials into substrates for synergistic wear reduction (Ref 26). However, another opinion is that implanted ions do not migrate below the surface. Four mechanisms have been identified as contributing to the reduction of sliding wear of ion implanted metals (Ref 27):

- Production of low-friction surfaces by altering surface chemistry
- Modification of microstructure to harden the surface
- Stabilization of microstructure against deleterious work-hardening effects
- Introduction of residual (compressive) stresses to combat fracture

Impact wear with and without sliding wear was also studied for various materials implanted with nitrogen (Ref 30). The materials used in this study were:

- *Soft*: aluminum alloy 2024, aluminum alloy 7075, and oxygen-free high-conductivity copper (CDA C10100)
- *Medium*: cartridge brass (CDA C26000) Cu-10Sn, aluminum bronze (Cu-18Al), and sintered iron (Fe-5Cu-1C)
- *Hard*: AISI 1018 carbon steel and AISI type 304 stainless steel

Impact wear improved significantly, but the combination of impact and sliding wear showed only a very small effect. Interestingly, one of the materials, cartridge brass, showed no significant improvement.

Critical Evaluation of Product Utility. Numerous examples have been given to define application areas that may benefit from the use of implantation techniques for materials processing. The advantages and disadvantages relate to the small dimension of implantation depth (nanometers). Significant niches are being created for materials processed by ion implantation and related techniques. These surface modification techniques for industrial applications are perhaps best appraised by considering the information presented in Table 1.

Table 1 Successful applications of ion implantation for industrial exploitation

Application	Material	Typical results
Orthopedic implants (artificial hip and knee joints)	Ti-6Al-4V	Significant (400 times) lifetime increase in laboratory tests
Bearings (precision bearings for aircraft)	M50 tool steel, 440C stainless steel, or 52100 bearing steel	Improved protection against corrosion, sliding wear, and rolling-contact fatigue
Extrusion tools (spinnerettes, nozzles, and dies)	Various alloys	Four to six times normal performance
Punching and stamping tools (pellet punches for nuclear fuel, scoring dies for cans)	D2 tool steel	Three to five times normal life and better end products

Source: Ref 2

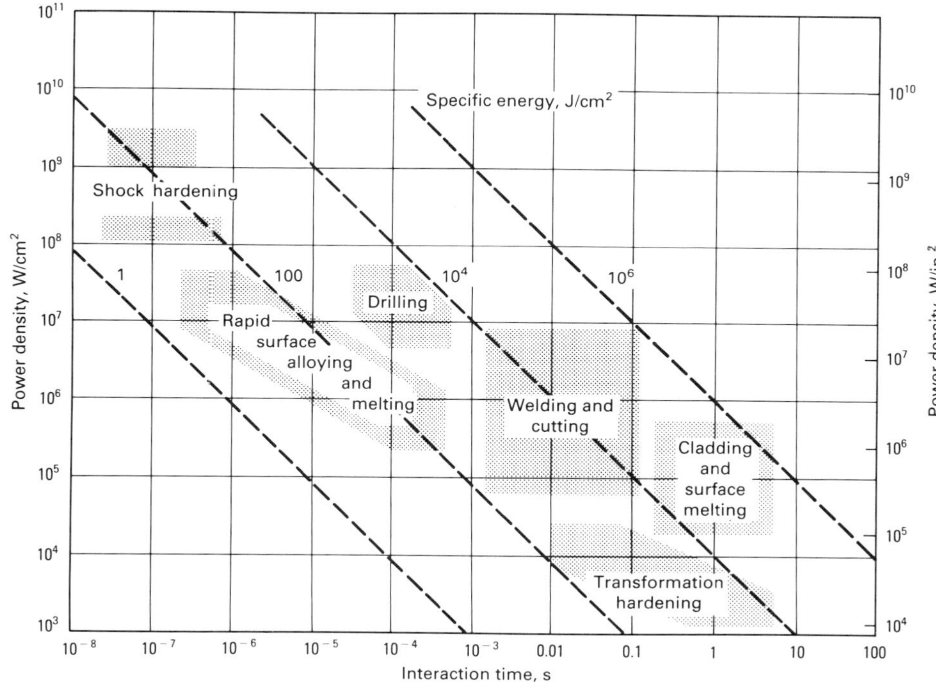

Fig. 4 Interaction times and power densities necessary for various laser surface modification processes

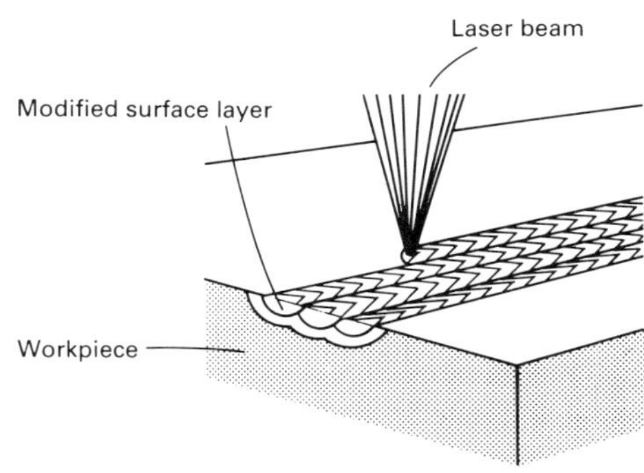

Fig. 5 Schematic of the procedure for the laser processing of metals

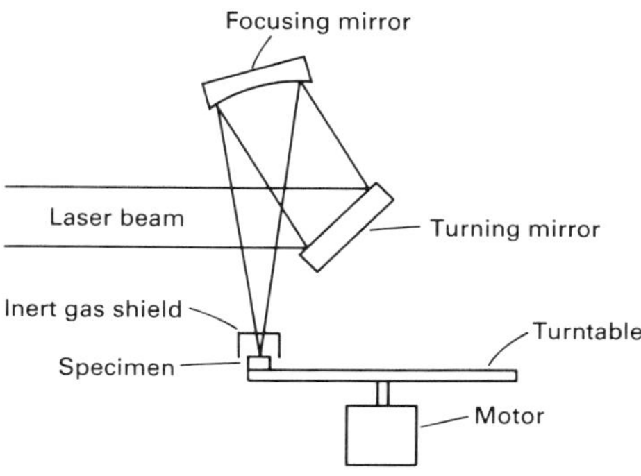

Fig. 6 Schematic of an apparatus for laser processing

Laser Surface Processing

Lasers with continuous outputs of 0.5 to 10 kW can be used to modify the metallurgical structure of a surface and to tailor the surface properties without adversely affecting the bulk properties. The surface modification can take the following three forms. The first is transformation hardening, in which a surface is heated so that thermal diffusion and solid-state transformations can take place (Ref 31-33). The second is surface melting, which results in a refinement of the structure due to the rapid quenching from the melt (Ref 32-37). The third is surface alloying, in which alloying elements are added to the melt pool to change the composition of the surface (Ref 38-44). The novel structures produced by laser surface melting and alloying can exhibit improved electrochemical behavior.

Materials Processing

The processing parameters necessary for laser surface processing depend on the type of modification to be done (for example, transformation hardening or surface alloying) and the properties of the material that are being treated (for example, thermal diffusivity, heat capacity, and transformation temperatures). Figure 4 shows typical ranges of conditions for various processes. The laser power, power density, and interaction time are the primary variables. Other variables, such as the composition of the atmosphere during treatment or the rate of material addition, are determined by the details of the processing—for example, the necessity of shielding against oxidation and the desired thickness, composition, and structure of the surface layer.

The laser beam modified layer can range in thickness from 0.01 to 5 mm (0.4 to 200 mils), depending on the processing variables, although thicknesses of 0.05 to 1 mm (2 to 40 mils) are more common. The longer the interaction time of the laser beam with the material, the deeper the processed layer will be. Of the processes shown in Fig. 4, the areas labeled "Cladding and surface melting" and "Transformation hardening" delineate process parameters that typically affect the material to depths from 0.5 to 5 mm (20 to 200 mils) and result in metallurgical structures similar to welded structures. The parameters designated "Rapid surface alloying and melting" affect a surface layer only 0.02 to 0.6 mm (0.8 to 24 mils) thick, but result in quench rates to 10^7 K/s and therefore allow for the production of novel metallurgical structures and alloys.

The procedure for laser beam surface modification is shown in Fig. 5. A high-power laser beam is brought to a focus near the surface of the workpiece. After a melt pool is established, the laser beam can be rastered across the surface, melting a strip of material with each sweep, until the entire surface has been processed. The processing is performed in an inert gas atmosphere or vacuum to protect the metal from oxidation. In practice, it is more convenient to move the workpiece rather than the laser beam.

Figure 6 shows a schematic of an apparatus for processing laboratory specimens. The specimen is passed through the laser beam by the rotation of the turntable, and the entire apparatus is translated a fraction of a melt width between passes to process a larger area. The process is varied by changing the power level of the laser beam, the power density (or spot diameter), or the sweep speed. The effects of these variables on the depth of melting for three materials are shown in Fig. 7. As can be seen, the melt depth can be varied by an order of magnitude simply by changing the sweep speed. The width of the melt pass is determined primarily by the diameter of the laser spot and the thermal properties of the material and is very slightly dependent on the sweep speed. The rate at which a surface can be processed is typically about 1 cm^2/s (0.16 $in.^2/s$) for a variety of conditions and materials.

Product Characteristics

Solidification Effects. The heat flow characteristics of such processing can reasonably be described by one-dimensional heat flow in which the rate at which energy enters the surface is balanced by the rate at which heat is conducted away by the cool substrate (Ref 45). A number of studies have been devoted to characterizing the temperature profiles that result from the laser and electron beam processing of materials (Ref 31, 33, 46). The character of the temperature profiles experienced by material at various depths below the surface of the metal is shown in Fig. 8. The three reference temperatures indicated represent the liquidus and solidus as well as a solid-state transformation temperature. As the laser beam sweeps across the surface of a metal sample, the surface begins to heat. When the melting temperature of the material is exceeded, a melt front is established and moves from the free surface of the sample toward the bulk. Shortly after the beam is turned off (or has moved away), the melt front slows and then stops. It then moves back toward the free surface as a solidification front. The unique characteristics of this solidification front are the high cooling rates (such as 10^7 K/s) and the steep temperature gradient (such as 10^5 K/cm). These characteristics promote the formation of metastable compositions and structures.

Because of this steep temperature gradient, there is only a thin layer of metal that is cool enough to solidify. The material above this layer

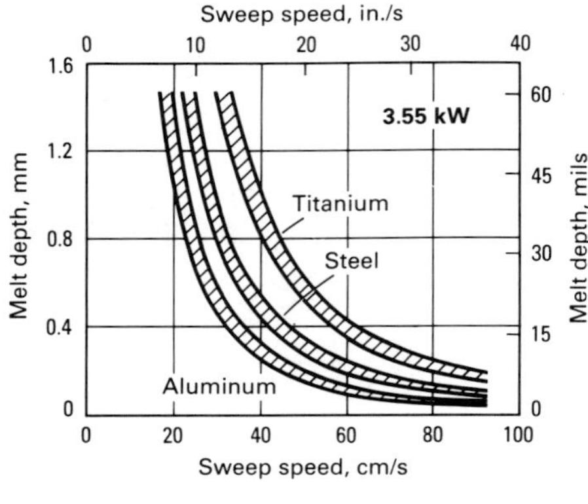

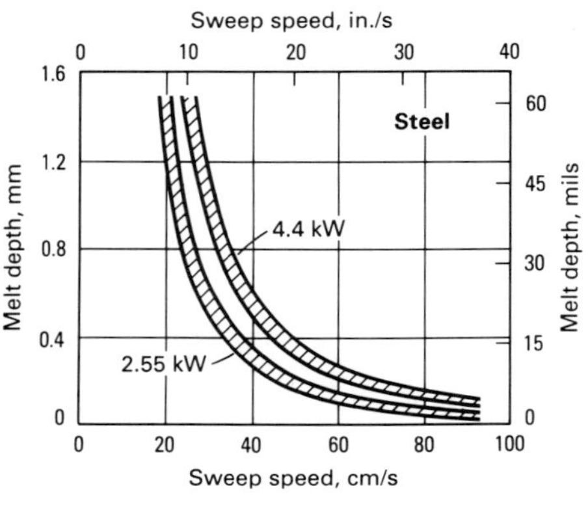

Fig. 7 Effects of laser-processing conditions on melt depth. (a) For three materials at one power level. (b) For steel at two power levels

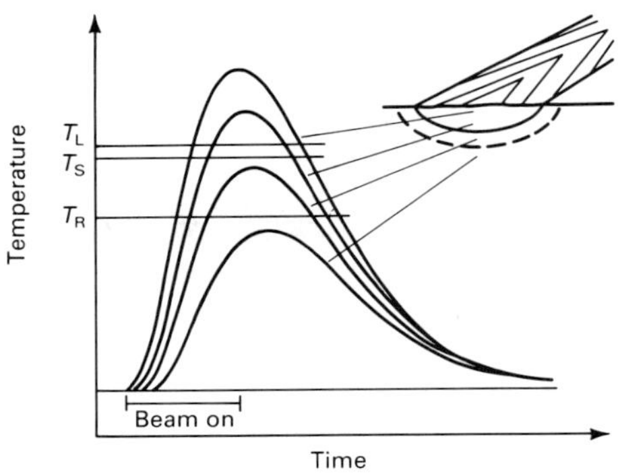

Fig. 8 View of laser-melted surface and temperature profiles experienced at different points on the surface during laser melting. Liquidus (T_L), solidus (T_S), and solid-state transformation (T_R) temperatures are indicated.

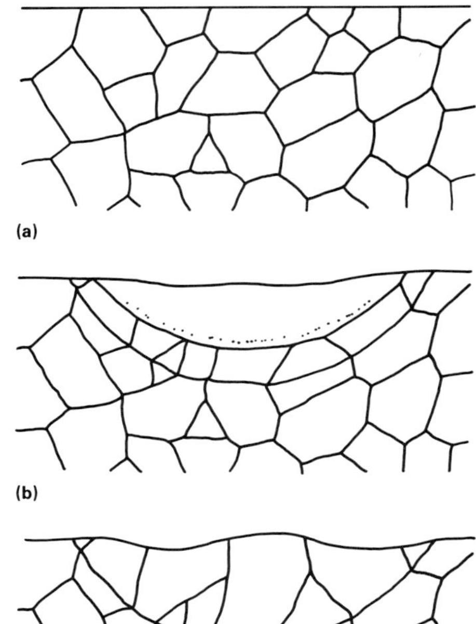

Fig. 9 Origins of epitaxial resolidification as a result of laser processing. (a) Before laser melting. (b) During resolidification. (c) After quenching

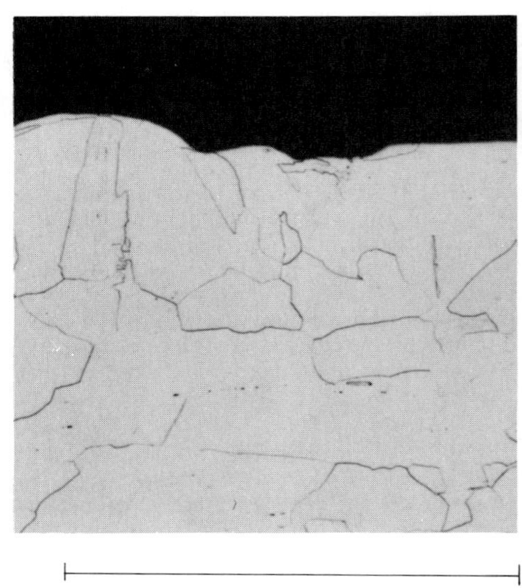

Fig. 10 Planar resolidification of a β-titanium alloy

is above the liquidus temperature, and the material below this layer has already solidified. There is a strong epitaxial character to the resolidification as a result of the steep temperature gradients (Ref 47, 48). This is illustrated in Fig. 9. The two effects that compete during the solidification are the regrowth of grains from the substrate and the heterogeneous nucleation and growth of new grains in this thin layer of chilled metal. For most alloys, grains from the substrate will grow into the chilled zone, and any grains that may have nucleated within the melt pool will either redissolve or be transported by the melt front to the free surface. As an example of this, Fig. 10 shows a β-titanium alloy that has been laser processed.

The epitaxial solidification can have a striking effect on the grain size of the surface layer and on the crystallographic texture. Because some crystal planes grow faster than others—for example, (100) for cubic crystals and (1010) for hexagonal close-packed crystals—those substrate grains having a large component of these orientations in the direction of the temperature gradient will grow at the expense of less favorably oriented grains. As these grains grow, they will tend to squeeze out less favorably oriented grains. This affects the crystallographic texture in two ways. First, there is a higher probability for grains at the free surface to be oriented with the fast-growing direction normal to the free surface, and second, low-angle grain boundaries will occur with a higher frequency than in the substrate. As shown in Fig. 11, when the grain size of the metal before processing is small compared to the depth of melting, there is a coarsening of the grain size, and the texture effects are more pronounced. Examples of these effects are shown in Fig. 9 for a large-grain material and in Fig. 12 for a fine-grain material.

During the solidification process, there is generally a redistribution of solute atoms away from the solidification front so that the interface can remain as close to equilibrium as possible (Ref 49). After an initial transient, this partitioning results in the geometrical destabilization of the solidification front, and dendritic solidification results. For a single-crystal alloy that undergoes rapid surface melting, the dendritic character of the resulting melt is shown in Fig. 13. This effect was documented by the laser melting of single crystals of a nickel-base superalloy (Ref 47). The high-temperature gradient affects the character of the dendritic growth in two ways. First, it restrains crystal growth to the directions of the fast-growing crystal planes as described above, and second, it minimizes the amount of segregation because of the short times allowed for lateral diffusion. An example of the dendritic character of laser-processed material is shown in Fig. 14. The diameter of the dendrites can be used to estimate the cooling rates experienced by the material as it resolidified. Higher cooling rates result in less lateral diffusion and smaller dendrites, while lower cooling rates result in more lateral diffusion and larger diameters.

To the extent that equilibrium cannot be maintained at the solidification front by partitioning as described in the previous paragraph, more than the equilibrium amount of solute atoms will be incorporated into the solidification structure. An increase in the cooling or solidification rates results in still greater deviations from the equilibrium concentrations (Ref 50, 51). In systems that exhibit solubility at all compositions, for example, the aluminum-manganese system, the solubility limits can be extended. Deviations from equilibrium can also result in the formation of metastable crystal structures when the system has several phases with comparable free energies of formation (Ref 52, 53). Amorphous or glassy surface layers can be produced by laser processing only when the system has several energetically and structurally similar crystalline phases; thus, atoms attaching to the solidification front will have a variety of closely spaced lattice sites with similar energies to choose from. This confuses the long-range order and destroys the crystalline character of the solid.

Changing the Chemical Composition. In addition to the structural modifications described in the previous section, the composition of the surface layer can also be changed to form surface alloys whose compositions can be tailored to suit surface requirements. Changes in composition

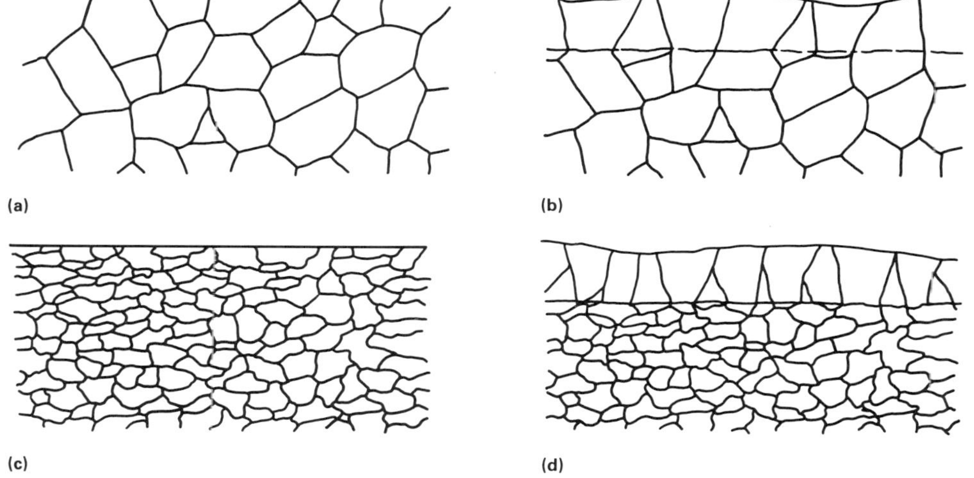

Fig. 11 Effect of epitaxial resolidification on grain size for a material with large original grain size (a and b) and a material with small original grain size (c and d) compared to the melt depth. (a) and (c) Before processing. (b) and (d) After processing

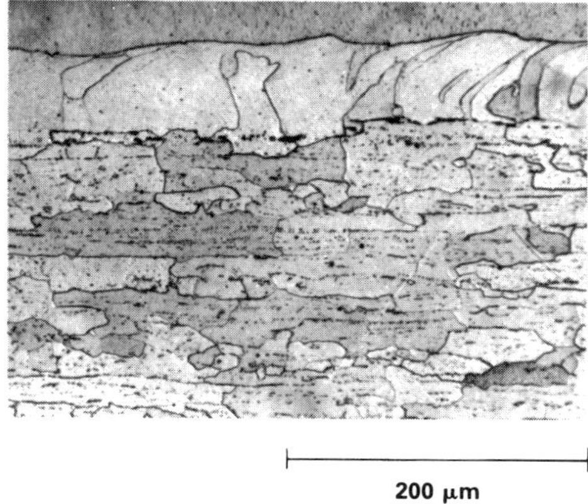

Fig. 12 Coarsening of grain size in a ferritic stainless steel by laser melting

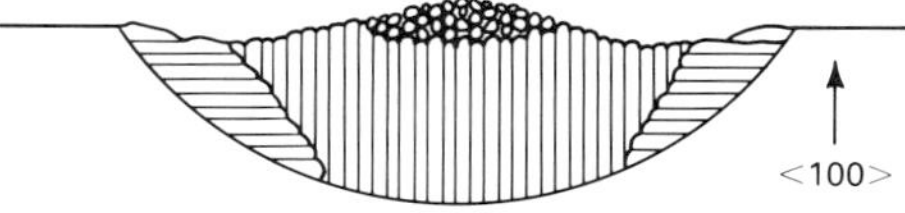

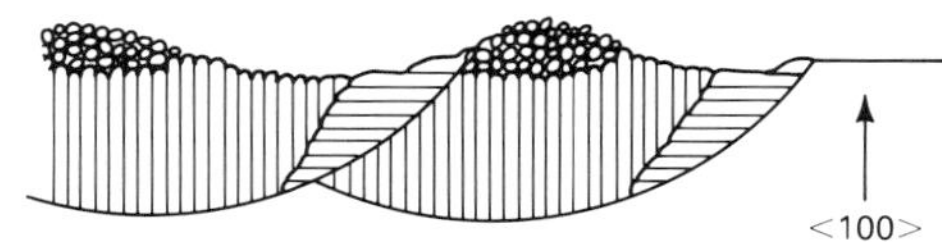

Fig. 13 Dendritic solidification of cubic single crystals. The high-temperature gradient resulting from laser melting causes dendrites to grow in the preferred ⟨100⟩ direction.

have been accomplished with several techniques (Ref 31, 39, 41, 42). For example, the specimen can be coated before laser processing with a thin layer of the desired alloying element or elements. Such coatings have been applied by vacuum deposition, electroplating, spraying, or the use of a slurry. The alloying elements can also be introduced directly into the melt pool during processing in the form of wire or as an injected powder. When the composition is changed, care must be taken to ensure that the alloying elements are distributed uniformly in the surface layer. Three mechanisms have been identified for facilitating the mixing of components: liquid-state diffusion, convection currents within the melt pool, and agitation of the melt pool by a laser-sustained plasma (Ref 42). A uniform composition can generally be obtained by processing two or more times, as shown in Fig. 15. The first set of passes disperses the alloying elements to the desired depth and determines the overall composition of the surface alloy. Subsequent passes (using processing conditions selected to obtain slightly shallower melt depths) further homogenize the surface.

Effects of Microstructure on Properties. The microstructural changes that result from laser processing can take several forms: a redistribution of major alloy components, a redistribution of any second phases or precipitates, a change in the crystalline character of the surface, or a change in the composition. The melting process results in a leveling of any large-scale composition variations, but the solidification transients can also result in a surface that is enriched in one or more of the alloy components. The high cooling rates during solidification and the accompanying extension of solid-solubility limits often result in either the elimination of second phases or a fine distribution of second phases or precipitates. Alloying may be necessary to take advantage of the higher solubility limits. The steep temperature gradients present during processing generally result in epitaxial resolidification, which introduces some crystallographic texture of the surface grains and grain boundaries (Ref 54). The modification of composition holds the most promise for improving electrochemical behavior because the surface chemistry can be adjusted so that a solidification structure is obtained upon quenching that exhibits more corrosion resistance than the base alloy (Ref 55).

Corrosion Properties

The corrosion resistance of laser-processed materials has been studied by several research groups. For example, laser beam processing was used in one study to melt and rapidly solidify AISI type 304 stainless steel and ferritic Fe-13Cr-xMo (x = 0, 2, 3.5, or 5) stainless steels (Ref 56). The critical current density for the passivation of laser-processed type 304 stainless steel in 1 N sulfuric acid decreased two orders of magnitude compared to unprocessed material. The width of the active peaks decreased for ferritic stainless steels versus unprocessed material. Pitting potentials increased by approximately 150 mV for type 304 stainless steel and 5% Mo ferritic stainless steel. Ferritic steels with 0, 2, or 3.5% Mo showed decreases in the pitting potentials.

In another case, the corrosion behavior of iron-aluminum bronzes was studied electrochemically in 5% H_2SO_4 (Ref 57). Substantial beneficial effects were observed and attributed to homogenization of the complex microstructure.

Lastly, laser surface alloying was performed to incorporate chromium in SAE 1018 steel and molybdenum in type 304 stainless steel (Ref 58). The SAE 1018-Cr materials passivated similarly to corresponding bulk alloys (5 to 80% Cr). The 304-3Mo material was similar in pitting resistance to type 316 stainless steel. The 304-9Mo material was superior to type 316 stainless steel and showed no pitting up to oxygen evolution potentials. Table 2 shows some of the results for the 304-Mo materials. Plasma-sprayed titanium coatings on steel were consolidated by laser processing to remove residual porosity, which is the major obstacle to enhanced corrosion resistance (Ref 59).

Critical Evaluation of Product Utility

Laser processing provides unique opportunities for producing corrosion-resistant surface layers. High-performance surface layers can be designed while conserving scarce, expensive, or critical materials. However, the major limitations associated with laser processing are the compatibility of substrates with thermal conduction requirements and the current restriction to planar substrates.

REFERENCES

1. E. McCafferty, P.G. Moore, J.D. Ayers, and G.K. Hubler, in *Corrosion of Metals Processed by Directed Energy Beams*, C.R. Clayton and C.M. Preece, Ed., The Metallurgical Society, 1982, p 1-21
2. P. Sioshansi, *Thin Solid Films*, Vol 118, 1984, p 61-71
3. B.G. Delves, in *Ion Implantation Into Metals*, V. Ashworth, W.A. Grant, and R.P.M. Proctor, Ed., Pergamon Press, 1982, p 126-134
4. G. Dearnaley, J.H. Freeman, R.S. Nelson, and J. Stephen, *Ion Implantation*, North-Holland, 1973
5. F.A. Smidt and B.D. Sartwell, *Nucl. Instr. Methods Phys. Res. B*, Vol B6, 1985, p 70-77
6. A.S. Denholm and A.B. Wittkower, *Nucl. Instr. Methods Phys. Res. B*, Vol B6, 1985, p 88-93
7. F.H. Scott, J.S. Punni, G.C. Wood, G. Dearnaley, *Ion Implantation Into Metals*, V. Ashworth, W.A. Grant, and R.P.M. Proctor,

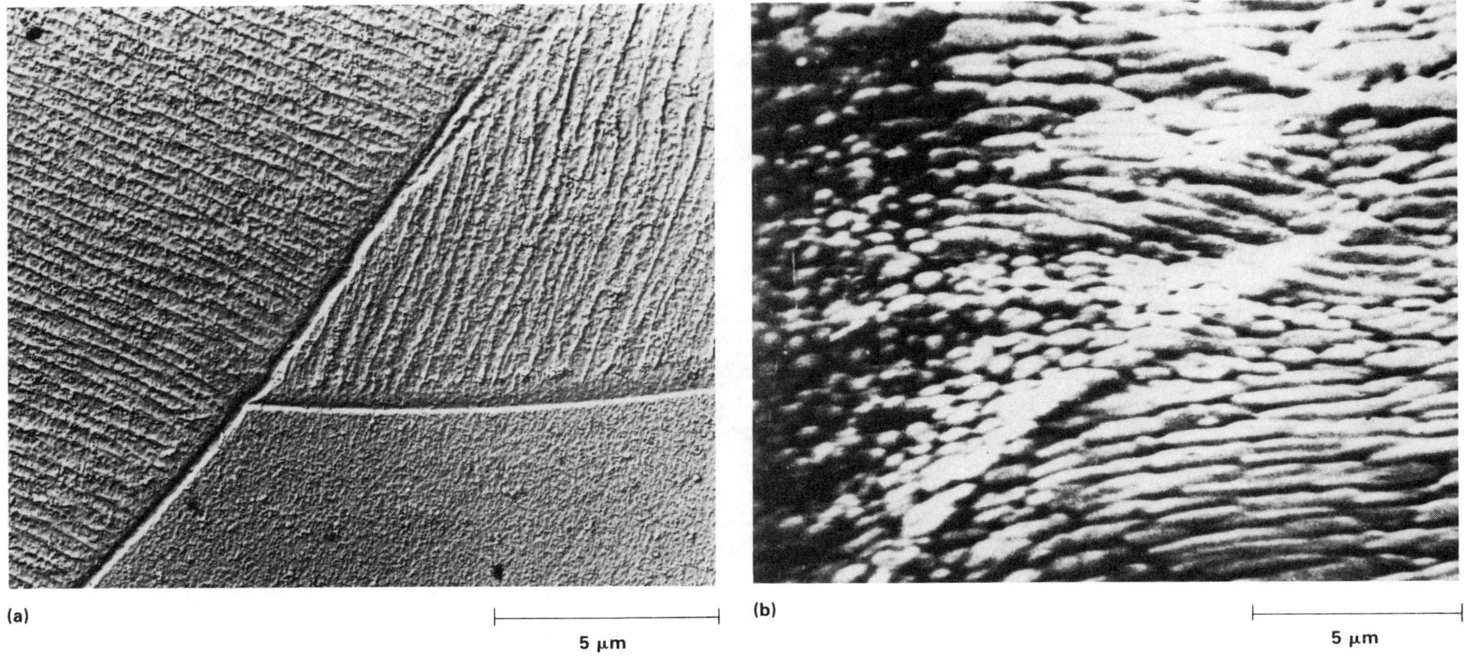

Fig. 14 Dendritic solidification in laser surface melted type 304 stainless steel. (a) Surface replica of a polished-and-etched cross section. (b) SEM micrograph of the free surface

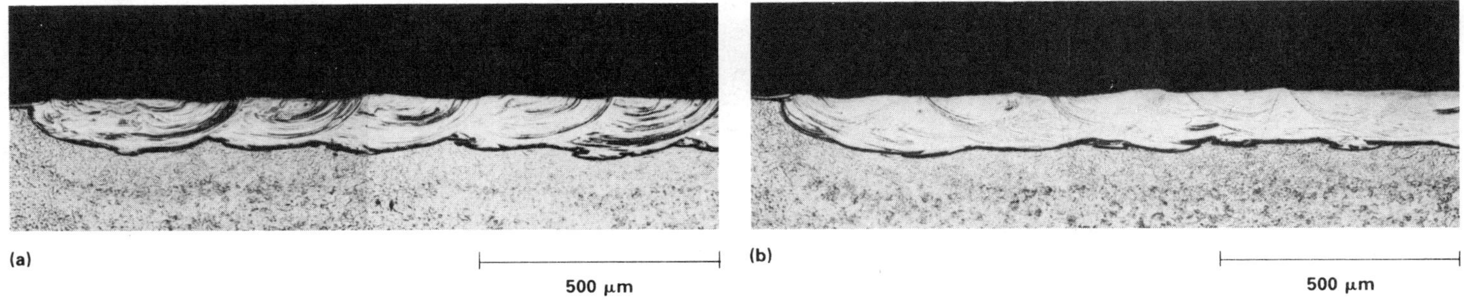

Fig. 15 Laser surface alloying of molybdenum into Ti-6Al-4V showing the effectiveness of multiple passes in homogenizing the chemical composition. (a) Single pass. (b) Double pass

Table 2 Effect of laser surface alloying with molybdenum on pitting potentials of austenitic stainless steels in 0.1 *M* NaCl

Sample	Composition, %			Pitting potential
	Cr	Ni	Mo	(E_{pit}), (V vs SCE)
Type 304	18–20	8–10	0	0.300
Type 316	16–18	10–14	2–3	0.550
304-3Mo	18.9	9.1	3.7	0.500
304-9Mo	19.2	11.7	9.6	Did not pit

Ed., Pergamon Press, 1982, p 245-254

8. M.J. Bennett, G. Dearnaley, M.R. Houlton, and R.W.M. Hawes, *Ion Implantation Into Metals*, V. Ashworth, W.A. Grant, and R.P.M. Proctor, Ed., Pergamon Press, 1982, p 264-276
9. U. Bernabai, M. Cavallini, G. Bombara, G. Dearnaley, and M.A. Wilkins, *Corros. Sci.*, Vol 20, 1980, p 19
10. A. Galerie and G. Dearnaley, *Nucl. Instr. Meth.*, Vol 209-210, 1983, p 823-829
11. G. Dearnaley, in *Fundamental Aspects of Corrosion Protection by Surface Modification*, E. McCafferty, C.R. Clayton, and J. Oudar, Ed., The Electrochemical Society, 1984, p 1-16
12. G.C. Bentini, M. Berti, A. Carnera, G. Della Mea, A.V. Drigo, S. Lo Russo, P. Mazzoldi, and G. Dearnaley, *Corros. Sci.*, Vol 20, 1980, p 27
13. R. Valori, D. Popgoshev, and G.K. Hubler, *J. Lubr. Technol. (Trans. ASME)*, Vol 105, 1985, p 534-541
14. P.D. Hicks and F.P.A. Robinson, *Corros. Sci.*, Vol 24 (No. 10), 1984, p 885-900
15. V. Ashworth, D. Baxter, W.A. Grant, and R.P.M. Proctor, *Corros. Sci.*, Vol 17, 1977, p 947
16. K. Hashimoto, in *Amorphous Metallic Alloys*, F.E. Luborsky, Ed., Butterworths, 1983, p 471
17. G.K. Wolf, J. Dressler, W. Ensinger, H. Ferber, A. Meger, and P. Munn, in *Fundamental Aspects of Corrosion Protection by Surface Modification*, E. McCafferty, C.R. Clayton, and J. Oudar, Ed., The Electrochemical Society, 1984, p 42-51
18. C.R. Clayton and C.M. Preece, Ed., *Corrosion of Metals Processed by Directed Energy Beams*, The Metallurgical Society, 1982
19. E. McCafferty, C.R. Clayton, and J. Oudar, Ed., *Fundamental Aspects of Corrosion Protection by Surface Modification*, The Electrochemical Society, 1984
20. C.M. Preece and J.K. Hirvonen, Ed., *Ion Implantation Metallurgy*, The Metallurgical Society, 1980
21. V. Ashworth, W.A. Grant, and R.P.M. Proctor, *Ion Implantation Into Metals*, Pergamon Press, 1982
22. J.K. Hirvonen, Ed., *Ion Implantation*, Vol 18, *Treatise on Materials Science and Technology*, Academic Press, 1980

23. F.A. Smidt, Ed., *Ion Implantation for Materials Processing*, Noyes Data Corporation, 1983
24. G.K. Hubler, O.W. Holland, C.R. Clayton, and C.W. White, Ed., *Ion Implantation and Ion Beam Processing of Materials: B*, Vol 27, Proceedings of the Materials Research Society, North-Holland, 1984
25. R.D. Granata, M.A. De Crosta, and H. Leidheiser, Jr., in *Proceedings of the Ninth National Congress on Metallic Corrosion*, Vol 1, National Research Council of Canada, 1984, p 264-268
26. G. Dearnaley, *Nucl. Instr. Methods Phys. Res. B*, Vol B7/8, 1985, p 158-165
27. I.L. Singer, in *Ion Implantation and Ion Beam Processing of Materials*, Vol 27, G.K. Hubler, O.W. Holland, C.R. Clayton, and C.W. White, Ed., Proceedings of the Materials Research Society, North-Holland, 1984, p 585-595
28. J.B. Pethica, in *Ion Implantation Into Metals*, V. Ashworth, W.A. Grant, and R.P.M. Proctor, Pergamon Press, 1982, p 147-156
29. R. Hutchings and W.C. Oliver, *Wear*, Vol 92, 1983, p 143
30. K.K. Shih, *Wear*, Vol 105, 1985, p 341-347
31. J.W. Hill, M.J. Lee, and I.J. Spalding, *Opt. Laser Technol.*, Vol 6, Dec 1974, p 276
32. E.M. Breinan, B.H. Kear, and C.M. Banas, *Phys. Today*, Vol 29, 1976, p 11, 44
33. H.E. Cline and T.R. Anthony, *J. Appl. Phys.*, Vol 48, 1977, p 3895
34. T.R. Anthony and H.E. Cline, *J. Appl. Phys.*, Vol 49, 1978, p 1248
35. P.R. Strutt, H. Nowotny, M. Tuli, and B.H. Kear, *Mater. Sci. Eng.*, Vol 36, 1978, p 217
36. G. Staehli and Ch. Sturzenegger, *Scr. Metall.*, Vol 12, 1978, p 617
37. P.L. Bonora, M. Bassoli, P.L. De Anna, B. Battaglin, G. Della Mea, P. Mazzoldi, and A. Miotello, *Electrochim. Acta*, Vol 25, 1980, p 1497
38. C.W. Draper, R.E. Woods, and L.S. Meyer, *Corrosion*, Vol 36, 1980, p 405
39. R.J. Pangborn and D.R. Beaman, *J. Appl. Phys.*, Vol 5, 1980, p 5992
40. P.G. Moore and E. McCafferty, *J. Electrochem. Soc.*, Vol 128, 1981, p 1391
41. D.S. Gnanamuthu, *Opt. Eng.*, Vol 19, 1980, p 783
42. P.G. Moore, in *Proceedings of the International Laser Processing Conference*, Laser Institute of America, 1981
43. C.W. Draper, *Appl. Opt.*, Vol 20, 1981, p 3093
44. H.W. Bergmann and B.L. Mordike, *Z. Werkstofftech.*, Vol 12, 1981, p 142
45. H.S. Carslaw and J.C. Jaeger, *Conduction of Heat in Solids*, Clarendon Press, 1959
46. L.E. Grenwald, E.M. Breinan, and B.H. Kear, in *Laser-Solid Interactions and Laser Processing—1978*, S.D. Ferris, E.J. Leany, and J.M. Poate, Ed., American Institute of Physics, 1979, p 189
47. S.L. Narasimhan, S.M. Copley, E.W. Van Stryland, and M. Bass, *ICEE J. Quant. Electr.*, Vol 13, 1977, p 2D
48. A. Munitz, *Metall. Trans. B*, Vol 11B, 1980, p 563
49. B. Chalmers, *Principles of Solidification*, Robert E. Krieger, 1977
50. C.W. Draper, S.P. Sharma, J.L. Yeh, and S.L. Bernasek, *Surf. Interface Anal.*, Vol 2 (No. 5), 1980, p 179
51. D.G. Beck, S.M. Copley, and M. Bass, *Metall. Trans. A*, Vol 13A, 1982, p 1879
52. P.T. Sarjeant and R. Roy, *Mater. Res. Bull.*, Vol 3, 1968, p 265
53. J.A. Dantzig and S.H. Davis, *Mater. Sci. Eng.*, Vol 32, 1978, p 199
54. L.S. Weinman, J.N. DeVault, and P.G. Moore, in *Applications of Lasers in Materials Processing*, E.A. Metzbower, Ed., American Society for Metals, 1979, p 259
55. L.E. Grenwald, E.M. Breinan, and B.H. Kear, in *Laser-Solid Interactions and Laser Processing—1978*, S.D. Ferris, E.J. Leany, and J.M. Poate, Ed., American Institute of Physics, 1979, p 239
56. J.B. Lumsden, D.S. Gnanamuthu, and R.J. Moores, in *Fundamental Aspects of Corrosion Protection by Surface Modification*, E. McCafferty, C.R. Clayton, and J. Oudar, Ed., The Electrochemical Society, 1984, p 122-129
57. J. Javadpour, C.R. Clayton and C.W. Draper, in *Corrosion of Metals Processed by Directed Energy Beams*, C.R. Clayton and C.M. Preece, Ed., The Metallurgical Society, 1982, p 135-145
58. E. McCafferty and P.G. Moore, *J. Electrochem. Soc.*, Vol 133 (No. 6), 1986, p 1090-1096
59. E. McCafferty, G.K. Hubler, P.M. Natishan, P.G. Moore, R.A. Kant, and B.D. Sartwell, Ion Beam or Laser Processing of Metal Surfaces for Improved Corrosion Resistance, *Mater. Sci. Eng.*, to be published

Specific Alloy Systems

Section Chairman:
Lawrence J. Korb, Rockwell International
Section Co-chairmen:
Ralph M. Davison, Avesta Stainless, Inc.
Aziz I. Asphahani, Haynes International
R.T. Webster, Teledyne Wah Chang Albany

Corrosion of Carbon Steels

Chairman: James H. Bryson, Inland Steel Company

CARBON STEEL, the most widely used engineering material, accounts for over 64 million tons, or approximately 88%, of the annual steel production in the United States (Tables 1 and 2). Despite its relatively limited corrosion resistance, carbon steel is used in large tonnages in marine applications, nuclear power and fossil fuel power plants, transportation, chemical processing, petroleum production and refining, pipelines, mining, construction, and metal-processing equipment. The specific steps taken in these industries to mitigate the corrosion of carbon steel are discussed in detail in the Section "Specific Industries and Environments" in this Volume and will not be repeated here.

This article consists of five major sections:

- Carbon Steels
- Weathering Steels
- Protection of Steel From Corrosion
- Metallic Coated Steels
- Organic Coated Steels

Where applicable, reference will be made to a specific article in this Handbook in which more detailed information can be found. In addition, previous 9th Edition Volumes of *Metals Handbook* should be consulted for data related to carbon steels. Of particular note are Volume 1, *Properties and Selection: Irons and Steels*; Volume 4, *Heat Treating*; and Volume 9, *Metallography and Microstructures*.

Carbon Steels

A.G. Preban
Inland Steel Company

The cost of metallic corrosion to the total economy must be measured in hundreds of millions of dollars per year. Because carbon steels represent the largest single class of alloys in use, both in terms of tonnage and total cost, it is easy to understand that the corrosion of carbon steels is a problem of enormous practical importance. This, of course, is the reason for the existence of entire industries devoted to providing protective systems for irons and steels. Aspects of corrosion control will be discussed later in this article. In this section, only the intrinsic aspects of the corrosion system, primarily the environmental and compositional factors, will be addressed. This is not meant to imply that design is not important. Indeed, design changes are often the most efficient manner of dealing with a particular corrosion problem. This aspect of corrosion control is discussed in the article "Design Details to Minimize Corrosion" in this Volume.

Carbon, or mild, steels are by their nature of limited alloy content, usually less than 2% by weight for the total of all additions. Unfortunately, these levels of addition do not generally produce any remarkable changes in general corrosion behavior. One possible exception to this

Table 1 Net shipments of U.S. steel mill products—all grades

Steel products	1985 Tons(a)	1985 %	1984 Tons(a)	1984 %
Ingots and steel castings	256	0.4	275	0.4
Blooms, slabs, and billets	1 118	1.5	1 031	1.4
Skelp	9	. . .	11	. . .
Wire rods	2 962	4.0	3 090	4.2
Structural shapes (≥8 cm, or 3 in.)	4 373	6.0	3 868	5.2
Steel piling	326	0.5	288	0.4
Plates	4 327	5.9	4 339	5.9
Rails				
standard (over 60 lbs)	679	0.9	933	1.3
all other	25	. . .	32	. . .
Joint bars	6	. . .	7	. . .
Tie plates	87	0.1	124	0.2
Track spikes	34	0.1	51	0.1
Wheels (rolled and forged)	68	0.1	59	0.1
Axles	33	0.1	33	. . .
Bars				
hot rolled	5 698	7.8	6 070	8.2
bar-size light shapes	1 329	1.8	1 184	1.6
reinforcing	4 325	5.9	4 432	6.0
cold finished	1 255	1.7	1 484	2.0
Tool steel	60	0.1	61	0.1
Pipe and tubing				
standard	855	1.1	743	1.0
oil country goods	1 299	1.8	1 406	1.9
line	775	1.1	775	1.0
mechanical	812	1.1	940	1.3
pressure	87	0.1	89	0.1
structural	219	0.3	270	0.4
stainless	49	0.1	52	0.1
Wire				
drawn	874	1.2	963	1.3
nails and staples	170	0.2	147	0.2
barbed and twisted	38	0.1	52	0.1
woven wire fence	20	. . .	26	. . .
bale ties and baling wire	34	0.1	35	0.1
Black plate	215	0.3	286	0.4
Tin plate	2 611	3.6	2 765	3.7
Tin free steel	889	1.2	945	1.3
Tin mill products—all other	58	0.1	66	0.1
Sheets				
hot rolled	12 952	17.7	13 133	17.8
cold rolled	13 574	18.6	13 664	18.5
Sheets and strip, galvanized				
hot dipped	6 850	9.4	6 100	8.3
electrolytic	697	0.9	659	0.9
all other metallic coated	1 122	1.5	1 109	1.5
electrical	413	0.6	490	0.7
Strip				
hot rolled	586	0.8	650	0.9
cold rolled	876	1.2	1 002	1.3
Total steel mill products	**73 043**	**100.0**	**73 739**	**100.0**
Carbon	64 360	88.1	65 009	88.2
Stainless and heat resisting	1 251	1.7	1 248	1.7
Alloy (other than stainless)	7 432	10.2	7 482	10.1

(a) Thousands of net tons. Source: American Iron and Steel Institute statement would be the weathering steels, in which small additions of copper, chromium, nickel, and/or phosphorus produce significant reductions in corrosion rate in certain environments. These steels will be discussed more fully in the following section of this article. At the levels present in low-alloy steels, the usual impurities have no significant effect on corrosion rate in the atmosphere, neutral waters, or soils. Only in the case of acid attack is an effect observed. In this latter case, the presence of phosphorus and sulfur markedly increase the rate of attack. Indeed, in acid systems, the pure irons appear to exhibit the best resistance to attack.

In solving a particular corrosion problem, a dramatic change in attack rate can often be attained by altering the corrosive environment. The deaeration of water and the addition of corrosion inhibitors are two examples that have broad application in the area of aqueous corrosion. These and other methods that may be useful in some cases are discussed in the Section "Corrosion Protection Methods" in this Volume.

Because corrosion is such a multifaceted phenomenon, it is generally useful to attempt to categorize the various types. This is usually done on an environmental basis. Therefore, in this section, atmospheric corrosion, aqueous corrosion, and some other corrosion types of interest, such as corrosion in soils, concrete, and boilers and heating plants, will be addressed.

Table 2 Net shipments of U.S. steel mill products by market classification—all grades

Market classifications	1985 Tons	1985 %	1984 Tons	1984 %
Steel for converting and processing	5 484	7.5	5 136	7.0
Independent forgers (not elsewhere classified)	708	1.0	775	1.1
Industrial fasteners	386	0.5	455	0.6
Steel service centers and distributors	18 439	25.2	18 364	24.9
Construction, including maintenance	7 900	10.8	7 522	10.2
Contractors' products	3 330	4.6	2 631	3.6
Total	11 230	15.4	10 153	13.8
Automotive				
Vehicles, parts, etc.	12 689	17.3	12 571	17.1
Independent forgers	261	0.4	311	0.4
Total	12 950	17.7	12 882	17.5
Rail transportation				
Freight cars, passenger cars, and locomotives	280	0.4	347	0.4
Rails and all other	781	1.0	1 091	1.5
Total	1 061	1.4	1 438	1.9
Shipbuilding and marine equipment	337	0.5	471	0.6
Aircraft and aerospace	39	...	37	0.1
Oil and gas industry	2044	2.8	2003	2.7
Mining, quarrying, and lumbering	298	0.4	298	0.4
Agricultural				
Agricultural machinery	442	0.7	480	0.6
All other	187	0.2	193	0.3
Total	629	0.9	673	0.9
Machinery, industrial equipment and tools	2 271	3.1	2 886	3.9
Electrical equipment	1 869	2.6	2 365	3.2
Appliances, utensils, and cutlery	1 466	2.0	1 635	2.2
Other domestic and commercial equipment	1 215	1.7	1 339	1.8
Containers, packaging and shipping materials				
Cans and closures	3044	4.2	3268	4.4
Barrels, drums, and shipping pails	465	0.6	507	0.7
All other	580	0.8	577	0.8
Total	4 089	5.6	4 352	5.9
Ordnance and other military	267	0.4	242	0.3
Export (reporting companies only)	494	0.7	428	0.6
Nonclassified shipments	7 767	10.6	7 807	10.6
Total shipments	**73 043**	**100.0**	**73 739**	**100.0**

Source: American Iron and Steel Institute

Atmospheric Corrosion

Atmospheres are often classified as being rural, industrial, or marine in nature. Such a classification is of course a gross oversimplification of the situation. It is easy to list locations along the seacoast that have heavy industrial pollution in the atmosphere. Such locations are both marine and industrial. Furthermore, two decidedly rural environments can differ widely in average yearly temperature and rainfall and can therefore have considerably different corrosive tendencies. Industrial expansion into formerly rural areas can easily change the aggressiveness of a particular location. Finally, long-term trends in the environment, such as changes in rainfall patterns, mean temperature, and perhaps acid rain, can make extrapolations from past behavior less reliable.

Other factors that limit the usefulness of atmospheric-exposure data are the general nonlinearity of weight loss due to corrosion with time and the fact that most atmospheric-corrosion data are presented as an average over the entire test panel surface. Most atmospheric exposure data for steels show a decrease in the rate of attack with time of exposure so that extrapolations of such data to times longer than those covered by the exposure data can lead to an overdesign in cross section. Finally, in many cases, the average weight loss per unit area is of less concern than the time to perforation. This factor is more related to localized attack, which can be masked by the averaging of data, as is done in weight loss determinations.

Given all the caveats mentioned above, the design engineer is well justified in using atmospheric-corrosion data as more indicative than quantitative. With all of this in mind, some of the corrosion data for various carbon steels in some representative environments will be discussed. More detailed information is available in Ref 1 to 3.

The effect of various environments on the corrosion rate is indicated in Table 3. Table 3 is a compilation of weight loss measurements for cold-rolled carbon steels after 2 years of exposure. The most startling feature of Table 3 is the extreme range of corrosion rates existing. Galeta Point Beach, Panama, is more than 450 times as aggressive than the site at Norman Wells, N.W.T., Canada. This difference in corrosion rate is easily greater than any effect that can be produced by small changes in composition of the steel. Again, this underscores the fact that in dealing with the corrosion of carbon steels the alteration of design or environmental factors is usually more effective than changing the grade of steel.

Further examination of Table 3 shows that the marine environments tend to be near the aggressive end of the list and that cold environments are generally less aggressive than warm sites. The average yearly temperature cannot, in general, be isolated from the moisture effect, because most of the more tropical exposure sites are also in regions with high humidity. One exception is arid Phoenix, AZ.

Corrosion Film Formation and Breakdown. The corrosion of carbon steel in the atmosphere and in many aqueous environments is best understood from a film formation and breakdown standpoint. It is an inescapable fact that iron in the presence of oxygen and/or water is thermodynamically unstable with respect to its oxides. Thus, the question is never whether the steel will rust, but rather at what rate. In the absence of film formation and with a constant environment, one would expect the oxidation rate to be constant. Conversely, if the corrosion product film that forms isolates the steel from the corrosive environment, one would expect a zero corrosion rate after the initial film formation period. A tightly adherent film that permits only diffusion transfer of the reactants would be characterized by a corrosion rate that decreases with the square root of the exposure time.

Because the above idealizations are rarely encountered in the corrosion of carbon steels, it is obvious that other factors that tend to disrupt stable film formation must be operative. These factors can be external, such as erosion by wind or rain, or they may be internal to the film itself, such as stresses caused by the different specific volumes of metal and oxide. It is the complexity of these various breakdown processes that makes a quantitative prediction of corrosion so difficult, makes extrapolations of data from short-term behavior to longer times so difficult, and makes precise predictions of behavior at a particular test site based on observations at a different site virtually impossible.

The corrosion of iron in the atmosphere proceeds by the formation of hydrated oxides. The half-cell reactions can be expressed as:

Table 3 Comparative rankings of 45 locations based on steel loss of weight (grams) of 10- × 15-cm (4- × 6-in.) specimens of steel

Ranking to State College, PA	Location	2-year exposure, grams lost
1	Norman Wells, N.W.T., Canada	0.73
2	Phoenix, AZ	2.23
3	Saskatoon, Sask., Canada	2.77
4	Esquimalt, Vancouver Island, Canada	6.50
5	Detroit, MI	7.03
6	Fort Amidor Pier, Panama, C.Z.	7.10
7	Morenci, MI	9.53
8	Ottawa, Ont., Canada	9.60
9	Potter County, PA	10.00
10	Waterbury, CT	11.00
11	State College, PA	11.17
12	Montreal, P.Q., Canada	11.44
13	Melbourne, Australia	12.70
14	Halifax (York Redoubt), N.S.	12.97
15	Durham, NH	13.30
16	Middletown, OH	14.00
17	Pittsburgh, PA	14.90
18	Columbus, OH	16.00
19	South Bend, PA	16.20
20	Trail, B.C. Canada	16.90
21	Bethlehem, PA	18.3
22	Cleveland, OH	19.0
23	Miraflores, Panama, C.Z.	20.9
24	London (Battersea), England	23.0
25	Monroeville, PA	23.8
26	Newark, NJ	24.7
27	Manila, Philippine Islands	26.2
28	Limon Bay, Panama, C.Z.	30.3
29	Bayonne, NJ	37.7
30	East Chicago, IN	41.1
31	Cape Kennedy, 0.8 km (0.5 mile) from ocean	42.0
32	Brazos River, TX	45.5
33	Pilsey Island, England	50.0
34	London (Stratford), England	54.3
35	Halifax (Federal Building), N.S.	55.3
36	Cape Kennedy, 55 m (60 yd) from ocean, 60-ft elevation	64.0
37	Kure Beach, NC, 250-m (800-ft) lot	71.0
38	Cape Kennedy, 55 m (60 yd) from ocean, 9-m (30-ft) elevation	80.2
39	Daytona Beach, FL	144.0
40	Widness, England	174.0
41	Cape Kennedy, 55 m (60 yd) from ocean, ground level	215.0
42	Dungeness, England	238.0
43	Point Reyes, CA	244.0
44	Kure Beach, NC, 25-m (80-ft) lot	260.0
45	Galeta Point Beach, Panama, C.Z.	336.0

Source: Ref 4

$\frac{1}{2}O_2 + H_2O + 2e \rightarrow 2(OH)^-$ (cathodic)

$Fe \rightarrow Fe^{2+} + 2e$ (anodic)

Further reactions can then occur, such as:

$Fe^{2+} + 2OH^- \rightarrow Fe(OH)_2$

$2Fe(OH)_2 + H_2O + \frac{1}{2}O_2 \rightarrow 2Fe(OH)_3$

The hydrated oxides can lose water during dry periods and revert to the anhydrous ferrous and ferric oxides. In addition, a layer of magnetite (Fe_3O_4) or $FeO{\cdot}Fe_2O_3$ often forms between iron oxide (FeO) and hematite (Fe_2O_3). Actually, the various oxides and hydroxides of iron form a rather complicated system of compounds. The compound FeOOH has been found to exist in three different crystal forms plus an amorphous form. The occurrence of the various oxide species is dependent on pH, oxygen availability, various atmospheric pollutants, and the composition of the steel, as in weathering steels containing copper and phosphorus (Ref 5). The actual nature of the rust film is important because FeO and FeOOH seem to be more adherent than Fe_3O_4 and Fe_2O_3, and therefore more likely to slow the corrosive attack, but the higher oxides and oxy-hydroxides are more prone to spallation.

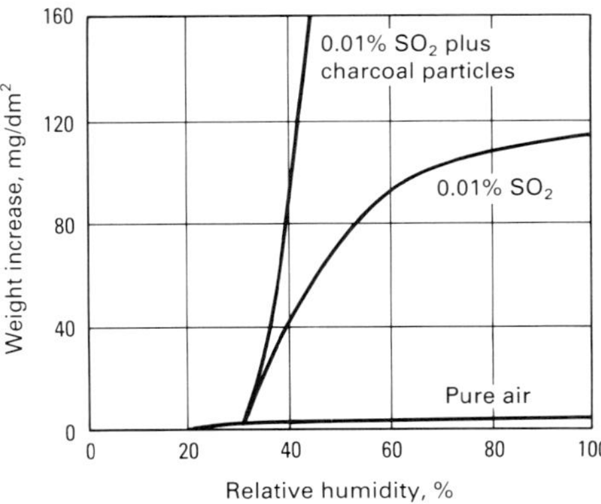

Fig. 1 Relationship between the corrosion rate of iron and the relative humidity in an environment containing 0.01% SO_2. Exposure period: 55 days. The corrosion rate of iron exposed for 55 days at 100% relative humidity is shown for comparison. Source: Ref 6

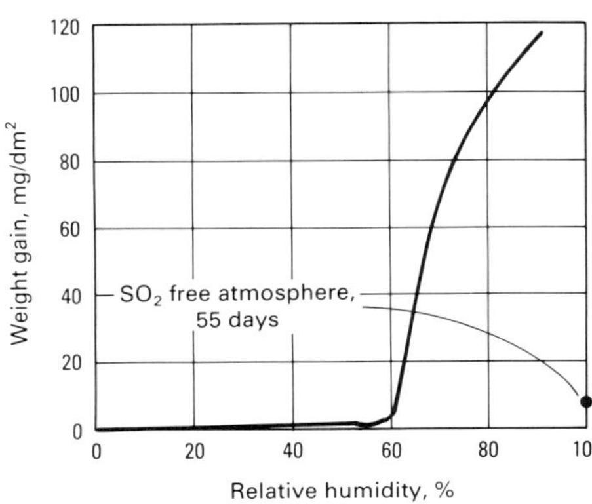

Fig. 2 Effect of relative humidity and atmospheric pollution on the rusting of iron. A critical humidity near 60% relative humidity is observed. Source: Ref 18

Atmospheric Factors in Atmospheric Corrosion

Because there is a substantial variation in the corrosion rates of carbon steels at different atmospheric-test locations, it is only logical to ask which factors contribute to these differences. Although the prediction of corrosivity is still not possible, it appears that humidity, temperature, and the levels of chloride, sulfate, and probably other atmospheric pollutants present each exert an influence on the corrosion rate of carbon steels.

Humidity and Atmospheric Pollutants. Because atmospheric corrosion is an electrolytic process, the presence of an electrolyte is required. This should not be taken to mean that the steel surface must be awash with water; a very thin adsorbed film of water is all that is required. During an actual exposure, the metal spends some portion of the time awash with water because of rain or splashing and a portion of the time covered with a thin adsorbed water film. The portion of time spent covered with the thin water film depends quite strongly on relative humidity at the exposure site (Fig. 1). This fact has led many corrosion scientists to investigate the influence of the time of wetness on the corrosion rate (Ref 7-17).

These studies have shown that time of wetness, although an important factor, cannot be considered in isolation when estimating corrosion rates. An excellent example of this fact is demonstrated in Fig. 2, in which the weight gain of iron is plotted as a function of relative humidity for an exposure of 55 days in an atmosphere containing 0.01% sulfur dioxide (SO_2). In the lower right-hand corner of Fig. 2 is the measured corrosion rate for iron exposed for the same time

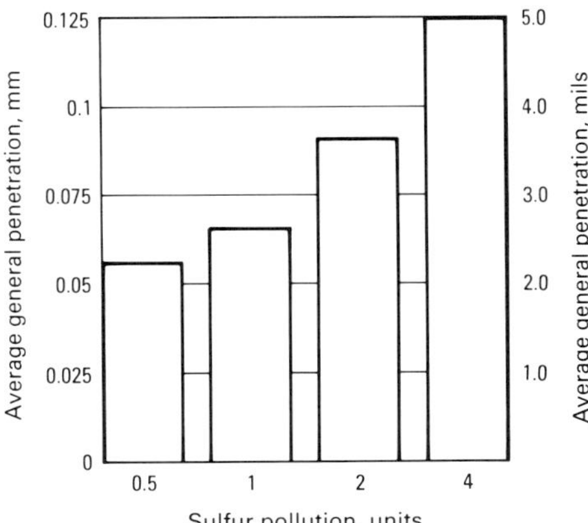

Fig. 3 Effect of atmospheric pollution on rusting. Exposure sites include Godalming, Teddington, U.K. (0.5 sulfur pollution unit); Hornchurch, Leicester, U.K. (1 sulfur pollution unit); Barking, Salford, U.K. (2 sulfur pollution units); and Manchester, Sheffield, U.K. (4 sulfur pollution units). The steel tested contained 0.28% Cu. Test duration: 1 year. Source: British Iron and Steel Research Association

in an SO_2-free atmosphere at 99% relative humidity. The increase in corrosion rate produced by the addition of SO_2 to the atmosphere is substantial, as demonstrated in Fig. 3, which shows actual exposure data from four different exposure sites with varying levels of sulfur pollution in the atmosphere. The accelerating effect of sulfur is obvious.

Another feature of interest is the apparent existence of a critical humidity level below which the corrosion rate is small. As shown in Fig. 2, the critical humidity in an SO_2-containing atmosphere is approximately 60%. This behavior contrasts with that of steel in contact with particles of sea salt, as illustrated in Fig. 4. In Fig. 4, the corrosion rate shows a steady increase with increasing humidity. Although there is a scarcity of data, it seems reasonable that oxides of nitrogen in the atmosphere would also exhibit an accelerating effect on the corrosion of steel. Indeed, any gaseous atmospheric constituent capable of strong electrolytic activity should be suspected as being capable of increasing the corrosion rate of steel. Figure 5 shows the corrosion rate per day of wetness at six test sites and demonstrates the accelerating effect of temperature on corrosion rate. From Fig. 5(e) and (f), one can infer the accelerating effect of chloride ions on atmospheric corrosion.

Effects of Alloying Additions. Because carbon steels are by definition not very highly alloyed, it is not surprising that most grades do not exhibit large differences in atmospheric-corrosion rate. Nevertheless, alloying can make changes in the atmospheric-corrosion rate of carbon steel. The elements generally found to be most beneficial in this regard are copper, nickel, silicon, chromium, and phosphorus. Of these, the most striking example is that of copper; increases from 0.01 to 0.05% have been shown to decrease the corrosion rate by a factor of two to three (Ref 20, 21). Additions of the above elements in combination are generally more effective than when added singly, although the effects are not additive. The effects from one study are shown in Fig. 6 to 10. The effectiveness of these elements in retarding corrosion also appears to depend on the corrosive environment, with the most benefit appearing in industrial atmospheres (Ref 23).

Kinetics of Atmospheric Corrosion

The rate of atmospheric corrosion of steels is not constant with time but usually decreases as the length of exposure increases. This fact indicates the difficulty in using most of the published atmospheric-corrosion data in any quantitative way. Much of the published data consists of weight loss due to corrosion averaged over the time of exposure, which is itself often variable. Such corrosion rate calculations are misleading, especially when the exposure time is short, because the ensuing rate of attack can be considerably lower. Several authors have shown that the amount of corrosion occurring as a function of time can be expressed as (Ref 24-30):

$$W = Kt^n \qquad \text{(Eq 1)}$$

where W is the weight loss of metal due to corrosion, t is the exposure time in years, and K and n are empirical constants. The goodness of fit seems to be excellent, but until more data from various atmospheric-exposure sites are analyzed, Eq 1 is of limited value. Because the values of K and n depend on both the alloy system and the exposure site, a great deal of work must be done before general use of Eq 1 can be made in real applications. The one possible exception would be when data for the exposure situation and alloy system are already known. In this case, Eq 1 can be very useful in estimating long-term corrosion behavior from as little as 2 years of data (Ref 24), although 3- to 4-year data give a better estimate.

Perhaps most important, Eq 1 points out that it is impossible to describe either the extent or rate of corrosion under atmospheric conditions with a single parameter, which is what much of the reported corrosion data persists in doing. When the results of a several-year exposure test are condensed to a single value, such as the average loss per year or the total loss for the exposure period, one cannot estimate the values of the kinetic parameters governing the system. Without the values of these parameters, extrapolation of results to longer exposure periods becomes quite unreliable. When good estimates for the kinetic parameters are available, extrapolations to 7- or 8-year performance from 1- and 2-year data have been found to agree within 5% of observed performance (Ref 24).

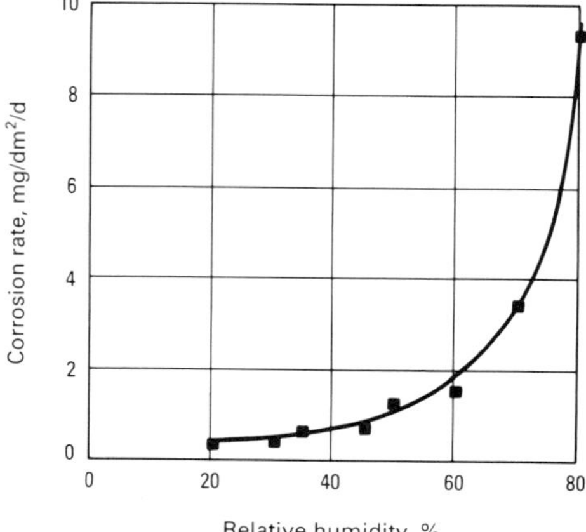

Fig. 4 Dependence on relative humidity of the initial corrosion rate of steel in contact with particles of sea salt. (0.7 g/cm). Exposure period: 13 days

Aqueous Corrosion

Carbon steel pipe and vessels are often required to transport water or are submerged in water to some extent during service. This exposure can be under conditions of varying temperature, flow rate, pH, and so on, all of which can alter the rate of corrosion. The relative acidity of the solution is probably the most important factor to be considered. At low pH, the evolution of hydrogen tends to eliminate the possibility of protective film formation so that steel continues to corrode, but in akaline solutions, the formation of protective films greatly reduces the corrosion rate. The greater the alkalinity, the slower the rate of attack becomes. In neutral solutions, other factors, such as aeration, become rate determining so that generalization becomes more difficult. All of these factors can be graphically demonstrated, as in Fig. 11.

The corrosion of steels in aerated seawater is about the same overall as in aerated freshwater, but this is somewhat misleading because the improved electrical conductivity of seawater can lead to increased pitting. Furthermore, because of the improved conductance, the concentration cells can operate over long distances, and this leads to a more nonuniform attack than in freshwater. It is also well documented that alternate cycling through immersion and exposure to air produces more pitting attack than continuous immersion (Ref 32-34).

The effect of various alloying additions and exposure conditions on the corrosion behavior of bars exposed for a period of 15 years in England is summarized in Table 4. Although this study showed a beneficial effect of both copper and nickel additions, other studies reported no significant benefit (Ref 35, 36). Interestingly, the corrosion rates of specimens completely immersed in seawater do not appear to depend on the geographical location of the test site; therefore, by inference, the mean temperature does not appear to play an important role (Ref 37).

This constancy of the corrosion rate in seawater has been attributed to the more rapid fouling of the exposed steel by marine organisms, such as barnacles and algae, in warmer seas (Ref 38). It is further speculated that this fouling offsets the increases expected from the temperature rise. For example, it has been demonstrated that under laboratory conditions of rapidly flowing seawater where fouling is suppressed a rise of 18 °C (32 °F) will approximately double the attack rate. The effect of velocity of the seawater on the corrosion rate is illustrated in Fig. 12. In actual marine exposures, periods of rapid flow from tidal motion may not be effective, because the slack periods at reversal may allow marine organisms to attach themselves to the metal surface. If these organisms can survive the subsequent high flow, then a growth on the exposed surface can develop. This effectively reduces the velocity of seawater at the metal/water interface so that bulk flow rates are no longer rate determining.

Additional information on the corrosion of steels in marine environments can be found in the article "Marine Corrosion" in this Volume. The effect of marine organisms on corrosion is discussed in the articles "General Corrosion" and "Localized Corrosion" in this Volume. Information on corrosion in freshwater—which includes all nonsaline natural waters, polluted or unpolluted, found in inland bodies such as streams, rivers, ponds, and lakes—is available in the article "Corrosion in Fresh Water" in Volume 1 of the 9th Edition of *Metals Handbook*.

Soil Corrosion

The response of carbon steel to soil corrosion depends primarily on the nature of the soil and certain other environmental factors, such as the availability of moisture and oxygen. These factors can lead to extreme variations in the rate of attack. For example, under the worst conditions, a buried vessel may perforate in less than 1 year, although archaeological digs in arid desert regions have uncovered iron tools dating back hundreds of years. Because of this intrinsic variability, any cumulative weight loss data or kinetic information is virtually worthless for design use. Some general rules can be formulated, however. Soils with high moisture content, high electrical conductivity, high acidity, and high dissolved salts will be most corrosive. At the other extremes of character, the soil will be virtually inert to carbon steel. The effect of aeration on soils is somewhat different from the effect of aeration on water because poorly aerated conditions in water can lead to accelerated attack by sulfate-reducing anaerobic bacteria.

The effect of low levels of alloying additions on the soil corrosion of carbon steels is modest at best, with most common additions showing no significant benefit. Some data (Fig. 13) seem to show a small benefit of 1% Cu + 2.5% Ni over plain carbon steel; however, it is debatable whether this improved performance could justify the added alloy cost. This decision must involve other factors, such as the cost of a failure and the cost of replacement.

Interestingly, it has been demonstrated that both the weight loss and maximum pit depth in soil corrosion can be represented by an equation of the form (Ref 40):

$$Z = at^m \qquad \text{(Eq 2)}$$

where Z is either the weight loss of maximum pit depth, t is time of exposure, and a and m are constants that depend on the specific soil corrosion situation. Equation 2 is of the same form as

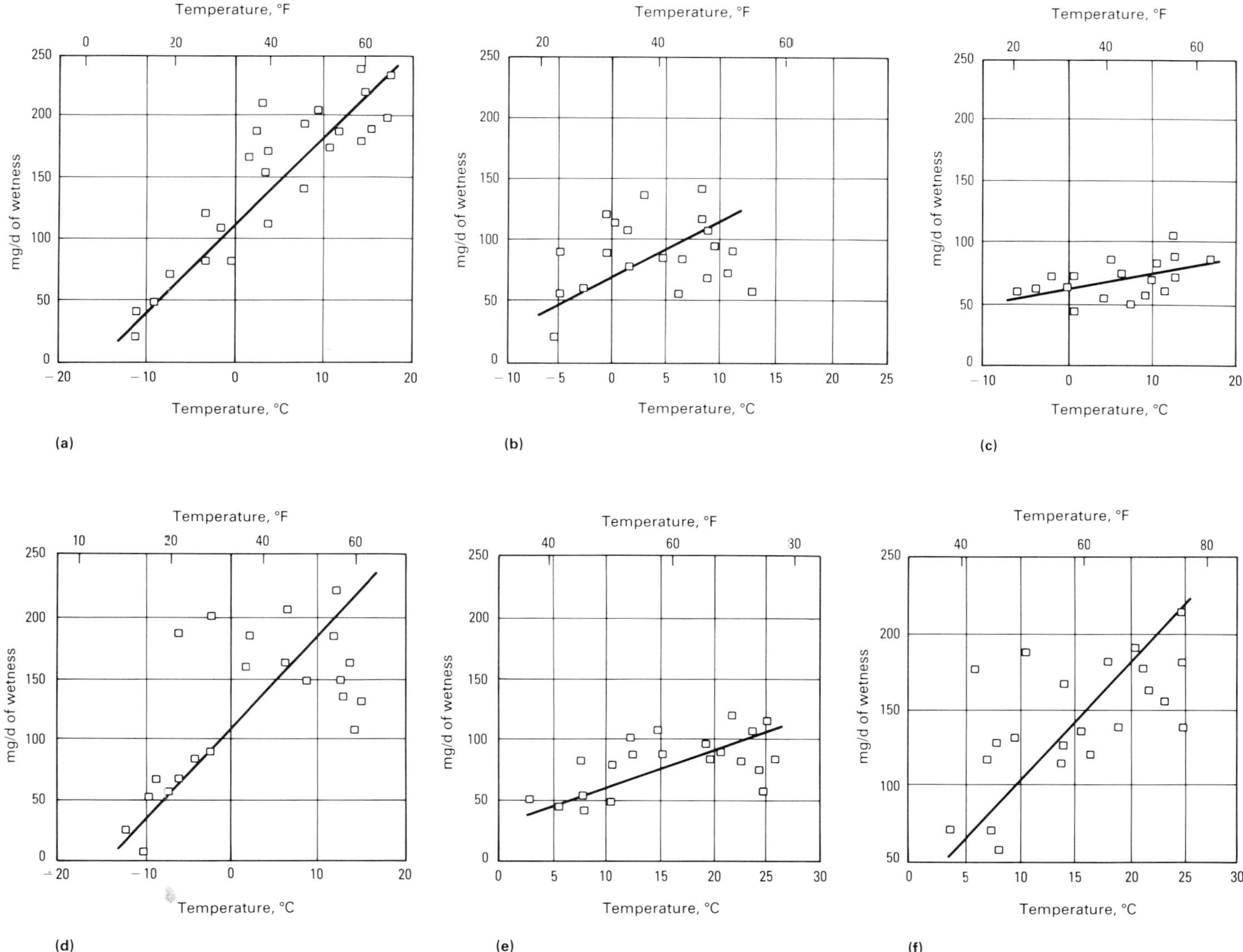

Fig. 5 Corrosion rates (in mg/d of wetness) for steel panels exposed at Cleveland, OH (a); Trail, B.C. (b); South Bend, PA (c); Ottawa, Canada (d); Kure Beach, NC, 250 m (800 ft) from ocean (e); Kure Beach, NC, 24 m (80 ft) from ocean. Source: Ref 19

Eq 1. The similarities between Eq 1 and 2 may be due to the general applicability of a power law relationship, but because both types of corrosion are fundamentally electrolytic, there may be some underlying connection. A more detailed discussion of underground corrosion is available in Ref 40. Tests for evaluating corrosion in soils are discussed in the article "Simulated Service Testing" (see the Section "Corrosion Testing in Soils") in this Volume.

Corrosion in Concrete

The corrosion process in concrete is such that it tends to create conditions that increase the rate of attack. This phenomenon is related to the fact that the various corrosion products of iron and steel have a larger specific volume than the steel itself. The increase in volume of the corrosion products causes stresses that can lead to cracks in the concrete. These cracks allow easier access for the attacking medium and therefore more rapid attack. When the cracks are open to the exterior, corrosion products can be washed out and often lead to cosmetically objectionable stains. Of greater concern is the case of prestressed concrete. In this material, the corrosion process can lead to loss of structural strength and eventual failure. Of the cases studied, the failures of prestressed structures are not associated with stress corrosion, but are merely due to the loss of load-bearing area of steel (Ref 41).

The presence of chloride ions appears to be the principal cause of steel corrosion in concrete (Ref 42). Various attempts at reducing or eliminating the corrosion problem have focused on protective coatings for the steel members (galvanized, paints, and so on), decreasing the concrete permeability, increasing the depth of concrete cover, or eliminating the chloride ion through the use of sealants, and so on. Although many of these approaches have shown some degree of success, the application of cathodic protection has been the most successful in arresting corrosion (Ref 43).

There does not appear to be any significant body of data relating the severity of corrosion to the composition of the steel reinforcing member. It is likely that alloying could reduce the overall rate of attack, but whether the reduction would be significant in light of the added cost is problematic. Additional information on the corrosion and corrosion protection of steels in concrete can be found in Ref 44 to 48 and in the articles "Corrosion in Structures" and "Cathodic Protection" in this Volume.

Boiler Service

Corrosion in boilers is a special case of aqueous corrosion that involves elevated temperatures. The availability of oxygen appears to be the rate-determining step insofar as general attack is concerned. In closed-loop systems, the initial oxygen supply of the water is rapidly consumed in the early stages of film formation so that corrosion rates are usually not a problem. In nonclosed-loop systems, deaeration is usually adequate for eliminating general corrosion problems.

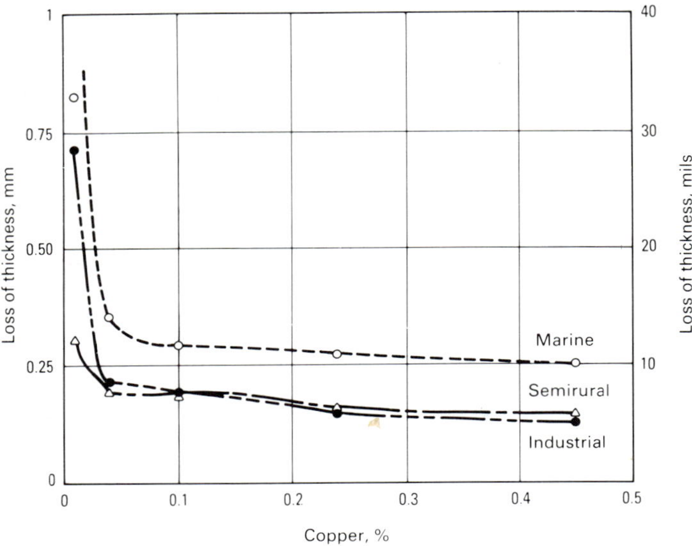

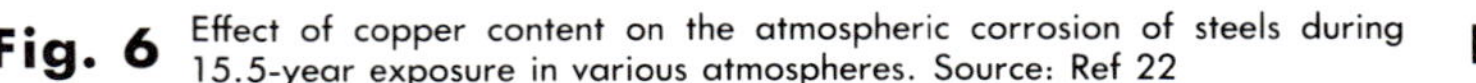

Fig. 6 Effect of copper content on the atmospheric corrosion of steels during 15.5-year exposure in various atmospheres. Source: Ref 22

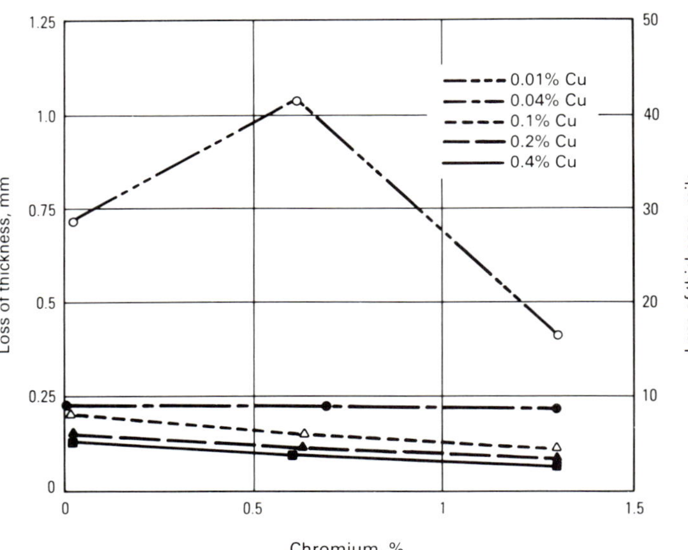

Fig. 7 Effect of chromium content on the atmospheric corrosion of five copper-bearing steels during 15.5-year exposure in an industrial atmosphere (Kearny, NJ). Source: Ref 22

Of more concern in boiler systems is the occurrence of pitting. In pitting corrosion, both dissolved oxygen and carbon dioxide (CO_2) promote attack. Deaeration is useful in stopping the oxygen attack, but CO_2 pitting is more effectively handled by maintaining an alkaline pH in the water. Surface deposits of corrosion products, mill scale, or even oil films have occasionally been implicated in the pitting attack of boilers. The treatment of boiler water is described in Ref 49 and in the article "Corrosion in Steam Systems" (see the discussion of feedwater) in Volume 1 of the 9th Edition of *Metals Handbook*.

Steel boilers are also susceptible to a form of stress-corrosion cracking (SCC) termed caustic cracking. This can occur even when the bulk concentration of caustic is below the danger level because of a variety of processes that can lead to localized accumulations of caustic well in excess of average concentrations. Because alkalinity is desired to avoid pitting attack, it is obvious that methods for preserving the benefit of pit suppression without incurring caustic cracking are necessary. Fortunately, such procedures have been devised; they are described in Ref 49. As a guide, Fig. 14 is presented for boiler water treatment with trisodium phosphate as a function of pH level necessary for crack suppression. Additional information on high-temperature aqueous corrosion can be found in the article "Corrosion in Fossil Fuel Power Plants" in this Volume.

Liquid-Metal Corrosion

Liquid metals can attack plain carbon steels in several ways. First, if iron is soluble in the liquid metal, dissolution will occur until the liquid metal is saturated. At this point, dissolution will cease in an isothermal system. Unfortunately, real systems almost always have temperature gradients. The presence of these gradients leads not only to dissolution in the high-temperature portions of the system but also to a precipitation or plating out in the colder areas. This precipitation decreases the solute content below the level that was present at the higher temperature, thus enabling additional dissolution to occur at the high temperature. The end result of this material transfer process is either the complete dissolution of the container on the high-temperature side of the system or, in the case in which the liquid metal is contained in channels or pipes, a blockage on the cold side from the continued precipitation.

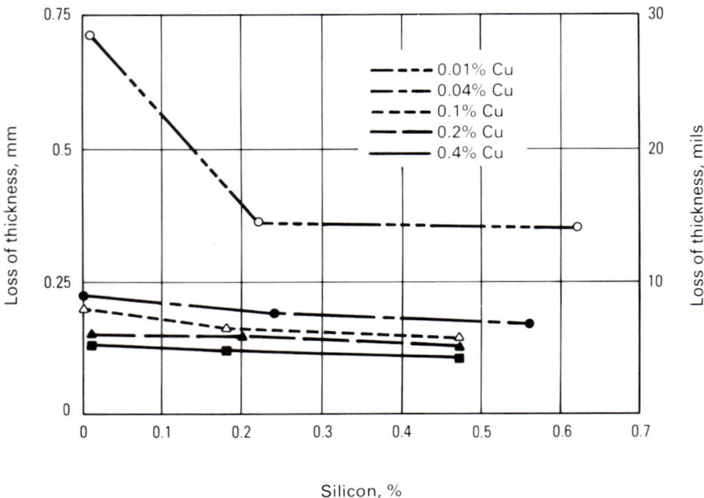

Fig. 8 Effect of silicon content on the atmospheric corrosion of five copper-bearing steels during 15.5-year exposure in an industrial atmosphere (Kearny, NJ). Source: Ref 22

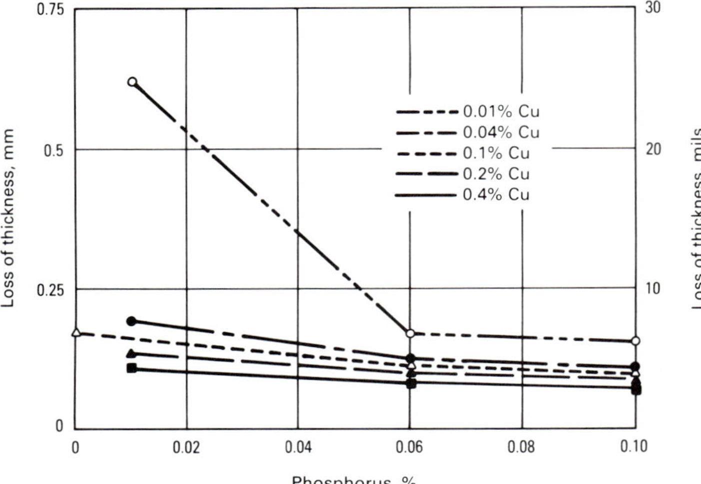

Fig. 9 Effect of phosphorus content on the atmospheric corrosion of five copper-bearing steels during 15.5-year exposure in an industrial atmosphere (Kearny, NJ). Source: Ref 22

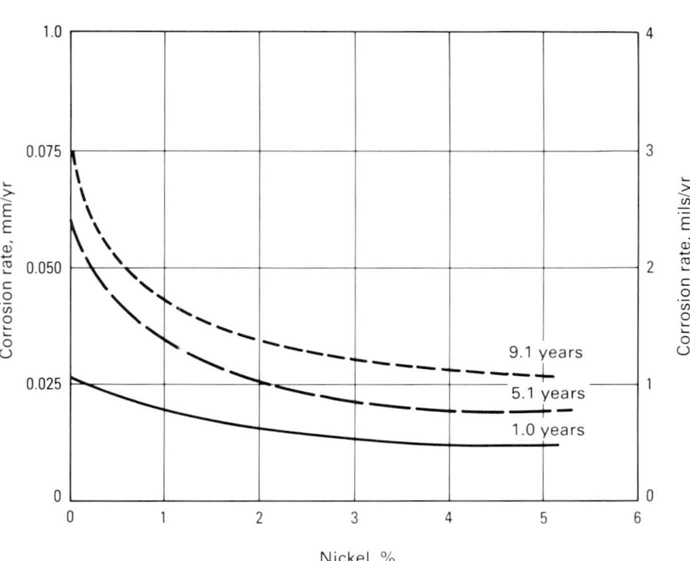

Fig. 10 Effect of nickel on the atmospheric corrosion of steel at Bayonne, NJ. Source: Ref 22

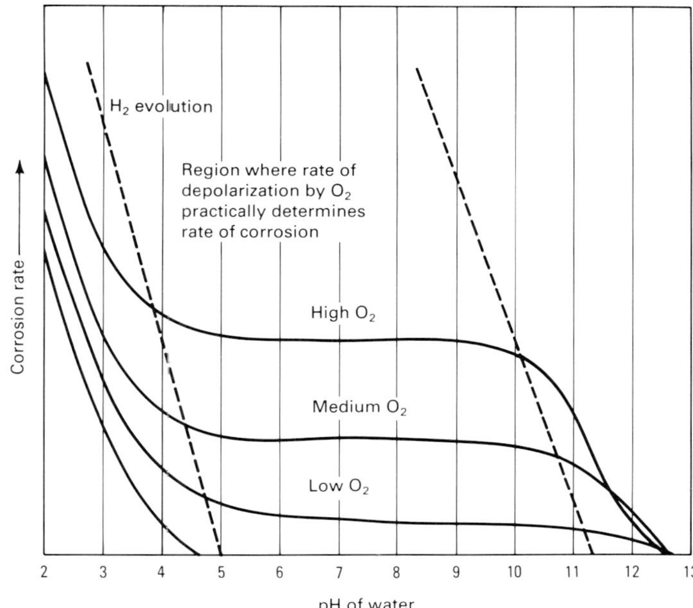

Fig. 11 Corrosion of iron by aqueous solutions. Source: Ref 31

A second mode of attack involves the formation of intermetallic alloys of the steel with the liquid metal. If the particular intermetallic is poorly adherent or if it can be spalled off by mechanical or thermal stresses (this exposes fresh surface to form more alloy), perforation can eventually result. On the other hand, if the intermetallic alloy that forms is tightly adherent and resistant to cracking, it can act as a barrier to diffusion and can slow the rate of attack.

Another form of attack that can occur and is more pernicious is intergranular penetration and/or dissolution by liquid metal. The unfortunate aspect of this mode of attack is that it can result in a loss of strength without any large weight loss or change in appearance. In this respect, it resembles the more familiar aqueous intergranular corrosion. More detailed information on dissolution by liquid metals can be found in the articles "Fundamentals of High-Temperature Corrosion in Liquid Metals" and "General Corrosion" (see the section "Corrosion in Liquid Metals") in this Volume.

Finally, there is the phenomenon of liquid-metal embrittlement, which requires the presence of both stress and a liquid metal. The cracks that occur during the embrittlement process may be intergranular or transgranular. The process seems to be similar in many ways to SCC. Steels have been reported to undergo embrittlement by lithium, indium, cadmium, zinc, tellurium, and various lead-tin solders. Additional information can be found in the article "Environmentally Induced Cracking" (see the Section "Liquid Metal Embrittlement") in this Volume.

Resistance of Plain Carbon Steels to Liquid-Metal Corrosion

Sodium and Sodium-Potassium Alloys. Plain carbon and low-alloy steels are generally suitable for long-term use in these media at temperatures to 450 °C (840 °F). Beyond these temperatures, stainless steels are required.

Lithium is somewhat more aggressive to plain carbon steels than sodium or sodium-potassium. As a result, low-alloy steels should not be considered for long-term use above 300 °C (570 °F). At higher temperatures, the ferritic stainless steels show better results at higher temperatures.

Cadmium. Low-alloy steels exhibit good serviceability to 700 °C (1290 °F).

Zinc. Most engineering metals and alloys show poor resistance to molten zinc, and carbon steels are no exception.

Antimony. Low-carbon steels have poor resistance to attack by antimony.

Mercury. Although plain carbon steels are virtually unattacked by mercury under nonflowing or isothermal conditions, the presence of either a temperature gradient or liquid flow can lead to drastic attack. The corrosion mechanism seems to be one of dissolution, with the rate of attack increasing rapidly with temperature above 500 °C (930 °F). Alloy additions of chromium, titanium, silicon, and molybdenum, alone or in combination, show resistance to 600 °C (1110 °F). Where applicable, the attack of ferrous alloys by mercury can be reduced to negligible amounts by the addition of 10 ppm Ti to the mercury; this raises the useful range of operating temperatures to 650 °C (1200 °F). Additions of metal with a higher affinity for oxygen than titanium, such as sodium or magnesium, may be required to prevent oxidation of the titanium and loss of the inhibitive action.

Aluminum. Plain carbon steels are not satisfactory for long-term containment of molten aluminum.

Gallium is one of the most aggressive of all liquid metals and cannot be contained by carbon or low-alloy steels at elevated temperatures.

Indium. Carbon and low-alloy steels have poor resistance to molten indium.

Lead, Bismuth, Tin, and Their Alloys. Low-alloy steels have good resistance to lead up to 600 °C (1110 °F), to bismuth up to 700 °C (1290 °F), and to tin only up to 150 °C (300 °F). The various alloys of lead, bismuth, and tin are more aggressive.

Weathering Steels

S.K. Coburn
Corrosion Consultants, Inc.

Yong-Wu Kim
Inland Steel Company

The weathering steels had their origin in the early studies of D.M. Buck. After a decade of effort, Buck established the efficacy of copper as a means of enhancing the atmospheric-corrosion resistance of unpainted carbon steel in a variety of environments (Ref 50-52). While this work was going on, a large study was initiated in 1916 by the American Society for Testing and Materials (ASTM) to evaluate the atmospheric performance of a variety of ferrous materials.

By 1929, United States Steel Corporation had initiated studies to enhance the performance of copper-bearing steel further through the addition of a number of alloying elements. By 1933, the first commercially available high-strength low-alloy steel was introduced into the railroad industry for coal hopper car use in the unpainted condition. These high-strength low-alloy steels were capable of resisting the leachates from sulfur-bearing coals better than the existing carbon and copper-bearing steel cars. Since that time, the original architectural grade has been covered by ASTM A 242 (Ref 53). When the heavier structural grades of high-strength low-alloy steels became available, they were covered by ASTM A 588 (Ref 54).

Through the exposure of small test panels in various atmospheres, the performance of the steel composition as well as the aggressiveness of

Table 4 Comparison of results under different types of exposure

Effects of alloy selection, chemical composition, and alloy additions	Sea air	Freshwater	Alternately wet with seawater or spray and dry	Continuously wet with seawater
Characteristic type of corrosion for ferrous alloys	Pockmarked	Vermiform on cleaned bars	Pitting, particularly on bars with scale	Pitting, particularly on bars with scale
Wrought iron versus carbon steel	Steel superior to wrought and ingot irons	Iron and steel equal in low-moor areas	Low-moor iron superior to carbon steel	Low-moor iron superior to carbon steel
Effect of scale	Cleaned steels lost less in weight and were less deeply pitted	Cleaned steels lost more in weight but were less deeply pitted	As in freshwater	Cleaned steels lost only slightly more in weight and were less deeply pitted
Sulfur and phosphorus contents	Best results when sulfur and phosphorus are low	Best results when sulfur and phosphorus are low, but effect less pronounced than in aerial tests	Best results when sulfur and phosphorus are low, but effect even less pronounced than in freshwater	Apparently little influence
Addition of copper	Very beneficial; effect increasing with the copper content	Very beneficial; 0.635% Cu almost as good as 2.185%	Beneficial; 0.635% and 2.185% Cu much the same	0.635% Cu slightly beneficial; 2.185% Cu somewhat less so
Addition of nickel	3.75% Ni superior even to 2% Cu; 36% Ni steel almost perfect even after 15-year exposure	3.75% Ni superior even to 2% Cu; 36% Ni steel offered excellent resistance	3.75% Ni beneficial; usually more so than Cu; 36% Ni steel the best metal in the set	3.75% Ni slightly beneficial and slightly superior to Cu; 36% Ni steel the best metal in the set
Addition of 13.5% Cr	Excellent resistance to corrosion; almost perfect after 15-year exposure; equal to 36% Ni steel	Excellent resistance to corrosion; equal to 36% Ni steel	Subject to severe localized corrosion that virtually destroys the metal	Subject to severe localized corrosion that virtually destroys the metal
Behavior of cast irons	Excellent resistance to corrosion; cold blast metal superior to hot; no graphitic corrosion	Undergoes graphitic corrosion	Undergoes graphitic corrosion	Undergoes graphitic corrosion

Source: Ref 31

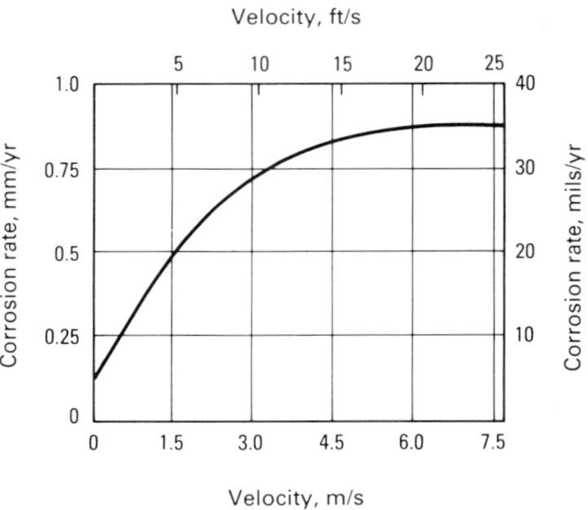

Fig. 12 Effect of velocity on corrosion of steel by seawater at ambient temperature

the particular location were both calibrated. Other characteristics of the high-strength low-alloy steels were studied, such as the ability to develop the protective oxide film under sheltered conditions in the atmosphere, in the soil, and immersed in freshwater and seawater. Studies were conducted on the staining characteristics and the ability to perform in contact with other materials. Finally, studies were performed to determine the manner in which protective coatings would function.

Copper-Bearing Steel

To appreciate the nature of the current composition of the weathering steels, it is useful to recall some of Buck's findings. In the early days, scrap steel was not in common use. Thus, Buck accidentally noticed that one test sheet outperformed the others. Upon examination, the copper level of this sheet was found to exceed the 0.01 to 0.02% common to the remainder of the test sheets. This finding resulted in Buck's initiating a series of studies to identify the minimum amount of copper (found to be 0.20%) necessary to effect an improvement in performance and to determine the relationship of the copper content to the sulfur content of the steel. From 1929 to 1933, much effort was expended toward developing compositions with superior atmospheric-corrosion resistance to the accepted 0.20% Cu-containing steels.

High-Strength Low-Alloy Steels

In 1962, the results of a comprehensive 15.5-year study were published in which some 270 different steels were exposed in three atmospheres beginning in the late 1940s (Ref 22). The sites were at Kearny, NJ (industrial); South Bend, PA (semirural); and Kure Beach, NC (250 m, or 800 ft, from the ocean). Table 5 lists the performance of 18 representative compositions in which the different levels of copper are combined with one of the four alloying elements (nickel, chromium, silicon, and phosphorus) to show their respective influences on corrosion in the industrial and marine sites. In addition, this group contains seven compositions in which one of the alloying elements is omitted, the purpose being to demonstrate how the remaining elements are capable of contributing to a satisfactory performance in the various atmospheres. Figures 6 to 10 in the previous section, "Carbon Steels," in this article also show the effects of the above-mentioned alloying elements on steel corrosion.

The significance of copper levels is shown in compositions 1, 2 and 3 in Table 5. Compositions 4 and 5 show how nickel can compensate for a low copper level. In contrast, compositions 6 and 7 show that chromium requires copper except in the marine environment. In compositions 8 and 9, silicon shows useful properties in the absence of copper. Phosphorus also contributes to this effect (compositions 10 and 11). Compositions 12 through 18 show the results of combining all of the alloying elements or omitting one of the elements. Additional research revealed that lower concentrations of the alloying elements were still effective. The current compositions of two of the major suppliers are shown in Table 6. Other proprietary compositions can be found in Ref 55.

Corrosion Behavior Under Different Exposure Conditions

The standard method for developing typical corrosion rates for comparative purposes is to expose 100- × 150-mm (4- × 6-in.) panels on test racks at an inclination of 30° from the horizontal facing south. An exposure rack with a sulfur dioxide candle in a louvered box is shown in Fig. 15. More detailed information on atmospheric-corrosion testing can be found in the article "Simulated Service Testing" in this Volume.

The panel performance is judged by the loss in weight sustained after varying exposure periods. The relationship between the loss in weight sustained by the skyward surface and the groundward surface by exposing test panels in semirural South Bend, PA, and industrial Kearny, NJ, for 4 years was first reported in Ref 56. In this test, carbon steel, copper-bearing steel, and USS COR-TEN steel panels were exposed. In both environments, the contribution to the total weight loss of the test panels was essentially the same. The skyward surface that was washed by the rain and warmed by the wind and sun contributed 37% to the weight loss, while the groundward surface that was never washed by the rain nor dried as much by the sun contributed 63% to the weight loss. This is significant because the sheltered surface has a coarse granular oxide film due to the loosely attached initial oxide film that tends

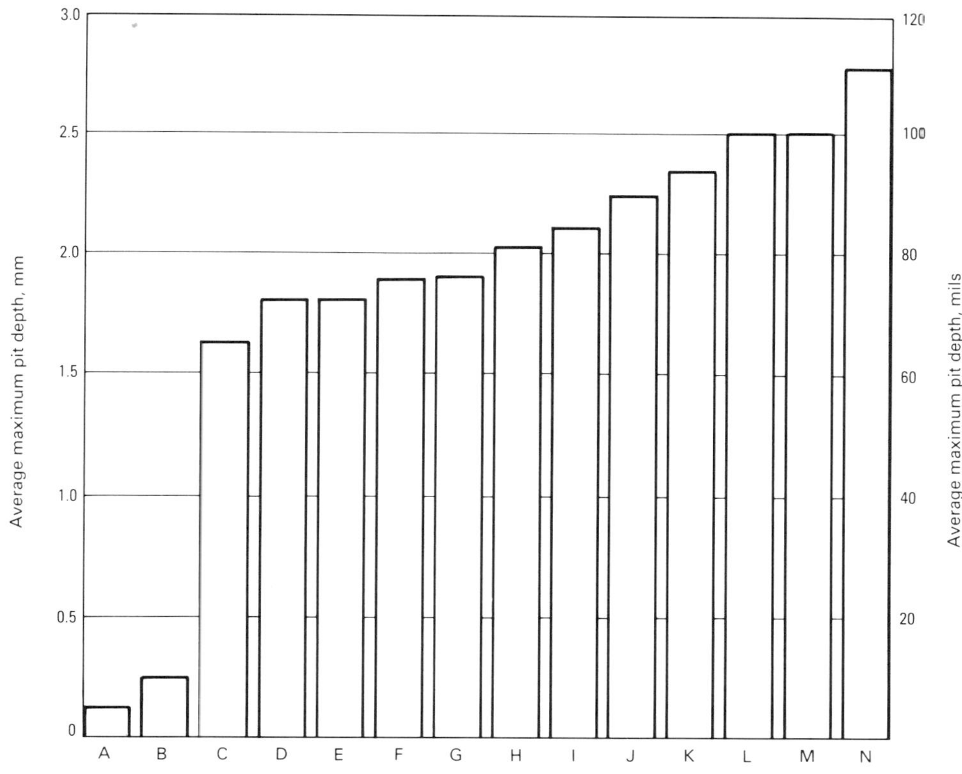

Alloy	Nominal composition, wt %
A	18% Cr, 8% Ni
B	18% Cr
C	Special cast iron (2.5C-1.4Si-0.3Mn-0.08S-0.13P-0.5Cu)
D	High-alloy cast iron (3.0C-2.1Si-1.0Mn-2.6Cr-15Ni-6.6Cu)
E	Copper-nickel steel (2.5Ni-1.0Cu)
F	Special cast iron (2.9C-2.0Si-0.8Mn-0.06S-0.25P-0.6Cu)
G	Cast iron (3.6C-1.6Si-0.5Mn-0.07S-0.7P)
H	Cast iron (3.6C-1.6Si-0.5Mn-0.07S-0.8P)
I	Low-carbon steel + 5% Cr
J	Low-alloy cast iron (3.5C-2.5Si-0.7Mn-0.05S-0.4P-0.3Cu-0.15Ni)
K	Wrought iron (0.02C-0.1Si-0.3Mn-0.02S-0.16P)
L	Low-carbon steel (0.15C-0.49Mn-0.03S-0.013P)
M	Wrought iron (0.02C-0.13Si-0.4Mn-0.02S-0.11P)
N	Copper-molybdenum iron (0.52Cu-0.15Mo)

Fig. 13 Averages of maximum pit depths on ferrous pipes removed from Merced clay adobe after an exposure of 5 years. Source: Ref 39

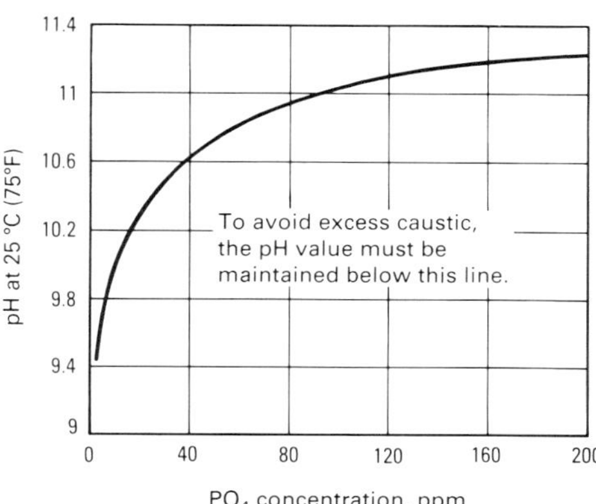

Fig. 14 Relationship of pH values to trisodium phosphate concentration in boiler water

to retain dampness and to promote additional corrosion.

Another study was conducted to compare the loss in weight between a vertical panel and an inclined panel in three different environments using copper-bearing steel and copper-free steel (Ref 57). It is apparent from the data discussed in Table 7 that a vertical surface that is only occasionally washed by rain is likely to corrode approximately 20 to 25% more than an inclined surface.

Because some weathering steels are licensed in many countries in Europe and the Far East, comparisons sometimes are made in which it is stated that poor results may occur because of some reduced performance level or excessive time interval before the protective oxide coating has matured in these different countries. What is often overlooked is that most European countries are located in latitudes that range from the equivalent Canadian border to the Arctic Circle. The sun is lower and the hours of sunlight are fewer than in the continental United States; thus, the time of wetness is longer (Table 8).

Characteristics of the Protective Oxide Film

It is likely that rusting occurs more during the night when atmospheric moisture condenses on metal surfaces when the dew point is reached. On average, in the Midwest and middle Atlantic states up through Canada, a 5.5- to 8.3-°C (10- to 15-°F) drop in temperature when the daytime relative humidity is around 75% is sufficient to result in condensation at night. The moisture absorbs the gaseous contaminants of the local atmosphere and nucleates around a dust particle and thus brings an acidic droplet to the metal surface. The thin acidic moisture film is capable of solubilizing the steel and initiating the rusting or oxidizing process. By morning, the sun and the moving air dry the gelatinous ferric hydroxide compounds that form, and the oxide begins to consolidate itself on the surface. This alternate wetting and drying cycle produces the protective oxide film on weathering steel compositions, but it produces the more porous nonprotective oxide film on carbon steel. Repeated cycles of this type ultimately result in complete coverage of the surface and a slowing of the corrosion rate. This behavior can best be expressed by time/corrosion curves, as shown in Fig. 16.

It should be noted that the formation of a protective rust film can result in a deceleration, although not necessarily a cessation, of corrosion. However, as the alloying increases, the quality of protection increases, as evident with the weathering steel composition. There are a number of ways of demonstrating this improving condition. A practical test is that demonstrated with rusted test panels that are evaluated for their wicking tendency after various exposure periods. Table 9 contains the results of a wicking test in which the edge of a test panel contacts the surface of a dish containing water. The height of wicking is an indication of the compactness of the protective oxide film.

It is easy to understand the sealing action that occurs as the oxide film ages by virtue of the behavior of a drop of water applied to the surface of a rusted test panel. The water droplet on a weathering steel panel retains its spherical form because of the sealed surface (Fig. 17). In contrast, the porous rust film on the carbon steel panel permits the droplet to wet out and penetrate the rust film.

As indicated earlier, the key to the development of the protective oxide film is the alternate wetting and drying cycle typified by the normal night and day exposure. Under conditions of long-term immersion in freshwater or seawater, the corrosion rate is the same as that for carbon steel—about 0.13 mm/yr (5 mils/yr). Similarly, burial in soil having varying moisture levels will result in behavior similar to that of carbon steel. The lack of a drying cycle inhibits the formation of the characteristic oxide film. The implication then is to avoid features in any structure, such as pockets, that can retain water for lengthy periods and to paint any portion of a structure, such as a column, that will be in the soil subject to rain and snow drainage.

Table 5 Average reduction in thickness of steel specimens after 15.5-year exposure in different atmospheres

Specimen No.	Composition, wt% Cu	Ni	Cr	Si	P	Thickness reduction: Kearny, NJ (industrial) μm	mils	Kure Beach, NC, 250-m (800-ft) lot (moderate marine) μm	mils
1	0.012	...	...	...	...	731	28.8	1321	52.0
2	0.04	...	...	...	...	223	8.8	363	14.3
3	0.24	...	...	...	...	155	6.1	284	11.2
4	0.008	1	...	...	...	155	6.1	244	9.6
5	0.2	1	...	...	...	112	4.4	203	8.0
6	0.01	...	0.61	...	...	1059	41.7	401	15.8
7	0.22	...	0.63	...	...	117	4.6	229	9.0
8	0.01	...	...	0.22	...	373	14.7	546	21.5
9	0.22	...	...	0.20	...	152	6.0	251	9.9
10	0.02	...	...	...	0.06	198	7.8	358	14.1
11	0.21	...	...	...	0.06	124	4.9	231	9.1
12	...	1	1.2	0.5	0.12	66	2.6	99	3.9
13	0.21	...	1.2	0.62	0.11	48	1.9	84	3.3
14	0.2	1	...	0.16	0.11	84	3.3	145	5.7
15	0.18	1	1.3	...	0.09	48	1.9	97	3.8
16	0.22	1	1.3	0.46	...	48	1.9	94	3.7
17	0.21	1	1.2	0.48	0.06	48	1.9	84	3.3
18	0.21	1	1.2	0.18	0.10	48	1.9	97	3.8

Source: Ref 22

Table 6 Representative compositions of A588 high-strength low-alloy steel

Proprietary grade	Composition, wt%(a) C	Mn	P	S	Si	Cu	Ni	Cr	V
USS COR-TEN	0.19(a)	0.80–1.25	0.04(a)	0.05(a)	0.30–0.65	0.25–0.40	0.40(a)	0.40–0.65	0.02–0.10
Mayari R	0.20(a)	0.75–1.35	0.04(a)	0.05(a)	0.15–0.50	0.20–0.40	0.50(a)	0.40–0.70	0.01–0.10

(a) Maximum. Source: Ref 55

Fig. 15 Typical exposure rack with sulfur dioxide candle in louvered box

To illustrate the results of constant dampness, a sheet of copper-bearing steel was exposed on a 30° test rack (Ref 26). The bottom of the sheet was turned up to serve as a trough to retain rainwater. The top of the sheet was turned down on the back of the rack to serve as a vertical wall facing north. The failure time in months in terms of severe rusting is shown in Table 10. It is clearly evident from the service life of the bottom and the sides of the trough that long-term dampness has a deleterious effect on the steel. It is also significant that the lack of the drying effect of the sun and the lack of the washing effect of the rain have combined to limit the service life of the vertical portion facing the north. The ideal exposure conditions are those in which the surface is washed frequently to remove contaminants and the sun is present to dry the surface.

Case Histories and Design Considerations

Thus, from the data presented in this section, the working rules for creating optimum conditions for the formation of the protective oxide film have evolved. The following case histories illustrate both the violations of these rules and suggestions on how to avoid certain maintenance problems.

Example 1: Assessing the Influence of Location. The Gulf Coast, where onshore breezes are the rule, experiences considerable penetration of salt air because there are no forested areas or concentration of tall buildings to provide a snow fence effect to deflect the incoming breezes upward. Thus, structures are prone to impact by salt-laden air, permitting a buildup of a salt residue that can inhibit formation of the characteristic oxide film.

To assess a location properly, one should expose a small test rack for 18 to 24 months with panels of weathering steel and copper-bearing steel or carbon steel with less than 0.02% Cu. Two or three removals for weight loss determination will indicate whether a protective oxide is forming on the weathering steel. If proximity to the ocean is a question, then exposure of a chloride candle, either at ground level or preferably at an elevation comparable to the height of the structure, should be made, and the monthly chloride determinations should be performed for at least 12 months in order to assess the influence of the seasons.

Example 2: Storage and Stacking of Weathering Steels. Electrical utilities store their tower angles at a central or regional location and along a power line site. In either case, angles should be stored facedown rather than nesting faceup. This eliminates the possibility of retaining water in between nested members. They should be stored on small steel angles with one end of the bundles elevated to facilitate drainage. When angles or channels are nested so that they can retain water, a loose voluminous rust scale develops, as seen in Fig. 18. Fortunately, such a scale can be readily removed by hammering, brushing, or with a power-driven wire wheel.

Large girders, H-beams, and channels are often nested for economical stacking. These should also be stacked with one end elevated and resting on steel angles. All configurations should facilitate drainage and minimize retention of rainwater. Draping with a cover cloth is preferred when space restrictions require that nesting be done.

Before heavy girders and columns are erected, they should be inspected by hammering to ensure that a laminated sheet of rust has not formed during the storage period. If this inspection is not performed, the rusted slab will begin to delaminate once in place, and this will raise questions as to whether the steel is truly of the weathering composition.

Example 3: Galvanic Corrosion Problems. Care must be exercised in preventing the mixing of carbon steel with the weathering steel stock.

Table 7 Relative corrosivity of panels in vertical and inclined positions in different environments

Location and environment	Position	Reduction in thickness after 2 years: Copper-free, µm	Copper-free, mils	Ratio	Copper-bearing, µm	Copper-bearing, mils	Ratio
Kearny, NJ (industrial)	Vertical	105	4.12	1.25	88	3.45	1.25
	Inclined	84	3.30		70	2.76	
Vandergrift, PA (industrial)	Vertical	116	4.55	1.22	102	4.02	1.25
	Inclined	95	3.73		81	3.20	
South Bend, PA (semirural)	Vertical	57	2.26	1.16	55	2.18	1.20
	Inclined	50	1.95		46	1.81	

Source: Ref 57

Table 8 Latitudes of North American and European countries

Latitude	U.S.	Canada(a)	Italy	France	Germany	England	Scandinavia
From	30°	45°	40°	45°	46°	52°	55°
To	45°	55°	45°	50°	53°	57°	70°

(a) Most densely populated area

When a missing member is encountered, the erection crews may substitute a carbon steel member. This may go unnoticed for several years and then result in excessive deterioration, such as that shown in Fig. 19.

One of the more vivid examples of galvanically coupled metals is the use of the hanger pin detail shown in Fig. 20 to facilitate girder movement during expansion and contraction. In this case, a bronze washer is part of the assembly. When such a device is used in the snow-belt states, it can create a strong galvanic cell with the steel when deicing salt solution drains from the deck through the expansion joint and through the crevice created by the connection. The outcome can be corrosion of the steel, with the resulting rust formation freezing and therefore immobilizing the joint. The resulting corrosion is evident in Fig. 21.

Example 4: Packout Rust Formation, Bolting, and Sealing. One of the major differences between a galvanized steel bolted structure, such as a transmission tower, and a weathering steel tower is the ability to tolerate loose bolts. In the case of a galvanized structure, moisture draining between a gusset plate and structural angle because of a loose bolt will cause little or no harm. In contrast, such retained drainage can initiate corrosion and rust formation in weathering steel joints, such as the condition of packout evident in Fig. 22. This condition can be overcome through the use of high-strength bolts; use of such bolts expands the area of clamping power over that of conventional bolts.

To overcome the possibility of packout formation, it is necessary to seal a joint effectively by an appropriate distribution of bolts. This is necessary for overcoming any tendency toward wicking action through capillary openings. The working guidelines for bolting, as discussed in Ref 58 and 59, deal with the establishment of bolt spacings and bolt-to-edge distances to provide adequate joint stiffness in order to avoid distortion due to packout corrosion products. Briefly, these guidelines provide that the pitch (spacing on a line of fasteners adjacent to a free edge of plates or shapes in contact with one another) should not exceed 14 times the thickness of the thinnest part and in any event should not exceed 180 mm (7 in.). The distance from the center of any bolt to the nearest free edge of plates or shapes in contact with one another should not exceed eight times the thickness of the thinnest part and in any event should not exceed 130 mm (5 in.). These factors are illustrated in Fig. 23.

At times, it would be wise to apply a caulk or sealant to the edges to ensure that an effective means for preventing the entry of moisture is used. Such a case is that shown in Fig. 24.

Example 5: Protection of Buried Members. When columns are located on concrete footers several feet below grade, they must not be installed in the unpainted condition. Such an action can result in moisture wicking upward from beneath the concrete pad and can create a condition of lamellar corrosion above grade. This condition is shown in Fig. 25. To avoid this, one must prepare the surface by blast cleaning or power brushing and apply a coal tar epoxy coating to extend several inches above grade. Arranging a grill work and drainage system is a desirable means of drawing off rainwater drainage and melted snow.

Table 9 Height of wicking of exposed steel panels

Exposure period, years	USS COR-TEN steel, mm	USS COR-TEN steel, in.	Carbon steel, mm	Carbon steel, in.
0.5	25.4	1.0	48.0	1.89
1.0	19.0	0.75	41.5	1.63
2.0	1.6	0.063	31.7	1.25
4.0	1.6	0.063	14.2	0.56

Fig. 17 Water droplet wetted out and penetrated carbon steel rust (left) but failed to penetrate sealed rust of weathering steel (right)

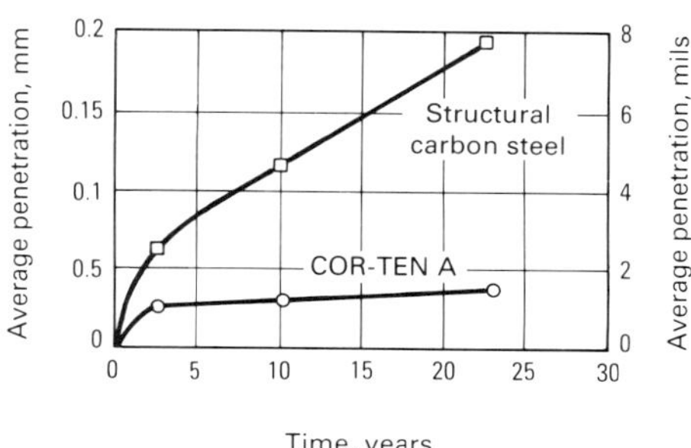

Fig. 16 Time/corrosion curves showing relative performance in a semi-industrial environment

Example 6: Contact With Fire-Retardant Wood Panels. A condition is often encountered in which a weathering steel curtain wall is placed over plywood panels that are treated with preservatives or fire retardants (Fig. 26). Because most fire-retardant compositions consist of inorganic salts capable of being leached from the panels if they become wet either through entry of water or through the maintenance of high relative humidities, it becomes necessary to insert a vapor barrier or effective sealant sheet, such as polyethylene. Alternatively, the interior face of the steel must be painted with a system capable of resisting the presence of water and the resulting salt leachate.

The major cause of failure of weathering steel curtain walls is inside-out corrosion due to the intrusion of moisture. A primary reason is that the quality of the interior protective coating is inadequate for resisting the destructive effects of long-term or frequent contact with liquid water rather than moisture vapor. Another cause is the failure of certain types of foamed-in-place insulation to adhere completely over the entire interior surface of the curtain wall.

Example 7: Painted Weathering Steels. Experience has demonstrated that paint, regardless of composition, will adhere better and give longer service when applied to an appropriately prepared weathering steel surface as compared to a carbon steel surface. This was demonstrated by exposing ten different paint systems over blast-cleaned panels of USS COR-TEN steel and carbon steel in the 25-m (80-ft) lot at Kure Beach, NC, for 15 years (Fig. 27).

The paint systems that failed on the (from left to right) fourth, seventh, and tenth carbon steel panels continued to function effectively on the weathering steel panels. The paint system that failed on the sixth carbon steel panel reached its true service life on the weathering steel panel; this permitted the exposed steel to develop its

Table 10 Failure time for copper-bearing steel sheet

Sheet face	Months to severe rust	Performance ratio
Bottom of trough	25	1
Sides of trough	25	1
Vertical portion facing north	130	5.2
Inclined skyward surface facing south	170	6.8

Fig. 18 View of loosely attached rust scale that formed among nested angles in a utility storage yard

protective oxide film to resist the environment further. In addition, it can be seen that the integrity of the paint, regardless of its composition, is retained on all but the sixth and seventh weathering steel panels. From this test and other similar exposure tests, it is suggested, conservatively, that paint life over a weathering steel surface can be doubled.

Example 8: Steel Thickness for Curtain Walls. Experience has demonstrated that if there is a desire to use a weathering steel as a curtain wall, the minimum thickness specified should be 18 gage (1.2141 mm, or 0.0478 in.). Thinner sections result in oil canning, which leads to irregular weathering, as noted in Fig. 28.

Example 9: Removing or Avoiding Stains. Staining is due to the small particles of dried rust becoming detached as condensed dew or rain drains down the sides of a structure. The particles of rust will dry on window panes or become trapped in the surface roughness of concrete columns and sidewalks. These deposits can be readily removed from windows with household abrasives. Such stains can be removed from concrete using typical building supplier's products. These concrete stain removers are generally acidic in nature and eliminate the stain by removing an extremely thin layer of concrete.

There are several ways to avoid staining of adjacent materials. To avoid staining of building entrance walks, one can install an anodized aluminum channel or gutter to retain and divert any drainage products, as shown in Fig. 29. Another method is to install a firm plastic sheet beneath a structural member to act as a deflector to protect lower walls, as shown in Fig. 30. Where horizontal and vertical structural members project beyond vertical members, condensate drippage can be retained through the use of shrubbery beds.

Example 10: Protection of Tower Legs and Lighting Standards. A very important form of protection for transmission tower legs and lighting standards at ground level is to maintain a clean area free of grass, bushes, and field crops. All of these forms of plant life tend to maintain a damp environment for long periods and interfere with the development of the protective oxide film. They are especially damaging when covering bolts and nuts at the base of these towers around concrete footers; the bolts and nuts in these areas can lose section and weaken in just a few years.

Summary

Weathering steels, used within the limitations outlined in this section, are useful structural materials, falling midway between painted carbon steel and the stainless steels. Depending on environment conditions, they can be used in the unpainted or the painted condition. In the painted condition, weathering steels contribute synergistically to extending the service life of the protective coating and therefore to lessening maintenance costs.

The primary limitations involve frequent and long-term contact with water caused by the inadvertent creation of pockets and crevices that trap and retain moisture. Another limitation is that found on bridge structures in which insufficient attention is paid to preventing attack of the below-deck structural members by deicing salt solution leaking through poorly maintained expansion joint devices. Also, bridge structures located close to the Gulf Coast shoreline are subject to constant onshore breezes, which contact the structure and leave sea salt residues in places on the structure where they cannot be washed away by the rain.

Like any below-ground carbon steel structure, the weathering steels require a protective coating, as they do when constantly immersed in freshwater or seawater. The protective oxide coating can develop only under conditions of alternate wetting and drying that occur in normal day and night exposure.

To avoid the staining that results from the drainage of moisture that contains particles of rust, one must resort to the techniques of retention and diversion. Finally, care must be taken to

Fig. 19 Results of mixing carbon steel angle in a weathering steel structure

Fig. 20 Typical hanger pin assembly with bronze washer

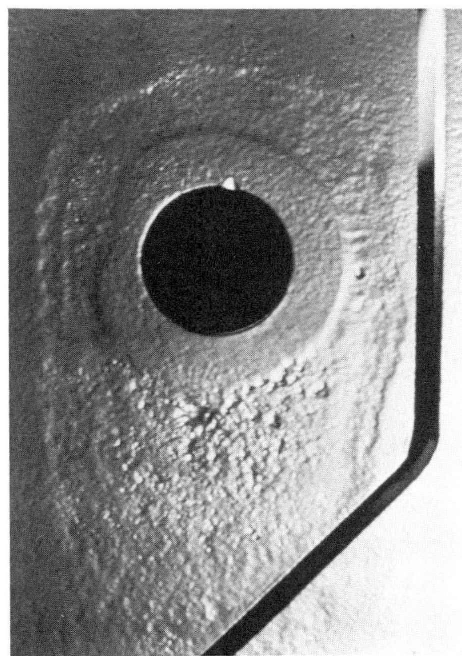

Fig. 21 View of blast-cleaned assembly showing effects of corrosion due to crevice attack and galvanic activity

protect field installations at ground level from the destructive effects of damp shrubbery, grass, and field crops. Clear space is necessary so that the structure can maintain a dry state, except for the usual periods of rain and snow.

Protection of Steel From Corrosion*

Corrosion protection is often an essential consideration in selecting carbon or alloy steel for a given structural application. Corrosion can reduce the load-carrying capacity of a component either by generally reducing its size (cross section) or by pitting, which not only reduces the effective cross section in the pitted region but also introduces stress raisers that may initiate cracks. Obviously, any measure that reduces or eliminates corrosion will extend the life of a component and increase its reliability.

Overall economics, environmental conditions, degree of protection needed for the projected life of the part, consequences of unexpected service failure, and importance of appearance are the chief factors that determine not only whether a steel part needs to be protected against corrosion but also the most effective and economic method of achieving that protection. There are two methods of minimizing the corrosion of steels. The first is to separate the reacting phases, and the second is to reduce the reactivity of the reacting phases. The separation of the reacting phases can be accomplished by metallic, inorganic or organic coatings, and film-forming inhibitors. Reactivity

Fig. 22 Distortion caused by packout rust formation because excessive spacing between bolts permitted entry of moisture into joint

can be reduced by alloying, anodic or cathodic protection, and chemical treatment of the environment. Some methods of protection combine two or more forms. For example, a baked epoxy paint applied over a chromate conversion coating on a galvanized steel part is, in reality, a combination of three forms. The epoxy paint provides a physical barrier to the corroding medium, the chromate conversion coating provides an inhibitor if the medium somehow penetrates the paint, and the galvanized coating (zinc) is an effective sacrificial anode (galvanic device) that diverts corrosion from the underlying steel part. A guide to the corrosion protection of carbon steels is provided in Table 11.

In addition to the active forms of corrosion prevention described above, design and material selection are important to the overall corrosion performance of a part. Many exterior steel structures—highway guard rails and electric transmission towers, for example—have been designed so that rain does not collect on horizontal surfaces in depressions, or at joints, but washes freely from all exposed surfaces. When such structures are made from a weathering steel containing small amounts of certain alloying elements (ASTM A242 and ASTM A588 structural steels, for example), the need for a protective coating is eliminated. Normal atmospheric conditions cause the steel to develop an adherent protective oxide that has a pleasing color. Maintenance is almost totally unnecessary because any mars or scratches in the oxide coating are naturally and automatically repaired by normal oxidation of the damaged area. Only physical damage affecting the function of the structure must be corrected by maintenance or repair personnel.

Galvanic couples usually are not serious considerations for steel exposed to the atmosphere, but become very important when a strong electrolyte such as seawater is involved. The effects of galvanic couples are described in the articles "Corrosion of Alloy Steels" and "General Corrosion" (see the Section "Galvanic Corrosion") in this Volume.

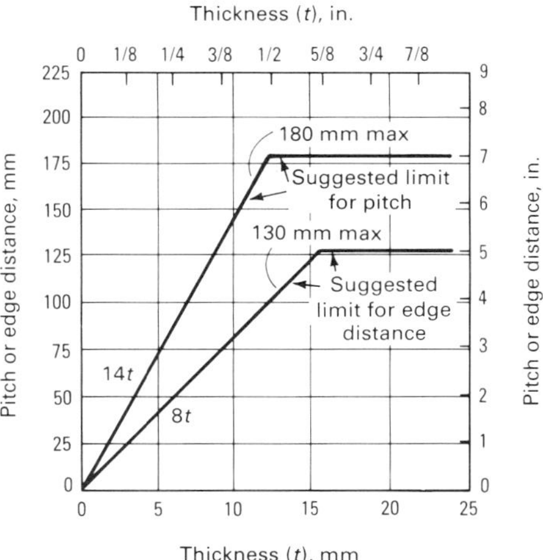

Fig. 23 Suggested spacing limits for joints in bolted weathering steel structures

Fig. 24 Gusset plate that should be strip caulked to prevent entry of moisture. Note the possibility for wicking action.

Fig. 25 Formation of lamellar rust due to moisture wicking upward from beneath concrete pad. The buried portion must be painted.

*Revised by H.H. Lawson, Armco Inc. Research & Technology, from Volume 1 of the 9th Edition of *Metals Handbook*.

Fig. 26 Example of fire-retardant treated plywood panels that must be separated from the weathering steel cladding by application of a paint system capable of resisting ingredients moisture-leached from the plywood

Fig. 27 Exposure test of ten paint systems applied to carbon steel panels (top) and weathering steel panels 25 m (80 ft) from ocean after 15 years. Courtesy of the LaQue Center for Corrosion Technology

Coatings

Many types of coatings are applied to enhance the corrosion resistance of carbon steels. Coating practices range from oiling for low-cost, temporary protection to vapor deposition for long-term corrosion, heat, and wear resistance. For economic reasons, the desired degree of protection must be determined before a coating is selected.

Effective temporary protection during shipment or storage can be obtained by coating the steel with mineral oil, solvents combined with inhibitors, emulsions of petroleum-base coatings, or waxes. These types of coatings are applied after acid pickling or between coating sequences, and they are not expected to provide long-term corrosion protection.

Surface preparation is important for all coating processes. Any oxide on the steel surface must be removed by pickling or blasting. Degreasing is necessary after oxide removal or when the steel has been given a temporary coating, and it can be accomplished by several means. More information on the cleaning and degreasing of steels is available in the article "Cleaning for Surface Conversion" in this Volume and in Volume 5 of the 9th Edition of *Metals Handbook*. Ideally, the first step in the coating process should be started immediately after cleaning.

Hot-dip coating processes are used to apply coatings of zinc, aluminum, lead, tin, and some alloys of these metals to carbon steels. The hot-dip process consists of immersing the steel in a molten bath of the coating metal. Zinc coating (galvanizing) protects steel galvanically because

Fig. 28 Clad columns with less than 18-gage cover, resulting in oil canning

Fig. 29 Protection offered by anodized aluminum channel to retain and divert dew condensate drippage

Fig. 30 Use of stiff rubberized sheeting beneath beam to divert drainage beyond stone wall

Table 11 Guide to corrosion prevention for carbon steels in various environments

Preventive method	Atmosphere	Soil	Freshwater	Seawater	Steam systems	Acids and pickling baths
Metal coatings: electroplating, galvanizing	Galvanizing very effective; plating with other metals used for both decorative appearance and corrosion protection	Galvanized steel widely used in drainage pipe and storm sewers	Galvanizing used in potable water	Not recommended	Not recommended	Not recommended
Painting: chemical treatment, priming, and painting	Economical and effective corrosion prevention	Seldom used	Fairly effective	Special paint systems used	Not recommended	Not recommended
Cathodic protection	Not recommended	Most economical and effective method, especially with organic coatings other than paint	Fairly effective with organic coatings	Very effective	Not recommended	Effective under special conditions
Inhibitors: liquid and vapor	Effective in closed areas	Not recommended	Effective in some applications, especially cooling waters	Fairly effective in some applications	Very effective	Very effective
Alloying additions to steel	Very effective, especially copper-bearing and HSLA steels	Not effective	Not effective	Only effective with much alloying	Chromium-molybdenum steels are very effective	Only effective with much alloying
Removal of oxygen from environment	Not applicable	Not recommended	Seldom used	Very effective, especially in desalination and hot seawater	Very effective	Not recommended
Removal of more noble metals; elimination of galvanic couples	Usually not necessary	Fairly effective	Effective	Necessary	Advisable	Not effective
Organic coatings other than paint	Sometimes used to replace painting	Used to advantage with cathodic protection	Fairly effective with cathodic protection	Used to advantage with cathodic protection	Not recommended	Have been used

the zinc is anodic to the steel base metal and therefore corrodes preferentially in most environments. Hot-dip galvanizing affords adequate atmospheric-corrosion protection to steel (see the section "Metallic Coated Steels" in this article and the articles "Hot Dip Coatings" and "Corrosion of Zinc" in this Volume). Aluminum hot-dip coatings (aluminizing) provide carbon steels with resistance to both corrosion and heat. In many environments, aluminum protects steel galvanically in much the same way as zinc.

Zinc-aluminum and aluminum-zinc alloys are also applied to steel by hot dipping. Heating aluminized steel results in the formation of an iron-aluminum intermetallic compound that resists oxidation at temperatures up to about 800 °C (1500 °F). Aluminized steel is often used where heat resistance is required—for example, in automotive exhaust systems (see the section "Metallic Coated Steels" in this article and the articles "Hot Dip Coatings" and "Corrosion in the Automotive Industry" in this Volume).

Hot-dip tin coatings provide a decorative and nontoxic barrier coating. Tin does not galvanically protect the steel substrate; for this reason, lacquers or other organic coatings are frequently used to fill pores in the tin coating and provide enhanced barrier protection. More information on tin coatings for steel is available in the article "Corrosion of Tin and Tin Alloys" in this Volume.

Hot-dip lead coatings are sometimes used on steel that will be exposed to sulfuric acid fumes or other aggressive chemical environments. Terne plate, a lead-tin alloy coating, gives more protection than pure lead coatings and is solderable (see the article "Corrosion of Lead and Lead Alloys" in this Volume).

Electroplated coatings are applied to steel for corrosion resistance, appearance, solderability, or other special requirements. A wide variety of materials are electroplated on steel, including zinc, aluminum, chromium (see the article "Corrosion of Electroplated Hard Chromium" in this Volume), copper, cadmium, tin, and nickel. Multilayer coatings can also be applied by electroplating; an example is the copper-nickel-chromium plating system used for bright automotive trim. More information on the electroplating process is available in Volume 5 of the 9th Edition of *Metals Handbook*. The article "Electroplated Coatings" in this Volume also contains information on the corrosion resistance of various plating materials.

Clad Metals. Carbon steels can be bonded to more corrosion-resistant materials, such as copper and stainless steels, by cold roll bonding, hot roll bonding, hot pressing, explosion bonding, and extrusion bonding. The resulting lamellar composite material has specific properties not obtainable in a single material. Examples of various clad metal systems and guidelines for designing with clad metals are enumerated in the article "Corrosion of Clad Metals" in this Volume.

Thermal spray coatings provide effective long-term corrosion protection for steels in a wide range of corrosive environments. They are applied by one of several processes, including wire flame spraying, powder flame spraying, and electric arc spraying. Zinc, aluminum, and zinc-aluminum alloys are the most common coating materials applied by thermal spray techniques; austenitic stainless steels, aluminum bronzes, and MCrAlY coating materials have also been used for specific applications. For maximum corrosion resistance, thermal sprayed coatings are sealed with an organic topcoat. Thermal spray coatings are often used for corrosion protection in marine applications. More information on thermal spray coatings is available in the articles "Thermal Spray Coatings" and "Marine Corrosion" in this Volume.

Vapor-deposited coatings are sometimes used for the protection of steel, although the cost of such coatings can be prohibitive. In vapor deposition, whether it be physical vapor deposition or chemical vapor deposition, the coating material is transported to the substrate in the form of individual atoms or molecules. A wide range of coating materials can be applied by vapor deposition. If applied to a sufficient thickness, the coating is essentially pore free and dense, thus providing excellent barrier protection. A well-known application for vapor-deposited coatings on steel is ion vapor deposited aluminum coatings on steel aircraft and aerospace components (see the article "Corrosion in the Aircraft Industry" in this Volume). More information on materials for and applications of vapor-deposited coatings is available in the article "CVD/PVD Coatings" in this Volume.

Phosphate or chromate conversion coatings are used to enhance the corrosion resistance of steels. By themselves, they provide slightly better corrosion resistance than bare steel; more often, they are used in conjunction with another coating system. Conversion coatings are applied after hot-dip galvanizing and provide good corrosion protection when topcoated with an organic coating system. More information on conversion coatings for the protection of steel is available in the articles "Chromate Conversion Coatings" and "Phosphate Conversion Coatings" in this Volume.

Organic coatings (paints) are used more often for corrosion protection of steels than any other type of coating. Properly applied, they provide excellent protection at a relatively low cost. A wide variety of coating materials and application methods are available; these are discussed in detail in the section "Organic Coated Steels" in this article and in the article "Organic Coatings and Linings" in this Volume.

Ceramic coatings used to protect steel include silicate cements and porcelain enamels. Monolithic cement linings provide good resistance to chemicals and thermal insulation. They can be applied by casting or spraying. More information on these materials is available in the article "Chemical-Setting Ceramic Linings" in this Volume. Porcelain enamels are glass coatings that are fused onto the steel surface at or above 425 °C (800 °F) to provide a glassy coating with good corrosion resistance and high hardness. The composition of the enamel can be varied to provide desired properties, such as improved resistance to alkalies. More information on porcelain enamels is available in the article "Porcelain Enamels" in this Volume.

Other nonmetallic materials are sometimes used as coatings or linings for steel in corrosion applications. These include rubbers (both natural and synthetic) and other elastomers and such plastic materials as epoxies, phenolics, and vinyls. A wide variety of properties and resistances to specific environments are available. Rubber linings have been used for many years in steel storage tanks for hydrochloric acid. Plastic linings are employed for plating tanks and similar applications. The articles "Corrosion in the Chemical Processing Industry," "Corrosion of Metal Processing Equipment," "Corrosion in the Pharmaceutical Industry," and "Corrosion in the Brewery Industry" in this Volume contain information on the use of these materials for the protection of steel.

Inhibitors

Inhibitors find their major uses in acid-pickling solutions, acidic service environments, steam systems, and neutral and near-neutral aqueous solutions. Inhibitors may be organic or inorganic compounds, and they are usually dissolved in aqueous environments. Inhibitors have been added to chemical conversion treatment baths and to paint primers. A few vapor-phase inhibitors are used in confined atmospheres (see the discussion "Packaging Applications" in this section) and in steam systems.

Some of the most effective inorganic inhibitors are chromates, nitrites, silicates, carbonates, phosphates, and arsenates (it should be noted that environmental concerns have significantly impacted the use of chromates). The organic inhibitors are many and include amines, heterocyclic nitrogen compounds, sulfur compounds (such as thioethers, thioalcohols, thioamides, thiourea, and hydrazine), some natural compounds (such as glue or proteins), and mixtures of two or more compounds. More detailed information on inhibitor types and applications can be found in the articles "Corrosion Inhibitors for Oil and Gas Production," "Corrosion Inhibitors for Crude Oil Refineries," and "Control of Environmental Variables in Water Recirculating Systems" in this Volume.

Corrosion Inhibitors in Acid Pickling. Mill scale is most often removed from steel by pickling in either sulfuric (H_2SO_4) or hydrochloric acid (HCl) solutions. Because pickling is essentially a chemical dissolution (corrosion) process, some of the underlying metal is inevitably removed from the surface along with the mill scale. Inhibitors are added to pickling baths to minimize metal loss, to minimize the extent of hydrogen pickup (which can lead to embrittlement), to protect the metal against pitting and poor surface quality, to reduce acid fumes, and to reduce acid consumption. Inhibitors prolong pickling time, but the benefits outweigh the increase in time. Table 12 gives mill scale removal rates in uninhibited H_2SO_4 and HCl. The removal rate is much higher for HCl but acid cost and acid consumption are lower for H_2SO_4.

Acid-pickling inhibitors should be stable in the acid at all operating temperatures and must not stain or contaminate the steel. The natural compounds that have been used include bran flour, gelatin, glue, sulfonated coal tar products, asphaltum, and wood tars. Synthetic compounds that have been used are nitrogen-base materials and their derivatives, pyridines, guanidines, aldehydes, and thioaldehydes and other sulfur-containing compounds.

Most inhibitors used for large-scale pickling are comparatively inexpensive technical grade chemicals that contain varying amounts and types of impurities. The inhibitive properties are often due to the presence of certain impurities rather than to the major constituent that determines the name of the product. When purified, many of these chemicals are ineffective as inhibitors. For example, various naphthalene sulfonic acids lose their inhibitive properties after purification by repeated recrystallization. Proteins and their breakdown products—the amino acids—can retard corrosion; however, purified proteins and amino acids are often found to be weaker inhibitors. The inhibitive properties of plant extracts are also due to the presence of natural nitrogen-containing bases known as alkaloids. Many of these—for example, brucine, strychnine, and papaverine—are very powerful corrosion inhibitors in sulfuric acid.

For a given metal, an inhibitor that is quite effective in one acid often has comparatively little influence in another acid of equal concentration. Thus, amines of high molecular weight retard corrosion of ferrous metals in hydrochloric acid far more effectively than in sulfuric acid. Proteins, on the other hand, are more effective in sulfuric acid.

Hydrogen Embrittlement. During pickling, hydrogen ions accumulate at the metal surface. The hydrogen ions combine to form molecules of hydrogen, which escape from the solution as bubbles. However, a few atoms penetrate the crystal lattice, where they may either diffuse through the metal and exit at the opposite surface or accumulate within the metal and embrittle it. Sometimes, as a result of high hydrogen pressures in subsurface pores, pickling blisters are formed on the surface.

Metal articles with polished surfaces absorb less hydrogen than those with rough, rusty, or scaled surfaces. Pickling in H_2SO_4 leads to much greater hydrogen pickup and more pronounced embrittlement than pickling in HCl of the same concentration. Inhibitors added to pickling solutions have a great effect on hydrogen pickup and its diffusion rate. Some inhibitors do not affect hydrogen pickup. In the presence of strong inhibitors, hydrogen absorption and diffusion is lowered to such an extent that quantitative results have no practical significance. It is important to recognize the inhibitors that strongly retard corrosion do not always lower the diffusion of hydrogen into the metal. For example, the comparatively weak inhibitor diethylaniline protects the steel effectively against hydrogen pickup when steel is pickled in 5 N H_2SO_4. On the other hand, when chemically pure pyridine is added to acids, the diffusion of hydrogen through the steel is appreciably enhanced despite effective inhibition against corrosion. Arsenic inhibitors also enhance the diffusion of hydrogen into steel.

Inhibitors act differently in nitric acid (HNO_3). Although the concentrated acid is a strong oxidizer, dilute solutions have more evident acidic properties. During pickling in uninhibited HNO_3, the HNO_3 is reduced to nitrous acid, which catalyzes further reduction to nitrous acid, which attacks ferrous metals. Hydrogen is consumed in producing nitrous acid and is not liberated at the metal surface; consequently, hydrogen embrittlement does not occur when steel is pickled in uninhibited HNO_3. A 0.007% solution of thiourea in a 1.5 N HNO_3 solution reduces the corrosion of carbon steel by 250 to 300 times. However, the mechanical properties after pickling are considerably impaired because thiourea hinders the reduction of HNO_3 to nitrous acid and causes hydrogen to be liberated at the metal surface along with other gases rather than being consumed within the acid solution.

Hydrazine, which is used as an oxygen scavenger, is an effective inhibitor because it is converted to hydrazoic acid, which decomposes the nitrous acid that forms. Potassium permanganate and potassium dichromate are also powerful inhibitors for HNO_3. The activity of these substances persists over a long period. For example, dichromate ions are active for several months in 1 N HNO_3 but they are persistent for only 10 h in 5 N HNO_3. Halides are also good inhibitors in HNO_3, potassium iodide being more effective than potassium bromide or potassium chloride. None of these inhibitors causes hydrogen embrittlement in HNO_3 solutions. More detailed information on hydrogen embrittlement is available in the articles "Environmentally Induced Cracking" and "Evaluation of Hydrogen Embrittlement" in this Volume.

Inhibitors in Acid Environments. The solution rate for steel in various acids depends on the surface finish of the steel, type and concentration of the acid, temperature, and the presence or absence of impurities that can either promote or inhibit corrosion. Generally, the corrosion rate of steel in HCl increases with acid concentration; however, in HNO_3 and H_2SO_4, the corrosion rate reaches a maximum and then decreases with

Table 12 Mill scale removal rates in uninhibited aqueous acids

Concentration of acid, %	Removal rate (a), g/h HCl	H_2SO_4
5	0.035	0.035
10	0.29	0.039
15	1.51	0.056

(a) At 18 °C (65 °F)

increasing acid concentration. With HNO_3, the peak is at 6 N; with H_2SO_4, the peak is at 12 N.

The corrosion rate of steel in all types and concentrations of acid solutions tends to increase with carbon content as long as there is little or no change in the concentration of other alloying elements in the steel. In 4 N acetic acid, for example, the corrosion rate of carbon steel is about 1.2 g/m²/h for carbon steel containing 0.05% C and 2.8 g/m²/h for 0.90% C. In 12 N H_2SO_4, the corrosion rate increases from 60 g/m²/h for steel with 0.05% C to 120 g/m²/h for steel with 0.90% C. In 16 N H_2SO_4, however, corrosion rates are the same regardless of carbon content, and the metal loss is less than 10 g/m²/h.

Inhibitors are often used in acidic process fluids to reduce the rate of corrosion to a tolerable level, thus prolonging the life of steel components in contact with the process fluid. Sometimes, it is not possible to use a particular inhibitor because of an adverse effect on the process. Then, a different inhibitor can sometimes be used, even though it is not as effective as the optimum choice. With some processes, the use of any inhibitor would have an adverse effect on the process. In such cases, the only alternative to high corrosion rates and expensive maintenance programs is the use of corrosion-resistant materials, either as coatings or linings or as basic materials of construction.

Volatile Inhibitors. The protection of metals against atmospheric corrosion and in closed steam systems by means of volatile (vapor phase) inhibitors is one of the more recent, significant methods for combating metallic corrosion. For atmospheric corrosion, the method consists of saturating an enclosed storage space with vapors that enhance the corrosion resistance of the metal surface. Volatile inhibitors protect metals primarily by changing the kinetics of the corrosion reactions much the same as inhibitors in solution. On the basis of general inhibition theory gained through experience, the kinds of groups necessary to effect protection can be predicted. For steel, they are chromate, dichromate, benzoate, nitrite, phosphate, hydroxyl, and ammonia. Such salts as sodium nitrite and potassium chromate (which are exceptionally effective as contact inhibitors or passivators in solution) could, in principle, be used as volatile inhibitors if their vapor pressures were suitable. Therefore, if an inorganic compound contains a desirable protective group, an organic radical can be substituted for the inorganic metal so that an organic salt is obtained that will now possess the two desirable properties, namely volatility and a protective group. Often, it may not be possible to vaporize a protective group at ordinary temperatures, and means must be taken to reduce the vapor pressure of the compound.

Amines were the first group of chemical compounds that attracted attention as volatile inhibitors. Amines are obtained by the substitution of an organic radical for one hydrogen atom in the ammonia molecule. An amine can combine with water to form organic cations and hydroxyl anions. Thus, the medium is made more alkaline, and the corrosion of iron is reduced. It was assumed that the principal passivating effect is produced by the hydroxyl ion; however, electrochemical investigations have shown that the organic cation, not the hydroxyl anion, plays the more significant role in the passivation process. Aliphatic, alicyclic, and heterocyclic amines are more effective than aromatic amines; for example, cyclohexylamine, dicyclohexylamine, and monoethanolamine are very effective. These amines have significant shortcomings, however. Because their vapor pressures are too high, they cannot provide lasting protection.

Vapor pressure is one of the basic characteristics of volatile inhibitors. The higher the vapor pressure, the sooner an airtight storage space is saturated. Because it is not economically feasible to produce airtight systems, a less volatile inhibitor is not consumed as quickly and can ensure more lasting protection. On the other hand, because more time is required to attain a protective vapor concentration, steel parts may corrode before a protective concentration is reached. Most of the amines used with steel have a vapor pressure that is too high to provide lasting protection; therefore, it is necessary to reduce the vapor pressure of the amine, which can be done by converting the amine to a salt.

Packaging Applications. The major application utilizing vapor-phase inhibition has been the packaging of steel articles ranging from small hand tools to large coils of steel strip. Paper can be impregnated with varying quantities of inhibitor—0.2 to 2 g/m²—depending on prevailing temperature and on the time a nonaggressive environment must be maintained. Packaging for long-term storage imposes volatility requirements that are different from those of packaging only for protection in transit. Paper impregnated with 0.2 g/m² of dicyclohexylamine nitrite is capable of conferring protection up to 10 years if the temperature does not exceed 23 °C (73 °F). However, protection is effective for only about 100 days at 75 °C (167 °F). It has been found that the logarithm of the effective service time of paper impregnated with inhibitor is inversely proportional to the temperature.

Various inhibitor compounds have been encapsulated and used in tablet form to protect various iron and steel articles in museum cabinets where humidity control was difficult. The same tablets are now being used in cardboard cartons employed for shipping numerous steel articles. Aqueous solutions of mixtures of these compounds are used to fog the holds of seagoing barges and to fog closed railroad cars and trucks carrying coiled steel, which supplements the impregnated paper wrapping. Paper wrapping can be eliminated by spraying the edges of lifts and coils of steel and fogging the atmosphere of the cargo container so as to offer short-term transit protection. Grease and oil films have been the conventional means for protecting coils of wire; this produces hazardous and unsightly oil residues on the floors of transit vehicles and warehouses. A dip into an aqueous solution of an inhibitor plus water-dispersable waxes results in a clean, dry product that is self-lubricating and merely requires dry warehouse storage.

Automotive Applications. Corrosion inhibitors are extensively used in that part of the automotive industry associated with the production and storage of parts. In addition, inhibitors are included in many automotive fluids, such as gasoline, antifreeze, and oil and brake fluids. The addition of inhibitors to rust-preventive compounds increases the protection provided by hydrophobic coatings. Organic compounds such as alcohol and ethylene glycol are the principal substances used to prevent engine cooling fluid from freezing in cold climates. Both substances undergo oxidation to form acidic products. The evaluation of corrosion inhibitors to prevent attack of metal parts by organic acids has become a complex problem. Many variables are involved, such as aeration, cooling system temperature, pressure, coolant velocity, impurities in the water, corrosion products, and galvanic couples, among other operating conditions. The concentration of inhibitor required to prevent corrosion is dependent on the impurities present in the water used. For example, ferrous metal corrosion can increase as the concentration of chloride ions is increased to 100 ppm and the concentration of sulfate ions to 300 ppm.

In both the United States and Europe, chromates were among the first corrosion inhibitors to be used. A significant portion of the inhibitor formulations presently used in the United States contains borates or combinations of borax and sodium silicate. However, the effectiveness of borates is reduced with increases in the concentrations of chloride and sulfate ions. Mixed inhibitor systems, such as benzoate-nitrite combinations, have been quite successful (5% sodium benzoate plus 0.3% sodium nitrite is effective in a 25% ethylene glycol solution). General Motors developed an antifreeze combination that included sodium nitrite, borax, sodium silicate, and trisodium phosphate, with each compound functioning in a special fashion. For example, nitrite was used to protect aluminum and solder; borate was used for its buffer action and reserve alkalinity, as well as for protection of ferrous metals. Silicate was used for general protection of all metals, phosphate for protection of ferrous metals, aluminum ions for their buffering action, sodium hydroxide for reserve alkalinity and mercapto-benzothiazole (MBT) for protection of copper and brass. There is also a need for cavitation protection in cooling systems made of aluminum; phosphates and silicates are helpful in this instance. Sodium silicate in amounts greater than 0.3% (by weight) added to undiluted ethylene glycol can produce gelling during storage. Additional information on prevention of automobile corrosion can be found in the article "Corrosion in the Automotive Industry" in this Volume.

Cathodic Protection

The most economical method for protecting underground or underwater steel structures from corrosion is usually cathodic protection. The use of cathodic protection for long-term corrosion prevention for underground pipelines, oil and gasoline tanks, offshore drilling rigs, well-head structures, steel piling, piers, bulkheads, offshore pipelines, gathering systems, drilling barges, and other underground and underwater structures is now a fairly standard procedure. In this method, a direct electrical current is applied to the steel structure. The direct current can be supplied by galvanic action from sacrificial anodes or by impressed current from a rectifier. Coatings of the coal tar or coal tar epoxy types or fusion-bonded epoxy powders are generally used on steel structures in conjunction with cathodic protection to lessen the amount of current needed. The current then must protect the steel only at holidays and flaws in the coatings. Detailed information on the principles of cathodic protection and the types of sacrificial anodes and impressed-current anodes can be found in the article "Cathodic Protection" in this Volume. Other articles that include discussions of the use of cathodic protection of steels include "Marine Corrosion," "Corrosion in Petroleum Production Operations," "Corrosion of Pipelines," and "Corrosion in Structures" in this Volume.

Metallic Coated Steels

Harvie H. Lee
Inland Steel Company

The main purpose for the application of a metallic coating to a steel substrate is for corrosion protection. Most metallic coatings are applied either by hot dipping into a molten bath of metal or by electroplating in an aqueous electrolyte. To a lesser extent, coatings are also applied by other methods, such as metal spraying, cementation, and metal cladding. A number of coating processes were reviewed in the section "Protection of Steel From Corrosion" in this article.

From the standpoint of corrosion protection of iron and steel, metallic coatings can be classified into two types—noble coatings and sacrificial coatings. Noble coatings such as lead, copper, or silver are noble in the galvanic series with respect to steel. For noble coatings, at areas with surface defects or porosity, the galvanic current accelerates attack of the base steel and eventually undermines the coating. Sacrificial coatings such as zinc or cadmium are anodic (more active) to steel. For sacrificial coatings at uncoated areas (pores), the direction of galvanic current through the electrolyte is from coating to the base steel; as a result, the base steel is cathodically protected. In general, the thicker the coating, the longer the duration of cathodic protection. This section will emphasize hot-dipped zinc and aluminum coatings. More detailed information is provided in the articles "Hot Dip Coatings," "Corrosion of Zinc," and "Corrosion in the Automotive Industry" in this article.

Zinc-Base Coatings

Types. Zinc-coated steels are generally produced by either hot dipping or electroplating. The main difference between these two types of zinc coatings is in their coating structure. The hot-dip galvanized steel coatings consist of a layer of zinc-iron intermetallics at the steel/zinc interface (Ref 60). Typical coating microstructures with various intermetallic layers are shown in Fig. 31. For galvanized sheet steel with better coating adhesion and good coating forming properties, the thickness of the intermetallic layers should be controlled below 20% of the total coating thickness. Control of intermetallic layer thickness can be achieved by adding 0.1 to 0.3% Al into the zinc bath; this addition retards the growth of the intermetallic layer.

In the case of electrogalvanized steels, the coatings display a smooth, uniform, and spangle-free surface and do not have a zinc-iron intermetallic layer. Currently, almost all galvanized sheet steel in the United States is produced by continuous process. The two commercial processes used in the continuous hot-dip galvanizing of sheet steel are the Sendzimir process and the Cook-Norteman process.

In the Sendzimir process, the steel strip is heated in a high-temperature furnace consisting of an oxidizing atmosphere to remove organic oils and surface contaminants, followed by heating in a reducing furnace with a hydrogen-rich atmosphere to reduce the surface oxide layer and to anneal the steel substrate. The discharge end of the reducing furnace is below the surface of the zinc bath; this allows the continuous sheet to enter the bath without passing through a contaminating atmosphere. Precise control of oxidizing and reducing temperature is critical in developing and maintaining the cleanliness of the steel surface.

In the Cook-Norteman process, an in-line furnace is not used. The sheet is chemically cleaned by alkaline degreasing and acid pickling. After cleaning, the sheet is coated with a film of zinc ammonium chloride, dried, and preheated to less than 260 °C (500 °F) before entering the galvanizing bath.

Aqueous Corrosion of Galvanized Steel. Zinc is an amphoteric metal that corrodes in acid and alkaline solutions. The hydrogen ion concentration in water and aqueous solutions has a significant effect on the corrosion rate of zinc. This effect is shown in Fig. 32, which plots the average overall corrosion rate versus the hydrogen ion concentration expressed in terms of pH value (Ref 61). In the pH range of 6 to 12.5, a protective film is formed on the zinc surface, and the zinc corrodes very slowly. At pH values below 4 and above 12.5, the major form of attack on zinc is hydrogen evolution, and zinc corrodes very rapidly. The pH values for natural water and mild alkaline soap-bearing water fall within the safe range and are not corrosive toward zinc coatings. The corrosion rate is higher in soft water than in hard water, which often forms a protective film. The corrosion rate of pure zinc in terms of weight losses ranges from 0.3 g/m^2/d in hard water to 2.7 g/m^2/d in distilled water. In aerated hot water, it has been observed that the polarity between the zinc coating and the base steel is reversed at 60 °C (140 °F) or above (Ref 62). In this case, zinc becomes a noble coating instead of a sacrificial coating and induces pitting of the bare steel. In seawater, zinc coatings corrode at approximately 0.025 mm/yr (1 mil/yr).

Atmospheric Corrosion of Galvanized Steel. The corrosion rate of zinc coatings in outdoor exposure depends on the type of outdoor atmosphere involved. Factors such as the frequency and duration of moisture contact, the rate of drying, and the extent of industrial pollution in the air have significant effect on the corrosion rate. In general, the corrosion rate of zinc coatings in a rural atmosphere is very low. Seacoast atmospheres are less corrosive to zinc coating than industrial atmospheres.

A large-scale long-term test program was conducted on galvanized steel wire (both hot-dipped and electroplated) by ASTM (Ref 63). Carbon steel wires with different coating weights were exposed at several testing sites, including Pittsburgh, PA (severe industrial); Sandy Hook, NJ (marine); Bridgeport, CT (industrial); State College, PA (rural); Lafayette, IN (rural); Ithaca, NY (rural); and Ames, IA (rural). After fifteen years of exposure at these sites, the average corrosion rates of the zinc coatings were obtained by dividing the coating weight (oz/ft^2) by the number of years of exposure before the first rust was observed. These rates are summarized in

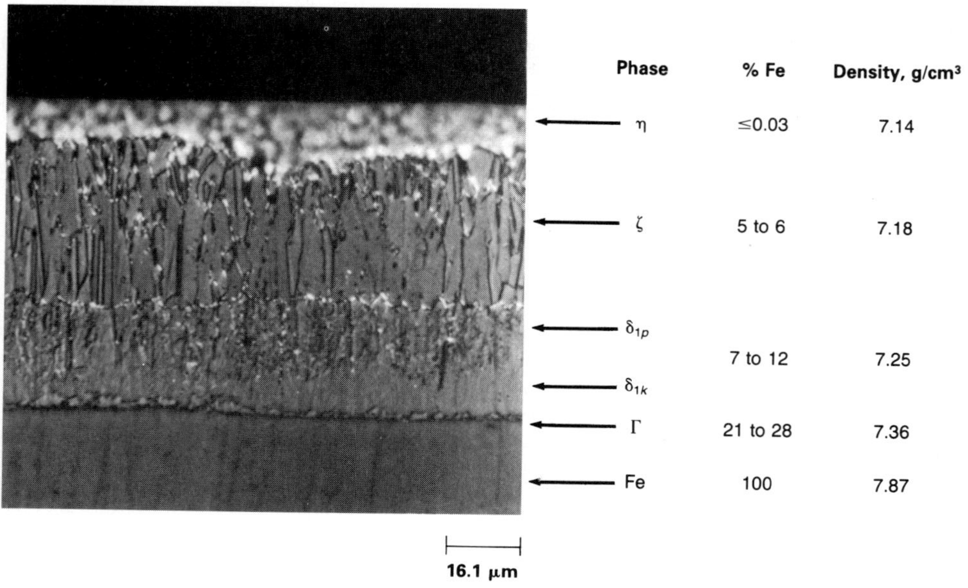

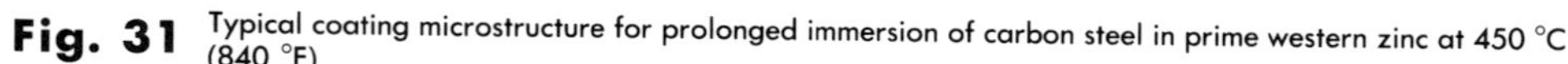

Fig. 31 Typical coating microstructure for prolonged immersion of carbon steel in prime western zinc at 450 °C (840 °F)

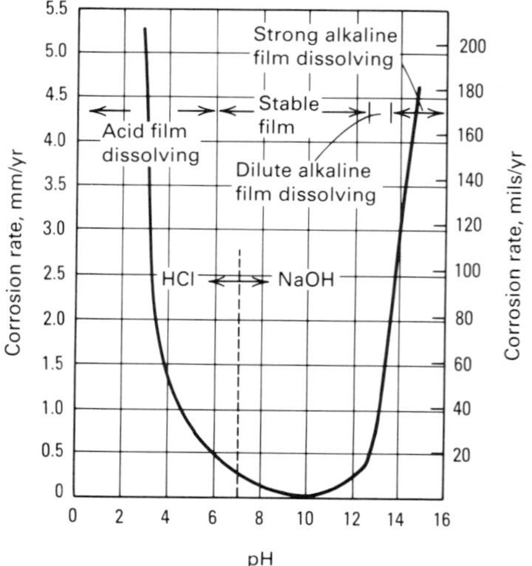

Fig. 32 Effect of pH value on the corrosion of zinc

Table 13. The service life of the zinc coating appears to be in direct proportion to the weight of the coating. As shown in Table 13, the corrosion rate can range from 12 g/m^2/yr (0.04 oz/ft^2/yr) in a rural atmosphere (Ames, IA) to 104 g/m^2/yr (0.34 oz/ft^2/yr) in a severe industrial atmosphere (Pittsburgh, PA). The gage of the wire or the type of zinc coating (either hot-dipped or electrodeposited) within the test limits seems to have no effect on the corrosion rate of the zinc coating.

In 1969, the atmospheric-corrosion behavior of hot-dip galvanized steel sheet was evaluated at three testing sites: a semi-industrial test site (Porter County, IN), a severe industrial test site (East Chicago, IN), and a marine test site (Kure Beach, NC, 250-m, or 800-ft, lot) (Ref 64). The galvanized steel used was 0.81 mm (0.032 in.) in thickness with an average coating weight of 168 g/m^2 (0.55 oz/ft^2). All test panels were 100 × 150 mm (4 × 6 in.) in size. Two panels were made into one sandwich-type test specimen for exposure so that the corrosion rate on the skyward and groundward side of each specimen could be evaluated independently. All specimens were exposed inclined 30° from the horizontal. Forty sandwich-type test specimens were placed at each of the three test sites. Specimens were removed at the conclusion of the exposure period of 6 months, 1 year, 2 years, 3 years, 4 years, and 5 years. The standard ASTM recommended practice G 1 was used in preparing, cleaning, and evaluating the specimens (Ref 65). The average weight loss data obtained from the skyward panels of each specimen were fitted to Eq 1.

Predictions of corrosion rates using Eq 1 are shown in Fig. 33. The correlation coefficients (R^2) shown demonstrate the high-quality fit of each curve in Fig. 33. These curves can be used to predict the service life of galvanized steel for a given coating thickness, or vice versa. For example, the predicted weight loss for a 10-year exposure, as shown in Fig. 33, is 121.6 g/m^2 (0.40 oz/ft^2) at the Porter County site, 290 g/m^2 (0.95 oz/ft^2) at the East Chicago site, and 103.3 g/m^2 (0.34 oz/ft^2) at the Kure Beach site.

Intergranular Corrosion of Galvanized Steel. It has been known since 1923 that zinc die casting alloys are susceptible to intergranular attack in an air-water environment (Ref 66). The adverse effect of intergranular corrosion of hot-dip galvanized steel was first observed in 1963, and was investigated at Inland Steel Company in 1972 (Ref 67). The observed effect associated with intergranular corrosion was termed delayed adhesion failure. Delayed adhesion failure is a deterioration in coating adhesion due to selective corrosion at grain boundaries. It was found that the small amount of lead normally added to commercial galvanizing spelters was a critical factor in the susceptibility of the zinc coating to intergranular attack. By using lead-free zinc spelter (<0.01% Pb), the damaging effect of intergranular corrosion was essentially eliminated.

Table 13 Atmospheric-corrosion rates of zinc-coated wire

Test site	Type of atmosphere	Average corrosion rate g/m^2/yr	oz/ft^2/yr
Pittsburgh, PA	Severe industrial	104	0.34
Sandy Hook, NJ	Marine (275 m, or 900 ft, from ocean)	40	0.13
Bridgepoint, CT	Industrial	40	0.13
State College, PA	Rural	18	0.06
Lafayette, IN	Rural	21	0.07
Ithaca, NY	Rural	18	0.06
Ames, IA	Rural	12	0.04

Source: Ref 63

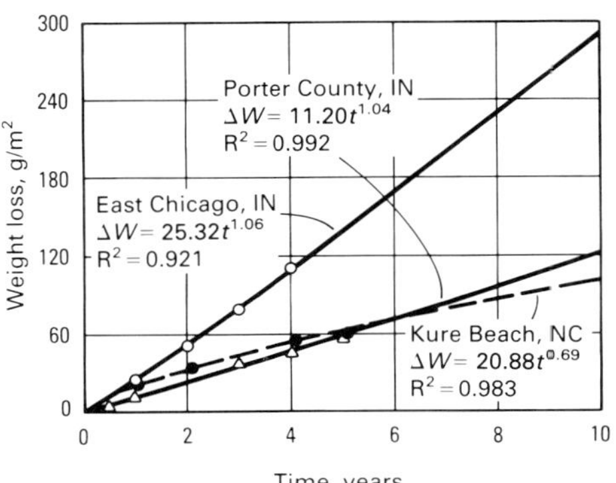

Fig. 33 Predictive equations for galvanized steel based on 5 years of exposure

For the continuous hot-dip galvanizing process, the main reason to add 0.07 to 0.15% Pb to zinc spelter was to produce a spangled coating and to lower the surface tension of the zinc bath in order to provide the necessary fluid properties to produce a ripple-free coating. It was found that by using antimony additions to the zinc spelter similar beneficial effects, such as those obtained by the use of lead additions, could be achieved without causing any detrimental effect with respect to intergranular corrosion. For galvanized coatings produced by electroplating process, no intergranular corrosion has been observed.

Aluminum-Base Coatings

Types. Aluminum coatings on steel are primarily produced by spraying or hot dipping. Spray coatings are mainly applied on structural steel by using a wire-type gun (see the article "Thermal Spray Coatings" in this Volume). Pure aluminum or aluminum alloy wires are continually melted in the oxygen-fuel gas flame and atomized by a compressed air blast that carries the melted metal particles to the prepared surface where they agglomerate to form a coating. The coating thickness is in the range of 0.08 to 0.2 mm (3 to 8 mils). Coatings are commonly sealed with organic lacquers or paints to delay the formation of visible surface rust.

Aluminum coatings on sheet steel are primarily produced by a continuous hot-dip process (the Sendzimir process). Molten baths of aluminum for hot dipping usually contain silicon in the range of 7 to 11% to retard the growth of a brittle iron-aluminum intermetallic layer. This alloy, which is one of the most fluid and easily cast aluminum alloys, forms a coating with a much thinner and more uniform alloy layer. This coated product, which has relatively good coating adhesion and forming properties, was commercially introduced in 1940; it is now identified as type I aluminized steel. A typical type I coating bath contains 9% Si, 87.5% Al, and 3.5% Fe.

Type II aluminized steel was commercially produced in 1954 with a coating consisting mainly of pure aluminum. This coating could withstand mild forming, such as corrugating and roll forming. A typical type II coating bath contains 97.5% Al, 2% Fe, and 0.5% Si.

Aqueous Corrosion of Aluminized Steels. In aqueous environments, pure aluminum or aluminum-silicon alloy coatings exhibit good general corrosive resistance. The coatings become passive in a pH range between 4 and 9 and corrode rapidly in acid or alkali solutions. These coatings tend to pit in environments containing chloride ions (Ref 68), particularly at crevice or stagnant areas where passivity breaks down through the action of a differential aeration cell.

In soft water, aluminum coatings exhibit a potential that is positive to steel; therefore, they act like a noble coating. In seawater or in aqueous environments containing Cl^- or SO_4^{2-}, the potential of aluminum coatings becomes active, and the polarity of aluminum-iron couples may reverse. Under these conditions, the aluminum coating is sacrificial and cathodically protects steel.

Atmospheric Corrosion of Aluminized Steel. The atmospheric-corrosion resistance of aluminum coatings is generally related to that of solid aluminum of the same thickness (see the article "Corrosion of Aluminum and Aluminum Alloys" in this Volume). The protection of steel by aluminum coating depends partly on cathodic protection and partly on the inert barrier layer of oxide film that forms on the metal surface. For thermally sprayed aluminum coatings, initial corrosion may produce slight superficial rust staining through pores in the coating. Subsequently, insoluble aluminum corrosion products block the pores and retard further corrosion of the coating.

When type I aluminized steel is exposed to the atmosphere, pitting corrosion can occur because of the difference in electrochemical potential between the silicon-rich phase and the aluminum matrix. The resulting corrosion product causes a red-brown blush discoloration on the metal surface (Ref 69). The corrosion product retards any further corrosion reaction. Type I panels have been exposed to a mild industrial atmosphere for over 40 years with no evidence of base metal corrosion.

For type II aluminum coating, the alloy layer is much thicker than that of the type I coating. Because of the protective nature of the oxide film formed on the coating surface, type II aluminum coatings have shown much better atmospheric-corrosion resistance than type I coatings.

In 1969, type I and type II aluminized steels were evaluated at three atmospheric-testing sites (Ref 64). This evaluation was part of the program outlined in the discussion "Zinc-Base Coatings" in this section. The weight loss data obtained from the skyward panels of each specimen were fitted to the same equation, $W = kt^n$, which has been used for hot-dipped zinc coatings. The results of this curve fitting, together with k and n values obtained for type I and type II aluminum coatings, are shown in Fig. 34 and 35 for all three test sites. Table 14 provides a summary of the predicted weight loss for type I and type II aluminized steel in comparison with hot-dip zinc coatings based on 5-year exposure data. As indicated in Table 14, the atmospheric-corrosion rate of aluminum coating (type I or type II) is equivalent to only 10 to 50% of the corrosion rate of zinc coating, depending on the type of atmosphere.

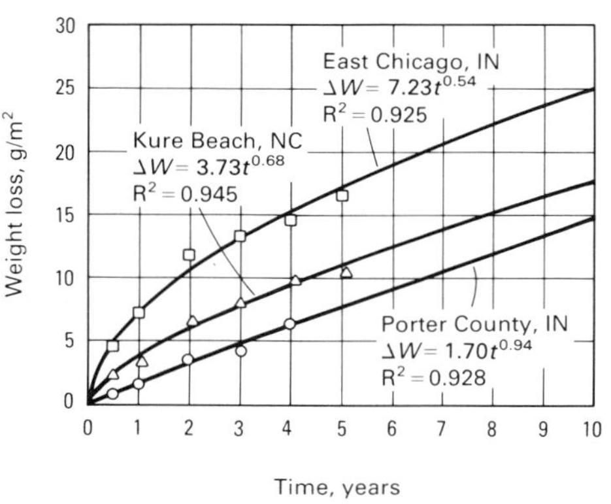

Fig. 34 Predictive equations for type I aluminized steel based on 5 years of exposure

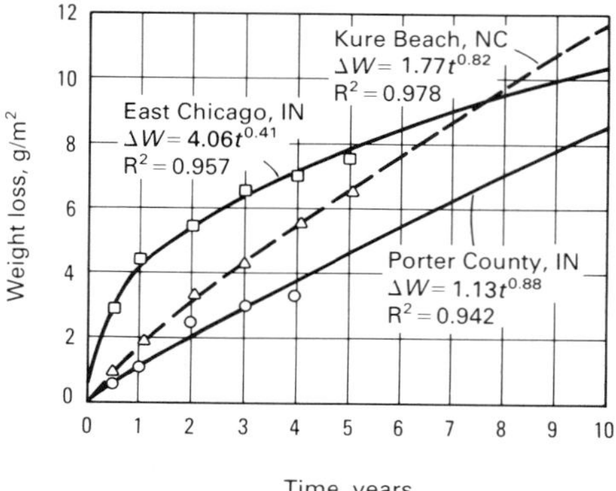

Fig. 35 Predictive equations for type II aluminized steel based on 5 years of exposure

Table 14 Predicted weight loss after 10-year atmospheric exposure of hot-dip galvanized steel and type I and type II aluminized steel based on 5-year exposure data

Test site	Weight loss — Hot-dip galvanized, g/m²	Hot-dip galvanized, mils	Type I aluminized, g/m²	Type I aluminized, mils	Type II aluminized, g/m²	Type II aluminized, mils
Porter County, IN (semi-industrial)	121.6	0.64	14.7	0.2	8.5	0.12
East Chicago, IN (semi-industrial)	290.1	1.53	25.2	0.35	10.5	0.15
Kure Beach, NC (250 m, or 800 ft, lot—marine)	103.3	0.54	17.8	0.25	11.6	0.16

Source: Ref 64

Organic Coated Steels

James H. Bryson
Inland Steel Company

Paint is applied to a steel sheet or a part for one or both of the following reasons: enhancement of the esthetic value of the product and/or preservation of structural integrity. The former goal is a consideration for the designer and will receive no further discussion in this section. The latter goal, however, will be discussed as it applies to prepainted steel.

This section will describe how paints deter corrosion, the prepainting process, the primary differences between prepaint paint formulations and postpaint formulations, considerations about part design, and selection criteria for the appropriate paint system. More detailed information on organic coating materials can be found in the article "Organic Coatings and Linings" in this Volume and in the article "Painting of Structural Steel" in Volume 5 of the 9th Edition of *Metals Handbook*.

How Paint Films Deter Corrosion

In the presence of water and oxygen, iron corrodes to form iron oxides and hydroxides (see the discussion "Corrosion Film Formation and Breakdown" in the section "Carbon Steels" in this article). The corrosion rate is accelerated when electrolytic solutes, such as the chloride or sulfate salts of alkali metals, are present. Of course, temperature also increases the corrosion rate, and where possible, decreasing the service temperature will increase the service life of the part. However, because little can usually be done to change service temperature, the exclusion of one or more of the principal reactants (oxygen, water, or electrolytes) from the steel surface will help deter corrosion. Such exclusion of reactants is the purpose of the paint film.

There are primarily three methods of protection: barrier, passivation, and galvanic. In barrier protection, the paint film retards the diffusion of water, oxygen, or salts to the steel substrate. It has been demonstrated that the flow of water and oxygen through the film is rapid; therefore, the contribution of the coating to lowering the corrosion rate is the addition of a high electrolytic resistance (Ref 70). Flake-shaped pigment particles that leaf (overlap) can increase the path length a reactant must traverse before reaching the surface; this increases the effectiveness of the barrier film. Some aluminum and stainless steel pigments protect in this fashion. The reactivity of the steel surface can be decreased when the paint film contains passivating pigments such as chromate salts. Paints are also often pigmented with zinc for both barrier and galvanic protection. The zinc loading must be sufficiently high for interparticle contact, a condition that requires that the critical pigment volume be exceeded; that is, the pigment particles are not completely wetted by the paint vehicle. Although some galvanic protection is afforded, most of the protection is provided by the barrier formed by zinc corrosion products (Ref 71).

Prepaint Processing

Much of the painted steel used today is prepainted in coil form (coil coated) before shipment to fabricators. Modern, high-speed paint lines can process bare or coated steel strip, and can be used to apply a wide variety of organic coatings.

After decoiling, the first step in the prepaint process is to clean the steel strip with an alkaline detergent. The steel strip is then brushed with an abrasive roll to remove mill oils and grime and to reduce the level of an amorphous form of surface carbon indigenous to steel strip processing. High levels of surface carbon lower corrosion resistance (Ref 72). Cleaning is usually more effective on flat strip than on a formed part. Next, the strip is rinsed and pretreated to improve paint adhesion and to reduce corrosion. A prepaint treatment may consist of phosphate coating or an organic pretreatment known as wash primer or etch primer. Such pretreatments are described in the article "Painting" in Volume 5 of the 9th Edition of *Metals Handbook*. Following pretreatment, a paint is then applied and cured in an oven. Depending on the paint formulation and the paint line, the dwell time in the oven is generally between 20 and 50 s. A second coat may then be applied and cured.

Differences Between Prepaint and Postpaint

In formulating a paint designed for a prepaint application, the forming step must be considered. Aside from steel considerations, successful forming of the part will depend on the flexibility and the abrasion resistance of the coating. The paint film must be flexible enough to withstand the strains induced from bending without crazing, which may compromise corrosion resistance. In addition, in the forming stages, the bend radii are often more severe than in the final part.

The coating must also withstand the abrasive forces of handling and forming. For a given coating type, the harder the coating, the more abrasion resistant the coating will be. Unfortunately, flexibility and hardness are inversely related; that is, the more flexible the coating, the softer the coating.

A method for overcoming the problems associated with coating flexibility is covered in the discussion "Part Design Considerations" in this section. Flexibility and hardness are considerations only for the end use of postpainted parts, while forming and handling are also factors of concern in the formulation of a paint designed for precoating a steel strip.

The final dried paint thickness, or dry-film thickness, on a prepainted steel strip is usually no more than 0.025 mm (1 mil). Plastisols and organosols are the major exceptions. Therefore, the prepaint dry-film thickness is much less than the typical dry-film thickness on a postpainted part. However, because of the method of application, the film is more evenly distributed; this results in significantly fewer areas of low dry-film thickness and the elimination of many of the appearance defects observed on finished postpainted parts. Many areas on postpainted parts will receive little or no paint because of the part shape. The formulations for prepaints are engineered to account for the lower dry-film thicknesses.

Part Design Considerations

When designing a part to be fabricated from prepainted steel, the maximum bend radius, the

Table 15 Relative rankings of various coatings in different performance categories

Category key: A, hardness; B, flexibility; C, humidity resistance; D, corrosion resistance to industrial atmospheres; E, salt spray; F, exterior durability, pigmented film; G, exterior durability, clear film; H, paint cure temperature, in °F; I, cost guide. Ratings key: 1, excellent; 2, good; 3, fair; 4, poor; H, high cost; M, moderate cost; L, low cost

Type	A	B	C	D	E	F	G	H	I
Silicone acrylic	1	3	2	2	2	2	1	450	H
Thermoset acrylic	2	2	1	2	1	2	2	430	M
Amine-alkyd	2	3	2	2	3	2	3	340	L
Silicone alkyd	2	3	2	2	2	1	2	420	H
Vinyl-alkyd	2	2	1	2	2	3	3	340	M
Straight epoxy	1	2	1	1	1	4	4	400	H
Epoxy-ester	2	2	1	2	1	4	4	400	M
Organosol	2	1	1	1	1	2	3	350	L
Plastisol	3	1	1	1	1	2	3	350	L
Polyester (oil-free)	1	2	1	2	1	2	3	400	M
Silicone polyester	2	2	1	2	1	1	2	450	H
Poly-vinyl fluoride	2	1	1	1	1	1	1	450	H
Poly-vinyl idene fluoride	2	1	1	1	1	1	1	450	H
Solution vinyl	2	1	1	2	1	2	3	300	M

forming equipment, and the joining method must be considered. As mentioned earlier, the maximum bend radius is often smaller than that specified on the blueprint of the part. This radius should be as generous as the structural and decorative criteria will allow. In considering the part shape, the avoidance of catchment areas, where possible, will decrease failures due to corrosion.

The forming equipment should be well maintained to avoid marring the surface. Where possible, roll forming is preferable to stamping. In cases in which hard finishes in conjunction with tight radii (high flexibility) are required, the prepainted strip can be warm formed. In warm forming, the paint is heated into or above its glass transition temperature range. In this temperature range, the paint is softer and more flexible, thus allowing tighter radii to be achieved during forming. After cooling, the paint becomes harder and more abrasion resistant.

Lastly, the part may require joining. Welding and mechanical fastening damage the paint film. Therefore, it is necessary to touch up the scars to restore corrosion resistance. Adhesive bonding eliminates the need for touch up of these damage areas. Taking these factors into account, prepainted steel has been successfully fabricated into finished or semifinished (requiring a post finish coat operation) parts in many automotive, appliances, or office furniture manufacturing plants. Prepainted parts have been produced on lines designed for their use and on existing lines, sometimes with no modification to the line.

Selection Guideline

As an aid to understanding the evaluation process, Table 15 compares various common coatings in several categories of performance. Changes in pigmentation and resin source for the vehicle can influence the rating by a factor of one or more. Table 15 is merely a guideline to the performance of these coatings. Comments from technical personnel at a paint company should be sought before making any decisions on paint selection.

The columns in Table 15 are self-explanatory, with the exception of those involving exterior durability and salt spray. Exterior durability is the resistance to weathering, particularly, the resistance to ultraviolet light. Ultraviolet light causes some coatings to chalk. Proper pigmentation will prevent this phenomenon for some coatings, and this can be determined by comparing the columns for pigmented and clear films.

Salt spray is not a predictor of service life and coatings cannot be compared for end use on this basis. However, salt spray does detect coating defects and can be put to good use for detecting induced flaws by comparing results for flat panels with those of panels with coating defects induced, for example, by forming or abrasion.

The first step in the evaluation is the selection of a steel mill and/or paint company that is willing and able to help evaluate the needs of the final product. These needs can be categorized as either preservice or service. The preservice conditions involve forming, handling, and joining. The service conditions are those to which the customer exposes the product: humidity, temperature, corrosive agents, sunlight, and abrasion. Of course, the preservice conditions can affect the service life of the final product, and these effects should be evaluated.

The next step is the experimental design. The test program compares candidate materials to the current product, if possible. Evaluation in actual service conditions is often not possible because of time limitations. Therefore, accelerated and laboratory tests are needed (see the article "Laboratory Testing" in this Volume). From these results, acceptable candidates are identified and are included in the next level of tests. A set of suitable parts is identified for testing the candidates. After the parts are fabricated, they are inspected to determine whether coating damage occurred and whether corrosion resistance was compromised. In general, one material will not be superior in all aspects. Therefore, the desirable properties must be prioritized.

Advantages of Prepainted Steels

Although the above evaluation sequence may seem formidable, many manufacturers have found the use of prepainted steel to be productive and economical. The use of prepainted steel reduces or eliminates the problems of the waste treatment of the emissions from paint lines. The postpainting line is often the slow step in the process, and using prepainted steel increases output. Although the material cost of prepainted steel is higher than the bare metal, the final part cost is lower because of increased productivity and the reduction of other costs, such as emissions control. Although prepainted steel cannot replace postpainted steel in every application, prepainted steel has demonstrated its productive and economic advantages.

REFERENCES

1. S.K. Coburn, Ed., *Atmospheric Factors Affecting the Corrosion of Engineering Metals*, STP 646, American Society for Testing and Materials, 1978
2. *Metal Corrosion in the Atmosphere*, STP 435, American Society for Testing and Materials, 1968
3. W.H. Ailor, Ed., *Atmospheric Corrosion*, Wiley-Interscience, 1982
4. Corrosiveness of Various Atmospheric Test Sites as Measured by Specimens of Zinc and Steel, in *Metal Corrosion in the Atmosphere*, STP 435, American Society for Testing and Materials, 1968, p 383
5. T. Misawa, K. Hashimoto, W. Suetaka, and S. Shimodaira, Mechanism of Formation of Iron Oxide and Oxyhydroxides in Aqueous Solutions, in *Proceedings of the Fifth International Congress on Metallic Corrosion*, National Association of Corrosion Engineers, 1972, p 775-779
6. L.L. Shreir, Ed., *Corrosion*, Vol 1, John Wiley & Sons, 1963, p 3.8
7. P.R. Grossman, Investigation of Atmospheric Exposure Factors That Determine Time-of-Wetness of Outdoor Structures, in *Atmospheric Factors Affecting the Corrosion of Engineering Metals*, STP 646, S.K. Coburn, Ed., American Society for Testing and Materials, 1978, p 5-16
8. K. Deaves, K.F. Meves, and E.H. Schultz, *Korros. Met.*, Vol 19, 1943, p 233
9. F.H. Haynie and J.B. Upham, *Mater. Prot. Perform.*, Vol 10-11, 1971, p 18
10. J. Michalczewski and T. Zak, *Powloki Ochr.*, Vol 4, 1974, p 20-24
11. T. Zak and G. Chojnacka-Kalinowski, *Powloki Ochr.*, Vol 27.5, 1977, p 2-4
12. J.B. Upham, *J. Air Poll. Control Assoc.*, Vol 17, 1967, p 398
13. P.J. Sereda, *Ind. Eng. Chem.*, Vol 52, 1960, p 157
14. K. Oma, T. Sugano, and Y. Harai, *Boshoku Gijutsu (Corros. Eng.)*, Vol 14 (No. 1), 1965, p 16
15. K. Tripathi, U.S. Agninotui, and J.N. Nanda, *Br. Corros. J.*, Vol 7, 1972, p 212
16. H. Guttman and J.P. Sereda, Measurement of Atmospheric Factors Affecting the Corrosion of Metals, in *Metal Corrosion in the Atmosphere*, STP 435, American Society for Testing and Materials, 1968, p 326
17. V. Kucera and J. Gullmann, "Practical Experience With an Electrochemical Technique for Atmospheric Corrosion Monitoring," Paper presented at ASTM symposium on Progress in Electro-chemical Corrosion Testing, San Francisco, CA, American Society for Testing and Materials, 1979

18. P.W. Brown and L.W. Masters, Factors Affecting the Corrosion of Metals in the Atmosphere, in *Atmospheric Corrosion*, W.J. Ailor, Ed., John Wiley & Sons, 1982, p 31-47
19. P.R. Grossman, Investigation of Atmospheric Exposure Factors That Determine Time-of-Wetness of Outdoor Structures, in *Atmospheric Factors Affecting the Corrosion of Engineering Metals*, STP 646, S.K. Coburn, Ed., American Society for Testing and Materials, 1978, p 14
20. A.P. Jahn, Report of Committee A-5 on Corrosion of Iron and Steel, in *Proceedings ASTM*, Vol 54, American Society for Testing and Materials, 1954
21. *Alloy Steels Pay Off*, Climax Molybdenum Company, 1953, p 67
22. C.P. Larrabee and S.K. Coburn, The Atmospheric Corrosion of Steels as Influenced by Changes in Chemical Composition, in *Proceedings of the First International Congress on Metallic Corrosion*, Butterworths, 1962, p 276-285
23. C.P. Larrabee, Corrosion Resistance of High Strength Low Alloy Steels as Influenced by Composition and Environment, *Corrosion*, Vol 9 (No. 8), 1953, p 259-271
24. R.A. Legault and A.G. Preban, Kinetics of the Atmospheric Corrosion of Low-Alloy Steels in an Industrial Environment, *Corrosion*, Vol 31 (No. 4), 1975, p 117-122
25. R.F. Passano, The Harmony of Outdoor Weathering Tests, in *Proceedings of the Thirty-Seventh Annual Meeting*, Vol 34, American Society for Testing and Materials, 1934, p 132
26. N.B. Pilling and W.A. Wesley, Atmospheric Durability of Steels Containing Nickel and Copper, in *Proceedings of the Forty-Third Annual Meeting*, Vol 40, American Society for Testing and Materials, 1940, p 643-647
27. J.B. Horton, "The Composition, Structure and Growth of Atmospheric Rust on Various Steels," Thesis, Lehigh University, 1964
28. R.A. Legault and V.P. Pearson, "Atmospheric Corrosion in Marine Environments," Paper OTC 2374, presented at the Offshore Technology Conference, Dallas, TX, 1975
29. R.A. Legault and V.P. Pearson, "The Atmospheric Corrosion of Galvanized and Aluminized Steel. A Comparison of the Skyward Side Corrosion Behavior of Three Different Coated Materials Exposed for a Five Year Period at Three Different Atmospheric Test Sites," Inland Steel Research Laboratory, 1976
30. K. Bohnenkamp, G. Burgmann, and W. Schwenk, EUR.5250d, (1 Teil), Untersuchengen iiber die Atmosphar-ishce Korrosion von un-und Niedriglegierten Stahlen in Meeres-Land-und Industrieluft, Recherche CECA 5A
31. C.P. Larrabee and W.L. Mathay, in *Corrosion Resistance of Metals and Alloys*, F.L. LaQue and H.R. Copson, Ed., Reinhold, 2nd ed., 1963, p 305-349
32. H.R. Ambler and A.J. Bain, *J. Appl. Chem.*, Vol 5, 1955, p 437
33. "Deterioration of Structures in Sea Water," Eighteenth Report of the Committee of the Institution of Civil Engineers, 1938
34. "Materials Corrosion Investigation of Eastport, ME," Fifth Interim Report, U.S. Engineers Office, 1946
35. H.J. French and F.L. LaQue, *Alloy Constructional Steels*, American Society for Metals, 1942
36. C.P. Larrabee, U.S. Patent 2,315,156, 1943
37. J.T. Crennell, in *Corrosion*, Vol 1, L.L. Shreir, Ed., John Wiley & Sons, 1963, p 2.34
38. F.L. LaQue, in *The Corrosion Handbook*, H.H. Uhlig, Ed., John Wiley & Sons, 1948, p 391
39. K.H. Logan, Corrosion by Soils, in *The Corrosion Handbook*, H.H. Uhlig, Ed., John Wiley & Sons, 1948, p 446-466
40. M. Romanoff, "Underground Corrosion," NBS Circular 579, National Bureau of Standards, 1957, p 38-47
41. G.J. Verbeck, Mechanisms of Corrosion of Steel in Concrete, in *Corrosion of Metals in Concrete*, Publication SP-49, American Concrete Institute, June 1975, p 22-38
42. C.E. Locke, "Mechanism of Corrosion of Steel in Concrete," Paper presented at NACE Seminar, Solving Rebar Corrosion Problems in Concrete, Chicago, IL, National Association of Corrosion Engineers, Sept 1982, p 2/2- 2/10
43. "Bridge Deck Deterioration, a 1981 Perspective," Federal Highway Administration Memorandum, Office of Research, Dec 1981
44. D.A. Lewis, Some Aspects of the Corrosion of Steel in Concrete, in *Proceedings of the First International Congress on Metallic Corrosion*, National Association of Corrosion Engineers, 1961, p 547-555
45. J. Ishikawa, B. Bresler, and I. Cornet, Electrochemical Study of the Corrosion Behavior of Galvanized Steel in Concrete, in *Proceedings of the Fourth International Congress of Metallic Corrosion*, National Association of Corrosion Engineers, 1972, p 556-559
46. I. Medgyesi, Problems Related to the Corrosion of Reinforcing Rods in Concrete, in *Proceedings of the Fourth International Congress on Metallic Corrosion*, National Association of Corrosion Engineers, 1972, p 591-593
47. B. Heuze, Cathodic Protection for Reinforced or Prestressed Concrete, in *Proceedings of the Fifth International Congress of Metallic Corrosion*, National Association of Corrosion Engineers, 1972, p 598-601
48. "Solving Rebar Corrosion Problems in Concrete," Seminar, Chicago, IL, National Association of Corrosion Engineers, Sept 1982
49. M. Hecht, W.C. Schroeder, E.P. Partridge, and S.F. Whirl, Boiler Corrosion, in *The Corrosion Handbook*, H.H. Uhlig, Ed., John Wiley & Sons, 1948, p 520-537
50. D.M. Buck, Copper in Steel—The Influence on Corrosion, *J. Ind. Eng. Chem.*, Vol 5, 1913, p 447
51. D.M. Buck and J.O. Handy, Research on the Corrosion Resistance of Copper Steels, *J. Ind. Eng. Chem.*, Vol 8, 1916, p 209
52. D.M. Buck, The Influence of Very Low Percentages of Copper in Retarding the Corrosion of Steel, in *Proceedings ASTM*, Vol 19, American Society for Testing and Materials, 1919, p 224
53. "Standard Specification for High-Strength Low-Alloy Structural Steel," A 242, *Annual Book of ASTM Standards*, American Society for Testing and Materials
54. "Standard Specification for High-Strength Low-Alloy Structural Steel with 50 ksi (345 MPa) Minimum Yield Point to 4 in. (100 mm) Thick," A 588, *Annual Book of ASTM Standards*, American Society for Testing and Materials
55. *Steel—Structural, Reinforcing, Pressure Vessel, Railway*, Vol 01.04, Section 1, *Annual Book of ASTM Standards*, American Society for Testing and Materials, 1987
56. C.P. Larrabee, The Effect of Specimen Position on Atmospheric Corrosion Testing on Steel, *Trans. Electrochem. Soc.*, 1945, p 297
57. E.S. Taylerson, Atmospheric Exposure Tests on Copper-Bearing and Other Irons and Steels in the United States, *J. Iron Steel Inst.*, Vol 143 (No. 1), 1941, p 287
58. R.L. Brockenbrough and R.J. Schmitt, Preprint C75 041-9, Paper presented at the Winter Meeting, New York, NY, Institute of Electrical and Electronics Engineers, Jan 1975
59. J.B. Vrable, R.T. Jones, and E.H. Phelps, The Application of High-Strength Low-Alloy Steels in the Chemical Industry, *Mater. Perform.*, Vol 18, 1979, p 39
60. R.P. Krepski, "The Influence of Bath Alloy Additions in Hot-Dip Galvanizing," St. Joe Minerals Corporation, 1980
61. B. Roetheli, G. Cox, and W. Littreal, *Met. Alloys*, Vol 3, 1932, p 73
62. G. Schikorr, *Trans. Electrochem. Soc.*, Vol 76, 1939, p 247
63. A.P. Jahn, Atmospheric Corrosion of Steel Wires, in *ASTM Proceedings*, Vol 52, American Society for Testing and Materials, 1952, p 987
64. R.A. Legault and V.P. Pearson, "The Atmospheric Corrosion of Galvanized and Aluminized Steel," Research Report, Inland Steel Company, 1980
65. "Standard Practice for Preparing, Cleaning, and Evaluating Corrosion Test Specimens," G 1, *Annual Book of ASTM Standards*, Vol 03.02, American Society for Testing and Materials
66. H.F. Brauer and W.M. Peirce, The Effect of Impurities on the Oxidation and Swelling of Zinc-Aluminum Alloys, *Trans. Am. Inst. Min. Metall. Eng.*, Vol 60, 1923, p 796
67. H.H. Lee, Galvanized Steel With Improved Resistance to Intergranular Corrosion, *Proc. Galvanized Committee*, Vol 69, 1977, p 17
68. H.H. Uhlig, *Corrosion and Corrosion Control*, John Wiley & Sons, 1971, p 335
69. J.H. Rigo, *Corrosion*, Vol 17 (No. 5), 1961, p 245
70. J.E.O. Mayne, *Anti-corros.*, Oct 1973, p 3-8
71. N.C. Fawcett, *Polymer Mater. Sci. Eng.*, Vol 53, 1985, p 855-859
72. R.A. Iezzi and H. Leidheiser, Jr., *Corrosion*, Vol 37, 1981, p 28

Corrosion of Alloy Steels

Thomas G. Oakwood, Inland Steel Research Laboratories

ALLOY STEELS comprise a category of ferrous materials that exhibit mechanical properties superior to those of ordinary carbon steels as the result of additions of such alloying elements as chromium, nickel, and molybdenum. Total alloy content can range from 0.5 to 1% and up to levels just below that of stainless steels. For many alloy steels, the primary function of the alloying elements is to increase hardenability in order to optimize mechanical properties and toughness after heat treatment. In some cases, however, alloy additions are used to reduce environmental degradation under certain specified service conditions.

Alloy steels are used in a broad spectrum of applications. In some cases, corrosion resistance is a major factor in alloy selection; in other applications, it is only a minor consideration. The information available on the corrosion resistance of alloy steels is end-use oriented and often addresses rather specialized types of corrosion. As a result, this article will emphasize those applications where corrosion resistance is either a major factor in steel selection or where available data have shown that variations in alloy content or steel processing affect resistance to corrosion.

Many applications use steels with a relatively low alloy content. The steels involved include those designated by the American Iron and Steel Institute (AISI), ASTM, and the Society of Automotive Engineers (SAE) as standard alloy steels, as well as modifications of these grades. In addition, potential standard grades, formerly SAE EX (experimental) grades, are applicable, along with high-strength low-alloy (HSLA) and structural alloy steels. The primary function of the alloying element additions is to enhance hardenability and/or mechanical properties. As will be discussed later, small additions of some alloying elements will enhance corrosion resistance in moderately corrosive environments. In severe environments, however, the corrosion resistance of this group of steels is often no better than that of carbon steel (see the article "Corrosion of Carbon Steels" in this Volume).

Other applications require more highly alloyed steels that, in addition to achieving the necessary mechanical properties, provide increased resistance to specific types of corrosion in certain environments. In this group of steels, corrosion resistance is an important factor in alloy design (see the article "Corrosion of Stainless Steels" in this Volume).

An extensive collection of data on alloy steel products, which encompasses compositions, mechanical and physical properties, applications, and service characteristics, can be found in Volume 1, Properties and Selection: Irons and Steels, of *Metals Handbook*. Information on the metallographic preparation and microstuctural interpretation of alloy steels is available in Volume 9, Metallography and Microstructures, of *Metals Handbook*. Finally, fracture characteristics of alloy steels are reviewed in Volume 12, Fractography, of *Metals Handbook*.

Corrosive Environments Encountered In The Use of Alloy Steels

Atmospheric corrosion is a factor in nearly all applications of alloy steels. It is the principal form of corrosion of concern in the automotive, off-highway equipment, machinery, construction, and aerospace industries. As will be discussed later, the atmospheric-corrosion resistance of various alloy steels, as well as the role of various alloying elements, depends on the severity of the environment. Corrosion rates vary as the environment is changed, for example, from rural to industrial.

Some industries that use alloy steels present certain specific corrosion problems. These include the production, refining, and distribution of oil and gas; energy conversion systems involving the combustion of fossil fuels; the chemical-process industries; and certain marine applications.

During the drilling and primary production of oil and gas, alloy steels are exposed to crude oil and gas formations containing varying amounts of hydrogen sulfide (H_2S), carbon dioxide (CO_2), water, and chloride compounds. High pressures and temperatures are also encountered in some cases. Refining operations subject low-alloy steels to environments containing both hydrogen and hydrocarbons. Transmission and distribution of oil and gas expose pipelines and piping systems to environments containing varying amounts of many of the constituents mentioned above.

In most cases, energy conversion systems involve consideration of the corrosion problems associated with the combustion of fossil fuels. Contaminants in coal, oil, and natural gas result in the accelerated attack of alloy steels at elevated temperatures. In steam-generating electric power plants, corrosion due to impurities in boiler feedwater and in high-pressure high-temperature steam needs to be addressed.

Alloy steels used in the construction of chemical-processing plants are subject to corrosion

Table 1 Effect of composition on 15.5-year atmospheric corrosion of high-strength low-alloy steels

Composition, %					Average reduction in thickness					
					Industrial (a)		Semirural (b)		Moderate marine (c)	
Cu	Ni	Cr	Si	P	μm	mils	μm	mils	μm	mils
0.01	...	...	...	...	731	28.8	312	12.3	1320	52
0.04	...	...	...	...	224	8.8	201	7.9	363	14.3
0.24	...	...	...	...	155	6.1	163	6.4	284	11.2
0.008	1	...	...	...	155	6.1	132	5.2	244	9.6
0.2	1	...	...	...	112	4.4	117	4.6	203	8.0
0.01	...	0.61	...	...	1060	41.7	419	16.5	401 (d)	15.8 (d)
0.2	...	0.63	...	...	117	4.6	145	5.7	229	9.0 (d)
0.1	...	1.3	...	...	419	16.5	287	11.3	465	18.3 (d)
0.22	...	1.3	...	...	89	3.5	114	4.5	...	...
0.012	...	...	0.22	...	373	14.7	257	10.1	546	21.5
0.22	...	...	0.20	...	152	6.0	155	6.1	251	9.9
0.02	...	...	...	0.06	198	7.8	175	6.9	358	14.1
0.21	...	...	...	0.06	124	4.9	130	5.1	231	9.1
0.01	1	0.62	0.26	0.08	86	3.4	89	3.5	130	5.1
0.2	1	0.61	0.17	0.1	58	2.3	71	2.8	102	4.0

(a) Kearny, NJ. (b) South Bend, PA. (c) Kure Beach, NC, about 250 m (800 ft) from ocean. (d) Estimated. Source: Ref 1

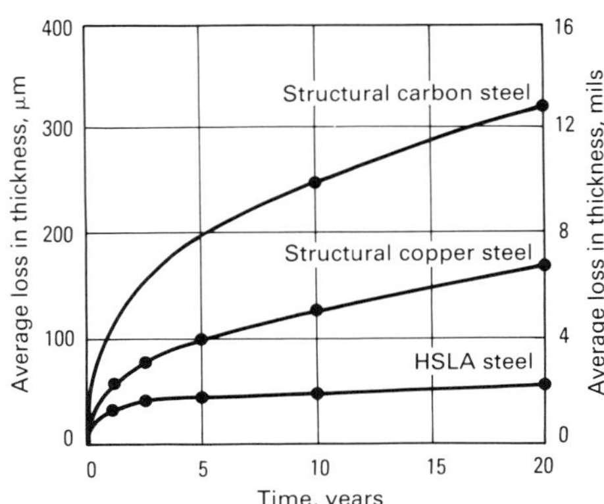

Fig. 1 Atmospheric corrosion versus time in a semi-industrial or industrial environment. Source: Ref 1

Table 2 Corrosion of structural steels in various environments

Type of atmosphere	Time, years	Average reduction in thickness											
		Structural carbon steel		Structural copper steel		UNS K11510(a)		UNS K11430(b)		UNS K11630(c)		UNS K11576(d)	
		μm	mils	μm	mils	μm	mils	μm	mils	μm	mils	μm	mils
Industrial (Newark, NJ)	3.5	84	3.3	66	2.6	33	1.3	46	1.8	36	1.4	56	2.2
	7.5	104	4.1	81	3.2	38	1.5	53	2.1	43	1.7	...	...
	15.5	135	5.3	102	4.0	46	1.8	...	...	53	2.1	...	...
Semi-industrial (Monroeville, PA)	1.5	56	2.2	43	1.7	28	1.1	36	1.4	30	1.2	41	1.6
	3.5	94	3.7	64	2.5	30	1.2	53	2.1	36	1.4	61	2.4
	7.5	130	5.1	81	3.2	36	1.4	61	2.4	43	1.7	...	...
	15.5	185	7.3	119	4.7	46	1.8	...	...	46	1.8	...	...
Semi-industrial (South Bend, PA)	1.5	46	1.8	36	1.4	25	1.0	33	1.3	25	1.0	38	1.5
	3.5	74	2.9	56	2.2	33	1.3	48	1.9	38	1.5	61	2.4
	7.5	117	4.6	81	3.2	46	1.8	69	2.7	48	1.9	...	...
	15.5	178	7.0	122	4.8	56	2.2	...	...	64	2.5	...	...
Rural (Potter County, PA)	2.5	...	...	33	1.3	20	0.8	30	1.2	...	...	...	...
	3.5	51	2.0	43	1.7	28	1.1	36	1.4	30	1.2	46	1.8
	7.5	76	3.0	64	2.5	33	1.3	38	1.5	38	1.5	...	...
	15.5	119	4.7	97	3.8	36	1.4	...	...	51	2.0	...	...
Moderate marine (Kure Beach, NC, 250 m, or 800 ft, from ocean)	0.5	23	0.9	20	0.8	15	0.6	20	0.8	18	0.7	25	1.0
	1.5	58	2.3	48	1.9	28	1.1	43	1.7	30	1.2	43	1.7
	3.5	124	4.9	84	3.3	46	1.8	64	2.5	48	1.9	56	2.2
	7.5	142	5.6	114	4.5	64	2.5	94	3.7	74	2.9	...	...
Severe marine (Kure Beach, NC, 25 m, or 80 ft, from ocean)	0.5	183	7.2	109	4.3	56	2.2	97	3.8	28	1.1	18	0.7
	2.0	914	36.0	483	19.0	84	3.3	310	12.2	...	...	53	2.1
	3.5	1448	57.0	965	38.0	...	...	729	28.7	99	3.9	99	3.9
	5.0	...	(e)		(e)	493	19.4	986	38.8	127	5.0	...	...

(a) ASTM A242 (type 1). (b) ASTM A588 (grade A). (c) ASTM A514 (type B) and A517 (grade B). (d) ASTM A514 (type F) and A517 (grade F). (e) Specimen corroded completely away. Source: Ref 1

from a wide variety of environments. Compounds of chlorine, sulfur, ammonia (HN_3), and acids and alkalies are typical.

Finally, alloy steels are often used in marine environments involving direct contact with seawater. Applications include ship construction and offshore drilling equipment.

Atmospheric-Corrosion Resistance of Alloy Steels

As noted earlier, atmospheric-corrosion resistance is of concern in most of the applications for alloy steels. The atmospheric-corrosion resistance of alloy steels is a function of the specific environment and steel composition. Some of the data available describe the effects of various alloying elements on corrosion resistance; in other cases, specific alloy steel grades are addressed. Thus, the available information provides a general guide for the selection of an alloy steel based on overall alloy content.

Table 1 lists some of the results of a systematic study of 270 alloy steels. Experimental heats of steel involving systematic combinations of chromium, copper, nickel, silicon, and phosphorus were tested to determine their individual and joint contributions to corrosion resistance. Table 1 summarizes test results for compositions whose performance established the value of specific concentrations of each of the five elements. These data were developed over 15.5 years in three environments: industrial (Kearny, NJ), semirural (South Bend, PA), and marine (Kure Beach, NC). The data show that the long-term atmospheric-corrosion of carbon steel can be reduced with a small addition of copper. Additions of nickel are also effective, and chromium in sufficient amounts is helpful if copper is present. The maximum resistance to corrosion was obtained in this study when alloy contents were raised to their highest levels.

Figure 1 summarizes some of the results from industrial environments. The carbon steel corrosion rate becomes constant after about 5 years. The corrosion rate of the copper steel levels off to a constant value after about 3 years, and the high-strength low-alloy steel, which utilizes several alloy elements, exhibits a constant rate after approximately 2 years. Corrosion of the high-strength low-alloy steel eventually ceases. Table 2 compares the corrosion behavior of carbon steel, a copper steel, and ASTM types A242, A588, A514, and A517 low-alloy steels in a variety of environments. It is evident that the low-alloy steels exhibit significantly better performance than either carbon steel or the structural copper steel.

Although the data given above provide good estimates of average corrosion behavior, it is important to note that corrosion rates can increase significantly in severe environments. Table 3 lists corrosion rates for several steels exposed to various atmospheres in chemical plants. Comparison of these data with the industrial atmosphere data shown in Table 2 illustrates the significant increase in corrosion rate associated with severe environments. Table 3 does, however, demonstrate the effectiveness of increased alloy content on corrosion resistance.

Protective coatings offer a means of providing significant additional protection from atmospheric corrosion. Well-cleaned, primed, and painted steel can give good service in many applications (see the article "Organic Coatings and Linings" in this Volume).

Galvanizing is also used to provide protection even under conditions in which the corrosive environment is quite severe. This zinc coating is anodic and corrodes preferentially; this protects exposed steel surfaces existing at cut edges or other areas where breaks in the coating are found. Corrosion resistance increases with coat-

Fig. 2 Sulfide stress corrosion cracking in a low-alloy steel. 100×

Table 3 Corrosion losses for high-strength low-alloy steels and carbon steel exposed to various atmospheres in chemical plants

Type of plant	Atmospheric constituents	Exposure period, months	Average reduction in thickness: Carbon steel, µm	Carbon steel, mils	A242 type 1 HSLA steel, µm	A242 type 1 HSLA steel, mils	A588 grade A HSLA steel, µm	A588 grade A HSLA steel, mils
Elastomers	Chlorine and sulfur compounds	6	33	1.3	20	0.8	229	0.9
		16	81	3.2	46	1.8	46	1.8
		24	122	4.8	51	2.0	48	1.9
Chlor-alkali	Moisture, lime, and soda ash	6	69	2.7	30	1.2	33	1.3
		12	119	4.7	43	1.7	46	1.8
		24	211	8.3	53	2.1	48	1.9
Chlor-alkali	Moisture, chlorides, and lime	6	104	4.1	61	2.4	69	2.7
		12	244	9.6	81	3.2	99	3.9
		24	478	18.8	145	5.7	188	7.4
Sulfur	Chlorides, sulfur, and sulfur compounds	6	394	15.5	188	7.4	239	9.4
		12	660	26.0	277	10.9	470	18.5
		24	1100	43.3	518	20.4	823	32.4
Petrochemical	Chlorides, hydrogen sulfide, and sulfur dioxide	6	51	2.0	23	0.9	30	1.2
		12	76	3.0	30	1.2	41	1.6
		24	86	3.4	30	1.2	48	1.9
Sulfuric acid	Sulfuric acid fumes	6	84	3.3	46	1.8	48	1.9
		12	114	4.5	53	2.1	56	2.2
		24	226	8.9	76	3.0	84	3.3
Chlorinated hydrocarbons	Chlorine compounds	6	137	5.4	46	1.8	46	1.8
		12	272	10.7	56	2.2	56	2.2
		24	1120	44.1	104	4.1	117	4.6
Petrochemical	Ammonia and ammonium acetate fumes	6	38	1.5	25	1.0	28	1.1
		12	58	2.3	33	1.3	48	1.9
		24	86	3.4	43	1.7	74	2.9
Detergent	Alkalies and organic compounds	6	20	0.8	15	0.6	15	0.6
		12	33	1.3	20	0.8	20	0.8
		24	48	1.9	23	0.9	25	1.0
Detergent	Sulfur compounds	6	30	1.2	15	0.6	23	0.9
		12	53	2.1	23	0.9	30	1.2
		24	81	3.2	23	0.9	30	1.2
Alkylation	Moisture, chlorides	8	460	18.1	292	11.5	297	11.7
		12	668	26.3	432	17.0	409	16.1
		36	1468	57.8	1016	40.0	1016	40.0
Hydrochloric acid	Chlorine, hydrochloric acid fumes	6	312	12.3	147	5.8	180	7.1
		12	640	25.2	345	13.6	396	15.6
		24	1265	49.8	640	25.2	803	31.6

Source: Ref 2

ing thickness. In mild environments, galvanized steels can be used with no further treatment. In more severe environments, galvanized steels can be painted. In some cases, a prior treatment is used to provide a zinc phosphate conversion coating over the zinc coating to improve paint adherence. Information on zinc-base coatings can be found in the articles "Hot Dip Coatings," "Organic Coatings and Linings," "Corrosion of Zinc," "Corrosion of Carbon Steels," and "Corrosion in the Automotive Industry" in this Volume.

Finally, electroplating, usually with chromium, can be used where decorative requirements must be met in addition to atmospheric-corrosion resistance. Detailed information is available in the articles "Electroplated Coatings" and "Corrosion of Electroplated Hard Chromium" in this Volume.

Corrosion of Alloy Steels in Specific End-Use Environments

As with carbon steels, alloy steels are used in a wide variety of industrial applications. This section will review four major industries that rely heavily on alloy steel products. These include oil and gas production, energy conversion systems, marine applications, and chemical processing. Additional information is provided in the Section "Specific Industries and Environments" in this Volume.

Oil and Gas Production

Drilling and Primary Production. A variety of corrosion problems can be encountered in the drilling and primary production of oil and gas. These include weight loss corrosion, pitting corrosion, corrosion fatigue, stress-corrosion crack-

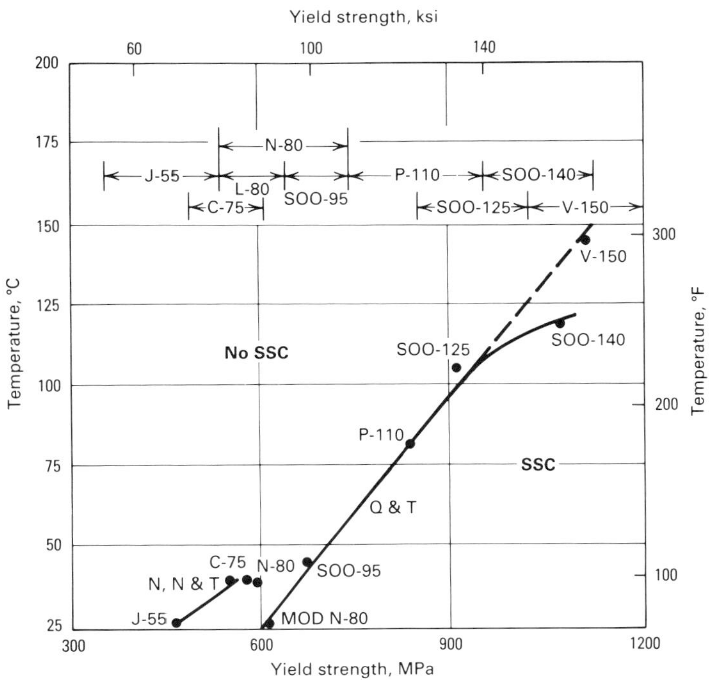

Fig. 3 Effect of temperature on SSC of high-strength steels. N, normalized; N & T, normalized and tempered; Q & T, quenched and tempered. Source: Ref 3

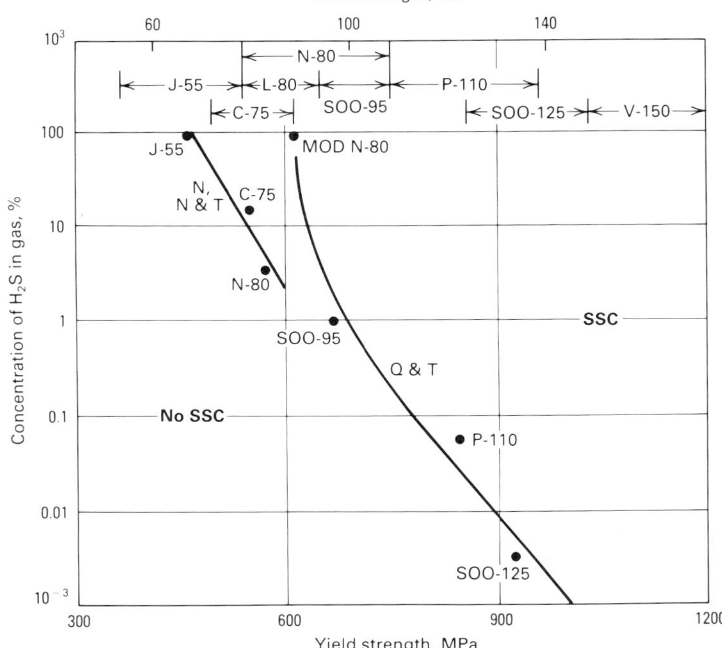

Fig. 4 Effect of H_2S concentration on SSC of high-strength steels. Source: Ref 3

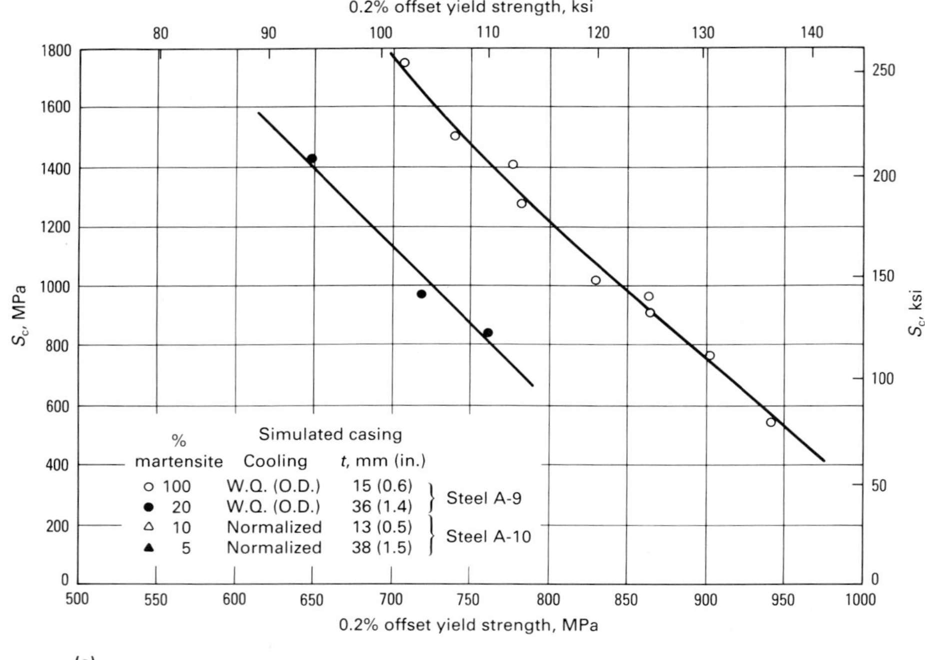

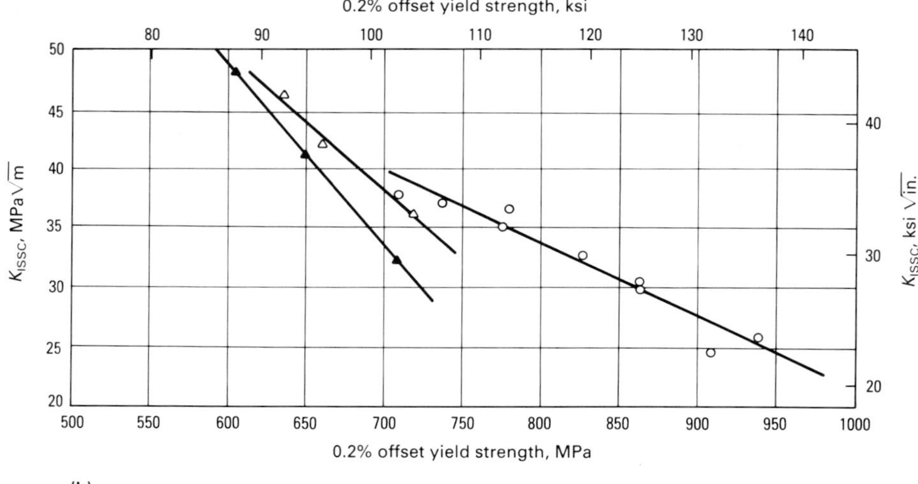

Fig. 5 Effect of yield strength on the critical stress, S_c, and sulfide fracture toughness, K_{ISSC}, of molybdenum-niobium modified 4135 steel cooled from the austenitizing temperature at different rates to produce a wide range of martensite contents and then tempered. W.Q. (O.D.), externally water quenched. (a) Bent-beam test. (b) Double-cantilever beam test (without salt). See Table 4 for steel compositions. Source: Ref 4

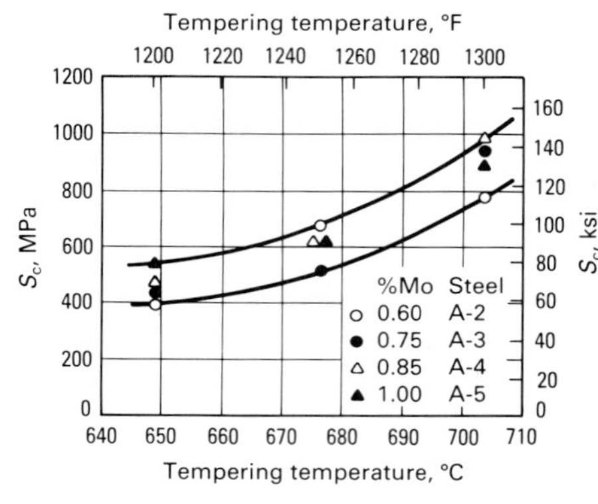

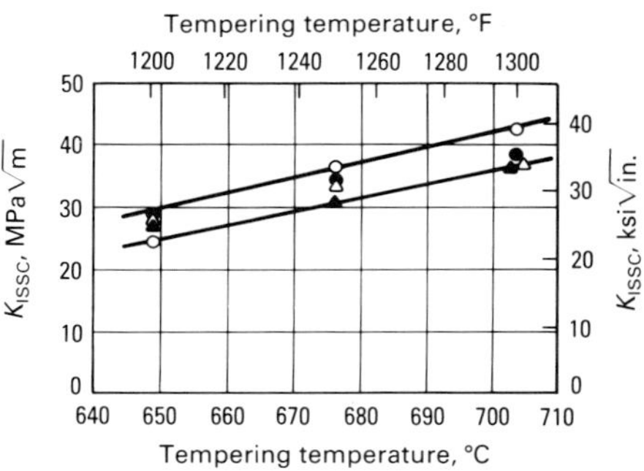

Fig. 6 Effect of temperature of a 1-h temper on the critical stress, S_c, and sulfide fracture toughness, K_{ISSC}, of molybdenum-niobium modified 4130 steels. (a) Bent-beam test. (b) Double-cantilever beam test. See Table 4 for steel compositions. Source: Ref 4

ing, sulfide stress cracking, and hydrogen-induced cracking.

In relatively shallow wells, lower-strength carbon or carbon-manganese steels can be employed in many of the components used in drilling and primary production. Oil and gas deposits are often such that corrosion is limited to weight loss corrosion, which can be effectively controlled by chemical inhibition. For the case of deep wells, however, high-strength alloy steels are usually required. Furthermore, very hostile environments are often encountered—in particular, high H_2S levels. In the case of deep gas wells, for example, formations have been encountered with H_2S concentrations ranging from 28 to 46%. In addition, temperatures to 200 °C (390 °F) can be found, along with pressures to 140 MPa (20 ksi). Also, H_2S is often found in combination with chloride-containing brines and CO_2, which add to the harshness of the environment.

Although chemical inhibition is used even in deep wells to control weight loss corrosion (see the article "Corrosion Inhibitors for Oil and Gas Production" in this Volume), the presence of H_2S can still result in the embrittlement of high-strength steels. This phenomena, known as sulfide stress cracking (SSC), depends on H_2S concentration and temperature. An example is shown in Fig. 2. Figures 3 and 4 illustrate typical SSC data for alloy steels used in oil field tubular components. The data shown are for high-strength steels covered under the following American Petroleum Institute specifications:

- 5A: "Welded or Seamless Steel Pipe for Oil or Gas Well Casing, Tubing, or Drill Pipe"
- 5AC: "Welded or Seamless Steel Pipe with Restricted Yield-Strength Range for Oil or Gas Well Casing or Tubing"
- 5AX: "High-Strength Seamless Steel Pipe for Oil or Gas Well Casing, Tubing, or Drill Pipe"

Certain proprietary grades are also included. As temperatures increase, some higher-strength steels can be used and resistance to SSC can be maintained. However, higher-strength steels are generally more susceptible to SSC than lower-strength steels.

Sulfide stress cracking resistance is influenced by steel microstructure, which in turn depends on steel composition and heat treatment. It has been observed that a martensitic structure provides better SSC resistance than other microstructures. Figure 5 illustrates this for a molybdenum-niobium modified AISI 4135 steel (compositions of the steels discussed in Fig. 5 to 7 are given in Table 4). The data in Fig. 5(a) were developed by using simple beam specimens strained in three-point bending for measuring a critical stress, S_c, and the data in Fig. 5(b) were obtained from double-cantilever beam specimens strained by wedge opening loading for measuring a threshold stress intensity, K_{ISSC}. Thus, it is important to

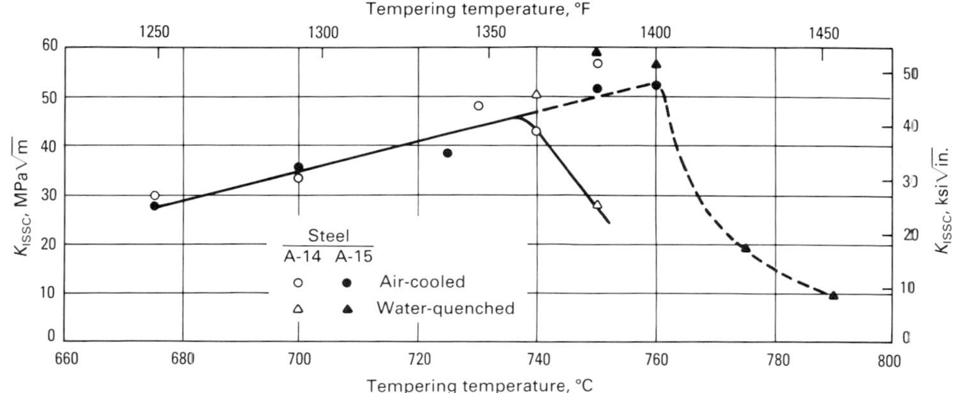

Fig. 7 Effect of tempering temperature on sulfide fracture toughness, K_{ISSC}, of molybdenum-niobium modified 4130 steels A-14 and A-15. See Table 4 for steel compositions. Source: Ref 4

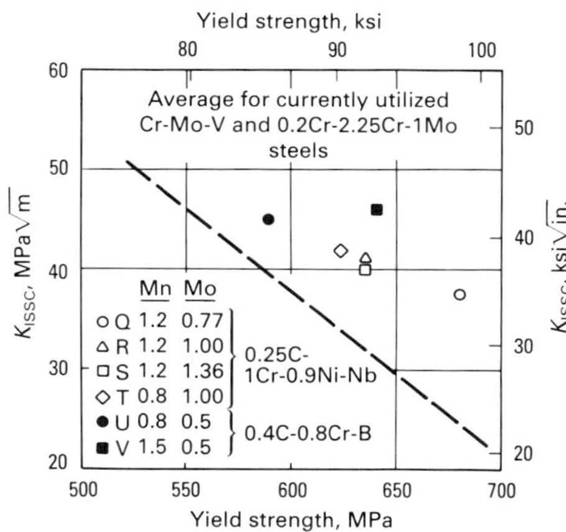

Fig. 8 Effects of molybdenum and manganese content on the SSC resistance of manganese-nickel-chromium-molybdenum-niobium and manganese-chromium-molybdenum-boron steels. Open symbols are 400-mm (16-in.) section thickness; closed symbols are 250-mm (10-in.). Source: Ref 5

select an alloy steel that has sufficient hardenability to achieve 100% martensite for a given application.

Furthermore, proper tempering of martensite is essential in order to maximize SSC resistance. Figure 6 illustrates the effects of tempering temperature on SSC behavior. It is evident that higher tempering temperatures improve SSC performance. The presence of untempered martensite, however, is extremely detrimental to SSC resistance. This is illustrated in Fig. 7, which shows the effect of tempering above the Ac_1 temperature for molybdenum-niobium modified 4130 steels containing two levels of silicon. Water quenching from above the Ac_1 temperature results in untempered martensite with a subsequent loss in SSC resistance. It has also been found that the development of a fine prior-austenite grain size and the use of accelerated cooling rates after tempering improve SSC resistance. The necessity for adequate hardenability is quite evident when considering alloy steels for heavy section wellhead components. Figure 8 shows how the SSC resistance of conventional steels used in wellhead equipment can be improved through modifications in composition, which increase hardenability.

With the advent of enhanced oil recovery techniques, additional corrosion problems must be considered. Carbon dioxide injection is one method of displacing crude oil from a formation for increased recovery (CO_2 injection is discussed in the section "Problems Encountered and Protective Measures" of the article "Corrosion in Petroleum Production Operations" in this Volume). This method involves development of CO_2 source wells, that is, those having large quantities of CO_2-containing gas. The gas from these wells is processed, transported to the production reservoir, and injected. Corrosion in source wells and in production wells results from the highly acidic environment created when CO_2 and water are present. The presence of chlorides and H_2S adds to the aggressiveness of the environment. Furthermore, corrosion rates change as temperatures change.

Figures 9(a) to (d) illustrate the complexities of corrosion in CO_2 environments. In Fig. 9(a), the effects of increasing CO_2 concentration on weight loss corrosion at a temperature of 65 °C (150 °F) are shown. The lower-alloy steels show a slight increase in corrosion rate with increasing CO_2 concentration, but the higher-alloy materials show little or no dependence on CO_2 level. As chromium content increases, corrosion resistance improves at a given CO_2 level. At a temperature of 175 °C (350 °F), however, the corrosion resistance of the lower-alloy steels improves, but that of the higher-alloy steels remains the same or decreases (Fig. 9b). With the addition of significant amounts of chloride at 65 °C (150 °F), some of the higher-alloyed steels begin to show an increase in corrosion rate with increasing CO_2 level (Fig. 9c). An increase in chloride concentration, along with an increase in temperature, results in a significant decrease in the corrosion resistance of the more highly alloyed steels (Fig. 9d). Finally, if H_2S is present in CO_2-brine environments, the corrosion rate of lower-alloy steels can be expected to increase. This is illustrated in Table 5.

The corrosion rates of various alloy steels in CO_2-brine-H_2S environments vary considerably with the specific environment encountered. As a result, control of the environment through chemical inhibition becomes an important tool, along with alloy selection, in reducing corrosion failures. Detailed information on corrosion and corrosion prevention in oil field operations is provided in the article "Corrosion in Petroleum Production Operations" in this Volume.

Petroleum Refining/Hydrocarbon Processing. A principal concern in petroleum refining and hydrocarbon processing is the problem of the interaction of hydrogen with the alloy steels used in these applications. Prolonged exposure to hydrogen, particularly at elevated temperatures, results in loss of ductility and premature failure. Figure 10 shows the delayed failure characteristics of an AISI 4340 steel resulting from cathodic charging of hydrogen. At higher tensile strengths, the effects of hydrogen become more severe.

The phenomenon often encountered in actual service is hydrogen attack. This involves the chemical reaction of hydrogen with metal carbides at elevated temperatures to form methane (CH_4). Because CH_4 cannot diffuse out of steel, an accumulation occurs, and this causes fissuring and blistering. The combined action of decarburization and fissuring results in loss of strength and ductility. The empirical limits on the use of alloy steels commonly used in a hydrogen environment are shown in Fig. 11. The Nelson diagrams shown are continually updated as additional service experience is acquired. Detailed information on corrosion problems and prevention in petroleum-refining operations is available in the articles "Corrosion in Petroleum Refining and Petrochemical Operations" and "Corrosion Inhibitors for Crude Oil Refineries" in this Volume.

Table 4 Chemical compositions of the molybdenum-niobium modified 4130/4135 test steels discussed in Fig. 5 to 7

Steel Code	Composition, wt%									
	C	Mn	Si	Cr	Mo	Nb	P	S	Al	N, ppm
A-2	0.34	0.74	0.39	1.06	0.60	0.035	0.030	0.022	0.12	208
A-3	0.31	0.73	0.39	1.05	0.75	0.036	0.034	0.021	0.16	166
A-4	0.32	0.74	0.39	1.04	0.85	0.035	0.027	0.026	0.16	138
A-5	0.32	0.74	0.40	1.05	0.98	0.036	0.027	0.025	0.17	148
A-9	0.34	0.68	0.38	1.00	0.75	0.034	0.025	0.027	ND	260
A-10	0.36	0.68	0.29	1.03	0.74	0.033	0.017	0.014	0.078	151
A-15	0.33	0.23	0.71	1.04	0.73	0.031	0.018	0.015	0.047	180

Source: Ref 4

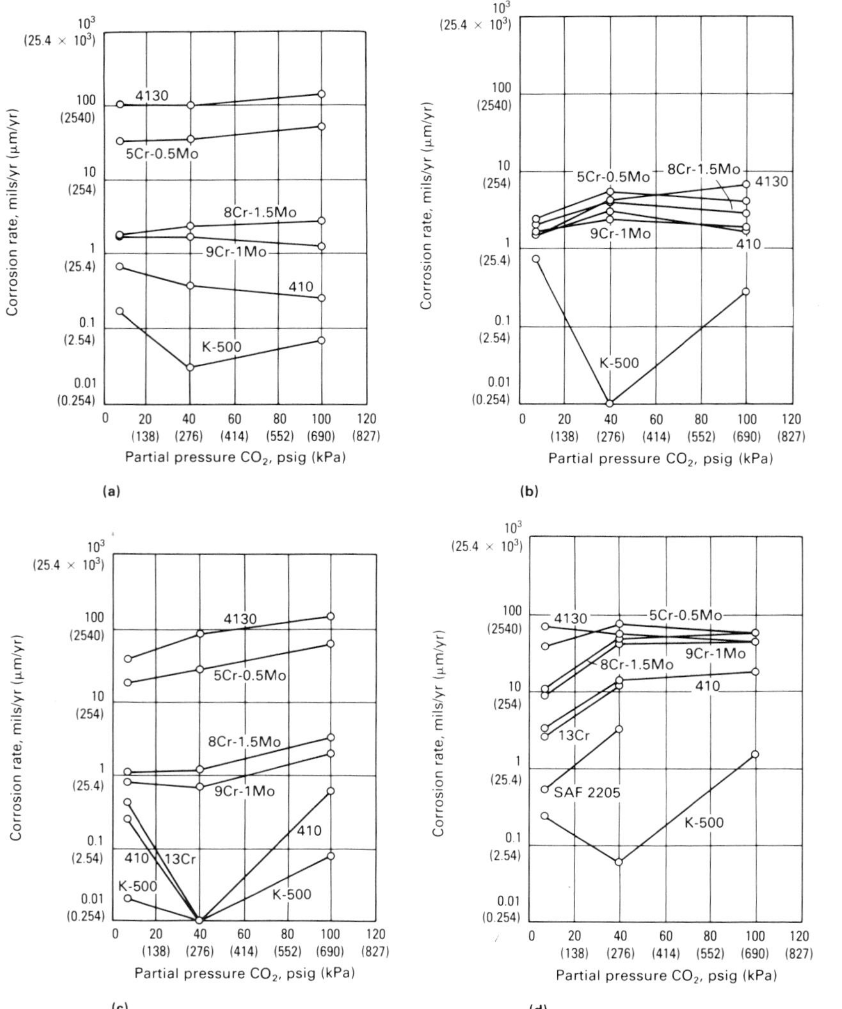

Fig. 9 Effect of partial pressure of CO_2 on the corrosion rates of various alloy steels. (a) 0% chlorides at 65 °C (150 °F). (b) 0% chlorides at 175 °C (350 °F). (c) 15.2% chlorides at 65 °C (150 °F). (d) 15.2% chlorides at 175 °C (350 °F). Source: Ref 6

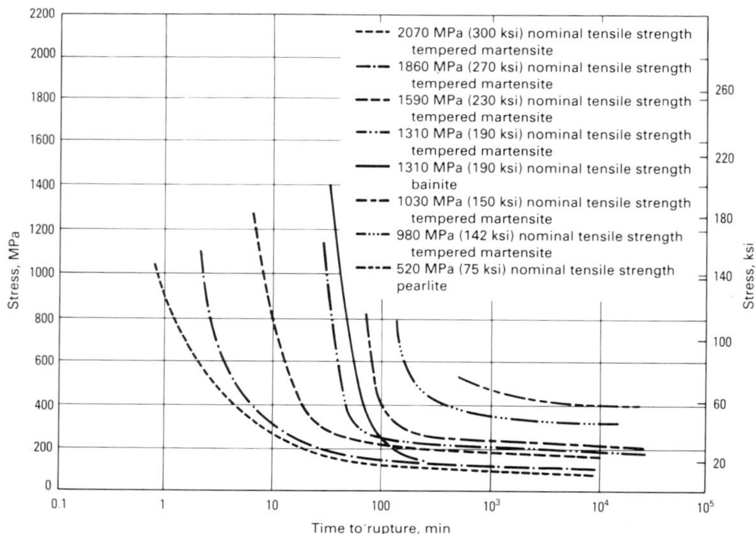

Fig. 10 Delayed-failure characteristics of unnotched specimens of an AISI 4340 steel during cathodic charging with hydrogen under standardized conditions. Electrolyte: 4% H_2SO_4 in water. Poison: 5 drops/liter of cathodic poison composed of 2 g phosphorus dissolved in 40 mL CS_2. Current density: 1.2 mA/cm^2 (8 mA/in.^{2}). Source: Ref 7

Table 5 Corrosion rate data for alloys exposed to seawater solutions at 175 °C (350 °F) with and without H_2S

Material	Corrosion rates: 1.93% Cl^-, 690 kPa (100 psig) CO_2, no H_2S		Corrosion rates: 1.92% Cl^-, 690 kPa (100 psig) CO_2, 0.1% H_2S	
	μm/yr	mils/yr	μm/yr	mils/yr
AISI 4130	890	35	2565	101
5Cr–1.5 Mo	330	13	1016	40
Type 410	36	1.4	30	1.2
13% Cr	...	...	25	1.0
Monel K-500	3	0.12	43	2.7

Source: Ref 6

Oil and Gas Transmission. The transmission of oil and gas involves consideration of the corrosion problems associated with linepipe steels. In addition to carbon steels, high-strength low-alloy steels are often used in pipeline service. Atmospheric corrosion, which has been discussed above, needs to be considered for exposed pipelines, and the corrosive actions of various soil formations must be addressed. A summary of an extensive study of the corrosion encountered by various alloy steels in several different types of soils is presented in Fig. 12(a) and (b). It is evident that factors such as soil pH, resistivity, degree of aeration, and level of acidity have more bearing on the severity of corrosion encountered than the alloy content of the steel. In some cases, increasing alloy content has a beneficial effect, but in other cases, it does not. In general, the use of protective coatings and cathodic protection offers the best means of reducing the level of corrosive attack.

Linepipe steels can be susceptible to a specialized form of hydrogen damage when H_2S is present in oil and gas. This type of embrittlement, known as hydrogen-induced cracking, results from the accumulation of hydrogen at internal surfaces within the steel. Examples include interfaces at nonmetallic inclusions and at microstructure constituents that differ significantly from the surrounding matrix. Martensite islands in a ferrite-pearlite matrix would be typical. Microcracks that form at these interfaces grow in a stepwise fashion toward the surface of the pipe, with the result being failure. Figure 13 illustrates an example of this type of cracking.

Very few failures due to hydrogen-induced cracking have been reported. However, they can be catastrophic, and considerable investigative work has been done to understand the nature of the problem and to develop preventive measures. Hydrogen-induced cracking can usually be prevented by control of the environment—for example, dehydration to remove water and through chemical inhibition. A number of metallurgical factors have also been identified that influence resistance to hydrogen-induced cracking and offer a means of reducing the susceptibility of linepipe steels to this form of embrittlement.

Two factors that influence the susceptibility of linepipe steels to hydrogen-induced cracking are steel cleanliness and degree of alloying element segregation. This might be expected, because the degree of steel cleanliness affects the volume fraction of nonmetallic inclusions present and therefore the number of interfaces available for the accumulation of hydrogen. Segregation of alloying elements can lead to the formation of

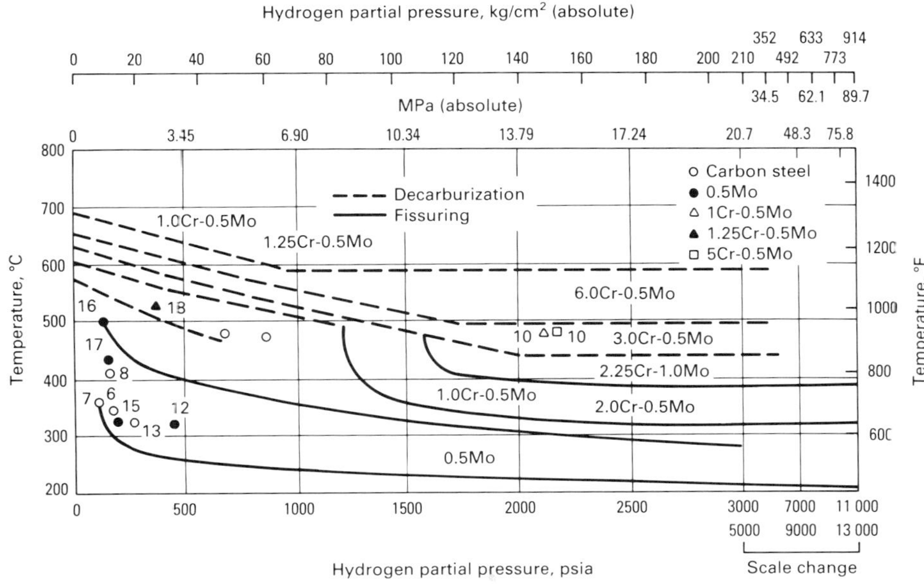

Fig. 11 Nelson curve defining safe upper limits for steels in hydrogen service. Source: Ref 7

low-temperature austenite decomposition products, thus providing additional sites for hydrogen accumulation.

Hydrogen-induced cracking has been found to be associated with manganese sulfide inclusions that have become elongated during hot rolling. Elongated silicate inclusions also provide interfaces for hydrogen accumulation. Laboratory tests have shown that reduction in the sulfur level of a linepipe steel reduces susceptibility to hydrogen-induced cracking. Reducing the sulfur content to levels of 0.002% or less can result in a significant improvement in resistance to hydrogen-induced cracking. It has also been observed that resistance to hydrogen-induced cracking can be improved through the use of sulfide shape control techniques. Calcium or rare-earth metals are added to the steel to form calcium or rare-earth sulfides. These inclusions are not plastic at hot working temperatures and therefore do not elongate during hot rolling.

The effects of various alloying elements on resistance to hydrogen-induced cracking are uncertain and somewhat controversial. The alloying element that has received the most attention is copper. Laboratory results have shown that cop-

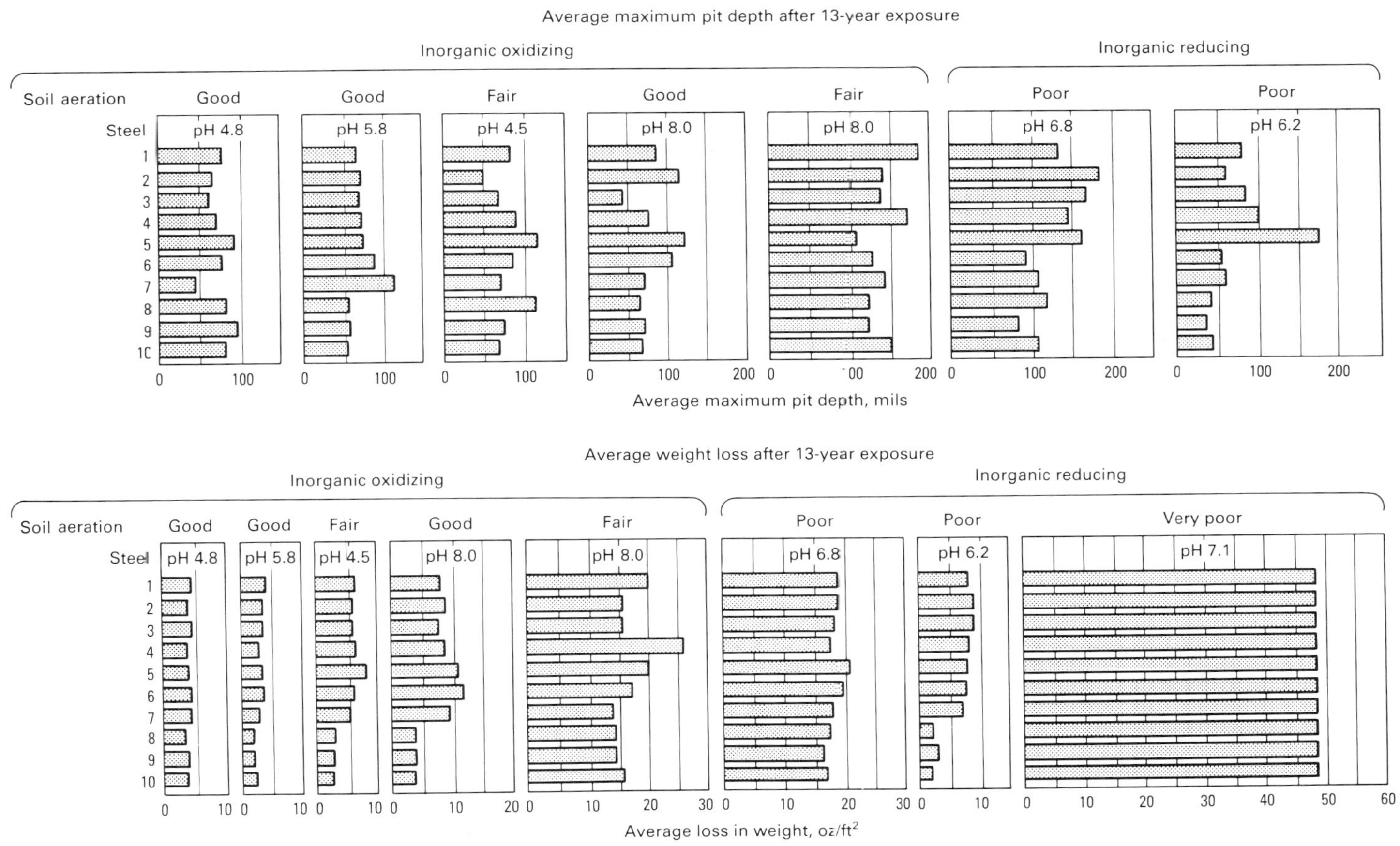

Element	Compositions, for steel numbers 1	2	3	4	5	6	7	8	9	10
Chromium	0.049	0.02	0.02	...	...	1.02	2.01	5.02	4.67	5.76
Nickel	0.034	0.15	0.14	0.52	1.96	0.22	0.07	0.09	0.09	0.17
Copper	0.052	0.45	0.54	0.95	1.01	0.428	0.004	0.008	0.004	0.004
Molybdenum	...	0.07	0.13	...	...	...	0.57	...	0.51	0.43

Fig. 12(a) Effect of composition on corrosion of low-alloy ferrous materials in various disturbed (backfilled) soils. Environmental data are given in Fig. 12(b).

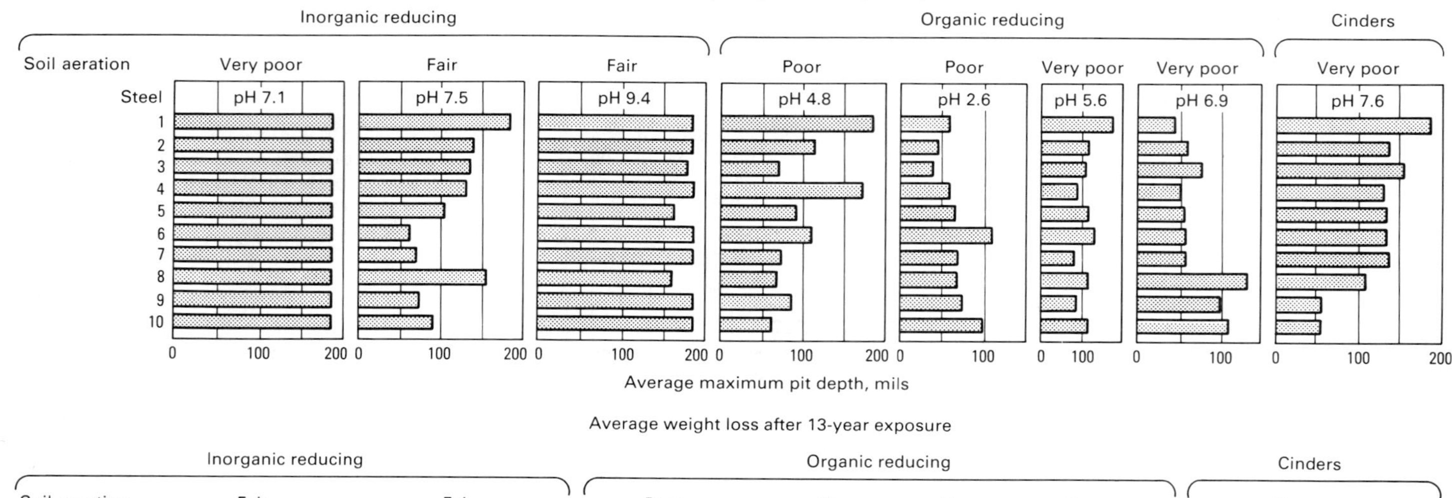

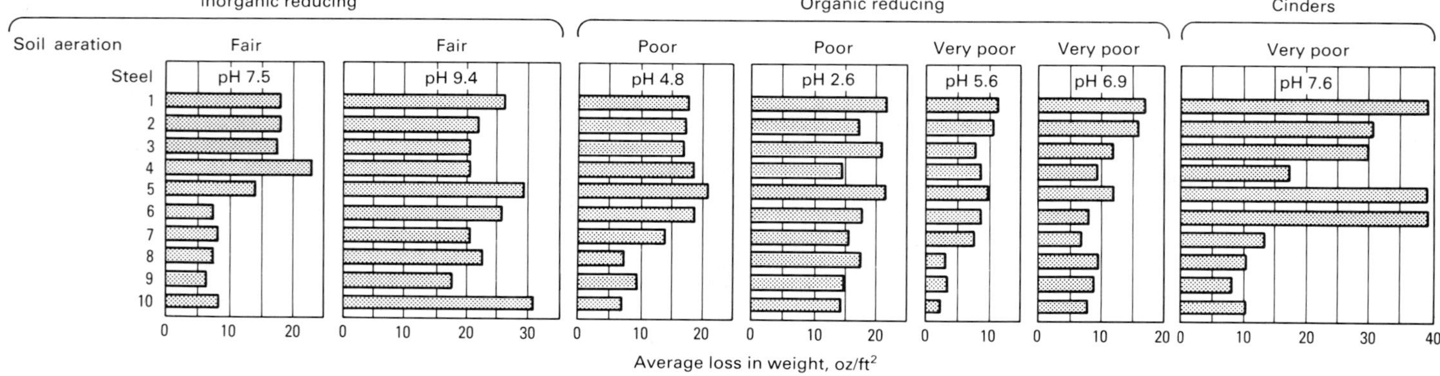

Environment	Inorganic oxidizing					Inorganic reducing					Organic reducing				Cinders
	Acid			Alkaline		Acid			Alkaline		Acid				Alkaline
Aeration	Good	Good	Fair	Good	Fair	Poor	Poor	Very poor	Fair	Fair	Poor	Poor	Very poor	Very poor	Very poor
Resistivity, Ω·m	177.9	52.1	69.2	1.48	2.32	1.9	9.43	4.06	0.62	2.78	7.12	2.18	16.6	0.84	4.55
pH	4.8	5.8	4.5	8.0	8.0	6.8	6.2	7.1	7.5	9.4	4.8	2.6	5.6	6.9	7.6

Fig. 12(b) Effect of composition on corrosion of low-alloy ferrous materials in various disturbed (backfilled) soils. Compositions are given in Fig. 12(a).

per can significantly reduce susceptibility to hydrogen-induced cracking. Apparently, the benefits of copper are realized only in environments with a pH of 4.5 to 4.8 and above. At pH levels less than this, copper has no effect on resistance to hydrogen-induced cracking. Detailed information on corrosion and corrosion prevention in linepipe steels can be found in the article "Corrosion of Pipelines" in this Volume.

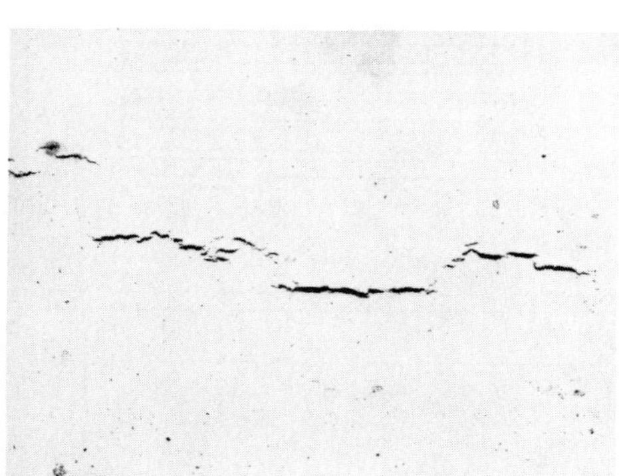

Fig. 13 Hydrogen-induced cracking in a linepipe steel. 30×

Energy Conversion Systems

Fossil Fuel Power Systems. As noted earlier, corrosion problems associated with energy conversion systems often concern the combustion of fossil fuels (these corrosion problems are discussed in detail in the article "Corrosion in Fossil Fuel Power Plants" in this Volume). The most typical applications involve the operation of coal- and oil-fired utility steam boilers. In these applications, corrosion due to boiler feedwater and high-temperature high-pressure steam must also be considered (see the section "Hot Corrosion in Coal- and Oil-Fired Boilers" in the article "Corrosion in Fossil Fuel Power Plants").

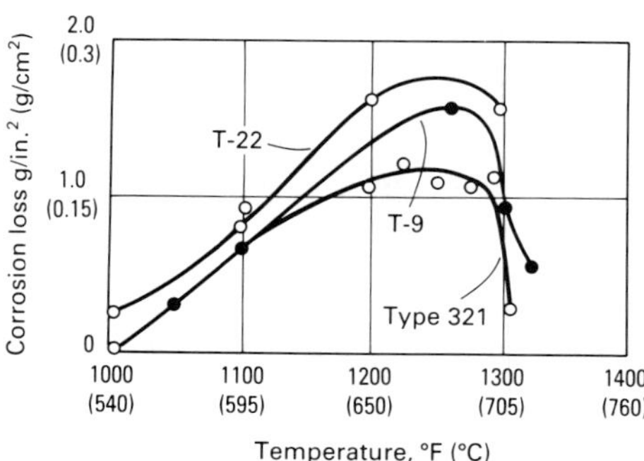

Fig. 14 Effect of alloying on corrosion rate of T-9 (8.0-10.0Cr, 0.90-1.10Mo), T-22 (1.9-2.6Cr, 0.87-1.13Mo), and type 321 (17-19Cr, 9-12Ni) alloy steels. Source: Ref 8

Combustion of fossil fuels can result in so-called fire-side corrosion, which is an elevated-temperature attack on metal surfaces stemming from the products of combustion. There are three general areas where external corrosion problems occur: the water wall or boiler tubes near the firing zone, the high-temperature superheater and reheater tubes, and the ductwork that handles the combustion flue gases.

Corrosion of the ductwork is a low-temperature attack that results mainly from acid condensation. Prevention of this corrosion depends primarily on maintaining flue gas temperatures and metal surface temperatures above acid dewpoints.

Corrosion on water wall, superheater, or reheater tubes results from fuel ash deposits at

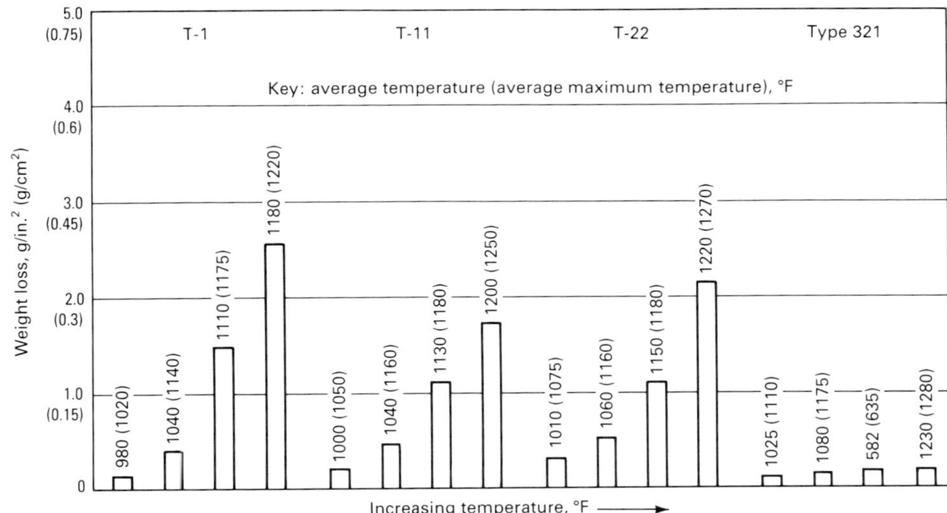

Fig. 15 Weight loss versus temperature data for corrosion probes made of alloy steels (T-1, T-11, T-22) and type 321 stainless steel. Source: Ref 9

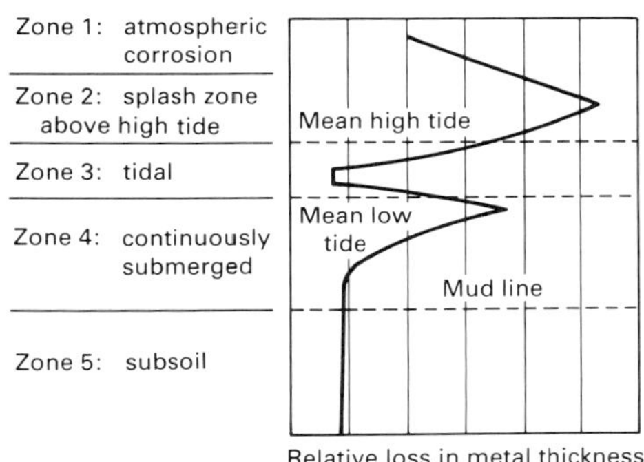

Fig. 16 Corrosion profile of steel piling after 5 years of exposure in seawater at Kure Beach, NC. Source: Ref 11

higher temperatures (see the section "Corrosion of Superheaters and Air Heaters" in the article "Corrosion in Fossil Fuel Power Plants" in this Volume). In these situations, the corrosive nature of fossil fuels varies considerably with the chemical composition of the fuel. It should be noted that many fuels are not especially corrosive. However, coals containing significant levels of sulfur and alkali metals are particularly damaging, as are oils that contain alkali metals, sulfur, and vanadium. These constituents have been identified as principal sources of corrosive attack in a number of studies involving the analysis of fuel ash deposits on boiler and superheater tubes.

In the case of coal combustion, corrosive attack results from complex chemical reactions involving sulfur and alkali metals (sodium and potassium) to form alkali sulfates. These alkali sulfates, along with sulfur trioxide (SO_3), react with the protective iron oxide. This reaction breaks down the iron oxide and forms a complex alkali iron sulfate. At temperatures of 425 to 480 °C (800 to 900 °F), this deposit can spall from the surface, exposing fresh iron for further attack. Such would be the case with water wall tubing. As temperatures increase to levels encountered by superheater or reheater tubing—for example, 565 to 705 °C (1050 to 1300 °F)—the complex sulfate becomes liquid and attacks the tubing directly. Figure 14 compares the corrosion behavior of two alloy steels and an austenitic stainless steel in this higher temperature range. The data were developed in a laboratory simulation that created the complex alkali iron sulfates. Corrosion rates increase with temperature until the sulfates become unstable, leading to a decrease in corrosion rate. Figure 15 illustrates weight loss data obtained from corrosion probes that were fabricated from various alloy steels and installed in a coal-fired steam boiler system. It is evident from both Fig. 14 and 15 that, although they can be used in these environments, alloy steels do not perform as well as stainless steels.

Several courses of action are taken to prevent fire-side corrosion in coal-fired facilities. At the lower temperatures encountered by water wall tubing, procedures are implemented to avoid spalling of combustion deposits. These procedures involve controlling fuel flow and combustion conditions to avoid impingement by particulate matter on critical metal surfaces. At higher temperatures, where liquid-phase attack can occur, protective shields have been used to maintain metal surfaces at temperatures above the corrosive range. The use of coal blending to counteract the corrosive nature of a given coal offers an additional means of corrosion prevention. Also, studies have shown that certain additives to coal are effective in reducing corrosion rates. Success has been achieved with kaolin, diatomaceous earth, and magnesium oxide or other alkaline earth oxides. These additives prevent the formation of complex alkali iron sulfates by forming stable compounds with one or more of their components. As noted earlier, the lower-alloy chromium-molybdenum steels have limited corrosion resistance to highly aggressive coals. Although some improvement can be achieved by using 9Cr-1Mo steels, such as T-9, maximum corrosion resistance requires the use of stainless steels.

In oil-fired boilers, the principal source of corrosion comes from a fluxing action of molten sodium-vanadium complexes with the protective oxide scale formed on metal surfaces. Although this can occur at lower temperatures if the current ratio of sodium to vanadium is present, this form of corrosion generally takes place at temperatures above 595 °C (1100 °F). Superheater and reheater tube corrosion rates of as much as 0.75 mm/yr (30 mils/yr) have been observed in field measurements.

An effective method of preventing oil ash corrosion is to remove vanadium, alkali metals, and sulfur chemically from the fuel. However this approach can be costly. Certain magnesium and calcium compounds have been found to be effec-

Table 6 Corrosion factors for carbon and alloy steel immersed in seawater

Factor in seawater	Effect on iron and steel
Chloride ion	Highly corrosive to ferrous metals. Carbon steel and common ferrous metals cannot be passivated (sea salt is about 55% chloride).
Electrical conductivity	High conductivity makes it possible for anodes and cathodes to operate over long distances; thus, corrosion possibilities are increased, and the total attack may be much greater than that for the same structure in freshwater.
Oxygen	Steel corrosion is cathodically controlled for the most part. Oxygen, by depolarizing the cathode, facilitates the attack; thus, a high oxygen content increases corrosivity.
Velocity	Corrosion rate is increased, especially in turbulent flow. Moving seawater may destroy rust barrier and provide more oxygen. Impingement attack tends to promote rapid penetration. Cavitation damage exposes the fresh steel surface to further corrosion.
Temperature	Increasing ambient temperature tends to accelerate attack. Heated seawater may deposit protective scale or lose its oxygen; either or both actions tend to reduce attack.
Biofouling	Hard-shell animal fouling tends to reduce attack by restricting access of oxygen. Bacteria can take part in corrosion reaction in some cases.
Stress	Cyclic stress sometimes accelerates failure of a corroding steel member. Tensile stresses near yield also promote failure in special situations.
Pollution	Sulfides, which are normally present in polluted seawater, greatly accelerate attack on steel. However, the low oxygen content of polluted waters could favor reduced corrosion.
Silt and suspended sediment	Erosion of the steel surface by suspended matter in the flowing seawater greatly increases the tendency toward corrosion.
Film formation	A coating of rust or of rust and mineral scale (calcium and magnesium salts) will interfere with the diffusion of oxygen to the cathode surface, thus slowing the attack.

Source: Ref 11

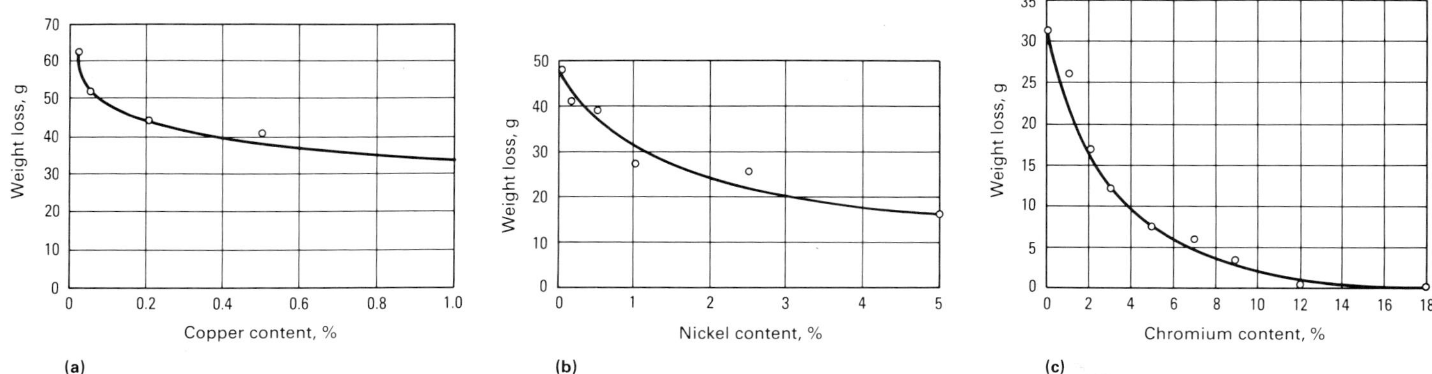

Fig. 17 Effect of alloying additions on the corrosion of steel in a marine atmosphere at Kure Beach, NC (90-month exposure). (a) Effect of copper (100 × 150 mm, or 4 × 6 in., specimen). (b) Effect of nickel (100 × 150 mm, or 4 × 6 in., specimen). (c) Effect of chromium (75 × 150 mm, or 3 × 6 in., specimen). Source: Ref 14

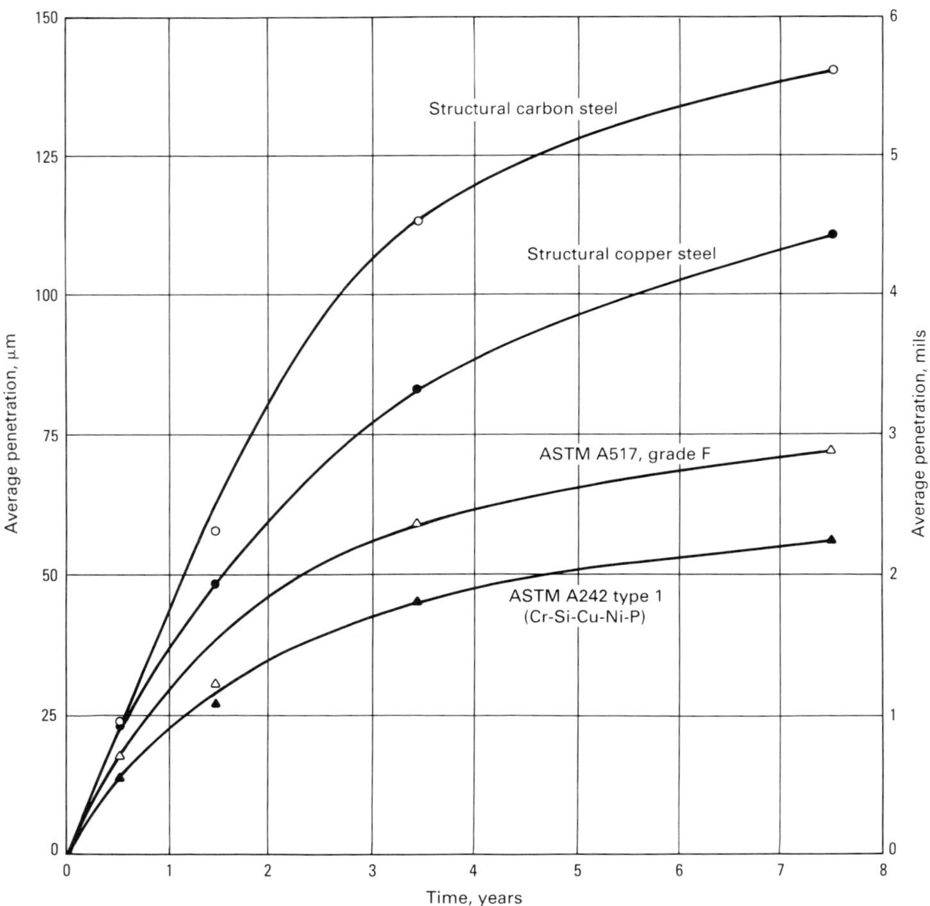

Fig. 18 Comparative corrosion performance of constructional steels exposed to moderate marine atmosphere at Kure Beach, NC. Source: Ref 19

tive in reducing corrosion rates. These compounds form high melting point complexes with oil ash constituents. In terms of alloy steel selection, it has been found that the 9Cr-1Mo alloys exhibit excellent corrosion resistance to oil ash corrosion. In addition, modifications of these alloys with additional molybdenum and/or vanadium provide high corrosion resistance and increased strength. Thus, for this application, alloy steels are available at a lower cost than stainless steels.

Steam/water side corrosion is another major problem encountered in fossil fuel power plants. In most cases, contaminant deposition reduces equipment efficiency and induces corrosion by a variety of mechanisms and is implicated in a variety of boiler tube failure mechanisms that are most common in water walls and economizers, which are often constructed from low-chromium ferritic steels such as ASTM A213 T11.

Water side deposits often begin as accumulations of corrosion products transported to the boiler from other parts of the system. The corrosion product deposit is porous, unlike the protective magnetite (Fe_3O_4) film. This porous deposit serves as a trap for corrosive impurities, such as caustic, chlorides, and acid sulfates.

Alloy steel boiler tube failures in steam-containing tubing have also almost exclusively been the result of contaminant entrainment within the steam. Chlorides, sulfates, and caustic are the most common contaminants. However, the growth of Fe_3O_4 on the inside tube surface can also be a secondary contributor to tube failure. If its rate of growth is excessive, this will act as a thermal barrier and cause the tube wall temperature to rise, sometimes above the point at which excessive creep damage will result in an overheating failure. Additional information is available in the section "Corrosion of Steam/Water Side Boilers" of the article "Corrosion in Fossil Fuel Power Plants" in this Volume.

Nuclear Power Systems. High-strength chromium-molybdenum and nickel-chromium-molybdenum alloy steels are also used in components for commercial light water reactors. For example, most modern light water reactor steam turbine rotors in the United States are made from 3.5NiCrMoV steel in conformance with the requirements of ASTM A471 (Class 1 through 9). A serious concern associated with steam turbine materials is that of stress-corrosion cracking (SCC), which has occurred in quenched-and-tempered and normalized-and-tempered low-alloy steels with a wide range of grain sizes. Metallurgical and environmental conditions that influence SCC behavior in low-alloy steels and methods of reducing or eliminating SCC are discussed in the section "Stress Corrosion Cracking in Steam Turbine Materials" of the article "Corrosion in the Nuclear Power Industry" in this Volume.

Wet steam erosion-corrosion of nuclear plant piping represents another serious problem that can lead to costly power outages and repairs. The most widely used material for U.S. nuclear plant wet steam piping has been carbon steel, which has shown a susceptibility to erosion-corrosion. With alloying additions of chromium, copper, and molybdenum, however, erosion-corrosion resistance can be significantly improved. In com-

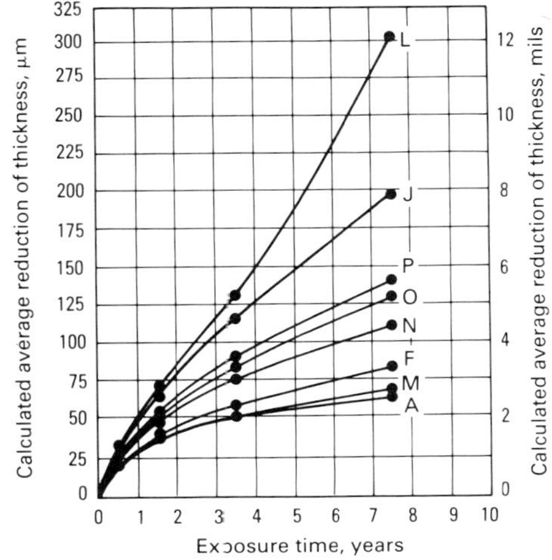

Steel	C	Mn	P	S	Si	Cu	Ni	Cr
	Composition, %							
A(a)	0.09	0.24	0.15	0.024	0.80	0.43	0.05	1.1
M(a)	0.06	0.48	0.11	0.030	0.54	0.41	0.51	1.0
F(a)	0.05	0.36	0.05	0.016	0.008	1.1	2.0	0.01
N(a)	0.11	0.55	0.08	0.026	0.06	0.55	0.28	0.31
O(a)	0.16	1.4	0.013	0.021	0.18	0.30	0.50	0.03
P(a)	0.23	1.5	0.018	0.021	0.19	0.29	0.04	0.08
J(b)	0.19	0.52	0.008	0.039	0.01	0.29	0.05	0.05
L(b)	0.16	0.42	0.013	0.021	0.01	0.02	0.02	0.01

(a) High-strength low-alloy steels. (b) Structural carbon and structural copper steels

Fig. 19 Effect of exposure time on corrosion of steels in marine atmosphere at Kure Beach, NC. Source: Ref 17

parison to ordinary carbon steel, erosion-corrosion rates can be reduced by three times with carbon-molybdenum steel and more than ten times with chromium-molybdenum steels. Field experience has shown that 1.25Cr, 0.5Mo, and 2.25Cr-1Mo steels are virtually immune to erosion-corrosion in nuclear power plant applications. Data comparing the erosion-corrosion behavior of carbon and alloy steel piping material can be found in the section "Erosion-Corrosion in Wet Steam Flow" of the article "Corrosion in the Nuclear Power Industry" in this Volume.

Nuclear Waste Disposal. The disposal of high-level nuclear waste in deep underground repositories requires the development of waste packages that will keep the radioisotopes contained for periods of up to 1000 years. The primary geologic media currently being considered in the United States for repository siting are salt, basalt, tuff, and granite. A number of alloy steel and other iron-base materials are being considered for the structural barrier members of waste packages. Their uniform and nonuniform (pitting and intergranular) corrosion behavior and

Table 7 Corrosion of low-alloy steels in a marine atmosphere

Data collected over 15.5 years at 250-m (800-ft) lot, Kure Beach, NC

Group	Description	C	Mn	Si	S	P	Ni	Cu	Cr	Mo	Approximate total alloy content, %	Weight loss, mg/dm² (a)
		Composition, %										
I	High-purity iron plus copper	0.020	0.020	0.003	0.03	0.006	0.05	0.020	...	...	...	...
		0.020	0.023	0.002	0.03	0.005	0.05	0.053	...	...	0.1	43
		0.02	0.07	0.01	0.03	0.003	0.18	0.10	...	...	0.4	29.8
II	Low-phosphorus steel plus copper	0.040	0.39	0.005	0.02	0.007	0.004	1.03	0.06	...	1.5	17.3
III	High-phosphorus steel plus copper	0.09	0.43	0.005	0.03	0.058	0.24	0.36	0.06	...	1.2	16.9
		0.095	0.41	0.007	0.05	0.104	0.002	0.51	0.02	...	1.0	16.5
IV	High-manganese and -silicon steels plus copper	0.17	0.67	0.23	0.03	0.012	0.05	0.29	0.14	...	1.4	16.6
V	Copper steel plus chromium and silicon	0.072	0.27	0.83	0.02	0.140	0.03	0.46	1.19	...	2.9	6.3
VI	Copper steel plus molybdenum	0.17	0.89	0.05	0.03	0.075	0.16	0.47	...	0.28	1.9	11.8
VII	Nickel steel	0.16	0.57	0.020	0.02	0.015	2.20	0.24	...	...	3.0	9.4
		0.19	0.53	0.009	0.02	0.016	3.23	0.07	...	...	3.9	9.2
		0.17	0.58	0.26	0.01	0.007	4.98	0.09	...	...	5.9	6.1
		0.13	0.23	0.07	0.01	0.007	4.99	0.03	0.05	...	5.4	7.5
VIII	Nickel steel plus chromium	0.13	0.45	0.23	0.03	0.017	1.18	0.04	0.65	0.01	2.6	10.5
IX	Nickel steel plus molybdenum	0.16	0.53	0.25	0.01	0.013	1.84	0.03	0.09	0.24	3.0	9.8
X	Nickel steel plus chromium and molybdenum	0.10	0.59	0.49	0.01	0.013	1.02	0.09	1.01	0.21	3.4	6.5
		0.08	0.57	0.33	0.01	0.015	1.34	0.19	0.74	0.25	3.4	7.6
XI	Nickel-copper steel	0.12	0.57	0.17	0.02	0.01	1.00	1.05	...	...	2.8	10.6
		0.09	0.48	1.00	0.03	0.055	1.14	1.06	...	...	3.8	5.6
		0.11	0.43	0.18	0.02	0.012	1.52	1.09	...	...	3.2	10.0
XII	Nickel-copper steel plus chromium	0.11	0.65	0.13	0.02	0.086	0.29	0.57	0.66	...	2.4	10.5
		0.11	0.75	0.23	0.04	0.020	0.65	0.53	0.74	...	2.9	9.3
		0.08	0.37	0.29	0.03	0.089	0.47	0.39	0.75	...	2.4	9.1
XIII	Nickel-copper steel plus molybdenum	0.03	0.16	0.01	0.03	0.009	0.29	0.53	...	0.08	1.1	18.2
		0.13	0.45	0.066	0.02	0.073	0.73	0.573	...	0.087	2.0	11.2

(a) A weight loss of 10 mg/dm²/15.5 years = 0.32 mil/yr. Source: Ref 13

Table 8 Average decrease in thickness of 6-m (20-ft) specimens after 5-year exposure to splash, seawater, and mud zones at Harbor Island, NC

Average distance from top		Decrease in thickness													
		Sheet steel piling		0.54Ni-0.52Cu-0.12P		0.55Ni-0.22Cu-0.17P		0.54Ni-0.20Cu-0.11P		0.55Ni-0.20Cu-0.14P		0.28Ni-0.20Cu-0.14P		0.28Ni-0.22Cu-0.17P	
m	ft	μm	mils	μm	mils	μm	mils	μm	mils	μm	mils	μm	mils	μm	mils
0.15	0.5(a)	229	9	279	11	305	12	229	9	610	24	229	9	254	10
0.46	1.5	2210	87	330	13	406	16	762	30	457	18	533	21	533	21
0.76	2.5	2490	98	432	17	660	26	1372	54	762	30	1143	45	1854	73
Approximate high-tide line															
1.1	3.5	1219	48	102	4	229	9	229	9	178	7	152	6	559	22
1.4	4.5	25	1	25	1	51	2	25	1	51	2	25	1	51	2
1.7	5.5	...	0	25	1	51	2	51	2	178	7	76	3	51	2
2.0	6.5	356	14	940	37	864	34	1041	41	737	29	711	28	610	24
Approximate low-tide line															
2.3	7.5	1422	56	1321	52	1321	52	1626	64	1346	53	1067	42	1168	46
2.6	8.5	1143	45	1041	41	1118	44	1245	49	1067	42	965	38	864	34
2.9	9.5	1321	52	965	38	1041	41	1245	49	1245	49	1092	43	813	32
3.2	10.5	1346	53	1219	48	1016	40	1245	49	1067	42	1041	41	813	32
3.5	11.5	1143	45	991	39	890	35	1245	49	1067	42	940	37	813	32
3.8	12.5	1168	46	940	37	965	38	1168	46	813	32	890	35	838	33
Approximate ground line															
4.1	13.5	1143	45	330	13	610	24	940	37	279	11	305	12	457	18
4.4	14.5	736	29	152	6	610	24	610	24	152	6	178	7	432	17
4.7	15.5	533	21	127	5	127	5	356	14	127	5	152	6	457	18
5.0	16.5	559	22	254	10	330	13	279	11	178	7	381	15	711	28
5.3	17.5	762	30	457	18	559	22	254	10	305	12	711	28	787	31
5.6	18.5	762	30	305	12	381	15	254	10	559	22	635	25	864	34
5.9	19.5	686	27	381	15	610	24	432	17	991	39	965	38	787	31

Note: Approximate mean high tide 0.6 to 0.9 m (2 to 3 ft) from tops of specimens; approximate mean low tide about 1.8 m (6 ft) from tops of specimens.
(a) Unrealistic values because of partial protection from top supporting member. Source: Ref 18

their resistance to SCC in aqueous environments relevant to salt media are under study at the Pacific Northwest Laboratory (Ref 10). These and the results of other studies on alternate nonferrous alloys are presented in the section "Corrosion of Containment Materials for Radioactive Waste Isolation" of the article "Corrosion in the Nuclear Power Industry" in this Volume.

Marine Applications

Carbon and alloy steels are extensively used for submerged or partly submerged structures—both in harbors (sea walls and piers, for example) and offshore (oil drilling platforms, for example). As will be shown in this section, however, low-alloy steels exhibit superior resistance to marine corrosion.

Marine structures exhibit five separate zones that are susceptible to corrosion at different rates, depending primarily on elevation above the tidal zone or depth of immersion in seawater. These zones are described below and are identified in Fig. 16, which also shows the usual relative corrosion rate associated with each zone:

- *Atmospheric zone:* The portion of the elevated structure subject to a marine atmosphere, including sea mist and high relative humidity, but without significant wetting by splash from waves
- *Splash zone:* The portion above the level of mean high tide that is subject to wetting by large droplets of seawater
- *Tidal zone:* The portion of the structure between mean high tide and mean low tide; it is alternately immersed in seawater and exposed to a marine atmosphere
- *Submerged zone:* The portion of the structure from about 0.3 to 1 m (1 to 3 ft) below mean low tide down to the mud line
- *Subsoil zone:* The portion below the mud line, where the structure has been driven into the ocean bottom

The effects of each of these zones on the corrosion behavior of alloy steels is reviewed below. More detailed information on marine atmospheres (zones 1 and 2, Fig. 16) and marine environments (zones 3 to 5) and the variables (such as salinity, temperature, pollutants, and biological activity) that influence corrosion can be found in the article "Marine Corrosion" in this Volume. A brief summary of some of the more influential variables is presented in Table 6.

Atmospheric-Zone Corrosion. Low-alloy steels demonstrate greatly improved resistance to marine atmospheres compared to the resistance of carbon steels. The first indication that composition affects corrosion behavior was the observation that copper-bearing steels exhibit improved endurance in industrial atmospheres (Ref 12). It was later found that copper-bearing steels also perform better than plain carbon steels at ocean sites (Ref 11).

A number of marine corrosion studies have evaluated the benefits of copper, nickel, chromium, and phosphorus additions to steel (Ref 1, 13-18). The benefit derived from the addition of copper to steel exposed to an industrial atmosphere has been attributed to the relatively insoluble basic sulfates from the SO_2 in the polluted air, which slowly develop a fine-grain, tightly adherent protective rust film (Ref 11). Additions of nickel, chromium, silicon, and phosphorus also promote relatively insoluble corrosion products (Ref 13). Chlorine has a deleterious effect on the protective rust layer on alloy steels, and the manner in which protective rust coats form in marine atmospheres is less understood than in the case of the industrial atmosphere. However, tests have shown that alloying additions do provide enhanced corrosion resistance in marine atmospheres. The effects of individual additions of copper, nickel, and chromium are shown in Fig. 17.

Tests performed at a 250-m (800-ft) lot at Kure Beach, NC, for 15.5 years indicated a corrosion rate of 7.6 μm (0.3 mil/yr) or less for copper-bearing and low-alloy steels (Ref 13). Table 7 identifies the compositions of the steels used in these tests and gives the weight losses determined. A wide range of compositions gave improved corrosion resistance. A comparison of marine atmosphere corrosion of plain carbon, copper-bearing steel and two low-alloy steels is shown in Fig. 18. Data for a series of low-alloy steels with total alloy additions up to 3.5% are shown in Fig. 19. Additional information on the effects of various alloying additions to carbon steels can be found in the section "Weathering Steels" of the article "Corrosion of Carbon Steels" in this Volume.

Splash- and Tidal-Zone Corrosion. Low-alloy steel undergoes decidedly less corrosion at the splash zone (zone 2, Fig. 16) than carbon steel (Ref 11) does. Some experimental results comparing carbon and alloy steel 6 m (20 ft) specimens after 5 years of exposure to splash, seawater, and mud zones are presented in Table 8. At the 0.45- and 0.75-m (1.5- and 2.5-ft) levels, the loss in thickness for the carbon steel was three to six times higher than that for the low-alloy steels. A graphical comparison of the 5-year results for a plain carbon and an Fe-0.54Ni-0.52Cu-0.12P alloy steel is shown in Fig. 20. Other experiences with low-alloy steels, especially in exposures in which the wave action is vigorous, also indicate

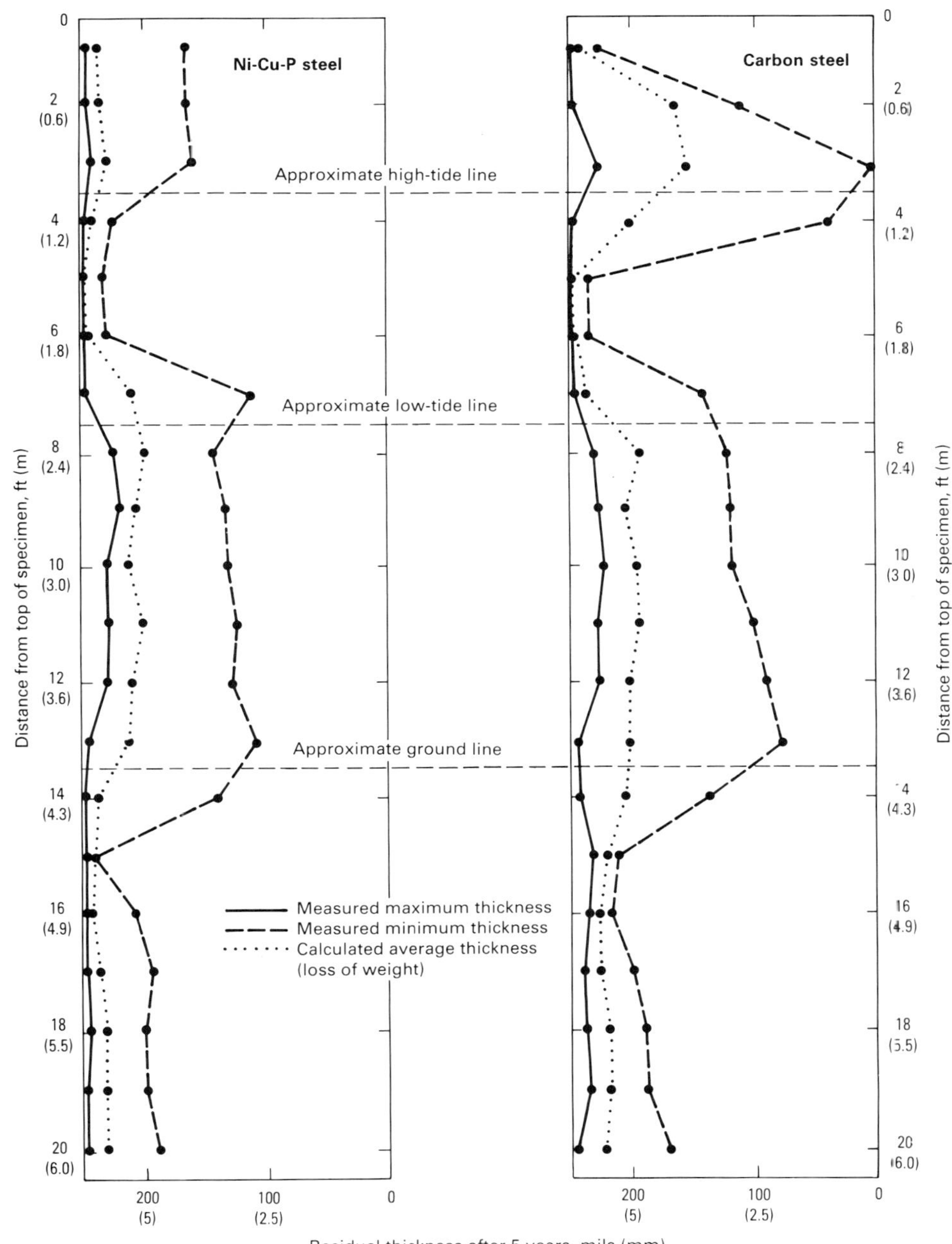

Fig. 20 Comparison of corrosion results for two steels in marine environments. Source: Ref 18

that they have considerable merit for splash-zone service (Ref 11).

Submerged Zone. Low-alloy steels exhibit corrosion rates in the range of about 65 to 125 μm/yr (2.5 to 5 mils/yr) when fully immersed in seawater (Ref 11). As such, low-alloy steels offer no particular advantage over carbon steel in applications involving submergence in the ocean. Examples of corrosion rates for plain carbon steel and low-alloy steels after 8 and 16 years in the Pacific Ocean near the Panama Canal are given in Table 9. The inferior corrosion performance of alloy steels in seawater is due to the fact that the special conditions in the atmosphere, which lead to the formation of the protective rust films, do not operate in the submerged condition.

Low-alloy steels also develop deeper pits in seawater than carbon steels do. This is demonstrated by the 8-year results from the Panama-Pacific exposures given in Table 10. The total penetration calculated from the weight loss (column 1) is compared with the average of the 20 deepest pits (column 2) and with the deepest pit (column 3). Assuming that the average of the 20 deepest pits is a more significant criterion than the deepest pit, this average pitting value can be compared with the weight loss penetration. The ratio of these two values for low-carbon steel at the 4.25-m (14-ft) depth is 2.6. The range for the alloy steels, some of which have higher weight loss penetrations to start with, is 1.6 to 3.7. At the mean tide level, the factor is lower, as is the pit depth for many of the steels involved in the comparison.

For a given strength, the designer would be tempted to specify a thinner wall for a low-alloy steel in a seawater application (Ref 11). Because the corrosion rate is higher, corrosion failure would be more rapid. Thus, from a design standpoint, the corrosion allowance for a low-alloy steel should be greater than that for a low-carbon steel. However, low-alloy steels have good strength characteristics, and if protective coatings were applied, these steels could be used to advantage. Cathodic protection must be applied with care for high-strength low-alloy steels because some tend to be more susceptible to hydrogen damage than carbon steel. Information on the cathodic protection of marine structures can be found in the article "Marine Corrosion" in this Volume and in Ref 21.

Burial Zone. Bottom conditions vary, but local attack is sometimes observed just above the mud zone or in the bottom mud itself (Ref 11). As in the soil, bottom mud is often aggressive to steel because of the presence of sulfate-reducing bacteria. For steel structures standing in the mud, the anodic and cathodic sites may be a considerable distance apart, and their locations may shift somewhat with time.

Galvanic corrosion in seawater is a matter of concern because the corroding medium has a fairly high conductivity. Service conditions can differ considerably because of solution composition, solute concentration, agitation, aeration, temperature, and purity of the metals, as well as corrosion product formation and biological growth, each of which can result in a different galvanic series.

In a structural joint, the ratio of the areas of two dissimilar metals has enormous influence on the corrosion rate of one of the members of the joint—the one that is more anodic in the galvanic series. The greater the ratio of the cathode to the anode, the greater the corrosion rate. A surprisingly small difference in solution potential can often result in a significant difference in corrosion rate. For example, tests were conducted in which carbon steel was coupled to itself and to ASTM A242 (type 1) high-strength low-alloy steel and type 410 stainless steel and in which ASTM A242 (type 1) high-strength low-alloy steel was coupled to itself and to type 410 stainless steel. The results of these tests after 6 months of immersion in seawater are given in Table 11. Coupling carbon steel to stainless steel in an anode-to-cathode ratio of 1:8 can result in an approximately eightfold greater corrosion loss for the carbon steel. Also important to design engineers is the significant increase in corrosion that occurs when carbon steel is coupled to high-strength low-alloy steel, despite the fact that their solution (galvanic) potentials are practically the same. An example of a carbon steel/alloy steel galvanic couple is shown in Fig. 21.

Ship and Submarine Applications. The selection of alloy steels for ship and submarine hulls, structures, and deck railings is based on toughness, ductility, and weldability rather than corrosion performance. Protection from corrosion is generally supplied by coatings and cathodic protection, as discussed in the article "Marine Corrosion" in this Volume. Compositions of high-strength alloy steels used for ship and submarine

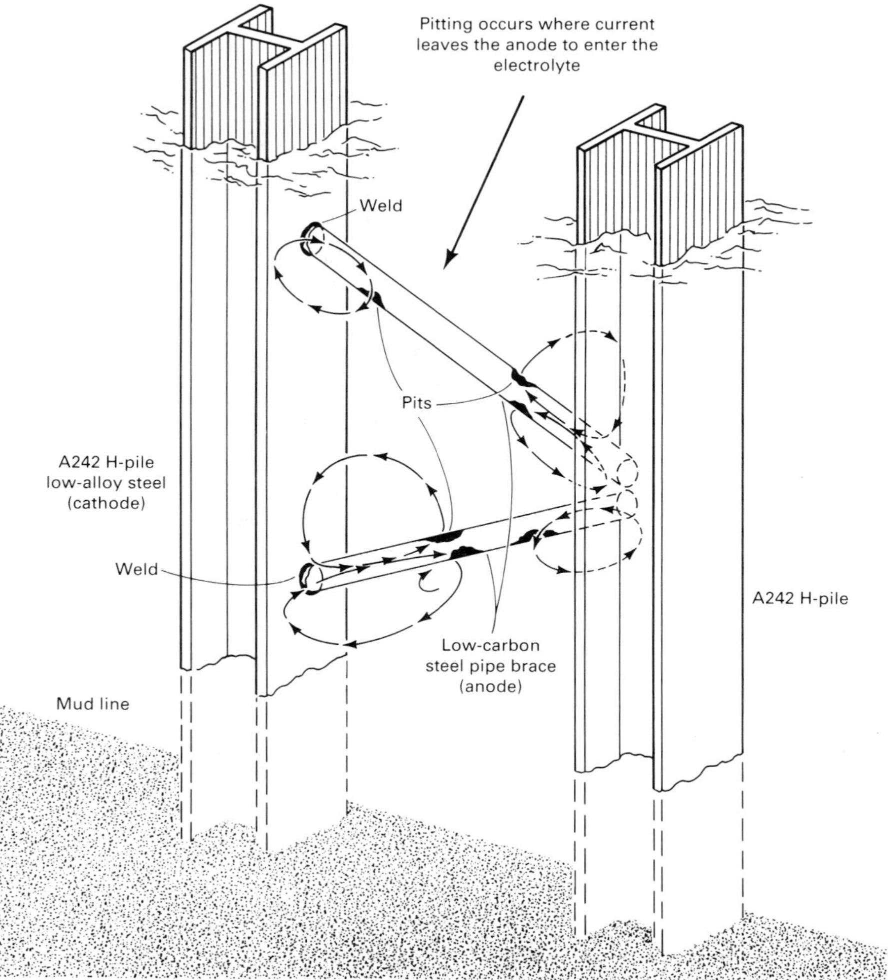

Fig. 21 Example of a carbon steel/alloy steel galvanic couple. Source: Ref 21

Table 9 Composition of structural steels and their corrosion rates immersed 4.25 m (14 ft) deep in the Pacific Ocean near the Panama Canal Zone

Type	Composition, %									Corrosion rate, 8 years		Corrosion rate, 16 years	
	C	Mn	P	S	Si	Cr	Ni	Cu	Mo	µm/yr	mils/yr	µm/yr	mils/yr
Unalloyed low carbon (A)	0.24	0.48	0.040	0.027	0.008	0.03	0.051	0.080	...	74	2.9	69	2.7
Copper bearing (D)	0.22	0.44	0.019	0.033	0.009	Trace	0.14	0.35	...	76	3.0	...	...
Nickel (2%)(E)	0.20	0.54	0.012	0.023	0.18	0.15	1.94	0.63	...	97	3.8	69	2.7
Nickel (5%)(F)	0.13	0.49	0.010	0.014	0.16	0.10	5.51	0.062	...	91	3.6	69	2.7
Chromium (3%)(G)	0.08	0.44	0.010	0.017	0.13	3.16	0.16	0.11	0.02	147	5.8	97	3.8
Chromium (5%)(H)	0.08	0.41	0.020	0.019	0.20	5.06	0.11	0.062	0.52	109	4.3	89	3.5
Low alloy (Cu-Ni)(I)	0.08	0.47	0.007	0.026	0.060	None	1.54	0.87	...	76	3.0	69	2.7
Low alloy (Cu-Cr-Si)(J)	0.15	0.45	0.113	0.026	0.47	0.68	0.49	0.42	...	135	5.3	122	4.8
Low alloy (K) (Cu-Ni-Mn-Mo)	0.078	0.75	0.058	0.022	0.04	Trace	0.72	0.61	0.13	69	2.7	64	2.5
Low alloy (Cr-Ni-Mn)(L)	0.13	0.60	0.089	0.021	0.15	0.55	0.30	0.61	0.059	140	5.5	127	5.0

Source: Ref 20

structural applications are given in Table 12. Of the compositions given, the corrosion resistance of ASTM A710 in both flowing and still seawater has been characterized (Ref 22). These results are given in Fig. 22(a) and (b) where ASTM A710 is compared with several other high-strength steels as well as with carbon steel. The conclusion drawn from this study is that ASTM A710 exhibits corrosion resistance comparable to other high-strength and carbon steels.

Chemical-Processing Industry

Many factors influence corrosion in the chemical-processing industry. Temperature, pressure, and velocity of the process stream all affect corrosion, and minute amounts of contaminants can result in large increases in corrosion rates. The use of alloy steels in such environments is generally limited to static or low-velocity applications, such as storage tanks or low-velocity piping. Applications for bare steel are particularly limited. More often, some form of protection is used both to protect the steel equipment and to maintain the purity of the product. Organic linings are commonly used for this purpose; the use of cathodic and anodic protection is also becoming more common. Some applications for alloy steels in the chemical-processing industry are listed below.

Sulfuric Acid. Steel tanks are used to store sulfuric acid at ambient temperatures at all concentrations to 101%. Corrosion can rapidly become catastrophic at these concentrations and at temperatures above 25 °C (75 °F). When product purity is of concern, anodic protection can be used to limit iron contamination over long storage periods (see the article "Anodic Protection" in this Volume). The addition of 0.1 to 0.5% Cu to steels used for sulfuric acid storage has been shown to reduce corrosion rates in acid concentrations to about 55%, but this beneficial effect has not been observed at concentrations greater than 60% (Ref 23).

Organic Acids. Alloy steel can be used for ambient-temperature storage of some high molecular weight organic acids, but steel is attacked rapidly by formic, acetic, and propionic acids.

Alkalies. Bare steel storage tanks are used for sodium hydroxide at concentrations to 50% and at temperatures to about 65 °C (150 °F). Where iron contamination of the product is of concern, spray-applied neoprene latex or phenolic-epoxy linings are used.

Anhydrous Ammonia. Low-alloy steel storage tanks have been used for many years for ammonia storage. Stress-corrosion cracking has been the primary corrosion problem in vessels used for ammonia storage. It has been shown in several investigations that high stresses and oxygen (air) contamination are the primary causes of such cracking and that the addition of 0.1 to 0.2% H_2O inhibits SCC in alloy steel storage vessels (Ref 24-29).

Chlorine. Steel is used to handle dry chlorine, and corrosion rates are generally low. Ignition can be a problem, however, and the recommended maximum service temperature in this application is 150 °C (300 °F) (Ref 30). Steel is also used to handle refrigerated liquid chlorine, but care must be taken at potential leak sites. Chlorine from small leaks can be trapped beneath ice formed on the equipment; this will form corrosive wet chlorine gas. More information on the materials used in the chemical-processing industry is

Table 10 Corrosion penetration of alloy steels immersed in the Pacific Ocean near the Panama Canal Zone after 8 years

See Table 9 for compositions.

		Penetration													
		Mean tide(a)							4.25 m (14 ft) Below Surface(a)						
		1		2		3			1		2		3		
Steel	Type	µm	mils	µm	mils	µm	mils	Ratio(b)	µm	mils	µm	mils	µm	mils	Ratio(b)
A	Low carbon	589	23.2	1016	40	1651	65	1.7	648	25.5	1676	66	2184	86	2.6
D	Copper bearing	615	24.2	1143	45	1600	63	1.9	704	27.7	1600	63	2743	108	2.3
E	Ni (2%)	582	22.9	991	39	1270	50	1.7	805	31.7	2388	94	4547	179	3.0
F	Ni (5%)	508	20.0	991	39	1905	75	2.0	813	32.0	2972	117	5436	214	3.7
G	Cr (3%)	653	25.7	2082	82	2362	93	3.2	1029	40.5	1651	65	1981	78	1.6
H	Cr (5%)	622	24.5	2235	88	2515	99	3.6	813	32.0	1600	63	2286	90	2.0
I	Low alloy (Cu-Ni)	1008	39.7	1778	70	3404	134	1.8	671	26.4	2083	82	3861	152	3.2
J	Low alloy (Cu-Cr-Si)	536	21.1	1194	47	1372	54	2.2	1097	43.2	2032	80	4445	175	1.8
K	Low alloy (Cu-Ni-Mn-Mo)	630	24.8	1016	40	2388	94	1.6	648	25.5	1422	56	3531	139	2.2
L	Low alloy (Cr-Ni-Mn)	521	20.5	991	39	1270	50	1.9	1115	43.9	2464	97	6579	259(c)	2.2

(a) 1, calculated from weight loss; 2, average of 20 deepest pits; 3, deepest pit.
(b) Ratio of average of 20 deepest pits to weight loss penetration. The higher the number, the greater is the pitting tendency in relation to the corrosion rate.
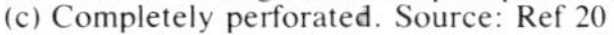
(c) Completely perforated. Source: Ref 20

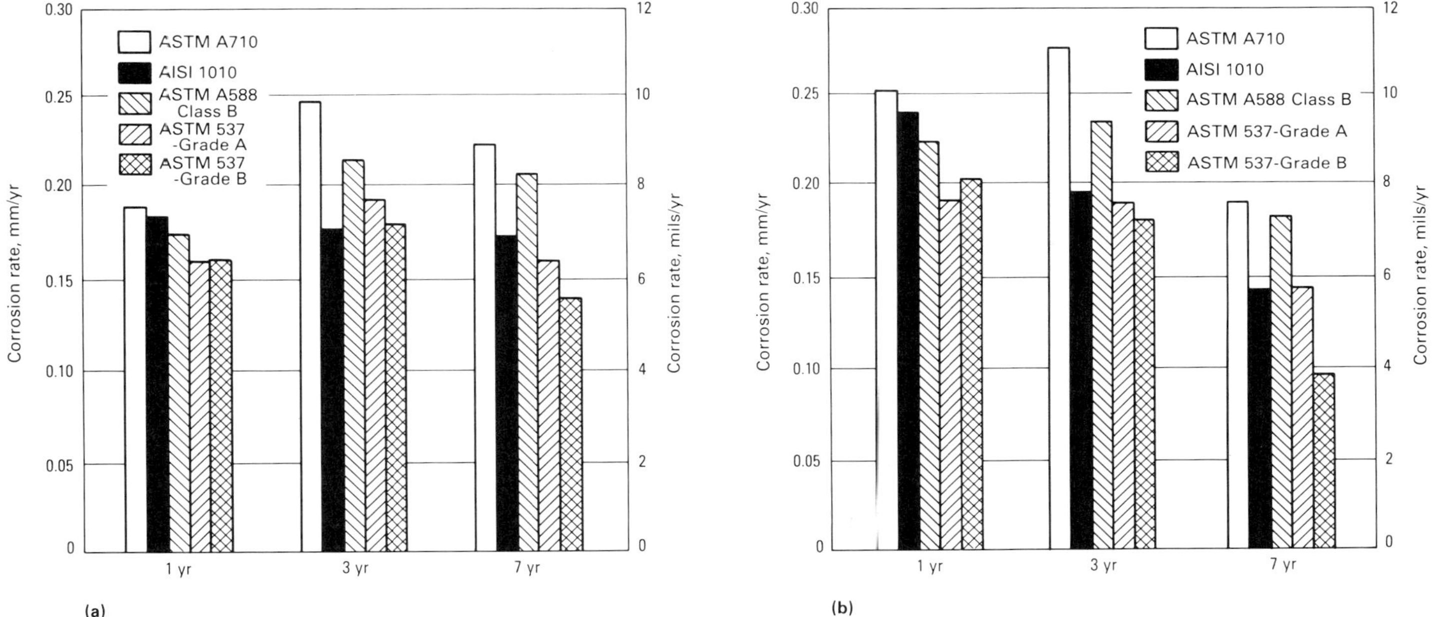

Fig. 22 Corrosion results for ASTM A710 and other steels exposed to (a) low velocity (0.5 m/s or 1.6 ft/s) seawater and (b) quiet (still) seawater. Source: Ref 22

Table 11 Corrosion of members of couples in seawater after 6 months

		Weight loss ($mg/m^2/d$) for area ratio (a) of					
		1:1		8:1		1:8	
Specimen 1	Couple specimen 2	1	2	1	2	1	2
Carbon steel	Carbon steel	5.5	...	...	...	...	...
Carbon steel	ASTM A242 (b)	8.2	2	6.7	2.7	17	3.2
Carbon steel	Type 410 stainless steel	13	0.03	7.0	...	47	0.04
ASTM A242 (b)	ASTM A242 (b)	4.5	...	...	...	...	...
ASTM A242 (b)	Type 410 stainless steel	9.5	0.03	6.2	0.04	35	0.02

(a) Area of specimen 1 to area of specimen 2. (b) Type 1, containing chromium, silicon, copper, nickel, and phosphorus. Source: Ref 14

available in the article "Corrosion in the Chemical Processing Industry" in this Volume.

REFERENCES

1. C.P. Larrabee and S.K. Coburn, The Atmospheric Corrosion of Steels as Influenced by Changes in Chemical Composition, in *Metallic Corrosion—First International Congress on Corrosion*, Butterworths, 1962, p 276-284
2. R.T. Jones, Carbon and Alloy Steels, in *Process Industries Corrosion*, National Association of Corrosion Engineers, 1986
3. R.D. Kane and W.K. Boyd, Materials Technology for Oil and Gas Production, in *Alloys for the Eighties*, Climax Molybdenum Com-

Table 12 High-strength alloy steels used for ship and submarine structural applications

Steel	Composition, %									
	C	Mn	S	P	Si	Ni	Cr	Mo	Cu	Other(max)
HY-80	0.12–0.18	0.10–0.40	0.020(max)	0.020(max)	0.15–0.35	2.00–3.25	1.0–1.80	0.20–0.60	0.25	0.02Ti, 0.03V, 0.025Al, 0.025Sb, 0.030Sn
HY-100	0.12–0.18	0.10–0.40	0.020(max)	0.020(max)	0.15–0.35	2.25–3.50	1.0–1.80	0.20–0.60	0.25	0.02Ti, 0.03V, 0.025Al, 0.025Sb, 0.030Sn
HY-130	0.12(max)	0.60–0.90	0.60–0.90	0.010(max)	0.15–0.35	4.75–5.25	0.4–0.7	0.30–0.65	0.25	0.02Ti, 0.05–0.10V
HY-180	0.12–0.15	0.30(max)	0.30(max)	0.010(max)	...	10.0	2.0	1.0	...	...
ASTM A710	0.035	0.44	0.015	0.010	0.28	0.89	0.68	0.21	1.16	0.045Nb

Data supplied by David Taylor, Naval Ship Research and Development Center.

pany, p 225-233

4. D.L. Sponseller, R. Garber, and J.A. Straatmann, Effect of Microstructure on Sulfide-Stress-Cracking Resistance of High-Strength Casing Steels, in *MiCon '82: Optimization of Processing, Properties, and Service Performance Through Microstructural Control*, STP 792, American Society for Testing and Materials, 1983, p 172-204
5. R. Garber, T. Woda, F.B. Fletcher, and T.B. Cox, *J. Mater. Energy Syst.*, Vol 7 (No. 2), 1985, p 91
6. J.R. Bryant and C.B. Chitwood, Paper 58, presented at Corrosion/83, National Association of Corrosion Engineers, 1983
7. G.R. Prescott, Material Problems in the Hydrocarbon Processing Industries, in *Alloys for the Eighties*, Climax Molybdenum Company, p 303-315
8. *Corrosion Problems in Coal Fired Boiler Superheater and Reheater Tubes—Fireside Corrosion*, Publication CS1653, Electric Power Research Institute, 1980
9. A.L. Plumley, J.A. Burnett, and V. Vaidya, Paper presented at CIM 21st Annual Conference of Metallurgists, 1982
10. R.E. Westerman *et al.*, "Corrosion and Environmental-Mechanical Characterization of Iron-Base Nuclear Waste Package Structural Barrier Materials," Pacific Northwest Laboratory, March 1986
11. M. Schumacher, Ed., Seawater Corrosion Handbook, Noyes Data Corporation, 1979
12. D.M. Buck, The Influence of Very Low Percentages of Copper in Retarding the Corrosion of Steel, in *Proceedings ASTM*, Vol 19, American Society for Testing and Materials, 1919, p 224
13. H.R. Copson, Long-Time Atmospheric Corrosion Tests on Low Alloy Steels, in *ASTM Proceedings*, Vol 60, American Society for Testing and Materials, 1960, p 650-665
14. F.L. LaQue, Corrosion Testing, in *ASTM Proceedings*, Vol 51, American Society for Testing and Materials, 1951, p 495-582
15. C.P. Larrabee, Steel Has Low Corrosion Rate During Long Seawater Exposure, *Mater. Prot.*, Vol 1 (No. 12), 1962, p 95-96
16. C.P. Larrabee, Corrosion of Steels in Marine Atmospheres and in Seawater, *Trans. Electrochem. Soc.*, Vol 87, 1945, p 161-182
17. C.P. Larrabee, Corrosion Resistance of High Strength Low-Alloy Steels as Influenced by Composition and Environment, *Corrosion*, Vol 9 (No. 8), 1953, p 259-271
18. C.P. Larrabee, Corrosion Resistant Experimental Steels for Marine Applications, *Corrosion*, Vol 14 (No. 11), 1958, p 501t-504t
19. R.J. Schmitt and E.H. Phelps, Corrosion Performance of Constructional Steels in Marine Applications, *J. Met.*, March 1970
20. C.R. Southwell and A.L. Alexander, "Corrosion of Structural Ferrous Metals in Tropical Environments—Sixteen Year's Exposure to Sea and Fresh Water," Paper 14, Preprint, NACE Conference, Cleveland, OH, National Association of Corrosion Engineers, 1968
21. H.S. Preiser, Cathodic Protection, in *Handbook of Corrosion Protection for Steel Pile Structures in Marine Environments*, American Iron and Steel Institute, 1981, p 67-100
22. D.G. Melton and D.G. Tipton, Corrosion Behavior of A710 Grade A Steel in Marine Environments, LaQue Center for Corrosion Technology Inc., Wrightsville Beach, NC, June 1983
23. H. Endo and S. Morioka, "Dissolution Phenomenon of Copper-Containing Steels in Aqueous Sulfuric Acid Solutions of Various Concentrations," Paper presented at the third symposium, Japanese Metal Association, April 1938
24. A.W. Loginow and E.H. Phelps, *Corrosion*, Vol 18 (No. 8), 1962, p 299-309
25. D.C. Deegan and B.E. Wilde, *Corrosion*, Vol 29 (No. 8), 1973, p 310-315
26. D.C. Deegan, B.E. Wilde, and R.W. Staehle, *Corrosion*, Vol 32 (No. 4), 1976, p 139-142
27. T. Kawamoto, T. Kenjo, and Y. Imasaka, *IHI Eng. Rev.*, Vol 10 (No. 4), 1977, p 17-25
28. F.F. Lyle and R.T. Hill, "SCC Susceptibility of High-Strength Steels in Liquid Ammonia at Low Temperatures," Paper 225, presented at Corrosion/78, National Association of Corrosion Engineers, 1978
29. K. Farrow, J. Hutchings, and G. Sanderson, *Br. Corros. J.*, Vol 16 (No. 1), 1981, p 11-19
30. W.Z. Friend and B.B. Knapp, *Trans. AIChE*, Section A, 25 Feb 1943, p 731

Corrosion of Stainless Steels

Ralph M. Davison, Avesta Stainless Inc.
Terry DeBold, Carpenter Technology Corporation
Mark J. Johnson, Allegheny Ludlum Steel Corporation

STAINLESS STEELS are iron-base alloys containing at least 10.5% Cr. With increasing chromium content and the presence or absence of some ten to fifteen other elements, stainless steels can provide an extraordinary range of corrosion resistance. Various grades have been used for many years in environments as mild as open air in architectural applications and as severe as the chemically active product streams in the chemical-processing industries. Stainless steels are categorized in five distinct families according to their crystal structure and strengthening precipitates. Each family exhibits its own general characteristics in terms of mechanical properties and corrosion resistance. Within each family, there is a range of grades that varies in composition, corrosion resistance, and cost.

Stainless steels are susceptible to several forms of localized corrosive attack. The avoidance of such localized corrosion is the focus of most of the effort involved in selecting stainless steels. Furthermore, the corrosion performance of stainless steels can be strongly affected by practices of design, fabrication, surface conditioning, and maintenance.

The selection of a grade of stainless steel for a particular application involves the consideration of many factors, but always begins with corrosion resistance. It is first necessary to characterize the probable service environment. It is not enough to consider only the design conditions. It is also necessary to consider the reasonably anticipated excursions or upsets in service conditions. The suitability of various grades can be estimated from laboratory tests or from documentation of field experience in comparable environments. Once the grades with adequate corrosion resistance have been identified, it is then appropriate to consider mechanical properties, ease of fabrication, the types and degree of risk present in the application, the availability of the necessary product forms, and cost.

Identification Systems for Stainless Steels

Grades of stainless steel are most commonly designated in one or more of the following ways: the American Iron and Steel Institute (AISI) numbering system, the Unified Numbering System (UNS), and proprietary designations. Other designations have been established by the national standards organizations of various major industrialized countries. These systems are generally similar to those of the United States, but there can be significant differences that must be taken into account when designing under these codes or using materials from these areas.

The AISI System. The most common designations are those of AISI, which recognizes grades as standard compositions on the basis of meeting criteria of total production and number of sources. Most of these grades have a three-digit designation in the 200, 300, or 400 series, and some have a one- or two-letter suffix that indicates a particular modification of the composition. There is a general association of the various microstructural families of grades with particular parts of the numbering series, but there are several significant exceptions to the system. Table 1 lists the AISI grades and their chemical analyses. Some proprietary designations are similar in structure to the AISI system, but are not standard grades. Also, commercial offerings of the standard grades may use the AISI number with some additional prefix or suffix to indicate the producer or a particular modification of the grade for a certain type of application.

The UNS system was introduced in the 1970s to provide a systematic and encyclopedic listing of metal alloys, including the stainless steels. Although not perfect, the UNS numbering system has been successful in maintaining a degree of order during a period when many new grades were introduced. In Table 1, the AISI grades also have UNS designations. Most stainless steels—those having more than 50% Fe—have a UNS number that consists of the letter S followed by five digits. For the AISI grades, the first three digits usually correspond to an AISI number. The basic AISI grades have 00 as the last two digits, while the modifications of the most basic grades show some other two digits. There are some significant exceptions in the UNS system, just as there are in the AISI system.

These designations are not normally a sufficient basis for specifying a stainless steel. To purchase a particular grade and product form, it is advisable to consult a comprehensive specification. American Society for Testing and Materials (ASTM) specifications, for example, are the most commonly used. These specifications usually define compositional limits; minimum mechanical properties; production, processing, and testing requirements; and in some cases particular corrosion performance requirements. Other standard specifications, such as those of the American Society of Mechanical Engineers (ASME), the National Association of Corrosion Engineers (NACE), the Technical Association of the Pulp and Paper Industry (TAPPI), or those of individual companies, may apply to certain types of equipment.

Proprietary Designations. In addition to the standard grades, there are well over 100 special grades that represent modifications, extensions, or refinements of the basic grades. In the early 1970s, the introduction of new stainless steel refining practices, most commonly argon-oxygen decarburization (AOD), greatly facilitated the production of stainless steels. In addition to permitting the use of lower-cost forms of alloy element additions, AOD also allows precise control of individual elements. This process also makes possible the economical removal of interstitial and tramp elements that are detrimental to corrosion resistance, mechanical properties, and processing.

Because of these capabilities, stainless steel producers have greatly extended the range of stainless steel grades. Very few of these grades have been accepted as AISI standards, but all were assigned UNS numbers when they were introduced into ASTM standards. Table 2 provides a representative sampling of these grades across the range of alloy content and corrosion resistance. Some of the grades are identified by common trade names or trademarks in order to facilitate understanding and to enhance the usefulness of this discussion. This listing is not intended to be exhaustive, and the omission of a grade does not indicate its disqualification from consideration.

Families of Stainless Steels

There are five major families of stainless steels, as defined by crystallographic structure. Each family is distinct with regard to its typical mechanical properties. Furthermore, each family tends to share a common nature in terms of resistance/susceptibility to particular forms of corrosion. However, within each family, it is possible to have a substantial range of composition. Therefore, each family is applicable to a broad range of corrosion environments.

Ferritic Stainless Steels. The simplest stainless steels contain only iron and chromium. Chromium is a ferrite stabilizer; therefore, the stability of the ferritic structure increases with chromium content. Ferrite has a body-centered cubic crystal structure, and it is characterized as magnetic and relatively high in yield strength but low in

Table 1 Compositions of AISI standard grades of stainless steels

UNS designation	AISI type	Composition, %(a) C	Mn	P	S	Si	Cr	Ni	Mo	Others
Austenitic grades										
S20100	201	0.15	5.60–7.50	0.06	0.03	1.00	16.00–18.00	3.50–5.50	. . .	0.25N
S20200	202	0.15	7.50–10.0	0.06	0.03	1.00	17.00–19.00	4.00–6.00	. . .	0.25N
S20500	205	0.12–0.25	14.00–15.50	0.03	0.03	0.50	16.50–18.00	1.00–1.75	. . .	0.32–0.40N
S30100	301	0.15	2.00	0.045	0.03	1.00	16.00–18.00	6.00–8.00	. . .	. . .
S30200	302	0.15	2.00	0.045	0.03	1.00	17.00–19.00	8.00–10.00	. . .	. . .
	302B	0.15	2.00	0.045	0.03	2.00–3.00	17.00–19.00	8.00–10.00	. . .	. . .
S30300	303	0.15	2.00	0.2	0.15	1.00	17.00–19.00	8.00–10.00	0.60	. . .
	303Se	0.15	2.00	0.2	0.06	1.00	17.00–19.00	8.00–10.00	. . .	0.15Se min
S30400	304	0.08	2.00	0.045	0.03	1.00	18.00–20.00	8.00–10.50	. . .	. . .
S30403	304L	0.03	2.00	0.045	0.03	1.00	18.00–20.00	8.00–12.00	. . .	. . .
S30430	S30430	0.08	2.00	0.045	0.03	1.00	17.00–19.00	8.00–10.00	. . .	3.00–4.00Cu
	304N	0.08	2.00	0.045	0.03	1.00	18.00–20.00	8.00–10.50	. . .	0.10–0.16N
S30500	305	0.12	2.00	0.045	0.03	1.00	17.00–19.00	10.50–13.00	. . .	. . .
S30800	308	0.08	2.00	0.045	0.03	1.00	19.00–21.00	10.00–12.00	. . .	. . .
S30900	309	0.2	2.00	0.045	0.03	1.00	22.00–24.00	12.00–15.00	. . .	. . .
	309S	0.08	2.00	0.045	0.03	1.00	22.00–24.00	12.00–15.00	. . .	. . .
S31000	310	0.25	2.00	0.045	0.03	1.50	24.00–26.00	19.00–22.00	. . .	. . .
	310S	0.08	2.00	0.045	0.03	1.50	24.00–26.00	19.00–22.00	. . .	. . .
S31400	314	0.25	2.00	0.045	0.03	1.50–3.00	23.00–26.00	19.00–22.00	. . .	. . .
S31600	316	0.08	2.00	0.045	0.03	1.00	16.00–18.00	10.00–14.00	2.00–3.00	. . .
	316F	0.08	2.00	0.2	0.10 min	1.00	16.00–18.00	10.00–14.00	1.75–2.50	. . .
S31603	316L	0.03	2.00	0.045	0.03	1.00	16.00–18.00	10.00–14.00	2.00–3.00	. . .
	316N	0.08	2.00	0.045	0.03	1.00	16.00–18.00	10.00–14.00	2.00–3.00	0.10–0.16N
S31700	317	0.08	2.00	0.045	0.03	1.00	18.00–20.00	11.00–15.00	3.00–4.00	. . .
S31703	317L	0.03	2.00	0.045	0.03	1.00	18.00–20.00	11.00–15.00	3.00–4.00	. . .
S32100	321	0.08	2.00	0.045	0.03	1.00	17.00–19.00	9.00–12.00	. . .	Ti:5×C min
	329	0.10	2.00	0.04	0.03	1.00	25.00–30.00	3.00–6.00	1.00–2.00	. . .
	330	0.08	2.00	0.04	0.03	0.75–1.50	17.00–20.00	34.00–37.00	. . .	0.10Ta, 0.20Nb
S34700	347	0.08	2.00	0.045	0.03	1.00	17.00–19.00	9.00–13.00	. . .	Nb:10×C min
S34800	348	0.08	2.00	0.045	0.03	1.00	17.00–19.00	9.00–13.00	. . .	Nb:10×C min
	384	0.08	2.00	0.045	0.03	1.00	15.00–17.00	17.00–19.00	. . .	. . .
Ferritic grades										
S40500	405	0.08	1.00	0.04	0.03	1.00	11.50–14.50	. . .	. . .	0.10–0.30Al
S40900	409	0.08	1.00	0.045	0.045	1.00	10.50–11.75	. . .	. . .	Ti:6×C–0.75
S42900	429	0.12	1.00	0.04	0.03	1.00	14.00–16.00	. . .	. . .	. . .
S43000	430	0.12	1.00	0.04	0.03	1.00	16.00–18.00	. . .	. . .	. . .
	430F	0.12	1.25	0.06	0.15	1.00	16.00–18.00	. . .	0.60	. . .
	430FSe	0.12	1.25	0.06	0.06	1.00	16.00–18.00	. . .	. . .	0.15Si min
S43400	434	0.12	1.00	0.04	0.03	1.00	16.00–18.00	. . .	0.75–1.25	. . .
S43600	436	0.12	1.00	0.04	0.03	1.00	16.00–18.00	. . .	0.75–1.25	Nb:5×C–0.70
S44200	442	0.20	1.00	0.04	0.03	1.00	18.00–23.00	. . .	. . .	. . .
S44400	444	0.25	1.00	0.04	0.03	1.00	17.50–19.50	. . .	. . .	(Ti+Nb): 0.2+ 4(C+N)–0.8
S44600	446	0.20	1.50	0.04	0.03	1.00	23.00–27.00	. . .	. . .	0.25N
Martensitic grades										
S40300	403	0.15	1.00	0.04	0.03	0.50	11.50–13.00	. . .	. . .	. . .
S41000	410	0.15	1.00	0.04	0.03	1.00	11.50–13.50	. . .	. . .	. . .
S41400	414	0.15	1.00	0.04	0.03	1.00	11.50–13.50	1.25–2.50	. . .	. . .
S41600	416	0.15	1.25	0.06	0.15 min	1.00	12.00–14.00	. . .	0.60	. . .
	416Se	0.15	1.25	0.06	0.06	1.00	12.00–14.00	. . .	. . .	0.15Se min
S42000	420	0.15 min	1.00	0.04	0.03	1.00	12.00–14.00	. . .	. . .	. . .
	420F	0.15 min	1.25	0.06	0.15 min	1.00	12.00–14.00	. . .	0.60	. . .
S42200	422	0.20–0.25	1.00	0.025	0.025	0.75	11.00–13.00	0.50–1.00	0.75–1.25	0.15–0.30V, 0.75–1.25W
S43100	431	0.20	1.00	0.04	0.03	1.00	15.00–17.00	1.25–2.50	. . .	. . .
	440A	0.60–0.75	1.00	0.04	0.03	1.00	16.00–18.00	. . .	0.75	. . .
	440B	0.75–0.95	1.00	0.04	0.03	1.00	16.00–18.00	. . .	0.75	. . .
	440C	0.95–1.20	1.00	0.04	0.03	1.00	16.00–18.00	. . .	0.75	. . .
Precipitation-hardening grades										
S13800	S13800	0.05	0.20	0.010	0.008	0.10	12.25–13.25	7.50–8.50	2.00–2.50	0.90–1.35Al, 0.01*N*
S15500	S15500	0.07	1.00	0.04	0.03	1.00	14.00–15.50	3.50–5.50	. . .	2.50–4.50Cu, 0.15–0.45Nb
S17400	S17400	0.07	1.00	0.04	0.03	1.00	15.50–17.50	3.00–5.00	. . .	3.00–5.00Cu, 0.15–0.45Nb
S17700	S17700	0.09	1.00	0.04	0.04	0.04	16.00–18.00	6.50–7.75	. . .	0.75–1.50Al

(a) Maximum unless otherwise indicated; all compositions include balance of iron.

ductility and work hardenability. Ferrite shows an extremely low solubility for such interstitial elements as carbon and nitrogen. The ferritic grades exhibit a transition from ductile to brittle behavior over a rather narrow temperature range. At higher carbon and nitrogen contents, especially at higher chromium levels, this ductile-to-brittle transition can occur above ambient temperature. This possibility severely limited the use of ferritic grades before the use of AOD. The ferritic family was then limited to AISI type 446 for oxidation-resistant applications and to AISI types 430 and 434 for such corrosion applications as automotive trim. The fact that these grades were readily sensitized to intergranular corrosion

Table 2 Compositions of some proprietary and nonstandard stainless steels

UNS designation	Common name	Composition, %(a) C	Mn	P	S	Si	Cr	Ni	Mo	Others
Austenitic grades										
S24100	18Cr-2Ni-12Mn	0.15	11.0–14.0	0.060	0.03	1.00	16.50–19.50	0.5–2.50	. . .	0.2–0.45N
S20910	Nitronic 50 (22-13-5)	0.06	4.0–6.0	0.040	0.03	1.00	20.50–23.50	11.50–13.50	1.50–3.00	0.1–0.3Nb, 0.2–0.4N, 0.1–0.3V
S30345	303Al MODIFIED	0.15	2.00	0.050	0.11–0.16	1.00	17.00–19.00	8.00–10.00	0.40–0.60	0.60–1.00Al
	303BV(b)	0.11	1.75	0.03	0.14	0.35	17.75	9.00	0.50	0.75Al
	302HQ-FM	0.06	2.00	0.04	0.14	1.00	16.00–19.00	9.00–11.00	. . .	1.3–2.4Cu
S30430	302HQ	0.10	2.00	0.045	0.03	1.00	17.00–19.00	8.00–10.00	. . .	3.0–4.0Cu
S30453	304LN	0.03	2.00	0.045	0.03	1.00	18.00–20.00	8.00–12.00	. . .	0.1–0.16N
S31653	316LN	0.03	2.00	0.045	0.03	1.00	16.00–18.00	10.00–14.00	2.00–3.00	0.1–0.16N
S31753	317LN	0.03	2.00	0.045	0.30	1.00	18.00–20.00	11.00–15.00	3.00–4.00	0.1–0.2N
S31725	317LM	0.03	2.00	0.045	0.03	0.075	18.00–20.00	13.00–17.00	4.00–5.00	0.1N, 0.75Cu
	317LMN	0.03	2.00	0.045	0.03	0.075	18.00–20.00	13.00–17.00	4.00–5.00	0.1–0.2N
N08904	904L	0.02	2.00	0.045	0.035	1.00	19.00–23.00	23.0–28.0	4.00–5.00	1.0–2.0Cu
N08700	JS700	0.04	2.00	0.04	0.03	1.00	19.00–23.00	24.0–26.0	4.3–5.0	0.5Cu, Nb:(8×C) −1.00, 0.005Pb, 0.035Sn
	JS777	0.025	1.70	0.03	0.03	0.50	19.00–23.00	24.0–26.0	4.00–5.00	2.10Cu, 0.25Nb
N08020	20Cb-3	0.07	2.00	0.045	0.035	1.00	19.00–21.00	32.00–38.00	2.00–3.00	3.0–4.0Cu, (8×C)Nb
N08024	20Mo-4	0.03	1.00	0.035	0.035	0.50	22.5–25.00	35.00–40.00	3.50–5.00	0.5–1.5Cu, 0.15–0.35Nb
N08026	20Mo-6	0.03	1.00	0.03	0.03	0.50	22.00–26.00	33.00–37.20	5.00–6.70	2.0–4.0Cu
N08028	Alloy 28	0.03	2.50	0.03	0.03	1.00	26.00–28.00	29.5–32.5	3.0–4.0	0.6–1.4Cu
N08367	AL-6×N	0.03	2.0	0.04	0.03	1.00	20.0–22.0	23.50–25.50	6.0–7.0	0.18–0.25N
S31254	254SMO	0.02	1.00	0.03	0.01	0.80	19.50–20.50	17.50–18.50	6.0–6.5	0.5–1.0Cu, 0.18–0.22N
Ferritic grades										
S44627	E-Brite	0.01	0.40	0.02	0.02	0.40	25.0–27.0	0.50	0.75–1.50	0.05–0.2Nb, 0.2Cu, 0.015N
S44635	MONIT	0.025	1.00	0.04	0.03	0.75	24.5–26.0	3.50–4.50	3.50–4.50	0.035N, Nb+Ti: 0.20 + 4(C+N)-0.80
S44660	Sea-Cure	0.025	1.00	0.04	0.03	1.00	25.0–27.0	1.50–3.50	2.50–3.50	0.035N, Nb + Ti: 0.20 + 4(C+N)-0.80
S44735	AL-29-4C	0.03	1.00	0.04	0.03	1.00	28.0–30.0	1.00	3.60–4.20	0.045N, Nb + Ti:6(C+N)
S4473S	Usinor 290 Mo	0.03	1.00	0.04	0.03	1.00	28.0–30.0	1.00	3.60–4.20	0.045N, Nb + Ti:6(C+N)
S44800	AL-29-4-2	0.01	0.30	0.025	0.02	0.20	28.0–30.0	2.0–2.5	3.50–4.20	0.15Cu, 0.02N, C+N:0.025 max
Duplex grades										
S31803	2205	0.03	2.00	0.03	0.02	1.00	21.0–23.0	4.50–6.50	2.50–3.50	0.08–0.2N
S31200	44LN	0.03	2.00	0.045	0.03	1.00	24.0–26.0	5.50–6.50	1.20–2.00	0.14–0.2N
S31260	DP-3	0.03	1.00	0.03	0.03	0.75	24.0–26.0	5.50–7.50	2.50–3.50	0.2–0.8Cu, 0.1–0.3N
S31500	3RE60	0.03	1.20–2.00	0.03	0.03	1.4–2.0	18.0–19.0	4.25–5.25	2.50–3.00	. . .
S32550	Ferralium 255	0.04	1.50	0.04	0.03	1.00	24.0–27.0	4.50–6.50	2.00–4.00	1.5–2.5Cu, 0.1–0.25N
S32950	7Mo-PLUS	0.03	2.00	0.035	0.01	0.60	26.0–29.0	3.50–5.20	1.00–2.50	0.15–0.35N
Martensitic grades										
S41040	XM-30	0.15	1.00	0.04	0.03	1.00	11.50–13.50	. . .	. . .	0.05–0.2Nb
S41610	XM-6	0.15	2.50	0.06	0.15	1.00	12.00–14.00	. . .	0.60	. . .
Precipitation-hardenable grades										
S31800	PH13-8Mo	0.05	0.20	0.01	0.008	0.10	12.25–13.25	7.50–8.50	2.00–2.50	0.90–1.35Al, 0.01N
S45000	Custom 450	0.05	1.00	0.03	0.03	1.00	14.00–16.00	5.00–7.00	0.5–1.00	1.25–1.75Cu, Nb:8 × C min
S45500	Custom 455	0.05	0.50	0.04	0.03	0.50	11.00–12.50	7.50–9.50	0.50	0.1–0.5Nb, 1.50–2.50Cu, 0.8–1.40Ti

(a) Maximum unless otherwise indicated; all compositions contain balance of iron. (b) Nominal composition

as a result of welding or thermal exposure further limited their use.

With AOD, it was possible to reduce the levels of carbon and nitrogen significantly. The activity of carbon and nitrogen could further be reduced by the use of stabilizers, which are highly reactive elements, such as titanium and niobium, that precipitate the remaining interstitials. This newer generation of ferritic stainless steels includes AISI type 444 and the more highly alloyed ferritic grades shown in Table 2. With control of interstitial elements, it is possible to produce grades with unusually high chromium and molybdenum contents. At these low effective carbon levels, these grades are tougher and more weldable than the first generation of ferritic stainless steels. Nevertheless, their limited toughness generally restricts use of these grades to sheet or lighter-gage tubulars.

Ferritic stainless steels are highly resistant and are in some cases immune to chloride stress-corrosion cracking (SCC). These grades are frequently considered for thermal transfer applications.

The same properties and advantages are also responsible for the extraordinary development of the lowest-alloyed grade of the ferritics, AISI type 409. This grade, developed for automotive muffler and catalytic converter service, has gained in technical sophistication. It is increasingly used in automotive exhaust systems and in other moderately severe atmospheric-exposure applications.

Austenitic Stainless Steels. The detrimental effects of carbon and nitrogen in ferrite can be overcome by changing the crystal structure to austenite, a face-centered cubic crystal structure. This change is accomplished by adding austenite stabilizers—most commonly nickel but also manganese and nitrogen. Austenite is characterized as nonmagnetic, and it is usually relatively low in yield strength with high ductility, rapid work-hardening rates, and excellent toughness. These desirable mechanical properties, combined with ease of fabrication, have made the austenitic grades, especially AISI type 304, the most common of the stainless grades. Processing difficulties tend to limit increases in chromium content; therefore, improved corrosion resistance is usually obtained by adding molybdenum. The use of nitrogen as an intentional alloy addition stabilizes the austenite phase, particularly with regard to the precipitation of intermetallic compounds. With the nitrogen addition, it is possible to produce austenitic grades with up to 6% Mo for improved corrosion resistance in chloride environments. Other special grades include the high-chromium grades for high-temperature applications and the high-nickel grades for inorganic acid environments.

The austenitic stainless steels can be sensitized to intergranular corrosion by welding or by longer-term thermal exposure. These thermal exposures lead to the precipitation of chromium carbides in grain boundaries and to the depletion of chromium adjacent to these carbides. Sensitization can be greatly delayed or prevented by the use of lower-carbon L-grades (<0.03% C) or stabilized grades, such as AISI types 321 and 347, which include additions of carbide-stabilizing elements (titanium and niobium, respectively).

The common austenitic grades, AISI types 304 and 316, are especially susceptible to chloride SCC. All austenitic stainless steels exhibit some degree of susceptibility, but several of the high-nickel high-molybdenum grades are satisfactory with respect to stress-corrosion attack in most engineering applications.

Martensitic Stainless Steels. With lower chromium levels and relatively high carbon levels, it is possible to obtain austenite at elevated temperatures and then, with accelerated cooling, to transform this austenite to martensite, which has a body-centered tetragonal structure. Just as with plain carbon and low-alloy steels, this strong, brittle martensite can be tempered to favorable combinations of high strength and adequate toughness. Because of the ferrite-stabilizing character of chromium, the total chromium content, and thus the corrosion resistance, of the martensitic grades is somewhat limited. In recent years, nitrogen, nickel, and molybdenum additions at somewhat lower carbon levels have produced martensitic stainless steels of improved toughness and corrosion resistance.

The duplex stainless steels can be thought of as chromium-molybdenum ferritic stainless steels to which sufficient austenite stabilizers have been added to produce steels in which a balance of ferrite and austenite is present at room temperature. Such grades can have the high chromium and molybdenum responsible for the excellent corrosion resistance of ferritic stainless steels as well as the favorable mechanical properties of austenitic stainless steels. In fact, the duplex grades with about equal amounts of ferrite and austenite have excellent toughness and their strength exceeds either phase present singly.

First-generation duplex grades, such as AISI type 329, achieved this phase balance primarily by nickel additions. These early duplex grades have superior properties in the annealed condition, but segregation of chromium and molybdenum between the two phases as re-formed after welding often significantly reduced corrosion resistance. The addition of nitrogen to the second generation of duplex grades restores the phase balance more rapidly and minimizes chromium and molybdenum segregation without annealing. The newer duplex grades combine high strength, good toughness, high corrosion resistance, good resistance to chloride SCC, and good production economy in the heavier product forms.

The precipitation-hardening stainless steels are chromium-nickel grades that can be hardened by an aging treatment at a moderately elevated temperature. These grades may have austenitic, semiaustenitic, or martensitic crystal structures. Semiaustenitic structures are transformed from a readily formable austenite to martensite by a high-temperature austenite-conditioning treatment. Some grades use cold work to facilitate transformation. The strengthening effect is achieved by adding such elements as copper and aluminum, which form intermetallic precipitates during aging. In the solution-annealed condition, these grades have properties similar to those of the austenitic grades and are therefore readily formed. Hardening is achieved after fabrication within a relatively short time at 480 to 620 °C (900 to 1150 °F). The precipitation-hardened grades must not be subjected to further exposure to elevated temperature by welding or environment, because the strengthening can be lost by overaging of the precipitates. The precipitation-hardened grades have corrosion resistance generally comparable to that of the chromium-nickel grades.

Mechanism of Corrosion Resistance

The mechanism of corrosion protection for stainless steels differs from that for carbon steels, alloy steels, and most other metals. In these other cases, the formation of a barrier of true oxide separates the metal from the surrounding atmosphere. The degree of protection afforded by such an oxide is a function of the thickness of the oxide layer, its continuity, its coherence and adhesion to the metal, and the diffusivities of oxygen and metal in the oxide. In high-temperature oxidation, stainless steels use a generally similar model for corrosion protection. However, at low temperatures, stainless steels do not form a layer of true oxide. Instead, a passive film is formed. One mechanism that has been suggested is the formation of a film of hydrated oxide, but there is not total agreement on the nature of the oxide complex on the metal surface. However, the oxide film should be continuous, nonporous, insoluble, and self-healing if broken in the presence of oxygen.

Passivity exists under certain conditions for particular environments. The range of conditions over which passivity can be maintained depends on the precise environment and on the family and composition of the stainless steel. When conditions are favorable for maintaining passivity, stainless steels exhibit extremely low corrosion rates. If passivity is destroyed under conditions that do not permit restoration of the passive film, then stainless steel will corrode much like a carbon or low-alloy steel.

The presence of oxygen is essential to the corrosion resistance of a stainless steel. The corrosion resistance of stainless steel is at its maximum when the steel is boldly exposed and the surface is maintained free of deposits by a flowing bulk environment. Covering a portion of the surface—for example, by biofouling, painting, or installing a gasket—produces an oxygen-depleted region under the covered region. The oxygen-depleted region is anodic relative to the well-aerated boldly exposed surface, and a higher level of alloy content in the stainless steel is required to prevent corrosion.

With appropriate grade selection, stainless steel will perform for very long times with minimal corrosion, but an inadequate grade can corrode and perforate more rapidly than a plain carbon steel will fail by uniform corrosion. Selection of the appropriate grade of stainless steel is then a balancing of the desire to minimize cost and the risk of corrosion damage by excursions of environmental conditions during operation or downtime.

Confusion exists regarding the meaning of the term passivation. It is not necessary to chemically treat a stainless steel to obtain the passive film; the film forms spontaneously in the presence of oxygen. Most frequently, the function of passivation is to remove free iron, oxides, and other surface contamination. For example, in the steel mill, the stainless steel may be pickled in an acid solution, often a mixture of nitric and hydrofluoric acids (HNO_3-HF), to remove oxides formed in heat treatment. Once the surface is cleaned and the bulk composition of the stainless steel is exposed to air, the passive film forms immediately.

Effects of Composition

Chromium is the one element essential in forming the passive film. Other elements can influence the effectiveness of chromium in forming or maintaining the film, but no other element can, by itself, create the properties of stainless steel. The film is first observed at about 10.5% Cr, but it is rather weak at this composition and affords only mild atmospheric protection. Increasing the chromium content to 17 to 20%, as typical of the austenitic stainless steels, or to 26 to 29%, as possible in the newer ferritic stainless steels, greatly increases the stability of the passive film. However, higher chromium may adversely affect mechanical properties, fabricability, weldability, or suitability for applications involving certain thermal exposures. Therefore, it is often more efficient to improve corrosion resistance by altering other elements, with or without some increase in chromium.

Nickel, in sufficient quantities, will stabilize the austenitic structure; this greatly enhances mechanical properties and fabrication characteristics. Nickel is effective in promoting repassivation, especially in reducing environments. Nickel is particularly useful in resisting corrosion in mineral acids. Increasing nickel content to about 8 to 10% decreases resistance to SCC, but further increases begin to restore SCC resistance. Resistance to SCC in most service environments is achieved at about 30% Ni. In the newer ferritic grades, in which the nickel addition is less than that required to destabilize the ferritic phase, there are still substantial effects. In this range, nickel increases yield strength, toughness, and resistance to reducing acids, but it makes the ferritic grades susceptible to SCC in concentrated magnesium chloride ($MgCl_2$) solutions.

Manganese in moderate quantities and in association with nickel additions will perform many of the functions attributed to nickel. However, total replacement of nickel by manganese is not practical. Very high manganese steels have some unusual and useful mechanical properties, such as resistance to galling. Manganese interacts with sulfur in stainless steels to form manganese sulfides. The morphology and composition of these sulfides can have substantial effects on corrosion resistance, especially pitting resistance.

Molybdenum in combination with chromium is very effective in terms of stabilizing the passive film in the presence of chlorides. Molybdenum is especially effective in increasing resistance to the initiation of pitting and crevice corrosion.

Carbon is useful to the extent that it permits hardenability by heat treatment, which is the basis of the martensitic grades, and that it provides strength in the high-temperature applications of stainless steels. In all other applications, carbon is detrimental to corrosion resistance through its reaction with chromium. In the ferritic grades, carbon is also extremely detrimental to toughness.

Nitrogen is beneficial to austenitic stainless steels in that it enhances pitting resistance, retards the formation of the chromium-molybdenum σ phase, and strengthens the steel. Nitrogen is essential in the newer duplex grades for increasing the austenite content, diminishing chromium and molybdenum segregation, and for raising the corrosion resistance of the austenitic phase. Nitrogen is highly detrimental to the mechanical properties of the ferritic grades and must be treated as comparable to carbon when a stabilizing element is added to the steel.

Effects of Processing, Design, Fabrication, and External Treatments

Corrosion failures in stainless steels can often be prevented by suitable changes in design or process parameters and by use of the proper fabrication technique or treatment. The solution to a corrosion problem is not always to upgrade the stainless steel. It is of course very important to establish the types of corrosion that may occur in a given service environment, and if failure does occur, it is important to establish the type of corrosion that caused the failure in order that the proper preventive measures can be implemented.

Heat Treatment

Improper heat treatment can produce deleterious changes in the microstructure of stainless steels. The most troublesome problems are carbide precipitation (sensitization) and precipitation of various intermetallic phases, such as σ, χ, and laves (η).

Sensitization, or carbide precipitation at grain boundaries, can occur when austenitic stainless steels are heated for a period of time in the range of about 425 to 870 °C (800 to 1600 °F). Time at temperature will determine the amount of carbide precipitation. When the chromium carbides precipitate in grain boundaries, the area immediately adjacent is depleted of chromium. When the precipitation is relatively continuous, the depletion renders the stainless steel susceptible to intergranular corrosion, which is the dissolution of the low-chromium layer or envelope surrounding each grain. Sensitization also lowers resistance to other forms of corrosion, such as pitting, crevice corrosion, and SCC.

Time-temperature-sensitization curves are available that provide guidance for avoiding sensitization and illustrate the effect of carbon content on this phenomenon (Fig. 1). The curves shown in Fig. 1 indicate that a type 304 stainless steel with 0.062% C would have to cool below 595 °C (1100 °F) within about 5 min to avoid sensitization, but a type 304L with 0.030% C could take about 20 h to cool below 480 °C (900 °F) without becoming sensitized. These curves are general guidelines and should be verified before they are applied to various types of stainless steels.

Another method of avoiding sensitization is to use stabilized steels. Such stainless steels contain titanium and/or niobium. These elements have an affinity for carbon and form carbides readily; this allows the chromium to remain in solution even for extremely long exposures to temperatures in the sensitizing range. Type 304L can avoid sensitization during the relatively brief exposure of welding, but it will be sensitized by long exposures.

Annealing is the only way to correct a sensitized stainless steel. Because different stainless steels require different temperatures, times, and quenching procedures, the user should contact the material supplier for such information. A number of tests can detect sensitization resulting from carbide precipitation in austenitic and ferritic stainless steels. The most widely used tests are described in ASTM standards A 262 and A 763 (Ref 1, 2). More detailed information on sensitization of stainless steels can be found in the article "Metallurgically Influenced Corrosion" in this Volume.

Precipitation of Intermetallic Phases. Sigma-phase precipitation and precipitation of other intermetallic phases also increase susceptibility to corrosion. Sigma phase is a chromium-molybdenum-rich phase that can render stainless steels susceptible to intergranular corrosion, pitting, and crevice corrosion. It generally occurs in higher-alloyed stainless steels (high-chromium high-molybdenum stainless steels). Sigma phase can occur at a temperature range between 540 and 900 °C (1000 and 1650 °F). Like sensitization, it can be corrected by solution annealing. Precipitation of intermetallic phases in stainless steels is also covered in detail in the article "Metallurgically Influenced Corrosion."

Cleaning Procedures. Any heat treatment of stainless steel should be preceded and followed by cleaning. Steel should be cleaned before heat treating to remove any foreign material that may be incorporated into the surface during the high-temperature exposure. Carbonaceous materials on the surface could result in an increase in the carbon content on the surface, causing carbide precipitation. Salts could cause excessive intergranular oxidation. Therefore, the stainless steel must be clean before it is heat treated.

After heat treatment, unless an inert atmosphere was used during the process, the stainless steel surface will be covered with an oxide film. Such films are not very corrosion resistant and must be removed to allow the stainless steel to form its passive film and provide the corrosion resistance for which it was designed. There are numerous cleaning methods that may be used before and after heat treating. An excellent guide is ASTM A 380 (Ref 3).

Welding

The main problems encountered in welding stainless steels are the same as those seen in heat treatment. The heat of welding (portions of the base metal adjacent to the weld may be heated to 430 to 870 °C, or 800 to 1600 °F) can cause sensitization and formation of intermetallic phases, thus increasing the susceptibility of stainless steel weldments to intergranular corrosion, pitting, crevice corrosion, and SCC. These phenomena often occur in the heat-affected zone of the weld. Sensitization and intermetallic phase precipitation can be corrected by solution annealing after welding. Alternatively, low carbon or stabilized grades may be used.

Another problem in high heat input welds is grain growth, particularly in ferritic stainless steels. Excessive grain growth can increase susceptibility to intergranular attack and reduce toughness. Thus, when welding most stainless steels, it is wise to limit weld heat input as much as possible. More detailed information on welding of stainless steels and the problems encountered can be found in the article "Corrosion of Weldments" in this Volume.

Cleaning Procedures. Before any welding begins, all materials, chill bars, clamps, hold down bars, work tables, electrodes, and wire, as well as the stainless steel, must be cleaned of all foreign matter. Moisture can cause porosity in the weld that would reduce corrosion resistance. Organic materials, such as grease, paint, and oils, may result in carbide precipitation. Copper contamination may cause cracking. Other shop dirt can cause weld porosity and poor welds in general. Information on cleaning is available in Ref 3.

Weld design and procedure are very important in producing a sound corrosion-resistant weld. Good fit and minimal out-of-position welding will minimize crevices and slag entrapment. The design should not place welds in critical flow areas. When attaching such devices as low-alloy steel supports and ladders on the outside of a stainless steel tank, a stainless steel intermediate pad should be used. In general, stainless steels with higher alloy content than type 316 should be welded with weld metal richer in chromium, nickel, and molybdenum than the base metal. Every attempt should be made to minimize weld spatter.

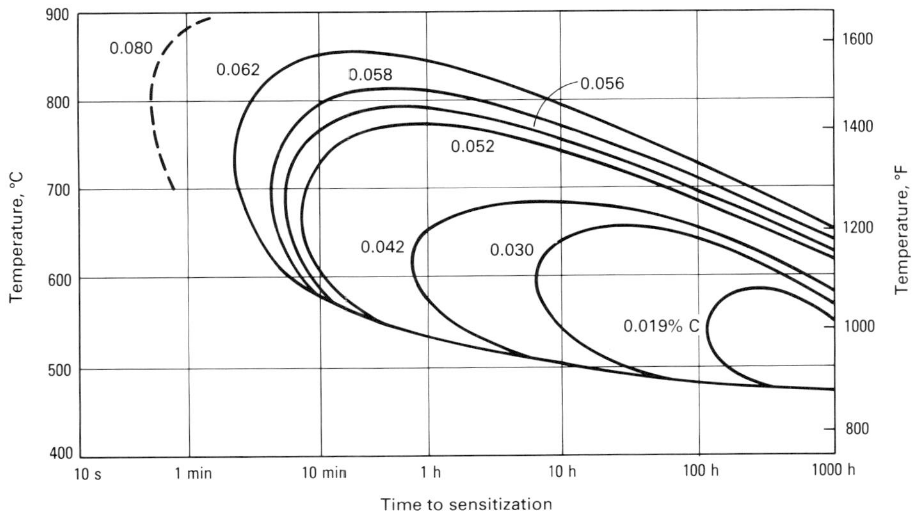

Fig. 1 Time-temperature-sensitization curves for type 304 stainless steel in a mixture of $CuSO_4$ and H_2SO_4 containing free copper. Curves show the times required for carbide precipitation in steels with various carbon contents. Carbides precipitate in the areas to the right of the various carbon content curves.

After welding, all weld spatter, slag, and oxides should be removed by brushing, blasting, grinding, or chipping. All finishing equipment must be free of iron contamination. It is advisable to follow the mechanical cleaning and finishing with a chemical cleaning. Such a cleaning will remove any foreign particles that may have been embedded in the surface during mechanical cleaning without attacking the weldment. Procedures for such cleaning or descaling are given in Ref 3, and more information on welding of stainless steels is available in Volume 6 of the 9th Edition of *Metals Handbook*.

Surface Condition

To ensure satisfactory service life, the surface condition of stainless steels must be given careful attention. Smooth surfaces, as well as freedom from surface imperfections, blemishes, and traces of scale and other foreign material, reduce the probability of corrosion. In general, a smooth, highly polished, reflective surface has greater resistance to corrosion. Rough surfaces are more likely to catch dust, salts, and moisture, which tend to initiate localized corrosive attack.

Oil and grease can be removed by using hydrocarbon solvents or alkaline cleaners, but these cleaners must be removed before heat treatment. Hydrochloric acid (HCl) formed from residual amounts of trichloroethylene, which is used for degreasing, has caused severe attack of stainless steels. Surface contamination may be caused by machining, shearing, and drawing operations. Small particles of metal from tools become embedded in the steel surface and, unless removed, may cause localized galvanic corrosion. These particles are best removed by the passivation treatments described below. Additional information on cleaning and descaling of stainless steel is available in Ref 3 and in Volume 5 of the 9th Edition of *Metals Handbook*.

Shotblasting or sandblasting should be avoided unless iron-free silica is used; metal shot, in particular, will contaminate the stainless steel surface. If shotblasting or shotpeening with metal grit is unavoidable, the parts must be cleaned after blasting or peening by immersing them in an HNO_3 solution, as noted above.

Passivation Techniques

During handling and processing operations, such as machining, forming, tumbling, and lapping, particles of iron, tool steel, or shop dirt may be embedded in or smeared on the surfaces of stainless steel components. These contaminants may reduce the effectiveness of the natural oxide (passive) film that forms on stainless steels exposed to oxygen at low temperatures (the formation of these passive films is discussed in the section "Mechanism of Corrosion Resistance" in this article). If allowed to remain, these particles may corrode and produce rustlike spots on the stainless steel. To prevent this condition, semifinished or finished parts are given a passivation treatment. This treatment consists of cleaning and then immersing stainless steel parts in a solution of HNO_3 or of HNO_3 plus oxidizing salts. The treatment dissolves the embedded or smeared iron, restores the original corrosion-resistant surface, and maximizes the inherent corrosion resistance of the stainless steel.

Cleaning. Each workpiece to be passivated must be cleaned thoroughly to remove grease, coolant, or other shop debris (Ref 4). A worker will sometimes eliminate the cleaning step based on the reasoning that the cleaning and passivation of a grease-laden part will occur simultaneously by immersing it in an HNO_3 bath. This assumption is mistaken. The grease will react with the HNO_3 to form gas bubbles, which collect on the surface of the workpiece and interfere with passivation. Also, contamination of the passivating solution (particularly with high levels of chlorides) can cause flash attack, which results in a gray or black appearance and deterioration of the surface.

To avoid such problems, each part should be wiped clean of any large machining chips or other debris. More tenacious deposits should be removed by brushing with a stainless steel wire brush, grinding, polishing with an iron-free abrasive, or sandblasting. Tools and materials used for these processes should be clean and used only for stainless steels. Machining, forming, or grinding oils must be removed in order for passivation to be effective. Cleaning should begin with solvent cleaning, which may be followed by alkaline soak cleaning and thorough water rinsing. Optimum results are obtained in passivation when the parts to be treated are as clean as they would have to be for plating. When large parts or bulky vessels are to be cleaned, it may be necessary to apply cleaning liquids by means of pressure spray; exterior surfaces may be cleaned by immersion or swabbing.

Passivating. After cleaning, the workpiece can be immersed in the passivating acid bath. As shown in Table 3, the composition of the acid bath depends on the grade of stainless steel. The 300-series stainless steels can be passivated in 20 vol% HNO_3. A sodium dichromate ($Na_2Cr_2O_7{\cdot}2H_2O$) addition or an increased concentration of HNO_3 is used for less corrosion-resistant stainless steels to reduce the potential for flash attack.

The procedure suggested for passivating free-machining stainless steels is somewhat different from that used for nonfree-machining grades (Ref 4). This is because sulfides of sulfur-bearing free-machining grades, which are totally or partially removed during passivation, create microscopic discontinuities in the surface of the machined part. Even normally efficient water rinses can leave residual acid trapped in these discontinuities after passivation. This acid can then attack the surface of the part unless it is neutralized or removed. For this reason, a special passivation process, referred to as the alkaline-acid-alkaline method, is suggested for free-machining grades.

Table 3 Passivating solutions for stainless steels (nonfree-machining grades)

Grade	Passivation treatment
Austenitic 300-series grades	20 vol% HNO_3 at 50–60 °C (120–140 °F) for 30 min
Grades with ≥17% Cr (except 440 series)	
Straight chromium grades (12–14% Cr)	20 vol% HNO_3 plus 22 g/L (3 oz/gal) $Na_2Cr_2O_7{\cdot}2H_2O$ at 50–60 °C (120–140 °F) for 30 min
High-carbon/high-chromium grades (440 series)	or
Precipitation-hardening grades	50 vol% HNO_3 at 50–60 °C (120–140 °F) for 30 min

Source: Ref 4

The following steps should be followed when passivating free-machining stainless steels with the alkaline-acid-alkaline technique:

- After degreasing, soak the parts for 30 min in 5 wt% sodium hydroxide (NaOH) at 70 to 80 °C (160 to 180 °F)
- Water rinse
- Immerse the part for 30 min in 20 vol% HNO_3 plus 22 g/L (3 oz/gal) $Na_2Cr_2O_7{\cdot}2H_2O$ at 50 to 60 °C (120 to 140 °F)
- Water rinse
- Immerse for 30 min in 5 wt% NaOH at 70 to 80 °C (160 to 180 °F)
- Water rinse and dry

Testing is often performed to evaluate the passivated surface. For example, 400-series, precipitation-hardening, and free-machining stainless steels are often tested in a cabinet capable of maintaining 100% humidity at 35 °C (95 °F) for 24 h. Material that is properly passivated will be virtually free of rust, although light staining may occur (Ref 1). Austenitic 300-series grades can be evaluated using a technique given in ASTM Standard Method A 380 (Ref 3). This test consists of swabbing the part with a copper sulfate ($CuSO_4{\cdot}5H_2O$)/sulfuric acid (H_2SO_4) solution; wetness should be maintained for 6 min (Ref 4). Free iron, if present, plates out the copper from the solution, and the surface develops a copper cast or color. Precautions for this procedure and details on additional tests for detecting the presence of iron on passivated surfaces are outlined in Ref 3 and 4. Information on passivation treatments for corrosion-resistant steels is also available in Federal Specification QQ-P-35B (Ref 5).

Design

Corrosion can often be avoided by suitable changes in design without changing the type of steel. The factors to be considered include joint design, surface continuity, and concentration of stress. Designs that tend to concentrate corrosive media in a small area should be avoided. For example, tank inlets should be designed such that concentrated solutions are mixed and diluted as they are introduced (Fig. 2). Otherwise, localized pockets of concentrated solutions can cause excessive corrosion.

Poor design of heaters can create similar problems, such as those that cause hot spots and thus accelerate corrosion. Heaters should be centrally located (Fig. 3). If a tank is to be heated externally, heaters should be distributed over as large

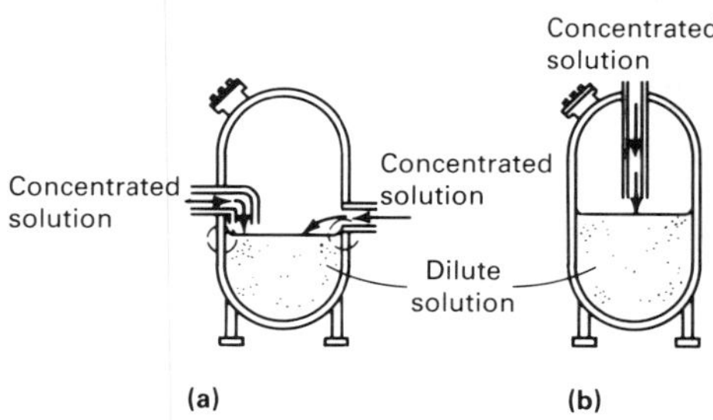

Fig. 2 Poor (a) and good (b) designs for vessels used for mixing concentrated and dilute solutions. Poor design causes concentration and uneven mixing of incoming chemicals along the vessel wall (circled areas). Good design allows concentrated solutions to mix away from vessel walls.

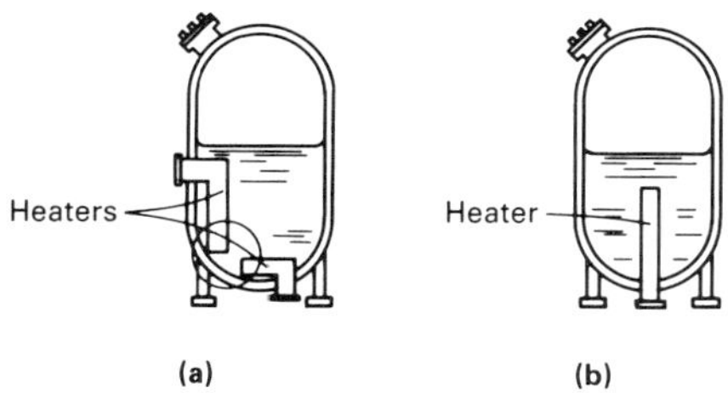

Fig. 3 Poor (c) and good (b) designs for heating of solutions. Poor design creates hot spots (circled area) that may induce boiling under the heater at the bottom of the vessel or may cause deposits to form between heaters and vessel walls. Good design avoids hot spots and pockets in which small volumes of liquid can become trapped between the heater and the vessel wall.

a surface area as possible, and circulation of the corrosive medium should be encouraged, if possible.

Hot gases that are not corrosive to stainless steel may form corrosive condensates on the cold portions of a poorly insulated unit. Proper design or insulation can prevent such localized cooling (Fig. 4). Conversely, vapors from noncorrosive liquids may cause attack; exhausts and overflows should be designed to prevent hot vapor pockets (Fig. 5). In general, the open ends of inlets, outlets, and tubes in heat exchangers should be flush with tank walls or tubesheets to avoid buildup of harmful corrodents, sludges, and deposits (Fig. 6). This is also true of tank bottom and drainage designs (Fig. 7).

Tanks and tank supports should be designed to prevent or minimize corrosion due to spills and overflows (Fig. 8). A tank support structure may not be as corrosion resistant as the tank itself, but it is a very important part of the unit and should not be made vulnerable to spilled corrodents.

Designs that increase turbulence or result in excessive flow rates should be avoided where erosion-corrosion may be a problem (Fig. 9). Gaskets in flanges should fit properly, intrusions in a flow stream should be avoided, and elbows should be given a generous radius. Finally, crevices should be avoided. Where crevices cannot be avoided, they should be sealed by welding, soldering, or the use of caulking compounds or sealants. Additional information is available in the article "Design Details to Minimize Corrosion" in this Volume.

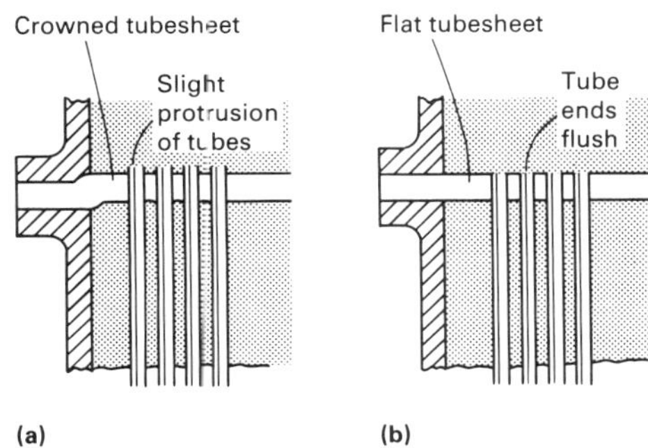

Fig. 6 Poor (a) and good (b) designs for tube/tubesheet assemblies. Crowned tubesheet and protruding tubes in (a) allow buildup of corrosive deposits; in (b), tubesheet is flat and tubes are mounted flush.

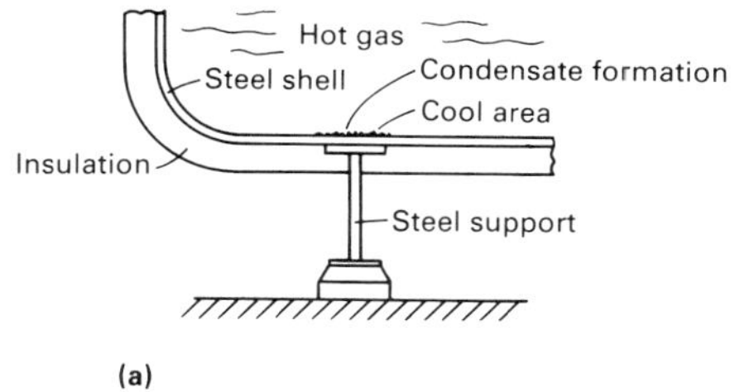

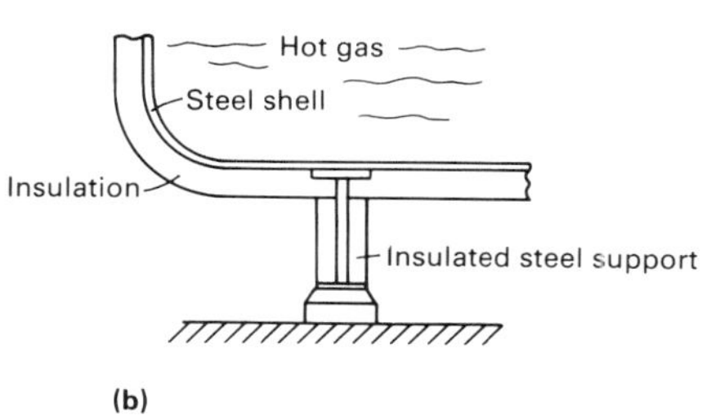

Fig. 4 Design to prevent localized cooling. In the poor design (a), the uninsulated steel support radiates heat, which causes a cool area on the steel shell. In (b), the steel support is insulated to prevent temperature decrease at the base of the shell.

Forms of Corrosion of Stainless Steels

The various forms of corrosive attack will be briefly discussed in this section. Detailed information on each of these forms of corrosion is available in the Section "Forms of Corrosion" in this Volume.

General (uniform) corrosion of a stainless steel suggests an environment capable of stripping the passive film from the surface and preventing repassivation. Such an occurrence could indicate an error in grade selection. An example is the exposure of a lower-chromium ferritic stainless steel to a moderate concentration of hot sulfuric acid (H_2SO_4).

Galvanic corrosion results when two dissimilar metals are in electrical contact in a corrosive medium. As a highly corrosion-resistant metal, stainless steel can act as a cathode when in contact with a less noble metal, such as steel. The corrosion of steel parts—for example, steel bolts in a stainless steel construction—can be a significant problem. However, the effect can be used in a beneficial way for protecting critical stainless steel components within a larger steel construction. In the case of stainless steel connected to a more noble metal, consideration must be given to the active-passive condition of the stainless steel. If the stainless steel is passive in the environment, galvanic interaction with a more noble metal is unlikely to produce significant corrosion. If the stainless steel is active or only marginally passive, galvanic interaction with a more noble metal will probably produce sustained rapid corrosion of the stainless steel without repassivation. The most important aspect of galvanic interaction for stainless steels is the necessity of selecting fasteners and weldments of adequate corrosion resistance relative to the bulk material, which is likely to have a much larger exposed area.

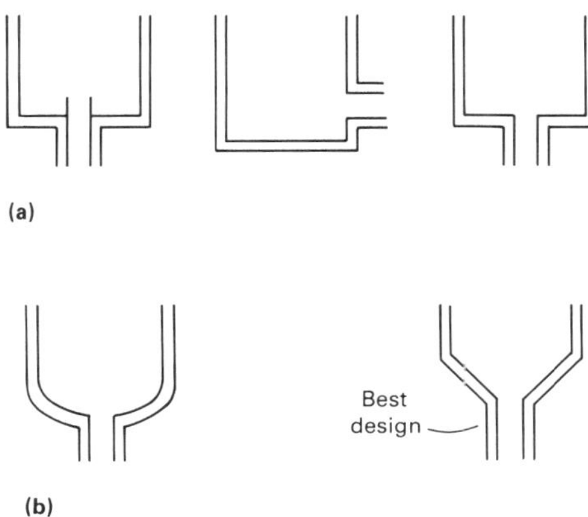

Fig. 7 Examples of poor (a) and good (b) designs for drainage, corners, and other dead spaces in vessels. Sharp corners and protruding outlet pipes in (a) can cause buildup of corrosive deposits and crevice corrosion; these design features are avoided in (b).

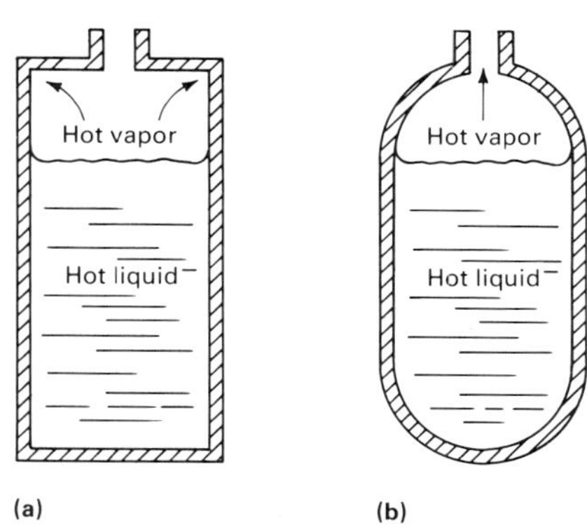

Fig. 5 Poor (a) and good (b) designs for vessels holding both liquid and vapor phases. Sharp corners and protruding outlet end in (a) allow hot gases to become trapped in the vapor space. This is avoided in (b) by using rounded corners and mounting the vessel outlet pipe flush.

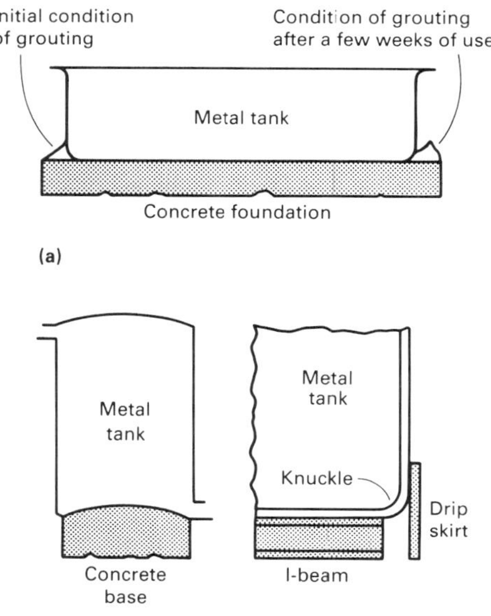

Fig. 8 Design for preventing external corrosion from spills and overflows. (a) Poor design. (b) Good designs

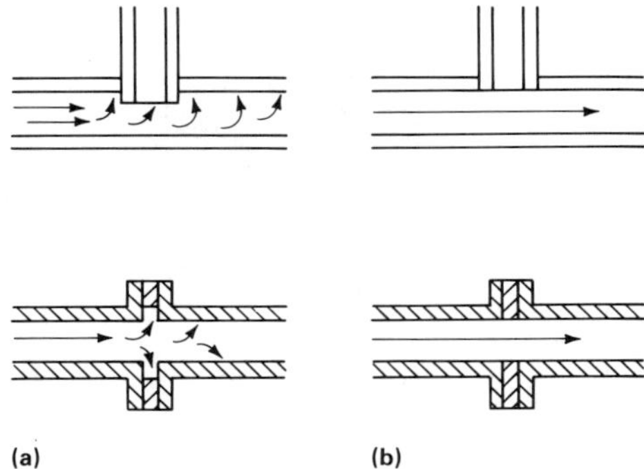

Fig. 9 Designs for preventing excessive turbulence. (a) Poor designs (both top and bottom figures). (b) Good designs (both top and bottom figures)

Pitting is a localized attack that can produce penetration of a stainless steel with almost negligible weight loss to the total structure. Pitting is associated with a local discontinuity of the passive film. It can be a mechanical imperfection, such as an inclusion or surface damage, or it can be a local chemical breakdown of the film. Chloride is the most common agent for initiation of pitting. Once a pit is formed, it in effect becomes a crevice; the local chemical environment is substantially more aggressive than the bulk environment. This explains why very high flow rates over a stainless steel surface tend to reduce pitting corrosion; the high flow rate prevents the concentration of corrosive species in the pit. The stability of the passive film with respect to resistance to pitting initiation is controlled primarily by chromium and molybdenum. Minor alloying elements can also have an important effect by influencing the amount and type of inclusions (for example, sulfides) in the steel that can act as pitting sites.

Pitting initiation can also be influenced by surface condition, including the presence of deposits, and by temperature. For a particular environment, a grade of stainless steel may be characterized by a single temperature, or a very narrow range of temperatures, above which pitting will initiate and below which pitting will not initiate. It is therefore possible to select a grade that will not be subject to pitting attack if the chemical environment and temperature do not exceed the critical levels. If the range of operating conditions can be accurately characterized, a meaningful laboratory evaluation is possible. Formation of deposits in service can reduce the pitting temperature.

Although chloride is known to be the primary agent of pitting attack, it is not possible to establish a single critical chloride limit for each grade. The corrosivity of a particular concentration of chloride solution can be profoundly affected by the presence or absence of various other chemical species that may accelerate or inhibit corrosion. Chloride concentration may increase where evaporation or deposits occur. Because of the nature of pitting attack—rapid penetration with little total weight loss—it is rare that any significant amount of pitting will be acceptable in practical applications.

Crevice corrosion can be considered a severe form of pitting. Any crevice, whether the result of a metal-to-metal joint, a gasket, fouling, or deposits, tends to restrict oxygen access, resulting in attack. In practice, it is extremely difficult to prevent all crevices, but every effort should be made to do so. Higher-chromium, and especially higher-molybdenum, grades are more resistant to crevice attack. Just as there is a critical pitting temperature for a particular environment, there is also a critical crevice temperature. This temperature is specific to the geometry and nature of the crevice and to the precise corrosion environment for each grade. The critical crevice temperature can be useful in selecting an adequately resistant grade for particular applications.

Intergranular corrosion is a preferential attack at the grain boundaries of a stainless steel. It is generally the result of sensitization. This condition occurs when a thermal cycle leads to grain-boundary precipitation of a carbide, nitride, or intermetallic phase without providing sufficient time for chromium diffusion to fill the locally depleted region. A grain-boundary precipitate is not the point of attack; instead, the low-chromium region adjacent to the precipitate is susceptible.

Sensitization is not necessarily detrimental unless the grade is to be used in an environment capable of attacking the region. For example, elevated-temperature applications for stainless steel can operate with sensitized steel, but concern for intergranular attack must be given to possible corrosion during downtime when condensation might provide a corrosive medium. Because chromium provides corrosion resistance, sensitization also increases the susceptibility of chromium-depleted regions to other forms of corrosion, such as pitting, crevice corrosion, and SCC. The thermal exposures required to sensitize a steel can be relatively brief, as in welding, or can be very long, as in high-temperature service.

Stress-corrosion cracking is a corrosion mechanism in which the combination of a susceptible alloy, sustained tensile stress, and a particular environment leads to cracking of the metal. Stainless steels are particularly susceptible to SCC in chloride environments; temperature and the presence of oxygen tend to aggravate chloride SCC of stainless steels. Most ferritic and duplex stainless steels are either immune or highly resistant to SCC. All austenitic grades, especially AISI types 304 and 316, are susceptible to some degree. The highly alloyed austenitic grades are resistant to sodium chloride (NaCl) solutions, but crack readily in $MgCl_2$ solutions. Although some localized pitting or crevice corrosion probably precedes SCC, the amount of pitting or crevice attack may be so small as to be undetectable. Stress corrosion is difficult to detect while in progress, even when pervasive, and can lead to rapid catastrophic failures of pressurized equipment.

It is difficult to alleviate the environmental conditions that lead to SCC. The level of chlorides required to produce stress corrosion is very low. In operation, there can be evaporative concentration or a concentration in the surface film on a heat-rejecting surface. Temperature is often a process parameter, as in the case of a heat exchanger. Tensile stress is one parameter that might be controlled. However, the residual stresses associated with fabrication, welding, or thermal cycling, rather than design stresses, are often responsible for SCC, and even stress-relieving heat treatments do not completely eliminate these residual stresses.

Erosion-Corrosion. Corrosion of a metal or alloy can be accelerated when there is an abrasive removal of the protective oxide layer. This form of attack is especially significant when the thickness of the oxide layer is an important factor in determining corrosion resistance. In the case of a stainless steel, erosion of the passive film can lead to some acceleration of attack.

Oxidation. Because of their high chromium contents, stainless steels tend to be very resistant to oxidation. Important factors to be considered in the selection of stainless steels for high-temperature service are the stability of the composition and microstructure of the grade upon thermal exposure and the adherence of the oxide scale upon thermal cycling. Because many of the stainless steels used for high temperatures are austenitic grades with relatively high nickel contents, it is also necessary to be alert to the possibility of sulfidation attack.

Corrosion in Specific Environments

Selection of a suitable stainless steel for a specific environment requires consideration of several criteria. The first is corrosion resistance. Alloys are available that provide resistance to mild atmospheres (for example, type 430) or to many food-processing environments (for example, type 304 stainless). Chemicals and more severe corrodents require type 316 or a more highly alloyed material, such as 20Cb-3. Factors that affect the corrosivity of an environment

Table 4 Atmospheric corrosion of austenitic stainless steels at two industrial sites

	New York City (industrial)		Niagara Falls (industrial-chemical)	
Type(a)	Exposure time, years	Specimen surface evaluation	Exposure time, years	Specimen surface evaluation
302	5	Free from rust stains	$<^2/_3$	Rust stains
302	26	Free from rust stains	...	...
304	26	Free from rust stains	<1	Rust stains
304	...	...	6	Covered with rust spots and pitted
347	26	Free from rust stains	...	...
316	23	Free from rust stains	$<^2/_3$	Slight stains
316	...	...	6	Slight rust spots, slightly pitted
317	...	...	$<^2/_3$	Slight stains
317	...	...	6	Slight stains
310	...	...	<1	Rust stains
310	...	...	6	Rust spots; pitted

(a) Solution-annealed sheet, 1.6 mm ($^1/_{16}$ in.) thick

include the concentration of chemical species, pH, aeration, flow rate (velocity), impurities (such as chlorides), and temperature, including effects from heat transfer.

The second criterion is mechanical properties, or strength. High-strength materials often sacrifice resistance to some form of corrosion, particularly SCC.

Third, fabrication must be considered, including such factors as the ability of the steel to be machined, welded, or formed. Resistance of the fabricated article to the environment must be considered—for example, the ability of the material to resist attack in crevices that cannot be avoided in the design.

Fourth, total cost must be estimated, including initial alloy price, installed cost, and the effective life expectancy of the finished product. Finally, consideration must be given to product availability.

Many applications for stainless steels, particularly those involving heat exchangers, can be analyzed in terms of a process side and a water side. The process side is usually a specific chemical combination that has its own requirements for a stainless steel grade. The water side is common in many applications. This section will discuss the corrosivity of various environments for stainless steels.

Atmospheric Corrosion

The atmospheric contaminants most often responsible for the rusting of structural stainless steels are chlorides and metallic iron dust. Chloride contamination may originate from the calcium chloride ($CaCl_2$) used to make concrete or from exposure in marine or industrial locations. Iron contamination may occur during fabrication or erection of the structure. Contamination should be minimized, if possible.

The corrosivity of different atmospheric exposures can vary greatly and can dictate application of different grades of stainless steel. Rural atmospheres, uncontaminated by industrial fumes or coastal salt, are extremely mild in terms of corrosivity for stainless steel, even in areas of high humidity. Industrial or marine environments can be considerably more severe.

Table 4 demonstrates that resistance to staining can depend on the specific exposure. For example, several 300-series stainless steels showed no rust during long-term exposures in New York City. On the other hand, staining was observed after much shorter exposures at Niagara Falls in a severe industrial-chemical environment near plants producing chlorine or HCl.

Although marine environments can be severe, stainless steels often provide good resistance. Table 5 compares several AISI 300-series stainless steels after a 15-year exposure to a marine atmosphere 250 m (800 ft) from the ocean at Kure Beach, NC. Materials containing molybdenum exhibited only extremely slight rust stain, and all grades were easily cleaned to reveal a bright surface. Type 304 stainless steel may provide satisfactory resistance in many marine applications, but more highly alloyed grades are often selected when the stainless is sheltered from washing by the weather and is not cleaned regularly.

Type 302 and 304 stainless steels have had many successful architectural applications. Type 430 stainless steel has been used in many locations, but there have been problems. For example, type 430 stainless steel rusted in sheltered areas after only a few months' exposure in an industrial environment. The type 430 stainless steel was replaced by type 302, which provided satisfactory service. In more aggressive environments, such as marine or severely contaminated atmospheres, type 316 stainless steel is especially useful.

Table 5 Corrosion of AISI 300-series stainless steels in a marine atmosphere

Based on 15-year exposures 250 m (800 ft) from the ocean at Kure Beach, NC

AISI type	Average corrosion rate, mm/yr	Average corrosion rate, mils/yr	Average depth of pits, mm	Average depth of pits, mils	Appearance(a)
301	$<2.5 \times 10^{-5}$	<0.001	0.04	1.6	Light rust and rust stain on 20% of surface
302	$<2.5 \times 10^{-5}$	<0.001	0.03	1.2	Spotted with rust stain on 10% of surface
304	$<2.5 \times 10^{-5}$	<0.001	0.028	1.1	Spotted with slight rust stain on 15% of surface
321	$<2.5 \times 10^{-5}$	<0.001	0.067	2.6	Spotted with slight rust stain on 15% of surface
347	2.5×10^{-5}	0.001	0.086	3.4	Spotted with moderate rust stain on 20% of surface
316	$<2.5 \times 10^{-5}$	<0.001	0.025	1.0	Extremely slight rust stain on 15% of surface
317	$<2.5 \times 10^{-5}$	<0.001	0.028	1.1	Extremely slight rust stain on 20% of surface
308	$<2.5 \times 10^{-5}$	<0.001	0.04	1.6	Spotted by rust stain on 25% of surface
309	$<2.5 \times 10^{-5}$	<0.001	0.028	1.1	Spotted by slight rust stain on 25% of surface
310	$<2.5 \times 10^{-5}$	<0.001	0.01	0.4	Spotted by slight rust stain on 20% of surface

(a) All stains easily removed to reveal bright surface. Source: Ref 7

Table 6 SCC of U-bend test specimens 25 m (80 ft) from the ocean at Kure Beach, NC

Alloy	Final heat treatment	Hardness, HRC	Specimen orientation	Time to failure of each specimen, days(a)
Custom 450	Aged at 480 °C (900 °F)	42	Transverse	NF, NF, NF, NF, NF
Type 410	Tempered at 260 °C (500 °F)	45	Longitudinal	379, 379, 471
	Tempered at 550 °C (1025 °F)	35	Longitudinal	4, 4
Alloy 355	Tempered at 540 °C (1000 °F)	38	Longitudinal	NF, NF, NF
15Cr-7Ni-Mo	Aged at 510 °C (950 °F)	49	Longitudinal	1, 1, 1
17Cr-4Ni	Aged at 480 °C (900 °F)	42	Longitudinal	93, 129, NF
	Aged at 620 °C (1150 °F)	32	Longitudinal	93, 129, NF
14Cr-6Ni	Aged at 480 °C (900 °F)	39	Longitudinal	93, 872, NF

(a) NF, no failure in over 4400 days for Custom 450 and 1290 days for the other materials. Source: Ref 9

Stress-corrosion cracking is generally not a concern when austenitic or ferritic stainless steels are used in atmospheric exposures. Several austenitic stainless steels were exposed to a marine atmosphere at Kure Beach, NC. Annealed and quarter-hard wrought AISI types 201, 301, 302, 304, and 316 stainless steels were not susceptible to SCC. In the as-welded condition, only type 301 stainless steel experienced failure. Following sensitization at 650 °C (1200 °F) for 1.5 h and furnace cooling, failures were obtained only for materials with carbon contents of 0.043% or more (Ref 8).

Stress-corrosion cracking must be considered when quench-hardened martensitic stainless steels or precipitation-hardening grades are used in marine environments or in industrial locations where chlorides are present. Several hardenable stainless grades were exposed as U-bends 25 m (80 ft) from the ocean at Kure Beach, NC. Most samples were cut longitudinally, and two alloys received different heat treatments to produce different hardness or strength levels. The results of the study (Table 6) indicated that Custom 450 stainless and stainless alloy 355 resisted cracking. Stainless alloy 355 failed in this type of test when fully hardened; resistance was imparted by the 540-°C (1000-°F) temper. Precipitation-hardenable grades are expected to exhibit improved corrosion resistance when higher aging temperatures (lower strengths) are used.

Resistance to SCC is of particular interest in the selection of high-strength stainless steels for fastener applications. Cracking of high-strength fasteners is possible and often results from hydrogen generation due to corrosion or contact with a less noble material, such as aluminum. Resistance to SCC can be improved by optimizing the heat treatment, as noted above.

Fasteners for atmospheric exposure have been fabricated from a wide variety of alloys. Type 430 and unhardened type 410 stainless steels have been used when moderate corrosion resistance is required in a lower-strength material. Better-than-average corrosion resistance has been obtained by using type 305 and Custom Flo 302HQ stainless steels when lower strength is acceptable.

Corrosion in Waters

Waters may vary from extremely pure to chemically treated water to highly concentrated chloride solutions, such as brackish water or seawater, further concentrated by recycling. This chloride content poses the danger of pitting or crevice attack of stainless steels. When the appli-

Table 7 Crevice corrosion indexes of several alloys in tests in filtered seawater

Mill-finished panels exposed for 30, 60, and 90 days to seawater at 30 °C (85 °F) flowing at <0.1 m/s (<0.33 ft/s); crevice washers tightened to 2.8 or 8.5 N · m (25 or 75 in. · lb)

Alloy	UNS Designation	Number of sides (S) attacked(a)	Maximum pit depth (D) mm	Maximum pit depth (D) mils	Crevice corrosion index (S × D)
AL-29-4C	S44735	0	nil		0
MONIT	S44635	3	0.01	0.4	0.03
Ferralium 255	S32550	1	0.09	3.5	0.09
Alloy 904L	N08904	3	0.37	14.6	1.1(b)
254SMO	S31254	6	0.19	7.5	1.1
Sea-Cure	S44660	14	0.11	4.3	1.5
AL-6X	N08366	8	0.34	13.4	2.7
JS777	...	6	2.3	90.6	14(b)
JS700	N08700	14	1.8	70.9	24
AISI type 329	...	17	1.6	63	28(c)
Nitronic 50	S20910	17	1.2	47.2	20

(a) Total number of sides was 18. (b) Also showed tunneling attack perpendicular to the upper edge, or attack at edges. (c) Perforated by attack from both sides. Source: Ref 11

Table 8 Corrosion of austenitic stainless steels in boiling glacial acetic acid

Data are from averaged results of 11-, 12-, and 21-day field tests.

AISI type	Corrosion rate mm/yr	Corrosion rate mils/yr
304	0.46	18
321	1.19	47
347	1.04	41
308	1.35	53
310	0.99	39
316	0.015	0.6

Source: Ref 17

cation involves moderately increased temperatures, even as low as 45 °C (110 °F), and particularly when there is heat transfer into the chloride-containing medium, there is the possibility of SCC. It is useful to consider water with two general levels of chloride content: freshwater, which can have chloride levels up to approximately 600 ppm, and seawater, which encompasses brackish and severely contaminated waters. The corrosivity of a particular level of chloride can be strongly affected by the other chemical constituents present, making the water either more or less corrosive.

Permanganate ion (MnO_4^-), which is associated with the dumping of chemicals, has been related to pitting of type 304 stainless steel. The presence of sulfur compounds and oxygen or other oxidizing agents can affect the corrosion of copper and copper alloys, but does not have very significant effects on stainless steels at ambient or slightly elevated temperatures (up to approximately 250 °C or 500 °F).

In freshwater, type 304 stainless steel has provided excellent service for such items as valve parts, weirs, fasteners, and pump shafts in water and wastewater treatment plants. Custom 450 stainless steel has been used as shafts for large butterfly valves in potable water. The higher strength of a precipitation-hardenable stainless steel permits reduced shaft diameter and increased flow. Type 201 stainless steel has seen service in revetment mats to reduce shoreline erosion in freshwater. Type 316 stainless steel has been used as wire for microstrainers in tertiary sewage treatment and is suggested for waters containing minor amounts of chloride.

Seawater is a very corrosive environment for many materials. Stainless steels are more likely to be attacked in low-velocity seawater or at crevices resulting from equipment design or attachment of barnacles. Type 304 and 316 stainless steels suffer deep pitting if the seawater flow rate decreases below about 1.5 m/s (5 ft/s) because of the crevices produced by fouling organisms. However, in one study, type 316 stainless steel provided satisfactory service as tubing in the heat recovery section of a desalination test plant with relatively high flow rates (Ref 10).

The choice of stainless steel for seawater service can depend on whether or not stagnant conditions can be minimized or eliminated. For example, boat shafting of 17Cr-4Ni stainless steel has been used for trawlers where stagnant exposure and the associated pitting would not be expected to be a problem. When seagoing vessels are expected to lie idle for extended periods of time, more resistant boat shaft materials, such as 22Cr-13Ni-5Mn stainless steel, are considered. Boat shafts with intermediate corrosion resistance are provided by 18Cr-2Ni-12Mn and high-nitrogen type 304 (type 304HN) stainless steels.

The most severe exposure conditions are often used in seawater test programs. In one example of such data, flat-rolled specimens of 11 commercially available alloys with several mill finishes were exposed to seawater (Table 7). Triplicate samples were prepared with plastic multiple-crevice washers, each containing 20 plateaus or crevices. These washers were affixed to both sides of each panel by using a torque of either 2.8 or 8.5 N · m (25 or 75 in. · lb). The panels were exposed for up to 90 days in filtered seawater flowing at a velocity of less than 0.1 m/s (<0.33 ft/s).

The results given in Table 7 show the number of sides that experienced crevice attack and the maximum attack depth at any crevice for that alloy. A crevice corrosion index (CCI) was calculated by multiplying the maximum attack depth times the number of sides attacked. This provided a ranking system that accounts for both initiation and growth of attack. Lower values of the CCI imply improved resistance.

Attack in the above test does not mean that materials with high CCIs cannot be used in seawater. For example, 22Cr-13Ni-5Mn stainless steel with a CCI of 20 has proved to be a highly resistant boat shaft alloy. Some of the more resistant materials in the above tests have been used for utility condenser tubing. These alloys include MONIT, AL-29-4C, 254SMO, Sea-Cure, and AL-6XN.

The possibility of galvanic corrosion must be considered if stainless steel is to be used in contact with other metals in seawater. Figure 10 provides corrosion potentials in flowing seawater for several materials. Preferably, only those materials that exhibit closely related electrode potentials should be coupled to avoid attack of the less noble material. Galvanic differences have been used to advantage in the cathodic protection of stainless steel in seawater. Crevice corrosion and pitting of austenitic type 302 and 316 stainless steels have been prevented by cathodic protection, but type 410 and 430 stainless steels develop hydrogen blisters at current densities below those required for complete protection.

Other factors that should be noted when applying stainless steels in seawater include the effects of high velocity, aeration, and temperature. Stainless steels generally show excellent resistance to high velocities, impingement attack, and cavitation in seawater. Also, stainless steels provide optimum service in aerated seawater because a lack of aeration at a specific site often leads to crevice attack. Very little oxygen is required to maintain the passive film on a clean stainless surface. Increasing the temperature from ambient to about 50 °C (120 °F) often reduces attack of stainless steels, possibly because of differences in the amount of dissolved oxygen, changes in the surface film, or changes in the resistance of the boldly exposed sample area (Ref 13). Further temperature increases can result in increased corrosion, such as SCC.

Corrosion in Chemical Environments

Selection of stainless steels for service in chemicals requires consideration of all forms of corrosion, along with impurity levels and degree of aeration. When an alloy with sufficient general corrosion resistance has been selected, care must be taken to ensure that the material will not fail by pitting or SCC due to chloride contamination. Aeration may be an important factor in corrosion, particularly in cases of borderline passivity. If dissimilar-metal contact or stray currents occur, the possibility of galvanic attack or hydrogen embrittlement must be considered.

Alloy selection also depends on fabrication and operation details. If a material is to be used in the as-welded or stress-relieved condition, it must resist intergranular attack in service after these thermal treatments. In chloride environments, the possibility of crevice corrosion must be considered when crevices are present because of equipment design or the formation of adherent deposits. Higher flow rates may prevent the formation of deposits, but in extreme cases may also cause accelerated attack due to erosion or cavitation. Increased operating temperatures generally increase corrosion. In heat transfer applications, higher metal wall temperatures result in higher rates than expected from the lower temperature of the bulk solution. These and other items may require consideration in the selection of stainless steels, yet suitable materials continue to be chosen for a wide variety of chemical plant applications (see the article "Corrosion in the Chemical Processing Industry" in this Volume).

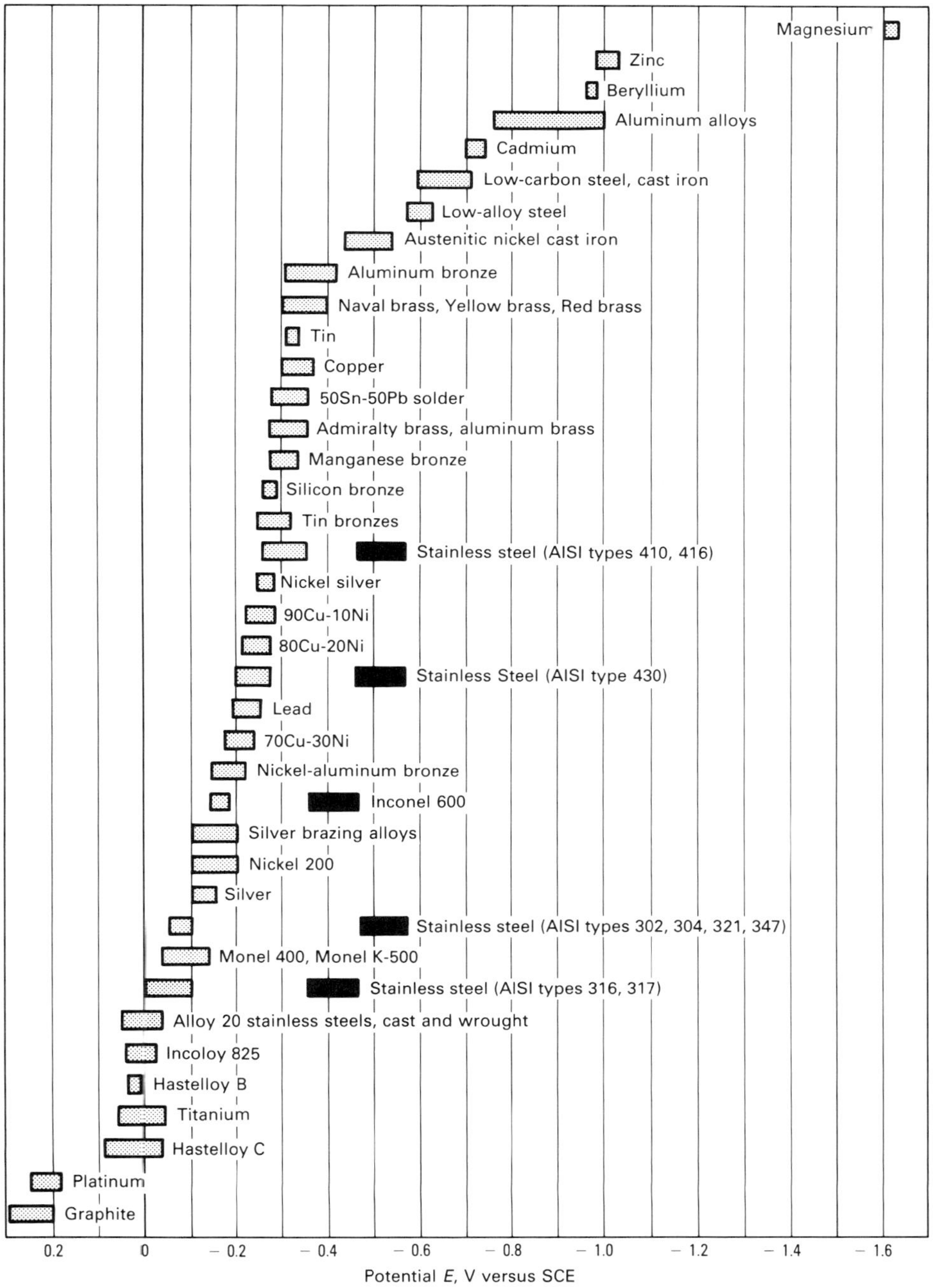

Fig. 10 Corrosion potentials of various metals and alloys in flowing seawater at 10 to 25 °C (50 to 80 °F). Flow rate was 2.5 to 4 m/s (8 to 13 ft/s); alloys are listed in order of the potential versus SCE that they exhibited. Those metals and alloys indicated by a black bar may become active and exhibit a potential near −0.5 V versus SCE in low velocity or poorly aerated water and in shielded areas. Source: Ref 12

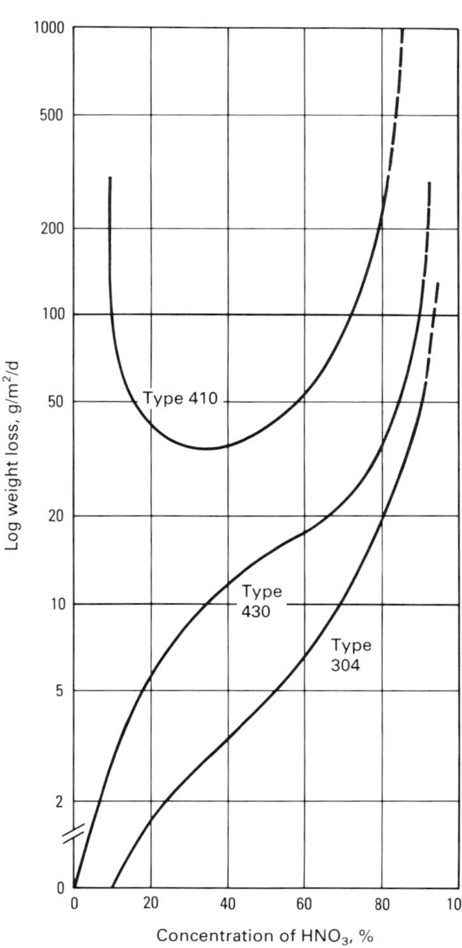

Fig. 11 Corrosion rates of various stainless steels in boiling HNO_3. Source: Ref 15

Some generalizations can be made regarding the performance of various categories of stainless steels in certain types of chemical environments. These observations relate to the compositions of the grades. For example, the presence of nickel and copper in some austenitic grades greatly enhances resistance to H_2SO_4 compared to the resistance of the ferritic grades. However, combinations of chemicals that are encountered in practice can be either more or less corrosive than might be expected from the corrosivity of the individual components. Testing in actual or simulated environments is always recommended as the best procedure for selecting a stainless steel grade. Additional information describing service experience is available from alloy suppliers.

Mineral Acids. The resistance of stainless steel to acids depends on the hydrogen ion (H^+) concentration and the oxidizing capacity of the acid, along with such material variables as chromium content, nickel content, carbon content, and heat treatment (Ref 14). For example, annealed stainless steel resists strong HNO_3 in spite of the low pH of the acid, because HNO_3 is highly oxidizing and forms a passive film due to the chromium content of the alloy. On the other hand, stainless steels are rapidly attacked by strong HCl because a passive film is not easily attained. Even in strong HNO_3, stainless steels can be rapidly attacked if they contain sufficient carbon and are sensitized. Oxidizing species, such as ferric salts, result in reduced general corrosion in some acids, but can cause accelerated pitting attack if chloride ions (Cl^-) are present.

Nitric Acid. As noted above, stainless steels have broad applicability in HNO_3 primarily because of their chromium content. Most AISI 300-series stainless steels exhibit good or excellent resistance in the annealed condition in concentrations from 0 to 65% up to the boiling point. Figure 11 illustrates the good resistance of type 304 stainless steel, particularly when compared with the lower-chromium type 410 stainless steel. More severe environments at elevated temperatures require alloys with higher chromium. In HNO_3 cooler-condensers, such stainless alloys as 7-Mo PLUS (UNS S32950) and 2RE10 (UNS S31008), are candidates for service.

In sulfuric acid, stainless steels can approach the borderline between activity and passivity. Conventional ferritic grades, such as type 430, have limited use in H_2SO_4, but the newer ferritic grades containing higher chromium and molybdenum (for example, 28% Cr and 4% Mo) with additions of at least 0.25% Ni have shown good resistance in boiling 10% H_2SO_4 (Ref 16), but corrode rapidly when acid concentration is increased.

The conventional austenitic grades exhibit good resistance in very dilute or highly concentrated H_2SO_4 at slightly elevated temperatures. Acid of intermediate concentration is more aggressive, and conventional grades have very limited utility. Resistance of several stainless steels in up to about 50% H_2SO_4 is shown in Fig. 12. Aeration or the addition of oxidizing species can significantly reduce the attack of stainless steels in H_2SO_4. This occurs because the more oxidizing environment is better able to maintain the chromium-rich passive oxide film.

Improved resistance to H_2SO_4 has been obtained by using austenitic grades containing high levels of nickel and copper, such as 20Cb-3 stainless steel. In addition to reducing general corrosion, the increased nickel provides resistance to SCC. Because of its resistance to these forms of corrosion, 20Cb-3 stainless steel has been used for valve springs in H_2SO_4 service.

Phosphoric Acid. Conventional straight-chromium stainless steels have very limited general corrosion resistance in phosphoric acid (H_3PO_4) and exhibit lower rates only in very dilute or more highly concentrated solutions. Conventional austenitic stainless steels provide useful general corrosion resistance over the full range of concentrations up to about 65 °C (150 °F); use at temperatures up to the boiling point is possible for acid concentrations up to about 40%.

In commercial applications, however, wet-process H_3PO_4 environments include impurities derived from the phosphate rock, such as chlorides, fluorides, and H_2SO_4. These three impurities accelerate corrosion, particularly pitting or crevice corrosion in the presence of the halogens. Higher-alloyed materials than the conventional austenitic stainless steels are required to resist wet-process H_3PO_4. Candidate materials include alloy 904L, alloy 28, 20Cb-3, 20Mo-4, and 20Mo-6 stainless steels.

Hydrochloric Acid. Stainless steels are generally not used for HCl service, except perhaps for very dilute solutions at room temperature. Stainless materials can be susceptible to accelerated general corrosion, SCC, and pitting in HCl environments.

Sulfurous Acid. Although sulfurous acid (H_2SO_3) is a reducing agent, several stainless steels have provided satisfactory service in H_2SO_3 environments. Conventional austenitic stainless steels have been used in sulfite digestors, and type 316, type 317, 20Cb-3, and cast Alloy Casting Institute alloys CF-8M and CN-7M stainless steels have seen service in wet sulfur dioxide (SO_2) and H_2SO_3 environments. Cast stainless steels are discussed in the article "Corrosion of Cast Steels" in this Volume. Service life is improved by eliminating crevices, including those from settling of suspended solids, or by using molybdenum-containing grades. In some environments, SCC is also a possibility.

Organic acids and compounds are generally less aggressive than mineral acids because they do not ionize as completely, but they can be corrosive to stainless steels, especially when impurities are present. The presence of oxidizing agents in the absence of chlorides can reduce corrosion rates.

Acetic Acid. Corrosion rates for several stainless steels in acetic acid are listed in Table 8. Resistance to pure acetic acid has been obtained by using type 316 and 316L stainless steels over all concentrations up to the boiling point. Type 304 stainless steel may be considered in all concentrations below about 90% at temperatures up to the boiling point. Impurities present in the manufacture of acetic acid, such as acetaldehyde, formic acid, chlorides, and propionic acid, are expected to increase the attack of stainless steels. Chlorides may cause pitting or SCC.

Formic acid is one of the more aggressive organic acids, and corrosion rates can be higher in the condensing vapor than in the liquid. Type 304 stainless steel has been used at moderate temperatures. However, type 316 stainless steel or higher alloys, such as 20Cb-3, are often preferred, and high-alloy ferritic stainless steels containing 26% Cr and 1% Mo or 29% Cr and 4% Mo also show some promise.

Other Organic Acids. The corrosivity of propionic and acrylic acids at a given temperature is generally similar to that of acetic acid. Impurities are important and may strongly affect the corrosion rate. In citric and tartaric acids, type 304 stainless steel has been used for moderate temperatures, and type 316 has been suggested for all concentrations up to the boiling point.

Organic Halides. Most dry organic halides do not attack stainless steels, but the presence of water allows halide acids to form and can cause pitting or SCC. Therefore, care should be exercised when using stainless steels in organic halides to ensure that water is excluded.

Other Organic Compounds. Type 304 stainless steel has generally been satisfactory in aldehydes, in cellulose acetate at lower temperatures, and in fatty acids up to about 150 °C (300 °F). At higher temperatures, these chemicals require type 316 or 317. Type 316 stainless steel is also used in amines, phthalic anhydride, tar, and urea service.

Stainless steels have been used in the plastic and synthetic fiber industries. Type 420 and 440C

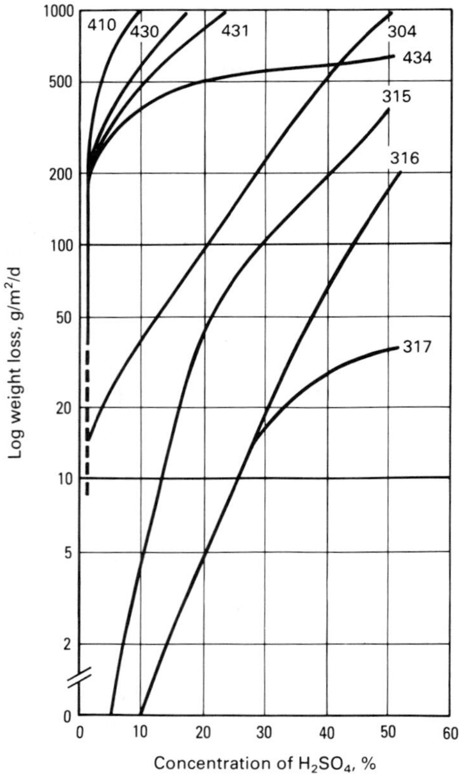

Fig. 12 Corrosion rates of various stainless steels in underaerated H_2SO_4 at 20 °C (70 °F). Source: Ref 15

Table 9 Generally accepted maximum service temperatures in air for AISI stainless steels

AISI type	Maximum service temperature: Intermittent service °C	°F	Continuous service °C	°F
Austenitic grades				
201	815	1500	845	1550
202	815	1500	845	1550
301	840	1545	900	1650
302	870	1600	925	1700
304	870	1600	925	1700
308	925	1700	980	1795
309	980	1795	1095	2000
310	1035	1895	1150	2100
316	870	1600	925	1700
317	870	1600	925	1700
321	870	1600	925	1700
330	1035	1895	1150	2100
347	870	1600	925	1700
Ferritic grades				
405	815	1500	705	1300
406	815	1500	1035	1895
430	870	1600	815	1500
442	1035	1895	980	1795
446	1175	2145	1095	2000
Martensitic grades				
410	815	1500	705	1300
416	760	1400	675	1250
420	735	1355	620	1150
440	815	1500	760	1400

Source: Ref 19

stainless steels have been used as plastic mold steels. More resistant materials, such as Custom 450, have been used for extruding polyvinyl chloride (PVC) pipe. Spinnerettes, pack parts, and metering pumps of Custom 450 and Custom 455 stainless steels have been used in the synthetic fiber industry to produce nylon, rayon, and polyesters.

Alkalies. All stainless steels resist general corrosion by all concentrations of sodium hydroxide (NaOH) up to about 65 °C (150 °F). Type 304 and 316 stainless steels exhibit low rates of general corrosion in boiling NaOH up to nearly 20% concentration. Stress-corrosion cracking of these grades can occur at about 100 °C (212 °F). Good resistance to general corrosion and SCC in 50% NaOH at 135 °C (275 °F) is provided by E-Brite and 7-Mo stainless steels (Ref 18). In ammonia (NH_3) and ammonium hydroxide (NH_4OH), stainless steels have shown good resistance at all concentrations up to the boiling point.

Salts. Stainless steels are highly resistant to most neutral or alkaline nonhalide salts. In some cases, type 316 is preferred for its resistance to pitting, but even the higher-molybdenum type 317 stainless steel is readily attacked by sodium sulfide (Na_2S) solutions.

Halogen salts are more corrosive to stainless steels because of the ability of the halide ions to penetrate the passive film and cause pitting. Pitting is promoted in aerated or mildly acidic oxidizing solutions. Chlorides are generally more aggressive than the other halides in their ability to cause pitting.

Gases. At lower temperatures, most austenitic stainless steels resist chlorine or fluorine gas if the gas is completely dry. The presence of even small amounts of moisture results in accelerated attack, especially pitting and possibly SCC.

Oxidation. At elevated temperatures, stainless steels resist oxidation primarily because of their chromium content. Increased nickel minimizes spalling when temperature cycling occurs. Table 9 lists generally accepted maximum safe service temperatures for wrought stainless steels. Maximum temperatures for intermittent service are lower for the austenitic stainless steels, but are higher for most of the martensitic and ferritic stainless steels listed.

Contamination of the air with water and CO_2 often increases corrosion at elevated temperatures. Increased attack can also occur because of sulfidation as a result of SO_2, H_2S, or sulfur vapor.

Carburization of stainless steels can occur in carbon monoxide (CO), methane (CH_4), and other hydrocarbons. Carburization can also occur when stainless steels contaminated with oil or grease are annealed without sufficient oxygen to burn off the carbon. This can occur during vacuum or inert gas annealing as well as open air annealing of oily parts with shapes that restrict air access. Chromium, silicon, and nickel are useful in combating carburization.

Nitriding can occur in dissociated NH_3 at high temperatures. Resistance to nitriding depends on alloy composition as well as NH_3 concentration, temperature, and pressure. Stainless steels are readily attacked in pure NH_3 at about 540 °C (1000 °F).

Liquid Metals. The 18-8 stainless steels are highly resistant to liquid sodium or sodium-potassium alloys. Mass transfer is not expected up to 540 °C (1000 °F) and remains at moderately low levels up to 870 °C (1600 °F). Accelerated attack of stainless steels in liquid sodium occurs with oxygen contamination, with a noticeable effect occurring at about 0.02% oxygen by weight (Ref 14).

Exposure to molten lead under dynamic conditions often results in mass transfer in common stainless alloy systems. Particularly severe corrosion can occur in strongly oxidizing conditions. Stainless steels are generally attacked by molten aluminum, zinc, antimony, bismuth, cadmium, and tin.

Corrosion in Various Applications

Every industry features a variety of applications encompassing a range of corrosion environments. This section will characterize the experience of each industry according to the corrosion problems most frequently encountered and will suggest appropriate grade selections.

Food and Beverage Industry. Stainless steels have been relied upon in these applications because of the lack of corrosion products that could contaminate the process environment and because of the superior cleanability of the stainless steels. The corrosion environment often involves moderately to highly concentrated chlorides on the process side, often mixed with significant concentrations of organic acids. The water side can range from steam heating to brine cooling. Purity and sanitation standards require excellent resistance to pitting and crevice corrosion.

Foods such as vegetables represent milder environments and can generally be handled by using type 304 stainless steel. Sauces and pickle liquors, however, are more aggressive and can pit even type 316 stainless steel. For improved pitting resistance, such alloys as 22Cr-13Ni-5Mn, 904L, 20Mo-4, 254SMO, AL-6XN, and MONIT stainless steels should be considered.

At elevated temperatures, materials must be selected for resistance to pitting and SCC in the presence of chlorides. Stress corrosion must be avoided in heat transfer applications, such as steam jacketing for cooking or processing vessels or in heat exchangers. Cracking may occur from

Table 10 Pitting of stainless steel spool test specimens in an FGD system

The slurry contained 7000 ppm dissolved Cl^-; test duration was 6 months, with 39 days on bypass.

		Maximum temperature		Maximum chloride concentration,	Maximum pit depth, mm (mils), and pit density								
Spool Location(a)	pH	°C	°F	ppm	Type 304	Type 316L	Type 317L	Type 317LM	Incoloy 825	JS700	JS777	904L	20Mo-6
Wet/dry line at inlet duct	1–2(b)	60–170	140–335	7000(b)	>1.24 (>49) Profuse	>0.91 (>36) Profuse	0.53 (21) Sparse	0.53 (21) Sparse	0.74 (29) Profuse	0.33 (13) Sparse	0.33 (13) Profuse	0.43 (17) Sparse	(c)
Quencher sump (submerged; 1.8-m, or 6 ft, level)	4.4	60	140	7000	>1.19 (>47) Sparse	>0.91 (>36) Sparse	0.28 (11)	0.1 (4) Single	<0.02 (<1)	nil	nil	nil	nil
Quencher sump (submerged; 3.4-m, or 11-ft, level)	4.4	60	140	7000	>1.2 (>48) Profuse	>0.9 (>36) Sparse	<0.03 (<1)	0.05 (2)	0.25 (10)	nil	nil	nil	nil
Quencher spray header, above slurry	4.4	60	140	100	>1.19 (>47) Profuse	0.58 (23) Profuse	0.61 (24) Profuse	0.46 (18) Profuse	0.66 (26) Profuse	0.33 (13) Sparse	0.61 (24) Profuse	0.25 (10) Sparse	0.15 (6) Sparse
Absorber, spray area	6.2	60	140	100	0.58 (23) Sparse	0.10 (4)	nil	nil	nil	nil	nil	nil	nil
Outlet duct	2–4(d) 1.5(e)	55 170	130(d) 335(e)	100(d) 82000(e)	>1.19 (>47) Profuse	>0.91 (>36) Profuse	0.58 (23) Profuse	0.58 (23) Profuse	0.48 (19) Profuse	0.18 (7) Single	0.51 (20) Profuse	0.53 (21) Profuse	0.36 (14) IG etch

(a) Slurry contained 7000 ppm dissolved Cl^-. Deposits in the quencher, inlet duct, absorber, and outlet ducting contained 3000–4000 ppm Cl^- and 800–1900 ppm F^-. (b) Present as halide gases. (c) Not tested. (d) During operation. (e) During bypass. Bypass condition gas stream contained SO_2, SO_3, HCl, HF and condensate. Source: Ref 25 the process or water side or may initiate outside the unit under chloride-containing insulation. Brewery applications of austenitic stainless steels have been generally successful except for a number of cases of SCC of high-temperature water lines. The use of ferritic or duplex stainless steels is an appropriate remedy for the SCC. More information on this subject is available in the article "Corrosion in the Brewery Industry" in this Volume.

Stainless steel equipment should be cleaned frequently to prolong its service life. The equipment should be flushed with freshwater, scrubbed with a nylon brush and detergent, then rinsed. On the other hand, consideration should be given to the effect of very aggressive cleaning procedures on the stainless steels, as in the chemical sterilization of commercial dishwashers. In some cases, it may be necessary to select a more highly alloyed stainless steel grade to deal with these brief exposures to highly aggressive environments.

Conventional AISI grades provide satisfactory service in many food and beverage applications. Type 304 stainless steel is widely used in the dairy industry, and type 316 finds application as piping and tubing in breweries. These grades, along with type 444 and Custom 450 stainless steels, have been used for chains to transfer food through processing equipment. Machined parts for beverage-dispensing equipment have been fabricated from type 304, 304L, 316, 316L, and 303Al MODIFIED, 302HQ-FM, and 303BV stainless steels. When the free-machining grades are used, it is important to passivate and rinse properly before service in order to optimize corrosion resistance.

Food-handling equipment should be designed without crevices in which food can become lodged. In more corrosive food products, extra-low-carbon stainless steels should be used when possible. Improved results have been obtained when equipment is finished with a 2B (general-purpose cold-rolled) finish rather than No. 4 (general-purpose polished) finish. Alternatively, an electropolished surface may be considered.

Pharmaceutical Industry. The production and handling of drugs and other medical applications require exceedingly high standards for preserving the sterility and purity of process streams. Process environments can include complex organic compounds, strong acids, chloride solutions comparable to seawater, and elevated processing temperatures. Higher-alloy grades, such as type 316 or higher, may be necessary instead of type 304 in order to prevent even superficial corrosion. Electropolishing may be desirable in order to reduce or prevent adherent deposits and the possibility of underdeposit corrosion. Superior cleanability and ease of inspection make stainless steel the preferred material.

The 18-8 stainless grades have been used for a wide variety of applications from pill punches to operating tables. However, care is required in selecting stainless steels for pharmaceutical applications because small amounts of contamination can be objectionable. For example, stainless steel has been used to process vitamin C, but copper must be eliminated because copper in aqueous solutions accelerates the decomposition of vitamin C. Also, stainless steel is not used to handle vitamin B_6 hydrochloride, even though corrosion rates may be low, because trace amounts of iron are objectionable (see the article "Stainless Steels in Corrosion Service" in Volume 3 of the 9th Edition of *Metals Handbook*).

The effects of temperature and chloride concentration must be considered. At ambient temperature, chloride pitting of 18Cr-8Ni stainless steel may occur, but SCC is unlikely. At about 65 °C (150 °F) or above, SCC of austenitic grades must be considered. Duplex alloys, such as 7-Mo PLUS, alloy 2205, and Ferralium 255 possess improved resistance to SCC in elevated-temperature chloride environments. Ferritic grades with lower nickel content, such as 18Cr-2Mo stainless steel, provide another means of avoiding chloride SCC. Additional information on the use of stainless steels in the pharmaceutical industry is available in the article "Corrosion in the Pharmaceutical Industry" in this Volume.

Stainless steels have also found application as orthopedic implants. Material is required that is capable of moderately high strength and resistance to wear and fretting corrosion, along with pitting and crevice attack. Vacuum-melted type 316 stainless steel has been used for temporary internal fixation devices, such as bone plates, screws, pins and suture wire. Higher purity improves electropolishing, and increased chromium (17 to 19%) improves corrosion resistance.

In permanent implants, such as artificial joints, very high strength and resistance to wear, fatigue, and corrosion are essential. Cobalt- or titanium-base alloys are used for these applications. More information on this subject is available in the article "Corrosion of Metallic Implants and Prosthetic Devices" in this Volume.

Oil and Gas Industry. Stainless steels were not frequently used in oil and gas production until the tapping of sour reservoirs (those containing hydrogen sulfide, H_2S) and the use of enhanced recovery systems in the mid-1970s. Sour environments can result in sulfide stress cracking (SSC) of susceptible materials. This phenomenon generally occurs at ambient or slightly elevated temperatures; it is difficult to establish an accurate temperature maximum for all alloys. Factors affecting SSC resistance include material variables, pH, H_2S concentration, total pressure, maximum tensile stress, temperature, and time. A description of some of these factors, along with information on materials that have demonstrated resistance to SSC, is available in Ref 20.

The resistance of stainless steels to SSC improves with reduced hardness. Conventional materials, such as type 410, 430, and 304 stainless steels, exhibit acceptable resistance at hardnesses below 22 HRC. Specialized grades, such as 22Cr-13Ni-5Mn, Custom 450, 20Mo-4, and some duplex stainless steels, have demonstrated resistance at higher hardnesses. Duplex alloy 2205 has been used for its strength and corrosion resistance as gathering lines for CO_2 gas before gas cleaning. Custom 450 and 22Cr-13Ni-5Mn stainless steels have seen service as valve parts. Other grades used in these environments include 254SMO and alloy 28, particularly for chloride and sulfide resistance, respectively.

In addition to the lower-temperature SSC, resistance to cracking in high-temperature environments is required in many oil field applications. Most stainless steels, including austenitic and duplex grades, are known to be susceptible to elevated-temperature cracking, probably by a mechanism similar to chloride SCC. Failure appears to be accelerated by H_2S and other sulfur compounds. Increased susceptibility is noted in material of higher yield strength, for example, because of the high residual tensile stresses imparted by some cold-working operations.

The above discussion is pertinent to the production phase of a well. However, drilling takes place in an environment of drilling mud, which usually consists of water, clay, weighting materials, and an inhibitor (frequently an oxygen scavenger). Chlorides are also present when drilling through salt formations. Austenitic stainless steels containing nitrogen have found use in this environment as nonmagnetic drill collars, as weight for the drill bit, and as housings for measurement-while-drilling (MWD) instruments. Nonmagnetic materials are required for operation of these instruments, which are used to locate the drill bit in directional-drilling operations. Nonstandard stainless steels used as drill collars or MWD components include type 316LN (high nitrogen), Staballoy, XM 19 H, Gammalloy (T286), Super Gammalloy (T287), AM20, Nipponel, 15-15LC, and SCF-19. More information on corrosion during petroleum production is available in the article "Corrosion in Petroleum Production Operations" in this Volume.

In refinery applications, the raw crude contains such impurities as sulfur, water, salts, organic acids, and organic nitrogen compounds. These and other corrosives and their products must be considered in providing stainless steels for the various refinery steps.

Raw crude is separated into materials from petroleum gas to various oils by fractional distillation. These materials are then treated to remove impurities, such as CO_2, NH_3, and H_2S, and to optimize product quality. Refinery applications of stainless steels often involve heat exchangers. Duplex and ferritic grades have been used in this application for their improved SCC resistance. Type 430 and type 444 stainless steel exchanger tubing has been used for resisting hydrogen, chlorides, and sulfur and nitrogen compounds in oil refinery streams. The article "Corrosion in Petroleum Refining and Petrochemical Operations" in this Volume contains detailed information on corrosion of materials in these applications.

Power Industry. Stainless steels are used in the power industry for generator components, feedwater heaters, boiler applications, heat exchangers, condenser tubing, flue gas desulfurization (FGD) systems, and nuclear power applications.

Generator blades and vanes have been fabricated of modified 12% Cr stainless steel, such as ASTM types 615 (UNS S41800) and 616 (UNS S42200). In some equipment, Custom 450 has replaced AISI type 410 and ASTM type 616 stainless steels.

Heat Exchangers. Stainless steels have been widely used in tubing for surface condensers and feedwater heaters. Both of these are shell and tube heat exchangers that condense steam from the turbine on the shell side. In these heat exchangers, the severity of the corrosion increases with higher temperatures and pressures. Stainless steels resist failure by erosion and do not suffer SCC in NH_3 (from decomposition of boiler feedwater additives), as do some nonferrous materials.

Stainless steels must be chosen to resist chloride pitting. The amount of chloride that can be tolerated is expected to be higher with higher pH and cleaner stainless steel surfaces, that is, the absence of deposits. For example, type 304 stainless steel may resist pitting in chloride levels of

1000 ppm or higher in the absence of fouling, crevices, or stagnant conditions. The presence of one or more of these conditions can allow chlorides to concentrate at the metal surface and initiate pits. Several high-performance stainless steels have been used to resist chloride pitting in brackish water or seawater. High-performance austenitic grades have been useful in feedwater heaters, although duplex stainless steels may also be considered because of their high strength. Ferritic stainless steels have proved to be economically competitive in exchangers and condensers. High-performance austenitic and ferritic grades have been satisfactory for seawater-cooled units. These grades include MONIT, AL-29-4C and Usinor 290 Mo, Sea-Cure, AL-6X, AL-6XN, and 254SMO stainless steels.

Compatibility of materials and good installation practice are required. Tubes of such materials as those listed above have been installed in tubesheets fabricated of alloy 904L, 20Mo-4, and 254SMO stainless steels. Crevice corrosion can occur when some tube materials are rolled into type 316 stainless steel tubesheets (Ref 21). Appropriate levels of cathodic protection have been identified (Ref 22).

Flue Gas Desulfurization. A wide variety of alloys have been used in scrubbers, which are located between the boiler and smokestack of fossil fuel power plants to treat effluent gases and to remove SO_2 and other pollutants. Typically, fly ash is removed, and the gas travels through an inlet gas duct, followed by the quencher section. Next, SO_2 is removed in the absorber section, most often using either a lime or limestone system. A mist eliminator is employed to remove suspended droplets, and the gas proceeds to the treated-gas duct, reheater section, and the stack.

Two important items for consideration in selecting stainless steels for resistance to pitting in scrubber environments are pH and chloride level. Stainless steels are more resistant to higher pH and lower chloride levels, as shown in Fig. 13 for type 316L stainless steel. Environments that cause pitting or crevice attack of type 316 stainless steel can be handled by using higher-alloy materials, for example, those with increased molybdenum and chromium.

Some of the materials being considered and specified for varying chloride levels are given in Ref 24. Other materials can also provide good resistance, as evidenced by the results given in Table 10 for samples exposed to several scrubber environments. The maximum depth of localized corrosion and pit density are given for the stainless steels tested. Exposure at the quencher spray header (above slurry) was more severe than expected, probably because of wet-dry concentration effects. Severe attack also occurred in the outlet duct. Samples in this area were exposed to high chlorides, high temperatures, and low pH during the 39 days on bypass operation. More information on corrosion in fossil fuel power generation is available in the article "Corrosion in Fossil Fuel Power Plants" in this Volume; corrosion in FGD systems is also discussed in the article "Corrosion of Emission-Control Equipment" in this Volume.

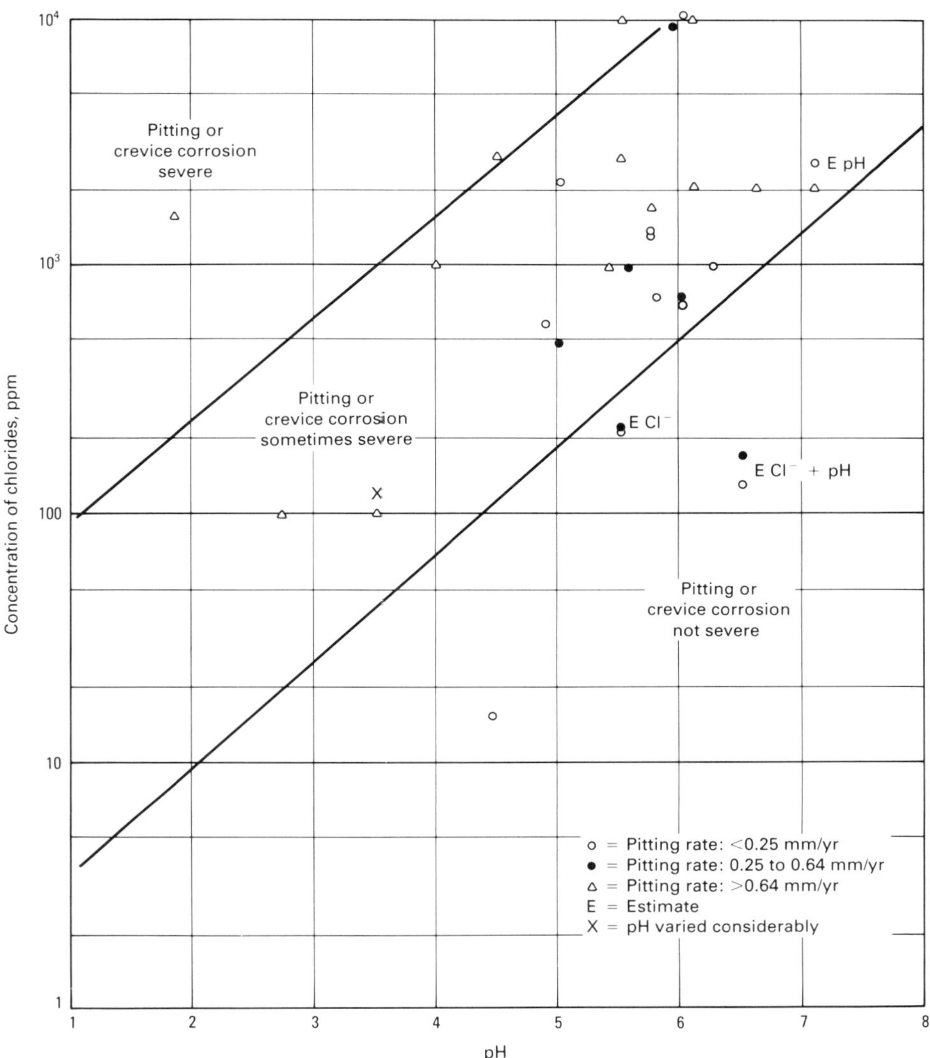

Fig. 13 Pitting of type 316L stainless steel in FGD scrubber environments. Solid lines indicate zones of differing severity of corrosion; because the zones are not clearly defined, the lines cannot be precisely drawn. Source: Ref 23

Nuclear Power Applications. Type 304 stainless steel piping has been used in boiling-water nuclear reactor power plants. The operating temperatures of these reactors are about 290 °C (550 °F), and a wide range of conditions can be present during startup, operation, and shutdown. Because these pipes are joined by welding, there is a possibility of sensitization. This can result in intergranular SCC in chloride-free high-temperature water that contains small amounts of oxygen, for example, 0.2 to 8 ppm. Nondestructive electrochemical tests have been used to evaluate weldments for this service (Ref 26).

Type 304 stainless steel with additions of boron (about 1%) has been used to construct spent-fuel storage units, dry storage casks, and transportation casks. The high boron level provides neutron-absorbing properties. More information on nuclear applications is available in the article "Corrosion in the Nuclear Power Industry" in this Volume.

Pulp and Paper Industry. In the kraft process, paper is produced by digesting wood chips with a mixture of Na_2S and NaOH (white liquor). The product is transferred to the brown stock washers to remove the liquor (black liquor) from the brown pulp. After screening, the pulp may go directly to the paper mill to produce unbleached paper or may be directed first to the bleach plant to produce white paper.

The digester vapors are condensed, and the condensate is pumped to the brown stock washers. The black liquor from these washers is concentrated and burned with sodium sulfate (Na_2SO_4) to recover sodium carbonate (Na_2CO_3) and Na_2S. After dissolution in water, this green liquor is treated with calcium hydroxide ($Ca(OH)_2$) to produce NaOH to replenish the white liquor. Pulp bleaching involves treating with various chemicals, including chlorine, chlorine dioxide (ClO_2), sodium hypochlorite (NaClO), calcium hypochlorite ($Ca(ClO)_2$), peroxide, caustic soda, quicklime, or oxygen.

The sulfite process uses a liquor in the digester that is different from that used in the kraft process. This liquor contains free SO_2 dissolved in water, along with SO_2 as a bisulfite. The compositions of the specific liquors differ, and the pH can range from 1 for an acid process to 10 for alkaline cooking. Sulfur dioxide for the cooking liquor is produced by burning elemental sulfur, cooling rapidly, absorbing the SO_2 in a weak alkaline solution, and fortifying the raw acid.

Table 11 ASTM standard tests for susceptibility to intergranular corrosion in stainless alloys

ASTM test method	Test medium and duration	Alloys	Phases detected
A 262, practice A	Oxalic acid etch; etch test	AISI types 304, 304L, 316, 316L, 317L, 321, 347 casting alloys	Chromium carbide
A 262, practice B	$Fe_2(SO_4)_3$-H_2SO_4; 120 h	Same as above	Chromium carbide, σ phase(a)
A 262, practice C	HNO_3 (Huey test); 240 h	Same as above	Chromium carbide, σ phase(b)
A 262, practice D	HNO_3-HF; 4 h	AISI types 316, 316L, 317, 317L	Chromium carbide
A 262, practice E	$CuSO_4$-16% H_2SO_4, with copper contact; 24 h	Austenitic stainless steels	Chromium carbide
A 708 (formerly A 393)	$CuSO_4$-16% H_2SO_4, without copper contact; 72 h	Austenitic stainless steels	Chromium carbide
G 28	$Fe_2(SO_4)_3$-H_2SO_4; 24-120 h	Hastelloy alloys C-276 and G; 20Cb-3; Inconel alloys 600, 625, 800, and 825	Carbides and/or intermetallic phases(c)
A 763, practice X	$Fe_2(SO_4)_3$-H_2SO_4; 24-120 h	AISI types 403 and 446; E-Brite, 29-4, 29-4-2	Chromium carbide and nitride intermetallic phases(d)
A 763, practice Y	$CuSO_4$-50% H_2SO_4; 96-120 h	AISI types 446, XM27, XM33, 29-4, 29-4-2	Chromium carbide and nitride
A 763, practice Z	$CuSO_4$-16% H_2SO_4; 24 h	AISI types 430, 434, 436, 439, 444	Chromium carbide and nitride

(a) There is some effect of σ phase in type 321 stainless steel. (b) Detects σ phase in AISI types 316, 316L, 317, 317L, and 321. (c) Carbides and perhaps other phases detected, depending on the alloy system. (d) Detects χ and σ phases, which do not cause intergranular attack in unstabilized iron-chromium-molybdenum alloys. Source: Ref 30

Various alloys are selected for the wide range of corrosion conditions encountered in pulp and paper mills. Paper mill headboxes are typically fabricated from type 316L stainless steel plate with superior surface finish and are sometimes electropolished to prevent scaling, which may affect pulp flow. The blades used to remove paper from the drums have been fabricated from type 410 and 420 stainless steels and from cold-reduced 22Cr-13Ni-5Mn stainless steel.

Evaporators and reheaters must deal with corrosive liquors and must minimize scaling to provide optimum heat transfer. Type 304 stainless steel ferrite-free welded tubing has been used in kraft black liquor evaporators. Cleaning is often performed with HCl, which attacks ferrite. In the sulfite process, type 316 (≥2.75% Mo) and type 317 stainless steels have been used in black liquor evaporators. Digester liquor heaters in the kraft and sulfite processes have used 7-Mo stainless for resistance to caustic or chloride SCC.

Bleach plants have used type 316 and 317 stainless steels and are upgrading to austenitic grades containing 4.5 and 6% Mo in problem locations. Tightening of environmental regulations has generally increased temperature, chloride level, and acidity in the plant, and this requires grades of stainless steel that are more highly alloyed than those used in the past. Tall oil units have shifted from type 316 and 317 stainless steels to such candidate alloys as 904L or 20Mo-4 stainless steels and most recently to 254SMO and 20Mo-6 stainless steels.

Tests including higher-alloyed materials have been coordinated by the Metals Subcommittee of the TAPPI Corrosion and Materials Engineering Committee. Racks of test samples, which included crevices at polytetrafluoroethylene (PTFE) spacers, were submerged in the vat below the washer in the C (chlorination), D (chlorine dioxide), and H (hypochlorite) stages of several paper mills. The sum of the maximum attack depth on all samples for each alloy—at crevices and remote from crevices—is shown in Fig. 14. It should be noted that the vertical axes are different in Fig. 14(a), (b), and (c). Additional information on corrosion in this industry is available in the article "Corrosion in the Pulp and Paper Industry" in this Volume.

Transportation Industry. Stainless steels are used in a wide range of components in transportation that are both functional and decorative. Bright automobile parts, such as trim, fasteners, wheel covers, mirror mounts, and windshield wiper arms, have generally been fabricated from 17Cr or 18Cr-8Ni stainless steel or similar grades. Example alloys include type 430, 434, 304, and 305 stainless steels. Type 302HQ-FM remains a candidate for such applications as wheel nuts, and Custom 455 stainless has been used as wheel lock nuts. Use of type 301 stainless steel for wheel covers has diminished with the weight reduction programs of the automotive industry.

Stainless steels also serve many nondecorative functions in automotive design. Small-diameter shafts of type 416 and, occasionally, type 303 stainless steels have been used in connection with power equipment, such as windows, door locks, and antennas. Solenoid grades, such as type 430FR stainless steels, have also found application. Type 409 stainless steel has been used for mufflers and catalytic converters for many years, but it is now being employed throughout the exhaust system. The article "Corrosion in the Automotive Industry" in this Volume contains detailed information on corrosion in the automotive environment.

In railroad cars, external and structural stainless steels provide durability, low-cost maintenance, and superior safety through crashworthiness. The fire resistance of stainless steel is a significant safety advantage. Modified type 409 stainless steel is used as a structural component in buses. Types 430 and 304 are used for exposed functional parts on buses. Type 304 stainless steel has provided economical performance in truck trailers. For tank trucks, type 304 has been the most frequently used stainless steel, but type 316 and higher-alloyed grades have been used where appropriate to carry more corrosive chemicals safely over the highways.

Stainless steels are used for seagoing chemical tankers, with types 304, 316, 317, and alloy 2205 being selected according to the corrosivity of the cargoes being carried. Conscientious adherence to cleaning procedures between cargo change-

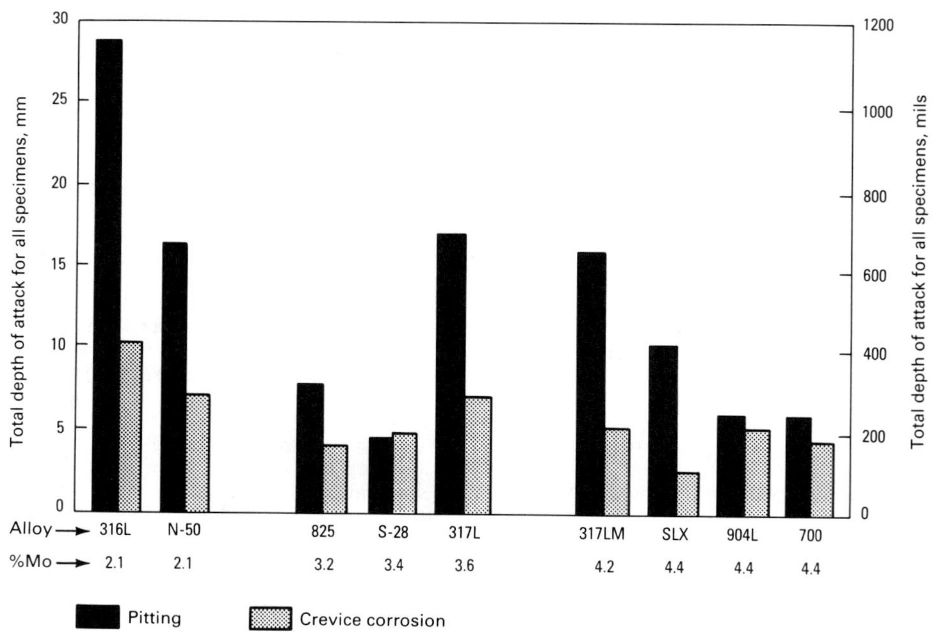

Fig. 14(a) Resistance of austenitic stainless steels containing 2.1 to 4.4% Mo to localized corrosion in paper mill bleach plant environment. Total depth of attack has been divided by 4 because there were four crevice sites per specimen. See also Fig. 14(b) and 14(c). Source: Ref 27

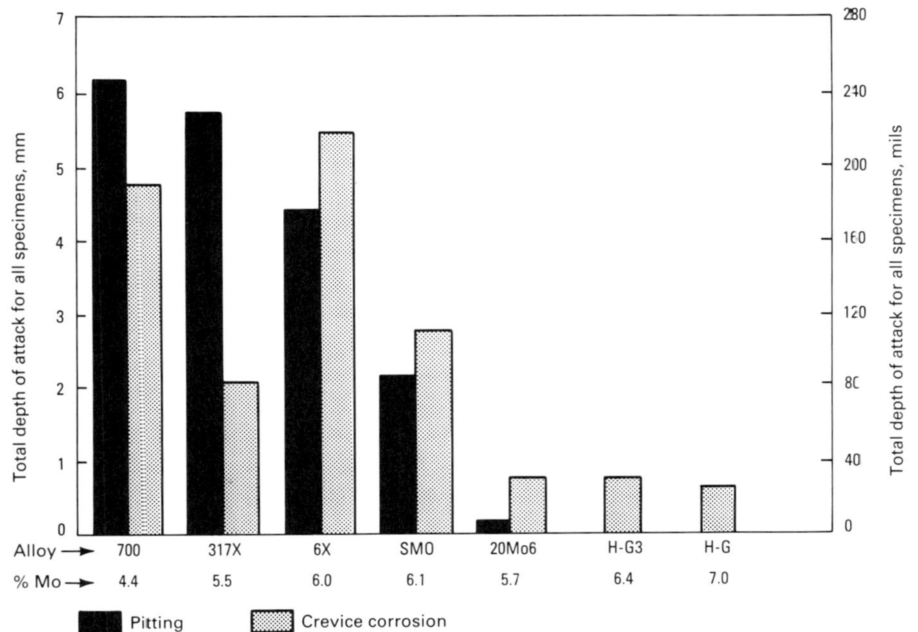

Fig. 14(b) Resistance of austenitic stainless steels containing 4.4 to 7.0% Mo to localized corrosion in paper mill bleach plant environment. Total depth of attack has been divided by 4 because there were four crevice sites per specimen. See also Fig. 14(a) and 14(c). Source: Ref 27

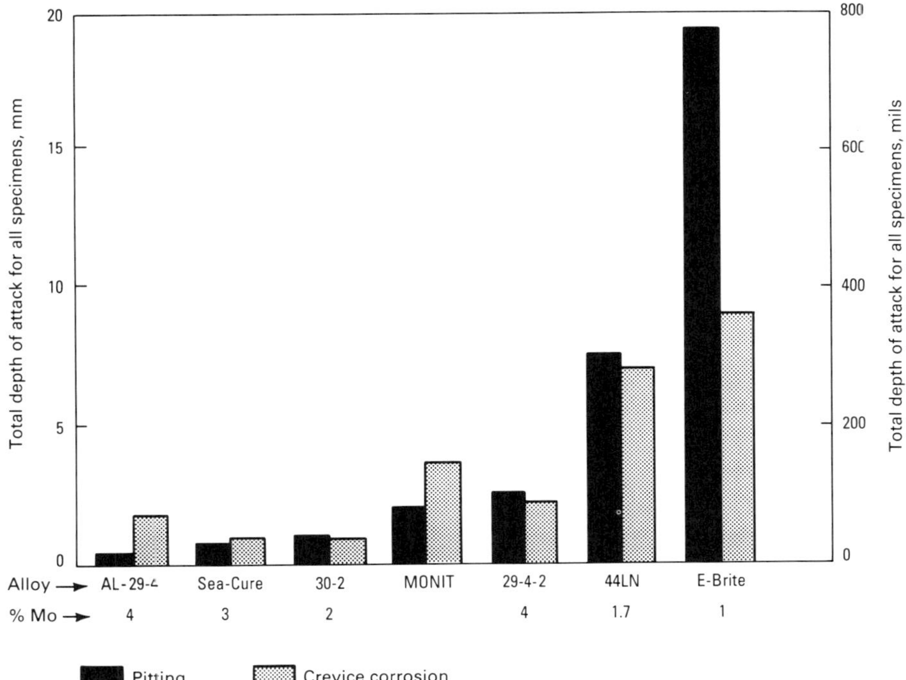

Fig. 14(c) Resistance of ferritic and duplex stainless steels to localized corrosion in paper mill bleach plant environment. Total depth of attack has been divided by 4 because there were four crevice sites per specimen. See also Fig. 14(a) and 14(b). Source: Ref 27

overs has allowed these grades to give many years of service with a great variety of corrosive cargoes.

In aerospace, quench-hardenable and precipitation-hardenable stainless steels have been used in varying applications. Heat treatments are chosen to optimize fracture toughness and resistance to SCC. Stainless steel grades 15-5PH and PH13-8Mo have been used in structural parts, and A286 and PH3-8Mo stainless steels have served as fasteners. Parts in cooler sections of the engine have been fabricated from type 410 or A286 stainless steel. Custom 455, 17-4PH, 17-7PH, and 15-5PH stainless steels have been used in the space shuttle program (see the articles "Corrosion in the Aircraft Industry" and "Corrosion in the Aerospace Industry" in this Volume).

Architectural Applications. Typically, type 430 or 304 has been used in architectural applications. In bold exposure, these grades are generally satisfactory; however, in marine and industrially contaminated atmospheres, type 316 is often suggested and has performed well (see the article "Corrosion in Structures" in this Volume).

In all applications, but particularly in these cases where appearance is important, it is essential that any chemical cleaning solutions be thoroughly rinsed from the metal.

Corrosion Testing

The physical and financial risks involved in selecting stainless steels for particular applications can be reduced through consideration of corrosion tests. However, care must be taken when selecting a corrosion test. The test must relate to the type of corrosion possible in the application. The steel should be tested in the condition in which it will be applied. The test conditions should be representative of the operating conditions and all reasonably anticipated excursions of operating conditions.

Corrosion tests vary in their degree of simulation of operation in terms of the design of the specimen and the selection of medium and test conditions. Standard tests use specimens of defined nature and geometry exposed in precisely defined media and conditions. Standard tests can confirm that a particular lot of steel conforms to the level of performance expected of a standard grade. Standard tests can also rank the performance of standard and proprietary grades. The relevance of test results to performance in particular applications increases as the specimen is made to resemble more closely the final fabricated structure—for example, bent, welded, stressed, or creviced. Relevance also increases as the test medium and conditions more closely approach the most severe operating conditions. However, many types of failures occur only after extended exposures to operating cycles. Therefore, there is often an effort to accelerate testing by increasing the severity of one or more environmental factors, such as temperature, concentration, aeration, and pH. Care must be taken that the altered conditions do not give spurious results. For example, an excessive temperature may either introduce a new failure mode or prevent a failure mode relevant to the actual application. The effects of minor constituents or impurities on corrosion are of special concern in simulated testing.

Pitting and crevice corrosion are readily tested in the laboratory by using small coupons and controlled-temperature conditions. A procedure for such tests using 6% $FeCl_3$ (10% $FeCl_3 \cdot 6H_2O$) is described in ASTM G 48 (Ref 28). This procedure is performed in 3 days. The coupon may be evaluated in terms of weight loss, pit depth, pit density, and appearance. Several suggestions for methods of pitting evaluation are given in ASTM G 46 (Ref 29). Reference 28 also describes the construction of a crevice corrosion coupon (Fig. 15). It is possible to determine a temperature below which crevice corrosion is not initiated for a particular material and test envi-

Fig. 15 Assembled crevice corrosion test specimen. Source: Ref 30

Fig. 16 Multiple-crevice cylinders for use in crevice corrosion testing. Source: Ref 30

ronment. This critical crevice temperature (CCT) provides a useful ranking of stainless steels. For the CCT to be directly applicable in design, it is necessary to determine that the test medium and conditions relate to the most severe conditions to be encountered in service.

Figure 16 shows one of several frequently used specimens with a multiple crevice assembly. The presence of many separate crevices helps to deal with the statistical nature of corrosion initiation. The severity of the crevices can be regulated by means of a standard crevice design and the use of a selected torque in its application.

Laboratory media do not necessarily have the same response of corrosivity as a function of temperature as do engineering environments. For example, the ASTM G 48 solution is thought to be roughly comparable to seawater at ambient temperatures. However, the corrosivity of $FeCl_3$ increases steadily with temperature. The response of seawater to increasing temperature is quite complex, relating to such factors as concentration of oxygen and biological activity. Also, the various families of stainless steels will be internally consistent, but will differ from one another in response to a particular medium.

Pitting and crevice corrosion may also be evaluated by electrochemical techniques. When immersed in a particular medium a metal coupon will assume a potential that can be measured relative to a standard reference electrode. It is then possible to impress a potential on the coupon and observe the corrosion as measured by the resulting current. Various techniques of scanning the potential range provide extremely useful data on corrosion resistance. Figure 17 demonstrates a simplified view of how these tests may indicate the corrosion resistance for various materials and media.

The nature of intergranular sensitization has been discussed earlier in this article. There are many corrosion tests for detecting susceptibility to preferential attack at the grain boundaries. The appropriate media and test conditions vary widely for the different families of stainless steels. Table 11 summarizes the ASTM tests for intergranular sensitization. Figure 18 shows that electrochemical techniques may also be used, as in the electrochemical potentiostatic reactivation (EPR) test.

Stress-corrosion cracking covers all types of corrosion involving the combined action of tensile stress and corrodent. Important variables include the level of stress, the presence of oxygen, the concentration of corrodent, temperature, and the conditions of heat transfer. It is important to recognize the type of corrodent likely to produce cracking in a particular family of steel. For example, austenitic stainless steels are susceptible to chloride SCC (Table 12). Martensitic and ferritic grades are susceptible to cracking related to hydrogen embrittlement.

It is important to realize that corrosion tests are designed to single out one particular corrosion mechanism. Therefore, determining the suitability of a stainless steel for a particular application will usually require consideration of more than one type of test. No single chemical or electrochemical test has been shown to be an all-purpose measure of corrosion resistance. More information on corrosion testing is available in the Section "Corrosion Testing and Evaluation" in this Volume.

Table 12 Stress-corrosion cracking resistance of stainless steels

Grade	Stress-corrosion cracking test(a): Boiling 42% $MgCl_2$	Wick test	Boiling 25% NaCl
AISI type 304	F(b)	F	F
AISI type 316	F	F	F
AISI type 317	F	[P(c) or F](d)	(P or F)
Type 317LM	F	(P or F)	(P or F)
Alloy 904L	F	(P or F)	(P or F)
AL-6XN	F	P	P
254SMO	F	P	P
20Mo-6	F	P	P
AISI type 409	P	P	P
Type 439	P	P	P
AISI type 444	P	P	P
E-Brite	P	P	P
Sea-Cure	F	P	P
MONIT	F	P	P
AL 29-4	P	P	P
AL 29-4-2	F	P	P
AL 29-4C	P	P	P
3RE60	F	NT	NT
2205	F	NT	(P or F)(e)
Ferralium	F	NT	(P or F)(e)

(a) U-bend tests, stressed beyond yielding. (b) Fails, cracking observed. (c) Passes, no cracking observed. (d) Susceptibility of grade to SCC determined by variation of composition within specified range. (e) Susceptibility of grade to SCC determined by variation of thermal history.
Source: Ref 6

REFERENCES

1. "Standard Practices for Detecting Susceptibility to Intergranular Corrosion Attack in Austenitic Stainless Steels," A 262, *Annual Book of ASTM Standards*, American Society for Testing and Materials
2. "Standard Practices for Detecting Susceptibility to Intergranular Attack in Ferritic Stainless Steels," A 763, *Annual Book of ASTM Standards*, American Society for Testing and Materials

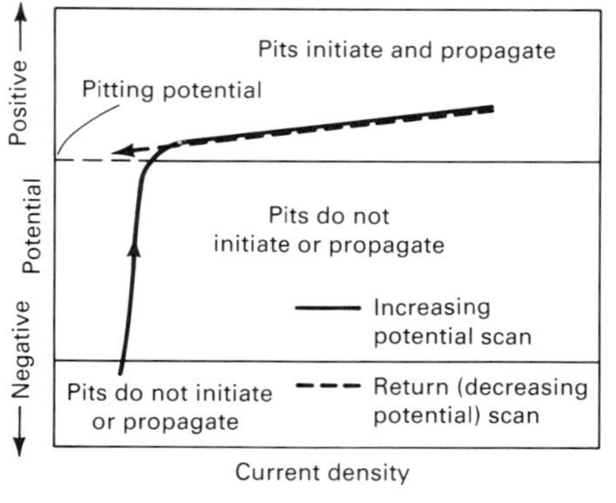

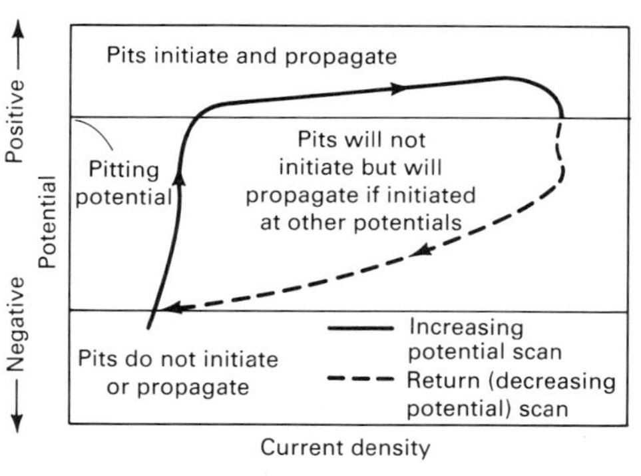

Fig. 17 Schematics showing how electrochemical tests can indicate the susceptibility to pitting of a material in a given environment. (a) Specimen has good resistance to pitting. (b) Specimen has poor resistance to pitting. In both cases, attack occurs at the highest potentials. Source: Ref 30

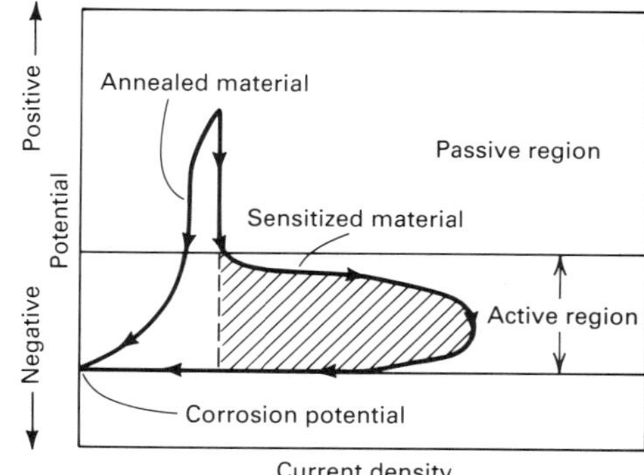

Fig. 18 Schematic showing the use of the EPR test to evaluate sensitization. The specimen is first polarized up to a passive potential at which the metal resists corrosion. Potential is then swept back through the active region, where corrosion may occur. Source: Ref 30

3. "Standard Recommended Practice for Cleaning and Descaling Stainless Steel Parts, Equipment, and Systems," A 380, *Annual Book of ASTM Standards*, American Society for Testing and Materials
4. T. DeBold, Passivating Stainless Steel Parts, *Mach. Tool Blue Book*, Nov 1986
5. "Passivation Treatments for Corrosion-Resisting Steels," Federal Specification QQ-P-35B, United States Government Printing Office, April 1973
6. R.M. Davison *et al.*, A Review of Worldwide Developments in Stainless Steels in Specialty Steels and Hard Materials, Pergamon Press, 1983, p 67-85
7. *Corrosion Resistance of the Austenitic Chromium-Nickel Stainless Steels in Atmospheric Environments*, The International Nickel Company, Inc., 1963
8. K.L. Money and W.W. Kirk, Stress Corrosion Cracking Behavior of Wrought Fe-Cr-Ni Alloys in Marine Atmosphere, *Mater. Perform.*, Vol 17, July 1978, p 28-36
9. M. Henthorne, T.A. DeBold, and R.J. Yinger, "Custom 450—A New High Strength Stainless Steel," Paper 53, presented at Corrosion/72, National Association of Corrosion Engineers, 1972
10. *The Role of Stainless Steels in Desalination*, American Iron and Steel Institute, 1974
11. M.A. Streicher, Analysis of Crevice Corrosion Data From Two Sea Water Exposure Tests on Stainless Alloys, *Mater. Perform.*, Vol 22, May 1983, p 37-50
12. A.H. Tuthill and C.M. Schillmoller, *Guidelines for Selection of Marine Materials*, The International Nickel Company, Inc., 1971
13. R.M. Kain, "Crevice Corrosion Resistance of Austenitic Stainless Steels in Ambient and Elevated Temperature Seawater," Paper 230, presented at Corrosion/79, National Association of Corrosion Engineers, 1979
14. F.L. LaQue and H.R. Copson, Ed., *Corrosion Resistance of Metals and Alloys*, Reinhold, 1963, p 375-445
15. J.E. Truman, in *Corrosion: Metal/Environment Reactions*, Vol 1, L.L. Shreir, Ed., Newness-Butterworths, 1976, p 352
16. M.A. Streicher, Development of Pitting Resistant Fe-Cr-Mo Alloys, *Corrosion*, Vol 30, 1974, p 77-91
17. H.O. Teeple, Corrosion by Some Organic Acids and Related Compounds, *Corrosion*, Vol 8, Jan 1952, p 14-28
18. T.A. DeBold, J.W. Martin, and J.C. Tverberg, Duplex Stainless Offers Strength and Corrosion Resistance, in *Duplex Stainless Steels*, R.A. Lula, Ed., American Society for Metals, 1983, p 169-189
19. L.A. Morris, in *Handbook of Stainless Steels*, D. Peckner and I.M. Bernstein, Ed., McGraw-Hill, 1977, p 17-1
20. "Material Requirements: Sulfide Stress Cracking Resistant Metallic Materials for Oil Field Equipment," MR-01-84, National Association of Corrosion Engineers
21. J.R. Kearns, M.J. Johnson, and J.F. Grubb, "Accelerated Corrosion in Dissimilar Metal Crevices," Paper 228, presented at Corrosion/86, National Association of Corrosion Engineers, 1986
22. L.S. Redmerski, J.J. Eckenrod, and K.E. Pinnow, "Cathodic Protection of Seawater-Cooled Power Plant Condensers Operating With High Performance Ferritic Stainless Steel Tubing," Paper 208, presented at Corrosion/85, National Association of Corrosion Engineers, 1985
23. E.C. Hoxie and G.W. Tuffnell, A Summary of INCO Corrosion Tests in Power Plant Flue Gas Scrubbing Processes, in *Resolving Corrosion Problems in Air Pollution Control Equipment*, National Association of Corrosion Engineers, 1976
24. *Effective Use of Stainless Steel in FGD Scrubber Systems*, American Iron and Steel Institute, 1978
25. G.T. Paul and R.W. Ross, Jr., "Corrosion Performance in FGD Systems at Laramie River and Dallman Stations," Paper 194, presented at Corrosion/83, National Association of Corrosion Engineers, 1983
26. A.P. Majidi and M.A. Streicher, "Four Non-Destructive Electrochemical Tests for Detecting Sensitization in Type 304 and 304L Stainless Steels," Paper 62, presented at Corrosion/85, National Association of Corrosion Engineers, 1985
27. A.H. Tuthill, Resistance of Highly Alloyed Materials and Titanium to Localized Corrosion in Bleach Plant Environments, *Mater. Perform.*, Vol 24, Sept 1985, p 43-49
28. "Standard Test Methods for Pitting and Crevice Corrosion Resistance of Stainless Steels and Related Alloys by the Use of Ferric Chloride Solution," G 48, *Annual Book of ASTM Standards*, American Society for Testing and Materials
29. "Standard Recommended Practice for Examination and Evaluation of Pitting Corrosion," G 46, *Annual Book of ASTM Standards*, American Society for Testing and Materials
30. T.A. DeBold, Which Corrosion Test for Stainless Steels, *Mater. Eng.*, Vol 2 (No. 1), July 1980

Corrosion of Cast Irons

Donald R. Stickle, The Duriron Company, Inc.

CAST IRON is a generic term that identifies a large family of ferrous alloys. Cast irons are primarily alloys of iron that contain more than 2% carbon and 1% or more silicon. Low raw material costs and relative ease of manufacture make cast irons the least expensive of the engineering metals. Cast irons can be cast into intricate shapes because of their excellent fluidity and relatively low melting points and can be alloyed for improvement of corrosion resistance and strength. With proper alloying, the corrosion resistance of cast irons can equal or exceed that of stainless steels and nickel-base alloys.

Because of the excellent properties obtainable with these low-cost engineering materials, cast irons find wide application in environments that demand good corrosion resistance. Services in which cast irons are used for their excellent corrosion resistance include water, soils, acids, alkalies, saline solutions, organic compounds, sulfur compounds, and liquid metals. In some services, alloyed cast irons offer the only economical alternative for constructing equipment.

Basic Metallurgy of Cast Irons

The metallurgy of cast irons is similar to that of steels except that sufficient silicon is present to necessitate use of the iron-silicon-carbon ternary phase diagram rather than the simple iron-carbon binary diagram. Figure 1 shows a section of the iron-iron carbide-silicon ternary diagram at 2% Si. The eutectic and eutectoid points in the iron-silicon-carbon diagram are both affected by the introduction of silicon into the system. In the 1 to 3% Si levels normally found in cast irons, eutectic carbon levels are related to silicon levels as follows:

$$\%C + \tfrac{1}{3}(\%Si) = 4.3 \qquad \text{(Eq 1)}$$

where %C is the eutectic carbon level, and %Si is the silicon level in the cast iron. The metallurgy of cast iron can occur in the metastable iron-iron carbide system, the stable iron-graphite system, or both. This causes structures of cast irons to be more complex than those of steel and more susceptible to processing conditions.

An appreciable portion of carbon in cast irons separates during solidification and appears as a separate constituent in the microstructure. The level of silicon in the cast iron has a strong effect on the manner in which carbon segregates in the microstructure. Higher silicon levels favor the formation of graphite, but lower silicon levels favor the formation of iron carbides. The form and shape in which the carbon occurs determine the type of cast iron (Table 1).

The structure of the metal matrix around the carbon-rich constituent establishes the class of iron within each type of iron. Four basic matrix structures occur in cast iron: ferrite, pearlite, bainite, and martensite.

Ferrite is generally a soft constituent, but it can be solid solution hardened by silicon. When silicon levels are below 3%, the ferrite matrix is readily machined, but exhibits poor wear resistance. Above 14% Si, the ferritic matrix becomes very hard and wear resistant, but is essentially nonmachinable. The low carbon content of the ferrite phase makes hardening difficult. Ferrite can be observed in cast irons upon solidification, but is generally present as the result of special annealing heat treatments. High silicon levels promote the formation of ferritic matrices in the as-cast condition.

Pearlite consists of alternate layers of ferrite and iron carbide (Fe_3C, or cementite). It is very strong and tough. The hardness, strength, machinability, and wear resistance of pearlitic matrices vary with the fineness of its laminations. The carbon content of pearlite is variable and depends on the analysis of the iron and its cooling rate.

Bainite is an acicular structure in cast irons that can be obtained by heat treating, alloying, or combinations of these. Bainitic structures provide very high strength at a machinable hardness.

Martensitic structures also occur in cast irons. These structures can be obtained by alloying, heat treating, or a combination of these practices. Martensitic microstructures are the hardest, most wear-resistant structures obtainable in cast irons. Molybdenum, nickel, manganese, and chromium can be used to produce martensitic or bainitic structures. Silicon has a negative effect on martensite formation, because it promotes the formation of pearlite or ferrite.

Influence of Alloying

Alloying elements can play a dominant role in the susceptibility of cast irons to corrosion attack. The alloying elements generally used to enhance the corrosion resistance of cast irons include silicon, nickel, chromium, copper, and molybdenum. Other alloying elements, such as vanadium and titanium, are sometimes used, but not to the extent of the five elements mentioned.

Silicon is the most important alloying element used to improve the corrosion resistance of cast irons. Silicon is generally not considered an alloying element in cast irons until levels exceed 3%. Silicon levels between 3 and 14% offer some increase in corrosion resistance to the alloy, but above about 14% Si, the corrosion resistance of the cast iron increases dramatically. Silicon levels up to 17% have been used to enhance the corrosion resistance of the alloy further, but silicon levels over 16% make the alloy extremely brittle and difficult to manufacture. Even at 14% Si, the strength and ductility of the material is low, and special design and manufacturing parameters are required to produce and use these alloys.

Alloying with silicon promotes the formation of strongly adherent surface films in cast irons. Considerable time may be required to establish these films fully on the castings. Consequently, in some services, corrosion rates may be relatively high for the first few hours or even days of exposure, then may decline to extremely low steady-state rates for the rest of the time the parts

Table 1 Summary of cast iron classification based on carbon form and shape

Type of cast iron	Carbon form and shape
White cast iron	Iron carbide compound
Malleable cast iron	Irregularly shaped nodules of graphite
Gray cast iron	Graphite flakes
Ductile cast iron	Spherical graphite nodules
Compacted graphite cast iron	Short, fat interconnected flakes (intermediate between ductile and gray cast iron)

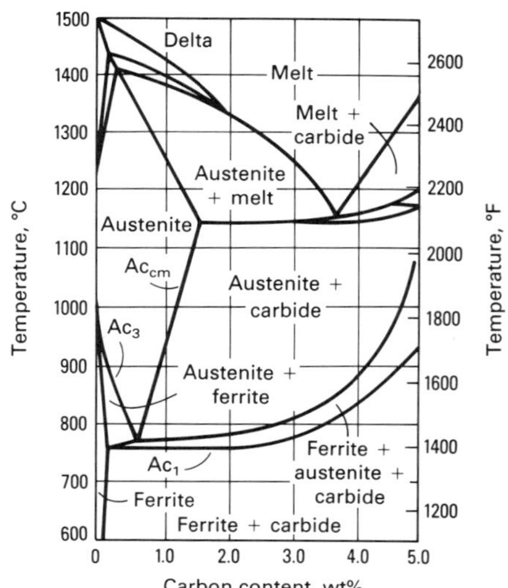

Fig. 1 Section of the iron-iron carbide-silicon ternary phase diagram at 2% Si

are exposed to the corrosive environment (Fig. 2).

Nickel is used to enhance the corrosion resistance of cast irons in a number of applications. Nickel increases corrosion resistance by the formation of protective oxide films on the surfaces of the castings. Up to 4% Ni is added in combination with chromium to improve both strength and corrosion resistance in cast iron alloys. The enhanced hardness and corrosion resistance obtained is particularly important for improving the erosion-corrosion resistance of the material. Nickel additions enhance the corrosion resistance of cast irons to reducing acids and alkalies. Nickel additions of 12% or greater are necessary to optimize the corrosion resistance of cast irons.

Nickel is not as common an alloying addition as either silicon or chromium for enhancing the corrosion resistance in cast irons. It is much more important as a strengthening and hardening addition.

Chromium is frequently added alone and in combination with nickel and/or silicon to increase the corrosion resistance of cast irons. As with nickel, small additions of chromium are used to refine graphite and matrix microstructures. These refinements enhance the corrosion resistance of cast irons in seawater and weak acids. Chromium additions of 15 to 30% improve the corrosion resistance of cast irons to oxidizing acids, such as nitric acid (HNO_3).

Chromium increases the corrosion resistance of cast iron by the formation of protective oxides on the surfaces of castings. The oxides formed will resist oxidizing acids, but will be of little benefit under reducing conditions. High chromium additions, like higher silicon additions, reduce the ductility of cast irons.

Copper is added to cast irons in special cases. Copper additions of 0.25 to 1% increase the resistance of cast iron to dilute acetic (CH_3COOH), sulfuric (H_2SO_4), and hydrochloric (HCl) acids as well as acid mine water. Small additions of copper are also made to cast irons to enhance atmospheric-corrosion resistance. Additions of up to 10% are made to some high nickel-chromium cast irons to increase corrosion resistance. The exact mechanism by which copper improves the corrosion resistance of cast irons is not known.

Molybdenum. Although an important use of molybdenum in cast irons is to increase strength structural uniformity, it is also used to enhance corrosion resistance, particularly in high-silicon cast irons. Molybdenum is particularly useful in HCl. As little as 1% Mo is helpful in some high-silicon irons, but for optimum resistance, 3 to 4% Mo is added.

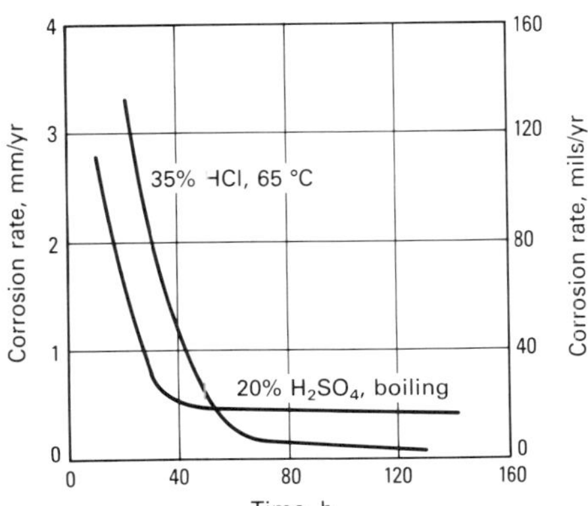

Fig. 2 Corrosion rates of high-silicon cast irons as a function of time and corrosive media

Other Alloying Additions. In general, other alloying additions to cast irons have a minimal effect on corrosion resistance. Vanadium and titanium enhance the graphite morphology and matrix structure and impart slightly increased corrosion resistance to cast irons. Few other additions are made to cast irons that have any significant effect on corrosion resistance.

Influence of Microstructure

Although the graphite shape and the amount of massive carbides present are critical to mechanical properties, these structural variables do not have a strong effect on corrosion resistance. Flake graphite structures may trap corrosion products and retard corrosion slightly in some applications. Under unusual circumstances, graphite may act cathodically with regard to the metal matrix and may accelerate attack.

The structure of the matrix has a slight influence on corrosion resistance, but the effect is small compared to that of composition. In gray irons, ferrite structures are generally the least resistant, and graphite flakes exhibit the greatest corrosion resistance. Pearlite and cementite show intermediate corrosion resistance.

Shrinkage or porosity can degrade the corrosion resistance of cast iron parts. The presence of porosity permits the corrosive medium to enter the body of the casting and can provide continuous leakage paths for corrosives in pressure-containing components.

Commercially Available Cast Irons

Based on corrosion resistance, cast irons can be grouped into five basic categories. Each will be discussed.

Unalloyed gray, ductile, malleable, and white cast irons represent the first and largest category. All of these materials contain carbon and silicon of 3% or less and no deliberate additions of nickel, chromium, copper, or molybdenum. As a group, these materials exhibit corrosion resistance that equals or slightly exceeds that of unalloyed steels, but they show the highest rates of attack for cast irons. These materials are available in a wide variety of configurations and alloys. Major ASTM standards that cover these materials are listed in Table 2.

Low and moderately alloyed irons constitute the second major class. These irons contain the iron and silicon of unalloyed cast irons plus up to several percent of nickel, copper, chromium, or molybdenum. As a group, these materials exhibit two to three times the service life of unalloyed cast irons. Major ASTM standards that include these materials are listed in Table 3.

High-nickel austenitic cast irons represent a third major class of cast irons for corrosion service. These materials contain large percentages of nickel and copper and are fairly resistant to such acids as concentrated H_2SO_4 and phosphoric (H_3PO_4) acid at slightly elevated temperatures, HCl at room temperature, and such organic acids as CH_3COOH, oleic, and stearic. When nickel levels exceed 18%, austenitic cast irons are nearly immune to alkali or caustics, although stress corrosion can occur. High-nickel cast irons can be nodularized to yield ductile irons. They can be purchased to the ASTM standards listed in Table 4.

High-chromium cast irons are the fourth class of corrosion-resistant cast irons. These materials are basically white cast irons alloyed with 12 to 30% Cr. Other alloying elements may also be added to improve resistance to specific environments. When chromium levels exceed 20%, high-chromium cast irons exhibit good resistance to oxidizing acids, particularly HNO_3. High-chromium irons are not resistant to reducing acids. They are used in saline solutions, organic acids, and marine and industrial atmospheres. These materials display excellent resistance to abrasion, and with proper alloying additions, they can also resist combinations of abrasives and liquids, including some dilute acid solutions. High-chromium cast irons are covered in ASTM standard A

Table 2 ASTM standards that include unalloyed cast irons

Standard	Materials/products covered
A 47	Malleable iron castings
A 48	Gray iron castings
A 74	Cast iron soil pipe and fittings
A 126	Gray iron castings for valves, flanges, and pipe fittings
A 159	Automotive gray iron castings
A 197	Cupola malleable iron
A 220	Pearlitic malleable iron castings
A 278	Gray iron castings for pressure-containing parts for temperatures up to 345 °C (650 °F)
A 319	Gray iron castings for elevated temperatures for nonpressure-containing parts
A 395	Ferritic ductile iron pressure-retaining castings for use at elevated temperatures
A 476	Ductile iron castings for paper mill dryer rolls
A 536	Ductile iron castings
A 602	Automotive malleable iron castings
A 716	Ductile iron culvert pipe
A 746	Ductile iron gravity sewer pipe

Table 3 ASTM standards that include low alloyed cast iron materials

Standard	Materials/products covered
A 159	Automotive gray iron castings
A 319	Gray iron castings for elevated temperatures for nonpressure-containing parts
A 532	Abrasion-resistant cast irons

Note: Because most cast iron standards make chemical composition subordinate to mechanical properties, many of the standards listed in Table 2 may also be used to purchase low alloyed cast iron materials.

Table 4 ASTM standards that include high-nickel austenitic cast iron materials

Standard	Materials/products covered
A 436	Austenitic gray iron castings
A 439	Austenitic ductile iron castings
A 571	Austenitic ductile iron castings for pressure-containing parts suitable for low-temperature service

532. In addition, some proprietary alloys not covered by national standards are produced for special applications.

High-silicon cast irons represent the fifth class of corrosion-resistant cast irons. The principal alloying element is 12 to 18% Si, with more than 14.2% Si needed to develop excellent corrosion resistance. Chromium and molybdenum are also used in combination with silicon to develop corrosion resistance to specific environments. High-silicon cast irons represent the most universally corrosion-resistant alloys available at moderate cost. When silicon levels exceed 14.2%, high-silicon cast irons exhibit excellent resistance to H_2SO_4, HNO_3, HCl, CH_3COOH, and most other mineral and organic acids and corrosives. These materials display good resistance in oxidizing and reducing environments and are not appreciably affected by concentration or temperature. Exceptions to universal resistance are hydrofluoric acid (HF), fluoride salts, sulfurous acid (H_2SO_3), sulfite compounds, strong alkalies, and alternating acid-alkali conditions. High-silicon cast irons are defined in ASTM standards A 518 and A 861. In addition, some proprietary compositions not included in these standards, such as alloy SD77 (Fe-4Cr-3Mo-16Si-1Mn-1C), are manufactured for high-temperature HCl service.

Forms of Corrosion

Cast irons exhibit the same general forms of corrosion as other metals and alloys. Examples of the forms of corrosion observed in cast irons include:

- Uniform or general attack
- Galvanic or two-metal corrosion
- Crevice corrosion
- Pitting
- Intergranular corrosion
- Selective leaching
- Erosion-corrosion
- Stress corrosion
- Corrosion fatigue
- Fretting corrosion

Graphite Corrosion. A form of corrosion unique to cast irons is a selective leaching attack commonly referred to as graphitic corrosion. Graphitic corrosion is observed in gray cast irons in relatively mild environments in which selective leaching of iron leaves a graphite network. Selective leaching of the iron takes place because the graphite is cathodic to iron and the gray iron structure establishes an excellent galvanic cell. This form of corrosion generally occurs only when corrosion rates are low. If the metal corrodes more rapidly, the entire surface, including the graphite, is removed, and more or less uniform corrosion occurs. Graphitic corrosion can cause significant problems because, although no dimensional changes occur, the cast iron loses its strength and metallic properties. Thus, without detection, potentially dangerous situations may develop in pressure-containing applications. Graphitic corrosion is observed only in gray cast irons. In both nodular and malleable iron, the lack of graphite flakes provides no network to hold the corrosion products together.

Fretting corrosion is commonly observed when vibration or slight relative motion occurs between parts under load. The relative resistance of cast iron to this form of attack is influenced by such variables as lubrication, hardness variations between materials, the presence of gaskets, and coatings. Table 5 compares the relative fretting resistance of cast iron under different combinations of these variables.

Pitting and Crevice Corrosion. The presence of chlorides and/or crevices or other shielded areas presents conditions that are favorable to the pitting and/or crevice corrosion of cast iron. Pitting has been reported in such environments as dilute alkylaryl sulfonates, antimony trichloride ($SbCl_3$), and calm seawater. Alloying can influence the resistance of cast irons to pitting and crevice corrosion. For example, in calm seawater, nickel additions reduce the susceptibility of cast irons to pitting attack. High-silicon cast irons with chromium and/or molybdenum offer enhanced resistance to pitting and crevice corrosion. Although microstructural variations probably exert some influence on susceptibility to crevice corrosion and pitting, there are few reports of this relationship.

Intergranular attack is relatively rare in cast irons. In stainless steels, in which this type of attack is most commonly observed, intergranular attack is related to chromium depletion adjacent to grain boundaries. Because only the high-chromium cast irons depend on chromium to form passive films for resistance to corrosion attack, few instances of intergranular attack related to chromium depletion have been reported. The only reference to intergranular attack in cast irons involves ammonium nitrate (NH_4NO_3), in which unalloyed cast irons are reported to be intergranularly attacked. Because this form of selective attack is relatively rare in cast irons, no significant references to the influence of either structure or chemistry on intergranular attack have been reported.

Erosion-Corrosion. Fluid flow by itself or in combination with solid particles can cause erosion-corrosion attack in cast irons. Two methods are known to enhance the erosion-corrosion resistance of cast irons. First, the hardness of the cast irons can be increased through solid-solution hardening or phase transformation induced hardness increases. For example, 14.5% Si additions to cast irons cause substantial solid-solution hardening of the ferritic matrix. In such environments as the sulfate liquors encountered in the pulp and paper industry, this hardness increase enables high-silicon iron equipment to be successfully used, while lower-hardness unalloyed cast irons fail rapidly by severe erosion-corrosion. Use of martensitic or white cast irons can also improve the erosion-corrosion resistance of cast irons as a result of hardness increases.

Second, better inherent corrosion resistance can also be used to increase the erosion-corrosion resistance of cast irons. Austenitic nickel cast irons can have hardnesses similar to unalloyed cast irons, but may exhibit better erosion resistance because of the improved inherent resistance of nickel-alloyed irons compared to unalloyed irons. Microstructure can also affect erosion-corrosion resistance slightly. Gray cast irons generally show better resistance than steels under erosion-corrosion conditions. This improvement is related to the presence of the graphite network in the gray cast iron. Iron is corroded from the gray iron matrix as in steel, but the graphite network that is not corroded traps corrosion products; this layer of corrosion products and graphite offers additional protection against erosion-corrosion attack.

Stress-corrosion cracking (SCC) is observed in cast irons under certain combinations of environment and stress. Because stress is necessary to initiate SCC and because design factors often limit stresses in castings to relatively low levels, SCC is not observed as often in cast irons as in other more highly stressed components. However, under certain conditions, SCC can be a serious problem. Because unalloyed cast irons are generally similar to ordinary steels in resistance to corrosion, the same environments that cause SCC in steels will likely cause problems in cast irons. Environments that may cause SCC in unalloyed cast irons include (Ref 2):

- Sodium hydroxide (NaOH) solutions
- NaOH-Na_2SiO_2 solutions
- Calcium nitrate ($Ca(NO_3)_2$) solutions
- NH_4NO_3 solutions
- Sodium nitrate ($NaNO_3$) solutions
- Mercuric nitrate ($Hg(NO_3)_2$) solutions
- Mixed acids (H_2SO_4-HNO_3)
- Hydrogen cyanide (HCN) solutions
- Seawater
- Acidic hydrogen sulfide (H_2S) solutions
- Molten sodium-lead alloys
- Acid chloride solutions
- Oleum

Graphite morphology can play an important role in SCC resistance in certain environments. In oleum (fuming H_2SO_4), flake graphite structures present special problems. Acid tends to penetrate along graphite flakes and corrodes the iron matrix. The corrosion products formed build up pressure and eventually crack the iron. This

Table 5 Relative fretting resistance of cast iron

Poor	Average	Good
Aluminum on cast iron	Cast iron on cast iron	Cast iron on cast iron with phosphate coating
Magnesium on cast iron	Copper on cast iron	Cast iron on cast iron with coating of rubber cement
Cast iron on chrome plate	Brass on cast iron	Cast iron on cast iron with coating of tungsten sulfide
Laminated plastic on cast iron	Zinc on cast iron	Cast iron on cast iron with rubber gasket
Bakelite on cast iron	Cast iron on silver plate	Cast iron on cast iron with Molykote lubricant
Cast iron on tin plate	Cast iron on copper plate	Cast iron on stainless with Molykote lubricant
Cast iron on cast iron with coating of shellac	Cast iron on amalgamated copper plate	
	Cast iron on cast iron with rough surface	

Source: Ref 1

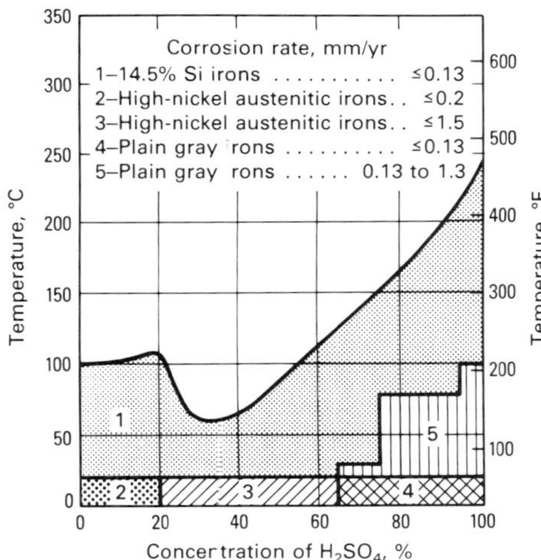

Fig. 3 Corrosion of high-nickel austenitic cast iron in H_2SO_4 as a function of acid concentration and temperature. Source: Ref 2

problem is found in both gray irons and high-silicon irons, which have flake graphite morphologies, but is not seen in ductile irons that have nodular graphite shapes.

Resistance to Corrosive Environments

No single grade of cast iron will resist all corrosive environments. However, a cast iron can be identified that will resist most of the corrosives commonly used in industrial environments. Cast irons suitable for the more common corrosive environments are discussed below.

Sulfuric Acid. Unalloyed, low-alloyed, and high-nickel austenitic as well as high-silicon cast irons are used in H_2SO_4 applications. Use of unalloyed and low-alloyed cast iron is limited to low-velocity low-temperature concentrated (>70%) H_2SO_4 service. Unalloyed cast iron is rarely used in dilute or intermediate concentrations, because corrosion rates are substantial. In concentrated H_2SO_4 as well as other acids, ductile iron is generally considered superior to gray iron, and ferritic matrix irons are superior to pearlitic matrix irons. In hot, concentrated acids, graphitization of the gray iron can occur. In oleum, unalloyed gray iron will corrode at very low rates. However, acid will penetrate along the graphite flakes, and the corrosion product will form and build up sufficient pressure to split the iron. Interconnecting graphite is believed to be necessary to cause this form of cracking; therefore, ductile and malleable irons are generally acceptable for this service. Some potential galvanic corrosion between cast iron and steel has been reported in 100% H_2SO_4.

High-nickel austenitic cast irons exhibit acceptable corrosion resistance in room temperature and slightly elevated temperature services. As shown in Fig. 3, they are adequate over the entire range of H_2SO_4 concentrations, but are a second choice compared to high-silicon cast irons.

High-silicon cast irons are the best choice among the cast irons and perhaps among the commonly available engineering material for resistance to H_2SO_4. The material resists the entire H_2SO_4 concentration range at temperatures to boiling (Fig. 4). Rapid attack occurs at

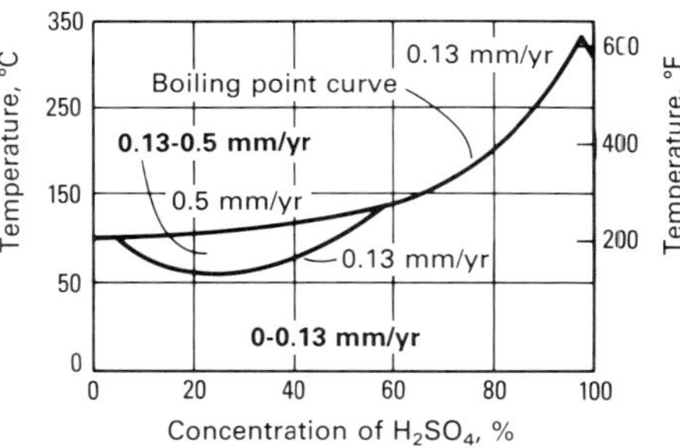

Fig. 4 Corrosion of high-silicon cast iron in H_2SO_4 as a function of acid concentration and temperature

concentrations over 100% and in services containing free sulfur trioxide (SO_3). High-silicon cast irons are relatively slow to passivate in H_2SO_4 services. Corrosion rates are relatively high for the first 24 to 48 h of exposure and then decrease to very low steady-state rates (Fig. 2).

Nitric Acid. All types of cast iron, except high-nickel austenitic iron, find some applications in HNO_3. The use of unalloyed cast iron in HNO_3 is limited to low-temperature low-velocity concentrated acid service. Even in this service, caution must be exercised to avoid dilution of acid because the unalloyed and low-alloyed cast irons both corrode very rapidly in dilute or intermediate concentrations at any temperature. High-nickel austenitic cast irons exhibit essentially the same resistance as unalloyed cast iron to HNO_3 and cannot be economically justified for this service.

High-chromium cast irons with chromium contents over 20% give excellent resistance to HNO_3, particularly in dilute concentrations (Fig. 5). High-temperature boiling solutions attack these grades of cast iron.

High-silicon cast irons also offer excellent resistance to HNO_3. Resistance is exhibited over essentially all concentration and temperature ranges with the exception of dilute, hot acids (Fig. 6). High-silicon cast iron equipment has been used for many years in the manufacture and handling of HNO_3 mixed with other chemicals, such as H_2SO_4, sulfates, and nitrates. Contamination of the HNO_3 with HF, such as might be experienced in pickling solutions, may accelerate attack of the high-silicon iron to unacceptable levels.

Hydrochloric Acid. Use of cast irons is relatively limited in HCl. Unalloyed cast iron is unsuitable for any HCl service. Rapid corrosion occurs at a pH of 5 or lower, particularly if appreciable velocity is involved. Aeration or oxidizing conditions, such as the presence of metallic salts, result in rapid destructive attack of unalloyed cast irons even in very dilute HCl solutions.

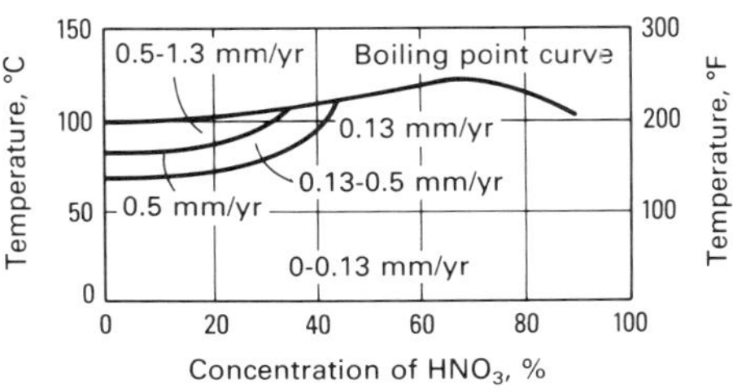

Fig. 6 Corrosion of high-silicon cast iron in HNO_3 as a function of concentration and temperature

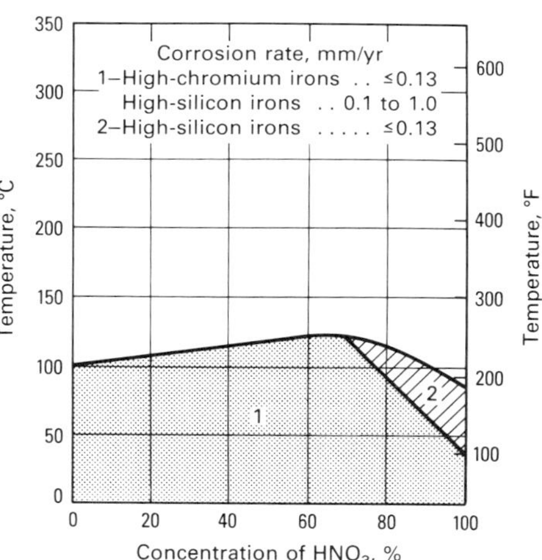

Fig. 5 Corrosion of high-chromium cast iron in HNO_3 as a function of acid concentration and temperature. Source: Ref 2

High-nickel austenitic cast irons offer some resistance to all HCl concentrations at room temperature or below. High-chromium cast irons are not suitable for HCl services.

High-silicon cast irons offer the best resistance to HCl of any cast iron. When alloyed with 4 to 5% Cr, high-silicon cast iron is suitable for all concentrations of HCl to 28 °C (80 °F). When high-silicon cast iron is alloyed with chromium, molybdenum, and higher silicon levels, the temperature for use can be increased (Fig. 7). In concentrations up to 20%, ferric ions (Fe^{3+}) or other oxidizing agents inhibit corrosion attack on high-silicon iron alloyed with chromium. At over 20% acid concentration, oxidizers accelerate attack on the alloy. As in H_2SO_4, corrosion rates in high-silicon cast iron are initially high in the first 24 to 48 h of exposure then decrease to very low steady-state rates (Fig. 2).

Phosphoric Acid. All cast irons find some application in H_3PO_4, but the presence of contaminants must be carefully evaluated before selecting a material for this service. Unalloyed cast iron finds little use in H_3PO_4, with the exception of concentrated acids. Even in concen-

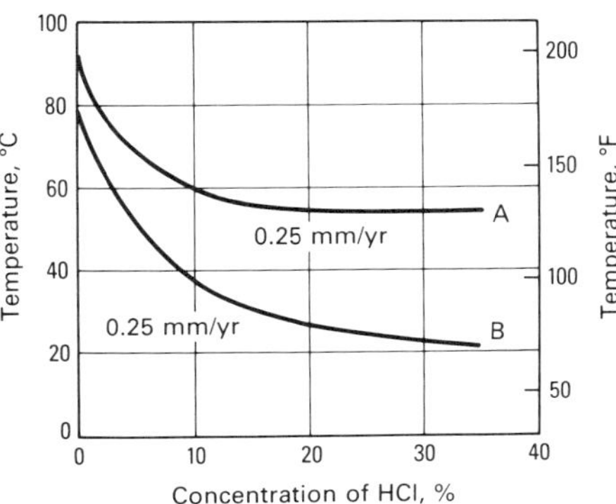

Fig. 7 Isocorrosion diagram for two high-silicon cast irons in HCl. A, Fe-14.3Si-4Cr-0.5Mo; B, Fe-16Si-4Cr-3Mo

trated acids, use may be severely limited by the presence of fluorides, chlorides, or H_2SO_4.

High-nickel cast irons find some application in H_3PO_4 at and slightly above room temperature. These cast irons can be used over the entire H_3PO_4 concentration range. Impurities in the acid may greatly restrict the applicability of this grade of cast iron.

High-chromium cast irons exhibit generally low rates of attack in H_3PO_4 up to 60% concentration. High-silicon cast irons show good-to-excellent resistance at all concentrations and temperatures of pure acid. The presence of fluoride ions (F^-) in H_3PO_4 makes the high-silicon irons unacceptable for use.

Organic acids and compounds are generally not as corrosive as mineral acids; consequently, cast irons find many applications in handling these materials. Unalloyed cast iron can be used to handle concentrated CH_3COOH and fatty acids, but will be attacked by more dilute solutions. Unalloyed cast irons are used to handle methyl, ethyl, butyl, and amyl alcohols. If the alcohols are contaminated with water and air, some discoloration of the alcohols may occur. Unalloyed cast irons can also be used to handle glycerine, although slight discoloration of the glycerine may result.

Austenitic nickel cast irons exhibit adequate resistance to CH_3COOH, oleic acid, and stearic acid. High-chromium cast irons are adequate for CH_3COOH, but will be more severely corroded by formic acid (HCOOH). High-chromium cast irons are excellent for lactic and citric solutions.

High-silicon cast irons show excellent resistance to most organic acids, including HCOOH and oxalic acid, in all temperature and concentration ranges. High-silicon cast irons also exhibit excellent resistance to alcohols and glycerine.

Alkali solutions require material selections that are distinctly different from those of acid solutions. Alkalies include NaOH, potassium hydroxide (KOH), sodium silicate (Na_2SiO_3), and similar chemicals that contain sodium, potassium, or lithium.

Unalloyed cast irons exhibit generally good resistance to alkalies—approximately equivalent to that of steel. These unalloyed cast irons are not attacked by dilute alkalies at any temperature. Hot alkalies at concentrations exceeding 30% attack unalloyed iron. Temperatures should not exceed 80 °C (175 °F) for concentrations up to 70% if corrosion rates of less than 0.25 mm/yr (10 mils/yr) are desired. Ductile and gray iron exhibit about equal resistance to alkalies; however, ductile iron is susceptible to cracking in highly alkaline solutions, but gray iron is not. Alloying with 3 to 5% Ni substantially improves the resistance of cast irons to alkalies. High-nickel austenitic cast irons offer even better resistance to alkalies than unalloyed or low-nickel cast irons.

High-silicon cast irons show good resistance to relatively dilute solutions of NaOH at moderate temperatures, but should not be applied for more concentrated conditions at elevated temperatures. High-silicon cast irons are usually economical over unalloyed and nickel cast irons in alkali solutions only when other corrosives are involved for which the lesser alloys are unsuitable. High-chromium cast irons have inferior resistance to alkali solutions and are generally not recommended for alkali services.

Atmospheric corrosion is basically of interest only for unalloyed and low-alloy cast irons. Atmospheric corrosion rates are determined by the relative humidity and the presence of various gases and solid particles in the air. The high humidity, sulfur dioxide (SO_2) or similar compounds found in many industrialized areas, and chlorides found in marine atmospheres increase the rate of attack on cast irons.

Cast irons typically exhibit very low corrosion rates in industrial environments—generally under 0.13 mm/yr (5 mils/yr)—and the cast irons are usually found to corrode at lower rates than steel structures in the same environment. White cast irons show the lowest rate of corrosion of the unalloyed materials. Pearlitic irons are generally more resistant than ferritic irons to atmospheric corrosion.

In marine atmospheres, unalloyed cast irons also exhibit relatively low rates of corrosion. Low alloy additions are sometimes made to improve corrosion resistance further. Higher alloy additions are even more beneficial, but are rarely warranted. Gray iron offers some added resistance over ductile iron in marine atmospheres.

Corrosion in Soils. Cast iron use in soils, as in atmospheric corrosion, is basically limited to unalloyed and low-alloyed cast irons. Corrosion in soils is a function of soil porosity, drainage, and dissolved constituents in the soil. Irregular soil contact can cause pitting, and poor drainage increases corrosion rates substantially above the rates in well-drained soils.

Neither metal structure nor graphite morphology has an important influence on the corrosion of cast irons in soils. Some alloying additions are made to improve the resistance of cast irons to attack in soils. For example, 3% Ni additions to cast iron are made to reduce initial attack in cast irons in poorly drained soils. Alloyed cast irons would exhibit better resistance than unalloyed or low-alloyed cast irons, but are rarely needed for soil applications, because unalloyed cast irons generally have long service lives.

Anodes placed in soils are frequently constructed from high-silicon cast iron. The high-silicon cast iron is not needed to resist the basic soil environment but rather to extend service life when subjected to the high electrical current discharge rates commonly used in protective anodes.

Corrosion in Water. Unalloyed and low-alloyed cast irons are the primary cast irons used in water. The corrosion resistance of unalloyed cast iron in water is determined by its ability to form protective scales. In hard water, corrosion rates are generally low because of the formation of calcium carbonate ($CaCO_3$) scales on the surface of the iron. In softened or deionized water, the protective scales cannot be fully developed, and some corrosion will occur.

In industrial waste waters, corrosion rates are primarily a function of the contaminants present. Acid pH waters increase corrosion, but alkaline pH waters lower rates. Chlorides increase the corrosion rates of unalloyed cast irons, although the influence of chlorides is small at a neutral pH.

Seawater presents some special problems for cast irons. Gray iron may experience graphitic corrosion in calm seawater. It will also be galvanically active in contact with most stainless steels, copper-nickel alloys, titanium, and Hastelloy C. Because these materials are frequently used in seawater structures, this potential for galvanic corrosion must be considered. In calm seawater, the corrosion resistance of cast iron is not greatly affected by the presence of crevices. However, intermittent exposure to seawater is very corrosive to unalloyed cast irons.

Use of high-alloy cast irons in water is relatively limited. High-nickel austenitic cast irons are used to increase the resistance of cast iron components to pitting in calm seawater. High-silicon cast iron is used to produce anodes for the anodic protection systems used in seawater and brackish water.

Corrosion in Saline Solutions. The presence of salts in water can have dramatic effects on the selection of suitable grades of cast iron. Unalloyed cast irons exhibit very low corrosion in such salts as cyanides, silicates, carbonates, and sulfides, which hydrolyze to form alkaline solutions. However, in salts such as ferric chloride ($FeCl_3$), cupric chloride ($CuCl_2$), stannic salts, and mercuric salts, which hydrolyze to form acid solutions, unalloyed cast irons experience much higher rates. In salts that form dilute acid solutions, high-nickel cast irons are acceptable. More acidic and oxidizing salts, such as $FeCl_3$, usually necessitate the use of high-silicon cast irons.

Chlorides and sulfates of alkali metals yield neutral solutions, and unalloyed cast iron experiences very low corrosion rates in these solutions. More highly alloyed cast irons also exhibit low rates, but cannot be economically justified for this application.

Unalloyed cast irons are suitable for oxidizing salts, such as chromates, nitrates, nitrites, and permanganates, when the pH is neutral or alkaline. However, if the pH is less than 7, corrosion rates can increase substantially. At the lower pH with oxidizing salts, high-silicon cast iron is an excellent material selection.

Ammonium salts are generally corrosive to unalloyed iron. High-nickel, high-chromium, and high-silicon cast irons provide good resistance to these salts.

Other Environments. Unalloyed cast iron is used as a melting crucible for such low-melting metals as lead, zinc, cadmium, magnesium, and aluminum. Resistance to molten metals is summarized in Table 6. Ceramic coatings and washes are sometimes used to inhibit metal attack on cast irons.

Cast iron can also be used in hydrogen chloride and chlorine gases. In dry hydrogen chloride, unalloyed cast iron is suitable to 205 °C (400 °F), while in dry chlorine, unalloyed cast iron is suitable to 175 °C (350 °F). If moisture is present, unalloyed cast iron is unacceptable at any temperature.

Coatings

Four general categories of coatings are used on cast irons to enhance corrosion resistance: metallic, organic, conversion, and enamel coatings. Coatings on cast irons are generally used to enhance the corrosion resistance of unalloyed and low-alloy cast irons. High-alloy cast irons are rarely coated.

Metallic coatings are used to enhance the corrosion resistance of cast irons. These coatings may either be sacrificial metal coatings, such as zinc, or barrier metal coatings, such as nickel-phosphorus. From a corrosion standpoint, these two classes of coatings have important differences. Sacrificial coatings are anodic when compared to iron, and the coatings corrode preferentially and protect the cast iron substrate. Small cracks and porosity in the coatings have a mini-

Table 6 Resistance of gray cast iron to liquid metals at 300 and 600 °C (570 and 1110 °F)

Liquid metal	Liquid metal melting point, °C	Resistance of gray cast iron(a) 300 °C (570 °F)	600 °C (1110 °F)
Mercury	−38.8	Unknown	Unknown
Sodium, potassium, and mixtures	−12.3 to 97.9	Limited	Poor
Gallium	29.8	Unknown	Unknown
Bismuth-lead-tin	97	Good	Unknown
Bismuth-lead	125	Unknown	Unknown
Tin	321.9	Limited	Poor
Bismuth	271.3	Unknown	Unknown
Lead	327	Good at 327 °C (621 °F)	Unknown
Indium	156.4	Unknown	Unknown
Lithium	186	Unknown	Unknown
Thallium	303	Unknown	Unknown
Cadmium	321	Good at 321 °C (610 °F)	Good
Zinc	419.5	...	Poor
Antimony	630.5	...	Poor at 630.5 °C (1167 °F)
Magnesium	651	...	Good at 651 °C (1204 °F)
Aluminum	660	...	Poor at 660 °C (1220 °F)

(a) Good, considered for long-time use, <0.025 mm/yr (<1.0 mil/yr); Limited, short-time use only, 0.025–0.25 mm/yr (1.0–10 mils/yr); Poor, no structural possibilities, >0.25 mm/yr (>10 mils/yr); Unknown, no data for these temperatures. Source: Ref 3

mal overall effect on the performance of the coatings. Barrier coatings are cathodic compared to iron, and the coatings can protect the cast iron substrate only when porosity or cracks are not present. If there are defects in the coatings, the service environment will attack the cast iron substrate at these imperfections, and the galvanic couple set up between the relatively inert coating and the casting may accelerate attack on the cast iron.

Several methods are used to apply metallic coatings to cast iron. Cast irons may be electroplated, hot dipped, flame sprayed, diffusion coated, or hard faced. Table 7 lists the metals that can be applied by these various techniques.

Zinc is one of the most widely used coatings on cast irons. Although zinc is anodic to iron, its corrosion rate is very low, and it provides relatively long-term protection for the cast iron substrate. A small amount of zinc will protect a large area of cast iron. Zinc coatings provide optimum protection in rural or arid areas.

Other metal coatings are also commonly used on cast irons. Cadmium provides atmospheric protection similar to that of zinc. Tin coatings are frequently used to improve the corrosion resistance of equipment intended for food handling,

Table 7 Summary of metallic coating techniques to enhance corrosion resistance of cast irons

Coating technique	Metals/alloys applied
Electroplating	Cadmium, chromium, copper, lead, nickel, zinc, tin, tin-nickel, brass, bronze
Hot dipped	Zinc, tin, lead, lead-tin, aluminum
Hard facing	Cobalt-base alloys, nickel-base alloys, metal carbides, high-chromium ferrous alloys, high-manganese ferrous alloys, high chromium and nickel ferrous alloys
Flame spraying	Zinc, aluminum, lead, iron, bronze, copper, nickel, ceramics, cermets
Diffusion coating	Aluminum, chromium, nickel-phosphorus, zinc, nitrogen, carbon

and aluminum coatings protect against corrosive environments containing sulfur fumes, organic acids, salts, and compounds of nitrate-phosphate chemicals. Lead and lead-tin coatings are primarily applied to enhance the corrosion resistance of iron castings to H_2SO_3 and H_2SO_4. Nickel-phosphorus diffusion coatings offer corrosion resistance approaching that obtainable with stainless steel.

Organic coatings can be applied to cast irons to provide short-term or long-term corrosion resistance. Short-term rust preventatives include oil, solvent-petroleum-base inhibitors and film formers dissolved in petroleum solvents, emulsified-petroleum-base coatings modified to form a stable emulsion in water, and wax.

For longer-term protection and resistance to more corrosive environments, rubber-base coatings, bituminous paints, asphaltic compounds, or thermoset and thermoplastic coatings can be applied. Rubber-base coatings include chlorinated rubber neoprene, and Hypalon. These coatings are noted for their mechanical properties and corrosion resistance but not for their decorative appearance. Bituminous paints have very low water permeability and provide high resistance in cast iron castings exposed to water. Use of bituminous paints is limited to applications that require good resistance to water, weak acids, alkalies, and salts. Asphaltic compounds are used to increase the resistance to cast irons to alkalies, sewage, acids, and continued exposure to tap water. Their application range is similar to that of bituminous paints. Cast irons are also lined with thermoset and thermoplastics, such as epoxy and polyethylene, to resist attack by fluids.

Fluorocarbon coatings offer superior corrosion resistance except in abrasion services. Fluorocarbon coatings applied to cast irons include such materials as polytetrafluoroethylene (PTFE), perfluoroalkoxy resins (PFA), and fluorinated ethylene polypropylene (FEP). Fluorocarbon coatings resist most common industrial services and can be used to 205 °C (400 °F). Cast iron lined with fluorocarbons is very competitive with stainless, nickel-base, and even titanium and zirconium materials in terms of range of services covered and product cost.

Conversion coatings are produced when the metal on the surface of the cast iron reacts with another element or compound to produce an iron-containing compound. Common conversion coatings include phosphate coatings, oxide coatings, and chromate coatings. Phosphate coatings enhance the resistance of cast iron to corrosion in sheltered atmospheric exposure. If the surface of the casting is oxidized and black iron oxide or magnetite is formed, the corrosion resistance of the iron can be enhanced, particularly if the oxide layer is impregnated with oil or wax. Chromate coatings are formed by immersing the iron castings in an aqueous solution of chromic acid (H_2CrO_4) or chromium salts. Chromate coatings are sometimes used as a supplement to cadmium plating in order to prevent the formation of powdery corrosion products. The overall benefits of conversion coatings are small with regard to atmospheric corrosion.

Enamel Coatings. In the enamel coating of cast irons, glass frits are melted on the surface and form a hard, tenacious bond to the cast iron substrate. Good resistance to all acids except HF can be obtained with the proper selection and application of the enamel coating. Alkaline-resistant coatings can also be applied, but they do not exhibit the same general resistance to alkalies as acids do.

Proper design and application are essential for developing enhanced corrosion resistance on cast irons with enamel coatings. Any cracks, spalling, or other coating imperfections may permit rapid attack of the underlying cast iron.

Selection of Cast Irons

Cast irons can provide excellent resistance to a wide range of corrosion environments when properly matched with the service environment for which they are intended. The basic parameters to consider before selecting cast irons for corrosion services include:

- Concentration of solution components in weight percent
- Contaminants, even at parts per million levels
- pH of solution
- Temperature and its potential range and rate of change
- Degree of aeration
- Percent and type of solids
- Continuous or intermittent operation
- Upset potential: maximum temperature and concentration
- Unusual conditions, such as high velocity and vacuum
- Materials currently used in the system and potential for galvanic corrosion

Although it is advisable to consider each of the parameters before ultimate selection of a cast iron, the information needed to assess all variables of importance properly is often lacking. In such cases, introduction of test coupons of the candidate materials into the process stream should be considered before extensive purchases of equipment are made. If neither test coupons nor complete service data are viable alternatives, consultation with a reputable manufacturer of the equipment or the cast iron with a history of applications in the area of interest should be considered.

REFERENCES

1. J.R. McDowell, in *Symposium on Fretting Corrosion*, STP 144, American Society for Testing and Materials, 1952, p 24
2. E.C. Miller, *Liquid Metals Handbook*, 2nd ed., Government Printing Office, 1952, p 144
3. R.I. Higgins, Corrosion of Cast Iron, *J. Res.*, Feb 1956, p 165-177

SELECTED REFERENCES

- *Corrosion Data Survey*, 6th ed., National Association of Corrosion Engineers, 1985
- M.G. Fontana, *Corrosion Engineering*, 3rd ed., McGraw-Hill, 1986
- "High Silicon Iron Alloys for Corrosion Services," Bulletin A12e, The Duriron Company, April 1972
- *Properties and Selection: Irons and Steels*, Vol 1, 9th ed., *Metals Handbook*, American Society for Metals, 1978
- C.F. Walton, Ed., *The Gray Iron Castings Handbook*, A.L. Garber, 1957
- C.F. Walton, Ed., *Gray and Ductile Iron Castings Handbook*, R.R. Donnelley & Sons, 1971

Corrosion of Cast Steels

Raymond W. Monroe, Maynard Steel Casting Company
Steven J. Pawel, University of Tennessee

STEEL CASTING COMPOSITIONS are generally divided into the categories of carbon, low-alloy, corrosion-resistant, or heat-resistant, depending on alloy content and intended service. Castings are classified as corrosion resistant if they are capable of sustained operation when exposed to attack by corrosive agents at service temperatures normally below 315 °C (600 °F).

The high-alloy ferrous-base compositions are usually given the name stainless steel, although this name has been questioned. Actually, they are widely referred to as cast stainless steels. Some of the high alloys, such as 12% Cr steel, exhibit many of the familiar physical characteristics of carbon and low-alloy steels, and some of their mechanical properties, such as hardness and tensile strength, can be altered by suitable heat treatment. The alloys of higher chromium content (20 to 30%)—chromium-nickel and nickel-chromium—do not show the changes in phase observed in ordinary steel when heated or cooled in the range from room temperature to the melting point. Consequently, these materials are nonhardenable, and their mechanical properties depend on their composition rather than heat treatment.

The high-alloy steels differ from carbon and low-alloy steels in other respects, such as their production and properties. Special consideration must be given to each grade with regard to casting design and foundry practice. For example, such elements as chromium, nickel, carbon, nitrogen, silicon, molybdenum, and niobium may exert a profound influence on the ultimate structure of these complex alloys; therefore, balancing of the alloy compositions is frequently required to obtain a satisfactory product. The chemical ranges used in the manufacture of wrought stainless alloys are not used to produce castings, because a different balance of alloying elements may be required to provide castability, desired mechanical properties, and optimum corrosion resistance.

Corrosion resistance is a relative term that depends on the particular environment to which a specific alloy is exposed. Carbon and low-alloy steels are considered resistant only to very mild corrosives, but the various high-alloy grades are applicable for varying situations from mild to severe services, depending on the particular conditions involved.

It is often misleading to list the comparative corrosion rates of different alloys exposed to the same corroding medium. In this article, no attempt will be made to recommend alloys for specific applications, and the data supplied should be used only as a general guideline. Alloy casting users will find it helpful to consult materials and corrosion specialists when selecting alloys for a particular application. The factors that must be considered in material selection include:

- The principal corrosive agents and their concentrations
- Known or suspected impurities, including abrasive materials and their concentration
- Average operating temperature, including variations even if encountered only for short periods
- Presence (or absence) of oxygen or other gases in solution
- Continuous or intermittent operation
- Fluid velocity

Each of these can have a vital effect on the service life of both cast and wrought equipment, and such detailed information usually must be provided. Many rapid failures are traceable to these details being overlooked—often when the information was available.

Selection of the most economical alloy is often made by the judicious use of corrosion data. However, discretion is suggested in evaluating the relative corrosion rates of various steels because of the uncertainties of the actual test or service conditions. Corrosion rates determined in controlled laboratory tests should be applied cautiously when considering actual service. The best information is obtained from equipment used under similar operating conditions. However, exposing samples to service conditions will also provide valuable information.

Corrosion of Carbon and Low-Alloy Cast Steels

Unless shielded by a protective coating, iron and steel will corrode in the presence of water and oxygen; therefore, steel will corrode when it is exposed to moist air. The rate at which corrosion proceeds in the atmosphere depends on the corroding medium, the conditions of the particular location in which the material is in use, and the steps that have been taken to prevent corrosion. The rate of corrosion also depends on the character of the steel as determined by its chemical composition and heat treatment. The probable rate of corrosion of a material in an environment can generally be estimated only from long-term tests.

Cast steel and wrought steel of similar analysis and heat treatment exhibit about the same corrosion resistance in the same environments. Plain carbon steel and some of the low-alloy steels do not ordinarily resist drastic corrosive conditions, although there are some exceptions, such as strong sulfuric acid (H_2SO_4). To increase the corrosion resistance of steel significantly, it is necessary to resort to extensive alloying. Small amounts of copper and nickel slightly improve the resistance of steel to atmospheric attack, but appreciably larger amounts of other elements, such as chromium or nickel, improve resistance significantly.

Atmospheric Corrosion. A 15-year research program compared the corrosion resistance of nine cast steels in marine and industrial atmospheres. Table 1 shows the compositions of the cast steels tested. The cast steel specimens exposed were 13-mm (½-in.) thick, 100- × 150-mm (4- × 6-in.) panels with beveled edges. The surfaces of half the specimens were machined. Specimens of each composition and surface condition were divided into three groups. One group was exposed to an industrial atmosphere at East Chicago, IN, and the other two groups were exposed to marine atmospheres 24 and 240 m (80 and 800 ft) from the ocean at Kure Beach, NC. The weight losses of the specimens during exposure were converted to corrosion rates in terms of millimeters (mils) per year. The results of this research are shown in Fig. 1 to 4.

Figure 5 shows the results of another portion of this project. Corrosion rates for a 3-year exposure of various cast steels, wrought steels, and malleable iron in both atmospheres are compared. The following conclusions can be drawn from these tests:

- The condition of the specimen surface has no significant effect on the corrosion resistance of cast steels. Unmachined surfaces with the casting skin intact have corrosion rates similar to those of machined surfaces regardless of the atmospheric environment
- The highest corrosion rate occurs in the marine atmosphere 24 m (80 ft) from the ocean, with lower but similar corrosion rates occurring in the industrial atmosphere and the marine atmosphere 240 m (800 ft) from the ocean
- The corrosion rate of cast steel decreases as a function of time, because corrosion products (scale and rust coating) build up and act as a protective coating on the cast steel surface. However, the corrosion rate of the most resistant cast steel (2% Ni) is always less than that of lesser corrosion-resistant cast steels
- Cast steels containing small percentages of nickel, copper, or chromium as alloying elements have corrosion resistance superior to that of cast carbon steels and those containing manganese when exposed to atmospheric environments
- Increasing the nickel and the chromium contents of cast steel increases the corrosion resis-

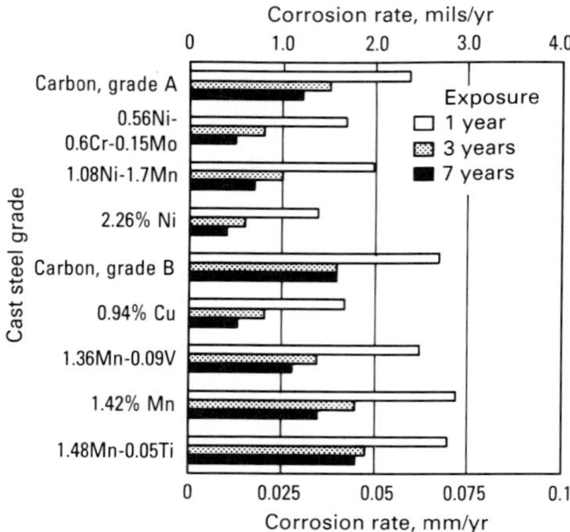

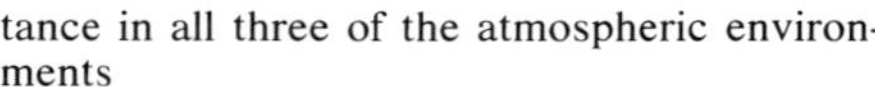

Fig. 1 Corrosion rates of various cast steels in a marine atmosphere. Nonmachined specimens were exposed 24 m (80 ft) from the ocean at Kure Beach, NC. Source: Ref 1

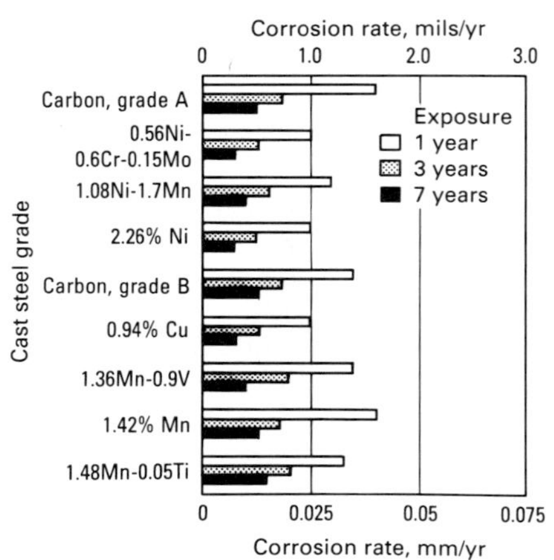

Fig. 2 Corrosion rates of various cast steels exposed at the 240-m (800-ft) site at Kure Beach, NC. Specimens were not machined. Source: Ref 1

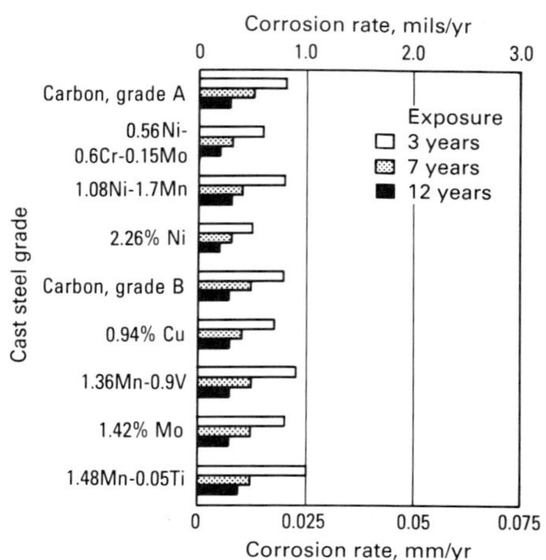

Fig. 3 Corrosion rates for cast steels in an industrial atmosphere. Nonmachined specimens were exposed at East Chicago, IN. Source: Ref 1

tance in all three of the atmospheric environments

- All cast steels have greater corrosion resistance than malleable iron in industrial atmospheres and are superior or equivalent to the wrought steels in this environment. The corrosion rate in the marine atmosphere depends primarily on the alloy content. The cast carbon steel is much superior to the AISI 1020 wrought steel, but is slightly inferior to malleable iron (Ref 1)

Other Environments. Several low- and high-alloy cast steels have been studied regarding their corrosion resistance to high-temperature steam. Test specimens 150 mm (6 in.) in length and 13 mm (½-in.) in diameter were machined from test coupons and then exposed to steam at 650 °C (1200 °F) for 570 h. The steel compositions and test results are given in Table 2. Table 3 shows the resistance of cast steels to petroleum corrosion, and Tables 4 and 5 supply similar data relating to water and acid attack. These data show the value of higher chromium content for improved corrosion resistance.

Corrosion of Cast Stainless Steels

Cast stainless steels are usually specified on the basis of composition by using the alloy designation system established by the Alloy Casting Institute (ACI). The ACI designations, such as CF-8M, have been adopted by the American Society for Testing and Materials (ASTM) and are preferred for cast alloys over the designations used by the American Iron and Steel Institute (AISI) for similar wrought steels.

The first letter of the ACI designation indicates whether the alloy is intended primarily for liquid corrosion service (C) or heat-resistant service (H). The second letter denotes the nominal chromium-nickel type, as shown in Fig. 6. As the nickel content increases, the second letter in the ACI designation increases from A to Z. The numerals following the two letters refer to the maximum carbon content (percent × 100) of the alloy. If additional alloying elements are included, they can be denoted by the addition of one or more letters after the maximum carbon content. Thus, the designation CF-8M refers to an alloy for corrosion-resistant service (C) of the 19Cr-9Ni (F) type, with a maximum carbon content of 0.08% and containing molybdenum (M).

Corrosion-resistant cast steels are also often classified on the basis of microstructure. The classifications are not completely independent, and a classification by composition often involves microstructural distinctions. Cast corrosion-resistant alloy compositions are listed in Table 6.

Composition and Microstructure

The principal alloying element in the high-alloy family is usually chromium, which, through the formation of protective oxide films, is the first step for these alloys in achieving stainless quality. For all practical purposes, stainless behavior requires at least 12% Cr. As will be discussed later, corrosion resistance further improves with additions of chromium to at least the 30% level. As shown in Table 5, nickel and lesser amounts of molybdenum and other elements are often added to the iron-chromium matrix.

Although chromium is a ferrite and martensite promoter, nickel is an austenite promoter. By varying the amounts and ratios of these two elements (or their equivalents), almost any desired combination of microstructure, strength, or other property can be achieved. Equally important is heat treatment. Temperature, time at temperature, and cooling rate must be controlled to obtain the desired results.

It is useful to think of the compositions of high-alloy steels in terms of the balance between austenite promoters and ferrite promoters. This is done on the widely used Schaeffler-type diagrams (Fig. 7). The phases shown are those that persist after cooling to room temperature at rates normally used in fabrication (Ref 2, 3).

The empirical correlations shown in Fig. 7 can be understood from the following. The field designated as martensite encompasses such alloys as CA-15, CA-6NM, and even CB-7Cu. These alloys contain 12 to 17% Cr, with adequate nickel, molybdenum, and carbon to promote high hardenability, that is, the ability to transform completely to martensite when cooled at even the moderate rates associated with the air cooling of heavy sections. High alloys have low thermal conductivities and cool slowly. To obtain the desired properties, a full heat treatment is required after casting; that is, the casting is austenitized by heating to 870 to 980 °C (1600 to 1800 °F), cooled

Table 1 Compositions of cast steels tested in atmospheric corrosion

Cast steel	Composition, %(a)									
	Ni	Cu	Mn	Cr	V	C	Mo	P	S	Si	Other
Carbon, grade A	0.10	0.13	0.61	0.21	0.03	0.14	trace	0.016	0.026	0.41	...
Nickel-chromium-molybdenum	0.56	0.13	0.80	0.60	0.04	0.26	0.15	...	...	0.44	...
1Ni–1.7Mn	1.08	0.08	1.70	0.08	0.04	0.27	...	0.02	0.023	0.42	...
2% Ni	2.26	0.12	0.77	0.19	0.03	0.17	trace	0.017	0.021	0.65	...
Carbon, grade B	0.03	0.03	0.65	0.10	0.04	0.25	...	0.011	0.021	0.51	...
1% Cu	0.04	0.94	0.87	0.11	0.07	0.28	...	...	...	0.42	...
1.36Mn–0.09V	0.01	0.15	1.36	0.08	0.09	0.37	...	0.031	0.038	0.34	...
1.42% Mn	0.01	0.13	1.42	0.16	0.04	0.37	...	0.027	0.022	0.38	...
1.5Mn–0.05Ti	0.01	0.11	1.48	0.04	0.03	0.33	...	0.016	0.025	0.40	0.05 Ti

(a) All compositions contain balance of iron. Source: Ref 1

Table 2 Corrosion of cast carbon and alloy steels in steam at 650 °C (1200 °F) for 570 h

Type of steel	Composition, %				Average penetration rate	
	C	Cr	Ni	Mo	mm/yr	mils/yr
Carbon	0.24	...	...	...	0.3	12
Carbon	0.25	...	...	...	0.28	11
Carbon-molybdenum	0.21	...	...	0.49	0.3	12
Carbon-molybdenum	0.20	...	...	0.49	0.25	10
Nickel-chromium-molybdenum	0.35	0.64	2.13	0.26	0.25	10
Nickel-chromium-molybdenum	0.28	0.73	2.25	0.26	0.25	10
5Cr-molybdenum	0.22	5.07	...	0.47	0.1	4
5Cr-molybdenum	0.27	5.49	...	0.43	0.1	4
7Cr-molybdenum(a)	0.11	7.33	...	0.59	0.05	2
9Cr-1.5Mo	0.23	9.09	...	1.56	0.025	1

(a) Not a cast steel. Source: Ref 1

Table 3 Petroleum corrosion resistance of cast steels

1000-h test in petroleum vapor under 780 N (175 lb) of pressure at 345 °C (650 °F)

Type of material	Weight loss	
	mg/cm²	mg/in.²
Cast carbon steel	3040	196
Cast steel, 2Ni-0.75Cr	2370	153
Seamless tubing, 5% Cr	1540	99.2
Cast steel, 5Cr-1W	950	61.5
Cast steel, 5Cr-0.5Mo	730	47
Cast steel, 12% Cr	6.4	100
Stainless steel, 18Cr-8Ni	2.1	30

Source: Ref 1

Table 4 Corrosion of cast steels in waters

Corrosive medium	Exposure time, months	Corrosion factor(a)		
		Fe-0.29C-0.69Mn-0.44Si	Fe-0.32C-0.66Mn-1.12Cr	Fe-0.11C-0.41Mn-3.58Cr
Tap water	2	100	85	58
	6	100	73	61
Seawater	2	100	60	26
	6	100	80	40
Alternate immersion and drying	2	100	93	30
	6	100	109	25
Hot water	1	100	100	64
0.05% H_2SO_4	2	100	71	68
	6	100	89	102
0.50% H_2SO_4	2	100	223	61

(a) Corrosion factor is the ratio of average penetration rate of the alloy in question to Fe-0.29C-0.69Mn-0.44Si steel. Source: Ref 1

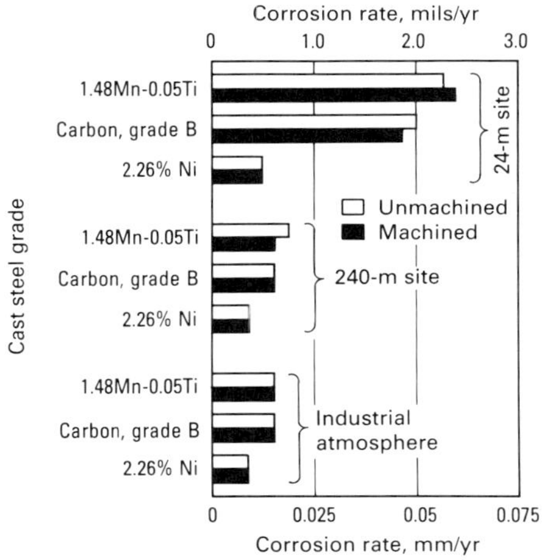

Table 5 Corrosion of cast chromium and carbon steels in mineral acids

Steel	Weight loss in 5 h					
	5% H_2SO_4		5% HCl		5% HNO_3	
	mg/cm²	mg/in.²	mg/cm²	mg/in.²	mg/cm²	mg/in.²
Carbon steel, 0.31% C	2.7	17.42	2.1	13.55	80.79	521.1
Chromium steel, 0.30C-2.42Cr	4.9	31.6	5.41	34.9	47.36	305.5

Source: Ref 1

Fig. 4 Corrosion rates of machined and nonmachined specimens of cast steels after 7 years in three environments. The effect of surface finish on corrosion rates is negligible. Source: Ref 1

to room temperature to produce the hard martensite, and then tempered at 595 to 760 °C (1100 to 1400 °F) until the desired combination of strength, toughness, ductility, and resistance to corrosion or stress corrosion is obtained (Ref 2, 3).

Increasing the nickel equivalent (moving vertically in Fig. 7) eventually results in an alloy that is fully austenitic, such as CC-20, CH-20, CK-20, or CN-7M. These alloys are extremely ductile, tough, and corrosion resistant. On the other hand, the yield and tensile strength may be relatively low for the fully austenitic alloys. Because these high-nickel alloys are fully austenitic, they are nonmagnetic. Heat treatment consists of a single step: water quenching from a relatively high temperature at which carbides have been taken into solution. Solution treatment may also homogenize the structure, but because no transformation occurs, there can be no grain refinement. The solutionizing step and rapid cooling ensure maximum resistance to corrosion. Temperatures between 1040 and 1205 °C (1900 and 2200 °F) are usually required (Ref 2, 3).

Adding chromium to the lean alloys (proceeding horizontally in Fig. 7) stabilizes the δ-ferrite that forms when the casting solidifies. Examples are CB-30 and CC-50. With high chromium content, these alloys have relatively good resistance to corrosion, particularly in sulfur-bearing atmospheres. However, being single-phase, they are nonhardenable, have moderate-to-low strength, and are often used as-cast or after only a simple solutioning treatment. Ferritic alloys also have poor toughness (Ref 2, 3).

Between the fields designated M, A, and F in Fig. 7 are regions indicating the possibility of two or more phases in the alloys. Commercially, the most important of these alloys are the ones in which austenite and ferrite coexist, such as CF-3, CF-8, CF-3M, CF-8M, CG-8M, and CE-30. These alloys usually contain 3 to 30% ferrite in a matrix of austenite. Predicting and controlling ferrite content is vital to the successful application of these materials. Duplex alloys offer superior strength, weldability, and corrosion resistance. Strength, for example, increases directly with ferrite content. Achieving specified minimums may necessitate controlling the ferrite within narrow bands. Figure 8 and Schoefer's equations are used for this purpose. These duplex alloys should be solution treated and rapidly cooled before use to ensure maximum resistance to corrosion (Ref 2, 3).

The presence of ferrite is not entirely beneficial. Ferrite tends to reduce toughness, although this is not of great concern given the extremely high toughness of the austenite matrix. However, in applications that require exposure to elevated temperatures, usually 315 °C (600 °F) and higher, the metallurgical changes associated with the ferrite can be severe and detrimental. In the low end of this temperature range, the reductions in toughness observed have been attributed to carbide precipitation or reactions associated with 475-°C embrittlement. The 475-°C embrittlement is caused by precipitation of an intermetallic phase with a composition of approximately 80Cr-20Fe. The name derives from the fact that this embrittlement is most severe and rapid when it occurs at approximately 475 °C (885 °F). At 540 °C (1000 °F) and above, the ferrite phase may transform to a complex iron-chromium-nickel-molybdenum intermetallic compound known as σ phase, which reduces toughness, corrosion resistance, and creep ductility. The extent of the reduction increases with time and temperature to about 815 °C (1500 °F) and may persist to 925 °C (1700 °F). In extreme cases, Charpy V-notch energy at room temperature may be reduced 95% from its initial value (Ref 2, 3). More information on the metallography and microstructures of these alloys is available in the article "Cast Stainless Steels" in Volume 9 of the 9th Edition of *Metals Handbook*.

Corrosion Behavior of H-Type Alloys

The ACI heat-resistant (H-type) alloys must be able to withstand temperatures exceeding 1095 °C (2000 °F) in the most severe high-temperature

Table 6 Compositions of ACI heat- and corrosion-resistant casting alloys

ACI designation	Wrought alloy type(a)	Composition, % (balance iron)(b)							
		C	Mn	Si	P	S	Cr	Ni	Other elements
CA-15	410	0.15	1.00	1.50	0.04	0.04	11.5–14	1	0.5Mo(c)
CA-15M	...	0.15	1.00	0.65	0.04	0.04	11.50–14.0	1.00	0.15–1.00Mo
CA-40	420	0.20–0.40	1.00	1.50	0.04	0.04	11.5–14	1	0.5Mo(c)
CA-6NM	...	0.06	1.00	1.00	0.04	0.03	11.5–14.0	3.5–4.5	0.4–1.0Mo
CA-6N	...	0.06	0.50	1.00	0.02	0.02	10.5–12.0	6.0–8.0	
CB-30	431	0.30	1.00	1.50	0.04	0.04	18–21	2	...
CB-7Cu-1	...	0.07	0.70	1.00	0.035	0.03	14.0–15.5	4.5–5.5	0.15–0.35Nb, 0.05N, 2.5–3.2Cu
CB-7Cu-2	...	0.07	0.70	1.00	0.035	0.03	14.0–15.5	4.5–5.5	0.15–0.35Nb, 0.05N, 2.5–3.2Cu
CC-50	446	0.50	1.00	1.50	0.04	0.04	26–30	4	
CD-4MCu	...	0.04	1.00	1.00	0.04	0.04	24.5–26.5	4.75–6.00	1.75–2.25Mo, 2.75–3.25Cu
CE-30	...	0.30	1.50	2.00	0.04	0.04	26–30	8–11	...
CF-3	304L	0.03	1.50	2.00	0.04	0.04	17–21	8–21	...
CF-8	304	0.08	1.50	2.00	0.04	0.04	18–21	8–11	...
CF-20	302	0.20	1.50	2.00	0.04	0.04	18–21	8–11	...
CF-3M	316L	0.03	1.50	1.50	0.04	0.04	17–21	9–13	2.0–3.0Mo
CF-8M	316	0.08	1.50	2.00	0.04	0.04	18–21	9–12	2.0–3.0Mo
CF-8C	347	0.08	1.50	2.00	0.04	0.04	18–21	9–12	3 × C min, 1.0 max Nb
CF-16F	303	0.16	1.50	2.00	0.17	0.04	18–21	9–12	1.5Mo, 0.2–0.35Se
CG-12	...	0.12	1.50	2.00	0.04	0.04	20–23	10–13	
CG-8M	317	0.08	1.50	1.50	0.04	0.04	18–21	9–13	3.0–4.0Mo
CH-20	309	0.20	1.50	2.00	0.04	0.04	22–26	12–15	...
CK-20	310	0.20	2.00	2.00	0.04	0.04	23–27	19–22	...
CN-7M	...	0.07	1.50	1.50	0.04	0.04	19–22	27.5–30.5	2.0–3.0Mo, 3.0–4.0Cu
CN-7MS	...	0.07	1.00	2.50–3.50	0.04	0.03	18–20	22–25	2.0–3.0Mo, 1.5–2.0Cu
CW-12M	...	0.12	1.00	1.50	0.04	0.03	15.5–20	bal	7.5Fe
CY-40	...	0.40	1.50	3.00	0.03	0.03	14–17	bal	11.0Fe
CZ-100	...	1.00	1.50	2.00	0.03	0.03	...	bal	3.0Fe, 1.25Cu
N-12M	...	0.12	1.00	1.00	0.04	0.03	1.0	bal	0.26–0.33Mo, 0.60V, 2.50Co, 6.0Fe
M-35	...	0.35	1.50	2.00	0.03	0.03	...	bal	28–33Cu, 3.5Fe
HA	...	0.20	0.35–0.65	1.00	0.04	0.04	8–10	...	0.90–1.20Mo
HC	446	0.50	1.00	2.00	0.04	0.04	26–30	4	0.5Mo(c)
HD	327	0.50	1.50	2.00	0.04	0.04	26–30	4–7	0.5Mo(c)
HE	...	0.20–0.50	2.00	2.00	0.04	0.04	26–30	8–11	0.5Mo(c)
HF	302B	0.20–0.40	2.00	2.00	0.04	0.04	18–23	8–12	0.5Mo(c)
HH	309	0.20–0.50	2.00	2.00	0.04	0.04	24–28	11–14	0.5Mo(c), 0.2N
HI	...	0.20–0.50	2.00	2.00	0.04	0.04	26–30	14–18	0.5Mo(c)
HK	310	0.20–0.60	2.00	2.00	0.04	0.04	24–28	18–22	0.5Mo(c)
HL	...	0.20–0.60	2.00	2.00	0.04	0.04	28–32	18–22	0.5Mo(c)
HN	...	0.20–0.50	2.00	2.00	0.04	0.04	19–23	23–27	0.5Mo(c)
HP	...	0.35–0.75	2.00	2.50	0.04	0.04	24–28	33–37	0.5Mo(c)
HP-50WZ	...	0.45–0.55	2.00	2.00	0.04	0.04	24–28	33–37	4.0–6.0W, 0.2–1.0Zr
HT	330	0.35–0.75	2.00	2.50	0.04	0.04	15–19	33–37	0.5Mo(c)
HU	...	0.35–0.75	2.00	2.50	0.04	0.04	17–21	37–41	0.5Mo(c)
HW	...	0.35–0.75	2.00	2.50	0.04	0.04	10–14	58–62	0.5Mo(c)
HX	...	0.35–0.75	2.00	2.50	0.04	0.04	15–19	64–68	0.5Mo(c)

(a) Cast alloy chemical composition ranges are not the same as the wrought composition ranges; buyers should use cast alloy designations for proper identification of castings. (b) Maximum, unless range is given. (c) Molybdenum not intentionally added. Source: Ref 3

service. An important factor pertaining to the corrosion behavior of these alloys is chromium content. Chromium imparts resistance to oxidation and sulfidation at high temperatures by forming a passive oxide film. Heat-resistant casting alloys must also have good resistance to carburization. More information on the corrosion of metals and alloys in high-temperature gases is available in the article "Fundamentals of Corrosion in Gases" in this Volume.

Oxidation. Resistance to oxidation increases directly with chromium content (Fig. 9). For the most severe service at temperatures above 1095 °C (2000 °F), 25% or more chromium is required. Additions of nickel, silicon, manganese, and aluminum promote the formation of relatively impermeable oxide films that retard further scaling. Thermal cycling is extremely damaging to oxidation resistance because it leads to breaking, cracking, or spalling of the protective oxide film. The best performance is obtained with austenitic alloys containing 40 to 50% combined nickel and chromium. Figure 10 shows the behavior of the H-type grades.

Sulfidation environments are becoming increasingly important. Petroleum processing, coal conversion, utility and chemical applications, and waste incineration have heightened the need for alloys resistant to sulfidation attack in relatively weak oxidizing or reducing environments. Fortunately, high chromium and silicon contents increase resistance to sulfur-bearing environments. On the other hand, nickel has been found to be detrimental to the most aggressive gases. The problem is attributable to the formation of low-melting nickel-sulfur eutectics. These produce highly destructive liquid phases at temperatures even below 815 °C (1500 °F). Once formed, the liquid may run onto adjacent surfaces and rapidly corrode other metals. The behavior of H-type grades in sulfidizing environments is represented in Fig. 11.

Carburization. High alloys are often used in nonoxidizing atmospheres in which carbon diffusion into metal surfaces is possible. Depending on chromium content, temperature, and carburizing potential, the surface may become extremely rich in chromium carbides, rendering it hard and possibly susceptible to cracking. Silicon and nickel are thought to be beneficial and enhance resistance to carburization.

Corrosion Behavior of C-Type Alloys

The ACI C-type (for liquid corrosion service) stainless steels must resist corrosion in the various environments in which they regularly serve. In this section, the general principles and important highlights of corrosion behavior will be discussed as influenced by the metallurgy of these materials. Topics include general corrosion, intergranular corrosion, localized corrosion, corrosion fatigue, and stress corrosion.

General Corrosion of Martensitic Alloys. The martensitic grades include CA-15, CA-15M, CA-6NM, CA-6NM-B, CA-40, CB-7Cu-1 and CB-7Cu-2. These alloys are generally used in applications requiring high strength and some corrosion resistance.

Alloy CA-15 typically exhibits a microstructure of martensite and ferrite. This alloy contains the minimum amount of chromium to be considered a stainless steel (11 to 14% Cr) and as such may not be used in aggressive environments. It does, however, exhibit good atmospheric-corrosion resistance, and it resists staining by many organic

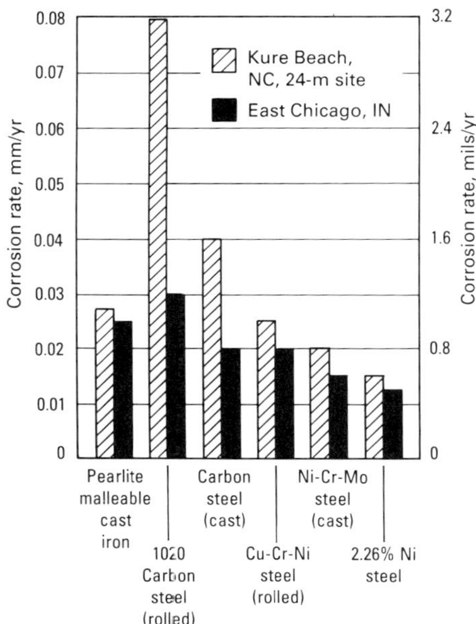

Fig. 5 Comparison of corrosion rates of cast steels, malleable cast iron, and wrought steel after 3 years of exposure in two atmospheres. Source: Ref 1

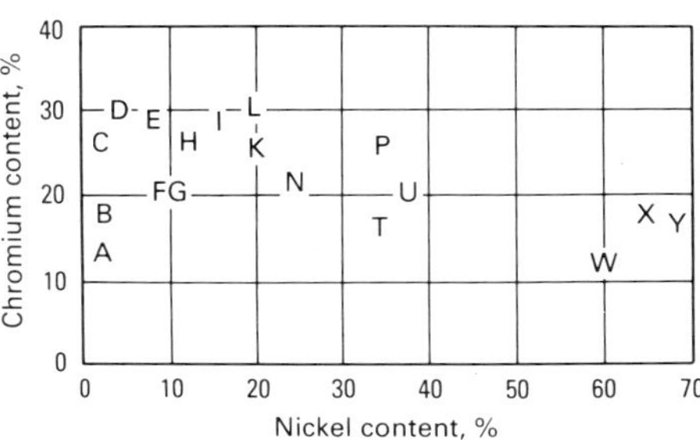

Fig. 6 Chromium and nickel contents in ACI standard grades of heat- and corrosion-resistant castings. See text for details. Source: Ref 2

environments. Alloy CA-15M may contain slightly more molybdenum than CA-15 (up to 1% Mo) and therefore may have improved general corrosion resistance in relatively mild environments. Alloy CA-6NM is similar to CA-15M except that it contains more nickel and molybdenum, which improves its general corrosion resistance. Alloy CA-6NM-B is a lower-carbon version of this alloy. The lower strength level promotes resistance to sulfide stress cracking. Alloy CA-40 is a higher-strength version of CA-15, and it also exhibits excellent atmospheric-corrosion resistance after a normalize and temper heat treatment. Microstructurally, the CB-7Cu alloys usually consist of mixed martensite and ferrite, and because of the increased chromium and nickel levels compared to the other martensitic alloys, they offer improved corrosion resistance to seawater and some mild acids. These alloys also have good atmospheric-corrosion resistance. The CB-7Cu alloys are hardenable and offer the possibility of increased strength and improved corrosion resistance among the martensitic alloys.

General Corrosion of Ferritic Alloys. Alloys CB-30 and CC-50 are higher-carbon and higher-chromium alloys than the CA alloys previously mentioned. Each alloy is predominantly ferritic, although a small amount of martensite may be found in CB-30. Alloy CB-30 contains 18 to 21% Cr and is used in chemical-processing and oil-refining applications. The chromium content is sufficient to have good corrosion resistance to many acids, including nitric acid (HNO_3). Figure 12 shows an isocorrosion diagram for CB-30 in HNO_3. Alloy CC-50 contains substantially more chromium (26 to 30%) and offers relatively high resistance to localized corrosion and high resistance to many acids, including dilute H_2SO_4 and such oxidizing acids as HNO_3.

General Corrosion of Austenitic and Duplex Alloys. Alloy CF-8 typically contains approximately 19% Cr and 9% Ni and is essentially the cast equivalent of AISI 304-type wrought alloys. Alloy CF-8 may be fully austenitic, but it more commonly contains some residual ferrite (3 to 30%) in an austenite matrix. In the solution-treated condition, this alloy has excellent resistance to a wide variety of acids. It is particularly resistant to highly oxidizing acids, such as boiling HNO_3. Figure 13 shows isocorrosion diagrams for CF-8 in HNO_3, phosphoric acid (H_3PO_4), and sodium hydroxide (NaOH). The duplex nature of the microstructure of this alloy imparts additional resistance to stress-corrosion cracking (SCC) compared to its wholly austenitic counterparts. Alloy CF-3 is a reduced-carbon version of CF-8 with essentially identical corrosion resistance except that CF-3 is much less susceptible to sensitization (Fig. 14). For applications in which the corrosion resistance of the weld heat-affected zone (HAZ) may be critical, CF-3 is a common material selection.

Alloys CF-8A and CF-3A contain more ferrite than their CF-8 and CF-3 counterparts. Because the higher ferrite content is achieved by increasing the chromium/nickel equivalent ratio, the CF-8A and CF-3A alloys may have slightly higher chromium or slightly lower nickel contents than the low-ferrite equivalents. In general, the corrosion resistance is very similar, but the strength increases with ferrite content. Because of the high ferrite content, service should be restricted to temperatures below 400 °C (750 °F) due to the possibility of severe embrittlement. Alloy CF-8C is the niobium-stabilized grade of the CF-8 alloy class. This alloy contains small amounts of niobium, which tend to form carbides preferentially over chromium carbides and improve intergranular corrosion resistance in applications involving relatively high service temperatures.

Alloys CF-8M, CF-3M, CF-8MA, and CF-3MA are molybdenum-bearing (2 to 3%) versions of the CF-8 and CF-3 alloys. The addition of 2 to 3% Mo increases resistance to corrosion by seawater and improves resistance to many chloride-bearing environments. The presence of 2 to 3% Mo also improves crevice corrosion and pitting resistance compared to the CF-8 and CF-3 alloys. Molybdenum-bearing alloys are generally not as resistant to highly oxidizing environments (this is particularly true for boiling HNO_3), but for weakly oxidizing environments and reducing environments, Mo-bearing alloys are generally superior.

Alloy CF-16F is a selenium-bearing free-machining grade of cast stainless steel. Because

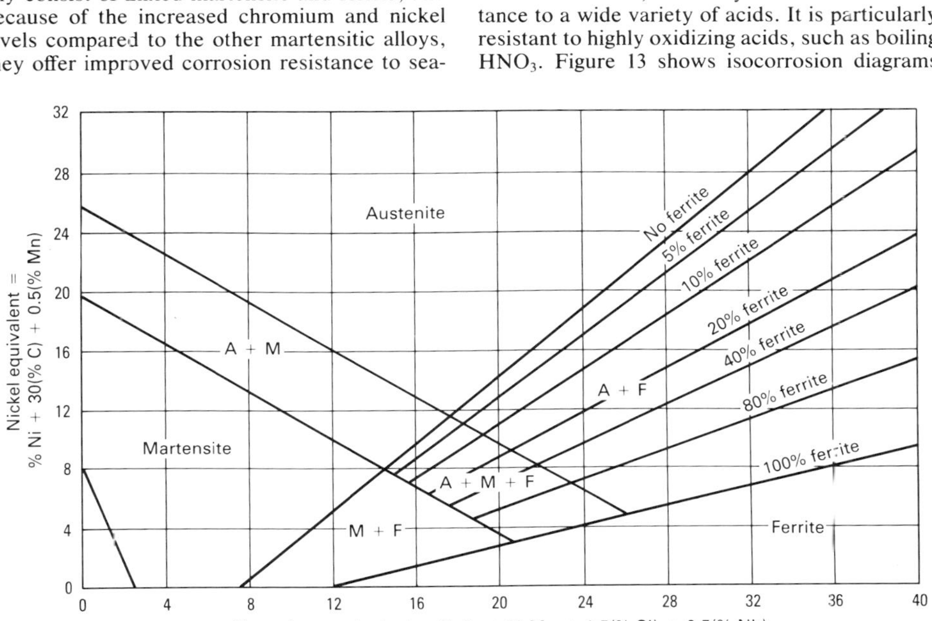

Fig. 7 Schaeffler diagram showing the amount of ferrite and austenite present in weldments as a function of chromium and nickel equivalents. Source: Ref 2

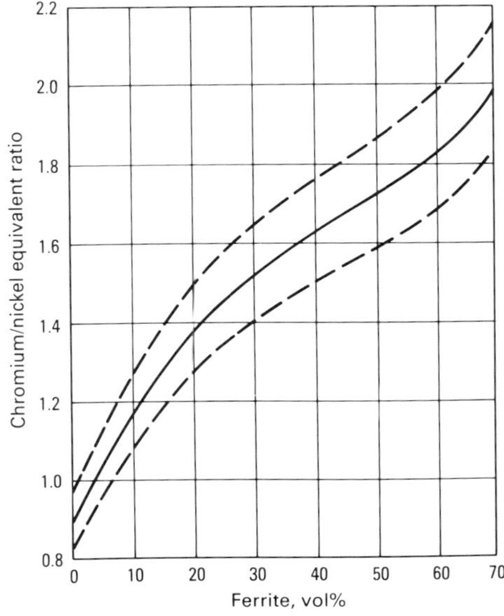

Fig. 8 Schoefer diagram for estimating the average ferrite content in austenitic iron-chromium-nickel alloy castings. Source: Ref 2

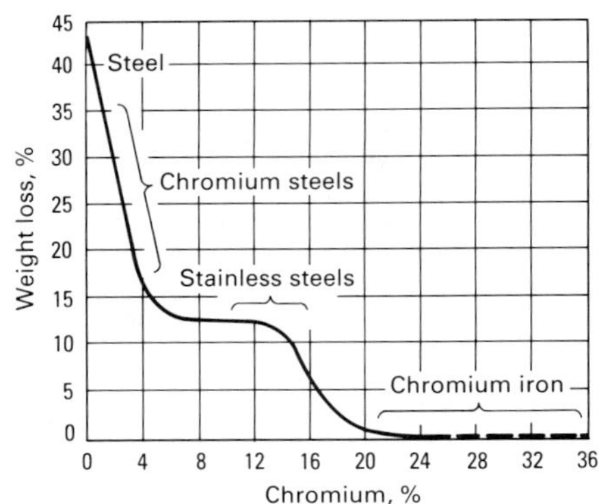

Fig. 9 Effect of chromium on oxidation resistance of cast steels. Specimens (13-mm, or 0.5-in., cubes) were exposed for 48 h at 1000 °C (1830 °F). Source: Ref 3

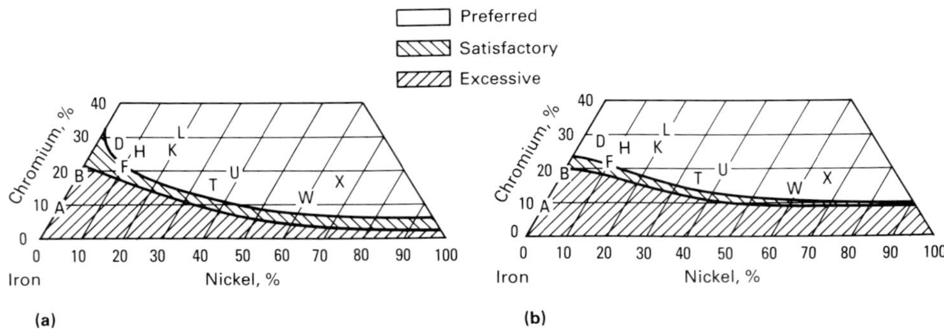

Fig. 10 Corrosion behavior of ACI H-type (heat-resistant) alloy castings in air (a) and in oxidizing flue gases containing 5 grains of sulfur per 2.8 m^3 (100 ft^3) of gas (b). Source: Ref 3

CF-16F nominally contains 19% Cr and 10% Ni, it has adequate corrosion resistance to a wide range of corrodents, but the large number of selenide inclusions makes surface deterioration and pitting definite possibilities.

Alloy CF-20 is a fully austenitic, relatively high-strength corrosion-resistant alloy. The 19% Cr content provides resistance to many types of oxidizing acids, but the high carbon content makes it imperative that this alloy be utilized in the solution-treated condition for environments known to cause intergranular corrosion.

Alloy CE-30 is a nominally 27Cr-9Ni alloy that normally contains 10 to 20% ferrite in an austenite matrix. The high carbon and ferrite contents provide relatively high strength. The high chromium content and duplex structure act to minimize corrosion because of the formation of chromium carbides in the microstructure. This particular alloy is known for good resistance to sulfurous acid and sulfuric acid, and it is extensively used in the pulp and paper industry (see the article "Corrosion in the Pulp and Paper Industry" in this Volume).

Alloy CG-8M is slightly more highly alloyed than the CF-8M alloys, with the primary addition being increased molybdenum (3 to 4%). The increased amount of molybdenum provides superior corrosion resistance to halide-bearing media and reducing acids, particularly H_2SO_3 and H_2SO_4 solutions. The high molybdenum content, however, renders CG-8M generally unsuitable in highly oxidizing environments.

Alloy CD-4MCu is the most highly alloyed material in this group of alloys; it has a nominal composition of Fe-26Cr-5Ni-2Mo-3Cu. The chromium/nickel equivalent ratio for this alloy is quite high, and a microstructure containing approximately equal amounts of ferrite and austenite is common. The low carbon content and high chromium content render the alloy relatively immune to intergranular corrosion. High chromium and molybdenum provide a high degree of localized corrosion resistance (crevices and pitting), and the duplex microstructure provides SCC resistance in many environments. This alloy can be precipitation hardened to provide strength and is also relatively resistant to abrasion and erosion-corrosion. Figures 15 and 16 show isocorrosion diagrams for CD-4MCu in HNO_3 and H_2SO_4, respectively.

Fully Austenitic Alloys. Alloys CH-10 and CH-20 are fully austenitic and contain 22 to 26% Cr and 12 to 15% Ni. The high chromium content minimizes the tendency toward the formation of chromium-depleted zones due to sensitization. These alloys are used for handling paper pulp solutions and are known for good resistance to dilute H_2SO_4 and HNO_3.

Alloy CK-20 contains 23 to 27% Cr and 19 to 22% Ni and is less susceptible than CH-20 to intergranular corrosion attack in many acids after brief exposures to the chromium carbide formation temperature range. Maximum corrosion resistance is achieved by solution treatment. Alloy CK-20 possesses good corrosion resistance to many acids and, because of its fully austenitic structure, can be used at relatively high temperature.

Alloy CN-7M, with a nominal composition of Fe-29Ni-20Cr-2.5Mo-3.5Cu, exhibits excellent corrosion resistance in a wide variety of environments and is often used for H_2SO_4 service. Figure 17 shows isocorrosion diagrams for CN-7M in H_2SO_4, HNO_3, H_3PO_4, and NaOH. Relatively high resistance to intergranular corrosion and SCC make this alloy attractive for very many applications. Although relatively highly alloyed, the fully austenitic structure of CN-7M may lead to SCC susceptibility for some environments and stress states.

Intergranular Corrosion of Austenitic and Duplex Alloys. The optimum corrosion resistance for these alloys is developed by solution treatment. Depending on the specific alloy in question, temperatures between 1040 and 1205 °C (1900 and 2200 °F) are required to ensure complete solution of all carbides and phases, such as

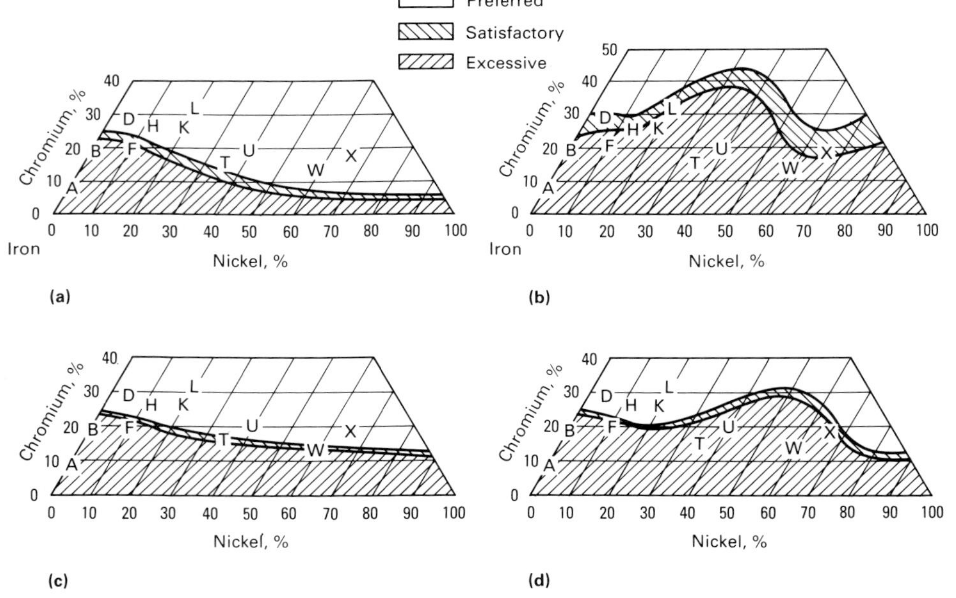

Fig. 11 Corrosion behavior of ACI H-type alloys in 100-h tests at 980 °C (1800 °F) in reducing sulfur-bearing gases. (a) Gas contained 5 grains of sulfur per 2.8 m^3 (100 ft^3) of gas. (b) Gas contained 300 grains of sulfur per 2.8 m^3 (100 ft^3) of gas. (c) Gas contained 100 grains of sulfur per 2.8 m^3 (100 ft^3) of gas; test at constant temperature. (d) Same sulfur content as gas in (c), but cooled to 150 °C (300 °F) each 12 h

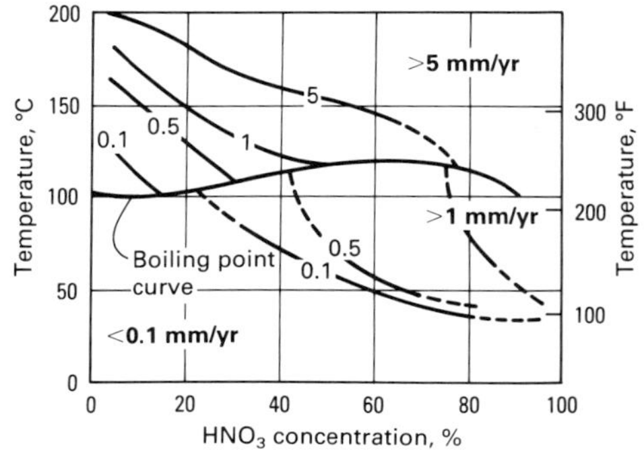

Fig. 12 Isocorrosion diagram for ACI CB-30 in HNO_3. Castings were annealed at 790 °C (1450 °F), furnace cooled to 540 °C (1000 °F), and then air cooled to room temperature.

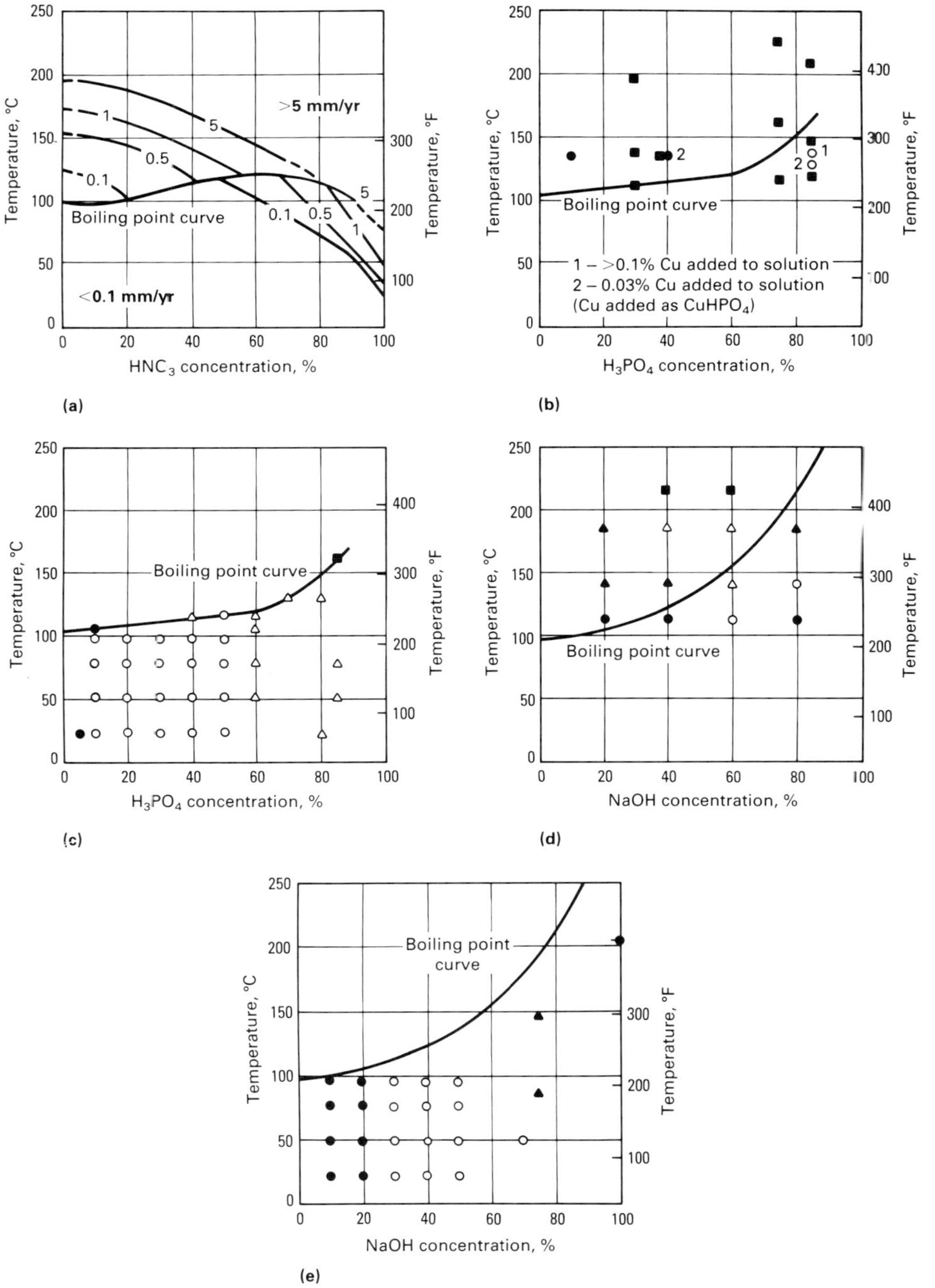

(a) (b) (c) (d) (e)

Fig. 13 Isocorrosion diagrams for ACI CF-8 in HNO_3 (a), H_3PO_4 (b and c), and NaOH solutions (d and e). (b) and (d) Tests performed in a closed container at equilibrium pressure. (c) and (e) Tested at atmospheric pressure

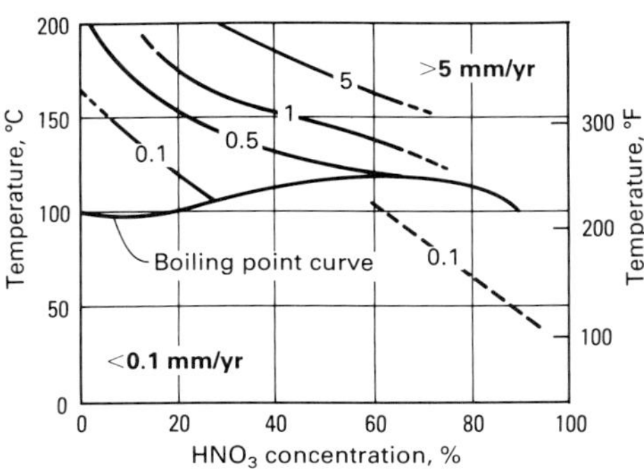

Fig. 14 Isocorrosion diagram for solution-treated quenched and sensitized ACI CF-3 in HNO_3

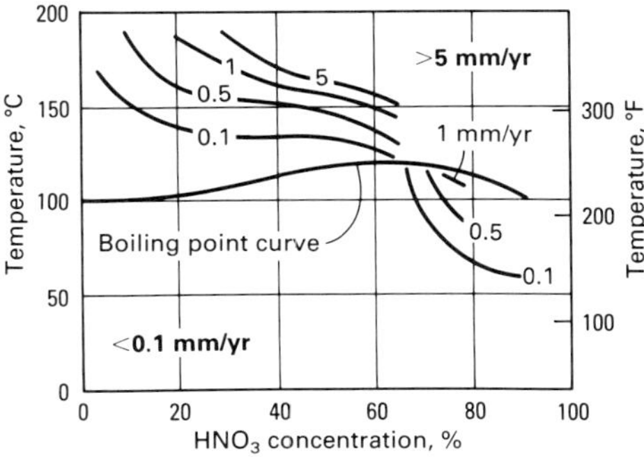

Fig. 15 Isocorrosion diagram for ACI CD-4MCu in HNO_3. The material was solution treated at 1120 °C (2050 °F) and water quenched.

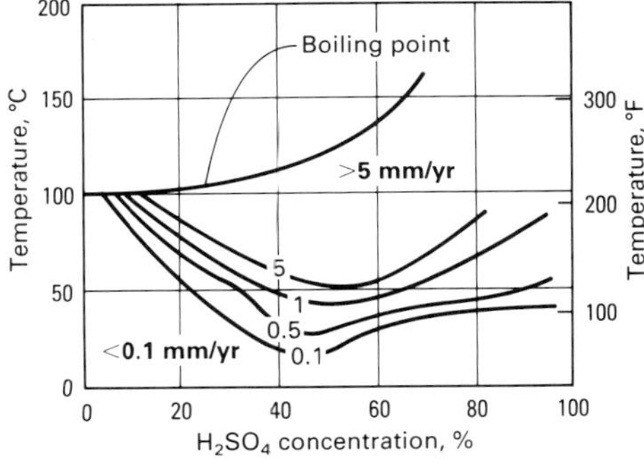

Fig. 16 Isocorrosion diagram for ACI CD-4MCu in H_2SO_4. The material was solution annealed at 1120 °C (2050 °F) and water quenched.

σ and χ, that sometimes form in highly alloyed stainless steels. Alloys containing relatively high total alloy content, particularly high molybdenum content, often require the higher solution treatment temperature. Water quenching from the temperature range of 1040 to 1205 °C (1900 to 2200 °F) normally completes the solution treatment.

Failure to solution treat a particular alloy or an improper solution treatment may seriously compromise the observed corrosion resistance in service. Inadvertent or unavoidable heat treatment in the temperature range of 480 to 820 °C (900 to 1500 °F)—for example, welding—may destroy the intergranular corrosion resistance of the alloy. When austenitic or duplex (ferrite in austenite matrix) stainless steels are heated in or cooled slowly through this temperature range, chromium-rich carbides form at grain boundaries in austenitic alloys and at ferrite/austenite interfaces in duplex alloys. These carbides deplete the surrounding matrix of chromium, thus diminishing the corrosion resistance of the alloy. An alloy in this condition of reduced corrosion resistance due to the formation of chromium carbides is said to be sensitized. In small amounts, these carbides may lead to localized pitting in the alloy, but if the chromium-depleted zones are extensive throughout the alloy or HAZ of a weld, the alloy may disintegrate intergranularly in some environments.

If solution treatment of the alloy after casting and/or welding is impractical or impossible, the metallurgist has several tools from which to choose to minimize potential intergranular corrosion problems. The low carbon grades CF-3 and CF-3M are commonly used as a solution to the sensitization incurred during welding. The low carbon content (0.03% C maximum) of these alloys precludes the formation of an extensive number of chromium carbides. In addition, these alloys normally contain 3 to 30% ferrite in an

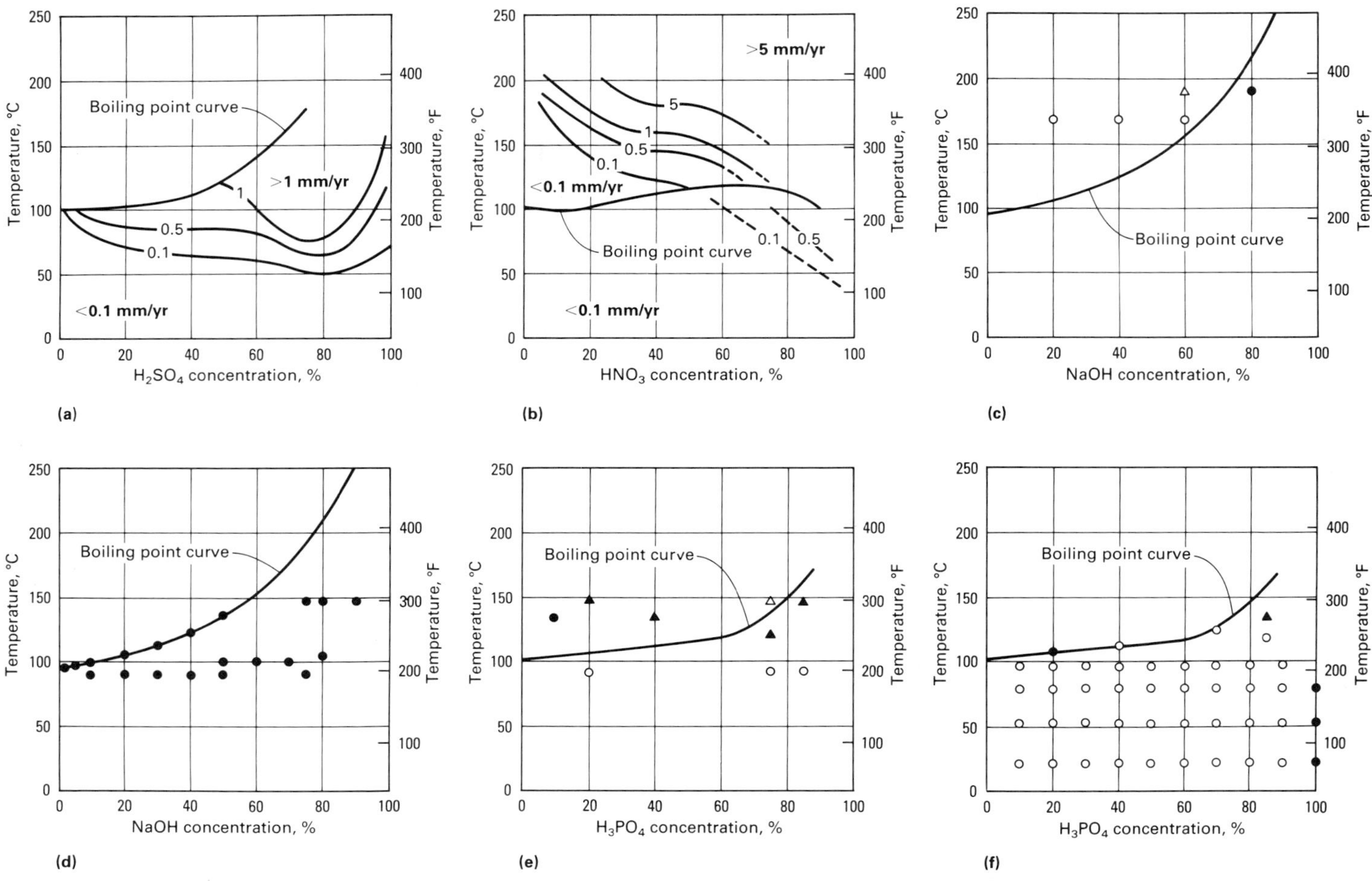

Fig. 17 Isocorrosion diagrams for solution-annealed and quenched ACI CN-7M in H_2SO_4, HNO_3, NaOH, and H_3PO_4. (a), (b), (d), and (f) Tested at atmospheric pressure. (c) and (e) Tested at equilibrium pressure in a closed container. See Fig. 13 for legend.

austenitic matrix. By virtue of rapid carbide precipitation kinetics at ferrite/austenite interfaces compared to austenite/austenite interfaces, carbide precipitation is confined to ferrite/austenite boundaries in alloys containing a minimum of about 3 to 5% ferrite (Ref 4, 5). If the ferrite network is discontinuous in the austenite matrix (depending on the amount, size, and distribution of ferrite pools), then extensive intergranular corrosion will not be a problem in most of the environments to which these alloys would be subjected.

An example of attack at the ferrite/austenite boundaries is shown in Fig. 18. These low-carbon alloys need not sacrifice significant strength compared to their high-carbon counterparts, because nitrogen may be added to increase strength. However, a large amount of nitrogen will begin to reduce the ferrite content, which will cancel some of the strength gained by interstitial hardening. Appropriate adjustment of the chromium/nickel equivalent ratio is beneficial in such cases. Fortunately, nitrogen is also beneficial to the corrosion resistance of austenitic and duplex stainless steels (Ref 6). Nitrogen seems to retard sensitization and improve the resistance to pitting and crevice corrosion of many stainless steels.

The standard practices of ASTM A 262 (Ref 7) are commonly implemented to predict and measure the susceptibility of austenitic and duplex stainless steels to intergranular corrosion. Table 7 indicates some representative results for CF-type alloys as tested according to practices A, B, and C of Ref 7 as well as two electrochemical tests described in Ref 10 and 11. Table 8 lists the compositions of the alloys investigated. The data indicate the superior resistance of the low-carbon alloys to intergranular corrosion. Table 7 also indicates that for highly oxidizing environments (represented here by A 262C-boiling HNO_3) the CF-3 and CF-3M alloys are equivalent in the solution-treated condition but that subsequent heat treatment causes the corrosion resistance of the CF-3M alloys to deteriorate rapidly for service in oxidizing environments (Ref 9). In addition, the degree of chromium depletion necessary to cause susceptibility to intergranular corrosion appears to increase in the presence of molybdenum (Ref 5). The passive film stability imparted by molybdenum may offset the loss of solid-solution chromium for mild degrees of sensitization.

Intergranular Corrosion of Ferritic and Martensitic Alloys. Ferritic alloys may also be sensitized by the formation of extensive chromium carbide networks, but because of the high bulk chromium content and rapid diffusion rates of chromium in ferrite, the formation of carbides can be tolerated if the alloy has been slowly cooled from a solutionizing temperature of 780 to 900 °C (1435 to 1650 °F). The slow cooling allows replenishment of the chromium adjacent to carbides. Martensitic alloys normally do not contain sufficient bulk chromium to be used in applications in which intergranular corrosion is likely to be of concern. Typical chromium contents for martensitic alloys may be as low as 11 to 12%.

Localized Corrosion. Austenitic and martensitic alloys display a tendency toward localized corrosion. The conditions conducive to this behavior may be any situation in areas where flow is restricted and an oxygen concentration cell may be established. Duplex alloys have been found to be less susceptible. Localized corrosion is particularly acute in environments containing chloride ion (Cl^-) and in acidic solutions.

Increasing the alloy content improves resistance to localized corrosion. Molybdenum has long been recognized as effective in reducing localized corrosion, although it is not a total solution. Excellent results have been obtained with CG-8M, but the CF-3M or CN-7M alloys are readily attacked. Nitrogen is also effective at retarding localized corrosion.

It has been suggested that resistance to pitting is good when a crevice factor (%Cr + 3(%Mo) + 15(%N)) exceeds 35 (Fig. 19). Another technique for comparing alloy composition resistance to localized corrosion is to ascertain the critical crevice temperature (CCT). This involves determining the maximum temperature at which no

Table 7 Intergranular corrosion test results for ACI casting alloys

Metallurgical condition	Test(a)	Alloy(b)/Test results(c) CF-8 (4)	CF-8 (11)	CF-8 (20)	CF-8M (5)	CF-8M (11)	CF-8M (20)	CF-3 (2)	CF-3 (5)	CF-3 (8)	CF-3M (5)	CF-3M (9)	CF-3M (16)
Solution treated	A 262A	P	P	P	P	P	P	P	P	P	P	P	P
	A 262B	P	P	P	P	P	P	P	P	P	P	P	P
	A 262C	P	P	P	P	P	P	P	P	P	P	P	P
	EPR	P	P	P	P	P	P	P*	P*	P*	P	P	P
	JEPR	P	P	P	P	P	P	P	P	P	P	P	P
Simulated weld repair	A 262A	X	X	X	X	X	X	P	P	P	P	P	P
	A 262B	X	X	X	X	X	X	P	P	P	P	P	P
	A 262C	X	X	X	X	X	X	P	P	P	P	P	P
	EPR	X	X	X	P	P	P	P*	P*	P*	P	P	P
	JEPR	X	X	X	P	P	P	P	P	P	P	P	P
Solution treated, held 1 h at 650 °C (1200 °F)	A 262A	X	X	X	X	X	X	X	X	X	X	X	X
	A 262B	X	X	X	X	X	X	P	P	P	P	P	P
	A 262C	X	X	X	X	X	X	P	P	P	X	X	X
	EPR	X	X	X	X	X	X	X/P*	X/P*	X/P*	X/P	P	P
	JEPR	X	X	X	P	X	X	P	P	P	P	P	P
As-cast	A 262A	X	X	X	X	X	X	X	X	X	X	X	X
	A 262B	X	X	X	X	X	X	P	P	P	P	X	P
	A 262C	X	X	X	X	X	X	P**	P**	P**	X	X	X
	EPR	X	X	X	X	X	X	X/P*	X/P*	X/P*	X/P	X/P	P
	JEPR	X	X	X	X	X	X	X/P	P	P	P	P	P

(a) See Ref 7 for details of ASTM A 262 practices. EPR, electrochemical potentiokinetic reactivation test; see Ref 10 for details. JEPR, Japanese electrochemical potentiokinetic reactivation test; see Ref 11 for details. (b) Parenthetical value is the percentage of ferrite. See Table 8 for alloy compositions. (c) P, pass; X, fail, based on the following criteria: A 262A ditching, <10% = pass; A 262B, penetration rate < 0.64 mm/yr (25 mils/yr) = pass; A 262C, penetration rate < 0.46 mm/yr (18 mils/yr) and not increasing = pass; EPR, peak current density <100 μA/cm² (645 μA/in.²) = pass; JEPR, ratio <1% = pass. P*, pass, but matrix pitting complicates test results. X/P, near pass. X/P*, likely pass; small EPR indication complicated by matrix pitting. P**, pass; actual heat treatment 4 h at 650 °C (1200 °F) after solution treatment rather than as-cast. Source: Ref 5, 8, 9

Table 8 Composition of research alloys

Material	Ferrite number(a)	Element, % C	Mn	Si	P	S	Cr	Ni	Mo	N
CF-8 LO	4	0.058	0.60	1.52	0.012	0.013	18.53	9.98	0.02	0.02
CF-8 INT	11	0.086	0.84	1.10	0.031	0.012	19.90	8.73	0.50	0.02
CF-8 HI	20	0.066	0.79	1.25	0.031	0.011	20.81	8.85	0.45	0.02
CF-8M LO	5	0.063	0.94	1.21	0.011	0.014	18.26	11.17	2.28	0.02
CF-8M INT	11	0.083	1.20	1.20	0.030	0.013	19.78	9.53	2.21	0.02
CF-8M HI	20	0.071	1.19	1.16	0.030	0.011	19.92	9.40	1.95	0.02
CF-3 LO	2	0.016	0.98	1.12	0.010	0.008	17.36	10.10	0.10	0.04
CF-3 INT	5	0.023	0.68	1.24	0.011	0.009	19.35	10.27	0.10	0.06
CF-3 HI	8	0.015	0.67	1.09	0.013	0.006	19.82	8.73	0.10	0.04
CF-3M LO	5	0.027	0.96	0.85	0.011	0.010	17.55	12.00	2.18	0.04
CF-3M INT	9	0.027	1.04	1.02	0.009	0.009	18.78	10.79	2.12	0.03
CF-3M HI	16	0.022	0.94	1.14	0.012	0.007	19.85	10.08	2.26	0.02

(a) This value is the percentage of ferrite.

Table 9 CCTs for several common cast and wrought alloys

Alloy	Structure	CCT °C	°F	Ref
Wrought AISI type 317L	Austenitic	2	35	13
Cast CF-3M	90% austenite, 10% ferrite	2	35	12
Cast CN-7M	Austenitic	−1.1	30	12
Cast CF-8M	90% austenite, 10% ferrite	−2.5	28	15
Wrought AISI type 316L	Austenitic	−2.5	28	14
Wrought AISI type 316	Austenitic	−3	27	13

Note: See text and Ref 12 for information on CCTs.

crevice attack occurs during a 24-h testing period. These tests have been conducted on a number of cast stainless alloys; the results are given in Table 9. Although the CCT has been shown to correlate well with tests in aerated seawater (Ref 16), it must not be used as the maximum operating temperature in seawater or other chloride-containing media. The ferric chloride ($FeCl_3$) test environment is a very severe, highly oxidizing environment containing about 39 000 ppm Cl^- at a pH of about 1.4. Therefore, the $FeCl_3$ CCT is lower than that normally found in aerated seawater (Ref 16), which contains about 20 000 ppm Cl^- with a pH of about 7.5 to 8.0.

Corrosion fatigue is one of the most destructive and unpredictable corrosion-related failure mechanisms. Behavior is highly specific to the environment and alloy. The martensitic materials are degraded the most in both absolute and relative terms. If left to corrode freely in seawater, they have very little resistance to corrosion fatigue. This is remarkable in view of their very high strength and fatigue resistance in air.

Properties can be protected if suitable cathodic protection is applied. However, because these materials are susceptible to hydrogen embrittlement, cathodic protection must be carefully applied. Too large a protective potential will lead to catastrophic hydrogen-stress cracking.

Austenitic materials are also severely degraded in corrosion fatigue strength under conditions conducive to pitting, such as in seawater. However, they are easily cathodically protected without fear of hydrogen embrittlement and perform well in fresh waters. The corrosion fatigue behavior of duplex alloys has not been widely studied.

Stress-Corrosion Cracking. The SCC of cast stainless steels has been investigated for only a limited number of environments, heat treatments, and test conditions. From the limited information available, the following generalizations apply.

First, SCC resistance seems to improve as the composition is adjusted to provide increasingly

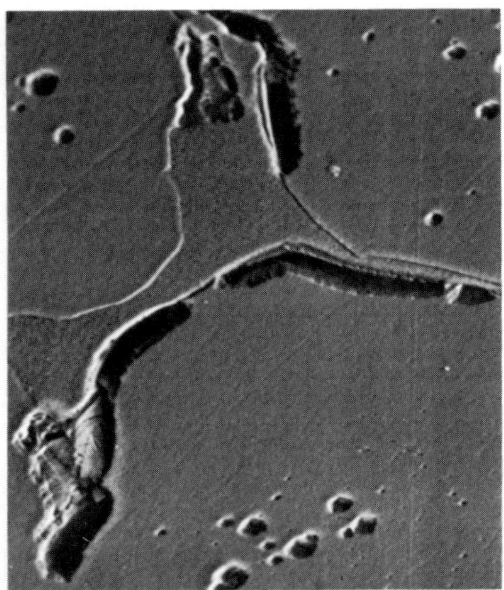

Fig. 18 Ferrite/austenite grain-boundary ditching in as-cast ACI CF-8. The specimen, which contained 3% ferrite, was EPR tested. SEM micrograph. 4550×. Source: Ref 5

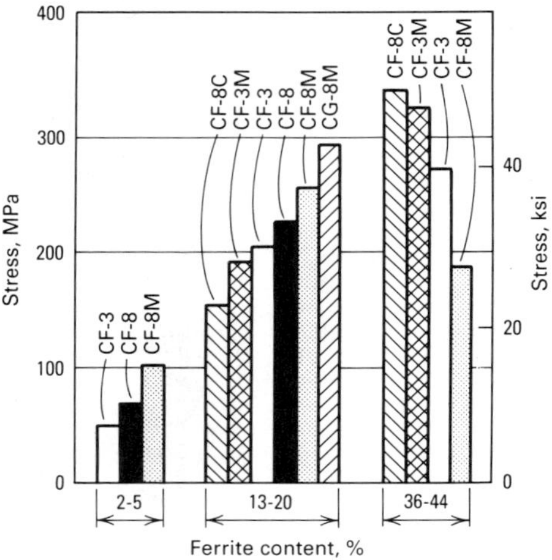

Fig. 20 Stress required to produce SCC in several ACI alloys with varying amounts of ferrite

greater amounts of ferrite in an austenitic matrix. This trend continues to a certain level, apparently near 50% ferrite (Fig. 20 and 21). Second, lower nickel contents tend to improve SCC resistance in cast duplex alloys, possibly because of its effect on ferrite content (Ref 17). Third, ferrite appears to be involved in a keying action in discouraging SCC. At low and medium stress levels, the ferrite tends to block the propagation of stress-corrosion cracks. This may be due to a change in composition and/or crystal structure across the austenite/ferrite boundary (Fig. 21). As the stress level increases, crack propagation may change from austenite/ferrite boundaries to transgranular propagation (Ref 17, 18). Finally, reducing the carbon content of cast stainless alloys—thus reducing the susceptibility to sensitization—improves SCC resistance. This is also true for wrought alloys (Ref 17, 19-21).

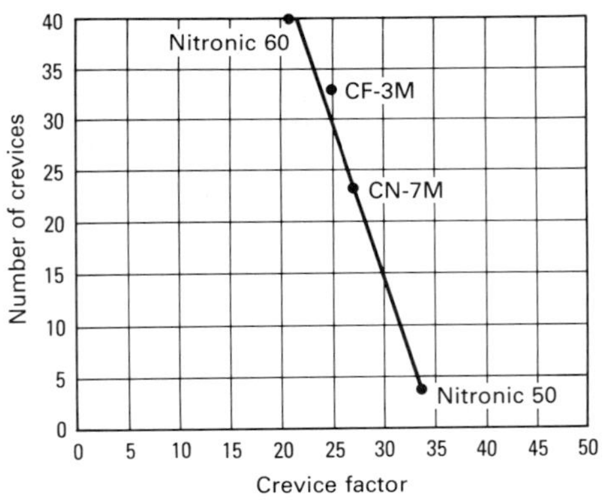

Fig. 19 Crevice corrosion resistance of various alloys in 5-day test in $FeCl_3$ at room temperature. See text and Ref 12 for explanation of crevice factor. Source: Ref 12

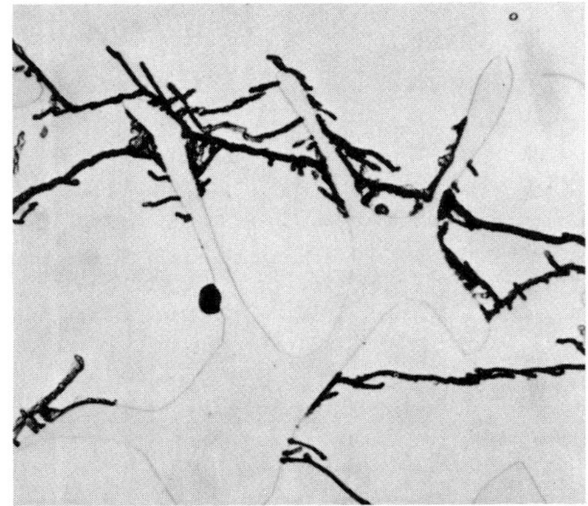

Fig. 21 Ferrite pools blocking the propagation of stress-corrosion cracks in a cast stainless steel

REFERENCES

1. C. Briggs, Ed., *Steel Casting Handbook*, 4th ed., Steel Founders' Society of America, 1970, 662-667
2. M. Prager, Cast High Alloy Metallurgy, in *Steel Casting Metallurgy*, J. Svoboda, Ed., Steel Founders' Society of America, 1984, 221-245
3. C.E. Bates and L.T. Tillery, *Atlas of Cast Corrosion-Resistant Alloy Microstructures*, Steel Founders' Society of America, 1985
4. T.M. Devine, Mechanism of Intergranular Corrosion and Pitting Corrosion of Austenitic and Duplex 308 Stainless Steel, *J. Electrochem. Soc.*, Vol 126 (No. 3), 1979, p 374
5. E.E. Stansbury, C.D. Lundin, and S.J. Pawel, Sensitization Behavior of Cast Stainless Steels Subjected to Simulated Weld Repair, in *Proceedings of the 38th SFSA Technical and Operating Conference*, Steel Founders' Society of America, 1983, p 223
6. S.J. Pawel, Literature Review on the Role of Nitrogen in Austenitic Steels, *Steel Founders' Res. J.*, Issue 5, 1st Quarter, 1984
7. "Standard Practices for Detecting Susceptibility to Intergranular Attack in Austenitic Stainless Steels," A 262, *Annual Book of ASTM Standards*, American Society for Testing and Materials
8. S.J. Pawel, "The Sensitization Behavior of Cast Stainless Steels Subjected to Weld Repair," MS thesis, University of Tennessee, June 1983
9. S.J. Pawel, E.E. Stansbury, and C.D. Lundin, Evaluation of Post Weld Repair Requirements for CF3 and CF3M Alloys—Exposure to Boiling Nitric Acid, in *First International Steel Foundry Congress Proceedings*, Steel Founders' Society of America, 1985, p 45
10. W.L. Clarke, R.L. Cowan, and W.L. Walker, Comparative Methods for Measuring Degree of Sensitization in Stainless Steel, in *Intergranular Corrosion of Stainless Alloys*, STP 656, R.F. Steigerwald, Ed., American Society for Testing and Materials, 1978, p 99
11. M. Akashi *et al.*, Evaluation of IGSCC Susceptibility of Austenitic Stainless Steels Using Electrochemical Methods, *Boshoku Gijutsu (Corros. Eng.)*, Vol 29, 1980, p 163 (BTSITS trans.)
12. J.A. Larson, 1984 SCRATA Exchange Lecture: New Developments in High Alloy Cast Steels, in *Proceedings of the 39th SFSA T & O Conference*, Steel Founders' Society of America, 1984, p 229-239
13. J.R. Maurer and J.R. Kearns, "Enhancing the Properties of a 6% Molybdenum Austenitic Alloy With Nitrogen," Paper 172, presented at Corrosion/85, National Association of Corrosion Engineers, 1985
14. A.P. Bond and H.J. Dundas, "Resistance of Stainless Steels to Crevice Corrosion in Seawater," Paper 26, presented at Corrosion/84, National Association of Corrosion Engineers, 1984
15. A. Poznansky and P.J. Grobner, "Highly Alloyed Duplex Stainless Steels," Paper 8410-026, presented at the International Conference on New Developments in Stainless Steel Technology, Detroit, MI, American Society for Metals, Sept 1984
16. A. Garner, Crevice Corrosion of Stainless Steels in Seawater: Correlation of Field Data With Laboratory Ferric Chloride Tests, *Corrosion*, Vol 37 (No. 3), March 1981, p 178-184
17. S. Shimodaira *et al.*, Mechanisms of Transgranular Stress Corrosion Cracking of Duplex and Ferrite Stainless Steels, in *Stress Corrosion Cracking and Hydrogen Embrittlement in Iron Base Alloys*, NACE Reference Book 5, National Association of Corrosion Engineers, 1977
18. P.L. Andresen and D.J. Duquette, The Effect of Cl^- Concentration and Applied Potential on the SCC Behavior of Type 304 Stainless Steel in Deaerated High Temperature Water, *Corrosion*, Vol 36 (No. 2), 1980, p 85-93
19. J.N. Kass *et al.*, Stress Corrosion Cracking of Welded Type 304 and 304L Stainless Steel Under Cyclic Loading, Corrosion, Vol 36 (No. 6), 1980, p 299-305
20. J.N. Kass *et al.*, Comparative Stress Corrosion Behavior of Welded Austenitic Stainless Steel Pipe in High Temperature High Purity Oxygenated Water, *Corrosion*, Vol 36 (No. 12), 1980, p 686-698
21. G. Cragnolino *et al.*, Stress Corrosion Cracking of Sensitized Type 304 Stainless Steel in Sulfate and Chloride Solutions at 250 and 100C, *Corrosion*, Vol 37 (No. 6), 1981, p 312-319

Corrosion of Aluminum and Aluminum Alloys

E.H. Hollingsworth (retired) and H.Y. Hunsicker (retired), Aluminum Company of America*

ALUMINUM, as indicated by its position in the electromotive force (emf) series, is a thermodynamically reactive metal; among structural metals, only beryllium and magnesium are more reactive. Aluminum owes its excellent corrosion resistance and its usage as one of the primary metals of commerce to the barrier oxide film that is bonded strongly to its surface and that, if damaged, re-forms immediately in most environments. On a surface freshly abraded and then exposed to air, the barrier oxide film is only 1 nm (10 Å) thick but is highly effective in protecting the aluminum from corrosion.

The oxide film that develops in normal atmospheres grows to thicknesses much greater than 1 nm (10 Å) and is composed of two layers (Ref 1). The inner oxide next to the metal is a compact amorphous barrier layer whose thickness is determined solely by the temperature of the environment. At any given temperature, the limiting barrier thickness is the same in oxygen, dry air, or moist air. Covering the barrier layer is a thicker, more permeable outer layer of hydrated oxide. Most of the interpretation of aluminum corrosion processes has been developed in terms of the chemical properties of these oxide layers.

The natural film can be visualized as the result of a dynamic equilibrium between opposing forces—those tending to form the compact barrier layer and those tending to break it down. If the destructive forces are absent, as in dry air, the natural film will consist only of the barrier layer and will form rapidly to the limiting thickness. If the destructive forces are too strong, the oxide will be hydrated faster than it is formed, and little barrier will remain. Between these extremes, where the opposing forces reach a reasonable balance, relatively thick (20 to 200 nm, or 200 to 2000 Å) natural films are formed (Ref 2).

The conditions for thermodynamic stability of the oxide film are expressed by the Pourbaix (potential versus pH) diagram shown in Fig. 1. As shown by this diagram, aluminum is passive (is protected by its oxide film) in the pH range of about 4 to 8.5. The limits of this range, however, vary somewhat with temperature, with the specific form of oxide film present, and with the presence of substances that can form soluble complexes or insoluble salts with aluminum. The relative inertness in the passive range is further illustrated in Fig. 2, which gives results of weight loss measurements for alloy 3004-H14 specimens exposed in water and in salt solutions at various pH values.

Beyond the limits of its passive range, aluminum corrodes in aqueous solutions because its oxides are soluble in many acids and bases, yielding Al^{3+} ions in the former and AlO_2^- (aluminate) ions in the latter. There are, however, instances when corrosion does not occur outside the passive range, for example, when the oxide film is not soluble or when the film is maintained by the oxidizing nature of the solution (Ref 4).

Pitting Corrosion

Corrosion of aluminum in the passive range is localized, usually manifested by random formation of pits. The pitting-potential principle establishes the conditions under which metals in the passive state are subject to corrosion by pitting (Ref 5-7). Simply stated, pitting potential E_p is that potential in a particular solution above which pits will initiate and below which they will not.

Four laboratory procedures have been developed to measure E_p—one based on fixed current and the other three on controlled potential (Ref 8). The most widely used is controlled potential, in which the potential of a specimen, usually immersed in a deaerated electrolyte of interest, is made more positive and the resulting current density from the specimen measured. The potential at which the current density increases sharply and remains high is called the oxide breakdown potential, E_{br}. With polished specimens in many electrolytes, E_{br} is a close approximation of E_p, and the two are used interchangeably.

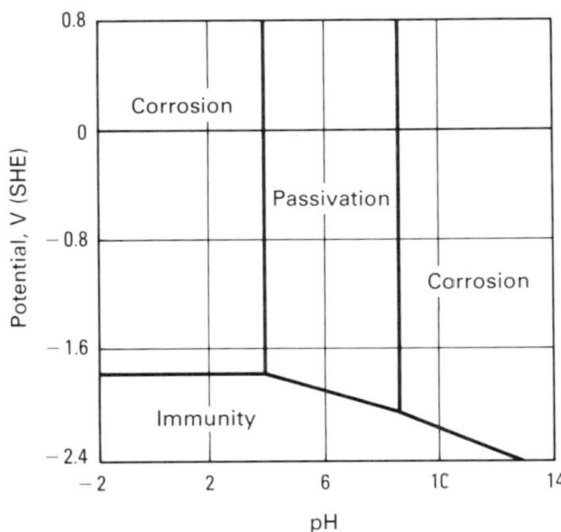

Fig. 1 Pourbaix diagram for aluminum with an $Al_2O_3 \cdot 3H_2O$ film at 25 °C (75 °F). Potential values are for the standard hydrogen electrode (SHE) scale. Source: Ref 3

An example is shown in Fig. 3. A specimen of aluminum alloy 1100 was immersed in a neutral deaerated sodium chloride (NaCl) solution, and the relationship between anode potential and current density was plotted (solid line, Fig. 3). At potentials more active than E_p, where the oxide layer can maintain its integrity, anodic polarization is easy, and corrosion is slow and uniform. Above E_p, anodic polarization is difficult, and the current density sharply increases. The oxide ruptures at random weak points in the barrier layer and cannot repair itself, and localized corrosion develops at these points.

Potential-current relationships for various cathodic reactions are indicated by the dashed lines in Fig. 3. Only when the cathodic reaction is sufficient to polarize the metal to its pitting potential will significant current flow and pitting corrosion start.

For aluminum, pitting corrosion is most commonly produced by halide ions, of which chloride (Cl^-) is the most frequently encountered in service. The effect of chloride ion concentration on the pitting potential of aluminum 1199 (99.99 + % Al) is shown in Fig. 4. Pitting of aluminum in halide solutions open to the air occurs because, in

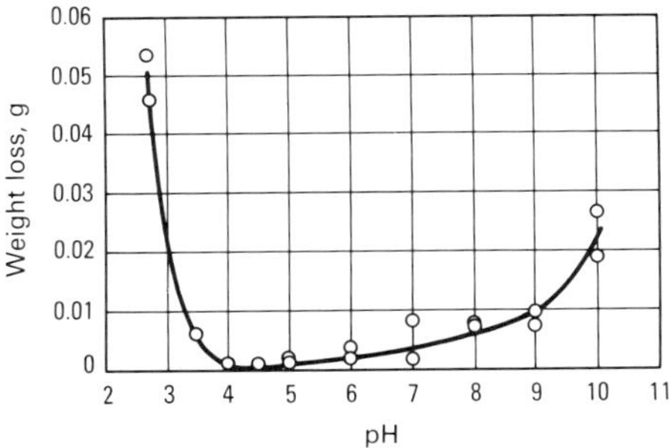

Fig. 2 Weight loss of alloy 3004-H14 exposed 1 week in distilled water and in solutions of various pH values. Specimens were 1.6 × 13 × 75 mm (0.06 × 0.5 × 3 in.). The pH values of solutions were adjusted with HCl and NaOH. Test temperature was 60 °C (140 °F).

*Revised by the ASM Committee on Corrosion of Aluminum: R.L. Horst (chairman), E.L. Colvin, and B.W. Lifka, Aluminum Company of America; S.C. Dexter, University of Delaware; F.N. Smith and T.E. Wright (retired), Alcan International Ltd.

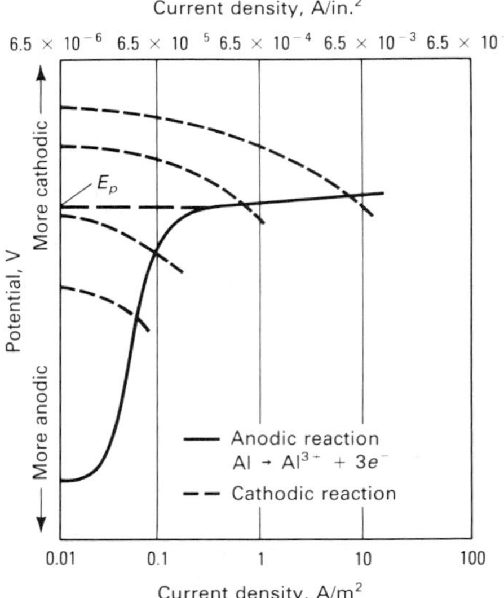

Fig. 3 Anodic-polarization curve for aluminum alloy 1100. Specimens were immersed in neutral deaerated NaCl solution free of cathodic reactant. Pitting develops only at potentials more cathodic than the pitting potential E_p. The intersection of the anodic curve for aluminum (solid line) with a curve for the applicable cathodic reaction (one of the representative dashed lines) determines the potential to which the aluminum is polarized, either by cathodic reaction on the aluminum itself or on another metal electrically connected to it. The potential to which the aluminum is polarized by a specific cathode reaction determines corrosion current density and corrosion rate.

the presence of oxygen, the metal is readily polarized to its pitting potential. In the absence of dissolved oxygen or other cathodic reactant, aluminum will not corrode by pitting because it is not polarized to its pitting potential. Generally, aluminum does not develop pitting in aerated solutions of most nonhalide salts because its pitting potential in these solutions is considerably more noble (cathodic) than in halide solutions, and it is not polarized to these potentials in normal service (Ref 7).

Pitting potentials for selected aluminum alloys in several electrolytes are reported in Ref 8. Examples of application of pitting-potential analysis to particular corrosion problems are given in Ref 9 and 10.

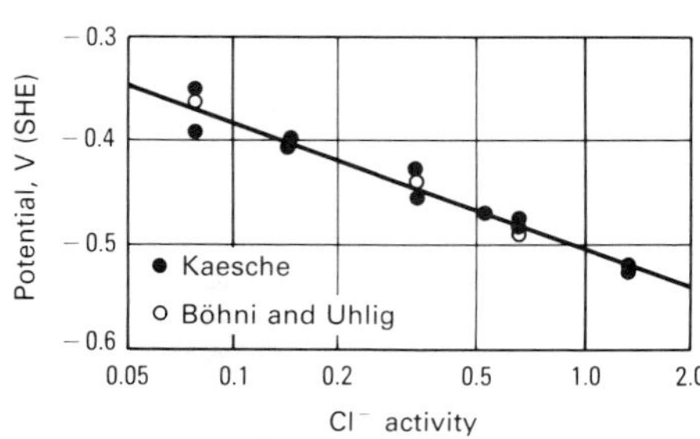

Fig. 4 Effect of chloride-ion activity on pitting potential of aluminum 1199 in NaCl solutions. Source: Ref 5 and 6.

Solution Potentials

Because of the electrochemical nature of most corrosion processes, relationships among solution potentials of different aluminum alloys, as well as between potentials of aluminum alloys and those of other metals, are of considerable importance. Furthermore, the solution-potential relationships among the microstructural constituents of a particular alloy significantly affect its corrosion behavior. Compositions of solid solutions and additional phases, as well as amounts and spatial distributions of the additional phases, may affect both the type and extent of corrosion.

The solution potential of an aluminum alloy is primarily determined by the composition of the aluminum-rich solid solution, which constitutes the predominant volume fraction and area fraction of the alloy microstructure (Ref 11). Solution potential is not affected significantly by second-phase particles of microscopic size, but because these particles frequently have solution potentials differing from that of the solid-solution matrix in which they occur, localized galvanic cells may be formed between them and the matrix.

The effects of principal alloying elements on solution potential of high-purity aluminum are shown in Fig. 5. For each element, the significant changes that occur do so within the range in which the element is completely in solid solution. Further addition of the same element, which forms a second phase, causes little additional change in solution potential.

Most commercial aluminum alloys contain additions of more than one of these elements; effects of multiple elements in solid solution on solution potential are approximately additive. The amounts retained in solid solution, particularly for more highly alloyed compositions, depend highly on fabrication and thermal processing so

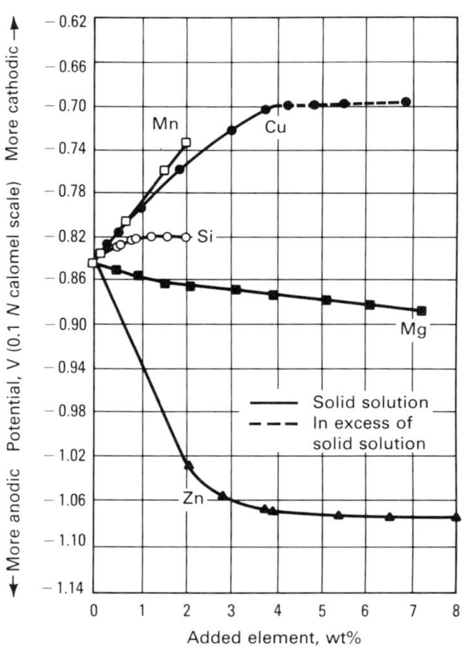

Fig. 5 Effects of principal alloying elements on the electrolytic-solution potential of aluminum. Potentials are for solution-treated and quenched high-purity binary alloys in a solution of 53 g/L NaCl plus 3 g/L H_2O_2 at 25 °C (75 °F).

that heat treatment and other processing variables influence the final electrode potential of the product. Tables 1a to d present representative solution potentials of commercial aluminum alloys and of several other metals and alloys.

The amounts of second phases present in aluminum and aluminum alloy products vary from nearly zero in those of aluminum 1199 and some others that also are nearly pure solid solutions to over 20% in hypereutectic aluminum-silicon casting alloys, such as 392.0 and 393.0. These phases

Table 1a Solution potentials of nonheat-treatable commercial wrought aluminum alloys

Values are the same for all tempers of each alloy.

Alloy	Potential(a), V
1060	−0.84
1100	−0.83
3003	−0.83
3004	−0.84
5050	−0.84
5052	−0.85
5154	−0.86
5454	−0.86
5056	−0.87
5456	−0.87
5182	−0.87
5083	−0.87
5086	−0.85
7072	−0.96

(a) Potential versus standard calomel electrode

Table 1b Solution potentials of heat-treatable commercial wrought aluminum alloys

Alloy	Temper	Potential(a), V
2014	T4	−0.69(b)
	T6	−0.78
2219	T3	−0.64(b)
	T4	−0.64(b)
	T6	−0.80
	T8	−0.82
2024	T3	−0.69(b)
	T4	−0.69(b)
	T6	−0.81
	T8	−0.82
2036	T4	−0.72
2090	T8E41	−0.83
6009	T4	−0.80
6010	T4	−0.79
6151	T6	−0.83
6351	T5	−0.83
6061	T4	−0.80
	T6	−0.83
6063	T5	−0.83
	T6	−0.83
7005	T6	−0.94
X7016	T6	−0.86
X7021	T6	−0.99
X7029	T6	−0.85
X7146	T6	−1.02
7049	T73	−0.84(c)
	T76	−0.84(c)
7050	T73	−0.84(c)
	T76	−0.84(c)
7075	T6	−0.83(c)
	T73	−0.84(c)
	T76	−0.84(c)
7475	T6	−0.83(c)
	T73	−0.84(c)
	T76	−0.84(c)
7178	T6	−0.83(c)

(a) Potential versus standard calomel electrode. (b) Varies ±0.01 V with quenching rate. (c) Varies ±0.02 V with quenching rate

Table 1c Solution potentials of cast aluminum alloys

Alloy	Temper	Type of mold(a)	Potential(b), V
208.0	F	S	−0.77
238.0	F	P	−0.74
295.0	T4	S or P	−0.70
	T6	S or P	−0.71
	T62	S or P	−0.73
296.0	T4	S or P	−0.71
308.0	F	P	−0.75
319.0	F	S	−0.81
	F	P	−0.76
355.0	T4	S or P	−0.78
	T6	S or P	−0.79
356.0	T6	S or P	−0.82
443.0	F	S	−0.83
	F	P	−0.82
514.0	F	S	−0.87
520.0	T4	S or P	−0.89
710.0	F	S	−0.99

(a) S, sand; P, permanent. (b) Potential versus standard calomel electrode

Table 1d Solution potentials of some nonaluminum base metals

Metal	Potential(a), V
Magnesium	−1.73
Zinc	−1.10
Cadmium	−0.82
Mild carbon steel	−0.58
Lead	−0.55
Tin	−0.49
Copper	−0.20
Bismuth	−0.18
Stainless steel(b)	−0.09
Silver	−0.08
Nickel	−0.07
Chromium	−0.40 to +0.18

(a) Potential versus standard calomel electrode. (b) Series 300, type 430

are generally intermetallic compounds of binary, ternary, or higher-order compositions, although some elements in excess of their solid solubility are present as elemental phases. Electrode potentials of some of the simpler second-phase constituents have been measured and are presented in Table 2. Measurement of the solution potentials of aluminum alloys has been standardized in ASTM G 69 (Ref 12).

Solution-potential measurements are useful for the investigation of heat-treating, quenching, and aging practices, and they are applied principally to alloys containing copper, magnesium, or zinc. In aluminum-copper and aluminum-copper-magnesium (2*xxx*) alloys, potential measurements can determine the effectiveness of solution heat treatment by measuring the amount of copper in solid solution. Also, by measuring the potentials of grain boundaries and grain bodies separately, the difference in potential responsible for intergranular corrosion, exfoliation, and stress-corrosion cracking (SCC) can be quantified. Solution-potential measurements of alloys containing copper also show the progress of artificial aging as increased amounts of precipitates are formed and the matrix is depleted of copper.

Table 2 Solution potentials of some second-phase constituents in aluminum alloys

Phase	Potential(a), V
Si	−0.26
Al_3Ni	−0.52
Al_3Fe	−0.56
Al_2Cu	−0.73
Al_6Mn	−0.85
Al_8Mg_5	−1.24

(a) Potential versus standard calomel electrode

Potential measurements are valuable with zinc-containing (7*xxx*) alloys for evaluating the effectiveness of the solution heat treatment, for following the aging process, and for differentiating among the various artificially aged tempers. These factors can affect corrosion behavior. In the magnesium-containing (5*xxx*) alloys, potential measurements can detect low-temperature precipitation and are useful in qualitatively evaluating stress-corrosion behavior. Potential measurements can also be used to follow the diffusion of zinc or copper in alclad products, thus determining whether the sacrificial cladding can continue to protect the core alloy (Ref 13).

Effects of Composition and Microstructure on Corrosion

1*xxx* Wrought Alloys. Wrought aluminums of the 1*xxx* series conform to composition specifications that set maximum individual, combined, and total contents for several elements present as natural impurities in the smelter-grade or refined aluminum used to produce these products. Aluminums 1100 and 1135 differ somewhat from the others in this series in having minimum and maximum specified copper contents. Corrosion resistance of all 1*xxx* compositions is very high, but under many conditions, it decreases slightly with increasing alloy content. Iron, silicon, and copper are the elements present in the largest percentages. The copper and part of the silicon are in solid solution. The second-phase particles present contain either iron or iron and silicon—Al_6Fe, Al_3Fe, and $Al_{12}Fe_3Si_2$ (Ref 14). The specific phase present or the relative amounts when more than one are present depend on the ratio of iron to silicon and on thermal history. The microstructural particles of these phases are cathodic to the aluminum solid solution, and exposed surfaces of these particles are covered by an oxide film thinner than that covering exposed areas of the solid solution (Ref 15). Corrosion may be initiated earlier and progress more rapidly in the aluminum solid solution immediately surrounding the particles. The number and/or size of such corrosion sites is proportional to the area fraction of the second-phase particles.

Not all impurity elements are detrimental to corrosion resistance of 1*xxx* series aluminum alloys, and detrimental elements may reduce the resistance of some types of alloys but have no ill effects in others. Therefore, specification limitations established for impurity elements are often based on maintaining consistent and predictable levels of corrosion resistance in various applications rather than on their effects in any specific application.

2*xxx* wrought alloys and 2*xx.x* casting alloys, in which copper is the major alloying element, are less resistant to corrosion than alloys of other series, which contain much lower amounts of copper. Alloys of this type were the first heat-treatable high-strength aluminum-base materials and have been used for more than 75 years in structural applications, particularly in aircraft and aerospace applications (Ref 16). Much of the thin sheet made of these alloys is produced as an alclad composite, but thicker sheet and other products in many applications require no protective cladding.

Electrochemical effects on corrosion can be stronger in these alloys than in alloys of many other types because of two factors: greater change in electrode potential with variations in amount of copper in solid solution (Fig. 5) and, under some conditions, the presence of nonuniformities in solid-solution concentration. However, that general resistance to corrosion decreases with increasing copper content is not primarily attributable to these solid-solution or second-phase solution-potential relationships, but to galvanic cells created by formation of minute copper particles or films deposited on the alloy surface as a result of corrosion. As corrosion progresses, copper ions, which initially go into solution, replate onto the alloy to form metallic copper cathodes. Reduction of copper ions and increased efficiency of O_2 and H^+ reduction reactions in the presence of copper increase the corrosion rate.

These alloys are invariably solution heat treated and are used in either the naturally aged or the precipitation heat-treated temper. Development of these tempers using good heat-treating practice can minimize electrochemical effects on corrosion resistance. The rate of quenching and the temperature and time of artificial aging can both affect the corrosion resistance of the final product.

2*xxx* Wrought Alloys Containing Lithium. Lithium additions decrease the density and increase the elastic modulus of aluminum alloys, making aluminum-lithium alloys good candidates for replacing the existing high-strength alloys, primarily in aerospace applications.

One of the earliest aluminum alloys containing lithium was 2020. This alloy in the T6 temper was commercially introduced in 1957 as a structural alloy with good strength properties up to 175 °C (350 °F). It has a modulus 8% higher and a density 3% lower than alloy 7075-T6, but was rarely used in aircraft because of its relatively low fracture toughness. It was used in the thrust structure of the Saturn S-II, the second stage of the Saturn V launch vehicle (Ref 17).

Two recently registered lithium-bearing alloys are 2090 and 8090. Alloy 2090, in T8-type tempers, has a higher resistance to exfoliation than that of 7075-T6, and the resistance to SCC is comparable (Ref 18). Alloy 8090 is being designed by various producers to meet other combinations of mechanical-property goals (Ref 19).

Although lithium is highly reactive, addition of up to 3% Li to aluminum shifts the pitting potential of the solid solution only slightly in the anodic direction in 3.5% NaCl solution (Ref 20). In an extensive corrosion investigation of several binary and ternary aluminum-lithium alloys, modifications to the microstructure that promote formation of the δ phase (AlLi) were found to reduce the corrosion resistance of the alloy in 3.5% NaCl solution (Ref 21). It was concluded that an understanding of the nucleation and growth of the δ phase is central to an understanding of the corrosion behavior of these alloys.

3*xxx* Wrought Alloys. Wrought alloys of the 3*xxx* series (aluminum-manganese and aluminum-manganese-magnesium) have very high resistance to corrosion. The manganese is present in the aluminum solid solution, in submicroscopic

Table 3 Relative ratings of resistance to general corrosion and to SCC of wrought aluminum alloys

Alloy	Temper	Resistance to corrosion General(a)	SCC(b)
1060	All	A	A
1100	All	A	A
1350	All	A	A
2011	T3, T4, T451	D(c)	D
	T8	D	B
2014	O	...	...
	T3, T4, T451	D(c)	C
	T6, T651, T6510, T6511	D	C
2017	T4, T451	D(c)	C
2018	T61	...	...
2024	O	...	...
	T4, T3, T351, T3510, T3511, T361	D(c)	C
	T6, T861, T81, T851, T8510, T8511	D	B
	T72	...	...
2025	T6	D	C
2036	T4	C	...
2117	T4	C	A
2218	T61, T72	D	C
2219	O	...	...
	T31, T351, T3510, T3511, T37	D(c)	C
	T81, T851, T8510, T8511, T87	D	B
2618	T61	D	C
3003	All	A	A
3004	All	A	A
3105	All	A	A
4032	T6	C	B
5005	All	A	A
5050	All	A	A
5052	All	A	A
5056	O, H11, H12, H32, H14, H34	A(d)	B(d)
	H18, H38	A(d)	C(d)
	H192, H392	B(d)	D(d)
5083	All	A(d)	B(d)
5086	O, H32, H116	A(d)	A(d)
	H34, H36, H38, H111	A(d)	A(d)
5154	All	A(d)	A(d)
5252	All	A	A
5254	All	A(d)	A(d)
5454	All	A	A
5456	All	A(d)	B(d)
5457	O	A	A
5652	All	A	A
5657	All	A	A
6053	O	...	...
	T6, T61	A	A
6061	O	B	A
	T4, T451, T4510, T4511	B	B
	T6, T651, T652, T6510, T6511	B	A
6063	All	A	A
6066	O	C	A
	T4, T4510, T4511, T6, T6510, T6511	C	B
6070	T4, T4511, T6	B	B
6101	T6, T63, T61, T64	A	A
6151	T6, T652	...	...
6201	T81	A	A
6262	T6, T651, T6510, T6511, T9	B	A
6463	All	A	A
7001	O	C(c)	C
7075	T6, T651, T652, T6510, T6511	C(c)	C
	T73, T7351	C	B
7178	T6, T651, T6510, T6511	C(c)	C

(a) Ratings are relative and in decreasing order of merit, based on exposure to NaCl solution by intermittent spraying or immersion. Alloys with A and B ratings can be used in industrial and seacoast atmospheres without protection. Alloys with C, D, and E ratings generally should be protected, at least on faying surfaces. (b) SCC ratings are based on service experience and on laboratory tests of specimens exposed to alternate immersion in 3.5% NaCl solution. A, no known instance of failure in service or in laboratory tests; B, no known instance of failure in service; limited failures in laboratory tests of short transverse specimens; C, service failures when sustained tension stress acts in short-transverse direction relative to grain structure; limited failures in laboratory tests of long tranverse specimens; D, limited service failures when sustained stress acts in longitudinal or long-transverse direction relative to grain structure. (c) In relatively thick sections, the rating would be E. (d) This rating may be different for material held at elevated temperatures for long periods.

particles of precipitate, and in larger particles of $Al_6(Mn,Fe)$ or $Al_{12}(Mn,Fe)_3Si$ phases, both of which have solution potentials almost the same as that of the solid-solution matrix (Ref 22). Such alloys are widely used for cooking and food-processing equipment, chemical equipment, and various architectural products requiring high resistance to corrosion.

4*xxx* Wrought Alloys and 3*xx.x* and 4*xx.x* Casting Alloys. Elemental silicon is present as second-phase constituent particles in wrought alloys of the 4*xxx* series, in brazing and welding alloys, and in casting alloys of the 3*xx.x* and 4*xx.x* series. Silicon is cathodic to the aluminum solid-solution matrix by several hundred millivolts and accounts for a considerable volume fraction of most of the silicon-containing alloys. However, the effects of silicon on the corrosion resistance of these alloys are minimal because of low corrosion current density resulting from the fact that the silicon particles are highly polarized.

Corrosion resistance of 3*xx.x* casting alloys is strongly affected by copper content, which can be as high as 5% in some compositions, and by impurity levels. Modifications of certain basic alloys have more restrictive limits on impurities, which benefit corrosion resistance and mechanical properties.

5*xxx* Wrought Alloys and 5*xx.x* Casting Alloys. Wrought alloys of the 5*xxx* series (aluminum-magnesium-manganese, aluminum-magnesium-chromium, and aluminum-magnesium-manganese-chromium) and casting alloys of the 5*xx.x* series (aluminum-magnesium) have high resistance to corrosion, and this accounts in part for their use in a wide variety of building products and chemical-processing and food-handling equipment, as well as applications involving exposure to seawater (Ref 23).

Alloys in which the magnesium is present in amounts that remain in solid solution, or is partially precipitated as Al_8Mg_5 particles dispersed uniformly throughout the matrix, are generally as resistant to corrosion as commercially pure aluminum and are more resistant to salt water and some alkaline solutions, such as those of sodium carbonate and amines. The wrought alloys containing about 3% or more magnesium under conditions that lead to an almost continuous intergranular Al_8Mg_5 precipitate, with very little precipitate within the grains, may be susceptible to exfoliation or SCC (Ref 24). Tempers have been developed for these higher-magnesium wrought alloys to produce microstructures having extensive Al_8Mg_5 precipitate within the grains, thus eliminating such susceptibility.

In the 5*xxx* alloys that contain chromium, this element is present as a submicroscopic precipitate, $Al_{12}Mg_2Cr$. Manganese in these alloys is in the form of $Al_6(Mn,Fe)$, both submicroscopic and larger particles. Such precipitates and particles do not adversely affect corrosion resistance of these alloys.

6*xxx* Wrought Alloys. Moderately high strength and very good resistance to corrosion make the heat-treatable wrought alloys of the 6*xxx* series (aluminum-magnesium-silicon) highly suitable in various structural, building, marine, machinery, and process-equipment applications. The Mg_2Si phase, which is the basis for precipitation hardening, is unique in that it is an ionic compound and is not only anodic to aluminum but also reactive in acidic solutions. However, either in solid solution or as submicroscopic precipitate, Mg_2Si has a negligible effect on electrode potential. Because these alloys are normally used in the heat-treated condition, no detrimental effects result from the major alloying elements or from the supplementary chromium, manganese, or zirconium, which are added to control grain structure. Copper additions, which augment strength in many of these alloys, are limited to small amounts to minimize effects on corrosion resistance. In general, the level of resistance decreases somewhat with increasing copper content.

When the magnesium and silicon contents in a 6*xxx* alloy are balanced (in proportion to form only Mg_2Si), corrosion by intergranular penetration is slight in most commercial environments (Ref 25). If the alloy contains silicon beyond that needed to form Mg_2Si or contains a high level of cathodic impurities, susceptibility to intergranular corrosion increases (Ref 26).

7*xxx* wrought alloys and 7*xx.x* casting alloys contain major additions of zinc, along with magnesium or magnesium plus copper in combinations that develop various levels of strength. Those containing copper have the highest strengths and have been used as construction-al materials, primarily in aircraft applications, for more than 40 years. The copper-free alloys of the series have many desirable characteristics: moderate-to-high strength; excellent toughness; and good workability, formability, and weldability.

Use of these copper-free alloys has increased in recent years and now includes automotive applications (such as bumpers), structural members and armor plate for military vehicles, and components of other transportation equipment.

The 7*xxx* wrought and 7*xx.x* casting alloys, because of their zinc contents, are anodic to 1*xxx* wrought aluminums and to other aluminum alloys. They are among the aluminum alloys most susceptible to SCC. However, SCC can be avoided by proper alloy and temper selection and by observing appropriate design, assembly, and application precautions (Ref 27). Stress-corrosion cracking of aluminum alloys is discussed in greater detail in a subsequent section in this article.

Resistance to general corrosion of the copper-free wrought 7*xxx* alloys is good, approaching that of the wrought 3*xxx*, 5*xxx*, and 6*xxx* alloys (Ref 28). The copper-containing alloys of the 7*xxx* series, such as 7049, 7050, 7075, and 7178 have lower resistance to general corrosion than those of the same series that do not contain copper. All 7*xxx* alloys are more resistant to general corrosion than 2*xxx* alloys, but less resistant than wrought alloys of other groups.

Although the copper in both wrought and cast alloys of the aluminum-zinc-magnesium-copper type reduces resistance to general corrosion, it is beneficial from the standpoint of resistance to SCC. Copper allows these alloys to be precipitated at higher temperatures without excessive loss in strength and thus makes possible the development of T73 tempers, which couple high strength with excellent resistance to SCC (Ref 29).

Composites. Aluminum alloys reinforced with silicon carbide (Ref 30), graphite (Ref 31), or boron (Ref 32) show promise as metal matrix composites for lightweight structural applications with increased modulus and strength and are potentially well suited to aerospace and military needs. The corrosion behavior of composites is governed by galvanic action between the aluminum matrix and the reinforcing material. When both are exposed to an aggressive environment, corrosion of the aluminum is accelerated. Silicon carbide, graphite, and boron are cathodic to aluminum and do not polarize easily.

For a useful service life, some form of corrosion protection is needed. Aluminum thermal spraying has been reported as a successful protection method for discontinuous silicon carbide/aluminum composites; for continuous graphite/aluminum or silicon carbide/aluminum, sulfuric acid (H_2SO_4) anodizing has provided protection, as have organic coatings or ion vapor deposited aluminum (Ref 33).

Effects of Additional Alloying Elements. In addition to the major elements that define the various alloy systems discussed above, commercial aluminum alloys may contain other elements that provide special characteristics. Lead and bismuth are added to alloys 2011 and 6262 to improve chip breakage and other machining characteristics. Nickel is added to wrought alloys 2018, 2218, and 2618, which were developed for elevated-temperature service, and to certain 3*xx.x* cast alloys used for pistons, cylinder blocks, and other engine parts subjected to high temperatures. Cast aluminum bearing alloys of the 850.0 group contain tin. In all cases, these alloying additions introduce constituent phases that are cathodic to the matrix and decrease resistance to corrosion in aqueous saline media. However, these alloys are often used in environments in which they are not subject to corrosion.

Table 4a Relative ratings of resistance to general corrosion and to SCC of aluminum sand casting alloys

Alloy	Temper	Resistance to corrosion: General(a)	SCC(b)
Sand castings			
208.0	F	B	B
224.0	T7	C	B
240.0	F	D	C
242.0	All	D	C
A242.0	T75	D	C
249.0	T7	C	B
295.0	All	C	C
319.0	F, T5	C	B
	T6	C	C
355.0	All	C	A
C355.0	T6	C	A
356.0	T6, T7, T71, T51	B	A
A356.0	T6	B	A
443.0	F	B	A
512.0	F	A	A
513.0	F	A	A
514.0	F	A	A
520.0	T4	A	C
535.0	F	A	A
B535.0	F	A	A
705.0	T5	B	B
707.0	T5	B	C
710.0	T5	B	B
712.0	T5	B	C
713.0	T5	B	B
771.0	T6	C	C
850.0	T5	C	B
851.0	T5	C	B
852.0	T5	C	B

(a) Relative ratings of general corrosion resistance are in decreasing order of merit, based on exposures to NaCl solution by intermittent spray or immersion. (b) Relative ratings of resistance to SCC are based on service experience and on laboratory tests of specimens exposed to alternate immersion in 3.5% NaCl solution. A, no known instance of failure in service when properly manufactured; B, failure not anticipated in service from residual stresses or from design and assembly stresses below about 45% of the minimum guaranteed yield strength given in applicable specifications; C, failures have occurred in service with either this specific alloy/temper combination or with alloy/temper combinations of this type; designers should be aware of the potential SCC problem that exists when these alloys and tempers are used under adverse conditions. (c) For electric motor rotors

Corrosion Ratings of Alloys and Tempers

Simplified ratings of resistance to general corrosion and to SCC for wrought and cast aluminum alloys are presented in Tables 3 and 4a and 4b. These ratings may be useful in evaluating and comparing alloy/temper combinations for corrosion service (more detailed ratings of resistance to SCC for high-strength wrought aluminum alloys are given in Table 6 and in Ref 34).

Galvanic Corrosion and Protection

The solution-potential values in Tables 1a to d, as measured against a standard calomel electrode (SCE), form a galvanic series for aluminum alloys and other metals. The galvanic relationships indicated by these values have wide applicability because of the similarity of the electrochemical behavior of these metals in the NaCl solution to that in marine and other saline environments. This galvanic series, however, is not necessarily valid in nonsaline solutions. For example, aluminum is anodic to zinc in an aqueous 1 *M* sodium chromate (Na_2CrO_4) solution and cathodic to iron in an aqueous 1 *M* sodium sulfate (Na_2SO_4) solution.

Under most environmental conditions frequently encountered in service, aluminum and its alloys are the anodes in galvanic cells with most other metals, protecting them by corroding sacrificially. Only magnesium and zinc are more anodic. Sacrificial corrosion of aluminum or cadmium is slight when these two metals are coupled in a galvanic cell, because of the small difference in electrode potential between them.

Contact of aluminum with more cathodic metals should be avoided in any environment in which aluminum by itself is subject to pitting corrosion. Where such contact is necessary, protective measures should be implemented to minimize sacrificial corrosion of the aluminum. In such an environment, aluminum is already polarized to its pitting potential, and the additional potential imposed by contact with the more cathodic metal greatly increases the corrosion current. In many environments, aluminum can be used in contact with chromium or stainless steels with only slight acceleration of corrosion; chromium and stainless steels are easily polarized cathodically in mild environments, so that the corrosion current is small despite the large differences in the open-circuit potentials between these metals and aluminum.

To minimize corrosion of aluminum wherever contact with more cathodic metals cannot be avoided, the ratio of the exposed surface area of the aluminum to that of the more cathodic metal should be as high as possible to minimize the current density at the aluminum and therefore the rate of corrosion. The area ratio may be increased by painting the cathodic metal or both metals, but painting only the aluminum is not effective and may even accelerate corrosion.

Corrosion of aluminum in contact with more cathodic metals is much less severe in solutions of most nonhalide salts, in which aluminum alone normally is not polarized to its pitting potential,

Table 4b Relative ratings of resistance to general corrosion and to stress-corrosion cracking of aluminum permanent mold, die casting, and rotor metal alloys

Alloy	Temper	Resistance to corrosion General(a)	SCC(b)
Permanent mold castings			
242.0	T571, T61	D	C
308.0	F	C	B
319.0	F	C	B
	T6	C	C
332.0	T5	C	B
336.0	T551, T65	C	B
354.0	T61, T62	C	A
355.0	All	C	A
C355.0	T61	C	A
356.0	All	B	A
A356.0	T61	B	A
F356.0	All	B	A
A357.0	T61	B	A
358.0	T6	B	A
359.0	All	B	A
B443.0	F	B	A
A444.0	T4	B	A
513.0	F	A	A
705.0	T5	B	B
707.0	T5	B	C
711.0	T5	B	A
713.0	T5	B	B
850.0	T5	C	B
851.0	T5	C	B
852.0	T5	C	B
Die castings			
360.0	F	C	A
A360.0	F	C	A
364.0	F	C	A
380.0	F	E	A
A380.0	F	E	A
383.0	F	E	A
384.0	F	E	A
390.0	F	E	A
392.0	F	E	A
413.0	F	C	A
A413.0	F	C	A
C443.0	F	B	A
518.0	F	A	A
Rotor metal(c)			
100.1		A	A
150.1		A	A
170.1		A	A

(a) Relative ratings of general corrosion resistance are in decreasing order of merit, based on exposures to NaCl solution by intermittent spray or immersion. (b) Relative ratings of resistance to SCC are based on service experience and on laboratory tests of specimens exposed to alternate immersion in 3.5% NaCl solution. A, no known instance of failure in service when properly manufactured; B, failure not anticipated in service from residual stresses or from design and assembly stresses below about 45% of the minimum guaranteed yield strength given in applicable specifications; C, failures have occurred in service with either this specific alloy/temper combination or with alloy/temper combinations of this type; designers should be aware of the potential SCC problem that exists when these alloys and tempers are used under adverse conditions. (c) For electric motor rotors

than in solutions of halide salts, in which it is. As shown in Fig. 3, increases in potential, as long as the value does not reach the pitting potential, have small effects on current density.

Galvanic current between aluminum and another metal also can be reduced by removing oxidizing agents from the electrolyte. Thus, the corrosion rate of aluminum coupled to copper in seawater is greatly reduced wherever the seawater is deaerated. In closed multimetallic systems, the corrosion rate of aluminum, although initially high, decreases to a low value whenever the cathodic reactant is depleted. Galvanic current is also low in solutions having high electrical resistivity, such as high-purity water, but some semiconductors, such as graphite and magnetite, are cathodic to aluminum, and when in contact with them, aluminum corrodes sacrificially.

In alclad products, the difference in solution potential between the core alloy and the cladding alloy is used to provide cathodic protection to the core (Ref 35). These products, primarily sheet and tube, consist of a core clad on one or both surfaces with a metallurgically bonded layer of an alloy that is anodic to the core alloy. The thickness of the cladding layer is usually less than 10% of the overall thickness of the product.

Cladding alloys are generally of the nonheat-treatable type, although heat-treatable alloys are sometimes used for higher strength. For mechanical-design calculations, such sacrificial claddings are treated as corrosion allowances and are not normally included in the determination of the strength of an alclad product.

Composition relationships of core and cladding alloys are generally designed so that the cladding is 80 to 100 mV anodic to the core. Table 5 lists several core alloy/cladding alloy combinations for common alclad products. Because of the cathodic protection provided by the cladding, corrosion progresses only to the core/cladding interface, then spreads laterally. This is highly effective in eliminating perforation of thin-wall products.

Table 5 Combinations of aluminum alloys used in some alclad products

Core alloy	Cladding alloy
2014	6003 or 6053
2024	1230
2219	7072
3003	7072
3004	7072 or 7013
6061	7072
7075	7072, 7008, or 7011
7178	7072

Surface Treatments. A process that produces an effect similar to that of conventional sacrificial cladding is called diffusion cladding. Aluminum products can be clad using this process, regardless of their shape (Ref 36). The process involves two steps: first, a thin film of zinc is deposited on the aluminum surface by chemical displacement from an alkaline zincate solution, then the zinc is diffused into the aluminum to produce a zone of zinc-enriched alloy that is anodic to the underlying aluminum. It was found that 3003 aluminum with a correctly balanced zinc diffusion treatment exhibited uniform corrosion and that the depth of corrosion was restricted to about one-half the thickness of the diffusion zone (Ref 37). These results suggest that a zinc diffusion treatment may be as effective as conventional alcladding for the prevention of localized pitting.

Another way to simulate alcladding is to apply a coating of an anodic alloy to an aluminum surface by thermal-spray techniques, such as flame or plasma spray. These coatings act in the same way as the cladding layer on an alclad product and corrode sacrificially to protect the core alloy (Ref 38, 39).

Cathodic Protection. In some applications, aluminum alloy parts, assemblies, structures, and pipelines are cathodically protected by anodes either made of more anodic metals or made anodic by using impressed potentials. In either case, because the usual cathodic reaction produces hydroxyl ions, the current on these alloys should not be high enough to make the solution sufficiently alkaline to cause significant corrosion (Ref 40).

The criterion for cathodic protection of aluminum in soils and waters has been published by the National Association of Corrosion Engineers (Ref 41). The suggested practice is to shift the potential at least -0.15 V but not beyond the value of -1.20 V as measured against a saturated copper sulfate ($Cu/CuSO_4$) reference electrode. In some soils, potentials as low as -1.4 V have been encountered without appreciable cathodic corrosion (Ref 42). Essentially the same criterion is followed in Eastern Europe (Ref 43).

Several examples of cathodic protection of aluminum equipment in chemical plants, as well as a preference for sacrificial anodes of zinc or aluminum-zinc alloy, are discussed in Ref 44. Such protection is most successful in electrolytes in the pH range of 4 to 8.5—the so-called neutral range. The cathodic protection of aluminum structures is reviewed in Ref 45, which supports general experience that cathodic protection is effective in preventing or greatly reducing several types of corrosion attack.

Buried aluminum pipelines are usually protected by sacrificial anodes—zinc for coated lines and magnesium for uncoated lines. It is generally accepted that such coatings as extruded polyethylene or a tape wrap should be applied to aluminum pipes for underground service. Because of the effectiveness and longevity of sacrificial anode systems and the need to avoid overprotection, impressed current (rectifier) systems generally are not used to protect aluminum pipelines.

The cathodic protection of aluminum alloys in seawater has been extensively studied (Ref 46, 47). Sacrificial anodes were found to be effective in reducing surface pitting and crevice corrosion without causing cathodic attack.

Deposition Corrosion

In designing aluminum and aluminum alloys for satisfactory corrosion resistance, it is important to keep in mind that ions of several metals have reduction potentials that are more cathodic than the solution potential of aluminum and therefore can be reduced to metallic form by aluminum. For each chemical equivalent of so-called heavy-metal ions reduced, a chemical equivalent of aluminum is oxidized. Reduction of only a small amount of these ions can lead to severe localized corrosion of aluminum, because the metal reduced from them plates onto the aluminum and sets up galvanic cells. The more important heavy metals are copper, lead, mercury, nickel, and tin. The effects of these metals on aluminum are of greatest concern in acidic solutions; in alkaline solutions, they have much lower solubilities and therefore much less severe effects.

Copper is the heavy metal most commonly encountered in applications of aluminum. A copper-ion concentration of 0.02 to 0.05 ppm in neutral or acidic solutions is generally considered to be the threshold value for initiation of pitting on aluminum. A specific value for the copper-ion threshold is normally not proposed because the pitting tendency also depends on the aluminum alloy; the pH of the water; concentrations of other ions in the water, particularly bicarbonate (HCO_3^-), chloride (Cl^-), and calcium (Ca^{2+}); and on whether the pits that develop are open or occluded (Ref 48). Copper contamination of solutions in contact with aluminum should be minimized or avoided. As discussed previously, the relatively low corrosion resistance of aluminum-copper alloys results from reduction of copper ions present in the corrosion product of the alloy.

Ferric (Fe^{3+}) ion can be reduced by aluminum, but does not form a metallic deposit. This ion is rarely encountered in service because it reacts preferentially with oxygen and water to form insoluble oxides and hydroxides, except in acidic solutions outside the passive range of aluminum. On the other hand, at room temperature, the most anodic aluminum alloys (those with a corrosion potential approaching −1.0 V versus the SCE) can reduce ferrous (Fe^{2+}) ions to metallic iron and produce a metallic deposit on the surface of the aluminum. The presence of (Fe^{2+}) ion also tends to be rare in service; it exists only in deaerated solutions or in other solutions free of oxidizing agents (Ref 49).

Mercury amalgamates with aluminum with difficulty because the natural oxide film on aluminum prevents metal-to-metal contact. However, after the two metals have been brought together, if the oxide film is broken by mechanical or chemical action, amalgamation occurs immediately, and in the presence of moisture, corrosion of the aluminum proceeds rapidly (Ref 50). Aluminum in contact with a solution of a mercury salt forms metallic mercury, which then readily amalgamates the aluminum. Of all the heavy metals, mercury can cause the most corrosion damage to aluminum (Ref 51). The effect can be severe when stress is present. For example, attack by mercury and zinc amalgam combined with residual stresses from welding caused cracking of the weldment (Fig. 6). The corrosive action of mercury can be serious with or without stress because amalgamation, once initiated, continues to propagate unless the mercury can be removed. If an aluminum surface has become contaminated with mercury, the mercury can be removed by treatment with 70% nitric acid (HNO_3) or by evaporation in steam or hot air (Ref 52). It is difficult to determine the safe level of mercury that can be tolerated on aluminum. In solutions, concentrations exceeding a few parts per billion should be viewed with suspicion; in atmospheres, any amount exceeding that allowed by Environmental Protection Agency regulations is suspect.

Intergranular Corrosion

Intergranular (intercrystalline) corrosion is selective attack of grain boundaries or closely adjacent regions without appreciable attack of the grains themselves. Intergranular corrosion is a generic term that includes several variations associated with different metallic structures and thermomechanical treatments (Fig. 7).

Intergranular corrosion is caused by potential differences between the grain-boundary region and the adjacent grain bodies (Ref 53). The location of the anodic path varies with the different alloy systems. In 2*xxx* series alloys, it is a narrow band on either side of the boundary that is depleted in copper; in 5*xxx* series alloys, it is the anodic constituent Mg_2Al_3 when that constituent forms a continuous path along a grain boundary; in copper-free 7*xxx* series alloys, it is generally considered to be the anodic zinc- and magnesium-bearing constituents on the grain boundary; and in the copper-bearing 7*xxx* series alloys, it appears to be the copper-depleted bands along the grain boundaries (Ref 54, 55). The 6*xxx* series alloys generally resist this type of corrosion, although slight intergranular attack has been observed in aggressive environments. The electrochemical mechanism for intergranular corrosion proposed by E.H. Dix has been verified (Ref 56) and related to the pitting potentials of aluminum (Ref 57).

Because intergranular corrosion is involved in SCC of aluminum alloys, it is often presumed to be more deleterious than pitting or general cor-

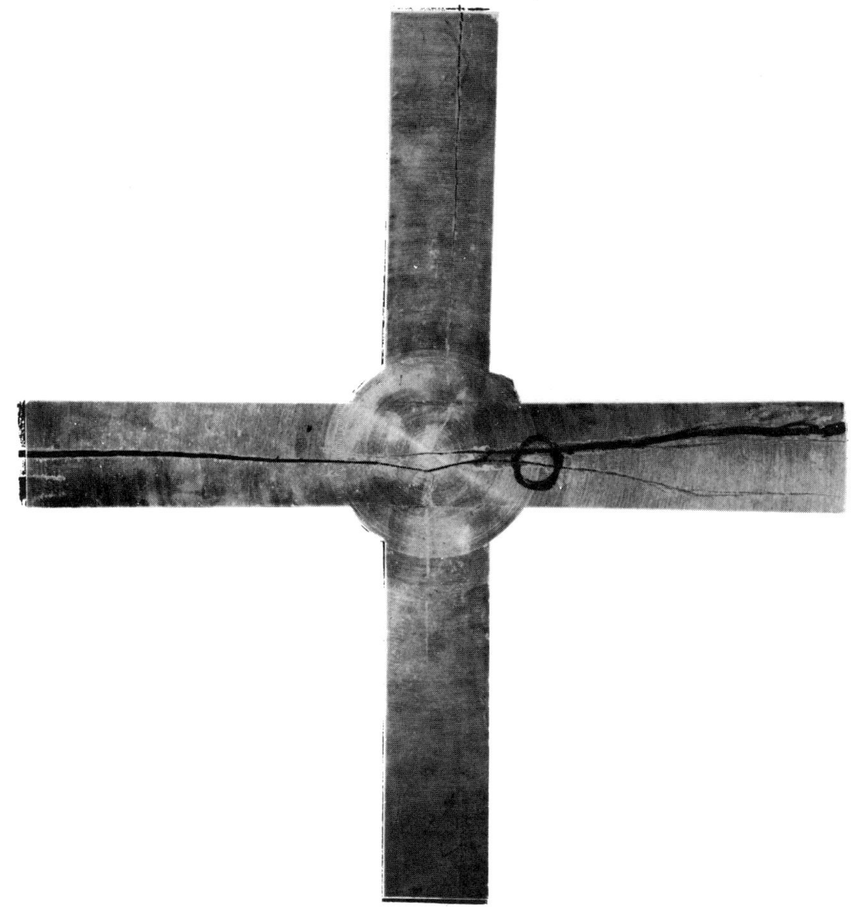

Fig. 6 Section through cruciform weldment of alloy 5083-H131 plate cracked by mercury. Attack was initiated by applying a few drops of mercury chloride ($HgCl_2$) solution and zinc amalgam to the sectioned surface at the circled area (right of center). 0.33×

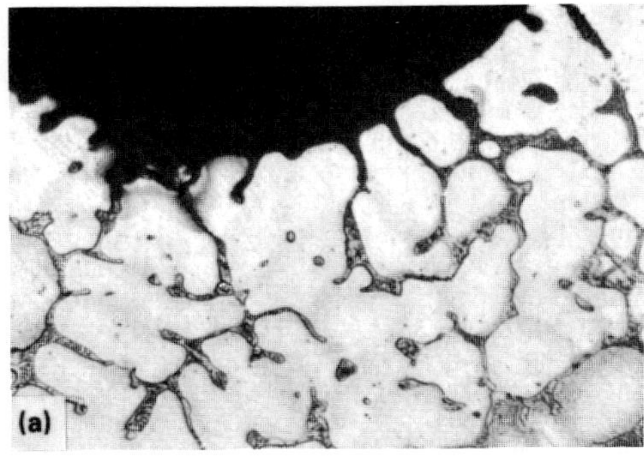

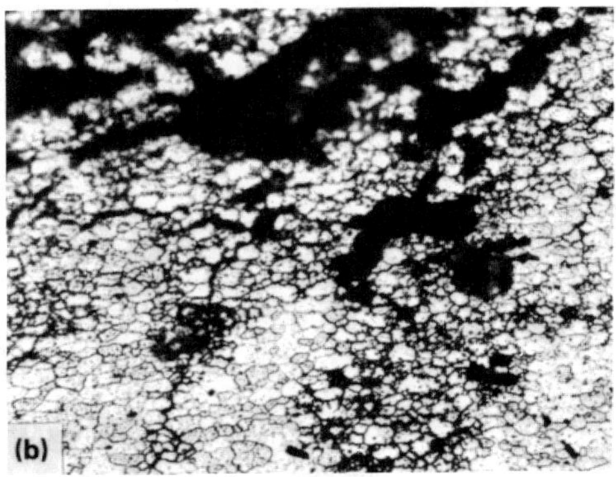

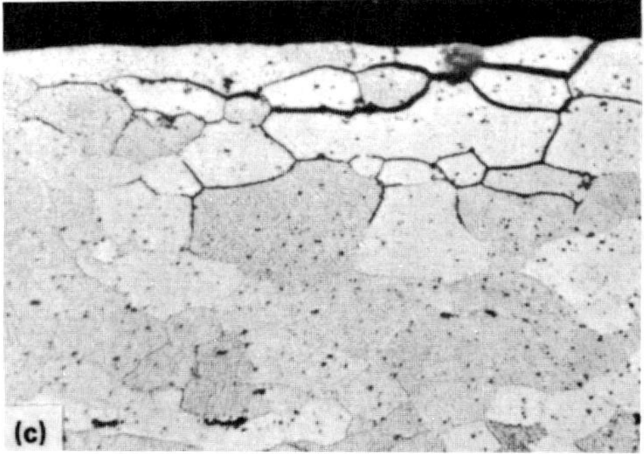

Fig. 7 Various types of intergranular corrosion. (a) Interdendritic corrosion in a cast structure. (b) Interfragmentary corrosion in a wrought, unrecrystallized structure. (c) Intergranular corrosion in a recrystallized wrought structure. All etched with Keller's reagent. 500×

rosion. However, in alloys that are not susceptible to SCC—for example, the *6xxx* series alloys—intergranular corrosion is usually no more severe than pitting corrosion, tends to decrease with time and, for equal depth of corrosion, its effect on strength is no greater than that of pitting corrosion, although fatigue cracks may be more likely to initiate at areas of intergranular corrosion than at random pits.

Evaluation of intergranular attack is more complex than evaluation of pitting. Visual observations are generally not reliable. For *5xxx* series alloys, a weight loss method has been accepted by The American Society for Testing and Materials (ASTM) (Ref 58). Electrochemical techniques provide some evidence of the susceptibility of a particular alloy or microstructure to intergranular corrosion, but such techniques should be accompanied by a metallographic examination of carefully prepared sections.

Stress-Corrosion Cracking

Only aluminum alloys that contain appreciable amounts of soluble alloying elements, primarily copper, magnesium, silicon, and zinc, are susceptible to SCC. For most commercial alloys, tempers have been developed that provide a high degree of immunity to SCC in most environments.

The electrochemical theory of stress corrosion, which was developed in about 1940, describes certain conditions required for SCC of aluminum alloys (Ref 53, 59, 60). Further research showed inadequacies in this theory, and the complex interactions among factors that lead to SCC of aluminum alloys are not yet fully understood (Ref 61). However, there is general agreement that for aluminum the electrochemical factor predominates and the electrochemical theory continues to be the basis for developing aluminum alloys and tempers resistant to SCC (Ref 62).

Stress-corrosion cracking in aluminum alloys is characteristically intergranular. According to the electrochemical theory, this requires a condition along grain boundaries that makes them anodic to the rest of the microstructure so that corrosion propagates selectively along them. Such a condition is produced by localized decomposition of solid solution, with a high degree of continuity of decomposition products, along the grain boundaries. The most anodic regions may be either the boundaries themselves (most commonly, the precipitate formed in them) or regions adjoining the boundaries that have been depleted of solute.

In *2xxx* alloys, the solute-depleted regions are the most anodic; in *5xxx* alloys, it is the Mg_2Al_3 precipitate along the boundaries. The most anodic grain-boundary regions in other alloys have not been identified with certainty. Strong evidence for the presence of anodic regions, and of the electrochemical nature of their corrosion in aqueous solutions, is provided by the fact that SCC can be greatly retarded, if not eliminated, by cathodic protection (Ref 60).

Figure 8 shows four different microstructures in an alloy containing 5% Mg. These microstructures represent degrees of susceptibility to SCC ranging from high susceptibility to high resistance, depending on heat treatment. The treatments that provide high resistance to cracking are those that produce microstructures either free of precipitate along grain boundaries (Fig. 8a) or with precipitate distributed as uniformly as possible within grains (Fig. 8d). In the latter case, corrosion along boundaries is minimized because the presence of precipitate or depleted regions throughout the microstructure increases the ratio of the total area of anodic regions to that of cathodic ones, thereby reducing the corrosion current on each anodic region. For alloys requiring microstructural control to avoid susceptibility, resistance is obtained by using treatments that produce precipitate throughout the microstructure, because precipitate always forms first along boundaries, and its formation there usually cannot be prevented.

According to electrochemical theory, susceptibility to intergranular corrosion is a prerequisite for susceptibility to SCC, and treatment of aluminum alloys to improve resistance to SCC also improves their resistance to intergranular corrosion. For most alloys, however, optimum levels of resistance to these two types of failure require different treatments, and resistance to intergranular corrosion is not a reliable indication of resistance to SCC.

In many cases, susceptibility to SCC of an aluminum alloy cannot be predicted reliably by examining its microstructure. Many observations have been made of the progressive changes in dislocation network, precipitation pattern, and other microstructural features that occur as an alloy is treated to improve its resistance to SCC, but these changes have not been correlated quantitatively with susceptibility.

Effect of Stress. Whether or not SCC develops in a susceptible aluminum alloy product depends on both magnitude and duration of tensile strength acting at the surface. The effects of the factors have been established most commonly by means of accelerated laboratory tests; results of one set of such tests are reflected in the shaded bands in Fig. 9. Despite introduction of fracture mechanics techniques capable of determining crack growth rates, such tests continue to be the basic tools used in evaluating resistance of aluminum alloys to SCC. These tests suggest a minimum (threshold) stress that is required for cracking to develop.

Although empirical in nature, the threshold value provides a valid measure of the relative susceptibilities of aluminum alloys to SCC under the specific conditions of a particular test or environment. Also, for some alloy/temper combinations, results of accelerated laboratory tests reliably predict stress-corrosion performance in service; for example, results of an 84-day alternate immersion test of alloy 7075 and alloy 7178 products correlated well with performance of these products in a seacoast environment.

Stress Relieving. Residual stresses are induced in aluminum alloy products when they are solution heat treated and quenched. Figure 10(a) shows the typical distribution and magnitude of residual stresses in thick high-strength material of constant cross section. Quenching places the surfaces in compression and the center in tension. If the compressive surface stresses are not disturbed by subsequent fabrication practices, the surface has an enhanced resistance to SCC because a sustained tensile stress is necessary to initiate and propagate this type of corrosion.

On the other hand, one of the most common practices associated with SCC problems is machining into the residual high tensile stress areas of material that has not been stress relieved. If the exposed tensile stresses are in a transverse direction or have a transverse component and if a susceptible alloy or temper is involved, the probability of SCC is present (Ref 63).

Aluminum products of constant cross section are stress relieved effectively and economically by mechanical stretching. The stretching operation must be done after quenching and, for most alloys, before artificial aging. Note the low magnitude of residual stresses after stretching (Fig. 10b) as compared to the as-quenched material in Fig. 10(a). Federal specifications for rolled and extruded products provide for stress relieving by stretching on the order of 1 to 3%.

Thus, the use of the stress-relieved temper for heat-treated mill products will minimize SCC problems related to quenching stresses. The stress-relieved temper for most alloys is identified by the designation T*x*5*x* or T*x*5*xx* after the alloy number, for example, 2024-T351 or 7075-T6511 (Ref 64).

Effects of Grain Structure and Stress Direction. Many wrought aluminum alloy products have highly directional grain structures (Fig. 11). Such products are highly anisotropic with respect

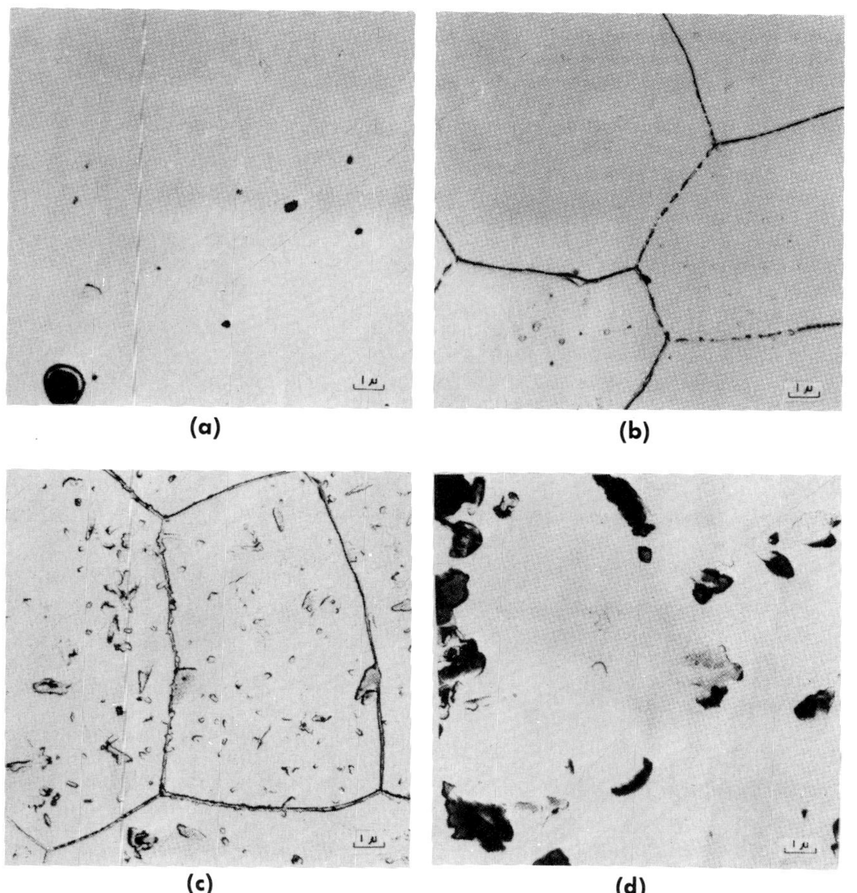

Fig. 8 Microstructures of alloy 5356-H12 after treatment to produce varying degrees of susceptibility to SCC. (a) Cold rolled 20%; highly resistant. (b) Cold rolled 20%, then heated 1 year at 100 °C (212 °F); highly susceptible. (c) Cold rolled 20%, then heated 1 year at 150 °C (300 °F); slightly susceptible. (d) Cold rolled 20%, then heated 1 year at 205 °C (400 °F); highly resistant

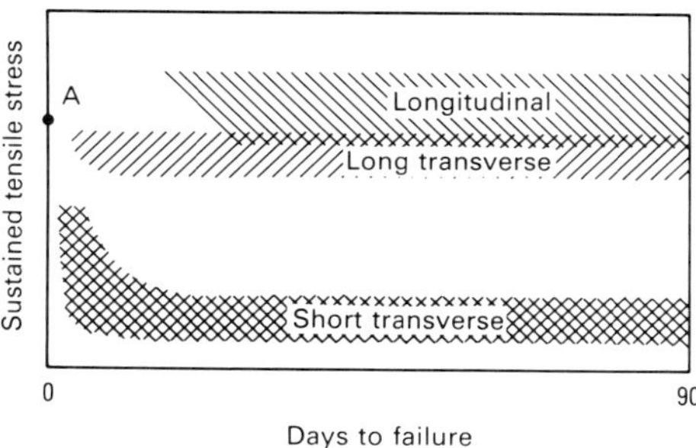

Fig. 9 SCC of alloy 7075-T651 plate. Shaded bands indicate combinations of stress and time known to produce SCC in specimens intermittently immersed in 3.5% NaCl solution. Point A is minimum yield strength in the long-transverse direction for a 75-mm (3-in.) thick plate.

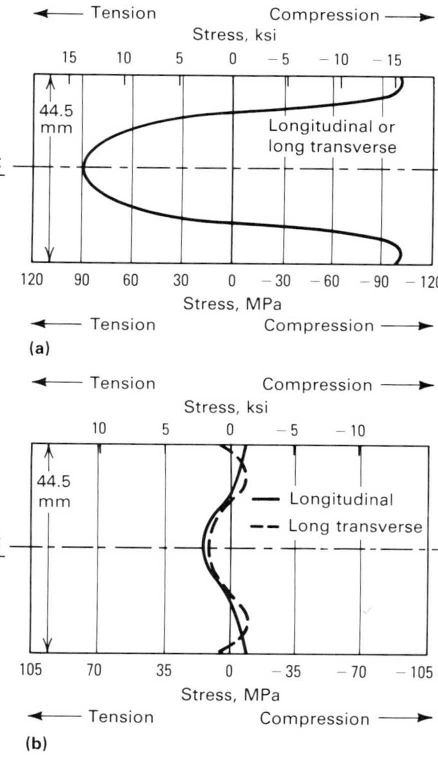

Fig. 10 Comparison of residual stresses in a thick, constant cross-section 7075-T6 aluminum alloy plate before and after stress relief. (a) High residual stresses in the solution-treated and quenched alloy. (b) Reduction in stresses after stretching 2%. Source: Ref 63

to resistance to SCC (Fig. 9). Resistance, which is measured by magnitude of tensile stress required to cause cracking, is highest when the stress is applied in the longitudinal direction, lowest in the short-transverse direction, and intermediate in other directions. These differences are most noticeable in the more susceptible tempers, but are usually much lower in tempers produced by extended precipitation treatments, such as T6 and T8 tempers for 2*xxx* alloys and T73, T736, and T76 tempers for 7*xxx* alloys.

Thus, direction and magnitude of stresses anticipated under conditions of assembly and service may govern alloy and temper selection. For products of thin section, applied in ways that induce little or no tensile stress in the short-transverse direction, resistance of 2*xxx* alloys in T3 or T4 tempers or of 7*xxx* alloys in T6 tempers may suffice. Resistance in the short-transverse direction usually controls application of products that are of thick section or are machined or applied in ways that result in sustained tensile stresses in the short-transverse direction. More resistant tempers are preferred in these cases.

Effects of Environment. Research indicates that water or water vapor is the key environmental factor required to produce SCC in aluminum alloys. Halide ions have the greatest effects in accelerating attack. Chloride is the most important halide ion because it is a natural constituent of marine environments and is present in other environments as a contaminant. Because it accelerates SCC, Cl^- is the principal component of environments used in laboratory tests to determine susceptibility of aluminum alloys to this type of attack. In general, susceptibility is greater in neutral solutions than in alkaline solutions and is greater still in acidic solutions.

Stress-Corrosion Ratings. A system of ratings of resistance to SCC for high-strength aluminum alloy products has been developed by a joint task group of ASTM and the Aluminum Association to assist alloy and temper selection, and has been incorporated into Ref 34. Definitions of these ratings, which range from A (highest resistance) to D (lowest resistance), are as follows:

- A: **Very high.** No record of service problems; SCC not anticipated in general applications
- B: **High.** No record of service problems; SCC not anticipated at stresses of the magnitude caused by solution heat treatment. Precautions must be taken to avoid high sustained tensile stresses (exceeding 50% of the minimum specified yield strength) produced by any combination of sources including heat treatment, straightening, forming, fit-up, and sustained service loading
- C: **Intermediate.** Stress-corrosion cracking not anticipated if total sustained tensile stress is maintained below 25% of minimum specified yield strength. This rating is designated for the short-transverse direction in products used primarily for high resistance to exfoliation corrosion in relatively thin structures, where appreciable stresses in the short-transverse direction are unlikely
- D: **Low.** Failure due to SCC is anticipated in any application involving sustained tensile stress in the designated test direction. This rating is currently designated only for the short-transverse direction in certain products

These stress levels are not to be interpreted as threshold stresses and are not recommended for

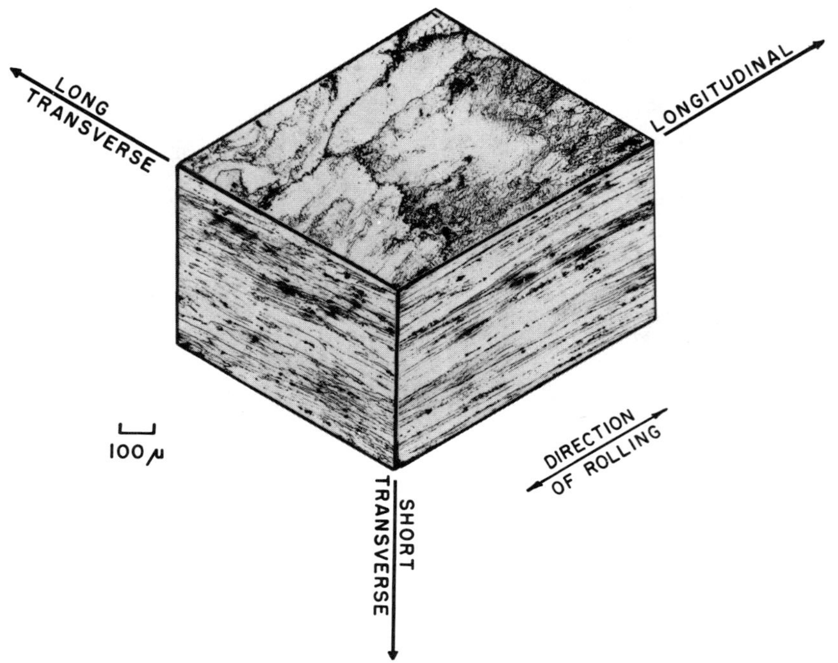

Fig. 11 Composite micrograph showing grain structure of a 38-mm (1.5-in.) alloy 7075-T6 plate

design. Documents such as MIL-HANDBOOK-5, MIL-STD-1568, NASC SD-24, and MSFC-SPEC-522A should be consulted for design recommendations.

The relative ratings of resistance to SCC for high-strength wrought aluminum alloys are presented in Table 6. These ratings, assigned primarily by alloy and temper, also distinguish among test directions and product types.

2*xxx* Alloys. Thick-section products of 2*xxx* alloys in the naturally aged T3 and T4 tempers have low ratings of resistance to SCC in the short-transverse direction. Ratings of such products in other directions are higher, as are ratings of thin-section products in all directions. These differences are related to the effects of quenching rate (largely determined by section thickness) on the amount of precipitation that occurs during quenching. If 2*xxx* alloys in T3 and T4 tempers are heated for short periods in the temperature range used for artificial aging, selective precipitation along grain or subgrain boundaries may further impair their resistance.

Longer heating, as specified for T6 and T8 tempers, produces more general precipitation and significant improvements in resistance to SCC. Precipitates are formed within grains at a greater number of nucleation sites during treatment to T8 tempers. These tempers require stretching, or cold working by other means, after quenching from the solution heat treatment temperature and before artificial aging. These tempers provide the highest resistance to SCC and the highest strength in the 2*xxx* alloy.

Some studies on aluminum-copper-lithium alloys indicate that these alloys have their highest resistance to SCC at or near peak-aged tempers (Ref 65-67). Underaging of these alloys (for example, 2090) is detrimental; overaging decreases resistance only slightly. The susceptibility of the underaged microstructure has been attributed to the precipitation of an intermetallic constituent, Al_2CuLi, on grain boundaries during the early stages of artificial aging. This constituent is believed to be anodic to the copper-rich matrix of an underaged alloy, causing preferential dissolution and SCC. As aging time increases, copper-bearing precipitates form in the interior of the grains, thus increasing the anode-cathode area ratio in the microstructure to a more favorable value that avoids selective grain-boundary attack. Similar studies of stress-corrosion behavior are being conducted on aluminum-lithium-copper-magnesium alloys (for example, 8090) (Ref 68).

5*xxx* alloys are not considered heat treatable and do not develop their strength through heat treatment. However, these alloys are processed to H3 tempers, which require a final thermal stabilizing treatment to eliminate age softening, or to H2 tempers, which require a final partial annealing. The H116 or H117 tempers are also used for high-magnesium 5*xxx* alloys and involve special temperature control during fabrication to achieve a microstructural pattern of precipitate that increases the resistance of the alloy to intergranular corrosion and SCC. The alloys of the 5*xxx* series span a wide range of magnesium contents, and the tempers that are standard for each alloy are primarily established by the magnesium content and the desirability of microstructures highly resistant to SCC and other forms of corrosion.

Although 5*xxx* alloys are not heat treatable, they develop good strength through solution hardening by the magnesium retained in solid solution, dispersion hardening by precipitates, and strain-hardening effects. Because the solid solutions in the higher-magnesium alloys are more highly supersaturated, the excess magnesium tends to precipitate out as Mg_2Al_3, which is anodic to the matrix. Precipitation of this phase with high selectivity along grain boundaries, accompanied by little or no precipitation within grains, may result in susceptibility to SCC.

The probability that a susceptible microstructure will develop in a 5*xxx* alloy depends on magnesium content, grain structure, amount of strain hardening, and subsequent time/temperature history. Alloys with relatively low magnesium contents, such as 5052 and 5454 (2.5 and 2.75% Mg, respectively), are only mildly supersaturated; consequently, their resistance to SCC is not affected by exposure to elevated temperatures. In contrast, alloys with magnesium contents exceeding about 3%, when in strain-hardened tempers, may develop susceptible structures as a result of heating or even after very long times at room temperature. For example, the microstructure of alloy 5083-O (4.5% Mg) plate stretched 1% (Fig. 12a) is relatively free of precipitate (no continuous second-phase paths), and the material is not susceptible to SCC. Prolonged heating below the solvus, however, produces continuous precipitate, which results in susceptibility (Fig. 12b).

6*xxx* Alloys. The service record of 6*xxx* alloys shows no reported cases of SCC. In laboratory tests, however, at high stresses and in aggressive solutions, cracking has been demonstrated in 6*xxx* alloys of particularly high alloy content, containing silicon in excess of the Mg_2Si ratio and/or high percentages of copper.

7*xxx* Alloys Containing Copper. The 7*xxx* series alloy that has been used most extensively and for the longest period of time is 7075, an aluminum-zinc-magnesium-copper-chromium alloy. Introduced in 1943, this aircraft construction alloy was initially used for products with thin sections, principally sheet and extrusions. In these products, quenching rate is normally very high, and tensile stresses are not encountered in the short-transverse direction; thus, SCC is not a problem for material in the highest-strength (T6) tempers. When 7075 was used in products of greater size and thickness, however, it became apparent that such products heat treated to T6 tempers were often unsatisfactory. Parts that were extensively machined from large forgings, extrusions, or plate were frequently subjected to continuous stresses, arising from interference misfit during assembly or from service loading, that were tensile at exposed surfaces and aligned in unfavorable orientations. Under such conditions, SCC was encountered in service with significant frequency.

This problem resulted in the introduction (in about 1960) of the T73 tempers for thick-section 7075 products. The precipitation treatment used to develop these tempers requires two-stage artificial aging, the second stage of which is done at a higher temperature than that used to produce T6 tempers. During the preliminary stage, a fine high-density precipitation dispersion is nucleated, producing high strength. The second stage is then used to develop resistance to SCC and exfoliation. Extensive accelerated and environmental testing has demonstrated that 7075-T73 resists SCC even when stresses are oriented in the least favorable direction, at stress levels of at least 300 MPa (44 ksi). Under similar conditions, the maximum stress at which 7075-T6 resists cracking is about 50 MPa (7 ksi). The excellent test results for 7075-T73 have been confirmed by extensive service experience in various applications.

The additional aging treatment required to produce 7075 in T73 tempers, which have high resistance to SCC, reduces strength to levels below those of 7075 in T6 tempers. Alloy 7175, a

Table 6 Relative SCC ratings for wrought products of high-strength aluminum alloys

Resistance ratings are as follows: A, very high; B, high; C, intermediate; D, low. See text for more detailed explanation of these ratings.

Alloy and temper(a)	Test direction(b)	Rolled plate	Rod and bar(c)	Extruded shapes	Forgings
2011-T3, -T4	L	(d)	B	(d)	(d)
	LT	(d)	D	(d)	(d)
	ST	(d)	D	(d)	(d)
2011-T8	L	(d)	A	(d)	(d)
	LT	(d)	A	(d)	(d)
	ST	(d)	A	(d)	(d)
2014-T6	L	A	A	A	B
	LT	B(e)	D	B(e)	B(e)
	ST	D	D	D	D
2024-T3, -T4	L	A	A	A	(d)
	LT	B(e)	D	B(e)	(d)
	ST	D	D	D	(d)
2024-T6	L	(d)	A	(d)	A
	LT	(d)	B	(d)	A(e)
	ST	(d)	B	(d)	D
2024-T8	L	A	A	A	A
	LT	A	A	A	A
	ST	B	A	B	C
2048-T851	L	A	(d)	(d)	(d)
	LT	A	(d)	(d)	(d)
	ST	B	(d)	(d)	(d)
2124-T851	L	A	(d)	(d)	(d)
	LT	A	(d)	(d)	(d)
	ST	B	(d)	(d)	(d)
2219-T3, -T37	L	A	(d)	A	(d)
	LT	B	(d)	B	(d)
	ST	D	(d)	D	(d)
2219-T6, -T8	L	A	A	A	A
	LT	A	A	A	A
	ST	A	A	A	A
6061-T6	L	A	A	A	A
	LT	A	A	A	A
	ST	A	A	A	A
7005-T53, -T63	L	(d)	(d)	A	A
	LT	(d)	(d)	A(e)	A(e)
	ST	(d)	(d)	D	D
7039-T63, -T64	L	A	(d)	A	(d)
	LT	A(e)	(d)	A(e)	(d)
	ST	D	(d)	D	(d)
7049-T73	L	A	(d)	A	A
	LT	A	(d)	A	A
	ST	A	(d)	B	A
7049-T76	L	(d)	(d)	A	(d)
	LT	(d)	(d)	A	(d)
	ST	(d)	(d)	C	(d)
7149-T73	L	(d)	(d)	A	A
	LT	(d)	(d)	A	A
	ST	(d)	(d)	B	A
7050-T736	L	A	(d)	A	A
	LT	A	(d)	A	A
	ST	B	(d)	B	B
7050-T76	L	A	A	A	(d)
	LT	A	B	A	(d)
	ST	C	B	C	(d)
7075-T6	L	A	A	A	A
	LT	B(e)	D	B(e)	B(e)
	ST	D	D	D	D
7075-T73	L	A	A	A	A
	LT	A	A	A	A
	ST	A	A	A	A
7075-T736	L	(d)	(d)	(d)	A
	LT	(d)	(d)	(d)	A
	ST	(d)	(d)	(d)	B
7075-T76	L	A	(d)	A	(d)
	LT	A	(d)	A	(d)
	ST	C	(d)	C	(d)
7175-T736	L	(d)	(d)	(d)	A
	LT	(d)	(d)	(d)	A
	ST	(d)	(d)	(d)	B
7475-T6	L	A	(d)	(d)	(d)
	LT	B(e)	(d)	(d)	(d)
	ST	D	(d)	(d)	(d)
7475-T73	L	A	(d)	(d)	(d)
	LT	A	(d)	(d)	(d)
	ST	A	(d)	(d)	(d)
7475-T76	L	A	(d)	(d)	(d)
	LT	A	(d)	(d)	(d)
	ST	C	(d)	(d)	(d)
7178-T6	L	A	(d)	A	(d)
	LT	B(e)	(d)	B(e)	(d)
	ST	D	(d)	D	(d)
7178-T76	L	A	(d)	A	(d)
	LT	A	(d)	A	(d)
	ST	C	(d)	C	(d)
7079-T6	L	A	(d)	A	A
	LT	B(e)	(d)	B(e)	B(e)
	ST	D	(d)	D	D

(a) Ratings apply to standard mill products in the types of tempers indicated and also in T*x*5*x* and T*x*5*xx* (stress-relieved) tempers and may be invalidated in some cases by use of nonstandard thermal treatments, or mechanical deformation at room temperature, by the user. (b) Test direction refers to orientation of direction in which stress is applied relative to the directional grain structure typical of wrought alloys, which for extrusions and forgings may not be predictable on the basis of the cross-sectional shape of the product: L, longitudinal; LT, long transverse; ST, short transverse. (c) Sections with width-to-thickness ratios equal to or less than two, for which there is no distinction between LT and ST properties. (d) Rating not established because product not offered commercially. (e) Rating is one class lower for thicker sections: extrusions, 25 mm (1 in.) and thicker; plate and forgings, 38 mm (1.5 in.) and thicker

variant of 7075, was developed for forgings. In the T736 temper, 7175 has strength nearly comparable to that of 7075-T6 and has better resistance to SCC. Other newer alloys—such as 7049 and 7475, which are used in the T73 temper, and 7050, which is used in the T736 temper—couple high strength with very high resistance and improved fracture toughness.

The T76 tempers, which also require two-stage artificial aging and which are intermediate to the T6 and T73 tempers in both strength and resistance to SCC, are developed in copper-containing 7*xxx* alloys for certain products. Comparative ratings of resistance for various products of all these alloys, as well as for products of 7178, are given in Table 6.

The microstructural differences among the T6, T73, and T76 tempers of these alloys are differences in size and type of precipitate, which changes from predominantly Guinier-Preston (GP) zones in T6 tempers to η', the metastable transition form of $\eta(MgZn_2)$, in T73 and T76 tempers. None of these differences can be detected by optical metallography. In fact, even the resolutions possible in transmission electron microscopy are insufficient for determining whether the precipitation reaction has been adequate to ensure the expected level of resistance to SCC. For quality assurance, copper-containing 7*xxx* alloys in T73 and T76 tempers are required to have specified minimum values of electrical conductivity and, in some cases, tensile yield strengths that fall within specified ranges. The validity of these properties as measures of resistance to SCC is based on many correlation studies involving these measurements, laboratory and field stress-corrosion tests, and service experience.

Copper-free 7*xxx* Alloys. Wrought alloys of the 7*xxx* series that do not contain copper are of considerable interest because of their good resistance to general corrosion, moderate-to-high strength, and good fracture toughness and formability. Alloys 7004 and 7005 have been used in extruded form and, to a lesser extent, in sheet form for structural applications. More recently introduced compositions, including 7016, 7021, 7029, and 7146, have been used in automobile bumpers formed from extrusions or sheet.

As a group, copper-free 7*xxx* alloys are less resistant to SCC than other types of aluminum alloys when tensile stresses are developed in the short-transverse direction at exposed surfaces. Resistance in other directions may be good, particularly if the product has an unrecrystallized microstructure and has been properly heat treated. Products with recrystallized grain structures are generally more susceptible to cracking as a result of stresses induced by forming or mechanical damage after heat treatment. When cold forming is required, subsequent solution heat treatment or precipitation heat treatment is recommended. Applications of these alloys must be carefully engineered, and consultation among designers, application engineers and product producers, or suppliers is advised in all cases.

Casting Alloys. The resistance of most aluminum casting alloys to SCC is sufficiently high that cracking rarely occurs in service. The

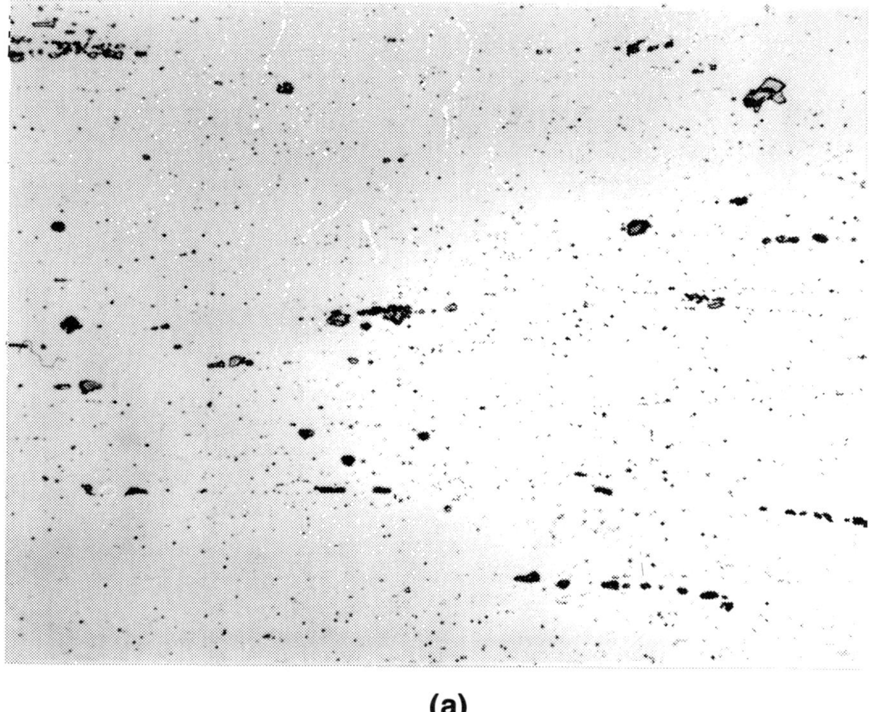

(a)

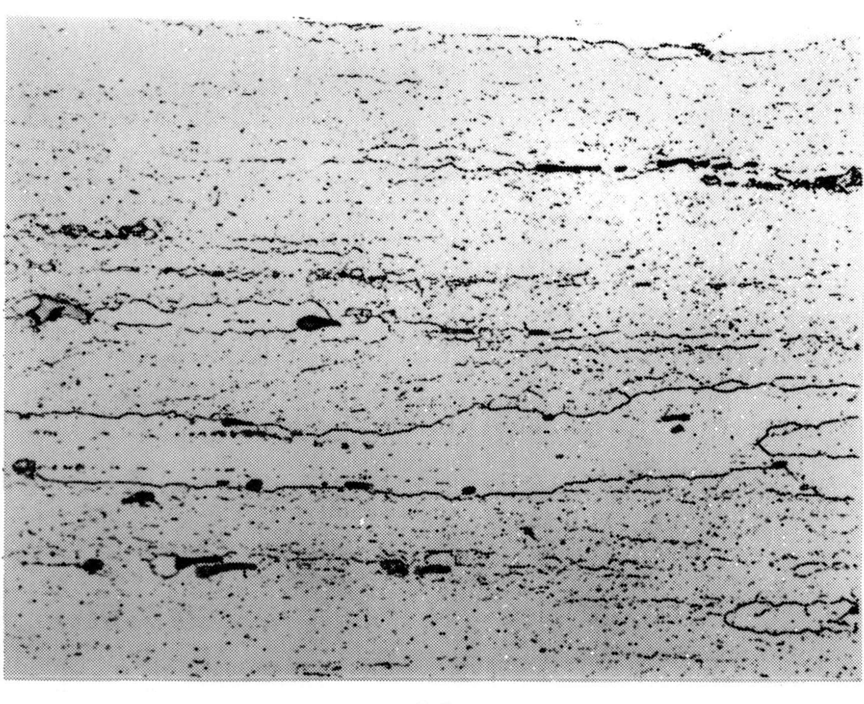

(b)

Fig. 12 Microstructures of alloy 5083-O plate stretched 1%. (a) As-stretched. (b) After heating 40 days at 120 °C (250 °F)

microstructures of these alloys are usually nearly isotropic; consequently, resistance to SCC is unaffected by orientation of tensile stresses.

Relative ratings of cast alloys, based primarily on accelerated laboratory tests, are listed in Table 4. It has been indicated by accelerated and natural-environment testing and verified by service experience that alloys of the aluminum-silicon 4*xx.x* series, 3*xx.x* alloys containing only silicon and magnesium as alloying additions, and 5*xx.x* alloys with magnesium contents of 8% or lower have virtually no susceptibility to SCC. Alloys of the 3*xx.x* group that contain copper are rated as less resistant, although the numbers of castings of these alloys that have failed by SCC have not been significant.

Significant SCC of aluminum alloy castings in service has occurred only in the highest-strength aluminum-zinc-magnesium 7*xx.x* alloys and in the aluminum-magnesium alloy 520.0 in the T4 temper. For such alloys, factors that require careful consideration include casting design, assembly and service stresses, and anticipated environmental exposure.

Specifications and Tests. Several aluminum alloy product specifications require defined levels of performance with respect to resistance to SCC. Standard tests used to measure such performance are described in methods standards and are referenced in materials specifications. Among these are tests for evaluating resistance to SCC of 2*xxx* alloys and of 7*xxx* alloys that contain copper by alternate immersion in 3.5% NaCl solution (Ref 69, 70). Lot acceptance criteria for products of 7*xxx* copper-containing alloys in T76, T73, and T736 tempers are based on combined requirements for tensile strength and electrical conductivity.

Exfoliation Corrosion

In certain tempers, wrought products of aluminum alloys are subject to corrosion by exfoliation, which is sometimes described as lamellar, layer, or stratified corrosion. In this type of corrosion, attack proceeds along selective subsurface paths parallel to the surface. As shown in Fig. 13(a), layers of uncorroded metal between the selective paths are split apart and pushed above the original surface by the voluminous corrosion product formed along the paths of attack. Because it can be detected readily at an early stage and is restricted in depth, exfoliation does not cause unexpected structural failure, as does SCC.

Exfoliation occurs predominantly in products that have markedly directional structures in which highly elongated grains form platelets that are thin relative to their length and width (Fig. 14). Susceptibility to this type of corrosion may result from the presence of aligned intergranular or subgrain boundary precipitates or from aligned strata that differ slightly in composition. The intensity of exfoliation increases in slightly acidic environments or when the aluminum is coupled to a cathodic dissimilar metal. Exfoliation is not accelerated by stress and does not lead to SCC.

Alloys most susceptible to exfoliation are the heat-treatable 2*xxx* and 7*xxx* alloys and certain cold-worked 5*xxx* alloys, such as 5456-H321 boat hull plates. Exfoliation problems with 5*xxx* alloys led to the development of special boat hull plate tempers, H116 and H117, for alloys 5083, 5086, and 5456. In these alloys, exfoliation is primarily caused by unfavorable distribution of precipitate. The processing to eliminate this form of attack promotes either more uniform precipitation within grains or a more advanced stage of precipitation. Thus, increases in the precipitation heat-treating time or temperature are as effective in reducing susceptibility to exfoliation as they are in reducing susceptibility to SCC.

During long-duration or high-temperature precipitation treatments, maximum resistance to exfoliation is usually achieved sooner than maximum resistance to SCC. Thus, precipitation treatments used to produce T76 tempers in 7*xxx*

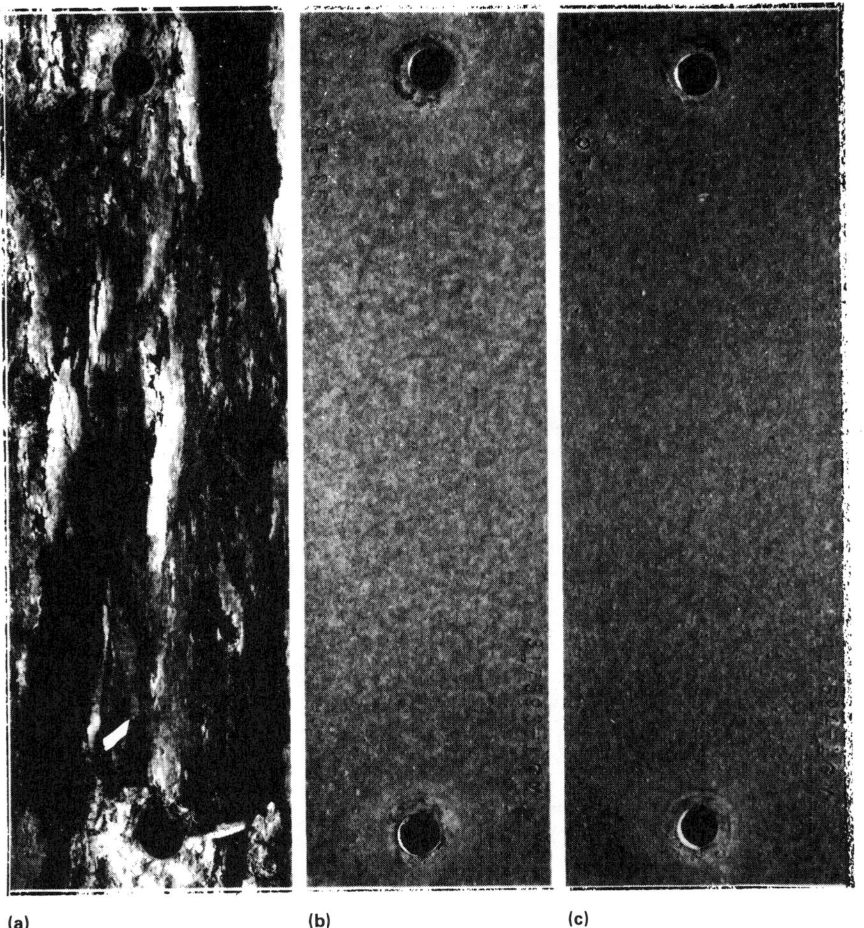

Fig. 13 Effect of temper on exfoliation resistance of an alloy 7075 extrusion exposed in a seacoast environment. Specimens were exposed for 4 years. (a) Specimen in the T6510 temper that developed exfoliation after only 5 months. (b) and (c) Specimens in the T76510 and T73510 tempers that were unaffected after 4 years

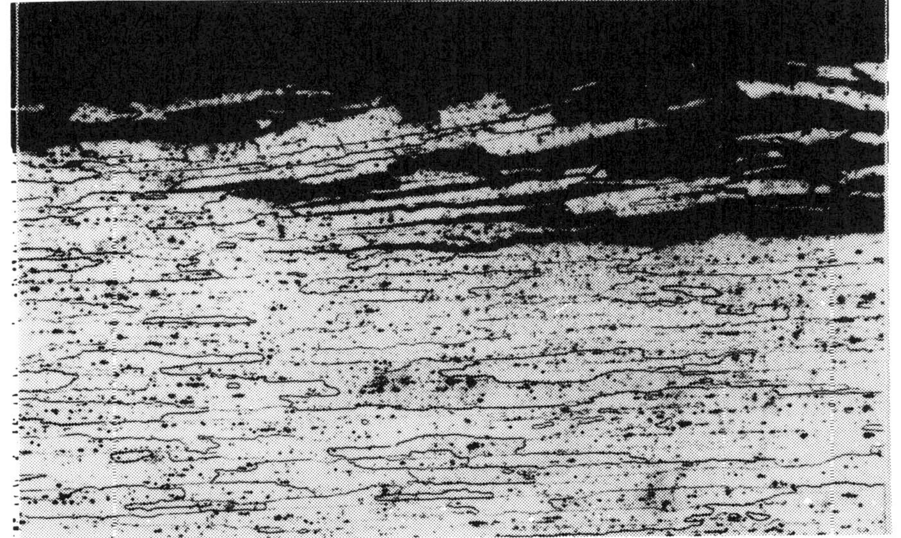

Fig. 14 Exfoliation corrosion in an alloy 7178-T651 plate exposed to a seacoast environment. Cross section of the plate shows how exfoliation develops by corrosion along boundaries of thin, elongated grains.

alloys, which use times and temperatures intermediate to those of T6 and T73 treatments, provide excellent resistance to exfoliation (Fig. 13b) but only intermediate resistance to SCC. The T73 tempers provide the highest resistance to both types of corrosion (Fig. 13c) but at a sacrifice in strength compared to T76 tempers.

Among the standard tests for evaluating resistance to exfoliation of 2*xxx*, 5*xxx*, and 7*xxx* alloys are those that require total immersion in aggressive acidified solutions of mixed salts or exposure to cyclic, acidified salt spray tests. Such tests are described in Ref 71 for 5*xxx* alloys and in Ref 72 to 74 for 2*xxx* and 7*xxx* alloys. Acceptability of aluminum-magnesium alloys 5083, 5086, and 5456 is based on a comparison of the microstructure disclosed by etching in a defined manner with a reference microstructure that is predominantly free from a continuous grain-boundary network of Al_8Mg_5 precipitate particles (Ref 75). Material containing such precipitate in amounts exceeding that shown by the reference standard is unacceptable unless it can be demonstrated by testing (Ref 71) that the material has acceptable resistance to exfoliation. References 76 to 80 compare the performance in these accelerated test methods to that in outdoor atmospheres.

Corrosion Fatigue

Fatigue strengths of aluminum alloys are lower in such corrosive environments as seawater and other salt solutions than in air, especially when evaluated by low-stress long-duration tests (Ref 81, 82). As shown in Fig. 15, such corrosive environments produce smaller reductions in fatigue strength in the more corrosion-resistant alloys, such as the 5*xxx* and 6*xxx* series, than in the less resistant alloys, such as the 2*xxx* and 7*xxx* series.

Like SCC of aluminum alloys, corrosion fatigue requires the presence of water. In contrast to SCC, however, corrosion fatigue is not appreciably affected by test direction, because the fracture that results from this type of attack is predominantly transgranular.

Erosion-Corrosion

In noncorrosive environments, such as high-purity water, the stronger aluminum alloys have

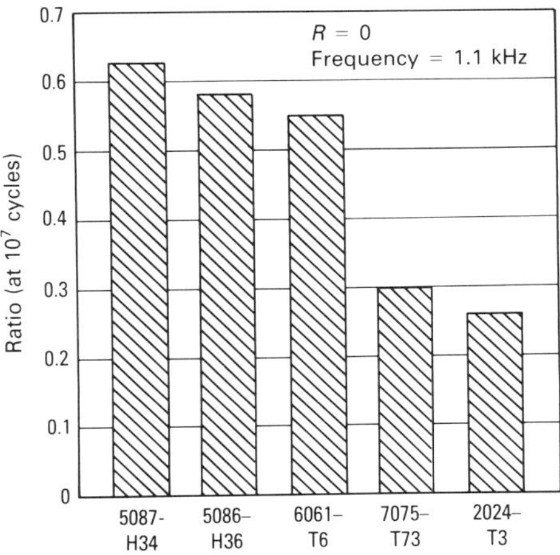

Fig. 15 Ratio of axial-stress fatigue strength of aluminum alloy sheet in 3% NaCl solution to that in air. Specimens were 1.6 mm (0.064 in.) thick.

the greatest resistance to erosion-corrosion because resistance is controlled almost entirely by the mechanical components of the system. In a corrosive environment, such as seawater, the corrosion component becomes the controlling factor; thus, resistance may be greater for the more corrosion-resistant alloys even though they are lower in strength. Corrosion inhibitors and cathodic protection have been used to minimize erosion-corrosion, impingement, and cavitation on aluminum alloys (Ref 83).

Atmospheric Corrosion

Most aluminum alloys have excellent resistance to atmospheric corrosion (often called weathering), and in many outdoor applications, such alloys do not require shelter, protective coatings, or maintenance. Aluminum alloy products that have no external protection and therefore depend critically on this property include electrical conductors, outdoor lighting poles, ladders, and bridge railings. Such products often retain a bright metallic appearance for many years, but their surfaces may become dull, gray, or even black as a result of pollutant accumulation. Corrosion of most aluminum alloys by weathering is restricted to mild surface roughening by shallow pitting, with no general thinning. However, such attack is more severe for alloys with higher copper contents, and such alloys are seldom used in outdoor applications without protection.

Corrosivity of the atmosphere to metals varies greatly from one geographic location to another, depending on such weather factors as wind direction, precipitation and temperature changes, amount and type of urban and industrial pollutants, and proximity to natural bodies of water. Service life may also be affected by the design of the structure if weather conditions cause repeated moisture condensation in unsealed crevices or in channels with no provision for drainage. Laboratory exposure tests, such as salt spray, total-immersion, and alternate-immersion tests, provide useful comparative information, but have limited value for predicting actual service performance and sometimes exaggerate differences among alloys that are negligible under atmospheric conditions (Ref 84). Consequently, extensive long-term evaluations of the effects of exposure in different industrial, chemical, seacoast, tropical, and rural environments have been made (Ref 85-88).

Data collected in these programs include measurements of maximum and mean depth of attack, weight loss, and changes in tensile properties. Because of the localized nature of the prevalent pitting corrosion, which leaves some (in many cases, most) of the original surface intact even after many years of weathering, weight loss or calculated average dimensional change based on weight loss may have limited significance. Changes in tensile strength, which reflect the effects of size, number, distribution, and acuity of pits, are generally most significant from a structural standpoint, while depth-of-attack determinations provide realistic measures of penetration rate.

Effect of Exposure Time. A very important characteristic of weathering of aluminum and of corrosion of aluminum under many other environmental conditions is that corrosion rate decreases with time to a relatively low, steady-state rate (Ref 85). This deceleration of corrosion (Fig. 16 to 18) occurs regardless of alloy composition, type of environment, or the parameter by which the corrosion is measured. However, loss in tensile strength, which is influenced somewhat by pit acuity and distribution but is basically a result of loss of effective cross section, decelerates more gradually than depth of attack (Fig. 16).

The decrease in rate of penetration of corrosion is dramatic. In general, rate of attack at discrete locations, which is initially about 0.1 mm/yr (4 mils/yr), decreases to much lower and nearly constant rates within a period of about 6 months to 2 years. For the deepest pits, the maximum rate after about 2 years does not exceed about 0.003 mm/yr (0.11 mil/yr) for severe seacoast locations and may be as low as 0.0008 mm/yr (0.03 mil/yr) in rural or arid climates. The dramatic deceleration in penetration is illustrated by the specimen cross sections shown in Fig. 17 and by the depth-of-attack curves shown in Fig. 18, both of which are from the same 30-year test program (Ref 89). Also shown in Fig. 18 are results (shown as vertical bars) from other test programs in which various articles made of aluminum alloys were continuously exposed for various periods and in different locations, many of which are less severe than the relatively aggressive industrial environment of New Kensington, PA.

Data for Wrought Alloys. Several major test programs have been conducted under the supervision of ASTM to investigate the weathering of aluminum alloy sheet. The first program, started in 1931, was limited in the variety of alloys tested, but included desert, rural, seacoast, and industrial exposures. Data obtained after 20 years of exposure are listed in Table 7. Corrosion rates were calculated from cumulative weight loss after 20 years, and average and maximum depths of attack were measured microscopically. In aggressive (seacoast and industrial) environments, the bare (nonalclad) heat-treated alloys—2017-T3 and, to a lesser extent, 6051-T4—exhibited more severe corrosion and greater resulting loss in tensile strength than the nonheat-treatable alloys. Alclad 2017-T3, although as severely corroded as the nonheat-treatable materials, did not show measurable loss in strength; in fact, some specimens of this alloy were 2 to 3% higher in strength after 20 years because of long-term natural aging.

Data from a comprehensive program initiated in 1958 were compiled from examinations and tests performed after 7 years of exposure (Ref 89). Thirty-four combinations of alloy and temper in the form of 1.27-mm (0.050-in.) thick sheet were exposed at four sites—two seacoast, one industrial, and one rural; Table 8 lists average values of measurements reported at two of the more aggressive sites. In another ASTM program, 10 years of weathering produced the changes in tensile strength reported in Table 9.

Data from these and other weathering programs (Ref 92, 93) demonstrate that differences in resistance to weathering among nonheat-treatable alloys are not great, that alclad products retain their strength well because corrosion penetration is confined to the cladding layer, and that corrosion and resulting strength loss tend to be greater for bare (nonalclad) heat-treatable 2*xxx* and 7*xxx* series alloys.

Data for Casting Alloys. The testing program that was the source of the strength change data for wrought alloys given in Table 9 also provided weathering data for casting alloys exposed for the same period of time and at the same sites. Specimens were separately sand-cast and permanent mold cast tensile bars, each with a reduced section 12.7 mm (0.5 in.) in diameter. Strength change data for these alloys are summarized in Table 10. Alloys with relatively high copper contents, such as 295.0-T6, 208.0-F, 319.0-T6, and 319.0-T61, showed the greatest losses. Alloys of the zinc-containing 7*xx.x* series generally exhibited larger strength losses than alloys having low zinc or copper contents. In all cases, as for wrought materials, severity of corrosion varied widely, depending on environmental conditions.

Comparison with Other Metals. Other metals were exposed to the same weathering environments over the same time periods used to evaluate corrosion of aluminum alloys. Comparative corrosion rates (average loss in thickness per side calculated from weight losses measured after exposures of 10 and 20 years) are listed in Table 11 for aluminum, copper, lead, and zinc panels. Figure 19 compares losses in tensile strengths at several weathering sites for unprotected low-carbon steel (0.09C, 0.07Cu) and for aluminum alloys.

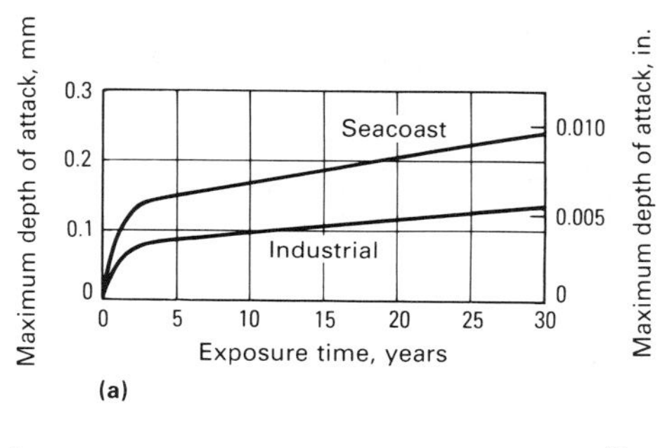

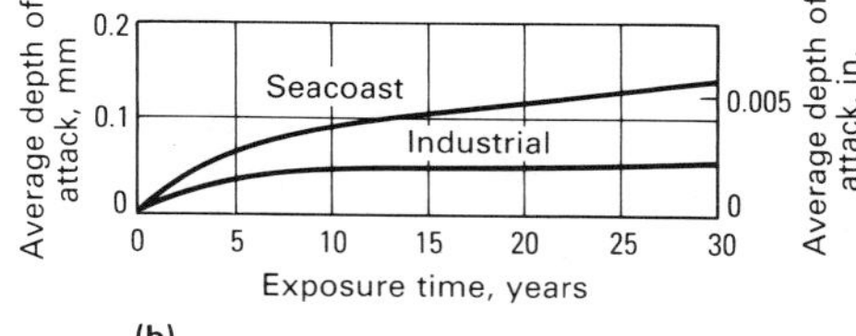

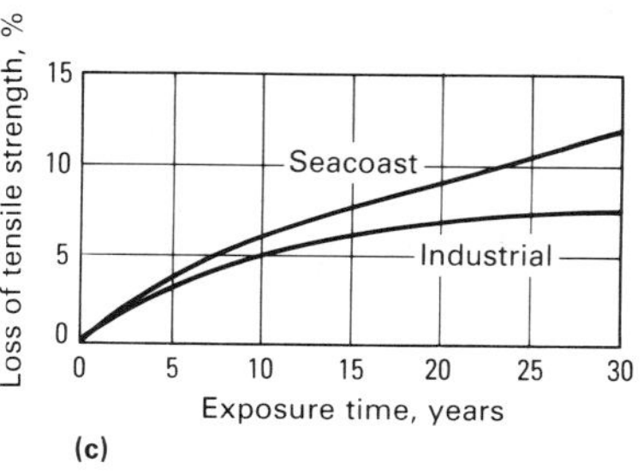

Fig. 16 Depth of corrosion and loss of tensile strength for alloys 1100, 3003, and 3004 (shown in graphs a, b, and c, respectively). Data are given for the average performance of the three alloys, all in the H14 temper. Seacoast exposure was at a severe location (Pt. Judith, RI); industrial exposure was at New Kensington, PA. Tensile strengths were computed using original cross-sectional areas, and loss in strength is expressed as a percentage of original tensile strength.

Filiform Corrosion

Filiform corrosion, sometimes termed worm-track corrosion, occurs on aluminum when it is coated with an organic coating and exposed to

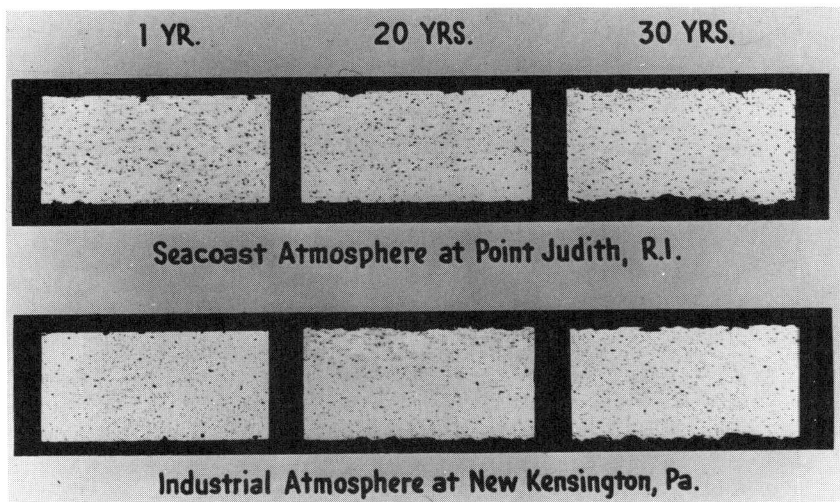

Fig. 17 Sectioned specimens cut from 1.6-mm (0.064-in.) thick alloy 3003-H14 panels after exposure in two environments

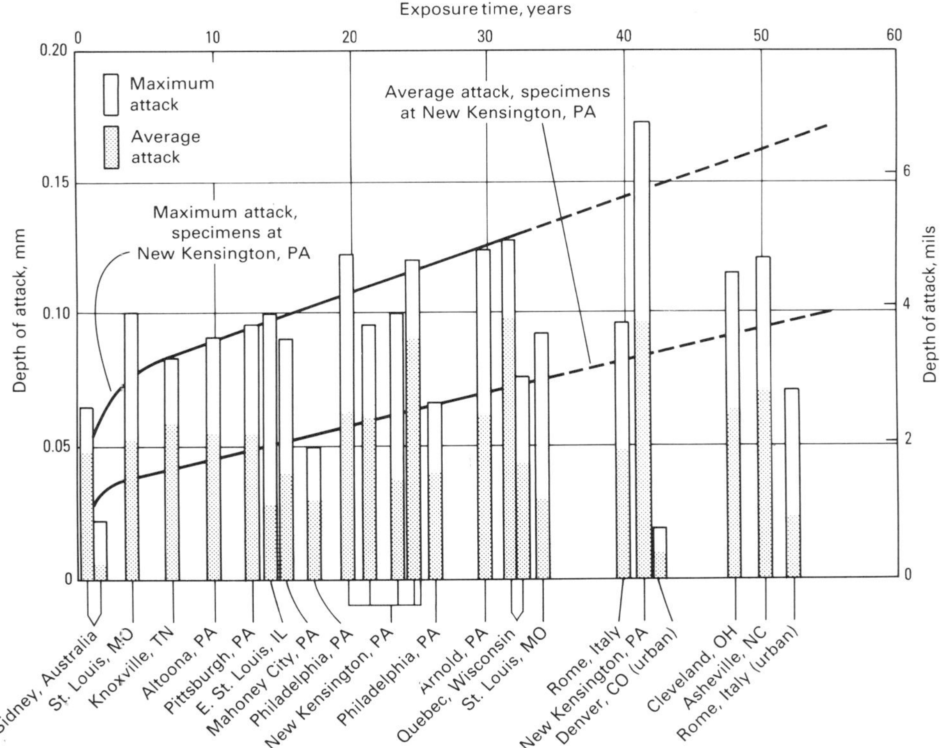

Fig. 18 Correlation of weathering data for specimens of alloys 1100, 3003, and 3004 (all in H14 temper) exposed to industrial atmosphere (curves) with service experience with aluminum alloys in various locations (bars)

warm, humid atmospheres. The corrosion appears as threadlike filaments that initiate at defects in the organic coating, are activated by chlorides, and grow along the metal/coating interface at rates to 1 mm/d (0.04 in./d). The moving end of the filament is called the head, and the remainder of the track is called the tail. It is not clear why this type of corrosion forms tracks instead of circular spots of increasing diameter.

Filiform corrosion occurs only in the atmosphere, and relatively humidity is the single most important factor. This type of attack is rare on aluminum below about 55% relative humidity or above 95%. In natural atmospheres, it occurs most readily on aluminum at relative humidities between 85 and 95%. Although temperature and the thickness of the organic coating are minor factors, elevating the temperature increases the rate of filament growth if the relative humidity stays within the critical range.

The presence of oxygen is fundamental because it supplies the primary reactant for the cathodic reaction. Essentially, filiform corrosion is a type of oxygen concentration cell in which the anodic area is the head of the filament and the cathode is the area surrounding it, including the tail (Ref 95).

Considerable acidity is generated at the leading edge of the head. Measurements of pH as low as 1.5 to 2.5 have been reported, with the electrolyte in the head containing large concentrations of chloride (Ref 96). Such acidic solutions can be highly corrosive to aluminum alloys if the filament head should stop moving.

Phosphate coatings or chromium-containing conversion coatings applied to the metal surface before the organic coating are widely used to protect against filiform corrosion, but they are not completely successful. Perfect coatings, the absence of chlorides, or relative humidity below 30 to 40% would also be beneficial, but these conditions are not likely to be encountered outside the laboratory.

Corrosion in Waters

High-Purity Water. Suitability of the more corrosion-resistant aluminum alloys for use with high-purity water at room temperature is well established by both laboratory testing and service experience (Ref 97). The slight reaction with the water that occurs initially ceases almost completely within a few days after development of a protective oxide film of equilibrium thickness. After this conditioning period, the amount of metal dissolved by the water becomes negligible.

Corrosion resistance of aluminum alloys in high-purity water is not significantly decreased by dissolved carbon dioxide or oxygen in the water or, in most cases, by the various chemicals added to high-purity water in the steam power industry to provide the required compatibility with steel. These additives include ammonia and neutralizing amines for pH adjustment to control carbon dioxide, hydrazine and sodium sulfate to control oxygen, and filming amines (long-chain polar compounds) to produce nonwettable surfaces. Somewhat surprisingly, the effects of alloying elements on corrosion resistance of aluminum alloys in high-purity water at elevated temperatures are opposite to their effects at room temperature; elements (including impurities) that decrease resistance at room temperature improve it at elevated temperatures.

At 200 °C (390 °F), high-purity aluminum of sheet thickness disintegrates completely within a few days by reaction with high-purity water to form aluminum oxide. In contrast, aluminum-nickel-iron alloys have the best elevated-temperature resistance to high-purity water of all aluminum metals; for example, alloy X8001 (1.0Ni-0.5Fe) has good resistance at temperatures as high as 315 °C (600 °F) (Ref 98).

Natural waters. Aluminum alloys of the 1*xxx*, 3*xxx*, 5*xxx*, and 6*xxx* series are resistant to corrosion by many natural waters (Ref 99-101). The more important factors controlling the corrosivity of natural waters to aluminum include water temperature, pH, and conductivity; availability of cathodic reactant; presence or absence of heavy metals; and the corrosion potentials and pitting potentials of the specific alloys. Various correlations of the corrosivity of natural waters to aluminum have been attempted (Ref 102), but none predicts the corrosivity of all natural waters reliably.

Seawater. Service experience with 1*xxx*, 3*xxx*, 5*xxx*, and 6*xxx* wrought aluminum alloys in

Table 7 Weathering data for 0.89-mm (0.035-in.) thick aluminum alloy sheet after 20-year exposure (ASTM program started in 1931)

Alloy and temper	Corrosion rate		Average depth of attack		Maximum depth of attack		Loss in tensile strength,
	nm/yr	µin./yr	µm	mils	m	mils	%
Phoenix, AZ (desert)							
1100-H14	76	3.0	8	0.3	18	0.7	0
2017-T3	76	3.0	23	0.9	51	2.0	0
2017-T3, alclad	13	0.5	10	0.4	23	0.9	0
3003-H14	13	0.5	5	0.2	10	0.4	0
6051-T4	13	0.5	28	1.1	74	2.9	0
State College, PA (rural)							
1100-H14	76	3.0	36	1.4	89	3.5	3
2017-T3	102	4.0	25	1.0	81	3.2	2
2017-T3, alclad	76	3.0	10	0.4	25	1.0	0
3003-H14	89	3.5	23	0.9	56	2.2	3
6051-T4	76	3.0	23	0.9	96	3.8	0
Sandy Hook, NJ (seacoast)							
1100-H14	279	11.0	96	3.8	231	9.1	3
2017-T3	...	...	43	1.7	132	5.2	10
2017-T3, alclad	...	...	23	0.9	33	1.3	...
3003-H14	356	14.0	36	1.4	84	3.3	...
6051-T4	343	13.5	58	2.3	137	5.4	9
La Jolla, CA (seacoast)							
1100-H14	584	23.0	102	4.0	356	14.0	8
2017-T3	2260	89.0	147	5.8	515	20.3	20
2017-T3, aclad	584	23.0	33	1.3	74	2.9	0
3003-H14	610	24.0	107	4.2	259	10.2	7
6051-T4	775	30.5	84	3.3	307	12.1	20
New York, NY (industrial)							
1100-H14	749	29.5	89	3.5	213	8.4	7
2017-T3	1260	49.6	51	2.0	180	7.1	7
2017-T3, alclad	762	30.0	28	1.1	36	1.4	0
3003-H14	965	38.0	51	2.0	163	6.4	8
6051-T4	914	36.0	74	2.9	170	6.7	12

Source: Ref 90

marine applications, including structures, pipeline, boats, and ships, demonstrates their good resistance and long life under conditions of partial, intermittent, or total immersion. Casting alloys of the 356.0 and 514.0 types also show high resistance to seawater corrosion, and these alloys are used widely for fittings, housings, and other marine parts.

Among the wrought alloys, those of the 5*xxx* series are most resistant and most widely used because of their favorable strength and good weldability. Alloys of the 3*xxx* series are also highly resistant and are suitable where their strength range is adequate. With the 3*xxx* and 5*xxx* series alloys, thinning by uniform corrosion is negligible, and the rate of corrosion based on weight loss does not exceed about 5 µm/yr (0.2 mil/yr), which is generally less than 5% of the rate for unprotected low-carbon steel in seawater. Corrosion is mainly of the pitting or crevice type, characterized by deceleration of penetration with time from rates of 3 to 6 µm/yr (0.1 to 0.2 mil/yr) in the first year to average rates over a 10-year period of 0.8 to 1.5 µm/yr (0.03 to 0.06 mil/yr).

The aluminum-magnesium-silicon 6*xxx* alloys are somewhat less resistant; although no general thinning occurs, weight loss may be two to three times that for 5*xxx* alloys. The more severe corrosion is reflected in larger and more numerous pits.

Alloys of the 2*xxx* and 7*xxx* series, which contain copper, are considerably less resistant to seawater than 3*xxx*, 5*xxx*, and 6*xxx* alloys and are generally not used unprotected. Protective measures, such as use of alclad products and coating by metal spraying or by painting, provide satisfactory service in certain situations.

Aluminum boats operating in salt water require antifouling paint systems because aluminum and its alloys do not inhibit growth of marine organisms. Aluminum is impervious to worms and borers, and the acids exuded from marine organisms are not corrosive to aluminum; but the accumulation of biofouling on the bottom of the boat impairs performance. Aluminum boats operating in both salt and fresh water, which alleviates fouling problems, have been able to leave underwater hull areas unpainted (Ref 103).

To make antifouling paint systems adhere properly to aluminum, careful surface preparation of the metal is necessary. A thorough precleaning and either a conversion coating or a washcoat primer are required, followed by a corrosion-inhibiting primer and a top coat. The antifouling paint is applied to the top coat. Primers containing red lead should not be used, because this substance may cause galvanic corrosion of the aluminum. For the same reason, copper-containing antifouling paints should not be used on aluminum hulls. The preferred antifouling paints for aluminum are those containing organic tin compounds.

The literature on corrosion testing of aluminum alloys in seawater is extensive. Summaries of information are provided in Ref 104 and 105, and in most of the selected references. Table 12 lists results of 10-year immersion testing of various alloys in the form of rolled plate exposed in three locations. The relationships among the types of alloys that have been discussed and a comparison with unprotected low-carbon steel are apparent. Similar data for extruded products of several 6*xxx* alloys and one 5*xxx* alloy are given in Table 13. Direct comparison of the data in Tables 12 and 13 is provided in Table 14, in which corrosion is expressed in terms of average weight loss, and in Fig. 20, which illustrates the deceleration of corrosion rate with time that is characteristic of aluminum alloys. Data on corrosion rates, maximum and average depth of pitting, and changes in tensile strength compiled during 10-year tidal and full-immersion exposure of seven 5*xxx* alloys and super-purity aluminum 1199 are summarized in Table 15. Full immersion generally resulted in more extensive corrosion than tidal exposure, although the reverse relationship has also been observed. Tensile-strength losses were 5% or less, and yield-strength losses were less than 5% in the panels completely immersed and generally lower in those exposed to tidal immersion.

The data in Table 16 illustrate the corrosion resistance of aluminum alloy plates, with and without riveted or welded joints, in flowing seawater. All assemblies and panels underwent only moderate pitting and retained most of their original strength.

The corrosion behavior of aluminum alloys in deep seawater, judging from tests at a depth of 1.6 km (1 mile), is generally the same as at the surface except that the rate of pit penetration may be higher and the effect of crevices somewhat greater (Ref 107). The corrosivity of unpolluted full-strength seawater depends on several factors: dissolved oxygen content; pH, temperature, and velocity of the water flow; and the presence or absence of heavy-metal ions, particularly copper (Ref 48). The corrosion rate tends to be increased by decreasing temperature, pH, and flow velocity and by increasing dissolved oxygen (Ref 48, 108-110). The higher corrosion rate in deep water is not caused by low dissolved oxygen, as stated in the older literature, but is caused by the combination of low pH and low temperature.

Surface-water conditions at various tropical locations are benign to aluminum alloys because of their high temperature, high pH, and high oxygen concentrations and the virtual absence of heavy-metal contamination (Ref 111-113). A variety of aluminum alloys in the form of heat-exchanger tubing have been tested for up to 3 years in surface water off Keahole Point, HI, with no significant pitting or crevice corrosion (Ref 111).

Experience with seawater desalination units has demonstrated the high degree of resistance of aluminum alloys to deaerated seawater at temperatures to 120 °C (250 °F). For example, a 11 355-L/day (3000-gal./day) multiflash aluminum unit at the Office of Saline Water Materials Test Center at Freeport, TX, operated at 99% efficiency and with minimal corrosion for more than 3 years under process conditions selected to match those of a commercial installation. Such experience has shown, however, that galvanic attack of aluminum alloys in contact with dissimilar metals is more severe at elevated temperatures than at room temperature.

Corrosion in Soils

Soils differ widely in mineral content, texture and permeability, moisture, pH and aeration, presence of organic matter and microorganisms, and electrical resistivity. Because of these varia-

Table 8 Weathering data for 1.27-mm (0.05-in.) thick aluminum alloy sheet after 7-year exposure (ASTM program started in 1958)

Average values from Kure Beach, NC, and Newark, NJ

Alloy and temper	Corrosion rate(a)		Maximum depth of attack in 7 years		Average depth of attack in 7 years		Loss in tensile strength in 7 years, %
	nm/yr	μin./yr	μm	mils	μm	mils	
Nonheat-treatable alloys							
1100-H14	345	13.6	70	2.6	29	1.1	0
1135-H14	321	12.6	83	3.3	37	1.5	0.4
1188-H14	250	9.8	121	4.8	46	1.8	0
1199-H18	205	8.1	96	3.8	57	2.2	3.9
3003-H14	295	11.6	86	3.4	52	2.0	1.1
3004-H34	414	16.3	119	4.7	44	1.7	1.1
4043-H14	335	13.2	105	4.1	34	1.3	2.8
5005-H34	373	14.7	76	3.0	27	1.1	0.9
5050-H34	349	13.7	107	4.2	58	2.3	0.5
5052-H34	362	14.3	62	2.4	43	1.7	0.8
5154-H34	326	12.8	91	3.6	65	2.6	0.9
5454-O	348	13.7	95	3.7	41	1.6	1.5
5454-H34	342	13.5	105	4.1	30	1.2	0.5
5456-O	381	15.0	104	4.1	37	1.5	0.4
5357-H34	292	11.5	138	5.4	102	4.0	0.4
5083-O	469	18.5	102	4.0	52	2.0	1.8
5083-H34	375	14.8	88	3.5	56	2.2	2.2
5086-H34	436	17.2	105	4.1	76	3.0	1.9
Heat-treatable alloys							
2014-T6	644	25.4	77	3.0	50	2.0	1.7
2024-T3	1022	40.2	76	3.0	67	2.6	2.0
2024-T81	725	28.5	97	3.8	76	3.0	6.0
2024-T86	806	31.7	77	3.0	58	2.3	6.2
6061-T4	378	14.9	57	2.2	38	1.5	0.4
6061-T6	422	16.6	98	3.9	42	1.7	0.7
7075-T6	688	27.1	119	4.7	71	2.8	1.7
7079-T6	635	25.0	65	2.6	37	1.5	0.5
Alclad alloys—heat treatable and nonheat treatable							
2014-T6	358	14.1	43	1.7	28	1.1	0
2024-T3	264	10.4	46	1.8	27	1.1	0
3003-H14	345	13.6	128	5.0	117	4.6	0
5155-H34	345	13.6	53	2.1	35	1.4	0
6061-T6	356	14.0	98	3.9	25	1.0	0.7
7075-T6	502	19.8	53	2.1	41	1.6	0.1
7079-T6	324	12.8	72	2.8	36	1.4	0

(a) Based on weight change. Source: Ref 89

tions, the corrosion performance of buried aluminum varies considerably, and a clear understanding of its behavior has depended on the accumulation of many field corrosion tests and actual case histories over an extended period of time (Ref 2, 114, 115).

Corrosion of the copper-containing 2*xxx* and 7*xxx* series alloys in moist low-resistivity soils, measured by weight loss and pitting depth, is several times greater than corrosion of the more resistant 1*xxx*, 3*xxx*, 5*xxx*, and 6*xxx* series alloys, and applications of the copper-bearing alloys for buried service is limited accordingly. Use of cathodic protection or alclad products effectively reduces corrosion or limits penetration.

Aluminum alloys 3003, 6061, and 6063 are most frequently used for surface and underground pipelines for irrigation, petroleum, and mining applications. Most early installations used uncoated pipe (Ref 116). Hundreds of miles of pipe were installed, ranging in wall thickness from 1.5 to 19 mm (0.06 to 0.75 in.). Some of these have been in service for over 40 years. When used, coatings are usually bituminous products or tape wraps. Unprotected sections exhibited corrosion attack ranging from almost none to deep pitting. Cathodically protected sections of some of the same pipes in corrosive soil showed either no attack or only mild etching. Cathodic protection of buried aluminum was standardized in 1963 (Ref 117). In addition to pipelines, extensive experience was gained with aluminum culverts in various soils (Ref 118).

Soil resistivity provides a useful guideline to soil corrosivity; corrosion problems are usually limited to soils having resistivities less than 1500 $\Omega \cdot$ cm (Ref 119). Experience has shown that soils, at least to the depth normally used to bury pipelines, are noncorrosive to aluminum over large areas of North America. However, noncorrosive soils can be rendered corrosive if they become contaminated with certain substances, such as cinders, and variability of soils along a long pipeline can lead to galvanic corrosion of portions of the line.

Techniques for installing buried aluminum pipelines have improved, including better joining methods and the ability to plow in long lengths of pipe directly from coils. A high-energy joining technique has replaced conventional field welding (Ref 119). The technique does not require filler metal and is sufficiently rapid that it does not produce heat-affected zones in the metal.

It was concluded from early field experience that buried aluminum pipelines should be coated because the risk of pitting could not be eliminated, even in high-resistivity soils. In addition, and in keeping with similar requirements for buried steel lines, buried aluminum lines should be cathodically protected. The current density requirement for protecting aluminum is roughly 10% of that required for similarly coated steel. Because of the risk of alkaline corrosion, applied cathodic voltages should not be more negative than −1.20 V versus the saturated $Cu/CuSO_4$ electrode.

Resistance of Anodized Aluminum

Anodizing is an electrolytic oxidation process that produces on an aluminum surface an integral coating of amorphous aluminum oxide that is much thicker than the natural barrier layer. The anodic coatings used for decoration and/or protection of aluminum have a thin nonporous barrier-type layer adjacent to the metal interface and a porous outer layer that can be sealed by hydrothermal treatment in water or in a metal salt solution to increase its protective value. The entire coating adheres tightly to the aluminum substrate, resists abrasion, and, when adequate in thickness, provides greatly improved protection against weathering and other corrosive conditions (Ref 120).

For outdoor applications of aluminum parts, a coating thickness of 5 to 7.6 μm (0.2 to 0.3 mil) is normally specified for bright automotive trim and 17 to 30 μm (0.7 to 1.2 mils) for architectural product finishes. Dichromate sealing affords added protection in severe saline environments. Because coatings can be attacked and stained by alkaline building materials (such as mortar, cement, and plaster), a clear, nonyellowing lacquer is often applied to anodized aluminum architectural parts to protect the finish during construction. An added advantage of lacquer coatings is that they minimize soil accumulation during service.

In general, chemical resistance of anodic coatings is greatest in approximately neutral solutions, but such coatings are usually serviceable and protective if the pH is between 4 and 8.5. More acidic and more alkaline solutions attack anodic coatings.

Under atmospheric weathering, the number of pits developed in the base metal decreases exponentially with increasing coating thickness (Fig. 21). The pits may form at minute discontinuities or voids in the coating, some of which result from large second-phase particles in the microstructure. The pit density was determined by dissolving the anodic coating in a stripping solution that does not attack the metal substrate. After the 8½-year exposure, the pits were of pin-point size and had penetrated less than 50 μm (2.0 mils). Specimens with coatings at least 22 μm (0.9 mil) thick were practically free of pitting.

Weathering of anodic coatings involves relatively uniform erosion of the coating by windborne solid particles, rainfall, and some chemical reaction with pollutants. The available information indicates that such erosion occurs at a reasonably constant rate, which averaged 0.33 μm/yr (0.013 mil/yr) for several alloys exposed to an industrial atmosphere for 18 years (Fig. 22).

A 3-year seacoast exposure of specimens of several alloys with 23-μm (0.9-mil) thick sulfuric acid coatings caused no visible pitting except in several alloys of the 7*xxx* series and in a 2*xxx* alloy (Table 17). Alloys that exhibited pitting were not protected any more effectively by 51-μm (2-mils) thick coatings. This confirms a general observation that optimum protection against atmospheric corrosion is achieved in the coating thickness range of 18 to 30 μm (0.7 to 1.2 mils)

Table 9a Loss in tensile strength for wrought aluminum alloys during various atmospheric exposures (ASTM program)

Exposed as 102- × 203-mm (4- × 8-in.) panels. Calculated from average tensile strength of several specimens (usually four)

Alloy and temper	Change in strength, %, during exposure of indicated length at — State College, PA — 6 mo	1 yr	3 yr	5 yr	10 yr	New York, NY — 6 mo	1 yr	3 yr	5 yr	10 yr	Kure Beach, NC — 6 mo	1 yr	3 yr	5 yr	10 yr
1.62-mm (0.064-in.) sheet															
2024-T3	8	1	2	0	1	2	−8	−7	−11(a)	−11(a)	6	−3	−4	−6	−4
3003-H14	6	0	2	0	1	4	−4	−5	−8	−6	5	0	−2	−4	0
3004-H34	6	−1	0	0	1	7	−2	−5	−5	−7	6	2	−2	−2	−1
5050-H34	6	0	−1	0	−1	4	−2	−1	−8	−4	5	−1	−1	−1	−2
5052-H34	9	0	−1	−1	0	...	−1	−6(a)	−5(a)	−7(a)	6	0	−2	−3(a)	−1
6061-T6	5	−2	−2	−3	0	...	−3	−7	−8	−11	4	−1	−1	−1	−4
7075-T6	5	−1	−3	0	−1	3	−1	−5	−6(a)	−8(c)	4	−2	−2	−4	−4
1.62-mm (0.064-in.) alclad sheet															
2014-T6	5	−1	−1	−2	2	4	1	−2	−4	−4	−2(a)	−1	−1	−4	−2
2024-T3	7	−1	1	1	0	8	−2	−1	−3	−3	6	1	0	0	−1
7075-T6	6	0	6	−2	−2	5	1	−2	−5	−5	6	2	2	−1	0
6.35-mm (0.25-in.) plate															
2014-T4	−3	0	0	0	0	−5	0	−2	−1	−4	−4	1	0	0	−12
2014-T6	0	−1	0	0	1	0	−2	−1	−1	−1	−2	−2	−1	−1	−1
6061-T6	−4	0	−2	−1	−5	7	−1	−2	4	3	−4	−1	0	−1	−8
6.35-mm (0.25-in.) alclad plate															
2014-T6	0	−1	0	1	−1	0	0	1	−1	−2	−1	0	0	0	0
2024-T3	0	0	0	−1	1	0	−2	−2	−2	−2	2	0	1	2	1
7075-T6	0	0	0	0	0	0	1	−1	0	1	0	1	0	0	−11(a)
6.35-mm (0.25-in.) extruded bar															
2014-T4	2	3	1	−1	−4	1	1	0	1	−2	−0	0	−1	−1	−13
2014-T6	−1	0	0	−1	0	−1	1	−2	−1	−2	−1	2	−1	−2	−1
6061-T6	0	0	0	−1	7	−2	−1	0	−3	−3	−1	−1	−2	−1	6
6063-T5	1	−1	−1	−1	1	1	−1	−2	9	11	−1	8	3	6	2
7075-T6	−1	−1	−3	−2	−3	−1	−2	−2	−1	−4	−2	−1	0	1	−2

(a) Average tensile strength values were below required minimum. Source: Ref 91

Table 9b Loss in tensile strength for wrought aluminum alloys during various atmospheric exposures (ASTM program)

Exposed as 102- × 203-mm (4- × 8-in.) panels. Calculated from average tensile strength of several specimens (usually four)

Alloy and temper	Change in strength, %, during exposure of indicated length at — Point Reyes, CA — 6 mo	1 yr	3 yr	5 yr	10 yr	Freeport, TX — 6 mo	1 yr	3 yr	5 yr	10 yr
1.62-mm (0.064-in.) sheet										
2024-T3		−13(a)	−19(a)	−19(a)	−23(a)	3	−2	−9(a)	−8	−13(a)
3003-H14		1	−3	−1	−4	3	0	−5	1	−4
3004-H34		−3	−1	−1	1	5	−1	−4	0	−2
5050-H34		2	−1	0	−2	5	0	−4	0	−3
5052-H34		−1	−2	0	−1	4	−1	−7(a)	0	−1
6061-T6		−3	−4	−5	−5	1	−3	−4	−1	−3
7075-T6		−3	−4	−4	−11(a)	1	−1	−5	−3	−8(a)
1.62-mm (0.064-in.) alclad sheet										
2014-T6		−3	−1	−4	−4	3	−1	−3	−3	−2
2024-T3		−1	−1	−1	−3	6	−1	−2	0	−3
7075-T6		3	−2	−3	−6	5	4	−1	−1	−2
6.35-mm (0.25-in.) plate										
2014-T4		−1	−3	−6	−5	1	...	−2	−1	−22(a)
2014-T6		−13(a)	−4	−8(a)	−8(a)	0	...	−2	0	−2
6061-T6		1	0	2	0	−4	0	−2	0	−2
6.35-mm (0.25-in.) alclad plate										
2014-T6		0	−1	0	−1	−1	2	−1	0	−2
2024-T3		2	0	−1	1	1	0	−1	0	0
7075-T6		1	−1	0	−1	0	2	−1	1	0
6.35-mm (0.25-in.) extruded bar										
2014-T4		3	−6	−3	−8	1	3	−2	2	−5
2014-T6		...	−4	−3	−7	1	1	−1	−2	−3
6061-T6		−1	−1	−1	...	0	0	−2	−1	−2
6063-T5		3	3	3	7	11	2	0	8	−1
7075-T6		−3	−3	−4	0	0	0	−1	−1	−4

(a) Average tensile strength values were below required minimum. Source: Ref 91

and that thicker coatings provide little additional protection.

Anodized aluminum exterior automotive parts, such as bright trim and bumpers, exhibit good resistance to deicing salts and other ingredients of road splash despite the limited thickness applied to maintain brightness and image clarity. Development of a hazy coating appearance is considered more of a problem than pitting during service in these applications. The hazy appearance results from scattering of light from a coating surface that has been microroughened as a result of inadequate sealing or use of excessively harsh alkaline cleaners.

Anodic coatings, unless used as part of a protective system that includes such other measures as shot peening or painting, are not reliable for protection against SCC of susceptible alloys. Data obtained with short-transverse direction specimens from plate of alloy 7075-T651 and other susceptible alloys show that the anodic coating may retard, have no effect, or even accelerate SCC, depending on the level of stress and, to some extent, on whether or not the stress was present before anodizing. High stresses applied after anodizing crack the coating. The effects of several applied protective measures on lifetimes of specimens in industrial and seacoast environments under relatively high elastic strain are shown in Fig. 23, in which the relatively small protective value of anodic coatings is apparent (Ref 122).

Effects of Nonmetallic Building Materials

Many nometallic building materials that contact aluminum during and after construction, either intentionally or accidentally, have been evaluated to determine their corrosive effects (Ref 123). Many of these materials that contain calcium or magnesium hydroxides are alkaline and, when wet, may cause overall surface attack of bare aluminum. This early reaction produces protective films of limited solubility that resist further corrosion. Such materials cause only superficial or mild surface attack, most of which occurs during initial stages of exposure.

Table 10a Loss in tensile strength for cast aluminum alloys during various atmospheric exposures (ASTM program)

Exposed as sepcrately cast tensile specimens. Calculated from average tensile strength of several specimens (usually six)

Alloy and temper	Change in strength, %, during exposure of indicated length at — State College, PA: 6 mo	1 yr	3 yr	5 yr	10 yr	New York, NY: 6 mo	1 yr	3 yr	5 yr	10 yr	Kure Beach, NC: 6 mo	1 yr	3 yr	5 yr	10 yr
Sand castings															
208.0-F	−1	−2	−2	−1	−2	−1	−4	−4	−3	0	−2	−5	−7	−6	−4
295.0-T6	1	−3	−2	−4	−2	−2	−6	−6	−5	−5	−7	−9	−9	−10	−9
319.0-T6	0	−1	−3	0	−3	1	−2	−6	−8	−5	−1	−5	−7	−6	−4
355.0-T6	0	1	−2	1	−3	1	0	−3	−1	−3	2	2	0	−1	−3
356.0-T6	1	−1	0	−1	−1	1	−1	−2	−2	−3	1	−1	0	−2	−2
443.0-F	3	0	−2	−2	−2	0	3	−2	−4	−3	−2	0	0	−1	−2
520.0-T4	1	−5	−4	−6	...	2	−1	−1	−2	...	−2	−2	−5	−6	...
705.0-T5	1	−2	−6	−4	−1	0	0	−4	−3	−10	1	−2	−3	−3	−4
707.0-T5	1	−2	−1	−3	0	2	1	−5	−9	−15	2	−3	−9	−13	−18
710.0-T5	2	−2	−2	−1	−5	1	−3	−3	−2	−1	2	−1	−1	−2	−1
712.0-T5	0	−8	−3	−2	−7	0	−2	−4	−5	−2	−4	−3	−8	−2	−8
713.0-T5	1	3	−2	1	−1	−3	−4	−1	−1	−5	−5	−3	−8	−1	−3
Permanent mold castings															
319.0-T61	1	−2	−1	−2	−2	1	−3	0	−4	−4	−5	−3	−4	−7	−5
355.0-T6	3	0	7	2	−4	1	−2	8	−2	−7	2	−7	5	−1	−5
443.0-F	3	0	−1	−1	−2	1	−3	−1	1	0	−1	0	−6	2	0
705.0-T5	−1	−2	−3	−5	−3	−2	0	−2	−3	−7	−3	−3	−5	−9	−5
707.0-T5	2	−2	−3	−3	−4	0	−2	−1	−4	−7	1	−2	−4	−7	−12
711.0-T5	−8	−11	−7	−6	−8	2	−4	−5	−2	−6	−2	−6	−6	−6	−11
713.0-T5	−2	−2	0	−1	−2	−1	−11	−2	−7	−2	−11	−12	−6	−4	−1

(a) Average tensile strength values were below required minimum. Source: Ref 91

Table 10b Loss in tensile strength for cast aluminum alloys during various atmospheric exposures (ASTM program)

Exposed as separately cast tensile specimens. Calculated from average tensile strength of several specimens (usually six)

Alloy and temper	Change in strength, %, during exposure of indicated length at — Point Reyes, CA: 6 mo	1 yr	3 yr	5 yr	10 yr	Freeport, TX: 6 mo	1 yr	3 yr	5 yr	10 yr
Sand castings										
208.0-F		−11	−13	−11	−10	−4	−5	−5	−9	−6
295.0-T6		−13	−15	−17	−16	−2	−9	−10	−10	−12
319.0-T6		−9	−14	−11	−10	−2	−1	−7	−5	−4
355.0-T6		−4	−8	−7	−10	1	−1	−4	−3	−7
356.0-T6		0	−1	−2	−5	2	−3	0	−3	−4
443.0-F		−7	−10	−10	−10	0	−1	−2	−4	−6
520.0-T4	1	−3	−6	−7	...	1	−4	−7	−11	...
705.0-T5		3	−8	−6	−4	6	3	−5	−4	−8
707.0-T5		−5	−8	−7	−9	−1	−5	−15	−16	−32(a)
710.0-T5		−1	−3	−4	−3	4	−1	−1	0	−2
712.0-T5		−7	−7	−8	−14	1	−7	−6	−9	−9
713.0-T5		−3	−6	0	−3	−4	−6	−7	−6	−9
Permanent mold castings										
319.0-T61		−7	−15(a)	−14(a)	−16(a)	0	−7	−4	−5	−5
355.0-T6		−6	−2	−8	−13	4	−4	5	−2	−7
443.0-F		−7	−11	−8	−10	0	−1	−3	−2	−2
705.0-T5		−5	−6	−3	−4	−3	−5	−5	−8	−14
707.0-T5		−3	−2	−2	−9	1	−3	−6	−10	−24(a)
711.0-T5		−5	−9	−6	−9	1	−4	−3	−1	−8
713.0-T5		−9	−6	−4	−9	−6	−9	−2	0	−6

(a) Average tensile strength values were below required minimum. Source: Ref 91

Drainage from freshly applied concrete, plaster, mortar, or stucco is highly alkaline and causes slight attack and discoloration. This is most likely to occur during or shortly after construction, and leaching by subsequent rains, as well as conversion to carbonates, reduces the alkalinity and further attack. Staining can be effectively prevented by organic coatings.

Some insulating materials that are porous and absorbent may cause corrosion when wet. If more cathodic metals, such as steel or copper alloys, are electrically coupled with the aluminum through these materials, galvanic attack may occur. Protective paint films on the cathodic metal, moisture barriers, or chemical inhibition are required for optimum performance under these conditions.

Concrete, plaster, mortar, and cements also cause superficial etching of aluminum, most of which occurs during the curing period. The surface attack involves dissolution of the natural oxide film and some of the metal, but a new film is formed that prevents further corrosion. Coupling with more cathodic metals has little effect on aluminum embedded in these materials except in those that contain certain curing or antifreeze additives.

When partly embedded in concrete, some metals undergo accelerated corrosion where the met-

Table 11 Atmospheric corrosion rates for aluminum and other nonferrous metals at several exposure sites

Location	Type of atmosphere	Depth of metal removed per side(a), in µm/yr, during exposure of indicated length for specimens of — Aluminum(b): 10 yr	20 yr	Copper(c): 10 yr	20 yr	Lead(d): 10 yr	20 yr	Zinc(e): 10 yr	20 yr
Phoenix, AZ	Desert	0.000	0.076	0.13	0.13	0.23	0.10	0.25	0.18
State College, PA	Rural	0.025	0.076	0.58	0.43	0.48	0.30	1.07	1.09
Key West, FL	Seacoast	0.10	...	0.51	0.56	0.56	...	0.53	0.66
Sandy Hook, NJ	Seacoast	0.20	0.28	0.66	...	...	...	1.40	...
La Jolla, CA	Seacoast	0.71	0.63	1.32	1.27	0.41	0.53	1.73	1.73
New York, NY	Industrial	0.78	0.74	1.19	1.37	0.43	0.38	4.8	5.6
Altoona, PA	Industrial	0.63	...	1.17	1.40	0.69	...	4.8	6.9

(a) Calculated from weight loss, assuming uniform attack, for 0.89-mm (0.035-in.) thick panels. (b) Aluminum 1100-H14. (c) Tough pitch copper (99.9% Cu). (d) Commercial lead (99.92% Pb). (e) Prime western zinc (98.9% Zn). Source: Ref 94

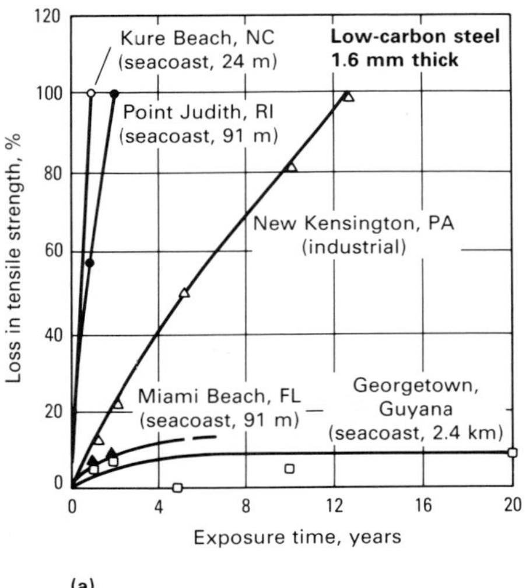

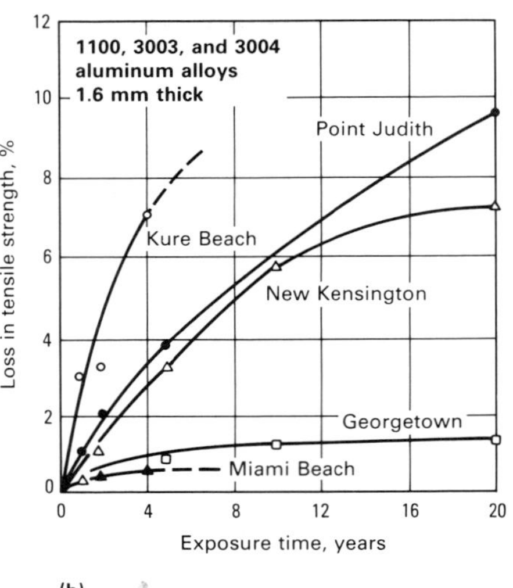

Fig. 19 Tensile-strength losses for low-carbon steel (a) and representative nonheat-treatable aluminum alloys (b) at several atmospheric exposure sites. Strength losses of the aluminum alloys are less than one-tenth that of the low-carbon steel.

al intersects the exposed surface of the concrete. This effect is usually not important for aluminum, but special consideration must be given to protection of faying surfaces or crevices between the aluminum and the concrete, which may entrap environmental contaminants. For example, highway railings and streetlight standards and stanchions are usually coated with a sealing compound where they are fastened to concrete in order to prevent entry of salt-laden road splash into crevices.

Contact with Foods, Pharmaceuticals, and Chemicals

The widespread use of aluminum in processing, handling, and packaging of foods, beverages, and pharmaceutical and chemical products is based on economic factors and the excellent compatibility of aluminum with many of these products (Ref 124). In addition to high corrosion resistance in contact with such products, many of these applications depend on the nontoxicity of aluminum and its salts, as well as its freedom from catalytic effects that cause product discoloration. Application of aluminum for packaging foods and pharmaceutical products has grown sharply since 1970; this application now accounts for about 26% of the aluminum marketed in the United States (Ref 125). The largest amount is used in beverage cans, and a smaller amount for foods. These cans generally have both internal and external organic coatings, primarily for decoration and for protection of product taste.

Large quantities of aluminum foil, either uncoated or with plastic coatings, are used in flexible packages. Coated foil is also used with fiber board in construction of rigid containers. The foil in such rigid containers, because of its extreme thinness, must be coated; only the slightest corrosion can be tolerated, and perforation must not occur even during long periods of storage. Packaging foils are produced from unalloyed aluminum corresponding to composition limits for aluminum 1230. Sheet for beverage can bodies is generally alloy 3004, food can bodies alloys 5352 or 5050, and can ends alloy 5182. These alloys have high corrosion resistance and are not normally subject to corrosion problems in such applications.

Aluminum alloy household cooking utensils, usually made of alloy 3003, have been used for many years. These utensils, as well as commercial food-processing equipment, do not require protective coatings; however, ceramic coatings are often applied to the exteriors of cooking utensils for aesthetic reasons, and polymeric coatings to the food-contacting surfaces for nonstick-ing characteristics. Alloys used in commercial food processing include alloy 3003, 5*xxx* alloys, and casting alloy 514.0. Unsatisfactory performance is sometimes caused by use of improper cleaners. Some alkaline cleaners cause excessive corrosion and should not be used unless they are inhibited effectively.

Aluminum alloys are used in processing, handling, and packaging a wide variety of chemical products (Ref 126, 127). Aluminum alloys are compatible with dry salts of most inorganic chemicals. Factors controlling compatibility of aluminum alloys with aqueous solutions have been discussed in earlier sections in this article. Within their passive pH range (about 4 to 9), aluminum alloys resist corrosion by solutions of most inorganic chemicals, but they are subject to pitting in aerated solutions, particularly halide solutions, in which they are polarized to their pitting potentials.

Figure 24 illustrates the corrosion behavior of aluminum in several acids and bases. Aluminum alloys are not suitable for handling mineral acids, with the exception of HNO_3 in concentrations above 82 wt% and H_2SO_4 from 98 to 100 wt%. Aluminum alloys resist most alcohols; however, some alcohols may cause corrosion when extremely dry and at elevated temperatures (Fig. 24d). The same characteristics are associated with phenol (Fig. 24f). Aldehydes have little or no action on aluminum (Fig. 24e). Under most conditions, particularly at room temperature, aluminum alloys resist halogenated organic compounds, but under some conditions, they may react rapidly or violently with some of these chemicals. If water is present, these chemicals may hydrolyze to yield mineral acids that destroy the protective oxide film of aluminum. Such corrosion by mineral acids may in turn promote further reaction with the chemicals themselves, because the aluminum halides formed by this corrosion are catalysts for some such reactions. To ensure safety, service conditions should be ascertained before aluminum alloys are used with these chemicals, and the most stringent precautions should be exercised before they are used in finely divided form.

Reactivity of aluminum alloys with halogenated organic chemicals is inversely related to the chemical stability of these reagents. Thus, they are most resistant to chemicals containing fluorine and are decreasingly resistant to those containing chlorine, bromine, and iodine. Aluminum alloys resist highly polymerized halogenated chemicals, reflecting the high degree of stability of these chemicals.

Resistance of aluminum and its alloys to many foods and chemicals, representing practically all classifications, has been established in laboratory tests and, in many cases, by service experience. Data are readily available from handbooks, proprietary literature, and trade association publications. Reference 128 is especially useful.

Much of the data from laboratory tests are for chemicals of high purity. Caution should be exercised in using these data to predict performance of aluminum alloys with commercial grades of chemicals. Corrosion of aluminum alloys by inorganic chemicals is frequently caused by such impurities as copper, lead, mercury, and nickel, and corrosion by organic chemicals often results from the presence of other organic chemicals. The combined effect of impurities may exceed the sum of their individual effects.

Care of Aluminum

Handling and Storage. Because of the excellent corrosion resistance of the 1*xxx*, 3*xxx*, 4*xxx*, 5*xxx*, and 6*xxx* series alloys, users occasionally have not employed good practice in the handling and storage of these alloys. This can result in water stains or in pitting. Methods to avoid these unsightly surface effects are described in the article "Cleaning and Finishing of Aluminum and Aluminum Alloys" in Volume 5 of the 9th Edition of *Metals Handbook*.

Water stain is superficial corrosion that occurs when sheets of bare metal are stacked or nested in the presence of moisture. The source of moisture may be condensation from the atmosphere that forms on the edges of the stack and is drawn between the sheets by capillary action. Aluminum should not be stored at temperatures or under atmospheric conditions conducive to condensation. When such conditions cannot be avoided, the metal sheets or parts should be separated and coated with oil or a suitable corrosion inhibitor. Once formed, water stain can be removed by either mechanical or chemical means, but the original surface brightness may be altered.

Outdoor storage of aluminum, even under a tarpaulin, is generally not desirable for long periods of time; this varies with the alloy, the end product, and the local environment. Moisture can collect on the surface, sometimes at relative humidities below the dew point, because of the hygroscopic nature of the dust or particles

Table 12 Average weight loss and maximum depth of pitting for aluminum alloy plate specimens after immersion in seawater

Specimens were 6.35 × 305 × 305 mm (0.250 × 12 × 12 in.) and weighed approximately 1.6 kg (3.5 lb).

Test series	Alloy and temper	Harbor Island, NC(a)				Halifax, NS				Esquimalt, BC(a)			
		1 yr	2 yr	5 yr	10 yr	1 yr	2 yr	5 yr	10 yr	1 yr	2 yr	5 yr	10 yr
Weight loss, g													
1	1100-H14	4.4	5.4	10.3	11.1	1.9	3.5	5.3	12.7	0.0	2.4	1.3	2.3
	3003-H14	4.1	6.4	9.3	11.2	0.0	3.3	4.6	7.5	0.0	0.0	3.0	2.2
	5052-H34	4.5	6.5	9.0	14.9	2.8	3.3	...	14.2	1.7	0.0	0.0	0.6
	6051-T4	3.7	4.9	9.9	12.3	0.0	0.7	3.5	8.0	1.9	7.8	19.0	14.6
	6051-T6	4.4	5.7	10.3	13.1	2.1	5.5	6.1	19.5	22.5	13.8	19.9	27.3
	6061-T4	4.8	6.6	12.4	18.6	4.4	6.0	8.0	15.6	0.9	2.3	28.2	62.0
	6061-T6	5.5	7.7	14.0	21.5	4.3	7.3	12.7	22.8	6.7	7.1	11.1	44.3
	7072	...	...	...	10.2	...	...	...	15.9	...	...	...	3.1
	7075-T6	...	...	...	149.0	...	...	...	242.6	...	...	...	246.5
2	5083	2.5	3.7	...	7.3	2.8	0.0	6.1	8.5	1.3	1.9	2.7	3.3
	5083	4.7	3.4	5.7	8.1	2.6	3.2	5.2	7.5	15.3	16.3	36.3	31.1
	5056	3.7	4.7	6.0	9.2	2.5	3.3	5.7	10.4	10.7	16.5	19.5	28.9
	5056	4.5	5.2	...	16.7	4.0	4.1	5.5	11.1	7.0	6.0	11.0	11.4
	6051-T4	3.9	4.2	12.1	9.1	3.6	3.1	5.5	9.2	9.1	18.8	15.3	51.0
	6051-T6	4.1	4.5	7.7	10.6	5.3	4.1	8.4	18.6	17.2	23.3	30.6	33.5
	6053-T6	4.1	4.5	6.6	9.7	3.0	3.3	5.6	14.8	15.7	25.1	19.3	25.8
	6061-T6	7.6	13.4	29.4	51.6	9.8	11.2	33.2	48.5	12.3	26.8	48.7	48.0
	6061-T6	5.5	6.5	15.4	34.2	10.0	9.4	19.1	54.1	7.3	7.0	21.3	18.6
	Al-7 Mg	4.1	4.1	6.5	9.4	2.4	2.4	4.6	8.0	1.6	2.9	2.1	3.3
	Low-carbon steel(b)	219.0	294.0	471.3	979.8	208.0	292.6	761.1	1450.0	277.0	455.4	1012.4	2240.8
3	5154	2.8	5.2	6.0	...	2.4	2.6	3.8	...	1.4	2.1	2.6	...
	5083	3.5	4.6	6.0	...	2.0	2.8	3.6	...	0.2	2.2	2.8	...
	6053-T6	3.8	6.6	25.9	...	19.3	29.2	4.7	...	45.6	80.4	86.0	...
	7075-T6	60.4	49.3	74.8	...	44.8	66.1	116.0	...	50.9	71.3	153.5	...
	3003, alclad	4.3	12.0	...	...	1.6	2.3	...	...	...	1.9	...	...
	6061, alclad	4.3	3.9	5.7	...	8.4	3.3	6.5	...	20.8	15.8	34.3	...
	7075, alclad	4.4	5.2	6.1	...	2.8	3.6	6.8	...	8.5	14.5	16.6	...
Maximum depth of pitting, mils													
1	1100-H14	0	0	40	0	17	32	0	29	30	26	15	0
	3003-H14	0	0	13	21	13	15	21	22	5	20	0	10
	5052-H34	0	0	0	0	5	20	6	12	16	6	0	5
	6051-T4	0	0	5	0	0	10	0	62	10	65	51	37
	6051-T6	2	0	5	0	19	56	15	64	70	60	181	238
	6061-T4	0	13	2	14	12	18	21	33	15	50	20	28
	6061-T6	36	24	60	95	36	43	43	54	30	25	80	116
	7072	...	...	...	56	...	...	...	150	...	...	...	26
	7075-T6	...	...	...	66	...	...	...	(c)	...	...	...	(c)
2	5083	12	9	6	0	3	0	12	7	13	5	0	6
	5083	16	13	6	10	4	23	16	22	29	38	47	55
	5056	7	10	7	5	3	0	12	15	20	39	34	35
	5056	10	10	5	28	0	0	10	24	20	1	0	11
	6051-T4	16	4	7	15	3	0	9	35	25	47	109	170
	6051-T6	11	17	9	15	5	8	25	60	55	34	184	200
	6053-T6	30	15	14	58	28	34	66	95	93	126	165	105
	6061-T4	67	100	144	130	50	67	90	122	60	100	125	125
	6061-T6	15	27	36	40	38	47	58	67	35	48	60	55
	Al-7 Mg	12	7	8	8	3	0	12	14	8	12	0	7
3	5154	12	9	5	...	0	12	15	...	0	0	3	...
	5083	22	1	7	...	0	11	7	...	0	0	5	...
	6053-T6	28	150	186	...	93	91	34	...	81	118	118	...
	7075-T6	25	25	25	...	18	17	(d)	...	15	11	(d)	...
	3003, alclad	0	12	...	...	0	13	...	...	...	13	...	...
	6061, alclad	10	10	9	...	9	11	9	...	11	13	9	...
	7075, alclad	10	15	13	...	13	14	12	...	12	14	15	...

(a) Harbor Island is near Wilmington, NC; Esquimalt is near Victoria, BC. (b) Original weight about 4.8 kg (10.6 lb). (c) Plate was perforated. (d) Could not determine because no original surface left. Source: Ref 2

that deposit on the metal from the atmosphere. The resulting staining or localized pitting, although of little structural consequence in the 1*xxx*, 3*xxx*, 4*xxx*, 5*xxx*, and 6*xxx* alloys, is undesirable if the aluminum will be used for an end product for which surface finish is critical. The 2*xxx* and 7*xxx* bare alloys are susceptible to intergranular attack under these conditions, and for these alloys, use of strippable coatings, protective wrappers, papers, or inhibited organic films is advisable when adverse conditions cannot be avoided.

Mechanical damage can be easily avoided by good housekeeping practices, proper equipment, and proper protection during transportation. When transporting flat sheets or plates, the aluminum should be oiled or interleaved with approved paper to prevent traffic marks, where fretting action at points of contact causes surface abrasion. Practices to avoid these defects are described in Ref 129.

Cleaning and Deactivation of Corrosion. For many applications, minor surface corrosion is of little consequence, and no cleaning is necessary. Where corrosion occurs that is detrimental to strength or appearance if allowed to continue, aluminum can be cleaned by a number of methods (Ref 129). Removal of corrosion products can be followed by deoxidizing or brightening cleaners, if desired. Specifications for all cleaners should state that they are suitable for aluminum. For architectural aluminum products, aggressive or heavy-duty cleaners should be avoided in favor of more frequent use of mild cleaners. Other preventive and maintenance procedures are discussed in Ref 130.

Table 13 Average weight loss and maximum depth of pitting for aluminum alloy extruded specimens after immersion in seawater

Specimens were 6.35 mm (0.250 in.) thick, 0.170 m^2 (1.83 ft^2) in area, and weighed approximately 1.2 kg (2.6 lb).

Test series	Alloy and temper	Harbor Island, NC(a)				Halifax, NS				Esquimalt, BC(a)			
		1 yr	2 yr	5 yr	10 yr	1 yr	2 yr	5 yr	10 yr	1 yr	2 yr	5 yr	10 yr
Weight loss, g													
1	6051-T4	2.9	0.0	8.0	8.2	7.8	1.5	4.5	6.3	2.8	0.0	...	...
	6051-T6	6.2	0.0	...	14.6	9.0	10.5	12.8	23.4	15.4	40.7	29.7	83.0
	6061-T4	4.7	10.9	...	7.4	0.0	0.4	4.3	7.4	0.2	10.0	29.8	38.3
	6061-T6	3.0	6.9	8.1	16.4	3.8	5.0	7.1	15.5	15.5	14.1	25.2	59.2
2	5056	3.2	2.7	6.3	9.9	6.3	2.8	4.2	6.2	5.0	1.9	3.6	4.8
	6051-T4	3.0	3.4	6.5	8.0	13.1	9.2	7.8	12.0	16.0	16.7	18.0	35.4
	6051-T6	4.9	11.1	5.7	9.4	19.9	...	23.0	78.9	23.0	30.2	41.3	122.8
	6053-T6	3.3	4.9	6.9	10.6	2.6	3.0	4.5	8.9	4.5	43.5	35.3	99.9
3	5056	2.0	5.2	5.1	...	1.3	1.3	2.1	...	3.5	9.4	2.4	...
	6063-T5	2.6	3.3	6.5	...	2.4	3.5	4.9	...	6.6	13.4	13.1	...
	6053-T6	2.8	3.0	5.3	...	25.7	12.0	30.6	...	43.5	29.9	77.1	...
Maximum depth of pitting, mils													
1	6051-T4	0	0	27	20	27	27	14	32	35	23	65	72
	6051-T6	70	40	46	67	52	68	125	(b)	70	(b)	160	(b)
	6061-T4	23	23	27	12	25	20	12	32	33	45	56	70
	6061-T6	13	13	10	15	15	9	14	27	20	30	46	45
2	5056	13	7	35	32	60	0	16	41	30	17	99	50
	6051-T4	57	5	20	15	34	65	30	74	66	65	90	115
	6051-T6	58	>100	34	45	100	...	84	(b)	64	85	107	>200(c)
	6053-T6	13	25	93	46	0	0	7	34	80	110	175	210
3	5056	28	68	17	...	0	3	15	...	37	72	63	...
	6063-T5	42	35	45	...	27	25	30	...	70	66	136	...
	6053-T6	28	1	20	...	185	90	(b)	...	178	(b)	(b)	...

(a) Harbor Island is near Wilmington, NC; Esquimalt is near Victoria, BC. (b) Plate was perforated. (c) In thick web of angle. Source: Ref 2

Table 14 Average weight loss (mg/m^2) for aluminum alloys in seawater (from Tables 12 and 13)

Test series	Harbor Island, NC				Halifax, NS				Esquimalt, BC			
	1 yr	2 yr	5 yr	10 yr	1 yr	2 yr	5 yr	10 yr	1 yr	2 yr	5 yr	10 yr
Series 1: plate(a)												
1100, 3003, 5052	22	32	49	64	9	18	26	60	3	4	7	9
6051, 6061	24	32	60	85	12	25	39	85	41	40	101	191
Series 1: extrusions(b)												
6051, 6061	25	26	47	68	30	26	42	78	50	95	166	354
Series 2: plate(a)												
Al-Mg	20	22	32	52	15	13	28	47	37	39	74	81
Al-Mg-Si	26	34	75	119	55	54	75	149	64	111	140	183
Series 2: extrusions(b)												
Al-Mg	19	19	38	62	25	16	21	37	18	11	15	26
Al-Mg-Si	22	38	38	41	70	24	70	196	85	177	185	506

(a) Plate surface area, 0.193 m^2 (2.08 ft^2). (b) Extrusion surface area, 0.170 m^2 (1.83 ft^2). Source: Ref 2

REFERENCES

1. M.S. Hunter and P. Fowle, Naturally and Thermally Formed Oxide Films on Aluminum, *J. Electrochem. Soc.*, Vol 103, 1956, p 482
2. H.P. Godard, W.B. Jepson, M.R. Bothwell, and R.L. Kane, *The Corrosion of Light Metals*, John Wiley & Sons, 1967
3. M. Pourbaix, *Atlas of Electrochemical Equilibria in Aqueous Solutions*, Pergamon Press, 1966, p 171
4. J.E. Hatch, Ed., Chapter 7, in *Aluminum: Properties and Physical Metallurgy*, American Society for Metals, 1984
5. H. Kaesche, Investigation of Uniform Dissolution and Pitting of Aluminum Electrodes, *Werkst. Korros.*, Vol 14, 1963, p 557
6. H. Bohni and H.H. Uhlig, Environmental Factors Affecting the Critical Pitting Potential of Aluminum, *J. Electrochem. Soc.*, Vol 116, 1969, p 906
7. J.R. Galvele, S.M. de Micheli, I.L. Muller, S.B. DeWexler, and I.L. Alanis, Critical Potentials for Localized Corrosion of Aluminum Alloys, in *Localized Corrosion*, B.F. Brown, J. Kruger, and R.W. Staehle, Ed., National Association of Corrosion Engineers, 1974, p 580
8. I.L. Muller and J.R. Galvele, Pitting Potentials of High Purity Binary Aluminum Alloys—Part I. Al-Cu Alloys, *Corros. Sci.*, Vol 17, 1977, p 179; Part II. Al-Mg and Al-Zn Alloys, *Corros. Sci.*, Vol 17, 1977, p 995
9. R.L. Horst and G.C. English, Corrosion Evaluation of Aluminum Easy-Open Ends on Tinplate Cans, *Mater. Perform.*, Vol 16 (No. 3), 1977, p 23
10. R.A. Bonewitz and E.D. Verink, Jr., Correlation Between Long Term Testing of Aluminum Alloys for Desalination and Electrochemical Methods of Evaluation, *Mater. Perform.*, Vol 14, 1975, p 16
11. R.H. Brown, W.L. Fink, and M.S. Hunter, Measurement of Irreversible Potentials as a Metallurgical Research Tool, *Trans. AIME*, Vol 143, 1941, p 115
12. "Standard Practice for Measurement of Corrosion Potentials of Aluminum Alloys," G 69, *Annual Book of ASTM Standards*, American Society for Testing and Materials
13. M.S. Hunter, A.M. Montgomery, and G.W. Wilcox, Microstructure of Alloys and Products, in *Aluminum*, Vol 1, K.R. Van Horn, Ed., American Society for Metals, 1967, p 77
14. J.E. Hatch, Ed., *Aluminum: Properties and Physical Metallurgy*, American Society for Metals, 1984, p 60
15. R.B. Mears and R.H. Brown, Causes of Corrosion Currents, *Ind. Eng. Chem.*, Vol 33, 1941, p 1001
16. J.D. Edwards *et al.*, *The Aluminum Industry: Aluminum Products and Their Fabrication*, McGraw-Hill, 1930, p 13
17. C.L. Burton, L.W. Mayer, and E.H. Spuhler, Aircraft and Aerospace Applications, in *Aluminum*, Vol II, K.R. Van Horn, Ed., American Society for Metals, 1967
18. P.E. Bretz and R.R. Sawtell, 'Alithilite' Alloys: Progress, Products and Properties, in *Proceedings of the Third Aluminum-Lithium Conference*, The Institute of Metals, 1986, p 47
19. C.J. Peel, B. Evans, and D. McDarmaid, Current Status of UK Lightweight Lithium-Containing Aluminum Alloys, in *Proceedings of the Third Aluminum-Lithium Conference*, 1986, p 26
20. H.F. DeJong and J.H.M. Martens, Investigation of the Pitting Potential of Rapidly Solidified Aluminum-Lithium Alloys, *Aluminium*, Vol 61 (No. 6), 1985, p 416
21. P. Niskanen, T.H. Sanders, Jr., J.G. Rinker, and M. Marek, Corrosion of Aluminum

Table 15 Summary of data from 10-year seawater exposures at Wrightsville Beach, NC

Alloy and temper	Mg, %	Thickness mm	Thickness in.	Corrosion rate based on weight change µm/yr	Corrosion rate based on weight change mil/yr	Maximum depth of attack in 10 years mm	Maximum depth of attack in 10 years mil	Average depth of attack in 10 years mm	Average depth of attack in 10 years mil	Change in tensile strength in 10 years, %
Half-tide exposure										
1199	...	1.27	0.050	0.91	0.036	0.99	0.039	0.07	0.003	0
5154-H38	3.5	1.27	0.050	0.94	0.037	0.50	0.020	0.13	0.005	−2.1
5454-H34	2.7	6.35	0.250	1.04	0.041	0.39	0.015	0.07	0.003	−0.7
5457-H34	1.0	1.02	0.040	0.91	0.036	0.56	0.022	0.03	0.001	−4.2
5456-O	5.1	6.17	0.243	0.36	0.014	1.74	0.069	0.32	0.013	−0.4
5456-H321	5.1	6.17	0.243	1.29	0.051	1.83	0.072	0.34	0.013	−4.5
5083-O	4.5	6.35	0.250	0.91	0.036	0.97	0.038	0.31	0.012	0
5086-O	4.0	2.03	0.080	0.89	0.035	0.69	0.027	0.06	0.002	−2.7
Full-immersion exposure										
1199	...	1.27	0.050	1.55	0.061	...	...	...	...	0
5154-H38	3.5	1.27	0.050	1.40	0.055	...	...	...	...	−5.1
5454-H34	2.7	6.35	0.250	1.50	0.059	0.51	0.020	0.10	0.004	−0.5
5457-H34	1.0	1.02	0.040	1.42	0.056	...	...	...	...	−5.2
5456-O	5.1	6.17	0.243	2.95	0.116	3.33	0.131	1.01	0.040	−3.0
5456-H321	5.1	6.17	0.243	1.62	0.064	1.12	0.044	0.31	0.012	−1.1
5083-O	4.5	6.35	0.250	1.50	0.059	0.61	0.024	0.03	0.001	0
5086-O	4.0	2.03	0.080	1.45	0.057	...	...	...	...	−3.7

Source: Ref 106

Alloys Containing Lithium, *Corros. Sci.*, Vol 22 (No. 4), 1982, p 283

22. M. Zamin, The Role of Mn in the Corrosion Behavior of Al-Mn Alloys, *Corrosion*, Vol 37, 1981, p 627
23. L.F. Mondolfo, *Aluminum Alloys: Structure and Properties*, Butterworths, 1976, p 812
24. E.H. Dix, W.A. Anderson, and M.B. Shumaker, "Development of Wrought Aluminum-Magnesium Alloys," Technical Paper 14, Alcoa Research Laboratories, 1958
25. H.P. Godard, W.B. Jepson, M.R. Bothwell, and R.L. Kane, *The Corrosion of Light Metals*, John Wiley & Sons, 1967, p 72
26. J. Zahavi and J. Yahalom, Exfoliation Corrosion of AlMgSi Alloys in Water, *J. Electrochem. Soc.*, Vol 129 (No. 6), 1982, p 1181
27. D.O. Sprowls and E.H. Spuhler, "Avoiding Stress-Corrosion Cracking in High Strength Aluminum Alloy Structures," Green Letter, Alcoa, Jan 1982
28. L.F. Mondolfo, *Aluminum Alloys: Structure and Properties*, Butterworths, 1976, p 851
29. P.L. Mehr, E.H. Spuhler, and L.W. Mayer, "Alcoa Alloy 7075-T73," Green Letter, Revision 1, Alcoa, Sept 1971
30. S.V. Nair, J.K. Tien, and R.C. Bates, SiC-Reinforced Aluminum Metal Matrix Composites, *Int. Met. Rev.*, Vol 30 (No. 6), 1985, p 275
31. D.M. Aylor, R.J. Ferrara, and R.M. Kain, Marine Corrosion and Protection for Graphite/Aluminum Metal Matrix Composites, *Mater. Perform.*, Vol 23 (No. 7), 1984, p 32
32. S.L. Fohlman, Corrosion and Electrochemical Behavior of Boron/Aluminum Composites, *Corrosion*, Vol 34 (No. 5), 1978, p 156
33. D.M. Aylor and P.J. Moran, "An Investigation of Corrosion Properties and Protection for Graphite/Aluminum and Silicon Carbide/Aluminum Metal Matrix Composites," Paper 202, presented at Corrosion/86, National Association of Corrosion Engineers, 1986
34. "Classification of Resistance to Stress-Corrosion Cracking of High-Strength Aluminum Alloys," G 64, *Annual Book of ASTM Standards*, American Society for Testing and Materials
35. R.H. Brown, Aluminum Alloy Laminates: Alclad and Clad Aluminum Alloy Products, in *Composite Engineering Laminates*, A.G.H. Dietz, Ed., M.I.T. Press, 1969
36. M.R. Bothwell, New Technique Enhances Corrosion Resistance of Aluminum, *Met. Prog.*, Vol 87, March 1985, p 81
37. H. Ikeda, Protection Against Pitting Corrosion of 3003 Aluminum Alloy by Zinc Diffusion Treatment, *Aluminium*, Vol 58 (No. 8), 1982, p 467
38. D.J. Scott, Aluminum Sprayed Coatings—Their Use for the Protection of Al Alloys and Steel, *Trans. IMF*, Vol 49, 1971, p 111
39. V.E. Carter and H.S. Campbell, Protecting Strong Aluminum Alloys Against Stress-Corrosion With Sprayed Metal Coatings, *Br. Corros. J.*, Vol 4, 1969, p 15
40. W.J. Schwerdtfeger, Effects of Cathodic Protection on the Corrosion of an Aluminum Alloy, *J. Res. Natl. Bur. Stand.*, Vol 68C (No. 4), 1964, p 283
41. Recommended Practice for Cathodic Protection of Aluminum Pipe Buried in Soil or Immersed in Water, *Mater. Protec.*, Vol 2 (No. 10), 1963, p 106
42. F.W. Hewes, Investigation of Maximum and Minimum Criteria for the Cathodic Protection of Aluminum in Soil, *Oil Week*, Vol 16 (No. 24-28), Aug-Sept 1965
43. M. Cerny, Present State of Knowledge About Cathodic Protection of Aluminum, *Prot. Met.*, Vol 11 (No. 6), 1975, p 645
44. R.B. Mears and H.J. Fahrney, Cathodic Protection of Aluminum Equipment, *Trans. AIChE*, Vol 37 (No. 6), 1941, p 911
45. B. Sandberg and A. Bairamov, "Cathodic Protection of Aluminum Structures," Report 1985:2, Swedish Corrosion Institute, 1985
46. T.J. Lennox, M.H. Peterson, and R.E. Groover, "Corrosion of Aluminum Alloys by Antifouling Paint Toxicants and Effects of Cathodic Protection," Paper 16, presented at NACE Conference, Cleveland, OH, National Association of Corrosion Engineers, 1968
47. R.E. Groover, T.J. Lennox, and M.H. Peterson, Cathodic Protection of 19 Aluminum Alloys Exposed to Seawater—Corrosion Behavior, *Mater. Protec.*, Vol 8 (No. 11), 1969, p 25
48. S.C. Dexter, Localized Corrosion of Aluminum Alloys for OTEC Heat Exchangers, *J. Ocean Sci. Eng.*, Vol 8 (No. 1), 1981, p 109
49. E.H. Cook and F.L. McGeary, Electrodeposition of Iron From Aqueous Solutions Onto an Aluminum Alloy, *Corrosion*, Vol 20 (No. 4), 1964, p 111t
50. J.D. Edwards, F.C. Frary, and Z. Jeffries, *The Aluminum Industry: Aluminum Products and Their Fabrication*, McGraw-Hill, 1930
51. M.H. Brown, W.W. Binger, and R.H. Brown, Mercury and its Compounds: A Corrosion Hazard, *Corrosion*, Vol 8 (No. 5), 1952, p 155
52. R.C. Plumb, M.H. Brown, and J.E. Lewis, A Radiochemical Tracer Investigation of the Role of Mercury in the Corrosion of Aluminum, *Corrosion*, Vol 11 (No. 6), 1956, p 277t
53. E.H. Dix, Acceleration of the Rate of Corrosion by High Constant Stresses, *Trans. AIME*, Vol 137, 1940, p 11
54. W.L. Fink and L.A. Willey, Quenching of 75S Aluminum Alloy, *Met. Technol.*, Vol 14 (No. 8), 1947, p 5
55. M.S. Hunter, G.R. Frank, and D.L. Robinson, in *Proceedings of Conference: Fundamental Aspects of Stress-Corrosion Cracking*, R.W. Staehle, Ed., National Association of Corrosion Engineers, 1969, p 497
56. H. Kaesche, Pitting Corrosion of Aluminum and Intergranular Corrosion of Aluminum Alloys, in *Localized Corrosion*, B.F. Brown, J. Kruger, and R.W. Staehle, Ed., National Association of Corrosion Engineers, 1974, p 516
57. J.R. Galvele and S.M. de Micheli, Mechanism of Intergranular Corrosion of Al-Cu Alloys, *Corros. Sci.*, Vol 10, 1970, p 795
58. "Standard Practice for Determining the Susceptibility to Intergranular Corrosion of

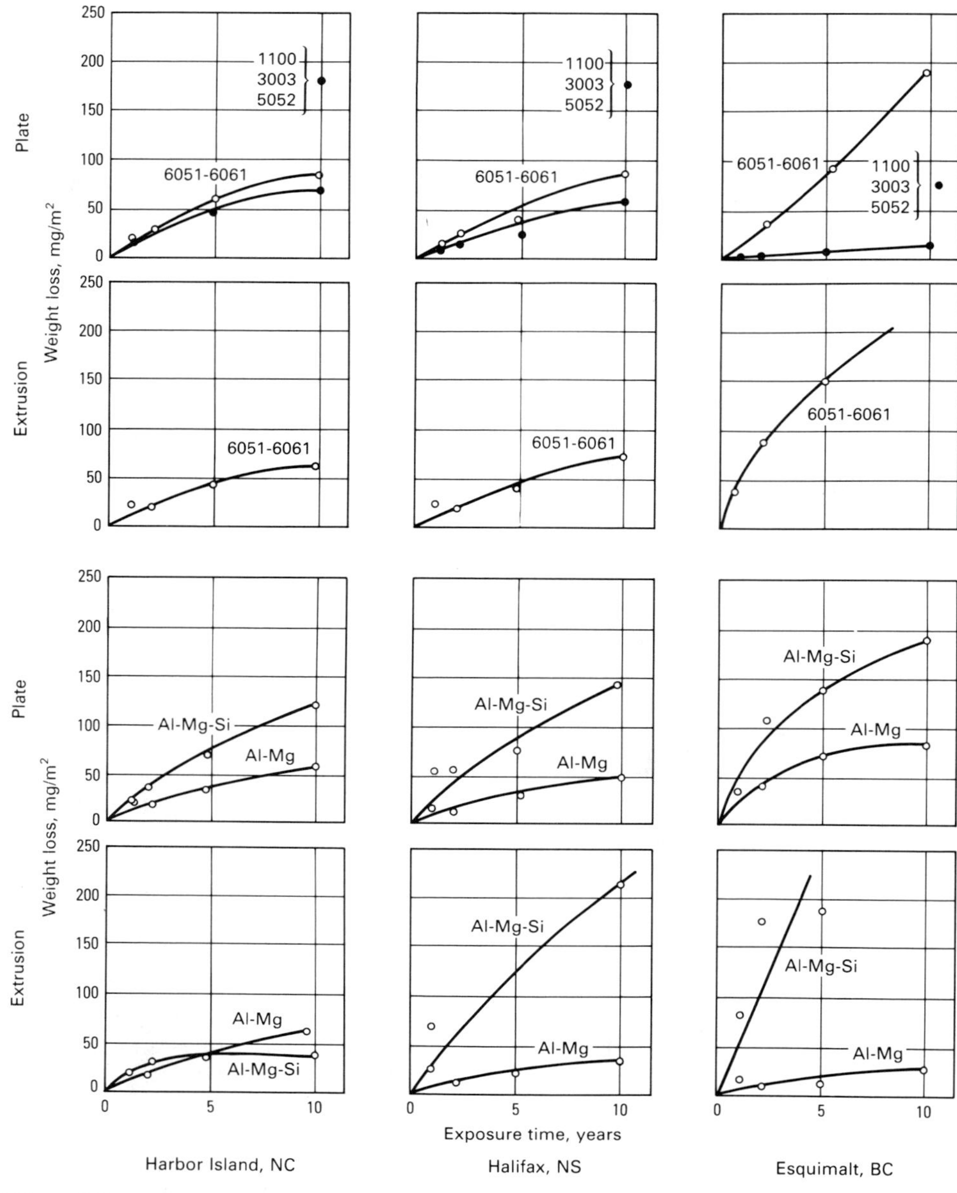

Fig. 20 Weight loss as a function of exposure time for three aluminum alloys in seawater

5*xxx* Series Aluminum Alloys by Weight Loss After Exposure to Nitric Acid (NAWLT Test)," G 67, *Annual Book of ASTM Standards*, American Society for Testing and Materials

59. R.B. Mears, R.H. Brown, and E.H. Dix, Jr., A Generalized Theory of the Stress-Corrosion Cracking of Alloys, in *Symposium on Stress-Corrosion Cracking of Metals*, American Society for Testing and Materials and American Institute of Mining and Metallurgical Engineers, 1945, p 323
60. D.O. Sprowls and R.H. Brown, Stress-Corrosion Mechanisms for Aluminum Alloys, in *Fundamental Aspects of Stress-Corrosion Cracking*, R.W. Staehle, A.J. Forty, and D. VanRooyen, Ed., National Association of Corrosion Engineers, 1969, p 466
61. M.O. Speidel, Hydrogen Embrittlement of Aluminum Alloys, in *Hydrogen in Metals*, L.M. Bernstein and A.W. Thompson, Ed., American Society for Metals, 1974, p 249
62. V.A. Marichev, The Mechanism of Crack Growth in Stress Corrosion Cracking of Aluminum Alloys, *Werkst. Korros.*, Vol 34, 1983, p 300
63. E.H. Spuhler and C.L. Burton, "Avoiding Stress-Corrosion Cracking in High Strength Aluminum Alloy Structures," Green Letter, Alcoa, 1970
64. *Aluminum Standards and Data*, The Aluminum Association, 1984, p 12
65. J.G. Rinker, M. Marek, and T.H. Sanders, Jr., Microstructure, Toughness and SCC Behavior of 2020, in *Aluminum-Lithium Alloys*, T.H. Sanders, Jr. and E.A. Starke, Jr., Ed, American Institute of Mining, Metallurgical, and Petroleum Engineers, 1983, p 597
66. A.K. Vasudevan, P.R. Ziman, S.C. Jha, and T.H. Sanders, Jr., Stress-Corrosion Resistance of Al-Cu-Li-Zr Alloys, in *Aluminum-Lithium Alloys*, Vol III, C. Baker, P.J. Gregson, S.J. Harris, and C.J. Peel, Ed., The Institute of Metals, 1986, p 303
67. E.L. Colvin, S.J. Murtha, and R.K. Wyss, The Effect of Aging Time on the Stress-Corrosion Cracking Resistance of 2090-T8E41, to be published in the proceedings of the International Conference on Aluminum Alloys, Charlottesville, VA, 15-20 June 1986
68. N.J.H. Holroyd, A. Gray, G.M. Scamans, and R. Herman, Environment-Sensitive Fracture of Al-Li-Cu-Mg Alloys, in *Aluminum-Lithium Alloys*, Vol III, C. Baker, P.J. Gregson, S.J. Harris, and C.J. Peel, Ed., The Institute of Metals, 1986, p 310
69. "Standard Recommended Practice of Alternate Immersion Stress Corrosion Testing in 3.5% Sodium Chloride Solution," G 44, *Annual Book of ASTM Standards*, American Society for Testing and Materials
70. "Standard Recommended Practice for Determining Susceptibility to Stress-Corrosion Cracking of High-Strength Aluminum Alloy Products," G 47, *Annual Book of ASTM Standards*, American Society for Testing and Materials
71. "Standard Test Method for Visual Assessment of Exfoliation Corrosion Susceptibility of 5*xxx* Series Aluminum Alloys (ASSET Test)," G 66, *Annual Book of ASTM Standards*, American Society for Testing and Materials
72. "Standard Test Method for Exfoliation Corrosion Susceptibility in 2*xxx* and 7*xxx* Series Aluminum Alloys (EXCO Test)," G 34, *Annual Book of ASTM Standards*, American Society for Testing and Materials
73. "Standard Method of Acidified Synthetic Sea Water (Fog) Testing," G 43, *Annual Book of ASTM Standards*, American Society for Testing and Materials
74. "Standard Practice for Modified Salt Spray (Fog) Testing," G 85, *Annual Book of ASTM Standards*, American Society for Testing and Materials
75. "Standard Specification for Aluminum and Aluminum-Alloy Sheet and Plate," B 209, *Annual Book of ASTM Standards*, American Society for Testing and Materials
76. "Exfoliation Corrosion Testing of Aluminum Alloys 5086 and 5456," Technical Report T1, Aluminum Association, circa 1972
77. D.O. Sprowls, J.D. Walsh, and M.B. Shumaker, Simplified Exfoliation Testing of Aluminum Alloys, in *Localized Corrosion—Cause of Metal Failure*, STP 516, American Society for Testing and Materials, 1972, p 38
78. T.J. Summerson, "Aluminum Association Task Group on Exfoliation and Stress-Corrosion Cracking of Aluminum Alloys for Boat Stock," Interim Report, in *Proceedings of the Tri-Service Conference on Corrosion of Military Equipment*, Technical Report AFML-TR-75-42, Vol II, 1975, p 193
79. S.J. Ketcham and P.W. Jeffrey, Exfoliation Corrosion Testing of 7178 and 7075 Aluminum Alloys, in *Localized Corrosion—Cause of Metal Failure*, STP 516, American Society for Testing and Materials, 1972, p 273
80. B.W. Lifka and D.O. Sprowls, Relationship of Accelerated Test Methods for Exfoliation Resistance in 7*xxx* Series Alloys with Expo-

Table 16 Corrosion resistance of aluminum alloy plate, with and without joints, partially immersed in flowing seawater at Kure Beach, NC

Alloy and temper	Type of joint	Exposure period, years	Maximum depth of attack, mils: Plate, Outside surface	Plate, Faying surface	Rivet or weld	Change in tensile strength due to corrosion(a), %
Continuously immersed						
6053-T6	Riveted(b)	6	1.4	3.0	8.4	0
6061-T6	Riveted(c)	1	1.4	2.8	2.8	0
6053-T6	Welded(d)	2	5.0	...	4.2	...
6061-T6	Welded(d)	1	5.0	...	9.8	...
6061-T4	None	3	2.1	...	...	2
6061-T6	None	3	1.4	...	...	1
2024-T4 alclad(e)	None	5	4.2	...	...	0
3004-H14 alclad(f)	None	5	1.4	...	...	−5
520.0-T4(g)	None	3	4.2	...	...	−4
Not immersed (atmospheric exposure)						
6053-T6	Riveted(d)	6	5.6	5.6	11.7	−1
6061-T6	Riveted(e)	1	5.6	2.1	8.5	0
6053-T6	Welded(f)	2	3.3	...	9.8	...
6061-T6	Welded(f)	1	7.0	...	9.8	...
6061-T4	None	3	2.1	...	...	1
6061-T6	None	3	4.2	...	...	1
2024-T4 alclad(e)	None	5	8.4	...	...	0
3003-H14 alclad(f)	None	5	7.0	...	...	−5
520.0T4(g)	None	3	1.4	...	...	4

(a) Results of testing 6.4-mm (0.25-in.) thick ASTM tensile specimens cut from indicated location in test plate (generally, two specimens were cut from each test plate and the results were averaged). (b) 6053-T6 rivets. (c) 6061-T43 rivets. (d) 4043 filler metal. (e) Average thickness of cladding on each surface, 297 μm (11.7 mils). (f) Average thickness of cladding on each surface, 307 μm (12.1 mils). (g) Sand cast. Source: Ref 4

Table 17 Results of 3-year seacoast exposure testing of anodized aluminum alloys(a)

Alloy and temper	Results
Sheet	
1100	No visible pitting
2024-T3, alclad	Edge pitting only
5456-H343	No visible pitting
5086-H34	No visible pitting
6061-T6	No visible pitting
7039-T6	No visible pitting
7075-T6	Edge pitting only
7075-F, alclad	Edge pitting only
7079-T6	Edge pitting only
Extrusions	
6351-T6	No visible pitting
6061-T6	No visible pitting
6063-T5	No visible pitting
6070-T6	No visible pitting
7039-T6	Scattered small pits

(a) H_2SO_4 anodic coatings 23 μm (0.9 mil) thick, sealed in boiling water on test panels 100 × 150 mm (4 × 6 in.) cut from sheet and extrusions

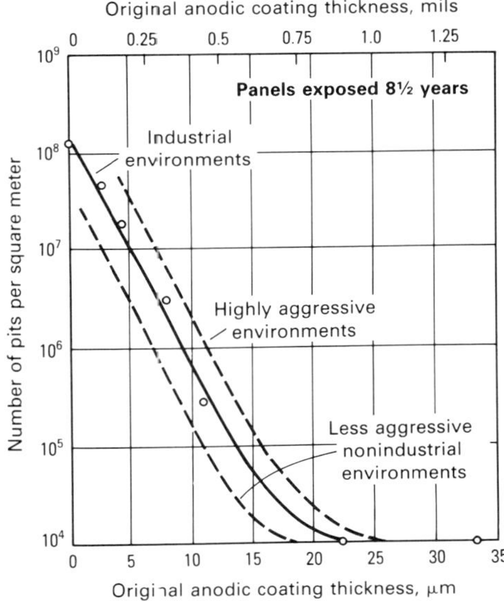

Fig. 21 Number of corrosion pits in anodized aluminum 1100 as a function of coating thickness. Source: Ref 121

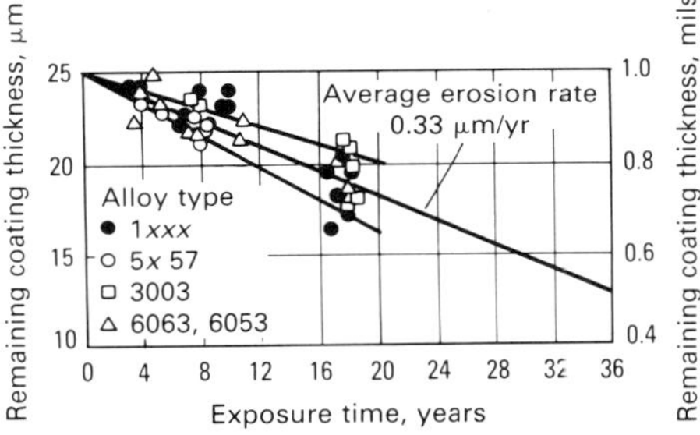

Fig. 22 Weathering data for anodically coated aluminum in an industrial atmosphere

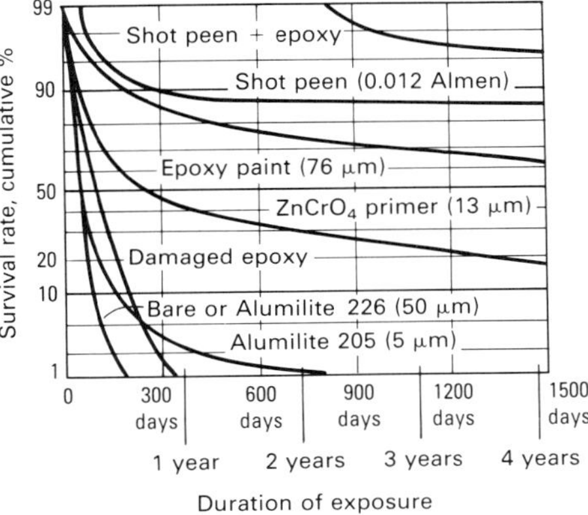

Fig. 23 Relative effectiveness of various protective systems in preventing SCC of susceptible aluminum alloys. Combined data for highly elastically strained specimens of alloys 2014-T651 and 7079-T651 exposed at Pt. Judith, RI; Comfort, TX; and New Kensington, PA

sure to a Seacoast Atmosphere, in *Corrosion in Natural Environments*, STP 558, American Society for Testing and Materials, 1974, p 306

81. O.F. Devereux, A.J. McEvily, and R.W. Staehle Ed., *Corrosion Fatigue: Chemistry, Mechanics, and Microstructure*, Part VII, *Aluminum Alloys*, National Association of Corrosion Engineers, 1972, p 451
82. H.L. Craig, T.W. Hooker, and D.W. Hoeppner, Ed., *Corrosion Fatigue Technology*, STP 642, American Society for Testing and Materials, 1978, p 51
83. J.E. Hatch, Ed., *Aluminum: Properties and Physical Metallurgy*, American Society for Metals, 1984
84. S.J. Ketcham and E.J. Jankowsky, Developing an Accelerated Test: Problems and Pitfalls, in *Laboratory Corrosion Tests and Standards*, G.S. Haynes and R. Babioan, Ed., STP 866, American Society for Testing and Materials, 1985, p 14
85. G. Sowinski and D.O. Sprowls, Weathering of Aluminum Alloys, in *Atmospheric Corrosion*, W.H. Ailor, Ed., John Wiley & Sons, 1982, p 297
86. M.A. Pelensky, J.J. Jaworski, and A. Galliccio, Corrosion Investigations at Panama Canal Zone, in *Atmospheric Factors Affecting the Corrosion of Engineering Materials*, S.K. Coburn, Ed., STP 646, American Society for Testing and Materials, 1976, p 58
87. C.J. Walton, D.O. Sprowls, and J.A. Nock, Jr., Resistance of Aluminum Alloys to Weathering, *Corrosion*, Vol 9 (No. 10), 1953, p 345
88. W.W. Binger, R.H. Wagner, and R.H. Brown, Resistance of Aluminum Alloys to Chemically Contaminated Atmospheres, *Corrosion*, Vol 9 (No. 12), 1953, p 440
89. F.L. McGeary, E.T. Englehart, and P.J. Ging, Weathering of Aluminum, *Mater. Protec.*, Vol 6 (No. 6), 1967, p 33
90. C.J. Walton and W. King, Resistance of Aluminum-Base Alloys to 20-Year Atmospheric Exposure, in *STP 174*, American Society for Testing and Materials, 1956, p 21
91. S.M. Brandt and L.H. Adams, Atmospheric Exposure of Light Metals, in *STP 435*, American Society for Testing and Materials, 1968, p 95
92. W.K. Boyd and F.W. Fink, "Corrosion of Metals in the Atmosphere," Report MCIC-74-33, Battelle Memorial Institute, 1974
93. S.C. Byrne and A.C. Miller, Effect of Atmospheric Pollutant Gases on the Formation of Corrosive Condensate on Aluminum, in *Atmospheric Corrosion of Metals*, S.W. Dean, Jr. and E.C. Rhea, Ed., STP 767, American Society for Testing and Materials, 1982, p 395
94. F. Mattsen and S. Lindgren, Hard-Rolled Aluminum Alloys, in *Metal Corrosion in the*

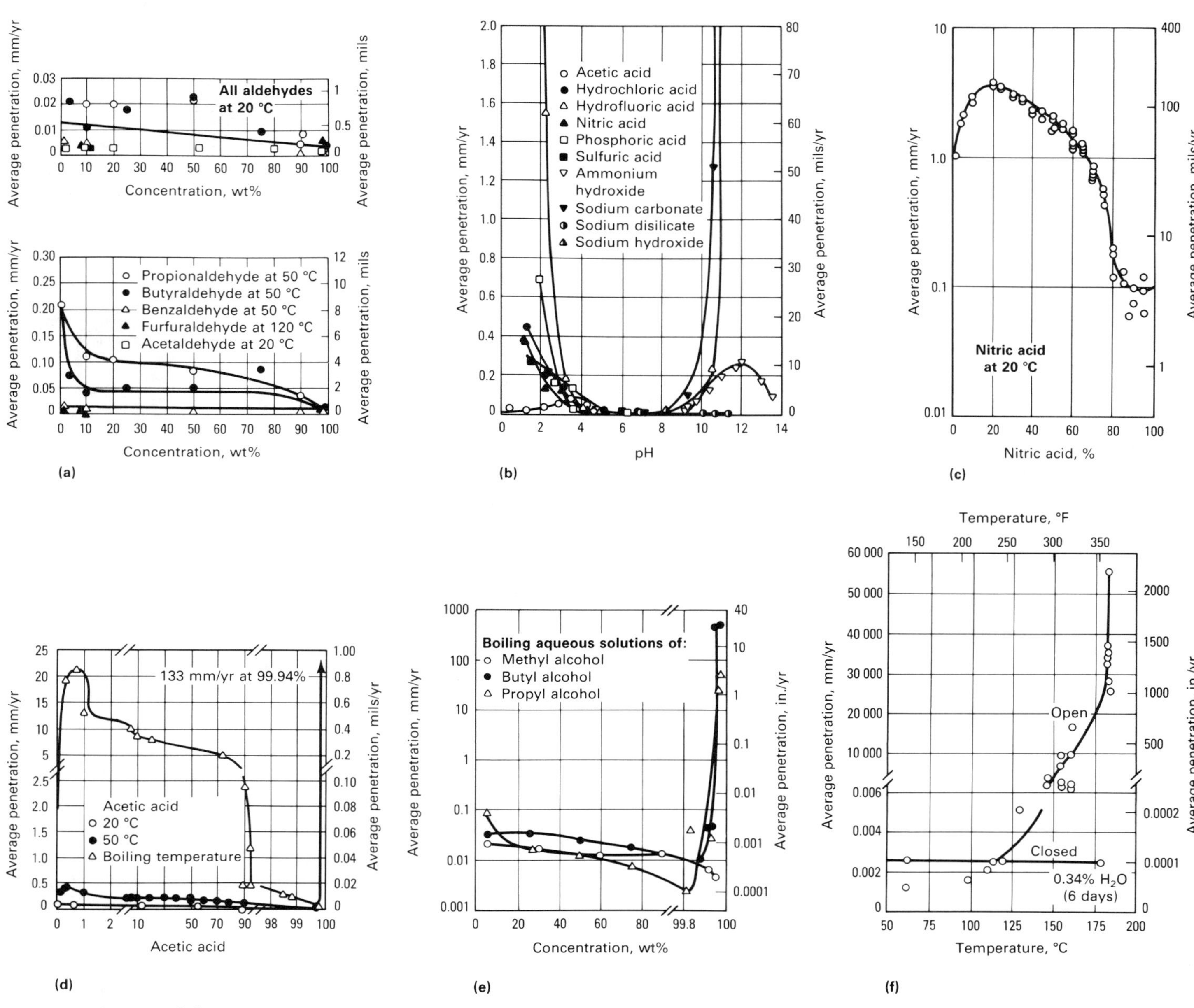

Fig. 24 Corrosion of aluminum 1100-H14 in various chemical solutions. Average penetration calculated from weight loss data in short-term tests. (a) Effect of concentration of aqueous solutions of several aldehydes. Rates of attack indicate that aluminum should be satisfactory for handling all these solutions. (b) Effect of pH. The concentration of all the solutions ranged from 0.00001 to 0.1 *N*, except acetic acid (0.00001 to 17.4 *N*), ammonium hydroxide (NH_4OH) (0.00001 to 15 *N*), and sodium disilicate ($Na_2Si_2O_5$) (0.00001 to 1 *N*). (c) Effect of concentration of HNO_3 solutions at room temperature. (d) Effect of concentration and temperature of acetic acid. (e) Effect of concentration of boiling aqueous solutions of three alcohols. (f) Effect of temperature of phenol. Rapid reaction above 120 °C (250 °F) can be stopped by small additions of steam or water.

Atmosphere, STP 435, American Society for Testing and Materials, 1968, p 240

95. T.P. Hoar, Discussion on Filiform Corrosion, *Chem. Ind.*, Nov 1952, p 1126
96. W.H. Slaybaugh, W. DeJager, S.E. Hoover, and L.L. Hutchinson, Filiform Corrosion of Aluminum, *J. Paint Technol.*, Vol 44 (No. 556), 1972, p 76
97. W.W. Binger and C.M. Marstiller, Aluminum Alloys for Handling High Purity Water, *Corrosion*, Vol 13 (No. 9), 1957
98. J.E. Draley and W.E. Ruther, Aqueous Corrosion of Aluminum, Part 2—Methods of Protection Above 200 °C, *Corrosion*, Vol 12 (No. 10), 1965, p 480t
99. D.W. Sawyer and R.H. Brown, Resistance of Aluminum Alloys to Fresh Waters, *Corrosion*, Vol 3 (No. 9), 1947, p 443
100. H.P. Godard, The Corrosion Behavior of Aluminum in Natural Waters, *Can. J. Chem. Eng.*, Vol 38, 1960, p 167
101. W.H. Ailor, Jr., A Review of Aluminum Corrosion in Tap Water, *J. Hydronautics*, Vol 3 (No. 3), 1969, p 105
102. B.R. Pathak and H.P. Godard, Equations for Predicting the Corrosivities of Natural Fresh Waters to Aluminum, *Nature*, Vol 218 (No. 5144), June 1968, p 893
103. W.A. Prey, N.W. Smith, and C.L. Wood, Jr., Marine Applications, in *Aluminum*, Vol II, K.R. Van Horn, Ed., American Society for Metals, 1967, p 389
104. K.G. Compton, Seawater Tests, in *Handook on Corrosion Testing and Evaluation*, W.H. Ailor, Ed., John Wiley & Sons, 1971, p 507
105. W.K. Boyd and F.W. Fink, "Corrosion of Metals in Marine Environments," Report MCIC-74-245R, Battelle Memorial Institute, 1975
106. W.H. Ailor, Jr., Ten-Year Seawater Tests on Aluminum, in *Corrosion in Natural Environments*, STP 558, American Society for Testing and Materials, 1974, p 117
107. F.M. Reinhart, "Corrosion of Metals and Alloys in the Deep Ocean," Report R834,

U.S. Naval Engineering Laboratory, 1976
108. S.C. Dexter, Effect of Variations in Seawater Upon the Corrosion of Aluminum, *Corrosion*, Vol 36 (No. 8), 1980, p 423
109. H.T. Rowland and S.C. Dexter, Effects of the Seawater Carbon Dioxide System on the Corrosion of Aluminum, *Corrosion*, Vol 36 (No. 9), 1980, p 458
110. S.C. Dexter, K.E. Lucas, J. Mihm, and W.E. Rigby, "Effect of Water Chemistry and Velocity of Flow on Corrosion of Aluminum," Paper 64, presented at Corrosion/83, Anaheim, CA, National Association of Corrosion Engineers, 1983
111. J. Larsen-Basse and S.H. Zaida, Corrosion of Some Aluminum Alloys in Tropical Surface and Deep Ocean Seawater, in *Proceedings of the International Congress on Metallic Corrosion*, Vol 4, June 1984, p 511
112. R.S.C. Munier and H.L. Craig, "Ocean Thermal Energy Conversion (OTEC) Biofouling and Corrosion Experiment (1977), St. Croix, U.S. Virgin Is., Part II, Corrosion Studies," Pacific Northwest Laboratory, Report PNL-2739, Feb 1978
113. D.S. Sasscer, T.O. Morgan, R. Ernst, T.J. Summerson, and R.C. Scott, "Open Ocean Corrosion Test of Candidate Aluminum Materials for Seawater Heat Exchangers," Paper 67, presented at Corrosion/83, Anaheim, CA, National Association of Corrosion Engineers, 1983
114. M. Romanoff, "Underground Corrosion," NBS 579, National Bureau of Standards, 1957
115. D.O. Sprowls and M.E. Carlisle, Resistance of Aluminum Alloys to Underground Corrosion, *Corrosion*, Vol 17, 1961, p 125t
116. T.E. Wright, New Trends in Buried Aluminum Pipelines, *Mater. Perform.*, Vol 15 (No. 9), 1976, p 26
117. Recommended Practice for Cathodic Protection of Aluminum Pipe Buried in Soil or Immersed in Water, *Mater. Protec.*, Vol 2 (No. 10), 1963, p 106
118. J.A. Apostolos and F.A. Myhres, "Cooperative Field Survey of Aluminum Culverts," Report FHWA/CA/TL80-12, California Department of Transportation, 1980
119. T.E. Wright, The Corrosion Behavior of Aluminum Pipe, *Mater. Perform.*, Vol 22 (No. 12), 1983, p 9
120. W.C. Cochran, Anodizing, in *Aluminum: Fabrication and Finishing*, Vol III, K.R. Van Horn, Ed., American Society for Metals, 1967, p 641
121. W.C. Cochran and D.O. Sprowls, "Anodic Coatings for Aluminum," Paper presented at Conference on Corrosion Control by Coatings, Lehigh University, Nov 1978
122. D.O. Sprowls *et al.*, "Investigation of the Stress-Corrosion Cracking of High Strength Aluminum Alloys," Final Report, Contract No. NAS-8-5340 for the period of May 1963 to Oct 1966, Accession No. NASA CR88110, National Technical Information Center, 1967
123. C.J. Walton, F.L. McGeary, and E.T. Englehart, The Compatibility of Aluminum with Alkaline Building Products, *Corrosion*, Vol 13, 1957, p 807t
124. *Aluminium in the Chemical and Food Industries*, The British Aluminium Company, Norfolk House, 1959
125. *Aluminum Statistical Review for 1984*, The Aluminum Association, 1984
126. E.H. Cook, R.L. Horst, and W.W. Binger, Corrosion Studies of Aluminum in Chemical Process Operations, *Corrosion*, Vol 17 (No. 1), 1961, p 97
127. R.L. Horst, Structures and Equipment for the Chemical, Food, Drug, Beverage and Atomic Industries, in *Aluminum: Design and Application*, Vol II, K.R. Van Horn, Ed., American Society for Metals, 1967, p 259
128. *Guidelines for the Use of Aluminum With Food and Chemicals*, 5th ed., The Aluminum Association, April 1984
129 *Care of Aluminum*, The Aluminum Association, 1977
130. E.T. Englehart, Cleaning and Maintenance of Surfaces, in *Aluminum*, Vol III, K.R. Van Horn, Ed., American Society for Metals, 1967, p 757

SELECTED REFERENCES

- J.D. Edwards, F.C. Frary, and Z. Jeffries, *The Aluminum Industry: Aluminum Products and Their Fabrication*, McGraw-Hill, 1930
- U.R. Evans, *The Corrosion and Oxidation of Metals: Scientific Principles and Practical Applications*, E. Arnold, 1960
- H.P. Godard, W.B. Jepson, M.R. Bothwell, and R.L. Kane, *The Corrosion of Light Metals*, John Wiley & Sons, 1967
- *Guidelines for the Use of Aluminum with Foods and Chemicals*, 5th ed., Aluminum Association, Inc., April 1984
- J.E. Hatch, Ed., *Aluminum: Properties and Physical Metallurgy*, American Society for Metals, 1984
- F.L. LaQue and H.R. Copson, *Corrosion Resistance of Metals and Alloys*, 2nd ed., Reinhold, 1963
- L.F. Mondolfo, *Aluminum Alloys: Structure and Properties*, Butterworths, 1976
- L.L. Shrier, *Corrosion*, Vol I and II, 2nd ed., Newnes-Butterworths, 1976
- H.H. Uhlig, Ed., *Corrosion Handbook*, John Wiley & Sons, 1948
- K.R. Van Horn, Ed., *Aluminum*, Vol I, II, and III, American Society for Metals, 1967

Corrosion of Copper and Copper Alloys

By the ASM Committee on Corrosion of Copper*
Chairman: Ned W. Polan, Olin Corporation

COPPER AND COPPER ALLOYS are widely used in many environments and applications because of their excellent corrosion resistance, which is coupled with combinations of other desirable properties, such as superior electrical and thermal conductivity, ease of fabricating and joining, wide range of attainable mechanical properties, and resistance to biofouling. Copper corrodes at negligible rates in unpolluted air, water, and deaerated nonoxidizing acids. Copper alloy artifacts have been found in nearly pristine condition after having been buried in the earth for thousands of years, and copper roofing in rural atmospheres has been found to corrode at rates of less than 0.4 mm (15 mils) in 200 years. Copper alloys resist many saline solutions, alkaline solutions, and organic chemicals. However, copper is susceptible to more rapid attack in oxidizing acids, oxidizing heavy-metal salts, sulfur, ammonia (NH_3), and some sulfur and NH_3 compounds. Resistance to acid solution depends mainly on the severity of oxidizing conditions in the solution. Reaction of copper with sulfur and sulfides to form copper sulfide (CuS or Cu_2S) usually precludes the use of copper and copper alloys in environments known to contain certain sulfur species.

Copper and copper alloys provide superior service in many of the applications included in the following general classifications:

- Applications requiring resistance to atmospheric exposure, such as roofing and other architectural uses, hardware, building fronts, grille work, hand rails, lock bodies, doorknobs, and kick plates
- Freshwater supply lines and plumbing fittings, for which superior resistance to corrosion by various types of waters and soils is important
- Marine applications—most often freshwater and seawater supply lines, heat exchangers, condensers, shafting, valve stems, and marine hardware—in which resistance to seawater, hydrated salt deposits, and biofouling from marine organisms is important
- Heat exchangers and condensers in marine service, steam power plants, and chemical process applications, as well as liquid-to-gas or gas-to-gas heat exchangers in which either process stream may contain a corrosive contaminant
- Industrial and chemical plant process equipment involving exposure to a wide variety of organic and inorganic chemicals
- Electrical wiring, hardware, and connectors; printed circuit boards; and electronic applications that require demanding combinations of electrical, thermal, and mechanical properties, such as semiconductor packages, lead frames, and connectors

Copper and its alloys are unique among the corrosion-resistant alloys in that they do not form a truly passive corrosion product film. In aqueous environments at ambient temperatures, the corrosion product predominantly responsible for protection is cuprous oxide (Cu_2O). This Cu_2O film is adherent and follows parabolic growth kinetics. Cuprous oxide is a *p*-type semiconductor formed by the electrochemical processes:

$$4Cu + 2H_2O \rightarrow 2Cu_2O + 4H^+ + 4e^- \text{ (anode)} \quad \text{(Eq 1)}$$

and

$$O_2 + 2H_2O + 4e^- \rightarrow 4(OH)^- \text{ (cathode)} \quad \text{(Eq 2)}$$

with the net reaction: $4Cu + O_2 \rightarrow 2Cu_2O$.

For the corrosion reaction to proceed, copper ions and electrons must migrate through the Cu_2O film. Consequently, reducing the ionic or electronic conductivity of the film by doping with divalent or trivalent cations should improve corrosion resistance. In practice, alloying additions of aluminum, zinc, tin, iron, and nickel are used to dope the corrosion product films, and they generally reduce corrosion rates significantly.

Effects of Alloy Compositions

Copper alloys are traditionally classified under the groupings listed in Table 1.

Coppers and high-copper alloys have similar corrosion resistance. They have excellent resistance to seawater corrosion and biofouling, but are susceptible to erosion-corrosion at high water velocities. The high-copper alloys are primarily used in applications that require enhanced mechanical performance, often at slightly elevated temperature, with good thermal or electrical conductivity. Processing for increased strength in the high-copper alloys generally improves their resistance to erosion-corrosion. A number of alloys in this category have been developed for electronic applications—such as contact clips, springs, and lead frames—that require specific mechanical properties, relatively high electrical conductivity, and atmospheric-corrosion resistance.

Brasses are basically copper-zinc alloys and are the most widely used group of copper alloys. The resistance of brasses to corrosion by aqueous solutions does not change markedly as long as the zinc content does not exceed about 15%; above 15% Zn, dezincification may occur. Quiescent or slowly moving saline solutions, brackish waters, and mildly acidic solutions are environments that often lead to the dezincification of unmodified brasses.

Susceptibility to stress-corrosion cracking (SCC) is significantly affected by zinc content; alloys that contain more zinc are more susceptible. Resistance increases substantially as zinc content decreases from 15 to 0%. Stress-corrosion cracking is practically unknown in commercial copper.

Elements such as lead, tellurium, beryllium, chromium, phosphorus, and manganese have little or no effect on the corrosion resistance of coppers and binary copper-zinc alloys. These elements are added to enhance such mechanical properties as machinability, strength, and hardness.

Tin Brasses. Tin additions significantly increase the corrosion resistance of some brasses, especially resistance to dezincification. Examples of this effect are two tin-bearing brasses: uninhibited admiralty metal (no active UNS number) and naval brass (C46400). Uninhibited admiralty metal was once widely used to make heat-exchanger tubes; it has largely been replaced by inhibited grades of admiralty metal (C44300, C44400, and C44500), which have even greater resistance to dealloying. Admiralty metal is a variation of cartridge brass (C26000) that is produced by adding about 1% Sn to the basic 70Cu-30Zn composition. Similarly, naval brass is the alloy resulting from the addition of 0.75% Sn to the basic 60Cu-40Zn composition of Muntz metal (C28000).

*Frank J. Ansuini, Consulting Engineer; Carl W. Dralle, Ampco Metal; Fraser King, Whiteshell Nuclear Research Establishment; W.W. Kirk, LaQue Center for Corrosion Technology, Inc.; T.S. Lee, National Association of Corrosion Engineers; Henry Leidheiser, Jr., Center for Surface and Coating Research, Lehigh University; Richard O. Lewis, Department of Materials Science and Engineering, University of Florida; Gene P. Sheldon, Olin Corporation

Table 1 Generic classification of copper alloys

Generic name	UNS numbers	Composition
Wrought alloys		
Coppers	C10100–C15760	>99% Cu
High-copper alloys	C16200–C19600	>96% Cu
Brasses	C205–C28580	Cu-Zn
Leaded brasses	C31200–C38590	Cu-Zn-Pb
Tin brasses	C40400–C49080	Cu-Zn-Sn-Pb
Phosphor bronzes	C50100–C52400	Cu-Sn-P
Leaded phosphor bronzes	C53200–C54800	Cu-Sn-Pb-P
Copper-phosphorus and copper-silver-phosphorus alloys	C55180–C55284	Cu-P-Ag
Aluminum bronzes	C60600–C64400	Cu-Al-Ni-Fe-Si-Sn
Silicon bronzes	C64700–C66100	Cu-Si-Sn
Other copper-zinc alloys	C66400–C69900	. . .
Copper-nickels	C70000–C79900	Cu-Ni-Fe
Nickel silvers	C73200–C79900	Cu-Ni-Zn
Cast alloys		
Coppers	C80100–C81100	>99% Cu
High-copper alloys	C81300–C82800	>94% Cu
Red and leaded red brasses	C83300–C85800	Cu-Zn-Sn-Pb (75–89% Cu)
Yellow and leaded yellow brasses	C85200–C85800	Cu-Zn-Sn-Pb (57–74% Cu)
Manganese and leaded manganese bronzes	C86100–C86800	Cu-Zn-Mn-Fe-Pb
Silicon bronzes, silicon brasses	C87300–C87900	Cu-Zn-Si
Tin bronzes and leaded tin bronzes	C90200–C94500	Cu-Sn-Zn-Pb
Nickel-tin bronzes	C94700–C94900	Cu-Ni-Sn-Zn-Pb
Aluminum bronzes	C95200–C95810	Cu-Al-Fe-Ni
Copper-nickels	C96200–C96800	Cu-Ni-Fe
Nickel silvers	C97300–C97800	Cu-Ni-Zn-Pb-Sn
Leaded coppers	C98200–C98800	Cu-Pb
Miscellaneous alloys	C99300–C99750	. . .

Cast brasses for marine use are also modified by the addition of tin, lead, and, sometimes, nickel. This group of alloys is known by various names, including composition bronze, ounce metal, and valve metal. These older designations are used less frequently, because they have been supplanted by alloy numbers under the UNS or Copper Development Association (CDA) system. The cast marine brasses are used for plumbing goods in moderate-performance seawater piping systems or in deck hardware, for which they are subsequently chrome plated.

Aluminum Brasses. An important constituent of the corrosion film on a brass that contains a few percent aluminum in addition to copper and zinc is aluminum oxide (Al_2O_3), which markedly increases resistance to impingement attack in turbulent high-velocity saline water. For example, the arsenical aluminum brass C68700 (76Cu-22Zn-2Al) is frequently used for marine condensers and heat exchangers in which impingement attack is likely to pose a serious problem. Aluminum brasses are susceptible to dezincification unless they are inhibited, which is usually done by adding 0.02 to 0.10% As.

Inhibited Alloys. Addition of phosphorus, arsenic, or antimony (typically 0.02 to 0.10%) to admiralty metal, naval brass, or aluminum brass effectively produces high resistance to dezincification. Inhibited alloys have been extensively used for such components as condenser tubes, which must accumulate years of continuous service between shutdowns for repair or replacement.

Phosphor Bronzes. Addition of tin and phosphorus to copper produces good resistance to flowing seawater and to most nonoxidizing acids except hydrochloric (HCl). Alloys containing 8 to 10% Sn have high resistance to impingement attack. Phosphor bronzes are much less susceptible to SCC than brasses and are similar to copper in resistance to sulfur attack. Tin bronzes—alloys of copper and tin—tend to be used primarily in the cast form, in which they are modified by further alloy additions of lead, zinc, and nickel. Like the cast brasses, the cast tin bronzes are occasionally identified by older, more colorful names that reflect their historic uses, such as G Bronze, Gun Metal, Navy M Bronze, and steam bronze. Contemporary uses include pumps, valves, gears, and bushings. Wrought tin bronzes are known as phosphor bronzes and find use in high strength wire applications, such as wire rope. This group of alloys has fair resistance to impingement and good resistance to biofouling.

Copper Nickels. Alloy C71500 (Cu-30Ni) has the best general resistance to aqueous corrosion of all the commercially important copper alloys, but C70600 (Cu-10Ni) is often selected because it offers good resistance at lower cost. Both of these alloys, although well suited to applications in the chemical industry, have been most extensively used for condenser tubes and heat-exchanger tubes in recirculating steam systems. They are superior to coppers and to other copper alloys in resisting acid solutions and are highly resistant to SCC and impingement corrosion.

Nickel Silvers. The two most common nickel silvers are C75200 (65Cu-18Ni-17Zn) and C77000 (55Cu-18Ni-27Zn). They have good resistance to corrosion in both fresh and salt waters. Primarily because their relatively high nickel contents inhibit dezincification, C75200 and C77000 are usually much more resistant to corrosion in saline solutions than brasses of similar copper content.

Copper-silicon alloys generally have the same corrosion resistance as copper, but they have higher mechanical properties and superior weldability. These alloys appear to be much more resistant to SCC than the common brasses. Silicon bronzes are susceptible to embrittlement by high-pressure steam and should be tested for suitability in the service environment before being specified for components to be used at elevated temperature.

Aluminum bronzes containing 5 to 12% Al have excellent resistance to impingement corrosion and high-temperature oxidation. Aluminum bronzes are used for beater bars and for blades in wood pulp machines because of their ability to withstand mechanical abrasion and chemical attack by sulfite solutions.

In most practical commercial applications, the corrosion characteristics of aluminum bronzes are primarily related to aluminum content. Alloys with up to 8% Al normally have completely face-centered cubic (fcc) α structures and good resistance to corrosion attack. As aluminum content increases above 8%, α-β duplex structures appear. The β phase is a high-temperature phase retained at room temperature upon fast cooling from 565 °C (1050 °F) or above. Slow cooling for long exposure at temperatures from 320 to 565 °C (610 to 1050 °F) tends to decompose the β phase into a brittle $\alpha + \gamma_2$ eutectoid having either a lamellar or a nodular structure. The β phase is less resistant to corrosion than the α phase, and eutectoid structures are even more susceptible to attack.

Depending on specific environmental conditions, β phase or eutectoid structure in aluminum bronze can be selectively attacked by a mechanism similar to the dezincification of brasses. Proper quench-and-temper treatment of duplex alloys, such as C62400 and C95400, produces a tempered β structure with reprecipitated acicular α crystals, a combination that is often superior in corrosion resistance to the normal annealed structures.

Iron-rich particles are distributed as small round or rosette particles throughout the structures of aluminum bronzes containing more than about 0.5% Fe. These particles sometimes impart a rusty tinge to the surface, but have no known effect on corrosion rates.

Nickel-aluminum bronzes are more complex in structure with the introduction of the κ phase. Nickel appears to alter the corrosion characteristics of the β phase to provide greater resistance to dealloying and cavitation-erosion in most liquids. For C63200 and perhaps C95800, quench-and-temper treatments may yield even greater resistance to dealloying. Alloy C95700, a high-manganese cast aluminum bronze, is somewhat inferior in corrosion resistance to C95500 and C95800, which are low in manganese and slightly higher in aluminum.

Aluminum bronzes are generally suitable for service in nonoxidizing mineral acids, such as phosphoric (H_3PO_4) sulfuric (H_2SO_4), and HCl; organic acids, such as lactic, acetic (CH_3COOH), or oxalic; neutral saline solutions, such as sodium chloride (NaCl) or potassium chloride (KCl); alkalies, such as sodium hydroxide (NaOH), potassium hydroxide (KOH), and anhydrous ammonium hydroxide (NH_4OH); and various natural waters including sea, brackish, and potable waters. Environments to be avoided include nitric acid (HNO_3); some metallic salts, such as ferric chloride ($FeCl_3$) and chromic acid (H_2CrO_4); moist chlorinated hydrocarbons; and moist HN_3. Aeration can result in accelerated corrosion in many media that appear to be compatible.

Exposure under high tensile stress to moist NH_3 can result in SCC. In certain environments, corrosion can lower the fatigue limit to 25 to 50% of the normal atmospheric value.

Types of Attack

Coppers and copper alloys, like most other metals and alloys, are susceptible to several forms of corrosion, depending primarily on environmental conditions. Table 2 lists the identifying characteristics of the forms of corrosion that commonly attack copper metals as well as the most effective means of combating each.

General Corrosion

General corrosion is the well-distributed attack of an entire surface with little or no localized penetration. It is the least damaging of all forms of attack. General corrosion is the only form of corrosion for which weight loss data can be used to estimate penetration rates accurately.

General corrosion of copper alloys results from prolonged contact with environments in which the corrosion rate is very low, such as fresh, brackish, and salt waters; many types of soil; neutral, alkaline, and acid salt solutions; organic acids; and sugar juices. Other substances that cause uniform thinning at a faster rate include oxidizing acids, sulfur-bearing compounds, NH_3, and cyanides. Additional information on this form of attack is available in the article "General Corrosion" in this Volume.

Galvanic Corrosion

An electrochemical potential almost always exists between two dissimilar metals when they are immersed in a conductive solution. If two dissimilar metals are in electrical contact with each other and immersed in a conductive solution, a potential results that enhances the corrosion of the more electronegative member of the couple (the anode) and partly or completely protects the more electropositive member (the cathode). Copper metals are almost always cathodic to other common structural metals, such as steel and aluminum. When steel or aluminum is put in contact with a copper metal, the corrosion rate of the steel or aluminum increases, but that of the copper metal decreases. The common grades of stainless steel exhibit variable behavior; that is, copper metals may be anodic or cathodic to the stainless steel, depending on conditions of exposure. Copper metals usually corrode preferentially when coupled with high-nickel alloys, titanium, or graphite. Additional information on this subject is available in the section "Galvanic Corrosion" of the article "General Corrosion" in this Volume.

Corrosion potentials of copper metals generally range from −0.2 to −0.4 V when measured against a saturated calomel electrode (SCE); the potential of pure copper is about −0.3 V. Alloying additions of zinc or aluminum move the potential toward the anodic (more electronegative) end of the range; additions of tin or nickel move the potential toward the cathodic (less electronegative) end. Galvanic corrosion between two copper metals is seldom a significant problem, because the potential difference is so small.

Table 3 lists a galvanic series of metals and alloys valid for dilute aqueous solutions, such as seawater and weak acids. The metals that are grouped together can be coupled to each other without significant galvanic damage. However, the connecting of metals from different groups leads to damage of the more anodic metal; the larger the difference in galvanic potential between groups, the greater the corrosion. Accelerated damage due to galvanic effects is usually greatest near the junction, where the electrochemical current density is the highest.

Another factor that affects galvanic corrosion is area ratio. An unfavorable area ratio exists when the cathodic area is large and the anodic area is small. The corrosion rate of the small anodic area may be several hundred times greater than if the anodic and cathodic areas were equal in size. Conversely, when a large anodic area is coupled to a small cathodic area, current density and damage due to galvanic corrosion are much less. For example, copper rivets (cathodic) used to fasten steel plates together lasted longer than 1.5 years in seawater, but steel rivets used to fasten copper plates were completely destroyed during the same period.

Five principal methods are available for eliminating or significantly reducing galvanic corrosion:

- Select dissimilar metals that are as close as possible to each other in the galvanic series
- Avoid coupling small anodes to large cathodes
- Insulate dissimilar metals completely wherever practicable
- Apply coatings and keep them in good repair, particularly on the cathodic member
- Use a sacrificial anode; that is, couple the system to a third metal that is anodic to both structural metals

Pitting

As with most commercial metals, corrosion of copper metals results in pitting under certain conditions. Pitting is sometimes general over the entire surface, giving the metal an irregular and roughened appearance. In other cases, pits are concentrated in specific areas and are of various sizes and shapes. Detailed information on this form of attack is available in the section "Pitting" in the article "Localized Corrosion" in this Volume.

Localized pitting is the most damaging form of corrosive attack because it reduces load-carrying capacity and increases stress concentration by creating depressions or holes in the metal. Pitting is the usual form of corrosive attack at surfaces on which there are incomplete protective films, nonprotective deposits of scale, or extraneous deposits of dirt or other foreign substances.

Copper alloys do not corrode primarily by pitting, but because of metallurgical and environmental factors that are not completely understood, the corroded surface does show a tendency toward nonuniformity. In seawater, pitting tends to occur more often under conditions of relatively low water velocity, typically less than 0.6 to 0.9 m/s (2 to 3 ft/s). The occurrence of pitting is somewhat random regarding the specific location of a pit on the surface as well as whether it will even occur on a particular metal sample. Long-term tests of copper alloys show that the average pit depth does not continually increase with extended times of exposure. Instead, pits tend to reach a certain limit beyond which little apparent increase in depth occurs. Of the copper alloys, the most pit resistant are the aluminum bronzes with less than 8% Al and the low-zinc brasses. Copper nickels and tin bronzes tend to have intermediate pitting resistance, but the high-copper alloys and silicon bronzes are somewhat more prone to pitting.

Crevice corrosion is a form of localized corrosion that occurs near a crevice formed either by

Table 2 Guide to corrosion of copper alloys

Form of attack	Characteristics	Preventive measures
General thinning	Uniform metal removal	Select proper alloy for environmental conditions based on weight loss data.
Galvanic corrosion	Corrosion preferentially near a more cathodic metal	Avoid electrically coupling dissimilar metals; maintain optimum ratio of anode to cathode area; maintain optimum concentration of oxidizing constituent in corroding medium.
Pitting	Localized pits, tubercles; water line pitting; crevice corrosion; pitting under foreign objects or dirt	Alloy selection; design to avoid crevices; keep metal clean.
Impingement		
Erosion-corrosion		
Cavitation	Erosion attack from turbulent flow plus dissolved gases, generally as lines of pits in direction of fluid flow	Design for streamlined flow; keep velocity low; remove gases from liquid phase; use erosion-resistant alloy.
Fretting	Chafing or galling, often occurring during shipment	Lubricate contacting surfaces; interleave sheets of paper between sheets of metal; decrease load on bearing surfaces.
Intergranular corrosion	Corrosion along grain boundaries without visible signs of cracking	Select proper alloy for environmental conditions based on metallographic examination of corrosion specimens.
Dealloying	Preferential dissolution of zinc or nickel, resulting in a layer of sponge copper	Select proper alloy for environmental conditions based on metallographic examination of corrosion specimens.
Corrosion fatigue	Several transgranular cracks	Select proper alloy based on fatigue tests in service environment; reduce mean or alternating stress.
SCC	Cracking, usually intergranular but sometimes transgranular, that is often fairly rapid	Select proper alloy based on stress-corrosion tests; reduce applied or residual stress; remove mercury compounds or NH_3 from environment.

Table 3 Galvanic series in seawater

Anodic End
Magnesium
Magnesium alloys
Zinc
Galvanized steel
Aluminum alloy 5052H
Aluminum alloy 3004
Aluminum alloy 3003
Aluminum alloy 1100
Aluminum alloy 6053
Alclad aluminum alloys
Cadmium
Aluminum alloy 2017
Aluminum alloy 2024
Low-carbon steel
Wrought iron
Cast iron
Ni-resist cast iron
AISI type 410 stainless steel (active)
50Pb-50Sn solder
AISI type 304 stainless steel (active)
AISI type 316 stainless steel (active)
Lead
Tin
Muntz metal (C28000)
Manganese bronze (C67500)
Naval brass (C46400)
Nickel (active)
Inconel (active)
Cartridge brass (C26000)
Admiralty metal (C44300)
Aluminum bronze (C61400)
Red brass (C23000)
Copper (C11000)
Silicon bronze (C65100)
Copper-nickel, 30% (C71500)
Nickel (passive)
Inconel (passive)
Monel
AISI type 304 stainless steel (passive)
AISI type 316 stainless steel (passive)
Silver
Gold
Platinum
Cathodic End

two metal surfaces or a metal and a nonmetal surface. Like pitting, crevice attack is a random occurrence, the precise location of which cannot always be predicted. Also, like pitting, the depth of attack appears to level off rather than to increase continually with time. This depth is usually less than that from pitting, and for most copper alloys, it will be less than 400 μm (15.8 mils).

For most copper alloys, the location of the attack will be outside but immediately adjacent to the crevice due to the formation of metal ion concentration cells. Classic crevice corrosion resulting from oxygen depletion and attack within crevices is less common in copper alloys. Aluminum- and chromium-bearing copper alloys, which form more passive surface films, are susceptible to differential oxygen cell attack, as are aluminum alloys and stainless steels. The occurrence of crevice attack is somewhat statistical in nature, with the odds of it occurring and its severity increasing if the area within a crevice is small compared to the area outside the crevice. Other conditions that will increase the odds of crevice attack are higher water temperatures or a flow condition on the surface outside the crevice.

Local cell action similar to crevice attack may also result from the presence of foreign objects or debris, such as dirt, pieces of shell, or vegetation, or it may result from rust, permeable scales, or uneven accumulation of corrosion product on the metallic surface. This type of attack can sometimes be controlled by cleaning the surfaces. For example, condensers and heat exchangers are cleaned periodically to prevent deposit attack.

Water line attack is a term used to describe pitting due to a differential oxygen cell functioning between the well-aerated surface layer of a liquid and the oxygen-starved layer immediately beneath it. The pitting occurs immediately below the water line.

Impingement

Various forms of impingement attack occur where gases, vapors, or liquids impinge on metal surfaces at high velocities, such as in condensers or heat exchangers. Rapidly moving turbulent water can strip away the protective films from copper alloys. When this occurs, the metal corrodes at a more rapid rate in an attempt to reestablish this film, but because the films are being swept away as rapidly as they are being formed, the corrosion rate remains constant and high. The conditions under which the corrosion product film is removed are different for each alloy and are discussed in the section "Corrosion of Copper Alloys in Specific Environments" in this article. Additional information on various types of impingement attack is available in the article "Mechanically Assisted Degradation" in this Volume.

Erosion-corrosion is characterized by undercut grooves, waves, ruts, gullies, and rounded holes; it usually exhibits a directional pattern. Pits are elongated in the direction of flow and are undercut on the downstream side. When the condition becomes severe, it may result in a pattern of horseshoe-shaped grooves or pits with their open ends pointing downstream. As attack progresses, the pits may join, forming fairly large patches of undercut pits. When this form of corrosion occurs in a condenser tube, it is usually confined to a region near the inlet end of the tube where fluid flow is rapid and turbulent. If some of the tubes in a bundle become plugged, the velocity is increased in the remaining tubes; therefore, the unit should be kept as clean as possible. Erosion-corrosion is most often found with waters containing low levels of sulfur compounds and with polluted, contaminated, or silty salt water or brackish water. The erosive action locally removes protective films, thus contributing to the formation of concentration cells and to localized pitting of anodic sites.

Cavitation is a phenomenon that occurs in moving water when the flow is disturbed so as to create a local pressure drop. Under these conditions, a vapor bubble will form and then collapse, applying a momentary stress of up to 1379 MPa (200 ksi) to the surface. The current theories of cavitation state that this repeated mechanical working of the surface creates a local fatigue situation that aids the removal of metal. This is in agreement with the observations that the harder alloys tend to have greater resistance to cavitation and that there is often an incubation period before the onset of cavitation attack. Of the copper alloys, aluminum bronze has the best cavitation resistance. Cavitation damage will be confined to the area where the bubbles collapse, usually immediately downstream of the low-pressure zone.

Impingement attack can be reduced, and the life of the unit extended, by decreasing fluid velocity, streamlining the flow, and removing entrained air. This is usually accomplished by redesigning water boxes, injector nozzles, and piping to reduce or eliminate low-pressure pockets, obstructions to smooth flow, abrupt changes in flow direction, and other features that cause local regions of high-velocity or turbulent flow. Condensers and heat exchangers are less susceptible to impingement attack if they are made of one of the aluminum brasses or copper nickels, which are more erosion resistant than the brasses or tin brasses. Erosion-resistant inserts at tube inlets and epoxy-type coatings are often effective repair methods in existing shell and tube heat exchangers. When contaminated waters are involved, filtering or screening the liquids and cleaning the surfaces can be very effective in minimizing impingement attack. The use of cathodic protection can lessen all forms of localized attack except cavitation.

Fretting

Another form of attack, called fretting or fretting corrosion, appears as pits or grooves in the metal surface that are surrounded or filled with corrosion product. Fretting is sometimes referred to as chafing, road burn, friction oxidation, wear oxidation, or galling.

The basic requirements for fretting are as follows:

- Repeated relative (sliding) motion between two surfaces must occur. The relative amplitude of the motion may be very small—motion of only a few tenths of a millimeter is typical
- The interface must be under load
- Both load and relative motion must be sufficient to produce deformation of the interface
- Oxygen and/or moisture must be present

Fretting does not occur on lubricated surfaces in continuous motion, such as axle bearings, but instead on dry interfaces subject to repeated, small relative displacements. A classic type of fretting occurs during shipment of bundles of mill products having flat faces. Fretting is not confined to coppers and copper alloys, but has been recognized on almost every kind of surface—steel, aluminum, noble metals, mica, and glass.

Fretting can be controlled, and sometimes eliminated, by:

- Lubricating with low-viscosity high-tenacity oils to reduce friction at the interface between the two metals and to exclude oxygen from the interface
- Separating the faying surfaces by interleaving an insulating material
- Increasing the load to reduce motion between faying surfaces; this may be difficult in practice, because only a minute amount of relative motion is necessary to produce fretting
- Decreasing the load at bearing surfaces to increase the relative motion between parts

Detailed information is available in the section "Fretting" of the article "Mechanically Assisted Degradation" in this Volume.

Intergranular Corrosion

Intergranular corrosion is an infrequently encountered form of attack that occurs most often in applications involving high-pressure steam. This type of corrosion penetrates the metal along grain boundaries—often to a depth of several grains—which distinguishes it from surface roughening. Mechanical stress is apparently not a factor in intergranular corrosion. The alloys that appear to be the most susceptible to this form of attack are Muntz metal, admiralty metal, aluminum brasses, and silicon bronzes. Additional information is provided in the section "Intergranular Corrosion" of the article "Metallurgically Influenced Corrosion" in this Volume.

Dealloying

Dealloying is a corrosion process in which the more active metal is selectively removed from an alloy, leaving behind a weak deposit of the more noble metal. Copper-zinc alloys containing more than 15% Zn are susceptible to a dealloying process called dezincification. In the dezincification of brass, selective removal of zinc leaves a relatively porous and weak layer of copper and copper oxide. Corrosion of a similar nature continues beneath the primary corrosion layer, resulting in gradual replacement of sound brass by weak, porous copper. Unless arrested, dealloying eventually penetrates the metal, weakening it structurally and allowing liquids or gases to leak through the porous mass in the remaining structure.

The term plug-type dealloying refers to the dealloying that occurs in local areas; surrounding areas are usually unaffected or only slightly corroded. In uniform-layer dealloying, the active component of the alloy is leached out over a broad area of the surface. Dezincification is the usual form of corrosion for uninhibited brasses in prolonged contact with waters high in oxygen and carbon dioxide (CO_2). It is frequently encountered with quiescent or slowly moving solutions. Slightly acidic water, low in salt content and at room temperature, is likely to produce uniform attack, but neutral or alkaline water, high in salt content and above room temperature, often produces plug-type attack.

Brasses with copper contents of 85% or more resist dezincification. Dezincification of brasses with two-phase structures is generally more severe, particularly if the second phase is continuous; it usually occurs in two stages: the high-zinc β phase, followed by the lower-zinc α phase.

Tin tends to inhibit dealloying, especially in cast alloys. Alloys C46400 (naval brass) and C67500 (manganese bronze), which are α-β brasses containing about 1% Sn, are widely used for naval equipment and have reasonably good resistance to dezincification. Addition of a small amount of phosphorus, arsenic, or antimony to admiralty metal (an all-α 71Cu-28Zn-1Sn brass) inhibits dezincification. Inhibitors are not entirely effective in preventing dezincification of the α-β brasses, because they do not prevent dezincification of the β phase.

Where dezincification is a problem, red brass, commercial bronze, inhibited admiralty metal, and inhibited aluminum brass can be successfully used. In some cases, the economic penalty of avoiding dealloying by selecting a low-zinc alloy may be unacceptable. Low-zinc alloy tubing requires fittings that are available only as sand castings, but fittings for higher-zinc tube can be die cast or forged much more economically. Where selection of a low-zinc alloy is unacceptable, inhibited yellow brasses are generally preferred.

Dealloying has been observed in other alloys. Dealloying of aluminum occurs in some copper-aluminum alloys, particularly with those having more than 8% Al. It is especially severe in alloys with continuous γ phase and usually occurs as plug-type dealloying. Nickel additions exceeding 3.5% or heat treatment to produce an α + β microstructure prevents dealloying. Dealloying of nickel in C71500 is rare, having been observed at temperatures over 100 °C (212 °F), low flow conditions, and high local heat flux. Dealloying of tin in cast tin bronzes has been observed as a rare occurrence in hot brine or steam. Cathodic protection generally protects all but the two-phase copper-zinc alloys from dealloying. Additional information on this form of attack is available in the section "Dealloying Corrosion" of the article "Metallurgically Influenced Corrosion" in this Volume.

Corrosion Fatigue

The combined action of corrosion (usually pitting corrosion) and cyclic stress may result in corrosion fatigue cracking. Like ordinary fatigue cracks, corrosion fatigue cracks generally propagate at right angles to the maximum tensile stress in the affected region. However, cracks resulting from simultaneous fluctuating stress and corrosion propagate much more rapidly than cracks caused solely by fluctuating stress. Also, corrosion fatigue failure usually involves several parallel cracks, but it is rare for more than one crack to be found in a part that has failed by simple fatigue. The cracks shown in Fig. 1 are characteristic of service failures resulting from corrosion fatigue.

Ordinarily, corrosion fatigue can be readily identified by the presence of several cracks emanating from corrosion pits. Cracks not visible to the unaided eye or at low magnification can be made visible by deep etching or plastic deformation or can be detected by eddy-current inspection. Corrosion fatigue cracking is often transgranular, but there is evidence that certain environments induce intergranular cracking in copper metals.

In addition to effective resistance to corrosion, copper and copper alloys also resist corrosion fatigue in many applications involving repeated stress and corrosion. These applications include such parts as springs, switches, diaphragms, bellows, aircraft and automotive gasoline and oil lines, tubes for condensers and heat exchangers, and fourdrinier wire for the paper industry.

Copper alloys that are high in fatigue limit and resistance to corrosion in the service environment are more likely to have good resistance to corrosion fatigue. Alloys frequently used in applications involving both cyclic stress and corrosion include beryllium coppers, phosphor bronzes, aluminum bronzes, and copper nickels. More information on corrosion fatigue is available in the section "Corrosion Fatigue" of the article "Mechanically Assisted Degradation" in this Volume.

Stress-Corrosion Cracking

Stress-corrosion cracking and season cracking describe the same phenomenon—the apparently spontaneous cracking of stressed metal. Stress-corrosion cracking is often intergranular (Fig. 2), but transgranular cracking may occur in some alloys in certain environments. Stress-corrosion cracking occurs only if a susceptible alloy is subjected to the combined effects of sustained stress and certain chemical substances.

Mechanism. Copper alloys crack in a wide variety of electrolytes. In some cases, the crack surfaces have the distinctive brittle appearance that is associated with SCC. In other cases, the threshold stress for cracking may be close to that observed in air, and the fracture surfaces resemble those of samples fractured in air. It is also clear in many systems that cracking occurs at low threshold stresses only when certain environmental conditions exist. Variables that control this threshold stress in a specific environment include pH, potential of the metal, temperature, extent of cold work before the test, and minor alloying elements in the copper alloy.

The best nonquantitative interpretation of SCC is the following. Stress-corrosion cracking occurs in those environmental/metal systems in which the rate of corrosion is low; the corrosion that does occur proceeds in a highly localized manner. Intergranular attack, selective removal of an alloy component, pitting, attack at a metal/precipitate interface, or surface flaws, when they occur in the presence of a surface tensile stress, may lead to a surface defect at the base of which the stress intensity factor, K_I, exceeds the threshold stress intensity for SCC K_{Iscc}, for that specific environment/alloy system under the conditions selected for the test or encountered in service. Whether or not a crack propagates depends on the specimen geometry and how the magnitude of the stress field at the crack tip changes as the crack develops. The critical factor is how the metal reacts at the crack tip. If the metallurgical structure or the kinetics of chemical corrosion at the crack tip is such that a small radius of curvature (sharp crack tip) is maintained at the crack tip, the crack will continue to propagate because the local stress at the crack tip is high. High rates of corrosion at the crack tip, which lead to a large radius of curvature (blunt), will favor pitting rather than crack growth.

A sharp crack tip is favored by:

- Selective removal of one component of an alloy with the resulting development of local voids that provide a brittle crack path
- Brittle fracture of a corrosion product coating at the base of a crack that continually reforms
- Attack along the interface of two discrete phases

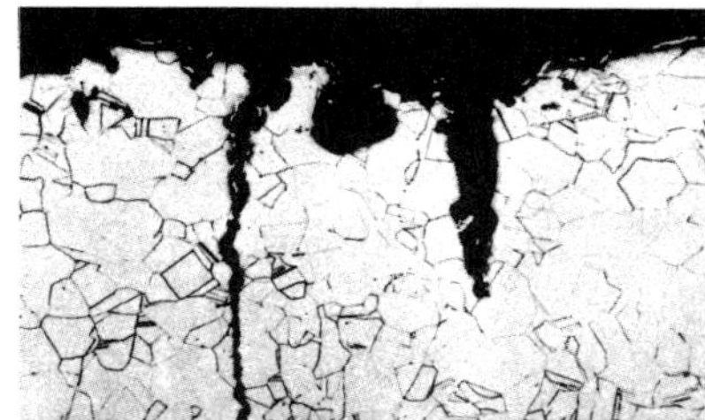

Fig. 1 Typical corrosion fatigue cracking of a copper alloy. Transgranular cracks originate at the base of corrosion pits on the roughened inner surface of a tube. Etched. About 150×

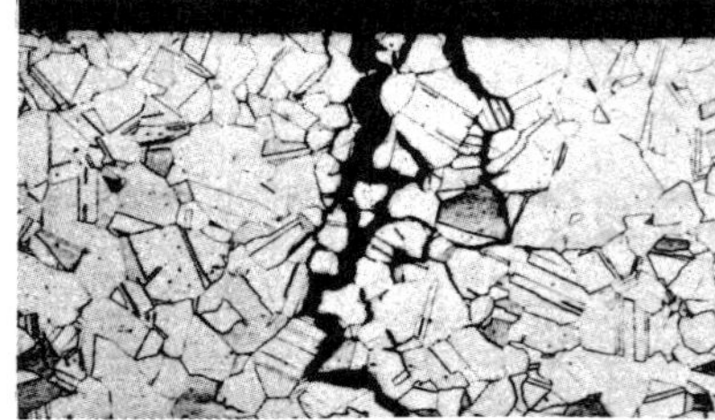

Fig. 2 Typical SCC in a copper alloy. Intergranular cracking in an etched specimen. About 60×

- Intergranular attack that does not spread laterally
- Surface energy considerations that encourage intrusion of the environment (a liquid metal in particular) into minute flaws

Since the discovery by E. Mattsson that a medium containing ammonium sulfate $[(NH_4)_2SO_4]$, NH_4OH, and copper sulfate $(CuSO_4)$ is an excellent one for studying the fundamentals of the SCC process caused by NH_3, many researchers have used this electrolyte, and the name Mattsson's solution has been given to this solution (Ref 1). Much of the knowledge of the specifics of SCC by NH_3 solutions has been obtained from brass exposed to this solution while under a tensile stress.

The chemistry and the electrochemistry of the brass-NH_3 system was recently reviewed and analyzed (Ref 2). Cupric (Cu^{2+}) ammonium complex was concluded to be necessary for the occurrence of SCC under open-circuit conditions in oxygenated NH_3 solutions. This complex becomes a component in the predominant cathodic reaction:

$$Cu(NH_3)_4^{2+} + e^- \rightarrow Cu(NH_3)_2^+ + 2NH_3 \quad \text{(Eq 3)}$$

Equation 3 permits cracking by cyclic rupture of a Cu_2O film generated at the crack tip (Ref 3) or by a mechanism involving dezincification (Ref 4). Cracking can also occur in deoxygenated solutions in the absence of significant concentrations of the Cu^{2+} ions provided the cuprous (Cu^+) complexes are available. It was suggested that the role of the Cu^+ complex is to provide a cathodic reaction, in this case allowing dezincification to occur. These findings are consistent with the recognition that SCC failures of brass are not limited to environments containing NH_3.

The most damaging evidence against the film rupture model is given in Ref 5. In this study, the tarnish film that formed on unstressed 70Cu-30Zn brass during exposure for 48 h to an NH_4OH-$(NH_4)_2SO_4$-$CuSO_4$ electrolyte at pH 7.2 was shown to fracture transgranularly when fractured in air. The reported film rupture mechanism predicts that these films should fracture intergranularly. The transgranular cracks do not propagate when a stressed specimen is immersed in the electrolyte; instead, very rapid intergranular SCC is observed. These facts are also difficult to reconcile with the repeated film rupture model.

It was first shown in 1972 that dezincification of 70Cu-30Zn brass occurs in the crack during SCC in an ammonium salt environment (Ref 4). More recently, mechanical strain was found to lead to dezincification of both 85Cu-15Zn and 70Cu-30Zn alloys in an NH_4OH-$(NH_4)_2SO_4$-$CuSO_4$ electrolyte (Ref 6). Unstressed samples of the same alloys did not show dezincification. Strain-induced dealloying was further shown to occur in both intergranular (copper-zinc) and transgranular (copper-zinc-nickel) (Ref 7). These observations indicated that stress corrosion of copper alloys is integrally related to strain-induced dealloying.

Conditions Leading to SCC. Ammonia and ammonium compounds are the corrosive substances most often associated with SCC of copper alloys. These compounds are sometimes present in the atmosphere; in other cases, they are in cleaning compounds or in chemicals used to treat boiler water. Both oxygen and moisture must be present for NH_3 to be corrosive to copper alloys; other compounds, such as CO_2, are thought to accelerate SCC in NH_3 atmospheres. Moisture films on metal surfaces will dissolve significant quantities of NH_3, even from atmospheres with low NH_3 concentrations.

A specific corrosive environment and sustained stress are the primary causes of SCC; microstructure and alloy composition may affect the rate of crack propagation in susceptible alloys. Microstructure and composition can be most effectively controlled by selecting the correct combination of alloy, forming process, thermal treatment, and metal-finishing process. Although test results may indicate that a finished part is not susceptible to SCC, such an indication does not ensure complete freedom from cracking, particularly where service stresses are high.

Applied and residual stresses can both lead to failure by SCC. Susceptibility is largely a function of stress magnitude. Stresses near the yield strength are usually required, but parts have failed under much lower stresses. In general, the higher the stress, the weaker the corroding medium must be to cause SCC. The reverse is also true: the stronger the corroding medium, the lower the required stress.

Sources of Stress. Applied stresses result from ordinary service loading or from fabricating techniques, such as riveting, bolting, shrink fitting, brazing, and welding. Residual stresses are of two types: differential-strain stresses, which result from nonuniform plastic strain during cold forming, and differential-thermal-contraction stresses, which result from nonuniform heating and/or cooling.

Residual stresses induced by nonuniform straining are primarily influenced by the method of fabrication. In some fabricating processes, it is possible to cold work a metal extensively and yet produce only a low level of residual stress. For example, residual stress in a drawn tube is influenced by die angle and amount of reduction. Wide-angle dies (about 32°) produce higher residual stresses than narrow-angle dies (about 8°). Light reductions yield high residual stresses because only the surface of the alloy is stressed; heavy reductions yield low residual stresses because the region of cold working extends deeper into the metal. Most drawing operations can be planned so that residual stresses are low and susceptibility to SCC is negligible.

Residual stresses resulting from upsetting, stretching, or spinning are more difficult to evaluate and to control by varying tooling and process conditions. For these operations, SCC can be prevented more effectively by selecting a resistant alloy or by treating the metal after fabrication.

Alloy Composition. Brasses containing less than 15% Zn are highly resistant to SCC. Phosphorus-deoxidized copper and tough pitch copper rarely exhibit SCC, even under severe conditions. On the other hand, brasses containing 20 to 40% Zn are highly susceptible. Susceptibility increases only slightly as zinc content is increased from 20 to 40%.

There is no indication that the other elements commonly added to brasses increase the probability of SCC. Phosphorus, arsenic, magnesium, tellurium, tin, beryllium, and manganese are thought to decrease susceptibility under some conditions. Addition of 1.5% Si is known to decrease the probability of cracking.

Altering the microstructure cannot make a susceptible alloy totally resistant to SCC. However, the rapidity with which susceptible alloys crack appears to be affected by grain size and structure. All other factors being equal, the rate of cracking increases with grain size. The effects of structure on SCC are not sharply defined, primarily because they are interrelated with effects of both composition and stress.

Control Measures. Stress-corrosion cracking can be controlled, and sometimes prevented, by selecting copper alloys that have high resistance to cracking (notably those with less than 15% Zn); by reducing residual stress to a safe level by thermal stress relief, which can usually be applied without significantly decreasing strength; or by altering the environment, such as by changing the predominant chemical species present or introducing a corrosion inhibitor.

Residual and assembly stresses can be eliminated by recrystallization annealing after forming or assembly. Recrystallization annealing cannot be used when the integrity of the structure depends on the higher strength of strain-hardened metal, which always contains a certain amount of residual stress. Thermal stress relief (sometimes called relief annealing) can be specified when the higher strength of a cold-worked temper must be retained. Thermal stress relief consists of heating the part for a relatively short time at low temperature. Specific times and temperatures depend on alloy composition, severity of deformation, prevailing stresses, and the size of the load being heated. Usually, time is from 30 min to 1 h and temperature is from 150 to 425 °C (300 to 795 °F). Table 4 lists typical stress-relieving times and temperatures for some of the more common copper alloys.

The exact thermal treatment should be established by examining specific parts for residual stress. If such examination indicates that a thermal treatment is insufficient, temperature and/or time should be adjusted until satisfactory results are obtained. Parts in the center of a furnace load may not reach the desired temperature as soon as parts around the periphery. Therefore, it may be necessary to compensate for furnace loading when set-

Table 4 Typical stress-relieving parameters for some common copper alloys

Common name	UNS number	Temperature °C	Temperature °F	Time, h
Commercial bronze	C22000	205	400	1
Cartridge brass	C26000	260	500	1
Muntz metal	C28000	190	375	½
Admiralty metal	C44300, C44400, C44500	300	575	1
Phosphor bronze, 5 or 10%	C51000, C52400	190	375	1
Silicon bronze	C65500	370	700	1
Aluminum bronze	C61300, C61400	400	750	1
Copper nickel, 30%	C71500	425	800	1

ting process controls or to limit the number of parts that can be stress relieved together.

Mechanical methods, such as stretching, flexing, bending, straightening between rollers, peening, and shot blasting, can also be used to reduce residual stresses to a safe level. These methods depend on plastic deformation to decrease dangerous tensile stresses or to convert them to less objectionable compressive stresses. Additional information on SCC is available in the section "Stress-Corrosion Cracking" of the article "Environmentally Induced Cracking" in this Volume.

Corrosion of Copper Alloys in Specific Environments

Selection of a suitably resistant material requires consideration of the many factors that influence corrosion. Operating records are the most reliable guidelines as long as the data are accurately interpreted. Some of the information in this article has been collected over a period of 20 years or more. Results of short-term laboratory and field testing are also described, but these data may not be as reliable for solving certain problems. Laboratory corrosion tests often do not duplicate such operating factors as stress, velocity, galvanic coupling, concentration cells, initial surface conditions, and contamination of the surrounding medium. If damage occurs by pitting, intergranular corrosion, or dealloying (as in dezincification) or if a thick adherent scale forms, corrosion rates calculated from a change in weight may be misleading. For these forms of corrosion, estimates of reduction in mechanical strength are often more meaningful. Corrosion fatigue and SCC are also potential sources of failure that cannot be predicted from routine measurements of weight loss or dimensional change.

Over the years, experience has been the best criterion for selecting the most suitable alloy for a given environment. The CDA has compiled much field experience in the form of the ratings shown in Table 5. Similar data for cast alloys are given in Table 6. These tables should be used only as a guide; small changes in the environmental conditions sometimes degrade the performance of a given alloy from "suitable" to "not suitable."

Whenever there is a lack of operating experience, whenever reported test conditions do not closely match the conditions for which alloy selection is being made, and whenever there is doubt as to the applicability of published data, it is always best to conduct an independent test. Field tests are the most reliable. Laboratory tests can be equally valuable, but only if operating conditions are precisely defined and then accurately simulated in the laboratory. Long-term tests are generally preferred because the reaction that dominates the initial stages of corrosion may differ significantly from the reaction that dominates later on. If short-term tests must be used as the basis for alloy selection, the test program should be supplemented with field tests so that the laboratory results can be reevaluated in light of true operating experience.

Erroneous conclusions based on laboratory results can also be reached by measuring corrosion damage inaccurately, especially when corrosion is slight. It is common practice to express test results in terms of penetration or average reduction in metal thickness, even when corrosion was actually measured by weight loss. Weight loss or average-penetration data are valid only when corrosion is uniform. When corrosion occurs predominantly by pitting or some other localized form or when corrosion is intergranular or involves the formation of a thick, adherent scale, direct measurement of the extent of corrosion provides the most reliable information. A common technique is to measure the maximum depth of penetration observed on a metallographic cross section through the region of interest. Statistical averaging of repeated measurements on one or more specimens may or may not be warranted. Despite the deficiencies in laboratory testing, information gained in this manner serves as a useful starting point for alloy selection. Operating experience may later indicate the need for a more discriminating selection.

Atmospheric Exposure

Comprehensive tests conducted over a 20-year period under the supervision of the American Society for Testing and Materials (ASTM), as well as many service records, have confirmed the suitability of copper and copper alloys for atmospheric exposure (Table 7). Copper and copper alloys resist corrosion by industrial, marine, and rural atmospheres except atmospheres containing NH_3 or certain other agents where SCC has been observed in high-zinc alloys (>20% Zn). The copper metals most widely used in atmospheric exposure are C11000, C22000, C23000, C38500, and C75200. Alloy C11000 is an effective material for roofing, flashings, gutters, and downspouts.

The colors of different copper alloys are often important in architectural applications, and color may be the primary criterion for selecting a specific alloy. After surface preparation, such as sanding or polishing, different copper alloys vary in color from silver to yellow to gold to reddish shades. Different alloys having the same initial color may show differences in color after weathering under similar conditions. Therefore, alloys having the same or nearly the same composition are usually used together for consistency of appearance in a specific structure.

Copper alloys are often specified for marine atmosphere exposures because of the attractive and protective patina they form during the exposure. In marine atmospheric exposures, this patina consists of a film of basic copper chloride or carbonate, sometimes with an inner layer of Cu_2O. The severity of the corrosion attack in marine atmospheres is somewhat less than that in industrial atmospheres but greater than that in rural atmospheres. However, these rates decrease with time.

Individual differences in corrosion rates do exist between alloys, but these differences are frequently less than the differences caused by environmental factors. Thus, it becomes possible to classify the corrosion behavior of copper alloys in a marine atmosphere into two general categories: those alloys that corrode at a moderate rate and include high-copper alloys, silicon bronze, and tin bronze and those alloys that corrode at a slower rate and include brass, aluminum bronze, nickel silver, and copper nickel. The average metal loss, d, of the former group can be approximated by $d = 0.1\ t^{2/3}$; the latter group can be approximated by $d = 0.1\ t^{1/3}$. In both equations, t is exposure time. These relationships are shown as solid lines in Fig. 3.

Environmental factors can cause this median thickness loss to vary by as much as 50% or more in a few extreme cases. Figure 3 shows the extent of this variation as a pair of dashed lines forming an envelope around the median. Those environmental factors that tend to accelerate metal loss include high humidity, high temperatures (either ambient or due to solar radiation), proximity to the ocean, long times of wetness, and the presence of pollutants in the atmosphere. The converse of these conditions would tend to retard metal loss.

Metallurgical factors can also affect metal loss. Within a given alloy family, those with a higher alloy content tend to corrode at a lower rate. Surface finish also plays a role in that a highly polished metal will corrode slower than one with a rougher surface. Finally, design details can affect corrosion behavior. For example, designs that allow the collection and stagnation of rainwater will often exhibit wastage rates in the puddle areas that are more typical of those encountered in seawater immersions.

Certain copper alloys are susceptible to various types of localized corrosion that can greatly affect their utility in a marine atmosphere. Brasses and nickel silvers containing more than 15% Zn can suffer from dealloying. The extent of this attack is greater on alloys that contain higher proportions of zinc. In addition, these same alloys are subject to SCC in the presence of small quantities of NH_3 or other gaseous pollutants. Inhibited grades of these alloys are available that resist dealloying but are susceptible to SCC.

Alloys containing large amounts of manganese tend to be somewhat prone to pitting in marine

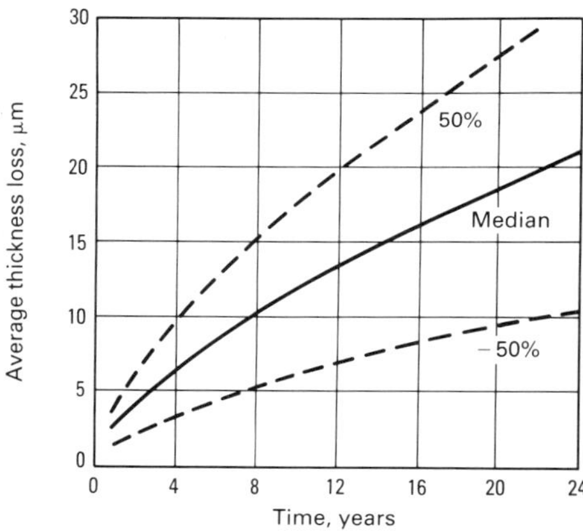

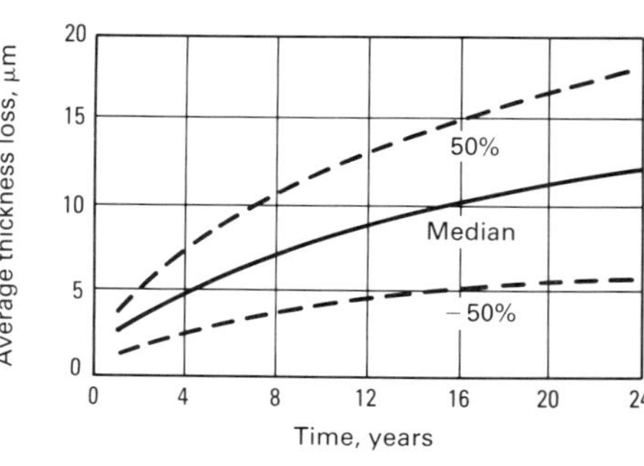

Fig. 3 Typical corrosion rates of representative copper alloys in a marine atmosphere. (a) Average data for copper, silicon bronze, and phosphor bronze. (b) Average data for brass, aluminum bronze, nickel silver, and copper-nickel

Table 5 Corrosion ratings of wrought copper alloys in various corrosive media

This table is intended to serve only as a general guide to the behavior of copper and copper alloys in corrosive environments. It is impossible to cover in a simple tabulation the performance of a material for all possible variations of temperature, concentration, velocity, impurity content, degree of aeration, and stress. The ratings are based on general performance; they should be used with caution, and then only for the purpose of screening candidate alloys.

The letters E, G, F, and P have the following significance:

E, excellent: resists corrosion under almost all conditions of service

G, good: some corrosion will take place, but satisfactory service can be expected under all but the most severe conditions.

F, fair: corrosion rates are higher than for the G classification, but the metal can be used if needed for a property other than corrosion resistance and if either the amount of corrosion does not cause excessive maintenance expense or the effects of corrosion can be lessened, such as by use of coatings or inhibitors.

P, poor: corrosion rates are high, and service is generally unsatisfactory.

Corrosive medium	Coppers	Low-zinc brasses	High-zinc brasses	Special brasses	Phosphor bronzes	Aluminum bronzes	Silicon bronzes	Copper nickels	Nickel silvers
Acetate solvents	E	E	G	E	E	E	E	E	E
Acetic acid(a)	E	E	P	P	E	E	E	E	G
Acetone	E	E	E	E	E	E	E	E	E
Acetylene(b)	P	P	(b)	P	P	P	P	P	P
Alcohols(a)	E	E	E	E	E	E	E	E	E
Aldehydes	E	E	F	F	E	E	E	E	E
Alkylamines	G	G	G	G	G	G	G	G	G
Alumina	E	E	E	E	E	E	E	E	E
Aluminum chloride	G	G	P	P	G	G	G	G	G
Aluminum hydroxide	E	E	E	E	E	E	E	E	E
Aluminum sulfate and alum	G	G	P	G	G	G	G	E	G
Ammonia, dry	E	E	E	E	E	E	E	E	E
Ammonia, moist(c)	P	P	P	P	P	P	P	F	P
Ammonium chloride(c)	P	P	P	P	P	P	P	F	P
Ammonium hydroxide(c)	P	P	P	P	P	P	P	F	P
Ammonium nitrate(c)	P	P	P	P	P	P	P	F	P
Ammonium sulfate(c)	F	F	P	P	F	F	F	G	F
Aniline and aniline dyes	F	F	F	F	F	F	F	F	F
Asphalt	E	E	E	E	E	E	E	E	E
Atmosphere:									
Industrial(c)	E	E	E	E	E	E	E	E	E
Marine	E	E	E	E	E	E	E	E	E
Rural	E	E	E	E	E	E	E	E	E
Barium carbonate	E	E	E	E	E	E	E	E	E
Barium chloride	G	G	F	F	G	G	G	G	G
Barium hydroxide	E	E	G	E	E	E	E	E	E
Barium sulfate	E	E	E	E	E	E	E	E	E
Beer(a)	E	E	G	E	E	E	E	E	E
Beet-sugar syrup(a)	E	E	G	E	E	E	E	E	E
Benzene, benzine, benzol	E	E	E	E	E	E	E	E	E
Benzoic acid	E	E	E	E	E	E	E	E	E
Black liquor, sulfate process	P	P	P	P	P	P	P	G	P
Bleaching powder (wet)	G	G	P	G	G	G	G	G	G
Borax	E	E	E	E	E	E	E	E	E
Bordeaux mixture	E	E	G	E	E	E	E	E	E
Boric acid	E	E	G	E	E	E	E	E	E
Brines	G	G	P	G	G	G	G	E	E
Bromine, dry	E	E	E	E	E	E	E	E	E
Bromine, moist	G	G	P	F	G	G	G	G	G
Butane(d)	E	E	E	E	E	E	E	E	E
Calcium bisulfate	G	G	P	G	G	G	G	G	G
Calcium chloride	G	G	F	G	G	G	G	G	G
Calcium hydroxide	E	E	G	E	E	E	E	E	E
Calcium hypochlorite	G	G	P	G	G	G	G	G	G
Cane-sugar syrup(a)	E	E	E	E	E	E	E	E	E
Carbolic acid (phenol)	F	G	P	G	G	G	G	G	G
Carbonated beverages(a)(e)	E	E	E	E	E	E	E	E	E
Carbon dioxide, dry	E	E	E	E	E	E	E	E	E
Carbon dioxide, moist(a)(e)	E	E	E	E	E	E	E	E	E
Carbon tetrachloride (dry)	E	E	E	E	E	E	E	E	E
Carbon tetrachloride (moist)	G	G	F	G	E	E	E	E	E
Castor oil	E	E	E	E	E	E	E	E	E
Chlorine, dry(f)	E	E	E	E	E	E	E	E	E
Chlorine, moist	F	F	P	F	F	F	F	G	F
Chloracetic acid	G	F	P	F	G	G	G	G	G
Chloroform, dry	E	E	E	E	E	E	E	E	E
Chromic acid	P	P	P	P	P	P	P	P	P
Citric acid(a)	E	E	F	E	E	E	E	E	E
Copper chloride	F	F	P	F	F	F	F	F	F
Copper nitrate	F	F	P	F	F	F	F	F	F
Copper sulfate	G	G	P	G	G	G	G	E	G
Corn oil(a)	E	E	G	E	E	E	E	E	E
Cottonseed oil(a)	E	E	G	E	E	E	E	E	E
Creosote	E	E	G	E	E	E	E	E	E
Dowtherm "A"	E	E	E	E	E	E	E	E	E
Ethanol amine	G	G	G	G	G	G	G	G	G
Ethers	E	E	E	E	E	E	E	E	E
Ethyl acetate (esters)	E	E	G	E	E	E	E	E	E
Ethylene glycol	E	E	G	E	E	E	E	E	E
Ferric chloride	P	P	P	P	P	P	P	P	P
Ferric sulfate	P	P	P	P	P	P	P	P	P
Ferrous chloride	G	G	P	G	G	G	G	G	G
Ferrous sulfate	G	G	P	G	G	G	G	G	G
Formaldehyde (aldehydes)	E	E	G	E	E	E	E	E	E
Formic acid	G	G	P	F	G	G	G	G	G
Freon, dry	E	E	E	E	E	E	E	E	E
Freon, moist	E	E	E	E	E	E	E	E	E
Fuel oil, light	E	E	E	E	E	E	E	E	E
Fuel oil, heavy	E	E	G	E	E	E	E	E	E
Furfural	E	E	F	E	E	E	E	E	E
Gasoline	E	E	E	E	E	E	E	E	E
Gelatin(a)	E	E	E	E	E	E	E	E	E
Glucose(a)	E	E	E	E	E	E	E	E	E
Glue	E	E	G	E	E	E	E	E	E
Glycerin	E	E	G	E	E	E	E	E	E
Hydrobromic acid	F	F	P	F	F	F	F	F	F
Hydrocarbons	E	E	E	E	E	E	E	E	E
Hydrochloric acid (muriatic)	F	F	P	F	F	F	F	F	F
Hydrocyanic acid, dry	E	E	E	E	E	E	E	E	E
Hydrocyanic acid, moist	P	P	P	P	P	P	P	P	P
Hydrofluoric acid, anhydrous	G	G	P	G	G	G	G	G	G
Hydrofluoric acid, hydrated	F	F	P	F	F	F	F	F	F
Hydrofluosilicic acid	G	G	P	G	G	G	G	G	G
Hydrogen(d)	E	E	E	E	E	E	E	E	E
Hydrogen peroxide up to 10%	G	G	F	G	G	G	G	G	G

(continued)

(a) Copper and copper alloys are resistant to corrosion by most food products. Traces of copper may be dissolved and affect taste or color of the products. In such cases, copper alloys are often tin coated. (b) Acetylene forms an explosive compound with copper when moisture or certain impurities are present and the gas is under pressure. Alloys containing less than 65% Cu are satisfactory; when the gas is not under pressure, other copper alloys are satisfactory. (c) Precautions should be taken to avoid SCC. (d) At elevated temperatures, hydrogen will react with tough pitch copper, causing failure by embrittlement. (e) Where air is present, corrosion rate may be increased. (f) Below 150 °C (300 °F), corrosion rate is very low; above this temperature, corrosion is appreciable and increases rapidly with temperature. (g) Aeration and elevated temperature may increase corrosion rate substantially. (h) Excessive oxidation may begin above 120 °C (250 °F). If moisture is present, oxidation may begin at lower temperatures. (j) Use of high-zinc brasses should be avoided in acids because of the likelihood of rapid corrosion by dezincification. Copper, low-zinc brasses, phosphor bronzes, silicon bronzes, aluminum bronzes, and copper nickels offer good resistance to corrosion by hot and cold dilute H_2SO_4 and to corrosion by cold concentrated H_2SO_4. Intermediate concentrations of H_2SO_4 are sometimes more corrosive to copper alloys than either concentrated or dilute acid. Concentrated H_2SO_4 may be corrosive at elevated temperatures due to breakdown of acid and formation of metallic sulfides and sulfur dioxide, which cause localized pitting. Tests indicate that copper alloys may undergo pitting in 90 to 95% H_2SO_4 at about 50 °C (122 °F), in 80% acid at about 70 °C (160 °F), and in 60% acid at about 100 °C (212 °F). (k) Wetting agents may increase corrosion rates of copper and copper alloys slightly to substantially when carbon dioxide or oxygen is present by preventing formation of a film on the metal surface and by combining (in some instances) with the dissolved copper to produce a green, insoluble compound.

Table 5 (continued)

Corrosive medium	Coppers	Low-zinc brasses	High-zinc brasses	Special brasses	Phosphor bronzes	Aluminum bronzes	Silicon bronzes	Copper nickels	Nickel silvers
Hydrogen peroxide over 10%	P	P	P	P	P	P	P	P	P
Hydrogen sulfide, dry	E	E	E	E	E	E	E	E	E
Hydrogen sulfide, moist	P	P	F	F	P	P	P	F	F
Kerosine	E	E	E	E	E	E	E	E	E
Ketones	E	E	E	E	E	E	E	E	E
Lacquers	E	E	E	E	E	E	E	E	E
Lacquer thinners (solvents)	E	E	E	E	E	E	E	E	E
Lactic acid(a)	E	E	F	E	E	E	E	E	E
Lime	E	E	E	E	E	E	E	E	E
Lime sulfur	P	P	F	F	P	P	P	F	F
Linseed oil	G	G	G	G	G	G	G	G	G
Lithium compounds	G	G	P	F	G	G	G	E	E
Magnesium chloride	G	G	F	F	G	G	G	G	G
Magnesium hydroxide	E	E	G	E	E	E	E	E	E
Magnesium sulfate	E	E	G	E	E	E	E	E	E
Mercury or mercury salts	P	P	P	P	P	P	P	P	P
Milk(a)	E	E	G	E	E	E	E	E	E
Molasses	E	E	G	E	E	E	E	E	E
Natural gas(d)	E	E	E	E	E	E	E	E	E
Nickel chloride	F	F	P	F	F	F	F	F	F
Nickel sulfate	F	F	P	F	F	F	F	F	F
Nitric acid	P	P	P	P	P	P	P	P	P
Oleic acid	G	G	F	G	G	G	G	G	G
Oxalic acid(g)	E	E	P	P	E	E	E	E	E
Oxygen(h)	E	E	E	E	E	E	E	E	E
Palmitic acid	G	G	F	G	G	G	G	G	G
Paraffin	E	E	E	E	E	E	E	E	E
Phosphoric acid	G	G	P	F	G	G	G	G	G
Picric acid	P	P	P	P	P	P	P	P	P
Potassium carbonate	E	G	E	E	E	E	E	E	E
Potassium chloride	G	G	P	F	G	G	G	E	E
Potassium cyanide	P	P	P	P	P	P	P	P	P
Potassium dichromate (acid)	P	P	P	P	P	P	P	P	P
Potassium hydroxide	G	G	F	G	G	G	G	E	E
Potassium sulfate	E	E	G	E	E	E	E	E	E
Propane(d)	E	E	E	E	E	E	E	E	E
Rosin	E	E	E	E	E	E	E	E	E
Seawater	G	G	F	E	G	E	G	E	E
Sewage	E	E	F	E	E	E	E	E	E
Silver salts	P	P	P	P	P	P	P	P	P
Soap solution	E	E	E	E	E	E	E	E	E
Sodium bicarbonate	E	E	G	E	E	E	E	E	E
Sodium bisulfate	G	G	F	G	G	G	G	E	E
Sodium carbonate	E	E	G	E	E	E	E	E	E
Sodium chloride	G	G	P	F	G	G	G	E	E
Sodium chromate	E	E	E	E	E	E	E	E	E
Sodium cyanide	P	P	P	P	P	P	P	P	P
Sodium dichromate (acid)	P	P	P	P	P	P	P	P	P
Sodium hydroxide	G	G	F	G	G	G	G	E	E
Sodium hypochlorite	G	G	P	G	G	G	G	G	G
Sodium nitrate	G	G	P	F	G	G	G	E	E
Sodium peroxide	F	F	P	F	F	F	F	G	G
Sodium phosphate	E	E	G	E	E	E	E	E	E
Sodium silicate	E	E	G	E	E	E	E	E	E
Sodium sulfate	E	E	G	E	E	E	E	E	E
Sodium sulfide	P	P	F	F	P	P	P	F	F
Sodium thiosulfate	P	P	F	F	P	P	P	F	F
Steam	E	E	F	E	E	E	F	E	E
Stearic acid	E	E	F	E	E	E	E	E	E
Sugar solutions	E	E	G	E	E	E	E	E	E
Sulfur, solid	G	G	E	G	G	G	G	E	G
Sulfur, molten	P	P	P	P	P	P	P	P	P
Sulfur chloride (dry)	E	E	E	E	E	E	E	E	E
Sulfur chloride (moist)	P	P	P	P	P	P	P	P	P
Sulfur dioxide (dry)	E	E	E	E	E	E	E	E	E
Sulfur dioxide (moist)	G	G	P	G	G	G	G	F	F
Sulfur trioxide (dry)	E	E	E	E	E	E	E	E	E
Sulfuric acid 80–95%(j)	G	G	P	F	G	G	G	G	G
Sulfuric acid 40–80%(j)	F	F	F	P	F	F	F	F	F
Sulfuric acid 40%(j)	G	G	P	F	G	G	G	G	G
Sulfurous acid	G	G	P	G	G	G	G	F	F
Tannic acid	E	E	E	E	E	E	E	E	E
Tartaric acid(a)	E	E	G	E	E	E	E	E	E
Toluene	E	E	E	E	E	E	E	E	E
Trichloracetic acid	G	G	P	F	G	G	G	G	G
Trichlorethylene (dry)	E	E	E	E	E	E	E	E	E
Trichlorethylene (moist)	G	G	F	G	E	E	E	E	E
Turpentine	E	E	E	E	E	E	E	E	E
Varnish	E	E	E	E	E	E	E	E	E
Vinegar(a)	E	E	P	F	E	E	E	E	G
Water, acidic mine	F	F	P	F	G	F	F	P	F
Water, potable	E	E	G	E	E	E	E	E	E
Water, condensate(c)	E	E	E	E	E	E	E	E	E
Wetting agents(k)	E	E	E	E	E	E	E	E	E
Whiskey(a)	E	E	E	E	E	E	E	E	E
White water	G	G	G	E	E	E	E	E	E
Zinc chloride	G	G	P	G	G	G	G	G	G
Zinc sulfate	E	E	P	E	E	E	E	E	E

(a) Copper and copper alloys are resistant to corrosion by most food products. Traces of copper may be dissolved and affect taste or color of the products. In such cases, copper alloys are often tin coated. (b) Acetylene forms an explosive compound with copper when moisture or certain impurities are present and the gas is under pressure. Alloys containing less than 65% Cu are satisfactory; when the gas is not under pressure, other copper alloys are satisfactory. (c) Precautions should be taken to avoid SCC. (d) At elevated temperatures, hydrogen will react with tough pitch copper, causing failure by embrittlement. (e) Where air is present, corrosion rate may be increased. (f) Below 150 °C (300 °F), corrosion rate is very low; above this temperature, corrosion is appreciable and increases rapidly with temperature. (g) Aeration and elevated temperature may increase corrosion rate substantially. (h) Excessive oxidation may begin above 120 °C (250 °F). If moisture is present, oxidation may begin at lower temperatures. (j) Use of high-zinc brasses should be avoided in acids because of the likelihood of rapid corrosion by dezincification. Copper, low-zinc brasses, phosphor bronzes, silicon bronzes, aluminum bronzes, and copper nickels offer good resistance to corrosion by hot and cold dilute H_2SO_4 and to corrosion by cold concentrated H_2SO_4. Intermediate concentrations of H_2SO_4 are sometimes more corrosive to copper alloys than either concentrated or dilute acid. Concentrated H_2SO_4 may be corrosive at elevated temperatures due to breakdown of acid and formation of metallic sulfides and sulfur dioxide, which cause localized pitting. Tests indicate that copper alloys may undergo pitting in 90 to 95% H_2SO_4 at about 50 °C (122 °F), in 80% acid at about 70 °C (160 °F), and in 60% acid at about 100 °C (212 °F). (k) Wetting agents may increase corrosion rates of copper and copper alloys slightly to substantially when carbon dioxide or oxygen is present by preventing formation of a film on the metal surface and by combining (in some instances) with the dissolved copper to produce a green, insoluble compound.

atmospheres, as are the cobalt-containing beryllium-coppers. A tendency toward intergranular corrosion has been observed in silicon bronzes and aluminum brass, but its occurrence is somewhat sporadic.

On the whole, however, even under somewhat adverse conditions, the average thickness losses for copper alloys in a marine atmosphere tend to be very slight, typically under 50 μm (Fig. 3). Thus, copper alloys can be safely specified for applications requiring long-term durability in a marine atmosphere. Design considerations for the atmospheric use of copper alloys include allowance for free drainage of structures, the possibility of staining from runoff water, and the use of smooth or polished surfaces.

Soils and Groundwater

Copper, zinc, lead, and iron are the metals most commonly used in underground construction. Data compiled by the National Bureau of Standards (NBS) compare the behavior of these materials in soils of the following four types: well-aerated acid soils low in soluble salts (Cecil clay loam), poorly aerated soils (Lake Charles clay), alkaline soils high in soluble salts (Docas clay), and soils high in sulfides (Rifle peat).

Corrosion data as a function of time for copper, iron, lead, and zinc exposed to these four types of soil are given in Fig. 4. Copper exhibits high

Table 6 Corrosion ratings of cast copper alloys in various media

The letters A, B, and C have the following significance: A, recommended; B, acceptable; C, not recommended

Corrosive medium	Copper	Tin bronze	Leaded tin bronze	High-leaded tin bronze	Leaded red brass	Leaded semi-red brass	Leaded yellow brass	Leaded high-strength yellow brass	High-strength yellow brass	Aluminum bronze	Leaded nickel brass	Leaded nickel bronze	Silicon bronze	Silicon brass
Acetate solvents	B	A	A	A	A	A	B	A	A	A	A	A	A	B
Acetic acid														
20%	A	C	B	C	B	C	C	C	C	A	C	A	A	B
50%	A	C	B	C	B	C	C	C	C	A	C	B	A	B
Glacial	A	A	A	C	A	C	C	C	C	A	B	B	A	A
Acetone	A	A	A	A	A	A	A	A	A	A	A	A	A	A
Acetylene(a)	C	C	C	C	C	C	C	C	C	C	C	C	C	C
Alcohols(b)	A	A	A	A	A	A	A	A	A	A	A	A	A	A
Aluminum chloride	C	C	C	C	C	C	C	C	C	B	C	C	C	C
Aluminum sulfate	B	B	B	B	B	C	C	C	C	A	C	C	A	A
Ammonia, moist gas	C	C	C	C	C	C	C	C	C	C	C	C	C	C
Ammonia, moisture-free	A	A	A	A	A	A	A	A	A	A	A	A	A	A
Ammonium chloride	C	C	C	C	C	C	C	C	C	C	C	C	C	C
Ammonium hydroxide	C	C	C	C	C	C	C	C	C	C	C	C	C	C
Ammonium nitrate	C	C	C	C	C	C	C	C	C	C	C	C	C	C
Ammonium sulfate	B	B	B	B	B	C	C	C	C	A	C	C	A	A
Aniline and aniline dyes	C	C	C	C	C	C	C	C	C	B	C	C	C	C
Asphalt	A	A	A	A	A	A	A	A	A	A	A	A	A	A
Barium chloride	A	A	A	A	A	C	C	C	C	A	A	A	A	C
Barium sulfide	C	C	C	C	C	C	C	C	B	C	C	C	C	C
Beer(b)	A	A	B	B	B	C	C	C	A	A	C	A	A	B
Beet-sugar syrup	A	A	B	B	B	A	A	A	B	A	A	A	B	B
Benzine	A	A	A	A	A	A	A	A	A	A	A	A	A	A
Benzol	A	A	A	A	A	A	A	A	A	A	A	A	A	A
Boric acid	A	A	A	A	A	A	A	B	A	A	A	A	A	A
Butane	A	A	A	A	A	A	A	A	A	A	A	A	A	A
Calcium bisulfite	A	A	B	B	B	C	C	C	C	A	B	A	A	B
Calcium chloride (acid)	B	B	B	B	B	B	C	C	C	A	C	C	A	C
Calcium chloride (alkaline)	C	C	C	C	C	C	C	C	C	A	C	A	C	B
Calcium hydroxide	C	C	C	C	C	C	C	C	C	B	C	C	C	C
Calcium hypochlorite	C	C	B	B	B	C	C	C	C	B	C	C	C	C
Cane-sugar syrups	A	A	B	A	B	A	A	A	A	A	A	A	A	B
Carbonated beverages(b)	A	C	C	C	C	C	C	C	C	A	C	C	A	C
Carbon dioxide, dry	A	A	A	A	A	A	A	A	A	A	A	A	A	A
Carbon dioxide, moist(b)	B	B	B	C	B	C	C	C	C	A	C	A	A	B
Carbon tetrachloride, dry	A	A	A	A	A	A	A	A	A	A	A	A	A	A
Carbon tetrachloride, moist	B	B	B	B	B	B	B	B	B	B	B	A	A	A
Chlorine, dry	A	A	A	A	A	A	A	A	A	A	A	A	A	A
Chlorine, moist	C	C	B	B	B	C	C	C	C	C	C	C	C	C
Chromic acid	C	C	C	C	C	C	C	C	C	C	C	C	C	C
Citric acid	A	A	A	A	A	A	A	A	A	A	A	A	A	A
Copper sulfate	B	A	A	A	A	C	C	C	C	B	B	B	A	A
Cottonseed oil(b)	A	A	A	A	A	A	A	A	A	A	A	A	A	A
Creosote	B	B	B	B	B	C	C	C	C	A	B	B	B	B
Ethers	A	A	A	A	A	A	A	A	A	A	A	A	A	A
Ethylene glycol	A	A	A	A	A	A	A	A	A	A	A	A	A	A
Ferric chloride, sulfate	C	C	C	C	C	C	C	C	C	C	C	C	C	C
Ferrous chloride, sulfate	C	C	C	C	C	C	C	C	C	C	C	C	C	C
Formaldehyde	A	A	A	A	A	A	A	A	A	A	A	A	A	A
Formic acid	A	A	A	A	A	B	B	B	B	A	B	B	B	C
Freon	A	A	A	A	A	A	A	A	A	A	A	A	A	B
Fuel oil	A	A	A	A	A	A	A	A	A	A	A	A	A	A
Furfural	A	A	A	A	A	A	A	A	A	A	A	A	A	A
Gasoline	A	A	A	A	A	A	A	A	A	A	A	A	A	A
Gelatin(b)	A	A	A	A	A	A	A	A	A	A	A	A	A	A
Glucose	A	A	A	A	A	A	A	A	A	A	A	A	A	A
Glue	A	A	A	A	A	A	A	A	A	A	A	A	A	A
Glycerin	A	A	A	A	A	A	A	A	A	A	A	A	A	A
Hydrochloric or muriatic acid	C	C	C	C	C	C	C	C	C	B	C	C	C	C
Hydrofluoric acid	B	B	B	B	B	B	B	B	B	A	B	B	B	B
Hydrofluosilicic acid	B	B	B	B	B	C	C	C	C	B	C	C	B	C
Hydrogen	A	A	A	A	A	A	A	A	A	A	A	A	A	A
Hydrogen peroxide	C	C	C	C	C	C	C	C	C	C	C	C	C	C
Hydrogen sulfide, dry	C	C	C	C	C	C	C	C	C	B	C	C	B	C
Hydrogen sulfide, moist	C	C	C	C	C	C	C	C	C	B	C	C	C	C

(continued)

(a) Acetylene forms an explosive compound with copper when moist or when certain impurities are present and the gas is under pressure. Alloys containing less than 65% Cu are satisfactory for this use. When gas is not under pressure, other copper alloys are satisfactory. (b) Copper and copper alloys resist corrosion by most food products. Traces of copper may be dissolved and affect taste or color. In such cases, copper metals are often tin coated.

resistance to corrosion by these soils, which are representative of most soils found in the United States. Where local soil conditions are unusually corrosive, it may be necessary to use some means of protection, such as cathodic protection, neutralizing backfill (limestone, for example), protective coating, or wrapping.

For many years, NBS has conducted studies on the corrosion of underground structures to determine the specific behavior of metals and alloys when exposed for long periods in a wide range of soils. Results indicate that tough pitch coppers, deoxidized coppers, silicon bronzes, and low-zinc brasses behave essentially alike. Soils containing cinders with high concentrations of sulfides, chlorides, or hydrogen ions (H^+) corrode these materials. In this type of contaminated soil, the corrosion rates of copper-zinc alloys containing more than about 22% Zn increase with zinc content. Corrosion generally results from dezincification. In soils that contain only sulfides, corrosion rates of the copper-zinc alloys decrease with increasing zinc content, and no dezincification occurs. Although not included in these tests, inhibited admiralty metals would offer significant resistance to dezincification.

Electric cables that contain copper are often buried underground. A recent study investigated the corrosion behavior of phosphorus-deoxidized copper (C12200) in four soil types: gravel, salt marsh, swamp, and clay (Ref 8). After 3 years of exposure, uniform corrosion rates were found to vary between 1.3 and 8.8 μm/yr (0.05 to 0.35 mil/yr). No pitting attack was observed. In general, the corrosion rate was highest for soils of lowest resistivity.

The possibility of disposing of nuclear waste in copper containers buried deep underground is currently under investigation. Except for the mining and oil industries, underground construction is usually limited to the first few tens of meters from the surface; an underground waste disposal vault would probably be located at a depth of 500 to 1000 m (1640 to 3280 ft) in stable bedrock. At these depths, the environment differs in several respects from that nearer the surface. With increasing depth, the natural groundwaters tend to become more saline and less oxidizing. In addition, the pressures exerted by hydrostatic and lithostatic forces become greater. These aspects affect the design and corrosion behavior of any metallic structure buried at such great depths.

A copper nuclear waste disposal container would be surrounded by a compacted claylike material. This serves a dual purpose: first it acts as a physical barrier, reducing the rate of transport of species to and from the container, and second, it provides some chemical buffering effects and effectively increases the pH of the environment. Both of these properties are beneficial in terms of the corrosion resistance of copper.

The clay most likely to be used is a montmorillonite clay, such as sodium bentonite. In the compacted form, this clay swells when wet and would effectively seal all cracks in the surrounding rock. The low permeability of the clay ensures that there would be no mass flow of groundwater and that transport of dissolved species would occur by diffusion only. The rate of diffusion in the clay is perhaps 100 times slower than in free solution. This slow rate of diffusion applies not only to the transport of oxidants, such as dissolved oxygen (O^2) or sulfide ions (S^{2-}), to the copper surface but also to the diffusion of soluble

Table 6 (continued)

Corrosive medium	Copper	Tin bronze	Leaded tin bronze	High-leaded tin bronze	Leaded red brass	Leaded semi-red brass	Leaded yellow brass	Leaded high-strength yellow brass	High-strength yellow brass	Aluminum bronze	Leaded nickel brass	Leaded nickel bronze	Silicon bronze	Silicon brass
Lacquers	A	A	A	A	A	A	A	A	A	A	A	A	A	A
Lacquer thinners	A	A	A	A	A	A	A	A	A	A	A	A	A	A
Lactic acid	A	A	A	A	A	C	C	C	C	A	C	C	A	C
Linseed oil	A	A	A	A	A	A	A	A	A	A	A	A	A	A
Liquors														
Black liquor	B	B	B	B	B	C	C	C	C	B	C	C	B	B
Green liquor	C	C	C	C	C	C	C	C	C	B	C	C	C	B
White liquor	C	C	C	C	C	C	C	C	C	A	C	C	C	B
Magnesium chloride	A	A	A	A	A	C	C	C	C	A	C	C	A	B
Magnesium hydroxide	B	B	B	B	B	B	B	B	B	A	B	B	B	B
Magnesium sulfate	A	A	A	A	B	C	C	C	C	A	C	B	A	B
Mercury, mercury salts	C	C	C	C	C	C	C	C	C	C	C	C	C	C
Milk(b)	A	A	A	A	A	A	A	A	A	A	A	A	A	A
Molasses(b)	A	A	A	A	A	A	A	A	A	A	A	A	A	A
Natural gas	A	A	A	A	A	A	A	A	A	A	A	A	A	A
Nickel chloride	A	A	A	A	A	C	C	C	C	B	C	C	A	C
Nickel sulfate	A	A	A	A	A	C	C	C	C	A	C	C	A	C
Nitric acid	C	C	C	C	C	C	C	C	C	C	C	C	C	C
Oleic acid	A	A	B	B	B	C	C	C	C	A	C	A	A	B
Oxalic acid	A	A	B	B	B	C	C	C	C	A	C	A	A	B
Phosphoric acid	A	A	A	A	A	C	C	C	C	A	C	A	A	A
Picric acid	C	C	C	C	C	C	C	C	C	C	C	C	C	C
Potassium chloride	A	A	A	A	A	C	C	C	C	A	C	C	A	C
Potassium cyanide	C	C	C	C	C	C	C	C	C	C	C	C	C	C
Potassium hydroxide	C	C	C	C	C	C	C	C	C	A	C	C	C	C
Potassium sulfate	A	A	A	A	A	C	C	C	C	A	C	C	A	C
Propane gas	A	A	A	A	A	A	A	A	A	A	A	A	A	A
Seawater	A	A	A	A	A	C	C	C	C	A	C	C	B	B
Soap solutions	A	A	A	A	B	C	C	C	C	A	C	C	A	C
Sodium bicarbonate	A	A	A	A	A	A	A	A	A	A	A	A	A	B
Sodium bisulfate	C	C	C	C	C	C	C	C	C	A	C	C	C	C
Sodium carbonate	C	A	A	A	A	C	C	C	C	A	C	C	C	A
Sodium chloride	A	A	A	A	A	B	C	C	C	A	C	C	A	C
Sodium cyanide	C	C	C	C	C	C	C	C	C	B	C	C	C	C
Sodium hydroxide	C	C	C	C	C	C	C	C	C	A	C	C	C	C
Sodium hypochlorite	C	C	C	C	C	C	C	C	C	C	C	C	C	C
Sodium nitrate	B	B	B	B	B	B	B	B	B	A	B	B	A	A
Sodium peroxide	B	B	B	B	B	B	B	B	B	B	B	B	B	B
Sodium phosphate	A	A	A	A	A	A	A	A	A	A	A	A	A	A
Sodium sulfate, silicate	A	A	B	B	B	B	C	C	C	A	C	C	A	B
Sodium sulfide, thiosulfate	C	C	C	C	C	C	C	C	C	B	C	C	C	C
Stearic acid	A	A	A	A	A	A	A	A	A	A	A	A	A	A
Sulfur, solid	C	C	C	C	C	C	C	C	C	A	C	C	C	C
Sulfur chloride	C	C	C	C	C	C	C	C	C	C	C	C	C	C
Sulfur dioxide, dry	A	A	A	A	A	A	A	A	A	A	A	A	A	A
Sulfur dioxide, moist	A	A	A	B	B	C	C	C	C	A	C	C	A	B
Sulfur trioxide, dry	A	A	A	A	A	A	A	A	A	A	A	A	A	A
Sulfuric acid														
78% or less	B	B	B	B	B	C	C	C	C	A	C	C	B	B
78% to 90%	C	C	C	C	C	C	C	C	C	B	C	C	C	C
90% to 95%	C	C	C	C	C	C	C	C	C	B	C	C	C	C
Fuming	C	C	C	C	C	C	C	C	C	A	C	C	C	C
Tannic acid	A	A	A	A	A	A	A	A	A	A	A	A	A	A
Tartaric acid	B	A	A	A	A	A	A	A	A	A	A	A	A	A
Toluene	B	B	A	A	A	B	B	B	B	B	B	B	B	A
Trichlorethylene, dry	A	A	A	A	A	A	A	A	A	A	A	A	A	A
Trichlorethylene, moist	A	A	A	A	A	A	A	A	A	A	A	A	A	A
Turpentine	A	A	A	A	A	A	A	A	A	A	A	A	A	A
Varnish	A	A	A	A	A	A	A	A	A	A	A	A	A	A
Vinegar	A	A	B	B	B	C	C	C	C	B	C	C	A	B
Water, acid mine	C	C	C	C	C	C	C	C	C	C	C	C	C	C
Water, condensate	A	A	A	A	A	A	A	A	A	A	A	A	A	A
Water, potable	A	A	A	A	A	A	B	B	B	A	A	A	A	A
Whiskey(b)	A	A	C	C	C	C	C	C	C	A	C	C	A	C
Zinc chloride	C	C	C	C	C	C	C	C	C	B	C	C	B	C
Zinc sulfate	A	A	A	A	A	C	C	C	C	B	C	A	A	C

(a) Acetylene forms an explosive compound with copper when moist or when certain impurities are present and the gas is under pressure. Alloys containing less than 65% Cu are satisfactory for this use. When gas is not under pressure, other copper alloys are satisfactory. (b) Copper and copper alloys resist corrosion by most food products. Traces of copper may be dissolved and affect taste or color. In such cases, copper metals are often tin coated.

corrosion products away from the surface. The net effect is reduction in the corrosion rate of copper compared with that in free solution. One study suggests that under such conditions uniform corrosion of oxygen-free electronic copper (C10100) would only amount to 1.1 mm (43.4 mils) in 10^6 years (Ref 9). Experimental results indicate that the clay may reduce the corrosion rate by about a factor of ten over that in bulk solution, although these results suggest a corrosion rate of about 1 μm/yr (0.04 mils/yr) (Ref 10).

Naturally occurring saline waters are also found deep underground. Although the composition and concentration of these groundwaters vary from site to site, the concentration of dissolved species generally increases with depth (Ref 11). Such groundwaters are encountered in mines, during oil drilling, and in deep boreholes. The waters have a complex composition, often being mixtures of sodium (Na^+), calcium (Ca^{2+}), magnesium (Mg^{2+}), chloride (Cl^-), sulfate (SO_4^{2-}) and bicarbonate (HCO_3^-) ions as well as trace amounts of other ions. Iron minerals in the bedrock react with dissolved oxygen in the groundwater and produce less oxidizing conditions than are found in waters nearer the surface. Additional information on the corrosion of nuclear waste containment materials is available in the section "Corrosion of Containment Materials for Radioactive Waste" of the article "Corrosion in the Nuclear Power Industry" in this Volume.

The corrosion rate of copper in quiescent groundwaters tends to decrease with time. This is due to the formation of a protective film, an example of which is shown in Fig. 5. The underlying layer consists of species from the groundwater as well as copper. This layer is brittle and is extensively cracked, permitting continued dissolution of copper ions into solution. In Fig. 5, some of these copper ions have precipitated on the underlying layer in the form of cupric hydroxychloride [$CuCl_2 \cdot 3(Cu(OH)_2)$] and copper oxide crystals. The corrosion layer is not truly passivating, and corrosion will continue, although at a reduced rate.

For both copper and copper alloys, corrosion rate depends strongly on the amount of dissolved oxygen present. The data in Table 8 illustrate this point for both pure copper and Cu-10Ni in various synthetic groundwaters. These data are derived from experiments lasting from 2 to 4 weeks; therefore, they include the high initial rates of corrosion and do not represent long-term corrosion rates. However, they do serve to show that deoxygenation of the solution results in at least an order of magnitude decrease in the short-term corrosion rate. It is also apparent from these data that, in aerated solutions at least, the addition of nickel decreases the uniform corrosion rate of copper. This is due to the formation of a more highly protective surface film.

The effects of salinity and temperature are less well understood. In general, increasing the total salinity of these groundwaters tends to increase their corrosiveness. However, it is not clear whether this is due to the sum effect of all the dissolved ions or of some of the species in particular. In open systems, it is difficult to distinguish the effect of temperature from that of dissolved oxygen, because the solubility of oxygen decreases with increasing temperature. The combination of these two opposing effects can lead to an apparent maximum in the corrosion rate at some intermediate temperature. Consequently, it is important that the rates refer to a

Table 7 Atmospheric corrosion of selected copper alloys

Alloy	Altoona, PA µm/yr	Altoona, PA mils/yr	New York, NY µm/yr	New York, NY mils/yr	Key West, FL µm/yr	Key West, FL mils/yr	La Jolla, CA µm/yr	La Jolla, CA mils/yr	State College, PA µm/yr	State College, PA mils/yr	Phoenix, AZ µm/yr	Phoenix, AZ mils/yr
C11000	1.40	0.055	1.38	0.054	0.56	0.022	1.27	0.050	0.43	0.017	0.13	0.005
C12000	1.32	0.052	1.22	0.048	0.51	0.020	1.42	0.056	0.36	0.014	0.08	0.003
C23000	1.88	0.074	1.88	0.074	0.56	0.022	0.33	0.013	0.46	0.018	0.10	0.004
C26000	3.05	0.120	2.41	0.095	0.20	0.008	0.15	0.006	0.46	0.018	0.10	0.004
C52100	2.24	0.088	2.54	0.100	0.71	0.028	2.31	0.091	0.33	0.013	0.13	0.005
C61000	1.63	0.064	1.60	0.063	0.10	0.004	0.15	0.006	0.25	0.010	0.51	0.002
C65500	1.65	0.065	1.73	0.068	...	...	1.38	0.054	0.51	0.020	0.15	0.006
C44200	2.13	0.084	2.51	0.099	...	...	0.33	0.013	0.53	0.021	0.10	0.004
70Cu-29Ni-1Sn(b)	2.64	0.104	2.13	0.084	0.28	0.011	0.36	0.014	0.48	0.019	0.10	0.004

Corrosion rates at indicated locations(a)

(a) Derived from 20-year exposure tests. Types of atmospheres: Altoona, industrial; New York City, industrial marine; Key West, tropical rural marine; La Jolla, humid marine; State College, northern rural; Phoenix, dry rural. (b) Although obsolete, this alloy indicates the corrosion resistance expected of C71500.

Table 8 Short-term corrosion rates of copper alloys in saline groundwaters

Alloy	Type of groundwater	Oxygen concentration, µg/g	Temperature °C	Temperature °F	Corrosion rate µm/yr	Corrosion rate mils/yr	Ref
C10100	Synthetic	<0.1	150	300	15	0.6	(b)
	55 g/L TDS(a)	6	150	300	340	13.4	
Copper	Brine A	<0.1	250	480	70	2.8	12
	306 g/L TDS	600	250	480	1200	47.2	
	Seawater	<0.1	250	480	50	2	
	35 g/L TDS	1750	250	480	5000	197	
Cu-10Ni (C70600)	Brine A	<0.1	250	480	140	5.5	12
		600	250	480	400	15.7	
	Seawater	<0.1	250	480	70	2.8	
		1750	250	480	700	27.6	

(a) TDS, total dissolved solids. (b) F. King and C.D. Litke, unpublished research, 1985

constant dissolved-oxygen concentration when considering the effects of temperature.

Water

Freshwater. Copper is extensively used for handling freshwater. Copper tubing in the K-gage range with flared fittings was designed for underground water service and, along with type L tubing, has now become standard for this application. The largest single application of copper tubing is for hot- and cold-water distribution lines in homes and other buildings, although considerable quantities are also used in heating lines (including radiant heating lines for homes), drain tubes, and fire safety systems.

Copper. Minerals in water combine with dissolved CO_2 and oxygen and react with copper to form a protective film. Therefore, the corrosion rate is low (5 to 25 µm/yr, or 0.2 to 1.0 mil/yr) in most exposures. In distilled water or very soft water, protective films are less likely to form; therefore, the corrosion rate may vary from less than 2.5 to 125 µm/yr (0.1 to 5 mils/yr) or more, depending on oxygen and CO_2 contents.

Copper-Zinc Alloys. The corrosion resistance of the brasses is good in unpolluted freshwater—normally 2.5 to 25 µm/yr (0.1 to 1.0 mil/yr). Corrosion rates are somewhat higher in nonscaling water containing CO_2 and oxygen. Uninhibited brasses of high zinc content (35 to 40% Zn) are subject to dezincification when used with stagnant or slowly moving brackish or slightly acid waters. On the other hand, inhibited admiralty metals and brasses containing 15% Zn or less are highly resistant to dezincification and are used very successfully in these waters. Inhibited yellow brasses are widely used in Europe and are gaining acceptance in North America. Alloy C68700 (arsenical aluminum brass, an inhibited 77Cu-21Zn-2Al alloy) has been successfully used for condenser and heat-exchanger tubes.

Copper nickels generally have corrosion rates under 25 µm/yr (1 mil/yr) in unpolluted water. They are sometimes used to resist impingement attack where severe velocity and entrained-air conditions cannot be overcome by changes in operating conditions or equipment design.

Copper-silicon alloys (silicon bronzes) also have excellent corrosion resistance, and for these alloys, the amount of dissolved oxygen in the water does not influence corrosion significantly. If CO_2 is also present, the corrosion rate will

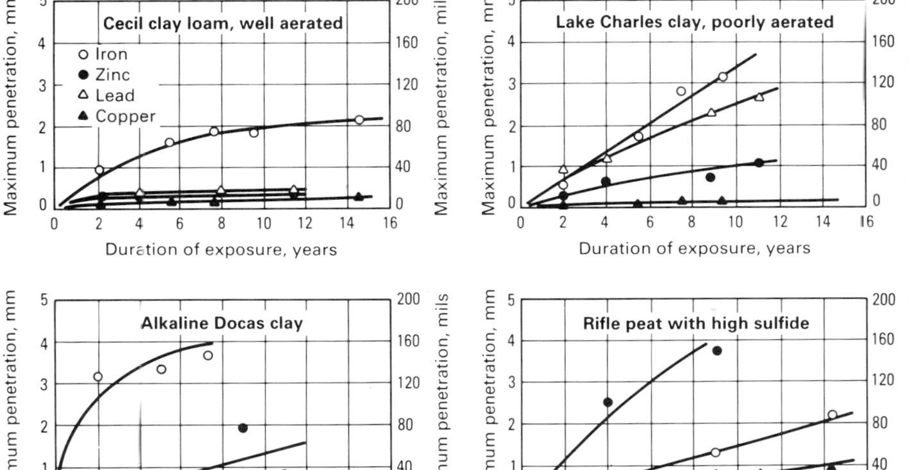

Fig. 4 Corrosion of copper, iron, lead and zinc in four different soils

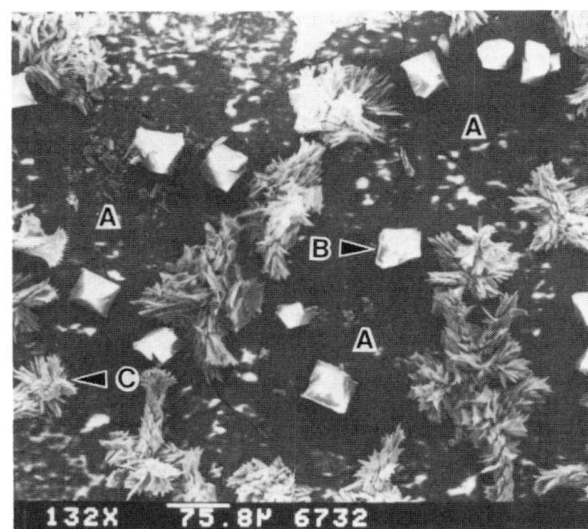

Fig. 5 Scanning electron micrograph of the corrosion product formed on C10100 in complex groundwater at 150 °C (300 °F). A, underlying film containing copper, silicon, calcium, chlorine, and magnesium; B, crystals of $CuCl_2 \cdot 3(Cu(OH)_2)$; C, crystals of CuO or Cu_2O. Courtesy of F. King and C.D. Litke

increase (but not excessively), particularly at temperatures above 60 °C (140 °F). Corrosion rates for silicon bronzes are similar to those for copper.

Copper-Aluminum Alloys. The aluminum bronzes have been used in many waters, from potable water to brackish water to seawater. Softened waters are usually more corrosive to these materials than hard waters. Alloys C61300 and C63200 are used in cooling tower hardware in which the makeup water is sewage effluent. Aluminum bronzes resist oxidation and impingement corrosion because of the aluminum in the surface film.

Steam. Copper and copper alloys resist attack by pure steam, but if much CO_2, oxygen, or NH_3 is present, the condensate is corrosive. Even though wet steam at high velocities can cause severe impingement attack, copper alloys are extensively used in condensers and heat exchangers. Copper alloys are also used for feedwater heaters, although their use in such applications is somewhat limited because of their rapid decline in strength and creep resistance at moderately elevated temperatures. Copper nickels are the preferred copper alloys for the higher temperatures and pressures.

Use of copper in systems handling hot water and steam is limited by the working pressures of tubes and joints. For example, copper tubing of 6.4 to 25 mm (¼ to 1 in.) nominal diameter joined with 50Sn-50Pb solder can be used at temperatures to 120 °C (250 °F) and pressures to 585 kPa (85 psi). The working pressure at this temperature in tubing of the same size can be increased to 1380 kPa (200 psi) when the system is joined with 95Sn-5Sb solder. When the joining material is a silver-base brazing alloy with a melting point above 540 °C (1000 °F), the working pressure at 120 °C (250 °F) for tubing in this size range can be increased to 2070 kPa (300 psi). A few copper alloys have shown a tendency to fail by SCC when they are highly stressed and exposed to steam. Alpha aluminum bronzes that do not contain tin are among the susceptible alloys.

Steam condensate that has been properly treated so that it is relatively free of noncondensate gases, as in a power-generating station, is relatively noncorrosive to copper and copper alloys. Rates of attack in most such exposures are less than 2.5 μm/yr (0.1 mil/yr). Copper and its alloys are not attacked by condensate that contains a significant amount of oil, such as condensate from a reciprocating steam engine.

Dissolved CO_2, oxygen, or both significantly increase the rate of attack. For example, condensate with 4.6 ppm O, and 14 ppm CO_2, and a pH of 5.5 at 68 °C (155 °F) caused an average penetration of 175 to 350 μm/yr (6.9 to 13.8 mils/yr) when in contact with C12200 (phosphorus-deoxidized copper), C14200 (arsenical copper),

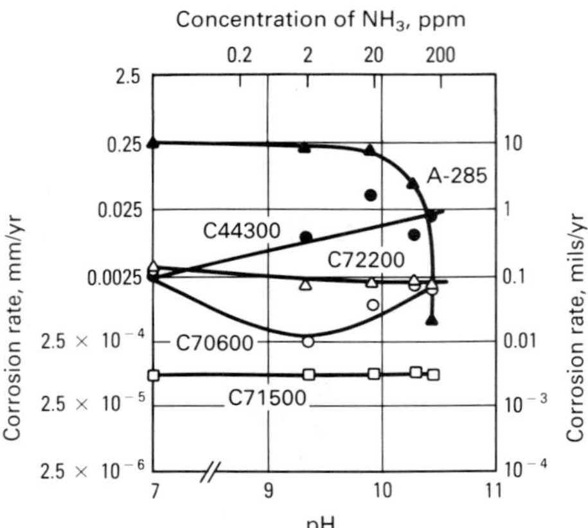

Fig. 6 Corrosion rates of copper alloys in aerated NH_3 solutions. Test duration: 1000 h. Source: Ref 13

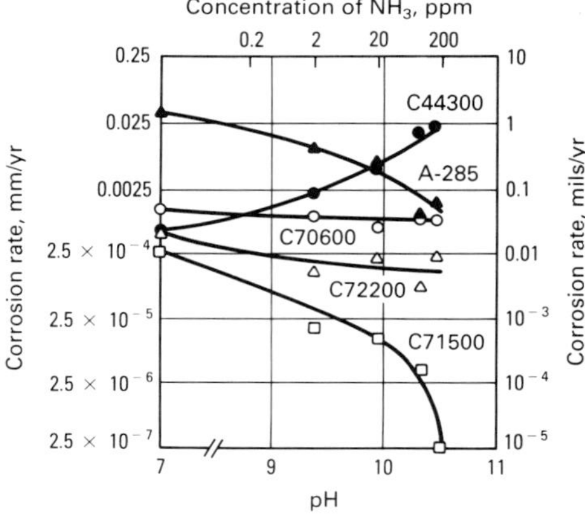

Fig. 7 Corrosion rates of copper alloys in deaerated NH_3 solutions. Test duration: 1000 h. Source: Ref 13

C23000 (red brass), C44300 to C44500 (admiralty metal), and C71000 (copper nickel, 20%). Steel tested under the same conditions was penetrated at about twice the rate given for the copper alloys listed above, but tin-coated copper proved to be much more resistant and was attacked at a rate of less than 25 μm/yr (1 mil/yr). To attain the optimum service life in condensate systems, it is necessary to ensure that the tubes are installed with enough slope to allow proper drainage, to reduce the quantity of corrosive agents (usually CO_2 and oxygen) at the source by mechanical or chemical treatment of the feedwater, or to treat the steam chemically.

Modern power utility boiler feedwater treatments commonly include the addition of organic amines to inhibit the corrosion of iron components of the system by scavenging oxygen and increasing the pH of the feedwater. These chemicals, such as morpholine, and hydrazine, decompose in service to yield NH_3, which can be quite corrosive toward some copper alloys. In the main body of well-monitored operating condensers, oxygen and NH_3 levels are quite low, and corrosion is usually mild. More aggressive conditions exist in the air removal section. Abnormal operating conditions, tube leakage, and shutdown-startup cycles may also increase the corrosivity of the steam-side environment by raising the oxygen concentration. The corrosion resistance in laboratory tests of a number of copper alloys and low-carbon steel in both aerated (8 to 12 ppm O_2) and deaerated (100 to 200 ppb O_2) NH_3 solutions are illustrated in Fig. 6 and 7. In these tests, NH_3 enhanced the corrosion resistance of the copper-nickel alloys, modifying surface oxides by increasing nickel content. Elevated oxygen levels are generally more deleterious than elevated NH_3 levels. However, C71500 was minimally affected by the elevated oxygen content. These laboratory data correlate well with field corrosion data from operating power plants (Table 9). Additional information on corrosion in power plant applications is available in the article "Corrosion of Fossil Fuel Power Plants" in this Volume.

Salt Water. An important use of copper alloys is in handling seawater in ships and tidewater power stations. Copper itself, although fairly useful, is usually less resistant to general corrosion than C44300 to C44500, C61300, C68700, C70600, or C71500. The superior performance of these alloys results from the combination of insolubility in seawater, erosion resistance, and biofouling resistance. The corrosion rates of copper and its alloys in relatively quiescent seawater are typically less than 50 μm/yr (2 mils/yr).

In the laboratory and in service, copper-nickel alloys C70600, C71500, C72200, and C71640 exhibit excellent corrosion resistance in seawater. Average corrosion rates for both C70600 and C71500 were shown to range from 2 to 12 μm/yr (0.08 to 0.5 mils/yr) (Ref 15). The long-term evaluations illustrated in Fig. 8 and 9 revealed corrosion rates under 2.5 μm/yr (0.1 mil/yr) for both alloys after 14 years of exposure to quiescent and low-velocity seawater (Ref 16). Sixteen-year tests confirmed this same low corrosion rate (Ref 17).

Pitting Resistance. Alloys C70600 and C71500 both display excellent resistance to pitting in seawater. The average depth of the 20 deepest pits in C71500 observed at the end of the 16-year tests was less than 127 μm (5 mils) (Ref 17).

Table 9 Comparison of field and laboratory condensate corrosion of copper alloys

Data are weight loss measured after total exposure time, expressed as penetration rates

Alloy	Corrosion rate, μm/yr (mils/yr)					
	Field tests(a)			Laboratory tests(b)		
	Plant A	Plant B	Plant C	0 ppm NH_3	2 ppm NH_3	20 ppm NH_3
C71500	0.2 (0.0083)	0.1 (0.004)	0.4 (0.0151)	0.3 (0.012)	0.05 (0.002)	0.025 (0.001)
C72200	0.4 (0.016)	0.4 (0.016)	0.38 (0.015)	0.61 (0.024)	0.2 (0.008)	0.18 (0.007)
C70600	0.48 (0.019)	0.36 (0.014)	0.46 (0.018)	1.3 (0.053)	1.1 (0.043)	0.94 (0.037)
C44300	1.27 (0.05)	0.79 (0.031)	0.61 (0.024)	0.61 (0.024)	2.3 (0.09)	5.6 (0.22)
A-285	6.2 (0.243)	10.4 (0.411)	2.6 (0.103)	38 (1.5)	8.3 (0.325)	4.6 (0.183)

(a) 2-year tests in hot wells at three plant sites (A, B, and C). Plant A, pH range of 8.8–9.7; typical pH of 9.1–9.3. Plant B, pH range of 9–10, typical pH of 9.3–9.6. (b) Laboratory data extrapolated from 1000-h tests in deaerated beakers. 0 ppm NH_3 solution, pH 7; 2 ppm NH_3 solution, pH 9.4; 20 ppm NH_3 solution, pH 10. Source: Ref 14

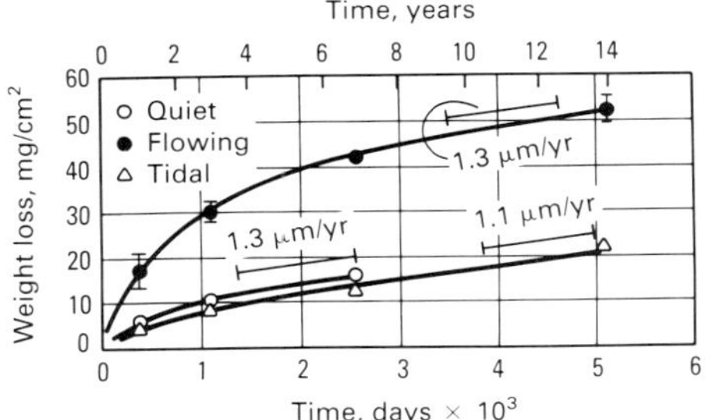

Fig. 8 Chronogravimetric curves for C70600 in quiet, flowing, and tidal seawater. Source: Ref 16

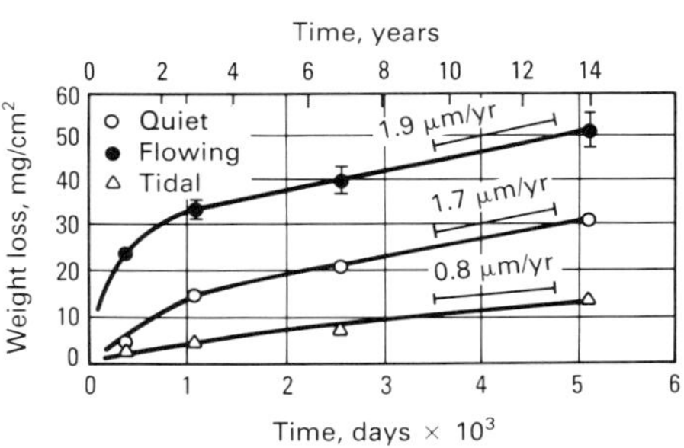

Fig. 9 Chronogravimetric curves for C71500 in quiet, flowing, and tidal seawater. Source: Ref 16

Chromium-modified copper-nickel alloys, developed for resistance to high-velocity seawater, were evaluated in both low- and high-velocity conditions. The quiescent and low-velocity performances of C72200, C70600, and C71500 were compared (Ref 18, 19); results showed uniform corrosion (5 to 25 μm/yr, or 0.2 to 1 mil/yr) on all three alloys. The chromium containing alloys, however, were slightly more susceptible to localized attack in quiet seawater. Another study reported that the pitting behavior of C72200 is influenced by the presence of iron and chromium in or out of solid solution (Ref 20). The fraction of iron plus chromium in solution in C72200 must be kept higher than 0.7 to avoid pitting corrosion.

Velocity Effects. The corrosion resistance of copper alloys in flowing seawater depends on the growth and maintenance of a protective film or corrosion product layers. These alloys typically exhibit velocity-dependent corrosion rates. The more adherent and protective the film on a particular alloy, the higher its breakaway velocity (the velocity at which there is a transition from low to high corrosion rate) and the greater its resistance to impingement attack or erosion-corrosion.

Some of the earliest work on copper-nickel alloys demonstrated the beneficial effects of iron additions on seawater impingement resistance. The graphical summary of the effects of iron shown in Fig. 10 qualitatively illustrates the balance between pitting resistance and impingement resistance that defines the optimum iron content for 90Cu-10Ni and 70Cu-30Ni at 1.5 and 0.5% Fe, respectively. The effects of manganese level in association with iron in copper-nickel alloys are also addressed in Ref 21. The relative beneficial effects of 2% Fe and 2% Mn in a 70Cu-30Ni alloy (C71640) are shown in Fig. 11, which indicates that the C71640 and C72200 alloys are markedly more resistant to erosion-corrosion than C70600 at velocities up to 9 m/s (30 ft/s). The chromium-modified copper-nickel alloys also provide increased resistance to impingement attack compared to copper-nickel-iron alloys. In jet impingement tests (Ref 19) on several copper-base alloys at impingement velocities as high as 10 m/s (33 ft/s), no measurable impingement attack was observed on alloys C72200 and C71900 at 4.6 m/s (15 ft/s) (Table 10).

The behaviors of several copper-nickel alloys, including C71640 and C72200, have been characterized under conditions simulating partial blockage of a condenser tube (Ref 22). In the 1-year natural seawater tests, enhanced erosion-corrosion resistance was observed for the C71640 and C72200 alloys as compared to C70600 and C71500. Some localized pitting and/or crevice corrosion associated with the nonmetallic blockage device was noted for C71640 and C72200, with no such attack occurring for the C70600 and C71500 alloys. Superior performance of the modified copper-nickel alloys C72200 and C71640 was also observed under severely erosive conditions in seawater containing entrained sand (Ref 23).

The combined results of laboratory impingement studies and service performance have produced maximum acceptable design velocities for condenser tube materials (Table 11). Erosion-corrosion was recently studied on the basis of fluid dynamics (Ref 24-26). Instead of defining the critical velocity for a material, which is difficult to relate to service conditions and which is specific to tubing diameter, the use of critical surface shear stress was advocated. This shear stress in a dynamic fluid system is a measure of the force applied by the moving fluid to the surface with which it interacts. It takes into account the changes in fluid density and kinematic viscosity with variations in temperature, specific gravity, and hydrodynamic parameters. Values of critical surface shear stress for several copper-base alloys are shown in Table 12.

Table 10 Summary of jet impingement test data for several copper alloys at three velocities

Test duration: 1–2 months; 10- to 26-°C (50- to 80-°F) seawater

Alloy	Impingement attack at velocity: 4.6 m/s (15 ft/s), mm/yr	4.6 m/s (15 ft/s), mils/yr	6.8 m/s (22 ft/s), mm/yr	6.8 m/s (22 ft/s), mils/yr	9.8 m/s (32 ft/s), mm/yr	9.8 m/s (32 ft/s), mils/yr
C44300	1.8–4.8	71–189	Not tested		Not tested	
C68700	0.36–3	14.2–118	Not tested		Not tested	
C70600	0.12–2.16	4.7–85	0.36–1.56	14.2–61.4	1.56	61.4
C71500	0.12–1.08	4.7–42.5	0.36–6.84	14.2–269	1.68–2.04	66–80.3
C71900	No attack		0.12–0.36	4.7–14.2	1.08–1.44	42.5–56.7
C72200	No attack		0.12	4.7	No attack	

Source: Ref 19

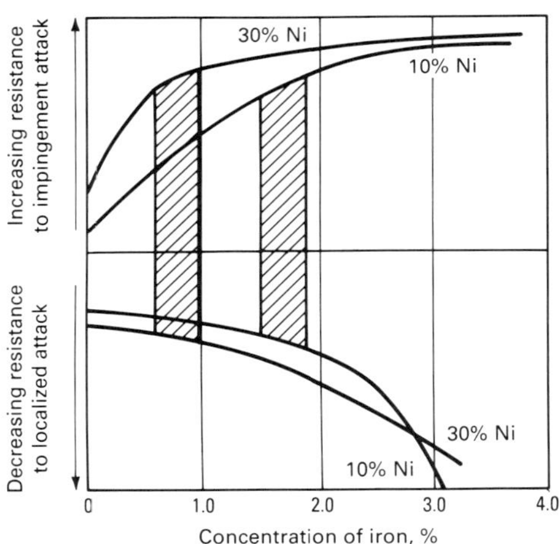

Fig. 10 Corrosion resistance of copper-nickel alloys as a function of iron content. Shaded areas indicate optimum iron contents for good balance between pitting resistance and impingement resistance. Source: Ref 21

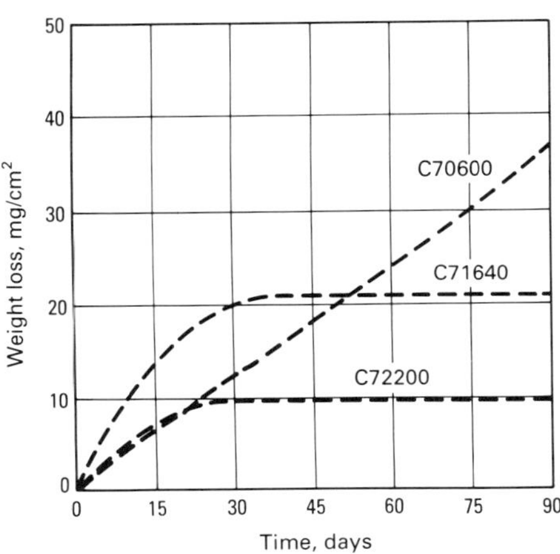

Fig. 11 Weight loss versus time curves for C70600, C71640, and C72200 exposed in seawater at a velocity of 9 m/s (40 ft/s). Source: Ref 21

Galvanic Effects. In general, the copper-base alloys are galvanically compatible with one another in seawater. The copper-nickel alloys are slightly cathodic (noble) to the nickel-free copper-base alloys, but the small differences in corrosion potential generally do not lead to serious galvanic effects unless unusually adverse anodic/cathodic area ratios are involved.

The data given in Table 13 demonstrate the increased attack of less noble carbon steel coupled to copper-nickel alloys, the increased attack on the copper-nickel alloys when coupled to more noble titanium, and the general compatibility of copper-nickel alloys with aluminum bronze. Coupling copper-nickel alloys to less noble materials affords protection to the copper-nickel that effectively reduces its corrosion rate, thus inhibiting the natural fouling resistance of the alloy.

Results of short-term galvanic couple tests between C70600 and several cast copper-base and ferrous alloys are listed in Table 14. The corrosion rate of cast 70Cu-30Ni was unaffected by coupling with an equal area of C70600, but

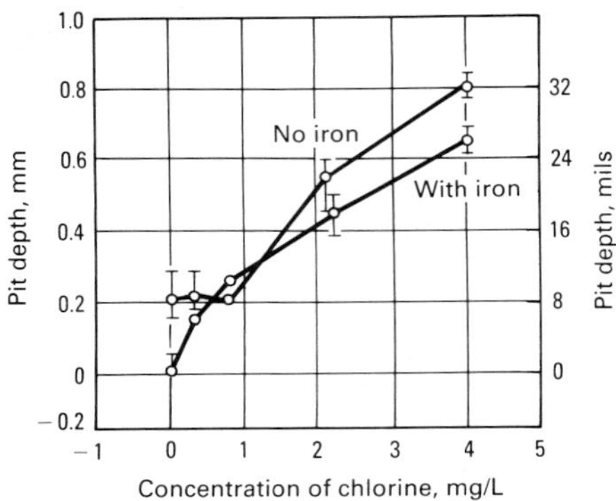

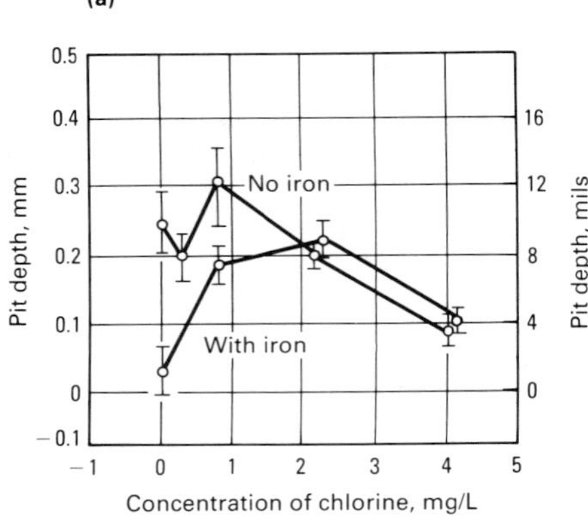

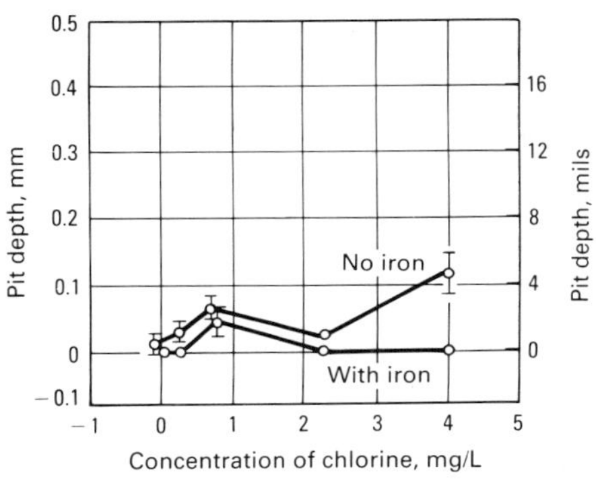

Fig. 12 Impingement attack versus chlorine levels for three copper alloys. (a) C70600. (b) C71500. (c) C71640

some increased corrosion of other cast copper-base alloys was noted. Corrosion rates of cast stainless steels were reduced, with a resultant increase in the corrosion of C70600. Gray iron displayed the largest galvanic effect, while the corrosion rates of Ni-Resist cast irons nominally doubled. Although some caution should be exercised in using absolute values from any short-term tests, the relative degree of acceleration of corrosion from galvanic coupling was shown to be unaffected by extending some tests with Ni-Resist/C70600 couples to 1 year.

Effect of Oxygen, Depth, and Temperature. The corrosion of copper and copper-base alloys in clean seawater is cathodically controlled by oxygen reduction, with H^+ reduction being thermodynamically unfavorable. Dissolved oxygen retards corrosion by the promotion of a protective film on the copper alloy surface, but increases the rate of corrosion by depolarizing cathodic sites and oxidizing Cu^+ ions to more aggressive Cu^{2+} ions. Other factors, such as velocity, temperature, salinity, and ocean depth, affect the dissolved oxygen content of seawater, thus influencing the corrosion rate. In general, oxygen concentration decreases with increasing salinity, temperature, and depth. These factors can vary with depth in a complex manner and also vary from location to location in the oceans of the world (Ref 27).

Although cathodic control by oxygen reduction suggests a strong dependence of corrosion rate on dissolved oxygen concentration, the growth of a protective oxide film on copper-nickel alloys minimizes the influence within the normally observed range of oxygen content found in seawater. Deep-ocean testing indicated that the corrosion rates of copper and copper-nickel alloys do not change significantly for dissolved oxygen contents between 1 and 6 mL/L of seawater and consequently were not significantly affected by variations in depth of exposure (Ref 27).

Short-term laboratory tests indicated only a small increase in corrosion rate with increasing temperature up to 30 °C (85 °F) (Ref 28). Long-term corrosion rate data from tests conducted at a coastal site near Panama (Ref 17) agree very well with long-term data for exposures in Wrightsville Beach, NC (Ref 16), where the seasonal temperature variation is 5 to 30 °C (40 to 85 °F). Final steady-state corrosion rates at both locations for C71500 ranged from 1 to 3 μm/yr (0.04 to 0.12 mils/yr).

Studies performed at higher temperatures relative to those in desalination plant environments show considerable disagreement in results (Ref 29-34). From 60 to 107 °C (140 to 225 °F), temperature may increase, decrease, or have no significant effect on the corrosion rate of copper-nickel alloys. Lower corrosion rates for C70600 over an intermediate temperature range were reported in seawater corrosion tests between 32 and 107 °C (90 and 225 °F) with controlled seawater chemistry; bicarbonate alkalinity, dissolved oxygen, and pH were noted as critical factors controlling corrosion (Ref 33). Other studies confirmed lower average corrosion rates at 40 °C (105 °F) than at lower temperatures (Ref 35). The variation in results reported in the literature can perhaps be explained by variations in seawater chemistry between test sites and/or control of operating conditions in desalination plants.

Effect of Chlorine. Coastal power plants that use seawater as a coolant have long used chlorine to control fouling and slime formation. The effect of chlorination, both continuous and intermittent, on the corrosion of copper-nickel alloys was studied (Ref 36, 37). Continuous chlorine additions increased the corrosion rate of C70600 by a factor of two. Intermittent chlorination at a higher level controlled fouling, yet had no apparent

Table 11 Accepted maximum tubular design velocities for some copper alloys

Alloy	Maximum design velocity m/s	ft/s
C12200	0.6–0.9	2–3
C44300	1.2–1.8	4–6
C60800, C61300	2.7	9
C68700	2.4	8
C65100, C85500	0.9	3
C70600	3.0–3.6	10–12
C71500	4.5–4.6	14.8–15
C72200	9.0	30

Table 12 Critical surface shear stress for copper-base alloys in seawater

Alloy	Critical shear stress Pa	psi
C12200	9.6	0.0014
C68700	19.2	0.028
C70600	43.1	0.0063
C71500	47.9	0.007
C72200	296.9	0.043

Source: Ref 24

Table 13 Galvanic couple data for C70600 and C71500 with other materials in flowing seawater

2-year exposures of equal area couples at a velocity of 0.6 m/s (2 ft/s)

Alloy	Corrosion rate μm/yr	mils/yr
Uncoupled		
C70600	31	1.2
C71500	20	0.8
C61400	43	1.7
Carbon steel	330	13
Titanium	2	0.08
Coupled		
C70600	25	1
C61400	43	1.7
C70600	3	0.12
Carbon steel	787	31
C70600	208	8.2
Titanium	2	0.08
C71500	18	0.7
C61400	64	2.5
C71500	3	0.12
Carbon steel	711	28
C71500	107	4.2
Titanium	2	0.08

Table 14 Galvanic corrosion data for C70600/cast alloy couples in seawater

32-day tests of equal area couples in seawater at 10 °C (50 °F). Velocity: 1.8 m/s (6 ft/s)

Alloy	Galvanic effect(a) C70600	Other alloy
C70600	1.0	...
Cast 90Cu-10Ni	0.8	1.6
Cast 70Cu-30Ni	0.9	1.0
85-5-5-5 (C83600)	0.9	1.5
M Bronze (C92200)	0.7	1.8
ACI CN7M	1.5	0.6
ACI CF8M	1.2	0.1
Gray iron	0.1	6.0
Ni-Resist type I(b)	0.4	2.1
Ni-Resist type II	0.3	2.6
Ni-Resist type D2	0.3	2.0

(a) Ratio of weight loss in couple to weight loss of an uncoupled control specimen. (b) Ni-Resist couple tests at 29 °C (85 °F)

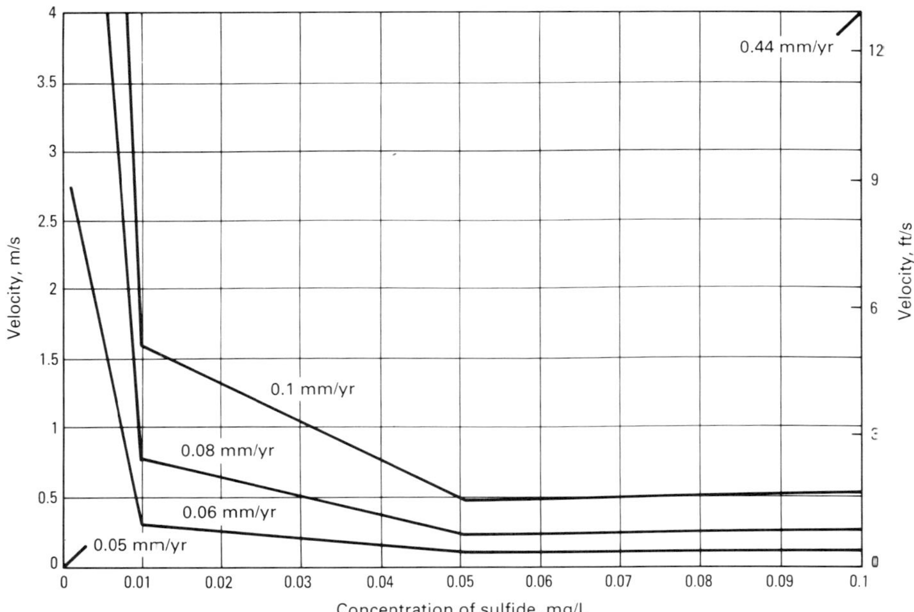

Fig. 13 Corrosion rates of C70600 as a function of seawater velocity and sulfide content

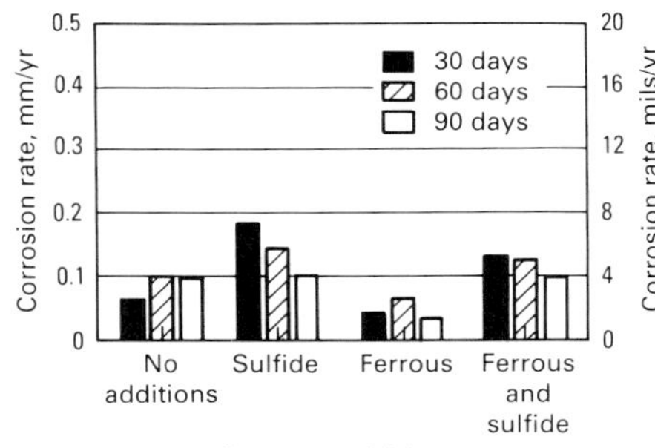

Fig. 14 Corrosion rates for C70600 exposed to seawater with additions of sulfide (0.05 mg/L) and/or Fe^{2+} (0.01 mg/L) ions. Source: Ref 46

effect on corrosion rates. A net reduction was noted in the corrosion rate of C71500 with continuous and most intermittent chlorine additions.

Seawater impingement tests were conducted on C70600, C71500, and C71640 with continuous additions of chlorine (and iron) (Ref 38). Additions of 0.5 to 4.0 mg/L of chlorine caused increased susceptibility to impingement attack on C70600 at a velocity of 9 m/s (30 ft/s). Addition of chlorine up to 4.0 mg/L had little effect on the impingement resistance of C71500. Figure 12 summarizes the results of these tests.

Polluted cooling waters, particularly in coastal harbors and estuaries, reportedly cause numerous premature failures of power station and shipboard condensers using copper-base alloys, including the copper-nickels. During the early 1950s, polluted waters were identified as the most important contributing factor in the failure of condenser tubes (Ref 39). Although enforcement of strict pollution standards has dramatically reduced pollution in many harbors in recent years, accelerated attack of condenser tubes and seawater piping materials by polluted waters is still reported.

The attack of copper-containing materials by polluted seawater has been addressed in numerous test programs. The primary causes of accelerated attack of copper-base alloys in polluted seawater are (1) the action of sulfate-reducing bacteria, under anaerobic conditions (for example, in bottom muds or sediments), on the natural sulfates present in seawater and (2) the putrefaction of organic sulfur compounds from decaying plant and animal matter within seawater systems during periods of extended shutdown (Ref 40). Partial putrefaction of organic sulfur compounds may also result in the formation of organic sulfides, such as cystine or glutathione, which can cause pitting of copper alloys in seawater (Ref 41).

Alloy C70600 has been found to be susceptible to sulfide-induced attack in aerated seawater containing sulfide concentrations as low as 0.01 mg/L (Ref 42). Recent work has demonstrated that, although the presence of 0.01 mg/L sulfide in aerated seawater can accelerate corrosion of copper-nickel alloys, the influence of seawater velocity is more significant (Ref 43). Figure 13 shows the rate of accelerated corrosion for C70600 as a function of sulfide and velocity.

Inhibition of Corrosion. In some applications, the corrosion resistance of copper alloys is further enhanced by adding iron to the seawater. This iron is introduced either through the addition of ferrous sulfate ($FeSO_4$) or by direct oxidation of a sacrificial iron anode either with or without an externally applied current.

The effectiveness of environmental iron additions against sulfide corrosion of copper-nickel alloys was evaluated (Ref 44, 45). Iron added continuously at a level of 0.2 mg/L by a stimulated iron anode was effective against low level (0.01 mg/L) sulfide corrosion of both C70600 and C71500, although some attack was still observed. Corrosion, already actively proceeding, was significantly reduced, and the effects of additional low-level sulfide exposure were nullified by ferrous ion (Fe^{2+}) treatment. Intermittent injection of $FeSO_4$ for 2 h per day at 1.0 to 5.0 mg/L was not found effective against high sulfide levels (0.2 mg/L), but was effective in reducing corrosion at lower sulfide levels (0.01 to 0.04 mg/L). Additional work demonstrated that continuous low-level additions of $FeSO_4$ could counteract sulfide-accelerated corrosion of copper-nickel alloys (Fig. 14).

In the use of $FeSO_4$ or stimulated iron anodes to counteract sulfide-induced corrosion, it should also be considered that iron additions affect heat-exchanger efficiency. The continued use of iron additions can result in a significant buildup of scale on the tube surface. At high enough levels of iron addition, sufficient sludge or precipitate may develop to result in complete blockage of the heat-exchanger tubes. At lower levels of iron addition, a bulky deposit will develop on the tube surface that may also interfere with heat transfer.

In a study of the increase in deposit formation and loss of heat transfer for aluminum brass in seawater with both intermittent and continuous Fe^{2+} ion dosing, it was recommended that some consideration be given to a gradual reduction in dosing levels after the initial film formation (Ref 47).

Other preventive measures can be taken to minimize the deleterious effects of sulfides (Ref 48-50). Elimination of decaying plant and animal life from inlet pipes and channels can alleviate the effects of sulfate-reducing bacteria. Initial design or operational procedures, such as eliminating stagnant legs in a piping system or careful use of screening and filtration systems, can yield a valuable return on investment. Aeration of the seawater, such as by the use of cooling towers or cascading systems, also helps to displace any dissolved hydrogen sulfide (H_2S). In one study, impingement tests were performed on C71500 in seawater containing 10 mg/L cystine (an organic sulfur compound) and varying amounts of an inhibitor, sodium dimethyldithiocarbamate (Ref 50). The results indicated a reduction in the depth of impingement attack. It was noted, however, that a 0.10% solution would be cost prohibitive on a once-through basis, but would be cost effective if circulated through the shipboard piping system upon first flooding and upon shutting down. It was further noted that inhibitor injection is necessary only when the cooling water source is polluted estuarine seawater.

Biofouling. Copper alloys, including the copper-nickels, have long been recognized for their inherent resistance to marine fouling. This fouling resistance is usually associated with macrobiological fouling, such as barnacles, mussels and marine invertebrates of corresponding size. Service experience with shrimp trawlers and private yachts fabricated with C70600 or C71500 hulls has demonstrated excellent resistance to hard-shell fouling and an accompanying reduction in hull maintenance costs (Ref 49). Copper-nickel alloys have also performed successfully as seawater intake screens by virtue of their mechanical strength, corrosion resistance, and resistance to biofouling (Ref 50).

Research demonstrated that fouling was not observed on copper-nickel alloys containing 80% or more copper and that only incipient fouling was noted on the 70Cu-30Ni alloy (Ref 51, 52). More recent evaluations indicated approximately equivalent fouling resistance for C70600 and C71500 in 14- and 5-year exposures, respectively (Ref 16, 53). One investigation concluded that the fouling resistances of pure copper, C70600, and C71500 were virtually identical (Ref 53).

Studies of copper-nickel alloys found that some minimum copper solution rate from the corrosion process is required to prevent fouling (Ref 52). It was not established whether the effect was due to toxicity of copper ions released from the metal surface or to a continual sloughing off of corrosion products. Fouling was minimal on C71500 exposed for 14 years, during which time the corrosion rate approached 1.0 μm/yr (0.04 mils/yr) (Ref 16). It was further demonstrated that copper ions released from a bare C70600 surface offered no fouling protection to an adjacent painted surface (Ref 53). This work concluded that the duplex nature of corrosion products on copper alloy surfaces is responsible for fouling resistance. The initial film formed on copper alloys exposed to seawater is Cu_2O. This inherently fouling-resistant material subsequently oxidizes to $CuCl_2 \cdot 3(Cu(OH)_2)$, which does not appear to be as toxic to marine organisms. The $CuCl_2 \cdot 3(Cu(OH)_2)$ periodically sloughs off from the material surface, carrying with it many marine organisms that may have attached. This reexposes the adherent, toxic Cu_2O film and renews fouling resistance.

Whatever the mechanism, the resistance to fouling is a result of corrosion of the alloy. If this is suppressed by galvanic effects or impressed cathodic protection, fouling will not be prevented.

Biofouling growth was studied on titanium and C70600 at 27 °C (80 °F) and at various velocities (Ref 54). Results (Fig. 15) indicated that the major fouling problem on titanium in the tests was silt particles bound by organic growths, while C70600 is fouled both by silt and corrosion products. Increasing velocity removes more of the silt and binding organisms, but not the corrosion products. Because titanium does not produce corrosion products, the change in the fouling rate with increasing velocity was more dramatic. The behavior of C70600 suggested the periodic sloughing off of portions of the fouling layer previously noted (Ref 53). At sufficient velocities (1.8 and 2.4 m/s, or 6 to 8 ft/s), macroorganisms did not adhere to the C70600 surface, and heat transfer resistance was due to corrosion products and entrapped particles. Fouling rates decrease by a factor of ten on titanium with an increase in velocity from 0.6 to 2.4 m/s (2 to 8 ft/s) and decrease by a factor of five on C70600 for the same velocity range.

Other studies demonstrated the excellent resistance to fouling and resulting retention of heat transfer efficiency in natural seawater of the copper alloys (Ref 55, 56). Figure 16 shows corrosion data for C70600 specimens. The relatively infrequent sponge ball mechanical cleaning did not increase corrosion of the C70600 compared to uncleaned controls. Mechanical cleaning was required much more frequently for the titanium in order to maintain a given level of heat transfer efficiency. Intermittent chlorination did increase the initial corrosion rates, although the rates were comparable to uncleaned controls after approximately 90 days. By contrast, in other tests in which excessive mechanical cleaning was used in natural seawater, a significant acceleration of corrosion occurred with daily sponge ball cleaning at a rate of 12 passes/h (Ref 57).

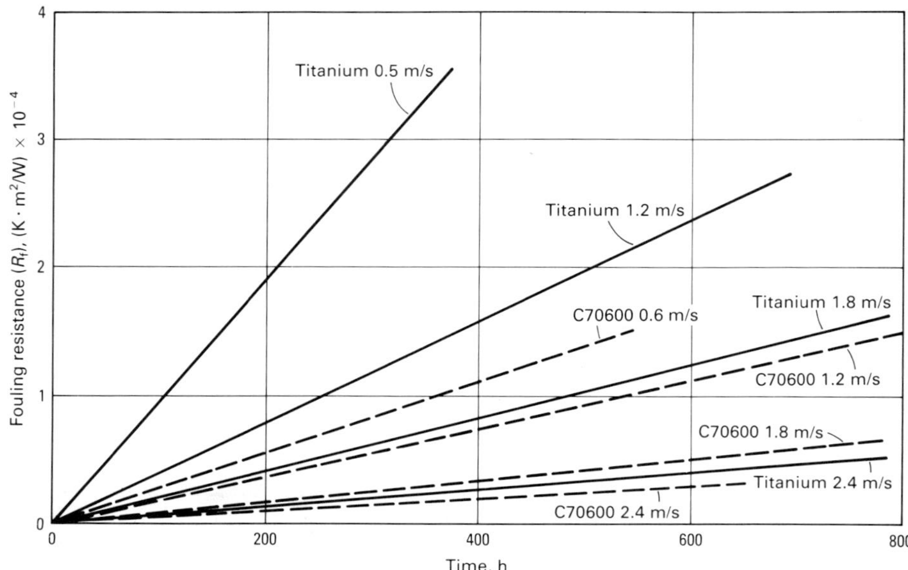

Fig. 15 Fouling rates of C70600 and titanium as a function of seawater velocity. Source: Ref 54

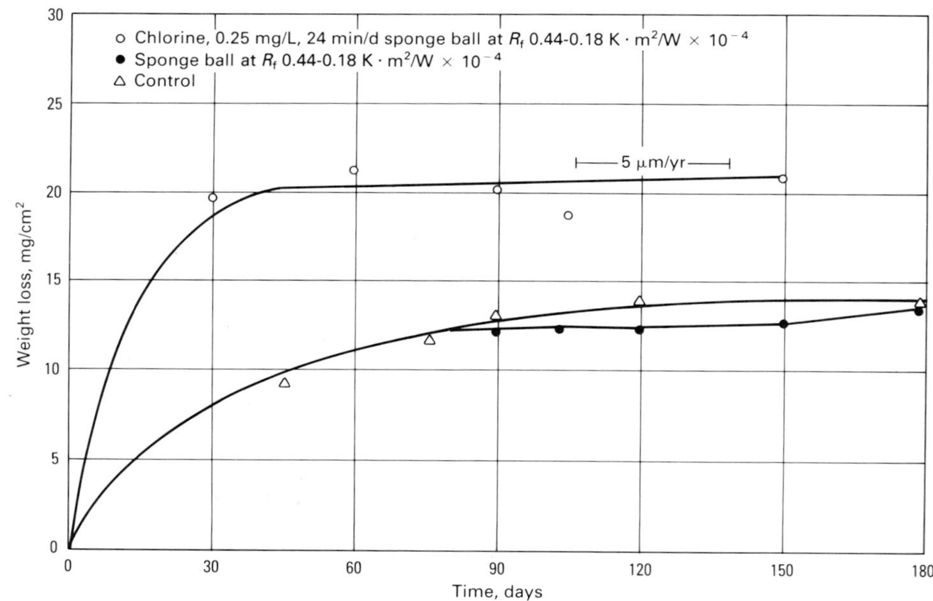

Fig. 16 Weight loss/corrosion data for C70600 cleaned by chlorinated sponge ball and sponge ball without chlorination

Heat Exchangers and Condensers. The selection of material for condenser and heat-exchanger tubes necessitates a survey of service conditions, an examination of tubes previously used and evaluation of its service life, and a review of the type, form, and location of corrosion experienced in the unit or in similar units. Types of water and operating conditions vary widely, and any estimate of probable tube performance must be based on specific operating factors. The tubes of the various alloys discussed in this section provide satisfactory and economical performance for the services described.

Inhibited admiralty metal (C44300, C44400, and C44500) has good corrosion resistance and is extensively used for tubing in various services, especially steam condensers cooled with fresh, salt, or brackish water. Admiralty metal tubes are also used for heat exchangers in oil refineries, in which corrosion from sulfur compounds and contaminated water may be very severe, and for feedwater heaters and heat-exchanger equipment as well as other industrial processes. Admiralty metal tubes are often used in equipment operating at temperatures of 200 °C (400 °F) or higher. Small amounts of phosphorus (0.02 to 0.06%) added to admiralty metal markedly increase dezincification resistance.

Inhibited aluminum brass (C68700) resists the action of high-velocity salt and brackish water and is commonly used for condenser tubes. The outstanding characteristic of C68700 is its high resistance to impingement attack. Tubes of this alloy are frequently recommended for use in marine and land power stations, in which cooling water velocities are high and inhibited admiralty metal tubes have failed because of impingement attack.

Aluminum Bronzes. Tube sheets made of C61300 and C63200 have been specified for coastal power station condensers. Alloy C61300 is also used for emergency raw seawater cooling system piping in coastal nuclear power plants. The aluminum bronzes of C61300, C63000, and C63200 in wrought form and C95400, C95500, and C95800 in cast form are extensively used in salt water environments. They are used in Navy seawater systems and submarine systems in pumps, valves, heat exchangers, and structural components for mounting electronic gear and propulsion units and are even more widely used in minesweepers, for which their nonmagnetic characteristics are important. They are used in cast or wrought form for tube sheets and water boxes in saltwater evaporators and in seawater cooling loops in fossil and nuclear power plants. Corrosion rates are of the order of 10 to 50 μm/yr (0.4 to 2 mils/yr) depending on temperature and velocity, and generally decrease with time. Temper annealing is particularly important in the cast forms of these alloys when used in seawater.

Copper nickel, 10% (C70600) exhibits excellent resistance to impingement attack; it appears to be inferior only to copper nickel, 30%. It is also highly resistant to SCC. This alloy is suitable for marine condenser tube installations in place of aluminum brass, especially where higher water velocities are encountered.

Copper nickel, 30% (C71500) has, in general, the best resistance of any of the copper alloys to impingement attack and to corrosion by most acids and waters. It is being used in increasing quantities under severely corrosive conditions for which service lives longer than those of other copper alloys are desired. It is used by the United States Navy for most shipboard condensers and heat exchangers.

Phosphorus-deoxidized coppers (C12000 to C12300) are extensively used in sugar refineries for condensers and evaporators. Deoxidized copper and standard materials in the refrigeration industry and for transferring heat from steam to water or air, because of their excellent resistance to corrosion by freshwater and their high thermal conductivities.

Bimetal tubes are sometimes used to meet severe corrosion problems not handled adequately by tubes of a single metal or alloy. Two tubes of different alloys, one inside the other, form one integral tube. Copper may be the inner or outer layer, depending on the application.

Drain Tubes. Copper is used for waste and vent lines in drains. The first such installations were made in the mid-1930s, and since then, many municipalities have approved the use of copper drain lines. Development of Sovent fittings now enables construction of a single-stack drain system in high-rise buildings instead of the two-stack system formerly used.

Corrosion in Acids

Copper is widely employed for industrial equipment used to handle acid solutions. A fairly definite separation exists between those acids that can be handled by copper and those that cannot. In general, copper alloys are successfully used with nonoxidizing acids, such as

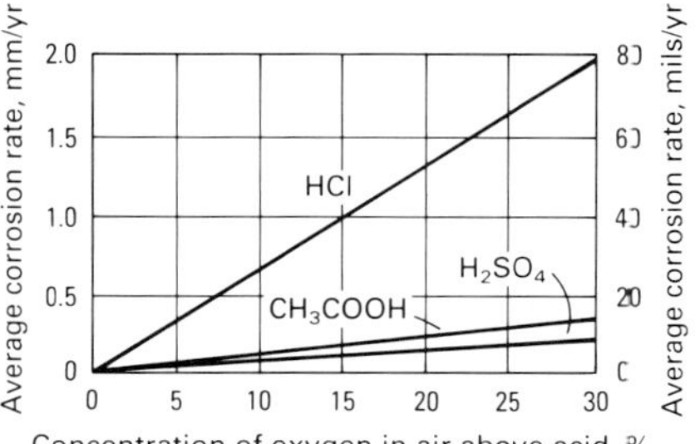

Fig. 17 Effect of oxygen on corrosion rates for copper in 1.2 *N* solutions of nonoxidizing acids. Specimens are immersed for 24 h at 24 °C (75 °F). Oxygen content of the solutions varied from test to test, depending on the concentration of oxygen in the atmosphere above the solutions.

CH_3COOH, H_2SO_4, HCl, and H_3PO_4, as long as the concentration of oxidizing agents, such as dissolved oxygen (air) and ferric (Fe^{3+}) or dichromate ions, is low. Broadly speaking, a thoroughly agitated or stirred solution or one into which a stream of air has been bubbled approaches air saturation and is therefore not a suitable acid medium for copper. Acids that are oxidizing agents in themselves, such as HNO_3; sulfurous (H_2SO_3); hot, concentrated H_2SO_4; and acids carrying such oxidizing agents as Fe^{3+} salts, dichromate ions, or permanganate (MnO_4^-) ions, cannot be handled in equipment made of copper or its alloys.

The corrosive action of a dilute (up to 1% acid) nonoxidizing acid on copper is relatively low; corrosion rates are usually less than 6 g/m²/d (equivalent penetration rate: 250 μm/yr, or 10 mils/yr). This is true only of oxidizing acids when the concentration does not exceed 0.01%. At such low acid concentrations, aeration has little effect in either oxidizing or nonoxidizing acids.

Nonoxidizing acids with near-zero aeration have virtually no corrosive effect. Rates in 1.2 *N* H_2SO_4, HCl, and CH_3COOH are less than 0.1 g/m²/d (4 μm/yr, or 0.15 mils/yr) in the absence of air. Figure 17 shows the general effect of various concentrations of oxygen on the corrosion rate of copper in these acids.

Except for HCl, nonoxidizing acids that contain as much air as is absorbed in quiet contact with the atmosphere are weakly corrosive. Rates generally range from 0.5 to 6 g/m²/d (approximately 20 to 250 μm/yr, or 0.8 to 10 mils/yr).

Air-saturated solutions of nonoxidizing acids are likely to be strongly corrosive, with corrosion rates of 5 to 30 g/m²/d (0.2 to 1.25 mm/yr, or 8 to 50 mils/yr). This rate is higher for HCl. The actual corrosion in any aerated acid depends on acid concentration, temperature, and other factors that are difficult to classify. Except in very dilute solutions, oxidizing acids corrode copper rapidly—usually at rates above 50 g/m²/d (2.1 mm/yr, or 85 mils/yr). The reaction is independent of aeration.

The corrosion rates of three common acids are compared below (temperature and aeration are not specified):

Acid	Corrosion rate g/m²/d	mm/yr	mils/yr
32% HNO_3	5700	240	9450
Concentrated HCl	18	0.75	30
17% H_2SO_4	2	0.1	4

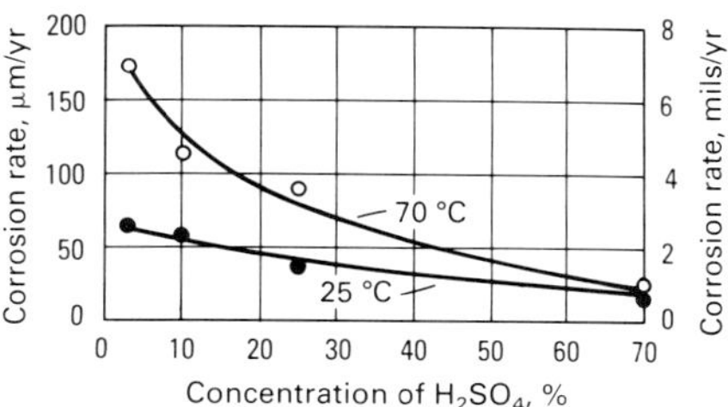

Fig. 18 Corrosion of C65500 in H_2SO_4 solutions. Specimens were immersed for 48 h at the indicated temperatures. The solution was not agitated or intentionally aerated.

Phosphoric, CH_3COOH, tartaric, formic, oxalic, malic, and similar acids normally react comparably to H_2SO_4. Many of the copper alloys can be brazed with brazing rod of the same composition, which provides a joint that is about as corrosion resistant in acids as the base metal.

Factors that may accelerate corrosion vary from one plant to another, and it is advisable to conduct preliminary service or field tests under actual operating conditions before purchasing large quantities of an alloy. Corrosion-accelerating factors can then be evaluated. Selection of the most suitable material for use in a chemical process depends not only on corrosion resistance but also on such factors as continuing availability of the alloy in the desired form and size (which should be ensured before any alloy is given serious consideration).

The following corrosion data were obtained in tests made under various conditions for handling different acids and acid solutions. Because of the variety of factors affecting all chemical reactions, the values shown cannot be taken as absolute and should be considered only as trends.

Sulfuric Acid. The corrosion rate of C65500 (3% silicon bronze) in H_2SO_4 indicates that this alloy can be successfully used with solutions of 3 to 70% H_2SO_4 (by weight) at temperatures of 25 to 70 °C (75 to 160 °F). Laboratory test results are shown in Fig. 18.

Rate of attack by H_2SO_4 varies with concentration (Table 15). The presence of copper or iron salts in acid solutions accelerates the corrosion rate of copper (Table 16).

Aluminum bronze C61300 (wrought), as well as C95200 and C95800 (cast), are used extensively in dilute (10 to 20%) H_2SO_4 service, particularly in steel-pickling acids. Because these alloys have good corrosion resistance and high mechanical properties, thinner sections can withstand the required loads. In general, the copper alloys are quite resistant to the environment, but when in contact with the steel being pickled, they are galvanically protected and in turn accelerate the cleaning action of the acid on the steels. In time, the iron salts are changed from Fe^{2+} to Fe^{3+} (oxidizing) form, and there is increased corrosion; therefore, filtering or elimination of the salts is beneficial. Also, open tanks made of copper for this medium will have a higher corrosion rate at the liquid level line because of higher oxygen concentration. Hydrochloric acid added to H_2SO_4 greatly increases the corrosion rate of copper alloys compared to that in either acid individually.

Phosphoric Acid. Copper and copper alloys are used in heat-exchanger tubes, pipes, and fittings for handling H_3PO_4, although the corro-

Table 15 Corrosion of copper alloys completely immersed in H_2SO_4 of various strengths

Alloy	Average penetration for H_2SO_4 concentration of: 30%, μm/yr	30%, mils/yr	40%, μm/yr	40%, mils/yr	50%, μm/yr	50%, mils/yr
Exposure time 24–48 h, boiling at a pressure of 13.3 kPa (100 torr)						
C11000	670–700	26.4–27.6	487–700	19.2–27.6	660–792	26.0–31.2
C14200	640–670	25.2–26.4	487–548	19.2–21.6	610	24.0
C51000	640	25.2	395–457	15.6–18.0	915	36.0
C26000	...	...	...	...	...	...
Exposure time: 16–24 h, solution agitated						
C11000	60–245	2.4–9.6	18–60	0.7–2.4	60	2.4
C14200	92–335	3.6–13.2	nil	nil	50–60	2.0–2.4

Alloy	Average penetration for H_2SO_4 concentration of: 60%, μm/yr	60%, mils/yr	70%, μm/yr	70%, mils/yr	80%, μm/yr	80%, mils/yr
Exposure time: 24–48 h, boiling at a pressure of 13.3 kPa (100 torr)						
C11000	2195–2255	86.4–88.8	853–1067	33.6–42.0	39 630–166 420	1560–6552
C14200	2285–2377	90.0–93.6	945	37.2	67 310–527 300	650–20 760
C51000	2957–3385	116.4–133.2	945–1067	37.2–42.0	60 660–62 080	2388–2444
C26000	...	...	580–793	22.8–31.2	72 850–206 050	2868–8112
Exposure time: 16–24 h, solution agitated						
C11000	60–92	2.4–3.6	1830–2745	72.0–108.0	39 370–40 890	1550–1610
C14200	15–60	0.6–2.4	2135	84.0	39 370–50 550	1550–1990

Table 16 Corrosion of copper in boiling 30% H_2SO_4 containing copper and iron salts

Copper, ppm	Average penetration, μm/yr	mils/yr	Iron, ppm	Average penetration, μm/yr	mils/yr	Iron and copper, ppm	Average penetration, μm/yr	mils/yr
0	60	2.4	0	122	4.8	0	13	0.5
20	183	7.2	28	122	4.8	20Cu + 28Fe	152	6.0
40	213	8.4	58	245	9.6	40Cu + 56Fe	244	9.6
80	243	9.6	112	427	16.8	80Cu + 112Fe	457	18.0
200	335	13.2	196	782	30.8	200Cu + 196Fe	730	28.8
280	360	14.2	280	975	38.4	280Cu + 280Fe	1005	39.6
360	427	16.8	364	1097	43.2	360Cu + 364Fe	1250	49.2
440	457	18.0	447	1280	50.4	440Cu + 447Fe	1525	60.0

sion rates of some of these alloys may be comparatively high. Laboratory tests were performed on eight groups of copper alloys in aerated and unaerated acid, with specimens at the water line, in quiet immersion and totally submerged. Acid concentrations ranged from 5 to 90%, and temperatures ranged from 20 to 85 °C (70 to 185 °F) except for the copper-aluminum-silicon alloy, which was tested only in 6.5% H_3PO_4 at 20 °C (70 °F) with specimens at the water line and in quiet immersion. Corrosion rates for the eight alloy groups were as follows:

Alloy type	Corrosion rate, mm/yr	mils/yr
Copper	0.55–3.7	22–148
Copper-zinc (70% Cu min)	0.13–7.0	5–280
Copper-tin	0.025–1.30	1–52
Copper-nickel	0.025–0.63	1–25
Copper-silicon	0.13–0.93	5–37
Copper-aluminum-iron	0.13–0.25	5–10
Copper-aluminum-silicon	0.28–2.4	11–97

In general, copper and copper alloys provide satisfactory service in handling pure H_3PO_4 solutions in various concentrations. The acid concentration seems to have less effect on the corrosion rate than the amount of impurities. The impure H_3PO_4 produced by the H_2SO_4 process may contain a markedly higher concentration of Fe^{3+}, SO_4^{2-}, sulfite (SO_3^{2-}), Cl^-, and fluoride (F^-) ions than acid produced by the electric furnace process. These ions increase the corrosion rate up to 150 times, which limits the service lives of copper alloys.

Pure H_3PO_4 produced by the electric furnace process contains only small quantities of impurities and is therefore only slightly corrosive to copper and its alloys. Inhibited admiralty metals C44300, C44400, and C44500 are suggested for solutions of pure H_3PO_4.

Accumulation of corrosion products on metal surfaces may also increase both the rate of corrosion and the possibility of pitting. Low-copper alloys, such as C46400 (naval brass), appear to form thin, adherent films of corrosion products. Copper, copper-silicon alloys, and other high-copper alloys form more voluminous, porous films or scales beneath which roughened or pitted surfaces are likely to be found.

The H_3PO_4 vapors that condense in electrostatic precipitators at about 120 °C (250 °F) are noticeably more corrosive than solutions of pure H_3PO_4 at the same or lower temperatures. The corrosion rates encountered in precipitators are so high that copper alloy wires will not give satisfactory service as electrodes. The high rate of corrosion is probably caused by an abundant supply of oxygen.

Although the corrosion rates of copper cooling tubes in H_3PO_4 condensation chambers are high (about 10 mm/yr, or 400 mils/yr), the rates are lower than those of some other materials. Therefore, the use of copper tubes is feasible for this application.

Table 17 Corrosion of C65800 totally submerged in HCl

Size of specimens, 50 × 25 × 1.3 mm (2 × 1 × 0.050 in.); surface condition, pickled; velocity of solution, natural convection; aeration, none; duration of test, 48 h

HCl concentration, wt%	Corrosion rate, g/m²/d	μm/yr	mils/yr
At 25 °C (75 °F)			
3	2.3	99	3.9
10	2.3	99	3.9
20	1.8	79	3.1
35	12.3	526	20.7
At 70 °C (160 °F)			
3	18.3	780	30.7
10	13.7	508	20.0
20	23.8	102	40.1
35	160.8	6860	270.1

The above discussion on the effect of H_3PO_4 on copper and its alloys emphasizes the value of keeping service records. Such records are valuable for anticipating repairs, making changes to minimize the effect of various factors, and selecting materials for replacement parts.

Hydrochloric acid is one of the most corrosive of the nonoxidizing acids when in contact

Table 18 Corrosion of wrought copper alloys in anhydrous HF

Temperature, °C	Temperature, °F	C51000, μm/yr	C51000, mils/yr	C44400, μm/yr	C44400, mils/yr	C71500, μm/yr	C71500, mils/yr
16–27	60–80	510	20	255	10	180	7
27–38	80–100	480	18.8	480	18.8	...	...
82–88	180–190	1525	60	510	20	255	10

Corrosion rate(a)

(a) These values are representative of results on copper alloys having high copper content, such as copper, aluminum bronze, silicon bronze, and inhibited admiralty metal. Corrosion rates for C23000 are between those for C44400 and C51000.

Table 19 Corrosion of copper in CH_3COOH-$(CH_3CO)_2O$ mixtures

Copper alloy	Exposure time, h	Test conditions	Average penetration rate, μm/yr	Average penetration rate, mils/yr
C11000	1115	CH_3COOH-$(CH_3CO)_2O$-acetone mixture, 110 to 140 °C (230 to 285 °F)	483	19.0
	2952	Same as above	66–70	2.6–2.8
C65500	1115	Same as above	213	8.4
	2952	Same as above	70–90	2.7–3.6
C11000	1115	1:1 CH_3COOH-$(CH_3CO)_2O$ mixture, 130 to 145 °C (265 to 295 °F)	120–533	4.7–21.0
C65500	1115	Same as above	116–236	4.6–9.3
C11000	865	95% CH_3COOH-5% $(CH_3CO)_2O$, liquid phase, 120 °C (250 °F)	97–116	3.8–4.4
C11000 coupled to type 316 stainless steel	865	Same as above	102–216	4.0–8.5
C11000	865	95% CH_3COOH-5% $(CH_3CO)_2O$, vapor phase, 120 °C (250 °F)	102–104	4.0–4.1
C11000 coupled to type 316 stainless steel	865	Same as above	94–213	3.7–8.4
C11000	2448	50:50 CH_3COOH-$(CH_3CO)_2O$, 150 °C (300 °F)	84–90	3.3–3.6
C11000	2448	Essentially pure CH_3COOH	5	0.2

Table 20 Corrosion of C11000 in isopropyl ether-CH_3COOH mixtures

Concentration, %: Isopropyl ether	Concentration, %: CH_3COOH	Average penetration rate, μm/yr	Average penetration rate, mils/yr
Exposed 72 h at 60–65 °C (140–150 °F)			
93	7	40–50	1.6–2.0
85	15	18–20	0.7–0.8
Exposed 328 h at 20 °C (70 °F)			
93	7	100	4.0
85	15	13	0.5

with copper and its alloys and is successfully handled only in dilute concentrations. The rates for C65800 in HCl of various concentrations are listed in Table 17. The corrosion rates for two nonstandard silicon bronzes were about the same as those for C65800.

The corrosion rate of copper nickels in 2 *N* HCl at 25 °C (75 °F) may range from 2.3 to 7.6 mm/yr (90 to 300 mils/yr), depending on the degree of aeration and other factors. Specimens of C71000 (copper nickel, 20%) in stagnant 1% HCl solutions at room temperature corrode at a rate of 305 μm/yr (12 mils/yr); in 10% HCl, 790 μm/yr (31 mils/yr).

Hydrofluoric acid (HF) is less corrosive than HCl and can be successfully handled by C71500 (copper nickel, 30%), which has good resistance to both aqueous and anhydrous HF. Unlike some other copper alloys, C71500 is not sensitive to velocity effects. The data given in Table 18 were generated from laboratory tests in conjunction with the HF alkylation process in anhydrous acid.

Acetic Acid and Acetic Anhydride [$(CH_3CO)_2O$]. Copper and copper alloys are successfully used in commercial processes involving exposure to CH_3COOH and related chemical compounds or in the manufacture of this acid. One plant kept records concerning the corrosion rate of C11000 used in two different CH_3COOH still systems. One still operated at 115 to 140 °C (240 to 285 °F) and handled a solution containing 50% CH_3COOH and about 50% $(CH_3CO)_2O$, with some esters also present. After operating for 663 h, the kettle showed an average penetration rate of 210 μm/yr (8.4 mils/yr). The rate was lower (60 μm/yr, or 2.4 mils/yr) for the bottom column and was lower yet (30 μm/yr, or 1.2 mils/yr) for the middle and top columns. A second still operating at 60 to 140 °C (140 to 285 °F) contained a 70% solution of CH_3COOH, the remainder being anhydride, esters, and ketones. After 1464 h, the kettle showed a corrosion rate of 120 μm/yr (4.8 mils/yr). The rate was only 30 μm/yr (1.2 mils/yr) for the middle and top columns.

In another field test, C11000 and C65500 coupons were placed in an CH_3COOH storage tank at ambient temperature. The stored solution contained 27% CH_3COOH, 1% butyl acetate, 70% H_2O, and small amounts of acetates, aldehydes, and other acids. During the 3984-h exposure, the specimens were immersed in the liquid phase 80% of the time and were in the vapor phase 20% of the time. The C11000 specimens showed a corrosion rate of 38 to 53 μm/yr (1.5 to 2.1 mils/yr); the C65500 specimens, 30 to 45 μm/yr (1.2 to 1.8 mils/yr).

The results of other field tests for C11000 and C65500 exposed in CH_3COOH mixtures are given in Tables 19 to 21. Test conditions involved various temperatures, concentrations, exposure times, locations in equipment as well as the presence of other chemicals.

In laboratory tests at room temperature, C61300 and C62300 exhibited typical corrosion rates of 65 to 80 μm/yr (2.5 to 3.2 mils/yr) in 10 to 40% CH_3COOH. The copper-aluminum alloys are suitable for use in CH_3COOH and the range of aliphatic and aromatic organic acids. The addition of chlorine atoms to the organic molecule will not increase the tendency toward pitting or crevice corrosion. Alloy C61300 is extensively used for pressure and valve castings.

Hydrocyanic acid (HCN) can be successfully handled by copper and copper alloys. Results of field tests for C11000 and C65500 are given in Tables 22 and 23.

Fatty Acids. Under severe service conditions, fatty acids attack copper alloys at somewhat higher rates than other organic acids, such as CH_3COOH or citric. Tests were conducted for 400 h in a copper-lined wooden splitting tank containing a mixture of about 60% fatty acids, 39% H_2O, and 1.17% H_2SO_4 heated to 100 °C (212 °F) and agitated violently with an open steam jet. Specimens of C71000 (copper nickel, 20%) showed a corrosion rate of 64 μm/yr (2.6 mils/yr); specimens of C71500 (copper nickel, 30%), 59 μm/yr (2.4 mils/yr) when submerged just below the liquid level in the tank. Similar specimens submerged 150 mm (6 in.) from the bottom of the tank showed corrosion rates of 178 and 185 μm/yr (7.0 and 7.3 mils/yr) for C71000 and C71500, respectively.

Oleic Acid. Copper and copper-zinc alloys are highly resistant to attack by pure oleic acid. However, oleic acid will attack these alloys when air and water are present. Temperature also influences the rate of attack. Copper and several copper alloys were tested in oleic acid at 25 °C (75 °F); C51000 and C61300 corroded at less than 50 μm/yr (2 mils/yr) compared with about 500 μm/yr (20 mils/yr) for C26000 and C65500.

Stearic acid, like all other fatty acids, attacks copper and copper alloys when moisture and air are present. Temperature and impurities also influence the rate of attack. Tests made at 25 to 100 °C (75 to 212 °F) in stearic acid showed corrosion rates of C11000, C26000, and C65500 to be in the range of 500 to 1250 μm/yr (20 to 50 mils/yr).

Tartaric Acid. Copper and its alloys corrode rather slowly when exposed to various concentrations of tartaric acid, as indicated by the laboratory test data given in Table 24.

Corrosion in Alkalies

Copper and its alloys resist alkaline solutions, except those containing NH_4OH or compounds that hydrolyze to NH_4OH or cyanides. Ammonium hydroxide reacts with copper to form soluble complex copper cations, but the cyanides react to form soluble complex copper anions. The rate of attack for copper-zinc alloys exposed to alkalies other than those specified above is about 50 to 500 μm/yr (2 to 20 mils/yr) at room temperature under stagnant conditions, but is about 500 to 1750 μm/yr (20 to 70 mils/yr) in aerated boiling solutions.

Alloy C71500 corrodes at less than 5 μm/yr (0.2 mil/yr) in 1 *N* to 2 *N* NaOH solutions at room temperature, and the degree of aeration usually has no significant effect. This rate is two to three times as great as the rate in boiling solutions. Copper-tin alloys (phosphor bronzes) corrode at less than 250 μm/yr (10 mils/yr) in 1 *N* to 2 *N* NaOH solutions at room temperature and are apparently unaffected by aeration.

Copper and two grades of silicon bronze were tested in a 50% NaOH solution at 60 °C (140 °F) for 4 weeks. The specimens were bright rolled and degreased sheet measuring about 25 × 50 × 1.3 mm (1 × 2 × 0.05 in.). The solution was exposed to air (no additional aeration), and velocity was limited to natural convection. Alloy C11000 showed a corrosion rate of 1.7 g/m²/d (70 μm/yr, or 2.8 mils/yr); C65100, 1.5 g/m²/d (63 μm/yr, or 2.5 mils/yr); and C65500, 1.1 g/m²/d (47 μm/yr, or 1.85 mils/yr).

Ammonium Hydroxide. Strong NH_4OH solutions attack copper and copper alloys rapidly, as compared with the rates of attack by metallic hydroxides because of the formation of a soluble complex copper-ammonium compound. However, in some applications, the corrosion of copper exposed to dilute solutions of NH_4OH is low. For example, copper specimens submerged in 0.01 *N* NH_4OH solution at room temperature for 1 week experienced weight loss of 1.5 m/m²/d (60 μm/yr, or 2.5 mils/yr).

Ammonium hydroxide solutions also attack copper-zinc alloys. Alloys containing more than 15% Zn are susceptible to SCC when exposed to NH_4OH. The stress may be due to applied service loads or to unrelieved residual stresses. In quiescent 2 *N* NH_4OH solutions at room temperature, copper-zinc alloys corrode at 1.8 to 6.6 mm/yr (70 to 260 mils/yr), copper-nickel alloys at 0.25 to 0.50 mm/yr (10 to 20 mils/yr), copper-tin alloys at 1.3 to 2.5 mm/yr (50 to 100 mils/yr), and copper silicon alloys at 0.75 to 5 mm/yr (30 to 200 mils/yr).

Anhydrous NH_3. Copper and its alloys are suitable for handling anhydrous NH_3 if the NH_3 remains anhydrous and is not contaminated with water and oxygen. In one test conducted for 1200 h, C11200 and C26000 each showed an average penetration of 5 μm/yr (0.05 mil/yr) in contact with anhydrous NH_3 at atmospheric temperature and pressure. Tests showed the rates of corrosion to be low in the presence of small amounts of water, but oxygen was probably excluded. Table 25 lists data on exposure for 1600 h. For any new installation, tests simulating the expected conditions are recommended.

Corrosion in Salts

Copper metals are widely used in equipment for handling saline solutions of various kinds, particularly those that are nearly neutral. Among these are the nitrates, sulfates, and chlorides of sodium and potassium. Chlorides are usually more corrosive than the other salts, especially in strongly agitated, aerated solutions.

The nonoxidizing acid salts, such as the alums and certain metal chlorides (magnesium and calcium chlorides) that hydrolyze in water to produce an acidic pH, exhibit essentially the same behavior as dilute solutions of the corresponding acids. Corrosion rates generally range from 2.5 to 1500 μm/yr (0.1 to 60 mils/yr) at room temperature, depending on the degree of aeration and the acidity. Table 26 lists test data for corrosion of copper in 30% calcium chloride ($CaCl_2$) refrigeration brine with and without inhibitors.

Neutral saline solutions can be successfully handled by copper alloys. Consequently, these alloys are used in heat-exchanger and condenser equipment exposed to seawater. Corrosion rates of copper in NaCl brine are given in Table 27. These rates are not necessarily the same as those in seawater.

Such alkaline salts as sodium silicate (Na_2SiO_3), sodium phosphate (Na_3PO_4), and sodium carbonate (Na_2CO_3) attack copper alloys at low but different rates at room temperature. On the other hand, alkali cyanide is aggressive and attacks copper alloys fairly rapidly because it forms a soluble complex copper anion. Table 28 provides specific corrosion rates.

Oxidizing salts corrode copper and copper alloys rapidly; therefore, copper metals should not be used with oxidizing saline solutions except those that are very dilute. Aqueous sodium dichromate ($Na_2Cr_2O_7$) solutions can be safely handled by copper alloys, but the presence of a highly ionized acid, such as H_2CrO_4 or H_2SO_4, may increase the corrosion rate several hundred times, because the dichromate acts as an oxidizing agent in acidic solutions. In one test, a copper-nickel corroded at 2.5 to 250 μm/yr (0.1 to 10 mils/yr) and a copper-tin alloy (phosphor bronze) at 5 μm/yr (0.2 mil/yr) when handling an aqueous $Na_2Cr_2O_7$ solution. The rate increased 200 to 300 times for both metals when H_2CrO_4 was added to the solution. In solutions containing Fe^{3+}, mercuric (Hg^{2+}) or stannic (Sn^{4+}) ions, a copper-nickel showed a corrosion rate of 27.4 mm/yr (1080 mils/yr), while copper-zinc and copper-tin alloys showed a still greater rate of 228 mm/yr (8980 mils/yr).

Salts of metals more noble than copper, such as the nitrates of mercury and silver, corrode copper alloys rapidly, simultaneously plating out the noble metal on the copper surface. Temperature and acidity influence the rate of attack. A film of mercury on high-zinc brass (more than 15% Zn) may cause intergranular cracking by liquid-metal embrittlement (LME) if the alloy is under tensile stress, either residual or applied.

Table 21 Corrosion of copper alloys in CH_3COOH

Alloy	Exposure time, h	Average penetration rate, μm/yr	Average penetration rate, mils/yr
$(CH_3CO)_2O$(a)			
C11000	2448(b)	60	2.4
	2448(c)	915–1100	36.0–43.2
C65500	2448(b)	60	2.4
	2448(c)	488–732	19.2–28.8
90% CH_3COOH(d)			
C11000, annealed	672	60	2.4
	816	30	1.2
C11000, cold worked	672	90	3.6
	792	90	3.6
Copper joint(e)	1512	183	7.2
	4000	120	4.8
Copper joint(f)	1512	183	7.2
	4000	120	4.8
45% CH_3COOH(g)			
C11000	1038	30 max	1.2 max
C65500	1038	30 max	1.2 max
Copper joint(f)	1038	30 max	1.2 max
25% CH_3COOH(h)			
C11000	432	274	10.8
	792	152	6.0

(a) Test specimens were exposed in stills separating CH_3COOH from $(CH_3CO)_2O$. (b) Top of column. (c) Kettle. (d) Test specimens were exposed in cycle feed lines at 30–50 °C (85–120 °F). (e) Joint brazed with BCuP-5 filler metal. (f) BAg filler metal. (g) Test specimens were exposed in the CH_3COOH recovery column, in which concentration of the acetic acid was 45% max. (h) Test specimens were exposed to crude by-product CH_3COOH (approximately 25% concentration) in pump suction line from storage tank.

Table 22 Corrosion of copper alloys in production of HCN

Alloy	Exposure time, h	Stripping still, μm/yr	Stripping still, mils/yr	Top of HCN refining still, μm/yr	Top of HCN refining still, mils/yr	Base of HCN stripping still, μm/yr	Base of HCN stripping still, mils/yr	Base of partial condenser, μm/yr	Base of partial condenser, mils/yr
		Average penetration rate(a)							
C11000	573	173–218	6.8–8.6	54–60	2.1–2.4	1033–1186	40.7–46.7	1534–14 170	60.4–558
	671	155–609	6.1–24.0	18–25	0.7–1.0	nil	nil	478	18.8
C65500	573	229–244	9.0–9.6	18–25	0.7–1.0	777–1145	30.6–45.1	1138–5385	44.8–212
	671	137–503	5.4–19.8	...	...	275	10.8	343	13.5

(a) All data from separate specimens; differences at similar locations imply expected variability.

Table 23 Corrosion of C11000 and C65500 in HCN solutions

Alloy	Exposure time, h	Test conditions	Average penetration rate µm/yr	mils/yr
C11000	3144	Ethylene cyanohydrin residues, 70 °C (160 °F)	5–35	0.2–1.4
C11000	2232	Ethylene cyanohydrin residues, 30 to 90 °C (85 to 195 °F)	13	0.5
C65500	2232	Same as above	40	1.6
C11000	1621	Cyanohydrin stripping still products (kettle)	690	27
C65500	1621	Same as above	35	1.4

Table 25 Corrosion of copper and brass in NH_3

Alloy	Average penetration rate(a): Liquid µm/yr	Liquid mils/yr	Vapor µm/yr	Vapor mils/yr
Anhydrous NH_3				
C11000	2.5	0.1	<2.5	<0.1
C26000	<2.5	<0.1	<2.5	<0.1
Anhydrous NH_3 plus 1% H_2O(b)				
C11000	<2.5	<0.1	<2.5	<0.1
C26000	2.5	0.1	<2.5	<0.1
Anhydrous NH_3 plus 2% H_2O(b)				
C11000	2.5	0.1	2.5	0.1
C26000	5.0	0.2	2.5	0.1

(a) Atmospheric temperature and pressure of 345 to 1035 kPa (50 to 150 psi) for 1600-h exposure. Specimens were placed at the top and bottom of 2-L bombs that were charged with NH_3. Pressure varied throughout the test, depending on temperature. Water was added to two of the bombs before charging with NH_3. (b) Any air present was probably depleted rapidly during initial stages of test.

Table 24 Corrosion of copper alloys in contact with tartaric acid at 25 °C (75 °F)

Acid concentration, %	Corrosion rate µm/yr	mils/yr
C26000 and C23000		
10	50 max	2 max
30	500–1250	20–50
50	500–1250	20–50
100	50 max	2 max
C71000		
5	25 max	1 max
C71300		
2	40	1.6

Table 26 Corrosion of C11000 in 30% $CaCl_2$ refrigeration brine

Inhibitor	Corrosion rate µm/yr	mils/yr
None(a)	10	0.4
$K_2Cr_2O_7$(b)	6	0.23
Aluminum foil(c)	119	4.7

(a) Exposed for 325 days at −12 °C (10 °F). (b) Exposed for 372 days, cold. (c) Exposed for 50 days with slight agitation in brine with a pH of 9

Table 27 Corrosion of C11000 in a NaCl brine refrigeration system

Field test; 98 days at −15 °C (4 °F)

Location in equipment	Corrosion rate µm/yr	mils/yr
With $Na_2Cr_2O_7$ inhibitor; pH 6.0 to 6.5		
Brine tank for near main outlet	5	0.2
Top of brine pump, high agitation	10	0.4
Inside cooler tube	15	0.6
Return line to storage tank	2.5	0.1
Brine tank near agitator	2.5	0.1
Without inhibitor; pH 10.5		
Open brine tank	160	6.3
Brine cooler outlet, rapid flow	360	14.2
Cooler inlet	157	6.2
Cooler outlet	250	9.8

Corrosion in Organic Compounds

Copper and many of its alloys resist corrosive attack by most organic solvents and by organic compounds, such as amines, alkanolamines, esters, glycols, ethers, ketones, alcohols, aldehydes, naphtha, and gasoline. Although the corrosion rates of copper and copper alloys in pure alkanolamines and amines are low, they can be significantly increased if these compounds are contaminated with water, acids, alkalies, salts, or combinations of these impurities, particularly at high temperatures. Tables 29 to 35 list the results of corrosion testing of copper and a limited but representative variety of copper alloys in contact with various organic compounds under many conditions.

Gasoline, naphtha, and other related hydrocarbons in pure form will not attack copper or any of the copper alloys. However, in the manufacture of hydrocarbon materials, process streams are likely to be contaminated with one or more of such substances as water, sulfides, acids, and various organic compounds. These contaminants attack copper and its alloys. Corrosion rates for C44300 and C71500 exposed to gasoline are low (Table 36), and these two alloys are successfully used in equipment for refining gasoline. Table 37 lists corrosion rates for copper and for alloys exposed to contaminated naphtha in two different environments.

Creosote. Copper and copper alloys are generally suitable for use with creosote, although creosote attacks some high-zinc brasses. Alloys C11000, C23000, C26000, C51000, and C65500 typically corrode at rates less than 500 µm/yr (20 mils/yr) when exposed to creosote at 25 °C (75 °F).

Linseed Oil. Copper and its alloys are fairly resistant to corrosion by linseed oil. All of the alloys show some attack, but none exhibits corrosion severe enough to make it unsuitable for this application. Alloys C11000, C51000, and C65500 showed corrosion rates less than 500 µm/yr (20 mils/yr) in linseed oil at 25 °C (75 °F). Alloy C26000 had a rate of 500 to 1250 µm/yr (20 to 50 mils/yr).

Benzol and Benzene. Alloys C11000, C23000, C26000, C51000, and C65500 tested in these materials at 25 °C (75 °F) had corrosion rates under 500 µm/yr (20 mils/yr).

Sugar. Copper is successfully used for vacuum-pan heating coils, evaporators, and juice extractors in the manufacture of both cane and beet sugar. Inhibited admiralty metals, aluminum brass, aluminum bronzes, and copper nickels are also used for tubes in juice heaters and evaporators. Bimetal tubes of copper and steel have been used by manufacturers of beet sugar to counteract SCC of copper tubes caused by NH_3 from beets grown in fertilized soil. Table 38 lists the results of tests conducted on copper and copper alloys in a beet-sugar refinery.

Beer. Copper is extensively used in the brewing of beer. In one installation, the wall thickness of copper kettles thinned from an original thickness of 16 mm (5/8 in.) to 10 mm (3/8 in.) in a 30-year period. Brazing with BAg (copper-silver) filler metals eliminates the possibility that the alkaline compounds used for cleaning copper equipment will destroy joints by attacking tin-lead solders. Steam coils require more frequent replacement than any other component in brewery equipment. They have service lives of 15 to 20 years. The service lives of other copper items exposed to process streams in a brewery range from 30 to 40 years. Additional information on metals and alloys for this application is available in the article "Corrosion in the Brewery Industry" in this Volume.

Sulfur compounds free to react with copper, such as H_2S, sodium sulfide (Na_2S), or potassium sulfide (K_2S), form CuS. Reaction rates depend on alloy composition; the alloys of highest resistance are those of high zinc content.

Strip tensile specimens of seven copper alloys were exposed in a fractionating tower in which oil containing 1.4% S was being processed. The results of this accelerated test are given in Table 39. These data show the suitability of the higher-zinc alloys for use with sulfur-bearing compounds. Alloy C28000 (60Cu-40Zn) showed good corrosion resistance, but C23000 (85Cu-15Zn) was completely destroyed.

Inhibited admiralty metals are also excellent alloys for use in heat exchangers and condensers that handle sulfur-bearing petroleum products and use water as the coolant. Alloys C44300, C44400, and C44500, which are inhibited toward dezincification by the addition of arsenic, antimony, or phosphorus to the basic 70Cu-29Zn-1Sn composition, offer good resistance to corrosion from sulfur as well as excellent resistance to the water side of the heat exchanger.

Table 28 Corrosion of copper alloys in alkaline saline solutions

Alloy family	Common name	Corrosion rate µm/yr	Corrosion rate mils/yr
Na_2SiO_3, Na_3PO_4, or Na_2CO_3			
Copper-zinc	Brasses	50–125	2–5
Copper-tin	Phosphor bronzes	<50	<2
Copper-nickel	Copper nickels	2.5–40	0.1–1.5
NaCN			
Copper-zinc	Brasses	250–500	10–20
Copper-tin	Phosphor bronzes	875	35
Copper-nickel	Copper nickels	500–2500	20–100

Table 29 Corrosion of copper alloys in amine system service

Alloy	Exposure time, h	Test conditions	Average penetration rate µm/yr	Average penetration rate mils/yr
C11000	1622	Coupons exposed in ethylenediamine refining still	nil–180	nil–7
C11000	1580	Aqueous ethylenediamine	25	1
C71500	1580	Same as above	75	3
C11000	806	Liquid vapor containing NH_3 and mono-, di-, and triethanolamines; 90–156 °C (195–315 °F)	760	30
C65500	806	Same as above	790	31.2
C11000	1437	Liquid vapor containing NH_3 and mono-, di-, and triethanolamines; 180–195 °C (355–385 °F)	28	1.1
C65500	1437	Same as above	48	1.9
C11000	2622	Vapor phase of diethanolamine still containing mono-, di-, and triethanolamines; 180–195 °C (355–385 °F)	28	1.1
C26000	887	Denuded monoethanolamine (20%)	nil	nil
C26000 coupled to carbon steel	168	20% monoethanolamine (4 mol CO_2 per mol MEA); 60 °C (140 °F)	4 550	179
C44200	900	Lean solution of diethanolamine containing impurities	50	2
C26000	1440	Rich solution of monoethanolamine	330	13
C11000	1440	Same as above	Dissolved	Dissolved
C26000	1440	Lean solution of monoethanolamine	3 000	118
C11000	1440	Same as above	11 500	454

Corrosion in Gases

Carbon dioxide and carbon monoxide (CO) in dry forms are usually inert to copper and its alloys, but some corrosion takes place when moisture is present. The rate of reaction depends on the amount of moisture. Because some alloy steels are attacked by CO, the high-pressure equipment used to handle this gas is often lined with copper or copper alloys.

Sulfur Dioxide (SO_2). Gases containing SO_2 attack copper in a manner similar to oxygen. The dry gas does not corrode copper or copper alloys, but the moist gas reacts to produce a mixture of oxide and sulfide scale. Table 40 lists the corrosion rates of some copper alloys in hot paper mill vapor that contains SO_2.

Hydrogen Sulfide. Moist H_2S gas reacts with copper and copper-zinc alloys to form CuS. Alloys containing more than 20% Zn have considerably better resistance than lower-zinc alloys or copper. Hot, wet H_2S vapors corrode C26000, C28000, or C44300 at a rate of only 50 to 75 µm/yr (2 to 3 mils/yr), but the rate for C11000 and C23000 under the same conditions is 1250 to 1625 µm/yr (50 to 65 mils/yr).

Halogen Gases. When they are dry, fluorine, chlorine, bromine, and their hydrogen compounds are not corrosive to copper and its alloys. However, they are aggressive when moisture is present. The corrosion rates of copper metals in wet hydrogen compounds are comparable to those given for HF and HCl in Tables 17 and 18.

Hydrogen. Copper and its alloys are not susceptible to attack by hydrogen unless they contain copper oxide. Tough pitch coppers, such as C11000, contain small quantities of Cu_2O. Deoxidized coppers with low residual deoxidizer contents—C12000, for example—may contain Cu_2O, but will contain less than the tough pitch coppers. These deoxidized coppers are not immune to hydrogen embrittlement. Deoxidized coppers with high residual deoxidizer contents, however, are not susceptible to hydrogen embrittlement, because the oxygen is tied up in complex oxides that do not react appreciably with hydrogen.

When oxygen-bearing copper is heated in hydrogen or hydrogen-bearing gases, the hydrogen diffuses into the metal and reacts with the oxide to form water, which is converted to high-pressure steam if the temperature is above 375 °C (705 °F). The steam produces fissures, which decrease the ductility of the metal. This condition is generally known as hydrogen embrittlement. Any degree of embrittlement can lead to catastrophic failure and therefore should be avoided; there is no safe depth of attack.

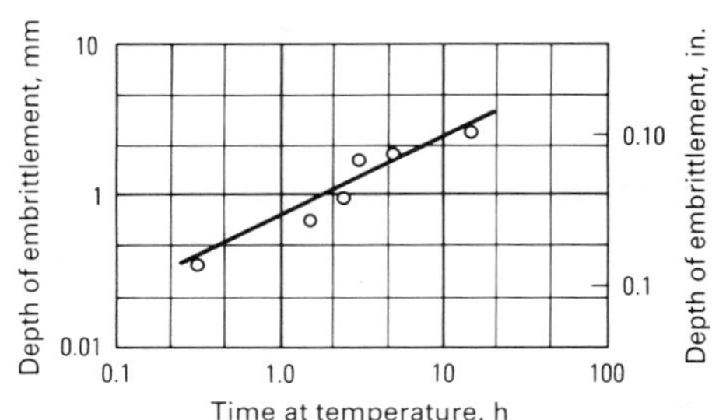

Fig. 19 Hydrogen embrittlement of tough pitch coppers heated in pure hydrogen at 600 °C (1100 °F)

Figure 19 shows the depth of damage, or embrittlement, of C11000 after it has been heated in hydrogen at about 600 °C (1100 °F) for varying times. The reaction is especially important when oxygen-containing copper is bright annealed in reducing atmospheres containing relatively small amounts of hydrogen (1 to 1.5%). Annealing of tough pitch coppers in such atmospheres at temperatures much above 475 °C (900 °F) may lead to severe embrittlement, especially when annealing times are long. In fact, tough pitch coppers should not be exposed to hydrogen at any temperature if they will subsequently be exposed to temperatures above 370 °C (700 °F).

When tough pitch coppers are welded or brazed, the possibility of hydrogen embrittlement must be anticipated, and hydrogen atmospheres must not be used. Where copper must be heated in hydrogen atmospheres, an oxygen-free copper or deoxidized copper with high residual deoxidizer content should be selected. No hydrogen embrittlement problems have been encountered with these materials. Additional information on hydrogen damage in metals is available in the section "Hydrogen Damage" of the article "Environmentally Induced Cracking" in this Volume.

Dry Oxygen. Copper and copper alloy tubing is used to convey oxygen at room temperature, as in hospital oxygen service systems. When heated in air, copper develops a Cu_2O film that exhibits a series of interference tints (temper colors) as it increases in thickness. The colors associated with different oxide film thicknesses are:

Color	Film thickness, nm
Dark brown	37–38
Very dark purple	45–46
Violet	48
Dark blue	50–52
Yellow	94–98
Orange	112–120
Red	124–126

Black cupric oxide (CuO) forms over the Cu_2O layer as the film thickness increases above the interference color range.

Scaling results when copper is used at high temperatures in air or oxygen. At low temperatures (up to 100 °C, or 212 °F), the oxide film increases in thickness logarithmically with time.

Table 30 Corrosion of copper alloys in ester solutions

Alloy	Exposure time, h	Test conditions	Average penetration rate, µm/yr	Average penetration rate, mils/yr
Acetates				
C11000	400	Alkenyl acetate plus H_2SO_4	6100	240
C65500	400	Same as above	3050	120
	257	Allylidene diacetate; 110 °C (230 °F)	183–213	7.2–8.4
C11000	240	Butyl acetate plus 1% H_2SO_4	1625–4090	64–161
C65500	240	Same as above	2870	113
C11000	2328	2-chloroallylidene diacetate	5	0.2
C11000	250	Crude vinyl acetate; 110–150 °C (230–300 °F)	25	1.0
C71500	250	Same as above	7.5–125	0.3–5
C11000	550	Ethyl acetate plus 1.0% H_2SO_4	483	19
C65500	550	Same as above	400	16
C11000	991	Ethyl acetate reaction mixture; liquid; 90 °C (195 °F)	550	21.6
C62300	991	Same as above	395	15.6
C65500	991	Same as above	518	20.4
C11000	991	Ethyl acetate reaction mixture; vapor; 90 °C (195 °F)	5	0.2
C62300	991	Same as above	15	0.6
C65500	991	Same as above	13	0.5
C11000	2976	Ethyl acetoacetate	10	0.4
C65500				
Cold-worked	216	Isopropyl acetate	6700	264
Annealed	216	Isopropyl acetate	6100	240
	480	Isopropyl acetate process; liquid; 120 °C (250 °F)	300	12
C11000	519	Methylamyl acetate process; batch still coils; 115 °C (240 °F)	500–685	22–27
C65500	519	Same as above	280–300	11–12
C11000	519	Methylamyl acetate process; batch still down pipe; 115 °C (240 °F)	330	13
C65500	51	Same as above	300	12
C11000	1345	Methylamyl acetate process; batch still condenser; 30 °C (85 °F)	840–940	33–37
C65500	1345	Same as above	1400–1575	55–62
C63600	3312	Methylamyl acetate process; batch still coils; 95 °C (205 °F)	483	19
C51000	3312	Same as above	430–483	17–19
C60800	3312	Same as above	330	13
C51000	3312	Methylamyl acetate process; batch still downpipe; 95 °C (205 °F)	380–483	15–19
C63600	3312	Same as above	400–460	16–18
C60800	3312	Same as above	280	11
C11000	217	Refined isopropenylacetate; 98 °C (210 °F)	60	2.4
	2784	Vinyl acetate, inhibited	2.5	0.1
C11000	250	Vinyl acetate, process; 150–190 °C (300–375 °F)	355–400	14–16
C71500	250	Same as above	685–1250	27–49
C11000	768	Vinyl acetate process; batch still kettle	685–1170	27–46
C65500	768	Same as above	150–483	6–19
C11000	864	Same as above	2290–3500	90–138
C65500	864	Same as above	660–2160	26–85
Acrylates				
C11000	240	Acidified sodium acrylate containing 5% H_2SO_4; 49 °C (120 °F)	945	37.2
	254	Ethyl acrylate process; 130 to 150 °C (265 to 300 °F)	1220	48
C65500	254	Same as above	430	16.8
C11000	240	Isopropyl ether solution of acrylic acid (18%); 49 °C (120 °F)	18	0.7
	240	Sodium acrylate solution containing 1% NaOH; 49 °C (120 °F)	5	0.2
	240	Washings from isopropyl ether solution of acrylic acid; 49 °C (120 °F)	210	8.3
	240	Wet calcium acrylate	240	9.4
	504	2-ethylhexylacrylate process; 95 °C (205 °F)	230–275	9.0–10.8
C65500	504	Same as above	220–275	8.6–10.8
C11000	566	2-ethylhexyl acrylate process; condensate tank; 30 °C (85 °F)	66–74	2.6–2.9
C51000	566	Same as above	60–86	2.4–3.4
C65500	566	Same as above	114–122	4.5–4.8
C11000	566	2-ethylhexyl acrylate process; 120 °C (250 °F)	236–239	9.3–9.4
C51000	566	Same as above	264	10.4
C65500	566	Same as above	328–360	12.9–14.2
Benzoates				
C11000	1680	Butyl benzoate	nil	nil
	1296	Butyl benzoate process; circulating line; 40 °C (100 °F)	800–1025	31.4–40.4
C60800	1296	Same as above	1060	41.8
C65500	1296	Same as above	843–1090	33.2–42.8
C23000	1296	Same as above	790–1085	31.2–42.7
C22000	1296	Same as above	900–985	35.6–38.8
C11000	1296	Butyl benzoate process; 40 °C (100 °F)	280	11.1
C65500	1296	Same as above	350–400	13.7–15.7
C11000	1296	Butyl benzoate process; batch still kettle; 185 °C (365 °F)	7.5–38	0.3–1.5
C65500	1296	Same as above	7.5–25	0.3–1.0
C11000	1680	Methyl benzoate (refined)	2.5	0.1
C11000	1680	Methyl benzoate (copper-free)	7.5	0.3

Scaling rate increases irregularly with further increases in temperature and rises rapidly with pressure up to 1.6 kPa (12 torr). Above 20 kPa (150 torr) the rate of increase is steady. Beyond the interference color range, the growth rate of the oxide film is approximately defined by:

$$W^2 = kt \qquad \text{(Eq 4)}$$

where W is weight gain (or increase in equivalent thickness) per unit area, t is time, and k is a constant of proportionality. Values for k are given in Table 41. Different investigators report different oxidation rates, but those given in Ref 58 appear to be reliable.

Low concentrations of lead, oxygen, zinc, nickel, and phosphorus in copper have little influence on oxidation rate. Silicon, magnesium, beryllium, and aluminum form very thin insulating (nonconductive) oxide films on copper, which protect the metal surface and retard oxidation.

SCC of Copper Alloys in Specific Environments

Properly selected copper alloys possess excellent resistance to SCC in many industrial and chemical environments; nevertheless, cracking has been identified in a significant number of environments. In some cases, the conditions for cracking are very limited and exist only within a narrow range of pH values or a narrow range of potentials. In many cases, the experimental data are limited to a single alloy, and it is not known if the environment is generally deleterious to many copper alloys or to a restricted group of alloys. Data are summarized below for environments in which cracking has been recognized. Additional information is available in the references cited in this section; they should be consulted when selecting a copper alloy for a specific application.

Acetate Solution. Pure copper wire stressed beyond the yield strength was observed to crack in 0.05 N cupric acetate ($Cu(C_2H_3O_2)_2$) (Ref 59). Alloy C26000 is susceptible to cracking in the same solution, and the cracking rate under slow strain rate conditions is a function of both pH and applied potential (Ref 60).

Amines. Alloy C26000 is susceptible to cracking in solutions of methyl amine, ethyl amine, and butyl amine when dissolved copper is present in the solution (Ref 61). Susceptibility is a maximum at a potential approximately 50 mV anodic from the rest potential. Tubing fabricated from C68700 exhibited cracks from the steam side of a condenser system after 3048 h of service in a desalination plant. The most likely cause of the cracking was an amine used as a water treatment chemical (Ref 62).

Ammonia. All copper-base alloys can be made to crack in NH_3 vapor, NH_3 solutions, ammonium ion (NH_4^+) solutions, and environments in which NH_3 is a reaction product. The rate at which cracks develop is critically dependent on many variables, including stress level, specific alloy, oxygen concentration in the liquid, pH, NH_3 or NH_4^+ concentration, copper ions concentration, and potential.

Early work on the stress corrosion of brass in NH_3 provided the following summary of findings (Ref 63):

- Stress-corrosion cracking occurs in a great variety of brasses that differ widely in composition, degree of purity, and microstructure

Table 31 Corrosion of copper alloys in ethers

Alloy	Exposure time, h	Test conditions	Average penetration rate µm/yr	mils/yr
C11000	2784	γ-methylbenzyl ether, N_2 atmosphere	2.5 max	0.1 max
C11000	2784	γ-methylbenzyl ether, air atmosphere	2.5 max	0.1 max
C11000	288	Recovered butyl ether	nil	nil
C65500	288	Same as above	2.5	0.1
C11000	94	Dichloro ethyl ether residues, 80 °C (175 °F)	183–915	7.2–36
C65500	94	Same as above	61–245	2.4–9.6
C11000	71	Crude dichloro ethyl ether, 80 °C (175 °F)	2130–3050	84–120
C65500	71	Same as above	1220–3050	48–120
C11000	70	Dichloro ethyl ether, 80 °C (175 °F)	150	6
C65500	70	Same as above	120	4.8
C11000	70	Dichloro ethyl ether, 100 °C (212 °F)	610	24
C65500	70	Same as above	245	9.6
C11000	70	Dichloro ethyl ether, boiling	183	7.2
C65500	70	Same as above	213	8.4

Table 32 Corrosion of copper alloys in ketones

Alloy	Exposure time, h	Test conditions	Average penetration rate µm/yr	mils/yr
C11000	138	Phenylxylol ketone mixture	41–43	1.6–1.7
C65500	138	Same as above	76	3.0
C11000	163	Pentanedione mixture	46–91	1.8–3.6
C65500	163	Same as above	33–84	1.3–3.3
C12000	43	Diethyl ketone, 30 °C (86 °F)	nil	nil
C12000	42	Diethyl ketone, boiling	nil–7.6	nil–0.3
C12000	43	Methyl *n*-propyl ketone, 30 °C (85 °F)	nil	nil
C12000	42	Methyl *n*-propyl ketone, boiling	nil	nil
C11000	216	Methylamyl ketone, boiling	2.5	0.1
C11000	353	Methyl ethyl ketone, boiling	12.7	0.5
C11000	409	Phenylxylol ketone containing NaOH	457–518	18–20.4
C65500	409	Same as above	701–823	27.6–32.4
C11000	165	Acetone dispersion of cellulose acetate, 56 °C (135 °F)	10.2	0.4
C26000	165	Same as above	5.1	0.2

Table 33 Corrosion of copper alloys in aldehydes

Alloy	Exposure time, h	Test conditions	Average penetration rate µm/yr	mils/yr
C11000	49	Boiling 2-ethylbutyraldehyde	33	1.3
C11000	112	Boiling butyraldehyde	33	1.3
C11000	1752	2-hydroxyadipaldehyde	20–23	0.8–0.9
C11000	168	Diethyl acetal mixture, 45 °C (115 °F)	60–120	2.4–4.8
C65500	168	Same as above	90–150	3.6–6.0
C26000	168	Same as above	90–150	3.6–6.0
C11000	70	2-ethyl-3-propylacrolein, 98 °C (210 °F)	33	1.3
C11000	168	Diacetoxybutyraldehyde, 160 °C (320 °F)	230–240	9.0–9.4
C65500	168	Same as above	75	3.0
C51000	168	Same as above	18–20	0.7–0.8
C11000	540	Propionaldehyde	1420–1550	56.0–61.0
C11000	216	Propionaldehyde, 190 °C (375 °F)	610–1220	24.0–48.0
C11000	443	Butylaldehyde	310	12.2
C51000	443	Same as above	360	14.2
C11000	2374	Same as above	165	6.5
C65500	2374	Same as above	20	0.8
C51000	2374	Same as above	10	0.4

- Cracking occurs only in objects that are subjected to external or internal stresses
- Visible corrosion is frequently associated with the effect, but the corrosion may often be superficial
- Lacquer coatings do not offer complete protection against SCC
- Sufficient and continuous coatings of a metal, such as nickel, confer complete protection
- Highly stressed articles may be kept for years in a clean air atmosphere without developing cracks
- Ammonia and ammonium NH_4^+ salts induce cracking
- Surface defects which localize stresses, do not appear to contribute to the development of cracks in the absence of an essential corroding agent, such as NH_3
- Severe corrosion and pitting do not of themselves lead to cracking
- Cracks often follow an intercrystalline path
- Traces of NH_3 in the environment are an important agent in inducing SCC in atmospheric exposure
- Ammonia has a specific and selective action on the material in the grain boundaries of brass
- Cracking always begins in surface layers that are under tension
- The behavior of a copper alloy subjected to the combined effect of tensile stress and NH_3 is an index of susceptibility to SCC
- Susceptibility to SCC diminishes as the copper content of the brass is increased
- Protracted heating of 70Cu-30Zn brass at 100 °C (212 °F) does not develop cracks and does not reduce the internal stress appreciably

Table 42 provides a ranking of various copper alloys according to their relative SCC susceptibility in NH_3 environments.

Atmosphere. Many natural environments contain pollutants that, in the presence of moisture, may cause stress-corrosion problems (Ref 64). Sulfur dioxide, oxides of nitrogen, and NH_3 are known to induce SCC of some copper alloys. Chlorides may also cause problems. Atmospheric-exposure test data are summarized in Table 43. In these tests, 150- × 13-mm (6- × ½-in.) U-bend samples were stressed in the long-transverse direction. The bend was produced by bending around a 19-mm (¾-in.) diam mandrel, and the legs of each specimen were held in nonconductive jigs during the test. The stress on the specimens was not determined. The stressed specimens were exposed in two industrial locations in New Haven, CT, and Brooklyn, NY, and in one marine location at Daytona Beach, FL.

Chlorate Solutions. Brass was observed to crack intergranularly and transgranularly when immersed in 0.1 to 5 *M* sodium chlorate ($NaClO_3$) solutions at pHs from 3.5 to 9.5 when subjected to slow straining (Ref 66). Crack velocities in 1 *N* $NaClO_3$ at pH 6.5 were 10^{-7} m/s at a crosshead speed of 10^{-4} cm/min (4×10^{-5} in./min) and 10^{-6} m/s at a crosshead speed of 10^{-3} cm/min (4×10^{-4} in./min).

Chloride Solutions. The service lives of copper alloys under cyclic stress are shorter in chloride solutions than in air. Slow strain rate experiments have also shown that C26000 (Ref 67) and C44300 (Ref 68) have lower fracture stresses in NaCl solutions when the metal is anodically polarized. The changes in fracture stress are insignificant relative to those in air in the absence of an applied potential.

Citrate Solutions. Alloy C72000 is sensitive to cracking in citrate solutions containing dissolved copper in the pH range of 7 to 11. The U-bend test specimens exhibited intergranular cracking (Ref 69).

Formate Solutions. Brass is susceptible to SCC in sodium formate ($NaCHO_2$) solutions at pHs exceeding 11 over a considerable range of applied potentials (Ref 60).

Hydroxide Solutions. Brass exhibits increased crack growth rates under slow strain rate conditions when it is exposed to NaOH at pHs of 12 and 13. The rate of crack growth is a function of the applied potential (Ref 60).

Mercury and Mercury Salt Solutions. Stressed alloys and alloys with internal stress crack readily when exposed to metallic mercury or mercury salt solutions that deposit mercury on the surface of the alloy. This high sensitivity to mercury is the basis of an industry test for the detection of internal stresses in which the alloy is immersed in a solution of mercurous nitrate. Cracking in mercury is the result of LME, not stress corrosion. It does not indicate the SCC susceptibility of an alloy.

Nitrate Solutions. Transgranular cracking was observed on C44300 specimens immersed in

Table 34 Corrosion of copper alloys in ethylene glycol solutions

Alloy	Exposure time, h	Test conditions	Average penetration rate µm/yr	mils/yr
C11000	1344	Triethylene glycol solution, aerated; room temperature	nil	nil
C11000	2560	Triethylene glycol air-conditioning system; 175 °C (345 °F)	40	1.6
C26000	2560	Same as above	50	2.0
C11000	3320	Same as above	10	0.4
C26000	3320	Same as above	15	0.6
C11000	8328	Same as above	25	1.0
C26000	8328	Same as above	35	1.4
C11000	2880	Triethylene glycol air-conditioning system(a); 160 °C (320 °F)	7.5	0.3
C26000	2880	Same as above	7.5	0.3
C11000	5760	Same as above	2.5	0.1
C26000	5760	Same as above	2.5	0.1
C51000	2880	Ethylene glycol solution(b) plus 0.03% H_2SO_4; 99 °C (210 °F)	7.5–10	0.3–0.4
C60800	2880	Same as above	2.5–7.5	0.1–0.3
C63000	2880	Same as above	2.5–18	0.1–0.7
C65500	2880	Same as above	20–25	0.8–1.0
C11000	2400	Ethylene glycol solution(b) plus 0.03–0.05% H_2SO_4; second run; 99 °C (210 °F)	580	23
C61800	2400	Same as above	380	15
C70600	2400	Same as above	480	19
C71500	2400	Same as above	460	18
C11000	305	Glycol maleate, 79 °C (175 °F)	20	0.8

(a) 87–95% glycol. (b) 15% glycol, 85% H_2O

Table 35 Corrosion of copper alloys in alcohols

Alloy	Exposure time, h	Test conditions	Average penetration rate µm/yr	mils/yr
C11000	503	Crude C-5 alcohols; 126–140 °C (260–285 °F)	7.5	0.3
C11000	210	Crude decyl alcohol; 175 °C (345 °F)	3–5	0.1–0.2
C11000	288	Primary decyl alcohol; 175 °C	15–45	0.6–1.8
C65500	288	Same as above	20–60	0.8–2.4
C44400	8160	Isopropanol and water; 118–145 °C (245–295 °F)	10–38	0.4–1.5
C23000	8160	Same as above	10–56	0.4–2.2
C11000	8160	Same as above	8–75	0.3–3.0
C65500	8160	Same as above	10–63	0.4–2.5
C11000	264	Allyl alcohol; refluxed at 88 °C (190 °F)	25	1
C11000	94	Methanol; boiling	nil	nil
C11000	46	Denaturing grade ethanol; boiling	25	1
C23000	165	2-ethyl-2-butyl-1,3 propanediol; 45 °C (115 °F)	5	0.2

Table 36 Corrosion of C44300 and C71500 exposed to gasoline in a refinery

Service condition(a)	Temperature °C	°F	Average penetration rate µm/yr	mils/yr
C44300				
Straight-run (untreated)				
Tower liquid(b)	121	250	1270 min	50 min
Storage(c)	4–27	40–80	63	2.5
Distilled tops from straight-run gasoline(d)	35	95	1270	50
Cracked gasoline (top tray in tower)(e)	204	400	15	0.6
Sweet gasoline vapor(f)	177	350	7.5	0.3
C71500				
Straight-run (untreated)				
Tower liquid(b)	121	250	180	7
Storage(c)	4–27	40–80	180	7
Distilled tops from straight-run gasoline(d)	35	95	1140	45
Cracked gasoline (top tray in tower)(e)	204	400	200	8
Sweet gasoline vapor(f)	177	350	10	0.4
Aviation gasoline (top of column)	121	250	2.5	0.1

(a) Gasoline or related hydrocarbons will not attack copper or its alloys. Attack depends on the type and amount of impurities in the gasoline, such as water, sulfides, mercaptans, aliphatic acids, naphthenic acids, phenols, nitrogen bases, and dissolved gases. (b) 100 lb of H_2S present per 1000 bbl of gasoline. (c) 0.02–0.03 g H_2S per liter of gasoline. (d) pH controlled by NH_3. (e) H_2S and HCl present. (f) Vacuum operation

naturally aerated 1 *N* sodium nitrate ($NaNO_3$) at pH 8 and a potential of 0.15 V versus standard hydrogen electrode (SHE). The fracture stress relative to air was 0.34 (Ref 68).

Copper alloy (Cu-23Zn-12Ni) wires measuring 0.6 mm (0.023 in.) in diameter and normally under a 6-g load and a positive potential in telephone equipment were observed to undergo SCC within 2 years (Ref 70). Laboratory tests suggested that nitrate salts were the cause. The phenomenon was duplicated in the laboratory by exposing the wires to such nitrate salts as zinc nitrate ($Zn(NO_3)_2$), ammonium nitrate (NH_4NO_3), calcium nitrate ($Ca(NO_3)_2$), and cupric nitrate ($Cu(NO_3)_2$) at high humidity; a potential was applied such that the wires were anodic to the normal corrosion potential. The wires were tested under a constant load of 386 MPa (56 ksi). Cracking also occurred in the absence of an applied potential when the nitrate concentration of the surface was high. Cracking did not occur in the presence of $(NH_4)_2SO_4$ and ammonium chloride (NH_4Cl) salts. Wires of Cu-20Ni did not crack under similar conditions.

Nitrite Solutions. Copper, 99.9 and 99.996% pure, exhibited transgranular cracking when subjected to a strain rate of 10^{-6} s^{-1} while immersed in 1 *M* sodium nitrite ($NaNO_2$) at a pH of 8.2 (Ref 71). The 99.9% Cu tested in solution showed an ultimate tensile strength of 160 MPa (23 ksi) and 25% elongation, as opposed to the 196 MPa (28.5 ksi) and 55% elongation obtained in air. Cracking in 1 *M* $NaNO_2$ was also observed in C26000, Admiralty brasses, and C70600.

Solder. In one investigation of the susceptibility to cracking of copper alloys by various solders, a U-shaped tube was coated with solder at 400 °C (750 °F) and then immediately flattened between steel tools in a hand press (Ref 72). The sample was then examined for cracks. The data are given in Table 44.

Sulfur Dioxide. Brass is susceptible to SCC in moist air containing 0.05 to 0.5 vol% SO_2. In addition, pre-exposure of the brass to a solu-

Table 37 Corrosion of copper alloys in contaminated naphtha

Alloy	Corrosion rate µm/yr	mils/yr
At 21 °C (70 °F)(a)		
C23000	230	9
C46400	50	2
C28000	75	3
C44200	200	8
C11000	1270	50
At 177 °C (350 °F)(b)		
C23000	2030	80
C46400	10	0.4
C28000	10	0.4
C44200	200	8

(a) The naphtha contained H_2S, H_2O, and HCl. (b) The naphtha contained H_2S, mercaptans, and naphthenic acids

Table 38 Corrosion of copper alloys in beet-sugar solution

Alloy	Decrease in tensile strength, %, for test rack number(a) 1	2	3	4
C11000	0	4.0	3.5	0
C44300	2.0	9.5	11.5	2.5
C44400	0	3.0	6.0	0
C44500	4.5	9.0	12.5	5.5
C71000	1.0	4.5	7.0	0
C71500	0	5.0	8.0	0

(a) Corrosion specimens (0.8-mm, or 0.032-in., thick strips) were exposed in contact with beet-sugar solution for 100 days in normal refinery operations. Test racks 1 and 4 were at the finishing pan containing Steffen's filtrate; rack 2 was in the first-effect thin-juice evaporator; rack 3 was at the third body of the triple-effect evaporator.

Table 39 Corrosion of selected copper alloys in cracked oil containing 1.4% S

Alloy type	UNS number	Exposure time, days	Loss in tensile strength(a), % 360 °C (680 °F)	315 °C (600 °F)	285 °C (545 °F)	255 °C (490 °F)
Red brass, 85%	C23000	27	100(b)	100(c)	100	100
Muntz metal	C28000	27	12(b)	7.5(d)	1	1.5
Naval brass	C46400	24	...	1.5	0	2
Uninhibited admiralty metal	...	27	13(b)	6(c)	3	2
Antimonial admiralty metal	C44400	27	16.5(b)	6(c)	4	2.5
Aluminum brass	...	24	...	7	16	10
Copper nickel, 30%	C71500	24	...	100	100	57
Silicon bronze, 3%	...	34	...	100	100	100

(a) Specimens 0.8 × 13 mm (0.032 × 0.50 in.) in cross section were exposed at different locations within a high-pressure fractionating column, each location having a characteristic average temperature. (b) 115-day exposure. (c) 26-day exposure. (d) Length of exposure unavailable

Table 40 Corrosion of copper alloys in hot paper mill vapor containing SO_2

Temperature, 200–220 °C (390–430 °F); atmosphere, 17–18% SO_2 plus 1–2% O_2; test duration, mainly 30 days, but some longer

Alloy type	UNS number	Weight loss, $g/m^2/d$
Bronze (90Cu-10Sn)	...	22.0
Aluminum bronze	C61800	26.4
Phosphor bronze	C51100	28.6
Nickel silver, 75–20	C73200	35.6
Phosphor bronze, 8% C	C52100	39.4
Silicon bronze	C65800	50.2
Nickel silver, 55–18	C77000	63.8
Nickel silver, 65–18	C75200	67.4
Nickel bronze (88.5Cu-5Sn-5Ni-1.5Si)	...	70.5

tion of benzotriazole inhibits the cracking (Ref 73).

Sulfate Solutions. Stress-corrosion cracking of C26000 was observed in a solution of 1 *N* sodium sulfate (Na_2SO_4) and 0.01 *N* H_2SO_4 when the alloy was polarized at a potential of 0.25 V versus SHE and subjected to a constant strain (Ref 74).

Sulfide Solutions. National Association of Corrosion Engineers (NACE) committee T-1F issued a report on the acceptability of various materials for valves for production and pipeline service (Ref 75). Bronze and other copper-base alloys are generally not acceptable for highly stressed parts in sour service. Some nickel-copper alloys are considered satisfactory.

Tungstate Solutions. Mild transgranular cracking of C44300 was observed in 1 *N* sodium tungstate (Na_2WO_4) at pH 9.4 and a corrosion potential of 0.080 V versus SHE. The fracture stress relative to that in air was 0.89, and the crack growth velocity was 2×10^{-9} m/s when a strain rate of 1.5×10^{-5} s^{-1} was used (Ref 68).

Water. Several cases of the SCC of admiralty brass heat-exchanger tubing are documented in Ref 76. The environments in which such SCC was observed included stagnant water, stagnant water contaminated with NH_3, and water accidentally contaminated with a nitrate. No cases were noted of SCC of the following alloys when used in heat-exchanger service: C70600, C71500, arsenical copper, C19400, and aluminum bronze.

Service data for various copper alloys used as condenser tubing are given in Ref 77. Information on six different alloys used in freshwater and in seawater service is summarized in Table 45.

An instance of the SCC of a Cu-7Al-2Si stud from an extraction pump exposed to wet steam is discussed in Ref 78. Also in Ref 78 are examples of SCC failures of copper alloys in marine service. These include tubing, a lifeboat keel pin, brass bolts and screws, a brass propeller, a flooding valve, and aluminum bronze valve parts. Some of the failures were attributed to bird excreta that provided a source of NH_3.

Protective Coatings

Copper metals resist corrosion in many environments because they react with one or more constituents of the environment upon initial exposure, thus forming an inert surface layer of protective reaction products. In certain applications, the corrosion resistance of copper metals may be increased by applying metallic or organic protective coatings. If the coating material is able to resist corrosion adequately, service life may depend on the impermeability, continuity, and adhesion to the basis metal of the coating. The electropotential relationship of the coating to the basis metal may be important, especially with metallic coatings and at uncoated edges. Tin, lead, and solder, used extensively as coatings, are ordinarily applied by hot dipping. Electroplating is also used.

Tin arrests corrosion caused by sulfur; it is most effective as a coating for copper wire and cable insulated by rubber that contains sulfur. Lead-coated copper is primarily used for roofing applications, in which contact with flue gases or other products that contain dilute H_2SO_4 is likely. Tin or lead coatings are sometimes applied to copper intended for ordinary atmospheric exposure, but this is done primarily for architectural effect; the atmospheric-corrosion resistance of bare copper is excellent in rural, urban, marine, and most industrial locations. Additional information on the use of tin for corrosion resistance is available in the article "Corrosion of Tin and Tin Alloys" in this Volume.

Electroplated chromium is used for decoration, for improvement of wear resistance, or for reflectivity. Because it is somewhat porous, it is not effective for corrosion protection. Where corrosion protection is important, electroplated nickel is most often used as a protective coating under electroplated chromium. Additional information on the corrosion resistance of chromium plate is available in the article "Corrosion of Hard Chromium Platings" in this Volume.

Various organic coatings are applied to copper alloys to preserve a bright metallic appearance. These are discussed in detail in the section "Organic Coatings" of the article "Protection by Coatings" in this Volume and in the articles "Painting" and "Cleaning and Finishing of Copper and Copper Alloys" in Volume 5 of the 9th Edition of *Metals Handbook*.

Table 41 Values of rate constant for oxide growth on unalloyed copper

Temperature °C	°F	Rate constant k(a) Pure O_2	Air
400	750	4.4×10^{-8}	...
500	950	4.4×10^{-7}	...
600	1100	3.24×10^{-6}	...
700	1300	1.6×10^{-5}	8.03×10^{-6}
800	1475	8.69×10^{-5}	7.97×10^{-5}
900	1650	3.49×10^{-4}	3.36×10^{-4}
950	1750	7.30×10^{-4}	...
1000	1850	1.78×10^{-3}	1.35×10^{-3}

(a) For calculation of weight gain in g/m^2 from Eq 4 when time is measured in seconds

Corrosion and Stress-Corrosion Testing

Aqueous Corrosion Testing. As with other alloys, static tests are used to examine the corrosion behavior of copper alloys in both natural and artificial environments. Tests common to many alloys systems are described in the Section "Corrosion Testing and Evaluation" in this Volume. General procedures are discussed in Ref 79 and 80.

One specific procedure that has been applied to copper alloys in closed-container tests is the determination of the partitioning of the major alloying elements between the corrosion product and the solution (Ref 13, 81). In this procedure, the samples are exposed to the test solution for some time period, after which the sample is removed and the solution filtered to remove any particulate. The collected particulate is dissolved in an acidified solution and quantitatively analyzed for copper and other alloying elements of interest; a similar analysis is performed on the filtered solution. The corrosion product is then stripped from the copper alloy using an inhibited HCl solution and analyzed. The results indicate which alloying elements contribute to film forma-

Table 42 Relative susceptibility to SCC of some copper alloys in NH_3

Alloy	Susceptibility index(a)
C26000	1000
C35300	1000
C76200	300
C23000	200
C77000	175
C66400	100
C68800	75
C63800	50
C75200	40
C51000	20
C11000	0
C15100	0
C19400	0
C65400	0
C70600	0
C71500	0
C72200	0

(a) 0, essentially immune to SCC under normal service conditions; 1000, highly susceptible to SCC as typified by C26000

Table 43 SCC of wrought copper alloys in three atmospheres

UNS Number	Temper, % cold rolled	Time to failure, years: New Haven, CT	Brooklyn, NY	Daytona Beach, FL	Crack morphology(a): New Haven, CT	Brooklyn, NY	Daytona Beach, FL
C11000	37	NF(b), 8.5	NF, 8.5	NF, 8.8	...	...	...
C19400	37	NF, 8.5	NF, 8.5	NF, 8.8	...	...	...
C19500	90	NF, 3.2	NF, 3.1	NF, 3.1	...	...	...
C23000	40	NF, 8.5	NF, 8.5	NF, 8.8	...	...	...
C26000	50	35–47 days	0–23 days	NF, 2.7	I	I	...
C35300	50	51–136 days	70–104 days	NF, 2.7	T+(I)	T+(I)	...
C40500	50	NF, 2.7	NT(c)	NT	...	...	...
C41100	50	NF, 2.7	NT	NT	...	...	...
C42200	37	NF, 8.5	NF, 8.5	NF, 8.8	...	...	...
C42500	50	NF, 2.7	NT	NT	...	...	...
C44300	10	NF, 2.7	NF, 2.7	NF, 2.7	...	...	...
	40	51–95 days	41–70 days	NF, 2.7	T	T	...
	40% + ordered(d)	51–67 days	33–49 days	NF, 2.7	T	T	...
C51000	37	NF, 8.5	NF, 8.5	NF, 8.8	...	...	...
C52100	37	NF, 5.7	NF, 5.7	NF, 5.7	...	...	...
C61900	40%, 9% β phase(e)	NF, 8.5	NF, 8.5	NF, 8.8	...	...	...
	40%, 95% β phase	NF, 8.5	NF, 8.5	NF, 8.8	...	...	...
C63800	50	NF, 5.7	NF, 5.7	NF, 5.7	...	...	...
C67200	annealed	0–30 days	0–134 days	NF, 3.1	I	I	...
	50	0–30 days	0–22 days	18–40 days	I	I	I
C68700	10	517–540 days	2.3-NF 2.7	NF, 2.7	T	T	...
	40	221–495 days	311–362 days	NF, 2.7	T	T	...
	40% + ordered(d)	216–286 days	143–297 days	NF, 2.7	T	T	...
C68800	10	NF, 2.7	NF, 2.7	NF, 2.7	...	...	...
	40	4.7-NF 6.4	2.7-NF 6.4	NF, 6.4	T	T	...
	40% + ordered(d)	NF, 2.7	NF, 2.7	NF, 2.7	...	...	...
C70600	50	NF, 2.2	NF, 2.3	NF, 2.2	...	...	...
C72500	40	NF, 2.2	NF, 2.3	NF, 2.2	...	...	...
C75200	annealed	NF, 3.2	NF, 3.1	NF, 3.1	...	...	...
	25	NF, 3.2	NF, 3.1	NF, 3.1	...	...	...
	50	NF, 3.2	NF, 3.1	NF, 3.1	...	...	...
C76200	annealed	171-NF 3.2	672-NF 3.1	NF, 3.1	T	T	...
	25	142–173 days	236–282 days	NF, 3.1	T	T	...
	50	142–270 days	236–282 days	NF, 3.1	T	T	...
C76600	38	127–966 days	197–216 days	754-NF 8.8	T	T	T
C77000	annealed	731–1003 days	337–515 days	NF, 3.1	T	T	...
	38	137–490 days	196–518 days	596–1234 days	T	T	T
	50	153–337 days	489–540 days	692–970 days	T	T	T
C78200	50	23–48 days	26–216 days	236–300 days	T+(I)	T+(I)	T

(a) I, intergranular; T, transgranular. Parentheses indicate minor mode. (b) NF, no failures in time specified. (c) NT, not tested. (d) Heated at 205 °C (400 °F) for 30 min. (e) Normal structure for this alloy.
Source: Ref 65

Table 44 Susceptibility of copper alloy tubes to cracking by solder

Solder applied at 400 °C (750 °F); specimens were immediately deformed and examined for cracks.

Solder	Alloy: 80Cu–20Ni	97Cu–3Zn	70Cu–30Zn
Lead	Shattered	Cracked	Cracked
97.5Pb–2.5Ag	Shattered	Uncracked	Cracked
95Pb–5Sn	Cracked	Uncracked	Cracked
80Pb–20Sn	Cracked	Uncracked	Uncracked
Grade B solder	Cracked	Uncracked	Uncracked
95Sn–5Sb	Uncracked	Uncracked	Uncracked

Source: Ref 72

Table 45 Experience with copper alloys in condenser tubing service

Medium	Alloy	Susceptibility to SCC
Freshwater	Admiralty brass	Very susceptible
	C70600	Resistant
	Arsenical Copper	Resistant
	C19400	Low Susceptibility
Seawater	C70600	Resistant
	C71500	Resistant
	Arsenic-aluminum brass bronze	Susceptible

Source: Ref 78

tion and whether the element is more prone to go into solution rather than into the film. In addition, the amount of copper that has entered solution and the amount that is actually particulate that spalled off of the surface can be determined. These data are of significance with regard to heavy-metal ion contamination of water sources.

Dynamic Corrosion Tests. One of the major uses of copper alloys is the transport of aqueous solutions; consequently, a significant number of tests have been designed to examine the effects of dynamic conditions on the corrosion behavior of the materials in these environments. The tests, which range in complexity from simple recirculating loops to jet impingement apparatus, examine the effects of such variables as flow rate, heat transfer conditions, and blockages, as well as various solution conditions. Of the systems developed, the flow loop is probably the most widely used test because it is easily constructed, requiring only a pump, ducting, and valves, and can incorporate a wide variety of test variables. Because of their simplicity, flow loops can be constructed on-site and tapped into process flow systems so that the actual operating environment can be used as the test environment. Descriptions of test loops are available in Ref 24, 43, and 82 to 85.

Tubular samples are the most easily tested in this system because they can be directly incorporated into the loop. As with any other corrosion test, the tube samples must be separated by insulating connectors to avoid galvanic effects; tube union fittings of plastic or flexible plastic hose clamped to the tubes are generally adequate. Flat samples can also be tested in flow loops by using special sample holders, such as those described in ASTM D 2688 (Ref 86) and in Ref 24, 82, 83, and 87.

A major variable that affects the corrosion behavior of copper alloys is solution velocity. The effect of flow rate on copper alloys has been examined by placing various diameters of the same tube material in series within a loop and pumping the solution through the loop at a constant pump speed (Ref 83). Velocity effects have also been studied in a parallel flow system with orifice size and header pressure controlled to produce various velocities simultaneously (Ref 24, 43). The effects of local velocity changes and crevices, conditions that arise in power plant condenser tubes because of lodged debris, have been examined by introducing artificial blockages into tubes (Ref 82). The blockage reduces the cross section of the tube, increasing local flow rate, and produces crevice corrosion conditions where it contacts the tube.

Heat transfer effects have been studied by running test tubes through small steam condensers to ovens and pumping the test solution through the tubes. It should be noted that the conditions provided by this type of test are unlike those obtained when the bulk solution is heated before pumping it through the tubes. Heating the bulk solution may change the concentration of components throughout the solution, such as decreasing the oxygen concentration or promoting precipitation. Under heat transfer conditions, these changes may only occur locally, resulting in different corrosion behavior. Corrosion behavior can also be affected by the temperature gradient that exists between the tube wall and the solution under heat transfer conditions, which is much larger than that of a heated solution passing through a tube surrounded by ambient air.

Loop tests are generally used to evaluate the corrosion rates of materials based on their weight loss over a period of time. Test duration depends in large measure on the aggressiveness of the solution and the sample thickness. However, for copper alloys in most aqueous solutions, the test duration should be at least 120 days in order to ensure attainment of steady-state corrosion rates.

When evaluating the samples that have been exposed to flowing systems, more than just the weight loss should be considered. Evidence of erosion should be sought, especially at leading edges and obstructions, and the depth of erosion should be monitored with respect to time. Evidence of pitting should also be looked for, and the depth of pitting as a function of time should be determined. Depth of crevice attack should be noted in samples with crevices, for example, at clamp sites. With regard to crevice corrosion in copper alloys, the attack usually occurs adjacent to the contact site; therefore, the contact site will generally be at the original thickness and can be used as a reference point when measuring the depth of attack.

Each alloy should also be examined for evidence of dealloying. This can generally be determined by metallographic examination of the cross section to see if a copper-rich layer at the sample surface is present. The material can also be mechanically tested to determine whether the mechanical properties have deteriorated with respect to a control sample. This type of testing, however, is generally performed only on materials that have not suffered from severe corrosion, which would obviously degrade the properties of the material.

Other dynamic systems, in addition to flow loops, have been developed primarily to evaluate the maximum flow rate that materials can withstand before erosion-corrosion occurs (Ref 88). An example of such a system is the jet impingement test (Ref 89). In this test, a high-velocity stream of solution is sprayed onto the specimen for some period of time, after which the depth of attack and the amount of surface area attacked are determined. Based on this evaluation, the relative erosion-corrosion resistance of various materials can be ranked.

The spinning-disk test is used to define the velocity that causes erosion in a material (Ref 89, 90). In this test, a disk of the material is immersed in the solution and rotated at a specific rate around the disk axis perpendicular to the plane of the disk. At the conclusion of the test, the sample is examined to determine the distance from the center of the disk, and therefore the velocity, at which erosion occurs.

One other test is used to examine the relative resistance of various materials to erosion by entrained particles in solution (Ref 23). In this test, silica sand of controlled size is introduced to the solution in which L-shaped samples are mounted on the periphery of a rotating disk.

Although any solution can be used in these dynamic test systems, most tests are conducted with seawater or freshwater. Natural waters, such as from the sea, rivers, or lakes, are used as test solutions, but their use is generally restricted by the location of the test facility. In addition, the compositions of natural waters vary not only with location but also with time, making a standardized test procedure difficult. To circumvent this problem for seawaters, substitute seawater (Ref 91) and a 3.4% NaCl solution have both been used. In general, these solutions are slightly more aggressive than natural seawaters; as a result, predictions of corrosion lifetimes based on data from these solutions are generally conservative with respect to actual performance.

A significant amount of work has recently been done on the behavior of copper alloys in sulfide-contaminated seawaters. An extensive bibliography is given in Ref 92. Sulfides are added to the seawater by either bubbling H_2S gas through the solution or adding a Na_2S solution. In general, sulfide concentrations of the order of 1 ppm are sufficient to cause accelerated attack. For rapid corrosion to occur, the copper alloy must be exposed to a solution that contains oxygen as well as sulfide or must be alternately exposed to sulfide-bearing deaerated solutions, followed by exposure to sulfide-free aerated solutions.

Because of the transient nature of sulfides in water, it is necessary to monitor the sulfide level in solution with time. Titration techniques are available for measuring the sulfide concentration, but these are generally time consuming and tedious if continual monitoring is required. An alternative is the use of a sulfide-specific ion electrode, which provides accurate sulfide readings in substitute ocean water in much less time.

Another environment of interest is the freshwater cooling tower environment typically found in power plant applications. A simulated cooling tower water environment has been developed based on the analysis of Ohio River water for its major constituents and their concentrations. The solution (Table 46) corresponds to a sixfold concentration of typical river water chemistry, simulating the concentrating effect of the cooling tower. Other cooling water solutions have also been used (Ref 93).

Atmospheric Testing. In a variety of applications, such as electrical and architectural components, the behavior of copper alloys when fully immersed in solution is not relevant with regard to their performance under various atmospheric conditions. Constant humidity and temperature chambers are used to evaluate the relative atmospheric behavior of the materials. The design and typical test environments are described in Ref 94 to 96. As with aqueous tests under artificial conditions, the corrosion behavior determined in these tests generally cannot be used to ensure the behavior of the material in the actual service environment. This is the case primarily because many variables in the service environment cannot be incorporated into the test or are overlooked and because the environment changes constantly. Such tests do, however, provide approximate data and allow ranking of the test materials.

Evaluation of tested specimens involves typical corrosion parameters, such as weight loss, depth of pitting, and crevice corrosion. In addition, patina (oxide film formation) is evaluated with regard to color, continuity, and film tenacity. After the specimen has been cleaned, evidence of dealloying should also be sought by examination of a metallographic cross section or by loss of mechanical properties (as compared to an untested control sample). In Military Standard 853C Method 1009.4 for electronic materials, materials exposed to the salt fog test are evaluated to determine the amount of corrosion over the surface and to determine whether pitting or corrosion contributed to failure of the bend test. Materials are also evaluated with regard to the effect of the oxide film on solderability and the corrosion resistance of the solder base metal.

Other environmental tests are Military Standard 202 Method 2080, in which candidate materials are aged in the steam from boiling distilled water and then solder dipped to evaluate their solderability. Ammonium sulfide ($(NH_4)_2S$) is used to determine the effectiveness of tarnish inhibitors in protecting the material. In this test, the specimen is held for a short period of time (usually 30 s) over an open beaker containing $(NH_4)_2S$ solution and then examined. If the surface has become blackened, the protective layer is considered to be inadequate.

Atmospheric testing of copper alloys in natural environments is conducted to evaluate the behavior of the materials in industrial, rural, and marine atmospheres. The procedure most widely used is given in Ref 97, which describes sample preparation, types of test racks, typical locations and orientation. This long-term test may last up to 20 years or more and therefore requires careful recordkeeping.

Stress-Corrosion Testing. Much of the early knowledge of the SCC tendencies of copper alloys was based on service experience. Such data were assimilated at laboratories involved in the development of copper alloys and were used to design alloys with greater resistance to SCC in specific environments. Some of this information reached the open literature. In other cases, researchers concerned with specific objectives, such as designing a desalination plant or operating power station, occasionally wrote summary articles in which they cited their experience with different alloy compositions. Such information is useful but qualitative, and the environmental constituents or conditions that led to the cracking are unknown.

In the past several decades, the study of SCC has been greatly accelerated, and the causes and mechanisms for the behavior have been addressed by materials scientists, physicists, chemists, metallurgists, and mechanical engineers. Laboratory studies under controlled conditions have been expanded, ASTM has developed standardized tests, and laboratories have compared data. As a result, considerable quantitative information is now available in the literature. In some cases, this information is obtained with full knowledge

Table 46 Composition of simulated cooling tower water

Component	Concentration, mg/L
CuO	700
$MgSO_4$ (anhydrous)	693
NaCl	614
KCl	59
KNO_3	24
Na_2CO_3	167
$SiO_2 \cdot XH_2O$	21
H_2SO_4	0.75 mL/L

of fracture mechanics principles. The methods of generating SCC data are numerous and include both static and dynamic tests.

In the static tests, the sample is put under tension by bending and restraining the sample or by mounting it in a tensile-testing machine. The data thus generated include time to first crack, time to fracture, or time to relax to a certain fraction (for example, 50 or 80%) of the unrestrained distance between the ends of the bent specimen. The data generated in this fashion allow comparison among different alloys, among different pretreatments, and among other experimental variables. The data are comparative within one data set but yield no absolute information.

Various NH_3 environments are widely used to test copper alloys, the most common being Mattsson's solution of pH 4.0, 7.2, and 10. Two other NH_3-base environments that produce very aggressive stress-corrosion conditions are a NH_3-0.5 *M* copper solution of pH 14, and a moist NH_3 test. The pH 14 solution is made by dissolving 3.18 g of copper powder in 1 L of 29.5% NH_4OH solution (typical reagent-strength NH_4OH). The moist NH_3 test requires the construction of a chamber in which 100% relative humidity and a constant NH_3 gas concentration are maintained (Ref 98).

One of the simplest laboratory stress-corrosion tests that provides a significant amount of information is the U-bend test, in which the springback of the sample is measured over time in the test solution. Two sample sets of each material are produced in a manner similar to that described in Ref 99. One is placed in the test solution, and the other remains in the room environment as a control. A variety of test jigs are described in Ref 99; however, the legs of the jig must be compressed the same distance when the sample is removed and then replaced in the jig. A typical example of this type of jig is given in Ref 99.

The samples are placed in the jig, removed, and the springback between the legs measured; this is also done for the control samples. The samples are reinserted in the test jig and placed in the test solution. At periodic intervals, the samples are removed from the solution, taken from the jig, and the springback distance between the legs remeasured. Similar measurements are made on the air control samples. The test continues until either physical failure occurs—that is, if the sample breaks or if it no longer has enough tension to hold it in the jig—or some predetermined performance criteria are met, for example, 1000 h elapsed time or springback reduction to 80% of its initial value. At the conclusion of the test, the average change in percent springback for each material at each time is determined, taking into account the loss in springback that occurred as a result of stress relaxation based on springback measurements of the air control samples.

A constant percent springback versus time indicates that the material is not susceptible to SCC in the test solution over that time period. A decrease in percent springback with time indicates the SCC has occurred. This should be verified by optical examination for cracking as well as metallographic examination of the sample to determine the mode of cracking. An increase in percent springback indicates that the tension side of the sample dissolved at a faster rate than the compressive side due to stress-assisted dissolution. Examination will reveal that the specimen has thinned and that failure occurred because of overload, not cracking. This result indicates that the solution is too aggressive for SCC to occur and that another solution should be used to compare stress-corrosion behavior.

Dynamic Tests. During the past decade, there has been a major swing toward the use of dynamic tests, which yield values that can be quantitatively applied to the proposed mechanisms of SCC. Primary among these is the slow strain rate technique. The application of this technique to the understanding of SCC began in the early 1960s (Ref 100). An excellent summary of the slow strain rate technique and its applications to SCC is given in Ref 101.

This method uses tensile-test specimens mounted in stiff-frame machines and strained at the rate of 10^{-7} to 10^{-5} s^{-1} in the presence of a specific environment. Strain rates in this range promote SCC, but the absence of cracking is no assurance of immunity to SCC. Various methods are used to assess the results when SCC is observed. These include the area under the stress-elongation curve, time to failure, crack velocity, and ratio of fracture stress in a medium to fracture stress in air.

REFERENCES

1. E. Mattsson, *Electrochim Acta*, Vol 3, 1961, p 279
2. U. Bertocci and E.N. Pugh, in *Proceedings of the Ninth International Congress on Metallic Corrosion* (Toronto, Canada), Vol 1, National Research Council of Canada, June 1984, p 144
3. H.L. Logan, *J. Res. Natl. Bur. Stand.*, Vol 48, 1952, p 99
4. H. Leidheiser, Jr. and R. Kissinger, *Corrosion*, Vol 28, 1972, p 218
5. R.P.M. Procter and M. Islam, *Corrosion*, Vol 32, 1976, p 267
6. N.W. Polan, J.M. Popplewell, and M.J. Pryor, *J. Electrochem. Soc.*, Vol 126, 1979, p 1299
7. A. Parthasarathi and N.W. Polan, *Metall. Trans. A*, Vol 13A, 1982, p 2027
8. G.S. Haynes and R. Baboian, A Comparative Study of the Corrosion Resistance of Cable Shielding Materials, *Mater. Perform.*, Vol 18, 1979, p 45-56
9. "Final Storage of Spent Nuclear Fuel," Publication KBS3, Swedish Nuclear Fuel Supply Company, 1983
10. J.P. Simpson, "Experiments on Container Materials for Swiss High-Level Waste Disposal Projects," Part II, Technical Report 84-01, National Cooperative for the Storage of Radioactive Waste, 1984
11. S.K. Frape, P. Fritz, and R.H. McNutt, Water-Rock Interaction and Chemistry of Groundwaters From the Canadian Shield, *Geochim. Cosmochin. Acta*, Vol 48, 1984, p 1617-1627
12. J.W. Braithwaite and M.A. Molecke, Nuclear Waste Canister Corrosion Studies Pertinent to Geologic Isolation, *Nuc. Chem. Waste Man.*, 1980, p 37-50
13. N.W. Polan, G.P. Sheldon, and J.M. Popplewell, "The Effect of NH_3 and O_2 Levels on the Corrosion Characteristics and Copper Release Rates of Copper Base Condenser Tube Alloys Under Simulated Steam Side Conditions," Paper No. 81-JPGC-Pwr-9, Joint ASME/IEEE Power Generation Conference, Oct 1981
14. G.P. Sheldon and N.W. Polan, Field Testing of Power Utility Condenser Tube Alloys, *J. Mater. Energy Syst.*, Vol 6 (No. 4), March 1985
15. A.H. Tuthill and C.M. Schillmoller, "Guidelines for Selection of Marine Materials," Paper presented at the Ocean Science and Ocean Engineering Conference, Marine Technology Society, June 1965
16. K.D. Efird and D.B. Anderson, *Mater. Perform.*, Vol 14 (No. 11), 1975
17. C.R. Southwell, J.D. Bultman, and A.L. Alexander, *Mater. Perform.*, Vol 15 (No. 7), 1976
18. D.B. Anderson and F.A. Badia, *Trans. ASME*, Vol 95 (No. 4), 1973
19. D.B. Anderson and K.D. Efird, *Proceedings of the Third International Congress on Marine Corrosion and Fouling* (Gaithersburg, MD), National Bureau of Standards, Oct 1972
20. R.D. Schelleng, "Heat Treatment and Corrosion Resistance of Cr-modified Cu-Ni," Technical Publication 949-OP, International Nickel Company, Nov 1976
21. C. Pearson, *Br. Corros. J.*, Vol 7, March 1972
22. N.W. Polan, M.A. Heine, J.M. Popplewell, and C.J. Gaffoglio, Paper 58, presented at Corrosion/82, National Association of Corrosion Engineers, 1982
23. N.W. Polan, M.A. Heine, and C.J. Gaffoglio, "Erosion Corrosion Resistance of Copper Alloys C72200 in Seawater Containing Entrained Sand," Paper No. 82-JPGC-Pwr-4, Joint ASME/IEEE Power Generation Conference, 1982
24. K.D. Efird, *Corrosion*, Vol 33 (No. 1), 1977
25. S. Sato and K. Nagata, *Sumitomo Light Met. Tech. Rep.*, Vol 19, July 1978
26. G. Bianchi, G. Fiori, P. Longhi, and F. Mazza, *Corrosion*, Vol 34 (No. 11), 1978
27. F.M. Reinhart, *Corrosion in Natural Environments*, STP 558, American Society for Testing and Materials, 1974
28. "Corrosion Resistance of Wrought 90/10 Copper-Nickel-Iron Alloy in Marine Environments," Marine Corrosion Bulletin 1, The International Nickel Co. Inc., 1975
29. C.L. Bulow, *Nav. Eng. J.*, Vol 77 (No. 3), 1965
30. R.N. Orava, Final Report DRI-2475, Contract No. 14-01-0001-1440, U.S. Department of the Interior, Office of Saline Water, Oct 1968
31. D.B. Anderson, *Mater. Prot. Perform.*, Vol 10 (No. 11), 1971
32. T.P. May, E.G. Holmberg, and J. Hinde, "Sea Water Corrosion at Atmospheric and Elevated Temperatures," Deckema-Monographian Bank 47, International Nickel Company, 1962
33. R.W. Ross, *Mater. Perform.*, Vol 18 (No. 7), 1979
34. C.F. Schrieber, "Seawater Corrosion Test Program: Part IV," Contract No. 14-30-311, U.S. Department of the Interior, Office of Water Research and Technology, April 1975
35. F.P. Ijsseling, L.J.P. Drolenga, and B.H. Kolster, *Br. Corros. J.*, Vol 17 (No. 4), 1982
36. W.C. Stewart and F.L. LaQue, *Corrosion*, Vol 8 (No. 8), 1952
37. F.L. LaQue, *Corrosion*, Vol 6 (No. 4), 1950
38. R. Francis, "Effects of Cooling Water Treatment on Ships' Condenser Tubes," Report A.1945, British Non-Ferrous Technology

Center, Jan 1979

39. P.T. Gilbert, *Trans. IME*, Vol 66 (No. 1), 1954
40. K.D. Efird and T.S. Lee, *Corrosion*, Vol 35 (No. 2), 1979
41. J.F. Bates and J.M. Popplewell, Paper 100, presented at Corrosion/74, National Association of Corrosion Engineers, 1974
42. J.P. Gudas and H.P. Hack, Paper 93, presented at Corrosion/77, National Association of Corrosion Engineers, 1977
43. T.S. Lee, H.P. Hack, and D.G. Tipton, in *Proceedings of the Fifth International Congress on Marine Corrosion and Fouling* (Barcelona, Spain), Orsi, May 1980
44. H.P. Hack and J.P. Gudas, Paper 23, presented at Corrosion/78, National Association of Corrosion Engineers, 1978
45. H.P. Hack and J.P. Gudas, Paper 234, presented at Corrosion/79, National Association of Corrosion Engineers, 1979
46. H.P. Hack and T.S. Lee, *Shell and Tube Heat Exchangers*, American Society for Metals, 1982
47. S. Sato, Corrosion and its Prevention in Copper Alloy Condenser Tubes under Modern Conditions, *Rev. Coat. and Corr.*, Vol 1 (No. 2) 1973, p 139
48. T.H. Michels, W.W. Kirk, and A.H. Tuthill, *J. Mater. Energy Syst.*, Vol 1 (No. 12), 1979
49. J.L. Manzolillo, E.W. Thiele, and A.H. Tuthill, "CA-706 Copper-Nickel Alloy Hulls: The Copper Mariner's Experience and Economics," Paper presented at the Society of Naval Architects and Marine Engineers, Nov 1976
50. F.J. Ansuini and K.L. Money, "Fouling Resistant Screens for OTEC Plants," Paper presented at fifth Ocean Thermal Energy Conversion Conference, Miami Beach, FL, Feb 1978
51. C.L. Bulow, *Trans. Electrochem. Soc.*, Vol 87, 1945
52. F.L. LaQue and W.F. Clapp, *Trans. Electrochem. Soc.*, Vol 87, 1945
53. K.D. Efird, *Mater. Perform.*, Vol 15 (No. 4), 1976
54. R.B. Ritter and J.W. Suitor, "Fouling Research on Copper and Its Alloys—Seawater Studies," Progress Report, Project 214B, International Copper Research Association, March 1976 to Feb 1978
55. R.O. Lewis, Paper 54, presented at Corrosion/82, National Association of Corrosion Engineers, 1982
56. G.P. Sheldon and N.W. Polan, The Heat Transfer Resistance of Various Heat Exchanger Tubing Alloys in Natural and Synthetic Seawaters, *J. Mater. Energy Syst.*, Vol 5 (No. 4), March 1984
57. D.G. Tipton, "Effect of Mechanical Cleaning on Seawater Corrosion of Candidate OTEC Heat Exchanger Materials," ANL/OTEC-BCM-018, National Bureau of Standards, June 1981
58. N.B. Pilling and R.E. Bedworth, *J. Inst. Met.*, Vol 29, 1923, p 529-582
59. E. Escalante and J. Kruger, *J. Electrochem. Soc.*, Vol 118, 1971, p 1062
60. R.N. Parkins and N.J.H. Holroyd, *Corrosion*, Vol 28, 1982, p 245
61. S.C. Sircar, U.K. Chatterjee, S.K. Roy, and S. Kisku, *Br. Corros. J.*, Vol 9, 1974, p 47
62. A.J. Fiocco, Annual Report, 1967-1968, Office of Saline Water, Denver, CO, 1969, p 241
63. H. Moore, S. Beckinsale, and C.E. Jallinson, *J. Inst. Met.*, Vol 25, 1921, p 35
64. W.H.J. Vernon, *Trans. Faraday Soc.*, Vol 23, 1927, p 162; Vol 27, 1931, p 255
65. J.M. Popplewell and T.V. Gearing, *Corrosion*, Vol 31, 1975, p 279
66. A.V. Bobylev, "Intercrystalline Corrosion and Corrosion of Metals Under Stress," I.A. Levin, Ed., Consultants Bureau, 1962
67. V.K. Gouda, H.A. El-Sayed, and S.M. Sayed, in *Eighth International Congress on Metallic Corrosion* (Frankfurt, West Germany), Dechema, Frankfurt am Main, 1981, p 479
68. A. Kawashima, A.K. Agrawal, and R.W. Staehle, in *Stress Corrosion Cracking—The Slow Strain Rate Technique*, STP 665, American Society for Testing and Materials, 1979, p 266
69. S.P. Nayak and A.K. Lahire, *Indian J. Technol.*, Vol 10, 1972, p 322
70. N. McKinney and H.W. Hermance, in *Stress Corrosion Cracking*, STP 425, American Society for Testing and Materials, 1967, p 274
71. S.P. Pednekar, A.K. Agrawal, H.E. Chaung, and R.W. Staehle, *J. Electrochem. Soc.*, Vol 126, 1979, p 701
72. R. Chadwick, *J. Inst. Met.*, Vol 97, 1969, p 93
73. J.B. Cotton, in *Second International Congress on Metallic Corrosion* (New York, NY), National Association of Corrosion Engineers, 1963, p 590
74. H.W. Pickering and P.J. Byrne, *Corrosion*, Vol 29, 1973, p 325
75. J.B. Greer and M.R. Chance, *Mater. Prot. Perform.*, Vol 12 (No. 3), 1973, p 41
76. S.D. Reynolds, Jr. and F.W. Pement, *Mater. Perform.*, Vol 13 (No. 9), 1974, p 2128
77. J. Papamarcos, *Power Eng.*, Vol 89, July 1973, p 24
78. B.F. Peters, J.A.H. Carson, and R.D. Barer, *Mater. Prot.*, Vol 4 (No. 5), 1965, p 24
79. "Standard Recommended Practice for Laboratory Immersion Corrosion Testing of Metals," G 31, *Annual Book of ASTM Standards*, American Society for Testing and Materials
80. B.W. Lifka and F.L. McGeary, Chapter 15, in *NACE Basic Corrosion Course*, National Association of Corrosion Engineers, 1970, p 15-1 to 15-35
81. A.J. Brock and J.M. Popplewell, Paper 105, presented at Corrosion/79, Atlanta, GA, National Association of Corrosion Engineers, U.S. Government Printing Office, March 1979
82. F.W. Fink and T.P. May, *Proceedings of the First International Symposium on Water Desalination*, Vol 1, 1965, p 432-438
83. W. Wolfe, Jr. and S.F. Hager, Paper G7-PWR-7, presented at Joint ASME/IEEE Power Generation Conference, Detroit, MI, Sept 1967
84. J.M. Popplewell and E.A. Thiele, Paper 30, presented at Corrosion/80, Chicago, IL, National Association of Corrosion Engineers, March 1980
85. G.A. Gehring, Jr., R.L. Foster, and B.C. Syrett, Paper 76, presented at Corrosion/83, Anaheim, CA, National Association of Corrosion Engineers, April 1983
86. "Standard Test Methods for Corrosivity of Water in the Absence of Heat Transfer (Weight Loss Methods)," D 2688, *Annual Book of ASTM Standards*, American Society for Testing and Materials
87. J.F. Bates and J.M. Popplewell, *Corrosion*, Vol 31 (No. 8), Aug 1975, p 267-275
88. B.C. Syrett, *Corrosion*, Vol 32 (No. 6), June 1976, p 242-252
89. F.L. LaQue and J.F. Mason, Jr., "The Behavior of Iron Modified 70-30 Copper-Nickel Alloy in Salt Water and in Some Petroleum Industry Environments," American Petroleum Institute, Division of Refining, May 1950
90. W.C. Stewart and F.L. LaQue, *Corrosion*, Vol 8 (No. 8), Aug 1952, p 259-277
91. "Standard Specification for Substitute Ocean Water," D 1141, *Annual Book of ASTM Standards*, American Society for Testing and Materials
92. P.T. Gilbert, *Mater. Perform.*, Vol 21 (No. 2), Feb 1982, p 47-53
93. O. Hollander and R.C. May, *Corrosion*, Vol 41 (No. 1), Jan 1985, p 39-45
94. "Standard Method of Salt Spray (Fog) Testing," B 117, *Annual Book of ASTM Standards*, American Society for Testing and Materials
95. "Standard Method of Acetic Acid-Salt Spray (Fog) Testing," B 287, *Annual Book of ASTM Standards*, American Society for Testing and Materials
96. "Standard Method for Copper-Accelerated Acetic Acid-Salt Spray (Fog) Testing (CASS Test)," B 368, *Annual Book of ASTM Standards*, American Society for Testing and Materials
97. "Recommended Practice for Conducting Atmospheric Corrosion Tests on Metals," G 50, *Annual Book of ASTM Standards*, American Society for Testing and Materials
98. J.M. Popplewell, *Corros. Sci.*, Vol 13, 1973, p 593
99. "Recommended Practice for Making and Using U-Bend Stress Corrosion Test Specimens," G 30, *Annual Book of ASTM Standards*, American Society for Testing and Materials
100. R.N. Parkins, in *Stress Corrosion Cracking—The Slow Strain Rate Technique*, STP 665, American Society for Testing and Materials, 1979, p 5
101. G.M. Ugiansky and J.H. Payer, Ed., *Stress Corrosion Cracking—The Slow Strain Rate Technique*, STP 665, American Society for Testing and Materials, 1979

Corrosion of Nickel-Base Alloys

Chairman: Aziz I. Asphahani, Haynes International, Inc.

Characteristics of Nickel and Nickel-Base Alloys

D.L. Klarstrom
Haynes International, Inc.

NICKEL AND NICKEL-BASE ALLOYS are vitally important to modern industry because of their ability to withstand a wide variety of severe operating conditions involving corrosive environments, high temperatures, high stresses, and combinations of these factors. There are several reasons for these capabilities. Pure nickel is ductile and tough because it possesses a face-centered cubic (fcc) crystal structure up to its melting point. Therefore, nickel and its alloys are readily fabricated by conventional methods, and they offer freedom from the ductile-to-brittle behavior of most body-centered cubic (bcc) and noncubic metals. Nickel has good resistance to corrosion in the normal atmosphere, in natural freshwaters and in deaerated nonoxidizing acids, and it has excellent resistance to corrosion by caustic alkalies (Ref 1). Therefore, nickel offers very useful corrosion resistance itself and it is an excellent base on which to develop specialized alloys. Its atomic size and nearly complete 3*d* electron shell enable it to receive large amounts of alloying additions before encountering phase instabilities (Ref 2). This allows a wide variety of alloys to be fashioned in a manner that can adequately capitalize on the unique properties of specific alloying elements. Finally, unique intermetallic phases can form between nickel and some of its alloying elements; this enables the formulation of very high strength alloys for both low- and high-temperature service. These and other important characteristics of the nickel alloy family will be discussed below.

Effects of Major Alloying Elements

The roles of the major alloying elements used to promote corrosion resistance in nickel-base alloys can be summarized as follows.

Copper. Additions of copper provide improvement in the resistance of nickel to nonoxidizing acids (Ref 3). In particular, alloys containing 30 to 40% Cu offer useful resistance to nonaerated sulfuric acid (H_2SO_4) and offer excellent resistance to all concentrations of nonaerated hydrofluoric acid (HF). Additions of 2 to 3% Cu to nickel-chromium-molybdenum-iron alloys have also been found to improve resistance to hydrochloric acid (HCl), H_2SO_4, and phosphoric acid (H_3PO_4) (Ref 4).

Chromium additions impart improved resistance to oxidizing media such as nitric (HNO_3) and chromic (H_2CrO_4) acids (Ref 3). Improved resistance to hot H_3PO_4 has also been shown (Ref 5). Chromium also improves resistance to high-temperature oxidation and to attack by hot sulfur-bearing gases (Ref 1). Although alloys have been formulated containing up to 50% Cr, alloying additions are usually in the range of 15 to 30%.

Iron is typically used in nickel-base alloys to reduce costs, not to promote corrosion resistance. However, iron does provide nickel with improved resistance to H_2SO_4 in concentrations above 50% (Ref 1). Iron also increases the solubility of carbon in nickel; this improves resistance to high-temperature carburizing environments (Ref 6).

Molybdenum in nickel substantially improves resistance to nonoxidizing acids (Ref 1, 3). Commercial alloys containing up to 28% Mo have been developed for service in nonoxidizing solutions of HCl, H_3PO_4, and HF, as well as in H_2SO_4 in concentrations below 60% (Ref 1, 7). Molybdenum also markedly improves the pitting and crevice corrosion resistance of nickel-base alloys (Ref 8). In addition, it is an important alloying element for imparting strength in metallic materials designed for high-temperature service (Ref 2).

Tungsten behaves similarly to molybdenum in providing improved resistance to nonoxidizing acids and to localized corrosion (Ref 9). However, because of its atomic weight, approximately twice as much tungsten as molybdenum must be added by weight to achieve atomically equivalent effects. Because of the negative impact this would have on alloy density and because of the typically higher cost and lower availability of tungsten, additions of molybdenum are generally preferred. However, additions of tungsten of the order of 3 to 4% in combination with 13 to 16% Mo in a nickel-chromium base result in alloys with outstanding resistance to localized corrosion (Ref 10).

Silicon is typically present only in minor amounts in most nickel-base alloys as a residual element from deoxidation practices or as an intentional addition to promote high-temperature oxidation resistance. In alloys containing significant amounts of iron, cobalt, molybdenum, tungsten, or other refractory elements, the level of silicon must be carefully controlled because it can stabilize carbides and harmful intermetallic phases. On the other hand, the use of silicon as a major alloying element has been found to improve greatly the resistance of nickel to hot, concentrated H_2SO_4 (Ref 9). Alloys containing 9 to 11% Si are produced for such service in the form of castings.

Cobalt. The corrosion resistance of cobalt is similar to that of nickel in most environments (Ref 9). Because of this and because of its higher cost and lower availability, cobalt is not generally used as a primary alloying element in materials designed for aqueous corrosion resistance. On the other hand, cobalt imparts unique strengthening characteristics to alloys designed for high-temperature service (Ref 2). Cobalt, like iron, increases the solubility of carbon in nickel-base alloys, and this increases resistance to carburization. Further, the melting point of cobalt sulfide is higher than that of nickel sulfide; therefore, alloying with cobalt also tends to improve high-temperature sulfidation resistance.

Niobium and Tantalum. In corrosion-resistant alloys, both niobium and tantalum were originally added as stabilizing elements to tie up carbon and prevent intergranular corrosion attack due to grain-boundary carbide precipitation (Ref 1, 4). However, the advent of argon-oxygen decarburization melting technology made it possible to achieve very low levels of residual carbon, and such additions of niobium and tantalum are no longer necessary. In high-temperature alloys, both elements are used to promote high-temperature strength through solid-solution and precipitation-hardening mechanisms (Ref 2). Additions of these elements are also considered to be beneficial in reducing the tendency of nickel-base alloys toward hot cracking during welding (Ref 11).

Aluminum and titanium are often used in minor amounts in corrosion-resistant alloys for the purpose of deoxidation or to tie up carbon and/or nitrogen, respectively (Ref 1). When added together, these elements enable the formulation of age-hardenable high-strength alloys for low- and elevated-temperature service. Additions of aluminum can also be used to promote the formation of a tightly adherent alumina scale at high temperature that resists attack by oxidation, carburization, and chlorination (Ref 12).

Carbon and Carbides. There is evidence that nickel forms a carbide of the formula Ni_3C at elevated temperatures, but it is unstable and decomposes into a mixture of nickel and graphite at low temperatures (Ref 13). Because this phase mixture tends to have low ductility, low-carbon forms of nickel are usually preferred in corrosion-resistant applications. This problem is also alleviated to some extent when nickel is alloyed with copper (Ref 1). In other nickel alloys, the carbides that form depend on the specific alloying elements present, their amounts, and the level of carbon present.

In corrosion-resistant alloys, many types of carbides are considered harmful because they can precipitate at grain boundaries during heat treatment or weld fabrication and subsequently promote intergranular corrosion or cracking in service. This results from the depletion of matrix elements essential to corrosion resistance during the carbide precipitation process. In high-temperature alloys, the presence of carbides is generally desired to control grain size and to enhance elevated-temperature strength and ductility. However, careful attention must be paid to the carbide types and morphologies after solution heat treatment or postfabrication heat treatment in order to avoid cracking during component manufacture or loss of strength and/or ductility in service.

In discussing carbides, it is necessary to distinguish between primary and secondary types. Primary carbides form during the solidification process. They are interdendritic and form from the last liquid to freeze, which is generally enriched in alloying elements. These carbides are typically metastable and would dissolve if given sufficient time at elevated temperatures. However, during metal manufacture, this is usually not the case; therefore, they can persist in the final product as stringers in the direction of predominant metal flow. Some level of carbide stringers usually must be tolerated because they cannot be economically avoided in complex alloy systems. However, large amounts of such stringers should be avoided, because they can adversely affect formability, weld fabrication, and service performance characteristics.

Secondary carbides are those that precipitate as the result of thermal exposures during metal manufacture and component fabrication operations or during component service life. These carbides precipitate preferentially on grain boundaries and internal structural defects, such as twin boundaries and dislocations. The quantity of secondary carbides that precipitate depends on the amount of carbon in solution, the exposure temperature, and the time at temperature. Therefore, conditions that generate a supersaturated solution of carbon followed by slow cooling or thermal arrests below carbide solvus temperatures will produce heavy secondary carbide precipitation. It is also possible to generate aligned or stringer-type structures of secondary carbides. This can occur if an alloy with secondary precipitates is mechanically worked and annealed below the solvus temperature. Heavy secondary carbide precipitation generally reduces ductility and toughness, and this adversely affects fabrication and service performance.

The carbide types occurring in nickel-base alloys can be separated into two broad categories: those that are chromium rich and those that are rich in refractory alloying elements. The chromium-rich carbides are of the forms Cr_7C_3 and $M_{23}C_6$. The Cr_7C_3 carbide forms only in a few simple alloys that are low in chromium as well as reactive and refractory alloying elements (Ref 14). It occurs as a blocky, intergranular precipitate, and it is stable only at temperatures of the order of 1050 to 1150 °C (1920 to 2100 °F) (Ref 15). It decomposes into $M_{23}C_6$ at lower temperatures.

The $M_{23}C_6$ carbide can range in chemistry from $Cr_{23}C_6$ in simple nickel-chromium alloys to $Cr_{21}(Mo,W)_2C_6$ in alloys containing molybdenum and tungsten (Ref 14). Other elements, such as nickel, iron, and cobalt, can partially substitute for chromium. The $M_{23}C_6$ carbide can assume a variety of morphologies, such as discrete, globular particles, continuous grain-boundary films, or a cellular grain-boundary structure. In some alloys, the latter two forms can seriously reduce ductility. Depending on alloy chemistry, $M_{23}C_6$ carbides can be stable to temperatures of 1150 °C (2100 °F) or above. Precipitation kinetics are particularly rapid in the temperature range of 760 to 980 °C (1400 to 1800 °F), which may be encountered during heat treatment or in-service exposure.

The common refractory metal carbides take on the forms of MC, M_6C, and $M_{12}C$. The MC carbides are formed by the reactive metal elements, such as titanium, zirconium, and hafnium, and by the refractory metal elements, such as vanadium, niobium, and tantalum. The MC carbides can contain mixtures of these elements (for example, (Ti,Nb)C) as well as less refractory elements, such as molybdenum and tungsten (for example, (TiMo)C) (Ref 14). If nitrogen is present in the alloy, carbonitride forms, such as Nb(C,N), are also possible. When present, these carbonitrides exhibit yellow to orange coloration as opposed to the steel gray appearance of the simple monocarbide. The MC carbides are usually formed as primary carbides. In wrought alloys, they are present as discrete, angularly shaped particles within grains and at grain boundaries. In cast alloys, they are usually present in a script morphology at interdendritic locations. The MC carbides are extremely stable and do not break down easily unless weakened by molybdenum or tungsten substitution (Ref 13). Decomposition of the MC carbide can yield the $M_{23}C_6$ or M_6C carbide forms.

The M_6C carbides, also known as the eta (η) carbides, are usually formed between the refractory elements molybdenum and tungsten and the major matrix elements of nickel, iron, cobalt, and chromium. Carbide chemistries can vary widely from forms such as $(Ni,Co)_3Mo_3C$ to $(Ni,Co)_2W_4C$, and carbon contents may range above and below that required for stoichiometry (Ref 14, 16). Silicon is also known to enter the carbide in large amounts (Ref 14). The M_6C carbides can form both as primary and secondary precipitates in a globular morphology. Their solvus temperatures are high, typically of the order of 1175 °C (2150 °F) or higher. In the temperature range of 760 to 980 °C (1400 to 1800 °F), they can decompose into the $M_{23}C_6$ or $M_{12}C$ carbide forms, depending on the alloy system.

The $M_{12}C$ carbides are closely related to the M_6C carbides. They usually form between the refractory elements molybdenum and tungsten and the matrix elements nickel, iron, cobalt, and chromium. However, unlike M_6C, the $M_{12}C$ carbides exhibit a very narrow chemistry range, with a composition such as Ni_6Mo_6C representing a typical example (Ref 16). Silicon is also known to enter the carbide preferentially in large quantities (Ref 17). The $M_{12}C$ carbide is usually observed as a secondary carbide after the dissolution or decomposition of primary M_6C carbides. Once formed, the $M_{12}C$ carbides are very stable, with solvus temperatures of 1175 °C (2150 °F) and above.

Intermetallic Phases

The occurrence of intermetallic phases in nickel-base alloys carries both good and bad connotations. On the positive side, the nickel-base system has been the most widely and successfully exploited of any alloy base in the development of high-strength high-temperature alloys because of the occurrence of unique intermetallic phases. On the negative side, the precipitation of certain intermetallic phases can seriously degrade ductility and corrosion resistance. This latter effect results from the fact that intermetallics, like carbides, can rob the matrix of elements vital to service performance.

In the case of corrosion-resistant alloys, especially the solid-solution type, the formulation of alloy chemistry to avoid intermetallic precipitation altogether is not necessary, because service temperatures are usually well below those at which precipitation kinetics become important. In such cases, it is necessary only to restrict alloy composition sufficiently to ensure successful manufacturing, fabrication, and use capabilities. For high-temperature alloys, the precipitation of undesired intermetallics can be a major concern, especially for applications requiring a long service life or ease of repair. Therefore, much effort has been devoted to understanding intermetallics and their effects on properties and to determining how to avoid their occurrence through alloy design.

Most high-strength nickel-base alloys depend on the precipitation of an A_3B-type compound known as gamma prime (γ'). In simple alloys, it takes on the form $Ni_3(Al,Ti)$. In complex alloys, other elements can substitute for nickel on the "A" side or for aluminum and titanium on the "B" side. This compound has an ordered fcc derivative, $L1_2$, crystal structure. Its low lattice mismatch (0 to 1%) allows the precipitate to nucleate homogeneously with low surface energy and long-term stability (Ref 14). The γ' phase is unique in a number of ways (Ref 2):

- Because it is an ordered compound, its antiphase boundary energy contributes to alloy strengthening
- It possesses the unusual characteristic of having increased strength with increasing temperature
- It is inherently ductile, unlike most other intermetallic compounds

The strength of γ' alloys can be increased by increasing the Al + Ti content to obtain a higher γ' volume fraction. However, alloys with high Al + Ti levels are difficult to manufacture in wrought forms and to fabricate, and they are best exploited as castings. Additions of refractory metals can also be used to increase strength by altering lattice mismatch and antiphase boundary energy and by solid-solution strengthening γ' and the matrix (Ref 2). Additions of cobalt are also effective by increasing the γ' solvus temperature (Ref 2).

Another important intermetallic phase that can be used to strengthen nickel-base alloys is a metastable form of Ni_3Nb known as gamma double prime (γ''). For the most part, it has been exploited in alloys containing significant amounts of iron (Ref 18). Gamma double prime has a body-centered tetragonal crystal structure and usually precipitates in the form of a disk. At temperatures of 705 °C (1300 °F) and above, it overages rapidly and transforms into the orthorhombic form of Ni_3Nb (Ref 19). Because of the sluggish nature of the precipitation reaction, alloys strengthened by γ'' can possess excellent weldability (Ref 11).

The most common harmful intermetallic phases that occur in nickel-base alloys are the topologically closed-packed (tcp) phases: sigma (σ), mu (μ), and Laves. These are complex layered structure phases that often nucleate at grain-boundary carbides (Ref 14). The σ phase has a complex tetragonal structure that can exist over a broad range of compositions involving the common transition alloying elements (Ref 20). In nickel-base alloys, σ is typically rich in chromium and often contains molybdenum and/or tungsten. It usually nucleates on $M_{23}C_6$ carbides that contain these elements and that are structurally related to σ through a slight shift in atomic arrangements (Ref 21). Typically, σ occurs in a platelike morphology that gives rise to a needle-shaped appearance when viewed two dimensionally (Ref 21).

The μ phases are complex rhombohedral compounds involving molybdenum and/or tungsten and the matrix elements nickel, chromium, iron, and cobalt. They occur in binary systems of iron and cobalt in forms such as Fe_7Mo_6, but they do not occur in nickel binary systems. However, a μ phase does form in the nickel-chromium-molybdenum ternary system (Ref 20). The μ phase tends to nucleate on M_6C or $M_{12}C$ carbides that have a compositional and structural similarity; this is analogous to σ-phase precipitation on $M_{23}C_6$ carbides (Ref 21, 22). The μ phase also occurs with a platelike morphology that is very distinctively faulted when observed by transmission electron microscopy.

The Laves phases are complex hexagonal compounds that, in nickel-base alloys, generally involve iron and the refractory elements molybdenum, tungsten, niobium, and tantalum. The chemical form is A_2B with iron on the "A" side. Laves phases occur because of atomic size effects; in contrast, the σ and μ phases form because of electronic effects. Unlike σ and μ, the Laves phases are not known to nucleate at carbide particles (Ref 21). However, their morphology is also platelike.

Because of the detrimental effects on properties produced by the tcp phases, much effort has been devoted to finding ways to avoid their occurrence. The early work of Beck and others led to the development of the electron vacancy (Nv) theory, which was based on Pauling's analysis of the transition metal *3d* electronic structure (Ref 21). This eventually resulted in a computerized calculation known as PHACOMP (phase computation), which makes a prediction for tcp phase precipitation after taking into account the matrix element depletion caused by the precipitation of γ', borides, and carbides (Ref 2, 21). Another approach is based on the computer calculation of phase diagrams called SIGMA-SAFE (Ref 23). Finally, an approach based on theoretical electronic structure calculations has also been proposed (Ref 24). Thus, powerful tools are available for designing new alloy chemistries that are free of tcp phases.

Behavior of Nickel-Base Alloys in Corrosive Environments

Narasi Sridhar
Haynes International, Inc.

Nickel and its alloys, like the stainless steels, offer a wide range of corrosion resistance. However, nickel can accommodate larger amounts of alloying elements, chiefly chromium, molybdenum, and tungsten, in solid solution than iron. Therefore, nickel-base alloys, in general, can be used in more severe environments than the stainless steels. In fact, because nickel is used to stabilize the austenite (fcc) phase of some of the highly alloyed stainless steels, the boundary between these and nickel-base alloys is rather diffuse. The nickel-base alloys range in composition from commercially pure nickel to complex alloys containing many alloying elements.

A distinction is usually made between those alloys that are primarily used for high-temperature strength, commonly referred to as superalloys, and those that are primarily used for corrosion resistance. Again, the distinction is not sharp, because some of the former class of alloys are used in corrosion service and some of the latter in high-temperature service. Many of the alloys that have high-temperature strength are multiphase alloys with precipitation-strengthening elements such as aluminum, titanium, and niobium. They also have higher carbon levels. Most of the corrosion-resistant alloys are primarily single-phase alloys that can be strengthened mainly by cold working. However, in applications in which geometry or size prevents strengthening by cold work, precipitation-strengthened alloys are used. A partial list of nickel-base alloys and their compositions is given in Table 1. The corrosion applications that are the topic of this section involve both aqueous and nonaqueous environments at relatively low temperatures (<260 °C, or 500 °F). High-temperature applications of nickel-base alloys involving hot gases or molten salts are discussed in the articles "Corrosion of Metal Processing Equipment," "Corrosion in Fossil Fuel Power Plants," and "Corrosion in the Nuclear Power Industry" in this Volume.

The types of corrosion of greatest importance in the nickel-base alloy system are uniform corrosion, pitting and crevice corrosion, intergranular corrosion, and galvanic corrosion (Ref 25). Stress-corrosion cracking (SCC), corrosion fatigue, and hydrogen embrittlement are also of great importance and are discussed in another section of this article. In order to estimate the performance of a set of alloys in any environment, it is of paramount importance to ascertain the composition and, for liquid environments, the electrochemical interaction of the environment with an alloy. A case in point is the nickel-molybdenum alloy B-2. The alloy performs exceptionally well in pure, deaerated H_2SO_4 and HCl, but it deteriorates rapidly when oxidizing impurities, such as oxygen (air) and ferric ions (Fe^{3+}), are present.

Electrochemical Interaction and Corrosion

This discussion will introduce some basic concepts necessary for understanding the corrosion behavior of metals. Although the electrochemical response of a metal in an electrolyte is a vast and complex subject, many of the corrosion phenomena observed in nickel-base alloys can be qualitatively understood in terms of the schematic diagram shown in Fig. 1. Generally, two types of behavior are exhibited by metals in solution:

- Type I, in which the corrosion or anodic current increases monotonically with potential
- Type II, in which the anodic current initially increases with potential (active behavior), then decreases to a small constant value (passive behavior), and finally increases again (transpassive)

These two types of behavior are not intrinsic properties of an alloy, such as the modulus of elasticity, but are the result of the interaction of the alloy with a given environment. Thus, alloy B-2, which exhibits Type I behavior in dilute H_2SO_4 and HCl, exhibits Type II behavior in concentrated H_2SO_4.

The potential of a metal in an environment is influenced by the balance of reduction-oxidation processes taking place at the solution/metal interface. Thus, in dilute H_2SO_4, the reduction (also termed cathodic) reaction is the conversion of hydrogen ion to hydrogen gas ($H^+ + e^- \rightarrow {}^1\!/_2H_2$). In a ferric chloride ($FeCl_3$) solution, the cathodic reaction is the conversion of Fe^{3+} to ferrous (Fe^{2+}) ion ($Fe^{3+} + e^- \rightarrow Fe^{2+}$). These cathodic reactions are represented by a downward-sloping line in Fig. 1, indicating that, as the potential is reduced, the reduction reaction takes place at a faster rate. Because no net current flows when a metal is dipped in a solution, the metal attains a potential that is the intersection of the cathodic and anodic lines. The higher the cathodic line, the greater the oxidizing power of the solution. For example, the reaction $Fe^{3+} + e^- \rightarrow Fe^{2+}$ has a potential that is more positive than the reaction $H^+ + e^- \rightarrow {}^1\!/_2H_2$. Thus, in a more oxidizing solution, the resultant potential of the metal is higher. The absolute metal-solution potential cannot, in theory, be measured, but a relative potential with respect to an unchanging reference potential can be measured.

With regard to the corrosion behavior of alloys, if an alloy exhibits Type I behavior, increasing the corrosion potential will result in an increased corrosion rate. Alloy B-2 in dilute, deaerated H_2SO_4 exhibits a low corrosion rate (Type I behavior) because the corrosion potential determined by the hydrogen reduction reaction is low. If Fe^{3+} impurities are present in the acid (this is quite common in acid streams that contact steel parts elsewhere in the system), the corrosion potential, and thus the corrosion rate, is increased (Fig. 2). The presence of oxygen can result in a similar behavior. In contrast, a high-chromium alloy such as alloy G-30 exhibits Type II behavior, and the presence of Fe^{3+} shifts the potential into the passive region, which results in a low corrosion rate.

A third possibility is that the solution is oxidizing enough to shift the potential into the transpassive region. In some solutions, this can lead to pitting corrosion.

Corrosion Resistance of Nickel-Base Alloys in Acid Media

Sulfuric acid is the most ubiquitous environment in the chemical industry. The electrochemical nature of the acid varies a great deal, depending on the concentration of the acid and the impurity content (Ref 25). The pure acid is considered to be a nonoxidizing acid up to a concentration of about 50 to 60% by weight, beyond which it is generally considered to be oxidizing (Ref 25, 26). The corrosion rates of nickel-base alloys, in general, increase with acid concentra-

Table 1 Nominal chemical compositions of some typical nickel-base alloys

Common alloy designation	UNS designation	Chemical composition, %: C(a)	Nb	Cr	Cu	Fe	Mo	Ni	Si(a)	Ti	W	Other
Nickel												
200	N02200	0.1	...	...	0.25 max	0.4 max	...	99.2 min	0.15	0.1 max	...	...
201	N02201	0.02	...	...	0.25 max	0.4 max	...	99.0 min	0.15	0.1 max	...	...
Nickel-copper												
400	N04400	0.15	...	...	31.5	1.25	...	bal	0.5	...	...	...
R-405	N04405	0.15	...	...	31.5	1.25	...	bal	0.5	...	...	0.0435
Nickel-molybdenum												
B-2	N10665	0.01	...	1.0 max	...	2.0 max	28	bal	0.1	...	...	...
B	N10001	0.05	...	1.0 max	...	5.0	28	bal	1.0	...	...	...
Nickel-chromium-iron												
600	N06600	0.08	...	16.0	0.5 max	8.0	...	bal	0.5	0.3 max	...	...
601	N06601	...	...	23.0	...	14.1	...	bal	...	...	...	1.35Al
800	N08800	0.1	...	21.0	0.75 max	44.0	...	32.5	1.0	0.38	...	...
800H	N08810	0.08	...	21.0	0.75 max	44.0	...	32.5	1.0	0.38	...	...
Nickel-chromium-iron-molybdenum												
825	N08825	0.05	...	21.5	2.0	29.0	3.0	42	0.5	1.0	...	...
G	N06007	0.05	2.0	22.0	2.0	19.5	6.5	43	1.0	...	1.0 max	...
G-2/2550	N06975	0.03	...	24.5	1.0	20.0	6.0	48	1.0	1.0	...	...
G-3	N06985	0.015	0.8	22.0	2.0	19.5	7.0	44	1.0	...	1.5 max	...
H	...	0.03	...	22.0	...	19.0	9.0	42	1.0	...	2.0	...
G-30	N06030	0.03	0.8	29.5	2.0	15.0	5.5	43	1.0	...	2.5	...
Nickel-chromium-molybdenum-tungsten												
N	N10003	0.06	...	7.0	0.35 max	5.0 max	16.5	71	1.0	0.5 max	0.5 max	...
W	N10004	0.12	...	5.0	...	6.0	24.0	63	1.0	...	...	...
625	N06625	0.1	4.0	21.5	...	5.0 max	9.0	62	0.5	...	...	...
690	N06690	0.02	...	29.0	...	10.0	...	61	...	0.3	...	...
C-276	N10276	0.01	...	15.5	...	5.5	16.0	57	0.08	...	4.0	...
C-4	N06455	0.01	...	16.0	...	3.0 max	15.5	65	0.08	...	...	...
C-22	N06022	0.015	...	22.0	...	3.0 max	13.0	56	0.08	...	3.0	...
ALLCORR	N06110	0.15	2.0 max	30.0	...	...	10.0	53	...	1.5 max	4.0 max	...
Nickel-silicon												
D	...	0.12	...	1.0 max	3.0	2.0 max	...	86	9.5	...	...	...
Precipitation-hardening												
K-500	N05500	0.25	...	...	29.0	2.0 max	...	63	0.5 max	0.6	...	2.7Al
R-41	N07041	0.09	...	19.0	...	5.0 max	10.0	52	0.5 max	3.1	...	1.5Al
718	N07718	0.05	5.0	18.0	...	19	3.0	53	...	0.4 max	...	...
X-750	N07750	...	0.9	15.5	...	7.0	...	bal	...	2.5	...	...
925	N09925	0.02	...	21.0	2.0	28	3.0	43.0	...	2.1	...	...

(a) Maximum

tion up to about 90% weight. Higher concentrations of the acid are generally less corrosive.

A broad view of the relative performance of various alloys in reagent grade H_2SO_4 is illustrated in Fig 3. In this graph, the temperatures at which the corrosion rate of an alloy in a given concentration of the acid exceeds 0.5 mm/yr (20 mils/yr) are plotted as isocorrosion curves. At low acid concentrations, the nickel-chromium-molybdenum alloys show a significantly higher resistance than type 316 stainless steel. Alloy 20, a high-nickel stainless steel, shows similar behavior. These high-nickel-chromium-molybdenum alloys can be used only to moderate temperatures in the intermediate and high concentrations of H_2SO_4. The nickel-molybdenum alloys B and B-2 can be used to higher temperatures for all concentrations of acid. However, in the presence of oxidizing species and in aqueous acid systems, alloys B and B-2 suffer serious corrosion. This effect is shown in Fig. 2 for alloy B-2.

The presence of oxidizing impurities can be beneficial to nickel-chromium-molybdenum alloys shown in Fig. 3 because these impurities can aid in the formation of passive films that retard corrosion. Another important consideration is the presence of chlorides (Cl^-). Chlorides generally accelerate the attack; the extent of acceleration differs for various alloys. This is shown in Fig. 4 in terms of the 0.13-mm/yr (5-mils/yr) isocorrosion curves for alloys C-276 and G.

Hydrochloric Acid. The corrosion resistance of several nickel-base alloys in HCl and related environments is reviewed in Ref 30. Commercially pure nickel (200 and 201) and nickel-copper alloys (alloy 400) have room-temperature corrosion rates below 0.25 mm/yr (10 mils/yr) in air-free HCl at concentrations up to 10%. In HCl concentrations of less than 0.5%, these alloys have been used at temperatures up to about 200 °C (390 °F) (Ref 30). Oxidizing agents, such as cupric (Cu^{2+}), Fe^{3+}, and chromate (CrO_4^{2-}) ions or aeration, raise the corrosion rate considerably. Under these conditions nickel-chromium-molybdenum alloys such as 625 or C-276 offer better corrosion resistance. They can be passivated by the presence of oxidizing agents.

The nickel-chromium-molybdenum alloys also show higher resistance to uncontaminated HCl (Fig. 5). For example, alloys C-276, 625, and C-22 show very good resistance to dilute HCl at elevated temperatures and to a wide range of HCl concentrations at ambient temperature. The corrosion resistance of these alloys depends on the molybdenum content. The alloy with the highest molybdenum content—Hastelloy alloy B-2—shows the highest resistance in HCl of all the nickel-base alloys. Accordingly, this alloy is used in a variety of processes involving hot HCl or nonoxidizing chloride salts hydrolyzing to produce HCl. In Fig. 5, alloy B-2 shows two isocorrosion curves—one at high temperatures near the boiling point and one at low temperature. This is due to variations in the oxygen content of the solutions. At the higher temperatures, the oxygen solubility is lower, and therefore the corrosion rate is lower. The major weakness of the alloy is that its corrosion resistance decreases dramatically under oxidizing conditions (Fig. 2). Chromium-containing alloys such as C-276 or 625 can be passivated when oxidizers are present, and thus display much lower corrosion rates.

Pitting Corrosion in Chloride Environments. The nickel-chromium-molybdenum alloys, such as alloys C-276, 625, ALLCORR, and C-22, exhibit very high resistance to pitting in oxidizing chloride environments. The critical pitting temperatures of various nickel-chromium-molybdenum alloys in an oxidizing chloride solution are shown in Table 2. Pitting corrosion is most prevalent in chloride-containing environments, al-

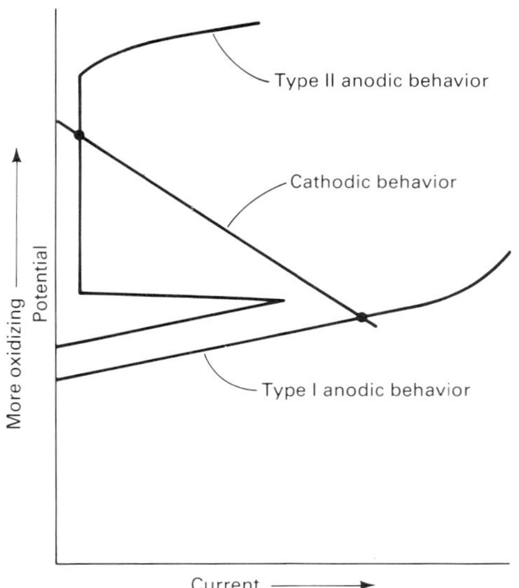

Fig. 1 Schematic showing the two basic types (Type I and Type II) of electrochemical response exhibited by nickel-base alloys. These responses are not intrinsic properties, but are interactions between alloys and environments.

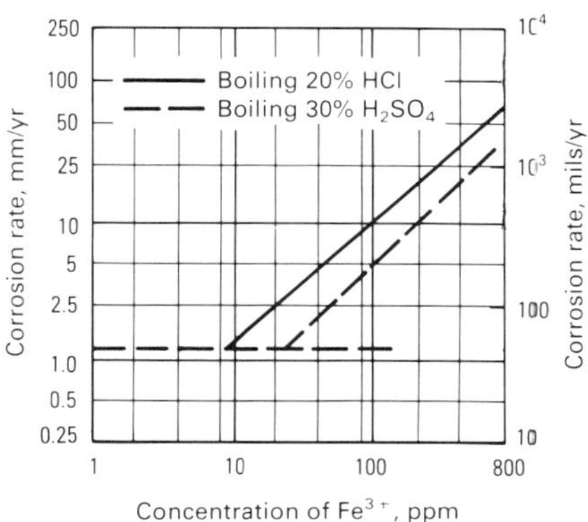

Fig. 2 Effect of Fe^{3+} in boiling H_2SO_4 and HCl on the corrosion rate of alloy B-2

though other halides and sometimes sulfides have been known to cause pitting. There are several techniques that can be used to evaluate resistance to pitting (Ref 31, 32). Critical pitting potential and pitting protection potential indicate the electrochemical potentials at which pitting can be initiated and at which a propagating pit can be stopped, respectively. These values are functions of the solution concentration, pH, and temperature for a given alloy; the higher the potentials, the better the alloy. The critical pitting temperature (below which pitting does not initiate) is often used as an indicator of resistance to pitting, especially in the case of highly corrosion-resistant alloys (Table 2). Chromium and molybdenum have been shown to be extremely beneficial to pitting resistance (Ref 8).

Nitric Acid. Chromium is an essential alloying element for corrosion resistance in HNO_3 environments because it readily forms a passive film in these environments (Ref 33). Thus, the higher chromium alloys show better resistance in HNO_3 (Fig. 6). In these types of environments, the highest chromium alloys, G-30 and 690, seem to show the highest corrosion resistance. Molybdenum is generally detrimental to corrosion resistance in HNO_3. For example, alloy C-22 (with 13% Mo) is not as good as alloy G-3 (with 7% Mo). In pure HNO_3, stainless steels find the greatest application. In concentrated (98%) HNO_3, stainless steels containing both chromium and silicon are finding greater use (Ref 34). Because of its oxidizing nature, HNO_3 streams are too severe for nickel-molybdenum alloys. The behavior of nickel-silicon alloys in HNO_3 seems to depend on their microstructure and composition (Ref 9). Cast Hastelloy alloy D, which consists of two phases (the nickel-silicon solid solution and the eutectic), does not show good resistance to any concentration of HNO_3. The Ni-9.5Si-2.5Cu-3Mo-2.75Ti cast alloy, which consists of a single phase, showed high resistance to boiling concentrated HNO_3 above 60% and to fuming (99%) HNO_3 up to 80 °C (175 °F). However, the corrosion rates in more dilute acid were very high.

Phosphoric acid is obtained by two different routes. The wet-process acid is obtained by reacting phosphate rock with concentrated H_2SO_4 and concentrating the resulting dilute acid by evaporation. The furnace acid is obtained by calcining the phosphate rock to produce elemental phosphorus, which is then oxidized and reacted with water to produce H_3PO_4. This latter acid is very pure and is used as reagent grade. The wet-process acid is used to make phosphatic fertilizers and usually contains a number of impurities, such as HF, H_2SO_4, and SiO_2. The percentage of these impurities depends on the source of the rock, the process of reaction with H_2SO_4, and the stage of concentration of the H_3PO_4.

The corrosion rates of various alloys in pure and wet-process acids are compared in Table 3. It is interesting to note the variation in corrosion rate of the same alloy in wet-process acid from two different manufacturers. Because of these differences, it is imperative that any comparison between alloys in this type of acid be made from tests conducted in the same batch of acid from the same source. It can also be seen that the wet-process acid can be considerably more corrosive than the reagent grade acid. It has been shown that in wet-process acid a high chromium content, such as in alloy G-30, is beneficial (Ref 35). In addition, Cl^- is inevitably present in these acids; therefore, molybdenum and tungsten additions are beneficial.

The plant test results shown in Table 4 indicate that alloys 20Cb3, 825, and C have the highest corrosion resistance in dilute H_3PO_4 process solutions. In very high-concentration H_3PO_4 at high temperatures, alloy B-2 shows the highest resistance (Table 5). The corrosion resistance in these acids seems to depend on molybdenum content.

Hydrofluoric acid is commercially available in concentrations ranging from 30 to 70% and as anhydrous HF. Plain carbon steel can withstand the anhydrous acid and is extensively used in rail cars carrying anhydrous HF. Of the nickel-base alloys, alloy 400 has been the most extensively examined (Ref 37). The alloy possesses good corrosion resistance in all concentrations of HF up to a temperature of about 120 °C (250 °F). However, the presence of oxygen in the solution is detrimental (Table 6).

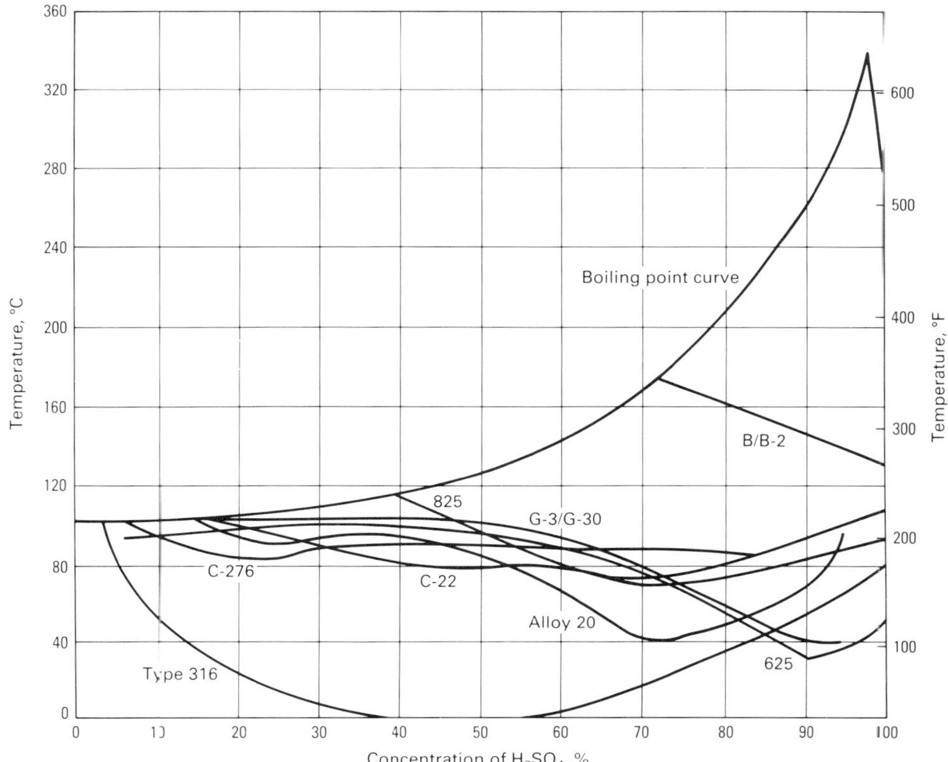

Fig. 3 Comparative behavior of several nickel-base alloys in pure H_2SO_4. The isocorrosion lines indicate a corrosion rate of 0.5 mm/yr (20 mils/yr). Source: Ref 27-29

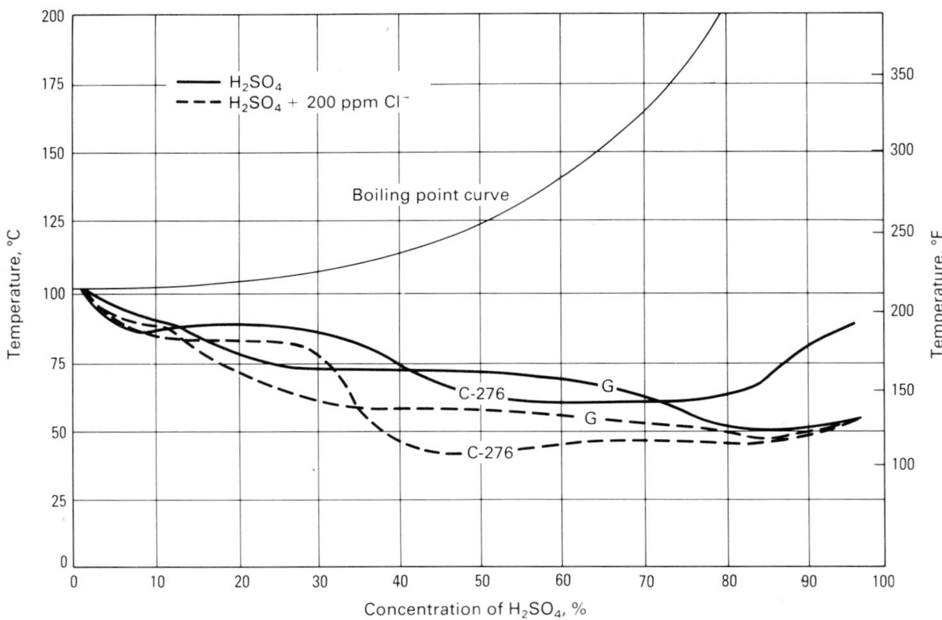

Fig. 4 Effect of dissolved chlorides on the corrosion resistance of nickel-base alloys in H_2SO_4. The isocorrosion lines indicate a corrosion rate of 0.13 mm/yr (5 mils/yr).

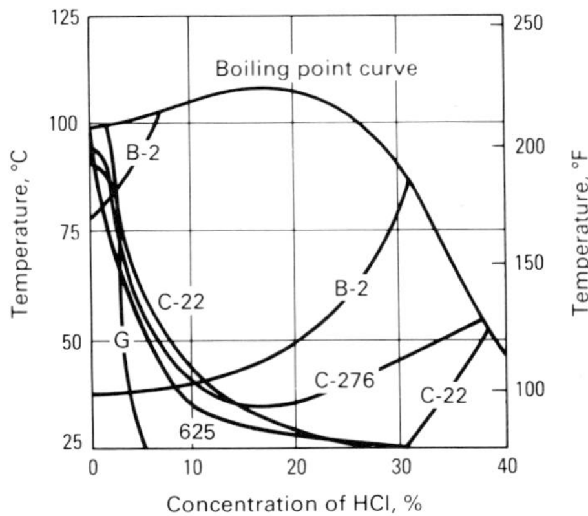

Fig. 5 Comparative isocorrosion plots of various nickel-base alloys in HCl. The lines indicate a corrosion rate of 0.13 mm/yr (5 mils/yr).

In another investigation, alloy 400 showed transgranular SCC in the vapor phase of dilute HF solutions (up to 0.5% HF) at temperatures up to 95 °C (200 °F) (Ref 38). The cracking susceptibility did not depend on the presence of oxygen, and no cracking was found in the liquid phase. The nickel-molybdenum alloys B and B-2 show good corrosion resistance at low temperatures. However, in one case, increasing the temperature from 25 to 95 °C (75 to 200 °F) in a 0.5% HF solution resulted in an increase in the corrosion rate of alloy B from 0.025 to 0.05 mm/yr (1 to 2 mils/yr) to 7.1 mm/yr (280 mils/yr). Alloy C-276 showed a low corrosion rate at all temperatures in HF up to 0.5% (Ref 38). The corrosion resistance of a variety of nickel-base alloys in various concentrations of HF is shown in Table 7. No control of aeration was attempted in these tests. The presence of aeration may be detrimental to the corrosion resistance of alloy C-276, as shown in Table 8 for 70% HF solution.

Table 2 Critical pitting temperatures for nickel alloys evaluated in 6% $FeCl_3$ for 24-h periods

Alloy	Critical pitting temperature °C	°F
825	0.0, 0.0	32.0, 32.0
904L	2.5, 5.0	36.5, 41
Type 317LM stainless steel	2.5, 2.5	36.5, 36.5
G	23.0, 25.0	73.5, 77
G-3	25.0, 25.0	77, 77
C-4	37.5, 37.5	99.5, 99.5
625	35.0, 40.0	95, 104
ALLCORR	52.5, 52.5	126.5, 126.5
C-276	60.0, 65.0	140, 149
C-22	70.0, 70.0	158, 158

Source: Ref 33

The organic acids are generally not as corrosive as the mineral acids because of their lower acidity. Of the organic acids, acetic and formic acids form the bulk of the corrosion data on nickel-base alloys. Corrosion data on other organic acids can be found in Ref 39 and 40.

Acetic Acid. The corrosion rates of Nickel 200, alloy 400, and alloy 600 are shown in Table 9. The detrimental effects of aeration can be seen for alloy 400 and Nickel 200. The corrosion rates of a variety of alloys in boiling 10 and 99% acetic acids are given in Tables 10 and 11. No deliberate aeration or deaeration was done in these tests. The corrosion rates of all the alloys are quite low in these environments. Even though pure acetic acid is not very aggressive, the addition of contaminants can increase the corrosion rates. This is shown for Nickel 200 and alloy 400 in Table 12. The detrimental effects of aeration and oxidizing ions are indicated, especially in the dilute acid.

Formic acid is generally more corrosive than acetic acid. This is partly because the dielectric constant of formic acid (56.1 at 25 °C, or 75 °F) is higher than that of acetic acid (6.2 at the same temperature); therefore, the ionic dissociation is higher in the former. Formic acid is also much more acidic than acetic acid. The corrosion rates of a variety of nickel-base alloys in boiling 40 and 88% formic acid are given in Tables 13 and 14. Among the nickel-chromium-molybdenum-iron alloys, the higher-molybdenum alloys generally tended to show higher resistance to corrosion. The isocorrosion curves for alloys C-276, B, and 400 and AISI type 316 stainless steel are shown in Fig. 7. For alloys 400 and B, the data shown in Fig. 7 and in boiling acids (Tables 13 and 14) do not agree. It is possible that these discrepancies are due to differences in aeration, because boiling acids are expected to have low aeration after a certain period of time.

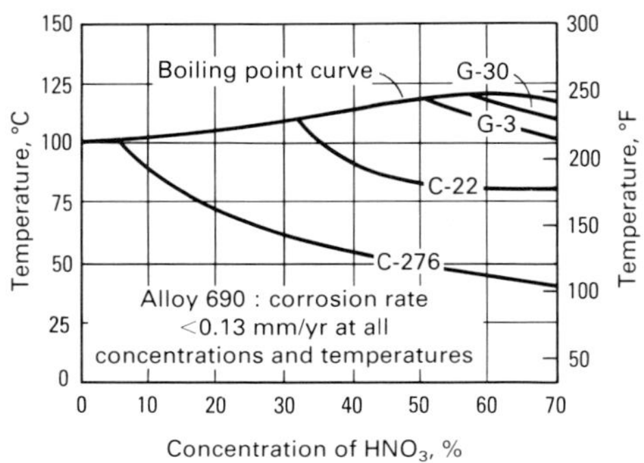

Fig. 6 Comparative behavior of nickel-base alloys in HNO_3. Isocorrosion curves indicate a corrosion rate of 0.13 mm/yr (5 mils/yr).

Propionic Acid. The corrosion rates of several alloys in boiling propionic acid are shown in Fig. 8. Lowering the temperature from the boiling point can increase the corrosion rates of those alloys that are sensitive to the degree of aeration (Table 15).

Other Organic Acids. The higher organic acids, such as acrylic acid, and the fatty acids, such as lauric and stearic acids, are generally not very corrosive to nickel-base alloys (Ref 39). The corrosion rates in these acids, as in other organic acids, are determined by the inorganic impurities present in the acids (for example, chlorides and oxidizing salts).

Corrosion in Acid Mixtures. In many processes, mixtures of several acids or acids and salts are encountered. Corrosion resistance in these types of environments is sometimes predictable qualitatively. In some cases, anomalous effects can be produced. Although it is impossible to list the corrosion rates of alloys in a wide range of acid mixtures within the constraints of this section, an attempt will be made to present those mixtures that are of greatest importance or that show surprising features.

Sulfuric + Nitric Acid Mixtures. In alloys that contain chromium and exhibit active-passive behavior, addition of HNO_3 or nitrates to H_2SO_4 will reduce the corrosion rate (Fig. 9). In nonchromium-containing alloys (for example, alloys B-2 or 400), addition of HNO_3 will increase the corrosion rates. The nitrate reduction reaction increases the redox potential in H_2SO_4 solution; the redox potential of H_2SO_4 solution is normally controlled by the H^+ reduction reac-

Table 3 Corrosion of various alloys in reagent grade and wet-process H_3PO_4

Alloy	Corrosion rate: Reagent grade 75% H_3PO_4, 115 °C (240 °F) mm/yr	mils/yr	Wet-process 75% H_3PO_4 (54% P_2O_5), 115 °C (240 °F): Source A mm/yr	mils/yr	Source B mm/yr	mils/yr
Type 316 stainless steel	0.76	30	29	1140	1.7	67
825	...	...	14	550	0.64	25
C-276	0.38	15	1.9	74	0.71	28
C-22	0.3	12	0.84	33	0.28	11
625	0.3	12	0.91	36	0.3	12
690	0.13	5	0.5	20	0.18	7
G-30	0.13	5	0.46	18	0.15	6

Table 4 Results of plant corrosion test in a H_3PO_4 evaporator

Specimens were exposed for 42 days to 53% H_3PO_4 containing 1 to 2% H_2SO_4 and 1.2 to 1.5% fluoride at 120 °C (250 °F). No pitting was observed.

Alloy	Corrosion rate mm/yr	mils/yr
20Cb	0.12	4.7
C	0.13	5
825	0.16	6.2
Type 317 stainless steel	0.26	10.4
400	0.64	25
Type 316 stainless steel	1.1	44
600	>33(a)	>1300(a)
B	>33(a)	>1300(a)

(a) Specimen was destroyed. Source: Ref 36

Table 5 Corrosion of various nickel-base alloys in pure H_3PO_4

Acid was nitrogen purged before the test, which was conducted at 280 °C (535 °F).

Alloy	Molybdenum content, %	Corrosion rate: 108% H_3PO_4 mm/yr	mils/yr	112% H_3PO_4 mm/yr	mils/yr
G-30	5.5	9.4	372	32.5	1280
G-3	7	7.9	310	24.4	961
625	9	4.6	180	14.9	590
C-22	13	2.6	102	3.9	152
C-276	16	1.4	54	1.8	72
B-2	28	0.23	9	0.05	2

Table 6 Effect of oxygen on corrosion of alloy 400 in HF solutions

The acids were purged with nitrogen containing various amounts of oxygen.

Concentration of oxygen in purge gas, ppm	Corrosion rate: Boiling (112 °C, or 234 °F) 38% HF: Liquid phase mm/yr	mils/yr	Vapor phase mm/yr	mils/yr	Boiling (108 °C, or 226 °F) 48% HF: Liquid phase mm/yr	mils/yr	Vapor phase mm/yr	mils/yr
<5	0.24	9.5	0.17	6.8	0.28	11	0.076	3
<500	0.43	17	0.3	12	0.56	22	0.1	4
1500	0.79	31	1.24	49	0.7	28	0.61	24
2500	0.74	29	0.46	18	0.69	27	0.23	9
3500	0.86	34	1.37	54	0.86	34	0.74	29
4700	1.3	53	2.7	107	1.1	43	2.1	83
10 000	1.2	46	0.64	25	1.2	48	1.9	75

Source: Ref 36

tion. In nonpassivating alloys, for example, alloy B-2, in which the corrosion current increases monotonically with potential, an increase in potential will increase the corrosion rate. In passivating alloys, an increase in potential can move the alloy from the active state to the passive state, thus reducing the corrosion rate. For high nitrate concentrations, the passive current density itself can increase, or the corrosion potential can move into the transpassive region, which results in an increase in corrosion rate. In the case of alloy C-276 in Fig. 9, only an increase in corrosion rate is observed with the HNO_3 addition. It is possible that for lower concentrations (<10%) of HNO_3 a decrease in corrosion rate could have been observed.

Sulfuric + Hydrochloric (or chloride) Mixtures. The addition of alkali chloride salts or HCl to H_2SO_4 increases the corrosion rates of all alloys. The behavior of three of the nickel-base alloys is shown in Fig. 4 and Fig. 10. For deaerated conditions, alloys B and B-2 are the most versatile, with alloys C-276 and C-22 following in that order. Generally, the higher the molybdenum, the better the performance of the alloy in H_2SO_4 + HCl mixtures.

Nitric + Hydrochloric Acid Mixtures. The effects of HNO_3 additions to HCl are similar to the effects of HNO_3 additions to H_2SO_4 (Fig. 11). However, in HNO_3 + HCl mixtures, pitting is the mechanism of corrosion, rather than uniform corrosion, which occurs in H_2SO_4 + HNO_3 mixtures. In addition, small changes in HCl concentration can result in enormous changes in corrosion rates (Fig. 12).

Nitric + Hydrofluoric Acid Mixtures. The addition of HNO_3 to HF reduces the corrosion rate initially, but beyond 10% HNO_3, the corrosion rate increases. Increasing the HF concentration results in an increasing corrosion rate (Fig 13). However, unlike the case of HCl addition, the higher-chromium alloys generally showed lower rates, irrespective of molybdenum level. Intergranular corrosion was also observed in many of these alloys. In all of the above cases, increasing temperature caused increased corrosion rates.

Corrosion Resistance of Nickel-Base Alloys in Alkalies

Nickel 200 and 201 are used extensively in caustic production and processes involving all concentrations and temperatures of sodium hydroxide (NaOH) and potassium hydroxide (KOH). This resistance prevails even when these alkalies are molten. The lower-carbon Nickel 201 is used at temperatures above 315 °C (600 °F) to avoid graphite precipitation. Isocorrosion curves for Nickel 200 and 201 are shown in Fig. 14. Nickel 200 is used extensively in caustic evaporator service where dilute caustic is concentrated up to 50%. In this service contaminants such as chlorates, hypochlorites and chlorides can cause increased corrosion rates, especially under velocity conditions (Ref 43-46). Though nickel is resistant to most alkalies, it is not resistant to ammonium hydroxide (NH_4OH) solutions.

The resistance of nickel alloys to general corrosion and SCC increases with increasing nickel content. In some caustic applications where higher strength or resistance to other corrodents is required, alloys 400, 600, C-276, and others are used. These alloys are highly resistant to general corrosion and SCC, but can be attacked at high caustic concentrations and temperatures.

Alloys 600 and 800 have been used extensively in nuclear steam generator service at about 300 °C (570 °F). In this service resistance to caustic, which can be formed and concentrated at tube sheets, is of concern (Ref 47). The higher-chromium alloy 690 is more resistant to SCC under some conditions.

Environmental Embrittlement of Nickel-Base Alloys

Juri Kolts
Conoco Inc.

Nickel-base alloys are frequently used because of their improved resistance to environmental embrittlement over steels and stainless steels. However, nickel-base alloys can exhibit environmental embrittlement under the combined action of tensile stresses (either residual or applied) and specific environmental conditions. In the most severe cases, cracking or failure may result after an incubation period in which no apparent damage has occurred. These incubation periods may be of the order of minutes, days, months, or years.

The embrittlement of nickel-base alloys by the combined action of tensile stress and a suitable environment is thought to occur by two phenomena: hydrogen embrittlement and SCC. No inference is made as to the mechanisms of embrittle-

Table 7 Corrosion of various nickel-base alloys in HF

24-h tests with no control of aeration.

	Corrosion rate, mm/yr (mils/yr)			
	2% HF		5% HF	
Alloy	70 °C (160 °F)	Boiling	70 °C (160 °F)	Boiling
C-276	0.24 (9.5)	0.076 (3)	0.25 (10)	0.1 (4)
C-22	0.23 (9.4)	0.94 (37)	0.34 (13.5)	0.84 (33)
625	0.5 (20)	. . .	0.4 (16)	. . .
C-4	0.43 (17)	. . .	0.38 (15)	. . .
200	. . .	. . .	0.46 (18)	. . .
600	. . .	. . .	0.23 (9)	. . .
B-2	. . .	. . .	0.38 (15)	. . .
G-3	. . .	. . .	0.5 (20)	. . .
G-30	(10)	. . .	0.76 (30)	. . .

Table 8 Effect of aeration on corrosion of two nickel-base alloys in 70% HF

Welded samples were tested from 14–36 days.

	Corrosion rate			
	Nitrogen blanket		Oxygen blanket	
Alloy	mm/yr	mils/yr	mm/yr	mils/yr
C-276	0.008	0.3	0.94	37
400	0.013	0.5	0.58	23

Table 9 Corrosion of high-nickel alloys in acetic acid

			Corrosion rate, mm/yr (mils/yr)					
	Temperature		Alloy 400		Nickel 200		Alloy 600	
Concentration of acetic acid, %	°C	°F	Aerated	Unaerated	Aerated	Unaerated	Aerated	Unaerated
2	30	86	. . .	0.0008 (0.03)	. . .	. . .	. . .	0.0013 (0.05)
5	116	240	. . .	0.0008 (0.03)	. . .	0.007 (0.28)	. . .	0.002 (0.08)
10	30	86	0.008 (0.33)	0.002 (0.08)	. . .	0.0025 (0.1)	. . .	0.0005 (0.02)
25	30	86	0.01 (0.41)	0.002 (0.08)	. . .	. . .	. . .	. . .
30	30	86	. . .	. . .	0.084 (3.3)	. . .	. . .	. . .
50	30	86	0.019 (0.74)	0.0025 (0.1)	0.11 (4.3)	0.006 (0.25)	. . .	. . .
75	30	86	0.009 (0.36)	0.0013 (0.05)	. . .	. . .	. . .	. . .
99.9	30	86	0.006 (0.23)	0.002 (0.08)	. . .	0.003 (0.13)	. . .	. . .
99.9	116	240	0.004 (0.15)	. . .	0.009 (0.36)	. . .	. . .	. . .

Source: Ref 38

Table 10 Corrosion of various mill-annealed alloys in boiling 10% acetic acid

Results are based on four 24-h test periods.

	Corrosion rate	
Alloy	mm/yr	mils/yr
Type 316L stainless steel	0.015–0.018	0.58–0.72
Alloy 825	0.0152–0.016	0.60–0.63
Alloy G	0.011–0.014	0.43–0.54
Alloy 625	0.01–0.019	0.39–0.77
Alloy C-276	0.011–0.0114	0.41–0.45
Alloy B-2	0.0112–0.013	0.44–0.50

ment or to what extent hydrogen is involved in SCC. Phenomenologically, hydrogen embrittlement is distinguished from SCC in this section by the influence of two parameters (environmental temperature and anodic/cathodic polarization) on the susceptibility of alloys to embrittlement. Increasing the temperature from ambient generally results in increasing susceptibility to SCC and decreasing susceptibility to hydrogen embrittlement. Cathodic polarization often results in increasing hydrogen embrittlement and decreasing SCC susceptibility.

Hydrogen embrittlement and SCC have only recently been recognized as important phenomena in nickel-base alloys. However, because these alloys are often used specifically to solve problems of hydrogen embrittlement and SCC, these topics are of great importance.

Stress-Corrosion Cracking

The nickel-base alloys are generally used to combat SCC where austenitic stainless steels have failed because of SCC. However, two events have recently occurred that require increased knowledge of the SCC resistance of nickel-base alloys. First, a large number of alloys have been developed and included in the market; this has resulted in an almost continuous change in performance (alloy content) between stainless steels and the numerous nickel-base alloys. Second, the nickel-base alloys have been historically considered to be immune to SCC in all but a few environments, but the increased requirements for current processes have extended the use of materials to temperatures at which the SCC of nickel-base alloys must be considered.

Stress-corrosion cracking of nickel-base alloys has been found to occur in three types of environments: high-temperature halogen ion solutions, high-temperature waters, and high-temperature alkaline environments. In addition, SCC has been detected in liquid metals, near-ambient-temperature polythionic acid solutions, and environments containing acids and hydrogen sulfide (H_2S).

High-Temperature Halogen Ion Solutions

Although the nickel-base alloys are specifically used for resistance to SCC in high-temperature Cl^- solutions, virtually every nickel-base alloy is susceptible to SCC in chloride solutions if the proper conditions exist. Table 16 lists the alloy-environment combinations in which SCC has been detected. Unlike the common stainless steels, which show SCC at low temperatures, low chloride contents (ppm ranges), and near-neutral solutions, the conditions that promote SCC in nickel-base alloys are much more severe. Stress-corrosion cracking is promoted by the following parameters in aqueous halide systems:

- Elevated temperatures, especially above 205 °C (400 °F)
- High Cl^- contents in the percent range
- Acidity, usually in the range pH <4
- Aeration or presence of other oxidizing species
- Presence of H_2S
- High stress and/or high-strength materials

Although Table 16 lists the alloy-environment combinations that exhibit SCC, this does not suggest that only these alloys exhibit SCC in the respective solutions. Other nickel-base alloys that are less resistant than the reported alloys may not have been tested in a particular investigation.

Because of the range of alloy compositions available, large differences in performance have been shown among the various nickel-base alloys. In the nickel-chromium-molybdenum-iron alloy systems, increasing the molybdenum and nickel contents (correspondingly decreased iron contents) appears to improve performance. Some evidence suggests that precipitation-hardenable

Table 11 Corrosion of various mill annealed nickel-base alloys in boiling 99% acetic acid

Results are based on four 24-h tests.

	Corrosion rate	
Alloy	mm/yr	mils/yr
200	0.11	4.5
400	0.015	0.6
G	0.03	1.2
G-2	0.005	0.2
G-3	0.015	0.6
625	0.01	0.4
C-4	0.0005	0.02
C-276	0.0076	0.3
B-2	0.03	1.2

Table 12 Effect of oxidizing ions on corrosion of Nickel 200 and alloy 400 in boiling acetic acid

		Corrosion rate					
		No air		Air sparge		3200 ppm Cu^{2+}(a)	
Alloy	Concentration of acetic acid, %	mm/yr	mils/yr	mm/yr	mils/yr	mm/yr	mils/yr
Nickel 200	100	0.036	1.4	0.025	1	0.81	32
Alloy 400	100	0.0025	0.1	0.05	2	2.97	117
Nickel 200	50	0.076	3	1.6	63	0.71	28
Alloy 400	50	0.025	1	2.1	84	0.9	36

(a) Added as acetate. Source: Ref 39

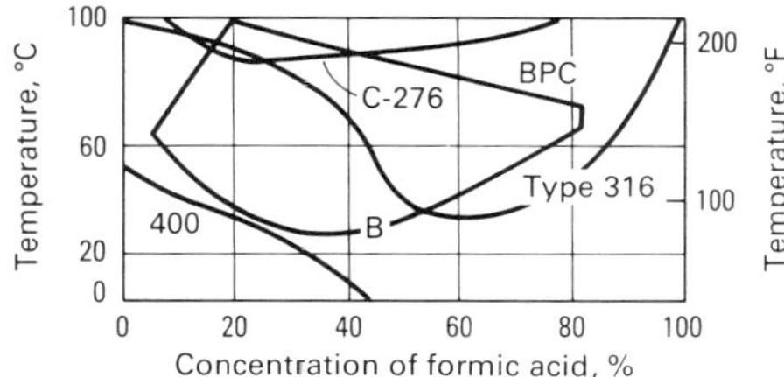

Fig. 7 Isocorrosion curves (0.1 mm/yr, or 4 mils/yr) of various nickel-base alloys in formic acid. BPC, boiling point curve. Source: Ref 39

alloys, such as X-750 and IN-718, have lower SCC resistance than nonprecipitation-hardenable alloys at comparable strength levels with otherwise similar compositions. Many of the test results have been developed under severe conditions. Stress-corrosion cracking may not necessarily occur in field applications with environments similar to those in Table 16, because all of the above parameters, along with tensile stress, must be present for SCC to occur.

Factors that may reduce susceptibility to SCC include:

- Lower stress level
- Presence of naturally occurring or intentionally added inhibitors
- Changes in alloy potential by galvanic coupling
- Lower material strength
- Time at temperature
- Surface treatments
- Concentration of environmental species
- Increasing pH

Fracture modes cannot be used to distinguish the mode of cracking unambiguously; however, there are general guidelines. Cracks in high-temperature chloride solutions are generally branching and transgranular. The precipitation-hardened nickel-base alloys, however, may frequently show either transgranular or intergranular cracking. The cracking in alloys B and B-2 tends to be intergranular in halogen ion solutions, although transgranular cracking can also occur. Stress-corrosion cracking in nickel-base alloys often occurs with the presence of a large number of secondary cracks. Of course, cracking occurs in the presence of tensile, rather than compressive, stresses. Excessive uniform corrosion or pitting need not be associated with SCC.

The application of nickel-base alloys in the oil and gas industry has made halogen ion SCC an area of special interest. Much research has been generated in developing materials selection criteria for production fluids with temperatures up to 230 °C (450 °F). Hydrogen sulfide has been found to play a significant role in decreasing resistance to SCC in oil and gas applications. However, because most produced fluids are fully deaerated, the main oxidant that must be considered is elemental sulfur. Pressure, temperature, and pH are important environmental parameters. Although there is little agreement regarding specific alloy selection criteria, the SCC resistance of nickel-base alloys has become a major basis for alloy selection for deep, hot sour gas wells.

Table 13 Corrosion of various mill-annealed nickel-base alloys in 40% formic acid

Results are based on four 24-h test periods.

Alloy	Corrosion rate, mm/yr	Corrosion rate, mils/yr
825	0.2	7.9
200	0.26–0.27	10.3–10.5
400	0.038–0.068	1.5–2.7
600	0.25	10.0
G	0.013–0.0132	5.0–5.2
G-3	0.046–0.05	1.8–2.1
625	0.17–0.19	6.8–7.8
C-4	0.07–0.076	2.9–3.0
C-276	0.07–0.074	2.8–2.9
B-2	0.008–0.01	0.31–0.40

Table 14 Corrosion of various mill-annealed nickel-base alloys in boiling 88% formic acid

Results are based on four 24-h test periods.

Alloy	Corrosion rate, mm/yr	Corrosion rate, mils/yr
825	0.064–0.08	2.5–3.1
200	0.31–0.34	12.2–13.2
400	0.024–0.028	0.97–1.1
G	0.099–0.12	3.9–4.6
G-2	0.05–0.067	2.0–2.6
G-3	0.14–0.15	5.4–5.9
625	0.236–0.238	9.3–9.4
C-4	0.05–0.076	2.0–3.0
C-276	0.043–0.048	1.7–1.9
B-2	0.00025–0.001	0.01–0.04
C-22	0.023	0.9

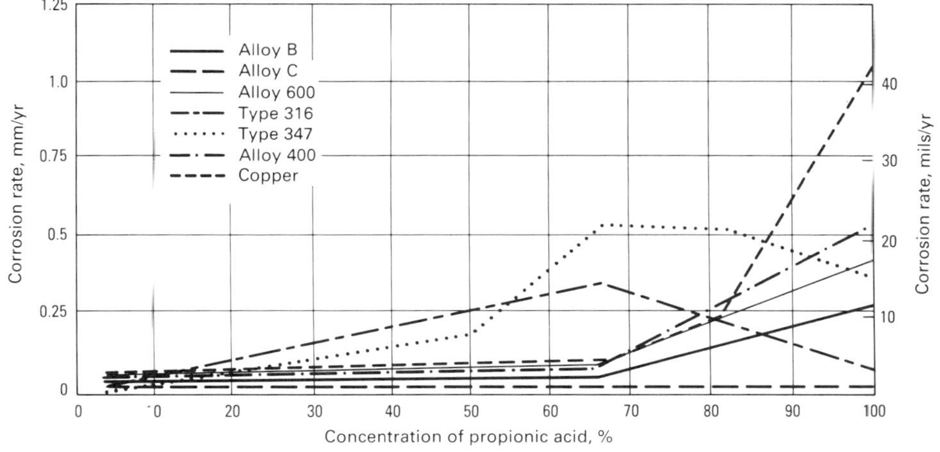

Fig. 8 Corrosion of various alloys in boiling propionic acid. Source: Ref 41

Table 15 Effect of test temperature on corrosion of various nickel-base alloys in propionic acid

No attempt was made to control aeration.

Alloy	Acid concentration, %	50 °C (120 °F), mm/yr	50 °C (120 °F), mils/yr	75 °C (165 °F), mm/yr	75 °C (165 °F), mils/yr	Boiling, mm/yr	Boiling, mils/yr
Alloy 400	50	0.28	11	0.13	5	0.076	3
	80	0.4	16	0.15	6	0.25	10
	99	0.48	19	1.19	47	0.53	21
Alloy B	50	0.38	15	0.1	4	0.05	2
	80	0.61	24	0.3	12	0.13	5
	99	0.15	6	0.64	25	0.28	11
Alloy C	50	nil		nil		0.025	1
	80	nil		nil		0.025	1
	99	nil		nil		0.025	1

High-Temperature Waters

Because of the commercial importance of nuclear energy, the intergranular SCC of alloy 600 in high-temperature pressurized water has been extensively investigated. Alloys 600, 800, and 690

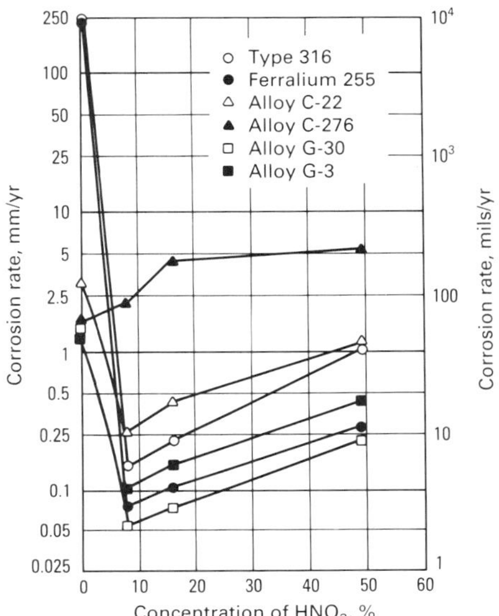

Fig. 9 Effect of HNO_3 on the corrosion of various alloys in a boiling 30% H_2SO_4 solution. Source: Ref 25

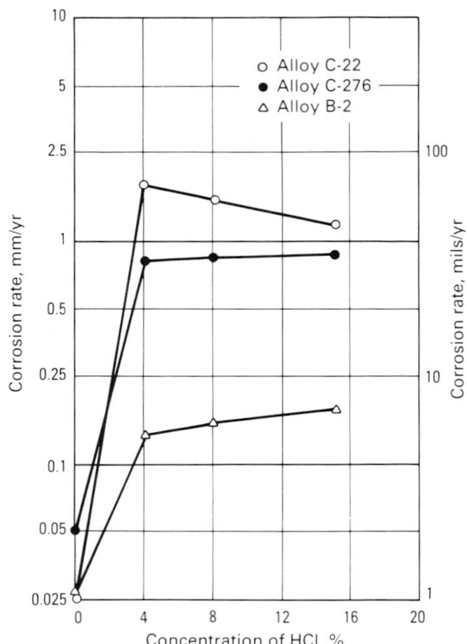

Fig. 10 Effect of HCl on the corrosion of various alloys in 15% H_2SO_4 at 80 °C (175 °F). Source: Ref 25

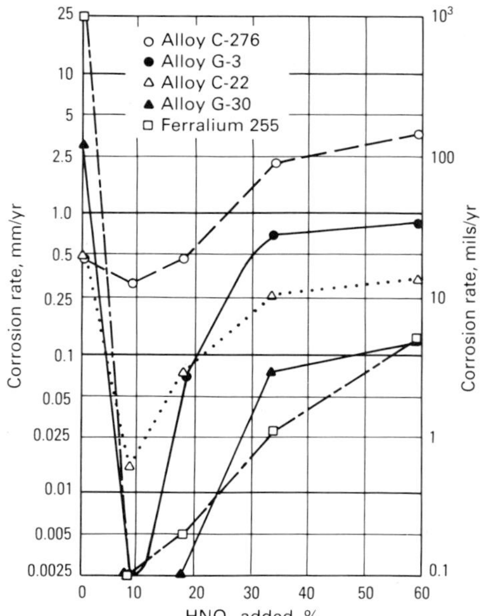

Fig. 11 Effect of HNO_3 on the corrosion of various alloys and Ferralium 255 in a 4% HCl solution at 80 °C (175 °F). Source: Ref 25

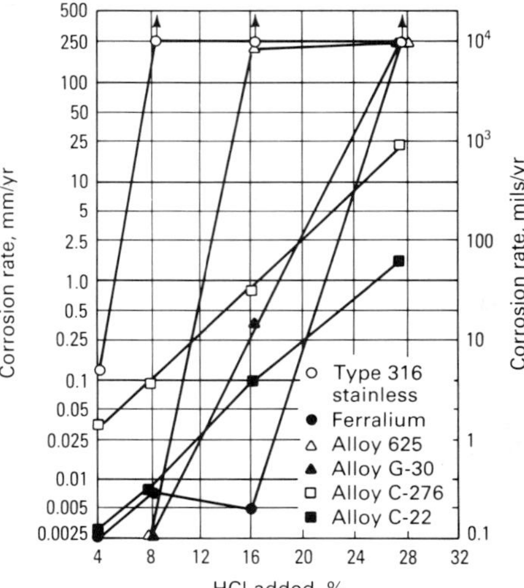

Fig. 12 Effect of HCl additions on the corrosion rates of various alloys in 9% HNO_3 at 52 °C (125 °F)

are used in steam generator tubing in pressurized water reactors (Ref 59). Stress-corrosion cracking has occurred in this environment and is generally intergranular. Although much of the literature deals with SCC of alloy 600, this is not the only alloy that exhibits SCC in high-temperature waters. Table 17 shows this phenomenon to extend to stainless steels and other nickel-base alloys. Although alloy 600 is susceptible to cracking in steam generators, it is much more resistant than type 304 stainless steel (Ref 48).

Heat treatments that affect grain-boundary structure and composition can be important in determining susceptibility to SCC. Control of such elements as phosphorus, carbon, nitrogen, and niobium can play a significant role in controlling crack behavior (Ref 60, 66). Many investigations have been conducted on the effect of various heat treatments of alloy 600 (Ref 67, 68), usually in the range of 595 to 705 °C (1100 to 1300 °F), on the susceptibility to SCC in waters. Aging in a temperature range that causes semicontinuous precipitation of carbides at grain boundaries and partitioning of solute elements at these boundaries is beneficial. The heat treatments are selected to reduce significant chromium depletion in the grain-boundary region.

Alloy selection and control of metallurgy have not been the only methods of controlling SCC in high-temperature waters. Operating procedures, water quality, design, and control of manufacturing have also been important. Control of steam purity is of primary importance with limits set not only on total purity level but also on the concentrations of individual impurities. Generally, all-volatile water treatments are specified.

Reduction in both residual stresses and operating stresses is the third major method of controlling SCC of nickel-base alloys in nuclear applications. Design considerations include minimizing carryover and reducing temperatures of superheated hot spots that permit concentration by evaporation of wet regions.

Caustic Environments

Many investigators believe that SCC in oxygenated water and high-temperature caustic SCC are related. Consequently, a number of alloy evaluations have been made of alloys 600, 800, and 690 in caustic environments at temperatures near 315 °C (600 °F). Stress-corrosion cracking in caustic solutions is highly dependent on applied potential, with a number of potential regions present where SCC can occur (Ref 69).

In austenitic stainless steels, the caustic cracking is often transgranular and not easily distinguished from chloride SCC. Alloys 600 and 800 generally exhibit intergranular SCC in high-temperature caustic environments, although alloy 800 can exhibit transgranular cracking.

Table 18 outlines some of the alloy-environment combinations that have caused SCC in alkaline environments. All major classes of nickel-base alloys exhibit SCC in alkaline environments. Of the alloys considered for nuclear applications, alloy 690 appears to possess better resistance to caustic SCC than alloys 600 or 800.

Other Environments

Stress-corrosion cracking of nickel-base alloys can occur in polythionic acids at or near ambient temperatures (Table 19). Polythionic acids are often produced by partial oxidation of wet H_2S when exposed to air. Intergranular cracking generally occurs only in highly sensitized alloys after extended exposure to sensitizing temperature ranges. Usually alloys that are not stabilized and contain high carbon (for example, alloys 600 and 800) are the most readily susceptible to this form of SCC.

Stress-corrosion cracking has been found in near-ambient-temperature HCl environments containing H_2S. The cracking is transgranular

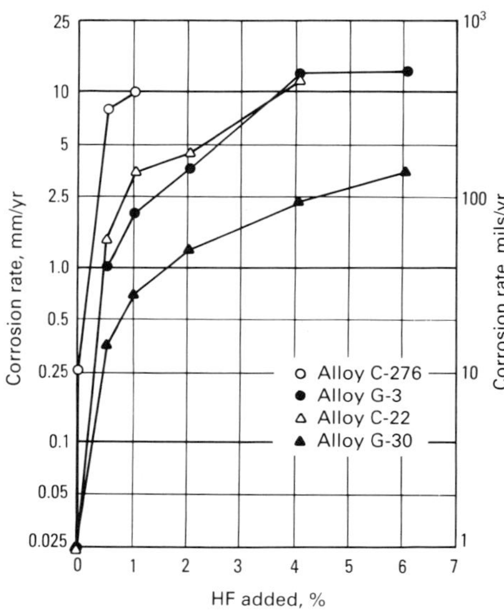

Fig. 13 Effect of HF additions on the corrosion of various alloys in 20% HNO_3 solutions at 80 °C (175 °F)

and does not occur without H_2S. Corrosion inhibitors can mitigate this form of SCC.

Hydrogen Embrittlement

Hydrogen embrittlement of nickel-base alloys is exemplified by three forms: brittle (usually intergranular) delayed fracture, a loss in reduction of area while often retaining a microvoid coalescent fracture, or a reduction in properties such as fatigue strength. Although cleavage-type cracks have been reported in nickel-base alloys, they are not the predominant mode of fracture. Table 20 lists alloys and environments in which hydrogen embrittlement is known to occur.

Delayed Failure. The most severe form of embrittlement by hydrogen generally occurs as a

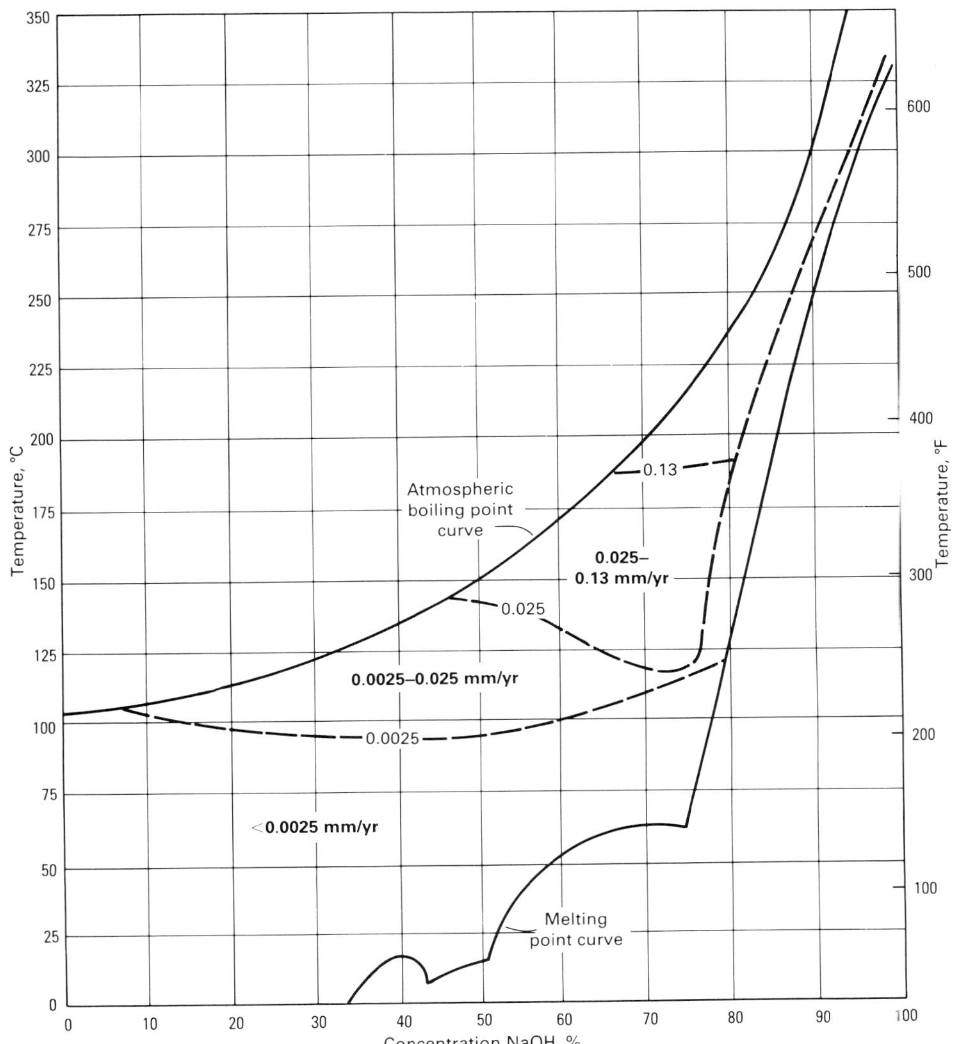

Fig. 14 Isocorrosion diagram for Nickel 200 and Nickel 201 in NaOH. Source: Ref 42

delayed failure or a reduction of K_{ISCC} to values at which sustained crack growth can occur. Under these circumstances, fracture can occur after an alloy has been placed in service under stress. With few exceptions, fully annealed nickel-base alloys are essentially immune to hydrogen-induced delayed failure. Only when the strength level of the alloys is increased by cold working or by heat treatment is the incidence of hydrogen embrittlement significant. Also, only the low-iron-containing alloys (generally less than 10% Fe) are significantly affected.

The susceptibility of precipitation-hardenable alloys, such as alloy X-750, alloy K-500, and alloy IN-718, has been reported from laboratory tests and from field exposures. The susceptibility to embrittlement decreases with decreasing strength level and correspondingly with increasing aging temperatures. However, at similar strength levels, increased aging temperatures appear to be beneficial (Ref 86). Some of the solid-solution alloys, such as alloy C-276 and alloy 625, are susceptible to hydrogen embrittlement under conditions of hydrogen charging when cold worked to high strength levels. Aging at temperatures (540 °C, or 1000 °F) at which ordering and/or grain-boundary segregation can occur greatly increases this susceptibility to hydrogen embrittlement (Ref 87-89). Alloy B-2 is susceptible to hydrogen embrittlement in H_2S-containing environments even in the annealed condition.

Some nickel alloys do not show significant susceptibility to hydrogen embrittlement. These alloys include alloys 825, G, and G-3, and alloy 20Cb3, as well as austenitic nickel-base alloys with correspondingly high iron contents.

Reductions in Ductility or Tensile Strength. Reductions in ductility have been observed during simultaneous deformation of nickel-base alloys and charging in gaseous hydrogen-containing atmospheres. Although these conditions generally do not pose a risk of failure, they have been used as a method of comparing the relative hydrogen embrittlement susceptibilities of alloys.

Annealed nickel has been shown to be susceptible to intergranular hydrogen embrittlement. This phenomenon appears to be related to the sulfur content/segregation in nickel.

Large reductions in the ductility of nickel-base alloys have been observed at strain rates near 10^{-4} to 10^{-6} s^{-1}. Although the fracture mode may remain microvoid coalescence, reductions in the dimple size ratio occur in air (Ref 90). In some alloys, intergranular fractures may occur. The embrittlement during slow strain rate testing is often associated with the dislocation transport of hydrogen (Ref 91).

Table 16 SCC of nickel-base alloys in halogen ion containing environments

Alloy	Environment	Ref
800, 825, 718	Boiling (155 °C, or 310 °F) 42% $MgCl_2$	48
800, 825, G-3, G, X, C-276	20% $MgCl_2$, 230 °C (450 °F)	49
800, C-276, MP35N	Boiling 85–89% $ZnCl_2$	50, 51
825, G, G-3, C-276	47% $ZnBr_2$, 205 °C (400 °F)	52
400, 600	Seawater, 285 °C (550 °F)	53
400, alloy 20	Salton sea brine, deaerated, 230 °C (450 °F)	54, 55
625, C-276	Salton sea brine, aerated, 230 °C (450 °F)	54
825, G, G-3	Aerated NaCl, $CaCl_2$, 230 °C (450 °F)	52
825, G-3	25% NaCl + H_2S + CH_3COOH	8
400, X-750, 600, K500	HF and H_2SiF_6, 50 °C (120 °F)	53
804, 825	H_2O + 100 ppm Cl + 50 ppm O_2, 300 °C (570 °F)	53
825, G-3	10% HCl + H_2S, 20 °C (70 °F)	52
825, 625, C-276, 718, G-3	1% HCl, 205 °C (400 °F)	56
B	1% HBr, 230 °C (450 °F)	56
B	1% HI, 205 °C (400 °F)	56
625, 718, C-276, G, 825	HAc + Cl^- + H_2S, 205 °C (400 °F)	57
B-2	HCl + H_2S, 175 °C (350 °F)	58

Table 17 SCC of nickel-base alloys in high-temperature waters

Alloy	Environment	Ref
B-2/B	Deionized H_2O + O_2, 205 °C (400 °F)	56
X-750	Pressurized and boiling water reactors (150–175 °C, or 300–350 °F)	60
600	Pressurized water reactors	60, 61
901	Steam turbines	62
800	Pressurized water reactors	63
718	Light water reactors	64
Type 304 stainless steel	Pressurized water reactors	65

Table 18 Alloy-alkaline environment combinations found to cause SCC of nickel-base alloys

Alloy	Environment	Ref
6B, B, C-276, MP35N	NaOH, 175 °C (350 °F)	70
600, 800, 690	NaOH, 325 °C (620 °F)	71
722	50% NaOH, 315 °C (600 °F)	72
600	LiOH, 335 °C (635 °F)	53
600	10% NaOH, 285 °C (550 °F)	60
800	10% NaOH, 285 °C (550 °F)	69
400	10% NaOH, 300 °C (570 °F)	73

Table 19 SCC in other alloy-environment combinations

Alloy	Environment	Ref
200	Room-temperature-mercury	74
400, 405, K500, 625, 718, X-750, 200, 600	Room-temperature-mercury	75
200, 600	Room-temperature polythionic acids	56
600 or 800, sensitized	Room-temperature polythionic acids	53, 59, 76

Increasing the temperature above room temperature often results in decreasing susceptibility to hydrogen embrittlement (Ref 52). Reducing the temperature much below ambient may also reduce susceptibility (Ref 80, 83).

The nickel-base alloys often exhibit such high corrosion resistance that the corrosion currents do not generate sufficient hydrogen to cause embrittlement, even in the most susceptible alloy conditions. However, the application of external cathodic currents, as in cathodic protection, or the generation of hydrogen on susceptible nickel-base alloys by coupling to active metals that do corrode in an environment may charge hydrogen into the nickel-base alloys, thus causing embrittlement in susceptible alloys. Alternatively, hydrogen can be charged into alloys at elevated temperatures at which corrosion rates may be high (Ref 92). Subsequent slow strain rate testing can then exhibit ductility losses.

Table 20 Nickel-base alloys exhibiting hydrogen embrittlement

Alloy	Environment	Ref
600	Cathodic charge/tensile test	77
C-276, C-4, MP35N, 625, 718		78, 79
Pure nickel	Cathodic charge H_2SO_4	80
K500	Sour oil and gas environment	81
X750	5% NaCl + 5% CH_3COOH + H_2S	82
718, A286, 625	Hydrogen gas (34 MPa, or 5000 psi) charged at room temperature	83
Pure nickel, 600	Hydrogen gas charged at 723 K, tested at room temperature	84
K500	Cathodic charge, seawater	85

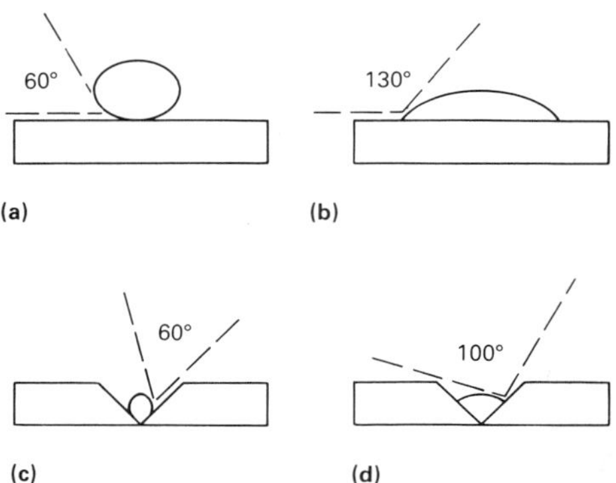

Fig. 15 Acceptable (b and d) and unacceptable (a and c) wetting profiles for nickel alloy weldments. (a) and (b) Overlay welds. (c) and (d) Butt welds

Fabrication and Weldability of Nickel Alloys

Samuel Dwight Kiser
Inco Alloys International, Inc.

Nickel-base alloys can be fabricated with most conventional welding processes, such as gas tungsten arc, gas metal arc, and shielded metal arc welding using coated electrodes. As the nickel content of an alloy increases, several general trends can be observed:

- Toleration for contamination decreases
- Puddle viscosity during welding increases
- Coefficient of expansion decreases as iron is replaced by nickel
- Penetration during welding decreases

Because of these differences between high-nickel materials and iron-base alloys, alternate techniques must sometimes be used for nickel-base alloys to produce acceptable weldments. For example, greater care in precleaning and cleanliness during welding are required to prevent the occurrence of contamination-induced cracking. The increased puddle viscosity (lack of fluidity) can result in poor wetting or in lack-of-fusion defects along bead edges. Also, if no weave is used, rollover or a very poor wetting profile is sometimes encountered. Because penetration is naturally low with high-alloy materials, the angle formed between the base alloy and the weld bead edge should be greater than 100° to facilitate good tie ins with subsequent beads. Figure 15 shows acceptable and unacceptable wetting profiles for nickel alloy weldments.

In Fig. 15(a) and (c), the highly convex weld with poor wetting can be caused by too slow a travel speed, too low a voltage, lack of weave, or the wrong polarity. Oxidized or dirty base metal can also inhibit wetting. It is very difficult to fuse these profiles into subsequent beads. Entrapped slag or lack-of-fusion defects often result from attempts to tie into these profiles.

Figures 15(b) and (d) illustrate good wetting profiles that easily accommodate subsequent beads. These profiles are produced by a proper balance of travel speed, parameters, weave, and base plate preparation. Also, greater viscosity usually requires increased included angles for butt joints (75 ± 5° is recommended for nickel-base alloys).

With regard to the subject of bead contour, it should be noted that each bead produced should be slightly convex. Highly stressed flat or concave weldments are prone to centerline cracking. Particular care should also be devoted to producing slightly convex stops. Convex welds can be produced by reducing voltage or by reducing travel speed; convex stops are made by back filling the crater area.

The relatively low coefficient of expansion of nickel-base alloys, as compared to that of stainless steels, gives nickel-base austenitic alloys better resistance to cracking during welding than the fully austenitic stainless steels. Although both fully austenitic alloy families (stainless and nickel alloys) have low tolerances to contamination, the lower coefficient of expansion of nickel-base alloys allows the joining of nickel to take place under lower shrinkage stresses and thus results in a decreased tendency toward shrinkage cracks.

The tendency for reduced penetration in high-nickel alloys has been discussed briefly. It can be rendered harmless by using increased root gaps, decreased lands, and greater included angles.

Welding Corrosion-Resistant Composites

To reduce material costs in corrosion applications, a relatively thin layer of corrosion-resistant material is sometimes applied over a less resistant, less expensive structural material. Welding is employed in such fabrications, and special considerations are necessary to achieve adequate corrosion resistance in the weldments.

Weld overlay cladding is commonly applied to steel, and the typical items that are overlaid include tubesheets and flanges. Common problems encountered in overlaying are excessive iron dilution and irregular penetration patterns. Excessive iron dilution usually does not result in corrosion problems, because multilayer overlays are most often specified. However, most overlay procedures are qualified by bend tests, and excessive iron dilution can cause failure of such tests. Irregular penetration patterns also result in bend test difficulties.

Excessive iron dilution and irregular penetration patterns can both be improved by maintaining approximately 50% bead overlap. Excessive iron dilution can be reduced by selecting smaller-diameter electrodes, by implementing lower heat input procedures, and by dissipating much of the arc energy on each previous bead.

Welded Metal Linings and Cladding. In the installation of applied linings and the fabrication of bonded clad plate, the primary concern is to produce weldments of equal or greater corrosion resistance than that of the lining or cladding material. The added limitation of only one or two layers of weldment sometimes prohibits the normal choice of consumable. For example, if it is known that certain weldments that contain some iron dilution must be exposed to the same corrosion media as the lining or cladding, then overalloyed filler metals are often selected.

Overalloyed consumables should be selected so that after dilution the weldment is still of slightly higher alloy content than the lining or cladding. For example, many alloys containing 4 to 6% Mo and clad materials have been welded with welding alloy 112 or alloy 625 filler metals, which contain 9% Mo. The consumable contains about 60% Ni, which accepts iron dilution well, and when the 9% Mo is diluted with moderate amounts of iron, there remains enough molybdenum to impart greater pitting and crevice corrosion resistance in the weld than in the cladding or lining.

Corrosion of Welds and Heat-Affected Zones

As was briefly addressed in the discussion "Welding Corrosion-Resistant Composites" in this article, the corrosion resistance of weldments is often very important in the fabrication of nickel alloys and their linings and clad plates. For general corrosion resistance, the selection of a welding consumable is straightforward; the companion wires and electrodes designed for use with the particular base alloy are usually used. If the fabrication is destined for service in which intergranular attack may occur, low-carbon or stabilized alloys and welding consumables are usually selected. In this way, the danger of grain-boundary precipitation of carbides and intermetallic phases (sensitization) is minimized. The phenomenon of sensitization, along with the resulting intergranular corrosion, is most prevalent in certain stainless steels (see the articles "Corrosion of Stainless Steels" and "Corrosion of Weldments" in this Volume), but nickel-base alloys are subject to similar mechanisms.

Intergranular attack may be counteracted by three methods. The first method involves the selection of alloy and welding consumables that contain extra-low carbon such that carbide precipitation is minimized because of insufficient carbon.

The second method involves the selection of stabilized alloys and welding consumables, which contain such carbide-stabilizing elements as titanium and niobium. Such alloys are rendered insensitive to intergranular attack by the formation of harmless titanium and niobium carbides, rather than chromium carbides. Appropriate annealing treatments are also used to promote the formation of these harmless carbides and to allow the rediffusion of chromium to depleted areas when slight chromium carbide precipitation may have occurred.

Third, if the fabrication is small enough and simple enough, a postweld heat treatment may be performed after all welding has been completed. This heat treatment will redissolve any precipitated carbides. The fabrication is then rapidly cooled (quenched) from the treatment to prevent reprecipitation of carbides.

In applications in which pitting or crevice corrosion attack is expected, alloys must be selected with great care. Furthermore, welding consumables and procedures must be chosen with even greater care.

The localized nature of pitting and crevice corrosion dictates that materials that have been developed and fabricated to resist these types of attack must have reasonably continuous surfaces, must be fabricated without the introduction of crevices and laps, and must have reasonable chemical homogeneity on a microscopic scale. These requirements are not difficult to achieve in wrought alloy forms (hot-rolled plate, rod and bar, and so on). The initial requirements of chemical composition can be met with weldments, but homogeneity within an as-deposited weld cannot be achieved. A weld is a casting, and dendrite formation and growth occur first from the highest melting point constituents as the weld puddle solidifies. As dendrite growth continues, lower melting point materials are typically relegated to the interdendritic spaces, which causes chemical segregation within the weld.

It is commonly known that molybdenum imparts pitting resistance to chromium-containing alloys of iron and nickel. It is not as widely known that molybdenum also serves as an effective melting range depressant. Therefore, the dendrites (those areas that solidify first) characteristically contain lower levels of molybdenum than interdendritic areas (the last to solidify). This means that base metals must be selected to have adequate corrosion resistance in the service environment and that overalloyed filler metals, as described earlier, must be used to ensure that weld metal dendrites have a molybdenum content at least equal to that of the base metal.

Major Applications and Corrosion Performance of Nickel Alloys

J.R. Crum
Inco Alloys International, Inc.

Nickel-base alloys are used for corrosion resistance or for combined corrosion resistance and high-temperature strength in a wide range of commercial applications. These various applications may demand resistance to aqueous corrosion mechanisms, such as general corrosion, localized attack, and SCC, or resistance to elevated-temperature oxidation, sulfidation, and carburization. Many nickel-base alloys have been developed to resist these and other forms of attack. The alloys often find application in areas outside the specific industry or process for which they were designed. A list of the nickel-base alloys discussed in this section and their nominal chemical compositions is given in Table 1. Alloys containing 30% or more nickel are considered.

Chemical-Processing Applications

Caustic Soda. The chemical-processing industry involves a great variety of corrosive environments. Thus, a variety of nickel alloys are used in this industry. A major use for Nickel 200 is in the production of NaOH. Nickel 200 exhibits outstanding corrosion resistance to NaOH at concentrations up to anhydrous at boiling or molten temperatures. Caustic soda (NaOH) is normally produced at 11 to 15% concentration and further concentrated by evaporation to 50% or higher. As NaOH concentration and temperature increase during the evaporation process or during other chemical-processing conditions, the corrosivity increases dramatically. Similarly, increasing the nickel content of nickel-base alloys produces increasing resistance to general corrosion and SCC in caustic. Thus, a number of nickel-base alloys can be used for handling NaOH, depending on solution concentration and temperature.

At temperatures above 315 °C (600 °F), Nickel 200 is replaced by Nickel 201 because of its low carbon content. In the handling of brines in NaOH production as well as the commercial production of salt, alloy 400 has performed well in a variety of components, such as heat exchangers, vacuum pans, heater tubes, rotary dryers, and transfer piping. In acidic salt environments, however, high-molybdenum nickel alloys, such as 825, 625, G-3, G-30, C-22, and C-276, are used.

Mineral Acids. The production of mineral acids requires the use of nickel alloys. In the production and handling of H_2SO_4, alloys 825 and 400 and alloy 20-type materials are used. When the acid becomes contaminated with halides, high-molybdenum alloys are used, such as alloys G, 625, C-22, C-4, C-276, and ALLCORR. These alloys are also specified where H_2SO_4 is used in the production of other chemicals, such as H_3PO_4, HF, titanium dioxide (TiO_2), ammonium sulfate [$(NH_4)_2SO_4$], and in the refining of copper and nickel ores. Hydrochloric acid is best handled by alloy B-2, especially at elevated temperatures. Some alloys suitable for handling various concentrations of H_3PO_4 are 825, G-3, G-30, and 625.

The extreme reactivity of HF excludes many materials from equipment for making HF and from equipment for processes that employ HF as a catalyst or reagent. Alloy 400 is used in valves in HF alkylation, storage tanks, and retorters. Where HF and H_2SO_4 are used, alloys 825, 625, G-3, and C-276 are used.

Nickel alloys are extensively used in the production of HNO_3. Alloy 617 is used for its high-temperature strength and corrosion resistance in the catalyst-support grids in high-pressure plants. In older plants, alloys 600 and 601 are used because of lower pressure. Alloy 800 is used in the heat-exchanger train. Where reboiling conditions exist, alloy 690 has given exceptional performance.

There are many more areas in the chemical-processing industry in which nickel alloys are used than can be covered in this brief summary (see the article "Corrosion in the Chemical Processing Industry" in this Volume). A few other important applications are the use of alloy 600 or Nickel 200 in chlorination processes, such as the Kroll process for production of zirconium and titanium. The production of organic acids and compounds (acetic and formic acids, phenol, fertilizers, urea, pesticides, and plastics) are other examples. All use the alloys discussed above in various chemical-handling equipment, such as pumps, piping, valves, bellows, evaporators, heat exchangers, scrubbers, linings, and tanks.

Water and Seawater Applications

Nickel and nickel-base alloys generally have very good resistance to corrosion in distilled water and freshwater. Typical corrosion rates for Nickel 200 (commercially pure nickel) in a distilled water storage tank at ambient temperature and domestic hot water service are <0.0025 mm/yr (<0.1 mil/yr) and <0.005 mm/yr (<0.2 mil/yr), respectively. Nickel-copper alloys such as 400 and R-405 also have very low corrosion rates and are used in freshwater systems for valve seats and other fittings. Because of the cost of nickel alloys, less expensive stainless steels or other materials are usually specified for pure or freshwater applications unless increased resistance to SCC or pitting is required. Alloys 600 and 690, for example, are used for increased SCC resistance in high-purity water nuclear steam generators.

In steam-hot water systems, such as condensers, appreciable corrosion of Nickel 200 and alloy 400 may occur if noncondensables (CO_2 and air) in the steam exist in certain proportions. Deaeration of the feedwater or venting of the noncondensable gases will prevent this attack. Alloy 600 is resistant to all mixtures of steam, air, and CO_2 and is particularly useful in contact with steam at high temperatures.

Table 21 Resistance of various alloys to stagnant seawater

3-year exposure tests

Alloy	Maximum pit depth — mm	Maximum pit depth — mils
625	nil	nil
825	0.025	0.98
K-500	0.864	34
400	1.067	42
AISI type 316	1.575	62

Nickel 200 and alloy 400 and nickel-base alloys containing chromium and iron are very resistant to flowing seawater, but in stagnant or very low velocity seawater, pitting or crevice corrosion can occur, especially under fouling organisms or other deposits. In moderate- and high-velocity seawater or brackish water, alloy 400 is frequently used for pump and valve trim and transfer piping. It has excellent resistance to cavitation erosion and exhibits corrosion rates less than 0.025 mm/year (1 mil/yr). Alloy 400 sheathing also provides economical seawater splash zone protection to steel offshore oil and gas platforms, pilings, and other structures. Although pitting can occur in alloy 400 under stagnant conditions, such pitting tends to slow down after fairly rapid initial attack and rarely exceeds 1.3 mm (50 mils) in depth. Age-hardened alloy K-500, with corrosion resistance similar to that of alloy 400, is frequently used for high-strength fasteners and pump and propeller shafting in freshwater and seawater applications.

Other nickel-base alloys containing chromium and molybdenum offer increased resistance to localized corrosion in stagnant seawater. Corrosion resistance of some nickel-base alloys and type 316 stainless steel in ambient temperature seawater is compared in Table 21. In hot seawater applications, such as heat exchangers, highly alloyed materials such as alloys 625 or C-276 may be required. In addition, alloys 625, 400, and K-500 are frequently specified for U.S. Naval wetted components in contact with seawater. Specific high tensile strength requirements are met with alloy 718.

Atmospheric Applications

Nickel and nickel-base alloys have very good resistance to atmospheric corrosion. Corrosion rates are typically less than 0.0025 mm/yr (0.1 mil/yr), with varying degrees of surface discoloration depending on the alloy (Table 22). Nickel 200 will become dull and acquire a thin adherent corrosion film, which is usually a sulfate. A greater tarnish will result in industrial sulfur-containing atmospheres than in rural or marine atmospheres.

Corrosion of alloy 400 is negligible in all types of atmospheres, although a thin gray-green patina will develop. In sulfurous atmospheres, a brown patina may be produced. Because of its low corrosion rate and pleasing patina, alloy 400 has been used for architectural service, such as roofs, gutters, and flashings, and for outdoor sculpture.

Nickel alloys containing chromium and iron, such as alloys 600 and 800, also have very good atmospheric corrosion resistance, but may develop a slight tarnish after prolonged exposure, especially in industrial atmospheres. Nickel-chromium-molybdenum materials such as alloys 825, 625, G, C-276, and C-22 develop very thin and protective passive oxide films that prevent even significant tarnishing. A mirror finish can be maintained after extended exposure to the atmosphere.

Table 22 Atmospheric corrosion and pitting of nickel-base alloys

Results of a 20-year exposure 24.4 m (80 ft) from the ocean at Kure Beach, NC

Alloy	Average weight loss, mg/dm²	Average corrosion rate(a) mm/yr	Average corrosion rate(a) mils/yr
Nickel 200	468.6	<0.0025	<0.1
Alloy 800	27.9	<0.0025	<0.1
Alloy 600	19.7	<0.0025	<0.1
Alloy 400	644.7	<0.0025	<0.1
Alloy 825	8.7	<0.0025	<0.1

(a) No pitting recorded for Nickel 200 and alloy 600. The average of the four deepest pit depths for the other three alloys was less than 0.025 mm (0.001 in.).

Although alloy 400 has been used for atmospheric service in the past, atmospheric exposures requiring nickel alloys are now relatively infrequent. Less costly low-alloy stainless steels or plated materials are normally used.

Applications in Pulp and Paper Mills

Nickel alloys are used in pulp and paper mills generally where conditions are the most corrosive. Alloys 600 and 800 have been utilized for over 25 years for digester liquor heater tubing because their high nickel content provides excellent resistance to chloride SCC. In the disposal of organic wastes in unevaporated black liquor, alloy 600 has been used for the reactor vessel, transfer lines, and piping.

A major problem in this industry has been the cracking of steel digesters in the weld area. A practical solution to the problem is to weld overlay the weld area with alloy 600 or 625 filler metals. Other methods are the use of alloy 600 or 625 sheet liners to cover repaired weld areas or the use of alloy clad steel plate in the initial digester fabrication.

One of the oldest applications of nickel alloys in the pulp and paper industry is alloy K-500 doctor blades. Alloy K-500 provides high corrosion resistance with the abrasive and wear resistance needed for good service even in wet creping operations.

The bleaching circuit and pollution control areas are the most demanding corrosive environments in the pulp and paper mill process. Alloys needed in this area must be resistant to oxidizing conditions of high-temperature low-pH liquors containing chlorine, chlorine dioxide, oxygen, hypochlorate, peroxides, and chlorides. Only the high-molybdenum alloys 625, C-276, and C-22 have the required corrosion resistance to resist these aggressive liquors. In equipment used to clean off-gas from the recovery boiler, alloys 825, 625, G, and C-276 have been used, depending on the severity of the conditions. Typical conditions found are acid mists, low-pH liquors containing chlorides and organics, and sometimes high temperatures. Alloy G scrubbers have given over 10 years of good service at one pulp and paper mill. Other plants have used alloys 625 and C-276 for internal components of scrubbers, fans, ducting, and so on. Detailed information on corrosion in pulp and paper mills is available in the article "Corrosion in the Pulp and Paper Industry" in this Volume.

Flue Gas Desulfurization Applications

Nickel alloys were first used in flue gas desulfurization (FGD) systems in wet-induced draft fans in a particulate scrubber system in 1972. Alloy 625 replaced type 316 stainless steel because of the inadequate mechanical properties of type 316. Construction of FGD equipment increased greatly after the Clean Air Act Amendments were enacted in 1977. Nickel alloys saw limited use during the early years because of their high initial cost. Today, there are over 125 FGD units operating in the United States, all having some nickel alloy applications.

During the period of rapid growth in the utilization of FGD equipment, the industry experienced severe corrosion problems at nearly every operating FGD system. Catastrophic failures were reported for coatings and linings on carbon steel, nonmetallics, stainless steel, and occasionally of nickel alloys. The corrosion problems were due to hot acids (nitric, sulfurous, sulfuric, hydrochloric, and hydrofluoric), chlorides, fluorides, and crevice conditions.

A wide variety of nickel alloys are available for use in FGD applications. Alloys 825 and G are significantly more corrosion resistant in sulfurous and sulfuric acids than austenitic stainless steels. With increasing molybdenum levels, alloys 625, ALLCORR, G, and C-276 are used to combat pitting and crevice corrosion. These nickel alloys have been successfully used in FGD components, such as quenchers, absorber towers, dampers, stack gas reheaters, wet ID fans and fan housings, outlet ducting, and stack liners. Some innovative FGD applications include the use of alloy 625 in a unique heat recovery system, all-alloy scrubber towers constructed with alloy 625, the application of thin-gage nickel alloy clad steel in new construction, and nickel alloy sheet liners to repair existing FGD components. Severe corrosion problems in FGD systems can be prevented and long life achieved economically by the proper application of nickel alloys in FGD systems. Additional information is available in the selected references at the end of this article.

Sour Gas Applications

Sour gas is defined by National Association of Corrosion Engineers (NACE) Material Recommendation MR-01-75 as gas being handled at an absolute pressure in excess of 0.45 MPa (65 psi) with a partial pressure of H_2S greater than 345 Pa (0.05 psi). This combination can cause sulfide stress cracking. Sulfide stress cracking can be controlled by using resistant alloys, by controlling the environment, or by isolating the component from the sour environment. The sulfide stress cracking problem is complicated because the environment normally also contains brackish water and CO_2 and is found at 65 to 245 °C (150 to 475 °F), depending on the depth and location of the well.

The best nickel alloys to use in these environments are those containing a minimum of 42% Ni (to resist chlorides), high chromium, and at least 3% Mo. Alloys 825, 925, 2550, 28, G-3, and C-276 are good examples. These are normally furnished for production tubing with minimum yield strengths of 758 or 862 MPa (110 or 125 ksi). Yield strengths up to 1030 MPa (150 ksi) minimum have been used. These strengths are obtained through cold work.

Similar strengths are required for tools and components. These can be much larger in cross section and therefore cannot be cold worked to derive strength. In these cases, precipitation-hardenable alloys are used. The same basic guidelines for composition—high nickel, high chromium, and at least 3% Mo—hold true for these components. The best alloy for use in each environment is usually determined by corrosion testing in that environment. The NACE specification MR-01-75 includes a list of alloys that have passed their initial corrosion screening tests. A list is also provided of the maximum hardness that an alloy can have and still resist sulfide stress cracking. More information on corrosion in the oil patch is available in the article "Corrosion in Petroleum Production Operations" in this Volume.

Applications in Fused Salts

Most applications for nickel alloys in fused salts occur in the heat treating of metals with salt baths. Common heat-treating applications and temperature ranges are:

- *Cyaniding*: 760 to 870 °C (1400 to 1600 °F)
- *Austempering*: 205 to 595 °C (400 to 1100 °F)
- *Martempering*: 205 to 370 °C (400 to 700 °F)
- *Tempering*: 165 to 760 °C (325 to 1400 °F)
- *Heat Treating of tool steels, stainless steels, aluminum, and copper alloys*: 205 to 1095 °C (400 to 2000 °F)

At temperatures in the range of 205 to 815 °C (400 to 1500 °F), cyanides, nitrates, and nitrites are employed. As the temperature requirements increase beyond 705 °C (1300 °F), various chloride mixtures of sodium, potassium, and barium are utilized. It is in the higher-temperature chloride mixtures that high-nickel alloys find significant applications. Increasing the temperatures and basic oxide contents normally increase corrosion rates.

Alloy 600 has proved to be one of the most resistant alloys to high-temperature chloride salt attack. Alloys 601, 617, 690, RA330, 800, and 825 also provide useful resistance.

Heating and Heat-Treating Applications

Cast and wrought nickel-iron-chromium and nickel-chromium alloys are commonly used for heating and heat treating. To be useful in these applications, an alloy must possess a high degree of environmental resistance, thermal fatigue resistance, high-temperature strength and stability, and, in most uses, good workability and weldability.

In heating and heat treating, the alloys of construction may encounter oxidation, sulfidation, carburization, nitridation, carbonitridation, and molten salt corrosion. Alloy selection will depend, in part, on which of these high-temperature corrosion environments are encountered in given service. Resistance to high-temperature oxidation is generally an essential criteria for alloy selection because most heating elements, furnace muffles, retorts, and radiant tubes are exposed to air on at least one side.

Those alloys that develop tight adherent oxide films, such as alloys 601, 617, and 214, generally provide the best long service life, particularly under cyclic temperature conditions. In intermediate-temperature environments that are alternately oxidizing and carburizing, the nickel-iron-chromium alloys are used. In carburizing and carbonitriding applications, nickel-iron-chromium and nickel-chromium alloys have all proved their resistance over the years in many types of service. Alloy 600 has shown exceptional resistance to both carburization and nitridation. The higher iron versions of the nickel-iron-chromium alloys offer the best sulfidation resistance.

The nickel-iron-chromium and the nickel-chromium alloys, such as RA330, 800, 601, 617, and X, have good overall high-temperature strength and microstructural phase stability. These alloys can be readily fabricated with sound welds by using weld metals specifically developed for them. These alloys can also be repair welded by a variety of joining processes.

These alloys have been used for years to manufacture heating elements, recuperators, retorts, radiant tubes, belts, baskets, and other furnace hardware. Heating and heat-treating operations include annealing, quenching and tempering, carburizing, carbonitriding, nitriding, powder metallurgy sintering, brazing, and enameling. More information on materials for and corrosion of heating and heat treating equipment is available in the article "Corrosion of Metal Processing Equipment" in this Volume.

Petrochemical and Refining Applications

The petrochemical and refining industries are defined as those involved in the production of basic organic chemicals and fuels from petroleum-base feedstocks. Examples include fuels and lubricants, ethylene and ethylene derivatives, steam hydrocarbon reforming, NH_3, methanol, and the many organic chemicals. In general, most petrochemical and refining applications are endothermic, requiring temperatures from 425 to 1000 °C (800 to 1830 °F) and higher. Typical environmental conditions include oxidizing, carburizing, nitriding, sulfidizing, and halogen gases (chlorine, hydrogen chloride, fluorine, and so on). Materials used in these applications must exhibit corrosion resistance, metallurgical stability, and strength in these high-temperature environments.

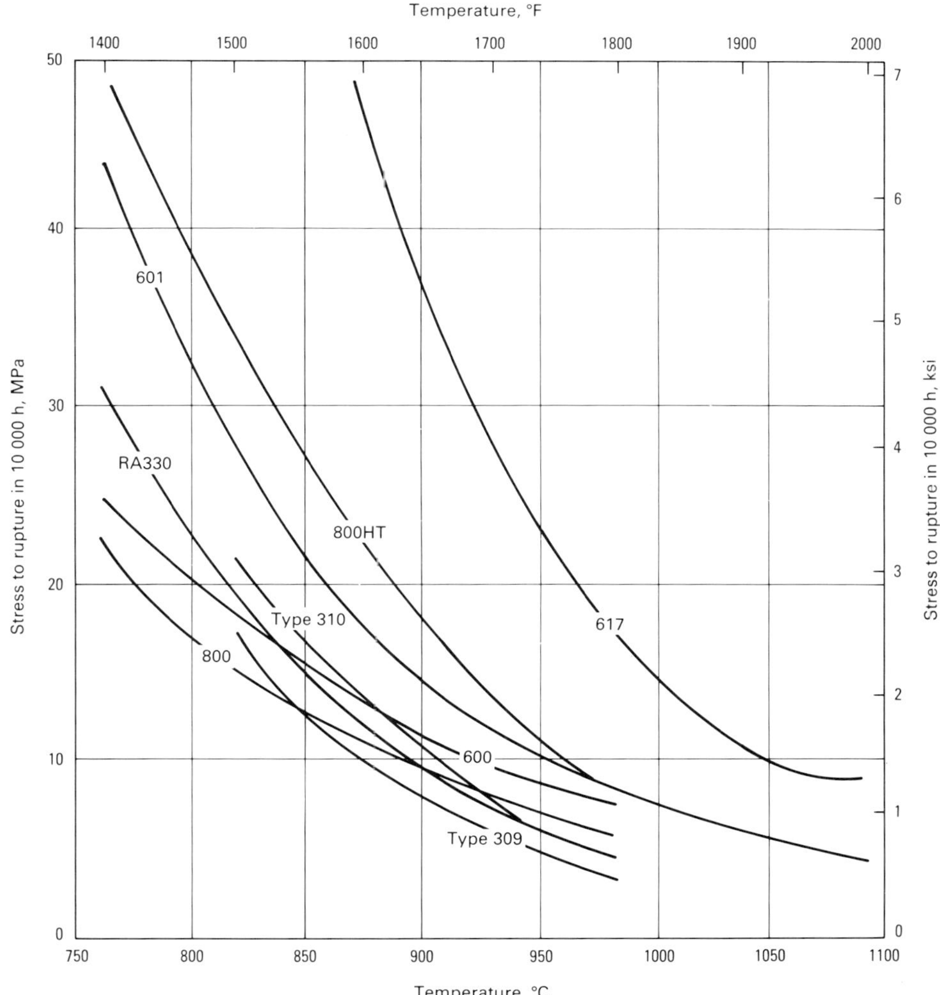

Fig. 16 High-temperature stress rupture data for various nickel-base and stainless alloys. Test duration was up to 10 000 h. Data are from manufacturers' published data.

Alloys 800H and 800HT are the standard materials for intermediate temperatures (620 to 925 °C, or 1150 to 1700 °F). These alloys combine corrosion resistance with strength and metallurgical stability (Fig. 16). Major applications include hydrogen reformer manifolds and pigtail piping, ethylene dichloride furnace tubing, transfer piping, and high-temperature pressure vessels.

As the environments become more aggressive and high-temperature strength requirements more demanding, alloys 600, 601, 617, 625, and X are utilized. High-temperature halogenations require alloys 600 and 625, while high-temperature oxygen requires alloy 601. Aggressive carburizing and nitriding environments utilize alloys 600 and 617. Expansion joints for high-temperature applications require alloys with good fatigue resistance such as alloys 617 or 625. More information on corrosion of petroleum refining equipment is available in the article "Corrosion in Petroleum Refining and Petrochemical Operations" in this Volume.

Nuclear Fuel Cycle Applications

Nickel 200 and 201 can be found in gaseous diffusion enrichment equipment because of their excellent corrosion resistance to HF-UF_x. For similar corrosion resistance reasons, Nickel 200 and 201 and alloys 400 and 600 are used in nuclear fuel reprocessing fluorination and hydrofluorination reactors.

The nickel-chromium-iron alloys are very popular materials for critical applications in pressurized and boiling water reactors because of their excellent corrosion resistance in steam and water environments and because of their resistance to chloride SCC. The age-hardenable alloy X-750 is used as a spring material for fuel pellet hold down springs, fuel element divider plates, and reactor scram springs, and for bolting. Another age-hardenable alloy, alloy 718, because of its high strength and spring characteristics, is also used for fuel assembly divider plates. It also has demonstrated excellent wear resistance in sodium fast breeder reactor environments. The single most significant application of nickel-base alloys in the nuclear industry has been the use of alloy 600 for steam generator tubing in pressurized water reactors. An alternate material for this application is alloy 690. Because of its high chromium content, alloy 690 has provided acceptable corrosion resistance to molten glass in nuclear fuel waste vitrification furnaces.

Alloy 825, because of its resistance to H_2SO_4 corrosion, and alloys 625 and C-276, because of their corrosion resistance to chloride-contaminated H_2SO_4, have been shown to be excellent materials of construction for radioactive waste disposal evaporators. Detailed information on corrosion in these types of applications is available in the article "Corrosion in the Nuclear Power Industry" in this Volume.

ACKNOWLEDGMENT

J.R. Crum wishes to thank the following contributors to the section "Major Applications and Corrosion Performance of Nickel Alloys": W.G. Lipscomb, E.L. Hibner, J.M. Martin, T.F. Lemke, R.W. Ross, Jr., R.H. Moeller, J.A. Harris, and G.D. Smith, all of Inco Alloys International, Inc.

REFERENCES

1. W.Z. Friend, chapter 2, in *Corrosion of Nickel and Nickel-Base Alloys*, John Wiley & Sons, 1980
2. R.F. Decker, "Strengthening Mechanisms of Nickel-Base Superalloys," Paper presented at Climax Molybdenum Company Symposium, Zurich, 1969
3. H.H. Uhlig, *Corrosion and Corrosion Control*, chapter 22, John Wiley & Sons, 1963
4. W.O. Binder, U.S. Patent 2,777,766, 1957
5. A.I. Asphahani *et al.*, U.S. Patent 4,410,489, 1983
6. W. Betteridge, chapter 4, in *Nickel and Its Alloys*, Ellis Horword Ltd., 1984
7. F.G. Hodge and R.W. Kirchner, An Improved Ni-Mo Alloy for Hydrochloric Acid Service, *Mater. Perform.*, Vol 15 (No. 8), 1976, p 40
8. J. Kolts, J.B.C. Wu, and A.I. Asphahani, Highly Alloyed Austenitic Materials for Corrosion Service, *Met. Prog.*, Sept 1983
9. W.Z. Friend, chapter 10, in *Corrosion of Nickel and Nickel-Base Alloys*, John Wiley & Sons, 1980
10. A.I. Asphahani, Advanced Materials Technologies of Interest to the Process Industries, *Werkst. Korros.*, Vol 36, 1985, p 501-510
11. W. Betteridge and J. Heslop, chapter 16, in *The Nimonic Alloys*, Crank, Russak & Company, 1974
12. R.B. Herchenroeder, G.Y. Lai, and K.V. Rao, A New, Wrought, Heat-Resistant Ni-Cr-Al-Fe-Y Alloy, *J. Met.*, Vol 35 (No. 11), Nov. 1983, p 16-22
13. W. Betteridge, chapter 3, in *Nickel and Its Alloys*, Ellis Horword Ltd., 1984
14. R.F. Decker and C.T. Sims, The Metallurgy of Nickel-Base Alloys, in *The Superalloys*, John Wiley & Sons, 1972
15. W. Betteridge and J. Heslop, chapter 5, in *The Nimonic Alloys*, Crank, Russak & Company, 1974
16. A.C. Fraker and H.H. Stadelmaier, The Eta Carbides of Molybdenum-Iron, Molybdenum-Cobalt, and Molybdenum-Nickel, *Trans. AIME*, Vol 245, April 1969, p 847-850
17. J.M. Leitnaker *et al.*, The Composition of Eta Carbide in HASTELLOY N After Aging 10,000 h at 815 °C, *Metall. Trans. A*, Vol 9A, March 1978, p 397-400
18. D.R. Muzyka, The Metallurgy of Nickel-Iron Alloys, in *The Superalloys*, John Wiley & Sons, 1972
19. D.F. Paulonis, J.M. Oblak, and D.S. Duvall, Precipitation in Nickel-Base Alloy 718, *Trans. ASM*, Vol 62, 1969, p 611-622
20. H.J. Wernick, Topologically Closed-Packed Structures, in *Intermetallic Compounds*, John Wiley & Sons, 1967
21. C.T. Sims, The Occurrence of Topologically Closed-Packed Phases, in *The Superalloys*, John Wiley & Sons, 1972
22. H.M. Tawancy, Long-Term Aging Characteristics of HASTELLOY alloy X, *J. Mater. Sci.*, Vol 18, 1983, p 2976-2986
23. E.S. Machlin and J. Shao, SIGMA-SAFE: A Phase Diagram Approach to the Sigma Problem in Ni Base Superalloys, *Metall. Trans. A*, Vol 9A, April 1978, p 561-568
24. M. Morinaga *et al.*, New PHACOMP and Its Applications to Alloy Design, in *Superalloys 1984*, The Metallurgical Society, 1984, p 523-532
25. N. Sridhar, Paper 182, presented at Corrosion/86, Houston, TX, National Association of Corrosion Engineers, 1986
26. T.W. Evans and A.C. Hart, *Electrochim. Acta*, Vol 16, 1971, p 1955-1970
27. Data Brochures H2002, H2006, H2007, H2009, H2019, H2028, and H2043, Haynes International, Inc.
28. *Materials of Construction for Handling Sulfuric Acid*, Publication 5A151, National Association of Corrosion Engineers
29. J.R. Crum and M.E. Atkins, *Mater. Perform.*, Vol 25 (No. 2), 1986, p 27-32
30. Corrosion Engineering Bulletin 3, International Nickel Company, Inc.
31. R.J. Brigham and E.W. Tozer, *Corrosion*, Vol 29 (No. 1), 1973, p 33-36
32. P.E. Manning, Paper 176, presented at Corrosion/82, Houston, TX, National Association of Corrosion Engineers, 1982
33. E. Hibner, Paper 181, presented at Corrosion/86, Houston, TX, National Association of Corrosion Engineers, 1986
34. H. Tischner *et al.*, *Werkst. Korros.*, Vol 37, 1986, p 119-129
35. G. Berglund and S.D. Bernhardson, *Phosphorus Potassium*, July/Aug 1982, p 29
36. Corrosion Engineering Bulletin 4, International Nickel Company, Inc.
37. Corrosion Engineering Bulletin 5, International Nickel Company, Inc.
38. S.W. Ciaraldi, M.R. Berry, and J.M. Johnson, Paper 98, presented at Corrosion/82, Houston, TX, National Association of Corrosion Engineers, 1982
39. Corrosion Engineering Bulletin 6, International Nickel Company, Inc.
40. D.L. Graver, Ed., *Corrosion Data Survey—Metals Section*, National Association of Corrosion Engineers, 1985
41. G.B. Elder, in *Process Industries Corrosion*, National Association of Corrosion Engineers, 1975, p 247
42. Corrosion Engineering Bulletin 2, International Nickel Company, Inc.
43. B.M. Barkel, Paper 13, presented at Corrosion/79, Atlanta, GA, National Association of Corrosion Engineers, 1979
44. J.R. Crum and W.G. Lipscombe, *Mater. Perform.*, Vol 25 (No. 4), 1986, p 9-12
45. J.R. Crum and W.G. Lipscombe, Paper 23, presented at Corrosion/83, Anaheim, CA, National Association of Corrosion Engineers, 1983
46. M. Yasuda, F. Takeya, and F. Hine, *Corrosion*, Vol 39 (No. 10), 1983, p 399-405
47. A.R. McIlree and H.T. Michels, *Corrosion*, Vol 33 (No. 2), 1977, p 60-67
48. A.J. Sedriks, Stress Corrosion Cracking of Stainless Steels and Nickel Alloys, *J. Inst. Met.*, Vol 101, 1973, p 225
49. J. Kolts, "Temperature Limits for Stress Corrosion Cracking of Selected Stainless Steels and Nickel Base Alloys in Chloride Containing Environments," Paper 241, presented at Corrosion/82, National Association of Corrosion Engineers, 1982
50. R.D. Kane, J.B. Greer, D.F. Jacobs, H.R. Hanson, B.J. Berkowitz, and G.A. Vaughn, "Stress Corrosion Cracking of Nickel and Cobalt Base Alloys in Chloride Containing Environments," Paper 174, presented at Corrosion/79, National Association of Corrosion Engineers, 1979
51. A.J. Sedriks, Comparative Stress Corrosion

Cracking Behavior of Austenitic Iron-Base and Nickel Base Alloys, *Corrosion*, Vol 31 (No. 9), Sept 1975, p 339
52. J. Kolts, "Laboratory Evaluation of Corrosion Resistant Alloys for the Oil and Gas Industry," Paper 323, presented at Corrosion/86, National Association of Corrosion Engineers, 1986
53. W.K. Boyd and W.E. Berry, Stress Corrosion Cracking Behavior of Nickel and Nickel Alloys, in *Stress Corrosion Cracking of Metals—A State of the Art*, American Society for Testing and Materials, 1972, p 58
54. P.B. Needham, S.D. Cramer, J.P. Carter, and F.X. McCawley, "Corrosion Studies in High Temperature Hypersaline Geothermal Brines," Paper 59, presented at Corrosion/79, National Association of Corrosion Engineers, 1979
55. S.D. Cramer and J.P. Carter, "Laboratory Corrosion Studies in Low and High Salinity Geobrines of the Imperial Valley, California," Report RI 8415, United States Bureau of Mines, 1980
56. Corrosion Data Bank, Haynes International, Inc.
57. G.A. Vaughn and H.-E. Chaung, "Wireline Materials for Sour Service," Paper 182, presented at Corrosion/81, National Association of Corrosion Engineers, 1981
58. J. Kolts and S.M. Corey, "Corrosion and Stress Corrosion Cracking of High Performance Alloys in Simulated Acidizing Environments to 350 °F," Paper 217, presented at Corrosion/84, National Association of Corrosion Engineers, 1984
59. P.H. Berge and J.R. Donatu, Materials Requirements for Pressurized Water Reactor Steam Generator Tubing, *Nucl. Technol.*, Vol 55, Oct 1981, p 88
60. J.T.A. Roberts, Metallurgical Improvements for Enhanced Stress Corrosion Performance of Light Water Reactor Structural Materials, *J. Mater. Energy Syst.*, Vol 4 (No. 3), p 142
61. J.A. Board, Stress Corrosion Cracking in the Power Industry, *J. Inst. Met.*, Vol 101, Sept 1973, p 241
62. R.J. Lindinger and R.M. Curran, "Experience With Stress Corrosion Cracking in Large Steam Turbines," Paper 7, presented at Corrosion/81, National Association of Corrosion Engineers, 1981
63. S.J. Green and J.P.N. Paine, Materials Performance in Nuclear Pressurized Water Reactor Steam Generators, *Nucl. Technol.*, Vol 55, Oct 1981, p 10
64. J. Prybylowski and R. Ballinger, "The Influence of Microstructure on Environmentally Assisted Cracking of Alloy 718," Paper 244, presented at Corrosion/86, National Association of Corrosion Engineers, 1986
65. A. Carlson and W.K. Kratzer, Nuclear Steam Generator Performance, *Nucl. Technol.*, Vol 28, March 1976, p 383
66. Y. Fujiwara, "Influence of C, N, Nb Content on IGC, SCC, and Mechanical Strength of Alloy 600," Nippon Yakin Kogyo, Ltd., p 1
67. J. Blanchet, H. Coriou, L. Groll, C. Mahieu, C. Otter, and G. Turluer, Historical Review of the Principal Research Concerning the Phenomenon of Cracking of Nickel Base Austenitic Alloys, in *Stress Corrosion Cracking and Hydrogen Embrittlement of Iron Base Alloys*, National Association of Corrosion Engineers, 1973, p 1149
68. K.H. Lee, G. Cragnolino, and D.D. MacDonald, "Effect of Heat Treatment on the Stress Corrosion Cracking Susceptibility of Inconel 600 in Boiling 25 Molal NaOH Solution," Paper 242, presented at Corrosion/82, National Association of Corrosion Engineers, 1982
69. G.J. Theus, Caustic Stress Corrosion Cracking of Inconel 600, Incoloy 800, and Type 304 Stainless Steel, *Nucl. Technol.*, Vol 28, March 1976, p 388
70. A.I. Asphahani, Slow Strain Rate Technique and Its Application to the Environmental Stress Cracking of Nickel-Base and Cobalt-Base Alloys, in *Stress Corrosion Cracking: The Slow Strain Rate Technique*, STP 665, American Society for Testing and Materials, 1979, p 279
71. F.W. Pement, I.L.W. Wilson, and R.G. Aspden, "Stress Corrosion Cracking Studies of High Nickel Austenitic Alloys in Several High Temperature Aqueous Solutions"
72. H.T. Michels and S. Floreen, The Relationship Between Microstructure, Deformation Behavior, and Stress Corrosion Cracking Resistance of an Age Hardened Ni Base Alloy, *Metall. Trans. A*, Vol 8A, April 1977, p 617
73. R.S. Pathania, "Caustic Cracking of Steam Generator Tube Materials," Paper 98, presented at Corrosion/76, National Association of Corrosion Engineers, 1976
74. S.P. Lynch, Hydrogen Embrittlement and Liquid Metal Embrittlement in Nickel Single Crystals, *Scr. Metall.*, Vol 13, 1979, p 1051
75. C.E. Price and J.K. Good, The Tensile Fracture Characteristics of Nickel, Monel, and Selected Superalloys Broken in Liquid Mercury, *Trans. ASME*, Vol 106, 1984, p 184
76. C.D. Stephens and R.C. Scarberry, "The Relation of Sensitization of Polythionic Acid Cracking of Incoloy Alloys 800 and 801," Paper presented at the 25th Conference, National Association of Corrosion Engineers, March 1969
77. M. Cornet, C. Bertrand, and M. Da Sunhabelo, Hydrogen Embrittlement of Ultra-Pure Alloys of the Inconel 600 Type Influence of the Additions of Element (C, P, Sn, Sb), *Metall. Trans. A*, Vol 13A, 1982, p 141
78. R.D. Kane and J.B. Greer, "Embrittlement of High Strength, High Alloy Tubular Materials in Sour Environments," Paper 6798, presented at the Society of Petroleum Engineers Symposium, American Institute of Mining, Metallurgical, and Petroleum Engineers, 1977
79. R.D. Kane, Accelerated Hydrogen Charging of Nickel and Cobalt Base Alloys, *Corrosion*, Vol 34 (No. 12), 1978, p 443
80. R.M. Latanision and H. Opperhauser, Jr., The Intergranular Embrittlement of Nickel by Hydrogen, The Effect of Grain Boundary Segregation, *Metall. Trans.*, Vol 5, Feb 1974, p 483
81. P.R. Rhodes, "Stress Cracking Risks in Corrosive Oil and Gas Wells," Paper 322, presented at Corrosion/86, National Association of Corrosion Engineers, 1986
82. J.W. Kochera, A.K. Dunlop, and J.P. Tralmer, "Experience With Stress Corrosion Cracking of Nickel Base Alloys in Sour Hydrocarbon Production"
83. R.J. Walter and W.T. Chandler, "Influence of Gaseous Hydrogen on Metals," NASA-25579, National Aeronautics and Space Administration, Oct 1973
84. M. Hasegawa and M. Osawa, Hydrogen Damage of Nickel Base Heat Resistant Alloys, *Trans. Iron Steel Inst. Jpn.*, Vol 21, 1981, p 25
85. K.O. Efird, Failure of Monel Ni-Cu-Al Alloy K500 Bolts in Seawater, *Mater. Perform.*, April 1985, p 37
86. R.P. Jewett, R.J. Walter, W.T. Chandler, and R.P. Frohmberg, "Hydrogen-Environment Embrittlement of Metals," NASA CR-2163, National Aeronautics and Space Administration
87. S. Hinotani, Y. Ohmori, and F. Terasaki, Effect of Phosphorus Segregation and Ni_2 Cr Formation on Hydrogen Embrittlement in 70 Ni-30 Cr Alloys, *Mater. Sci. Technol.*, Vol 1, April 1985, p 297
88. B.J. Berkowitz and C. Miller, Effect of Ordering on Hydrogen Embrittlement Susceptibility of Ni_2Cr, *Metall. Trans. A*, Vol 11A, 1980, p 1877
89. A.I. Asphahani, "High Performance Alloys for Deep Sour Gas Wells," Paper 42, presented at Corrosion/78, National Association of Corrosion Engineers, 1978
90. C.G. Rhodes and A.W. Thompson, Microstructure—Hydrogen Performance of Alloy 903, *Metall. Trans.*, 1977
91. M. Kurkela and R.M. Latanision, The Effect of Plastic Deformation on the Transport of Hydrogen in Nickel, *Scr. Metall.*, Vol 13, 1979, p 927
92. G. Herbsleb, The Stress Corrosion Cracking of Sensitized Austenitic Stainless Steels and Nickel Base Alloys, *Corros. Sci.*, Vol 20, 1980, p 243

SELECTED REFERENCES

- Corrosion Engineering Bulletins 1 through 6, International Nickel Company, Inc.
- W.Z. Friend, *Corrosion of Nickel and Nickel-Base Alloys*, Wiley-Interscience, 1980
- D.L. Graver, Ed., *Corrosion Data Survey—Metals Section*, National Association of Corrosion Engineers, 1985
- *Proceedings of CORROSION/83 Symposium on Performance of Construction Materials in Flue Gas Desulfurization Systems—CORROSION/84 Symposium on Materials Evaluation and Environmental Effects on Corrosion in Flue Gas Desulfurization Systems*, National Association of Corrosion Engineers, 1984
- *Resolving Problems in Air Pollution Control Equipment*, National Association of Corrosion Engineers, 1976
- *Solving Corrosion Problems in Air Pollution Control Equipment*, National Association of Corrosion Engineers, 1981, 1984

Corrosion of Cobalt-Base Alloys

Chairman: Aziz I. Asphahani, Haynes International, Inc.

Introduction

Paul Crook
Haynes International, Inc.

MANY OF THE COMMERCIAL cobalt alloys that are currently in use stem from the work of Elwood Haynes at the turn of the century. He discovered the high strength and stainless nature of the binary cobalt-chromium alloy, and he later identified tungsten and molybdenum as powerful strengthening agents within the cobalt-chromium system. When he discovered these alloys, Haynes suggested that they might be used for cutting tools, cutlery, and surgical instruments. He named them the Stellite alloys because of their permanent, starlike luster (the Latin for star is *stella*).

Because of their high strength over a wide temperature range and their resistance to many environments, cobalt-chromium alloys have found two major uses within industry. First, they are used to resist wear, particularly in hostile environments, and second, they are used as structural materials at high temperatures. Typically, the alloys used to resist wear contain higher carbon levels (0.25 to 2.5%) for carbide formation, and they are normally cast or applied to critical surfaces by welding (a process known as hardfacing). Alloys used for structural purposes at high temperatures are normally low in carbon, contain appreciable quantities of nickel, and are available as wrought products. The carbides in the wear-resistant alloys enhance abrasion resistance but reduce ductility.

Physical Metallurgy

Table 1 gives the compositions of commercial cobalt-base alloys. As can be deduced from Table 1, the chief difference among the individual Stellite hardfacing alloys is carbon content and thus carbide volume fraction in the material.

Chromium has a dual function in the Stellite alloys. It is both the predominant carbide former (that is, most of the carbides are chromium rich) and the most important alloying element in the matrix, where it provides added strength (as a solute) and resistance to corrosion and oxidation. The most common carbide in the Stellite alloys is a chromium-rich M_7C_3 type, although chromium-rich $M_{23}C_6$ carbides are abundant in low-carbon alloys such as Stellite alloy 21.

Tungsten and molybdenum in the Stellite alloys serve to provide additional strength to the matrix. They do so by virtue of their large atomic size (that is, they impede dislocation flow when present as solute atoms). When present in large quantities (for example, in Stellite alloy 1), they participate in the formation of carbides during alloy solidification and promote the precipitation of M_6C. They also improve general corrosion resistance of the alloys. Although these alloying elements are critical to the performance of the Stellite alloys in service, the main reason for the commercial success of the Stellite alloys is the cobalt.

Cobalt imparts to its alloys an unstable, face-centered cubic crystal (fcc) structure with a very low stacking fault energy. The instability arises from the fact that elemental cobalt, if cooled extremely slowly, transforms from an fcc to hexagonal close-packed (hcp) crystal structure at 417 °C (782.6 °F). In most cobalt alloys, the transformation temperature is somewhat higher.

Because of the sluggish nature of the transformation, the fcc structure in cobalt and its alloys is usually retained to room temperature, and hcp formation is triggered only by mechanical stress or time at elevated temperature. The unstable fcc structure, as well as its associated low stacking fault energy, are believed to result in:

- High yield strengths
- High work-hardening rates (due to the interaction between stacking faults)
- Limited fatigue damage under cyclic stresses (due to the lack of cell walls within plastically deformed material)
- The ability to absorb stresses (through transformation of the structure to hcp)

The first three of these attributes are believed to be important in preventing metallic damage during sliding wear. The last two are believed to be responsible for the outstanding resistance to cavitation and erosion-corrosion of the cobalt alloys.

The size and shape of the carbide particles within the Stellite alloys are strongly influenced by cooling rate and subtle chemistry changes. This is illustrated in Fig. 1 and 2, which show typical overlay microstructures as applied by different welding processes. Such changes markedly affect abrasion resistance, because there is a distinct relationship among the size of abrading species, the size of the structural hard particles, and the abrasive wear rate.

The success of the structural cobalt alloys can be attributed to both their inherent strength (over a wide temperature range) and their resistance to the environment. Their strength stems largely from the unusual microstructural properties of cobalt and from the strengthening solutes, such as chromium and tungsten. Their corrosion properties will be discussed in subsequent sections of this article.

The structural alloys generally contain significant quantities of nickel. This serves to stabilize the fcc structure with a view toward improved ductility during service. With sufficient nickel, the structural cobalt alloys tend to exhibit twinning during deformation.

Although the structural cobalt alloys are low in carbon as compared to most of the wear-resistant alloys, they nevertheless depend on carbide precipitation for additional strength. The most abundant carbide in the structural cobalt alloys is chromium-rich $M_{23}C_6$, although M_6 and MC carbides are common, depending on the type and level of other alloying additions, such as tungsten and tantalum. Table 1 lists the compositions of several structural cobalt alloys.

Behavior of Cobalt-Base Alloys in Corrosive Environments

Narasi Sridhar
Haynes International, Inc.

A variety of cobalt-containing alloys, ranging in cobalt content from 19 to 60% will be discussed in this section (see Table 1). Because the corrosion properties of these alloys have thus far not been examined systematically with respect to such variables as alloying elements, the aim in this discussion is to describe the corrosion resistance of commercially available cobalt-base alloys in various environments. Stress-corrosion interactions will not be considered. This section will also exclude hardfacing alloys, which are discussed in the section "Fabrication and Weldability of Cobalt-Base Alloys" in this article.

General Corrosion

The corrosion behavior of pure cobalt has not been documented as extensively as that of nickel. The behavior of cobalt is similar to that of nickel, although cobalt possesses lower overall corrosion resistance (Ref 1). For example, the passive behavior of cobalt in 1 *N* sulfuric acid (H_2SO_4) has been shown to be similar to that of nickel, but the critical current density necessary to achieve passivity is 14 times higher for the former (Ref 2). Several investigations have been carried out on binary cobalt-chromium alloys. In cobalt-base alloys, it has been found that as little as 10% Cr is sufficient to reduce the anodic current density necessary for passivation from 500 to 1 mA/cm^2 (3225 to 6.5 mA/in.^{2}) (Ref 2). For nickel, about

Table 1 Compositions of a variety of commercial cobalt-base alloys

Compositions of cobalt-base alloys for orthopedic implants are given in Table 12.

Alloy	UNS designation	Composition, % C	Co	Cr	Fe	Mo	Ni	Si	W	Other
Wrought alloys										
Elgiloy	R30003	0.15(a)	40	20	16	7	15.5	...	...	...
Havar	R30004	0.20	42.5	20	19	2.5	13.0	...	3	...
Haynes alloy 6B	...	1.2	57	30	3(a)	1.5(a)	3	1.5(a)	4.5	...
Haynes alloy 21	R30021	0.25	60	27	3(a)	5.5	3	1.0(a)	...	...
MP35N	R30035	0.025(a)	35	20	1(a)	10	35	0.15(a)	...	...
MP159	R30159	0.04(a)	36	19	9	7	25	0.2(a)	...	3.0Ti 0.5Nb
Haynes alloy 188	R30188	0.15(a)	39	22	3(a)	...	22	0.5(a)	14	0.15La(a)
Duratherm 600	R30600	0.05(a)	42	12	8.5	4	26	0.6(a)	4	2.0Ti 0.6Al
Haynes alloy 25	...	0.1	42	20	3	...	10	...	15	...
Haynes alloy 556	R30556	0.1	19	22	29	3.5	21	0.8(a)	3	1.25Ta(a) 0.2N
Hardfacing alloys										
Stellite alloy 3	...	2.43	50.5	30.5	1.5	0.5	1.5	0.5	12.5	0.1B
Stellite alloy 6	R30006	1.15	61.5	28.3	1.5	0.5	1.5	1.1	4.5	...
Stellite alloy 21	...	0.25	62.7	27.3	1.0	5.5	2.5	0.5	0.25	...
Tribaloy alloy T-400	...	0.06	57.5	8.5	...	28 5	...	2.6	...	3.0(Ni + Fe)
Tribaloy alloy T-800	...	0.06	50.0	17.5	...	28 5	...	3.4	...	3.0(Ni + Fe)

(a) Maximum

14% Cr is needed to reduce the passivating anodic current density to the same level. The corrosion rates of a variety of cobalt-chromium alloys in several oxidizing and reducing media are shown in Table 2. In another study, researchers measured the open-circuit potentials in a dilute salt solution (Ringers solution) of a number of cast binary cobalt-chromium alloys having 10 to 40% Cr (Ref 4). They found that the behavior of the alloy was controlled not only by chromium content but also by microstructural changes produced by the addition of chromium to cobalt. Addition of up to 25% Cr made the alloys more noble. However, above 25% Cr, precipitation of chromium-rich σ phase made the alloys more active because of local chromium depletion around the σ-phase precipitates. Unfortunately, they did not measure corrosion rates of these alloys.

Another study examined the effect of alloying elements in cobalt-base alloys in H_2SO_4 and phosphoric acid (H_3PO_4) (Ref 5). These alloys contained 2% C for abrasive wear resistance. The data are given in Table 3. It can be seen that addition of nickel and copper lowered the corrosion rates significantly. Addition of molybdenum or nickel plus molybdenum did not have a great effect, possibly because of intermetallic precipitation. These researchers also conducted tests in an H_3PO_4 solution under abrasive conditions and found molybdenum, nickel, and copper to be highly beneficial (Ref 5).

The corrosion resistance of a number of commercial cobalt-base alloys in H_2SO_4 and hydrochloric acid (HCl) is given in Tables 4 and 5. It should be noted that all of these alloys, regardless of their chromium and molybdenum contents, exhibit similar corrosion resistance in dilute H_2SO_4 (Table 4). Thus, the high-chromium Haynes alloys 6B and 21 show approximately the same corrosion rates as the lower-chromium Haynes alloys 188 and 556. Similar behavior has been observed in the nickel-iron-chromium-molybdenum alloys (Ref 6). In H_2SO_4 and HCl, the nickel and cobalt contents govern the behavior of the alloy as long as minimum amounts of chromium and molybdenum or tungsten are present. The corrosion resistance of wrought cobalt-base alloys in HCl solutions is given in Table 5. Again, as with the nickel-iron-chromium-molybdenum alloys, the corrosion resistance of cobalt-

(a)

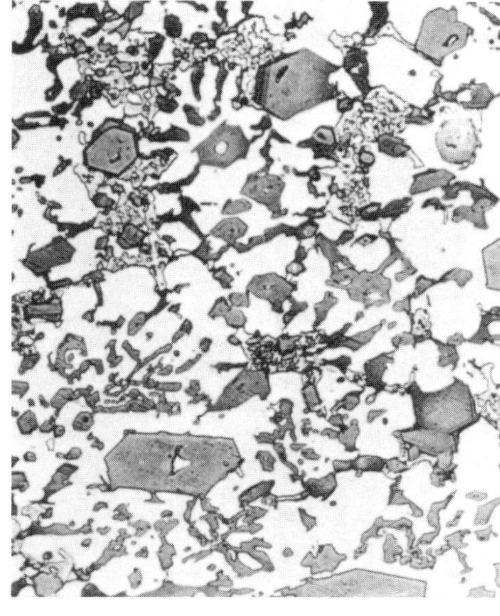

(b)

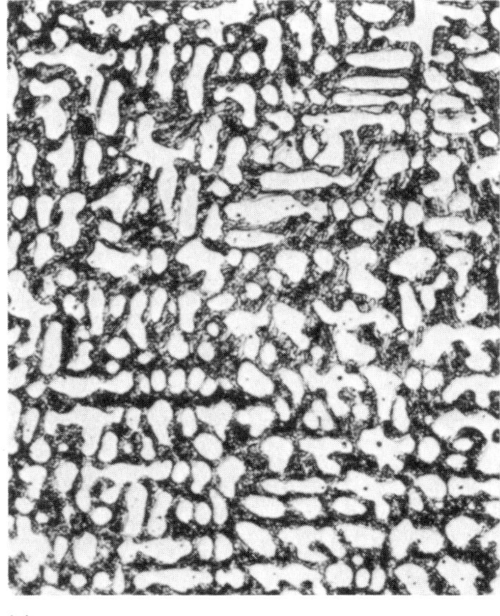

(c)

Fig. 1 Typical overlay microstructures of Stellite 1 alloy applied by different weld processes. (a) Three-layer gas tungsten arc. (b) Three-layer oxyacetylene. (c) Three-layer shielded metal arc. See also Fig. 2. All 500×

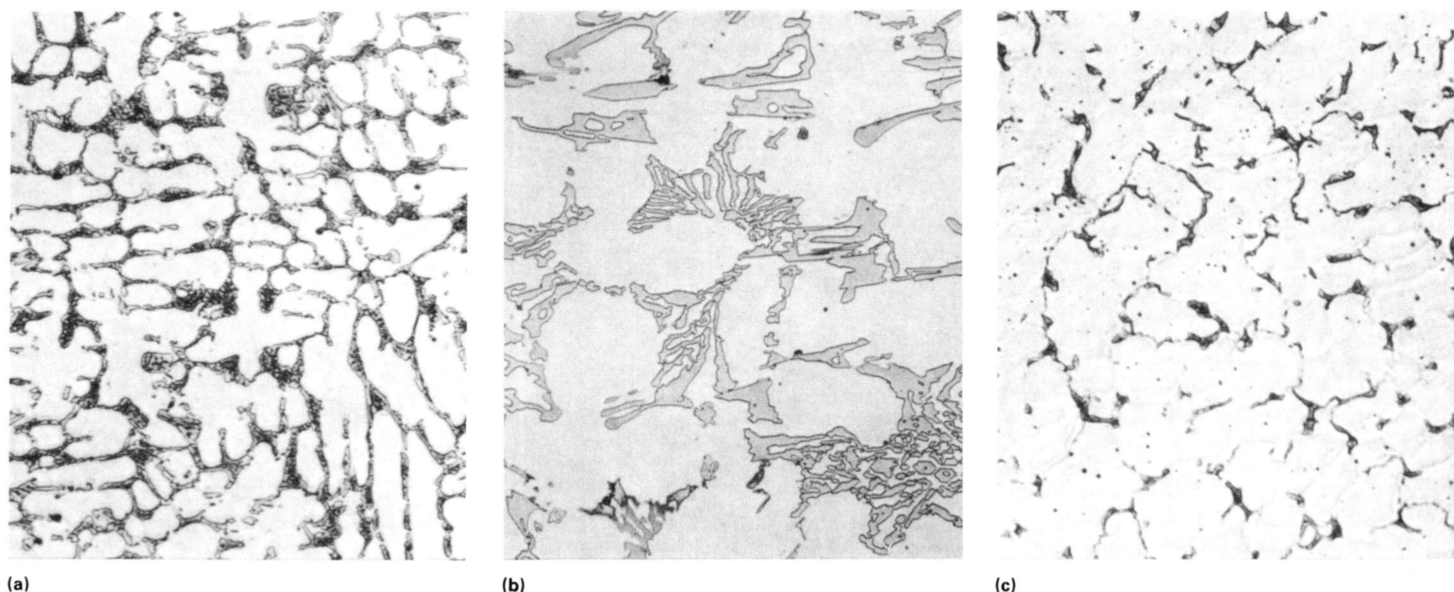

Fig. 2 Typical overlay microstructures of Stellite 6 alloy applied by different weld processes. (a) Three-layer gas tungsten arc. (b) Three-layer oxyacetylene. (c) Three-layer shielded metal arc. See also Fig. 1. All 500×

Table 2 Corrosion of cobalt- and nickel-base alloys in aqueous solutions

Alloy	Solution and temperature	Corrosion rate, mm/yr	Corrosion rate, mils/yr	Ref
Cobalt	1 *N* HNO_3, 25 °C (75 °F)	223	8900	2
Nickel	1 *N* HNO_3, 25 °C (75 °F)	19	770	2
Ni-20Cr	11% HNO_3, hot	nil		3
Co-20Cr	11% HNO_3, boiling	nil		3
Co-20Cr	11% H_2SO_4, boiling	110	4400	3
Co-20Cr	10% HCl, boiling	250	10 000	3

Table 3 Effect of alloying additions on the corrosion of cobalt-base alloys in acid media

Data obtained in 192-h tests at 25 °C (75 °F)

Alloy composition	Weight loss, mg/cm², 10% H_2SO_4	Weight loss, mg/cm², 28% P_2O_5 + 13% H_2SO_4 + 2% HF
Co-2.1C-31Cr-12W	0.03	0.09
Co-2.1C-32Cr-12.5W-19.5Ni	0.02	0.03
Co-2.2C-32Cr-18.8Ni-14W	0.04	0.06
Co-2.3C-32Cr-14W-9.4Mo	0.06	0.06
Co-2.1C-32Cr-17.6Ni-13W	nil	nil
Co-2.2C-31Cr-18Ni-15W-9Mo-4Cu	1.8	1.0
Co-2.1C-32Cr-13.6W-4.2Cu	Not determined	Not determined

Source: Ref 5

Table 4 Corrosion of various wrought cobalt-base alloys in H_2SO_4 solutions

Tests were conducted in four 24-h periods.

Alloy	2% boiling, mm/yr	2% boiling, mils/yr	10% 66 °C (150 °F), mm/yr	10% 66 °C (150 °F), mils/yr	10% boiling, mm/yr	10% boiling, mils/yr	20% 66 °C (150 °F), mm/yr	20% 66 °C (150 °F), mils/yr	20% boiling, mm/yr	20% boiling, mils/yr	80% boiling, mm/yr	80% boiling, mils/yr
Haynes alloy 6B	0.8	31	0.0005	0.02	3.99	157	...	...	9.2	361	250	10 000
Haynes alloy 25	0.86	34	0.2	8	4.4	174	0.5	20	10	395	>250	10 000
Haynes alloy 188	0.89	35	nil		3.02	119	0.25	10(a)	5.6	220	...	...
Haynes alloy 556	0.76	30	nil		3.45	136	0.33	13	8	316	...	...
Haynes alloy 21	...	...	...	...	1.83	72	...	...	...	...	...	...
Havar	...	...	...	...	3	120	...	...	...	...	...	...
MP35N	...	...	...	...	1.3	51	...	...	...	...	...	...

(a) Corrosion rate varied within test period.

base alloys is not good except in very dilute HCl.

Because many of the commercial alloys contain appreciable amounts of chromium, their corrosion resistance to dilute nitric acid (HNO_3) is quite good (Table 6). However, in concentrated HNO_3, Haynes alloy 6B exhibited high corrosion rates, while Haynes alloys 188 and 25, which have lower chromium, showed lower corrosion rates. This could have been a consequence of the high carbon content of alloy 6B. High carbon and chromium are present intentionally in this alloy to provide chromium carbides for abrasion resistance. However, concentrated HNO_3 is known to attack high-chromium phases such as σ phase and chromium-rich carbides.

The corrosion resistance of some cobalt-base alloys in a variety of reducing and oxidizing environments is given in Table 7. In highly oxidizing chromic acid (H_2CrO_4), the chromium-containing alloys, whether cobalt- or nickel-base, do not show good resistance, because the passive chromium oxide film is unstable in this acid. However, high amounts of tungsten seem to decrease the corrosion rates of Haynes alloys 188 and 25. In H_3PO_4, the corrosion rates of the cobalt-base alloys are very high. Although there are no compositionally equivalent nickel-base

Table 5 Corrosion of various wrought cobalt-base alloys in HCl solutions

Tests were conducted in four 24-h periods except for the 5% HCl test, which was for one 24-h period.

	Corrosion rate							
	1% boiling		2.5% boiling		5% boiling		10% room temperature	
Alloy	mm/yr	mils/yr	mm/yr	mils/yr	mm/yr	mils/yr	mm/yr	mils/yr
Haynes alloy 6B	...	...	96.5	3800	>250	10 000	...	...
Haynes alloy 25	0.56	22	57.4	2260	188	7400	nil	
Haynes alloy 188	nil(a)		61	2400	140	5500	0.008–0.8	0.3–31(b)
Haynes alloy 556	nil(a)		63.5	2500	167.6	6600	0.6	23

(a) Corrosion decreased from a high value. (b) Corrosion rate oscillated during the test period.

Table 6 Corrosion resistance of various wrought cobalt-base alloys in HNO_3 solutions

	Average corrosion rate					
	10% boiling		65% boiling		70% boiling	
Alloy	mm/yr	mils/yr	mm/yr	mils/yr	mm/yr	mils/yr
Haynes alloy 188	0.02	0.8(a)	0.56	22(b)	...	...
Haynes alloy 25	0.02	0.8(a)	0.99	39(b)	0.94	37(b)
Haynes alloy 6B	0.023	0.9(a)	...	...	96.5	3800(a)
Haynes alloy 21	0.025	1.0(a)	...	...	...	...
Haynes alloy 556	...	...	0.28	11(b)	...	...
Havar	0.066	2.6	...	...	...	...
MP35N	0.056	2.2	...	...	...	...

(a) One 24-h test period. (b) Five 24-h test periods.

alloys, the corrosion rates of many of the commercial nickel-chromium-molybdenum-iron alloys in H_3PO_4 are lower (Ref 7). In acetic acid, the corrosion rates of all of the alloys are quite low. Nickel is widely used in caustic service. The nickel-chromium-molybdenum alloys offer lower corrosion resistance than nickel. In comparison, the cobalt-base alloys do not offer very high resistance to corrosion in caustic environments. Indeed, the data in Table 7 seem to indicate that the corrosion rates of the three cobalt-base alloys tested in caustic decreased with increasing nickel content (alloy 6B < 25 < 188).

Addition of oxidizing agents to H_2SO_4 results in lower corrosion rates for the cobalt-chromium-molybdenum-tungsten alloys, just as in the case of the nickel-base alloys. This is shown in Table 8 for the 50% H_2SO_4 + 2.5% ferric sulfate ($Fe_2(SO_4)_3$) mixture recommended by the American Society For Testing and Materials (ASTM) as an intergranular corrosion test solution.

Localized Corrosion

The resistance to localized corrosion (pitting and crevice corrosion) of the cobalt-base alloys is generally determined by the chromium, molybdenum, and tungsten contents. Generally, localized corrosion resistance is measured by immersion tests in oxidizing chloride solutions or by electrochemical tests in chloride solutions. Crevice corrosion resistance is usually measured by attaching blocks to the sample in order to create a crevice between the sample and the block. The test methods are outlined in Ref 7 to 10.

An example of the localized corrosion resistance of several cobalt-base alloys in an oxidizing chloride-containing test solution is given in Table 9. This is an extremely acidic chloride solution containing ferric and cupric salts to increase the oxidizing potential of the solution. As the results indicate, there is a wide variation in alloy performance. Havar alloy and Haynes alloy 556 start to corrode at a rapid rate at about 70 °C (160 °F). Alloys MP35N, 188, and 21 begin to corrode rapidly near the boiling point of the solution. Alloy 25 exhibits the highest resistance of these cobalt-base alloys. Alloy 6B is not listed in Table 9, because it exhibits a high corrosion rate even in less severe solutions, such as 10% ferric chloride ($FeCl_3$). The corrosion rates of the alloys tested are given in Table 9. In the 20 to 22% Cr alloys, the alloys containing the lowest molybdenum and tungsten (alloys Havar and 556) show the highest corrosion rates. The alloy with the intermediate molybdenum level (alloy MP35N) shows lower corrosion rates, while the alloys with high tungsten (alloy 25) and high chromium plus molybdenum (alloy 21) show even lower rates. The exception to this case is alloy 188, which shows higher corrosion rates than alloy 25 in spite of similarity in composition. Unfortunately, no data are available at intermediate temperatures, so a more accurate comparison of alloys cannot be made.

Results of corrosion tests conducted in a milder localized test environment of $FeCl_3$ are given in Table 10. Again, the trend is similar to that shown in Table 9 in that Haynes alloys 188 and 25 are superior in pitting resistance to Haynes alloy 556. In summary, it is believed that the effects of alloying elements (chromium, molybdenum, and tungsten) on pitting resistance are qualitatively similar in cobalt-base alloys to those observed in nickel-base alloys. Their effects in cobalt-base alloys cannot be predicted precisely, however, because of the lack of extensive data.

Environmental Embrittlement of Cobalt-Base Alloys

Juri Kolts
Conoco Inc.

Cobalt-base alloys are primarily used in high-temperature applications. In such uses, hydrogen embrittlement and stress-corrosion cracking (SCC) are generally not thought to be important. However, in applications in which cobalt-base alloys are used for aqueous corrosion service, both of these modes of fracture may become important.

Hydrogen Embrittlement

Cobalt-base alloys can be used to combat hydrogen embrittlement where steels have failed by this mechanism. The literature on the hydrogen embrittlement of cobalt-base alloys is not very extensive. Most of the experimental work has been performed on Haynes alloys 188 and 25 and on alloy MP35N. The performance of these alloys is in many ways similar to that of many nickel-base alloys (Ref 11, 12).

Effects of Strength Level. The cobalt-base alloys can be processed to achieve much higher yield/tensile strengths than most of the nickel-base alloys. Susceptibility to hydrogen embrittlement is closely related to yield strength; therefore, the potential for hydrogen embrittlement in cobalt-base alloys may be higher than in nickel-base alloys, not because of a higher inherent susceptibility to embrittlement, but because of the higher strength levels achievable in cobalt-base alloys.

Annealed cobalt-base alloys do not show significant susceptibility to hydrogen embrittlement, even in the most severe hydrogen-charging conditions. When cold worked to levels exceeding 1380 MPa (200 ksi) yield strength, the cobalt-base alloys may not exhibit embrittlement (Ref 13).

Table 7 Corrosion of cobalt-base alloys in a variety of environments at boiling temperature

	Corrosion rate									
	99% acetic acid		85% phosphoric acid		60% formic acid		10% chromic acid		50% sodium hydroxide	
Alloy	mm/yr	mils/yr	mm/yr	mils/yr	mm/yr	mils/yr	mm/yr	mils/yr	mm/yr	mils/yr
Haynes alloy 6B	0.0008	0.03	15.5	610	1.22	48	...	...	2.74	108
Haynes alloy 21	0.017	0.67	17.3	680	...	...	...	...	...	...
Haynes alloy 25	0.0056	0.22	19.2	754	0.61	24	1.0	40	0.53	21
Haynes alloy 188	0.005	0.20	13.5	530	...	...	1.37	54	0.43	17
Haynes alloy 556	0.005	0.20	0.84	33	...	...	2.8	110	...	...
MP35N	0.015	0.6	12.7	500	...	...	...	...	...	...
Havar	0.078	3.1	>100	4000	...	...	...	...	...	...

Table 8 Corrosion of cobalt-base alloys in a boiling oxidizing mixture of 50% H_2SO_4 and 2.5% $Fe_2(SO_4)_3$

Alloy	Corrosion rate mm/yr	Corrosion rate mils/yr
Havar	0.72	28.5
MP35N	0.41	16.2
Haynes alloy 21	0.36	14.0
Haynes alloy 25	0.66	26.0
Haynes alloy 188	0.43	17.0
Haynes alloy 556	0.36	14.0

Table 9 Results of pitting corrosion tests in an aqueous mixture of 11.5% H_2SO_4 + 1.2% HCl + 1% $CuCl_2$ + 1% $FeCl_3$ at various temperatures

Test duration: 24 h

Alloy	Corrosion rate, mm/yr (mils/yr) Room	70 °C (160 °F)	Boiling
Havar	nil	15 (600)	5.6 (2200)
MP35N	nil	nil–0.015 (0.6)	0.076–50 (3–2000)
Haynes alloy 21	nil	nil	0.76 (30)
Haynes alloy 188	nil–0.01 (0.4)	0.005–0.2 (0.2–8)	2.5–24.1 (100–950)
Haynes alloy 25	0.0076–0.025 (0.3–1.0)	nil–0.01 (0.4)	nil–0.05 (2)
Haynes alloy 556	0.0025–0.023 (0.1–0.9)	1.3–19.3 (50–760)	14.2–53 (560–2100)

Two factors—impressed hydrogen charging and low-temperature aging (as is the case for nickel-base alloys)—are often required to produce hydrogen embrittlement.

External Hydrogen Charging. Because of the high corrosion resistance of many cobalt-base alloys, the alloys may not corrode at a sufficient rate even in acid systems to charge hydrogen into the alloys. Hydrogen charging will often be achieved through the cathodic protection systems used to protect other components in a system or through the galvanic coupling of cobalt-base alloys to more actively corroding metals. Hydrogen entry may be enhanced by the presence of hydrogen recombination poisons, such as sulfides, cyanides, or other substances commonly found to be effective hydrogen recombination poisons on steels. Under such conditions, cobalt-base alloys with yield strengths over 1380 MPa (200 ksi), and especially greater than 1724 MPa (250 ksi), can exhibit hydrogen embrittlement. The time to failure (usually by an intergranular mode) is strongly dependent on the charging current (Ref 14).

Effect of Thermal Treatments. The aging of cobalt alloys has two opposite effects on hydrogen embrittlement, depending on the temperature range. At temperatures ranging from 205 to about 650 °C (400 to 1200 °F), aging is highly detrimental to hydrogen embrittlement resistance (Ref 15, 16). Aging above about 760 °C (1400 °F) is highly beneficial (Ref 17). The most susceptible microstructures are produced by a combination of cold work and aging near 540 to 650 °C (1000 to 1200 °F). Aging at temperatures above 790 °C (1450 °F) produces microstructures that are extremely resistant to embrittlement.

Long-term aging treatments that take place for periods of years at temperatures as low as 205 °C (400 °F) may have a noticeable detrimental effect (Ref 18, 19). Very high strength levels and hydrogen charging are both required to produce this effect.

Effects of Other Factors. Anisotropy in mechanical properties may play an important role in hydrogen embrittlement (Ref 16). Crack path sensitivity is greater in directions transverse to the direction of mechanical working. Other factors, such as stress level, environmental temperature, and environmental corrosivity, have effects similar to those seen in steels and nickel-base alloys (see the articles "Environmentally Induced Cracking," "Corrosion of Alloy Steels," and "Corrosion of Nickel-Base Alloys" in this Volume).

Table 10 Corrosion of cobalt-base alloys in 3.8% $FeCl_3$ solutions

24-h test; all alloys in the mill-annealed condition

Alloy	Corrosion rate, mm/yr (mils/yr) Room	70 °C (160 °F)	Boiling
Haynes alloy 188	nil	nil	nil
Haynes alloy 25	0.013 (0.5)	nil	nil
Haynes alloy 556	0.033 (1.3)	14 (550)	36 (1419)

Stress-Corrosion Cracking

Only a limited number of environments have been reported to cause SCC of cobalt-base alloys. These include two general classes of environments: acid chlorides and strong alkalies. Both of these environments produce SCC only at temperatures exceeding about 150 to 175 °C (300 to 350 °F) (Ref 20), although useful lives can be expected even above 205 °C (400 °F) (Ref 21). The fracture mode in sodium hydroxide (NaOH) was intergranular on Stellite alloy 6B (Ref 22). In acid chlorides containing hydrogen sulfide (H_2S), the SCC of Haynes alloy 188 was intergranular; cracking of alloy MP35N was reported to be transgranular (Ref 20).

Minor surface SCC has been reported for MP35N in hypersaline geothermal brine at 200 to 230 °C (390 to 445 °F) (Ref 23). In some conditions, cobalt-base alloys have been shown to be susceptible to SCC in boiling magnesium chloride ($MgCl_2$) at 155 °C (310 °F) (Ref 24), although other alloys are resistant at this temperature (Ref 25). No data have been obtained for $MgCl_2$ solutions at higher temperatures.

The SCC behavior discussed above is similar to that found for nickel-base alloys where much more data are available. In view of these similarities, the article "Corrosion of Nickel-Base Alloys" in this Volume may provide further insight into the possible environments in which precautions must be taken or additional data developed for cobalt-base alloy applications.

Fabrication and Weldability of Cobalt-Base Alloys

S.J. Matthews
Haynes International, Inc.

The weldability of wrought cobalt-base alloys, especially those with greater than 10% Ni, is very similar to the weldability of nickel-base alloys (see the article "Corrosion of Nickel-Base Alloys" in this Volume). Conventional fusion-welding processes can be used, although oxyacetylene welding is not recommended for low-carbon cobalt-base alloys. Gas tungsten arc welding generally produces the most satisfactory results.

Table 11 Advantages and disadvantages of commonly used hardfacing processes and consumables

Process	Consumables(a)	Advantage	Disadvantage
Gas tungsten arc	CR	High-quality deposits	Relatively slow process
Shielded metal arc (coated electrode)	CR, TW	Portability (field repair)	Slag removal and low deposit efficiency
Open arc	TW	High deposition rate	Spatter and rough deposits
Submerged arc	TW	High deposition rate and efficiency	High base metal dilution
Gas metal arc	TW	Good quality and good deposition rate	Relatively high dilution
Oxyacetylene	CR, TW, P	Low dilution	Slow process
Plasma arc	CR, TW, P	Very smooth high-quality deposits	Some overspray (powder loss) with plasma-transferred arc process
Flame spray	P	Very smooth deposit (after fusing)	Maximum thickness of 3.2 mm (1/8 in.)
Plasma spray	P	No dilution, no distortion	Only thin coatings, not 100% dense
Jet Kote II surfacing system	P	High-quality, dense coatings	Higher gas consumption than flame spray
Laser	P	High volume production capability	Very expensive equipment

(a) CR, cast rod; TW, tube wire; P, powder

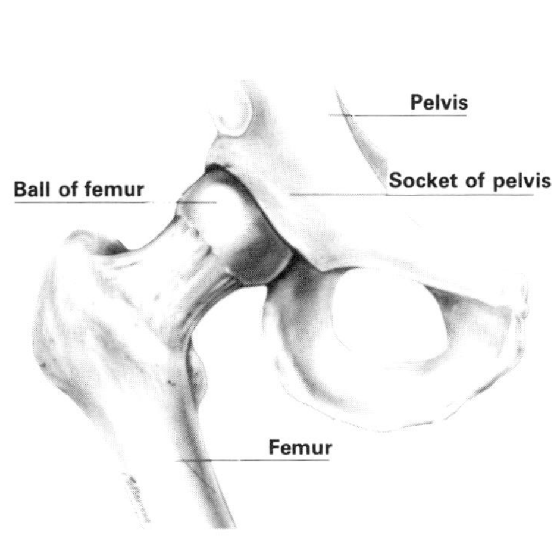

(a)

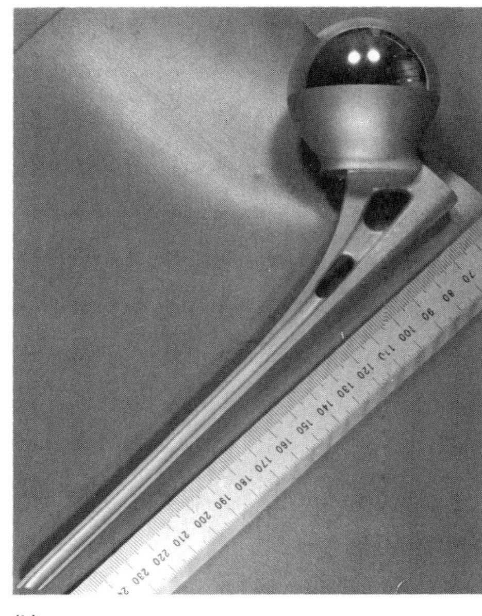

(b)

(c)

Fig. 3 The natural hip joint (a) and two types of artificial joints (b and c). The cast cobalt-chromium-molybdenum endoprosthesis (b) has a large diameter head that articulates in the acetabulum (the natural socket of the pelvis). The total hip replacement (c) replaces both the femoral head and the socket of the pelvis. The prosthesis shown in (c) uses a modular cobalt-chromium-molybdenum head that fits on a forged cobalt-chromium-molybdenum stem; the socket component is ultrahigh molecular weight polyethylene with a metal backing.

Sound welds are readily obtained with good joining practices. Joint preparations and cleaning before welding are the most important joining practices. For low-carbon cobalt-base alloys (<0.15% C), preheat is not required, and weld interpass temperatures should be below 95 °C (200 °F).

Cobalt alloys are highly susceptible to weld cracking at high temperatures when they are contaminated with copper. Molten copper will initiate liquid-metal embrittlement in the heat-affected zone (Ref 26). For example, a minute amount of copper that is inadvertently transferred onto the surface of a sheet from a backing bar can cause severe cracking if the copper melts during welding. This may be prevented by plating the copper backing with nickel or chromium to prevent accidental copper transfer. Copper or brass wire brushes can also contaminate the metal and should not be used for cleaning.

Hardfacing and Wear

A sizeable amount of cobalt alloy used for corrosion environments is in the form of hardfacing deposits, for which the primary need is wear protection as well as corrosion resistance. Hardfacing is a term that describes the application of a material to the surface of a component by welding for the main purpose of reducing wear. The article "Hardfacing" in Volume 6 of the 9th Edition of *Metals Handbook* contains detailed information on hardfacing materials and processes. Wear can be defined as the loss of material by abrasion, sliding wear, or erosion (solid particle/liquid droplet/cavitation).

Abrasion accounts for the largest share of industrial wear problems. It occurs when hard particles are forced against and move along a solid surface. The basic mechanisms of abrasion include cutting, plowing, and chipping. The predominant mechanism will depend on the size, shape, and hardness of the abrasive relative to the hardness of the wearing surface. Cobalt hardfacing alloys designed to resist abrasion generally contain large amounts of chromium carbide. Their abrasion resistance depends on the volume fraction, size, and morphology of the carbide structure. As a general rule, the higher the abrasion resistance of a cobalt-base hardfacing alloy, the lower its corrosion resistance.

Sliding wear (also called metal-to-metal wear) occurs when two contacting metal surfaces slide under load. At a low contact load, it is common to encounter a mild, oxidative wear mode (controlled by surface films and generating oxide wear debris). A more severe wear that generates metallic wear debris is encountered at higher contact loads, that is, galling. Unlike abrasion, hard phases play only a secondary role in controlling the galling resistance of an alloy. The factors that contribute to the sliding wear resistance of an alloy are largely determined by matrix composition and deformation characteristics. Cobalt-base hardfacing alloys have an industry-accepted reputation for outstanding resistance to sliding wear and galling.

Erosion is another form of surface degradation. It differs from the previous modes of wear in that the energy required to cause damage originates from momentum transfer rather than large-scale traction forces. Solid-particle erosion results from the impingement of gas-borne hard particles onto a solid surface. Liquid droplet erosion occurs when high-speed liquid drops strike a solid surface. Cavitation erosion occurs when gas bubbles (formed by high fluid flow rates or rapid pressure drops) in a liquid collapse on a solid surface, striking it with minute liquid jets.

Like abrasion, solid-particle erosion mechanisms include cutting and plowing. The mechanisms of liquid droplet erosion and cavitation erosion are primarily fatigue and fracture. Because some mechanisms of solid-particle erosion are similar to abrasion, cobalt alloys with high volume fractions of hard phases are also effective in resisting solid-particle erosion. However, al-

Table 12 Compositions of cobalt-base alloys for orthopedic implants

Alloy type	Alloy designation/ applicable ASTM standard	Composition, %(a) C	Co	Cr	Fe	Mn	Mo	N	Ni	P	S	Si	Ti
Cobalt-chromium-molybdenum	F 75	0.35	bal	27–30	0.75	1.0	5–7	...	1.0	...	...	1.0	...
Cobalt-chromium-molybdenum, thermomechanically processed	F 799	0.35	bal	26–30	1.5	1.0	5–7	0.25	1.0	...	...	1.0	...
Cobalt-nickel-chromium-molybdenum	F 562	0.025	bal	19–21	1.0	0.15	9–10.5	...	33–37	0.015	0.01	0.15	1.0

(a) Maximum un ess range is given

loys that can withstand large plastic strains without fracturing are also useful under some service conditions. The alloy characteristics that provide resistance to liquid droplet and cavitation erosion are more difficult to define, although the ability to accommodate strain through slip or transformation, combined with good fatigue resistance, appears to be important. It is generally recognized that cobalt-base wear alloys, specifically in wrought form, have excellent liquid droplet and cavitation erosion resistance.

Hardfacing Welding Processes

With regard to the subject of hardfacing, it should be noted that the operation involves the deposition of an overlay, usually by welding or allied process (that is, thermal spray). A wide variety of welding and thermal spray processes are available. Table 11 lists the commonly used processes, along with their advantages and disadvantages. Part size and shape, accessibility considerations, base material composition, dilution requirements, area to be surfaced, and the number of parts to be hardfaced usually influence process selection. Cobalt-base alloys are available in a wide variety of different product forms for depositing hardfacing overlays (bare cast rod, coated electrodes, tubular wires, solid wires, and powder).

Most welding processes are readily adaptable to hardfacing if proper techniques are implemented to prevent stress cracking and to minimize base metal dilution. Stress cracking generally occurs as a result of thermally induced stresses. The occurrence of thermally induced stress can be minimized by the liberal use of preheat, high interpass temperature, and very slow cooling. The hardfacing of transformation-hardenable steels, such as type 410 stainless steel, can compound the stress operating on the hardfacing deposit during cooling and can require special precautions to minimize cracking. Specifically, preheat and interpass temperatures should be maintained above the M_s temperature (temperature at which austenite begins to transform to martensite) of the steel. Depending on the size and mass of the part, a postweld heat treatment immediately after hardfacing may also be required to minimize cracking.

Three welding processes that have been extensively used for hardfacing with cobalt-base alloys are oxyacetylene, gas tungsten arc, and plasma-transferred arc.

The oxyacetylene process is characterized by the lowest achievable base metal dilution (<5%); unfortunately, the process is relatively slow and time consuming, depositing only about 1 kg/h (2.2 lb/h) of hardfacing deposit. Furthermore, the low-carbon cobalt-chromium-molybdenum alloys such as Stellite alloy 21 intrinsically do not have good oxyacetylene weldability. Stellite alloy 6 can be deposited by oxyacetylene. However, proper melting practice and chemistry control are required during the manufacture of the cast rod in order to produce consumables that do not generate porosity during oxyacetylene deposition.

Gas Tungsten Arc Welding. Oxyacetylene methods have given way in many cases to the gas tungsten arc processes, especially when hardfacing austenitic stainless steels, which will sensitize if exposed to a carburizing oxyacetylene flame. Because gas tungsten arc is a more intense heat source, more base metal dilution (~20%) can be expected. However, overall dilution can usually be minimized by using two or more layers of hardfacing deposit. Hot cracking can be a potential problem in gas tungsten arc hardfacing. Hot cracking may be caused by high levels of deleterious elements, such as sulfur. Attempts to hardface a free-machining steel, such as type 303 or 303Se, may result in hot cracking because the presence of harmful elements will be introduced to the deposit through base metal dilution.

The plasma-transferred arc process is characterized by the ideal combination of relatively low base metal dilution (~10%) and a relatively high deposition rate (up to 5 kg/h, or 11 lb/h). In the plasma-transferred arc process, powder is used as the consumable, rather than a cast welding rod.

The plasma-transferred arc process is a mechanized rather than a manual process. It can be described as follows. A tungsten electrode, recessed into a torch body, generates a transferred arc to the workpiece. Plasma gas (usually argon) is ionized within the torch and exits through a constricted orifice. At this location, the hardfacing filler material is introduced in powder form through powder injection ports assisted by an argon carrier gas. The powder particles melt completely and resolidify as a fusion welded overlay. The weldability of cobalt-base alloy powders for plasma-transferred arc hardfacing is very good, and they will usually produce clean, smooth, sound deposits.

Hardfacing by Thermal Spray

Hardfacing techniques that use welding as the method of deposition should always attempt to minimize base metal dilution for the obvious reason that excessive base metal dilution will compromise the metallurgical effectiveness of the hardfacing alloy. The deposition of cobalt-base alloys by thermal spray methods offers the advantage of no base metal dilution because most spray processes do not melt the substrate material. Cobalt hardfacing alloys in powder form can be deposited by the conventional flame spray

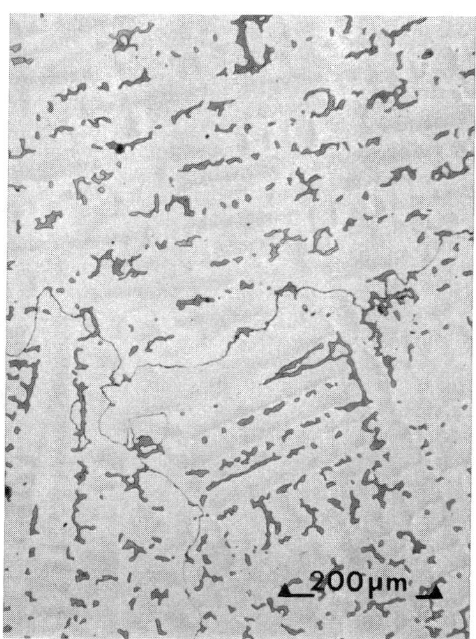

Fig. 4 Microstructure of investment cast cobalt-chromium-molybdenum alloy. Note the relatively coarse carbides and the large grain size (ASTM macrograin size 7.5). Compare with Fig. 5.

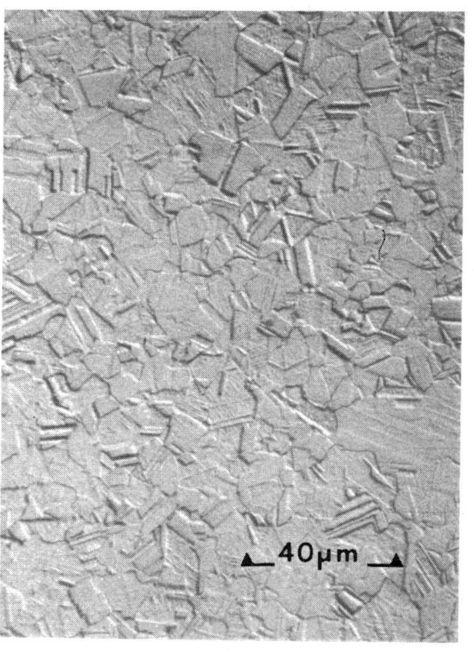

(a)

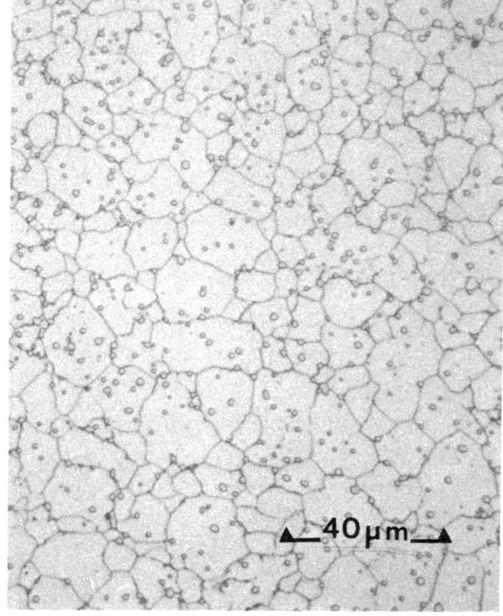

(b)

Fig. 5 Microstructures of high-strength fine-grain cobalt-chromium-molybdenum alloy. (a) Forged. (b) HIP powder. Note the decrease in grain size from Fig. 4.

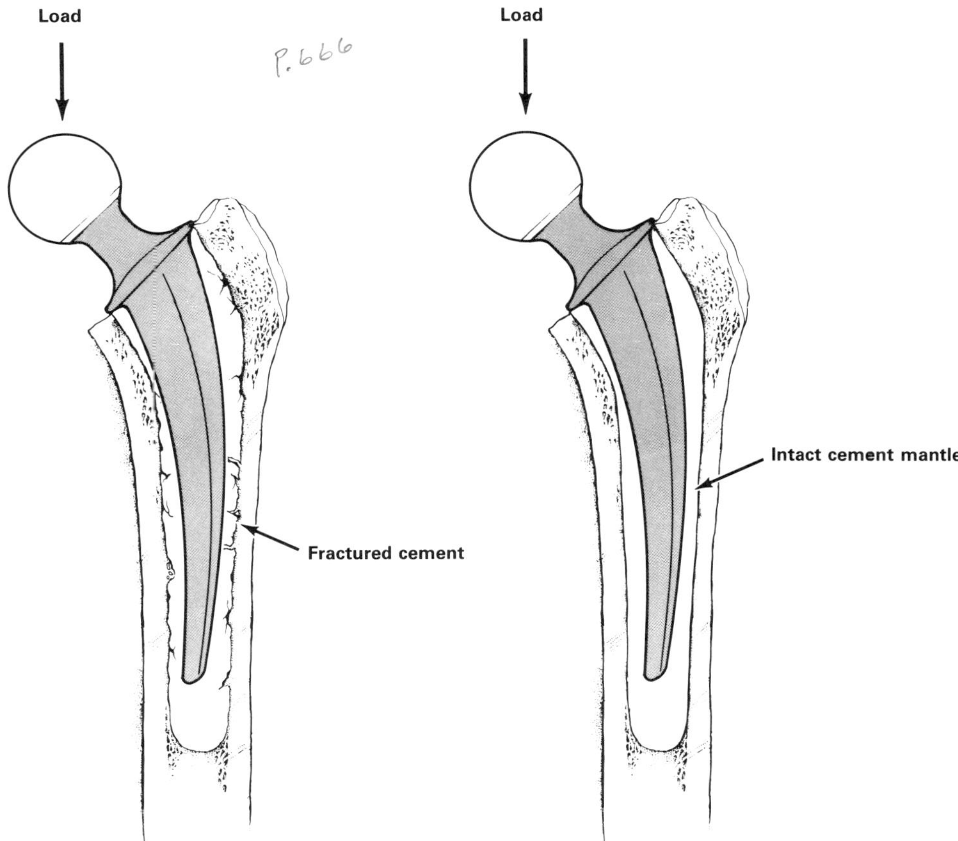

Fig. 6 Fracture of the cement mantle around a femoral hip stem results in increased stresses in the distal region of the stem. Loads applied to hips with fractured cement (a) act over a longer lever arm; and on a small cross section. This can lead to fatigue failure of the distal portion of the stem. The load applied to a hip with intact cement (b) acts over a short lever arm and is distributed over a large area.

process. The flame spray process is usually followed by a second fusing operation with an oxyacetylene torch. For this reason, cobalt-base alloys intended for spray and fuse deposition are modified with intentional additions of boron to lower the melting point and to allow for good fusing.

Cobalt-base alloy hardfacing powders that are not modified with boron can be deposited by other thermal spray processes, but in these cases, higher-energy processes are used, such as plasma spray, detonation gun, or Jet Kote. All three of these processes are designed to achieve an extremely high velocity gas stream into which the hardfacing alloy powders are introduced. Powder particles pick up heat and kinetic energy from the high-velocity stream and are driven against the substrate surface; this produces an extremely dense coating. Coating densities, however, never achieve full theoretical density, and some degree of porosity is intrinsic to this type of hardfacing.

For maximum effectiveness in corrosion environments, thermal spray coatings are generally sealed with a suitable sealer, such as epoxy. More information on thermal spray materials and processes is available in the articles "Thermal Spray Coatings" in this Volume, "Metal Powders Used for Hardfacing" in Volume 7 of the 9th Edition of *Metals Handbook*, and "Thermal Spray Coating" in Volume 5 of the 9th Edition of *Metals Handbook*.

Applications*

Phillip J. Andersen
Zimmer
Paul Crook
Haynes International, Inc.

Cobalt-base alloys are used in applications that require good corrosion and wear resistance. These alloys generally possess good high-temperature strength and therefore are used in such applications as jet engine turbines and gas turbine generators. The good wear resistance, fatigue strength, and biocompatibility of cobalt-base alloys have prompted their use as orthopedic implants. Cobalt-base alloys are also used in the nuclear power and chemical-processing industries. These latter three applications will be discussed in more detail in this article.

Orthopedic Implants

Cobalt-base alloys are widely used for the fabrication of various devices that are surgically implanted in the body. These devices must be highly reliable. The failure of certain implants (for example, heart valves) can be fatal, and the failure of other implants can necessitate further surgery. The longevity of implants is therefore a critical requirement. The lifetime of a device is influenced by a variety of factors, including patient variables (weight, activity level, compliance with physician instructions), surgical technique, and device variables (design, materials, manufacturing practice).

Implant materials must meet exacting requirements if the device is to perform its function successfully for extended periods of time. These requirements include biocompatibility, mechanical strength, corrosion and wear resistance, and approval by various regulatory agencies. Only a few materials meet these requirements at this time. The vast majority of metallic implants are made from type 316L stainless steel (covered by ASTM F 138), cobalt-chromium-molybdenum alloys (ASTM F 75, F 799), MP35N (cobalt-nickel-chromium-molybdenum alloy) (ASTM F 562), unalloyed titanium, and Ti-6Al-4V (ASTM F 67 and F 136). More information on materials for orthopedic implants and corrosion in biomedical applications is available in the article "Corrosion of Metallic Implants and Prosthetic Devices" in this Volume.

The use of cobalt-base alloys as orthopedic implants dates back at least as far as 1924, when Stellite alloy was implanted in dogs by A.R. Zierold (Ref 27). C.S. Venable and W.G. Stuck implanted cobalt-chromium bone screws in animals in 1937 and observed positive results (Ref 28). Similar experimental results were observed in subsequent experiments. Dr. Smith-Petersen implanted "moulds" made from the cobalt-base alloy Vitallium in humans in 1938. These moulds were basically a shell that was placed over the femoral head to reduce pain and to increase mobility (Ref 29). Investment cast cobalt-base alloys were subsequently used for endoprostheses in 1950 (Ref 30). An endoprosthesis consists of a large ball that replaces the femoral head attached to a shaft that extends down into the intermedullary canal of the femur. Endoprostheses remain in wide use today.

More recently, total hip replacement has become common. In this surgery, a plastic socket is placed into the acetabulum, while a metallic hip stem consisting of a shaft and ball is placed in the femur. The shaft is often held in place by a poly (methyl methacrylate) (PMMA) cement that acts like a grout between the bone and the stem. This ball and socket arrangement replaces a diseased joint to provide mobility and pain relief. These types of devices are shown in Fig. 3.

Cobalt-base alloys are also widely used in total knee replacements and, to a lesser extent, as implants that fix bone fractures, that is, bone screws, staples, and plates. The support structures for heart valves are also often fabricated from cobalt-base alloys. A variety of dental implants have also been produced from cobalt alloys (see the article "Tarnish and Corrosion of Dental Alloys" in this Volume).

Alloys Used

Most of the cobalt-base alloys currently in use as implants meet the requirements of ASTM F 75, F 799, or F 562. Standards F 75 and F 799 describe requirements for cobalt-chromium-molybdenum alloys, while F 562 describes requirements for a cobalt-chromium-nickel-molybdenum

*The section on orthopedic implants was written by Phillip J. Andersen; the sections on nuclear and chemical-processing applications were written by Paul Crook.

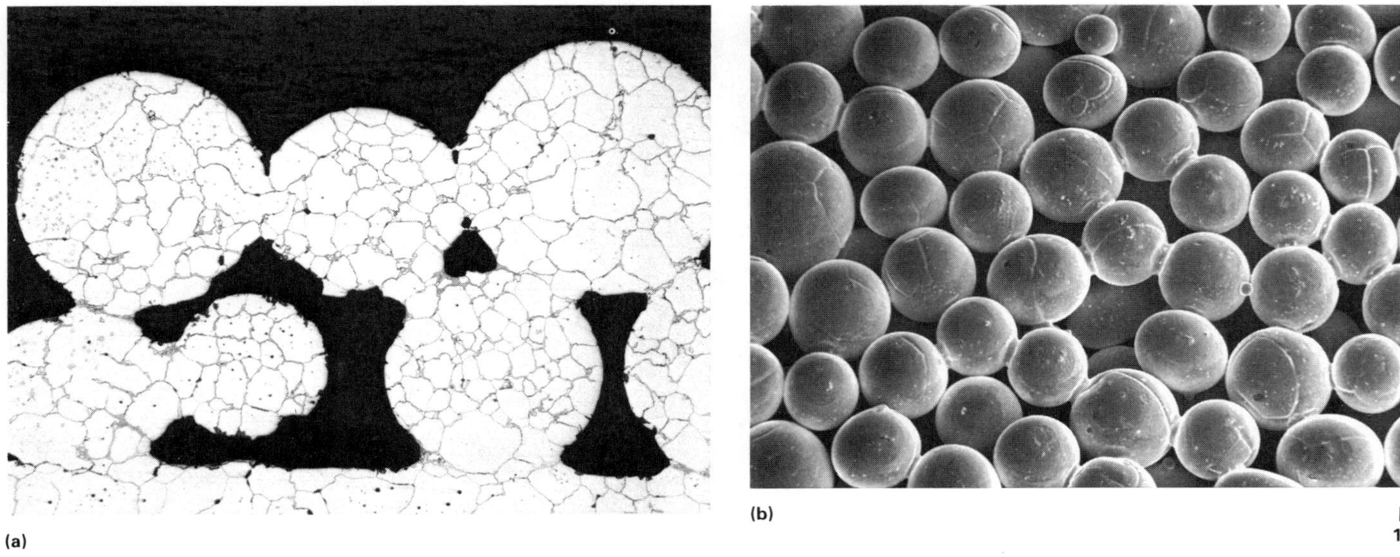

Fig. 7 Micrographs of a porous-coated cobalt-chromium-molybdenum implant material. (a) Cross section of coating; voids within the coating and the substrate are caused by incipient melting of carbides during the sintering process. (b) SEM micrograph of the porous layer. Note the interparticle necking.

Table 13 Specified minimum mechanical properties of cobalt-base alloys used for orthopedic implants

Alloy designation	Tensile strength MPa	Tensile strength ksi	Yield strength, 0.2% offset MPa	Yield strength, 0.2% offset ksi	Elongation, %	Reduction of area, %
F 75	655	95	450	65	8	8
F 799	1172	170	827	120	12	12
F 562						
Solution annealed(a)	793–1000	115–145	241–448	35–65	50	65
Cold-worked and aged(b)	1793 min	260 min	1586 min	230 min	8	35

(a) Annealed 1 to 2 h at 1050 ± 15 °C (1925 ± 25 °F), air cooled, and water quenched. (b) Cold worked 53%, aged 4 h at 540 to 650 ± 15 °C (1000 to 1200 ± 25 °F), and air cooled

Table 14 Reported fatigue strengths of cobalt-base alloys

Alloy	Test technique	10^7 cycle endurance limit MPa	10^7 cycle endurance limit ksi	Ref
Cast F 75	Rotating beam	267	38.7	36
		310	45	31
Forged F 799	Rotating beam	793–966	115–140	31
HIP P/M F 799	Flat plate bending	765	111	32
Forged F 562	Rotating beam	515	75	37
Cold-worked and aged F 562	Tension-tension	862	125	38
Cold-worked and aged F 562	Rotating beam	669	97	38

alloy. The compositions of these alloys are listed in Table 12.

The cobalt-chromium-molybdenum alloy is available with either carbide strengthening or nitrogen solid-solution strengthening. Investment cast cobalt-chromium-molybdenum alloys are carbide strengthened; a typical investment cast microstructure is shown in Fig. 4. The grains in this microstructure are quite coarse, and the carbides are large. Higher-strength fine-grain cobalt-chromium-molybdenum devices have been made by forging of nitrogen-strengthened bar stock or by hot isostatic pressing (HIP) of carbide-strengthened powders (Ref 31, 32). Microstructures of these materials are shown in Fig. 5.

The cobalt-nickel-chromium-molybdenum alloy MP35N has also been successfully used as an implant. This alloy can be strengthened by work hardening and by a precipitation reaction. The alloy was widely used in the mid-1970s to early 1980s, but its popularity in the United States has declined recently. This appears to be at least partially due to concerns about nickel release from the alloys, which can cause metal sensitivity reactions in some patients.

Manufacturing Techniques

The earliest cobalt-base implants were made by classical investment-casting techniques. This technology is well suited to a variety of medical device applications. Unfortunately, the fatigue failure of a small percentage of cast cobalt-chromium-molybdenum total hip implants has been observed.

Fatigue failure typically occurs after the PMMA cement layer around the implant degrades, leaving only the tip of the stem fixed. Under these conditions, there is a long moment arm acting on a relatively small implant cross section. This is shown schematically in Fig. 6. A loose implant is generally cause for reoperation prior to mechanical failure, but occasionally, the fatigue limit is exceeded and the device fails.

Several approaches are available for improving the mechanical properties of cobalt-chromium-molybdenum alloys in order to provide a greater margin of safety. Finer grain size materials that are free of residual casting porosity and inclusions have dramatically improved high-cycle fatigue strength compared to the as-cast alloy. These structures can be achieved by closed die forging or by powder metallurgy techniques such as HIP. The fatigue limit of these alloys is two to three times higher than that of cast cobalt-chromium-molybdenum. Closed die forgings are currently the most common approach to producing total hip implants with high fatigue strength. These techniques have been applied since the mid-1970s. Only one failure has been reported out of the thousands of hip stems fabricated by these methods (Ref 33).

The most recent development in total hip implants is the application of porous coatings to certain regions of the implant. The porous coating can be used either to enhance fixation with the PMMA cement or to provide sites that bone or fibrous tissue can grow into. In the latter case, cement is not used; this eliminates the potential breakdown of cement over long periods of time. In cobalt-base devices, relatively large spherical powders (250 to 1000 μm in diameter) are sintered onto the implant to create a porous layer. Porous-coated implants are typically investment cast; forged implants are not used, because the sintering cycles cause grain growth and a dramatic reduction in fatigue strength. Figure 7 shows an SEM micrograph and a cross-sectional view of a porous-coated cobalt-chromium-molybdenum component.

Properties of Cobalt Alloy Implants

The properties of interest for orthopedic implant materials include biocompatibility, corro-

Table 15 Some applications for cobalt-base alloys in the chemical and food processing industries

Application	Environment	Alloy	Form	Wear Mode
Mixer for glass production (linings and bearings)	Dry ingredients (Sand, soda ash, feldspar, borax, barium)	Haynes alloy 6B	Wrought plates and half sleeves	Abrasion
Hot asphalt agitator bearings	Hot asphalt at 480 °C (900 °F)	Haynes alloy 6B	Wrought half sleeves	Abrasion/erosion
Catalytic reactor valves (butadiene production)	Butane at 650 °C (1200 °F)	Stellite alloy 6	Weld overlay	Galling
Scraper blades on rotary filters for TiO_2 production	Muds containing dilute H_2SO_4	Haynes alloy 6B	Wrought plate	Abrasion
Pug mill paddles and extrusion screws in catalyst manufacture	Kaolin clay and H_2SO_4	Stellite alloy 1	Weld overlay	Abrasion
Screw conveyor bearings in household cleaner production	Silica	Haynes alloy 6B	Wrought half sleeves	Abrasion
Cutters for synthetic fibers	Polyesters	Haynes alloy 6B	Wrought plate	...
Meat grinder blades	Meat	Stellite alloy 19PM	P/M	...
Extrusion cooker screw flights	Cereal doughs at up to 205 °C (400 °F)	Stellite alloy 12	Weld overlay	Abrasion
Screws for oil extraction	Vegetable oils, seeds	Stellite alloy 12	Weld overlay	Abrasion
Homogenizer valves	Various liquids	Stellite alloy 6	Weld overlay	Cavitation, erosion

sion resistance, wear resistance, and such mechanical properties as strength, ductility, and high-cycle fatigue behavior.

Biocompatibility is a measure of the interaction between the living host and the implanted material. All implants cause some response because the body recognizes them as foreign objects. The good biocompatibility of cobalt-chromium-molybdenum components was a major factor in the early interest in this material. Cobalt-chromium-molybdenum is well tolerated by the body. Typically, a thin, fibrous layer is formed when these alloys are implanted in direct apposition to bone; more severe reactions to these materials are extremely rare.

There is some controversy regarding the biocompatibility of porous-coated cobalt-base implants (Ref 34, 35). This concern is due to the greater surface area of the porous layer and the fact that the porous-coated implant can be in direct contact with bone. However, in traditional cemented implants, most of the device is isolated from bone by the cement layer. Thus, porous-coated implants can result in potentially higher release of metal ions (especially chromium and nickel) into the surrounding tissues.

Corrosion resistance is a key aspect of the biocompatibility of a material. The corrosion resistance of all implanted metals depends on the formation of passive films. In the cobalt-base alloys, a Cr_2O_3 film is formed. More information on the formation of passive films in implant materials is available in the article "Corrosion of Metallic Implants and Prosthetic Devices" in this Volume.

Wear behavior is important because artificial joints are subject to millions of cycles during their lifetimes. Total knee and hip components have a highly polished metal surface that articulates on an ultrahigh molecular weight polyethylene bearing surface. Retrieved cobalt-chromium-molybdenum implants generally show very low amounts of wear.

The mechanical properties of cobalt-base alloys depend on processing history. The tensile properties required by ASTM specifications for cast and wrought cobalt-base implant alloys are shown in Table 13. Fatigue properties are extremely important in such applications as total hip replacements. High-cycle fatigue endurance limits measured by several investigators using various test methods are shown in Table 14. The wrought products have fatigue strengths that are two to three times higher than those of cast or porous-coated cobalt-chromium-molybdenum.

Nuclear Power Applications

Despite the maintenance hazards imposed by irradiated cobalt (which, under neutron bombardment, is transformed to a radioactive isotope with a long half-life), wear alloys of the cobalt-chromium type have been extensively used to protect the sealing surfaces of steam control and safety valves within the nuclear industry. Their use in these valves arises from their outstanding resistance to galling, steam erosion, and, to a lesser extent, oxidation/corrosion. Stellite alloys 21 and 6 are the most popular for valve facing. Application techniques include tungsten inert gas (with cast rod) and plasma-transferred arc welding (with powder).

The wrought version of Stellite alloy 6 has been used for control shaft guide bushings in sodium-cooled reactors. Resistance to galling and the corrosive effects of liquid sodium (at operating temperatures up to 900 °C, or 1650 °F) were the factors involved in the selection of this alloy.

Away from the reactor core, the cobalt-chromium high-temperature alloy Haynes alloy 25 has been used in the form of tubing to construct nitrogen gas heaters as part of a closed-loop gas turbine system. In this case, the cobalt alloy was selected for its high-temperature strength and resistance to nitriding at elevated temperatures. More information on materials and corrosion in nuclear power plants is available in the article "Corrosion in the Nuclear Power Industry" in this Volume.

Chemical-Processing Industry

Within the chemical- (and food) processing industry, cobalt alloys play a vital role in protecting critical components from the combined effects of wear and corrosion. In particular, they are extensively used to protect control valve sealing faces and pump seals and bearings. Normally, they are applied as weld overlay coatings, but they are also used in the form of small castings, wrought products, and small powder metallurgy parts. Examples of applications and environments in which the cobalt alloys have been successfully used in the chemical-processing industry are given in Table 15. Detailed information on materials used and corrosion in chemical-processing applications is available in the article "Corrosion in the Chemical Processing Industry" in this Volume.

REFERENCES

1. H.H. Uhlig and A.I. Asphahani, *Mater. Perform.*, Vol 18 (No. 11), 1979, p 9
2. A.P. Bond and H.H. Uhlig, *J. Electrochem. Soc.*, Vol 107, 1960, p 488
3. F. Wever and V. Hashimoto, *Die Korros. Metall. Werkst.*, Vol 2, 1938, p 745
4. A. Acharya, E. Freise, and E.H. Greener, *Cobalt*, Vol 47, June 1970, p 75
5. A. Davin and D. Coutsouradis, *Cobalt*, Vol 52, Sept 1971, p 160
6. N. Sridhar, Paper 19, presented at Corrosion/87, San Francisco, CA, National Association of Corrosion Engineers, March 1987
7. "Standard Recommended Practice for Laboratory Immersion Corrosion Testing of Metals," G 31, *Annual Book of ASTM Standards*, American Society for Testing and Materials
8. "Standard Test Methods for Pitting and Crevice Corrosion Resistance of Stainless Steels and Related Alloys by the Use of Ferric Chloride Solution," G 48, *Annual Book of ASTM Standards*, American Society for Testing and Materials
9. "Standard Practice for Conducting Cyclic Potentiodynamic Polarization Measurements for Localized Corrosion," G 61, *Annual Book of ASTM Standards*, American Society for Testing and Materials
10. "Standard Test Method for Pitting at Crevice Corrosion of Metallic Surgical Implant Materials," F 746, *Annual Book of ASTM Standards*, American Society for Testing and Materials
11. E.P. Whelan, in *Hydrogen Effect in Metals*, I.M. Bernstein and A.W. Thompson, Ed., The Metallurgical Society, 1981, p 979
12. A.I. Asphahani, Paper presented at the Second International Congress on Hydrogen in Metals, Paris, 1977
13. J.P. Stroup, A.H. Bauman, and A. Simkovich, *Mater. Perform.*, Vol 15, June 1976, p 43
14. R.D. Kane, *Corrosion*, Vol 34 (No. 12), 1978, p 442
15. R.D. Kane and B.J. Berkowitz, *Corrosion*, Vol 36 (No. 1), 1980, p 29
16. R.D. Kane, M. Watkins, N.F. Jacobs, and G.L. Hancock, *Corrosion*, Vol 33 (No. 9), 1977, p 309
17. J. Kolts, Paper 407, presented at Corrosion/86, National Association of Corrosion Engineers, 1986

18. R.D. Kane and J.B. Greer, Paper SPE 6798, presented at Society of Petroleum Engineers, 1977
19. J. Kolts, Paper 323, presented at Corrosion/86, National Association of Corrosion Engineers, 1986
20. R.D. Kane *et al.*, Paper 174, presented at Corrosion/79, National Association of Corrosion Engineers, 1979
21. G.A. Vaughn and H.-E. Chaung, Paper 182, presented at Corrosion/81, National Association of Corrosion Engineers, 1981
22. A.I. Asphahani, in *Stress Corrosion Cracking: The Slow Strain Rate Technique*, STP 665, American Society for Testing and Materials, 1979, p 279
23. A. Goldberg and R.P. Kershaw, "Evaluation of Materials Exposed to Scale Control/Nozzle-Exhaust Experiments at the Salton Sea Geothermal Field," Report UCRL 52664, Lawrence Livermore Laboratory, Feb 1979
24. A.I. Asphahani, Paper 42, presented at Corrosion/78, National Association of Corrosion Engineers, 1978
25. E. Taylor, *Mater. Prot.*, March 1970, p 29
26. S.J. Matthews, M.O. Maddock, and W.F. Savage, How Copper Surface Contamination Affects Weldability of Cobalt Superalloys, *Weld. J.*, Vol 51 (No. 5), May 1972, p 326-328
27. A.R. Zierold, Reaction of Bone to Various Metals, *Arch. Surg.*, Vol 9, 1924, p 365-412
28. C.S. Venable and W.G. Stuck, Electrolysis Controlling Factor in the Use of Metals in Treating Fractures, *JAMA*, Vol 111 (No. 15), 1938, p 1349-1352
29. O.E. Aufranc, Constructive Hip Surgery With the Vitallium Mold, *J. Bone Joint Surg.*, Vol 39A (No. 2), 1957, p 237-248
30. D.C. Mears, *Materials and Orthopaedic Surgery*, Williams & Wilkins, 1979, p 17
31. S. Weisman, Vitallium FHS Forged High-Strength Alloy, in *Current Concepts of Internal Fixation of Fractures*, H. Uhthoff, Ed., Springer Verlag, 1980, p 118
32. D.I. Bardos, High Strength Co-Cr-Mo Alloy for Prostheses, in *Current Concepts of Internal Fixation of Fractures*, H. Uhthoff, Ed., Springer Verlag, 1980, p 111
33. E.H. Miller, R. Shastri, and C.I. Shih, Fracture Failure of a Forged Vitallium Prosthesis, *J. Bone Joint Surg.*, Vol 64-A (No. 9), 1982, p 1360-1363
34. J.L. Woodman, R.M. Urban, K. Lim, and J.O. Galante, Cobalt, Chromium and Nickel Release From Porous Coated Cast Cobalt-Chromium Alloy, in *Transactions of the 30th Annual Meeting of the Orthopaedic Research Society*, Vol 9, 1984, p 150
35. L.C. Jones and D.S. Hungerford, Urinary Metal Ion Levels in Patients Implanted with Porous Coated Total Hip Prostheses, in *Transactions of the 33rd Annual Meeting of the Orthopaedic Research Society*, Vol 12, 1987, p 317
36. F.S. Georgette and J.A. Davidson, Effect of HIPing on the Fatigue and Tensile Strength of a Cast Porous Coated Co-Cr-Mo Alloy, *J. Biomed. Mater. Res.*, Vol 20 (No. 8), Oct 1986, p 1236
37. M. Semlitsch, Metallurgical and Clinical Experience With Cast and Forged Cobalt/Chromium Base Implant Metals of Compound Construction for Artificial Joint Endoprostheses, *Reconstruction Surgery and Traumatology*, C. Chapchal, Ed., Verlag S. Karger, 1976, p 82-101
38. Multiphase MP35N Technical Data, Latrobe Steel Company

SELECTED REFERENCES

- S.J. Matthews and P. Crook, Hardfacing Materials and Processes for Valve Applications, in *Proceedings of the International Conference on Welding Technology for Energy Applications*, ORNL CONF-820544, Oak Ridge National Laboratory, 1982, p 127-135
- S.J. Matthews, R.D. Zordan, and P. Crook, Laboratory Engineers Solutions to Unlubricated Wear Problems, *Met. Prog.*, July 1984, p 57-67

Corrosion of Titanium and Titanium Alloys

Ronald W. Schutz, TIMET Corporation
David E. Thomas, RMI Company

TITANIUM ALLOYS were originally developed in the early 1950s for aerospace applications, in which their high strength-to-density ratios were especially attractive. Although titanium alloys are still vital to the aerospace industry for these properties, recognition of the excellent resistance of titanium to many highly corrosive environments, particularly oxidizing and chloride-containing process streams, has led to widespread nonaerospace (industrial) applications. Because of decreasing cost and the increasing availability of titanium alloy products, many titanium alloys have become standard engineering materials for a host of common industrial applications (Ref 1). In fact, a growing trend involves the use of high-strength aerospace-founded titanium alloys for industrial service in which the combination of strength to density and corrosion resistance properties is critical and desirable.

The objectives of this article are severalfold and include:

- Describing and characterizing the relevant modes of corrosion observed on titanium alloys
- Providing a comprehensive overview of the available corrosion data base on titanium alloys
- Providing basic explanations, relevant summaries, and guiding comments relative to the data base, where applicable
- Offering practical strategies for expanding the useful corrosion resistance of titanium in corrosive environments

References are provided as sources of additional information.

The designations and nominal compositions of commercial titanium alloys addressed in this article are listed in Table 1. The first eight alloys listed are titanium alloys that are commonly used in industrial applications in which corrosion resistance is often of primary concern. It is for this reason that most of the corrosion data presented in this article applies to this group of alloys. With the exception of the Ti-6Al-4V alloy, these alloys consist of single α-phase (hexagonal close-packed crystal structure) or near-α alloys containing relatively small amounts of β phase (body-centered cubic crystal structure) in an α matrix.

The rest of the titanium alloys were developed for aerospace purposes, in which significantly increased strengths are achieved by solid-solution alloying and stabilization of two-phase structures. These alloys, which contain varying amounts of stabilized β phase, are generally heat treatable (aged) to high strength levels. Titanium alloy metallurgy (Ref 2) and the mechanical and physical properties of titanium alloys have been extensively reviewed in the literature (Ref 3-7).

Mechanism of Corrosion Resistance

The excellent corrosion resistance of titanium alloys results from the formation of very stable, continuous, highly adherent, and protective oxide films on metal surfaces. Because titanium metal itself is highly reactive and has an extremely high affinity for oxygen, these beneficial surface oxide films form spontaneously and instantly when fresh metal surfaces are exposed to air and/or moisture. In fact, a damaged oxide film can generally reheal itself instantaneously if at least traces (that is, parts per million) of oxygen or water (moisture) are present in the environment. However, anhydrous conditions in the absence of a source of oxygen may result in titanium corrosion, because the protective film may not be regenerated if damaged.

The nature, composition, and thickness of the protective surface oxides that form on titanium alloys depend on environmental conditions. In most aqueous environments, the oxide is typically TiO_2, but may consist of mixtures of other titanium oxides, including TiO_2, Ti_2O_3, and TiO (Ref 8). High-temperature oxidation tends to promote the formation of the chemically resistant, highly crystalline form of TiO_2 known as rutile, whereas lower temperatures often generate the more amorphous form of TiO_2, anatase, or a mixture of rutile and anatase (Ref 8). Although these naturally formed films are typically less than 10 nm thick (Ref 9) and are invisible to the eye, the TiO_2 oxide is highly chemically resistant and is attacked by very few substances, including hot, concentrated HCl, H_2SO_4, NaOH, and (most notably) HF. This thin surface oxide is also a highly effective barrier to hydrogen, as discussed in a later section in this article.

Furthermore, the TiO_2 film, being an *n*-type semiconductor, possesses electronic conductivity. As a cathode, titanium permits electrochemical reduction of ions in an aqueous electrolyte. On the other hand, very high resistance to anodic current flow through the passive oxide film can be expected in most aqueous solutions. Because the

Table 1 Designations and nominal compositions of titanium alloys

Common alloy designation	UNS designation	Nominal composition, %	ASTM grade	Alloy type
Grade 1	R50250	Unalloyed titanium	1	α
Grade 2	R50400	Unalloyed titanium	2	α
Grade 3	R50550	Unalloyed titanium	3	α
Grade 4	R50700	Unalloyed titanium	4	α
Ti-Pd	R52400/ R52250	Ti-0.15Pd	7/11	α
Grade 12	R53400	Ti-0.3Mo-0.8Ni	12	Near-α
Ti-3-2.5	...	Ti-3Al-2.5V	9	Near-α
Ti-6-4	R56400	Ti-6Al-4V	5	α-β
Ti-6-2-1-.8	...	Ti-6Al-2Nb-1Ta-0.8Mo	...	Near-α
Ti-5Ta	...	Ti-5Ta	...	Near-α
Ti-5-2.5	...	Ti-5Al-2.5Sn	...	α
Ti-8-1-1	...	Ti-8Al-1V-1Mo	...	Near-α
Ti-6-2-4-2	...	Ti-6Al-2Sn-4Zr-2Mo	...	Near-α
Ti-4-3-1	...	Ti-4Al-3Mo-1V	...	α-β
Ti-550	...	Ti-4Al-2Sn-4Mo-0.5Si	...	α-β
Ti-6-6-2	...	Ti-6Al-6V-2Sn-0.6Fe-0.6Cu	...	α-β
Corona 5	...	Ti-4.5Al-1.5Cr-5Mo	...	α-β
Ti-6-2-4-6	R56260	Ti-6Al-2Sn-4Zr-6Mo	...	α-β
Ti-10-2-3	...	Ti-10V-2Fe-3Al	...	Near-β
Transage 129	...	Ti-2Al-11.5V-2Sn-10Zr	...	Near-β
Transage 207	...	Ti-2.5Al-2Sn-9Zr-8Mo	...	Near-β
Ti-15-3-3-3	...	Ti-15V-3Sn-3Cr-3Al	...	β
Ti-3-8-6-4-4	R58640	Ti-3Al-8V-6Cr-4Zr-4Mo	...	β
Ti-13-11-3	...	Ti-3Al-13V-11Cr	...	β
Ti-8-8-2-3	...	Ti-8V-8Mo-3Al-2Fe	...	β
Ti-15-5	...	Ti-15Mo-5Zr	...	β

passivity of titanium stems from the formation of a stable oxide film, an understanding of the corrosion behavior of titanium is obtained by recognizing the conditions under which this oxide is thermodynamically stable. The Pourbaix (potential-pH) diagram for the titanium-water system at 25 °C (75 °F) is shown in Fig. 1 and depicts the wide regime over which the passive TiO_2 film is predicted to be stable, based on thermodynamic (free energy) considerations. Oxide stability over the full pH scale is indicated over a wide range of highly oxidizing to mildly reducing potentials, whereas oxide film breakdown and the resultant corrosion of titanium occur under reducing acidic conditions. Under strongly reducing (cathodic) conditions, titanium hydride formation is predicted.

Thus, successful use of titanium alloys can be expected in mildly reducing to highly oxidizing environments in which protective TiO_2 and Ti_2O_3 films form spontaneously and remain stable. On the other hand, uninhibited, strongly reducing acidic environments may attack titanium, particularly as temperature increases. However, shifting the alloy potential in the noble (positive) direction by various means can induce stable oxide film formation, often overcoming the corrosion resistance limitations of titanium alloys in normally aggressive reducing media.

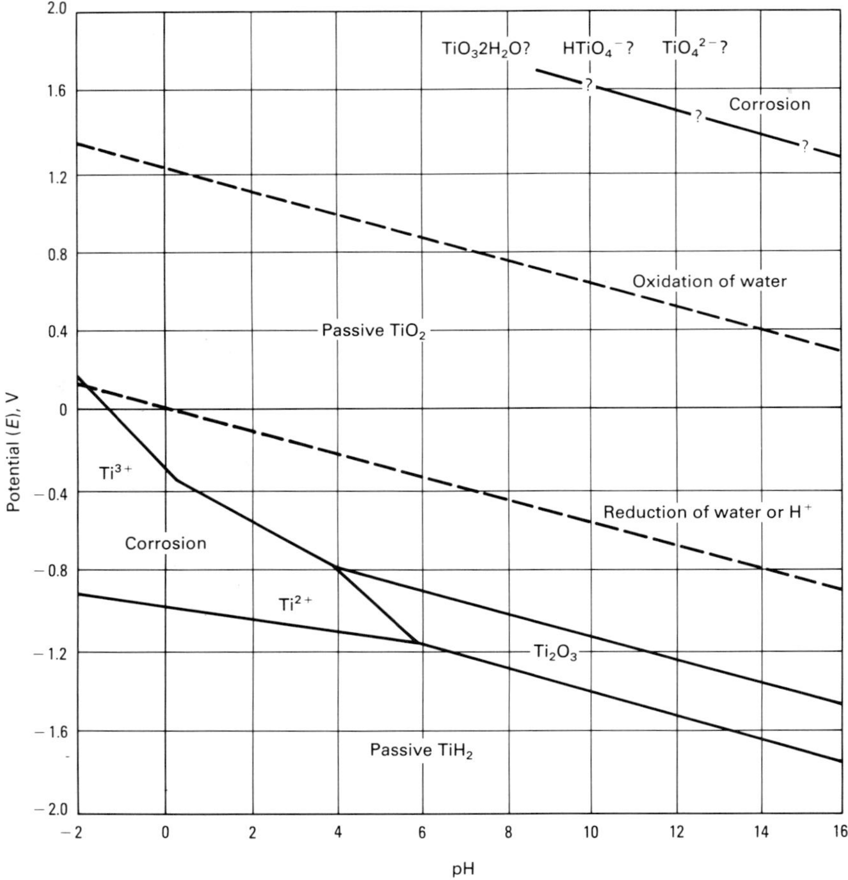

Fig. 1 Pourbaix (potential-pH) diagram for the titanium-water system at 25 °C (77 °F). Source: Ref 10

Enhancing Corrosion Resistance

The methods of expanding the corrosion resistance of titanium into reducing environments include:

- Increasing the surface oxide film thickness by anodizing or thermal oxidation
- Anodically polarizing the alloy (anodic protection) by impressed anodic current or galvanic coupling with a more noble metal in order to maintain the surface oxide film
- Applying precious metal (or certain metal oxides) surface coatings
- Alloying titanium with certain elements
- Adding oxidizing species (inhibitors) to the reducing environment to permit oxide film stabilization

Of these five methods, the last two have been very practical, effective, and most widely used in actual service.

Alloying titanium with precious metals (such as palladium), nickel, and/or molybdenum or coating with certain precious metals (or their oxides) facilitates cathodic depolarization by providing sites of low hydrogen overvoltage on alloy surfaces and by shifting alloy potential in the noble (positive) direction. For example, this is the basis for the significantly improved crevice corrosion resistance of titanium grades 7 and 12 in reducing acids and hot brines as compared to that of unalloyed titanium.

Various dissolved reducible (oxidizing) species in normally reducing media also serve to depolarize cathodic reactions on titanium alloy surfaces; this passivates the alloy by shifting the alloy potential in the noble direction. Many of these species, which include a host of multivalent transition metal ions, are very potent inhibitors and may be effective at concentrations of 100 ppm or less (see the section "Inhibition of Reducing Acid Corrosion" in this article). These inhibiting species often occur as natural process stream constituents or contaminants and need not be intentionally added to achieve complete titanium alloy passivation. More information on the five methods listed above are provided in the section "Expanding and Enhancing the Corrosion Resistance of Titanium" in this article.

Alloy Composition Effects. The nature of the oxide film on titanium alloys basically remains unaltered in the presence of minor alloying constituents; thus, small additions (<2 to 3%) of most commercially used alloying elements or trace alloy impurities generally have little effect on the basic corrosion resistance of titanium in normally passive environments. For example, despite small differences in interstitial element (carbon, oxygen, and nitrogen) and iron content, all unalloyed grades of titanium possess the same useful range of resistance in environments in which corrosion rates are normally very low (Ref 11, 12). However, under active conditions in which titanium exhibits significant general corrosion, certain alloying elements may accelerate corrosion. Increasing the alloy iron and sulfur content, for example, increases corrosion rates when corrosion rates exceed 0.13 mm/yr (5 mils/yr) (Ref 11, 12). Thus, minor variations in alloy chemistry may be of concern only under conditions in which the passivity of titanium is borderline or when the metal is fully active. On the other hand, minor nickel and palladium additions are highly effective in expanding the corrosion resistance of titanium alloys under reducing conditions.

The influence of certain major alloying elements on the general and crevice corrosion behavior of various commercial titanium alloys has been determined in reducing aqueous acid media (Ref 13). Results indicate that vanadium and, especially, molybdenum additions (≥4% Mo) improve corrosion resistance but that increasing the aluminum content appears to be detrimental. The influence of alloying elements on the resistance of titanium alloys to pitting and stress-corrosion cracking (SCC) is addressed in subsequent sections of this article.

Weldments (Ref 12) and castings (Ref 14, 15) of the first eight titanium alloys listed in Table 1 generally exhibit corrosion resistance similar to that of their unwelded, wrought counterparts. These titanium alloys contain so little alloy content and second phase that metallurgical instability and thermal response are not significant. Therefore, titanium weldments and associated heat-affected zones generally do not experience corrosion limitations in welded components when normal passive conditions prevail for the base metal. However, under marginal or active conditions (for corrosion rates ≥0.10 mm/yr, or 4 mils/yr), weldments may experience accelerated corrosion attack relative to the base metal, depending on alloy composition (Ref 12). The increasing impurity (iron, sulfur, oxygen) content associated with the coarse, transformed-β microstructure of weldments appears to be a factor. Few published data are available concerning the corrosion resistance of other α-β and β titanium alloy weldments and castings.

Forms of Corrosion and Related Test Methods

Titanium alloys, like other metals, are subject to corrosion in certain environments. The primary forms of corrosion that have been observed on these alloys include general corrosion, crevice corrosion, anodic pitting, hydrogen damage, and SCC. In any contemplated application of titanium, its susceptibility to degradation by any of these forms of corrosion should be considered. In order to understand the advantages and limitations of titanium alloys, each of these forms of corrosion will be explained. Although they are not common limitations to titanium alloy performance, galvanic corrosion, corrosion fatigue, and erosion-corrosion are included in the interest of completeness. More information on the mechanisms of all these forms of attack is available in the Section "Forms of Corrosion" in this Volume.

General Corrosion

General corrosion is characterized by a relatively uniform attack over the exposed surface of the metal. At times, general corrosion in aqueous media may take the form of mottled, severely roughened metal surfaces that resemble localized attack. This often results from variations in the corrosion rates of localized surface patches due to localized masking of metal surfaces by process scales, corrosion products, or gas bubbles; such localized masking can prevent true uniform surface attack. When titanium is in the fully passive condition, corrosion rates are typically much lower than 0.04 mm/yr (1.5 mils/yr)—well below the 0.13-mm/yr (5-mils/yr) maximum corrosion rate commonly accepted by designers. This very small, acceptable corrosion is attributable to the finite oxidation (typically TiO_2 film growth) of titanium alloy surfaces. As a result, titanium is most often designed with a zero corrosion allowance in normal passive environments. In many environments in which titanium is fully resistant, slight surface oxide growth may occur; this oxide growth manifests itself as colored surfaces and very slight weight gain by test coupons.

General corrosion becomes a concern in reducing acid environments, particularly as acid concentration and temperature increase. In strong and/or hot reducing acids (in the absence of inhibitors), the oxide film of titanium can deteriorate and dissolve, and the unprotected metal is oxidized to the soluble trivalent ion ($Ti \rightarrow Ti^{3+} + 3e^-$). This ion has a characteristic violet color in acid solutions. If dissolved oxygen or other oxidizing species are present in hot acid, the Ti^{3+} ion is readily oxidized to the less soluble (pale yellow) Ti^{4+} ion, which may subsequently hydrolyze to form insoluble TiO_2 precipitates (scales). Titanium ion hydrolysis often produces highly colored metal surfaces, involving thin titanium oxide films that may inhibit subsequent corrosion. Gray-matte or dull silver surface finishes can also be observed in reducing acid exposures involving severe corrosion attack. These are titanium hydride surface films, which are typically of the order of 0.05 mm (2 mils) thick.

General Corrosion Testing. General corrosion rates for titanium alloys can be determined from weight loss data, dimensional changes, and electrochemical methods. Electrochemical anodic and cathodic polarization testing is often used to supplement weight loss testing. Polarization testing can identify whether the alloy is truly fully passive or possibly metastable; this is often not discernible from weight loss tests alone. The immersion test procedures described in ASTM G 1 and G 31 apply, provided several modifications are observed (Ref 16). These modifications focus on test sample surface preparation and posttest sample-cleaning procedures.

The type of surface finish tested should resemble the one expected in service. For titanium alloys, this will often be the pickled finish, although sandblasted or ground surfaces are also common. The initial degreasing of test samples should avoid chlorinated organic solvents (with higher-strength titanium alloys), anhydrous methanol, or hot alkaline cleaners, if possible. Acceptable cleaning solvents include methyl ethyl ketone (MEK), acetone, most alcohols, benzene, and most detergent solutions. The pickled finish can be prepared by pickling the metal in a 35 vol% HNO_3-5 vol% HF (balance water) solution at 20 to 55 °C (70 to 130 °F) for several minutes or more. Typically, 0.02 to 0.05 mm (0.8 to 2 mils) of surface is removed in this process, depending on surface requirements. More dilute solutions, such as 12 vol% HNO_3-1 vol% HF, can also be used if slower pickling rates are desired. In any case, a minimum 7/1 HNO_3/HF vol% ratio should be maintained to avoid excessive uptake of hydrogen in titanium alloys during pickling. After pickling, a quick rinse in deionized water leaves a shiny specimen that is ready for weighing after air drying. Blasted and abraded surfaces are prepared by procedures similar to those used for other metals.

After laboratory or *in situ* test exposure, titanium samples can be coated with tenacious, insoluble corrosion product (TiO_2) films or scales, which require removal before final weighing. Because titanium oxides are not soluble in common mineral acids, very light (<5-s exposure) sandblasting has been found to be most effective. If scaling consists of silicaceous, carbonaceous, sulfate, or other typical process stream deposits, then acids or alkaline solutions that are properly inhibited with oxidizing species must be used; common amine inhibitors are not effective on titanium. Recommended cleaning solutions for these scales are discussed in detail in Ref 1. Serious consideration must be given to the presence of test medium contaminants that may significantly affect the corrosion rate of titanium. Metal ion contaminants or dissimilar-metal corrosion products can promote the passivity of titanium (see the section "Expanding and Enhancing the Corrosion Resistance of Titanium" in this article). The degree of aeration and other background chemistry variables in the test media that influence alloy passivity in service must be taken into account in order to avoid false test indicators.

Immersion testing will generate weight loss data, or corrosion current measurements can be obtained from electrochemical polarization tests (ASTM G 5). Corrosion rates in millimeters per year for titanium alloys can be calculated from weight loss data as follows:

$$\text{Corrosion rate} = \frac{(8.76 \times 10^4)\,(W)}{(d)(A)(t)} \qquad \text{(Eq 1)}$$

where d is the titanium alloy density (in grams per cubic centimeter), A is the sample surface area (in square centimeters), t is the exposure time (in hours), and W is the weight change (in grams).

Corrosion rates (mm/yr) can be calculated from electrochemical measurements by using Eq 2:

$$\text{Corrosion rate} = \frac{(0.0033)(i_{corr})(\text{EW})}{d} \qquad \text{(Eq 2)}$$

where i_{corr} is the measured corrosion current (in milliamps per square centimeter), d is alloy density (in grams per cubic centimeter), and EW is the equivalent weight for titanium. The equivalent weight for titanium is approximately 16 under reducing acid conditions and 12 under oxidizing conditions. The value of i_{corr} is typically determined from Tafel slope extrapolation or linear polarization methods (Ref 17, 18). More information on general corrosion testing is available in the article "Evaluation of Uniform Corrosion" in this Volume.

Crevice Corrosion

Titanium alloys may be subject to localized attack in tight crevices exposed to hot (>70 °C, or 160 °F) chloride, bromide, iodide, fluoride, or sulfate-containing solutions. Crevices can stem from adhering process stream deposits or scales, metal-to-metal joints (for example, poor weld joint design or tube-to-tubesheet joints), and gasket-to-metal flange and other seal joints.

The mechanism for crevice corrosion of titanium is similar to that for stainless steels, in which oxygen-depleted reducing acid conditions develop within tight crevices. The model for crevice corrosion is illustrated in Fig. 2. Dissolved oxygen or other oxidizing species in the bulk solution are depleted in the restricted volume of solution in the crevice. Finite surface oxidation in crevices consumes these species faster than diffusion from the bulk solution can replenish them (Ref 20). As a result, metal potentials in crevices become active (negative) relative to metal surfaces exposed to the bulk solution. This creates an electrochemical cell in which the crevices become anodic and corrode, and the surrounding metal surface is cathodic.

Titanium chlorides formed within the crevice are unstable and tend to hydrolyze, forming hydrochloric acid (HCl) and titanium oxide/hydroxide corrosion products. Because of the small, restricted volumes of solution in these crevices, crevice pH levels as low as 1 or below can develop. These local reducing acidic conditions can result in severe and rapid localized active corrosion within crevices, depending on alloy resistance and temperature.

Although dissolved oxidizing species such as oxygen, chlorine, ferric ion (Fe^{3+}), and cupric ion (Cu^{2+}) tend to inhibit effectively the general corrosion of exposed titanium surfaces, most of these species tend to accelerate the onset and propagation of titanium alloy crevice corrosion. These species are excellent cathodic depolarizers and thus accelerate cathodic reduction kinetics, which often are rate controlling. These cationic oxidizing species will not diffuse into the active crevice to effect passivation. On the other hand, certain anionic oxidizing species, such as NO_3^-, ClO_3^-, OCl^-, CrO_3^{2-}, ClO_4^-, and MnO_4^-, can migrate into the crevice and inhibit crevice attack when present in halide solutions.

Crevice corrosion on titanium typically generates irregularly shaped pits (Fig. 3). Microstructural examination of hand-polished and etched sections of crevices often reveals a surrounding layer of precipitated titanium hydride in α alloys. These are a by-product of hydrogen reduction at cathodic sites surrounding the crevice.

Although frequently interpreted as a pitting phenomenon, smeared surface iron pitting of unalloyed titanium in hot brines appears to be a special case of crevice corrosion (Ref 21). It

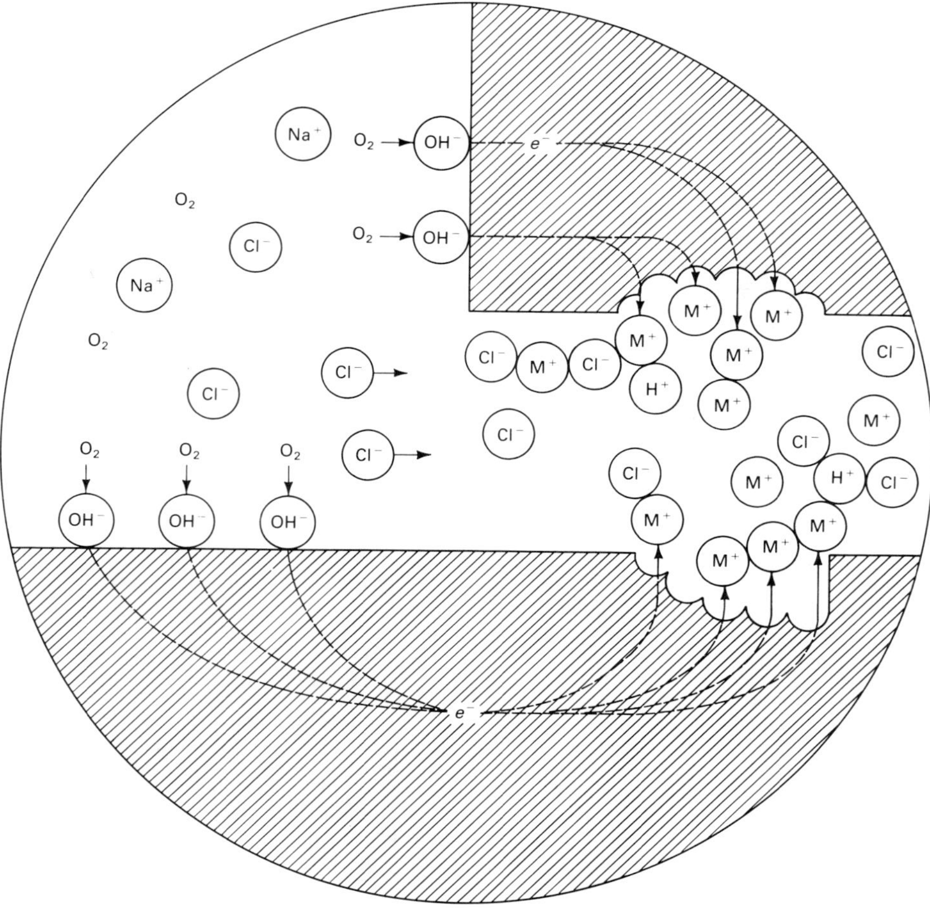

Fig. 2 Schematic showing the mechanism of crevice corrosion for titanium in aqueous chloride media. Source: Ref 19

results when iron, carbon steel, or low-alloy steel is gouged, scratched, smeared, and embedded into a titanium surface, breaching the titanium oxide film. During hot (>80 °C, or 175 °F) brine exposure, the embedded iron can either corrode off the surface and permit repassivation or develop local acidic conditions if occluded by titanium metal smears or laps. Localized attack initiated by this mechanism creates a very characteristic circular pit morphology (Fig. 4) and can involve local hydrogen absorption. Pit initiation has not been observed with copper, nickel, or austenitic stainless steel alloys smeared into titanium surfaces. Titanium grades 7 and 12 appear to be much more resistant to this form of localized attack.

Several highly effective strategies are available for preventing titanium alloy crevice corrosion and smeared iron pitting, as discussed in a subsequent section of this article. In all cases, the basic remedy aims at maintaining creviced metal surfaces at sufficiently noble potentials where titanium alloy passivity is assured.

Crevice corrosion testing of titanium alloys generally aims at determining go/no-go performance information. The rate of crevice corrosion is of little practical interest, because crevice attack is generally insidious and very rapid. Thus, crevice corrosion cannot be tolerated in any form. Many crevice test assemblies have been used, including the multiple-crevice washer (Ref 22). Unfortunately, crevice corrosion initiation is not very reproducible, especially when pH and temperature conditions are marginal and few or no cathodic depolarizers are present.

Because the susceptibility of titanium to crevice corrosion increases dramatically as creviced surface area increases and crevice gap decreases, the larger sandwich-type crevice test assembly illustrated in Fig. 5 has proved to be most effective (Ref 16). This approach employs an assembly consisting of 25- × 25-mm (1- × 1-in.) minimum flat sheet or plate specimen, with thin gasket sheets (typically Teflon) interspersed to provide the desired number of metal-to-metal and metal-to-gasket crevices per assembly. This assembly is fastened together with a centerline titanium bolt and nut and tightened with a controlled torque (typically 2.8 N · m, or 2 ft · lb). Titanium assembly bolts are covered with an insulating Teflon sleeve to avoid galvanic interactions between coupons. Coupons must be flat and surfaces must be smooth to ensure tight, uniform crevice gaps. The preparation of crevice test coupons involves considerations similar to those described for general corrosion test coupons.

Although crevice corrosion can actually initiate after several hours of exposure, test duration should be a minimum of 2 to 4 weeks in order to develop a sufficient degree of attack for detection and measurement purposes. Posttest evaluation of creviced coupons may require removal of titanium oxide/hydroxide corrosion products if crevice attack has occurred. Common mineral acids and reagents will not dissolve these tenacious deposits. A light (<5 to 8 s) sandblasting of the coupon will readily remove these scales and facilitate visual examination of pitted surfaces (ASTM G 46). Creviced coupon surfaces are often highly colored after exposure, but reveal no visible pits after sandblasting. This is indicative of the very slight growth of a protective surface titanium oxide film, which is considered quite normal and acceptable from a performance standpoint. Slight coupon weight gain is often measured in this situation. In addition to visual examination and weight change measurements, monitoring of creviced specimen potential (Ref 23) and current (Ref 24) has been used to a limited extent to identify the initiation of crevice corrosion. More information on testing of crevice corrosion specimens is available in the article "Evaluation of Crevice Corrosion" in this Volume.

Pitting

Pitting is defined as localized corrosion attack occurring on openly exposed metal surfaces in the absence of any apparent crevices. This pitting occurs when the potential of the metal exceeds the anodic breakdown potential of the metal oxide film in a given environment. When the anodic breakdown (pitting) potential of the metal is equal to or less than the corrosion potential under a given set of conditions, spontaneous pitting can be expected.

Because of its protective oxide film, titanium exhibits anodic pitting potentials, E_b, that are very high (≫1 V); thus, pitting corrosion is generally not of concern for titanium alloys. For example, pitting potentials exceed +80 V versus the saturated calomel electrode (SCE) in sulfate and phosphate solutions and are typically in the +5- to +10-V range for chlorides. Although pitting is normally not a limiting factor in titanium performance, pitting potential values provide useful guidelines for titanium for anode applications in which impressed anodic potentials may be high.

The anodic pitting potential of titanium is dependent on alloy content, medium chemistry, temperature, potential scan rate, and, especially, surface condition. A more intrinsic alloy property is the repassivation (protection) potential, which is defined as the minimum potential at which pitting can be maintained (Ref 25). This pitting parameter is not sensitive to surface condition or measuring technique artifacts, and it represents a more conservative design guideline than the anodic breakdown potential. The repassivation potentials of titanium alloys are also very high relative to the alloy corrosion potentials, and this explains why titanium alloys are generally resistant to pitting attack. Anodic pitting and repassivation potential values for titanium alloys under various conditions are presented in the section "Anodic Pitting" in this article.

Pitting Potential Testing. The potentiostatic (constant potential) and potentiodynamic (potential scan) electrochemical techniques used on other metals to measure anodic pitting potential apply to titanium alloys as well (Ref 18). Guidelines are described in ASTM G 3 and G 5. Determination of anodic pitting potential requires slow scan rates (−0.5 mV/s) and consideration of the surface condition tested. For example, abraded or sandblasted sample finishes will exhibit significantly lower pitting potentials than as-pick-

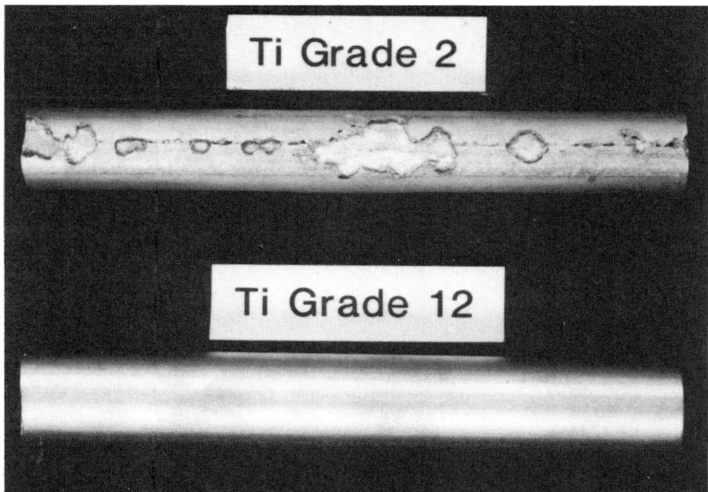

Fig. 3 Under-deposit (inside-diameter surface initiated) crevice corrosion of grade 2 titanium tubing in saturated NaCl brine. Grade 12 titanium is resistant in the same service.

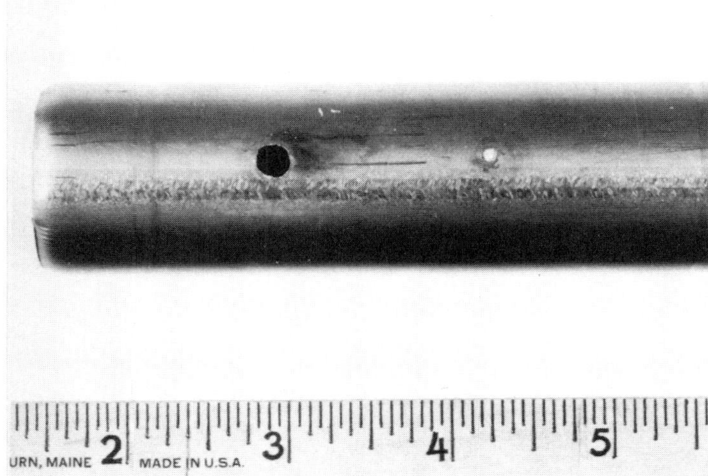

Fig. 4 Smeared surface iron pitting of unalloyed titanium tubing in hot brine service. Source: Ref 21

led surfaces. Also, because of the relatively high pitting potential of titanium, a potentiostat with a potential scan range of at least −2 to +10 V (SCE) is generally required. Recommended auxiliary electrodes include high-density graphite, glassy carbon, or platinum.

Repassivation potentials are readily determined by using the galvanostatic method (Ref 25) or the constant potential-surface scratch test (Ref 18, 25). The galvanostatic method involves impressing an anodic current density of approximately 200 mA/cm² (1290 mA/in.²) on the specimen for at least several minutes before measuring the repassivation potential of the sample. Reproducible, unambiguous repassivation potentials are more difficult to derive by using reverse scan potentiodynamic techniques.

Hydrogen Damage

Titanium alloys are widely used in hydrogen-containing environments and under conditions in which galvanic couples or cathodic charging (impressed current) causes hydrogen to be evolved on metal surfaces. Although excellent performance is revealed for these alloys in most cases, hydrogen embrittlement has been observed.

The surface oxide film of titanium is a highly effective barrier to hydrogen penetration. Traces of moisture or oxygen in hydrogen gas containing environments very effectively maintain this protective film, thus avoiding or limiting hydrogen uptake (Ref 26-29). On the other hand, anhydrous hydrogen gas atmospheres may lead to absorption, particularly as temperatures and pressures increase.

In α and α-β alloys, excessive hydrogen uptake can induce the precipitation of titanium hydride in the α phase. These acicular-appearing hydride platelets (Fig. 6) are brittle and have been well characterized in the literature (Ref 30-32). Small amounts of hydride precipitates are not detrimental from an engineering standpoint in most cases, but cause severe reduction in alloy ductility and toughness when present in greater amounts. For example, hydride precipitates can be observed in grade 2 titanium microstructures at hydrogen concentrations above approximately 100 ppm, depending on the amount of β phase present, but these precipitates do not result in gross embrittlement of grade 2 titanium until levels in excess of 500 to 600 ppm are achieved.

Although uniaxial tensile properties may experience little effect from increasing hydrogen levels, biaxial or triaxial stress properties, such as bend ductility, cup (cold-drawing) formability, and impact toughness, in α and near-α alloys are very sensitive to hydrogen levels (Ref 31-36). In α and, especially, α-β alloys, hydrogen contents above critical levels can result in sustained-load cracking, which dramatically reduces useful maximum service loads in notched or cracked components under slow strain rate or constant tensile load situations (Ref 31-37).

Beta titanium alloys have a very high solubility for hydrogen such that embrittlement is generally not associated with hydride precipitation (Ref 31, 36). Significant losses in ductility or formability may not occur below levels of several thousand parts per million of hydrogen (Ref 31). The tolerance to hydrogen decreases somewhat in the aged (high-strength) condition. This increased tolerance of the β alloys must be weighed against the significantly higher hydrogen uptake rates that result from the much larger hydrogen diffusion coefficient for β titanium (Ref 36, 38).

Factors that can lead to hydrogen uptake and possible embrittlement of α and near-α titanium alloys in aqueous media have been identified from field and laboratory experience. The three general conditions that must exist simultaneously for the hydrogen embrittlement of α alloys are (Ref 26, 29):

- A mechanism for generating nascent (atomic) hydrogen on a titanium surface. This may be from a galvanic couple, an impressed cathodic current, corrosion of titanium, or severe continuous abrasion of the titanium surface in an aqueous medium
- Metal temperature above approximately 80 °C (175 °F), where the diffusion rate of hydrogen into α titanium is significant (Ref 39-71)
- Solution pH less than 3 or greater than 12, or impressed potentials more negative than −0.70 V (SCE) (Ref 29, 41-44)

The key to preventing hydrogen embrittlement is simply to avoid one or more of these conditions.

Galvanic couples between titanium and certain active metals and excessive cathodic charging from impressed-current cathodic protection systems are the usual causes of excessive hydrogen absorption. In near-neutral electrolytes such as seawater, active metals such as zinc, magnesium, and aluminum can lead to hydrogen uptake and eventual embrittlement when coupled to titanium above 80 °C (175 °F) (Ref 29, 45). A similar problem occurs when titanium is in galvanic contact with carbon steels or active stainless steels in aqueous sulfide media above 80 °C (175 °F) (Ref 29). Like arsenic, antimony, and cyanide species, the sulfide acts as a hydrogen recombination poison (that is, prevents the recombination of atomic hydrogen) and enhances hydrogen uptake in this situation.

No hydrogen uptake and embrittlement problems occur when titanium is coupled to fully passive materials in a given environment. These compatible materials may include other titanium alloys, resistant stainless steels, copper alloys, and nickel-base alloys, depending on conditions.

Cathodic charging of hydrogen onto unalloyed titanium surfaces is not recommended when temperatures exceed 80 °C (175 °F). At metal temperatures below 80 °C (175 °F), thin surface hydride films may form on α titanium alloys (Fig. 7); these are usually not detrimental from the standpoint of corrosion or mechanical properties (Ref 46). However, very high cathodic current densities may lead to enhanced hydride film growth and eventual wall penetration and embrittlement even at room temperature. Practically speaking, impressed cathodic potentials should remain more noble than −1.0 V (SCE) in ambient-temperature seawater applications. Surface thermal oxides on titanium appear to inhibit hydrogen uptake effectively under moderate cathodic charging conditions, but can break down at high current densities (Ref 47).

High-temperature alkaline conditions may also result in excessive hydrogen uptake and embrittlement of titanium alloys. The nascent hydrogen generated on titanium surfaces from small but finite general corrosion in hot (>80 °C,

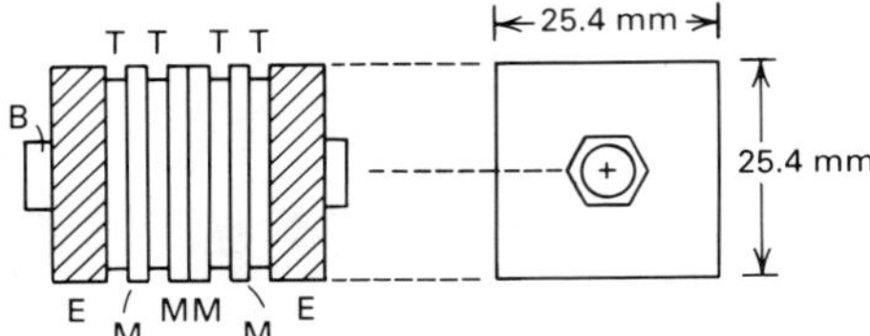

Fig. 5 Schematic of typical crevice corrosion test assembly used for titanium alloy sheet and plate samples. E, assembly and plates; M, alloy test coupons; T, Teflon sheet spacers; B, titanium bolt

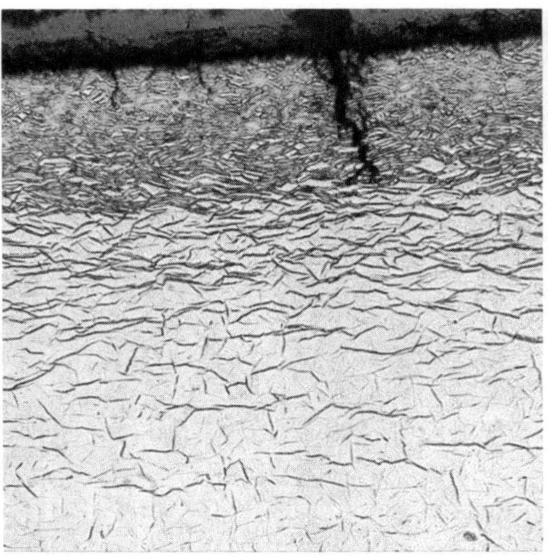

Fig. 6 Photomicrograph of severely hydrided unalloyed titanium. Approximately 200×

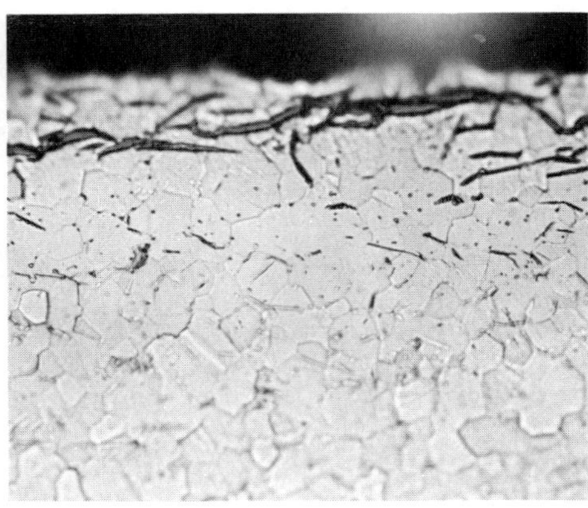

Fig. 7 Photomicrograph of unalloyed titanium sheet revealing a very thin, innocuous surface layer of titanium hydrides. Approximately 500×

or 175 °F), strongly alkaline (pH ≥ 12) media appears to be responsible.

Hydrogen Testing. Testing to determine the susceptibility of a titanium alloy to hydrogen uptake and embrittlement should simulate conditions expected in service. In hydrogen gas atmospheres, simple coupon exposures for as long as is deemed practical (1 week minimum) are recommended to ensure significant uptake, if it occurs. The gas atmosphere must duplicate exact gas chemistry, particularly with respect to water and oxygen content. Mere traces of moisture, for example, will effectively inhibit hydrogen absorption by titanium in dry hydrogen gas and possibly cause test interference.

Galvanic coupling tests or cathodic charging tests can also be conducted to evaluate susceptibility to hydrogen uptake. For a given environment, an active metal (iron, aluminum, etc.) sample is galvanically coupled to the titanium alloy sample such that a specific anode-to-cathode surface area is established. Impressed cathodic charging tests are performed in electrolytic cells containing a specific electrolyte. A power supply (potentiostat or galvanostat) impresses a constant potential or current on the cell such that the titanium is cathodic relative to an inert counterelectrode such as graphite or platinum. A reference electrode can also be used to control or to measure the polarization potential of the test cathode. A typical cathodic charging cell and procedures are presented in Ref 16.

The surface condition of the coupon is a critical variable in all hydrogen uptake tests. Studies have shown that abraded or sandblasted surfaces absorb hydrogen more readily than as-pickled surfaces. Thickening of the surface oxide film by anodizing or thermal oxidation further retards absorption. The actual surface finish anticipated in service should be evaluated.

After test exposure, sample evaluation may involve tensile, notched tensile, bend, ductility (for example, drawn cup), Charpy V-notch impact, and/or hydrogen analysis. Uniaxial, smooth-specimen tensile testing is generally of little value in diagnosing the subtle embrittling effects of hydrogen. Titanium alloys tend to exhibit susceptibility under biaxial or triaxial stress states; therefore, bend tests, cup tests, or notched tensile tests are generally more sensitive to hydrogen effects. Impact toughness testing will be particularly sensitive to hydrogen effects. Impact toughness testing can be an especially sensitive indicator of hydrogen effects in α alloys, whereas slow strain rate methods are very suitable for α-β alloys (Ref 31, 34, 35). Because hydrogen content has relatively little effect on alloy hardness, hardness testing is generally not used.

Hydrogen analysis of coupons is performed by the hot vacuum extraction method. In the hot vacuum extraction apparatus, a small sample is heated to 1100 to 1400 °C (2010 to 2550 °F) for several minutes to reversibly release the absorbed hydrogen, followed by evolved gas measurements. More information on testing for hydrogen damage in metals is available in the article "Evaluation of Hydrogen Embrittlement" in this Volume.

Stress-Corrosion Cracking

Stress-corrosion cracking is a fracture, or cracking, phenomenon caused by the combined action of tensile stress, a susceptible alloy, and a corrosive environment. The metal normally shows no evidence of general corrosion attack, although slight localized attack in the form of pitting may be visible. Usually, only specific combinations of metallurgical and environmental conditions cause SCC. This is important because it is often possible to eliminate or reduce SCC sensitivity by modifying either the metallurgical characteristics of the metal or the makeup of the environment.

Another important characteristic of SCC is the requirement that tensile stress be present. These stresses may be provided by cold work, residual stresses from fabrication, or externally applied loads.

The key to understanding SCC of titanium alloys is the observation that no apparent corrosion, either uniform or localized, usually precedes the cracking process (Ref 48, 49). As a result, it can sometimes be difficult to initiate cracking in laboratory tests by using conventional test techniques.

It is also important to distinguish between the two classes of titanium alloys. The first class, which includes ASTM grades 1, 2, 7, 11, and 12, is immune to SCC except in a few specific environments. These specific environments include anhydrous methanol/halide solutions, nitrogen tetroxide (N_2O_4), red fuming HNO_3, and liquid or solid cadmium. The second class of titanium alloys, including the aerospace titanium alloys, has been found to be susceptible to several additional environments, most notably aqueous chloride solutions. However, this susceptibility is almost always associated with high stress concentrations typical of laboratory testing with loaded, precracked specimens, and generally is not observed with loaded smooth or notched specimens. In fact, the SCC susceptibility identified for these alloys is seldom observed in actual field applications.

Over the years, a variety of mechanisms or models have been proposed to explain SCC phenomena in titanium alloys (Ref 50-54). In general, the mechanisms fall into two broad categories. The first category is the anodic electrochemical dissolution in highly localized areas that, aided by the applied tensile stress, propagates cracks into the metal. The second category involves the absorption of a species (often hydrogen) that embrittles the metal just ahead of the advancing crack, thereby promoting further crack growth.

The first mechanism, anodic-assisted cracking, generally begins when a corrosion pit is formed. In the presence of a tensile stress, the pit will produce a stress concentration that deepens the pit. If corrosion is not so rapid as to allow the advancing crack tip to blunt, the crack will continue to advance into the metal and eventually lead to failure. Once a crack initiates, the balance among the crack tip corrosion rate, the crack tip environment, and the crack tip stress state is critical to crack propagation.

The second mechanism, hydrogen-assisted cracking, is said to occur by absorption of hydrogen near the crack tip. Hydrogen absorption leads to embrittlement of the metal and promotes crack formation. The source of hydrogen is normally associated with anodic dissolution (that is, from the concurrent reduction reaction) at freshly exposed metal at the crack tip. As a result, anodic dissolution in the vicinity of the crack tip is normally required for this mechanism to operate. Obviously, this confuses the nature of the true mechanism, because the second mechanism relies to a certain extent upon at least a portion of the first mechanism to generate the embrittling species.

As might be expected, this situation has led to a great deal of discussion in the literature for many years. It is quite likely, given the tremendous diversity of cracking observations, that no single mechanism exists to explain SCC in titanium alloys. Stress-corrosion cracking mechanisms put forth for titanium alloys generally fall into the two categories discussed above. However, there is greater emphasis on the metallurgical aspects of SCC than is typical for other metal systems in the literature. It is sufficient to say that the makeup of the environment, the metallurgical characteristics of the metal, and the stress state of the system all influence the cracking mechanism. If one of these characteristics is altered, it should not be unexpected that a different mechanism may be necessary to explain the SCC phenomenon adequately.

Stress-Corrosion Testing. Because there is likely no universal, predictive mechanism for SCC, one must rely upon experience for successful application of metals in corrosive environments. Experience may come from actual service problems involving SCC or from laboratory tests designed to reveal susceptibility to SCC. It is often the case that service experience is not available, and laboratory tests are the sole basis for material selection.

In testing metals for SCC resistance, two principal variables must be considered: the environment to be employed and the specimen configuration to be selected. An environment must be selected that is representative of that expected in service. More often than not, this choice is fixed by the intended application. Artificial environments selected to accelerate the test must be recognized as compromises and will not produce wholly reliable information, although they can be used for assessing the relative susceptibilities of various alloys.

Selection of specimen configuration is somewhat different from choosing the test environment in that the investigator is free to choose from a multitude of previously designed configurations. Unfortunately, these configurations do not always produce the same result, nor do they even evaluate the same properties.

Most specimen configurations used for SCC testing fall into three categories:

- *Category 1:* smooth, statically loaded specimens, such as U-bends, C-rings, bent beams, and dead-load tensile bars (ASTM G 30, G 35, G 36, G 38, G 47)
- *Category 2:* notched and precracked specimens, such as cantilever beams, compact tension specimens, and double-cantilever beams (ASTM E 399)
- *Category 3:* smooth, dynamically loaded specimens, primarily the slow strain rate tensile specimens

Each of these categories is used to evaluate a different characteristic of the SCC process, so comparisons across specimen types can lead to incorrect conclusions.

Category 1 specimens are used to evaluate the susceptibility of a material to both initiation and propagation of SCC. Because the samples are smooth and subjected to a static load, they represent the most favorable conditions a material would experience in service. These samples are least effective when used in metal/environment combinations in which localized corrosion (for example, pitting) is unlikely, because SCC initiation may not occur for a long time. Thus, the test duration chosen is of paramount importance.

This configuration also provides limited quantitative information because the test gives simply all-or-nothing results. Once the specimen is stressed and exposed to an environment, it either cracks or it does not. If it does not crack, it is not assured that SCC cannot occur; the test duration may have been too short, or the stress too low, or the orientation incorrect. Data typically reported in addition to pass or fail usually consist of time to failure at a given stress level or something similar. Obviously, this information is of limited value to the person with the responsibility for selecting materials.

Category 2 specimens are quite different in that the initiation step is avoided by creating a crack in the material prior to the test. The specimen may be statically or dynamically loaded, depending on the information required, but is normally statically loaded. Because a crack is assumed to be present from the start, this configuration approaches worst-case conditions and is quite conservative. In addition, when precracked (fracture mechanics) specimens are used, quantitative information is produced that a designer can use to determine the most appropriate material for the application. For this reason, it is often the case than one notched, precracked specimen can provide far more information than dozens of smooth, statically loaded specimens.

With the category 2 specimen, it is often important that the fatigue precrack be initiated in the environment in which the SCC susceptibility test is to be performed. This is especially true for titanium, because the highly passive nature of the metal may repair the precracked area before the environment of interest can be introduced. Individual evaluations are required to determine whether precracking should be performed in the corrosive environment.

Category 3 specimens were developed from category 1 specimens in order to remove the uncertainty associated with SCC initiation. Because the smooth specimen is dynamically loaded to failure, the investigator is assured of two things. First, the specimen will always fail, although not always as the result of SCC. Second, at some point during the test, a crack will be mechanically produced that may serve as an initiation site for SCC. This specimen configuration also represents a worst-case situation because most materials are not intentionally subjected to mechanical fracture during their real-life exposure. The major difficulty with this configuration stems from the data produced; such data are far from quantitative. The quantitative data derived from these tests, such as the ratio of time to failure in air versus time to failure in the test environment, provide some comparative information, but are of limited value to the materials selector. Therefore, testing with category 3 specimens resolves some of the deficiencies of testing with the category 1 specimen, although it creates several new problems. Unfortunately, like category 1 specimens, the category 3 specimens are also pass/fail and produce essentially no quantifiable data.

Given the obvious advantages of category 2 specimens and the difficulty of initiating SCC on titanium category 1 specimens, it is not surprising that almost all of the laboratory testing on titanium alloys has been performed with category 2 specimens. In the following sections, metallurgical, environmental, and stress effects on the SCC susceptibility of titanium alloys will be discussed. The data presented are primarily derived from fracture mechanics test procedures in corrosive environments and, as such, are quantitative in nature. In many cases, results are compared with those attained in noncorrosive environments (often air) to give the reader some appreciation for the degree to which exposure to the environment reduces fracture toughness.

Most fracture toughness data are presented in one of two general forms. The first form is designed to determine the threshold stress intensity required for SCC, K_{ISCC}, and it plots initial stress intensity factor, K_I, versus time to failure. The second form plots crack velocity versus applied stress intensity. The second form can also indicate K_{ISCC} if one exists.

Figure 8 shows several methods of presenting SCC data. Method a is typical of category 1 specimens, and methods b and c are typical of category 2 specimens. For category 3 specimens, data are usually presented in tabular form. Detailed information on SCC testing is available in the article "Evaluation of Stress-Corrosion Cracking" in this Volume.

Table 2 Galvanic series in flowing seawater

Velocity: 4 m/s (13 ft/s); temperature: 24 °C (75 °F)

Material	Steady-state electrode potential, V versus SCE
Graphite	+0.25
Platinum	+0.15
Zirconium	−0.04
AISI type 316 stainless steel (passive)	−0.05
AISI type 304 stainless steel (passive)	−0.08
Monel alloy 400	−0.08
Hastelloy alloy C	−0.08
Titanium	−0.10
Silver	−0.13
AISI type 410 stainless steel (passive)	−0.15
AISI type 316 stainless steel (active)	−0.18
Nickel	−0.20
AISI type 430 stainless steel (passive)	−0.22
Copper alloy C71500 (70Cu-30Ni)	−0.25
Copper alloy C70600 (90Cu-10Ni)	−0.28
Copper alloy 442 (admiralty brass)(a)	−0.29
G bronze	−0.31
Copper alloy	−0.32
Copper	−0.36
Copper alloy C46400 (uninhibited naval brass)	−0.40
AISI type 410 stainless steel (active)	−0.52
AISI type 304 stainless steel (active)	−0.53
AISI type 430 stainless steel (active)	−0.57
Carbon steel	−0.61
Cast iron	−0.61
Aluminum alloy 3003-H	−0.79
Zinc	−1.03

(a) No longer listed by the Copper Development Association; admiralty brasses are now inhibited with small additions of arsenic (CDA C44300), antimony (C44400), or phosphorus (C44500). Source: Ref 57

Galvanic Corrosion

The coupling of titanium with dissimilar metals usually does not accelerate the corrosion of titanium. The exception is in strongly reducing environments in which titanium is severely corroding and not readily passivated. In this uncommon situation, accelerated corrosion may occur when titanium is coupled to more noble metals. In its normal passive condition, titanium is beneficially influenced by materials that exhibit more noble (positive) corrosion potentials. In this regard, graphite and various precious metals (such as platinum, palladium, ruthenium, iridium, and gold) provide anodic protection when coupled to titanium by further stabilizing the oxide film of titanium at more noble potentials.

As shown in Table 2, the corrosion potential of titanium under normally passive conditions is quite noble, but similar to stainless steel or nickel-base alloys in the passive condition. The small potential difference between these passive engineering alloys generally mean negligible galvanic interactions and good galvanic compatibility as long as passive conditions prevail for the alloys involved.

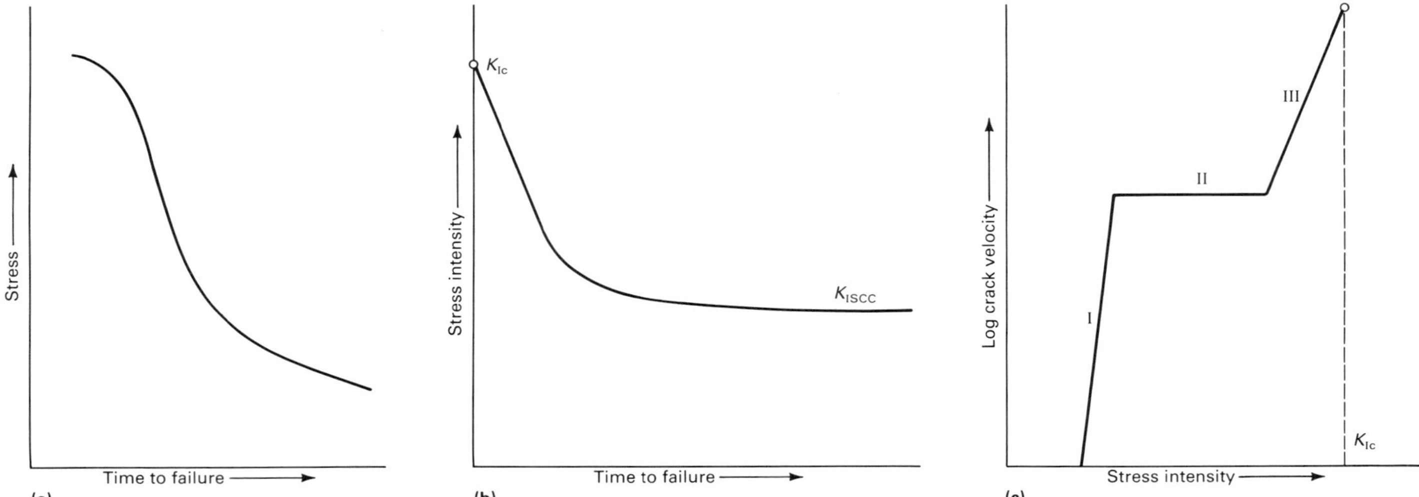

Fig. 8 Typical data representations for SCC using category 1 and category 2 specimens. (a) Stress versus time to failure for category 1 specimens. (b) and (c) Stress intensity versus time to failure and crack velocity versus stress intensity for category 2 specimens

However, when titanium is coupled to a metal that is active in an environment, accelerated anodic attack of the active metal may result. The rate of accelerated attack depends on many factors, including the cathode-to-anode surface area ratio, concentration of dissolved cathodic depolarizers (for example, oxygen, or atomic hydrogen), temperature, solution flow velocity, and medium chemistry. Depending on environmental conditions, active metals may include carbon or low-alloy steels, aluminum, zinc, copper alloys, or stainless steels that are active (depassivated) or pitting. When galvanic corrosion is unacceptably high, consideration should be given to all-titanium component design, coupling to more compatible (passive) alloys, use of dielectric (insulating) joints, or cathodic protection of the active metal.

As discussed in the section "Hydrogen Damage," attention should be given to possible excessive hydrogen uptake by titanium when it is galvanically coupled to active metals. This is of concern in α titanium alloys when temperatures exceed 80 °C (175 °F) in conductive aqueous media, especially when hydrogen recombination poisons, such as sulfide ion (S^{2-}), arsenic, or cyanide ion (CN^-), are present.

Information on the evaluation of galvanic attack is available in the article "Evaluation of Galvanic Corrosion" in this Volume.

Erosion-Corrosion

Erosion-corrosion is defined as the acceleration in metal corrosion rate because of relative movement between a corrosive fluid and a metal surface. This form of attack is highly dependent on fluid velocity and is favored in areas where high local turbulence, impingement, or cavitation of the fluid occur on metal surfaces. Suspended solids in fluid can also result in abrasion, which can drastically accelerate metal removal.

In normal passive environments, the hard, tenacious TiO_2 surface film of titanium provides a superb barrier to erosion-corrosion. For this reason, titanium alloys can withstand flowing water or seawater velocities as high as 30 m/s (100 ft/s) with insignificant metal loss. The ability of the oxide film to repair itself when damaged and the intrinsic hardness of titanium alloys both contribute to their excellent resistance to erosion-corrosion. Therefore, inlet turbulence in shell and tube heat exchangers, entrained gas bubble impingement, and pump cavitation effects are generally not of concern in titanium tubing, piping, and other components.

Titanium alloys also exhibit relatively high resistance to fluids containing suspended solids. Critical velocities for excessive metal removal depend on the concentration, shape, size, and hardness of the suspended particles, in addition to fluid impingement angle (Ref 55), local turbulence, and titanium alloy properties. The typically low concentrations of silt entrained in seawater are generally of little consequence, but continuous exposure to high-velocity slurries of hard particles can lead to finite metal removal. The harder higher-strength titanium alloys, such as grade 4 or 5, may offer improved erosion resistance when marginal erosion of the softer unalloyed titanium grades is observed. When abrasive conditions are severe, application of hard surface coatings should be considered. In potential applications involving high-velocity slurries or suspended solids, it is advisable to conduct erosion tests whenever possible.

Corrosion Fatigue

Corrosion fatigue refers to the reduction in fatigue resistance of a metal due to the presence of a corrosive medium. Because of the protective oxide film of titanium, the smooth or notched fatigue strength of the more common titanium alloys (Table 1) and their weldments is not significantly affected by water, seawater, and many other aqueous chloride media (Ref 56-60). These alloys typically exhibit smooth fatigue run-out stress to tensile strength ratios in the range of 0.5 to 0.6, which remain unchanged in 3.5% sodium chloride (NaCl) solutions and in seawater (Ref 58, 61).

Corrosion in Specific Media

The corrosion rate data reviewed in the following sections were generated from both laboratory and field tests and, in some cases, from actual equipment service. Most of the data are derived from weight change measurements involving varying exposure periods. Samples of plate and sheet in the fully annealed condition were tested, most frequently in the as-pickled surface condition.

These corrosion rate values should be used as general indicators of corrosion resistance, but it must be recognized that corrosion rates can change with exposure time. For example, corrosion rates often decrease with exposure time in normal passive environments; this reflects the finite growth of the protective oxide film. In addition, these values do not always reveal whether the alloy is in a fully passive or stable condition. Therefore, supplemental electrochemical testing may be necessary when borderline passivity, such as in reducing environments, is suspected. Finally, it should be recognized that titanium alloy corrosion behavior is highly sensitive to small, even trace, concentrations of various oxidizing or complexing species in normally aggressive, reducing environments. Therefore, it is necessary to take all trace species or background contaminants in the anticipated service (degree of aeration, and so on) into account when deciding alloy suitability based on pure or simplified simulated medium chemistry. In cases in which the simulation of complex or variable media chemistry is difficult, *in situ* testing is recommended if possible. Supplemental corrosion data not presented in these sections can be found in the Appendices to this article.

General Corrosion in Specific Media

Water and Seawater. Titanium and its alloys are fully resistant to water, all natural waters, and steam to temperatures in excess of 315 °C (600 °F) (Ref 62). Slight weight gain is usually experienced in these benign environments, along with some surface discoloration at higher temperatures from finite passive film thickening. The immunity to attack of α alloys is observed regardless of oxygen level or in high-purity water, such as that normally used in nuclear reactor coolant

Table 3 Corrosion of titanium in ambient seawater

Alloy	Ocean depth, m	Ocean depth, ft	Corrosion rate, mm/yr	Corrosion rate, mils/yr
Unalloyed titanium	Shallow		8×10^{-7}	0.00003
Unalloyed titanium	720–2070	2360–6800	$<2.5 \times 10^{-4}$	<0.1
Unalloyed titanium	2–2070	6.5–6800	nil	
Unalloyed titanium	1720	5640	4×10^{-5}	0.0015
Ti-6Al-4V	2–2070	6.5–6800	$<2.5 \times 10^{-4}$	<0.01
Ti-6Al-4V	1720	5640	8×10^{-6}	0.0003
Ti-6Al-4V	1720	5640	$<1 \times 10^{-3}$	<0.04

Source: Ref 67–71

Table 4 Corrosion of unalloyed titanium in chromic acid solutions

Concentration of CrO_3, %	Temperature, °C	Temperature, °F	Corrosion rate, mm/yr	Corrosion rate, mils/yr
10	Boiling		0.003	0.12
15	24	75	0.005	0.2
15	82	180	0.015	0.6
36.5	90	195	0.046	1.8
50	24	75	0.013	0.5
50	82	180	0.025	1.0

Source: Ref 80

systems (Ref 63-67). The typical contaminants encountered in natural water streams, such as iron and manganese oxides, sulfides, sulfates, carbonates, and chlorides, do not compromise the passivity of titanium. In media containing chloride levels greater than 1000 ppm (for example, seawater) at temperatures about 75 °C (165 °F), consideration should be given to possible crevice corrosion when tight crevices exist in service (see the section "Crevice Corrosion" in this article).

Titanium alloys exhibit negligible corrosion rates in seawater to temperatures as high as 260 °C (500 °F). As shown in Table 3, extremely low corrosion rates of unalloyed titanium and Ti-6Al-4V after up to 3 years of exposure in ambient seawater are recorded. Pitting and crevice corrosion are totally absent in ambient seawater, even if marine deposits form and biofouling occurs. Titanium tubing exposed for 16 years to polluted and sulfide-containing seawater shows no evidence of corrosion (Ref 72). Similar reports of nil corrosion rate in ambient seawater have been reported for unalloyed titanium (Ref 73, 74) and titanium alloys, such as Ti-5Al-2Sn, Ti-13V-11Cr-3Al, Ti-6Al-2Nb-1Ta-0.8Mo, and titanium-palladium (Ref 74-77). Exposure of titanium to marine atmospheres (Ref 74, 78), splash or tide zone, and soils also does not cause corrosion (Ref 68-71, 79). As indicated in the section "Erosion-Corrosion" in this article, the excellent resistance of titanium to seawater is relatively unaffected by velocity.

Unalloyed titanium has provided more than 20 years of outstanding service in seawater for the chemical, oil refining, desalination, and power industries. As result of its immunity to ambient seawater corrosion, titanium is considered to be the technically correct material for many critical marine applications, including many naval and offshore components. More information on the applications of titanium in these industries is available in the articles "Corrosion in the Chemical Processing Industry," "Corrosion in Petroleum Refining and Petrochemical Operations," "Corrosion in the Nuclear Power Industry," and "Corrosion in Fossil Fuel Power Plants" in this Volume.

Oxidizing Media. Titanium alloys are generally highly resistant to oxidizing media and oxidizing acids over a wide range of concentrations and temperatures. Common chemicals in this category include chromic, nitric, perchloric, and hypochlorous acids and salts of these acids. Other oxidizing salts include thiosulfates, vanadates, permanganates, and molybdates. Corrosion rates at and below the boiling point of these aqueous salt solutions over the full range of concentration will typically be less than 0.03 mm/yr (1.2 mils/yr). Data for unalloyed titanium in chromic acid solutions are given in Table 4.

Nitric Acid. Unalloyed titanium has been extensively used for handling and producing nitric acid in applications in which stainless steels have experienced significant uniform or intergranular attack (Ref 67, 81-83). Titanium offers excellent resistance over the full concentration range at subboiling temperatures. As temperatures exceed 80 °C (175 °F), however, the corrosion resistance becomes highly dependent on nitric acid purity.

In hot, very pure solutions or vapor condensates of nitric acid, significant uniform corrosion rates may occur, particularly as temperatures increase. The data plotted in Fig. 9 and 10 show that the mid-range HNO_3 concentrations (20 to 70 wt%) are most aggressive when full inhibition to attack is not achieved in pure refreshed solutions. In these situations, semiprotective oxide surface films form that do not fully retard continued oxidation of the metal surface.

As the impurity levels increase in hot HNO_3 solutions, the resistance of unalloyed titanium improves dramatically. In particular, relatively small amounts of certain dissolved metallic species, including Si^{4+}, Cr^{6+}, Fe^{3+}, Ti^{4+}, or various precious metal ions, can effectively inhibit the high-temperature corrosion of titanium in nitric acid (Ref 67, 82-89). This inhibitive effect is very potent, as shown in Fig. 11 and 12 for Cr^{6+} and Ti^{4+}, respectively, and in Table 5 for Ti^{4+}. Thus, titanium exhibits excellent resistance to recirculating nitric acid process streams, such as stripper reboiler loops (Table 6) in which steady-state levels of dissolved Ti^{4+} inhibitor are achieved. Hold tanks and stripper sumps are also good applications for similar reasons. Another good example of this inhibitive effect is the excellent performance of unalloyed titanium in evaporator reboilers and other components in the high-temperature metal-contaminated process streams used for U_3O_8 recovery (Ref 89-92).

The significant discrepancies and variations in titanium corrosion rates in hot HNO_3 media reported by investigators over the years appear to be the result of these inhibitive metal ion effects. Because titanium corrosion is inhibited by its own corrosion product (Ti^{4+}), the titanium surface area to acid volume ratio, the test duration, and the rate of solution replenishment will all be critical to the rate obtained. The container material and acid purity (chemistry) will also be influential. These factors are reviewed in more detail in Ref 93). Based on this information, it is clear that all design and operating factors must be taken into account when evaluating titanium for high-temperature concentrated HNO_3 service.

Limited corrosion testing of α-β and β titanium alloys in boiling HNO_3 indicates that increasing aluminum and β alloying elements tend to decrease corrosion resistance. The corrosion data for various titanium alloys listed in Table 7 show that α alloys are generally most resistant to hot HNO_3. Other studies have shown that high-purity (low iron, sulfur, and so on) unalloyed titanium does not experience the significant accelerated weldment attack in high-temperature HNO_3 exhibited at times by the less pure unalloyed grades and the near-α alloys.

Fuming Nitric Acid. Titanium alloys exhibit good resistance to white fuming nitric acid. However, dangerous and violent pyrophoric reactions may occur with titanium alloys exposed to red fuming nitric acid or to nitrogen tetroxide (see the section "Gases" in this article). The attack is intergranular and results in a surface residue of finely divided titanium particles that are highly reactive. The critical variables are the nitrogen dioxide (NO_2) and water contents of the acid (Fig. 13). Fuming nitric acid containing less than 1.4 to 2.0% water or more than 6% NO_2 may cause this rapid impact-sensitive reaction to occur (Ref 48, 94-98). Both water and NO are effective inhibitors for this attack, but increasing oxygen and NO_2 are detrimental in this situation. Corrosion rate data in red fuming nitric acid for various alloys as a function of NO_2 and water content also can be found in Ref 8.

Other Oxidizing Media. Titanium alloys exhibit outstanding resistance to solutions of oxidizing chlorine compounds over the full range of concentrations and to relatively high temperatures. Various data for unalloyed titanium are presented in Table 8. Titanium is highly resistant to wet chlorine gas, although a minimum water content must be present, depending on temperature, to maintain full passivity. This is discussed in more detail in the section "Gases" in this article. Titanium is unique among the common engineering alloys in its immunity to general and pitting corrosion in oxidizing chloride environments. These comments also apply to bromine and iodine-containing media. Halide salts of oxidizing cationic species also enhance the passivity of titanium alloys such that negligible corrosion rates can be expected. Examples include $FeCl_3$, $CuCl_2$, and $NiCl_2$ solutions and their bromide counterparts.

Although peroxides are generally oxidizing, titanium alloys can experience general corrosion in aqueous peroxide solutions, depending on concentration, temperature, and pH. As shown in Table 9, corrosion rates are minimal in dilute near-neutral hydrogen peroxide solutions, but increase dramatically under alkaline conditions because of the formation of soluble titanium-peroxyl (complex) compounds (Ref 99, 100). However, corrosion is effectively inhibited by small additions of calcium, strontium, or barium ions (Ref 100). Sodium silicate and sodium hexametaphosphate additions have also been shown to reduce corrosion rates substantially (Ref 99). Significant attack may occur in highly concentrated (90%) H_2O_2 solutions.

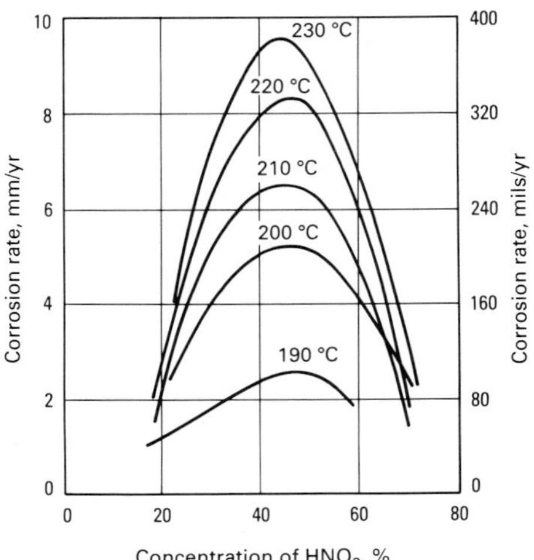

Fig. 9 Corrosion of unalloyed titanium in high-temperature HNO_3 solutions. Source: Ref 80, 86

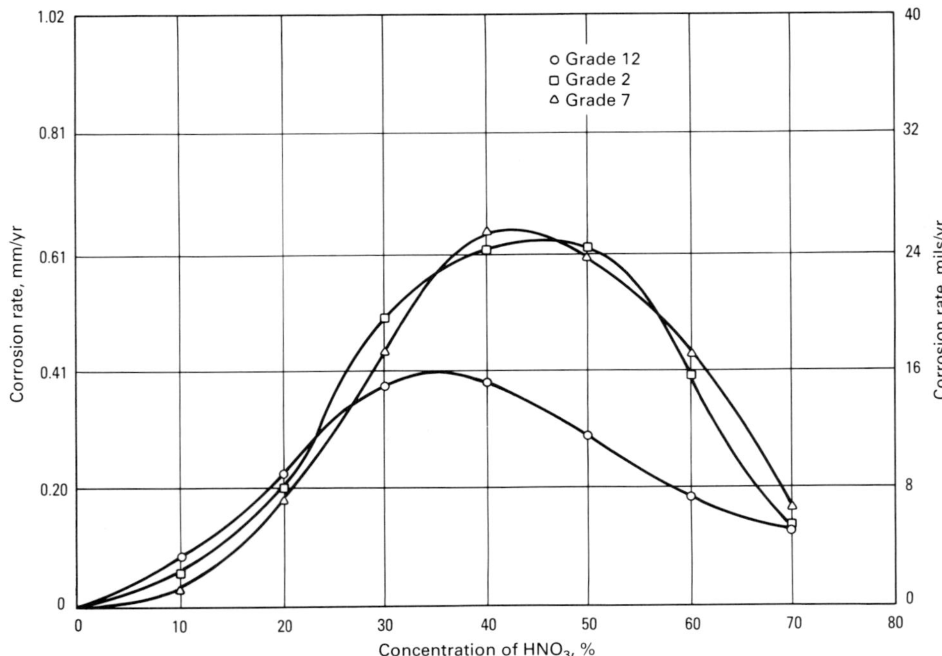

Fig. 10 Corrosion of titanium alloys in boiling, uninhibited HNO_3 solutions. Acid solutions were refreshed every 24 h.

Reducing Acids. The corrosion resistance of titanium alloys in reducing acid media is very sensitive to acid concentration, temperature, background chemistry, and purity of the acid solution, in addition to titanium alloy composition. When the temperature and/or concentration of pure (uncontaminated) reducing acid solutions exceed certain values, the protective oxide film of titanium may break down, which would result in severe general corrosion. Included in this category are hydrochloric, sulfuric, hydrobromic, hydriodic, hydrofluoric, phosphoric, sulfamic, oxalic, and trichloroacetic acids.

Temperature-acid concentration guidelines for titanium grades 2, 7, and 12 in naturally aerated but pure (uninhibited) HCl, H_2SO_4, and H_3PO_4 solutions are presented in Fig. 14, 15, and 16, respectively. Supplemental corrosion data in these acids are provided in Tables 10 to 12 and in the Appendices to this article. Table 13 lists data for several alloys in HBr and HI solutions. Corrosion data for unalloyed titanium in highly concentrated and SO_3-fuming H_2SO_4 media can be found in Ref 8. It should be noted that the grade 7 (or 11) titanium alloy exhibits substantially greater resistance to dilute uninhibited reducing acids than any other commercial titanium alloy. General corrosion rates for as-annealed high-strength titanium alloys in uninhibited boiling HCl solutions are plotted in Fig. 17 to 19 (Ref 14). Figure 20 suggests that aging has little effect on the general resistance of a titanium alloy to HCl (Ref 14). The beneficial effect of molybdenum alloying additions on alloy corrosion resistance is apparent from these results.

Hydrofluoric acid solutions can aggressively attack titanium alloys over the full range of concentrations and temperatures, because the fluoride ion (F^-) forms highly stable, soluble complexes with titanium. Although the addition of oxidizing species, such as HNO_3, will tend to reduce corrosion and retard hydrogen uptake in HF solutions, significant rates of attack still pre-

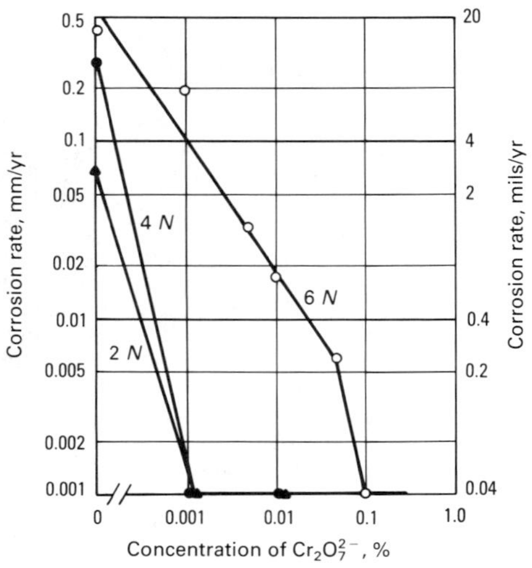

Fig. 11 Inhibitive effect of dissolved $Cr_2O_7^{2-}$ ions on the corrosion of commercially pure titanium after 65 h in boiling HNO_3 solutions. Source: Ref 88

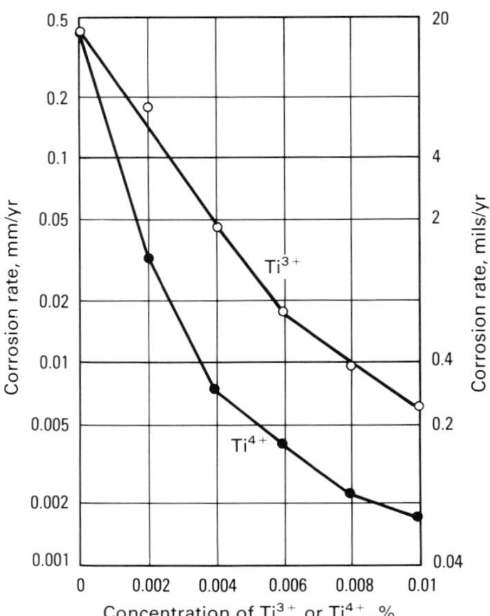

Fig. 12 Inhibitive effect of dissolved Ti^{2+} or Ti^{4+} ions on the corrosion of titanium after 65 h in boiling 6 *N* HNO_3 solutions. Source: Ref 88

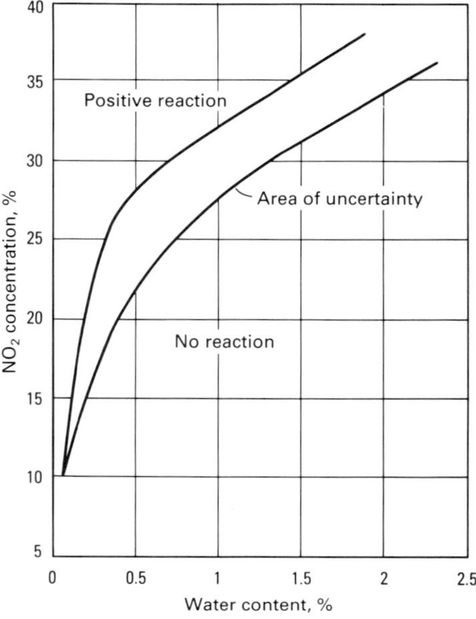

Fig. 13 Acid composition limits for avoiding rapid, pyrophoric reactions of titanium with red fuming HNO_3. Source: Ref 94

Table 5 Effect of dissolved Ti^{4+} on the corrosion rate of unalloyed titanium in boiling HNO_3 solutions

Titanium ion added, mg/L	Corrosion rate: 40% HNO_3 mm/yr	40% HNO_3 mils/yr	68% HNO_3 mm/yr	68% HNO_3 mils/yr
0	0.75	29.5	0.81	32
10	...	...	0.02	0.8
20	0.22	8.7	0.06	2.4
40	0.05	2	0.01	0.4
80	0.02	0.8	0.01	0.4

Source: Ref 85

Table 6 Corrosion of titanium grade 2 and type 304L stainless steel heating surfaces exposed to boiling 90% HNO_3

Metal temperature °C	°F	Corrosion rate: Grade 2 mm/yr	Grade 2 mils/yr	Type 304L mm/yr	Type 304L mils/yr
116	240	0.03–0.17	1.2–6.7	3.8–13.2	150–520
135	275	0.04–0.15	1.6–6	17.2–73.7	675–2900
154	310	0.03–0.06	1.2–2.4	18.3–73.7	720–2900

Source: Ref 82

Table 7 Corrosion of various titanium alloys in boiling HNO_3 solutions

Test duration: 196 h

Alloy	Corrosion rate at indicated HNO_3 concentration: 25% mm/yr	25% mils/yr	45% mm/yr	45% mils/yr	70% mm/yr	70% mils/yr
Grade 1	0.15	6	0.39	15	0.08	3.1
Grade 7	0.17	6.7	0.38	14.9	0.07	2.8
Grade 12	0.18	7	0.27	10.6	0.06	2.4
Ti-6-2-1-0.8	0.39	15	0.73	28.7	0.21	8.3
Grade 9	0.18	7	0.54	21.3	0.10	4
Ti-550	0.83	32.6	1.14	44.9	0.30	12
Grade 5	0.67	26.4	0.86	33.8	0.02	0.8
Transage 207	8.0	315	15.6	614	0.95	37.4
Ti-6-2-4-6	4.3	170	5.7	224	0.78	30.7
Ti-10-2-3	0.48	18 9	1.2	47.2	0.07	2.8
Ti-3-8-6-4-4	1.13	44 5	3.6	141.7	1.46	57.5
Ti-5Ta	0.04	1 6	0.08	3.1	0.03	1.2

Table 8 Corrosion of unalloyed titanium in solutions of oxidizing chlorine compounds

Reagent	Concentration, %	Temperature °C	°F	Corrosion rate mm/yr	mils/yr
Water saturated with chlorine	...	75	165	0.003	0.12
Water saturated with chlorine	...	88	190	0.002(a)	0.08
Water saturated with chlorine	...	97	207	0.07	2.8
NaOCl	6	25	77	nil	
ClO_2 + HOCl	15	43	110	nil	
ClO_2 + steam	5	100	212	0.005	0.2
$Ca(OCl)_2$	2	100	212	0.001	0.04
$Ca(OCl)_2$	6	100	212	0.001	0.04
$Ca(OCl)_2$	18	25	77	nil	
HOCl + ClO_2 + Cl_2	17	38	100	nil	

(a) Welded sample. Source: Ref 80

Table 9 General corrosion of grade 2 titanium in hydrogen peroxide solutions

Medium	pH(a)	Temperature °C	°F	Corrosion rate mm/yr	mils/yr
5% H_2O_2	1	23	73	0.064	2.5
5% H_2O_2	4.3	23	73	0.013	0.5
5% H_2O_2	1	66	150	0.152	6
5% H_2O_2	4.3	66	150	0.061	2.4
5% H_2O_2 + 500 ppm Ca^{2+}	1	66	150	nil	
20% H_2O_2	1	66	150	0.686	27
20% H_2O_2 + 500 ppm Ca^{2+}	1	66	150	nil	
10 g/L H_2O_2 + 20 g/L NaOH	...	60	140	55.9	2200
0.75 g/L H_2O_2	11	70	160	0.42	16.5
3.5 g/L H_2O_2 + 10 g/L NaOH + 10 g/L Na_2SiO_3 + 0.5 g/L Na_3PO_4	...	60	140	nil	

(a) Acid solutions were prepared with HCl additions.

vail. Inhibition of corrosion can be achieved in very dilute acid fluoride solutions when an excess of complexing metal ions (for example, Fe^{3+}, Al^{3+}, and Cr^{6+}) is present (Ref 101, 102). In the absence of these complexing metal ions, solutions containing more than 20 ppm F^- may attack titanium when solution pH falls below 6 to 7.

Titanium alloys exhibit good resistance to most mildly reducing acid solutions whether they are inhibited or not. These environments include sulfurous acid, aqueous hydrogen sulfide solutions, boric acid, or carbonic acid. Near-nil corrosion rates can be expected over the full concentration range to temperatures well beyond their boiling points (Ref 26, 80, 103).

Inhibition of Reducing Acid Corrosion. Although strong reducing acids may seriously corrode titanium alloys in pure form, the presence of certain oxidizing species (cathodic depolarizers) in these acids can effectively inhibit general corrosion; this expands the useful range of application of these alloys. These species may be prime constituents of a process stream, naturally occurring background contaminants (such as dissolved oxygen), intentionally added inhibitors, or ferrous corrosion products.

The inhibitors commonly encountered in industrial service are listed in Table 14. Many of these effective inhibitors are multivalent metal ions in their highest valence states. Inhibitor potency at levels as low as 100 ppm or below is typically observed, as indicated by the data presented in Table 15 and those listed under the respective acids in the Appendices to this article. The beneficial influence of minute ferric ion concentrations on the useful resistance of titanium grades 2, 7, and 12 in HCl media is shown in Fig. 21. Nitric acid or nitrate additions effectively inhibit titanium alloy corrosion on HCl media, but more limited inhibition is observed in H_2SO_4 and H_3PO_4 solutions containing nitric acid or nitrates.

Because the performance of titanium alloys in reducing acids is highly influenced by the presence of many inhibiting species, the nature and background chemistry of a reducing acid environment should be thoroughly examined before determining alloy suitability. Titanium is often selected for normally aggressive reducing acid solutions, such as the hydrometallurgical acid-leaching process streams for metallic ores, because of the beneficial effect of these inhibitive ions (Ref 106).

Anodic protection is also an effective means of passivating and protecting titanium alloys in reducing acids. Table 16 shows that impressed anodic potentials can significantly reduce the corrosion rate of unalloyed titanium in various hot, concentrated acids. Generally, an increase in anodic potential will decrease the corrosion rate as long as the anodic pitting potential is not exceeded for titanium in the electrolyte. Information on the mechanism of and apparatus for anodic protection is available in the article "Anodic Protection" in this Volume.

Salt Solutions. Titanium alloys are highly resistant to practically all salt solutions over the pH range of 3 to 11 and to temperatures well in excess of boiling. Titanium withstands exposure to solutions of chlorides (Ref 26, 109), bromides, iodides, sulfites, sulfates, borates, phosphates, cyanides, carbonates, bicarbonates, and ammonium compounds. Corrosion rate values for titanium alloys in these various salt solutions are generally less than

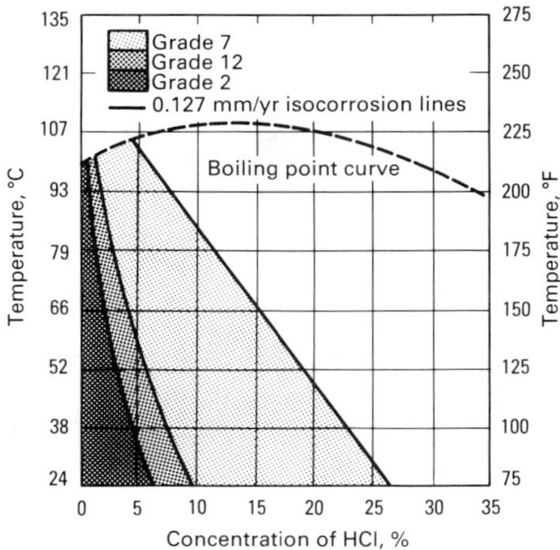

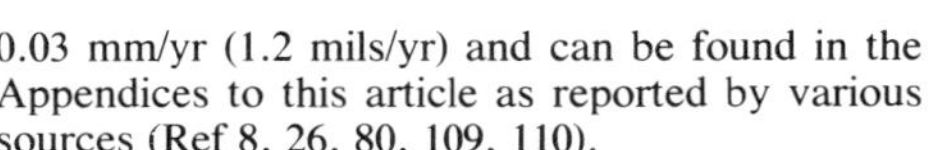

Fig. 14 Isocorrosion diagram for titanium alloys in pure, naturally aerated HCl solutions. 0.127-mm/yr (5-mils/yr) isocorrosion lines are shown.

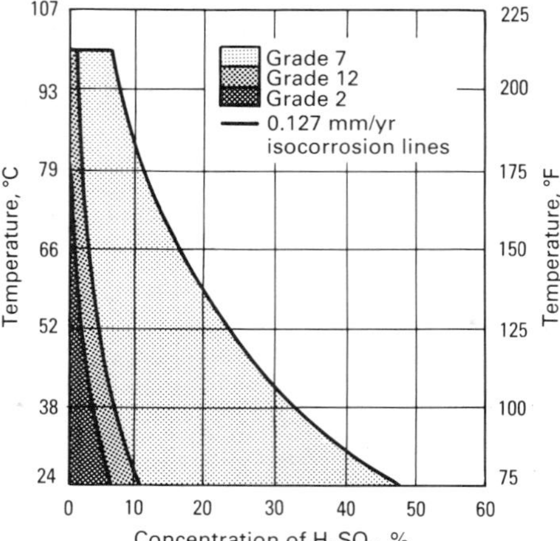

Fig. 15 Isocorrosion diagram for titanium alloys in pure, naturally aerated H_2SO_4 solutions. 0.127-mm/yr (5-mils/yr) isocorrosion lines are shown.

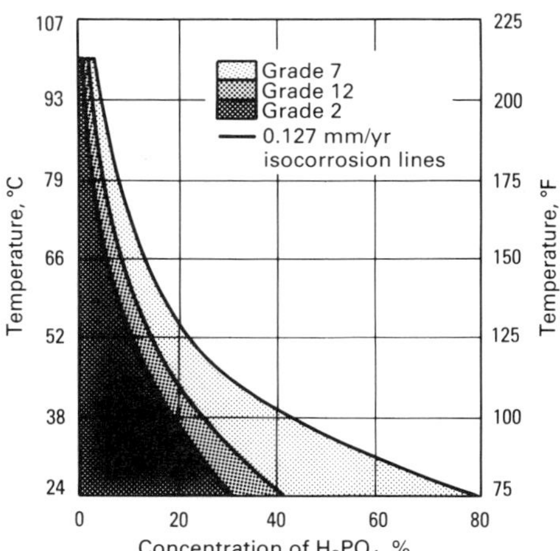

Fig. 16 Isocorrosion diagram for titanium alloys in pure, naturally aerated H_3PO_4 solutions. 0.127-mm/yr (5-mils/yr) isocorrosion lines are indicated.

0.03 mm/yr (1.2 mils/yr) and can be found in the Appendices to this article as reported by various sources (Ref 8, 26, 80, 109, 110).

Oxidizing anionic salts such as nitrates, hypochlorites, chlorites, chlorates, perchlorates, molybdates, chromates, permanganates, and vanadates further extend titanium alloy passivity into stronger acidic and alkaline solutions. Similar beneficial effects on the low-pH (acidic) side can be expected from oxidizing cationic salts, such as ferric, cupric, and nickelous chlorides or sulfates. In fact, titanium is often the most practical metal for handling hot, oxidizing, acidic chloride conditions.

Titanium alloys are frequently selected because of their superior resistance to the chlorides typically found in many process streams, brines, and seawater. In hot chloride media, susceptibility to pitting is usually not an issue, but crevice corrosion may be possible, depending on pH and temperature (see the section "Crevice Corrosion" in this article). Special attention must be given to nonoxidizing acidic or hydrolyzable salt solutions as temperatures and concentrations increase. To avoid general or localized HCl attack resulting from salt hydrolysis, special concentration-temperature guidelines for titanium should be observed for concentrated $AlCl_3$ (Ref 8, 109), $ZnCl_2$ (Ref 111), $MgCl_2$ (Ref 112), and $CaCl_2$ solutions. Guidelines for titanium grades 2, 7, and 12 in hot concentrated $MgCl_2$ and in concentrated sea salt and NaCl slurries are presented in Fig. 22 and 23, respectively, and in Fig. 24 for concentrated $CaCl_2$ brines. Grades 12 and 7 and higher molybdenum-containing titanium alloys exhibit superior resistance to general and localized corrosion in these high-temperature acid salt solutions.

Alkaline Media. Titanium alloys are generally very resistant to alkaline media, including solutions of NaOH, KOH, $Ca(OH)_2$, $Mg(OH)_2$, and NH_4OH (Ref 26). Near-nil corrosion rates can be expected in boiling solutions of the latter three alkalies up to saturation. As shown in Table 17, titanium exhibits low corrosion rates in NaOH and KOH solutions at subboiling temperatures. However, significant increases in corrosion are noted as the concentrations of these two strong alkalies increase at higher temperatures. Potassium hydroxide tends to be more aggressive than sodium hydroxide under these conditions.

Although corrosion rates are relatively low in alkaline media, titanium alloys may experience excessive hydrogen pickup and eventual embrittlement under certain conditions. For α and near-α alloys, hydrogen embrittlement is possible when temperatures exceed 80 °C (175 °F) and pH is 12 or more. The presence of dissolved oxidizing species in hot caustic solutions, such as chlorate, hypochlorite, or nitrate compounds, can extend resistance to hydrogen uptake to somewhat higher temperatures.

Organic Compounds. As indicated in Table 18, titanium alloys are highly resistant to most organic compounds, including alcohols, ketones, ethers, aldehydes, and hydrocarbons. The traces of moisture (ppm levels) normally present in industrial organic process streams are sufficient to maintain the protective oxide film of titanium. Totally anhydrous organic streams may prevent oxide film repair and should be avoided. In the special case of absolute methanol, at least 1.5% H_2O must be added to prevent depassivation and SCC (Ref 113-116). Higher molecular weight alcohols are generally quite benign toward titanium alloys.

Chlorinated hydrocarbons generally do not pose any problems for most titanium alloys (Table 18). A few high-strength alloys may be susceptible to SCC under specific circumstances (see the section "Stress-Corrosion Cracking" in this article). If significant quantities of water are also present, many chlorinated hydrocarbons may undergo hydrolysis to form HCl at higher temperatures. Titanium alloy performance will depend on the temperature and the extent of HCl formation and concentration in the aqueous phase.

Titanium alloys are generally very resistant to organic acids. Nil corrosion can be expected in concentrated solutions of very weak organic acids such as adipic, hydroxyacetic, acetic, terephthalic, tannic, stearic, maleric, tartaric, benzoic, butyric, and succinic acids to tempera-

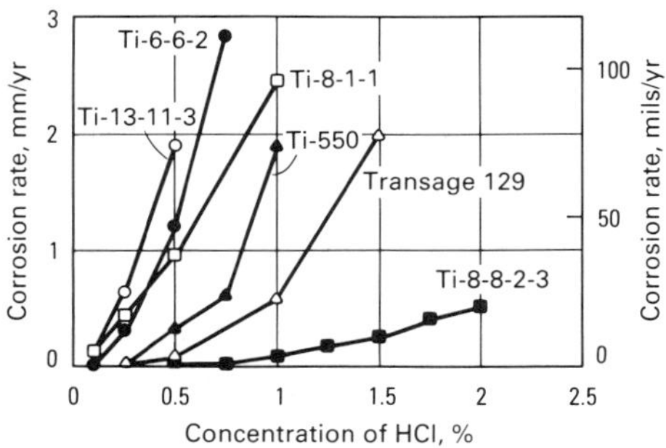

Fig. 17 General corrosion of annealed titanium alloys in naturally aerated HCl solutions. See also Fig. 18 and 19.

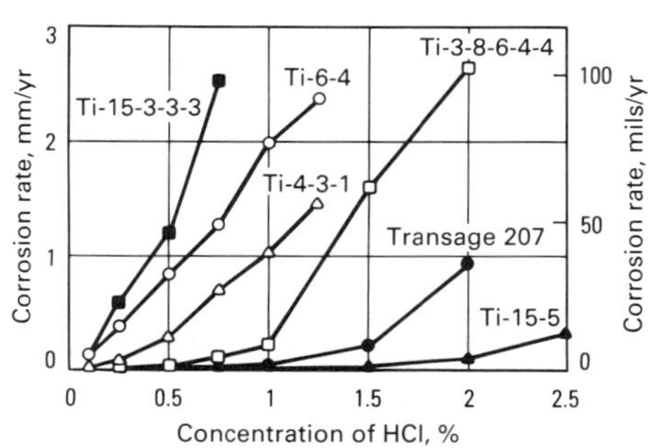

Fig. 18 General corrosion of annealed titanium alloys in naturally aerated HCl solutions. See also Fig. 17 and 19.

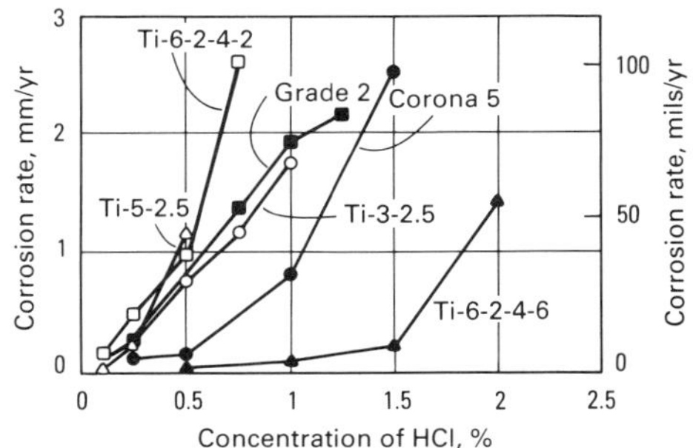

Fig. 19 General corrosion of annealed titanium alloys in naturally aerated HCl solutions. See also Fig. 17 and 18.

Table 10 Corrosion of grade 2 titanium in naturally aerated HCl solutions

Concentration of HCl, %	Temperature °C	Temperature °F	Corrosion rate mm/yr	Corrosion rate mils/yr
5	24	75	nil	
6	24	75	0.07	2.8
8	24	75	0.2	8
9	24	75	0.25	1
17.3	24	75	0.51	2
26	24	75	2.59	102
1.4	52	125	0.02	0.8
5.8	52	125	0.51	2
6	52	125	0.68	26.8
7	52	125	1.27	50
11.5	52	125	3.07	121
1	66	150	0.01	0.4
1.5	66	150	0.02	0.8
1.7	66	150	0.13	5
2	66	150	0.61	24
3	66	150	1	40
4.7	66	150	7.08	279
0.05	Boiling		0.02	0.8
0.1	Boiling		0.1	4
0.2	Boiling		0.23	9
0.4	Boiling		0.53	21
0.5	Boiling		0.84	33.1
1	Boiling		1.83	72

Table 11 Corrosion of grade 12 titanium in naturally aerated HCl solutions

Concentration of HCl, %	Temperature °C	Temperature °F	Corrosion rate mm/yr	Corrosion rate mils/yr
6	24	75	0.008	0.31
8	24	75	0.008	0.31
10	24	75	1.40	55
12	24	75	2.54	100
28.5	24	75	5.58	220
3	52	125	nil	
4	52	125	0.001	0.04
5.9	52	125	0.51	20
7	52	125	5.30	209
2.4	66	150	0.01	0.4
3.6	66	150	0.03	1.2
5.9	66	150	0.51	20
7	66	150	8.98	354
0.6	Boiling		0.025	1
1.7	Boiling		0.16	6.3
2	Boiling		0.51	20
2.5	Boiling		6.85	270

Table 12 Corrosion of grade 7 titanium in naturally aerated HCl solutions

Concentration of HCl, %	Temperature °C	Temperature °F	Corrosion rate mm/yr	Corrosion rate mils/yr
9	24	75	nil	
18	24	75	nil	
20	24	75	0.01	0.4
26.5	24	75	0.02	0.8
27	24	75	0.70	27.6
9	52	125	0.008	0.3
11.5	52	125	0.02	0.8
14.7	52	125	0.03	1.2
16.8	52	125	0.06	2.4
19	52	125	0.08	3.1
21.9	52	125	0.41	16.1
6	66	150	0.01	0.4
9.6	66	150	0.03	1.2
11.5	66	150	0.04	1.6
16.8	66	150	0.13	5
17	66	150	0.39	15.4
20.9	66	150	0.51	20
2	Boiling		0.025	1
3	Boiling		0.05	2
4	Boiling		0.10	4
6	Boiling		0.23	9
9	Boiling		0.51	20
16.8	Boiling		2.97	117

tures in excess of 100 °C (212 °F). Table 19 provides a selection of corrosion rates derived from several sources. Corrosion rates can become significant in the stronger, nonaerated organic acids as acid concentration and temperature increase. These acids include formic, lactic, citric, trichloroacetic, and, especially, oxalic acid (see data in the Appendices to this article). Grades 7 and 12 exhibit much greater resistance to these acids than unalloyed titanium.

Solution aeration is often sufficient to inhibit corrosion on unalloyed titanium in the stronger organic acids, such as formic, lactic, and citric (Ref 8, 26, 80). Addition of oxidizing species (Table 14) may also effectively inhibit corrosion in the more aggressive organic acids, such as oxalic.

Gases. The oxide film on titanium alloys provides an effective barrier to attack by most gases in wet or dry condition, including oxygen, nitrogen, dry HCl, SO_2, NH_3, HCN, CO_2, CO, and H_2S (Ref 117). This protection extends to temperatures in excess of 150 °C (300 °F). The outstanding resistance of titanium alloys to rural, marine, and urban atmospheric exposure has been documented (Ref 78).

Titanium alloys experience no significant corrosion degradation in oxygen or sulfur-bearing gases below 300 °C (570 °F). Excessive surface oxidation and eventual interstitial embrittlement may occur above 340 to 370 °C (645 to 700 °F) after prolonged, continuous exposure to air, depending on alloy composition. Embrittlement is the result of the enhanced diffusion rate of interstitial oxygen into the metal at higher temperatures such that time to failure will also depend on metal section thickness and state of stress.

Titanium alloys exhibit reasonably good oxidation resistance to temperatures of approximately 550 °C (1020 °F) in short-term (less than several hours) air exposures. Studies have shown that aluminum, silicon, and niobium improve the oxidation resistance of titanium but that vanadium, iron, and certain other β-stabilizing elements may increase oxidation rates (Ref 8, 118). Other sources suggest that oxidation rates (kinetics) depend on gas composition, temperature, and exposure time (Ref 8). Below approximately 200 °C (390 °F), a logarithmic dependence with time is typical, but a parabolic rate law (Ref 8, 118) and a logarithmic rate law (Ref 8, 119, 120) have both been measured between 200 and 700 °C (390 and 1290 °F). The growth of thermal oxide films on unalloyed titanium in the air at 400 to 700 °C (750 to 1290 °F) is plotted in Fig. 25 as a function of time. Above 700 °C (1290 °F), a linear dependence of rate with time is typical, indicating little protective effect by the surface oxide.

Although the ignition of titanium alloys in normal air is generally not of concern in typical mill product forms (except powders), ignition is possible in enriched oxygen atmospheres. Figure 26 shows that ignition may occur when oxygen content exceeds 35 vol%. Thresholds for the ignition of unalloyed titanium in pure oxygen gas are plotted in Fig. 27 as a function of pressure and temperature. Ignition thresholds in low-temperature gaseous oxygen are provided in Fig. 28. Titanium alloys exhibit impact sensitivity in liquid oxygen (Ref 123). Ignition is not easily achieved unless the oxide film is mechanically damaged and fresh metal surfaces are exposed.

Ignition and burning of titanium alloys can be avoided in oxygen-rich atmospheres by proper equipment design, surface coatings, and/or avoidance of mechanical damage to exposed titanium surfaces (Ref 106). Recommended guidelines for preventing ignition of titanium powders have also been established (Ref 124).

Titanium alloys exhibit very low reaction rates in pure nitrogen atmospheres below 650 °C (1200 °F) because of the formation of a protective nitride surface film. Reaction and diffusion rates are significantly lower than those observed in oxygen (Ref 8).

Although titanium is the preferred metallic material for handling wet chlorine and bromine gas environments, rapid, dangerous, exothermic halogenation reactions may occur with titanium in dry chlorine and bromine gas environments. A minimum water content (or oxygen content) in these cases is necessary to maintain total alloy passivity, as indicated in Fig. 29 and 30. The critical water content depends on gas temperature and flow rate. Mechanical damage to metal surfaces to expose fresh metal facilitates reaction with dry chlorine, but thicker oxide films (thermal oxides) tend to retard initiation of the reaction. Titanium alloys cannot be fully passivated in liquid bromine, because of the extremely low solubility of water in this medium.

Rapid, pyrophoric reactions with titanium alloys are also possible in anhydrous N_2O_4 gas atmospheres (Ref 94-96). Small water additions (Ref 127) or the presence of 0.6 to 1.0 wt% nitric acid effectively inhibits metal attack.

Liquid Metals and Fused Salts. Titanium exhibits good resistance to many liquid metals at moderately elevated temperatures, at which corrosion rate increases with temperature and flow rate (Ref 8, 67, 80, 109). As shown in Table 20, these metals include molten aluminum, sodium, potassium, sodium-potassium mixtures, magnesium, tin, and lead. In contrast, useful performance of titanium in molten lithium, bismuth, zinc, gallium, cadmium, and mercury is limited to relatively low temperatures. Liquid mercury below 150 °C (300 °F) does not appear to affect titanium unless wetting of freshly exposed (mechanically damaged) surfaces occurs (Ref 67). As discussed in the earlier section on SCC, liquid cadmium, silver, and mercury may cause SCC of titanium alloys.

Titanium exhibits relatively high rates of attack in molten chloride salts, increasing with both temperature and the presence of oxygen. Aggressiveness of attack follows the order: KCl > NaCl > LiCl (Ref 8, 109). High rates of metal dissolution have also been reported in fused sodium carbonate, sodium hydroxide, sodium peroxide, and sodium bisulfate (Ref 109).

Crevice Corrosion in Specific Media

Titanium alloys generally exhibit superior resistance to crevice corrosion as compared to stainless steel and nickel-base alloys (Ref 128). Nevertheless, the susceptibility of titanium alloys to crevice corrosion should be considered when tight crevices exist in hot aqueous chloride, bromide, iodide, or sulfate solutions. Crevice test results indicate that the initiation of crevice cor-

Table 13 Corrosion of titanium alloys in naturally aerated HBr and HI solutions

Acid	Concentration, %	Temperature, °C (°F)	Alloy	Corrosion rate, mm/yr	Corrosion rate, mils/yr
HBr	0.3	Boiling	Grade 2	nil	
HBr	0.6	Boiling	Grade 2	0.003	0.12
HBr	0.9	Boiling	Grade 12	0.008	0.32
HBr	3.0	Boiling	Grade 2	1.45	57
HBr	3.0	Boiling	Grade 12	0.013	0.5
HBr	3.0	Boiling	Grade 7	0.010	0.4
HBr	8.0	Boiling	Grade 7	0.094	3.7
HBr	40	25 (75)	Grade 2	nil	
HI	10	Boiling	Grade 2	nil	
HI	57	25 (75)	Grade 2	0.15	6

Table 14 Species that inhibit the corrosion of titanium alloys in reducing acids

Inhibitor category	Species	Relative inhibitor potency
Oxidizing metal cations	Ti^{4+}, Fe^{3+}, Cu^{2+}, Hg^{4+}, Ce^{4+}, Sn^{4+}, VO_2^+, Te^{4+}, Te^{6+}, Se^{4+}, Se^{6+}, Ni^{2+}	High, High, Low
Oxidizing anions	ClO_4^{2-}, $Cr_2O_7^{2-}$, MoO_4^{2-}, MnO_4^{2-}, WO_4^-, IO_3^-, VO_4^{3-}, VO_3^-, NO_3^-, NO_2^-, $S_2O_3^{2-}$	Very high, Very high, Moderate
Precious metal ions	Pt^{2+}, Pt^{4+}, Pd^{2+}, Ru^{3+}, Ir^{3+}, Rh^{3+}, Au^{3+}	High, High
Oxidizing organic compounds	Picric acid, *o*-dinitrobenzene, 8-nitroquinoline, *m*-nitroacetanilide, trinitrobenzoic acid, and certain other nitro, nitroso, and quinone organics	Moderate–high
Others	O_2, H_2O_2, ClO_3^-, OCl^-	Moderate

Source: Ref 104, 105

Table 15 Effect of certain multivalent metal ions on the corrosion of titanium in boiling reducing acids

Inhibiting ion	Concentration of inhibiting ion, ppm	Boiling 5% HCl, mm/yr	Boiling 5% HCl, mils/yr	Boiling 10% H_2SO_4, mm/yr	Boiling 10% H_2SO_4, mils/yr
Fe^{3+}	0	29	1142	>76.2	>3000
	100	0.025	1	0.208	8.2
	500	0.02	0.8	0.069	2.7
Cu^{2+}	0	29	1142	>76.2	>3000
	100	0.033	1.3	0.419	16.5
	500	nil		0.361	14.2
Mo^{6+}	0	29	1142	>76.2	>3000
	100	nil		0.001	0.04
	500	nil		nil	
Cr^{6+}	0	29	1142	>76.2	>3000
	100	nil		0.001	0.04
	500	nil		0.001	0.04
V^{5+}	0	29	1142	>76.2	>3000
	100	0.02	0.8	0.005	0.2
	500	0.008	0.3	0.005	0.2

Source: Ref 106

rosion often lacks reproducibility, consistency, and regularity. These test data must be judged relative to their statistical significance (that is, number of data points). Factors that affect crevice attack significantly include alloy composition, pH, temperature, halide concentration, presence of oxidizing species (cathodic depolarizers), sample surface condition, type of gasket, type of crevice (gasket-to-metal, metal-to-metal, deposit-to-metal), and the crevice geometry, particularly crevice gap (tightness) (Ref 1, 23, 26, 129-131).

Chlorides. The susceptibility of titanium alloys to crevice corrosion in hot, concentrated chloride solutions increases significantly as temperatures increase and pH decreases. Figure 31 shows pH-temperature limits for the crevice corrosion of titanium grades 2, 7, and 12. These guidelines have been found to be applicable to most chloride salt solutions, including seawater, over a wide range of chloride concentrations (>0.1%). The limits for grade 2 are applicable to all unalloyed grades as well.

Crevice attack of titanium alloys will generally not occur below a temperature of 70 °C (160 °F) regardless of solution pH or chloride concentration or when solution pH exceeds 10 regardless of temperature. As indicated in Fig. 31 and Table 21, grade 12 provides crevice corrosion resistance when brine pH falls between 3 and 11 to temperatures as high as 300 °C (570 °F). The

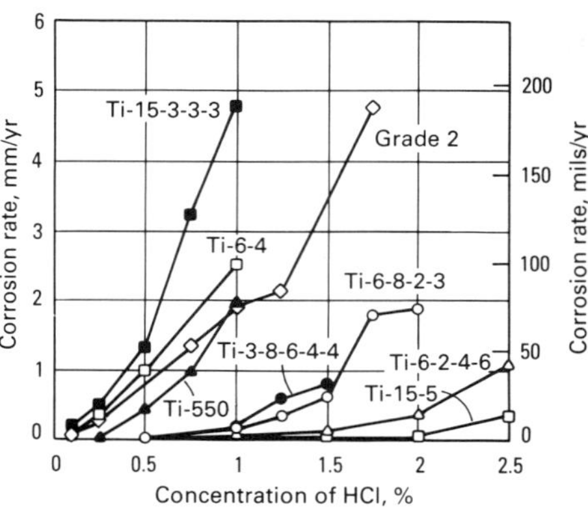

Fig. 20 General corrosion of aged titanium alloys in naturally aerated HCl solutions

grade 7 (or 11) alloy extends this resistance to brine pH values as low as 0.6 to 0.7, depending on brine composition and temperature. The grade 7 alloy is considered to be the most crevice corrosion resistant titanium alloy available commercially, and it is often preferred for hot low-pH salt solutions.

The crevice corrosion resistance of grade 12 titanium was extensively tested in concentrated near-neutral NaCl, NaCl-$MgCl_2$, and NH_4Cl brines at high temperatures. Studies were conducted to assess this alloy for potential application in hypersaline geothermal brine (Ref 132), high-level nuclear waste storage (Ref 134-136), oil refineries (Ref 137), and salt evaporator brine heaters (Ref 133). In all cases, crevice testing involving Teflon gasket-to-metal and metal-to-metal crevices for extended periods revealed no evidence of significant attack to temperatures as high as 250 °C (480 °F). In saturated NH_4Cl solution, no gasket-to-metal or under-salt-deposit attack was noted to 177 °C (350 °F) and at pH 3 to 7. Similar performance is noted for grade 7 titanium under these conditions.

It should be cautioned that deviations from normal crevice corrosion guidelines can be expected in certain acidic salts that may hydrolyze to form HCl at high temperatures when highly concentrated (Ref 111, 112). These salts include $MgCl_2$, $CaCl_2$, $ZnCl_2$, and $AlCl_3$. The grade 7 alloy is generally most resistant in these situations (Ref 112).

Teflon gasket-to-metal crevices of various high-strength titanium alloys were also tested in hot NaCl brines. The data, detailed in Ref 13, produced the ranking of alloy resistance shown in Table 22. Relative alloy crevice corrosion resistance generally parallels alloy resistance in reducing acid media. Titanium alloys containing at least 4 wt% Mo exhibit significantly increased crevice corrosion in high-temperature NaCl media down to relatively low pHs. The beneficial effect of higher molybdenum content is confirmed in Teflon gasket-to-metal crevice tests performed on the Ti-3-8-6-4-4 alloy under simulated deep oil well (downhole) conditions (Ref 138). Testing to 300 °C (570 °F) in a pH 3 solution consisting of 250 g/L NaCl, 1 g/L S, and 103-kPa (15-psi) H_2S revealed no susceptibility to crevice attack and indicated superior resistance to grade 12 titanium under these conditions.

The crevice corrosion resistance of grade 9 titanium appears to be essentially the same as

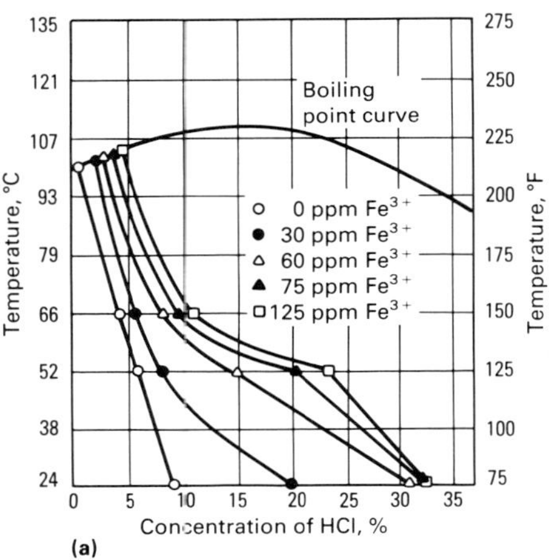

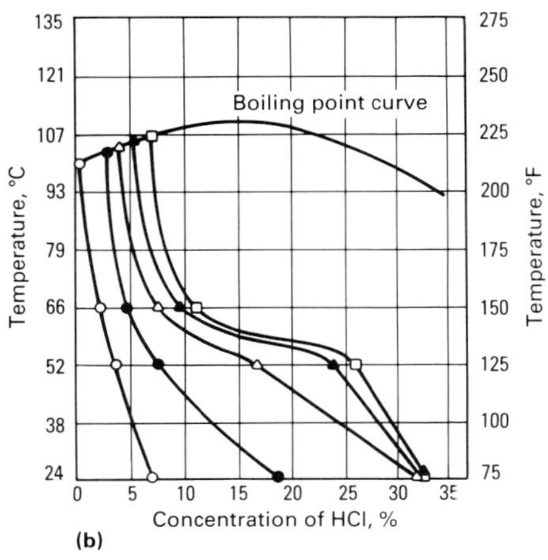

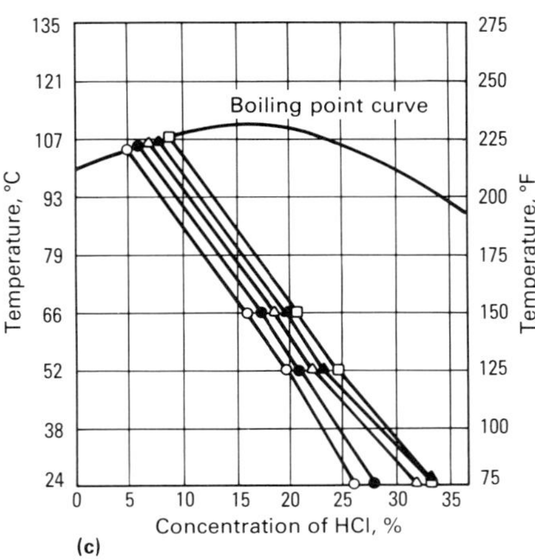

Fig. 21 Effect of minute Fe^{3+} ion concentrations on the useful corrosion resistance of grades 2 (a), 12 (b), and 7 (c) titanium in naturally aerated HCl solutions. 0.127-mm/yr (5-mils/yr) isocorrosion lines are shown.

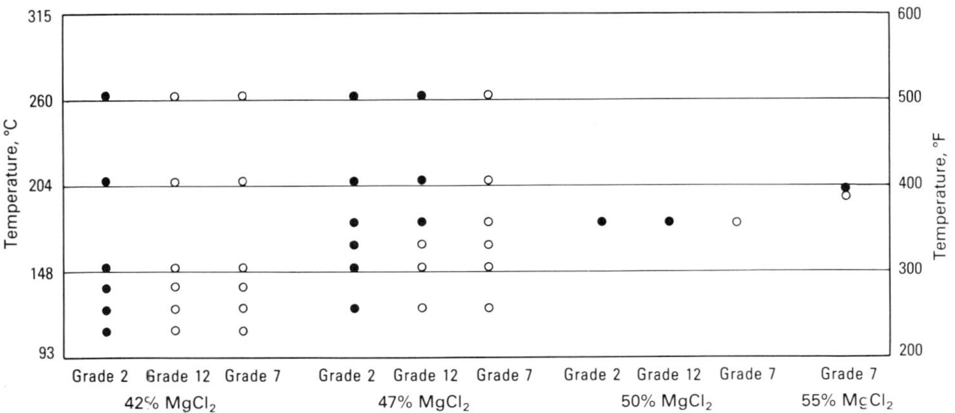

Fig. 22 Temperature guidelines for avoiding localized attack of grades 2, 7, and 12 titanium in concentrated $MgCl_2$ solutions in the absence of crevices. ● denotes susceptibility to attack. Source: Ref 112

Table 16 Effect of impressed anodic potentials on the corrosion of unalloyed titanium in hot reducing acids

Acid	Concentration, %	Temperature °C	Temperature °F	Applied potential, V versus SHE	Corrosion rate mm/yr	Corrosion rate mils/yr	Reduction in corrosion rate
Sulfuric	40	60	140	+2.1	0.005	0.2	11 000×
	40	90	195	+1.4	0.07	2.8	896×
	40	114	237	+2.6	1.8	71	189×
	60	60	140	+1.7	0.035	1.4	662×
	60	90	195	+3.0	0.10	4	163×
Hydrochloric	37	60	140	+1.7	0.068	2.7	2080×
Phosphoric	60	60	140	+2.7	0.018	0.7	307×
	60	90	195	+2.0	0.5	20	100×
Formic	50	Boiling		+1.4	0.083	3.3	70×
Oxalic	25	Boiling		+1.6	0.25	10	350×
Sulfamic	20	90	195	+0.7	0.005	0.2	2710×

Source: Ref 80, 107, 108

that for unalloyed titanium in hot NaCl brines. On the other hand, the greater aluminum content of the grade 5 alloy results in slightly reduced crevice resistance as compared to unalloyed grades.

Studies have addressed the effect of chloride concentration on crevice corrosion initiation on unalloyed titanium (Ref 130). Threshold temperatures of 250, 200, and 150 °C (480, 390, and 300 °F) were indicated for chloride concentrations of 0.01, 0.1, and 1%, respectively, in neutral (pH 7) NaCl brine. Unpublished results suggest that 0.01% Cl^- at 90 °C (195 °F) and 0.10% Cl^- at 70 °C (160 °F) may be threshold conditions for crevice attack in aerated, pH 3 to 5 solutions given tight Teflon gasket-to-metal crevices.

Tests have shown that crevice tightness and type of gasket are critical to crevice corrosion initiation. Teflon gasket-to-metal crevices are generally more susceptible to crack initiation than silicone rubber, neoprene rubber, asbestos, and polyvinyl chloride (PVC) gasket-to-metal crevices (Ref 23, 129). Certain polymeric sealants, such as styrol-acrylic copolymer or methacrylate polymers, may significantly increase susceptibility to crevice attack, especially when the sealant contains chloride salts (Ref 23, 129). On the other hand, metal-to-metal crevices are generally least susceptible to attack. In addition, it has been found that the incubation time for crevice corrosion in NaCl brine may be reduced, and attack aggravated, by impressed anodic potentials (Ref 23, 129). As reviewed in the previous section on crevice corrosion, certain dissolved oxidizing species (cathodic depolarizers) in the brine may similarly decrease crevice attack incubation period while increasing attack rate.

Bromides and Sulfates. Unpublished test results suggest that the pH-temperature guidelines for crevice corrosion of titanium in saturated NaCl (Fig. 31) are applicable in saturated NaBr solutions. However, the rate of crevice attack is measurably lower than that in chlorides at corresponding pH and temperature values.

For example, crevice testing of unalloyed titanium in saturated Na_2SO_4 solutions revealed attack in neutral solutions at temperatures above 110 °C (230 °F) and only below pH 6 at 104 °C (220 °F) with no attack at 93 °C (200 °F) as low as pH 3. Teflon gasket-to-metal crevices were also exposed to a simulated acid sulfate galvanizing solution for 38 days at temperatures of 66, 82, and 100 °C (150, 180, and 212 °F). The pH 1.3 solution consisted of 100 g/L Na_2SO_4, 120 g/L $ZnSO_4$, 0.6 g/L Fe^{3+} (as $Fe_2(SO_4)_3$), and 0.008% Cl^-. Grade 2 titanium proved to be fully resistant, except at boiling (100 °C, or 212 °F). Grade 7 and 12 titanium alloys resisted crevice attack under all conditions tested. All results indicate that threshold temperatures for crevice corrosion in sulfate solution are measurably higher than those in chloride brines.

Anodic Pitting in Specific Media

Anodic Breakdown Pitting. Titanium exhibits its relatively high anodic breakdown potentials, E_b, in aqueous solution as compared to most engineering metals. This is the basis for its use as

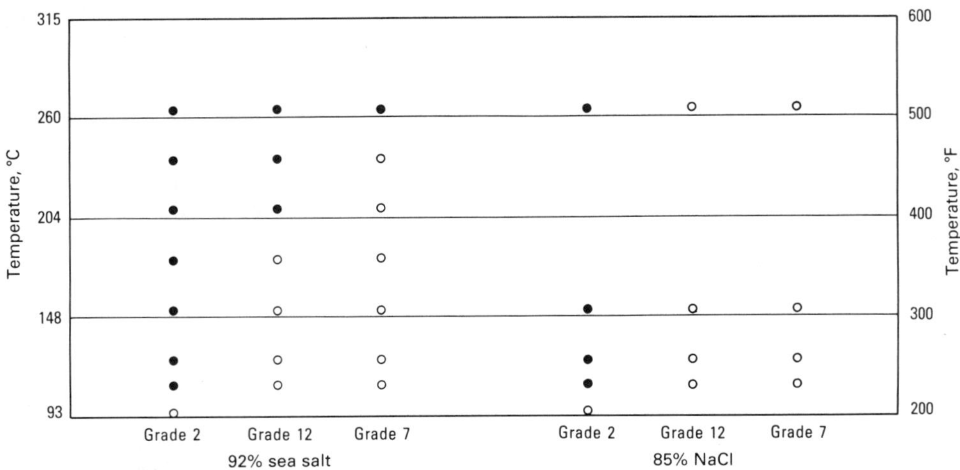

Fig. 23 Temperature guidelines for avoiding localized attack of grades 2, 7, and 12 titanium in pressure-bled tests in concentrated sea salt and NaCl slurries in the absence of crevices. ● denotes susceptibility to localized attack.

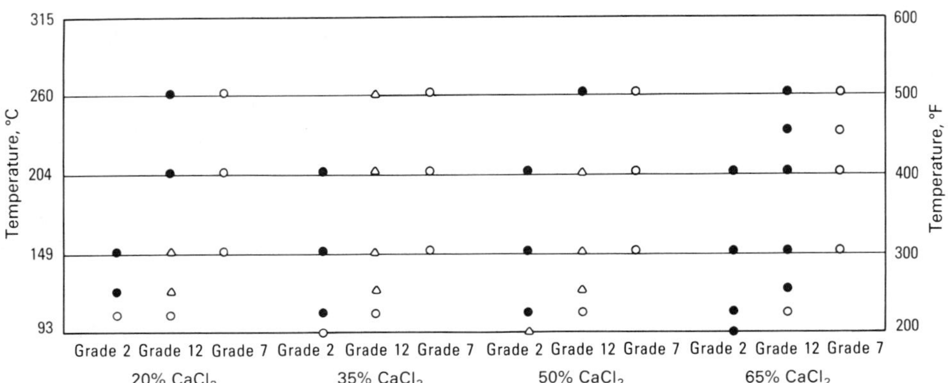

Fig. 24 Temperature guidelines for avoiding localized attack of grades 2, 7, and 12 titanium in concentrated $CaCl_2$ solutions in the absence of crevices. ● denotes susceptibility to localized attack; △ indicates incipient edge attack.

Table 17 Corrosion of unalloyed titanium in highly alkaline solutions

Medium	Concentration, %	Temperature °C	Temperature °F	Corrosion rate mm/yr	Corrosion rate mils/yr
Ammonium hydroxide	28	26	79	0.002	0.08
Ammonium hydroxide	70	Boiling		nil	
Sodium carbonate	20	Boiling		nil	
Sodium hydroxide	28	25	75	0.003	0.12
Sodium hydroxide	10	Boiling		0.02	0.8
Sodium hydroxide	40	66	150	0.038	1.5
Sodium hydroxide	40	93	200	0.064	2.5
Sodium hydroxide	40	121	250	0.13	5
Sodium hydroxide	50	66	150	0.018	0.7
Sodium hydroxide	50–73	188	370	>1.1	>43.3
Sodium hydroxide	73	110	230	0.05	2
Sodium hydroxide	73	Boiling		0.13	5
Potassium hydroxide	10	Boiling		0.13	5
Potassium hydroxide	25	Boiling		0.3	12
Potassium hydroxide	50	25	75	0.010	0.4
Potassium hydroxide	50	Boiling		2.7	106

Source: Ref 8, 26, 67, 80

dimensionally stable anodes for chlor-alkali cells, anodes for recovery of metals or metal oxides from solutions, zinc and nickel plating anode baskets, aluminum anodizing racks, and platinum anode substrates for impressed cathodic protection systems. In sulfate and phosphate media, anodic pitting potentials of titanium alloys are typically in the range of +80 to +100 V (versus Ag/AgCl reference electrode). For this reason, dilute sulfuric and phosphoric acid solutions (and their salts) are typical electrolytes for anodizing titanium to grow protective surface oxides and/or produce colored surfaces.

In halide salt solutions, titanium alloys exhibit somewhat lower but yet reasonably high pitting potentials. Table 23 reveals that the anodic breakdown pitting potential decreases measurably with increasing temperature. Values of +9 to +10.5 V (versus Ag/AgCl) can be expected in room-temperature chloride solutions, decreasing to approximately +1.2 V at 175 to 250 °C (345 to 480 °F). These values are dependent on sample surface condition. For example, abraded or sandblasted surfaces exhibit somewhat lower values than as-pickled surfaces. In one study, increasing alloy aluminum content was shown to reduce anodic pitting potential in titanium alloys (Ref 139, 140). Pitting potentials were determined for a 20% NH_4Cl solution under stagnant and flowing (1 m/s, or 3.3 ft/s) conditions (Ref 143). Values were measured at +9 to +10 V at 20 to 40 °C (70 to 100 °F), decreasing into the range of +4.0 to +5.5 V at 120 °C (250 °F). Increased flow rate indicated a minor negative effect on pitting potential. A relatively minor effect by pH and chloride concentration is also observed in NaCl brine (Ref 142).

It was shown that pitting potentials of titanium can be raised in chloride solutions by addition of sulfate ions (Ref 144). Above a critical sulfate concentration, titanium exhibits pitting potential values similar to those obtained in pure sulfate media (Table 23).

As shown in Fig. 32, anodic pitting potential values are significantly lower in bromide solutions, and they decrease with increasing temperature. The dependence of pitting potential on bromide concentration at room temperature is given by (Ref 142):

$$E_b \text{ (versus SCE)} = 1.1 - 0.43 \log [\text{Br}]$$

At room temperature, anodic pitting potentials of +0.90 to +1.4 V have been reported for titanium grades 2 and 5 (Ref 139, 142). One study has reported values for grade 1, 2, and 3 titanium ranging between +1.8 to +2.2 V in 1% NaBr (pH 6) solution at room temperature, decreasing to +1.0 to +1.2 V at 100 °C (212 °F) (Ref 145). As Fig. 32 suggests, pitting potentials may fall to potentials as low as +0.6 to +0.8 V at temperatures above 130 to 150 °C (265 to 300 °F). Thus, pitting of titanium alloys may be possible in pure bromide solutions at higher temperature if highly oxidizing conditions prevail.

However, additions of various oxidizing anions may inhibit pitting in NaBr solutions by significantly raising anodic pitting potentials (Ref 142). Critical concentrations of these inhibitive anions have been determined, and the relative efficiency of inhibition decreases in the order $SO_4^{2-} > NO_3^- > CrO_4^{2-} > PO_4^{3-} > CO_3^{2-}$.

Studies in room-temperature iodide solutions have revealed anodic pitting potentials of +1.7 to +1.8 V, with little effect of acidification indicated (Ref 139, 146). Above 40 to 50 °C (100 to 120 °F), values near +0.5 V (versus SCE) are reported.

Repassivation potentials, E_p, represent conservative measures of anodic pitting tendency because they represent minimum potentials below which pitting cannot be sustained. The values presented in Table 24 show that titanium alloys will not experience spontaneous pitting attack in aqueous chloride media, even if oxidizing species are present. Unalloyed titanium ex-

Table 18 Corrosion of unalloyed titanium in organic media

Medium	Concentration, %	Temperature, °C (°F)	Corrosion rate mm/yr	Corrosion rate mils/yr
Acetic anhydride	99–99.5	20 (70) to boiling	<0.13	<5
Adipic acid	0–67	204 (400)	<0.05	<2
Adipic acid + 20% glutaric acid + 5% acetic acid	25	200 (390)	nil	
Aniline hydrochloride	5–20	35–100 (95–212)	<0.001	<0.04
Benzene + HCl + NaCl	Vapor + liquid	80 (175)	0.005	0.2
Carbon tetrachloride	99	Boiling	0.003	0.12
Carbon tetrachloride	100	Boiling	0.003	0.12
Chloroform	100	Boiling	nil	
Chloroform + water	50	Boiling	0.12	4.7
Cyclohexane + traces formic acid	...	150 (300)	0.003	0.12
Ethyl alcohol	95	Boiling	0.013	0.5
Ethylene dichloride	100	Boiling	<0.13	<5
Formaldehyde	37	Boiling	<0.13	<5
Tetrachloroethylene	100	Boiling	nil	
Tetrachloroethylene + water	...	Boiling	0.13	5
Tetrachloroethane	100	Boiling	nil	
Trichloroethylene	99	Boiling	<0.13	<5

Source: Ref 8, 26, 67, 109

Table 19 Corrosion of titanium alloys in various organic acids

Acid	Concentration, %	Alloy	Temperature, °C (°F)	Corrosion rate mm/yr	Corrosion rate mils/yr
Acetic	0–99.5	Grades 2, 7, 12	Boiling	nil	
Adipic	67	Grade 2	240 (465)	nil	
Citric, aerated	10–50	Grade 2	100 (212)	0.01	0.4
Citric	50	Grade 2	Boiling	0.35	13.8
Citric	50	Grades 7, 12	Boiling	0.01	0.4
Di- and mono-chloroacetic	100	Grade 2	Boiling	<0.013	<0.5
Formic, aerated	25–90	Grade 2	100 (212)	0.001	0.04
Formic, aerated	25	Grade 2	Boiling	2.4	94.5
Formic	45	Grade 2	Boiling	11.0	433
Formic	45	Grades 7, 12	Boiling	nil	
Formic	10	Grade 2	Boiling	nil	
Lactic, aerated	10	Grade 2	Boiling	0.014	0.55
Lactic	10	Grade 2	100 (212)	0.048	1.9
Lactic	25	Grade 2	Boiling	0.028	1.1
Lactic	85–100	Grade 2	Boiling	0.01	0.4
Oxalic	0.5	Grade 2	60 (140)	2.4	94.5
Oxalic	1	Grade 2	35 (95)	0.15	6
Oxalic	10	Grade 7	Boiling	32.3	1272
Stearic	100	Grade 2	180 (355)	0.003	0.12
Tartaric	10–50	Grade 2	100 (212)	≤0.013	<0.5
Terephthalic	77	Grade 2	225 (437)	nil	
Trichloroacetic	100	Grade 2	Boiling	14.6	575

Source: Ref 8, 26, 80, 103, 109

hibits the highest E_p value, which decreases as alloy aluminum content increases. Increasing iron content over the range of 0.02 to 0.20% results in a minor (several tenths of a volt) decrease in E_p values in unalloyed titanium (Ref 11).

Like anodic pitting potentials, repassivation potentials are significantly lower in bromide and iodide media. Room-temperature E_p values of +1.2 and +0.95 V are measured for grades 2 and 5 titanium, respectively (Ref 139), whereas values of +0.9 V in dilute KBr solutions have been reported (Ref 146). Repassivation potentials for grades 2 and 5 titanium in dilute room-temperature iodide solutions have been measured to be +1.8 and +1.5 V, respectively (Ref 139, 146).

Hydrogen Damage in Specific Media

Gaseous Hydrogen. Absorption of hydrogen by titanium alloys in gaseous hydrogen is highly dependent on temperature, gas pressure, gas moisture or oxygen content, metal surface condition, alloy composition, and the nature of the surface oxide formed on the metal (Ref 148). In the absence of surface oxide film interferences, the hydrogen concentration in titanium is directly proportional to the square root of the hydrogen gas partial pressure (Ref 8, 149). Hydrogen absorption is a fully reversible process in that vacuum may remove absorbed hydrogen given sufficient temperatures to promote kinetics and overcome oxide film diffusion barriers. The ability of anodic and thermal oxide films on titanium to retard hydrogen absorption significantly in high-temperature hydrogen gas has been demonstrated (Ref 148-151).

As shown in Table 25, absorption of hydrogen by grade 2 titanium in dry hydrogen gas dramatically increases as both pressure and temperature increase. Table 25 also reveals the protective effect of surface oxides (as-pickled and anodized surfaces) and the detrimental effect of iron surface contamination. Similar results for unalloyed titanium are given in Table 26, in which hydrogen gas pressures of 4.1 to 6.2 MPa (600 to 900 psi) and temperatures of 100 to 300 °C (212 to 570 °F) were tested. These results also demonstrate the beneficial effects of gas moisture content. It has been shown that at least 2% H_2O content in

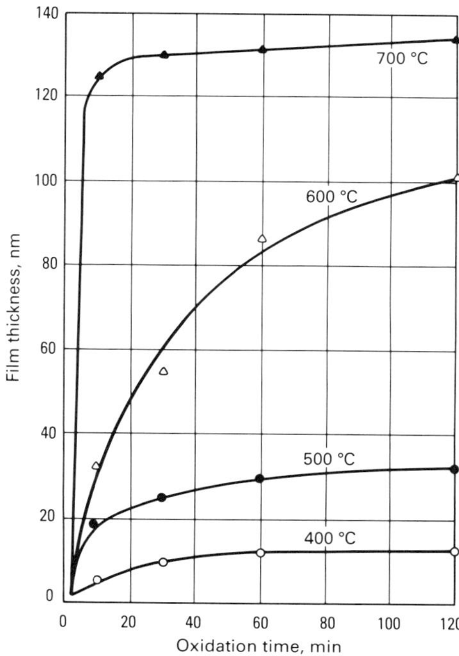

Fig. 25 Growth of thermal oxide films on unalloyed titanium in air. Source: Ref 120

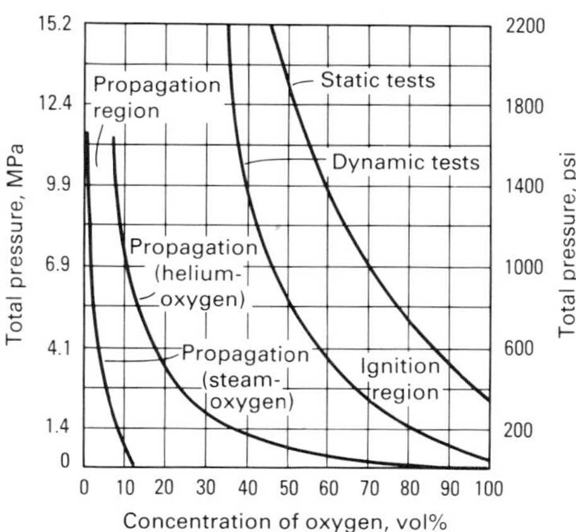

Fig. 26 Ignition and crack propagation limits for unalloyed titanium in various oxygen gas mixtures. Source: Ref 121, 122

5.5-MPa (800-psi) hydrogen gas at 315 °C (600 °F) effectively retards hydrogen uptake in commercially pure titanium (Ref 150).

Cathodic Hydrogen Uptake. As discussed in the section "Hydrogen Damage" in this article, titanium may absorb hydrogen if there is a mechanism for generating atomic hydrogen on the metal surface. This mechanism may involve an impressed cathodic current, galvanic coupling to active metals, or severe, continuous mechanical damage of titanium surfaces. In these situations, the factors of potential, current density, metal temperature, solution pH and chemistry, and alloy composition and surface condition all significantly influence the rate at which hydrogen is absorbed.

Figure 33 shows that unalloyed titanium may absorb hydrogen in near-neutral brines (seawater) at 25 and 100 °C (75 and 212 °F) when

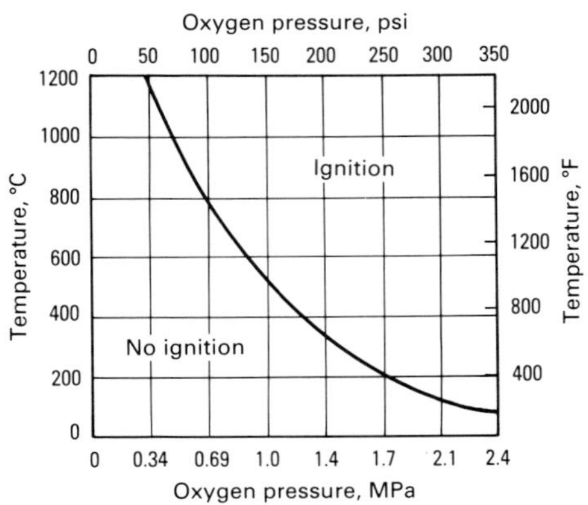

Fig. 27 Ignition limits for ruptured unalloyed titanium in pure oxygen gas atmospheres. Source: Ref 122

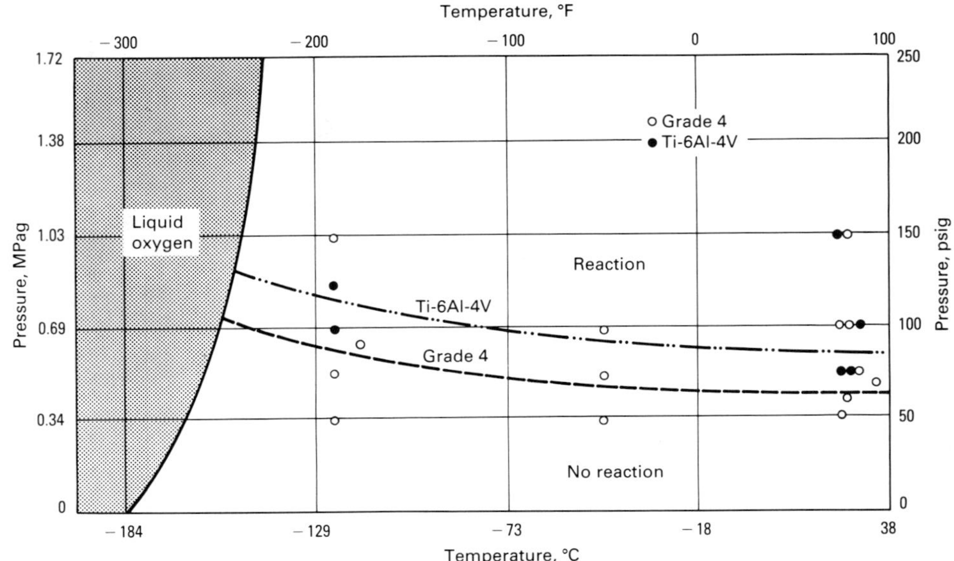

Fig. 28 Ignition limits for grade 4 titanium and Ti-6Al-4V alloy in low-temperature pure oxygen atmospheres. Source: Ref 67, 121

cathodic potentials are more active (negative) than −0.75 V versus SCE. This charging produces only thin, innocuous surface hydride films (Fig. 7) on unalloyed titanium as long as potentials remain more noble than −1.0 V and temperatures are below 80 °C (175 °F). For example, grade 2 titanium galvanically coupled to aluminum alloy 6063 in a process water electrolyte at 65 °C (150 °F) generated surface hydrides only. However, accelerated hydriding of unalloyed titanium may occur at impressed potentials cathodic (negative) to −1.0 V even at room temperature. The potential threshold indicated in Fig. 33 applies to α-β and β titanium alloys as well; however, significantly higher rates of hydrogen absorption and penetration can be expected at any given temperature under hydrogen charging conditions. This results from the significantly higher solubility and diffusion coefficients for hydrogen in the β phase (Ref 36, 38).

Increasing temperature and decreasing pH have both been shown to accelerate significantly the rate of hydrogen absorption of titanium alloys during cathodic charging (Ref 39, 44, 46, 152-154). These studies indicate that the rate of hydrogen absorption and surface hydride layer growth initially follow a parabolic rate law; this suggests that hydrogen diffusion is rate controlling at temperatures less than and equal to 100 °C (212 °F) (Ref 155). Linear rate behavior has been observed in long-term charging exposures (Ref 46). Dramatic increases in hydrogen absorption rate during cathodic charging are noted when electrolyte pH levels are reduced to 2 or below (Ref 29, 39, 46, 150, 152).

The presence of hydrogen recombination poisons, such as sulfide, arsenate, or antimony species, substantially increases hydrogen uptake during charging (Ref 152). This effect of sulfide has been observed in a few oil refinery heat exchangers in which grade 2 titanium tubes were galvanically coupled to carbon steel tubesheets and components in hot (≥80 °C, or 175 °F) sulfide-containing aqueous process streams. Severe hydriding (Fig. 6) and eventual embrittlement of the tubes occurred. Elimination of the detrimental couple with the active metal (steel) has been achieved by designing with more galvanically compatible passive alloys (including all-titanium design) or use of dielectric joints. In comparison, grade 2 titanium galvanically coupled to carbon steel in neutral sulfide-free electrolytes results in very minor increases in hydrogen uptake to temperatures as high as 120 °C (250 °F) (Ref 29, 44).

The surface condition of titanium also influences hydrogen absorption rates during cathodic charging. Studies consistently reveal that as-pickled and as-received surfaces are much less amenable to hydrogen uptake than abraded, vapor-blasted, or sandblasted surfaces (Ref 29, 39, 45). Furthermore, anodized and, particularly, thermally oxidized surfaces are highly effective barriers to hydrogen uptake in titanium during charging (Ref 26, 42, 47).

Stress-Corrosion Cracking in Specific Media

In the following sections, SCC of titanium alloys in a variety of environments will be discussed. Most of the discussion will concentrate on environments that were identified through the use of precracked, notched test specimens (cate-

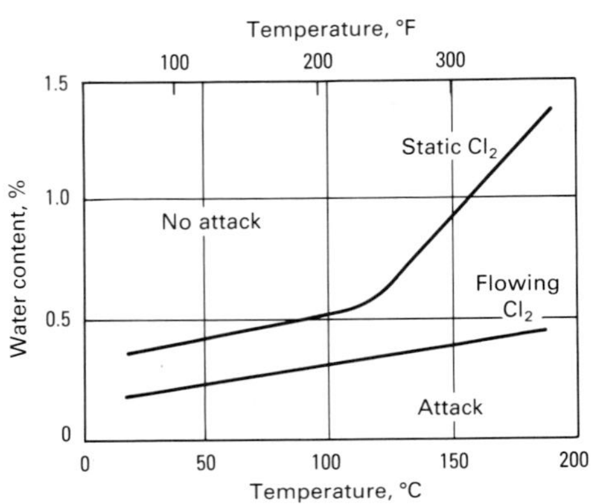

Fig. 29 Effect of temperature and gas flow on the critical water content required to passivate titanium in pure chlorine gas. Source: Ref 125

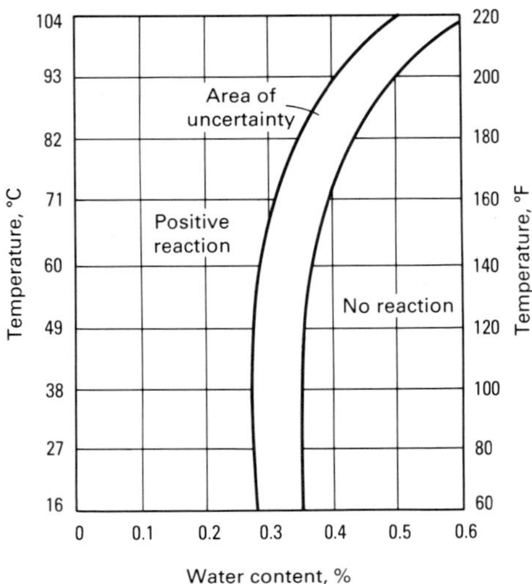

Fig. 30 Water content necessary to maintain passivity of unalloyed titanium in static chlorine gas atmospheres. Source: Ref 126

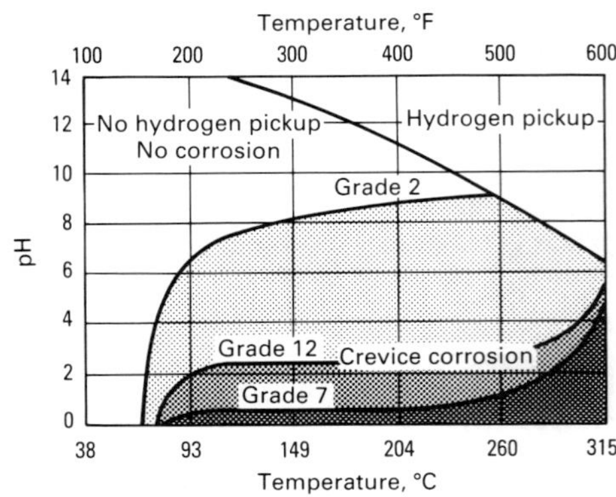

Fig. 31 Temperature-pH limits for crevice corrosion of grades 2, 7, and 12 titanium in saturated NaCl brines. Crevice corrosion will occur in the shaded area. Source: Ref 26, 132

Table 20 Corrosion of unalloyed titanium in liquid metals

Liquid metal	Temperature °C	Temperature °F	Corrosion rate mm/yr	Corrosion rate mils/yr
Bismuth-lead	300	570	<0.1	<4
Bismuth-lead	600	1110	0.13–1.3	5–50
Gallium	400	750	0.1	4
Gallium	450	840	>1.0	>40
Lithium	850	1560	0.1–1.0	4–40
Magnesium	750	1380	0.1	4
Magnesium	850	1560	0.1–1.0	4–40
Lead	400	750	<0.13	<5
Lead	600–950	1110–1740	0.1–1.0	4–40
Mercury	150	300	<0.1	<4
Mercury	150–300	300–570	0.1–1.0	4–40
Sodium, potassium	600	1110	<0.1	<4
Sodium, potassium	800	1470	0.1–1.0	4–40
Sodium, potassium	600	1110	<0.1	<4
Tin	350	660	<0.1	<4
Tin	600	1110	0.1–1.0	4–40
Aluminum	750	1380	<0.1	<4
Aluminum	850	1560	≥0.1	≥4
Cadmium	500	930	>1.0	>40
Zinc	445	830	≥1.0	≥40

Table 21 Resistance of titanium alloys to crevice corrosion in boiling salt solutions

Tight metal-to-gasket crevices were tested.

Solution	pH	Alloy/resistance(a) Grade 2	Grade 12	Grade 7
Saturated $ZnCl_2$	3.0	F	R	R
10% $MgCl_2$	4.2	F	R	R
10% $CaCl_2$	3.0	F	R	R
10% KCl	3.0	F	R	R
Saturated NaCl	3.0	F	R	R
Saturated NaCl + Cl_2	1–2	F	F	R
10% NH_4Cl	4.1	F	R	R
10% $FeCl_3$	0.6	F	F	R
10% Na_2SO_4	2.0	F	R	R

(a) F, failed; R, resistant. Source: Ref 133

Table 22 Ranking of titanium alloy crevice corrosion resistance in NaCl brines at temperatures ≥90 °C (195 °F)

Alloy	Maximum pH at which attack occurred
Least resistant	
Ti-6-4	≥5
Grade 2	≥5
Ti-4-3-1	5
Ti-550	3
Grade 12	2
Transage 210	2
Ti-6-2-4-6	2
Ti-3-8-6-4-4	1
Ti-15-5	<1
Most resistant	

gory 2), as in Ref 48. Mention also will be made of the few environments in which smooth specimens are sufficient for the identification of SCC.

Although the protective oxide film of titanium exhibits a high degree of stability and integrity in many aggressive environments, even to the extent that essentially no corrosion occurs, SCC has been identified in several environments. The SCC literature for titanium alloys is summarized in Table 27. Table 27 lists the environments that are known to cause SCC and the alloys that have been found to be susceptible. Each of these environments and the metallurgical factors that influence SCC susceptibility are discussed in the following sections.

Red Fuming Nitric Acid. The first reported observation of titanium SCC occurred in red fuming nitric acid (Ref 156). Cracking was observed for commercially pure titanium in a room-temperature environment containing 20% NO_2. Later investigators indicated that SCC occurred in as little as 6.5% NO_2 with less than 0.7% H_2O. Intergranular cracking was observed on smooth specimens, which indicates the high sensitivity to SCC in this medium. In addition to commercially pure titanium, cracking was observed on Ti-8Mn and Ti-6Al-4V, even in red fuming HNO_3 without free NO_2. Work on inhibitors showed that 1.5 to 2.0% H_2O completely inhibits SCC. It was also shown that 1% NaBr inhibits SCC (Ref 156). The work on water as an inhibitor is especially interesting because it is of importance in a number of other environments that promote SCC.

Nitrogen Tetroxide. As titanium use in aerospace increased, it was found that titanium alloys were highly resistant to corrosion attack in nitrogen tetroxide (N_2O_4), an oxidizer used with hydrazine rocket fuels. Unfortunately, SCC was rather dramatically revealed in an explosion during proof testing of a Ti-6Al-4V storage vessel. The vessel that exploded had been exposed to N_2O_4 at 40 °C (100 °F) at a stress level of 620 MPa (90 ksi). Testing revealed that titanium alloys would crack in NO-free N_2O_4 (that is, N_2O_4 with excess dissolved oxygen, or red N_2O_4) but would not crack in NO-containing N_2O_4 (oxygen free, or green N_2O_4) (Ref 127).

Methanol. Prior to the discovery of SCC in N_2O_4, methanol was found to cause stress cracking of titanium alloys. Methanol and NaBr were shown to be extremely corrosive to titanium, and in some cases, they promoted intergranular SCC of smooth specimens (Ref 157). Soon after this discovery, it was shown that methanol/HCl and methanol/H_2SO_4 mixtures also caused SCC of commercially pure titanium—once again on smooth specimens (Ref 158). Stress-corrosion cracking in methanol was dramatically rediscovered when a Ti-6Al-4V pressure vessel exploded during proof testing with methanol. This led to a flurry of research to discover the nature of this cracking phenomenon and the metallurgical factors that promoted cracking.

Two types of SCC are observed in methanol solutions and are characterized by the failure mode exhibited. In the first type, intergranular fracture is evident. This type of fracture is common for commercially pure titanium and β titanium alloys, such as Ti-13V-11Cr-3Al, that are exposed to methanol containing a halide ion, such as Cl^- or Br^-. Susceptibility to SCC measured as time to failure of smooth samples indicates that:

- Increasing the halide content decreases time to failure
- Water additions to a critical level decrease time to failure
- Higher halide concentrations increase the critical level of water for maximum susceptibility to SCC
- Water levels beyond the critical level reduce and can inhibit cracking susceptibility

Details of this failure mode are presented in Ref 113-116 and 159-161.

The effects of both cathodic and anodic polarization have also been investigated. Anodic polarization increases the susceptibility of titanium to SCC in methanol/halide mixtures. On the other hand, cathodic polarization dramatically reduces SCC susceptibility, as shown in Fig. 34. Potentials more negative than −250 mV versus Ag/AgCl prevent cracking in methanol (Ref 161).

Metal ion additions have also been examined and found to affect methanol SCC. In general, additions that have altered the cathodic reaction, such as palladium, chromium, iron, and gold, have accelerated cracking. This is shown in Fig. 35 for palladium.

The other fracture mode, typical of highly alloyed titanium, is characterized by transgranular α-phase cleavage. Alloys such as Ti-8Al-1Mo-1V typify this failure mode. In tests using precracked specimens, cracking changes from intergranular in stage I cracking (SCC initiation, Fig. 8) to transgranular in stage II (SCC propagation). Most of the α and α-β alloys susceptible to SCC in neutral aqueous solutions (discussed later) exhibit this mode of failure. In contrast, β titanium alloys, such as Ti-11.5Mo-6Zr-4.5Sn, exhibit intergranular fracture in stage II. Stage II is the region in which crack velocity is essentially independent of stress intensity (SCC propagation).

In general, halide additions increase crack velocity in stage II. It has also been reported that additions of sulfuric acid and acetic acid accelerate cracking. Application of cathodic potential between −1.5 and −1.0 V versus SCE has been shown to prevent crack initiation (stage I). Anodic potentials appear to increase crack velocity. As indicated earlier, small water additions prevent methanol SCC initiation. This effect is shown in Fig. 36. Little work has been performed on the effect of temperature on methanol SCC; however, the data available indicate that crack velocities increase with temperature.

Titanium alloys known to be susceptible to aqueous SCC (discussed later) are also susceptible to methanol SCC, and the alloys most susceptible to seawater are also most susceptible to methanol. In addition, the alloys that are susceptible to SCC in distilled water are not beneficially affected by the inhibiting effect of water (Fig. 37).

Other Alcohols. Very little work has been performed on SCC in alcohols other than metha-

Table 23 Anodic breakdown pitting potentials, E_b, for titanium alloys in chloride solutions

Alloy	Solution	pH	Temperature °C	Temperature °F	E_b, V(a)	Ref
Grade 2	1 N NaCl	7	25	75	+11.0	139
Grade 5	1 N NaCl	7	25	75	5.2	139
Grade 2	Saturated NaCl(b)	1, 7	25	75	9.6	...
Grade 12	Saturated NaCl(b)	1, 7	25	75	9.6	...
Grade 7	Saturated NaCl(b)	1, 7	25	75	9.6	...
Grade 5	Saturated NaCl(b)	1, 7	25	75	8.9	...
Grade 2	Saturated NaCl	1, 7	95	200	5.0–6.5	...
Grade 12	Saturated NaCl	1, 7	95	200	5.0–5.7	...
Grade 7	Saturated NaCl	1, 7	95	200	5.2–7.0	...
Grade 5	Saturated NaCl	1, 7	95	200	2.5–3.4	...
Grade 2	1 N NaCl	7	125	255	~4.4	140
Grade 2	1 N NaCl	7	150	300	~2.2	140
Grade 2	1 N NaCl	7	175	345	~1.2	140
Grade 2	1 N NaCl	7	200	390	~1.2	140
Grade 12	Seawater	8	245	475	2.3	141
Grade 12	O_2-saturated seawater	8	245	475	3.3	141
Grade 2	1 N KCl + 0.2 M H_2SO_4	...	25	75	80.0	142

(a) Measured versus Ag/AgCl reference electrode. (b) Similar values were obtained in synthetic seawater (pH 8).

Table 24 Repassivation potentials of as-annealed titanium alloys in boiling chloride media

Alloy	Repassivation potential, V(a): 5% NaCl (pH 3.5)	3% HCl	Saturated NaCl
Grade 1	...	...	+7.0
Grade 2	+6.7	+5.8	+5.7
Ti-6-4	+2.3	+1.7	...
Ti-550	+2.8	+2.3	...
Ti-6-2-4-6	+3.0	+2.4	...
Ti-3-8-6-4-4	+3.2	+2.6	...
Ti-8-8-2-3	+2.6	+2.4	...
Ti-15-5	+6.3	+5.6	...
Grade 12	...	...	+5.9
Grade 7	...	...	+5.6

(a) Measured versus Ag/AgCl reference electrode. Source: Ref 147

nol. The studies that have been reported on alloys other than commercially pure titanium show that certain α-β alloys, such as Ti-6Al-4V, may be susceptible to cracking in anhydrous ethanol. Cracking in ethylene glycol has also been reported for Ti-8Al-1Mo-1V. Other work indicates that cracking susceptibility significantly diminishes as the number of carbon atoms in the alcohol increases.

Halogenated Hydrocarbons. No testing of commercially pure titanium has been performed in common hydrocarbons. However, widespread use of titanium alloys in the aerospace industry has prompted considerable study of SCC in halogenated hydrocarbons common to aerospace processing. Stress-corrosion cracking of certain titanium alloys has been identified in the following hydrocarbons:

- Carbon tetrachloride
- Methylene chloride
- Methylene iodide
- Trichloroethylene
- Trichlorofluoromethane
- Trichlorofluoroethane
- Octafluorocyclobutane

In most of these environments, precracked specimens (category 2) are required to identify SCC.

Carbon tetrachloride (CCl_4) SCC was first noted in Ti-8Al-1Mo-1V (Ref 162-165). The threshold stress intensity was approximately the same as that observed for SCC in 3.5% NaCl. Crack velocities in CCl_4 were approximately ten times faster than velocities in methanol. Studies on dynamically loaded smooth specimens (category 3) also showed that Ti-5Al-2.5Sn was susceptible to SCC in CCl_4 at stresses approaching the tensile strength of the alloy.

The other hydrocarbons identified were found to cause cracking in Ti-8Al-1Mo-1V and Ti-5Al-2.5Sn, alloys known to be susceptible to SCC in distilled water (Ref 165, 166). No other alloys were found to be similarly affected.

Freons include any of a number of fluorinated hydrocarbons commonly used as refrigerants. Titanium alloys Ti-8Al-1Mo-1V and Ti-5Al-2.5Sn have been found to exhibit threshold stress intensities in commercial freons below air threshold stress intensities (Ref 166). The alloy Ti-6Al-4V was also identified as susceptible when exposed in the solution-treated and aged condition.

Hot Salts. In the late 1950s, cracking of a titanium alloy was discovered during routine creep testing. The failure was eventually traced to chlorides from fingerprints on the creep specimen. These findings were reproduced in several laboratory studies for a host of titanium alloys. Nearly all titanium alloys were found to be susceptible to this cracking phenomenon (termed hot salt cracking) with the exception of the commercially pure grades of titanium. With this discovery, a great deal of concern was expressed with regard to the multitude of existing applications similar to this laboratory environment. However, after much investigation, it was found that no failure in the field could be attributed to hot salt cracking.

Several complete descriptions of hot salt cracking can be found in the literature (Ref 167-175). Hot salt cracking is primarily influenced by temperature, stress, time, and the alloy itself. Cracking is observed in the temperature range from 285 to 425 °C (545 to 800 °F). In general, susceptibility increases with stress and/or temperature and does not occur below 260 °C (500 °F) or above 540 °C (1000 °F).

Cracking is normally characterized by extensive branching and is not necessarily associated with the regions of highest stress intensity; therefore, category 2 specimens are not required to initiate cracking. Indeed, it is often difficult to initiate cracks in precracked notches. Statically loaded beam specimens (category 1) have been used in most of the laboratory investigations.

The alloys that are most susceptible to hot salt cracking are α alloys with more than 3% Al, such as Ti-5Al-2.5Sn. Commercially pure titanium is apparently immune (Ref 176). Alpha-beta alloys are less susceptible to cracking, although alloys with high aluminum contents are most susceptible. Apparently, the least resistant titanium alloy is Ti-8Al-1Mo-1V. Alloys with higher molydenum content, such as Ti-4Al-3Mo-1V, are most resistant (Ref 169). The combined effect of time, temperature, and stress is shown in the Larsen-Miller diagram in Fig. 38 for several alloys. From

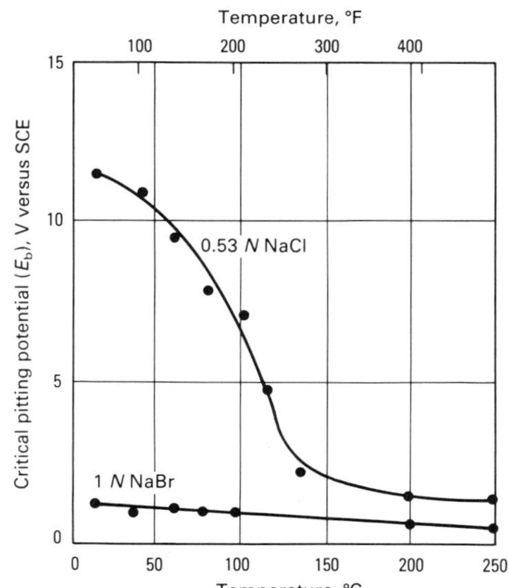

Fig. 32 Effect of temperature on the anodic pitting potential E_b of grade 2 titanium in dilute NaCl and NaBr solutions. Source: Ref 142

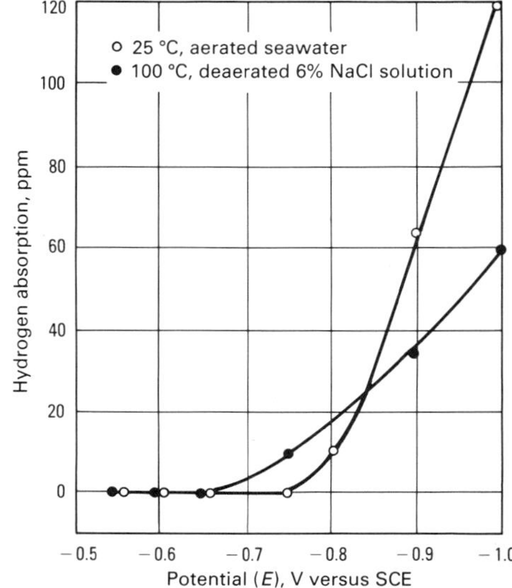

Fig. 33 Cathodic potential thresholds for absorption of hydrogen by unalloyed titanium in neutral brines. Source: Ref 41, 42

Table 25 Hydrogen absorption in grade 2 titanium after 96-h exposures in pure, dry hydrogen

Temperature °C	Temperature °F	Hydrogen pressure MPa	Hydrogen pressure psi	Hydrogen absorption, ppm/specimen surface preparation: Freshly pickled	Pickled + 2-month exposure	Iron-contaminated	Anodized
149	300	Atmospheric		0	0	0	0
149	300	2.76	400	58	20	174	0
149	300	5.52	800	28	0	117	0
315	600	Atmospheric		0	0	0	0
315	600	2.76	400	2586	6911	5951	516
315	600	5.52	800	4480	10 550	13 500	10 000

Source: Ref 150

Table 26 Hydrogen absorption in unalloyed titanium after 14-day exposures

Sample condition	Hydrogen absorption, ppm: Dry hydrogen	Moist hydrogen	Hydrogen + CO_2
As-received	58	1	4
Iron-contaminated	780	21	125
Copper-contaminated	27	5	5
Platinum-contaminated	...	...	770
Anodized	9	2	3

Source: Ref 151

Fig. 38, it is clear that alloy type and microstructural condition are important.

Oxygen has been reported as necessary for hot salt cracking. At least one study has shown that cracking will not occur in Ti-5Al-2.5Sn when the environmental pressure is reduced below 10 μm (Ref 169). Although the role of water (moisture) has not been clearly established, it appears that water is also a necessary environmental component in the cracking process (Ref 170, 171).

Chloride, bromide, and iodide salts have all been shown to produce similar cracking. Fluoride and hydroxide salts have not. The cation associated with the salt has also been reported to affect cracking susceptibility. The severity of attack has been shown to increase as follows (Ref 170, 171):

$$MgCl_2 > SrCl_2 > CsCl > CaCl_2 > KCl > BaCl_2 > NaCl > LiCl$$

Table 28 lists titanium alloys in order of their susceptibility to hot salt cracking. This list is taken from Ref 177 and has not met with unanimous agreement.

Cracking is normally intergranular in nature, but it depends largely on alloy type. Alpha alloys exhibit both transgranular and intergranular fracture, depending on whether the material was annealed above or below the β transus, respectively. Alpha-beta alloys exhibit predominantly intergranular fracture (Ref 171, 173, 176, 178).

From a practical standpoint, hot salt cracking appears to be a phenomenon that is restricted to the laboratory. As indicated earlier, no in-service failure has been attributed to hot salt cracking. The likely reason for this is the critical relationship among environment, stress level, and alloy type. Unless all of the conditions are met simultaneously and for extended time, cracking will not occur.

Molten Salts. It would appear that Ti-8Al-1Mo-1V is the only titanium alloy tested for SCC in molten salt environments. Cracking has been observed in pure chloride and bromide eutectic melts at temperatures between 300 and 500 °C (570 and 930 °F). In general, increasing temperature increases crack velocity. Cathodic protection has been observed to inhibit or stop cracking.

Nitrate salts below 125 °C (255 °F) do not induce cracking even when Cl^-, Br^-, or I^- anions are present. At higher temperatures in pure molten nitrates, cracking can occur only when halides are present (Ref 164).

Liquid/Solid-Metal Embrittlement. Several metals, both in liquid and solid form, have been found to induce cracking in contact with titanium alloys (Ref 127, 164, 179-190). The first reported incidence stemmed from a cracked compressor disk in contact with cadmium-plated steel bolts (Ref 179). Initial speculation hinted that the exposure temperature may have been above the melting point of cadmium, leading to liquid-metal embrittlement. However, later work found that cracking would occur well below the melting point of cadmium (Ref 186), such as at room temperature for Ti-6Al-4V.

Those metals known to cause cracking of titanium alloys include cadmium, mercury, zinc, and certain silver brazing alloys. The titanium alloys that are known to be susceptible to cracking in cadmium include commercially pure titanium (ASTM grade 3) with more than 0.2% oxygen, Ti-4Al-4Mn, Ti-8Mn, Ti-31V-11Cr-3Al, Ti-6Al-4V, and Ti-8Al-1Mo-1V. It is likely that most other titanium alloys are susceptible, but have not been tested.

Alloys tested and found to crack in mercury include commercially pure titanium (ASTM grade 4, ~0.3% oxygen), Ti-8Mn, Ti-13V-11Cr-3Al, Ti-6Al-4V, and Ti-8Al-1Mo-1V. As with cadmium, other alloys are probably susceptible, but have not been tested.

Zinc, in both solid and liquid form, has been reported to cause cracking of titanium alloys. However, there is conflicting evidence in the literature as to whether this is actually the case.

Silver and silver brazing alloys have been shown to cause cracking in titanium alloys that are particularly sensitive to SCC. These alloys include Ti-8Al-1Mo-1V, Ti-5Al-2.5Sn, and Ti-7Al-4Mo (Ref 187, 188). As with cadmium, both solid and liquid forms of silver may produce cracking. Susceptibility for Ti-6Al-4V is considered to be above 345 °C (650 °F).

Aqueous Environments. Under certain metallurgical conditions, several titanium alloys have been shown to be susceptible to SCC in distilled water. These include Ti-8Al-1Mo-1V, Ti-5Al-2.5Sn, and Ti-11.5Mo-6Zr-4.5Sn. Microstructural variation for each alloy affects the degree of susceptibility. For example, mill-annealed Ti-8Al-1Mo-1V is less susceptible than step-cooled Ti-8Al-1Mo-1V. Testing has been performed with category 2 type specimens where crack velocity and threshold stress intensity are determined. In these alloys, the degree of susceptibility is highly dependent on heat treatment.

Test results in neutral-pH environments with category 2 type specimens indicate that titanium alloys exhibit a threshold stress intensity, K_{ISCC}, below which cracks will not propagate. The individual effects of ionic species, concentration, potential, pH, and so on, have been extensively studied and are discussed below.

Ionic Species. The anions Cl^-, Br^-, and I^- are the only species shown to promote and/or induce SCC in titanium alloys. The few alloys susceptible to cracking in distilled water become more susceptible, while alloys that are not susceptible in distilled water may become susceptible when these species are present. Anions such as NO_3^-, SO_4^{2-}, OH^-, F^-, $Cr_2O_4^{2-}$, and PO_4^{3-} may reduce sensitivity.

In general, cations do not alter SCC sensitivity. However, oxidizing cations, such as Fe^{3+} or Cu^{2+}, can increase K_{ISCC} in more susceptible alloys. This effect is analogous to anodic polarization, which is discussed in the section "Potential and pH" in this article.

Concentration. As shown in Fig. 39, increasing the concentration of anions that promote SCC generally increases crack velocity (Ref 164, 191, 192) and decreases K_{ISCC}. Additions of SO_4^{2-} and NO_3^- to distilled water can completely inhibit SCC in alloy/heat-treatment combinations that are moderately susceptible, such as mill-annealed Ti-8Al-1Mo-1V. More susceptible combinations, such as step-cooled Ti-8Al-1Mo-1V, are not similarly affected.

Potential and pH. It is difficult to discuss the effect of potential and pH on SCC independently. In neutral brines containing halides, K_{ISCC} varies significantly with potential. Both cathodic and anodic polarization increase K_{ISCC}. Within a narrow potential region, K_{ISCC} reaches a minimum. Both the start and breadth of this range are dependent on the alloy and its metallurgical condition. In general, the range of maximum susceptibility (minimum K_{ISCC}) occurs at approximately −500 mV versus SCE. Examples of this behavior are shown in Fig. 40.

Stress-corrosion crack velocity is also affected by potential. For alloys such as Ti-13V-11Cr-3Al, crack velocity increases linearly with anodic polarization (Ref 192). The slope of the crack velocity curve depends on the alloy type and metallurgical condition, and application of sufficient cathodic potential can halt a stress-corrosion crack.

In acidic solutions, potential effects are somewhat different (Ref 193). First, cathodic polarization will not stop a propagating crack. Second, lowering pH usually decreases K_{ISCC} and increases crack velocity at a constant potential, as shown in Fig. 41. In alkaline solutions, SCC is similar to that found in neutral solutions (Ref 194, 195).

Temperature. Limited published data were available on the effect of temperature on SCC in aqueous environments (Ref 164, 191). Within a narrow temperature range (0 to 93 °C, or 32 to 200 °F), K_{ISCC} in NaCl solution was found to be independent of temperature for Ti-8Al-1Mo-1V. However, crack velocity was found to be strongly dependent on temperature.

Experimentation with category 1 and category 2 specimens has shown that the β alloy Ti-3Al-

Table 27 Environments known to promote SCC of commercial titanium and titanium alloys

Medium	Temperature °C	Temperature °F	Susceptible materials
Oxidizers			
Nitric acid, RFNA	Room		Unalloyed titanium, Ti-8Mn, Ti-6Al-4V, Ti-5Al-2.5Sn, Ti-2Fe-2Cr-2Mo
Nitrogen tetroxide (no excess NO)	30–75	85–165	Ti-6Al-4V
Organic compounds			
Methanol (chloride, bromide)	Room		Ti-6Al-4V
			Ti-6Al-4V, Ti-8Al-1Mo-1V, grade 4
			Ti-6Al-4V, Ti-8Al-1Mo-1V, Ti-5Al-2.5Sn, Ti-4Al-3Mo-1V
Methyl chloroform, inhibited	370	700	Ti-8Al-1Mo-1V, Ti-6Al-4V, Ti-5Al-2.5Sn, Ti-13V-11Cr-3Al
Ethyl alcohol	Room		Ti-8Al-1Mo-1V
			Ti-8Al-1Mo-1V, Ti-5Al-2.5Sn
Ethylene glycol	Room		Ti-8Al-1Mo-1V
Trichloroethylene	370, 620,	700, 1150,	Ti-8Al-1Mo-1V, Ti-5Al-2.5Sn
	815	1500	Ti-5Al-2.5Sn
Trichlorofluoroethane	788	1450	Ti-8Al-1Mo-1V, Ti-5Al-2.5Sn, Ti-6Al-4V, Ti-13V-11Cr-3Al
Chlorinated diphenyl	315–370	600–700	Ti-5Al-2.5Sn
Hot salt			
Chloride salts—various, residues	288–426	550–800	All commercial alloys except grades 1, 2, 7, 11, and 12
Metal embrittlement			
Cadmium	Room		Ti-6Al-4V
	329–400	625–750	Ti-8Mn
Mercury	Room		Grade 4 Ti, Ti-6Al-4V
	370	700	Ti-13V-11Cr-3Al
Silver			
Plate	470	875	Ti-7Al-4Mo, Ti-5Al-2.5Sn
AgCl	370–480	700–900	Ti-7Al-4Mo, Ti-5Al-2.5Sn
Ag-5Al-2.5Mn	340	650	Ti-6Al-4V, Ti-8Al-1Mo-1V
Seawater	Ambient		Unalloyed titanium (with high oxygen content)
			Ti-8Mn
			Ti-2.5Al-1Mo-11Sn-5Zr-0.2Sc
			Ti-3Al-11Cr-13V
			Ti-5Al-2.5Sn
			Ti-6Al-4V
			Ti-6Al-6V-2Sn
			Ti-8Al-1Mo-1V
			Ti-3Al-8V-6Cr-4Zr-4Mo
Miscellaneous			
Chlorine	288	550	Ti-8Al-1Mo-1V
Hydrochloric acid			
10%	35	95	Ti-5Al-2.5Sn
...	340	650	Ti-8Al-1Mo-1V

8V-6Cr-4Mo-4Zr does not crack in NaCl brines up to 205 °C (400 °F) (Ref 196). Tests above that temperature in NaCl brines containing H_2S may produce cracking with category 3 specimens.

Hydrogen Sulfide. Tests in acidic NaCl brine environments containing hydrogen sulfide employing category 1 specimens indicate that titanium alloys are immune to SCC at stress levels up to their yield strength (Ref 138). It is, however, unclear whether the category 1 type specimens are capable of determining SCC susceptibility in these environments. As noted in the previous section, elevated-temperature tests in H_2S brines indicate that Ti-3Al-8V-6Cr-4Mo-4Zr is susceptible to SCC. Unfortunately, no similar tests in environments with H_2S have been reported. Therefore, H_2S may not be the species responsible for SCC in this alloy.

Metallurgical Effects. In addition to environmental effects, the metallurgical condition of a particular titanium alloy will influence its susceptibility to SCC. Both chemistry and heat treatment (microstructure) play a part in these effects.

In α alloys, the two most important alloying elements are aluminum and oxygen. Binary titanium-aluminum alloys have been extensively studied. In this investigation, it has been established that 5% Al is necessary for SCC to occur in aqueous environments. As the aluminum level is increased, K_{ISCC} decreases and crack velocity increases (Ref 193, 197). The most susceptible heat treatments in aluminum-containing alloys are those that produce the α_2 phase.

Binary titanium-oxygen alloys also exhibit a critical level of oxygen below which SCC does not occur, because of the transition from wavy to planar slip (Ref 198). This level, generally taken as less than 0.20 to 0.25%, represents the break between ASTM grade 2 and ASTM grade 3. Titanium-aluminum-oxygen alloys generally suffer from the cumulative effects of both aluminum and oxygen on SCC (Ref 198). Tin additions to titanium-aluminum alloys generally cause a loss in SCC resistance. This is clearly demonstrated by Ti-5Al-2.5Sn, one of the more susceptible titanium alloys.

The α-β titanium alloys are considerably more difficult to generalize with regard to SCC behavior. This stems from the wide variety of microstructures that can be produced and the number of alloying elements involved.

Stress-corrosion cracking of Ti-8Al-1Mo-1V has been extensively studied because of its sensitivity to microstructure. The martensitic structures produced by quenching from high-temperature solution treatment are immune to SCC (Ref 190, 191). Lower-temperature solution treatment produces an equiaxed α-β structure that is susceptible to SCC. The degree of susceptibility is determined by the grain size, volume fraction, and mean free path of the susceptible α phase. Tempered martensitic structures, produced by annealing a martensitic microstructure, are also susceptible to SCC (Ref 190). Basket-weave or Widmanstätten microstructures produced by working and/or heat treatment above the β transus generally exhibit better toughness both in and out of an aqueous environment.

In general, titanium alloys with higher aluminum, oxygen, and tin contents are the most susceptible to SCC; the effect of aluminum is shown in Fig. 42. Molybdenum is usually beneficial in increasing SCC resistance. Microstructural effects in these alloys are similar to those discussed for Ti-8Al-1Mo-1V.

With the exception of Ti-13V-11Cr-3Al, all of the commercial β titanium alloys are immune to SCC in the β-phase condition (Ref 199-202). However, aging decomposes the β phase and produces a variety of phases. The ω phase, produced by low-temperature aging of many β alloys, does not induce SCC susceptibility. Aging at higher temperatures produces the α phase, which nearly always leads to SCC. The degree to which a given β alloy in the $\beta + \alpha$ condition is susceptible appears to be related to the alloy chemistry and to the quantity and morphology of the α phase.

Aging at higher temperatures produces a coarser α, which is less susceptible to SCC than the finer α. Alloys containing molybdenum are less susceptible to SCC, especially those without tin.

Galvanic Corrosion in Specific Media

In their normal passive condition, titanium alloys are most often the cathode when galvanically coupled to most common engineering alloys in service. As a result, galvanic corrosion of titanium is very rare and occurs only under very unusual conditions. This rare situation could occur in a medium such as a reducing acid, in which titanium is actively corroding. In this case, coupling to a more active metal (Ref 203) or a more noble metal could accelerate titanium alloy corrosion only if full passivation is not achieved. Coupling titanium alloys to a more noble material is most often very beneficial, resulting in establishing passivity when titanium is marginally active (corroding) or further maintaining passivity of titanium at more noble potentials. This form of anodic protection explains the excellent galvanic compatibility of titanium with noble metals (precious metals) and graphite composites in most environments.

Titanium alloys exhibit relatively noble corrosion potentials in the many environments in which full passivity is achieved. This is evident from Table 2. Similar data in natural seawater from other sources indicate that corrosion potentials for titanium and its alloys fall in the range of +0.1 to −0.3 V versus SCE (Ref 41, 44, 204-206).

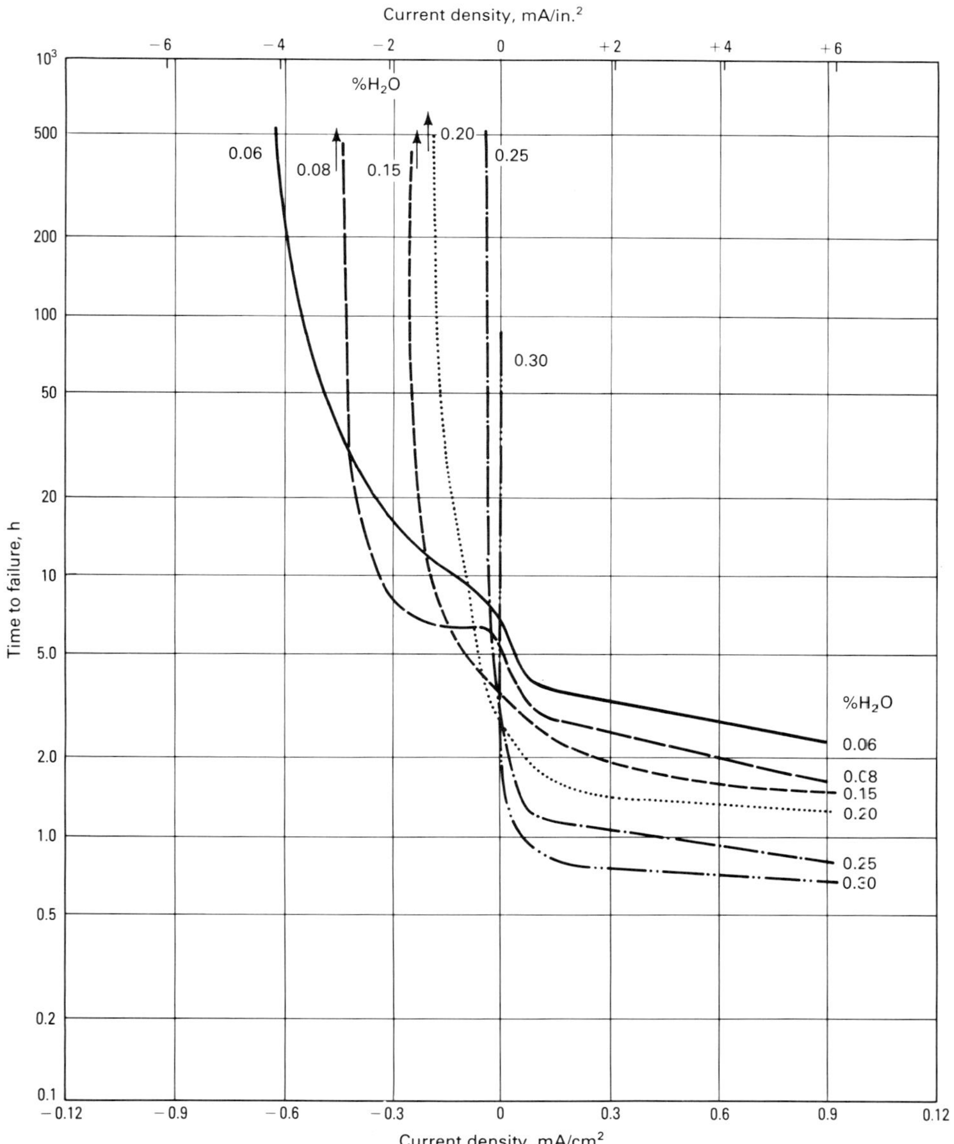

Fig. 34 Time to failure versus applied current and water content for cold-rolled and annealed Ti-6Al-4V stressed to 75% of yield strength in a methanol/HCl mixture. For 0.08% H_2O, 0.15% H_2O, and 0.20% H_2O, there was no failure in time shown.

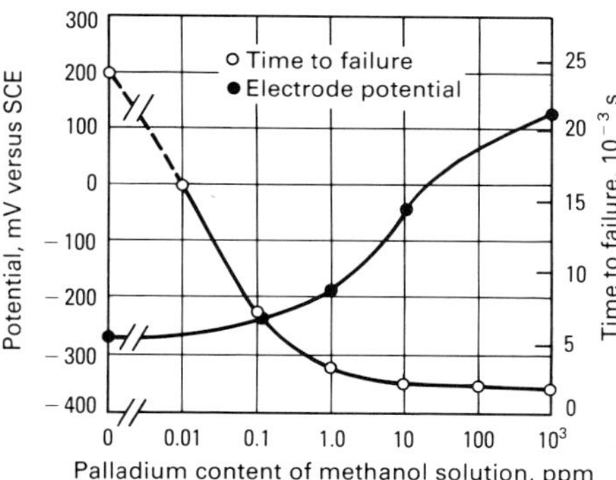

Fig. 35 Effect of palladium on time to failure and electrode potential for commercially pure titanium in methanolic solutions

Deaeration, increasing temperature (Ref 206), and sunlight (Ref 207) were all shown to cause a slight shift in titanium corrosion potential in the active direction.

In all cases, these referenced sources reveal that titanium exhibits corrosion potentials that are very similar to those of other more resistant alloys in the passive condition, including the stainless steels and nickel-base alloys. The minor potential differences between these resistant alloys result in very small, benign galvanic interactions as long as passive alloy conditions exist. Thus, galvanic compatibility can be expected in most environments when a titanium alloy is coupled to another resistant alloy, assuming both exist in a fully passive condition. The data in Table 29 for stainless steel and nickel-base alloys coupled to titanium in marine environments support this point.

However, when a titanium alloy is coupled to a metal that is active (corroding or pitting) in an environment, accelerated anodic attack of the active metal may ensue. Depending on the environment, active metals may include carbon steel, aluminum, zinc, copper alloys, or stainless steels that are active or pitting.

The effect of enhanced galvanic corrosion of various copper alloys coupled to commercially pure titanium in ambient temperature seawater is illustrated in Fig. 43 and 44. These sources and others reveal a wide variation in the extent of galvanic corrosion of copper alloys and steel (Ref 43, 69, 208-210). The degree of galvanic attack depends on many (often interacting) factors, including cathode-to-anode surface area ratio, concentration of dissolved cathodic depolarizers such as oxygen and atomic hydrogen, temperature (Ref 44, 208), medium flow velocity and flow characteristics (that is, turbulence, angle of impingement), and medium chemistry. The presence of sulfide, increasing cathode-to-anode area, oxygen content, and flow velocity all aggravate the galvanic attack of copper alloys in seawater. As shown in Fig. 45, deaeration of NaCl brines drastically reduces galvanic attack (compare to Fig. 43). Discussions of galvanic interactions between titanium and a wide array of engineering alloys in flowing seawater are presented in Ref 74.

As a result of the relatively high polarization resistance ($2.6 \times 10^6\ \Omega \cdot cm^2$) and hydrogen overvoltage of titanium in ambient seawater, the cathodic behavior of titanium is similar to that of the 18-8 stainless steels (Ref 211). Therefore, the galvanic effects of titanium on active metals are quite similar to those for 18-8 stainless, as observed in salt spray tests (Ref 212). Although these cathode characteristics tend to mitigate galvanic current, they also result in increased cathodic current throwing power in conductive electrolytes; therefore, the effective surface area of titanium involved in a galvanic couple may be quite large. For example, studies show that the effective length of titanium (and stainless steel) condenser tubing involved in the galvanic attack of copper alloy tubesheets in seawater is in excess of 6 m (20 ft) (Ref 208, 211, 213). This differs significantly from the two-tube diameter effective length rule used for copper alloys in seawater. Thus, significant cathodic polarization of titanium may occur in nonoxidizing media, such as seawater, when coupled to a more active metal. This often leaves the galvanic couple in cathodic control.

Effective design strategies for limiting or avoiding galvanic attack of active metals include:

- Coupling to more compatible (passive) alloys (including all-titanium design)
- Use of dielectric (insulating) joints between dissimilar metals
- Cathodic protection of the active metal by either impressed current or sacrificial anode means (Ref 44, 208, 210)

Coating of titanium cathode surfaces may also mitigate galvanic current, assuming that the galvanic couple is in cathodic control.

Erosion-Corrosion in Specific Media

Unalloyed titanium and grade 5 titanium have both been shown to withstand silt-free flowing

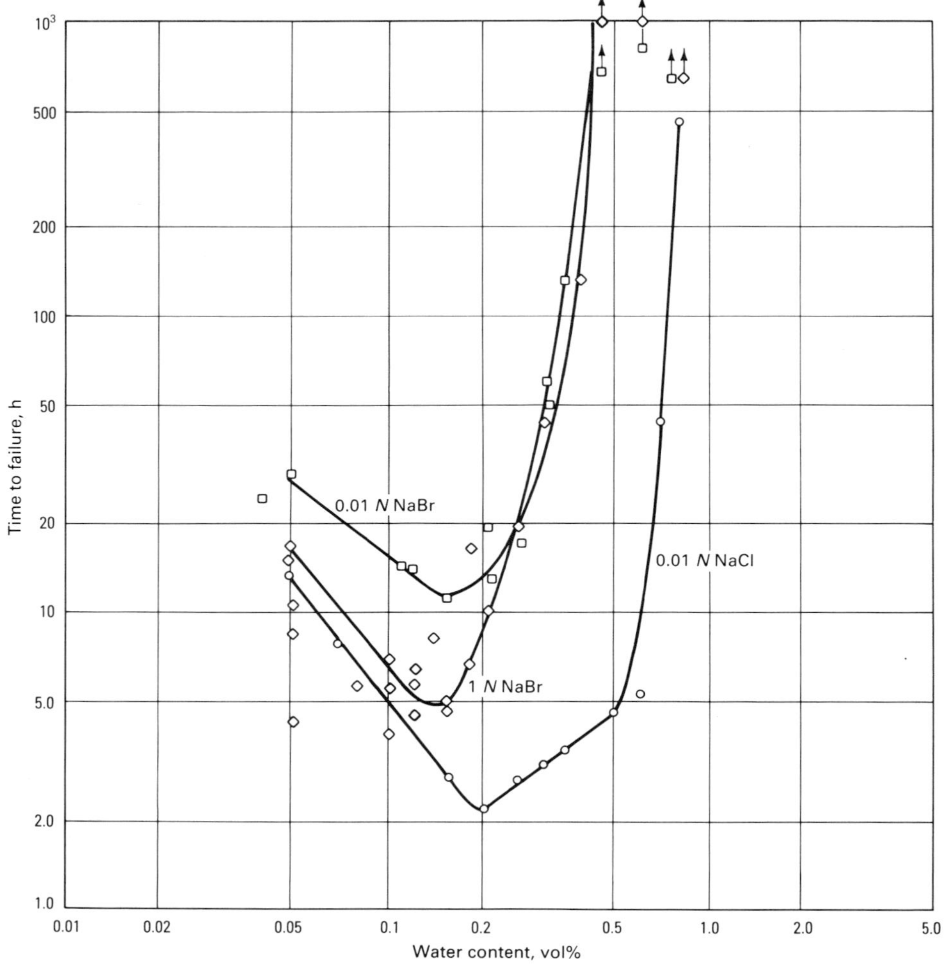

Fig. 36 Effect of bromide and chloride additions on SCC of cold-rolled and annealed commercially pure titanium stressed to 75% of yield strength in methanol/water solutions. Arrows indicate no failure in time shown.

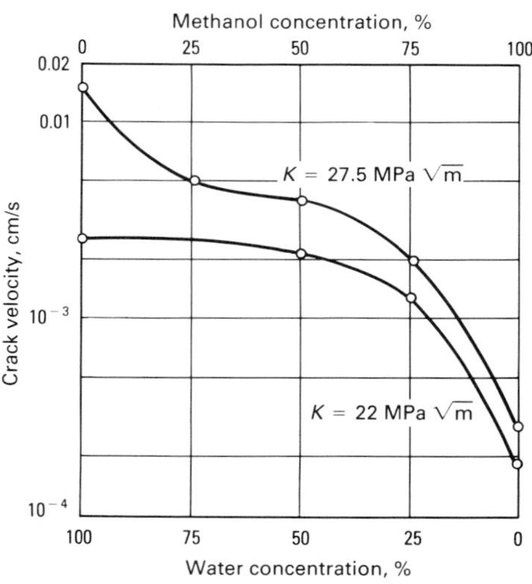

Fig. 37 Effect of water on crack velocity for Ti-8Al-1Mo-1V in methanolic solutions

Table 28 Relative resistance of titanium alloys to hot salt cracking

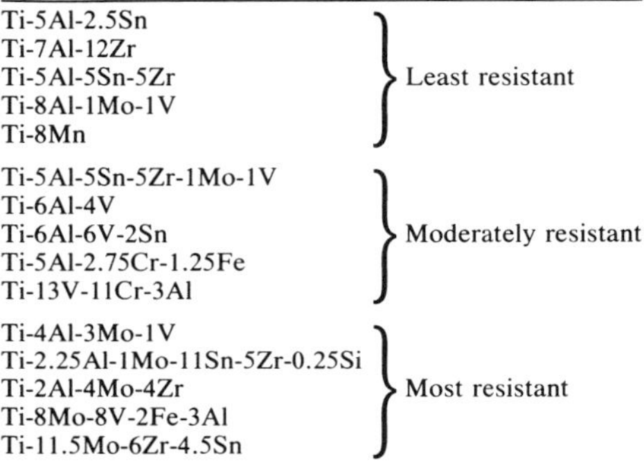

Alloy	Resistance
Ti-5Al-2.5Sn	Least resistant
Ti-7Al-12Zr	
Ti-5Al-5Sn-5Zr	
Ti-8Al-1Mo-1V	
Ti-8Mn	
Ti-5Al-5Sn-5Zr-1Mo-1V	Moderately resistant
Ti-6Al-4V	
Ti-6Al-6V-2Sn	
Ti-5Al-2.75Cr-1.25Fe	
Ti-13V-11Cr-3Al	
Ti-4Al-3Mo-1V	Most resistant
Ti-2.25Al-1Mo-11Sn-5Zr-0.25Si	
Ti-2Al-4Mo-4Zr	
Ti-8Mo-8V-2Fe-3Al	
Ti-11.5Mo-6Zr-4.5Sn	

Source: Ref 177

seawater to velocities as high as 30 m/s (100 ft/s) (Ref 41, 57, 214-218). In fact, high-speed water wheel tests in seawater indicate erosion rates for grade 5 titanium of approximately 0.013 mm/yr (5 mils/yr) at 46 m/s (150 ft/s) (Ref 216). Jet impingement tests also involving seawater velocities of 46 m/s (150 ft/s) reveal rates of 0.03 to 0.06 mm/yr (1.2 to 24 mils/yr) for commercially pure titanium, with values of approximately 0.03 mm/yr (1.2 mils/yr) for welded and unwelded grade 5 titanium samples alike (Ref 218). Extremely low erosion rates are also reported for grade 2 titanium at various seawater locations, as shown in Table 30. The superior erosion-corrosion resistance of titanium has also been reported in other media (Ref 103).

Studies involving sand and emery particle-laden seawater indicate satisfactory erosion-corrosion resistance to flow rates of approximately 6 m/s (20 ft/s) (Ref 57). Data generated from rotating disk tests are presented in Table 31. The immunity of titanium to erosion-corrosion in silt-laden seawater flowing at approximately 2 m/s (6.5 ft/s) has been demonstrated in more than 20 years of power plant surface condenser tube service (Ref 69, 72). The outstanding resistance of titanium alloys to cavitation damage has also been documented (Ref 217, 219, 220), and it has been confirmed that the harder higher-strength titanium alloys are more resistant to cavitation (Ref 221).

Testing in a hypersaline geothermal brine further demonstrated the superior erosion-corrosion characteristics of titanium alloys (Ref 132, 222, 223). Expanded brine (104 °C, or 220 °F; pH 2.5 to 4.5) impinged on high-strength ferrous, nickel, cobalt, and titanium alloy samples at a velocity of 240 m/s (800 ft/s). Grade 5 titanium uniquely experienced no detectable erosive wear after 120 h of exposure.

The relative erosion resistance of unalloyed titanium was investigated in two special versions of a rotating drum test involving erosive wear by wet TiO_2 filter cake solids and feed slurry (Ref 224). Titanium exhibited minor metal loss, being measurably superior to the steels, stainless steels, and nickel alloys tested. The superior erosion-corrosion resistance of titanium compared to Hastelloy alloy C was noted in a high-velocity gas scrubber venturi device in which a chloride-rich solution was used to scrub hot (315 °C, or 600 °F) process gas (Ref 224).

Erosion-corrosion testing in coal-water slurries representative of those associated with coal-washing plants also affirmed the superior performance of unalloyed titanium as compared to carbon steel, Ni-hard cast iron, and stainless steel alloys (Ref 225). Test results revealed that titanium was inert to attack to slurry velocities of 5 m/s (16 ft/s) but that types 304 and 316, alloy 904L, and type 440C stainless steels exhibited significant wear above 2 m/s (6.5 ft/s). Unalloyed titanium that was thermally oxidized at 700 °C (1290 °F) provided full wear resistance to velocities as high as 8 m/s (26 ft/s) in these rotary tests.

Extensive erosion-corrosion testing of Ti-6Al-4V, Ti-5Al-2.5Sn, and Ti-7Al-4Mo alloys has been conducted in high-velocity wet steam environments for application in low-pressure steam turbine blading in power plants. These alloys have demonstrated superior resistance to 403 stainless steel (12 to 13% Cr steel) in operating turbines and in water droplet erosion and water jet impingement tests (Ref 220). Full erosion resistance of Ti-6Al-4V blades to velocities of 440 to 530 m/s (1450 to 1740 ft/s) at 10% steam moisture has been noted in turbines. In fact, these studies suggest useful erosion resistance of Ti-6Al-4V in approximately 8% steam moisture

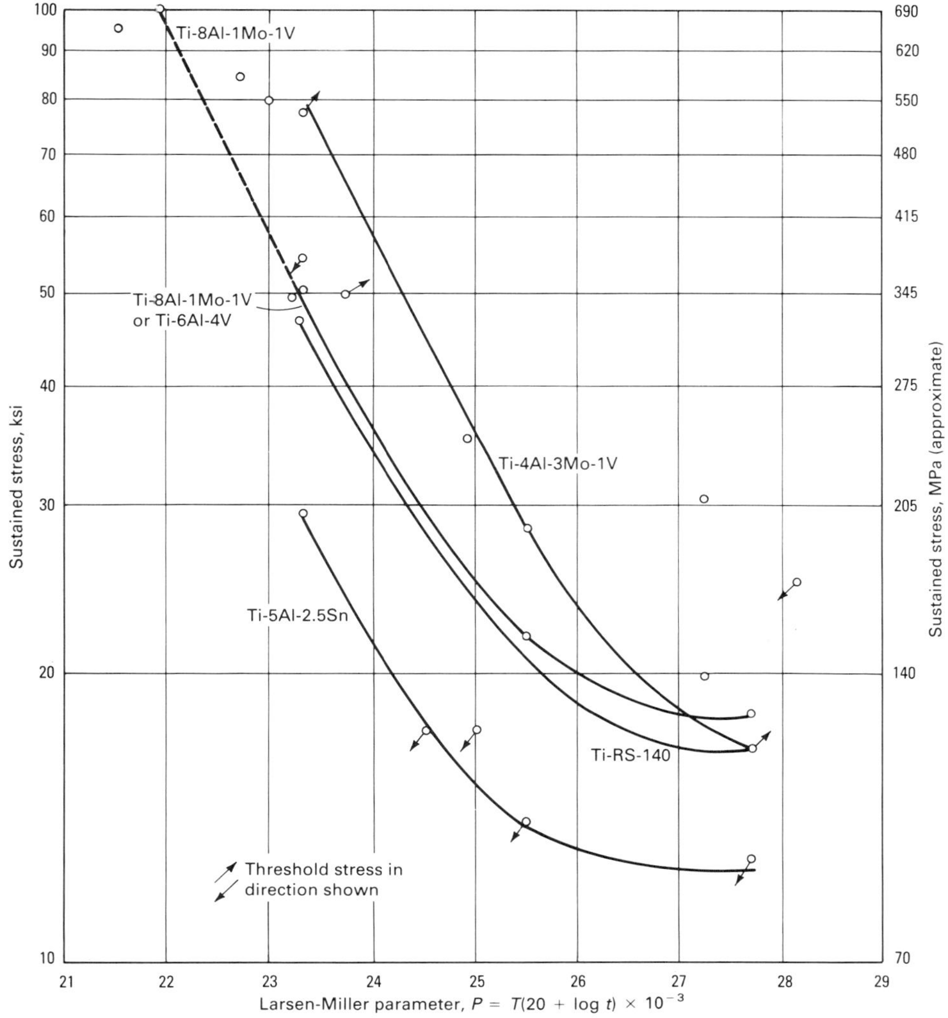

Fig. 38 Larsen-Miller plot for hot salt cracking of several annealed α-β titanium alloys. T is temperature (°R), and t is exposure time (hours).

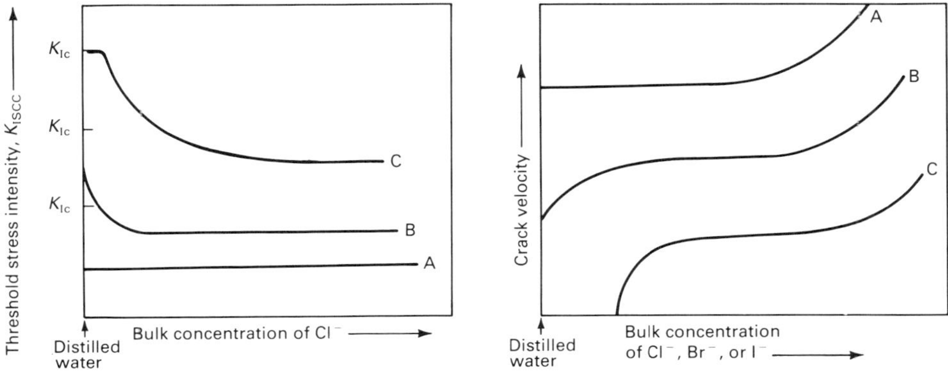

Fig. 39 Effect of halide ion concentration on K_{ISCC} and crack velocity in aqueous solutions. A, highly susceptible; B, moderately susceptible; C, slightly susceptible

to 549 m/s (1800 ft/s), and in 11% steam moisture to 488 m/s (1600 ft/s) (Ref 220). Single-shot water jet impingement testing has shown that annealed Ti-7Al-4Mo alloy is significantly more erosion resistant than 12% Cr steel, type 303 stainless steel, or Stellite alloy 6 at jet velocities of 610 and 915 m/s (2000 and 3000 ft/s).

Expanding and Enhancing the Corrosion Resistance of Titanium

The general corrosion resistance of titanium can be improved or expanded by one or a combination of the following strategies:

- Alloying
- Inhibitor additions to the environment
- Precious metal surface treatments
- Thermal oxidation
- Anodic protection

Alloying. Perhaps the most effective and preferred means of extending resistance to general corrosion into reducing environments has been by alloying titanium with certain elements. Beneficial alloying elements include precious metals (>0.05 wt% Pd) (Ref 226-229), nickel (≥0.5 wt%) (Ref 226, 229, 230-232), and/or molybdenum (≥4 wt%) (Ref 13, 229, 233, 234). These additions facilitate cathodic depolarization by providing sites of low hydrogen overvoltage, which shifts alloy potential in the noble direction where oxide film passivation is possible. Relatively small concentrations of certain precious metals (of the order of 0.1 wt%) are sufficient to expand significantly the corrosion resistance of titanium in reducing acid media.

These beneficial alloying additions have been incorporated into several commercially available titanium alloys, including the titanium-palladium alloys (grades 7 and 11), Ti-0.3Mo-0.8Ni (grade 12), Ti-3Al-8V-6Cr-4Zr-4Mo, Ti-15Mo-5Zr, and Ti-6Al-2Sn-4Zr-6Mo. As shown by the corrosion data in this article, these alloys all offer expanded application into hotter and/or stronger HCl, H_2SO_4, H_3PO_4, and other reducing acids as compared to unalloyed titanium. The high-molybdenum alloys offer a unique combination of high strength, low density, and superior corrosion resistance.

Inhibitor Additions. Various oxidizing species can effectively inhibit the corrosion of titanium in reducing acid environments when present in very small concentrations. Typical potent inhibitors for titanium alloys in aggressive reducing acids are listed in Table 14. Many of these inhibitors are effective at levels as low as 20 to 100 ppm, depending on acid concentration and temperature. If not normally present in a given corrosive acid stream, minute additions of a process-compatible inhibitive species may be considered to protect titanium components. These can be especially practical when process streams are recycled.

Minute additions of water may be required to maintain titanium alloy passivity in certain anhydrous environments. This has been highly effective in certain anhydrous organic compounds, absolute methanol, red fuming nitric acid, dry hydrogen or chlorine gas, and nitrogen tetroxide.

Precious Metal Surface Treatments. Precious metals such as platinum and palladium have been ion plated, ion implanted, or thermal diffused into titanium alloy surfaces to achieve improved resistance to reducing acids (Ref 235). This approach has not been used commercially for industrial components because of high cost, coating application limitations, and the limitations (mechanical and corrosion damage) normally associated with very thin surface films. However, ion plated platinum or gold surface films impart significant improvements in titanium alloy oxidation resistance at temperatures up to 650 °C (1200 °F) (Ref 236, 237).

Thermal Oxidation. Protective thermal oxide films can form when titanium is heated in air at temperatures of 600 to 800 °C (1110 to 1470 °F) for 2 to 10 min. The rutile TiO_2 film formed measurably improves resistance to dilute reducing acids as well as absorption of hydrogen under cathodic charging (Ref 47) or gaseous hydrogen conditions. Corrosion studies in hot, dilute HCl solu-

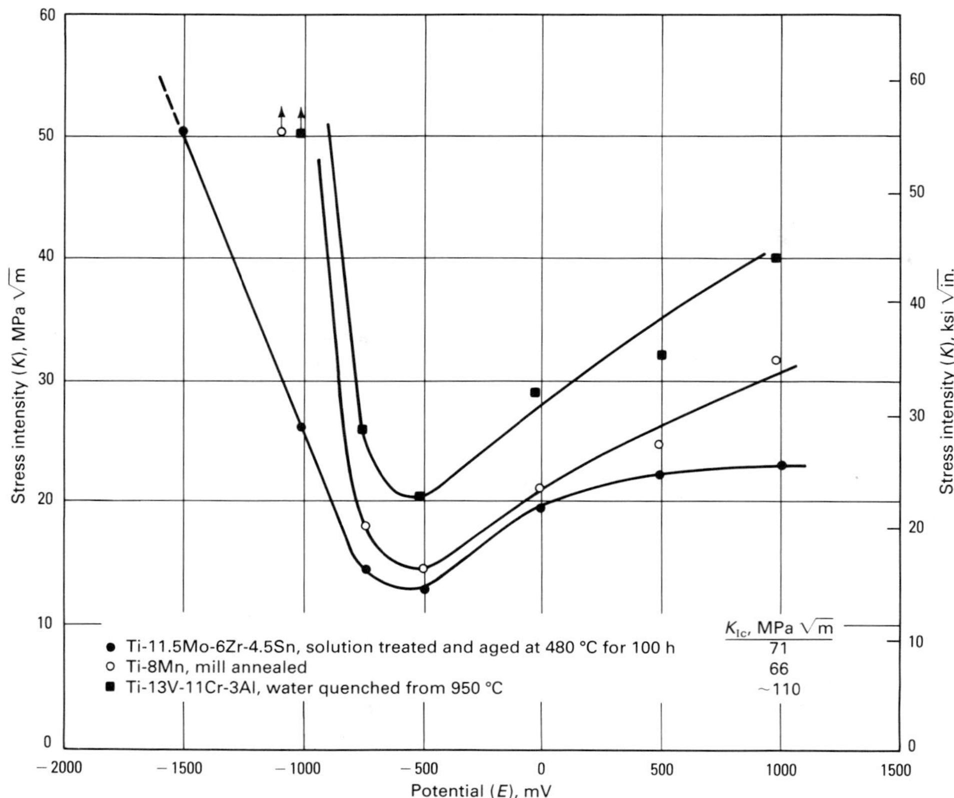

Fig. 40 Stress-corrosion cracking of β titanium alloys as a function of potential in 0.6 *M* KCl at 24 °C (75 °F). Source: Ref 164

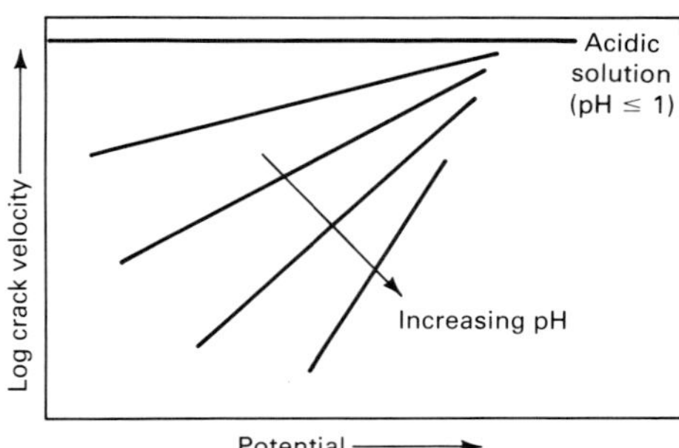

Fig. 41 Stage II crack velocity as a function of pH and potential in aqueous solutions

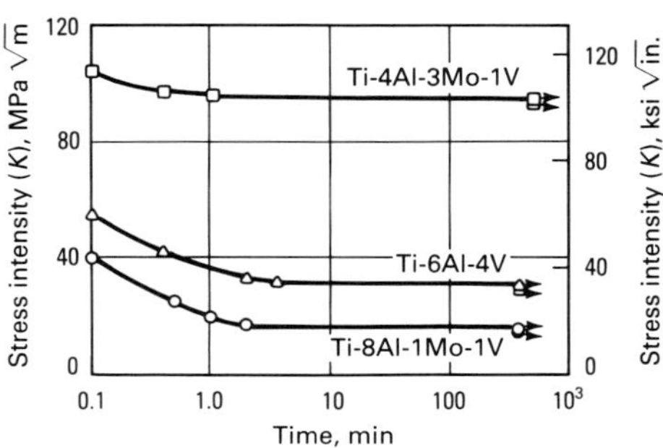

Fig. 42 Effect of alloy composition on SCC resistance of mill-annealed titanium alloys in aqueous 3.5% NaCl solution at 24 °C (75 °F)

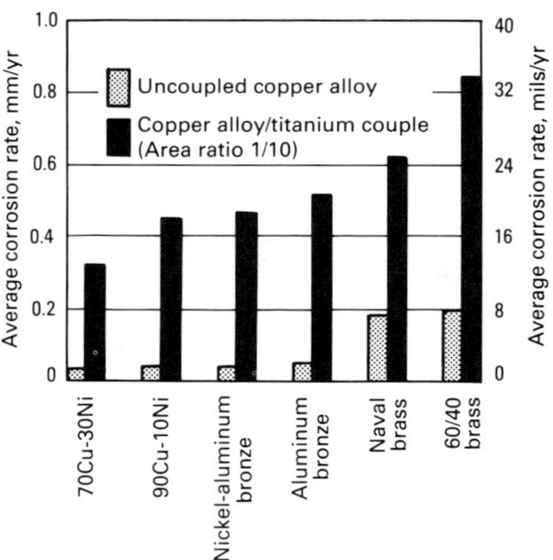

Fig. 43 Corrosion of various copper alloys that were galvanically coupled to titanium in aerated seawater at 25 °C (77 °F). Compare with Fig. 45. Source: Ref 41, 44

tions have confirmed its superior protective benefits as compared to as-pickled, polished, or anodized surfaces on unalloyed titanium (Ref 47, 238). Corrosion and hydrogen uptake resistance was afforded by thermal oxidation in molten urea at 200 °C (390 °F) (Ref 238). Enhanced protection from dry chlorine attack can also be expected. Like anodizing, thermal oxidation offers no improvements in titanium resistance in highly alkaline or oxidizing aqueous media.

Although the thermal oxide has proved to be protective in relatively short-term tests in dilute reducing acids, long-term performance has not been fully demonstrated. Mechanical damage and plastic strain of thermally oxidized components must be avoided for effective protection. The oxide has been successfully applied on tubing and small components, but may be impractical for large components or where component distortion may occur during heating.

Anodic Protection. Titanium alloys can be effectively protected in reducing acid media by impressed anodic (direct current) potentials. Sustained impressed potentials in the range of +1 to +4 V versus the standard hydrogen electrode (SHE) are usually adequate to measure full passivation of titanium in many acids, as indicated in Table 16. Limited use of anodic protection by impressed currents has been made in concentrated H_2SO_4 and H_3PO_4 solutions in which a very wide range of impressed potentials can be applied. The added cost of impressed current systems, challenges with protecting complex component geometries, and stray current problems have inhibited its application. Also, titanium surfaces exposed to alternating wet/dry or vapor-phase conditions are not protected by this method.

Other Surface Treatments. Surface films of titanium nitrides and carbides are highly resistant to reducing acids. Studies have shown that the dense adherent nitride films produced by reactive plasma ion plating provided superior protection in deaerated H_2SO_4 solutions when compared to several other film-forming methods (Ref 239). Methods of applying nitride surface films to titanium include ion implantation (Ref 240), ion plating, sputter deposition, or thermal diffusion (nitrogen gas or molten cyanide bath). Because of the cost and limitations of film application and the inherent thin film performance limitations, these films are generally not used for corrosion resistance only. The improved water resistance offered by these hard films is generally the primary incentive.

The crevice corrosion resistance of titanium alloys can be enhanced by the following strategies:

- Alloying titanium
- Precious metal surface treatments
- Other metallic coatings
- Thermal oxidation
- Noble alloy contact
- Surface pickling (for smeared surface iron)

Alloying. The crevice corrosion resistance of titanium alloys tends to parallel general corrosion resistance in reducing acids. Thus, alloying with certain precious metals, such as palladium (Ref 131, 226, 241), nickel (Ref 23, 133, 226, 231, 232), and/or molybdenum (Ref 13, 138), also significantly improves the resistance of titanium to crevice corrosion (Ref 242).

The commercially available titanium alloys that exhibit superior crevice corrosion resistance include grades 7, 11, and 12. In addition, high-strength titanium alloys containing at least 4 wt% Mo offer an excellent combination of improved crevice corrosion resistance, low density, and high strength (Ref 13). Proper alloy selection depends largely on pH, temperature, and other conditions, as discussed previously in this article.

Selection of a more resistant titanium alloy is generally preferred from a long-term reliability

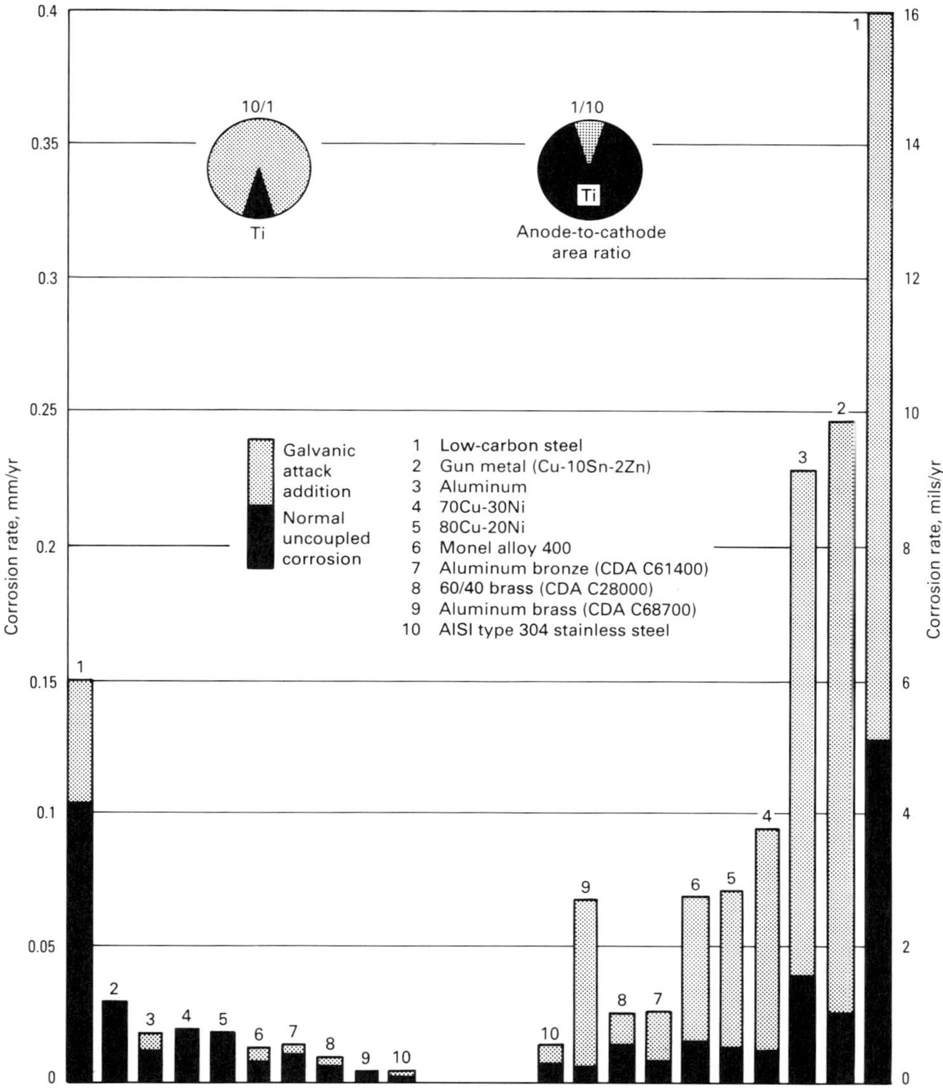

Fig. 44 Corrosion of dissimilar metals coupled to titanium in flowing ambient-temperature seawater. Source: Ref 56

Table 29 Corrosion rates of various metals galvanically coupled to titanium in marine exposures

Coupled material	Corrosion rate after indicated exposure, mm/yr (mils/yr) — 193 days at half tide: Uncoupled	Coupled(a)	Coupled(b)	56 months in sea air: Uncoupled	Coupled(b)
Alclad 2024-T3	0.015 (0.6)	0.03 (1.2)	0.043 (1.7)	0.001 (0.04)	0.007 (0.28)
Copper	0.013 (0.5)	0.023 (0.9)	0.025 (1)	0.002 (0.08)	0.006 (0.24)
Low-carbon steel	0.15 (6)	0.31 (12.2)	0.43 (17)	0.156 (6.1)	. . .
Monel alloy 400	0.025 (1)	0.003 (0.12)	0.003 (0.12)	nil	0.001 (0.04)
Inconel alloy 600	nil	nil	nil	nil	nil
AISI type 302 stainless steel	0.002 (0.08)	nil	0.003 (0.12)	nil	nil
AISI type 316 stainless steel	nil	nil	nil	nil	nil

(a) Area ratio of titanium to other metal: 1 to 7. (b) Area ratio of titanium to other metal: 7 to 1. Source: Ref 69

and cost standpoint over surface treatment options for less resistant titanium alloys. On the other hand, localized surface treatments may be cost-effective when very heavy alloy wall sections are involved or when it is necessary to upgrade existing equipment components.

Surface films of many precious metals and/or their oxides offer significant improvements in the crevice corrosion resistance of titanium. In fact, these treatments can offer crevice resistance approaching that of the grade 7 or 11 titanium alloy. The most common precious metals applied to titanium surfaces are palladium, platinum, and ruthenium, and their oxides. Thermal diffusion coatings of palladium (Ref 235), platinum, and ruthenium are readily applied by firing in an air furnace after coating titanium with modified metal chloride solutions (Ref 243). The resultant coating is typically a mixture of titanium and precious metal oxides, depending on firing temperature (Ref 235). Thermal palladium surface treatments have been successfully used in oil refinery tubular exchangers (Ref 244).

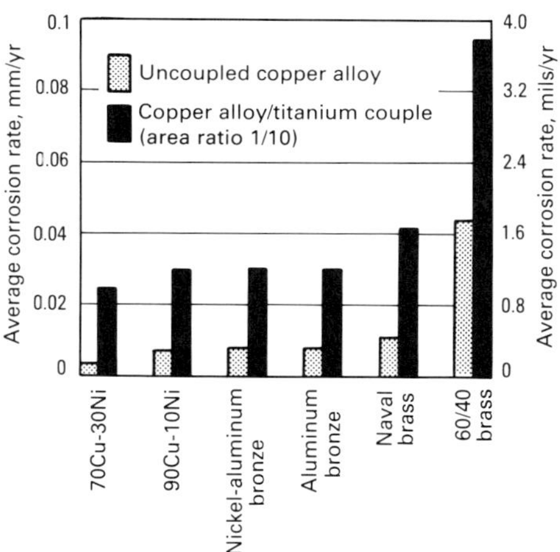

Fig. 45 Corrosion of copper alloys that were galvanically coupled to titanium in boiling, deaerated 6% NaCl at 100 °C (212 °F). Compare with Fig. 43, which shows corrosion rates in aerated seawater. Source: Ref 44

Other methods of applying these precious metals on titanium surfaces include electroplating, brush plating, and ion implantation. Electroplating or brush plating provides precious metal layer thicknesses of the order of 0.01 mm (0.4 mils). Brush plating is a special technique for achieving localized electroplating of surfaces (Ref 245). Palladium implantation generates palladium-rich surface layers of the order of 0.5 μm (0.02 mil) or less. These surface layers have been shown to be effective in hot, concentrated chloride brines (Ref 246).

Other Metallic Coatings. The application of certain metals and their oxides within titanium alloy crevices can effectively inhibit crevice corrosion initiation in hot chloride media. These metals include nickel and copper, their oxides, and the Fe_2O_3 and MoO_3 oxides (Ref 247, 248). In finely powdered form, these materials can be painted on creviced titanium surfaces in slurry form. Alternatively, these powders can be formulated into sealants, such as silicone or ethylene propylene diene monomer (EPDM) rubber. For example, a 5 wt% addition of nickel metal/NiO powder (50/50) blended into EPDM gaskets for chlor-alkali cell anode components is known to prevent the crevice corrosion of unalloyed titanium in hot, low-pH, chlorine-saturated NaCl brines. It has been shown that unalloyed titanium anodized in a molybdate solution also provides crevice corrosion resistant metal surfaces (Ref 248).

Thermal Oxidation. Crevice corrosion resistance is afforded from thermal oxide films formed when titanium alloys are heated in air at 500 to 800 °C (930 to 1470 °F) for 2 to 10 min. Increasing the temperature (or time) within this range results in thicker, more protective oxide films. In hot

Table 30 Erosion-corrosion of grade 2 titanium in seawater at various locations

Location	Flow rate m/s	Flow rate ft/s	Type of test	Test duration, months	Erosion-corrosion rate mm/yr	Erosion-corrosion rate mils/yr
Brixham Sea	9.8	32	Model condenser	12	0.003	0.12
Kure Beach, NC	1	3.3	Ducting	54	7.5×10^{-7}	0.00003
Kure Beach, NC	8.5	28	Rotating disk	2	1.3×10^{-4}	0.005
Kure Beach, NC	9	29.5	Micarta wheel	2	2.8×10^{-4}	0.01
Kure Beach, NC	7.2	23.6(a)	Jet impingement	1	5×10^{-4}	0.02
Wrightsville Beach, NC	1.3	4.3	...	6	1×10^{-4}	0.004
Wrightsville Beach, NC	9	29.5	Micarta wheel	2	1.8×10^{-4}	0.007
Mediterranean Sea	7.2	23.6(a)	Jet impingement	0.5	0.5 mg/day	...
Dead Sea	7.2	23.6(a)	Jet impingement	0.5	0.2 mg/day	...

(a) Included air. Source: Ref 57

Table 31 Erosion-corrosion of titanium grade 2 in seawater containing suspended solids

Flow rate m/s	Flow rate ft/s	Seawater suspension	Test duration, h	Erosion-corrosion rate mm/yr	Erosion-corrosion rate mils/yr
7.2	23.6	No solids	10 000	nil	
2.0	6.5	40 g/L of 60-mesh sand	2000	0.0025	0.1
2.0	6.5	40 g/L of 10-mesh emery	2000	0.0125	0.5
3.5	11.5	1% 80-mesh emery	17.5	0.0037	0.15
4.1	13.5	4% 80-mesh emery	17.5	0.083	3.3
7.2	23.6	40% 80-mesh emery	1	1.5	60

Source: Ref 57

NaCl brines, thermally oxidized titanium has proved its superior resistance to anodized or as-pickled titanium in both metal-to-metal and metal-to-gasket crevices (Ref 47, 129). Avoidance of mechanical damage or plastic metal strain is necessary for good protection. Thermal oxide films may exhibit limitations at higher temperatures and in low-pH brine, and they are better considered for situations in which only borderline crevice conditions exist for unalloyed titanium.

Noble Alloy Contact. Crevice corrosion of unalloyed titanium or other titanium alloys can be averted in metal-to-metal crevice situations. If a more noble alloy is one member of the metal-to-metal crevice, the titanium metal surface can be anodically protected by the galvanic couple achieved in the crevice. More noble metals include the precious metals, more resistant titanium alloys such as grade 7 or 12, or copper alloys. For example, it is well known that unalloyed titanium tubes that are roll expanded into copper alloy tubesheets resist tube joint crevice corrosion in high-temperature seawater (Ref 41, 44). Similarly, the more noble grade 7 titanium alloy will protect unalloyed titanium with which it is in direct contact; this provides resistance similar to that of grade 7 titanium. For example, a grade 7-grade 2 titanium sheet crevice will resist crevice attack in boiling chlorine-saturated NaCl brines, in which a grade 2-grade 2 joint would undergo crevice corrosion. This principle also applies to treated titanium surfaces in contact with untreated unalloyed titanium.

Surface Pickling. If the presence of galled or smeared surface iron is suspected on titanium equipment, it should be removed to prevent possible pitting or hydrogen uptake in titanium exposed to hot chloride brines. This surface iron contamination can be removed with a light (~5 min) pickle in near-ambient temperature 35 vol% HNO_3-5 vol% HF or 12-1 HNO_3-HF solution, followed by water flushing. This procedure will remove less than 0.03 mm (1.2 mils) from the titanium alloy surface. Studies have shown that pickling in dilute HNO_3-HF solutions is much more effective in removing surface iron contamination than either anodizing (Ref 47) or pure HNO_3 exposures.

REFERENCES

1. R.W. Schutz, Titanium, in *Process Industries Corrosion—The Theory and Practice*, National Association of Corrosion Engineers, 1986, p 503
2. "Titanium and Its Alloys," Course 27, Lesson 3, Metals Engineering Institute, American Society for Metals
3. *Aerospace Structural Metals Handbook*, Mechanical Properties Data Center, U.S. Department of Defense, 1985
4. F. Schwartzberg, F. Holden, H. Ogden, and R. Jaffee, "The Properties of Titanium Alloys at Elevated Temperatures," TML Report 82, Titanium Metallurgical Laboratory, Battelle Memorial Institute, Sept 1957
5. M. Mote, R. Hooper, and P. Frost, "The Engineering Properties of Commercial Titanium Alloys," TML Report 92, Titanium Metallurgical Laboratory, Battelle Memorial Institute, June 1958
6. *Titanium Alloys Handbook*, MCIC-HB-02, Metals and Ceramics Information Center, Department of Defense Information Analysis Center, Dec 1972
7. *Handbook on Materials for Superconducting Machinery*, MCIC-HB-04, Metals and Ceramics Information Center, Battelle-Columbus Laboratories, and Advanced Research Projects Agency and Cryogenics Division, National Bureau of Standards, Nov 1974
8. N.D. Tomashov and P.M. Altovskii, *Corrosion and Protection of Titanium*, Government Scientific-Technical Publication of Machine-Building Literature (Russian translation), 1963
9. V.V. Andreeva, *Corrosion*, Vol 20, 1964, p 35
10. M. Pourbaix, *Atlas of Electrochemical Equilibria in Aqueous Solutions*, National Association of Corrosion Engineers, 1974, p 217
11. R.W. Schutz, J.S. Grauman, and J.A. Hall, Effect of Solid Solution Iron on the Corrosion Behavior of Titanium, in *Titanium—Science and Technology*, Proceedings of the Fifth International Conference on Titanium, Deutsche Gesellschaft fur Metallkunde E.V., 1985, p 2617-2624
12. L.C. Covington and R.W. Schutz, Effects of Iron on the Corrosion Resistance of Titanium, in *Industrial Applications of Titanium and Zirconium*, STP 728, American Society for Testing and Materials, 1981, p 163-180
13. R.W. Schutz and J.S. Grauman, Fundamental Corrosion Characterization of High-Strength Titanium Alloys, in *Industrial Applications of Titanium and Zirconium: Fourth Volume*, STP 917, American Society for Testing and Materials, 1986, p 130-143
14. J. Newman, Fighting Corrosion With Titanium Castings, *Chem. Eng.*, June 4, 1979
15. J.P. Dippel, Manufacturing Titanium Castings for the Pump Industry, *World Pumps*, Dec 1982
16. R.W. Schutz and L.C. Covington, Guidelines for Corrosion Testing of Titanium, in *Industrial Applications of Titanium and Zirconium*, STP 728, American Society for Testing and Materials, 1981, p 59-70
17. S.W. Dean, Jr., Electrochemical Methods of Corrosion Testing, in *Electrochemical Techniques for Corrosion*, National Associaton of Corrosion Engineers, 1977, p 52-60
18. E.L. Liening, Electrochemical Corrosion Testing Techniques, in *Process Industries Corrosion*, National Association of Corrosion Engineers, 1986, p 85-122
19. M.G. Fontana and N.D. Greene, *Corrosion Engineering*, McGraw-Hill, 1967
20. J.C. Griess, Jr., Crevice Corrosion of Titanium in Aqueous Salt Solutions, *Corrosion*, Vol 24 (No. 4), April 1968, p 96-109
21. L.C. Covington, Pitting Corrosion of Titanium Tubes in Hot Concentrated Brine Solutions, in *Galvanic and Pitting Corrosion—Field and Laboratory Studies*, STP 576, American Society for Testing and Materials, 1976, p 147-154
22. B.J. Moniz, Field Coupon Corrosion Testing, in *Process Industries Corrosion—The Theory and Practice*, National Association of Corrosion Engineers, 1986, p 67-83
23. M. Kobayashi *et al.*, Study on Crevice Corrosion of Titanium, in *Titanium '80—Science and Technology*, Vol 4, The Metallurgical Society, 1980, p 2613-2622
24. R.B. Diegle, "Electrochemical Cell For Monitoring Crevice Corrosion in Chemical Plants," Paper 154, presented at Corrosion/81, Toronto, Canada, National Association of Corrosion Engineers, April 1980
25. L. Szklarska-Smialowska and M. Janik-Czachor, *Corros. Sci.*, Vol 11, 1971, p 901-914
26. L.C. Covington and R.W. Schutz, "Corrosion Resistance of Titanium," TIMET Corporation, 1982
27. J.B. Cotton, *Chem. Eng. Prog.*, Vol 66 (No. 10), 1970, p 57
28. L.C. Covington, "Factors Affecting the Hydrogen Embrittlement of Titanium," Paper presented at Corrosion/75, Toronto, Canada, National Association of Corrosion En-

gineers, April 1975
29. L.C. Covington, *Corrosion*, Vol 35 (No. 8), Aug 1979, p 378-382
30. V.A. Livanov *et al.*, *Hydrogen in Titanium*, Israel Program for Scientific Translation Ltd., Catalog No. 2163, Daniel Davey & Company, Inc., 1965
31. N.E. Paton and J.C. Williams, Effect of Hydrogen on Titanium and Its Alloys, in *Titanium and Titanium Alloys—Source Book*, American Society for Metals, 1982, p 185-207
32. R.R. Boyer and W.F. Spurr, Characteristics of Sustained-Load Cracking and Hydrogen Effects in Ti-6Al-4V, *Metall. Trans. A*, Vol 9A, Jan 1978, p 23-29
33. R. Bourcier and D. Koss, *Acta Metall.*, Vol 32 (No. 11), 1984, p 2091-2099
34. G.A. Lenning *et al.*, Effect of Hydrogen on Alpha Titanium Alloys, *Trans. AIME*, Oct 1956, p 1235
35. C.M. Craighead *et al.*, Hydrogen Embrittlement of Beta-Stabilized Titanium Alloys, *Trans. AIME*, Aug 1956, p 923
36. J.J. DeLuccia, "Electrolytic Hydrogen in Beta Titanium," Report NADC-76207-30, Air Vehicle Technical Department, Naval Air Development Center, June 1976
37. D.A. Meyn, Effect of Hydrogen on Fracture and Inert-Environment Sustained Load Cracking Resistance of Alpha-Beta Titanium Alloys, *Metall. Trans.*, Vol 5, Nov 1974, p 2405
38. W.R. Holman *et al.*, Hydrogen Diffusion in a Beta Titanium Alloy, *Trans. AIME*, Vol 233, Oct 1965, p 1836
39. I.I. Phillips, P. Pool, and L.L. Shreir, Hydride Formation During Cathodic Polarization of Ti.-II. Effect of Temperature and pH of Solution on Hydride Growth, *Corros. Sci.*, Vol 14, 1974, p 533-542
40. R.L. Jacobs and J.A. McMaster, Titanium Tubing: Economical Solution to Heat Exchanger Corrosion, *Mater. Prot. Perform.*, Vol 11 (No. 7), July 1972, p 33-38
41. "Get More Advantages By Applying Titanium Tubing Not Only For Power Plants But Also For Desalination Plants!!," Technical Brochure, Japan Titanium Society, May 1984
42. H. Satoh, T. Fukuzuka, K. Shimogori, and H. Tanabe, "Hydrogen Pickup by Titanium Held Cathodic in Seawater," Paper presented at the Second International Congress on Hydrogen in Metals, Paris, June 1977
43. S. Sato, K. Nagata, and M. Nagayama, "Experiences of Welded Titanium Condenser Tubes in Japan," Technical Research Laboratory, Sumitomo Light Metal Industries Ltd.
44. T. Fukuzuka, K. Shimogori, H. Satoh, and F. Kamikubo, "Corrosion Problems and Countermeasures in MSF Desalination Plant Using Titanium Tube," Kobe Steel Ltd., 1985
45. L.A. Charlot and R.E. Westerman, Low-Temperature Hydriding of Zircaloy-2 and Titanium in Aqueous Solutions, *Electrochem. Technol.*, Vol 6 (No. 3-4), March/April 1968
46. J. Lee and P. Chung, "A Study of Hydriding of Titanium in Seawater Under Cathodic Polarization," Paper 259, Presented at Corrosion/86, Houston, TX, National Association of Corrosion Engineers, March 1986
47. R.W. Schutz and L.C. Covington, *Corrosion*, Vol 37 (No. 10), Oct 1981, p 585-591
48. B.R. Brown, "Stress Corrosion Cracking in High Strength Steels and in Titanium and Aluminum Alloys," Naval Research Laboratory, 1972
49. I.R. Lane Jr., J.L. Cavallaro, and A.G.S. Morton, in *Stress Corrosion Cracking of Titanium*, STP 397, American Society for Testing and Materials, 1966
50. T.R. Beck, Electrochemical Aspects of Titanium Stress-Corrosion Cracking, in *Proceedings of Conference—Fundamental Aspects of Stress-Corrosion Cracking*, National Association of Corrosion Engineers, 1969, p 605
51. T.R. Beck and E.A. Grens, An Electrochemical Mass-Transport-Kinetic Model for Stress-Corrosion Cracking of Titanium, *J. Electrochem. Soc.*, Vol 116 (No. 2), 1969, p 117
52. D.T. Powell and J.C. Scully, Stress-Corrosion Cracking of Alpha Titanium Alloys at Room Temperature, *Corrosion*, Vol 24 (No. 6), 1968, p 151
53. G. Sanderson, D.T. Powell, and J.C. Scully, The Stress-Corrosion Cracking of Ti Alloys in Aqueous Chloride Solutions at Room Temperature, *Corros. Sci.*, Vol 8, 1968, p 473
54. R.J.H. Wanhill, Aqueous Stress-Corrosion in Titanium Alloys, *Br. Corros. J.*, Vol 10, 1975, p 69
55. N. Gat and W. Tabakoff, Effects of Temperature on the Behavior of Metals Under Erosion by Particulate Matter, *J. Test. Eval.*, Vol 8 (No. 4), 1980, p 177-186
56. J.B. Cotton and B.P. Downing, Corrosion Resistance of Titanium to Seawater, *Trans. Inst. Marine Eng.*, Vol 69 (No. 8), 1957, p 311
57. "Titanium Heat Exchangers for Service in Seawater, Brine and Other Natural Aqueous Environments: The Corrosion, Erosion and Galvanic Corrosion Characteristics of Titanium in Seawater, Polluted Inland Waters and in Brines," Titanium Information Bulletin, Imperial Metals Industries (Kynoch) Ltd., May 1970
58. J.P. Doucet *et al.*, Corrosion Fatigue Behavior of Ti-6Al-4V For Marine Application, in *Titanium 1986—Products and Applications*, Vol 1, Proceedings of the Technical Program from the 1986 International Conference, Titanium Development Association, 1986
59. D.R. Mitchell, "Fatigue Properties of Ti-50A Welds in 1-Inch Plate," TMCA Case Study W-20, Titanium Metals Corporation of America, March 1969
60. A.G.S. Morton, "Mechanical Properties of Thick Plate Ti-6Al-4V," MEL Report 266/66, U.S. Navy Marine Engineering Laboratory, Jan 1967
61. R. Ebara *et al.*, Corrosion Fatigue Behavior of Ti-6Al-4V in NaCl Aqueous Solution, in *Corrosion Fatigue: Mechanics, Metallurgy, Electrochemistry, and Engineering*, STP 801, American Society for Testing and Materials, 1983, p 135-146
62. P.C. Hughes and I.R. Lamborn, Contamination of Titanium by Water Vapour, *J. Inst. Met.*, Vol 89, 1960-1961, p 165
63. C.R. Breden, *Met. Prog.*, Vol 64, 1953, p 194
64. S.C. Datski, Report ANL 5354, U.S. Atomic Energy Commission, 1954
65. D. Schlain, "Corrosion Properties of Titanium," Bulletin 619, U.S. Bureau of Mines, 1964
66. J.D. Tkach and R.H. Meservey, "Corrosion of Thermocouple Sheath Materials in a Pressurized Water Reactor Environment," Aerojet Nuclear Company, 1971
67. R.L. Kane, The Corrosion of Titanium, in *The Corrosion of Light Metals*, The Corrosion Monograph Series, John Wiley & Sons, 1967
68. F.M. Reinhart, "Corrosion of Materials in Hydrospace, Part III, Titanium and Titanium Alloys," Technical Note N-921, U.S. Naval Civil Engineering Laboratory, Sept 1967
69. H.B. Bomberger, P.J. Cambourelis, and G.E. Hutchinson, Corrosion Properties of Titanium in Marine Environments, *J. Electrochem. Soc.*, Vol 101, 1954, p 442
70. W.L. Wheatfall, "Metal Corrosion in Deep-Ocean Environments," Research and Development Phase Report 429/66, U.S. Navy Marine Engineering Laboratory, Jan 1967
71. M.A. Pelensky, J.J. Jawarski, and A. Gallaccio, Air, Soil, and Sea Galvanic Corrosion Investigation at Panama Canal Zone, in *Galvanic and Pitting Corrosion—Field and Laboratory Studies*, STP 576, American Society for Testing and Materials, 1967, p 94
72. L.C. Covington, W.M. Parris, and D.M. McCue, "The Resistance of Titanium Tubes to Hydrogen Embrittlement in Surface Condensers," Paper 79, presented at Corrosion/79, Houston, TX, National Association of Corrosion Engineers, March 1976
73. K.O. Gray, *Mater. Prot.*, Vol 3 (No. 7), 1964, p 46
74. J.A. Beavers *et al.*, chapter 3, in *Corrosion of Metals in Marine Environments*, MCIC-86-50, Metals and Ceramics Information Center, Battelle-Columbus Division, July 1986
75. F.M. Reinhart, "Corrosion of Materials in Hydrospace," R-504, U.S. Naval Civil Engineering Laboratory, Dec 1966, p 118
76. F.M. Reinhart and J.F. Jenkins, "Corrosion of Alloys in Hydrospace—189 Days at 5,900 Feet," Final Report NCEL-TN-1224, U.S. Naval Civil Engineering Laboratory, April 1972, p 4
77. F.M. Reinhart and J.F. Jenkins, "The Relationship Between the Concentration of Oxygen in Seawater and the Corrosion of Metals," U.S. Naval Civil Engineering Laboratory, 1971, p 562-577
78. L.C. Covington and R.W. Schutz, "Resistance of Titanium to Atmospheric Corrosion," Paper 113, presented at Corrosion/81, Toronto, Ontario, National Association of Corrosion Engineers, April 1981
79. B. Sanderson and M. Romanoff, Performance of C.P. Titanium in Corrosive Soils, *Mater. Prot.*, April 1969, p 29-32
80. *Corrosion Resistance of Titanium*, Technical Handbook, Imperial Metals Industries (Kynoch) Ltd., Birmingham, UK
81. C.R. Bishop, Corrosion Tests at Elevated Temperatures and Pressures, *Corrosion*, Vol 19, Sept 1963, p 308-314
82. T.F. Degnan, Materials for Handling Hydrofluoric, Nitric and Sulfuric Acids, in *Process Industries Corrosion*, National Association of Corrosion Engineers, 1975, p 229
83. E.E. Millaway, Titanium: Its Corrosion Behavior and Passivation, *Mater. Prot. Perform.*, Jan 1965, p 16-21
84. A. Takamura, K. Arakawa, and Y. Moriguchi, Corrosion Resistance of Titani-

um and Titanium-5% Tantalum Alloys in Hot Concentrated Nitric Acid, in *The Science, Technology and Applications of Titanium*, R.I. Jaffee and N.E. Promisel, Ed., Pergamon Press, 1970, p 209
85. S.H. Weiman, *Corrosion*, Vol 22, April 1966, p 98-106
86. H. Keller and K. Risch, The Corrosion Behavior of Titanium in Nitric Acid at High Temperatures, *Werkst. Korros.*, Vol 9, 1964, p 741-743
87. T. Furuya *et al.*, "Corrosion Resistance of Zirconium and Titanium Alloy in HNO_3 Solutions," Paper presented at the International Meeting, Jackson Hole, WY, American Nuclear Society, Aug 1984
88. H. Satoh, F. Kamikubo, and K. Shimogori, Effect of Oxidizing Agents on Corrosion Resistance of CP Titanium in Nitric Acid Solution, in *Titanium—Science and Technology,* Proceedings of the Fifth International Conference on Titanium, Deutsche Gesellschaft fur Metallkunde, E.V., 1985, p 2649-2655
89. M.W. Wilding and B.F. Paige, "Survey on Corrosion of Metals and Alloys in Solutions Containing Nitric Acid," Report ICP-1107, Allied Chemical Corporation, Idaho Chemical Programs, Dec 1976
90. C.M. Slansky, "Review of Corrosion and Materials Selection in Radioactive Waste Handling," Allied Chemical Corporation, Idaho Chemical Programs, 1977
91. C.E. Stevenson, "Idaho Chemical Processing Plant—Technical Progress Report for January thru March 1958," IDO-14443, Allied Chemical Corporation, Sept 1958
92. R. Villemez and C. Millet, Evaluation of Alloys for Nuclear Waste Evaporators, *Mater. Perform.*, July 1980, p 19-25
93. D.E. Thomas, Titanium Alloy Corrosion Resistance in Nitric Acid Solutions, *Titanium 1986—Products and Applications*, Vol 1, Proceedings of the Technical Program from the 1986 International Conference, San Francisco, CA, Titanium Development Association, 1986
94. L.L. Gilbert and C.W. Funk, Explosions of Titanium and Fuming Nitric Acid Mixtures, *Met. Prog.*, Nov 1956, p 93-96
95. H.B. Bomberger, Titanium Corrosion and Inhibition in Fuming Nitric Acid, *Corrosion*, Vol 13 (No. 5), May 1957, p 287-291
96. R.L. Wallner *et al.*, *Mater. Prot.*, Jan 1965, p 55-56
97. W.K. Boyd, "Stress Corrosion Cracking of Titanium Alloys—An Overview," Paper presented at the International Symposium on Stress Corrosion Mechanisms in Titanium Alloys, Georgia Institute of Technology, Jan 1971
98. J.B. Rittenhouse and C.A. Popp, Inhibition of Corrosion in Fuming Nitric Acid, *Corrosion*, Vol 14, June 1958, p 283-284
99. T.M. Sigulovskaya *et al.*, Corrosion-Electrochemical Behavior of Titanium and Its Alloys in Alkaline Solutions of Hydrogen Peroxide, UDC 620.193.01, *Zashch. Met.*, Vol 12 (No. 4), July/Aug 1976, p 363-367
100. L. Clerbois and L. Plumet, Process for Inhibiting the Corrosion of Equipment Made of Titanium, U.S. Patent 4,372,813, 1983
101. D.E. Thomas and E.B. Bomberger, The Effect of Chlorides and Fluorides on Titanium Alloys in Simulated Scrubber Environments, *Mater. Perform.*, Nov 1983, p 29-36
102. E.G. Koch, N.G. Thompson, and J.L. Means, "Trace Elements in FGD Environments and Their Effect on Corrosion of Alloys," Paper 297, presented at Corrosion/84, New Orleans, LA, National Association of Corrosion Engineers, April 1984
103. D.W. Stough, F.W. Fink, and R.S. Peoples, "The Corrosion of Titanium," Report 57, Titanium Metallurgical Laboratory, Battelle Memorial Institute, 1956
104. J.A. Petit *et al.*, *Corros. Sci.*, Vol 21 (No. 4), 1981, p 279-299
105. V.P. Gupta, Process for Decreasing the Rate of Titanium Corrosion, U.S. Patent 4,321,231, 1982
106. R.W. Schutz and L.C. Covington, Hydrometallurgical Applications of Titanium, in *Industrial Applications of Titanium and Zirconium: Third Conference*, STP 830, American Society for Testing and Materials, 1984, p 29-47
107. J.C. Cotton, *Chem. Eng. Prog.*, Vol 66 (No. 10), 1970, p 57
108. J.B. Cotton, *Chem. Ind.*, Vol 3, Jan 1958, p 68-69
109. R.L. LaQue and H.R. Copson, *Corrosion Resistance of Metals and Alloys*, 2nd ed., ACS Monograph, Reinhold, 1963, p 646-661
110. F.W. Fink and W.K. Boyd, "The Corrosion of Metals in Marine Environments," DMIC Report 245, Defense Materials Information Center, Battelle Memorial Institute, May 1970
111. R.E. Smallwood, Corrosion of Titanium and Zirconium Alloys in Zinc Chloride Solutions, in *Industrial Applications of Titanium and Zirconium*, STP 728, American Society for Testing and Materials, 1981, p 147-162
112. R.W. Schutz and J.S. Grauman, Selection of Titanium Alloys for Concentrated Seawater, NaCl and $MgCl_2$ Brines, in *Titanium 1986—Titanium Products and Applications*, Proceedings of the Technical Program from the 1986 International Conference, San Francisco, CA, Titanium Development Association, 1986
113. E.G. Haney, G. Goldberg, R.E. Emsberger, and W.T. Brehm, "Investigation of Stress Corrosion Cracking of Titanium Alloys," Second Progress Report, NASA Grant N6R-39-008-014, Mellon Institute, May 1967
114. C.M. Chem, H.B. Kirkpatrick, and H.L. Gegel, "Cracking of Titanium Alloys in Methanolic and Other Media," Paper presented at the International Symposium on Stress Corrosion Mechanisms in Titanium Alloys, Georgia Institute of Technology, Jan 1971
115. E.G. Haney and W.R. Wearmouth, Effect of Pure Methanol on the Cracking of Titanium, *Corrosion*, Vol 25 (No. 2), Feb 1969, p 87
116. A.J. Sedriks and J.S.A. Green, Stress Corrosion of Titanium in Organic Liquids, *J. Met.*, April 1971, p 48
117. B.J. Hanson, Behavior of C.P. Titanium in Hydrogen Sulfide Atmospheres at Elevated Temperatures, in *Industrial Applications of Titanium and Zirconium: Third Conference*, STP 830, American Society for Testing and Materials, 1984, p 19-28
118. C. Coddet *et al.*, Oxidation of Titanium Base Alloys for Application in Turbines, in *Titanium '80—Science and Technology*, Vol 4, The Metallurgical Society, 1980, p 2755-2764
119. D. David *et al.*, A Structural and Analytical Study of Titanium Oxide Thin Films, in *Titanium '80—Science and Technology*, Vol 4, The Metallurgical Society, 1980, p 2811-2817
120. T. Fukuzuka *et al.*, On the Beneficial Effect of the Titanium Oxide Film Formed by Thermal Oxidation, in *Titanium '80—Science and Technology*, Vol 4, The Metallurgical Society, p 2783-2792
121. J.D. Jackson, W.K. Boyd, and P.D. Miller, "Reactivity of Metals With Liquid and Gaseous Oxygen," DMIC Memorandum 163, Defense Materials Information Center, Battelle Memorial Institute, Jan 1963
122. F.E. Littman and F.M. Church, "Reactions of Metals With Oxygen and Steam," Final Report AECU-4092, Stanford Research Institute to Union Carbide Nuclear Company, Feb 1959
123. J.D. Jackson *et al.*, Technical Report 60-258, Wright Air Development Center, Battelle Memorial Institute, June 1960
124. "Standard for the Production, Processing, Handling, and Storage of Titanium," NFPA 481-1982, National Fire Protection Association
125. H.B. Bomberger, in *Industrial Applications of Titanium and Zirconium: Third Conference*, STP 830, American Society for Testing and Materials, 1984, p 143-158
126. E.E. Millaway and M.H. Klineman, Factors Affecting Water Content Needed to Passivate Titanium in Chlorine, *Corrosion*, Vol 23 (No. 4), 1972, p 88
127. J.D. Jackson and W.K. Boyd, "Corrosion of Titanium," DMIC Memorandum 218, Defense Materials Information Center, Battelle Memorial Institute, Sept 1966
128. R.W. Schutz and J.S. Grauman, *Mater. Perform.*, Vol 25 (No. 4), April 1986, p 35-42
129. H. Satoh *et al.*, Effect of Gasket Materials on Crevice Corrosion of Titanium, in *Titanium—Science and Technology*, Proceedings of the Fifth International Conference on Titanium, Deutsche Gesellschaft fur Metallkunde E.V., 1985, p 2633-2639
130. K. Shimogori and Mitarbeiter, Crevice Corrosion of Titanium in NaCl Solutions in the Temperature Range 100 to 250 °C, *J. Jpn. Inst. Met.*, Vol 44 (No. 6), 1978, p 567-572
131. J.C. Griess Jr., *Corrosion*, Vol 24, 1968, p 96-109
132. R.W. Schutz, *Mater. Perform.*, Vol 24 (No. 1), Jan 1985, p 39-47
133. R.W. Schutz, J.A. Hall, and T.L. Wardlaw, "TI-CODE 12, An Improved Industrial Alloy," Paper presented at the Japan Titanium Society 30th Anniversary International Symposium, Japan Titanium Society, Aug 1982
134. J.W. Braithwaite and M.A. Molecke, *Nucl. Chem. Waste Mgmt.*, Vol 2, 1980, p 37-50
135. J.A. Ruppen, R.S. Glass, and M.A. Molecke, Titanium Utilization in Long-Term Nuclear Waste Storage, in *Titanium for Energy and Industrial Applications*, The Metallurgical Society, 1981, p 355-369
136. R.W. Schutz and J.A. Hall, "Optimization of Mechanical/Corrosion Properties of TI-CODE 12 Plate and Sheet, Part I: Compositional Effects, Stage I Final Report," SAND83-7438, Sandia National Laboratories, 1984
137. W.J. Neill, Experience With Titanium Tubing in Oil Refinery Heat Exchangers, *Mater.*

Perform., Sept 1980, p 57-63
138. D.E. Thomas *et al.*, Beta-C: An Emerging Titanium Alloy For the Industrial Marketplace, in *Industrial Applications of Titanium and Zirconium: Fourth Volume*, STP 917, American Society for Testing and Materials, 1986, p 144-163
139. H.J. Raetzer-Scheive, *Corrosion*, Vol 34 (No. 12), Dec 1978, p 437-442
140. F.A. Posey and E.G. Bohlmann, *Desalination*, Vol 3, 1967, p 268
141. J.W. Braithwaite, N.J. Magnani, and J.W. Munford, "Titanium Alloy Corrosion in Nuclear Waste Environments," Paper 213, presented at Corrosion/80, Chicago, IL, National Association of Corrosion Engineers, March 1980
142. T. Koizumi and S. Furuya, in *Titanium—Science and Technology*, Vol 4, Proceedings of the Second International Conference, Plenum Press, 1973, p 2383-2393
143. F. Kamikubo, H. Satoh, and K. Shimogori, Corrosion of Titanium and Its Prevention in a Fertilizer Plant, in *Titanium—Science and Technology*, Proceedings of the Fifth International Conference on Titanium, Deutsche Gesellschaft fur Metallkunde E.V., 1985, p 1173
144. I. Dugdale and J.B. Cotton, *Corros. Sci.*, Vol 4, 1964, p 397
145. F. Kamikubo *et al.*, Effects of a Small Amount of Impurity Elements on Pitting Potential of C.P. Titanium in Sodium Bromide Solutions, in *Metallic Corrosion*, Vol 2, Proceedings of the Eighth International Congress on Metallic Corrosion, Deutsche Gesellschaft fur Chemisches Apparatewesen E.V., 1981, p 1378-1383
146. T.R. Beck, *J. Electrochem. Soc.*, Vol 120, 1973, p 1310
147. R.W. Schutz and J.S. Grauman, "Compositional Effects on Titanium Alloy Repassivation Potential in Chloride Media," Paper presented at the International Conference on Localized Corrosion, Orlando, FL, National Association of Corrosion Engineers, June 1987
148. G.R. Caskey Jr., The Influence of a Surface Oxide Film on Hydriding of Titanium, in *Hydrogen in Metals*, I.M. Bernstein and A.W. Thompson, Materials/Metalworking Technology Series, American Society for Testing and Materials, 1974, p 465-474
149. E.A. Gulbransen and K.F. Andrew, *Trans. Am. Inst. Mining Met. Engrs.*, Vol 185, 1949, p 174
150. L.C. Covington, "Factors Affecting the Hydrogen Embrittlement of Titanium," Paper 59, presented at Corrosion/75, Toronto, Ontario, National Association of Corrosion Engineers, April 1975
151. J.B. Cotton and J.G. Hines, Hydriding of Titanium Used in Chemical Plant and Protective Measures, in *The Science, Technology and Application of Titanium*, Pergamon Press, 1970, p 150-170
152. Z.A. Foroulis, Factors Influencing Absorption of Hydrogen in Titanium From Aqueous Electrolytic Solutions, in *Titanium '80—Science and Technology*, Vol 4, The Metallurgical Society, 1980, p 2705-2711
153. L.C. Covington and N.G. Feige, A Study of Factors Affecting the Hydrogen Uptake Efficiency of Titanium in Sodium Hydroxide Solutions, in *Localized Corrosion—Cause of Metal Failure*, STP 516, American Society for Testing and Materials, 1972, p 222-235
154. I. Phillips *et al.*, *Corros. Sci.*, Vol 12, 1972, p 855-866
155. R. Gruner, B. Streb, and E. Brauer, Hydrogen in Titanium, in *Titanium—Science and Technology*, Proceedings of the Fifth International Conference on Titanium, Deutsche Gesellschaft fur Metallkunde E.V., 1985, p 2571
156. G.C. Kiefer, *Iron Age*, Vol 169, 1952, p 170
157. N.D. Tomashov, R.M. Altovskiy, and V.B. Vladimirov, Study of the Corrosion of Titanium and Its Alloys in Methyl Alcohol Solutions of Bromine, Trans. FTD-TT 63-672/1 + 2, Translation Division, Foreign Technology Division WPAFB, *Korroz. Zashch. Konstruktsionnykh Metallichoskikh Materialov*, 1961, p 221-233a
158. L.K. Mori, A. Takamura, and T. Shimose, Stress-Corrosion Cracking of Ti and Zr in HCl-Methanol Solutions, *Corrosion*, Vol 22 (No. 2), Feb 1966, p 29-31
159. A.J. Sedriks and J.A.S. Green, Stress-Corrosion Cracking and Corrosion Behavior of Titanium in Methanol Solutions: Effect of Metal Ions in Solution, *Corrosion*, Vol 25 (No. 8), 1969, p 324
160. E.G. Haney and W.R. Wearmouth, "Investigation of Stress-Corrosion Cracking of Titanium Alloys," Report 6, Research Grant NGR-39-008-014, National Aeronautics and Space Administration, May 1969
161. B.S. Hickman, J.C. Williams, and H.L. Marcus, Transgranular and Intergranular Stress-Corrosion Cracking of Titanium Alloys, *Aust. Inst. Met.*, Vol 14 (No. 3), 1969, p 138
162. K.E. Weber, J.S. Fritzen, D.S. Cowgill, and W.C. Gillchriest, Similarities in Titanium Stress-Corrosion Cracking Processes in Salt Water and in Carbon Tetrachloride, in "Accelerated Crack Propagation of Titanium by Methanol, Halogenated Hydrocarbons, and Other Solutions," DMIC Memorandum 228, Defense Metals Information Center, Battelle Memorial Institute, March 1967, p 39
163. H.R. Herrigel, Titanium U-Bends in Organic Liquids: Effect of Inhibitors, in "Accelerated Crack Propagation of Titanium by Methanol, Halogenated Hydrocarbons, and Other Solutions," DMIC Memorandum 228, Defense Metals Information Center, Battelle Memorial Institute, March 1967, p 16
164. T.R. Beck, M.J. Blackburn, W.H. Smyrl, and M.O. Speidel, "Stress-Corrosion Cracking of Titanium Alloys: Electrochemical Kinetics, SCC Studies With Ti: 8-1-1, SCC and Polarization Curves in Molten Salts, Liquid Metal Embrittlement, and SCC Studies With Other Titanium Alloys," Quarterly Progress Report 14, Contract NAS 7-489, Boeing Scientific Research Laboratories, Dec 1969
165. T.R. Beck and M.J. Blackburn, Stress-Corrosion Cracking of Titanium Alloys, *AIAA J.*, Vol 6 (No. 2), 1968, p 326
166. C.C. Seastrom and R.A. Gorski, The Influence of Fluorocarbon Solvents on Titanium Alloys, in "Accelerated Crack Propagation of Titanium by Methanol, Halogenated Hydrocarbons, and Other Solutions," DMIC Memorandum 228, Defense Metals Information Center, Battelle Memorial Institute, March 1967, p 20
167. J.D. Jackson and W.K. Boyd, "The Stress-Corrosion and Accelerated Crack Propagation Behavior of Titanium and Titanium Alloys," DMIC Technical Note, Defense Metals Information Center, Battelle Memorial Institute, Feb 1966
168. *Stress-Corrosion Cracking of Titanium*, STP 397, American Society for Testing and Materials, 1965
169. A.J. Hatch, H.W. Rosenberg, and E.F. Erbin, Effect of Environment on Cracking in Titanium Alloys, in *Stress-Corrosion Cracking of Titanium*, STP 397, American Society for Testing and Materials, 1965
170. H.L. Logan, Studies of Hot-Salt Cracking of the Titanium-8% Al-1% Mo-1% V Alloy, in *Proceedings of Conference—Fundamental Aspects of Stress-Corrosion Cracking*, National Association of Corrosion Engineers, 1969, p 662
171. H.L. Logan, M.J. McBee, G.M. Ugiansky, C.J. Bechtoldt, and B.T. Sanderson, Stress-Corrosion Cracking of Titanium, in *Stress-Corrosion Cracking of Titanium*, STP 397, American Society for Testing and Materials, 1965, p 215
172. S.P. Rideout, R.S. Ondrejcin, M.R. Louthan, and D.E. Rawl, The Role of Moisture and Hydrogen in Hot-Salt Cracking of Titanium Alloys, in *Proceedings of Conference—Fundamental Aspects of Stress-Corrosion Cracking*, National Association of Corrosion Engineers, 1969, p 650
173. R.V. Turley and C.H. Avery, Elevated Temperature Static and Dynamic Sea-Salt Stress Cracking of Titanium Alloys, in *Stress-Corrosion Cracking of Titanium*, STP 397, American Society for Testing and Materials, 1965, p 1
174. S.P. Rideout, R.S. Ondrejcin, and M.R. Louthan, Hot-Salt Stress-Corrosion Cracking of Titanium Alloys, in *The Science, Technology and Application of Titanium*, Pergamon Press, 1970
175. G. Sanderson and J.C. Scully, The Stress Corrosion of Titanium Alloys in Aqueous Magnesium Chloride Solution at 154 °C, *Corrosion*, Vol 24 (No. 3), 1968, p 75
176. M.A. Donachie, W.P. Danesi, and A.A. Pinkowish, Effect of Salt Atmosphere on Crack Sensitivity of Commercial Titanium Alloys at 600 °F-900 °F, in *Stress Corrosion Cracking of Titanium*, STP 397, American Society for Testing and Materials, 1965, p 179
177. W.K. Boyd, Stress-Corrosion Cracking of Titanium and Its Alloys, in *Proceedings of Conference—Fundamental Aspects of Stress-Corrosion Cracking*, National Association of Corrosion Engineers, 1969, p 593
178. D.E. Piper and D.N. Fager, The Relative Stress-Corrosion Susceptibility of Titanium Alloys in the Presence of Hot-Salt, in *Stress-Corrosion Cracking of Titanium*, STP 397, American Society for Testing and Materials, 1965, p 31
179. "Examination of Cracks in Titanium-Alloy Compressor Disc, Westinghouse Electric Corporation, Aviation Gas Turbine Division, Caused by Molten Cadmium," Memorandum Report, Titanium Metallurgical Laboratory, Battelle Memorial Institute, Feb 1956
180. H.A. Johnson, "Stress Cracking of Titanium," Technical Memorandum WCRT TM 56-97, Wright Air Development Center,

Wright-Patterson Air Force Base, 1956
181. W.M. Robertson, Embrittlement of Titanium by Liquid Cadmium, *Metall. Trans.*, Vol 2, 1970, p 68
182. A.R.C. Westwood, C.M. Preece, and M.H. Kamdar, Adsorption-Induced Brittle Fracture in Liquid Metal Environments, in *A Treatise on Brittle Fracture*, H. Liebouritz, Ed., Academic Press, 1970
183. J. Bingham, "Effect of Temperature on Preloaded Cadmium-Plated Titanium Fasteners," Hi-Shear Corporation, unpublished research, June 1969
184. D.N. Fager and W.F. Spurr, "Solid Cadmium Embrittlement: Titanium Alloys," *Corrosion*, Vol 26, 1970, p 40
185. "Nuclear Fuels and Materials Development," 2nd ed., Report TID-11295, U.S. Atomic Energy Commission, Sept 1962
186. J.B. Hollowell, J.G. Dunleavy, and W.K. Boyd, "Liquid-Metal Embrittlement," DMIC Technical Note, Defense Metals Information Center, Battelle Memorial Institute, April 1965
187. R.E. Duttweiler, R.R. Wagner, and K.C. Antony, An Investigation of Stress-Corrosion Failures in Titanium Compressor Components, in *Stress-Corrosion Cracking of Titanium*, STP 397, American Society for Testing and Materials, 1965, p 152
188. G. Martin, Investigation of Long-Term Exposure Effects Under Stress of Two Titanium Structural Alloys, in *Stress-Corrosion Cracking of Titanium*, STP 397, American Society for Testing and Materials, 1965, p 95
189. W. Rostocket, J.M. McCaughey, and H. Marcus, *Embrittlement by Liquid Metals*, Reinhold, 1960
190. D.N. Fager and W.F. Spurr, Some Characteristics of Aqueous Stress Corrosion in Titanium Alloys, *Trans. ASM*, Vol 61, 1968, p 283
191. J.D. Boyd, P.J. Moreland, W.K. Boyd, R.A. Wood, D.N. Williams, and R.I. Jaffee, "The Effect of Composition on the Mechanism of Stress-Corrosion Cracking of Titanium Alloys in N_2O_4 and Aqueous and Hot-Salt Environments," Contract NASr-100(09), Battelle Memorial Institute, Aug 1969
192. T.R. Beck, Stress-Corrosion Cracking of Titanium Alloys: II. An Electrochemical Mechanism, *J. Electrochem. Soc.*, Vol 115, 1968, p 890
193. M.J. Blackburn and J.C. Williams, Metallurgical Aspects of Stress-Corrosion Cracking of Titanium Alloys, in *Proceedings of Conference—Fundamental Aspects of Stress-Corrosion Cracking*, National Association of Corrosion Engineers, 1969, p 620
194. T.R. Beck, "Stress-Corrosion Cracking of Titanium Alloys. Preliminary Report on Ti-8Al-1Mo-1V and Proposed Electrochemical Mechanism," D1-82-0554, The Boeing Company, July 1965
195. D.A. Litvin and B. Hill, Effect of pH on Sea-Water Stress-Corrosion Cracking of Ti-7Al-2Cb-1Ta, *Corrosion*, Vol 26 (No. 3), 1970, p 89
196. D.E. Thomas, unpublished research, 1985
197. H.A. Johanson, G.B. Adams, and P. Van Rysselberghe, *J. Electrochem. Soc.*, Vol 104, 1957, p 339
198. S.R. Seagle, R.R. Seeley, and G.S. Hall, The Influence of Composition and Heat Treatment on Aqueous Stress Corrosion of Titanium, *Applications Related Phenomena in Titanium Alloys*, STP 432, American Society for Testing and Materials, 1967, p 170
199. B.S. Hickman, J.C. Williams, and H.L. Marcus, "Stress-Corrosion Cracking of High-Beta-Phase Content Titanium Alloys," Paper presented at Spring Meeting, Las Vegas, NV, American Institute of Mining, Metallurgical, and Petroleum Engineers, May 1970
200. T.R. Beck and M.J. Blackburn, "Stress-Corrosion Cracking of Titanium Alloys: SCC of Titanium: 8%Mn Alloy; Pitting Corrosion of Aluminum and Mass-Transport-Kinetic Model for SCC of Titanium," Progress Report 7, Contract NAS 7-489, Boeing Scientific Research Laboratories, April 1968
201. T.R. Beck, M.J. Blackburn, and M.O. Speidel, "Stress-Corrosion Cracking of Titanium Alloys: SCC of Aluminum Alloys, Polarization of Titanium Alloys in HCl and Correlation of Titanium and Aluminum SCC Behavior," Quarterly Progress Report 11, Contract NAS 7-489, Boeing Scientific Research Laboratories, March 1969
202. F.A. Crossley, C.J. Reichel, and C.R. Simcoe, "The Determination of the Effects of Elevated Temperature on the Stress-Corrosion Behavior of Structural Materials," Technical Report 60-191, WADD, Armour Research Foundation of Illinois Institute of Technology, May 1960
203. D. Schlain *et al.*, Galvanic Corrosion Behavior of Titanium and Zirconium in Sulfuric Acid Solutions, *J. Electrochem. Soc.*, Vol 102 (No. 3), March 1955, p 102-109
204. T.S. Lee, Preventing Galvanic Corrosion in Marine Environments, *Chem. Eng.*, April 1985, p 89
205. T.S. Lee *et al.*, *Corrosion*, Nov 1984, p 44-46
206. C.A. Smith and K.G. Compton, *Corrosion*, Vol 31 (No. 9), Sept 1975, p 320-326
207. J. Symonds, "The Influence of Sunlight on the Behavior of Galvanic Couples Between Ti and Cu-Base Alloys in Seawater," Oceanic Engineering Report 82-26, Westinghouse Electric Company, Oceanic Divisions, Feb 1982
208. G.A. Gehring, Jr. and R.J. Kyle, "Galvanic Corrosion in Steam Surface Condensers Tubed with Either Stainless Steel or Titanium," Paper 60, presented at Corrosion/82, Houston, TX, National Association of Corrosion Engineers, March 1982
209. G.A. Gehring, Jr. and J.R. Maurer, "Galvanic Corrosion of Selected Tubesheet/Tube Couples Under Simulated Seawater Condenser Conditions," Paper 202, presented at Corrosion/81, Toronto, Canada, National Association of Corrosion Engineers, April 1981
210. H.T. Hack and W.L. Adamson, "Analysis of Galvanic Corrosion Between a Titanium Condenser and a Copper-Nickel Piping System," Report 4553, David W. Taylor Naval Ship Research and Development Center, Jan 1976
211. G.A. Gehring, Jr. *et al.*, "Effective Tube Length—A Consideration on the Galvanic Corrosion of Marine Heat Exchanger Materials," Paper presented at Corrosion/80, Chicago, IL, National Association of Corrosion Engineers, 1980
212. D.W. Stough, F.W. Fink, and R.S. Peoples, "The Galvanic Corrosion Properties of Titanium and Titanium Alloys in Salt-Spray Environments," TML Memorandum, Battelle Memorial Institute, Oct 1957
213. L.C. Covington and G.A. Gehring Jr., "Experimental Verification of the Effective Tube Length Tending to Galvanically Corrode Copper Alloy Tubesheet," Paper presented at ASTM meeting, San Francisco, CA, American Society for Testing and Materials, May 1979
214. G.J. Danek, Jr., The Effect of Seawater Velocity on the Corrosion Behavior of Metals, *Naval Eng. J.*, Vol 78 (No. 5), 1966, p 763
215. C.F. Hanson, *Titanium—Science and Technology*, Vol 1, Pergamon Press, 1973, p 145
216. J.A. Davis and G.A. Gehring, Jr., *Mater. Perform.*, Vol 14 (No. 4), 1975, p 32-39
217. D.F. Hasson and C.R. Crowe, Titanium For Offshore Oil Drilling, *J. Met.*, Vol 34 (No. 1), 1982, p 23-28
218. A.E. Hohman and W.L. Kennedy, *Mater. Prot.*, Vol 2 (No. 9), Sept 1963, p 56-68
219. W.L. Williams, *J. Am. Soc. Naval Eng.*, Vol 62, Nov 1950, p 865-869
220. R.A. Wood, "Status of Titanium Blading For Low Pressure Steam Turbines," EPRI AF-445, Final Report, Electric Power Research Institute, Feb 1977
221. J.Z. Lichtman, *Corrosion*, Vol 17, 1961, p 119
222. A. Goldberg and R. Kershaw, "Evaluation of Materials Exposed to Scale-Control/Nozzle-Exhaust Experiments at the Salton Sea Geothermal Field," VCRL-52664, Lawrence Livermore Laboratory, Feb 1979
223. A. Goldberg and R. Garrison, "Materials Evaluation for Geothermal Applications: Turbine Materials," VCRL-79360, Lawrence Livermore Laboratory, 1977
224. R.P. Lee, *Mater. Perform.*, July 1976, p 26-32
225. G. Hoey and J. Bednar, *Mater. Perform.*, Vol 22 (No. 4), April 1983, p 9-14
226. H.B. Bomberger and L.F. Plock, Methods Used to Improve Corrosion Resistance of Titanium, *Mater. Prot.*, June 1969, p 45-48
227. M. Stern and H. Wissenberg, The Influence of Noble Metal Alloy Additions on the Electrochemical and Corrosion Behavior of Titanium, *J. Electrochem. Soc.*, Vol 106 (No. 9), Sept 1959, p 759
228. M. Stern and C.R. Bishop, The Corrosion Behavior of Titanium-Palladium Alloy, *Trans. ASM*, Vol 52, 1960, p 239
229. N.D. Thomashov *et al.*, Corrosion and Passivity of the Cathode-Modified Titanium-Based Alloys, in *Titanium and Titanium Alloys—Scientific and Technological Aspects*, Vol 2, Plenum Press, 1982, p 915-925
230. A.J. Sedriks, Further Observations on the Electrochemical Behavior of Ti-Ni Alloys on Acidic Chloride Solutions, *Corrosion*, Vol 29 (No. 2), 1973, p 64
231. J.C. Griess, Jr., *Corrosion*, April 1968, p 99-103
232. L.C. Covington and H.R. Palmer, "A New Corrosion Resistant Titanium Alloy Ti-38A for High Temperature Brine Service," Paper presented at the AIME Titanium Committee Session on Corrosion and Biomedical Applications of Titanium, Detroit, MI, American Institute of Mining, Metallurgical, and Petroleum Engineers, Oct 1974
233. M. Stern and C. Bishop, The Corrosion Resistance and Mechanical Properties of Titanium-Molybdenum Alloys Containing Noble Metals, *Trans. ASM*, Vol 54, Sept 1961, p 286-298
234. N. Thomashov, R. Al'tovskii, and G.

Chernova, Passivity and Corrosion Resistance of Titanium and Its Alloys, *J. Electrochem. Soc.*, Vol 108, Feb 1961, p 113-119
235. T. Fukuzuka, K. Shimogori, H. Satoh, and F. Kamikubo, Protection of Titanium Against Crevice Corrosion by Coating With Palladium Oxide, in *Titanium '80—Science and Technology*, The Metallurgical Society, 1980, p 2631-2638
236. S. Fujishiro and D. Eylon, *Thin Solid Films*, Vol 54, 1978, p 309-315
237. S. Fujishiro and D. Eylon, *Metall. Trans. A*, Vol 11A, Aug 1980, p 1261-1263
238. T. Fukuzuka *et al.*, in *Industrial Applications of Titanium and Zirconium*, STP 728, American Society for Testing and Materials, 1981, p 71-84
239. A. Erdemir *et al.*, *Mater. Sci. Eng.*, Vol 69, 1985, p 89-93
240. P. Sioshansi, Surface Modification by Ion Implantation, *Mach. Des.*, 20 March 1966
241. A. Takamura, Corrosion Resistance of Ti and a Ti-Pd Alloy in Hot, Concentrated Sodium Chloride Solutions, *Corrosion*, Oct 1967, p 306-313
242. H.B. Bomberger, "Alloying to Improve Crevice Corrosion Resistance of Titanium," Paper presented at the AIME Titanium Committee Session on Corrosion and Biomedical Applications, Detroit, MI, American Institute of Mining, Metallurgical, and Petroleum Engineers, Oct 1974
243. K. Shimogori *et al.*, Chemical Apparatus Free From Crevice Corrosion, U.S. Patent 4,154,897, 1979
244. K. Suzuki and Y. Nakamoto, *Mater. Perform.*, June 1981, p 23-26
245. S. Senderoff, Brush Plating, *Prod. Finish.*, Dec 1955
246. P. Munn and G. Wolf, *Mater. Sci. Eng.*, Vol 69, 1985, p 303-310
247. L.C. Covington, The Role of Multivalent Metal Ions in Suppressing Crevice Corrosion of Titanium, in *Titanium—Science and Technology*, Vol 4, Proceedings of the Second International Conference, Plenum Press, 1973, p 2395-2403
248. T. Moroishi and H. Miyuki, Effect of Several Ions on the Crevice Corrosion of Titanium, in *Titanium '80—Science and Technology*, The Metallurgical Society, 1980, p 2623-2630

Appendix 1: General Corrosion Data for Unalloyed Titanium

This appendix is a compilation of general corrosion rate values for unalloyed titanium (ASTM grades 1 to 4). These values were derived from various published sources (Ref 8, 26, 67, 80, 85, 86, 109, 127) and from unpublished in-house laboratory tests. These data should be used only as a guideline for alloy performance. Rates may vary depending on changes in medium chemistry, temperature, length of exposure, and other factors. Also, total alloy suitability cannot be assumed from these values alone, because other forms of corrosion, such as localized attack, may be limiting. The text should be consulted to assess overall alloy suitability more thoroughly for a given set of environmental conditions. In complex, variable, and/or dynamic environments, *in situ* testing may provide more reliable data. In the following table, temperatures are given only in centigrade, and corrosion rates are reported only in millimeters per year.

Medium	Concentration, %	Temperature, °C	Corrosion rate, mm/yr
Acetaldehyde	75	149	0.001
	100	149	nil
Acetate, *n*-propyl	· · ·	87	nil
Acetic acid	5–99.7	124	nil
	33–vapor	Boiling	nil
	99	Boiling	0.003
	65	121	0.003
	58	130	0.381
	99.7	124	0.003
Acetic acid + 3% acetic anhydride	Glacial	204	1.02
Acetic acid + 1.5% acetic anhydride	Glacial	204	0.005
Acetic acid + 109 ppm Cl	31.2	Boiling	0.259
Acetic acid + 106 ppm Cl	62.0	Boiling	0.272
Acetic acid + 5% formic acid	58	Boiling	0.457
Acetic anhydride	100	21	0.025
	100	150	0.005
	99.5	Boiling	0.013
Adipic acid + 15–20% glutaric + 2% acetic acid	25	199	nil
Adipic acid	67	240	nil
Adipylchloride and chlorobenzene solution	· · ·	· · ·	nil
Adiponitrile	Vapor	371	0.008
Aluminum chloride, aerated	10	100	0.002
	25	100	3.15
Aluminum chloride	10	100	0.002
	10	150	0.03
	25	60	nil
	25	100	6.55
Aluminum	Molten	677	164.6
Aluminum fluoride	Saturated	Room	nil
Aluminum nitrate	Saturated	Room	nil
Aluminum sulfate	Saturated	Room	nil
	10	80	0.05
	10	Boiling	0.12
Aluminum sulfate + 1% H_2SO_4	Saturated	Room	nil
Ammonium acid phosphate	10	Room	nil
Ammonium aluminum chloride	Molten	350–380	Very rapid attack
Ammonia, anhydrous	100	40	<0.127
Ammonia, steam, water	· · ·	222	11.2
Ammonium acetate	10	Room	nil
Ammonium bicarbonate	50	100	nil
Ammonium bisulfite, pH 2.05	Spent pulping liquor	71	0.015
Ammonium carbamate	50	100	nil
Ammonium chloride	Saturated	100	<0.013
Ammonium chlorate	300 g/L	50	0.003
Ammonium fluoride	10	Room	0.102
Ammonium hydroxide	28	Room	0.003
	28	100	nil
Ammonium nitrate	28	Boiling	nil
Ammonium nitrate + 1% nitric acid	28	Boiling	nil
Ammonium oxalate	Saturated	Room	nil
Ammonium perchlorate	20	88	nil
Ammonium sulfate	10	100	nil
Ammonium sulfate + 1% H_2SO_4	Saturated	Room	0.010
Aniline	100	Room	nil
Aniline + 2% $AlCl_3$	98	158	>1.27
Aniline hydrochloride	5	100	nil
	20	100	nil
Antimony trichloride	27	Room	nil
Aqua regia	3:1	Room	nil
	3:1	80	0.86
	3:1	Boiling	1.12
Arsenous oxide	Saturated	Room	nil
Barium carbonate	Saturated	Room	nil
Barium chloride	5	100	nil
	20	100	nil
	25	100	nil
Barium hydroxide	Saturated	Room	nil
Barium nitrate	10	Room	nil
Barium fluoride	Saturated	Room	nil
Benzaldehyde	100	Room	nil
Benzene (traces of HCl)	Vapor and liquid	80	0.005
	Liquid	50	0.025
Benzene	Liquid	Room	nil
Benzoric acid	Saturated	Room	nil
Bismuth	Molten	816	High
Bismuth/lead	Molten	300	Good resistance
Boric acid	Saturated	Room	nil
	10	Boiling	nil
Bromine	Liquid	30	Rapid attack
Bromine, moist	Vapor	30	<0.003
Bromine gas, dry	· · ·	21	Dissolves rapidly
Bromine-water solution	· · ·	Room	nil
Bromine in methyl alcohol	0.05	60	0.03 (cracking possible)

(continued)

Medium	Concentration, %	Temperature, °C	Corrosion rate, mm/yr
N-butyric acid	Undiluted	Room	nil
Calcium bisulfite	Cooking liquor	26	0.001
Calcium carbonate	Saturated	Boiling	nil
Calcium chloride	5	100	0.005
	10	100	0.007
	20	100	0.015
	55	104	0.001
	60	149	<0.003
	62	154	0.406
	73	175	0.80
Calcium hydroxide	Saturated	Room	nil
	Saturated	Boiling	nil
Calcium hypochlorite	2	100	0.001
	6	100	0.001
	18	21	nil
	Saturated	21	nil
Carbon dioxide	100	. . .	Excellent
Carbon tetrachloride	99	Boiling	0.005
	Liquid	Boiling	nil
	Vapor	Boiling	nil
Carbon tetrachloride + 50% H_2O	50	25	0.005
Chlorine gas, wet	>0.7 H_2O	Room	nil
	>0.95 H_2O	140	nil
	>1.5 H_2O	200	nil
Chlorine saturated water	Saturated	97	nil
Chlorine gas, dry	<0.5 H_2O	Room	May react
Chlorine dioxide	5	82	<0.003
Chlorine dioxide + HOCl, H_2O + Cl_2	15	43	nil
Chlorine dioxide in steam	5	99	nil
Chlorine dioxide	10	70	0.03
Chlorine monoxide (moist)	Up to 15	43	nil
Chlorine trifluoride	100	30	Vigorous reaction
Chloracetic acid	30	82	<0.127
	100	Boiling	<0.127
Chlorosulfonic acid	100	Room	0.312
Chloroform	Vapor and liquid	Boiling	0.000
Chloroform + 50% H_2O	50	25	0.000
Chloropicrin	100	95	0.003
Chromic acid	10	Boiling	0.003
	15	24	0.006
	15	82	0.015
	50	24	0.013
	50	82	0.028
Chromic acid + 5% nitric acid	5	21	<0.003
Citric acid	10	100	0.009
	25	100	0.001
	50	60	0.000
	50	Boiling	0.127–1.27
	672	149	Corroded
Citric acid (aerated)	50	100	<0.127
Copper nitrate	Saturated	Room	nil
Copper sulfate	50	Boiling	nil
Copper sulfate + 2% H_2SO_4	Saturated	Room	0.018
Cupric carbonate + cupric hydroxide	Saturated	Ambient	nil
Cupric chloride	20	Boiling	nil
	40	Boiling	0.005
	55	118	0.003
Cupric cyanide	Saturated	Room	nil
Cuprous chloride	50	90	<0.003
Cyclohexylamine	100	Room	nil
Cyclohexane (plus traces of formic acid)	. . .	150	0.003
Dichloroacetic acid	100	Boiling	0.007
Dichlorobenzene + 4–5% HCl	. . .	179	0.102
Diethylene triamine	100	Room	nil
Ethyl alcohol	95	Boiling	0.013
	100	Room	nil
Ethylene dichloride	100	Boiling	0.005–0.127
Ethylene dichloride + 50% water	50	25	0.005
Ethylene diamine	100	Room	nil
Ferric chloride	10–20	Room	nil
	1–30	100	0.004
	10–40	Boiling	nil
	1–30	Boiling	nil
	50	150	0.003
Ferric chloride	10	Boiling	0.00
Ferric sulfate	10	Room	nil
Ferrous chloride + 0.5% HCl	30	79	0.006
Ferrous sulfate	Saturated	Room	nil
Fluoboric acid	5–20	Elevated	Rapid attack
Fluorine, commercial	Gas–liquid	Gas-109	0.864
Fluorine, HF free	Liquid	-196	0.011
	Gas	-196	0.011
Fluorosilicic acid	10	Room	47.5
Formaldehyde	37	Boiling	nil
Formamide vapor	. . .	300	nil
Formic acid, aerated	10	100	0.005
	25	100	0.001
	50	100	0.001
	90	100	0.001
Formic acid, nonaerated	10	100	nil
	25	100	2.44
	50	Boiling	3.20
	90	100	3.00
Formic acid	9	50	<0.127
Furfural	100	Room	nil
Gluconic acid	50	Room	nil
Glycerin	. . .	Room	nil
Hydrogen chloride, gas	Air mixture	25–100	nil
Hydrochloric acid, aerated	1	60	0.004
	2	60	0.016
	5	60	1.07
	1	100	0.46
	5	35	0.01
	10	35	1.02
	20	35	4.45
Hydrochloric acid	0.1	Boiling	0.10
	1	Boiling	1.8
Hydrochloric acid + 4% $FeCl_3$ + 4% $MgCl_2$	19	82	0.51
Hydrochloric acid + 4% $FeCl_3$ + 4% $MgCl_2$ + Cl_2 saturated	19	82	0.46
Hydrochloric acid, chlorine saturated	5	190	<0.025
	10	190	28.5
Hydrochloric acid, + 200 ppm Cl_2	36	25	0.432
Hydrochloric acid			
+1% HNO_3	5	40	nil
+1% HNO_3	5	95	0.091
+5% HNO_3	5	40	0.025
+5% HNO_3	5	95	0.030
+10% HNO_3	5	40	nil
+10% HNO_3	5	95	0.183
+3% HNO_3	8.5	80	0.051
+5% HNO_3	1	Boiling	0.074
Hydrochloric acid			
+2.5% $NaClO_3$	10.2	80	0.009
+5.0% $NaClO_3$	10.2	80	0.006
Hydrochloric acid			
+0.5% CrO_3	5	38	nil
+0.5% CrO_3	5	95	0.031
+1% CrO_3	5	38	0.018
+1% CrO_3	5	95	0.031
Hydrochloric acid			
+0.05% $CuSO_4$	5	38	0.040
+0.05% $CuSO_4$	5	93	0.091
+0.5% $CuSO_4$	5	38	0.091
+0.5% $CuSO_4$	5	93	0.061
+1% $CuSO_4$	5	38	0.031
+1% $CuSO_4$	5	93	0.091
+5% $CuSO_4$	5	38	0.020
+5% $CuSO_4$	5	93	0.061
+0.05% $CuSO_4$	5	Boiling	0.064
+0.5% $CuSO_4$	5	Boiling	0.084
Hydrochloric acid			
+0.05% $CuSO_4$	10	66	0.025
+0.20% $CuSO_4$	10	66	nil
+0.5% $CuSO_4$	10	66	0.023
+1% $CuSO_4$	10	66	0.023
+0.05% $CuSO_4$	10	Boiling	0.295
+0.5% $CuSO_4$	10	Boiling	0.290
Hydrochloric acid + 0.1% $FeCl_3$	5	Boiling	0.01
Hydrochloric acid + 1 g/L Ti^{4+}	10	Boiling	0.000
Hydrochloric acid + 5.8 g/L Ti^{4+}	20	Boiling	0.000

(continued)

Medium	Concentration, %	Temperature, °C	Corrosion rate, mm/yr
Hydrochloric acid + 18% H_3PO_4 + 5% HNO_3	18	77	0.000
Hydrofluoric acid	1	26	127
Hydrofluoric acid, anhydrous	100	Room	0.127–1.27
Hydrofluoric-nitric acid 5 vol% HF-35 vol% HNO_3	. . .	25	452
Hydrofluoric-nitric acid 5 vol% HF-35 vol% HNO_3	. . .	35	571
Hydrogen peroxide	3	Room	<0.127
	6	Room	<0.127
	30	Room	<0.305
Hydrogen peroxide + 2% NaOH	1	60	55.9
Hydrogen peroxide			
pH 4	5	66	0.061
pH 1	5	66	0.152
pH 1	20	66	0.69
pH 11	0.08	70	0.42
Hydrogen sulfide (water saturated)	. . .	21	<0.003
Hydrogen sulfide, steam, and 0.077% mercaptans	7.65	93–110	nil
Hydroxy-acetic acid	. . .	40	0.003
Hypochlorous acid + ClO and Cl_2 gases	17	38	0.000
Iodine, dry or moist gas	. . .	25	0.1
Iodine in water + potassium iodide	. . .	Room	nil
Iodine in alcohol	Saturated	Room	Pitted
Lactic acid	10–85	100	<0.127
	10	Boiling	<0.127
Lead	. . .	816	Attacked
	. . .	324–593	Good
Lead acetate	Saturated	Room	nil
Linseed oil, boiled	. . .	Room	nil
Lithium, molten	. . .	316–482	nil
Lithium chloride	50	149	nil
Magnesium	Molten	760	Limited resistance
Magnesium chloride	5–20	100	<0.010
	5–40	Boiling	0.005
Magnesium hydroxide	Saturated	Room	nil
Magnesium sulfate	Saturated	Room	nil
Manganous chloride	5–20	100	nil
Maleic acid	18–20	35	0.002
Mercuric chloride	1	100	0.000
	5	100	0.011
	10	100	0.001
	Saturated	100	0.001
Mercuric cyanide	Saturated	Room	nil
Mercury	100	Up to 38	Satisfactory
	100	Room	nil
	. . .	371	3.03
Methyl alcohol	91	35	nil
	95	100	<0.01
Mercury + iron	. . .	371	0.079
Mercury + copper	. . .	371	0.063
Mercury + zirconium	. . .	371	0.033
Mercury + magnesium	. . .	371	0.083
Monochloracetic acid	30	80	0.02
	100	Boiling	0.013
Nickel chloride	5	100	0.004
	20	100	0.003
Nickel nitrate	50	Room	nil
Nitric acid, aerated	10	Room	0.005
	30	Room	0.004
	40	Room	0.002
	50	Room	0.002
	60	Room	0.001
	70	Room	0.005
	10	40	0.003
	20	40	0.005
	30	50	0.015
	40	50	0.016
	50	60	0.037
	60	60	0.040
	70	70	0.040
	40	200	0.610
	70	270	1.22
	20	290	0.305
Nitric acid	35	80	0.051–0.102
	70	80	0.025–0.076
	17	Boiling	0.076–0.102
	35	Boiling	0.127–0.508
	70	Boiling	0.064–0.900
Nitric acid, not refreshed	5–60	35	0.002–0.007
	5–60	60	0.01–0.02
	30–50	100	0.10–0.18
	5–20	100	0.02
	30–60	190	1.5–2.8
	70	270	1.2
	20	290	0.4
	70	290	1.1
Nitric acid, white fuming	Liquid or vapor	Room	nil
	. . .	82	0.152
	. . .	122	<0.127
	. . .	160	<0.127
Nitric acid, red fuming	<About 2% H_2O	Room	Ignition sensitive
	>About 2% H_2O	Room	Not ignition sensitive
Nitric acid	40	Boiling	0.63
+0.01% $K_2Cr_2O_7$	40	Boiling	0.01
+0.01% CrO_3	40	Boiling	0.01
+0.01% $FeCl_3$	40	Boiling	0.68
+1% $FeCl_3$	40	Boiling	0.14
+1% $NaClO_3$	40	Boiling	0.31
+1% $NaClO_3$	40	Boiling	0.02
+1% $Ce(SO_4)_2$	40	Boiling	0.10
+0.1% $K_2Cr_2O_7$	40	Boiling	0.016
Nitric acid, saturated with zirconyl nitrate	33–45	118	nil
Nitric acid + 15% zirconyl nitrate	65	127	nil
Nitric acid + 179 g/L $NaNO_3$ and 32 g/L NaCl	20.8	Boiling	0.127–0.295
Nitric acid + 170 g/L $NaNO_3$ and 2.9 g/L NaCl	27.4	Boiling	0.483–2.92
Oxalic acid	1	35	0.03
	5	35	0.13
	1	Boiling	107
	25	60	11.9
	Saturated	Room	0.508
Perchloroethylene + 50% H_2O	50	25	nil
Perchloryl fluoride + liquid ClO_3	100	30	0.002
Perchloryl fluoride + 1% H_2O	99	30	Liquid 0.290
	. . .	. . .	Vapor 0.003
Phenol	Saturated solution	25	0.102
Phosphoric acid	10–30	Room	0.020–0.051
	30–80	Room	0.051–0.762
	5.0	66	0.005
	6.0	66	0.117
	0.5	Boiling	0.094
	1.0	Boiling	0.266
	12	25	0.005
	20	25	0.076
	50	25	0.19
	9	52	0.03
	10	52	0.38
	5	Boiling	3.5
	10	80	1.83
Phosphoric acid + 3% nitric acid	81	88	0.381
Phosphorus oxychloride	100	Room	0.004
Phosphorus trichloride	Saturated	Room	nil
Photographic emulsions	. . .	. . .	<0.127
Phthalic acid	Saturated	Room	nil
Potassium bromide	Saturated	Room	nil
Potassium chloride	Saturated	Room	nil
	Saturated	60	nil
Potassium dichromate	Saturated	Room	nil
Potassium ethyl xanthate	10	Room	nil
Potassium ferricyanide	Saturated	Room	nil
Potassium hydroxide + 13% potassium chloride	13	29	nil
Potassium hydroxide	50	29	0.010
	10	Boiling	<0.127
	25	Boiling	0.305

(continued)

Medium	Concentration, %	Temperature, °C	Corrosion rate, mm/yr
Potassium hydroxide	50	Boiling	2.74
	50 anhydrous	241–377	1.02–1.52
Potassium iodide	Saturated	Room	nil
Potassium permanganate	Saturated	Room	nil
Potassium perchlorate	20	Room	0.003
	0–30	50	0.003
Potassium sulfate	10	Room	nil
Potassium thiosulfate	1	Room	nil
Propionic acid	Vapor	190	Rapid attack
Pyrogallic acid	355 g/L	Room	nil
Salicylic acid	Saturated	Room	nil
Seawater	· · ·	24	nil
Seawater, 4 1/2-year test	· · ·	Ambient	nil
Sebacic acid	· · ·	240	0.008
Silver nitrate	50	Room	nil
Sodium	100	To 1100 (593)	Good
Sodium acetate	Saturated	Room	nil
Sodium aluminate	25	Boiling	0.091
Sodium bifluoride	Saturated	Room	Rapid
Sodium bisulfate	Saturated	Room	nil
	10	66	1.83
Sodium bisulfite	10	Boiling	nil
	25	Boiling	nil
Sodium carbonate	25	Boiling	nil
Sodium chlorate	Saturated	Room	nil
Sodium chlorate + NaCl 80–250 g/L	0–721 g/L	40	0.003
Sodium chloride	Saturated	Room	nil
pH 7	23	Boiling	nil
pH 1.5	23	Boiling	nil
pH 1.2	23	Boiling	0.71
pH 1.2, some dissolved chlorine	23	Boiling	nil
Sodium citrate	Saturated	Room	nil
Sodium cyanide	Saturated	Room	nil
Sodium dichromate	Saturated	Room	nil
Sodium fluoride	Saturated	Room	0.008
pH 7	1	Boiling	0.001
pH 10	1	Boiling	0.001
pH 7	1	204	0.000
Sodium hydrosulfide + sodium sulfide and polysulfides	5–12	110	<0.003
Sodium hydroxide	5–10	21	0.001
	10	Boiling	0.021
	28	Room	0.003
	40	80	0.127
	50	57	0.013
	50	Boiling	0.051
	73	129	0.178
	50–73	188	>1.09
	50	38	0.023
Sodium hypochlorite	6	Room	nil
Sodium hypochlorite + 15% NaCl + 1% NaOH	1.5–4	66–93	0.030
Sodium nitrate	Saturated	Room	nil
Sodium perchlorate	900 g/L	50	0.003
Sodium phosphate	Saturated	Room	nil
Sodium silicate	25	Boiling	nil
Sodium sulfate	10–20	Boiling	nil
	Saturated	Room	nil
Sodium sulfide	10	Boiling	0.027
	Saturated	Room	nil
Sodium sulfite	Saturated	Boiling	nil
Sodium thiosulfate	25	Boiling	nil
Sodium thiosulfate + 20% acetic acid	20	Room	nil
Soils, corrosive	· · ·	Ambient	nil
Stannic chloride	5	100	0.003
	24	Boiling	0.045
Stannic chloride, molten	100	66	nil
Stannic chloride	100	35	nil
	Saturated	Room	nil
Steam + air	· · ·	82	nil
Steam + 7.65% hydrogen sulfide	· · ·	93–110	nil
Stearic acid, molten	100	180	0.003
Succinic acid	100	185	nil
	Saturated	Room	nil
Sulfanilic acid	Saturated	Room	nil
Sulfamic acid	3.75 g/L	Boiling	nil
	7.5 g/L	Boiling	2.74
Sulfamic acid + 0.375 g/L $FeCl_3$	7.5 g/L	Boiling	0.030
Sulfur, molten	100	240	nil
Sulfur monochloride	· · ·	202	>1.09
Sulfur dioxide, dry	· · ·	21	nil
Sulfur dioxide, water saturated	Near 100	Room	0.003
Sulfur dioxide gas + small amount SO_3 and approximately 3% O_2	18	316	0.006
Sulfuric acid, aerated	1	60	0.008
	3	60	0.013
	5	60	4.83
	10	35	1.27
	40	35	8.64
	75	35	1.07
	75	Room	10.8
	1	100	0.005
	3	100	23.4
	Concentrated	Room	1.57
	Concentrated	Boiling	5.38
	1	100	7.16
	3	100	21.1
Sulfuric acid	1	Boiling	17.8
	5	Boiling	25.4
Sulfuric acid + 0.25% $CuSO_4$	5	95	nil
	30	38	0.061
	30	95	0.088
Sulfuric acid + 0.5% $CuSO_4$	30	38	0.067
	30	95	0.823
Sulfuric acid + 1.0% $CuSO_4$	30	38	0.020
	30	95	0.884
Sulfuric acid + 0.5% CrO_3	5	95	nil
	30	95	nil
Sulfuric acid + 1.0% $CuSO_4$	30	Boiling	1.65
Sulfuric acid vapors	96	38	nil
	96	66	nil
	96	200–300	0.013
Sulfuric acid + 10% HNO_3	90	Room	0.457
Sulfuric acid + 50% HNO_3	50	Room	0.635
Sulfuric acid + 70% HNO_3	30	Room	0.102
Sulfuric acid + 90% HNO_3	10	Room	nil
Sulfuric acid + 90% HNO_3	10	60	0.011
Sulfuric acid + 95% HNO_3	5	60	0.005
Sulfuric acid + 50% HNO_3	50	60	0.399
Sulfuric acid + 20% HNO_3	80	60	1.59
Sulfuric acid saturated with chlorine	45	24	0.003
	62	16	0.002
	5, 10	190	<0.025
	82	50	>1.19
Sulfuric acid + 4 g/L Ti^{4+}	40	100	nil
Sulfurous acid	6	Room	nil
Tannic acid	25	100	nil
Tartaric acid	10–50	100	<0.127
	10	60	0.003
	25	60	0.003
	50	60	0.001
	10	100	0.003
	25	100	nil
	50	100	0.0121
Terephthalic acid	77	218	nil
Tetrachloroethane, liquid and vapor	100	Boiling	0.001
Tetrachloroethylene + H_2O	· · ·	Boiling	0.127
Tetrachloroethylene	100	Boiling	nil
Tetrachloroethylene, liquid and vapor	100	Boiling	0.001
Titanium tetrachloride	99.8	300	1.57
Trichloroacetic acid	100	Boiling	14.6
Trichloroethylene	99	Boiling	0.003–0.127
Trichloroethylene + 50% H_2O	50	25	0.001
Uranium chloride	Saturated	21–90	nil
Uranyl ammonium phosphate filtrate + 25% chloride + 0.5% fluoride + 1.4% ammonia + 2.4% uranium	20.9	165	<0.003
Uranyl nitrate containing 25.3 g/L Fe^{3+}, 6.9 g/L Cr^{3+}, 2.8 g/L Ni^{2+}, 4.0 M HNO_3 + 1.0 M Cl	120 g/L	Boiling	nil
Uranyl sulfate + 3.1 M Li_2SO_4 + 100–200 ppm O_2	3.1 M	250	<0.020
Uranyl sulfate + 3.6 M Li_2SO_4, 50 psi oxygen	3.8 M	350	0.006–0.432

(continued)

Medium	Concentration, %	Temperature, °C	Corrosion rate, mm/yr
Urea + 32% ammonia + 20.5% H_2O, 19% CO_2	28	182	0.079
Water, degassed	...	316	nil
Water, river, saturated with chlorine	...	93	nil
X-ray developer solution	...	Room	nil
Zinc chloride	5	Boiling	nil
	20	104	nil
	50, 75	150	nil
	75	150	0.06
	75	200	Rapid pitting
	80	173	2.1
Zinc sulfate	Saturated	Room	nil

Source: Ref 8, 26, 67, 80, 85, 86, 109, 127

Appendix 2: General Corrosion Data for Titanium Alloys

This appendix is a compilation of general corrosion rate values for commercial titanium alloys other than the unalloyed grades. These values were derived from various published sources (Ref 13, 26, 68, 80, 109, 133, 138) and from unpublished in-house laboratory tests. These data should be used only as a guideline for alloy performance. Rates may vary depending on changes in medium chemistry, temperature, length of exposure, and other factors. Total alloy suitability cannot be assumed from these values alone, because other forms of corrosion, such as localized attack, may be limiting. The text should be consulted to assess overall alloy suitability more thoroughly for a given set of environmental conditions. In complex, variable, and/or dynamic environments, *in situ* testing may provide more reliable data. In the following table, temperatures are given only in centigrade, and corrosion rates are reported only in millimeters per year.

Medium	Alloy	Concentration, %	Temperature, °C	Corrosion rate, mm/yr
Acetic acid	Grade 9	99.7	Boiling	nil
Acetic acid + 5% formic acid	Grade 12	58	Boiling	nil
Ammonium hydroxide	Grade 12	30	Boiling	nil
Aluminum chloride	Grade 12	10	Boiling	nil
	Grade 7	10	100	<0.025
	Grade 7	25	100	0.025
Ammonium chloride	Grade 12	10	Boiling	nil
Ammonium hydroxide	Grade 9	8, 28	150	nil
Aqua regia	Grade 7	3:1	Boiling	1.12
	Grade 12	3:1	Boiling	0.61
	Grade 9	3:1	Boiling	1.29
	Grade 9	3:1	25	0.015
Calcium chloride	Grade 7	62	150	nil
	Grade 7	73	177	nil
Chlorine, wet	Grade 7	...	25	nil
Chromic acid	Grade 7	10	Boiling	nil
	Grade 9	10	Boiling	0.008
	Grade 9	30	Boiling	0.053
	Grade 9	50	Boiling	0.26
Citric acid	Grade 7	50	Boiling	0.025
	Grade 12	50	Boiling	0.013
	Grade 9	50	Boiling	0.38
Ferric chloride	Grade 7	10	Boiling	nil
	Grade 12	10	Boiling	nil
	Ti-5Ta	10	Boiling	nil
	Grade 7	30	Boiling	nil
	Ti-6-4	10	Boiling	nil
	Ti-3-8-6-4-4	10	Boiling	nil
	Ti-10-2-3	10	Boiling	nil
	Ti-6-2-4-6	10	Boiling	0.06
	Transage 207	10	Boiling	0.19
	Ti-550	10	Boiling	nil
	Grade 9	10	Boiling	nil
	Ti-6-2-1-.8	10	Boiling	nil
Formic acid	Grade 9	25	88	<0.13
Formic acid, nitrogen-sparged	Grade 9	25	35	<0.13
Formic acid	Grade 9	50	Boiling	5.08
	Grade 7	45	Boiling	nil
	Grade 12	45, 50	Boiling	nil
	Grade 7	50	Boiling	0.01
	Ti-6-4	50	Boiling	7.92
	Transage 207	50	Boiling	0.90
	Ti-6-2-4-6	50	Boiling	0.62
	Ti-3-8-6-4-4	50	Boiling	0.98
	Ti-5Ta	50	Boiling	3.16
	Ti-550	50	Boiling	0.02
	Grade 12	90	Boiling	0.56
	Grade 7	90	Boiling	0.056
Hydrochloric acid	Ti-550	0.5	Boiling	0.056
	Ti-550	1.0	Boiling	0.64
	Transage 207	0.5	Boiling	0.005
	Transage 207	1.0	Boiling	0.025
	Ti-6-2-4-6	0.5	Boiling	nil
	Ti-6-2-4-6	1.0	Boiling	0.03
Hydrochloric acid, aerated	Ti-6-2-4-6	pH 1	Boiling	0.01
Hydrochloric acid	Ti-10-2-3	0.5	Boiling	1.10
	Ti-3-8-6-4-4	0.5	Boiling	0.003
	Ti-3-8-6-4-4	1.0	Boiling	0.058
	Ti-3-8-6-4-4	1.5	Boiling	0.26
Hydrochloric acid, aerated	Ti-3-8-6-4-4	pH 1	Boiling	nil
Hydrochloric acid	Ti-5Ta	0.5	Boiling	0.013
	Ti-5Ta	1.5	Boiling	2.10
	Ti-6-4	1.0	Boiling	2.52
Hydrochloric acid, aerated	Ti-6-4	pH 1	Boiling	0.60
Hydrochloric acid	Grade 9	0.5	Boiling	1.08
	Grade 9	1	88	0.009
	Grade 9	3	88	3.10
Hydrochloric acid, deaerated	Grade 7	3	82	0.013
	Grade 7	5	82	0.051
	Grade 7	10	82	0.419
Hydrochloric acid	Grade 9	1	Boiling	2.79
Hydrochloric acid, aerated	Grade 9	5	35	0.001
Hydrochloric acid, nitrogen saturated	Grade 9	5	35	0.185
Hydrochloric acid	Ti-6-2-1-.8	0.5	Boiling	0.020
	Ti-6-2-1-.8	1.0	Boiling	1.07
	Grade 7	0.5	Boiling	nil
	Grade 7	1.0	Boiling	0.008
	Grade 7	1.5	Boiling	0.03
	Grade 7	5.0	Boiling	0.23
	Grade 12	0.5	Boiling	nil
	Grade 12	1.0	Boiling	0.04
	Grade 12	1.5	Boiling	0.25
Hydrochloric acid, hydrogen saturated	Grade 7	1–15	25	<0.025
	Grade 7	20	25	0.102
	Grade 7	5	70	0.076
	Grade 7	10	70	0.178
	Grade 7	15	70	0.33
	Grade 7	3	190	0.025
	Grade 7	5	190	0.102
	Grade 7	10	190	8.9
Hydrochloric acid, oxygen saturated	Grade 7	3, 5	190	0.127
	Grade 7	10	190	9.3
Hydrochloric acid, chlorine saturated	Grade 7	3, 5	190	<0.03
	Grade 7	10	190	29.0
Hydrochloric acid, aerated	Grade 7	1, 5	70	<0.03
	Grade 7	10	70	0.05
	Grade 7	15	70	0.15
Hydrochloric acid + 4% $FeCl_3$ + 4% $MgCl_2$	Grade 7	19	82	0.49
Hydrochloric acid + 4% $FeCl_3$ + 4% $MgCl_2$, chlorine saturated	Grade 7	19	82	0.46
Hydrochloric acid +5 g/L $FeCl_3$	Grade 7	10	Boiling	0.279
+16 g/L $FeCl_3$	Grade 7	10	Boiling	0.076
+16 g/L $CuCl_2$	Grade 7	10	Boiling	0.127

(continued)

Medium	Alloy	Concentration, %	Temperature, °C	Corrosion rate, mm/yr
Hydrochloric acid				
+2 g/L $FeCl_3$	Grade 12	4.2	91	0.058
+0.2% $FeCl_3$	Grade 9	1	Boiling	0.005
+0.2% $FeCl_3$	Grade 9	5	Boiling	0.033
+0.2% $FeCl_3$	Grade 9	10	Boiling	0.305
+0.1% $FeCl_3$	Grade 9	5	Boiling	0.008
+0.1% $FeCl_3$	Ti-550	5	Boiling	0.393
+0.1% $FeCl_3$	Transage 207	5	Boiling	0.048
+0.1% $FeCl_3$	Ti-6-2-4-6	5	Boiling	0.068
+0.1% $FeCl_3$	Ti-10-2-3	5	Boiling	0.008
+0.1% $FeCl_3$	Ti-3-8-6-4-4	5	Boiling	0.018
+0.1% $FeCl_3$	Ti-5Ta	5	Boiling	0.020
+0.1% $FeCl_3$	Ti-6-4	5	Boiling	0.015
+0.1% $FeCl_3$	Ti-6-2-1.-.8	5	Boiling	0.051
+0.1% $FeCl_3$	Grade 7	5	Boiling	0.013
+0.1% $FeCl_3$	Grade 12	5	Boiling	0.020
Hydrochloric acid +				
18% H_3PO_4 + 5% HNO_3	Grade 7	18	77	nil
Hydrogen peroxide				
pH 1	Grade 7	5	23	0.062
pH 4	Grade 7	5	23	0.010
pH 1	Grade 7	5	66	0.127
pH 4	Grade 7	5	66	0.046
+500 ppm Ca^{2+}, pH 1	Grade 7	5	66	nil
+500 ppm Ca^{2+}, pH 1	Grade 7	20	66	0.76
Hydrogen peroxide,				
pH 1 + 5% NaCl	Grade 7	20	66	0.008
Magnesium chloride	Grade 7	Saturated	Boiling	nil
Methyl alcohol	Grade 9	99	Boiling	nil
Oxalic acid	Grade 7	1	Boiling	1.14
Nitric acid	Grade 9	10	Boiling	0.084
	Grade 9	30	Boiling	0.497
Phosphoric acid,				
naturally aerated	Grade 12	25	25	0.019
	Grade 12	30	25	0.056
	Grade 12	45	25	0.157
	Grade 12	8	52	0.02
	Grade 12	13	52	0.066
	Grade 12	15	52	0.52
	Grade 12	5	66	0.038
	Grade 12	7	66	0.15
	Grade 12	0.5	Boiling	0.071
	Grade 12	1.0	Boiling	0.14
	Grade 7	40	25	0.008
	Grade 7	60	25	0.07
	Grade 7	15	52	0.036
	Grade 7	23	52	0.15
	Grade 7	8	66	0.076
	Grade 7	15	66	0.104
	Grade 7	0.5	Boiling	0.050
	Grade 7	1.0	Boiling	0.107
	Grade 7	5.0	Boiling	0.228
Potassium hydroxide	Grade 9	50	150	9.21
Seawater	Grade 9	...	Boiling	nil
Sodium chloride, pH 1	Grade 9	Saturated	93	nil
Sodium fluoride				
pH 7	Grade 12	1	Boiling	0.001
pH 7	Grade 7	1	Boiling	0.002
Sodium hydroxide	Grade 9	50	150	0.49
Sodium sulfate, pH 1	Grade 7	10	Boiling	nil
Sulfamic acid	Grade 12	10	Boiling	11.6
	Grade 7	10	Boiling	0.37
Sulfuric acid,				
naturally aerated	Grade 12	9	24	0.003
	Grade 12	9.5	24	0.006
	Grade 12	10	24	0.38
	Grade 12	3.5	52	0.013
	Grade 12	3.75	52	1.73
	Grade 12	2.75	66	0.015
	Grade 12	3.0	66	1.65
	Grade 12	0.75	Boiling	0.003
	Grade 12	1.0	Boiling	0.91
	Grade 7	1.0	204	0.005
	Grade 7	2.0	204	nil
	Grade 12	1.0	204	0.91
	Grade 9	0.5	Boiling	8.48
Sulfuric acid,				
nitrogen saturated	Grade 7	5	70	0.15
	Grade 7	10	70	0.25
	Grade 7	1, 5	190	0.13
	Grade 7	10	190	1.50
Sulfuric acid,				
oxygen saturated	Grade 7	1–10	190	0.13
Sulfuric acid,				
chlorine saturated	Grade 7	10	190	0.051
	Grade 7	20	190	0.38
Sulfuric acid,				
nitrogen saturated	Grade 7	10	25	0.025
	Grade 7	40	25	0.23
Sulfuric acid, aerated	Grade 9	5	35	0.025
Sulfuric acid,				
nitrogen saturated	Grade 9	5	35	0.405
Sulfuric acid,				
naturally aerated	Ti-3-8-6-4-4	1	Boiling	nil
	Ti-3-8-6-4-4	5	Boiling	1.85
Sulfuric acid, aerated	Grade 7	10	70	0.10
	Grade 7	40	70	0.94
Sulfuric acid +				
5 g/L $Fe_2(SO_4)_3$	Grade 7	10	Boiling	0.178
Sulfuric acid +				
16 g/L $Fe_2(SO_4)_3$	Grade 7	10	Boiling	<0.03
Sulfuric acid +				
16 g/L $Fe_2(SO_4)_3$	Grade 7	20	Boiling	0.15
Sulfuric acid +				
15% $CuSO_4$	Grade 7	15	Boiling	0.64
Sulfuric acid +				
3% $Fe_2(SO_4)_3$	Ti-3-8-6-4-4	50	Boiling	<0.03
Sulfuric acid +				
1 g/L $FeCl_3$	Ti-3-8-6-4-4	10	Boiling	0.15
Sulfuric acid +				
50 g/L $FeCl_3$	Ti-3-8-6-4-4	10	Boiling	0.05
Sulfuric acid +				
1% $CuSO_4$	Grade 7	30	Boiling	1.75
Sulfuric acid +				
100 ppm Cu^{2+} +				
1% thiourea (deaerated)	Grade 7	1	100	nil
Sulfuric acid +				
100 ppm Cu^{2+} +				
1% thiourea (deaerated)	Grade 12	1	100	0.23
Sulfuric acid +				
1000 ppm Cl^-	Grade 7	15	49	0.015

Source: Ref 13, 26, 68, 80, 109, 133, 138

Corrosion of Zirconium and Hafnium

T.L. Yau and R.T. Webster, Teledyne Wah Chang Albany

ZIRCONIUM was first identified by Klaproth in 1789. In 1824, Berzelius made the first impure metal by reducing potassium fluorozirconate with potassium. In 1925, van Arkel and de Boer prepared the first high-purity zirconium by using an iodide decomposition process. The commercial Kroll process was developed in 1946 at the Bureau of Mines in Albany, OR.

Although zirconium is sometimes described as an exotic or rare element, it is in fact plentiful. It is ranked 19th in abundance of the chemical elements occurring in the earth's crust, and it is more abundant than many common metals, such as nickel, chromium, and cobalt. The most important source for zirconium is zircon ($ZrSiO_4$), which occurs in several regions throughout the world in the form of beach sand.

In 1940, Gillett discovered the excellent corrosion resistance of zirconium in a large number of acids and alkalies. This property was confirmed by Kroll in 1946 when ductile zirconium became available. Kroll predicted that zirconium would find uses in hydrochloric acid (HCl) applications. Hydrochloric acid is regarded as the most corrosive of the common acids. Indeed, one of the earliest applications for zirconium was in the handling of HCl.

About the time of Kroll's work, Kaufman and Utermeyer found that the early measurements of the thermal neutron cross section of zirconium were incorrect, because the metal that was tested contained hafnium. Hafnium occurs naturally with zirconium in ores (the corrosion of hafnium is discussed in section "Corrosion Resistance of Hafnium" in this article). When the hafnium was removed, zirconium was found to have a very low thermal neutron cross section. This high transparency to thermal neutrons, coupled with excellent corrosion resistance and good mechanical properties, makes zirconium very useful in nuclear power applications, especially as cladding for uranium fuel and for other reactor internals.

Nuclear applications account for a large portion of all the zirconium consumed. The excellent corrosion resistance of zirconium to strong acids and alkalies, salts, seawater, and other agents has attracted increasing attention for applications in chemical-processing equipment. Zirconium is used as a getter in vacuum tubes, as an alloying element, and in the manufacture of such diverse items as surgical appliances, photoflash bulbs, and explosive primers. Along with niobium, zirconium is superconductive at low temperatures and is used to make superconductive magnets.

Table 1 Typical physical and mechanical properties of zirconium

Property	Value
Physical Properties	
Atomic number	40
Atomic weight, amu	91.22
Atomic radius, Å	
0 charge	1.60–1.62
4+ charge	0.80–0.90
Density, g/cm^3 ($lb/in.^3$)	6.510 (0.235)
Crystal structure	
α phase	Hexagonal close-packed (below 865 °C, or 1590 °F)
β phase	Body-centered cubic (above 865 °C or 1590 °F)
α + β phase	...
Melting point, °C (°F)	1852 (3365)
Boiling point, °C (°F)	4377 (7910)
Coefficient of thermal expansion per °C at 25 °C (75 °F)	5.89×10^{-6}
Thermal conductivity (300–800 K)	
$Btu \cdot ft/h \cdot ft^2 \cdot °F$	13
W/m·K	22
Specific heat, J/kg·K (Btu/lb·°F)	285 (0.068)
Vapor pressure, mm Hg	
2000 °C (363 °F)	0.01
3600 °C (651 °F)	900.0
Electrical resistivity, μΩ·cm at 20 °C (70 °F)	39.7
Temperature coefficient of resistivity per °C 20 °C (70 °F)	0.0044
Latent heat of fusion, cal/g	60.4
Latent heat of vaporization, cal/g	1550
Mechanical Properties	
Modulus of elasticity, MPa (ksi)	9.9×10^4 (14.4×10^3)
Shear modulus, MPa (ksi)	3.6×10^4 (5.25×10^3)
Poisson's ratio (ambient temperature)	0.35

Physical and Mechanical Properties of Zirconium

Typical physical and mechanical properties of zirconium are given in Table 1 for comparison with the properties of other structural metals. First, the density of zirconium is lower than that of iron or nickel. Second, zirconium has a low coefficient of thermal expansion. The coefficient of thermal expansion of zirconium is about two-thirds that of titanium, about one-third that of AISI type 316 stainless steel, and about one-half that of Monel. Third, zirconium has high thermal conductivity—about 18% better than that of type 316 stainless steel.

Zirconium forms intermetallic compounds with most metallic elements, and only a limited number of alloys have been developed. For nuclear service, it is desirable to have zirconium alloys with improved strength and corrosion resistance in high-temperature water or steam. The most common alloys—Zircaloy-2 and Zircaloy-4—contain the strong α stabilizers tin and oxygen, as well as the β stabilizers iron, chromium, and nickel. The other alloys of commercial importance are Zr-2.5Nb and Zr-1Nb. In zirconium, niobium is a mild β stabilizer.

Zirconium ores generally contain a few percent of its sister element, hafnium. Hafnium has chemical and metallurgical properties similar to those of zirconium, although its nuclear properties are markedly different. Hafnium is a neutron absorber, but zirconium is not. As a result, there are nuclear and non-nuclear grades of zirconium and zirconium alloys. The nuclear grades are essentially hafnium free, and the non-nuclear grades may contain up to 4.5% Hf. Properly speaking, the alloy names Zircaloy, Zr-2.5Nb, and Zr-1Nb apply to nuclear grade materials. American Society for Testing and Materials (ASTM) specifications for non-nuclear grades list UNS R60704 as the alloy corresponding closely to Zircaloy-4 and UNS R60705 and R60706 as the alloys corresponding closely to Zr-2.5Nb. Properties and design specifications for zirconium alloys are given in Tables 2 to 5.

Corrosion Resistance of Zirconium

Zirconium is a reactive metal, as evidenced by its redox potential of −1.53 V versus the normal hydrogen electrode at 25 °C (75 °F). It has a high affinity for oxygen. When zirconium is exposed to an oxygen-containing environment, an adherent, protective oxide film forms on its surface. This film is formed spontaneously in air or water at ambient temperature and below. Moreover, this film is self-healing and protects the base metal from chemical and mechanical attack at temperatures to 300 °C (570 °F). As a result, zirconium is very resistant to corrosive attack in most mineral and organic acids, strong alkalies, saline solutions, and some molten salts. Zirconium is not attacked by oxidizing media unless halides are present.

There are a few media that will attack zirconium. Among them are hydrofluoric acid (HF),

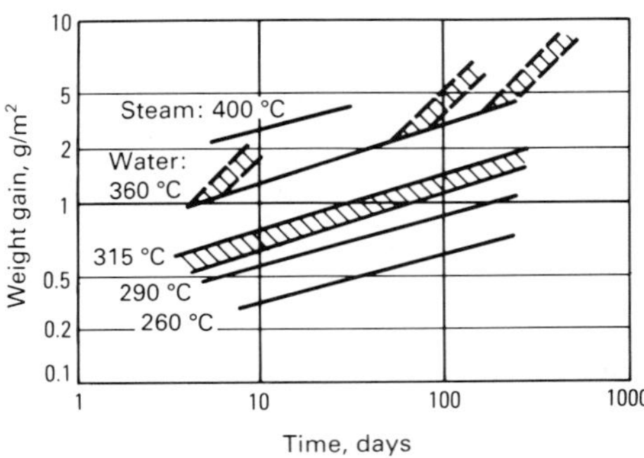

Fig. 1 Corrosion of arc-melted crystal bar zirconium in pressurized water and in steam. Source: Ref 2

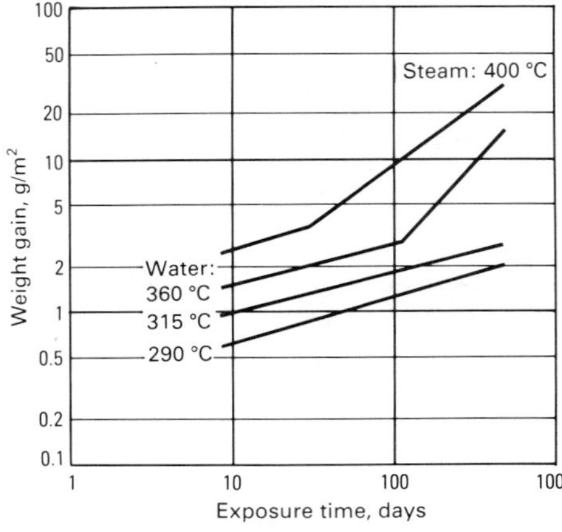

Fig. 2 Corrosion of Zircaloy-2 in pressurized water and in steam. Source: Ref 2

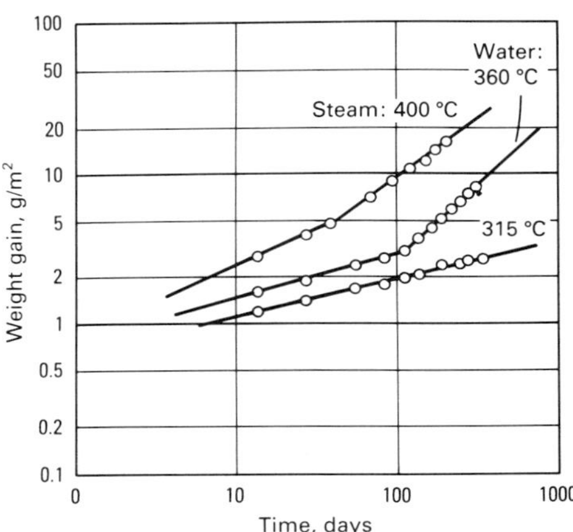

Fig. 3 Corrosion of Zircaloy-4 in pressurized water and in steam. Source: Ref 2

ferric chloride ($FeCl_3$), cupric chloride ($CuCl_2$), aqua regia, concentrated sulfuric acid (H_2SO_4), and wet chlorine gas. Table 6 lists media for which corrosion test data have been reported for zirconium and its alloys. The data in Table 6 should be viewed as a guide to application of zirconium in chemical process media. Corrosion resistance should be determined *in situ* if possible because the process medium may differ greatly from the reported media.

Specific Media

Water and Steam. Corrosion and oxidation of unalloyed zirconium in water and steam are reported to be irregular (Ref 1, 2). This behavior is probably caused by variations in the impurity content in the metal. Nitrogen and carbon impurities are particularly harmful. The corrosion rate of zirconium increases markedly when nitrogen and carbon concentrations exceed 40 and 300 ppm, respectively (Ref 1, 2).

Test results for a large number of heats of high-quality crystal bar zirconium are plotted in Fig. 1. The irregular behavior of unalloyed zirconium corrosion can be seen in the curves for 315 and 360 °C (600 and 680 °F). The data for corrosion resistance at 315 °C (600 °F) must be plotted as a band because there is too much scatter in the data. The curve at 360 °C (680 °F) has three bands extending upward from it; each band represents a change in corrosion rate from the basic rate indicated by the single line and represents data from a different set of test specimens. The irregular corrosion behavior of unalloyed zirconium stimulated alloy development programs. Zircaloy-2, Zircaloy-4, Zr-2.5Nb and Zr-1Nb are the most important alloys used in water-cooled nuclear reactors because they have the most reliable corrosion resistance in high-temperature water and steam.

Zircaloy-2 is superior to unalloyed zirconium in high-temperature water and steam. A tightly adherent oxide film forms on this alloy at a rate that is at first quasi-cubic but after an initial period undergoes a transition to linear behavior. Unlike the oxide film on unalloyed zirconium, the oxide film on Zircaloy-2 remains dark and adherent throughout transition and in the post-transition region. Test results for Zircaloy-2 in water and steam are plotted in Fig. 2.

Zircaloy-4 differs in composition from Zircaloy-2 only in having no nickel and a slightly greater iron content. Both variations are intended to reduce hydrogen pickup in reactor operation. The corrosion behavior of Zircaloy-4 is very similar to that of Zircaloy-2 (compare Fig. 3 and Fig. 2). However, hydrogen pickup for Zircaloy-4 is significantly lower, particularly when the alloy is exposed to water at 360 °C (680 °F). At this temperature, hydrogen pickup for Zircaloy-4 is about 25% of theoretical, or less than half that for Zircaloy-2. In addition, hydrogen pickup for Zircaloy-4 is less sensitive to hydrogen overpressure than that for Zircaloy-2 (Fig. 4). For both Zircaloys, hydrogen pickup is markedly decreased when dissolved oxygen is present in the corrosion medium (Ref 1).

Alloy Zr-2.5Nb is considered to be somewhat less resistant to corrosion than the Zircaloys. Nevertheless, Zr-2.5Nb is acceptable for many applications. An example is the use of Zr-2.5Nb pressure tubes in the primary loops of some reactors. The corrosion resistance of Zr-2.5Nb can be substantially improved by heat treatments (Ref 3, 4). Also, Zr-2.5Nb is superior to Zircaloys in steam at temperatures above 400 °C (750 °F) (Ref 5).

Salt Water. Zirconium has excellent corrosion resistance to seawater, brackish water, and polluted water. The corrosion properties of zirconium grades Zr702 and Zr704 in natural seawa-

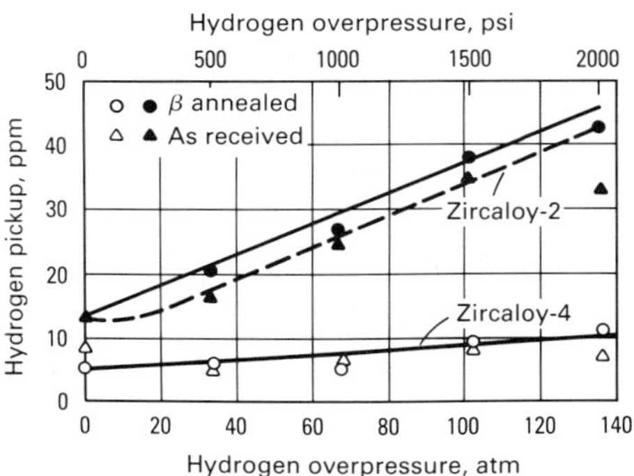

Fig. 4 Hydrogen pickup versus hydrogen overpressure for Zircaloy-2 and Zircaloy-4. Specimens were immersed for 14 days in water pressurized by admitting compressed hydrogen into the vapor space above the waterline. Water temperature: 343 °C (650 °F). Source: Ref 2

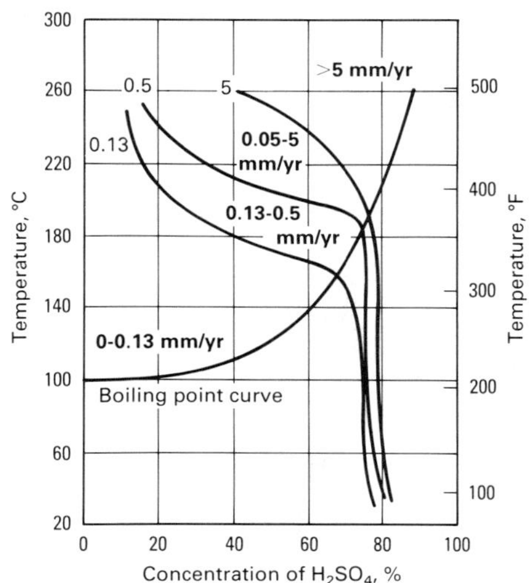

Fig. 5 Isocorrosion diagram for zirconium in H_2SO_4

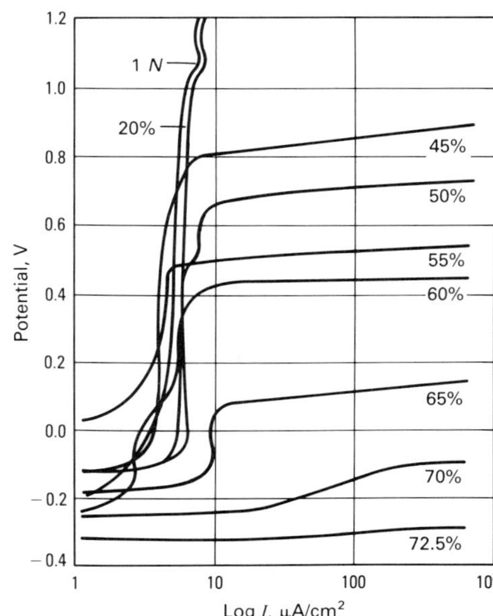

Fig. 6 Anodic polarization curves for zirconium in various concentrations of H_2SO_4 (wt%). Applied potential is given in volts versus the saturated calomel electrode. Temperature: near boiling point

Table 2 Chemical compositions of zirconium alloys

Alloy	Composition, % — Zr + Hf, min	Hf, max	Fe + Cr	Sn	H, max	N, max	C, max	Nb	O, max
Zr702	99.2	4.5	0.20	. . .	0.005	0.025	0.05	. . .	0.16
Zr704	97.5	4.5	0.2–0.4	1–2	0.005	0.025	0.05	. . .	0.18
Zr705	95.5	4.5	0.2 max	. . .	0.005	0.025	0.05	2–3	0.18
Zr706	95.5	4.5	0.2 max	. . .	0.005	0.025	0.05	2–3	0.16

Table 3 Minimum ASTM requirements for the room-temperature mechanical properties of zirconium alloys

Alloy	Minimum tensile strength MPa	Minimum tensile strength ksi	Minimum yield strength MPa	Minimum yield strength ksi	Minimum elongation (0.2% offset), %	Bend test radius(a)
Zr702	380	55	207	30	16	5*T*
Zr704	414	60	240	35	14	5*T*
Zr705	552	80	380	55	16	3*T*
Zr706	510	74	345	50	20	2.5*T*

(a) Bend tests are not applicable to material more than 4.75 mm (0.187 in.) thick. *T* is the thickness of the bend test specimen.

Table 4 Densities of zirconium alloys at 20 °C (70 °F)

Alloy	Density g/cm³	Density lb/in.³
Zr702	6.51	0.235
Zr704	6.57	0.237
Zr705	6.64	0.24
Zr706	6.64	0.24

ter are given in Ref 6. Specimens with or without crevice attachment of Zr702 were placed in the Pacific Ocean at Newport, OR, for up to 129 days. All welded and nonwelded specimens exhibited negligible corrosion rates. Marine biofouling was observed; however, no corrosion was found beneath the marine organisms or within the crevices. Laboratory tests were performed on Zr702 and Zr704 in boiling seawater for 275 days and in 200 °C (390 °F) seawater for 29 days. Both alloys were resistant to general, pitting, and crevice corrosion.

Tests of U-bend specimens with or without steel coupling of Zr702, nickel-containing Zr704, and nickel-free Zr704 were conducted in boiling seawater for 365 days. Results are reported in Table 7. No cracking was observed during the testing period. Overstressing of the tested U-bend indicated that all specimens were still ductile except for the welded nickel-containing Zr704 with steel coupling. Steel-coupled nickel-containing Zr704 showed much higher hydrogen and oxygen absorption and formed hydrides, particularly in the weld heat-affected zone. Chemical analyses and metallographic examinations on other U-bends did not show evidence of hydride formation.

Sulfuric acid is a corrosive and complicated acid. It has a wide range of strengths, changing from the reducing character of dilute solutions to the oxidizing character of concentrated solutions. The resistance of most engineering metals and alloys depends greatly on the acid concentration and temperature. The usefulness of many materials is restricted to some specific conditions, such as low concentrations, high concentrations, or low temperatures.

The corrosion of zirconium in H_2SO_4 solutions is rather straightforward (Fig. 5). It can be seen that zirconium resists attack by H_2SO_4 at all concentrations up to 70% and at temperatures to boiling and above. In 70 to 80% H_2SO_4, the corrosion resistance of zirconium depends strongly on temperature. In highly concentrated H_2SO_4, the corrosion rate of zirconium increases rapidly with concentration.

In the range in which zirconium shows corrosion resistance in H_2SO_4, a protective film is formed on zirconium that is predominantly cubic zirconium oxide (ZrO_2) with only traces of the monoclinic phase (Ref 7). Zirconium corrodes in highly concentrated H_2SO_4 (for example, 80%) because loose films are formed that prove to be zirconium disulfate tetrahydrate ($Zr(SO_4)_2 \cdot 4H_2O$) (Ref 8). Also, at the higher acid concentrations, films that flake off are formed, and they are probably partly zirconium hydride (Ref 8).

The two most attractive regions for the use of zirconium in H_2SO_4 can be seen in Fig. 5. The first region is the concentration range of 40 to 60% H_2SO_4, and the second is the dilute H_2SO_4 at elevated temperatures. In these two regions, iron- and nickel-base alloys corrode rapidly, or their resistance depends strongly on concentration, temperature, or aeration. The comparative corrosion rates of zirconium and several iron- and nickel-base alloys in H_2SO_4 are given in Table 8. As demonstrated in Table 8, stainless steels have very poor corrosion resistance in hot H_2SO_4. In general, stainless steels provide useful service only at low temperatures in dilute or highly concentrated H_2SO_4. Nickel-base alloys have higher corrosion resistance than stainless steels in H_2SO_4 (Table 8). However, the corrosion resistance of Alloy B-2 (UNS N10665) in H_2SO_4 decreases greatly at elevated temperatures. For example, the corrosion rate of Alloy B-2 in 10% H_2SO_4 increases from 0.025 to 26 mm/yr (1 to 1022 mils/yr) when temperature changes from 102 to 225 °C (215 to 435 °F). Furthermore, the corrosion resistance of Alloy C-276 (UNS N10276) is sensitive to temperature and concentration.

The corrosion resistance of zirconium in H_2SO_4 can be further understood from electrochemical measurements. The anodic polarization curves of zirconium in 4.9 to 72.5% H_2SO_4 at near-boiling temperatures are shown in Fig. 6. All polarization curves in this article were obtained according to ASTM G 5 (Ref 9). As indicated in Fig. 6, zirconium experiences a passive-to-transpassive transition in H_2SO_4 with an increasing potential. Zirconium does not have the active region in H_2SO_4, as is the case with common metals and alloys.

Figure 6 shows that the transpassive (breakdown) potential of zirconium in H_2SO_4 decreases with increasing concentration. As shown in Fig. 7, in less than 65% H_2SO_4, zirconium can tolerate some amounts of strong oxidizing agents, such as ferric (Fe^{3+}), cupric (Cu^{2+}), and nitrate ions (NO_3^-), without a reduction in corrosion resis-

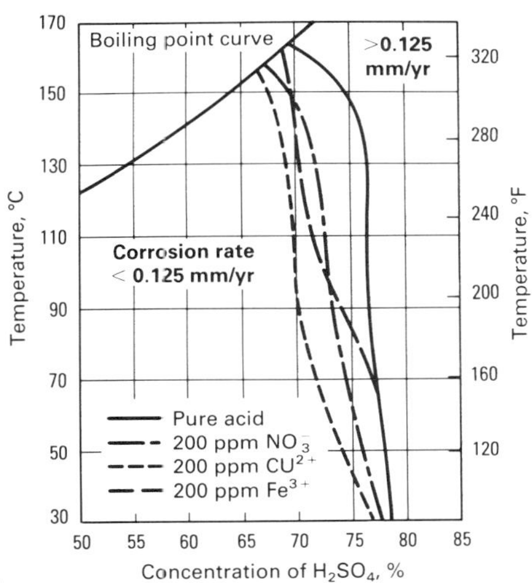

Fig. 7 Effect of 200 ppm of various impurities on the 0.125-mm/yr isocorrosion line for zirconium in H_2SO_4

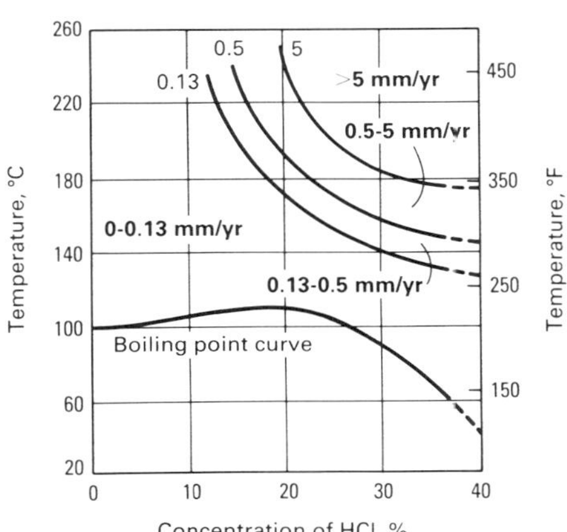

Fig. 8 Isocorrosion diagram for zirconium in HCl

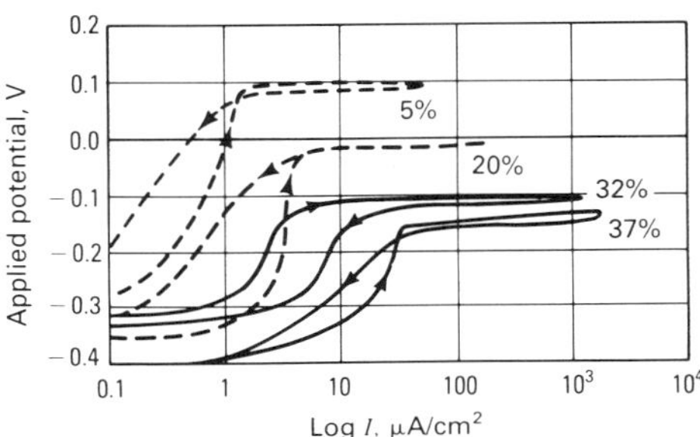

Fig. 9 Potentiodynamic curves for zirconium in various concentrations of HCl. Applied potential is given in volts versus the saturated calomel electrode.

Table 5 ASME mechanical requirements for Zr702 and Zr705 used for unfired pressure vessels

Material form and condition	ASME specification number	Alloy grade	Tensile strength MPa	Tensile strength ksi	Minimum yield strength MPa	Minimum yield strength ksi	Maximum allowable stress in tension for metal temperature not exceeding °C (°F): 40 (100) MPa	40 (100) ksi	95 (200) MPa	95 (200) ksi	150 (300) MPa	150 (300) ksi	205 (400) MPa	205 (400) ksi	260 (500) MPa	260 (500) ksi	315 (600) MPa	315 (600) ksi	370 (700) MPa	370 (700) ksi
Flat-rolled products	SB 551	702	359	52	207	30	90	13	76	11	64	9.3	48	7	42	6.1	41	6	33	4.8
		705	552	80	379	55	138	20	115	16.6	98	14.2	86	12.5	78	11.3	72	10.4	62	9.9
Seamless tubing	SB 523	702	359	52	207	30	90	13	76	11	64	9.3	48	7	42	6.1	41	6	33	4.8
		705	552	80	379	55	138	20	115	16.6	98	14.2	86	12.5	78	11.3	72	10.4	62	9.9
Welded tubing(a)	SB 523	702	359	52	207	30	77	11.1	65	9.4	55	7.9	41	6	36	5.2	35	5.1	28	4.1
		705	552	80	379	55	117	17	97	14.1	83	12	73	10.6	66	9.6	59	8.5	52	7.6
Forgings	SB 493	702	359	52	207	30	90	13	76	11	64	9.3	48	7	42	6.1	41	6	33	4.8
		705	552	80	379	55	138	20	115	16.6	98	14.2	86	12.5	78	11.3	72	10.4	62	9.9
Bar	SB 550	702	359	52	207	30	90	13	76	11	64	9.3	48	7	42	6.1	41	6	33	4.8
		705	552	80	379	55	138	20	115	16.6	98	14.2	86	12.5	78	11.3	72	10.4	62	9.9

(a) 85% joint efficiency was used to determine the allowable stress value for welded tube. Filler metal shall not be used in the manufacture of welded tube.

tance, as confirmed by immersion test results (Ref 10). Moreover, in 20% or less H_2SO_4, zirconium can tolerate a great amount of strong oxidizing agents. Consequently, zirconium equipment is often used in steel pickling. In more than 65% H_2SO_4, zirconium becomes sensitive to the presence of oxidizing agents (Ref 7).

Zirconium weld metal may corrode preferentially when H_2SO_4 concentration is approximately 55% and higher. Heat treatment at 775 ± 15 °C (1425 ± 25 °F) for 1 h per 25.4 mm (1 in.) of thickness was found to restore the corrosion resistance to the same high resistance of the parent metal (Ref 11-13).

The acid concentration limit is very important when zirconium is used to handle H_2SO_4 at elevated temperatures in the marginal concentration region, such as 60% or more. When the limit is exceeded, zirconium may corrode rapidly. In less than 65% H_2SO_4, the vapor phase is almost entirely water vapor (Ref 14). Also, the concentration change is negligible when the system is under a pressurized condition (Ref 15). Consequently, acid concentration can change significantly because of, for example, imperfect sealing of a system. In a nonpressurized system, the acid concentration can exceed the concentration limit. Acid concentration can easily change when the system is under vacuum because the water vapor is continuously taken away.

The results of the impurity effect study can be found in Ref 10 and 16. General conclusions include:

- Ferric, Cu^{2+}, and NO_3^- ion impurities in H_2SO_4 degrade at acid concentrations above 65%
- Chloride ions (Cl^-) alone do not change the corrosion resistance of zirconium in H_2SO_4
- When heavy-metal ions and halide ions coexist in H_2SO_4 (for example, when $FeCl_3$ is present), the optimum acid concentration range for zirconium is 60 to 65%, as demonstrated in Table 9
- Zirconium can tolerate only very small amounts of fluoride ions (F^-) in H_2SO_4 even at low acid concentrations. Fluoride ions must be complexed using inhibitors—such as silica (SiO_2), aluminum nitrate ($Al(NO_3)_3$), and zirconium sponge—when zirconium equipment is used to handle H_2SO_4-F^- solutions

When the corrosion resistance limits of zirconium in H_2SO_4 are exceeded, a pyrophoric surface layer may be formed on zirconium under some specific conditions (Ref 17, 18). The pyrophoric surface layer on zirconium formed in 77.5% H_2SO_4 + 200 ppm Fe^{3+} at 80 °C (175 °F) consisted of γ-hydride, ZrO_2, zirconium sulfate, and fine metallic particles (Ref 18). The combination of γ-hydride and metallic particles is suggested to be responsible for the pyrophoricity. Treating in hot air or steam can be used to eliminate this tendency (Ref 18).

Hydrochloric acid is regarded as the most difficult to handle of the common acids for metallic materials (Ref 19). The presence of even a small amount of HCl in a medium may cause pitting and stress-corrosion cracking (SCC) of common metals and alloys.

Zirconium is totally resistant to attack in all concentrations of HCl to temperatures well above boiling (Ref 10, 20, 21). The isocorrosion diagram for zirconium in HCl is shown in Fig. 8. Moreover, zirconium is not as susceptible to hydrogen embrittlement in HCl as tantalum is (Ref 22, 23). As discussed in Ref 23, tantalum lost 33% and 18% of its ductility after 1000 h in 11 *M* HCl and 11 *M* HCl + 7% gallium chloride ($GaCl_3$), respectively, at 70 °C (160 °F). Under the same testing conditions, zirconium remained unattacked and retained 100% of its ductility.

The comparative corrosion rates of zirconium and several iron- and nickel-base alloys in boiling 20% HCl are given in Table 10. Table 10 shows that zirconium outperforms other tested alloys. Normally, stainless steels and high-performance alloys can be considered only for handling very dilute and/or low-temperature HCl (Ref 24, 25).

Although HCl is highly reducing, the anodic polarization curves of zirconium still do not have the active region (Fig. 9). This corrosion property explains the resistance of zirconium to crevice corrosion in chloride-containing environments.

However, Fig. 9 shows that zirconium can suffer pitting and/or SCC when it is anodically polarized to a potential at or exceeding the pitting potentials. The same types of corrosion problems can be developed in HCl when strong oxidizing ions, such as Fe^{3+}, are present. Figure 10 illustrates the detrimental effect of Fe^{3+} in 20% HCl at 100 °C (212 °F). It can be seen that the presence of Fe^{3+} polarizes the zirconium surface to a potential exceeding the pitting potential. Thus, local breakdown of the passive surface at preferred sites occurs, and a condition develops that favors pitting and/or cracking. Maintaining zirconium at a potential in its passive region, which is arbitrarily set at 50 to 100 mV below the corrosion potential, can counteract the detrimental effects resulting from the presence of Fe^{3+} (Ref 26).

Zirconium resists other halogen acids except HF. Zirconium has corrosion rates of less than 0.13 mm/yr (<5 mils/yr) in boiling 20, 45, and 48% hydrobromic acid (HBr) and less than 0.025 mm/yr (<1 mil/yr) in boiling 47% hydroiodic acid (HI). The results of pitting potential measurements indicate that zirconium has lower pitting tendency in 1 *N* HBr or 1 *N* HI than in 1 *N* HCl:

Solution	Pitting potential, mV versus SCE
1 *N* HCl	140
1 *N* HBr	420
1 *N* HI	670

Nitric acid (HNO_3), due to its passivating power, is not considered to be a difficult acid for metallic materials to handle. However, HNO_3 becomes highly corrosive when its temperature is high or when impurities, such as heavy-metal ions, are present.

The excellent corrosion resistance of zirconium in HNO_3 has been known for over 30 years (Ref 20, 27, 28). Below the boiling point and at 98% HNO_3 and up to 250 °C (480 °F) and at 70% HNO_3, the corrosion rate of zirconium is less than 0.13 mm/yr (<5 mils/yr) (Fig. 11). Recent autoclave tests showed that the corrosion rates of zirconium were less than 0.025 mm/yr (<1 mil/yr) in 80% HNO_3 and 90% HNO_3 at 120 and 150 °C (250 and 300 °F) (Ref 29). Moreover, the corrosion rates were still under 0.025 mm/yr (<1 mil/yr) when zirconium was tested in boiling 30% to 70% HNO_3 with up to 1% $FeCl_3$, 1% NaCl, 1% seawater, 1% Fe^{3+}, or 1.45% stainless steel at 205 °C (400 °F) (Ref 29). These results indicated that the presence of heavy-metal ions and Cl^- in HNO_3 has little effect on the corrosion resistance of zirconium.

Zirconium is normally susceptible to pitting in acidic oxidizing chloride solutions. However, NO_3^- ions effectively inhibit the pitting of zirconium (Ref 30-32). The minimum ratio of NO_3^-/Cl^- required to inhibit pitting of zirconium was determined to be 1 (Ref 30, 31) or 5 (Ref 32). Nevertheless, the presence of appreciable amounts of HCl should be avoided, because zirconium is not resistant to aqua regia.

In the production of HNO_3, ammonia (NH_3) is oxidized with air over platinum catalysts. The resulting nitric oxide (NO) is further oxidized into

Table 6 Corrosion resistance of zirconium alloys in various media

Medium	Concentration, %	Temperature		Corrosion rate						Remarks
				Zr702		Zr704		Zr705		
		°C	°F	mm/yr	mils/yr	mm/yr	mils/yr	mm/yr	mils/yr	
Acetaldehyde	100	Boiling		<0.05	<2	...	...	...	...	...
Acetic acid	5–99.5	35	95 to boiling	<0.025	<1	...	...	<0.025	<1	...
Acetic acid anhydride	99	Room-boiling		<0.025	<1	...	...	<0.025	<1	...
Acetic acid (glacial)	99.7	Boiling		<0.13	<5	...	...	...	...	...
Acetic acid	100	160	320	<0.025	<1	...	...	...	...	...
Acetic acid + 50 ppm I^-	100	160,200	320,390	<0.025	<1	...	...	...	...	...
Acetic acid + 1% I^- + 100 ppm Fe^{3+}	99	200	390	<0.025	<1	...	...	<0.025	<1	...
Acetic acid + 2% HI	80	100	212	<0.025	<1	...	...	<0.025	<1	...
Acetic acid + 2% HI + 1000 ppm iron added as powder	80	100	212	<0.025	<1	...	...	...	...	...
Acetic acid + 2% HI, 1% methanol, 500 ppm formic acid, 100 ppm Cu	80	150	300	<0.025	<1	...	...	<0.025	<1	...
Acetic acid + 2% HI, 1% methanol, 500 ppm formic acid, 100 ppm Fe	80	150	300	<0.025	<1	...	...	<0.025	<1	...
Acetic acid + 2% HI	98	150	300	<0.025	<1	...	...	<0.025	<1	...
Acetic acid + 2% HI + 200 ppm Cl^-	80	100	212	<0.025	<1	...	...	<0.025	<1	...
Acetic acid + 2% HI + 200 ppm Fe^{3+}	80	100	212	<0.025	<1	...	...	<0.025	<1	...
Acetic acid + 2% I^-	98	150	300	<0.025	<1	...	...	<0.025	<1	...
Acetic acid + 2% HI + 1% CH_3OH + 500 ppm formic acid	80	150	300	<0.025	<1	...	...	<0.025	<1	...
Acetic acid + 2% HI + 200 ppm Cl^-	80	100	212	<0.025	<1	...	...	<0.025	<1	...
Acetic acid + 50% acetic anhydride	50	Boiling		<0.025	<1	...	...	<0.025	<1	...
Acetic acid + 50% 48% HBr	50	115	240	<0.025	<1	...	...	<0.025	<1	...
Acetic acid + saturated gaseous HCl and Cl_2	100	Boiling		>5	>200	...	...	>5	>200	...
Acetic acid + saturated, gaseous HCl and Cl_2	100	40	100	<0.025	<1	...	...	...	...	...
Acetic acid + 10% CH_3OH	90	200	390	<0.025	<1	...	...	...	...	...
Aluminum chlorate	30	100	212	<0.05	<2	...	...	...	...	...
Aluminum chloride	5, 10, 25	35–100	95–212	<0.025	<1	...	...	...	...	...
	25	Boiling		<0.025	<1	...	...	<0.025	<1	...
	40	100	212	<0.05	<2	...	...	...	...	...
Aluminum chloride (aerated)	5, 10	60	140	<0.05	<2	...	...	...	...	...
Aluminum fluoride	20	Room		>1.3	>50	...	...	...	...	pH 3.2
Aluminum potassium sulfate	10	Boiling		nil		...	...	nil		pH 3.2
Aluminum sulfate	25	Boiling		nil		...	...	nil		...
	60	100	212	<0.05	<2	...	...	...	...	...
Ammonia (wet)	$+H_2O$	38	100	<0.13	<5	...	...	...	...	...
Ammonium carbamate		193	380	<0.025	<1	...	...	...	...	58.4% urea, 16.8% ammonia, 14.8% CO_2, 9.9% H_2O at 22–24 MPa (3200–3500 psi)
Ammonium chloride	1, 10, saturated	35–100	95–212	<0.025	<1	...	...	...	...	...
Ammonium hydroxide	28	Room to 100	212	<0.025	<1	...	...	...	...	...
Ammonium fluoride	20	28	80	>1.3	>50	...	...	...	...	pH 8
	20	98	210	>1.3	>50	...	...	...	...	pH 8
Ammonium oxalate	100	100	212	<0.05	<2	...	...	...	...	...
Ammonium sulfate	5, 10	100	212	<0.13	<5	...	...	...	...	...
Aniline hydrochloride	5, 20	35–100	95–212	<0.025	<1	...	...	...	...	...
	5, 20	100	212	<0.05	<2	...	...	...	...	...
Aqua regia	3:1	Room		>1.3	>50	...	...	...	...	3 parts HCl/1 part HNO_3
Barium chloride	5.20	35–100	95–212	<0.025	<1	...	...	...	...	...
	25	Boiling		0.13–0.25	5–10	...	...	...	...	...
Bromine	100-liquid	20	70	<0.25	<10	...	...	0.5–1.3	20–50	Pitting
	vapor	20	70	...	...	...	...	>1.3	>50	Pitting
Bromochloromethane	100	100	212	<0.05	<2	...	...	...	...	...
Cadmium chloride	100	Room		<0.05	<2	...	...	...	...	...
Calcium bromide	100	100	212	<0.05	<2	...	...	...	...	...
Calcium chloride	5, 10, 25	35–100	95–212	<0.025	<1	...	...	...	...	...
	70	Boiling		<0.025	<1	...	...	<0.025	<1	B.P. = 162 °C (324 °F)
	75	Boiling		<0.13	<5	...	...	...	...	...
	Mixture	79	175	<0.025	<1	...	...	...	...	14% CaCl, 8% NaCl, 0.2% $Ca(OH)_2$
Calcium fluoride	Saturated	28	80	nil		...	...	...	...	pH 5
	Saturated	90	195	nil		...	...	...	...	pH 5
Calcium hypochlorite	2, 6, 20	100	212	<0.13	<5	...	...	...	...	...
Carbonic acid	Saturated	100	212	<0.13	<5	...	...	...	...	...

(continued)

Table 6 (continued)

Medium	Concentration, %	Temperature °C	Temperature °F	Corrosion rate Zr702 mm/yr	Zr702 mils/yr	Zr704 mm/yr	Zr704 mils/yr	Zr705 mm/yr	Zr705 mils/yr	Remarks
Carbon tetrachloride	0–100	Room to 100	212	<0.13	<5	...	...	...	...	...
Chlorine (water saturated)		Room		>1.3	>50	...	...	...	...	...
		75	165	>1.3	>50	...	...	...	...	...
Chlorine gas (more than 0.13% H_2O)	100	94	200	>1.3	>50	...	...	...	...	...
Chlorine gas (dry)	100	Room		<0.13	<5	...	...	...	...	...
Chlorinated water		100	212	<0.05	<2	...	...	...	...	...
Chloroacetic acid	100	Boiling		<0.025	<1	...	...	...	...	...
Chromic acid	10–50	Boiling		<0.025	<1	...	...	...	...	...
Citric acid	10–50	35–100	95–212	<0.025	<1	...	...	...	...	...
	10, 25, 50	100	212	<0.025	<1	...	...	...	...	...
	50	Boiling		<0.13	<5	...	...	...	...	...
Chromium plating solution		66	150	>1.3	>50	...	...	>1.3	>50	M + T chemicals CR-100
Cupric chloride	5, 10, 20	35–100	95–212	>1.3	>50	>1.3	>50	>1.3	>50	...
	20, 40, 50	Boiling		>1.3	>50	>1.3	>50	>1.3	>50	...
Cupric cyanide	Saturated	Room		>1.3	>50	...	...	...	...	...
Cupric nitrate	40	Boiling		Weight gain		...	...	Weight gain		B.P. = 115 °C (239 °F)
Dichloroacetic acid	100	Boiling		<0.5	<20	...	...	...	...	...
Ethylene dichloride	100	Boiling		<0.13	<5	...	...	...	...	...
Ferric chloride	0–50	Room to 100	212	>1.3	>50	>1.3	>50	>1.3	>50	...
	0–50	Boiling		>1.3	>50	>1.3	>50	>1.3	>50	...
Ferric sulfate	10	0–100	32–212	<0.05	<2	...	...	...	...	...
Formaldehyde	6–37	Boiling		<0.025	<1	...	...	<0.025	<1	...
	0–70	Room to 100	212	<0.05	<2	...	...	...	...	...
Fluoboric acid	5–20	Elevated		>1.3	>50	...	...	...	...	...
Fluosilicic acid	10	Room		>1.3	>50	...	...	...	...	...
Formic acid	10–90	35	95 to boiling	<0.13	<5	...	...	...	...	...
Formic acid (aerated)	10–90	Room to 100	212	<0.13	<5	...	...	...	...	...
Hydrazine	Mixture	109	230	<0.025	<1	...	...	...	...	2% hydrazine + saturated NaCl + 6% NaOH
	Mixture	130	265	nil		...	...	...	...	2% hydrazine + saturated NaCl + 6% NaOH
Hydrobromic acid	48	Boiling		<0.13	<5	...	...	<0.13	<5	B.P. = 125 °C (257 °F); shallow pits
	Mixture	Boiling		<0.025	<1	...	...	<0.025	<1	24% HBr + 50% acetic acid (glacial)
Hydrochloric acid	2	225	435	<0.025	<1	...	...	<0.025	<1	...
	5	Room		<0.025	<1	...	...	...	...	...
	10	35	95	<0.025	<1	...	...	...	...	...
	20	35	95	<0.025	<1	...	...	...	...	...
	32	30	85	<0.025	<1	...	...	...	...	...
	32	82	180	<0.025	<1	...	...	...	...	...
20% HCl + Cl_2 gas		58	135	0.13–0.25	5–10	...	...	...	...	Pitting
37% HCl + Cl_2 gas		58	135	<0.13	<5	...	...	...	...	...
10% HCl + 100 ppm $FeCl_3$		30	85	<0.025	<1	<0.05	<2	<0.025	<1	SCC observed
10% HCl + 100 ppm $FeCl_3$		105	220	<0.13	<5	...	...	...	...	Pitting rate
20% HCl + 100 ppm $FeCl_3$		105	220	<0.13	<5	...	...	...	...	...
37% HCl + 100 ppm $FeCl_3$		53	125	0.13–0.25	5–10	...	...	...	...	SCC observed
Hydrochloric acid	Mixture	Room		Dissolved		...	...	...	...	20% HCl + 20% HNO_3
	Mixture	Room		Dissolved		...	...	...	...	10% HCl + 10% HNO_3
Hydrofluoric acid	0–100	Room		>1.3	>50	...	...	...	...	...
Hydrogen peroxide	50	100	212	<0.05	<2	...	...	...	...	...
Hydroxyacetic acid		40	104	<0.13	<5	...	...	...	...	...
Lactic acid	10–100	148	298	<0.025	<1	...	...	...	...	...
	10–85	35	95 to boiling	<0.025	<1	...	...	...	...	...
Magnesium chloride	5–40	Room to 100	212	<0.05	<2	...	...	...	...	...
	47	Boiling		nil		...	...	nil		...
Manganese chloride	5, 20	Room to 100	212	<0.025	<1	...	...	...	...	...
Mercuric chloride	1-saturated	35–100	95–212	<0.025	<1	...	...	...	...	...
	Saturated	Boiling		<0.025	<1	...	...	<0.025	<1	...
Nickel chloride	5, 20	35–100	95–212	<0.025	<1	...	...	...	...	...
	5–20	100	212	<0.025	<1	...	...	...	...	...
	30	Boiling		nil		...	...	nil		...
Nitric acid	20	103	215	<0.025	<1	<0.025	<1	<0.025	<1	...
	70	121	250	<0.025	<1	<0.025	<1	<0.025	<1	...
	10–70	Room to 260	500	<0.025	<1	...	...	...	...	...
	70–98	Room-boiling		<0.025	<1	...	...	...	...	SCC observed
Nitric acid + 1% Fe	65	120	248	<0.025	<1	...	...	...	...	...
Nitric acid + 1% Fe	65	204	400	<0.025	<1	...	...	...	...	...
Nitric acid + 1.45% 304 stainless steel	65	204	400	nil		...	...	...	...	...
Nitric acid + 1% Cl^-	70	120	248	nil		...	...	...	...	...
Nitric acid + 1% seawater	70	120	248	nil		...	...	...	...	...
Nitric acid + 1% $FeCl_3$	70	120	248	nil		...	...	...	...	...
Oxalic acid	0–100	100	212	<0.025	<1	...	...	...	...	...
Perchloric acid	70	100	212	<0.05	<2	...	...	...	...	...

(continued)

Table 6 (continued)

Medium	Concentration, %	Temperature °C	°F	Corrosion rate: Zr702 mm/yr	Zr702 mils/yr	Zr704 mm/yr	Zr704 mils/yr	Zr705 mm/yr	Zr705 mils/yr	Remarks
Sodium sulfate	0–20	Room to 100	212	<0.05	<2	...	...	...	...	...
Sodium sulfide	33	Boiling		nil		...	...	nil		...
Stannic chloride	5	100	212	<0.025	<1	...	...	...	...	...
	24	Boiling		<0.025	<1	...	...	...	...	...
Succinic acid	0–50	100	212	<0.05	<2	...	...	...	...	...
	100	150	300	<0.05	<2	...	...	...	...	...
Sulfuric acid	0–75	20	70	<0.025	<1	<0.025	<1	<0.025	<1	...
	80	20	70	<0.13	<5	>1.3	>50	...	...	...
	80	30	85	0.5–1.3	20–50	>1.3	>50	>1.3	>50	...
	77.5	60	140	0.25–0.5	10–20	...	...	<0.25	<10	...
	75	50	120	<0.025	<1	...	...	...	...	...
	77	50	120	0.13–0.25	5–10	>1.3	>50	...	...	...
	80	50	120	>1.3	>50	>1.3	>50	>1.3	>50	...
	75	80	125	<0.13	<5	...	...	<0.13	<5	...
	65	100	212	...	...	<0.025	<1	<0.13	<5	...
	70	100	212	<0.05	<2	...	...	<0.13	<5	...
	75	100	212	<0.13	<5	...	...	<0.13	<5	...
	76	100	212	<0.25	<10	...	...	...	...	...
	77	100	212	<0.5	<20	...	...	...	...	...
	77.5	100	212	>1.3	>50	>1.3	>50	>1.3	>50	...
	60	130	265	...	...	...	...	<0.13	<5	...
	65	130	265	<0.025	<1	...	...	...	...	...
	70	140	285	<0.13	<5	...	...	<0.25	<10	...
	58	Boiling		...	...	<0.025	<1	<0.13	<5	B.P. = 140 °C (284 °F)
	62	Boiling		<0.13	<5	...	...	0.5–1.3	10–20	B.P. = 146 °C (295 °F)
	64	Boiling		<0.13	<5	...	...	0.5–1.3	20–50	B.P. = 152 °C (306 °F)
	68	Boiling		<0.13	<5	...	...	...	...	B.P. = 165 °C (329 °F)
	69	Boiling		<0.13	<5	...	...	...	...	B.P. = 167 °C (333 °F)
	71	Boiling		<0.13	<5	...	...	...	...	B.P. = 171 °C (340 °F)
	72–74	Boiling		0.13–0.25	5–10	>1.3	>50	...	...	...
	75	Boiling		0.25–0.5	10–20	>1.3	>50	...	...	B.P. = 189 °C (372 °F)
Sulfuric acid										
+ 1000 ppm Fe^{3+}	60	Boiling		<0.025	<1	...	...	...	...	B.P. = 138–142 °C (280–288 °F)
+ 10 000 ppm Fe^{3+}	60	Boiling		<0.13	<5	...	...	...	...	Added as $Fe_2(SO_4)_3$
Sulfuric acid										
+200–1000 ppm Fe^{3+}	65	Boiling		<0.13	<5	...	...	...	...	B.P. = 152–155 °C (306–311 °F)
+10 000 ppm Fe^{3+}	65	Boiling		0.13–0.25	5–10	...	...	...	...	Added as $Fe_2(SO_4)_3$
Sulfuric acid										
+14 ppm–141 ppm Fe^{3+}	70	Boiling		0.13–0.25	5–10	...	...	...	...	B.P. = 167–171 °C (333–340 °F)
+200 ppm Fe^{3+}	70	Boiling		0.25–0.5	10–20	...	...	...	...	Added as $Fe_2(SO_4)_3$
+1410 ppm–10 000 ppm Fe^{3+}	70	Boiling		>1.3	>50	...	...	...	...	...
Sulfuric acid										
+1000 ppm $FeCl_3$	60	Boiling		<0.13	<5	<0.13	<5	<0.5	<20	B.P. = 138–142 °C (280–288 °F)
+10 000 ppm $FeCl_3$	60	Boiling		<0.13	<5	<0.5	<20	0.5–1.3	20–50	...
+20 000 ppm $FeCl_3$	60	Boiling		0.5–1.3	20–50	0.5–1.3	20–50	>1.3	>50	...
Sulfuric acid										
+200 ppm $FeCl_3$	65	Boiling		<0.13	<5	<0.13	<5	<0.5	<20	B.P. = 152–155 °C (306–311 °F)
+1000 ppm $FeCl_3$	65	Boiling		<0.13	<5	<0.13	<5	<0.5	<20	...
+10 000 ppm $FeCl_3$	65	Boiling		<0.13	<5	<0.13	<5	<0.5	<20	...
Sulfuric acid										
+10 ppm $FeCl_3$	70	Boiling		<0.5	<20	<0.5	<20	>1.3	>50	B.P. = 167–171 °C (333–340 °F)
+100 ppm $FeCl_3$	70	Boiling		<0.5	<20	<0.5	<20	>1.3	>50	...
+200 ppm $FeCl_3$	70	Boiling		<0.5	<20	<0.5	<20	>1.3	>50	...
+1000 ppm $FeCl_3$	70	Boiling		<0.5	<20	<0.5	<20	>1.3	>50	...
+10 000 ppm $FeCl_3$	70	Boiling		0.5–1.3	20–50	>1.3	>50	>1.3	>50	...
Sulfuric acid										
+200 ppm Cu^{2+}	60	Boiling		<0.13	<5	...	...	...	...	Added as $CuSO_4$
+1000–10 000 ppm Cu^{2+}	60	Boiling		<0.025	<1	...	...	...	...	...
Sulfuric acid										
+200–10 000 ppm Cu^{2+}	65	Boiling		<0.13	<5	...	...	...	...	Added as $CuSO_4$
Sulfuric acid										
+3 ppm Cu^{2+}	70	Boiling		0.13–0.25	5–10	...	...	...	...	Added as $CuSO_4$
+27–226 ppm Cu^{2+}	70	Boiling		>1.3	>50	...	...	...	...	...
Sulfuric acid										
+1000–10 000 ppm NO_3^-	60	Boiling		<0.13	<5	...	...	...	...	Added as $NaNO_3$
+50 000 ppm NO_3^-	60	Boiling		>1.3	>50	...	...	...	...	...
Sulfuric acid										
+200–1000 ppm NO_3^-	65	Boiling		<0.13	<5	...	...	...	...	Added as $NaNO_3$
+10 000 ppm NO_3^-	65	Boiling		0.25–0.5	10–20	...	...	...	...	...
+50 000 ppm NO_3^-	65	Boiling		>1.3	>50	...	...	...	...	...
Sulfuric acid										
+200 ppm NO_3^-	70	Boiling		0.13–0.25	5–10	...	...	...	...	Added as $NaNO_3$
+6000 ppm NO_3^-	70	Boiling		0.5–1.3	20–50	...	...	...	...	...
Sulfuric acid										
+1000 ppm NO_3^-	60	Boiling		<0.13	<5	...	...	...	...	Added as HNO_3

(continued)

Table 6 (continued)

Medium	Concentration, %	Temperature °C	Temperature °F	Corrosion rate Zr702 mm/yr	Zr702 mils/yr	Zr704 mm/yr	Zr704 mils/yr	Zr705 mm/yr	Zr705 mils/yr	Remarks
Phenol	Saturated	Room		<0.13	<5	...	...	...	...	...
Phosphoric acid	5–30	Room		<0.13	<5	...	...	...	...	...
	5–35	60	140	<0.13	<5	...	...	...	...	...
	5–50	100	212	<0.13	<5	...	...	...	...	...
	35–50	Room		<0.13	<5	...	...	...	...	...
	45	Boiling		<0.13	<5	...	...	...	...	...
	50	Boiling		<0.13	<5	0.13–0.25	5–10	0.25–0.38	10–15	B.P. = 108 °C (226 °F)
	65	100	212	0.13–0.25	5–10	...	...	<0.5	<20	...
	70	Boiling		>1.3	>50	...	...	>1.3	>50	B.P. = 123–126 °C (253–259 °F)
	85	38	100	0.13–0.5	5–20	...	...	...	...	...
	85	80	175	0.5–1.3	20–50	...	...	0.5–1.3	20–50	...
	85	Boiling		>1.3	>50	...	...	>1.3	>50	B.P. = 156 °C (313 °F)
	Mixture	Room		nil		...	...	...	...	88% H_3PO_4 + 0.5% HNO_3
	Mixture	Room		Weight gain .		...	...	...	...	88% H_3PO_4 + 5% HNO_3
	Mixture	89	190	>1.3	>50	...	...	>1.3	>50	85% H_3PO_4 + 4% HNO_3
Potassium chloride	Saturated	60	140	<0.025	<1	...	...	...	...	...
	Saturated	Room		<0.025	<1	...	...	...	...	...
Potassium fluoride	20	28	82	nil		...	...	...	...	pH 8.9
	20	90	195	>1.3	>50	...	...	...	...	pH 8.9
	0.3	Boiling		<0.025	<1	...	...	...	...	...
Potassium hydroxide	50	27	80	<0.025	<1	...	...	...	...	...
	10	Boiling		<0.025	<1	...	...	...	...	...
	25	Boiling		<0.025	<1	...	...	...	...	...
	50	Boiling		<0.13	<5	...	...	...	...	...
	50-anhydrous	241–377	465–710	>1.3	>50	...	...	...	...	...
	Mixture	29	85	<0.025	<1	...	...	...	...	13% KOH, 13% KCl
Potassium iodide	0–70	Room to 100	212	<0.05	<2	...	...	...	...	...
Potassium nitrite	0–100	Room to 100	212	<0.05	<2	...	...	...	...	...
Seawater (Pacific)		Boiling		nil		...	...	nil		...
		200	390	nil		...	...	...	...	pH 7.6
Silver nitrate	50	Room		<0.13	<5	...	...	...	...	...
Sodium bisulfate	40	Boiling		<0.025	<1	...	...	<0.025	<1	B.P. = 107 °C (225 °F)
Sodium chloride	3-saturated	35	95 to boiling	<0.025	<1	...	...	<0.025	<1	...
	29	Boiling		<0.025	<1	...	...	...	...	...
	Saturated	Room		<0.025	<1	...	...	...	...	...
	Saturated	Boiling		<0.025	<1	...	...	<0.025	<1	Adjusted to pH 1
	Saturated	107	225	nil		...	...	...	...	Adjusted to pH 0
Sodium chloride + saturated SO_2	3.5	80	175	nil		...	...	...	...	...
Sodium chloride + saturated SO_2	25	80	175	nil		...	...	...	...	...
Sodium chloride + saturated SO_2	Saturated	80	175	nil		...	...	...	...	...
Sodium chloride	Mixture	215	420	nil		nil		nil		25% NaCl + 0.5% acetic acid + 1% S + saturated H_2S
Sodium fluoride	Saturated	28	82	nil		...	...	...	...	...
	Saturated	90	195	>1.3	>50	...	...	...	...	...
Sodium formate	0–80	100	212	<0.05	<2	...	...	...	...	...
Sodium hydrogen sulfite	40	Boiling		<0.025	<1	...	...	<0.025	<1	...
Sodium hydroxide	5–10	21	70	<0.025	<1	...	...	...	...	...
	28	Room		<0.025	<1	...	...	...	...	...
	10–25	Boiling		<0.025	<1	...	...	...	...	...
	40	100	212	<0.025	<1	...	...	...	...	...
	50	38–57	100–135	<0.025	<1	...	...	...	...	...
	50–73	188	370	0.5–1.3	20–50	...	...	...	...	...
	73	110–129	230–265	<0.05	<2	...	...	...	...	...
	73 to anhydrous	212–538	415–1000	0.5–1.3	20–50	...	...	...	...	...
	Mixture	82	180	<0.025	<1	...	...	...	...	9–11% NaOH, 15% NaCl
	Mixture	10–32	50–90	<0.025	<1	...	...	...	...	10% NaOH, 10% NaCl, and wet $CoCl_2$
	Mixture	129	265	<0.025	<1	...	...	...	...	0.6% NaOH, 2% $NaClO_3$ + trace of NH_3
	Mixture	191	375	<0.025	<1	...	...	...	...	7% NaOH, 53% NaCl, 7% $NaClO_3$, 80–100 ppm NH_3
	Mixture	138	280	<0.13	<5	...	...	...	...	52% NaOH + 16% NH_3
Sodium hydroxide (suspended salt, violent boiling)	20	60	140	0.25–0.5	10–20	...	...	...	...	...
Sodium hydroxide + 750 ppm free Cl_2	50	38	100	<0.025	<1	...	...	...	...	...
	50	38–57	100–135	<0.025	<1	...	...	...	...	...
Sodium hypochlorite	6	100	212	<0.13	<5	...	...	...	...	...
	6	50	120	nil		...	...	nil		...
Sodium iodide	0–60	100	212	<0.05	<2	...	...	...	...	...
Sodium peroxide	0–100	Room to 100	212	<0.025	<2	...	...	...	...	...
Sodium silicate	0–100	Room to 100	212	<0.025	<2	...	...	...	...	...

(continued)

Table 6 (continued)

Medium	Concentration, %	Temperature °C	Temperature °F	Corrosion rate Zr702 mm/yr	Zr702 mils/yr	Zr704 mm/yr	Zr704 mils/yr	Zr705 mm/yr	Zr705 mils/yr	Remarks
$+10\ 000$ ppm NO_3^-	60	Boiling		0.25–0.5	10–20	...	...	...	...	...
$+50\ 000$ ppm NO_3^-	60	Boiling		>1.3	>50	...	...	...	...	...
Sulfuric acid										
$+1000$ ppm NO_3^-	65	Boiling		<0.13	<5	...	...	...	...	Added as HNO_3
$+10\ 000$–$50\ 000$ ppm NO_3^-	65	Boiling		>1.3	>50	...	...	...	...	...
Sulfuric acid	Mixture	Room to 100	212	<0.025	<1	...	...	...	...	1% H_2SO_4, 99% HNO_3
	Mixture	Room to 100	212	nil		...	...	...	...	10% H_2SO_4, 90% HNO_3
	Mixture	Boiling		<0.025	<1	...	...	...	...	14% H_2SO_4, 14% HNO_3
	Mixture	100	212	>1.3	>50	>1.3	>50	>1.3	>50	25% H_2SO_4, 75% HNO_3
	Mixture	Room		<0.025	<1	...	...	...	...	50% H_2SO_4, 50% HNO_3
	Mixture	Boiling		>1.3	>50	>1.3	>50	>1.3	>50	68% H_2SO_4, 5% HNO_3
	Mixture	Boiling to 135	275	0.25–0.5	10–20	0.25–0.5	10–20	>1.3	>50	68% H_2SO_4, 1% HNO_3
	Mixture	Room		>1.3	>50	>1.3	>50	>1.3	>50	75% H_2SO_4, 25% HNO_3
	Mixture	Boiling		<0.025	<1	...	...	...	...	7.5% H_2SO_4, 19% HCl
	Mixture	Boiling		<0.025	<1	...	...	...	...	34% H_2SO_4, 17% HCl
	Mixture	Boiling		<0.025	<1	...	...	...	...	40% H_2SO_4, 14% HCl
	Mixture	Boiling		0.025–0.13	1–5	...	...	...	...	56% H_2SO_4, 10% HCl
	Mixture	Boiling		<0.025	<1	...	...	...	...	60% H_2SO_4, 1.5% HCl
	Mixture	Boiling		<0.13	<5	...	...	...	...	69% H_2SO_4, 1.5% HCl
	Mixture	Boiling		0.25–0.5	10–20	...	...	...	...	69% H_2SO_4, 4% HCl
	Mixture	Boiling		<0.5	<20	...	...	...	...	72% H_2SO_4, 1.5% HCl
	Mixture	Boiling		>1.3	>50	...	...	>1.3	>50	20% H_2SO_4, 7% HCl with 50 ppm F^- impurities
Sulfurous acid	6	Room		<0.13	<5	...	...	...	...	...
	Saturated	192	380	0.13–1.3	5–50	...	...	...	...	...
Sulfamic acid	10	Boiling		nil		...	...	nil		B.P. = 101 °C (214 °F)
Tannic acid	25	35–100	95–212	<0.025	<1	...	...	...	...	...
Tartaric acid	10–50	35–100	95–212	<0.025	<1	...	...	...	...	...
Trichloroacetic acid	10–40	Room		<0.05	<2	...	...	...	...	...
	100	Boiling		>1.3	>50	...	...	...	...	...
	100	100	212	>1.3	>50	...	...	...	...	B.P. = 195 °C (383 °F)
Tetrachloroethane	100	Boiling		<0.13	<5	...	...	...	...	B.P. = 146 °C (295 °F) symmetrical; B.P. = 129 °C (264 °F) unsymmetrical
Trichloroethylene	99	Boiling		<0.13	<5	...	...	...	...	B.P. = 87 °C (189 °F)
Trisodium phosphate	5–20	100	212	<0.13	<5	...	...	...	...	...
Urea reactor mixture	Mixture	193	380	0.025	<1	...	...	...	...	58% urea, 17% NH_3, 15% CO_2, 10% H_2O
Zinc chloride	70	Boiling		nil		...	...	nil		...
	5–20	35	95 to boiling	<0.025	<1	...	...	...	...	...
	40	180	355	<0.025	<1	...	...	<0.025	<1	...

Table 7 Hydrogen and oxygen analyses of zirconium-steel coupled U-bend test specimens

365-day test in boiling seawater

Specimen	Condition	Hydrogen/oxygen content, ppm Hydrogen	Oxygen
Zr702-steel	Unwelded	6	1350
Zr704 (nickel-containing)-steel	Unwelded	8	1480
Zr704 (nickel-free)-steel	Unwelded	9	1440
Zr702-steel	Welded	8	1250
Zr704 (nickel-containing)-steel	Welded	450	5000
Zr704 (nickel-free) steel	Welded	5	1480

Source: Ref 6

nitrogen dioxide (NO_2), then absorbed in water to form HNO_3. Acid of up to 70% concentration is produced at temperatures to 204 °C (400 °F) by the process. As indicated in Table 11, all tested alloys except zirconium corrode rapidly in 65% HNO_3 at 204 °C (400 °F).

The polarization curves of zirconium in HNO_3 are shown in Fig. 12. A passive-to-active transition similar to that which occurs in H_2SO_4 takes place with increasing acid concentration. However, corrosion potentials are very noble because of the oxidizing nature of HNO_3. Common oxidizing agents, such as oxygen and Fe^{3+} ions, will not affect the corrosion resistance of zirconium. The polarization curves do suggest that, although corrosion rates are low, zirconium may be sensitive to stress in concentrated HNO_3. This is consistent with the observation of SCC in U-bend specimens in more than 70% HNO_3 (Ref 33). The slow strain rate technique can reveal cracking of zirconium in less than 70% HNO_3 (Ref 34).

The primary concern in the use of zirconium for HNO_3 service is cracking in concentrated HNO_3. Results of C-ring tests indicate that zirconium specimens will have a long life when they are stressed below the yield point (Ref 33). Cracking can be prevented by avoiding high sustained tensile stresses or by applying other preventive measures (Ref 35).

Other concerns include the accumulation of chlorine gas in the vapor phase and the presence of noncomplexed F^- ions (Ref 29). Chlorine gas can be generated by the oxidation of chlorides by HNO_3. Areas that can trap chlorine gas should be avoided for zirconium equipment when Cl^- is present in HNO_3. The corrosion of zirconium in HNO_3-F^- solutions can be controlled by adding an inhibitor, such as zirconium sponge, zirconium nitrate ($Zr(NO_3)_4 \cdot 5H_2O$), or phosphorus pentoxide (P_2O_5), to convert F^- ions into noncorrosive complex ions.

Phosphoric acid (H_3PO_4) is less corrosive than other mineral acids. Many materials demonstrate useful resistance in H_3PO_4 at least at low temperatures. As usual, corrosion rates increase with temperature, concentration, and impurities in the acid. Areas such as the liquid-level line or the condensing zone are particularly vulnerable to attack.

Zirconium resists attack in H_3PO_4 at concentrations up to 55% and temperatures exceeding the boiling point. Above 55% H_3PO_4, the corrosion rate could increase greatly with temperature (Fig. 13). The most interesting area for zirconium would be dilute H_3PO_4 at elevated temperatures. As indicated in Table 12, zirconium outperforms tested stainless alloys in 20% H_3PO_4 at 150 °C (300 °F).

Figure 14 shows the anodic polarization curves of zirconium in H_3PO_4 at near-boiling temperatures. As the concentration increases, the passive range diminishes gradually, and the passive current increases progressively. It appears that zir-

Table 8 Comparative corrosion rates of Zr702 and stainless alloys in H_2SO_4

Concentration of H_2SO_4, %	Temperature °C	Temperature °F	Zr702 mm/yr	Zr702 mils/yr	type 304 mm/yr	type 304 mils/yr	type 310 mm/yr	type 310 mils/yr	type 316 mm/yr	type 316 mils/yr	Hastelloy B-2 mm/yr	Hastelloy B-2 mils/yr	Hastelloy C-276 mm/yr	Hastelloy C-276 mils/yr
10	102	216(a)	0.0025	<0.1	52	2053	3.7	145	14.2	561	0.025	1.0	0.18	7.1
30	108	226(a)	0.0025	<0.1	>127	>5000	28.9	1138	>127	>5000	0.053	2.1	1.4	55
55	132	269(a)	0.0025	0.1	$>23 \times 10^4$	$>9 \times 10^5$	>8890	$>3.5 \times 10^5$	$>10^4$	4×10^5	0.048	1.9	7.5	295
55	168	335	0.5	2.0	...	...	...	...	...	...	0.94	37	5.4	214
2	225	435	nil		...	...	...	...	...	...	0.38	15	1	40
5	232	450	0.0025	0.1	...	...	...	...	...	...	2.8	111	3.9	152
10	225	435	0.018	0.7	...	...	...	...	...	...	26	1022	16.8	660
15	225	435	0.06	2.4	...	...	...	...	...	...	...	...	...	...

(a) Boiling point

Table 9 Inhibiting effect of SO_4^{2-} on the corrosion of zirconium in boiling H_2SO_4 containing $FeCl_3$

Medium	Temperature °C	Temperature °F	Corrosion rate(a) mm/yr	Corrosion rate(a) mils/yr
40% H_2SO_4 + 2% $FeCl_3$	115	240	5(b)	199(b)
50% H_2SO_4 + 2% $FeCl_3$	124	255	2	80
55% H_2SO_4 + 2% $FeCl_3$	131	270	2.1	83
60% H_2SO_4 + 2% $FeCl_3$	141	285	0.66	26
70% H_2SO_4 + 2% $FeCl_3$	166	330	0.84	33

(a) Calculated from four 1-day cycles. (b) Localized corrosion

Table 10 Corrosion rates of zirconium and stainless alloys in boiling 20% HCl

Alloy	Test duration, h	Corrosion rate mm/yr	Corrosion rate mils/yr
Zr702	28	0.018	0.7
Type 304	0.5	>8890	$>3.5 \times 10^5$
Type 310	0.5	>2795	$>1.1 \times 10^5$
Type 316	0.5	1145	$>4.5 \times 10^4$
Hastelloy B-2	28	0.43	17
Hastelloy C-276	28	6.9	273

Table 11 Corrosion rates of zirconium and stainless alloys in 65% HNO_3 at 204 °C (400 °F)

Alloy	Test duration, h	Corrosion rate mm/yr	Corrosion rate mils/yr
Zr702	48	0.008	0.3
Type 304	3	>380	$>1.5 \times 10^4$
Type 310	20	>305	$>1.2 \times 10^4$
Type 316	20	>380	$>1.5 \times 10^4$
Hastelloy B-2	0.25	$>1.4 \times 10^4$	$>5.5 \times 10^5$
Hastelloy C-276	0.25	>660	$>2.6 \times 10^4$

conium passivates more slowly in H_3PO_4 than in other mineral acids.

If H_3PO_4 contains more than a trace of F^- ion, attack on zirconium may occur. Because fluoride compounds are usually present in H_3PO_4, the use of zirconium has always been questioned. However, because P_2O_5 is an effective fluoride inhibitor for zirconium (Ref 36) and a large amount of P_2O_5 is often present in H_3PO_4 processes, tests should be performed to determine the suitability of zirconium in the actual H_3PO_4 medium.

Alkalies. Zirconium resists attack in almost all alkalies, either fused or in solution (Ref 37-39). This makes zirconium distinctly different from some other highly corrosion-resistant materials, such as tantalum, glass, graphite, and polytetrafluoroethylene (PTFE), which are attacked by strong alkalies. Moreover, steels and stainless alloys, which are resistant to alkalies, are subject to cracking at certain concentrations and temperatures.

Zirconium U-bend specimens were tested in boiling, concentrated sodium hydroxide (NaOH) (Ref 40). During the test period, the concentration changed from 50% to about 85%, and temperature increased from 150 to 300 °C (300 to 570 °F). The PTFE washers and tubes used to make the U-bend dissolved. However, the zirconium U-bends specimens remained ductile and did not show any cracks after 20 days.

Zirconium coupons were tested in a white liquor paper pulping solution, which contained NaOH and sodium sulfide (Na_2S), at 120, 175, and 225 °C (250, 345, and 435 °F) (Ref 40). All coupons had corrosion rates of less than 0.025 mm/yr (<1 mil/yr). In the same solution, graphite and glass both corroded rapidly at 100 °C (212 °F).

Salts. Zirconium is highly resistant to corrosion by most saline solutions. Corrosion rates are usually very low at all temperatures to the boiling point. Ferric chloride and $CuCl_2$ are examples of the few exceptions. Although zirconium has good corrosion resistance in sodium fluoride (NaF) and potassium fluorides (KF) at low temperatures, resistance decreases rapidly with increasing temperature. Consequently, zirconium is not recommended for handling any fluoride-containing solutions unless F^- ions are complexed.

Zirconium is considerably more resistant to chloride cracking than stainless steels are. No failure was observed in U-bend tests conducted in boiling 42% magnesium chloride ($MgCl_2$). The other attractive corrosion property of zirconium is its high crevice corrosion resistance. Zirconium is not subject to crevice corrosion even in

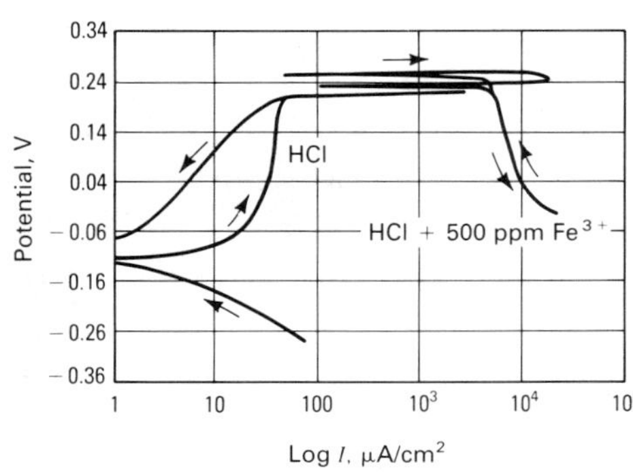

Fig. 10 Anodic polarization curves for zirconium in 20% HCl with and without 500 ppm Fe^{3+}. Test temperature: 100 °C (212 °F). Potential is given in volts versus the standard hydrogen electrode.

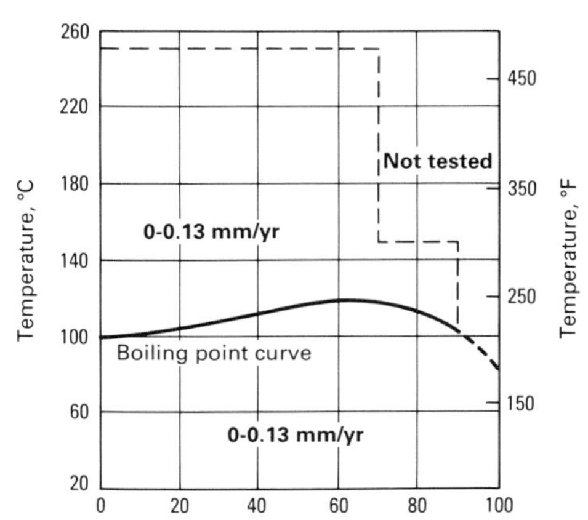

Fig. 11 Isocorrosion diagram for zirconium in HNO_3

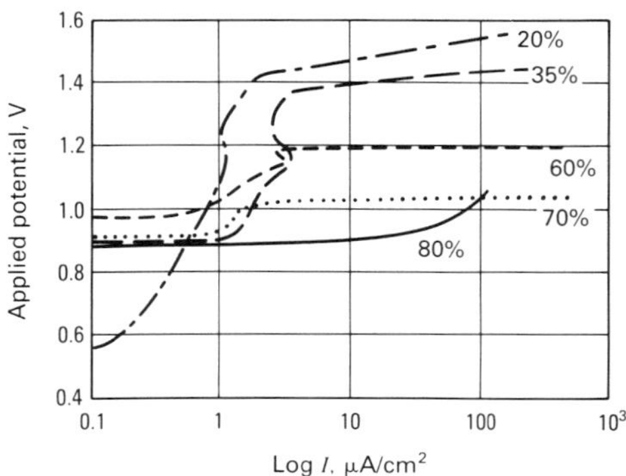

Fig. 12 Anodic polarization curves for zirconium in various concentrations of HNO_3. Applied potential is given in volts versus the saturated calomel electrode.

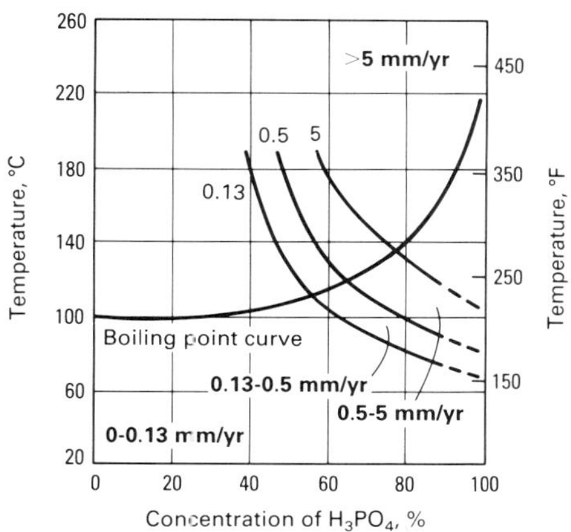

Fig. 13 Isocorrosion diagram for zirconium in H_3PO_4

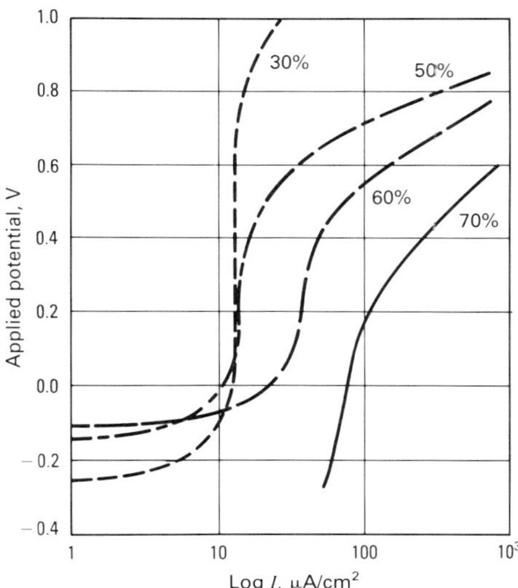

Fig. 14 Anodic polarization curves for zirconium in various concentrations of H_3PO_4. Applied potential is given in volts versus the saturated calomel electrode.

acidic chloride solutions at elevated temperatures.

Unlike many common metals, zirconium has very little affinity with sulfur. Zirconium-sulfur compounds were found to form only at temperatures above 500 °C (930 °F) (Ref 41). Furthermore, there is no instance of zirconium-sulfur bonds forming in aqueous systems (Ref 42). Consequently, the highly corrosive hydrogen sulfide (H_2S) will play a neutral role in the corrosion reactions of zirconium in sulfide-containing solutions. Zirconium coupons and U-bends were tested in numerous NaCl-H_2S solutions at temperatures to 205 °C (400 °F) (Ref 6, 43, 44). No pitting, crevice corrosion, or cracking was observed.

Zirconium is susceptible to pitting in acidic oxidizing chloride solutions. Cupric ions are more detrimental than Fe^{3+} ions in promoting the pitting and general corrosion of zirconium in acidic chloride solutions. The test results of zirconium in NaCl + $CuCl_2$ given in Tables 13 and 14 demonstrate the effects of pH surface condition, and heat treatment. The pitting and general corrosion of zirconium in NaCl + $CuCl_2$ can be controlled by adjusting the pH to 6 or higher (Table 13) or by heat treatments (Table 14). Also, resistance to pitting and general corrosion can be improved by surface conditioning. As usual, the weld metal is more susceptible to pitting than the base metal.

Organic Media. Zirconium resists corrosion in a wide range of organic compounds, including acetic acid, acetic anhydride, formic acid, urea, ethylene dichloride, formaldehyde, citric acid, lactic acid, oxalic acid, tannic acid, and trichloroethylene.

Gases. Zirconium will react with oxygen in the air at temperatures above 540 °C (1000 °F), producing a white ZrO_2 film that is brittle and porous. At temperatures above 700 °C (1290 °F), zirconium will absorb oxygen and become embrittled after prolonged exposure.

The oxide film on zirconium provides an effective barrier to hydrogen absorption up to 760 °C (1400 °F). In an all-hydrogen atmosphere, hydrogen absorption will begin at 310 °C (590 °F), and the metal will ultimately become embrittled. Hydrogen can be removed from zirconium by prolonged vacuum annealing at temperatures above 750 °C (1380 °F). Zirconium is stable in NH_3 up to about 1000 °C (1830 °F), in most gases (CO, CO_2, SO_2, C_3H_8, and N_2) up to about 300 to 400 °C (570 to 750 °F), and in halogens up to about 200 °C (390 °F).

Molten Salts and Metals. Zirconium resists attack in some molten salts. It is very resistant to corrosion by molten NaOH to temperatures above 1000 °C (1830 °F). It is also fairly resistant to potassium hydroxide (KOH). The oxidation properties of zirconium in nitrate salts are similar to those in air.

Zirconium resists some types of molten metals, but the corrosion rate is affected by trace impurities, such as oxygen, hydrogen, or nitrogen, in the specific molten metal. Zirconium has a corrosion rate of less than 0.025 mm/yr (1 mil/yr) in liquid lead to 600 °C (1110 °F), lithium to 800 °C (1470 °F), mercury to 100 °C (212 °F), and sodium to 600 °C (1110 °F). The molten metals known to attack zirconium are zinc, bismuth, and magnesium.

Forms of Corrosion

Galvanic Corrosion. Because of the protective oxide film that forms on zirconium in air, zirconium assumes a noble potential similar to that of silver. When this protective film is damaged, zirconium can become activated and can therefore corrode when in contact with other more noble metals (Table 15).

Other less noble metals will corrode when in contact with zirconium when its oxide film is intact. For example, in seawater or acid solutions, corrosion of steels, aluminum, and zinc is accelerated in electrical contact with zirconium. Therefore, care should be taken to keep them electrically insulated.

Crevice Corrosion. Of all the corrosion-resistant structural metals, zirconium and tantalum are the most resistant to crevice corrosion. In low-pH chloride solutions or chlorine gas, for example, zirconium is not subject to crevice attack.

Pitting Corrosion. Zirconium is quite resistant to pitting corrosion in chloride, bromide, and iodide solutions. Electrochemical measurements, however, can reveal the pitting tendency of zirconium in these halide solutions. Zirconium does not pit in most halide environments, because the corrosion potential is lower than the pitting potential. The presence of oxidizing ions, such as Fe^{3+} or Cu^{2+}, in halide solutions increases the corrosion potential, and pitting will begin when the corrosion potential exceeds the pitting potential. Zirconium will also pit in perchlorate solutions if oxidizing impurities are present. Nitrate and sulfate ions (SO_4^{2-}) will inhibit the pitting of zirconium in certain concentrations.

Fretting corrosion takes place when vibrational contact is made at the interface of tight-fitting highly loaded surfaces, such as between the leaves of a spring or the parts of ball and roller bearings. Fretting of zirconium occurs when its protective oxide coating is damaged or removed. The best way to overcome this type of corrosion (if it cannot be eliminated mechanically) is to apply a heavy oxide coating on the zirconium. This coating reduces friction drastically and prevents the removal of the passive protective oxide.

Table 12 Corrosion rates of zirconium and stainless alloys in 20% H_3PO_4 at 150 °C (300 °F)

Alloy	Test duration, h	Corrosion rate mm/yr	Corrosion rate mils/yr
Zr702	24	0.023	0.9
Type 304	24	60.4	2376
Type 310	24	41.2	1623
Type 316	24	10.4	410
Hastelloy B-2	24	1.1	43
Hastelloy C-276	24	0.99	39

Table 13 Corrosion of zirconium in boiling NaCl solutions containing 500 ppm Cu^{2+}

Test consisted of seven 1-day runs.

Concentration of NaCl, %	pH	Average corrosion rate: Unwelded specimens mm/yr	Unwelded specimens mils/yr	Welded specimens mm/yr	Welded specimens mils/yr
3.5	1	0.053(a)	2.1(a)	0.59(a)	23.6(a)
25	1	0.04	1.6	0.55(a)	21.7(a)
3.5	4.8	0.009(a)	0.38(a)	0.6(a)	23.8(a)
25	4	0.025	1	0.56(a)	21.9(a)
3.5	5	0.018	0.7	0.64(a)	25.3(a)
25	5	nil		nil	
3.5	6	nil		nil	
25	6	nil		nil	
3.5	7.5	nil		nil	
25	7.5	nil		nil	

(a) Pitting

Table 14 Effects of heat treatment on the corrosion of sandblasted and pickled zirconium in boiling NaCl solutions containing 500 ppm Cu^{2+}

Test consisted of seven 1-day runs; Cu^{2+} was added as $CuCl_2$.

Specimen type	Metallurgical condition	Average corrosion rate — 3.5% NaCl, mm/yr	3.5% NaCl, mils/yr	25% NaCl, mm/yr	25% NaCl, mils/yr
Unwelded	Sandblasted and pickled	0.007(a)	0.27(a)	0.024(a)	0.98(a)
Welded	Sandblasted and pickled	0.011(a)	0.45(a)	0.033(a)	1.3(a)
Welded	Heated to 760 °C (1400 °F), air cooled	0.0025(a)	0.1(a)	0.006(a)	0.23(a)
Welded	Heated to 760 °C (1400 °F), water quenched	0.0025(a)	0.1(a)	0.004(a)	0.17(a)
Welded	Heated to 870 °C (1600 °F), air cooled	0.0033	0.13	0.006(a)	0.23(a)
Welded	Heated to 870 °C (1600 °F) water quenched	0.0033	0.13	0.004(a)	0.17(a)
Welded	Heated to 980 °C (1800 °F), air cooled	0.005	0.2	0.007	0.27
Welded	Heated to 980 °C (1800 °F), water quenched	0.005	0.2	0.007	0.27

(a) Pitting

Table 15 Galvanic series in seawater

Cathodic (Noble)

Platinum
Gold
Graphite
Titanium
Silver
Zirconium
AISI type 316, 317 stainless steels (passive)
AISI type 304 stainless steel (passive)
AISI type 410 stainless steel (passive)
Nickel (passive)
Silver solder
Copper nickel (70–30)
Bronzes
Copper
Brasses
Nickel (active)
Naval brass
Tin
Lead
AISI type 316, 317 stainless steels (active)
AISI type 304 stainless steel (active)
Cast iron
Steel or iron
Aluminum alloy 2024
Cadmium
Aluminum alloy 1100
Zinc
Magnesium and magnesium alloys

Anodic (active)

Stress Corrosion. Zirconium and its alloys resist SCC in many environments, such as NaCl, HCl, $MgCl_2$, NaOH, and H_2S, that would induce SCC in other alloys. Zirconium service failures due to SCC are few in chemical applications. The high SCC resistance of zirconium can probably be attributed to its high repassivation rate. Any break in the surface film will be quickly healed if sufficient oxygen is present. Even in dehydrated systems, sufficient oxygen is generally present for repassivation.

The environments known to cause SCC of zirconium alloys include $FeCl_3$ or $CuCl_2$ solutions, mixtures of methanol (CH_3OH) and HCl or methanol and iodine, concentrated HNO_3, and liquid mercury or cesium. Stress-corrosion cracking in zirconium alloys can be prevented by:

- Avoiding high sustained tensile stress
- Modifying the environment
- Achieving a crystallographic texture with the hexagonal basal planes perpendicular to the cracking path
- Maintaining a high-quality surface film, that is, one low in impurities, defects, and mechanical damage
- Using electrochemical protection techniques

Corrosion Protection of Zirconium

Oxide Film Formation. One of the unique properties of zirconium that makes it attractive for chemical use is the protective nature of its oxide film. This film gives excellent corrosion resistance in spite of the reactive nature of the metal. If it is mechanically destroyed, this impervious oxide barrier will regenerate itself in many environments. Aluminum and titanium exhibit similar characteristics, but the film on zirconium is considerably more corrosion resistant.

Several methods of oxide formation are possible, depending on the properties desired. Normal methods include anodizing, autoclaving in high-temperature water or steam, and formation in air.

Anodizing forms a very thin film, but may be useful in some applications. Because it is formed near room temperature, the film does not have the adhesion to the underlying metal of thermally produced coatings.

Autoclave film formation is a practice common to the nuclear reactor industry. In this process, the uniform film of high integrity that is formed is superior to the films formed at lower temperatures. In addition to the slower corrosion rate, the rate of hydrogen absorption is drastically reduced. Films can be formed in 14 days in pressurized (19 MPa, or 2700 psi) deionized water at 360 °C (680 °F) or in 1 to 3 days at 400 °C (750 °F) and 10 MPa (1500 psi) in high-purity steam.

Film Formation in Air. The most common film used in the chemical industry is one formed in air. This film is often formed during the final stress relief of a component (30 min to 4 h at 550 °C, or 1020 °F). This film will range in color from a straw yellow to an iridescent blue or purple to a powdery tan or light grey. Such films need not be taken as a sign of metal contamination. This low-temperature stress relief does not cause significant penetration of oxygen into the metal, but it does form an oxide layer that is diffusion bonded to the base metal.

In addition to the corrosion resistance provided by the oxide film, a properly formed film serves as an excellent bearing surface against a variety of materials in several environments and over a broad temperature range (Ref 45). This treatment consists of cleaning the surface, followed by 4 to 6 h in air at 550 °C (1020 °F). The resultant oxide layer, approximately 5 μm (0.2 mil) thick, is equivalent to sapphire in hardness and is diffusion bonded to the base metal. The oxide layer can be damaged by a striking action, but it serves as an excellent bearing surface for sliding contacts. Oxidized zirconium pump shafts are an example of a common application.

Electrochemical Protection. Zirconium experiences a passive-to-transpassive transition with increasing potential in all mineral acids except HF (Ref 46). The commonly observed active nose in many metal-acid systems is not observed for zirconium. Consequently, zirconium prefers a more reducing condition, which produces more stable passive films. In fact, zirconium is one of the best metals for handling reducing acids.

By impressing a potential that is arbitrarily 50 to 100 mV below its corrosion potential, zirconium can be used to handle oxidizing HCl solutions (Ref 26). Tables 16 and 17 demonstrate the benefits of electrochemical protection in greatly reducing pitting and SCC. The uniform corrosion rates of unprotected coupons are usually low. However, the penetration rates are much greater than the uniform corrosion rates and increase with immersion time. Electrochemical protection eliminates this local attack. Similarly, unprotected welded U-bends cracked in all but one case shortly after exposure, while protected U-bends resisted cracking for the 32-day test interval in all but one acid concentration. Thus, electrochemical protection offers a very definite improvement to the corrosion properties of zirconium in oxidizing acid chlorides.

This electrochemical protection technique can also be used to combat the SCC of zirconium in concentrated HNO_3 (Ref 35). In HNO_3, because of the difference between the corrosion potential and the critical potential that can cause SCC, it is desirable to control the potential of zirconium a few hundred millivolts below the corrosion potential, or at 500 mV versus SCE.

Applications

Nuclear Industry. The development of water-cooled nuclear power reactors brought about the use of zirconium for uranium fuel cladding and for structural components. As a result of these developments in the nuclear industry, the cost of zirconium and its alloys decreased considerably. Zirconium and its alloys have emerged as engineering materials rather than laboratory curiosities. Materials for fuel cladding and structural components in nuclear reactors are restricted because of the following crucial requirements:

- Low absorption cross section for thermal neutrons
- Adequate strength, creep resistance, and ductility after prolonged irradiation in reactor coolant
- Excellent corrosion and oxidation resistance
- Absence of interactions with the fuel material and fission products

Table 16 Corrosion of protected and unprotected Zr702 in HCl solutions containing 500 ppm Fe^{3+}

Test duration: 32 days

Medium	Acidity	Temperature °C	Temperature °F	Corrosion rate: Unprotected mm/yr	Unprotected mils/yr	Protected(a) mm/yr	Protected(a) mils/yr
10% HCl	3 *N*	60	140	0.18	7.1	<0.0025	0.1
		102	215	1.3	51	<0.0025	0.1
Spent acid (15% Cl^-)	5 *N*	65	150	0.9	36	<0.0025	0.1
		80	175	0.9	36	<0.0025	0.1
20% HCl	6 *N*	60	140	0.09	3.6	<0.0025	0.1
		107	225	1.5	59	<0.0025	0.1

(a) Protected means specimens were protected electrochemically; see text for details.

Table 17 Time to failure of welded Zr702 U-bend specimens in HCl solutions containing 500 ppm Fe^{3+}

Medium	Acidity	Temperature °C	Temperature °F	Time to failure, days(a) Unprotected	Protected
10% HCl	3 *N*	60	140	0.1	NF
		102	215	0.1	NF
Spent acid (15% Cl^-)	5 *N*	65	150	0.3	NF
20% HCl	6 *N*	60	140	NF	NF
		107	225	0.1	NF
28% HCl	9 *N*	60	140	2	NF
		94	200	0.1	NF
32% HCl	10 *N*	53	125	1	32
		77	170	0.1	20
37% HCl	12 *N*	30	85	0.3	NF
		53	115	1	NF

(a) NF, no failure

Zirconium alloys, such as the Zircaloys and Zr-2.5Nb, have been developed to meet these requirements. In water-cooled reactors, zirconium alloys have found extensive use for fuel cladding and as pressure tubes. In systems in which the first requirement listed above is of overriding importance for reasons of neutron physics, the choice is virtually restricted to zirconium or one of its alloys.

Zirconium and its alloys also find application in other nuclear reactor systems, such as gas-cooled or organic coolant cooled reactors. More information on corrosion in nuclear power plants is available in the article "Corrosion in the Nuclear Power Industry" in this Volume.

Chemical-Processing Industry. The earliest application for zirconium was in the production of hydrogen peroxide (H_2O_2) using H_2SO_4. Zirconium shell and tube heat exchangers replaced graphite exchangers at one plant where they were used to condense a 75% concentrated H_2SO_4 medium. The average maintenance-free life of the heat exchanger was 12 years. The tubes were thinned at the liquid/gas interface area because of bubble collapse, but no other corrosion occurred.

Because of this experience, the company replaced the graphite heat exchangers with zirconium shell and tube exchangers used in the manufacturing of acrylic films and fibers. In this application, the H_2SO_4 concentration was as high as 60% at 150 °C (300 °F).

Another major application in H_2SO_4 concerns the manufacture of methyl methacrylate. The system at one company includes pressure vessels, columns, heat exchangers, piping systems, pumps and valves made from zirconium.

Zirconium is also widely used for column internals and reboilers in the manufacture of butyl alcohol. The operating conditions are 60 to 65% H_2SO_4 at temperatures to boiling and slightly above. Zirconium has corroded under upset conditions of elevated concentrations and when such impurities as $FeCl_3$ are present. Zirconium is also used in H_2SO_4 recovery and recycle systems in which fluorides are not present and the acid concentration does not exceed 65%.

Zirconium has many applications in HCl, such as the production of concentrated HCl and polymers. Zirconium heat exchangers, pumps, and agitators have been used for over 15 years in an azo dye coupling reaction. In addition to being very corrosion resistant in this environment, zirconium does not plate out undesirable salts that would change the color and stability of the dyes. Other applications in HCl include the breaking down of cellulose in the food industry and the polymerization of ethylene chloride, which is carried out in HCl and chlorinated solvents.

Zirconium and its alloys were identified to offer the best prospects from a cost standpoint as materials for an HI decomposer in hydrogen production (Ref 47). Tantalum, molybdenum, Nb-1Zr, and zirconium alloys showed good corrosion resistance in HI_x media (as liquid or gas) from room temperature to 300 °C (570 °F), but most nickel- and iron-base alloys and some nonmetallic materials showed corrosion resistance only at low temperatures.

There is an increasing interest in the use of zirconium for HNO_3 service. For example, because of the high degree of concern over safety, zirconium is chosen as a major structural material for the critical equipment used to reprocess spent nuclear fuels. Another application is the production of HNO_3.

Zirconium columns and reboilers are being used at one company to produce 67% HNO_3 from 57% HNO_3. Previously, such materials as AISI type 304L stainless steel, titanium, and glass-lined steel were used. Titanium and type 304L stainless steel had relatively short service lives, while glass-lined steel presented maintenance problems.

A 27-ton zirconium heat exchanger is being used by one company to produce 65% HNO_3 at 205 °C (400 °F). Before the use of zirconium, the company was faced with such problems as frequent replacement costs and downtime. In service since October, 1984, the zirconium heat exchanger has already outperformed the stainless steel predecessor.

With proper design, zirconium can be used to handle a highly concentrated HNO_3. For example, an Israeli chemical plant uses zirconium tubes in a U-tube cooler that processes bleached HNO_3 at concentrations between 98.5 and 99%. The unit cools the acid from 70 to 75 °C (160 to 170 °F) to 35 to 40 °C (95 to 100 °F). Previously, U-tube coolers were made from aluminum, which failed in 2 to 12 weeks. The zirconium has been in service for about 2 years, operating 24 h a day, six days a week.

Another very important application for zirconium is in processes that cycle between HCl or H_2SO_4 and alkaline solutions. One company replaced a lead and brick lined carbon steel reactor vessel with zirconium because the reaction alternated between hot H_2SO_4 and caustic. The vessel has been in use for several years with no corrosion problems.

Zirconium reactor vessels and heat exchangers have been used in urea production for over 30 years. The use of zirconium equipment allows the reaction to take place at higher temperatures and pressures, permitting greater conversion of CO_2 to urea.

Zirconium is commonly used in the production of acetic acid and acetic anhydride. Its corrosion rate is less than 0.05 mm/yr (<2 mils/yr) in all concentrations and temperatures.

A zirconium distillation column proved to be safe and economical in a chlorinated hydrocarbon environment at more than 150 °C (300 °F). The column was built as a replacement for an old brick-lined column. Corrosion of the old column necessitated frequent renovations, with resultant high maintenance costs and plant downtime.

In the zirconium extraction process, several corrosive chemicals are used, including methyl isobutyl ketone, HCl, ammonium thiocyanate, H_2SO_4, and zirconyl chloride. Zirconium has been found to withstand the rigors of this manufacturing process, as in the following five examples.

First, the pumps used to transport the process chemicals are primary candidates for failure. Various materials, including stainless alloys, plastics, and high-silicon cast iron had been tried, but each had some limitation in this severe service. All wetted pump parts were converted to zirconium. The oldest one has been operating for more than 5 years and shows no sign of corrosion.

Second, for heat exchanger service in the same process chemicals, zirconium tubes replaced impervious graphite tubes that were prone to failure. The oldest zirconium heat exchanger has been in service more than 7 years and shows no signs of corrosion.

Third, steam stripper columns were converted from a furan resin to zirconium. At the 105 °C (220 °F) operating temperature of the columns, the plastic would embrittle and crack. Failure of these columns often resulted in expensive spills. By contrast, zirconium columns require no maintenance due to material failure.

Table 18 Average corrosion rates of hafnium in various boiling solutions

Medium	Exposure time, days	Corrosion rate, mm/yr (mils/yr)
Seawater	10	nil
Saturated NaCl (pH 1)	21	<0.025 (<1)
70% $CaCl_2$	10	nil
40% HBr	10	0.025 (1.0)(a)
40% Cu $(NO_3)_2$	10	nil
40% $NaHSO_4$	10	1.1 (44)
10% NH_2SO_3H	10	nil
25% $Al_2(SO_4)$	10	nil
70% $ZnCl_2$	10	nil
20% HCl	8	0.005 (0.2)
60% H_2SO_4	8	0.005 (0.2)
60% H_3PO_4	8	0.22 (8.5)
30% HNO_3	8	nil
30% HNO_3 + 1% NaCl	8	nil
50% HNO_3	8	nil
50% HNO_3 + 1% NaCl	8	nil
70% HNO_3	8	nil
70% HNO_3 + 1% NaCl	8	nil
50% KOH	2	0.013 (0.5)
50% NaOH	2	0.39 (15.3)

(a) Pitting

Table 19 Corrosion of hafnium-zirconium alloys in boiling 20% HCl + 200 ppm Fe^{3+}

Fe^{3+} added as $FeCl_3$

Alloy	Average corrosion rate: First test (2 days) mm/yr	mils/yr	Second test (4 days) mm/yr	mils/yr	Pitting Ranking(a)
Hf-2.9Zr	0.19	7.4	0.058	2.3	2
Hf-17.3Zr	0.18	7.0	0.053	2.1	5
Hf-47.4Zr	0.22	8.8	0.053	2.1	4
Hf-59.5Zr	0.09	3.7	0.036	1.4	1
Hf-81.4Zr	0.13	5.1	0.058	2.3	3

(a) 1, best; 5, worst

Table 20 Composition of hafnium-zirconium alloys in boiling water saturated with chlorine gas

Alloy	Average corrosion rate: First test (2 days) mm/yr	mils/yr	Second test (4 days) mm/yr	mils/yr	Pitting ranking(a)
Hf-2.9Zr	0.015	0.6	0.005	0.2	3–4
Hf-17.3Zr	<0.0025	<0.1	0.005	0.2	3–4
Hf-47.4Zr	<0.0025	<0.1	<0.0025	<0.1	2
Hf-59.5Zr	nil	nil	<0.0025	<0.1	1
Hf-81.4Zr	<0.0025	<0.1	<0.0025	<0.1	5

(a) 1, best; 5, worst

Fourth, zirconium was selected for use in an electrostatic precipitator installed to scrub corrosive gases emitted from rotary kilns. The emitting electrodes and all fixtures exposed to the ammonium-sulfate and chloride-gas process stream were constructed from zirconium. The emitting electrodes are charged with 45 000 V.

Fifth, several parts of a crude chlorination scrubber were converted from plastic to zirconium. The plastic parts were replaced approximately every 6 months. The zirconium replacement is expected to last at least 25 years.

Zirconium is not toxic and has very low corrosion rates in many media. It is an excellent material for equipment used in food processing, the manufacture of electronic grade chemicals, and pharmaceutical preparations.

Potential applications also include heat exchangers exposed to seawater, cladding for high-strength or lightweight alloys, parts subjected to thermal and mechanical stresses in a marine environment, and nuclear waste packages.

Finally, zirconium is an excellent material for laboratory equipment, such as crucible and autoclave. Preoxidized zirconium can be used as insulating washers and measuring devices exposed to high-temperature corrosives. More information on corrosion in the chemical industry is available in the article "Corrosion in the Chemical Processing Industry" in this Volume.

Corrosion Resistance of Hafnium

Hafnium occurs naturally in zirconium ores and is separated from zirconium for use in nuclear reactors. Hafnium shares many properties with zirconium, especially its high corrosion resistance to many media. Hafnium has a high thermal neutron absorption cross section in contrast to the very low absorption cross section of zirconium. This capability has made it a primary material for nuclear reactor control rods. The unique combination of high neutron absorption and excellent corrosion resistance to mineral acids also makes hafnium an excellent material for nuclear fuel reprocessing applications.

Hafnium is superior to zirconium and Zircaloy alloys in corrosion resistance in water and steam, molten alkali metals, and air. For example, weight gains in one of the standard corrosion tests (water at 360 °C, or 680 °F, and 18.5 MPa, or 2690 psi) for these nuclear grade metals are compared as follows:

Material	Weight gain, mg/dm^2: 28 days	56 days
Zircaloy	20–22	25–28
Hafnium	3–6	5–7(a)

(a) Estimated range; test not run on routine basis.

However, at higher temperatures, hafnium begins to react appreciably with oxygen, nitrogen, and hydrogen. Hafnium begins to react slowly with air or oxygen to form hafnium oxide at approximately 400 °C (750 °F); with nitrogen to form nitrides at approximately 900 °C (1650 °F); and rapidly with hydrogen at about 700 °C (1290 °F) to form hydrides. The oxides and nitrides formed at these elevated temperatures tend to remain at the surfaces; however, hydrogen diffuses rapidly and forms hydrides throughout the metal. Oxides, nitrides, and hydrides all lead to impaired ductility.

Like other reactive metals, hafnium resists attack by many chemicals because of the thin, tenacious layer of oxide that forms naturally on the surface of the metal. The corrosion properties are unaffected as long as this thin layer is not penetrated by reactants at increasing temperatures.

In aqueous solutions, hafnium is soluble in HF and concentrated H_2SO_4. It is resistant to dilute HCl and H_2SO_4 and is unaffected by HNO_3 in all concentrations. Aqua regia dissolves hafnium and, with the addition of small amounts of soluble fluoride salts, the reaction with other acids is appreciably increased. Hafnium is also very resistant to alkalies. The corrosion resistance of hafnium in various media is given in Table 18.

Hafnium-Zirconium Alloys. The Hf-Zr system is one of the few metallic systems in which thermochemical properties are almost ideal. That is, hafnium and zirconium can form isomorphous alloys for all ratios of the components. Moreover, hafnium and zirconium exhibit similar excellent corrosion-resistant properties, although they differ greatly in neutron absorption. Therefore, hafnium-zirconium alloys offer a broad range of neutron absorption for special applications in which corrosion resistance and the neutron absorption are both important.

Hafnium alloys with 2.9, 17.3, 42.4, 59.5, and 81.4% Zr were evaluated for their corrosion resistance in various media. All of the alloys exhibited low corrosion rates (<0.0025 mm/yr, or 0.1 mil/yr) in the following boiling solutions: 30% HNO_3 with or without 1% NaCl, 50% HNO_3 with or without 1% NaCl, and 70% HNO_3 with or without 1% NaCl. Transverse-cut U-bend specimens of these alloys were tested in 90% HNO_3 at room temperature for 60 days. No cracking was observed.

The pitting resistance of these alloys was evaluated in boiling 20% HCl containing 200 ppm Fe^{3+} (as $FeCl_3$) and in chlorine gas saturated water. Test results are given in Tables 19 and 20. Although all alloys showed some pitting in these oxidizing chloride solutions, Hf-59.5Zr appeared to have the highest pitting resistance.

REFERENCES

1. S. Kass, *The Development of the Zircaloys in Corrosion of Zirconium Alloys*, STP 368, American Society for Testing and Materials, 1964
2. D.E. Thomas, Corrosion in Water and Steam, in *Metallurgy of Zirconium*, B. Lustman and F. Kerze, Jr., Ed., McGraw-Hill, 1955
3. S.B. Dalgaard, Corrosion and Hydriding Be-

havior of Some Zr-2.5 wt.% Nb Alloys in Water, Steam and Various Gases at High Temperature, in *Proceedings of the Conference on Corrosion Reactor Materials*, Vol 2, International Atomic Energy Association, 1962
4. J.E. LeSurf, The Corrosion Behavior of 2.5Nb Zirconium Alloy, in *Symposium on Applications—Related Phenomena in Zirconium and Its Alloys*, STP 458, American Society for Testing and Materials, 1969
5. H.H. Klepfer, Zirconium-Niobium Binary Alloys for Boiling Water Reactor Service: Part 1—Corrosion Resistance, *J. Nucl. Mater.*, Vol 9, 1963, p 65
6. T.L. Yau, *Fourth Asian-Pacific Corrosion Control Conference*, Tokyo, Japan, 26-31 May 1985, Vol 2, 1985, p 136
7. B. Cox, *J. Electrochem. Soc.*, Vol 117 (No. 5), 1970, p 654
8. T. Smith, *J. Electrochem. Soc.*, Vol 107 (No. 2), 1960, p 82
9. "Practice for Standard Reference Method for Making Potentiostatic and Potentiodynamic Anodic Polarization Measurements," G 5, *Annual Book of ASTM Standards*, American Society for Testing and Materials
10. "Zircadyne Corrosion Properties," Teledyne Wah Chang Albany, 1981, p 7
11. T. Gunter, *Werkst. Korros.*, Vol 30, 1979, p 308
12. B.F. Frechem, J.G. Morrison, and R.T. Webster, in *Industrial Applications of Titanium and Zirconium*, STP 728, American Society for Testing and Materials, 1981, p 85
13. T.L. Yau and R.T. Webster, *Corrosion*, Vol 39, 1983, p 218
14. O.T. Fasullo, *Sulfuric Acid*, McGraw-Hill, 1965, p 290
15. T.L. Yau and R.T. Webster, in *Laboratory Corrosion Tests and Standards*, STP 866, American Society for Testing and Materials, 1985, p 36
16. T.L. Yau, in *Industrial Applications of Titanium and Zirconium: Third Conference*, STP 830, American Society for Testing and Materials, 1984, p 203
17. J.H. Schemel, in *Manual on Zirconium and Hafnium*, STP 639, American Society for Testing and Materials, 1977
18. T.L. Yau, in *Industrial Applications of Titanium and Zirconium: Third Conference*, STP 830, American Society for Testing and Materials, 1984, p 124
19. M.G. Fontana and N.D. Greene, *Corrosion Engineering*, 2nd ed., McGraw-Hill, 1978, p 250
20. L.B. Golden, I.R. Lane, and W.L Acherman, *Ind. Eng. Chem.*, Vol 44, 1952, p 1930; Vol 45, 1953, p 782
21. W.E. Kuhn, *Corrosion*, Vol 15, 1953, p 103; Vol 16, 1960, p 136
22. M.N. Fokin, R.L. Barn, and M.M. Kurtepox, "Corrosion and Metal Protection," J.L. Rosenfeld, Ed., Indian National Scientific Documentation Letter, 1975, p 113
23. L.L. Migai *et al.*, Nauchnye Trudy Nauchno—Issledovatel'shii, Proekt. IN-T Redkomet, Prom-Sti, Vol 106, 1981, p 123, in Russian
24. R.C. Scarberry, D.L. Graver, and C.D. Stephens, *Mater. Prot.*, Vol 6 (No. 6), 1967, p 54
25. "Corrosion Resistance of Hastelloy Alloys," Cabot Corporation, High Technology Materials Division, 1980
26. T.L. Yau and M. Maguire, *Corrosion*, Vol 40, 1984, p 289; Vol 41, 1985, p 397
27. V.V. Andreeva and A.I. Glukhova, *J. Appl. Chem.*, Vol 12, 1962, p 457
28. C.R. Bishop, *Corrosion*, Vol 19, 1963, p 308t
29. T.L. Yau, in *Industrial Applications of Titanium and Zirconium: Fourth Volume*, STP 917, American Society for Testing and Materials, 1986, p 57
30. V.V. Andreeva and A.I. Glukhova, *J. Appl. Chem.*, Vol 11, 1961, p 390
31. G. Jangg, R.T. Webster, and M. Simon, *Werkst. Korros.*, Vol 29, 1978, p 15
32. M. Maraghini *et al.*, *J. Electrochem. Soc.*, Vol 101, 1954, p 400
33. T.L. Yau, *Corrosion*, Vol 39, 1983, p 167
34. J.A. Beavers, J.C. Griess, and W.K. Boyd, *Corrosion*, Vol 36, 1981, p 292
35. T.L. Yau, "Factors Affecting the Stress Corrosion Cracking Susceptibility of Zirconium in 90% Nitric Acid," Paper No. 170, presented at Corrosion/87, National Association of Corrosion Engineers, 1987
36. T.L. Yau, Performance of Zirconium and Columbium in Simulated FGD Scrubber Solutions, in *Corrosion Flue Gas Desulfurization Systems*, National Association of Corrosion Engineers, 1984
37. S.M. Shelton, "Zirconium: Its Production and Properties," Bulletin 561, Bureau of Mines, 1956
38. P.G. Gegner and W.L. Wilson, *Corrosion*, Vol 15, 1959, p 342t
39. C.M. Graighead, L.A. Smith, and R.I. Jaffee, "Screening Tests on Metals and Alloys in Contact with Sodium Hydroxide at 1000 and 1500 °F," U.S. Energy Commission Report BMI-706, Battelle Memorial Institute, Nov 1951
40. Tests Conducted at Teledyne Wah Chang Albany
41. H. Bloom, Ph.D. thesis, University of London, 1947
42. W.B. Blumenthal, "Zirconium Compounds," National Lead Company, TAM Division, Sept 1969, p 65
43. T.L. Yau, *Corrosion*, Vol 38, 1982, p 615
44. T.L. Yau, "Zirconium Versus Corrosive Species in Geothermal Fluids," Paper 140, presented at Corrosion/84, National Association of Corrosion Engineers, 1984
45. R.D. Watson, "Oxidized Zirconium as a Bearing Material in Water Lubricated Mechanisms," Report CRE-996, Atomic Energy of Canada, 1960
46. M.A. Maguire and T.L. Yau, "Corrosion-Electrochemical Properties of Zirconium in Mineral Acids," Paper 265, presented at Corrosion/86, National Association of Corrosion Engineers, 1986
47. D.H. Krikorian, in *Proceedings: Materials and Corrosion Problems in Energy Systems*, W.J. Lochman and M. Indig, Ed., National Association of Corrosion Engineers, 1980

Corrosion of Niobium and Niobium Alloys

T.L. Yau and R.T. Webster, Teledyne Wah Chang Albany

NIOBIUM AND NIOBIUM ALLOYS are used in several corrosion-resistant applications, principally rocket and jet engines, nuclear reactors, sodium vapor highway lighting, and chemical-processing equipment. Niobium has many of the same properties of tantalum, its sister metal, but has one-half the density of tantalum (see the article "Corrosion of Tantalum" in this Volume). A common property of niobium and tantalum is the interaction with the reactive elements hydrogen, oxygen, nitrogen, and carbon at temperatures above 300 °C (570 °F). These reactions will cause severe embrittlement. Consequently, at elevated temperatures, the metal must be protectively coated or used in vacuum or inert atmospheres. Niobium resists a wide variety of corrosive environments, including concentrated mineral acids, organic acids, liquid metals (particularly sodium and lithium), metal vapors, and molten salts.

Mechanism of Corrosion Resistance

Niobium, like other reactive metals, derives its corrosion resistance from a readily formed, adherent, passive oxide film. The corrosion properties of niobium are similar to those of tantalum, but niobium is less resistant in aggressive media, such as hot concentrated mineral acids. Table 1 lists some typical corrosion data for niobium in aqueous media. Like tantalum, niobium is susceptible to hydrogen embrittlement if cathodically polarized by either galvanic coupling or by impressed potential. In addition to being very stable, the anodic niobium oxide film has a high dielectric constant and a high breakdown potential. These properties, coupled with its good electrical conductivity, have led to the use of niobium as a substrate for platinum-group metals in impressed-current cathodic protection anodes.

Corrosion in Specific Media

Acid Solutions. Niobium is resistant to most organic acids and mineral acids, except hydrofluoric acid (HF), at all concentrations and temperatures below 100 °C (212 °F). This list of acids includes the halogen acids hydrochloric (HCl), hydroiodic (HI), and hydrobromic (HBr); nitric acid (HNO_3); sulfuric acid (H_2SO_4); and phosphoric acid (H_3PO_4). Niobium is especially resistant under oxidizing conditions—for example, in concentrated H_2SO_4 containing ferric (Fe^{3+}) or cupric (Cu^{2+}) ions. At room temperature, niobium is resistant to H_2SO_4 at all concentrations up to 95%. The corrosion rate increases rapidly with temperature and concentration. Figure 1 shows the corrosion rates for niobium and tantalum at elevated temperatures versus concentration in H_2SO_4.

Niobium is completely resistant in dilute sulfurous acid (H_2SO_3) at 100 °C (212 °F). In concentrated acid at the same temperature, it has a corrosion rate of 0.25 mm/yr (10 mils/yr). Niobium is completely resistant in HNO_3, having a corrosion rate of 0.025 mm/yr (1 mil/yr) in 70% HNO_3 at 250 °C (480 °F). In chromium plating solutions, niobium experiences only a slight weight change, and in the presence of small amounts of fluoride (F^-) catalyst, it exceeds the corrosion resistance of tantalum.

Niobium is inert in mixtures of HNO_3 and HCl. It has a corrosion rate of less than 0.025 mm/yr (1 mil/yr) in aqua regia at 55 °C (130 °F). In boiling 40 and 50% H_3PO_4 with small amounts of F^- impurity (5 ppm), niobium has a corrosion rate of 0.25 mm/yr (10 mils/yr). In mixtures of HNO_3 and H_2SO_4, niobium can dissolve readily.

Alkaline Solutions. In ambient aqueous alkaline solutions, niobium has corrosion rates of less than 0.025 mm/yr (1 mil/yr). At higher temperatures, even though the corrosion rate does not seem excessive, niobium is embrittled even at low concentrations (5%) of sodium hydroxide (NaOH) and potassium hydroxide (KOH). Like tantalum, niobium is embrittled in salts that hydrolyze to form alkaline solutions. These salts include sodium and potassium carbonates and phosphates.

Salt Solutions. Niobium has excellent corrosion resistance in salt solutions, except those that hydrolyze to form alkalies. It is resistant to chloride solutions even in the presence of oxidizing agents. It does not corrode in 10% ferric chloride ($FeCl_3$) at room temperature, and it is resistant to attack in seawater. Niobium exhibits resistance similar to tantalum in salt solutions.

Gases. Niobium is easily oxidized. It will oxidize in air above 200 °C (390 °F). The reaction, however, does not become rapid until above red heat (about 500 °C, or 930 °F). At 980 °C (1795 °F), the oxidation rate is 430 mm/yr (17 in./yr). In pure oxygen, the attack is catastrophic at 390 °C (735 °F). Oxygen diffuses freely through the metal, and this causes embrittlement. Niobium reacts with nitrogen above 350 °C (660 °F), with water vapor above 300 °C (570 °F), with chlorine above 200 °C (390 °F), and with carbon dioxide, carbon monoxide, and hydrogen above 250 °C (480 °F). At a temperature of 100 °C (212 °F), niobium is inert in most common gases, for example, bromine, chlorine, nitrogen, hydrogen, oxygen, carbon dioxide, argon monoxide, and sulfur dioxide (wet or dry).

Liquid Metals. Niobium resists attack in many liquid metals to relatively high temperatures. These include bismuth below 510 °C (950 °F), gallium below 400 °C (750 °F), lead below 850 °C (1560 °F), lithium below 1000 °C (1830 °F), mercury below 600 °C (1110 °F), sodium, potassium, and sodium-potassium alloys below 1000 °C (1830 °F), thorium-magnesium eutectic below 850 °C (1560 °F), uranium below 1400 °C (2550 °F), and zinc below 450 °C (840 °F). The presence of excessive amounts of nonmetallic impurities (for example, gases) may reduce the resistance of niobium to these liquid metals.

Because liquid metals are excellent heat transfer media, they can be used in very compact thermal systems, such as the fast breeder reactor, reactors for space vehicles, and fusion reactors. Niobium is a serious candidate as a material for high-efficiency reactors.

Niobium resists attack by sodium vapor at high temperatures and pressures. The Nb-1Zr alloy is in use as the end caps in high-pressure sodium vapor lamps used for highway lighting.

Galvanic Effects. Niobium is susceptible to hydrogen embrittlement if it is polarized cathodically by galvanic coupling or by chemical attack. For this reason, niobium cannot be protected by cathodic protection. However, if niobium is polarized anodically, it forms a very stable passive film that protects the metal from corrosion. This property, combined with good electrical conductivity (13% that of copper) and good mechanical properties, has led to the use of niobium as a substrate metal for platinum in impressed-current cathodic protection anodes. Its anodic breakdown potential in chloride solution is about 115 V; titanium has a breakdown potential of about 10 V.

Niobium platinized anodes are used in high-resistivity waters and other environments that require high driving potential to obtain good current spread. In this application, niobium has an advantage over tantalum because it is less expensive. The cost can be further decreased by using a composite electrode with a copper core, which increases the conductivity of the anodes.

Localized Corrosion. Niobium is highly resistant to localized attacks, such as pitting, crevice corrosion, and stress-corrosion cracking (SCC). The presence of heavy-metal ions in reducing acids will improve the corrosion resistance of niobium (Table 2). Also, as shown in

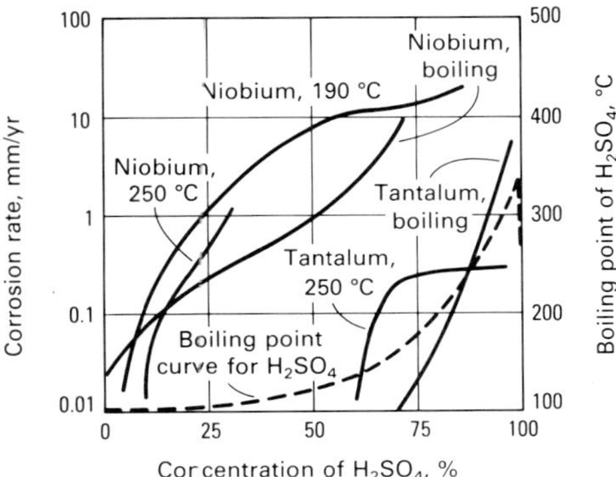

Fig. 1 Corrosion of niobium and tantalum in H_2SO_4 at various concentrations and temperatures. Source: Ref 1

Table 2, niobium does not suffer pitting or crevice corrosion in simulated scrubber solutions (Ref 2).

In one study, niobium did not exhibit SCC in 90% HNO_3 at room temperature using the slow strain rate technique or in liquid bromine using the U-bend method (Ref 3). However, with particular types of surface contamination and preparation, niobium, like tantalum, was found to be sensitive to crevice corrosion at anodic potentials below those normally regarded as safe (Ref 4).

Corrosion in Waters. Niobium reacts with water to form niobium oxide. There is a direct transition from immunity to passivity without an intermediate region where corrosion occurs. Figure 2 shows the Pourbaix (potential-pH) diagram of niobium in water at 25 °C (75 °F).

Applications

Chemical and Metal Processing. The high anodic breakdown potential of niobium (about 100 V in seawater), coupled with good electrical conductivity and mechanical properties, has led to its use as a substrate for platinum in impressed-current cathodic protection anodes. Niobium is an important metal in the chromium plating application. It is particularly attractive when fluoride ions are also present. It has the highest resistance to attack by fluoride ions among reactive and refractory metals.

Niobium retains excellent corrosion resistance in liquid bromine even when saturated with moisture. It can be used to make equipment designed to produce or transport liquid bromine.

Niobium would be well suited for handling hot, concentrated HNO_3. It can be considered for highly oxidizing solutions expected in chemical wastes, scrubber environments, and mining solutions.

The corrosion resistance and mechanical properties of niobium can now be more easily utilized because of advancements in cladding techniques. Thin layers (for example, 0.25 mm, or 0.01 in.) of niobium sheets can be cladded to common metals, such as steel, by using various cladding techniques. More information on corrosion in chemical- and metal-processing applications is available in the articles "Corrosion in the Chemical Processing Industry" and "Corrosion of Metal Processing Equipment" in this Volume.

Nuclear Applications. Niobium is used in certain fast breeder reactors because of its compatibility with uranium and liquid sodium/potassium.

Table 1 Corrosion of niobium in aqueous media

Medium	Concentration, %	Temperature, °C (°F)	Corrosion rate mm/yr	Corrosion rate mils/yr
Mineral acids				
Hydrochloric acid	1	Boiling	nil	
Hydrochloric acid (aerated)	15	Room-60 (140)	nil	
Hydrochloric acid (aerated)	15	100 (212)	0.025	1.0
Hydrochloric acid (aerated)	30	35 (95)	0.025	1.0
Hydrochloric acid (aerated)	30	60 (140)	0.05	2.0
Hydrochloric acid (aerated)	30	100 (212)	0.125	5.0
Hydrochloric acid	37	Room	0.025	1.0
Hydrochloric acid	37	60 (140)	0.25	10
Hydrochloric acid	37% with Cl_2	60 (140)	0.5	20
Hydrochloric acid	10% with 0.1% $FeCl_3$	Boiling	0.025	1.0
Hydrochloric acid	10% with 0.6% $FeCl_3$	Boiling	0.125	5.0
Hydrochloric acid	10% with 35% $FeCl_2$ and 2% $FeCl_3$	Boiling	0.05	2.0
Nitric acid	65	Room	nil	
Nitric acid	70	250 (480)	0.025	1.0
Phosphoric acid	60	Boiling	0.5	20
Phosphoric acid	85	Room	0.0025	0.1
Phosphoric acid	85	88 (190)	0.05	2.0
Phosphoric acid	85	100 (212)	0.125	5.0
Phosphoric acid	85	Boiling	3.75	150
Phosphoric acid	85% with 4% HNO_3	88 (190)	0.025	1.0
Phosphoric acid	40–50% with 5 ppm F^-	Boiling	0.25	10
Sulfuric acid	5–40	Room	nil	
Sulfuric acid	98	Room	Embrittlement	
Sulfuric acid	10	Boiling	0.125	5.0
Sulfuric acid	25	Boiling	0.25	10
Sulfuric acid	40	Boiling	0.5	20
Sulfuric acid	40% with 2% $FeCl_3$	Boiling	0.25	10
Sulfuric acid	60	Boiling	1.25	50
Sulfuric acid	60% with 0.1–1% $FeCl_3$	Boiling	0.5	20
Sulfuric acid	20% with 7% HCl and 100 ppm F^-	Boiling	0.25	10
Sulfuric acid	50% with 20% HNO_3	50–80 (120–175)	nil	
Sulfuric acid	50% with 20% HNO_3	Boiling	0.25	10
Sulfuric acid	72% + 3% CrO_3	100 (212)	0.025	1.0
Sulfuric acid	72% + 3% CrO_3	125 (255)	0.125	5.0
Sulfuric acid	72% + 3% CrO_3	Boiling	3.75	150
Organic acids				
Acetic acid	5–99.7	Boiling	nil	
Citric acid	10	Boiling	0.025	1.0
Formaldehyde	37	Boiling	0.0025	0.1
Formic acid	10	Boiling	nil	
Lactic acid	10–85	Boiling	0.025	1.0
Oxalic acid	10	Boiling	1.25	50
Tartaric acid	20	Room-boiling	nil	
Trichloroacetic acid	50	Boiling	nil	
Trichloroethylene	99	Boiling	nil	
Alkalies				
NaOH	1–40	Room	0.125	5.0
NaOH	1–10	98 (208)	Embrittlement	
KOH	5–40	Room	Embrittlement	
KOH	1–5	98 (208)	Embrittlement	
NH_4OH	...	Room	nil	
Salts				
$AlCl_3$	25	Boiling	0.005	0.2
$Al_2(SO_4)_3$	25	Boiling	nil	
$AlK(SO_4)_2$	10	Boiling	nil	
$CaCl_2$	70	Boiling	nil	
$Cu(NO_3)_2$	40	Boiling	nil	
$FeCl_3$	10	Room-boiling	nil	
$HgCl_2$	Saturated	Boiling	0.0025	0.1
K_2CO_3	1–10	Room	0.025	1.0
K_2CO_3	10–20	98 (208)	Embrittlement	
K_3PO_4	10	Room	0.025	1.0
$MgCl_2$	47	Boiling	0.025	1.0
NaCl	Saturated; pH = 1	Boiling	0.025	1.0
Na_2CO_3	10	Room	0.025	1.0
Na_2CO_3	10	Boiling	0.5	20
Na_2HSO_4	40	Boiling	0.125	5.0
NaOCl	6	50 (120)	1.25	50
Na_3PO_4	5–10	Room	0.025	1.0
Na_3PO_4	2.5	98 (208)	Embrittlement	
NH_2SO_3H	10	Boiling	0.025	1.0
$NiCl_3$	30	Boiling	nil	
$ZnCl_2$	40–70	Boiling	nil	

(continued)

Table 1 (continued)

Medium	Concentration, %	Temperature, °C (°F)	Corrosion rate mm/yr	Corrosion rate mils/yr
Miscellaneous				
Bromine	Liquid	20 (70)	nil	
Bromine	Vapor	20 (70)	0.025	1.0
Chromium plating solution	25% CrO_3, 12% H_2SO_4, H_2O	92 (198)	0.125	5.0
Chromium plating solution	17% CrO_3, 2% Na_5SiF_6, trace H_2SO_4, H_2O	92 (198)	0.125	5.0
H_2O_2	30	Room	0.025	1.0
H_2O_2	30	Boiling	0.5	20

Table 2 Corrosion of niobium in simulated scrubber solutions

Solution	Temperature °C	Temperature °F	Test duration, days	Average corrosion rate mm/yr	Average corrosion rate mils/yr	Pitting or crevice corrosion
7 vol% H_2SO_4 + 3 vol% HCl + 1% $FeCl_3$ + 1% $CuCl_2$	103	217	8	0.018	0.7	None
30 vol% H_2SO_4 + 3 vol% HCl + 1% $FeCl_3$ + 1% $CuCl_2$	110	230	8	0.12	4.8	None

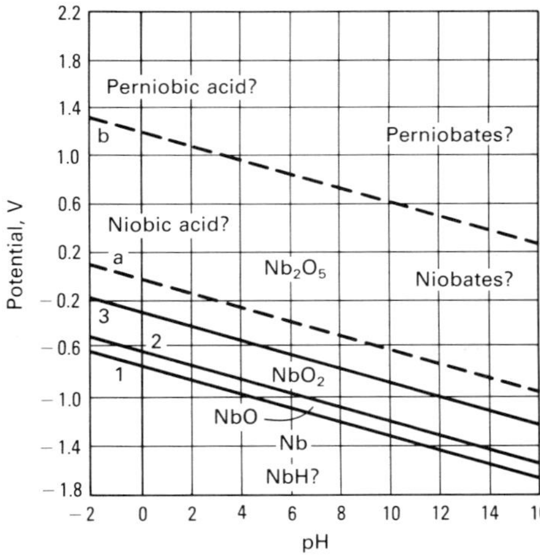

Fig. 2 Pourbaix (potential-pH) diagram for niobium in water at 25 °C (75 °F). See Ref 6 for details. Source: Ref 5 and 6

Miscellaneous Applications. Niobium is finding use in components for high-pressure sodium lamps, in satellite launch vehicles and spacecraft, and in the production of superconducting devices. Niobium alloys are used for rocket nozzles on satellite launch vehicles and spacecraft. Because niobium is easily oxidized at elevated temperatures, it must be coated for this use. Aluminide and silicide coatings are used for this purpose. A major application is in superconducting devices. Niobium alloys also are used in aircraft fasteners.

REFERENCES

1. C.R. Bishop, *Corrosion*, Vol 14, 1963, p 308
2. T.L. Yau, Performance of Zirconium and Columbium in Simulated FGD Scrubber Solutions, in *Corrosion in Desulfurization Systems*, National Association of Corrosion Engineers, 1984
3. Teledyne Wah Chang Albany, unpublished research
4. P.C.S. Hayfield, "Electrochemical Properties of Niobium in Impressed Current Cathodic Protection," Paper 103, presented at Corrosion/81, National Association of Corrosion Engineers, 1981
5. J. van Muylder and M. Pourbaix, Technical Report 53, Centre Belge d'Etude de la Corrosion, 1959
6. M. Pourbaix, *Atlas of Electrochemical Equilibria in Aqueous Solutions*, Pergamon Press, 1966

Corrosion of Tantalum*

Mortimer Schussler (retired) and Charles Pokross, Fansteel Inc.

TANTALUM is one of the most versatile corrosion-resistant metals known. It combines the inertness of glass with the strength and ductility of low-carbon steel and has a much higher heat transfer capability than glass. The relatively high cost of tantalum has been a limiting factor in its use, but where corrosion resistance is important, the economics are changing. Fabrication techniques, in which thin linings of tantalum are used in chemical-processing equipment, result in equipment that has the acid corrosion resistance provided by tantalum but at a much lower cost than an all-tantalum construction.

The long life and reliability of tantalum equipment in severe-corrosion applications often more than offsets its higher initial costs. Therefore, a new situation has been created for utilizing the benefits of tantalum products. When tantalum is properly applied in chemical applications—one of its primary uses—it can often be justified not only on a field replacement basis but also on initial installation. Table 1 lists numerous applications for tantalum in the chemical processing industry and in other industries.

Mechanism of Corrosion Resistance

The outstanding corrosion resistance and inertness of tantalum are attributed to a very thin, impervious, protective oxide film that forms upon exposure of the metal to slightly anodic or oxidizing conditions (Ref 1). Although tantalum pentoxide (Ta_2O_5) is the usual oxide form, suboxides may also exist in transition between the base metal and the outer film (Ref 2). It is only when these oxide films react with or are penetrated by a chemical reagent that attack occurs on the underlying metal substrate. A visible, continuous oxide film can be formed on tantalum by electrolytic anodizing in an acid solution such as dilute phosphoric acid (H_3PO_4). This film has a high dielectric constant, which prevents the flow of direct current from tantalum to an electrolyte when the metal is made anodic. The high stability of the oxide film makes tantalum valuable for capacitor and rectifier applications.

Tantalum occupies a position toward the electropositive end of the electromotive force (emf) series and thus tends to become cathodic in the galvanic cell circuit formed by contact with almost all other metals. Because of this cathodic behavior, atomic hydrogen, which may be liberated, can be absorbed by the tantalum and result in hydrogen embrittlement. Stray voltages can also cause this undesired effect. Therefore, when used in chemical-processing equipment, tantalum must be protected from becoming cathodic so that the material will not become embrittled. Hydrogen embrittlement and methods for minimizing its effect are discussed in the section "Hydrogen Embrittlement, Galvanic Effects, and Cathodic Protection of Tantalum" in this article.

Corrosion of Tantalum in Specific Media

Because of the wide ranges of type, concentration, and temperature of the media to which it exhibits excellent corrosion resistance, tantalum has found a primary application as a material of construction in the chemical-processing industry. This section will discuss the corrosion resistance of tantalum in many of the important media in which it is used (or could be used) commercially. Additional information is available in the article "Corrosion in the Chemical Processing Industry" in this Volume.

Water. Tantalum is not attacked by fresh water, mine waters (which are usually acidic), or seawater, either cold or hot. Tantalum shows no corrosion in deionized water at 40 °C (100 °F) (Ref 3).

For tantalum equipment exposed to boiler waters and condensates, the alkalinity must be controlled. The pH should be less than 9 and preferably no more than 8. No failures caused by exposure of tantalum to steam condensate have ever been recorded. Tantalum is used in many cases at saturated steam pressures above 1035 kPa (150 psi) at temperatures of 185 °C (365 °F) and is considered resistant to saturated steam below 250 °C (480 °F) at a pressure of 3.9 MPa (560 psi).

Acids. The chemical properties of tantalum are similar to those of glass. Like glass, tantalum is immune to attack by almost all acids except HF (see Table 2 for compatibility data on tantalum in numerous acids). Reactions conducted in the laboratory in glass equipment can be transferred to plant operations in tantalum equipment with complete assurance that the equipment will be free from corrosion, that the product will not become contaminated, and that undesired side reactions will not occur.

Tantalum is not attacked by such agents as sulfuric acid (H_2SO_4), nitric acid (HNO_3), hydrochloric acid (HCl), aqua regia, perchloric acid ($HClO_4$), chlorine, bromine, hydrobromic acid (HBr) or any of the bromides, phosphoric acid (H_3PO_4) when free of the F^- ion, nitric oxides, chlorine oxides, hypochlorous acid (HClO), organic acids, and hydrogen peroxide (H_2O_2) at ordinary temperatures. It is attacked, even at room temperature, by strong alkalies, HF, and free sulfur trioxide (SO_3) (as in fuming H_2SO_4). Table 3 compares the corrosion resistance of tantalum and other resistant metals in various acids.

Sulfuric Acid. Tantalum is highly resistant to corrosion by H_2SO_4 in all concentrations up to about 98%. It is inert to dilute acid even at boiling temperatures and is not attacked by concentrated acids at temperatures below 150 °C (300 °F). A slow, uniform attack by concentrated H_2SO_4 begins on tantalum at about 175 °C (345 °F). However, the corrosion by hot, concentrated H_2SO_4 is uniform, and at a temperature as high as 200 °C (390 °F), tantalum can be successfully used with 98% H_2SO_4 (Fig. 1).

Fuming H_2SO_4 (oleum) attacks the metal much more rapidly than the concentrated acid does (Fig. 1), but the attack on tantalum by either concentrated H_2SO_4 or oleum is uniform over the surface. The presence of impurities does not increase the corrosion rate of tantalum in H_2SO_4. No failures due to hydrogen embrittlement have been reported for tantalum chemical-processing equipment used in H_2SO_4 service. However, under special laboratory test conditions, hydrogen embrittlement can be produced in tantalum (Ref 5). This behavior is discussed in the section "Hydrogen Embrittlement, Galvanic Effects, and Cathodic Protection of Tantalum" in this article.

Phosphoric Acid. Figure 2 shows the corrosion rate of tantalum exposed to various concentrations of reagent grade (chemically pure) H_3PO_4 at the boiling temperature and at 190 °C (375 °F). Figure 2 also shows the boiling point of H_3PO_4 as a function of concentration. Tantalum exhibits superior resistance to boiling H_3PO_4 at all concentrations. At temperature in excess of boiling, the superiority of tantalum is evident. However, if the H_3PO_4 contains more than a few parts per million of F^- ion, as is frequently the case with commercial acid, attack on the tantalum may occur.

In one study, the corrosion resistance of commercially pure tantalum to a mixture of H_3PO_4, potassium chloride (KCl), and water initially containing 60 to 260 ppm F^- was evaluated at 120 °C (250 °F) and at atmospheric pressure (Ref 7). Corrosion rates calculated from the test data were of the order of 0.0005 to 0.15 mm/yr (0.02 to 6 mils/yr), indicating good corrosion resistance.

Hydrochloric Acid. Specific corrosion tests and many industrial applications show that tantalum is completely inert to HCl in all concentrations under atmospheric pressure to at least 90 °C (195 °F). This has been demonstrated by long industrial experience. For example, bayonet heaters

*Adapted with permission from M. Schussler and C. Pokross, *Corrosion Data Survey on Tantalum*, 2nd ed., Fansteel Inc., 1985

Table 1 Applications for tantalum equipment

Product	Operation	Industry	Equipment	Remarks
Acetic acid, crude	Recovery from wood distillate	Chemical	Bayonet heaters, condensers	...
Aluminum chloride	Concentration	Chemical	Bayonet heaters, condensers	Condensers for HCl recovery
Amino acids	Digesting proteins in HCl	Chemical, breweries, distilleries, food	Bayonet heaters, condensers, complete plants	See glutamic acid, monosodium glutamate
Ammonium chloride, crude	Concentration from gas house liquors	Gas and coke	Bayonet heaters, heat exchangers	...
Ammonium chloride, pure	Concentration before crystallizing	Chemical	Heat exchangers	High pressure
Ammonium nitrate	Concentration	Chemical	Heat exchangers	...
Aqua regia	Ore solution, stainless steel pickling	Chemical, steel	Bayonet heaters, heat exchangers, pickling tank coils	...
Aviation gasoline	Butane isomerization	Petroleum	Anhydrous HCl plants	...
Benzoic acid	Hydrolysis from benzyl chloride	Chemical, pharmaceutical	HCl absorbers	Recovery of by-product HCl
Benzyl chloride	Chlorination of toluene	Chemical	HCl absorbers	Recovery of by-product HCl
Bromine, crude	Steam and bromine condensation	Chemical	Condensers	...
Bromine, pure	Purification from chlorine and organics	Chemical	Boilers, Condensers, complete purification plants	...
Bromides, organic	Bromination	Pharmaceutical	Condensers	Both reflux and product condensers
Chloral	HCl absorption	Chemical	HCl absorbers	Recovery of by-product HCl
Chlorine	Brine cooling	Chemical	Heat exchangers	...
Chloroacetic acid	HCl absorption	Chemical	HCl absorbers	Recovery of by-product HCl
Chlorobenzene, also monochlorobenzene and paradichlorobenzene	Chlorinator operation HCl absorption	Chemical	Condensers, HCl absorbers	Recovery of by-product HCl
Chromic acid	Heating solutions	Electroplating	Coils, heat exchangers	...
DDT	HCl absorption	Chemical	HCl absorbers	...
Diphenyl chloride	HCl absorption	Chemical	HCl absorbers	Recovery of by-product HCl
Detergents (chlorinated)	HCl absorption	Chemical	HCl absorbers	Recovery of by-product HCl
Ethyl ether	Heating alcohol reactor	Pharmaceutical, chemical	Bayonet heaters, single and multiple	...
Ethyl bromide	Alcohol bromination	Chemical	Special HBr reactor; anhydrous HBr plant	...
Ethyl chloride	Alcohol chlorination	Chemical	Special HCl reactor; anhydrous HCl plant	...
Ethyl gasoline	Tetraethyl lead production	Chemical, petroleum	Heaters, coolers, condensers, anhydrous HCl plant	...
Ethylene dibromide	Ethyl gas stabilizer, ethylene bromination	Chemical	Condensers, bromide plants	...

(continued)

fabricated from tantalum with a wall thickness as thin as 0.33 mm (0.013 in.) have been in continuous industrial use in HCl distilling units for over 20 years without being attacked.

Figure 3 shows the corrosion rate of tantalum in HCl over the concentration range from 1 to 35%. Figure 4 shows a plot of the corrosion rate of tantalum in 20, 32, and 37% HCl as a function of temperature. Additions of HNO_3 and ferric or cupric chlorides ($FeCl_3$ or $CuCl_2$) to HCl tend to improve corrosion resistance.

Tests indicate that tantalum resists HCl at all temperatures and concentrations to 190 °C (375 °F). At concentrations near 30% and at 190 °C (375 °F), some tendency toward hydrogen embrittlement has been noted. This possibility should be considered when handling concentrated solutions of acid above the boiling point. The tendency was not noted at or below the boiling point of HCl solutions.

Nitric Acid. Tantalum is inert to HNO_3 solutions in all concentrations and at all temperatures to boiling. The presence of Cl^- in HNO_3 does not reduce the corrosion resistance of the metal to this acid. Figure 5 shows the corrosion rate of tantalum in HNO_3 concentrations ranging from 1 to 70%.

The corrosion rate of tantalum to HNO_3 at sub-boiling temperature is less than 0.4 μm/yr (0.015 mils/yr) for most concentrations and temperatures. In general, the use of tantalum at these temperatures would not be economical, considering the resistance offered by stainless steels. At temperatures near and above the normal boiling point of HNO_3, the superior resistance of tantalum becomes pronounced. Corrosion testing of tantalum for equipment to be used at these temperatures is recommended. Tantalum has been successfully used for years to handle fuming HNO_3 at service conditions up to 5.5 MPa (800 psig) and 315 °C (600 °F) in chemical-processing equipment.

Hydrofluoric acid is the best solvent for tantalum. The rate of attack varies from slow for dilute acid to rapid for concentrated solutions. The rate of attack by HF can be greatly accelerated by the addition of HNO_3 and other oxidizing agents, such as H_2O_2. Embrittlement of the metal due to the absorption of atomic (nascent) hydrogen can occur when the metal is attacked by HF. However, when sufficient HNO_3 is present, hydrogen embrittlement does not occur, even after nearly all of the tantalum is dissolved—for example, by severe pickling. The rate of hydrogen absorption is greatly reduced in dilute HF if the tantalum is made positive by impressing 2 to 10 V on the material in an electrolytic cell. Embrittlement by hydrogen does not occur when tantalum is made positive. In solutions of HF, which prevent the formation of the protective oxide film, tantalum is less noble than zinc, manganese, aluminum, and zirconium (Ref 8).

Attack on tantalum apparently does not occur in chromium plating baths containing F^- (Ref 9). In one test, a corrosion rate of 0.0005 mm/yr (0.02 mils/yr) was observed on a sample placed in a chromium plating bath for 2½ months. The solution contained 40% chromium trioxide (CrO_3) and 0.5% F^- ion at a temperature of 55 to 60 °C (130 to 140 °F). In another 4-day exposure, a corrosion rate of 0.005 mm/yr (0.2 mils/yr) occurred in a bath containing 250 g/L (33 oz/gal) CrO_3, 1.5 g/L (0.20 oz/gal) H_2SO_4, and 3 g/L (0.40 oz/gal) of F^- ion at 55 °C (130 °F). Complex ion formation between chromium and F^- ions is thought to explain the reduced activity of F^-.

Corrosion rates were determined on several construction materials, including tantalum, in HNO_3 plus HF mixtures (Ref 10). These studies were related to tests of various candidate materials for fuel element processing. Tantalum showed some promise but was not entirely satis-

Table 1 (continued)

Product	Operation	Industry	Equipment	Remarks
Ethylene dichloride	Chlorination	Chemical	Condensers	. . .
Ethylene glycol	Reactor, sulfuric acid concentrator	Chemical	Bayonet heaters, multiple	. . .
Ferric chloride	Dissolving concentration	Chemical	Bayonet heaters, single and multiple	. . .
	Sewage treatment	Sanitary districts	Bayonet heaters, flow regulator parts, thermometer wells	Heaters to warm storage tanks
Formic acid	Distillation	Chemical	Condensers	. . .
Fuming HNO_3	Distillation	Chemical	Multiple bayonet heaters, condensers	. . .
Glutamic acid	Digesting gluten in HCl, vacuum evaporation of HCl, recovery of HCl	Chemical	Bayonet heaters, single and multiple; condensers; synthetic HCl plants	. . .
Halogens (except fluorine)	Chlorine, bromine, iodine generators and recovery systems	Chemical, pharmaceutical	Bayonet heaters, coils, condensers, regulator parts, thermometer wells	. . .
Hydrobromic acid	Generation	Chemical	Bayonet heaters, coils, condensers, synthetic HBr plants	. . .
Hydrochloric acid	Production, purification, recovery, processing	Chemical, pharmaceutical, food	Bayonet heaters, heat exchangers, coils, condensers, HCl absorbers, synthetic HCl plants, acid coolers, gas coolers, chlorine burners, strippers, thermometer wells	. . .
Hydrochloric acid C.P.	Distillation	Chemical	Bayonet heaters, coils, condensers, complete stills	. . .
Hydrochloric acid, anhydrous	Production	Chemical	Absorbers, coolers, strippers, chlorine burners, complete plants	. . .
Hydriodic acid	Generation and recovery	Chemical, pharmaceutical	Bayonet heaters, coils, condensers	. . .
Hydrogen chloride	Production	Chemical	Absorbers, coolers, strippers, chlorine burners, complete plants	. . .
Hydrogen peroxide	Hydrolysis, concentration	Chemical	Bayonet heaters, heat exchangers	. . .
Iodine	Recovery from sour brines	Chemical	Coils	. . .
Isobutane	Isomerization for aviation gasoline	Petroleum	Anhydrous HCl plants	. . .
Isopropyl alcohol	Concentration of H_2SO_4	Chemical, petroleum	Bayonet heaters, multiple	. . .
Lactic acid	Distillation purification	Chemical, pharmaceutical	Bayonet heaters, condensers	. . .
Magnesium bromide	Concentration	Chemical	Bayonet heaters, coils	. . .
Magnesium chloride	Concentration	Chemical	Heat exchangers	. . .
Methyl chloride	Methanol chlorination	Chemical	Coils, condensers, complete units	. . .
Methyl ethyl ketone	Recovery of by-product HCl	Chemical	HCl absorbers	. . .
Mono-sodium glutamate	Glutamic hydrochloride production	Chemical, food	Bayonet heaters, condensers, complete plants	. . .

(continued)

factory. Its possible use was considered restricted to low fluoride concentrations and temperatures below boiling.

Acid Mixtures. Tests were conducted to determine the suitability of tantalum to either hot or cool mixed solutions of HCl, H_2SO_4, and potassium hydrogen sulfate ($KHSO_4$). The initial temperature was the boiling point of the solution, 115 °C (235 °F). Because of a loss of fluid over a weekend, the boiling point elevated so that the temperature at the completion of the test was 120 °C (250 °F). The corrosion rate of the tantalum based on 4 days of exposure was 0.001 mm/yr (0.04 mils/yr). Examination of the sample after the test indicated no apparent discoloration. There were no visible signs of attack, and the sample remained ductile after the test. The very slight weight loss reported was believed to have been caused by physical scraping on the sample. Thus, tantalum exhibited excellent corrosion resistance to this mixture.

Corrosion rates were also determined on tantalum exposed to a mixture of concentrated H_2SO_4, HNO_3, and HCl at temperatures ranging from 200 to 270 °C (390 to 520 °F) (Ref 11). Corrosion rates were reduced by a factor of three compared to the rate in H_2SO_4 alone. When only HCl or water was added, the corrosion rate was reduced only slightly. When HNO_3 was added alone, the corrosion rate was the same as when HCl and HNO_3 were both added. Thus, the oxidizing effect of HNO_3 reduced the tantalum corrosion rate.

Other Acids and Reagents. Over the temperature range commonly used in solution processes, tantalum is inert to sulfur and phosphorous chlorides, diphenyl and diphenyl oxide, and hydrogen sulfide (H_2S).

Hypochlorites do not affect tantalum unless they are strongly alkaline. The same is true of all chlorides, bromides, and iodides. Fluorides, however, will attack tantalum.

A detailed corrosion study was conducted in a chlorine dioxide (ClO_2) plant for pulp bleaching in an effort to secure useful information on the performance of various metals and alloys in this service (Ref 12). Tantalum was among the materials tested, and it showed no corrosion attack in any of the tests.

Salts. Tantalum is not attacked by dry salts or by salt solutions at any concentration or temperature unless HF is liberated when the salt dissolves or unless a strong alkali is present. Salts that form acidic solutions have no effect on tantalum. However, fused sodium hydrosulfate ($NaHSO_4$) or $KHSO_4$ dissolves tantalum (Ref 9). Table 4 gives compatibility data for tantalum in numerous salts.

Alkalies. Sodium hydroxide (NaOH) and potassium hydroxide (KOH) solutions do not dissolve tantalum, but tend to destroy the metal by formation of successive layers of surface scale. The rate of the destruction increases with concentration and temperature. Damage to tantalum equipment has been experienced unexpectedly when strong alkaline solutions are used during cleaning and maintenance.

Tantalum is attacked, even at room temperature, by concentrated alkaline solutions and is dissolved by molten alkalies. However, tantalum is fairly resistant to dilute alkaline solutions. In one long-term exposure test in a paper mill, tantalum suffered no attack in a solution with a pH of 10.

A study using change in electrical resistivity to measure corrosion rates found that tantalum wire totally immersed in 10% NaOH solution at room temperature for 210 days corroded at the rate of 0.24 μm/yr. A similar rate occurred in 10% NaOH at 100 °C (212 °F). In the latter case, there was some local effect at the points where the wire left the solution and entered submerged rubber stoppers in the sides of the corrosion vessel; this accounted for most of the weight loss.

Tantalum has been used as anode baskets in a number of silver cyanide barrel platers for several years of service life, and although the solutions are quite alkaline with free KOH, the tantalum has remained bright and ductile, with no failures

Table 1 (continued)

Product	Operation	Industry	Equipment	Remarks
Muriatic acid pickling	Heating, pickling tanks	Steel, metal working	Coils	...
Nitric acid	Distillation, recovery	Chemical	Bayonet heaters, single and multiple; condensers	...
Nitroglycerin	Nitration, HNO_3 recovery	Chemical, explosive	Condensers, thermometer wells	...
Nitrosyl bromide	Recovery, handling	Chemical	Condensers, pipe and fittings	...
Nitrosyl chloride	Generation, recovery	Chemical	Pipe and fittings	...
Oakite (tri-sodium phosphate)	Cleaning and degreasing metals	Metal working	Bayonet heaters, coils	...
Pentachlor phenol	Chlorination	Chemical	HCl absorbers	Recovery of by-product
Perchloric acid	Generation, concentration	Chemical	Coils, condensers	...
Persulfuric acid	Electrolysis, recovery	Chemical	Bayonet heaters, electrode supports	...
Phenol (carbolic acid)	Chlorination, hydrolysis	Chemical, plastic	Bayonet heaters, HCl absorbers	Tantalum equipment used in Raschig process
Phosphoric acid	Concentration	Chemical, food	Bayonet heaters	Tantalum can be used only when fluorine content is below 10 ppm
Phosgene	Generation	Chemical	Bayonet heaters, condensers	...
Pickling liquors	Heating, pickling tanks	Metal working, steel	Coils	Tantalum used with HCl, HNO_3, or mixtures of them
Plating	(See chromic acid)	...	...	...
Polystyrene	Recovery of by-product HCl	Chemical, plastic	HCl absorbers	...
Rayon (viscose process)	...	...	Bayonet heaters, spinneret cups, thermometer wells	...
Silver nitrate	Recovery	Photographic	Heat exchangers	...
Smokeless powder	Nitration, HNO_3 and H_2SO_4 recovery	Chemical equipment	Bayonet heaters, condensers, thermometer wells	...
Sulfuric acid	Concentration	Chemical, petroleum	Bayonet heaters	...
Styrene	Recovery of by-product HCl	Chemical, plastic	HCl absorbers	...
Sulfuric acid	Concentration, recovery	Chemical, petroleum, pharmaceutical	Bayonet heaters, single and multiple	...
Tartaric acid	Reactor concentrator	Chemical, pharmaceutical, food	Bayonet heaters, multiple	...
Tetraethyl lead	Production	Chemical, petroleum	Heaters, coolers, condensers, anhydrous HCl plant	...
T.N.T. (Tri-nitrotoluene)	Nitration, HNO_3 and H_2SO_4 recovery	Chemical, explosive	Bayonet heaters, condensers, thermometer wells	...
Tri-sodium phosphate	Cleaning and degreasing metals	Metal working	Bayonet heaters, coils	...
Vinyl chloride	Chlorination	Chemical, plastic	Anhydrous HCl plants	...

occurring. The tantalum anode basket is protected by the positive voltage of the cell itself.

Organic Compounds. In general, tantalum is completely resistant to organic compounds and is used in heat exchangers, spargers, and reaction vessels in several important organic reactions, particularly when corrosive inorganics are involved. These include solutions of phenol and of acetic, lactic, and oxalic acids. Table 5 lists the compatibility of tantalum in numerous corrosive environments, including many organic compounds.

Most organic salts, gases, alcohols, ketones, alkaloids, and esters have no effect on tantalum. Specific exceptions, however, should be made for reagents that may hydrolyze to HF or contain free SO_3 or strong alkalies. One important exception was noted in Ref 13, which states that mixtures of anhydrous methanol with chlorine, bromine, or iodine cause pit-type corrosion on tantalum at 65 °C (150 °F). This is of particular interest because tantalum is not attacked individually by methanol, the halogens involved, or the product, methyl halide, even at somewhat higher temperatures. Also, pit-type corrosion on tantalum is rare.

Fine Chemicals, Foods, and Pharmaceuticals. The immunity of tantalum to corrosion also ensures freedom from product contamination and undesired side reactions in the processing of fine chemicals, foods, and pharmaceuticals.

Body Fluids and Tissues. Tantalum is completely inert to body fluids and tissues. Bone and tissue do not recede from tantalum, which makes it attractive as an implant material for the human body; however, the superior strength and rigidity of stainless steel and the castability of the high-cobalt alloys have led to their much greater use as prosthetic materials. Tantalum has been used for bone replacement and repair, suture wire, cranial repair plates, and wire gauze for abdominal-muscle support in hernia-repair surgery (Ref 14). More information on this subject is available in the article "Corrosion of Metallic Implants and Prosthetic Devices" in this Volume.

Carbon, Boron, and Silicon. Tantalum reacts at elevated temperatures (not stated) directly with carbon, boron, and silicon to form Ta_2C and TaC, TaB and TaB_2, and $TaSi_2$, respectively, although other binary compounds of these elements have been reported (Ref 9). These compounds are characterized by metallic appearance and properties, high melting point, and high hardness.

Phosphorus. Tantalum phosphides, TaP and TaP_2, are formed by heating tantalum filings in phosphorus vapor at 750 to 950 °C (1380 to 1740 °F) (Ref 9).

Sulfur. Tantalum reacts with sulfur or H_2S at red heat to form tantalum sulfide (Ta_2S_4). Tantalum sulfide is also formed when Ta_2O_5 is heated in H_2S or carbon disulfide (CS_2). Little or no data are available on the effect of solutions of sulfides, such as those of the alkali and alkaline earth metals, but the highly alkaline nature of these compounds indicates that they probably corrode tantalum to some degree.

Selenium and Tellurium. Tantalum is attacked by selenium and tellurium vapors at temperatures of 800 °C (1470 °F) and higher. In contrast, there is little or only slight attack on the metal by liquid selenides and tellurides of yttrium, the rare earths, and uranium at temperatures of 1300 to 2100 °C (2370 to 3810 °F), and tantalum is considered to be a satisfactory material in which to handle these intermetallic compounds.

Gases. The solubility of oxygen in tantalum, as determined by an x-ray technique, is shown by curve B in Fig. 6. There is some disagreement among investigators, but oxygen solubility may be accepted as being close to values based on electrical resistivity and hardness measurements (Ref 15).

Table 2 Effects of acids on tantalum

Acid	Concentration, %	Temperature, °C (°F)	Code(a)
Acetic acid	5–99.5	Room to boiling	E
Acetic acid, glacial	99.7	Room to boiling	E
Acetic acid vapor	0–100	Room to boiling	E
Acetic anhydride	99	Room	E
Aqua regia	3 HCl, 1 HNO_3	Room to 77 (170)	E
Arsenic acid	90	Room	E
Benzoic acid	5	Room	E
	Saturated	Room	E
Boric acid	5	Room to boiling	E
	10	Room to boiling	E
	Saturated	Room	E
Butyric acid	5	Room	E
Carbolic acid	Saturated	Room	E
	...	...	E
Chloroacetic acid	30	Room to 82 (180)	E
	100	Room to boiling	E
Chloric acid	...	Room	E
Chlorosulfonic acid	10	...	E
Chromic acid	5–50	Room to boiling	E
Citric acid	5	Room	E
	10–25	Room to boiling	E
Citric acid (nonaerated)	50	Room to 100 (212)	E
Citric acid (aerated)	50	Room to 100 (212)	E
Citric acid	Concentrated	Room to boiling	E
Dichloroacetic acid	100	Room to 100 (212)	E
	100	Room to boiling	E
Fatty acids	...	...	E
Fluoboric acid	5–20	Elevated	NR
Fluorosilicic acid	10	Room	NR
Formic acid (still)	5	Room to 66 (150)	E
Formic acid (nonaerated)	10–50	Room to boiling	E
Formic acid (aerated)	10–90	Room to 100 (212)	E
Gallic acid	5	Room to boiling	E
Hydrobromic acid	0–100	Room to boiling	E
Hydrochloric acid (nonaerated)	5–20	Room to 35 (95)	E
Hydrochloric acid (aerated)	5–20	Room to 35 (95)	E
Hydrochloric acid	All	Room to 71 (160)	E
Hydrochloric acid fumes	Concentrated	Room to 38 (100)	E
Hydrocyanic acid	...	...	E
Hydrofluoric acid	5–48	Room	NR
Hydrofluoric acid (anhydrous)	100	Room	NR
Hydrofluoric acid vapors	...	Room	NR
Hydrofluoric-nitric acid	1 HF:15 HNO_3	Room	NR
Hydrofluosilicic acid	5	Room	V
Hydrofluosilicic acid vapors	...	100 (212)	NR
Hydroxyacetic acid	...	Room to 40 (105)	E
Lactic acid	5	Room to 66 (150)	E
	10–100	Room to boiling	E
Malic acid	...	Room and hot	E
Methyl-sulfuric acid	0–100	Room to 150 (300)	E
Molybdic acid	5	Room	E
Muriatic acid	...	Room	E
Nitric acid	5	Room	E
	10–40	Room to 100 (212)	E
	50–65	Room to boiling	E
	69.5	Room to 100 (212)	E
Nitric acid (white fuming)	90	Room to 82 (180)	E
Nitric acid	95	Room	E
Nitric acid	Concentrated	Room to boiling	E
Nitric acid	Fuming	Room	E
Nitrous acid	5	Room	E
Oleic acid	...	Room	E
Oxalic acid	1	Room to 38 (100)	E
	5	Room to 35 (95)	E
	10	Room to boiling	E
	0.5–25	Room to 60 (140)	E
	Saturated	Room to 93 (200)	E
Perchloric acid	0–100	Room to 150 (300)	E
Phenol (carbolic acid)	Saturated	Room	E
Phosphoric acid	1	Room	E
Phosphoric acid	5	Room to 100 (212)	E
Phosphoric acid (still)	10	Room to 175 (350)	E
Phosphoric acid (agitated)	10	Room	E
Phosphoric acid (aerated)	10	Room	E
Phosphoric acid	5–30	Room	E
Phosphoric acid	35–85	Room	E
Phosphoric acid	85	Room to 38 (100)	E
Phosphoric-sulfuric + $CuSO_4$	$15H_3PO_4$-$10H_2SO_4$	Room to 66 (150)	E
Picric acid	Concentrated	...	E
Propionic acid vapor	...	190 (375)	E
Pyrogallic acid	...	...	E
Salicylic acid	...	Room	E
Stearic acid	Concentrated	Room to 93 (200)	E
Succinic acid	...	Molten	E
Sulfuric acid	1–5	Room to 60 (140)	E
	5	Room to 60 (140)	E
	10	Room to boiling	E
	15	Room	E
	50	Room to boiling	E
	Concentrated	Room to 150 (300)	E
	Concentrated	Boiling	NR
	Fuming	Room	NR
Sulfuric acid vapors	96	Room to 150 (300)	E
Sulfuric anhydride	Dry	Room	NR
Sulfuric-nitric acid	90–10	Room	E
	70–30	Room	E
	50–50	Room to 60 (140)	E
	30–70	Room	E
	10–90	Room to 60 (140)	E
Sulfurous acid	6	Room	E
	Saturated	Room to 190 (375)	E
Sulfurous spray	...	Room	E
Tannic acid	10	Room to 66 (150)	E
	25	Room to 100 (212)	E
Tartaric acid	10	Room to 100 (212)	E
	25	Room to 100 (212)	E
	50	Room to 100 (212)	E

(a) E, no attack; V, variable depending on temperature and concentration; NR, not resistant

The conversion of tantalum into oxide was shown to occur by the nucleation and growth of small plates along the {100} planes of the bcc metal (Ref 16, 17). The kinetics of the oxidation of tantalum were studied at temperatures to 1400 °C (2550 °F) and at pressures ranging from less than 1 to over 40 atm (100 to 4050 kPa) (Ref 15). The reaction was found to be initially parabolic, with a transformation to a linear rate after a period of time.

Increasing the temperature not only increases the rate of oxidation of tantalum but also decreases the time before the reaction changes from parabolic to linear behavior. At a moderately high temperature (about 500 °C, or 930 °F), this transition occurs almost at once.

Under a pressure of 1 atm of oxygen, tantalum oxidizes rapidly and catastrophically at a temperature of 1300 °C (2370 °F). At slightly lower temperatures (1250 °C, or 2280 °F), the specimen oxidizes linearly at a high rate for a short period of time, then oxidizes catastrophically.

The effects of pressure on the oxygen-tantalum reaction have been established and can be summarized as follows: tantalum oxidizes linearly from 500 to 1000 °C (930 to 1830 °F) at all pressures between 1.3 and 4137 kPa (0.19 and 600 psi). From 600 to 800 °C (1200 to 1480 °F), the oxidation rate shows a pronounced increase in rate with oxygen pressures above 100 kPa (14.7 psi). Another investigator found that at a higher temperature (1000 °C, or 1830 °F), the rate tends to vary as the square root of pressure below 100 kPa (14.7 psi). At any one temperature between 500 and 800 °C (930 and 1480 °F), the rate of oxidation approaches a limiting rate as the pressure increases.

The presence of a few atomic percent of oxygen in tantalum increases electrical resistivity, hardness, tensile strength, and modulus of elasticity, but decreases elongation and reduction of area, magnetic susceptibility, and corrosion resistance to HF (Ref 18). When one realizes that 1 at.% O in tantalum equals 892 ppm, the effect of

Table 3 Corrosion resistance of tantalum and other metals to acids

Solution	Temperature °C	Temperature °F	Test duration, days	Corrosion rate: Tantalum mm/yr	Tantalum mils/yr	Niobium mm/yr	Niobium mils/yr	Zirconium mm/yr	Zirconium mils/yr	Titanium mm/yr	Titanium mils/yr
HCl, 18%	19–26	65–80	36	nil		nil		2.3 μm	0.09(a)	0.11	4.5
HCl, 37%	19–26	65–80	36	nil		0.003	0.12	2 μm	0.08	17.7	698
HCl, 37%	110	230	7	nil		0.1	4(b)	0.48	18.75	Not tested	
HNO_3, conc	19–26	65–80	36	nil		nil		nil		1.3 μm	0.05
$1HNO_3 \cdot 2HCl$	19–26	65–80	35	nil		0.5 μm	0.02	Very soluble		5.3 μm	0.21
$1HNO_3 \cdot 2HCl$	50–60	110–140	1	nil		0.025	1	Very soluble		Not tested	
H_2SO_4, 20%	95–100	205–212	4	nil		0.5 μm	0.02	4.6 μm	0.18	Not tested	
H_2SO_4, 50%	19–26	65–80	35	nil		Not tested		nil		0.053	2.1
H_2SO_4, 98%	19–26	65–80	36	nil		0.5 μm	0.02	Very soluble		1.2	46.8
H_2SO_4, 98%	21	70	. . .	nil		Not tested		Not tested		Not tested	
H_2SO_4, 98%	145	295	30	nil		4.6	180(b)	Very soluble		Very soluble	
H_2SO_4, 98%	175	345	30	0.25 μm	0.01	Not tested		Not tested		Not tested	
H_2SO_4, 98%	200	390	30	0.04	1.5	Not tested		Not tested		Not tested	
H_2SO_4, 98%	250	480	6 h	0.74	29	Not tested		Not tested		Not tested	
H_2SO_4, 98%	300	570	Not stated	8.7	342	Not tested		Not tested		Not tested	
H_3PO_4, 85%	19–26	65–80	36	nil		0.5 μm	0.02	0.5 μm	0.02(b)	0.17	6.75
$FeCl_3$, 10%	19–26	65–80	36	nil		nil		0.01	0.42(c)	0.76 μm	0.03

(a) Became brittle. (b) Tarnished. (c) Uneven corrosion

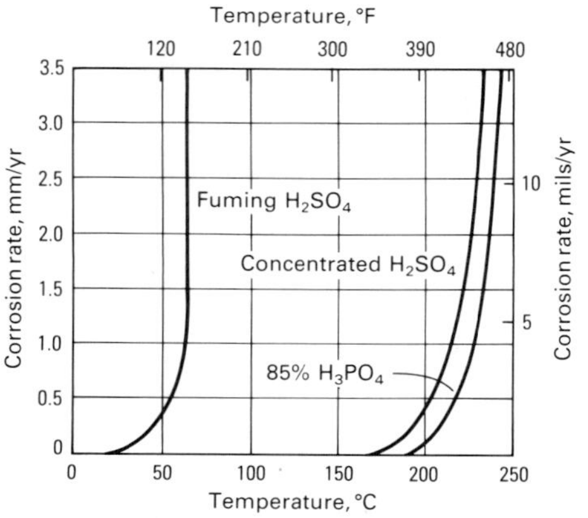

Fig. 1 Corrosion rates of tantalum in fuming H_2SO_4 (oleum) and concentrated H_2SO_4. Source: Ref 4

very small contents of oxygen on the properties is evident.

Nitrogen. The solubility of nitrogen in tantalum is shown by curve B in Fig. 7. It was found that tantalum dissolves 4 at.% N at 1000 °C (1830 °F) and that the solubility decreases rapidly with decreasing temperature. Other researchers indicated that the room-temperature solubility limit was 5.5 at.% for a stable solid solution and 7.7 at.% for a metastable solid solution (Ref 9, 15).

Below 800 °C (1470 °F), the tantalum-nitrogen reaction was reported to be cubic by one investigator, or parabolic by another (Ref 15). In the latter case, the nonlinear behavior of the Arrhenius plot of parabolic rate constants below 600 °C (1200 °F) suggests that the reaction does not obey a parabolic rate law. In contrast, a plot of cubic rate constants, although showing a wide range of scatter, was stated to indicate linear behavior (Ref 18).

Above 800 °C (1470 °F), the reaction is parabolic, with rate constants increasing uniformly as temperature increases. Although both investigators agree on the rate law followed in this temperature region, neither their initial weight gain versus time data nor their rate constants are in agreement.

As in the case of oxygen in tantalum, the presence of nitrogen in only a few atomic percent concentration increases hardness, tensile strength, and electrical resistivity and decreases elongation and density. One atomic percent N_2 in tantalum equals 780 ppm N_2.

Air. The kinetics of the reaction of tantalum with air can be regarded as an extension of the reaction of tantalum with oxygen, because tantalum forms oxides preferentially over nitrides. The rate law governing the initial period of the reaction has not been established, but by analogy to the oxygen-tantalum reaction, it may be linear (with a rate constant differing from that of subsequent linear oxidation rates), parabolic, or logarithmic, depending on temperature and pressure conditions. Tantalum is quite stable in air at 250 °C (480 °F) and below; at 300 °C (570 °F), it shows a tarnish after 24 h of exposure. The rate of corrosion, as measured by weight gain, increases rapidly at higher temperatures (see Fig. 8). At 500 °C (930 °F), the white oxide, Ta_2O_5, begins to form. Figure 8 shows plot of weight gain versus temperature (Ref 9, 15).

After a certain amount of time, ranging from over 6 h at 400 °C (750 °F) to less than 2 min at 900 °C (1650 °F), the reaction was said to become linear (Ref 19). The possibility was expressed that, at high temperatures or after extended exposure at low temperature, the linear oxidation data may also fit one of the other rate laws.

Protection Against Oxidation. Although proposed environments for tantalum and tantalum alloy applications involve gases ranging from carbon dioxide (CO_2) to halides, most of the interest centers on applications that will expose the metals to either oxygen or air at an elevated temperature. A great deal of work has been done in an effort to understand the reactions involved and to develop methods to protect the metals against attack. Although nitrogen in air may in some cases be deleterious, oxygen attack is usually seen as the mechanism of failure under low loads at elevated temperatures. Consequently, most attempts to protect tantalum against elevated-temperature gas attack have focused on imparting resistance to the base metal.

If any degree of oxidation resistance is to be imparted to tantalum, two approaches are:

- Form a denser, more adherent oxide film by alloy additions to the tantalum that alter and modify the oxide phase
- Provide a protective coating to inhibit oxygen attack on the tantalum. Coatings include silicides, aluminides, and noble metals (Ref 20)

The use of coatings to protect tantalum substrates is also described in the article "Cleaning and Finishing of Reactive and Refractory Al-

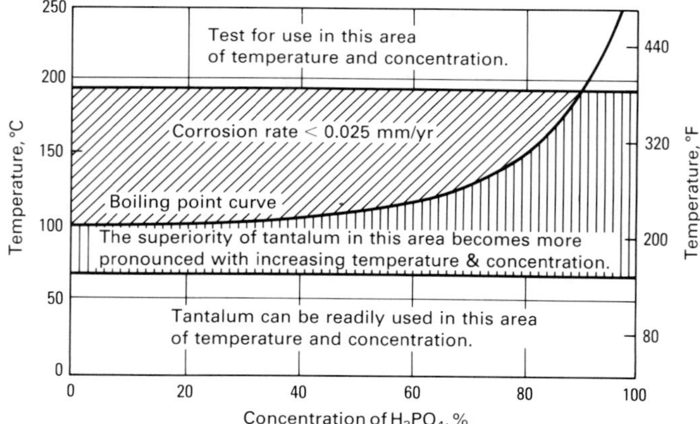

Fig. 2 Corrosion resistance of tantalum in H_3PO_4 at various concentrations and temperatures. Source: Ref 5, 6

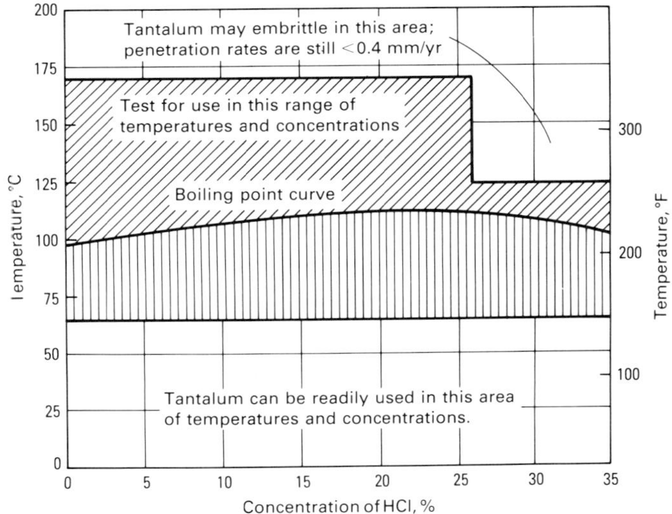

Fig. 3 Corrosion of tantalum in HCl at various concentrations and temperatures. Source: Ref 5, 6

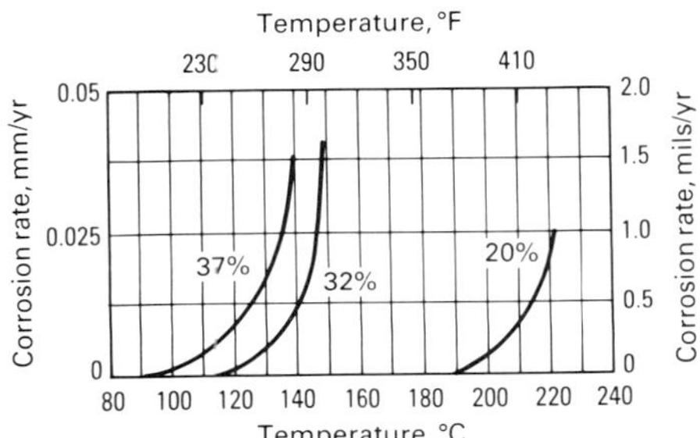

Fig. 4 Corrosion rates of tantalum in HCl of various concentrations

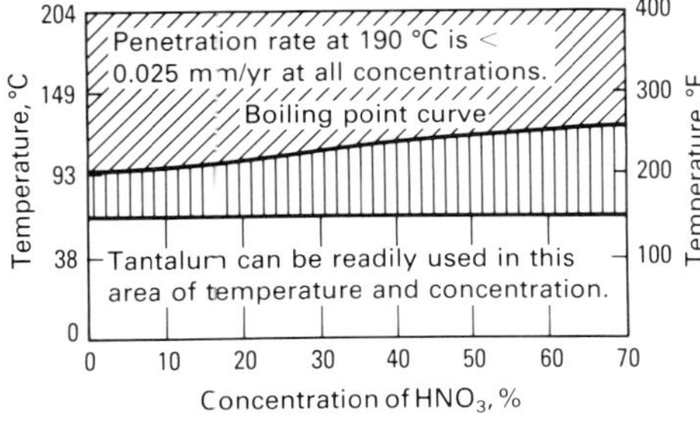

Fig. 5 Corrosion resistance of tantalum in HNO_3 at various concentrations and temperatures. Source: Ref 5, 6

loys'' in Volume 5 of the 9th Edition of *Metals Handbook*.

Other than by oxidation, tantalum structures may also fail in high-temperature service because of embrittlement by diffusion of an environmental gas into a subsurface layer of the base metal. Protection against this phenomenon, known as contamination, is achieved by the same techniques as used against surface oxidation, although the actual agents are often different. Because oxidation and contamination are both symptoms of the same basic problem, most work on contamination has been done as part of a larger effort on oxidation.

Hydrogen. Tantalum dissolves a considerable amount of hydrogen at comparatively low temperatures (Ref 9, 15). The maximum limit of solubility is 50 at.%.

Although tantalum generally does not react with molecular hydrogen below 250 °C (480 °F), it can absorb 740 times its own volume of hydrogen at red heat. However, it was reported that tantalum containing dissolved oxygen can absorb molecular hydrogen at room temperature when it is undergoing deformation. Tantalum containing more than 150 volumes of hydrogen loses its ductility (Ref 14).

Atomic hydrogen can be absorbed by tantalum, even at room temperature. Therefore, chemical equipment should be insulated from stray currents. In a few commercial cases, the metal is made electrolytically positive to prevent the liberation and absorption of atomic hydrogen on its surface.

Absorption of hydrogen is accompanied by an expansion of the bcc crystal lattice. When such material is heated to about 800 °C (1470 °F) or more in a high vacuum, it loses essentially all of its hydrogen. When permanent damage to the metal has not occurred, annealing or degassing at 800 °C (1470 °F) or higher temperature restores the metal to its original condition. In addition to decreasing the ductility, strength, and density of tantalum, the presence of hydrogen increases the hardness and the electrical resistivity.

Water Vapor. Although few published data have been found on the attack of steam on tantalum at high temperatures, it is known from industrial practice that tantalum is not affected adversely when heated with steam at pressures to 1380 kPa (200 psig), corresponding to a temperature of 200 °C (390 °F). It has been reported that at temperatures above 1125 °C (2240 °F), water is decomposed by tantalum with adsorption of oxygen by the metal and evolution of hydrogen. At 925 °C (1700 °F) and lower temperatures, the reaction is negligibly slow.

Halogens. Fluorine attacks tantalum at room temperature. Tantalum is totally inert to wet or dry chlorine, bromine, and iodine to 150 °C (300 °F), and these elements dissolved in solutions of salts or acids likewise have no effect. Chlorine begins to attack tantalum at about 250 °C (480 °F). The reaction is rapid at 450 °C (840 °F), and it occurs instantaneously at 500 °C (930 °F). The presence of water vapor sharply decreases corrosion by chlorine so that the maximum temperatures at which tantalum is sufficiently resistant to chlorine containing 1.5 to 30% H_2O for chemical equipment applications are 375 and 400 °C (705 and 750 °F), respectively. Bromine attacks tantalum at 300 °C (570 °F), and iodine begins to attack at about the same temperature (Ref 4, 9).

Carbon Dioxide and Carbon Monoxide. Tantalum is corroded by dry CO_2 at 810 kPa (8 atm) and 500 °C (930 °F). Tantalum reacts with CO_2 at 1100 °C (2010 °F) to form Ta_2O_5, and with carbon monoxide (CO) at the same temperature to form TaO. The latter reverts to Ta_2O_5 when exposed to oxygen (Ref 9, 15).

The oxidation rates of tantalum in various partial pressures of CO_2 in the temperature range of 700 to 950 °C (1290 to 1740 °F) were measured with a gravimetric balance (Ref 21). Oxidation involved a surface-controlled reaction associated with the formation of a nonprotective layer of Ta_2O_5. Below 720 °C (1330 °F), there was a change in the rate-determining mechanism, and above 830 °C (1525 °F), the linear rate was preceded by a complex region of nonlinear oxidation behavior. The linear oxidation of tantalum has been explained quantitatively in terms of CO_2 absorption, followed by a rate-controlling surface reaction. The initial absorption has been expressed both as an equilibrium process and as a steady-state reaction.

Nitrogen Monoxide and Nitrous Oxide. Below about 1125 °C (2050 °F), the reaction rate of nitrogen monoxide (NO) (as a 5% mixture in argon) with tantalum cannot be detected (Ref 9). As the temperature is increased, the reaction takes place with increasing rapidity, rising from 0.0065% area loss per second at 1195 °C (2180 °F) to 0.076% at 1457 °C (2670 °F) (Ref 22).

The oxidation by nitrous oxide (N_2O) on evaporated films of niobium, tantalum, and titanium was studied in the temperature range of 195 to 435 K (Ref 23). For tantalum, fast dissolution absorption of N_2O occurred at 195 K, accompanied by N_2 evolution. Some incorporation of N_2O also occurred. The rate of N_2O absorption was found to depend on the pressure of the reacting gas.

Other Reactive Gases. Although little published data appear to exist on the subject, it is expected that tantalum would react at some elevated temperature with oxygen-containing gaseous compounds, such as sulfur dioxide (SO_2) and nitrogen dioxide (NO_2) (Ref 19). With such hydrocarbons as benzene or naphthalene, tantalum reacts at temperatures from 1700 to 2500 °C (3090 to 4530 °F) to form tantalum carbide. Tantalum is used as a getter in vacuum tubes at temperatures of 650 to 1000 °C (1200 to 1830 °F) to absorb gases and maintain a high vacuum.

Inert Gases. Pure helium and argon do not react with tantalum, even at its melting point. These gases are used as an inert atmosphere for arc melting or welding the metal.

Liquid Metals. Tantalum and some tantalum-base alloys exhibit good resistance to many liquid metals (Ref 5, 9, 24). Table 6 lists the effect of various liquid metals on tantalum. Such tantalum materials exhibit remarkable resistance to several liquid metals even to high temperatures (900 to 1100 °C, or 1650 to 2010 °F) in the absence of oxygen or nitrogen.

Table 4 Effects of salts on tantalum

Salt	Concentration, %	Temperature, °C (°F)	Code(a)
Aluminum acetate	Saturated	Room	E
Aluminum chloride	5	Room	E
Aluminum chloride (aerated)	5–10	Room to 60 (140)	E
Aluminum chloride	25	Room to 100 (212)	E
Aluminum fluoride	5	Room	NR
	Saturated	Room	NR
Aluminum hydroxide	Saturated	. . .	E
Aluminum potassium sulfate (alum)	2	Room	E
	10	Room to boiling	E
Aluminum sulfate	10–saturated	Room to boiling	E
Ammonium acid phosphate	10	Room	E
Ammonium alum	. . .	. . .	E
Ammonium alum (slightly ammoniacal)	. . .	. . .	E
Ammonium bicarbonate	50	Room to 100 (212)	E
Ammonium bromide	5	Room	E
Ammonium carbonate	50	Room to 100 (212)	E
Ammonium carbonate (aqueous)	50	Room to boiling	E
Ammonium carbonate	All	Room to hot	E
Ammonium chloride	1	Room	E
	10–50	Room to boiling	E
Ammonium fluoride	10	Room	NR
Ammonium hydroxide	. . .	. . .	V
Ammonium monosulfate	. . .	. . .	E
Ammonium nitrate	0–100	Room to 150 (300)	E
Ammonium oxalate	5	Room	E
Ammonium persulfate	5	Room	E
Ammonium phosphate	5	Room	E
Ammonium sulfate (aerated)	1	Room	E
	5	Room to 100 (212)	E
Ammonium sulfate	10	Room to 100 (212)	E
Ammonium sulfate	Saturated	Room to boiling	E
Ammonium sulfite	Saturated	Room to boiling	E
Amyl acetate	. . .	. . .	E
Aniline hydrochloride	5	Room	E
	20	Room to 38 (100)	E
Antimony trichloride	. . .	Room	E
Barium carbonate	Saturated	Room	E
Barium chloride	5 to saturated	Room	E
	5	Room to 100 (212)	E
	20	Room to 100 (212)	E
	25	Room to boiling	E
Barium hydroxide	Saturated	. . .	EE
Barium hydroxide·$8H_2O$	Saturated	Room	E
Barium nitrate	Aqueous solution	Room to hot	E
Barium sulfate	. . .	Room	E
Butyl acetate	. . .	Room	E
Calcium bisulfite	. . .	Room	E
Calcium carbonate	Saturated	Room to boiling	E
Calcium chlorate	Dilute	Room to hot	E
Calcium chloride	5–20	Room to 100 (212)	E
	28	Room to boiling	E
	Concentrated	Room	E
Calcium hydroxide	10–saturated	Room to boiling	E
Calcium hypochlorite	2–saturated	Room to boiling	E
Calcium sulfate	Saturated	Room	E
Copper acetate	Saturated	Room	E
Copper carbonate	Saturated	. . .	E
Copper chloride (agitated, aerated)	1	Room	E
Copper chloride (agitated)	5	Room	E
Copper chloride (aerated)	5	Room	E
Copper cyanide (electroplating solution)	. . .	Room	E
Copper cyanide	Saturated	Room to boiling	E
Copper nitrate	1–saturated	Room	E
Copper sulfate	5	Room	E
	Saturated	Room to boiling	E
Cupric carbonate-cupric hydroxide	Saturated	Room	E
Cupric chloride	20–50	Room to boiling	E
Cupric cyanide	Saturated	Room	E
Cupric nitrate	. . .	Room to 40 (105)	E
Cuprous chloride	50	Room to 90 (195)	E
Ferric chloride (still)	1–50	Room to boiling	E
Ferric chloride (agitated)	5	Room	E
Ferric chloride (aerated)	5	Room	E
Ferric hydroxide	. . .	Room	E
Ferric nitrate	1–5	Room	E
Ferric sulfate	1–saturated	Room	E
Ferrous chloride	. . .	Room	E
Ferrous sulfate	Dilute	Room	E
Ferrous ammonium citrate	. . .	. . .	E
Fluoride salts	Variable	Variable	V
Hydrogen bromide	. . .	. . .	E
Hydrogen peroxide	3–30	Room	E
	. . .	Room to boiling	E
Hydrogen iodide	. . .	. . .	E
Hydrogen sulfide	Dry	Room	E
	Saturated H_2O	Room	E
Hyposulfite soda (hypo)	. . .	. . .	E
Lactic acid salts	. . .	Room	E
Lead acetate	Saturated	Room	E
Magnesium carbonate	. . .	. . .	E
Magnesium chloride (still)	1–5	Room to hot	E
Magnesium chloride	5–40	Room to boiling	E
Magnesium hydroxide	Saturated	Room	E
	Thick suspension	Room	E
Magnesium nitrate	. . .	. . .	E
Magnesium sulfate	5	Room to hot	E
	Saturated	Room	E
Manganese carbonate	. . .	. . .	E
Manganese chloride	10–50 (Aqueous)	Room to boiling	E
Manganous chloride	5–20	Room to 100 (212)	E
Mercuric bichloride	0.07	Room	E
Mercuric chloride	1–saturated	Room to 100 (212)	E
Mercuric cyanide	Saturated	. . .	E
Mercurous nitrate	. . .	. . .	E
Nickel chloride	5–20	Room to 100 (212)	E
Nickel nitrate	10	Room	E
Nickel nitrate plus 6% H_2O	50	Room	E
Nickel sulfate	10	Room	E
Phosphoric anhydride	Dry	Room	E
Phosphorus trichloride		. . .	. . .
	Saturated	Room	E
Phthalic anhydride	. . .	. . .	E
Potassium bichromatic (neutral)	. . .	Room	E
Potassium bromide	5–saturated	Room	E
Potassium carbonate	1	Room	E
Potassium chlorate	. . .	. . .	E
Potassium chloride	1–36	Room to boiling	E
	Saturated	Room	E
Potassium cyanide	. . .	. . .	E
Potassium dichromate (neutral)	. . .	. . .	E
Potassium ferricyanide	5–saturated	Room	E
Potassium ferricyanide plus 5% NaCl	0.5	Room	E
Potassium ferrocyanide	5	Room	E
Potassium hydrate	. . .	. . .	E
Potassium hydroxide	5	Room	E
	27–50	Boiling	NR
Potassium iodide	Saturated	Room	E
Potassium iodide—iodine	. . .	. . .	E
Potassium iodide plus 0.1% Na_2CO_3	Saturated	Room	E
Potassium nitrate	5	Room	E
Potassium oxalate	. . .	. . .	E
Potassium permanganate (neutral)	. . .	. . .	E
Potassium pyrosulfate	. . .	Molten	NR
Potassium sulfate	1–5	Room to hot	E
	10	Room	E
Potassium sulfide	. . .	. . .	E
Potassium thiosulfate	1	Room	E
Silver bromide	. . .	. . .	E

(continued)

(a) E, no attack; V, variable, depending on temperature and concentration; NR, not resistant

Table 4 (continued)

Salt	Concentration, %	Temperature, °C (°F)	Code(a)
Silver chloride	...	...	E
Silver cyanide	...	...	E
Silver nitrate	50	Room	E
Sodium acetate (moist)	5	Room	E
Sodium acetate	Saturated	Room	E
Sodium aluminate	25	Room to boiling	E
Sodium benzoate	...	...	E
Sodium bicarbonate	All	Room to 66 (150)	E
Sodium bichromate (neutral)	...	...	E
Sodium bisulfate	Solution	...	E
	10–25	Room to boiling	E
	Saturated	Room	E
	...	Molten	NR
Sodium borate	...	...	E
Sodium bromide	5	Room	E
Sodium carbonate	10–25	Room to boiling	E
	All	Room	E
Sodium chlorate	10–25	Room	E
	Saturated	Room	E
Sodium chloride (still)	5	Room to 40 (105)	E
Sodium chloride (aerated)	20	Room	E
Sodium chloride	29	Room to boiling	E
	Saturated	Room to boiling	E
Sodium citrate	Saturated	Room	E
Sodium cyanide	Saturated	Room	E
Sodium dichromate	Saturated	Room	E
Sodium ferricyanide	...	...	E
Sodium ferrocyanide	...	...	E
Sodium fluoride	5–saturated	Room	NR
Sodium hydrosulfite	...	...	E
Sodium hydroxide	10–saturated	Room	NR
	10	Boiling	NR
	25	Room to boiling	NR
	40	80 (175)	NR
Sodium hypochlorite	6	Room	E
Sodium hyposulfite	Dilute	Room	E
Sodium lactate	...	...	E
Sodium nitrate	All	Room	E
Sodium nitrite	...	...	E
	Saturated	Room	E
Sodium peroxide	...	100 (212)	V
Sodium phosphate	5–saturated	Room	E
Sodium pyrosulfate	...	Molten	NR
Sodium silicate	...	...	E
	25	Room to boiling	E
Sodium sulfate (still)	5	Room	E
Sodium sulfate	10–20	Room to boiling	E
Sodium sulfate	Saturated	Room	E
Sodium sulfide	10	Room to boiling	E
	Saturated	Room	E
Sodium sulfite	5	Room	E
	10–saturated	Room to boiling	E
Sodium thiosulfate	10–25	Room to boiling	E
Sodium thiosulfate—acetic acid	20	Room	E
Stannic chloride	5	Room to 100 (212)	E
	24	Room to boiling	E
	100	Molten	E
Stannous chloride	5–saturated	Room	E
Sulfur chloride	Dry	...	E
Sulfuryl chloride	...	...	E
Thionyl chloride	...	...	E
Tin salts	...	...	E
Titanium tetrachloride	...	...	E
Zinc chloride (still)	5	Room to boiling	E
Zinc chloride	10	Room to boiling	E
	20	Room to 100 (212)	E
	Saturated	Room	E
Zinc sulfate	5–saturated	Room	E
	25	Room to boiling	E

(a) E, no attack; V, variable, depending on temperature and concentration; NR, not resistant

The severity of attack on tantalum by liquid metals may be markedly increased by increasing temperature. Because operating temperatures somewhat in excess of 700 °C (1290 °F) are desirable in many cases, refractory metals, including tantalum and niobium, seem to be particularly promising materials of construction for containing liquid metals.

Liquid aluminum reacts rapidly with tantalum to form the stable compound aluminum (Al_3Ta) (Ref 9).

Liquid bismuth has little action on tantalum at temperatures below 1000 °C (1830 °F) and exerts no detrimental effects on the stress rupture properties of tantalum at 815 °C (1500 °F); but it causes some intergranular attack at 1000 °C (1830 °F) (Ref 5, 25).

Calcium. Tantalum is only slightly attacked by calcium at 1200 °C (2190 °F). A crucible with a wall thickness of 0.15 mm (5.8 mils) was reduced to 0.13 mm (5.3 mils) after 12 days of exposure to calcium at 1200 °C (2190 °F) (Ref 9).

Cesium. Similar lack of corrosion resistance to cesium was found for tantalum as reported for niobium. Refluxing capsule tests indicated surface dissolution and severe attack after 720 h at 980 and 1370 °C (1800 and 2500 °F) (Ref 26).

Gallium. The resistance of tantalum to molten gallium is considered to be good at temperatures to 450 °C (840 °F), but poor at temperatures above 600 °C (1110 °F).

Lead. Tantalum is highly resistant to liquid lead at temperatures to 1000 °C (1830 °F), with a rate of attack of less than 0.025 mm/yr (1 mil/yr). It exhibits no detrimental effects when stress rupture tests are conducted in molten lead at 815 °C (1500 °F) (Ref 5).

Lithium. Tantalum possesses good resistance to molten lithium at temperatures to 1000 °C (1830 °F) (Ref 5, 24, 27, 28). Tantalum capability with lithium is similar to that of niobium in that corrosion resistance depends on oxygen concentration. Tantalum metal will exhibit good corrosion resistance to lithium as long as the oxygen concentration of the tantalum is maintained below 100 to 200 ppm.

Specimens of T-111 and T-222 (Ta-9.6W-2.4Hf-0.01C) alloys, oxygen contaminated to 500 ppm and welded in argon, were exposed to lithium at 750 and 1200 °C (1380 and 2190 °F) for 100 h. Evaluation indicated no attack in the weld areas; however, intergranular penetration was observed in the base metal of both alloys. Heat treatment at 1315 °C (2400 °F) eliminated the attack. In addition, a method of inhibiting the corrosion of tantalum by liquid lithium at temperatures above 1000 °C (1830 °F) by the addition of 0.15 to 1.5 at.% Si to the lithium is discussed in Ref 29.

Lithium and Uranium Mononitride Fuel. In one investigation, simulated nuclear-fuel element specimens, consisting of uranium mononitride (UN) fuel cylinders clad with tungsten-lined T-111 alloy were exposed in a pumped lithium loop operated at 1040 °C (1905 °F) (Ref 30) for up to 7500 h. The lithium flow velocity was 1.5 m/s (5 ft/s) in the specimen test section. A cladding crack was simulated in one specimen exposed for 50 000 h by an axial slot machined through both the cladding and the tungsten liner.

All of the fuel element specimens appeared to be in excellent condition after the tests. No evidence of any chemical compatibility problems between the specimens and the flowing lithium was found. Except for a slight reduction in the oxygen content of the T-111, very little change in chemistry was observed in the T-111 or the UN. No microstructural changes were observed in the UN, but bands of fine precipitates were seen in the T-111 after the lithium exposure. These precipitates were thought to be the result of thermal aging, not lithium exposure.

Direct exposure of the UN to the lithium through the simulated cladding crack resulted in some erosion of the UN and in some nitrogen contamination of the T-111 cladding in the area of the defect. The T-111 in the fuel element clad specimens was ductile after the long-term lithium exposure. Thermal aging at 1040 °C (1905 °F), however, resulted in the T-111 becoming sensitive to hydrogen embrittlement during post-test handling and testing.

Magnesium and Magnesium Alloys. Tantalum is unattacked by molten magnesium at 1150 °C (2100 °F) (Ref 9.)

Mercury. The limited amount of corrosion testing of refractory metals in mercury is summarized below. The results for tantalum are consistent with the solubility information (Ref 31).

In static tests, tantalum exhibited good resistance to mercury at temperatures to 600 °C (1110 °F). Refluxing capsule tests showed no attack of tantalum up to 760 °C (1400 °F). The corrosion resistance of tantalum to mercury was further documented in a two-phase natural-circulation loop test that ran for 19 975 h with a boiling temperature of 650 °C (1200 °F) and superheat temperature of 705 °C (1300 °F).

Post-test evaluation of the loop revealed no corrosion. As a result of the inertness of tantalum to mercury attack demonstrated in this long-term experiment, tantalum was evaluated as a replace-

Table 5 Effects of miscellaneous corrosive reagents on tantalum

Medium	Concentration, %	Temperature, °C (°F)	Code(a)
Acetone	...	Boiling	E
Air	...	Below 300 (570)	E
	...	Above 300 (570)	NR
Amines	...	...	E
Aniline	Concentrated	Room	E
Baking oven gases	...	...	E
Beer	...	...	E
Benzene	...	Room	E
Benzol	...	Hot	E
Bleaching powder	Solution	Hot	V
Blood (meat juices)	...	Cold	E
Body fluids	...	...	E
Borax	...	Fused	NR
Bromine	Dry	Below 300 (570)	E
	Wet	...	E
Bromine water	...	Room	E
Buttermilk	...	Room	E
Carbon bisulfide	...	Room	E
Carbon tetrachloride	99	Boiling	E
	Liquid	Boiling	E
	Pure	Room	E
	5–10 aqueous solution	Room	E
Chlorinated brine	...	...	E
Chlorinated hydrocarbons	...	...	E
Chlorinated water	Saturated	Room	E
Chlorine dioxide	...	180 (355)	E
Chlorine gas	Dry	Up to 250 (480)	E
	Moist (1.5% H_2O)	Up to 375 (705)	E
	Moist (30% H_2O)	Up to 400 (750)	E
Chloroform	...	Room	E
Chromium plating bath	...	Room	E
Cider	...	Room	E
Coffee	...	Boiling	E
Copal varnish	...	...	E
Cream of tartar	...	...	E
Creosote (coal tar)	...	Hot	E
Crude oil	...	...	E
Developing solutions	...	Room	E
Distillery wort	...	...	E
Dyewood, liquor	...	Room	E
Ether	...	Room	E
Ethyl acetate	...	...	E
Ethyl chloride	5	Room	E
Ethyl sulfate	...	...	E
Ethylene chloride	...	Room	E
Ethylene dibromide	...	...	E
Ethylene dichloride	100	Boiling	E
Flue gases	...	...	E
Fluorine	...	Room	NR
Food pastes	...	...	E
Formaldehyde	...	Room	E
Formaldehyde plus 2.5% H_2SO_4	50	158 (315)	E
Fuel oil	...	Hot	E
Fuel oil (containing H_2SO_4)	...	Hot	E
Fruit juices	...	Room	E
Furfural	...	...	E
Gasoline	...	...	E
Glauber's salt	Solution	Hot	E
Glue, dry	...	Room	E
Glue, solution acid	...	Hot	E
Glycerine	...	Room	E
Gypsum	...	...	E
Hydrocarbons	...	...	E
Hydrogen	...	Up to 300 (570)	E
Ink	...	...	E
Iodine	...	Up to 300 (570)	E
Idoform	...	...	E
Kerosene	...	Room	E
Ketchup	...	Room	E
Lard	...	Room	E
Linseed oil	...	...	E
Lye (caustic)	34	110 (230)	NR
Lysol	...	100 (212)	E
Mayonnaise	...	Hot and cold	E
Meats (unsalted)	...	Room	E
Mash	...	Hot	E
Methylene chloride	40	Room to boiling	E
Milk	Fresh or sour	Hot or cold	E
Mine water, acid	...	...	E
Molasses	...	...	E
Mustard	...	Room	E
Naphtha	...	...	E
Nitre cake	...	Fused	NR
Nitric oxides	...	...	E
Nitrosyl chloride	...	...	E
Nitrous oxide	Dry	...	E
Oils, crude	...	Hot and cold	E
Oils, mineral, vegetable	...	Hot and cold	E
Organic chlorides	...	...	E
Oxygen	...	Up to 300 (570)	E
Paraffin	...	Molten	E
Paraffin	...	Molten	E
Paregoric compound	...	...	E
Petroleum ether	...	...	E
Phenol	...	...	E
Phenolic resins	...	...	E
Pine tar oil	...	...	E
Potash	Solution	Hot	NR
Quinine bisulfate (dry)	...	...	E
Quinine sulfate (dry)	...	...	E
Rosin	...	Molten	E
Sal ammoniac	20	Boiling	E
Salt	Saturated	Room	E
Salt brine	Saturated	Hot	E
Salt water	...	...	E
Sewage	...	...	E
Soaps	...	Room	E
Soy bean oil	...	...	E
Starch	Solution	...	E
Steam	...	...	E
Sugar juice	...	...	E
Sulfur, dry	...	Molten	E
Sulfur, wet	...	...	E
Sulfur dioxide	Dry	Room	E
	Moist	Room	E
Sulfur trioxide	Dry	Room	NR
Tomato juice	...	Room	E
Turpentine oil	...	...	E
Tung oil	...	...	E
Varnish	...	...	E
Vegetable juices	...	...	E
Vegetable oil	...	Hot and cold	E
Vinegar	Still	Room	E
	Agitated	Room	E
	Aerated	Room	E
	Fumes	...	E
Vinegar and salt	...	...	E
Water	...	...	E
	...	Hot	E
	Salt	...	E
	Sea	...	E
Whiskey	...	...	E

(a) E, no attack; V, variable, depending on temperature and concentration; NR, not resistant

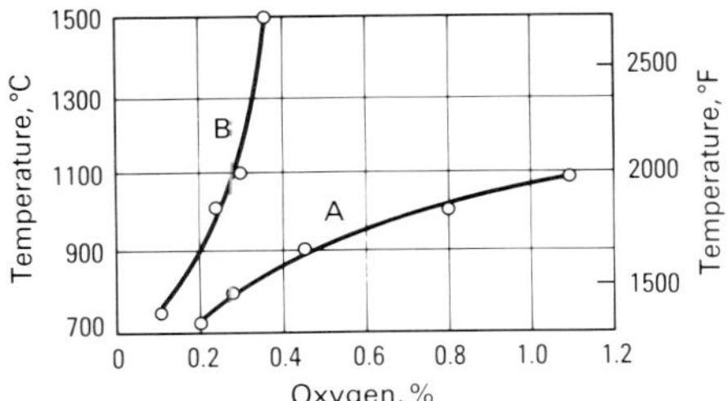

Fig. 6 Solubility of oxygen in niobium and tantalum. Curve A, niobium; Curve B, tantalum. Source: Ref 15

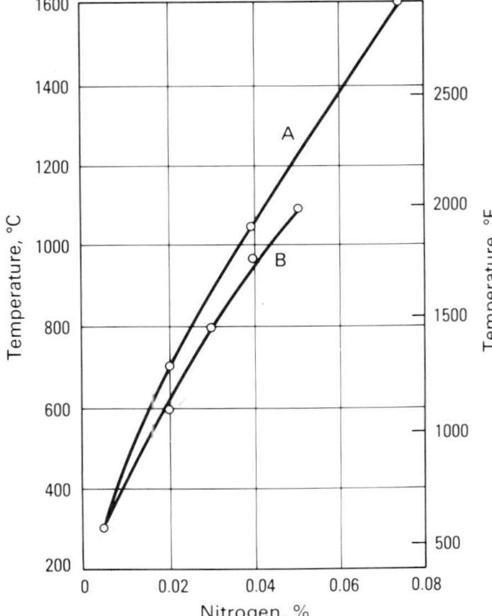

Fig. 7 Solubility of nitrogen in niobium and tantalum. Curve A, niobium; Curve B, tantalum. Source: Ref 15

ment material for Croloy 9M steel in a mercury boiler (Ref 24). Results of other tests of tantalum in mercury are described in Ref 31 and 32.

Potassium. The compatibility of tantalum and potassium at 600, 800, and 1000 °C (1110, 1470, and 1830 °F) was studied in static capsule tests (Ref 33). As the oxygen concentration of potassium was increased, the amount of tantalum in the potassium was also found to increase. The results indicated the formation of an unidentified ternary oxide phase that is either nonadherent or dissolved when the potassium is dissolved for chemical analysis. When the tantalum specimens contained oxygen above a certain threshold concentration, potassium penetrated the tantalum, and intergranular, as well as transgranular, attack was observed. The threshold levels for intergranular attack at the test temperatures were found to be 500, 700, and 1000 ppm O, respectively. The mechanism of attack was believed to be the formation of a ternary oxide phase.

Silver. Tantalum is only slightly attacked by silver at 1200 °C (2190 °F); a tantalum crucible tested in silver at this temperature for 35 days showed a loss in wall thickness of 0.02 mm (0.8 mil).

Liquid sodium, potassium, or alloys of these elements have little effect on tantalum at temperatures to 1000 °C (1830 °F), but oxygen contamination of sodium causes increases in corrosion. Sodium does not alloy with tantalum (Ref 5).

The presence of oxygen in liquid sodium leads to slight weight loss of tantalum in flowing systems. In addition, extensive intergranular and transgranular attack of tantalum by sodium was observed. This attack was attributed to the high (390 ppm) oxygen concentration of the tantalum before exposure to the sodium.

The compatibility of tantalum alloys T-111 and Ta-10W with static sodium was demonstrated in capsules tested at 1315 °C (2400 °F) for 6271 h and 300 h, respectively. No corrosion was found in either alloy.

Tellurium. Corrosion of candidate construction materials for stills to extract radioactive polonium-210 from bismuth by distillation at temperatures of 450 to 950 °C (840 to 1740 °F) was investigated (Ref 27, 34). Tellurium, which is chemically similar to polonium, was used as a nonradioactive substitute for polonium. Of the materials investigated, tantalum appeared to be the most satisfactory from the standpoint of fabricability and long-term corrosion resistance. Tantalum was corroded at rates up to 0.5 μm/h (0.02 mils/h) during the initial 100 to 200 h of exposure; the rate decreased to less than 0.05 μm/h (0.002 mils/h) after 400 h for concentrations of tellurium of less than 30% in bismuth.

Thorium-Magnesium. In static tests, the thorium-magnesium eutectic had no appreciable effect on tantalum at 1000 °C (1830 °F). No measurable corrosion of tantalum by the thorium-magnesium eutectic was noted in dynamic tests for 28 days with a temperature range of 700 to 840 °C (1290 to 1545 °F) (Ref 5). Extensive tests on components for molten-metal fuel reactors revealed that tantalum is a satisfactory material for several thousand hours of service in high-temperature circulating loops containing a molten magnesium-thorium alloy having a composition in the range of the magnesium-rich eutectic (Ref 9).

Uranium and Plutonium Alloys. Short-term tests indicated that the practical upper limit for tantalum as a container material for uranium is about 1450 °C (2640 °F). However, attack below this temperature is also significant because a tantalum crucible with a wall thickness of 1.5 mm (0.06 in.) was completely corroded within a test period of 50 h at 1275 °C (2325 °F) (Ref 5). Other investigations showed that tantalum is not attacked by uranium-magnesium and plutonium-magnesium alloys at 1150 °C (2100 °F) (Ref 9). Extensive tests on components for molten-metal fuel reactors revealed that tantalum is a satisfactory material for several thousand hours of service in several liquid-metal environments (Ref 9).

Zinc is reported to wet and attack tantalum, the surface of which is abraded in zinc at 440 °C (825 °F); also, molten zinc attacks tantalum at significant rates at temperatures above 450 °C (840 °F) (Ref 9). Tantalum showed appreciable attack from molten zinc at 750 °C (1380 °F) (Ref 15). However, one industrial zinc producer observed excellent corrosion resistance at 500 °C (930 °F) (Ref 9). The maintenance of the oxide film on the tantalum may account for the latter result.

Other Molten Metals. The intermetallic compounds YSb, ErSb, LaSb, and YBi have little effect on tantalum at 1800 to 2000 °C (3270 to 3630 °F), but antimony vapor severely attacks tantalum at temperatures of 1000 °C (1830 °F) and higher (Ref 9).

Hydrogen Embrittlement, Galvanic Effects, and Cathodic Protection of Tantalum

The hydrogen embrittlement experienced in corrosion tests on tantalum exposed to HCl in sealed capsules (Ref 5) was mentioned in the section "Corrosion of Tantalum in Specific Media" in this article. Also, hydrogen embrittlement of tantalum was observed when it was exposed to concentrated H_2SO_4 at 250 °C (480 °F) or to concentrated HCl at 150 °C (300 °F) (Ref 35).

Failures due to hydrogen embrittlement have occurred in some severe aqueous acid media in chemical industry applications where tantalum was, or became, electrically coupled to a less noble material, such as low-carbon steel. Under these conditions, tantalum became the cathode in the galvanic cell thus created.

Because of the presence of stray currents, tantalum may become a cathode in the system and consequently may absorb and become embrittled by atomic hydrogen in the electrolytic cell (galvanic cell). The presence of stray current can result from induction from adjacent lines, leakages, variable ground voltages, and other sources. Although stray voltages may be transient, absorbed hydrogen is cumulative in its effect on producing hydrogen embrittlement of tantalum.

For applications of pure tantalum in aggressive acids at high temperatures, hydrogen embrittlement rather than uniform corrosion attack is the crucial concern (Ref 5, 35). References cited in Ref 35 propose a specific mechanism for hydrogen embrittlement of tantalum based on stress-induced hydride precipitation.

Several methods have been used or proposed to reduce hydrogen embrittlement of tantalum (Ref 5, 36-38):

- Complete electrical insulation of tantalum from all metals in system. For additional protection, the insulated tantalum may be connected to the positive pole of a DC source (about 15 V), while the negative pole is connected to some other metallic part, which is exposed to the solution in the vessel or to ground
- Addition of a selected oxidizing agent to the solution
- Coupling the tantalum surface to a noble metal
- Anodizing the tantalum

Galvanic Effects. If tantalum is the cathode in a galvanic couple, hydrogen embrittlement can prove disastrous (Ref 5). If tantalum is the anode in such a cell, anodization occurs so readily that no damage occurs and galvanic current quickly drops to a very low value. In couples of tantalum with platinum, silver, copper, bismuth, antimony, molybdenum, nickel, lead, tin, zinc, and aluminum, tantalum was initially the more electronegative member of the couple (except for zinc and aluminum). However, galvanic current rapidly decayed as the tantalum spontaneously anodized. In HF, tantalum was again more noble than zinc and aluminum and was more active than platinum, silver, copper, antimony, nickel, and lead. The latter six couples result in high steady-state currents, because tantalum dissolves rather than anodizes in fluoride solution. The anomalous behavior of tantalum when coupled with bismuth or iron in HF is apparently attributable to the formation of insoluble

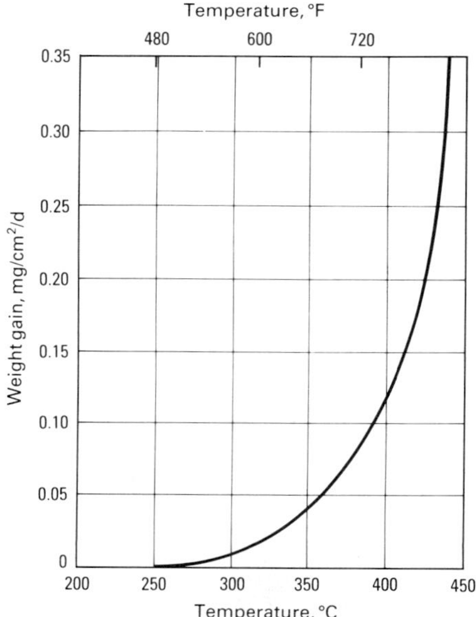

Fig. 8 Corrosion rate of tantalum in air as a function of temperature

fluorides on the surfaces of the bismuth or iron electrodes.

Time is an important factor in determining whether tantalum will be damaged by galvanic effects (Ref 39). In practice, it is dangerous to depend on laboratory tests to provide information on whether tantalum is anodic in a given galvanic couple. In results reported in Ref 39, tantalum was cathodic to aluminum in dilute NaCl, HCl, and NaCl after times of 0.5 and 60 min. When tantalum was coupled to certain of the other metals—notably Hastelloy B, nickel, and lead—the polarity of the tantalum reversed with time in some electrolytes from anodic to cathodic. When tantalum is cathodic, or becomes cathodic, hydrogen embrittlement can occur. In probably all of the couples, tantalum would become cathodic if given sufficient time (Ref 39). The conclusion is drawn that it is highly desirable to prevent galvanic cell formation by providing adequate electrical insulation.

Cathodic Protection. Applying low-overvoltage elements, such as platinum and other noble metals, to tantalum and other metals has received further attention in efforts to develop improved anodes for cathodic protection (Ref 40). The first of the precious metals containing anodes to become commercial was platinum or platinum alloy on a titanium substrate. The uniqueness of this concept was not that of platinum, which has been used as an electrode material for many years, but the nature of titanium (or tantalum) itself. Titanium and tantalum have a very adherent, nonporous inert oxide film that will not transfer current unless the electrode-to-electrode potential reaches a certain value. With proper treatment, this oxide film can be removed, and a very thin layer of platinum can be applied to the titanium (or tantalum) surface. Any unplatinized surface will reoxidize, leaving a surface that will pass current from the platinized areas but not from the oxidized areas.

The commonly used precious metal and anode materials are lead-platinum bielectrodes, platinized titanium, platinized niobium, and platinized tantalum. The latter two substrates allow greater applied voltage to the anode system. Like platinized titanium anodes, the mechanism for successful operation of the platinized niobium or platinized tantalum anodes is the valve-like nature of the oxide film on the niobium or tantalum substrates. Niobium in such anodes operates to a breakdown voltage of 40 to 50 V, but tantalum will operate up to 200 V. For both materials, the resistivity of the alloys used is approximately one-third that of titanium, which is said to give better anode characteristics (Ref 40). More information on this subject is available in the article "Cathodic Protection" in this Volume.

Table 6 Effects of molten metals on tantalum

Metal	Remarks	Temperature, °C (°F)	Code(a)
Aluminum	Forms Al_3Ta	Molten	NR
Antimony	...	to 1000 (1830)	NR
Bismuth	...	to 900 (1650)	E
Cadmium	...	Molten	E
Gallium	...	to 450 (840)	E
Lead	...	to 1000 (1830)	E
Lithium	...	to 1000 (1830)	E
Magnesium	...	to 1150 (2100)	E
Mercury	...	to 600 (1110)	E
Potassium	...	to 900 (1650)	E
Sodium	...	to 900 (1650)	E
Sodium-potassium alloys	...	to 900 (1650)	E
Zinc	...	to 500 (930)	E/V
Tin	...	...	V
Uranium	...	...	V
Mg-37Th	In helium	to 800 (1470)	S
Bi-5 to 10U	In helium	to 1100 (2010)	S
Bi-5U-0.3Mn	In helium	to 1050 (1920)	S
Bi-10U-0.5Mn	In helium	to 1160 (2120)	S
Al-18Th-6U	Failed	to 1000 (1830)	NR
U-10Fe	Failed	to 900 (1650)	NR
U-Cr (eutectic)	Failed	to 900 (1650)	NR
YSb-intermetallic compound	...	1800–2000 (3270–3630)	S
YBi-intermetallic compound	...	1800–2000 (3270–3630)	S
ErSb-intermetallic compound	...	1800–2000 (3270–3630)	S
LaSb-intermetallic compound	...	1800–2000 (3270–3630)	S
Plutonium-cobalt-cerium	...	to 650 (1200)	V

(a) E, no attack; S, satisfactory; V, variable, depending on temperature and concentration; NR, not resistant

Corrosion Resistance of Tantalum-Base Alloys

Most of the preceding discussion has concerned the corrosion resistance of unalloyed tantalum. Limited corrosion data have been obtained on tantalum-base alloys. Some of the data on the more promising tantalum alloys are cited in this section, especially on Tantaloy 63 (Ta-2.5W-0.15Nb). In addition, the outstanding corrosion resistance found for certain low substitutional alloy content tantalum-base alloys, notably tantalum-molybdenum, is described (Ref 35). Data found in the literature on the corrosion resistance of other tantalum-base alloys are also summarized.

Tantalum-Tungsten Alloys. Samples of 0.18-to 0.75-mm (0.007-to 0.03-in.) thick strip or sheet of the following materials were exposed for selected times in concentrated (95.5 to 98%) H_2SO_4 at chosen temperatures ranging from 175 to 200 °C (345 to 390 °F):

- Tantalum, electron beam melted
- Tantalum, powder metallurgy
- Tantaloy 63 (Ta-2.5W-0.15Nb)
- Fansteel 65 metal (Ta-5W)
- Fansteel 60 metal (Ta-10W)

Some tests on electron beam melted tantalum and Tantaloy 63 were conducted on materials in the as-rolled, stress-relieved, and fully recrystallized conditions. The average corrosion rates observed in these tests are listed in Table 7. Corrosion rate as a function of tungsten content is plotted in Fig. 9 for additional tests on electron beam melted tantalum, Tantaloy 63 (Ta-2.5W-0.15Nb), Ta-5W, and Ta-10W exposed to concentrated H_2SO_4 at 180 and 210 °C (360 and 405 °F).

The corrosion behavior of substitutional tantalum-molybdenum, tantalum-tungsten, tantalum-niobium, tantalum-hafnium, tantalum-zirconium, tantalum-rhenium, tantalum-nickel, tantalum-vanadium, tantalum-tungsten-molybdenum, tantalum-tungsten-niobium, tantalum-tungsten-hafnium, and tantalum-tungsten-rhenium alloys was studied in various corrosive mediums, including concentrated H_2SO_4 at 200 and 250 °C (390 and 480 °F) (Ref 35). Figure 10 gives corrosion data in 95% H_2SO_4 for several binary tantalum alloys. Tantalum became embrittled in concentrated H_2SO_4 at 250 °C (480 °F). Additions of tungsten reduced the corrosion rate and hydrogen absorp-

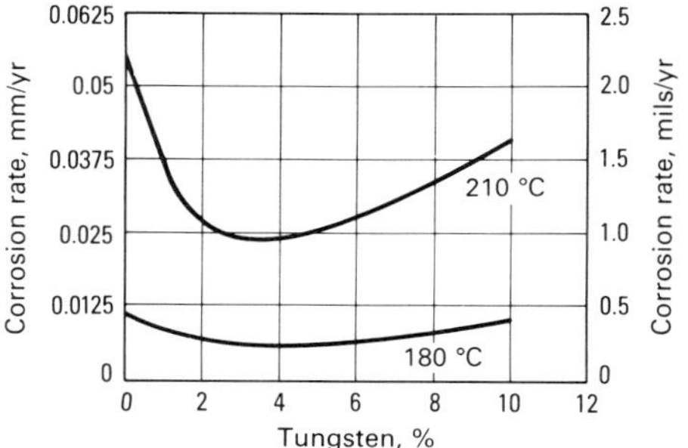

Fig. 9 Corrosion rate versus tungsten content for tantalum-tungsten alloys exposed to concentrated H_2SO_4 at 180 °C (360 °F) and 210 °C (405 °F)

Table 7 Corrosion rates for tantalum materials exposed to concentrated H_2SO_4 at 175 to 200 °C (345 to 392°F)

Material	Metallurgical condition	Temperature °C	Temperature °F	Exposure, days	Corrosion rate mm/yr	Corrosion rate mils/yr
Ta, electron beam melted	Recrystallized	175	345	60	0.005	0.189
Ta, P/M	Recrystallized	175	345	60	0.0055	0.217
Tantaloy 63	Recrystallized	175	345	60	0.0058	0.229
Tantaloy 63	As-rolled	181	360	7	0.0026	0.104
Tantaloy 63	Stress-relieved	181	360	7	0.0022	0.087
Tantaloy 63	Recrystallized	181	360	7	0.0022	0.087
Ta, electron beam melted	As-rolled	199	390	3	0.018	0.72
Ta, electron beam melted	Recrystallized	199	390	3	0.024	0.96
Tantaloy 63	As-rolled	199	390	3	0.0048	0.19
Tantaloy 63	Stress-relieved	199	390	3	0.0043	0.17
Tantaloy 63	Recrystallized	199	390	3	0.0045	0.18
Ta, electron beam melted	Recrystallized	200	392	32	0.057	2.24
Ta, (P/M)	Recrystallized	200	392	32	0.058	2.27
Tantaloy 63	Recrystallized	200	392	32	0.029	1.15
Tantaloy 63	Recrystallized	200	392	13	0.03	1.24
Ta-5W	Recrystallized	200	392	13	0.034	1.34
Ta-10W	Recrystallized	200	392	13	0.05	1.98

tion, but additions of molybdenum and rhenium were more effective in reducing both effects. Additions of niobium and vanadium had only a slight influence on the corrosion rate of tantalum; lower-valence elements, such as hafnium, increased the corrosion rate (Ref 35).

Hydrochloric Acid. Samples of recrystallized sheet of electron beam melted tantalum and Tantaloy 63 were exposed in the same test in a tantalum autoclave to concentrated (37 to 38%) HCl at 100 °C (212 °F) for 24 h. The corrosion rates of these materials were:

Alloy	Corrosion rate mm/yr	Corrosion rate mils/yr
Tantalum	0.04	1.6
Tantaloy 63	0.023	0.9

Corrosion rates for Tantaloy 63 in three metallurgical conditions exposed to concentrated HCl at 100 °C (212 °F) for 24 h in another autoclave test were about tenfold lower:

Condition	Corrosion rate μm/yr	Corrosion rate mils/yr
As-rolled	0.8	0.032
Stress relieved	1.0	0.039
Fully recrystallized	1.0	0.039

Researchers in other tests also found that tantalum and substitutional tantalum-base alloys became hydrogen embrittled in concentrated HCl at 150 °C (300 °F) (Ref 35).

Nitric Acid. Samples of unalloyed, electron beam melted tantalum and Tantaloy 63 were exposed to concentrated (70%) HNO_3 for 3 days at about 200 °C (390 °F). Following this test, neither material showed any measurable weight loss. Additional tests on Tantaloy 63 tested in 70% HNO_3 at about 200 °C (390 °F) for 72 h in an autoclave reactor gave the following corrosion rates for material in three metallurgical conditions:

Condition	Corrosion rate μm/yr	Corrosion rate mils/yr
As-rolled	0.1	0.0038
Stress-relieved	0.04	0.0016
Fully recrystallized	0.038	0.0015

Other Aqueous Media. The Ta-10W binary solid-solution alloy, also known as Fansteel 60 Metal, has been used in some applications, such as pump and valve parts, for which a material appreciably harder and stronger than the pure metal is desired. For example, Ta-10W alloy is used as an insert in the plug of a tantalum-lined split-body valve to give a hard plug to soft seat combination when used with a tantalum seat.

A search of the literature indicated that very little corrosion data were available on the Ta-10W alloy (Ref 41). This alloy is of interest as a repair metal for glass-lined steel equipment because it is of much higher strength than unalloyed tantalum. Consequently, corrosion tests were conducted on the Ta-10W alloy in various environments (Ref 41).

Figure 11 shows the corrosion rate as a function of H_2SO_4 concentration for tests at 205 and 230 °C (400 and 450 °F). These data indicate that either unalloyed tantalum or Ta-10W alloy can be used at 230 °C (450 °F) to handle H_2SO_4 in concentrations below 90%. Although the corrosion rates at 205 °C (450 °F) are similar for the two materials, on the basis of this graph, the corrosion weight loss of the Ta-10W alloy is about twice that of unalloyed tantalum at 230 °C (450 °F) in H_2SO_4 over the concentration range of 70 to 90%.

Little or no weight loss occurred below 175 °C (350 °F) in HCl, even at 30% concentration. At 190 °C (375 °F), a small amount of attack was detected, with that for the Ta-10W alloy being more severe.

In the HNO_3 corrosion tests, even up to 175 °C (350 °F) in concentrations to 60%, it was concluded that neither the tantalum nor the Ta-10W alloy showed any perceptible loss in weight (Ref 48). No comments were made regarding the small weight gains observed in most cases.

Tests were also conducted in 5% NaOH solution at 100 °C (212 °F). Considerable weight losses occurred on both the unalloyed tantalum and the alloy under caustic conditions. The difference, if any, in the corrosion rates of the two materials appears small.

The NaOH corrosion test was conducted because tantalum is known to be susceptible to caustic embrittlement; therefore, it was desired to determine whether the Ta-10W alloy suffered embrittlement also. Unalloyed tantalum showed approximately a 25% increase in yield strength and a 10% increase was attributed to a pickup of interstitial elements (oxygen, nitrogen, and hydrogen), although chemical analyses of the materials before and after exposure were not conducted. With the Ta-10W alloy, the exposure to 5% NaOH at 100 °C (212 °F) produced embrittlement as evidenced by the premature fracture in the tensile test. Reportedly, such embrittlement was not evident on the sample of the Ta-10W alloy to which a platinum spot had been welded before the test. The corrosion resistance of tantalum-tungsten alloys was also studied in 50% KOH at 30 and 80 °C (85 and 175 °F), 20% HF at 20 °C (70 °F), and $KOH{:}3K_3Fe(CN)_6$ mixture (concentration not given) (Ref 9).

In the hydroxide solutions, a maximum corrosion rate was obtained at about 60 at.% tantalum. Although the alloy system reportedly represents a continuous series of solid solutions, a maximum electrical resistance also was found at the same composition. In 20% HF solution, the tantalum-tungsten alloy system essentially exhibits the relatively low corrosion rates associated with tungsten, except when the tantalum concentration exceeds 80 at.%, at which concentration the corrosion increases markedly. Alloys containing more than 18% W show no corrosion in 20% HF, thus offering an advantage over pure tantalum. Other tests have been conducted on tantalum and tantalum-tungsten alloys having tungsten contents ranging from 8.7 to 20.4% in 20% HF. This work showed that tantalum and tantalum-tungsten alloys containing less than about 20% W were more susceptible to attack by 20% HF than was previously reported. Tantalum-tungsten alloys showed little improvement over tantalum when tested in the $KOH{:}3Fe(CN)_6$ mixture.

Combined Reagents. In a proposed dehydrator application, the exposure was to be to an environment containing 61.5% HNO_3 plus 7% magnesium nitrate ($Mg(NO_3)_2$) at 115 °C (238 °F). Although no significant corrosion on tantalum would be expected upon exposure to the separate reagents, tests were conducted on the mixture to determine corrosion rates. Base metal and weldment specimens of tantalum and Tantaloy 63 were used. Within the precision of the weight measurements, no specimen of either tantalum or

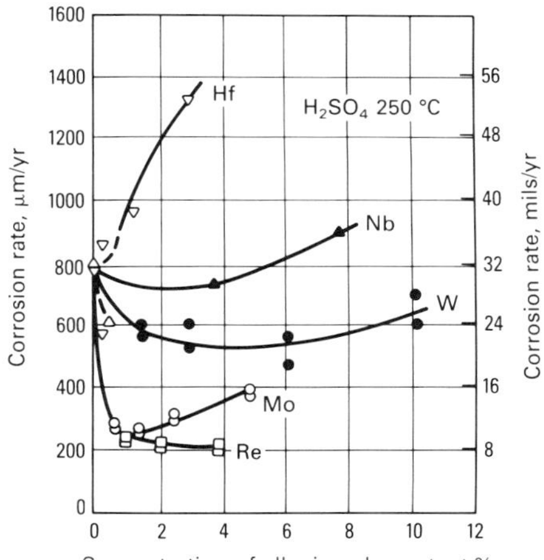

Fig. 10 Influence of alloying elements on the corrosion rate of binary tantalum alloys exposed 3 days to 95% H_2SO_4 at 250 °C (480 °F). Source: Ref 35

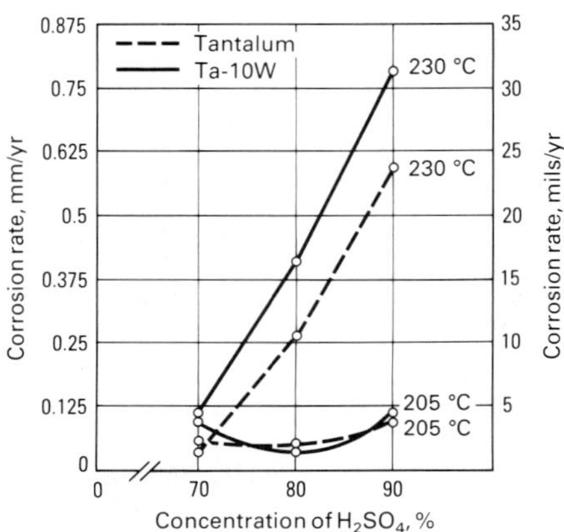

Fig. 11 Corrosion rate versus concentration for tantalum and Ta-10W alloy exposed to H_2SO_4 at various temperatures. Source: Ref 41

Tantaloy 63 showed any weight loss in these corrosion tests.

Phosphoric Acid Plus Residual Hydrofluoric Acid. The use of tantalum in H_3PO_4, other than the high-purity Food Grade acid, has always been questioned because fluoride compounds are usually present. The observation has been made in several publications and bulletins that the presence of a small amount of HF (<5 ppm) in commercial H_3PO_4 causes severe corrosion of tantalum (Ref 42). Therefore, tantalum has not been considered resistant to H_3PO_4 containing more than a trace of F^- ion.

In one test, a sample of Tantaloy 63 was exposed to a wet-process H_3PO_4 at an average temperature of 155 °C (310 °F) for a total of 283.5 h. Following the corrosion exposure, the sample (0.64 mm, or 0.025 in., thick) was fully ductile and withstood bending flat on itself with no evidence of cracking. There was no indication of corrosion on the surface of the specimen, and metallographic examination of a cross section of the specimen showed no evidence of corrosion attack. The weight loss data for the exposure equated to a corrosion penetration rate of 0.004 mm/yr (0.15 mils/yr). These data suggested that the tantalum alloy showed good corrosion resistance to the hot-wet process H_3PO_4.

Tantalum-Molybdenum Alloys. Corrosion resistance was studied on tantalum-molybdenum alloys that form a continuous series of solid solutions (Ref 5). The entire alloy system is extremely corrosion resistant, and the corrosion resistance of tantalum is generally retained when the alloy contains more than about 50% Ta. Table 8 gives weight loss data for tantalum-molybdenum alloys in concentrated H_2SO_4 and concentrated HCl at 150 °C (300 °F).

Table 8 Corrosion rates of tantalum-molybdenum alloys in concentrated H_2SO_4 at 150 °C (300 °F)

Solutions were saturated with oxygen.

Tantalum in molybdenum, at. %	Average corrosion rate, mg/cm²/d: Concentrated H_2SO_4 (98%)	Concentrated HCl (37%)
0	0.008	0.018
10.1	0.009	0.017
20.1	0.008	0.018
30.0	0.010	0.009
40.0	0.009	0.010
50.0	0.000	0.010
61.2	0.000	0.000
71.5	0.000	. . .
82.8	0.000	0.000
91.4	0.000	0.000
100.0	0.000	0.000

Tantalum-Niobium Alloys. Corrosion tests were conducted in hot and cold concentrated HCl and H_2SO_4 on alloys having various proportions of tantalum and niobium (Ref 43). The corrosion rates increased roughly in proportion to the niobium content in the alloy. Even though the 95Ta-5Nb alloy showed excellent resistance in all exposures, the attack was three times that obtained on pure tantalum. Additional corrosion data on binary tantalum-niobium alloys are given in Ref 44 and 45.

Other data were reported on corrosion tests of binary tantalum-niobium alloys and ternary alloys based on the Ta-Nb system (Ref 46). Tests on the materials were carried out in 75% H_2SO_4 at 185 °C (360 °F), in 70% H_2SO_4 at 165 °C (330 °F), in 75% H_2SO_4 at room temperature, and in 20% HCl at room temperature. The tantalum-niobium alloys containing approximately 60% or more tantalum appeared promising for boiling 70% H_2SO_4. Ternary alloys containing elements of groups IVA, VA, and VIA with tantalum and niobium did not offer any advantages in fabricability, and addition of zirconium, hafnium, chromium, and vanadium lowered the corrosion resistance.

One manufacturer developed an alloy designated WC-640 for corrosion-resistant applications (Ref 47). Analysis of a sheet sample showed that the chemical composition actually corresponded to a ternary tantalum-niobium-tungsten alloy, nominally Ta-40Nb-0.5W. Tensile properties of WC-640 were found to be similar to those of Tantaloy 63. The alloy exhibited excellent resistance to all test environments except concentrated H_2SO_4 at 200 °C (390 °F) and 40% NaOH at 100 °C (212 °F).

Tantalum-Titanium Alloys. Considerable data have been accumulated on the corrosion resistance of tantalum-titanium alloys (Ref 5). Dilution of tantalum with titanium shows considerable promise for the possibility of providing a lower-cost alloy with corrosion resistance almost comparable to that of tantalum in some selected environments. In addition to dilution with a lower-cost material, the resulting marked reduction in density is particularly advantageous because corrosion applications generally require materials on a volume rather than a weight basis. Corrosion tests in 10 to 70% HNO_3 at the boiling point and at 190 °C (375 °F) (in sealed glass tubes) were conducted on tantalum-titanium alloys ranging from pure tantalum to Ta-90Ti. All of these materials showed excellent behavior, with corrosion rates less than 0.025 mm/yr (1 mil/yr) and no indication of embrittlement.

Hydrogen embrittlement may occur when this alloy system is exposed to reducing corrosive conditions in tests conducted in sealed capsules. The tendency for hydrogen damage is markedly decreased as the tantalum concentration is increased (Ref 9).

Ternary Alloys. The corrosion behavior of alloys in the Ti-Ta-Nb system in 5% HCl at 100 °C (212 °F) was investigated (Ref 48). The corrosion rate dropped by a factor of 1.5 to 2 times less than that of titanium with a total of 15 to 20% Nb or Ta in the alloy. At 20 to 30% additions of these elements, the corrosion rate dropped by a factor of 10 to 70 times less than that of titanium. Tantalum increased the corrosion resistance of the alloys more effectively than niobium did.

The role of the structural factor in increasing the corrosion resistance of the titanium-tantalum-chromium and titanium-chromium alloys in a 5% HCl solution at 100 °C (212 °F) was demonstrated (Ref 49). The corrosion rate of quenched alloys was 2 to 10 times or more lower than that of annealed alloys. As the tantalum content increases, the corrosion rate decreases for both quenched and annealed alloys. The ternary alloys with a tantalum to chromium ratio of 3:1 and binary titanium-tantalum alloys with 20% more tantalum were found to have good corrosion resistance.

Other Tantalum Alloys. It has been observed that the presence of a small amount of iron or nickel, for example, in a tantalum weld makes that site subject to about the same acid attack as would be experienced by iron or nickel alone (Ref 9). Galvanic action, as well as simple chemical attack, is undoubtedly involved.

REFERENCES

1. J. Chelius, Use of Refractory Metals in Corrosive Environment Service, *Mater. Eng. Quart.*, Aug 1957, p 57-59
2. R.B. Flanders, Try Tantalum for Corrosion Resistance, *Chem. Eng.*, 17 Dec 1979
3. B.A. Johnson, "Corrosion of Metals in Deionized Water at 38 C," Technical Memorandum TM X-1791, National Aeronautics and Space Administration, 10 March 1969
4. D.F. Taylor, "Tantalum: Its Resistance to Corrosion," Paper presented at the Chicago Section, the Electrochemicals Society, 4 May 1956
5. M. Stern and C.R. Bishop, Corrosion and Electrochemical Behavior, in *Columbium and Tantalum*, F.T. Sisco and E. Epremian, Ed., John Wiley & Sons, 1963
6. "Tantalum, Corrosion Data, Comparative Charts and Coating Characteristics," General Technologies Corporation
7. A. Alon, M. Schor, and A. Vromen, Corrosion Resistance of Ta to Mixture of Phosphoric Acid and Potassium Chloride at 120 C, *Corrosion*, Vol 22 (No. 1), Jan 1966
8. Corrosionomics, Vol 5 (No. 5) Fansteel Inc., 1960, p 5
9. C.A. Hampel, Ed., *Tantalum in Rare Metals Handbook*, 2nd ed., Reinhold, 1967
10. W.T. Edwards, Aqueous Corrosion in the Atomic Energy Industry, in *First International Congress on Metallic Corrosion*, 1962
11. J. Vehlow and H. Geisert, "Tantalum Corrosion Under Wet Incineration Conditions—Influence of the Dosing Components and Study of Welded Specimens, "Paper presented at International Corrosion Conference, 6 Sept 1981
12. H.O. Peeple and R.L. Adams, Jr., A Corrosion Study in a Chlorine Dioxide Pulp Bleaching Plant, reprinted from *TAPPI*, Vol 38 (No. 1), Jan 1955, copyright 1955 by Technical Association of the Pulp and Paper Industry, reprinted with permission
13. E. Rabald, *Werkst. Korros.*, Vol 12, 1961, p 695-698
14. D.F. Taylor, Tantalum and Tantalum Com-

pounds, in *Encyclopedia of Chemical Technology*, Vol 19, 2nd ed., John Wiley & Sons, 1969, p 630-652

15. F.E. Bacon and P.M. Moanfeldt, Reaction with the Common Gases, in *Columbium and Tantalum*, F.T. Sisco and E. Epremian, Ed., John Wiley & Sons, 1963
16. E. Gebhardt and H.D. Seghezzi, *Z. Metallkd.*, Vol 50, 1959, p 248
17. R. Bakish, *J. Electrochem. Soc.*, Vol 105, 1958, p 71
18. F.F. Schmidt, W.D. Klopp, W.M. Albrecht, F.C. Holden, H.R. Ogden, and R.I. Jafee, Technical Report WADD-TR-59-13, United States Air Force, 1959
19. W.M. Albrecht and W.D. Goode, Jr., Publication BMI-1360, Battelle Memorial Institute, 6 July 1959
20. J.B. Hallowell, D.J. Maykuth, and H.R. Ogden, Silicide Coatings for Tantalum and Tantalum-Base Alloys, in *Refractory Metals and Alloys III: Applied Aspects*, Vol 30, Part 2, American Institute of Mining, Metallurgical, and Petroleum Engineers, Dec, 1963
21. K.J. Richards and M.E. Wadsworth, Oxidation Kinetics of Tantalum in Carbon Dioxide, *Trans. AIME*, Vol 230, Feb 1964, p 33-38
22. M. Farber, A.J. Darnell, and D.M. Ehrenberg, *J. Electrochem. Soc.*, Vol 102, 1955, p 446-453
23. J.M. Saleh and M.H. Matloob, Oxidation of Titanium, Tantalum, and Niobium Films by Oxygen and Nitrous Oxide, *J. Phys. Chem.*, Vol 76 (No. 24), 1974, p 2486-2489
24. E.E. Hoffman and R.W. Harrison, The Compatibility of Refractory Metals with Liquid Metals, in *Refractory Metals and Alloys: Metallurgy and Technology*, I. Machlin, R.T. Begley, and E.D. Weisert, Ed., Proceedings of Metallurgy and Technology of Refractory Metals Symposium, Washington, DC, 25 and 26 April 1968, Plenum Press, 1968
25. E.C. Miller, chapter 4, in *Liquid Metals Handbook*, Atomic Energy Commission, Department of the Navy, 1952, p 144-183
26. W.D. Manly, *Corrosion*, Vol 12, 1956, p 336t-342T
27. P. Cybulskis, "Review of Metals Technology, Liquid Metals," Metals and Ceramics Information Center, 21 Dec 1973
28. R.H. Cooper and E.E. Hoffman, *Refractory Alloy Technology for Space Nuclear Power Applications*, Oak Ridge National Laboratory, 10-11 Aug 1983
29. J.Y.N. Wang, Method of Inhibiting the Corrosion of Tantalum by Liquid Lithium at High Temperatures, U.S. Patent 3,494,805, 1970
30. G.W. Watson, "Evaluation of Tantalum-Alloy-Clad Uranium Mononitride Fuel Specimens from 7500-Hour 1040 C Pumped-Lithium-Loop Test," Technical Note TN D-7619, National Aeronautics and Space Administration, April 1974
31. J.R. Weeks, "Liquidus Curves and Corrosion of Fe, Cr, Ni, Co, V, Cb, Ta, Ti, Zr in 500-750 C Mercury," revised and expanded version of a paper presented at the 20th NACE Conference, National Association of Corrosion Engineers, 9-13 March 1974
32. L.B. Engel, Jr. and R.W. Harrison, "Corrosion Resistance of Tantalum, T-111, and Cb-1Zr to Mercury at 1200 F," Contractor Report CR-1811, National Aeronautics and Space Administration, May 1971
33. R.L. Klueh, "Effect of Oxygen on the Corrosion of Niobium and Tantalum by Liquid Lithium," Report ORNL TM-4069, Oak Ridge National Laboratory, March 1973
34. W.R. Kanne, Jr., Corrosion of Metals by Liquid Bismuth-Tellurium Solutions, *Corrosion*, Vol 29, Feb 1973, p 75-82
35. L.A. Gypen, M. Brabers, and A. Deruyttere, Corrosion Resistance of Tantalum Base Alloys, Elimination of Hydrogen Embrittlement in Tantalum by Substitutional Alloying, *Werkst. Korros.*, Vol 35, 1984, p 37-46
36. "Tantalum and Galvanic Action," Corrosionomics, Fansteel Inc., March 1957
37. T. Fukuzuka, K. Shimogori, H. Satok, and F. Kamikulo, Inhibition of Corrosion and Hydrogen Embrittlement of Tantalum in Concentrated Sulfuric Acid Solutions at Elevated Temperatures by Additions of N-O Compounds, *Boshoku Gyutsu*, Vol 30 (No. 6), June 1981, p 327-336
38. N. Kagarva, K. Yamanmoto, R. Sasano, T. Kusakabe, and Y. Moriya, Method for Preventing Corrosion and Hydrogen Embrittlement of Tantalum-Made Equipment Handling Hot Concentrated Sulphuric Acid Therein, U.S. Patent 4,356,148, 1982
39. R. Wehrmann, "Oxidation-Reduction Potential of Tantalum," Corrosionomics, Fansteel Inc., Sept 1956
40. E.W. Dreyman, Precious Metal Anodes: State of the Art, *Mater. Protect. Perform.*, Vol 11 (No. 9), Sept 1972, p 17-20
41. B.G. Staples and W.S. Galloway, Jr., Recent Results of Testing Corrosion Performance of 90-10 Tantalum-Tungsten Alloy, Pfaudler Reprint 581, reprinted from *Mater. Protect.*, Vol 7 (No. 7), July 1968, p 34-39
42. Corrosionomics, Vol 3 (No. 1), Fansteel Inc., 1958
43. G.L. Miller, *Tantalum and Columbium*, Academic Press, 1959
44. A.V. Mosolov, I.D. Nefedova, and I.Y. Klinov, Corrosion and Electrochemical Behavior of Niobium-Tantalum Alloys in HCl at Elevated Temperatures and Pressures, *Z. Metallov.*, Vol 4 (No. 3), May-June 1968, p 248-251 (in English)
45. A.P. Gulyaev and I.Y. Georgieva, Corrosion Resistance of Binary Niobium Alloys, *Z. Metallov.*, Vol 1 (No. 6) Nov-Dec 1965, p 652-657 (in English)
46. D. Lupton and F. Aldinger, Possible Substitutes for Tantalum in Chemical Plant Handling Mineral Acids, in *Trends in Refractory Metals, Hard Metals, and Special Materials and Their Technology*, Proceedings of the 10th Plansee Seminar, Reutte, Austria, 1981, p 101-130
47. "Columbium (niobium)," TWCA-8204 Cb Teledyne Wah Chang Albany
48. N.D. Tomashov, T.V. Chukalovskaya *et al.*, "Investigation of the Corrosion Resistance of Alloys of the Ti-Ta-Nb Systems," translation of the Foreign Technology Division, Air Force Systems Command, FTD-HT-149-74
49. N.D. Tomashov, G.P. Chernova *et al.*, "Investigation of the Structure and Corrosion Behavior of Alloys in the Ti-Ta-Cr Systems," Translation of the Foreign Technology Division, Air Force Systems Command, FTD-HT-150-74
50. "Corrosion Prevention Products," Fansteel Bulletin, Fansteel Inc.

Corrosion of Magnesium and Magnesium Alloys

Allan Froats, Chromasco/Timminco, Ltd.; Terje Kr. Aune, Norsk Hydro; David Hawke, Amax Magnesium; William Unsworth, Magnesium Elektron, Ltd.; and James Hillis, Dow Chemical Company

THE CORROSION RESISTANCE of a magnesium or a magnesium alloy part depends on many of the same factors that are critical to other metals. However, because of the electrochemical activity of magnesium (Table 1), the relative importance of some factors is greatly amplified. This article will discuss the effects of heavy-metal impurities, the type of environment (rural atmosphere, marine atmosphere, elevated temperatures, and so on), the surface condition of the part (such as as-cast, treated, and painted), and the assembly practice. In some environments, a magnesium part can be severely damaged unless galvanic couples are avoided by proper design or surface protection. Therefore, this aspect will be discussed in detail in the section "Galvanic Corrosion" in this article.

Unalloyed magnesium is not extensively used for structural purposes. Consequently, the corrosion resistance of magnesium alloys is of primary concern. Two major magnesium alloy systems are available to the designer. The first includes alloys containing 2 to 10% Al, combined with minor additions of zinc and manganese. These alloys are widely available at moderate cost, and their mechanical properties are good to 95 to 120 °C (200 to 250 °F). Beyond this, the properties deteriorate rapidly with increasing temperature. The second group consists of magnesium alloyed with various elements (rare earths, zinc, thorium, silver, and so on) except aluminum, all containing a small but effective zirconium content that imparts a fine grain structure (and thus improved mechanical properties). These alloys generally possess much better elevated-temperature properties, but their more costly elemental additions, combined with the specialized manufacturing technology required, result in significantly higher costs. Table 2 lists the compositions commonly available in both systems.

If a magnesium part is to perform satisfactorily in a particular application, it must not only be designed to meet the mechanical requirements but environmental factors, finishing, and assembly methods must also be properly assessed. This article will describe in detail the causes of past corrosion failures and the measures available to prevent such failures in future applications.

Metallurgical Factors

Chemical Composition. Figure 1 shows the effects of 14 elements on the saltwater corrosion performance of magnesium in binary alloys with increasing levels of the individual elements. Six of the elements included in Fig. 1 (aluminum, manganese, sodium, silicon, tin, and lead) plus thorium, zirconium, beryllium, cerium, praseodymium, and yttrium are known to have little if any deleterious effect on the basic saltwater corrosion performance of pure magnesium when present at levels exceeding their solid solubility or up to a maximum of 5% (Ref 2). Four elements in Fig. 1 (cadmium, zinc, calcium, and silver) have mild-to-moderate accelerating effects on corrosion rates, whereas four others (iron, nickel, copper, and cobalt) have extremely deleterious effects because of their low solid-solubility limits and their ability to serve as active cathodic sites for the reduction of water at the sacrifice of elemental magnesium. Although cobalt is seldom encountered at detrimental levels and cannot be introduced even through the long immersion of cobalt steels in magnesium melts, iron, nickel, and copper are common contaminants that can be readily introduced through poor molten-metal handling practices. These elements must be held to levels under their individual solubility limits (or their activity moderated through the use of alloying elements such as manganese or zinc) to obtain good corrosion resistance.

Figure 2 illustrates the effect of increasing iron, nickel, and copper contamination on the standard ASTM salt spray performance of die-cast AZ91 test specimens as compared to the range of performance observed for cold-rolled steel and die-cast aluminum alloy 380 samples. Such results have led to the definition of the critical contaminant limits for two magnesium-aluminum alloys in both low- and high-pressure cast form and the introduction of improved high-purity versions of the alloys. Table 3 lists some of the critical contaminant limits defined to date. The iron tolerance for the magnesium-aluminum alloys depends on the manganese present, a fact suggested many years ago but only recently proved. For AZ91 with a manganese content of 0.15%, this means that the iron tolerance would be 0.0048% (0.032 × 0.15%) (Ref 7).

It should also be noted that the nickel tolerance depends strongly on the cast form, which influences grain size, with the low-pressure cast alloys showing just a 10-ppm tolerance for nickel in the as-cast (F) temper. Therefore, alloys intended for low-pressure cast applications should be of the lowest possible nickel level (Ref 4). The low tolerance limits for the contaminants in AM60 alloy when compared to AZ91 alloy can be related to the absence of zinc. Zinc is thought to improve the tolerance of magnesium-aluminum alloys for all three contaminants, but it is limited to 1 to 3% because of its detrimental effects on microshrinkage porosity and its accelerating effect on corrosion.

For the rare-earth, thorium, and zinc alloys containing zirconium, the normal saltwater corrosion resistance is only moderately reduced when compared to high-purity magnesium-aluminum alloys—0.5 to 0.76 mm/yr (20 to 30 mils/yr) as opposed to less than 0.25 mm/yr (10 mils/yr) in 5% salt spray—but contaminants must again be controlled. The zirconium alloying element is effective in this case because it serves as a strong grain refiner for magnesium alloys and it precipitates the iron contaminant from the alloys before casting. However, if alloys containing more than 0.5 to 0.7% Ag or more than 2.7 to 3% Zn are used, a sacrifice in corrosion resistance should be

Table 1 Standard reduction potentials

Electrode	Reaction	Potential, V
Li,Li^+	$Li^+ + e^- \rightarrow Li$	−3.02
K,K^+	$K^+ + e^- \rightarrow K$	−2.92
Na,Na^+	$Na^+ + e^- \rightarrow Na$	−2.71
Mg,Mg^{2+}	$Mg^{2+} + e^- \rightarrow Mg$	−2.37
Al,Al^{3+}	$Al^{3+} + e^- \rightarrow Al$	−1.71
Zn,Zn^{2+}	$Zn^{2+} + e^- \rightarrow Zn$	−0.76
Fe,Fe^{2+}	$Fe^{2+} + e^- \rightarrow Fe$	−0.44
Cd,Cd^{2+}	$Cd^{2+} + e^- \rightarrow Cd$	−0.40
Ni,Ni^{2+}	$Ni^{2+} + e^- \rightarrow Ni$	−0.24
Sn,Sn^{2+}	$Sn^{2+} + e^- \rightarrow Sn$	−0.14
Cu,Cu^{2+}	$Cu^{2+} + e^- \rightarrow Cu$	0.34
Ag,Ag^+	$Ag^+ + e^- \rightarrow Ag$	0.80

Table 2 Typical magnesium alloy systems and nominal compositions

Alloy	Element, %(a) Al	Zn	Mn	Ag	Zr	Th	Re	Product form(b)
AM60	6	...	0.2	...	...	...	...	C
AZ31	3	1	0.2	...	...	...	...	W
AZ61	6	1	0.2	...	...	...	...	W
AZ63	6	3	0.2	...	...	...	...	C
AZ80	8	0.5	0.2	...	...	...	...	C, W
AZ91	9	1	0.2	...	...	...	...	C
EZ33	...	2.5	...	...	0.5	...	2.5	C
ZM21	...	2	1	...	...	...	...	W
HK31	...	0.1	...	...	0.5	3	...	C, W
HZ32	...	2	...	...	0.5	3	...	C
QE22	...	...	...	2.5	0.5	...	2	C
QH21	...	...	...	2.5	0.5	1	1	C
ZE41	...	4.5	...	...	0.5	...	1.5	C
ZE63	...	5.5	...	...	0.5	...	2.5	C
ZK40	...	4.0	...	...	0.5	...	...	C, W
ZK60	...	6.0	...	...	0.5	...	...	C, W

(a) For details, see alloying specifications. (b) C, castings; W, wrought products

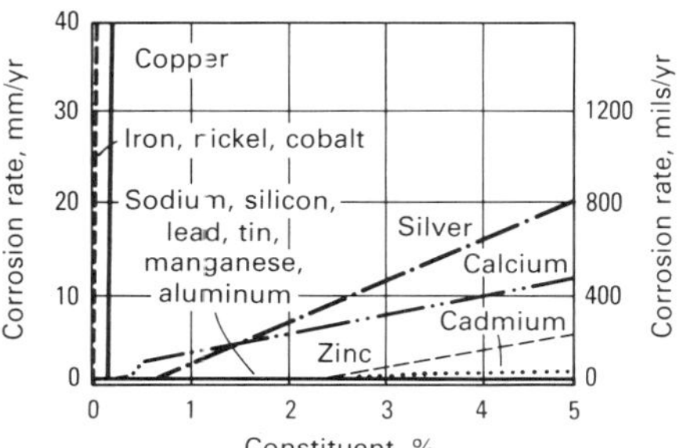

Fig. 1 Effect of alloying and contaminant metals on the corrosion rate of magnesium as determined by alternate immersion in 3% NaCl solution. Source: Ref 1

expected (Fig. 1). Nevertheless, when properly finished, these alloys provide excellent service in harsh environments.

Heat-Treating, Grain Size, and Cold-Work Effects. Using controlled-purity AZ91 alloy cast in both high- and low-pressure forms, the contaminant tolerance limits have been defined as summarized in Table 4 for the as-cast (F), the solution-treated (T4, held 16 h at 410 °C, or 775 °F, and quenched), and the solution-treated and aged (T6, held 16 h at 410 °C, or 775 °F, quenched, and aged 4 h at 215 °C, or 420 °F) (Ref 4, 5).

Table 5 compares the average 5% salt spray corrosion performance of sand-cast samples produced in a standard AZ91C and a high-purity AZ91E composition. The alloys were cast with and without standard grain-refining practices used to evaluate physical and compositional effects. The cast samples were then tested in the F, T4, T6, and T5 (aged 4 h at 215 °C, or 420 °F) tempers. In the case of the high-iron-containing AZ91C, none of the variations tested significantly affected the poor corrosion performance resulting from an iron level 2 to 3 times the alloy tolerance. In the case of the high-purity alloy, however, the T5 and T6 tempers consistently gave salt spray corrosion rates under 0.25 mm/yr (10 mils/yr), whereas the as-cast and solution-treated samples exhibited an inverse response to grain size and/or the grain-refining agents (Ref 4, 5). Welds on aluminum-zinc alloys should be aged or should be solution treated and aged to obtain good corrosion resistance in harsh environments and to reduce the risk of failure due to stress-corrosion cracking (SCC).

Cold working of magnesium alloys, such as stretching or bending, has no appreciable effect

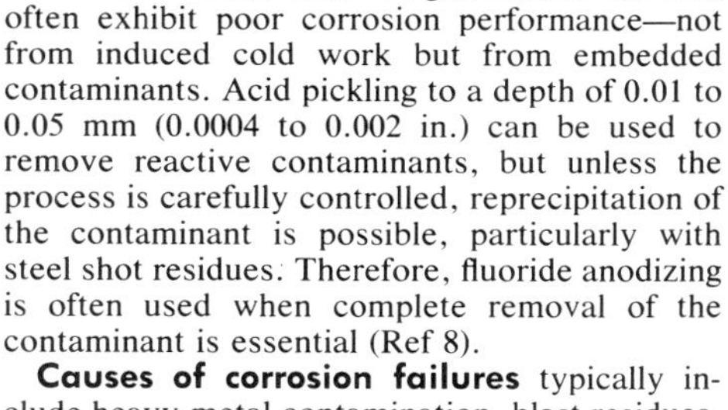

on corrosion rate. Shot- or grit-blasted surfaces often exhibit poor corrosion performance—not from induced cold work but from embedded contaminants. Acid pickling to a depth of 0.01 to 0.05 mm (0.0004 to 0.002 in.) can be used to remove reactive contaminants, but unless the process is carefully controlled, reprecipitation of the contaminant is possible, particularly with steel shot residues. Therefore, fluoride anodizing is often used when complete removal of the contaminant is essential (Ref 8).

Causes of corrosion failures typically include heavy-metal contamination, blast residues, flux inclusions, and galvanic attack. The characteristics of each will be discussed.

Heavy-metal contamination often results in general pitting attack that is unassociated with fasteners or dissimilar-metal attachments. The rate of attack on unpainted surfaces will be essentially unaltered by surface condition, that is, freshly sanded or machined, acid pickled, or chrome treated. The iron, nickel, or copper content will analyze in excess of the tolerance limit for one or more of the elements. Figure 3 illustrates the effect of heavy-metal contamination on the ASTM salt spray corrosion performance of low-pressure cast AZ91.

Blast residues can cause general pitting attack in saline environments. Attack is normally limited to unmachined surfaces of sand castings. Sanded or acid-etched (2% sulfuric acid (H_2SO_4) for 15 to 30 s) samples will show vastly improved performance in saltwater immersion or salt spray tests because of removal of the contaminant. Scanning electron microscopy and energy-dispersive x-ray analysis samples cleaned in chromic acid (H_2CrO_4) can be used to confirm and identify the presence of the contaminant, which is usually iron (from steel shot blasting) or silica (from sand blasting). Preventive measures are discussed in the section "Protective Coating Systems" in this article.

Flux inclusions result in localized attack that is clustered or distributed randomly on machined surfaces of castings. Freshly machined surfaces exposed to 70 to 90% relative humidity will develop active corrosion sites overnight. Scanning electron microscopy/energy-dispersive x-ray analysis of a freshly machined surface (free of fingerprints or other sources of contamination) will reveal pockets of magnesium and potassium chloride, as well as possible traces of calcium, barium, and sulfur. In zirconium-bearing alloys, elemental zirconium and zirconium-iron compounds may also be associated with the deposits. Chromic acid pickling followed by chemical treatment and surface sealing can alleviate the problem of inclusions in finished castings. With the use of sulfur hexafluoride (SF_6) rapidly replacing fluxes for the protection of melts during casting, this problem should be eliminated in the future.

Galvanic attack is usually observed as heavy localized attack on the magnesium, normally within 3.2 to 4.8 mm (1/8 to 3/16 in.) of fasteners or an interface with other parts of dissimilar metal. Proper design and assembly methods can minimize galvanic attack, as detailed in subsequent sections in this article.

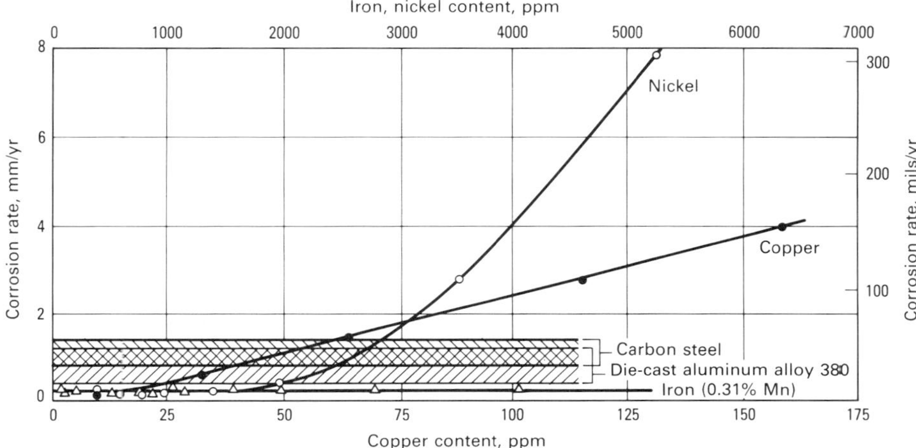

Fig. 2 Effect of nickel and copper contamination on the salt spray corrosion performance of die-cast AZ91 alloy. Source: Ref 3

Environmental Factors

The information in this section is based on the behavior of unprotected metal. Improved resistance is obtained by using proper protection systems. The alloy composition as summarized

Table 3 Known contaminant tolerance limits in high- and low-pressure cast forms

Alloy/form	Grain size, µm	Critical contaminant limit, % Fe	Ni	Cu	Ref
Unalloyed magnesium	...	0.015	0.0005	0.1	1
AZ91/high pressure	5-10	0.032 Mn(a)	0.0050	0.040	4
AZ91/low pressure	100-200	0.032 Mn(a)	0.0010	0.040	4
AM60/high pressure	5-10	0.021 Mn(b)	0.0030	0.010	3
AM60/low pressure	100-200	0.021 Mn(b)	0.0010	0.010	5
AZ63/low pressure	...	0.003(c)	0.0040	>0.45	1
K1A/low pressure	...	>0.003	0.003	...	6

(a) Iron tolerance equals manganese content of alloy times 0.032. (b) Iron tolerance equals manganese content of alloy times 0.021. (c) Magnesium content of AZ63 reported as 0.2%

Table 4 Contaminant tolerance limits versus temper and cast form for AZ91 alloy

High-pressure die cast, 5-10 µm average grain size; low-pressure cast, 100-200 µm average grain size

Contaminant, %	Critical contaminant limit(a) High pressure F	Low pressure F	T4	T6
Iron	0.032 Mn	0.032 Mn	0.035 Mn	0.046 Mn
Nickel	0.0050	0.0010	0.001	0.001
Copper	0.040	0.040	<0.010	0.040

(a) Tolerance limits expressed in wt% except for iron, which is expressed as the fraction of the manganese content (for example, the iron tolerance of 0.2% Mn alloy = 0.0064% Fe in F temper)

Table 5 Typical corrosion rates versus temper and grain size for two magnesium alloys

ASTM B 117 salt spray test

Alloy	Grain size, µm	Mn, %	Fe(a)	Temper corrosion rate F mm/yr	F mils/yr	T4 mm/yr	T4 mils/yr	T6 mm/yr	T6 mils/yr	T5 mm/yr	T5 mils/yr
AZ91C (untreated)	187	0.18	0.087	18	700	15	600	15	600	...	...
AZ91C (degassed and grain refined)	66	0.16	0.099	17	690	18	700	15	600	...	...
AZ91E(b) (untreated)	146	0.23	0.008	0.64	25	4	160	0.15	6	0.12	5
AZ91E (degassed and grain refined)	78	0.26	0.008	2.2	90	1.7	70	0.12	5	0.12	5
AZ91E (untreated)	160	0.33	0.004	0.35	14	3	120	0.22	9	0.12	5
AZ91E (degassed and grain refined)	73	0.35	0.004	0.72	29	0.82	33	0.1	4	0.1	4

(a) Iron is expressed as a fraction of analyzed manganese content. (b) AZ91E alloy pending ASTM approval. Source: Ref 4, 6

above does affect corrosion behavior. Table 6 indicates the suitability of magnesium for exposure to a wide range of substances. This information is intended as an indication; further testing is recommended before final conclusions are drawn concerning specific alloys and exposure conditions.

Atmospheres. A clean, unprotected magnesium alloy surface exposed to indoor or outdoor atmospheres free from salt spray will develop a gray film that protects the metal from corrosion while causing only negligible losses in mechanical properties. Chlorides, sulfates, and foreign materials that hold moisture on the surface can promote corrosion and pitting of some alloys unless the metal is protected by properly applied coatings.

The surface film that ordinarily forms on magnesium alloys exposed to the atmosphere gives limited protection from further attack. Unprotected magnesium and magnesium alloy parts are resistant to rural atmospheres and moderately resistant to industrial and mild marine atmospheres, provided they do not contain joints or recesses that entrap water in association with an active galvanic couple.

Compositions of the corrosion products that form on magnesium vary from one location to another and from indoor to outdoor exposure. X-ray diffraction analysis of corrosion products scraped from magnesium ingots after 18 months of exposure in a rural atmosphere have shown the presence of various hydrated carbonates of magnesium, including $MgCO_3 \cdot H_2O$, $MgCO_3 \cdot 5H_2O$, and $3MgCO_3 \cdot Mg(OH)_2 \cdot 3H_2O$. In an industrial atmosphere, hydrated and basic carbonates were found, together with magnesium sulfite ($MgSO_3 \cdot 6H_2O$) and magnesium sulfate ($MgSO_4 \cdot 7H_2O$).

These analyses, in addition to similar analyses after shorter periods of exposure, indicate that the primary reaction in corrosion of magnesium is the formation of magnesium hydroxide ($Mg(OH)_2$), followed by a secondary reaction with carbonic acid to convert the hydroxide to a hydrated carbonate. In atmospheres contaminated with sulfur compounds, sulfites or sulfates may also be present in the corrosion product. The sulfates may be formed by the reaction of acidic sulfur-bearing gases with $Mg(OH)_2$ or $MgCO_3$.

Corrosion of magnesium alloys increases with relative humidity. At 9.5% humidity, neither pure magnesium nor any of its alloys exhibit evidence of surface corrosion after 18 months. At 30% humidity, only minor corrosion may occur. At 80% humidity, the surface may exhibit considerable corrosion. In marine atmospheres heavily loaded with salt spray, magnesium alloys require protection for prolonged survival.

Fresh Water. In stagnant distilled water at room temperature, magnesium alloys rapidly form a protective film that prevents further corrosion. Small amounts of dissolved salts in water, particularly chlorides or heavy-metal salts, will break down the protective film locally, which usually results in pitting.

Dissolved oxygen plays no major role in the corrosion of magnesium in either freshwater or saline solutions. However, agitation or any other means of destroying or preventing the formation of a protective film leads to corrosion. When magnesium is immersed in a small volume of stagnant water, its corrosion rate is negligible. When the water is constantly replenished so that the solubility limit of $Mg(OH)_2$ is never reached, the corrosion rate may increase.

The corrosion of magnesium alloys by pure water increases substantially with temperature. At 100 °C (212 °F), the AZ alloys corrode typically at 0.25 to 0.50 mm/yr (10 to 20 mils/yr). Pure magnesium and alloy ZK60A corrode excessively at 100 °C (212 °F) with rates up to 25 mm/yr (1000 mils/yr). At 150 °C (302 °F), all alloys corrode excessively (Ref 9).

Salt Solutions. Severe corrosion may occur in neutral solutions of salts of heavy metals, such as copper, iron, and nickel. Such corrosion occurs when the heavy metal, the heavy-metal basic salts, or both plate out to form active cathodes on the anodic magnesium surface.

Chloride solutions are corrosive because chlorides, even in small amounts, usually break down the protective film on magnesium. Fluorides form insoluble magnesium fluoride and consequently are not appreciably corrosive. Oxidizing salts, especially those containing chlorine or sulfur atoms, are more corrosive than nonoxidizing salts, but chromates, vanadates, phosphates, and many others are film forming and thus retard corrosion, except at elevated temperatures.

Acids and Alkalis. Magnesium is rapidly attacked by all mineral acids except hydrofluoric acid (HF) and H_2CrO_4. Hydrofluoric acid does not attack magnesium to an appreciable extent, because it forms an insoluble, protective magnesium fluoride film on the magnesium; however, pitting develops at low acid concentrations. With increasing temperature, the rate of attack increases at the liquid line, but to a negligible extent elsewhere.

Pure H_2CrO_4 attacks magnesium and its alloys at a very low rate. However, traces of chloride ion in the acid will markedly increase this rate. A boiling solution of 20% H_2CrO_4 in water is widely used to remove corrosion products from magnesium alloys without attacking the base metal. Magnesium resists dilute alkalis, and 10% caustic solution is commonly used for cleaning at temperatures up to the boiling point.

Organic Compounds. Aliphatic and aromatic hydrocarbons, ketones, ethers, glycols, and higher alcohols are not corrosive to magnesium and its alloys. Ethanol causes slight attack, but anhydrous methanol causes severe attack. The rate of attack in the latter is reduced by the presence of water. Gasoline-methanol fuel blends in which the water content equals or exceeds about 0.25 wt% of the methanol content do not attack magnesium (Ref 10).

Pure halogenated organic compounds do not attack magnesium at ambient temperatures. At elevated temperatures or if water is present, such

Fig. 3 Effect of heavy-metal contamination on the salt spray (ASTM B 117) performance of sand-cast AZ91 samples in the T6 temper. The samples, containing less than 10 ppm Ni and less than 100 ppm Cu, were simultaneously exposed for 240 h. The sample at left contained 160 ppm Fe and had a corrosion rate of 15 mm/yr (591 mils/yr). The sample at right contained 19 ppm Fe, and the corrosion rate was 0.15 mm/yr (5.9 mils/yr).

compounds may cause severe corrosion, particularly those compounds having acidic final products.

Dry fluorinated hydrocarbons, such as the freon refrigerants, usually do not attack magnesium alloys at room temperature, but when water is present, they may stimulate significant attack. At elevated temperatures, fluorinated hydrocarbons may react violently with magnesium alloys.

In acidic foodstuffs, such as fruit juices and carbonated beverages, attack of magnesium is slow but measurable. Milk causes attack, particularly when souring.

At room temperature, ethylene glycol solutions produce negligible corrosion of magnesium that is used alone or galvanically connected to steel; at elevated temperatures, such as 115 °C (240 °F), the rate increases, and corrosion occurs unless proper inhibitors are added. Measures should be taken to control the galvanic corrosion.

Gases. Dry chlorine, iodine, bromine, and fluorine cause little or no corrosion of magnesium at room or slightly elevated temperature. Even when it contains 0.02% H_2O, dry bromine causes no more attack at its boiling temperature (58 °C, or 136 °F) than at room temperature. The presence of a small amount of water causes pronounced attack by chlorine, some attack by iodine, and negligible attack by fluorine. Wet chlorine, iodine, or bromine below the dew point of any aqueous phase causes severe attack of magnesium. Dry, gaseous sulfur dioxide causes no attack at ordinary temperatures. If water vapor is present, some corrosion may occur. Wet (below dew point) sulfur dioxide gas is severely corrosive to magnesium due to the formation of sulfurous and sulfuric acids. Ammonia, wet or dry, causes no attack at ordinary temperatures. Dry, gaseous sulfur dioxide (SO_2) or ammonia causes no attack at ordinary temperatures; some corrosion may occur if water vapor is present.

Water vapor in air or in oxygen sharply increases the rates of oxidation of magnesium and its alloys above 100 °C (212 °F), but boron trifluoride (BF_3), SO_2, and SF_6 are effective in reducing oxidation rates. The presence of BF_3 or SF_6 in the ambient atmosphere is particularly effective in suppressing high-temperature oxidation up to and including the temperature at which the alloy normally ignites.

The oxidation rate of magnesium in oxygen increases with temperature. At elevated temperature (approaching melting), the oxidation rate is a linear function of time. Cerium, lanthanum, calcium, and beryllium in the metal reduce the oxidation rate below that of pure magnesium. Beryllium additions have the most striking effects, protecting some alloys at temperatures up to the melting point over extended periods of time. Structural applications of magnesium alloys at elevated temperature are usually limited by creep strength rather than by oxidation.

Soils. Except when used as galvanic anodes, magnesium alloys have good corrosion resistance in clay or nonsaline sandy soils, but have poor resistance in saline sandy soils.

Corrosion in Real and Simulated Environments

Structural applications of magnesium alloys usually involve environmental conditions ranging from indoor atmospheres (computer disk drive components) to intermittent salt splash (automotive clutch housings). Allowable corrosion in these applications may range from zero for disk drive parts or critical military and aerospace components to substantial for some automotive castings, provided there is no interference with function for a specified period. Magnesium alloys of controlled high purity have demonstrated the capability of withstanding these environments in properly designed assemblies. Designers, however, must have advance information about the type and magnitude of corrosion that can occur under particular service conditions.

The best sources of such information would be previous service experience or long-term data from tests in the actual environment. When these are lacking, accelerated corrosion tests, such as alternate (intermittent) immersion in salt water or salt spray, are often used to compare the corrosion resistance of magnesium alloys to each other and to other metals. In addition, such tests are used to examine the relative merits of protective chemical treatments and coatings and the galvanic compatibilities of dissimilar metals with magnesium. The ASTM B 117 salt spray test (Ref 11) in particular is widely used and is firmly established as an acceptance test in many specifications. Magnesium is also subjected to various saltwater-based accelerated corrosion tests developed for special purposes. An important example is the proving ground cycle tests used by automobile manufacturers. Correlation has been established between proving ground cycle performance and vehicle corrosion in long-term service.

Specific Effects of Chloride Environments. Salt-based accelerated tests can provide useful information on the performance of magnesium alloys and assemblies, both bare and coated, in saline environments. The results of such tests must be interpreted with care, however, and no attempt should be made to relate them to the behavior of magnesium in rural, urban, or industrial atmospheres in which chloride is not the controlling constituent of the environment.

The corrosion of magnesium alloys in saline solutions is governed by the concentration and distribution of the critical impurities nickel, iron, and copper, whose presence in precipitated alloy phases creates active cathodic sites of low hydrogen overvoltage. The chloride ion further stimulates corrosion through its interference with protective-film formation and through the high solubility and acidic nature of magnesium chloride which accumulates at local anodic sites. The high conductivity of the chloride electrolyte also promotes the flow of corrosion current. If an adverse galvanic couple is introduced, for example, by attaching a steel bolt to the magnesium, and if the junction is bridged by salt water, corrosion of magnesium is greatly accelerated, and alloy purity provides no defense. In a salt spray test, galvanic corrosion of magnesium in such an assembly would be excessive compared to that which would occur in a marine atmosphere or even under roadsalt splash conditions.

Rural, Urban, and Industrial Atmospheres. In contrast to wet chloride exposure, a rural atmosphere would probably cause negligible corrosion damage to the magnesium assembly referred to above. Some corrosion, still minor compared to that caused by salt water, would be expected in urban or industrial atmospheres containing significant acidic gas pollution, principally SO_2. These gases convert the insoluble hydroxide-carbonate films that form naturally on magnesium alloys into soluble bicarbonates, sulfites, and sulfates that can be washed away by rain. This increases the rate of weathering and can provide conductive electrolytic paths, allowing some galvanic corrosion to occur.

To simulate atmospheric exposure where chloride does not exert a controlling influence, various humidity tests have been used or proposed. Water fog and water immersion tests are also used, primarily to evaluate the adhesion and blister resistance of organic coatings. Table 7 lists some simulated environmental tests and their intended purposes.

Table 6 Magnesium suitability for testing in various substances

Follow precautions in text.

Chemical	Concentration, %	Service test warranted
Acetaldehyde	Any	No
Acetic acid	Any	No
Acetone	Any	Yes
Acetylene	100	Yes
Alcohol, butyl	100	Yes
Alcohol, ethyl	100	Yes
Alcohol, isopropyl	100	Yes
Alcohol, methyl	100	No
Alcohol, propyl	100	Yes
Ammonia (gas or liquid)	100	Yes
Ammonium salts (most)	Any	No
Ammonium hydroxide	Any	Yes
Aniline	100	Yes
Anthracene	100	Yes
Arsenates (most)	Any	Yes
Benzaldehyde	Any	No
Benzene	100	Yes
Bichromates	Any	Yes
Boric Acid	1–5	No
Brake Fluids (most)	100	Yes
Bromides (most)	Any	No
Bromobenzene	100	Yes
Butter	100	No
Butylphenols	100	Yes
Calcium arsenate	Any	Yes
Calcium carbonate	100	Yes
Calcium chloride	Any	No
Calcium hydroxide	100	Yes
Camphor	100	Yes
Carbon bisulfide	100	Yes
Carbon dioxide (dry)	100	Yes
Carbon monoxide	100	Yes
Carbon tetrachloride	100	Yes
Carbonated water	Any	No
Castor oil	100	Yes
Cellulose	100	Yes
Cement	100	Yes
Chlorides (most)	Any	No
Chlorine	100	No
Chlorobenzenes	100	Yes
Chloroform	100	Yes
Chlorophenols	Any	No
Chlorophenylphenol	100	Yes
Chromates (most)	Any	Yes
Chromic acid	Any	Yes
Citronella oil	100	Yes
Cod liver oil (crude)	...	Yes
Copals	100	Yes
Coumarin	100	Yes
Cresol	100	Yes
Cyanides (most)	Any	Yes
Dichlorohydrin	100	Yes
Dichlorophenol	100	Yes
Dichromates (see bichromates)		
Diethanolamine	100	Yes
Diethyl aniline	100	Yes
Diethyl benzene	100	Yes
Diethylene glycol solutions	Any	Yes, may need inhibitors
Diphenyl	100	Yes
Diphenylamine	100	Yes
Diphenylmethane	100	Yes
Diphenyl oxide	100	Yes
Dipropylene glycol	100	Yes
Divinylbenzene	100	Yes
Dry cleaning fluids	100	Yes
Ethers	100	Yes
Ethanolamine (mono)	100	Yes
Ethyl acetate	100	Yes
Ethyl benzene	100	Yes
Ethyl bromide	100	No
Ethylcellulose	100	Yes
Ethyl chloride	100	Yes
Ethyl salicylate	100	Yes
Ethylene (gas)	100	Yes
Ethylene dibromide	100	Yes
Ethylene glycol solutions	Any	Yes, may need inhibitors
Fats, cooking (acid-free)	100	Yes
Fatty acids	Any	No
Ferric chloride	Any	No
Fluorides (most)	Any	Yes
Fluosilicic acid	Any	No
Formaldehyde	Any	Yes
Fruit juices and acids	Any	No
Fuel oil	100	Yes
Gasohol (10% ethanol)	100	Yes, if inhibited
Gasohol (10% methanol)	100	Yes, if inhibited
Gasoline (lead-free)	100	Yes, if inhibited
Gasoline (leaded)	100	Yes, if inhibited
Gelatine	Any	Yes
Glycerine C.P.	100	Yes
Grease (acid-free)	100	Yes
Heavy metal salts (most)	Any	No
Hexamine	3	Yes
Hydrochloric acid	Any	No
Hydrofluoric acid	5–60	yes
Hydrogen peroxide	Any	No
Hydrogen sulfide	100	Yes
Iodides	Any	No
Iodine crystals (dry)	100	Yes
Isopropyl acetate	100	Yes
Isopropyl benzene	100	Yes
Isopropyl bromide	Any	No
Kerosene	100	Yes
Lanolin	100	Yes
Lard	100	Yes
Lead arsenate	Any	Yes
Lead oxide	Any	No
Linseed oil	100	Yes
Magnesium arsenate	Any	Yes
Magnesium carbonate	100	Yes
Magnesium chloride	Any	No
Mercury salts	Any	No
Methane (gas)	100	Yes
Methyl bromide	Any	No
Methyl cellulose	100	Yes
Methyl chloride	100	Yes
Methylene chloride	100	Yes
Methyl salicylate	100	Yes
Milk (fresh and sour)	100	No
Mineral acids	Any	No
Monobromobenzene	100	Yes
Monochlorobenzene	100	Yes
Naphtha	100	Yes
Naphthalene	100	Yes
Nicotine sulfate	40	Yes
Nitrates (all)	Any	No
Nitrous gases	100	No
Nitric acid	Any	No
Nitroglycerin	Any	No
Oil, animal (acid- and chloride-free)	Any	Yes
Oil, mineral (chloride-free)	100	Yes
Oil, vegetable (chloride-free)	100	Yes
Oleic acid	100	Yes
Olive oil	100	Yes
Organic acids (most)	Any	No
Orthochlorophenol	100	No
Orthodichlorobenzene	100	Yes
Orthophenylphenol	100	Yes
Oxygen	100	Yes
Paraphenylphenol	100	Yes
Paradichlorobenzene	100	Yes
Pentachlorophenol	100	Yes
Perchloroethylene	100	Yes
Permanganates (most)	Any	Yes
Phenol	100	Yes
Phenyl ethyl acetate	100	Yes
Phenylphenols	100	Yes
Phosphates (most)	Any	Yes
Phosphoric acid	Any	No
Polypropylene glycols	100	Yes
Potassium fluoride	Any	Yes
Potassium hydroxide	Any	Yes
Potassium nitrite	Any	No
Potassium permanganate	Any	Yes
Propylene glycol U.S.P.	100	Yes
Propylene oxide	100	Yes, may need inhibitors
Pyridine (acid free)	100	Yes
Pyrogallol	Any	No
Rubber and rubber cements	100	Yes
Seawater	100	No
Sodium bromate	Any	No
Sodium bromide	Any	No
Sodium carbonate	Any	Yes
Sodium chloride	Any	No
Sodium cyanide	Any	Yes
Sodium dichromate	Any	Yes
Sodium fluoride	Any	Yes
Sodium hydroxide	Any	Yes
Sodium phosphate (tribasic)	Any	Yes
Sodium silicate	Any	Yes
Sodium sulfide	3	Yes
Sodium tetraborate	3	Yes
Steam	100	No
Stearic acid (dry)	100	Yes
Styrene polymer	100	Yes
Sugar solutions (acid-free)	Any	Yes
Sulfates (most)	Any	No
Sulfur	100	Yes
Sulfur dioxide (dry)	100	Yes
Sulfur chloride	Any	No
Sulfuric acid	Any	No
Sulfurous acid	Any	No
Tannic acid	3	No
Tanning solutions	Any	No
Tar, crude and its fractions	100	Yes
Tartaric acid	Any	No
Tetrahydronaphthalene	100	Yes
Titanium tetrachloride	100	Yes
Toluene (toluol)	100	Yes
Trichlorbenzene	100	Yes
Trichloroethylene	100	Yes
Trichlorophenol	100	Yes
Tung oil	100	Yes
Turpentine	100	Yes
Urea	100	Yes
Urea in aqueous solution (cold)	Any	Yes
Urea in aqueous solution (warm)	Any	No
Vinegar	Any	No
Vinylidine chloride	100	Yes
Vinyl toluene	100	Yes
Water, boiling	100	No
Water, distilled	100	Yes
Water, rain	100	Yes
Waxes (acid-free)	100	Yes
Xylol	100	Yes

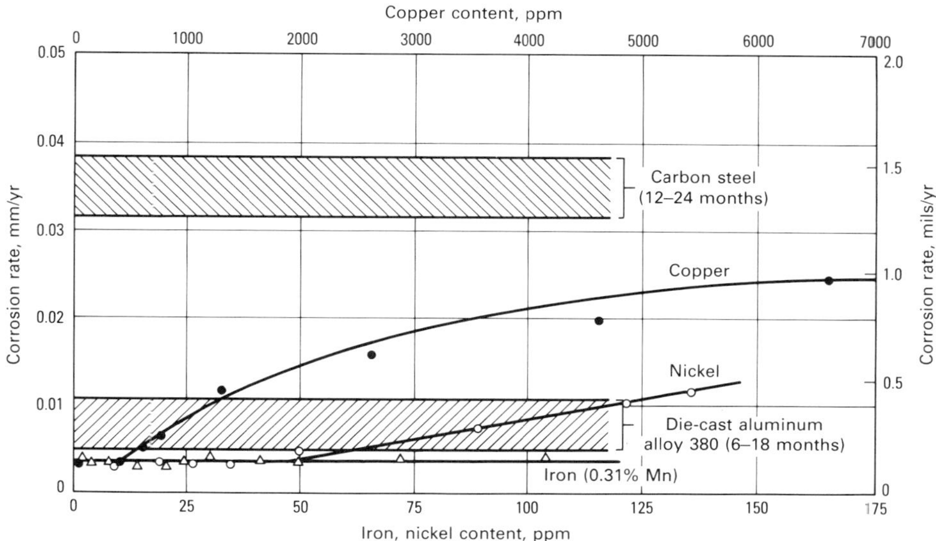

Fig. 4 Effect of copper and nickel contamination on the atmospheric corrosion of die-cast AZ91 alloy exposed for 2 years on the Texas Gulf Coast

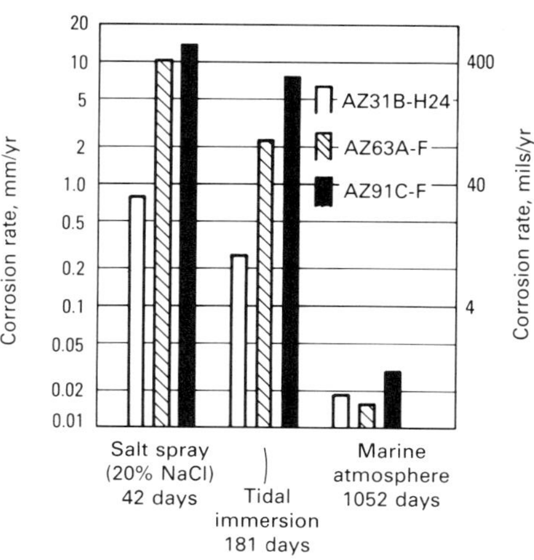

Fig. 5 Relative performance of magnesium alloys in saltwater exposures

Tests for Stress-Corrosion Cracking. Magnesium alloys containing more than about 1.5% Al are susceptible to stress-corrosion cracking, and the tendency increases with aluminum content. Alloys in wrought form appear to be more susceptible than castings. The stress sources likely to promote cracking are weldments and inserts. Welded structures of these alloys require stress-relief annealing. While there is little documented record of SCC failures of castings in service, magnesium castings have been shown to fail in laboratory tests under tensile loads as low as 50% of yield strength in environments causing negligible general corrosion. The apparent low incidence of SCC service failures of castings is attributable to low stresses actually applied or to stress relaxation by yielding or creep when a fixed deflection is imposed.

Although laboratory tests of the type described are useful in encouraging conservative design of magnesium alloy structures, principal reliance has been placed on long-term atmospheric tests of tensile-loaded specimens. Short-term accelerated tests, such as the sodium chloride/potassium chromate ($NaCl/K_2CrO_4$) tests, do not indicate SCC behavior reliably in practice.

Corrosion Fatigue. Substantial reductions in fatigue strength of magnesium alloys are shown in laboratory tests using NaCl spray or drops. Such tests are useful for comparing alloys, heat treatments, and protective coatings. Effective coatings, by excluding the corrosive environment, provide the primary defense against corrosion fatigue.

Corrosion Rates in Salt Spray, Salt Immersion, and Natural Marine Environments. Figures 2 and 4 show the effects of nickel, iron, and copper content on the corrosion of controlled-purity AZ91 die-casting alloy in a 10-day salt spray (5% NaCl, ASTM B 117) test and in a 2-year atmospheric exposure on the Texas Gulf Coast. Although the salt spray corrosion scale is 200 times larger than the marine atmospheric corrosion scale, some parallels can be drawn:

- The breakpoints relative to nickel and copper contamination are the same in both exposures (50 and 500 ppm, respectively)

Table 7 Accelerated or simulated environmental corrosion tests for magnesium alloys

Test condition	Used to evaluate
Humidity tests	
95% relative humidity, 38 °C (100 °F)	Flux inclusions, indoor tarnishing, filiform corrosion
100% relative humidity, 38 °C (100 °F), condensing (ASTM D 2247)	Paint adhesion and blistering, corrosion in rural (uncontaminated) atmospheres
Polluted atmosphere (DIN-50018-1960), 100% relative humidity, 40 °C (105 °F), Air + SO_2 + CO_2 for 8 h, then 16 h in air at room temperature	Corrosion and coating performance in industrial atmospheres
Water tests	
Water fog (ASTM D 1735), deionized water, 38 °C (100 °F)	Paint adhesion and blistering (roughly equivalent to condensing humidity)
Water immersion (ASTM D 870), deionized water, 38 °C (100 °F)	Paint adhesion and blistering (severe test)
Salt tests	
Salt spray (ASTM B 117), 5% NaCl, pH 6.5–7.2	Corrosion of magnesium alloys, impurity effects, surface treatments and coatings on the same substrate alloy, galvanic compatibilities of other materials with magnesium. Valid for chloride environments, with careful interpretation. A severe accelerated test
Salt immersion, 5% NaCl, 25 °C (75 °F), pH 10.5, intermittent or continuous immersion with mild air agitation	Same as salt spray
Copper-accelerated acetic acid salt spray (ASTM B 368), 5% NaCl, 1 g $CuCl_2 \cdot 2H_2O$ per 3.8 L of solution, 49 °C (120 °F), pH 3.1–3.3	Plated coatings on magnesium
Salt spray-SO_2, 5% NaCl + SO_2, 35 °C (95 °F), pH 2.5–3.2, naval air development center	Naval aircraft materials (simulates sea spray plus ship stack gases)
Automotive proving ground test, repeated cycles of salt-mud splash, partial drying, and high-humidity storage	Magnesium alloys, bare or coated, for exposed automobile and truck parts. Galvanic compatibilities. Simulates severe road deicing salts exposure

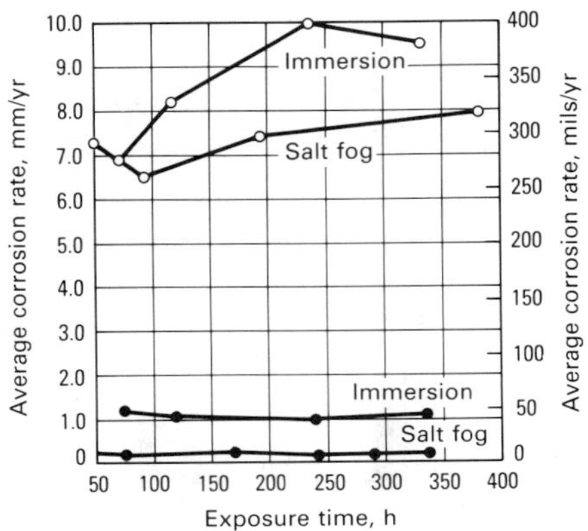

Analysis of die-cast plates, %		
	AM60A (○)	AZ91D (●)
Aluminum	6.2	9.7
Zinc	0.09	0.74
Manganese	0.22	0.19
Nickel	0.003	0.0018
Iron	0.005	0.006
Copper	0.03	0.0067

Fig. 6 Corrosion rates of die-cast magnesium in 5% NaCl salt spray and continuous-immersion exposures. Source: Ref 12

- AZ91 corrosion remains low in both exposures over the iron contamination range studied (0.31% Mn alloy), confirming that the critical iron/manganese ratio (0.032 for this alloy) was not significantly exceeded
- High-purity AZ91 alloy is shown in both exposures to have lower overall corrosion rates than carbon steel or die-cast aluminum 380

Figure 5 shows average corrosion rates found for three unprotected magnesium alloys in salt spray (20% NaCl), tidal-immersion, and marine atmospheres. The principal alloy and impurity constituents of the specimens tested were as follows:

Element	AZ31B-H24 sheet	AZ63-F sand cast	AZ91C-F sand cast
Aluminum	2.6	5.8	8.8
Zinc	1.0	2.9	0.68
Manganese	0.51	0.25	0.22
Silicon	0.0017	<0.05	<0.05
Nickel	0.0005	<0.001	<0.01
Iron	0.0007	0.005	0.006
Copper	0.0019	0.015	0.013
Fe/Mn =	0.0014	0.020	0.027

The marine atmosphere and tidal-immersion exposures were conducted at the Naval Air Station, Norfolk, VA. The atmospheric panels were suspended directly over the water, 3 m (10 ft) above mean tide level, at an angle of 45° from the horizontal and facing east-southeast. The tidal-immersion panels were mounted vertically so that they were totally immersed in the water at high tide and totally exposed to atmosphere at low tide.

The salt spray and tidal-immersion results are in good agreement in rating the three alloys in order of actual corrosion rates. The atmospheric panels, even though directly above the water, sustained only a small fraction of the corrosion damage seen in the salt spray and tidal tests.

No fundamental differences exist between spray and immersion tests that would lead to serious misjudgments about the relative corrosion resistance of magnesium alloys or the relative value of protective coating systems. (Caution should be exercised, however, when comparing the relative performance of magnesium alloys of different families, for example, QE22 and ZE41, or when comparing magnesium with other metals.) Figure 6 shows average corrosion rates over various exposure times for two die-cast alloys in salt spray and in 5% NaCl continuous immersion.

Influence of Galvanic Couples in Salt Spray and Marine Atmospheres. The combination of salt spray and dissimilar metals coupled to magnesium represents an extreme in accelerated corrosion testing. The gap between such a test and a marine atmosphere exposure is illustrated in Fig. 7 and in Table 8. Figure 7 shows magnesium die-cast plates that were assembled with steel cap screws and nuts and exposed to salt spray (5% NaCl) for 10 days. Table 8 lists losses of tensile strength of AZ31B-H24 sheet with an original thickness of 4.8 mm (0.188 in.), coupled with various dissimilar-metal cleats and exposed for 1.1 years at Kure Beach, NC, at 24.4 m (80 ft) from the mean tide point. Comparative data are shown for rural and an urban-industrial test site. Marine exposure of steel magnesium couples will produce nearly complete perforation failure of 4.8 mm (0.188 in.) magnesium panels in 10 days. However, it should be noted that the steel panels retained 88% of their tensile strength after 13 months.

Atmospheric Corrosion Rates in Rural, Industrial, and Marine Environments. Table 9 lists the average corrosion rates reported for a number of alloys after 2.5 to 3 years of exposure at three different atmospheric test sites: a rural site (Midland, MI) an industrial site (Madison, IL), and a severe marine site (the Kure Beach, NC, site mentioned earlier). These data reveal that the alloy with the best corrosion resistance in all three environments is AZ91-T6 (10 ppm Fe). The alloys with the worst performance are ZE10A-0, EZ33A-T5/ZH62A-T5, and ZH62A-T5 in the rural, industrial, and marine exposures, respectively. Based on the average site corrosion rates for all alloys, the relative severity of the exposures is 1:2.1:1.7. Therefore, the industrial, SO_2-rich atmosphere is the more severe of the three sites.

Effect of SO_2 Pollution on Atmospheric Corrosion. Figure 8, prepared from data of the National Research Council of Canada, shows the effects of SO_2 pollution on the corrosion of ZK61A, AZ80A, and low-carbon steel. These effects may cause industrial atmospheres to corrode magnesium at average rates that are somewhat higher than those found in marine atmospheres (Table 9). The marine atmosphere poses a greater threat, however, for two reasons:

- The greater sensitivity of the marine corrosion rate to critical impurity content
- Greatly increased susceptibility to galvanic-corrosion damage by incompatible coupling metals in a wet chloride environment

These principles are demonstrated in Tables 10 and 11. Table 10 lists the corrosion rates of high- and low-iron samplings of AZ31 sheet and AZ91C-T6 castings at the three test sites represented in Table 9 plus a 20% NaCl salt spray test for the AZ91 castings. Table 11 assesses the relative galvanic damage to AZ31B-H24 caused by various cleat metals in rural, industrial, and marine exposures.

Indoor Atmospheres. Before the computer age, reaction of magnesium alloys with indoor atmospheres was of concern primarily from the standpoint of appearance, not function. The widespread introduction of magnesium die castings into the computer disk drive environment has imposed strict new standards of surface stability on the metal because of the need to maintain a clean particle-free atmosphere at the disk/head interface. The corrosion of magnesium alloys in indoor atmospheres increases with relative humidity. At relative humidities to about 90%, corrosion is very minor, resulting in the formation of a nearly invisible film of amorphous $Mg(OH)_2$. As humidity increases beyond this level, heavier tarnish films develop, the principal corrosion product now being crystalline

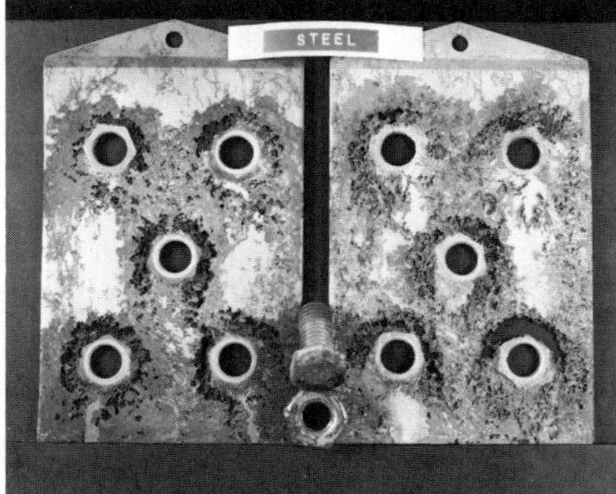

Fig. 7 Galvanic corrosion of AZ91D caused by bare steel fasteners during a 10-day exposure to 5% NaCl salt spray. Source: Ref 12

Table 8 Loss in tensile strength of galvanically coupled AZ31 sheet versus coupled metal cleat and exposure

Cleat metal(a)	Loss of tensile strength, % — State College, PA (rural)	Newark, NJ (urban-industrial)	Kure Beach, NC, 24.4-m (80-ft) site (marine)
AZ31B	(0.1 gain)	0.7	0.1
Aluminum alloy 6061	(0.2 gain)	0.8	1.0
Aluminum alloy 5052	0.5	1.2	1.6
AlClad alloy 7075	0.5	2.0	5.1
AISI type 304 stainless steel	0.9	3.4	9.0
Monel	1.1	4.4	10.7
Low-carbon steel	1.5	6.8	12.4
85-15 brass	1.8	7.1	15.2

(a) Cleats bolted in register to both sides of 4.8-mm (0.188-in.) thick AZ31B-H24 sheet; tensile specimens cut so that stress is at right angle to the four corrosion bands produced. Source: Ref 2

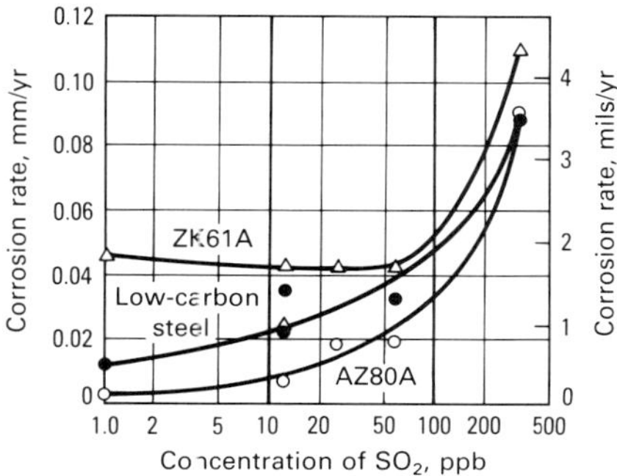

Fig. 8 Corrosion rates versus SO_2 pollution levels at six exposure sites. Rainfall at the sites ranged from 533 to 965 mm/yr (21 to 38 in./yr). Source: Ref 14

$Mg(OH)_2$. In die castings, small traces of chloride residues from cover or refining fluxes may serve as nuclei for corrosion spots in humid air.

The strict requirements of these applications are being met by a combination of the following factors:

- Constant efforts by alloy producers and die casters to provide metal that is extremely low in nonmetallic inclusions
- Use of selected conversion coatings and proprietary protective coatings for supplementary surface stabilization

Accelerated tests involving cyclic humidity and temperature variation provide useful information for these applications, both on coating performance and by detection of any flux inclusions that might be present.

Galvanic Corrosion

Insufficient attention to galvanic corrosion has been one of the major obstacles to the growth of structural applications of magnesium alloys. Serious galvanic problems occur mainly in wet saline environments for the reasons described earlier.

Outstanding improvements in general saltwater corrosion resistance of magnesium alloys have been achieved by reducing the internal corrosion currents through strict limitations on the critical impurities nickel, iron, and copper, as well as on the iron-to-manganese ratio. These improvements have no significant effect on galvanic corrosion, because the electromotive force (emf) for corrosion now comes from an external source, the dissimilar metal coupled to magnesium. Prevention of galvanic damage thus requires consideration of a combination of measures involving:

- Design to prevent access and entrapment of salt water at the dissimilar-metal junction
- Selection of the most compatible dissimilar metals
- Introduction of high resistance into the metallic portion of the circuit through insulators or into the electrolytic portion of the circuit by increasing the length of the path the electrolytic current must follow
- Protective coating of the full assembly

Details of these measures are treated in the section "Protection of Assemblies" in this article. It is useful, however, to review the basic principles governing the galvanic corrosion of magnesium and the relative compatibility of dissimilar metals with magnesium.

Relative Compatibility of Metals. All structural metals are cathodic to (more noble than) magnesium. The degree to which the corrosion of magnesium is accelerated under a given set of exposure conditions depends partly on the relative positions of the two metals in the emf series, but more importantly on how rapidly the effective potential of the couple is reduced by polarization as galvanic current flows. The principal polarization mechanism in a magnesium couple in salt water is the resistance to the formation and liberation of hydrogen gas at the cathode. Therefore, metals of low hydrogen overvoltage, such as nickel, iron, and copper, constitute efficient cathodes for magnesium and cause severe galvanic corrosion. Metals that combine active potentials with higher hydrogen overvoltages, such as aluminum, zinc, cadmium, and tin, are much less damaging, although not fully compatible with magnesium.

Data were compiled in tests at Kure Beach, NC, in which sheets of dissimilar metals were fastened to panels of AZ31B and AZ61A. The dissimilar metals were divided into five groups

Table 9 Corrosion rates of magnesium alloys during atmospheric exposure for 2.5 to 3 years

Alloy and temper	Rural mm/yr	Rural mils/yr	Industrial mm/yr	Industrial mils/yr	Marine-rural mm/yr	Marine-rural mils/yr
Sheet						
AZ31B-H24 (10 ppm Fe)	0.013	0.52	0.025	1.0	0.017	0.69
AZ31C-O (70 ppm Fe)	0.012	0.46	0.025	1.0	0.038	1.5
HK31A-H24	0.018	0.73	0.030	1.2	0.016	0.64
HM21A-T8	0.020	0.80	0.032	1.3	0.022	0.88
ZE10A-O	0.022	0.88	0.030	1.2	0.028	1.1
Extrusions						
AZ31B-F	0.013	0.53	0.025	1.0	0.019	0.77
HM31A-F	0.018	0.70	0.035	1.4	0.020	0.80
ZK60A-T5	0.017	0.66	0.032	1.3	0.025	1.0
Castings						
AZ63A-T4	0.0086	0.34	0.022	0.88	0.019	0.76
AZ91C-T6 (350 ppm Fe)	0.0043	0.17	0.015	0.62	0.022	0.88
AZ91C-T6 (10 ppm Fe)	0.0027	0.11	0.014	0.57	0.0064	0.25
AZ92A-T6	0.0094	0.37	0.020	0.80	0.025	1.0
EZ33A-T5	0.020	0.79	0.040	1.6	0.028	1.1
HK31A-T6	0.017	0.67	0.035	1.4	0.028	1.1
HZ32A-T5	0.015	0.61	0.038	1.5	0.028	1.1
ZH62A-T5	0.015	0.58	0.040	1.6	0.041	1.6
ZK51A-T5	0.014	0.57	0.035	1.4	0.025	1.0
Site average	**0.014**	**0.56**	**0.030**	**1.2**	**0.024**	**1.0**

Source: Ref 13

Table 10 Corrosion rates versus exposure and iron content for AZ31 sheet and AZ91 castings

Alloy(a)	Iron, ppm	Rural mm/yr	Rural mils/yr	Industrial mm/yr	Industrial mils/yr	Marine mm/yr	Marine mils/yr	20% NaCl spray mm/yr	20% NaCl spray mils/yr
		Exposure and average corrosion rates							
(1) AZ31	70	0.012	0.5	0.025	1.0	0.038	1.5	...	...
(2) AZ31	10	0.013	0.51	0.025	1.0	0.018	0.7	...	...
Ratio of (1):(2)	7	0.9	0.9	1.0	1.0	2.2	2.2	...	...
(3) AZ91	350	0.0043	0.17	0.016	0.6	0.022	0.87	95	3740
(4) AZ91	10	0.0028	0.11	0.014	0.55	0.0064	0.25	0.71	27.9
Ratio of (3):(4)	35	1.5	1.5	1.1	1.1	3.5	3.5	134	134

(a) (1) AZ31C sheet, (2) AZ31B-H24 sheet, (3) AZ91C-T6 cast plate (Fe/Mn = 0.15), (4) AZ91C-T6 cast plate (Fe/Mn = 0.007). Source: Ref 13

Table 11 Severity of galvanic attack on AZ31 sheet versus exposure and coupled metal

Cleat metal	Relative loss of tensile strength(a): Rural (State College, PA)	Urban-industrial (Newark, NJ)	Marine (Kure Beach, NC, 24.4-m, or 80-ft, site)
Aluminum alloy 5052	1.0	2.4	3.2
Aluminum alloy 7075	1.0	4.0	10.2
AISI type 304 stainless steel	1.8	6.8	18.0
Monel	2.2	8.8	21.4
Low-carbon steel	3.0	13.6	24.8
85-15 brass	3.6	14.2	30.4

(a) See data in Table 8; aluminum alloy 5052 tested at State College, PA, = 1.00. Source: Ref 2

Table 12 Relative effects of various metals on galvanic corrosion of magnesium alloys AZ31B and AZ61A exposed at the 24.4- and 244-m (80- and 800-ft) stations, Kure Beach, NC

Group 1 (least effect)	Group 2	Group 3	Group 4	Group 5 (greatest effect)
Aluminum alloy 5052	Aluminum alloy 6063	AlClad alloy 2024	Zinc-plated steel	Low-carbon steel
Aluminum alloy 5056	AlClad alloy 7075	Aluminum alloy 2017	Cadmium-plated steel	Stainless steel
Aluminum alloy 6061	Aluminum alloy 3003	Aluminum alloy 2024		Monel
	Aluminum alloy 7075	Zinc		Titanium
				Lead
				Copper
				Brass

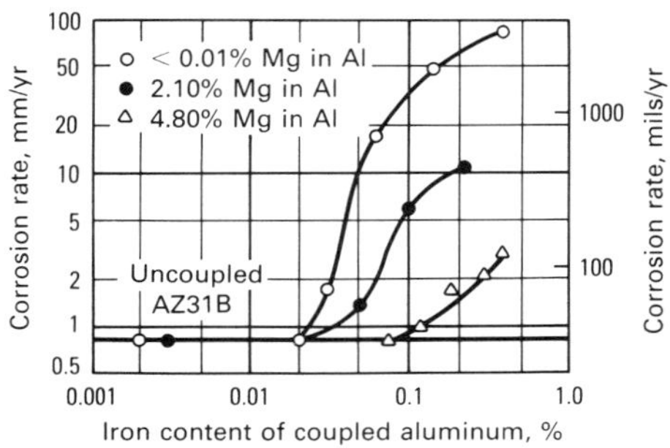

Fig. 9 Corrosion rates in 3% NaCl solution of magnesium alloy AZ31B coupled with aluminum containing varying amounts of iron and magnesium. The corrosion rate of uncoupled AZ31B is shown for comparison.

based on observed gradations of galvanic damage to magnesium. These ratings are summarized in Table 12.

Effects of Anode and Cathode Areas. The relative areas of the magnesium anode and the dissimilar-metal cathode have an important effect on the corrosion damage that occurs. A large cathode coupled with a small area of magnesium results in rapid penetration of the magnesium, because the galvanic current density at the small magnesium anode is very high, and anodic polarization in chloride solutions is very limited. This explains why painted magnesium should not be coupled with an active cathodic metal if the couple will be exposed to saline environments. A small break in the coating at the junction results in a high concentration of galvanic current at that point unmitigated by any polarization. Unfavorable area effects can also be seen in the behavior of some proprietary coatings using aluminum or zinc powder. When used as a coating on a steel bolt attached to magnesium, the metallic pigment can present a very large effective surface area, which may be more detrimental than bare steel. Galvanic action is further accelerated if the metallic pigment contains such impurities as iron.

Effects of Minor Constituents on Compatibility of Aluminum With Magnesium. Aluminum alloys containing small percentages of copper (7000 and 2000 series and 380 die-casting alloy) may cause serious galvanic corrosion of magnesium in saline environments. Very pure aluminum is quite compatible, acting as a polarizable cathode; but when iron content exceeds 200 ppm, cathodic activity becomes significant (apparently because of the depolarizing effect of the intermetallic compound $FeAl_3$), and galvanic attack of magnesium increases rapidly with increasing iron content. The effect of iron is diminished by the presence of magnesium in the alloy (Fig. 9). This agrees with the relatively compatible behavior of aluminum alloys 5052, 5056, and 6061 shown in Table 11.

Cathodic Corrosion of Aluminum. Judgments of compatibility of aluminum alloys with magnesium alloys are complicated by the fact that aluminum can be attacked by the strong alkali generated at the cathode when magnesium corrodes sacrificially in static NaCl solutions. Such attack destroys compatibility in alloys containing significant iron contamination, apparently by exposing fresh, active sites with low overvoltage. The aluminum alloys having substantial magnesium content (5052 and 5056) are more resistant, to this effect, but not completely so. The essential requirement for a fully compatible aluminum alloy, as indicated in Fig. 9, would be met by a 5052 alloy with a maximum of 200 ppm Fe or a 5056 alloy with a maximum of 1000 ppm Fe. Commercially produced 5052 alloy is permitted by specification to have a maximum (iron + silicon) content of 0.45% and may typically contain 0.3% Fe. In a severe exposure such as 5% NaCl immersion, this iron content, combined with the cathodic corrosion caused by the current from the magnesium, can render the 5052 alloy incompatible with magnesium. In most real situations, however, this extreme condition would not exist, and a 5052 washer under the head of a plated steel bolt in a magnesium assembly would reduce galvanic attack of the magnesium. For maximum effect of the washer, the linear distance along the aluminum from the bolt should be about 4.8 mm (3/16 in.)

Cathodic corrosion of aluminum is much less severe in seawater than in NaCl solution, because the buffering effect of magnesium ions reduces the equilibrium pH from 10.5 to about 8.8 (Ref 15). The compatibility of aluminum with magnesium is accordingly better in seawater and is less sensitive to iron content.

Cathodic Damage to Coatings. Hydrogen evolution and strong alkalinity generated at the cathode can damage or destroy organic coatings applied to fasteners or other accessories coupled to magnesium. Alkali-resistant resins are necessary, but under severe conditions such as salt spray or salt immersion, the coatings may be simply blown off by hydrogen, starting at small voids or pores. Because of its severity, the salt spray test can lead to rejection of some fastener coatings that may provide useful benefits in real service environments. Salt spray should not be relied upon exclusively to evaluate these coatings.

Compatibility of Plated Steel. Zinc, cadmium, or tin plating on steel all reduce galvanic attack of magnesium substantially compared to bare steel. This agrees with the more compatible potentials and/or the higher hydrogen overvoltages of the plated deposits. The relative merit of the three electroplates is generally considered to be (in decreasing order) tin, cadmium, zinc. The salt spray test is biased against zinc because zinc is rapidly removed from the steel substrate in this test medium due to general corrosion as well as cathodic attack when coupled with magnesium. This does not occur in many natural environments, and the failure of the salt spray test to rate zinc and cadmium plating properly in marine atmospheres is well known (Fig. 10).

Protection of Assemblies

A dissimilar metal in contact with magnesium will not by itself result in galvanic corrosion. For corrosion to occur, both surfaces must also be wetted by a common electrolyte. The degree to which precautions against galvanic corrosion are taken will depend on many factors, of which the operating environment is of primary importance.

For indoor use, where condensation is not likely, no protection is necessary. Even in some

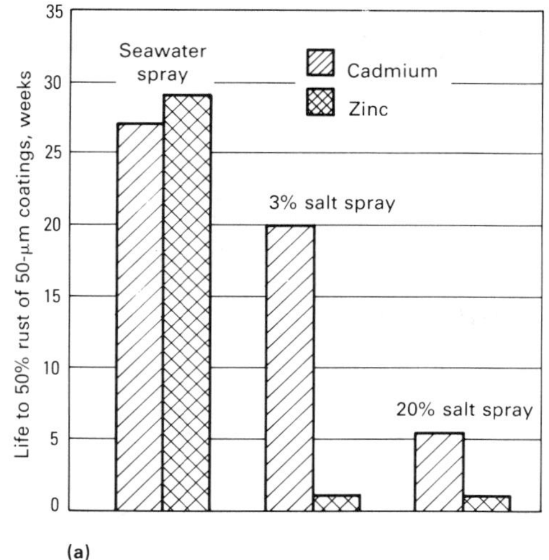

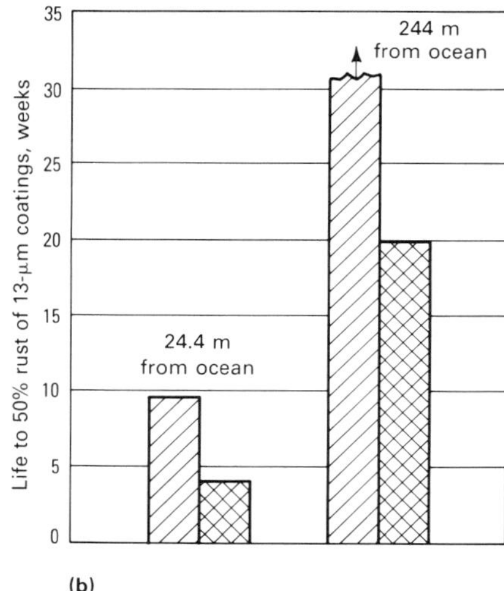

Fig. 10 Comparison of cadmium and zinc plate on steel in salt spray tests and in marine atmospheres. (a) 200-μm coating. (b) 50-μm coating. Source: Ref 16

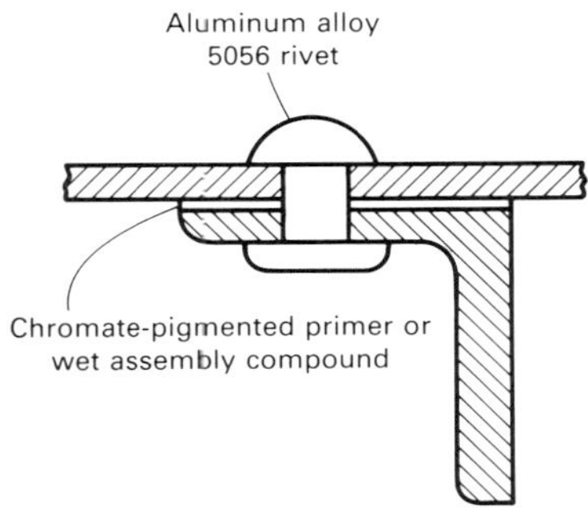

Fig. 11 Schematic of the proper method of protecting faying surfaces in magnesium-to-magnesium assemblies

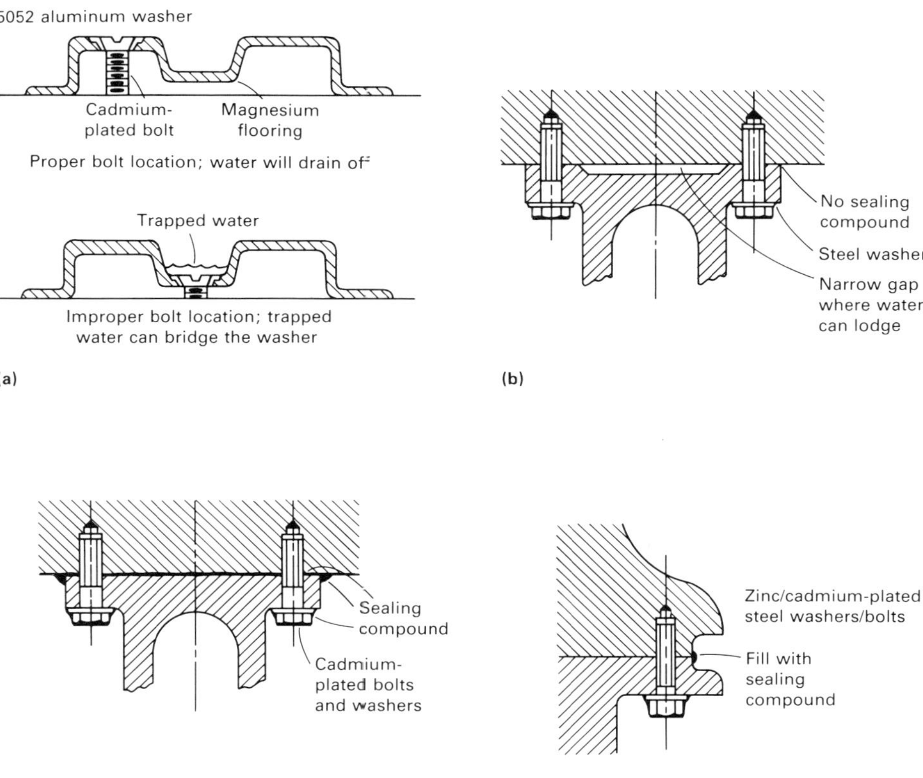

Fig. 12 Design considerations for reducing galvanic corrosion. (a) Proper bolt location. (b) Poor practice. (c) Good with no gap. (d) For use when direct metal-to-metal contact is required for electrical reasons

sheltered outdoor environments, magnesium components can give good service lives without special precautions against galvanic attack provided other mitigating factors are present. These might include design elimination of water traps, good ventilation, component warmth, or the presence of an oil film.

For continuous outdoor use, during which magnesium assemblies may be wetted or subjected to salt splash or spray, precautions against galvanic attack must be taken. Although corrosive attack from any source can jeopardize the satisfactory performance of magnesium components, attack resulting from galvanic corrosion is probably the most detrimental. Under corrosive conditions, use of high-purity magnesium alloys will have no significant influence in reducing the effects of galvanic corrosion.

Magnesium-to-Magnesium Assemblies. For all practical purposes, galvanic corrosion between magnesium alloys is negligible. However, because joining two magnesium components almost invariably involves use of dissimilar-metal fasteners and the formation of a crevice at the joint, good assembly practice dictates that in corrosive conditions some precautions should be taken (Fig. 11). Magnesium faying or mating surfaces should be assembled using wet assembly techniques. Chromate-inhibited primers or sealing compounds are placed between the surfaces at the time of assembly. Sealing/jointing compounds of the polymerizing or nonpolymerizing type are preferred because they remain flexible and resist cracking. Polymerizing-type compounds are also used for caulking operations. In bolted assemblies, the retorquing of bolts shortly after assembly will help eliminate any joint relaxation problems. For additional protection, mating surfaces should be primed before assembly and painted after assembly.

Magnesium-to-Nonmetallic Assemblies. Although the joining of most nonmetallic materials, such as plastics and ceramics, to magnesium will not result in any potential corrosion hazard, there are some notable exceptions. Magnesium-to-wood assemblies present an unusual problem because of the water absorbency of wood and the tendency of the assemblies to leach out natural acids. To protect magnesium from attack, the wood should first be sealed with paint or varnish, and the faying surface of the magnesium treated as described above for magnesium-to-magnesium assemblies. The joining of magnesium to carbon fiber reinforced plastics is another exception that, in the presence of a common electrolyte, could result in corrosion of the magnesium unless similar assembly precautions are observed.

Magnesium-to-Dissimilar-Metal Assemblies. Several techniques can be implemented to minimize or eliminate galvanic corrosion or, if breakdown occurs, to reduce its effect in magnesium-to-dissimilar-metal couples. These include:

- Elimination of the common electrolyte
- Reduction of the relative area of dissimilar metal present
- Minimization of the potential difference of the dissimilar metal
- Protection of the dissimilar metal and the magnesium from the common electrolyte

Good design can play a vital role in reducing the threat of galvanic corrosion (Fig. 12). Elimination of a common electrolyte may be possible by the provision of a simple drain hole or shield to prevent liquid entrapment at the dissimilar-metal junction. Alternatively, the location of screws or bolts on raised bosses may also help avoid common electrolyte contact, as would use of nylon washers, spacers, or similar moisture-impermeable gaskets. The use of studs in place of bolts will reduce the area of dissimilar metal exposed by up to 50% provided the captive ends of the studs are located in blind holes.

The degree of attack resulting from galvanic corrosion is, among other things, proportional to the potential difference between the metals involved. Consequently, this should be reduced to a minimum by careful material selection or the use of selected plating or coating of metals brought into contact with magnesium.

Dissimilar metals that are relatively compatible with magnesium are the aluminum-magnesium (5000 series) or aluminum-magnesium-silicon (6000 series) aluminum alloys, which should be used for washers, shims, fasteners (rivets and special bolts), and structural members, where possible. Other aluminum alloys, steels, titanium, copper, brass, monel, and so on, will corrode magnesium when coupled with it under corrosive conditions, and protection is therefore required.

Aluminum, zinc, cadmium, and tin are used to coat steel or brass components to reduce the galvanic couple with magnesium. Reducing the potential difference or plating using materials with high hydrogen overvoltages will help reduce galvanic corrosion under mildly corrosive environments, but will have minimal effect in severe environments; additional precautions are required for corrosion protection.

Use of wet assembly techniques, as discussed previously, will eliminate galvanic corrosion in crevices. Caulking the metal junctions will increase the electrical resistance of the galvanic couple by lengthening the electrolytic path and thus reduce the degree of attack should it occur (Fig. 13). Vinyl tapes have also been used to separate magnesium from dissimilar metals or a common electrolyte and thus prevent galvanic attack (Fig. 14). Finally, painting the magnesium and, more important, the dissimilar metal after assembly will effectively insulate the two materials externally from any common electrolyte.

Protective Coating Systems

Inorganic Surface Treatments. A full range of chemical and electrochemical cleaning and surface pretreatments before application of paint finishes is available for magnesium. Whichever

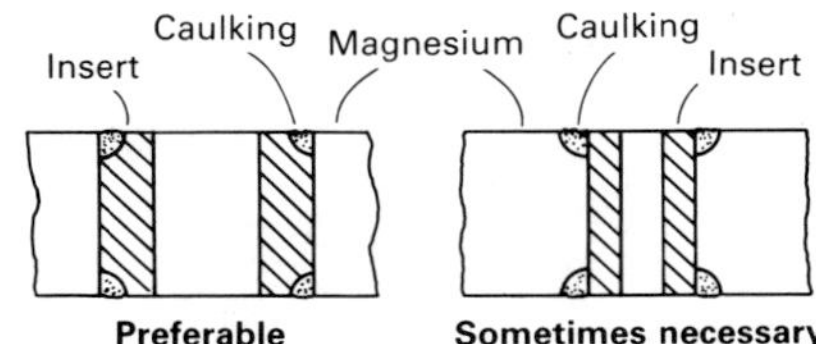

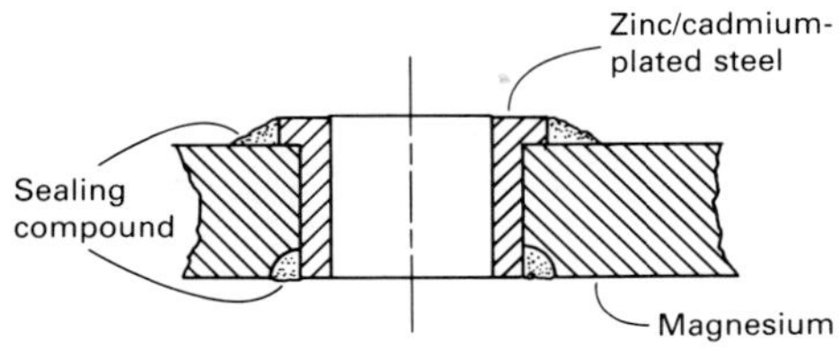

Fig. 13 Examples of good practice for bushing installations

pretreatment is selected, it must be applied to a clean metal surface. In the case of magnesium, this implies the removal of oil, dirt, or grease, and more important, a surface free of other contaminants. Heavy-metal contamination arising from blasting, brushing, tumbling, lapping, and other abrasive operations is particularly detrimental, as is contamination from graphite-containing die-forming lubricants. The use of abrasive materials compatible with magnesium—for example, high-purity alumina, silicon carbide, and glass—will help ensure that heavy metal pickup is kept to a minimum.

Oil, dirt, and grease are removed by conventional solvent immersion or vapor-degreasing techniques using chlorinated solvents. Alkali cleaning in high-pH cleaners is also suitable. Oxides, die-forming compounds, and other surface contaminants are removed by a range of acid-pickling solutions. Details are given in Table 13. In addition, an electrochemical process known as fluoride anodizing will more effectively remove sand or heavy-metal contamination. This process is also applicable to finished work when dimensional losses cannot be tolerated.

The primary function of dip or anodic coatings on magnesium is to provide a suitable surface to promote the adhesion of subsequent organic coatings. Conversion coatings should not be regarded as protective treatments in their own right unless they are to be exposed only to noncorrosive environments. Under these conditions, they will delay the onset of natural surface oxidation and may provide a more visually attractive surface appearance.

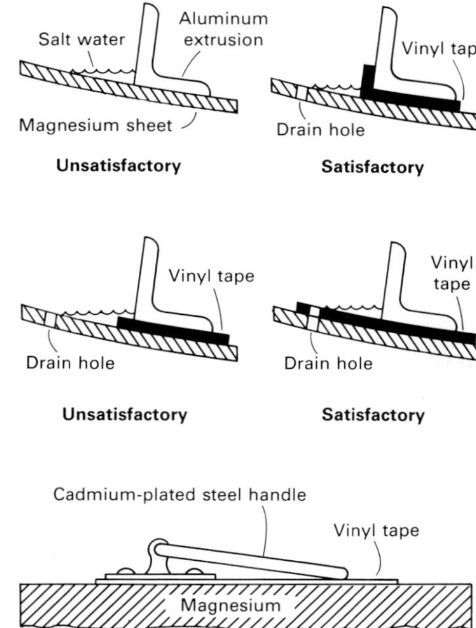

Fig. 14 Use of insulating tapes to avoid galvanic corrosion

Table 13 Some chemical cleaning treatments for magnesium alloys

Type	Composition(a)	Operating conditions: Time, min	Temperature, °C	Temperature, °F	Typical metal removal, mm/surface	Typical metal removal, in./surface	Comments
Nitric acid	50–100 mL 70% HNO_3, to 1 mL H_2O	1/2–1 1/2	21–27	70–80	0.01–0.05	0.0004–0.002	General cleaning of rough castings, forgings, etc.
Acetic-nitrate	200 mL glacial acetic acid, 50 g $NaNO_3$ (sodium nitrate), to 1 L H_2O	1/2–1	21–27	70–80	0.012–0.025	0.0005–0.0009	Removal of mill scale and other surface contamination from wrought products
Chromic-nitrate	180 g CrO_3, 30 g $NaNO_3$, to 1 L H_2O	2–20	21–32	70–90	0.012–0.025	0.0005–0.0009	Removal of mill scale and graphite lubricants from wrought or founded products
Hydrofluoric-sulfuric	250 mL 60% HF, 31 mL 96% H_2SO_4, to 1 L H_2O	2–5	21–32	70–90	0.003	0.0001	Brightens die castings. Improves response to chemical pretreatment
Chromic-nitric-hydrofluoric	280 g CrO_3, 8 mL 60% HF, 25 mL 70% HNO_3, to 1 L H_2O	1/2–2	21–32	70–90	0.012–0.025	0.0005–0.0009	Removal of surface segregation from die castings to leave a smut-free surface. Improves response to chemical treatment
Chromic acid	100–200 g CrO_3, to 1 L H_2O	1–15	90–100	195–212	Negligible		Removal of oxides, corrosion product, and conversion coatings. Negligible metal removal providing bath uncontaminated by chlorides, sulfates, etc. Use silver chromate addition to control chloride contamination.
Chromic-sulfuric	100 g CrO_3, 10 mL 96% H_2SO_4, to 1 L H_2O	Swab until clean	21–32	70–90	Negligible		Local removal of superficial corrosion product
Nitric-sulfuric	80 mL 70% HNO_3, 20 mL 96% H_2SO_4, to 1 L H_2O	10–15 s	21–32	70–90	0.05	0.002	Preliminary treatment for sand castings to remove surface- contaminating effects of blast cleaning
Nitric acid/hydrofluoric acid treatment	(1) 50–100 mL 70% HNO_3, to 1 L H_2O (2) 100 mL 60% HF, to 1 L H_2O	Up to 2 min in (1), rinse, then 15 min in (2)	21–27	70–80	0.05	0.002	Removal of heavy metal contamination from surface of rough castings. Second-stage immersion removes reprecipitated contaminants remaining after first- stage cleaning.
Fluoride, anodizing	150–250 g $NH_4F{\cdot}HF$ (ammonium bifluoride), to 1 L H_2O	AC anodize at 200 A/m^2 (0.13 A/in.^{2}) up to 120 V	30	85 max	Negligible		Super cleaning of heavy metal contaminated surfaces. Fluoride film formed requires removal in H_2CrO_4 before chemical pretreatment.

(a) Whenever water is specified, use deionized water.

Table 14 Some chemical conversion coating treatments for magnesium alloys

Name	Bath composition(a)	Procedure	Appearance	Typical metal removal	Comments
Chrome pickle (acid chromate)	180 g $Na_2Cr_2O_7{\cdot}2H_2O$ (sodium dichromate), 187 mL 70% HNO_3, to 1 L H_2O	1/2- to 2-min immersion at room temperature; allow to drain for 5–30 s; rinse in cold water, then hot water to aid drying	Golden yellow, often with iridescence	Up to 0.015 mm (0.0006 in.)	Applicable to all alloys and forms; mainly applied to wrought and die castings; good paint base
Modified chrome pickle	15 g $NaHF_2$ (sodium acid fluoride), 180 g $Na_2Cr_2O_7{\cdot}2H_2O$, 10 g $Al_2(SO_4)_3{\cdot}14H_2O$ (aluminum sulfate), 125 mL 70% HNO_3, to 1 L H_2O	1/2- to 2-min immersion at room temperature; allow to drain for 5 s; rinse in cold water, then hot water to aid drying	Yellow-red iridescence to gold	Up to 0.012 mm (0.0005 in.)	Particularly suited to treatment of die castings; prepickle in $HF{\cdot}H_2SO_4$ mixture or hot alkaline clean; good paint base
Galvanic dichromate	(1) 50 g $NH_4F{\cdot}HF$ (ammonium bifluoride) (or sodium or potassium bifluoride), to 1 L H_2O	Immerse in activator for 5 min and rinse	Dark brown to black	Negligible	Applicable to all alloys and forms; matte black film useful for optical applications; good paint base
	(2) 30 g $(NH_4)_2SO_4$ (ammonium sulfate), 30 g $Na_2Cr_2O_7{\cdot}2H_2O$, 2.6 mL 0.880 NH_4OH, to 1 L H_2O	10- to 30-min immersion at 50–60 °C (120–140 °F) with parts coupled to low-carbon steel cathode; bath pH 5.6–6.2			
Dichromate	(1) 50 g $NH_4F{\cdot}HF$, or 187 mL 60% HF, to 1 L H_2O	5-min immersion in activator at room temperature, except for AZ31 alloy, which should only be immersed for 1/2–1 min if HF activator is used; rinse thoroughly	Brassy to dark brown	Negligible	Applicable to most alloys and all forms; as-cast die-cast surfaces should be prepickled to remove skin segregation; excellent paint base
	(2) 180 g $Na_2Cr_2O_7{\cdot}2H_2O$, 2.5 g CaF_2 or MgF_2 (calcium or magnesium fluoride), to 1 L H_2O	Immersion for 30 min in boiling solution (95 °C, or 205 °F, min); maintain pH 4.0–5.5; rinse and dry			
Chrome-manganese	100 g $Na_2Cr_2O_7{\cdot}2H_2O$, 50 g $MnSO_4{\cdot}5H_2O$ (manganese sulfate), 50 g $MgSO_4{\cdot}7H_2O$ (magnesium sulfate), to 1 L H_2O	Up to 2-h immersion at room temperature, proportionately less at higher temperatures, e.g. 10 min at boiling; maintain pH 4.0–6.0; rinse and dry	Dark brown to black	Negligible	Applicable to most alloys and all forms; as-cast die-cast surfaces should be prepickled to remove skin segregation; excellent paint base
Dilute chromic acid	10 g CrO_3, 7.5 g $CaSO_4{\cdot}2H_2O$ (calcium sulfate), to 1 L H_2O	Immerse or swab for 1–2 min; rinse and hot air dry	Brassy to brown	Negligible	For touch-up use or as complete treatment; moderate paint base; solution should be stirred or shaken vigorously before use
Iridite mag-coat	37.5 g proprietary Iridite 15 chromate compound, 58.5 mL 37% HCl, 0.26 mL proprietary wetting agent, to 1 L H_2O	Immerse or swab for 15–30 s at 21–32 °C (70–90 °F); maintain pH 0.2–0.6; rinse thoroughly in cold water and hot air dry	Brown to dark brown	Slight, up to 0.003 mm (0.0001 in.)	For touch-up use or as complete treatment; as-cast surfaces should be prepickled in $H_2CrO_4{\cdot}HNO_3{\cdot}HF$ to remove skin segregation; good paint base
Parker phosphate	(1) Proprietary Parco Coater 2557 (2) Proprietary Parcolene "Dilute"	3- to 5-min immersion at 55–70 °C (130–160 °F) followed by hot water rinse, then immerse for 15–45 s in hot supplementary treatment and hot air dry	Gray to matte silver	Negligible	Paint base for mild environments only when applied to high-purity alloys; not recommended for corrosive environments; nonchromate-containing treatment
Amchem phosphate	(1) Proprietary, Prep-n-Cote 978 (2) Deoxylite 41 (optional)	1- to 5-min immersion or spray at 38 °C (100 °F); water rinse; final rinse at 60 °C (140 °F) with Deoxylite	Light gray to beige	Negligible	Paint base for mild environments when applied to high-purity alloys; provides option of a chromium-free treatment
Dilute chromate; NH35	2.5 g $NaHF_2$, 2.5 g $Na_2Cr_2O_7{\cdot}2H_2O$, 3 g $MgSO_4{\cdot}7H_2O$, 32 mL 65% HNO_3, to 2 L H_2O	Clean (alkaline bath/or solvent); water rinse; NH35 treat 20–30 s; water rinse and dry	Yellow to yellow-brown low-pressure castings; yellow-brown to matte gray on high-pressure die castings	0.002 mm (0.00008 in.)	A dilute chromate treatment developed primarily for high-pressure die-cast alloys; is near equivalent of standard chromates in both shelf life and paint base performance

(a) Whenever water is specified, use deionized water.

Table 15 Details of two hard-anodizing treatments for magnesium alloys

Name	Bath composition(a)	Anodizing conditions: Time, min	Temperature °C	Temperature °F	Coating appearance	Coating buildup mm	Coating buildup in.	Comments
HAE	135–165 g KOH, 34 g $Al(OH)_3$ (aluminum hydroxide), 34 g KF (potassium fluoride), 34 g Na_3PO_4 (trisodium phosphate), 20 g K_2MnO_4 (potassium manganate), to 1 L H_2O	8 min at 200 A/m^2 (0.13 A/in.^{2}) for thin coating (70 V)	15–30 max (cooling required)	60–85	Light tan (thin)	0.005	0.0002	Applicable to all alloys and forms. Thin coating provides excellent paint base. Thick coating provides excellent wear resistance, and if sealed with organic resins, provides superior corrosion protection as well. Process has good throwing power.
		60 min at 250 A/m^2 (0.16 A/in.^{2}) for thick coating (90 V) (AC anodize)			Dark brown (thick)	0.040	0.0016	
No. 17	240 g $NH_4F{\cdot}HF$, 100 g $Na_2Cr_2O_7{\cdot}2H_2O$, 90 mL 85% H_3PO_4, to 1 L H_2O	5 min at 200 A/m^2 (0.13 A/in.^{2}) for thin coating (70 V)	70–80	160–175	Light green (thin)	0.006	0.0002	Applicable to all alloys and forms. Thin coating provides excellent paint base. Thick coating gives good wear resistance and, if sealed with organic resins, provides superior corrosion protection. Process has excellent throwing power.
		25 min at 200 A/m^2 (0.13 A/in.^{2}) for thick coating (90 V) (AC anodize)			Dark green (thick)	0.030	0.0012	

(a) Whenever water is specified, use deionized water.

Table 14 lists some of the treatments used worldwide, together with brief application details and uses. Some treatments are quick and inexpensive to use, but may not provide as good a paint base as others. Consequently, their use should be restricted to mildly corrosive environments.

Several hard-anodizing treatments are available for magnesium, but the most commonly used are the No. 17 and HAE treatments (Table 15). Both may be applied as thin (0.005 mm, or 0.0002 in.) or thick (0.038 mm, or 0.0015 in.) coatings, with the thicker treatments imparting wear and abrasion resistance. These coatings are porous and provide excellent bases for subsequent painting. However, particularly for the thicker films, conventional painting may not completely seal the anodic pores. To prevent the risk of subsurface lateral corrosion spread from a point of damage, resin impregnation is used for maximum serviceability in aggressive corrosive environments. Inorganic chemical post-treatments are sometimes used to impregnate the anodic film with corrosion inhibitors, but these treatments can be detrimental to subsequently applied organic coatings and are not as effective as resin impregnation.

Organic Coatings. Adhesion and subsequent corrosion protection to cleaned and pretreated magnesium surfaces are enhanced by the use of alkali-resistant paint systems. Paints based on epoxy, epoxy ester, phenolic, polyurethane, vinyl, acrylic, polyester, silicone, and epoxy silicone systems are generally suitable. Those based on linseed, soya or other oils, alkyds, or nitrocellulose are best avoided unless applied for decorative purposes only.

Inhibiting pigments, such as strontium or zinc chromates, are often incorporated into primer systems for magnesium. These slightly soluble compounds will release chromate ions to retard subsequent corrosion should the paint film be damaged. Chromate inhibitors are, however, not as effective on magnesium as on aluminum alloys, particularly under corrosive conditions. Primers containing metallic zinc, lead, or any other metallic pigmentation should not be used on magnesium.

The full range of application techniques can be used. These include brushing, dipping, solvent or electrostatic powder spray, and electrophoretic techniques. Multilayer compatible coatings of primer, filler, and top coat systems will provide optimum protection in corrosive environments. Use of high-temperature stoving (baking) systems is also beneficial in developing maximum resistance to moisture permeability.

General Recommendations. Selection of a suitable protective scheme depends on many factors, especially the expected operational environment, design life, inspection and maintenance costs, the component cost, and of course the cost of original surface protection. For new applications, it is advisable to err on the side of overprotection until enough experience is gained to enable a more valued judgment to be made.

For indoor and similar noncorrosive environments, surface protection requirements are min-

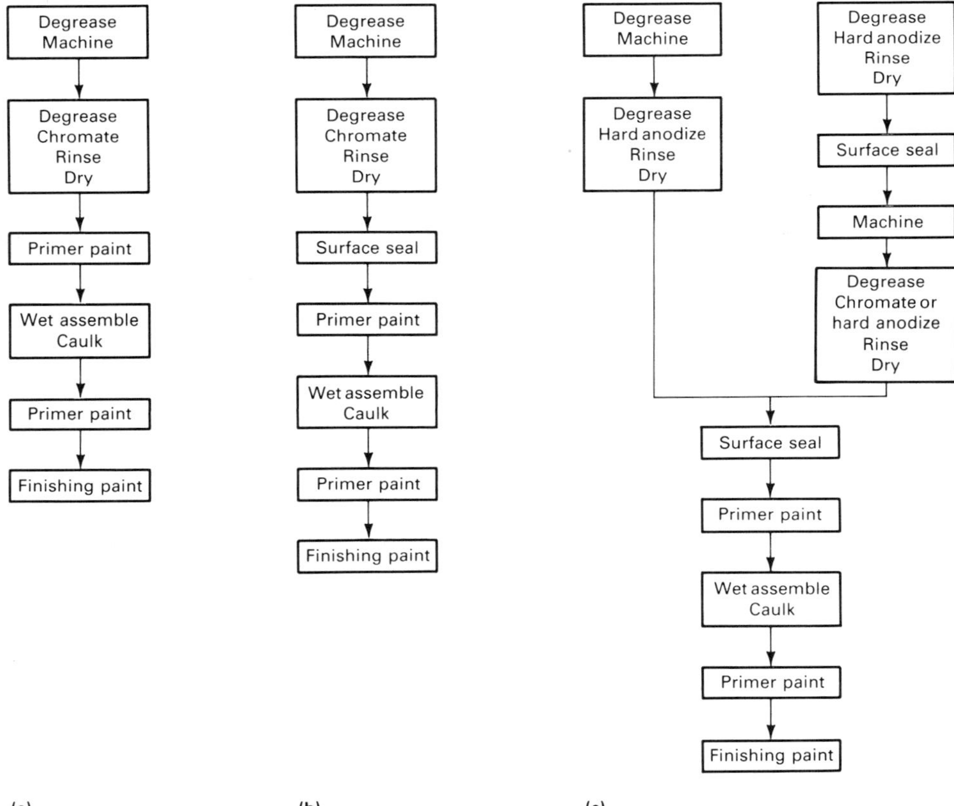

Fig. 15 Diagram of protection schemes for critical applications in corrosive environments. (a) For moderately corrosive environments. (b) For severely corrosive environments. (c) For severely corrosive environments with risk of abrasion or damage

imal and may range from none to simple chromate or phosphate conversion coatings with primer only or a decorative paint finish. Even under apparently more corrosive conditions, other mitigating factors, such as the use of high-purity alloys, good ventilation, component warmth, good design, and oil films, will enable magnesium components to be used with little or no protection.

For mildly or moderately corrosive environments, a chromate pretreatment followed by one coat of suitable primer and one or more coats of compatible finish should be applied. Under these conditions, the effect of galvanic couples, if present, should be considered, and some precautions taken, as previously outlined. It is expected that most commercial applications would be covered by this and the preceding general recommendation. (See the section "Proven Protection Systems" in this article.)

For moderately to severely corrosive environments, good-quality chromate conversion coatings or thin anodic pretreatments should be used. Chromate-inhibited epoxy primer systems, careful wet assembly procedures, and painting after assembly with primer and top coat are recommended. Use of low temperature baking paints is beneficial for improving humidity resistance (Fig. 15a).

For severely corrosive environments, for which maximum chemical, salt spray, and humidity resistance are required (Fig. 15b), specialized paints and coating techniques are used. Good-quality chromate or anodic pretreatments are required. Use of thick anodic coatings will also impart a measure of abrasion and damage resistance (Fig. 15c). Pretreatments should then be sealed with high temperature baking organic resins. This is achieved by a process known as Surface Sealing (MIL-M-3171 Section 3.9.3, MIL-M-46080), in which three coats of thinned resin are applied to a preheated component by spraying or, preferably, by dipping. Epoxy resin systems are preferred (MIL-C-46079), although the technique would be beneficial for other high temperature baking resins, such as phenolics and epoxy silicones.

After this foundation treatment, full wet assembly procedures, including caulking of joints, should be performed before application of a cold or low temperature curing chromate-inhibited primer and a compatible top coat system. For additional protection, a high temperature baking paint system should be maintained throughout. The above recommendations are necessarily very general in scope, but represent the various protection schemes in worldwide use on magnesium.

Proven Protection Systems

Aerospace Applications. Even within this specialized area, many surface protection schemes that reflect the differing operational environments encountered are in use.

For military and helicopter applications, comprehensive protection schemes are required to achieve extended component life and to reduce maintenance costs. Such schemes, as recommended for severely corrosive environments in the previous section, are required, and indeed, are mandatory in certain countries. Full wet assembly procedures and the coating of all exposed surfaces are essential. These schemes have given total protection for thousands of hours in various accelerated testing programs and will minimize corrosion spread.

For civil and other less aggressive aerospace applications, the protection schemes outlined for moderately corrosive environments are applicable. Good-quality aerospace paint systems should be used with wet assembly techniques.

Automotive Applications. The most notable automotive use of magnesium has undoubtedly been for the engine and transmission casings on the Volkswagen Beetle. For this application, no protective treatment was required to obtain long and reliable service life because of the mitigating environmental factors previously outlined. More modern automotive applications include air cleaner covers, engine compartment grills, retractable headlight assemblies, clutch and brake pedal supports, and clutch and transmission housings. For some applications, no protection is required. For others, where aesthetic appearance or corrosion protection is needed, the protective schemes outlined for mildly corrosive environments are suitable.

Cathodic electrophoretic epoxy primers applied to chromic acid pickled or dichromate pretreatments have proved effective, particularly in conjunction with high-purity AZ91D die castings. Electrostatic powder spray is used for top coating. Specially designed fasteners incorporating nylon or plastic washers, sleeves, shims, and so on, help reduce the risk of galvanic corrosion on exposed parts.

Some evaluations (Ref 17, 18), however, demonstrated that for structural underbody and road wheel applications, in which frequent exposure to water splash, stone impact damage, and the absence of mitigating environmental factors are problems, more comprehensive protection schemes may be required, together with protection against galvanic corrosion. Where applicable, use of underbody wax-type coatings can provide additional protection. It is expected that use of high-purity alloys will enable designers to specify many more magnesium automotive applications in the future.

Electronic and Computer Applications. Die-cast, investment-cast, and wrought magnesium components are used by computer and computer peripheral manufacturers for several applications where lightweight, low inertia, rigidity, and heat sink requirements preclude the use of other metals or plastics. The noncorrosive operational environment within the computer eliminates the need for galvanic-corrosion precautions.

For exterior housings, decorative surface treatments consisting of chromate pretreatment followed by textured epoxy powder coatings have proved satisfactory. Within disk drive units, where minute dust or other particles could cause disk or head failure, very thin (0.003 mm, or 0.0001 in.) specialized conformal coatings are applied to protect against atmospheric oxidation. Other similar commercial applications include housing or support frames in portable video equipment and a range of optical and medical electronic equipment.

Other Applications. Some magnesium applications may require surface protection against wear and corrosion. The use of resin-sealed hard-

Fig. 16 Magnesium-bodied atmospheric deep-sea diving suit

anodizing treatments, particularly thick HAE treatments, will provide excellent abrasion and corrosion resistance, but in some applications—for example, pulley wheels—will themselves cause excessive wear on other components. Under these conditions, dry-lubricant coatings are required. Nylon coatings applied by electrostatic powder or fluidized bed techniques onto a chromated and suitably primed magnesium surfaces have proved effective for use in aircraft pulley control systems. Thick nylon coatings also offer good damage and erosion resistance. Other systems, based on fluorocarbon (Teflon) impregnation of anodic pretreatments or resin-bonded solvent-base fluoropolymer coatings, are also effective in providing combined corrosion resistance and lubricity.

For high-temperature (above 200 °C, or 390 °F) applications, silicone-, epoxy-silicone-, and polyimide-based paints will provide effective corrosion-resistant coatings on magnesium. In environments where lubricants may be present, polyimide coatings are preferred, particularly when applied onto No. 17 anodic pretreatment.

There are a few applications in which magnesium components may be subjected to prolonged immersion or contact with corrosive electrolyte. In some systems, it may be possible to add corrosion-inhibiting agents to the electrolyte. Maintaining electrolyte pH above 10.5 or adding soluble chromates or neutral fluorides is effective in reducing magnesium-base metal corrosion. In other applications, comprehensive surface protection schemes and good maintenance are essential to achieve satisfactory service life. One such example is an atmospheric pressure deep-sea diving suit, the body and helmet of which are in magnesium alloy (Fig. 16). Surface protection consists of a thick HAE anodic film that is surface sealed with high temperature stoving epoxy resin, full wet assembly procedures, and final painting with primer and top coat. Coupled with good maintenance, this protection scheme has given satisfactory service between major overhaul intervals of 4 years.

Some special applications may require the metal plating of a magnesium component. Zinc and nickel can be directly plated onto magnesium from special electroless baths. Other metals may be electroplated from standard plating baths after surface cleaning, activation, zinc immersion coating, and a copper strike. It should be emphasized that metal plating of magnesium is not a corrosion-protective coating and that plated components should not be exposed to severely corrosive environments. Plated magnesium die castings are suitable, however, for such applications as interior automotive door handles and window cranks, where it is necessary to combine weight reduction with strength durability and good appearance.

REFERENCES

1. J.D. Hanawalt, C.E. Nelson, and J.A. Peloubet, *Trans. AIME*, Vol 147, 1942, p 273
2. M.R. Bothwell, in *The Corrosion of Light Metals*, John Wiley & Sons, 1967, p 269
3. J.E. Hillis and K.N. Reichek, Paper 860288, Society of Automotive Engineers, 1986
4. K.N. Reichek, K.J. Clark, and J.E. Hillis, Paper 850417, Society of Automotive Engineers, 1985
5. Dow Chemical Company, unpublished research
6. E.F. Emley, *Principles of Magnesium Technology*, Pergamon Press, 1966, p 685
7. I. Lunder, T.K. Aune, and K. Nisancioglu, Paper 382, presented at Corrosion/85 Conference, National Association of Corrosion Engineers, 1985
8. E.F. Emley, *Principles of Magnesium Technology*, Pergamon Press, 1966, p 692-695
9. M.R. Bothwell, in *The Corrosion of Light Metals*, John Wiley & Sons, 1967, p 84
10. D.L. Hawke, Paper 860285, Society of Automotive Engineers, 1986
11. "Standard Method for Salt Spray (Fog) Testing," B 117, *Annual Book of ASTM Standards*, American Society for Testing and Materials
12. AMAX Magnesium, unpublished research
13. M.R. Bothwell, in *The Corrosion of Light Metals*, John Wiley & Sons, 1967, p 292
14. M.R. Bothwell, in *The Corrosion of Light Metals*, John Wiley & Sons, 1967, p 291
15. M.R. Bothwell, *J. Electrochem. Soc.*, Vol 106, 1959, p 1021
16. F.L. LaQue, Corrosion Testing, in *Proceedings of the American Society for Testing and Materials*, Vol 51, 1951, p 557
17. F. Kaumle, N.C. Toemmeraas, and J.A. Bolstad, Paper 850420, Society of Automotive Engineers, 1985
18. A. Wickberg and R. Ericsson, Paper 850418, Society of Automotive Engineers, 1985

Corrosion of Zinc

Dale C.H. Nevison, Zinc Information Center, Ltd.

CORROSION LOSSES in the United States amount to approximately $167 billion annually. It is estimated that about 35% of this loss is avoidable through the application of corrosion control techniques already developed and available for use. More important, much greater savings could be realized if corrosion were more thoroughly understood and if proper specifications for dealing with it were used at the initial stage of design.

Most of the money spent on corrosion control is used to protect steel. Approximately 40% of domestic steel production is devoted to replacing steel that has failed because of corrosion. Although steel is strong and easy to form and fabricate, it corrodes rapidly if left unprotected.

Therefore, it would appear that the simplest method of hindering corrosion on a metal surface is to seal it with a coating that prevents the access of moisture and oxygen. Such coatings need to be strongly resistant to the corrosive influence of their environment and sufficiently impermeable to prevent oxygen and water vapor from penetrating the metal surface. They also need to be tough, abrasion resistant, and strongly adherent to the surface they are required to protect.

Coatings of metallic zinc are generally regarded as the most economical means of protecting steel against corrosion. Zinc offers threefold protection in that it:

- Provides a rugged, lasting sheath that seals the underlying metal from contact with its corrosive environment
- Provides galvanic (sacrificial) protection when zinc-coated steels are subjected to mechanical damage and when the base metal is subsequently exposed to the environment
- Provides further protective action at minor discontinuities in the coating as a result of the selective dissolution of the coating, because comparatively bulky corrosion products accumulate in the damaged area and tend to form a barrier to further electrochemical action

It is not the purpose of this article to explicate the electrochemical reactions that occur at the surfaces of the anode and cathode in an electrolytic or galvanic cell. However, some discussion of the theory is necessary to provide information on corrosion principles that will illustrate why zinc is useful in commercial cathodic protection systems.

Galvanic Series for Metals

All metals and alloys can be arranged in decreasing order of activity, as shown in Table 1. When a metal near the anodic (tendency to lose electrons) end of the series is electrically connected to a metal below it and when both are immersed in an electrolyte, a galvanic cell is established, and electrical current flows from the more anodic metal and through the electrolyte to the other more cathodic (tendency to gain electrons) metal; this shifts the potential of the latter metal in a more electronegative direction. Current flows because of the natural potential difference that exists between different metals. When two metals of a galvanic cell are widely separated in the galvanic series, there is a greater flow of current because of the larger potential difference that exists between them than when metals closer to one another are involved.

Figure 1 illustrates a typical galvanic, or sacrificial anode, corrosion cell that contains the four elements necessary for corrosion to proceed: an anode, a cathode, an electrolyte, and an electron path. Zinc ions that are positively charged go into solution, and the electrons migrate to the cathode through the electron path. When the electrolyte is water, the reduction of dissolved oxygen occurs first at the surface:

$$O_2 + 2H_2O + 4e^- \rightarrow 4OH^- \quad \text{(Eq 1)}$$

Once the dissolved oxygen has become depleted or current limited (that is, insufficient diffusion of oxygen to support an upper current limit) in the electrolyte, hydrogen ions react with electrons to form hydrogen gas:

$$2H^+ + 2e^- \rightarrow H_2 \quad \text{(Eq 2)}$$

The evolution of hydrogen gas at the cathode is generally visible evidence that a reaction (reduction) at the cathode has occurred. The evolution of hydrogen at the cathode increases the cell resistance because of hydrogen gas bubbles on the cathode, with a resultant decrease in current flow from the anode. However, dissolved oxygen may react with excess hydrogen ions (to form water) at the cathode and increase the reaction rate. Under these conditions, greater current will flow from the anode.

Thus, when the zinc metal ions of the anode enter the solution, a sacrificial corrosion of the anode occurs and the steel cathode is protected. The corroding areas on a metal surface are at the anode, where an electric current flows out of the metal, and the protected areas are at the cathode, where the current flows into the metal. This is what occurs in a cathodic (galvanic) protection system. This type of sacrificial or cathodic protection also occurs when zinc coatings on steel surfaces are subjected to mechanical damage in which the continuity of the zinc coating is broken and the steel surface is exposed.

The scale indicates that magnesium, aluminum, and cadmium should also protect steel. In most normal applications magnesium is highly reactive and is too rapidly consumed. Aluminum forms a resistant oxide coating, and its effectiveness in providing cathodic protection is limited. Cadmium provides the same cathodic protection for steel as zinc, but its applications are limited for technical and economic reasons to small electroplated parts, such as fasteners.

The galvanic series is typical of those for corrosive solutions, such as seawater. However, in the presence of certain electrolytes, the relative positions of some of the metals listed can be altered; an important example occurs when tinned steel is in contact with fruit juice, in which case tin is anodic to the base metal. In general, however, the protection afforded by tin or nickel

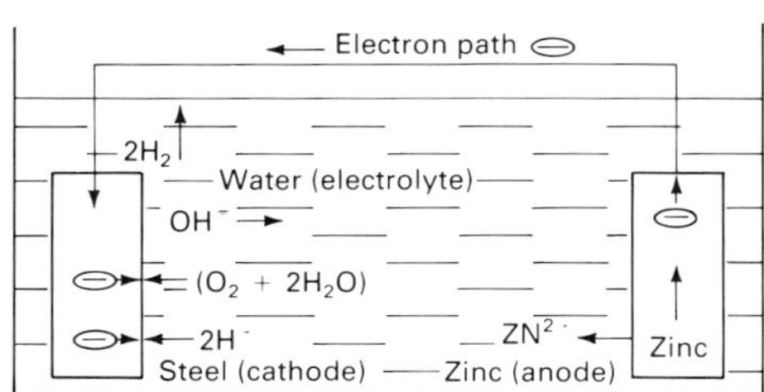

Fig. 1 Typical galvanic corrosion cell showing mechanism of sacrificial behavior of zinc in protecting steel against corrosion

Table 1 Galvanic series of metals

Corroded End—Anodic or less noble (Electronegative)
Magnesium
Zinc
Aluminum
Cadmium
Iron or steel
Stainless steels (active)
Soft solders
Tin
Lead
Nickel
Brass
Bronzes
Nickel-copper alloys
Copper
Stainless steels (passive)
Silver solder
Silver
Gold
Platinum
Protected End—Cathodic or most noble (Electropositive)

Note: Any one of these metals and alloys will theoretically corrode while protecting any other that is lower in the series as long as both form part of an electric circuit.

coatings on steel is effective only as long as the coating is continuous and free from pores or other discontinuities. When there are breaks in the coating, electrolytic action between the coating and the basis metal accelerates the dissolution of the steel; such dissolution can proceed at the steel surface beneath the cathodically protected metallic coating until breakdown occurs over a large area.

Upon exposure to environments other than highly acidic atmospheric pollution, zinc forms a self-protecting film of fairly impermeable basic corrosion products. This protects the metal from further attack. If the coating is continuous, the rate at which the coating is attacked is the only factor. If the coating is not continuous or if it becomes discontinuous because of weathering, pores, flaws and cracks, then the electrochemical property or anodic nature of zinc becomes effective in providing protection.

Life of Zinc Coatings

There are different concepts regarding what constitutes the useful life of a zinc coating. These include the first visible evidence of rust, time to perforation, and time to breakage. Each concept has its relevance. In the latter two, the thickness and corrosion resistance of the base metal are factors in the useful life of the zinc-coated items. Another generally accepted criterion of the useful life of a protective coating, especially where appearance is of minor importance, is the period of time during which the coating prevents failure of the base metal.

Basically, the degree of corrosion protection or the life of a zinc coating can be evaluated in terms of the rate at which the coating loses weight or thickness in a given environment. The most practical way of expressing the corrosion resistance (the response of zinc coatings to different exposure conditions) is in terms of the area on which rust appears as a result of exposed steel.

The protective value of zinc coatings in American Society for Testing and Materials (ASTM) tests is expressed in terms of the percentage of the surface showing rust. This method of evaluation is realistic for several reasons. Corrosion takes place initially at the most vulnerable spots and then spreads, because of such influences as nonuniform coating thickness, the nonuniformity of wetting and drying of different parts of a specimen, the presence of bare metal where the steel has been sheared, and the presence in the coating of an alloy layer that may corrode at a rate different from that of the relatively pure zinc surface. The curves representing the progress of rust with time are S shaped.

Coating Thickness and Uniformity. Although the corrosion rate of zinc in atmospheres is low, it must be remembered that galvanized coatings are normally quite thin. The weight of a zinc coating is always expressed in terms of ounces per square foot of surface. For galvanized sheet, the weight of coating is given in ounces per square foot of sheet; because the sheet is coated on both sides, this is twice the average weight of coating per square foot of surface. Thus, a coating of 1.25 oz/ft^2 (380 g/m^2) has a total of 1.25 oz of zinc on the front and back surfaces of each square foot of sheet. Therefore, the average weight of coating per square foot of each surface of the sheet is 0.625 oz (17.7 g), or slightly more than 1 mil (0.025 mm) thick. It is important to specify a coating that is thick enough for the intended service. Uniformity of coating thickness is also significant, especially in light coatings. When coatings are uneven, the thinner areas fail even though there is ample zinc elsewhere on the surface. With a heavy coating, variations in thickness are less important, since the sacrificial protection of the exposed steel is more effective because of the increased zinc layer.

Comparison of the protective value of different zinc coatings may be expressed in terms of the average thickness or weight per unit area of zinc removed in a given area. Suggested units are millimeters per square decimeter per day, inches per year, and mils per year. This last unit is convenient because its reciprocal represents the life in years of a coating that is 1 mil (0.025 mm) thick.

To provide more complete information, these weight figures should be accompanied by equivalent thickness values. One ounce of zinc per square foot of surface is equivalent to a coating thickness of 1.7 mils (0.0017 in., or 0.043 mm). Any other coating thickness can be calculated from this value if the weight is given, or vice versa. The weights of coatings for many galvanized products are covered by various ASTM specifications. They do not represent the maximum weight of coating desirable for every application or the maximum obtainable. Specifications from ASTM should be consulted for a description of the methods of measuring properly the thickness of the coating.

In continuous or hand-dip galvanizing, the zinc coating consists of a relatively pure layer of zinc on the outside, followed by three layers of iron-zinc alloy phases with decreasing proportions of zinc and correspondingly higher percentages of iron from surface to interface. The aggregate alloy layer also provides corrosion protection to the base metal; when the pure zinc layer has finally corroded away, the reddish corrosion product of the exposed layer sometimes encountered should not be mistaken for rust of the underlying steel. On the contrary, this is the optimum stage at which paint should be applied.

Atmospheric Corrosion

The behavior of zinc and zinc coatings during atmospheric exposure has been closely examined in tests conducted throughout the world. The performance of zinc in a specific atmospheric environment can be predicted within reasonable limits.

Precise comparison of corrosion behavior in atmospheres is complex because of the many factors involved, such as prevailing wind direction, type and intensity of corrosive fumes, the amount of sea spray, and the relative periods of moisture or condensation and dryness. However, it is generally accepted that the corrosion rate of zinc is low; it ranges from 0.13 μm/yr (0.005 mil/yr) in dry rural atmospheres to 0.013 mm/yr (0.5 mil/yr) in more moist industrial atmospheres. Zinc is more corrosion resistant than steel in most natural atmospheres, the exceptions being ventilated indoor atmospheres where the corrosion of both steel and zinc is extremely low and certain highly corrosive industrial atmospheres. For example, in seacoast atmospheres, the corrosion rate of zinc is about 1/25 that of steel.

Zinc owes its high degree of resistance to atmospheric corrosion to the formation of insoluble basic carbonate films. Environmental conditions that interfere with the formation of such films may attack zinc quite rapidly. The most important factors that control the rate at which zinc corrodes in atmospheric exposures are:

- The duration and frequency of moisture contact
- The rate at which the surface dries
- The extent of industrial pollution of the atmosphere

The latter is the most important because the formation of basic corrosive films is prevented when the zinc is attacked by acidic moisture. This effect of industrial pollutants is well illustrated by considering that in highly industrial environments a 2-oz/ft^2 (610-g/m^2) zinc coating will begin to exhibit rusting after 4 years and may be 80% rusted in 10 years. A similar coating would not be expected to display rusting after 30 to 40 years of exposure in rural atmospheres and after 15 to 25 years in marine environments.

In dry air, zinc is slowly attacked by atmospheric oxygen. A thin, dense layer of oxide is formed on the surface of the zinc, and a porous outer layer then forms on top of it. Although the outer layer breaks away occasionally, the thin under layer remains and protects the metal by restricting its interaction with the oxygen. Under these conditions, which occur in some inland tropical climates, the zinc oxidizes very slowly.

The rate of drying is also an important factor because a thin moisture film with higher oxygen concentration promotes corrosion. For normal exposure conditions, the films dry quite rapidly, and only in sheltered areas are drying times slow enough to accelerate the attack of the zinc significantly.

In coastal districts and industrial areas, the air is contaminated with water containing considerable amounts of dissolved salts. Sodium chloride (NaCl) is of course the main salt present in marine atmospheres, and oxides of sulfur are the most important pollutants in industrial and urban atmospheres. The oxides of sulfur dissolve in water to form sulfuric acid (H_2SO_4), which in turn reacts with the zinc hydroxide ($Zn(OH)_2$) or zinc carbonate ($ZnCO_3$) formed on the surface to produce zinc sulfate ($ZnSO_4$). Because this salt is very soluble, it is easily removed by rainwater, and this leaves the zinc surface relatively unprotected. This solubility of the zinc corrosion products is one of the most important factors affecting the rate of corrosion of zinc in industrial atmospheres, and it primarily accounts for the fact that this rate may be as much as four or five times greater than that in rural or marine atmospheres, depending on the amount of sulfur present.

If a fresh zinc surface is allowed to stand with large drops of dew on it, as may easily happen if it is stored in a closed space in which the temperature varies periodically, it will be attacked by the oxygen dissolved in the drops of water. These drops of water conduct electricity slightly, and because the oxygen concentration is greater at the outer edges of the drops than at their centers, some electrochemical action results. Therefore, bulky deposits of porous zinc oxide (ZnO) or $Zn(OH)_2$ form on the surface instead of forming a protective layer all over it. This oxide quickly takes up carbon dioxide (CO_2) to form a basic carbonate, commonly known as wet-storage stain. This may form on any new zinc surface unless care is taken to store the metal in a dry airy place until the protective layer has formed over the entire surface. The formation of wet-storage stain may also be prevented by a simple chromate treatment.

Influence of Atmospheric Variables

Atmospheric corrosion has been defined to include corrosion by air temperatures between −18 to 70 °C (0 to 160 °F) in the open and in enclosed spaces of all kinds. Deterioration in the atmosphere is sometimes called weathering. This definition encompasses a great variety of environments of differing corrosivities. The factors that determine the corrosivity of an atmosphere include industrial pollution, marine pollution, humidity, temperature (especially the spread between daily highs and lows that influences condensation and evaporation of moisture) and rainfall. The atmosphere, as far as corrosion is concerned, is not a simple invariant environment. The influence of these factors on the corrosion of zinc is related to their effect on the initiation and growth of protective films.

Relative Humidity and Rainfall. In relatively dry air, the initial film formed on zinc surfaces is ZnO from the reaction of the zinc with atmospheric oxygen. This will be converted to a hydroxide in the presence of moisture. These films have a relatively minor protective effect. The $Zn(OH)_2$ reacts further with CO_2 in the atmosphere, which forms a basic $ZnCO_3$. This film is very protective and is mainly responsible for the excellent resistance of zinc to ordinary atmospheres.

The significance of atmospheric humidity in the corrosion of zinc is related to the conditions that may cause condensation of moisture on the metal surface and to the frequency and duration of the moisture contact. If the air temperature drops below the dew point, moisture will be deposited. Similarly, lowering the temperature of a metal surface below the air temperature in a humid atmosphere will cause moisture to condense on the metal. If the water evaporates quickly, corrosion is usually not severe, and a protective film is formed on the surface. On the other hand, water that remains in contact with zinc at high humidities, and particularly under poorly ventilated conditions, may cause severe corrosion.

Rainfall is not considered to be a source of serious corrosion unless it remains in contact with the zinc for some time, particularly if access to air is limited. The weight losses sometimes observed after a heavy rainfall indicate the washing away of soluble corrosion products that formed before the rainfall.

In some short-term tests, zinc kept under a roof, but otherwise exposed to the atmosphere, corroded at about the same rate as completely exposed zinc. Sulfur dioxide (SO_2), not rain, is the significant factor in the corrosive attack. In long-term tests, however, the specimens protected from falling rain corroded at a rate considerably slower than that for completely exposed specimens. This is attributed to the formation of a $ZnSO_4$ film on the samples under the roof that is not washed away by the rain. The surface is more acid and absorbs less SO_2 from the air. The higher corrosion rate in the open is not attributed directly to rainfall, but is attributed indirectly to the washing away of the protective film by the rain.

Results of tests on the effect of atmospheric pollution and rainfall on the corrosion of zinc conducted by the Corrosion Committee of the Iron and Steel Institute proved that the pH of the rainwater and the total solids contents of the air have some significance (Ref 1). In highly industrialized localities, the rainwater is contaminated by acidic sulfur compounds and becomes acidic enough to interfere with the formation of a protective film. This pH value is very significant because it interferes with the zinc formation of protective coatings of carbonates, sulfates, or oxides.

In acid rain environments, such coatings are never stabilized, because they are in a constant flux of dissolution. The United States and Canadian governments, as well as the Electric Power Research Institute, began investigations in the early 1980s on the acid rain corrosion of zinc and galvanized steel. These studies will provide detailed information on the long-term effects of acid rain on these metals.

Industrial and Marine Pollutants. In industrial or marine locations, condensed dew is likely to be contaminated with impurities that are corrosive to zinc. In such circumstances, the corrosivity of the contaminant may be more important than the degree of moisture condensation. Sulfur dioxide is one of the most harmful pollutants in the atmosphere, and it plays a major part in the corrosion of steel and zinc. Exposure tests showed that the correlation between sulfur pollution and corrosion is high for copper-bearing steel and for zinc, and these tests demonstrated that the SO_2 concentration in the air is the determining factor for the intensity of the corrosion of these metals.

Another series of tests found that even at test sites situated far from industrial towns the corrosion products contained a strong sulfate component derived from atmospheric sulfur compounds. This indicates that the effects of this type of pollution are far reaching (Ref 2).

Atmospheric Exposure Tests on Zinc

Several atmospheric exposure programs have been conducted throughout the world to obtain corrosion rate data for zinc exposed to representative natural atmospheres. These programs have provided quantitative evidence of the excellent resistance of zinc over a wide range of atmospheric conditions. Although there is considerable spread in terms of percentage in the corrosion rates observed, the actual corrosion rate rarely exceeded about 8 μm/yr (0.3 mil/yr) in average metal loss, even under the more severe conditions. This is well within all standards of acceptable corrosion performance.

A quantitative assessment of the relative corrosivity of different atmospheres can be seen from the results of the exposure tests reported by ASTM (Ref 3). The results calculated from ASTM data provide a comparison of the corrosion of steel and zinc on the basis of 45 worldwide locations (Table 2). From a cross-section of these exposure sites, it can be noted that the corrosion rates of steel and zinc vary significantly with different locations and climates.

The amount of zinc corrosion with increasing time of exposure depends on the type of atmosphere. In semi-industrial and rural atmospheres, the corrosion rate is approximately constant. On the other hand, the corrosion rate decreases with time in marine atmospheres and increases with time in industrial atmospheres. In all cases, the yearly change in corrosion rate decreases with time and approaches a steady rate. The equilibrium corrosion rate also depends on the environment. The rate in a rural atmosphere is only about one-fifth that in an industrial atmosphere and one-third that in a marine atmosphere.

Although neither iron nor zinc corrodes appreciably in the rural climate of Norman Wells Northwest Territory (N.W.T.), the atmosphere at the Geleta Point Beach site in the Panama Canal zone appears to be the most corrosive to these metals (Ref 4). Corrosivity can vary appreciably over very short distances. For example, a comparison of the corrosion rates for two Kure Beach, NC, sites shows that the corrosion rate for zinc at the 25-m (80-ft) site is roughly three times that of the 250-m (800-ft) site, although the sites were only 225 m (720 ft) apart.

The corrosion rate data for steel in the same atmosphere show that steel, in general, corrodes at a greater rate than zinc in all atmospheres. The progress of corrosion with time, however, may be quite different for the two metals; therefore, direct comparison of corrosion rates should specify the duration of exposure for which the comparisons are made. For example, the corrosion rate of zinc decreases with time in a severe seacoast environment (Kure Beach: 25-m, or 80-ft, site), but that of steel increases with time. On the other hand, the reverse is true in an industrial environment. The effect of the nature of the atmosphere on the relative corrosion rates as influenced by duration of exposure is indicated in Table 3.

Similar results were obtained in the very comprehensive series of tests conducted by ASTM (Ref 3). Table 4 presents a summary of the corrosion behavior of zinc based on 20-year ASTM exposure data.

Another important result was revealed in a series of tests by the Swedish Committee on Corrosion (Ref 2). A zinc-coated surface was found to be much less likely to suffer pitting attack than unprotected steel. The depth of pits on unprotected steel was up to six times the average loss of metal, but for zinc-coated steel, the ratio was only two or three.

Atmospheric Exposure Tests on Zinc Coatings. Measurements of the corrosion rates of zinc and zinc coatings have been made simultaneously in many parts of the world, and it has been found that the rates are substantially the same. Therefore, the same figures can be used to calculate the estimated lives of zinc coatings of a given weight that have been exposed in a particular environment.

The life of a zinc coating is difficult to define. Quantitative weight loss data are not always meaningful, because of the complicating effects of interface alloy layers and perforation of the coatings. Furthermore, the criterion for failure may vary with the type of coating and its intended applications. If a steel specimen were coated with a perfectly uniform zinc coating and if all parts were exposed to the same corrosive influences, the zinc coating would suddenly disappear from the entire surface, after which the exposed steel would corrode at its normal rate. This situation obviously does not exist in practice, because of the nonuniform distribution of the zinc coating over the surface and the resulting irregular penetration of the coating by corrosion. Therefore, arbitrary values, such as time to initial rusting or to some percentage of the total surface showing rust, have been used as a measure of the life of zinc-coated components.

In general, the underlying ferrous metal is protected as long as it can fulfill the intentions of the designer. Therefore, a galvanized roof has not

Table 2 Comparative rankings of 45 locations based on steel and zinc losses

Ranking: Zinc	Ranking: Steel	Location	2-year exposure, grams lost: Zinc	2-year exposure, grams lost: Steel	Steel:zinc loss ratio
1	1	Norman Wells, N.W.T., Canada	0.07	0.73	10.3
2	2	Phoenix, AZ	0.13	2.23	17.0
3	3	Saskatoon, Sask., Canada	0.13	2.77	21.0
4	4	Esquimalt, Vancouver Island, Canada	0.21	6.50	31.0
5	6	Fort Amidor Pier, Panama, Canal Zone (C.Z.)	0.28	7.10	25.2
6	13	Melbourne, Australia	0.34	12.70	37.4
7	8	Ottawa, Ontario, Canada	0.49	9.60	19.5
8	23	Miraflores, Panama, C.Z.	0.50	20.9	41.8
9	31	Cape Kennedy, 0.8 km (0.5 mile) from ocean	0.50	42.0	84.0
10	11	State College, PA	0.51	11.17	22.0
11	7	Morenci, MI	0.53	7.03	18.0
12	16	Middletown, OH	0.54	14.00	26.0
13	9	Potter County, PA	0.55	10.00	18.3
14	21	Bethlehem, PA	0.57	18.3	32.4
15	5	Detroit, MI	0.58	7.03	12.2
16	27	Manila, Philippine Islands	0.66	26.2	39.8
17	43	Point Reyes, CA	0.67	244.0	364.0
18	20	Trail, B.C., Canada	0.70	16.90	24.2
19	15	Durham, NH	0.70	13.30	19.0
20	14	Halifax (York Redoubt), N.S.	0.70	12.97	18.5
21	19	South Bend, PA	0.78	16.20	20.8
22	30	East Chicago, IN	0.79	41.1	52.1
23	32	Brazos River, TX	0.81	45.4	56.0
24	25	Monroeville, PA	0.84	23.8	28.4
25	39	Daytona Beach, FL	0.88	144.0	164.0
26	37	Kure Beach, NC, 250-m (800-ft) lot	0.89	71.0	80.0
27	18	Columbus, OH	0.95	16.00	16.8
28	12	Montreal, Quebec, Canada	1.05	11.44	10.9
29	24	London (Battersea), Eng.	1.07	23.0	21.6
30	17	Pittsburgh, PA	1.14	14.90	13.1
31	10	Waterbury, CT	1.12	11.00	9.8
32	28	Limon Bay, Panama, C.Z.	1.17	30.3	25.9
33	22	Cleveland, OH	1.21	19.0	15.7
34	42	Dungeness, Eng.	1.60	238.0	148.0
35	26	Newark, NJ	1.63	24.7	15.1
36	38	Cape Kennedy, 55 m (60 yds) from ocean, 9-m (30-ft) elevation	1.77	80.2	45.5
37	41	Cape Kennedy, 55 m (60 yds) from ocean, ground level	1.83	215.0	117.0
38	36	Cape Kennedy, 55 m (60 yds) from ocean, 18-m (60-ft) elevation	1.94	64.0	33.0
39	29	Bayonne, NJ	2.11	37.7	17.9
40	33	Pilsey Island, Eng.	2.50	50.0	20.0
41	44	Kure Beach, NC, 25-m (80-ft) lot	2.80	260.0	93.0
42	34	London (Stratford), Eng.	3.06	54.3	17.8
43	35	Halifax (Federal Building), N.S.	3.27	55.3	17.0
44	40	Widness, Eng.	4.48	174.0	39.0
45	45	Galeta Point Beach, Panama, C.Z.	6.80	336.0	49.4

Source: Ref 3

failed as long as it keeps rain out of a building, nor has a fence failed as long as its strands are strong enough to keep animals from straying. However, in other applications, slight discoloration of the surface may be unacceptable, even though the zinc layer is intact.

The behavior of zinc coatings of different thicknesses and applied by different methods was examined in tests under actual outdoor service conditions. Results showed that for a particular exposure condition the life of a zinc coating is approximately proportional to the weight of zinc and is independent of the method by which it is applied (Fig. 2).

This is a very important result that should always be kept in mind when protective coatings are specified for a particular service application. If a steel structure or component is expected to have a long service life, it is always more economical to apply a sufficiently heavy zinc coating at the start than to renew the coating later on because the initial zinc coating was of inadequate weight. The amount of surface preparation required is nearly always less with the new structure than with an old one and is the same whether a thick or thin coating is to be applied. On the other hand, there is no point in applying a heavy zinc coating to an article that will be discarded for other reasons after a short period of service under mildly corrosive conditions.

The thickness of coating that can be obtained depends on the process and to some extent on the part to be coated. Table 5 lists typical coating thicknesses applied by normal practice. There are variations in thickness, uniformity, density, and composition of the coatings applied by different methods, and these variations should be considered in evaluating the corrosion performance of coatings. As long as the coatings remain continuous and intact, the corrosion behavior is generally equivalent to that of metallic zinc, but galvanic action begins when the zinc layer is penetrated to the steel interface. The original thickness, density, and uniformity of the zinc coating determine the progress of sacrificial corrosion and the protection of the steel.

Table 3 Comparison of corrosion rates of zinc and steel in various environments

Location	Ratios of corrosion of steel to corrosion of zinc after indicated exposure: 1 year	8 years
State College, PA	20:1	13:1
New York, NY (spring start)	17:1	4:1
New York, NY (autumn start)	32:1	6:1
Middletown, OH	28:1	12:1
Kure Beach, NC (25-m, or 80-ft, site)	47:1	168:1 (4 years)
Norman Wells, N.W.T.	2:1	1.6:1 (2 years)

Table 4 Corrosion of zinc in various environments based on 20-year ASTM exposure data

Environment	Years to corrode 0.025 mm (1 mil)
Semiarid	100+
Rural	22
Southwestern seacoast	14
Heavy industrial seacoast	4.6
Heavy industrial	3.4

Source: Ref 3

Atmospheric Exposure Tests on Zinc Die Castings. The corrosion behavior of zinc die castings in various natural atmospheres was investigated by ASTM in a program extending over a 20-year period (Ref 5). Because mechanical properties have an important effect on the practical applications of die castings, changes in these properties were taken as a measure of corrosion damage rather than weight loss. Specimens of alloys AG41A (1% Cu) and AG40A (copper free) were exposed at several locations, and their mechanical properties were determined after 5, 10, and 20 years of exposure. The average properties after the 20-year exposure at two industrial, one rural, and two indoor locations are given in Table 6.

Round specimens measuring 6.4 mm (1/4 in.) in diameter were used for tensile tests, and 6.4-mm (1/4-in.) square bars were used as impact specimens. The percentage changes from the original values are included. As in all previous tests on zinc sheet, the industrial atmospheres appeared to be the most harmful. The differences between the rural and the mild indoor atmospheres were relatively minor. In the summary of the original publication, the data for all locations were averaged, and no distinction was made between the atmospheres.

Table 6 demonstrates that a rather large decrease in the impact strength of both alloys occurred in the interval between the 10- and 20-year exposures in the industrial atmospheres. This also occurred for alloy AC41A exposed to the indoor atmospheres. Such decreases in mechanical properties are probably caused by intergranular corrosion, to which die casting alloys are often very susceptible. Intergranular attack can reduce cross-sectional areas and create stress-raising notches, but it does not reduce the overall specimen dimensions significantly.

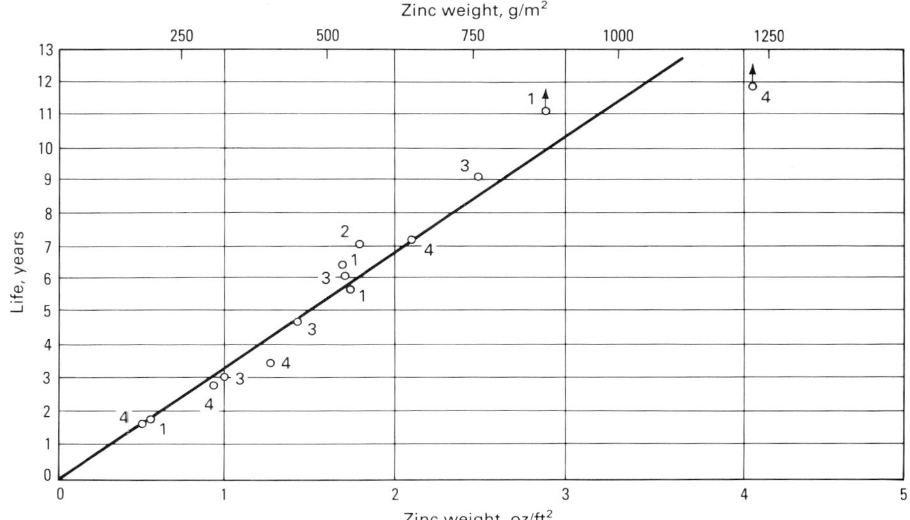

Fig. 2 Service lives of various zinc coatings according to the weight of the zinc present. Results are for exposure in a very aggressive industrial atmosphere. 1, electrodeposited; 2, electrodeposited (passivated with chromate solution); 3, hot-dip galvanized; 4, sprayed

Table 5 Typical zinc coating thicknesses applied by various processes

Method	Process	Specification	Coating thickness µm	Coating thickness mils	Application
Electrogalvanizing	Electrolysis	ASTM A 591	0.5–4	0.02–0.165(a)	Interior; appliance panels, studs, acoustical ceiling members
Zinc plating	Electrolysis	ASTM B 633	39	1.53	Interior or exterior; fasteners and hardware items
Mechanical plating	Peening	ASTM B 695	2.5–145	0.098–5.75	Interior or exterior; fasteners and hardware items
Thermal spraying	Hot-zinc spray	AWS C2.2	84–210	3.32–8.33	Interior or exterior; items that cannot be galvanized because of size or must be performed on site
Continuous galvanizing	Hot dip	ASTM A 525	50	2.0(a)	Interior or exterior; roofing, gutters, culverts, automobile bodies
Hot dip galvanizing	Hot dip	ASTM A 123 ASTM A 386 ASTM A 153	35–115	1.4–4.6	Interior or exterior; nearly all shapes and sizes, ranging from nails, nuts, and bolts to large structural assemblies
Zinc-rich painting	Spray, roller, or brush	SSPC-SP Guide 12.00, 22.00 SSPC-SP Paint 20 SSPC-SP 12.01	15–125 µm/coat	0.6–5.0 mils/coat	Interior or exterior; items that cannot be galvanized because of size or must be performed on-site, large structural assemblies, aesthetic requirements

(a) Total for both sides of sheet

Effect of Alloying Additions

Hot-dip galvanized zinc coatings seldom consist of high-purity zinc. Lead is often added to the plating bath to produce a smoother finish, and aluminum is included to reduce the thickness of the iron-zinc diffusion zone. Numerous other elements may be present in the bath as additions or impurities.

Improved coating performance over normal galvanized zinc coatings has been realized with Zn-55Al-1.5Si compositions. Also, Zn-5Al containing 0.05% mischmetal has proved effective.

Electrodeposited coatings may also contain impurities, which originate in the plating solution. Die casting alloys often include several percent each of aluminum and copper as strengtheners. These alloying elements, whether added deliberately or not, can have a considerable effect on the corrosion of the resulting zinc-base alloys. Composition specifications for slab zinc, rolled zinc, and zinc-base die castings are given in Tables 7, 8, and 9, respectively.

The mechanisms behind the effects of various alloying elements on zinc corrosion are complex and not completely understood. In some cases, the alloying elements may become concentrated in the coating surface as zinc dissolves away, leaving behind a more corrosion-resistant coating. Alloying elements may affect the corrosion scales and their corresponding electronic and diffusion properties, as is known to happen during oxidation of several metals at very high temperature. Most of the recent literature, however, points toward the effects of alloying elements on the intergranular corrosion of zinc. Although general surface corrosion and intergranular attack both contribute to coating performance, severe intergranular corrosion can clearly lead to the deterioration of thin zinc coatings during weathering. Intergranular attack creates sites for pitting corrosion and opens pathways for the environment to reach the substrate.

Several elements have various effects on the corrosion resistance of galvanized coatings. The effects of aluminum, lead, copper, magnesium, and other elements are discussed below.

Aluminum. Small additions of aluminum decrease the corrosion resistance of zinc, especially if lead is also present. The maximum decrease occurs at 0.3% Al (in an alloy containing 0.2% Pb). Aluminum appears to increase intergranular attack in zinc alloys. Large amounts of aluminum (5 and 55%) improve the performance of the coating.

Lead by itself is not particularly harmful. However, in conjunction with any concentration of aluminum, it appears to accelerate corrosion.

Copper improves atmospheric-corrosion resistance and retards intergranular attack. This effect may be only temporary and may not make any significant difference in long-term corrosion.

Magnesium improves the corrosion resistance of zinc somewhat. It is especially useful for counteracting the adverse effects of aluminum. Again, this effect may be only temporary.

Other Alloying Additions. Small amounts of chromium, nickel, or titanium may be beneficial to the corrosion resistance of zinc.

Corrosion of Zinc in Water

The corrosion of zinc in water is largely controlled by the impurities present in the water. Naturally occurring waters are seldom pure. Even rainwater, which is distilled by nature, contains nitrogen, oxygen, CO_2, and other gases, as well as entrained dust and smoke particles. Water that runs over the ground carries with it eroded soil, decaying vegetation, living microorganisms, dissolved salts, and colloidal and suspended matter. Water that seeps through soil contains dissolved CO_2 and becomes acidic. Groundwater also contains salts of calcium, magnesium, iron, and manganese. Seawater contains many of these salts in addition to its high NaCl content.

All of these foreign substances in natural waters affect the structure and composition of the resulting films and corrosion products on the surface, which in turn control the corrosion of zinc. In addition to these substances, such factors as pH, time of exposure, temperature, motion, and fluid agitation influence the aqueous corrosion of zinc.

As in the atmosphere, the corrosion resistance of a zinc coating in water depends on its initial ability to form a protective layer by reacting with the environment. In distilled water, which cannot form a protective scale to reduce the access of oxygen to the zinc surface, the attack is more severe than in most types of domestic or river water, which do contain some scale-forming salts.

The scale-forming ability of water depends principally on three factors: the hydrogen ion concentration (pH value), the total calcium con-

Table 6 Change in mechanical properties of zinc die cast alloys upon exposure to several natural atmospheric environments

Results are from 20-year exposure tests at the following sites: outdoor/industrial (New York, NY, and Altoona, PA), outdoor/rural (State College, PA), and indoor

Alloy	Property	Original value	After 20-year exposure: Outdoor/industrial(a)		Outdoor/rural(a)		Indoor(a)	
AC41A	Tensile strength	305 MPa (44.3 ksi)	211 MPa (30.6 ksi)	[−31%]	259 MPa (37.6 ksi)	[−15%]	254 MPa (36.9 ksi)	[−17%]
	Elongation in 50 mm (2 in.), %	7	3.7	[−50%]	9	[+28%]	12	[+71%]
	Hardness, HRE	91	78	[−14%]	80	[−10%]	83	[−8%]
	Charpy impact	56J (41 ft·lb)	12J (9 ft·lb)	[−78%](b)	45J (33 ft·lb)	[−20%]	27J (20 ft·lb)	[−52%](b)
AG40A	Tensile strength	254 MPa (36.9 ksi)	206 MPa (29.9 ksi)	[−19%]	225 MPa (32.6 ksi)	[−12%]	228 MPa (33.1 ksi)	[−10%]
	Elongation in 50 mm (2 in.), %	15	7	[−47%]	18	[+20%]	21	[+40%]
	Hardness, HRE	83	72	[−13%]	67	[−19%]	74	[−11%]
	Charpy impact	53J (39 ft·lb)	16J (12 ft·lb)	[−69%]	51J (38 ft·lb)	[−3%]	40	[+3%]

(a) Bracketed figures are the percentage changes from the original values caused by exposure. (b) Large change between 10 and 20 years of exposure. Source: Ref 5

Table 7 ASTM composition specifications for slab zinc

Grade	Impurities, %: Pb(a)	Fe(a)	Cd(a)	Zn (minimum by difference)
Special high grade	0.003	0.003	0.003	99.990
High grade	0.07	0.02	0.03	99.90
Intermediate	0.20	0.03	0.40	99.5
Brass special	0.6	0.03	0.50	99.0
Prime Western	1.6	0.05	0.50	98.0

(a) Maximum composition. Source: Ref 6

Table 8 Typical compositions of rolled zinc

Compositions (remainder zinc), %(a): Pb	Fe	Cd	Cu	Mg	Ti
0.05	0.010	0.005	0.001	...	...
0.05–0.12	0.012	0.005	0.001	...	...
0.30–0.65	0.020	0.20–0.35	0.005	...	...
0.50–0.12	0.012	0.005	0.65–1.25	...	...
0.05–0.12	0.015	0.005	0.75–1.25	0.007–0.02	...
0.20	0.015	0.01	0.5–0.8	...	0.08–0.16

(a) Maximum composition

Table 9 Standard composition specifications for zinc-base alloy die castings

Element	Composition, %: Alloy AG40A	Alloy AC41A
Copper	0.25 max(a)	0.75–1.25
Aluminum	3.5–4.3	3.5–4.3
Magnesium	0.020–0.05	0.03–0.08
Iron(a)	0.100	0.100
Lead(a)	0.005	0.005
Cadmium(a)	0.004	0.004
Tin(a)	0.003	0.003
Zinc	rem	rem

(a) Maximum composition. Source: Ref 7

Table 10 Effect of oxygen on the corrosion of zinc in distilled water

Test condition	Temperature: °C	°F	Corrosion rate(a): $mg/dm^2/d$	mils/yr
Boiled distilled water; specimens immersed in sealed flasks	Room	Room	5.0	1.0
Boiled distilled water; specimens immersed in sealed flasks	40	104	9.4	1.9
Boiled distilled water; specimens immersed in sealed flasks	65	149	16.5	3.3
Oxygen bubbled slowly through the water	Room	Room	43.0	8.6
Oxygen bubbled slowly through the water	40	104	68.6	13.7
Oxygen bubbled slowly through the water	65	149	62.0	12.4

(a) High-grade zinc specimens, in duplicate, immersed for 7 days. The corrosion rate was calculated after removal of corrosion products.

tent, and the total alkalinity. If the pH value is below that at which the water would be in equilibrium with calcium carbonate ($CaCO_3$), the water will tend to dissolve rather than to deposit scale. Waters with a high content of free CO_2 also tend to be aggressive toward zinc.

Water hardness is an important variable in zinc corrosion. The corrosion rate of zinc in hard water may be 15 μm/yr (0.6 mil/yr), but in soft water, it can be 150 μm/yr (6 mils/yr). Hard waters are usually less corrosive toward zinc because they deposit protective scales on the metallic surface, but softer waters would not deposit these scales. Similarly, seawater also deposits protective scales on zinc and is less corrosive than soft water.

Softer waters, with their higher content of dissolved oxygen and CO_2, generally attack zinc more vigorously than the fairly hard waters. River waters have been found to deposit scale more easily than well waters. The normal corrosion product on zinc in water is $ZnCO_3$.

Corrosion of Zinc by Distilled Water

Agitation, Aeration, and CO_2. Under conditions in which the oxygen content cannot be replaced as quickly as it is consumed by the corrosion process, such as in stagnant water, zinc is attacked rapidly at local areas, and this causes pitting. As more oxygen is made available, the corrosion becomes more uniform. With further increases in the oxygen content of the water, the corrosion rate increases. For example, when thin films of moisture condense on a zinc surface, the concurrent rapid supply of oxygen at the corroding surface has a decided accelerating effect on the corrosion rate.

Both this and the stagnant-water types of attack can be minimized by the use of chromate films. Experimentation has shown that with testpieces immersed in water through which oxygen was bubbled, corrosion occurred about eight times as fast as with specimens in water that was boiled to remove gases and then cooled out of contact with air. Effects of concentration of dissolved oxygen and temperature on corrosion in distilled water are shown in Table 10. Although the corrosion rate increased when oxygen was bubbled through the water, the attack was uniform. The presence of oxygen in the water accelerates the corrosion by depolarization of the cathodic areas. The rate of corrosion is then controlled by diffusion of oxygen through the film of $Zn(OH)_2$ corrosion products.

Temperature. In practical applications, the temperature of the water has been shown to be a very important factor affecting the rate of corrosion of zinc in water. In one study, a marked increase in corrosion rate was found to occur at a temperature of about 60 °C (140 °F), followed by a decrease in corrosion at higher temperature (Ref 8). At temperatures near 70 °C (160 °F), a reversal in potential may occur where zinc coatings become cathodic to iron. Low oxygen and high bicarbonate contents favor reversal, but the presence of oxygen, sulfates, and chlorates tends to maintain the natural anodic state of the zinc.

The results of this study were obtained on 99.9% pure zinc immersed for 15 days in aerated distilled water and rotated at 56 rpm. The water was continuously aerated by air bubbling through it, and no attempt was made to remove CO_2 from the air. Also, the specimens were supported on a wooden disk that was rotated through the test water at 56 rpm. Thus, all of the factors that cause increased corrosion as discussed above were present in these tests. The temperature, however, seems to have been the controlling factor in this experiment.

Table 11 lists metal loss and penetration and describes the appearance of corrosion products on the zinc surfaces. There was a direct relationship between the character of the corrosion film and the extent of corrosion. At the temperatures of maximum corrosion, the film is granular and nonadherent, but at the lower temperatures, it is

Table 11 Effect of temperature on the corrosion of zinc in distilled water
Rolled high-grade zinc immersed for 15 days in water aerated by air bubbles

Temperature, °C	Temperature, °F	Corrosion rate, mg/dm²/d	Corrosion rate, mils/yr	Appearance of corrosion film
20	68	3.9	0.78	Gelatinous, very adherent
50	122	13.7	2.74	Less gelatinous, adherent
55	131	76.2	15.2	Mostly granular, nonadherent
65	149	577.0	115.4	Granular to flaky, nonadherent
75	167	460.0	92.0	Granular, flaky, nonadherent
95	203	58.7	11.7	Compact, dense, nonadherent
100	212	23.5	4.7	Very dense and adherent

Table 12 Corrosion of zinc in distilled water and tap water

Alloy	Distilled water, mg/dm²/d	Distilled water, mils/yr	Tap water, mg/dm²/d	Tap water, mils/yr
High-purity grade	27	5.4	3.5	0.7
Die cast alloy (AC41A)	28.5	5.7	11.5	2.3

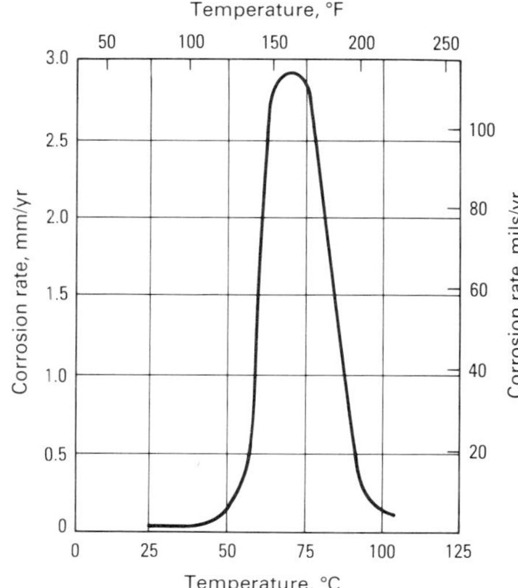

Fig. 3 Effect of temperature on corrosion of zinc in distilled water. Source: Ref 9

gelatinous and adherent. At higher temperatures, it is compact and adherent. From this it might be concluded that the granular coating is more permeable to oxygen, which would account for the increased corrosion.

Tests showed that the observed corrosion peak at approximately 65 °C (150 °F) occurs both in waters under a pure oxygen atmosphere and under a CO_2-free air atmosphere (Fig. 3). In contrast, under an oxygen-free nitrogen atmosphere, the peak disappears completely. Additional experiments showed that as the partial pressure of oxygen over the water was reduced from that in air the peak decreased in magnitude and shifted to lower temperatures that involve the decreased solubility of oxygen in water at elevated temperature. This theory holds that the corrosion of zinc proceeds by the reaction:

$$Zn + 2H_2O \rightarrow Zn(OH)_2 + H_2 \quad \text{(Eq 3)}$$

Oxygen, when present, depolarizes and accelerates the reaction by combining with the evolved hydrogen:

$$H_2 + \frac{1}{2}O_2 \rightarrow H_2O \quad \text{(Eq 4)}$$

As the temperature increases, the reaction rate initially increases; but the increasing temperature decreases the O_2 supply, and the depolarizing effect of the oxygen is decreased. Thus, at elevated temperature, the reaction polarizes and its rate slows. The stable film at elevated temperature is ZnO, not $Zn(OH)_2$. If ZnO is formed from $Zn(OH)_2$ by this reaction:

$$Zn(OH)_2 \rightarrow ZnO + H_2O \quad \text{(Eq 5)}$$

then the process is still limited by the rate of $Zn(OH)_2$ formation, which is limited by the rate of Eq 3 and 4, which depend on the oxygen supply.

This is consistent with the observation that decreased oxygen supplies (caused by reduced oxygen pressure over the water) lower the corrosion peak temperature (Ref 10). As the oxygen partial pressure is lowered, the equilibrium oxygen content in the water is lowered for any temperature. The reduced oxygen content of the water thus lowers the temperature at which depolarization (Eq 5) becomes rate limiting, because the critical oxygen level for depolarization is reached at a lower temperature.

Corrosion Inhibitors

When zinc is used in contact with water in a closed system, inhibitors are frequently employed to minimize corrosion. Various inorganic inhibitors are available for use with zinc, such as sodium dichromate, sodium silicate, borax, and hexametaphosphate. Mechanical exclusion films provided by adsorptive-type organic compounds, such as lanolin, are also useful for the corrosion inhibition of zinc. For most purposes, the adjustment of the pH to the mildly alkaline range and the addition of sodium dichromate are preferred.

There is some danger of intensified pitting when too little of this type of inhibitor is added. If zinc-coated parts are stored for significant periods of time under conditions in which moisture may be entrapped between closely adjacent surfaces, the part may become defaced by the appearance of a whitish reaction product powder on the surface. This is known as wet-storage stain. Although it affects appearance, this product is generally not harmful to the zinc coating. Unchecked, wet-storage stain can shorten the service life before the underlying ferrous metal rusts.

To avoid the occurrence of wet-storage stain, it is essential to store zinc-coated material under conditions of free circulation of air and to avoid the large periodic temperature change that causes condensation of moisture on the metal. The reaction causing this white staining does not occur in the normal use of zinc-coated products once they have aged at the surface, but new parts must be stored carefully to avoid this problem.

Corrosion of Zinc in Domestic Waters

Cold Water. Galvanized pipe is widely used in handling domestic water supplies, and the results have been satisfactory. Therefore, quantitative corrosion rates are not of primary interest, and very few data are found in the literature. Hard waters contain dissolved salts that may affect the corrosion of zinc. Carbonates and bicarbonates tend to deposit protective films that stifle corrosion, and it is generally agreed that soft or distilled waters are more corrosive than hard water. For example, the corrosion rates of two grades of zinc in distilled and tap water of moderate hardness are given in Table 12.

Hot Water. In domestic water systems, for which zinc is widely adopted as a protective coating, the sacrificial dissolution of zinc at discontinuities in the coating in the presence of calcium bicarbonate (a normal constituent of hard water supplies) leads to the deposition of an insoluble layer of $CaCO_3$ on the exposed surfaces. Because this layer is impervious to the passage of ions and electrons, it inhibits any further corrosive action. This reaction, which depends on the presence of dissolved calcium bicarbonate in the water, cannot occur in systems in soft water areas, and it is generally agreed that soft or distilled water is more corrosive than hard water. Other constituents in natural waters, such as nitrates, sulfates, and chlorides, may tend to increase corrosion, but their effect is usually overcome by the carbonates that form films of relatively low solubility in close contact with the zinc surface.

Changing the temperature of a solution can influence the corrosion tendency. For example, historically, household water heater tanks were made of galvanized steel. The zinc coating on the carbon steel base offered some cathodic protection to the underlying steel, and the service life (usually judged by how long it took to produce red, or rusty, water) was considered to be adequate. Water tanks were seldom operated above 60 °C (140 °F).

With the development of automatic dishwashers and automatic laundry equipment, the average water temperature was increased so that temperatures of approximately 80 °C (175 °F) are not unusual in household hot-water tanks. Coinciding with the widespread use of automatic dishwashers and laundry equipment was a sudden upsurge of complaints about the short service lives of galvanized steel water heater tanks. Electrochemical measurements showed that in many cases iron was anodic to zinc at 77 °C (170 °F), but zinc was anodic to iron at temperatures below 60 °C (140 °F). This explained why zinc offered no cathodic protection at 77 °C (170 °F) and why red water and premature perforation of galvanized water tanks had occurred. The problem was reduced by the use of magnesium anodes and protective coatings as well as the development of new alloys.

In 1974, the International Lead Zinc Research Organization (ILZRO) patented a method for protecting galvanized steel used in fabricating domestic hot-water systems, electric water heat-

Table 13 Relationship of soil corrosion to electrical resistivity

Soil Class	Corrosion resistance	Electrical resistivity, Ω·cm
Sandy	Excellent	6000–10 000
Loams	Good	4500–6000
Clays	Fair	2000–4500
Peat/muck	Bad	0–2000

ers and distribution piping (Ref 11). This method is based on the formation of a passivating coat of zinc pyrophosphate ($Zn_2P_2O_7$), which is partly formed at the expense of the zinc coating. This compound is insoluble and a good electrical insulator because of its covalent property. It prevents the formation of electrochemical cells and thus prevents attack on the zinc coating. The $Zn_2P_2O_7$ coating can be applied by a chemical or an electrochemical method, depending on the application.

Corrosion of Zinc in Natural Water and Seawater

The factors influencing the corrosion of zinc by tap or supply waters discussed in the preceding section also apply to natural waters, such as lake or river waters. Few data have been reported on the corrosion rate of zinc in natural freshwaters. In the case of seawater, however, dissolved salts, principally chlorides and sulfates, determine the corrosion behavior of zinc. The high chloride content of seawater would normally tend to increase corrosion, but the presence of magnesium and calcium ions inhibits the attack. The effect of time of exposure on the corrosion rate in natural waters indicates that the corrosion rate in seawater initially exceeds that in freshwater, but after about 2 years of exposure, the rate in seawater decreases so that it is approximately the same as that in freshwater.

The frequency with which seaside piers and ship hulls require repainting testifies to the corrosivity of seawater and the importance of taking adequate protective measures. The salts present in the North Sea amount to about 3.5% of the total, mainly NaCl with smaller amounts of magnesium and calcium salts. This produces a solution with a pH of about 8.

The corrosion resistance of zinc compares very favorably with that of other coating materials when totally immersed in seawater, possibly because of the slightly inhibitive action of the magnesium salts present. In tests, aluminum-, cadmium-, lead-, tin-, and zinc-coated specimens were immersed for 2 years. All except the zinc-coated specimens failed in this time. The zinc-coated specimens were then transferred to another test site and immersed for another 4 years. At the end of this time—a total of 6 years—the 3-oz/ft^2 (915-g/m^2) coatings were just ceasing to give complete protection to the basis steel. Therefore, it was concluded that the consumption of zinc is approximately 0.5 oz/ft^2 (150 g/m^2) of steel surface protected per year.

In most cases, zinc coatings would not be used alone when applied to steel immersed in seawater, but would form the first layer of a more elaborate protective system. Conditions within a few hundred yards of the surf line on beaches are intermediate between total immersion in seawater and normal exposure to a marine atmosphere. Such conditions have been studied in a series of tests carried out at Lighthouse Beach in Nigeria.

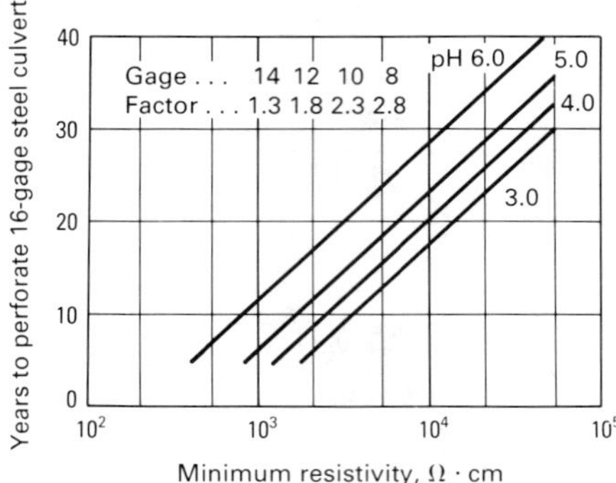

Fig. 4 Method of estimating service life as developed by the California Division of Highways. This correlates pH with the electrical resistivity of soils to determine years to perforate a steel sheet. Local durability records are used for confirmation or control. Multiply years to perforation by factor for increase in metal gage. This figure is based on 16-gage galvanized steel pipe with a coating thickness of 1.6 mm (0.064 in.). Source: Ref 12

At the high-water mark, about 45 m (50 yd) from the surf line on the tropical beach, conditions were extremely severe, and iron and zinc both corroded more rapidly than if they had actually been immersed in the sea.

Upon moving inland, however, it was found that both the salt content of the atmosphere and the rates of corrosion of iron and zinc fell off very rapidly. The values obtained only 1 km (3/4 mile) from the sea were similar to those obtained at inland sites.

Corrosion Resistance of Zinc in Soils

Galvanized pipe is frequently used to provide underground water service for farms, hot houses, and so on. Inspection of the plain galvanized pipe used in culverts and in storm drains will show a long service life and durability that is often surprising to the engineer unfamiliar with such structures. Pipe with only a galvanized coating can often be used to handle runoffs from rainstorms for more years than the facility above the pipe will be used.

More than 200 soil series have been categorized in the United States according to texture, color, and natural drainage. The physical properties of importance in corrosion are those that determine the permeability of the soil to air and water. Soils with a coarse texture, such as sands and gravels, permit free circulation of air. The corrosion under such conditions approaches that occurring in the atmosphere. Clay and silt soils are generally characterized by a fine texture and high water-holding capacity, resulting in poor aeration and drainage. The aeration characteristics of a soil depend on particle size and distribution and on the apparent specific gravity. The latter is a measure of the compactness of the soil.

Numerous chemical elements are present in soils, but the ones of interest in corrosion studies are those that are soluble in water. Analyses are usually made for base-forming elements, such as sodium, potassium, calcium, and magnesium, and for the acid-forming groups, such as carbonate, bicarbonate, chloride, nitrate, and sulfate.

However, corrosion engineers have determined that the best measure of the potential corrosivity of a given soil is its electrical resistance measured in ohms per cubic centimeter. The electrical resistivity of the soils is determined by the nature and concentration of the ions formed by the chemical salts dissolved in the soil moisture. These ions may also determine the progress of corrosion. If the primary product of corrosion is relatively insoluble and deposits as a film on the metal surface, further corrosion may be reduced or completely stifled. Numerous inspections of existing structures have established that in sandy and loamy soils with resistivity of 4500 Ω · cm and higher the engineer need not be concerned with the corrosion on the soil side of galvanized steel pipe (Table 13).

When the engineer wishes to confirm his judgment of service life requirements or when the corrosion and abrasion aspects of the proposed site may be in question, reference to the California Division of Highways method for determining service life is recommended. This comprehensive method uses measurable environmental corrosion criteria and is applicable to all parts of the country. The California method correlates pH with the electrical resistivity of soils to determine the number of years necessary to perforate a galvanized steel sheet. For example, an engineer considering a typical site environment with a resistivity of 10 000 Ω · cm and pH of 7 can refer to Fig. 4 and determine that it would take 40 years to perforate 16-gage (~1.5-mm, or 0.06-in.) galvanized steel pipe.

The engineer can determine the years to perforation for other gages of pipe by using the appropriate multiplying factor. For example, years to perforation for 8-gage (~4.2-mm, or 0.16-in.) pipe would be 40 times 2.8, or 112 years. Moreover, perforations resulting from the pitting action of galvanic cells will not affect the useful service of a storm drainage structure, as compared to a pressure conduit from which a valuable or dangerous product would be lost. With proper maintenance, the life of corrugated steel pipe can be extended indefinitely.

Except in the most corrosive soils, the maximum depth of pitting in steel specimens exposed for approximately 12½ years was more than 11 times that in zinc specimens, although the ratio for the rates of corrosion was only approximately one-half that figure (Ref 13). This resistance to pitting, combined with the fact that rusting does not appear to start until nearly all of the zinc and zinc-alloy layers have corroded away, reduces the risk of premature failure in galvanized piping. It cannot be overemphasized that, although these results serve as a useful guide to the performance of zinc coatings in particular soils, local experience should always be sought.

The Corrosion of Zinc in Chemical Environments

As indicated previously, the use of zinc for corrosion-resistant applications accounts for about one-half of the total consumption of the metal. However, the available information reveals that in most applications the resistance of zinc to the atmosphere and in various waters is the primary corrosion criterion. In one sense, these may be considered chemical environments, because the chemical reactions between zinc and the constituents of pure or polluted atmospheres and waters determine the life of the zinc exposed to air or water.

Zinc is usually not considered to be a useful metal in the acidic or strongly alkaline chemical environments encountered, for example, in the chemical-processing industries (see the article "Corrosion in the Chemical Processing Industry" in this Volume). The corrosion of zinc increases in aqueous chemical solutions on either side of the 6 to 12 pH range. This should not be considered a fixed rule, because many other factors, such as agitation, aeration, temperature, polarization, and, in some cases, the presence of inhibitors, may have considerable influence on the corrosion.

There is considerable interest in the use of zinc in milder chemical environments. For example, zinc is used in contact with many organic chemical and chemical specialties, such as detergents, insecticides, and agricultural chemicals. In most cases, zinc comes in contact with such chemicals during the handling, packaging, and storage of the commercial products.

The corrosion resistance of zinc to the chemicals is usually the primary consideration, but in some cases, the effect of zinc corrosion on a consumer product or chemical is of greater concern than the actual corrosion rate of zinc. For example, zinc in contact with certain organic chemicals may in rare cases cause polymerization or catalyze some other undesirable change that would alter the original composition of properties of the product. In other cases, some change in the appearance or texture of a consumer product may be caused by the relatively slight corrosion of zinc. Thus, zinc would be considered incompatible, although it does not corrode excessively.

There are many situations in the chemical industry in which zinc serves a useful purpose. Zinc-coated tanks and cylinders are widely used in oil refineries and other plants for storing oil and petroleum products, chlorine, CO_2, and other industrial gases (see the articles "Corrosion in Petroleum Production Operations" and "Corrosion in Petroleum Refining and Petrochemical Operations" in this Volume). Refrigerating plants and cooling equipment, as well as degreasing plants, are almost universally protected by zinc coatings. Zinc-coated steel is also used on structural steelwork around chemical plants, where it is exposed to high humidity and a variety of chemical fumes. Galvanized steel is extensively used for roofing and siding on pulp and paper processing buildings (see the article "Corrosion in the Pulp and Paper Industry" in this Volume).

Other uses of zinc-coated steel in the chemical industry include applications in floating roof type storage tanks for volatile liquids as well as galvanized wire cloth and mesh belts for the movement of chemicals through various production stages. Galvanized steel containers are used to store strategic chemicals in outdoor locations. From these representative examples it is apparent that zinc and zinc-coated steel are definitely useful in the chemical industry.

Corrosion in Dissolved Salts, Acids, and Bases

Zinc is not used in contact with acid and strong alkaline solutions, because it corrodes rapidly in such media. The section "Corrosion of Zinc in Water" in this article indicates the safe range in which it may be used.

Very dilute concentrations of acids accelerate corrosion rates beyond the limits of usefulness. Alkaline solutions of moderate strength are much less corrosive than corresponding concentrations of acid, but are still corrosive enough to impair the usefulness of zinc.

Zinc-coated steel is used in handling refrigeration brines that may contain calcium chloride ($CaCl_2$). In this case, the corrosion rate is kept under control by adding sufficient alkali to bring the pH into the mildly alkaline range and by the addition of inhibitors, such as sodium chromate (Na_2CrO_4). Certain salts, such as the dichromates, borates, and silicates, act as inhibitors to the aqueous corrosion of zinc.

Nonaqueous Corrosion

Organic Compounds. Many organic liquids that are nearly neutral in pH and substantially free from water do not attack zinc. Therefore, zinc and zinc-coated products are commonly used with gasoline, glycerine, and inhibited trichlorethylene. The presence of free water may cause local corrosion because of the lack of access to oxygen. When water is present, zinc may function as a catalyst in the decomposition of such solutions as trichlorethylene, with acid attack as the result. Some organic compounds that contain acidic impurities, such as low-grade glycerine, attack zinc. Although neutral soaps do not attack zinc, there may be some formation of zinc soaps in dilute soap solutions.

Gases. Zinc may be safely used in contact with most common gases at normal temperatures if water is absent. Moisture content stimulates attack. Dry chlorine does not affect zinc. Hydrogen sulfide (H_2S) is also harmless because insoluble zinc sulfide (ZnS) is formed. On the other hand, SO_2 and chlorides have a corrosive action because water-soluble and hygroscopic salts are formed.

Indoor Exposure. Zinc corrodes very little in ordinary indoor atmospheres of moderate relative humidity. In general, a tarnish film begins to form at spots where dust particles are present on the surface; the film then develops slowly. This attack may be a function of the percentage of relative humidity at which the particles absorb moisture from the air. However, moisture has little effect on the tarnish formation up to 70% relative humidity. The degree of corrosion is related to the relative humidity at and above this point because the zinc corrosion products absorb enough moisture to stimulate the attack to a perceptible rate.

Rapid corrosion can occur where the temperature decreases and where visible moisture that condenses on the metal dries slowly. This is related to the ease with which such thin moisture films maintain a high oxygen content because of the small volume of water and large water/air interface area. Considerably accelerated corrosion can then take place with the formation of a film that is too thick. Chromate protective films are used to a considerable extent to prevent attack where accidental or limited contact with water is expected. Atmospheres inside industrial buildings can be corrosive, particularly where heated moisture and gases, such as SO_2, condense near a cool room.

Contact With Food Products. Zinc should not be used in contact with acidic foodstuffs unless they can be expected to remain dry. Otherwise, the zinc must be adequately protected by copper-nickel-chromium plating or another satisfactory impervious coating. The slight acidity present in many foodstuffs can attack the zinc, and this may give the food a metallic taste. For the same reason, the zinc die castings used in any equipment to hold or dispense drinks should also be plated or otherwise protected. Consumption of food contaminated with zinc may cause nausea but is not dangerous, and this rarely arises because of the taste. A summary of the compatibility of untreated zinc with various media is presented in Table 14.

Stress-Corrosion Cracking and Corrosion Fatigue

Stress-Corrosion Cracking (SCC). The effects of corrosion and stress on the performance of a material are often treated as separate concerns. However, in conjunction, the two can cause the phenomenon of SCC, which can destroy a metallic component faster than either stress or corrosion separately. In essence, SCC is a process in which cracks in the metal grow under the combined effects of tensile stress and a corrosive environment. When the cracks become sufficiently large, the component fractures. Often, but not always, the stress-corrosion resistance of an alloy is very dependent on its microstructure. Information on the mechanisms of SCC is provided in the article "Environmentally Induced Cracking" in this Volume. Testing and interpretation of SCC data can be found in the article "Evaluation of Stress-Corrosion Cracking" in this Volume.

Corrosion Fatigue. When metal is subjected to vibration or cyclic stress, fatigue strength is more important than ultimate tensile strength. Under corrosive conditions, even the advantages of alloy additions and heat treatments that considerably increase the tensile strength and fatigue limit of steel have only a minor effect on corrosion fatigue behavior. Corrosion fatigue is the weakening that occurs in a metal after it has been stressed repeatedly in the presence of corrosive agents. Generally, the corrosion fatigue characteristics of low-alloy steels are little or no better than those of ordinary low-carbon steel unless the metal has a protective coating.

The plastic deformation that occurs in the metal at the bottom of a crack during repeated stressing raises its energy and thus increases its susceptibility to chemical attack. The strengthened material that would otherwise prevent the cracks from spreading is therefore destroyed by corrosion in preference to the surrounding metal, and the crack continues to grow. This process continues by further plastic deformation of metal, which is in turn destroyed.

In addition to this basic mechanism of corrosion fatigue, certain subsidiary effects also play a part. For example, the strains occurring in the surface of the stressed metal tend to disrupt such protective films as would otherwise be formed by the corrosion products, with the damage to the protective film often taking the form of small cracks or crevices. The small areas of metal exposed by these cracks are anodic to the surrounding film and corrode at an accelerated rate with the formation of pits; this sets the mechanism of corrosion fatigue in motion. The total amount of corrosion involved in corrosion fatigue is often extremely small. It is important to prevent corrosion from the start because once it has begun cyclic stressing may lead to early failure even though further attack is prevented.

These facts clearly point to the necessity of sacrificial, electrochemical protection as a means of preventing corrosion fatigue. Steel compo-

Table 14 Compatibility of untreated zinc with various media

Medium	Media descriptor	Compatibility
Aerosol propellants	. . .	Excellent
Acid solutions	Weak, cold, quiescent	Fair
	Strong	Not recommended
Alcohols	Anhydrous	Good
	Water mixtures	Not recommended
	Beverages	Not recommended
Alkaline solutions	Up to pH 12.5	Fair
	Strong	Not recommended
Carbon tetrachloride	. . .	Excellent
Cleaning solvents	Chlorofluorocarbon	Excellent
Detergents	Inhibited	Good
Diesel oil	Sulfur free	Excellent
Fuel oil	Sulfur free	Excellent
Gas(a)	Towns, natural, propane, butane	Excellent
Glycerine	. . .	Excellent
Inks	Printing	Excellent
	Aqueous writing	Not recommended
Insecticides	Dry	Excellent
	In solution	Not recommended
Lubricants	Mineral, acid free	Excellent
	Organic	Not recommended
Paraffin	. . .	Excellent
Perchlorethylene	. . .	Excellent
Petroleum(a)	. . .	Excellent
Refrigerants	Chlorofluorocarbon	Excellent
Soaps	. . .	Good
Trichloroethylene	. . .	Excellent

(a) Chromate passivation treatment recommended because of the possible presence of moisture traces.

nents can be protected throughout their life by a sacrificial zinc coating to give them complete protection. If a coating with no sacrificial properties is used, a surface fault may well lead to fatal pitting.

Steel and Corrosion Fatigue. Detailed investigations into the corrosion fatigue resistance of steels in moist air and saline solutions have given some surprising results. The ultimate tensile strength and indeed the pure fatigue limits of steels are increased considerably by alloying or special heat treatments, but these have only a minor effect on corrosion fatigue behavior.

Thus, under practical conditions the behavior of a special steel may be little better than that of ordinary carbon steel unless it has a protective zinc coating. These are significant considerations in determining what preventive measures to adopt in practice, and that explain why sacrificial or electrochemical protection is of major importance. Because corrosion fatigue cracks may be difficult to detect until they have reached dangerous proportions, a practical safeguard is to protect steel components throughout their service lives by a sacrificial coating of zinc, which gives them complete cathodic protection. When zinc coatings are used, small to medium-sized imperfections in the coating are of relatively little importance, and corrosion is prevented even at the growing points of the microscopically small crevices that cause corrosion fatigue.

Any type of coating that covers the surface completely and retards the onset of corrosion may help, but if it is not anodic to steel, it cannot exert cathodic protection at coating defects. In many general uses, the fact that nonanodic coatings cannot prevent corrosion at small defects may be unimportant. When fatigue failure is not a consideration, a very small amount of corrosion has no significant effect.

Wherever corrosion fatigue can occur, however, the situation is very different. Coatings that cannot protect sacrificially may, as a result of discontinuities, cause serious pitting, and trouble may be accentuated by concentrating the corrosion in small areas, thus simulating the growth of fatigue cracks. Additional information on corrosion fatigue can be found in the articles "Mechanically Assisted Degradation" and "Evaluation of Corrosion Fatigue" in this Volume.

Corrosion Protection With Zinc Anodes

Historical Development of Cathodic Protection. Historically, zinc was one of the very first metals to be used as a galvanic anode; early in the 19th century, Sir Humphrey Davy secured pieces of zinc to the copper sheathing on wood hulls of British Navy vessels to prevent severe underwater corrosion of the copper. The latter was used as a barrier shield to stop penetration and destruction of the wood hulls by marine borers and to prevent attachment of barnacles to hulls.

Sir Humphrey thus became the first practitioner of cathodic protection and is credited with developing the concept of the electrochemical series of the elements. He concluded from actual sea trials aboard ship that corrosion of the copper sheathing could be arrested by using zinc protectors, but the fouling problems by marine organisms were not entirely resolved. Further experiments demonstrated that by varying the area of the protector metal in relation to the copper the latter would corrode at a low, tolerable rate that was insufficient to cause perforation but adequate to exhibit antifouling characteristics. The electrolytic protection of metals from corrosion in a bulk electrolyte, as demonstrated by Sir Humphrey, has developed into a universally accepted method of corrosion control known as cathodic protection.

Cathodic Protection Systems. In practice, cathodic protection is applied to potentially corrodible metals underground or underwater by galvanic anodes (self-generated current) or power-impressed systems. Galvanic anode systems do not require an external source of current, because the protective current is self-generated when the galvanic anode is electrically connected to the structure to be protected in a bulk electrolyte.

The most commonly used power-impressed systems employ anodes with low anodic corrosion rates that are electrically connected to a rectifier that converts alternating current (ac) to direct current (dc). The negative terminal of the rectifier is grounded to the structure to be cathodically protected, and the positive terminal is connected to the relatively inert anodes. When the system is activated, current flows from the rectifier to the anodes, through the electrolyte to the structure (cathode) to be protected, and back to the rectifier through the return path. The anodes commonly used in impressed-current systems include, but are not limited to, high-silicon cast iron, graphite, platinized titanium, and lead-silver alloys (seawater only). The cathode, or structure to be protected, does not recognize the source of protective current, because it is possible to polarize most structures by using galvanic anodes or impressed-current systems.

Other power sources used in impressed-current systems include, but are not limited to, the following: engine generator sets, wind-powered generators, thermoelectric generators, gas turbines, solar cells and fuel cells. Although the sources of power for supplying protective current in cathodic protection systems are numerous, zinc is the one that is most frequently used; therefore, the rest of this section will discuss the details relating to the function of zinc as a galvanic anode.

Applications of Zinc Anodes. The use of zinc in commercial cathodic protection systems involves environments as diverse as seawater, brackish water, freshwater, and a wide variety of soils. Because the environment plays a significant role in determining the success or failure of zinc galvanic anode systems, seawater and soils will be treated separately.

The application of cathodic protection to reduce or prevent corrosion damage occurring on the steel hulls of marine craft, such as ships, launches, barges, floating docks, buoys, and pontoons, makes it possible to increase substantially the interval between dry dockings and to reduce the amount of maintenance work to be accomplished during each dry docking period. The material and labor costs of installing a zinc anode cathodic protection system are usually small compared to such maintenance work as chipping, painting, and replacing hull plates.

The economic benefits that accrue from marine cathodic protection depend to a large extent on the duty cycle of the craft under consideration. Maximum savings are obtained when cathodic protection is applied to vessels that are dry-docked only for reasons of anticorrosion maintenance. Included in this category would be most barges, dredges, buoys, pontoons, and stored ships. It is not unusual under these conditions for a cathodic protection system, designed for a 3-year life, to pay for itself within 1 year. Even in the case of merchant ships, which must be dry-docked annually for removal of marine growths and reapplication of antifouling paint, cathodic protection, if properly applied, can prove economical because it reduces or eliminates corrosion damage to exposed steel propeller shafts and to areas on the hull where the paint has become damaged.

Zinc is an ideal metal for cathodic protection in seawater because it does not subject the adjacent

Table 15 Zinc anode composition specifications for seawater use

Element	MIL-A-18001H composition, %	ASTM B 418 Type I composition, %
Aluminum	0.10–0.50	0.10–0.4
Cadmium	0.025–0.15	0.03–0.10
Iron(a)	0.005	0.005
Lead(a)	0.006	. . .
Copper(a)	0.005	. . .
Silicon(a)	0.125	. . .
Zinc	rem	rem

(a) Maximum composition

painted surfaces of the hull to high potentials, which are injurious to many commonly used paints. Nevertheless, good marine paints, application practice, and reasonable maintenance are of course necessary. Zinc has a high ampere-hour capability per unit volume (530 A · yr/m^3, or 15 A · yr/ft^3). This means that the total volume of zinc required is not large; therefore, the effect of the installation on the speed of the ship is generally very small if the anodes are correctly installed. The installation of zinc anodes on marine craft is straightforward, particularly if it is accomplished when the vessel is in dry dock.

Table 15 lists the chemical composition and impurity limits of the current United States Government (MIL-A-18001H) and ASTM (B 418) zinc anode specifications. The basic zinc anode specification containing aluminum and cadmium as alloying elements has become the worldwide standard for zinc anodes used in cathodic protection systems in seawater and brackish water at ambient temperatures. Committee B-2 of ASTM cautioned that the threshold level for intergranular corrosion of type I composition zinc-aluminum-cadmium alloy is particularly severe above a temperature of about 50 °C (120 °F). For additional information on anode selection, see the articles "Marine Corrosion," "Corrosion of Magnesium and Magnesium Alloys," "Corrosion of Aluminum and Aluminum Alloys," and "Cathodic Protection" in this Volume.

Prevention of Intergranular Corrosion. The zinc industry has long been aware of the susceptibility of zinc alloys containing aluminum to intergranular corrosion when exposed to elevated temperatures. Committee B-2 of ASTM on nonferrous metals and alloys recognized this potential problem when zinc alloy anodes were exposed to service conditions involving elevated temperatures and recommends the use of unalloyed low-iron zinc anodes, ASTM Type II (updated zinc anode specification, ASTM B 418), shown in Table 16 to avoid intergranular corrosion at elevated temperatures.

Underground Zinc Anodes. The cost of installing cathodic protection as a means of stopping the corrosion of coated steel distribution piping is usually small when compared to the cost of repairing leaks and making replacements. The perforation of the pipe wall as a result of corrosion may occur sooner on a well-coated pipe than on a bare pipe, because corrosion current is concentrated at holidays or damaged areas in the coating. The total metal loss, however, of a coated pipe will usually be negligible compared to that which would occur on a bare pipe, and the danger of developing general structural weakness is therefore greatly reduced.

Fortunately, it is relatively easy and inexpensive to apply cathodic protection to a coated

Table 16 Zinc anode composition specification for elevated-temperature exposure

Element	ASTM B 418 Type II composition, %(a)
Aluminum	0.005
Cadmium	0.003
Iron	0.0014
Zinc	rem

(a) Maximum composition

pipeline as compared to a bare pipeline, because the current requirements are only a small fraction of that required for the bare pipeline. The combination of a good pipe coating plus cathodic protection has proved to be both economical and successful to such an extent that the current practice is to provide both a coating and cathodic protection on virtually all new transmission pipelines. This practice is often carried out even in areas where there is no certainty that severe corrosion would occur.

Because the environment around an anode is significant when determining the type and nature of anodic films or coatings, successful anode performance is related to the presence of a friendly environment in contact with the zinc. Soils containing significantly more dissolved sulfates and chlorides than carbonates, bicarbonates, nitrates, and phosphates are often compatible with zinc anodes. However, by packing a prepared backfill consisting of hydrated gypsum, bentonite clay, and sodium sulfate (Na_2SO_4) around an anode, a friendly environment is created, and such a packaged anode can be used in almost all soils. The most popular consists of 75% gypsum, 20% bentonite clay, and 5% Na_2SO_4; this environment provides a relatively low resistivity and thus permits high current output levels. In time, however, the Na_2SO_4 tends to leach out of the backfill, and higher resistivity occurs.

The principal zinc specification (ASTM B 418 Type II) used for underground cathodic protection systems given in Table 16 is the unalloyed high-purity zinc with iron controlled to 0.0014% maximum. The literature does not show any long-term field tests comparing the unalloyed Type II materials, with the zinc-aluminum-cadmium composition of MIL-A-18001H, or the ASTM B 418 Type I given in Table 15. Although the latter alloy anodes prepackaged in prepared backfill have been installed, comparative results are lacking.

Zinc Coating Processes

Seven methods of applying a zinc coating to iron and steel are in general use: hot dip galvanizing, continuous-line galvanizing, electrogalvanizing, zinc plating, mechanical plating, zinc spraying, and painting with zinc-bearing paints. This section and the section "Painting With Zinc-Bearing Paints" in this article contain brief descriptions of each process, the nature of the coating formed, and the practical advantages and limitations of each method. There is usually at least one process that is applicable to any specific purpose. Because the processes are complementary, there are rarely more than two processes to be seriously considered as the best choice for a particular application. Additional information on zinc coating processes can be found in the articles "Hot Dip Coatings" (continuous and batch processes are described), "Electroplated Coatings," "Thermal Spray Coatings," "Corrosion of Carbon Steels," and "Corrosion in the Automotive Industry" in this Volume. Reference to Volume 5 of the 9th Edition of *Metals Handbook* is also recommended.

Hot Dip Galvanizing

In hot dip galvanizing, the steel or iron to be zinc coated is usually completely immersed in a bath of molten zinc. It is by far the most widely used of the zinc coating processes and has been practiced commercially for almost two centuries. The modern hot dip galvanizing process is conducted in carefully controlled plants by applying the results of scientific research, and it is far removed from that of years ago, although it is still dependent on the same basic principles.

The process is primarily applied to finished parts and to semifabricated materials, such as sheet, strip, wire, and tube, on the continuous automated lines of the steel producers. There is an obvious advantage in galvanizing after fabrication in that the zinc completely seals edges, rivets, and welds so that there are no uncovered parts at which rusting can begin.

Continuous Galvanizing. In 1936, a revolutionary new process for continuously coating coils of sheet steel by hot dipping was introduced in the United States. This process, known as the Sendzimir process, uses a small amount of aluminum in the zinc bath and produces a coating with essentially no iron-zinc alloy and with sufficient ductility to permit deep drawing and folding without damage to the coating. Other processes for continuous zinc coating of sheet steel without alloy layer formation were later developed and joined the Sendzimir process. Today, nearly all hot dip galvanized sheet steel is produced by continuous methods.

Machines measuring 150 m (500 ft) or more in length that galvanize the sheet at speeds frequently exceeding 90 m/min (300 ft/min) are producing at the rate of more than 9 000 000 Mg/yr (10 000 000 tons/yr) in the United States. A variety of coating weights and types are produced that vary from as little as 0.5 oz/ft^2 (150 g/m^2) of sheet to a maximum of 2.75 oz/ft^2 (840 g/m^2) of sheet. The standard product is 1.25 oz/ft^2 (380 g/m^2) of sheet. The most dramatic advance in the use of galvanized steel sheet is in the automotive industry. Corrosion problems, intensified by accelerated usage of salt on the highways in the winter months, have resulted in an increase to current levels of over 900 000 Mg (1 000 000 tons).

The annual production of galvanized pipe is being increasingly mechanized and should in the near future approximate the speeds of, and apply coatings comparable to, those of the continuous lines for sheet. The product is broadly used for fencing, sign poles, playground equipment, plumbing, and other construction purposes.

The steel industry also galvanizes steel wire for bridge cables, fencing, armored conductor cable, and general steel wire strand—for hundreds of uses. The product is woven into wire used in armored cable with aluminum or copper conductors and manufactured into many products.

In addition, continuous lines were developed to apply heat-cured organic finishes to metal strips. Prepainted galvanized strip products are now used for roofing and siding, interior partitions, rain goods, furniture, appliances, and other applications.

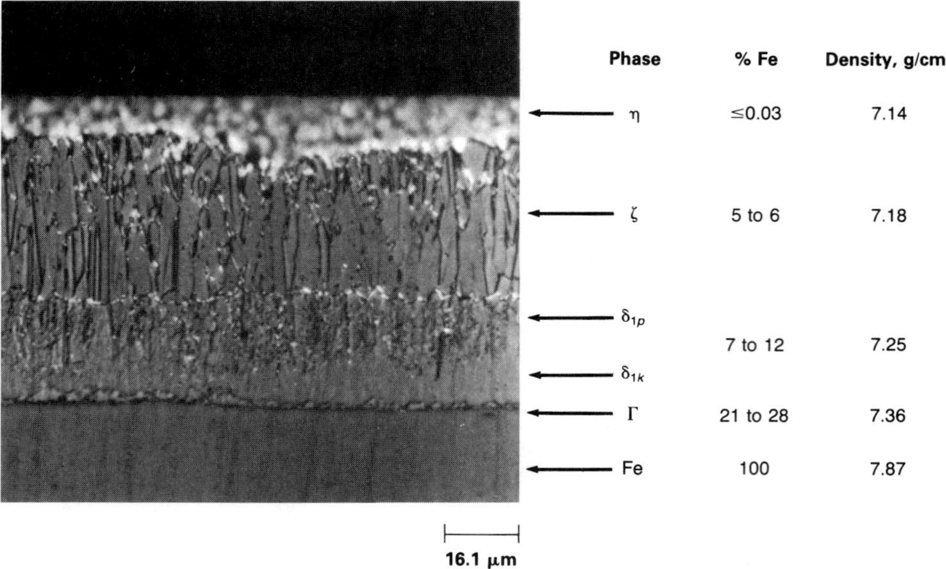

Fig. 5 Typical hot dip galvanized coating. Note the gradual transition from layer to layer, which results in a strong bond between base metal and coating.

The Galvanizing Process. Before the iron or steel parts are dipped in the molten zinc, it is necessary to remove all scale and rust. This is usually done by pickling in an inhibited acid. To remove molding sand and surface graphite from iron castings, shot- or gritblasting is generally used, usually followed by a brief pickling operation.

In the dry galvanizing process, the work is prefluxed by dipping in a flux solution of zinc ammonium chloride, then passed to a low-temperature drying oven. It is then ready for dipping into the molten zinc bath, the surface of which is kept relatively clear of flux. In contrast, in wet galvanizing, the pickled articles are dipped into the molten zinc bath through a substantial flux blanket.

In all processes, as the work enters the bath, the protective layer that has prevented oxidation of the freshly pickled surface peels off, and the work is immediately wetted by the molten zinc. This involves interpenetration of the iron and zinc with the formation of the alloy layers.

Current galvanizing baths are adapted to close thermostatic control. In some of these, hot gases are circulated around the sides of the bath; in others, direct gas or electric heating systems are used. The usual operating temperature is 445 to 460 °C (830 to 860 °F).

Developments in galvanizing resulted in improved quality through better fluxes, better temperature control, and better cleaning practices. Larger kettles and improved materials made it possible to galvanize large structural pieces. Methods were developed for continuously coating wire and pipe in semicontinuous lines.

Nature of the Hot Dip Galvanized Coating. Figure 5 shows a photomicrograph of a typical hot dip galvanized coating consisting of a series of layers. Starting from the base steel at the bottom of the section, each successive layer contains a higher proportion of zinc until the outer layer, which is relatively pure zinc, is reached. Therefore, there is no real line of demarcation between the iron and the zinc; instead, there is a gradual transition through the series of iron-zinc intermetallics, which provide a powerful bond between the base metal and the coating.

The structure of the coating (that is, the number and extent of the alloy layers) and its thickness depend on the composition and physical condition of the steel being treated and on a number of factors within the control of the galvanizer. For example, heavier coatings tend to be deposited on rough-surface and coarse-grain steel, and the total thickness of the alloy layer tends to be slightly greater at corners than at hollows, or shallow areas.

The total thickness of the coating may be controlled by varying the time for which the work is immersed in the molten zinc and the speed at which it is removed. If only a thin coating is required, as is sometimes the case in sheet or wire galvanizing, the work is mechanically wiped upon being withdrawn from the bath in order to remove excess zinc. The temperature of the bath has little effect on the nature of the coating if it is kept between 430 and 470 °C (805 and 880 °F). Small or threaded parts are often centrifuged after being hot dipped galvanized in order to remove excess zinc and to produce a more uniform coating.

The ratio of the total thickness of the alloy layers to that of the outer zinc coating is also affected by varying the time of immersion and the speed of withdrawal of the work from the molten zinc bath; the rate of cooling of the steel after withdrawal is another factor to be taken into account. Sheet galvanizers operating continuous-strip processes usually suppress the formation of alloy layers by adding 0.1 to 0.2% Al to the bath; this increases the ductility of the coating and makes the sheet more amenable to fabrication.

Other elements may be added to galvanizing baths to improve the characteristics and appearance of the coating. Tin and antimony give rise to well-defined spangle effects, and the presence of some lead in the bath is generally considered desirable. The addition of 1% Pb reduces the surface tension by more than 40% when compared with pure zinc. This reduction in surface tension tends to help the drainage of the bath metal as the workpiece exits the bath. Aluminum also improves the appearance of the coating and the corrosion resistance.

Advantages and Limitations. An important advantage of hot dip galvanizing is that, unless zinc is removed by mechanical devices (a practice confined to the automated galvanizing of sheet, strip, wire, and tube) or by centrifuging, the work will probably be thoroughly covered and will carry a thick coating usually weighing 1.8 to 2.2 oz/ft^2 (550 to 670 g/m^2). All edges, rivets, seams, and welds are thus sealed by the hot dip process. Furthermore, it will often be found to be the most economical process when large amounts of steel are to be treated. The size of galvanizing baths does of course limit the size of articles that may be treated, but large parts can be protected by suitable double dipping (one end at a time).

Heating fabricated parts to 450 °C (840 °F) by immersion in molten zinc baths occasionally has undesirable effects, but these can usually be overcome. For example, warping can be eliminated by paying careful attention to welding techniques so as to balance stresses. Possible embrittlement of malleable cast iron, which is likely only if its phosphorus content exceeds 0.07%, can be avoided by quenching in water from 650 °C (1200 °F) before galvanizing. The danger of embrittlement in galvanizing articles that have previously undergone severe localized cold working can be overcome by suitable stress relieving.

Electrogalvanizing

The additional development of continuous electrogalvanizing lines added another dimension to zinc-coated steel, that is, very thin, formable coatings ideally suited to deep drawing or painting. Zinc is electrodeposited on a variety of mill products by the steel industry: sheet, wire, and, in some cases, pipe. Electrogalvanizing at the mill produces a thin, uniform coat of pure zinc with excellent adherence. The coating is smooth, readily prepared for painting by phosphatizing, and free of the spangle that is characteristic of some other zinc coatings.

Electrogalvanized steel is produced by electrodepositing an adhering zinc film on the surface of sheet steel or wire. These coatings are not as thick as those produced by hot dip galvanizing and are mainly used as a base for paint.

The coating produced on strip coils or sheets has a coating weight in the range of less than about 0.06 to 0.2 oz/ft^2 (18 to 60 g/m^2), or 1.3 to 4.3 μm (0.05 to 0.17 mil) thick on each side. A small amount carries considerably less—approximately 0.025 oz/ft^2 (7.6 g/m^2), or 0.5 μm (0.21 mil) on each side.

Zinc is usually electrodeposited on steel wire in the range of 0.3 to 3 oz/ft^2 (90 to 915 g/m^2). The diameter of plated wire (including wire that is cold drawn after plating) usually ranges from 0.23 to 4.9 mm (0.009 to 0.192 in.). Steel carbon contents range from 0.08 to 0.85%. Tensile strengths range from 345 to 2070 MPa (50 to 300 ksi). Heat-treated and coated wire can be cold drawn to approximately 95% reduction in area, depending on chemical composition, heat treatment, and diameter.

Nature of the Electrogalvanized Coating. The pure zinc coating deposited is highly ductile. Because of its excellent adhesion, electrogalvanized steel strip and wire have good working properties, and the coating remains intact after severe deformation.

Electrodeposited zinc coatings are simpler in structure than hot dip galvanized coatings. They are composed of pure zinc and have a homogeneous structure. Surfaces have a smooth texture whose appearance can be varied by additives and special treatments in the plating bath. They can be used where a fine finish is needed.

Electrogalvanizing provides adequate protection for many types of mild exposures. These coatings are frequently treated with chromate conversion solutions to improve appearance, reduce staining, and retard the formation of white corrosion products under high-humidity conditions.

Advantages and Limitations. In electrogalvanizing, steel strip or wire is continuously fed through suitable entry equipment, a series of washes and rinses, and a plating bath. Either an acid sulfate zinc or cyanide zinc bath is used as the plating bath. Both produce even, adhering zinc deposits. Although brighteners are not used for electrogalvanizing, grain refiners are usually added to help produce a fine, tightknit zinc surface on the steel.

Zinc electrodeposits are considered to have the best adhesion of any metallic coating. Good adhesion depends on very close physical conformity of the coating with the basis metal. Therefore, particular care must be taken during initial cleaning. Electrodeposition affords a continuous process for applying zinc coatings to parts that cannot be hot dipped. They are especially useful where a high processing temperature could damage a part.

Applications. Electrogalvanized sheets are produced in various tempers suitable for simple bending or forming, for curving, and for rolling into cylinders without fluting. Spot welding is easily accomplished if care is taken.

Electrogalvanized steel is easily prepared to receive decorative finishes. Much of it is produced with a phosphate treatment or an organic coating. The phosphate treatment provides an adequate surface for a good bond with organic finishing materials. Organic coating applied over electrozinc thus treated maintains good adhesion in adverse conditions, such as sudden changes in temperature and high humidity. Phosphated electrogalvanized steel is used for parts subject to atmospheric corrosion or salt spray and for parts that will be lacquered or painted. Phosphate treatment increases corrosion resistance markedly, particularly in atmospheres with a high sulfur content.

Electrogalvanized sheet is used for manufacturing water cooler housings, exterior panels of ranges, freezers, dryers, washers, air conditioners, and other major appliances. It is used for deep-drawn parts for kitchen cabinets, refrigerators, and allied products instead of plain cold-rolled sheet because zinc holds better in the dies and reduces breakage significantly. Bakery goods and other merchandizing display cases, stud systems for steel building construction, acoustical ceiling members, and television antennas are also made of electrogalvanized steel.

A new application that should greatly benefit the automotive industry is one-side electrogalvanizing. The galvanized side protects against corrosion, and the bare side can take the baked enamel finish required by the outer automobile surface (see the article "Corrosion in the Automotive Industry" in this Volume).

Electrogalvanized wire is especially useful in applications in which the wire must be bent, twisted, or wrapped around its own diameter. When formed, the coating does not crack, peel, or flake. Many chain link fences are made from zinc-electrocoated wire because it is not rough and therefore is safe to handle. The wire is used for conveyor belts, twisted wire brushes, chains, baskets, kitchen utensils, staples, cages, bobby pins, clotheslines, and telephone and transmission wire.

Zinc Plating

Coatings of zinc may also be applied to iron and steel surfaces by electroplating. The article to be plated is made the negative electrode in an electrolytic cell containing a solution of a zinc salt through which an externally generated dc current is passed. Zinc is supplied to the cell as expendable positive electrodes or as zinc salts added directly to the plating solution.

Surface Preparation. The adherence of electrodeposited zinc coatings depends on the metal-to-metal bond between the plated coating and the underlying steel surface. Therefore, particular attention must be given to the preparation of the surface before plating to obtain a coating in true physical contact with the entire steel surface.

The usual method of removing all rust, scale, and grease from the steel surface involves cleaning the surface thoroughly in a hot alkaline bath by soaking the parts for a short period of time. This is often followed by use of an electrolytic alkaline cleaner and a spray alkaline cleaner. An acid dip is then carried out to remove oxides and scale. There must be adequate rinsing between the alkaline/acid baths and the acid/plating baths to avoid contamination of the plating bath by carryover from the cleaning baths.

Zinc Plating Baths. Zinc plating is done in an acid or an alkaline bath. Although the alkaline-cyanide zinc baths are the most efficient and have the best throwing power, they do create a serious pollution problem. Therefore, more alkaline, hydrochloric acid (HCl), and H_2SO_4 zinc baths are being used. The H_2SO_4 baths are primarily used in the tubing, wire, and sheet electrogalvanized areas. All of the other baths are used for barrel plating. Barrel plating is used to plate a large volume of small articles. These are placed in a suitably constructed rotating barrel that is immersed in the plating solution.

Various brightening agents may be added to the baths to give a deposit that is more lustrous than that obtained from normal zinc plating baths. The amount of brightening agent requires very careful control, and the bath and the zinc anode must both be kept particularly pure when brighteners are used.

Nature of the Electroplated Zinc Coating. The normal electroplated zinc coating is dull gray with a matte finish, but whiter and more lustrous deposits can be produced by adding special agents. The coating consists of pure zinc and is of uniform composition throughout. It adheres by metal-to-metal bonds.

Advantages and Limitations. Electroplating is the most precise of all zinc coating processes and therefore is particularly suitable for coating delicate articles, such as instrument parts, on which a fine finish is essential. Furthermore, the parts need never be heated to a temperature above the boiling point of water.

Because the pure zinc coating is extremely ductile, it is quite easy to form zinc-plated sheets without damaging the coating. On simply shaped articles, it is possible to control the thickness of electroplated zinc coatings within fine limits. Barrel-plated articles have coatings that are more uniform in thickness than those normally obtained.

Zinc electrodeposits expand very slightly during plating; therefore, if the deposit does not adhere properly, blistering may develop. The effect develops slowly and takes several hours or even days for completion; it is most marked in acid-sulfate deposits.

Mechanical Plating

This process is used to deposit a specified thickness of zinc on a steel surface in a uniform manner, even in threaded areas. Parts plated by the process do not encounter hydrogen embrittlement. The plate thickness can be varied from 2.5 to 125 μm (0.0001 to 0.005 in.).

In the plating of fasteners, a large, lined, steel tumbling container is used for the complete process. A predetermined amount of parts, glass beads, and water are added to the tumbler, and the following system is used. First, an inhibited acidic powder cleaner and an acid copper salt are added and maintained at a pH of 1.8. The parts are then tumbled to achieve proper cleaning and an immersion coating of copper. Second, the catalyst (promoter chemical) and zinc powder (8 to 20 μm) are added and tumbled.

The resulting zinc deposit is cold welded to the copper by the peening action of the glass beads. By regulating the amount of zinc, it is possible to predetermine the film thickness of zinc that will result.

Zinc Thermal Spraying

Zinc thermal spraying consists of projecting atomized particles of molten zinc onto a prepared surface. Two types of spray guns are in commercial use today: the powder gun and the wire gun.

Surface Preparation. The surface preparation of the work is the same for each process and involves cleaning and then roughening the surface to be sprayed. The usual method of roughening is coarse gritblasting.

Whatever method of surface preparation is used, the sprayed zinc coating should be applied as soon as possible after the surface has been prepared in order to reduce the possibility of oxidation and thus increase the effectiveness of the metal-to-metal bond. The time lag is reduced to a minimum in mechanized plants that have been built for production line work and for treating construction steelwork before assembly.

Two Methods of Spraying Zinc. Two methods of spraying zinc produce coatings of comparable quality. They are the powder process and the wire process.

The Powder Process. Zinc particles are suspended in a gaseous medium and driven through a blowpipe flame. The stream of powder-laden gas is heated by the flame surrounding the nozzle, and outside this, a cone of compressed air gives impetus to the stream of molten droplets.

There are four feeds to the gun: combustible gas, oxygen, compressed air, and zinc powder suspended in air or a gas. A wide choice of fuel gases is available because the gas does not have to function as an atomizer.

The Wire Process. Zinc wire is fed axially into the center of a blowpipe flame issuing from a ring of small gas ports. A stream of compressed air disintegrates the film of molten metal as it forms and sprays it out of the nozzle. The oxygen and fuel gas supplied for the flame are at the same pressure so that they may be simply mixed in a small chamber.

In hand tools, such as those used for zinc coatings, a compressed air turbine is used to drive the wire feed through worm reduction gearing. Nozzles often consist of a sleeve of hard steel surrounded by copper to disperse the heat generated by the blowpipe flame. Various types of wire-spraying guns have been developed in which electricity is used in place of gas for heating.

The Nature of the Thermally Sprayed Zinc Coating. The sprayed coating is slightly rough and slightly porous. The specific gravity of a typical zinc coating is about 6.35 as compared to 7.1 for cast zinc. This slight porosity does not affect the protective value of the coating, because the zinc is anodic to steel. The zinc corrosion products that form when the coating is in service fill up the pores to produce a solid coating. The slight roughness of the surface makes it an ideal base for paint.

The mechanism by which the zinc adheres to the underlying surface has been the subject of many experiments and much speculation. Although no alloy layer is formed and the bond is purely mechanical, the adhesion of zinc to a properly prepared iron or steel surface is excellent. Recently, it has been suggested that preheating the base metal gives even greater values for adhesion. In the ordinary process, the impact of the particles of molten zinc on the surface causes only a very slight increase in temperature of the base metal.

Advantages and Limitations. Zinc spraying has a significant advantage over most other methods of zinc coating in that it can be applied to work of almost any shape or size on the site. When applied to finished parts, the welds, ends, and rivets receive adequate coverage. Moreover, it is the only satisfactory method of depositing unusually heavy zinc coatings with thicknesses of 0.25 mm (0.01 in.) and greater.

Zinc spraying is usually not suitable for depositing coatings inside cavities, although special types of nozzles are available for applying coatings inside short lengths of tube. The process is seldom economical for treating open structures, such as wire mesh, because of the large loss of metal that would result.

The mechanized plants, such as those used for treating construction steelwork before assembly, make it possible to deposit completely uniform coatings. In hand spraying, the degree of uniformity achieved depends entirely on the skill of the operator.

Painting With Zinc-Bearing Paints

Although zinc coatings can be applied in several ways, the size of a structure or piece of equipment places limitations on the method used. On very large structures, painting is often the only practical method. Two types of zinc pigment coatings are available for corrosion control: zinc dust/zinc oxide and zinc-rich coatings.

Zinc Dust/Zinc Oxide Coatings

Zinc dust/zinc oxide paints, also known as metallic zinc paints, contain a pigmentation of approximately 80% zinc dust, 20% ZnO, and approximately 80% pigment by weight. These paints offer excellent rust-inhibitive properties, adhesion, film distensibility, and abrasion resistance. Because they adhere tightly to zinc and other metals, they are ideal for prime and finish coat applications and may be used as a primary first coat even over partially rusted surfaces.

Zinc dust/zinc oxide paints are used for the protection of many types of steel structures under a variety of service conditions. They are particularly well suited to use on galvanized steel; are highly satisfactory for priming steel for atmospheric and underwater exposure; and can be used on many outdoor structures, such as bridges, water tanks, and dams, where rusting must be prevented.

In accordance with Federal Specification TT-P-641 which covers primer paints and zinc dust/zinc oxides for galvanized surfaces, there are three types of zinc dust/zinc oxide paints:

- *Type I*: zinc dust/zinc oxide linseed oil for outdoor exposure, recommended as a primer or finish coat for broad, general use, especially when there is widespread rusting of the steel surfaces; should be air dried only
- *Type II*: zinc dust/zinc oxide alkyd resin paint, a heat-resistant paint sometimes sold as a stack paint, may also be used for outdoor exposures where rust is not severe; quick drying; can be air dried or baked at temperatures to 150 °C (300 °F)
- *Type III*: zinc dust/zinc oxide phenolic resin paint; used for water immersion and other severe moisture conditions; may be air dried or baked at temperatures to 150 °C (300 °F)

These paints, when properly formulated and prepared, can be applied by brushing, dipping, or spraying. Although the presence of ZnO prevents rapid or hard settling, adequate agitation of the paint in the dip tank is necessary to ensure the coating homogeneity necessary for maximum metal protection. Pressure equipment should be used when spraying, and the distance between the paint reservoir and the spray gun should be as short as possible to ensure the proper rate of feed to the nozzle. Again, some agitation of the paint in the reservoir is recommended.

Zinc dust/zinc oxide paints possess high covering power and can hide backgrounds of almost any color when spread at the rate of approximately 20 m^2/L (800 ft^2/gal). However, because the protection afforded by a paint coating is directly related to its thickness, the necessary protection cannot be guaranteed unless the dry film is thick enough for the specific environmental conditions. Therefore, care must be taken to avoid spreading the paints too thin. The natural blue-gray color of zinc dust/zinc oxide paints provides an aesthetic appearance, but if another color is desired, red, buff (orange-yellow), and green can be obtained by varying the pigment.

To prepare surfaces for zinc dust/zinc oxide paints, rust (or scale) and any accumulation of leaves, dirt or other foreign materials should be removed. This may be accomplished on large structures by sandblasting and on small structures or areas with a deck or wire brush.

Zinc-Rich Coatings

In recent years, a number of paints have been developed that will deposit a film of metallic zinc having many properties in common with zinc coatings applied by hot-dip galvanizing, electroplating, metal spraying, and mechanical plating. Such paint films will protect the underlying steel sacrificially if they contain 92 to 95% metallic zinc in the dry film and if the film is in electrical contact with the steel surface at a sufficient number of points. They are effective where steel is subjected to high humidity and water immersion. Under normal conditions, zinc-rich coatings are long lasting and most effective where a regular maintenance program may be difficult. In applications in which steel is immersed in brackish or salt water, zinc-rich coatings, along with a suitable top coat, should be used. Most zinc-rich paints are of the air-drying type, although oven-cured primers containing a high content of zinc dust are available.

The type of zinc dust used is a heavy powder, light blue-grey in color, with spherically shaped particles having an average diameter of approximately 4 μm. Such powder normally contains 95 to 97% free metallic zinc with a total zinc content exceeding 99%.

Surface Preparation. Zinc-rich primers must be applied over clean steel surfaces to provide the metal-to-metal contact essential to successful performance of the coating. Abrasive blasting is the most effective method of cleaning steel. Although white metal blast-cleaning (NACE No. 1) is preferred, near-white SSPC-SP-10 or Commercial Blast Cleaning SSPC-SP-6 is acceptable (Ref 14).

Where the zinc is supplied as a separate component, it should be added slowly to the vehicle with constant agitation. After a homogeneous mix is obtained, the primer may be applied with air spray. Airless spray may also be used, but the nozzles may wear quickly. Because zinc settles rapidly, continuous agitation of the paint is essential during application, and fluid lines should be kept as short as possible.

To obtain a wet coat, the gun should be kept within 30 cm (1 ft) of the surface. Uneven film thickness due to brushing or rolling may result in mudcracking in the thick portions. Zinc-rich primers should be applied at a dry film thickness of 0.06 to 0.08 mm (2.5 to 3.5 mils).

The Nature of the Zinc-Rich Coating. Depending on the binder, zinc-rich coatings fall into two classes: organic and inorganic. The inorganic solvent-base types are derived from organic alkyl silicates, which become totally inorganic upon curing. Each offers particular protection characteristics, and each requires different preparation of the steel surface. The following comparisons should be helpful in selecting the most useful binder system.

The organic zinc-rich coatings are formed by using zinc dust as a pigment in an organic binder. This binder may be any of the well-known coating vehicles, such as chlorinated rubber and epoxy. The zinc dust must be in sufficient concentration so that the zinc particles are in particle-to-particle contact throughout the film. Thus, zinc provides cathodic protection. In the case of the organic binder, there is no reaction with the underlying surface other than for the organic vehicle to wet the steel surface thoroughly and to obtain mechanical adhesion.

Organic zinc-rich coatings do not require a white blast preparation of the steel surface, although a commercial blast should be included if the application is heavy service. For mild-service applications, the organic coating can be applied to a well-hand-cleaned surface, even if traces of rust are present.

Some proponents feel that maintaining proper humidity during surface preparation, application, and curing is not necessary. Because this type of coating is more flexible than inorganic coatings, exacting surface preparation for bonding to a substrate is not required. Finally, although organic coatings are more compatible with top coats, they are somewhat less abrasion resistant than the inorganic types.

As to the advantages of these coatings, organic zinc-rich coatings require less critical surface preparation, allow greater variation in application techniques, are less sensitive to varying climatic conditions during application and curing, and are more flexible and more resistant to chemical environments. Their disadvantages include flammability, blistering, harmful solvent effects, sensitivity to atmospheric influences, and relatively low heat resistance.

For better resistance against continuous exposure to salt water and to acid or alkali chemical fumes, zinc-rich coatings should be top coated with organic topcoats to provide a totally organic system, with optimum intercoat compatibility. A top coating may also be applied to provide color or to prevent gradual erosion of the zinc coating. Although zinc-rich coatings vary in application characteristics, they can be applied by brush or spray, and depending on the specific formulation, one coat can vary in thickness from 2 to 7 mils.

Inorganic Zinc-Rich Coatings. Many inorganic zinc-rich coatings use water solutions of alkali silicates as vehicles. Others use phosphates, silicones, and modifications of these groups.

Self-cured coatings are two-package materials consisting of zinc dust and a vehicle; they are mixed immediately before application. Postcured coatings are three-package materials that consist of zinc dust, the vehicle to be mixed with it before application, and a curing agent that is applied on top of the coating.

The inorganic zinc coating forms its film and its adhesion to the steel surface by methods quite different from those of the organics. The coating system is a chemically reactive system, and the chemical activity is similar for either the water- or the solvent-base inorganic. Zinc is the principal reactive element in the inorganic coating systems and is primarily responsible for the development of initial insolubility. Depending on the formulation, other metal ions may be present in the system that also react and aid in the insolubilization of the coating. The silicate vehicle can also react with underlying iron surface to form a chemical bond with the iron or steel substrate.

Inorganic zinc-rich coatings commonly require a white metal blast as preparation for the steel surface. Because inorganic coatings generally have limited flexibility and tend to break or crack upon bending or impact, careful preparation of the steel surface is required to ensure a good bond between the coating and the steel. However, despite the difficulties of preparation, these inorganic coatings are unaffected by solvents, oils, petroleum products, aliphatics, aromatics, ketones, and alcohols. They do not chalk, peel, or lose thickness over long periods of time. Also, they are easier to weld through and have excellent abrasion resistance and surface hardness.

Inorganic zinc-rich coatings offer good conductivity; good adhesion to clean steel; excellent resistance to weather, sunlight, and variations in temperature; resistance to radiation, heat, and abrasion; and reduced undercutting. Conversely, these coatings require unusually good surface preparation, display a lack of distensibility and adhesion to some metals other than steel and zinc, require moderate temperatures and atmospheric humidity for cure, and exhibit unsatisfactory durability under conditions of continuous immersion in electrolytes and a lack of resistance to strong acids and alkalies.

Zinc Dust/Zinc Oxide Paint Versus Zinc-Rich Coating

Whether to use a zinc dust/zinc oxide paint or a zinc-rich coating depends on a number of factors, including cost of surface preparation, paint application, and anticipated length of surface. Zinc dust/zinc oxide coatings are ideal for rural or semi-industrial atmospheres. They are particularly effective on galvanized surfaces.

The widely used zinc dust/zinc oxide primers based on ordinary drying oil media do not give general electrolytic protection against corrosion and therefore do not fall in the category of zinc-rich paints.

Zinc-rich coatings are preferred for the protection of steel or galvanized steel structures exposed to marine environments or immersed in seawater. Applications include interiors of floating roof tanks, cooling tower piping, pipe racks and exterior piping in refineries, stacks, chemical plant maintenance, offshore drilling platforms, aboveground pipelines, structural steel before erection, exterior of pressure vessels, ammonia tanks, ship holds, and air conditioning equipment.

A top coat finish may be necessary in aggressive atmospheres. The top coat must adapt to the environment and must guarantee compatibility with, and adhesion to, the zinc-rich primer.

Advantages and Limitations of Zinc-Rich Paints. Zinc-rich primers offer a more versatile form for applying zinc to steel than galvanization; large, continuous complex shapes and fabricated new or existing structures can be easily coated at manufacturing shops or in the field. Their performance has earned them a prominent place in the field of corrosion protection coatings. However, the limitations of zinc-rich paints include cost, difficulty in applying, and the requirement of clean steel surfaces. They must be top coated in severe environments (pH under 6.0 and over 10.5).

Zinc Chromate Paints

Zinc chromate pigments are unique in that they are useful as corrosion inhibitors for ferrous and nonferrous metals. Zinc chromate was developed just before World War II and was the major pigment in marine primers. It is yellow in color, but can be tinted green. Zinc chromate paints are used as an after-pickling coating on steel and as a primer for steel and aluminum. Federal Specification TT-P-645 covers zinc chromate paints.

REFERENCES

1. R.M. Burns and W.W. Bradley, *Protective Coatings for Metals*, 2nd ed., Reinhold, 1955, p 66-147
2. *Zinc Coatings for Corrosion Protection*, Zinc Institute, 1978, p 20
3. E.A. Anderson, The Atmospheric Corrosion of Rolled Zinc, in *Symposium on Atmospheric Corrosion of Nonferrous Metals*, STP 175, American Society for Testing and Materials, 1955, p 126-134
4. W. Machu, Corrosion of Metals and Metal Coatings in Tropical and Sub-Tropical Climates, *Werkst. Korros.*, Vol 5, 1954, p 395-398
5. Report of Subcommittee V on Atmospheric Exposure Tests of Zinc Alloy Die Castings, Committee B-6, in *Proceedings ASTM*, Vol 61, American Society for Testing and Materials, 1961, p 273-281
6. "Specification for Zinc (Slab Zinc)," B 6, *Annual Book of ASTM Standards*, American Society for Testing and Materials
7. "Specification for Zinc Alloy Die Castings," B 86, *Annual Book of ASTM Standards*, American Society for Testing and Materials
8. G.L. Cox, Effect of Temperature on the Corrosion of Zinc, *Ind. Eng. Chem.*, Vol 23, 1931, p 902-904
9. H. Grubitsch and O. Illi, The Hot Water Corrosion of Zinc II, *Korros. Metall.*, Vol 16, 1940, p 197
10. H. Grubitsch and H. Huemer, Scanning Electron Microscope Studies of the Hot Water Corrosion of Zinc, *Werkst. Korros.*, Vol 24 (No. 1), 1973, p 1-7
11. R. Rosset and A. Jardy, Protection of Galvanized Steel Against Cold and Hot Water Corrosion by a Pyrophosphate Coating, in *Proceedings of Intergalva/82*, May 1982; French patent 7402178, 1974; French patent 7637425, 1976
12. *Handbook of Steel Drainage and Highway Construction Products*, American Iron and Steel Institute, 1971, p 214-215
13. *Zinc Coatings for Corrosion Protection*, Zinc Institute, 1978, p 25
14. *Zinc Coatings for Corrosion Protection*, Zinc Institute, 1978, p 14

SELECTED REFERENCES

- *Good Painting Practices: Steel Structures Painting Manual*, Vol 1 and 2, 2nd ed., Steel Structures Painting Council, 1982
- C.J. Slunder and W.K. Boyd, *Zinc: Its Corrosion Resistance*, 2nd ed., International Lead Zinc Research Organization, 1983

Corrosion of Tin and Tin Alloys

Daniel J. Maykuth and William B. Hampshire, Tin Research Institute, Inc.

TIN is a soft, brilliant white, low-melting metal that is most widely known and characterized in the form of coating for steel, that is, tinplate. In the molten state, it reacts with and readily wets most of the common metals and their alloys. Because of its low strength, the pure metal is not regarded as a structural material and is rarely used in monolithic form. Rather, the metal is most frequently used as coating for other metals and in alloys to impart corrosion resistance, enhance appearance, or improve solderability. It also finds wide use in alloys, the most important of which are tin-base soft solders and bearing alloys and copper-base bronzes.

Pure Tin

Pure tin is subject to two phenomena that are sometimes confused with the corrosion process in the ordinary atmosphere. These are its low-temperature allotropic modification and its susceptibility to whisker growth. To avoid this confusion, these processes are discussed below.

Allotropic Modification. At temperatures from 13.2 °C (55.8 °F) to its melting point of 232 °C (449.6 °F), tin exists in a body-centered tetragonal (bct) structure commonly known as β-tin. Below 13.2 °C (55.8 °F), the β form can change to a diamond cubic structure known as α-tin, which lacks cohesion and appears as a friable gray powder. This is sometimes called the tin pest. This transformation does not occur spontaneously unless the tin is of extremely high purity and is exposed to subzero temperatures. The transformation can be accelerated by inoculating the β with α crystals or by deforming the β-tin at low temperatures (Fig. 1). Some details of the mechanisms and kinetics of this process are discussed in Ref 1.

The transformation is inhibited by the presence of small amounts of bismuth, antimony, or lead. Hot-dipped tin coatings and most electrodeposited coatings seem to be immune to this phenomenon, probably because of impurity effects. Thus, no traces of transformation were evident on hot-dipped tinplate cans after burial for 46 years in arctic snow or on electroplated tin coatings on refrigerator parts (Ref 2). However, transformation has occurred with thicker deposits; when such low-temperature exposure is anticipated, the incorporation of about 0.1% Bi is recommended to avoid the problem (Ref 3).

Tin Whiskers. Tin is subject to a form of recrystallization at room temperature that manifests itself as a growth of thin (1- to 2-μm, or 0.04- to 0.08-mils, diam) single-crystal filaments from the surface of tin coatings. These can begin to form in as little as 5 weeks and may grow at a rate up to 1 mm/mo (0.04 in./mo). Although the mechanism is not clearly understood, formation of tin whiskers appears to be favored by residual or applied stress, by the presence of a brass substrate, and by high-purity electrodeposited tin (Ref 4-6). The potential for whisker growth can be minimized if not completely eliminated by reflowing the tin coating or by incorporating 2 to 10% Pb into the electrodeposited tin.

Atmospheric Corrosion. In clean dry air, tin retains a bright appearance for many days. In one study, a light dulling was observed after 100 days, and noticeable, faint yellow-gray tarnish film was seen after 150 days (Ref 7). However, it was also reported that the reflectivity of tin remains practically unchanged over long periods when the tin is washed with soap and water (Ref 8). Thus, at ordinary temperatures, the surface oxide film on tin is very thin and exhibits a very slow rate of growth. The rate of oxidation increases with temperature. Above 190 °C (375 °F), a film thickness sufficient to produce interference colors is reportedly produced in a few hours; at 210 °C (410 °F), this film thickness is produced in 20 min (Ref 2).

The results of a comprehensive 20-year study of the atmospheric corrosion resistance of bulk tin were reported by an ASTM Committee (Ref 9-13). Sheets of commercial 99.85% purity tin, measuring 230 × 300 mm (9 × 12 in.) were exposed at seven sites in the United States, including industrial, seacoast, and rural atmospheres. Results are listed in Table 1.

Ancient tin coins from Malaysia were found to be covered with successive layers of brown and gray scale that were principally stannic oxide (SnO_2) that contained sulfate plus traces of silica and iron (Ref 14). Examination of seventeenth and eighteenth century sarcophagi in Vienna revealed some evidence of deterioration that was suspected to be the tin pest. It was found, however, that the casting was porous and that air and moisture produced corrosion products of stannous oxide (SnO) and SnO_2, causing the observed swelling, blistering, and cracking.

Oxidation. At extremely low temperatures, the oxidation of tin is very superficial. In one investigation, resistivity measurements were used on tin condensation films formed at 1.5 to 300 K; in all cases, a step function indicating that the growth of tin oxide first began at 23 K was found (Ref 15). No further growth of the oxide was detected at 50 to 150 K.

The most comprehensive studies of interactions between tin and oxygen were those discussed in Ref 16 to 19, in which 99.994% pure foils and a vacuum microbalance were used to measure oxidation rates at oxygen pressures between 10^{-3} and 500 torr (0.13 and 6.7×10^4 Pa) and temperatures from 150 to 220 °C (300 to 430 °F). The essential features of oxidation behavior were found to be explainable in terms of the microstructure of the oxide. With oxygen pressure under 1 torr (133 Pa), dendritic α-SnO crystallites grew at an increasing rate, with the rate-determining factor apparently being the dis-

Fig. 1 Gray tin transformation on pure tin. Both samples were stored at −20 °C (−4 °F), but the sample on the left was bent at this temperature and the other was left undisturbed.

Table 1 Corrosion of tin exposed in different environments for 10 and 20 years

Sample location	Average corrosion rate(a) 10 years mm/yr	10 years mils/yr	20 years mm/yr	20 years mils/yr
Heavy industrial	...	...	0.0017	0.067
Marine heavy industrial	...	...	0.0013	0.051
Marine (New Jersey)	0.0019	0.075	...	...
Marine (Florida)	0.0023	0.09	...	...
Marine (California)	...	...	0.0029	0.11
Semiarid	...	...	0.00044	0.017
Rural	0.00049	0.019	...	...

(a) Converted from weight loss data, assuming a tin density of 7.29 g/cm^3. Source: Ref 13

sociation of oxygen. Above 1 torr (133 Pa), the oxidation rate curves had a characteristic sigmoid shape, in which the initial stages corresponded to the lateral spread of oxide from numerous nuclei to form α-SnO platelets. Subsequent growth followed a logarithmic law and was consistent with control by tin diffusion through an oxide film under a parabolic or cubic law, while the formation of cavities in the oxide film progressively reduced the area through which diffusion could take place. For long oxidation times, the thick oxide film was subject to random fracture, leading to erratic results.

The oxidation of tin containing 0.17% Pb and 0.024% Sb was examined at 168 to 211.5 °C (335 to 413 °F) and oxygen pressures of 4 to 9 torr (533 to 1200 Pa) (Ref 20). These oxidation rate data were not significantly different from those given in Ref 16 to 19.

The effects of impurities on the oxidation rate of tin were also studied by using microbalance techniques under conditions similar to those described in Ref 16 to 19 (Ref 21). The results are summarized in Table 2. These results were later rationalized in terms of the relative thermodynamic stability of the oxides formed, as follows. If the oxide of the alloying element is less thermodynamically stable than SnO, the oxidation rate of the alloy remains unchanged for additions whose ions have the same valence as the tin. However, when the formal ionic charge of the alloying element exceeds that of the tin—for example, antimony, bismuth, iron, and titanium—then the oxidation rate of the tin increases. Those alloying elements forming an oxide more stable than SnO—for example, zinc, indium, phosphorus, and germanium—undergo preferential oxidation at the surface, thus inhibiting the oxidation of tin (Ref 22).

The oxidation rate of molten tin was studied at 400 to 800 °C (750 to 1470 °F) with oxygen pressures of 50 to 500 torr (6.7 to 67 kPa) (Ref 23). The rate varied greatly from specimen to specimen at any one temperature, but was apparently linear under all conditions. The variability was attributed to crystal orientation in the oxide film, which was in apparent agreement with the results of other investigations (Ref 24). One researcher commented that another possibility is the continuous conversion of SnO to a nonprotective SnO_2 (Ref 25). In one experiment conducted at 800 °C (1470 °F), the initially formed jet-black film of SnO became incandescent at one end of a boat, and the incandescence traveled rapidly to the other end of the boat, leaving an orange coating.

Another study investigated the effects of alloying additions at levels of 0.01, 0.1, and 1% on the oxidation of molten tin (Ref 26). Antimony, lead, bismuth, and copper had negligible effects, while

Table 2 Effect of alloy additions of 0.1 at.% on the oxidation rate of tin at 190 °C (375 °F) and an oxygen pressure of 10 torr (1330 Pa)

Alloying element	Increase in weight after 1000 min, μg/cm^2
Manganese	2.7
Antimony	2.5
Thorium	2.1
Bismuth	1.7
Iron	1.6
Lead	1.3
Nothing added	1.0
Cadmium	1.0
Phosphorus (0.5 at.%)	0.3
Zinc	0.2
Indium	0.1

Source: Ref 21

higher concentrations of lead increased the temperature at which significant oxidation occurs. Magnesium, lithium, and sodium significantly increased the oxidation rate, but zinc, phosphorus, indium, and aluminum decreased the rate. The oxidation of an alloy containing 0.01% Al was about the same as that of pure tin at 425 °C (795 °F).

Other laboratory oxidation studies were concerned with tin in contact with air. The formation of an oxide film was shown in Ref 27 and 28, and weight increment curves were developed in Ref 29. In another study, the oxidation rate was determined to be linear after the first few days and was nonprotective (Ref 6). Lastly, the oxidation of tin and tinplate was investigated by using coulometric and x-ray techniques (Ref 30, 31). Up to 130 °C (265 °F), the oxidation followed a logarithmic rate law that tended to become parabolic at higher temperatures. At room temperature, the oxide film appeared to be amorphous, but at higher temperatures, α-SnO was detected, possibly with some SnO_2.

One study found that SnO forms on tin immediately above its melting point and that SnO_2 forms at higher temperatures (Ref 32). This effect was demonstrated by spot heating a piece of tinfoil (Ref 33). Stannic oxide was found at the center and was surrounded by SnO, which was in turn ringed with an amorphous oxide. According to other researchers, the disproportionation of SnO to tin and SnO_2 is a slow process, even at 300 °C (570 °F) (Ref 34). The need for extreme care in oxidation studies, especially with regard to surface preparation, was emphasized in Ref 22. This was demonstrated by using cathodic cleaning to show the effects of humidity (Ref 35).

Minor impurities in tin also affect its oxidation behavior in air. Small amounts of indium, phosphorus, or zinc were found to slow the oxidation (Ref 30). In addition, traces of aluminum were shown to cause embrittlement as a result of intercrystalline attack (Ref 36). Antimony additions, however, counteracted this effect.

Reaction With Other Gases. Tin does not react with hydrogen or nitrogen below its melting point, nor is it reactive with dry ammonia (NH_3). Molten tin reacts with carbon dioxide (CO_2) according to:

$$Sn + 2C_2 \rightarrow SnO_2 + 2CO \qquad \text{(Eq 1)}$$

Above 650 °C (1200 °F), molten tin reacts with water vapor to form SnO_2 and hydrogen.

From 25 to 100 °C (75 to 212 °F), hydrogen sulfide (H_2S) has little apparent effect on tin, but above 100 °C (212 °F), stannous sulfide (SnS) forms. Stannous sulfide and stannic sulfide (SnS_2) are also formed by reacting tin with sulfur at high temperatures. Tin also reacts readily with SCl_2, S_2Cl_2, NOF, and hydrofluoric acid (HF). Tin is readily attacked by chlorine, bromine, and iodine at room temperature, but fluorine reactions become significant only above 100 °C (212 °F).

Water. In hot or cold distilled water, the only action of tin is the slow growth of an oxide film, with a negligible amount of metal entering solution. Water that was freshly distilled in a tin was found to have less than 1 ppb Sn in solution (Ref 2). Storage in tin-lined or tinned copper tanks for 24 h produced, in the worst instances, only a few ppb, but in some cases, the tin content remained below 1 ppb.

In tap water of 7.2 pH at 25 °C (75 °F), specimens of 99.99% cold-rolled tin showed a weight gain of 0.023 mg/dm^2/d (1.2×10^{-4} mm/yr, or 0.04 mils/yr) in 50 days and the formation of an insoluble film (Ref 37). With harder tap waters of 7.4 and 8.6 pH, weight losses of the order 0.046 and 0.01 mg/dm^2/d (2.3×10^{-4} and 5×10^{-5} mm/yr, or 0.09 and 0.02 mils/yr), respectively, were incurred in 50 days. Precipitated carbonate was mainly responsible for localized water line attack with hot and cold hard waters because no attack occurred without the precipitate. Addition of 5% Sb to the tin prevented localized attack by hard water.

The results of corrosion test data on tin and several tin alloys in seawater under conditions of total immersion are shown in Table 3. It was also observed that application of a fairly thick 60Pb-40Sn alloy coating over copper will protect it from erosion by seawater at high velocity (Ref 38).

Acids. Tin may be corroded by acidic aqueous solutions of pH less than 3 or by less acidic solutions containing compounds that form stable complex ions with tin. The corrosion rate is also highly influenced by oxygen concentration and the presence of metallic impurities in the tin or by the acid that can concentrate on the surface and facilitate the cathodic half reaction.

Table 4 compares the corrosion rates for tin samples exposed vertically in various acids open to the air at 30 °C (85 °F). The greater weight loss over the 96-h period was largely attributed to the access of oxygen to the solutions.

The following general comments concern the effects of other acids (Ref 25). Hot hydrobromic (HBr) and hydroiodic (HI) acids rapidly attack tin, but the rate of attack is slow with HF. Tin is

Table 3 Corrosion of tin and tin alloys totally immersed in seawater

Material	Form	Exposure time, years	Penetration rate(a) mm/yr	Penetration rate(a) mils/yr	Test location
99.75 tin	Cast bar	4	0.0022	0.087	Bristol Channel
99.2 tin	Cast bar	4	0.0008	0.03	Bristol Channel
Babbitt alloy (Sn-7.4Sb-3.7Cu)	Cast plate	1.4	0.060	2.4	Kure Beach, NC
Solder (Sn-50Pb)	Sheet	0.5	0.075	2.95	Bogue Inlet, NC
Solder (Sn-60Pb on copper)	Plate	2.1	0.011	0.43	Kure Beach, NC

(a) Converted from weight loss data, assuming cast densities of 7.29 g/cm³ for tin, 7.39 g/cm³ for babbitt, 8.90 g/cm³ for 50–50 solder, and 9.28 g/cm³ for 40–60 solder. Source: Ref 38

Table 4 Corrosion rate of tin in 0.1 *N* acids at 30 °C (85 °F) exposed vertically in solutions open to air

Acid	Average penetration rate(a) 24-h test mm/yr	24-h test mils/yr	96-h test mm/yr	96-h test mils/yr
Hydrochloric	0.40	15.7	0.30	11.8
Sulfuric	0.32	12.6	0.29	11.4
Phosphoric	0.03	1.2	0.01	0.4
Formic	0.34	13.4	0.25	9.8
Acetic	0.29	11.4	0.24	9.4
Oxalic	0.17	6.7	0.17	6.7
Citric	0.25	9.8	0.21	8.3
Malic	0.22	8.7	0.22	8.7
Lactic	0.24	9.4	0.21	8.3

(a) Converted from weight loss data, assuming a tin density of 7.29 g/cm³. Source: Ref 39

slowly attacked by $HClO_2$ and is readily attacked by $HClO_3$. Sulfurous acid (H_2SO_3) attacks tin, but sodium acid sulfite ($NaHSO_3$) is noncorrosive. Pyrosulfuric acid ($H_2S_2O_7$) and chlorosulfonic acid (SO_2ClOH) react rapidly with tin; nitric acid (HNO_3) reacts rapidly with tin over a wide range of concentrations, and the reaction is complex.

Bases. Tin may be dissolved by alkaline solutions, with the production of soluble stannates or stannites. Corrosion will usually follow if the surface oxide layer can be dissolved; this will occur with pH greater than 12 and may occur at pH values down to 10. When corrosion is possible, its rate is governed by the temperature and the rate of arrival of oxygen or other oxidizing agents to the initial surface and is not greatly affected by the character of the alkali in long periods of immersion. However, in intermittent immersion, the corrosion rate is affected by the nature of the alkali and its concentration because these affect the time for removal of the oxide film. The corrosion rates of tin in various alkaline solutions exposed to air at 60 °C (140 °F) are summarized in Table 5.

Hydrogen evolution does not occur on a tin surface in alkaline solutions. Thus, exclusion of oxidizing agents, including air, can provide complete protection unless the tin is in contact with another metal on which hydrogen evolution can occur. Additions of oxygen absorber can prevent corrosion even without the exclusion of air, but they must be replenished. Small additions of oxidizing agents to alkalies stimulate corrosion, but sufficiently large additions can be completely effective. Soluble chromates are particularly effective in this way. Saturated NH_3 solutions do not attack tin, but more dilute solutions behave like other alkaline solutions of comparable pH.

Other Liquid Media. Milk and milk products are usually nonreactive with tin, although a long period of stagnant contact may produce local corrosion (Ref 41). Sulfide solutions and materials containing sulfur dioxide (SO_2) as a preservative produce sulfide stains, but the rate of metal loss is low. Beer dissolves a trace of tin from freshly exposed metal. Although this may cause an objectionable haze in the beverage, the action usually ceases within a short period. To avoid this effect, the tin surfaces can be passivated by using alkaline chromate solutions.

Most organic liquids, including ethers, alcohols, ketones, esters, hydrocarbons, and chlorinated hydrocarbons, are inert toward tin in the absence of water (Ref 25). However, a reaction was reported between tin and lower alcohols at elevated temperatures, and when mineral acidity can arise, as with chlorinated hydrocarbons containing water, there may be some corrosion (Ref 42). Animal, vegetable, or mineral oils and fatty acids are also essentially inactive, and the absence of any catalytic action of tin on their oxidation makes tin or tin-coated vessels suitable for these products.

Galvanic Behavior. When immersed in electrical contact with a more noble metal, such as copper or nickel, tin is much more likely to be corroded, and any loss of metal will be faster, with an increase in the number of locally corroded spots in conditions favorable to local corrosion. However, contact with such metals as aluminum or zinc can prevent corrosion of tin entirely, and a tin coil or vessel can be protected by joining it to a strip of one of these metals. The galvanic-corrosion behavior of tin and tin-lead alloys in contact in seawater with numerous alloy steels and other structural materials is summarized in Table 6.

Passivation of Tin. Tin can be readily passivated with or without an applied potential. The solutions most frequently used are the strongly oxidizing chromate solutions, which produce a thin, tenacious oxide layer that is quite protective. This film is 40 to 50 Å (16×10^{-8} to 20×10^{-8} in.) thick when prepared by immersion in an alkaline chromate solution at 80 to 90 °C (175 to 195 °F) for 15 min (Ref 43). Anodic passivation with a current density of 500 A/dm² (32 A/in.²) for 5 s in 0.5% sodium hydroxide (NaOH) forms a 300-Å (12×10^{-7}-in.) thick film. In 0.005 *M* potassium chromate (K_2CrO_4) solution, SnO is oxidized to SnO_2 above a potential of 0.2 V versus an Ag/AgCl electrode, and the oxide continues to thicken even after the oxygen evolution potential is reached. The passivation behavior of tin in solutions of phosphoric acid (H_3PO_4) (Ref 44), NaOH (Ref 45), sodium borate ($NaBO_2$), and sodium carbonate (Na_2CO_3) (Ref 46) has also been studied, and is reviewed in detail in Ref 25.

Soft Solders

Most soft solders contain from 2 to 100% Sn, with the balance consisting of lead, although some special-purpose solders substitute silver or antimony for some or all of the lead. Two features are particularly relevant to the corrosion behavior of solders with regard to their function as a joining material. First, fluxes are usually used, and, second, the solder exposure areas are usually much smaller than the area of the materials being joined.

By nature, fluxes function as oxide removers and may contain hygroscopic products that, if not removed, will promote corrosion. A mild flux, such as pure natural resin, is inactive at normal (room) temperatures and therefore has a harmless residue.

More powerful fluxes may consist of natural resin with additions—for example, chlorides and bromides—or mixtures of chlorides, H_3PO_4, and derivatives. Residues from such fluxes must usually be completely removed by mechanical wiping or with solvents.

The area effect can be minimized by coating the joined metals with tin or tin-lead alloys. However, the suitable design of joints and the formation of protective corrosion products over the solder often permit the satisfactory use of soldered joints in conditions that may at first appear hostile.

Simple binary tin-lead solders consist essentially of eutectic mixtures, and their corrosion behavior is similar to that for either metal, with the overall behavior similar to that of the predominant metal. Both metals are attacked by acids and alkalies, but the presence of lead, which forms many more insoluble compounds than tin, creates further possibilities for the formation of protective layers in near-neutral aqueous media. The addition of other elements has not been found to affect the corrosion resistance of tin-lead alloys appreciably (Ref 2). Also, the behavior of lead-free solders containing silver or antimony

Table 5 Corrosion rate of tin in alkaline solutions exposed to air at 60 °C (140 °F)

Concentration of solution, %	Penetration rate(a) Na_3PO_4 mm/yr	Na_3PO_4 mils/yr	Na_2CO_3 mm/yr	Na_2CO_3 mils/yr	Na_2SiO_3 mm/yr	Na_2SiO_3 mils/yr	NaOH mm/yr	NaOH mils/yr
0.005	0.015	0.6	0.030	1.2	0.030	1.2	0.21	8.3
0.02	0.015	0.6	0.045	1.8	0.045	1.8	0.24	9.4
0.05	0.21	8.3	0.24	9.4	nil		0.21	8.3
0.10	0.23	9.1	0.26	10.2	0.015	0.6	0.20	7.9
0.15	0.24	9.4	0.27	10.6	0.075	2.95	0.20	7.9
0.20	0.26	10.2	0.27	10.6	0.090	3.5	0.21	8.3
0.25	0.26	10.2	0.27	10.6	0.12	4.7	0.24	9.4

(a) Converted from weight loss data, assuming a tin density of 7.29 g/cm³. Source: Ref 40

Table 6 Seawater corrosion of galvanic couples

□ The corrosion of the metal under consideration will be reduced considerably in the vicinity of the contact.
○ The corrosion of the metal under consideration will be reduced slightly.
△ The galvanic effect will be slight with the direction uncertain.
■ The corrosion of the metal under consideration will be increased slightly.
▲ The corrosion of the metal under consideration will be increased moderately.
● The corrosion of the metal under consideration will be increased considerably.

S Exposed area of the metal under consideration is small compared with the area of the metal with which it is coupled.
E Exposed area of the metal under consideration is approximately equal to that of the metal with which it is coupled.
L Exposed area of the metal under consideration is large compared to that of the metal with which it is coupled.

Metal considered		Magnesium	Magnesium alloys	Zinc	Galvanized steel	Aluminum 5052	Aluminum 3004	Aluminum 1100	Alclad	Aluminum 3003	Aluminum 6053	Aluminum 6061	Cadmium	Aluminum 2017	Aluminum 2117	Aluminum 2024	Low-carbon steel	Wrought iron	Low-alloy steels	Cast iron	Low-alloy cast irons	4–6% Cr	Ni cast iron	12–14% Cr	Lead-tin solders	16–18% Cr steel	Lead	Tin	Muntz metal	Manganese bronze	Naval brass	Nickel	Yellow brass	Admiralty brass	Aluminum bronze	Red brass	Copper	Silicon bronze	Nickel silver	70–30 Copper-nickel	Composition G bronze	Composition M bronze	Inconel	Silver solder	70–30 nickel-copper	25–30% Cr steel	Cr-Ni stainless steel	Cr-Ni-Mo stainless steel	Graphite
Low-carbon steel	S	□	□	□	□	□	□	□	□	□	□	□	□	□	□	□	△	▲	●	●	●	●	●	●	●	●	●	●	●	●	●	●	●	●	●	●	●	●	●	●	●	●	●	●	●	●	●	●	●
	E	□	□	○	○	□	□	□	□	□	□	□	□	□	□	□	△	△	■	■	■	■	▲	▲	▲	▲	▲	▲	▲	▲	▲	▲	▲	▲	▲	▲	▲	▲	▲	▲	▲	▲	▲	▲	▲	▲	▲	▲	▲
	L	○	○	○	○	○	○	○	○	○	○	○	○	○	○	○	△	△	■	■	■	■	■	■	■	■	■	■	■	■	■	■	■	■	■	■	■	■	■	■	■	■	■	■	■	■	■	■	■
Wrought iron	S	□	□	□	□	□	□	□	□	□	□	□	□	□	□	□	△	△	●	●	●	●	●	●	●	●	●	●	●	●	●	●	●	●	●	●	●	●	●	●	●	●	●	●	●	●	●	●	●
	E	□	□	○	○	□	□	□	□	□	□	□	□	□	□	□	△	△	■	■	■	■	▲	▲	▲	▲	▲	▲	▲	▲	▲	▲	▲	▲	▲	▲	▲	▲	▲	▲	▲	▲	▲	▲	▲	▲	▲	▲	▲
	L	○	○	○	○	○	○	○	○	○	○	○	○	○	○	○	○	△	■	■	■	■	■	■	■	■	■	■	■	■	■	■	■	■	■	■	■	■	■	■	■	■	■	■	■	■	■	■	■
Low-alloy steels	S	□	□	□	□	□	□	□	□	□	□	□	□	□	□	□	□	□	△	□	△	▲	●	●	●	●	●	●	●	●	●	●	●	●	●	●	●	●	●	●	●	●	●	●	●	●	●	●	●
	E	□	□	○	○	□	□	□	□	□	□	□	□	□	□	□	○	○	△	○	△	△	▲	▲	▲	▲	▲	▲	▲	▲	▲	▲	▲	▲	▲	▲	▲	▲	▲	▲	▲	▲	▲	▲	▲	▲	▲	▲	▲
	L	○	○	○	○	○	○	○	○	○	○	○	○	○	○	○	○	○	△	○	△	△	■	■	■	■	■	■	■	■	■	■	■	■	■	■	■	■	■	■	■	■	■	■	■	■	■	■	■
Cast iron	S	□	□	□	□	□	□	□	□	□	□	□	□	□	□	□	△	△	●	△	●	▲	●	●	●	●	●	●	●	●	●	●	●	●	●	●	●	●	●	●	●	●	●	●	●	●	●	●	●
	E	□	□	○	○	□	□	□	□	□	□	□	□	□	□	□	○	○	■	△	■	■	▲	▲	▲	▲	▲	▲	▲	▲	▲	▲	▲	▲	▲	▲	▲	▲	▲	▲	▲	▲	▲	▲	▲	▲	▲	▲	▲
	L	○	○	○	○	○	○	○	○	○	○	○	○	○	○	○	○	○	■	△	■	■	■	■	■	■	■	■	■	■	■	■	■	■	■	■	■	■	■	■	■	■	■	■	■	■	■	■	■
Low-alloy cast iron	S	□	□	□	□	□	□	□	□	□	□	□	□	□	□	□	□	□	△	□	△	▲	●	●	●	●	●	●	●	●	●	●	●	●	●	●	●	●	●	●	●	●	●	●	●	●	●	●	●
	E	□	□	○	○	□	□	□	□	□	□	□	□	□	□	□	○	○	△	○	△	▲	▲	▲	▲	▲	▲	▲	▲	▲	▲	▲	▲	▲	▲	▲	▲	▲	▲	▲	▲	▲	▲	▲	▲	▲	▲	▲	▲
	L	○	○	○	○	○	○	○	○	○	○	○	○	○	○	○	○	○	△	○	△	△	■	■	■	■	■	■	■	■	■	■	■	■	■	■	■	■	■	■	■	■	■	■	■	■	■	■	■
4–6% Cr steel	S	□	□	□	□	□	□	□	□	□	□	□	□	□	□	□	□	□	▲	□	▲	△	●	●	●	●	●	●	●	●	●	●	●	●	●	●	●	●	●	●	●	●	●	●	●	●	●	●	●
	E	□	□	○	○	□	□	□	□	□	□	□	□	□	□	□	○	○	△	○	△	△	▲	▲	▲	▲	▲	▲	▲	▲	▲	▲	▲	▲	▲	▲	▲	▲	▲	▲	▲	▲	▲	▲	▲	▲	▲	▲	▲
	L	○	○	○	○	○	○	○	○	○	○	○	○	○	○	○	○	○	△	○	△	△	■	■	■	■	■	■	■	■	■	■	■	■	■	■	■	■	■	■	■	■	■	■	■	■	■	■	■
Nickel cast iron	S	□	□	□	□	□	□	□	□	□	□	□	□	□	□	□	□	□	□	□	□	□	△	●	●	●	●	●	●	●	●	●	●	●	●	●	●	●	●	●	●	●	●	●	●	●	●	●	●
	E	□	□	○	○	□	□	□	□	□	□	□	□	□	□	□	○	○	○	○	○	○	△	■	■	■	■	■	■	■	■	■	■	■	■	■	■	■	■	■	■	■	■	■	■	■	■	■	▲
	L	○	○	○	○	○	○	○	○	○	○	○	○	○	○	○	○	○	○	○	○	○	■	■	■	■	■	■	■	■	■	■	■	■	■	■	■	■	■	■	■	■	■	■	■	■	■	■	■
12–14% Cr steel	S	□	□	□	□	□	□	□	□	□	□	□	□	□	□	□	□	□	□	□	□	□	□	△	▲	●	△	△	●	●	●	●	●	●	●	●	●	●	●	●	●	●	●	●	●	●	●	●	●
	E	□	□	□	□	□	□	□	□	□	□	□	□	□	□	□	□	□	□	□	□	□	○	△	■	■	■	■	▲	▲	▲	▲	▲	▲	▲	▲	▲	▲	▲	▲	▲	▲	▲	▲	▲	▲	▲	▲	●
	L	○	○	○	○	○	○	○	○	○	○	○	○	○	○	○	○	○	○	○	○	○	○	△	■	■	■	■	■	■	■	■	■	■	■	■	■	■	■	■	■	■	■	■	■	■	■	■	●
Lead-tin solders	S	□	□	□	□	□	□	□	□	□	□	□	□	□	□	□	□	□	□	□	□	□	□	△	△	△	△	△	■	■	■	■	■	■	■	■	■	■	■	■	■	■	■	■	■	■	■	■	●
	E	□	□	□	□	□	□	□	□	□	□	□	□	□	□	□	□	□	□	□	□	□	□	△	△	△	△	△	△	△	△	△	△	△	△	△	△	△	△	△	△	△	△	△	△	△	△	△	▲
	L	○	○	○	○	○	○	○	○	○	○	○	○	○	○	○	○	○	○	○	○	○	○	△	△	△	△	△	△	△	△	△	△	△	△	△	△	△	△	△	△	△	△	△	△	△	△	△	▲
16–18% Cr steel	S	□	□	□	□	□	□	□	□	□	□	□	□	□	□	□	□	□	□	□	□	□	□	□	■	△	■	■	●	●	●	●	●	●	●	●	●	●	●	●	●	●	●	●	●	●	●	●	●
	E	□	□	□	□	□	□	□	□	□	□	□	□	□	□	□	□	□	□	□	□	□	○	○	△	△	△	△	▲	▲	▲	▲	▲	▲	▲	▲	▲	▲	▲	▲	▲	▲	▲	▲	▲	▲	▲	▲	●
	L	○	○	○	○	○	○	○	○	○	○	○	○	○	○	○	○	○	○	○	○	○	○	○	△	△	△	△	■	■	■	■	■	■	■	■	■	■	■	■	■	■	■	■	■	■	■	■	●
25–30% Cr steel	S	□	□	□	□	□	□	□	□	□	□	□	□	□	□	□	□	□	□	□	□	□	□	□	△	□	△	△	□	□	□	□	●	●	●	●	●	●	●	●	●	●	●	●	●	△	□	△	●
	E	□	□	□	□	□	□	□	□	□	□	□	□	□	□	□	□	□	□	□	□	□	□	□	△	□	△	△	○	○	○	○	■	■	■	■	■	■	■	■	■	■	■	■	■	△	○	△	●
	L	○	○	○	○	○	○	○	○	○	○	○	○	○	○	○	○	○	○	○	○	○	○	○	△	○	△	△	○	○	○	△	△	△	△	△	△	△	△	△	△	△	△	△	△	△	△	△	●
Austenitic Cr-Ni stainless steel	S	□	□	□	□	□	□	□	□	□	□	□	□	□	□	□	□	□	□	□	□	□	□	□	△	□	△	△	□	□	□	□	●	●	●	●	●	●	●	●	●	●	●	●	●	△	△	●	●
	E	□	□	□	□	□	□	□	□	□	□	□	□	□	□	□	□	□	□	□	□	□	□	□	△	□	△	△	○	○	○	○	■	■	■	■	■	■	■	■	■	■	■	■	■	■	△	■	●
	L	○	○	○	○	○	○	○	○	○	○	○	○	○	○	○	○	○	○	○	○	○	○	○	△	○	△	△	○	○	○	△	△	△	△	△	△	△	△	△	△	△	△	△	△	△	△	△	●
Austenitic Cr-Ni-Mo stainless steel	S	□	□	□	□	□	□	□	□	□	□	□	□	□	□	□	□	□	□	□	□	□	□	□	△	□	△	△	□	□	□	□	●	●	●	●	●	●	●	●	●	●	●	●	●	△	□	△	●
	E	□	□	□	□	□	□	□	□	□	□	□	□	□	□	□	□	□	□	□	□	□	□	□	△	□	△	△	○	○	○	○	■	■	■	■	■	■	■	■	■	■	■	■	■	△	○	△	●
	L	○	○	○	○	○	○	○	○	○	○	○	○	○	○	○	○	○	○	○	○	○	○	○	△	○	△	△	○	○	○	△	△	△	△	△	△	△	△	△	△	△	△	△	△	△	△	△	▲
Lead	S	□	□	□	□	□	□	□	□	□	□	□	□	□	□	□	□	□	□	□	□	□	□	■	△	■	△	△	■	■	■	■	■	■	■	■	■	■	■	■	■	■	■	■	■	■	■	■	●
	E	□	□	□	□	□	□	□	□	□	□	□	□	□	□	□	□	□	□	□	□	□	□	△	△	△	△	△	△	△	△	△	△	△	△	△	△	△	△	△	△	△	△	△	△	△	△	△	▲
	L	○	○	○	○	○	○	○	○	○	○	○	○	○	○	○	○	○	○	○	○	○	○	△	△	△	△	△	△	△	△	△	△	△	△	△	△	△	△	△	△	△	△	△	△	△	△	△	▲
Tin	S	□	□	□	□	□	□	□	□	□	□	□	□	□	□	□	□	□	□	□	□	□	□	■	△	■	△	△	■	■	■	■	■	■	■	■	■	■	■	■	■	■	■	■	■	■	■	■	●
	E	□	□	□	□	□	□	□	□	□	□	□	□	□	□	□	□	□	□	□	□	□	□	△	△	△	△	△	△	△	△	△	△	△	△	△	△	△	△	△	△	△	△	△	△	△	△	△	▲
	L	○	○	○	○	○	○	○	○	○	○	○	○	○	○	○	○	○	○	○	○	○	○	△	△	△	△	△	△	△	△	△	△	△	△	△	△	△	△	△	△	△	△	△	△	△	△	△	▲

Source: Ref 38

with tin does not differ greatly from that of pure tin.

Atmospheric Corrosion. Even small additions of lead to tin impair the retention of its bright reflective surface in common atmospheres. With increasing lead content, the appearance of soldered joints becomes increasingly dull, like that of lead. However, destructive corrosion (except effects from flux residues) is highly unusual. On rare occasions, within enclosed spaces, condensed pure water may extract lead, but more common causes of trouble are volatile organic acids. Acetic acid (CH_3COOH) vapors from wood or insulating materials and formic acid (HCOOH) or other acids that may come from insulating materials may attack lead-containing solders to produce a white incrustation and cause serious destruction of metal. Where such attack occurs, substitution of a solder with a higher tin content may eliminate the problem.

Contact of solder with other metals can impose a serious risk in conditions of exposure to sea spray or where pockets or crevices can trap moisture or flux residues. In most atmospheric conditions, the formation of lead sulfate ($PbSO_4$) protects the solder. However, in chloride pollution conditions, nickel, copper, and their alloys are likely to be cathodic to solder. Zinc tends to be strongly anodic to soft solders, but correctly designed zinc roof coverings appear to suffer no deterioration at the soldered joints (Ref 2).

Immersion. Natural waters and commercial treated waters that are aggressive to lead are likely to corrode solder at a rate that increases slowly, in proportion to its lead content, up to about 70% Pb, then more rapidly at higher lead contents. Selective dissolution of lead can also occur in distilled, demineralized, or naturally soft waters, causing serious weakening of joints (Ref 2). In the general run of commercial waters, the ability of lead to form insoluble oxides, sulfates, and carbonates usually protects solders against serious attack. Although rare, selective dissolution of tin has been reported during prolonged contact of solders with solutions of anionic surface-active agents.

When freshly exposed to water, solders are anodic to copper, but soldered joints in copper pipes are widely used without trouble in conventional commercial and domestic cold- and hot-water systems. Despite this generally good corrosion resistance, it has been demonstrated that, under adverse conditions, lead may be leached from the commonly used 50Sn-50Pb plumbing solder into water traveling through the pipe; this is a cause of increasing concern (Ref 47, 48). The lead content of water passing through soldered copper pipes is usually less than that recommended by various regulatory authorities, although higher values may be found in new installations and in some soft water areas (Ref 48). Public concern about all sources of lead in the human diet is well documented in numerous publications, and in some countries, including the United States, legislative action has been undertaken to prohibit the use of lead-containing solders and to tighten existing water quality standards (Ref 49).

Soldered joints in brass usually perform well in domestic waters, but good joint design is imperative. In automobile radiators in which there are no inhibitors, ethylene glycol, although not directly aggressive, does appear able to detach protective deposits that may form on soldered joints. Properly tested and approved inhibitors avoid this problem. Sodium nitrite ($NaNO_2$), which is used as an inhibitor for some metals, will attack solders and must be used in conjunction with sodium benzoate ($NaC_7H_5O_2$).

In seawater or uninhibited brines, the high conductivity and predominance of chloride makes galvanic action at a soldered joint more likely to continue destructively, and soldered joints in copper, nickel, and their alloys may need protection by coatings. Although tin or tin-coated metals can be used in contact with aluminum alloys even in salt water, the soldering process introduces sufficient aluminum to the solder to render it susceptible to intergranular corrosion. If tin-zinc solders are used, the zinc can prevent the serious embrittling action, although some corrosion will still occur under moist conditions.

Pewter

By definition, modern pewter is an alloy that contains 90 to 98% Sn, 1 to 8% Sb, 0.25 to 3% Cu, and a maximum of 0.05% Pb and As (Ref 50). Material that conforms to these standards has about the same degree of corrosion resistance to ordinary atmospheres as pure tin. Alloys within this range are widely used for decorative items, containers, and flatware. Indoors, they retain a bright, white luster in the same manner as pure tin. Because contamination from fabrication residues can deteriorate the protective oxide, care should be exercised in finishing to remove residues from soldering fluxes and cleaning solutions. Regular, simple washing with a mild soap solution will ensure that the surface remains in good condition.

Pewter tankards and plates also have about the same degree of corrosion resistance to foods and drinks as tin does. With the normal contact time, the amount of tin dissolved by beer is insufficient to cause a haze. However, citrus juices or vinegar will etch a pewter surface if contact is maintained for more than an hour. Undisturbed neutral salt solutions may produce black spots and, later, local pitting. Strong alkaline cleaning agents may also etch the surface.

In years past, pewter alloys contained lead in sufficient quantities to affect its corrosion resistance significantly, for example, by producing a dark patina during atmospheric exposure. Modern pewter can be chemically treated to reproduce this patina. Several proprietary processes are available, including those based on immersion in iron chloride ($FeCl_3$) or sodium nitrate (Na_2NO_3) solutions or acidic solutions of copper and arsenic (Ref 51, 52).

Bearing Alloys

The most widely used babbitt bearing alloys are usually classified as tin- or lead-base and have composition ranges within the following limits:

Alloy addition	Composition, % Tin-base	Composition, % Lead-base
Tin	65–91	0–20
Lead	0.35–18	63 (min)
Antimony	4.5–15	10–15
Copper	2–8	1.5 (max)

The tin-base alloys are much more corrosion resistant against the action of the acids contained or formed in lubricating oils (see the article "Materials for Sliding Bearings" in Volume 3 of the 9th Edition of *Metals Handbook*). An addition of as little as 3% Sn in lead appears to prevent corrosion from the development of oil acidity (Ref 53).

In some instances of marine use, the formation of a hard, crusty oxidation product has been observed on tin-rich bearings (Ref 54). When free access of salt water to a bearing is possible, the cathodic relationship of the babbitt alloys to steel renders them unsuitable, and bearing alloys such as Zn-70Sn-1.5Cu are preferred (Ref 2).

Some aluminum-base alloys containing 5 to 40% Sn and 0.7 to 1.3% Cu have also found use as bearing alloys in automobiles. These alloys are manufactured using thermal treatments designed to produce structures that avoid a continuous network of the tin in order to obviate the risk of susceptibility to corrosion by the presence of moisture (Ref 55). With normal lubrication, the aluminum-tin alloys appear to be as fully resistant to corrosion as the tin-base babbitt alloys. The aluminum-tin alloys, however, are not suitable for exposure to wet conditions.

Other Tin Alloys

Tin-Copper. Alloys in this group are all copper-base and consist mainly of bronzes, gunmetal, and brass that contains tin additions. Understandably, their corrosion behavior in air is based on the behavior of copper, which tends to develop a layer of basic green salts (mainly sulfates), that is adherent, protective, and has a pleasing appearance. More information on the corrosion resistance of copper alloys is available in the article "Corrosion of Copper and Copper Alloys" in this Volume.

Atmospheric Corrosion. Early studies were conducted on Cu-6.3Sn-0.08P wire and Cu-6.3Sn-0.08P-0.5Zn sheet in rural, suburban, urban, industrial, and marine environments for 1 year (Ref 56). Evaluations included weight gains as well as changes in tensile strength and electrical resistance. The bronze samples ranked consistently high among the materials tested, as indicated by the tensile strength data shown in Table 7.

A more extensive study covering 20 years and seven sites compared the behavior of a variety of alloys, including phosphor bronze (Cu-7.85Sn-0.03P), admiralty brass (Cu-29.01Zn-1.22Sn), and a nickel-tin bronze (Cu-28.6Ni-1.04Sn-0.55Zn) (Ref 58). Weight changes were used to assess corrosion behavior, along with changes in electrical resistance and tensile strength. Some representative data are given in Table 8. Small tin additions also impart dezincification resistance to brass.

Table 7 Tensile strength loss in copper alloys after exposure for 1 year in various environments

Environments included industrial, marine, rural, suburban, and urban locations; data are averages for all five environments.

Alloy	Strength loss, %
Tin bronze (6% Sn)	1.2
High-conductivity copper	2.4
Aluminum bronze (3.5% Al)	2.1
70–30 nickel-copper	3.2
60–40 copper-zinc	18.4
70–30 copper-zinc	8.6

Source: Ref 57

Table 8 Tensile strength loss in copper alloys after exposure for 10 years at four sites

Exposure site	Strength loss, %			
	Copper	Tin bronze (8% Sn)	70–30 copper-zinc	70Cu-29Zn-1Sn
Heavy industrial	5.9	7.2	30.9	9.0
Marine, heavy industrial	6.3	8.0	28.2	7.9
Severe marine	7.6	5.7	8.0	2.5
Rural	3.1	3.1	3.2	2.2

Source: Ref 58

A similar study involved exposure of screen wire cloth at four sites for up to 9 years (Ref 59). A Cu-2Sn bronze was found to exhibit the lowest strength losses at all sites from a group of alloys that included brasses, aluminum bronze, and nickel-copper. Outstanding corrosion resistance of a Cu-2Sn bronze exposed to sulfur-bearing gases in railway tunnels was also reported (Ref 60).

Another investigation compared the behavior of five stainless steels and a low-alloy steel with that of a Cu-4.38Sn-0.36P bronze exposed at tropical inland and seacoast sites for 8 years (Ref 61). The coastal site was more aggressive toward the bronze, which showed higher weight losses at both sites than the stainless steels, but the low-alloy steel was more severely attacked. However, the bronze was free of pitting and suffered no loss in strength, which was not the case with some of the stainless steels. In Ref 62, these researchers summarized the results of 16-year exposures on three tin-containing alloys (Cu-4.38Sn-0.36P, Cu-39Zn-0.84Sn, and Cu-40Zn-1Fe-0.65Sn) exposed at marine, inland semirural, and two tropical sites. In general, the copper alloys resisted corrosion in the tropical zones, although less so at coastal sites as compared to inland sites. The tin-containing alloys were as good as, or slightly superior to, the other alloys.

More recent work by the same investigators included previous data plus additional information on the following cast bronzes: Cu-5Sn-5Pb-5Zn, Cu-6Sn-2Pb-3Zn-1Ni, Cu-9Sn-3Zn-1Ni, and Cu-3Sn-2Zn-6Ni (Ref 63). The conclusions were much the same as before. The later work included a study of the effect of coupling phosphor bronze to equal areas of numerous other metals, and this work indicated that the coastal sites were 4 to 8 times more aggressive than the inland sites. Evaluation of the effect of corrosion on the solderability of a Cu-2Sn-9Ni alloy was reported by workers at Bell Telephone, who found this material to be superior to both nickel-silver and an 8% Sn phosphor bronze (Ref 64).

Alloys in the Cu-Sn-Al system were evaluated, and those alloys containing at least 5% each of tin and aluminum were found to have good corrosion resistance in rural, urban, and industrial environments (Ref 65). The most promising material was Cu-5Sn-7Al. Another researcher noted that such alloys could be brittle, but that the addition of 1% Fe and 1% Mn overcame this difficulty without detracting from the corrosion resistance of the alloy (Ref 66).

Tin-Silver. In the mid-1930s, tin-silver alloys were assessed as potential replacements for sterling silver (silver-copper alloy) in decorative applications (Ref 67). In this work, an Ag-7.5Sn alloy was found to show improved corrosion resistance over pure silver in several environments. In a later extension of this work, alloys with up to 10% Sn were tested in atmospheres containing H_2S, SO_2, and indoor air as well as for resistance to salt and oxidation upon heating in air (Ref 68). Comparison to sterling silver showed the tin-silver alloys to be at least as good as the sterling alloys, and in some cases even better. Specifically, their resistance to chloride attack was considerably better, and less discoloration occurred upon heating in air. Also, preoxidation of the tin-silver alloys improved resistance to attack by sulfur-containing atmospheres.

Tin and Tin-Alloy Coatings

Tin coatings can be applied by various processes, including hot dipping, electrodeposition, spraying, and chemical displacement. Electrodeposits can be matte or bright as plated, and matte deposits less than 8 μm (0.3 mils) thick can be brightened by momentary fusion. The latter can be effected by conductive or resistive heating in air or by immersion in a suitable oil.

In the standard electrodeposition process, alkaline stannate, acid sulfate, or fluoborate solutions are all widely used. The alkaline solutions give smooth, matte deposits, but the acid solutions usually require organic addition agents to produce smooth, coherent coatings. If improperly controlled, these agents can increase the risk of dewetting during soldering or flow melting.

The ranges of coating thicknesses that are practical for the various processes are as follows (Ref 2):

Process	Thickness	
	μm	mils
Chemical replacement	Trace–2.5	Trace–0.1
Flow-melted electrodeposition	0.4–7.5	0.02–0.3
Electrodeposition, general	2.5–75	0.1–2.9
Hot dipping	1.5–25	0.06–1
Spraying	75–350	2.9–13.6

Coatings applied by any method may contain pores that will expose the base metal. Porosity should be minimal for electrodeposited and hot-dipped coatings thicker than 15 μm (0.6 mils) (Ref 2). However, the behavior of the coating will be strongly influenced by the relative polarity of tin and substrate, by the nature of any intermetallic layers formed by reactions between these, and by the extremely low rate of corrosion of tin in alkaline and mildly acidic media in the absence of oxygen or other cathode depolarizers.

Because deposits less than about 12 μm (0.5 mils) thick are not likely to be pore free, the heaviest practical deposits should be used when tin is specified for corrosion resistance. Table 9 lists recommended tin coating thicknesses for quality tin coatings for various service conditions.

Tests conducted by the Metal Finishing Supplies Association (MFSA) showed that bright acid tin deposits generally perform better than the matte tin deposits in salt spray corrosion tests (Ref 69). However, no published specifications recognize any difference between the corrosion performance of these processes. Similarly, because tin is cathodic to almost all of the commonly used base metals and undercoating metals, the MSFA recommends that the same tin coating thicknesses be applied to any of the common base metals. Also, the use of a copper or nickel undercoating does not justify the use of thinner tin deposits (Ref 69).

Tin Coatings on Steel. Tin on steel is widely used in packaging. The single most important product of this type is tinplate. Modern tinplate is a highly developed, sophisticated product that is produced at high speeds to yield a coiled, thin, low-carbon steel strip carrying a very thin (0.1 to 2 μm, or 0.004 to 0.08 mils) tin coating on each side. Because of the importance of tinplate, its preparation and properties as well as its performance as a container for food and food products will be discussed in the section "Tinplate" in this article.

This section will primarily deal with heavier tin coatings that are usually applied to individual components by batch processing for nonpackaging applications, such as food-processing equipment, electrical and electronic components, wire, and fasteners. Unless otherwise stated, these coatings, unlike tinplate, have not been subjected to fusion or reflow treatments and are therefore free of the iron-tin intermetallic layer, which can exert profound effects.

One study compared the behavior of 25-μm (1-mil) thick plated layers of tin, 80Sn-20Zn, and zinc on steel at three sites in Nigeria for 2 years (Ref 70). Samples were exposed at 30° to the horizontal, about 1.2 m (47 in.) above ground, facing south and in sheltered exposure where they were supported vertically inside a ventilated box. The test results are given in Table 10.

An evaluation of various protective coatings based on many years of testing is summarized in Ref 71, in which a 12-μm (0.5-mil) thick coating is concluded to be a practical minimum for reasonable protection of steel in a mild indoor exposure; for outdoors, the minimum coating thickness should be 50 μm (2 mils).

The Protective Coatings (Corrosion) Subcommittee of the Corrosion Committee of the British Iron and Steel Research Association reported test results after 12 years of exposure in an industrial area (Sheffield), two marine atmospheres (Colshot and Congella, South Africa), and a rural area with heavy rainfall (Flanwryted Falls) (Ref 72, 73). These data, listed in Table 11, indicated that tin deposited by any of several methods appeared more protective in the industrial area than at the other sites. This behavior was attributed to the production of protective corrosion products in the pores of the coatings. Similar observations have been reported (Ref 74), and similar evaluations have been conducted using accelerated corrosion tests and outdoor exposure in urban Berlin (Ref 75). One conclusion, based on 1 year of exposure, was that reflowing of tin coatings improved their corrosion resistance, except in salt spray exposure. Also, deposits from an acid electrolyte were said to be better than those from a stannate bath.

Additional atmospheric corrosion test results have been reported (Ref 76-79). In one study, data were summarized from 10 years of exposure for tin-plated steel in industrial, marine, and rural atmospheres that included estimates of the added cost of SO_2 pollution. Another study included tropical exposures of samples in China in two environments. In the first environment, samples were mounted at 45° outdoors facing south. In the second, the racks were sheltered from solar radiation, wind, and rain. Recommendations based on 58 months of testing were that matte tin

Table 9 Recommended tin coating thicknesses for typical applications

Service condition	Thickness range μm	mils	Typical applications
Very mild (little or no exposure to atmospheric conditions)	1.3–2.5	0.05–0.1	Insulated copper wire; pistons and other lubricated machine components
Mild (exposure to relatively clean indoor atmospheres)	2.5–5.0	0.1–0.2	Connectors, wires, etc., plated primarily for immediate solderability or where storage periods are short
Moderate (exposure to average shop and warehouse atmospheres)	3.8–7.6	0.15–0.3	This range is considered best for parts that must be reflowed: connectors, circuit boards, wire, busbars; deposits heavier than 7.5 μm (0.3 mil) may dewet.
	7.6–12.7	0.3–0.5	Connectors, fasteners, busbars, wire, transformer cans, chassis frames; adequate for good shelf life and in service
Severe (exposure to humid air, mildly corrosive industrial environments)	12.7–25.4	0.5–1.0	Connectors, wire, gas meter components, automotive air cleaners; adequate as a nitride stop-off
Very severe (exposure to seacoast atmospheres; contact with certain chemical corrosives)	25.4–127	1.0–5.0	Water containers; oil-drilling pipe couplings

Source: Ref 69

coatings 25 μm (1 mil) thick should not be exposed to either environment for more than 1 year and that the life of similar coatings 32 μm (1.2 mils) thick would be less than 2 years (Ref 79).

The general conclusion, based on results of most of the above outdoor studies, was that the corrosion resistance of tin coatings, that is, their protection of steel, was not very high (Ref 22). In addition, this is reflected in the international standard ISO 2093-1973 covering tin coatings, which carries the following tin coating thickness recommendations:

Type of service	Minimum tin thickness On steel μm	mil	On nonferrous metals(a) μm	mil
Exceptionally severe	30	1.2	30	1.2
Severe	20	0.8	15	0.6
Moderate	12	0.5	8	0.3
Mild	4	0.2	4	0.2

(a) Except brass

Tin Coatings on Nonferrous Metals. Tin coatings are widely used on nonferrous substrates, usually for one or more of the following reasons:

- Improvement and retention of solderability
- Excellent compatibility (low toxicity) with foods
- Prevention of galvanic effects between dissimilar metals
- Low electrical resistance

Not surprisingly, copper and copper-base alloys are the most frequently tinned nonferrous materials. Tin tends to be anodic to copper and copper alloys, including the intermetallic tin-copper compounds. Therefore, accelerated corrosion of the tin coating might be expected in aqueous environments. Indeed, this is sometimes evidenced by black spots on a tin coating that result from localized corrosion around discontinuities. Although normally associated with total aqueous immersion, these black spots can also appear on outdoor exposure involving cyclic condensation (Ref 22).

Deterioration of the solderability of tinned copper during aging has been studied by many reseachers, and accelerated test procedures have been devised to simulate the effect (Ref 80-82). Similarly, changes in the contact resistance of tin coatings have been related to increases in the thickness of the oxide film on its surface (Ref 83).

Special mention should be made of the corrosion behavior of tin coatings on brass in ordinary atmospheres. Zinc diffuses through tin coatings fairly rapidly; significant zinc levels are reached on the surface of a coating thickness of 7.5 μm (0.3 mil) in about 1 year (Ref 34). Zinc at the surface oxidizes readily to form white corrosion products that adversely affect its solderability and contact resistance. To avoid such problems, a 2.5-μm (0.1-mil) thick barrier layer of either copper or nickel is recommended over the brass (Ref 84).

Immersion Tin Coating. Contrary to the standard electromotive force (emf) series of metals, tin can be applied by immersion (chemical displacement) on copper. This is done by using a cyanide or a thiourea type of solution.

An outstanding application is tinning of the inside of copper tubing. Such tubing in coil form is used in water coolers. The tin prevents delivery of greenish water from new coolers and eventually disappears. By then, the copper surface has become conditioned to deliver water appearing as it did when it entered the cooler.

Tin-cadmium alloy coatings for the corrosion protection of tin were first studied by plating duplex coatings of tin on cadmium and then heat treating. These and later electrodeposited surfaces (Ref 85) were found to have phenomenal resistance to salt spray tests, and they were successfully used for some time to protect the engine components of naval aircraft (Ref 86).

Tin-cadmium coatings resemble tin-zinc coatings in appearance and behavior. Because cadmium is less effective at sacrificially protecting steel exposed at pores, the optimum cadmium content in the coating ranges from 25 to 50%. The initial electrolyte development discussed in Ref 85 was followed by an investigation of a range of alloys; it was concluded that the alloys performed better than cadmium alone in marine environments (Ref 87). Another study found that the attack on tin-cadmium coatings by organic vapors was less than for pure cadmium (Ref 88). In addition, tin-cadmium was found to be superior to tin-zinc when in contact with jet fuels or in hot synthetic oils. This work was supplemented by that described in Ref 89, which suggests that tin-cadmium alloys, particularly with a chromate surface treatment, performed better than cadmium coatings of the same thickness.

More recently, tin-cadmium alloy coatings were shown to provide better corrosion resistance to steel than duplex coatings of tin and cadmium (Ref 90). Lastly, zinc or tin-zinc coatings were found to be more protective to steel in industrial atmospheres than tin on cadmium or cadmium on tin, but this behavior was reversed in a marine environment (Ref 91).

Tin-Cobalt Coatings. As expected, the properties of tin-cobalt electrodeposits are similar to those for tin-nickel. Intermetallic deposits of SnCo (Ref 92, 93) or SnCo mixed with Sn_2Co (Ref 94) have been produced, and proprietary plating systems have been patented. These deposits are bright and are similar to chromium plate; most studies of their performance have concerned systems of steel coated by nickel, with a thin film of tin-cobalt applied to obtain a bright finish.

An evaluation of tin-cobalt coatings for their resistance to salt spray, NH_3, and in copper-accelerated salt spray (CASS) tests revealed that the deposit was resistant to all of these environments and was more ductile than tin-nickel electrodeposits (Ref 95).

Two researchers also tested systems of nickel plus tin-cobalt in CASS and outdoor exposure tests (Ref 96, 97). Their conclusions were similar even though different baths were used and minor differences in the deposits were obtained. Thus, their corrosion resistance was comparable to that for a nickel-chromium system in all but the more severe conditions.

Table 10 Corrosion rate of 25-μm (1-mil) thick coatings on steel at three tropical sites after 2 years of exposure

Coating	Material loss, full exposure test: Jungle μm/yr	Jungle mils/yr	Town μm/yr	Town mils/yr	Coast μm/yr	Coast mils/yr
Tin	0.18	0.007	1.02	0.04	3.02	0.12
80Sn-20Zn	0.46	0.018	1.35	0.053	2.87	0.113
Zinc	0.53	0.021	1.45	0.057	2.90	0.114

Coating	Weight loss, sheltered exposure test mg/dm²: Jungle	Town	Coast
Tin	1.6	11.3	40
80Sn-20Zn	9.3	15.5	23
Zinc	16.5	10.7	18.1

Source: Ref 70

Table 11 Summary of atmospheric corrosion tests on tin-coated steel at four exposure sites

Coating method	Sheffield		Flanwryted Falls		Colshot		Congella	
	T(a)	L(b)	T	L	T	L	T	L
Electrodeposited from stannate bath	0.076	>11.9	0.077	2.4	0.063	1.0	...	...
Hot-dipped	0.015	5.9	...	...	...	...	...	...
Sprayed by molten-metal pistol	0.023	1.5(c)	...	...	...	...	...	...
	0.031	5.9(c)	0.034	0.6	0.037	0.7(c)	0.041	0.9(c)
	0.067	>11.9	...	...	...	...	...	...
Sprayed by powder pistol	0.096	3.0	...	...	0.102	0.8	...	...

(a) *T*, Coating thickness in mm. (b) *L*, Lifetime in years of coating as determined by rust appearing on more than 5% of the specimen. (c) Average of duplicate results that did not agree well. Source: Ref 72, 73

Corrosion tests on coatings of 0.2-μm (0.8-mil) tin-cobalt over duplex bright nickel were compared with the same thickness of chromium (Ref 98). The tin-cobalt appeared markedly inferior to chromium in outdoor exposure and wear resistance, but was reasonably satisfactory as a substitute for decorative chromium for indoor use.

Tin-Copper Coatings. Tin alloys close to the Cu_3Sn intermetallic composition (40 to 45% Sn) were once used as a material for mirrors; hence the name speculum. These alloys resemble silver in brightness and appearance; they find some use as tableware and on bathroom fixtures, but are not used outdoors, where they rapidly turn dull and gray. However, even the indoor corrosion resistance of the alloy is seriously impaired if the composition is not optimum (~42% Sn), and the subsequent need for close control of plating conditions has prevented large-scale development of the coating.

Tin-bronze deposits containing about 12% Sn were reported to be superior to copper as an undercoat for nickel-chromium coatings with regard to weathering behavior (Ref 99, 100). Some results were also reported with tin-copper coatings over steel in industrial and marine environments (Ref 101).

Tin-lead coatings with a wide range of composition are applied by hot dipping or electrodeposition. Steel strip coated with tin-lead alloys by hot dipping and sold as sheet or coil carries the general designation of terneplate. The tin content varies from 2 to 20%. In general, the higher the tin content, the lower the porosity of the terneplate and therefore the greater the protection afforded to the substrate. Like tin coatings, tin-lead does not offer any galvanic protection to steel in the atmosphere; protection against rusting depends on coating continuity and on the formation of protective corrosion products. A comparison of the behavior of a Pb-12Sn coating with pure tin and lead revealed that both lead-containing coatings developed white films believed to be $PbSO_4$ (Ref 72).

In a more comprehensive study, a range of electrodeposited tin-lead coatings obtained with different bath additives was evaluated (Ref 57). The performance of a Pb-5.5Sn coating in salt spray and outdoor testing was found to be superior to pure lead and lead-tin alloys containing tin additions of 7, 10, or 15%. The Pb-10Sn and Pb-15Sn alloys were comparable in behavior to pure tin. These results were partially supported by those of another study, which consisted of atmospheric exposures at three sites on electrodeposited lead and coatings of Pb-5Sn and Pb-14Sn (Ref 102). Superior protection was achieved with the tin-containing alloys at all sites, which included severe industrial, rural, and marine environments.

The results of tests on a number of commercial terneplate compositions in accelerated corrosion tests, SO_2, humidity, and salt spray as well as outdoor exposure in both industrial and marine environments are given in Ref 103. Performance was assessed largely on the degree of rusting of the underlying steel after 12 months of exposure. Lead-tin alloys showed greater resistance to chloride attack than lead-antimony alloys. It was also noted that coverage of the steel increased with the tin content of the alloy and that resistance of the coating to attack appeared to increase in both chloride-rich and humid conditions.

Tin-Nickel Coatings. Alloys containing 18 to 25% Ni can be deposited from a cyanide-stannate bath to give bright coatings with good resistance to HNO_3 (Ref 104). However, because of their high hardness and brittleness, no interest has been shown in these coatings. Similar results have been reported with a complex pyrophosphate bath (Ref 105). Primary commercial interest has centered on the intermetallic compound NiSn (containing about 67% Sn), which is readily deposited from mixed chloride/fluoride electrolytes (Ref 106, 107).

The NiSn intermetallic is metastable and does not transform to a mixture of other intermetallics unless it is heated (Ref 108). The deposit is hard, bright, and has reasonable solderability. It also has good wear resistance and remarkable resistance to attack by a wide range of solutions. For these reasons, tin-nickel coatings have found use as decorative corrosion-resistant finishes for balance weights, drawing instruments, pistons in automobile braking systems, and some food contact applications. Recommended coating thicknesses for this alloy coating have been specified in ISO 2179 1972 as follows:

Intended duty	Thickness μm	mils
Severe environments	25	1
Moderate environments	15	0.6
Mild environments	8	0.3

For coatings on steel intended for moderate or severe service, an undercoat of copper, tin, or bronze with a minimum thickness of 8 μm (0.3 mils) is also specified, and porosity tests are required.

Studies of tin-nickel coatings showed them to be unaffected by atmospheres containing SO_2 or H_2S (Ref 109-111). This work also indicated that these coatings retained their brilliance more readily than nickel-chromium in positions sheltered from rain. Similar conclusions were reached in other investigations (Ref 112, 113), taking into consideration the undercoatings used for both alloy deposits on steel. On the other hand, nickel-chromium deposits were reported as superior to tin-nickel in marine environments (Ref 114).

Within the past 10 years, studies of the corrosion resistance of tin-nickel deposits have centered on their effect on the electrical contact resistance of this alloy, either alone or with a thin coating of gold. The contact resistance of tin-nickel is sufficiently low to merit consideration for moderate-voltage applications (about 50 V), but too high for low-voltage uses. Several extensive studies have been reported in this field (Ref 115, 116). This effect on contact resistance is largely a result of the insulating passive film that forms on SnNi and the high hardness of the material. An excellent review of the work in this field is available in Ref 22. It is generally agreed that tin tends to concentrate at the surface of tin-nickel electrodeposits, but no adequate explanation of the oxidation behavior of this alloy is currently available.

The resistance to attack of the coating by various acids and chemicals has been studied, and the results are given in Table 12, which also compares these with coatings of tin and nickel (Ref 117). Generally, the results show that tin-nickel has a high resistance to attack by acids, alkalies, and several neutral salt solutions. This behavior is attributed to the presence of a passive air-formed film.

Tin-Zinc Coatings. The general shortage of cadmium after World War II led to an interest in the possibilities of tin-zinc coatings for the protection of steel. One of the first studies to explore this possibility compared the behavior of tin-zinc alloy coatings containing 8 to 72% Zn with that of electrodeposited coatings of tin, zinc, and cadmium and with hot-dipped zinc (Ref 118). Coatings 8 to 25 μm (0.3 to 1 mil) thick were compared to exposures to a humidity cabinet, salt spray, and hot water. Coating failure occurred as a result of zinc dissolution such that deposits containing low percentages of zinc did not protect the steel from rusting at pores, but coatings with more than 40% Zn soon developed voluminous white corrosion products at the surface. The best overall results were indicated for tin-zinc coatings with compositions near Sn-25Zn, which were superior to zinc and cadmium in salt spray and were superior to zinc but about equal to cadmium in the humidity test. This work also suggested that chromate passivation treatments improved the overall performance of tin-zinc coatings, making them less susceptible to staining by finger or grease marks. This study was followed by a number of others that reached the same general conclusion about the usefulness of tin-zinc coatings for protecting steel against atmospheric corrosion (Ref 71, 75, 119, 120).

Another comprehensive study compared coating corrosion resistance in urban and marine environments (Ref 121). The order of merit (best to worst performance) in urban exposures was zinc, 50Sn-50Zn, 80Sn-20Zn, and cadmium. For marine exposures, 50Sn-50Zn and zinc were superior, while 80Sn-20Zn and cadmium performed about the same. Details on the results of the marine exposures are given in Table 13.

The effectiveness of tin-zinc coatings in protecting steel nuts and screws was studied by

Table 12 Corrosion resistance in various media of tin, nickel, and tin-nickel alloy

Solution(a)	Weight losses, mg — Tin	Nickel	Tin-nickel	Change in appearance of tin-nickel coating
1 *M* hydrochloric acid	61.5	41.5	24.8	Covered by adherent brown film
0.5 *M* sulfuric acid	19.5	25.6	14.5	Slightly darkened
1 *M* nitric acid	205.0	97.6	1.1	None
0.05 *M* sulfurous acid	0.2	725.0	0.5	None
1 *M* formic acid (pH 1.8)	22.1	35.5	nil	None
1 *M* acetic acid (pH 2.4)	22.2	43.6	0.6	Very slightly darkened
0.5 *M* oxalic acid (pH 1.1)	12.3	16.4	12.0	Etched on immersed area; dark stain at waterline
1 *M* lactic acid (pH 1.9)	18.0	17.8	2.1	None
0.5 *M* tartaric acid (pH 1.7)	10.6	10.0	0.5	None
0.3 *M* citric acid (pH 1.9)	12.0	19.2	0.4	None
0.5 *M* phenol (pH 2.3)	nil	0.3	nil	None
1 *M* sodium chloride	0.5	1.0	0.8	None
Seawater	1.4	0.3	1.0	
0.3 *M* ferric chloride (pH 1.5)	290.0	303.0	2.3 and 6.2	None, except slight local action on edge
Sodium hypochlorite (40 g/L available chlorine)	1.3	625.0	22.0 and 67.0	Bottom edge badly etched; none elsewhere
Sodium hypochlorite (0.1 g/L available chlorine)	1.8	1.8	0.8	None
1 *M* sodium hydroxide	36.6	0.2	0.7	None

(a) Specimens (75 × 25 mm, or 3 × 1 in.) vertically suspended in solutions at 30 °C (85 °F) with a length of 58 mm (2.3 in.) immersed for 24 h. Source: Ref 117

Table 13 Relative ability of different coatings to prevent the rusting of steel in a marine atmosphere

Coating	Months to first appearance of rust for coating thicknesses indicated, μm — 7.5	12.5	25
Zinc	18	33	36
Passivated zinc	18	18	36
50-50 tin-zinc	25	35	>48
Passivated 50-50 tin-zinc	29	35	>48
80-20 tin-zinc	9	18	36
Passivated 80-20 tin-zinc	13	21	36
Cadmium	8	21	34
Passivated cadmium	13	21	25
Tin	1	1	1

Source: Ref 121

exposing these coatings to suburban, industrial, and marine environments in contact with aluminum plates (Ref 122). Although failure of the 80Sn-20Zn coating was indicated by rusting of the steel more quickly than with zinc or cadmium, the presence of the 80Sn-20Zn was observed to prevent rapid attack of the aluminum. Another advantage was the absence of hygroscopic products on the tin-zinc; that is, rings of moisture tended to form around the corrosion products on nuts and screws coated with zinc and cadmium, but this behavior was not noted with the 80Sn-20Zn.

Tin-zinc alloys have also been used to solder aluminum (Ref 123). Again, it was found that corrosion resistance of the solders in a tropical atmosphere was a function of zinc content. After 9 months of exposure, the 80Sn-20Zn alloy appeared to be the most resistant to attack and had the best retention of strength.

In another study, contact resistance measurements were used to follow the progress of corrosion on binary alloys of tin with zinc, lead, antimony, and cadmium on steel in a rural outdoor atmosphere (Ref 124). It was noted that tin-zinc and tin-cadmium coatings maintained a lower contact resistance than equal thicknesses of tin, tin-lead, or tin-antimony alloys after 2 months of exposure.

The most recent studies with tin-zinc coatings explored the effects of four passivation treatments on the resistance to attack of a Sn-25Zn coating by a salt fog and in cyclic humidity (Ref 125). An electrolytic treatment using sodium dichromate ($Na_2Cr_2O_7$) was found to be superior to the others and was particularly outstanding in the salt fog. The other treatments, in decreasing order of merit, used passivation based on electrolytic molybdate, electrolytic tungstate, and nonelectrolytic chromate solutions.

Another investigation concluded that tin-zinc solders exhibit a significant decrease in shear strength after immersion in 3% sodium chloride (NaCl)-0.1% hydrogen peroxide (H_2O_2) solutions (Ref 123). However, other soft solders, including tin-cadmium and tin-antimony alloys, also behaved in the same manner.

Tinplate

As noted earlier, the term tinplate is reserved for a low-carbon steel strip product coated on both sides with a thin layer of tin. For almost 200 years, tinplate has been the primary material used to make containers (tin cans) for the long-term storage of food. Most of the tinplate manufactured is used to make food cans, and nearly all food cans are made of tinplate.

Modern tinplate is much more sophisticated than a simple coating of tin on steel. To achieve the demanding deep-drawing properties necessary for the production of can bodies for two-piece can manufacture, the steel base for tinplate is often continuously cast using the most current technology. Inclusions or other defects in the steel may otherwise cause breakage in the can-body drawing operation. Because the economics of canmaking depend on high-speed operation using a continuous coiled strip, such breakage cannot be tolerated due to the lost production time; therefore, the steel must be as clean as possible.

In preparing the base steel, the metal is processed to strip form, the final step being a cold reduction that brings the strip to a thickness that is typically from 0.15 to 0.50 mm (6 to 20 mils). Next, the strip is annealed and then temper rolled to obtain the desired mechanical properties. At the final stage of temper rolling, textured rolls can be used to produce a special surface finish for particular applications. A cold reduction in place of temper rolling yields a product that is termed double reduced.

The coiled steel is now ready for the tinplate line. It is first welded to the end of the previous coil to form a continuous strip for processing. The strip passes through cleaning and pickling sections to prepare it for plating, then immediately through the plating cells, where up to 11.2 g/m^2 (1 g/ft^2) of tin is deposited. Any one of three different electrolytes can be used, depending on the other details of the installation, and strip speeds typically approach 600 m/min (1970 ft/min). More details of tinplate production and commerce are available in Ref 126 to 129.

The production steps that typically follow plating create additional layers in the tinplate structure that significantly affect corrosion properties. Upon exiting the plating cells, the tinplate has a matte surface that is usually reflowed by momentarily melting the tin coating in a resistance or induction heating unit. In doing so, a thin layer of tin-iron intermetallic compound is formed at the tin/steel interface. Next, an extremely thin passivation film based on chromium oxide is created by immersion or spraying of chromic acid (H_2CrO_4) on the tinplate surface or by passing the tinplate through a solution of $Na_2Cr_2O_7$, with or without the simultaneous application of electrical current. Finally, a very thin, uniform layer of lubricant, usually either dioctyl sebacate or acetyl tributyl citrate, is electrostatically applied.

Therefore, as supplied to the canmaker, the typical tinplate product consists of five layers, the innermost being a steel sheet about 200 to 300 μm (7.8 to 11.7 mils) thick. This steel is covered on each side with perhaps 0.08 μm (0.004 mils) of tin-iron intermetallic compound. The next layer is free tin that is perhaps 0.3 μm (0.012 mils) thick, with a passivation film of about 0.002 μm (0.00008 mils) and an oil film also about 0.002 μm (0.00008 mils) thick. All five layers affect corrosion behavior.

General Properties. Although the unique corrosion properties of tinplate have kept it the material of choice for food cans, other useful properties should be mentioned. Until a few years ago, all food cans were soldered, and tin coatings were used quite often for their excellent solderability. Can welding has recently replaced soldering, and the favorable electrical contact properties of tinplate have made it very amenable to high-speed resistance welding.

The strength of the steel base gives tin cans the durability to withstand the filling, sterilization, and transportation phases of processing. Because a wide range of mechanical properties is possible, the steel base properties can be adjusted, for example, to maximize strength and stiffness, to maximize ductility and elongation, or to minimize directional properties.

Corrosion Resistance in Sealed Cans. A cursory glance at a seawater galvanic series would lead one to expect tin to be cathodic to steel. Therefore, with a very thin coating of tin, the rapid dissolution of iron at any plating pores, scratches, or other breaks in the coating would be anticipated, resulting in pitting corrosion and eventual perforation of the tinplate. Fortunately, inside the sealed can of food, the situation is very different.

The good performance of the tinplate food can begins with the ability of tin to form chemical complexes with a variety of organic liquids, especially those found in foodstuffs. This fact reverses the situation described in the preceding paragraph; therefore, tin becomes a sacrificial anode, greatly diminishing the rate of dissolution of the iron. Once in solution, the tin ions have a very strong inhibiting effect on iron dissolution because the tin may actually plate out on the exposed steel to form a thin surface layer of tin-iron intermetallic that would be more noble than the steel surface it covers (Ref 25).

With the available atmospheric oxygen limited to only that in the headspace (the volume between the top of the contents and the bottom of the can lid), dissolution of tin is usually rapid only at the very beginning of storage, that is, until the cathodic reaction involving this small amount of oxygen has gone to completion. This effect is desirable because it provides quick protection of the steel. Too large a headspace, however, might allow too much tin to be dissolved, which would allow iron dissolution and probably hydrogen evolution. Hydrogen evolution usually leads to swells, or bulging can ends, a condition the consumer has come to recognize as the sign of can failure. A similar situation applies if a leak in the can allows the free entry of oxygen; the contents are spoiled by iron dissolution, but swelling is not expected to occur.

The small amount of tin dissolved in the food passes easily and quickly through the human body with no known effect, although there is circumstantial evidence that tin is an essential element for human well-being. Too much tin, that is, of the order of 1000 ppm, in the food may cause gastric distress in sensitive individuals. This distress lasts only as long as the irritant is present, and no permanent damage is expected. As a precaution, most governmental regulations limit the tin content of food containers to well below this threshold, typically at a level of about 250 ppm. This lower level is not known to have an effect on any individuals and appears to provide a substantial margin of safety while assuring the effectiveness of tin in preserving the food.

Until a decade ago, food cans were typically of three-piece soldered construction. Recent concerns over the lead content of foods, however, resulted in the abandonment of soldering in favor of welded construction or of two-piece (no side seam) fabrication, even though this source of lead was probably a small component in the overall human intake of lead. Both of these newer techniques appear to produce improved can integrity in general, although the soldered tinplate container is capable of many years of safe, stable shelf life. Soldered construction is still used for dry packs, in which the absence of a liquid food component eliminates the migration of lead into the packed product, and of course for nonfood items, because there is no reason to change from a successful time-tested production method.

Can Corrosion Problems. Even with nearly 200 years of experience and some of the strictest quality control programs imaginable, the canning industry is not perfect. Although not at all common, failures tend to be serious when they occur, because the economics of efficient food production depend on large, rapid production runs. The slightest misjudgment can result in a problem during long-term storage.

The first consideration in matching a given food product to the appropriate can has always been the thickness of the tin coating, as indicated above. Recent years have seen more extensive use of lacquers to provide an inert barrier against any metal dissolution (see the section "Lacquers" in this article). However, the inevitable defects in lacquer coverage make the tin coating the last defense against corrosion.

Tin is necessary for pale (uncolored or yellow) fruits and many vegetables to preserve the taste and color of the product. Suitable cans are made of plain (unlacquered) tinplate with a coating weight (the usual way of expressing the thickness) of 8.4 to 11.2 g/m^2 (0.8 to 1 g/ft^2). A special grade of tinplate called grade K uses an altered tin-iron intermetallic layer (probably thicker and more continuous over the tin/steel interface) to improve corrosion resistance with these products (Ref 2). Grade K tinplate then allows the use of tin coatings at the lower end of the thickness range.

Parenthetically, it might be mentioned that inorganic tin chemicals have been used as intentionally added preservatives for some food products that have been packed in containers other than tin cans. This fact emphasizes the importance of tin in preserving the organoleptic properties of foodstuffs.

Certain other vegetable packs, such as asparagus, green beans, and tomato-base products, would benefit from tin availability, but are strong detinners. This rapid dissolution of tin may produce an unsightly interior can surface or tin levels above the regulatory limit. The usual remedy is to use lacquer as an inert barrier, but this practice involves the possibility of even more rapid and concentrated attack at any interior scratch or other defect in coverage. Double lacquering reduces this possibility. Dark fruit packs, such as cherries, behave similarly in that strong detinning may produce undesirable color changes, which can be controlled by double lacquering.

Beer and other beverages cannot tolerate iron or tin dissolution, because of taste degradation or cloudiness, respectively; therefore, these cans are fully lacquered with a two-coat system. Baby foods are similarly treated because of special concerns regarding metal pickup in such cases.

Dairy products can be stored in plain tinplate cans, but it is advisable to use a weak passivation film because dark stains may be produced. Cathodic (only) $Na_2Cr_2O_7$ and a Na_2CO_3 treatment have each proven successful for these packs.

A stronger-than-normal passivation film (or, more precisely, one with a higher metallic chromium content) is used to prevent staining from sulfur, which is a naturally occurring contaminant of some meat products and soups, for example. Tin sulfide stains may be unsightly on the can interior surface or may be protective enough of the tin surface to reverse the tin-iron polarity and cause rapid iron dissolution. Too thick a passivation film may adversely affect lacquerability; therefore, again, correct specification and quality control are needed to produce the required product. An alternative is special lacquers containing zinc compounds that react with the sulfur to form less objectionable, nonstaining sulfides.

Sulfur can also have a deleterious effect if present as an impurity in the base steel, as can copper, phosphorus, and silicon. In these cases, impurity control in steelmaking provides a usable base material. Various tests are used to provide the suitability of the complete tinplate product (see the section "Corrosion Testing of Coatings" in this article).

Still another source of sulfur is as a residual chemical contaminant in the foodstuff. Nitrate contamination is also possible as plant uptake from fertilizers. Nitrates and certain other naturally occurring and additive organic compounds can act as cathode depolarizers, which, by increasing cathodic activity, require an increase in anodic activity, that is, tin dissolution, probably leading to early pack failure.

The wide variations in natural food products, even the same product produced in different locales, make it impossible to be more specific regarding container-product interactions. Tinplate makers and users always find it necessary to perform pack testing, that is, preparing larger samples of the canned product and observing their performance over several months of storage. Although this testing is expensive, there is no alternative, given the wide variations in product chemistry. A laboratory simulation test has recently been developed that may help in screening variables for subsequent pack testing (Ref 130).

Lacquers, also called enamels, are combinations of various resins modified with various additives (Table 14). These formulations (only the main ingredients are listed in Table 14) must produce an inert, protective film on the tinplate surface at a reasonable cost. To keep costs low, the lacquers must wet well over a surface that has received minimal preparation. In fact, cost is often one of the primary reasons for using lacquers, because they tend to substitute for the use of thicker tin coatings. In covering up so much of the sacrificial anode, however, the integrity of the lacquer film becomes extremely important in determining can performance due to the risk of more concentrated attack at defects.

Difficulties in lacquer application appear as eye holing, which consists of roundish areas of uncovered tinplate surface in the cured coating. These eyeholes may occur singly or as parts of larger affected areas. They may be caused by incompatibility between the lacquer and the surface to be coated, either the oil that must be displaced or the passivation film to which the lacquer must adhere. Dust contamination or improper tinplate surface temperature are other possible causes. Heating the tinplate before lacquer is applied usually alleviates all of these potential problems.

Once successfully applied, the lacquer must adhere to the tinplate surface through processing and storage. It must not crack during mechanical deformation, as in a beading process in which circumferential expansions in the sidewall are used for strengthening. The lacquer must also adhere through the heat-processing steps. If the lacquer cracks, there may be rapid attack in the crack. The tin layer under the crack is corroded away (undermining corrosion), causing collapse

Table 14 Main types of internal can lacquer

General type of resin and components blended to produce it	Flexibility	Sulfide-stain resistance	Typical uses	Comments
Oleo-resinous (drying oil and natural or synthetic resins)	Good	Poor	Acid fruits	Good general-purpose range at relatively low cost
Sulfur-resistant oleo-resinous (added ZnO)	Good	Good	Vegetables and soups (especially can ends or as topcoat over epoxy phenolic)	Not for use with acid products; possible intense green color with such vegetables as spinach
Phenolic (phenol or substituted phenol with formaldehyde)	Moderate-poor	Very good	Meat, fish, vegetables, and soups	Good at relatively low cost, but film thickness is restricted by flexibility
Epoxy-phenolic (epoxy resins with phenolic resins)	Good	Poor	Meat, fish, vegetables, soups, beer, and beverages (first coat)	Wide range of properties can be obtained by modifications
Epoxy-phenolic with ZnO (ZnO added)	Good	Good	Vegetables and soups (especially can ends)	Not for use with acid products; possible color change with some green vegetables
Aluminized epoxy-phenolic (metallic aluminum powder added)	Good	Very good	Meat products	Clean but rather dull appearance
Vinyl, solution (vinyl chloride-vinyl acetate copolymers)	Excellent	Not applicable	Spray on can bodies, roller coating on ends, as topcoat for beer and beverages	Free from flavor taints; sensitive to soldering heat and not usually suitable for direct application to tinplate
Vinyl, organosol, or plastisol (high molecular weight vinyl resins suspended in a nonsolvent)	Good	Not applicable	Beer and beverage topcoat on ends, bottle closures, drawn cans for sweets, pharmaceuticals, and tobacco	Same as for vinyl solutions, but giving a thicker, tougher layer
Acrylic (acrylic resin, usually pigmented white)	Very good in some ranges	Very good when pigmented	Vegetables, soups, and prepared foods containing sulfide stainers	Attractive, clean appearance of opened cans
Polybutadiene (hydrocarbon resins)	Moderate-poor	Very good if zinc-oxide is added	Beer and beverages first coat; vegetables and soups if with ZnO	Cost and, therefore, popularity depend on country

of the lacquer, widening of the crack, and exposure of the steel. Undermining corrosion may also cause failure after initiation at scratches or other lacquer defects that may occur during processing, transportation or storage. Corrosion proceeds rapidly in such cracks because of their small relative areas. Therefore, lacquers are usually applied as two coats; additional coats are applied to particularly sensitive corrosion areas, such as the side seam in three-piece can technology.

Undermining corrosion becomes more severe with increased tin coating weights because there is more of the readily dissolvable tin under the defect. Therefore, lacquer detachment is a greater possibility, leading to consumer complaints. As outlined above, the use of thinner tin coatings to combat undermining corrosion also increases the possibility of pack failure due to iron dissolution.

External corrosion of tinplate cans follows more closely the classical galvanic behavior described above. The ready availability of oxygen and the lack of complexing agents make the tin coating cathodic, and the unprotected steel will readily rust at plating pores and scratches in the coating that expose the underlying base steel. Coils or sheets of tinplate in transit from tinplate production to canmaking, as well as unfilled cans, are equally susceptible to this kind of deterioration.

A thicker tin coating will provide improved protection by reducing the steel exposure through plating pores. The passivation film also helps to provide corrosion protection. Because both of these factors are usually limited by other considerations, special precautions are taken to minimize moisture and pollution exposure. Where water is an essential part of the processing, as in steam retorting of cans or the subsequent cooling, the water is deoxygenated and treated with corrosion control agents. Minimizing the time of exposure to water is also practiced, as is the use of wetting agents to promote drying of the cans.

The paper to be used for labels and the materials to be used for shipping containers must be carefully selected. The presence of chloride or sulfate compounds in the paper, for example, may create a serious corrosion problem during storage. Wooden shipping cases might release corrosive organic vapors, and these and other materials can promote moisture exposure during transit. Again, it is difficult to be specific, because various foods are packed and then shipped literally all around the world.

A special type of corrosion may occur during transportation or handling of the tinplate. Fretting corrosion results from the intimate contact of two tinned surfaces combined with small relative movements between the two. The rubbing together of the surface asperities produces erosion, and the increased surface area yields more oxidation. Oxide particles are formed, and they act as very effective abrasives to cause additional damage to the surface. Fretting corrosion, therefore, typically features fine spots of dark, embedded tin oxide particles in areas of the tinplate that were subjected to pressure, such as from steel strapping used to secure a bundle of tinplate sheets. The most effective preventive measure is to pack the tinplates so as to minimize the relative movement of the tinplate surfaces. One purpose of the lubricant applied to the tinplate is to reduce fretting corrosion.

Corrosion Testing of Coatings

The preceding sections point out the excellent corrosion resistance of tin and tin alloys and how this property is used to advantage by coatings on stronger structural metals, typically steel or a copper alloy. Because the tin is cathodic to the base metal in normal environments, complete coverage of the base metal is important for preventing rapid attack at any pores in the coating. Therefore, porosity testing is valuable in predicting corrosion performance; however, because porosity tends to decrease with increasing coating thickness, thickness measurements often provide a more convenient indication of suitability.

In fact, most international and national specifications stipulate certain minimum coating thicknesses for anticipated service conditions. For example, ASTM B 545 specifies a minimum tin thickness of 5 μm (0.2 mils) for mild service conditions or where solderability is a primary concern (Ref 131). For exceptionally severe service conditions, such as where abrasion is combined with corrosion, the specification calls for a minimum tin thickness of 30 μm (1.2 mils). Between these extremes, for so-called normal conditions, 20 μm (0.8 mils) on steel or 8 μm (0.3 mils) on copper alloys is the specified minimum (Table 9).

The test methods described below are suited only to the purpose for which they were designed. Corrosion resistance can only be defined relative to a metal and to a particular environment; it is not an absolute property (Ref 2).

Coating Thickness Measurements. Several commercial instruments are available for measuring tin and tin alloy coating thicknesses. They have been developed to satisfy the need for a nondestructive test, and each has advantages and disadvantages.

Perhaps the simplest are the magnetic methods, which require the base metal to be magnetic or ferromagnetic. The tests determine how the coating alters the strength of a magnetic field that is passed through the coating when a magnet is positioned on the coating surface. The force required to remove the magnet is proportional to the tin thickness. The accuracy of the determination is enhanced by the use of accurate standards of known coating thickness to which comparisons can be made. The surface condition of the test sample is obviously very important for accurate measurements because the magnet is brought into contact with that surface.

The β backscatter method directs a beam of electrons (β particles) at the surface to be measured and detects particles scattered back in the direction of the source. This backscatter is in proportion to the coating thickness, and the instrument reads thickness after careful calibration to standards. The accuracy of the method relies on proper alignment of the source, sample, and backscatter collector, and periodic recalibration is important to allow for the gradual depletion of the radioactive source that generates the electron beam. Because the method does not require contact with the surface to be measured, it is particularly useful for high-speed continuous plating operations.

Somewhat similar is the x-ray fluorescence method, which directs x-rays onto the test sample and records secondary emissions that are caused by the excitation of the coating and/or base metal. The secondary emissions are not only in proportion to the coating thickness but are also characteristic of the coating alloy. X-ray fluorescence, therefore, provides a quick alloy analysis in addition to the coating thickness measurement. It also has the advantage of being noncontacting.

Among the destructive tests are the coulometric and microscopic methods. The coulometric test involves essentially a controlled stripping or deplating of the coated surface within an accurately determined area. An electrical current is applied through an electrolyte, resulting in anodic dissolution of the coating. When the coating is removed, a voltage change signals the end of the test. The amount of current passed through this small test cell is proportional to the amount of material removed and therefore implies a coating thickness.

In the microscopic method, the surface to be measured is simply sectioned and then examined optically. A direct measurement of the coating thickness is possible without using plated standards, but there is the potential problem of subjective evaluation. The other metallurgical observations that can be made on a cross section often constitute the reason for using this method.

A few other methods are available for coating thickness measurements. A micrometer check before and after application of the coating can be used for thicker coatings. Gravimetric methods involve weighing parts before and after coating or before and after coating dissolution and then calculating an average thickness based on the weight difference (coating weight) and the surface area from which it was dissolved. A drop test involves a stream of droplets of, for example, trichloroacetic acid solution, applied to a specific spot at a certain rate until the coating is penetrated, with the time to penetration being an indication of the coating thickness. In all test methods, the thinness of the typical tin alloy coating makes it important to perform the test carefully.

Porosity and Rust Resistance Testing. Although thickness testing can be performed quickly to provide process control feedback, common industrial practice includes porosity testing of statistically selected samples. This testing can predict performance more accurately by revealing the extent of variations in coating coverage that might go unobserved in thickness testing, because the latter tends to give an average over an area. The porosity test often uses simulated service conditions, usually with some accelerating factor to provide results more quickly.

Although techniques have been devised for automated test evaluation, most porosity tests rely on a visual assessment of the results. Therefore, the corrosive medium is often selected to give a readily visible corrosion product. This choice facilitates the reporting of both the quantity and distribution of porosity, along with any abnormalities in coverage.

For tin coatings on steel, SO_2, ferricyanide, or ammonium thiocyanate (NH_4CNS) tests have been used, and either of the first two is suitable for testing tin-lead coatings on steel. Tin coatings on copper alloys can be evaluated by using SO_2 or ammonium persulfate ($(NH_4)_2S_2O_8$) tests. Tin-nickel coatings are tested in a manner similar to tin coatings.

The sample of the coated metal is exposed to the corrosive medium for a specified time, perhaps at an elevated temperature to accelerate the corrosive action. After exposure, the test panel is removed and examined for rust or corrosion product. The number of attacked pores per unit area or the percentage of attacked area compared to the total area is the criterion for evaluation.

Solderability. Because tin and tin alloy coatings are so widely used to provide long-term protection to a solderable surface, much effort has recently been devoted to developing solderability tests, including accelerated aging techniques to predict shelf life. Indeed, if the coating is applied molten or if a plated coating is reflowed (also called flow melting), a form of solderability test has already been done in the sense that wetting difficulties will have been revealed during processing.

The simplest solderability test is a vertical dip of the properly fluxed sample into a solder pot. After a typical dwell time of about 3 s, the sample is withdrawn, the coating is allowed to solidify, and the surface is visually inspected for evidence of good wetting.

The other solderability tests tend to be variations of this dip test. In the rotary dip test, the sample is fixed to the end of a rotating arm, which is aligned so that the sample is passed through the upper surface of a solder bath. This test physically simulates the relative motion of surface and bath as it might happen in a wave-soldering operation. By testing a series of samples, a minimum time for complete wetting can be established and compared to the required standards.

Another popular solderability test, the surface tension balance test, is also a variation of the vertical dip test. The most significant difference is that the sample is suspended from an instrumented test rig that accurately records the forces that act on the sample during the dip into the solder pot. If wetting occurs, a force develops that attempts to pull the sample into the bath. The speed with which this force develops and its magnitude are two of the sensitive parameters often specified in standards for this test.

As indicated above, accelerated aging techniques are under investigation. The problem is that the current solderability tests may give a good correlation to actual soldering behavior in the short term, but none can predict soldering performance weeks or months from the time of testing; the latter is almost always more important. Solderability tends to maintain a certain level over the shelf life of the part and then to degrade very quickly to an unacceptable level. Aging the samples in steam for 16 to 24 h seems to give tin or tin-lead coatings the correct temperature and moisture exposure to cause poor samples to degrade. This procedure has won some acceptance for estimating the effects of 6 months to 1 year of normal storage, but only for tin-base coatings.

Special Tests for Tinplate. Because of its commercial importance, tinplate is subjected to several special tests. Coating weights (thicknesses) are often determined coulometrically, although some installations use β backscatter or x-ray fluorescence methods to obtain a quick continuous evaluation for process control. Porosity tests may involve the SO_2 or NH_4CNS tests mentioned in the section "Porosity and Rust Resistance Testing" in this article. Other tests determine the presence of tin oxides, the composition of the passivation film, and the coverage of the oil film, all of which are important for good corrosion performance. Several of the special tinplate tests are outlined below.

The Iron Solution Test (Ref 2). The tinplate sample is exposed to a solution containing sulfuric acid (H_2SO_4), H_2O_2, and NH_4CNS under controlled conditions. The amount of iron (in micrograms) dissolved during the fixed test period is termed the iron solution value (ISV). This value reflects to some extent the continuity of the coating; however, it may also be influenced by the quality of the steel, as the test solution was devised as one in which exposed steel is on the threshold of protection by tin.

The Pickle-Lag Test (Ref 2). The steel base of the tinplate is exposed to 6 *M* hydrochloric acid (HCl) under defined conditions, and the time before hydrogen is steadily evolved is measured. This time period (in seconds) is the pickle-lag value; the lower the value, the better. A high value is associated with subsurface oxidation during annealing, and it seems likely that this defect may influence both the continuity of the coating and the continuity of the tin-iron alloy layer.

The Alloy-Tin Couple (ATC) Test (Ref 2). A sample of the tinplate from which the free tin layer has been removed, but with the tin-iron intermetallic layer intact, is coupled to a relatively large electrode of pure tin in deoxygenated grapefruit juice. The current flowing between the test sample and the tin electrode is measured; its value after 23 h is termed the ATC value. The purpose of this test is to assess the restraining effect of the tin-iron compound layer on the cathodic efficiency of the metal exposed when the free tin layer is dissolved from part of the surface.

Thus, test results are affected by the continuity of the compound layer and by the characteristics of the steel.

The Tin Grain Size Test (Ref 2). The tin coating is lightly etched, and the size of the crystals revealed is expressed on the ASTM Standard Scale for the grain size of nonferrous metals. Increased grain size is considered beneficial. This is based on experience without, as yet, support from experimental measurements, but the effect of this factor is most likely to be seen in the initial rate of tin dissolution. It has been shown that different crystal faces of tin have differing dissolution and oxidation rates, and perhaps the effects of crystal orientation and crystal size are associated. It is also possible that impurities segregated at the grain boundaries produce a change related to boundary length and thus to grain size.

Other Tests. A cysteine hydrochloride ($C_3H_7O_2NS \cdot HCl$) staining test measures the tendency toward sulfide staining. A heating test simulates stoving (baking) to reveal any tendency toward discoloration during that operation. Finally, a series of tests can be used to evaluate lacquerability and lacquer adhesion to the tinplate. Additional information on these specialized tests is available in Ref 2, 128, and 129.

REFERENCES

1. W.B. Burgers and L.J. Groen, *Faraday Soc. Discuss.*, Vol 23, 1957, p 183
2. S.C. Britton, *Tin Versus Corrosion*, Publication 510, Tin Research Institute, 1975
3. E.S. Hedges, *Tin and Its Alloys*, Edward Arnold, Ltd., 1960
4. S.C. Britton and M. Clarke, *Trans. IMF*, Vol 40, 1963, p 205
5. S.C. Britton, *Trans. IMF*, Vol 52, 1974, p 95
6. N.A.J. Sabbagh and H.J. McQueen, *Met. Finish.*, 27 March 1975
7. L. Kenworthy, *Trans. Faraday Soc.*, Vol 31, 1935, p 1331
8. L. Kenworthy and J.M. Waldram, *J. Inst. Met.*, Vol 55, 1934, p 247
9. W. Finkeldy, *Symposium on Outdoor Weathering of Metals*, American Society for Testing and Materials, 1934, p 69-87
10. Report of Committee B3, *Proc. ASTM*, Vol 34 (No. 1), 1934, p 221
11. Report of Committee B3, *Proc. ASTM*, Vol 35 (No. 1), 1935, p 1
12. G.O. Hiers, *Symposium on Outdoor Weathering of Metals*, American Society for Testing and Materials, 1946
13. G.O. Hiers and E. Minarcik, *Symposium on Atmospheric Corrosion of Nonferrous Metals*, STP 175, American Society for Testing and Materials, 1956, p 135
14. H.J. Plenderleith and R.M. Organ, *Studies Conserv.*, Vol 1, 1953, p 63
15. W. Ruehl, *Z. Phys.*, Vol 176, 1963, p 409
16. W.E. Boggs, R.G. Kachik, and G.E. Pellissier, *J. Electrochem. Soc.*, Vol 108 (No. 1), 1961
17 W.E. Boggs, P.S. Trozzo, and G.E. Pellissier, *J. Electrochem. Soc.*, Vol 108 (No. 1), 1961, p 13
18. W.E. Boggs, *J. Electrochem. Soc.*, Vol 108 (No. 2), 1961
19. W.E. Boggs, R.H. Kachik, and G.E. Pellissier, *J. Electrochem. Soc.*, Vol 111 (No. 6), 1964, p 636
20. C. Luner, *Trans. Met. Soc. AIME*, Vol 218 (No. 3), 1960, p 572
21. W.E. Boggs, R.H. Kachik, and G.E. Pellissier, *J. Electrochem. Soc.*, Vol 110 (No. 1), 1963, p 4
22. M.E. Warwick, *Atmospheric Corrosion of Tin and Tin Alloys*, Publication 602, Tin Research Institute, 1980
23. L.L. Bircumshaw and G.D. Preston, *Philos. Mag.*, Vol 21, 1936, p 686
24. J.H. Bilbrey, D.A. Wilson, and M.J. Spendlove, Publication 5181, *US Bur. Mines Rep. Invest.*, 1955
25. H. Leidheiser, Jr., *The Corrosion of Copper, Tin, and Their Alloys*, R.E. Krieger, 1979
26. W. Gruhl and V. Gruhl, *Metall*, Vol 6, 1952, p 177
27. A. Kutzelnig, *Z. Anorg. Allg. Chem.*, Vol 202, 1931, p 418
28. A. Kutzelnig, *Z. Electrochem.*, Vol 41, 1935, p 450
29. D.J. MacNaughton and E.S. Hedges, *Proceedings of the International Congress on Mining, Metallurgy, and Applied Geology*, 1935
30. S.C. Britton and K. Bright, *Metallurgia*, Vol 56, 1957, p 163
31. S.C. Britton and J.C. Sherlock, *Br. Corros. J.*, Vol 9, 1974, p 96
32. R.D. Jenkins, *Proc. Phys. Soc.*, Vol 47, 1935, p 107
33. R.K. Hart, *Proc. Phys. Soc.*, Vol 65B, 1952, p 955
34. J.C. Platteuw and G. Meyer, *Trans. Faraday Soc.*, Vol 52, 1956, p 1066
35. S.N. Shah and D.E. Davies, *First International Congress on Metallic Corrosion*, Butterworths, 1961, p 232
36. H.S. Rawdon, *Ind. Eng. Chem.*, Vol 19, 1927, p 613
37. T.P. Hoar, *J. Inst. Met.*, Vol 55, 1934, p 135-145
38. F.L. LaQue, Corrosion by Seawater, Behavior of Metals and Alloys in Seawater, in *The Corrosion Handbook*, H.H. Uhlig, Ed., John Wiley & Sons, 1948, p 383-430
39. S.C. Britton, *Anti-Corrosion Manual*, Scientific Surveys, Ltd., 1958
40. C.L. Baker, *Ind. Eng. Chem.*, Vol 27, 1935, p 1358
41. O.F. Hunziker, W.H. Cordes, and B.H. Nissen, *J. Dairy Sci.*, Vol 12, 1929, p 140
42. A. Guillemin, *Ann. Chim.*, Vol 19, 1944, p 145
43. A.I. Levin, M.E. Prostakov, and V.P. Kochergin, *Zh. Prikl. Khim.*, Vol 33, 1960, p 2102
44. A. Ragheb and L.A. Kamel, *Corrosion*, Vol 18, 1962, p 153t
45. A.M.S. El Din and F.M.A. El Wahab, *Electrochim. Acta*, Vol 9, 1964, p 883
46. D.E. Davies and S.N. Shah, *Electrochim. Acta*, Vol 8, 1963, p 703
47. Lead and Health, in *Report of a DHSS Working Party on Lead in the Environment*, Her Majesty's Stationary Office, 1980
48. *Toxic Metals in Drinking Water*, Sierra Club, 1984
49. EEC Council Directive 15.7.80 Relating to the Quality of Water Intended for Human Consumption, *J. Eur. Comm.*, Vol L229, p 11
50. "Standard Specification for Modern Pewter Alloys," B 560-79, *Annual Book of ASTM Standards*, American Society for Testing and Materials
51. *Working with Pewter*, Publication 566, International Tin Research Institute, 1979
52. H. Schmidt and A. Niessen, *Blasberg-Mitteilungen*, Vol 22, 1971, p 10
53. Symposium on Lead-Base Babbitt Alloys, *Met. Prog.*, Vol 69, 1956, p 174
54. J.B. Bryce and T.G. Roehner, *Trans. IME*, Vol 73, 1961, p 377 and 393
55. *Aluminum-Tin Alloy Bearings*, Publication 463, International Tin Research Institute, 1972
56. J.C. Hudson, *Trans. Faraday Soc.*, Vol 25, 1929, p 177
57. A.H. DuRose, *J. Electrochem. Soc.*, Vol 89, 1946, p 101
58. A.W. Tracey, *Symposium on Outdoor Weathering of Metals*, American Society for Testing and Materials, 1955, p 67
59. G.W. Quick, *J. Res. Natl. Bur. Stand.*, Vol 14, 1935, p 175
60. S.C. Britton, *J. Inst. Met.*, Vol 47, 1941, p 119
61. A.L. Alexander, C.R. Southwell, and B.W. Forgeson, *Corrosion*, Vol 17 (No. 7), 1961, p 97
62. C.W. Hummer, C.R. Southwell, and A.L. Alexander, *Mater. Protect.*, Vol 7 (No. 1), 1968, p 41
63. C.R. Southwell, J.D. Bultman, and A.L. Alexander, *Mater. Perform.*, Vol 15 (No. 7), 1976, p 9
64. D.M. Ward and P.A. Lovett, *Electron*, 15 Jan 1970
65. L. Habraken, C. Rogister, A. Davin, and D. Cousouradis, *Metall*, Vol 23 (No. 11), 1969, p 1148
66. Z. Ahmad, *Anti-Corros. Methods Mater.*, Vol 24 (No. 1), 1977, p 8
67. L.E. Price and G.J. Thomas, *J. Inst. Met.*, Vol 63, 1938, p 29
68. R. Duckett, D.A. Robins, and S.C. Britton, *Metallurgia*, Vol 65, 1962, p 291
69. "Quality Metal Finishing Guide—Tin and Tin Alloy Coatings," Vol 1 (No. 1), Metal Finishing Suppliers Association, 1977
70. S.C. Clarke and E. Longhurst, *First International Congress on Metallic Corrosion*, Butterworths, 1961, p 70
71. K.G. Compton, *Corrosion*, Vol 4, 1948, p 112
72. J.C. Hudson and T.H. Banfield, *J. Iron Steel Inst.*, Vol 154, 1946, p 229
73. J.C. Hudson and J.F. Stanners, *J. Iron Steel Inst.*, Vol 175, 1953, p 381
74. S.C. Britton, *Metall. Ital.*, Vol 46, 1954, p 89
75. M. Dettner, *Plating*, Vol 46 (No. 5), 1959, p 469
76. P.L. Speddin, *Australas. Corros. Eng.*, Vol 15 (No. 8), 1971, p 27
77. T. Biestek, *Pr. Inst. Mech.*, Vol 7 (No. 26), 1959, p 89
78. T. Biestek, *Pr. Inst. Mech.*, Vol 11, 1963, p 11
79. T. Biestek, *Met. Finish.*, Oct 1974, p 39
80. W.G. Bader and R.G. Baker, *Plating*, March 1973, p 242
81. M.L. Ackroyd, *A Survey of Accelerated Aging Techniques for Solderable Substrates*, Publication 531, International Tin Research Institute, 1976
82. C.A. MacKay, *Surface Finishes and Their Solderability*, Publication 561, International

Tin Research Institute, 1979
83. H.B. Gibson, *Prod. Eng.*, Nov 1956, p 1
84. M.L. Ackroyd and C.A. MacKay, *Solders, Solderable Finishes, and Reflowed Coatings*, Publication 529, International Tin Research Institute, 1977
85. B.E. Scott and R.D. Gray, *Iron Age*, Vol 167 (No. 3), 1951, p 59
86. N.E. Promisel and G.S. Mustin, *Corrosion*, Vol 7, 1951, p 377
87. S.C. Britton and R.W. de Vere Stacpoole, *Trans. IMF*. Vol 32, 1955, p 211
88. B. Cohen, *Plating*, Vol 44 (No. 9), 1957, p 963
89. I.T. Turner, *Plating*, Vol 52, (No. 7), 1965, p 677
90. W. Beck and E.J. Janowsky, *J. Electrochem. Soc.*, Vol 109, 1962, p 496
91. F. Cook, J.K. Cosslet, R.W. Scott, and C.E.A. Shanahan, *Br. Corros. J.*, Vol 1 (No. 7), 1966, p 283
92. V. Sree and T.L. Rama Char, *Metalloberflaĉhe*, Vol 15, 1961, p 301
93. M. Clarke, R.G. Elbourne, and C.A. MacKay, *Trans. IMF*, Vol 50 (No. 4), 1972, p 160
94. M. Clarke and R.G. Elbourne, *Electrochim. Acta*, Vol 16 (No. 11), 1971, p 1949
95. H. Miyashita and S. Kurihara, *J. Met. Finish. Soc. Jpn.*, Vol 21, 1970, p 79
96. J. Hyner, *Plat. Surf. Finish.*, Vol 64, (No. 2), 1977, p 33
97. J.D.C. Hemsley and M.E. Roper, *Trans. IMF*, Vol 57 (No. 2), 1979, p 77
98. J.W. Price, *Tin and Tin Alloy Plating*, Electrochemical Publications Ltd., 1983
99. F.A. Lowenheim, *Proc. Am. Electropl. Soc.*, Vol 44, 1957, p 1
100. W.H. Safranek and C.L. Faust, *Proc. Am. Electropl. Soc.*, Vol 41, 1954, p 201
101. H. Laub, *Werkst. Korros.*, Vol 15 (No. 6), 1964, p 437
102. A.K. Graham and H.C. Pinkerton, *Plating*, Vol 54, (No. 4), 1967, p 367
103. R. Smith, *Sheet Met. Ind.*, Vol 49 (No. 12), 1972, p 761; Vol 50 (No. 2), 1973, p 92
104. R.G. Monk and J.H.T. Ellingham, *Trans. Faraday Soc.*, Vol 31, 1935, p 1460
105. J.L. Rama Char and J. Vaid, *Electropl. and Met. Finish.*, Vol 14, 1961, p 367
106. *Electroplated Tin-Nickel Alloy*, Publication 235, International Tin Research Institute, 1968
107. J.W. Cuthbertson, N. Parkinson, and H.P. Rooksby, *J. Electrochem. Soc.*, Vol 100, 1953, p 107
108. R.F. Smart and D.A. Robins, *Trans. IMF*, Vol 37, 1960, p 108
109. S.C. Britton and R.M. Angles, *Trans. IMF*, Vol 29, 1953, p 26
110. S.C. Britton, D.G. Michael, and R.M. Angles, *Trans. IMF*, Vol 29, 1953, p 40
111. S.C. Britton, *Met. Ind.*, Vol 87 (No. 25), 1955, p 510
112. J. Chadwick, *Electroplating*, Vol 6 (No. 12), 1953, p 451
113. F.A. Lowenheim, W.W. Sellers, and F.X. Carlin, *J. Electrochem. Soc.*, Vol 105 (No. 6), 1958, p 338
114. R.M. Angles, *Tin-Nickel Alloy Plating*, International Nickel Company (Mond) Ltd., 1964
115. M. Antler, *Proceedings of the Conference on Corrosion Control by Coatings*, Lehigh University, 1978
116. M. Antler, M. Feder, C.F. Hornig, and J. Bohland, *Plat. Surf. Finish.*, Vol 63 (No. 7), 1976, p 30
117. S.C. Britton and R.M. Angles, *J. Electrodep. Tech. Soc.*, Vol 27, 1951, p 293
118. R.M. Angles and R. Kerr, *Engineering*, Vol 161, 1946, p 289
119. H. Heinemann, *Metalloberflaĉhe*, Vol 6B, 1954, p 33
120. *Auto. Eng.*, 1955, p 75
121. S.C. Britton and R.M. Angles, *Metallurgia*, Vol 44, 1951, p 185
122. S.C. Britton and R.W. de Vere Stacpoole, *Metallurgia*, Vol 52, 1955, p 64
123. S.V. Lashko-Avakyan and N.F. Lashko, *Savarochnoe Proivodstvo*, Vol 5, 1961, p 13
124. S.L. Phillips and C.E. Johnson, *J. Electrochem. Soc.*, Vol 117, 1970, p 827
125. D.R. Cowieson and A.R. Schofield, *Passivation of Tin-Zinc Alloy Coated Steel*, Publication 661, International Tin Research Institute, 1985
126. W.E. Hoare, E.S. Hedges, and B.T.K. Barry, *The Technology of Tinplate*, Edward Arnold, 1965
127. *Steel Products Manual: Tin Mill Products*, American Iron and Steel Institute, 1979
128. *Guide to Tinplate*, Publication 662, International Tin Research Institute, 1983
129. E. Morgan, *Tinplate and Modern Canmaking Technology*, Pergamon Press, 1985
130. M.E. Warwick and W.B. Hampshire, *Laboratory Corrosion Tests and Standards*, STP 866, American Society for Testing and Materials, 1985, p 48
131. "Standard Specification for Electrodeposited Coatings of Tin," B 545, *Annual Book of ASTM Standards*, American Society for Testing and Materials

Corrosion of Lead and Lead Alloys*

Jerome F. Smith, Lead Industries Association, Inc.

LEAD has such a successful record of service in exposure to the atmosphere and to water that its resistance to corrosion by these media is often taken for granted. Underground, thousands of kilometers of lead-sheathed cable and lead pipe give reliable long-term performance all over the world. In the chemical industry, lead is used in the corrosion-resistant equipment necessary for handling many chemicals (see the article "Corrosion in the Chemical Processing Industry" in this Volume). For information on the corrosion performance of lead and lead alloys used in lead-acid storage batteries, see the article "Corrosion of Batteries and Fuel-Cell Power Sources" in this Volume. General information on compositions, properties, and applications, can be found in the articles "Lead and Lead Alloys" and "Properties of Lead and Lead Alloys" in Volume 2 of the 9th Edition of *Metals Handbook*.

The Nature of Lead Corrosion

The corrosion of lead in aqueous electrolytes is an electrochemical process. The metal either enters the solution at anodic sites as metallic cations or is converted anodically to solid compounds. Both corrosion reactions can be represented by the reaction:

$$Pb - 2e^- \rightarrow Pb^{2+} \quad \text{(Eq 1)}$$

This oxidation reaction, which takes place at anodic sites, is accompanied by a reduction of some constituent in the electrolyte at cathodic sites. In neutral salt solutions, the cathodic reaction is the reduction of dissolved oxygen:

$$\tfrac{1}{2}O_2 + H_2O + 2e^- \rightarrow 2OH^- \quad \text{(Eq 2)}$$

In acid solutions free of oxygen, the corresponding cathodic reaction is:

$$2H^+ + 2e^- \rightarrow H_2 \quad \text{(Eq 3)}$$

The rate of corrosion is a function of the current flowing between the anodes and cathodes of the corrosion cell. Many factors and conditions can initiate or influence this flow of current. In the corrosion of a single metal, such as lead, local anodes and cathodes may be set up as a result of inclusions, inhomogeneities, stress variations, and differences in temperature. In bimetallic (galvanic) corrosion, the anodic and cathodic sites are on different metals, with the less noble metal (anode) corroding in preference to the more noble metal (cathode).

In most environments, lead is cathodic to steel, aluminum, zinc, cadmium, and magnesium and therefore will accelerate the corrosion of these metals. With titanium and passivated stainless steels, lead is the anode of the cell and suffers accelerated attack. In either case, the rate of corrosion is governed by the difference in potential between the two metals, the ratio of their areas, and their polarization characteristics.

The corrosion rate of lead is usually under anodic control, because the most important determinant generally is the solubility and other physical characteristics of the corrosion products formed at anodic sites. Most of these products are relatively insoluble lead salts that are deposited on the lead surface as impervious films, which tend to stifle further attack. The formation of such insoluble protective films is responsible for the high resistance of lead to corrosion by sulfuric (H_2SO_4), chromic (H_2CrO_4), and phosphoric (H_3PO_4) acids.

In general, anything that damages the protective film increases the corrosion rate. Factors that help create or strengthen the film reduce the corrosion rate. Therefore, the life of the lead-protected equipment can be extended, for example, by washing it with film-forming aqueous solutions containing sulfates, carbonates, or silicates. This procedure is suggested for protecting lead when it will be in contact with corrosives that do not form protective films.

Forms of Corrosion. The corrosion of lead can take many forms. Lead exposed to the usual type of atmospheric attack will corrode uniformly. Pitting will occur under conditions of partial passivity or cavitation, which is the formation and collapse of gas bubbles at a liquid/metal interface.

In some cases, a combination of corrosion and other forms of deterioration, such as erosion, fatigue, and fretting, will cause damage much more severe than that caused by each form of attack working independently. Another type of accelerated corrosion can occur when lead is in contact with a corrosive environment and is subjected to a continuous load exceeding its creep strength. The process of creep will continually expose fresh surface to the corroding environment.

Intergranular corrosion is another form of attack on lead. It occurs at grain boundaries of lead generally in the cast form and can cause a significant loss in strength.

It is evident that the specific rate and form of corrosion that occur in a particular situation depend on many complex variables. However, in each of the four major environments discussed below—water, atmosphere, underground, and chemical—certain factors have a determining influence on what form and rate lead corrosion will have.

Corrosion in Water

Distilled water free of oxygen and carbon dioxide (CO_2) does not attack lead. Distilled water containing CO_2 but not oxygen also has little effect on lead. The corrosion behavior of lead in distilled water containing dissolved CO_2 and dissolved oxygen depends on CO_2 concentration. This dependency, which causes many different reactions to take place in a narrow range of concentration, explains the contradictory nature of much of the corrosion data reported in the literature.

For example, lead steam coils that handle pure water condensate are not severely corroded in systems in which all condensate is returned to the boiler and negligible makeup water is used. However, if makeup water is used, dissolved oxygen can be introduced to the condensate, and corrosion can be severe. Carbon dioxide can also be generated from the breakdown of carbonates and bicarbonates in boiler water, decreasing the severity of corrosion of lead. The oxygen level in the makeup water is usually controlled by adding oxygen scavengers, such as hydrazine or sodium sulfite.

In general, the corrosion rate in natural and domestic waters depends on the degree of water hardness. Water hardness is primarily caused by calcium and magnesium salts in the water. These salts, if present in at least moderate amounts ($>$125 ppm), form films on lead that adequately protect it against corrosive attack. Silicate salts present in the water increase both the hardness and the protective value of the film. In contrast, nitrate and chloride ions either interfere with the formation of the protective film or penetrate it; thus they increase corrosion.

In soft, aerated natural and domestic waters, the corrosion rate depends on both the hardness and the oxygen content of the water. When water hardness is less than 125 ppm, corrosion rate, like the rate in distilled water, depends on the relative proportions of dissolved CO_2 and dissolved oxygen. Potable waters, in which lead content is not permitted to exceed 0.05 ppm, often have hard-

*Portions of this article are based on Chapters 5 and 6 in *Lead for Corrosion Resistant Applications—A Guide*, Lead Industries Association, Inc.

Table 1 Corrosion of chemical lead in industrial and domestic waters

Total immersion

Type of water	Temperature °C	°F	Aeration	Agitation	Corrosion rate µm/yr	mils/yr
Condensed steam, traces of acid	21–38	70–100	None	Slow	21.59	0.85
Mine water						
pH 8.3, 110-ppm hardness	20	68	Yes	Slow	6.60	0.26
160-ppm hardness	19	67	Yes	Slow	7.11	0.28
110-ppm hardness	22	72	Yes	Slow	6.35	0.25
Cooling tower water, oxygenated, from Lake Erie	16–29	60–85	Complete	None	134.6	5.3
Los Angeles aqueduct water, treated with chlorine and copper sulfate	Ambient		. . .	150 mm/s (0.5 ft/s)	9.65	0.38
Spray cooling water, chromate treated	16	60	Yes	. . .	9.4	0.37

Table 2 Corrosion of lead in natural waters

Location and type of water	Type of test	Agitation	Corrosion rate µm/yr	mils/yr	Ref
Bristol Channel; seawater	Immersion about 93% of the time	. . .	12.7	0.50	1
Southhampton Docks; seawater	Half tide level	. . .	2.79	0.11	2
Gatun Lake, CZ; tropical freshwater	Immersion	None	2.03	0.08	3
Fort Amador, CZ; tropical Pacific Ocean	Immersion	Flowing(a)	9.14	0.36	
Fort Amador, CZ; tropical Pacific Ocean	Mean tide level	Flowing(a)	5.08	0.20	
San Francisco Harbor; seawater	Mean tide level	Flowing	10.67	0.42	
Port Hueneme Harbor, CA; seawater	Immersion	Flowing(b)	5.59	0.22	4
Kure Beach, NC; seawater	Immersion	. . .	15.24	0.60	

(a) At 150 mm/s (0.5 ft/s). (b) At 60 mm/s (0.2 ft/s)

Table 3 Corrosion of lead in various natural outdoor atmospheres

Location	Type of atmosphere	Duration of test, years	Type of lead	Corrosion rate µm/yr	mils/yr	Ref
Altoona, PA	Industrial	10	Chemical	0.737	0.029	5, 6
			Pb-1Sb	0.584	0.023	5, 6
New York City	Industrial	20	Chemical	0.381	0.015	5, 6
			Pb-1Sb	0.330	0.013	5, 6
Sandy Hook, NJ	Seacoast	20	Chemical	0.533	0.021	5, 6
			Pb-1Sb	0.508	0.020	5, 6
Key West, FL	Seacoast	10	Chemical	0.584	0.023	5, 6
			Pb-1Sb	0.559	0.022	5, 6
LaJolla, CA	Seacoast	20	Chemical	0.533	0.021	5, 6
			Pb-1Sb	0.584	0.023	5, 6
State College, PA	Rural	20	Chemical	0.330	0.013	5, 6
			Pb-1Sb	0.356	0.014	5, 6
Phoenix, AZ	Semiarid	20	Chemical	0.102	0.004	5, 6
			Pb-1Sb	0.308	0.012	5, 6
Kure Beach, NC (25-m, or 80-ft site)	East coast, marine	2	Chemical	1.321	0.052	7
			Pb-6Sb	1.041	0.041	7
Newark, NJ	Industrial	2	Chemical	1.473	0.058	7
			Pb-6Sb	1.067	0.042	7
Point Reyes, CA	West coast, marine	2	Chemical	0.914	0.036	7
			Pb-6Sb	0.660	0.026	7
State College, PA	Rural	2	Chemical	1.397	0.055	7
			Pb-6Sb	0.991	0.039	7
Birmingham, England	Urban	7	99.96%Pb	0.939	0.037	8
			Pb-1.6Sb	0.102	0.004	8
Wakefield, England	Industrial	1	99.995%Pb	1.879	0.074	8
Southport, England	Marine	1	99.995%Pb	1.778	0.070	8
Bourneville, England	Suburban	1	99.995%Pb	1.956	0.077	8
Cardington, England	Rural	1	99.995%Pb	1.422	0.056	8
Cristobal, CZ	Tropical, marine	8	Chemical	1.346	0.053	3
Miraflores, CZ	Tropical, marine	8	Chemical	0.762	0.030	3

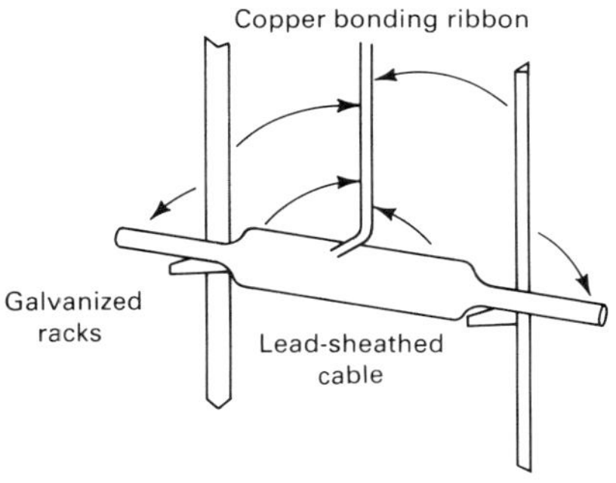

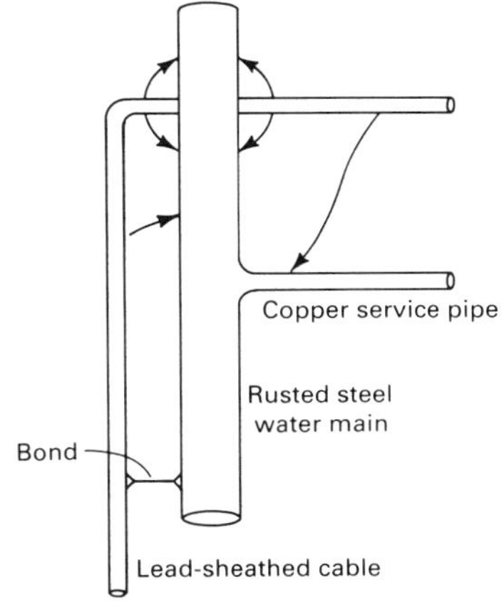

Fig. 1 Corrosion caused by galvanic coupling. Arrows indicate direction of current flow. Source: Ref 2

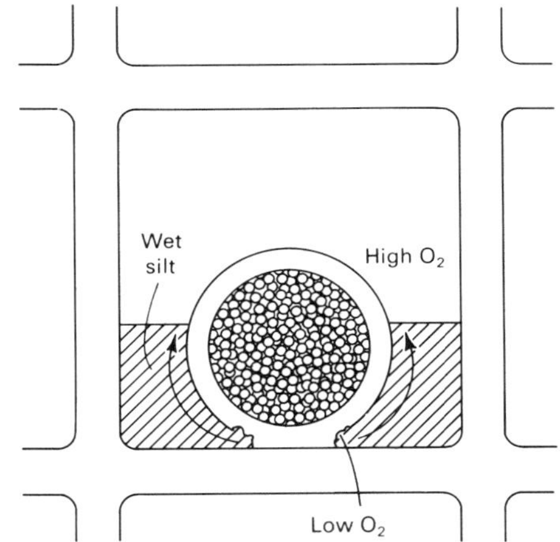

Fig. 2 Corrosion caused by differential aeration in a duct. Arrows indicate direction of current flow. Source: Ref 2

nesses below 125 ppm and often contain considerable amounts of CO_2 and oxygen; thus, lead frequently cannot be used for pipe or containers that handle potable waters. This problem of contamination limits the use of lead in such applications, even though from a service point of view, corrosion rate is negligible.

The corrosion rates of chemical lead (99.9% Pb) in several industrial and domestic waters are presented in Table 1. It should be noted that corrosion rate is relatively low, even where water hardness is below 125 ppm. A corrosion rate for a freshwater is also included among the data for seawater in Table 2.

The corrosion of lead in seawater is relatively slight and may be retarded by incrustations of lead salts. Data on the performance of lead in seawater at several locations are given in Table 2. Comparison of two of the entries in this table shows that at the same tropical location (Panama Canal Zone), the corrosion rate of lead in freshwater is about one-fourth the rate in seawater.

Extensive service experience and laboratory testing have indicated that the corrosion rate of

Table 4 Corrosion of lead alloys in various soils

Maximum exposure time: 11 years

Type of soil	Chemical lead(a) Corrosion rate, μm/yr	Chemical lead(a) Corrosion rate, mils/yr	Chemical lead(a) Max pit depth, μm	Chemical lead(a) Max pit depth, mils	Tellurium lead(b) Corrosion rate, μm/yr	Tellurium lead(b) Corrosion rate, mils/yr	Tellurium lead(b) Max pit depth, μm	Tellurium lead(b) Max pit depth, mils	Antimonial lead(c) Corrosion rate, μm/yr	Antimonial lead(c) Corrosion rate, mils/yr	Antimonial lead(c) Max pit depth, μm	Antimonial lead(c) Max pit depth, mils
Cecil clay loam	<2.54	<0.1	457	18	<2.54	<0.1	406	16	<2.54	<0.1	229	9
Hagerstown loam	<2.54	<0.1	787	31	<2.54	<0.1	762	30	<2.54	<0.1	406	16
Lake Charles clay	7.62	0.3	2540	100	10.16	0.4	2718	107	10.16	0.4	2642	104
Muck	7.62	0.3	1321	52	7.62	0.3	1346	53	7.62	0.3	1295	51
Carlisle muck	5.08	0.2	508	20	5.08	0.2	533	21	2.54	0.1	305	12
Rifle peat	<2.54	<0.1	838	33	<2.54	<0.1	584	23	<2.54	<0.1	711	28
Sharkey clay	7.62	0.3	1778	70	7.62	0.3	1854	73	10.16	0.4	2261	89
Susquehanna clay	<2.54	<0.1	864	34	2.54	0.1	1016	40	2.54	0.1	356	14
Tidal marsh	<0.25	<0.01	305	12	<0.25	<0.01	203	8	<0.25	<0.01	152	6
Docas clay	<2.54	<0.01	635	25	<2.54	<0.1	432	17	<2.54	<0.1	483	19
Chino silt loam	<2.54	<0.1	381	15	<2.54	<0.1	508	20	<2.54	<0.1	178	7
Mohave fine gravelly clay	<2.54	<0.1	610	24	<2.54	<0.1	584	23	2.54	<0.1	406	16
Cinders	7.62	0.3	2159	85	7.62	0.3	1549	61	10.16	0.4	1168	46
Merced silt loam	<2.54	<0.1	610	24	<2.54	<0.1	406	16	<2.54	<0.1	229	9

(a) 0.056 Cu, 0.002 Bi, 0.001 Sb, (b) 0.08 Cu, 0.01 Sb, 0.043 Te. (c) 0.036 Cu, 5.3 Sb, 0.016 Bi. Source: Ref 15

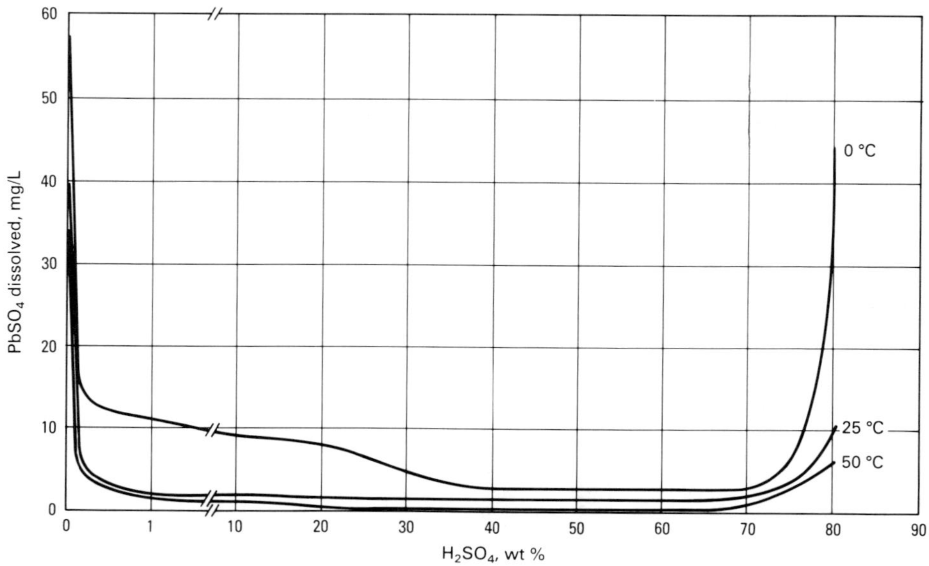

H_2SO_4, wt %	$PbSO_4$, dissolved, mg/L, at: 0 °C (32 °F)	25 °C (75 °F)	50 °C (120 °F)
0	33.0	44.5	57.7
0.005	8.0	10.0	24.0
0.01	7.0	8.0	21.0
0.10	4.6	5.2	13.0
1.0	1.8	2.2	11.3
10.0	1.2	1.6	9.6
20.0	0.5	. . .	8.0
30.0	0.4	1.2	4.6
60.0	0.4	1.2	2.8
70.0	1.2	1.8	3.0
75.0	2.8	3.0	6.6
80.0	6.5	11.5	42.0

Fig. 3 Solubility of lead sulfate in sulfuric acid

Table 5 Solubility of lead compounds

Lead compound	Formula	Temperature °C	Temperature °F	Solubility (a), Kg/m³
Acetate	$Pb(C_2H_3O_2)_2$	20	68	433
Bromide	$PbBr_2$	20	68	8.441
Carbonate	$PbCO_3$	20	68	0.0011
Basic carbonate	$2PbCO_3$, $Pb(OH)_2$	. . .	. . .	Insoluble
Chlorate	$Pb(ClO_3)_2$, H_2O	18	64	0.513
Chloride	$PbCl_2$	20	68	9.9
Chromate	$PbCrO_4$	25	77	0.000058
Fluoride	PbF_2	18	64	0.64
Hydroxide	$Pb(OH)_2$	18	64	0.155
Iodide	PbI_2	18	64	0.63
Nitrate	$Pb(NO_3)_2$	18	64	565
Oxalate	PbC_2O_4	18	64	0.0016
Oxide	PbO	18	64	0.017
Orthophosphate	$Pb_3(PO_4)_2$	18	64	0.00014
Sulfate	$PbSO_4$	25	77	0.0425
Sulfide	PbS	18	64	0.1244
Sulfite	$PbSO_3$	. . .	. . .	Insoluble

(a) In water at room temperature. Source: Ref 17

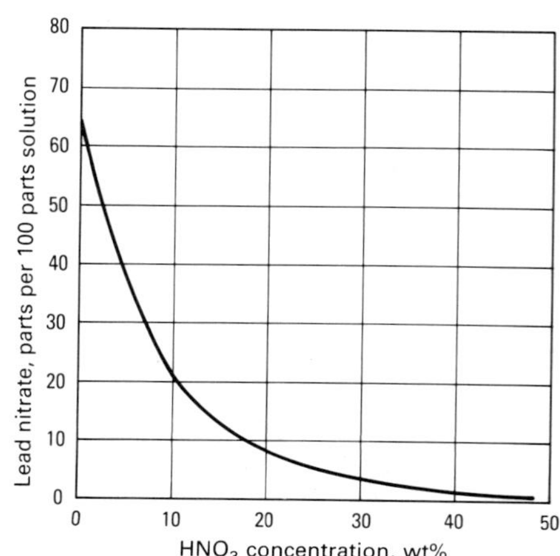

Fig. 4 Solubility of lead nitrate in nitric acid

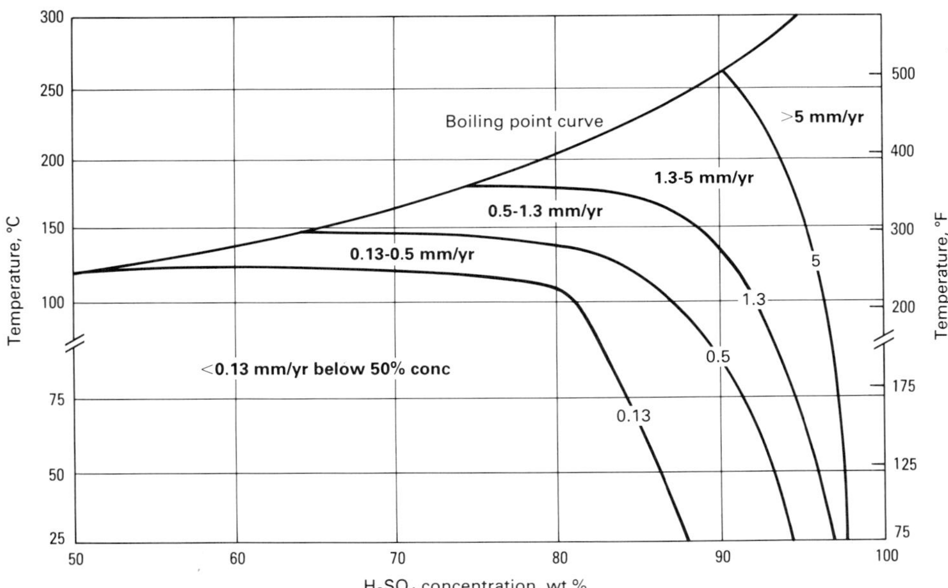

Fig. 5 Corrosion rate of lead in H_2SO_4. Source: Ref 18

lead is generally quite low in a wide variety of waters. The only major applications in which lead cannot be used are those involving some pure waters containing oxygen and soft natural waters, especially if contamination is of concern. In contrast, as discussed above, addition of calcium and magnesium salts further enhances the resistance of lead to corrosion by water.

Atmospheric Corrosion

In most of its forms, lead exhibits consistent durability in all types of atmospheric exposure, including industrial, rural, and marine (Table 3). These three atmospheric environments are distinct because each involves different factors that promote corrosion. In rural areas, which are relatively free of pollutants, the only important environmental factors influencing corrosion rate are humidity, rainfall, and air flow. However, near or on the sea, chlorides entrained in marine air often exert a strong effect on corrosivity. In industrial environments, sulfur oxide gases and the minerals in solid emissions change the patterns of corrosion behavior considerably. However, the protective films that form on lead and its alloys are so effective that corrosion is insignificant in most natural atmospheres. The extent of this protection is demonstrated by the survival of lead roofing and auxiliary products after hundreds of years of atmospheric exposure. In fact, the metal is preserved permanently if these films are not damaged (Ref 9).

Antimonial lead, such as UNS 52760 (Pb-2.75Sb-0.2Sn-0.18As-0.075Cu), exhibits approximately the same corrosion rate in atmospheric environments as chemical lead (99.9% commercial-purity lead). However, the greater hardness, strength, and resistance to creep of antimonial lead often make it more desirable for use in specific chemical and architectural applications. The ability of some antimonial leads to retain this greater mechanical strength in atmospheric environments has been demonstrated in exposure tests in which sheets containing 4% Sb and smaller amounts of arsenic and tin were placed in semirestricted positions for 3 years. They showed less tendency to buckle than chemical lead, indicating that their greater resistance to creep had been retained.

Painting of lead coatings, especially terne metal (a coating containing 8 to 12% Sn, rem Pb), further raises their resistance to corrosion in outdoor environments. Terne metal has such good paint retention that one coat will far outlast two separate coats on plain steel. The use of terne metal is discussed in the article "Corrosion in the Automotive Industry" in this Volume.

Table 7 Corrosion of lead in hydrochloric acid at 24 °C (75 °F)

HCl concentration, %	Chemical lead µm/yr	Chemical lead mils/yr	6% antimonial lead µm/yr	6% antimonial lead mils/yr
1	610	24	840	33
5	410	16	510	20
10	560	22	1090	43
15	790	31	3810	150
20	1880	74	4060	160
25	4830	190	5080	200
35(a)	8890	350	13 720	540

(a) Commercially concentrated HCl

Table 8 Corrosion of lead in hydrochloric acid-ferric chloride mixtures at 24 °C (75 °F)

Solution	Chemical lead µm/yr	Chemical lead mils/yr	6% antimonial lead µm/yr	6% antimonial lead mils/yr
5% HCl + 5% $FeCl_3$	711	28	940	37
10% HCl + 5% $FeCl_3$	1041	41	1930	76
15% HCl + 5% $FeCl_3$	2235	88	4064	160
20% HCl + 5% $FeCl_3$	3810	150	4826	190

Table 6 Corrosion of chemical lead in commercial phosphoric acid at 21 °C (70 °F)

Solution	Corrosion rate µm/yr	Corrosion rate mils/yr
20% H_3PO_4	86.4	3.4
30% H_3PO_4	124.5	4.9
40% H_3PO_4	144.8	5.7
50% H_3PO_4	162.6	6.4
85% H_3PO_4	40.6	1.6
80% H_3PO_4(a)	325.1	12.8

(a) Pure grade

Corrosion in Underground Ducts

Lead is extensively used in the form of sheathing for power and communications cables because of its impermeability to water and its excellent resistance to corrosion in a wide variety of soil conditions (the use of lead-sheathed cables is discussed in the article "Corrosion in Telephone Cable Plants" in this Volume). Cables are either buried directly in the ground or installed in ducts or conduits. In the United States, the preferred method is to lay cable in ducts or conduits made of such materials as cement, vitrified clay, or wood.

Severe corrosion of lead in underground service (in ducts or directly in the soil) is the exception rather than the rule. However, because repair or replacement of underground components is difficult and expensive, proper corrosion protection is recommended in any underground service. Although the discussion that follows is based on preventive methods used for lead-sheathed cables, it is directly applicable in many ways to the underground behavior of other lead products, such as chemical service pipe.

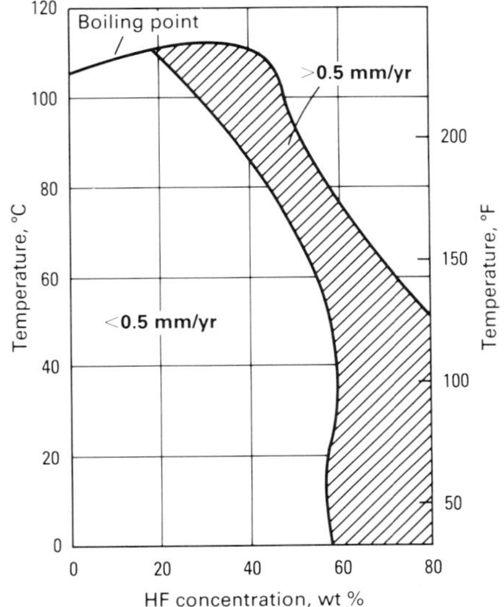

Fig. 6 Resistance of lead to corrosion in air-free hydrofluoric acid

Table 9 Corrosion of lead in nitric acid

HNO3 solution, %	Corrosion rate: 24 °C (75 °F) µm/yr	24 °C (75 °F) mils/yr	50 °C (122 °F) µm/yr	50 °C (122 °F) mils/yr
1	3556	140	15 240	600
5	41 910	1650	46 990	1850
10	86 360	3400	88 646	3490

Table 10 Effect of nitric acid in sulfuric acid on the corrosion of lead at 118 °C (245 °F)

Solution	Chemical lead µm/yr	Chemical lead mils/yr	6% antimonial lead µm/yr	6% antimonial lead mils/yr
78% H_2SO_4 + 0% HNO_3	188	7.4	356	14
78% H_2SO_4 + 1% HNO_3	150	5.9	559	22
78% H_2SO_4 + 5% HNO_3	213	8.4	2896	114

Table 11 Corrosion of chemical lead with sulfuric-nitric mixed acids

Solution	Corrosion rate: 24 °C (75 °F) µm/yr	24 °C (75 °F) mils/yr	50 °C (122 °F) µm/yr	50 °C (122 °F) mils/yr
78% H_2SO_4 + 0% HNO_3	25.4	1	50.8	2
78% H_2SO_4 + 1% HNO_3	76.2	3	304.8	12
78% H_2SO_4 + 3.5% HNO_3	91.4	3.6	457.2	18
78% H_2SO_4 + 7.5% HNO_3	101.6	4	889	35

The environment within ducts is often quite complex (Ref 9). It can include combinations of highly humid manhole and soil atmospheres, free lime leached from concrete, and alkalies formed by the electrolysis of salts in the water that seeps into ducts. Some of the factors involved in the corrosion of lead cable sheathing and how they relate to cable assembly and installation are discussed in Ref 10. Their influence in initiating or accelerating corrosion is described in Ref 10, with simple sketches used for illustration. Two of these factors—galvanic coupling and differential aeration—are discussed below.

Galvanic Coupling. Figure 1 illustrates two typical examples of contact between lead and other metals. In the presence of an electrolyte, such a dissimilar-metal couple forms a galvanic cell in which the more anodic metal is corroded. A difference in potential sufficient to cause corrosion may also arise when the surface of the lead is scratched to expose bright, active metal. In such cases, the exposed metal is the anode and is attacked.

Differential Aeration. Figure 2 illustrates differential-aeration corrosion. In this type of corrosion cell, areas exposed to low oxygen concentration tend to become anodic to areas exposed to higher oxygen concentrations. As shown, the amount of air able to penetrate the silt and reach the crevice where the cable sheath and the duct meet is less than the amount available at the upper surface of the sheath; this results in corrosion.

An actual example of differential-aeration corrosion is described in Ref 11. Lead-sheathed cable was pressed tightly against the inner surface of a tile duct, and water formed a meniscus extending from the sheathing surface to the tile. The area that was pressed against the tile did not corrode. However, an adjacent area, where the water was farthest from contact with air, corroded severely. The lead surface in contact with water closer to the air in the duct was the cathode.

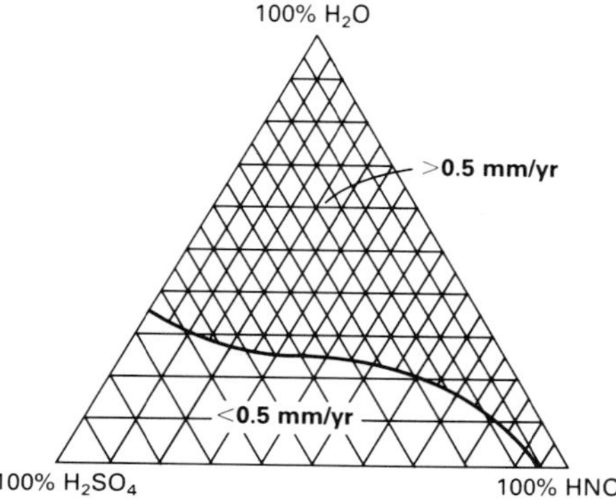

Fig. 7 Corrosion rates of lead in H_2SO_4-H_2O_3-H_2O mixtures

Alkalinity is another factor that causes the corrosion of cable sheathing (Ref 12). Sheathing on cable installed in continuous concrete or asbestos cement ducts in concrete tunnels under waterways was found to be severely corroded. Analysis of water samples from these locations revealed that the corrosion had resulted from the presence of up to 1000 ppm of hydroxides. These alkaline water samples (pH 10.9 to 12.2) contained mainly calcium hydroxide. Sodium hydroxide was also found in some tunnels.

The source of the calcium hydroxide was incompletely cured concrete. Electrolysis of solutions of deicing salts that had seeped into the tunnels was believed to be the source of sodium hydroxide. The buildup in concentration occurred because seepage water was not being removed (the ducts had been designed to function without removal of seepage water). Proper drainage and use of completely cured, impervious concrete were suggested as corrective measures.

Stray currents can cause severe corrosion of lead pipe or lead cable sheathing. Stray currents are those that follow paths outside intended circuits. They may also be minor earth currents. Stray currents cause corrosion at the point where they leave the metal. Sources of stray currents include electric railway systems, grounded electric dc power, electric welders, cathodic protection systems, and electroplating plants. Stray alternating currents are much less damaging than stray direct currents.

It has also been found that corrosion of lead cable sheathing in manhole waters depends more on the magnitude and polarity of the potential between the ground and the lead sheathing than it does on the natural dissolved salts in the water. Corrosion is at a minimum when the sheathing is

Table 12 Corrosion of lead in hydrochloric acid–sulfuric acid mixtures

Solution	Chemical lead, 24 °C (75 °F) µm/yr	Chemical lead, 24 °C (75 °F) mils/yr	Chemical lead, 66 °C (150 °F) µm/yr	Chemical lead, 66 °C (150 °F) mils/yr	6% antimonial lead, 24 °C (75 °F) µm/yr	6% antimonial lead, 24 °C (75 °F) mils/yr	6% antimonial lead, 66 °C (150 °F) µm/yr	6% antimonial lead, 66 °C (150 °F) mils/yr
1% HCl + 9% H_2SO_4	130	5	230	9	130	5	300	12
3% HCl + 7% H_2SO_4	360	14	810	32	530	21	1040	41
5% HCl + 5% H_2SO_4	360	14	1070	42	530	21	1650	65
7% HCl + 3% H_2SO_4	410	16	1140	45	560	22	1880	74
9% HCl + 3% H_2SO_4	460	18	1190	47	760	30	2130	84
5% HCl + 25% H_2SO_4	250	10	560	22	560	22	860	34
10% HCl + 20% H_2SO_4	430	17	1070	42	2030	80	1470	58
15% HCl + 15% H_2SO_4	1040	41	1880	74	2290	90	4570	180
20% HCl + 10% H_2SO_4	2180	86	3050	120	2790	110	4570	180
25% HCl + 5% H_2SO_4	3560	140	4060	160	3810	150	5330	210
5% HCl + 45% H_2SO_4	1580	62	...	...	1350	53	...	...
10% HCl + 40% H_2SO_4	1650	65	...	...	2130	84	...	...
15% HCl + 35% H_2SO_4	1680	66	...	...	3050	120	...	...
20% HCl + 30% H_2SO_4	2130	84	...	...	3300	130	...	...
25% HCl + 25% H_2SO_4	3050	120	...	...	5330	210	...	...

Table 13 Effect of sulfuric acid on the corrosion of lead by fluosilicic acid at 45 °C (113 °F)

Solution	Chemical lead µm/yr	Chemical lead mils/yr	6% Sb lead µm/yr	6% Sb lead mils/yr
5% H_2SiF_6	1346	53	1956	77
5% H_2SiF_6 + 5% H_2SO_4	229	9	356	14
10% H_2SiF_6	1626	64	2921	115
10% H_2SiF_6 + 1% H_2SO_4	2235	88	1930	76
1% H_2SiF_6 + 10% H_2SO_4	102	4	229	9

Table 14 Corrosion of lead in chemical process fluids

Fluids	Temperature °C	Temperature °F	Corrosion rate μm/yr	Corrosion rate mils/yr
Sulfation of oils with 25% sulfuric acid (66° Bé) at 60 °C (140 °F)				
Castor	...	...	76.2	3
Tallow	...	...	304.8	12
Olive	...	...	76.2	3
Cod Liver	...	...	152.4	6
Neatsfoot	...	...	279.4	11
Fish	...	...	279.4	11
Vegetable	...	...	584.2	23
Peanut	...	...	457.2	18
Sulfonation with 93% sulfuric acid (66° Bé)				
Naphthalene	166	330	1143	45
Phenol	120	248	76.2	3
Washing and neutralization of sulfated and sulfonated compounds				
Sulfated vegetable oil + water wash-neutralized with sodium hydroxide	60	140	228.6	9
Naphthalene sulfonic acid + water wash-neutralized with caustic soda pH 3	70	158	990.6	39
Washing tallow with 2% by wt 60° Bé sulfuric acid	121	250	127	5
Storage of liquid alkyl detergent	...	...	7.62	0.3
Storage of 50% chlorosulfonic acid-50% sulfur trioxide	...	...	15.24	0.6
Mixing tank and crystallizer-saturated ammonium sulfate-5% sulfuric acid solution	47	116	25.4–127	1–5
Splitting				
Olive oil and 0.5% sulfuric acid (66° Bé)	88	190	279.4	11
Storage of split fatty acids	...	...	Liquid 20.32	Liquid 0.8
Storage of split fatty acids	...	...	Liquid level 304.8	Liquid level 12
Extraction of aluminum sulfate from alumina				
Bauxite + sulfuric acid-boiling	...	...	Liquid 406.4	Liquid 16
Bauxite + sulfuric acid-boiling	...	...	Vapor 127	Vapor 5
Alum evaporator	116	240	76.2	3
Tank for dissolving alum paper mill	49	120	406.4	16
Storage of 24% alum solution	...	...	15.24	0.6
Dorr settling tank				
19.5 sulfuric acid, 20% ferrous sulfate, 10% titanium oxide as $TiSO_4$	70	158	254	10
Evaporator				
Nickel sulfate solution	100	212	152.4	6
Zinc sulfate solution	107	225	152.4	6
Ammonium sulfate production				
Solution-saturated ammonium sulfate + 5% sulfuric acid	47	116	Mixing tank 25.4	Mixing tank 1
Solution-saturated ammonium sulfate + 5% sulfuric acid	47	116	Crystallizer 127	Crystallizer 5
Acid washing				
Lube oil-treatment with 25% sulfuric acid	104	220	635	25
Sludge oil + 15% sulfuric acid-steam treatment	...	...	508	20
Benzol (crude)-treatment with 3% sulfuric acid washed with water, neutralized with lime	60	140	152.4	6
Tar oil-treatment with 25% sulfuric acid, washed with water, neutralized with sodium hydroxide	77	170	609.6	24
Wet acid gases from regeneration of sulfuric acid	121	250	152.4	6
Polymerization				
Polymerization of butenes with 72% sulfuric acid	80	175	12.7	0.5
Polymerization of butenes with 72% sulfuric acid	80	175	356 pits	14 pits
Viscose rayon spinning bath				
Evaporator—6% sulfuric acid, 17% sodium sulfate, 30% other inorganic sulfates	40	104	127	5
Evaporator—concentrated bath of 20% sulfuric acid, 30% sodium sulfate	55	130	101.6	4

(continued)

cathodic to the ground (Ref 13, 14). Grounding prevents this type of corrosion.

Other factors that can initiate the corrosion of lead sheathing include contact with acetic acid (in wooden ducts), microorganisms, and corroded steel-tape armor. Bacterial corrosion usually occurs when aeration is poor and mud, water, or organic matter is present. Bacteria capable of reducing sulfates to sulfides are the principal cause of attack. Microbial decomposition of the hydrocarbons present in cable coatings may also produce organic acids corrosive to lead. Corrosion of lead by corroded steel-tape armor can occur when the oxide coating formed on the steel is cathodic to lead.

Corrosion in Soil

Soils vary widely in physical and chemical characteristics and, consequently, in corrosive effect. More than 200 varieties of soil in the U.S. have been classified according to texture, color, and natural drainage. The physical properties of soils that most influence the corrosion of lead in underground service are those that affect the permeability of the soil to air and water, because good drainage tends to minimize corrosion. Soils with coarse textures, such as sands and gravels, permit free circulation of air. Corrosion in such soils is approximately the same as in the atmosphere. Clays and silty soils generally exhibit fine texture and high water-holding capacity and therefore poor aeration and drainage.

Numerous chemical compounds are present in soils, but only those soluble in water play important roles in the corrosion of metals. For example, the calcareous nature of some Indiana soils influences corrosion through alkaline attack or the promotion of bacterial activity.

Considerable corrosion testing of lead and lead alloys in numerous soils has indicated that corrosion rate decreases with increasing particle size and that the distribution of anodic and cathodic areas depends on soil particle size, water activation value of the soil, soil pH, and duration of exposure. Test results also show that lead tends to become passive in soils regardless of water content; however, addition of sodium bicarbonate reactivates it.

The data in Table 4 show that in most soils the average corrosion rate of lead is low—from less than 2.5 to 10 μm (0.1 to 0.4 mil) per year. It should be noted, however, that the depth of pitting is often a more important measure of underground corrosion behavior than corrosion rate.

The most comprehensive investigation of corrosion of metals buried in soils was conducted by the National Bureau of Standards from 1910 to 1955 (Ref 15). This investigation included lead alloy pipe of three different compositions buried in 14 soils. Specimens were removed periodically; maximum exposure time was 11 years (Table 4).

Analysis of the data in Table 4 indicates that, in general, weight loss and maximum pit depth decrease with increasing aeration of the soil. For example, poor aeration caused severely deep pitting of the lead buried in Sharkey clay, in Lake Charles clay, and in cinders, but pitting of pipe buried in the well-aerated Cecil clay loam was shallow.

Table 14 (continued)

Fluids	Temperature °C	Temperature °F	Corrosion rate µm/yr	Corrosion rate mils/yr
Vapors from spin bath evaporator	49	120	127	5
Spinning bath drippings	46	115	203.2	8
Storage-reclaimed spinning bath liquor	...	...	50.8	2
Pickling solution				
Brass and copper-sulfuric acid + 5% cupric sulfate	71	160	127	5

Source: Ref 19

Resistance to Chemicals

The excellent resistance of lead and lead alloys to corrosion by a wide variety of chemicals is attributed to the polarization of local anodes caused by the formation of a relatively insoluble surface film of lead corrosion products (Ref 16). The extent of protection depends on the compactness, adherence, and solubility of these films.

Solubilities of various lead compounds in water at room temperature are given in Table 5. These data are general indicators of the behavior of lead in solutions that promote the formation of these compounds. The solubility of a lead corrosion product, however, depends on the solution in which the lead is immersed. Therefore, the solubility of that corrosion product in water is not always an adequate indicator of its behavior in another solution. This fact is illustrated by the variation in solubility of lead sulfate ($PbSO_4$) in H_2SO_4 as acid concentration and temperature change (Fig. 3). The $PbSO_4$ film is less soluble in H_2SO_4 solutions than it is in water. Solubility drops to a minimum value at acid concentrations of 30 to 60% and then increases at higher concentrations. At intermediate concentrations, the sulfate film is so insoluble that corrosion is negligible.

Another example of the importance of the solubility relationship of the lead film to its environment is shown in Fig. 4. Lead nitrate ($Pb(NO_3)_2$) is quite soluble in dilute and intermediate-strength solutions of nitric acid (HNO_3) at room temperature. Lead is not resistant to corrosion under such conditions. However, above an HNO_3 concentration of 50%, $Pb(NO_3)_2$ is only slightly soluble, and lead is quite resistant to attack.

Increases in temperature generally increase corrosion rate (Fig. 3). This effect is primarily due to increases in film solubility.

Galvanic Corrosion. When lead is anodic to a metal to which it is coupled and a firm film develops on the lead, galvanic corrosion of the lead will be negligible. For example, when lead is galvanically connected to a copper or a copper alloy in an H_2SO_4, H_2CrO_4, or H_3PO_4 solution, the lead is protected by a firm film even though it is the anode in the galvanic cell. However, when the other metal is the anode, galvanic corrosion of the lead may occur, and it is sometimes severe. Aluminum or magnesium will be severely corroded if coupled with lead in the presence of an electrolyte. If lead is coupled with Monel in a 6% H_2SO_4 solution, corrosion of the Monel will be accelerated.

A factor to be kept in mind is that environmental changes can reverse the galvanic positions of the two metals. For example, iron is anodic to lead in acids and cathodic to it in alkalies.

In general, galvanic corrosion of lead is significant only when the lead is coupled with a metal to which it is anodic, when an electrolyte is present, and when a firm film cannot be maintained; corrosion of metals anodic to lead seldom occurs. When galvanic corrosion of either lead or the dissimilar metal does occur, it is unlikely to be severe, because lead occupies a central position in the galvanic series.

Physical variables also influence the corrosion rate of lead in many situations. For example, if the flow velocity of a solution is above a critical point, it can completely erode the protective film, leaving only a clean lead surface exposed for continued attack. This is demonstrated by the rapid increase in corrosion of lead in 20% H_2SO_4 at velocities greater than 1.5 m/s (300 ft/min). The presence of foreign insoluble matter further aggravates this condition.

Fatigue stresses may also break the protective film on lead, repeatedly exposing lead to environmental attack. However, applied fatigue stresses no greater than the creep strength of lead do not significantly affect corrosion properties.

Quantitative Corrosion Data. It is important to remember when evaluating quantitative corrosion data that lead weighs more per unit of volume and is normally used in greater thicknesses than most other metals. The effects of these two factors should be considered in evaluating the data presented in this section.

Lead has high corrosion resistance to H_2CrO_4, H_3PO_4, H_2SO_4, and sulfurous (H_2SO_3) acids and is widely used in their manufacture and handling. Lead satisfactorily resists all but the most dilute solutions of H_2SO_4. It performs well at acid concentrations up to 95% at ambient temperatures, up to 85% at 220 °C (428 °F), and up to 93% at 150 °C (302 °F) (Fig. 5). Below a concentration of 5%, the corrosion rate increases, but it is still relatively low. In the lower range of concentration, antimonial lead is recommended.

Lead exhibits the same excellent corrosion resistance to higher concentrations of H_2CrO_4, H_2SO_3, and H_3PO_4 at elevated temperatures. Lead is also generally resistant to solutions of salts formed by each of these salts. However, the reaction to mixtures of these salts is much more complex.

Lead finds especially wide application in the manufacture of H_3PO_4 from phosphate rock when H_2SO_4 is used in the process. Corrosion rates are low for all acid concentrations up to 85% (Table 6). The corrosion rate of 6% antimonial lead has been reported to be lower than that of chemical lead in a plant test using a solution containing 32% H_3PO_4, 0.4% H_2SO_4, and 1% chlorides at 88 °C (190 °F). In pure acid manufactured from elemental phosphor, lead corrodes at a higher rate because of the absence of sulfates.

Lead has fair corrosion resistance to dilute hydrochloric acid (HCl) (up to 15%) at 24 °C (75 °F); the corrosion rate increases at higher concentrations and at higher temperatures (Table 7). The presence of 5% ferric chloride also accelerates corrosion (Table 8).

The resistance of lead to corrosion by hydrofluoric acid (HF) is only fair. However, lead is used to handle HF because it is the only low-priced metal that has adequate corrosion resistance. The corrosion rate in this acid (if it is free of air) is less than 510 µm/yr (20 mils/yr) for a wide range of temperatures and concentrations (Fig 6).

Nitric, acetic, and formic acids in most concentrations corrode lead at rates high enough to preclude its use in these acids. However, although HNO_3 rapidly attacks lead when dilute, it has little effect at strengths of 52 to 70%. The same is true of HF, acetic acid, and acid sodium sulfate.

Addition of H_2SO_4 to acids corrosive to lead often lowers the corrosion rate. For example, although HNO_3 in concentrations less than 50% is quite corrosive to lead (Table 9), in the presence of 54% H_2SO_4 the corrosion rate in 1% and 5% HNO_3 is quite low even at 118 °C (245 °F) (Table 10). Other concentrations of H_2SO_4 also lower the corrosion rate in HNO_3 (Table 11). The composition range of mixed H_2SO_4 and HNO_3 solutions for which chemical lead has a corrosion rate of less than 500 µm (20 mils) per year is shown in Fig. 7. Chemical lead is preferred over 6% antimonial lead for handling these mixtures of acids.

The corrosion rates of chemical lead and 6% antimonial lead in HCl and in fluosilicic acid are retarded by the presence of H_2SO_4 (Tables 12 and 13). Data on the corrosion of lead in chemical process fluids containing H_2SO_4 or closely related compounds are presented in Table 14.

Qualitative corrosion data serve to provide guidelines for screening suitable metals for chemical equipment. Laboratory test environments may not always simulate actual plant conditions, and there may be significant variations among plants manufacturing the same product. Therefore, it is often more helpful to be less specific when categorizing the corrosion rates of lead in various chemicals. Table 15 presents such less specific information and should be used only as a guide for determining whether further tests are warranted. Most of the data in Table 15 are for chemical lead. The corrosion rates of different grades of lead in the same chemical all normally fall within the same category. Therefore, no mention is made of variations in corrosion rate for other grades of lead.

Tin-Lead Solder Alloys

Solders in the tin-lead system are the most widely used of all joining materials. Industrial solder alloys are in use that contain a combination of materials from 100% Pb to 100% Sn, as demanded by the particular application. Each alloy has unique characteristics. In general, properties are influenced by the melting characteris-

Table 15 Corrosion rate of lead in chemical environments

For corrosion rate information on lead in other chemical environments, see the more extensive tables contained in Lead for Corrosion Resistant Applications—A Guide, Lead Industries Association, Inc.

Chemical	Temperature °C	Temperature °F	Concentration, %	Corrosion class(a)
Acetic acid	24	75	Glacial	B
Acetic anhydride	24	75	. . .	A
Acetone	24–100	75–212	10–90	A
Alcohol, ethyl	24–100	75–212	10–100	A
Alcohol, methyl	24–100	75–212	10–100	A
Aluminum chloride	24	75	0–10	B
Aluminum potassium sulfate	24–100	75–212	10–20	A
Aluminum potassium sulfate	24–100	75–212	20–100	B
Ammonia	24–100	75–212	10–30	B
Ammonium chloride	24	75	0–10	B
Ammonium hydroxide	27	80	3.5–40	A
Ammonium nitrate	24–49	75–120	10–30	D
Ammonium sulfate	24	75	. . .	B
Amyl acetate	24	75	80–100	B
Aniline	20	68	. . .	A
Antimony chloride	24	75	. . .	C
Arsenic trichloride	100–149	212–300	. . .	B
Barium chloride	24–100	75–212	10	B
Benzaldehyde	24	75	10–100	D
Benzene	24	75	. . .	B
Benzoic acid	24	75	. . .	D
Benzyl alcohol	24–100	75–212	. . .	B
Benzyl chloride	24–100	75–212	. . .	B
Beryllium chloride	100	212	. . .	D
Boric acid	24–149	75–300	10–100	B
Bromine	24	75	. . .	B
Butyric acid	24	75	10–100	D
Cadmium sulfate	24–100	75–212	10–30	A
Calcium bicarbonate	24	75	. . .	C
Calcium chloride	24	75	20	A
Calcium fluoride	24–100	75–212	. . .	B
Calcium nitrate	24	75	10	D
Calcium sulfate	24–100	75–212	10	B
Carbon disulfide	24–100	75–212	. . .	A
Carbonic acid	24	75	. . .	D
Cellulose acetate	24	75	. . .	A
Cellulose nitrate	24–100	75–212	. . .	B
Chloroacetic acid	24	75	. . .	B
Chloric acid	24	75	10	D
Chlorine	38	100	. . .	B
Chloroform	24–62	75–143	. . .	B
Chromic acid	24	75	. . .	B
Copper chloride	24	75	10–40	D
Creosote	24	75	90	D
Dichlorobenzene	24–100	75–212	10–100	B
Diethyl ether	24	75	. . .	B
Dioxane	24–100	75–212	. . .	B
Ethyl acetate	24–79	75–175	. . .	B
Ferric ammonium sulfate	24–100	75–212	10–20	A
Ferric chloride	24	75	20–30	D
Ferric sulfate	24–79	75–175	10–20	A
Ferrous chloride	24	75	10–30	C
Ferrous sulfate	24–100	75–212	10	B
Fluosilicic acid	45	113	10	D
Formaldehyde	24–52	75–125	20–100	B
Formic acid	24–100	75–212	10–100	D
Glycerol	24	75	. . .	B
Hydrazine	24	75	20–100	D
Hydriodic acid	24	75	10–50	D
Hydrobromic acid	24	75	10–70	D
Hydrochloric acid	24	75	0–10	C
Hydrogen peroxide	24	75	10–30	D
Isopropanol	24	75	. . .	A
Lead acetate	24	75	10–30	D
Lead chloride	24–100	75–212	. . .	B
Lithium hydroxide	24	75	. . .	D
Magnesium chloride	24	75	10–100	D
Magnesium sulfate	24–100	75–212	10–60	B
Mercury	24	75	100	D
Methyl ethyl ketone	24–100	75–212	10–100	B
Nitrobenzene	24–52	75–125	. . .	B

(continued)

(a) The four categories of the Table are: A, <50 μm/yr (<2 mils/yr): negligible corrosion—lead recommended for use. B, <500 μm/yr (<20 mils/yr): practically resistant—lead recommended for use. (When the only information available is that "lead is resistant" to a certain chemical, that chemical was arbitrarily placed in this category.) C, 500–1270 μm/yr (20–50 mils/yr): lead may be used where this effect on life can be tolerated. D, >1270 μm/yr (>50 mils/yr): corrosion rate too high to merit any consideration of lead.

tics of the alloys, which in some measure are related to their load-carrying and temperature capabilities. Table 16 gives the melting temperatures of some tin-lead compositions and their respective applications. More detailed information on the compositions and melting characteristics of tin-lead solders can be found in the article "Soldering" in Volume 6 of the 9th Edition of *Metals Handbook*. Corrosion-related data of these alloys are also presented in the article "Corrosion of Tin and Tin Alloys" in this Volume.

Applications. Solder alloys containing less than 5% Sn are used for joining tin-plated containers and for automobile radiator manufacture. For automobiles, a small additional amount of silver is usually added to provide extra joint strength at automobile radiator operating temperatures. Solder alloys of 10Sn-90Pb and 20Sn-80Pb are also used in radiator joints. With compositions between 10Sn-90Pb and 25Sn-75Pb, care must be taken to avoid any kind of movement during the solidification phase to prevent hot tearing in solders with a wide freezing range.

Higher tin content solders at the 25Sn-75Pb and 30Sn-70Pb compositions have lower liquidus temperatures and can be used for joining materials with sensitivity to high temperature or where the wetting characteristics of the tin are important to providing sound soldering joints. Solder alloys in the composition range described above are usually applicable to industrial products and generally are used in conjunction with inorganic fluxing materials.

The widely used general-purpose solder alloys contain 40 to 50% Sn. These solders are used for plumbing applications, electrical connections, and general soldering of domestic items.The 60Sn-40Pb and 63Sn-37Pb alloys are used most extensively in the electronic industries for both hand soldering and wave or dip applications. Sometimes silver additions are made to alloys used in the electronics industries to reduce dissolution of silver-based coatings. The corrosion performance of tin-lead solders for electronic applications is discussed in the article "Corrosion in the Electronics Industry" in this Volume.

REFERENCES

1. J.N. Friend, The Relative Corrodibilities of Ferrous and Nonferrous Metals and Alloys—Part 1—The Results of Four Years of Exposure in Bristol Channel, *J. Inst. Met.*, Vol 39, 1928, p 111-143
2. J.N. Friend, The Relative Corrodibilities of Ferrous and Nonferrous Metals and Alloys—Part III—Results of Three Years Exposure at Southampton Docks, *J. Inst. Met.*, Vol 48, 1932, p 109-120
3. B.W. Forgeson *et al.*, Corrosion of Metals in Tropical Environments, *Corrosion*, Vol 14, 1958, p 73t-81t
4. C.V. Brouilette, Corrosion Rates in Port Hueneme Harbor, *Corrosion*, Vol 14, 1958, p 352t-356t
5. G.O. Hiers and E.J. Minarcik, The Use of Lead and Tin Outdoors, in *Symposium on Atmospheric Corrosion of Nonferrous Metals*, STP 175, American Society for Testing and Materials, 1955, p 135-140
6. Report of Subcommittee VI of ASTM Committee B-3 on Atmospheric Corrosion Tests

Table 15 (continued)

Chemical	Temperature °C	Temperature °F	Concentration, %	Corrosion class(a)
Oxalic acid	24	75	20–100	D
Phenol	24	75	90	B
Phosphoric acid	24–93	75–200	...	B
Potassium chloride	8	47	0.25–8.0	B
Potassium hydroxide	24–60	75–140	0–50	B
Pyridine	24–100	75–212	10	B
Sodium acetate	25	77	4	B
Sodium bicarbonate	24	75	10	B
Sodium chloride	25	77	0.5–24	A
Sodium hydroxide	26	79	0–30	B
Sodium nitrate	24	75	10	D
Sodium sulfate	24	75	2–20	A
Stannous chloride	24	75	10–50	D
Zinc sulfate	35	95	...	B
Zinc chloride	79	175	25	B

(a) The four categories of the Table are: A, <50 μm/yr (<2 mils/yr): negligible corrosion—lead recommended for use. B, <500 μm/yr (<20 mils/yr): practically resistant—lead recommended for use. (When the only information available is that "lead is resistant" to a certain chemical, that chemical was arbitrarily placed in this category.) C, 500–1270 μm/yr (20–50 mils/yr): lead may be used where this effect on life can be tolerated. D, >1270 μm/yr (>50 mils/yr): corrosion rate too high to merit any consideration of lead.

Table 16 Melting characteristics and applications of tin-lead solders

Composition, % Tin	Composition, % Lead	Solidus °C	Solidus °F	Liquidus °C	Liquidus °F	Uses
2	98	270	518	312	594	Side seams for can manufacturing
5	95	270	518	312	594	Coating and joining metals
10	90	268	514	299	570	
15	85	226	440	288	550	
20	80	183	361	277	531	Coating and joining metals, or filling dents or seams in automobile bodies
25	75	183	361	266	511	Machine and torch soldering
30	70	183	361	255	491	
35	65	183	361	247	477	General purpose and wiping solder
40	60	183	361	238	460	Wiping solder for joining lead pipes and cable sheaths. For automobile radiator cores and heating units
45	55	183	361	227	441	Automobile radiator cores and roofing seams
50	50	183	361	216	421	General purpose. Most popular of all
60	40	183	361	190	374	Primarily used in electronic soldering applications in which low soldering temperatures are required
63	37	183	361	183	361	Lowest-melting (eutectic) solder for electronic applications

of Nonferrous Metals and Alloys, in *ASTM Proceedings*, Vol 44, American Society for Testing and Materials, 1944, p 224

7. Report of Subcommittee VI of ASTM Committee B-3 on Atmospheric Corrosion Tests of Nonferrous Metals and Alloys, in *ASTM Proceedings*, Vol 62, American Society for Testing and Materials, 1962, p 216
8. J.N. Friend, The Relative Corrodibilities of Ferrous and Nonferrous Alloys—Part II—The Results of Seven Years Exposure to Air at Birmingham, *J. Inst. Met.*, Vol 42, 1929, p 149-155
9. R.M. Burns, Corrosion of Metals II—Lead and Lead Alloy Cable Sheathing, *Bell Syst. Tech. J.*, Vol 15, 1936, p 603-625
10. K.G. Compton, Factors Involved in Corrosion of Lead Cable Sheath, *Corrosion*, Vol 17, 1961, p 409t-412t
11. NACE Task Group T-4B-1, Cell Corrosion on Lead Cable Sheaths, *Corrosion*, Vol 12, 1956, p 257t-259t
12. R.I. Perry, Preventing Corrosion of Lead-Sheathed Power Cables in Concrete Tunnels, *Corrosion*, Vol 12, 1956, p 207t-212t
13. Y. Yamaguchi *et al.*, Studies on Corrosion of Communication Cable Lead Sheath by Manhole Water, *Corros. Eng. (Jpn.)*, Vol 5, 1956, p 302-306
14. NACE Technical Unit Committee T-4B, Corrosion of Lead Sheath in Manhole Water, *Corrosion*, Vol 14, 1958, p 85t-87t
15. M. Romanoff, "Underground Corrosion," NBS 579, National Bureau of Standards, April 1957, p 227
16. E.L. Littauer and H.C. Wesson, Lead and Lead Alloys, in *Corrosion*, Vol 1, L.L. Shrier, Ed., John Wiley & Sons, 1963, p 4:68-4:85
17. R.C. Weast, Ed., *Handbook of Chemistry and Physics*, 45th ed., CRC Press, 1964
18. M.G. Fontana, *Ind. Eng. Chem.*, Vol 43, 1951, p 105A
19. G.A. Nelson, *Corrosion Data Survey*, National Association of Corrosion Engineers, 1967

Corrosion of the Noble Metals

Gaylord D. Smith, Inco Alloys International, Inc.
Edward Zysk, Englehard Corporation

THIS ARTICLE seeks to characterize the corrosion resistance of the eight noble metals: ruthenium, rhodium, palladium, silver, osmium, iridium, platinum, and gold. The elements are listed in order of their atomic number as found in periods 5 and 6 (Group VIII and Ib) of the periodic table. These metals are unique in their nobility and for the most part offer industry corrosion resistance that is unmatched in base metals and their alloys. The available information on the corrosion resistance of each element varies widely. Generally, more data are available for the more abundant, more easily fabricated elements. Silver and platinum have been evaluated in more environments than the other elements. Conversely, very little data are available for the intractable elements, osmium and ruthenium.

The information is presented on an element by element basis. For each element, the general fabricability of the metal is briefly defined. More detailed information on fabricability is available in Ref 1 to 3 and Volume 2 of the 9th Edition of *Metals Handbook*. This will aid in assessing the relevance of any particular noble metal to the immediate problem at hand. Atomic, structural, and physical properties can also play an important role in the use of any element or alloy in a given environment. Table 1 lists selected physical properties for all of the noble metals. Similarly, where data exist, typical properties and the effect of working are given for each element. Corrosion data are divided into several categories, including high-temperature oxidation. These data follow a general summary of the performance of each element. For comparative purposes, certain tables have common corrosive environments in each table for each element. This should aid the reader in finding the optimal element or alloy for any given type of common corrosive environment. Much of these data, as found in the literature, were originally presented in a myriad of different corrosion units. This information has been standardized in units of millimeters per year (mils per year). Similarly, virtually all the corrosion results are based on short-duration tests, and as is true of virtually all corrosion data, the information is intended to serve principally as a guide of expected results, because short-term results do not always reliably predict longer-term corrosion rates. Relatively little data are available on alloying effects. Most alloying, however, is done to vary mechanical and physical properties. Therefore, the corrosion resistance of the principal elements of most noble metal alloys is still relevant to the performance of many of their available commercial alloys.

Corrosion data for the noble metals have been summarized for the various elements in the past (Ref 1, 3, 4-10). Some of these data are in conflict and some are in question. A best effort has been made to present the likely corrosion rate. Data have been taken from many sources, and it is impossible to reference each data point. However, the references cited will frequently contain additional detail in specific incidences.

Because of the high cost of the noble metals, a corrosion rate of approximately 0.05 mm/yr (2 mils/yr) was selected as an arbitrary practical upper limit of acceptable corrosion. Negligible corrosion is defined as a corrosion rate of less than 0.25 mm/yr (10 mils/yr) but generally greater than 0.05 mm/yr (2 mils/yr). Most of the corrosion data are presented as falling in one of these classifications. For some applications, more rigorous limits and knowledge of the actual corrosion rate are required as a result of product purity or economics. For these applications, a more rigorous search of the literature or actual laboratory/field testing may be warranted.

Silver

Fabrication. Next to gold, silver is the most easily fabricated metal in the periodic table. It is very soft and ductile in the annealed condition. Silver is available in a large variety of product forms, including sheet, strip, foil, bar, wire, and shaped extrusions. Although silver work hardens, it is easily annealed in air above about 300 °C (570 °F). Silver is readily fusion welded—preferably by argon arc welding, because an oxy-hydrogen flame can embrittle silver through gas absorption. Silver can be soldered by a number of silver- or tin-base solders. It can also be hammer peened to form strong joints.

Silver is commonly clad to copper, nickel, and steel alloys by either brazing or solid-phase bonding. Silver-lined tubing can be fabricated by expanding silver tubing inside the base metal tubing.

Atomic, Structural, and Physical Properties. There are a number of interesting features about silver that are defined in Table 1. For example:

- Silver has the lowest density of the noble metal group of elements
- Silver has the lowest melting point of all the noble metals
- The thermal conductivity of silver is the highest of all the elements near room temperature
- The room-temperature electrical resistance of silver is the lowest of all the elements

Mechanical Properties. Annealed silver has relatively low yield and tensile strengths, but appreciable strength can be induced through cold working. Tensile strength can be increased from 124 to 186 MPa (18 to 27 ksi) for annealed material to approximately 310 MPa (45 ksi) following 50 to 80% cold work. Silver loses strength rapidly above room temperature and, accordingly, is used with caution if high-temperature service is contemplated. Silver cladding is being successfully used to extend the useful temperature range of silver and its alloys.

Corrosion Resistance. Extensive reviews of the corrosion resistance of silver are available in the literature (Ref 1, 4-6, 8). The principal objectives of this article are to draw a comparison between silver and the other noble metals with respect to corrosion resistance and to highlight some of the commercial corrosive environments in which silver can or cannot be used.

The corrosion resistance of silver depends on purity; much of the data presented here is for 999 fine silver (the fineness of silver is the silver content in parts per thousand). Silver can be used in hydrochloric acid (HCl), but results can be unsatisfactory under strongly aerating conditions when the concentration of acid and the temperature are increased (Table 2). The halide acids, with the exception of hydrofluoric acid (HF), passivate silver under favorable conditions by forming a stable protective film. Silver-lined vessels are used for the bromination of organic materials, but hydrobromic acid (HBr) exposure is limited to room temperature and a maximum of 14% acid. Similarly, silver is restricted to room-temperature exposure in dilute hydroiodic acid (HI). Nitric acid (HNO_3) that contains traces of nitrous acid attacks silver vigorously. Hot, concentrated (60%) sulfuric acid (H_2SO_4) also attacks silver rapidly, as does 95% acid at room temperature. Phosphoric acid (H_3PO_4) can be handled at all concentrations and temperatures of 160 to 200 °C (320 to 390 °F). The corrosion resistance of silver in acids is given in Table 3. The Pourbaix diagram for silver is given in Ref 11. Silver is not attacked by water or steam to 600 °C (1110 °F).

Silver is resistant to dry and moist chlorine and dry bromine, but is not resistant to iodine and fluorine unless it is cathodically protected (Table 4). Silver is resistant to many hydroxide, sulfate, carbonate, nitrate-bisulfate, and halide salt solutions, but is attacked by others (Table 5).

Table 6 lists corrosion rate data for silver in nearly 100 organic compound environments. Silver is satisfactory for use in nearly 90% of these environments. This is an impressive account, given the myriad of environments presented. Several of the organics for which silver is not recommended did not attack silver; rather, silver affected the organic compound, such as certain

Table 1 Selected properties of the noble metals

Property	Metal: Platinum	Palladium	Iridium	Rhodium	Osmium	Ruthenium	Gold	Silver
Atomic number	78	46	77	45	76	44	79	47
Atomic weight, amu	195.09	106.4	192.2	102.905	190.2	101.07	196.967	107.87
Crystal structure	fcc	fcc	fcc	fcc	hcp	hcp	fcc	fcc
Electronic configuration (ground state)	$5d^9 6s$	$4d^{10}$	$5d^7 6s^2$	$4d^8 5s$	$5d^6 6s^2$	$4d^7 5s$	$5d^{10} 6s$	$4d^9 5s^2$
Chemical valence	2,4	2,4	3,4	3	4,6,8	3,4,6,8	1,3	1,2,3
Density at 20 °C (70 °F), g/cm^3 (lb/in.^{3})	21.45	12.02	22.65	12.41	22.61	12.45	19.32	10.49
	(0.774)	(0.434)	(0.818)	(0.448)	(0.816)	(0.449)	(0.697)	(0.378)
Melting point, °C (°F)	1769	1554	2447	1963	3045	2310	1064.4	961.9
	(3216)	(2829)	(4437)	(3565)	(5513)	(4190)	(1948)	1763.4
Boiling point, °C (°F)	3800	2900	4500	3700	5020 ± 100	4080 ± 100	2808	2210
	(6870)	(5250)	(8130)	(6690)	(9070 ± 180)	(7375 ± 180)	(5086)	(4010)
Electrical resistivity, μΩ·cm at 0 °C (32 °F)	9.85	9.93	4.71	4.33	8.12	6.80	2.06	1.59
Linear coefficient of thermal expansion, μin./in./°C	9.1	11.1	6.8	8.3	6.1	9.1	14.16	19.68
Electromotive force versus Pt-67 electrode at 1000 °C (1830 °F), mV	...	−11.457	12.736	14.10	...	9.744	12.34(a)	10.70(b)
Tensile strength, MPa (ksi)								
As-worked wire	207–241	324–414	2070–2480(c)	1379–1586(c)	...	496	207–221	290
	(30–35)	(47–60)	(300–360)	(200–230)		(72)(c)	(30–32)	(42)
Annealed wire	124–165	145–228	1103–1241	827–896	...	...	124–138	125–186
	(18–24)	(21–33)	(160–180)	(120–130)			(18–20)	(18.2–27)
Elongation in 50 mm (2 in.), %								
As-worked wire	1–3	1.5–2.5	15–18(c)	2	...	3(c)	4	3–5
Annealed wire	30–40	29–34	20–22	30–35	...	...	39–45	43–50
Hardness, HV								
As-worked wire	90–95	105–110	600–700(c)	...	...	...	55–60	...
Annealed wire	37–42	37–44	200–240	120–140	300–670	200–350	25–27	25–30
As-cast	43	44	210–240	...	800	170–450	33–35	...
Young's modulus at 20 °C (70 °F), GPa (10^6 psi)								
Static	171	115	517	319	558	414	77	74
	(24.8)	(16.7)	(75)	(46.5)	(81)	(60)	(11.2)	(10.8)
Dynamic	169	121	527	378	...	476	...	...
	(24.5)	(17.6)	(76.5)	(54.8)		(69)		
Poisson's ratio	0.39	0.39	0.26	0.26	...	...	0.42	0.37(d)

(a) At 800 °C (1470 °F). (b) At 700 °C (1290 °F). (c) Hot worked. (d) Annealed. Source: Engelhard Industries Division, Engelhard Corporation

Table 2 Corrosion of silver in hydrochloric acid

Concentration of HCl, %	Corrosion rate — Limited aeration: 20 °C (70 °F) mm/yr	20 °C (70 °F) mils/yr	100 °C (212 °F) mm/yr	100 °C (212 °F) mils/yr	Strong aeration: 20 °C (70 °F) mm/yr	20 °C (70 °F) mils/yr	100 °C (212 °F) mm/yr	100 °C (212 °F) mils/yr
5	...	...	0.035	1.4	0.04	1.6	...	...
15	0.007	0.28	...	...	0.085	3.3	...	...
25	0.14	5.5	...	...	0.36	14.2	...	...
36	0.07	2.8	...	...	...	...	2.5	100

fats and the essential oils. This is also true for hydrogen peroxide and hydrazine. Silver is attacked by organic amines and by ammoniacal solutions of copper acetate. The inertness of silver in many organic environments makes it the material of choice, especially where high product purity is essential.

Silver is attacked by most fused bisulfates, cyanides, halides, phosphates, and peroxides, but is not attacked at temperatures below 500 °C (930 °F) by sodium hydroxide and at slightly lower temperatures by potassium, cesium, and rubidium hydroxides. All the low-melting molten metals attack silver, including mercury, sodium, potassium, lead, tin, bismuth, and indium. The corrosion resistance of silver in various gases is given in Table 7. The standard electrode potential for silver ($Ag \rightleftharpoons Ag^+ + e^-$) is +0.79 V and is exceeded only by gold and the platinum group metals.

Oxidation. Silver is resistant to dry and moist air at ordinary temperatures. At slightly elevated temperatures, silver may form a thin film of silver oxide. Above about 455 °C (850 °F), any silver oxide present will dissociate at atmospheric pressure, leaving the surface clean once again. However, molten silver dissolves appreciable quantities of oxygen, which can result in considerable internal porosity upon resolidification as the oxygen is rejected from the lattice. Oxygen diffuses more freely through solid silver than through any metal. This fact has made it feasible to internally oxidize certain alloying elements, such as magnesium, in order to dispersion strengthen silver in, for example, the Ag-10MgO alloy used for electrical contacts. The information on oxygen in silver is reviewed in Ref 12. Similarly, the causes and characteristics of sulfur-induced tarnish on silver and silver-rich alloys are reviewed in Ref 5. Surprisingly, the extensive research aimed at improving the tarnish resistance of silver by alloying has been largely unsuccessful.

Silver is blackened by ozone, with the maximum rate of attack at 220 to 250 °C (430 to 480 °F); the black tarnish disappears above about 455 °C (850 °F). Silver can be exposed to carbon monoxide to 300 °C (570 °F), to hydrogen to 700 °C (1290 °F), and to nitrogen to 500 °C (930 °F). At red heat, sulfur dioxide and sulfur trioxide rapidly attack silver. This attack becomes progressively worse with temperatures beginning at or near room temperature (Table 7).

The relatively poor strength of silver at elevated temperatures has minimized its use at high temperatures. However, cladding has extended the useful temperatures at which silver can be used.

Alloying of Silver for Corrosion Applications. Sterling silver, by definition, must contain a minimum of 92.5% Ag. Although the remainder is unrestricted, it is usually copper. Because copper is soluble in silver to the extent of about 4%, sterling silver is commonly a duplex alloy. The copper in solid solution has little effect on corrosion resistance, but the duplex microstructure may be a source of galvanic attack in strong electrolytes, such as seawater. However, this effect is rare in most environments. At slightly elevated temperatures, the copper is selectively oxidized. In this regard, copper does adversely affect the tarnish resistance of silver. However, copper additions to silver improve the strength of the metal, an attribute that is exploited in making cooling coils for beer and other foodstuff manufacturers. In general, base metal alloying does little to improve the corrosion resistance of silver. Most alloying of silver with base metals is

Table 3 Corrosion of silver in acids

Acid	Temperature, °C (°F)	Corrosion rate mm/yr	Corrosion rate mils/yr
Acetic, all concentrations	Boiling	<0.05	2
Acetylsalicylic, all concentrations	Boiling	<0.05	2
Aqua regia	Room	Potential dissolution(a)	
Arsenic	Room	Dissolution	
Ascorbic, all concentrations	Room	<0.05	2
Benzoic, all concentrations	130 (265)	<0.05	2
Boric, salt	Boiling	<0.05	2
Butyric	Boiling	<0.05	2
Carbonic, all concentrations	Room	<0.05	2
Chloric, all concentrations	Room	Attacked	
Chlorotoluene-sulfonic	Room	<0.05	2
Chromic, all concentrations	100 (212)	<0.05	2
Citric, to 30% concentration	Boiling	<0.05	2
Crotonic	Boiling	<0.05	2
Fatty acids	400 (750)	<0.05	2
Fluorosilicic	65 (150)	<0.05	2
Formic, pure	Boiling	<0.05	2
Gluconic, all concentrations	Boiling	<0.05	2
Glycerophosphoric, to 50%	Boiling	<0.05	2
Hydrogen selenide	Room	Attacked	
Hydrogen sulfide	Room	Attacked	
Hydrobromic, below 14%	Room	<0.05	2
Hydrochloric		—See Table 2—	
Hydrofluoric, below 50%	Boiling	<0.05	2
Hydroiodic, dilute	Room	<0.25	10
Hypochlorous	Room	Attacked	
Isovaleric, all concentrations	Boiling	<0.05	2
Lactic	Boiling	<0.05	2
Laevulinic, all concentrations	Boiling	<0.05	2
Monochloroacetic, all concentrations	Boiling	<0.05	2
Nitric	Room	Rapid dissolution	
Nitrous	Room	Dissolution	
Oxalic	Boiling	<0.05	2
Phenylacetic, all concentrations	Boiling	<0.05	2
Phosphoric, %			
5	102 (215)	0.003	0.12
45	60 (14)	nil	
45	110 (230)	0.007	0.28
67	60 (140)	0.004	0.16
67	125 (255)	0.02	0.8
85	60 (140)	0.002	0.08
85	140 (285)	0.048	1.9
85	160 (320)	0.306	12
Phthalic, pure	Boiling	<0.05	2
Picric, pure	125 (255)	<0.05	2
Propionic	Boiling	<0.05	2
Pyridine-carboxylic, pure	Room	<0.05	2
Salicylic, all concentrations	Boiling	<0.05	2
Stearic, pure	160 (320)	<0.05	2
Sulfuric, %			
10	Boiling	0.003	0.12
50	Boiling	0.034	1.3
60	Boiling	0.88	34.6
95	Room	0.14	5.5
Sulfurous, all concentrations	90 (195)	<0.05	2
Tartaric, all concentrations(b)	100 (212)	<0.05	2

(a) Attack will occur whenever silver chloride film is ruptured. (b) Oxygen increases attack in dilute tartaric acid at room temperature.

done to change mechanical properties and melting point.

Some electrical contacts have been made with a 60 to 70% Ag alloy containing up to 30% Au. A ternary Au-24Ag-6Pt alloy is extensively used in telephone equipment. Silver-gold binary alloys are used for jewelry and dental applications as well (see the article "Tarnish and Corrosion of Dental Alloys" in this Volume). Alloys of silver containing 3 to 10% Pd are also used for electrical contact materials. It should be noted that the alloy Pd-23Ag, because of its dimensional stability and high hydrogen permeability, has found acceptance as a diffusion membrane for the production of ultrapure hydrogen. Alloys in the silver-gold-palladium ternary field possess excellent high-temperature oxidation resistance, favorable brazing characteristics, and a wide range of fusion temperatures from the melting point of silver (961.9 °C, or 1763.4 °F) to that of palladium (1554 °C, or 2829 °F).

Corrosion Applications of Silver. The use of silver and silver-rich alloys in jewelry, coinage, and sterling ware is well known and documented. Also of commercial significance are the uses of silver in manufacturing and processing equipment within the food, chemical, and pharmaceutical industries. Silver has been used for decades within the food industry to maintain purity and freedom from metallic taste. Silver is resistant to many organic acids, salts, compounds, and foodstuffs and is used to manufacture, concentrate, and evaporate these products. Silver equipment has been used for handling syrups, fruit juices, beer, cider, white vinegar, jellies, gelatin, fatty acids, and essential oils, as well as autoclaves, mixing vessels, boiling pans, storage vats, and transfer equipment. Similarly, in the pharmaceutical industry, silver equipment is used to produce certain fine chemicals, hormones, and vitamins (see the article "Corrosion in the Pharmaceutical Industry" in this Volume).

The chemical industry frequently uses silver-lined process equipment to minimize corrosion. Silver is used in apparatus to synthesize urea from ammonia and carbon dioxide. Dye manufacturers use silver to handle aniline, ethyl chloride, boric acid, iodine, formaldehyde, magnesium chloride, magnesium sulfate, and zinc oxide. Silver-lined equipment is used to dehydrate glacial acetic acid and to process and store acetic anhydride in the rayon industry and phenol in the synthetic coatings industry. Highly refined acid products, such as phthalic acid made from hardwood distillation, can be processed in silver-lined stills. Silver vessels are often used to produce highly corrosive writing inks. Silver condensers are used to produce high-purity methyl acetone. High-purity sodium and potassium hydroxide are concentrated to the anhydrous melt in silver pans and commercial grades of these products are also concentrated in silver-lined continuous vacuum evaporators operating at over 315 °C (600 °F). Silver-tubed condensers are used for condensing 70% HF from the hot hydrogen fluoride vapors formed during the hydrofluorination of uranium. Silver metering equipment is employed to disperse wet chlorine gas in water purification installations. The chemical industry uses equipment made from silver to produce fluorophosphoric acid and salts. Rupture disks for emergency relief of pressure in chemical-processing equipment are frequently made of silver.

The use of silver to produce mirrors is based on the relative inertness of silver in air as well as its high and uniform reflectivity in the visible light range. Silver-containing electrical contacts and batteries derive their utility from the chemical nobility and stability of silver, although in relay contacts the formation of silver sulfide, whose resistance increases as the current decreases, imposes restraints on the use of silver and its alloys in this field.

Table 4 Corrosion of silver in halogens

Halogen	Temperature, °C (°F)	Corrosion rate, mm/yr (mils/yr)
Chlorine, dry(a)	100 (212)	Slight >0.05 (2)
Chlorine, moist or in solution	100 (212)	<0.05 (2)
Saturated chlorine in water	Room	<0.05 (2)
Bromine, dry(a)	Room	<0.05 (2)
Bromine, in methanol	Room	>0.05, <0.5 (2, 20)
Bromine, in glacial acetic acid	50 (120)	<0.05 (2)
Iodine, dry	Room	Dissolution
Iodine, moist or in solution	Room	Dissolution
Fluorine, dry(b)	Room	Very slight attack

(a) Halogen attack accelerates with increasing temperature, especially with moisture present. (b) Can be used as cathode to 250 °C (480 °F) for electrolysis of potassium fluoride plus HF to produce fluorine

Table 5 Corrosion of silver in salts and other environments

Environment	Temperature, °C (°F)	Corrosion rate mm/yr	mils/yr
Alum, all concentrations	Boiling	<0.05	2
Aluminum chloride, all concentrations(a)	Boiling	<0.05	2
Aluminum fluoride, all concentrations	Boiling	<0.05	2
Aluminum sulfate, all concentrations	Boiling	<0.05	2
Ammonium chloride, all concentrations	Boiling	<0.05	2
Ammonium hydroxide(b)	Room	<0.05	2
Ammonium nitrate, <20%	Room	<0.05	2
Ammonium phosphate, all concentrations	Boiling	<0.05	2
Ammonium sulfate, all concentrations	Boiling	<0.05	2
Ammonium thiocyanate, pure	100 (212)	<0.05	2
Antimony pentachloride, pure	90 (195)	<0.05	2
Barium chloride, all concentrations	Boiling	<0.05	2
Barium chloride, all concentrations	Room	<0.05	2
Barium peroxide, all concentrations	Room	Attacked	
Bismuth oxide, all concentrations	Room	Slight attack	
Calcium bisulfite, pure	Boiling	<0.05	2
Calcium carbonate, all concentrations	Room	Slight attack	
Calcium chloride, all concentrations	100 (212)	<0.05	2
Calcium hydroxide, all concentrations	100 (212)	<0.05	2
Calcium sulfate, all concentrations	100 (212)	<0.05	2
Calcium sulfide, all concentrations	Room	Blackens	
Cesium hydroxide, all concentrations	500 (930)	<0.05	2
Cupric chloride, all concentrations	100 (212)	Attacked	
Cupric nitrate, all concentrations	Room	<0.05	2
Cupric sulfate, all concentrations	Room-boiling	<0.05	2
Cupric sulfate in sodium chloride	100 (212)	Attacked	
Cuprous chloride, all concentrations	100 (212)	Attacked	
Cuprous nitrate, all concentrations	100 (212)	Attacked	
Cuprous sulfate, all concentrations	100 (212)	Attacked	
Dyes, acid chromium	Boiling	<0.05	2
Ferric alum, all concentrations	100 (212)	Attacked	
Ferric chloride, <5%	Room	<0.05	2
Ferrous sulfate, all concentrations(c)	Room	<0.05	2
Fluorosilicate, all concentrations	100 (212)	<0.05	2
Hydrogen peroxide, all concentrations	Room	Peroxide decomposed	
Hydrogen sulfide, all concentrations	Room	Blackened	
Lithium chloride, all concentrations	Boiling	<0.05	2
Magnesium chloride, all concentrations	120 (250)	<0.05	2
Magnesium chloride, melt	710 (1310)	Attacked	
Mercuric chloride, all concentrations	Room	Not recommended	
Nitrosyl chloride, dry	Room	<0.05	2
Phosphorus chlorides, pure	Boiling	<0.05	2
Potassium bisulfate, all concentrations	Boiling	<0.05	2
Potassium bromide, all concentrations	200–400 (390–750)	<0.05	2
Potassium carbonate, all concentrations	Boiling	<0.05	2
Potassium chlorate, all concentrations	Boiling	<0.05	2
Potassium cyanide, concentrated, in air	Room	Attacked	
Potassium dichromate, all concentrations	Boiling	<0.05	2
Potassium ferrocyanide, all concentrations	Room	Attacked	
Potassium hydroxide, all concentrations(b)	300 (570)	<0.05	2
Potassium hydroxide, melt(b)	350 (680)	<0.05	2
Potassium nitrate, all concentrations	Boiling	<0.05	2
Potassium nitrate, melt	335 (635)	Attacked	
Potassium perborate, all concentrations(d)	50 (120)	<0.05	2
Potassium permanganate, all concentrations	Boiling	Attacked	
Potassium peroxide, melt	100 (212) above melting point	Attacked	
Potassium persulfate, all concentrations	Room	Attacked	
Potassium sulfate, all concentrations	Boiling	<0.05	2
Sodium bisulfate, melt	400 (750)	Attacked	
Sodium bisulfites, all concentrations	100 (212)	<0.05	2
Sodium carbonate	Boiling	<0.05	2
Sodium chloride, all concentrations	Boiling	<0.05	2
Sodium chromate, all concentrations	Boiling	<0.05	2
Sodium cyanide, all concentrations	Room	Attacked	
Sodium fluorosilicate, pure	100 (212)	<0.05	2
Sodium hydroxide, <95%	Boiling	<0.05	2
Sodium hydroxide, melt(b)(e)	500 (930)	<0.05	2
Sodium hypochlorite, all concentrations	Room	<0.05	2
Sodium hypochlorite plus sodium chloride, saturated solution	Room	<0.05	2
Sodium nitrate, all concentrations	Boiling	<0.05	2
Sodium perborate, all concentrations	50 (120)	<0.05	2
Sodium perchlorate, all concentrations	Boiling	<0.05	2
Sodium perchlorate, melt	480 (900)	Attacked	
Sodium peroxide, melt	400 (750)	Attacked	
Sodium phosphates, all concentrations	Boiling	<0.05	2
Sodium silicates, all concentrations	Boiling	<0.05	2
Sodium sulfate, all concentrations	Boiling	<0.05	2
Sodium sulfide, all concentrations	Room	Slight attack	
Sodium thiosulfate, all concentrations	Room	<0.05	2
Stannic ammonium chloride, all concentrations	Boiling	<0.05	2
Stannic chloride, all concentrations	Boiling	<0.05	2
Sulfuryl chloride, dry and wet	300 (570)	<0.05	2
Thionyl chloride, dry or wet	Boiling	<0.05	2
Uranyl nitrate, all concentrations	Boiling	<0.05	2
Zinc chloride, all concentrations	Boiling	<0.05	2

(a) Provided oxidizing agents are not present. (b) Air must be excluded. (c) Attacked upon heating. (d) Causes deterioration of potassium perborate. (e) Mass transfer possible above 600 °C (1110 °F)

Gold

Fabrication. Gold is a highly malleable element that does not work harden significantly at room temperature. It is available for industrial use in all product forms, including thin strip and leaf, tubing, and fine wire. It is readily fabricated into finished-product forms and claddings. Gold is easily fusion or hammer welded. Soft soldering is not recommended.

The common unit of gold content in an alloy is the karat. This is the proportion of gold in alloy in twenty-fourths. Karat weight K is calculated on the following basis:

$$K = \frac{\text{wt\% Au} \times 24}{100}$$

In jewelry, gold-clad stock is designated by the ratio of weight of gold to the total weight of the material.

Some atomic, structural, and physical properties of gold are given in Table 1.

Mechanical Properties. Annealed gold is extremely soft (25 to 27 HV) at room temperature and is relatively low in tensile strength (125 to 138 MPa, or 18 to 20 ksi). Cold-worked wire cannot be strengthened much above 207 MPa (30 ksi). Because of the low strength of gold, the use of gold cladding on stronger base metals is common. Gold cannot be used as a structural material at elevated temperatures because of creep of the metal. Gold rupture disks usually are not used above 80 °C (175 °F).

Corrosion Resistance. Pure gold essentially owes its corrosion resistance to the low chemical affinity of the element. Passive film protection, such as occurs for silver in halide environments, is rare.

Table 8 gives the corrosion resistance of gold in acids. Gold is very resistant to H_2SO_4 to 250 °C (480 °F), and attack above this temperature may be primarily dependent on available oxygen. Gold is also resistant to concentrated HCl to its boiling point and to HNO_3 concentrations of up to 50% at the boiling point. However, hot mixtures of HNO_3 and H_2SO_4 will rapidly attack gold, as will aqua regia and hydrogen cyanide (with oxygen present). Mixtures of HCl, HBr, and HI with HNO_3 are extremely corrosive to gold. Mixtures of HF and HNO_3 are not corrosive to gold. Gold is resistant to most other acids.

Gold is generally attacked by the gaseous halogens, wet or dry. Only dry fluorine and wet or dry iodine can be handled by gold and then only within limitations (Table 9). Gold, second only to platinum, is resistant to dry hydrogen chloride to 870 °C (1600 °F), to oxygen to its melting point, to ozone to 100 °C (212 °F), to phosgene at room temperature, and to sulfur dioxide, dry and wet, to 600 °C (1110 °F) (Table 10).

Gold is not attacked by most inorganic salts in solution for their typical conditions of use (Table 11). Exceptions are the solutions of alkali cyanide and peroxide. Most alkali and alkaline melts as nitrates, hydroxides, and bisulfates will eventually attack gold, but only at elevated temperatures. The corrosion resistance of gold in over 40 organic compounds is given in Table 12. In all cases, gold was found to be resistant enough for commercial use.

Gold is attacked by all low-melting alloys, including mercury, sodium, potassium, lead, tin, bismuth, and iridium. The standard electrode potential for gold ($Au \rightleftharpoons Au^+ + e^-$) is +1.68 V.

Oxidation Resistance. Gold does not oxidize at any temperature up to its melting point, but may possess an absorbed layer of oxygen on the surface. Gold and its alloys greater than 14 karat are not susceptible to tarnish by hydrogen sulfide, sulfur, or other nonoxidized sulfur com-

Table 6 Corrosion resistance of silver in organic compounds

Environment	Temperature, °C (°F)	Corrosion rate mm/yr	Corrosion rate mils/yr
Acetaldehyde, pure	200–400 (390–750)	<0.05	2
Acetic anhydride, all concentrations	Boiling	<0.05	2
Acetone, pure	Boiling	<0.05	2
Acetylene dichloride, wet and acid	Boiling	<0.05	2
Ethyl alcohol, all concentrations	Boiling	<0.05	2
Amyl acetate, pure	Boiling	<0.05	2
Amyl alcohol, pure	Boiling	<0.05	2
Aniline, pure	Boiling	<0.05	2
Benzaldehyde, pure and aqueous	Boiling	<0.05	2
Benzene, pure	Boiling	<0.05	2
Benzotrifluoride, pure	Boiling	<0.05	2
Benzyl chloride, pure	180 (355)	<0.05	2
-bromoisovaleryl bromide, pure	100 (212)	<0.05	2
-bromoisovaleryl urea, pure	Melting point	<0.05	2
Butyl acetate, pure	Boiling	<0.05	2
Butyl alcohol, pure	Boiling	<0.05	2
Carbon tetrachloride, dry and wet	Boiling	<0.05	2
Chlorobenzene, pure	Boiling	<0.05	2
Chlorocresols, all concentrations	Boiling	<0.05	2
Chloroform, dry or wet	Boiling	<0.05	2
Chlorohydrins, pure	Boiling	<0.05	2
Chloronitrobenzenes, pure	Boiling	<0.05	2
Chlorotoluene, pure	Boiling	<0.05	2
Coniferyl alcohol, all concentrations	80 (175)	<0.05	2
Copals, pure and wet	400 (750)	<0.05	2
Copper acetate, neutral solutions	100 (212)	<0.05	2
Copper acetate, ammoniacal solutions	Room	Attacked	
Coumarin, pure	100 (212)	<0.05	2
Cresols, pure	Boiling	<0.05	2
Dextrose, all concentrations	Boiling	<0.05	2
Dialkyl sulfates, pure	Boiling	<0.05	2
Dibutyl phthalate, pure	Boiling	<0.05	2
Dimethylaniline, pure	Boiling	<0.05	2
Diphenyl, pure	400 (750)	<0.05	2
Essential oils, pure(a)	Boiling	<0.05	2
Ether, pure	Boiling	<0.05	2
Ethyl acetate, pure	Boiling	<0.05	2
Ethyl benzene, pure	136 (277)	<0.05	2
Ethylene dibromide, wet and acid products	Boiling	<0.05	2
Ethylene dichloride, wet and acid products	Boiling	<0.05	2
Fats, pure	300 (570)	<0.05	2
Fatty acids, pure	400 (750)	<0.05	2
Formaldehyde, all concentrations	Boiling	<0.05	2
Furfural, wet and slightly acid	Boiling	<0.05	2
Gelatin, pure	Boiling	<0.05	2
Glycerol, pure	Boiling	<0.05	2
Guanidine nitrate, all concentrations	Room	Not recommended	
Guinolines, pure	Boiling	<0.05	2
Guinone, inorganic solvent and pure	100 (212)	<0.05	2
Hexachloroethane, dry and moist	187 (369)	<0.05	2
Hexamethylene tetramine, all concentrations(b)	Room	<0.05	2
Hydrazine, pure	Room	Not recommended	
Hydroguinone, pure	Boiling	<0.05	2
Isoborneol acetate, pure	Boiling	<0.05	2
Isobutyl chloride, dry and wet	Boiling	<0.05	2
Limonene, pure	Boiling	<0.05	2
Methyl alcohol, pure	Boiling	<0.05	2
Methylamines, aqueous	Room	Attacked	
Methyl chloride, dry and wet	300 (570)	<0.05	2
Methylene chloride, dry and wet	Boiling	<0.05	2
Methylglycol, pure	Boiling	<0.05	2
Milk, pure(c)	Boiling	<0.05	2
Nitrobenzene, pure	Boiling	<0.05	2
Nitrocellulose, in water or alcohol	Room	<0.05	2
Nitrophenols, pure	Boiling	<0.05	2
Nitrotoluenes, pure	Boiling	<0.05	2
Pentachloroethane, wet, dry, and acid	Boiling	<0.05	2
Phenol, all concentrations	Boiling	<0.05	2
Phthalic anhydride, pure	Boiling	<0.05	2
Potassium acetate, all concentrations	Boiling	<0.05	2
Quinine sulfate, all concentrations	70 (160)	<0.05	2
Sodium acetate, all concentrations	Boiling	<0.05	2
Sodium acetate, melt	400 (750)	<0.05	2
Sodium bisulfate, all concentrations	Boiling	<0.05	2
Sodium formate, all concentrations	Boiling	<0.05	2
Sodium isovalerate, all concentrations	Boiling	<0.05	2
Sodium isovalerate, melt with sodium hydroxide	290 (555)	<0.05	2
Sodium methylate, all concentrations in alcohol or ether	100 (212)	<0.05	2
Sodium pentachlorophenolate, all concentrations	Boiling	<0.05	2
Sodium phenolate, all concentrations	Boiling	<0.05	2
Sodium salicylate, all concentrations	Boiling	<0.05	2
Sodium tartrates, all concentrations	Boiling	<0.05	2
Sorbital, all concentrations	Boiling	<0.05	2
Sorbose, all concentrations	Boiling	<0.05	2
Toluene, pure	Boiling	<0.05	2
Toluenesulfonyl chlorides, pure	Boiling	<0.05	2
Triethanolamine, mixture with diethylene glycol	Room	<0.05	2
Vinyl chloride, pure	200 (390)	<0.05	2

(a) Silver may taint the flavor of fats. (b) Solutions must be free of air and ammonia. (c) Silver may impart metallic taste

pounds. However, because gold does not support its own mass at elevated temperatures, high-temperature applications are minimal.

Corrosion Applications. Gold is primarily used for its decorative appearance in jewelry, coinage, dentistry, and gold leaf. Because of its softness and lack of resistance to halogens, its use in chemical applications, even as gold-lined apparatus, is somewhat limited. Gold is resistant to nonoxidizing H_3PO_4 and phosphates; therefore, it is used for lining autoclaves handling phosphate mixtures to 500 °C (930 °F). In the production of zirconium by the iodide process, gold closure gaskets are used to handle dry iodine vapors at 500 °C (930 °F). The use of gold-lined equipment to perform hydrochlorinations and hydrofluorinations of organic compounds in the chemical industry has been established. Laboratory ware fabricated from an Au-10Pt alloy is frequently used in place of platinum.

The stability of gold, coupled with its ability to reflect infrared radiation, has made it a popular glass window coating for energy conservation in tall buildings. Gold-base brazing alloys are used in critical aerospace and chemical equipment applications. Electroplated gold is used for its stability in electrical applications, such as connectors, relay switches, and contacts. Gold-silver-copper alloys are used for pen nibs and for slip rings and brushes in electrical instruments. Gold-palladium-iron alloys are used for potentiometer wire. A gold-palladium-iron alloy having a high resistivity and a negative temperature coefficient of resistance has been used as a resistor and potentiometer wire.

Platinum

Fabrication. Platinum is a soft, ductile, white metal, and like other face-centered cubic (fcc) metals, it can be readily hot or cold worked. Hot working is typically begun at 1000 °C (1830 °F), with reductions per pass as great as 50% being possible. Cold work is usually performed with reductions of 10% per pass and intermediate annealing after 75% reduction in thickness. Wire processing is similarly performed. Platinum can be processed bare into leaf and wire as small as 2.5 μm (0.1 mil). Pure platinum can be annealed in a short time at 600 to 700 °C (1110 to 1290 °F). Alloys require higher temperatures. Platinum can be annealed in air, but other atmospheres, such as nitrogen, argon, and helium, can also be used without damage. Platinum is readily fusion or resistance welded. Platinum can be electrodeposited on base metals.

The principal atomic, structural, and physical properties of platinum are listed in Table 1.

Mechanical Properties. Platinum work hardens at about the same rate as copper or silver; the mechanical properties obtained are strongly influenced by the purity of the platinum. The hardness and tensile strength are in the range of 37 to 42 HV and 124 to 165 MPa (18 to 24 ksi) for annealed material. After a 50% cold reduction, these values increase to 90 to 95 HV and 207 to 241 MPa (30 to 35 ksi). Elongation decreases from 30 to 40% to about 3%. The tensile strength of annealed platinum wire versus temperature is given in Table 13. The 100-h stress rupture life of annealed platinum at 925 °C (1700 °F) is with a

Table 7 Corrosion of silver in gases

Gas	Temperature, °C (°F)	Corrosion rate mm/yr	mils/yr
Acetylene, dry	Room	Risk of explosion	
Ammonia, pure	190 (375)	<0.05	2
Ammonium chloride, vapor	200 (390)	Attacked	
Carbon dioxide, pure	Room	<0.05	2
Carbon monoxide, pure	300 (570)	<0.05	2
Hydrogen, pure	700 (1290)	<0.05	2
Hydrogen chloride, dry(a)	430 (805)	<0.05	2
Nitric oxide, pure	Room	Attacked	
Nitric tetroxide, pure	Room	<0.05	2
Nitrogen, pure	500 (930)	<0.05	2
Oxygen, pure(b)	100 (212)	<0.05	2
Ozone, with 98% oxygen	Room	<0.05	2
Steam, pure(c)	600 (1110)	<0.05	2
Sulfur dioxide, pure	Red heat	Attacked	
Sulfur trioxide, pure	Red heat	<0.05	2

(a) Silver is protected by a layer of silver chloride that forms rapidly on the surface. (b) Attack becomes appreciable at 200 °C (390 °F). (c) Without pressure

Table 8 Corrosion of gold in acids

Acid	Temperature, °C (°F)	Corrosion rate mm/yr	mils/yr
Acetic, glacial	100 (212)	<0.05	2
Aqua regia	Room	Rapid dissolution	
Arsenic, all concentrations	Room	<0.05	2
Chlorosulfonic, all concentrations	Boiling	<0.05	2
Chlorotoluene-sulfonic, all concentrations	Boiling	<0.05	2
Citric, 20%	Boiling	<0.05	2
Citric, 30%	Boiling	<0.05	2
Crotonic, all concentrations	Boiling	<0.05	2
Fatty acids, pure	Boiling	<0.05	2
Glycerophosphoric, 1 to 50%	Boiling	<0.05	2
Hydrobromic, specific gravity 1.7	Room	<0.05	2
Hydrochloric, 36%	Room-100 (212)	<0.05	2
Hydrofluoric, 40%	Room	<0.05	2
Hydrogen sulfide, moist	Room	<0.05	2
Hydroiodic, specific gravity 1.75	Room	<0.05	2
Isovaleric, all concentrations	Boiling	<0.05	2
Lactic, all concentrations	Boiling	<0.05	2
Laevulinic, all concentrations	Boiling	<0.05	2
Nitric, %			
1–50	Boiling	<0.05	2
70	Room	>0.05	2
70	Boiling	0.15	6
Oxalic, all concentrations	Boiling	<0.05	2
Phenol-2,4-disulfonic, all concentrations	100 (212)	<0.05	2
Phthalic, pure	Boiling	<0.05	2
Picric, pure	125 (255)	<0.05	2
Propionic, all concentrations	Boiling	<0.05	2
Pyridine, all concentrations	Boiling	<0.05	2
Pyridine-carboxylic, all concentrations	150 (300)	<0.05	2
Salicylic, all concentrations	Boiling	<0.05	2
Stearic, pure	Boiling	<0.05	2
Sulfuric, all concentrations	250 (480)	<0.05	2
Sulfurous, all concentrations	100 (212)	<0.05	2
Tartaric, all concentrations	Boiling	<0.05	2

stress of 14 MPa (2000 psi) and at 1300 °C (2370 °F) is with a stress of 1.7 MPa (245 psi).

Corrosion Resistance. The exceptional corrosion resistance of platinum, one of the more familiar of its characteristics, is widely acknowledged and is extensively covered in the literature (Ref 1, 7, 9, 10). Platinum is one of the few metals that is unaffected by atmospheric exposure, even in sulfur-bearing industrial atmospheres.

Platinum is resistant to corrosion by single acids (Table 14), alkalies, aqueous solutions of common salts (Table 15), and organic materials (Table 16). The potential-pH diagram for platinum as defined by Pourbaix shows that platinum at 25 °C (75 °F) is immune to attack at all but the lowest pH levels and high redox potentials (Ref 11). Even at elevated temperatures, platinum is resistant to dry hydrogen chloride and sulfurous gases (Table 17). Platinum is resistant to most halogen gases at room temperature, with dry and moist bromine being the exception (Table 18). Platinum is also essentially inert to many molten salts, and it resists the action of fused glasses if oxidizing conditions are maintained.

Aqua regia and mixtures of HCl and oxidizing agents will attack platinum, as will free halogens and selenic acid to some degree at elevated temperatures. A number of low-melting metals, including lead, tin, antimony, zinc, and arsenic, will readily alloy with and attack platinum at their melting temperatures. Low-melting phases are formed with silicon, phosphorus, bismuth, and boron, and salts or compounds of these metals can be detrimental at high temperatures under reducing conditions.

As an anode, platinum will resist attack in a wide variety of alkaline, neutral, and mild acid solutions. There is attack in strong HCl. If an alternating current is applied to platinum electrodes, attack will occur in cyanide and in some acid solutions. The standard electrode potential ($Pt \rightleftharpoons Pt^+ + 2e^-$) is approximately +1.2 V at 25 °C (75 °F).

Effect of Alloying on Corrosion Resistance. Alloys containing up to 25% Pd have essentially the same corrosion resistance as platinum and are not discolored by heating in air. The addition of palladium up to 25% raises the annealed hardness and tensile strength of platinum.

The corrosion resistance of the entire binary series of rhodium-platinum alloys is excellent, with corrosion resistance tending to improve with higher rhodium contents. For example, a 10% addition of rhodium to platinum reduces the corrosion rate in 36% HCl at 100 °C (212 °F) from 0.2 mm/yr (50 mils/yr) to 0 and the attack of 100 g/L ferric chloride ($FeCl_3$) at 100 °C (212 °F) from 16.7 to 0.2 mm/yr (660 to 50 mils/yr). Alloys containing less than about 20% Rh can be hot or cold worked, while those containing between 20 to 40% Rh must be hot worked prior to cold working. The practical limit for workability is about 40% Rh.

Iridium and ruthenium additions to platinum result in corrosion resistance similar to that obtained through rhodium additions. However, the ranges for working are slightly more restrictive for iridium and much more limiting for ruthenium. All the alloys of the gold-platinum binary system remain quite corrosion resistant. Alloys containing more than 60% Ag are rapidly attacked by HNO_3 and $FeCl_3$ and are tarnished by exposures to industrial atmospheres.

Nickel additions rapidly harden platinum and gradually reduce the nobility of platinum. Up to 50% Cu can be added to platinum while still retaining its resistance to HNO_3.

Platinum can be used as a minor alloying element in the base metals titanium and chromium. Platinum additions as small as 0.1% greatly increase the resistance of these base metals to HCl and H_2SO_4.

Oxidation Resistance. Platinum is outstanding in its resistance to oxidation, remaining untarnished upon heating in air at all temperatures and retaining its metallic luster up to the melting point. At temperatures above about 750 °C (1380 °F), an extremely small but measurable weight loss occurs because of the formation of a volatile oxide, probably PtO_2, and because of volatilization of the metal. The loss due to oxide

Table 9 Corrosion of gold in halogens

Halogen	Temperature, °C (°F)	Corrosion rate mm/yr	mils/yr
Bromine, all concentrations, dry and wet	Room	Attacked	
Chlorine, dry	Room	0.3	12
	120 (250)	0.7	27.6
	150 (300)	1.5	60
	175 (345)	3.0	120
	205 (400)	30.5	1200
Chlorine, wet	Room	Attacked	
Fluorine, dry	<500 (930)	<0.05	2
Iodine, dry and wet	Room	<0.05	2
Iodine, dry and wet	>50 (120)	Attacked	
Iodine, 5% in alcohol	Room	Attacked	

Table 10 Corrosion of gold in various gases

Gas	Temperature °C	°F	Corrosion rate, mm/yr	mils/yr
Hydrogen chloride, dry	870	1600	<0.05	2
Ozone, with 98% oxygen	100	212	<0.05	2
Phosgene	Room		<0.05	2
Steam	800	1470	<0.05	2
Sulfur dioxide, dry and wet	600	1110	<0.05	2

Table 11 Corrosion of gold in salts

Salt	Temperature, °C (°F)	Corrosion rate mm/yr	Corrosion rate mils/yr
Aluminum sulfate, 10%	100 (212)	<0.05	2
Ferric chloride in HCl solutions	Room	<0.25	10
Magnesium chloride, all concentrations	Boiling	<0.05	2
Mercuric chloride, 10%	100 (212)	50.0	2000
Nitrosyl chloride, dry	Room	<0.05	2
Potassium bisulfate, all concentrations	Boiling	<0.05	2
Potassium bromide, all concentrations	Boiling	<0.05	2
Potassium carbonate, all concentrations	Boiling	<0.05	2
Potassium chlorate, all concentrations	Boiling	<0.05	2
Potassium dichromate, all concentrations	Boiling	<0.05	2
Potassium hydroxide, all concentrations	300 (570)	<0.05	2
Potassium hydroxide, melt	360 (680)	<0.05	2
Potassium iodide, with iodine	Room	Attacked	
Potassium nitrate, all concentrations	Boiling	<0.05	2
Potassium permanganate, all concentrations	Boiling	<0.05	2
Potassium peroxide, melt	380 (715)	Attacked	
Potassium sulfate, all concentrations	Boiling	<0.05	2
Sodium bisulfate, all concentrations	Boiling	<0.05	2
Sodium bisulfate, melt	400 (750)	<0.05	2
Sodium bisulfites, all concentrations	100 (212)	<0.05	2
Sodium carbonate, all concentrations	Boiling	<0.05	2
Sodium chloride, all concentrations	Boiling	<0.05	2
Sodium chromate, all concentrations	Boiling	<0.05	2
Sodium cyanide, all concentrations	Room	Attacked	
Sodium hydroxide, <90%	Boiling	<0.05	2
Sodium nitrate, all concentrations	Boiling	<0.05	2
Sodium perborate, all concentrations	50 (120)	<0.05	2
Sodium phosphates, all concentrations	Boiling	<0.05	2
Sodium silicates, all concentrations	Boiling	<0.05	2
Sodium sulfate, all concentrations	Boiling	<0.05	2
Sodium sulfide, all concentrations	Boiling	<0.05	2
Sodium sulfite, all concentrations	Boiling	<0.05	2
Stannic ammonium chloride, all concentrations	Boiling	<0.05	2
Stannic chloride, all concentrations	Boiling		
Strontium nitrate, all concentrations	Boiling	<0.05	2
Sulfur monochloride, pure	Boiling	<0.05	2
Sulfuryl chloride, dry and wet	300 (570)	<0.05	2
Thionyl chloride, dry or wet	Boiling	<0.05	2
Uranyl nitrate, all concentrations	Boiling	<0.05	2
Zinc sulfate, 10%	100 (212)	<0.05	2

formation is the greater of the two effects as higher temperatures are reached, and it is influenced by such factors as oxygen pressure, atmosphere flow rate, degree of saturation of the atmosphere with the oxide, and the geometry of the system (Ref 13, 14). A comparison of the oxidation resistance of platinum with that of a number of other high-melting metals is given in Fig. 1.

Alloys with more than 5% Rh or Ir are slowly oxidized in air at temperatures between 750 °C (1380 °F) for rhodium, 900 °C (1650 °F) for iridium, and 1150 °C (2100 °F) with the formation of a superficial blue-black film. At higher temperatures, the film decomposes, and the alloy becomes bright again. Above about 800 to 900 °C (1470 to 1650 °F), both alloy systems lose mass because of the volatilization of oxides formed on the surface and the preferential volatilization of the alloying element. Iridium-containing alloys have greater volatilization rates at a given temperature than rhodium-containing platinum alloys. When heated in air, ruthenium is selectively oxidized from ruthenium-platinum alloys, although less vigorously than osmium.

Corrosion Applications. Pure platinum, as well as platinum containing small amounts of rhodium, gold (5%), or iridium, is used for crucibles and other laboratory ware. The first major use of platinum was in laboratory ware and, to a certain extent, in chemical equipment, particularly for stills and condensers in the concentration of H_2SO_4. Today, laboratory apparatus and ware made from platinum and other noble metals are extensively used in analytical control work (fusions, ignitions, ashing) and research demanding corrosion resistance at room and elevated temperatures. Platinum crucibles are used for the production of large synthetic crystals. The chemical industry uses platinum for constructing heating and cooling coils, evaporation apparatus, stills, and autoclaves. Corrosion-resistant platinum rupture disks are used for the protection of pressure vessels handling corrosive materials. Platinum electrodes are used for the electrolytic production of hydrogen peroxide and per-salts, such as perchlorates, because they have a high oxygen overvoltage and do not corrode or effect the purity of the product. In addition, platinum electrodes can be used for cathodic protection in seawater and in chlorine production.

Platinum and, preferably, Pt-5Rh or Pt-10Rh are used in woven screen form to catalytically oxidize ammonia to HNO_3. Platinum, platinum-rhodium, and other platinum-base alloys are used in large quantities by the glass industry for molds, nozzles, and spinnerets.

The platinum resistance thermometer and the 90%Pt-Pt thermocouple are used to define the International Practical Temperature Scale between 13.81 K and 630.74 °C and between 630.74 °C and 1064.43 °C (freezing point of gold), respectively. In commercial practice, the rhodium platinum-platinum thermocouple is regularly used to 1400 °C (2370 °F) and under special conditions to as high as 1600 °C (2910 °F).

Because of its tarnish resistance, platinum is used in the jewelry, dental, medical implant, and electrical contact fields.

Palladium

Fabrication. In many respects, the fabricability of palladium resembles that of platinum and gold, which are closely associated with palladium in the periodic table. Like other fcc metals, palladium is ductile and can be readily hot or cold worked. Palladium can withstand drastic working and forming operations and, like gold, can be processed into leaf as thin as 0.1 μm (4 μin.). Many wrought and fabricated forms of palladium and its commercial alloys are readily available, including sheet and strip, bar and wire, and shaped extrusions. Palladium and some alloys are easily electroplated.

Physical Properties. Palladium, with a density of 12.02 g/cm^3 (0.434 lb/in.^{3}), is the lightest of the platinum-group metals. It also has the lowest melting point of the platinum metal group (1554 °C, or 2829 °F). The principal physical properties of palladium are listed in Table 1.

Mechanical Properties. Palladium work hardens at about the same rate as platinum; the annealed hardness increases from 40 HV to about 105 HV after a 50% reduction in thickness. Similarly, the tensile strength increases from 207 MPa (30 ksi) annealed to about 379 MPa (55 ksi) after a 50% reduction in thickness to 448 MPa (65 ksi) following a 75% reduction. The tensile strength of annealed palladium wire decreases progressively with increasing temperature, as shown in Table 19. The 100-h stress rupture life of palladium at 925 °C (1700 °F) is with a stress of 4.1 MPa (595 psi) and at 1300 °C (2370 °F) is with a stress of 1.7 MPa (247 psi) (Ref 15, 16).

Corrosion Resistance. Palladium is generally resistant to corrosion by most single acids, alkalies, and aqueous solutions of many common salts (Tables 20 and 21). It is not attacked at room temperature by H_2SO_4, HCl, HF, acetic, or oxalic acids, but may be attacked at 100 °C (212 °F) or when air is present. Nitric and hot H_2SO_4 attack palladium, as do $FeCl_3$ and hypochlorite solutions, chlorine, bromine, and, to a negligible extent, iodine. Table 22 lists the resistance of palladium to halogens at room temperature. The Pourbaix diagram for palladium, although not complete, can be found in Ref 11. Palladium is readily corroded anodically in HCl or acid chloride solutions. As an electrode for cathodic protection use in seawater, palladium corrodes at a rate of 8.6 g/A · yr at a current density of 540 A/m^2 (50 A/ft^2). The current efficiency is 0.05%. In binary palladium-platinum alloys, the corrosion rate in this application becomes equal to that of platinum if the palladium content is less than 20%. Cyanide solutions containing an oxidizing agent are useful for metallographic etching.

At high temperatures, molten salts such as sodium peroxide, hydroxide, and carbonate attack palladium, but the molten nitrate does not. Hydrogen sulfide at temperatures above 600 °C (1110 °F) attacks the metal and produces a low-melting phase. The standard electrode potential ($Pd \rightleftharpoons Pd + 2e^-$) is approximately +0.83 V at 25 °C (75 °F).

Effect of Alloying Elements on Corrosion Resistance. The additions of 2% Pt to palladium makes the alloy resistant to the jewelers' HNO_3 drop test used to determine equivalency with gold alloys, and the addition of 10% Pt to palla-

Table 12 Corrosion of gold in organic compounds

Compound	Temperature, °C (°F)	Corrosion rate mm/yr	Corrosion rate mils/yr
Acetylene dichloride, wet and acid	Boiling	<0.05	2
Aniline, pure	Boiling	<0.05	2
C-bromoisovalerty bromide, pure	100 (212)	<0.05	2
Chloronitrobenzenes, pure	Boiling	<0.05	2
Copper acetate, neutral solutions	100 (212)	<0.05	2
Cresols, pure	Boiling	<0.05	2
Dextrose, all concentrations	Boiling	<0.05	2
Dimethylaniline, pure	Boiling	<0.05	2
Diphenyl, pure	400 (750)	<0.05	2
Essential oils, pure	Boiling	<0.05	2
Ether, pure	Boiling	<0.05	2
Ethylene dibromide, wet and acid products	Boiling	<0.05	2
Ethylene dichloride, wet and acid products	Boiling	<0.05	2
Glycerol, pure	Boiling	<0.05	2
Guinine sulfate	Boiling	<0.05	2
Guinolines, pure	Boiling	<0.05	2
Hydroguinone, all concentrations	Boiling		
Isoborneol acetate, pure	Boiling	<0.05	2
Isobutyl chloride, dry and wet	Boiling	<0.05	2
Limonene, pure	Boiling	<0.05	2
Methyl alcohol, pure	Boiling	<0.05	2
Methyl chloride, dry and wet	300 (570)	<0.05	2
Methylglycol, pure	Boiling	<0.05	2
Milk, pure	Boiling	<0.05	2
Nitrobenzene, pure	Boiling	<0.05	2
Nitrotoluenes, pure	Boiling	<0.05	2
Pentachloroethane, wet, dry and acid	Boiling	<0.05	2
Phenol, all concentrations	Boiling	<0.05	2
Phenylhydrazine, all concentrations	100 (212)	<0.05	2
Phthalic anhydride, pure	Boiling	<0.05	2
Potassium acetate, all concentrations	Boiling	<0.05	2
Pyridine, all concentrations	Boiling	<0.05	2
Sodium acetate, all concentrations	Boiling	<0.05	2
Sodium bisulfate, all concentrations	Boiling	<0.05	2
Sodium formaldehyde sulphoxylate, all concentrations	90 (195)	<0.05	2
Sodium formate, all concentrations	Boiling	<0.05	2
Sodium formate, melt	260 (500)	<0.05	2
Sodium isovalerate, all concentrations	Boiling	<0.05	2
Sodium isovalerate, melt with sodium hydroxide	290 (555)	<0.05	2
Sodium phenolate, all concentrations	Boiling	<0.05	2
Sorbital, all concentrations	Boiling	<0.05	2
Sorbose, all concentrations	Boiling	<0.05	2
Toluene, pure	Boiling	<0.05	2
Triethanolamine, pure and all concentrations	Boiling	<0.05	2
Vinyl chloride, pure	500 (930)	<0.05	2

dium makes it completely resistant to HNO_3. In $FeCl_3$ solution (100 g/L), 10% Pt decreases the room-temperature corrosion rate of palladium from 11.9 to 8.6 mm/yr (469 to 339 mils/yr). A 30% Pt addition further decreases the corrosion rate to 1.8 mm/yr (71 mils/yr).

Both iridium and rhodium are quite effective in improving the corrosion and tarnish resistance of palladium. Palladium alloys with 2% Ir or Rh are resistant to the HNO_3 drop test, and alloys with 10% of either element are untarnished by industrial sulfur-bearing atmospheres. The addition of up to 10% Ru only slightly improves the corrosion resistance of palladium. Alloys containing more than 10% Au are resistant to tarnish by industrial sulfur-bearing environments, and those with more than 20% Au are resistant to HNO_3 and HCl.

The corrosion resistance of binary palladium-nickel or cobalt alloys is intermediate between that of the component metals and can be raised to levels above those of gold alloys by the addition of platinum, rhodium, or iridium in quantities of from 5 to 20%. Small additions of palladium (0.15 to 0.20%) to titanium may have been found to be effective in improving its corrosion resistance. In such corrosive media as boiling HCl, an almost hundredfold increase in corrosion resistance is obtained. Similar effects in chromium have also been observed.

Oxidation Resistance. Palladium is not tarnished by dry or moist air at ordinary temperatures. At temperatures in the range of 400 to 790 °C (750 to 1450 °F), a thin oxide film forms in air. At higher temperatures, the superficial oxide decomposes to give off oxygen, leaving a clean metal surface.

At temperatures above about 1000 °C (1830 °F), the behavior of palladium in air or oxygen is complicated by the interplay of two phenomena: the solution of oxygen in palladium (which increases the weight) and the loss of metal by volatilization (which decreases the weight). The mass change data for palladium oxidized in air over the temperature range 700 to 1400 °C (1290 to 2550 °F) is given in Fig. 1. The oxidation of palladium (thin oxide film formation) can be eliminated by alloying with 75% Pt (Ref 17, 18). Because palladium lacks an oxide film above 1000 °C (1832 °F), the volatilization losses will depend on rate of flow of the gas stream over the surface of the metal (Ref 19, 20).

Corrosion Applications. Pure wrought or electroplated palladium is used for electrical contacts, and large quantities are currently in use in relays for telephone service. Palladium and palladium-rich alloys, because of their freedom from tarnish, perform with high reliability and low electrical noise even with low contact forces. Many alloys containing palladium have been developed to meet specific contact requirements. A list of some of these alloys is given in Table 23. The palladium content provides the noble metal characteristics; other metals are added for improved hardness, electrical resistance, or economy.

Table 13 Effect of temperature on the tensile strength of annealed platinum wire

1.3-mm (0.05-in.) diam wire annealed at 1100 °C (2010 °F) for 5 min

Test temperature °C	Test temperature °F	Tensile strength MPa	Tensile strength ksi
20	70	138	20
200	390	117	17
400	750	90	13
600	1110	76	11
800	1470	55	8
1000	1830	28	4

Palladium does not stain or discolor porcelain that is fired on it, and palladium-rich alloys are used as supports in porcelain overlay dental restorations. In the jewelry field, palladium hardened with a few percent of ruthenium or rhodium provides a light, white, tarnish-free alloy suitable for such articles as watch cases, brooches, and settings for gems. Palladium is an exceptionally effective whitener for gold; the addition of as little as 15% produces a satisfactory white color. In contrast to the great increase in hardness and working difficulties encountered when gold is whitened by other metals, the white gold-palladium alloys are ductile and readily worked.

Palladium alloys are well suited for the construction of fine wire precision resistors because of their ductility, corrosion resistance, and the range of electrical resistivities available. Alloys with exceptionally low-temperature coefficients of electrical resistivity have been developed for applications as potentiometer resistor wire. An alloy that is widely used for this is Pd-40Ag, which has a resistivity of 42 $\mu\Omega\cdot$cm in the as-worked condition and a temperature coefficient of only 0.00003/°C (0.00005/°F) in the temperature range 0 to 100 °C (32 to 212 °F).

Metallic films fired on glass or ceramic substrates are of interest for thin film resistors. For this application, palladium-gold resistance films can be produced by the thermal decomposition of resinate mixtures. In another process, a mixture of finely ground low-melting glass and palladium and silver powders is fired on the substrate. Thin film resistors made in this manner do not exhibit a critical resistance/composition dependence.

A palladium-alloy, noble metal thermocouple combination has been developed that approximates the high electromotive force (emf) developed by one of the commonly used base metal thermocouples. Designed to operate at temperatures up to about 1300 °C (2375 °F), the Platinel thermocouple also exhibits a superior stability and life over the base metal thermocouples. Consisting of a gold-palladium negative leg and a palladium-platinum-gold positive leg, the thermocouple produces an emf approximating that of base metal thermocouples over a wide temperature range. Because of the close match of emf, the palladium alloy thermocouple can be substi-

Table 14 Corrosion of platinum in acids

Acid	Temperature, °C (°F)	Corrosion rate mm/yr	mils/yr
Acetic, all concentrations	Boiling	<0.05	2
Acetylsalicylic, all concentrations	Boiling	<0.05	2
Aqua regia	Room	Rapid dissolution	
Ascorbic, all concentrations	Boiling	<0.05	2
Benzoic, all concentrations	130 (265)	<0.05	2
Benzene sulfonic, pure	Room	<0.05	2
Boric, saturated	Boiling	<0.05	2
Butyric, all concentrations	Boiling	<0.05	2
Carbonic, pure	1400 (2550)	<0.05	2
Chloric, all concentrations	Room	<0.05	2
Chlorosulfonic, all concentrations	Boiling	<0.05	2
Chlorotoluene-sulfonic, all concentrations	Boiling	<0.05	2
Citric, <20% concentrations	Boiling	<0.05	2
Citric, 30% concentrations	Boiling	<0.05	2
Crotonic, all concentrations	Boiling	<0.05	2
Fatty, pure	400 (750)	<0.05	2
Fluorosilicic (10% hydrofluoric, 5% fluorosilicic)	Boiling	<0.05	2
Formic, pure	Boiling	<0.05	2
Gluconic, all concentrations	Boiling	<0.05	2
Glycerophosphoric, 1–50% solution	Boiling	<0.05	2
Hydrobromic, fuming	Room	<0.25	10
	100 (212)	4.8	189
Hydrochloric, 36%	Room	nil	
	100 (212)	<0.25	10
Hydrofluoric, 40%	Room	nil	
Hydrogen sulfide, pure	1000 (1830)	<0.05	2
Hydroiodic, specific gravity 1.75	Room	<0.25	10
	100 (212)	13.7	539
Isovaleric, all concentrations	Boiling	<0.05	2
Lactic, all concentrations	Boiling	<0.05	2
Laevolinic, all concentrations	Boiling	<0.05	2
Monochloroacetic, all concentrations	Boiling	<0.05	2
Nitric, 70%	Room	<0.25	10
Nitric, 95%	Room-100 (212)	nil	
Nitrosyl-sulfuric, pure	100 (212)	<0.05	2
Oxalic, all concentrations	Boiling	<0.05	2
Phenol-2,4-disulfonic, all concentrations	100 (212)	<0.05	2
Phenylacetic, all concentrations	Boiling	<0.05	2
Phosphoric, 100 g/L	100 (212)	nil	
Phthalic, pure	Boiling	<0.05	2
Picric, pure	125 (255)	<0.05	2
Propionic, all concentrations	Boiling	<0.05	2
Pyridine, all concentrations	Boiling	<0.05	2
Pyridine-carboxylic, all concentrations	150 (300)	<0.05	2
Salicylic, all concentrations	Boiling	<0.05	2
Stearic, pure	Boiling	<0.05	2
Sulfuric	Room-100 (212)	nil	
Sulfurous, all concentrations	100 (212)	<0.05	2
Tartaric, all concentrations	Boiling	<0.05	2

tuted for the base metal thermocouple in many instances without the need to modify the existing instrument.

In general, brazing alloys containing palladium exhibit exceptional wettability, good flow and gap-filling characteristics, freedom from the tendency to attack and erode base metals, good ductility, and other desirable brazing qualities. The palladium-containing brazing alloys have proved useful for dissimilar-metal or metal-to-ceramic joints, for extremely thin sheet metal assemblies, and for assemblies that must withstand extreme service temperatures. The alloys can be used to braze a wide range of base metals, including low-alloy and stainless steels; nickel-, cobalt-, and copper-base alloys; and refractory metals such as tungsten and molybdenum. They are consequently gaining a significant place in the gas turbine, jet engine, air frame, missile, nuclear, and electronic industries.

Gold-palladium alloys are used for thermal fuses to prevent temperature override in furnaces. The range of melting temperatures possible with the different gold-palladium combinations (1063 to 1552 °C, or 1954 to 2825 °F), the narrow melting range of the alloys, and their corrosion and oxidation resistance make them well suited for this service.

Rhodium

Fabrication. Rhodium powder can be consolidated either by powder metallurgy techniques or by melting. Powder compacts are sintered by air, inert atmospheres, hydrogen, or vacuum at about 1200 °C (2190 °F).

The melting of rhodium requires careful control of atmospheric conditions because liquid rhodium will dissolve a large quantity of oxygen, which is rejected upon solidification. If conditions are reducing, the rhodium may become contaminated through reduction of the refractory crucibles. A satisfactory technique for the production of small rhodium ingots is electron beam or argon arc melting on a water-cooled copper hearth.

Sintered or cast rhodium ingots can be worked down to thin strip or fine wire. The fabrication of these wrought forms requires an initial hot working at a temperature of 1200 °C (2190 °F) or higher, but subsequently, the temperature can be dropped. At thinner gages, the metal can be cold worked. During cold work, the rhodium should be given stress-relief anneals at 600 to 800 °C (1110 to 1470 °F), but complete annealing should be avoided because recrystallized metal is less ductile than metal with a fibrous structure. Moderate amounts of cold work (up to 40 to 50%) can be given between stress-relief anneals.

The principal atomic, structural, and physical properties for rhodium are listed in Table 1. Rhodium is characterized by:

- High specular reflectivity
- The highest electrical conductivity of all the platinum-group metals
- The highest thermal conductivity of all the platinum-group metals
- A high melting point: 1963 °C (3565 °F)
- A density of 12.41 g/cm^3 (0.448 lb/in.^{3}), approximately 58% that of platinum

Mechanical Properties. The hardness of wrought rhodium in the annealed condition averages about 130 HV. Therefore, rhodium is harder than platinum or palladium (each about 40 HV).

Rhodium work hardens rapidly. A 15% reduction by cold rolling increases hardness to 300 HV. The hardness of annealed rhodium decreases progressively with increasing temperature; hardness falls from about 120 HV at room temperature to 80 HV at 600 °C (1110 °F) and to less than 50 HV at 1100 °C (2010 °F). The tensile strength of annealed rhodium is 827 to 896 MPa (120 to 130 ksi), while a tensile strength of 1379 to 1517 MPa (200 to 220 ksi) is obtainable in cold-drawn wire.

Corrosion Resistance. Rhodium is resistant to corrosion by nearly all aqueous solutions at room temperature, including concentrated acids, but it is slowly attacked by solutions of sodium hypochlorite. In wrought or cast form, it is unattacked at 100 °C (212 °F) by concentrated HCl, HNO_3, and aqua regia, but it is attacked slowly by concentrated H_2SO_4 and by HBr (Table 24). Table 25 lists corrosion rates of rhodium in other environments. Rhodium can be used for insoluble anodes in electrolytic processes in which evolution of oxygen and chlorine occurs. When alternating current is used, however, rhodium is dissolved fairly readily in a number of electrolytes.

Rhodium is unattacked by chlorine at room temperature, but it may be attacked at elevated temperatures. It is generally more resistant to the halogens than platinum, although it is less resistant than iridium (Table 26). In the massive form, it is attacked slowly by molten sulfur, but finely divided metal may react violently. Rhodium is resistant to wet or dry gaseous sulfur dioxide vapors to 1000 ° C (1830 °F) if elemental sulfur is not present.

Rhodium is resistant to attack by some molten salts. Rhodium crucibles have been used at 1620 °C (2950 °F) for growing calcium tungstate single crystals by the Czochralski technique.

Rhodium is attacked to varying extents by fused alkalies under oxidizing conditions; the rate of corrosion is comparable to that of platinum. It is corroded fairly rapidly by alkali cyanides and fused sodium bisulfate.

At temperatures 200 °C (360 °F) above their melting points, gold, silver, mercury, cesium, potassium, sodium, and gallium have negligible corrosive action on rhodium, but unlike iridium and ruthenium, rhodium is rapidly dissolved by lead and bismuth. The standard electrode poten-

Table 15 Corrosion of platinum in salts

Salt	Temperature, °C (°F)	Corrosion rate mm/yr	Corrosion rate mils/yr
Alum, all concentrations	Boiling	<0.05	2
Aluminum chloride, all concentrations	Boiling	<0.05	2
Aluminum fluoride, all concentrations	Boiling	<0.05	2
Aluminum sulfate, 100 g/L	Room-100 (212)	nil	
Aluminum sulfate, all concentrations	Boiling	<0.05	2
Ammonium chloride, all concentrations	Boiling	<0.05	2
Ammonium nitrate, all concentrations	Boiling	<0.05	2
Ammonium persulfate, all concentrations	60 (140)	<0.05	2
Ammonium phosphate, all concentrations	Boiling	<0.05	2
Ammonium sulfate, all concentrations	Boiling	<0.05	2
Ammonium thiocyanate, pure	Boiling	<0.05	2
Antimony pentachloride, pure	100 (212)	<0.05	2
Barium chloride, all concentrations	Boiling	<0.05	2
Calcium hypochlorite, all concentrations	Room	<0.05	2
Calcium bisulfite, pure	Boiling	<0.05	2
Calcium chloride, all concentrations	100 (212)	<0.05	2
Calcium sulfate, pure	To red heat	<0.05	2
Calcium sulfide, all concentrations	100 (212)	<0.05	2
Calcium tungstate, pure	800 (1470)	<0.05	2
Calcium tungstate, all concentrations	Boiling	<0.05	2
Carnallite, pure	500 (930)	<0.05	2
Carnallite, all concentrations	Boiling	<0.05	2
Carnallite, saturated solution	Boiling	<0.05	2
Cupric chloride, 100 g/L	Room	nil	
Cupric sulfate, 100 g/L	100 (212)	nil	
Ferric chloride, 100 g/L	Room	<0.25	10
	100 (212)	16.7	657
Ferrous sulfate, all concentrations	Room	<0.05	2
Fluorosilicate, all concentrations	100 (212)	<0.05	2
Lithium chloride, all concentrations	Boiling	<0.05	2
Magnesium chloride, all concentrations	Boiling	<0.05	2
Magnesium sulfate, all concentrations	100 (212)	<0.05	2
Mercury chloride, all concentrations	Boiling	<0.05	2
Nitrosyl chloride, dry	Room	<0.05	2
Phosphorus chlorides, pure	Boiling	<0.05	2
Potassium bisulfate, all concentrations	Boiling	<0.05	2
Potassium bisulfate, melt	200–400 (390–750)	<0.05	2
Potassium bromide, all concentrations	Boiling	<0.05	2
Potassium bromide, melt	760 (1400)	<0.05	2
Potassium carbonate, all concentrations	Boiling	<0.05	2
Potassium carbonate, melt(a)	900 (1650)	<0.05	2
Potassium chlorate, all concentrations(b)	Boiling	<0.05	2
Potassium cyanide, 50 g/L	Room	<0.25	10
	100 (212)	1.4	55
Potassium dichromate, all concentrations	Boiling	<0.05	2
Potassium ferricyanide, all concentrations	Boiling	<0.05	2
Potassium ferrocyanide, all concentrations	Boiling	<0.05	2
Potassium hydroxide, all concentrations	300 (570)	<0.05	2
Potassium hydroxide, melt(a)	300 (570)	<0.05	2
Potassium nitrate, all concentrations	Boiling	<0.05	2
Potassium nitrate, melt	335 (635)	Attacked	
Potassium permanganate, all concentrations	Boiling	<0.05	2
Potassium peroxide, all concentrations	100 (212)	<0.05	2
Potassium peroxide, melt	380 (715)	Attacked	
Potassium persulfate, all concentrations	60 (140)	Attacked	
Potassium sulfate, all concentrations(c)	Boiling	<0.05	2
Potassium sulfate, melt	Melting point	<0.05	2
Sodium bisulfate, all concentrations	Boiling	<0.05	2
Sodium bisulfate, melt	400 (750)	<0.05	2
Sodium bisulfites, all concentrations	100 (212)	<0.05	2
Sodium carbonate, all concentrations	Boiling	<0.05	2
Sodium carbonate, melt	860 (1580)	<0.05	2
Sodium chloride, all concentrations	Boiling	<0.05	2
Sodium chloride, melt(d)	800 (1470)	<0.05	2
Sodium chromate, all concentrations	Boiling	<0.05	2
Sodium cyanide, all concentrations	Room	<0.05	2
Sodium formaldehyde sulfoxylate, all concentrations	90 (195)	<0.05	2
Sodium formate, all concentrations	Boiling	<0.05	2
Sodium formate, melt	260 (500)	<0.05	2
Sodium fluorosilicate, all concentrations	100 (212)	<0.05	2
Sodium hydroxide, <90% pure	Boiling	<0.05	2
Sodium hydroxide, melt	350 (660)	<0.05	2
Sodium hypochlorite, all concentrations	100 (212)	<0.05	2
Sodium hypochlorite + sodium chloride, saturated solution	100 (212)	<0.25	10
Sodium nitrate, all concentrations	Boiling	<0.05	2
Sodium perborate, all concentrations	50 (120)	<0.05	2
Sodium percarbonate, all concentrations	50 (120)	<0.05	2
Sodium perchlorate, all concentrations	Boiling	<0.05	2
Sodium perchlorate, melt	480 (900)	Attacked	
Sodium peroxide, all concentrations	Boiling	<0.05	2
Sodium peroxide, melt	400 (750)	<0.05	2
Sodium phosphates, all concentrations	Boiling	<0.05	2
Sodium silicates, all concentrations	Boiling	<0.05	2
Sodium sulfate, all concentrations	Boiling	<0.05	2
Sodium sulfide, all concentrations	Boiling	<0.05	2
Sodium sulfide, melt	700 (1290)	<0.05	2
Sodium sulfite, all concentrations	Boiling	<0.05	2
Sodium thiocyanate, all concentrations	Boiling	<0.05	2
Sodium thiocyanate, melt	300 (570)	<0.05	2
Sodium thiosulfate, all concentrations	Boiling	<0.05	2
Stannic ammonium chloride, all concentrations	Boiling	<0.05	2
Stannic chloride, all concentrations	Boiling	<0.05	2
Strontium nitrate, all concentrations	Boiling	<0.05	2
Sulfite cooking liquor, pH 1.3	Boiling	<0.05	2
Sulfur monochloride, pure	Boiling	<0.05	2
Sulfuryl chloride, dry and wet	300 (570)	<0.05	2
Thionyl chloride, dry or wet	Boiling	<0.05	2
Uranyl nitrate, all concentrations	Boiling	<0.05	2

(a) Platinum is attacked if strong oxidizers are present. (b) Platinum-iridium anodes used to electrolytically manufacture potassium chlorate. (c) Provided reducing agents are not present. (d) Provided no ammonia is present

tial of rhodium ($Rh \rightleftharpoons Rh^{2+} + 2e^-$) is about +0.6 V at 25 °C (75 °F).

Oxidation Resistance. Rhodium does not tarnish in air at room temperature, even in the most severe atmospheric conditions, but heating in air at temperatures above about 600 °C (1110 °F) will produce a thin oxide film that is visible as a dark discoloration. The weight change due to oxidation at these temperatures is negligible even after prolonged heating.

The oxide decomposes at about 1100 °C (2010 °F) in air at normal pressure and at slightly higher temperatures in oxygen. At higher temperatures, rhodium dissolves some oxygen and simultaneously reacts with it to form a volatile oxide. The volatile oxide is formula rhodium dioxide (Ref 21).

The vapor pressure of rhodium dioxide is directly proportional to the partial pressure of oxygen in the atmosphere. It is slightly less than that of the corresponding platinum oxide at temperatures below 1200 °C (2190 °F), but at higher temperatures the reverse is true.

Corrosion Applications. Rhodium is principally used as an alloying element in conjunction with platinum and palladium. Rhodium-containing platinum alloys are used for crucibles, furnace windings, thermocouple elements, linings for glass extrusion nozzles and spinnerets, and woven screen oxidation catalysts for production of HNO_3 from ammonia. Rhodium is used to harden palladium in jewelry applications. It also increases the corrosion resistance. Similarly, rhodium increases the hardness and corrosion resistance of nickel.

As the element, rhodium is electroplated for a number of decorative and nontarnishing uses, including jewelry, reflective mirrors, and electrical contacts. Rhodium is also vacuum deposited on glass to produce mirrors with high reflectivity and chemical stability. Thin coatings of rhodium on glass make an excellent gray filter.

Iridium

Fabrication. Pure iridium is not readily amenable to conventional fabrication and is generally used as an alloying element. However, iridium powder can be consolidated by conventional powder metallurgy techniques or by melting. Powder compacts are preferably sintered in vacuum at 1500 °C (2730 °F) before forging, rolling, swaging, or other hot-working operations for consolidation. Melting can be carried out in an argon arc furnace on a water-cooled copper hearth or by induction heating in a zirconia crucible, again in an argon atmosphere. In both cases, iridium powder is the raw material, which is preferably first briquetted and vacuum sintered into a partially consolidated material.

Sintered or cast iridium has working characteristics that are similar to those of tungsten and therefore requires considerable care in the early stages of processing. Initial breakdown of cast or

Table 16 Corrosion of platinum in organic compounds

Environment	Temperature, °C (°F)	Corrosion rate mm/yr	Corrosion rate mils/yr
Acetaldehyde, pure	200–400 (390–750)	<0.05	2
Acetic anhydride, all concentrations(a)	Boiling	<0.05	2
Acetone, pure	Boiling	<0.05	2
Acetylene, dry	600 (1110)	Becomes spongy	
Acetylene dichloride, wet and acid	Boiling	<0.05	2
Ethyl alcohol, all concentrations	Boiling	<0.05	2
Amyl acetate, pure	Boiling	<0.05	2
Amyl alcohol, pure	Boiling	<0.05	2
Aniline, pure	Boiling	<0.05	2
Benzaldehyde, pure and aqueous	Boiling	<0.05	2
Benzene, pure	Boiling	<0.05	2
Benzotrifluoride, pure	Boiling	<0.05	2
Benzyl chloride, pure	180 (355)	<0.05	2
Butyl acetate, pure	Melting point	<0.05	2
Butyl alcohol, pure	Boiling	<0.05	2
Carbon bisulfide, pure	Boiling	<0.05	2
Carbon tetrachloride, dry and wet	Boiling	<0.05	2
Chloramine(s), all concentrations	Boiling	<0.05	2
Chlorobenzene, pure	Boiling	<0.05	2
Chlorocresols, all concentrations	Boiling	<0.05	2
Chloroform, dry or wet	Boiling	<0.05	2
Chlorohydrins, pure	Boiling	<0.05	2
Chloronitrobenzenes, pure	Boiling	<0.05	2
Chlorotoluene, pure	Boiling	<0.05	2
Copper acetate, neutral solutions	100 (212)	<0.05	2
Cresols, pure	Boiling	<0.05	2
Dextrose, all concentrations	Boiling	<0.05	2
Dibutyl phthalate, pure	Boiling	<0.05	2
Dichlorodifluoromethane, pure	600 (1110)	<0.05	2
Dimethylaniline, pure	600 (1110)	<0.05	2
Diphenyl, pure	400 (750)	<0.05	2
Essential oils, pure	Boiling	<0.05	2
Ether, pure	Boiling	<0.05	2
Ethyl acetate, all concentrations	Boiling	<0.05	2
Ethyl benzene, pure	135 (275)	<0.05	2
Ethylene dibromide, wet and acid products	Boiling	<0.05	2
Ethylene dichloride, wet and acid products	Boiling	<0.05	2
Fatty acids, pure	400 (750)	<0.05	2
Formaldehyde, all concentrations	500 (930)	<0.05	2
Furfural, wet and slightly acid	Boiling	<0.05	2
Gelatin, pure	Boiling	<0.05	2
Glycerol, pure	Boiling	<0.05	2
Guanidine nitrate, all concentrations	Room	<0.05	2
Guinine sulfate, all concentrations	Boiling	<0.05	2
Guinolines, pure	Boiling	<0.05	2
Guinone, inorganic solvent and pure	100 (212)	<0.05	2
Hexachloroethane, dry and moist	187 (370)	<0.05	2
Hydrazine, <50% solution	Room	<0.05	2
Hydroquinone, all concentrations	Boiling	<0.05	2
Isoborneol acetate, pure	Boiling	<0.05	2
Isobutyl chloride, dry and wet	Boiling	<0.05	2
Limonene, pure	Boiling	<0.05	2
Methyl alcohol, pure	Boiling	<0.05	2
Methylamines, all solutions and gaseous	Room	<0.05	2
Methyl chloride, dry and wet	300 (570)	<0.05	2
Methylene chloride, dry and wet	Boiling	<0.05	2
Methylglycol, pure	Boiling	<0.05	2
Milk, pure	Boiling	<0.05	2
Nitrobenzene, pure	Boiling	<0.05	2
Nitrocellulose, in water or alcohol	Room	<0.05	2
Nitrotoluenes	Boiling	<0.05	2
Pentachloroethane, wet, dry, and acid	Boiling	<0.05	2
Phenol, all concentrations	Boiling	<0.05	2
Phenyl hydrazine, all concentrations	100 (212)	<0.05	2
Phenylmercuric acetate, pure and all concentrations	Melting point	<0.05	2
Phthalic anhydride, pure	Boiling	<0.05	2
Potassium acetate, all concentrations	Boiling	<0.05	2
Pyridine, all concentrations	Boiling	<0.05	2
Sodium acetate, all concentrations	Boiling	<0.05	2
Sodium acetate, melt	400 (750)	<0.05	2
Sodium bisulfate, all concentrations	Boiling	<0.05	2
Sodium formaldehyde sulphoxylate, all concentrations	90 (195)	<0.05	2
Sodium formate, all concentrations	Boiling	<0.05	2
Sodium formate, melt	260 (500)	<0.05	2
Sodium isovalerate, all concentrations	Boiling	<0.05	2
Sodium isovalerate, melt with sodium hydroxide	290 (555)	<0.05	2
Sodium methylate, all concentrations in alcohol or ether	100 (212)	<0.05	2
Sodium pentachlorophenolate, all concentrations	Boiling	<0.05	2
Sodium phenolate, all concentrations	Boiling	<0.05	2
Sodium salicylate, all concentrations	Boiling	<0.05	2
Sodium tartrates, all concentrations	Boiling	<0.05	2
Sorbital, all concentrations	Boiling	<0.05	2
Sorbose, all concentrations	Boiling	<0.05	2
Toluene, pure	Boiling	<0.05	2
Toluenesulfonyl chlorides, all concentrations	Boiling	<0.05	2
Triethanolamine, pure and all concentrations	Boiling	<0.05	2
Vinyl chloride, pure	500 (930)	<0.05	2

(a) Platinum-gold alloys perform better than pure platinum.

sintered shapes is done in the temperature range of 1200 to 1500 °C (2190 to 2730 °F).

Subsequent drawing to wire is performed by warm working at 600 to 750 °C (1110 to 1380 °F), which is below the recrystallization temperature. Such wire has a fibrous structure, a hardness of 600 to 700 HV, and useful tensile strength and ductility. Drawing at lower temperatures leads to a rapid increase in hardness and splitting of the wire. Drawing of material that has been fully recrystallized by annealing results in frequent breakage.

Iridium can also be rolled at 600 to 750 °C (1110 to 1380 °F) into strip with the fibrous structure characteristic of drawn wire. In addition, iridium can be rolled at higher temperatures (1200 to 1500 °C, or 2200 to 2730 °F) to yield a product with an equiaxed structure and hardness of about 400

Table 17 Corrosion of platinum in gases

Gas	Temperature °C	Temperature °F	Corrosion rate mm/yr	Corrosion rate mils/yr
Ammonia, with oxidant(a)	950	1740	<0.05	2
Ammonia, pure	Elevated		Nitridation	
Carbon dioxide, no reductant present	1400	2550	<0.05	2
Carbon monoxide, no reductant present	1400	2550	<0.05	2
Hydrogen, pure	1000	1830	<0.05	2
Hydrogen chloride, dry(b)	1200	2190	<0.1	4
Hydrogen sulfide, moist	Room		Blackened(c)	
Nitrogen dioxide			<0.05	2
Ozone, with 98% oxygen	100	212	<0.05	2
Steam	600	1110	<0.05	2
Sulfur dioxide, dry and wet	600	1110	<0.05	2

(a) Use of platinum-rhodium alloys is preferred for ammonia oxidation (loss is <250 mg of platinum/ton of nitric acid. (b) Corrosion rate is increased by the presence of steam or oxidizing agent. (c) Platinum is blackened but unattacked in hydrogen sulfide to 1000 °C (1830 °F).

Table 18 Corrosion of platinum in halogens at room temperature

Halogen	Corrosion rate mm/yr	Corrosion rate mils/yr
Chlorine, dry	<0.25	10
Chlorine, moist	<0.25	10
Saturated chlorine in water	nil	
Bromine, dry	3.5	138
Bromine, moist	2.0	80
Saturated bromine in water	nil	
Iodine, dry	<0.25	10
Iodine, moist	nil	
Iodine in alcohol, 50 g/L	nil	

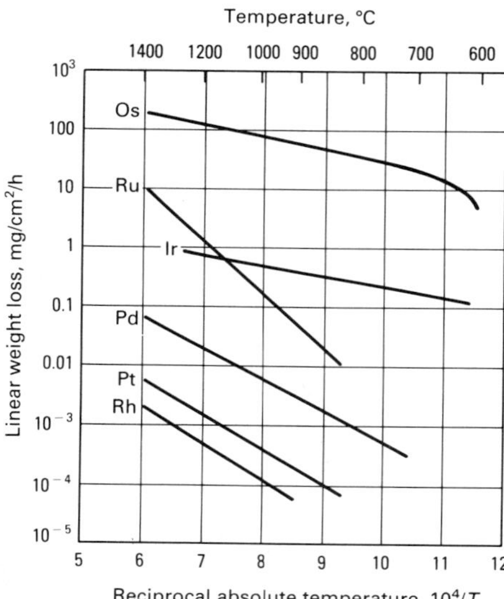

Fig. 1 Weight losses of platinum-group metals oxidized in air

HV. In general, worked material that has a fibrous structure is preferred because it has better ductility and strength.

Warmed-worked iridium does not exhibit a sharp recrystallization temperature. Some softening occurs upon heat treatment at 700 °C (1290 °F), but it is necessary to heat to a minimum of 1000 °C (1830 °F) or higher temperatures before full recrystallization occurs.

Atomic, Structural, and Physical Properties. A number of interesting features of the atomic, structural, and physical property data for iridium are listed in Table 1. For example:

- Iridium has the distinction of being the heaviest element known, with a density of 22.65 g/cm^3 (0.818 lb/in.^{3})
- Next to osmium, iridium has the highest melting point of the platinum-group metals, 2447 °C (4437 °F)

Table 21 Corrosion of palladium in common salts and other environments

Environment	Temperature °C	Temperature °F	Corrosion rate mm/yr	Corrosion rate mils/yr
Hydrogen sulfide, moist	Room		nil	
Sodium hypochlorite + sodium chloride, saturated solution	Room		1.8	71
	100	212	14.9	587
Ferric chloride, 100 g/L	Room		11.9	469
	100	212	Rapid dissolution	
Potassium cyanide, 50 g/L	Room		1.6	63
	100	212	62.7	2469
Mercuric chloride, solution	100	212	nil	
Cupric chloride, 100 g/L	Room		<0.25	10
Cupric sulfate, 100 g/L	100	212	nil	
Aluminum sulfate, 100 g/L	Room		nil	
	100	212	nil	

Table 19 Effect of temperature on the tensile strength of annealed palladium wire

1.3-mm (0.05-in.) diam wire annealed at 1100 °C (2010 °F) for 5 min

Test temperature °C	Test temperature °F	Tensile strength MPa	Tensile strength ksi
20	70	193	28
200	390	169	24.5
400	750	125	18
600	1110	88	13
800	1470	57	8
1000	1830	26	4

- The modulus of elasticity of iridium is one of the highest for an element, 517 GPa (75 000 ksi)

Mechanical Properties. Iridium shows a high degree of work hardening compared with other fcc metals. The hardness of annealed iridium increased by 250 HV with 20% cold reduction as compared to an increase of 30 HV for pure platinum worked to a similar degree. The very small amounts of the impurities segregated at the grain boundaries may be the cause of the rapid work-hardening behavior. The excellent high-temperature strength properties of iridium place it in the category of the refractory metals.

Corrosion Resistance. Iridium is the most corrosion-resistant metal known. It is completely unattacked by the common mineral acids at normal and high temperatures and by cold and boiling aqua regia (Table 27). It is the most resistant of the platinum metals to the halogens, as shown in Table 28. It is slightly attacked by fused sodium and potassium hydroxides and by fused sodium bicarbonate. Iridium is resistant to anodic corrosion in aqueous electrolytes, but may be attacked in aqueous potassium cyanide, HCl, and ammonium carbonate solutions under the action of an alternating current. The corrosion resistance of iridium in other environments is shown in Table 29. Iridium can be dissolved by aqua regia under pressure by heating to 250 to 300 °C (480 to 570 °F). Iridium is resistant to wet or dry gaseous sulfur dioxide to 1000 °C (1830 °F) if elemental sulfur is not present.

Iridium shows excellent resistance to attack by a wide range of molten metals. Iridium is unattacked by gallium, lithium, potassium, sodium, indium, mercury, and bismuth at temperatures up to 200 °C (360 °F) above their respective melting points under an atmosphere of argon. It is only slowly attacked by molten lead, tellurium, cadmium, antimony, tin, calcium, silver, and gold. On the other hand, the metal is readily attacked by molten copper, aluminum, zinc, and magne-

Table 22 Corrosion of platinum in halogens at room temperature

Halogen	Corrosion rate mm/yr	Corrosion rate mils/yr
Chlorine, dry	1.1	43
Chlorine, moist	14.0	551
Saturated chlorine in water	<0.25	10
Bromine, dry	24.5	965
Bromine, moist	25.0	984
Saturated bromine in water	27.7	1090
Iodine, dry	<0.25	10
Iodine, moist	<0.25	10
Iodine in alcohol, 50 g/L	<0.25	10

Table 20 Corrosion of palladium in acids

Acid	Temperature °C	Temperature °F	Corrosion rate mm/yr	Corrosion rate mils/yr
Aqua regia	Room		Rapid dissolution	
Hydroiodic, specific gravity 1.75	Room		65.7	2587
Hydrobromic, fuming	Room		161.2	6346
Hydrochloric, 36%	Room		<0.25	10
	100	212	1.3	51
Hydrofluoric, 40%	Room		nil	
Nitric, 70%	Room		61.3	2413
Nitric, 95%	100	212	Rapid dissolution	
Phosphoric, 100 g/L	100	212	<0.25	10
Hydrochloric, specific gravity 1.6	Room		nil	
	100	212	2.5	100
Sulfuric, concentrated	Room		<0.25	10
	100	212	1.6	63
Acetic acid, glacial	Room		<0.25	10

sium. The standard electrode potential of iridium ($Ir \rightleftharpoons Ir^{3+} + 3e^-$) is about +1.0 V at 25 °C (75 °F).

Oxidation Resistance. Iridium does not tarnish at room temperature, but heating in air at temperatures above about 600 °C (1110 °F) will produce a thin oxide tarnish. At temperatures of 1000 °C (1830 °F) and higher, iridium loses weight through the formation of a volatile oxide, which is reported to be either iridium sesquioxide, iridium trioxide, or iridium dioxide (Ref 21-24). The weight loss as a function of time is linear because base metal is continually exposed to the attacking air. The linear mass change for all the platinum metals is plotted in Fig. 1. Although the weight loss in air is much greater for iridium than for platinum, iridium is the only metal with a sufficiently high melting point that can be used unprotected in air at temperatures up to 2300 °C (4170 °F) without catastrophic failure.

Corrosion Applications. Except for the use of pure iridium as high-performance sparkplug electrodes and for very high-temperature crucibles used in the preparation of single crystals of certain optical and electronic glasses, iridium is largely used for hardening and increasing the corrosion resistance of platinum and palladium. In platinum, which in the fully annealed state has a tensile strength of 131 MPa (19 ksi), a 10% Ir addition increases tensile strength to 414 MPa (60 ksi); a 35% iridium addition increases tensile strength to 965 MPa (140 ksi). Additional hardening can be obtained in platinum alloys containing 10 to 30% Ir by heat treating in the temperature range of 500 to 800 °C (930 to 1470 °F). As a rather less potent hardener than ruthenium, iridium can be added to platinum in significantly higher concentration than ruthenium without impairing ductility and workability.

Iridium is unattacked by lithium, sodium, potassium, bismuth, gallium, lead, silver, and gold at up to 200 °C (360 °F) above the respective melting points of these metals (Ref 25). It is also unattacked by mercury up to 550 °C (1020 °F). Iridium is not wetted by the lithium, sodium, potassium, bismuth, gallium, and lead at these test temperatures, although at 400 °C (720 °F) above its melting point gallium does wet iridium and there is some surface attack. Gold and silver wet iridium and form continuous films over both inside and outside the sintered crucible.

Table 23 Properties and applications of typical palladium electrical contact materials

Material	Melting point °C	Melting point °F	Electrical conductivity, %IACS(a)	Hardness, HR15-T Annealed	Hardness, HR15-T Cold worked	Density g/cm³	Density lb/in.³	Typical applications
Palladium	1552	2825	16	65	80	11.99	0.43	Telephone relays
95.5Pd-4.5Ru	1590	2900	7	73	86	12.0	0.433	Heating pads
90Pd-10Ru	1650	3000	6.5	87	93	12.03	0.434	Voltage regulators
70Pd-30Ag	1375	2510	4.3	80	90	11.5	0.415	Relays
60Pd-40Ag	1340	2440	4.0	72	90	11.3	0.41	Relays
50Pd-50Ag	1290	2350	5.5	78	89	11.2	0.40	Relays
72Pd-26Ag-2Ni	1380	2520	4	80	90	11.5	0.415	Thermostats
60Pd-40Cu	1200	2190	5	83	95	10.6	0.38	Sliding contacts
45Pd-30Ag-20Au-5Pt	1370	2500	4.5	88	96	12.5	0.45	Sliding contacts
14Cu-10Pt-10Au-1Zn	1085	1985	5	90	97	11.9	0.429	Sliding contacts

(a) International Annealed Copper Standard

Tellurium, cadmium, and tin attack iridium at all temperatures above their melting points, along with the formation of intermediate compounds at the interface. These compounds have some protective value for the iridium, although they are brittle and do not adhere well to the iridium. Antimony also reacts with iridium to form an intermetallic compound, but this compound does not seem to adhere to the iridium and has no protective action.

Molten copper, calcium, and zinc at all temperatures penetrate iridium intergranularly; this is followed by the dissolution of iridium in these liquid metals. Copper, in particular, penetrates very rapidly into the iridium. Magnesium and aluminum attack iridium very rapidly by uniform diffusion at the interface; the solid solution so formed then dissolves in the liquid metal.

In addition to hardening platinum, iridium greatly enhances its corrosion resistance, particularly in environments involving nascent chlorine, aqua regia, and similar corrosives. Because of their exceptional resistance to corrosion and tarnish, platinum-iridium alloys containing up to 30% Ir have been used in:

- Chemical plants handling extremely corrosive materials
- Electrical contacts exposed to severe environments and where extreme reliability of chlorine is required
- Jewelry, surgical tools, and implants (catheters, microelectrodes, and pacemaker components)
- Primary standards of length and weight

Table 25 Corrosion of rhodium in common salts and other environments

Environment	Temperature °C	Temperature °F	Corrosion rate mm/yr	Corrosion rate mils/yr
Hydrogen sulfide, moist	Room		nil	
Sodium hypochlorite + sodium chloride, saturated solution	Room		<0.25	10
	100	212	<0.25	10
Ferric chloride, 100 g/L	100	212	nil	
Mercuric chloride, 100 g/L	100	212	nil	
Cupric sulfate, 100 g/L	100	212	nil	
Aluminum sulfate, 100 g/L	100	212	nil	

Iridium is also an effective hardener for palladium, and it imparts corrosion resistance. Although not so widely applied as their platinum counterparts, the palladium alloys have been used for jewelry and electrical contacts.

Iridium alloys containing up to 60% Rh have been proposed for high-temperature thermocouples. The couple iridium versus Ir-40Rh is regarded as one of the most satisfactory for use in oxidizing atmospheres at temperatures as high as 2100 °C (3810 °F).

The naturally occurring alloy osmiridium (30 to 65% Os balance iridium) has been widely used for the tipping of fountain pen nibs, for instrument pivots, and for similar applications requiring high hardness and extreme resistance to wear and corrosion. Recently, osmiridium has been replaced by a wide variety of complex synthetic alloys containing, among others, the refractory metals, together with iridium, ruthenium, osmium, rhodium, and platinum.

Iridium-tungsten alloys have been principally developed for springs required to operate at high temperatures. These springs have excellent relaxation properties at temperatures up to 800 °C (1470 °F).

Small additions of iridium are very effective in improving the corrosion resistance of titanium. As with the other platinum-group metals, addition of as little as 0.1% Ir increases the corrosion resistance of titanium to nonoxidizing acids a hundredfold. Similar improvements have been reported for chromium containing 0.5% Ir.

Ruthenium

Fabrication. Wrought forms of ruthenium are limited. Those products that are available are generally fabricated by powder metallurgy techniques. The ruthenium powder is obtained from the refining process. Hot working can be accomplished at 1150 to 1500 °C (2100 to 2730 °F), although rolling of bar to strip can be performed in the range of 1050 to 1250 °C (1920 to 2280 °F) to minimize grain growth and edge cracking due to large grain size. Cold rolling of ruthenium is very limited because of the low ductility of the metal. The metal work hardens rapidly and requires frequent intermediate anneals.

Single-crystal ruthenium, prepared by electron beam zone refining, shows a relatively high degree of ductility when compared to polycrystalline material. However, the material work hardens rapidly, and upon annealing, it reverts to a polycrystalline state and exhibits low ductility.

Atomic, Structural, and Physical Properties. A high melting point (2310 °C, or 4190 °F), a moderate density of 12.45 g/cm³ (0.45 lb/in.³) and a Young's Modulus of 414 GPa (60 000 ksi) characterize ruthenium. In relation to other metals of the platinum group, ruthenium bears closest resemblance to osmium and, like it, has a hexagonal close-packed (hcp) structure. The principal atomic, structural, and physical properties of the metal are given in Table 1. Some of these properties will vary according to the degree of preferred orientation present in the material.

Mechanical Properties. The hardness of ruthenium varies considerably according to the orientation of the hexagonal lattice. The hardness of sintered bar, swaged at 1500 °C (2730 °F) with a 45% reduction in area and subsequently annealed at the same temperature, is in the range 400 to 450 HV on longitudinal sections and 250 to 300 HV on transverse sections.

The tensile strength of a polycrystalline sintered bar, swaged with a 45% reduction in area at 1500 °C (2730 °F) and subsequently annealed at the same temperature, is 496 MPa (72 ksi), and the elongation is 3%. High-temperature tensile

Table 24 Corrosion of rhodium in acids

Acid	Temperature °C	Temperature °F	Corrosion rate mm/yr	Corrosion rate mils/yr
Aqua regia	Boiling		nil	
Hydroiodic, specific gravity 1.75	100	212	nil	
Hydrobromic, fuming	100	212	2.2	87
Hydrochloric, 35%	100	212	nil	
Hydrofluoric, 40%	Room		nil	
Nitric, 95%	100	212	nil	
Phosphoric, 100 g/L	100	212	nil	
Sulfuric, concentrated	100	212	<0.25	10
Acetic acid, glacial	100	212	nil	

Table 26 Corrosion of rhodium in halogens at room temperature

Halogen	Corrosion rate, mm/yr (mils/yr)
Chlorine, dry	nil
Chlorine, moist	nil
Saturated chlorine in water	nil
Bromine, dry	nil
Bromine, moist	nil
Saturated bromine in water	nil
Iodine, dry	nil
Iodine, moist	<0.25 (10)
Iodine in alcohol, 50 g/L	<0.25 (10)

Table 27 Corrosion of iridium in acids

Acid	Temperature, °C (°F)	Corrosion rate, mm/yr (mils/yr)
Aqua regia	Room-boiling	nil
Hydroiodic, specific gravity 1.75	Room-100 (212)	nil
Hydrobromic, specific gravity 1.7	100 (212)	nil
Hydrochloric, 36%	100 (212)	nil
Hydrofluoric, 40%	Room	nil
Nitric, 95%	100 (212)	nil
Phosphoric	100 (212)	nil
Sulfuric, concentrated	Room-100 (212)	nil
Acetic acid, glacial	100 (212)	nil

Table 28 Corrosion of iridium in halogens at room temperature

Halogen	Corrosion rate, mm/yr (mils/yr)
Chlorine, dry	nil
Chlorine, moist	nil
Bromine, dry	nil
Bromine, moist	<0.25 (10)
Saturated bromine in water	nil
Iodine, dry	nil
Iodine, moist	nil
Iodine in alcohol, 50 g/L	nil

Table 29 Corrosion of iridium in other environments

Environment	Temperature °C	Temperature °F	Corrosion rate, mm/yr (mils/yr)
Hydrogen sulfide, moist	Room		nil
Sodium hypochlorite + sodium chloride, saturated solution	100	212	<0.25 (10)
Ferric chloride, 100 g/L	100	212	nil
Mercuric chloride, solution	100	212	nil
Cupric sulfate, 100 g/L	100	212	nil
Aluminum sulfate, 100 g/L	100	212	nil

Table 30 Corrosion of ruthenium in acids

Acid	Temperature °C	Temperature °F	Corrosion rate, mm/yr (mils/yr)
Aqua regia	100	212	nil
Hydroiodic, 60%	100	212	nil
Hydrobromic, 62%	100	212	nil
Hydrochloric, 36%	100	212	nil
Hydrofluoric, 49%	Room		nil
Nitric, 95%	100	212	nil
Sulfuric, 95%	100	212	nil

tests carried out in air indicate a steady decrease in strength with increasing temperature and a maximum in elongation and reduction of area in the temperature range of 750 to 900 °C (1380 to 1650 °F).

Corrosion Resistance. Ruthenium is resistant to attack by cold and hot acid solutions, including aqua regia, and in this respect, is superior to platinum (Table 30). However, its resistance to attack under certain oxidizing conditions is not as high as that of platinum. Therefore, it is attacked fairly rapidly by sodium hypochlorite. Saturated aqueous solutions of chlorine and bromine and alcoholic solutions of iodine attack the metal slowly (Table 31). It is dissolved fairly rapidly as an anode in a large number of electrolytes and more rapidly by alternating current electrolysis. Ruthenium is attacked by fused alkaline hydroxides, carbonates, and cyanides and is attacked very rapidly by fused sodium peroxide. The corrosion resistance of ruthenium in other environments is given in Table 32.

Ruthenium exhibits good resistance to attack by molten lithium, sodium, potassium, copper, silver, and gold when it is heated in an atmosphere of argon. No solution attack by these metals occurs up to 100 °C (180 °F) above their melting points, although grain-boundary penetration is observed with sintered and unworked ruthenium. It is also resistant to attack by molten lead, and up to 700 °C (1290 °F), attack by liquid bismuth is extremely slight. The solubility of ruthenium in bismuth at this temperature is 0.029%, and at 1200 °C (2190 °F), it is 0.016%.

Oxidation Resistance. Ruthenium does not tarnish at room temperature even in heavily polluted atmospheres, but when heated in air or oxygen to temperatures approaching 800 °C (1470 °F), it oxidizes to form a surface film of ruthenium dioxide. Between this temperature and 1150 °C (2100 °F), there is simultaneous formation of a volatile oxide, probably ruthenium monoxide and ruthenium dioxide. At higher temperatures, only ruthenium monoxide is formed. The vapor pressure of the monoxide when formed on heating ruthenium in pure oxygen at temperatures in the range of 1200 to 1400 °C (2190 to 2555 °F) has been found to follow the approximate relationship:

$$\log_{10} P = \frac{11\,100}{T} + 4.83$$

where T is the absolute temperature, and P is the pressure in atmospheres (Ref 21). The actual rate of weight loss upon heating the metal in air will depend on a number of factors, including the geometric form of the sample under test and the degree of movement of the surrounding atmosphere. The mass change of ruthenium in air in comparison with the other platinum metals is given in Fig. 1

Corrosion Applications. Because of the lack of wrought forms of ruthenium that can be readily fabricated, the applications of ruthenium in corrosive environments are limited. High-ruthenium alloys containing other platinum metals or base metals have been used for electrical contacts and for severe wear resistance applications, such as tips for fountain pen nibs and for nonmagnetic instrument pivots. High hardness and excellent resistance to corrosion are the prime virtues of these alloys. Ruthenium dioxide coated titanium is an effective anode in chlorine-manufacturing cells.

Ruthenium is commercially added to platinum and palladium as a hardener. A 5% addition of ruthenium to annealed platinum will increase the hardness from 40 to 130 HV; similarly, the hardness of annealed palladium can be increased from 40 to 90 HV. The work-hardening rate of these metals is also increased by the addition of ruthenium.

Small additions of ruthenium have also been found to be effective in improving the corrosion resistance of titanium. As with other platinum metals, addition of as little as 0.01% Ru results in a hundredfold increase in resistance to corrosion in reducing acids.

Ruthenium is unattacked by lithium, sodium, potassium, gold, silver, copper, lead, bismuth, tin, tellurium, indium, cadmium, calcium, and gallium at temperatures up to 200 °C (360 °F) above the melting points of the respective metals (Ref 25). Gold, silver, and copper flow readily over the surface of ruthenium, but do not wet it. Ruthenium is also unattacked by mercury at temperatures to 550 °C (1020 °F). Ruthenium is apparently unattacked at lower temperatures by gallium, but there is some attack at temperatures 400 °C (720 °F) above the melting point of gallium. Similarly, bismuth dissolves ruthenium very slowly at 700 °C (1290 °F), with the dissolution occurring uniformly at the ruthenium surface.

Ruthenium is attacked by molten aluminum or zinc at all temperatures above their melting points. This attack appears to consist of uniform dissolution of the surface and does not result in the formation of intermetallic compounds or grain-boundary penetration. On the other hand, attack by magnesium and antimony occurs with the formation of an intermetallic compound at the interface, which appears to have some protective value.

Table 31 Corrosion of ruthenium in halogens at room temperature

Halogen	Corrosion rate, mm/yr (mils/yr)
Saturated chlorine in water	1.3 (51)
Chlorine, dry	nil
Chlorine, moist	nil
Saturated bromine in water	1.0 (40)
Bromine, dry	nil
Bromine, moist	nil
Iodine, dry	nil
Iodine, moist	nil
Iodine in alcohol, 50 g/L	1.0 (40)

Table 32 Corrosion of ruthenium in salts and other environments at 100 °C (212 °F)

Environment	Corrosion rate, mm/yr (mils/yr)
Hydrogen sulfide, moist, at room temperature	nil
Sodium hypochlorite + sodium chloride, saturated solution	Moderate attack
Ferric chloride, 100 g/L	nil
Mercuric chloride, 100 g/L	44.0 (1732)
Aluminum sulfate, 100 g/L	nil

Osmium

Fabrication. Osmium is an element that is essentially impossible to fabricate except by powder metallurgy, and even with powder metallurgy techniques, extreme care must be used. Arc melting must be done under vacuum or inert atmospheres.

Atomic, Structural, and Physical Properties. Osmium has the highest melting point (about 3045 °C, or 5513 °F) and the second-highest density (22.57 g/cm^3, or 0.815 lb/in.^{3}) of the platinum-group metals. Osmium is similar in

Table 33 Corrosion of osmium in acids

Acid	Temperature °C	Temperature °F	Corrosion rate, mm/yr (mils/yr)
Aqua regia	Boiling		Rapid dissolution
Hydroiodic, specific gravity 1.75	100	212	3.7 (148)
Hydrobromic, specific gravity 1.7	100	212	1.8 (72)
Hydrochloric, 36%	Room		nil
Hydrochloric, 36%	100	212	<0.25 (10)
Hydrofluoric, 40%	Room		nil
Nitric, 95%	100	212	Rapid dissolution
Phosphoric, concentrated	100	212	nil
Sulfuric, concentrated	100	212	nil

Table 34 Corrosion of osmium in halogens at room temperature

Halogen	Corrosion rate mm/yr	mils/yr
Chlorine, dry	<0.25	10
Bromine, dry	4.1	161
Iodine, dry	<0.25	10

Table 35 Corrosion of osmium in salts and in other environments

Environment	Temperature, °C (°F)	Corrosion rate mm/yr	mils/yr
Hydrogen sulfide, moist	Room	<0.25	10
Sodium hypochlorite + sodium chloride, saturated solution	Room	Rapid dissolution	
Ferric chloride, 100 g/L	100 (212)	3.0	120

many properties to ruthenium and, like ruthenium, has an hcp crystal structure. The principal physical properties of the element are given in Table 1. Some of these properties will vary according to the degree of preferred orientation present in the test specimen.

Mechanical property data on osmium are scarce. The annealed hardness of osmium is 300 to 670 HV (Ref 26). This makes osmium the hardest of the platinum-group metals.

Corrosion Resistance. Compared to other elements in the platinum-group metals, osmium has relatively modest corrosion resistance. The element is attacked by aqua regia and the oxidizing acids, but is resistant to HCl and H_2SO_4 (Table 33). Osmium is attacked by the halogens at room temperature (Table 34), and the attack becomes progressively worse upon heating. Osmium is dissolved fairly rapidly in sodium hypochlorite at room temperature and in $FeCl_3$ at 100 °C (212 °F) (Table 35). Osmium burns in the vapor of sulfur and phosphorus and is attacked by molten alkali hydrosulfates, potassium hydroxide, and oxidizing agents. Osmium powder readily absorbs considerable amounts of hydrogen.

Oxidation Resistance. Osmium powder will slowly oxidize even at room temperature to form osmium tetroxide. Osmium tetroxide boils at 130 °C (265 °F) and is extremely toxic.

Corrosion Applications. Osmium is the rarest of the platinum-group metals; annual worldwide production amounts to only a few thousand ounces. Much of this production is for medical applications. A principal use of osmium is in the manufacture of hard, nonrusting pivots for instruments, phonograph needles, tipping the nibs of fountain pens, and certain types of electrical contacts. Alloys containing at least 60% Os or osmium plus ruthenium in conjunction with another platinum-group metal are usually used in these applications.

Anodic Behavior of the Noble Metals

Under anodic conditions in nitrate and cyanide solutions, silver is readily dissolved. This behavior is the basis of numerous silver electroplating baths. Silver becomes passive in most halide and hydroxide solutions because of the formation of a silver halide layer and a silver oxide layer, respectively. In sulfate solution, silver first forms a passive film of sulfate, but upon application of a higher potential, the sulfate becomes an oxide that can be reduced again to silver sulfate at a potential even lower than that required for the initial sulfate film formation.

Gold dissolves readily when it is made anodic in chloride solution containing an oxidizing agent. Only under conditions of low acid concentration or high current densities will the surface gradually become passivated by an absorbed layer of oxygen. In H_2SO_4, gold initially dissolves as Au^+, but the surface gradually becomes passivated by a layer of gold hydroxide.

Platinum and iridium are capable of carrying high current densities in certain acidic and alkaline solutions, but can become passivated with difficulty by absorbed oxygen. This film formation is accelerated by the superimposition of an alternating current on the direct current. The low solubility of platinum has made it a valuable commercial anode material for the electrolytic production of persulfates and perchlorates, electroplating anodes, and cathodic protection. Platinum loses less than 0.6 mg/A · yr over a current density range of 540 to 5400 A/m^2 (50 to 500 A/ft^2) in flowing seawater, and less than 1 g per ton of chlorine produced in the electrolysis of brine. In strong HCl solutions, platinum may be attacked, particularly if the temperature is raised. Rhodium becomes anodically passive in HCl. Conversely, palladium is readily dissolved anodically in acid chloride solutions and under certain neutral chloride conditions. Ruthenium metal is dissolved anodically in both HCl and H_2SO_4.

Small additions of the noble metals, particularly palladium and platinum, can substantially lower the corrosion rate of such metals as titanium, stainless steel, and chromium. This is achieved through the noble metal additions by their cathodic reduction of dissolved oxygen or hydrogen ions, permitting larger current densities and higher positive potentials in the anodic regions of the surface. To be effective, this higher potential must move the metal into a passive condition; otherwise, the corrosion rate will be accelerated. The technique works well for many base metal alloys in H_2SO_4.

REFERENCES

1. G.W. Walkiden, The Noble Metals, in *Corrosion—Metal/Environmental Reactions*, Vol 1, 2nd ed., L.L. Sheir, Ed., Newnes-Butterworths, 1979, p 6.3-6.23
2. C.R. Marsland, The Fabrication of Silver and Silver-Base Alloys, in *Silver—Economics, Metallurgy and Use*, A. Butts and C.D. Cox, Ed., D. Van Nostrand, 1967, p 310-321
3. E.M. Wise, *Gold*, D. Van Nostrand, 1964
4. R.H. Leach, Silver and Silver Alloys, in *The Corrosion Handbook*, H.H. Uhlig, Ed., John Wiley & Sons, 1948, p 314-320
5. A. Butts and J.M. Thomas, Corrosion Resistance of Silver and Silver Alloys, in *Silver in Industry*, L. Addicks, Ed., Reinhold, 1940, p 357-400
6. A. Butts, The Chemical Properties of Silver, *Silver—Economics, Metallurgy and Use*, D. Van Nostrand, 1967, p 123-136
7. E.M. Wise, Platinum Group Metals and Alloys, in *The Corrosion Handbook*, H.H. Uhlig, Ed., John Wiley & Sons, 1948, p 299-313, 699-718
8. R.F. Vines, Noble Metals, in *Corrosion Resistance of Metals and Alloys*, 2nd ed., F.L. LaQue and H.R. Copson, Ed., ACS Monograph Series, No. 158, Reinhold, 1963, p 601-621
9. E.M. Wise, Corrosion Behavior of Palladium, in *Palladium—Recovery, Properties and Uses*, Academic Press, 1968, p 21-28
10. E. Rabald, *Corrosion Guide*, 2nd ed., Elsevier, 1968
11. M.J.N. Pourbaix, *Atlas of Electrochemical Equilibria in Aqueous Solutions*, Pergamon Press, 1966
12. J.C. Chaston, Oxygen in Silver, in *Silver—Economics, Metallurgy and Use*, A. Butts and C.D. Cox, Ed., D. Van Nostrand, 1917, p 304-309
13. J.S. Hill and H.J. Albert, Loss of Weight of Platinum, Rhodium and Palladium at High Temperatures, *Englehard Ind. Tech. Bull.*, Vol 4 (No. 2), 1963, p 59-63
14. J.C. Chaston, Reaction of Oxygen With the Platinum Metals: I-The Oxidation of Platinum, *Platinum Met. Rev.*, Vol 8 (No. 2), 1964, p 50-54
15. E.P. Sadowski, Stress-Rupture Properties of Some Platinum and Palladium Alloys, Part I, in *Metallurgical Society Conferences*, Vol 11, The Metallurgical Society, 1961, p 465-476
16. H.J. Albert, D.J. Accinno, and J.S. Hill, Stress-Rupture Properties of Some Platinum and Palladium Alloys, Part II, in *Metallurgical Society Conferences*, Vol II, The Metallurgical Society, 1961, p 476-482
17. E.D. Zysk, Noble Metals in Thermometry-Recent Developments, *Englehard Ind. Tech. Bull.*, Vol 5 (No. 3), 1964, p 69-99
18. E. Raub and W. Plate, The Solid State Reactions Between the Precious Metals or Their Alloys and Oxygen at High Temperatures, *Z. Metall.*, Vol 48, 1957, p 529-539
19. W. Betteridge and D.W. Rhys, High-Temperature Oxidation of Platinum Metals and Their Alloys, in *Proceedings of the First International Congress on Metallic Corrosion*, Butterworths, 1962, p 186-192
20. J.C. Chaston, Reaction of Oxygen With the Platinum Metals III-The Oxidation of Palladium, *Platinum Met. Rev.*, Vol 9 (No. 4), 1965, p 126-129
21. C.B. Alcock and G.W. Hooper, Thermodynamics of the Gaseous Oxides of the Platinum-Group Metals, *Proc. R. Soc. (London) A*, Vol 254, p 551-561
22. R.W. Douglass and R.J. Jaffee, Elevated-Temperature Properties of Rhodium, Iridium and Ruthenium, *Proc. ASTM*, Vol 62, 1962, p 627
23. H. Schafer and N.J. Heitland, Equilibrium Measurements in the System Iridium-Oxygen: Gaseous Iridium Trioxide, *Z. Anorg. Allg. Chem.*, Vol 304, 1960, p 249-265
24. W.L. Phillips, Jr., Oxidation of the Platinum Metals in Air, *Trans. ASM*, Vol 57, 1964, p 33-37
25. D.W. Rhys and E.G. Price, Resistance of Iridium and Ruthenium to Attack by Liquid Metals, *Englehard Ind. Tech. Bull.*, Vol V (No. 2), 1964, p 37-42
26. R.F. Vines, *The Platinum Metals and Their Alloys*, The International Nickel Company, Inc., 1941

Corrosion of Beryllium*

John J. Mueller, Battelle Columbus Laboratories
Donald R. Adolphson, Sandia Laboratories

THE CORROSION BEHAVIOR OF BERYLLIUM in aqueous and elevated-temperature oxidizing environments has been extensively studied for early-intended use of beryllium in nuclear reactors and in jet and rocket propulsion systems. Since that time, beryllium has been used as a structural material in less corrosive environments. Its primary applications include gyro systems, mirror and reentry vehicle structures, and aircraft brakes. Only a small amount of information has been published that is directly related to the evaluation of beryllium for service in the less severe or normal atmospheric environments associated with these applications.

Despite the lack of published data on the corrosion of beryllium in atmospheric environments, much can be deduced about its corrosion behavior from studies of aqueous corrosion and the experiences of fabricators and users in applying, handling, processing, storing, and shipping beryllium components. The methods of corrosion protection implemented to resist water and high-temperature gaseous environments provide useful information on methods that can be applied to protect beryllium for service in future long-term structural applications.

Numerous grades of beryllium are commercially available, and these grades provide a wide range of compositional variations. The grades of material included under the name beryllium contain other elements at low concentrations, all of which are considered impurities and are generally not introduced as alloying elements. The major controls applied in producing beryllium are the minimization of impurities and control of the ratio of certain impurities, that is, maintenance of a prescribed ratio of about 2:1 for iron and aluminum in some grades. The impurities in normal commercial grades vary from 1 to 4.5% and include oxygen, carbon, iron, chromium, nickel, silicon, aluminum, and copper (Ref 1). The impurities may be present in solid solution, as separate phases or compounds, or as particulate inclusions.

Reported corrosion results for beryllium often do not seem consistent or reproducible. This is perhaps because the investigators have used widely different grades of beryllium without properly characterizing the material involved or because a given grade has widely different compositions or distribution of impurities. Table 1 lists compositions of various grades of vacuum hot-pressed beryllium.

Corrosion of Beryllium in Air

Beryllium is commonly referred to as being self-protective against atmospheric oxidation; it resembles aluminum, titanium, and zirconium in this respect. Beryllium is unique among the alkaline earth metals in providing this characteristic. The protective nature of its oxide is partially due to the fact that temperature poses no reported corrosion problems for beryllium (Ref 2). On the other hand, air that contains water at a level at which condensation can occur with temperature cycling can cause beryllium to corrode under certain conditions. Because the atmospheric-corrosion problem can be regarded as an extension of the aqueous corrosion of beryllium, the corrosion behavior of beryllium in water and aqueous environments is important in understanding corrosion in air environments. The condensable water vapor in normal air environments provides the basis for this relationship.

Causes of Corrosion in Air. There are two primary situations in which beryllium has been found to be susceptible to corrosion in a water-bearing air atmosphere. One cause of corrosion in air is when the beryllium contains beryllium carbide (Be_2C) particles exposed at its surface by a machining operation. The water available from the surrounding air will slowly react with the carbides to form a flowery white buildup of beryllium oxide (BeO) according to the reaction:

$$Be_2C + 2H_2O = 2BeO + CH_4$$

with methane as a gaseous product. The reaction:

$$Be_2C + 4H_2O = 2Be(OH)_2 + CH_4$$

has been suggested as an intermediate step in arriving at the final BeO product (Ref 3). This is consistent with the observed formation of a white gelatinous product from the reaction of Be_2C in an aqueous environment (Ref 4).

This type of corrosion is commonly manifested by the bursting out of the beryllium at the edges of the carbide particles or by the formation of blisters as a result of the localized volume expansion during reaction.

The reactivity of carbides in an air atmosphere was first reported in 1955 (Ref 5). In work performed at Oak Ridge National Laboratory in 1947, white corrosion products and blisters were observed to form on the surface of extruded beryllium that had been exposed for about 6 months to the local humid air. The corrosion products were caused by carbide inclusions that were parallel to the extrusion axis. Similar corrosion attack was discovered during the shelf storage of finished precision-machined components (Ref 4, 6). The components were scrapped because of the disruption of the finely machined, unetched surfaces.

In one case, an instrument grade material corroded in service in a controlled gas environment containing a fixed value of relative humidity (Ref 3). The corrosion, which was in the form of spotty white buildup and disruption of the surface, occurred at carbide sites. The type of corrosion observed was experimentally duplicated with carbide-seeded beryllium subjected to the same type of environment. In addition, beryllium that exhibits flowery corrosion spots during extended air storage was also found to yield a similar pattern of corrosion attack within hours of immersion in an aqueous bath (Ref 4).

The carbon present in commercial beryllium may be distributed uniformly as fine carbide particles or randomly as a smaller number of bulk particles. Such a difference in distribution can lead to differing corrosion results. For example, one study investigated the effect of a large atyp-

Table 1 Compositions of four grades of vacuum hot-pressed beryllium

S, structural grade; I, instrument grade

Chemical composition	S-65B	S-200F	I-220A	I-400
Be, % (min)	99.0	98.5	98.0	94.0
BeO, % (max)	1.0	1.5	2.2	4.2(a)
Al, ppm (max)	600	1000	1000	1600
C, ppm (max)	1000	1500	1500	2500
Fe, ppm (max)	800	1300	1500	2500
Mg, ppm (max)	600	800	800	800
Si, ppm (max)	600	600	800	800
Other, ppm (max)	400	400	400	1000

(a) BeO specified is minimum in this instance.

*Adapted from J.J. Mueller and D.R. Adolphson, Corrosion, in *Beryllium Science and Technology*, Vol 2, D.R. Floyd and J.N. Lowe, Ed., Plenum, 1979, p 417-433 ical carbide inclusion corroding in an air environment during storage (Ref 7). The carbide inclusion was disk shaped, about 13 mm (1/2 in.) in diameter, in an approximately 76-mm (3-in.) diam beryllium component. The inclusion was exposed to the atmosphere at the machined surface. The volume expansion resulting from reaction with water in the air caused a large piece of beryllium to break out of the component. The volume expansion that occurred in this case from reaction with water vapor also causes the blistering and disruption of beryllium at the edges of smaller Be_2C inclusions.

In another study, during preparation of beryllium specimens with artificially added graphite inclusions, the formation in air of the typical white corrosion products was observed at particles exposed at the machined surfaces and fracture surfaces of tensile specimens (Ref 8). Metallographic examination of sections through graphite inclusions disclosed the presence of a shell of Be_2C around each graphite particle that formed during hot pressing of the beryllium.

The second major cause of corrosion in a humid air environment occurs when the beryllium surface contains contaminants in the form of chlorides and sulfates that remain on the beryllium after a final drying operation or are introduced after proper rinsing and drying. The use of tap water or a chloride-contaminated water bath without adequate cascade rinsing in deionized water prior to drying has been frequently reported as a cause of corrosion during subsequent processing or storage (Ref 9-12). The residual surface contaminants become concentrated upon drying and require very little water or condensation derived from temperature and humidity changes to provide a condition for corrosion at active sites. In one case, plastic piping, which is used to avoid metallic contamination of the rinse system, was found to be a source of chloride ions (Cl^-) that resulted in corrosion of components during storage (Ref 12).

Fingerprinting of clean, dried beryllium components is one of the most commonly encountered sources of surface contamination which can result in subsequent etching and corrosion of beryllium. The corrosive nature of fingerprints is apparent from the need to abrade and renew the affected surface mechanically before most chemical treatments (Ref 1, 13). All producers and users of beryllium strongly specify the avoidance of hand contact with semifinished and finished beryllium components.

The final cleaning treatment can be very important, particularly in view of the present-day trend of avoiding the chlorinated solvents that had often been used to remove oily surface contamination. Solvent and/or chemical cleaning is extremely important when the final machined surface will not be given a final acid-etching treatment to remove mechanically produced surface imperfections.

In one laboratory, a step has been taken toward developing cleaning procedures that are effective in removing surface contamination while avoiding the use of chlorinated solvents (Ref 14). An alkaline detergent-type bath with superimposed ultrasonic treatment has been found to be effective in removing all types of normal soils, with the exception of silicone greases. A suitable solvent for the silicone grease, followed by an alkaline detergent cleaning, should be an effective procedure in this instance.

A comparison was made of the results of cleaning in an alkaline detergent and by means of other solvents, including chlorinated solvents. The evaluation involved scanning electron microscopy of the cleaned surfaces and surface-outgassing measurements. The results showed that the alkaline detergent cleaning produced a surface that was cleaner than that obtained with the other cleaning methods. Surprisingly, the detergent-cleaned surface yielded a smaller amount of water in the outgassing experiments than the surface cleaned with chlorinated solvent. The higher yield from the chlorinated solvent cleaned surface was attributed to water held by surface contaminants that were not removed by this cleaning procedure.

An acid-etching treatment is commonly used on machined beryllium components to remove machining-damaged surface material. The material removed by etching may contain microcracks, which are open defects into which contaminants can penetrate and cause corrosion. Precision components that cannot be acid etched but are lapped to final size are susceptible to corrosion as a result of the unremoved microcracks (Ref 7). In the removal of only a few tenths of a mil of stock by lapping with an improper lapping agent, the microcracks can be smeared over with adjacent material. This results in entrapment of contaminants and corrosion of the component at a later time. A free-cutting diamond paste was recommended as the lapping agent.

Contamination From the Atmosphere. Rainwater has been recognized as a potential source of corrosive agents that can be inadvertently transported into a protective environment and cause deterioration of susceptible materials. Table 2 lists the corrosive agents commonly found in rainwater.

Inadvertent admission of atmospheric moisture into silos and launch control centers of missile systems has been cited as the cause of most of the general corrosion problems associated with the various metals contained in these structures (Ref 15). Despite the fact that corrosion problems have been encountered in missile silos, the maintenance of a humidity-controlled environment has generally been effective in protecting most components from corrosion. For example, one manufacturer has not found a single case of corrosion of beryllium components in rocket thrusters during periodic examinations of all operational units through years of continued service (Ref 16). These components were maintained under humidity-controlled conditions.

The open atmosphere above the ocean provides a site of unusually high salt content. Storms at sea and the mist generated by a moving vessel have been found to produce severe corrosion of beryllium brake shoe pads on military carrier bombers (Ref 17, 18). The brake shoes were typical of those that have performed well for years on the C5A military transport in less corrosive environments. The salt-laden environment of the aircraft carrier caused the brake pads to pit and to become covered with corrosion products.

One study demonstrated the noncorroding nature of pure water vapor when the beryllium surface is uncontaminated and free of exposed Be_2C inclusions (Ref 19). Polished, bare specimen coupons were subjected to an atmosphere of 95% relative humidity at 40 °C (100 °F) for 30 days with a result of no apparent corrosive attack. Neither microscopic examination nor weight gain measurements indicated corrosive attack on any of the specimens.

Aqueous Corrosion of Beryllium

Beryllium that is clean and free of surface impurities has very good resistance to attack in low-temperature high-purity water. The corrosion rate in good-quality water is typically less than 0.025 mm/yr (1 mil/yr) (Ref 20). Beryllium has been reported to perform without problems for 10 years in slightly acidified demineralized water in a nuclear test reactor (Ref 7, 21). This high-purity water environment also produced no evidence of accelerated corrosion with the beryllium galvanically coupled with stainless steel or aluminum.

Beryllium of normal commercial purity, however, is susceptible to attack, primarily in the form of localized pitting, when exposed to impure water (Ref 22). Chloride and sulfate ions (SO_4^{2-}) are the most critical contaminants in aqueous corrosion. Because these contaminants are present in tap water, most processing specifications warn against its use. In cases where tap water can be used, the final rinses are done in deionized water to ensure that the Cl^- or SO_4^{2-} impurities are not retained on the subsequently dried metal surface.

The pitting of beryllium in aqueous baths containing Cl^- and SO_4^{2-} ions has generally been attributed to attack in areas anodic to the bulk of the metal (Ref 23). In an investigation consisting of 12-month corrosion tests in water at 85 °C (185 °F), the pitting attack was attributed to a lack of metal purity (Ref 24). Years later, in a study prompted by rejection of beryllium components for the presence of hard inclusions, pitting corrosion was shown to be associated with points of silicon and aluminum segregation (Ref 25). In subsequent studies, an aqueous corrosive agent containing 200 ppm Cl^- ion, 400 ppm SO_4^{2-} ion,

Table 2 Corrosive agents in rainwater

Agent and source	Location	Concentration, ppm
Chloride (Cl^-) ions; major source is sea spray	Over sea or near coastlines	Average 2–20; with extreme winds, may increase up to 100
	500 miles or more inland	Average 0.1–0.2, but sometimes higher than 1.0
Sulfate (SO_4^{2-}) ions; major sources in industrial areas	Large cities, industrial areas	Average 10–50; higher under extreme conditions such as smog
	Other areas	1–5
Nitrate (NO_3^-) ions	Over land	0.5–5
Hydrogen ions (pH)	Over land	Average approximately 5; may decrease to 3 near industrial centers

and approximately 50 ppm carbonate ion (CO_3^{2-}) was used to produce rapid pitting attack of the beryllium. Electron microprobe analysis showed that the localized pitting occurred at sites rich in either silicon or aluminum.

On the basis of the examination of pitted specimen surfaces by microprobe analysis, the following general conclusions were drawn (Ref 26):

- Pitting corrosion is determined by the distribution of alloy impurities
- Sites that have high concentrations of iron, aluminum, or silicon (probably in solid solution) tend to form corrosion pits
- The corrosion sites are often characterized by the presence of alloy-rich particles, which are probably beryllides; however, the particles themselves do not appear to be corrosion active
- Pitting density is believed to be related to the number of segregated regions in the matrix containing concentrations of iron, aluminum, or silicon higher than concentrations present in the surrounding areas

Segregated impurities, such as Be_2C particles, intermetallic compounds, or alloy-rich zones, can contribute to localized attack when present in an exposed beryllium surface (Ref 7, 25). As previously noted, Be_2C forms a gelatinous corrosion product in an aqueous environment. As the purity of beryllium has improved during the years of its commercialization, the problem of localized corrosion at segregated particles and inclusions has significantly decreased (Ref 6, 27).

The effects of the surface finish of beryllium on aqueous corrosion were investigated in one study (Ref 28). The specimen conditions included (1) as-machined, (2) pickled in chromic-phosphoric (H_2CrO_4-H_3PO_4) acid after machining, and (3) machined, annealed beryllium reannealed in vacuum for 1 h at 825 °C (1515 °F). The results of long-term tests in demineralized water at 85 °C (185 °F) showed that the pickled specimens corroded at a higher rate than the others during the first 60 days of testing. The machined and annealed specimens initially corroded faster than the as-machined specimens. It was found, however, that with extended exposure times the magnitude of attack for all the types of surfaces reached about the same value: 0.0025 to 0.005 mm/yr (0.1 to 0.2 mil/yr).

Extensive information on the behavior of beryllium under the combined effects of stress and chemical environment is not readily available. The first reported work involved the use of extruded material in water containing 0.005 *M* hydrogen peroxide (H_2O_2) at pH 6 to 6.5 at 90 °C (195 °F) (Ref 29). No evidence of cracking was noted even though stresses up to 90% of the yield strength were used. Stress corrosion has been reported on cross-rolled sheet when synthetic seawater was used as the test medium (Ref 30). Studies of time to failure versus applied stress revealed that the time decreased from 2340 to 40 h as the applied stress was increased from 8.4 to 276 MPa (1220 to 40 000 psi, or 70% of yield strength). Failure appeared to be closely associated with random pitting attack. Certain pits appeared to remain active, which promoted severe localized attack.

In-Process Corrosion Problems and Handling and Storage Procedures

Many of the specified practices used by manufacturers and users result from knowledge of the problems experienced in processing and handling of components. One example of this is the common specification to avoid contact of beryllium by tap water. Early processors encountered serious problems of pitting or surface corrosion in beryllium that was allowed to stand in tap water, or was rinsed off with tap water and, after drying, was allowed to stand in a damp atmosphere where water condensation could easily occur on the contaminated surfaces. Improper handling, inadequate control of chemical and rinse baths, and improper storage or packaging procedures are the primary causes of in-process corrosion problems.

In-Process Problems

Inadvertent contact with water can be a damaging factor for beryllium. An airframe manufacturer experienced a major production problem when water vapor condensing on overhead piping dripped onto structures in the processing line (Ref 31).

Chloride ions from sources other than tap water can also lead to pitting attack in aqueous baths. Chlorinated solvents used to remove oils and greases have been found to be sources of Cl^- ions when the solvent is carried into a cleaning bath by an as-machined surface (Ref 14). As mentioned earlier, plastic piping, which is used to avoid metallic contamination in a chemical cleaning facility, provided a source of Cl^- ions (Ref 6).

Fingerprinting of in-process parts has been a commonly cited problem (Ref 9-11). Under certain atmospheric conditions, the salt in a fingerprint may seriously etch a beryllium surface. Such etching of a finished, critically sized component can cause rejection of the component (Ref 11, 12). Etching on a semifinished surface, if not seen and properly prepared, can lead to localized corrosion of a chemically treated surface.

Corrosion problems are often discovered well after their occurrence (Ref 12). One example of this is the use of a dye penetrant solution for nondestructive testing. If it is merely wiped off a surface, the dye penetrant solution can later seep out of crevices and cause localized attack. A second example is when localized galvanic attack is found on beryllium parts. The cause was traced back to a cutoff operation in which the beryllium contacted steel while wetted by the coolant.

Contact with apparently dry materials has also been found to be the cause of corrosion of beryllium. In one case, beryllium became etched when it was placed on an inspection bench on which a rubberized mesh was used to prevent nicking and scratching of finished parts (Ref 9). In another instance, a molded styrofoam support used for shipment of a complex satellite boom caused severe localized etching of the beryllium at points of pressure contact (Ref 9). The beryllium part was not shipped in direct contact with the styrofoam, but the part had been laid on the foam after being removed from its polyethylene bag for inspection. It can only be surmised that the etching resulted from attack by chemical compounds contained in the contacting materials. Lastly, small beryllium components that had been inserted into plastic shop-routing folders were attacked by the photocopied routing sheets, which directly contacted the components (Ref 12).

Handling and Storage Procedures

In handling and packaging beryllium components for shipment or storage, it is important to maintain an initially clean surface and to avoid unduly moist environments. The various practices employed by industry are discussed below.

Handling of beryllium parts with bare hands is often done and is not harmful when the part is in a rough or semifinished stage. In general, a finish-machined surface, even if it is to be acid etched to remove machine-damaged surface material, should not be handled with bare hands. A fingerprint left on the surface for some time, during which it may etch the beryllium, may retard the beneficial aspects of the acid-etching treatment or at least may provide a questionable artifact after etching. Fingerprints that have etched the beryllium adversely affect conversion or passivation coatings, disrupting the continuity of the coating and thus its effectiveness (Ref 11-13).

The common practice in handling finished and semifinished parts (or when handprinting is unacceptable) is to use protective gloves. White cotton, nylon, polyethylene, or rubber gloves are usually used.

Packaging for Shipment. Polyethylene bags are extensively used by industry as protective containment and/or barriers to contact between beryllium and its surroundings (Ref 10, 11, 27). A permeable container with a desiccant is often used inside the polyethylene bag to remove undesirable moisture (Ref 11, 27). An argon gas purge of the interior of the bag has been used as a means of displacing the moisture initially present (Ref 9). Because of the desire to minimize the availability of moisture to the beryllium, an attempt is generally made to provide an effective seal at the bag opening. An integrally molded interlocking seal or the practice of overlapping and taping provides an effective seal against free moisture access to the interior.

Molded foam, formed styrofoam, and contoured fiber-type insulation materials are typical means of supporting and protecting large beryllium structures or components against mechanical damage inside a shipping container (Ref 9, 10, 27). In some cases, when support is desired in all directions (if the container is overturned), a molded upper closure is used to encase the structure totally with supporting material. In almost every case, however, a polyethylene bag provides an effective separator between the beryllium and the contacting support material.

Storage of beryllium components while awaiting assembly or use can be safely accomplished under the conditions used for shipment, that is, in a compatible container made of metal or polyethylene with a desiccant to maintain a dry storage atmosphere. Heating a cabinet with a light bulb to keep the interior at about 50 °C (120 °F) has been found to be effective for long-term storage of gyro components (Ref 6). A room or chamber with humidity control, as well as a humidity-controlled missile silo, can be effective in preventing corrosion of beryllium rocket motor components during storage periods of several years (Ref 6, 11, 32). The following points should be considered in safely storing components for extended periods of time:

- Ensure that the component entering storage is free of corrosion-causing contaminants
- Maintain a moderately dry and noncondensing environmental atmosphere
- Ensure against contact with materials of nonproven compatibility

Corrosion-Protection Surface Treatments and Coatings

Extensive development and study of surface treatments and coatings for beryllium have been done to provide protection against many potentially severe environments. The status of these activities is summarized in Ref 7, 33, and 34. In addition to the treatment designed to provide normal storage and handling protection, many of the coatings and treatments developed for more severe applications are useful for less demanding applications. For example, those treatments designed to provide protection against seawater are effective in protecting against corrosion of beryllium during processing or use in air environments.

Chromate-type coatings are applied by a simple dip treatment. The coatings thus produced provide reasonable protection for beryllium during handling and storage and against attack by salt-containing environments for moderately long time periods. Some coatings are extremely thin, being virtually immeasurable by conventional techniques. One investigation demonstrated the ability of a chromate-type coating on S-100 beryllium to hold up well under a 100% relative humidity test involving 50 6-h cycles at 75 °C (170 °F) and under a 5% salt spray test for a period of 120 h (Ref 35).

Another study found that a passivation chromate treatment applied to a type of beryllium that showed susceptibility to white spot formation in moist air was very effective in improving resistance to such attack (Ref 36). The treatment involved a 30-min dip in a solution consisting of 25% H_3PO_4, 25% saturated solution of potassium chromate (K_2CrO_4), and 50% deionized water. Specimen life in 100% relative humidity at 70 °C (160 °F) increased from 1 day to 10 to 14 days because of the passivation treatment.

As further evidence of the protection afforded by a patented passivation treatment, an accelerated water immersion test consisting of 30 days in distilled water at 95 °C (205 °F) was used to evaluate the effectiveness of the treatment (Ref 13). The results showed a reduction in corrosion rate by a factor of 50. The protective chromate coating was less than 0.01 μm (0.0004 mil) thick. Also, another study found that a conversion coating on instrument grade beryllium containing Be_4C markedly reduced the corrosive attack by moisture in the service atmosphere (Ref 3).

In one case, a beryllium mirror that had been given a chromate passivation coating and had been stored on the roof of a California facility, unprotected except for a roof shelter, did not exhibit significant surface degradation after more than 6 months of exposure (Ref 32). The mirror surface appeared darker, but had retained 97 to 98% of its reflectivity and had lost only about 0.5% of its reflectivity to the 10-μm wavelength.

Commercial manufacturers of beryllium components are often required by drawing specifications to provide passivation-type coatings on finished parts. These coatings are effective in providing protection during shipment and short periods of storage under less-than-ideal conditions.

Fluoride coatings are produced by treating beryllium in fluorine above 520 °C (970 °F), which produces a glassy-appearing water-insoluble coating. The structure of the coating is of the rhombic tridymite type. This type of coating was shown to be very effective in resisting corrosion in chloride-containing water and in distilled water (Ref 37). Coatings 0.2 and 1.2 μm (0.008 and 0.05 mil) thick were unchanged after 3000 h in distilled water at room temperature. A 0.2-μm (0.008-mil) thick coating provided effective protection in water containing a 150-ppm concentration of Cl^- ion.

Anodic coatings that are similar in character to the well-known anodizing on aluminum have been shown to improve the resistance of beryllium to corrosion in normally corrosive aqueous solutions and to air oxidation at elevated temperature. There are a variety of formulations for anodizing solution compositions that provide differing qualities and results in various test environments. Nitric acid (HNO_3), H_2CrO_4, sodium dichromate ($Na_2Cr_2O_7$), and sodium hydroxide (NaOH) are common ingredients in many anodizing bath formations.

Anodized films vary in thickness from 2.5 μm (0.1 mil) to several mils, depending on solution type, applied voltage, current density, bath temperature, and time of treatment. The film qualities vary in accordance with descriptive terms, such as color, thickness, adherence uniformity, density, and electrical characteristics. It was concluded from a survey of many anodizing procedures that the following simple formulations produced the most uniform and adherent coatings (Ref 38):

- Solution of 50% HNO_3 with a current density of 2.15 A/m^2 (0.20 A/ft^2) for 5 min
- Solution of 7.5% NaOH with a current density of 108 A/m^2 10 A/ft^2 for 20 min.

The following experimental corrosion results on anodized coatings show the effectiveness of anodized coatings in resisting water and salt-containing environments. The results attest to the potential of anodized coatings in providing protection for beryllium in structural applications in which normal air environments are encountered.

Brush-produced anodic coatings were found to protect beryllium effectively for the following indicated times in the associated environments (Ref 3):

- 2400 h in a humidity cabinet
- 3 months at 40 °C (100 °F) in tap water
- 2000 h in ASTM salt spray test

In one study, approximately 5-μm (0.2-mil) thick anodic coatings produced in a 1% H_2CrO_4 solution and sealed in boiling water were evaluated (Ref 19). No corrosion was detected after exposure to 5% salt fog spray at 40 °C (100 °F) for 30 days.

Anodized coatings prepared in two distinctly different baths were evaluated as protection for beryllium experimentally seeded with Be_4C (Ref 3). When microanalysis techniques were used to determine gas generation in a humid gaseous environment, it was shown that the anodized coatings markedly reduced the corrosion rate.

The results of an evaluation of anodized coatings showed favorable performance in 5% salt spray tests at room temperature, but indicated poor performance at 50 °C (120 °F) (Ref 39). Only one specimen among five tested at room temperature showed attack after 24 h. Five similar specimens exposed at 50 °C (120 °F), however, were noticeably corroded within 24 h. In synthetic seawater, unprotected beryllium corroded at a rate 21 times that of the anodized beryllium. Pitting of the anodized material was first observed after 30 days in test. The performance of the coatings at 50 °C (120 °F) was much poorer than in other reference cases, but the results showed a definite improvement over uncoated material of similar quality. The difference in performance observed may reflect a difference in the quality of the coatings evaluated or of the beryllium used.

One prominent application of anodized coatings is in the mirror industry, in which beryllium is used for its light weight, high modulus of elasticity, and high specific heat and thermal conductivity. In this application, anodizing is used in conjunction with the good thermal properties of beryllium. The anodic coating is favored for its high thermal emissivity and absorptivity values. Its corrosion resistance is an added benefit. An anodic coating applied before lapping the spherical mirror surface has been found to provide corrosion protection against pitting corrosion at the unlapped edge of the mirror (Ref 27). The lapping compound tends to move to the rim of the lens structure, where it will readily attack bare beryllium if allowed to stand even for relatively short periods of time.

Plated Coatings. Beryllium is rarely plated to provide general corrosion protection but rather to comply with specific requirements. Plated coatings, whether applied by electroless or electrolytic plating techniques, are commonly used on beryllium to provide areas for electrical contact, to improve wear resistance, or to improve polishing characteristics in the case of mirrors.

Organic paint coatings are used when an electrically insulating barrier is required between the beryllium component and another metallic structure. Coatings used for this purpose provide protection against galvanic attack in the event of an electrolyte gaining access to the joint.

In one case, an electrodeposited paint coating added over a passivation coating was found to provide suitable protection for beryllium structures requiring long-term storage capability (Ref 31). A previously used black anodizing treatment yielded a coating that became troublesome because of its ease of scratching and chipping and the need for additionally protecting the uncoated sites of the electrical contact.

In one study, a polyvinyl butyrol phosphating primer was evaluated for the Navy for atmospheric protection of several metals, including beryllium (Ref 40). A two-coating system was recommended for storage for periods of 8 to 10 months in open exposure to the atmosphere.

An epoxy primer has been found to be effective for general protection of sheet and tubular components in communications satellite platform structures (Ref 12). The primer is used on all exposed surfaces, and all of the individually coated components are joined by adhesive bonding or mechanical fasteners. The primer is applied immediately after a flash etching of the surfaces to be coated and is cured by baking. The primer system of protection was selected to satisfy the requirement of years-duration storage capability for the fabricated structures.

REFERENCES

1. R.F. Bunshah, Future Trends in Beryllium Metallurgy Research, in *Beryllium: Its Metallurgy and Properties*, H. Hausner, Ed., University of California Press, 1965, p 279
2. S. Yamaguchi, A Study of Oxide Films on Light Metals by Electron Diffraction: Mg, Al and Be, *Sci. Pap. Inst. Phys. Chem. Res. (Jpn.)*, Vol 36, 1939, p 463
3. D. Beasley, Atomic Weapons Research Establishment, private communication
4. A. Brewer, Rocky Flats (Energy Research and Development Association Facility), private communication
5. J.L. English, *The Metal Beryllium*, American Society for Metals, 1955, p 530-532
6. L. Mellum and A. Polito, Honeywell Aerospace, private communication
7. A.J. Stonehouse and W.W. Beaver, Beryllium Corrosion and How to Prevent It, *Mater. Protec.*, Vol 4, 1965, p 24-28
8. D.B. King *et al.*, "Effect of Inclusions on the Mechanical Behavior of Beryllium," AFML-TR-76-33, Air Force Materials Laboratory, April 1976
9. L. Grant, Grant and Kamper, Inc., private communication
10. A. Stonehouse, Brush Wellman Company, private communication
11. G. Allen, American Beryllium Company, private communication
12. R. Dumphe and P. Smitz, Hughes Aircraft, private communication
13. Technical Data, File 302-704, Kawecki Berylco Industries, Inc.
14. J. Briggs, Rocky Flats (Energy Research and Development Association Facility), private communication RFP-2430
15. L.E. Gatzek, "Corrosion Control of Missiles in Long-Term Silo Environments," Paper presented at the National Aeronautics and Space Engineering Meeting, Los Angeles, CA, 5-9 Oct 1964
16. W. Kappen, Rocketdyne, private communication
17. B. King, Brush Wellman Company, private communication
18. T. McMacken, B.F. Goodrich, private communication
19. C.B. Gilpin and T.L. Mackay, "Corrosion Research Studies on Forged Beryllium," AFML-TR-66-294, Air Force Materials Laboratory, Jan 1967
20. J.N. Wanklyn and P.J. Jones, The Aqueous Corrosion of Reactor Metals, *J. Nucl. Mater.*, Vol 6 (No. 3), 1962, p 291-329
21. P.D. Miller and W.K. Boyd, Beryllium Deters Corrosion—Some Do's and Don'ts, *Mater. Eng.*, Vol 68 (No. 1), 1968, p 33-36
22. D.W. White, Jr. and J.E. Burke, *The Metal Beryllium*, American Society for Metals, 1955, p 533-548
23. T.L. Mackay and C.B. Gilpin, *Corros. Met.*, Vol 6, 1968, p 235-240
24. J.L. English, Report ORNL-772, United States Atomic Energy Commission, 1951
25. A.M. Kinan *et al.*, "The Significance of Inclusions/Precipitates in Beryllium," Paper presented at 1972 WESTEC Conference, Los Angeles, CA, March 1972
26. H.A. Moreen and A.G. Gross, Jr., "Pitting Corrosion in Beryllium," Unpublished Report, Autonetics Division
27. R. Paquin, Perkin-Elmer Corporation, private communication
28. A.R. Olsen, Oak Ridge National Laboratory, unpublished research, 1948
29. H.L. Logan and H. Hessing, "Summarizing Report of Stress Corrosion of Beryllium," NBS-6, National Bureau of Standards, Dec 1955
30. R.A. Miller *et al.*, *Corrosion*, Vol 23, 1967, p 11-14
31. R. Green, Lockheed Missile and Space Company, private communication
32. G. Speak and J. Heter, Hughes Aircraft, private communication
33. J.G. Beach, "Electrodeposited Electroless, and Anodized Coatings on Beryllium," Memorandum 197, Defense Materials Information Center, 1 Sept 1964
34. "Corrosion of Beryllium," Report 242, Defense Materials Information Center, 11 Dec 1967
35. J. Brooker and A.J. Stonehouse, Chemical Conversion Coatings Retard Corrosion of Beryllium, *Mater. Protec.*, Vol 8 (No. 2), 1969, p 43-47
36. S.J. Morana, "Surface Passivation Coating for Beryllium Metal," Paper presented at the Air Force Materials Laboratory Fifteenth Anniversay Corrosion of Military and Aerospace Equipment Technical Conference, 23-25 May 1967
37. P.M. O'Donnell, Beryllium Fluoride Coating as a Corrosion Retardant for Beryllium, *Corros. Sci.*, Vol 7, 1967, p 717-718
38. C.M. Packer, "Stress Corrosion Cracking of Beryllium," Report LMSC-288140, General Research in Materials and Propulsion, Section 6, Lockheed Corporation, Jan 1960; also LMSD-49735
39. F.E. Canpagna, "Properties of Beryllium for Rocket Engine Design," AFRPL-Tr-66-251-R-6718, Air Force Rocket Propulsion Laboratory, Nov 1966
40. G. Ya Terlo, "Protection of Steel and Light Alloys with Polyvinyl Butyral Phosphatizing Primers," VL-023, AD628185, Department of the Navy, 1966

Corrosion of Uranium and Uranium Alloys

Lawrence J. Weirick, Sandia National Laboratories

THE CORROSION of uranium and uranium alloys will be reviewed in this article. Included are sections on the oxidation of unalloyed uranium, the effect of alloying elements on oxidation response, the corrosion behavior of uranium alloys in aqueous solutions, the influence of microstructure on corrosion properties of the alloys, galvanic behavior, the stress-corrosion cracking (SCC) of uranium alloys, and protective coatings for uranium and uranium alloys, particularly metallic coatings of electroplated nickel and ion-plated aluminum. This article is not intended as a complete review of all the areas mentioned; additional information on a particular section is available in the cited references.

Oxidation of Unalloyed Uranium

The oxidation and corrosion of unalloyed uranium has been of interest to metallurgists and engineers since the early 1940s. However, only recently has there been agreement among researchers on reaction rates and mechanisms, particularly in environments containing both water vapor and oxygen.

Oxygen. Uranium reacts with oxygen according to:

$$U + \left(\frac{2 + x}{2}\right)O_2 \rightarrow UO_{2+x} \qquad \text{(Eq 1)}$$

where x is between 0.2 to 0.4 at temperatures to 200 °C (390 °F). The oxide thus formed is hyperstoichiometric uranium dioxide (UO_2). The most recent review of the literature on the reaction of uranium with oxygen or dry air at temperatures up to 300 °C (570 °F) is provided in Ref 1. The best estimate of the uranium-oxygen reaction rate constant (expressed in units of mg/cm²/h) for the temperature range of 40 to 300 °C (105 to 570 °F) is:

$$K = 9.4 \times 10^7 \exp\left(\frac{-18\,000}{RT}\right) \qquad \text{(Eq 2)}$$

where the weight gain (in milligrams per square centimeter) with time (in hours) is due to the quantity of oxygen consumed, R is a constant, and T is temperature. Thus, the reaction rates were found to be linear and independent of oxygen pressure above 133 Pa (1 torr). The reaction product was determined by x-ray diffraction analysis to be nominally $UO_{2.2}$, which has a face-centered cubic (fcc) crystal structure. The reaction product is very compact and adherent.

One of the most extensive studies of the oxidation of uranium covered the temperature range of 300 to 625 °C (570 to 1155 °F) and oxygen pressures from 133 to 667 kPa (100 to 500 torr) (Ref 2). No influence of oxygen pressure was observed in this range. The data were fitted to an equation of the form $w^n = kt$, where w is the quantity of oxygen consumed (in milligrams per square centimeter) in time t (in minutes), and k and n are constants. Values for n and k were functions of temperature T, and three rate equations were proposed:

$$w^{4/5} = 1.0 \times 10^5 t \exp\left(\frac{-16\,800}{RT}\right) \qquad \text{(Eq 3)}$$

$$w = 0.840 \qquad \text{(Eq 4)}$$

$$w^{6/5} = 1.8 \times 10^4 t \exp\left(\frac{-14\,300}{RT}\right) \qquad \text{(Eq 5)}$$

Equation 3 is for 300 °C $< T <$ 450 °C (570 °F $< T <$ 840 °F), Eq 4 is for T = 450 °C (840 °F), and Eq 5 is for $T >$ 450 °C (840 °F). The rate of oxidation accelerates with time up to 450 °C (840 °F), is constant at T = 450 °C (840 °F), and decreases with time at temperatures above 450 °C (840 °F). The data in Ref 2, along with those of several other investigators, are summarized in Fig. 1. The deviation in the curve from linearity at 450 °C (840 °F) was attributed to a change in the character of the oxide from a highly porous and nonprotective layer to a more compact scale, although all of these oxides were identified as nominally UO_2.

Water Vapor. Uranium reacts with water vapor according to:

$$U + (2 + x)H_2O \rightarrow UO_{2+x} + (2 + x)H_2 \qquad \text{(Eq 6)}$$

where x is between 0 and 0.1 in the absence of oxygen. In the most recent investigation of the uranium-water vapor reaction, Equation 6 was found to be linear between 30 and 80 °C (85 and 175 °F) (Ref 9). The reaction rate is proportional to the square root of the humidity and obeys Eq 7:

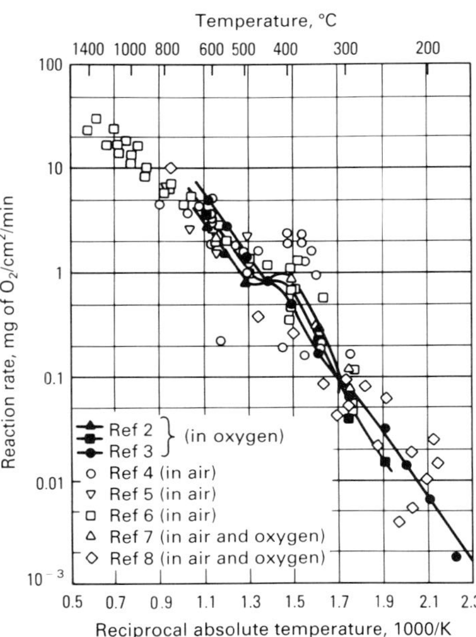

Fig. 1 Oxidation rates of uranium in air and in oxygen

$$K = 4.1 \times 10^8 r^{1/2} \exp\left(\frac{-15\,500}{RT}\right) \qquad \text{(Eq 7)}$$

where r is the fractional relative humidity. These data and Eq 7 agree very well with the results in Ref 10 and 11 (Fig. 2). Both investigators found a water vapor pressure dependency of 1:2.1. The crystal structure of the UO_2 is fcc. The reaction product forms loose flakes that readily spall from the base metal. The quantity of hydrogen produced is typically from 1 to 12% less than the 2 mol predicted for every mole of oxide produced.

Water Vapor-Oxygen Mixtures. Until the work done in the early 1980s, which is discussed in Ref 11, it was thought that uranium in a water vapor-oxygen mixture reacted with water vapor, forming UO_2 and releasing hydrogen that combined with the free oxygen to form more water vapor. This work, combining both thermogravimetric and gas spectrometric techniques, showed

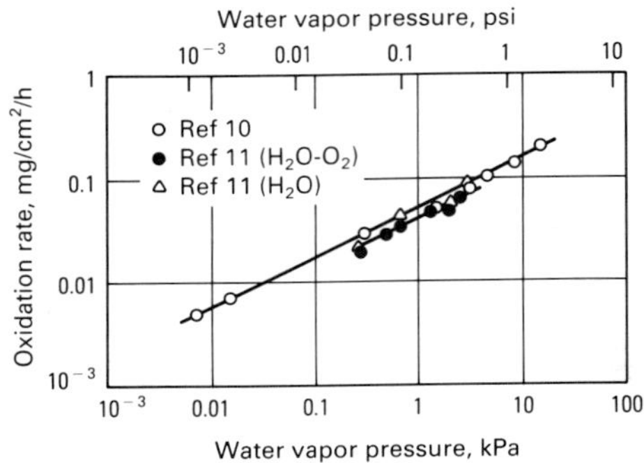

Fig. 2 Oxidation rate of uranium at 100 °C (212 °F) versus water vapor pressure

that uranium reacts with oxygen directly in a water vapor-oxygen mixture. The water vapor affects the uranium oxide structure formed, producing a more defective, or nonstoichiometric, oxide than that formed in the absence of water vapor. Thus, uranium reacts with oxygen in the water vapor-oxygen mixture according to:

$$4U + 9/2\ O_2 \rightarrow U_4O_9 \quad \text{(Eq 8)}$$

The most recent and extensive investigation of the uranium-oxygen-water vapor system measured the kinetics between 40 and 100 °C (105 and 212 °F), and between 11 and 75% relative humidity (Ref 12). The reaction rate is given by:

$$K = 7.6 \times 10^{13} \exp\left(\frac{-26\ 400}{RT}\right) \quad \text{(Eq 9)}$$

Equation 9 is linear and not proportional to either the water vapor pressure (relative humidity) or the oxygen pressure. However, at oxygen concentrations from 100 to 10 ppm, the reaction reverts to the uranium-water vapor reaction shown in Fig. 3. The reaction product, identified by x-ray diffraction analysis (Ref 11) as U_4O_9 or $UO_{2.25}$, is fcc and is compact and adherent. No hydrogen is produced.

Water. Although numerous papers and review articles have been written concerning the reactivity of uranium in water, that is, the corrosion of uranium, most of this work is decades old. The consensus from these investigations is that the rate of corrosion of uranium in water at 25 °C (75 °F) equals the rate of oxidation of uranium in water vapor at 100 °C (212 °F) as the relative humidity approaches 100%, that rate being 0.57 mg/cm²/h (Ref 13). The reaction product has been identified as $UO_3{\cdot}0.8H_2O$ (Ref 13).

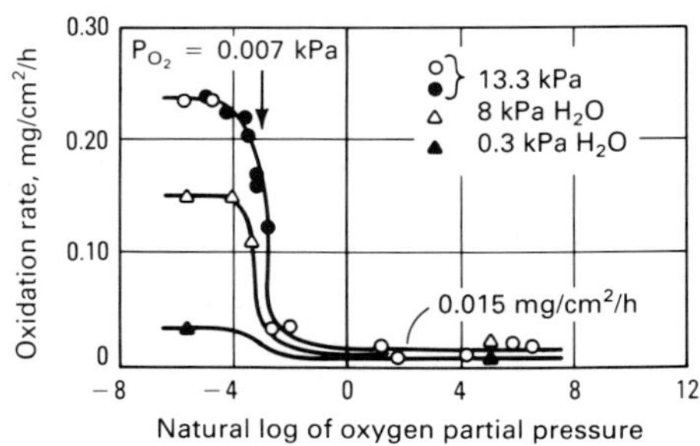

Fig. 3 Oxidation rate of uranium versus oxygen pressure in water vapor at 100 °C (212 °F). Source: Ref 10

Aqueous Solutions. The corrosion properties of uranium in aqueous solutions are highly dependent on the pH of the medium, owing to the participation of hydrogen ion (H^+) in the redox reactions involved in corrosion. General corrosion in caustic aqueous solutions in the presence of oxygen normally proceeds by coupling the anodic dissolution reaction to the oxygen to hydroxl ion (OH^-) cathodic reaction. The H^+ ion to hydrogen gas reaction does not seem to become significant until a low pH is reached. Factors in the specific corrosion mechanism that lead to the accumulation of hydrogen in the region of the surface film and film/metal interface cause increased general corrosion rates.

Effect of Alloying Elements on Oxidation Response

Most of the uranium alloys under study and in use have come from binary and higher combinations of the γ-miscible elements titanium, molybdenum, niobium, and zirconium. Addition of these elements stabilizes the γ phase and increases corrosion resistance.

Titanium. The most recent and extensive study involving the oxidation of a uranium-titanium alloy investigated the oxidation of a U-0.75Ti alloy in environments containing oxygen and/or water vapor at 140 and 100 °C (285 and 212 °F) (Ref 14). The reaction rate of U-0.75Ti with water vapor at 140 °C (285 °F) was found to be linear and proportional to the water vapor pressure ($K \propto P_{H_2O}^{1:2.1}$) for water vapor pressures between 267 and 2670 Pa (2 and 20 torr) (Fig. 4). Hydrogen was produced by the reaction at a rate of approximately 2 mol for every 1 mol of UO_2 formed. A pressure dependency of one-half suggests that a dissociative/adsorption equilibrium exists at the reaction interface.

In contrast, the behavior exhibited by U-0.75Ti when tested at 100 °C (212 °F) in water vapor showed no pressure dependency on water vapor after approximately 50 h of exposure (Fig. 4). The rate measured was 1.4×10^{-3} mg/cm²/h. This occurrence was believed to be due to the reaction rate being controlled by a solid-state diffusion process rather than a dissociative adsorption process.

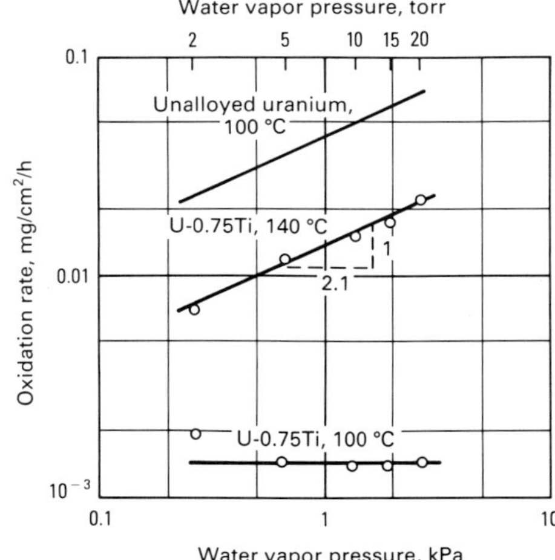

Fig. 4 Oxidation rate of U-0.75Ti versus water vapor pressure. Source Ref 14

The reaction rate of U-0.75Ti with oxygen was also linear and independent of the oxygen pressure for oxygen pressures between 67 Pa and 133 kPa (0.5 and 1000 torr) at both temperatures. The reaction rates at 140 and 100 °C (285 and 212 °F) were approximately 3.2×10^{-3} and 1.2×10^{-4} mg/cm²/h, respectively. The reaction rate was thought to be controlled by the reaction of the metal with oxygen ions (O^{2-}) at the metal/oxide interface.

The reaction rate of U-0.75Ti in environments containing both water vapor and oxygen was found to be linear and not dependent on either water vapor or oxygen pressure for the ranges investigated at both temperatures. The oxidation rates were 3.7×10^{-3} and 1.5×10^{-4} mg/cm²/h at 140 and 100 °C (285 and 212 °F), respectively. Thus, the addition of water vapor to a pure oxygen environment had only a small effect on the oxidation kinetics. Also, the mechanism for oxidation in the mixed environment was thought to be the same as that in oxygen, that is, the reaction of metal with O^{2-} ions at the metal/oxide interface.

The reaction products were identified as oxygen-rich variants of the form UO_{2+x}, where x is greater than 0 but less than 0.25. X-ray diffraction analyses of samples from each of the three environments showed that the crystal structure of the oxides was fcc.

The oxidation rates for this alloy were one to two orders of magnitude lower than the rates for uranium in equivalent environments. Otherwise, the alloy behaved similarly to unalloyed uranium in all three environments. In general, the oxidation resistance of uranium increases with increasing additions of titanium up to an addition level of 2 wt% (Ref 15).

Molybdenum. The published literature on the atmospheric corrosion of uranium-molybdenum alloys is very limited. The available literature on the corrosion of all uranium-molybdenum alloys in both moist air and moist nitrogen has been reviewed. The only conclusions that can be made are that these alloys have a corrosion resistance comparable to uranium-titanium alloys of the same atomic percentage and that the corrosion resistance increases with molybdenum content.

Niobium. Although a number of investigations of the oxidation of uranium-niobium binary alloys have been reported in recent years, the two studies that included the most alloys and controlled conditions are discussed in Ref 16 and 17. In the first study, the alloys were tested in water-saturated nitrogen or oxygen at 75 °C (165 °F), and in the second, specimens were exposed to pure water, water vapor plus oxygen, or water vapor plus air at 125 °C (255 °F). In both experiments, the alloys showed considerably reduced reaction rates compared to unalloyed uranium, with the resistance of the alloy to reaction with the moisture increasing with niobium content, and the presence of oxygen in the test environment reduced the rate of hydrogen evolution. In all cases, more hydrogen was absorbed by the alloy than was released into the atmosphere. The amount of hydrogen released into the atmosphere decreased significantly with increasing niobium content, but all of the alloys absorbed roughly the same amount of hydrogen.

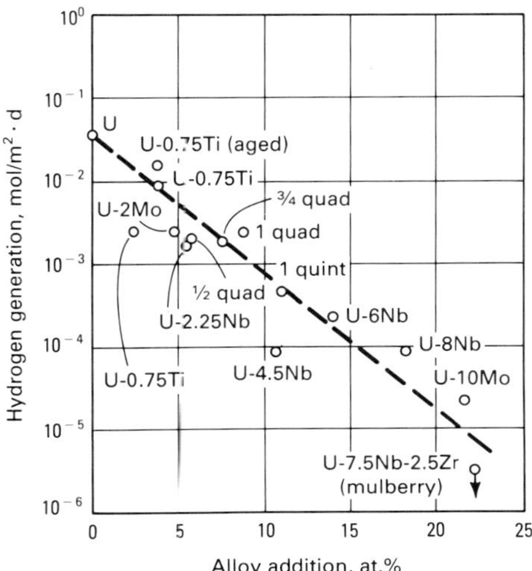

Fig. 5 Rate of hydrogen generation versus alloy addition of selected uranium alloys. ½ quad, U-0.5Nb-0.5Mo-0.5Zr-0.5Ti; ¾ quad, U-0.75Nb-0.75Mo-0.75Zr-0.5Ti; 1 quad, U-1Nb-1Mo-1Zr-0.5Ti; 1 quint, U-1Nb-1Mo-1Zr-0.5Ti-0.5V. Source: Ref 16

The most recent investigation of a uranium-niobium alloy involved the atmospheric corrosion of the U-6Nb alloy (Ref 18). The temperature range investigated was between 50 and 110 °C (120 and 230 °F), and the corresponding relative humidity ranged from 75 to 17%. The oxidation rate of U-6Nb in water vapor was so slow that it was not possible to derive an equation for the corrosion rate, even at 110 °C (230 °F). In moist air, the reaction was slightly faster than had been measured in moist nitrogen, but was still about 1/500 of the rate for unalloyed uranium.

Niobium-Zirconium. The uranium-niobium-zirconium alloy of most interest is a U-7.5Nb-2.5Zr alloy designated as mulberry, which was given the label of a "stainless" uranium alloy. In humid environments, mulberry reacts similarly to the binary U-6Nb alloy. Oxidation data for mulberry in humid environments are provided in Ref 19.

Corrosion Behavior of Uranium Alloys

The factors that control metallic corrosion can be categorized as thermodynamic or kinetic. The electrochemical principles of thermodynamics are those of reversible cells and standard potential; kinetics is governed by polarization techniques.

Humid Air. Figure 5 illustrates the relationship between the percent of alloying additions and corrosion response in hot, humid air. The data show the rate of hydrogen generation of selected uranium alloys exposed to 100% relative humidity in oxygen at 75 °C (165 °F). This relationship is shown as approximately linear. Thus, increasing the total amount of alloying additions, irrespective of the alloying element, increases the resistance to corrosion.

Water. In addition to work at Argonne National Laboratory, Battelle Columbus Laboratories, and the Army Materials and Mechanics Research Center (AMMRC), a series of Russian articles (Ref 20) discusses the relative resistance of numerous uranium alloys to boiling water at 100 °C (212 °F). It has been shown that the

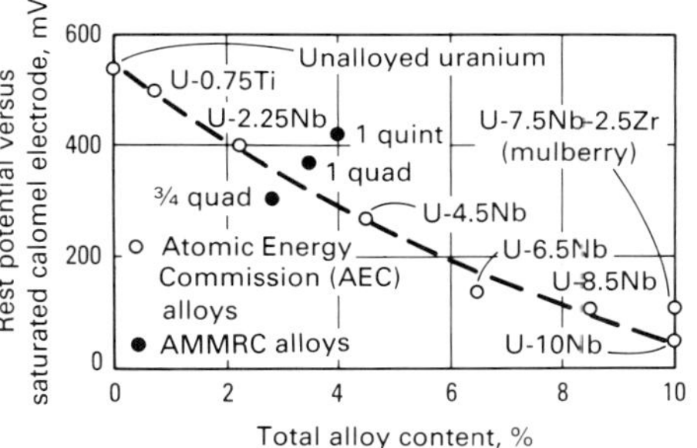

Fig. 6 Rest potential versus total alloy content for uranium alloys in 0.001 *M* KCl at 25 °C (75 °F). ½ quad, U-0.5Nb-0.5Mo-0.5Zr-0.5Ti; ¾ quad, U-0.75Nb-0.75Mo-0.75Zr-0.5Ti; 1 quad, U-1Nb-1Mo-1Zr-0.5Ti; 1 quint, U-1Nb-1Mo-1Zr-0.5Ti-0.5V

corrosion resistance of uranium alloys to boiling water is inversely proportional to the percentage of α-uranium present in the particular compositions. An increase in the corrosion resistance with alloying is due to the formation of the γ phase. The most promising alloying additives—molybdenum, niobium, and zirconium—have been extensively studied.

Dilute Salt Solutions. Because the corrosion resistance of uranium alloys in hot, moist air and in boiling water has been found to be proportional to the total alloying content, it is expected that this proportionality also holds true for corrosion in dilute salt solutions. Electrochemical measurements have indicated that this should be the case. In one study, standard or corrosion potentials of various uranium alloys were measured in a 0.001 *M* potassium chloride (KCl) solution against a standard calomel electrode (Ref 21). Figure 6 shows these measured potentials versus the total alloying content of the alloys. Although the relationship is not strictly linear between potential and alloy content, it is directly related. Kinetic data on the corrosion rates of uranium alloys in dilute salt solutions are scarce, but those that have been published confirm this relationship.

Ocean Water. As with the preceding corrosion environments, the corrosion susceptibility of uranium alloys in ocean water is expected to decrease with increasing alloy content. In one investigation, the corrosion potentials of a number of uranium alloys were measured in artificial seawater (Fig. 7). The observed trend of decreasing corrosion potential with increasing alloying content is very similar to that discussed previously for the dilute salt solution test. Thermodynamics again predicts a decreasing corrosion rate with increased alloy content. The kinetics of the uranium alloy-ocean water reaction were measured gravimetrically, the results showed a logarithmic relationship between the corrosion rate and the total alloy content (Fig. 8).

Acids and Bases. An indication of the possible anodic reactions can be obtained by examining the Pourbaix diagram for uranium shown in Fig. 9. In the potential range from −1.8 to 1.2 V at low pHs (0 to 2.0), uranium forms primarily soluble species. The uranium ion U^{3+} forms in the active region near the corrosion potential, and the uranyl ion (UO_2^{2+}) forms in the transpassive region. In the passive region, UO_2 undoubtedly forms. Anodic polarization techniques can be used to study the ease of transition from the active to the passive state as well as the dissolution behavior of the metal and its alloys. The transition from the active to the passive state is accompanied by a decrease in corrosion rate of the order of 10^4 to 10^6, which is extremely significant for many applications.

Anodic polarization techniques have been used to study the effects of alloying constituent, temperature, solution chemistry, solution concentration, pH, and presence of chloride on the corrosion response of uranium alloys (Ref 24). An example of the effect of alloying on anodic polarization is shown in Fig. 10; the passive current densities vary inversely with alloy content. Figure 11 shows an example of the effect of solution chemistry, particularly the addition of chloride, on the anodic polarization behavior. The uranium-molybdenum alloy passivates more easily in sodium sulfate (Na_2SO_4) than sulfuric acid (H_2SO_4), and the addition of chloride prevents passivation entirely.

Additional conclusions have been reached based on this work, as follows. Uranium binary alloys exhibit active-passive behavior in sodium hydroxide (NaOH), ammonium hydroxide (NH_4OH), sodium nitrate ($NaNO_3$), sodium chromate (Na_2CrO_4), and ammonium chromate ($[NH_4]_2CrO_4$). The critical current densities for passivity were inversely proportional to the H_2SO_4 concentration. Unlike those of most metals, the dissolution rates of uranium alloys decrease with increasing acid concentration. Chloride additions as small as 0.005 *M* affect the

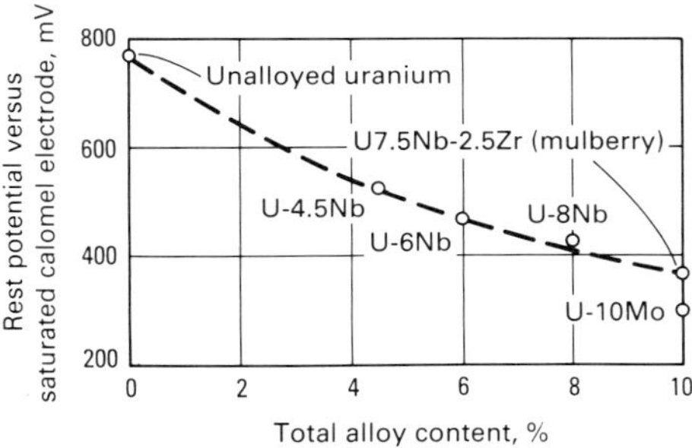

Fig. 7 Rest potential versus total alloy content for uranium alloys in ocean water at 20 °C (70 °F). Source: Ref 22

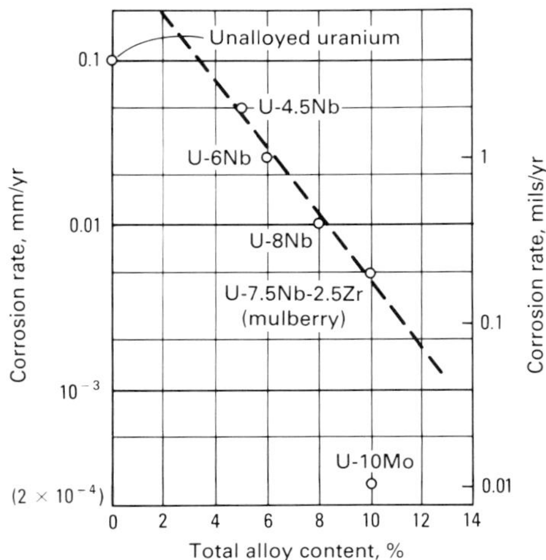

Fig. 8 Corrosion rate versus total alloy content for uranium alloys in ocean water at 20 °C (70 °F). Source: Ref 23

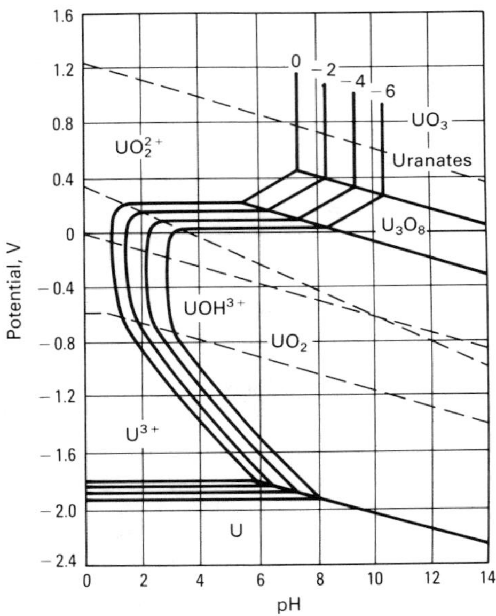

Fig. 9 Pourbaix diagram for uranium

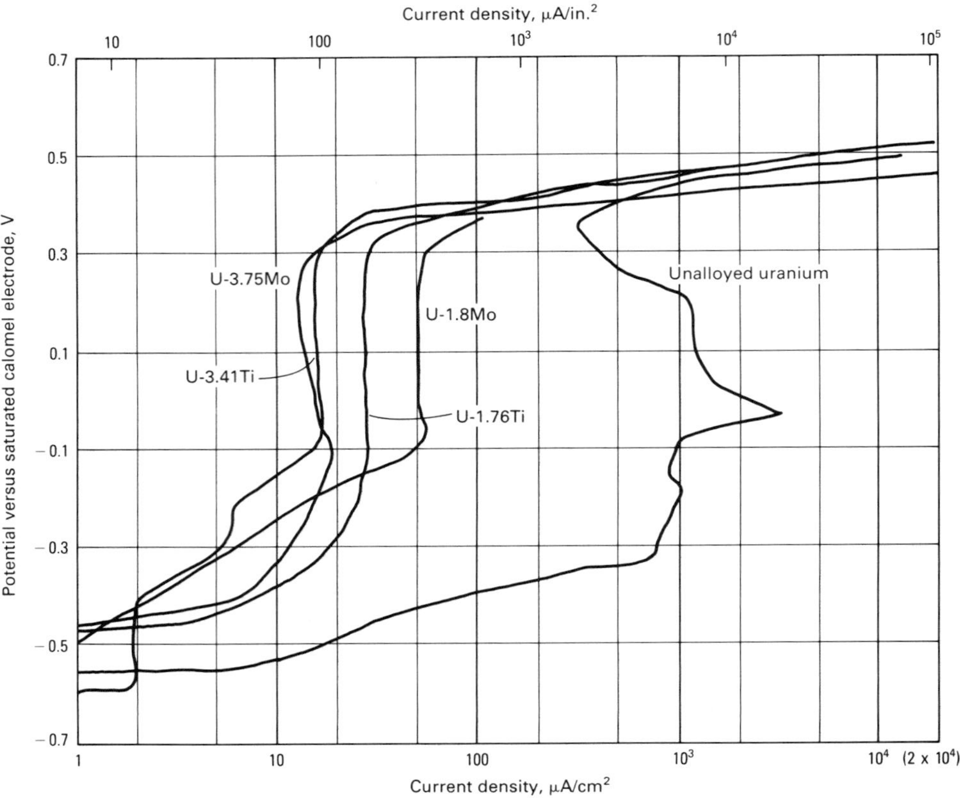

Fig. 10 Effect of alloying on the anodic polarization of uranium in H_2SO_4 at 25 °C (75 °F)

anodic polarization curve, but chromates, sulfates, and nitrates inhibit pitting at this low concentration of chloride. The uranium-titanium alloys were found to be more resistant to basic solutions than the uranium-molybdenum alloys.

Effect of Microstructure on Corrosion Properties

The influence of microstructure on the corrosion properties of uranium alloys is very much dependent on the particular uranium alloy. For example, the corrosion response of the U-0.75Ti alloy changes very little with process variables, but the U-6Nb alloy changes significantly.

U-0.75Ti. Research conducted for the United States Air Force measured the corrosion response of U-0.75Ti penetrators manufactured by various processes. The processing parameters included those in extrusion (temperature and rate), solutionizing (temperature, time, and media), aging (temperature, time, and atmosphere), and quenching (rate and media). In addition, U-0.75Ti penetrators made by such forming techniques as casting, swaging, forging, or grinding were tested. In all, more than 50 different combinations of processing techniques were investigated. The test environments chosen were a salt fog (to satisfy a military specification) and hot, moist nitrogen. The results showed that the largest difference in corrosion response to the salt fog environment between any two lots of material was a factor of two (Ref 25, 26). For hot, moist nitrogen testing, the difference in final weight loss between the most and least corroded material was a factor of ten. In that the weight loss in moist nitrogen for even the most susceptible material was very small, these differences were indeed small. Thus, the metal processing of the U-0.75Ti alloy did not significantly change the corrosion response.

U-6Nb. An investigation of the effect of cooling rate on the microstructure, mechanical behavior, corrosion resistance, and subsequent age hardenability of U-6Nb revealed that cooling rates exceeding 20 °C/s (36 °F/s) cause the parent γ phase to transform martensitically to a niobium-supersaturated variant of the α phase (Ref 27). This martensitic phase exhibits low hardness and strength, high ductility, good corrosion resistance, and substantial age hardenability. As cooling rates decrease from 10 °C/s (18 °F/s) to 0.2 °C/s (0.4 °F/s), fine-scale microstructural changes (consistent with spinodal decomposition) occur to an increasing extent. These changes were found to produce large increases in hardness and strength as well as large decreases in ductility, slight decreases in corrosion resistance, and slight changes in age hardenability. At cooling rates less than 0.2 °C/s (0.4 °F/s), the parent phase undergoes cellular decomposition to a coarse two-phase lamellar microstructure. This lamellar microstructure was seen to exhibit intermediate strength and ductility, substantially reduced corrosion resistance, and no age hardenability.

Figure 12 shows the corrosion behavior of U-6Nb as influenced by cooling rate. Very rapid cooling results in a low rest potential, as would be expected for a corrosion-resistant material. At low cooling rates, the rest potential is substantially higher than that for any of the intermediate or high cooling rates, which suggests that a substantial decrease in corrosion resistance occurs at low cooling rates. Corrosion tests done at 75 °C (165 °F) in 95% relative humidity air or nitrogen confirmed the inferences drawn from the rest potential measurements. Samples cooled at intermediate and high rates exhibited no measurable weight change or visual tarnishing, but the samples cooled at low rates formed a black surface film and had an average weight gain of 0.01 mg/cm²/d (equivalent to a penetration rate for UO_2 of 17 μm/yr, or (0.67 mils/yr). More information on the microstructures and metallography of uranium and uranium alloys is available in the article "Uranium and Uranium Alloys" in Volume 9 of the 9th Edition of Metals Handbook.

Galvanic Behavior

Some uses of uranium-base alloys require that they be in contact with other metals. When two dissimilar metals are electrically connected and immersed in a suitable electrolyte, a galvanic current flows between them. The driving force for the current flow is the potential difference between the two dissimilar metals. An electromotive force (emf) series is a list of metals and alloys arranged in order of the potentials generated when electrodes of each material are compared with one another in a specific environment. An emf series of a chloride-containing aqueous solution is often used for purposes of comparison for many applications.

Uranium-Niobium Alloys. The galvanic-corrosion behavior of uranium and uranium alloys, particularly uranium-niobium alloys, was investigated in ocean water and hydrochloric acid (HCl), and the emf series of selected materials generated in ocean water and in 0.1 *N* HCl are given in Tables 1 and 2. In addition, galvanic couples were immersed in 0.1 *N* HCl for 4 to 96 h, and the corrosion rates were determined for each

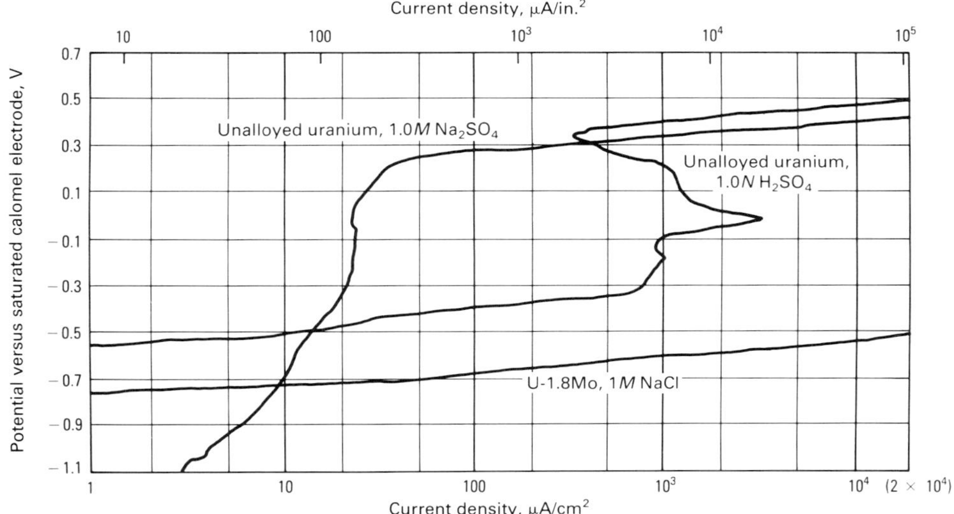

Fig. 11 Anodic polarization curves for uranium in differing aqueous solutions at 25 °C (75 °F)

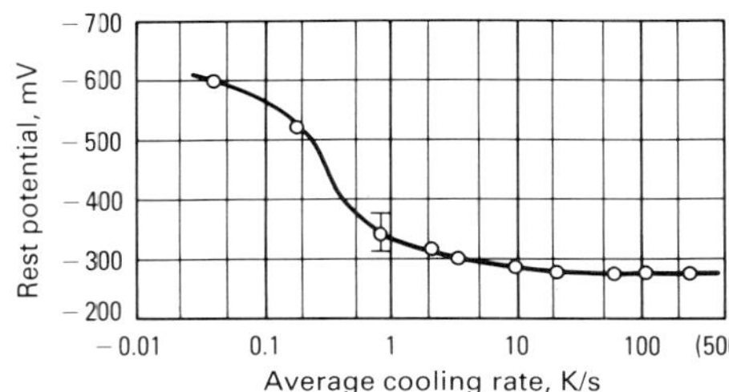

Fig. 12 Effects of cooling rate on rest potential of U-6Nb in 0.001 *M* KCl. The error bar indicates the uncertainty in the potential measurement. Source: Ref 27

couple. The emf series, based on electrode potentials in 0.1 *N* HCl at 25 and 70 °C (75 and 160 °F) and in oxygen-saturated ocean water, was found to be essentially the same as the series based on the gravimetric galvanic-corrosion tests. Therefore, the emf series based only on electrode potentials is useful for indicating potential galvanic-corrosion effects.

Polynary Uranium Alloys. The galvanic-corrosion behavior of polynary uranium alloys, particularly quaternary alloys, was studied in humid air and 4% salt spray; the results are summarized in Table 3. Galvanic corrosion became significant in water-saturated and condensing conditions, the different quad alloys behaved similarly, and aluminum alloy 6061-T6 and AISI type 304 stainless steel showed the best compatibility with the uranium alloys in resistance to humid atmospheres.

Stress-Corrosion Cracking of Uranium Alloys

One of the problems that has been encountered in the use of uranium alloys is SCC. Stress-corrosion cracking is defined as a cracking process that requires the simultaneous action of a corrosive environment and a sustained stress. In general, as the corrosion resistance and strength of an alloy increase, its susceptibility to SCC also increases. The conventional test methods use time to failure measurements on smooth specimens, incorporating both initiation and propagation processes, or precracked specimens in a fracture mechanics test, which tends to eliminate the initiation stages.

Uranium-Titanium Alloys. The U-0.75Ti alloy has been the only composition in the U-Ti system that has received a significant amount of stress-corrosion testing. Crack propagation tests using precracked specimens conducted in oxygen, hydrogen, water vapor, water, dry air, wet air, and chloride solutions have shown that water is the species responsible for cracking (Table 4). Oxygen apparently inhibits crack propagation in this alloy. The effect of the strength level on the susceptibility of U-0.75Ti to SCC has also been investigated (Ref 30). Figure 13 shows the relationship between ultimate tensile strength and K_{Iscc}, the threshold stress intensity for SCC. The data show that K_{Iscc} decreased linearly by a factor of two in the strength range investigated.

Table 1 Electromotive force series of selected materials in 0.1 *N* HCl

Alloy	Electrode potentials, mV: Oxygen-saturated, 25 °C (75 °F)	Air-equilibrated 25 °C (75 °F)	Air-equilibrated 70 °C (160 °F)
Aluminum alloy 7178	−740	−760	−805
Uranium	−740	−755	−790
U-4.5Nb	−465	−475	−600
4340 steel	−445	−460	−470
U-6Nb	−395	−420	−465
U-8Nb	−375	−400	−445
Ti-6Al-4V	−310	−385	−375
Mulberry	−305	−340	−410
U-10Mo	−170	−190	−240
Type 304 stainless steel	230	75	−230

Source: Ref 22

Table 2 Electromotive force series of selected materials in ocean water at 25 °C (75 °F)

Ocean water prepared according to ASTM D 1141

Alloy(a)	Electrode potential, mV: Oxygen-saturated	Air-equilibrated
Aluminum alloy 7178	−800	−800
Tuballoy (depleted uranium)	−770	−795
U-4.5Nb	−525	−530
4340 steel	−480	−540
U-6Nb	−470	−460
U-8Nb	−430	−415
Ti-6Al-4V	−390	−350
Mulberry	−370	. . .
U-10Mo	−300	. . .
Type 304 stainless steel(b)	−225	−250

(a) Alloys are listed in order of increasing mobility from top to bottom as determined by electrode potentials in oxygen-saturated ocean water. (b) Passive. Source: Ref 22

All of the tests conducted in this alloy system found that transgranular cracking was the SCC mode, and there was no evidence of cracking along prior γ-phase grain boundaries.

Uranium-Molybdenum Alloys. Uranium alloys with molybdenum concentrations from 0.6 to 12% have been found to be susceptible to SCC (Ref 29). Below 5% Mo, a series of metastable α-phase alloys is formed that has been found to be susceptible to cracking. Tests done in both wet and dry conditions on γ-phase alloys containing more than 5% Mo have shown that, contrary to the behavior observed with uranium-titanium alloys, oxygen is the species primarily responsible for SCC. Carbon content is important to the cracking behavior, alloys with higher carbon contents being more susceptible. The heat treating of quenched alloys can lead to significant improvements in SCC resistance in that alloys with equilibrium microstructures are much more resistant to cracking than metastable alloys. The predominant fracture mode observed in uranium-molybdenum alloys has been transgranular, particularly in nonaqueous environments. An intergranular mode has been observed in precracked tests conducted in aqueous environments (Ref 31).

Uranium-Niobium Alloys. The principal interest in the U-Nb system has centered on alloys with 2.3, 4.5, 6, and 8% Nb by weight, respectively. The SCC response is very much dependent on the particular alloy. The 2.3 and 4.5% Nb alloys are subject to SCC in water vapor, but the 6% Nb alloy is not. However, 6 and 8% Nb alloys will crack in environments containing oxygen, and water vapor will accelerate the SCC in these environments (Ref 32).

The U-4.5Nb alloy was shown to crack in either water vapor or oxygen and was more susceptible when both were present (Ref 32). Figure 14 shows that for the U-4.5Nb alloy the crack velocity decreases and K_{Iscc} increases as the oxygen pressure is decreased (Ref 32). The chloride ion (Cl^-) was also found to be very deleterious to the U-4.5Nb alloy in aqueous solutions (Ref 33). Specimens loaded at 33MPa$\sqrt{m}$ (30 ksi$\sqrt{in.}$) failed in 270 min in distilled water and in as little as 1 min in chloride solutions. However, another study revealed that the U-6Nb alloy was not susceptible to SCC initiation in chloride solutions when aged at temperatures below 200 °C (390 °F), but material aged between 250 and 400 °C (480 and 750 °F) was susceptible (Ref 34). When overaged (aged at temperatures above 600 °C, or 1110 °F), the alloy exhibited an increased resistance to cracking. In general, the data have shown for all of the uranium-niobium alloys that overaged specimens are more resistant to SCC than underaged specimens.

Table 3 Galvanic corrosion of uranium alloy couples

Couple	Test duration, days	Environment	Surface observation	Appearance of interface at 100×: Uranium	Appearance of interface at 100×: Other material
U-593/stainless steel(a)	90	4% NaCl	Heavy oxidation of U	Slight pitting	Normal
U-581/aluminum(b)	90	4% NaCl	Heavy oxidation of U	Corrosive attack	Slight pitting
U-581/stainless steel(b)	180	30% relative humidity	No apparent change	Normal	Normal
U-583/aluminum(c)	180	30% relative humidity	No apparent change	Normal	Normal
U-584/1042 steel(d)	76	30% relative humidity	No apparent change	Normal	Slight pitting
U-584/4340 steel(d)	76	30% relative humidity	No apparent change	Slight pitting	Slight pitting
U-584/1042 steel(d)	76	100% relative humidity	Oxidation	Normal	Corrosive attack
U-584/4340 steel(d)	76	100% relative humidity	Oxidation	Corrosive attack	Slight corrosive attack
U-584/1042 steel(d)	76	4% NaCl	Corrosion and heavy oxidation	Deep pitting	Slight corrosive attack
U-584/4340 steel(d)	76	4% NaCl	Corrosion and oxidation	Deep pitting	Slight corrosive attack
U-581/stainless steel(b)	180	100% relative humidity	Corrosion and oxidation	Slight corrosive attack	Slight corrosive attack
U-581/aluminum(b)	180	100% relative humidity	Corrosion and oxidation	Slight corrosive attack	Slight corrosive attack

Uranium alloy compositions: (a) U-0.99Mo–1.02Nb–1.27Zr–0.49Ti; (b) U-1.40Mo–1.50Nb–1.47Zr-0.52Ti; (c) U-1.80Mo–2.01Nb–1.90Zr–0.54Ti; (d) U-1.03Mo–1.04Nb–0.98Zr–0.62-Ti. Source: Ref 28

Table 4 SCC thresholds for U-0.75Ti in air and in aqueous environments

Heat treatment	Yield strength, MPa	Yield strength, ksi	Environment	K_{Iscc}, MPa$\sqrt{m}$	K_{Iscc}, ksi$\sqrt{in.}$
380 °C (715 °F), 6 h	986	143	Dry air	42	38
380 °C (715 °F), 6 h	986	143	100% relative humidity air	27.5	25
380 °C (715 °F), 6 h	986	143	50 ppm Cl^-	17.6	16
380 °C (715 °F), 6 h	986	143	3.5% NaCl	17.6	16
Air cooled	607	88	Water	23	21
Air cooled	607	88	3.5% NaCl	16.5	15
450 °C (840 °F)	...	...	Air	28.6	26
450 °C (840 °F)	...	...	Water	15.4	14
450 °C (840 °F)	...	...	3.5% NaCl	11	10

Source: Ref 29

Table 5 Threshold stress for SCC initiation of mulberry in dilute chloride solutions as a function of heat treatment

Heat treatment	K_{Iscc}, MPa$\sqrt{m}$	K_{Iscc}, ksi$\sqrt{in.}$
150 °C (300 °F), 1 h	<9	<8
350 °C (660 °F), 25 h	<13	<12
450 °C (840 °F), 5 h	~16.5	~15
550 °C (1020 °F), 9 h	>33	>30
600 °C (1110 °F), 8 h	>71	>65

Source: Ref 35

Fig. 13 Plane-strain threshold for SCC propagation versus ultimate tensile strength for U-0.75Ti in 50 ppm Cl^- solution at 25 °C (75 °F). Source: Ref 30

Both transgranular and intergranular cracking was observed in the U-Nb system, with the mode observed being dependent on the alloy composition, the environment, and the type of specimen. In chloride solutions, smooth-specimen tests showed that the U-2.3Nb alloys cracked transgranularly and that the U-6Nb and U-8Nb alloys cracked intergranularly. The U-4.5Nb alloy exhibited intergranular cracking in smooth specimens and transgranular cracking in precracked specimens.

Uranium-Niobium-Zirconium Alloys. The only uranium-niobium-zirconium alloy that has been studied to any significant extent is mulberry. The SCC behavior of mulberry is similar to that of U-6Nb. The SCC behavior of this alloy was extensively studied in the 1970s and was later reviewed in Ref 29 and 32. Two types of SCC are observed in the alloy, each with its own fracture mode. The intergranular cracking initiates easily, but requires oxygen, water, and chloride to do so. The transgranular SCC does not initiate easily, but can propagate in pure oxygen or in solutions containing strong oxidizers.

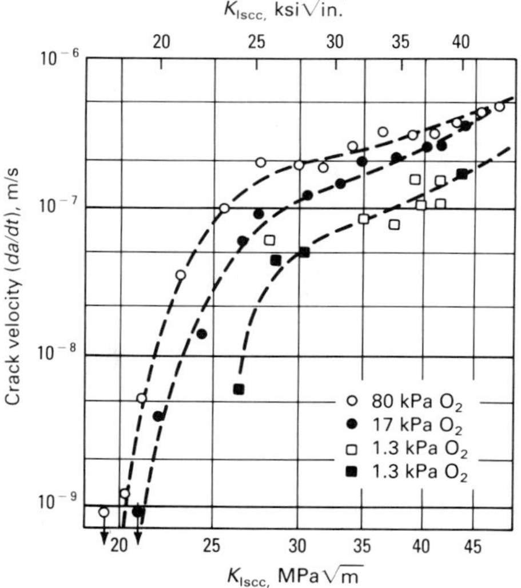

Fig. 14 Crack velocity versus SCC threshold as a function of oxygen pressure for U-4.5Nb. Source: Ref 32

An investigation of the effect of heat treatment on the SCC susceptibility of mulberry showed that the higher the aging temperature the greater the resistance to SCC initiation (Table 5). It was also found that crack propagation was fastest for material aged at intermediate temperature (350 to 450 °C, or 660 to 840 °F) and was slowest in the material aged at 150 °C (300 °F), the standard aging temperature for this alloy.

Polynary Uranium Alloys. The SCC behavior of a series of polynary uranium alloys containing small quantities of all the potent γ stabilizers (titanium, niobium, molybdenum, and zirconium) has been investigated, primarily by AMMRC (Ref 36). All of the SCC studies have been conducted on as-extruded material that has a partial microstructure of equilibrium phases in addition to the α phase from quenching. Precracked specimen tests have yielded K_{Iscc} values for polynary alloys tested in distilled water and in chloride solutions (Table 6). The data show that chloride increases the susceptibility to SCC. Tests done on smooth specimens have shown that these alloys are very resistant to crack initiation in humid air, but water promotes crack propagation in precracked specimens. All of the SCC failures have resulted in a transgranular stress-corrosion fracture.

Protective Coatings

The corrosion of unalloyed uranium under conditions of many applications was sufficiently fast to require the use of protective coatings. Most of the standard coatings for uranium alloys were initially developed and tested on unalloyed uranium. The following materials systems have been implemented in an attempt to develop the best coating system: ceramics for protective oxides, polymer science for organic films such as paints, and metallurgy for ion-plated and electroplated metallics. More information on protective coatings for metals is available in the Section "Corrosion Protection Methods" in this Volume.

Protective Oxides. An advantage of protective oxide coatings over other coatings is the

Table 6 SCC thresholds for polynary uranium alloys in aqueous solutions

Alloy	Yield strength MPa	Yield strength ksi	Environment	K_{Iscc} MPa$\sqrt{m}$	K_{Iscc} ksi$\sqrt{in.}$
U-0.75Nb-0.75Mo-0.75Zr-0.5Ti	772	112	Water	44	40
			3.5% NaCl	13	12
U-1Nb-1Mo-1Zr-0.5Ti	1172	170	Water	31	28
			50 ppm Cl^-	10	9
			3.5% NaCl	8	7
U-1Nb-1Mo-1Zr-0.5Ti-0.5V	1627	236	Water	10	9
			50 ppm Cl^-	5.5	5
			3.5% NaCl	5.5	5

small dimensional changes of the part that accompany the conversion of the metal at the surface to oxide. There have been two approaches to the application of protective oxides for uranium and uranium alloys: thermally grown and anodically grown. The first attempts at using thermally grown oxides were made in the late 1960s (Ref 37). In these investigations, the oxygen supply to the uranium surface was controlled while the material was heated at elevated temperatures (near 625 °C or, 1155 °F). Subsequent corrosion testing of the specimens showed that the onset of corrosion was delayed by the protective oxides but that once corrosion began the rate was the same as for unprotected uranium. Other efforts in this area, primarily at the Oak Ridge National Laboratory, were not any more successful, and this technique has not found a hardware application.

Another study examined films grown anodically on uranium in aqueous ammonium borate ($[NH_4]_2B_4O_7$), ethanol, ethylene glycol, and combinations of these liquids (Ref 38). Subsequent corrosion testing produced the same results shown for thermally grown films, that is, some initial, but not long-lasting, protection. In a third investigation, anodically grown films were applied to specimens of the U-0.75Ti alloy (Ref 39). Subsequent corrosion testing of these materials showed that these films offered no protection from corrosion in moist air (Ref 21). As with the thermally grown films, the anodically grown films have not found a hardware application.

Organic Films. To provide satisfactory inhibition of corrosion, an organic film must possess two essential properties: a very low permeability to water vapor and a relatively high oxygen permeability. The latter property is a safeguard because it is accepted that no organic coating would be completely impermeable to water vapor. A high water vapor concentration combined with a low oxygen concentration at the substrate/coating interface would lead to a high corrosion rate.

Most of the research done on organic coatings for uranium and uranium alloys was conducted at the Atomic Weapons Research Establishment. For example, in one study, eleven paint systems, representing the three major curing mechanisms, were examined in single and multiple coats (Ref 40). Specimens were exposed to hot, moist air, and the coating was given a merit ratio dependent on its response. The results obtained are given in Table 7. None of the coatings could be described as protective under the conditions of test. In light of these observations and results, it was apparent that the major reason for failure of coatings was that the water permeability of the materials was too high to become the rate-determining stage in the corrosion process. Further, because the reaction rate is independent of water vapor pressure in the presence of oxygen, even a reduction in water vapor pressure at the uranium/coating interface will have no effect until the level is reduced below about 1% relative humidity. In the same study, the water permeability of two coatings was reduced by the addition of leafing aluminum powder. Although this technique improved the protectiveness of the coatings, the water permeation rates were still high.

A number of investigators at various facilities have tried paint systems, particularly epoxies, on uranium alloys. For example, U-0.75Ti was coated with a monochlorotrifluoroethylene film (Ref 41), U-2Mo was coated with thermoplastic and thermosetting acrylic resin systems with both water and solvent vehicles (Ref 42), and U-4.2Nb was spray painted with a 75-μm (3-mil) thick epoxy primer with a 50-μm (2-mil) polyurethane cover coat (Ref 43). Corrosion tests were conducted on these coating-alloy systems in hot, moist air, in dilute salt solutions, and in salt spray. All gave results that showed a delay in the onset of corrosion of the base alloy but an eventual corrosion rate that was at least as high as uncoated material and, in some cases, even higher, particularly at flaws in the coating. Therefore, none of these coatings offered long-term protection, and their application on hardware was not pursued. However, it was found that if the monochlorotrifluoroethylene coating is applied as a top coat over an epoxy paint and the part is not subjected to a high-humidity environment, then very good protection is afforded (Ref 41).

Table 7 Weight loss of coated uranium rods exposed to water-saturated air at 70 °C (160 °F)

Coating applied	Number of coats	Corrosion rate, mg/cm³/h	Merit ratio (a)
Uncoated	1	0.028	...
Styrene-butadiene	1	0.026	1.07
Ethyl cellulose	1	0.033	0.84
Natural rubber	1	0.027	1.04
Long oil alkyd	1	0.038	0.75
Long oil alkyd	3	0.058	0.48
Short oil epoxy-ester	1	0.029	0.96
Styrenated epoxy-ester	1	0.029	0.96
Chlorinated rubber	1	0.074	0.37
Chlorinated rubber	3	0.125	0.22
Polyurethane	1	0.035	0.80
Polyurethane	3	0.102	0.27
Amine-cured epoxy	1	0.043	0.65
Amine-cured epoxy	3	0.160	0.17
Polyamide-cured epoxide	1	0.035	0.80
Polyamide-cured epoxide	3	0.097	0.28
Adduct-cured epoxide	1	0.036	0.77
Adduct-cured epoxide	3	0.115	0.24

(a) Ratio of corrosion rate of uncoated specimen to that of coated specimen. Source: Ref 40

Metallic Platings. Although metallic coatings for uranium and uranium alloys have been extensively studied, the two systems used almost exclusively are electroplated nickel and ion-plated aluminum. The most recent paper reviewing the electroplating of uranium and uranium alloys was done in 1982 (Ref 44). A complementary review paper was later written on the ion-plating of uranium alloys (Ref 45).

Electroplating of Uranium. The first promising results for the electroplating of nickel on unalloyed uranium were obtained at the Los Alamos Scientific Laboratory (Ref 46) and the Battelle Memorial Institute (Ref 47). The Los Alamos process, which involved chemical pretreatment of the uranium, resulted in a very coarse etched surface that provided a good bond through mechanical interlocking between the coating and the substrate. The corrosion protection was favorable. The main disadvantage, however, was that metallic tin produced by a displacement reaction in the pretreatment process had to be removed mechanically. An important part of this process, which was retained in subsequent process modifications, was the chemical etch that produced the good interlocking. The Battelle process, involving numerous electrochemical pretreatments, also resulted in a good bond and adequate corrosion protection, but removed a large amount of uranium (about 75 μm, or 3 mils, per surface) and required an outgassing step for best results.

Investigators at Sandia Laboratories developed a chemical pretreatment by first using a stannous chloride/nitric acid ($SnCl_2/HNO_3$) solution and finally a ferric chloride/nitric acid ($FeCl_3/HNO_3$) solution; this eliminated any mechanical removal steps and gave superior coatings (Ref 48, 49). The 16 steps necessary to electroplate nickel onto uranium are:

- Vapor degrease in trichloroethylene
- Caustic soak for 5 min at 70 to 80 °C (160 to 175 °F)
- Water rinse
- Scrub surfaces with pumice
- Water rinse
- Pickle in 35% solution of HNO_3 at room temperature for 2 min
- Water rinse
- Etch in 1400 g/L $FeCl_3$ solution for 10 to 15 min at 55 °C (130 °F)
- Water rinse
- Pickle in 35% solution of HNO_3 at room temperature for 2 min
- Water rinse
- Caustic soak for 5 min at 70 to 80 °C (160 to 175 °F)
- Water rinse
- Pickle in 35% solution of HNO_3 at room temperature for 2 min
- Water rinse
- Plate in nickel sulfamate solution

The first few steps are cleaning processes, such as a vapor degrease to remove organic contaminants and a caustic soak and pumice scrub to remove the oxide film. The HNO_3 pickle removes any remaining traces of oxide. The primary etchant for metal removal is $FeCl_3$. After the etchant step, the cleaning steps of pickling and the caustic soaks are used immediately before electroplating. Table 8 gives the composition of the nickel sulfamate bath from which the nickel is electroplated.

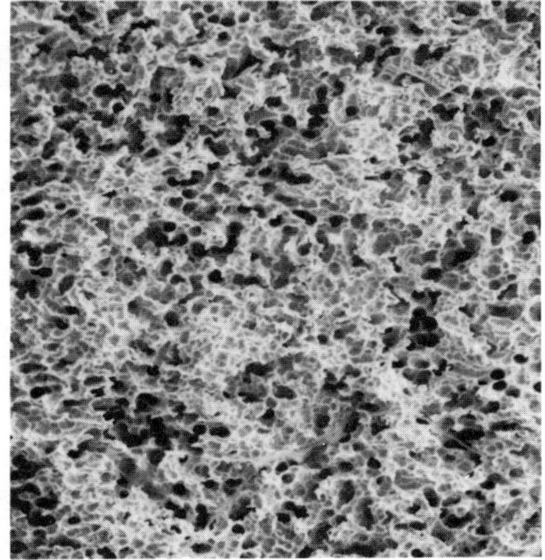

(a)

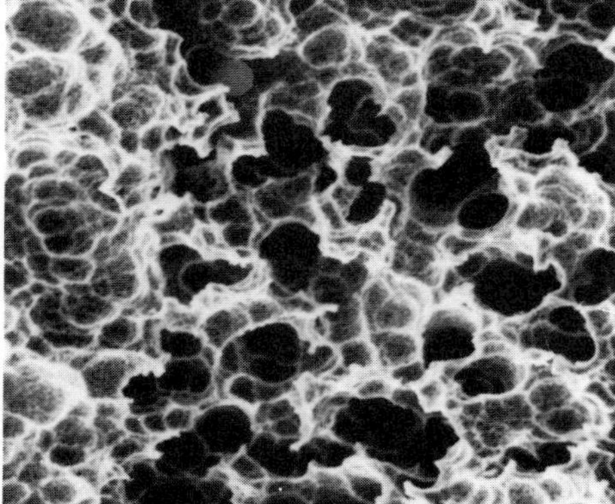

(b)

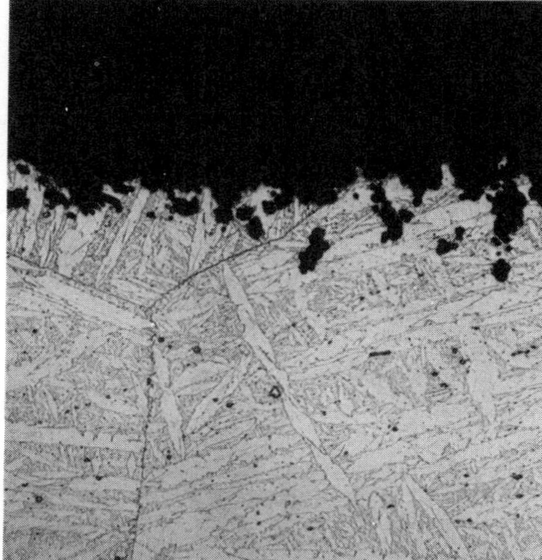

(c)

Fig. 15 Surface morphology (a and b) and cross section (c) of U-0.75Ti after chemical etching. The recesses extending into the base metal facilitate adherence of the electroplate, providing enhanced corrosion protection. (a) and (c) 300×. (b) 1000×

Table 8 Solution compositions and operating parameters for electroplating uranium and uranium alloys

Nickel sulfamate solution	
Nickel sulfamate	450 g/L
Boric acid	30 g/L
Surface tension of solution	0.034–0.038 N/m
pH of solution	3.8–4.0
Temperature of solution	48–50 °C (118–122 °F)
Anodes	Sulfur-depolarized nickel
Filtration	Continuous
Current density	270 A/m² (25 A/ft²)
Zinc plating solution	
Zinc	22.5 g/L
Sodium cyanide	56.5 g/L
NaOH	80 g/L
Cyanide/zinc ratio	2.5/1
Temperature	22 °C (72 °F)
Current density	270 A/m² (25 A/ft²)
Chromate conversion coating	
Proprietary	Concentration 10% by volume

On the surface of an etched uranium piece, the tunnels and fingers extending in the metal are the key to the good adherence of the subsequent coating (Fig. 15). The coating is thrown down into the recesses of the fingers and tunnels. This mechanical interlocking causes the nickel electroplate to be very adherent and results in good corrosion protection.

The effect of the thickness of the plating on the resistance to a corrosive environment is shown in Table 9. Because one of the corrosion products formed from the reaction of uranium with water in hydrogen, the measurement of hydrogen evolved during a corrosion test gives a measure of the amount of corrosion. The data in Table 9 indicate that at least 50 μm (2 mils) of nickel on uranium were required to stop the reaction between the water vapor and the uranium.

Electroplating of Uranium Alloys. The process of electroplating nickel onto lean uranium alloys, such as U-0.75Ti, is the same as that used for uranium. In addition to the nickel plate, two more coatings are typically applied over the nickel for lean uranium alloys. These are electroplated zinc and a zinc chromate conversion coating applied by a chemical dip (solution composition and operating conditions given in Table 8). In aqueous chloride electrolytes, a breached coating may behave in a protective, neutral, or destructive manner toward the substrate, depending on whether its electrochemical potential is negative, equivalent, or positive, respectively, to the substrate. In one investigation, kinetic measurements were made of couples of coating material and U-0.75Ti in a KCl electrolyte (Ref 50). The results, plotted in Fig. 16, show that both cadmium and zinc coupled to U-0.75Ti produce a negative corrosion current, which means that they corrode sacrificially and protect the uranium alloy. However, when nickel and aluminum were coupled to the U-0.75Ti, a positive corrosion current was produced, signifying that the uranium alloy corroded. Therefore, any flaws in a nickel or aluminum coating on U-0.75Ti could result in an increased local corrosion rate.

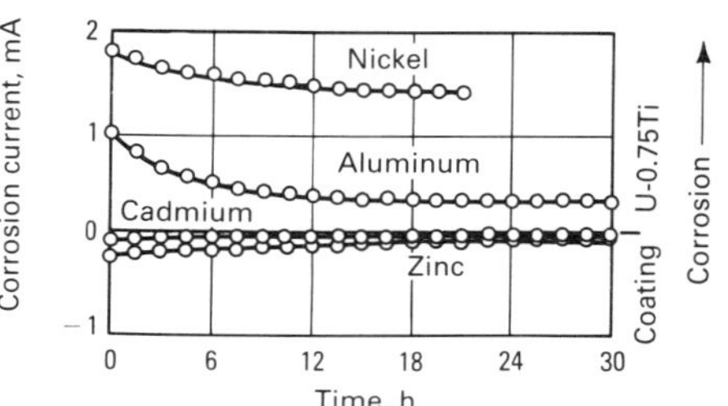

Fig. 16 Current developed by coating material/U-0.75Ti galvanic couples in 0.001 *M* KCl at 25 °C (75 °F). Source: Ref 50

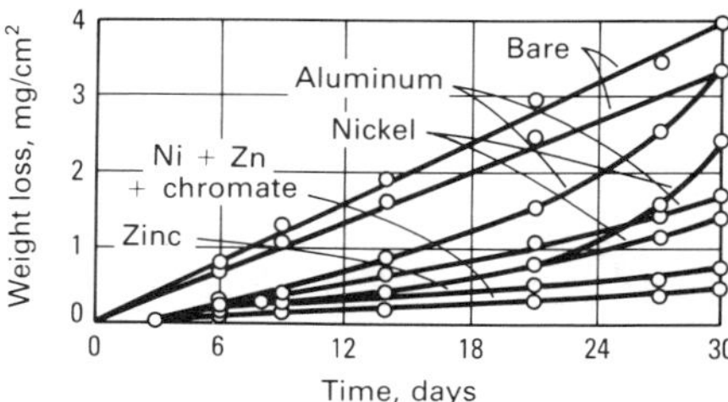

Fig. 17 Relative protection of U-0.75Ti by various coating materials in 50 ppm Cl^- solutions at 25 °C (75 °F)

Coated and uncoated samples of U-0.75Ti were tested in a chloride solution at room temperature for 30 days (Fig. 17). Four important results were obtained from these tests, as follows. All coatings in the unflawed condition gave some protection to the U-Ti alloy. When the electroplated nickel and ion-plated aluminum coatings began to fail, the corrosion rates, as shown by the slopes of the weight loss curves, accelerated until they became faster than that of uncoated material. In spite of the poor adherence of the ion-plated zinc coatings, they imparted the predicted galvanic protec-

Table 9 Corrosion of nickel-plated uranium in moist nitrogen

Specimens were plated in nickel sulfamate solution after etching in $FeCl_3$. Surface area of each specimen was approximately 3870 mm² (6 in.²). Nitrogen contained 2.8% water vapor and 97.2% N_2 at 70 °C (160 °F). Specimens were sealed in glass tubes, evacuated for a minimum of 3 h under hard vacuum, backfilled with the gas mixture, sealed, and placed in the oven for testing.

Plating thickness, μm	Plating thickness, mils	Test duration	Amount of hydrogen evolved, ppm
Unplated		6 weeks	180 000
13	0.5	8 days	7000
13	0.5	6 weeks	35 000
25	1.0	6 weeks	15
50	2.0	7 weeks	10
75	3.0	7 weeks	0
75	3.0	30 weeks	0

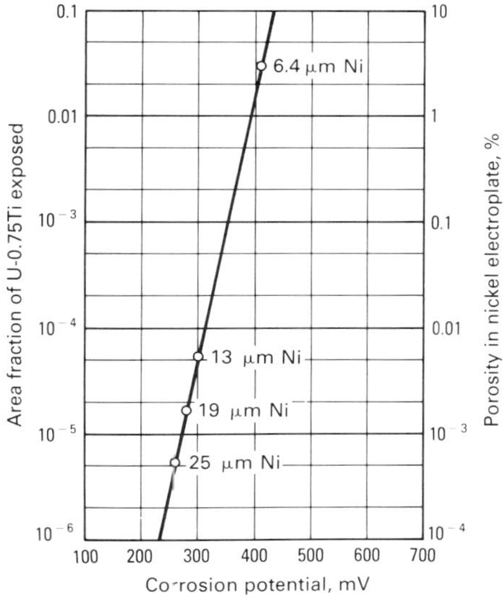

Fig. 18 Porosity in electroplated nickel on U-0.75Ti as a function of nickel thickness. Tests were performed in 0.001 *M* KCl at 25 °C (75 °F).

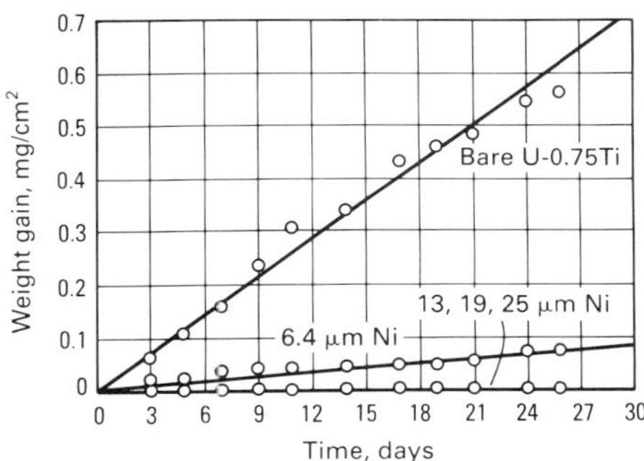

Fig. 19 Corrosion kinetics of U-0.75Ti as a function of nickel thickness in 10% relative humidity air at 105 °C (220 °F)

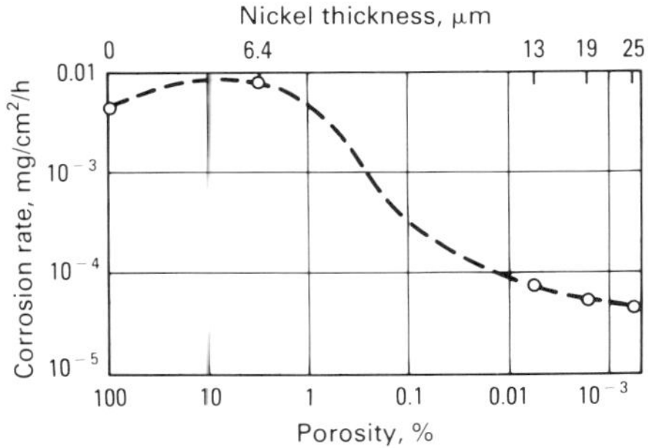

Fig. 20 Corrosion rates of U-0.75Ti as a function of porosity in nickel coatings. Tested in a 50-ppm Cl^- solution at 25 °C (75 °F)

tion to the uranium alloy. The duplex coating with the chromate finish provided the best long-term protection.

Increasing the nickel plating thickness from 6.4 to 13 µm (0.25 to 0.5 mil) significantly increases the coverage on U-0.75Ti. A technique was developed to measure quantitatively the porosity in a nickel plating on U-0.75Ti (Ref 51). The data obtained using this electrochemical porosity technique for coatings at various thicknesses are given in Fig. 18. The area fraction exposed, or pore content, decreases significantly when coating thickness is increased from 6.4 to 13 µm (0.25 to 0.5 mil). Corrosion tests reflected these porosity observations (Ref 52). A set of specimens with varying coating thicknesses was exposed to 10% relative humidity at 105 °C (220 °F) for 26 days; results are plotted in Fig. 19. The 6.4-µm (0.25-mil) coating did give some protection to the U-0.75Ti material as compared to uncoated. However, a 13-µm (0.5-mil) coating, or thicker, prevented any measurable corrosion. Another set of coated specimens was exposed to a more severe environment, namely a 50-ppm chloride solution (Fig. 20). In this case, the specimen with a 6.4-µm (0.25-mil) coating exhibited a higher corrosion rate than the uncoated material did. This increase in corrosion rate was due to the noble electrochemical coupling of the nickel to the U-0.75Ti, which increased the corrosion rate of the uranium alloy at pores in the nickel electroplate. The large decrease in corrosion rate due to increasing nickel thickness reflected the decreasing porosity in the nickel electroplate with increasing thickness. In summary, for lean uranium alloys, electroplated nickel, particularly with a zinc and chromate overplate, provides good corrosion protection in moist air and in chloride solution environments.

Ion Plating of Uranium. The development of an ion-plating process provided a means of coating uranium with an adherent thin protective coating without the presence of an intermediate oxide film (Ref 53). In this process, the sample is sputter cleaned by positive-ion bombardment and maintained in contamination-free state before and during film deposition. In addition, the high energy flux to the surface promotes chemical reaction and diffusion in the interfacial region. A study was performed to determine which process parameters influenced the ion-plated aluminum coating properties and to optimize processing conditions for maximum corrosion protection (Ref 54). The parameters that influenced corrosion protection were substrate precleaning, amount of gaseous contamination in the glow discharge, sputter cleaning voltage and pressure, filament contamination, deposition rate, and amount of aluminum-uranium interdiffusion, either during plating or in subsequent vacuum heat treatment. Vacuum heat treatment after plating promotes the formation of intermetallic compounds at the uranium/aluminum interface that provides the best corrosion protection.

Ion Plating of Uranium Alloys. The ion plating of aluminum has been primarily applied to the U-0.75Ti alloy. The basic procedure developed for the ion plating of uranium is also used for uranium alloys. This procedure consists of precleaning with a degreaser to remove organic contaminants, removing the surface oxides by grit blasting or electropolishing, ion cleaning in the vacuum chamber with an argon glow discharge, then ion plating with aluminum. The advantages of the ion-plating process are:

- No surface charging of hydrogen (hydrogen being extremely detrimental to the ductility of U-0.75Ti
- Ability to plate uniform thicknesses
- Very thin coatings are required for corrosion protection (coatings of 5 to 10 µm, or 0.2 to 0.4 mils, are sufficient for U-0.75Ti
- Excellent coating adherence
- Ability to plate uranium with aluminum which electroplating cannot presently accomplish

The disadvantages of the ion plating process are:

- The surface of the part will undergo heating (overheating of U-0.75Ti will cause a deleterious change in mechanical properties)
- The geometry of the part significantly affects the plating thickness (particularly holes and corners)
- The ion-plating process is predominantly a line-of-sight application (parts must be surrounded by sources or rotated)
- There is no quantitative test to measure the adhesion of the coating (the sticky tape test is typically used)

REFERENCES

1. A.G. Ritchie, A Review of the Rates of Reaction of Uranium with Oxygen and Water Vapor at Temperatures up to 300 C, *J. Nucl. Mater.*, Vol 102, 1981, p 170
2. L. Baker, Jr. and J.D. Bingle, The Kinetics of Oxidation of Uranium Between 300 and 625 C, *J. Nucl. Mater.*, Vol 20, 1966, p 11
3. L. Leibowitz, J.G. Schnizlein, J.D. Bingle, and R.C. Vogel, The Kinetics of Oxidation of Uranium Between 125 and 250 °C, *J. Electrochem. Soc.*, Vol 108, 1961, p 1155-1160
4. J.W. Isaacs and J.N. Wanklyn, "The Reaction of Uranium with Air at High Temperatures," Report No. AERE-R-3559, Atomic Energy Research Establishment, Harwell, England, Dec 1960
5. W.J. Megaw, R.C. Chadwick, A.C. Wells, and J.E. Bridges, The Oxidation and Release of Iodine-131 from Uranium Slugs Oxidizing in Air and CO_2, *J. Nucl. Energy*, Parts A/B, Vol 15, 1961, p 176-184
6. R.K. Hilliard, "Oxidation of Uranium in Air at High Temperatures," U.S. Atomic Energy Commission Contract Report No. HW-58022, General Electric Co., Hanford Atomic Products Operation, 1958
7. A.F. Bessonov and V.G. Vlasov, Oxidation Mechanism of Metallic Uranium, *Phys. Metals Metallogr.*, Vol 12, 1961, p 89-94 (translation of *Fix. Metal. Metalloved.*, Vol 12, 1961, p 403-408)
8. J.E. Antill and P. Murray, Reactions Between Fuel Elements and Gaseous Coolants, in *Progress in Nuclear Energy, Ser. IV, Technology, Engineering and Safety*, Vol 3, C.M. Nicholls, Ed., Pergamon Press, 1960, p 65-86
9. A.G. Ritchie, R.C. Greenwood, S.J. Randles, D.R. Netherton, and J. Whitehorn, "Measurement of the Rate of the Uranium-Water Vapor Reaction," Report O 4/86, Atomic Weapons Research Establishment, June 1986
10. C.C. Colmenares *et al.*, "Oxidation of Uranium Studied by Gravimetric and Positron Annihilation Techniques," UCRL-85549, Lawrence Livermore National Laboratories, April 1981
11. L.J. Weirick, "The Oxidation of Uranium in Low Partial Pressures of Oxygen and Water Vapor at 100 C," SAND83-0618, Sandia National Laboratories, June 1984
12. A.G. Ritchie, R.C. Greenwood, and S.J. Randles, "The Kinetics of the Uranium-Oxygen-Water Vapor Reaction Between 40 and 100 C," Report O 10/85, Atomic Weapons

Research Establishment, Oct 1985
13. S. Orman, "Oxidation of Uranium in Water and Water Vapor," Report O 25/64, Atomic Weapons Research Establishment, 1964
14. C.J. Greenholt and L.J. Weirick, "The Oxidation of Uranium-0.75 Weight Percent Titanium in Environments Containing Oxygen and/or Water Vapor at 140 C," SAND85-1688, Sandia National Laboratories, Sept 1985, to be published in *J. Nucl. Mater.*
15. A.M. Ammons, Oak Ridge Y-12 Plant, private communication, March 1975
16. N.J. Magnani, "The Reaction of Uranium and Its Alloys with Water Vapor at Low Temperatures," SAND-74-0145, Sandia National Laboratories, Aug 1974
17. S. Orman, Oxidation of Uranium and Uranium Alloys, in *Physical Metallurgy of Uranium Alloys*, J.J. Burke, D.A. Colling, A.E. Gorum, and J. Greenspan, Ed., Brook Hill, 1974
18. A.G. Ritchie, R.C. Greenwood, S.J. Randles, D.R. Netherton, and J. Whitehorn, "The Atmospheric Corrosion of Uranium-6% Niobium Alloy," Preprint, Atomic Weapons Research Establishment, Nov 1985
19. N.J. Magnani and H. Romero, "The Reaction of Water Vapor with U-7½wt%Nb-2½wt%Zr and U-4½wt%Nb," SC-RR-72-0635, Sandia National Laboratories, Sept 1972
20. V.B. Kishinevski, L.I. Gomozov, and O.S. Ivanov, Corrosion Resistance in Water of Some Alloys of Uranium with Zirconium, Niobium and Molybdenum, in *Physical Chemistry of Alloys and Refractory Compounds of Thorium and Uranium*, Report AEC-tr-7212, O.S. Ivanov, Ed., U.S. Atomic Energy Commission, 1972, translated from Russian
21. L.J. Weirick, Protective Coatings for Uranium Alloys, in *Physical Metallurgy of Uranium Alloys*, J.J. Burke, D.A. Colling, A.E. Gorum, and J. Greenspan, Ed., Brook Hill, 1974
22. J.M. Macki and R.L. Kochen, "The Galvanic Corrosion Behavior of Uranium Alloys in Hydrochloric Acid and Ocean Water," RFP-1592, Rocky Flats Plant, Feb 1971
23. J.M. Macki and R.L. Kochen, "The Corrosion Behavior of Uranium-Base U-Nb, U-Nb-Zr, and U-Mo Alloys in Hydrochloric Acid and Ocean Water," RFP-1586, Rocky Flats Plant, Feb 1971
24. M. Levy and C.V. Zabielski, Electrochemical Behavior of Some Binary and Polynary Uranium Alloys, in *Physical Metallurgy of Uranium Alloys*, J.J. Burke, D.A. Colling, A.E. Gorum, and J. Greenspan, Ed., Brook Hill, 1974
25. L.J. Weirick, "Corrosion Testing of the General Electric Mantech GAU 8/A Penetrator," SAND76-8055, Sandia National Laboratories, Feb 1977
26. H.R. Johnson and L.J. Weirick, "Corrosion Testing of the General Electric Mantech II GAU 8/A Penetrator," SAND78-8009, Sandia National Laboratories, May 1978
27. K.H. Eckelmeyer, A.D. Romig, Jr., and L.J. Weirick, The Effect of Quench Rate on the Microstructure, Mechanical Properties, and Corrosion Behavior of U-6 wt pct Nb, *Metall. Trans.* Vol 15A, July 1984, p 1319
28. C. Levy, "Uranium Alloys for XM-673 Projectile, Engineering Program," AMMRC SP 72-17, Army Materials and Mechanics Research Center, Oct 1972
29. N.J. Magnani, Stress Corrosion Cracking of Uranium Alloys, in *Physical Metallurgy of Uranium Alloys*, J.J. Burke, D.A. Colling, A.E. Gorum, and J. Greenspan, Ed., Brook Hill, 1974
30. N.J. Magnani, The Effect of Environment, Orientation and Strength Level on the Stress Corrosion Behavior of U-0.75wt.%Ti, *J. Nucl. Mater.*, Vol 54, 1974, p 108
31. N.J. Magnani, The Effect of Environment on the Cracking Behavior of Selected Uranium Alloys," SCR-72-2661, Sandia National Laboratories, March 1972
32. N.J. Magnani, Hydrogen Embrittlement and Stress-Corrosion Cracking of Uranium and Uranium Alloys, in *Advances in Corrosion Science and Technology*, Vol 6, M.G. Fontana and R.W. Staehle, Plenum Press, 1976
33. N.J. Magnani, The Effects of Chloride Ions on the Cracking Behavior of U-7.5wt%Nb-2.5wt%Zr and U-4.5wt%Nb, *J. Nucl. Mater.*, Vol 42, 1972, p 271
34. J.W. Koger, "Stress-Corrosion Cracking of Uranium Alloys," Y-DA-5624, Y-12 Plant, 1973
35. L.J. Weirick, The Effect of Heat Treatment Upon the Stress-Corrosion Cracking of Mulberry (U-7.5Nb-2.5Zr), *Corrosion*, Vol 31, 1975, p 5
36. W.F. Czyrklis and M. Levy, Stress Corrosion Cracking Behavior of Uranium Alloys, *Corrosion*, Vol 30, 1974, p 181
37. G.S. Petit, R.R. Wright, C.A. Keinberger, and C.W. Weber, "Formation of Corrosion-Resistant Oxide Film on Uranium," K-1778, Oak Ridge Gaseous Diffusion Plant, 1969
38. O. Flint, J.J. Polling, and A. Charlesby, The Anodic Oxidation of Uranium, *Acta Metall.*, Vol 2, 1954, p 696
39. T.S. Prevender, Sandia National Laboratories, private communication
40. S. Orman and P. Walker, The Corrosion of Uranium and Its Prevention by Organic Coatings, *J. Oil Colour Chem. Assoc.*, Vol 48, 1965, p 233
41. C.A. Colmenares, "Aluminum and Polymeric Coatings for Protection of Uranium," UCID-19970, Lawrence Livermore National Laboratory, Dec 1983
42. C.E. Miller, "Producibility Study, Cartridge 20MM, DS Mk 149," Final Report, Olin Energy Systems Operations, Olin Corporation, Dec 1974
43. J.M. Macki and R.L. Kochen, "The Corrosion and Stress-Corrosion Cracking of Painted U-4.2wt.%Nb," RFP-1891, Rocky Flats Plant, Aug 1972
44. L.J. Weirick, Electroplating and Corrosion Protection, in *Surface Metallurgy of Uranium and Uranium Alloys*, Vol 3, *Metallurgical Technology of Uranium and Uranium Alloys*, American Society for Metals, 1981
45. G.W. Vest, Ion Plating in Corrosion Protection, in *Surface Metallurgy of Uranium and Uranium Alloys*, Vol 3, *Metallurgical Technology of Uranium and Uranium Alloys*, American Society for Metals, 1981
46. J.K. Gore and R. Seegmiller, Surface Treatments and Electro-plated Coatings on Uranium, *Plating*, Vol 50, 1963, p 215
47. J.G. Beach and C.L. Faust, "Electroplates on Thorium and Uranium for Corrosion Protection and to Aid Joining," BMI-1537, Battelle Memorial Institute, Aug 1961
48. P.D. Anderson, P.R. Coronado, and L.M. Berry, U.S. Patent 3,341,350, granted Sept 1967, Method of Preparing a Uranium Artide for a Protective Coating
49. J.W. Dini and P.R. Coronado, Preparation of Uranium for Electro-plating with Nickel, *Trans. IMF*, Vol 47, 1969, p 1
50. L.J. Weirick, "Evaluation of Metallic Coatings for the Corrosion Protection of a Uranium-3/4 Weight Percent Titanium Alloy," SLL-73-5024, Sandia National Laboratories, Feb 1974
51. L.J. Weirick, Electrochemical Determination of Porosity in Nickel Electroplates on a Uranium Alloy, *J. Electrochem. Soc.*, Vol 47 (No. 1), 1975, p 937
52. L.J. Weirick and D.L. Douglass, Effect of Thin Electrodeposited Nickel Coatings on the Corrosion Behavior of U-0.75Ti, *Corrosion*, Vol 36 (No. 6), 1976, p 209
53. D.M. Mattox, Film Deposition Using Accelerated Ions, *Electrochem. Technol.*, Vol 2, 1964, p 295
54. R.D. Bland, A Parametric Study of Ion-Plated Aluminum Coatings on Uranium, *Electrochem. Technol.*, Vol 6, 1968, p 272

Corrosion of Powder Metallurgy Materials

Erhard Klar, SCM Metal Products

POWDER METALLURGY (P/M) is a branch of metallurgy that may be defined as the technology and art of producing metal powders and using them to construct massive materials and shaped objects. Although the commercial use of P/M dates back to the beginning of the 20th century, the major classes of P/M materials developed during the first four decades (cemented carbides, composite electrical contacts, self-lubricating bronze bearings, and composite friction materials) did not pose any major problems with regard to corrosion resistance in that the service lives of these materials were normally not limited by their corrosion properties. With the appearance of iron-base sintered structural parts in the 1940s, the P/M method began to compete with other metal-forming techniques, such as casting, machining, and forging. The rapid increase in the utilization of iron and copper powders from 1940 to 1970 is linked to the advent of mass production in the automotive industry, and the successful use of P/M technology was based mainly on labor and material savings. With the growth of the industry and the development of stronger sintered parts came the demand for improved corrosion resistance, which led to the development of several surface treatments for sintered iron and steel parts (steam treatment; impregnation with oils, plastics, and waxes; metallization; tumbling with fillers; and electroplating) and to the development of sintered stainless steels in the 1950s.

With the development of fully dense P/M materials in the 1950s and 1960s, and particularly with the appearance of rapid solidification technology in recent years, it became increasingly clear that P/M processing was capable not only of cost savings from reduced labor and reduced material usage (net shape and near net shape processing) but also of producing superior materials. The superior microstructures possible with P/M technology were responsible for significant improvements in mechanical and magnetic properties, workability, and corrosion resistance. Two classes of materials in which such improved properties are of utmost importance are fully dense P/M superalloys and P/M aluminum alloys. Powder metallurgy superalloys are currently used in advanced turbine engines. Although wrought P/M aluminum alloys are still in the evaluation stage, they have been used in aerospace structures on a limited trial basis in non-critical applications.

The corrosion properties of the above-mentioned major classes of P/M materials will be described in the following sections. Corrosion resistance will be compared with that of similar conventional alloys where possible. Brief accounts will be given for the manufacture and use of these materials. More detailed descriptions of all aspects of the P/M process, such as powder production, compaction, sintering, and other state-of-the art consolidation methods, are available in Volume 7 of the 9th Edition of *Metals Handbook*. In addition to these major classes of P/M materials, there are many P/M specialty materials in which corrosion resistance may also play an important role (see, for example, the article "Corrosion of Cemented Carbides," in this Volume). Information on such materials is also provided in Volume 7 of the 9th Edition of *Metals Handbook*.

Fig. 1 P/M parts steam treated using the cycle shown in Fig. 2

Sintered Iron Base P/M Parts

Powder metallurgy iron and carbon steel parts exhibit little resistance to corrosion. Therefore, steam treatment and/or impregnation with oils is widely used to prevent or minimize rusting in mild environments. Steam treatment will be described in this section.

Oxidation of small sintered iron and steel parts (Fig. 1) with superheated steam has been in use for over 40 years as an economical process to improve their hardness, compressive strength, wear, and corrosion resistance. This treatment can be combined with tempering and is then referred to as steam tempering. It leaves a tightly adhering bluish to black oxide that is usually acceptable as a final finish. The pores become coated or filled with the oxide as a result of the lower density of the oxide. This provides a certain pressure tightness and minimizes the entry of contaminants.

The Process. A typical procedure for the steam treatment (Fig. 2) of clean (degreased) parts involves heating to a temperature above 100 °C (212 °F) but below the critical point of air oxidation and discoloration (425 °C, or 800 °F). A temperature of 315 °C (600 °F) is typically used for the first stage. Parts are soaked for 15 min or until the center of the load is above 100 °C (212 °F). Steam is then introduced into the chamber at a high flow rate (4.5 to 165 kg/h, or 10 to 360 lbs/h, depending on furnace size) to purge air from the furnace through the relief valve.

Once the purge has been completed, steam flow is halved and temperature is increased to

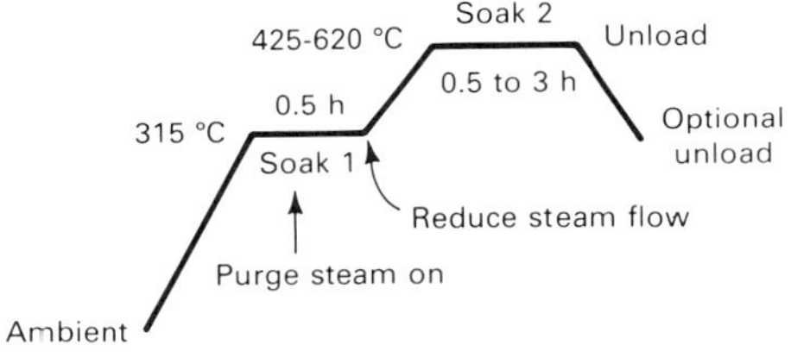

Fig. 2 Typical steam-treating cycle for iron P/M parts

Table 1 Applications for P/M stainless steels

Part	Alloy
Aerospace	
Seatback tray slides	316L
Galley latches	316L
Jet fuel refueling impellers	316L
Foam generators	316L
Agriculture	
Fungicide spray equipment	316L
Appliances	
Automatic dishwasher components	304L
Automatic washer components	304L
Garbage disposal components	410L
Pot handles	316L
Coffee filters	316L-Si
Electric knives	316L
Blenders	303L
Can opener gears	410L
Automotive	
Rearview mirror mounts	316L, 434L
Brake components	434L
Seat belt locks	304L
Windshield wiper pinions	410L
Windshield wiper arms	316L
Manifold heat control valves	304L
Building and construction	
Plumbing fixtures	303L
Spacers and washers	316L
Sprinkler system nozzles	316L
Shower heads	316L
Window hardware	304L, 316L
Thermostats	410L
Chemical	
Filters	304L-Si, 316L
High corrosion resistance filters	830
Cartridge assemblies	316L-Si
Electrical and electronic	
Limit switches	410L
G-frame motor sleeves	303L
Rotary switches	316L
Magnetic clutches	410L, 440A
Battery nuts	830
Electrical testing probe jaws	316L
Hardware	
Lock components	304L, 316L
Threaded fasteners	303L
Fasteners	316L
Quick-disconnect levers	303L, 316L
Industrial	
Water and gas meter parts	316L
Filters, liquid and gas	316L-Si
Recording fuel meters	303L
Fuel flow meter devices	410L
Pipe flange clamps	316L
High polymer filtering	316L-Si
Jewelry	
Coins, medals, medallions	316L
Watch cases	316L
Watch band parts	316L
Marine	
Propeller thrust hubs	316L
Cam cleats	304L
Medical	
Centrifugal drive couplings	316L
Dental equipment	304L
Hearing aids	316L
Anesthetic vaporizers	316L
Office equipment	
Nonmagnetic card stops	316L
Dictating machine switches	316L
Computer knobs	316L
Recreation and leisure	
Fishing rod guides	304L, 316L
Fishing rod gear ratchets	316L
Photographic equipment	316L
Soft drink vending machines	830, 316L
Travel trailer water pumps	316L

Table 2 Compositions of commercial P/M stainless steels

Alloy	Composition, % Cr	Ni	Si	Mo	Cu	Su	Mn	C	S	P	Fe	O(ppm)
Austenitic grades												
303	17–18	12–13	0.6–0.8	...	...	...	0.3(a)	0.03(a)	0.1–0.3	0.03(a)	rem	...
304L	18–19	10–12	0.7–0.9	...	...	...	0.3(a)	0.03(a)	0.03(a)	0.03(a)	rem	1000–2000
304LSC	18–20	10–12	0.8–1.0	...	2(b)	1(b)	0.3(a)	0.03(a)	0.03(a)	0.03(a)	rem	...
316L	16.5–17.5	13–14	0.7–0.9	2–2.5	...	...	0.3(a)	0.03(a)	0.03(a)	0.03(a)	rem	1000–2000
Martensitic grade												
410L	12–13	...	0.7–0.9	...	...	...	0.1–0.5	0.05(a)	0.03(a)	0.03(a)	rem	1500–2500
Ferritic grades												
430L	16–17	...	0.7–0.9	...	...	...	0.3(a)	0.03(a)	0.03(a)	0.03(a)	rem	...
434L	16–18	...	0.7–0.9	0.5–1.5	...	...	0.3(a)	0.03(a)	0.03(a)	0.03(a)	rem	...

(a) Maximum. (b) Typical

between 425 and 620 °C (800 and 1150 °F). The load is then soaked for 30 min to 3 h, depending on the amount of oxide required. When the soak has been completed and the load removed, the furnace is cooled to below 425 °C (800 °F) for the next load (Ref 1).

With correct processing, only two types of iron oxide form: FeO and Fe_3O_4. The furnace is purged of air before admitting the steam in order to avoid the formation of ferric oxide (Fe_2O_3) (red rust). Furthermore, the coldest part of the furnace charge must be above 100 °C (212 °F) before admission of steam to prevent the formation of iron hydroxide ($Fe(OH)_3$) and Fe_2O_3. Dark blue or light gray oxides form, depending on the temperature of the second soak.

The disadvantages of the treatment include some losses in tensile strength, impact resistance, and ductility (Ref 2). Details on the kinetics of oxidation, thermochemical equilibria, permeability and dimensional changes, oxide adherence, and hardness increases can be found in Ref 3 to 6.

Corrosion Resistance. Phase composition and other structural characteristics of oxidation products depend on the oxidation technique and the cooling conditions (Ref 4). For example, the corrosion resistance of a low alloy sintered steel [0.12% C, 0.01% Si (max), 0.24% Mn, 0.010% P, 0.025% S, 0.02% Cr, 0.01% Ni (max), 0.019% Al, 1.86% Cu, 0.01% Ti (max)] with a density of 6.4 g/cm^3 in 3% aqueous sodium chloride (NaCl) (by immersion) was inferior at low (580 °C, or 1075 °F) and high (630 °C, or 1165 °F) oxidation temperatures. The latter was attributed to the lower adherence of the oxides to the metallic phase.

In one study, potentiostatic anodic polarization measurements of steam-treated carbon steel in dilute (0.5%) sulfuric acid (H_2SO_4) and aqueous NaCl solutions (400 ppm chloride ion Cl^-), revealed significant reduction of corrosion rates for the Cl^- environment, while H_2SO_4 testing showed an increase in corrosion rate (Ref 7). The latter was attributed to the solubility of the oxides in dilute H_2SO_4 and their increased attack compared to the unoxidized steel.

Sintered (Porous) P/M Stainless Steels

In its first decade of rapid commercialization, sintered stainless steel use increased at a compound annual rate of nearly 20% to reach a consumption of 2000 tons per year in 1973 for the North American market. During this period, the major developmental efforts were aimed at improving the compacting properties of water-atomized stainless steel powder. Interest in corrosion resistance increased after the market had peaked at a consumption of about 3000 tons per year.

Although powder and processing requirements for improving corrosion resistance were better defined and improved stainless steel powders became available, the basic understanding of corrosion as well as corrosion data development for sintered stainless steels are both still in their infancy. There are many factors that distinguish sintered stainless steel from cast and wrought stainless steel. Complicating the issue is the fact that the corrosion resistance of sintered stainless steels depends as much on the sintering process as it does on the properties of the powder. Neither guidelines nor standards relating to corrosion behavior exist for most of the critical composition and process parameters. Furthermore, much of the published literature on specific corrosion data of sintered stainless steels is nearly obsolete because of the lack of information on process conditions.

Application and Selection of Sintered P/M Stainless Steels. In the absence of detailed and reliable corrosion data, tentative selection of a P/M stainless steel for a specific application is made by following the same principles developed for cast and wrought stainless steels. Thus, for

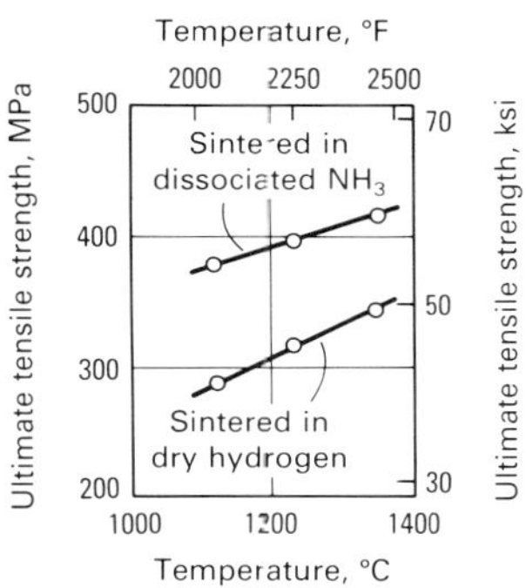

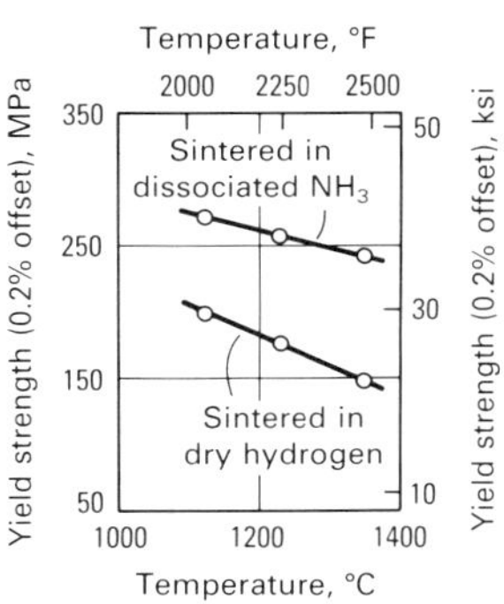

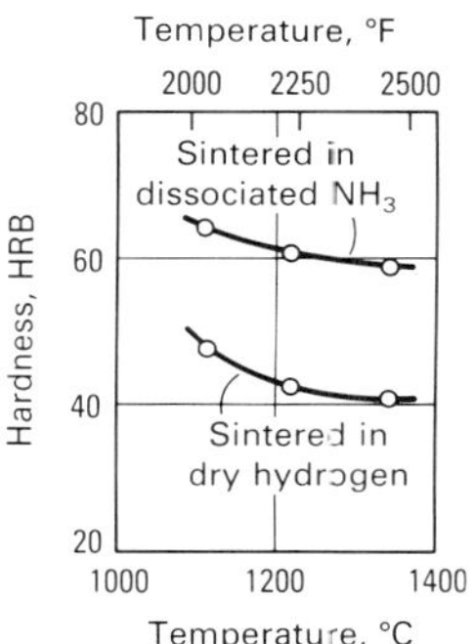

Fig. 3 Effect of sintering temperature on tensile and yield strengths and apparent hardness of type 316L stainless steel. Parts (density: 6.85 g/cm^3) were sintered for 30 min in various atmospheres.

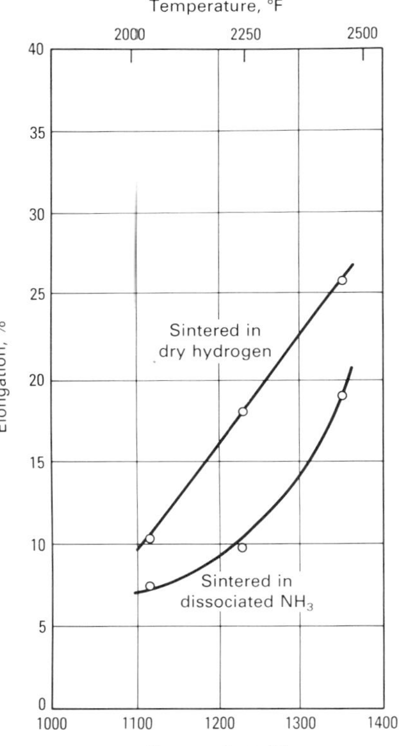

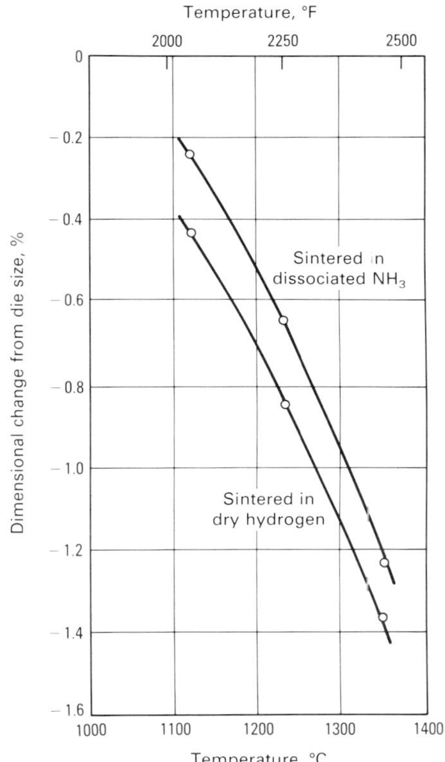

Fig. 4 Effect of sintering temperature on elongation and dimensional change during sintering of type 316L stainless steel. Parts (density: 6.85 g/cm^3) were sintered for 30 min in various atmospheres.

Table 3 Typical mechanical properties of medium-density P/M stainless steels

All materials sintered in dissociated NH_3

MPIF designation	Composition, % Cr	Ni	Mo	Si	Fe	Tensile strength MPa	ksi	0.2% yield strength MPa	ksi	Elongation in 25 mm (1 in.), %	Density, g/cm^3
SS-303	17	12	...	0.7	rem	241	35	220	32	1	6.2
	17	12	...	0.7	rem	358	52	324	47	2	6.6
SS-316	16	13	2	0.7	rem	262	38	220	32	2	6.2
	16	13	2	0.7	rem	372	54	275	40	4	6.6
SS-410	12	...	...	0.8	rem	289	42	283	41	<1	5.8
	12	...	...	0.8	rem	379	55	372	54	<1	6.2

Source: Ref 8

better corrosion resistance, the austenitic grades are preferred. However, type 410L stainless steel is often used for its good abrasion resistance. Because the corrosion resistance of sintered stainless steels depends so much on powder quality and parts-processing details, appropriate field testing is advisable to ensure compliance with specifications.

Table 1 provides an overview of market segments and applications for sintered stainless steels. The 300-series austenitic grades account for about two-thirds of total usage, and among the austenitic grades, type 316L is the most important. In terms of market distribution, automotive applications constitute the largest volume, followed by hardware and tools, filters, appliances, office machines, and a large segment of miscellaneous uses.

Powder Production and Compositions. Stainless steel powders suitable for cold compaction are produced by water atomization. Table 2 lists the typical powder compositions of the important commercial grades of stainless steel powders.

Properties of Sintered Parts. Sintered properties depend not only on powder characteristics but also on processing and sintering conditions. The effects of the most important processing parameters are shown in Fig. 3 to 5. Sintering in a nitrogen-containing atmosphere results in the absorption of considerable amounts of nitrogen, along with an increase in strength and a decrease in ductility. Additional mechanical properties are summarized in Table 3.

Corrosion Resistance of Sintered Stainless Steels

As noted above, reliable information on the corrosion behavior of sintered stainless steels remains scarce. Therefore, emphasis in the following discussion will be placed on summarizing recent basic and practical information on powder selection and on sintering process control for maximizing corrosion resistance.

Table 4 compares the corrosion resistance of sintered type 316L stainless steel of 85% of theoretical density with that of wrought stainless steel of similar composition in a 5% aqueous solution of NaCl after a 100-h 5% aqueous NaCl salt spray test in 10% ferric chloride ($FeCl_3$) and

Table 4 Corrosion resistance of wrought and sintered type 316L stainless steel

Corrodent	Wrought type 316L	P/M type 316L(a)
5% aqueous NaCl salt spray, 100 h(b)	Clean	Profuse, voluminous corrosion product and pitting
5% aqueous NaCl immersion(c)..........	>1000 h(d)	5 to 500 h(d)
10% aqueous $FeCl_3$, 25 °C (75 °F)(e)........	70 g/m^2/h	70 g/m^2/h
10% aqueous HNO_3, 20 °C (70 °F)	<0.1 g/m^2/h	0.001 to 0.4 g/m^2/h

(a) 85% of theoretical density. (b) Per ASTM B 117. (c) Per ASTM G 31. (d) Time in hours at which 90% of specimens had 1% of surface covered by stain. (e) Per ASTM G 48. Source: Ref 9

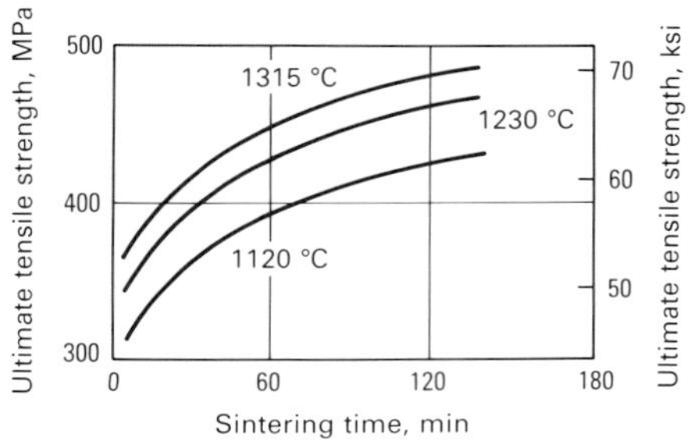

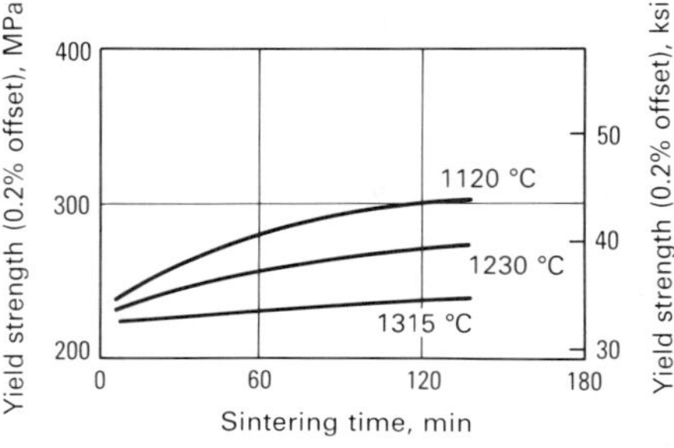

Fig. 5 Effect of sintering time on tensile and yield strengths of type 316L stainless steel. Parts were pressed to 6.85 g/cm^3 and sintered at various temperatures in dissociated NH_3.

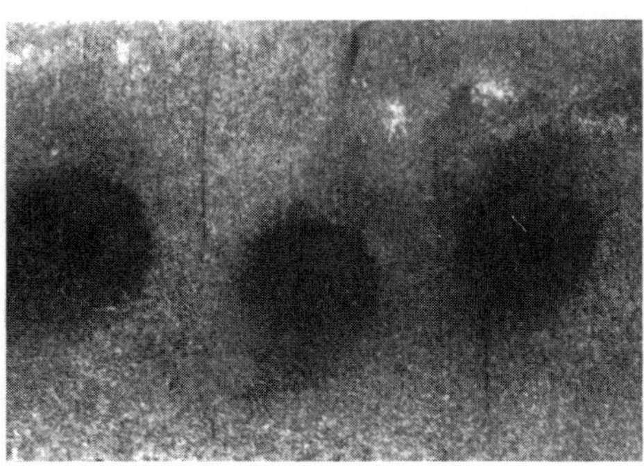

Fig. 7 Small circles of rust around iron particles embedded in the surface of sintered type 316L stainless steel after testing in 5% aqueous NaCl. 35×. Source: Ref 9

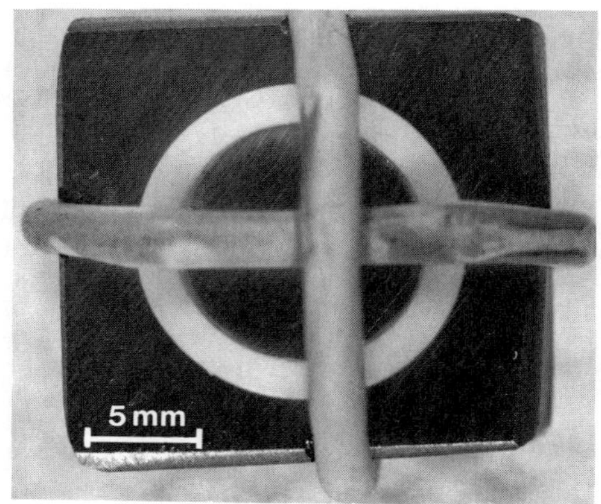

(a)

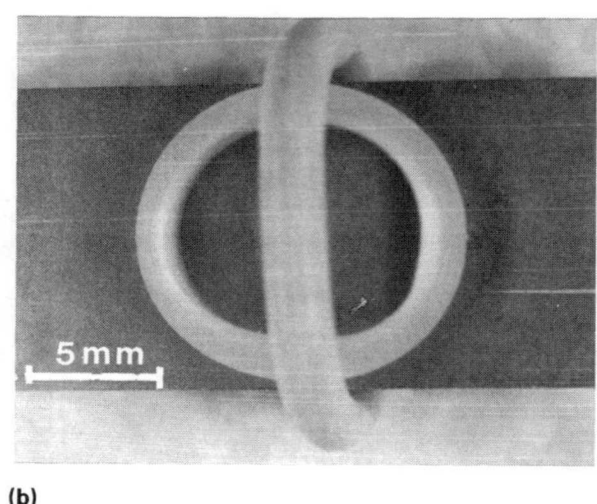

(b)

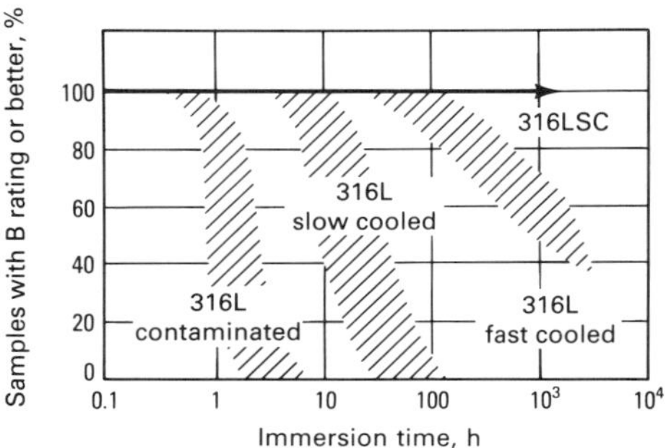

Fig. 8 Typical corrosion behavior of regular and copper-tin modified (type 316LSC) sintered type 316L stainless steel sintered in dissociated NH_3 under various conditions of cooling and contamination. B rating indicates that <1% of the specimen surface is covered by stain. Testing of the modified stainless steel was terminated after 1500 h. Source: Ref 9

(c)

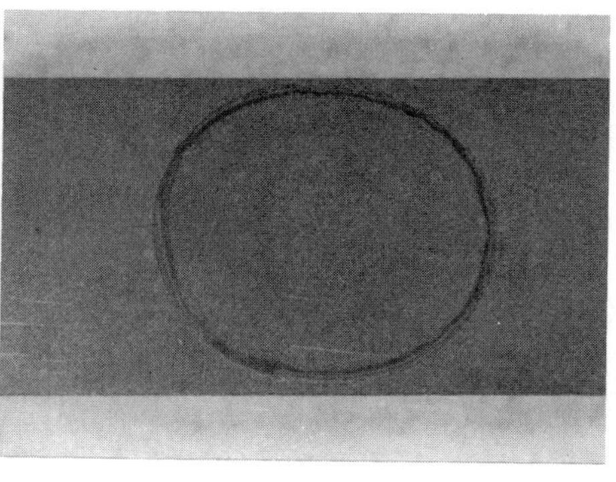

(d)

Fig. 6 Comparison of wrought and sintered type 316L stainless steels before and after testing in 10% aqueous $FeCl_3$. (a) Assembled crevice corrosion test specimen of wrought type 316L (100% dense). (b) Assembled crevice corrosion test specimen of sintered type 316L (85% dense). (c) Wrought specimen after test showing severe attack at four crevices under rubber bands and under teflon ring. (d) Sintered specimen after test showing slight attack under teflon ring. Source: Ref 10

in 10% nitric acid (HNO_3) solutions (Ref 9). For both dilute $FeCl_3$ and HNO_3, sintered type 316L performed similarly to wrought type 316L. For chloride environments, however, the P/M product is inferior. Because NaCl is probably the single most important corrodent for sintered stainless steels, a summary of recent investigations is presented below. Most of the data are for type 316L, but the general conclusions should also be valid for other stainless steel grades of the 300 and 400 series.

The limited corrosion resistance of sintered stainless steels, particularly in a chloride environment, is commonly thought to derive from the presence of residual pores that give rise to crevice corrosion as a result of oxygen depletion within the pores. There is evidence, however, that factors other than porosity often determine corrosion life. For example, despite similar pore volumes, pore sizes, and pore shapes, type 316L parts prepared from various powder lots and sintered under varying conditions had corrosion resistances in 5% aqueous NaCl that varied between 5 and 500 h for a specified degree of corrosion (Table 4) (Ref 9, 10). Furthermore, a comparison of wrought and sintered (85% of theoretical density) type 316L for susceptibility to crevice corrosion in 10% $FeCl_3$ showed that the wrought part was even more severely attacked than the porous P/M part (Fig. 6). Lastly, surface analyses of water-atomized stainless steel powders showed the presence of large amounts of oxidized silicon concurrent with a severe depletion of chromium (Ref 10). The surface composition of a sintered part depended on its sintering conditions.

It is clear that the factors of critical importance for the corrosion resistance of sintered stainless steels include control of iron contamination, carbon, nitrogen, oxygen, and sintered part density (Ref 9-13). The effects of these variables are discussed in the following sections and are summarized in Table 5. In addition, the precautions necessary for maximizing corrosion resistance differ with the sintering atmosphere.

Effect of Iron Contamination. Contamination of stainless steel powder with iron or iron-base powder may originate at the powder produc-

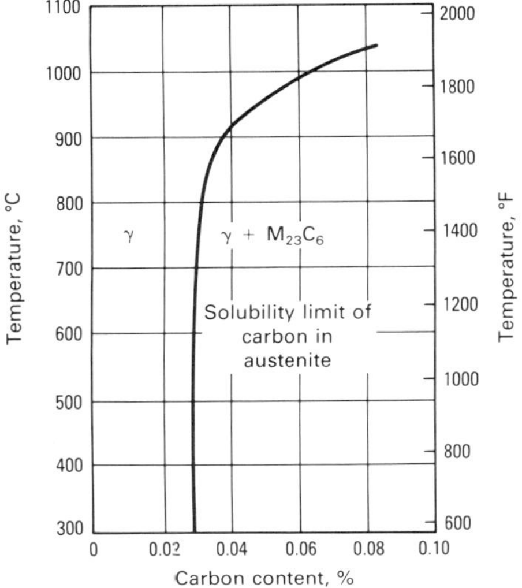

Fig. 9 Solid solubility of carbon in an austenitic stainless steel. Source: Ref 18

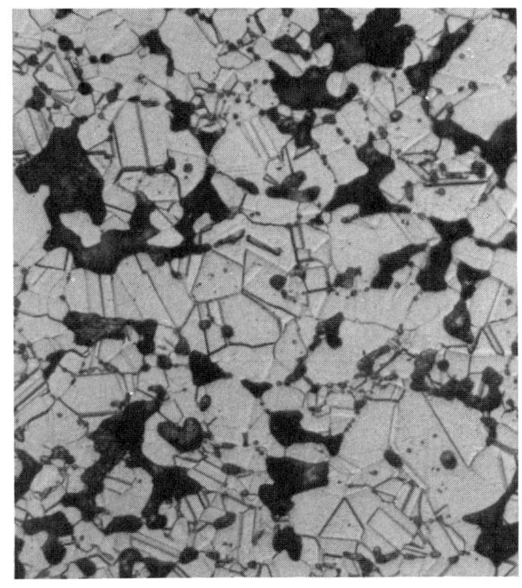

(a)

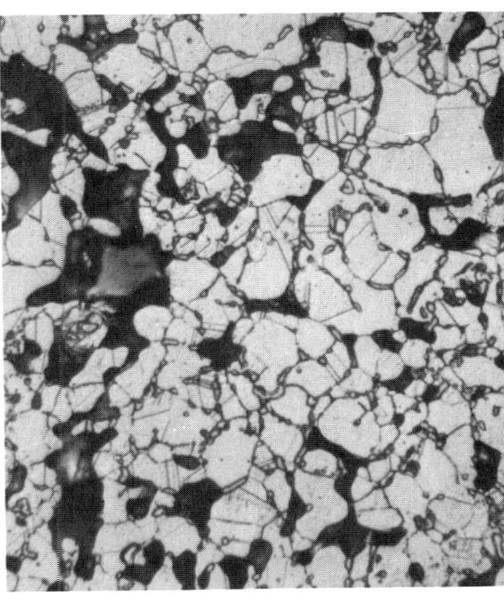

(b)

Fig. 10 Microstructures of type 316L stainless steel sintered in hydrogen at 1150 °C (2100 °F). (a) Low carbon content. (b) Excessive carbon content. Both 400×

er or the part manufacturer. Even extremely small amounts of iron contamination have a disastrous effect on the corrosion resistance of sintered parts in a saline environment. Utmost cleanliness, for example, through the use of separate production facilities and dedicated equipment, is mandatory. In saline solutions, active iron or iron-base powder particles form galvanic couples with the passive stainless steel and corrode anodically in preference to the stainless steel. Figure 7 shows this type of corrosion for iron particles embedded in the surface of a pressed and sintered type 316L part. Rusting occurs within minutes after exposure. The buildup of the initial corrosion product forms a crevice in which oxygen depletion causes acidification of the solution inside the part and further corrosion.

Because of its severity, iron contamination overshadows the other factors that affect corrosion resistance (Fig. 8). Active iron or iron alloy particles present in stainless steel powder or on the surface of a sintered stainless steel part will be revealed by placing the powder or part in a concentrated aqueous solution of copper sulfate ($CuSO_4$). The dissolved copper plates out on the iron particles within minutes, making them easy to identify with a low-magnification microscope. The powder must be tested in the unlubricated condition because lubricant will prevent the solution from wetting the powder. Experiments with very fine iron powder particles combined with high-temperature (>1260 °C, or 2300 °F), sintering have shown that this type of corrosion can be avoided if the sintering conditions result in complete alloying of the iron particles with the stainless steel matrix (Ref 9).

Effect of Carbon. Water-atomized austenitic and ferritic grades of stainless steel powders have low (<0.03%) carbon contents (Table 2) in order to resist intergranular corrosion from sensitization during cooling of the sintered part. However, in P/M sintering, there are additional sources of potential carbon contamination, and because stabilization with titanium, niobium, and tantalum is not practiced due to the excessive oxidation of

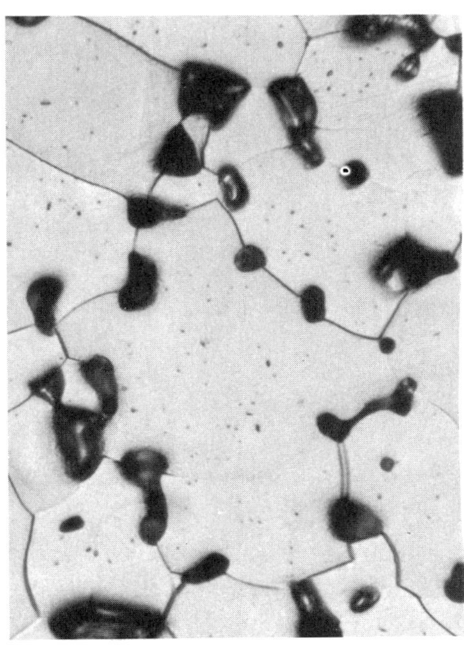

(a)

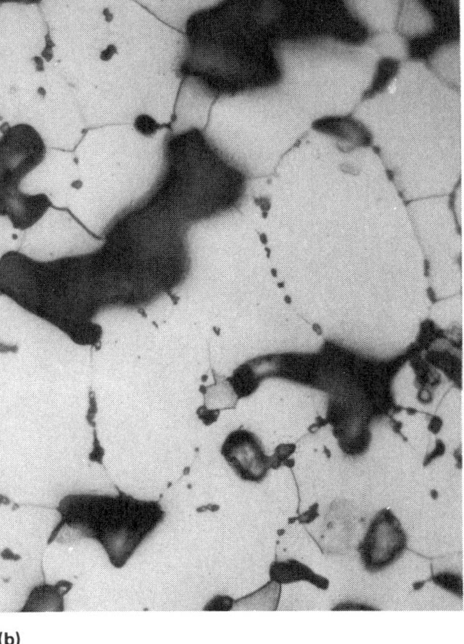

(b)

Fig. 11 Cross sections of vacuum-sintered (30 min at 1330 °C, or 2430 °F) type 430L stainless steel. (a) No oxides are present in grain boundaries after addition of 0.2% C. (b) Small, gray, rounded oxide particles in grain boundaries

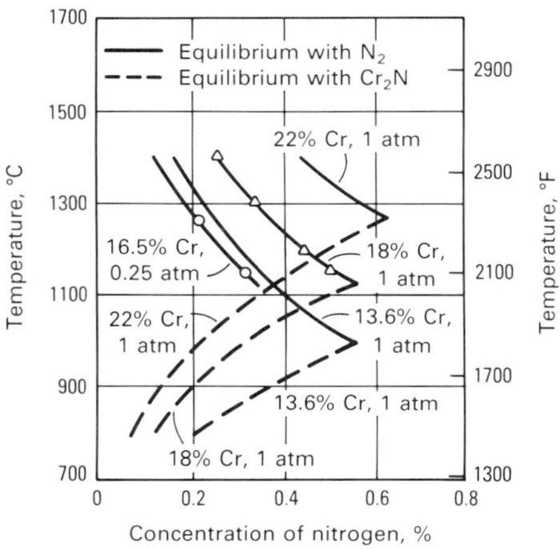

Fig. 12 Solubility of nitrogen in austenitic stainless steel in equilibrium with gaseous nitrogen or Cr_2N. Source: Ref 9

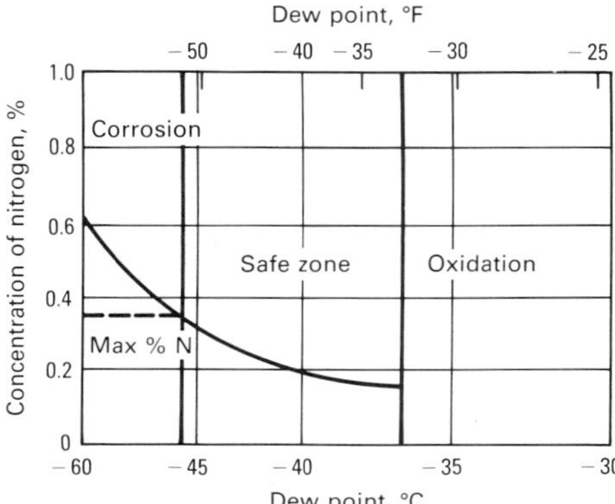

Fig. 13 Safe operating parameters with respect to dew point can be developed for a specific set of operating conditions and quality requirements. The safe zone here is for sintering in an atmosphere of 30% H_2-70% N_2 at 1035 °C (1900 °F). Source: Ref 14

these elements during water atomization, the phenomenon of sensitization will be discussed as it applies to sintered stainless steels.

Because of the decreasing carbon solubility with decreasing temperature (Fig. 9), grain-boundary precipitation of carbon as chromium-rich carbide ($M_{23}C_6$) with attendant chromium depletion occurs during cooling in the sintering furnace for materials with carbon contents exceeding 0.03%. The chromium-depleted regions will exhibit inferior corrosion resistance.

Figure 10 shows the microstructures of two sintered type 316L parts with carbon contents below and above the critical concentration of 0.03%. The low-carbon material has clean, thin grain boundaries, but the high-carbon material has grain boundaries with heavy precipitates of chromium-rich carbides.

The two main sources of carbon contamination of sintered stainless steels are (1) the organic lubricant present in prelubricated powder or added by the part manufacturer (typically 0.5 to 1%) to minimize die wear during powder compaction and (2) carbon (soot) contaminated sintering furnaces. Adequate lubricant removal is accomplished by a so-called dewaxing, or lubricant burnoff, process (Ref 15), which consists of heating the green parts in air or nitrogen to 425 to 540 °C (800 to 1000 °F). Length of heating should take into consideration the size (mass) of the parts. The oxygen absorbed during dewaxing in air will be reduced during sintering in a low dew point reducing atmosphere but not when sintering is done in a vacuum furnace.

Soot-containing sintering furnaces, that is, when a furnace was used with a low dew point endothermic atmosphere, can carburize stainless steel parts when soot that adheres loosely to furnace walls falls onto the stainless steel parts. Moisture that is from the sintering atmosphere or is formed by reduction of oxides can react with soot to form carbon monoxide (CO) and carburize the stainless steel.

For vacuum sintering of stainless steels, it may actually be beneficial to use powders with carbon contents exceeding 0.03%. With correct processing, the excessive carbon will be used up for the reduction of some of the oxides of the water-atomized stainless steel powders (typically 0.2 to 0.3% O) and thus improve the mechanical strength, ductility, and corrosion resistance of the sintered part. This phenomenon is illustrated in Fig. 11 for a vacuum-sintered ferritic stainless steel. Both parts were processed identically except for an addition of 0.2% graphite to one of the powders for the purpose of reducing its oxygen content. As expected, the graphite-containing material (Fig. 11a) produced clean grain boundaries, indicating the absence of both carbides and oxides, but the graphite-free powder (Fig. 11b) showed a heavy decoration of the grain boundaries with oxides, with an accompanying deterioration of corrosion resistance. Grain-boundary corrosion due to carbide sensitization, although somewhat less severe than the type of corrosion discussed above, still severely limits the attainment of adequate corrosion resistance of sintered P/M stainless steels.

Remedies such as heat treatment to rediffuse chromium back into the chromium-depleted austenite and the use of carbide stabilizers (titanium, vanadium, and niobium) are not practiced in the P/M industry. Such remedies are unneces-

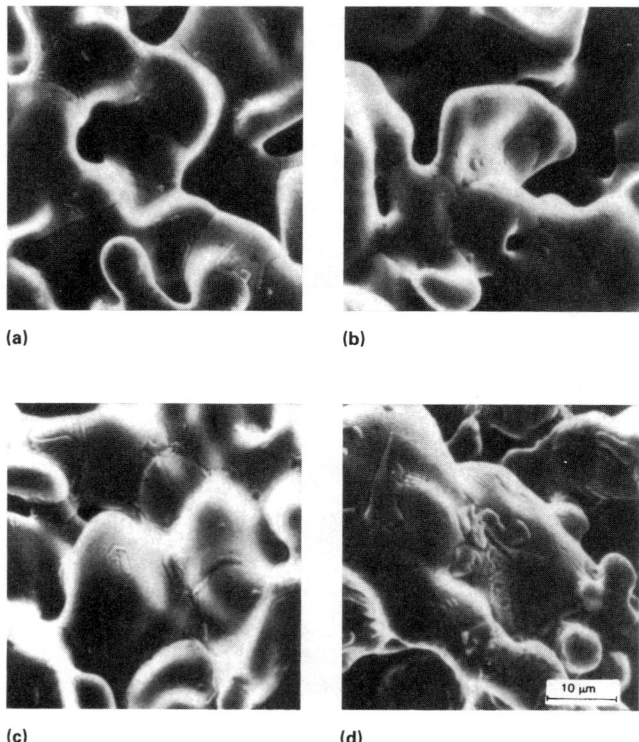

Fig. 14 Scanning electron micrographs of type 316L stainless steel. (a) Sintered 45 min in 100% H_2 at 1350 °C (2460 °F); 66 ppm N. (b) Sintered 45 min in 75% H_2 at 1350 °C (2460 °F); 3100 ppm N. (c) Sintered 45 min in 25% H_2 at 1350 °C (2460 °F); 4300 ppm N. (d) Sintered 45 min in 25% H_2 at 1150 °C (2100 °F); 6650 ppm N. The amount of intergranular precipitate increases with nitrogen content. Source: Ref 13

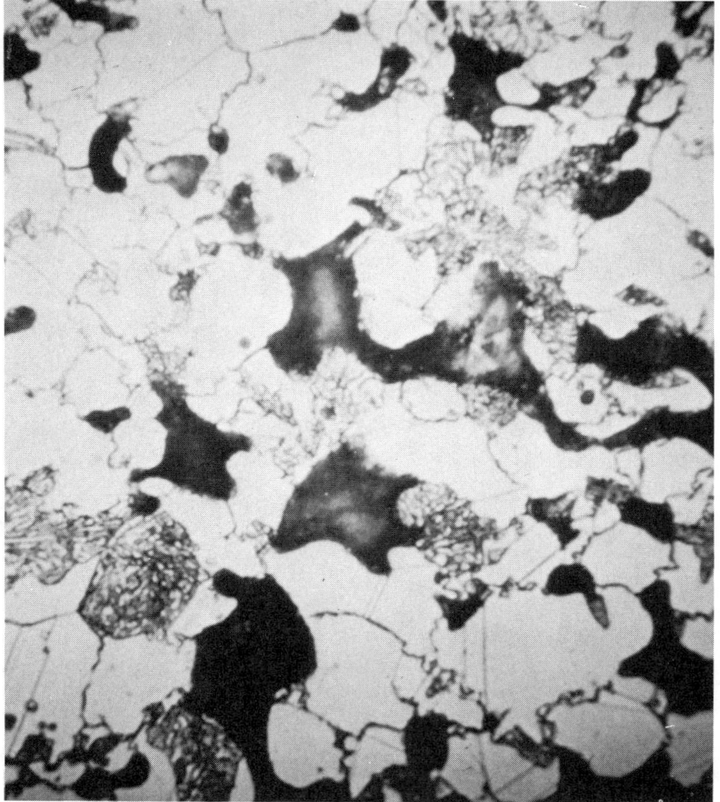

Fig. 15 Micrograph showing the lamellar structure of Cr_2N and low-chromium austenite in sintered type 316L that was slowly cooled in dissociated NH_3. Etched with Marble's reagent. 700×. Source: Ref 9

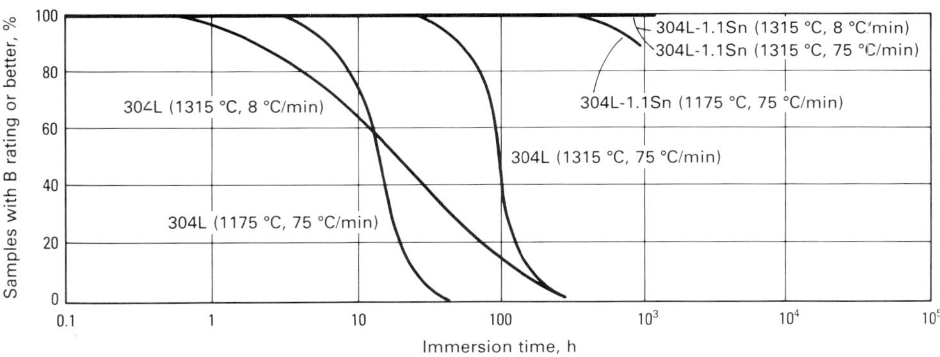

Fig. 16 Effect of composition, cooling rate, and sintering temperature on corrosion resistance of type 304L and tin-modified type 304L P/M stainless steels (sintered density: 6.5 g/cm^3; sintering atmosphere: dissociated NH_3) in 5% aqueous NaCl. B rating indicates <1% of specimen surface stained. Parenthetical values designate sintering temperature and cooling rate. Source: Ref 19

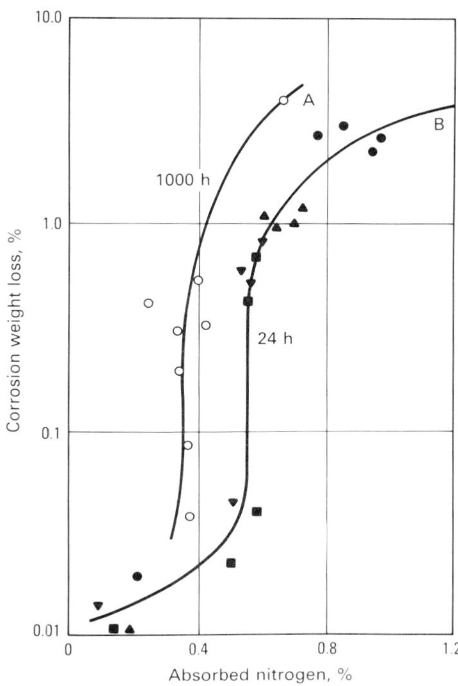

Fig. 17 Weight loss of austenitic stainless steel in 10% aqueous HNO_3 as a function of absorbed nitrogen content. Curve A: sintered in dissociated NH_3 at 1150 °C (2100 °F) with a dew point of −43 °C (−45 °F). Density: 5.10 to 5.20 g/cm^3. Curve B: sintered in various atmospheres with different dew points. Density: 5.2 to 5.8 g/cm^3. Source: Ref 9

sary if carbon pickup during sintering is avoided, because the various powders are available as low-carbon (L) grades.

Effect of Nitrogen. As with carbon, the key to understanding the effect of nitrogen on the corrosion resistance of sintered stainless steels is the solubility of nitrogen in the stainless steel matrix. Under certain conditions, dissolved nitrogen will precipitate as chromium nitride (Cr_2N), with accompanying chromium depletion and deterioration of corrosion resistance. Under industrial conditions of sintering in dissociated ammonia (NH_3), synthetic nitrogen-base atmospheres (typically 5 to 10% H_2, rem N_2) or in a vacuum with a partial pressure of nitrogen, the stainless steel part absorbs nitrogen in accordance with known phase equilibria (Fig. 12). The amount of nitrogen absorbed during sintering decreases with increasing sintering temperature and with decreasing chromium concentration of the stainless steel. Also, absorption follows Sievert's law; that is, absorption is proportional to the square root of the partial pressure of nitrogen in the sintering atmosphere. This nitrogen absorption provides significant strengthening (Fig. 3). Upon completion of sintering, when the part enters the cooling zone of the furnace, the solubility of nitrogen decreases sharply with temperature (Fig. 12). As a result, Cr_2N begins to precipitate at the temperature at which the nitrogen content crosses the solubility limits. More important, below about 1150 °C (2100 °F), additional nitrogen is absorbed from the sintering atmosphere, leading to more Cr_2N precipitation and chromium depletion along the grain boundaries. The net result is inferior corrosion resistance due to grain-boundary corrosion.

The rate of this detrimental nitrogen absorption increases with decreasing part density and with decreasing dew point. A high dew point, however, leads to the problem of excessive oxidation. The basic relationship of this phenomenon is shown in Fig. 13. The data in Fig. 13, which were developed for the bright annealing of stainless steel in dissociated NH_3 atmospheres, show the extent of nitrogen and oxygen absorption as a function of dew point. At high dew points (higher than about −37 °C, or −35 °F, depending on part size), the rate of oxidation is severe enough to produce a dull surface. At dew points of about −45 °C (−50 °F) or lower, nitrogen absorption increases so much that the corrosion resistance deteriorates because of excessive Cr_2N formation. Thus, optimum bright annealing of austenitic stainless steels must be done within a narrow dew-point range. Although the authors (Ref 14) caution against applying these findings to sintered stainless steels based on the unexplained higher nitrogen contents found for their parts sintered in dissociated NH_3, it should be noted that such

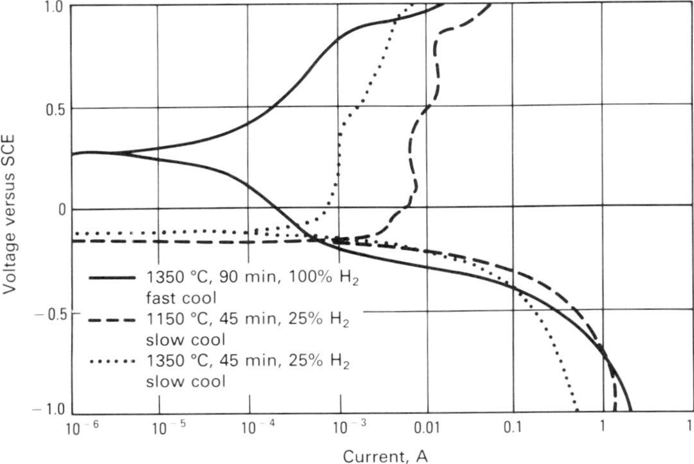

Fig. 18 Forward scan potentiodynamic corrosion curves for external surfaces of three sintered type 316L stainless steel samples in 10% HNO_3 at 25 °C (75 °F). Note the increasing corrosion currents in the 0- to 1-V range and the decreasing corrosion potential with nitrogen additions to the atmosphere, slow cooling, and lower sintering temperatures. SCE, saturated calomel electrode. See also Fig. 19. Source: Ref 20

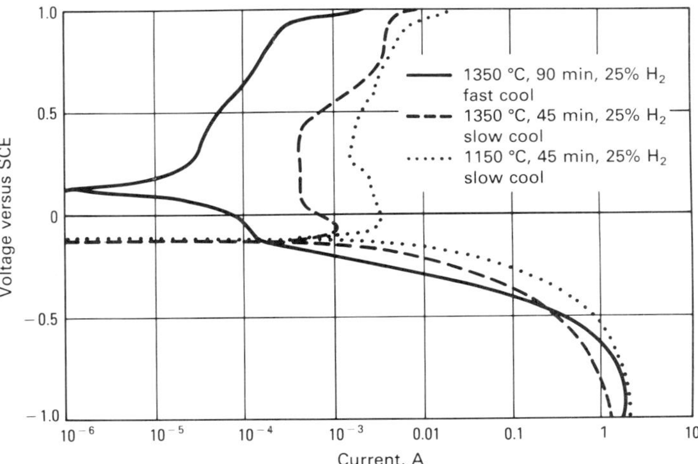

Fig. 19 Forward scan potentiodynamic corrosion curves of the internal microstructure (metallographic cross section) for type 316L stainless steel samples sintered in 25% H_2. Corrosion susceptibility in 10% HNO_3 at 25°C (75 °F) increases with a lower sintering temperature and slow cooling. Cr_2N precipitation is most severe on the surface of a sintered part. See also Fig. 18. Source: Ref 20

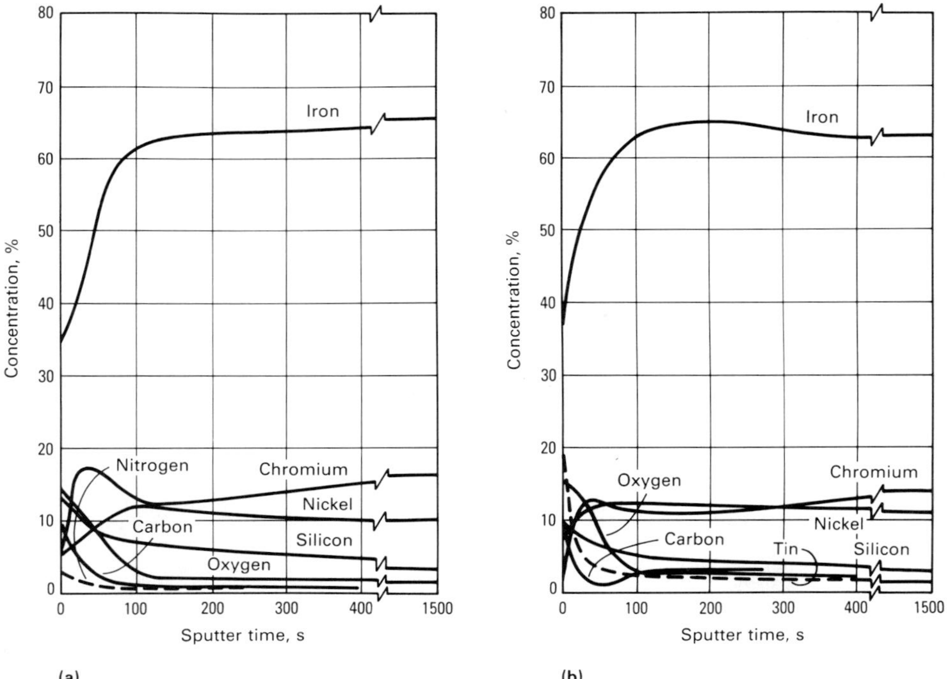

Fig. 20 Auger composition depth profiles of P/M type 316L stainless steel parts sintered in dissociated NH_3 at 1175 °C (2150 °F). (a) Type 316L. (b) Tin-modified type 316L. Source: Ref 9

higher nitrogen contents are expected on the basis of known solubility data for nitrogen in type 316L (Fig. 12) considering the differing methods of nitrogen analysis used.

Chromium nitride sensitization may in some cases be limited to a very shallow surface depth of the part. With very slow cooling, however, absorption and precipitation proceed toward the interior of the porous part. Figure 14 shows Cr_2N precipitates in the grain boundaries of parts that were sintered under conditions that produced nitrogen contents from 55 to 6650 ppm. Increasing nitrogen content correlates with increasing amounts of precipitation and increasing localized corrosion (Fig. 14). Figure 15 shows the microstructure of a type 316L part that was sintered in dissociated NH_3 and cooled very slowly. Slow cooling produced a lamellar structure of Cr_2N and low-chromium austenite of very poor corrosion resistance.

Corrosion resistance data for sintered types 304L and 316L in NaCl solutions and in 10% HNO_3, reflecting the effect of Cr_2N precipitation, are shown in Fig. 8, 16, and 17. Figures 8 and 16 show that a higher sintering temperature, fast cooling rates (75 °C/min, or 135 °F/min, versus 8 °C/min, or 14 °F/min), and the use of type 316L rather than type 304L provide better corrosion resistance. That these measures are beneficial follows directly from the austenite-nitrogen phase diagram (Fig. 12).

Figure 18 shows potentiodynamic corrosion curves for sintered type 316L in 10% HNO_3. The corrosion current density in the passive range increases and the corrosion potential decreases under conditions that promote Cr_2N precipitation, that is, lower sintering temperature, slower cooling rate, and high nitrogen concentration of the sintering atmosphere. Figure 19 is similar to Fig. 18 except that internal rather than external cross sections were used. The significantly lower corrosion currents of the internal surface confirm that Cr_2N precipitation is most severe on the surface of a sintered part.

Recently developed tin-containing grades of type 304L (Table 2) and 316L stainless steels have shown less sensitivity to nitride precipitation and correspondingly improved corrosion resistance (Fig. 8 and 16). The beneficial effect of tin has been confirmed in several studies (Ref 9, 10, 16, 19-22) and has been attributed to an enrichment of the surfaces of both the water-atomized powder and the sintered part with tin, presumably as a result of the low solubility of tin in solid stainless steel (Ref 10). Tin may also form stable acid-resistant passive films in a crevice and may cause cathodic surface poisoning, but its major beneficial effect is believed to lie in its formation of an effective barrier to nitrogen (and possibly also to oxygen) diffusion. This reduces the rate at which nitrogen is absorbed on the surface of the sintered part as it enters the cooling zone of the furnace. Auger composition depth profiles of regular and (1.5%) tin-containing type 316L parts sintered in dissociated NH_3 (Fig. 20) show that the presence of tin on the surface effectively suppresses nitrogen absorption. In addition, on the basis of potentiodynamic polarization tests in 10% HNO_3 and 5 N H_2SO_4, improvement in corrosion resistance has also been reported due to the presence of tin (Ref 20, 22).

The effect of oxygen on the corrosion resistance of sintered stainless steels is probably the most complex and least understood variable for several reasons. First, commercial water-atomized compactible stainless steel powders have typical oxygen contents of about 2000 ppm or more. Although much of this oxygen resides on the surfaces of individual powder particles as oxidized silicon (Fig. 21a), the exact nature and distribution of the oxides depends on atomizing conditions. Second, with typical industrial sintering practice, the reduction of these oxides remains incomplete and depends on many process parameters. Lastly, as a sintered part enters the cooling zone of the furnace, certain elements will oxidize upon reaching the temperature for the oxide-metal equilibrium of the high oxygen affinity elements (Fig. 22). Thus, a sintered part still reflects the history of its powder-making process, compaction, and sintering. Figure 21(b) shows the Auger composition depth profile of a type 316L part after sintering in hydrogen at 1260 °C (2300 °F). It is apparent that much of the oxidized silicon present in the green part has become reduced and that severely depleted chromium has been replenished.

An empirical correlation between the saltwater corrosion resistance of sintered type 316L and the oxygen content of the sintered parts suggests that sintering conditions resulting in lower oxygen contents provide better corrosion resistance (Fig. 23). With excessive dew points (>−34 °C, or −30 °F), the oxygen content of a sintered part may increase considerably. The microstructure (Fig. 24) of such a part shows a lack of particle bonding (compare with Fig. 10a for low oxygen content), and its mechanical strength and corrosion resistance are both inferior.

For optimum corrosion resistance, it appears that the following precautions are beneficial:

- Use of a powder with low oxygen content
- Sintering conditions that ensure a high degree of oxide removal
- Fast cooling through the high-temperature range after sintering

Cooling in a hydrogen atmosphere should be done with a water vapor content of less than 50 ppm (Ref 12). Cooling in a nitrogen-containing atmosphere should be done with a dew point between about −37 and −45 °C (−35 and −50 °F) (Ref 14).

Effect of Sintered Density. Applications of sintered stainless steels cover a wide density spectrum. Low densities of about 5 g/cm^3 may be typical of filters, but densities of 6.5 g/cm^3 or greater are typical of structural parts. It is therefore of interest to know the effect of density on corrosion resistance. Corrosion studies of sintered austenitic stainless steels have shown that the corrosion resistance improves significantly with increasing density in acidic environments, such as dilute H_2SO_4, HCl, and HNO_3. Figure 25 illustrates this behavior for three austenitic stainless steels (18Cr-11Ni to 18Cr-14Ni) that were vacuum sintered 1 h at 1150 and 1250 °C (2100 and 2280 °F) and tested in boiling 40% HNO_3.

For saline solutions, some investigators have found the effect of increasing density to be beneficial (Ref 20) while others have found it to be detrimental (Ref 10, 16, 23). This lack of agreement is perhaps not surprising considering that concentration changes in several of the critical variables, such as oxygen, carbon, and nitrogen, also depend on the density of a part. It should be noted, however, that the positive relationship between density and corrosion resistance was derived from short-term potentiodynamic polarization measurements (Ref 20), whereas the negative relationships were all derived from longer-term salt immersion tests.

Table 6 summarizes recent results on the effect of density on the salt corrosion resistance (im-

Table 5 Effect of iron, carbon, nitrogen, and oxygen on corrosion resistance of sintered austenitic stainless steels in NaCl

Variable	Origin of problem	Effect on corrosion resistance	Suggested solutions	Ref
Iron	Contamination of prealloyed powder with iron or iron-base powder at powder or parts producer's facility	Lowering of corrosion resistance by more than 99% due to galvanic corrosion	Utmost cleanliness at both powder and parts producer's manufacturing facilities, preferably separate and dedicated equipment and facilities	9
Carbon	Inadequate lubricant removal; carburizing sintering atmosphere; soot in sintering furnace; high carbon powder	Inferior resistance to intergranular corrosion	Use L-grade designation of stainless steel powder. Ensure adequate lubricant removal (before sintering). Use clean soot-free sintering furnace and carbon-free sintering atmospheres; carbon content of sintered part should be ≤ 0.03%.	9, 10
Nitrogen	Sintering in dissociated NH_3 or other nitrogen-containing atmosphere combined with slow cooling	Inferior resistance to intergranular corrosion	Reduce percentage of nitrogen in sintering atmosphere. Use fast cooling of parts preferably >150 to 200 °C/min (270 to 360 °F/min) through critical temperature range (700 to 1000 °C, or 1290 to 1830 °F). Use higher sintering temperature. Use intermediate dew points (−37 to −45 °C, or −35 to −50 °F) in cooling zone of furnace. Use tin-modified stainless steel powders.	9, 11, 12, 13, 14
Oxygen	Excessive oxygen in powder; excessive dew point of sintering atmosphere; slow cooling after sintering	Inferior resistance to general corrosion	Use low oxygen content powder, preferably <2000 ppm. Control dew point within sintering furnace to ensure reducing conditions. Fast cooling, preferably >200 °C/min (360 °F/min) For nitrogen-containing atmospheres, use dew point of −37 to −45 °C (−35 to −50 °F) in cooling zone. For sintering in H_2, ensure that water vapor content of atmosphere is below 50 ppm.	9, 11, 14
Density of sintered part	High sintered density	Inferior resistance to crevice corrosion	Use lower density to increase pore size and circulation of corrodent. In acidic environments, corrosion resistance improves with increasing density due to a decrease of specific surface area.	9, 11, 15, 16, 17

Table 6 Effect of density on corrosion performance of vacuum-sintered type 316L stainless steel

Compacting pressure, MPa	Compacting pressure, tons/in.²	Sintering temperature, °C	Sintering temperature, °F	Sintering time, min	Sintered density, g/cm³	Median pore size of pore volume(a)	Corrosion rating(b) for four specimens, each immersed in 5% aqueous NaCl, h: 1	3	19	27	46	91	140	210	314	380	558	525
276	20	1205	2200	45	5.67	9	A	A	A	A	A	B	B	B	A	B	B	B
							A	A	A	B	B	B	B	B	B	B	B	C
							A	A	B	B	B	B	B	B	B	B	C	C
							A	A	B	B	B	B	B	B	B	C	C	C
552	40	1205	2200	45	6.53	5	A	B	B	B	C	C	D	...	...	...	...	...
							A	B	...	B	B	C	C	C	D	...	...	...
							A	A	B	B	C	C	C	C	D	...	...	...
							A	A	B	C	C	C	C	D	...	...	...	...
276	20	1315	2400	45	5.86	8	A	A	A	A	B	B	B	B	B	B	B	B
							A	A	A	A	B	B	B	B	B	B	B	B
							A	A	A	A	B	B	B	B	B	B	B	B
							A	A	A	A	B	B	B	B	B	B	B	B
552	40	1315	2400	45	6.57	5	A	A	B	B	B	B	B	C	C	C	D	...
							A	A	B	C	C	C	C	D	...	...	...	...
							A	A	B	C	C	C	C	D	...	...	...	...
							A	A	A	A	B	B	B	C	C	D	...	...

(a) Determined by mercury porosimetry. (b) A, sample free from any corrosion; B, ≤1% of surface covered by stain; C, 1 to 25% of surface covered by stain with slight corrosion product; D, >25% of surface covered by stain with heavy corrosion product

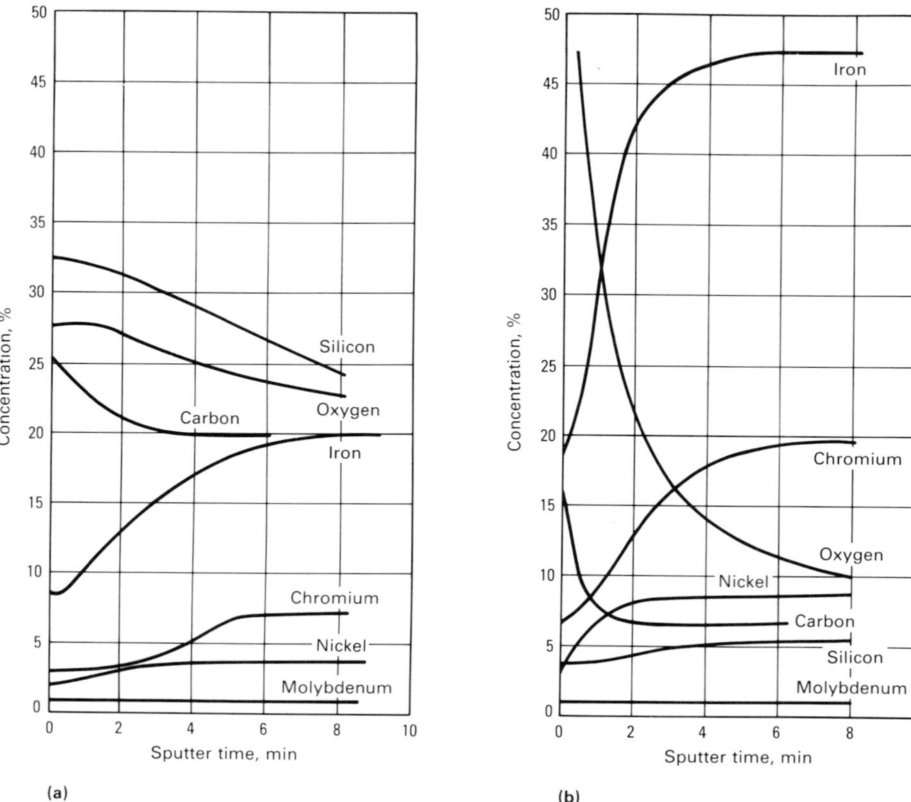

Fig. 21 Auger composition depth profiles of P/M type 316L stainless steel. (a) Green part. (b) Sintered part

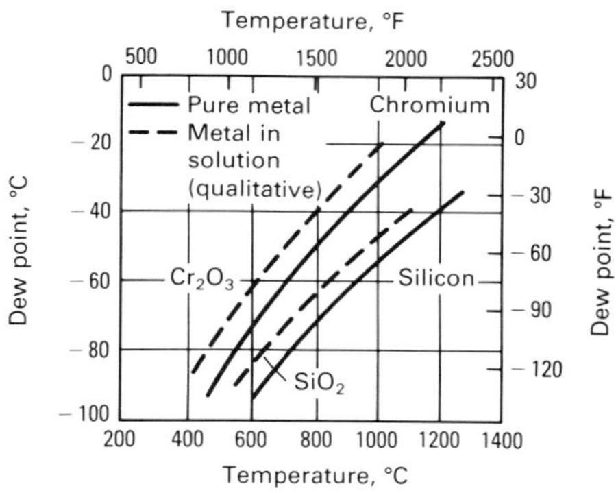

Fig. 22 Redox curves for chromium and silicon alone and in solution. Source: Ref 9

mersion in 5% aqueous NaCl) of vacuum-sintered type 316L parts. Unlike sintering in a reducing atmosphere, vacuum sintering does not lower the oxygen content with decreasing density. Thus, an improvement in corrosion resistance with decreasing density, as shown in Table 6, should not be attributed to a lower oxygen content, but is perhaps better explained in terms of reduced crevice corrosion as a result of the improved circulation of the corrodent through large pores (Ref 9). The average pore diameters of the parts pressed at the lower compacting pressures (Table 6), as measured by mercury porosimetry, were 60 to 80% larger than the size of the pores of the high-density parts. The standard deviations of the pore size distributions were similar and were around 2. Therefore, sintered stainless steel parts with densities from about 60 to 90% of theoretical have average pore sizes from about 10 to 2 or 3 μm that are likely to affect the circulation of the corrodent and thus its resistance to crevice corrosion.

Effect of Copper Additions to Type 304L. One study found that the corrosion resistance of copper-containing type 304L vacuum-sintered parts (1 h at 1200 °C, or 2190 °F; 88% dense) improved with increasing copper content (Ref 17). Figure 26 shows the weight loss of the parts kept for 6 h in boiling 5% H_2SO_4. Higher nickel content is also beneficial. Salt spray testing for 24 h with 5% NaCl solution resulted in almost no pitting. The effect of copper in P/M stainless steels is said to be identical to that observed in cast stainless steels.

Oxidation Resistance. Sintered stainless steels are not widely used for elevated-temperature service. Thus, information on elevated-temperature oxidation resistance is scarce.

Figure 27 shows the weight gain in air at 700 °C (1290 °F) for type 310L stainless steel parts that were vacuum sintered 1 h at 1250 °C (2280 °F) as a function of sintered density (circular plates), mesh size of powder used, and sintering temperature. This initial weight gain did not always show a parabolic course of oxidation. Within the density range studied, oxidation increased almost exponentially with decreasing density. Silicon-modified (4.06% Si) type 310L stainless steel showed weight gains that were less than 50% of those of regular type 310L. The increased oxidation of the parts made from the finer powder fraction is due to their large internal surface area. Higher sintering temperature and higher compacting pressure (higher densities) reduce surface porosity and specific pore surface area, thus lessening internal oxidation through early pore closure. The maximum recommended operating temperature for sintered austenitic stainless steels is 700 °C (1290 °F).

Higher-Alloyed Stainless Steels. Although the common stainless steel grades used in industry have maximum chromium and nickel contents of 20 and 14%, respectively (Table 2), higher-alloyed stainless steels have been used in the past to obtain improved corrosion resistance. Such steels are available from powder producers. In one investigation, a high-nickel/chromium/molybdenum austenitic stainless steel P/M material (SS-100) performed comparably to wrought type 216, 316, or 317 in 16-h salt solution immersion tests (Ref 25).

Other Approaches to Improving the Corrosion Resistance of Sintered Stainless Steels. If the corrosion resistance of sintered stainless steel parts remains inadequate after composition and process optimization, passivation and coating treatments are sometimes used. Chemical and thermal passivation treatments for sintered type 316L, effective in dilute H_2SO_4, are described in Ref 26. Chemical passivation with HNO_3 solutions similar to those applied to wrought stainless steels is not suitable for every material. On the basis of rest potential measurements of sintered type 316L, thermal passivation by heating the sintered parts for 20 to 30 min in air at temperatures of 400 to 500 °C (750 to 930 °F) is recommended.

In another study, the corrosion resistance of vacuum-sintered type 304L (6.9 g/cm^3) in 5% H_2SO_4 was improved by activating the parts in a mixture of 13 to 15% HNO_3, 2% hydrofluoric acid (HF), and 0.3% hydrochloric acid (HCl), followed by passivation for 30 min in 30% HNO_3 at 70 °C (160 °F) (Ref 27). After testing for 2 h in 5% H_2SO_4 (Fig. 28), the passivated specimens showed no weight loss, whereas the as-sintered specimens rapidly lost weight and turned the solution green. In addition, Ref 28 describes a phosphate-base passivating treatment for sintered stainless steels that is effective in acetic acid.

Improvement of the corrosion resistance of sintered stainless steel through the chemical vapor deposition (CVD) of chromium is discussed in Ref 29. The chemical vapor deposition of chromium onto sintered stainless steel parts was applied by pack cementation. Considerable infilling of the pores with chromium takes place; 50-μm thick pores with diameters of up to 50 μm may become sealed. Immersion of coated and uncoated specimens in 5 and 10% H_2SO_4 solutions for 168 h at room temperature showed significant attack of the uncoated specimens and no noticeable attack of the coated specimens. Electrochemical testing in 5% H_2SO_4 gave similar results, and a 3% salt spray test at room temperature showed many local sites of corrosion for the uncoated specimen and no corrosion after 250 h for the coated specimen. Sealing or coating of the pores of a sintered stainless steel part with an organic resin (Ref 25) or with water glass is sometimes recommended, but performance data proving the effectiveness of this treatment are lacking.

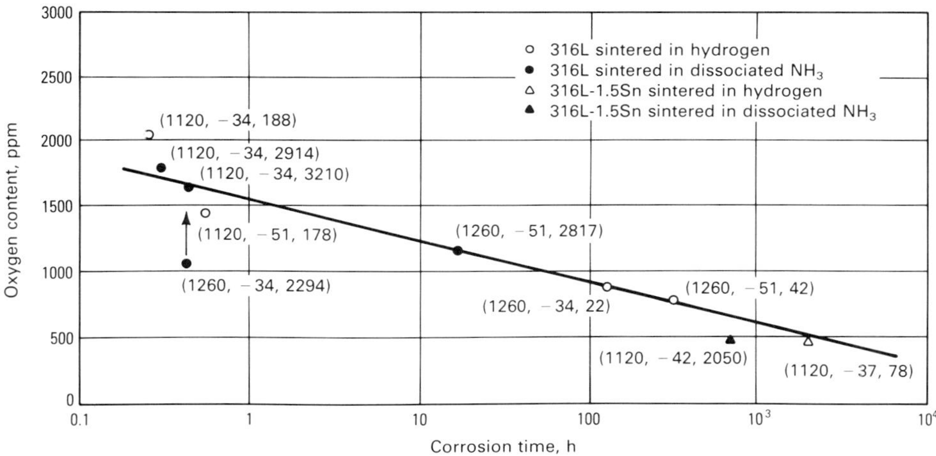

Fig. 23 Effect of oxygen content on corrosion resistance of sintered type 316L and tin-modified type 316L (sintered density: 6.65 g/cm^3; cooling rate: 75 °C/min, or 135 °F/min). Parenthetical values are sintering temperature (°C), dewpoint (°C), and nitrogen content (ppm), respectively. Time indicates when 50% of specimens showed first sign of corrosion in 5% aqueous NaCl. Source: Ref 10

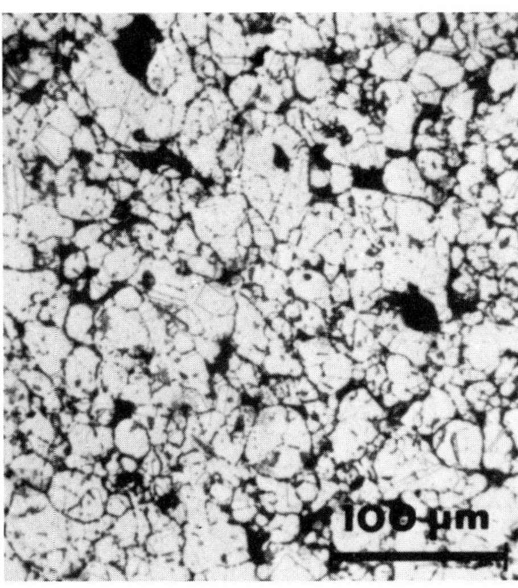

Fig. 24 Microstructure of type 316L stainless steel sintered in a high dew point atmosphere. Oxygen content: 5100 ppm; sintered density: 7.5 g/cm^3. Etched with Marble's reagent. 200×. Source: Ref 9

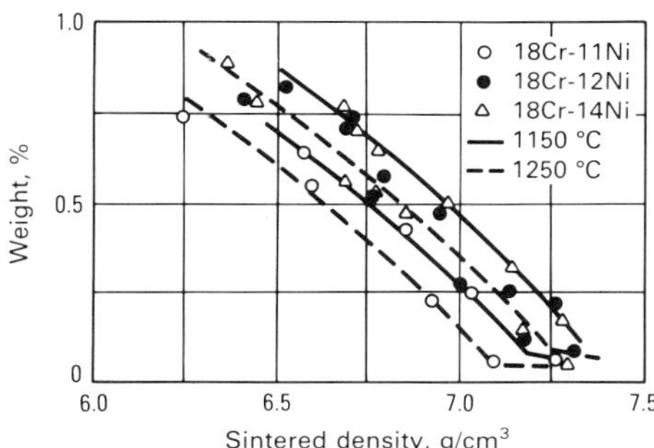

Fig. 25 Relationship between sintered density and weight decrease of three austenitic stainless steels in 40% HNO_3 solution. Source: Ref 23

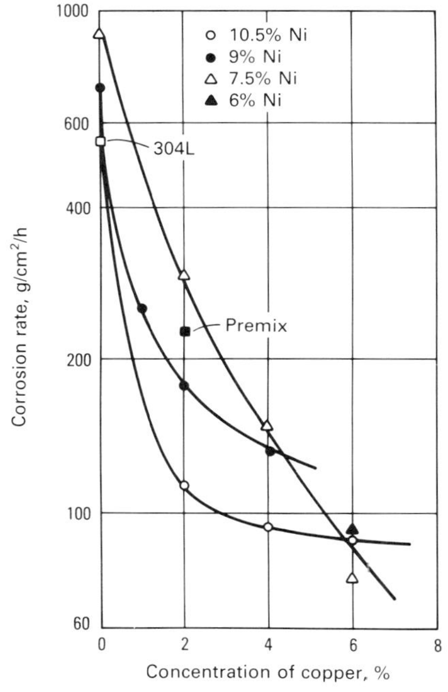

Fig. 26 Effect of nickel and copper additions on the corrosion rate of sintered austenitic stainless steel compacts exposed to boiling H_2SO_4 for 6 h. Relative sintered density is 88%. Source Ref 17

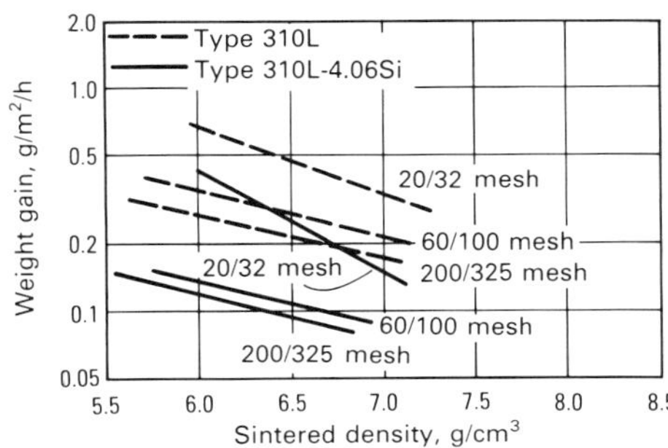

Fig. 27 Weight gain versus sintered density curves for materials prepared from powders of various particle sizes. Parts were sintered 1 h at 1250 °C (2280 °F). Source: Ref 24

Fully Dense P/M Stainless Steels

In the fully dense category of P/M stainless steels, parts made from water-atomized powders must be distinguished from those made from inert gas atomized powders.

Water-atomized powders, because of their irregular particle shape, are cold compactible and permit the pressing of complex parts, which, at temperatures approaching the melting point of the material, can be sintered to nearly full density. However, water-atomized stainless steel powders typically have oxygen contents of 2000 ppm or more, and sintering to full density usually does not reduce the oxygen content to the low level of the corresponding ingot material. The commercial production of such parts is still in its infancy, and corrosion data are not yet available.

Inert gas or centrifugally atomized powders are spherical and noncompactible. They have low oxygen contents (about 50 to 200 ppm) and are consolidated to full density by such processes as hot isostatic pressing (HIP), hot forging, and extrusion.

One company has manufactured seamless stainless steel tubes from gas-atomized powder since 1980. The P/M method is said to offer a competitive alternative to conventional production methods due to:

- Efficient use of raw materials
- Low energy consumption
- Short total production time
- High flexibility (less material in process; short delivery times)
- The ability to make difficult compositions (Ref 30)

The process consists of cold isostatic compaction of the encapsulated nitrogen-atomized powder, followed by heating to the extrusion temperature and hot extrusion. The capsule material is removed by decladding. Standard grades include most of the common austenitic stainless steels as well as some special austenitic, ferritic-austenitic, and ferritic stainless steels, together with nickel-base alloys.

In comparison to conventional material, the extruded P/M products possess a more homogeneous structure with reduced microsegregation due to the rapid cooling of the powder particles. Also, the grain size is somewhat finer, slag inclusions (particularly sulfides) are smaller, and the nitrogen content is somewhat higher (900 versus 500 ppm for wrought type 316).

Mechanical Properties and Corrosion Resistance. Attributed to the above differences are slightly higher yield and tensile strength (Table 7) without a loss in elongation. Mechanical properties at elevated temperatures are practically identical to conventionally produced materials. The impact toughness of the P/M material, although good, is lower than that of conventional material when tested in the longitudinal direction. Creep strength is similar to that of conventional material.

No difference between P/M and conventional material has been found regarding the resistance to intergranular corrosion according to practice C and practice E of ASTM A 262 (Ref 31). As shown in Fig. 29, the resistance to pitting attack,

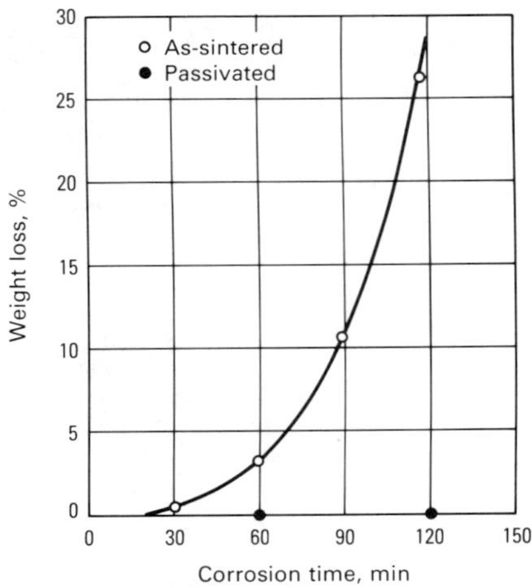

Fig. 28 Relationship between weight loss and corrosion time of vacuum-sintered type 304L stainless steel in 5% H_2SO_4

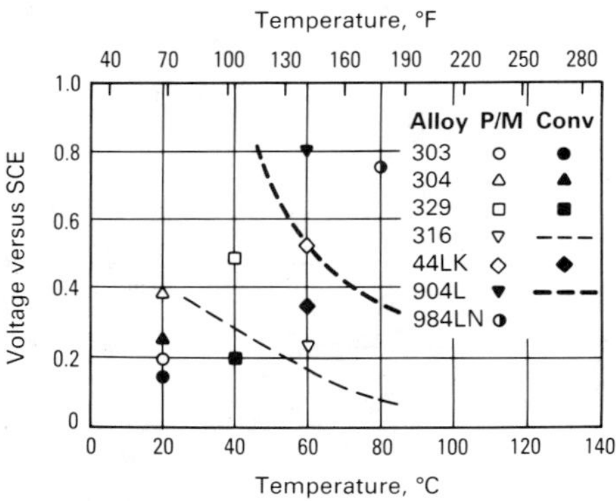

Alloy	Nominal composition, % C(max)	Fe	Cr	Ni	Mo	Others
303	0.10	rem	18	8.5	...	...
304	0.05	rem	18.5	8.5	...	...
329	0.05	rem	26	5	1.5	...
316	0.05	rem	17	11.5	2.2	...
44LK	0.03	rem	25	6	1.6	...
904L	0.02	rem	20	25	4.5	Cu
984LN	0.05	rem	20	33	2.2	Cu,N

Fig. 29 Comparison of pitting resistance of P/M and conventional stainless steels. Source: Ref 30

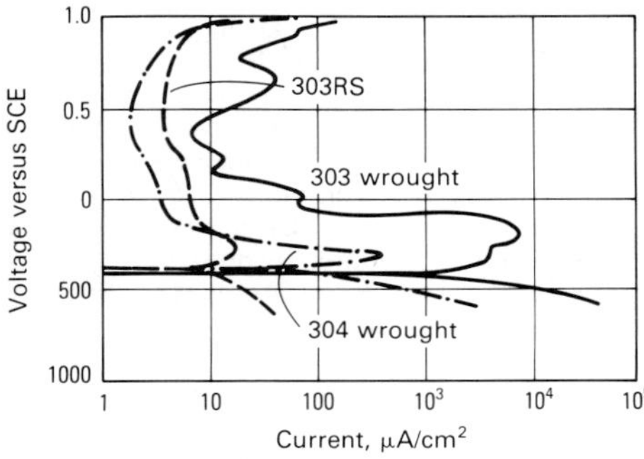

Fig. 30 Potentiodynamic polarization curves for conventional type 303 and 304 stainless steels and for rapidly solidified type 303 in deaerated 1 M H_2SO_4 at 30 °C (85 °F). Source: Ref 33

as measured by the pitting corrosion breakthrough potential, is superior for several P/M grades compared to the corresponding conventional grades. Table 8 gives the general and selective corrosion information from tests according to ASTM A 262, practice C (Ref 31) for two austenitic P/M grades. The improved corrosion resistance of the P/M grades is attributed to their lower segregation rate, their finer and more uniform distribution of inclusions, and their finer grain size.

The enhancement of the corrosion resistance of stainless steel parts made from rapidly solidified powders has been confirmed by several investigators. For example, the significantly superior oxidation resistance of type 303 stainless steel, made by extrusion of rapidly solidified powder, was attributed to the elevated-temperature grain growth inhibiting effect of uniformly dispersed manganese sulfide (MnS) particles (Ref 32). Figure 30 shows that this material maintains its good corrosion performance in aqueous environments, and potentiodynamic polarization curves in 1 M H_2SO_4 indicate that the P/M material exhibits the lowest corrosion rate at the corrosion potential. Finally, although wrought type 303 was highly susceptible to pitting, the P/M alloy showed no obvious pits on the surface and only a low pit density within the material. The pits were related to the presence of sulfide stringers in the wrought material, from which it was concluded that P/M steels with lower sulfur contents and with spherical sulfide morphology, such as type 304 and 316, might exhibit improved pitting resistance.

Injection molding technology (see the article "Injection Molding" in Volume 7 of the 9th Edition of *Metals Handbook*) is currently used to manufacture small and nearly fully dense stainless steel parts from fine powders. However, information on the corrosion performance of such parts is unavailable.

P/M Superalloys

Development of P/M superalloys began in the 1960s with the search by the aerospace industry (and later the electric power industry) for stronger high-temperature alloys in order to operate engines at higher temperatures and thus improve fuel efficiency. Figure 31 illustrates the great advances achieved since the 1940s by the introduction of new processes and alloys, such as vacuum melting, directional solidification of eutectics, development of alloys with high volume fractions of γ' phase, and P/M processing with and without oxide dispersions.

Initially, lower production costs were a major objective in exploring the P/M approach. Figure 32 illustrates the material savings possible with two different P/M methods due to their near net shape capabilities. Later, specific advantages linked to the P/M approach, such as the use of more complex and greater volume fractions of dispersoids, reduced segregation, and improved workability, led to the development of stronger alloys and to the use of these alloys not only in turbine disks but also in the higher-temperature turbine blades.

Much effort is currently being directed toward reducing the cost of consolidating superalloy powders, particularly of oxide dispersion strengthened (ODS) superalloys, through the development of suitable forging techniques (Ref 34). Efforts are also underway to exploit the advantages of microcrystallinity and extended solid solutions of rapid solidification technology.

Uses and State of Commercialization. P/M superalloys were first used in military engines in the mid-1970s. Table 9 summarizes

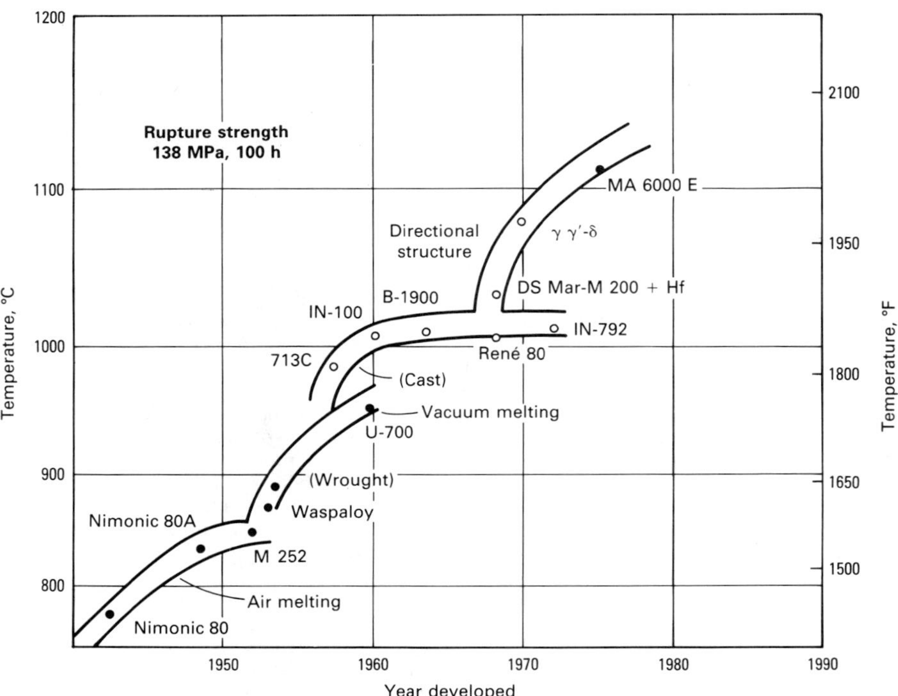

Fig. 31 Trends in alloy processing and development. Source: Ref 34

Table 7 Typical mechanical properties of cold-worked and annealed stainless steel tubes extruded from powders

Grade	Type(a)	Number of samples	Yield strength, 0.2% offset MPa	ksi	Tensile strength MPa	ksi	Elongation, %
Type 304L	C	84	302	44	582	84	57
	P/M	18	325	47	609	88	58
Type 304	C	133	321	46	600	87	57
	P/M	72	350	51	660	96	55
Type 316L	C	90	319	46	604	88	53
	P/M	128	336	49	632	92	52
Type 316	C	134	306	44	584	85	54
	P/M	125	346	50	649	94	51
Type 904L	C	49	334	48	651	94	45
	P/M	112	382	55	681	99	43

(a) C, conventional production; P/M, powder metallurgy. Source: Ref 30

Table 8 Huey test (ASTM A 262, practice C) corrosion data for two P/M extruded stainless steels

Grade	Number of samples	Corrosion rate, μm/48 h Average	Specific	Selective attack, μm Average	Specific
Type 725LN	14	0.57–0.69	1.5 max	<50	100 max
Type 724L	14	1.48–1.79	3.3 max	<30	200 max

Source: Ref 27

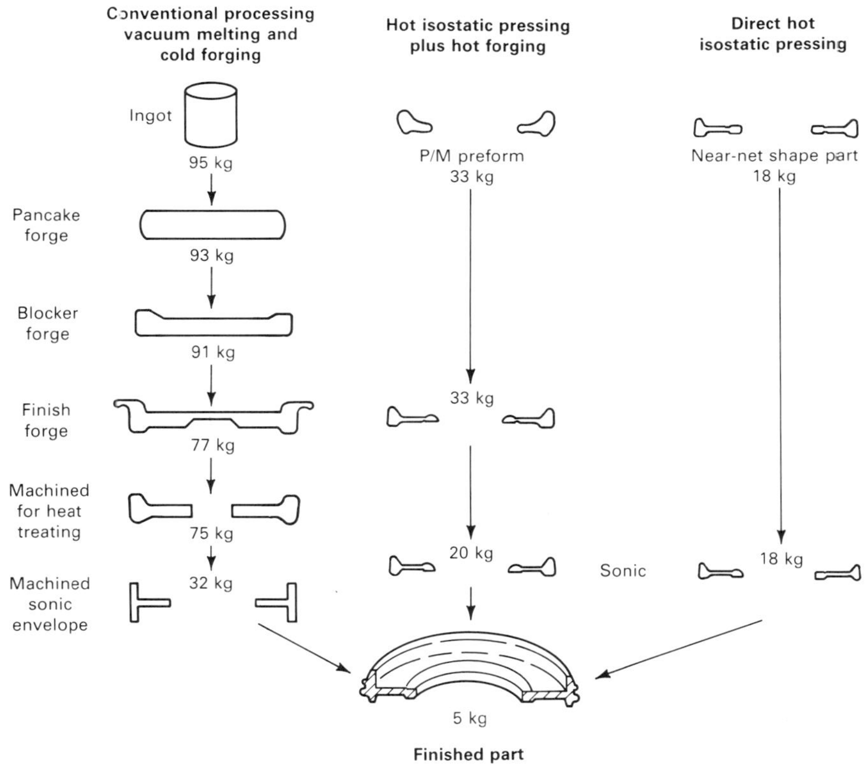

Fig. 32 Processing sequence in the production of jet engine compressor disks

the uses of P/M superalloys in terms of components, engine use, and reasons for using P/M technology. Other uses of superalloys include nuclear reactors, heat exchangers, furnaces, sour gas well equipment, and other high-temperature applications.

Manufacturing of P/M Superalloys. An important prerequisite for making P/M superalloys that possess reliable dynamic properties is the use of clean powders. Years of intensive work were spent in identifying and controlling the problems related to unclean powders. Today, argon and vacuum (also known as soluble gas process) atomization, as well as atomization by the rotating electrode process, are known to be suitable for producing powders with the required low oxygen content and low degree of contamination (details on these processes are available in the article "Atomization" in Volume 7 of the 9th Edition of *Metals Handbook*). The so-called prior particle boundary (PPB) problem, that is, the presence of carbides segregated at PPBs, was solved through the development of low-carbon alloys. Special equipment is used for removing ceramic particles and particles containing entrapped argon. Some of these problems are minimized or avoided in ODS alloys made by mechanical alloying. In mechanical alloying, elemental and master alloy powders as well as refractory compounds are mechanically alloyed by high-energy milling (Ref 36, 37).

Two established powder consolidation techniques for P/M superalloys are hot isostatic pressing (HIP) and isothermal forging. Figure 33 illustrates schematically the steps of the P/M processes in comparison to conventional processing. Both P/M methods permit the manufacture of so-called near net shape parts with attendant improved material use and reduced machining costs. Powder metallurgy forging exploits the improved forgeability deriving from the higher incipient melting temperature and reduced grain size of P/M material. Hot compaction by extrusion leads to very fine grain size, improved hot ductility, and superplasticity.

Depending on the application of a superalloy part, the powder consolidation process can be controlled in order to yield either a fine or a coarse grain size. Fine grain size is preferred for intermediate temperatures (up to about 700 °C, or 1290 °F) because of its higher strength and ductility at these temperatures. For high-temperature blade and vane applications, however, a large grain size (ASTM 1 to 2) provides superior creep strength due to reduced grain-boundary sliding. Grain coarsening of ODS alloys is achieved through special heat treatments after consolidation (Ref 34).

Compositions and Properties. Table 10 shows the compositions of the best known P/M superalloys. Many have the same compositions as cast alloys but are manufactured similarly to wrought alloys. The important P/M superalloys—IN-100, René 95, and Astroloy—were adapted to the P/M process by reducing their carbon content and by adding stable carbide formers to eliminate the problem of PPB carbides. To facilitate HIP, alloy compositions were modified to increase the temperature gap between the γ' solvus (above which HIP has to be carried out for increasing grain size) and the solidus temperature.

Oxidation. Nickel-, cobalt-, and iron-base superalloys use the selective oxidation of aluminum or chromium to develop oxidation resistance (Ref 38). These alloys are therefore often referred to as aluminum oxide (Al_2O_3) or chromium oxide (Cr_2O_3) formers, depending on the composition of the oxide scale that provides protection. Alloy composition, surface conditions, gas environment, and cracking of the oxide scale affect the selective-oxidation process (Ref 38). Figure 34 shows the development of superalloys in terms of the progress achieved against high-temperature oxidation.

Cyclic oxidation causes protective scales of Al_2O_3 and Cr_2O_3 to crack and spall. Regeneration of the scales will eventually result in the complete depletion of chromium and aluminum. The length of time for which superalloys are Al_2O_3 or Cr_2O_3 formers under given conditions is very important because of the subsequent appearance of less protective oxides. The importance of chromium

Table 9 Aerospace applications of P/M superalloys

P/M superalloy	Component	Engine	Aircraft/ manufacturer	Reasons for using P/M technology: Cost reduction	Reasons for using P/M technology: Improved properties
IN-100	Turbine disks, seals, spacers	F-100	Pratt & Whitney	X	X
René 95	Turbine disks, cooling plate	T-700	Helicopter/G.E.	...	...
René 95	Turbine disks, compressor shaft	F-404	F-18 Fighter	X	...
René 95	Vane	F-404	G.E.	...	...
René 95	High-pressure turbine blade retainer, disks, forward outer seals	F-101	...	X	...
Astroloy	High-pressure turbine disks	JT8D-17R Turbofan	...	X	...
Merl 76	Turbine disks	Turbofan	...	X	X
Inconel MA-754	Turbine nozzle vane	F-404	F-18 Fighter	...	X
Inconel MA-754	High- and low-pressure turbine vanes	Selected engines	...	...	X
Stellite 31	Turbine blade dampers	TF 30-P100	USAF F-111F	X	...
Inconel MA-6000E	Turbine blades	TFE 731	...	...	X

Source: Ref 35

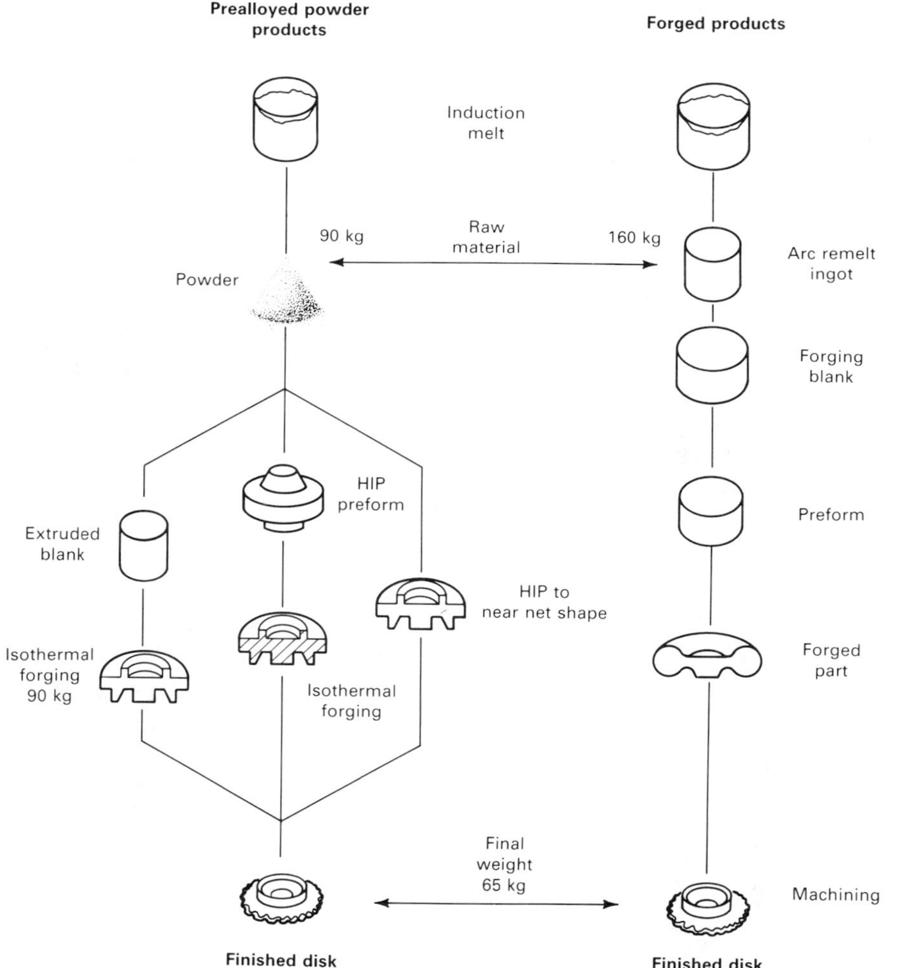

Fig. 33 Comparison of conventional processing and P/M processing for the fabrication of superalloy disks. Source: Ref 35

for imparting oxidation resistance is demonstrated in Table 11, which lists fatigue crack growth rates for different alloys.

Figures 35 and 36 show comparisons of the oxidation resistances of superalloys with and without oxide dispersions. Many studies have confirmed the beneficial effect of dispersed oxides on oxidation resistance. The lower oxidation rates of ODS alloys have been attributed to the reduced time required to form a continuous Cr_2O_3 scale due to the presence of dispersed oxides, which act as nuclei for oxidation (Ref 41). Based on marker studies with platinum, one investigation attributed the beneficial effect of oxide dispersions to the predominant, inward diffusion of oxygen ion (O^{2-}) and a slowdown of the chromic ion (Cr^{3+}) diffusion (Ref 42). The latter may be caused by the blocking of the dispersions in the Cr_2O_3 scale. With the dispersed oxides becoming dissolved in the scale, it also appears possible that trivalent ions, such as yttrium (Y^{3+}) and lanthanum (La^{3+}), will reduce the number of vacant cation sites, thus lowering the diffusivity of Cr^{3+}.

Dispersed oxides may also improve scale adhesion because of the thinner scale or because of increased porosity or smaller grain size in the oxide scale (Ref 40). It was reported that at 1300 °C (2370 °F) the outer regions of the Al_2O_3 film of MA 956 (Fe-20Cr-4.5Al-0.5Ti-0.5Y_2O_3) became enriched with titanium, giving rise to a continuous layer of titanium-rich oxide (Ref 43). Pegging of the oxide by titanium carbide particles and the irregular metal/oxide interface is said to contribute to the good spalling resistance of the alloy.

In oxidation tests in air and in an inert atmosphere at 1260 °C (2300 °F) for MA 754 (Ni-20Cr-0.5Ti-0.5Y_2O_3-0.3Al-0.05C), Ni-20Cr (cast/wrought), and an ODS nickel-chromium alloy, subsurface porosity was attributed to the oxidation of chromium and aluminum (Kirkendall porosity), and thermally induced porosity was excluded as a cause (Ref 44). This type of porosity decreases with improving oxidation resistance of the alloy.

Results of cyclic oxidation tests at 1100 °C (2010 °F) for MA 956, TD-NiCr, and Hastelloy X are given in Table 12. The superior resistance of MA 956 is attributed to a very stable Al_2O_3 film and parabolic oxidation for over 500 h. Tables 13 and 14 list sulfidation and carburization resistance data for MA 956. As in the case of oxidation, the alloy shows marked superiority to the other alloys tested. Tables 15 and 16 provide similar data for alloy MA 6000E (Ni-15Cr-4.5Al-4W-2Mo-2Ta-2.5Ti-1.1Y_2O_3) (Ref 45). The functions of the various alloying elements are as follows:

- Aluminum, titanium, and tantalum for γ' hardening
- Y_2O_3 for high-temperature strength and stability
- Aluminum and chromium for oxidation resistance
- Titanium, tantalum, chromium, and tungsten for sulfidation resistance
- Tungsten and molybdenum for solid-solution strengthening

Hot Corrosion. The requirement of hot corrosion resistance of superalloys derives from the use of sodium- and sulfur-containing fuels and the presence of salt in the air necessary for combustion. Under such conditions, combustion gases often leave deposits of sulfates or chlorides with metallic constituents of sodium, calcium, magne-

Table 10 Nominal compositions of several P/M superalloys

Alloy	Composition, %														
	C	Cr	Mo	W	Ta	Ti	Nb	Co	Al	Hf	Zr	B	Ni	Fe	V	Y_2O_3
IN-100	0.07	12.5	3.2	...	...	4.3	...	18.5	5.0	...	0.04	0.02	rem	...	0.75	...
René 95	0.07	13.0	3.5	3.5	...	2.5	3.5	8.0	3.5	...	0.05	0.01	rem	...	...	...
MERL 76	0.02	12.4	3.2	...	...	4.3	1.4	18.5	5.0	0.4	0.06	0.02	rem	...	...	...
AF 115	0.05	10.5	2.8	6.0	...	3.9	1.7	15.0	3.8	2.0	...	...	rem	...	...	...
PA101	0.1	12.5	...	4.0	4.0	4.0	...	9.0	3.5	1.0	...	...	rem	...	...	...
Low-carbon Astroloy	0.04	15.0	5.0	...	...	3.5	...	17.0	4.0	...	0.4	0.025	rem	...	...	...
MA 754	0.05	20.0	...	...	...	0.5	...	...	0.3	...	...	...	rem	...	...	0.6
MA 956	...	20.0	...	...	...	0.5	...	...	4.5	...	...	...	...	rem	...	0.5
MA 6000	0.05	15.0	2.0	4.0	2.0	2.5	...	...	4.5	...	0.15	0.01	rem	...	...	1.1
Stellite 31	0.5	25.5	...	7.5	...	...	...	rem	...	...	...	...	10.5	2.0	...	...

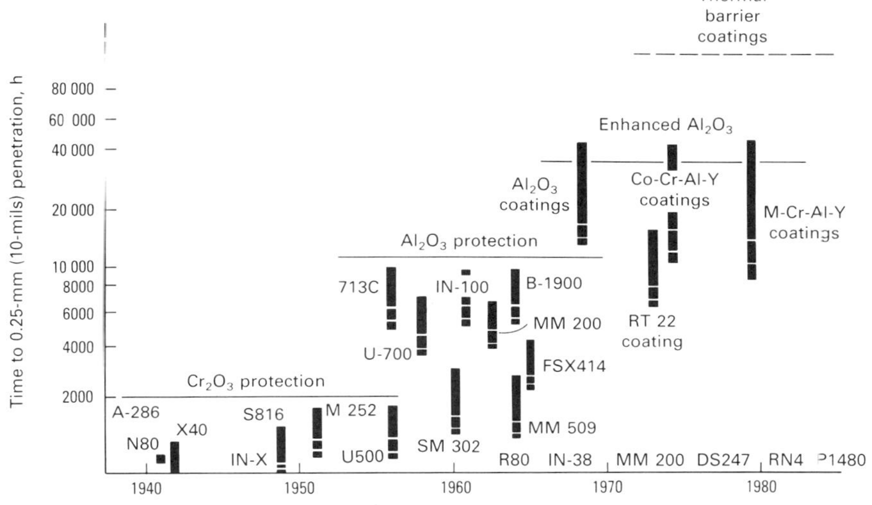

Fig. 34 Advancing steps in the protection of superalloys against oxidation at high temperatures showing life (in hours) to 0.25-mm (10-mils) penetration at 980 °C (1800 °F). Source: Ref 39

Table 11 Relative increase in fatigue crack growth rates after 15 min at 650 °C (1200 °F)

ΔK = 30 MPa$\sqrt{m}$ (27 ksi$\sqrt{in.}$)

Alloy	Relative increase in crack growth rates	Grain size, μm	Chromium content, %
René 95 after HIP + forge	242	50–70	12.8
IN-100	43.4	4–6	12.0
HIP-consolidated MERL 76	41.5	15–20	12.0
NASA II B-7	335	4–6	8.9
HIP-consolidated Astroloy	3.3	50–70	15.1
Waspaloy	4.0	40–150	19.3
Astroloy after HIP + forge	3.5	50–100	14.7

Source: Ref 40

sium, or potassium on the surfaces of superalloys. The resulting corrosion problems are particularly severe if these condensed phases are liquid. Hot corrosion may occur in gas turbines, boiler tubes, and incinerators. Typically (and similar to what happens in pure oxidation), hot corrosion of superalloys occurs in two stages: a slow rate initiation stage, followed by a propagation stage of rapid degradation. The difference, compared to oxidation, is that the conditions causing hot corrosion simply shorten the time in which superalloys form protective Al_2O_3 or Cr_2O_3 scales by selective oxidation (Ref 38). Factors affecting the length of the initiation stage (at the end of the initiation stage, the superalloy must be removed from service because of the start of excessive corrosion) include alloy composition, alloy fabrication conditions, gas composition and velocity, deposit composition and its physical state, amount of deposit, temperature, temperature cycles, erosion, and specimen geometry.

When a protective scale dissolves into a liquid deposit, so-called fluxing reactions can occur with the appearance of other basic or acidic nonprotective reaction products. Propagation may also be caused by components from the deposit that can accumulate in the deposit or the alloy and thus cause a nonprotective scale to form. Chlorine and sulfur produce such effects, and hot corrosion caused by the latter is known as sulfidation (see the section "High-Temperature Oxidation/Sulfidation" in the article "General Corrosion" in this Volume). Figure 37 shows the temperature ranges over which the various hot corrosion propagation modes are important. Additional information on the effects of individual elements on corrosion resistance is available in Ref 38 and in the articles "Corrosion of Nickel-Base Alloys" and "Corrosion of Cobalt-Base Alloys" in this Volume.

Some superalloys corrode in several modes. For example, hot corrosion of IN-738 proceeds by alloy-induced acidic fluxing, but is preceded by other propagation modes, including a basic fluxing mode. The higher chromium content alloys IN-738 and IN-939 were developed to improve the hot corrosion resistance of land-based gas turbines. Carbide stabilization through tungsten and tantalum and delay of $M_{23}C_6$ formation in service were expected to allow the large chromium content to impart improved hot corrosion resistance. Increasing the chromium and decreasing the Al_2O_3, however, lowered γ' solution temperatures and strength, which necessitated the use of coatings. The use of coatings led to the current use of enhanced aluminum, that is, carefully balanced coating alloys (based on nickel, iron, or cobalt with chromium, aluminum, and other active elements). Generally, all superalloy load-bearing parts used at very high temperatures under dynamic conditions are coated (Ref 39). Nevertheless, coatings generally last longer on more corrosion-resistant base materials.

In a model study, IN-738 was used to demonstrate the effect of grain size and Y_2O_3 dispersions on hot corrosion behavior (Ref 46). Under gas turbine simulated hot gas corrosion test conditions at 850 and 950 °C (1560 and 1740 °F) (Fig. 38), the presence of a Y_2O_3 dispersion lowered the corrosion rate. At 950 °C (1740 °F), a finer grain size further reduced the corrosion rate, which was thought to be mainly due to a higher diffusion rate of chromium and aluminum. The effect of the dispersion was predominant at 850 °C (1560 °F). Reduced sulfate formation at 850 °C (1560 °F) was attributed to the likely formation of yttrium oxysulfide.

In a study of the oxidation and hot corrosion resistance of P/M LC Astroloy and IN-100, isostatically pressed samples were found to be moderately attacked in a sulfate-chloride environment and heavily corroded by pure sodium sulfate (Na_2SO_4) (Ref 47). Heat treatment and the use of coarse powder (62 to 150 μm for Astroloy and 88 to 200 μm for IN-100) lowered the susceptibility to catastrophic corrosion. Additions of yttrium to IN-100 improved the corrosion resistance in pure sulfate, but were detrimental when NaCl was present. Therefore, yttrium additions to IN-100 cannot be recommended for marine turbines. It was concluded that in many cases impregnation coatings must be considered for components made of IN-100 alloys.

As a part of an evaluation of improved alloys for use in oil and gas drilling at depths of 6100 m (20 000 ft), HIP nickel-base alloy Inconel 625 was

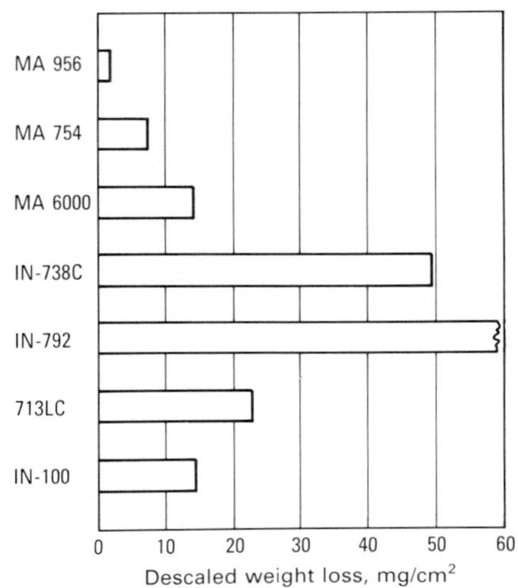

Fig. 35 Comparison of the oxidation resistance of ODS alloys MA 956, MA 754, and MA 6000 with that of other superalloys. Testing conditions: 504 h at 1100 °C (2010 °F) in air containing 5% H_2O. Temperature was cycled between test temperature and room temperature every 24 h. Source: Ref 40

studied in a simulated deep, hot, sour cell environment (Ref 48). The alloy demonstrated resistance to pitting and crevice corrosion, sulfide stress cracking, chloride stress-corrosion cracking (SCC), and elevated-temperature anodic stress cracking. Hot isostatic pressed Inconel 625 exhibited essentially the same corrosion resistance as wrought Inconel 625.

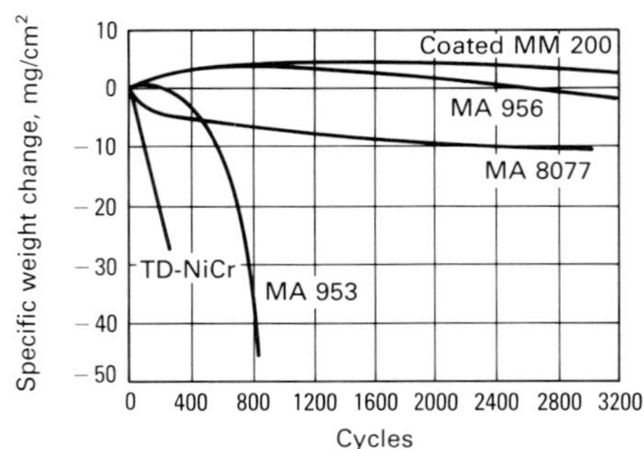

Fig. 36 Cyclic oxidation of ODS alloys MA 956, MA 8077, MA 953, and TD-NiCr compared to that of coated alloy MM 200. Testing conditions: held at 1100 °C (2010 °F) for 1 h and cooled by a 3-min air blast. Source: Ref 40

Fatigue and Creep Crack Growth. Fatigue crack growth rates of nickel-base superalloys measured at frequencies above 0.1 Hz, at intermediate temperatures, and at an intermediate stress intensity range, ΔK, were found to be several times higher than those measured in inert atmospheres (Ref 49). The buildup of corrosion products with decreasing ΔK, however, was thought to enhance crack closure, thus reducing the effective stress intensity range and leading to fatigue thresholds higher than those in inert environments.

Table 17 shows creep crack growth rates of IN-750 with various grain-boundary carbide microstructures (Ref 40, 49). Aggressive environments (helium + 3% sulfur dioxide, SO_2, and air) produce order of magnitude increases over the

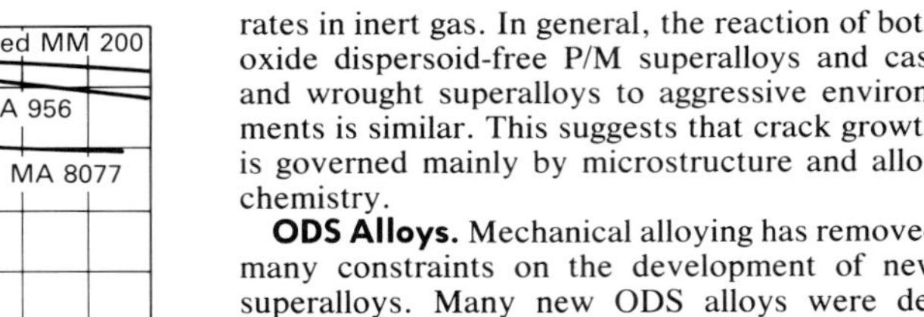
rates in inert gas. In general, the reaction of both oxide dispersoid-free P/M superalloys and cast and wrought superalloys to aggressive environments is similar. This suggests that crack growth is governed mainly by microstructure and alloy chemistry.

ODS Alloys. Mechanical alloying has removed many constraints on the development of new superalloys. Many new ODS alloys were designed specifically for corrosion resistance because alloying requirements for precipitation strengthening can be greatly reduced. Superior creep, corrosion, and erosion resistance at high temperatures have been claimed to enable the use of lower-grade fuels (Ref 50). Figures 39 to 42 show the corrosion resistances of ODS alloys MA 956, MA 6000, and MA 754 compared to several conventional superalloys.

Coatings for ODS alloys. As mentioned above, for extended high-temperature service, superalloys require additional protection through coatings. The use of aluminide coatings appears to be unsatisfactory due to the development of subsurface Kirkendall porosity and early spalling of the protective scale. Kirkendall porosity decreases with increasing aluminum content of the substrate alloy as well as with decreasing grain size (Ref 50). Only limited information exists on the properties of chromium-aluminum-yttrium coatings (Ref 40) and on diffusion barrier coatings (Ref 51-54).

P/M Aluminum Alloys

Although low- to medium-strength sintered (porous) P/M aluminum alloy parts were reported over 40 years ago and then commercialized in recent years, interest in the corrosion properties of P/M aluminum alloys paralleled the development of fully dense P/M aluminum products, which dates back some 25 years. Early work showed that heat-treated extrusions of alloy powders higher in zinc, magnesium, and copper than the conventional, 7000-series ingot metallurgy (I/M) wrought alloys provided higher tensile strength (Ref 55). In addition, it was found that high strength and resistance to SCC superior to that of I/M alloys could be obtained (Ref 56). Subsequently, rapid solidification processing and mechanical alloying were further exploited with additions of iron, nickel, cobalt, and oxides for grain refinement and stabilization of the structure

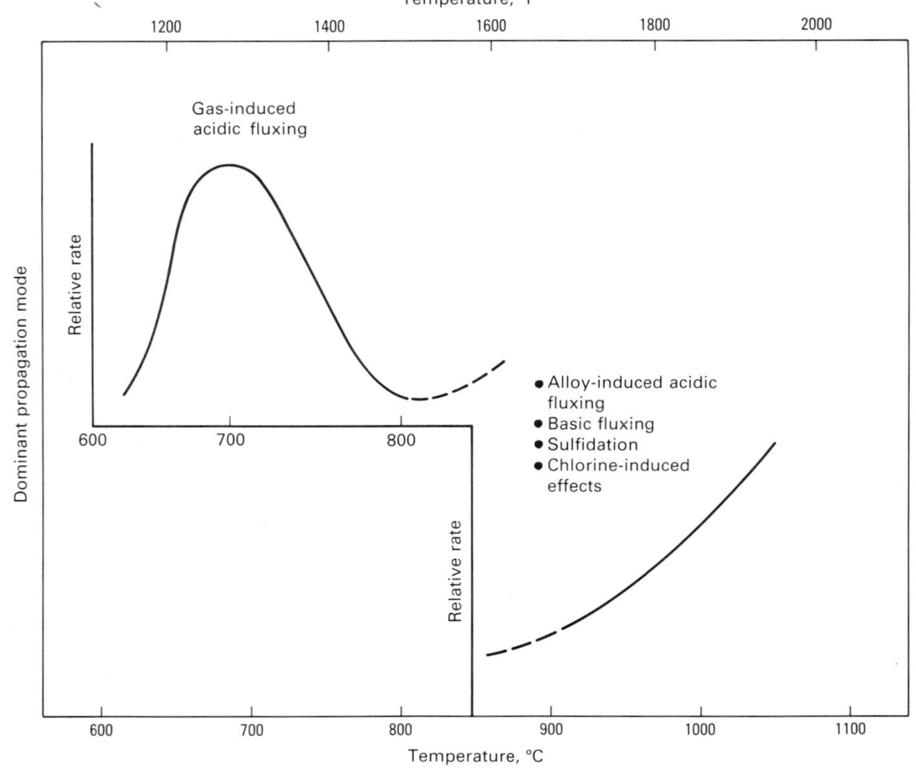

Fig. 37 Schematic showing the temperature regimes over which different propagation modes are most prevalent

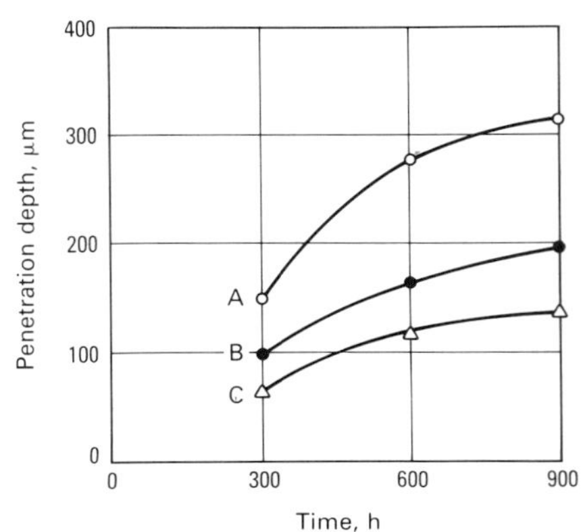

Fig. 38 Corrosion resistance of alloy IN-738LC in hot (850 °C, or 1560 °F) gases. A, IN-738LC; B, IN-738LC with Y_2O_3 dispersion, annealed at 1270 °C (2320 °F); C, IN-738LC with Y_2O_3 dispersion, annealed at 1100 °C (2010 °F). Source: Ref 34

Table 12 Cyclic oxidation resistance of superalloys at 1100 °C (2010 °F)

504-h test in atmosphere of air containing 5% H_2O; temperature cycled between 1100 °C (2010 °F) and room temperature every 24 h.

Alloy	Weight change, mg/cm² Undescaled	Weight change, mg/cm² Descaled	Metal loss, μm	Maximum attack, μm
MA 956	0.99	−1.57	2	15
TD-NiCr	−4.66	−12.52	20	33
Hastelloy X	−11.81	−20.62	50	256

Source: Ref 45

Table 13 Sulfidation resistance of superalloys at 925 °C (1700 °F)

312-h test in burner rig with air/fuel ratio of 30:1; fuel contained 0.3% S and 5 ppm seawater. Temperature cycle: 58 min at temperature, followed by 2 min cool to room temperature.

Alloy	Weight change, mg/cm² Undescaled	Weight change, mg/cm² Descaled	Metal loss, μm	Maximum attack, μm
MA 956	1.04	−0.17	5	18
TD-NiCr	−1.69	−11.57	25	129
Hastelloy X	−2.25	−6.83	33	132

Source: Ref 45

Table 14 Carburization resistance of superalloys at 1095 °C (2000 °F)

100-h test in atmosphere of hydrogen containing 2% methane.

Alloy	Weight change, mg/cm² Undescaled	Weight change, mg/cm² Descaled	Metal loss, μm	Maximum attack, μm
MA 956	0.07	−0.42	10	10
Incoloy 800	33.74	29.89	132	7615
Alloy 814	0.82	−0.73	13	363

Source: Ref 45

without the deleterious segregation effects that occur when I/M alloys are overalloyed (Ref 57).

The large, active constituents often found in conventional alloys are absent in these P/M alloys. Also absent are impurities and grain-boundary depletion of alloying elements, both of which can cause localized attack. Elements that are insoluble in the solid state are often soluble in the liquid state and may be uniformly dispersed in the powder particles by rapid quenching; metastable phases may also be formed (Ref 58). Figure 43 illustrates the differences in appearance between P/M and I/M microstructures. Additional information on the development of high-strength aluminum P/M alloys is available in Ref 59-64.

Because of higher solute contents, the P/M aluminum alloys can be aged to considerably higher yield strength than conventional alloys, with accompanying improvements in fatigue and toughness and without loss in SCC resistance. Rapid solidification processing and mechanical alloying also permit increased alloying contents. For example, increased amounts of lithium are possible (supersaturation), which results in lower density (7 to 20% weight savings) and higher specific elastic modulus.

Uses, Market, and State of Commercialization. Depending on alloy type, the attractiveness of these alloys in comparison to conventional alloys lies in their superior room-temperature strength combined with excellent corrosion and stress-corrosion resistance, their improved elevated-temperature properties, and their lower density and higher elastic modulus. High-performance P/M aluminum alloys are expected to find increasing use in aerospace, military, and marine applications and to replace heavier and more costly titanium alloys—for example, in major airframe primary load-carrying structural members, such as upper wing skins and landing gear components; helicopter rotors; low-temperature fan and compressor cases, vanes, and blades for gas-turbine engines; and fins, winglets, and rocket motor cases in missiles.

The addition of silicon carbide (SiC) to various aluminum alloy matrices by P/M or I/M processing results in lightweight, high-modulus composites. Potential uses include antennae yolks, torpedo hulls, mobile bridges, gyroscope supports, tractor tread shoes, and helicopter landing skids (Ref 60). The limited ductility and high price of these composite alloys, however, are impeding commercialization for use in sports equipment.

According to a recent market assessment, the targets of the high-performance P/M aluminum alloys are the high-strength 2*xxx* and 7*xxx* alloys (Ref 65). About 50 000 short tons (45 000 Mg), out of a total of 100 000 to 200 000 short tons (90 000 to 180 000 Mg) per year, are used by the United States aerospace industry. If the functional price can be made comparable to that of conventional alloys, a large potential also exists in the automotive market.

Of the many P/M alloys under development, only the aluminum-zinc-magnesium-cobalt alloys X7090 and X7091, which were introduced in 1981, are being produced commercially. Alloy X7090 is expected to be used for the landing gear support beam and the landing gear door actuator for the Boeing 757 aircraft. According to Ref 65, the following barriers must be overcome to achieve market introduction:

- Lack of exposure and experience
- Inadequate industry standards and the lack of reliable, repeatable, nondestructive tests for P/M parts
- The rate of technology change, which can inhibit major capital investment

The manufacture of high-performance fully dense P/M aluminum alloys is based on the use of either rapidly solidified materials, powders and particulates or mechanically attrited powders. Detailed descriptions of how these powders are produced are available in Ref 66 and the articles "Atomization" and "Milling of Brittle and Ductile Materials" in Volume 7 of the 9th Edition of *Metals Handbook*. Both methods can be used to produce compositions that are not practical with I/M and that have very fine and uniform dispersions of intermetallic particles.

The powders are usually degassed and then consolidated through the application of heat and pressure, that is, by forging, extrusion, hot pressing, or hot rolling, into a billet or near net shape part. Additional information on the critical steps of degassing and consolidation of these alloys is available in Ref 59.

Classes of High-Performance Aluminum P/M Alloys

As noted above, from an applications point of view, it is convenient to distinguish among three classes of high-performance P/M aluminum alloys (Ref 59, 60, 62, 63): high room-temperature strength, SCC/corrosion-resistant alloys; low-density high-stiffness alloys; and elevated-temperature alloys.

High Room-Temperature Strength SCC/Corrosion-Resistant Alloys. Table 18 shows the composition of six P/M aluminum alloys characterized by high room-temperature strength and high SCC resistance. Their room-temperature properties are given in Tables 19 and 20.

Table 15 Sulfidation resistance of superalloys at 925 °C (1700 °F)

Tested in burner rig with air/fuel ratio of 30:1; fuel contained 0.3% S and 5 ppm seawater. Temperature cycle: 58 min at temperature, followed by 2 min cool to room temperature.

Alloy	Exposure time, h	Descaled weight loss, mg/cm²	Maximum attack, μm
MA 6000E	312	−11.11	24
IN-100	48	−367.36	169
Alloy 713LC	168	−488.63	328
IN-738C	312	−9.73	28

Source: Ref 45

Table 16 Cyclic oxidation resistance of superalloys at 1100 °C (2010 °F)

504-h test in air containing 5% H_2O; temperature cycled from 1100 °C (2010 °F) to room temperature every 24 h.

Alloy	Descaled weight change, mg/cm²
MA 6000E	−14.12
IN-100	−7.27
Alloy 713LC	−22.08
IN-738C	−49.51

Source: Ref 45

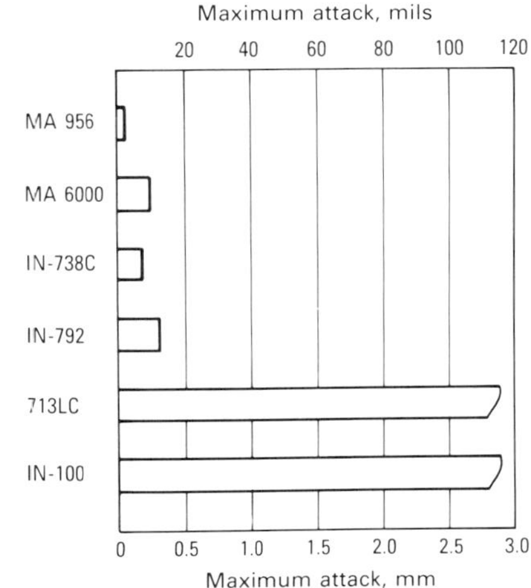

Fig. 39 Comparison of the corrosion resistance of MA 6000 and MA 956 with that of other superalloys. Tested in a burner rig for 312 h using a 30:1 air to fuel ratio. Fuel contained 0.3% S and 5 ppm seawater, and specimens were held at temperature for 58 min of each hour, then cooled 2 min with an air blast. Source: Ref 40

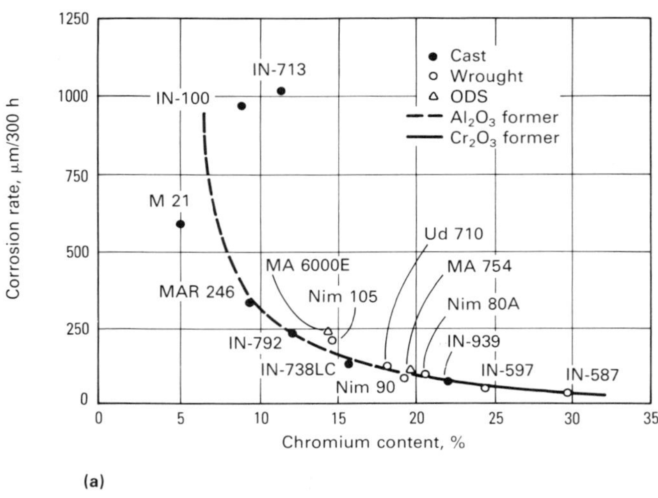

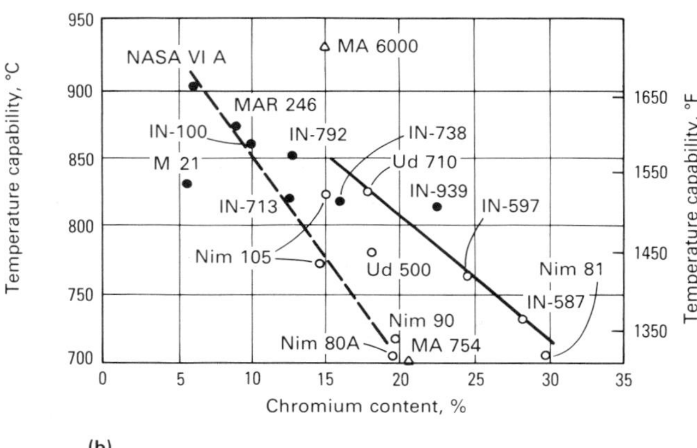

Fig. 40 Corrosion rate (a) and temperature capability (b) of MA 6000, MA 754, and non-ODS superalloys as a function of chromium content. A stress of 200 MPa (29 ksi) was applied to the specimens during the 10 000-h test. Source: Ref 40

Table 17 Dependence on carbide microstructure of creep crack growth rates of alloy IN-750 in four environments

Grain-boundary carbide microstructure	Crack growth rate (da/dt), mm/min(a)			
	Helium	Air	Helium + 4% methane	Helium + 3% SO_2
Blocky	7×10^{-4}	1.95×10^{-2}	7×10^{-4}	1.5×10^{-1}
Cellular	1.15×10^{-3}	2.6×10^{-3}	3.1×10^{-3}	1.3×10^{-1}
None	7×10^{-4}	6.3×10^{-2}	7×10^{-4}	1.6×10^{-1}

(a) $K = 35$ MPa $\sqrt{m}$ (32 ksi $\sqrt{in.}$). Source: Ref 40, 49

Table 18 Nominal compositions of high strength corrosion-resistant P/M aluminum alloys

Alloy	Composition, %								
	Zn	Mg	Cu	Co	Zr	Cr	O	C	Al
X7090	7.3–8.7	2.0–3.0	0.6–1.3	1.0–1.9	...	...	0.20–0.60	...	bal
X7091	6.8–7.1	2.0–3.0	1.1–1.8	0.20–0.60	...	...	0.20–0.50	...	bal
IN-9052	...	4.0	...	...	...	...	0.8	1.1	bal
IN-9021	...	1.5	4.0	...	...	...	0.8	1.1	bal
PM 64	6.8–8.0	1.9–2.9	1.8–2.4	0.1–0.4	0.1–0.35	0.08–0.25	<0.05	<0.5	bal
1519 B	6.8–8.0	1.9–2.9	1.8–2.4	...	...	0.1–0.5	<0.05	<0.05	bal

Source: Ref 59

Alloy CW67 (Table 20), a recent second-generation P/M product in the 7*xxx* series, is a zirconium- and nickel-containing aluminum-zinc-magnesium-copper alloy. It is still in the laboratory evaluation stage. Strength increases of 20 to 30% over wrought aluminum alloy 7075 translate into weight savings of 10% or more in aerospace structures. In the damage-tolerant T7 temper, CW67 demonstrates superior combinations of strength, fracture toughness, and resistance to fatigue and SCC compared to other I/M and P/M alloys (Ref 67, 68). Of the alloys shown in Table 18, alloys X7090 and X7091 are farthest along in the development of P/M high-strength aluminum alloys. In these alloys, cobalt is present as an intermetallic dispersoid of the composition Co_2Al_9 (in iron- and nickel-containing alloys, the analogous dispersoid is $(Fe,Ni)_2Al_9$) and is beneficial to both strength and SCC resistance. Figure 44 shows a transmission electron micrograph of a P/M MA 67 (the experimental precursor of alloy X7090) heat-treated die forging showing the uniformly distributed and small (0.05-μm) Co_2Al_9 particles as dark spheroids present on grain boundaries and within grains. In the T7*x* tempers, there were generally no precipitate-free zones surrounding the Co_2Al_9 particles. The magnesium- and aluminum-base oxide particles, another distinctive P/M feature, appear as clusters of dark particles, lighter and much finer than Co_2Al_9 (0.02 to 0.004 μm) and predominantly at grain boundaries.

In 1975, it was proposed that the superior SCC resistance of P/M aluminum-zinc-magnesium-copper alloys containing iron and nickel or cobalt may be related to the following factors: grain morphology (SCC fracture is intergranular and the crack path is tortuous because of the presence of fine dispersoids); grain-boundary movement restriction from the dispersoids (including oxides); hydrogen recombination at the cathodic Co_2Al_9 dispersoids; blunting of the leading edge of cracks by dispersoid particles; and shifting of the chemistry within cracks to more alkaline conditions at grain boundaries containing corroding oxide particles (Ref 58). The effect of different cobalt concentrations on the SCC behavior of several aluminum powder alloys was recently investigated in more detail (Ref 69-71). It was concluded that the increased SCC resistance was not due to the smaller grain size but to a direct effect of cobalt attributable to the cathodic behavior of the Co_2Al_9 particles that serve as sites for hydrogen recombination, thus reducing both the absorption of atomic hydrogen into the grain boundaries and hydrogen embrittlement.

Based on precracked double-cantilever beams in chromate inhibited brine, alloys X7090 and X7091 exhibited the greatest susceptibility to SCC in their undertempers (Ref 72). Susceptibility decreased in the peak-aged tempers and was absent in the overaged tempers after 500 h of immersion. In comparison, conventional alloy 7075, which uses chromium and manganese as dispersoid-forming elements, has a lower strength in its peak-aged temper and exhibits similar SCC propagation behavior to X7091 under the same test conditions.

Figure 45 shows time to failure curves for unnotched (reflecting both initiation and propagation behavior) specimens of alloys 7075 and X7091 in acetic acid brine. For a given strength, the P/M alloys exhibit superior SCC resistance. An investigation of the effect of varying amounts of dispersoid contents of Co_2Al_9 and $(Fe,Ni)_2Al_9$ found that the SCC resistances of Co_2Al_9 and $(Fe,Ni)_2Al_9$ dispersoid-containing alloys are similar in their T7 tempers (Ref 73). The influence of dispersoids on SCC was evident in the T6 but not the T7 tempers. Electrochemical tests showed a tendency for high cathodic reaction rates in the dispersoid-containing alloys. Anodic polarization tests revealed several distinct breakdown potentials.

Based on constant-amplitude ($R = 0.5$) fatigue crack growth rates measured in an inert and an aggressive environment, it was found that the corrosion fatigue crack growth rates for I/M 7075-T651 alloy are ten times faster than those for P/M X7091-T7470 (Ref 74). The results were explained in terms of crack closure of the P/M alloy with an effective load reduction of up to 40% of the applied load amplitude. Exfoliation

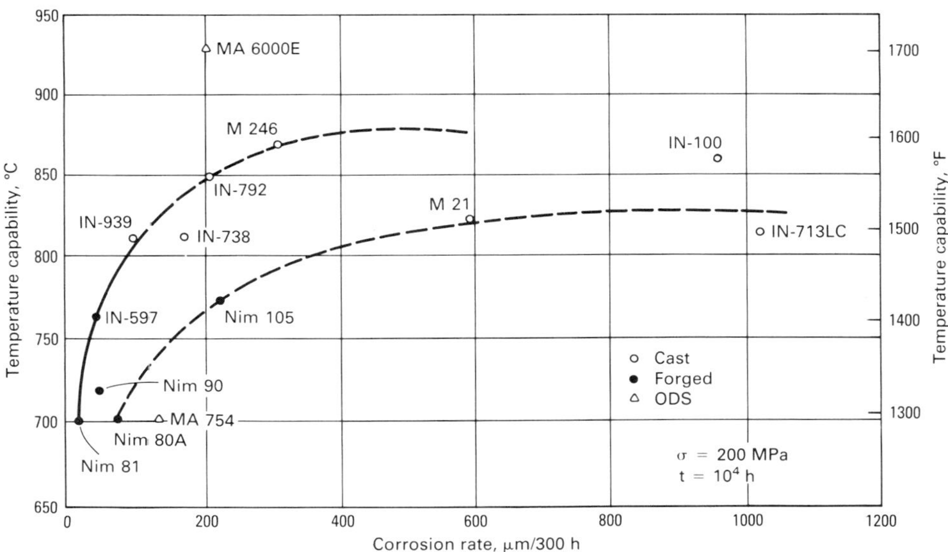

Fig. 41 Temperature capability as a function of corrosion rate for various superalloys. Same data as in Fig. 40. Source: Ref 40

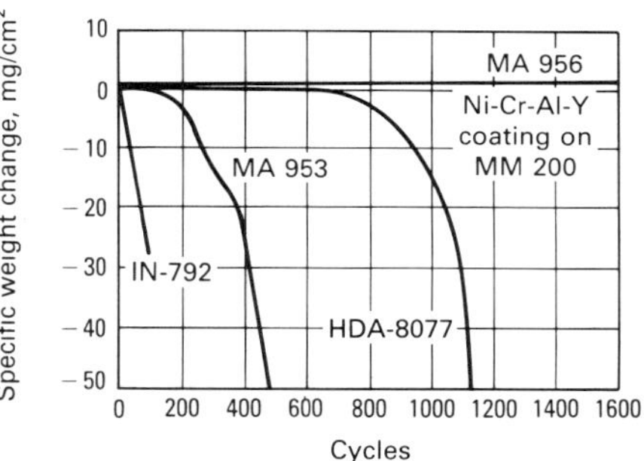

Fig. 42 Hot corrosion of alloys MA 953, HDA 8077, and MA 956 compared to that of some non-ODS alloys. Test conditions: 900 °C (1650 °F), 1 h, followed by a 3-min air blast, 5 ppm sea salt. Source: Ref 40

Table 19 Typical room-temperature tensile properties of high strength P/M aluminum alloys

Alloy	Temper	Product form	Ultimate tensile strength MPa	Ultimate tensile strength ksi	Yield strength MPa	Yield strength ksi	Elongation, %
X7090	T7E71	Extrusion	627	91	586	85	10
		Forging	614	89	579	84	10
X7091	T7E69	Extrusion	593	86	545	79	12
		Forging	579	84	531	77	13
1519 B	T76	Forging	558	81	510	74	10
	T73	Forging	510	74	448	65	11
PM 64	T76	Forging	600	87	552	80	6
	T73	Forging	558	81	496	72	9
IN-9052	F	Forging	595	86(a)	560	81	6
		Forging	410	59	380	55	7
IN-9021	T4	Forging	625	91	597	87	14

(a) For selected applications only. Note: these typical properties are subject to change as alloy development continues and should not be used for design. Potential users should contact manufacturers for the most up-to-date properties. Source: Ref 59

Table 20 Typical room-temperature properties of I/M and P/M high strength aluminum alloys

Alloy	SCC resistance MPa	SCC resistance ksi	Exfoliation rating(a)	Ultimate tensile strength MPa	Ultimate tensile strength ksi	Yield strength(b) MPa	Yield strength(b) ksi	Elongation, in 50 mm (2 in.), %	Fracture toughness (c) MPa √m	Fracture toughness (c) ksi √in.
Ingot metallurgy										
7075-T76	172	25	P—EA	524	76	462	67	12	29	26
7075-T73	290	42	P—EA	503	73	434	63	12	32	29
Powder Metallurgy										
X7090-T7E71	310	45	N—EA	621	90	579	84	9	26	24
X7091-T7E69	310	45	N—EA	593	86	545	79	11	46	42
CW67-T7	310	45	P—EA	614	89	579	84	12	47	43

(a) ASTM exfoliation ratings: N, no appreciable attack; P, pitting—either discrete or blistering; EA, visible lifting of surface; EB, thicker surface attack; EC, more severe surface attack; ED, most severe surface attack. (b) Longitudinal orientation. (c) Longitudinal-transverse orientation per ASTM E 399. Source: Ref 67

tests of P/M X7091 (T7E73 temper) using EXCO (Ref 75), salt spray, and marine exposure tests showed that this alloy resisted exfoliation because of its fine equiaxed grain structure (Ref 76). The dispersoid is beneficial to the stabilization of this structure.

The aluminum-magnesium alloys IN-9052 and IN-9021 are made by mechanical alloying. Uniform dispersions of MgO, Al_2O_3, and Al_4C_3 effectively pin the boundaries of the very fine grains (0.3 to 2 μm). In general corrosion tests in accordance with ASTM G 44 (Ref 77) (alternate immersion in NaCl solution), IN-9052 had a corrosion rate over two orders of magnitude lower than that of conventional 7075-T73 (Ref 78, 79). Figure 46 shows a comparison of crack velocity versus stress intensity factor of IN-9052 with alloys 7090, 7091, and conventional alloy 7075 in their highest-strength conditions for double-cantilever beam specimens in a chromate-inhibited brine solution.

Table 21 compares the mechanical and SCC properties of P/M alloy IN-9052 with those of I/M alloy 5083, which is of similar composition. The high increase in strength of IN-9052 is obtained without sacrificing the excellent corrosion resistance and at a marked improvement of SCC resistance after sensitization. Alloy IN-9052 is also superior in exfoliation resistance when tested in accordance with EXCO and Asset (Ref 80) and did not exhibit pitting. The factors responsible for the good SCC resistance of IN-9052 compared to conventional alloy 5083 are believed to be its fine grain size, the absence of Mg_2Al_3 grain boundary precipitates, and the small degree of segregation of magnesium to grain boundaries (Ref 71, 72). In another study using short-term exposure in a 3.5% NaCl solution, IN-9021 was found to be very susceptible to pitting and to exhibit a high general corrosion rate (three times the corrosion rates of conventional alloys 7075 and 2024) (Ref 81).

The low-density high-stiffness P/M aluminum alloys contain lithium, which decreases the density and increases the elastic modulus. Binary aluminum-lithium alloys, however, contain the ordered metastable δ′ precipitate (Al_3Li), which causes strain localization and accompanying low ductility and toughness. Alloy development work concentrates on the addition of dispersoids to overcome this problem.

With I/M, the maximum lithium content is about 2.5% due to difficulties in direct-chill casting at higher concentrations. With rapid solidification technology and mechanical alloying, lithium concentrations in ternary alloys can be as high as 4%. Alloys with ultimate tensile and yield strengths exceeding 600 MPa (87 ksi) and an elongation of 5% have been made and are under evaluation. A recent patent (Ref 82) claimed a UTS of >758 MPa (110 ksi) for an experimental mechanically alloyed aluminum-lithium binary alloy. Preliminary baseline corrosion studies on binary aluminum-lithium alloys have indicated that susceptibility to SCC is possible when the lithium concentration exceeds the room-temper-

Table 21 Typical longitudinal properties of equivalent P/M and I/M aluminum alloys

Alloy	Ultimate strength MPa	Ultimate strength ksi	0.2% yield strength MPa	0.2% yield strength ksi	Elongation, in 50 mm (2 in.), %	Fracture toughness MPa $\sqrt{m}$	Fracture toughness ksi $\sqrt{in.}$	Elastic modulus GPa	Elastic modulus psi × 10^6	Density, Mg/m^3	Threshold for SCC(a)
P/M IN-9052	448	65	379	55	13	44	40	71	10.3	2.66	55
I/M 5083-H111	290	42	152	22	14	...	...	70	10.2	2.66	Susceptible if sensitized at 65°C (150°F)

(a) 90-day alternate immersion in 3.5% NaCl; sensitized at 95 °C (200 °F) for 7 days and tested in the short-transverse direction. Source: Ref 72

Table 22 Nominal compositions of elevated-temperature P/M aluminum alloys

Alloy	Composition, % Fe	Ce	V	Mo	Zr	Al
CU-78	8.0	4.0	...	...	...	bal
Allied 1	9.3	...	3.5	...	...	bal
Allied 2	10.1	...	2.3	...	3.2	bal
Pratt & Whitney	8.0	...	...	2.0	...	bal

Source: Ref 59

Table 23 Mechanical properties of elevated-temperature P/M aluminum alloys

Alloy	Temperature °C	Temperature °F	Ultimate tensile strength MPa	Ultimate tensile strength ksi	Yield strength (0.2% offset) MPa	Yield strength (0.2% offset) ksi	Elongation, in 50 mm (2 in.), %
CU 78	20	70	565	82	448	65	5
	230	450	424	61	391	57	5
	345	650	165	24	124	18	7
Al-8Fe-2Mo	20	70	519	75	414	60	3
	230	450	370	54	331	48	NA
	345	650	207	30	172	25	NA
Allied 1	20	70	621	90	596	86	6.3
	230	450	433	63	425	62	6.8
	345	650	270	39	253	37	11.4
Allied 2	20	70	645	94	632	91	8.7
	230	450	420	61	414	60	7.3
	345	650	252	37	240	35	10.2

(a) NA, not available. Source: Ref 59

ature solid solubility limit (Ref 83). Overaged tempers have been found to be nonsusceptible, but peak-aged tempers have given the highest susceptibility.

In one investigation, the SCC behavior of two aluminum-lithium-copper P/M alloys (2.6% Li) was studied with and without magnesium additions by using three test methods: threshold stress tests on tuning fork specimens, slow crack growth tests on fracture mechanics specimens, and slow strain rate tests on electrically isolated tensile coupons (Ref 84). Each of these test methods, together with other experimental parameters was found to yield important information on the SCC behavior of the alloys.

Another study compared the SCC susceptibility of the aluminum-lithium alloys (Ref 85). It was found that alloys exposed to aqueous 3.5% NaCl exhibited time-dependent fracture and that SCC susceptibility was composition dependent. Alloy X2020 and mechanically alloyed Al-3Li-2.1Mg appeared very resistant to SCC. The lower limit of threshold stress necessary to cause fracture in the most susceptible P/M alloy (Al-2.6Li-1.4Cu-1.6Mg) was about 355 MPa (51.5 ksi). The P/M aluminum-lithium-copper alloys were susceptible to intergranular crack initiation in aqueous NaCl, but were resistant to sustained subcritical crack growth for stress intensity factors up to 15 MPa$\sqrt{m}$ (13.7 ksi$\sqrt{in.}$). Intergranular corrosion began at active pitting sites correlated to oxide particles strung along the extrusion direction of the alloys.

Elevated-Temperature Alloys. Table 22 shows the compositions and Table 23 shows the elevated-temperature properties of P/M aluminum alloys exhibiting superior elevated-temperature strength. These alloys are currently under development, but preliminary results indicate that they have good corrosion and SCC resistance. In these ternary and quaternary alloys, iron functions as the major dispersoid-forming element. The solute levels of iron and other intermetallic compound forming transition metals are considerably higher than is acceptable in conventional alloys (Ref 86).

Figure 47 shows the superior high-temperature (230 to 340 °C, or 450 to 650 °F) strength of some of these alloys as well as of aluminum-iron-vanadium-silicon alloys made by the planar flow cast process in comparison to conventional aluminum alloys. Figure 48 shows the much improved saline environment (ASTM B 117) weight loss data for several high-temperature aluminum-iron-zirconium-vanadium alloys compared to some conventional alloys (Ref 87).

REFERENCES

1. D.J. Orzeske, Steam Treatment Improves P/M Part Quality, *Ind. Heat.*, Vol 53, May 1986, p 34-37
2. D. Feinberg, *Surface Finishing of Powder Metallurgy Parts*, Metal Powder Industries Federation, 1968
3. K. Razavizadeh and B.L. Davies, Influence of Powder Type and Density on Pore Closure and Surface Hardness Changes Resulting from Steam Treatment of Sintered Iron, *Powder Metall.*, Vol 22 (No. 4), 1979, p 187-192
4. K. Volenik, H. Volrabova, J. Neid, M. Sebenji, and J. Cirak, Structure of Oxidation Products of Sintered Steel in Superheated Steam, *Powder Metal.*, Vol 21 (No. 3), 1978, p 149-154
5. P. Franklin and B.C. Davies, The Effects of Steam Oxidation on Porosity in Sintered Iron, *Powder Metall.*, Vol 20 (No. 1), 1977, p 11-16
6. G.F. Bocchini, A. Gallo, and I. Montevecchi, *Treatment of Sintered Steel Parts in Steam Atmosphere*, Proceedings of the 2nd International Congress on Heat Treatment of Materials: 1st National Congress on Metallurgical Coatings, Florence, Italy, Associazone Italiana di Metallurgia, 1982, p 1091-1102
7. B.J. Santer and B.D. Cash, Effect of Surface Treatment on the Corrosion Properties of Ferrous Powder Metallurgy Parts, *Powder Metall.*, Vol 17 (No. 34), 1974, p 319-330
8. "P/M Standards and Specifications," MPIF 35, Metal Powder Industries Federation
9. M.A. Pao and E. Klar, Corrosion Phenomena in Regular and Tin-modified P/M Stainless Steels, in *Progress in Powder Metallurgy*, H.S. Nayar, S.M. Kaufman, and K.E. Meiners, Ed., Metal Powder Industries Federation, 1984, p 431-444
10. D. Ro and E. Klar, Corrosive Behavior of P/M Austenitic Stainless Steels, in *Modern Developments in Powder Metallurgy*, Vol 13, H.H. Hausner and P.W. Taubenblat, Ed., Metal Powder Industries Federation, 1980, p 247-287
11. R.L. Sands, G.F. Bidmead, and D.A. Oliver, The Corrosion Resistance of Sintered Stainless Steels, in *Modern Developments in Powder Metallurgy*, Vol 2, H.H. Hausner, Ed., Plenum Press, 1966, p 73-85
12. H.S. Nayar, R.M. German, and W.R. Johnson, The Effect of Sintering on the Corrosion Resistance of 316L Stainless Steel, in *Progress in Powder Metallurgy*, Vol 37, 1981, p 255-265
13. G. Lei, R.M. German, and H.S. Nayar, Influence of Sintering Variables on the Corrosion Resistance of 316L Stainless Steel, *Powder Metall. Int.*, Vol 15 (No. 2), 1983, p 70-76
14. R.H. Shay, T.L. Ellison, and K.R. Berger, Control of Nitrogen Absorption and Surface Oxidation of Austenitic Stainless Steels in H-N Atmospheres, in *Progress in Powder Metallurgy*, Vol 39, H.S. Nayar, S.M. Kaufman, and K.E. Meiners, Ed., Metal Powder Industries Federation, 1983, p 411-430
15. K.H. Moyer, The Burn-Off Characteristics of Common Lubricants in 316L Powder Compacts, *Int. J. Powder Metall.*, Vol 7 (No. 3), 1971, p 33-43
16. S.K. Chatterjee, M.E. Warwick, and D.J. Maykuth, The Effect of Tin, Copper, Nickel, and Molybdenum on the Mechanical Properties and Corrosion Resistance of Sintered Stainless Steel (AISI 304L), in *Modern De-*

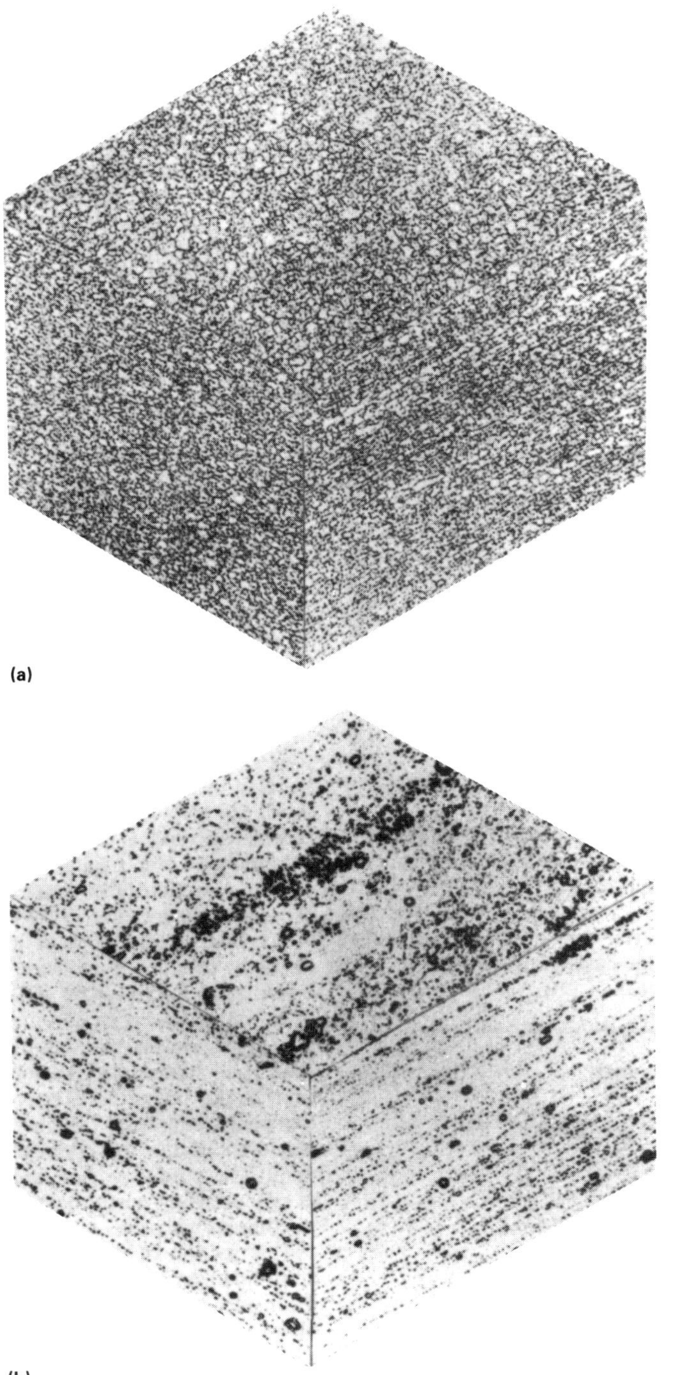

Fig. 43 Comparison of grain structure in P/M and I/M aluminum alloys. (a) P/M alloy X7090. (b) I/M alloy 7178. Both 500×

velopments in Powder Metallurgy, Vol 16, E.N. Aqua and C.I. Whitman, Ed., Metal Powder Industries Federation, 1984, p 277-293

17. T. Kato, K. Kusaka, and T. Hisada, Influence of Cu Addition on Some Properties of SUS 304L Stainless Steel Powders, *Denki Seiko (Electro. Furn. Steel)*, Vol 51 (No. 4), Nov 1980, p 252-263
18. A.J. Sedriks, *Corrosion of Stainless Steels*, John Wiley & Sons, 1979, p 15
19. M.A. Pao and E. Klar, *On the Corrosion Resistance of P/M Austenitic Stainless Steels*, Proceedings of the International Powder Metallurgy Conference, Florence, Italy, Associazone Italiano di Metallurgia, 1982
20. G. Lei, R.M. German, and H.S. Nayar, Corrosion Control in Sintered Austenitic Stainless Steels, in *Progress in Powder Metallurgy*, Vol 39, H.S. Nayar, S.M. Kaufman, and K.E. Meiners, Ed., Metal Powder Industries Federation, 1984, p 391-410
21. K. Kusaka, T. Kato, and T. Hisada, Influence of S, Cu, and Sn Additions on the Properties of AISI 304 L Type Sintered Stainless Steel, in *Modern Developments in Powder Metallurgy*, Vol 16, E.N. Aqua and C.I. Whitman, Ed., Metal Powder Industries Federation, 1984, p 247-259
22. D. Itzhak and S. Harush, The Effect of Sn Addition on the Corrosion Behaviour of Sintered Stainless Steel in H So, *Corros. Sci.*, Vol 25 (No. 10), 1985, p 883-888
23. F.M.F. Jones, The Effect of Processing Variables on the Properties of Type 316L Powder Compacts, in *Progress in Powder Metallurgy*, Vol 30, Metal Powder Industries Federation, 1970, p 25-50
24. T. Kato and K. Kusaka, On Some Properties of Sintered Stainless Steels at Elevated Temperatures, *Powder Metall.*, Vol 27 (No. 5), July 1980, p 2-8
25. O.W. Reen and G.O. Hughes, Evaluating Stainless Steel Powder Metal Parts, *Precis. Met.*, Vol 35 (No. 8), Aug 1977, p 53-54
26. M.H. Tikkanen, Corrosion Resistance of Sintered P/M Stainless Steels and Possibilities for Increasing It, *Scand. J. Metall.*, Vol 11, 1982, p 211-215
27. T. Takeda and K. Tamura, Compacting and Sintering of Chrome-Nickel Austenitic Stainless Steel Powders, H.B. Transl 8311 from Powder and Powder Metallurgy (Japan), Vol 17 (No. 2), 1970, p 70-76
28. T.J. Treharne, U.S. Patent 4,536,228, granted 20 Aug 1985, Corrosion Inhibition in Sintered Stainless Steel
29. A. Kempster, J.R. Smith, and C.C. Hanson, Chromium Diffusion Coatings on Sintered Stainless Steel, Metal Powder Report, MPR Publishing Services Ltd., England, June 1986, p 455-460
30. C. Tornberg, "The Manufacture of Seamless Stainless Steel Tubes from Powder," Paper 8410-013, presented at the 1984 ASME International Conference on New Developments in Stainless Steel Technology, Detroit, MI, American Society of Mechanical Engineers, 1984, p 1-6
31. "Standard Practices for Detecting Susceptibility to Intergranular Attack in Austenitic Stainless Steels," A 262, *Annual Book of ASTM Standards*, American Society for Testing and Materials
32. G.S. Yurek, D. Eisen, and A.J. Garrat-Reed, *Metall. Trans. A*, Vol 13A, 1982, p 473
33. P.C. Searson and R.M. Latanision, The Corrosion and Oxidation Resistance of Iron- and Aluminum-Based Powder Metallurgical Alloys, *Corros. Sci.*, Vol 25 (No. 10), 1985, p 947-968
34. G.H. Gessinger, Recent Developments in Powder Metallurgy of Superalloys, *Powder Metall. Int.*, Vol 13 (No. 2), 1981, p 93-101
35. R.F. Singer, Recent Developments and Trends in High Strength P/M Materials, *Powder Metall. Int.*, Vol 17 (No. 6), 1985, p 284-288
36. L.R. Curwick, The Mechanical Alloying Process: Powder to Mill Product, in *Frontiers of High Temperature Materials*, J.S. Benjamin, Ed., Proceedings of the International Conference on Oxide Dispersion Strengthened Superalloys by Mechanical Alloying, Inco Alloy Products Company, 1981, p 3-10
37. J.S. Benjamin and T.E. Volin, The Mecha-

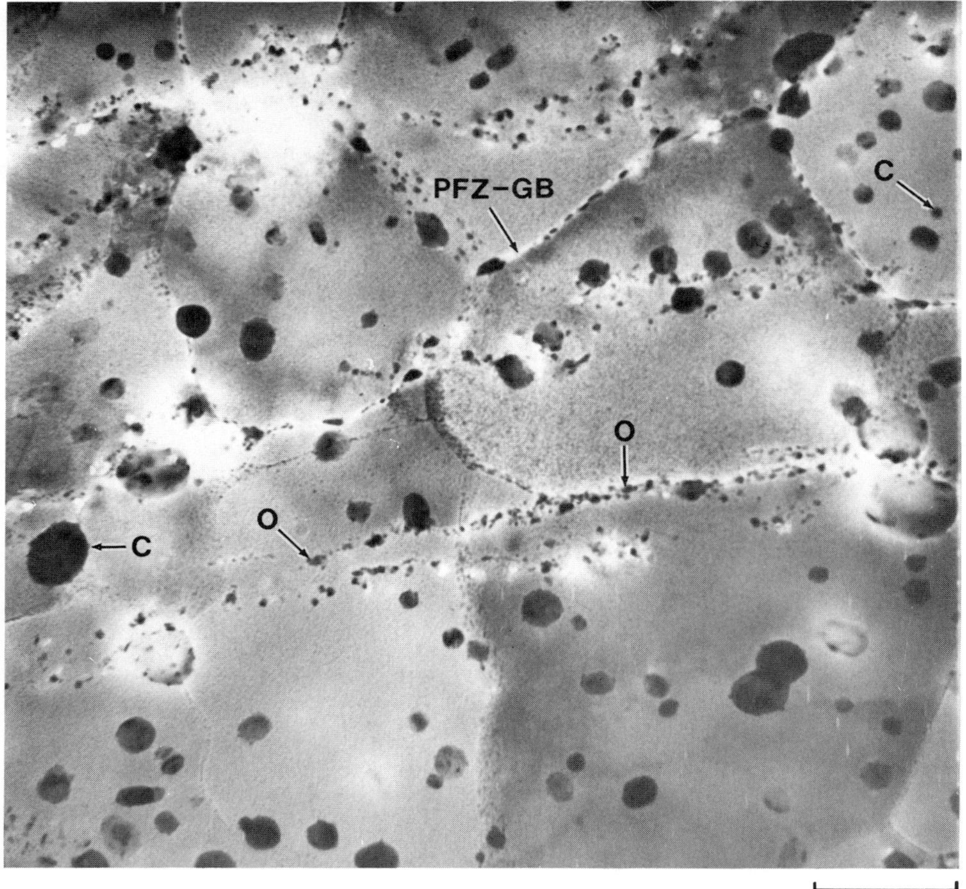

Fig. 44 Structure of P/M MA 67 die forging that was solution heat treated 2 h at 495 °C (920 °F), water quenched, and aged 7 days at room temperature plus 24 h at 120 °C (250 °F), plus 12 h at 165 °C (325 °F). Transverse section showing Co_2Al_9 (arrow C), oxide cluster (arrow O), and a precipitate-free zone (arrow PFZ-GB) at a grain boundary. 18 000×. Source: Ref 58

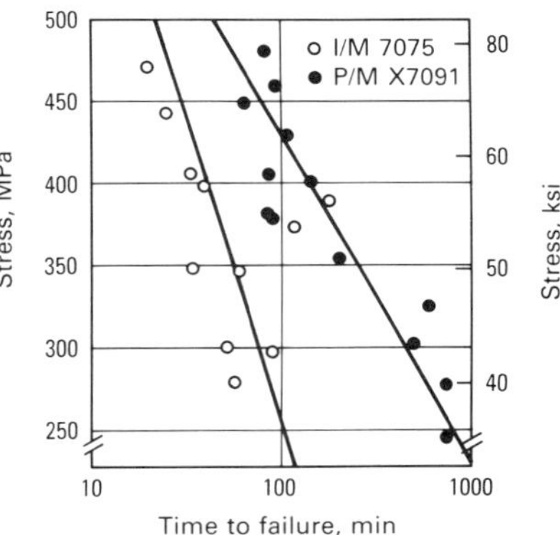

Fig. 45 Stress versus time to failure curves for unnotched specimens of conventional alloy 7075 and powder alloy X7091 at the same level in acetic acid brine. Source: Ref 33

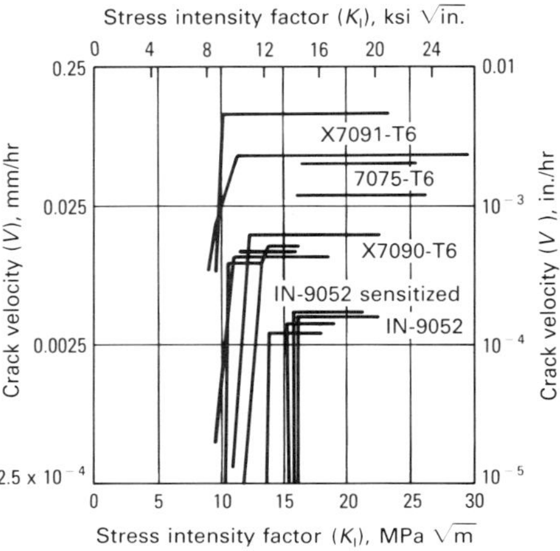

Fig. 46 Crack velocity versus stress intensity factor for P/M alloys IN-9052, X7090, and X7091 as well as conventional alloy 7075. All alloys were in their highest-strength conditions. Source: Ref 71, 72

nism of Mechanical Alloying, *Metall. Trans.*, Vol 5, 1974, p 1929-1934

38. F.S. Pettit and G.H. Meier, Oxidation and Hot Corrosion of Superalloys, in *Superalloys, 1984*, M. Gell *et al.*, Ed., Proceedings of the Fifth International Symposium on Superalloys, The Metallurgical Society, 1984, p 651-687
39. C.T. Sims, A History of Superalloy Metallurgy for Superalloy Metallurgists, in *Superalloys, 1984*, M. Gell *et al.*, Ed., Proceedings of the Fifth International Symposium on Superalloys, The Metallurgical Society, 1984, p 309-419
40. G.H. Gessinger, *Powder Metallurgy of Superalloys*, Butterworths, 1984
41. J. Stringer, B.A. Wilcox, and P.I. Jaffee, *Oxid. Met.*, Vol 5, 1972, p 11
42. C.S. Giggins and F.S. Pettit, The Oxidation of TD Ni Cr (Ni-20 Cr-2 Vol. pct ThO_2) Between 900 and 1200 C, *Metall. Trans.*, Vol 2, 1971, p 1071-1078
43. F. Perry, Oxide-Dispersion-Strengthened P/M Alloys Produced for Severe Service Applications, *Ind. Heat.*, Vol 49 (No. 5), May 1982, p 22-25
44. J.H. Weber and P.S. Gilman, Environmentally Induced Porosity in Ni-Cr Oxide Dispersion Strengthened Alloys, *Scr. Metall.*, Vol 18, 1984, p 479-482
45. J.H. Weber, *High Temperature Oxide Dispersion Strengthened Alloys*, Proceedings of the 25th National SAMPE Symposium and Exhibition, Society for the Advancement of Material and Process Engineering, 1980, p 752-763
46. G.H. Gessinger, *High Temperature Alloys for Gas Turbines*, D. Coutsouradis *et al.*, Ed., Applied Science, 1978, p 817
47. P.L. Antona, A. Bennani, P. Cavalloti, and O. Ducati, Heat Treatments and Oxidation Behavior of Some P/M Ni-Base Superalloys, in *European Symposium on Powder Metallurgy*, Vol 1, Jernkontoret Activity Group B, 1978, p 137-142
48. W.K. Uhl, M.R. Pendley, and S. McEvoy, "Evaluation of HIP Nickel-Base Alloys for Extreme Sour Service," Paper 219, presented at Corrosion/84, New Orleans, LA, National Association of Corrosion Engineers, April 1984
49. S. Floreen, Effects of Environment on Intermediate Temperature Crack Growth in Superalloys, in *Proceedings of the AIME Symposium* (Louisville, KY), American Society of Mining, Metallurgical, and Petroleum Engineers, 1981
50. G.A.J. Hack, Inconel Alloy MA 6000—A New Material for High Temperature Turbine Blades, *Met. Powder Rep.*, Vol 36 (No. 9), Sept 1981, p 425-429
51. D.H. Boone, D.A. Crane, and D.P. Whittle, *Thin Solid Films*, Vol 84, 1981, p 39
52. F.R. Wermuth and A.R. Stetson, Report NASA CR-120852, National Aeronautics and Space Administration, 1971
53. M.A. Gedwill, T.K. Glasgow, and R.S. Levine, "A New Diffusion Inhibited Oxidation Resistant Coating for Superalloys," NASA TM 82687, National Aeronautics and Space Administration, 1981
54. T.K. Glasgow and G.J. Santoro, Oxidation and Hot Corrosion of Coated and Bare Oxide Dispersion Strengthened Superalloy MA 755E, *Oxid. Met.*, Vol 15 (No. 314), April 1986, p 251-276
55. S.G. Roberts, An Exploratory Investigation of Prealloyed Powders of Aluminum, in *Powder Metallurgy*, W. Leszynski, Ed., 1961, p 799-817
56. A.P. Haarr, Final Report, Section III, Contract No. DA-36-04-ORD-3559RD, DDC, Alexandria, VA, 31 May 1966
57. H. Jones, *Rapid Solidification of Metals and Alloys*, Monograph 8, Institution of Metallur-

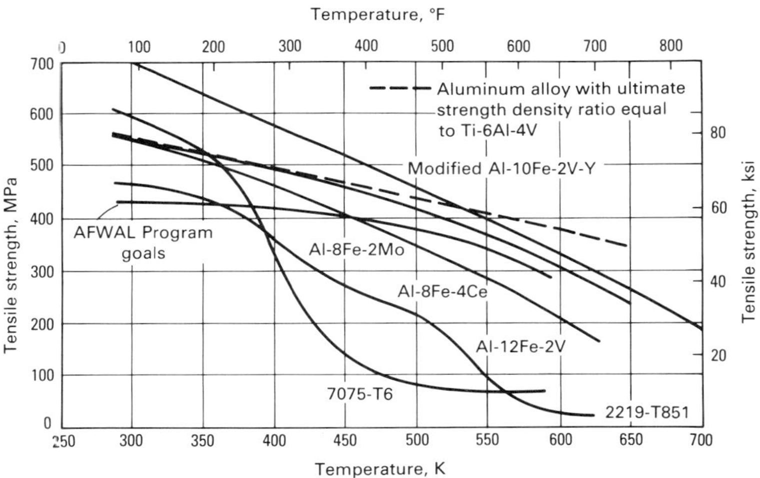

Fig. 47 Elevated-temperature strength of advanced aluminum alloys. Source: Ref 86

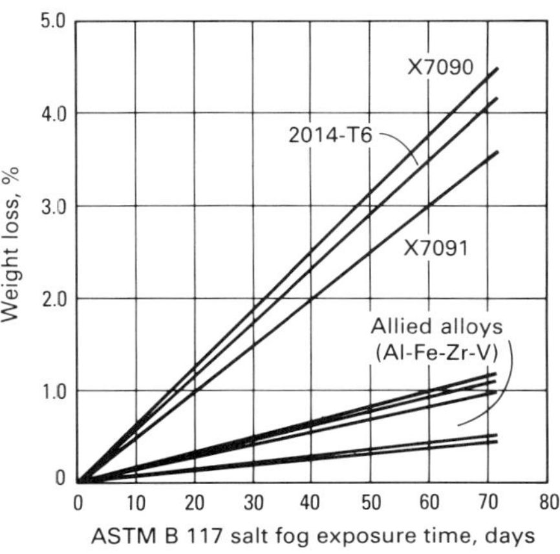

Fig. 48 Saline environment weight loss data for several aluminum alloys. Source: Ref 86

gists, 1982

58. J.P. Lyle and W.S. Cebulak, Powder Metallurgy Approach for Microstructure and Properties in High Strength Aluminum Alloys, *Metall. Trans. A*, Vol 6A, 1975, p 685-699
59. J.R. Pickens, Review: Aluminum Powder Metallurgy Technology for High-Strength Applications, *J. Mater. Sci.*, Vol 16, 1981, p 1437-1457
60. J.R. Pickens, High Strength Aluminum Alloys Made by Powder Metallurgy: A Brief Overview, in *Rapidly Quenched Metals*, S. Steeb and H. Warlimont, Ed., Elsevier, 1985, p 1711-1718
61. T.E. Tietz and I.G. Palmer, Advanced P/M Aluminum Alloys in *Proc. 1981 ASM Mater. Sci. Sem., Advances in Powder Metallurgy*, 1981
62. J.R. Pickens and E.A. Starke, Jr., The Effect of Rapid Solidification on the Microstructures and Properties of Aluminum Powder Metallurgy Alloys, in *Rapid Solidification Processing: Principle and Technologies, III*, R. Mehrabian, Ed., National Bureau of Standards, p 150-170
63. F.H. Froes and J.R. Pickens, Powder Metallurgy of Light Metal Alloys for Demanding Applications, *J. Metals*, Vol 36 (No. 1), 1984, p 14-28
64. J.R. Pickens, K.S. Kumar, and T.J. Langan, High Strength Aluminum Alloy Development, to be published in *The Proceedings of the 33rd Sagamore Army Materials Research Conference, Corrosion Prevention and Control*, Vol 33, AMMRC
65. E. Lavernia, B. Poggiali, I. Servi, J. Clark, F. Katrak, and N. Grant, Rapidly Solidified Aluminum Alloys: A Market Assessment, *J. Met.*, Vol 37 (No. 11), Nov 1985, p 35-38
66. N.J. Grant, in *Proceedings of the International Conference on the Rapid Solidification Process*, Claitor's Publishing, p 230
67. A. Hafeez, "Corrosion Resistance of Wrought P/M Alloys," Tech Brief 2 3/85, Aluminum Company of America, Pittsburgh, PA
68. G.J. Hildeman, L.C. Labarre, A. Hafeez, and L.M. Angers, "Microstructural, Mechanical Property and Corrosion Evaluations of 7xxx P/M Alloy CW 67," Internal Report, Aluminum Company of America
69. L. Christodoulou, J.R. Gordon, and J.R. Pickens, Effect of Co Content on the Stress-Corrosion Cracking Behavior of 7091-Type Aluminum Powder Alloys, *Metall. Trans. A*, Vol 16A, 1985, p 945-951
70. L. Christodoulou, T.J. Langan, D. Venables, J.A.S. Green, and J.R. Pickens, Crack-Tip Chemistry of Stress-Corrosion Cracks in 7xxx Aluminum Powder Alloys, accepted for publication in *Corrosion NACE*
71. J.R. Pickens and L. Christodoulou, The Stress-Corrosion Cracking Behavior of High-Strength Aluminum Powder Metallurgy Alloys, accepted for publication in *Metall. Trans. A*
72. J.R. Pickens, L. Christodoulou, and T.J. Langan, Report DAA 929-81-C-0031, Part I, Martin Marietta, 1983
73. A. Hafeez, Aluminum Corporation of America, private communication, Sept 1986
74. J.S. Santner and M. Kumar, Corrosion-Fatigue Crack Propagation Rates in Conventional 7075 and P/M X7091 Aluminum Alloys, in Corrosion Fatigue: Mechanics, Metallurgy, Electrochemistry and Engineering, STP 801, T.W. Crooker and B.N. Leis, Ed., American Society for Testing and Materials, 1983, p 229-255
75. "Standard Test Method for Exfoliation Susceptibility in 2xxx and 7xxx Series Aluminum Alloys (EXCO Test)," G 34, *Annual Book of ASTM Standards*, American Society for Testing and Materials
76. D.L. Erich and S.J. Donachie, Benefits of Mechanically Alloyed Aluminum, *Met. Prog.*, Vol 121 (No. 2), Feb 1982
77. "Standard Recommended Practice for Alternate Immersion Stress Corrosion Testing in 3.5% Sodium Chloride Solution," G 44, *Annual Book of ASTM Standards*, American Society for Testing and Materials
78. J.R. Pickens, R.D. Schelleng, S.J. Donachie, and T.J. Nichol, High Strength Aluminum Alloy and Process, U.S. Patent 4,292,079, 1981
79. J.R. Pickens, Techniques for Assessing the Corrosion Properties of Aluminum Powder Metallurgy Alloys, in *Proceedings of Rapidly Solidified Powder Aluminum Alloys Conference*, American Society for Testing and Materials, 1984
80. "Standard Test Method for Visual Assessment of Exfoliation Corrosion Susceptibility of 5xxx Series Aluminum Alloys (ASSET Test)," G 66, *Annual Book of ASTM Standards*, American Society for Metals
81. W.J.D. Shaw, Surface Corrosion Comparisons of Some Aluminum Alloys in 3.5% NaCl Solution, in *Microstructural Science*, Vol 12, Elsevier, p 243-261
82. J.R. Pickens, Mechanically Alloyed Dispersion Strengthened Aluminum-Lithium Alloy, U.S. Patent 4,532,106, 1985; European Examiner Application No. 45622A, 1982; Japanese Examiner Application No. 57857/82, 1982
83. L. Christodoulou, L. Struble, and J.R. Pickens, Stress-Corrosion Cracking in Al-Li Binary Alloys, in *Proceedings of the Second International Aluminum-Lithium Conference* (Monterey, CA), The Metallurgical Society, 1983, p 561-579
84. P.P. Pizzo, R. Galvin, and H.G. Nelson, Utilizing Various Test Methods to Study the Stress Corrosion Behavior of Al-Li-Cu Alloys, in *Environment-Sensitive Fracture: Evaluation and Comparison of Test Methods*, STP 821, S.W. Dean, E.N. Pugh, and G.M. Ugiansky, Ed., American Society for Testing and Materials, 1984, p 173-201
85. P.P. Pizzo, R. Galvin, and H.G. Nelson, Stress-Corrosion Behavior of Aluminum-Lithium Alloys in Aqueous Salt Environments, in *Aluminum-Lithium Alloys II*, The Metallurgical Society, 1984, p 627-656
86. C.M. Adam and R.E. Lewis, High Performance Aluminum Alloys, in *Rapidly Solidified Crystalline Alloys*, S.K. Das, B.H. Kear, and C.M. Adam, Ed., Proceedings of TMS-AIME Northeast Regional Meeting, The Metallurgical Society, 1985
87. "Standard Method of Salt Spray (Fog) Testing, "B 117, *Annual Book of ASTM Standards*, American Society for Testing and Materials

Corrosion of Cemented Carbides

Herbert S. Kalish, Adamas Carbide Corporation

CEMENTED CARBIDES consist of hard refractory metal compounds that have a lower-melting ductile metal binder or cement (internationally, the term hardmetal is used in preference to the term cemented carbides, which is used almost exclusively in the United States). Figure 1 shows microstructures of both the basic tungsten carbide-cobalt (WC-Co) materials and materials containing titanium carbide (TiC) and tantalum carbide (TaC). Table 1 shows the physical properties of the commonly available refractory metal or hard metal carbides used to make cemented carbides. Only two—WC and TiC—are used as true base compound materials that comprise over 50% of the composition. Tungsten carbide base materials are by far the most predominant and have been in widespread use for more than 50 years. They were originally used as early as 1916 (Ref 3-5). During this time, it was found that WC could be combined with cobalt to make a high-hardness, wear-resistant, strong material. This material was initially used for wire drawing dies instead of diamond dies.

The first key to the successful development of cemented carbides was that these refractory metal compounds, particularly WC, are best produced as powders. In fact, the only logical way to produce tungsten is the hydrogen reduction of WO_3 or ammonium paratungstate powder into tungsten metal powder. The carburization of tungsten to WC also results in a fine powder. The second key was the discovery of the eutectic system WC-Co (Fig. 2). Liquid-phase sintering is possible well below the melting point of the WC and even below the melting point of cobalt.

Cemented WC is produced by mixing from 3 wt% or less up to as much as 30 wt% of cobalt metal powder with a balance of WC powder. The mixed powders are ball milled, generally in volatile solvents, for times ranging from a few hours to as long as 7 days. Alternatively, the powders are milled in an attritor for 1 to 10 h.

A suitable transient binder is added to the powder, which is then pelletized and pressed to form the shape. Finally, the part is sintered at temperatures between 1300 and 1600 °C (2370 and 2910 °F), most often in vacuum. Because a liquid phase is formed during sintering, virtually 100% density is achieved. More information on the production of cemented carbides is available in the articles "Production of Tungsten, Molybdenum, and Carbide Powders" and "Production Sintering Practices for P/M Materials" in Volume 7 of the 9th Edition of *Metals Handbook*.

Effect of Composition on Properties

The two most common variables in cemented carbides are the cobalt or binder content and the grain size. As shown in Fig. 3, increased grain size decreases hardness, and increased cobalt content also decreases hardness (Ref 6). Increased contents of cobalt or other binders, however, are necessary to increase strength. As shown in Fig. 4, strength increases with increased cobalt content; although a maximum appears to occur at about 15 to 18% Co, this is true only for transverse rupture strength (Ref 6). Very high impact strength requires very high cobalt contents (up to 25 or 30 wt%) and coarse-grain carbide. In corrosion applications, however, the binder content ranges from virtually nil (there are some so-called "binderless" compositions that actually contain 1 to 2% binder) up to about 10%, with exceptions running up to 15% binder.

Cemented carbides are not selected for corrosion applications *per se*. They are extremely important in corrosion conditions in which high hardness, wear resistance, or abrasion resistance is required. When this is the case and the selection of a cemented carbide is logical, the corrosion-resistant properties are examined. For ordinary corrosion resistance, many metals and ceramics are better choices, but when wear resistance is also a requirement, the cemented carbide is needed.

Binder Composition and Content. The corrosion resistance of cemented carbides is based on the two very different components. The cobalt binder has very poor corrosion and oxidation resistance, and the WC has excellent corrosion resistance and good oxidation resistance. Alternate binders, such as nickel, have better corrosion resistance than cobalt and are used in spite of their lower hardness and strength. Nickel is a superior binder for cemented TiC and therefore is used in all cemented TiC materials regardless of the need for corrosion resistance. In some applications, cemented TiC shows superior corrosion resistance, and in other applications, cemented WC is better.

The addition of nickel to the usual cobalt binder used for WC, or the substitution of it entirely for cobalt, always improves corrosion resistance. There is, however, a sacrifice in strength, hardness, and wear resistance. A chromium addition also enhances corrosion resistance.

The most important variable in the corrosion of cemented carbides is the binder content. Because the binder corrodes more than the carbide, the smaller the amount of binder the better. On the other hand, decreasing the binder decreases the strength.

Carbides. Additions of TaC and TiC to the WC-Co materials are common for the compositions used for machining steel. These additives give the carbide crater resistance. Cratering on the top of a metal-cutting insert is the result of a physicochemical reaction. The addition of TaC and/or TiC will slow this reaction; indeed, it has been found that TaC also enhances the outright chemical corrosion resistance of these materials.

Other additives, such as chromium carbide (Cr_2C_3), molybdenum carbide (Mo_2C), niobium carbide (NbC), and vanadium carbide (VC), are often added in small quantities as grain growth inhibitors. Little has been published about their effect on corrosion, but chromium has been shown to be a beneficial binder additive to WC-Ni binder compositions (Ref 7). Vanadium carbide and Mo_2C will probably have a weakening effect on the strength of a WC-base hardmetal.

For TiC-base hardmetals, Mo_2C is invariably added to the composition, but there are no known studies of the effect of molybdenum on corrosion resistance. The molybdenum has always been added to enhance the liquid-phase sintering of the TiC-base compositions. In general, these compositions have been made for their hardness and strength characteristics, with corrosion resistance being a secondary consideration. Most recent TiC-base compositions have titanium nitride

Table 1 Physical properties of carbides used in the manufacture of cemented carbides

Carbide	Microhardness, kg/mm^2	Melting point °C	Melting point °F	Density, g/cm^3
TiC	3200	3200	5790	4.94
VC	2950	2830	5125	5.71
HfC	2700	3890	7030	12.76
ZrC	2560	3530	6385	6.56
NbC	2400	3500	6330	7.80
Cr_2C_3	2280	1895	3440	6.66
WC	2080	2600	4710	15.67
Mo_2C	1950	2675	4850	9.18
TaC	1790	3780	6835	14.50

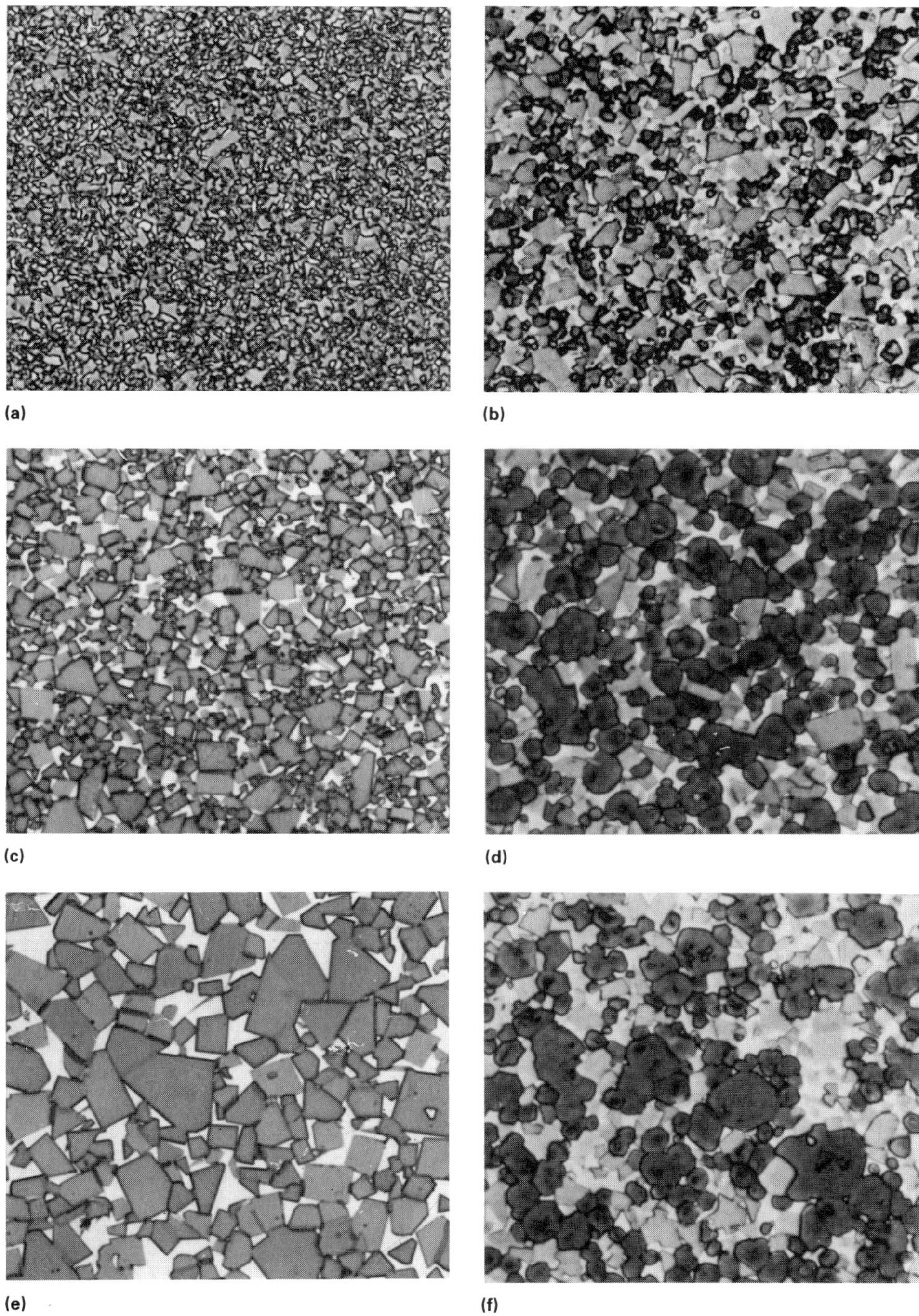

Fig. 1 Microstructures of WC-Co (a, c, and e) and WC-TaC-TiC-Co (b, d, and f) cemented carbides. In a, c, and e, the white areas are cobalt binder phase. In b, d, and f, the darker, more rounded grains are the $W_xTa_yTi_zC$ cubic solid-solution γ phase. (a) and (b) Fine grain structures. (c) and (d) Medium grain structures. (e) and (f) Coarse grain structures. All 1500×. Source: Ref 1 and 2

(TiN) added, and this has been shown to improve the corrosion resistance (Ref 8).

Perhaps it is not surprising that compositions developed primarily for machining should show improved corrosion resistance. In machining, there is heat with resultant oxidation and often corrosionlike mechanisms. Thus, some of the improved machining compositions also show better corrosion behavior. On the other hand, optimum corrosion resistance is obtained by tailoring the composition and amount of the binder phase. This can result in lower-strength materials with limited usefulness in machining applications.

Because carbon is the basis of cemented carbides, its variation within a given composition is very important to properties and corrosion resistance. Figure 5 shows the range of carbon content allowable in the simple WC-Co compositions as cobalt content is varied (Ref 9-11). Corrosion-resistant compositions have three problems:

- The lower the cobalt or binder content, the better the resistance to corrosion, but this limits the safe zone, in which neither carbon porosity nor η phase (hard, brittle M_6C or $M_{12}C$ intermetallics) exist
- The lower the carbon content, the better the corrosion resistance, but falling into the η-phase zone results in embrittlement of the material
- The addition of alternate binders, such as nickel, decreases the safe zone

In making corrosion-resistant cemented carbides, manufacturers must be aware of these problems and limitations. Information on the metallography and microstructures of these materials is available in the article "Cemented Carbides" in Volume 9 of the 9th Edition of *Metals Handbook*.

Applications of Cemented Carbides

The major applications of cemented carbides actually involve environments that are inherently corrosive. For example, the major use of cemented carbides is for metal-cutting (machining) applications. In these applications, extreme heat is generated whether or not coolants are used, and in those cases in which coolants are used, the corrosive attack of the coolant is a factor in the performance of the cutting tool. In general, however, very little heed is paid to this factor; cemented carbides are more often chosen for their wear resistance in such applications as mining and oil well drilling. In actuality, there is a corrosive environment to be contended with in mining (Ref 12) and oil well drilling; the natural waters and other fluids involved are often very corrosive. Other well-known examples in which cemented carbide is performing in a corrosive environment include balls for ball point pens and dental drills. In both of these examples, the corrosion resistance of the most frequently used WC-6Co composition was serendipitous. The material was selected for its wear resistance. It just happens to have good corrosion resistance in the saline and ink solutions. The dulling of cemented carbide saw tips used for sawing green or unseasoned wood is a corrosive as well as a wear phenomenon (see the section "Saw Tips and Corrosion" in this article).

Examples of the use of cemented carbide in true corrosion applications include the following:

- Ball point pen balls
- Dental drills and burrs
- Surgical and orthodontic tweezers, pliers, and clamps
- Valve seats
- Valve balls and valve stems
- Valve and shaft seals (seal rings)
- Spray nozzles
- Pulverizing hammers
- Compressor plungers
- Bearings
- Cage mills
- Ball mill linings and balls
- Internal parts in industrial meters

The article "Cemented Carbides" in Volume 7 of the 9th Edition of *Metals Handbook* contains more information on applications for cemented carbides.

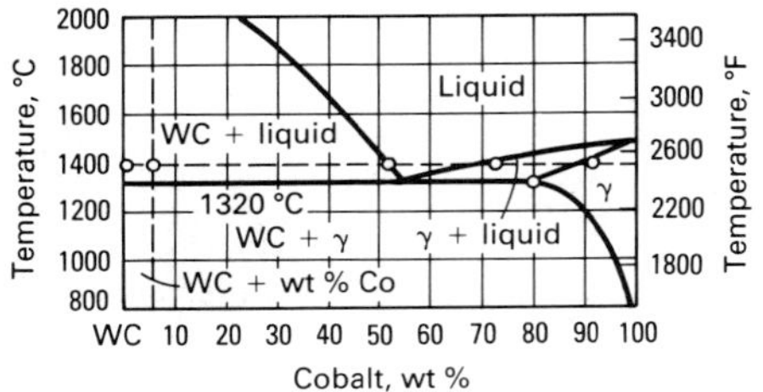

Fig. 2 Quasi-binary phase diagram for the WC-Co system

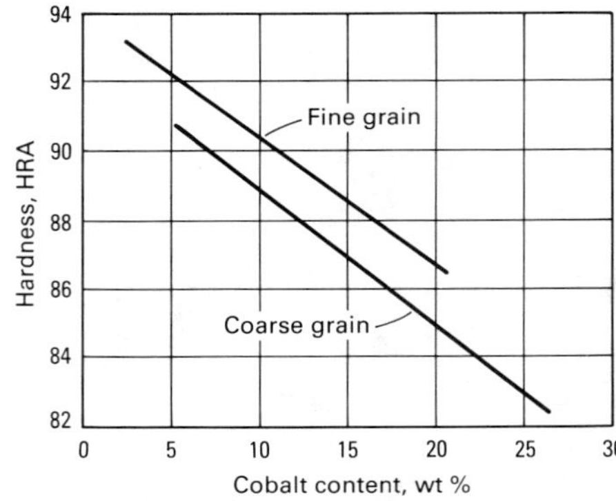

Fig. 3 Effect of cobalt content and grain size on the hardness of WC-Co cemented carbides

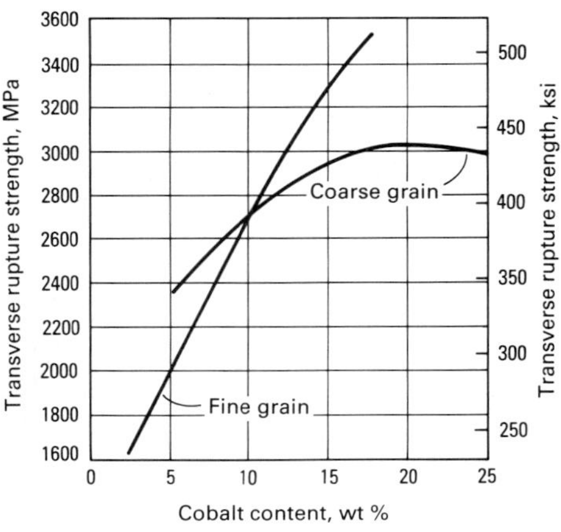

Fig. 4 Effect of cobalt content and grain size on the transverse rupture strength of WC-Co cemented carbides

Selection of Cemented Carbides for Corrosion Applications

The selection of cemented carbides is a very difficult problem for the user. There has been a lack of standardization on the part of the producers, and this lack has not been answered by any national or international standards organization. Some attempts have been made to standardize with regard to metal-cutting applications. There is International Organization for Standardization (ISO) standard 513 for metal-cutting applications for carbide (Ref 13). It is widely used in Europe and most other parts of the industrialized world, but it is not recognized in the United States (Ref 11). In addition, there is no ISO standard for cemented carbides used for wear, mining, or corrosion applications, and if any exist in other industrialized countries, the producers choose to ignore them, or they may be so broad that a given producer can have three or more grades falling into one category (Ref 14, 15).

The producers also tend to disregard attempts at standardization in the hope of having a unique product. Even in the established WC-6Co grades, the producers offer several different varieties based on different grain size or different minor element additions. For example, the company that developed the WC-6Co composition about 70 years ago offers five different grades of this composition, and two of them have identical published properties. They are not alone. In

Table 2 Some physical properties of corrosion-resistant cemented carbide grades

Properties of a carbon steel, a tool steel, and a cast cobalt alloy are included for comparison.

Special attributes	Proprietary designation	Nominal composition, wt%: WC	Co	TaC	TiC	Ni	Cr	Mo_2C	Hardness, HRA	Density, g/cm³	Transverse rupture strength, MPa	ksi	Abrasion resistance factor(a)	Coefficient of thermal expansion, μm/m·K	Thermal conductivity, W/m·K	cal/cm·s·°C
Abrasion-resistant, wear, and structural grades																
Maximum abrasion resistance	GU-2(b)	96.5	3	0.5	...	...	...	...	93.3	15.30	1655	240	1.8	4.9	125.5	0.30
	PWX(b)	94.0	5.5	0.5	...	...	...	...	92.5	15.05	2137	310	2.1	5.2	108.8	0.26
	A(b)	94.0	6.0	...	...	...	...	...	91.8	15.00	2206	320	3.4	5.5	104.6	0.25
	B(b)	91.0	9.0	...	...	...	...	...	90.8	14.70	2758	400	6.8	5.5	96.2	0.23
	BB(b)	87.0	13.0	...	...	...	...	...	89.5	14.28	3103	450	17	6.2	87.9	0.21
Toughness	GU-1(b)	81.5	18.0	0.5	...	...	...	...	88.4	13.84	3448	500	32	6.8	83.7	0.20
Gall resistance	474(b)	79.0	12	9	...	...	...	...	89.6	14.29	2241	325	16.5	5.8	87.9	0.21
	GG(b)	60.0	12	28	...	...	...	...	89.0	14.09	2069	300	18	7.1	83.7	0.20
Oxidation resistance	Titan 80(b)	...	...	...	74	12.5	...	13.5	93.0	5.63	1379	200	22	7.8	16.7	0.04
	Titan 60(b)	...	...	...	70.5	17.5	1.0	11.0	91.7	5.71	1724	250	28	8.4	16.7	0.04
	Titan 50(b)	...	...	...	66.5	22.5	1.0	10.0								
Special corrosion resistance	K602(c)	88.2	1.8	10.0	...	...	...	...	94.3	15.6	759	110	...	4.9	...	...
	K701(c)	85.8	10.1	...	...	...	4.1	...	92.0	14.0	1138	165	...	6.5	62.8	0.15(d)
	K703(c)	93.3	5.8	...	...	...	0.9	...	91.5	14.7	1931	280	...	4.5	...	...
	K714(c)	88.4	6.1	4.5	1.0	...	...	...	92.5	13.1	1827	265	1.8(d)	4.0	...	...
	K801(c)	93.7	...	0.3	...	6.0	...	...	90.0	14.8	2103	305	17(d)	5.6	96.2	0.23(d)
	K803(c)	89.0	...	...	1.0	10.0	...	...	91.0	14.4	2000	290	...	5.6	...	...
Grades for heading and forming dies																
Impact resistance	HD-15(b)	85.0	15	...	...	...	...	...	87.4	14.10	3172	460	30	6.5	83.7	0.20
	HD-20(b)	80.0	20	...	...	...	...	...	85.3	13.60	3103	450	45	6.8	83.7	0.20
	HD-25(b)	75.0	25	...	...	...	...	...	83.5	13.15	2965	430	65	7.5	83.7	0.20
Gall resistance	HD-20T(b)	75.0	20	5	...	...	...	...	85.3	13.55	2896	420	46	7.1	83.7	0.20
	HD-25T(b)	70.0	25	5	...	...	...	...	83.5	13.15	2827	410	67	7.8	83.7	0.20
Mining grades																
Strength and impact resistance	575(b)	94.0	6	...	...	...	...	...	90.8	15.00	2413	350	8.1	4.9	104.6	0.25
	569(b)	90.0	10	...	...	...	...	...	88.6	14.51	2930	425	13	5.8	104.6	0.25
	783(b)	89.0	11	...	...	...	...	...	88.1	14.41	3103	450	19	5.8	104.6	0.25
	502(b)	88.0	12	...	...	...	...	...	87.6	14.31	2965	430	21	6.2	104.6	0.25
Noncarbide metals																
Carbon steel	...	...	...	...	...	...	...	...	To 79	7.8	To 1379 (tensile strength)	200	>140	14.8	50.2	0.12
T1 tool steel	...	...	...	...	...	...	...	...	To 87	8.7	3448	500	70	12.6	...	...
Cast Co-Cr-W alloy	...	...	...	...	...	...	...	...	To 83	8.6	2069	300	110	13–16	...	...

(a) Determined in accordance with ASTM B 657 (Ref 2). The lower the number, the better the resistance to abrasion. (b) Adamas designation. (c) Kennametal designation. (d) Values estimated from available data. Source: Ref 16, 17

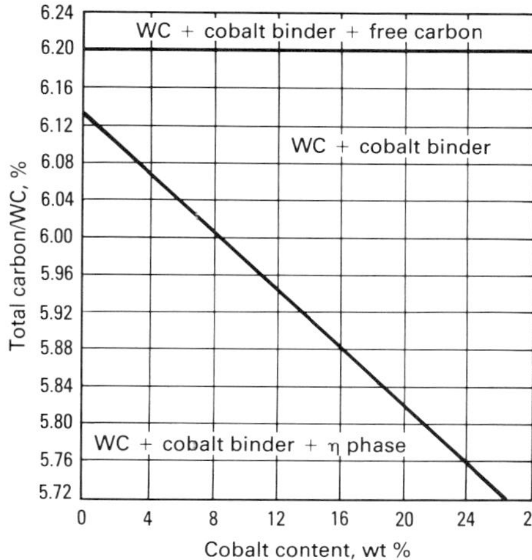

Fig. 5 Effect of cobalt content and carbon content on the phases present in WC-Co cemented carbides

some cases, three grades are shown with the same composition and properties. A good example is the nine differently designated 6% Co compositions of one company. Five of the nine are indeed different because of small TaC additions or grain size, but one of the compositions has three designations, and two of them have two designations. More often, the reason for the multiple designations of the same composition is that one designation is for cutting tools, another for wear parts or dies, and another for mining. Another problem area is the selection of composition by the manufacturer. For example, if one producer establishes a WC-25Co composition, another producer will make and market a grade with 24% Co, and another a product with 26% Co. Despite these problems, Table 2 lists the properties of various representative grades for corrosion applications; Table 3 lists approximate compositions and proprietary designations for a number of corrosion-resistant grades.

Table 3 includes grades from 16 manufacturers worldwide. These are meant to be representative only in a general sense. There are well over 100 manufacturers throughout the world (over 25 in the United States alone); therefore, it is not feasible to include all. In addition, cross comparisons are not precisely possible. For example, a grade listed with an approximate composition of WC-25Co may be cross referenced with a comparable grade that contains only 24% Co. Reference 15 contains more complete data on any grade, and manufacturers can be consulted for more information.

In addition to small differences in cobalt content from one manufacturer to another, there are small differences in minor additives and in grain size. For example, with the 6% Co grades, there are two basic grain size classes—fine and coarse—but these two are not precisely the same from one manufacturer to another. Some have a slightly finer or coarser size within the defined category of fine and coarse. Again, precise standards are lacking. References 1, 2, and 18 are the attempts at standardization, but they are useful only in a general sense; moreover, no producer

Table 3 Representative compositions and proprietary designations of corrosion-resistant cemented carbide grades

Representative composition, %							Proprietary designations							
WC	Co	TaC	TiC	Ni	Cr	Mo_2C	Adamas Carbide	Anderson Strathclyde	Carbidie	Carmet	Danit	General Carbide	General Electric Carboloy	GTE Valenite
96.5	3.0	0.5	...	...	...	...	GU2	CA	CD20	CA8	K04	GC003	999	VC3
94.0	5.5	0.5	...	...	...	...	PWX	CF	CD24	CA306	K10	GC005	895	VC2
94.0	6.0	...	...	...	...	...	A	CG	CD30	CA4	K20	GC106	883	...
91.0	9.0	...	...	...	...	...	B	...	CD35F	CA12	K30	GC009	...	VC152
87.0	13.0	...	...	...	...	...	BB	...	CD40	CA10	...	GC313	258	VC11
81.5	18.0	0.5	...	...	...	...	GU1	...	CD650	...	...	...	...	...
79.0	12.0	9.0	...	...	...	...	474	...	...	...	...	...	...	VC047
...	...	...	74.0	12.5	...	13.5	Titan 80	...	...	CA100	...	...	...	VC83
...	...	...	70.5	17.5	1.0	11.0	Titan 60	...	...	...	...	...	...	...
...	...	...	66.5	22.5	1.0	10.0	Titan 50	...	...	...	...	...	...	...
89.0	...	...	1.0	10.0	...	...	...	...	...	R10	...	...	...	VC099
85.0	15.0	...	...	...	...	...	HD15	CPM	CD50	CA11	DG30	GC315	268	VC12
80.0	20.0	...	...	...	...	...	HD20	...	...	CA20	DG40	GC320	...	VC13
75.0	25.0	...	...	...	...	...	HD25	...	...	CA225	DG50	GC325	190	VC14
75.0	20.0	5.0	...	...	...	...	HD20T	...	CD60	...	...	...	...	...
70.0	25.0	5.0	...	...	...	...	HD25T	...	CD70	...	...	...	...	...
94.0	6.0	..	...	...	...	...	575	CR	...	CA3	B030	GC206	44A	...
90.0	10.0	..	...	...	...	...	569	CM	...	2102	B050	GC410	90	...
89.0	11.0	..	...	...	...	...	783	...	CD337	CA411	B055	GC411	115	...
88.0	12.0	...	...	...	...	...	502	CT	...	CA412	B060	GC412	120	...

Representative composition, %							Proprietary designations							
WC	Co	TaC	TiC	Ni	Cr	Mo_2C	Kennametal	Krupp Widia	Mefasa	Metallwerk Plansee	Mitsubishi	Sandvik	Sumitomo	Teledyne Firth Sterling
96.5	3.0	0.5	...	...	...	...	K11	THF	...	H03T	...	CS05	...	HF
94.0	5.5	0.5	...	...	...	...	K68	GT05	K1	H10T	GTi05	CS10	H1	HA
94.0	6.0	...	...	...	...	...	K6	GT10	K2	H16T	GTi10	HML	G10E	H6
91.0	9.0	...	...	...	...	...	K9	GT15	MK30	H30T	GTi15	H10F	G3	H8
87.0	13.0	...	...	...	...	...	...	GT3H	...	H40T	GTi20	R4	G5	H81
81.5	18.0	0.5	...	...	...	...	...	...	...	...	GTi40	...	...	...
...	...	...	74.0	12.5	...	13.5	K165	...	...	F05T	NX33	CN02	...	...
...	...	...	70.5	17.5	1.0	11.0	...	...	...	F10T	NX55	...	T12A	...
...	...	...	66.5	22.5	1.0	10.0	...	TTF	...	...	...	...	T12B	...
88.2	1.8	10.0	...	...	...	...	K602	...	...	...	...	...	...	...
85.8	10.1	...	...	...	4.1	...	K701	...	...	...	...	...	...	...
93.3	5.8	...	...	...	0.9	...	K703	...	...	...	...	...	...	...
88.4	6.1	4.5	1.0	...	...	...	K714	...	...	...	...	...	...	...
93.7	...	0.3	...	6.0	...	...	K801	...	...	WC6Ni	...	...	...	...
89.0	...	...	1.0	10.0	...	...	K803	...	...	TCR30	...	...	...	...
85.0	15.0	...	...	...	...	...	SP212	BT40	G3	B50T	...	CT60	G6	MPD160
80.0	20.0	...	...	...	...	...	...	...	G4	H60T	GTi40	CT75	G7	...
75.0	25.0	...	...	...	...	...	...	GT55	G5	H70T	...	CT85	G8	...
75.0	20.0	5.0	...	...	...	...	K91	...	...	...	...	...	...	ND20
70.0	25.0	5.0	...	...	...	...	K90	...	...	...	...	...	...	ND25
94.0	6.0	...	...	...	...	...	K3404	BT10	K3	B10T	...	CT30	...	HAN6
90.0	10.0	...	...	...	...	...	K3070	BT25	MK35	B30T	...	CT45	G3	MPD10
89.0	11.0	...	...	...	...	...	K3047	...	MK40	B36T	...	CT50	...	MPD11
88.0	12.0	...	...	...	...	...	K3030	BT30	...	B40T	...	...	G5	...

Table 4 Selected mechanical properties of corrosion-resistant cemented carbide grades

Proprietary designation	Poisson's ratio	Charpy V-notch impact resistance(a) J	Charpy V-notch impact resistance(a) in.·lb	Tensile strength MPa	Tensile strength ksi	Compressive strength MPa	Compressive strength ksi	Modulus of elasticity GPa	Modulus of elasticity 10^6 psi
GU-2(b)	0.21	1.24	11	1034	150	6068	880	662	96
PWX(b)	0.21	1.36	12	1241	180	5929	860	652	94.5
A(b)	0.23	1.47	13	1310	190	5516	800	648	94
B(b)	0.26	1.69	15	1586	230	4482	650	607	88
BB(b)	0.28	2.71	24	1793	260	4137	600	545	79
GU-1(b)	0.27	3.38	30	1862	270	4068	590	510	74
474(b)	0.26	1.58	14	1586	230	4206	610	538	78
GG(b)	0.26	1.47	13	1517	220	4137	600	524	76
Titan 80(b)	0.20	0.90	8	1103	160	3448	500	448	65
Titan 60(b)	0.22	1.02	9	1172	170	3275	475	414	60
K602(c)	0.21	0.23	2	...	...	5653	820	586	85
K701(c)	0.24	0.28	2.5	...	...	...	...	531	77
K703(c)	...	0.45	4	...	...	6033	875	627	91
K714(c)	0.20	0.79	7	...	...	5998	870	552	80
K801(c)	0.25	0.90	8	...	...	5275	765	621	90
K803(c)	0.23	1.36	12	...	...	5447	790	552	80
HD15(b)	0.30	2.82	25	1862	270	3965	575	531	77
HD20(b)	0.30	3.05	27	1793	260	3723	540	496	72
HD25(b)	0.30	3.28	29	1724	250	3516	510	462	67
HD20T(b)	0.27	2.94	26	1724	250	3792	550	483	70
HD25T(b)	0.27	3.16	28	1655	240	3620	525	455	66
575(b)	0.27	1.36	12	1517	220	5171	750	641	93
569(b)	0.29	1.92	17	1793	260	4309	625	579	84
783(b)	0.29	2.15	19	1862	270	4240	615	572	83
502(b)	0.29	2.37	21	1862	270	4137	600	565	82

(a) Values extrapolated from available data. Not necessarily based on actual Charpy V-notch impact tests. (b) Adamas designation. (c) Kennametal designation. Source: Ref 16, 17

Table 5 Relative solubilities in acids and bases of the basic constituents of cemented carbides

Constituent	Medium and solubility(a) Dilute HNO_3	HCl	H_2SO_4	$20HNO_3$-$60HCl$-$20H_2O$	$25HNO_3$-$25HF$-$50H_2O$	Alkali solutions	Salt solutions
Cobalt	V	Sl	Sl	V	V	I	I
Nickel	V	Sl	Sl	V	V	I	I
WC	I	I	I	S	S	I	I
TaC	I	I	Sl?(b)	S	S	I	I
TiC	S	I	I	S	S	I	I

(a) Solubility: V, very soluble; Sl, slightly soluble; I, insoluble; S, soluble. (b) Data from Ref 21 and Ref 22 are contradictory. Source: Ref 21, 22

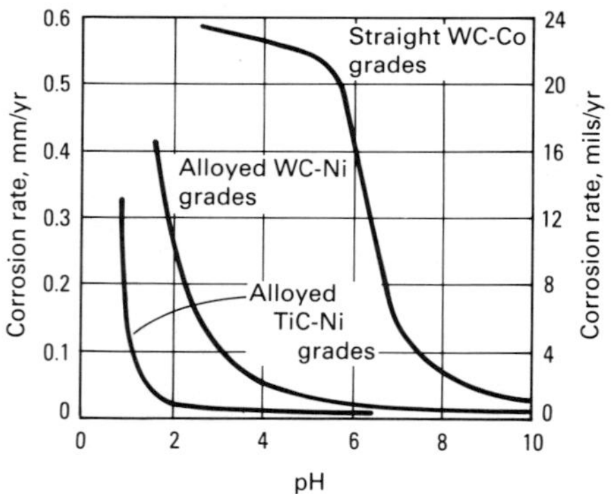

Fig. 6 Corrosion rate of various cemented carbide grades as a function of pH. Source: Ref 23

ever refers to the specifics of these standards in designating the cemented carbide it produces. Another factor is the intentional addition of minor elements such as tantalum, titanium, vanadium, chromium, and molybdenum as grain growth inhibitors or the inadvertent introduction of minor amounts of these and other elements in the raw materials or through recycling. These elements affect hardness and strength and cannot be discounted in the selection of a cemented carbide for corrosion applications.

Table 6 Corrosion resistance of cemented carbides in various media at room temperature

Medium	Corrosion resistance(a) WC-Co cemented carbides	TiC-Ni cemented carbides
Acid salts in water	E	E
Neutral salts in water	V	E
Alkalies		
KOH in water	F-G	F-P
NaOH in water	V	E
NH_3 in water	F	E
Weak acids	G	G-E
Distilled water	E	E
Seawater	V	E
Organic solvents, including acetone, alcohols, gasoline, benzene, carbon tetrachloride, and ethylene glycol	E	E

(a) Corrosion resistance: E, excellent; V, very good; F, fair; G, good; P, poor

Other processing variables also affect properties and performance. Among the important results of processing variables is the amount of porosity in the final cemented carbide product. In some cases, the porosity is negligible, and theoretical density is achieved. In other cases, porosity is present. This can be rated in accordance with ISO 4505 (Ref 19) or ASTM B 276 (Ref 20), both of which are based on the same standard photomicrographs. The ultimate in freedom from porosity is achieved by hot isostatic pressing. This operation, when carried out properly at about 138 MPa (20 000 psi) and at temperatures of 1200 to 1400 °C (2190 to 2550 °F), has no detrimental or beneficial effect on the cemented carbide except for the removal of the last vestiges of porosity. Table 4 shows some typical values of selected mechanical properties of cemented carbide grades that may also be helpful in selecting a proper composition for a particular application.

Corrosion in Aqueous Media

The corrosion of cemented carbides is based on the solubility of the key ingredients used in the various compositions. Although some alloying occurs, the solubility of the WC or TiC in cobalt or nickel is very limited. The main alloying in the WC-Co compositions is primarily based on the addition of TiC, TaC, and NbC, which form cubic-phase solid solutions with WC.

Table 5 shows the relative solubilities of the chief constituents of cemented carbides in various media. Tungsten carbide is insoluble in most acids as well as in basic and salt solutions. It is soluble only in very strong mixtures of nitric acid plus hydrochloric acid (HNO_3 + HCl) and HNO_3 plus hydrofluoric acid (HF). Cobalt and nickel show the same significant solubility in all acids. Even so, the nickel binder compositions show somewhat less attack in some acid solutions than the cobalt binder alloys. From this elementary information, it is obvious that the lower the binder content, the less the corrosion.

Corrosion of cemented carbides, therefore, is generally based on the surface depletion of the

Table 7 Corrosion resistance of cemented carbides in various media

Data for two AISI austenitic stainless steels are included for comparison.

Medium	Chemical designation	Concentration, %	Temperature, °C (°F)	pH	Type of cemented carbide/corrosion resistance(a)					AISI stainless steels(b)	
					WC-Co	TiC-NiMo	WC-Ni	WC-CoCr	WC-TaC-Co	Type 302	Type 316
Acetic acid, unaerated	CH_3COOH	4	Room	...	C	B	B	B	A	...	...
Acetic acid (glacial), unaerated	CH_3COOH	99.8	Room	...	C	C	B	A	A	...	...
Acetone	$(CH_3)_2CO$	...	Room	...	A	A	A	A	A	A	A
Alcohols	...	...	Room	...	A	A	A	A	A	...	...
Ammonia, anhydrous	NH_3	...	...	...	B	B	B	B	A	...	...
Argon gas	Ar	...	...	...	A	A	A	A	A	...	...
Benzene, liquid	C_6H_6	...	Room	...	A	A	A	A	A	...	...
Carbon tetrachloride	CCl_4	Pure	Room	...	A	A	A	A	A	...	...
Chlorine gas, dry	Cl	...	Room	...	C	C	C	C	B	...	...
Chlorine gas, wet	$Cl{\cdot}H_2O$	...	Room	...	D	C	C	D	B	...	...
Citric acid	$C_3H_4(OH)(COOH)_3$	5	Room	1.7	C	A	A	...	...	A	A
Citric acid	$C_3H_4(OH)(COOH)_3$	5	60 (140)	1.7	D	A	B	...	...	A	A
Copper sulfate solution	$CuSO_4$	0.01	Room	6	C	A	A	...	...	A-C	A-C
Copper sulfate solution	$CuSO_4$	0.01	70 (160)	6	D	A	A	...	...	A-C	A-C
Digester liquor, black	...	...	66 (150)	...	B	B	B	B	A	...	...
Esters	...	...	Room	...	A	A	A	A	A	...	...
Ethanol	C_2H_5OH	96	Room	...	A	A	A	...	...	A	A
Ethylene glycol	$C_2H_6O_2$	...	Room	...	A	A	A	A	A	...	...
Ferrous sulfide	FeS	Slurry in water	Room	...	C	C	C	C	A	...	...
Fluorine, liquid	F	...	−188 (−305)	...		B					
50% formaldehyde, 50% alcohol	...	...	Room	...	C	Uncoupled B Coupled C(c)	C	C	A	...	...
Formic acid	HCOOH	5	Room	...	C	A	C	...	...	A	A
Formic acid	HCOOH	5	60 (140)	1.8	D	A	...	...	...	B	A
Freon gas	$C_2Cl_3F_3/CH_2Cl_3$	...	Room	...	A	A	A	A	A	...	...
Gasoline	...	...	Room	...	A	A	A	A	A	...	...
Helium, liquid	He	...	−269 (−450)	...	A	A	A	A	A	...	...
Hydrochloric acid	HCl	0.5	Room	1	D	C	C	...	...	C	A
Hydrochloric acid	HCl	0.5	60 (140)	1	D	C	C	...	...	D	A
Hydrochloric acid	HCl	10	Room	...	D	D	D	...	...	D	C
Hydrochloric acid	HCl	37	Room	...	D	D	D	D	A	...	...
Hydrochloric acid	HCl	37	100 (212)	...	D	D	D	D	B	...	...
Hydrofluoric acid, anhydrous	HF	...	Room	...	B	B	B	B	A	...	...
Hydrofluoric acid	HF	1–60	Room	...	D	D	D	D	D	...	...
Hydrogen, liquid	H	...	253 (488)	...	A	A	A	A	A	...	...
Kerosene	...	...	Room	...	A	A	A	A	A	...	...
Magnesium bisulfite digester liquor	$MgHSO_3$	...	Room	...	B	B	B	B	A	...	...
Methane, liquid	CH_4	...	162 (324)	...	A	A	A	A	A	...	...
Methanol, anhydrous	CH_3OH	...	Room	...	A	A	A	A	A	...	...
Methanol, 20% water	CH_3OH/H_2O	...	Room	...	A	A	A	A	A	...	...
Nitric acid	HNO_3	0.5	Room	1.1	D	C	A	...	...	A	A
Nitric acid	HNO_3	5	Room	...	D	D	D	D	B	...	...
Nitric acid	HNO_3	...	100 (212)	...	D	D	D	D	B	...	...
Nitric acid	HNO_3	10	Room	...	D	B	C	...	...	A	A
Nitrogen, liquid	N	...	196 (385)	...	A	A	A	A	A	...	...
Oil, crude (Sand, salt water, high in sulfur)	...	...	Room	...	C	C	C	C	A	...	...
Oxalic acid	$(COOH)_2{\cdot}2H_2O$	5	Room	1	A-B	A	A	...	...	A	A
Oxalic acid	$(COOH)_2{\cdot}2H_2O$	5	60 (140)	1	B-C	A	...	...	...	B	A
Oxygen, liquid	O	...	183 (361)	...	A	A	A	A	A	...	...
Perchloric acid	$HClO_4$	0.5	Room	1.3	C-D	A	C	...	...	D	...
Perchloric acid	$HClO_4$	0.5	60 (140)	1.3		D	A	D	...	D	D
Phosphoric acid	H_3PO_4	5	Room	1.2	D	B	C	...	...	A	A
Phosphoric acid	H_3PO_4	85	Room	...	D	C	C	D	A	...	...
Crude phthalic acid and anhydride	C_6H_4-1,2 $(COOH)_2$/ C_6H_4-1,2 $(CO)_2O$	...	250–280 (480–535)	...	C	C	B	C	A	...	...
Sodium carbonate	Na_2CO_3	5	Room	12	A	A	A	...	...	A	A
Sodium carbonate	Na_2CO_3	5	60 (140)	12	A	A	A	...	...	A	A
Sodium chloride	NaCl	3	Room	7	A-B	A	A	...	...	A	A
Sodium chloride	NaCl	3	60 (140)	7	A-B	A	A	...	...	A	A
Sodium cyanide	NaCN	10	Room	...	D	D	D	D	A	...	...
Sodium hydrogen sulfate	$NaHSO_4$	5	Room	1.2	C-D	A	A-B	...	...	D	A
Sodium hydrogen sulfate	$NaHSO_4$	5	60 (140)	1.2	D	C	C-D	...	...	D	A
Sodium hydroxide	NaOH	5	Room	14	A	A	A	...	...	A	A
Sodium hydroxide	NaOH	5	60 (140)	14	B	A	A	...	...	A	A
Sodium hydroxide	NaOH	40	Room	16	A	A	A	...	...	A	A
Sodium hydroxide	NaOH	40	60 (140)	16	A	A	A	...	...	A	A
Steam, superheated	H_2O	...	600 (1110)	...	A	A	A	A	A	...	...
Sulfuric acid	H_2SO_4	0.5	Room	1.2	C-D	A	B-C	...	...	C	A
Sulfuric acid	H_2SO_4	0.5	60 (140)	1.2	D	D	D	...	...	D	A

(continued)

(a) A, highly resistant, negligible attack; B, resistant, light attack; C, poor resistance, medium attack; D, not resistant, not suitable. This table should be used only as a guide. Many factors, such as temperature variations, changes in chemical environment, purity of solutions, and stress or loading conditions, may invalidate these recommendations. Tests under operating conditions should be made. (b) Results were obtained under laboratory conditions in pure solutions and are classified with reference to corrosion resistance only. (c) Coupled to brass. Source: Ref 23, 24

Table 7 (continued)

Medium	Chemical designation	Concentration, %	Temperature, °C (°F)	pH	Type of cemented carbide/corrosion resistance(a): WC-Co	TiC-NiMo	WC-Ni	WC-CoCr	WC-TaC-Co	AISI stainless steels(b): Type 302	Type 316
Sulfuric acid	H_2SO_4	5	Room	...	C	B	C	C	A	...	...
Sulfuric acid	H_2SO_4	5	100 (212)	...	D	C	C	D	A	...	...
Sulfuric acid	H_2SO_4	10	Room	0	D	D	B	...	...	D	A
Sulfuric acid	H_2SO_4	10	60 (140)	0	D	D	...	...	...	D	D
Sulfur, liquid	S	100	130 (265)	...	A	A	...	...	...	...	A
Water, boiler feed	H_2O	...	66 (150)	...	B	C	A	A	A	...	...
Water, fresh, distilled, purified	H_2O	...	Room	...	A	A	A	A	A	...	...
Water, tap	H_2O	...	Room	...	B	A	B	B	A	...	...
Water, sea	...	...	Room	...	B	B	B	...	A	...	...

(a) A, highly resistant, negligible attack; B, resistant, light attack; C, poor resistance, medium attack; D, not resistant, not suitable. This table should be used only as a guide. Many factors, such as temperature variations, changes in chemical environment, purity of solutions, and stress or loading conditions, may invalidate these recommendations. Tests under operating conditions should be made. (b) Results were obtained under laboratory conditions in pure solutions and are classified with reference to corrosion resistance only. (c) Coupled to brass. Source: Ref 23, 24

Table 8 Corrosion of WC-Co cemented carbides in mineral acids

Corrosion rates for AISI type 304 stainless steel are shown for comparison.

Cobalt content, wt%	Weight loss mg/mm^2: 37% HCl, Room temperature, 10 h	37% HCl, Room temperature, 100 h	37% HCl, 100 °C (212 °F), 10 h	5% HCl, 10% H_2SO_4, 100 °C (212 °F), 20 h	5% HCl, 10% H_2SO_4, Room temperature, 200 h	5% H_2SO_4, 100 °C (212 °F), 20 h	10% HNO_3, Room temperature, 20 h	5% HNO_3, 100 °C (212 °F), 20 h
5.5	0.001	0.015	0.05	0.01	0.020	0.10	0.02	0.02
6	0.003	0.02	0.01	0.02	0.030	0.20	0.10	0.15
9	0.005	0.03	0.2	0.08	0.033	0.25	0.20	Destroyed
13	0.01	0.05	0.04	0.12	0.036	0.35	Destroyed	Destroyed
15	0.015	0.13	1.8	0.15	0.040	0.40	Destroyed	Destroyed
Type 304 stainless steel	1.2	Destroyed	Destroyed	1.2	0.18	Destroyed	None	None

Source: Ref 25

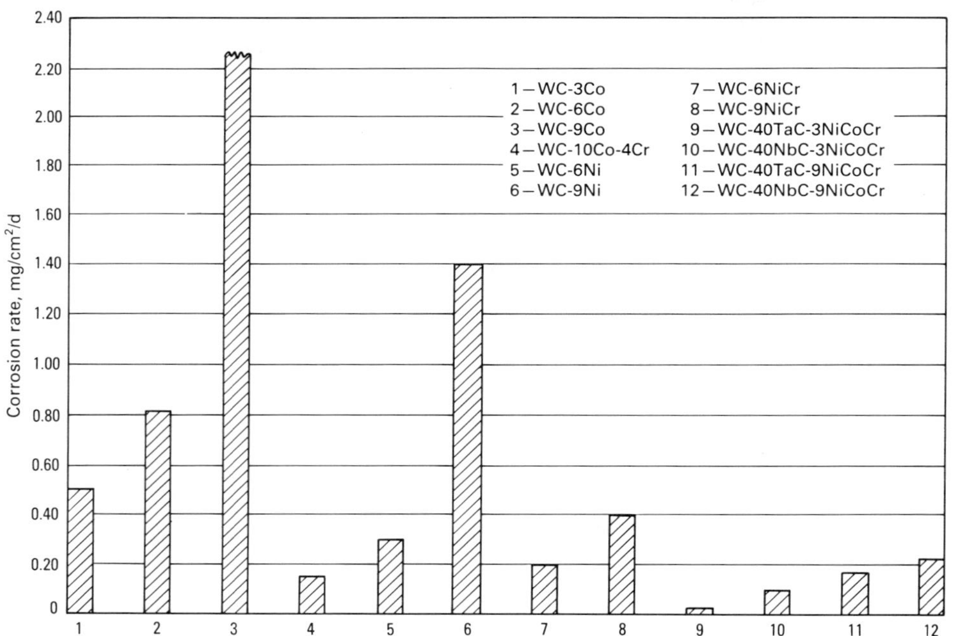

Fig. 7 Corrosion resistance of cemented carbides in 22% HCl at room temperature. See Table 9 for properties of these grades, and Fig. 8 to 14 for corrosion resistance in other media. Source: Ref 7 and 28

binder phase such that at the surface region only a carbide skeleton remains; because the applications are invariably for wear or abrasion, this skeleton is rapidly worn away. At low binder phase contents, the rate of attack is diminished, and in conditions in which the corrosion is not too severe, the reduced binder content will be beneficial. In more severe corrosion, however, the use of a cobalt binder is prohibited, and the WC-Co grade is simply not resistant enough. In these cases, certain corrosion-resistant grades should be used.

The most common of these are WC with nickel alloy binders and TiC-Ni-Mo_2C-base cemented carbide. Figure 6 shows the corrosion rate as a function of pH for these different types of cemented carbides tested in buffered solutions. These tests included a final surface wear treatment by tumbling in order to obtain a true value of the depth of the corroded surface.

As can be seen in Fig. 6, straight WC-Co grades are resistant down to pH 7. This is also valid for WC-Co grades containing cubic carbides such as TiC, TaC, and NbC. The highest corrosion resistance is obtained for certain alloyed TiC-Ni grades, which are resistant down to about pH 1, but compared to the straight WC-Co grades, they are less tough and have lower thermal conductivity. They also have the disadvantages of being difficult to grind and braze; therefore, they are used only in specific applications.

In many corrosion-wear situations, the proper choice is specially alloyed WC-Ni grades, which are resistant down to pH 2 to 3. Even in certain solutions with pH values less than 2, they have proved to be resistant to corrosion. Because WC is the hard principal constituent and because nickel and cobalt are similar metals in many respects, their mechanical and thermal properties are comparable to those of the straight WC-Co grades.

The pH value is one of the most important parameters when determining the corrosivity of a medium, but other factors such as temperature and electrical conductivity also have a great influence. The latter is dependent on the ion concentration, that is, the amount of dissolved salts in the solution. Thus, one cannot define the corrosivity of a certain medium in a simple way, and accordingly, no general rules that are valid in all situations can be given. However, Table 6 gives general guidelines for the corrosion resistance of WC-Co and TiC-Ni cemented carbides in various room-temperature media. Table 7 gives compatibility data for several types of cemented carbides in aqueous media at various temperatures, and Table 8 lists weight loss as a function of cobalt content for cemented carbides in mineral acids.

In general, it can be stated that the corrosion of cemented WC is fair to good in a limited way in all acids except HNO_3. The corrosion resistance of cemented TiC is excellent in phosphoric acid (H_3PO_4), boric acid, and picric acid and is somewhat better than cemented WC in HCl or sulfuric acid (H_2SO_4). Cemented TiC is poor in HNO_3. As expected, increasing the cobalt content to increase strength significantly decreases the corro-

Table 9 Properties of corrosion-resistant cemented carbide grades

See Fig. 7 to 14 for the corrosion resistance of 12 of these grades in various media.

Proprietary designation	Grade number(a)	Composition symbol(a)	Composition, wt% WC	Ni	Co	Cr	Others	Hardness HV (30-gf load)	Converted to HRA	Density, g/cm³	Transverse rupture strength MPa	ksi
H03T(b)	1	WC-3Co	96.7	...	3	...	0.3TaC	1850	92.9	15.3	1400	203
H10T(b)	2	WC-6Co	94.5	...	5.5	...	(c)	1730	92.4	15.0	1900	276
H30T(b)	3	WC-9Co	90.4	...	9	...	0.2TiC, 0.4TaC	1450	90.7	14.6	2000	290
H40T(b)	...	...	88	...	12	...	(c)	1340	89.7	14.3	2600	377
K701(d)	4	WC-10Co-4Cr	85.8	...	10.1	4.1	...	1645	92.0	14.0	1140	165
WC6Ni(b)	5	WC-6Ni	94	6	...	...	...	1400	90.2	15.0	1500	218
WC9Ni(b)	6	WC-9Ni	91	9	...	...	...	1150	87.6	14.6	1800	261
TCR10(b)	7	WC-6NiCr	94	5.7	...	0.3	...	1520	91.2	14.8	2000	290
TCR30(b)	8	WC-9NiCr	91	8.5	...	0.5	...	1420	90.4	14.4	2500	363
H032(e)	...	WC-10TaC-3NiCoCr	87	1.5	1	0.5	10TaC	2000	93.3	15.3	1300	189
H031(e)	13	WC-20TaC-3NiCoCr	77	1.5	1	0.5	20TaC	1940	93.1	14.9	1430	207
V492(e)	9	WC-40TaC-3NiCoCr	57	1.5	1	0.5	40TaC	2000	93.3	14.9	1400	203
H035(e)	10	WC-40NbC-3NiCoCr	57	1.5	1	0.5	40NbC	1870	92.9	11.1	1280	186
V455(e)	11	WC-40TaC-9NiCoCr	50	4	4	1	41TaC	1450	90.7	14.2	1850	268
V473(e)	12	WC-40NbC-9NiCoCr	50	4	4	1	41NbC	1400	90.2	14.5	1750	254
TWF18(b)	...	...	...	18	...	...	66TiC, 16Mo$_2$C	1470	90.8	6.0	1270	184

(a) Used to refer to grades in Fig. 7 to 14 (b) Metallwerk Plansee grade designation. (c) Depending on the reference used, these grades are sometimes shown with small additions of TaC and TiC. (d) Kennametal designation. (e) Metal werk Plansee experimental designation. Source: Ref 7 and 28

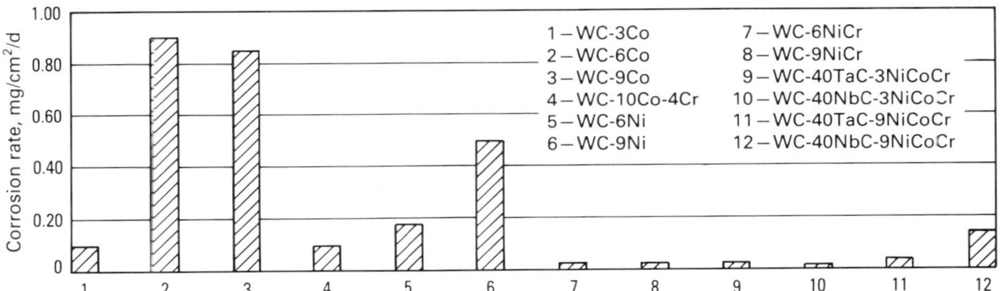

Fig. 8 Corrosion resistance of cemented carbides in 37.8% HNO_3 at room temperature. See Fig. 7 for key to identification and compositions. See also Table 9 and Fig. 9 to 14. Source: Ref 7 and 28

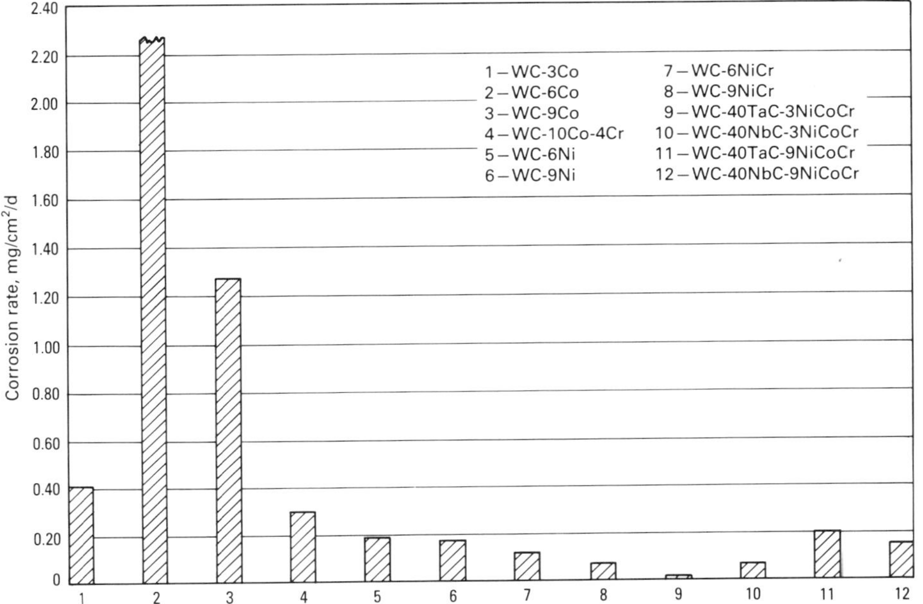

Fig. 9 Corrosion resistance of cemented carbides in 9.8% H_2SO_4 at room temperature. See also Table 9 and Fig. 7, 8, and 10 to 14. Source: Ref 7 and 28

sion resistance (Table 8). The same situation exists in virtually all corrosive environments, and because the same effect is seen for abrasion resistance, it is recommended that the minimum cobalt content be used for all wear and corrosion applications. This means that for a given application the hardest grade that will give adequate strength, impact resistance, and resistance to chipping should be chosen.

Special Corrosion-Resistant Grades. To obtain corrosion resistance above and beyond that available with the regular WC-Co and TiC-Ni grades, the special corrosion-resistant grades are used. These always result in a sacrifice in strength, hardness, and/or abrasion resistance, as shown in Table 2. On the other hand, the corrosion-resistant grades do offer significant benefits in corrosion resistance in many media (Table 7). These grades include the WC + Ni binder, the WC + Co-Cr binder, and the so-called binderless WC, which generally contains about 10% TaC and between 1 and 2% Co. In addition, there are other special grades, such as the 0.1 to 1.0% Pt addition patented as an improvement toward ink corrosion resistance in ballpoint pen balls (Ref 26).

Sintered cemented carbide compositions based on more than 50% Cr_2C_3, for corrosion resistance are also mentioned in patents (Ref 27) and the literature (Ref 3, 4, 5). These are generally not commercially viable and are brittle materials; therefore, they cannot compete with the ceramic materials, such as silicon carbide, silicon nitride, aluminum oxide, boron nitride, and the whisker-reinforced ceramics, which have superb corrosion resistance. Where impact and chipping are not problems, these ceramic materials are a better choice than the cemented carbides. The cemented carbides have the advantage, however, in strength, impact resistance, thermal conductivity, and often greater ease of manufacture.

The best recent work showing the performance of the special corrosion-resistant compositions compared to the standard compositions and even some experimental compositions is that done at

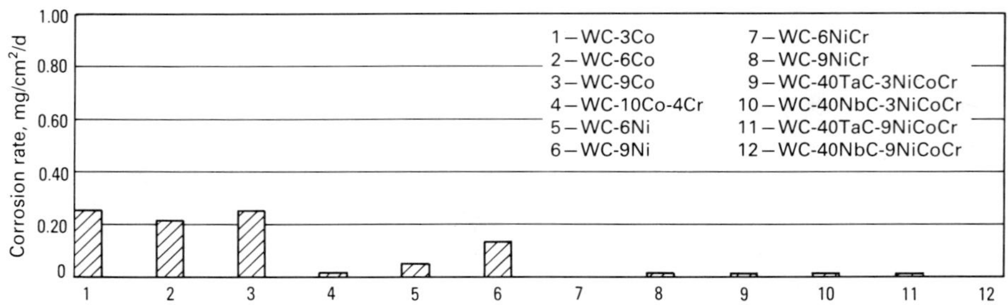

Fig. 10 Corrosion resistance of cemented carbides in 6% acetic acid at room temperature. See Fig. 9 for key to identification and compositions. See also Table 9, Fig. 7, and Fig. 11 to 14. Source: Ref 7 and 28

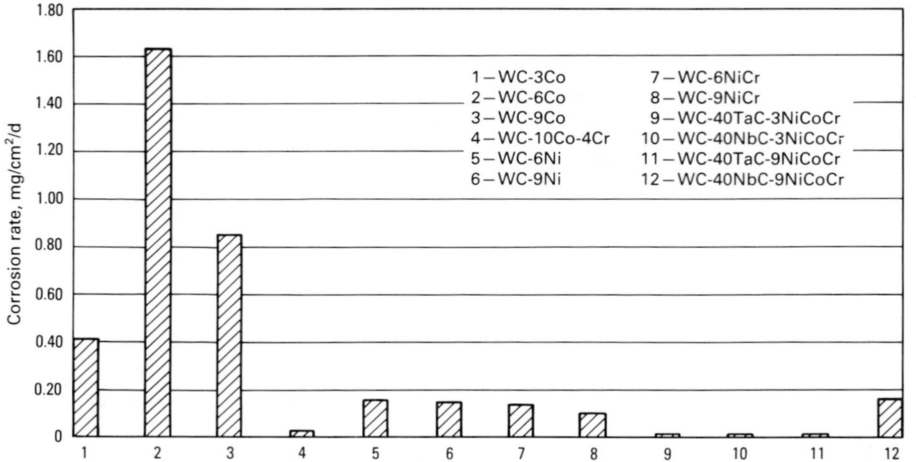

Fig. 11 Corrosion resistance of cemented carbides in 6.5% H_3PO_4 at room temperature. See Fig. 9 for key to identification and compositions. See also Table 9, Fig. 7 to 10, and Fig. 12 to 14. Source: Ref 7 and 28

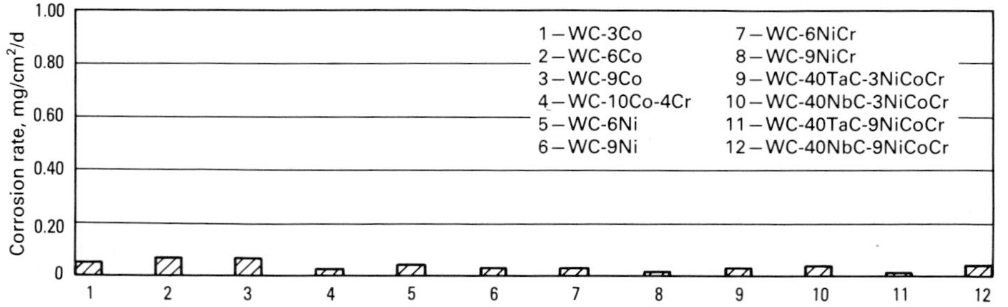

Fig. 12 Corrosion resistance of cemented carbides in 4% NaOH at room temperature. See Fig. 9 for key to identification and compositions. See also Table 9, Fig. 7 to 11, and Fig. 12 and 13. Source: Ref 7 and 28

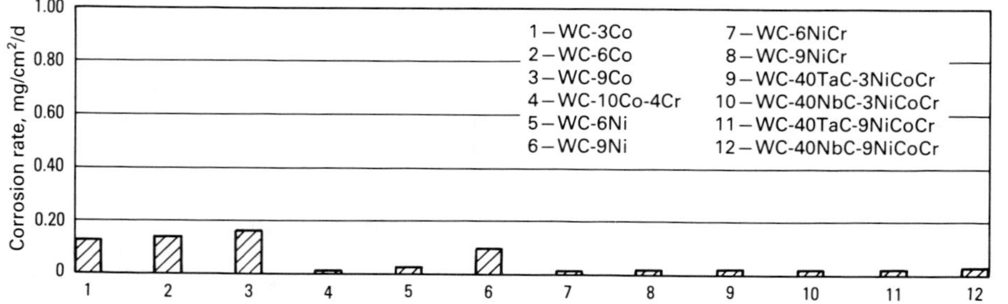

Fig. 13 Corrosion resistance of cemented carbides in 2.9% NaCl at room temperature. See Fig. 9 for key to identification and compositions. See also Table 9, Fig. 7 to 12, and Fig. 14. Source: Ref 7 and 28

Metallwerk Plansee (Ref 7, 28). Table 9 lists the properties of these grades; for convenience, the proprietary designations are given, and the grades are also noted by composition, such as WC-10Co-4Cr. Grades are also listed by a grade number that can be used when referring to Fig. 7 to 14.

Figure 7 shows the corrosion of the 12 different compositions listed in Table 9 in 22% HCl at room temperature. Grades 1 to 3 (WC-3Co, WC-6Co, and WC-9Co, respectively) illustrate the increase in corrosion rate that results from increasing cobalt binder content. The nickel binders (grades 5 and 6; WC-6Ni and WC-9Ni, respectively) are an improvement, but again, the increase in binder content increases the corrosion rate. Of the more exotic compositions, grades 4 (WC-10Co-4Cr) and 7 (WC-6NiCr) are viable choices for limited use in HCl at room temperature. The best of the experimental compositions is grade 9 (WC-40TaC-3NiCoCr); it has greater strength and higher hardness. If additional strength is needed above grade 9, grade 11 (WC-40TaC-9NiCoCr) is a good choice with the increased binder content, but as is generally the case, this results in a loss of corrosion resistance.

Figure 8 shows the same type of information for 38% HNO_3 at room temperature. In general, corrosion is lower, but again, the higher-cobalt WC-Co compositions (grades 2 and 3) are not suitable, nor is the WC-9Ni composition (grade 6). Grade 5 (WC-6Ni) is marginal in HNO_3, but grades 1 and 4 are still better. On the other hand, the commercially available grades 7 and 8 (WC-6NiCr and WC-9NiCr, respectively) show very limited corrosion attack that is virtually equal to that of three of the four experimental grades; the commercial alloys in this case have better strength.

The basic cemented carbides are attacked most severely by H_2SO_4 (Fig. 9). Some of the WC-Ni or WC-NiCr commercial compositions can tolerate limited use. However, the experimental grade 7 (WC-40TaC-3NiCoCr) provides exceptional corrosion resistance.

Figure 10 shows that many compositions are available for use in acetic acid with little corrosion. Attack in H_3PO_4 is relatively rapid only on the WC-Co compositions (Fig. 11).

Figures 12 and 13 show the suitability of all of the compositions listed in Table 9 in sodium hydroxide (NaOH) and sodium chloride (NaCl). In NaCl, there is significant benefit in choosing a nickel binder cemented carbide (for example, grade 5, WC-6Ni) if the loss in strength can be tolerated.

Figure 14 shows the resistance to erosion-corrosion of different cemented carbide compositions in a slurry of artificial seawater and sand. It follows the pattern of benefit for the use of nickel binders in saline applications. The best of the WC-Co compositions is obviously the one with the lowest binder content (WC-3Co; grade 1). It shows a rate, however, more than 10 times greater than the experimental grade 9 (WC-40TaC-3NiCoCr), and both have the same transverse rupture strength and equivalent hardness. For a commercial composition, the grade 9 (WC-9NiCr) cemented carbide shows excellent performance, with one-half the rate of attack of the low-cobalt composition (grade 1, WC-3Co) and much higher transverse rupture strength.

Some of the same data are shown in Fig. 15 and 16 to compare the relative corrosion of the different compositions in various media. These tests

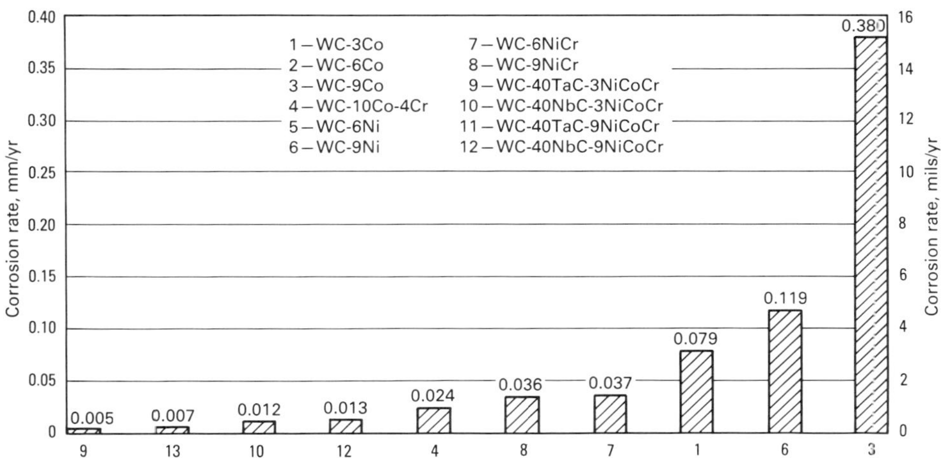

Fig. 14 Resistance to erosion-corrosion of cemented carbides in a room-temperature slurry of artificial seawater and sand. See Fig. 9 for key to identification and compositions. See also Table 9 and Fig. 7 to 13. Source: Ref 7 and 28

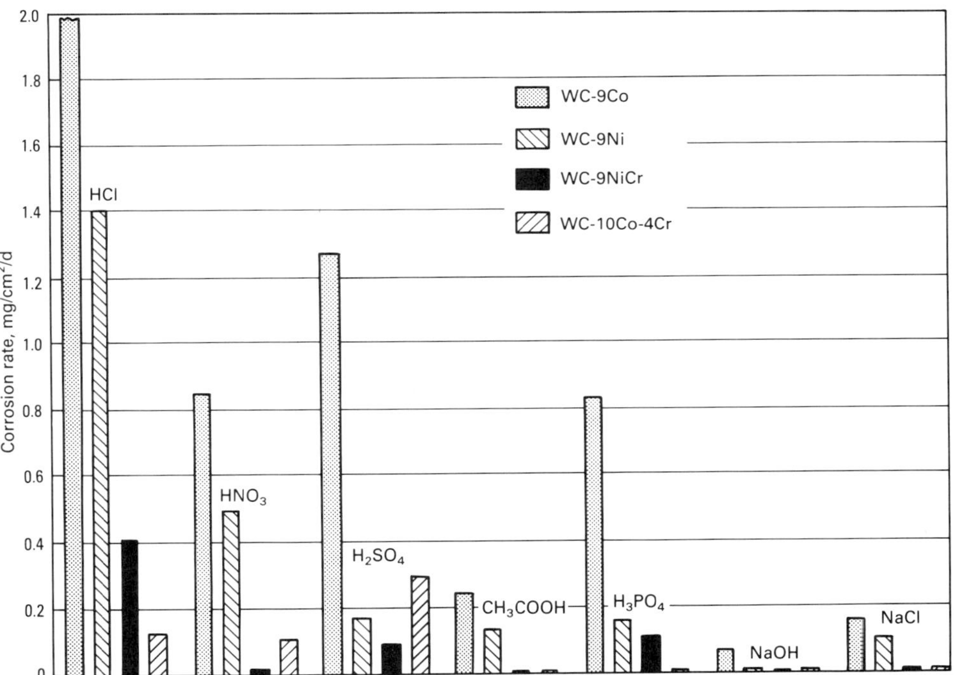

Fig. 15 Corrosion resistance of four commercial cemented carbide compositions in aqueous media at room temperature. Source: Ref 28

were performed at room temperature, and solution concentrations are the same as those in Fig. 7 to 14. In Fig. 16, the cemented carbides are also compared to an Fe-20Cr-32Ni alloy; the superiority of the experimental WC-40TaC-3NiCoCr cemented carbide is evident. As with all corrosion test data, care must be taken not to extrapolate these data to different solution concentrations and temperatures. It would be logical to assume, for example, that the WC-40TaC-3NiCoCr alloy would always outperform WC-3Co in these media at different concentrations and temperatures, but the validity of this assumption must be verified through further testing.

Corrosion in Warm Acids and Bases. The corrosion rate of various cemented carbide compositions in warm (50 °C, or 120 °F) acids is shown in Table 10. The straight WC-Co compositions show rapid attack in dilute H_2SO_4 and HNO_3, and little attack in those concentrated acids. Although the corrosion rate is lower in HCl, it is obvious that these compositions are not suitable for use in warm or hot acid solutions. The TiC-6.5Ni-5Mo composition is quite good in H_2SO_4, moderately good in HCl, and very poor in HNO_3. Several of the binderless compositions and the TaC-base cemented carbide show very acceptable corrosion resistance in these warm acids. These results are to be expected, because the cobalt and nickel binders are completely soluble in these acids.

The corrosion rates of various cemented carbides in basic solutions at 50 °C (120 °F) is quite a different matter, as shown in Table 11. Although corrosion does proceed, it is slow enough to demonstrate the utility of even the WC-Co compositions in such applications as seal rings in these basic solutions.

Galvanic Corrosion. The resistance to galvanic corrosion of various cemented carbides coupled to AISI type 316 stainless steel has been investigated (Ref 30). Immersion testing of uncoupled specimens was also performed for comparison. Compositions of the materials tested are given in Table 12.

The apparatus used for the galvanic-corrosion testing is shown in Fig. 17. Figure 18 shows the corrosion rates of the materials in the immersion test. The binderless WC-3TiC-2TaC alloy performed the best, followed by the TiC-base cermet, the WC-Ni-CrMo alloy, the sintered cobalt-base alloy, and the WC-6Co alloy. The logarithm of weight loss plotted against the logarithm of time yielded the linear weight loss curves in this test. Based on this, it was postulated that the movement of electrons between cemented carbide and stainless steel is the rate-determining factor in galvanic corrosion. Table 13 compares the corrosion rates of the materials in the immersion and galvanic tests. For most of the alloys tested, the rate of galvanic corrosion is greater than the corrosion rate in the simple immersion test. It is thought that the larger the potential difference between the cemented carbide and the stainless steel, the greater the difference between the corrosion rates obtained in the immersion test and in the galvanic-corrosion test.

Figure 19 shows cross sections of specimen rings after the galvanic corrosion test. Corrosion proceeded inward from the surface that contacted the seawater in the WC-6Co alloy (Fig. 19a). The investigators postulated that the electrode potential is large and that electrons would move smoothly between the cemented carbide and the contacting stainless steel; therefore, attack proceeded according to the galvanic-corrosion mechanism. In the case of the binderless WC-3TiC-2TaC alloy (Fig. 19b), corrosion is very slight even after 1 year. The TiC-base cermet (Fig. 19c) shows corrosion only on the inner side surface (the side contacting the teflon; see Fig. 17). In the sintered cobalt-base alloy and the WC-Ni-CrMo alloy (Fig. 19c and d), corrosion proceeded from the corner that contacted both the seawater and the stainless steel. It was postulated that the electrode potential and the distance of electron movement were smaller than those for the WC-6Co alloy. Based on the results of these tests, either the binderless alloy or the TiC-base alloy should be acceptable for this type of application.

Crevice Corrosion. The same investigators also reported on the crevice corrosion resistance of cemented carbides in seawater with specimens of type 316 stainless steel, teflon, and silicon carbide adjacent to the cemented carbide specimens (Ref 30). Of the five compositions tested, only the WC-6Co specimen showed any significant attack after 1 year. The attack was moderate and progressed the least against the silicon carbide and the most against the stainless steel (Ref 30).

Oxidation Resistance of Cemented Carbides

The ordinary WC-Co cemented carbides are reasonably resistant to oxidation in air up to about 650 to 700 °C (1200 to 1290 °F). The constituent affected the fastest is WC, which will oxidize to WO_3. In oxygen, the temperature limit

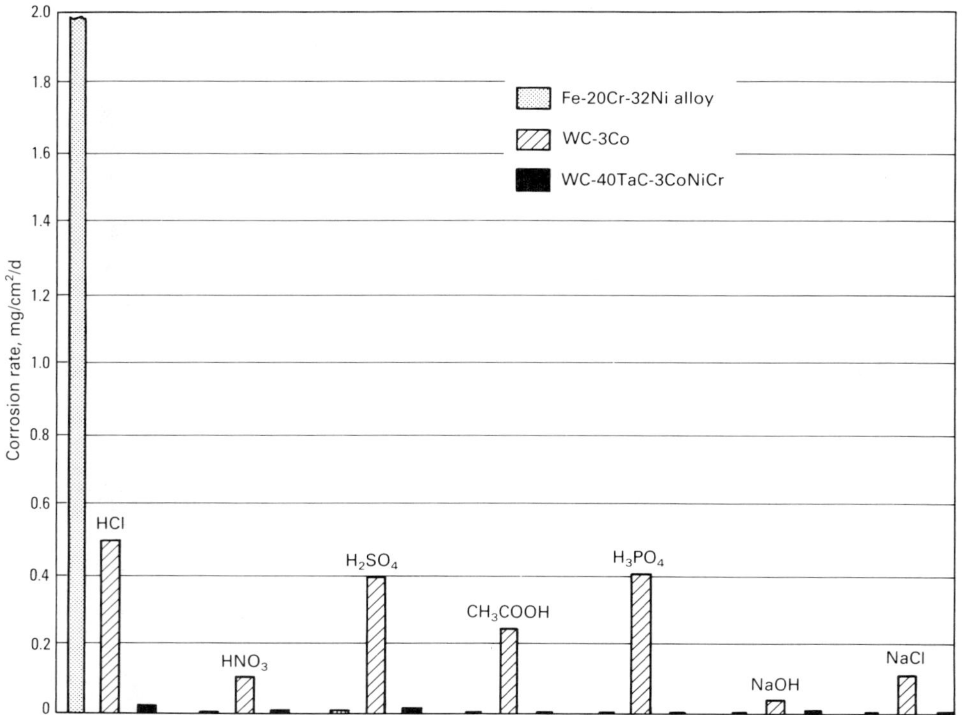

Fig. 16 Comparison of the corrosion resistance of a commercial WC-3Co cemented carbide and two experimental compositions in aqueous media. Source: Ref 28

is lower, and rapid deterioration will occur at about 500 °C (930 °F). Even in air, however, the practical temperature limit for WC-Co compositions for any length of time is 500 to 600 °C (930 to 1110 °F). Nonetheless, these compositions do stand up, for example, in cutting tools in which localized higher temperatures at the cutting tip will be encountered. The addition of both or either TiC or TaC to the WC-Co compositions increases the oxidation resistance somewhat and is undoubtedly also related to the improvement found for these additions for machining steel. In applications in which oxidation resistance combined with wear resistance is required, as in hot glass forming and shearing tools, the addition of TiC and/or TaC to the basic WC-Co is of little benefit. The TiC-Mo_2C-Ni compositions have clearly superior oxidation resistance and can be used at temperatures up to 900 °C (1650 °F), at which point they start to oxidize fairly rapidly. At the lower temperature, the TiC-base compositions form a tight adherent oxide film that tends to resist rapid attack. This behavior difference is analogous to the behavior difference between cobalt and nickel alone, but WC is also more readily oxidized than TiC.

Saw Tips and Corrosion

Cemented carbides are in widespread use in slitter saws, which are used to saw all types of metals, composites, lumber, and many other materials. Small saw blades are sometimes manufactured from a single piece of carbide; larger blades, which may run up to 2 m (6 ft) in diameter, more commonly use cemented carbide tips brazed onto the steel saw body. The heavy-duty chain saws used in the lumber industry also have carbide teeth. Selection of cemented carbides for these applications is invariably based on the need for excellent wear resistance and toughness. Basic WC-Co compositions are almost always used.

The rapid dulling of saws in such applications, however, is attributable to corrosive as well as abrasive conditions. For example, one investigation studied the corrosion of WC cutting tools used to cut western red cedar (Ref 31). Tests were performed to determine the relative rates of attack of WC and cobalt in substances extracted from western red cedar, which has a higher content of such substances than other commercial lumber species. Because the WC was not attacked, it was concluded that the cobalt binder content should be reduced to minimize attack. Alternatively, the cobalt binder could be replaced with another binder material, such as nickel; however, such a substitution would result in a serious loss of strength. Thus, the solution to this particular problem is not a simple one, and western red cedar is still being sawed primarily with WC-Co cemented carbide compositions.

It was also suggested that the carbide be coated with TiC, TiN, or Al_2O_3 (or a combination of these). To date, these coatings are not used in such applications because of the need for resharpening and because of the difficulties of brazing a coated tip.

Coating of Cemented Carbides

This widely used process has been primarily applied to metal cutting tools. Certain special applications can be cited, such as the coating of cemented WC watch cases with TiN to form a hard, corrosion-resistant gold-colored watch case (bezel). Clearly, the potential exists to utilize these thin (2 to 10 μm, or 0.08 to 0.4 mil) coatings on wear- and corrosion-resistant parts. The limitation is that the coating must be very thin to avoid spalling or chipping. In addition, because the use of a cemented carbide in a corrosion application will only be in a very high-integrity, relatively costly application, the potential danger of a coating failing or being locally breached rules out consideration in most applications. Despite this, coating of cemented carbides is an important state of the art that must be considered in special applications.

Coating is most commonly done by chemical vapor deposition (CVD), and this process gives a wide range of possible coating materials. In addition to the common TiN, TiC, Al_2O_3, perfectly feasible coating materials include hafnium carbide (HfC), hafnium nitride (HfN), zirconium carbide (ZrC), zirconium nitride (ZrN), TaC, and NbC. The state of the art includes all combinations of TiC, TiN, Al_2O_3, and titanium carbonitride (TiCN), with limited commercial use of HfN and TaC as coating materials. Chemical vapor deposition is generally performed at 900 to 1100 °C (1650 to 2010 °F). Titanium nitride is coated at lower temperatures, down to perhaps 700 °C (1290 °F), in less used commercial apparatus.

Physical vapor deposition (PVD) has the advantage of being done at lower temperatures,

Table 10 Weight losses of cemented carbides immersed in various acids at 50 °C (120 °F) for 72 h

Composition	Weight loss, mg/cm²/d								
	HCl, %			H_2SO_4, %			HNO_3, %		
	5	10	37	10	50	98	5	10	50
WC-6Co	2.29	2.43	0.79	8.72	2.82	0.72	13.50	1.45	0.16
WC-9Co	2.55	1.96	1.92	12.70	5.05	0.72	25.60	6.48	0.18
WC-8Ni-2Mo-3Cr	0.07	0.01	+0.01	5.01	0.76	0.01	5.71	1.23	0.11
WC-5TaC	+0.02	nil	0.05	+1.03	0.23	0.02	+0.02	0.03	0.15
WC-2TaC-3TiC	0.06	nil	+0.02	+1.02	0.33	0.06	0.35	0.08	0.12
WC-47NbC-15TiC-9Ni-4Mo	1.06	0.98	0.14	5.31	0.51	0.52	8.07	1.35	0.24
TaC-4Co-3Ni-1Cr	0.09	0.02	0.22	2.03	0.41	0.51	0.09	0.05	0.01
TaC-23TiC-3Co-2Ni-1Cr	0.26	0.59	1.03	1.98	0.48	0.43	8.04	5.12	6.35
TiC-6.5Ni-5Mo	0.59	1.73	4.91	0.17	0.35	0.39	35.20	19.80	68.2

Source: Ref 29

Table 11 Weight changes of cemented carbides immersed in NaOH, KOH, and NaOCl at 50 °C (120 °F) for 72 h

Composition	Weight loss, mg/cm²/d — NaOH, % 5	NaOH, % 10	KOH, % 5	KOH, % 10	NaOCl
WC-6Co	+0.75	+0.85	0.39	0.30	1.44
WC-9Co	+0.83	+0.88	0.24	0.28	2.35
WC-8Ni-2Mo-3Cr	+0.09	+0.11	0.08	0.09	1.12
WC-5TaC	+0.89	+0.92	0.18	0.18	+0.15
WC-2TaC-3TiC	+0.87	+0.90	0.20	0.20	+0.13
TaC-4Co-3Ni-1Cr	+0.71	+0.68	0.11	0.14	0.05

Source: Ref 29

Table 12 Compositions and properties of galvanic corrosion test specimens

Specimen	Composition, wt%	Hardness, HRA	Transverse rupture strength, MPa	ksi
WC-6Co	WC-6Co	91.0	2400	348
WC-3TiC-2TaC alloy	WC-3TiC-2TaC	92.9	1200	174
TiC-base cermet	TiC-10TiN-2.5Mo$_2$C-15Ni	91.5	1500	218
Sintered cobalt-base alloy	Co-Cr-W-C	85.5	1400	203
WC-NiCrMo alloy	WC-3TiC-1.5(Cr$_3$C$_2$Mo$_2$C)-15Ni	89.0	2100	305

Source: Ref 30

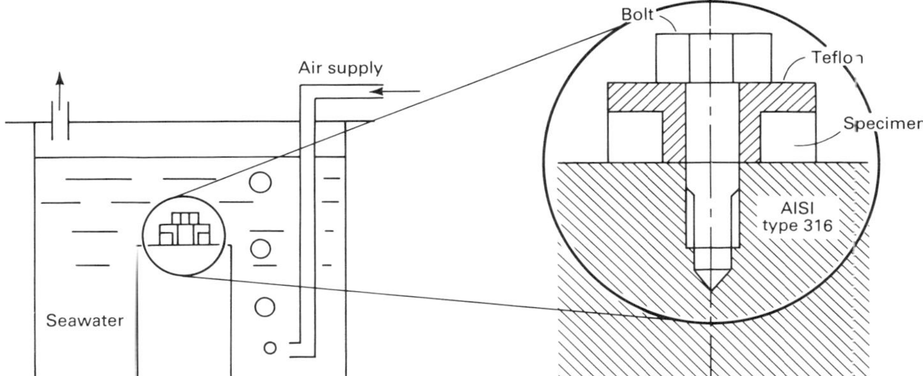

Fig. 17 Schematic of experimental apparatus used to study galvanic corrosion of cemented carbides in seawater. Source: Ref 30

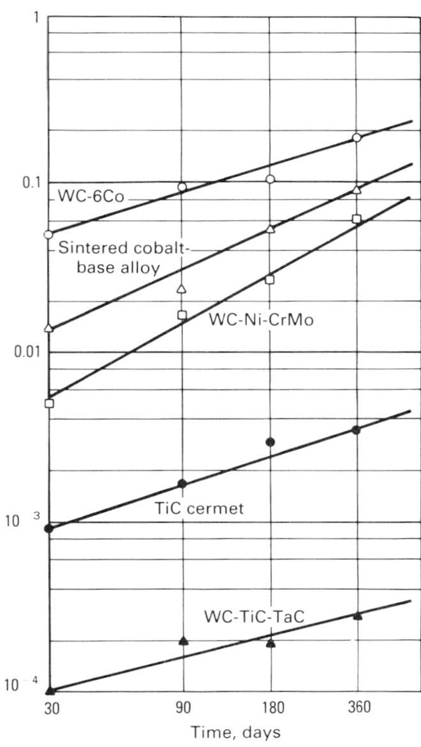

Fig. 18 Corrosion weight loss as a function of time for uncoupled test specimens from Ref 30

Table 13 Corrosion rates of immersion and galvanic corrosion test specimens

Specimen	Weight loss, g/m²/d — Immersion test	Galvanic test
WC-6Co alloy	4.2	16.67
WC-3TiC-2TaC alloy	0.6	0.03
TiC-base cermet	0.2	0.58
Sintered cobalt-base alloy	0.3	4.77
WC-NiCrMo alloy	0.2	1.71

Source: Ref 30

down to perhaps 500 °C (930 °F), but it is a line-of-sight process that generally requires rotation of the parts being coated. Deposition rates for PVD are much lower than those of CVD, and PVD equipment is more expensive. Physical vapor deposited coatings have also been limited commercially to TiN, usually at thicknesses of 3 μm (0.12 mil) or less. More information on the corrosion and wear resistance of coatings applied by these methods is available in the article "CVD/PVD Coatings" in this Volume.

Although there are few applications in which cemented carbides are used solely for corrosion resistance, it is essential to recognize the availability of the coated carbides. Coatings of TiN, TiC, or Al_2O_3 can impart very important corrosion and oxidation resistance to cemented carbides.

Special Surface Treatments

Considerable work has been done to enhance the surface properties of cemented carbides (Ref 32-34), but it generally has been derived from surface modification processes developed for other metals. These surface treatments include boriding, nitriding, and ion implantation. Most of the treatments have been used to enhance resistance to wear, abrasion, or erosion. The benefits, if any, of such treatments in increasing resistance to oxidation and corrosion are not yet well documented. Nevertheless, these processes may have potential in special applications. The article "Surface Modification" in this Volume contains information on the ion implantation and laser surface modification processes and their effects on the surface properties of metals.

REFERENCES

1. "Hardmetals—Metallographic Determination of Microstructure," ISO 4499, International Organization for Standardization, 1978
2. "Standard Method for Determination of Microstructure in Cemented Carbides," B 657, *Annual Book of ASTM Standards*, American Society for Testing and Materials
3. P. Schwarzkopf and R. Kieffer, *Cemented Carbides*, Macmillan, 1960
4. R. Kieffer and F. Benesovsky, *Hartmetalle* (Hard Metals), Springer Verlag, 1965
5. C. Goetzel, *Treatise on Powder Metallurgy*, Vol I to III, Interscience, 1949
6. H.S. Kalish, Some Plain Talk About Carbides, *Mfg. Eng. Mgmt.*, Vol 71 (No. 1), July 1973
7. E. Kny, T. Bader, Ch. Hohenrainer, L. Schmid, and R. Glätzle, Korrosionsresistente, hochverschlei β feste Hartmetalle, *Werkst. Korros.*, May 1986
8. H. Suzuki *et al.*, *Choukougoukin to Syouketu Koushitu Goukin*, Tokyo, Maruzen, 1986, p 514
9. J. Gurland and P. Bardzil, Relation of Strength, Composition, and Grain Size of Sintered WC-Co Alloys, *J. Met.*, Feb 1955, p 311-315
10. H. Suzuki, Variation in Some Properties of Sintered Tungsten Carbide-Cobalt Alloys With Particle Size and Binder Composition, *Trans. Jpn. Inst. Met.*, Vol 7, 1966, p 112
11. H.S. Kalish, Carbide Grade Classifications—What They Mean, *Mfg. Eng. Mgmt.*, Vol 76 (No. 1), 1976

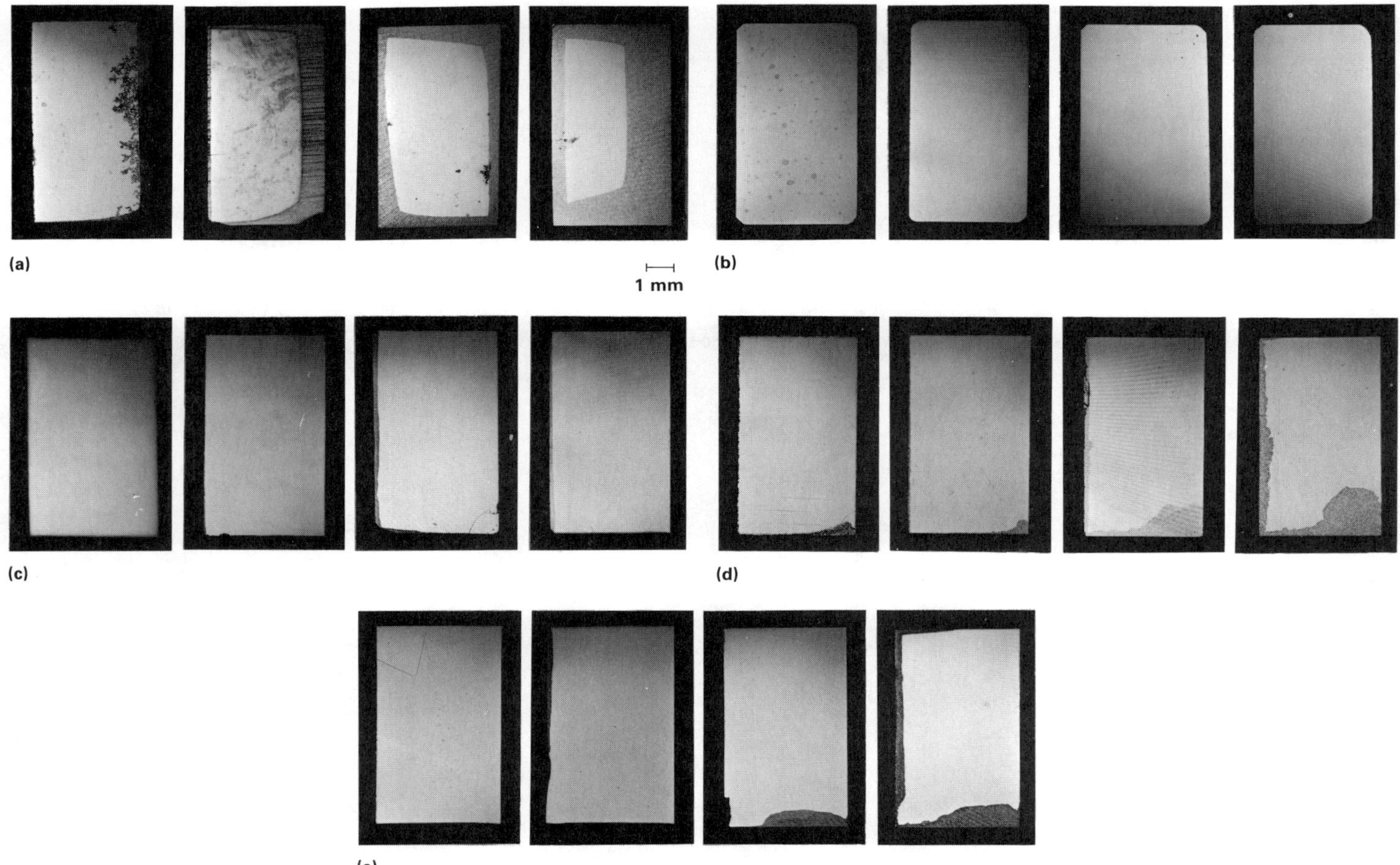

Fig. 19 Cross sections of galvanic-corrosion test specimens after (left to right) 1 month, 3 months, 6 months, and 12 months. (a) WC-6Co alloy. (b) WC-3TiC-2TaC binderless alloy. (c) TiC-base cermet. (d) Sintered cobalt-base alloy. (e) WC-NiCrMo alloy. Source: Ref 30

12. R.S. Montgomery, The Mechanism of Percussive Wear of Tungsten Carbide Composites, *Wear*, 1968, p 309-329
13. "Application of Carbides for Machining by Chip Removal," ISO R 513, 1st ed., UDC 621.9.027, International Organization for Standardization, Nov 1986
14. K.J.A. Brookes, "Metalworking Production's Guide to Hardmetals for Machining," Application Code, International Organization for Standardization, 1984
15. K.J.A. Brookes, *World Directory and Handbook of Hardmetals*, 3rd ed., 1982
16. Published and unpublished data, Adamas Carbide Corporation
17. *Properties and Proven Uses of Kennametal Hard Carbide Alloys*, Kennametal, Inc., 1976
18. "Evaluating Apparent Grain Size and Distribution of Cemented Tungsten Carbide," B 390, *Annual Book of ASTM Standards*, American Society for Testing and Materials
19. "Hardmetals—Metallographic Determination of Porosity and Uncombined Carbon," ISO 4505, International Organization for Standardization, 1978
20. "Standard Test Method for Apparent Porosity in Cemented Carbides," B 276, *Annual Book of ASTM Standards*, American Society for Testing and Materials
21. R.C. Weast, Ed., *Handbook of Chemistry & Physics*, 67th ed., CRC Press, 1986
22. E.K. Storms, *Refractory Carbides*, Academic Press, 1967
23. Published data, Sandvik Coramant
24. Published data, Kennametal, Inc.
25. A. Hara and Y. Saito, Corrosion and Oxidation Resistance of "Igetalloy," *Sumitomo Elec. Tech. Rev.*, No. 13, Jan 1970
26. Corrosion Resistant Binder for Tungsten Carbide Materials and Titanium Carbide Materials, U.S. Patent 3,628,921, 1971
27. Sintered Alloys of a Chromium Carbide-Tungsten Carbide-Nickel System, U.K. Patent 1202844, 1970
28. E. Kny and L. Schmid, *New Hardmetal Alloys With Improved Erosion and Corrosion Resistance*
29. K. Takao and K. Terasaki, Chemical Resistance of Various Cemented Carbides, *Nippon Tungsten Rev.*, Vol 10, 1977
30. Y. Masumoto, K. Takechi, and S. Imasato, Corrosion Resistance of Cemented Carbide, *Nippon Tungsten Rev.*, Vol 19, 1986
31. E. Kirbach and S. Chow, Chemical Wear of Tungsten Carbide Cutting Tools by Western Redcedar, *Forest Prod. J.*, Vol 26 (No. 3)
32. H. Ito and Y. Mohashi, Corrosion-Resistant Cemented Carbides by Chromium Diffusion Methods, *Nippon Tungsten Rev.*, Vol 6, Sept 1973
33. H.S. Kalish, Method of Forming a Hard Surface on Cemented Carbides and Resulting Article, U.S. Patent 3,744,979, 1973
34. Materials Development Corporation, Medford, Mass.

SELECTED REFERENCES

- O.A. Drobysheva and V.N. Latyshev, Interaction of a Hard Alloy and a Cutting Fluid, *Fiz. Khim. Mekh. Mater.*, Vol 8 (No. 3), May-June 1972, p 38-40
- A.L. Echtenkamp, "Combating Corrosion/Wear With the Hard Carbide Alloys," Paper presented at the ASLE/ASME Lubrication Conference, Minneapolis, MN, American Society of Lubrication Engineers, Oct 1978
- E. Suganuma, Electrochemical Behaviour of Cemented Carbide, *Bull. Yagamata Univ. Eng.*, Vol 11 (No. 2), March 1971
- G. VerWeyst, Corrosion-Resistant Tooling for Metal-Forming Operations, *Lubr. Eng.*, Vol 41 (No. 6), June 1985, p 370-374
- V.A. Zhilin and V.M. Druzhinin, Corrosion-Induced Erosion of Hard Alloy Tools During Machining of High-Strength Steels, *Fiz. Khim. Mekh. Mater.*, Vol 6 (No. 4), July-Aug 1970, p 57-62
- V.A. Zhilin and V.M. Durzhinin, Electrochemical Corrosion of Hard Alloys, *Poroshk. Metall.*, Aug 1970, p 68-71

Corrosion of Metal Matrix Composites

Denise M. Aylor, David Taylor Naval Ship Research and Development Center

METAL MATRIX COMPOSITE (MMC) materials have developed substantially over the past 20 years. Developmental efforts have involved aluminum-, copper-, magnesium-, titanium-, and lead-base MMCs, but the primary emphasis has been on aluminum-base materials. Interest in the use of these composites is due to the higher attainable strength and stiffness properties as compared to materials prepared by conventional alloying. However, successful application of MMCs in the marine environment requires adequate corrosion resistance.

This article will discuss the ambient-temperature corrosion characteristics of MMCs that have potential application in marine environments, with primary emphasis on aluminum-base composites. Structural characteristics, design criteria, and coatings for optimum protection of MMCs will also be discussed. More information on various types of MMCs is available in Volume 1 of the 1st Edition of the *Engineered Materials Handbook*.

Structural Characteristics

Metal matrix composites basically consist of a nonmetallic reinforcement incorporated into a metallic matrix. Reinforcements, characterized as either continuous or discontinuous fibers, typically constitute 20 vol% or more of the composite. Reinforcements in continuous-fiber composites include graphite (Gr), silicon carbide (SiC), boron (B), or aluminum oxide (Al_2O_3). Fabrication techniques for these composites vary from chemical vapor deposition (CVD) coating of the fibers, liquid-metal infiltration, and diffusion bonding to liquid-metal infiltration and direct casting to near-net shape. Discontinuous-fiber composites consist almost exclusively of SiC in whisker (w) or particulate (p) form. These MMCs are produced by using modified powder metallurgy techniques (Ref 1). Figure 1 shows cross sections of typical continuous and discontinuous reinforced MMCs.

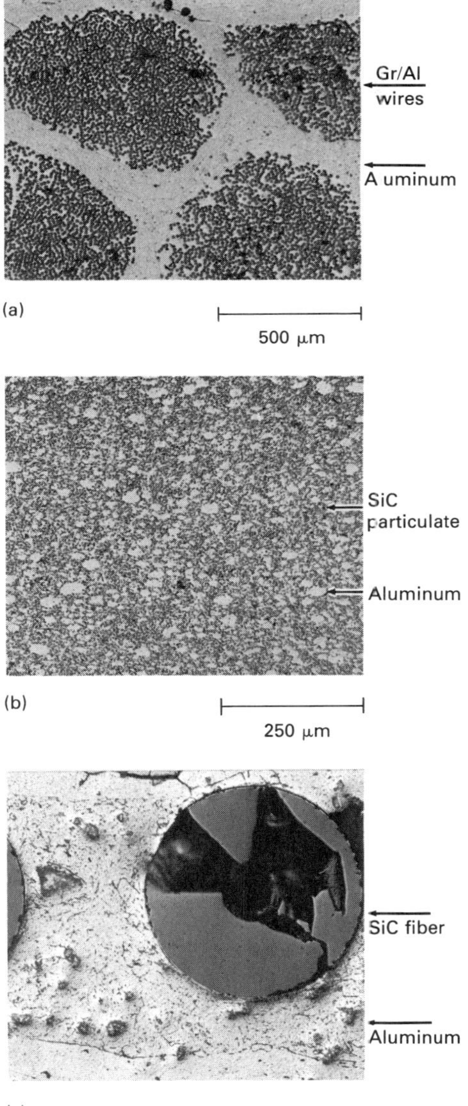

Fig. 1 Cross sections of typical fiber-reinforced MMCs. (a) Continuous-fiber reinforced graphite/aluminum composite. (b) Discontinuous silicon carbide(p)/aluminum composite. (c) Continuous-fiber silicon carbide/aluminum composite

Corrosion Behavior of Aluminum-Base MMCs

Graphite/aluminum composites exhibit accelerated corrosion in marine environments when graphite fibers and aluminum are simultaneously exposed. Assuming that the edges of the graphite/aluminum composite are masked off to prevent exposure of both the graphite and the aluminum, only the aluminum surface foils will initially be exposed to the environment. The aluminum surface foils will pit at an average rate of 0.025 to 0.035 mm/yr (1.0 to 1.4 mils/yr) in seawater and at 0.5 to 0.76 μm/yr (0.02 to 0.03 mils/yr) in the marine atmosphere (1100, 6061, and 5000 series aluminum alloys). Pits may also be present with depths much greater than the average rates reported (Ref 2). Crevice corrosion of the aluminum foils may also occur at the edges because of the crevice formed between the aluminum surface foil and the masking material.

The pitting and crevice corrosion processes eventually penetrate the foils and result in exposure of the graphite/aluminum composite matrix below, at which point the corrosion rate becomes extremely accelerated. Corrosion has been shown to proceed preferentially along foil/foil, wire/wire, and wire/foil interfaces in the composite (Ref 3). Severe exfoliation occurs because of wedging of the hydrated alumina ($Al_2(OH)_3$) corrosion products within the composite. Figure 2 shows an example of severe graphite/aluminum corrosion (known as catastrophic failure). This catastrophic condition can occur within 30 days in seawater after exposure of the graphite-aluminum matrix. Catastrophic failure in the marine atmosphere and in splash/spray environments is less rapid than in seawater, but can occur within 6 months (Ref 4). This accelerated corrosion is believed to result from the aluminum carbides that are formed at the reinforcement/matrix interface during fabrication, which alter the properties of the aluminum surface film at these locations and render the composite more susceptible to breakdown (Ref 3, 5).

The aluminum surface foils alone provide reasonably good corrosion protection to the composites. Marine exposure tests of graphite/aluminum MMCs with 6061, 5056, and 1100 aluminum alloy surface foils (graphite/aluminum edges masked) revealed no pitting penetration through the foils to expose the graphite/aluminum composite wires below during a 20-month exposure (Ref 4). Pitting of the foils, which occurred on most of the graphite/aluminum panels, ranked as light pitting in the splash/spray zone and marine atmosphere and as localized pitting in filtered seawater.

In summary, graphite/aluminum composites undergo extremely severe corrosion in marine environments when the graphite and the aluminum are mutually exposed. Aluminum surface foils have provided 20 months of protection to MMCs, assuming there is no graphite-aluminum exposure. However, the composite will start to fail upon foil penetration by the deepest pit. Service life can be extended by applying corro-

Fig. 2 Catastrophic failure of a graphite/aluminum MMC after 6 months in a marine atmosphere

sion-resistant coatings. Primary emphasis should be placed on preventing exposure of both the graphite and the aluminum, and the graphite/aluminum composite should be frequently inspected while the component is in service.

Silicon Carbide/Aluminum Composites. Marine corrosion of silicon carbide/aluminum composites is much less severe than that observed on graphite/aluminum MMCs. Discontinuous silicon carbide/aluminum MMCs, however, are susceptible to localized corrosion. Mild-to-moderate pitting has been reported on SiC whisker- and particulate-reinforced composites containing 6061 and 5000 series aluminum matrices exposed for a maximum of 42 months in splash/spray and marine atmospheric environments. The degree of corrosion present on the composites is slightly accelerated compared to that on unreinforced aluminum alloys.

Silicon carbide/aluminum composites immersed in natural seawater are susceptible to significantly more severe corrosion than is typical for silicon carbide/aluminum MMCs in the aforementioned environments (splash/spray and atmosphere). Silicon carbide/aluminum panels in seawater undergo pitting, both localized at the edges and distributed uniformly across the surface. The extent of pitting varies from minimal attack through 33 months of exposure to extensive corrosion that is equivalent to a rate as high as 0.25 mm/yr (9.8 mils/yr).

Corrosion rates for silicon carbide/aluminum MMCs in seawater are also generally higher than is typical for unreinforced aluminum alloys. This was documented in Ref 4 for discontinuous SiC in 6061 and 5000 series aluminum matrices and in Ref 6, which reports that silicon carbide/2024 aluminum corroded approximately 40% faster than 2024 aluminum in sodium chloride (NaCl) solution. Contrary to these findings, in another study, little difference in weight loss measurements was noted between silicon carbide/6061 aluminum and 6061 aluminum in NaCl (Ref 7).

Discontinuous silicon carbide/aluminum MMCs are believed to corrode at the silicon carbide/aluminum interfaces (Ref 4, 6, 7). Concentration of the corrosion at these interfaces is presumably due to the crevices formed there, which are preferential sites for pitting. Evidence of the pitting concentrated at the silicon carbide/aluminum interfaces in both whisker and particulate composites is shown in Fig. 3.

Electrochemical studies of discontinuous silicon carbide/aluminum MMCs containing 6061 and 5000 series aluminum alloy matrices demonstrated that the presence of the SiC does not increase the susceptibility of the composite to pit initiation (Ref 5, 8). Research on silicon carbide/2024 aluminum did show a more electropositive pitting potential for the composite relative to the 2024 aluminum (Ref 8); however, this difference in pitting potential may be due to the difference in microstructure between the composite matrix and the 2024 aluminum (Ref 1).

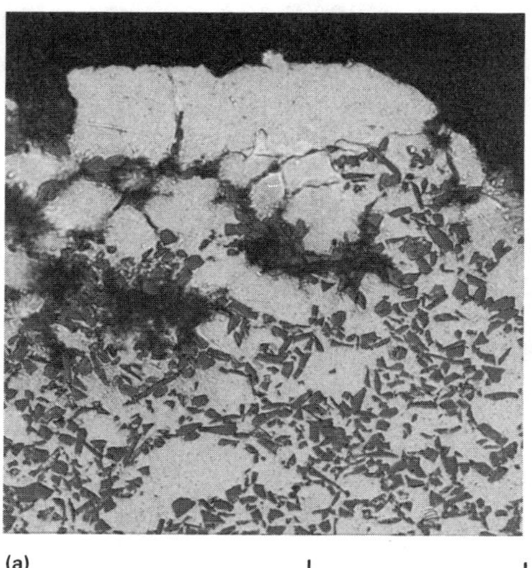

(a)

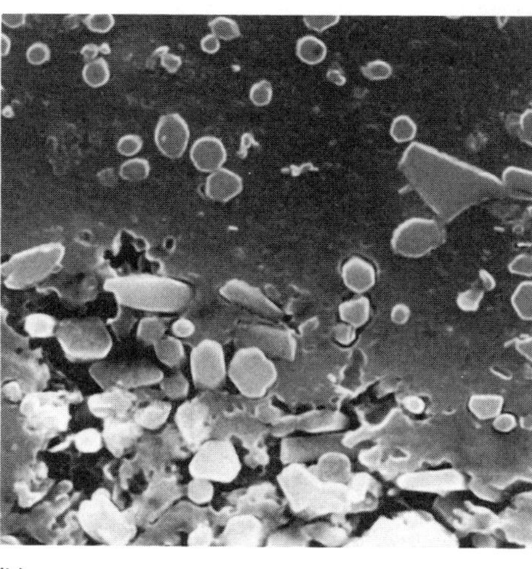

(b)

Fig. 3 Cross sections of discontinuous silicon carbide/aluminum MMC panels. (a) Silicon carbide(p)/6061 aluminum MMC after a 230-day, tidal-immersion exposure. (b) Silicon carbide(w)/6061 aluminum MMC after a 60-day filtered-seawater exposure

Continuous-fiber silicon carbide/aluminum composites also undergo localized corrosion (Ref 4). These composites are susceptible to both crevice corrosion and pitting. Seawater entry into the silicon carbide/aluminum composite matrix will result in crevice corrosion at the fiber/matrix interfaces, which accelerates the corrosion rate and eventually results in delamination of the aluminum surface foils. However, the rate of silicon carbide/aluminum corrosion is much less severe than is typical for graphite/aluminum. Figure 4 contrasts the extent of corrosion evident on the silicon carbide/aluminum panels described above.

In summary, silicon carbide/aluminum MMCs are susceptible to localized corrosion in marine environments. Generally, the susceptibility to pit initiation is similar for composites and unreinforced alloys; however, the rate of pit propagation is higher for composites. Silicon carbide/aluminum corrosion in seawater is increasingly more severe than in the other marine environments. Corrosion-resistant coatings are recommended for these composites to enhance their service lives.

Boron/Aluminum Composites. The corrosion properties of boron/aluminum composites are extensively reviewed in Ref 1. This section will summarize the significant findings.

Boron/aluminum MMCs experience severe corrosion in chloride environments and are significantly less corrosion resistant than unreinforced aluminum alloys. The concentration of corrosion in these composites has been found at fiber/matrix interfaces and at the bonds between foils (Ref 9, 10). The accelerated corrosion at these sites has been attributed to imperfect bonding and fissures in the composite and emphasizes the need for eliminating fabrication flaws to reduce corrosion of boron/aluminum MMCs in chloride environments (Ref 9). Corrosion at the fiber/matrix interfaces has also been attributed to the presence of aluminum boride formed during fabrication (Ref 10).

Aluminum Oxide/Aluminum Composites. The corrosion properties of aluminum oxide/aluminum composites are reviewed in Ref 1. The significant findings are summarized below.

To obtain good wettability and bonding in aluminum oxide/aluminum MMCs, the aluminum matrix is alloyed to form a bonding compound between the fiber and the matrix. Corrosion studies of Al_2O_3/Al-2Li MMCs (containing a Li_2O-$5Al_2O_3$ bond layer) in NaCl solutions indicated no severe attack at the fiber/matrix interfaces. The corrosion rate of the MMCs (based on weight loss measurements) was only slightly higher than for aluminum alloy 6061-T6 (Ref 11).

Corrosion evaluations of Al_2O_3/Al-2Mg MMCs identified pitting at the fiber/matrix interfaces, presumably due to the Mg_5Al_8 precipitated there during fabrication (Ref 12). Research on aluminum oxide/6061 aluminum MMCs also reported preferential corrosion at the fiber/matrix interface (Ref 1). These findings suggest that the corrosion resistance of aluminum oxide/aluminum composites is highly dependent on the bonding compound formed at the fiber/matrix interface. To date, no severe corrosion problems have been identified with Al_2O_3/Al-Li composites.

The stress-corrosion cracking (SCC) properties of graphite/aluminum MMCs are discussed in Ref 13 and 14. Based on evaluation of a limited number of specimens, an initial stress-dependent corrosion mechanism was reported for graphite/

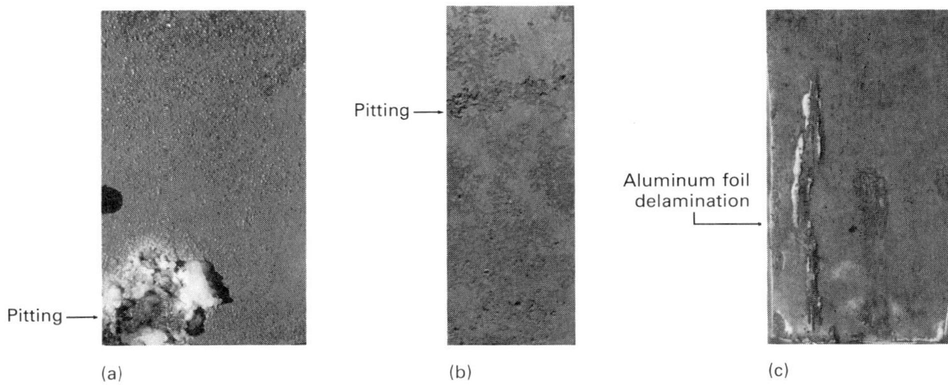

Fig. 4 Silicon carbide/aluminum MMC panels after exposure to filtered seawater. (a) Silicon carbide(w)/6061 aluminum after a 4-month exposure. (b) Silicon carbide(p)/6061 aluminum after a 24-month exposure. (c) Silicon carbide (continuous fiber)/6061 aluminum after a 33-month exposure

aluminum, which then shifted to a corrosion-dominated failure as the exposure in seawater increased (>100 h) (Ref 13). In another study, a corrosion-dominated mechanism was also noted at longer exposure times, but it was suggested that the failures were creep related as well (Ref 14).

Stress-corrosion cracking testing of boron/aluminum MMCs at lower stress intensities (<80% of overload fracture toughness) identified no failures within the 1000-h test limit (Ref 14). Delayed-time failures were reported at high stress intensities for boron/aluminum MMCs evaluated in air and seawater. It was suggested that the failures resulted from room-temperature creep.

Corrosion Fatigue. The seawater and air fatigue properties of graphite/6061 aluminum MMCs are superior to those of aluminum alloy 6061-T6; however, composites and unreinforced aluminum alloys both exhibit a degradation in seawater fatigue properties as compared to the corresponding air fatigue properties (Ref 13). Discontinuous silicon carbide/6061 aluminum MMCs also retain improved fatigue properties over unreinforced aluminum alloy 6061 in chloride environments (Fig. 5) (Ref 15). It has been suggested that the improved corrosion fatigue properties of silicon carbide/aluminum MMCs are due to an increased resistance to crack initiation (Ref 16).

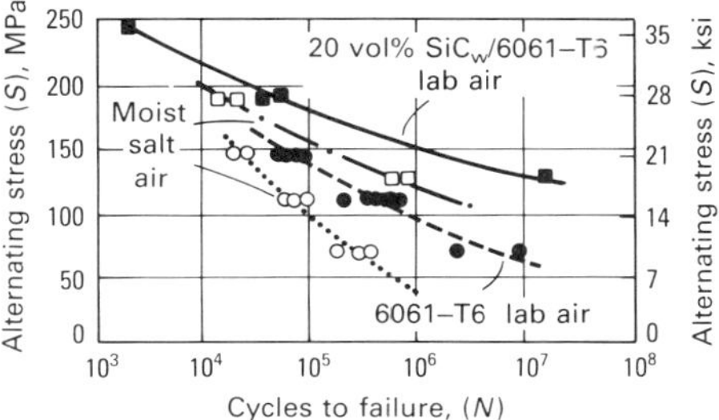

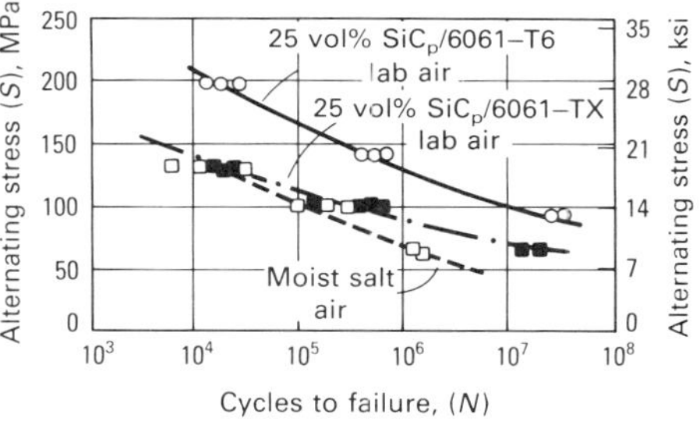

Fig. 5 Corrosion fatigue of discontinuous silicon carbide/aluminum in moist salt air. Source: Ref 15

Corrosion Behavior of Copper-Base MMCs

To date, very limited corrosion research has been conducted on copper-base MMCs. These composites offer substantial potential in marine applications because of their antifouling characteristics as well as their potentially higher strength and stiffness and lower density than that attainable in conventional copper alloys.

A corrosion evaluation of copper-base MMCs was recently completed by the author. The composites evaluated were some of the first copper MMCs ever produced and were not optimized materials. Many of them contained porosity, inconsistent sizes of reinforcement, and an uneven reinforcement distribution. The specific composites evaluated included continuous-fiber graphite/copper, silicon carbide/copper-tin, and silicon carbide/copper-titanium as well as discontinuous titanium carbide(p)/copper, titanium carbide(p)/copper-aluminum, silicon nitride(w)/copper, boron carbide(w)/copper, silicon carbide(w)/copper, and aluminum oxide(chopped fiber)/copper-titanium.

Overall, copper MMCs displayed minimal evidence of corrosive attack. Filtered-seawater specimens developed a green corrosion product film on their surfaces, which presumably was copper hydroxide chloride ($Cu_2(OH)_3Cl$) (Ref 17). Localized corrosion was commonly present on edges, surfaces, and adjacent to crevice sites. Localized corrosion outside the crevice area is typical for copper alloys (Ref 2). The localized corrosion present on the surfaces and edges was possibly due to porosity in the as-fabricated composites, which provide continuing corrosion paths during seawater exposure. Figure 6 shows the typical corrosion found on copper-base MMCs and controls in filtered seawater.

In the marine atmosphere, copper-base MMCs developed a thin, green corrosion product film on their surfaces. This film was much thinner than the film formed in seawater, and it developed more slowly. Graphite/copper MMCs showed evidence of surface foil blistering within 4 months of exposure in this environment; however, this degradation may have been due to imperfect bonding of the composite.

In general, copper-base MMCs display a marine corrosion resistance that is similar to that of copper alloys. Porosity in some of the composites could have contributed to increased corrosion. Further developmental efforts are required to improve the overall structure and the fiber-matrix bonding in these materials.

Corrosion Behavior of Magnesium-Base MMCs

Very limited research has been conducted on graphite/magnesium composites because of the known extreme reactivity of magnesium; graphite/magnesium MMCs are susceptible to extremely severe corrosion in chloride-containing solutions. Research on graphite/magnesium with magnesium alloy AZ91C wires and AZ31B foils identified severe corrosion after only 5 days of immersion (Ref 18). Corrosion was especially severe in areas where both the graphite fibers and the magnesium matrix were exposed; this can be attributed to galvanic-corrosion processes.

As the volume percentage of graphite is increased, the MMC corrosion rate also increases. The corrosion rate of composites containing only 10 vol% graphite (1.8 mg/cm^2 day, or 4×10^{-4} oz/in.^{2} day) is still unacceptable for component applications. Therefore, to use graphite/magnesium MMCs effectively in chloride environments, graphite fiber exposure to the environment must be eliminated and the corrosion resistance of the magnesium matrix improved. Use of low impurity magnesium alloys and a fiber surface treatment have been suggested to improve corrosion resistance (Ref 19).

Coatings for Corrosion Control

The coatings recommended in this section are applicable to aluminum-base MMCs. To date, no research has been published on protective coatings for copper or magnesium-base MMCs. Figures 7 and 8 show photographs illustrating the corrosion evident on select coated continuous- and discontinuous-fiber MMCs.

Continuous-Fiber Composites. Various coatings have been evaluated for protecting continuous-fiber graphite/aluminum (Ref 4, 20). Organic coatings have been identified as providing excellent corrosion protection for graphite/aluminum MMCs. In service, composites protected with an organic coating must be frequently inspected because this coating provides only barrier protection and any coating holidays are potential sites for corrosion attack of the composite.

Noble metal coatings, such as nickel and titanium, applied by CVD, physical vapor deposition (PVD), and electroplating methods also provide barrier protection; however, graphite/aluminum corrosion at coating flaw sites is much worse for these coatings than for an organic coating. This is due to the highly unfavorable anodic (aluminum): cathodic (noble coating) area ratio formed at a flaw site. Consequently, noble metal coatings are not recommended for protecting aluminum-base MMCs.

Sulfuric acid (H_2SO_4) anodizing offers good marine corrosion protection for graphite/aluminum MMCs. Earlier studies in filtered seawater have shown that the anodized layer thins as a function of exposure time (Ref 21). However,

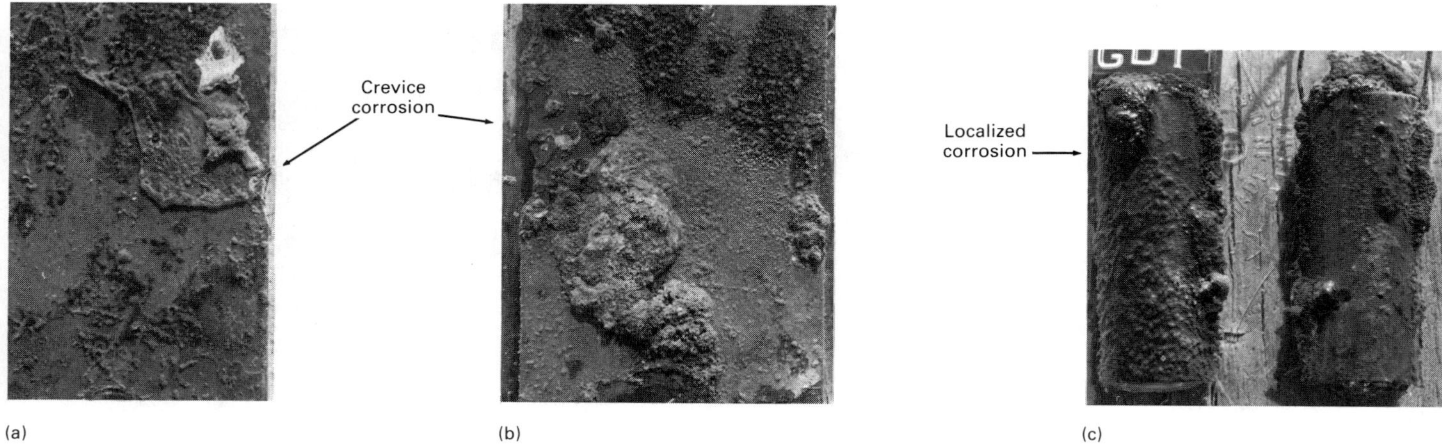

Fig. 6 Copper-base MMC and control specimens after exposure to filtered seawater. (a) Unreinforced 99.95% Cu after a 24-month exposure. (b) Titanium carbide(p)/copper MMC after a 21-month exposure. (c) Silicon nitride(w)/copper MMC after a 21-month exposure

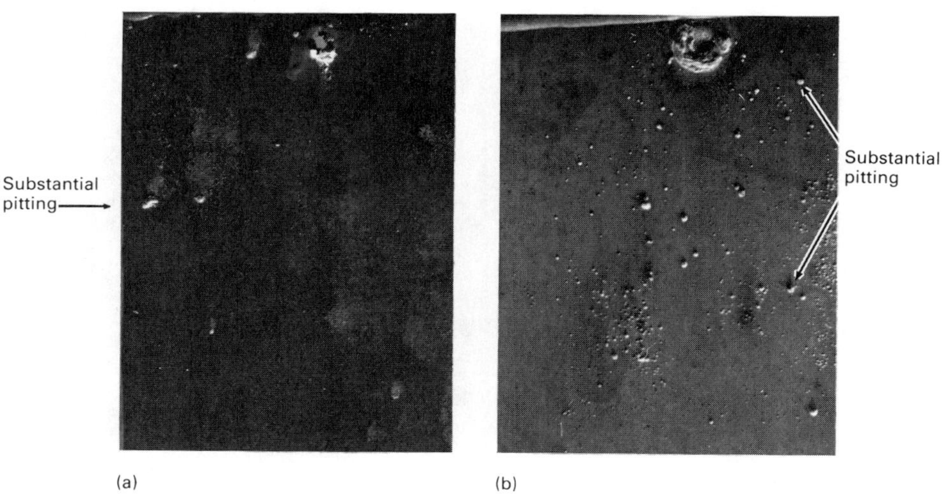

Fig. 7 Coated continuous-fiber graphite/aluminum MMCs after 1 month in filtered seawater. (a) Chromate/phosphate conversion coating. (b) Electrodeposited aluminum/manganese coating

current results indicate that anodizing has provided at least 28 months of protection for graphite/aluminum MMCs. Chromate/phosphate conversion and electrodeposited aluminum/manganese coatings on graphite/aluminum are less corrosion resistant than H_2SO_4 anodizing. Graphite/aluminum MMCs with these coatings are susceptible to substantial pitting and/or blistering within the first 3 months of marine exposure.

Aluminum arc spray coatings exhibit excellent corrosion resistance in marine environments, but are not recommended for use on conventional graphite/aluminum MMCs. The reduced thickness of the surface foils on these composites (0.13 to 0.18 mm, or 5 to 7 mils) prohibits grit blast surface preparation without the occurrence of severe warpage. Thermal spraying can be used for protecting graphite/aluminum only if the surface foils are thick enough (≥0.51 mm, or 20 mils) to prevent warpage.

In summary, sulfuric acid anodizing (0.025 mm, or 1 mil, thick), organic coatings, (0.13 mm, or 5 mils, thick), or preferably a combination of both are recommended for corrosion protection of continuous-fiber graphite/aluminum MMCs. These coatings will provide a minimum of 28 to 33 months of protection in the marine environment, assuming there are no substantial coating defects to expose the graphite/aluminum MMC. These coatings should also provide a similar degree of protection for continuous-fiber silicon carbide/aluminum, boron/aluminum, and aluminum oxide/aluminum MMCs.

Discontinuous-Fiber Composites. Both metallic and ceramic coatings have been investigated for protecting discontinuous silicon carbide/aluminum MMCs (Ref 4). Aluminum flame- and arc-sprayed coatings exhibit excellent corrosion resistance for a minimum of 33 months. Zinc arc-sprayed coatings are less corrosion resistant than aluminum thermal-sprayed coatings. Zinc coatings protect the silicon carbide/aluminum substrate by zinc dissolution. Although zinc coatings can provide more effective cathodic protection than aluminum, their mechanism of protection restricts their useful life. Zinc-coated composites immersed in seawater have exhibited pitting of the silicon carbide/aluminum substrate between 4 and 9 months of exposure, indicating a loss of protection by the zinc coating.

Aluminum oxide plasma-sprayed coatings applied to silicon carbide/aluminum MMCs exhibited varied results in the marine environment. The performance variations were attributable to differences in the quality of coating application rather than to poor corrosion resistance of the coating. Aluminum oxide coating performance ranged from excellent after 33 months of exposure to coating blistering that was extensive enough to expose the substrate after 4 months.

Aluminum oxide coatings will provide only barrier protection to the silicon carbide/aluminum composites. Aluminum thermal-sprayed coatings, on the other hand, do have some sacrificial-protection capabilities. The aluminum coatings will protect exposed silicon carbide/aluminum areas to a greater extent than a barrier coating.

Ion vapor deposited aluminum provides good corrosion protection to silicon carbide/aluminum MMCs in the marine atmosphere. This same coating tends to blister within 1 month of exposure in filtered seawater. The blistering is assumed to occur at silicon carbide particle sites because these particles project from the surface and interfere with a uniform surface for coating.

Aluminum, applied by flame or arc spraying, is recommended for protection of silicon carbide/aluminum MMCs in the marine environment. A coating thickness of 0.13 to 0.20 mm (5 to 8 mils) is optimum. Organic topcoats may also be applied for added protection. Proper thermal-spray and organic-coating surface preparation and application procedures must be followed to avoid premature failure in service.

Design for Corrosion Prevention

For long-term use of MMC components in service, effective coating protection must be employed. Consequently, MMC design should take into consideration the ease of initial coating application as well as coating maintenance (Ref 22). A simple component design is optimum for ensuring effective coating application; the more complicated the design, the more difficult it is to obtain an adherent, uniform coating. Areas that are difficult to coat, such as sharp edges and corners, overlaps, rivets, fasteners, and welds, should be eliminated as much as possible during

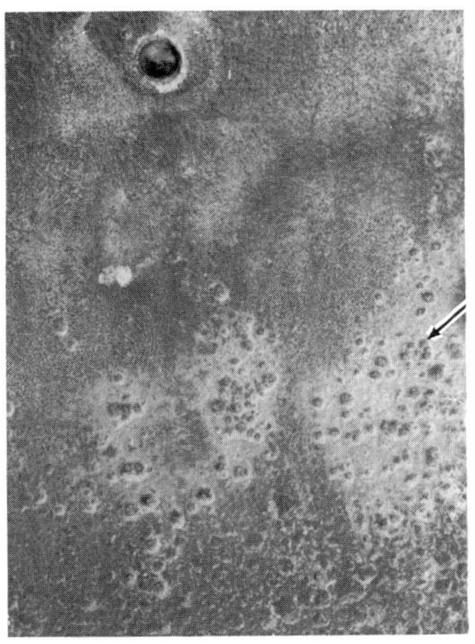

(a)

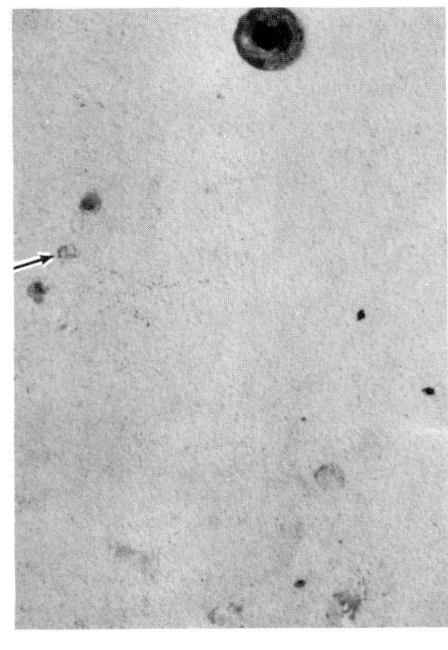

(b)

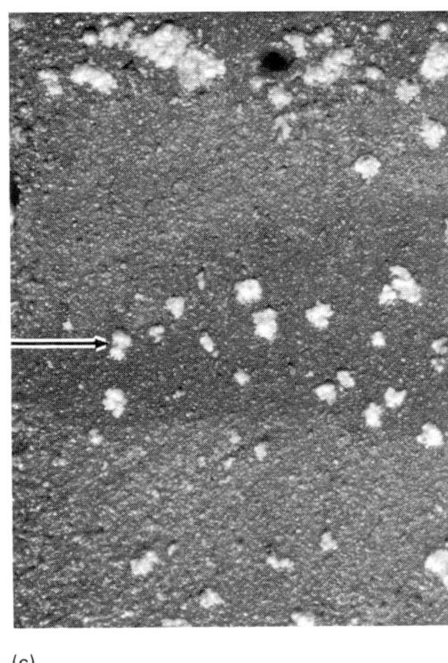

(c)

Fig. 8 Coated discontinuous silicon carbide(p)/aluminum MMCs after seawater exposure. (a) Coated with ion vapor deposited aluminum; 4-month exposure. (b) Coated with plasma-sprayed aluminum oxide; 18-month exposure. (c) Coated with arc-sprayed zinc; 9-month exposure

design. Also, recesses or low spots should be avoided, because these areas will collect water and lessen the corrosion resistance of the coating. For maintenance considerations, it is imperative that all areas to be coated be readily accessible.

REFERENCES

1. M. Metzger and S.G. Fishman, *Ind. Eng. Chem., Prod. Res. Dev.*, Vol 22, 1983, p 296
2. W.K. Boyd and F.W. Fink, *Corrosion of Metals in Marine Environments*, Metals and Ceramics Information Center, 1978, p 44, 57-67, 85-87
3. W.H. Pfeifer, in *Hybrid and Select Metal Matrix Composites: A State of the Art Review*, W.J. Renton, Ed., American Institute of Aeronautics and Astronautics, 1977, p 231-252
4. D.M. Aylor and P.J. Moran, Preprint 202, *Proceedings of the Corrosion/86 Symposium*, National Association of Corrosion Engineers, 1986
5. D.M. Aylor and P.J. Moran, *J. Electrochem. Soc.*, Vol 132, 1985, p 1277
6. H.M. DeJarnette and C.R. Crowe, Naval Surface Weapons Center, unpublished research, 1982
7. K.D. Lore and J.S. Wolf, Abstract 154, The Electrochemical Society, Oct 1981
8. P.P. Trzaskoma, E.M. McCafferty, and C.R. Crowe, *J. Electrochem. Soc.*, Vol 130, 1983, p 1804
9. A.J. Sedriks, J.A.S. Green, and D.L. Novak, *Metall. Trans.*, Vol 2, 1971, p 871
10. S.L. Pohlman, *Corrosion*, Vol 34, 1978, p 156
11. A.R. Champion, W.H. Krueger, H.S. Hartmann, and A.K. Dhingra, in *Proceedings of the Second International Conference on Composite Materials*, B. Noton *et al.*, Ed., The Metallurgical Society of AIME, 1978, p 883
12. J.Y. Yang and M. Metzger, University of Illinois, unpublished research, 1980
13. D.A. Davis, M.G. Vassilaros, and J.P. Gudas, *Mater. Perform.*, Vol 21, 1982, p 38
14. W.L. Phillips, "Sharp Notch SCC of B/Al and Gr/Al Composites," Report 3616, Naval Research Laboratory, Oct 1977
15. C.R. Crowe and D.F. Hasson, in *Proceedings of the 6th International Conference on the Strength of Metals and Alloys*, Vol 2, 1982, p 859
16. S-S Yau, Ph.D. dissertation, North Carolina State University, 1983
17. J.M. Popplewell, Marine Corrosion of Copper Alloys: An Overview, in *Proceedings of the Corrosion/78 Symposium*, National Association of Corrosion Engineers, 1978
18. P.P. Trzaskoma, "The Corrosion Behavior of a Graphite Fiber/Magnesium Metal Matrix Composite in Aqueous Chloride Solution," Report 5640, Naval Research Laboratory, Sept 1985
19. W.A. Ferrando, "Corrosion Resistant Mg-Based Materials for Naval Applications," NSWC TR 85-88, Naval Surface Weapons Center, April 1985
20. M.J. Snyder and J.H. Payer, "The Engineering Development of Graphite Fiber Reinforced Aluminum Composites," Report 74-4312A, Launch Vehicle Materials Technology Program, Battelle Columbus Laboratories, Dec 1976
21. D.M. Aylor and R.M. Kain, *Mater. Perform.*, Vol 23, 1984, p 32
22. C.G. Munger, *Corrosion Prevention by Protective Coatings*, National Association of Corrosion Engineers, 1984, p 173-191

Corrosion of Amorphous Metals

N. Robert Sorensen and Ronald B. Diegle, Sandia National Laboratories

AMORPHOUS, or glassy, metal systems have been extensively studied since their introduction for their unique structural, mechanical, electronic, magnetic, and corrosion properties. Glassy metals have seen practical applications that exploit many of these properties, but they are also interesting as a tool for probing the influence of atomic structure and chemical composition on the corrosion process. They contain none of the classical crystalline or chemical defects found in crystalline solids, such as grain boundaries and second-phase particles, and they are chemically and structurally homogeneous.

Glassy alloys can be produced from two or more transition metals, for example, copper-zirconium and nickel-niobium, or from a combination of metals and metalloids, such as iron, nickel, and chromium containing boron, phosphorus, silicon, or carbon. They are produced by such techniques as liquid quenching, molecular deposition, or external action, and they retain a disordered structure resembling the liquid. As such, they have no long-range order. There is, however, short-range order over a few atomic distances (as found in the liquid state); this has prompted the use of the terms glassy or vitreous to distinguish them from materials that are truly amorphous on the atomic scale.

Very high cooling rates ($>10^5$ K/s) are required to retain the highly metastable glassy state. Although the temperature decrease in quenching from the liquid to the solid is not large, the rate of heat extraction is very high and requires at least one dimension of the resulting alloy to be very thin. Because of this requirement, glassy metals produced by liquid quenching are typically in the form of ribbons, wires, and filaments.

Techniques for producing glassy metals can be divided into three main groups (Ref 1). The first, which is termed quenching, involves rapid solidification from the melt under a set of conditions, such as cooling rate and sample dimensions, that precludes the formation of the stable equilibrium structure.

The second technique, termed atomic or molecular deposition, involves growth from the vapor phase, such as thermal evaporation and sputtering, or from a liquid phase, such as electroless deposition and electrodeposition, to form the desired alloy. These techniques have higher effective cooling rates than liquid quenching processes; therefore, they allow the formation of glassy alloys that cannot be produced by rapid liquid quenching.

The third set of techniques is classified as external action techniques, and they rely on such procedures as solid deformation and irradiation to form the metastable glassy alloy. Ion implantation and ion beam mixing, for example, can produce amorphous surface layers on bulk crystalline substrates. The latter two groups of techniques—molecular deposition and external action techniques—have the advantage of being able to produce considerably thicker alloys, but they typically require considerably more time for completion of the process.

Some glassy metals exhibit extremely good corrosion resistance because of several factors. Glassy metals are free from such defects as the grain boundaries and second-phase particles that are present in crystalline metals. Corrosion often occurs preferentially at such sites; therefore, glassy metals might be expected to exhibit better corrosion resistance than crystalline alloys. The galvanic corrosion associated with chemical inhomogeneities, such as secondphase particles, is also impossible in glassy metals. In addition, the passive films responsible for corrosion resistance in crystalline alloys also play a role in glassy metal corrosion. Thus, the effect of the amorphous structure, chemical homogeneity, and unique chemical composition on the formation and stability of the passive film must also be considered.

Any corrosion reaction involves two or more partial anodic and cathodic half-cell reactions (Fig. 1). Two sets of curves are shown in Fig. 1. The bottom curves represent metal dissolution and plating, and the top curves are for proton reduction and hydrogen oxidation. The reversible potential, $E°$, represents equilibrium between the oxidized and reduced species. For the metal, the reaction is $M \rightarrow M^{n+} + ne^-$. The current density $M \rightarrow M^{n+} + ne^-$ at $E°$ is termed the exchange current density, i_o, and is a characteristic of the metal. The cathodic half cell in acid solutions is typically proton reduction, which is represented by the reaction $2H^+ + 2e^- \rightarrow H_2$. In the case of the cathodic reaction, i_o is a function of the surface on which the proton reduction reaction occurs. The corrosion rate of a metal in deaerated acid is represented by the intersection of the metal dissolution and hydrogen reduction curves. This intersection establishes the corrosion potential, E_{corr}, and the corrosion current, i_{corr}, because the anodic and cathodic partial currents are equal at that point. The net result is that the metal corrodes and hydrogen is evolved simultaneously at the metal surface.

There are several possibilities for explaining the difference in corrosion behavior between amorphous and crystalline metals. A metal-metalloid glassy metal typically contains of the order of 20 at.% metalloids. The exchange current density for the hydrogen reduction reaction may be lower on the metalloid surface than on the metal (Fig. 1b). If it is assumed that the anodic kinetics do not change, the Evans diagram predicts a decrease in i_{corr} and a more active E_{corr}.

A more likely possibility is that the anodic kinetics do in fact change with structure and composition. Because dissolution occurs preferentially at active sites, including kinks, ledges, steps, and grain boundaries, and because there are fewer long-range defects in an amorphous alloy, a lower exchange current density might be expected for metal dissolution in the case of amorphous alloys (Fig. 1c). In this case, the decrease in i_{corr} is accompanied by a shift in E_{corr} in the noble direction.

If amorphous alloys are compared with crystalline alloys with different chemical compositions (~20% metalloids for the glassy metal versus none for the conventional crystalline alloy), a shift in the reversible potential for metal oxidation would be expected. If this shift is in the noble direction and if all other factors remain constant, a decrease in i_{corr} would be expected, but a shift in the active direction would tend to increase the corrosion rate.

The above analysis is for a metal that is dissolving under activation control and at open-circuit potential. With systems containing film formers, the effect of the passive film and the interaction between metalloids and the film formers must be included, and the above analysis may not be appropriate.

Corrosion Behavior: A Historical Review

The first published information on the corrosion behavior of metallic glasses appeared in 1974 (Ref 3), and it concerned the Fe-Cr-P-C alloy system. Figure 2 shows the corrosion rates of $Fe_{70}Cr_{10}P_{13}C_7$ and $Fe_{65}Cr_{10}Ni_5P_{13}C_7$ metallic glasses and a typical AISI type 304 stainless steel in hydrochloric acid (HCl) of various concentrations at 30 °C (85 °F). It also includes data from Ref 4 obtained under similar test conditions. The corrosion rates, calculated from gravimetric measurements, were relatively large for the stainless steel because of pitting attack, but the rates for the metallic glasses were so low that they could not be detected even after immersion for 168 h. This early work illuminated the distinct differences in corrosion behavior between crystalline stainless steel and iron-base metal-metalloid glasses.

Other early research includes work that appeared in 1976 (Ref 5, 6). Figure 3 shows the relative corrosion rates in 1 N sodium chloride (NaCl) of crystalline iron-chromium alloys as compared to those of glassy $Fe\text{-}Cr_xP_{13}C_7$ alloys, where x ranges between 2 and 10 at.%. The corrosion rates of the crystalline alloys were about 0.6 mm/yr (24 mils/yr) and were largely unaffected by chromium content, which is the result of pitting corrosion. Conversely,

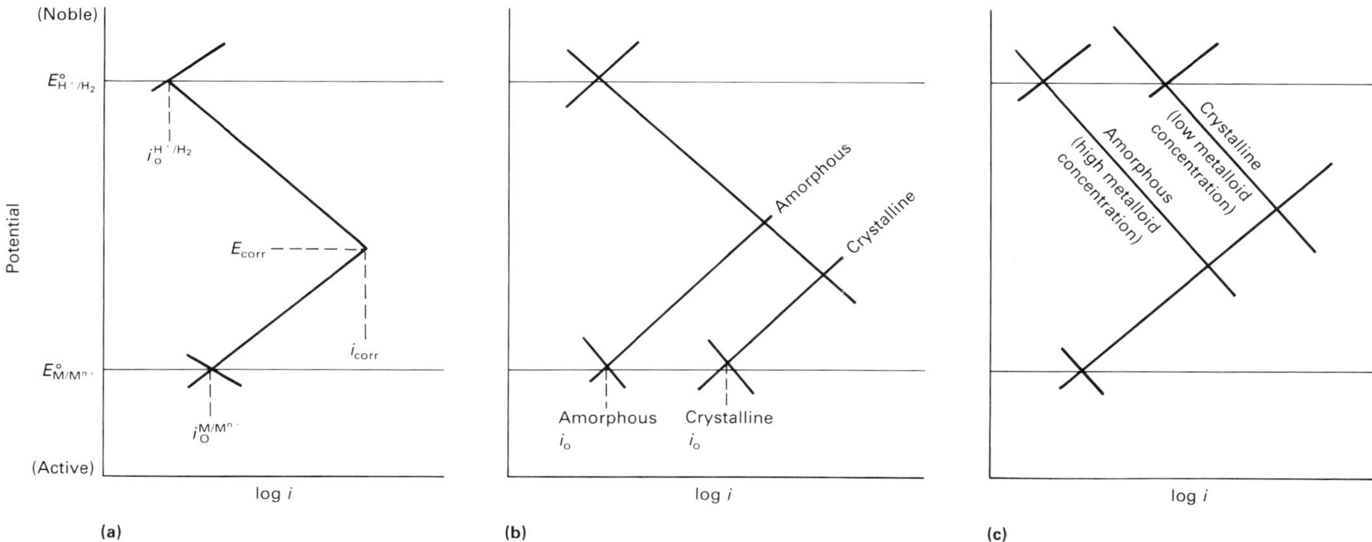

Fig. 1 Schematic Evans diagrams showing the possible influence of alloy structure and composition on the corrosion rate, i_{corr}, and corrosion potential, E_{corr}. See text for discussion. Source: Ref 2

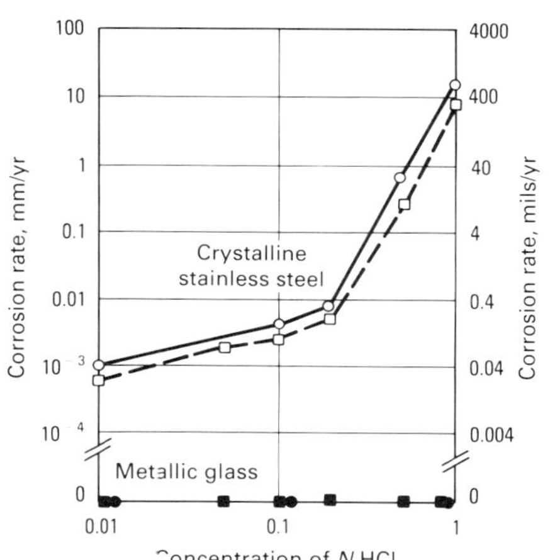

Fig. 2 Comparison of the corrosion rates of metallic glasses and crystalline stainless steel as a function of HCl concentration at 30 °C (85 °F). No weight changes of the metallic glasses of $Fe_{70}Cr_{10}P_{13}C_7$ were detected by a microbalance after immersion for 200 h. Open/closed circles (Ref 3); open/closed squares (Ref 4)

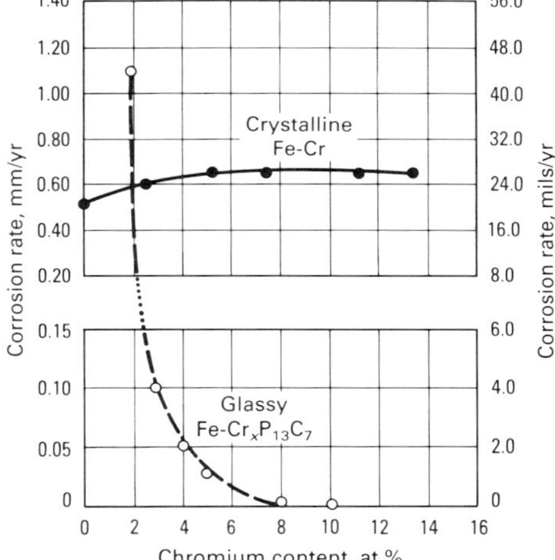

Fig. 3 Comparison of the corrosion rates of glassy $Fe\text{-}Cr_xP_{13}C_7$ alloys and crystalline iron-chromium alloys in 1 *N* NaCl solution at 30 °C (85 °F). Source: Ref 5

the glassy alloys exhibited a sharp decrease in corrosion rate with increasing chromium content, with an undetectable rate occurring above 8 at.% Cr. Pitting did not occur on the glassy alloys, even those with only a few atomic percent of chromium.

In another study, the anodic polarization behavior of glassy $Fe_{25}Ni_{40}Cr_{15}P_{16}B_4$ and $Fe_{40}Ni_{40}P_{16}B_4$ alloys was compared in sulfuric acid (H_2SO_4) with and without NaCl additions. The presence of chromium facilitated passivation over a broad potential range. Thermal crystallization of the metallic glasses caused the corrosion rates during anodic polarization to increase sharply, especially in the presence of chloride ion (Cl^-). It was concluded that crystallization probably decreased corrosion resistance by introducing chemical and structural heterogeneities into the alloys.

Studies such as those summarized above emphasized the excellent resistance to uniform and localized corrosion that could be obtained with certain types of metallic glasses. Results of these studies stimulated additional research into broader compositional ranges. Research during the late 1970s focused primarily on the transition metal-metalloid compositions, although some work was also initiated on metal-metal systems, such as copper-zirconium. Regarding the former compositional class of glassy alloys, research addressed the effects of phosphorus, boron, silicon, and carbon, which are added to stabilize the glassy structure.

These additive elements strongly influence the corrosion behavior of glassy alloys, as shown in Fig. 4 for $Fe_{70}Cr_{10}B_{13}X_7$ and $Fe_{70}Cr_{10}P_{13}X_7$ alloys. Specifically, in acidic solutions, the corrosion rates of the alloy system containing phosphorus as the major metalloid are two orders of magnitude lower than those of the alloy system with boron as the major metalloid. In addition, the corrosion rate of the glassy iron-chromium alloy progressively decreased by the addition of silicon, boron, carbon, and phosphorus in 0.1 *N* H_2SO_4. The addition of chromium without phosphorus to the glassy alloys is relatively ineffective in reducing corrosion rates, as evident from the $Fe_{70}Cr_{10}B_{13}C_7$ and $Fe_{70}Cr_{10}B_{13}Si_7$ alloys. Thus, phosphorus was identified as the single most effective metalloid element among phosphorus, carbon, silicon, and boron for improving the corrosion resistance of glassy iron-base alloys containing chromium. The combination of metalloids that is most effective in providing corrosion resistance in glassy iron-chromium alloys is phosphorus and carbon.

It was also recognized that the corrosion behavior of glassy alloys is strongly influenced by additions of metallic elements, especially those that form films on the alloy surface, that is, film former additions. Figure 3 shows an early example of the strong beneficial effect of chromium additions to an iron-base glassy alloy. The effect of chromium content on the corrosion rates of glassy $Ni\text{-}Cr\text{-}P_{15}B_5$ alloys in 10% ferric chloride ($FeCl_3$) is apparent in Fig. 5, which shows that an undetectably small corrosion rate was attained with 7 at.% Cr. A large variety of other metal additions have also been investigated. For example, such elements as titanium, manganese, niobium, vanadium, tungsten, and molybdenum can benefit the corrosion resistance of $Fe\text{-}Cr_3P_{13}C_7X$ alloys in 1 *N* HCl (Ref 9).

Corrosion research involving metal-metalloid systems soon led to research with metal-metal glasses. One study characterized the corrosion behavior of copper-zirconium and copper-titanium alloys in H_2SO_4, HCl, nitric acid (HNO_3), and sodium hydroxide (NaOH) (Ref 10). In all of the solutions except NaOH, the crystalline and glassy copper-titanium alloys exhibited corrosion rates lower than those of pure copper, and in all cases, the corrosion resistance of the glassy alloy was better than that of the crystalline alloy. The glassy alloys in these compositional systems are not unusually corrosion resistant; in fact, neither the crystalline nor glassy forms of the alloys were more corrosion resistant than pure titanium or

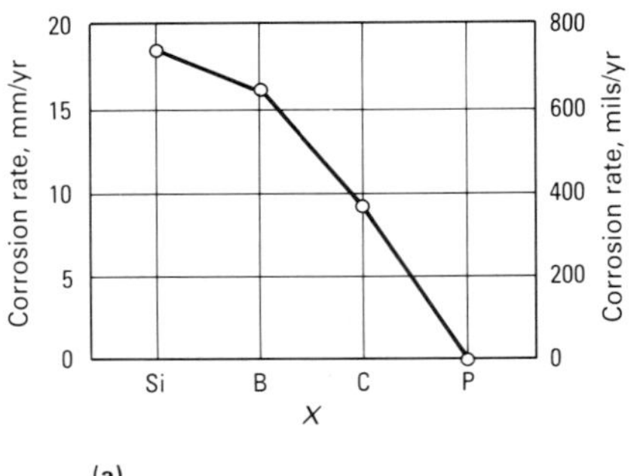

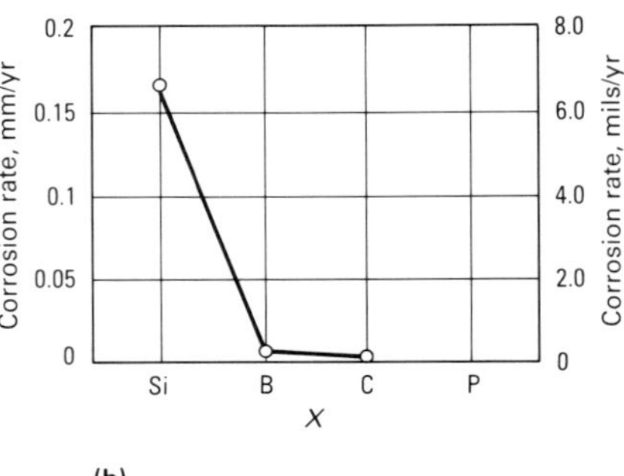

Fig. 4 Average corrosion rates estimated from the weight loss of amorphous $Fe_{70}Cr_{10}B_{13}X_7$ and $Fe_{70}Cr_{10}P_{13}X_7$ alloys in 0.1 *N* H_2SO_4 at 30 °C (85 °F), where (a) *X* is silicon, boron, carbon, and phosphorus, and (b) *X* is silicon, boron, and carbon. Source: Ref 7

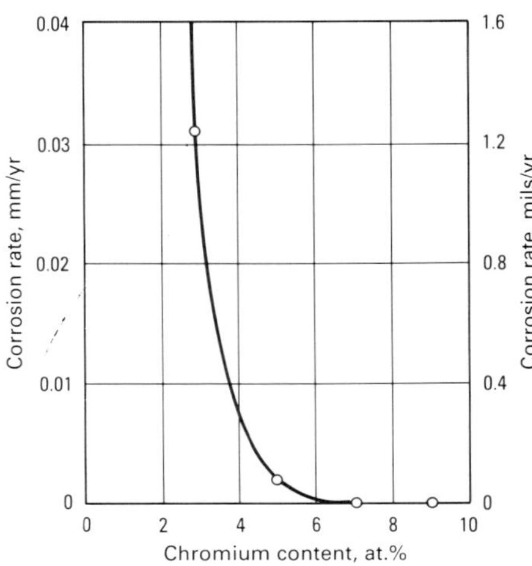

Fig. 5 Effect of the chromium content on the corrosion rates of amorphous Ni-Cr-$P_{15}B_5$ alloys in 10% $FeCl_3 \cdot 6H_2O$ at 30 ± 1 °C (85 ± 2 °F). The corrosion rate was estimated from weight loss during immersion for 168 h. Source: Ref 8

pure zirconium. This fact suggests that the corrosion resistance of the glassy alloys is the result of the presence of the passivating element (titanium or zirconium), not the presence of the glassy state.

In another investigation, several alloys in the Cu-Zr system were examined in H_2SO_4 electrolyte (Ref 11). It was shown that the copper-zirconium alloys remained resistant to corrosion regardless of whether they were devitrified to a single-phase or a multiphase equilibrium microstructure; however, the glassy state was about 20% more corrosion resistant than the devitrified state.

Since the early work with iron-base metal-metalloid glasses, the field of study has been extended to include many alloy systems. Results with nickel-, titanium-, copper-, and cobalt-base alloy systems, among others, have been reported in the literature. The effect of metalloid additions on corrosion behavior is reasonably well characterized, and theories have been proposed to explain the beneficial effect of phosphorus on corrosion (Ref 7). The influence on corrosion behavior of a wide variety of elemental additions has been evaluated, and many such additions increase corrosion resistance; those with the strongest effect are the classical film formers, such as chromium, titanium, and molybdenum. Research on glassy alloy corrosion in the past few years has expanded to include developing means for using these alloys in practical applications. Therefore, research has accelerated in such areas as laser surface remelting, ion implantation, sputtering, electrodeposition, and chemical vapor deposition. See, for example, the articles "Surface Modification" (ion implantation and laser surface processing are discussed) and "CVD/PVD Coatings" in this Volume.

General Corrosion Behavior

Glassy alloys can be grouped into two major categories with intrinsically different corrosion behaviors. The first group includes the transition metal-metal binary alloy systems, such as Cu-Zr, Ni-Ti, W-Si, and Ni-Nb. The second class consists of transition metal-metalloid alloys. These alloys are usually iron-, nickel-, or cobalt-base systems, may contain film formers (such as chromium and titanium), and normally contain approximately 20 at.% P, B, Si, and/or C as the metalloid component.

Transition Metal-Metal Binary Alloys. Research in the transition metal-metal systems indicates that corrosion resistance is primarily determined by the behavior of the more corrosion-resistant component of the alloy. For example, in an investigation of the corrosion behavior of copper-zirconium alloys, the potentiodynamic polarization behavior of the alloys exhibited characteristics of both components, but it was not superior to the more passive material (zirconium) (Ref 2).

This work also examined the effect of alloy structure. By choosing the proper alloy composition ($Cu_{60}Zr_{40}$), the researchers devitrified the glassy alloy to a single-phase $Cu_{10}Zr_7$ and found a slight improvement in corrosion resistance when the material was in the glassy state. They worked with a series of copper-zirconium alloys and found that whether or not the alloy forms a single-phase or multiphase alloy upon devitrification has very little effect on corrosion resistance. This lead to the conclusion that, at least in this alloy system, structure plays a secondary role in establishing corrosion resistance.

Other alloys, including copper-titanium, copper-zirconium, and nickel-titanium, have demonstrated that the corrosion behavior of the transition metal-metal class of glassy metals is determined almost completely by the more corrosion-resistant component. Thus, the corrosion resistance in the transition metal-metal alloy systems is apparently the result of the presence of a passivating element, not the glassy structure.

Transition Metal-Metalloid Alloys. The second class of glassy alloys consists of transition metal-metalloid elements. This family includes iron-, nickel-, and cobalt-base alloys containing combinations of phosphorus, boron, carbon, and silicon as the metalloid constituents. These alloys can be formed as binary systems, such as Ni-P and Fe-B, or they may be considerably more complex, such as Fe-Ni-P-B quaternary systems. In addition to the base metal, they often contain appreciable concentrations of film formers to promote passivity—for example, the Fe-Ni-Cr-P-B system. They derive their corrosion resistance from the same type of processes as crystalline alloys, namely the development of a passive film. The significant difference between corrosion-resistant glassy alloys and their crystalline counterparts, such as stainless steels, is that the level of chromium necessary to promote passivity can be considerably less in the glassy alloys.

Figure 3 shows a comparison between the corrosion rates of crystalline iron-chromium alloys and amorphous iron-chromium-phosphorus-carbon alloys as a function of chromium concentration. At low chromium levels, the amorphous alloy corrodes at a higher rate than the crystalline material. However, at slightly higher chromium levels (4 at.%), there is a significant decrease in the corrosion rate of the glassy alloy, but the crystalline material is essentially unchanged. At an intermediate level of 8 at.% Cr, no corrosion of the glassy alloy was detected by weight loss experiments after immersion for 168 h. It was also found that the concentration of HCl, which has a profound effect on corrosion behavior of crystalline alloys, had no effect on corrosion of the glassy Fe-Cr-P-C or Fe-Ni-Cr-P-C alloy systems, which exhibited no weight loss after exposure for 168 h (Ref 5).

Figure 6 shows the effect of chromium concentration on the corrosion behavior of iron-, nickel-, and cobalt-base alloys. In all cases, the corrosion rate decreases with increasing chromium concentration and becomes vanishingly small at some level of chromium. In addition, the Fe-Cr-P-C alloy system, which exhibits the highest corrosion rate at low chromium contents, exhibits no weight loss in immersion tests with a chromium concentration of as little as 8 at.%. This behavior supports the theory that, when an alloy contains a strong film former, the higher the initial reactivity of the alloy, the more rapidly the film former can be accumulated at the interface and the more rapid the rate of passivation (Ref 9).

The corrosion behavior of nickel-base (Ref 8) and cobalt-base (Ref 13) glassy alloys is very similar to that of the iron-base systems, and it is also a strong function of chromium concentration. Figure 7 shows the corrosion rate for a nickel-base glassy alloy as a function of chromium concentration. These data, which are very similar to the data for the iron-base system, show that at a concentration of about 7 at.% Cr the alloy is extremely resistant to corrosion in a $FeCl_3$ solution.

In one investigation, ion implantation was used to make glassy iron-chromium-phosphorus alloys in which the chromium level varies from 6 to 18 at.% (Ref 14). An interaction between chromium and phosphorus was observed, which suggests a mechanism for the passivation of these amor-

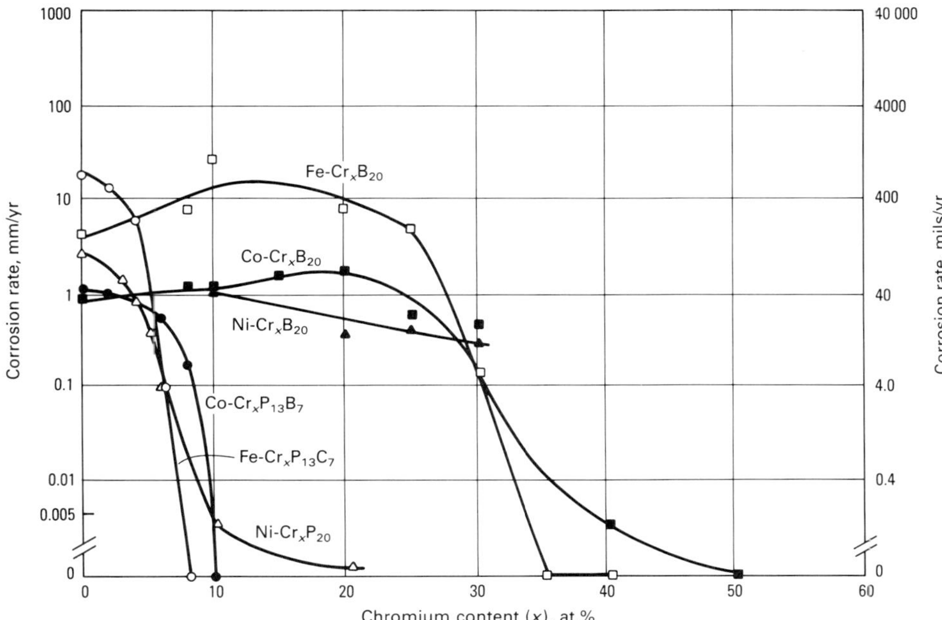

Fig. 6 The influence of chromium content on the corrosion rates of iron-, cobalt-, and nickel-base alloys in 1 *N* HCl. Source: Ref 12

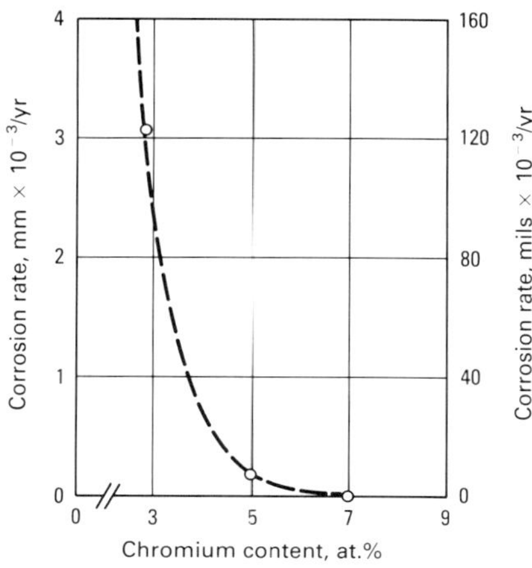

Fig. 7 Plot of corrosion rates of metallic glasses of Ni-Cr-$P_{15}B_5$ in 10 wt% $FeCl_3 \cdot 6H_2O$ at 30 °C (85 °F) versus chromium content

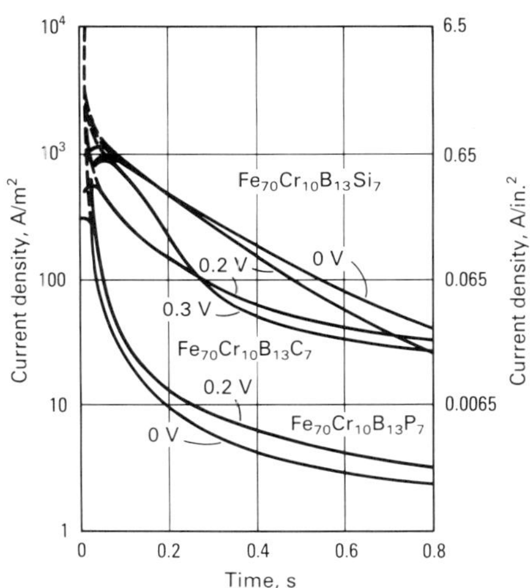

Fig. 8 Current density transients for glassy $Fe_{70}Cr_{10}B_{13}X_7$ alloys following mechanical abrasion of specimen surfaces during anodic polarization at constant potentials in 0.1 *N* H_2SO_4. *X* denotes minor metalloid content, and potentials (SCE) are indicated in the figure. Source: Ref 12

phous alloys. Specifically, at low chromium levels, phosphorus implantation degrades passivity and induces pitting. At high chromium concentrations, there is a slight improvement in passivation, although the crystalline and amorphous alloys are both spontaneously passive and exhibit current densities of the order of 1 uA/cm^2. However, at intermediate chromium levels (8 to 10 at.%), there is a profound benefit from the phosphorus implantation. At these intermediate levels, there is a decrease in current density relative to the unimplanted alloy, namely four orders of magnitude for the Fe-Cr_{10}P alloy. In fact, the Fe-Cr_{10}P alloy exhibits current decay characteristics similar to an Fe-18Cr crystalline alloy.

This research indicates that there is a critical chromium concentration required to provide passivity and that below this level the combination of phosphorus and the amorphous structure increases the initial dissolution rate. Also, below their critical chromium level, there is insufficient chromium for passivation; therefore, pitting is observed. Above this critical chromium concentration, phosphorus and the glassy structure increase the initial dissolution rate, cause a rapid accumulation of chromium in the passive film, and result in an increased rate of passivation.

One study examined the influence of alloying elements on the corrosion behavior of iron-chromium base alloys (Ref 12). Figure 8 shows current decay transients for glassy $Fe_{70}Cr_{10}B_{13}X_7$ alloys (*X* is silicon, carbon, or phosphorus) that were potentiostated in the passive region and abraded under potentiostatic control to produce the repassivation transients. The glassy alloy containing phosphorus exhibited the highest initial current, the most rapid repassivation kinetics, and the lowest passive current density.

Thus, the most effective elements in providing corrosion resistance are chromium and phosphorus. Published concepts concerning the interactions between chromium and phosphorus appear to be consistent with existing data, but the relative effects of structure and composition on corrosion behavior remain to be quantified.

Localized Corrosion Behavior

One of the most outstanding characteristics regarding the corrosion behavior of certain metallic glasses is their ability to resist localized corrosion. In this article, the term localized corrosion refers to pitting and crevice attack (stress-assisted forms of corrosion, such as stress-corrosion cracking and hydrogen embrittlement, are discussed in the section "Environmental Cracking Behavior" in this article). In many cases, this resistance to localized attack extends over wide ranges of oxidizing potential and pH and to alloy compositions that would be considered lean in film former elements compared to conventional crystalline stainless steels.

Effect of Chromium on Pitting. The work summarized in Fig. 3 and described previously in this article indicates that iron-base glasses with only several atomic percent of chromium very effectively resist pitting in chloride-containing solutions. Polarization curves of glassy alloys obtained in 1 *N* NaCl do not show a characteristic pitting potential; rather, they exhibit stable passivity until the onset of transpassivity. In addition, results from the study discussed in Ref 6 with $Fe_{25}Ni_{40}Cr_{15}P_{16}B_4$ showed that the passive range in 1 *N* H_2SO_4 plus 0.1 *N* NaCl is not interrupted by pitting but extends to transpassivity.

In another study, increasing the chromium content from 0 to 16 at.% in a series of Fe-Ni-Cr-P-B alloy systems facilitated passivation in acidified 1 *N* NaCl, but pitting was not observed on any alloy polarized below the transpassive potential region (Ref 15). Polarization at transpassive potentials caused numerous pits to form that penetrated the filament and were noncrystallographic in shape.

Chromium was shown to be very effective in conferring pitting resistance, such as for the glassy alloys Fe-$Cr_xB_{13}C_7$ and Fe-$Cr_xB_{13}Si_7$ in 3% NaCl (Ref 16). With chromium levels of 2 and 5 at.%, both alloy types pitted at potentials slightly anodic to the free corrosion potential of about −0.6 V (saturated calomel electrode, SCE). The addition of 8 at.% Cr extended the pitting resistance to about 1 V(SCE), which is an extremely aggressive condition for alloys containing such a low level of chromium. By contrast, type 304 stainless steel contains about 18 wt.% Cr, yet its pitting potential is several hundred millivolts less positive than that of these glassy alloys.

About 7 at.% Cr was sufficient to prevent pitting of Ni-Cr-P-B alloy systems in 10% $FeCl_3 \cdot H_2O$ at 30 °C (85 °F) (Ref 17). Glassy $Fe_{73}Cr_7P_{15}B_5$ passivated spontaneously in 1 *N* HCl. Surface analysis by x-ray photoelectron spectroscopy showed that chromium and phosphorus were enriched and that nickel depleted in the alloy substrate beneath the passive film. Figure 9 compares the effect on corrosion rate of adding chromium and titanium to Ni-X-P_{20} glasses. Chromium is more effective than titanium in conferring corrosion resistance, and the chromium-containing alloys exhibited a stronger tendency for spontaneous passivation. The corrosion rate decreased approximately logarithmically with increasing chromium or titanium up to about 10 and 7 at.%, respectively.

Effect of Molybdenum on Pitting. Molybdenum benefits the pitting resistance of glassy alloys and crystalline steels. The addition of molybdenum to glassy Fe-$Mo_xP_{13}C_7$ alloys suppressed pitting and decreased the critical current density for passivation and the passive current density (Ref 19). As little as 4 at.% Mo prevented pitting in 1 N HCl, and small additions of molybdenum were more effective than chromium in decreasing corrosion rates. Molybdenum has been shown to facilitate the formation of a passive hydrated chromium or iron oxy-hydroxide film through its enrichment in the corrosion product layer during active dissolution (Ref 20). The enrichment assists the accumulation of the passivating species in the film by lowering the dissolution rate of the species; the molybdenum-rich product subsequently dissolves and thus leaves little molybdenum behind in the film.

Effect of Other Alloying Elements on Pitting. Titanium, tantalum, molybdenum, and tungsten were incorporated by high-rate sputter deposition into alloys of the general composition T_1-T_2, where T_1 = titanium, tantalum, molybdenum, or tungsten and T_2 = rhenium, iron, cobalt, nickel, or copper (Ref 21). Tungsten-iron and titanium-copper resisted pitting corrosion up to 2.5 V(SCE) in chloride solutions of pH 1 and 7. Addition of tungsten to Fe-$W_xP_{13}C_7$ increased the critical pitting potential, E_{crit}, to above 2 V(SCE) at x = 6 at.%, but x = 10 at.% caused transpassive dissolution at 1 V(SCE). Addition of tungsten to commercial type 304 stainless steel by sputtering stabilized the glassy structure and increased E_{crit} in chloride electrolyte (Ref 22).

One study investigated the effects of the alloying additions titanium, zirconium, vanadium, niobium, chromium, molybdenum, tungsten, manganese, cobalt, nickel, copper, ruthenium, rhodium, palladium, and platinum in the glassy alloy Fe-X-$P_{13}C_7$ (Ref 23). All elements except manganese decreased the corrosion rate in H_2SO_4, HCl, HNO_3, and NaCl solutions. Although the base alloy, Fe-$P_{13}C_7$, did not passivate, additions of any of the preceding elements at levels from 0.5 to several atomic percent enabled passivation to occur during anodic polarization in 0.1 N H_2SO_4. Chromium was most effective, and molybdenum and titanium also were very beneficial. Pitting was not observed in 3% NaCl for those alloys that passivated. The alloys that did not passivate, such as Fe-Co-$P_{13}C_7$, also did not pit, but they dissolved uniformly.

Passivation. It has been proposed that the excellent resistance of certain glassy alloys to uniform and localized corrosion results from their enhanced chemical reactivity relative to conventional stainless alloys (Ref 24). Transient repassivation experiments with glassy $Fe_{70}Cr_{10}P_{13}C_7$ and crystalline type 304 stainless steel in acidified chloride electrolyte showed a higher initial reactivity on the glassy alloy after abrasion and a more rapid rate of repassivation. These experiments demonstrated that there is a synergistic effect between chromium and phosphorus in transition metal-metalloid glasses such that maximum

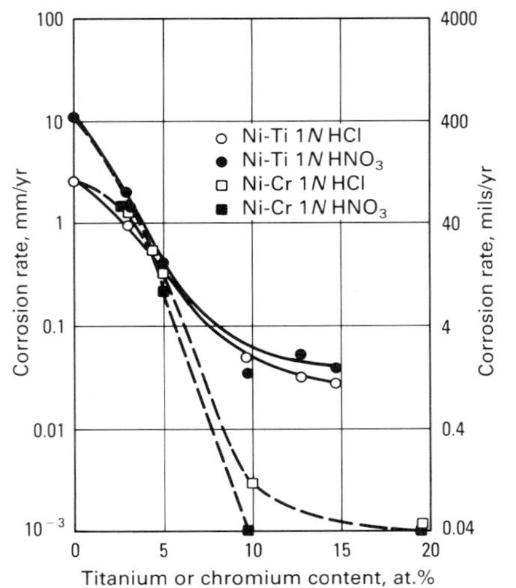

Fig. 9 Changes in corrosion rates of glassy Ni-Ti-P_{20} and Ni-Cr-P_{20} alloys measured in 1 N HCl and 1 N HNO_3 at 30 ± 1 °C (85 ± 2 °F) as a function of the titanium or chromium content. Source: Ref 18

resistance to localized corrosion exists when these two elements are both present. The excellent resistance to localized corrosion may result from the rapid re-formation of a passive film at regions where it is damaged by mechanical or electrochemical means, combined with enrichment of Cr^{3+} species in the film.

Research cited earlier in this article concerning the use of ion implantation to make iron-chromium-phosphorus alloys (Ref 14) demonstrated that the synergistic effect of chromium and phosphorus is a strong function of the chromium concentration. Potentiostatic polarization experiments in acidic chloride solutions showed behavior that was a strong function of chromium concentration. At low chromium levels, phosphorus implantation induced pitting, but at high chromium levels (18 at.%), a slight improvement in passivation was observed. At intermediate levels (10 to 12 at.%), substantial improvement in the passive film was obtained through phosphorus implantation. It was proposed that when the alloy contains small amounts of chromium, there is not enough chromium to passivate the alloy, but when the phosphorus stimulates the initial dissolution, the alloy becomes susceptible to pitting. As the chromium concentration increases, there is sufficient chromium for passivation, and the phosphorus promotes the accumulation of chromium and a very protective passive film.

Crevice Corrosion Resistance. One investigation examined the resistance of glassy alloys to crevice corrosion in acidic solutions containing Cl^- ion. Crevice corrosion was studied as a means of circumventing the need for initiating localized corrosion; that is, crevice corrosion behavior was considered to represent more a measure of the resistance to propagation, rather than initiation, of localized corrosion (Ref 25). Introducing microcracks into Fe-Ni-Cr-P-B glassy alloy filaments by cold rolling (Ref 26) was found to cause susceptibility to a transient form of crevice attack; however, the crevices widened into pit-shaped cavities and then passivated spontaneously.

Subsequent research with an electrochemical cell consisting of a prepared crevice instrumented with microreference and pH electrodes showed that the glassy alloy possessed a strong tendency to passivate, even under the aggressive conditions of low pH, low dissolved oxygen concentration, and oxidizing potential that prevail within crevices (Ref 27). The conclusion was that the alloy resisted crevice attack because of its strong ability to passivate, which in turn stifled propagation. This resistance to crevice corrosion could be expected to extend to other glassy transition metal-metalloid compositions containing both a film former and phosphorus.

Glassy nickel-phosphorus is another alloy system that has been recently investigated and that appears to resist chloride-induced corrosion (Ref 27). In fact, the potentiodynamic polarization curves are virtually identical in both chloride-containing and chloride-free electrolytes. A form of chemical passivity has been proposed to explain the corrosion behavior. Passivation in this system is due to the formation of an ionic barrier layer, not to the formation of a classical passive oxide film. This barrier layer consists of hypophosphite ion adsorbed on the nickel-phosphorus surface, which may in turn be hydrogen bonded to an outer layer of water molecules. This barrier layer inhibits the transport of water to the surface and thus prevents hydration of nickel, which is the first step in the nickel dissolution process.

Environmental Cracking Behavior

The environmentally induced fracture of glassy alloys, namely hydrogen embrittlement and stress corrosion cracking (SCC), will be discussed in this section. Details on the mechanisms of these phenomena can be found in the article "Environmentally Induced Cracking" in this Volume.

Stress-Corrosion Cracking. One of the first reported experiments on the SCC of glassy alloys concerned $Ni_{49}Fe_{29}P_{14}B_6Al_2$ (Ref 28). Loading to 75% of the fracture stress in air in 3.5 N NaCl solution resulted in a slow, presumably SCC fracture region and a final, fast fracture region. However, another researcher suggested that the fracture of this alloy was actually induced by hydrogen (Ref 29).

In another case, the SCC behavior of a glassy $Fe_{32}Ni_{36}Cr_{14}P_{12}B_6$ alloy in boiling magnesium chloride ($MgCl_2$) at 125 °C (255 °F) was studied by means of constant extension rate tensile tests and constant strain tests (Ref 30). Stress-corrosion cracking occurred at the corrosion potential and anodic overpotentials, and slight cathodic polarization prevented SCC. Examination of the fracture surfaces led to the conclusion that localized corrosion enhances hydrogen entry and subsequent embrittlement.

The SCC behavior of glassy Fe-Cr-Ni-P-C alloy systems in acidic chloride solutions was investigated with constant extension rate tensile tests (Ref 31). Hydrogen embrittlement occurred at cathodic polarizations up to −300 mV relative to the corrosion potential. In the passive potential region, no cracking occurred in neutral NaCl solutions and in acidic solutions at low Cl^- concentrations. Fracture stress decreased only when the specimens were strained in strong acidic solutions containing Cl^-, and this phenomenon was also attributed to hydrogen embrittlement.

The tendency of glassy $Fe_{40}Ni_{40}P_{14}B_6$ to undergo SCC and hydrogen embrittlement in acidic electrolytes was also studied (Ref 32). Cathodic polarization of elastically stressed specimens in 1 M HCl resulted in failure by hydrogen embrittlement. Specimens immersed in aqueous $FeCl_3$ solution at the free corrosion potential failed by SCC, as did those that were anodically polarized in 1 M HCl. These specimens were covered by an iron oxide film, and selective leaching (dealloying) of nickel from pits and cracks occurred.

Hydrogen Embrittlement. Although classical SCC (defined as cracking caused directly by anodic dissolution at the crack tip) of glassy alloys apparently occurs, hydrogen embrittlement is a more common mode of environmentally assisted failure in aqueous electrolytes. Hydrogen embrittlement of glassy alloys has been observed during bending or tensile tests during or after hydrogen charging in the following alloys:

- $Fe_{80}P_{13}C_7$ and $Fe_{70}Cr_{10}P_{13}C_7$ (Ref 33)
- $Fe_{32}Ni_{36}Cr_{14}P_{12}C_6$ (Ref 34)
- $Fe_{49.5}Cr_{7.5}Ni_{23}P_{13}C_7$ and $Fe_{53}Cr_7Ni_{20}P_{14}C_6$ (Ref 35)

As shown in Fig. 10, the stress-strain curve of an $Fe_{49.5}Cr_{7.5}Ni_{23}P_{13}C_7$ glassy alloy exhibits almost completely elastic behavior in air and in various acid chloride solutions. The fracture strain decreased to about 30% of that in air as a result of charging the specimen with hydrogen. It was proposed that local corrosion at the open-circuit and even passive potentials can produce hydrogen and thus create embrittlement; fractographic evidence and the return of ductility by baking after corrosion were cited as evidence for this claim (Ref 29).

Although it appears that hydrogen embrittlement is more common than SCC as the environmentally assisted failure mode for glassy alloys, the detailed mechanism of the hydrogen embrittlement of transition metal-metalloid alloys is uncertain. At cathodic potentials in deaerated solutions, the principal cathodic reaction produces hydrogen by the following reaction sequence: $2H^+ + 2e \rightarrow H_2$, which can be separated into a proton reduction step and a hydrogen adatom-adatom combination step. Elements such as phosphorus, sulfur, arsenic, and antimony poison the reaction $H_{ads} + H_{ads} \rightarrow H_2$; thus, they increase the concentration of adsorbed (ads) hydrogen on the electrode surface and consequently the flux of atomic hydrogen absorbed through the surface into the bulk alloy. Because phosphorus is commonly found in transition metal-metalloid glassy alloys, it would seem likely that these phosphorus-containing alloys might have a large tendency toward absorbing hydrogen from the electrolyte and consequently a significant tendency toward hydrogen embrittlement. Augmenting this tendency would be the very high strength and limited ductility characteristic of this compositional class of glassy alloys.

In this regard, an investigation of hydrogen permeation through glassy phosphorus-containing nickel-base alloys concluded that phosphorus increases the rate of hydrogen absorption relative to that for pure nickel (Ref 2). It was proposed that internal voids in the alloys act as traps for the atomic hydrogen and that this atomic hydrogen may subsequently combine to form molecular hydrogen, which ultimately produces internal pressure that can shatter the specimen.

Another researcher characterized the effects of metalloid additions on the susceptibility to hydrogen embrittlement of glassy Fe-$Cr_5Mo_{12}X$ and Fe-$Cr_{10}Mo_{12}X$ (X = 18C, 20B, or 13P-7C) in 1 N HCl, 0.5 N NaCl, and 1 N H_2SO_4 (Ref 29).

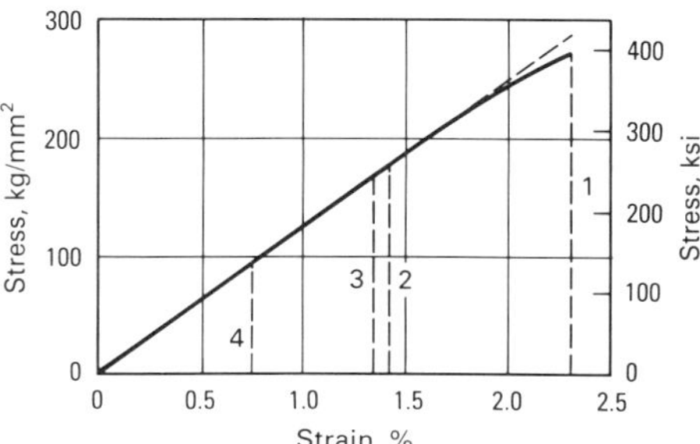

Fig. 10 Stress-strain behavior of a glassy $Fe_{49.5}Cr_{7.5}Ni_{23}P_{13}C_7$ alloy at various potentials in air and in solution at a strain rate of 4.2×10^{-6} s^{-1}. Line 1: in air; line 2: in 5 N H_2SO_4 + 0.1 N NaCl, E_{corr} = −20 mV; line 3: 5 N H_2SO_4 + 0.1 N NaCl, E = +500 mV; line 4: 5 N H_2SO_4 + 0.1 N NaCl, E = −500 mV. Source: Ref 31

Although phosphorus is an effective hydrogen recombination poison, alloys containing this element showed a lower susceptibility to hydrogen embrittlement. (The alloys containing carbon were the most susceptible.) This lower susceptibility was ascribed to the higher rate of repassivation of phosphorus-containing alloys; because the corrosion rate was decreased, the amount of hydrogen produced by the open-circuit corrosion reaction and that absorbed into the alloys should also be lowered.

The specific effects of each metalloid on hydrogen embrittlement susceptibility are still uncertain. However, the large concentrations of metalloids present in transition metal-metalloid alloys almost certainly influence the high susceptibility of these materials to hydrogen embrittlement.

Applications

Although metallic glasses are interesting from a research standpoint, there are several engineering applications in which their unique properties may be important. In considering possible applications of glassy metals, two obvious limitations are apparent. First, all alloys in this class of materials are metastable. If they are subjected to elevated temperatures, devitrification will occur, which normally results in loss of the properties of interest. The glass transition temperature is of course a function of the alloy composition; therefore, some alloys are suitable for use at temperatures substantially above ambient. The second limitation concerns the physical dimensions of the material produced. Because of the high cooling rate required, at least one dimension of the alloy must be very thin. Therefore, the most common forms of glassy metals include filaments, wires, and ribbons.

An obvious application of glassy alloys is that of corrosion-resistant coatings or barriers. In certain applications, a thin, highly corrosion-resistant coating may be sufficient, and these coatings permit the use of less-expensive base materials.

To make commercial use of glassy alloys, advances are required in several areas. Glassy coatings are very appealing in that they can be applied to engineering parts of complex geometry. However, most of the deposition processes used to make glassy alloys are slow, expensive, or both. Techniques are needed that enable the uniform and rapid coating of large areas. In addition, because most currently used deposition techniques tend to be expensive, replacement of corrosion-resistant bulk alloys with glassy metal coatings must await the development of cost-competitive procedures. Additional information on the applications of amorphous materials, such as tool applications, aluminum die-casting mold inserts, metal-bonded abrasive wheels, and hardfacing coatings, can be found in the article "Amorphous Powder Metals" in Volume 7 of the 9th Edition of *Metals Handbook*.

REFERENCES

1. H. Jones, *Proceedings of the Second International Conference on Rapidly Quenched Metals*, N.J. Grant and B.C. Giessen, Ed., Elsevier, 1976
2. R.M. Latanision, J.C. Turn, and C.R. Compeau, in *Proceedings of the Third International Conference on Mechanical Behavior of Metals*, Vol 2, 1979, p 475
3. M. Naka, K. Hashimoto, and T. Masumoto, *J. Jpn. Inst. Met.*, Vol 38, 1974, p 38
4. Y. Waseda and K.T. Aust, *J. Mater. Sci.*, Vol 16, 1981, p 2337
5. M. Naka, K. Hashimoto, and T. Masumoto, *Corrosion*, Vol 32, 1976, p 146
6. R.B. Diegle and J.E. Slater, *Corrosion*, Vol 32, 1976, p 155
7. M. Naka, K. Hashimoto, and T. Masumoto, *J. Non-cryst. Solids*, Vol 28, 1978, p 403
8. K. Hashimoto, M. Kasaya, K. Asami, and T. Masumoto, *Corros. Eng.*, Vol 26, 1977, p 445
9. K. Hashimoto, M. Naka, J. Noguchi, K. Asami, and T. Masumoto, in *Proceedings of the Fourth International Conference on Passivity*, R.P. Frankenthal and J. Kruger, Ed., The Electrochemical Society, 1978, p 156-169
10. M. Naka, K. Hashimoto, and T. Masumoto, *J. Non-cryst. Solids*, Vol 30, 1978, p 29
11. J.C. Turn and R.M. Latanision, *Corrosion*, Vol 39, 1983, p 271
12. K. Hashimoto, "Supplement to the Scientific Report of the Research Institutes of Tohoku University," A 201-216, Tohoku University, 1980
13. M. Naka, K. Hashimoto, K. Asami, and T. Masumoto, *Proceedings of the Third International Conference on Rapidly Quenched Metals*, B. Cantor, Ed., The Metals Society, 1978, p 449
14. N.R. Sorensen, R.B. Diegle, and S.T. Picraux, *J. Mater. Res.*, Vol 1 (No. 6), 1986, p 752
15. R.B. Diegle, *Corrosion*, Vol 35, 1979, p 250
16. M. Naka, K. Hashimoto, and T. Masumoto, "Scientific Report of the Research Institutes of Tohoku University, A-26, Tohoku University, 1979, p 283
17. K. Hashimoto, M. Kasaya, K. Asami, and T. Masumoto, *Boshoku Gijutsu*, Vol 26, 1977, p 445
18. M. Naka, K. Hashimoto, and T. Masumoto, 1703rd Report, Research Institute for Iron, Steel, and Other Metals, 1980, p 156
19. M. Naka, K. Hashimoto, and T. Masumoto, *J. Non-cryst. Solids*, Vol 29, 1978, p 61
20. K. Asami, M. Naka, K. Hashimoto, and T. Masumoto, *J. Electrochem. Soc.*, Vol 127, 1980, p 2130
21. R. Wang, Abstract 80-2, in *Extended Abstracts*, The Electrochemical Society, 1980, p 620
22. R. Wang, Paper presented at the Fall Meeting, Pittsburgh, PA, The Metallurgical Soci-

ety, Oct 1980
23. M. Naka, K. Hashimoto, and T. Masumoto, *J. Non-cryst. Solids*, Vol 31, 1979, p 355
24. K. Hashimoto, M. Naka, and T. Masumoto, "Scientific Report of the Research Institutes of Tohoku University, A-26, Tohoku University, 1976, p 48
25. R.B. Diegle, *Corrosion*, Vol 35, 1979, p 250
26. T.M. Devine, *J. Electrochem. Soc.*, Vol 124, 1977, p 38
27. R.B. Diegle, *Corrosion*, Vol 36, 1980, p 362
28. C.A. Pampillo, *J. Mater. Sci.*, Vol 10, 1975, p 1194
29. A. Kawashima, K. Hashimoto, and T. Masumoto, *Corrosion*, Vol 36, 1980, p 577
30. R.F. Sandenbergh and R.M. Latanision, *Corrosion*, Vol 41, 1985, p 369
31. A. Kawashima, K. Hashimoto, and T. Masumoto, *Corros. Sci.*, Vol 16, 1976, p 935
32. M.D. Archer and R.J. McKim, *J. Mater. Sci.*, Vol 18, 1983, p 1125
33. M. Nagumo and T. Takahashi, *Mater. Sci. Eng.*, Vol 23, 1976, p 257
34. R.K. Viswanadham, J.A.S. Green, and W.G. Montague, *Sci. Metall.*, Vol 10, 1976, p 229
35. A. Kawashima, K. Hashimoto, and T. Masumoto, *Corros. Sci.*, Vol 16, 1976, p 935

Corrosion of Electroplated Hard Chromium

Allen R. Jones, M&T Chemicals, Inc.

HARD CHROMIUM plated parts have more than about 1.2 μm (0.05 mils) of chromium. Parts that are plated with less than this amount are referred to as decorative applications. Electroplated chromium protects substrates by means of a barrier coating as opposed to a sacrificial coating, such as zinc. Chromium is more electrochemically active than steel; however, it forms a dense self-healing oxide layer on its surface. Chromium can be passivated, or the oxide layer can be formed by exposure to air or by immersion in room-temperature oxidizing acids. Electroplated chromium is chemically resistant to most compounds and offers excellent corrosion protection in various environments. It is especially useful in applications that also require wear resistance.

Electrodeposition Parameters

Chromium Plating Baths. Commercially, hard chromium is deposited from four types of baths: conventional, fluoride, and the two high-efficiency etch-free baths. All of the baths contain chromic acid (CrO_3) and sulfate (SO_4^{2-}). The SO_4^{2-} acts as a catalyst. Chromium cannot be electrodeposited from an aqueous CrO_3 solution unless one or more catalysts are present. Depending on which catalysts are present and the plating parameters, between 10 and 45% of the cathodic current will be used to reduce hexavalent chromium (Cr^{6+}) to chromium metal. The properties of the electrodeposits are influenced by the ratio of CrO_3 to the catalysts, plating temperature, and current density.

Stress and Microcracks. The tensile stress in most electroplated chromium deposits increases until microcracks are formed (Ref 1, 2). The microcracks decrease the stress in the deposit as the thickness of the deposit increases. Stress is inversely proportional to the number of microcracks. The number of microcracks is more important in controlling stress than the type of bath chemistry. Crack-free deposits are highly stressed.

Microcracks are present in most electroplated hard chromium deposits. Figure 1 shows a typical microcrack structure. The density of microcracks in chromium deposits varies from 0 to more than 1200 cracks/cm (3000 cracks/in.), depending on bath chemistry, current density, and temperature. The number of microcracks increases with the concentration of catalyst in the plating bath. The depth of a microcrack is less than about 8 μm (0.3 mils) on a deposit that is 130 μm (5 mils) thick with crack counts of about 800 cracks/cm (2000 cracks/in.).

Because chromium protects substrates by forming a barrier, the coatings must be thicker than the microcracks to provide good corrosion resistance. Thin coatings may not form microcracks and can offer as much corrosion resistance as thicker coatings (see the section "Coating Thickness" in this article). Chromium electrodeposits that are about 25 μm (1 mil) thick with crack counts of about 400 cracks/cm (1000 cracks/in.) are as resistant to corrosion as deposits with crack counts of about 100 cracks/cm (250 cracks/in.). Deposits with very low crack counts have deeper microcracks than deposits with higher crack counts. Therefore, highly microcracked deposits are as resistant to corrosion as sparsely microcracked deposits.

Microcracks are not as detrimental to corrosion resistance as might be expected. There are two reasons for this. First, the microcracks are not voids, but are areas with a structure and composition that are different from those of the bulk. Second, because the microcracks are very narrow (about 0.1 μm wide) and because water does not wet chromium, the water does not readily enter the microcracks.

Microcrack-free thick chromium deposits can be plated from baths at low current densities and high temperatures. These microcrack-free deposits provide better corrosion protection than microcracked chromium. However, these deposits are highly stressed and are not as hard as microcracked chromium. Crack-free deposits can be used when corrosion protection is the only requirement for the deposit. Postplating grinding or cutting may cause pickout (chromium fracturing from chromium) in highly stressed deposits. The conditions under which some crack-free coatings are deposited will result in a deposition efficiency that is lower than that normally observed for the plating bath.

Additional Deposit Properties Influencing Corrosion. Hardness is related to microcracking, which is related to corrosion. Chromium coatings have hardnesses between 850 and 1050 HK (100-gf load). Microcrack-free deposits can have hardnesses as low as 600 or 300 HK (Ref 2). According to one study, as deposit hardness increases or crystal size decreases, the rate of attack by sulfuric acid (H_2SO_4), hydrochloric acid (HCl), and CrO_3 decreases (Ref 3).

Table 1 Corrosion of chromium-plated parts in salt spray

Five different carburetor parts were plated with between 1.3 and 1.9 μm (0.05 and 0.075 mils) of chromium; four parts of each type were used in the test.

Condition of parts before plating	Average corrosion rating(a) After 24 h	Average corrosion rating(a) After 150 h
Unfinished	1.6	3.0
Roto-finished(b)	1.3	2.6
Buffed	0.9	2.1

(a) Corrosion ratings: 0, none; 1, very light initial corrosion; 2, light localized corrosion; 3, light general corrosion (b) Parts were prepared by tumbling in a barrel containing granite chips for 1.5 h.

Coating Thickness

Figure 2 shows that the corrosion resistance of hard chromium plated steel in salt spray undergoes a maximum and a minimum, then increases with the chromium thickness (Ref 4). Figure 2 also shows the average of two panels in a salt spray exposure. Maximum corrosion resistance occurred at a chromium thickness of about 5 μm (0.2 mils). As the thickness increased above 5 μm (0.2 mils), microcracking occurred and corrosion resistance decreased. When the chromium thickness increased to about 10 μm (0.4 mils), the initial cracks were covered by more chromium, there were fewer corrosion paths to the substrate, and the corrosion resistance of the deposit increased. These deposits were plated from a conventional bath containing 250 g/L of CrO_3 and 2.5 g/L of H_2SO_4 at 31 A/dm^2 (2 A/in.^{2}); no temperature was specified.

Figure 3 shows additional data on corrosion resistance and chromium thickness. The electrodeposits were prepared from a conventional bath containing 295 g/L of CrO_3 and 3 g/L of H_2SO_4. Data are given for two plating conditions: 30 °C (85 °F) at 20 A/dm^2 (1.3 A/in.^{2}) and 60 °C (140 °F) at 43 A/dm^2 (2.8 A/in.^{2}). The first condition produced cold chromium that was crack free and soft. The second condition produced conventional microcracked hard chromium. The cold chromium deposit showed excellent corrosion resistance at thicknesses of 4.8, 9.1, and 12.4 μm (0.2, 0.36, and 0.49 mils), but the corrosion resistance was very poor at a thickness of 15.5 μm (0.6 mils). The 15.5-μm (0.6-mil) coating was not porous, and no reason was given for its poor corrosion resistance. The high stress in the coating and the poor adhesion of cold chromium may have resulted in coating failure. The thinner ($<$15 μm, or 0.59 mils) cold chromium coatings performed much better than the conventional chromium deposits.

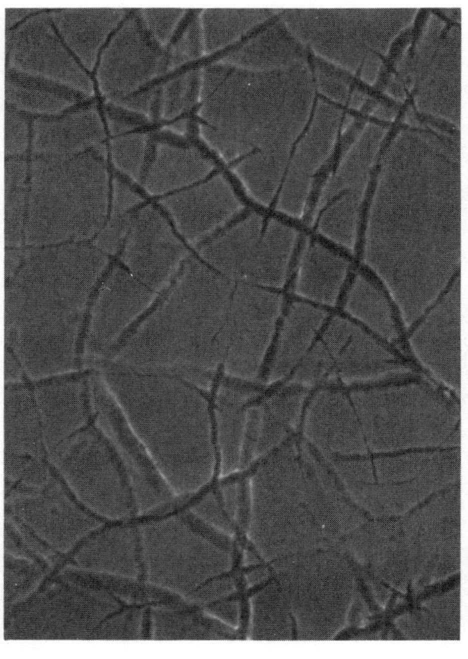

(a)

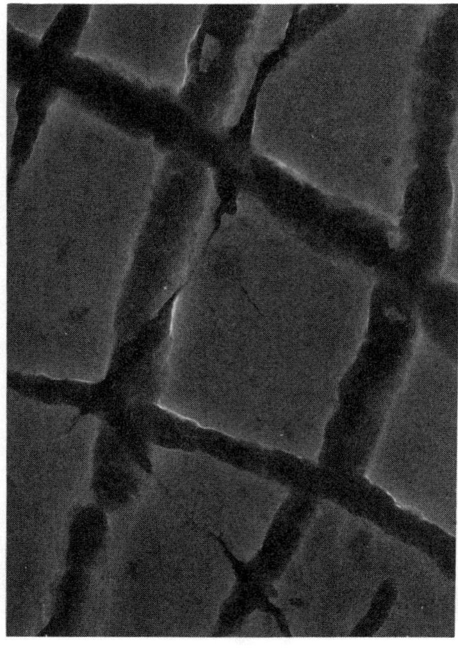

(b)

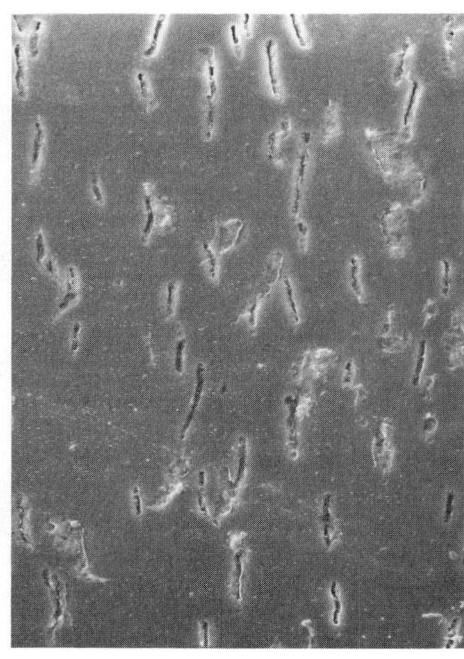

(c)

Fig. 1 Photomicrographs of chromium deposits (plated in a high-efficiency etch-free bath) after etching. (a) and (b) Deposit plated at 78 A/dm^2 (5 A/in.^{2}) and at 55 °C (130 °F). (a) 540×. (b) 2300×. (c) Cross section of a chromium deposit plated at 93 A/dm^2 (6 A/in.^{2}) and at 58 °C (135 °F). The specimen was polished before etching. 880×. Both deposits contain 800 microcracks/cm (2000 microcracks/in.).

The conventional deposits shown in Fig. 2 do not exhibit the trend observed in Fig. 3. The thickest deposit (18.8 μm, or 0.74 mils) did show a slight decrease in corrosion resistance relative to the 14.4 μm (0.57 mil) deposit. Statistically, however, the corrosion resistance was the same for the two coating thicknesses.

Fig. 2 Chromium corrosion in salt spray versus thickness of deposit. Curve A shows time to general rust; curve B is for time to initial corrosion. Parts were plated in a conventional bath (250 g/cL CrO_3 and 2.5 g/L SO_4^{2-} at 31 A/dm^2, or 2 A/in.^{2}). Source: Ref 4

Basis Metal Preparation

The importance of basis metal preparation and its effect on corrosion are discussed in Ref 4. Table 1 shows that corrosion resistance improves as the surface finish of the basis metal improves. The improvement is significant for both 24- and 150-h salt spray exposures. Superfinishing the parts before plating produces a smooth finish after plating that has fewer nodules. Large nodules are plated at a higher localized current density and may lead to the presence of local cells on the surface.

Defects in the basis metal that cause pits or atypical cracking expose the basis metal to corrosive media (Ref 6). Gas pits are usually caused by particles in or on the substrate surface that have a low hydrogen overvoltage. These pits can occur on overetched cast iron parts and on reworked parts that have not been baked after stripping.

The composition of the basis metal will significantly improve the corrosion resistance of thin deposits if the basis metal is corrosion resistant. On thicker deposits, the substrate may affect corrosion by influencing adhesion or the stress of the deposit (microcracks).

Postplating Treatment

Chromium is often plated on hardened steel parts, and these parts will retain some of the hydrogen that is codeposited with the chromium. It is necessary to bake these parts after plating in order to reduce hydrogen embrittlement. Baking

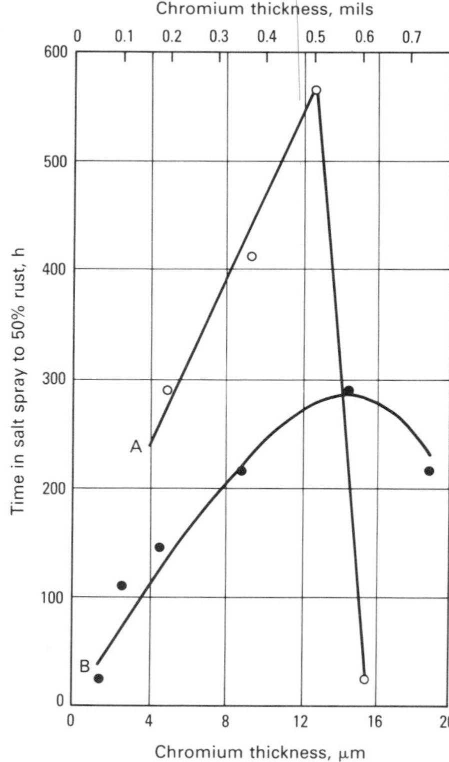

Fig. 3 Chromium corrosion in salt spray versus thickness of deposit. The electrodeposits were plated from a conventional bath (295 g/L CrO_3 and 3.0 g/L SO_4^{2-}) at two conditions: 20 A/dm^2 (1.3 A/in.^{2}) at 30 °C (85 °F) (curve A) and 43 A/dm^2 (2.8 A/in.^{2}) at 60 °C (140 °F) (curve B). Source: Ref 5

Table 2 Corrosion resistance of electroplated chromium deposits immersed in various acids

Acid	Concentration, %	Temperature °C	Temperature °F	Corrosion rate, mm/yr (mils/yr)
Acetic	10	12	55	nil
		58	135	0.38 (15)
	100	12	55	nil
		58	135	0.20 (8)
Anthranilic	Saturated	12	55	nil
		58	135	nil
Anthraquinone 2-sulfonic	10	12	55	nil
		58	135	0.03 (1.2)
Arsenic	10	12	55	nil
		58	135	0.28 (11)
Benzene, sulfonic	10	12	55	nil
		58	135	0.05 (2)
Benzoic	Saturated	12	55	nil
		58	135	nil
Butyric	10	12	55	nil
		58	135	0.15 (6)
Carbolic (phenol)	Saturated	12	55	nil
		58	135	nil
Chloric	10	12	55	0.40 (16)
Chlorine Acid	...	...	...	Attacked
Chromic	...	...	...	Slow attack
Cinnamic	Saturated	12	55	nil
		58	135	nil
Citric	10	12	55	nil
		58	135	0.18 (7)
Dichloroacetic	10	12	55	nil
		58	135	1.57 (62)
Dinitrobenzoic (3,5)	Saturated	12	55	nil
		58	135	0.03 (1.2)
Formic	10	12	55	nil
		58	135	30.48 (1200)
Fumaric	Saturated	12	55	nil
		58	135	nil
Furoic (pyromucic)	10	12	55	nil
		58	135	nil
Gluconic	10	12	55	nil
		58	135	nil
Glycollic	10	12	55	nil
		58	135	0.58 (23)
Hydrobromic	10	12	55	0.03 (1.2)
		58	135	4.72 (186)
Hydrochloric	...	...	...	Rapid attack
Hydrofluoric	10	12	55	25.4 (1000)
Hydroiodic	10	12	55	nil
		58	135	0.38 (15)
Lactic	10	12	55	nil
		58	135	0.15 (6)
Maleic	10	12	55	nil
		58	135	0.46 (18)
Malic	10	12	55	0.05 (2)
		58	135	0.23 (9)
Malonic	10	12	55	0.03 (1.2)
		58	135	0.36 (14)
Mandelic (amygdalic)	Saturated	12	55	nil
		58	135	0.03 (1.2)
Mixed acid 36% HNO_3, 61% H_2SO_4, 3% H_2O	100	12	55	nil
		58	135	0.03 (1.2)
Monochloroacetic	10	12	55	nil
		58	135	0.08 (3)
Mucic	Saturated	12	55	nil
		58	135	nil
Naphthalene 2,7 disulfonic	10	12	55	nil
		58	135	0.03 (1.2)
Naphthionic	Saturated	12	55	nil
		58	135	nil
Nitric	10	12	55	nil
		58	135	0.30 (12)
	100	12	55	nil
		58	135	0.13 (5)
Nitrobenzoic (meta)	Saturated	12	55	nil
		58	135	nil
Nitrocinnamic (meta)	Saturated	12	55	nil
		58	135	nil
Oleic	100	12	55	nil
		58	135	nil
Oxalic	10	12	55	nil
		58	135	0.03 (1.2)
Palmitic	100	12	55	nil
		58	135	nil
Perchloric	10	12	55	0.03 (1.2)
		58	135	1.07 (42)
Phenolsulfonic (ortho)	10	12	55	nil
		58	135	0.66 (26)
Phenylacetic	Saturated	12	55	nil
		58	135	0.03 (1.2)
Phosphoric	10	12	55	0.03 (1.2)
		58	135	0.86 (34)
	85	12	55	nil
		58	135	0.05 (2)
Phosphoric (crude)	28	81	180	Attacked
	60	81	180	Slight attack
Phthalic	Saturated	12	55	nil
		58	135	0.08 (3)
Picric	Saturated	12	55	nil
		58	135	nil
Propionic	10	12	55	nil
		58	135	0.13 (5)
Pyrogallic	10	12	55	nil
		58	135	nil
Pyruvic	10	12	55	nil
		58	135	nil
Salicylic	Saturated	12	55	nil
		58	135	0.05 (2)
Stearic	100	12	55	nil
		58	135	nil
		340	645	Resistant
Succinic	10	12	55	0.03 (1.2)
		58	135	0.25 (10)
Sulfanilic	Saturated	12	55	0.03 (1.2)
		58	135	0.20 (8)
Sulfobenzoic (ortho)	10	12	55	nil
		58	135	3.28 (129)
Sulfuric	10	12	55	0.28 (11)
		58	135	254 (10 000)
	100	12	55	0.76 (30)
		58	135	1.75 (69)
Tannic	10	12	55	nil
		58	135	nil
Tartaric	10	12	55	nil
		58	135	0.10 (4)
Toluene, sulfonic (para)	10	12	55	nil
		58	135	nil
Trichloroacetic	10	12	55	0.03 (1.2)
		58	135	2.62 (103)
Uric	...	...	...	Resistant

Source: Ref 14, 15

will affect corrosion in two ways. First, it will develop (open up) the microcracks. Immersion in boiling water is used to make cracks visible so that they can be counted after the part is reverse etched. Second, if the temperature is high enough, it will reduce the stress in the deposit (Ref 1). This reduced stress will decrease stress-corrosion cracking (SCC). Stress-corrosion cracking occurs at pits or grooves in the coating caused by differences in potentials at the bottom of the groove. The bottom of the groove will be dissolved anodically and will deepen the groove (Ref 7).

In one study, diamond smoothing (compacting) and baking in oil after plating improved the corrosion resistance of electrodeposited chromium (Ref 8). Diamond smoothing at forces of 10 and 15 kgf (22 and 33 lbf) after baking improved the corrosion resistance of 9- to 25-μm (0.35- to 1-mil) thick and 50- to 80-μm (2- to 3-mil) thick chromium coatings relative to samples that were not smoothed. Chromium-plated samples baked in oil were much more corrosion resistant than samples baked in air. Both methods sealed pores or cracks in the chromium, which improved the corrosion resistance. Superfinishing techniques may also seal pores and cracks and enhance corrosion resistance. In addition, mechanical recompression of electrodeposited chromium improves corrosion resistance (Ref 9).

Table 3 Corrosion resistance of electroplated chromium deposits immersed in salt solutions

Salt	Concentration, %	Temperature °C	Temperature °F	Corrosion rate, mm/yr (mils/yr)
Alum	...	...	...	Resistant
Aluminum chloride	10	12	55	nil
		58	135	0.08 (3)
Aluminum sulfate	10	12	55	nil
		58	135	0.20 (8)
Amino G salt	Saturated	12	55	0.03 (1.2)
		58	135	0.38 (15)
Ammonium chloride	10	12	55	nil
		58	135	0.10 (4)
Barium chloride	10	12	55	nil
		58	135	0.03 (1.2)
Calcium chloride	10	12	55	nil
		58	135	nil
Calcium hypochlorite	10	12	55	0.05 (2)
		58	135	0.89 (35)
Chromic chloride	10	12	55	nil
		58	135	0.08 (3)
Cupric chloride	10	12	55	0.38 (15)
Cupric nitrate	10	12	55	0.05 (2)
		58	135	0.18 (7)
Cupric sulfate	...	...	...	Resistant
Ferric chloride	10	12	55	nil
		58	135	0.41 (16), pitting
Ferrous chloride	10	12	55	nil
		58	135	0.14 (5.5)
Magnesium chloride	10	12	55	nil
		58	135	nil
Manganese chloride	10	12	55	nil
		58	135	nil
Mercuric chloride	10	12	55	2.01 (79), pitting
Potassium chloride	10	12	55	nil
		58	135	nil
Scheaffer salt	Saturated	12	55	nil
		58	135	0.15 (6)
Sodium benzene sulfonate	10	12	55	nil
		58	135	nil
Sodium carbonate	...	...	...	Resistant
Sodium chloride	10	12	55	nil
		58	135	nil
Sodium formate	10	12	55	nil
		58	135	nil
Sodium hydrosulfite	10	12	55	nil
		58	135	nil
Sodium hydroxide	10	12	55	nil
		58	135	nil
Sodium phenol sulfonate	10	12	55	nil
		58	135	0.03 (1.2)
Sodium sulfate	10	12	55	nil
		58	135	0.05 (2)
Stannous chloride	10	12	55	nil
		58	135	0.89 (35)
Strontium chloride	10	12	55	nil
		58	135	nil
Zinc chloride	10	12	55	nil
		58	135	0.03 (1.2)

Source: Ref 14, 15

Table 4 Corrosion resistance of electroplated chromium deposits in miscellaneous environments

Environment	Concentration, %	Temperature °C	Temperature °F	Corrosion rate, mm/yr (mils/yr)
Acid green	10	12	55	nil
		58	135	0.08 (3)
Aminophenol (meta)	Saturated	12	55	nil
		58	135	0.03 (1.2)
Aniline hydrochloride	10	12	55	0.03 (1.2)
		58	135	0.58 (23)
Bakelite during molding	...	...	...	Resistant
Beer and wort	...	...	...	Resistant
Beet sugar juice	...	...	...	Resistant
Benzyl chloride	Saturated	12	55	nil
		58	135	nil
	100	12	55	nil
		58	135	nil
Biscuit dough	...	...	...	Resistant
Brass, molten	...	...	...	Resistant
Brine, neutral	...	...	...	Resistant
Bronze, aluminum, molten	...	...	...	Resistant
Carbonaceous material, hot	...	...	...	Resistant
Carbon tetrachloride	Saturated	12	55	nil
		58	135	nil
	100	12	55	nil
		58	135	nil
Chlorobenzene	Saturated	12	55	nil
		58	135	nil
	100	12	55	nil
		58	135	nil
Chloroform	Saturated	12	55	nil
		58	135	nil
	100	12	55	nil
		58	135	nil
Chlorohydroquinone	10	12	55	nil
		58	135	0.003
Chlorophenol (ortho)	Saturated	12	55	nil
		58	135	0.03 (1.2)
Cyanides, fused	...	...	...	Resistant
Ebonite during molding	...	...	...	Resistant
Fruit acids	...	...	...	Generally resistant
Glass, molten	...	...	...	Resistant
Glue, hot	...	...	...	Resistant
Milk	...	...	...	Resistant
Nitrophenol (para)	Saturated	12	55	nil
		58	135	0.03 (1.2)
Oil, crude	...	...	...	Resistant
Oils, essential	...	...	...	Resistant
Paper pulp suspension	...	12	55	nil
		58	135	nil
Phthalimide	Saturated	12	55	nil
		58	135	0.03 (1.2)
Printing ink	...	...	...	Resistant
Resins, synthetic	...	...	...	Resistant
Thiourea (during molding)	...	...	...	Resistant
Vinyl (during molding)	...	...	...	Resistant
Rubber (during vulcanizing)	...	...	...	Resistant
Soap	...	...	...	Resistant
Steam	...	...	...	Resistant
Steam (superheated)	...	...	...	Resistant
Succinimide	Saturated	12	55	nil
		58	58	0.03 (1.2)
Sugar	...	...	...	Resistant
Sulfite liquors	...	Below boiling		Resistant
Sulfur (in petrol)	...	...	...	Resistant
Tar	...	...	...	Resistant
Tartrazine	10	12	55	nil
		58	135	0.15 (6)
Tetrachlorobenzene	Saturated	12	55	nil
		58	135	nil
Vegetable oil acid	...	...	...	Generally resistant
Water, deep well	...	...	...	Resistant
Water, sea	...	...	...	Resistant
Zinc, molten	...	...	...	Attacked

Source: Ref 14, 15

Grinding of a chromium-plated part to finished dimensions should be performed with adequate cooling and lubrication. Excessive forces and heat may cause macrocracking of the chromium down to the basis metal. Coarse grinding will promote SCC.

The passivation of chromium by exposure to air or by immersion in oxidizing acid will significantly improve its corrosion resistance (Ref 10). In one study, nitrogen ion implantation of a chromium-plated molding tool improved the corrosion resistance of the part to molding gases and fluids and increased its service life by more than four times (Ref 11). Postplating treatments are extensively discussed in Ref 12.

Corrosion Resistance Data

Corrosion of hard chromium deposits usually begins at microcracks or intersections of microcracks (Ref 10). After moderate acid attack, the corrosion reveals the microcrack pattern. Attack will continue on all of the chromium, and the microcrack pattern will no longer be visible. The corrosion of chromium in sodium chloride (NaCl) solutions will produce mounds of corrosion products. When these mounds are removed, concentric rings define the attacked area. The center and outside area are unattacked.

Chromium-plated steel with and without diffusion treatment at 1000 °C (1830 °F) resists corrosion by sodium polysulfides (Na_2S_4, Na_2S_5) and sulfur at temperatures to 440 °C (825 °F). In a 12-month static test, chromium-diffused samples (plating thickness: 50 to 200 μm, or 2 to 8 mils) performed better than the as-plated samples (Ref 13). Tables 2 to 5 provide detailed data on the corrosion of hard chromium in various media.

Applications

Electrodeposited chromium is used in a wide variety of applications and environments. Reference 15 includes a listing of 38 industrial categories with 315 specific applications of electroplated chromium. The corrosion resistance of electroplated chromium is important in wear applications. The wear resistance of a part will decrease if corrosion occurs on a wearing surface—for example, shocks and struts. Industrial applications in which chromium is not exposed directly to aggressive chemicals may expose chromium coatings to elevated temperature and corrosive environments, such as combustion products.

Electroplated chromium for atmospheric-corrosion applications should be between 20 and 30 μm (0.8 and 1 mil) thick. For corrosion resistance in chemical exposures, electroplated chromium should be 50 to 75 μm (2 to 3 mils) thick. Electroplated chromium is attacked at 58 °C (135 °F) in formic acid (HCOOH), hydrobromic acid (HBr), HCl, perchloric acid ($HClO_4$), H_2SO_4, and trichloroacetic acid (CCl_3COOH) and is attacked at 12 °C (55 °F) in hydrofluoric acid (HF). Hot (58 °C, or 135 °F) solutions of ferric chloride ($FeCl_3$), mercuric chloride ($HgCl_2$), and stannous chloride ($SnCl_2$) attack electroplated chromium more severely than most other salt solutions.

Table 5 Corrosion resistance of electroplated chromium in various gases

Gas	Temperature, °C (°F)	Corrosion rate
Air, gas works	...	Resistant
Air, hot oxidizing	...	Resistant
Air, hot reducing	...	Resistant
Air, nitric acid works	...	Resistant
Air, normal	...	Resistant
Ammonia	...	Resistant
Carbon monoxide	...	Resistant
Carbon dioxide	...	Resistant
Chlorine, dry	<300 (<570)	Resistant
Chlorine, wet	...	Attacked
Coal gas	...	Resistant
Hydrogen sulfide	...	Resistant
Oxygen	>1200 (>2180)	Oxidizes
Petroleum and diesel fuel combustion products, hot	...	Resistant
Steam	...	Resistant
Steam, superheated	...	Resistant

Source: Ref 14, 15

REFERENCES

1. J.E. Stareck, E.J. Seyb, and A.C. Tulumello, Stress in Chromium Deposits, *Plating*, Vol 41, 1954, p 1171-1182
2. A. Bremmer, P. Burkhead, and C. Jennings, Physical Properties of Electrodeposited Chromium, *J. Res. Natl. Bur. Stand.*, Vol 40, 1948, p 31-59 RP, 1954
3. M. Cymboliste, The Structure and Hardness of Electrochemical Chromium, *Trans. Electrochem. Soc.*, Vol 73, 1938, p 353-363
4. W.E. Moline, Corrosion Resistance of Chromium Plated and Surface Conditioned 13 Per Cent Chromium Steel, *Mon. Rev. Am. Electroplat. Soc.*, No. 4, April 1946, p 401-408
5. R. Kausalya and N.V. Parhasaradhy, Chromium Electrodeposits With Improved Corrosion Resistance, *Plating*, Vol 57 (No. 12), 1970, p 1238-1249
6. H. Chessin, E.C. Knill, and E.J. Seyb, Jr., Defects in Hard Chromium Deposits, *Plat. Surf. Finish.*, Vol 70, 1983, p 24-29
7. A.K. Graham, Ed., *Electroplating Engineering Handbook*, 3rd ed., Van Nostrand Reinhold, 1971, p 408-409
8. E.Y. Beider, E.V. Plaskeev, G.N. Petrova, and S.M. Pankratov, Enhancing the Protective Properties of Chromium Coatings, *Prot. Met.*, Vol 20 (No. 1), 1984, p 127-129
9. K. Schreck, "Producing a Corrosion-Proof Surface on Workpieces," U.S. Patent PCT Int. Appl. WO 85/3090, 1985
10. N. Hackerman and D.I. Marshall, Corrosion Studies on Electrolytic Chromium, *Trans. Electrochem. Soc.*, Vol 89, 1946, p 195-205
11. Combined Ion Implantation and Hard Chrome Plating Gives Ten Times Tool Life, *Prod. Finish.*, Vol 39 (No. 5), 1986, p 14
12. F.A. Lowenheim, Ed., *Modern Electroplating*, 3rd ed., John Wiley & Sons, 1974, p 112-114
13. A. Wicker, "Corrosion of Chromium-Coated Steel in Sodium Polysulfide Environments," Report EPRI-EM-2947, Electric Power Research Institute, 1983
14. H.H. Uhlig, *The Corrosion Handbook*, John Wiley & Sons, 1948, p 825-828
15. P. Morisset, *Chromium Plating*, Robert Draper, 1954, p 179-183, 260-271

SELECTED REFERENCES

- J.P. Greenwood, *Hard Chromium Plating*, Robert Draper, 1964
- F.A. Lowenheim, Ed., *Modern Electroplating*, 3rd ed., John Wiley & Sons, 1974
- P. Morisset, *Chromium Plating*, Robert Draper, 1954

Corrosion of Brazed Joints

M.M. McDonald, Rockwell International

CORROSION is a concern for many reasons when materials are brazed. Brazing involves the joining of dissimilar metals; therefore, galvanic corrosion or preferential attack is a possibility. Also, many of the fluxes used in brazing are corrosive; therefore, if improper cleaning does not remove all of the flux or if poor joint design results in flux entrapment, corrosion may result. Poor joint design may also contribute to corrosion by the introduction of stresses.

The corrosion behavior of brazed joints is a complicated phenomenon. The interacting factors include the brazing procedure, the materials, and the environment. The pertinent steps of the brazing procedure include joint design, heating process selection, flux selection, and cleaning. Material factors include selection of the base materials and the brazing alloy. The corrosiveness of an environment depends primarily on the relative concentration of water and impurities, the ability of electric currents to exist, and the presence or absence of oxidizing agents (Ref 1). The velocity and temperature of the environment are often important factors as well.

This article consists of four main sections. The first section provides background information on brazing. The second section briefly describes the major types of corrosion that are of importance in brazing, and the third section discusses the role of proper brazing procedures in minimizing corrosion. The fourth and major section reviews the work done to date on the corrosion behavior of the major classes of filler alloys. Additional information on brazing methods and practices can be found in Volume 6 of the 9th Edition of *Metals Handbook*.

Background on Brazing

The American Welding Society (AWS) defines brazing as a group of welding processes that produce coalescence of materials by heating them to a suitable temperature and using a filler metal having a liquidus above 450 °C (840 °F) and below the solidus of the base materials; the filler metal is distributed between closely fitted surfaces of the joint by capillary attraction (Ref 2). Brazing processes are generally classified according to the sources or methods of heating, such as torch, furnace, induction, dip, or infrared. The general procedure for brazing is discussed in the following section.

Brazing offers certain advantages over other joining processes. These advantages, which stem mainly from the lower heating temperatures (Ref 2, 3, 4), include the elimination of warping; increased joining speed; high production rates (due to simultaneous brazing); reduction of the effects of expansion, contraction, and distortion; reduced chance of locked-in stress; and reduced expense. The metals that are commonly brazed include aluminum, magnesium, copper, and their alloys; stainless steels; low-carbon and low-alloy steels; cast iron; nickel-base and cobalt-containing alloys; and alloys containing the reactive metals titanium, zirconium, and beryllium.

Major Types of Corrosion Important in Brazing

All types of corrosion can be observed in brazed joints. The types of corrosion that affect materials in a brazed joint are similar to those that affect the materials in service individually. Certain types of corrosion are seen more frequently in brazed joints because of conditions that are inherent in brazing. For example, because brazed joints are bonds of dissimilar metals, galvanic corrosion is always a concern. Interfacial corrosion is also of particular concern in brazed joints. In the following sections, the forms of corrosion will be briefly discussed as they pertain to brazed joints.

General Attack. All metals will exhibit general corrosion in some corrosive medium. Because two or three different types of metals are part of a brazed joint, the behavior of each different metal in the service environment must be considered with respect to the possibility of general attack. This type of corrosion is uniform; therefore, failure by this mechanism can be predicted on the basis of relatively simple tests in which the rate of dissolution is determined. Additional information is available in the articles "General Corrosion" and "Evaluation of Uniform Corrosion" in this Volume.

Galvanic Corrosion. A potential difference usually exists between two dissimilar metals immersed in an electrolyte. This potential difference produces electron flow between the metals, leading to increased attack of the less corrosion-resistant material and decreased attack of the more resistant material as compared to the behavior of these metals when they are not in contact (Ref 5). Because brazing is often performed when metallurgical incompatibility precludes the use of other welding processes, many dissimilar-metal combinations are brazed. Therefore, galvanic corrosion is always a concern, and the galvanic series for the service environment should be consulted. Table 1 is the galvanic series for seawater.

Table 1 Galvanic series of some commercial metals and alloys in seawater

Noble or cathodic
Platinum
Gold
Graphite
Titanium
Silver
  Chlorimet 3 (62Ni-18Cr-18Mo)
  Hastelloy C (62Ni-17Cr-15Mo)
  18-8 Mo stainless steel (passive)
  18-8 stainless steel (passive)
  Chromium stainless steel 11-30% Cr (passive)
  Inconel (passive) (80Ni-13Cr-7Fe)
  Nickel (passive)
Silver solder
  Monel (70Ni-30Cu)
  Cupronickels (60–90% Cu, 40–10% Ni)
  Bronzes (copper-tin)
  Copper
  Brasses (copper-zinc)
  Chlorimet 2 (66Ni-32Mo-1Fe)
  Hastelloy B (60Ni-30Mo-6Fe-1Mn)
  Inconel (active)
  Nickel (active)
Tin
Lead
Lead-tin solders
  18-8 Mo stainless steel (active)
  18-8 stainless steel (active)
Ni-Resist (high-nickel cast iron)
Chromium stainless steel, 13% Cr (active)
  Cast iron
  Steel or iron
2024 aluminum (4.5Cu-1.5Mg-0.6Mn)
Cadmium
Commercially pure aluminum (1100)
Zinc
Magnesium and magnesium alloys
Active or anodic

Source: Ref 5

In determining the potential for galvanic corrosion in a brazed joint, environmental, distance, and area effects must be considered, as well as position in the galvanic series (Ref 5). In brazing, moisture is generally the most important environmental factor. Moisture enables the electrolyte to carry current between the two electrode areas; therefore, the corrosion rate generally increases with humidity (Ref 6), and keeping the materials dry inhibits corrosion. A critical humidity level exists for most metals above which corrosion proceeds rapidly—for example, 100% for copper, 85% for certain nickel alloys, and 65% for iron (Ref 7). Galvanic corrosion is readily recognized by the somewhat localized attack near the junction and by the decreasing evidence of attack with increasing distance from the junction. The distance over which galvanic corrosion is evident depends on the conductivity of the solution. An unfavorable area ratio consists of a large cathode (the more corrosion-resistant material) and a small anode (the less corrosion-resistant material); for a given current flow, the current density is

(a)

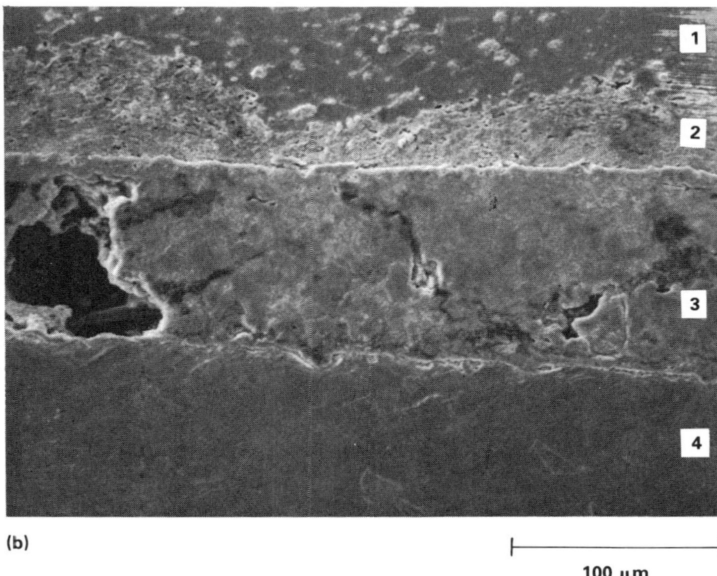

(b)

Fig. 1 Example of selective leaching. (a) Macrograph of a brass tube soldered to a copper tube. The arrow indicates the region of dezincification. 6.5×. (b) Scanning electron micrograph of the region of dezincification. Region 1, brass base metal; 2, dezincification zone; 3, lead-tin solder; 4, copper base metal. 275×. Source: Ref 9

greater for the smaller electrode than for the larger one.

Knowledge of the factors that influence galvanic corrosion enables determination of a practice that would minimize the occurrence of this phenomenon. Practices that accomplish this include selecting combinations of metals as close together as possible in the galvanic series, avoiding unfavorable area effects, and reducing moisture as much as possible. Additional information is available in the articles "General Corrosion" (in particular, the discussion of galvanic corrosion) and "Evaluation of Galvanic Corrosion" in this Volume.

Crevice Corrosion and Pitting. Within a crevice or other shielded area on a metal surface, stagnant solutions may lead to localized attack, or crevice corrosion (Ref 8). Lap brazed joints are a potential site for crevice corrosion. The geometry of lap joints creates conditions that are favorable for stagnant-solution buildup. The possibility of crevice corrosion can be minimized by making sure that the braze is continuous along the joint.

Pitting is an extremely localized form of corrosion. A pit may form at imperfections in the metal surface, such as grain boundaries, or at crevices or other areas of stagnant electrolyte (Ref 5). Additional information is provided in the articles "Localized Corrosion" (in particular, the discussions of crevice and pitting corrosion), "Evaluation of Crevice Corrosion," and "Evaluation of Pitting Corrosion" in this Volume.

Dealloying corrosion, or selective leaching, is the preferential removal of one component from an alloy. A classic example of selective leaching is the dezincification of brass. During dezincification, the brass actually dissolves, and the copper redeposits on the remaining brass (Ref 8). An example of dezincification of a brass tube soldered to a copper tube is shown in Fig. 1.

Because selective leaching involves only one alloy, brazing does not increase the possibility of the occurrence of selective leaching. If an alloy susceptible to selective leaching is used in a brazed joint, however, the possibility of its occurrence should be considered. Additional information is available in the article "Metallurgically Influenced Corrosion" (in particular, the discussion of dealloying corrosion) in this Volume.

Interfacial Corrosion. Three conditions are necessary for interfacial corrosion (Ref 10). First, one member of the joint must be stainless steel. Second, the brazing alloy must be susceptible to this form of attack; that is, it must contain a copper-zinc rich phase. Third, the finished joint must be exposed to damp or wet conditions. The exact nature of this form of corrosion is under dispute. Several mechanisms have been proposed to explain the failure of brazed joints under these circumstances. Two of the more popular theories are (1) crevice corrosion based on an oxygen concentration and potential gradient across the brazing alloy and base metal (Ref 11-15) and (2) selective dissolution of the copper-zinc rich phase at the braze/steel interface and redeposition of the copper, that is, selective leaching or dezincification (Ref 16-18). Figure 2 schematically shows crevice formation and deposition of corrosion products on the metal surface. Other proposed mechanisms include oxygen deficiency (Ref 19), electrochemical action between the base and filler materials in the electrolyte (Ref 20), and chromium diffusion into the brazing alloy with the creation of a chromium-depleted zone at the base metal (Ref 21).

The susceptibility of a brazed joint to interfacial corrosion depends on the type of filler alloy, the type of stainless steel, the service environment, and to some extent even the brazing technique (Ref 10). Polishing the finished joint has also been found to affect the likelihood of attack (Ref 10). Unfortunately, the most common brazing techniques, those using a silver-containing filler alloy and a flux, create the conditions under which joints are most susceptible to failure. Table 2 lists the available information related to the susceptibility of various alloys and base metal combinations to interfacial corrosion.

All types of stainless steel are susceptible to attack by interfacial corrosion. Low-nickel fer-

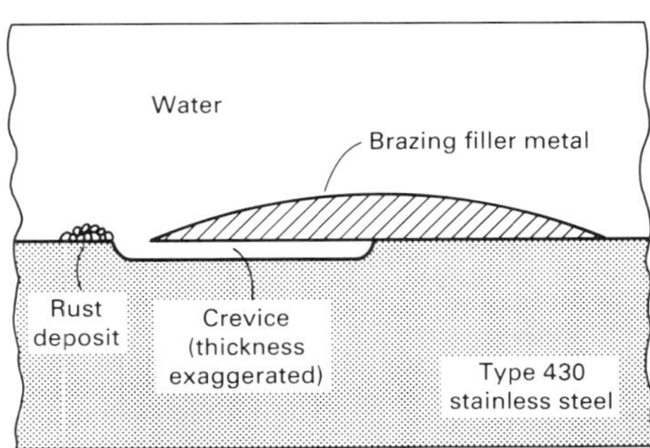

Fig. 2 Crevice formation and deposition of corrosion products on the stainless steel surface

ritic and martensitic stainless steels are more susceptible to attack and suffer a more rapid rate of corrosion than the austenitic grades (Ref 10). In the case of such austenitic stainless steels as types 302, 304, and 316, the use of a brazing filler metal containing a small amount of nickel apparently eliminates interfacial corrosion completely (Ref 15). The use of a nickel-containing filler alloy also retards interfacial corrosion in type 430 stainless steel (Ref 15). The presence of nickel in the filler alloy is beneficial because it is the major constituent of an interfacial layer that forms between the steel and filler alloy when these filler alloys are used (Ref 10).

Brazing in the absence of flux, that is, in dry hydrogen, inert gas, or vacuum, appears to eliminate interfacial corrosion in straight chromium stainless steels and chromium-nickel stainless steels (Ref 15). A notable exception is when certain brazing filler metals containing manganese are used (Ref 22). Evidently, manganese diffuses from the filler alloy into the surface of the stainless steel and produces a vulnerable manganese-rich layer.

Erosion-corrosion is the increase in rate of attack on a metal because of relative motion between a corrosive fluid and the metal surface. Impingement, fretting, and cavitation are the three

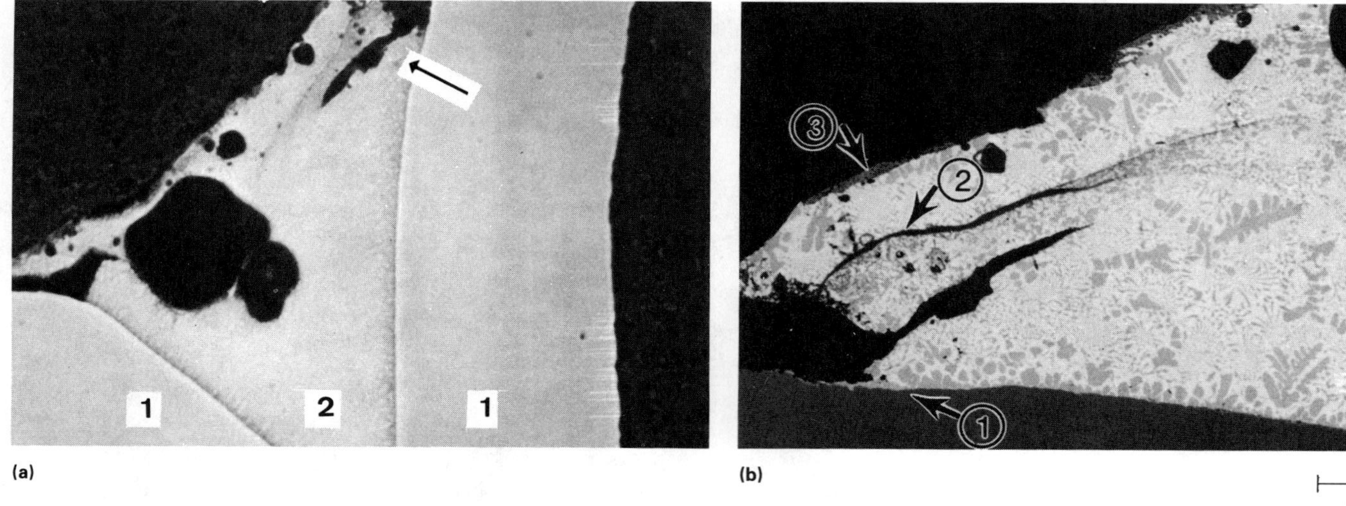

Fig. 3 Example of high-temperature corrosion. (a) Optical micrograph showing a stainless steel joint brazed with a silver-copper-palladium filler alloy that exhibited high-temperature corrosion. Region 1 is the stainless steel base metal; region 2 is the filler alloy. 80×. (b) Scanning electron micrograph of upper region of braze alloy in (a) (indicated by the arrow). Oxidation products occurred in regions 1 to 3. 165×. Source: Ref 25

most important types of erosion-corrosion. Erosion-corrosion may be equally detrimental to braze filler alloys and base metals because much of the effect is mechanical in nature (Ref 1). Inappropriate brazing practices, particularly designs that have rapid changes in section, have been shown to result in failure (perforation) at brazed joints due to erosion-corrosion (Ref 23). Additional information is available in the articles "Mechanically Assisted Degradation" (in particular, the discussion on erosion-corrosion) and "Evaluation of Erosion and Cavitation" in this Volume.

High-Temperature Corrosion. Corrosion may occur at elevated temperatures by mechanisms similar to those at lower temperatures. The most common type of high-temperature corrosion is oxidation. Oxidation can be defined as the loss of electrons by a constituent in a chemical reaction (Ref 24). Therefore, the phenomenon of oxidation includes attack by carbon oxides, sulfur compounds, nitrogen, and hydrogen, as well as oxygen. Corrosion testing of brazed joints often involves high-temperature exposure of the joints to various environments. The brazing filler alloys that are most resistant to oxidation are those containing a high percentage of noble elements and those that form tenacious protective films (Ref 1). Examples of such filler alloys are the gold- and nickel-base alloys. In addition, if the flux used during the brazing process is not completely removed from a joint, its presence may lead to corrosion of the joint at elevated temperatures.

An example of high-temperature corrosion is given in Fig. 3. In this example, the copper in the silver-copper-palladium filler alloy oxidized. The alloy was used to braze stainless steel. Additional information is available in the articles "Fundamentals of Corrosion in Gases" and "General Corrosion" (in particular, the discussion of high-temperature oxidation/sulfidation) in this Volume.

Stress-Corrosion Cracking (SCC). Unfortunately, there is considerable confusion and controversy concerning the classification of SCC phenomena. In this context, SCC refers strictly to the situation in which a metal that is subject to a constant tensile stress and exposed simultaneously to a specific corrosive environment cracks immediately or after a given time (Ref 26). The stress may be residual in the metal or may be externally applied. Failure may be either intergranular or transgranular. Of particular interest in brazing is the susceptibility of copper-base alloys to SCC in the presence of ammonia.

Few documented failures of brazed joints have been attributed to SCC, probably because they were not identified as such (Ref 1). Some results of SCC tests on brazed joints have been reported in the literature. For example, one study examined the stress-corrosion behavior of Ti-6Al-4V butt joints brazed with aluminum-copper filler alloys and smooth Ti-6Al-4V base metal specimens in 3.5% sodium chloride (NaCl) at room temperature under a tensile stress of 714 MPa (102 ksi) (Ref 27). The joints and specimens were found to be essentially immune to SCC after 1000 h of exposure. In another study, all of the titanium joints tested were found to be unsusceptible to SCC in the environments to which they were exposed (Ref 28). These tests involved titanium joints brazed with aluminum-base filler alloys exposed to aqueous 3.5% NaCl at room temperature and to dry NaCl at 150 °C (300 °F) under a load of 91 to 105 MPa (13 to 15 ksi) for 500 h. Additional information is available in the articles "Environmentally Induced Cracking" (in particular, the discussion of SCC) and "Evaluation of Stress-Corrosion Cracking" in this Volume.

Liquid-Metal Embrittlement (LME). Many materials (including stainless steels, nickel alloys, and copper-nickel alloys, particularly age-hardenable ones) will crack during brazing if they are stressed while in contact with the molten brazing metal (Ref 2). The cracking occurs virtually instantaneously during the brazing operation, and the molten filler metal follows and fills in the cracks, making them easily visible. This type of failure is known as LME. Although LME is not, strictly speaking, a corrosion phenomenon, it is often discussed in the context of SCC. This is presumably because both mechanisms are often categorized as forms of stress cracking.

An example of LME of TZM (Mo-0.5Ti-0.08Zr-0.015C) brazed with a gold-base filler alloy is shown in Fig. 4. The backscattered electron micrograph taken in the scanning electron microscope highlights the filler alloy present in the cracks of the base metal. Similar behavior has been observed in molybdenum brazed with gold-base filler alloys (Ref 30).

Liquid-metal embrittlement is generally believed to be specific; that is, a particular solid will be significantly embrittled only by certain specific liquid metals (Ref 26, 31). The problem of determining which solid metal/liquid metal combinations will exhibit LME is complicated by the fact that this phenomenon is strongly dependent on temperature and strain rate. The solid iron-base alloys/liquid copper combination has received considerable attention. Many investigators have shown that this combination is susceptible to embrittlement (Ref 32-37). In a series of studies, the mechanism of cracking in a number of stainless steels, steels, and superalloys was identified as LME by molten copper (Ref 38-40). Another investigation showed that the austenitic stainless

Table 2 Susceptibility of brazing alloy/stainless steel types to interfacial corrosion

Brazing alloy type	Susceptibility of steel type(a) Ferritic	Austenitic
Ag-Cu-Zn-Cd	C	D
Ag-Cu-Zn	C	D
Ag-Cu-Zn-Cd-Ni	D	B
Ag-Cu-Zn-Ni	D	B
Ag-Cu-Zn-Ni-Mn	D	B
Ag-Cu-Zn-Sn	D	B
Ag-Cu-In-Ni	A	A
Ag-Cu-Sn-Ni	A	A
Ag-Cu (silver copper eutectic)	C	C
Ag-Mn (15% Mn-Ag)	C	C
Palladium-containing alloys	A	A
Gold-containing alloys	A	A
Nickel-base alloys	A	A
High-copper alloy	A	A

(a) The designations are as follows: A, should be resistant under most conditions; B, may offer limited or complete protection, depending on conditions; C, rapid joint failure can be expected in a very short period; D, joint failure likely to occur after some time (less rapid than above). Source: Ref 10

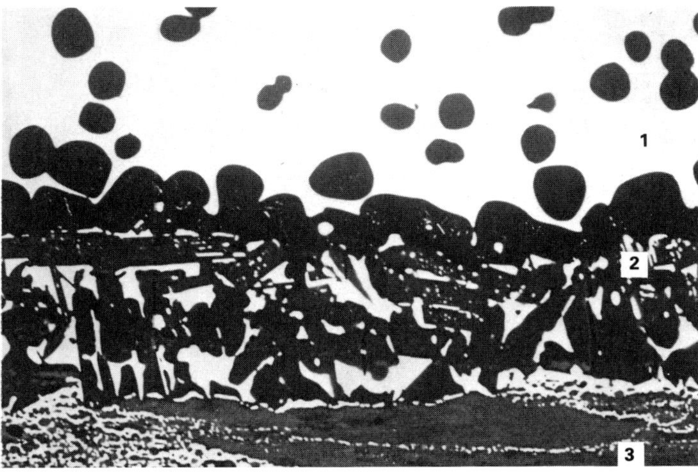

Fig. 4 Backscattered electron micrograph of TZM molybdenum-base metal that exhibited stress cracking due to thermal stresses and the presence of molten gold-base filler alloy. Region 1 is the filler alloy, region 2 is an intermetallic phase formed by interaction between base metal and braze alloy, and region 3 is the TZM base metal. The cracks in the base metal are quite evident because of the penetration of the filler alloy. Source: Ref 29

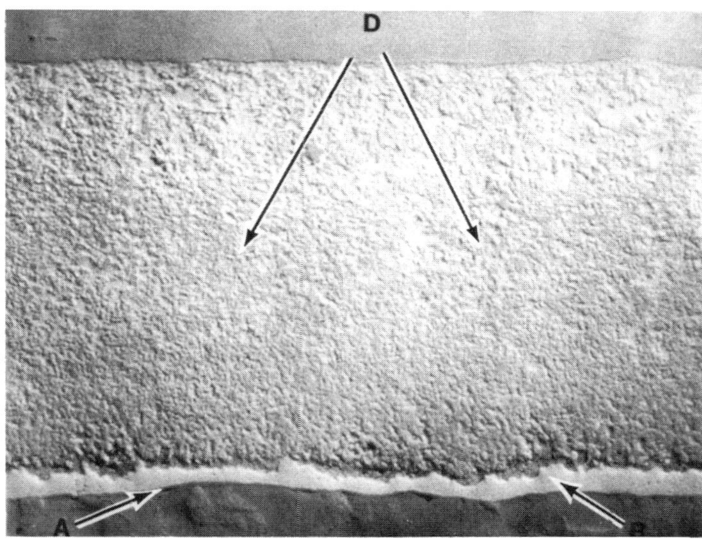

Fig. 5 Optical micrograph of a cross section of a beryllium-to-Monel joint brazed with a silver-copper filler alloy. Arrow A points to a beryllium-copper intermetallic phase, arrow B points to a copper-beryllium intermetallic phase, and arrow D points to the copper-nickel-beryllium reaction matrix. 330×. Source: Ref 42

steels are severely embrittled by copper and high-copper braze alloys and that the base metals are nearly unaffected by silver-base braze alloys (Ref 31). In addition, it was found that a type 430 ferritic stainless steel was not embrittled by any of the several liquid braze alloys to which it was exposed, namely copper and several copper-base alloys, several silver-base alloys, aluminum, and gold (Ref 31). Liquid-metal embrittlement has been eliminated by (Ref 2):

- Removing the source of an externally applied stress as determined from a critical analysis of the brazing process
- Redesigning parts or revising joint design
- Selecting a brazing filler metal that is less likely to cause LME in the substrate

Solid-phase embrittlement refers to the phenomenon of brittle-phase formation in the region of the brazing alloy/base metal interface. The presence of this brittle phase can lead to cracking. Like SCC, solid-phase is not, strictly speaking, a corrosion phenomenon. However, it is discussed here because it is also often categorized as a stress-cracking phenomenon. Embrittlement of base metals containing more than a few percent nickel or iron is known to occur when filler alloys containing free phosphorus are used for brazing (Ref 41). Another example of solid-phase embrittlement is sulfide formation in nickel-base alloys by reaction of the base metal with a brazing atmosphere containing sulfur compounds or from contamination from sulfurized cutting oil (Ref 41).

Figure 5 shows the presence of β′, a brittle beryllium-copper intermetallic phase that formed at the interface of beryllium and a silver-copper filler alloy in a braze between beryllium and Monel. In this case, the occurrence of the brittle intermetallic phase was suppressed by coating the beryllium with titanium hydride prior to brazing (Ref 43).

Intergranular corrosion refers to the phenomenon of localized attack at and adjacent to grain boundaries, with relatively little corrosion of the grains. Sensitization of stainless steels is a classic example of intergranular corrosion. Sensitization occurs because of chromium depletion near the grain boundaries caused by precipitation of chromium carbides at the grain boundaries. The chromium-depleted zone near the grain boundaries may then corrode because it does not contain sufficient corrosion resistance to resist attack in many corrosive environments. Figure 6 shows sensitization of a stainless steel joint that was overheated during brazing with a silver-copper-palladium filler alloy (Ref 25). In general, sensitization is not a concern when brazing properly with silver-base filler alloys, because of the low brazing temperatures. When brazing stainless steels with copper-base filler alloys, however, care should be taken to cool quickly through the 510- to 790-°C (950- to 1450 °F) temperature range in which sensitization occurs. Additional information is available in the articles "Metallurgically Influenced Corrosion" (in particular, the discussion of intergranular corrosion) and "Evaluation of Intergranular Corrosion" in this Volume.

The Role of Proper Brazing Procedures In Minimizing Corrosion

Efforts to minimize the occurrence of corrosion can be made at each step of the brazing process. These steps include joint design, selection of materials, brazing process and flux, and final cleaning.

Joint Design. In designing a joint that is to be brazed, attention should be given to geometries that may make a joint more susceptible to corrosion. For example, if an assembly is to be pickled, the internal geometry of the entire assembly should allow complete exchange of pickling and cleaning solutions (Ref 41). If a lap joint is to be fabricated, the possibility of the geometry leading to the buildup of stagnant solution should be considered. If a butt joint is to be fabricated, the efficiency of the joint is maximized by making sure the joint has no defects, such as entrapped flux, voids, incomplete joining, and porosity (Ref 1, 44). These defects, in addition to impairing the structural integrity of the joint, may increase the susceptibility of the joint to corrosion. Entrapped flux may lead to high-temperature corrosion, and incomplete joining may lead to crevice corrosion.

The concept of distributing the stress in the joint as evenly as possible is also important in joint design. If the stress is concentrated, the conditions for SCC may exist. Reference 3 provides suggestions for distributing stress.

Materials Selection. In addition to the material properties normally considered, such as strength, the selection of the materials for a brazed joint should involve consideration of the potential for corrosion. In particular, the appropriate galvanic series should be consulted (Table 1). Also, attention should be paid to the susceptibility of certain alloys to other types of corrosion, such as selective leaching and general attack. The service environment often dictates the potential for corrosion. Alloy systems are discussed in more detail later in this article and in the Section "Specific Alloy Systems" in this Volume.

Brazing Process Selection. Brazing processes are generally classified according to the sources or methods of heating, such as torch, furnace, induction, dip, or infrared (Ref 2). Table 3 summarizes the major advantages and disadvantages of these common methods of brazing. The intensity of the heat source can play a role in the distribution of stress and the formation of phases. If the material is known to be susceptible to SCC, the distribution of thermal stresses should be considered. If an alloy system is susceptible to brittle-phase formation, the temperature and time at temperature the alloy is subjected to should be considered in choosing the brazing process. The susceptibility of alloy systems to brittle-phase formation may be determined by

Table 3 Advantages and disadvantages of some common brazing methods

Method	Advantages	Disadvantages
Torch	Inexpensive; lower temperature; small parts	Need flux; oxidation
Vacuum furnace	Fluxless; prevents oxidation; good temperature control 41)	Higher temperature (because fluxless); expensive; slow heating
Induction	Localized heating; atmosphere or vacuum conditions; fast heating	Poorer temperature control
Dip	Good temperature control; small parts	Part cleanup
Infrared	Able to do large parts fluxless	Slow heating

examining phase diagrams or consulting the literature.

Torch brazing is a method of brazing that may involve the use of a low-temperature, air-gas, large flame to minimize the possibility of localized overheating (Ref 41). Localized overheating may overstress a filler alloy as it solidifies if the metal is sufficiently restrained. Cracking may then result from the overstressing. Induction heating can be used to keep a uniform temperature distribution around the joint to prevent the stress concentrations from thermal expansion and contraction that may lead to stress cracking. Vacuum brazing has the advantage of thoroughly degasing the material and protecting the materials from oxidation and other forms of high-temperature corrosion. The cost of vacuum brazing, however, may preclude its use.

The Brazing Procedure. To achieve a good joint by any method of brazing, it is necessary to follow certain guidelines (Ref 2, 45). The parts must be properly cleaned before brazing, and they must be protected by fluxing or by the use of an inert atmosphere during the heating process to prevent excess oxidation or other forms of high-temperature corrosion. Also, as previously discussed, the proper filler alloy, flux, and joint design must be chosen, and a heating process must be selected that will provide the proper brazing temperature and heat distribution. Guidelines for the selection of filler alloy and flux are available in Ref 46 and the Section "Brazing" in Volume 6, 9th Edition, of *Metals Handbook*. Finally, the braze joint must be cleaned to remove flux residue and any oxide scale formed during the brazing process. Flux removal is particularly important because flux residues are often chemically corrosive and, if not removed, could possibly weaken certain joints. Figure 7 shows a copper to stainless steel joint that exhibited corrosion products as a result of improper flux removal.

Corrosion Resistance of Brazed Joints

The corrosion resistance of brazed joints involves the individual and collective corrosion resistance of the base metals and the filler alloys. This section will summarize the work done to date on characterizing the behavior of the main categories of braze filler alloys in corrosive environments. A more detailed discussion of this topic is provided in Ref 1. The articles in this Volume on the corrosion of specific metals and alloys should also be consulted. American Welding Society designations for the classes of filler alloys discussed in this section are contained in Ref 2.

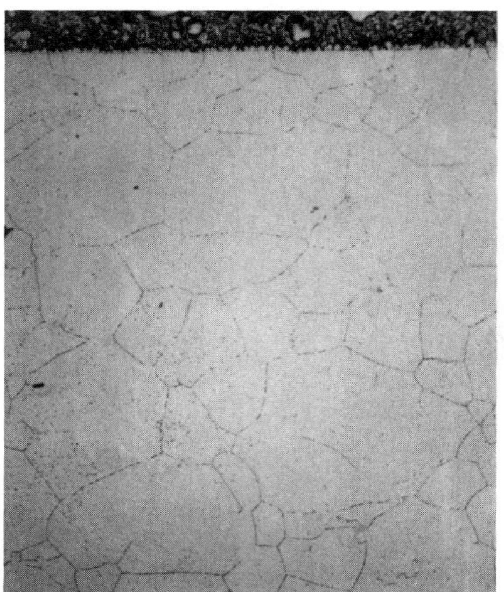

Fig. 6 Optical micrograph of a cross section of a stainless steel to stainless steel joint brazed with a silver-copper-palladium filler alloy, and etched with oxalic acid. Sensitization is evident in the stainless steel base metal. 300×

Silver-Base Filler Alloys. The class of silver-base filler alloys is used to braze most types of base metals. Examples of alloy systems brazed with silver-base alloys referred to in this section include titanium to titanium, zirconium to zirconium, aluminum to titanium, copper to copper, and stainless steel to stainless steel.

In general, the corrosion resistance of silver-base filler alloys is similar to that of other nonferrous metals used for their low-temperature corrosion-resistant properties (Ref 1). These brazing alloys are unsuitable for use with strong mineral acids, although the attack is slow enough to allow pickling in sulfuric (H_2SO_4) or hydrochloric (HCl) acids for cleaning (Ref 24).

Many corrosion tests involving silver-base filler alloys have been performed in saltwater environments. For example, titanium joints, brazed with silver-aluminum-manganese filler alloys, that were exposed to 50 h of salt spray and were flexure tested failed sooner if the filler alloy contained a lower amount or no manganese (Ref 48, 49). However, in another test, aluminum brazed to titanium with silver or silver-aluminum alloys exhibited very limited damage after salt spray exposure of over 250 days (Ref 50). In addition, a researcher found that the silver-copper-zinc filler alloy of stainless steel joints corroded out by interfacial corrosion after 1 month of exposure to a NaCl solution (Ref 21). Another researcher obtained similar results for exposure to NaCl solutions and related the decreased interfacial corrosion resistance to increasing zinc content of the filler alloy and increasing nickel plus chromium content of the base metal (Ref 51).

Many corrosion tests involving silver-base filler alloys have also been performed in water not containing salt. In one study, Zircaloy-2 joints brazed with silver-copper alloys were tested in pressurized water at 360 °C (680 °F) (Ref 52). Straight silver-copper alloys and silver-copper alloys with a small amount of silicon failed (separated) in 154 h or less. Silver-copper alloys containing 10% Sn did not fail in 1250 h; the filler alloy, however, was completely oxidized. Another researcher tested copper joints brazed with a silver-tin alloy for 1½ to 2 years (Ref 53). The filler alloy with the highest amount of tin exhibited the least amount of corrosion, but the corrosion was heavy after 22 months for this alloy.

Fig. 7 Macrograph of a copper to stainless steel braze joint that exhibited corrosion products as a result of improper flux removal. 1×. Source: Ref 47

A problem with galvanic corrosion would be expected when aluminum is coupled with titanium. However, no evidence of galvanic corrosion was found in aluminum-to-titanium joints brazed with silver-base alloys tested in simulated and actual aircraft service environments (Ref 50). Another researcher tested aluminum brazed to titanium with silver-base (silver-aluminum) alloys under weathering conditions (Ref 54). Although no galvanic corrosion was found, there was appreciable corrosion of the filler alloy after 3 months and total disappearance of the filler after 1 year.

Silver-base filler alloys have also been tested in other environments. For example, stainless steel joints brazed with silver-copper-zinc alloys were found to exhibit rapid declines in shear strength in a matter of days with exposure to HNO_3, HCl, and H_2SO_4 (Ref 21). In another test, stainless steel joints brazed with silver-copper-palladium and silver-manganese filler alloys were exposed to carbon dioxide (CO_2) + 5% carbon monoxide (CO) at 900 °C (1650 °F) for 5000 h and steam at 700 °C (1290 °F) for 5000 h (Ref 55). In both environments, most of the filler alloy disappeared because of corrosion. A third investigation involved testing of type 410 stainless steel joints brazed with a silver-copper-palladium alloy for oxidation resistance at temperatures ranging from 315 to 650 °C (600 to 1200 °F) for 500 to 1000 h (Ref 56). The filler alloy exhibited less than 0.025 mm (1 mil) of penetration after exposure to 315 °C (600 °F) for 500 h, and it exhibited a completely oxidized filler after exposure to 540 °C (1000 °F) for 1000 h. Lastly, Inconel joints brazed with a silver-palladium-manganese alloy were exposed to air at elevated temperatures (Ref 57). At 815 °C (1500 °F), the oxidation penetrated more than 0.13 mm (5 mils) after 200 h, and the filler was completely destroyed after exposure at 925 °C (1700 °F) for 200 h.

Table 4 Brazing temperatures, corrosion tests, and wettability observations of candidate brazing filler metals on Zircaloy-2 sheet

Brazing filler metal, wt%(a)	Braze temperature		Wettability(b)	500 h corrosion rate (c), mg/dm²		
	°C	°F		Zircaloy-2 control	Braze specimens 1	Braze specimens 2
Zirconium binary alloys						
Zr-33Ag	1330	2425	0	+5.8	−2310	−9810
Zr-40Ag	1220	2230	+	+39.0	−3355	−3144
Zr-50Ag	1420	2590	+	+33.8	−6	0
Zr-8.5Al	1455	2650	0	+22.5	−16 634	−19 371
Zr-11.5Al	1380	2515	0	+25.4	−25 535	−28 137
Zr-14Al	1420	2590	+	+37.7	−17 835	−15 277
Zr-12Au	1680	3055	0	+45.2	−1312	+556
Zr-24.5Au	1505	2740	0	+45.2	−16 843	−15 608
Zr-35Au	1170	2140	0	+42.6	−27 500	...
Zr-3Be	1230	2245	+	+18	−6	+235
Zr-5Be	995	1823	+	+15	+11	+145
Zr-8Be	1030	1885	+	+19	−27	+36
Zr-12Co	1230	2245	0	+35.6	−500	+125
Zr-15Co	1030	1890	−	+38.3	+878	−33
Zr-15Co(d)	1030	1885	−	+10	+184	+188
Zr-20Co	980	1795	0	+39.3	−16 852	+27
Zr-20Co(d)	1115	2040	−	+15	+79	+104
Zr-15Cr	1230	2245	+	+42.9	−155	−111
Zr-18Cr	1480	2695	0	+31.9	−200	−94
Zr-24Cr	1330	2425	0	+23.2	0	...
Zr-15Cu	1410	2570	−	+44.5	+577	+459
Zr-21Cu	1330	2425	−	+33.7	+206	+1428
Zr-25Cu	1130	2065	−	+45.2	+1153	...
Zr-54Dy	1530	2785	+	+28.2	−9702	−13 700
Zr-70Dy	1310	2390	+	+33.8	−10 166	−3864
Zr-65Er	1630	2965	+	+30.7	−11 038	−12 870
Zr-75Er	1480	2695	+	+49.3	−20 925	−25 763
Zr-12Fe	1150	2100	0	+34.4	−111	+256
Zr-16Fe	964	1167	−	+30.7	+461	...
Zr-21Fe	1130	2065	+	+62.9	+240	+16
Zr-15Ga	1430	2605	+	+41.9	−17 274	−12 444
Zr-18Ga	1280	2335	0	+22.9	−19 377	−18 281
Zr-25Ga	1330	2425	0	+83.3	−8760	−17 724
Zr-6Ge	1600	2910	+	+18.0	+170	+510
Zr-7.7Ge	1565	2850	+	+58.3	+623	...
Zr-7.7Ge(d)	1565	2850	+	+16	+467	−12 203
Zr-12Ge	1580	2875	+	+48.2	+554	...
Zr-16Mn	Did not melt up to 1800 °C (3270 °F); test discontinued		...	...	...	...
Zr-21.5Mn	1165	2130	−	+5.8	+83	−43
Zr-29Mn	1280	2335	0	+6.8	+207	−108
Zr-21Mo	1730	3145	0	+11.4	−15 625	−15 176
Zr-31Mo	1550	2820	0	+25.3	−22 238	−24 417
Zr-40Mo	1730	3145	Eliminated(f)	...	...	...
Zr-14Ni	1150	2100	0	+12.6	+240	+605
Zr-14Ni(d)	1150	2100	0	+18	+787	+709
Zr-18Ni	990	1815	+	0	+2394	+1720
Zr-22Ni	1145	2095	−	−3.0	+754	+820
Zr-22Pd	1230	2245	0	+21.9	−25 970	−34 563
Zr-27.5Pd	1070	1960	−	0	−41 393	−28 093
Zr-33Pd	1110	2030	0	+18.1	−21 258	−37 704
Zr-31Pt	1330	2425	0	+20.5	+794	+447
Zr-37Pt	1215	2220	0	+17.8	+1243	+469
Zr-40Pt	1330	2425	+	+17.3	+635	+231
Zr-1.5Si	1730	3145	0	+15.3	−166	−172
Zr-3Si	1640	2985	0	+6.1	+73	−103
Zr-4.5Si	1650	3000	0	+15.3	0	0
Zr-17Sn	1730	3145	+	+6.0	+414	+440
Zr-24Sn	1730	3145	+	+9.8	+71	−7
Zr-29Sn	1620	2950	+	+9.6	+392	+290
Zr-24V	1260	2300	−	+3.5	−31 000	−19 371
Zr-30V	1430	2605	−	+12.3	−17 692	−20 098
Zr-37V	1280	2335	−	+14.5	−26 951	−33 500
Zr-29Y	1530	2785	−	+8.9	−26 121	−16 742
Zr-59Y	1390	2535	−	+6.4	−14 233	−11 792
Zr-23Nb	Did not melt up to 1800 °C (3270 °F); test discontinued		...	...	...	...
Zr-14Ta	Did not melt up to 1800 °C (3270 °F); test discontinued		...	...	...	...
Zr-47.5Ti	1650	3000	0	+2.9	−1765	−26 782

(continued)

(a) The filler metal compositions are given in weight percent before arc melting. Some alterations in composition may have resulted during melting and subsequent brazing. (b) + = excellent; 0 = good; − = poor. (c) Weight change of the brazing filler metal is computed by subtracting the $\Delta W/A$ of the braze coupon on the control specimen times the unbrazed area of the braze coupon from the total ΔW of the braze coupon and dividing the balance by the area of the brazing filler metal: $C_3 = \Delta W_1 - (\Delta W_2/A_2)(A_1 - A_3)/A_3$, where subscripts are 1, brazed specimen; 2, Zircaloy-2 control; and 3, brazing filler metal. (d) Repeat runs. (e) Later formulation did not melt at this temperature. (f) Severely attacked Zircaloy-2 tab; test discontinued. (g) These filler metals were evaluated by mixing particles of Zr-Be and Zr-Cu binary systems. The braze temperature does not necessarily represent 30 °C (85 °F) superheat for an actual ternary melt. Source: Ref 87

Table 4 (continued)

Brazing filler metal, wt%(a)	Braze temperature °C	Braze temperature °F	Wettability(b)	500 h corrosion rate (c), mg/dm²: Zircaloy-2 control	Braze specimens 1	Braze specimens 2
Nonzirconium binary alloys						
Au-9Co	995	1825	+	+11.2	−10 000	−8980
100Au	1053	1945	+	+15.5	−330 000	−30 120
Co-29Ge	1145	2095	−	+6.4	−1344	−974
Co-77Ge	840	1545	0	+3.3	−22 143	−5871
Co-34Sn	1025	1875	Eliminated(f)	...	...	...
Cr-62Sb	1130	2065	−	−16.8	−30 354	−56 500
Cu-20Ir	970	1780	Eliminated(f)	...	...	...
Fe-20Si	1160	2120	−	+9.6	−6050	...
Ge-30Fe	890	1635	0	+5.9	−1815	−2022
Pd-3.5B	950	1740	+	+6.7	−17 140	−7970
Pd-23Sb	1035	1895	+	+13.8	−36 240	...
Pd-42Sb	840	1545	0	+3.1	−28 000	−43 670
Pd-3Si	1200	2190	Eliminated(f)	...	...	...
Pd-5Si	870	1600	0	+3.0	+1935	+4020
Pt-3.5B	950	1740	+	+5.8	+26	+5
Pt-3Si	900	1650	+	+7.4	+225	+185
Pt-4.2Si	890	1635	+	+9.7	−65 140	−11
Pt-4.2Si(d)	890	1635	+	+6.0	−21	−15
Pt-29Sn	1100	2010	0	+6.0	+238	0
Ta-1Al	1700+	3090+	Eliminated(f)	...	...	...
Ta-3.0Au	1700+	3090+	−	+13.6	−65 110	−38 000
Ta-48Au	1250	2280	...	+13.7	−65 710	−38 000
Ta-68Co	1275	2325	Eliminated(f)	...	...	...
Zirconium-base ternary alloys						
Zr-10Ag-10Si	1590	2895	Eliminated(f)	...	...	...
Zr-15Ag-15Si	1555	2830	Eliminated(f)	...	...	...
Zr-20Ag-20Si	1575	2865	Eliminated(f)	...	...	...
Zr-10Al-20Mo	1295	2365	+	+38.9	−1800	−12 955
Zr-20Al-10Mo	1340	2445	+	+25.1	−9810	−8660
Zr-20Al-20Mo	1325	2415	+	+10.7	−4700	−10 645
Zr-15Au-15Mo	1420	2590	+	+25.7	−100	−920
Zr-20Au-20Mo	1320	2410	+	+25.9	−16 190	−34 070
Zr-30Au-20Mo	1200	2190	+	+33.1	−13 850	−16 800
Zr-10Au-10Si	1550	2820	Eliminated(f)	...	...	...
Zr-15Au-15Si	1560	2840	Eliminated(f)	...	...	...
Zr-20Au-20Si	1515	2760	Eliminated(f)	...	...	...
Zr-25Au-25Ta	1320	2410	+	+12.9	−23 210	−35 210
Zr-10Co-20Mo	1160	2120	+	+34.0	−845	−1858
Zr-20Co-10Mo	970	1780	+	+31.7	+120	+635
Zr-20Co-20Mo	1120	2050	+	+33.5	+40	−11 655
Zr-20Co-20Mo(d)	1120	2050	0	+6	−43	+214
Zr-25Co-25Ta	1030	1890	Eliminated(f)	...	...	...
Zr-33Co-33Ta	1060	1940	Eliminated(f)	...	...	...
Zr-50Co-25Ta	1095	2005	Eliminated(f)	...	...	...
Zr-33Co-33Y	960	1760	+	+18.0	−40 470	−38 990
Zr-10Cr-10Fe	1190	2175	0	+7	+33	−40
Zr-15Cr-15Fe	1295	2365	Eliminated(f)	...	...	...
Zr-20Cr-20Fe	1310	2390	Eliminated(f)	...	...	...
Zr-49Cr-1In	1310	2390	Eliminated(f)	...	...	...
Zr-25Cr-25Mn	1320	2410	Eliminated(f)	...	...	...
Zr-25Cr-25Ni	1000	1830	Eliminated(f)	...	...	...
Zr-25Cr-25Si	1420	2590	Eliminated(f)	...	...	...
Zr-10Cr-10Sn	1315	2400	0	+7	+68	...
Zr-15Cr-15Sn	1330	2425	+	+8	+137	+156
Zr-20Cr-20Sn	1360	2480	+	+16	+97	+112
Zr-25Cr-25Sn	1340	2445	+	+12.4	−57	+42
Zr-12.5Cu-1.5Be(g)	1230	2245	0	+14	−397	−1963
Zr-12.5Cu-2.5Be(g)	1130	2065	+	+12	−6	+8
Zr-12.5Cu-4Be(g)	1130	2065	+	+16	−23	+11
Zr-10Ga-20Mo	1280	2335	+	+32.4	−13 220	−24 460
Zr-20Ga-10Mo	1305	2380	+	+29.6	−6680	−30 330
Zr-20Ga-20Mo	1295	2565	Eliminated(f)	...	...	...
Zr-10Ga-40Ta	1350	2460	+	+21.6	−32 290	−20 870
Zr-20Ga-30Ta	1445	2635	Eliminated(f)	...	...	...
Zr-49Mn-1In	1190	2175	Eliminated(f)	...	...	...
Zr-49Ni-1In	985	1805	Eliminated(f)	...	...	...
Zr-25Ni-25Si	1460	2660	Eliminated(f)	...	...	...
Zr-10Pt-10B	1495(e)	2723(e)	0	+30	+1	−21 600

(continued)

(a) The filler metal compositions are given in weight percent before arc melting. Some alterations in composition may have resulted during melting and subsequent brazing. (b) + = excellent; 0 = good; − = poor. (c) Weight change of the brazing filler metal is computed by subtracting the $\Delta W/A$ of the braze coupon on the control specimen times the unbrazed area of the braze coupon from the total ΔW of the braze coupon and dividing the balance by the area of the brazing filler metal: $C_3 = \Delta W_1 - (\Delta W_2/A_2)(A_1 - A_3)/A_3$, where subscripts are 1, brazed specimen; 2, Zircaloy-2 control; and 3, brazing filler metal. (d) Repeat runs. (e) Later formulation did not melt at this temperature. (f) Severely attacked Zircaloy-2 tab; test discontinued. (g) These filler metals were evaluated by mixing particles of Zr-Be and Zr-Cu binary systems. The braze temperature does not necessarily represent 30 °C (85 °F) superheat for an actual ternary melt. Source: Ref 87

Table 4 (continued)

Brazing filler metal, wt%(a)	Braze temperature °C	Braze temperature °F	Wettability(b)	500 h corrosion rate (c), mg/dm² Zircaloy-2 control	Braze specimens 1	Braze specimens 2
Zr-10Pt-10B(d)	1595	2905	0	+16	−16 264	+35
Zr-15Pt-15B	1505	2740	Eliminated(f)	...	...	...
Zr-20Pt-20B	1505	2740	Eliminated(f)	...	...	...
Zr-10Pt-10Si	1525	2775	Eliminated(f)	...	...	...
Zr-15Pt-15Si	1545	2815	Eliminated(f)	...	...	...
Zr-20Pt-20Si	1545	2815	Eliminated(f)	...	...	...
Zr-25Si-25Mn	1545	2815	Eliminated(f)	...	...	...
Zr-25Si-25Sn	1050	1922	Eliminated(f)	...	...	...
Zr-47.5Ti-5Be	940	1725	+	+18	+360	+694
Nonzirconium ternary alloys						
Cr-15Pd-15B	Did not melt up to 1800 °C (3270 °F); test discontinued		...	...	...	...
Cr-20Pd-10Si	1255	2290	Eliminated(f)	...	...	...
Cr-35Pt-10Si	1250	2280	Eliminated(f)	...	...	...
Cr-12.6Pt-15.4Sn	Did not melt up to 1800 °C (3270 °F); test discontinued		...	...	...	...
Fe-20Cr-20Ge	1270	2320	Eliminated(f)	...	...	...
Fe-30Cr-30Sb	1325	2415	Eliminated(f)	...	...	...
Ge-15Cr-15Fe	925	1695	Eliminated(f)	...	...	...
Ni-10Si-19Cr	1120	2050	Eliminated(f)	...	...	...
Pd-15Cr-15B	1105	2020	Eliminated(f)	...	...	...
Pd-15Cr-20Sb	1095	2005	Eliminated(f)	...	...	...
Pd-15Cr-10Si	1170	2140	Eliminated(f)	...	...	...
Pd-35Cr-10Si	1500	2730	Eliminated(f)	...	...	...
Pt-15Cr-15Sn	1280	2335	Eliminated(f)	...	...	...
Pt-20Cr-10Si	1205	2200	Eliminated(f)	...	...	...
Pt-25Cr-25Sb	1280	2335	Eliminated(f)	...	...	...
Pt-35Cr-10Si	1310	2390	Eliminated(f)	...	...	...
Ta-33Co-33Y	895	1645	Eliminated(f)	...	...	...

(a) The filler metal compositions are given in weight percent before arc melting. Some alterations in composition may have resulted during melting and subsequent brazing. (b) + = excellent; 0 = good; − = poor. (c) Weight change of the brazing filler metal is computed by subtracting the $\Delta W/A$ of the braze coupon on the control specimen times the unbrazed area of the braze coupon from the total ΔW of the braze coupon and dividing the balance by the area of the brazing filler metal: $C_3 = \Delta W_1 - (\Delta W_2/A_2)(A_1 - A_3)/A_3$, where subscripts are 1, brazed specimen; 2, Zircaloy-2 control; and 3, brazing filler metal. (d) Repeat runs. (e) Later formulation did not melt at this temperature. (f) Severely attacked Zircaloy-2 tab; test discontinued. (g) These filler metals were evaluated by mixing particles of Zr-Be and Zr-Cu binary systems. The braze temperature does not necessarily represent 30 °C (85 °F) superheat for an actual ternary melt. Source: Ref 87

Silver-base braze joints are generally discolored by sulfur-containing substances (Ref 24). However, some steel members brazed with silver-base brazing alloys have been successfully used in handling sulfite pulps in paper mills (Ref 58).

Tests have shown that joints brazed with silver-base filler alloys have good long-term oxidation resistance at temperatures to 425 °C (800 °F) (Ref 1). In fact, joints with silver-base filler alloys are being used successfully in engine components having service environments to 425 °C (800 °F) (Ref 56). Additional information on the corrosion of silver and silver alloys is available in the article "Corrosion of Noble Metals" in this Volume.

Copper-Base Filler Alloys. As a brazing filler alloy, copper is stronger and less expensive than silver. Copper brazing is performed at higher temperatures (1100 °C, or 2000 °F) than silver brazing and therefore requires an exceptional high-temperature flux or controlled atmosphere to prevent oxidation (Ref 59).

As was the case with silver-base filler alloys, many of the studies of the corrosion resistance of copper-base filler alloys have been done on joints involving stainless steel. Copper filler alloys containing significant amounts of zinc (such as RBCuZn) or phosphorus (such as BCuP) are not recommended for joining stainless steels, because brittle compounds tend to form at the braze/base metal interface when these filler alloys are used (Ref 50). As discussed in the section "Liquid-Metal Embrittlement" in this article, the combination of copper-base brazing alloys and stainless steel base metals is particularly susceptible to LME. The intergranular penetration of stainless steel base metals by copper-base brazing alloys has been investigated (Ref 61, 62). It was found that intergranular penetration could be avoided by using a Ni-13Cr-10P filler alloy (Ref 61). Also, as discussed previously, when copper-base alloys are used to braze stainless steel, care should be taken to cool quickly through the carbide sensitization range.

Many investigators have examined the condition of joints brazed with copper-base filler alloys exposed to water. One study found that copper pipes joined with a copper-phosphorus alloy exhibited no corrosion of the filler alloy after up to 2 years of exposure; the base metal, however, was heavily corroded after as little as 17 months of exposure (Ref 53). In a second investigation, Zircaloy-2 joints brazed with copper-phosphorus alloys exhibited failure after exposure to 360 °C (680 °F) water for as little as 86 h (Ref 52). Similar joints brazed with copper-silver-tin alloys failed after 1200 h of exposure, and a Cu-20Pd-3In alloy appeared virtually unattacked after over 1400 h of exposure. However, in another study, Zircaloy-3 joints brazed with this same filler alloy were tested in the same environment. A marked decrease in the tensile strength and the formation of a black corrosion product at the extremely brittle joint were found (Ref 63). Finally, an examination of copper and cupro-nickel joints brazed with a copper-silver-phosphorus alloy and a copper-zinc-tin alloy found that the copper-zinc-phosphorus joints were susceptible to interfacial corrosion and that the copper-zinc-tin joint exhibited severe corrosion of the β phase in the copper joint (Ref 64).

The behavior of joints brazed with copper-base filler alloys in other environments has also been investigated. For example, Ti-6Al-4V joints diffusion brazed with copper were exposed to air at 425 °C (800 °F) for 100 h, and no degradation of tensile strength of the joint was found (Ref 65). These joints were also examined for corrosion or stress corrosion after exposure at 703 MPa (102 ksi) in 3.5% NaCl for 1000 h; there was no evidence of corrosion. However, the base metal and joints were mildly susceptible to corrosion fatigue, with the joints showing somewhat higher susceptibility than the base metal. In another case, stainless steel joints brazed with copper-base filler alloys exposed to CO_2 + 5% CO at 480 °C (900 °F) for 5000 h exhibited depths of penetration in the filler alloy from 60 μm (2.5 mils) for the 82Cu-18Pd alloy to 0.6 mm (25 mils) for the 52.5Cu-38.5Mn-9Ni alloy (Ref 55). Lastly, a systematic investigation of the oxidation resistance of stainless steel joints brazed with a copper-manganese-cobalt alloy found less than 25 μm (1 mil) of penetration after 1000 h at 425 °C (800 °F), from 50 to 125 μm (2 to 5 mils) of penetration after 1000 h at 540 °C (1000 °F), and more than 125 μm (5 mils) of penetration after 500 h at 650 °C (1200 °F) (Ref 56). Additional information is available in the article "Corrosion of Copper and Copper Alloys" in this Volume.

Nickel-Base Filler Alloys. Brazed joints produced with nickel-base filler alloys are used at temperatures ranging from −200 to above 1200 °C (−330 to above 2190 °F) (Ref 1). The characteristics of nickel make it a good starting point for developing brazing alloys to withstand high service temperatures; it is ductile, and with proper alloying additions, such as chromium, it has excellent strength and oxidation resistance as well as corrosion resistance at elevated temperatures (Ref 61, 66-67). One investigation involved examination of a series of nickel-chromium-silicon-boron-iron metallic glass filler alloys with varying amounts of chromium used to braze type

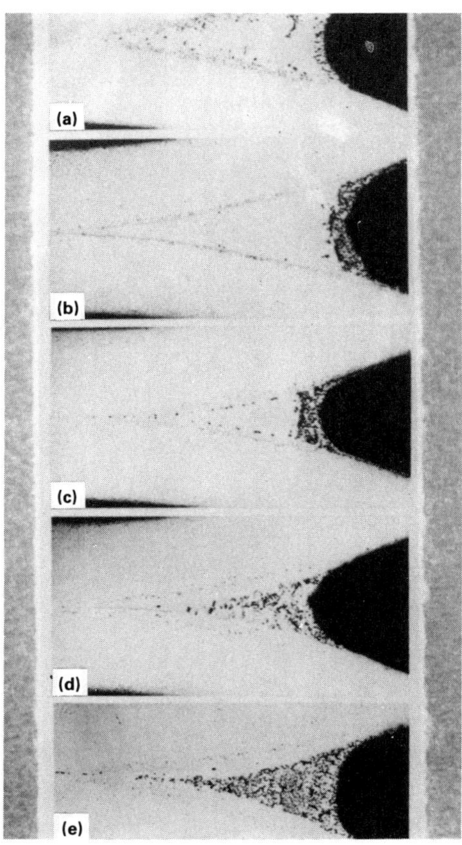

Fig. 8 Type 316 stainless steel brazed with (a) BNi-3, (b) BNi-5, (c) Ni-19Cr-10Si-10Mo, (d) BNi-7, (e) BNi-6, and exposed to potassium at 835 °C (1535 °F) for 6500 h. Source: Ref 79

410 stainless steel after exposure to saltwater vapor at 540 °C (1000 °F) for 336 h (Ref 68). A strong influence of chromium content on corrosion resistance was found. The chromium-free filler alloy was severely corroded, with both general corrosion and pitting being evident; the high-chromium alloy exhibited only minimal evidence of corrosion. In another study, similar results were obtained for titanium-stabilized austenitic stainless steel brazed with nickel-chromium filler alloys (for example, BNi-2 and BNi-7) and exposed to air at 800 and 850 °C (1470 and 1500 °F) (Ref 69).

Much of the work on the corrosion behavior of nickel-base filler alloys has been done with nickel-base metal joints. One researcher examined the effect of air exposure at 815 °C (1500 °F) for 1 week on a series of 25 nickel-base filler alloys used to braze thin nickel sheet (Ref 70). The nickel-manganese alloys exhibited excellent oxidation resistance, and the nickel-phosphorus alloys exhibited poor-to-good resistance, with resistance increasing with decreasing manganese content. In another study, Inconel joints brazed with nickel-chromium-germanium-lithium, nickel-chromium-germanium-indium, and nickel-chromium-silicon-indium exhibited significant oxidation at 900 °C (1650 °F) after 500 h (Ref 71). In a third investigation, a nickel-chromium-silicon alloy exhibited good resistance to hot corrosion and sufficiently good resistance to oxidation when used to braze nickel alloy vanes (Ref 72).

Lastly, nickel 201 joints that were vacuum brazed using a nickel-silicon-boron filler alloy according to two different vendor processes were compared (Ref 73). The processes resulted in significantly different thermal exposures, and field failures were found to be a function of the brazing process. The higher-temperature process caused rapid grain-boundary diffusion of boron from the brazing filler metal to the base metal, resulting in intergranular corrosion. The lower-temperature, faster cooling process resulted in less diffusion of the boron; therefore, failure was not intergranular. Crevice corrosion, involving attack of a needlelike nickel boride phase, led to failure.

The oxidation resistance of nickel-base filler alloys has received considerable attention. In addition to the influence of the chromium content of the filler alloy on corrosion, other factors in corrosion behavior have been examined. For example, in one study, there was no effect on austenitic/ferritic stainless steel joints brazed with a series of nickel-base filler alloys (BNi-2,3,6,7) exposed to air at 575 °C (1065 °F) for up to 10 000 h (Ref 74). Another researcher found a continuous oxide layer on the base metal of Inconel 718 joints, but there was no oxide penetration into the nickel-palladium filler alloy region after 200 h at 815 °C (1500 °F) (Ref 75). A third study found that a nickel-chromium-phosphorus and a nickel-chromium-palladium-silicon alloy had good oxidation resistance after 100 h at 800 °C (1470 °F) but that a nickel-manganese-palladium alloy had poor resistance under the same conditions (Ref 76). Finally, a nickel-tin alloy exhibited severe corrosion after exposure in air at 815 and 930 °C (1500 and 1700 °F) for up to 1300 h, but other nickel-base filler alloys, such as nickel-chromium-silicon-boron-iron alloys and nickel-chromium-germanium alloys, that were tested under these conditions exhibited only slight penetration (Ref 57).

Special tests are required for brazed joints that are put into service environments in which liquid-metal coolants, such as sodium or potassium, are used. One researcher recently examined the sodium compatibility of refractory metal alloy/type 304 stainless steel joints brazed with a series of nickel-base filler alloys (BNi-3,5,7) (Ref 77). No significant chemical corrosive attack was observed for any of the joints exposed to a sodium environment at 800 °C (1470 °F) for 130 h. Other researchers had previously conducted extensive studies of the corrosive behavior of nickel-base filler alloys in liquid sodium environments (Ref 57, 71, 78). In other studies, the corrosion behavior of nickel-base filler alloys used to join stainless steel was examined in potassium environments (Ref 79, 80). As shown in Fig. 8, the lower-melting filler alloys (for example, BNi-6,7) exhibited the least corrosion resistance in boiling potassium at 835 °C (1540 °F) for up to 6500 h, but the higher-melting filler alloys (for example, BNi-3,4) suffered minimal attack.

The corrosion behavior of a series of nickel-chromium-phosphorus alloys used to join stainless steel, Hastelloy X, and a nickel-base alloy was investigated in mixtures of steam and CO_2-CO at ambient and elevated temperatures (Ref 81, 82). All of the filler alloys showed generally good corrosion resistance under all conditions. Additional information is given in the article "Corrosion of Nickel-Base Alloys" in this Volume.

Aluminum-Base Filler Alloys. Aluminum-based brazing alloys are primarily used to braze aluminum. Corrosion in an electrolyte is minimized when using aluminum-silicon filler alloys because the electrolytic potential between the base and braze materials is minimal (Ref 1). Aluminum filler alloys that contain significant levels of zinc or copper exhibit less corrosion resistance, but their resistance is often adequate for all but extremely harsh environments. In addition, when brazing with aluminum-base filler alloys, it is important to remove the flux thoroughly after brazing (Ref 1, 83) because the highly active nature of these fluxes may cause corrosion even by simply absorbing moisture from the air (additional information is available in the article "Brazing of Aluminum Alloys" in Volume 6 of the 9th Edition of *Metals Handbook*). Removal may be accomplished by immersion in boiling water (Ref 23) or concentrated acid solutions (Ref 2).

The work of one researcher indicated that penetration of the filler alloy into the base metal was a problem when aluminum-silicon-yttrium alloys were used to braze aluminum alloys (Ref 84). In another study, the susceptibility of several aluminum alloys (6951, 6061, 6063, 3003, 3004, 3105) to penetration of aluminum-silicon filler alloys was investigated (Ref 85). Only the alloys 6951 and 6061 were found to exhibit a tendency toward excessive grain-boundary penetration. In studying the pitting corrosion properties of vacuum-brazed 7072 clad A3003 aluminum alloy, it was found that the potential difference between clad and core materials maintained the corrosion resistance of the joint (Ref 86).

Aluminum-base filler alloys are also used to join aluminum to titanium or titanium to titanium. The galvanic potential between aluminum and titanium alloys indicates that galvanic corrosion should occur in an electrolyte (Ref 50), but electropotential studies have shown that the natural oxide passivation films on the titanium and aluminum surfaces greatly reduce the danger of that couple (Ref 1). In one study, no susceptibility to SCC was observed in titanium alloys brazed with aluminum-base filler alloys (Ref 28). These corrosion tests on titanium joints indicated adequate corrosion resistance when an aluminum-silicon filler alloy was exposed to aqueous 3.5% NaCl at room temperature, but there was inadequate corrosion resistance when aluminum-copper-silver filler alloy was exposed. Other researchers found that weathered titanium joints brazed with aluminum-zinc filler metals showed no visible corrosion after 1 year (Ref 54). Additional information is available in the article "Corrosion of Aluminum and Aluminum Alloys" in this Volume.

Other Filler Alloy Systems. In one study, the corrosion behavior of more than 150 candidate filler alloys (mostly zirconium-base) was evaluated for joining Zircaloy alloys (Ref 87). The results of these corrosion tests in 3-L stainless steel autoclaves with exposure to boron shim water at 360 °C (680 °F) and 18.6 MPa (2705 psi) for 500 h are presented in Table 4. Another investigation examined the corrosion resistance of Zircaloy brazements with a wide variety of filler alloys, including many zirconium-, palladium-, and silver-base alloys, exposed to 360-°C (680-°F) water (Ref 52). Test results are given in Table 5. Additional information on the corrosion of Zircaloy alloys can be found in the articles "Corrosion of Zirconium and Hafnium" and "Corrosion in the Nuclear Power Industry" (in particular the discussion of corrosion of Zircaloy-clad LWR fuel rods) in this Volume.

In a study of the corrosion resistance of a series of iron-base filler alloys for brazing molybdenum, the alloys were found to exhibit outstanding corrosion resistance when exposed to liquid sodium and molten fluoride salts at 600 to 700 °C (1110 to 1290 °F) (Ref 88). Also, a titanium-vanadium-beryllium filler alloy was tested in potassium and in vacuum; the alloy performed satisfactorily (Ref 89).

Several hundred binary and ternary alloys (mainly based on zirconium and/or titanium systems) were investigated in an effort to find a suitable filler alloy for joining niobium (Ref 90). The results of these shear tests on joints aged 100 h at 815 °C (1500 °F) are summarized in Table 6.

An extensive investigation was conducted on the galvanic corrosion of metals and coatings when coupled to uranium in severe environments (Ref 91). This report also contains a comprehensive bibliography on this subject. The materials that were investigated included brazing alloys, namely silver-copper and silver. The environments in which the galvanic series was determined were as follows:

- 0.05 *M* H_2SO_4, pH 1.0, 29 °C (84 °F)
- deionized water, pH 5.8, 29 °C (84 °F)
- 5.0 wt% NaCl, pH 6.8, 29 °C (84 °F)
- 0.5 *M* sodium carbonate (Na_2CO_3), pH 11.3, 29 °C (84 °F)

Many other braze alloy systems have been examined for their corrosion behavior in various environments. Reference 1 provides an extensive review of the literature on this subject.

Table 5 Corrosion test results for Zircaloy specimens brazed without spacing shims, exposed to 360 °C (680 °F)

Filler metal composition (nominal), wt%	Total exposure time, h	Condition of specimen(a)
Ag-28Cu	154 failed	...
Ag-28Cu	24 failed	...
Ag-28Cu	29.5 failed	...
Ag-30Cu-10Si	29.5 failed	...
Ag-30Cu-10Sn	1250	Filler metal was completely oxidized.
Al-1Ni	1318	Center of filler metal was completely oxidized, although a corrosion-resistant alloy layer was found adjacent to the base metal.
Al (hot dip coating)	163 failed	...
Al-24Pd	113 failed	...
Al-12Si	1413	Considerable attack along centerline of filler metal
Cu-7P	86 failed	Was thick joint
Cu-7P	1318	Virtually complete failure
Cu-15Ag-5P-5Sn	1201	Failure virtually complete
Cu-15Ag-5P-10Sn	1201	Failure virtually complete
Cu-20Pd-3In	1413	Virtually unattacked
Ni-7P (Kanigen plate)	1367	Virtually unattacked
Ni-3.4P	1313	75% of joint destroyed
Ni-30Ge-13Cr	104 failed	...
Ni-20Pd-10Si	1367	Slight attack on filler metal
Ni-20Cr-10Si	1315	Extensive corrosion
Pd-6Si	1310	Very extensive attack of Zircaloy adjacent to filler metal
Pd-10Nb	1315	Very heavy attack of base metal
Pd-20Sn	1413	Extensive base metal attack extending into filler metal
Zr-21Cu	1311	80% of joint destroyed
Zr-17Ni	1200	Heavy attack
Zr-15Ni-10Sn	1200	Extensive attack throughout joint
Zr-5Be	1259	Unattacked
Zr-10Fe-10Sn	1200	50% of joint corroded
Zr-15Fe-15Mn	1200	20% of joint corroded
Gold	23 failed	...
Au-10Co	1450	Virtually complete failure

(a) Specimens were removed from test after a minimum of 1200 h and examined. Source: Ref 52

Table 6 Results of room-temperature shear tests on brazed niobium joints

Shear strength of the niobium specimen: 207 MPa (30 ksi)

Alloy, wt%	Brazement shear strength(a): As-brazed, MPa	As-brazed, ksi	Aged 100 h at 815 °C (1500 °F), MPa	Aged, ksi
48Ti-48Zr-4Be	230 (6)	33.3 (6)	169 (3)	24.5 (3)
46Ti-46Zr-4V-4Be	192 (6)	27.8 (6)	224 (4)	32.5 (4)
63Ti-27Fe-10V	187 (6)	27.088 (6)	155 (4)	22.5 (4)
68Ti-28V-4Be	185 (6)	26.8 (6)	154 (4)	22.3 (4)
63Ti-27Fe-10Mo	172 (5)	24.9 (5)	134 (4)	19.5 (4)
75Zr-19Nb-6Be	165 (6)	24.0 (6)	171 (4)	24.8 (4)
45Ti-40Zr-15Fe	155 (3)	22.5 (3)	121 (4)	17.5 (4)
60Zr-25V-15Nb	148 (4)	21.5 (4)	85 (3)	12.4 (3)
67Zr-29V-4Fe	139 (5)	20.2 (5)	101 (3)	14.6 (3)
80Zr-17Fe-3Be	117 (3)	16.2 (3)	(b)	(b)

(a) Numbers in parentheses indicate number of specimens. (b) Specimens broke during machining. Source: Ref 90

REFERENCES

1. N.C. Cole, Corrosion Resistance of Brazed Joints, *Weld. Res. Counc. Bull.*, No. 247, April 1979
2. *Brazing Manual*, American Welding Society, 1976
3. M.M. Schwartz, Ed., *Source Book on Brazing and Brazing Technology*, American Society for Metals, 1980
4. *The Brazing Book*, Handy and Harman, 1985
5. M.G. Fontana and N.D. Greene, *Corrosion Engineering*, McGraw-Hill, 1978
6. W.H. Vernon, *J. Trans. Faraday Soc.*, Vol 19, 1923-1924, p 839
7. F.L. LaQue and H.R. Copson, *Corrosion Resistance of Metals and Alloys*, 2nd ed., Reinhold, 1963
8. J.F. Bosich, *Corrosion Prevention for Practicing Engineers*, Barnes and Noble, 1970
9. J.R. Winkel and J. McAndrews, Rockwell International, unpublished research, 1986
10. J.A. Willingham, Interfacial Corrosion in Stainless Steel Brazed Joint, *Stainless Steel Ind.*, Vol 8 (No. 46), Nov 1980, p 17
11. G.H. Sistare, J.J. Halbig, and L.C. Grenell, Silver Brazing Alloys for Corrosion Resistant Joints in Stainless Steels, *Weld. J.*, Vol 33 (No. 2), 1954, p 137
12. G. Lewis, Short Communication: A Photo-Electron Spectroscopy Examination of Brazed Stainless Steel Joints, *Corros. Sci.*, Vol 20, 1980, p 1259
13. S. Takahashi, *Int. Dent. J. Lond.*, Vol 18 (No. 4), 1968, p 823
14. A.T. Kuhn and R.M. Trummer, Review of the Aqueous Corrosion of Stainless Steel-Silver Brazed Joints, *Br. Corros. J.*, Vol 17 (No. 1), 1982
15. Interface Corrosion in Brazed Joints in Stainless Steel, *Brazing Tech. Bull.*, No. T-9, 1958
16. R.A. Jarman, J.W. Myles, and J.C. Booker, Interfacial Corrosion of Brazed Stainless Steel Joints in Domestic Tap Water, *Br. Corros. J.*, Vol 8, 1973, p 33
17. I. Okamoto, T. Takemoto, and C. Funiwara, Corrosion Behavior of Silver Brazed Stainless Steel in Chloride Solutions, *Trans. Jpn. Weld. Res. Inst.*, Vol 8 (No. 1), 1979, p 59
18. P. Caille, C. Niney, and M.E. Vacille, Assembly of Welding Equipment for the Food and Drink Industry, *Int. Inst. Weld. Fnn.*, 1978
19. J. Hinde and E.R. Perry, *Weld. Met. Fabr.*, Vol 28, 1960, p 145
20. C.S. Churchill, Brazing and Soldering Techniques Part 2: Brazing, in *The Machinery Press*, 1963
21. I. Kawakatsu, Corrosion of B-Ag Brazed Joints in Stainless Steel, *Weld. J.*, Vol 52, 1973, p 233s
22. A.S. McDonald, Alloys for Brazing Thin Sections of Stainless Steel, *Weld. J.*, March 1957, p 131s
23. E. Kauczor, Erosion-Corrosion in Pipe Systems Due to Inappropriate Brazing and Welding Practices, *Praktiker*, Vol 32 (No. 4), April 1980, p 128
24. N.C. Cole, Corrosion Resistance of Brazed Joints, in *Source Book on Brazing and Braz-*

ing Technology, American Society for Metals, 1980, p 365
25. T.G. Glenn and S.E. Krause, Rockwell International, unpublished research, 1980
26. H.H. Uhlig, *Corrosion and Corrosion Control*, John Wiley & Sons, 1971
27. A.H. Freedman, Basic Properties of Thin-Film Diffusion Brazed Joints in Ti-6Al-4V, *Weld. J.*, Aug 1971, p 343s
28. R.R. Wells, Low Temperature Large-Area Brazing of Damage Tolerant Titanium Structures, *Weld. J.*, Oct 1975, p 348s
29. C.L. Quimby, M.M. McDonald, and E.L. Brown, unpublished research, 1986
30. M.M. McDonald and D.L. Keller, Wetting of Molybdenum by Various Commercial Brazing Alloys, in *International Trends in Welding Research Conference*, American Society for Metals, 1987
31. C. Heiple, W. Bennett, and T. Rising, Embrittlement of Several Stainless Steels by Liquid Copper and Liquid Braze Alloys, *Mater. Sci. Eng.*, Vol 52, 1982, p 277
32. R. Genders, *J. Inst. Met.*, Vol 37, 1927, p 215
33. V.H. Schottky, K. Schichtel, and R. Stolle, *Arch. Eisenhüttenwes.*, Vol 4, 1931, p 54
34. S.V. Williams, *J. Inst. Met.*, Vol 37, 1927, p 226
35. O. Paterman, *J. Inst. Met.*, Vol 56, 1935, p 254
36. D.A. Melford, *J. Iron Steel Inst.*, Vol 200, 1962, p 290
37. R.R. Hough and R. Rolls, Creep Fracture Phenomena in Iron Embrittled by Copper, *J. Mater. Sci.*, Vol 6, 1971, p 1493
38. W.F. Savage, E.F. Nippes, and R.P. Stanton, Intergranular Attack of Steel by Molten Copper, *Weld J.*, Vol 57, 1978, p 9s
39. W.F. Savage, E.F. Nippes, and M.C. Mushala, Copper Contamination Cracking in the Weld Heat-Affected Zone, *Weld. J.*, Vol 57, 1978, p 145s
40. W.F. Savage, E.F. Nippes, and M.C. Mushala, Liquid-Metal Embrittlement of the Heat-Affected Zone by Copper Contamination, *Weld. J.*, Vol 57, 1978, p 237s
41. H.H.H. Watson, Fluid-Tight Joints for Exacting Applications, in *Source Book on Brazing and Brazing Technology*, American Society for Metals, 1980, p 249
42. D.F. Averette, W.L. Johns, J.H. Doyle, and D.H. Riefenberg, Rockwell International, unpublished research, 1984
43. E.F. Westland, Vacuum Furnace Brazing of Beryllium, *The Metallurgy of Beryllium*, Chapman and Hall, 1963, p 843
44. Welding Workbook: Brazing Joint Designs and Filler Metal Placement, *Weld. J.*, Sept 1984, p 51
45. M.M. McDonald, The Metallurgy and Metallography of Braze Joints, in *International Metallographic Society Denver Symposium*, American Society for Metals, 1987
46. G.M.A. Blanc, J. Colbus, and C.G. Keel, Notes on the Assessment of Filler Metals and Fluxes, in *Source Book on Brazing and Brazing Technology*, American Society for Metals, 1980, p 138
47. D.L. Olson, Colorado School of Mines, unpublished research, 1977
48. W.T. Kaarlela and W.S. Margolis, Development of the Ag-Al-Mn Brazing Filler Metal for Titanium, *Weld. J.*, Vol 53 (No. 10), 1974, p 629
49. W.E. Johnson, W.O. Sanafrank, W.T. Kaarlela, C.J. Kastrop, and W.M. Slifer, "Titanium Sandwich Panel Research and Development," AMC TR59-7-618, Vol 111, Nov 1959
50. S.D. Elrod, D.T. Lovell, and R.A. Davis, Aluminum Brazed Titanium Honeycomb Sandwich Structure—a New System, *Weld. J.*, Oct 1973, p 425s
51. T. Takemoto and I. Okamoto, Effect of Composition on the Corrosion Behavior of Stainless Steels Brazed With Silver-Based Filler Metals, *Weld. J.*, Oct 1984, p 300s
52. J.B. McAndrew, H. Schwartzbart, and R. Necheles, Corrosion Resistance of Zircaloy-2 Brazements in High-Temperature Water, *Weld. J.*, June 1957, p 287s
53. K. Nielsen, Corrosion of Soldered and Brazed Joints in Tap Water, *Br. Corros. J.*, Vol 19 (No. 2), 1984, p 57
54. F. Bollenrath and G. Metzger, The Brazing of Titanium to Aluminum, *Weld. J.*, Oct 1963, p 442s
55. G.E. Sheward, Brazing Applied to the Development and Manufacture of Nuclear Fuel Elements, *Weld. J.*, July 1970, p 548
56. R.P. Schaefer, J.E. Flynn, and J.R. Doyle, Brazing Filler Metal Evaluation for an Aircraft Gas Turbine Engine Application, *Weld. J.*, Sept 1971, p 394s
57. G.M. Slaughter, C.F. Leitten, Jr., P. Patriarca, E.E. Hoffman, and W.D. Manly, Sodium Corrosion and Oxidation Resistance of High Temperature Brazing Alloys, *Weld. J.*, Vol 36 (No. 5), 1957, p 217s
58. C.H. Chatfield and A.W. Swift, Silver Alloy Brazing Stainless Steel, *Weld. J.*, Vol 28, 1949, p 1142
59. B.R. Williams, The Basics of Copper Brazing, in *Source Book on Brazing and Brazing Technology*, American Society for Metals, 1980, p 151
60. V.P. Weaver and J. Imperati, Copper and Copper Alloys for Pressure Vessels, *Weld. Res. Counc. Bull.*, Nov 1961
61. I. Amato, Some Developments in Rotating Heat Exchanger Brazing Technology, *Weld. J.*, Vol 49 (No. 4), April 1970, p 177s
62. N. Bredz and H. Schwartzbart, *Weld. J.*, Vol 38 (No. 8), 1959, p 305s
63. J.M. Gerken, "Brazing of Zircaloy-3 for Temperature Instrumented Probes, S3G Task 18," KAPL-M-JMG-9, May 1957
64. B. Upton, Brazing Alloys for Marine Service, *Br. Corros. J.*, Vol 1, 1966, p 134
65. A.H. Freedman, Basic Properties of Thin Film Diffusion Brazed Joints in Ti-6Al-4V, *Weld. J.*, Vol 50 (No. 8), 1971, p 343s
66. S. Lamb and F.M. Miller, The Effects of Aggression by Nickel-Base Brazing Filler Metals, *Weld. J.*, July 1969, p 283s
67. I. Amato, F. Baudrocco, and M. Ravizza, Spreading and Aggression Effects by Nickel-Base Brazing Filler Metals on the Alloy 718, *Weld. J.*, July 1972, p 341s
68. D. Bose, A. Datta, and N. DeCristofaro, Comparison of Gold-Nickel With Nickel-Base Metallic Glass Brazing Foils, *Weld. J.*, Oct 1981, p 29
69. V. Ruza, N. Lehka, and J.K. Malik, Oxidation Resistance of Braze Joints in Stainless Steel, *Met. Constr./Br. Weld. J.*, June 1974, p 183
70. W.A. Petersen, Brazing Thin Nickel Sheet, *Weld. J.*, April 1963, p 190s
71. D. Canonico and H. Schwartzbart, Development of Oxidation- and Liquid-Sodium-Resistant Brazing Alloys, *Weld. J.*, March 1960, p 122s
72. L.J. Malcolm, Developments in Vacuum Braze Coatings of Aero-Engine Nozzle Guide Vanes, *Weld. J.*, July 1972, p 483
73. E.I. Savage and J.J. Kane, Microstructural Characterization of Nickel Braze Joints as a Function of Thermal Exposure, *Weld. J.*, Oct 1984, p 316s
74. H.R. Heap and C.C. Riley, Development of Brazed Austenitic/Ferritic Steel Steam Pipe Joints for Turbines, *Weld. J.*, June 1971, p 253s
75. T.L. D'Silva, Nickel-Palladium Base Brazing Filler Metal, in *Source Book on Brazing and Brazing Technology*, American Society for Metals, 1980, p 209
76. J.T. Klomp, Heat Resistant Ceramic-to-Metal Seals, *Weld. J.*, Vol 50 (No. 2), 1971, p 88s
77. F.M. Hosking, Sodium Compatibility of Refractory Metal Alloy-Type 304L Stainless Steel Joints, *Weld. J.*, July 1985, p 181s
78. M. Soenen, "Compatibility of Different Brazing Alloys During Long Time Exposure in Sodium Loop," Centre D'Etude De L'EnergieNuckleaire,Mol,Belgium,Cen-EURATOM, Contract No. 006-60-5/BRAB, SM-85/2
79. N.C. Cole and G.M. Tolson, "Corrosion of Nickel-Based Filler Metals in Boiling Potassium," Paper presented at the 53rd Annual Meeting, Detroit, MI, American Welding Society, April 1972
80. N.C. Cole and J.W. Hendricks, Compatibility of Brazes With Liquid Metals and Molten Salts, *Trans. Am. Nucl. Soc.*, Vol 15, 1972, p 237
81. G.E. Sheward and G.R. Bell, Development and Evaluation of a Ni-Cr-P Brazing Filler Alloy, *Weld. J.*, Oct 1976, p 285
82. G.E. Sheward and G.R. Bell, Development and Evaluation of a Ni-Cr-P Brazing Filler Alloy, in *Source Book on Brazing and Brazing Technology*, American Society for Metals, 1980, p 213
83. W. Schultze and H. Schoer, Fluxless Brazing of Aluminum, *Weld. J.*, Oct 1973, p 644
84. W.J. Werner, G.M. Slaughter, and F.B. Gurtner, Development of Filler Metals and Procedures for Vacuum Brazing of Aluminum, *Weld. J.*, Feb 1972, p 64s
85. D.J. Schmatz, Grain Boundary Penetration During Brazing of Aluminum, *Weld. J.*, Oct 1983, p 267s
86. T. Hattori and A. Sakamoto, Pitting Corrosion Property of Vacuum Brazed 7072 Clad Aluminum Alloy, *Weld. J.*, Oct 1982, p 339s
87. R.E. Beal and Z.P. Saperstein, Development of Brazing Filler Metals for Zircaloy, *Weld. J.*, July 1971, p 275s
88. N.C. Cole, R.W. Gunkel, and J.W. Koger, Development of Corrosion Resistant Filler Metals for Brazing Molybdenum, *Weld. J.*, Oct 1973, p 466s
89. J.W. Hendricks, N.C. Cole, and G.M. Slaughter, Compatibility of Braze Joints With Potassium and Vacuum, *Weld. J.*, July 1972, p 329s
90. C.W. Fox, R.G. Gilliland, and G.M. Slaughter, Development of Alloys for Brazing Columbium, *Weld. J.*, Dec 1963, p 535s
91. J.R. Winkel and E.L. Childs, "Galvanic Corrosion of Metals and Coatings When Coupled to Uranium in Severe Environments," RFP-3486, Rockwell International, Rocky Flats Plant, Oct 1983

Corrosion of Clad Metals

Robert Baboian and Gardner Haynes, Texas Instruments, Inc.

CLAD METALS are metallurgical materials systems that are a part of a large group of materials termed composites. As shown in Fig. 1, clad metals are categorized as bonded metal-metal laminar composite systems that can be fabricated by several processes. They are also referred to as sandwich metals, metal laminates, and multimetals. Clad metals can be provided in plate, strip, tube, rod, and wire form.

The use of clad metal systems dates back to 3000 B.C., when gold was used to cover bronze. Hammering was the method usually used for cladding. Even today, clad metals—including platinum- and gold-clad (gold-filled) systems—are widely used in the jewelry industry. In about 300 B.C., laminated swords fabricated by hammer cladding were stronger, lighter, and more durable than the ones made with a monometal. Modern cladding processes originated in the early 1800s, when English craftsmen developed the Old Sheffield process for cladding silver (or gold) to another metal. This was the first use of the roll-bonding process.

The Cladding Process

The cladding process is generally differentiated from other bonding processes, such as brazing and welding, by the fact that none of the metals to be joined is molten when a metal-to-metal bond is achieved. Also, there are no intermediate layers, such as adhesives. The principle cladding techniques include cold roll bonding, hot roll bonding, hot pressing, explosion bonding, and extrusion bonding (Fig. 1). Regardless of the technique used, the bond is achieved by forcing clean oxide-free metal surfaces into intimate contact; this causes a sharing of electrons between the metals. Gaseous impurities diffuse into the metals, and nondiffusible impurities consolidate by spheroidization. All the techniques involve some form of deformation to break up surface oxides, to create metal-to-metal contact, and to heat in order to accelerate diffusion. The techniques differ in the amount of deformation and heat used to form the bond and in the method of bringing the metals into intimate contact. Cold and hot roll bonding apply primarily to sheet (less than 5 mm, or 0.2 in., thick), but explosion bonding is usually restricted to thicker gages (up to several inches).

Most engineering metals and alloys can be clad by using one or more of these techniques. As many as 100 different metal combinations with up to 15 layers have been cold roll bonded. Clad combinations that have been commercially produced on a large scale are shown in Fig. 2.

Certain combinations are more difficult to bond. Table 1 categorizes a number of cladding combinations by degree of difficulty in bonding. Properties that affect the bonding process include ductility, nature of oxide films, and tendency to form intermetallic compounds. Because bonding processes commonly use large amounts of reduction, alloys with greater ductility are more easily clad. Metals and alloys with virtually no ductility, such as beryllium and chromium, are impossible to bond. Metals and alloys that form tenacious oxide films are more difficult to bond. These include stainless steels, aluminum alloys, and the refractory metals. Finally, some combinations of metals are thermally unstable and form brittle intermetallic compounds above a certain temperature. These combinations can be clad, but require stringent control of process variables to avoid the formation of intermetallic compounds.

Designing With Clad Metals

The choice of a material for a particular application depends on such factors as cost, availability, appearance, strength, fabricability, electrical or thermal properties, mechanical properties, and corrosion resistance. Clad metals provide a means of designing into a composite material specific properties that cannot be obtained in a single material. The early use of clad metals in the jewelry industry combined the aesthetics of precious metals with the low-cost strength of base metals. These materials systems are currently being used for electrical and electronics applica-

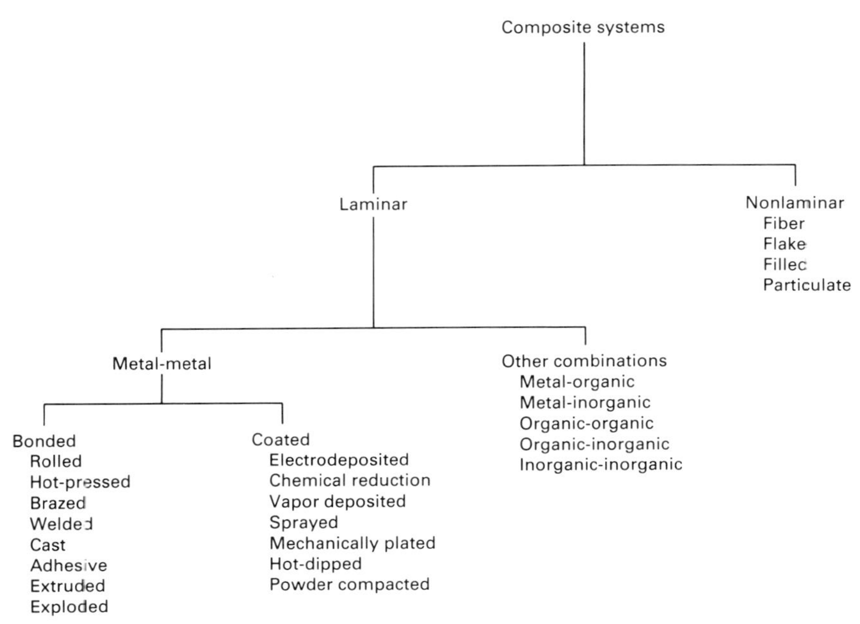

Fig. 1 Categorization of clad metals as bonded metal-metal laminar composite systems

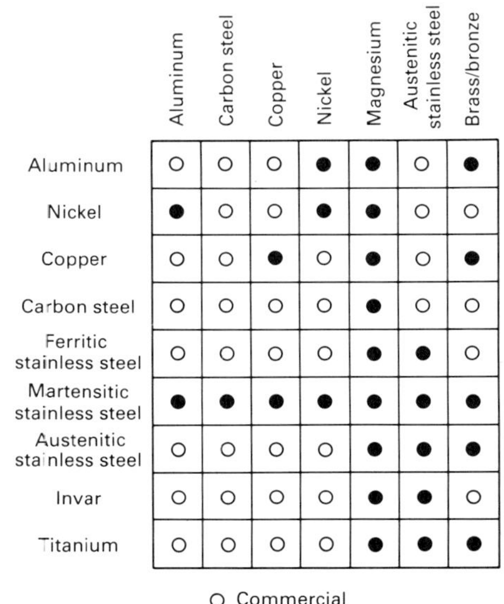

	Aluminum	Carbon steel	Copper	Nickel	Magnesium	Austenitic stainless steel	Brass/bronze
Aluminum	○	○	○	●	●	○	●
Nickel	●	○	○	●	●	○	○
Copper	○	○	●	○	●	○	●
Carbon steel	○	○	○	○	●	○	○
Ferritic stainless steel	○	○	○	○	●	●	○
Martensitic stainless steel	●	●	●	●	●	●	●
Austenitic stainless steel	○	○	○	○	●	●	●
Invar	○	○	○	○	●	●	○
Titanium	○	○	○	○	●	●	●

○ Commercial
● Requires development

Fig. 2 High-volume commercially available clad metals

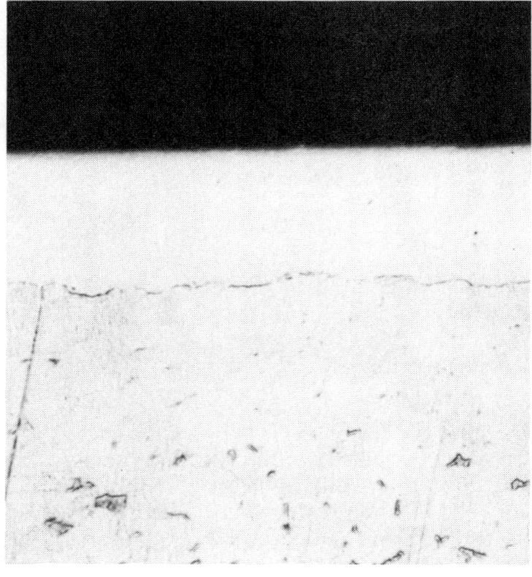

(a)

(b)

Fig. 3 Photomicrographs of cross sections of type 304 stainless steel clad carbon steel. (a) As-polished. 300×. (b) Polished and etched. 500×

tions, such as contacts and connectors with selectively clad (inlay) precious metals for low contact resistance and high reliability.

Clad metals can provide properties not available in a monolithic material at any cost. The best example is thermostat bimetals. An alloy with a high coefficient of thermal expansion is clad to an alloy with a low coefficient of thermal expansion. When heated, the resulting bimetal will bend about its neutral axis. Thermostat bimetals are used in various temperature-sensing devices, including motor protectors, circuit breakers, automotive chokes, vent dampers, and room thermostats.

Clad metals can be designed with unique coefficients of linear expansion by cladding two or more alloys in a symmetrical configuration. Specific examples are copper-clad invar (iron-nickel alloy) wire for glass-to-metal seals and copper-clad invar (Cu/invar/Cu), that matches the coefficient of expansion of ceramics for leadless semiconductor chip carriers. Different expansion rates can be obtained by varying the cladding ratio.

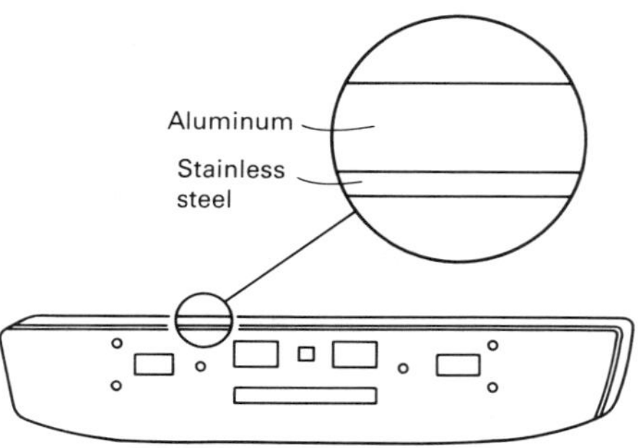

Fig. 4 Stainless steel clad aluminum truck bumper material that combines the corrosion resistance of stainless steel with lightweight aluminum

Self-brazing materials, such as copper-clad stainless steel (Cu/SS/Cu) and copper clad steel (Cu/steel/Cu), provide another example of the unique properties designed into a clad material. Multilayer heat exchangers are fabricated from these materials by simply stacking the layers of clad material and furnace brazing the entire assembly.

A final example of a unique clad system is the brass-clad steel used in bullet jackets. In this case, the brass cladding provides the drawability that allows steel to be used as a bullet jacket with improved strength and aerodynamics.

Equations for calculating the properties of clad systems are available in the literature. Basically, such properties as density, electrical conductivity, and thermal conductivity involve weighted averages of the components. Other properties, such as thermal expansion and modulus of elasticity, depend on the clad configuration, but can be calculated.

Designing Clad Metals for Corrosion Control

Clad metals designed for corrosion control can be categorized according to the following systems:

- Noble metal clad systems
- Corrosion barrier systems
- Sacrificial metal systems
- Transition metal systems
- Complex multilayer systems

Proper design is essential for providing maximum corrosion resistance with clad metals. This section will discuss the basis for designing clad metals for corrosion resistance.

Noble metal clad systems are materials having a relatively inexpensive base metal covered with a corrosion-resistant metal. Selection of the substrate metal is based on the properties required for a particular application. For example, when strength is required, steel is frequently chosen as the substrate. The cladding metal is chosen for its corrosion resistance in a particular environment, such as seawater, sour gas, high temperature, and automotive.

A wide range of corrosion-resistant alloys clad to steel substrates have been used in industrial applications. One example is type 304 stainless steel on steel. Figure 3 shows cross sections of this material. The uniformity of the bond interface is apparent in Fig. 3(a), and in the polished-and-etched condition (Fig. 3b), the metallographic structure of the stainless steel is clearly visible.

Table 1 Categorization of clad metals by degree of difficulty in bonding

In most cases, when a metal is named, alloys of that metal also apply.

Easy to bond
Copper/steel
Copper/nickel
Copper/silver
Copper/gold
Aluminum/aluminum alloys
Tin/copper
Tin/nickel
Gold/nickel
Difficult to bond
Copper/aluminum
Aluminum/carbon steel
Stainless steel/aluminum
Copper/tantalum, niobium, titanium
Titanium/carbon steel, stainless steel
Stainless steel/carbon steel
Aluminum/nickel
Nickel/steel
Copper/stainless steel
Manganese/nickel/copper
Copper/manganese
Silver/manganese
Silver/steel
Uranium/zirconium
Zirconium/copper, steel, stainless steel
Platinum/nickel, copper, steel
Tantalum/niobium
Impractical to bond
Gold/aluminum
Zirconium/aluminum
Cobalt/aluminum
Impossible to bond
Beryllium/anything
Chromium/anything

The grain structure is analogous to that of annealed stainless steel strip.

Clad metals of this type are typically used in the form of strip, plate, and tubing. The noble metal cladding ranges from commonly used stainless steels, such as type 304, to high-nickel alloys, such as Inconel 625. These clad metals find various applications in the marine, chemical-processing, power, and pollution control industries. Specific uses include heat exchangers, reaction and pressure vessels, furnace tubes, tubes and tube elements for boilers, scrubbers, and other systems involved in the production of chemicals.

Another group of commonly used noble metal clad metals uses aluminum as a substrate. For example, in stainless steel clad aluminum truck bumpers (Fig. 4), the type 302 stainless steel cladding provides a bright corrosion-resistant surface that also resists the mechanical damage (stone impingement) encountered in service. The aluminum provides a substrate with a high strength-to-weight ratio.

A wide range of precious metals are clad to lower-cost materials. Platinum-clad niobium consists of a thin layer of the precious metal bonded to a niobium substrate. This clad material, available in strip, wire, and rod form, is used as an anode for impressed-current cathodic protection (Fig. 5) and for such other applications as elec-

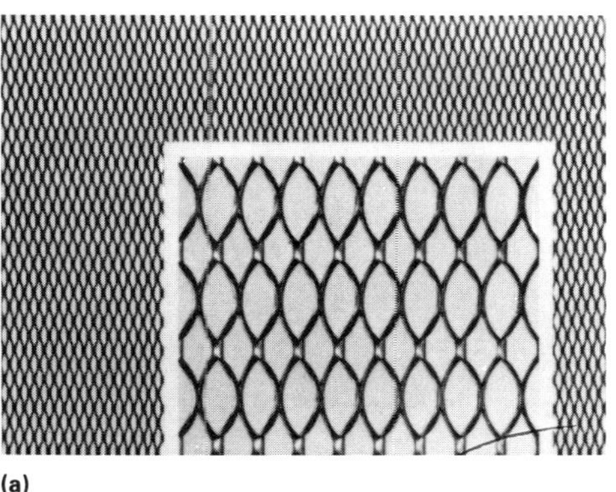

(a)

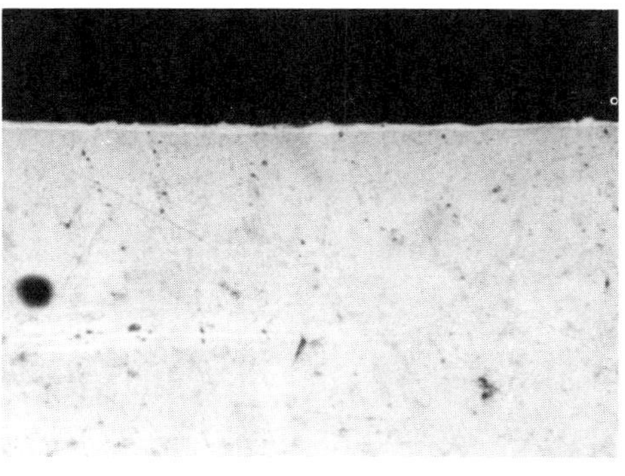

(b)

Fig. 5 Platinum-clad niobium, used widely as an anode material in electroplating and in impressed-current cathodic protection. (a) Expanded anode. (b) Cross section showing 1-μm (0.04-mil) thick platinum cladding on a niobium substrate. 500×

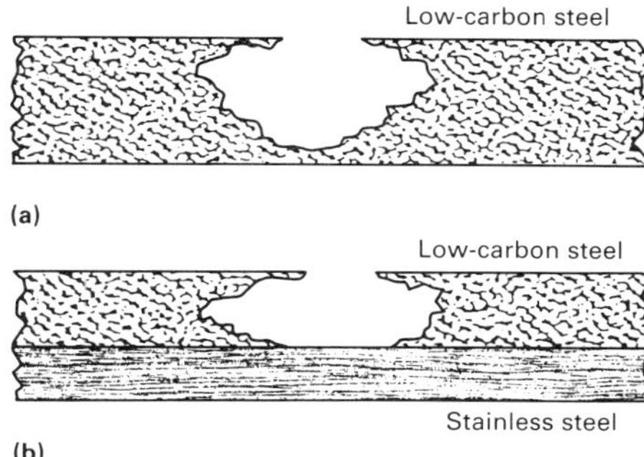

Fig. 6 Illustrations of the corrosion barrier principle. (a) Solid carbon steel. (b) Carbon steel clad stainless steel

troplating and desalination. Gold that is clad to such substrate materials as brass is widely used in wire, strip, and tube form. The most familiar example is the gold-filled writing instrument, which consists of gold clad to brass.

Corrosion Barrier Systems. The combination of two or more metals to form a corrosion barrier system is most widely used where perforation caused by corrosion must be avoided (Fig. 6). Low-carbon steel and stainless steel are susceptible to localized corrosion in chloride-containing environments and may perforate rapidly. When steel is clad over the stainless steel layer, the corrosion barrier mechanism prevents perforation. Localized corrosion of the stainless steel is prevented; the stainless steel is protected galvanically by the sacrificial corrosion of the steel in the metal laminate. Therefore, only a thin pore-free layer is required.

The example shown in Fig. 7 of carbon steel clad to type 304 stainless steel demonstrates how perforation is avoided in seawater compared to solid type 304 stainless steel. This material can be used for tubing and for wire in applications requiring strength and corrosion resistance.

Carbon steel cannot be used when increased general corrosion resistance of the outer cladding is required. A low-grade stainless steel with good resistance to uniform corrosion but poor resistance to localized corrosion can be selected. In seawater service, type 304 stainless steel that is clad to a thin layer of Hastelloy C-276 provides a substitute for solid Hastelloy C-276. In this corrosion barrier system, localized corrosion of the type 304 stainless steel is arrested at the C-276 alloy interface (Fig. 6).

The most widely used clad metal corrosion barrier material is copper-clad stainless steel (Cu/430 SS/Cu) for telephone and fiber optic cable shielding. In environments in which the corrosion rate of copper is high, such as acidic or sulfide-containing soils, the stainless steel acts as a corrosion barrier and thus prevents perforation, while the inner copper layer maintains high electrical conductivity of the shield.

Sacrificial metals, such as magnesium, zinc, and aluminum, are in the active region of the galvanic series and are extensively used for corrosion protection. The location of the sacrificial metal in the galvanic couple is an important consideration in the design of a system. By cladding, the sacrificial metal may be located precisely for efficient cathodic protection.

The single largest application for cold roll bonded materials is stainless-steel-clad aluminum for automotive trim (Fig. 8). The stainless steel exterior surface provides corrosion resistance, high luster, and abrasion and dent resistance, and the aluminum on the inside provides sacrificial protection for the painted auto body steel and for the stainless steel.

The largest application for hot roll bonded materials—Alclad aluminum—also falls into this category. In this case, a more active aluminum alloy is bonded to a more noble aluminum alloy. In service, the outer clad layer of aluminum corrodes sacrificially and protects the more noble aluminum substrate.

Transition Metal Systems. A clad transition metal system provides an interface between two incompatible metals. It not only reduces galvanic corrosion where dissimilar metals are joined but also allows welding techniques to be used when direct joining is not possible.

The principle of a clad transition metal is illustrated in Fig. 9. In this example, aluminum is joined to low-carbon steel through a steel-clad aluminum transition metal. Steel and aluminum form brittle intermetallic compounds and are difficult to weld directly. The transition metal insert allows steel to be welded to steel and aluminum to aluminum; the actual bond between the steel and the aluminum occurs in the clad transition. In addition, the dissimilar-metal crevice is eliminated, which reduces susceptibility to galvanic corrosion.

This concept has been applied commercially in a number of applications. Roll-bonded copper-clad aluminum has been deep drawn into transition tubes for joining aluminum and copper refrigeration tubing. Explosion-bonded steel-clad aluminum has been used to weld aluminum superstructures to steel ship hulls and aluminum bus bars to steel electrodes in aluminum smelting plants. Because of the increased use of aluminum on automobiles, steel-clad aluminum transition materials are being used for spot welding steel to aluminum in such applications as steering wheels and door components.

Complex Multilayer Systems. In many cases, materials are exposed to dual environments; that is, one side is exposed to one corrosive medium, and the other side is exposed to a different one. A single material may not be able to

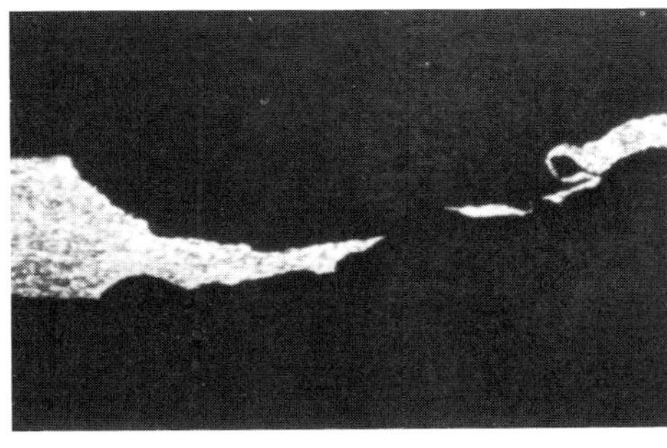

(a)

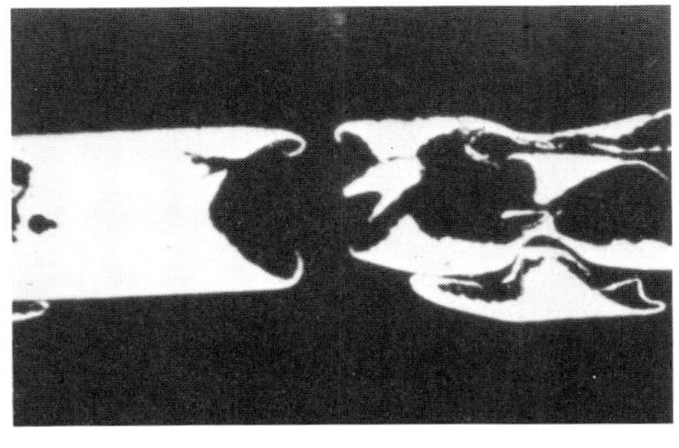

(b)

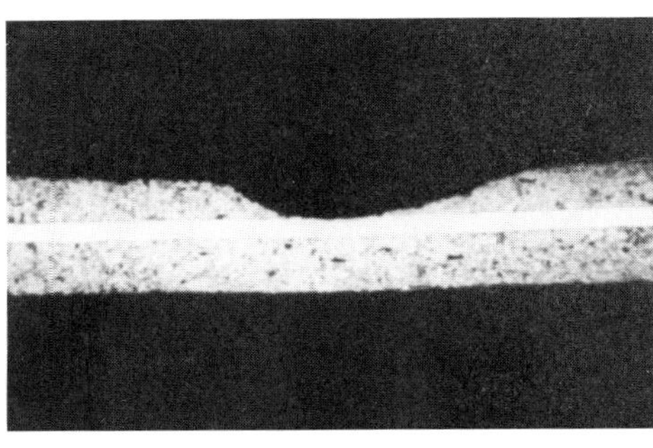

(c)

Fig. 7 Photomicrographs of cross sections of materials after 18 months of immersion in seawater at Duxbury, MA. (a) Low-carbon steel. (b) Type 304 stainless steel. (c) Carbon steel clad type 304 stainless steel

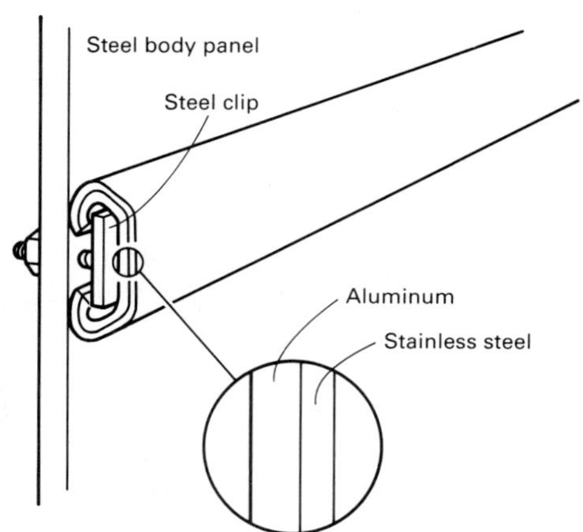

Fig. 8 Stainless steel clad aluminum automotive trim provides sacrificial corrosion protection to the auto body while maintaining a bright corrosion-resistant exterior surface.

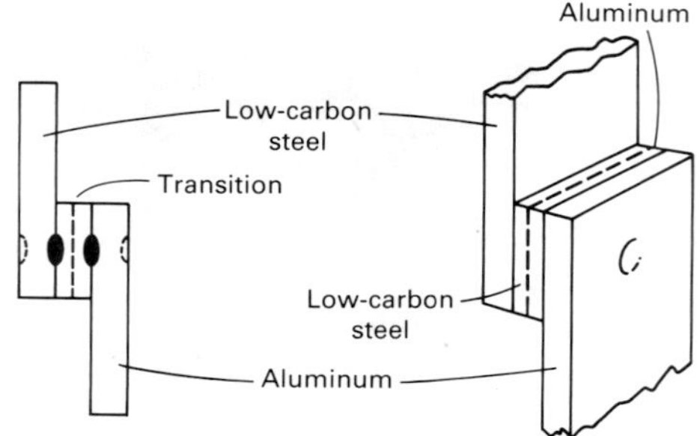

Fig. 9 Illustration of a steel-clad aluminum transition material insert used for joining aluminum to carbon steel

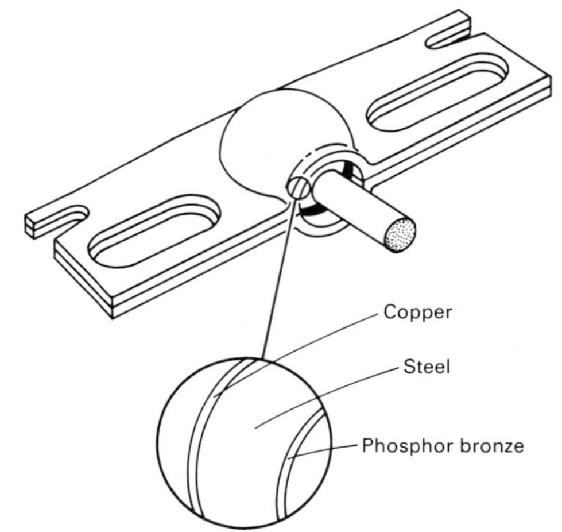

Fig. 10 Clad metal windshield wiper socket, which consists of copper-clad, steel-clad phosphor bronze

meet this requirement, or a critical material may be required in large quantity.

Clad metals provide an ideal solution to the materials problem of dual environments. For example, in automobile windshield wiper sockets (Fig. 10), wear resistance is required on the mating surface, atmospheric-corrosion resistance is required on the external surface, and high strength is incorporated into the design. Multilayer phosphor-bronze-clad, steel-clad copper is used in this application. The phosphor bronze provides the required bearing surface, copper provides atmospheric-corrosion resistance, and steel provides the required strength.

In the application of small battery cans and caps, copper-clad, stainless-steel-clad nickel (Cu/SS/Ni) is used where the external nickel layer provides atmospheric-corrosion resistance and low contact resistance. The copper layer on the inside provides the electrode contact surface as well as compatible cell chemistry. The stainless steel layer provides strength and resistance to perforation corrosion.

Other examples include the titanium-clad, copper-clad nickel (Ti/Cu/Ni) bipolar electrode used in fuel cells. Titanium is required on the anode side, nickel provides a hydrogen barrier on the cathode side, and copper provides electrical and thermal conductivity. Tantalum-clad, copper-clad nickel (Ta/Cu/Ni) is used for capacitor cans. Tantalum provides internal corrosion resistance, nickel provides external atmospheric-corrosion resistance and low contact resistance, and copper provides electrical and thermal conductivity.

SELECTED REFERENCES

- R. Baboian, Designing Clad Metals for Corrosion Control, *Trans. SAE*, Vol 81, 1973, p 1763
- R. Baboian, Conservation of Critical Materials With Clad Metal Systems, *Mater. Perform.*, Vol 33, 1984, p 13
- R.G. Delagi, Designing With Clad Metals, *Mach. Des.*, Vol 52 (No. 27), Nov 20, 1980
- A. Pocalyko, Explosion Clad Plate for Corrosion Service, *Mater. Prot.*, June 1965
- R.D. Sisson, Ed., *Coatings and Bimetallics for Aggressive Environments*, American Society for Metals, 1985

Specific Industries and Environments

Section Co-chairmen:
Robert S. Charlton, B.H. Levelton & Associates, Ltd.
James A. Hanck, Pacific Gas and Electric Company
Fred H. Meyer, Jr., Air Force Wright Aeronautical Laboratories
Lawrence J. Korb, Rockwell International

Marine Corrosion

Chairman: Robert H. Heidersbach, Metallurgical Engineering Department, California Polytechnic State University

MARINE CORROSION includes the deterioration of structures and vessels immersed in seawater, the corrosion of machinery and piping systems that use seawater for cooling and other industrial purposes, and corrosion in marine atmospheres. Although salt water is generally considered to be a corrosive environment, it is not widely understood how corrosive salt water is in comparison to other environments, such as fresh (salt-free) water.

Figure 1 shows the corrosion rate of iron in aqueous sodium chloride (NaCl) solutions of various concentrations. The maximum corrosion rate occurs near 3.5% NaCl—the approximate salt concentration of seawater.

Other variables in seawater and in the marine environment affect corrosion rates in different ways; these variables are discussed in the sections "Seawater" and "Marine Atmospheres" in this article. The remainder of this article is devoted to corrosion protection in the marine environment (see the sections "Metallic Coatings," "Organic Coatings," and "Cathodic Protection"). Corrosion of specific metals and alloys in the marine environment is discussed in the articles in the Section "Specific Alloy Systems" in this Volume.

Seawater

Stephen C. Dexter
College of Marine Studies
University of Delaware

The general marine environment includes a great diversity of subenvironments, such as full-strength open ocean water, coastal seawater, brackish and estuarine waters, bottom sediments, and marine atmospheres. Exposure of structural materials to these environments can be continuous or intermittent, depending on the application. Structures in shallow coastal or estuarine waters are often exposed simultaneously to five zones of corrosion. Beginning with the marine atmosphere, the structure then passes down through the splash, tidal, continuously submerged (or subtidal), and subsoil (or mud) zones. The relative corrosion rates often experienced on a steel structure passing through all of these zones are illustrated in Fig. 2.

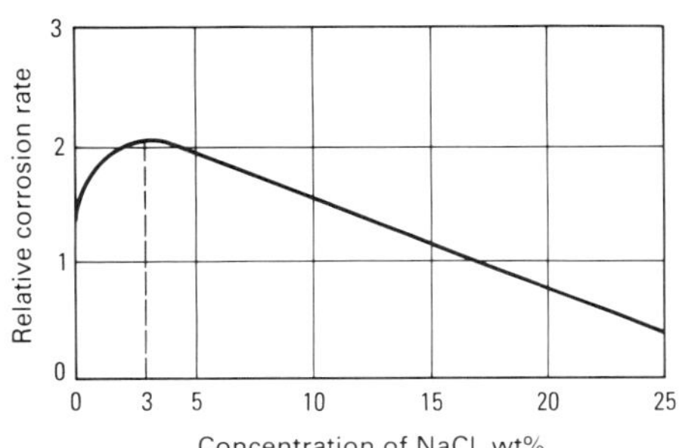

Fig. 1 Effect of NaCl concentration on the corrosion rate of iron in aerated room-temperature solutions. Data are compiled from several investigations. Source: Ref 1

As will be discussed in this section, the major chemical constituents of seawater are consistent worldwide. The minor constituents, however, vary from site to site and with season, storms, and tidal cycles. These minor constituents include dissolved trace elements and dissolved gases. In addition, seawater contains dissolved organic materials and living microscopic organisms. Frequently, the minor chemical constituents of seawater, together with the organic materials and living organisms, are the rate-controlling factor in the corrosion of structural metals and alloys.

Because of its variability, seawater is not easily simulated in the laboratory for corrosion-testing purposes. Stored seawater is notorious for exhibiting behavior as a corrosive medium that is different from that of the water mass from which it was taken. This is due in part to the fact that the minor constituents, including the living organisms and their dissolved organic nutrients, are in delicate balance in the natural environment. This balance begins to change as soon as a seawater sample is isolated from the parent water mass, and these changes often have a large effect on the types of corrosion experienced and the corrosion rate.

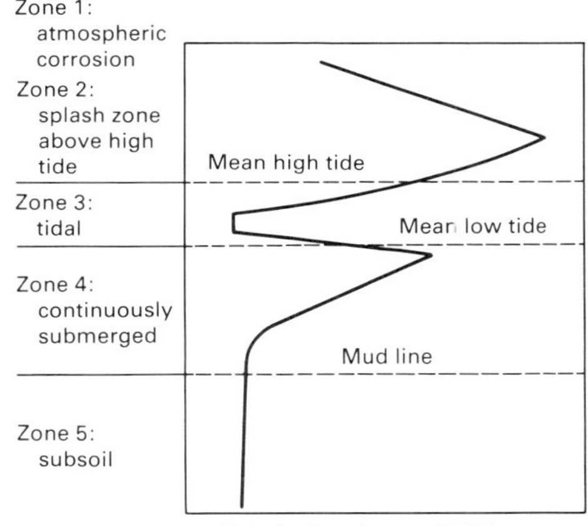

Fig. 2 Zones of corrosion for steel piling in seawater, and relative loss of metal thickness in each zone. Source: Ref 2

Variations in the chemistry of open ocean seawater tend to take place slowly (over time periods of 3 to 6 months) and over horizontal and vertical distances that are large in comparison to the dimensions of most marine structures. Such gradual changes may produce an equally gradual change in the corrosion rate of structural materials with season and location, but they are unlikely to produce sharp changes in either corrosion mechanism or rate. Moreover, such gradual changes are relatively easy to measure and monitor.

On the other hand, changes that take place over periods of hours to days and over distances of centimeters to meters can occur as the result of point inputs of various chemical pollutants or the attachment of micro- and macroscopic marine plants and animals to the surface of a structure. The chemical changes produced by the attachment of biological fouling organisms take place directly at the metal/water interface where the corrosion occurs, not in the bulk water. This means that the chemical environment in which the corrosion reactions occur in the presence of a fouling film may bear little resemblance to that of the bulk water.

It is these types of effects, which can be produced quickly and can lead to sharp chemical gradients over short distances, that often result in the onset of localized corrosion. Crevice corrosion beginning under the base of an isolated barnacle on stainless steel is a good example of this type of influence. Whether the fouling film is composed of microscopic bacteria or large sedentary fouling organisms is often less important than whether the film provides complete or spotty coverage of the metal surface. Almost invariably, a spotty film, or one that forms in discrete colonies of organisms with bare metal in between, will be more likely to induce structurally significant corrosion than a film that produces a continuous layer.

In this section, the general properties of ocean water and their effects on corrosion will be discussed. The major and minor features, including the effects of variability, pollutants, and fouling organisms, will be covered. Also included is a discussion of the factors that control the corrosivity of marine atmospheres.

Consistency and the Major Ions

The concentrations of the major constituents of full-strength seawater are shown in Table 1. Major constituents are considered to be those that have concentrations greater than 1 mg/L and are not greatly affected by biological processes. The behavior of these major ions and molecules is said to be conservative because their concentrations bear a relatively constant ratio to each other over a wide range of dilutions. Although most of the known elements can be found dissolved in seawater, the ions and molecules listed in Table 1 account for over 99.85% of the total dissolved solids. Moreover, the conservative nature of these major ions means that all of their concentrations can be calculated if the concentration of any one of them is measured.

Salinity and Chlorinity. The most commonly measured property of seawater is its salinity. Historically, salinity, S, in parts per thousand (‰), has been defined as the total weight in grams of inorganic salts in 1 kg of seawater when all bromides and iodides are replaced by an equivalent quantity of chlorides and all carbonates are replaced by an equivalent quantity of oxides. Salinity is usually determined by measuring either the chlorinity or the electrical conductivity of the seawater. Chlorinity, Cl, is defined as the mass in grams of silver required to precipitate the halogens in 0.3285234 kg of seawater. This is nearly equal to the mass of chloride in the seawater sample. Chlorinity is related to salinity by:

$$S = 1.80655\ \mathrm{Cl} \qquad \text{(Eq 1)}$$

where S and Cl are measured in parts per thousand.

If pure water is the only substance added to or removed from seawater, the concentration of any ion, χ, from Table 1 at a salinity other than 35‰ can be calculated from the following relationship:

$$[\chi] \text{ at salinity } S = [\chi] \text{ at } 35‰ \text{ salinity} \times \frac{S}{35‰} \qquad \text{(Eq 2)}$$

where the brackets denote concentration, and S is again given in parts per thousand. For example, using Eq 2 and Table 1, the concentration of sodium ion in seawater of 20‰ salinity is:

$$[Na^+] \text{ at } 20‰ = (0.468 \text{ mol/kg of seawater}) \times \frac{20‰}{35‰} = 0.2667 \text{ mol } Na^+\text{/kg of seawater}$$

This relationship may break down at very low salinities (<10‰ for Ca^{2+} and HCO_3^- and <5‰ for the other major ions). The relationship may also break down in grossly polluted seawater. In this latter case, the concentration of the ion of interest must be known in both seawater and in the solution being added to it.

In the short term, the only processes that affect the concentrations of the major ions are evaporation, precipitation, and river discharge. Because water is the only substance added to or removed from seawater by the first two of these processes, they affect the absolute concentrations of each ion but have no effect on the concentration ratios.

Effect of River Discharge. River discharge does add constituents other than pure water. Therefore, the conservative behavior of the major ions may not hold in the low-salinity regions near river outlets. To a first approximation, river water is a 0.4 millimolar solution of calcium bicarbonate, as shown in Table 2. At a salinity of 10‰, the calcium concentration calculated using Eq 2 and Table 1, but ignoring the river input, will be 10% too low. The error for bicarbonate ion will be even larger. Errors in concentrations of the other major ions calculated by ignoring the river input, however, will be much less because of their relatively high concentrations in seawater compared to those in river water.

Salinity Variations. The total salt content of open ocean seawater varies from 32 to 36‰. It can rise above that range in the tropics or in enclosed waterways, such as the Red Sea, where evaporation exceeds freshwater input. It will be lower than that range in estuaries and bays, where there is appreciable dilution from river input. Salinity variations in the surface waters of the Pacific Ocean are shown in Fig. 3. The data for these figures, as well as the surface water data for other variables to follow, were taken from the *Russian Atlas of the Pacific Ocean*. Several additional sources of similar information are given in the Selected References at the end of this article. Salinity variations with depth at given locations in the Atlantic and Pacific Oceans are shown in Fig. 4. The locations from which these and other data on variations with depth were taken are illustrated in Fig. 5. It should be noted that the open ocean salinity variations with horizontal location and depth are quite small.

Effect of Salinity on Corrosion. The main effects of salinity on corrosion result from its influence on the conductivity of the water and from the influence of chloride ions on the breakdown of passive films. Specific conductance varies with temperature and chlorinity, as indicated in Table 3. The high conductivity of seawater means that the resistance of the electrolyte plays a minor role in determining the rate of corrosion reactions and that surface area relations play a major role.

Two examples will serve to illustrate this point. First, galvanic corrosion in freshwater systems tends to be localized near the two-metal junction by the high resistivity of the electrolyte. In seawater, however, anodes and cathodes that are tens of meters apart can operate; therefore, the galvanic corrosion is much more spread out and is less intense at the junction. In the second example, a large area of cathodic metal, such as stainless steel, will produce more severe galvanic attack on an anodic metal in seawater than in freshwater, because high conductivity allows the entire area of stainless to participate in the reaction. Similarly, pitting corrosion tends to be more intense in seawater because large areas of boldly exposed cathode surface are available to support the relatively small anodic areas at which pitting takes place.

The second effect of salinity on corrosion in seawater is related to the role of chloride ions in the breakdown of passivity on active-passive metals such as stainless steels and aluminum alloys. The higher the salinity of the water, the more readily chloride ions succeed in penetrating the passive film and initiating pitting and crevice corrosion at localized sites on the metal surface.

The open ocean salinity changes shown in Fig. 4 and 5 have very little effect on the processes of galvanic, pitting, and crevice corrosion. The much larger salinity changes found in coastal and brackish waters can have a substantial effect on both the susceptibility to and the intensity of localized corrosion. These coastal salinity changes have undoubtedly contributed to the variability in reported pitting and crevice corrosion rates with season and location or with time at a given location. For alloys that corrode uniformly, variations in corrosion rate due to salinity changes are small compared to those caused by changes in oxygen concentration and temperature, as discussed in the following sections.

Variability and the Minor Ions

This section will consider how the variability of the minor constituents of seawater affects corrosion. Variabilities in seawater properties with horizontal location and depth will be presented for temperature, the dissolved gases oxygen and carbon dioxide, and pH. Each of these properties has a range over which it typically varies in the marine environment. The effect on corrosion of each property as it changes within this range will be considered.

Temperature. When all other factors are held constant, an increase in temperature increases the corrosivity of seawater. If the dissolved oxygen concentration is held constant, the corrosion rate of low-carbon steel in seawater will approximately double for each 30 °C (55 °F) increase in temperature.

Open ocean temperature variations are shown in Fig. 6 and 7. From these data alone, one would expect corrosion rates in tropical surface waters to be about twice those in the polar regions or in deep water. Corrosion rates are usually higher in warm surface waters than in cold deep waters, as illustrated in Fig. 8 and 9, but the picture is not nearly as simple as temperature alone would

Table 1 Concentrations of the most abundant ions and molecules in seawater of 35‰ salinity

Density of seawater: 1.023 g/cm³ at 25 °C (75 °F)

Ion or molecule	Concentration, m mol/kg of seawater	Concentration, g/kg of seawater
Na^+	468.5	10.77
K^+	10.21	0.399
Mg^{2+}	53.08	1.290
Ca^{2+}	10.28	0.412
Sr^{2+}	0.09	0.008
Cl^-	545.9	19.354
Br^-	0.84	0.067
F^-	0.07	0.0013
HCO_3^-	2.30	0.140
SO_4^{2-}	28.23	2.712
$B(OH)_3$	0.416	0.0257

Source: Ref 3

Table 2 Concentrations of the most abundant ions and molecules in average river water

Ion or molecule	Concentration, m moles/L of river water	Concentration, mg/L of river water
Na^+	0.274	6.3
K^+	0.059	2.3
Mg^{2+}	0.171	4.1
Ca^{2+}	0.375	15.0
Cl^-	0.220	7.8
HCO_3^-	0.958	58.4
SO_4^{2-}	0.117	11.2
NO_3^-	0.016	1.0
Fe^{2+}	0.012	0.67
$Si(OH)_4$	0.218	20.9

Source: Ref 4

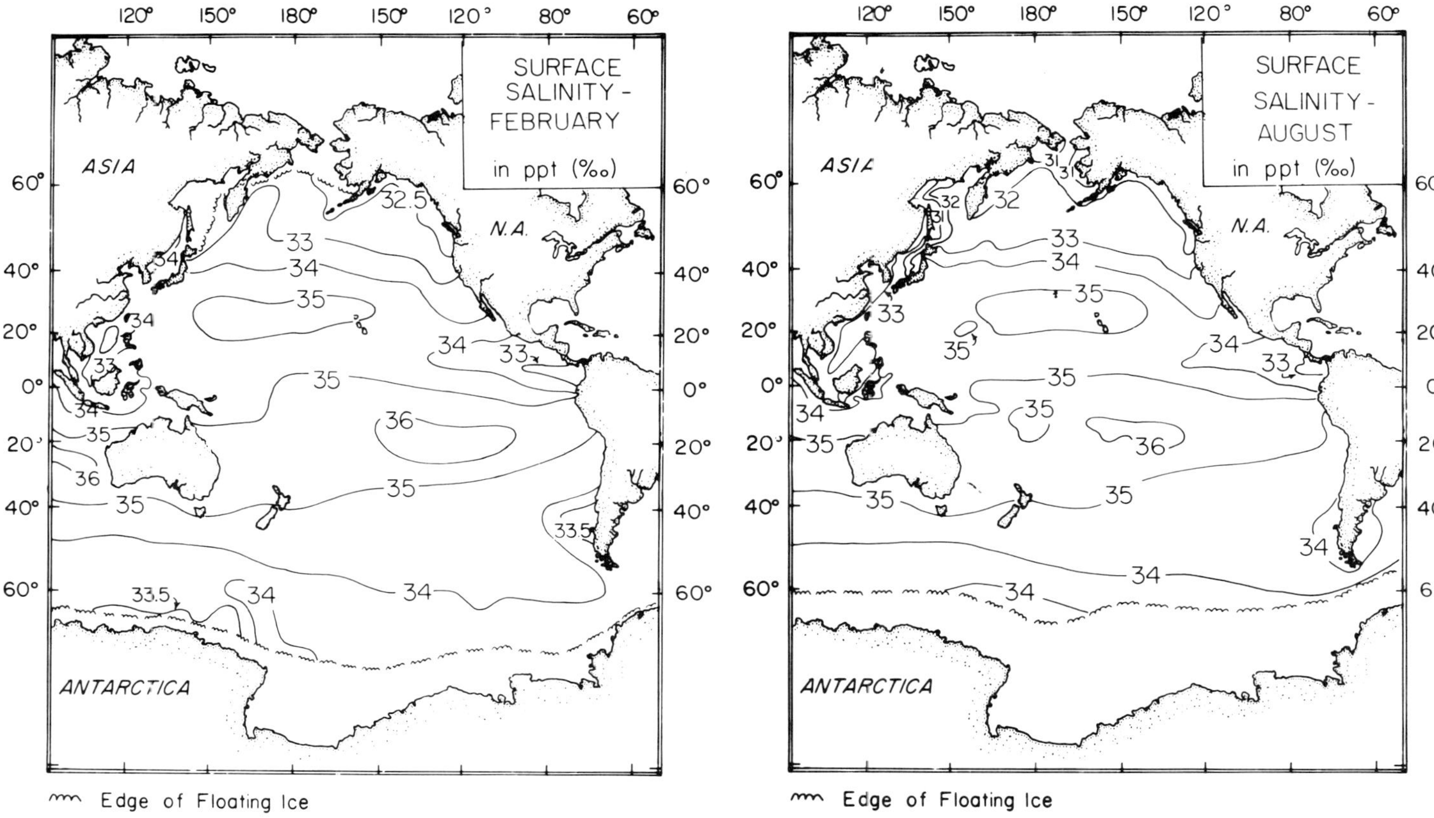

(a) (b)

Fig. 3 Pacific Ocean surface salinity (‰) for February (a) and August (b). Source: Ref 5

indicate. The saturation level of dissolved oxygen increases as the temperature decreases, and the effects of dissolved oxygen on the corrosion rate are often stronger than those of temperature, as will be discussed below. Short-term local temperature fluctuations and the effects of biofouling and scaling films must also be considered.

Data for the temperature and salinity variations with depth at the four coastal locations in Fig. 5 are shown in Fig. 10 to 13. The temperature and salinity profiles for Cook Inlet (station 1) and for the Oregon coast (station 3) given in Fig. 10 and 11(a), respectively, show only small differences with depth and season. No reliable data are available for the winter months at the Cook Inlet station. The differences shown here are inconsequential with regard to corrosion.

Larger differences are shown for the Gulf of Mexico (station 4) and the New Jersey coast (station 5) in Fig. 12 and 13. The temperature differences with depth and season shown for these two locations will be large enough to influence both corrosion rate and calcareous deposition. The salinity changes shown will have little influence. The seasonal differences in the surface waters of the Gulf of Mexico disappear at greater depths, while those off the New Jersey coast persist all the way to the bottom.

Dissolved Oxygen. Many of the minor constituents that are important to corrosion processes are dissolved gases such as carbon dioxide and oxygen. Their concentrations are not conservative (that is, constant), because they are influenced by air-sea exchange as well as by biochemical processes. The concentration of dissolved oxygen in surface waters will usually be within a few percent of the equilibrium saturation value with atmospheric oxygen at a given temperature. The solubility of oxygen in seawater varies inversely with both temperature and salinity, but the effect of temperature is greater. If the absolute temperature T (°K) and salinity S (‰) are known, the solubility of oxygen can be calculated from the relationship:

$$\ln [O_2] = A_1 + A_2 \left(\frac{100}{T}\right) + A_3 \ln \left(\frac{T}{100}\right) + A_4 \left(\frac{T}{100}\right) + S\left[B_1 + B_2\left(\frac{T}{100}\right) + B_3\left(\frac{T}{100}\right)^2\right] \quad \text{(Eq 3)}$$

where oxygen concentration is given in milliliters per liter (mL/L), and salinity S is again in parts per thousand (‰). The constants A_1 through B_3 are given in Table 4. Table 5 lists the equilibrium oxygen saturation levels in milliliters per liter as a function of temperature and salinity calculated from Eq 3 and Table 4.

Generally, the surface waters of the ocean are in equilibrium with the oxygen in the atmosphere at a specific temperature. Two sets of conditions, however, can lead to the waters becoming substantially supersaturated with oxygen. The first of these conditions is oxygen production due to photosynthesis by microscopic marine plants. During high growth periods, intense photosynthesis can produce concentrations as high as 200% saturation for periods of up to a few weeks. Such oxygen supersaturation is most often found in near-shore regions as a transient phenomenon.

The second condition that may cause oxygen supersaturation is the entrainment of air bubbles

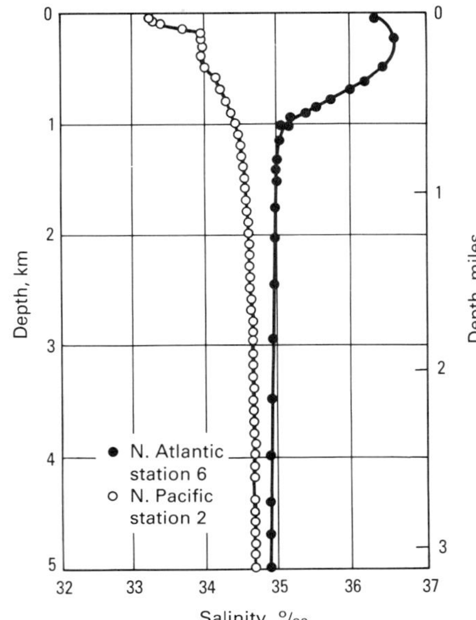

Fig. 4 Comparison of salinity-depth profiles for open ocean sites 2 and 6 (see Fig. 5 for site locations). Source: Ref 5

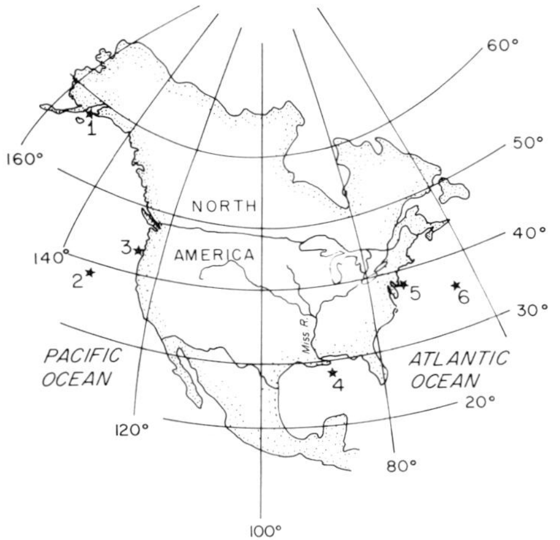

Station number	Location	Chart coordinates
1	Cook Inlet	58° 57′ N 152° 59′ W
2	Northeast Pacific	38° 21′ N 133° 38′ W
3	West coast, U.S.	44° 39′ N 124° 39′ W
4	Gulf of Mexico	28° 12′ N 88° 26′ W
5	East coast, U.S.	39° 20′ N 73° 40′ W
6	Northwest Atlantic	36° 44′ N 64° 28′ W

Fig. 5 Station positions for Fig. 4, 7, 10-13, 16, and 28. Source: Ref 5

Table 3 Specific conductance of seawater as a function of temperature and chlorinity

Conductance: $\Omega^{-1} \cdot cm^{-1}$

Chlorinity, ‰	Temperature, °C (°F) 0 (32)	5 (40)	10 (50)	15 (60)	20 (70)	25 (75)
1	0.001839	0.002134	0.002439	0.002763	0.003091	0.003431
2	0.003556	0.004125	0.004714	0.005338	0.005971	0.006628
3	0.005187	0.006016	0.006872	0.007778	0.008702	0.009658
4	0.006758	0.007845	0.008958	0.010133	0.011337	0.012583
5	0.008327	0.009653	0.011019	0.012459	0.013939	0.015471
6	0.009878	0.011444	0.013063	0.014758	0.016512	0.018324
7	0.011404	0.013203	0.015069	0.017015	0.019035	0.021121
8	0.012905	0.014934	0.017042	0.019235	0.021514	0.023868
9	0.014388	0.016641	0.018986	0.021423	0.023957	0.026573
10	0.015852	0.018329	0.020906	0.023584	0.026367	0.029242
11	0.017304	0.020000	0.022804	0.025722	0.028749	0.031879
12	0.018741	0.021655	0.024684	0.027841	0.031109	0.034489
13	0.020167	0.023297	0.026548	0.029940	0.033447	0.037075
14	0.021585	0.024929	0.028397	0.032024	0.035765	0.039638
15	0.022993	0.026548	0.030231	0.034090	0.038065	0.042180
16	0.024393	0.028156	0.032050	0.036138	0.040345	0.044701
17	0.025783	0.029753	0.033855	0.038168	0.042606	0.047201
18	0.027162	0.031336	0.035644	0.040176	0.044844	0.049677
19	0.028530	0.032903	0.037415	0.042158	0.047058	0.052127
20	0.029885	0.034454	0.039167	0.044114	0.049248	0.054551
21	0.031227	0.035989	0.040900	0.046044	0.051414	0.056949
22	0.032556	0.037508	0.042614	0.047948	0.053556	0.059321

Source: Ref 6

due to wave action. This factor usually will not cause supersaturations greater than about 10%, because vigorous wave action also promotes reequilibration with the atmosphere.

Oxygen Variability. The distribution of dissolved oxygen in the surface waters of the Pacific Ocean is shown in Fig. 14(a) for the months of January through March and in Fig. 14(b) for July through September. Comparison with Fig. 6 reveals that the highest concentrations of oxygen coincide with the lowest temperatures; this agrees with the oxygen solubility data given in Table 5.

As discussed previously, surface waters are either saturated or supersaturated with oxygen at atmospheric conditions. In contrast, deep waters are often undersaturated because of the consumption of oxygen during the biochemical oxidation of organic matter. Figure 15 shows horizontal maps of dissolved oxygen concentration in the Pacific Ocean at depths of 500 and 1000 m (1640 and 3280 ft). The level of oxygen is generally lower at these depths, especially in the northeastern Pacific.

This decrease in oxygen concentration with depth is shown more clearly in Fig. 16. The oxygen profiles for the open Atlantic and Pacific stations both go through a minimum at intermediate depths and increase again at great depths. In the Atlantic Ocean, the surface oxygen concentrations are usually lower, and the oxygen minimum is not as intense as in the Pacific. In addition, the oxygen concentrations in the deep Atlantic are higher than those in the deep Pacific, and they can be even higher than those in the Atlantic surface waters.

Figure 17 shows the depth of the dissolved oxygen minimum in the Pacific (solid contours) and the concentration of oxygen at that depth (dotted contours). The depth of the oxygen minimum ranges from 400 m (1310 ft) in the equatorial eastern Pacific to over 2400 m (7875 ft) in the central south Pacific. The concentration of oxygen at the depth of the minimum ranges from 0.01 to 0.40 mg · atm/L (1 mg · atm/L = 12.2 mL/L = 16 ppm at 25 °C or 75 °F). Oxygen profiles with depth for the four coastal stations are shown in Fig. 10 to 13.

Effect of Oxygen on Corrosion. The corrosion rate of active metals (for example, iron and steel) in aerated electrolytes such as seawater at constant temperature is a direct linear function of the dissolved oxygen concentration, as shown in Fig. 18. When oxygen and temperature vary together, as they do in the marine environment, the oxygen effect tends to predominate. This trend is illustrated by data for the corrosion rate of steel at various depths in the Pacific Ocean in Fig. 19. The corrosion rate decreases with dissolved oxygen down to the oxygen minimum, then increases again with oxygen at greater depths, despite a continuing decrease in temperature. The corrosion rates of nickel and nickel-copper alloys are somewhat less affected by oxygen concentration, as shown in Fig. 20. The effect of oxygen on copper alloys depends on the flow velocity. Figure 21 shows that there is very little effect of oxygen on copper alloys exposed in quiet open ocean water. At a flow velocity of 1.8 m/s (6 ft/s), however, increasing oxygen has a marked accelerating effect (Fig. 22).

In contrast, the effect of oxygen on the corrosion rates of active-passive metals, such as the aluminum and stainless alloys, can be quite variable (Fig. 23). In such alloy systems, high oxygen concentrations tend to promote healing of the passive film and thus retard initiation of pitting corrosion. On the other hand, high oxygen favors a vigorous cathodic reaction and tends to increase the rate of pit and crevice propagation after initiation.

For all alloy systems, the conditions most conducive to corrosion are those in which differences in dissolved oxygen are allowed to develop between two regions of the wetted metal surface. This can lead to an oxygen concentration cell, with potential differences as large as 0.5 V. The portion of metal surface on which the oxygen concentration is lowest becomes the anode and is subject to localized corrosion. Differences in dissolved oxygen concentration of this type are unlikely to be caused by variations in the water; instead, they are caused by localized deposits or structural design factors that create oxygen-shielded regions on the metal surface.

These effects also can lead to pitting corrosion of active metals such as carbon and low-alloy steels. The average uniform corrosion rates of carbon and low-alloy steels in a wide variety of marine environments are found to range from 50 to 125 μm/yr (2 to 5 mils/yr), slowly decreasing with time of exposure. Data from the Panama Canal zone showed that, although the average penetration rate was 68 μm/yr (2.7 mils/yr), the penetration by pitting was some five to eight times higher (Fig. 24).

Differences in dissolved oxygen from point to point along the metal surface caused by spotty biofouling films can contribute to the pitting rate. Dissolved oxygen differences are not the only factor in controlling pitting, however. Metallurgical factors can also be major contributors (see the article "Metallurgically Influenced Corrosion" in this Volume). In contrast to the effects of a spotty film, complete coverage of the surface by hard-shelled, sedentary fouling organisms can lead to a marked decrease in the overall corrosion rate by acting as a diffusion barrier against dissolved oxygen reaching the metal surface. In the case of aluminum and stainless alloys, point to point differences in oxygen concentration can lead to both pitting and crevice corrosion.

Dissolved Carbon Dioxide and pH. The concentration of carbon dioxide is less affected by air-sea interchange than the concentration of dissolved oxygen, because the carbon dioxide

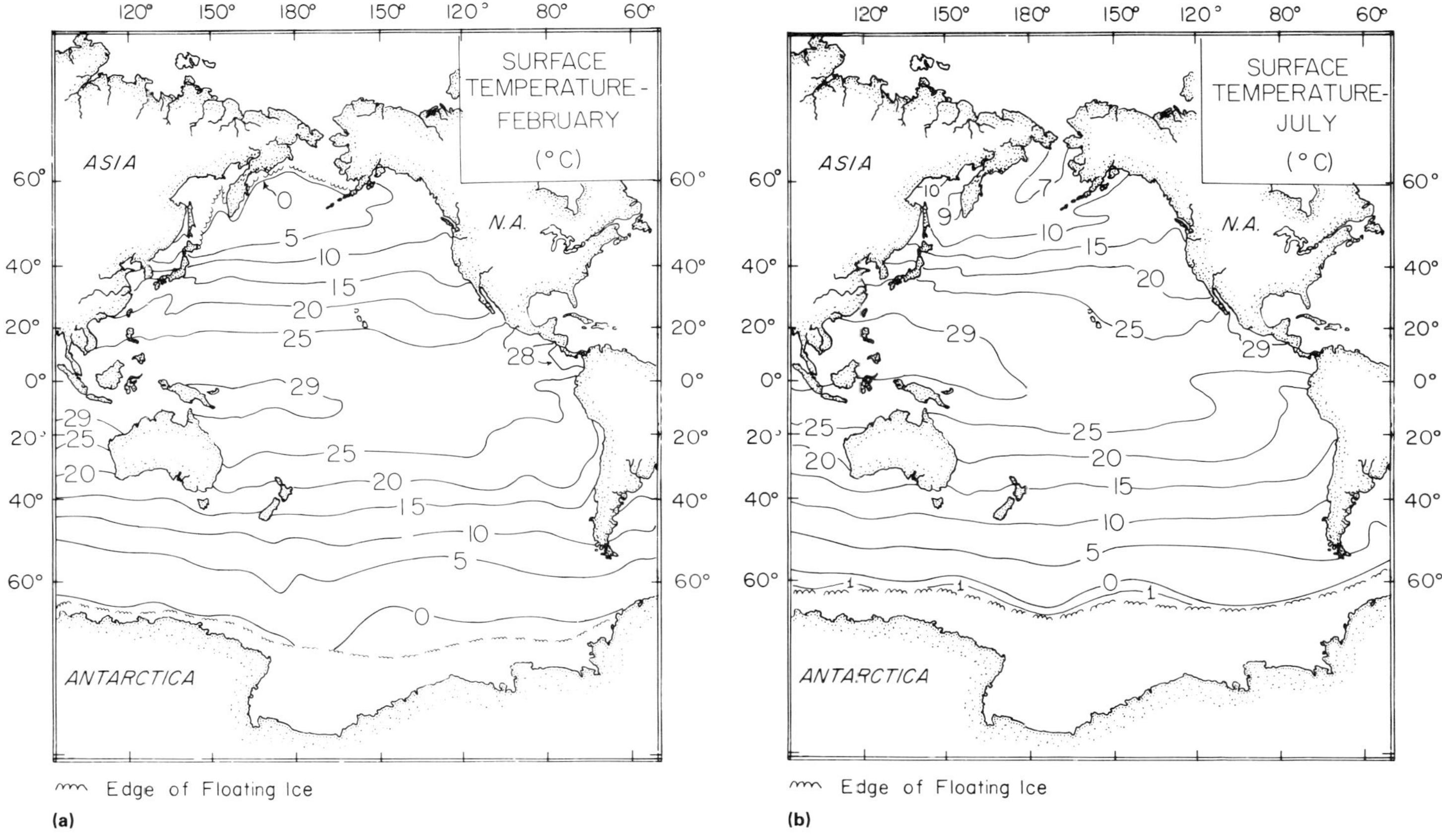

Fig. 6 Pacific Ocean surface temperature (°C) for February (a) and July (b). Source: Ref 5

system in seawater is buffered by the presence of bicarbonate and carbonate ions. Carbon dioxide is a weak acid and undergoes two ionizations in aqueous solutions:

$$CO_2 + H_2O = H^+ + HCO_3^- \quad \text{First ionization} \quad \text{(Eq 4)}$$

$$HCO_3^- = H^+ + CO_3^{2-} \quad \text{Second ionization} \quad \text{(Eq 5)}$$

Surface seawater usually has a pH value greater than 8 because of the combined effects of air-sea exchange and photosynthesis. At this pH, 93% of the total inorganic carbon is present as HCO_3^-, 6% as CO_3^{2-}, and 1% as CO_2. Bicarbonate ion accounts for at least 85% of the total inorganic carbon under all naturally occurring conditions. However, the relative concentrations of CO_2 and CO_3^{2-} vary greatly depending on pH. The CO_3^{2-} concentration is relatively high in surface waters, and surface waters are nearly always supersaturated with respect to the calcium carbonate phases calcite and aragonite. This supersaturation favors deposition of calcareous scales on metal surfaces undergoing cathodic protection, as will be discussed in the section "Effects of pH on Corrosion and Calcareous Deposition" in this article.

Relationship Among CO_2, Oxygen, and pH. The concentrations of carbon dioxide and oxygen are closely coupled and related to the pH of seawater through the processes of photosynthesis and biochemical oxidation, as represented in the following general reaction:

$$CH_2O + O_2 \underset{\text{Biochemical oxidation (respiration)}}{\overset{\text{Photosynthesis}}{\rightleftharpoons}} CO_2 + H_2O \quad \text{(Eq 6)}$$

where CH_2O represents a typical carbohydrate molecule. During decomposition of this organic material in seawater, Eq 6 proceeds from left to right, dissolved oxygen is consumed, and CO_2 is produced. Production of CO_2, in turn, makes the water more acidic (that is, lower pH) and decreases the saturation state with respect to carbonates.

Variability of pH. The pH of the Pacific surface waters ranges from 8.1 to 8.3, and its general distribution for the months of January to March is shown in Fig. 25. Distributions of pH at depths of 500 and 1000 m (1640 and 3280 ft) are shown in Fig. 26 and 27. In comparing these to the dissolved oxygen distributions at the same depths (Fig. 14 and 15), it should be noted that the trends for the two variables are similar. For example, at a depth of 500 m (1640 ft), the region of maximum dissolved oxygen, centered on 180° longitude between 20 and 40° north latitude in Fig. 15(a), is reproduced closely for pH in Fig. 26.

Profiles of pH with depth for the two open ocean locations are shown in Fig. 28. A comparison of the corresponding pH and oxygen profiles from Fig. 16 and 28 reveals the closely coupled nature of their relationship through the carbon dioxide system, as discussed above. The oxygen and pH minima are reached at the same depth for a given location, as was predicted. The deep north Pacific water is from 0.15 to 0.40 pH units more acidic than that in the North Atlantic, primarily because of the increased oxidation of organic matter in the North Pacific.

Profiles of pH for the coastal waters off Oregon and New Jersey are shown in Fig. 11 and 13, respectively. The close correlation between the shapes of the oxygen and pH profiles in both winter and summer for the Oregon data in Fig. 11 is particularly striking. Upon close examination, the oxygen and pH profiles in Fig. 13 do not appear to be closely related in the manner seen earlier. In March, the water column is well mixed down to the bottom, and the changes with depth of all four variables are small. In August, however, the dissolved oxygen profile is nearly independent of depth, while the pH and temperature profiles show substantial changes. Based on salinity and temperature, the oxygen saturation levels during August are about 5.2 mL/L in the surface waters and 6.5 mL/L in the deep water. The oxygen profile for August shows that the surface waters are nearly saturated, while in the deep waters, biological activity has used up enough oxygen and produced enough CO_2 to decrease the pH—but not enough to produce a strong oxygen minimum. This indicates the danger inherent in assuming that a pH minimum will always correspond to a similar minimum in oxygen. The two profiles may not correspond closely in shape when the biological demand for oxygen is not sufficiently intense to produce a strong oxygen minimum or when there is a strong temperature gradient.

Effects of pH on Corrosion and Calcareous Deposition. The pH of open ocean seawater ranges from about 7.5 to 8.3. Changes within this

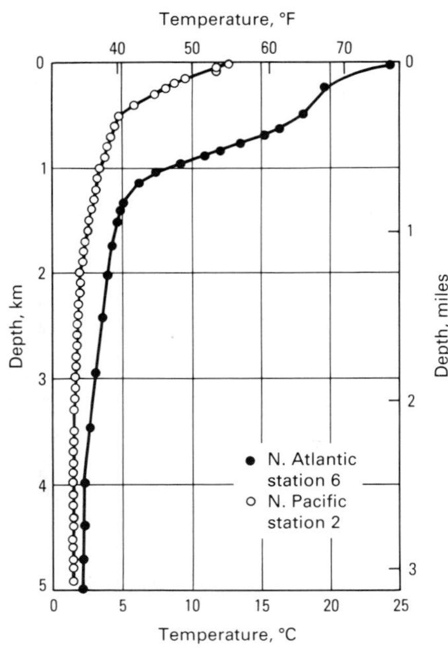

Fig. 7 Comparison of temperature-depth profiles for open ocean sites 2 and 6 (see Fig. 5 for site locations). Source: Ref 5

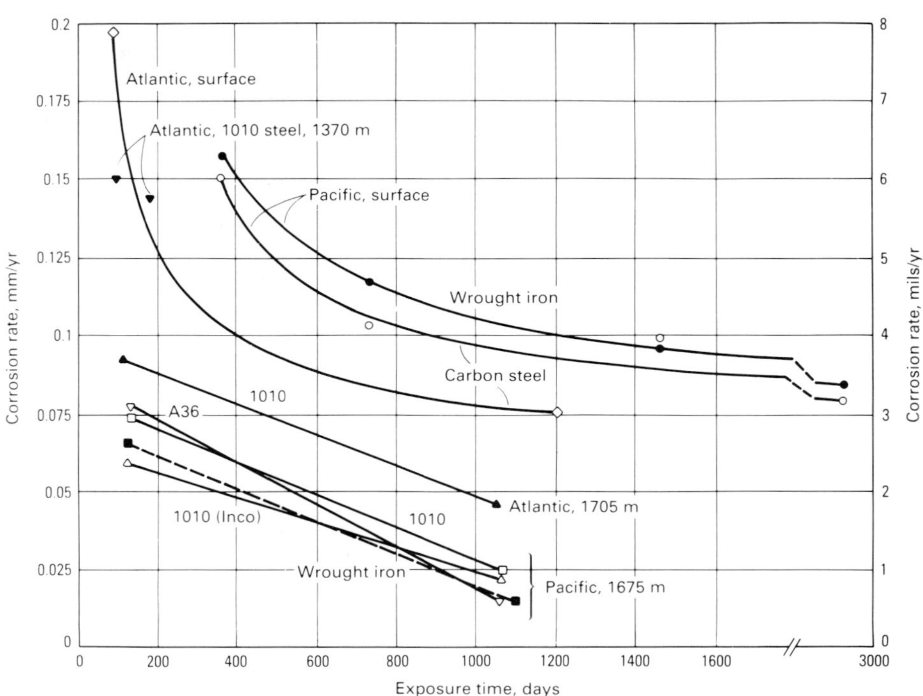

Fig. 8 Corrosion rates of carbon steels and wrought iron in the Atlantic and Pacific Oceans at various depths. Source: Ref 7

range have no direct effect on the corrosion of most structural metals and alloys. The one exception to this general statement is the effect of pH on aluminum alloys. A decrease in pH from the surface water value of 8.2 to a deep water value of 7.5 to 7.7 causes a marked acceleration in the initiation of both pitting and crevice corrosion. This effect accounts for the reported increase in corrosion of aluminum alloys in the deep ocean (see the article "Corrosion of Aluminum and Aluminum Alloys" in this Volume).

Although variations in seawater pH have little direct effect on corrosion, they do have an indirect effect through their influence on calcareous deposition. The surface waters of most of the oceans of the world are 200 to 500% supersaturated with respect to the calcium carbonate species calcite and aragonite. This means that precipitation of carbonate-type scales is likely to be an important part of any corrosion reaction in surface water at most locations. The predominant species precipitated in warm surface waters are aragonite and, at interface pH values above 9.3 as experienced in cathodic protection, brucite ($Mg(OH)_2$).

Scale precipitation is most likely to occur in the elevated-pH regime adjacent to cathodically protected surfaces, where OH^- ions are produced during reduction of dissolved oxygen. For many years, the corrosion protection industry has relied on the buildup of calcareous scales to make cathodic protection more economical. The higher the pH at the water/metal interface, the more brucite is favored and the lower the calcium-magnesium ratio of the deposit will be. A lower calcium-magnesium ratio, in turn, makes the scale less dense and less protective. Thus, a high level of cathodic protection applied in the early stages of immersion, as is sometimes done to accelerate scale buildup, can be counterproductive in terms of scale quality.

In deep waters, where the temperature and pH are both lower than at the surface, calcareous deposits do not form spontaneously under ambient conditions, and it has often been difficult to form deposits even under cathodic protection conditions. This is partly because the deep waters—below 300 m (985 ft) in the Atlantic and 200 m (655 ft) in the North Pacific—are undersaturated in carbonates because of low pH and high pressure. At the low temperatures of the deep water, calcite is the predominant calcium carbonate phase. At first, this would seem to be beneficial because calcite forms a dense, protective film. However, calcite formation is strongly inhibited by the free magnesium ions that are abundant in seawater. Therefore, only brucite, which is much less protective, tends to form in deep water, and even brucite forms only under cathodic protection conditions when the interface pH is greater than 9.7.

In the laboratory, fine-grain, dense, and protective deposits can be formed in cold water with elevated calcium and bicarbonate concentrations and decreased magnesium. An economical way to achieve these conditions on a large structure in the real environment does not yet exist.

Effect of Pollutants

The ratios of the major ions are not affected by pollution of the water as long as the salinity remains above 5 to 10‰. The relations between the major, conservative ions will hold constant, except perhaps in a confined waterway with poor tidal flushing in which a pollutant containing a large concentration of one of the major ions is introduced in quantities approaching that of the waterway itself.

In contrast, the concentrations of the minor constituents of seawater may be radically changed by pollution. This is an important fact because it is usually the minor ions and dissolved gases that determine the corrosion rate. Concentrations of heavy metals; nutrients such as nitrates and phosphates; dissolved organics; and dissolved gases such as oxygen, carbon dioxide, and hydrogen sulfide are particularly sensitive to pollution.

Effects Related to Dissolved Oxygen. Pollutants containing organic material usually increase the utilization of dissolved oxygen in the water. As the organics becomes oxidized, oxygen concentrations fall, carbon dioxide concentrations rise, and the water becomes more acidic. If the pH does not fall below 4, these conditions often result in a decrease in the corrosivity of the water toward carbon and low-alloy steels. During the first half of this century, for example, the upper Delaware estuary in the Chester-Philadelphia, PA, area was sufficiently polluted that the yearly mean dissolved oxygen in the Delaware River was nearly zero. Consequently, the corrosion rates of industrial steel structures in that waterway were very low during that period. As political pressure directed toward cleaning up the river mounted during the 1950s and 1960s, the yearly mean dissolved oxygen began to recover. By the mid-1970s, the oxygen concentrations had increased enough that "lace-paper" conditions were being noted on sheet steel and H-pilings in the area.

It is not always true, however, that a decrease in oxygen concentration will decrease corrosion rates. For active-passive metals such as aluminum and stainless alloys, a decrease in oxygen can produce either an increase or decrease in the corrosion rate. The corrosion rate can also increase for steel structures in low (or even no) oxygen waters if certain types of bacteria are active, as discussed below.

Sulfides. Hydrogen sulfide and various sulfates are frequent components of organic pollutants. Sulfates themselves are not particularly detrimental except that they can be reduced to sulfides by the action of sulfate-reducing bacte-

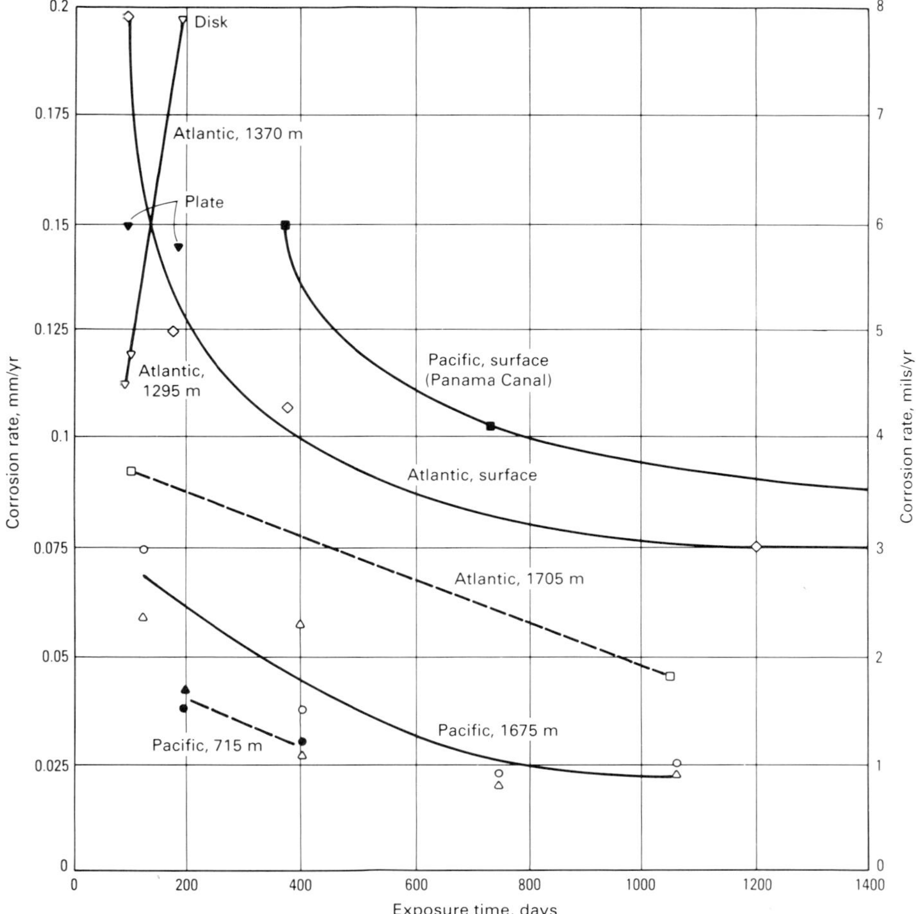

Fig. 9 Corrosion rates of low-carbon steels in the Atlantic and Pacific Oceans at various depths. Source: Ref 7

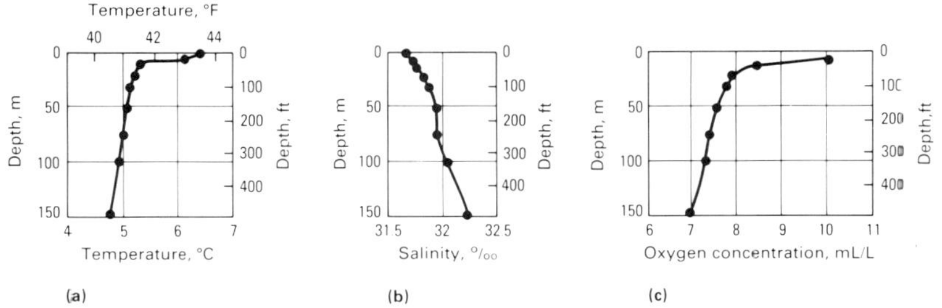

Fig. 10 Variation of temperature (a), salinity (b), and dissolved oxygen concentration (c) with depth at Cook Inlet (station 1, Fig. 5) for May 1968. Source: Ref 5

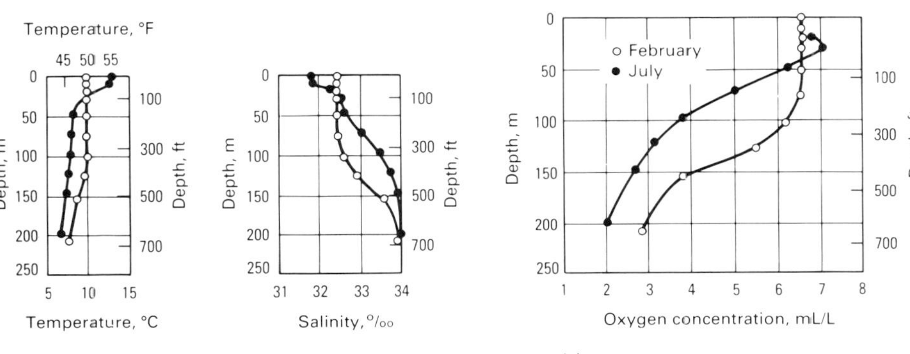

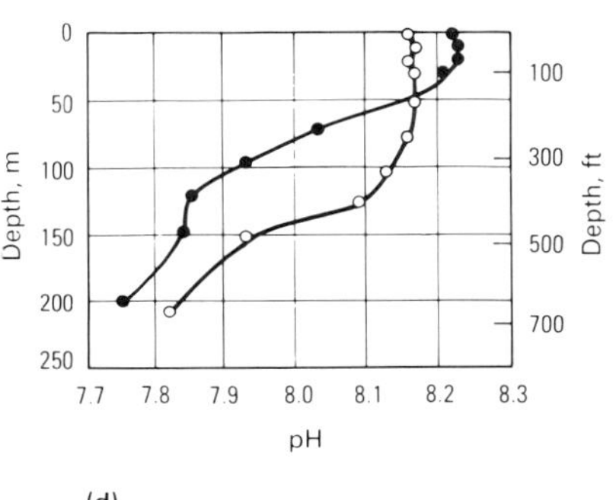

Fig. 11 Variation of temperature (a), salinity (b), dissolved oxygen (c), and pH (d) with depth and season off the Oregon coast (station 3, Fig. 5). Source: Ref 5

ria. The effects of these bacteria will be considered in the section "Influence of Biological Organisms" in this article. Hydrogen sulfide may reach levels of 50 ppm or higher in severely polluted estuarine or harbor waters. Bottom muds in harbors and salt marshes rich in decomposing organic matter may also have high sulfide concentrations.

Penetration by pitting corrosion of low-carbon steel panels in the polluted seawater of the San Diego harbor was several times higher than the uniform penetration rates usually experienced (Table 6). Similarly, the corrosion rate of several copper alloys used in condenser service was 3 to 10 times higher in polluted than in clean seawater (Table 7). It is not known exactly what role the sulfides played in accelerating the corrosion documented in Tables 6 and 7. It has been shown in other tests, however, that as little as 4 ppm of hydrogen sulfide can seriously increase the corrosion rate of copper alloys, as shown in Table 8.

Sulfide films are known to form on copper alloys in polluted waters. These films can be very harmful. Under most conditions, the sulfide film is itself cathodic to the bare copper alloy surface. This makes the film very effective in accelerating pitting corrosion at any break in the film, and the effects on corrosion are known to persist long after the polluted water has been removed. For this reason, it is important to remove sulfide films from copper alloys, even when the source of pollution has been eliminated. The sulfide film will continue to accelerate pitting corrosion as long as it remains on the metal surface, even in clean water. It is usually recommended that the first exposure of copper alloys be in clean, rather than sulfide-polluted, seawater whenever possible. Experience has shown that if sulfide films are allowed to form before other corrosion product films on the copper alloys of a marine condenser or piping system, they can be very difficult to remove. Moreover, even after cleaning, traces of the sulfide film are likely to plague that system throughout the service life.

Heavy Metals. Nominally unpolluted seawater contains nearly every known element, most of them in very small concentrations. For example, the copper concentration in clean seawater is about 0.2 ppb. This does not normally cause corrosion problems for any of the common marine structural alloys. At elevated copper concentrations, however, aluminum alloys can suffer accelerated corrosion. The copper concentration in the water can be elevated by copper-containing pollutants, by leaching from copper-base antifouling paints, or by corrosion of copper alloys.

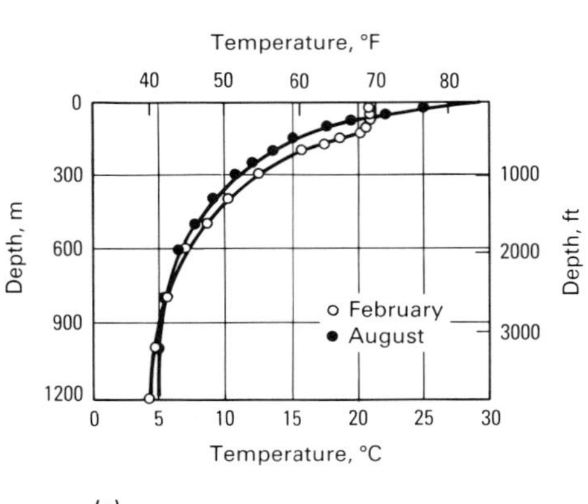

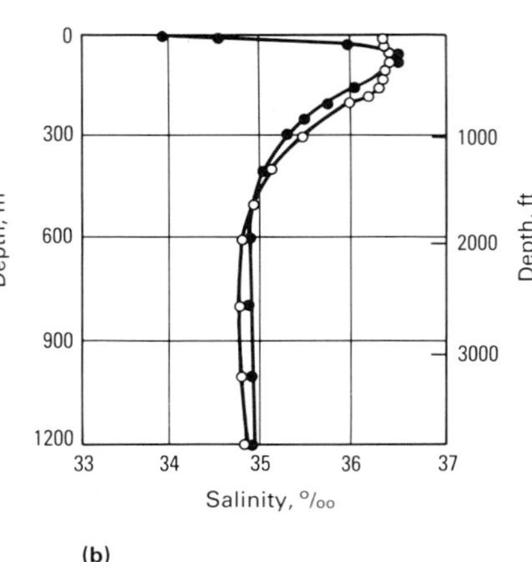

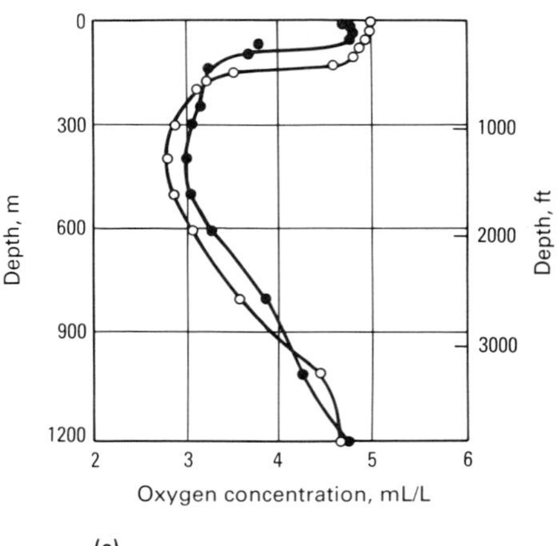

Fig. 12 Variation of temperature (a), salinity (b), and oxygen concentration (c) with depth and season in the Gulf of Mexico (station 4, Fig. 5). Source: Ref 5

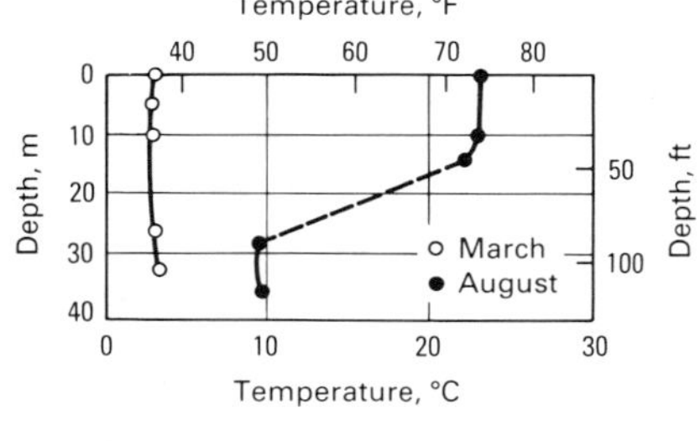

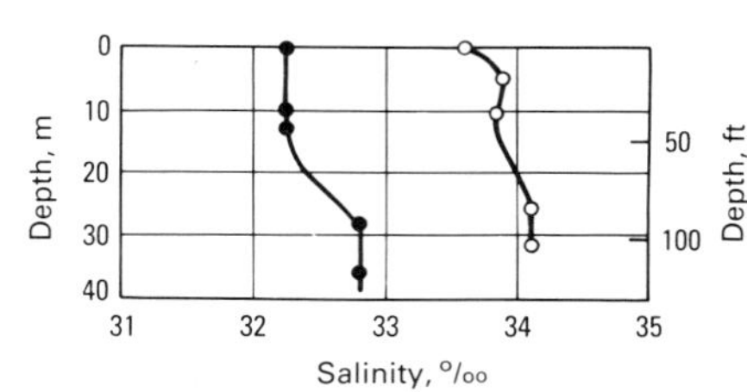

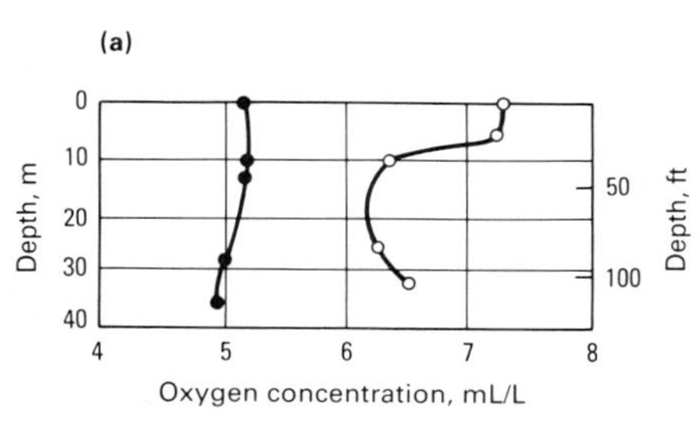

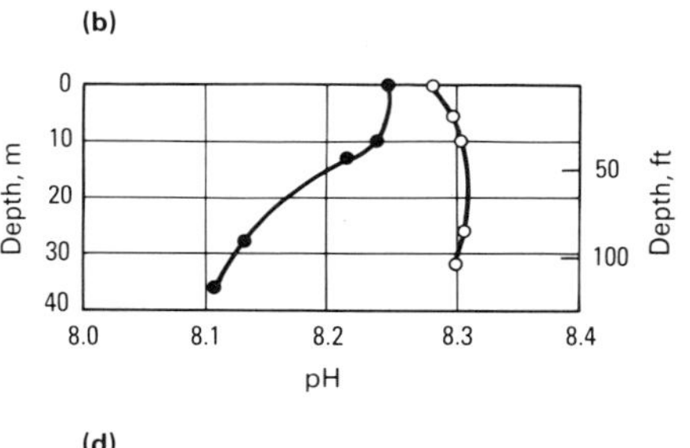

Fig. 13 Variation of temperature (a), salinity (b), dissolved oxygen (c), and pH (d) with depth and season off the New Jersey and Delaware coasts (station 5, Fig. 5). Source: Ref 5

Acceleration of aluminum corrosion by copper corrosion products has often been observed in seawater piping systems having copper alloy pumps, even when the aluminum piping is not in direct electrical contact with the copper alloy. In freshwater, copper concentrations as low as 0.05 ppm have been found to accelerate aluminum corrosion. In seawater, the threshold concentration below which copper contamination has no effect seems to be about 0.03 ppm, as shown in Fig 29. The copper accumulates on the aluminum surface by electrochemical deposition and provides an efficient cathode; this depolarizes the aluminum and can lead to the initiation of pitting corrosion.

A similar effect has sometimes been observed for iron and steel corrosion products generated upstream of aluminum components in desalination plants. However, the effect of iron contamination is not as strong or as consistent as that of copper.

Influence of Biological Organisms

Seawater is a biologically active medium that contains a large number of microscopic and macroscopic organisms. Many of these organisms are commonly observed in association with solid surfaces in seawater, where they form biofouling films. Because this subject has been dealt with in detail in the sections on biological corrosion in the articles "General Corrosion" and "Localized Corrosion" in this Volume, only a brief description will be given here.

Immersion of any solid surface in seawater initiates a continuous and dynamic process, beginning with adsorption of nonliving, dissolved organic material and continuing through the formation of bacterial and algal slime films and the settlement and growth of various macroscopic plants and animals. This process, by which the surfaces of all structural materials immersed in seawater become colonized, adds to the variability of the ocean environment in which corrosion occurs.

Bacterial Films. The process of colonization begins immediately upon immersion with the

Table 4 Constants for use with Eq 3

These values can be used with Eq 3 to calculate oxygen concentration relative to air at 1 atm total pressure and 100% relative humidity.

Constant	Value
A_1	−173.4292
A_2	249.6339
A_3	143.3483
A_4	−21.8492
B_1	−0.033096
B_2	0.014259
B_3	−0.0017000

Source: Ref 8

Table 5 Solubility of oxygen in seawater as a function of temperature and salinity

Solubility values were calculated using Eq 3.

Temperature		Oxygen solubility (mL/L) at indicated salinity (‰)					
°C	°F	0	8	16	24	31	36
0	32	10.22	9.70	9.19	8.70	8.27	7.99
5	41	8.93	8.49	8.05	7.64	7.28	7.04
10	50	7.89	7.52	7.14	6.79	6.48	6.28
15	60	7.05	6.72	6.40	6.10	5.83	5.65
20	70	6.35	6.07	5.79	5.52	5.29	5.14
25	75	5.77	5.52	5.27	5.04	4.84	4.70
30	85	5.28	5.06	4.84	4.63	4.45	4.33

Source: Ref 5

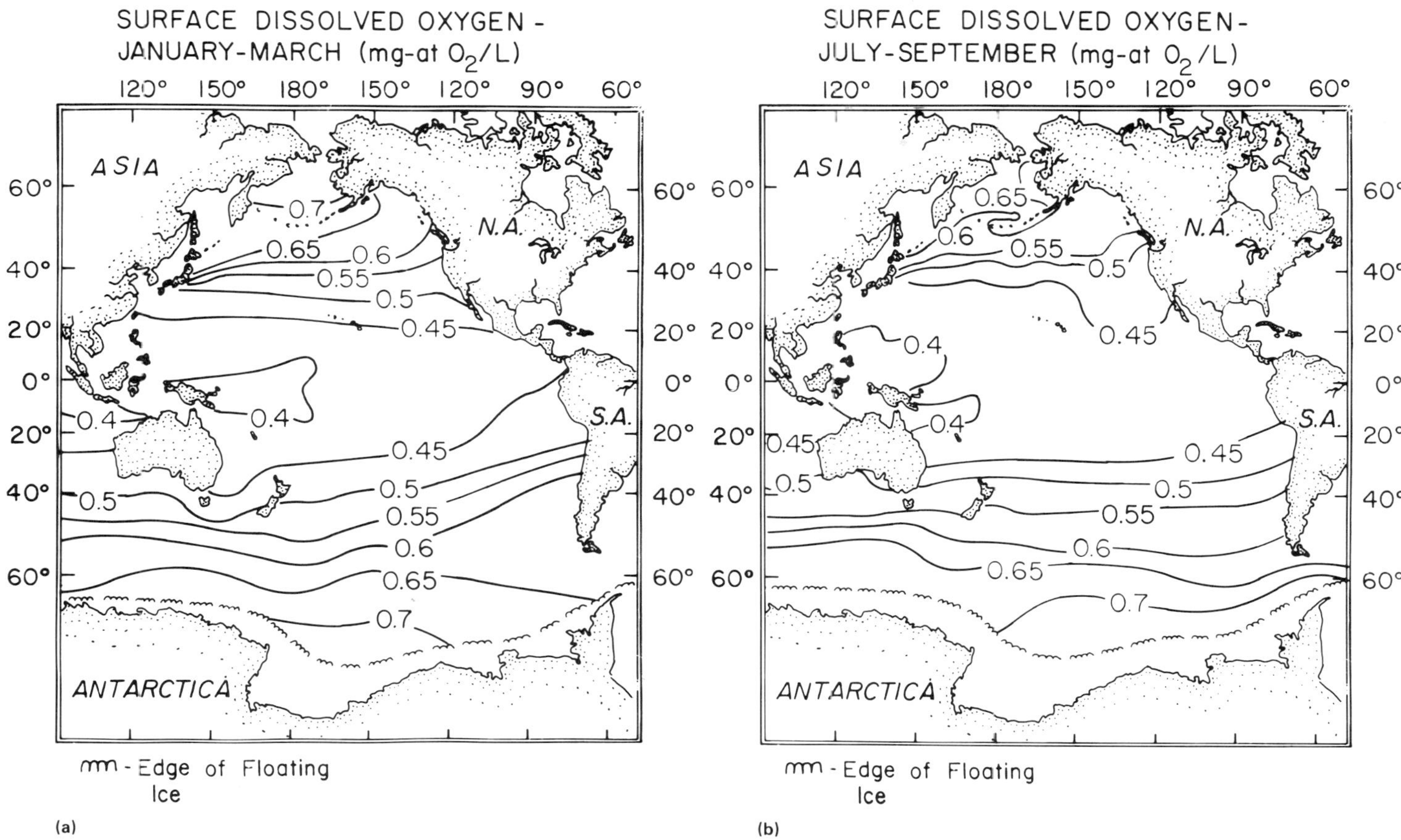

Fig. 14 Pacific Ocean surface dissolved oxygen (mg · atm/L) for January through March (a) and July through September (b). Source: Ref 5

adsorption of a nonliving organic conditioning film. This conditioning film is nearly complete within the first 2 h of immersion, at which time the initially colonizing bacteria begin to attach in substantial numbers. The bacterial, or primary, slime film develops over a period of 24 to 48 h in most natural seawaters, although further changes in the film can often be observed over more than a 2-week period. Additional information is available in the Selected References at the end of this article.

The bacterial film changes the chemistry at the metal/liquid interface in a number of ways that have an important bearing on corrosion. As the biofilm grows, the bacteria in the film produce a number of by-products. Among these are organic acids, hydrogen sulfide, and protein-rich polymeric materials commonly called slime. The first effect of the composite film of bacteria and associated polymer, an example of which is shown in Fig. 30, is to create a diffusion barrier between the metal/liquid interface and the bulk seawater. The barrier itself is over 90% water, so it does not truly isolate the interface; instead, it supports strong concentration gradients for various chemical species. Thus, the water chemistry at the interface may be different from that in the bulk water, although the two are closely coupled through diffusive processes.

Two chemical species, oxygen and hydrogen, that are often implicated (or even rate-controlling) in corrosion are also important in the metabolism of the bacteria. A given bacterial slime film can be either a source or a sink for either oxygen or hydrogen. Moreover, these films are rarely continuous. Usually, they provide only spotty coverage of the metal surface. Thus, they are capable of inducing oxygen (or other chemical) concentration cells. Bacterial action on decaying organic matter in the slime film can also result in the production of ammonia and sulfides. Ammonia causes stress-corrosion cracking of copper alloys, and sulfides have been implicated in accelerated localized and/or uniform corrosion of both copper alloys and steels.

Under anaerobic (no oxygen) conditions, such as those found in marshy coastal areas, in which all the dissolved oxygen in the mud is used in the decay of organic matter, the corrosion rate of steel is expected to be very low. Under these conditions, however, the sulfate-reducing bacteria of the genus *desulfovibrio* utilize the hydrogen produced at the metal surface in reducing sulfates from the decaying organic material to sulfides, including H_2S. The sulfides combine with iron from the steel to produce an iron sulfide (FeS) film, which is itself corrosive. The bacteria thus transform a benign environment into an aggressive one in which steel corrodes quite rapidly.

Even under open ocean conditions at air saturation, the presence of a bacterial slime film can result in anaerobic conditions at the metal surface. Oxygen-utilizing bacteria in the initial film may eventually increase sufficiently in numbers that they use all the oxygen diffusing through the film before it can reach the metal surface. This creates an anaerobic layer right next to the metal surface and provides a place where the sulfate-reducing organisms can flourish. In all of these examples, the biofilm is able to change the chemistry of the electrolyte substantially at the water/metal interface. Thus, the corrosion rate may depend as much on the details of the electrolyte chemistry at the interface as it does on the ambient bulk seawater chemistry. Additional details about many aspects of biological corrosion can be found in the Selected References at the end of this article and in the articles "General Corrosion," "Localized Corrosion," and "Evaluation of Microbiological Corrosion" in this Volume.

Macrofouling Films. Within the first 2 or 3 days of immersion, the solid surface, already having acquired both conditioning and bacterial films, begins to be colonized by the macrofouling organisms. A heavy encrustation of these organisms can have a number of undesirable effects on marine structures. Both weight and hydrodynamic drag on the structure will be increased by the fouling layer. Interference with the functioning of moving parts may also occur.

In terms of corrosion, the effects of the macrofouling layer are similar to those of the microfouling layer. If the macrofoulers form a continuous layer, they decrease the availability of dissolved oxygen at the metal/water interface and can reduce the corrosion rate. If the layer is discontinuous, they may induce oxygen or chemical concentration cells; this leads to various types of localized corrosion. Fouling films may also break down protective paint coatings by a combination of chemical and mechanical action.

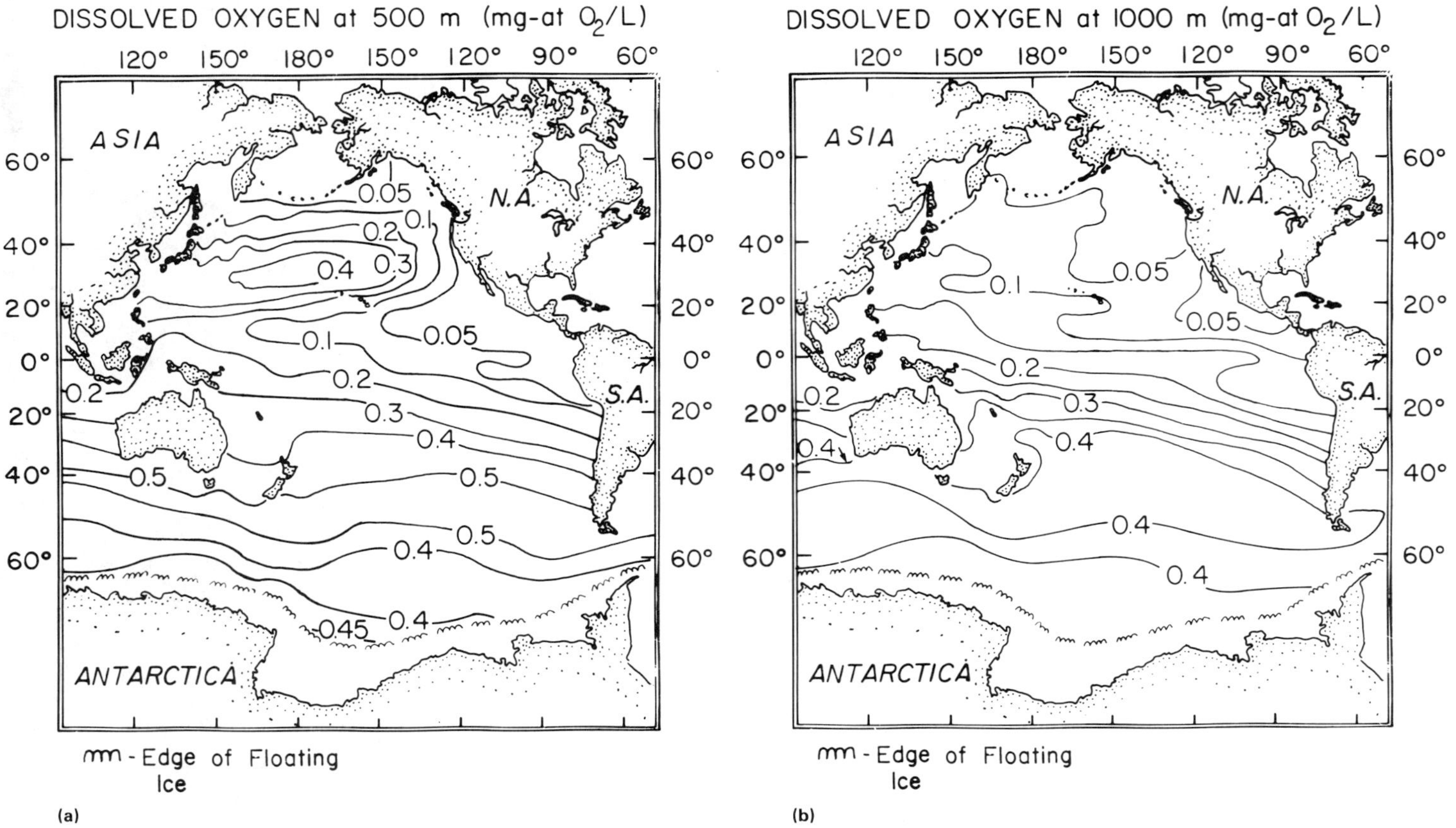

DISSOLVED OXYGEN at 500 m (mg-at O_2/L)

DISSOLVED OXYGEN at 1000 m (mg-at O_2/L)

(a)

(b)

Fig. 15 Pacific Ocean dissolved oxygen (mg · atm/L) at depth of 500 m (a) and 1000 m (b). Source: Ref 5

Additional information is available in the Selected References at the end of this article.

Marine Atmospheres

Richard B. Griffin
Department of Mechanical Engineering
Texas A&M University

The annual cost of corrosion in the United States has been estimated at $167 billion. A reasonable fraction of this amount is the result of atmospheric corrosion. The buildings, automobiles, bridges, storage tanks, ships, and other items that must be coated, repaired, or replaced represent only some of the problem areas of corrosion to the U.S. economy.

Typically, atmospheric corrosion is broken down into the types listed in Table 9. A variety of factors affect the atmospheric corrosion behavior of materials. These include the time of wetness, temperature, material, air contaminants, solar radiation, biological species, and the composition of the corrosion products. The particular location of a component is also important with respect to its corrosion behavior.

The marine or marine-industrial type is generally considered to be the most aggressive environment. This discussion of marine atmospheric corrosion will include atmospheric corrosion (zone 1, Fig. 2) and the splash zone above high tide (zone 2, Fig. 2). For carbon steels in marine exposure, the maximum corrosion rate occurs in the splash zone, in which the alloy is wet almost continually with well-aerated seawater. The atmospheric corrosion of low-carbon steel is in the range of 0.025 to 0.75 mm/yr (1 to 30 mils/yr). This section will discuss the specific details associated with the rates of corrosion in marine atmospheres.

Important Variables

A number of factors, such as moisture, temperature, winds, airborne contaminants, alloy content, location, and biological organisms, contribute to atmospheric corrosion. Each of these factors will be discussed with regard to its contribution to corrosion in the marine atmosphere.

Moisture. For corrosion to occur by an electrochemical process, there must be an electrolyte present. An electrolyte is a solution that will allow a current to pass through it by the diffusion of anions (negatively charged ions) and cations (positively charged ions). Water that contains ions is a very good electrolyte. Therefore, the amount and availability of moisture present is an important factor in atmospheric corrosion. For steel beyond a certain critical humidity, there will be an acceleration in the rate of corrosion in the atmosphere. An example of this is shown in Fig. 31, in which the critical humidity is 60% for iron in an atmosphere free of sulfur dioxide (Fig. 31a). For magnesium under similar conditions, the critical relative humidity is 90% (Fig. 31b). The critical relative humidity is not a constant value; it depends on the hygroscopicity (tendency to absorb moisture) of the corrosion products and the contaminants.

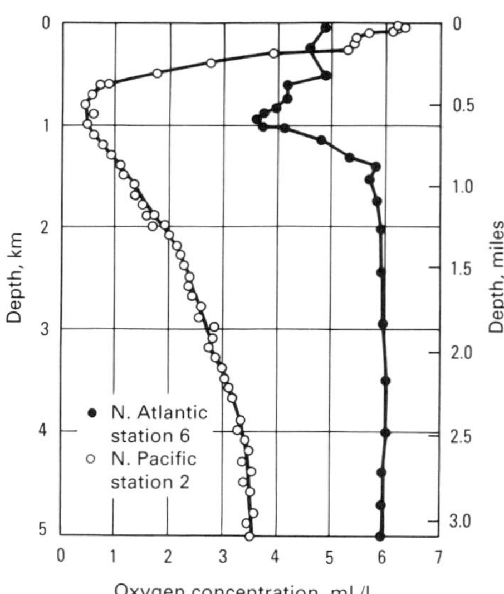

Fig. 16 Comparison of dissolved oxygen-depth profiles for open ocean stations 2 and 6 (see Fig. 5). Source: Ref 5

One of the measures of the effects of moisture is the time of wetness. As Fig. 32 shows, corrosion rate increases as time of wetness increases. In addition, Fig. 32 shows the importance of a contaminant. When the sulfur dioxide level in-

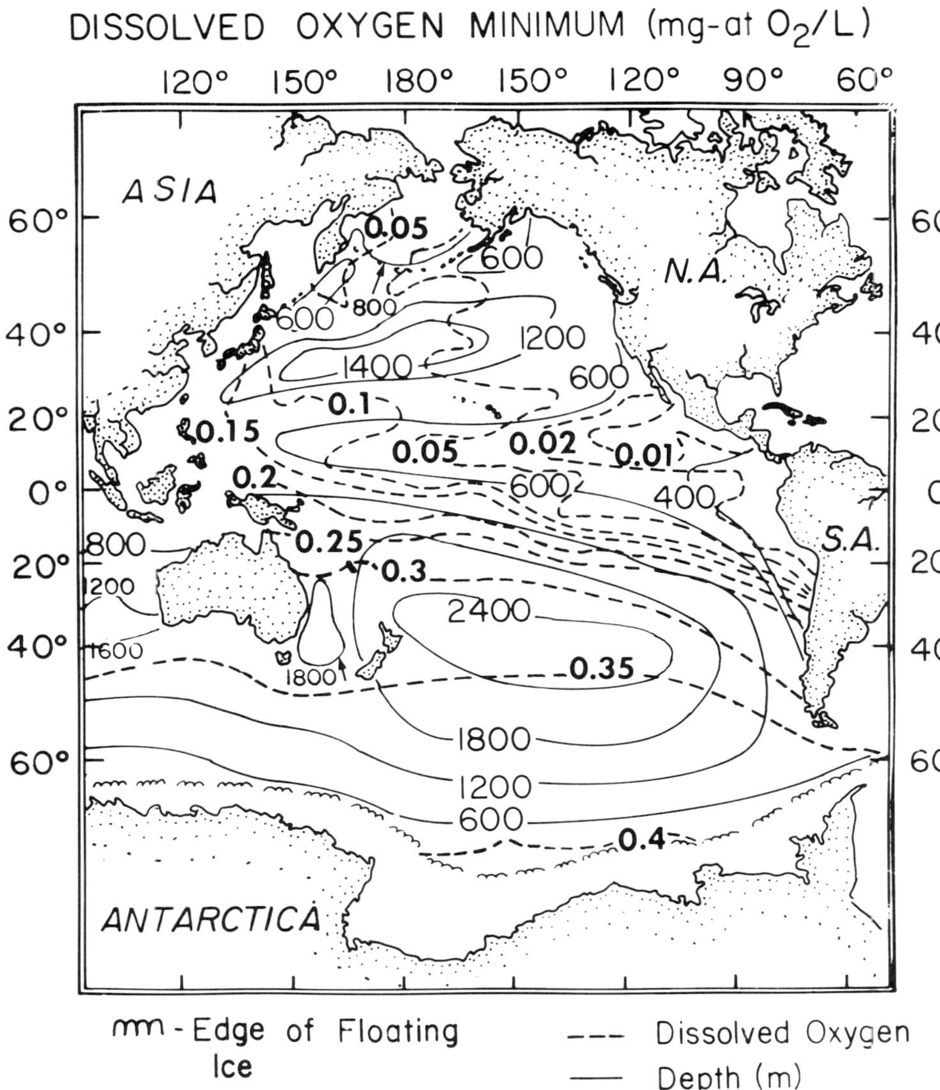

Fig. 17 Depth (meters) of the dissolved oxygen minimum in the Pacific (solid contours) and the value of the minimum in mg · atm/L (dashed contours). Source: Ref 5

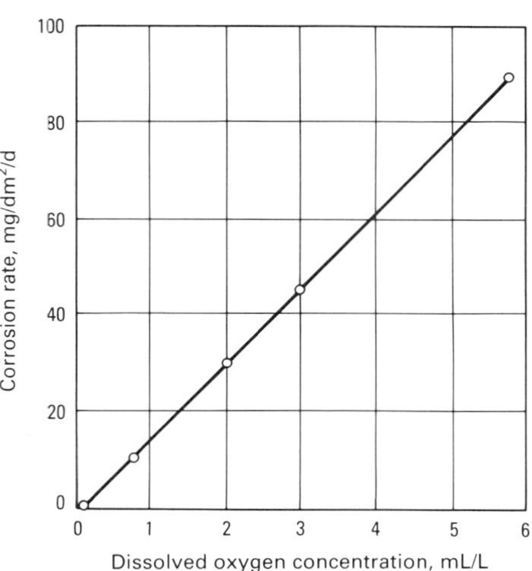

Fig. 18 Effect of oxygen concentration on the corrosion of low-carbon steel in slowly moving water containing 165 ppm $CaCl_2$. The 48-h test was conducted at 25 °C (75 °F). Source: Ref 1

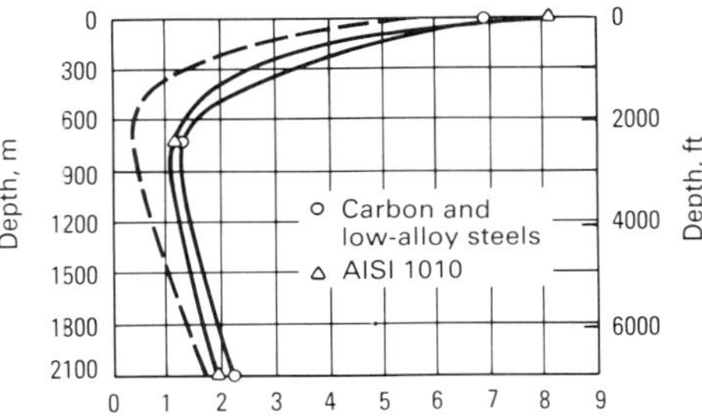

Fig. 19 Corrosion of steels versus depth after 1 yr of exposure compared to the shape of the dissolved oxygen profile (dashed line). Source: Ref 2

creases, there is a corresponding increase in the overall corrosion rate. However, the severity of the marine environment is related to the salt content of the sea spray or dew that contacts the material surface, which is usually more corrosive than rainfall.

For acid rain conditions, there appears to be no significant increase in corrosion rate. A study conducted in Sweden from October 1974 to November 1976 for carbon steel showed an increase in corrosion rates with increasing sulfur dioxide; however, the incidences were relatively infrequent. The study also showed that the corrosion rates measured for a longer time do not seem to be influenced by the incidences of acid rain. Similar results were obtained in a British study on the atmospheric-corrosion rate of zinc.

Airborne Contaminants. The second most important factor in atmospheric corrosion is the contaminants found in the air. These can be manmade or natural, such as airborne moisture carrying salt from the sea or sulfur dioxide put into the atmosphere by a coal-burning utility plant. Figure 32 illustrates the importance of the atmospheric sulfur dioxide level on the corrosion rate of zinc. The important contaminants are chlorides, sulfur dioxide, carbon dioxide, nitrogen oxides, and hard dust particles (for example, sand or minerals).

Chlorides. There is a direct relationship between atmospheric salt content and measured corrosion rates. The amount of sea salts measured off the coast of Nigeria illustrate this relationship between the salinity and the corrosion rate. This is shown in Fig. 33, in which salinity of 10 mg/m^2/d results in a corrosion rate of less than 0.1 g/dm^2/mo, while a salinity of 1000 mg/m^2/d results in a corrosion rate of almost 10 g/dm^2/mo. At the LaQue Center for Corrosion Technology test site at Kure Beach, NC, a similar effect has been observed for carbon steel. The corrosion rate at the site 25 m (80 ft) from the mean tide line was 1.19 mm/yr (47 mils/yr), while at the 250-m (800-ft) site, the corrosion rate for the same material was 0.04 mm/yr (1.6 mils/yr).

The average atmospheric chloride levels, as collected in rainwater for the United States, are shown in Fig. 34. The highest levels occur along the coast of the Atlantic Ocean, Pacific Ocean, and the Gulf of Mexico. The maximum corrosion rate is related to the maximum chloride in the atmosphere. This will of course be related to the distance inland, the height above sea level, and the prevailing winds. The chlorides of calcium and magnesium are hygroscopic and have a tendency to form liquid films on metal surfaces.

Sulfur Dioxide. The presence of SO_2 in the atmosphere lowers the critical relative humidity while increasing the thickness of the electrolyte film and increasing the aggressiveness of the environment. For carbon steel, the effect of SO_2 levels is shown in Fig 35, in which data are plotted from three Norwegian test sites. The data show that as SO_2 concentrations are increased the corrosion rate, measured as weight loss, increases. For example, at an SO_2 concentration of 25 μg/m^3, the corrosion rate is approximately 55 g/m^2/mo, while for an SO_2 concentration of 100 μg/m^3 the corresponding corrosion rate is approximately 170 g/m^2/mo. A summary of Scandinavian data for carbon steel and zinc showed the following relationships between corrosion rate and SO_2 concentration:

$$K_{steel} = 5.28\ [SO_2] + 176.6 \qquad \text{(Eq 7)}$$

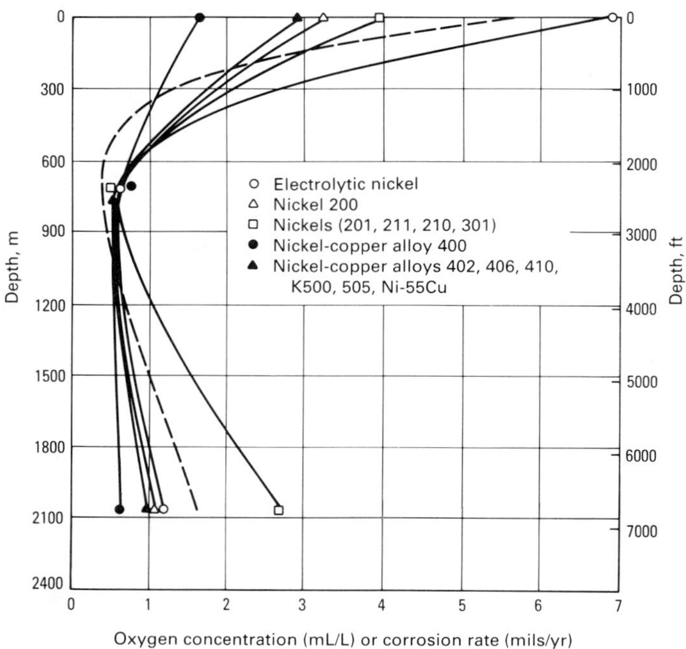

Fig. 20 Corrosion of nickels and nickel-copper alloys versus depth after 1 year of exposure compared to the shape of the dissolved oxygen profile (dashed line). Source: Ref 2

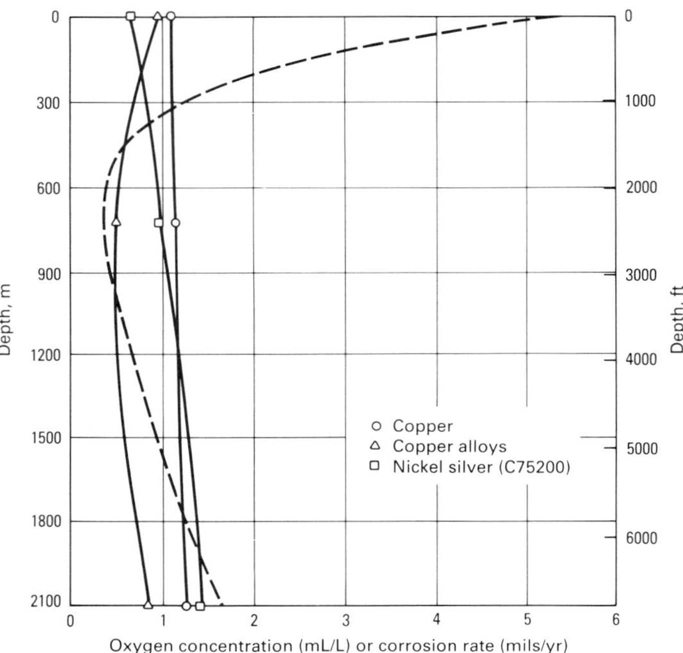

Fig. 21 Corrosion of copper alloys versus depth after 1 year of exposure compared to the shape of the dissolved oxygen profile (dashed line). Source: Ref 2

and

$$\dot{K}_{zinc} = 0.22\,[SO_2] + 6 \quad \text{(Eq 8)}$$

where $\dot{K}$ is the atmospheric corrosion rate in $g/m^2/yr$, and $[SO_2]$ represents the concentration of SO_2 in $\mu g/m^3$. Similar types of relationships have been shown for other alloy systems and locations.

Carbon Dioxide. The opinion of the majority of investigators is that carbon dioxide (CO_2) has an effect on the corrosion of metals. Carbon dioxide in the presence of water forms carbonic acid (Eq 4); a pH of 5.6 could be obtained with atmospheric CO_2 in equilibrium with pure water. Carbonates are found in the corrosion products on a number of metals. It is possible that for steel and copper the presence of CO_2 might lessen the corrosion effects of SO_2, because of the nature of the corrosion products formed. Carbon dioxide does not have nearly the same level of importance in atmospheric corrosion as SO_2 and chlorides do.

Location is a very important variable. The distance from the sea and the height above the ground are both significant.

Distance. Figure 33 shows the effect of moving inland along the coast of Nigeria from the 45-, 365-, and 1190-m (50-, 400-, and 1300-yd) sites at Lagos. From studies done on Barbados, the effect of distance is confirmed by the map of the island shown in Fig. 36. As can be seen, this represents one of the worst conditions: tropical beach, on-shore winds, and facing a large, uninterrupted stretch of ocean. Similarly, at a site in Aracaju, Brazil, low-carbon steel samples were tested at five sites from about 0.1 km (0.06 miles) to almost 4 km (2.5 miles) from the sea. There was, as Fig. 37 shows, a rapid falloff in the corrosion rate as the testing site was moved inland. By about 1.5 km (0.9 miles) inland, the corrosion rate had reached a value that shows it is basically independent of the marine atmosphere.

The height of the specimens above seawater is also important. This is illustrated in Fig. 38(a) in which the corrosion rate of the carbon steel specimens in the 25-m (80-ft) lot at Kure Beach, NC, varies from less than 400 μm/yr (16 mils/yr) at a height of 5 m (16.5 ft) to 600 μm/yr (24 mils/yr) at a height of about 8 m (26 ft). The corrosion rate is in direct proportion to the amount of chloride in the atmosphere. There is considerably less corrosion for the carbon steel at the Kure

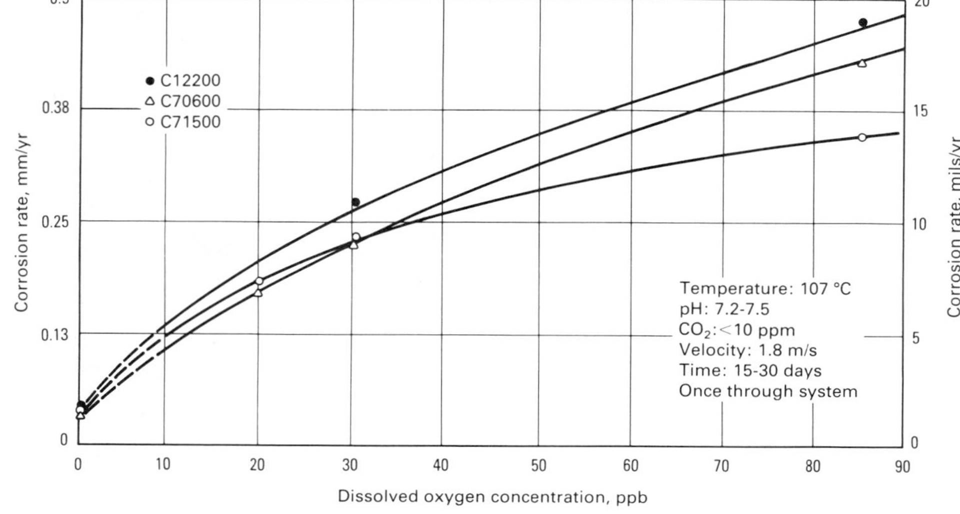

Fig. 22 Effect of dissolved oxygen in seawater on the corrosion rate of three CDA copper alloys. Source: Ref 7

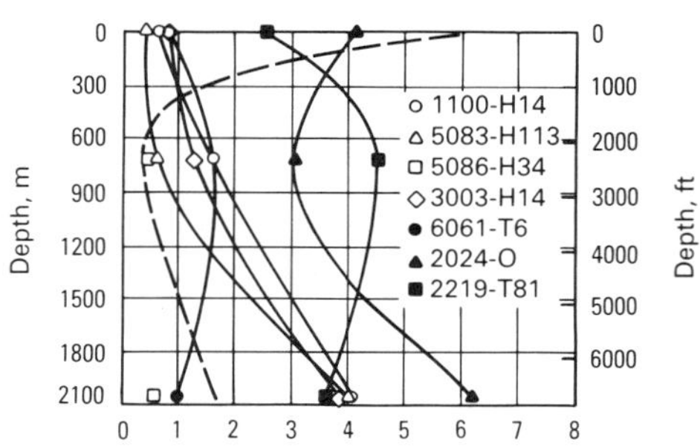

Fig. 23 Corrosion rates of aluminum alloys versus depth after 1 year of exposure compared to the shape of the dissolved oxygen profile (dashed line). Source: Ref 2

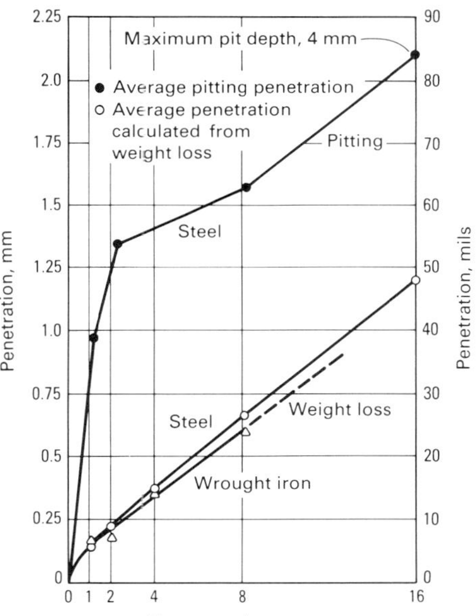

Fig. 24 Corrosion of carbon steel and wrought iron continuously immersed in seawater. Average penetration rate was 68 μm/yr (2.7 mils/yr) for steel; that of wrought iron was 61 μm (2.4 mils/yr). Source: Ref 7

Beach, NC, 250-m (800-ft) test site (Fig. 38b). The 25-m (80-ft) lot has an average chloride content of approximately 400 mg/m^2/d, while the 250-m (800-ft) lot has an average chloride value of approximately 100 mg/m^2/d.

In the splash zone, the effect of height above the sea is illustrated in Fig. 2, which shows that the corrosion rate is highest slightly above mean high tide. This zone would not only have a high chloride content, but would also be alternately wet and dry. As the height above the sea increases, the corrosion rate decreases because the specimen is not wet as often.

Orientation. Another important factor with regard to the atmospheric corrosion of a material

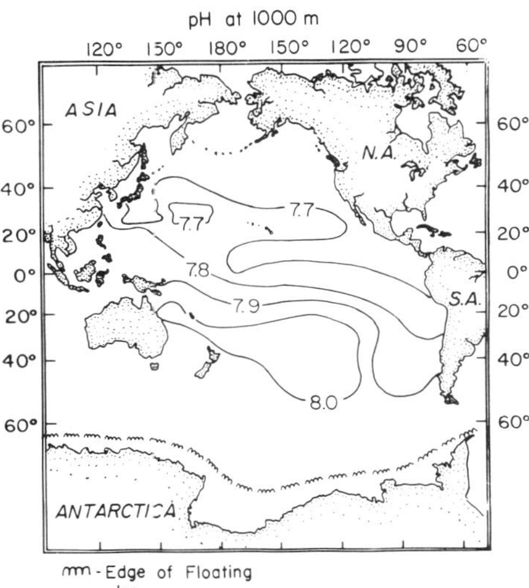

Fig. 27 Pacific Ocean pH at a depth of 1000 m (3280 ft). Source: Ref 5

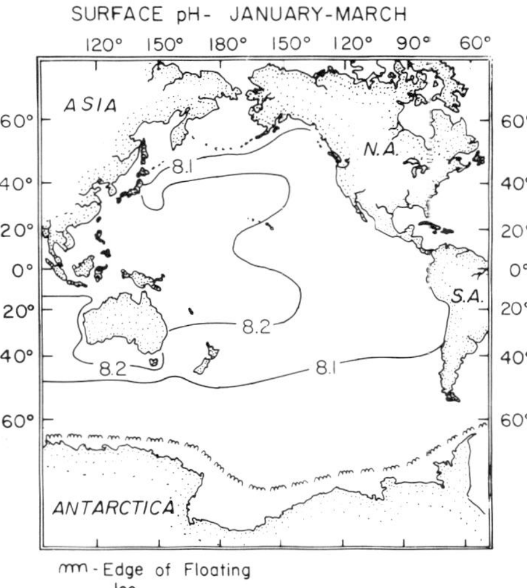

Fig. 25 Pacific Ocean surface pH for the period January to March. Source: Ref 5

is its orientation with respect to the earth's surface. Results for a 1-year exposure of iron specimens placed vertically and inclined at an angle of 30° with respect to the ground are shown in Fig. 39. The spread in the data is much greater for the Kure Beach 25-m (80-ft) test lot than for the 250-m (800-ft) test lot. In both cases, the vertical specimens showed a higher corrosion rate. This was attributed to the formation of a nonuniform, less protective oxide in the vertical position than in the 30° position. It is also possible that the 30° samples have the chloride deposits cleaned from their surfaces more easily than the vertical specimens. Ratios of the corrosion rate in the vertical position to that in the 30° position are given in Table 10 for five sites. In the vertical position, the corrosion rate is greater on the side facing the sea than on the side facing land. At the 25-m (80-ft)

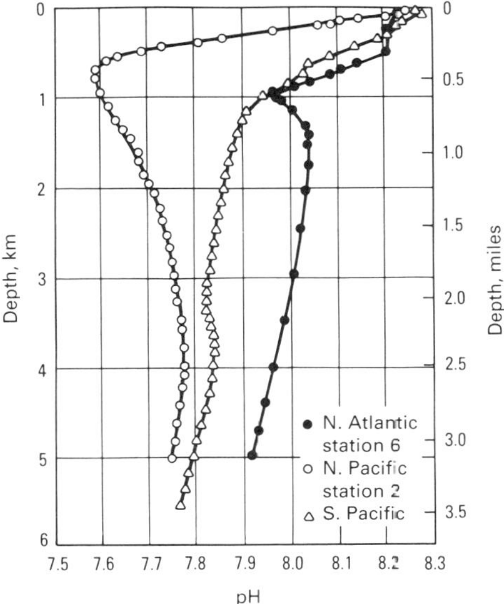

Fig. 28 Comparison of pH-depth profiles for open ocean sites 2 and 6 (see Fig. 5). Note that the data for the south Pacific are highest at the surface, but are intermediate at depths greater than 500 m (1640 ft). Source: Ref 5

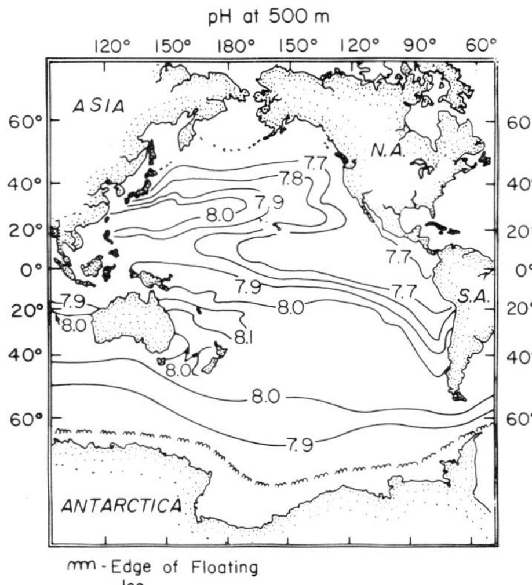

Fig. 26 Pacific Ocean pH at a depth of 500 m (1640 ft). Source: Ref 5

lot at Kure Beach, steel pipe specimens corroded at the rate of 850 μm/yr (33.5 mils/yr) facing the ocean, as compared to 50 μm/yr (2 mils/yr) facing away from the ocean, over a 4.5 year period.

The corrosion rate can also be measured on the skyward or groundward side of specimens that are parallel to the earth's surface. Tests conducted at Kure Beach showed that the skyward side corroded at a greater rate after 3 months. However, after 6 months of testing, the rates were identical. Similarly, for an AZ31 B magnesium alloy in a 30-day test, the skyward-facing specimens lost more material than the groundward-facing ones.

The temperature affects the relative humidity, the dew point, the time of wetness, and the kinetics of the corrosion process. For atmospheric corrosion, the presence of moisture, as determined by the time of wetness, is probably the most important role of temperature. Figure 40 illustrates the effect of temperature on iron, zinc, and copper. There are three distinct patterns with increasing temperature over the range of 20 to 40 °C (70 to 100 °F):

- Corrosion rate increases for iron
- Corrosion rate decreases for zinc
- Corrosion remains constant for copper

The temperature of interest may not be the average daily temperature. It may be more important to know the dew-point temperature or the test panel surface temperature. From an atmospheric corrosion standpoint, dry, hot conditions are preferable to cooler, moist conditions.

Sunlight influences the degree of wetness and affects the performance of coatings and plastics. Sunlight may also stimulate photosensitive corrosion reactions on such metals as copper and iron. In addition, it may stimulate biological reactions, such as the development of fungi. Ultraviolet (UV) light and photo-oxidation can cause embrittlement and surface cracks in polymers. This can be avoided by the addition of UV stabilizers (for example, carbon black).

Wind. The direction and velocity of the wind affect the rate of accumulation of particles on metal surfaces. Also, wind disperses the airborne contaminants and pollutants. Figure 36 illustrates the very severe corrosion that can occur on an

Table 6 Pitting of low-carbon steel submerged in the San Diego harbor (polluted seawater)

Penetration rate averaged 0.056 mm/yr (2.2 mils/yr) for this exposure.

		Penetration			
		Average of five deepest pits per panel		Deepest pit per panel	
Exposure time, days	Number of panels	mm	mils	mm	mils
155	6	0.33–0.61	13–24	0.46–0.75	18–30
361	12	0.5–1.34	20–53	0.74–1.5	29–60
552	6	0.81-1.04	32–41	0.66–1.3	26–50

Source: Ref 7

Table 7 Corrosion of copper alloy condenser tubes in polluted and clean seawater

Velocity: 2.3 m/s (7.5 ft/s). Test duration: 64 days

		Corrosion rate			
		Clean seawater		Polluted seawater(a)	
Alloy	CDA/UNS designation	mm/yr	mils/yr	mm/yr	mils/yr
90Cu-10Ni	C70600	0.075	3	0.86	34
70Cu-30Ni	C71500	0.13	5	0.66	26
2% Al brass	C68700	0.075	3	0.56	22
6% Al brass	C60800	0.13	5	0.53	21
Arsenical admiralty brass	C44300	0.33	13	0.89	35
Phosphorus deoxidized copper	C12200	0.36	14	2.7	105

(a) Contained 3 ppm hydrogen sulfide. Source: Ref 7

Table 8 Effect of hydrogen sulfide in seawater on corrosion of copper condenser tube alloys

64-day test in seawater flowing at 2.3 m/s (7.5 ft/s). Test temperature: 27 °C (80 °F)

	Corrosion rate			
	Clean seawater		Seawater plus 4 ppm H_2S	
Alloy	mm/yr	mils/yr	mm/yr	mils/yr
Phosphorus deoxidized copper	0.36	14	0.38	15
Admiralty brass	0.33	13	0.89	35
70Cu-30Ni	0.13	5	0.66	26

Source: Ref 7

ocean beach facing the prevailing wind. The effect caused by the chloride ions being carried inland is illustrated in Fig. 37, which shows an increased corrosion rate at 1 km (0.6 mile) inland. Stronger prevailing winds can carry the airborne contaminants even further inland. A marine site may be made even more aggressive by the prevailing winds bringing industrial pollutants, particularly SO_2, to the marine site.

Time. For many materials, there is a decrease in the corrosion rate as time increases. This decrease is associated with the formation of protective corrosion layers. Figure 41 provides an example of this for a low-carbon steel at eight sites in South Africa. There is an initial sharp increase in the atmospheric-corrosion rate, followed by a slowing down of the corrosion rate as corrosion products form on the alloy surface. This is particularly true for the sites C through G. For site B, the corrosion rate is sufficiently high to prevent the formation of a protective layer; therefore, a very high corrosion rate can be maintained. The effect of corrosion on tensile strength is shown in Fig. 42 for a low-carbon steel and aluminum. The rate of loss in ultimate tensile strength is initially large, but as time continues, the rate of loss decreases.

Starting Date. There can be a variation in the corrosion rate that depends on when the tests were started. Figure 43 compares the measured weight losses for iron and zinc in tests started at two different dates. Over a 60-day test, the variation in corrosion rate for zinc is much larger than that for iron. Similarly, for iron specimens at the Kure Beach, 25-m (80-ft) lot, there are variations of hundreds of microns per year in corrosion rates, as measured on samples exposed vertically for 1 and 2 years each. This is shown in Fig. 44 for iron calibration specimens tested from 1949 to 1979.

Site Variability. Large variations in atmospheric corrosion rate occur within a particular type of region. An example would be the various corrosion behaviors of steel and zinc in different tropical environments, especially in a tropical seacoast for 1 year. Figure 45 shows the average penetration for steel in a 1-year test at various tropical sites. For zinc under similar conditions, the average penetration varied from 3 to 15 μm (0.11 to 0.6 mil). Similarly, Fig. 42 illustrates the loss of tensile strength with increasing time in different temperate marine environments. As Fig. 42 shows, there is a wide variation in the loss of tensile strength between the four seacoast locations.

Temperate and tropical marine sites, along with inland sites, are compared in Fig. 46 for copper and zinc. Overall, the long-term rates are similar at both marine and inland sites.

However, a similar comparison for carbon steels and low-alloy steels (Fig. 47) illustrates that for these materials the tropical environment has a higher overall corrosion rate. Figure 47(a) compares the stabilized corrosion rate of carbon steel at Cristobal, Panama (20 μm/yr, or 0.8 mil/yr), to that at Kure Beach 250 m (800 ft) from the ocean (16 μm/yr, or 0.63 mil/yr). Low-alloy steels exhibit a similar increased rate of corrosion, as shown in Fig. 47(b) and a similar pattern is exhibited for carbon steels compared at inland sites (Fig. 47c).

Alloy Content. The selection of a particular alloy composition can make a significant difference in the corrosion rate of a material. For steels, a comparison can be made for carbon steels, low-alloy steels, and steels with 5% alloying elements (Fig. 48). Figure 48 also compares marine versus inland exposure, and in each case, the long-term corrosion rate is greater for the marine environment. Figure 48 also shows the more rapid corrosion that takes place in the first 1 to 3 years and the leveling off associated with long-term atmospheric corrosion. Very similar results have been reported for a study done in South Africa at eight sites that were classified as rural to severe marine.

The results of 15.5-year studies of low-alloy steels in 13 groups conducted at the Kure Beach, NC, 250-m (800-ft) lot are shown in Fig. 49, in which the mass loss per unit area is plotted as a function of the total alloy content. Alloy additions of about 2 wt% result in the mass loss per area being reduced from 40 mg/dm^2 to less than 12 mg/dm^2.

The significance of chromium as an alloying element is shown in Fig. 50 for atmospheric-corrosion conditions classified as moderate and severe marine. Above 12 or 12.5 wt% Cr, the atmospheric corrosion becomes negligible; lower chromium levels result in a rapid increase in the corrosion rate. This effect is also illustrated in Table 11. For AISI 300-series stainless steels tested at the Kure Beach, NC, 250-m (800-ft) lot over a 15-year period, the atmospheric-corrosion rates were equal to or less than 0.03 μm/yr (0.001 mil/yr) (see Table 11).

An excellent summary of marine atmospheric-corrosion data is given in Table 12. A wide variety of metals and alloys are listed. In addition, information on general corrosion, pitting, and loss of tensile strength is given for materials exposed at a seacoast site at Cristobal, Panama.

Exposure Time. One of the difficulties with marine atmospheric corrosion testing is the length of time required for the tests. For steels, a reasonable estimate of long-term corrosion performance can be made from short-term data. This is not always the case; any short-term result must be used very cautiously, and it is always best to have long-term test data available.

Atmospheric Corrosion Test Sites

There are a large number of atmospheric corrosion sites throughout the world; Table 13 lists some of them. Where available, the 2-year corrosion rates for low-carbon steel and zinc are given. Some of the sites have a marine corrosion index and/or an atmospheric corrosion index number after them. The higher the index number, the more aggressive the environment.

Metallic Coatings

Jean A. Montemarano
and Barbara A. Shaw
David Taylor Naval Ship R & D Center

Effective protection from the marine environment can be provided by metallic coatings, which include thermal spray, galvanizing, and, for certain applications, electroplating. In general, me-

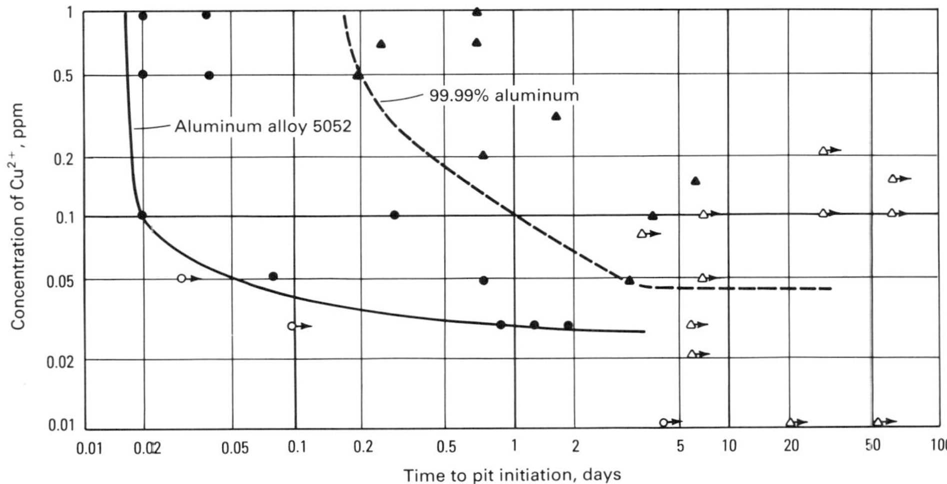

Fig. 29 Effect of adding Cu^{2+} ion to seawater on the time to pit initiation for aluminum alloy 5052 and 99.99% Al. Solid points represent conditions under which pitting started; open points indicate conditions under which no pitting occurred. Source: Ref 9

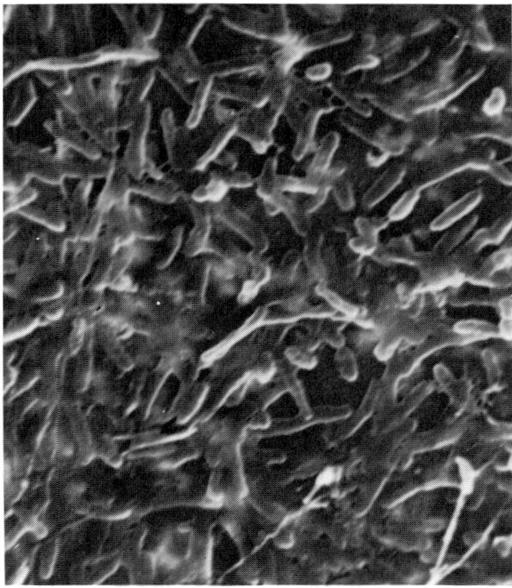

Fig. 30 Rod-shaped marine bacteria embedded in slime film growing on the surface of a copper-base antifouling paint after immersion in natural seawater at Woods Hole, MA, for 7 days. Depth of immersion was 5 m. 3000×

Table 9 Types of atmospheres and corrosion rates of low-carbon steel

Test duration: 2 years

Atmosphere	Location	Corrosion rate mm/yr	Corrosion rate mils/yr
Marine	Point Reyes, CA	0.5	19.71
Severe	25-m (80-ft) lot, Kure Beach, NC	0.53	21.00
Industrial	Brazos River, TX	0.093	3.67
Mild	250-m (800-ft) lot, Kure Beach, NC	0.146	5.73
Rural	Esquimalt, BC, Canada	0.013	0.53
Industrial	East Chicago	0.084	3.32
Marine	Bayonne, NJ	0.077	3.05
Urban	Pittsburgh, PA	0.03	1.20
Suburban (semi-industrial)	Middletown, OH	0.029	1.13
Rural	State College, PA	0.023	0.90
Marine	Esquimalt, BC, Canada	0.013	0.53
Desert	Phoenix, AZ	0.0046	0.18

Source: Ref 10

tallic coatings are two to three times more expensive than their traditional organic counterparts; therefore, their use is usually justified by longer service life and reduced maintenance. Metallic coatings function by providing a barrier, similar to organic coatings, for the bare metal from the marine environment and by corroding preferentially with respect to the substrate (normally, steel) when the coating is scratched or nicked. Aluminum and zinc are more active in seawater than steel and are the metals most widely used as protective coatings. Aluminum and zinc can be thermal sprayed. Zinc, aluminum, and zinc-aluminum alloy hot dip coatings are also used for the corrosion protection of steel in marine environments. Only the hot-dip zinc process, better known as galvanizing, is commercially available for the coating of fabricated articles. Lastly, the electroplating process is commonly used for the zinc or cadmium plating of fasteners for marine applications.

Metal Spray

Thermal spray has been used by industry and in European countries for corrosion protection since the 1940s. Bridges, structural steel work, boat holds and tanks, sluice gates and canal lock gates, and offshore drilling rigs are some of the items that have been metallized (Ref 13). Thermal spray for corrosion protection is normally applied by either the wire flame (combustion) or wire arc process. In either case, the metal wire is fed into a gun and melted either by a flame (normally oxyacetylene) or an electric arc. The atomized particles are propelled by means of compressed air onto the surface, where they cool, forming layers of splat-quenched particles (Ref 14). Both methods are portable and can be easily automated. Figures 2 and 4 in the article "Thermal Spray Coatings" in this Volume illustrate the wire flame and wire arc processes.

Surface preparation, as with all coatings, is an essential part of the coating process. Care must be taken to ensure that the surface is properly prepared and cleaned. Grit-blasting to white metal is used to achieve the necessary surface roughness. Typical grit media include aluminum oxide and chilled iron. The coating is largely mechanically bonded to the substrate in the flame or arc spray process used for corrosion control (Ref 15). If a component cannot be cleaned such that all rust and oil are removed, the thermal spray coating will not remain attached for long.

For example, field tests were conducted in Norway on steel piles coated with aluminum thermal spray followed by a wash primer, a coal tar vinyl paint, and then a topcoat. After 1 year or less, in spite of the organic coatings, blisters appeared in the coatings on all the piles in the splash zone. The failure analysis indicated that the major contributing factor was inadequate adhesion between the steel and aluminum thermal spray coating due to poor surface preparation (Ref 16). On the other hand, carefully prepared thermal sprayed steel specimens gave 19 years of complete base metal corrosion protection in seawater and marine atmosphere tests performed by the American Welding Society (AWS) (Ref 17).

The metal coating is normally applied to a thickness of 75 to 180 μm (3 to 7 mils) to provide adequate corrosion protection. The thickness of the coating is selected to limit interconnected porosity (too thin a coating) and to minimize thermal expansion mismatch (too thick a coating) with the substrate, which could result in bond-line separation (Ref 18). For marine applications, thermal sprayed aluminum coatings 180 to 250 μm (7 to 10 mils) thick are used in order to limit through porosity (Ref 19).

Zinc is widely used by industry and in European countries for thermal spray for corrosion control protection. In addition, a duplex coating of aluminum followed by zinc has also been used in Europe, although aluminum thermal spray coatings are being stressed for marine applications in the United States (Ref 20). The zinc thermal spray coating has a high electrochemical activity and therefore high corrosion rates; this results in depletion of the coating, although it affords excellent cathodic protection to steel. On the other hand, the aluminum coating forms a passive film of aluminum oxide, resulting in very low corrosion rates (Ref 21). In an attempt to combine the best features of both materials, zinc-aluminum coatings (85Zn-15Al) were examined for their marine corrosion protection. This combination of metals can be obtained either in a prealloyed wire or by spraying the two wires simultaneously in the dual wire arc process. Both coatings provided adequate cathodic protection to the substrate, but suffered severe coating degradation after just 6 months of exposure (Ref 14).

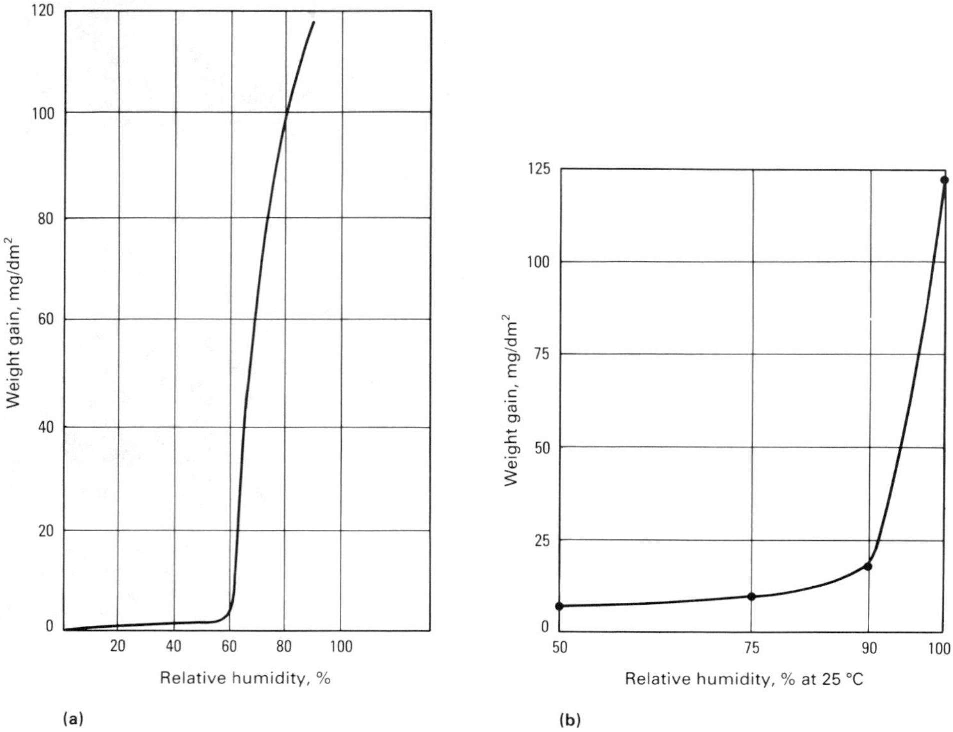

Fig. 31 Corrosion rates of iron and magnesium as a function of relative humidity. (a) For iron, the critical relative humidity is 60%. (b) For magnesium, corrosion rate increases significantly at a critical relative humidity of about 90%. Source: Ref 10

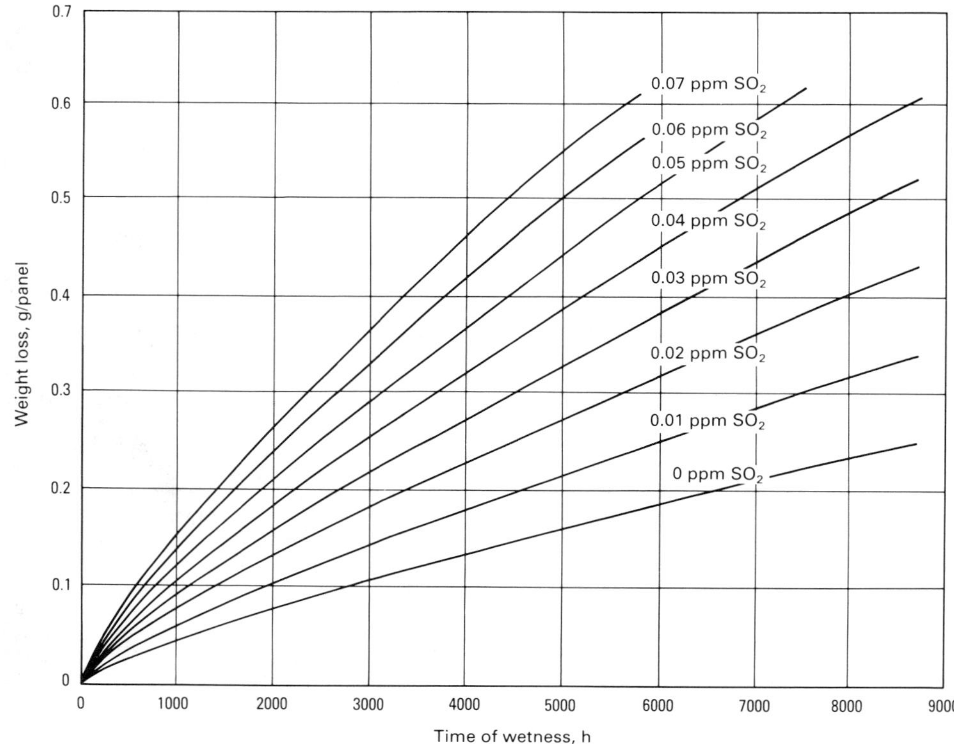

Fig. 32 The increase in corrosion rate of zinc as a function of time of wetness and SO_2 concentration. Source: Ref 10

Sealing. Standard practice is to seal the thermal spray coating with low-viscosity sealers because the metal coating is inherently porous (Fig. 51). The sealer is normally sprayed or brushed on, and it penetrates and fills the pores. Sealers should also be used in acidic or alkali environments (Ref 18). Vinyls and thinned epoxies are typical sealers. For high-temperature applications, a silicone alkyd sealer is used (Ref 22). Topcoats are normally used for cosmetic reasons. The roughness of the thermal spray coatings offers an excellent surface for paint adhesion. Rusting of the substrate underneath the paint film, which is the common mode of paint failure, is not the same mode of failure for thermal spray coatings (Ref 18). The topcoated thermal spray system provides excellent long-term maintenance-free corrosion protection (10 to 15 years).

For ship applications, thermal spray coatings offer corrosion protection for topside weather equipment, machinery spaces, and interior wet spaces (Ref 18). Specifically, these categories include auxiliary exhaust stacks; diesel headers; steam valves, piping, and traps; boiler skirts; stanchions, pipe hangers; rigging fittings; lighting fixtures; ladders; hatches and scuttles; boat davit machinery components; bilges; and machinery foundations. Figure 52 shows two topside applications. For marine atmospheric service, the use of thermal spray aluminum coatings is an outstanding method of corrosion control. However, use of thermal spray coatings for immersion service lacks extensive field experience. Underwater hulls of small boats have been thermal sprayed, followed by primer and antifouling topcoats. More recently, thermal spray coatings, sealed with a silicone sealer, were applied to the flare boom of a North Sea oil rig platform (Ref 23). In addition, tension-leg elements, the risers, and the flare tower of a North Sea tension-leg rig platform were coated using the same materials (Ref 23).

Comparative corrosion tests support the long-term performance of thermal spray in marine environments. The longest test (19 years), which was conducted by AWS, indicates that 75 to 150 μm (3 to 6 mils) of aluminum thermal spray coating, whether sealed or unsealed, provided

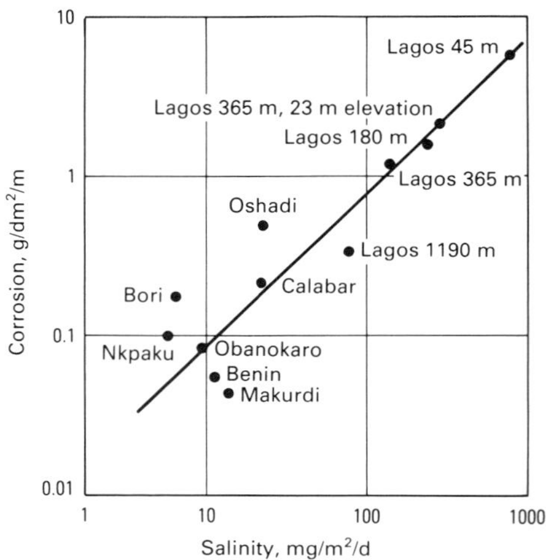

Fig. 33 Atmospheric corrosion as a function of salinity at various sites in Nigeria. Source: Ref 11

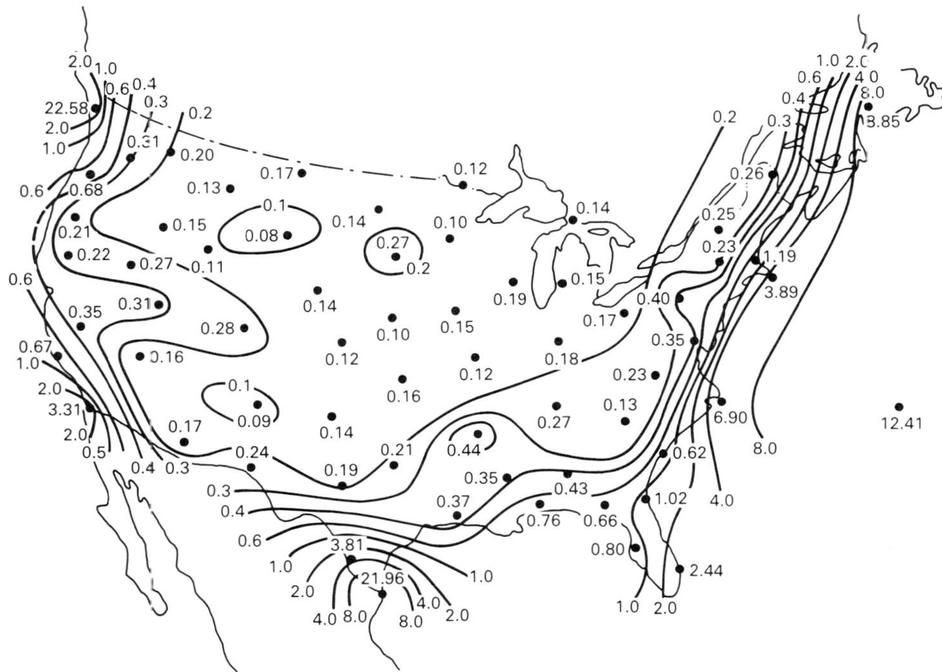

Fig. 34 Average chloride concentration (mg/L) in rainwater in the U.S. Source: Ref 10

long-term protection in total seawater immersion, splash and spray zone, and marine atmosphere tests. Although no significant base metal attack occurred, some blistering and rust staining of the aluminum coating on unsealed panels were noted (Ref 17). In this study, zinc thermal spray coatings gave equivalent performance in marine atmosphere; but for total seawater immersion and splash and spray zone tests, a minimum thickness of 300 μm (12 mils) was required to prevent base metal attack for unsealed panels, and a minimum of 230 μm (9 mils) was required for sealed panels. However, for the latter two marine environments, the unsealed zinc had completely converted to corrosion products (Ref 17).

Sealed aluminum thermal spray coatings again performed well in a 7-year study conducted by the National Bureau of Standards to ascertain what coating provided the best corrosion protection for steel piles. Sand abrasion, seawater immersion, splash and spray, and marine atmosphere test zones of the coated piles were evaluated for coating performance. Hot-dip zinc, sealed zinc thermal spray coatings, and unsealed aluminum thermal spray coatings performed equivalently and had at least two to three times the corrosion rate of sealed aluminum thermal spray coating (Ref 24).

A study of coated 3-m (10-ft) long panels designed to simulate coated piles also demonstrated the effectiveness of thermal spray coatings. After 20 years of field exposure, unpainted aluminum thermal spray coatings and painted zinc thermal spray coatings showed excellent corrosion protection (Ref 25). After 5 years, pipe columns, pumps, and oil flow lines were internally and externally protected against the marine atmosphere in offshore oil facilities by 200 μm (8 mils) of aluminum thermal spray, followed by a wash primer and two coats of aluminum sealer (Ref 26). Tensile steel links of a suspension bridge, which were coated with zinc flame spray, primed with red lead, and topcoated, showed no rusting after 44 years of marine exposure (Ref 27). After 1½ years of marine atmosphere and splash and spray exposure in Norway, flame-sprayed zinc and arc-sprayed and flame-sprayed aluminum steel panels showed good corrosion performance (Ref 28).

Laboratory tests, such as salt spray (fog), as well as electrochemical techniques, demonstrate the excellent corrosion performance of thermal spray coatings. For example, after 1600 h of exposure in tests, aluminum flame-sprayed panels showed no rust, although zinc flame-sprayed panels showed rust after 500 h of exposure (Ref 21). Other investigations indicated that zinc thermal spray panels rusted after 100 h of exposure (Ref 29). The results from the salt spray test are difficult to correlate with actual service performance; therefore, it is not a good accelerated test method for prediction (Ref 30, 31). Electrochemical methods have also been used to evaluate the corrosion behavior of thermal spray coatings (Ref 14, 31-34). For example, in one study, corrosion potential monitoring tests and potentiodynamic polarization measurements were conducted. The results from the latter test indicated that the aluminum flame spray coatings provide corrosion protection by passivation of the coating but that zinc and 85Zn-15Al alloy flame spray coatings operate solely through a galvanic mechanism (Ref 14). Another study, based on potential versus time curves, indicated that aluminum became electrochemically active after spraying (Ref 33). Duplicate studies by other researchers did not find this activation phenomenon (Ref 35). Other applications and materials for thermal spray coatings are discussed in the article "Thermal Spray Coatings" in this Volume.

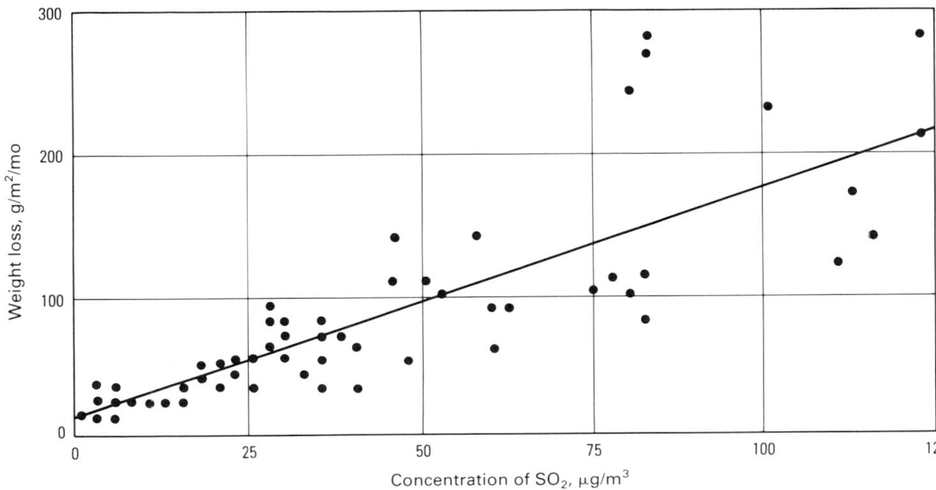

Fig. 35 Effect of SO_2 concentration on the corrosion rate of carbon steel at three Norwegian sites. Source: Ref 10

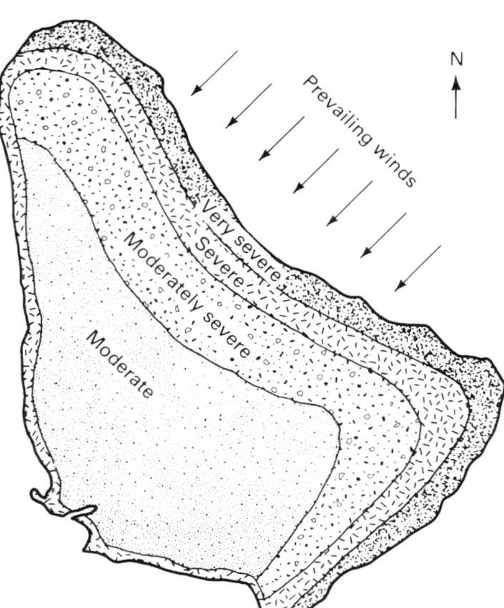

Fig. 36 Estimates of marine atmosphere corrosivity at various locations on the island of Barbados in the West Indies. Based on CLIMAT data. Source: Ref 10

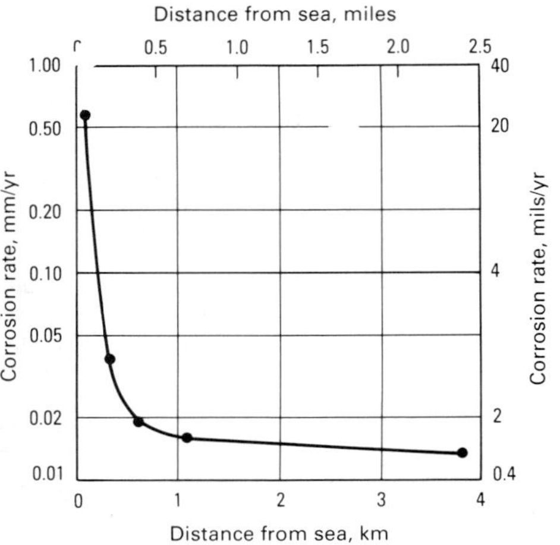

Fig. 37 Corrosion rate of carbon steel as a function of distance from the sea at Aracaju, Brazil. Source: Ref 10

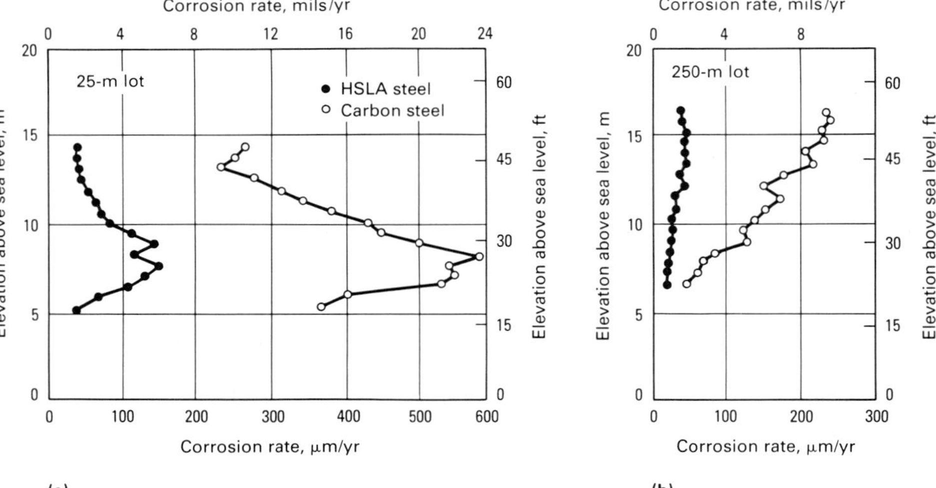

Fig. 38 Effect of elevation above sea level for carbon and HSLA steels at Kure Beach, NC. (a) 25-m (80-ft) lot; (b) 250-m (800-ft) lot. Source: Ref 12

Hot Dip Coatings

Galvanizing has been extensively used for protection against the marine environment. The advantages associated with applying zinc by flame spray versus galvanizing makes the former method attractive for certain applications (Ref 36). Hot dip galvanizing produces a fully dense coating that is metallurgically bonded to the substrate. In galvanizing, the size of part, heat distortion, ease of application, and the thickness and uniformity of coating are factors that must be considered. The thickness of galvanized coatings can vary from 75 to 200 μm (3 to 8 mils) and should be selected depending on the environment to be experienced and the desired lifetime. A 43-μm (1.7-mil) coating thickness is projected to give approximately 10 to 15 years of protection to steel in temperate to tropical marine atmospheric environments, but in service, this life may not be realized (see the article "Protection of Steel From Corrosion" in Volume 1 of the 9th Edition of *Metals Handbook*). Thick galvanizing and thermal spray were the only protection methods recommended by the British Standards Institution for providing long-term corrosion protection in a polluted marine atmosphere (Ref 37).

The American Society for Testing and Materials (ASTM) exposed galvanized sheet specimens in two marine environments—Sandy Hook, NJ, and Key West, FL—in 1926 and reported that panels with a coating weight of 760 g/m² (2.5 oz/ft²) of zinc first showed rust after 13.1 and 19.8 years of exposure, respectively (Ref 38). An extensive study on the atmospheric corrosion of galvanized steel at the 244-m (800-ft) lot at Kure Beach, NC, resulted in predicted weight losses after 10 years of 103 g/m² and 55 g/m² for skyward and groundward marine exposures, respectively (Ref 39). Most investigators agree that the life of a zinc coating is roughly proportional to its thickness in any particular environment and is independent of the method of application. Galvanizing is used for corrosion protection on cables of suspension bridges in Norway (Ref 16). One study indicated that galvanized steel panels were in good condition ater 1½ years (Ref 28). Hot-dip aluminized coated steel panels, which also were exposed, showed rusting after this time period.

Hot dip aluminum coatings, or aluminized coatings, are also used for the corrosion protection of steel in marine environments. Hot dip aluminizing of fabricated articles is no longer carried out in the United States or in Europe on a commercial basis (Ref 40). The coatings in use today are produced by a continuous strip process. An extensive comparative study was conducted on the atmospheric corrosion behavior of aluminized and galvanized steels (Ref 41, 42). Table 14 shows predicted 10-year weight losses of both of these coatings based on exposures conducted in the 250-m (800-ft) lot at Kure Beach, NC. A further comparison of the atmospheric-corrosion behavior of aluminized and galvanized panels was conducted by ASTM. After 20 years of marine atmospheric exposure (250-m, or 800-ft, lot, Kure Beach, NC), many of the galvanized steel panels were showing rust, but consistently good results were reported for the aluminized coating, which showed only minor pinholes of rust (Ref 43). Since 1972, a commercially produced aluminum-zinc (55Al-1.5Si-43.5Zn) hot dip coating has also been available for the corrosion protection of steel. One study reported that after 11 years of severe marine exposure (25-m, or 80-ft, lot, Kure Beach, NC) the 55Al-Zn coated panels were in good condition, with some corrosion products starting to creep inward on the faces of the panels from the cut edges (Ref 44). The advantages of hot dip aluminum coatings are discussed in Ref 45. The corrosion behavior of

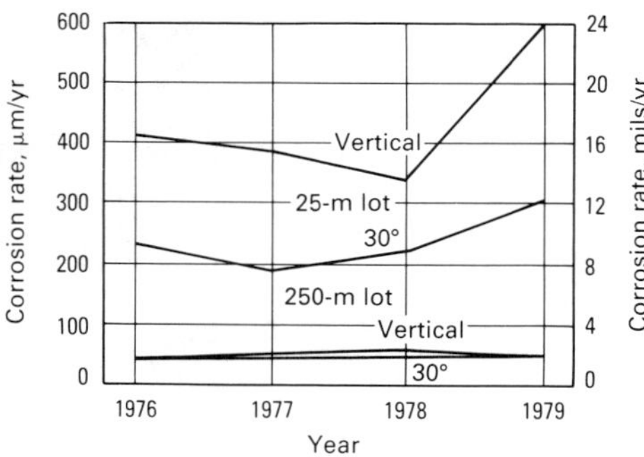

Fig. 39 Effect of specimen orientation on corrosion rates of iron specimens exposed vertically and at an angle of 30° to the horizontal. Results of 1-year test at Kure Beach, NC. Source: Ref 12

Table 10 Comparison of atmospheric-corrosion rates for specimens held vertically and inclined at 30° to the horizontal

Location	Corrosion rate ratio, vertical/30°
Kearny, NJ	1.25
Vandergrift, PA	1.26
South Bend, PA	1.20
25-m (80-ft) lot, Kure Beach, NC	1.41
250-m (800-ft) lot, Kure Beach, NC	1.25

Source: Ref 12

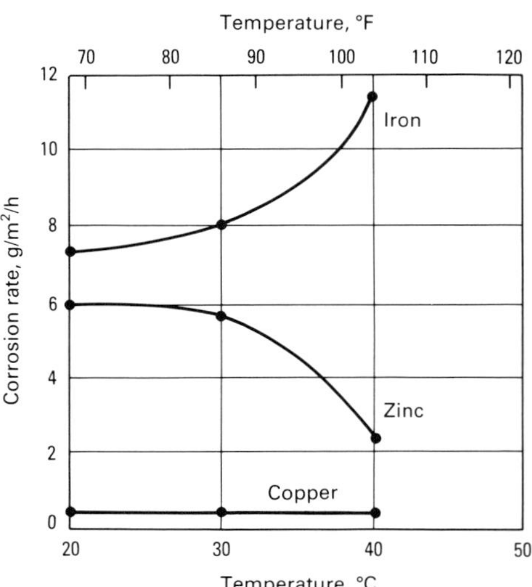

Fig. 40 The effect of temperature on the corrosion rates of iron, zinc, and copper. Source: Ref 10

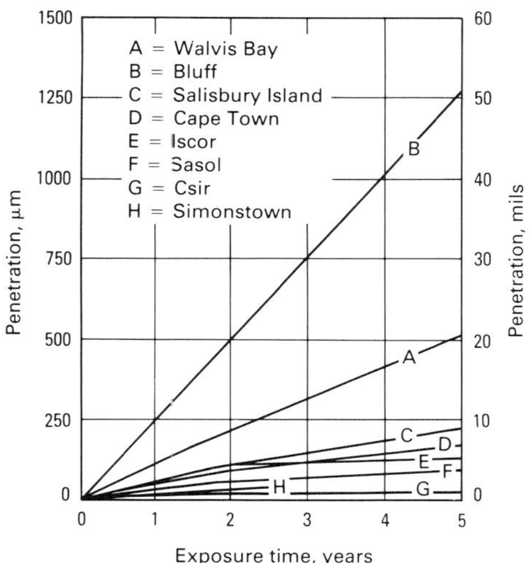

Fig. 41 Change in corrosion rate as a function of time for eight South African sites. Source: Ref 10

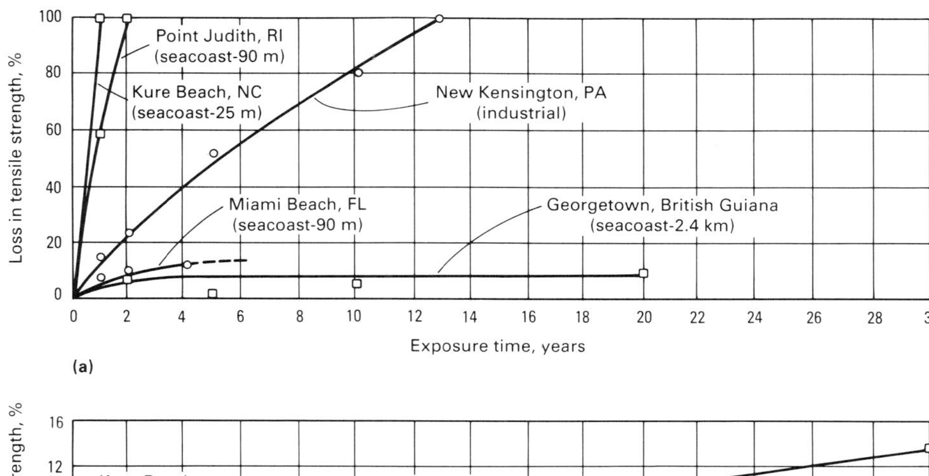

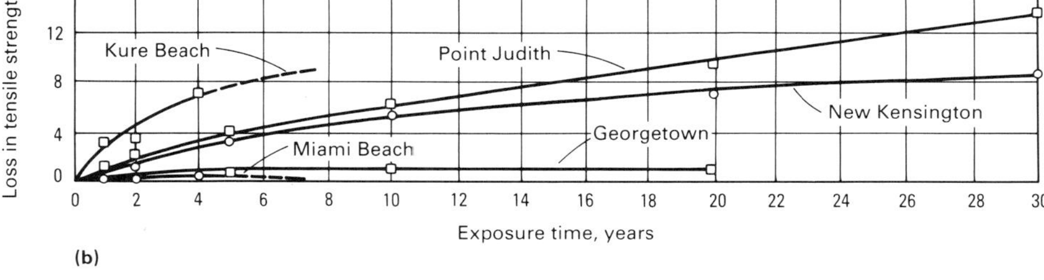

Fig. 42 Loss in tensile strength as a function of time for (a) 1.6 mm (1/16-in.) low-carbon steel and (b) aluminum alloys of the same thickness at five test sites. Data in (b) are averages for aluminum alloys 1100, 3003, and 3004. Source: Ref 10

aluminum coatings obtained from aluminizing baths of various compositions was studied in laboratory tests. Aluminized coatings containing manganese were suggested as possible candidates for corrosion protection for coastal structures and deep sea oil rigs. More information on hot dip galvanized and aluminized coatings is available in the article "Hot Dip Coatings" in this Volume.

Electroplating

Electroplated zinc or cadmium is the standard coating used to provide corrosion protection to steel fasteners in the marine environment. The cadmium coating is used because of its hardness, close dimensional tolerance, and barrier to hydrogen permeation into or out of steels (Ref 46). The disadvantages of cadmium plating are its short life (for example, 4 months) in the marine atmospheric environment and concerns about occupational health due to the toxicity in the plating process. Zinc plating also has a short service life. Alternatives for this application include ion vapor deposited aluminum and paints containing zinc or aluminum pigment in a ceramic binder. These coatings, including zinc with a

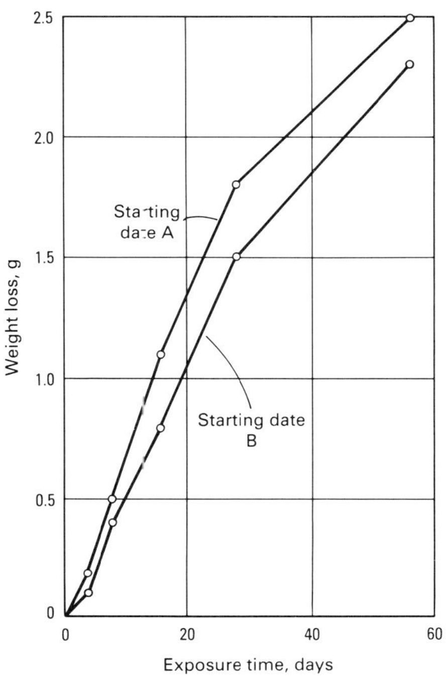

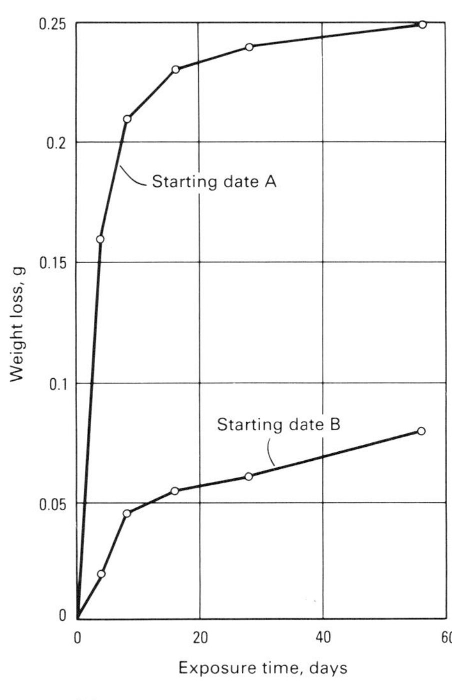

Fig. 43 Effect of different starting dates on the corrosion rate of iron (a) and zinc (b). Source: Ref 10

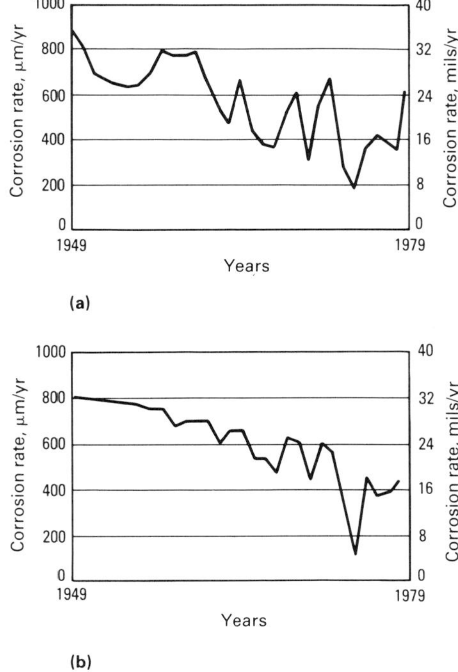

Fig. 44 Corrosion of iron calibration specimens tested for 1 year (a) and 2 years (b) at the 24.4-m (80-ft) lot at Kure Beach, NC. Source: Ref 12

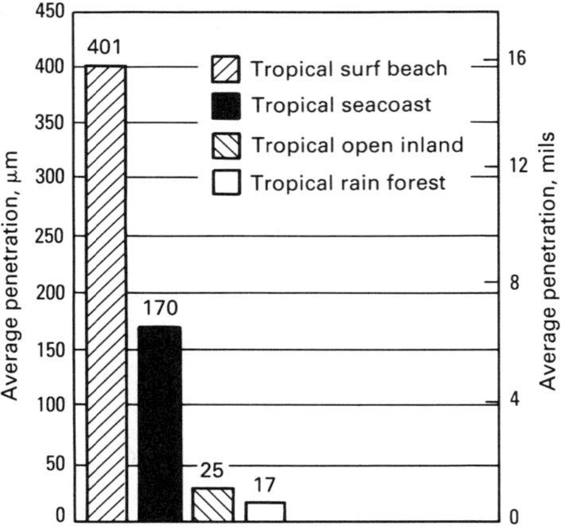

Fig. 45 Variation in corrosion rate after 1-year exposure of steel at four different tropical sites. Data are averaged from various investigations; see Ref 10 for details. Source: Ref 10

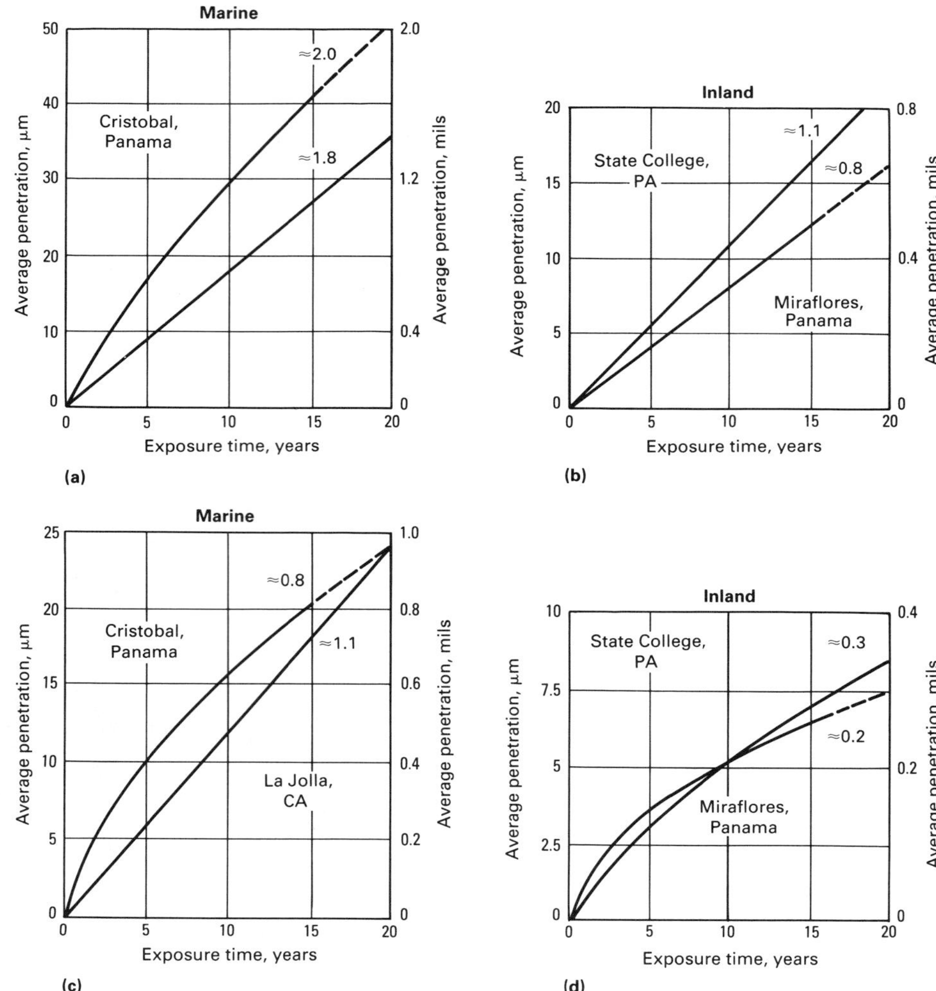

Fig. 46 Comparison of corrosion rates for zinc (a and b) and copper (c and d) at tropical and temperate exposure sites. Numbers on curves are stabilized corrosion rates in microns per year. Source: Ref 10

potassium silicate binder and aluminum with a phosphate-chromate binder, exhibit excellent corrosion protection for fasteners (minimum 1 year marine protection) (Ref 47). They are normally applied by conventional hand spraying.

Methods for electroplating aluminum are still in development, although plating using an organic aprotic solvent is a promising process (Ref 48). In laboratory polarization and galvanic tests, ion-deposited aluminum coatings performed well, indicating their potential for use on aircraft fasteners (Ref 49). Zinc and aluminum coatings for aircraft fastener applications showed variable results in corrosion tests (Ref 50). Aluminum and zinc pigmented paints performed better than electroplated zinc, ion vapor deposited aluminum, and electroplated cadmium on steel fasteners in laboratory seawater immersion tests (Ref 46). However, hydrogen permeability through the coating, as well as the corrosion performance of the coating, must be considered for a given fastener application (Ref 49).

Organic Coatings

J.S. Smart III
Amoco Production Company
R. Heidersbach
California Polytechnic State University

Organic coatings are the principal means of corrosion control for the hulls and topsides of ships and for the splash zones on permanent offshore structures. Most stationary offshore oil industry platforms are not painted below the waterline, and most marine pipelines are factory coated with special proprietary coatings (see the articles "Corrosion in Petroleum Production Operations" and "Corrosion of Pipelines" in this Volume).

Figure 53 shows the marine environments that are destructive to shipboard coatings. Similar environments are found on offshore oil production platforms (Fig. 54), lighthouses, docks, and other marine structures.

Before the 1960s, most marine coatings were fairly simple and could be applied by laborers such as seamen or maintenance personnel. Although the advent of high-performance marine coatings in the 1960s changed this, the performance of marine coatings has improved to such an extent that topside coating lives of 20 years have been experienced on some offshore oil production platforms.

Surface Preparation

Proper surface preparation is the most important consideration in determining the performance of organic coating systems. Surface cleanliness and proper surface profile are both important. Surface preparation frequently accounts for two-thirds of total painting costs for offshore structures.

The Steel Structures Painting Council (SSPC), the National Association of Corrosion Engineers (NACE), and standards groups in Sweden, Germany, the U.K., and Japan have all issued standards for surface preparation. These are listed in Table 15. Wet abrasive blast cleaning and waterblasting are not yet included in the standards, but are now being extensively used. Wet blasting is useful for dust control and for avoiding electrical sparking in Class I (explosive) areas. Generally, a small amount of nitrite inhibitor is added to the water to prevent re-rusting before priming.

Waterblasting can be used around rotating equipment, such as pumps, turbines, and generators; underwater; and in Class I areas. No grit is used; therefore, few solids result that could harm equipment. Waterblasting with 34.5 to 138 MPa (5000 to 20 000 psi) of pressure will remove all but the most adherent paint and oxide scale and, once the surface is blown dry, provides an excellent surface for painting. Waterblasting with detergent in the water at lower pressures is a good alternative to solvent washing for preparing oily and greasy surfaces for painting.

Gritblasting is usually used for surface preparation for marine coatings. The severe corrosion exposure conditions in offshore and coastal locations require the best possible surface preparation.

Inorganic zinc primers, which are frequently used in marine applications, require white metal gritblasting to remove all surface contamination because inorganic zinc has both a chemical bond and a mechanical bond to the surface. Epoxy primers can be applied over commercial grade surfaces for land-based exposures, but require near-white metal surfaces to maintain performance offshore.

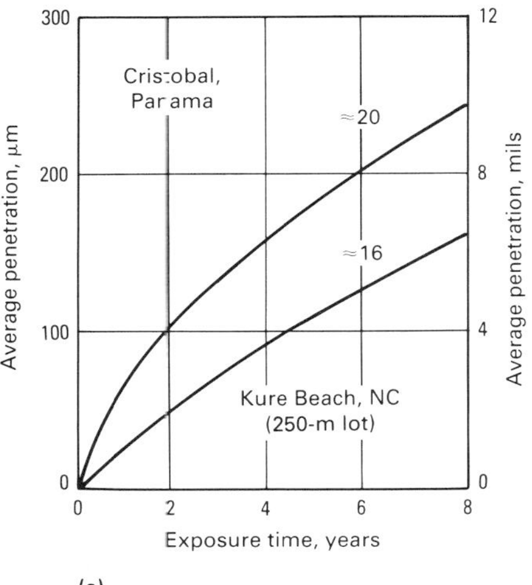

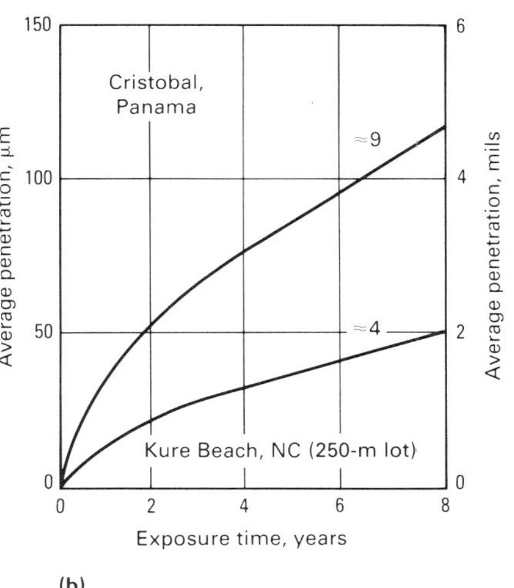

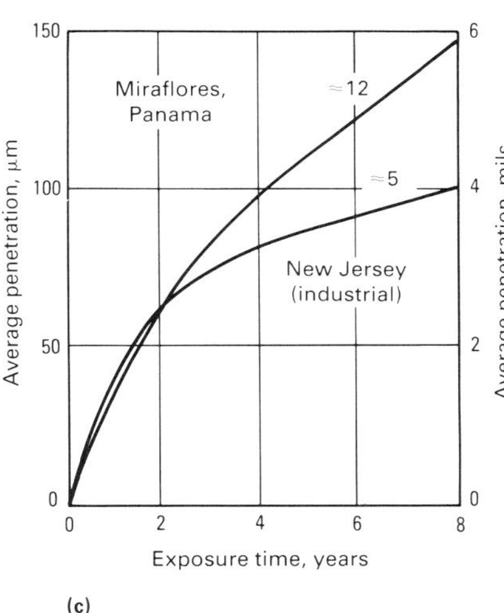

Fig. 47 Comparison of corrosion rates of steels at temperate and tropical exposure sites. Numbers on curve are stabilized corrosion rates in microns per year. (a) Carbon steel, marine exposure. (b) Low-alloy steel, marine exposure. (c) Carbon steel, inland exposure. Source: Ref 10

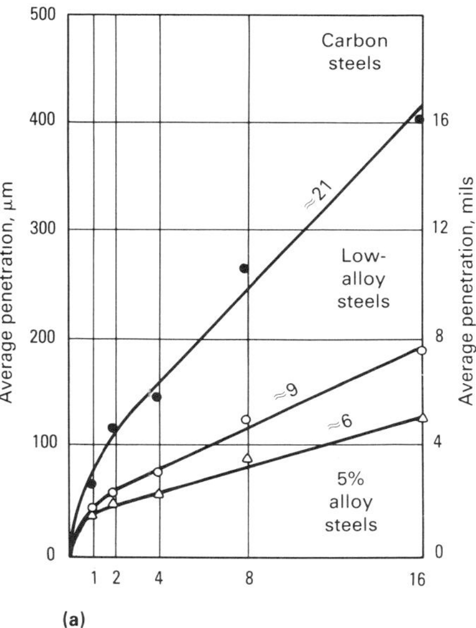

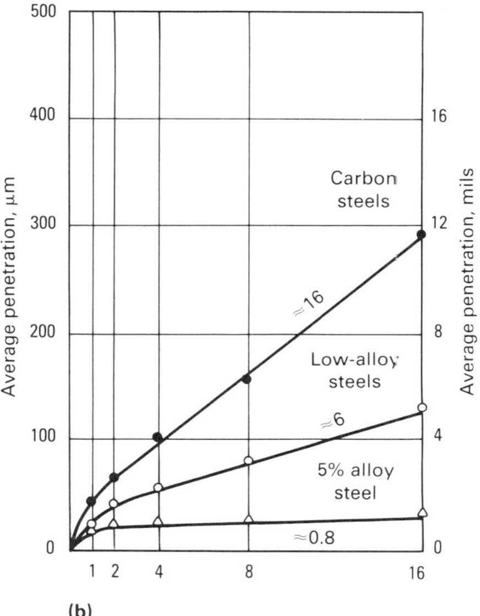

Fig. 48 Comparison of marine (a) and inland (b) corrosion rates for carbon steel, low-alloy steels, and 5% alloy steels at the Naval Research Laboratory test sites in Panama. Numbers on curves are stabilized corrosion rates in microns per year. Source: Ref 10

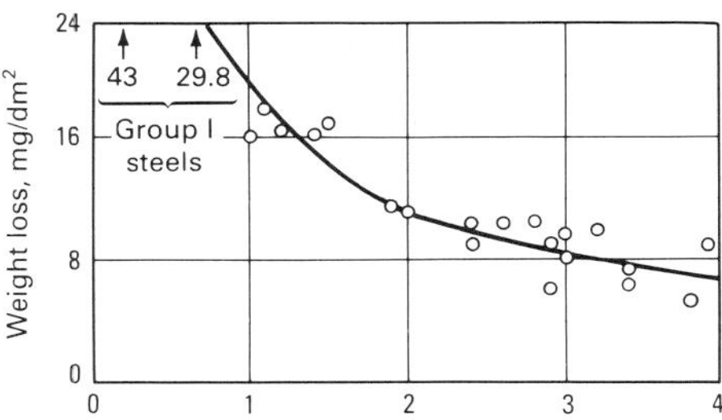

Fig. 49 Corrosion data for 25 low-alloy steels tested over a 15.5-year period at the Kure Beach, NC, 250-m (800-ft) lot. Source: Ref 11

Table 16 lists the characteristics of several types of grit. Grit that is used offshore is not recoverable; this limits the economical choices to either boiler slag or copper slag. Other types are too expensive unless they can be recycled. Silica sand is generally not used because of the possibility of silicosis, its friable nature, and its rounded shape, which is not conducive to high productivity.

Topside Coating Systems

Organic coatings are usually composed of three components: binders (resins), pigments, and solvents. Not all paints, however, have all three components. For example, solventless paints have been developed in response to environmental restrictions on the use of volatile solvents. Solventless paints can be applied at thicknesses to 13 mm (1/2 in.); such thick films would not be possible in a paint containing volatile solvents, because the thickness of the film would prevent solvent evaporation.

Paints can be classified by the type of binder or resin into the following categories.

- Air-drying oils (for example, linseed oil, alkyds)
- Lacquers (vinyls, chlorinated rubbers)
- Chemically cured coatings (epoxies, phenolics, and urethanes)
- Inorganic coatings (silicates)

The article "Organic Coatings and Linings" in this Volume contains detailed information on the formulation of all of these types of organic coatings.

Primers

The primer is by far the most important coat in the protection of steel substrates. The primary function of subsequent coats is to protect the primer and to give color and a pleasing appearance.

Inhibitive primers contain substances that resist the effects of contaminants on the steel, such as rust and salt, and that resist disbondment under corrosive conditions and cathodic protection. Formerly, lead pigments and chromates were used as inhibitors, but this is no longer the case, because of tightened environmental regulations that prohibit or severely restrict their use. The inhibitors currently in use include a number of proprietary compounds.

Zinc-Rich Primers. The introduction of coatings containing a high percentage of metallic zinc is probably the outstanding development in protective coatings in the last 30 years. Zinc dust is loaded into both organic and inorganic binders to form primers that are extremely effective in the prevention of corrosion, underfilm creepage, and coating system failure.

Inorganic zinc-rich primers are based on various silicate binders. There are several types of self-curing and postcuring primers. The binder serves as a strong, adherent matrix for the zinc metal. The zinc dust must be present in sufficient amounts to provide metal-to-metal contact between both the zinc particles and the steel surface. The zinc dust provides protection to the steel substrate in the same manner as in galvanizing. If a break develops in the coating, the zinc

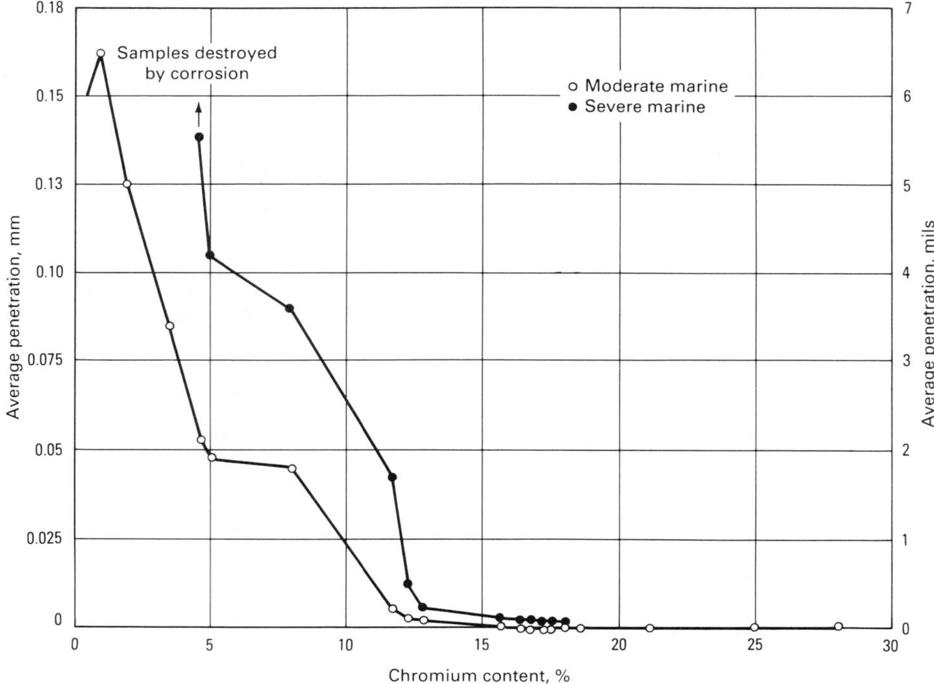

Fig. 50 Effect of chromium additions on the atmospheric corrosion of steels. Source: Ref 10

Table 11 Average corrosion rate and pit depth for ten austenitic stainless steels

Results of a 15-yr test at the Kure Beach, NC, 250-m (800-ft) lot

AISI type	Average corrosion rate, μm	Average corrosion rate, mils	Average depth of pits, μm	Average depth of pits, mils
301	<0.03	<0.001	40	1.6
302	<0.03	<0.001	30	1.2
304	<0.03	<0.001	30	1.2
321	<0.03	<0.001	70	2.8
347	<0.03	<0.001	90	3.5
316	<0.03	<0.001	30	1.2
317	<0.03	<0.001	30	1.2
308	<0.03	<0.001	40	1.6
309	<0.03	<0.001	30	1.2
310	<0.03	<0.001	10	0.4

Source: Ref 10

acts as a sacrificial anode and corrodes preferentially; this provides protection of the iron for long periods. Laboratory tests and field experience indicate that inorganic zinc-rich primers can at least double the life of a coating system and can often increase it tenfold. To be effective, however, inorganic zinc-rich primers must be applied to a clean surface.

Organic zinc-rich primers are alternatives to the inorganic zinc-rich coatings when conditions are not appropriate for inorganic zinc-rich coatings. Organic zinc-rich primers can be formulated with epoxy, urethane, vinyl, and chlorinated rubber binders. The most common binder used for marine applications is polyamide epoxy. Zinc-rich epoxies provide a lower degree of conductivity and cathodic protection than inorganic zinc but impart several other desirable characteristics:

- Organic zinc-rich primers frequently may be applied over old paint, which makes them a good choice for maintenance painting
- The good adhesion of the epoxy binder makes surface preparation requirements less stringent than those for inorganic zinc-rich coatings. Near-white metal surfaces are adequate for offshore applications, and commercial grade gritblasting can be used in less severe environments
- The epoxy binder provides some protection to the zinc, and this allows moderate exposure of the primer to the marine environment without corrosion of the zinc and formation of zinc corrosion products. Zinc corrosion products can cause intercoat adhesion problems and paint blistering

Topcoats

Topcoats for steel serve mainly to protect the primer and to add color and appearance. To serve this function, they must be:

- Barrier coatings impervious to moisture, salt, chemicals, solvents, and ion passage
- Strong and resistant to mechanical damage
- Of adequate color and gloss retention

Some of the most common topcoats in use are discussed below, and detailed information on each of these types is available in the article "Organic Coatings and Linings" in this Volume.

Alkyds are the most common and versatile coatings in existence, but they are seldom used in severe marine applications, because of their poor performance over steel. This poor performance is due to the oil base of the alkyd. As corrosion proceeds on steel, hydroxyl ions (OH^-) are generated at cathode sites. Hydroxyl ions saponify the oils in the coating, and this results in coating failure.

Alkyds are the product of the reaction of a polybasic acid, a polyhydric alcohol, and a monobasic acid or oil. The number of possible combinations is large; therefore, a wide range of performance is available. Alkyds are used in marine service in relatively mild applications, such as interior coatings for cabins, quarters, engine rooms, kitchens, heads, and some superstructure applications.

Vinyls also have a broad range of desirable properties. Most vinyl resins are the product of the polymerization of polyvinyl chloride (PVC) and polyvinyl acetate (PVA). Vinyls are solvent-base coatings that form a tight homogeneous film over the substrate. They are easy to apply by brush, roller, and spray. Intercoat adhesion is excellent because of the solvent-base nature of the coating. Vinyls do not oxidize or age, and they are inert to acids, alkalies, water, cement, and alcohols. They do soften slightly when covered with some crude oils. Vinyl coatings dry quickly and can be recoated in a short time (often in minutes), depending on the solvent used. Vinyl coatings are also flexible and can accommodate the motion of the steel beneath them, such as when a ship or platform is launched.

Vinyls were extensively used on ships and offshore platforms for many years, and they are still in use in many areas. However, they have given way to epoxies in most marine applications because vinyl coatings are relatively thin and are not very strong. Film thickness is typically only 50 μm (2 mils) per coat, and the coatings cannot withstand mechanical abuse. In addition, vinyls are not very effective for covering rough, previously corroded surfaces.

Chlorinated rubber coatings are based on natural rubber that has been reacted with chlorine to give a hard high-quality resin that is soluble in various solvents. Chlorinated rubbers have been used for many years as industrial-type paints because of their low moisture permeability, strength, resistance to UV degradation, and ease of application. Chlorinated rubbers have found application on ships and containers, railroads, and as traffic paint for road stripes. For many years, chlorinated rubber coatings were used to paint ships because of their ease of application and repairability, tolerance of poor surface preparation, fast drying characteristics, and relatively good wear and abrasion resistance. They are still used to a great extent on ships and are the standard paint system for containers. Modern fleet owners, however, have phased out chlorinated rubbers in favor of higher-quality coating systems, such as epoxy, for reasons to be discussed below.

Epoxies. The combination of excellent adhesion (some can be applied underwater), good impact and abrasion resistance, high film builds (up to 6.4 mm, or 1/4 in., on a wet, vertical surface), relatively low cost, and excellent chemical and solvent resistance has made epoxy coatings the workhorses of modern marine coatings. These properties result in service lives of 7 to 12 years on ships, offshore platforms, and coastal applications when epoxy topcoats are applied over inhibited epoxy or inorganic zinc-rich primers. Because epoxies are chemically cured, a wide range of properties can be achieved by varying the molecular weight of the resin, the type of curing agent, and the type of pigments or fillers used.

Table 12 Corrosion data for noncoupled metal panels exposed at the US Naval Research Laboratory tropical seacoast site at Cristobal, Panama

		General corrosion						Pitting			Loss in tensile strength (e), %	
		Average penetration (b), µm (mils)					Final corrosion rate(c), µm (mils)	Average deepest 20 pits (d), µm (mils)		Deepest pit, µm (mils)		
Metal or alloy	Surface(a)	1 year	2 years	4 years	8 years	16 years		8 years	16 years		8 years	16 years
Magnesium alloys												
AZ31X	...	28 (1.1)	48 (1.9)	91 (3.6)	201 (7.9)	381 (15)	23 (0.9)	178 (7)	559 (22)	864 (34)	25	47
AZ61X	...	12 (0.47)	33 (1.3)	...	157 (6.2)	304 (12)	19 (0.75)	177 (6.9)	466 (18.3)	533 21	28	32
Aluminum alloys												
1100	...	<0.3 (0.01)	1 (0.04)	<0.3 (0.01)	0.5 (0.02)	2.8 (0.11)	<0.3 (0.01)	<125 (4.9)	<125 (4.9)	<125 (4.9)	<1	<1
6061-T6	...	0.8 (0.3)	1.5 (0.06)	2.0 (0.08)	0.8 (0.03)	2.8 (0.11)	0.3 (0.01)	<125 (4.9)	<125 (4.9)	<125 (4.9)	1	<1
2024-T6	...	0.8 (0.3)	1.0 (0.04)	0.5 (0.02)	0.5 (0.02)	3.3 (0.13)	<0.3 (0.01)	125 (4.9)	125 (4.9)	125 (4.9)	1	1
Zinc (99.5%)	...	5.8 (0.23)	9.1 (0.36)	17 (0.67)	28 (1.1)	41 (1.6)	1.8 (0.07)	<125 (4.9)	<125 (4.9)	381 (15)	3	3
Iron												
Low copper ingot	Pickled	101 (4)	207 (8.1)	794 (31.2)	...	...	...	...	...	...	...	...
ASTM K	Pickled	52 (2)	79 (3.1)	128 (5)	210 (8.3)	...	19 (0.75)	762 (30)	...	...	...	...
Aston wrought	Pickled	70 (2.8)	99 (3.9)	177 (7)	281 (11.0)	475 (18.7)	24 (0.94)	737 (29)	1346 (53)	1549 (61)	...	...
Aston wrought	Mill scale	69 (2.7)	138 (5.4)	168 (6.6)	282 (11.1)	403 (15.9)	...	1041 (41)	1245 (49)	1549 (61)	...	...
Carbon steel												
0.24% C	Pickled	64 (2.5)	122 (4.8)	144 (5.7)	259 (10.2)	402 (15.8)	21 (0.83)	863 (33.9)	1295 (51)	3124 (123)	...	...
0.24% C	Mill scale	66 (2.6)	114 (4.5)	141 (5.6)	278 (10.9)	401 (15.78)	...	940 (37)	1321 (52)	3124 (123)	...	...
0.24% C	Machined	50 (2)	78 (3)	126 (5)	173 (6.8)	270 (10.6)	12 (0.47)	355 (13.9)	457 (17.9)	991 (39)	...	...
Copper-bearing	Pickled	55 (2.2)	78 (3)	116 (4.6)	222 (8.7)	345 (13.6)	19 (0.75)	787 (30.9)	762 (30)	1676 (66)	...	...
Low-alloy steel												
Cu, Ni	Pickled	44 (1.7)	60 (2.4)	79 (3.1)	127 (5)	198 (7.8)	10 (0.4)	301 (11.9)	356 (14)	432 (17)	...	...
Cu, Cr, Si	Pickled	43 (1.65)	57 (2.2)	79 (3.1)	130 (5.1)	204 (8)	10 (0.4)	305 (12)	457 (17.9)	889 (35)	...	...
Cu, Ni, Mn, Mo	Pickled	44 (1.7)	61 (2.4)	76 (3)	124 (4.9)	188 (7.4)	9.7 (0.38)	305 (12)	406 (16)	914 (36)	...	...
Cr, Ni, Mn	Pickled	42 (1.6)	57 (2.2)	71 (2.8)	115 (4.5)	160 (6.3)	7.9 (0.31)	305 (12)	330 (13)	737 (29)	...	...
Nickel steel (2% Ni)	Pickled	39 (1.5)	51 (2)	66 (2.6)	95 (3.7)	146 (5.7)	6.6 (0.26)	279 (10.9)	330 (13)	483 (19)	...	...
Nickel steel (5% Ni)	Pickled	34 (1.3)	47 (1.85)	58 (2.3)	90 (3.5)	136 (5.4)	6.4 (0.25)	305 (12)	305 (12)	381 (15)	...	...
Chromium steel (3% Cr)	Pickled	50 (2)	63 (2.5)	77 (3.03)	116 (4.6)	169 (6.7)	7.7 (0.3)	457 (17.9)	609 (24)	1600 (63)	...	...
Chromium steel (5% Cr)	Pickled	41 (1.6)	47 (1.85)	55 (2.2)	90 (3.5)	113 (4.4)	5.1 (0.2)	279 (10.9)	330 (13)	483 (19)	...	...
Cast steel (0.27% C)	Machined	44 (1.7)	63 (2.5)	90 (3.5)	140 (5.5)	217 (8.5)	11 (0.43)	356 (14)	432 (17)	914 (36)	...	...
Cast iron-gray (3.2% C)	Machined	39 (1.5)	56 (2.2)	88 (3.46)	133 (5.2)	196 (7.7)	8.1 (0.32)	356 (14)	457 (17.9)	940 (37)	...	...
Cast iron												
Austenitic (18% Ni)	Machined	25 (1)	34 (1.3)	44 (1.7)	113 (4.4)	233 (9.2)	15 (0.6)	558 (21.9)	1041 (41)	1499 (59)	...	...
Stainless steels												
Type 410	...	1.0 (0.04)	1.0 (0.04)	1.5 (0.06)	1.0 (0.04)	4.6 (0.18)	0.3 (0.01)	<125 (4.9)	<125 (4.9)	<125 (4.9)	<1	<1
Type 430	...	0.5 (0.02)	1.0 (0.04)	1.0 (0.04)	1.0 (0.04)	2.0 (0.08)	<0.3 (0.01)	<125 (4.9)	<125 (4.9)	<125 (4.9)	<1	<1
Type 301	...	0.3 (0.01)	<0.3 (0.01)	<0.3 (0.01)	0.3 (0.01)	0.5 (0.02)	0.3 (0.01)	<125 (4.9)	<125 (4.9)	<125 (4.9)	<1	<1
Type 321	...	<0.3 (0.01)	<0.3 (0.01)	<0.3 (0.01)	0.3 (0.01)	0.5 (0.02)	<0.3 (0.01)	<125 (4.9)	<125 (4.9)	<125 (4.9)	<1	<1
Type 316	...	<0.3 (0.01)	<0.3 (0.01)	<0.3 (0.01)	<0.3 (0.01)	<0.3 (0.01)	<0.3 (0.01)	<125 (4.9)	<125 (4.9)	<125 (4.9)	<1	<1

(continued)

(a) All specimens were degreased before exposure; any treatment prior to degreasing is listed.
(b) Average penetration over a 4.23-dm^2 (65.6-$in.^2$) exposed area; calculations based on weight loss and density.
(c) Rate after time-corrosion relation had stabilized; slope of the linear portion of the curve, usually after two to eight years.
(d) Averages obtained by measuring the five deepest measurable (>125 µm, or 5 mils) penetrations on each surface of duplicate panels.
(e) Percent loss in ultimate tensile strength for 1.59-mm (1/16-in.) thick metal. Source: Ref 10

Table 12 (continued)

Metal or alloy	Surface(a)	General corrosion: Average penetration (b), μm (mils): 1 year	2 years	4 years	8 years	16 years	General corrosion: Final corrosion rate(c), μm (mils)	Pitting: Average deepest 20 pits (d), μm (mils): 8 years	16 years	Pitting: Deepest pit, μm (mils)	Loss in tensile strength (e), %: 8 years	16 years
α-β Brass												
Muntz metal (1/4% As)	...	1.8 (0.07)	2.3 (0.091)	3.6 (0.14)	5.8 (0.23)	11 (0.43)	0.8 (0.03)	<125 (4.9)	<125 (4.9)	<125 (4.9)	4	8
Naval	...	1.5 (0.06)	2.0 (0.08)	3.3 (0.13)	5.3 (0.21)	9.9 (0.38)	0.5 (0.02)	<125 (4.9)	<125 (4.9)	<125 (4.9)	3	7
Manganese bronze	...	4.6 (0.18)	4.8 (0.19)	7.6 (0.3)	8.4 (0.33)	15 (0.6)	0.8 (0.03)	<125 (4.9)	<125 (4.9)	<125 (4.9)	6	8
α brass												
Cu-30Zn	...	1.3 (0.05)	1.8 (0.07)	2.8 (0.11)	4.6 (0.18)	8.4 (0.33)	0.5 (0.02)	<125 (4.9)	<125 (4.9)	<125 (4.9)	5	4
Cu-20Zn	...	2.0 (0.08)	2.8 (0.11)	4.1 (0.16)	5.8 (0.23)	9.4 (0.37)	0.5 (0.02)	<125 (4.9)	<125 (4.9)	<125 (4.9)	2	3
Cu-10Zn	...	3.0 (0.12)	3.6 (0.14)	5.6 (0.22)	7.8 (0.31)	12 (0.47)	0.5 (0.02)	<125 (4.9)	<125 (4.9)	<125 (4.9)	2	3
Bronze												
Aluminum (5%)	...	2.0 (0.08)	2.8 (0.11)	3.8 (0.15)	5.8 (0.23)	9.9 (0.38)	0.5 (0.02)	<125 (4.9)	<125 (4.9)	<125 (4.9)	1	2
Phosphor	...	5.1 (0.2)	7.4 (0.29)	10 (0.4)	15 (0.6)	24 (0.95)	1.0 (0.04)	<125 (4.9)	<125 (4.9)	<125 (4.9)	6	3
Silicon	...	7.9 (0.31)	10 (0.4)	17 (0.67)	28 (1.1)	48 (1.9)	2.3 (0.09)	<125 (4.9)	<125 (4.9)	<125 (4.9)	2	3
Cast bronze												
Tin (8%)	Machined	4.6 (0.18)	8.9 (0.35)	11 (0.43)	14 (0.55)	21 (0.83)	1.0 (0.04)	<125 (4.9)	<125 (4.9)	<152 (6)	...	...
Ni-Sn (6% Ni)	Machined	3.3 (0.13)	4.6 (0.18)	7.4 (0.29)	11 (0.43)	16 (0.63)	0.5 (0.02)	125 (4.9)	125 (4.9)	125 (4.9)	...	...
Copper (99.9%)	...	4.3 (0.17)	5.8 (0.23)	9.7 (0.38)	14 (0.55)	20 (0.78)	0.8 (0.03)	<125 (4.9)	<125 (4.9)	<125 (4.9)	4	5
Copper/nickel (70/30)	...	0.8 (0.03)	1.5 (0.06)	3.0 (0.1)	5.8 (0.23)	10 (0.4)	0.5 (0.02)	<125 (4.9)	<125 (4.9)	<125 (4.9)	<1	1
Monel 400	...	1.0 (0.04)	1.0 (0.04)	1.8 (0.07)	3.0 (0.1)	5.6 (0.22)	0.3 (0.01)	<125 (4.9)	<125 (4.9)	<125 (4.9)	<1	2
Nickel (99%)	...	0.2 (0.008)	0.5 (0.02)	0.8 (0.03)	1.5 (0.05)	5.0 (0.2)	<0.3 (0.01)	<125 (4.9)	<125 (4.9)	<125 (4.9)	<1	<1
Lead (99%)	...	1.5 (0.06)	3.4 (0.13)	6.3 (0.25)	11 (0.43)	20 (0.8)	1.3 (0.05)	<125 (4.9)	<125 (4.9)	<125 (4.9)	<1	<1
Coated steels												
Galvanized	...	6.6 (0.26)	...	15 (0.6)	24 (0.95)	...	...	<125 (4.9)	...	...	...	...
Zn sprayed	...	1.5 (0.06)	13 (0.51)	14 (0.55)	17 (0.67)	...	...	127 (5)	...	...	...	...
Pb coated	...	2.0 (0.08)	5.1 (0.2)	...	9.1 (0.36)	...	...	<125 (4.9)	...	...	...	...
Al sprayed	...	<0.3 (0.01)	<0.3 (0.01)	<0.3 (0.01)	<0.3 (0.01)	<0.3 (0.01)	<0.3 (0.01)	<125 (4.9)	<125 (4.9)	<125 (4.9)	...	...

(a) All specimens were degreased before exposure; any treatment prior to degreasing is listed.
(b) Average penetration over a 4.23-dm^2 (65.6-$in.^2$) exposed area; calculations based on weight loss and density.
(c) Rate after time-corrosion relation had stabilized; slope of the linear portion of the curve, usually after two to eight years.
(d) Averages obtained by measuring the five deepest measurable (>125 μm, or 5 mils) penetrations on each surface of duplicate panels.
(e) Percent loss in ultimate tensile strength for 1.59-mm (1/16-in) thick metal. Source: Ref 10

Immersion Coatings

Immersion coatings for marine service have far greater requirements than other organic coatings. They must resist moisture absorption, moisture transfer, and electroendosmosis (electrochemically induced diffusion of moisture through the coating). They also must be strong and have good adhesion.

Most ship hulls and many marine structures use cathodic protection to supplement the protection afforded by organic coatings (see the section "Cathodic Protection" in this article). This is desirable because it is virtually impossible to apply and maintain a defect-free organic coating system on a large structure.

Barrier Properties. To be effective in seawater immersion service, an organic coating must have a low moisture vapor transfer rate as well as low moisture absorption. Moisture absorption is the molecular moisture absorbed into and held within the molecular structure of the coating. This property is not important to the effectiveness of the coating unless the moisture absorption lowers the dielectric characteristics of the coating and increases the passage of electrical current. Moisture vapor transfer, on the other hand, is important, particularly when the coating is exposed to an external current (as in cathodic protection). Generally, the lower the moisture vapor transfer rate of a coating, the more effective the coating.

Where electroendosmosis may be encountered, adhesion is also very important. Most organic coatings are negatively charged, and under cathodic protection, the cathode has an excess of electrons, which makes it negatively charged. This being the case, coatings with a high moisture vapor transfer rate or questionable adhesion would be more subject to damage and blistering by cathodic potentials.

Mechanical Properties. Coatings used on marine structures must be strong. Most damage to marine coating systems is mechanical, not a breakdown of the coating from exposure to seawater. Immersion coatings must have good impact and abrasion resistance and must be able to flex well enough to maintain contact with the steel substrate when it is bent. Rubbing by mooring ropes, chains, and crane wire ropes, as well as impact from cargo handling, work parties, and berthing operations, are major causes of damage.

Types of Immersion Coatings. Many of the common paint formulations can be used for immersion service, but the most common coatings in use are coal tar epoxies and straight epoxies.

Coal tar epoxies were introduced in 1955 and are the most common coatings in use on fixed marine structures (Ref 52). These thermosetting materials are available with a variety of setting temperatures and chemical curing systems. Coal tar epoxies require near-white surface prepara-

Table 13 Some marine atmospheric-corrosion test sites around the world

Corrosion rates of steel and zinc are also listed for some sites.

Test site	Type of atmosphere	Distance from sea		Corrosivity index(a)		Corrosion rate from 2-year test			
						Steel		Zinc	
		km	miles	MCI	ACI	mm/yr	mils/yr	mm/yr	mils/yr
United States									
Cape Canaveral, FL									
0.8 km (1/2 miles) from ocean	Marine	0.8	0.5	...	...	0.086	3.39	0.0011	0.045
55 m (60 yd), 9 m (30 ft) elevation	Marine	0.055	0.035	...	...	0.165	6.48	0.004	0.158
55 m (60 yd), ground level	Marine	0.055	0.035	...	...	0.44	17.37	0.0041	0.163
55 m (60 yd), 18 m (60 ft) elevation	Marine	0.055	0.035	...	...	0.131	5.17	0.0044	0.173
Point Reyes, CA	Marine	0.400	0.25	11	0.183	0.50	19.71	0.0015	0.060
Brazos River, TX	Industrial marine	...	...	...	...	0.093	3.67	0.0018	0.072
Daytona Beach, FL	Marine	...	...	...	...	0.295	11.63	0.0022	0.079
Kure Beach, NC									
250-m (800-ft)	Marine	0.244	0.15	...	...	0.145	5.73	0.0022	0.079
25-m (80-ft)	Marine	0.0244	0.015	11.4	...	0.53	21.00	0.0064	0.250
Atlantic City, NJ	Marine	...	...	...	...	...	...	...	...
Annapolis, MD	Marine	...	...	...	...	...	...	...	...
Havre de Grace, MD	Marine	...	...	...	...	...	...	...	...
Key West, FL	Subtropical	...	...	...	...	...	...	...	...
La Jolla, CA	Marine	...	...	...	...	...	...	...	...
Miami, FL	Marine	4	2.5	5.9	0.04	...	...	...	...
Ormond Beach, FL	Marine	...	...	...	...	...	...	...	...
Point Judith, RI	Marine	...	...	...	...	...	...	...	...
Portsmouth, VA	Marine	...	...	...	...	...	...	...	...
Sandy Hook, NJ	Marine	...	...	...	...	...	...	...	...
Battelle, Sequin, WA	Marine	0.030	0.018	6.9	0.07	...	...	...	...
Hickham AFB, HI	Marine	0.150	0.09	8.7	1.4	...	...	...	...
Panama									
Fort Amidor	Marine	...	...	...	...	0.014	0.57	0.0011	0.045
Miraflores	Marine	...	...	...	...	0.043	1.69	0.0026	0.104
Limon Bay	Marine	...	...	...	...	0.062	2.45	0.0026	0.104
Galeta Point	Marine	...	...	...	...	0.69	27.14	0.015	0.607
Canada									
Esquimalt, Vancouver Island, BC	Rural marine	...	...	...	...	0.013	0.53	0.0005	0.019
Cape Beale, NC	Marine	0.025	0.015	12.4	0.20	...	...	...	...
Chebucto Head, NS	Marine	0.100	0.06	13.0	1.2	...	...	...	...
Estevan Point, BC	Marine	0.400	0.25	8.4	0.02	...	...	...	...
Daniels Harbor, NF	Marine	0.150	0.09	17.5	0.11	...	...	...	...
Sable Island, NS	Marine	...	...	13.9	0.99	...	...	...	...
St. Vincents, NF	Marine	0.150	0.09	14.7	0.18	...	...	...	...
Deadmans Bay, NF	Marine	0.030	0.018	11.9	0.12	...	...	...	...
England									
Dungeness	Industrial marine	...	...	...	...	0.49	19.22	0.0036	0.143
Pilsey Island	Industrial marine	...	...	...	...	0.103	4.04	0.0057	0.223
Cornwall	Industrial marine	0.400	0.25	12.9	0.83	...	...	...	...
Ghana									
Tema	...	0.030	0.018	77.5	3.4	...	...	...	...
Benin									
Cotonou	...	0.150	0.09	17.6	0.67	...	...	...	...
Togo									
Lome	...	0.100	0.06	23.6	0.27	...	...	...	...
South Africa									
Durban, Salisbury Island	Marine	0.010	0.006	64.0	5.7	0.056	2.20	0.015	0.607
Dyeban Bluff	Severe marine	...	...	...	...	0.26	10.22	0.0032	0.126
Cape Town docks	Mild marine	...	...	...	...	0.047	1.84	0.0032	0.126
Walvis Bay military camp	Severe marine	...	...	...	...	0.11	4.33	0.063	2.483
Simonstown	Marine	...	...	...	...	0.016	0.63	0.0032	0.126
Nigeria									
Lagos									
45-m (50-yd)	Severe marine	0.046	0.03	...	...	...	...	...	...
365-m (400-yd)	Marine	0.366	0.23	...	...	...	...	...	...
1190-m (1300-yd)	Mild marine	1.189	0.74	...	...	...	...	...	...

(continued)

(a) MCI, Marine Corrosivity Index: determined by the weight loss of an aluminum wire-mild steel bolt couple. ACI, Atmospheric Corrosivity Index: determined by the weight loss of an aluminum open helical coil specimen or an aluminum wire-plastic bolt specimen. Source: Ref 10

Table 13 (continued)

Test site	Type of atmosphere	Distance from sea km	Distance from sea miles	Corrosivity index(a) MCI	Corrosivity index(a) ACI	Corrosion rate from 2-year test: Steel mm/yr	Steel mils/yr	Zinc mm/yr	Zinc mils/yr
Bahrain									
Sadad	Marine	0.800	0.5	11.2	0.30	...	...	...	...
Iran									
Shapour	Marine	0.010	0.006	5.2	0.15	...	...	...	...
Pakistan									
Karachi	Marine	0.060	0.037	33.8	4.1	...	...	...	...
Yemen									
Rasketenib	Marine	0.100	0.06	14.3	1.1	...	...	...	...
Japan									
Hitachi	Marine	1.0	0.62	5.2	0.41	...	...	...	...
Okinawa	Marine	0.500	0.31	26.2	1.8	...	...	...	...
Zushi	Marine	0.016	0.01	2.6	1.4	...	...	...	...
Australia									
Sydney (beach)	Marine	0.010	0.006	6.4	2.0	...	...	...	...
Sydney (D.S.L.)	...	3	1.8	7.1	1.3	...	...	...	...
New Zealand									
Phia	Marine	0.2	0.12	15.8	2.4	...	...	...	...
Belgium									
Ostende	Marine	...	...	...	...	...	...	...	...
Greece									
Rafina	Marine	0.2	0.12	13.6	1.0	...	...	...	...
Rhodes	Marine	0.2	0.12	14.3	1.5	...	...	...	...
Netherlands									
Schagen	Marine	2.4	1.5	17.0	2.0	...	...	...	...
Den Helder	Marine	...	...	...	...	...	...	...	...
Spain									
Almeria	...	0.035	0.022	22.4	1.6	...	...	...	...
Cartagena	...	0.050	0.031	5.2	1.9	...	...	...	...
La Coruña	...	0.160	0.1	26.2	1.4	...	...	...	...
Germany									
Cuxhaven	Marine	...	...	...	...	...	...	...	...
France									
Biarritz	Marine	...	...	...	...	...	...	...	...
Italy									
Bari	Rural marine	...	...	...	...	...	...	...	...
Barbados									
Holetown	Marine	0.075	0.047	59	0.42	...	...	...	...
Dominican Republic									
El Macao	Marine	0.100	0.06	30	0.27	...	...	...	...
Colombia									
Barranquilla	Marine	0.010	0.006	12.6	3.3	...	...	...	...
Cartagena	Marine	0.010	0.006	16.3	3.1	...	...	...	...
Galera Zamba	Marine	0.190	0.12	48.0	8.8	...	...	...	...
Santa Marta	Marine	0.060	0.037	1.9	0.06	...	...	...	...
Guatemala									
Pacific Beach	Marine	...	...	17.2	2.7	...	...	...	...
Uruguay									
Punta Del Este	Marine	0.040	0.025	48.3	0.79	...	...	...	...
Venezuela									
Carmaine Chico	...	0.800	0.5	39.0	0.53	...	...	...	...

(a) MCI, Marine Corrosivity Index: determined by the weight loss of an aluminum wire-mild steel bolt couple. ACI, Atmospheric Corrosivity Index: determined by the weight loss of an aluminum open helical coil specimen or an aluminum wire-plastic bolt specimen. Source: Ref 10

tion and are very adherent and abrasion resistant. They tend to be brittle and should not be used on flexible structures. Straight epoxies have been commercially available longer than coal tar epoxies (Ref 53). Epoxies are usually applied in thinner coats and are more expensive than coal tar epoxies. Epoxies have become the material of choice for immersion service because of their superior performance (Ref 51, 53). They have replaced chlorinated rubbers for most ship hull applications, and they are available in a variety of polyamide- or amine-based formulations (Ref 52). Detailed information on these and other coating materials is available in the article "Organic Coatings and Linings" in this Volume.

Antifouling Topcoats. Most shipboard applications require antifouling topcoats. The formulations for these coatings are changing because of environmental legislation. In some parts of the world, copper-containing antifouling coatings are still popular, but in North America, these coatings have been replaced by organo-tin compounds.

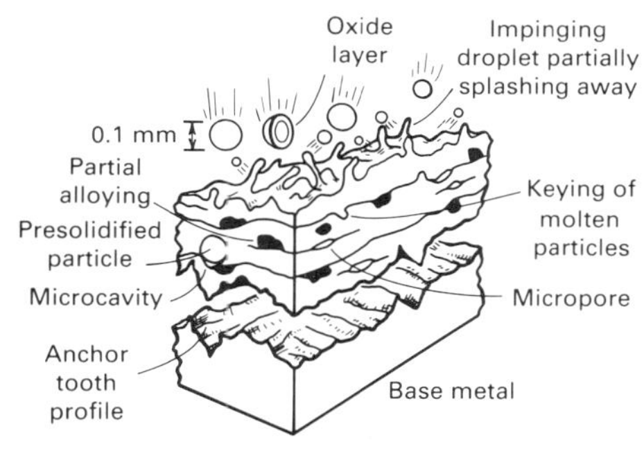

(a)

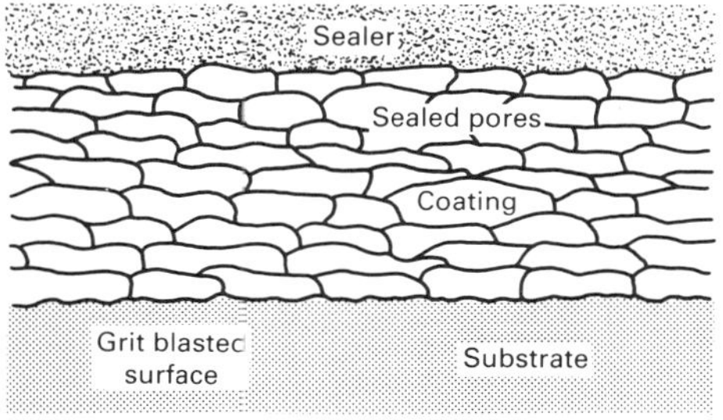

(b)

Fig. 51 Schematics of metal sprayed coating on steel. (a) As-sprayed coating. (b) Sealed coating

(a)

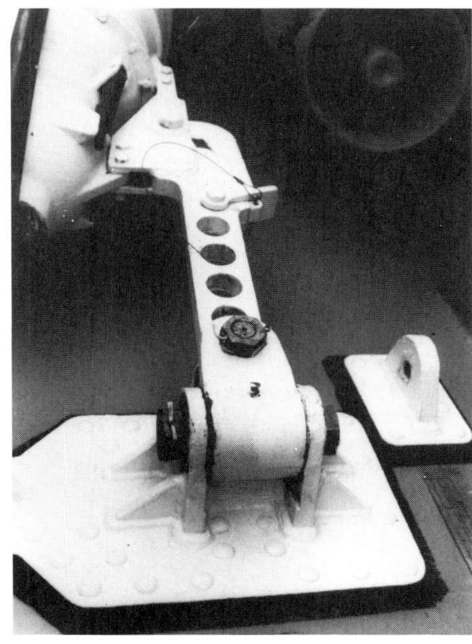

(b)

Fig. 52 Examples of aluminum flame sprayed topside weather equipment. (a) Stanchion. (b) Swivel arm assembly

Cathodic Protection

Robert H. Heidersbach
California Polytechnic State University
Richard Baxter
Deepwater Corrosion Services
John S. Smart III
Amoco Production Company
Michael Haroun
Oklahoma State University

Cathodic protection is an electrochemical means of corrosion control that is widely used in the marine environment. A detailed explanation of the principles of cathodic protection appears in the article "Cathodic Protection" in this Volume.

Cathodic protection can be defined as a technique of reducing or eliminating the corrosion rate of a metal by making it the cathode of an electrochemical cell and passing sufficient current through it to reduce its corrosion rate. All cathodic protection systems require the following components:

- Voltage source
- Anode
- Cathode
- Return circuit
- Electrolyte

Two types of cathodic protection systems are commonly used: impressed-current (active) systems and sacrificial anode (passive) systems. Both are common in marine applications. In recent years, hybrid systems—combinations of impressed-current and sacrificial anodes—have been used for

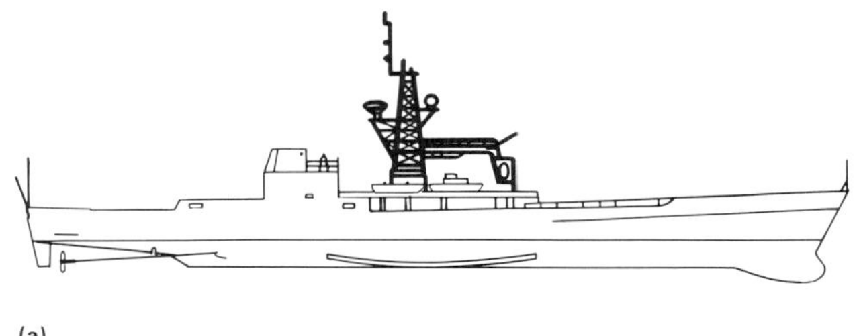

(a)

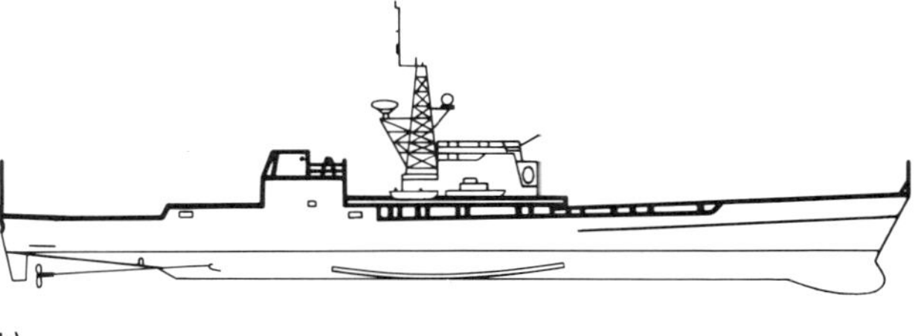

(b)

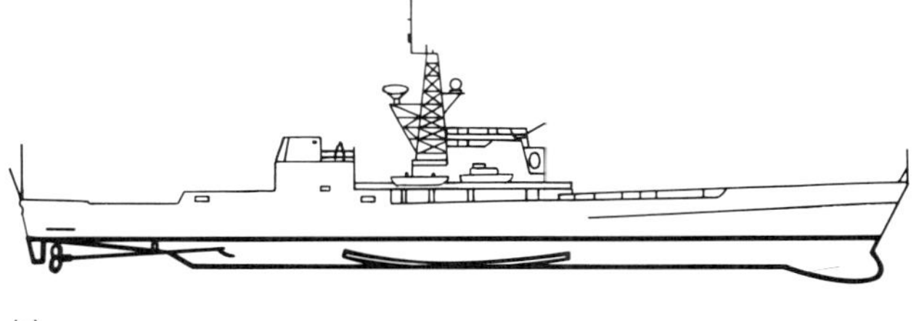

(c)

Fig. 53 Environments that are destructive to shipboard coatings. (a) Antennas and superstructures. (b) Deck areas. (c) Underwater hull

Table 14 Predicted 10-year corrosion rates for galvanized and aluminized steel panels

Tested 250 m (800 ft) from the ocean at Kure Beach, NC

Coating	Predicted weight loss, g/m² Skyward exposure	Groundward exposure
Galvanized	103.3	55.2
Type 1 aluminized (Al-Si)	17.8	20.1
Type 2 aluminized (pure aluminum)	11.6	17.9

Source: Ref 43

Table 15 Recommended surface preparation methods for various metallic substrates

Substrate	Recommended surface preparation method
Steel	Abrasive blast cleaning to near-white finish (SSPC SP10) for touchup and full-scale repainting Hand or power tool (preferably power) cleaning with wire brushes, chipping hammers, grinders, sanders, and so on, for maintenance
Galvanized steel	Solvent cleaning (degreasing) of new surfaces, followed by application of one coat of a wash primer
Aluminum	Anodizing or chromate conversion coating whenever possible. Otherwise, clean as for steel using only stainless steel wire brush.

very large marine structures. More information on land-based applications of cathodic protection is available in the articles "Cathodic Protection" and "Corrosion of Pipelines" in this Volume.

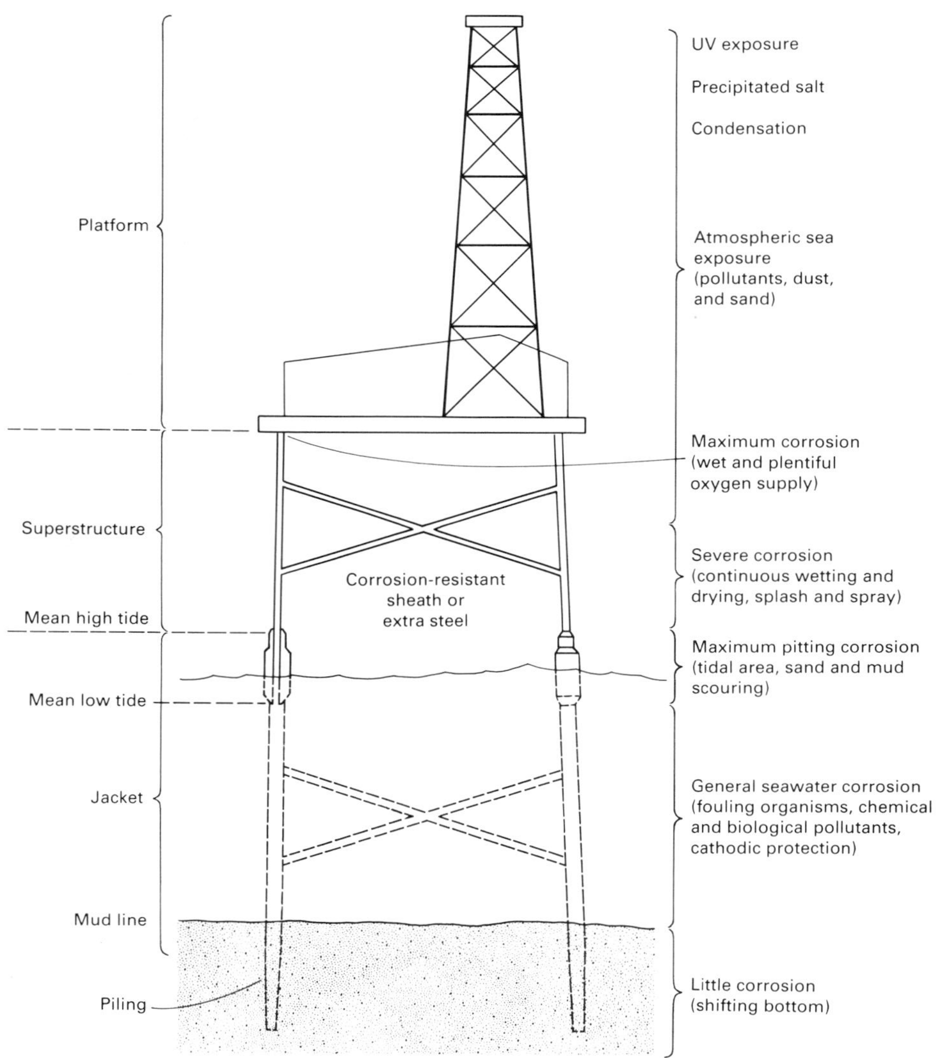

Fig. 54 Zones of severity of environment for a typical offshore drilling structure

Cathodic Protection Criteria

A number of criteria are used to determine whether or not a structure is cathodically protected. These criteria, which are covered in NACE RP-01-76 (Ref 54), include potential measurements, visual inspection, and test coupons.

Potential Measurements. Reference 46 specifies a negative (cathodic) voltage of at least 0.80 V between the platform and a silver-silver chloride reference electrode contacting the water. Normally, voltage is measured with the protective current applied. The 0.80 V standard includes the voltage drop across the steel/water interface, but does not include the voltage drop in the water.

Application of the protective current should produce a minimum negative (cathodic) voltage shift of 300 mV. The voltage shift is measured between the platform surface and a reference electrode contacting the water; it includes the voltage drop across the steel/water interface but not the voltage drop in the water.

Visual inspection should indicate no progression of corrosion beyond limits acceptable for platform life (Ref 54).

Corrosion test coupons must indicate a corrosion type and rate that is within acceptable limits for the intended platform life (Ref 54).

A number of other criteria are also possible, but in practice, −0.80 V versus Ag/AgCl is the most commonly used. Other reference electrodes can be used for marine applications. They are listed in Table 17.

Anode Materials

The choice of anode material depends on whether active (impressed-current) or passive (sacrificial anode) cathodic protection systems are under consideration. Sacrificial anodes must be naturally anodic to steel and must corrode reliably (avoid passivation) in the environment of interest. However, above all, sacrificial anodes should be inexpensive and durable. Impressed-current anodes rely on external voltage sources; therefore, they do not need to be naturally anodic to steel. They usually would be cathodic to steel if not forced to assume anodic potentials by the impressed current. Additional information on materials for sacrificial and impressed-current anodes is available in the article "Cathodic Protection" in this Volume.

Sacrificial Anodes. Commercial sacrificial anodes are either magnesium, aluminum, or zinc alloys. Table 18 lists the energy capabilities of sacrificial anode alloys.

Magnesium anodes have not been popular for offshore applications in recent years, because of developments in improved aluminum and zinc anodes. However, several operators have begun experimenting with composite sacrificial anode systems for offshore platforms. These designs use aluminum or zinc anodes for long-term performance and have magnesium anodes that are intended to provide an initially high current density and polarize the platform quickly to the desired protection potential. Results from the limited applications of this composite design are mixed, and this concept remains controversial.

Aluminum anodes and, more recently, aluminum-zinc alloys, have become the preferred sacrificial anodes for offshore platform cathodic protection. This is because aluminum anodes have reliable long-term performance when compared to magnesium, which may be consumed before the platform has served its useful life. Aluminum also has better current/weight characteristics than zinc. Weight can be a major consideration for large offshore platforms.

The major disadvantage of aluminum for some applications—for example, the protection of painted ship hulls—is that aluminum is too cor-

Table 16 Properties of abrasives

Abrasive	Moh's hardness	Shape	Bulk density kg/m³	Bulk density lb/ft³	Color	Free silica, wt%	Degree of dusting	Reuse
Naturally occurring abrasives								
Sand								
Silica	5	Rounded	1600	100	White	90+	High	Poor
Mineral	5–7	Rounded	2000	125	Variable	5	Medium	Good
Flint	6.7–7	Angular	1280	80	Light gray	90+	Medium	Good
Garnet	7–8	Angular	2320	145	Pink	nil	Medium	Good
Zircon	7.5	Cubic	2965	185	White	nil	Low	Good
Novaculite	4	Angular	1600	100	White	90+	Low	Good
By-product abrasives								
Slag								
Boiler	7	Angular	1360	85	Black	nil	High	Poor
Copper	8	Angular	1760	110	Black	nil	Low	Good
Nickel	8	Angular	1360	85	Green	nil	High	Poor
Walnut shells	3	Cubic	720	45	Brown	nil	Low	Poor
Peach pits	3	Cubic	720	45	Brown	nil	Low	Poor
Corn cobs	4.5	Angular	480	30	Tan	nil	Low	Good
Manufactured abrasives								
Silicon carbide	9	Angular	1680	105	Black	nil	Low	Good
Aluminum oxide	8	Blocky	1920	120	Brown	nil	Low	Good
Glass beads	5.5	Spherical	1600	100	Clear	67	Low	Good
Metallic abrasives								
Steel shot(a)	40–50(b)	Round	...	...	...	...	...	Excellent
Steel grit (made by crushing steel shot)(a)	40–60(b)	Angular	...	...	...	...	...	Excellent

(a) Steel shot produces a peened surface, while steel grit produces an angular, etched type of surface texture. (b) Rockwell C hardness. Source Ref 51

rosion-resistant in many environments. Aluminum alloys will not corrode reliably onshore or in freshwaters. In marine environments, the chloride content of seawater depassivates some aluminum alloys and allows them to perform reliably as anode materials. Unfortunately, it is necessary to add mercury, antimony, indium, tin, or similar metals to the aluminum alloy to ensure that this depassivation occurs. Heavy-metal pollution concerns have led to bans on the use of mercury alloys in some locations. Problems with alloy production result in aluminum alloys with supposedly adequate alloying additions that still corrode in an unreliable manner. This problem is the subject of ongoing research, but current indications are that alloy segregation may be the source of the problem.

Zinc anodes are similar to aluminum in many respects. They are used on ship hulls because aluminum passivates when ships enter the brackish water or freshwater of many harbors but zinc will continue to perform. Tankers with combination ballast/product tanks use zinc anodes because of their lower tendency to cause sparks if they fall from their supports and strike steel.

Zinc bracelet anodes are also used on offshore pipelines. Part of the reason for this is to ensure that the anode will corrode and provide protective current in bottom mud environments where aluminum may passivate. Because marine pipelines must be buoyancy compensated, the increased weight of zinc is an advantage, and zinc is also usually less expensive on a cost per delivered coulomb basis.

Impressed-Current Anodes. Impressed-current anodes need not be anodic to steel, and normally they are not. Most materials used as impressed-current anodes are insoluble and corrosion resistant, with very low rates of consumption (Table 19). Exchange current density—the ability of a material to sustain high current densities with lower power consumption—is an important consideration for some applications.

High-silicon cast iron is the most corrosion-resistant nonprecious alloy in commercial use. It is widely used as an impressed-current anode material, and the anodes are very strong, durable, and abrasion resistant. The major disadvantage in offshore applications is the high weight/current characteristics of high-silicon cast iron. Marine applications for high-silicon cast iron anodes include docks and similar coastal structures.

Precious metals have the advantage of high exchange current densities (Ref 54). The practical result is that a small precious metal surface is equivalent to thousands of times the surface area of other anode materials. Therefore, a small surface of platinum or palladium may be more economical than a much less expensive material. Early precious metal anodes were alloys that, after a short period of use, became enriched on the surface with the precious metal component. Anodes of this type are still used in harsh environments, such as Cook Inlet or the North Sea, where anode sleds weighing several tons are necessary to withstand high currents and storm conditions. For most other applications, smaller anodes consisting of precious metals clad to stronger substrates (for example, platinum bonded to a niobium substrate) are gaining acceptance. The extreme weight advantages of these anodes over other systems make them especially desirable for deep water structures.

Ceramic anodes consisting of metal oxides on conductive metal substrates are gaining acceptance for cathodic protection applications. Although these anodes have been used for years in the electrochemical-processing industry, their application to cathodic protection is recent. Ceram-

Table 17 Reference electrodes used for cathodic protection systems on offshore structures

Type of electrode	Protection potential of steel, V
Ag/AgCl	−0.80 (or more negative)
$Cu/CuSO_4$	−0.85 (or more negative)
Zinc	+0.25 (or less positive)

Table 18 Energy characteristics of materials used for sacrificial anodes

Material	Energy capability A·h/kg	Energy capability A·h/lb	Consumption rate kg/A·yr	Consumption rate lb/A·yr
Aluminum-zinc-mercury	2750–2840	1250–1290	3.1–3.2	6.8–7
Aluminum-zinc-indium	1670–2400	760–1090	3.6–5.2	8–11.5
Aluminum-zinc-tin	925–2600	420–1180	3.4–9.4	7.4–20.8
Zinc	815	370	10.8	23.7
Magnesium	1100	500	7.9	17.5

Table 19 Typical current densities and consumption rates of materials used for impressed-current anodes

Material	Typical anode current density A/m²	Typical anode current density A/ft²	Consumption rate kg/A·yr	Consumption rate lb/A·yr
Pb-6Sb-1Ag	160–215	15–20	0.045–0.09	0.1–0.2
Platinum (plated on substrate)	540–1080	50–100	6×10^{-6}	1.3×10^{-5}
Platinum (wire or clad)	1080–5400	100–500	10^{-5}	2.2×10^{-5}
Graphite	10.8–43	1–4	0.23–0.45	0.5–1.0
Fe-14Si-4Cr	10.8–43	1–4	0.23–0.45	0.5–1.0

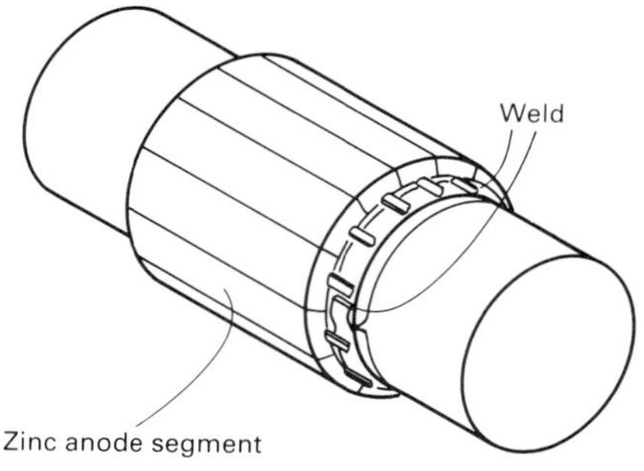

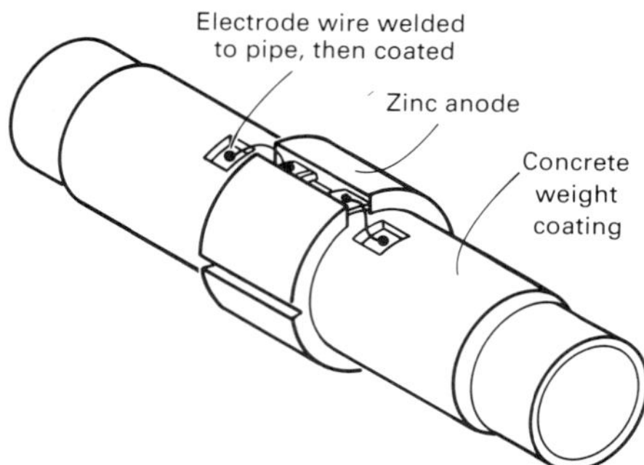

Fig. 55 Typical pipeline bracelet anodes

ic anodes have many of the same advantages as precious metal anodes (light weight, high current capacity).

Polymer Anodes. Several suppliers are marketing polymer anodes. These contain embedded graphite conductors and are used in applications requiring low current densities. Most offshore applications require the opposite—anodes with high current capacities. Coastal concrete structures that use cathodic protection to reduce the corrosion of reinforcing steel are possible marine applications for polymer anodes.

Comparison of Impressed-Current and Sacrificial Anode Systems

Sacrificial anode cathodic protection systems are simpler than impressed-current systems and require little or no maintenance except for periodic anode replacement. The capital cost of small systems is minimal, and they are often used for such applications as small pipelines and fishing boats.

The capital costs of impressed-current cathodic protection systems may be lower than those of sacrificial anode systems in certain applications (for example, long pipelines and large offshore platforms). Impressed-current systems are normally used for the protection of large structures or where the low conductivity of the electrolyte (freshwater, concrete, and so on) makes sacrificial anodes impractical.

Cathodic Protection of Marine Pipelines

Corrosion control of marine pipelines is usually achieved through the use of protective coatings and supplemental cathodic protection. A variety of organic protective coatings can be used. They are usually applied in a factory so that the only field-applied coatings are at joints in pipeline sections. Most marine pipelines have an outer "weight coating" of concrete. The cathodic protection system supplements these coatings and is intended to provide corrosion control at holidays (defects) in the protective coating.

Design Considerations. The average cathodic protection current density required to protect a marine pipeline will depend on the type of coating applied, the method used to coat field joints, the amount of damage inflicted on the coating during shipment and installation, whether or not burial is specified, and the location of the pipeline. Large-diameter pipelines can be protected by installing an impressed-cathodic protection system at one or both ends of the pipeline. Such a system would include a suitably sized transformer/rectifier unit and inert anodes, such as graphite or high-silicon cast iron.

Most marine pipelines are protected by the installation of bracelet-type zinc or aluminum alloy sacrificial anodes (Fig. 55). Electrical contact between the anode and the pipeline is made through an insulated copper cable bonded to the pipeline. Zinc anodes may either be high purity or alloyed. Aluminum anodes are usually fabricated from a proprietary aluminum-zinc-indium alloy.

Design Procedures. Typically, bracelet anodes are spaced at a maximum of 150 m (500 ft) on small-diameter pipelines (<355 mm, or 14 in.) and 300 m (1000 ft) on larger pipelines. The current required for a segment of pipeline is calculated by using the current density required for the given environment, the surface area of the pipe segment, and the fraction of steel assumed to be bare. Anodes are then sized to fit the conditions; that is, the anode must have adequate weight to satisfy the relationship:

$$\frac{W}{C} > IL \qquad \text{(Eq 9)}$$

where W is the anode weight in kilograms, C is the alloy consumption rate in kilograms per amp year (kg/A · yr), I is the anode current output in amps, L is the desired design life in years, and the anode nominal current output must exceed the required current. Anode current output I is determined from Ohm's law:

$$I = \frac{E}{R} \qquad \text{(Eq 10)}$$

where I is the current output in amps, E is the net driving voltage in volts, and R is the anode-to-electrolyte resistance in ohms.

Anode-to-electrolyte resistance R can be calculated using an empirical relationship, such as McCoy's equation:

$$R = \frac{0.315\,p}{\sqrt{A}} \qquad \text{(Eq 11)}$$

where p is the electrolyte resistivity in ohm centimeters (Ω · cm), and A is the anode area in square centimeters.

Data sheets from the manufacturer can also be consulted for information on the electrochemical properties of specific proprietary alloys. Anode geometry can be optimized by using a successive iteration technique, but most anode manufacturers offer a range of standard sizes, which can be optimized for specific applications. If the pipeline is to have a concrete weight coating for stabilization, the thickness of the anode should match the concrete thickness to facilitate installation. Anodes to be installed on nonweight-coated pipelines should have tapered ends so that the anodes do not hang up on the rollers of the lay barge.

Isolation. If the pipeline and the offshore production facilities are operated by separate parties, the pipeline is usually electrically isolated from the production facilities through the use of insulated flanges or monolithic isolation joints. This practice prevents loss of current from the pipeline cathodic protection system to the platform and facilitates recordkeeping. Pipelines are always electrically isolated from shore-based facilities, usually at the valve pit on the beach.

Cathodic Protection of Offshore Structures

Offshore oil production platforms are unusual because most platforms are not painted below the waterline. The cathodic protection system causes a pH shift in the water, which becomes more alkaline (higher pH). Most minerals are less soluble in alkalis than in near-neutral environments (neutral water has a pH of 7; the pH of seawater averages approximately 7.8). The higher pH near the cathode causes minerals to precipitate onto the steel surface and form a protective scale or calcareous deposit. Depending on such factors as water depth, temperature, and velocity, this protective scale may be calcium carbonate, magnesium hydroxide, or a mixture of these and other minerals (Ref 55; see also the section "Seawater" in this article).

The technology of offshore cathodic protection is rapidly changing. Many of these changes are required because water depths are increasing. Therefore, offshore structures are now being built in deeper, colder water in which mineral deposits are less likely to form. The formation of a mineral deposit in such conditions may require current densities as high as approximately 1000 to 2000 mA/m^2 (100 to 200 mA/ft^2).

Another problem associated with deep water platforms is that current density requirements change with depth. In recent years, several deep water platforms have been found to be underprotected. These North American platforms (Gulf of Mexico and Santa Barbara Channel) were located in deep water, but the inadequate protection was at intermediate depths. Gases are more soluble at depth, and carbonate scales (calcareous deposits) are harder to deposit in deep, cold waters. This difficulty is offset by the fact that many deep waters have little dissolved oxygen and should therefore be less corrosive than shallow waters. It should be noted however, that Cook Inlet, the North Sea, and other stormy waters may be oxygen saturated (and presumably carbon dioxide saturated) all the way to the bottom.

Many operators prefer to use sacrificial anodes on offshore platforms because the sacrificial anode systems are simple and rugged. In addition, they become effective as soon as the platform is launched and do not depend on external electric power supplies. Surveys of the reliability of impressed-current systems have led to the conclusion that they do not perform as well as sacrificial anode systems (Ref 56). One reason for this lack of performance may be the poor or fragile design of some early impressed-current systems.

Unfortunately, the weight of sacrificial anodes is becoming a serious consideration for deep water platforms. Impressed-current systems are gaining acceptance. Some operators have introduced hybrid designs. In these designs, the primary cathodic protection system uses impressed current, and a sacrificial anode system is used to protect the platform after launching and before the electrical system on the platform becomes operational. The Murchison Platform has the most widely publicized hybrid cathodic protection system (Ref 57).

In the past, the inefficiencies associated with cathodic protection design were not serious. Water depths were shallow, and cathodic protection systems were overdesigned to ensure satisfactory performance. This was justified based on economics. A typical cathodic protection system is only 1 to 2% of the total capital cost of a new platform, but a retrofit may cost as much as the platform itself. Early platforms in Cook Inlet and the North Sea were underdesigned, and the costs of retrofits led to the efforts that produced NACE RP-01-76 (Ref 54). References 58 and 59 detail some problems experienced with deep water platforms.

Design Procedures. A typical cathodic protection design procedure for an offshore platform might consist of the following sequence of steps:

- Selection of a proper maintenance current density; this will depend on the geographic location
- Calculation of surface areas of steel in mud and in seawater, and the addition of a safety factor
- Calculation of the total amount of anode material required to guarantee a desired life
- Selection of an anode geometry. Initial current density (calculated for a single anode from Dwight's equation; see the example below) should exceed 160 mA/m^2 (15 mA/ft^2), assuming a native potential of 0.45 V between bare steel and aluminum anodes
- Judicious distribution of anodes on the steel, assuming a throwing power of 7.6 m (25 ft) in line of sight and placing anodes within 3 m (10 ft) of all nodes

The criterion for complete cathodic protection is a steel structure potential more negative than −0.80 V (that is, −0.82, −0.85, etc.) at any point versus the Ag/AgCl reference electrode.

Example of System Design for an Offshore Platform. The following is a sample of an offshore cathodic protection design procedure. The platform is in the Gulf of Mexico, and the design parameters, which are similar to other Gulf platforms, include the following:

- Maintenance current density: 5 mA/ft^2
- Design life: 20 years
- Calculated surface area: 33 484 ft^2 (water zone) and 47 984 ft^2 (mud zone)
- Anode capacity: 1280 A · h/lb
- Safety factor: 25%
- Water resistivity: 20 Ω · cm
- Assumed anode parameters: 725-lb net aluminum weight; 8 ft long; 90.25 in.^{2} cross-sectional area

Given these numbers, the total current in the water zone and the mud zone can then be calculated:

$$\text{Current}_{\text{water}} = \frac{(5\ \text{mA/ft}^2)\ (33\ 484\ \text{ft}^2)}{1000} = 167.4\ \text{A}$$

$$\text{Current}_{\text{mud}} = \frac{(5\ \text{mA/ft}^2)\ (47\ 984\ \text{ft}^2)}{1000} = 239.9\ \text{A}$$

The next step is to evaluate the total weight, TW, of the anode material required:

$$\text{TW}_{\text{water}} = \frac{(167.4\ \text{A})\ (8760\ \text{h/yr})\ (20\ \text{yr})\ (1.25)}{(1280\ \text{A} \cdot \text{h/lb})} = 28\ 641\ \text{lb}$$

$$\text{TW}_{\text{mud}} = \frac{(239.9\ \text{A})\ (8760\ \text{h/yr})\ (20\ \text{yr})\ (1.25)}{(1280\ \text{A} \cdot \text{h/lb})} = 41\ 045\ \text{lb}$$

At this stage, the number of anodes needed in each section, N, can be evaluated:

$$\text{N}_{\text{water}} = \frac{28\ 641\ \text{lb}}{725\ \text{lb/anode}} = 39.5$$

$$\text{N}_{\text{mud}} = \frac{41\ 045\ \text{lb}}{725\ \text{lb/anode}} = 56.6$$

$$\text{N}_{\text{total}} = 40 + 57 = 97\ \text{anodes}$$

A total of 97 anodes are needed to prepolarize the steel or to provide the initial current density. The initial current density must be at least 160 mA/m^2 (15 mA/ft^2) to ensure the buildup of an adequate calcareous deposit on the steel members. The total initial current output per anode is calculated by using a potential of 0.45 V between bare, unpolarized steel (−0.60 V) and aluminum (−1.05 V) and an anode resistance calculated from Dwight's equation (Eq 12) for a single cylindrical anode, because experience shows no significant interference between the various numbers of a multiple anode design:

$$\text{Resistance} = \frac{P}{2\pi L}\left[\ln\left(\frac{4L}{R}\right) - 1\right] \qquad \text{(Eq 12)}$$

where L is the anode length in inches, π is 3.14159, p is water resistivity in Ω · cm, and R is equivalent radius in inches.

For anode shapes other than cylindrical:

$$R = \sqrt{\frac{\text{cross-sectional area}}{\pi}} \qquad \text{(Eq 13)}$$

In this case, the anode has a trapezoidal cross section, and the equivalent radius is calculated by using Eq 13, as follows:

$$R = \sqrt{\frac{90.25}{\pi}} = 5.36\ \text{in.}$$

Equation 12 can now be used to calculate the anode resistance:

$$\text{Resistance} = \frac{20\ \Omega \cdot \text{cm}}{2\pi\ (96\ \text{in.})}\left[\ln\left(\frac{4\ (96\ \text{in.})}{5.36\ \text{in.}}\right) - 1\right]\left(\frac{1\ \text{in.}}{2.54\ \text{cm}}\right) = 0.0427\ \Omega$$

Next, the initial current output, I, of the anode is determined:

$$I = \frac{\Delta E}{\text{Resistance}} \qquad \text{(Eq 14)}$$

where, ΔE is the potential difference (0.45 V), and the resistance = 0.0427 Ω.

$$I = \frac{0.45\ \text{V}}{0.0427\ \Omega} = 10.54\ \text{A}$$

The initial structure current density is determined as follows:

$$\text{Current density} = \frac{(10.54\ \text{A/anode})\ (97\ \text{anodes})}{33\ 484\ \text{ft}^2} = 0.0305\ \text{A/ft}^2 = 30.5\ \text{mA/ft}^2$$

A current of 328 mA/m^2 (30.5 mA/ft^2) is considered to be an adequate current density for the buildup of a calcareous deposit, which leads to a satisfactory polarization of steel.

Fully polarized steel develops a potential more negative than −0.80 V versus the Ag/AgCl reference electrode and will therefore establish a potential drop of 0.20 to 0.25 V versus aluminum anodes. Use of the 0.20 to 0.25 V potential drop for Ohm's law and calculation of the anode resistance with Dwight's equation (Eq 12) based on anode dimensions at 40 to 50% consumption will then result in a maintenance current density that is roughly one-half the value of the calculated initial current density. Therefore, if the steel is polarized to −0.80 V versus Ag/AgCl, the current output of the anode is (according to Eq 14):

$$I = \frac{\Delta E}{\text{Resistance}}$$

where ΔE = 1.05V − 0.8 V = 0.25 V, and resistance = 0.0441 Ω (the derated anode radius based on a 10% reduction of the initial radius).

$$I = \frac{0.25\ \text{V}}{0.0441\ \Omega} = 5.6\ \text{A}$$

The actual potential drop between structural steel cathode and aluminum anodes at some time after launching of the platform is estimated as the difference between the given aluminum versus Ag/AgCl potential and the average of polarized steel potentials as measured with Ag/AgCl reference electrodes at various locations of the structure. The true current output per anode is, therefore, available for predicting the life of the cathodic protection system given the capacity of 330-kg (725-lb) aluminum anodes, namely:

$$\text{Life} = \frac{(1280\ \text{A} \cdot \text{h/lb})\ (725\ \text{lb/anode})}{(8760\ \text{h/yr})\ (S\ \text{A/anode})}$$

which leads to the life in years for S amps of current output per anode as follows:

$$\text{Life} = \frac{106}{S}$$

Based on a current output per anode of 5.6 A, the design life of the cathodic protection system is:

$$\text{Life} = \frac{106}{5.6} = 18.9\ \text{years}$$

The required design life is 20 years, so another anode geometry should be chosen. Suppose that the new anode has the following anode characteristics: length = 10 ft (120 in.), radius = 6 in., weight = 1135 lb, and anode capacity = 1280

A · h/lb. The total number of anodes needed, based on the surface areas of the water and mud sections, is:

$$N_{water} = \frac{28\ 641\ \text{lb}}{1135\ \text{lb/anode}} = 25.2$$

$$N_{mud} = \frac{41\ 045\ \text{lb}}{1135\ \text{lb/anode}} = 36.2$$

$$N_{total} = 26 + 37 = 63\ \text{anodes}$$

The resistance of one anode using Dwight's equation (Eq 12) is:

Resistance

$$= \frac{20\ \Omega \cdot \text{cm}}{2\pi\ (120)} \left[\ln\left(\frac{(4)\ (120)}{(6)}\right) - 1\right] \left(\frac{1\ \text{in.}}{2.54\ \text{cm}}\right)$$

$$= 0.03532\ \Omega$$

The initial total current output I based on a potential drop of 0.45 V and using Eq 14 is:

$$I = \frac{0.45\ \text{V}}{0.03532\ \Omega} = 12.74\ \text{A}$$

The structure initial current density is now evaluated and is equal to:

$$\text{Current density} = \frac{(12.74\ \text{A/anode})\ (63\ \text{anodes})}{33\ 484\ \text{ft}^2}$$

$$= 0.0239\ \text{A/ft}^2 = 23.9\ \text{mA/ft}^2$$

An initial structure current density of 257 mA/m^2 (23.9 mA/ft^2) is acceptable. A more conservative design will allow for a higher number. Next, the anode resistance is calculated using Eq 12 and based on a derated radius of 5.4 in. (a 10% reduction):

Resistance

$$= \frac{20 \cdot \Omega\ \text{cm}}{2\pi\ (120)} \left[\ln\left(\frac{(4)\ (120)}{5.4}\right) - 1\right] \left(\frac{1\ \text{in.}}{2.54\ \text{cm}}\right)$$

$$= 0.03642\ \Omega$$

Using a potential drop of 0.25 V, the total maintenance current I available is:

$$I = \frac{0.25\ \text{V}}{0.03642\ \Omega} = 6.86\ \text{A}$$

Based on a current of 6.86 A, the design life in years of the cathodic protection system is:

$$\text{Life} = \frac{(1280\ \text{A} \cdot \text{h/lb})(1135\ \text{lb/anode})}{(8760\ \text{h/yr})(6.86\ \text{A/anode})}$$

$$= 24.2\ \text{years}$$

Therefore, the design is acceptable.

Anode Distribution. A final consideration concerns the positioning of anodes about the structure. They are placed within 3 m (10 ft) of nodes (welded or cast locations with the highest structural loading), but elsewhere are assumed to protect steel in line of sight within a 7.6-m (25-ft) radius. Thus, areas shadowed by other structural elements may not be fully protected by any particular anode.

Computer-aided cathodic protection designs for offshore structures have been tried by several organizations in recent years. These computer-aided designs are of two types. Small computers have been used to make the types of calculations (such as wetted surface area versus anode consumption) that have commonly been used for cathodic protection design. For these systems, the computer is a time saver that allows a greater number of alternatives to be considered but does not change the actual methodology of design.

An alternative approach is to use numerical techniques, such as finite element, finite difference, or boundary integral, to model the potential-current distribution field in the region of an offshore structure. Large computers can be programmed to generate complex analyses of various alternative designs (Ref 60). These computerized designs have found limited acceptance because of the expense associated with the designs and the time delay caused by communications difficulties among the operator, the cathodic protection designer, and the computer expert. Efforts are underway to develop software so that small computers can be used with these numerical programs. If these efforts are successful, design engineers will be able to compare a number of design alternatives quickly and inexpensively. The same basic types of computational techniques—finite element, finite difference, and boundary integral modeling—that are used for structural design can be used for cathodic protection design (Ref 60-63).

Cathodic Protection of Ship Hulls

Ships and pipelines normally have protective coatings as their primary means of corrosion control. Cathodic protection systems are then sized so that an adequate electric current will be delivered to polarize the structure to the desired level. This is done for new structures by estimating the percentage of bare steel that results from holidays in the protective coatings. Once the estimated amount of bare steel is determined, anodes are sized to provide adequate current densities for the design life of the system.

Ships are returned to drydocks; therefore, the size (and weight) of anodes can be reduced from what would be necessary for permanent anodes on offshore pipelines or platforms. Anodes are concentrated near the bow and stern, where coating damage is most likely to occur. The stern is also the location where galvanic couples (for example, propeller to hull) are possible. Relatively small anodes are placed on ships in these locations. Small anodes are desirable to minimize the drag effects caused by turbulence due to anode protrusions.

Anode Materials. Aluminum anodes are available for ship hulls, but they can passivate and become inactive on ships that enter rivers or brackish estuaries. For this reason, zinc anodes are almost universally used in commercial service.

Impressed-current cathodic protection systems are used on very large ships. The galvanic couple between the propeller, the shaft, and the hull of the ship can cause significant corrosion problems. Modeling of the current requirements for cathodic protection near tanker propellers was one of the first applications of the computer in cathodic protection design (Ref 64, 65).

Impressed-current cathodic protection systems can produce overprotection in some cases. Organic coatings can disbond because of the formation of hydrogen gas bubbles underneath coatings. Coating disbondment can produce increased surface areas that require more cathodic protection, and is controlled by placing dielectric shields between the impressed-current anode and the hull. Larger shields are sometimes fabricated from glass-reinforced epoxy, which is molded directly on the ship's hull. At one time, coating disbondment was a major concern, but most modern coatings are resistant to disbonding.

Hydrogen embrittlement of steel due to cathodic protection is sometimes a concern. This has been a problem on case-hardened shafts, bolts, and other high-strength attachments. Most structural steels have relatively low strength as well as minimum susceptibility to hydrogen embrittlement.

REFERENCES

1. H.H. Uhlig and R.W. Revie, *Corrosion and Corrosion Control*, 3rd ed., Wiley-Interscience, 1985, p 108
2. F.L. LaQue, *Marine Corrosion*, Wiley-Interscience, 1975
3. J.P. Riley and G. Skirrow, Ed., *Chemical Oceanography*, Vol 2, 2nd ed., Academic Press, 1975
4. D.A. Livingstone, Chemical Composition of Rivers and Lakes, in *Data of Geochemistry*, U.S. Geological Survey, Prof. Paper No. 440, Chapter G, M. Fleischer, Ed., 1963
5. S.C. Dexter and C.H. Culberson, Global Variability of Natural Sea Water, *Mater. Perform.*, Vol 19 (No. 19), 1980, p 16-28
6. B.D. Thomas, T.G. Thompson, and C.L. Utterback, *J. du conseil*, Vol 9, 1934, p 28-35
7. F.W. Fink and W.K Boyd, Corrosion of Metals in Marine Environments, MCIC Report No. 78-37, Metals and Ceramics Information Center, Battelle Columbus Laboratories, 1978
8. D.R. Kester, Dissolved Gases Other Than CO_2, in *Chemical Oceanography*, Vol 1, 2nd ed., J.P. Riley and G. Skirrow, Ed., Academic Press, 1975, p 498
9. S.C. Dexter, *J. Ocean Sci. and Eng.*, Vol 6 (No. 1), 1981, p 109-148
10. A.H. Ailor, Ed., *Atmospheric Corrosion*, John Wiley & Sons, 1982
11. M. Schumacher, Ed., *Seawater Corrosion Handbook*, Noyes Data Corporation, 1979
12. S.W. Dean and E.C. Rhea, Ed., *Atmospheric Corrosion of Metals*, STP 767, American Society for Testing and Materials, 1982
13. H. Herman and K. Altorfer, "Zinc Thermal Spray Metallizing for Corrosion Protection of Structural Steel," Paper presented at the ASM Metals Congress, St. Louis, MO, American Society for Metals, Oct 1982
14. B. Shaw and P. Moran, Characterization of the Corrosion Behavior of Zinc-Aluminum Thermal Spray Coatings, *Mater. Perform.*, Vol 24 (No. 11), Nov 1985, p 22-31
15. M. Thorpe, How to Spray With Molten Metal, *Chem. Technol.*, June 1982, p 364-372
16. R. Klinge, Sprayed Zinc and Aluminum Coatings for the Protection of Structural Steel in Scandinavia, in *Eighth International Ther-*

mal Spray Conference, American Welding Society, 1976, p 203-213
17. "Corrosion Tests of Flame-Sprayed Coated Steel—19-year Report," AWS C2.14-74, American Welding Society, 1974
18. "Metallized Coatings for Corrosion Control of Naval Ship Structures and Components," NMAB Report 409, National Academy Press, Feb 1983
19. "Metal Sprayed Coating Systems for Corrosion Protection Aboard Naval Ships," DOD-STD-2138(SH), Nov 1981
20. R.A. Parks, "Thermal Spray Coating Applications in the U.S. Navy," Paper presented at the ASM Thermal Spray Conference, Long Beach, CA, American Society for Metals, Oct 1984
21. E. Liberman, C. Clayton, and H. Herman, "Thermally-Sprayed Active Metal Coatings for Corrosion Protection in Marine Environments," Report SUSB-84-1, Department of Materials Science & Engineering, State University of New York, 1984
22. W. Cochran, Thermally Sprayed Aluminum Coatings on Steel, *Met. Prog.*, Dec 1982, p 37-40
23. W.H. Thomason, "Offshore Corrosion Protection with Thermal-Sprayed Aluminum," Paper presented at the Offshore-Shore Technology Conference, Houston, TX, May 1985
24. E. Escalante, W.P. Iverson, W.F. Gerhold, B.T. Sanderson, and R.L. Alumbaugh, *Corrosion Protection of Steel Piles in a Natural Seawater Environment*, Monograph 158, National Bureau of Standards, 1977
25. "Protective Coatings for Steel Piling: Results of Harbor Exposure on Ten-Foot Simulated Piling," Technical Report R4904, Naval Civil Engineering Laboratory, 1966
26. F.W. Garnter, Corrosion Protection of Iron and Steel by Thermal Spraying With Zinc and Aluminum, in *Eighth International Thermal Spray Conference*, American Welding Society, 1976, p 214-222
27. Corrosion Suspended on Menai Strait Suspension Bridge, *Anti-Corros.*, 1983, p 11
28. S. Haagenrud and R. Klinge, Atmospheric Corrosion Testing of Metallized and Painted Steel, in *Ninth International Thermal Spraying Conference*, 1980, p 385-391
29. H. Leclercq and R. Bensimon, Combined Paper: New Zinc-Based Alloy for Metallizing, in *Eighth International Thermal Spray Conference*, American Welding Society, 1976, p 417-429
30. F.L. La Que, A Critical Look at Salt Spray Tests, *Mater. Meth.*, Feb 1952, p 3-15
31. I.L. Rozenfeld, D.M. Kramarento, E.V. Antoshin, A.I. Nemkovskii, and Yu I. Sakharov, Electrochemical Study of the Protective Properties of Metal Spray Coatings, *Zashch. Met.*, Vol 8 (No. 2), March-April 1972
32. H.D. Steffens, Electrochemical Studies of Cathodic Protection Against Corrosion by Means of Sprayed Coatings, in *Seventh International Metal Spray Conference*, The Welding Institute, 1974, p 123-128
33. M. Magome, Y. Miam, K. Ueno, and G. Ueno, The Active State in Sprayed Metal Coatings. *Surfacing J.*, Vol 13 (No. 12), 1982, p 37-40
34. V. Vesely and J. Horkey, Electrochemical and Corrosion Properties of Thermally Sprayed Coatings of Aluminum Alloys, in *Eighth International Thermal Spray Conference*, American Welding Society, 1976, p 430-435
35. B. Shaw and P. Moran, The Active State in Sprayed Metal Coatings: A Discussion, *Surfacing J.*, Vol 16 (No. 3), 1985, p 66-69
36. L.J. Walters, Flame-Sprayed Protective Coatings for Iron and Steel, *Anti-Corros.*, Jan 1972, p 8-13
37. J.C. Bailey, U.K. Experience in Protecting Large Structures by Metal Spraying, in *Eighth International Thermal Spray Conference*, American Welding Society, 1976, p 223-231
38. R. Burns and W. Bradley, *Protective Coatings for Metals*, Reinhold, 1967
39. R. Legault and V. Pearson, Kinetics of the Atmospheric Corrosion of Galvanized Steel, in *Atmospheric Factors Affecting the Corrosion of Engineering Metals*, STP 646, K. Coburn, Ed., American Society for Testing and Materials, 1978, p 83-96
40. L. Shreir, Ed., *Corrosion Control*, Vol 2, *Corrosion*, Newnes-Butterworth, 1979
41. R. Legault and V. Pearson, *Corrosion*, Vol 34, 1978, p 349
42. R. Legault and V. Pearson, "Inland Steel Research Laboratories Report: The Atmospheric Corrosion of Galvanized and Aluminized Steel," Inland Steel Research Laboratories, Oct 1976
43. D.E. Tonini, Atmospheric Corrosion Test Results for Metallic Coated Steel Panels Exposed in 1960, in *Atmospheric Corrosion of Metals*, STP 767, S.W. Dean, Jr. and E.C. Rhea, Ed., American Society for Testing and Materials, 1982, p 163-185
44. J.C. Zoccola, H.E. Townsend, A.R. Borzillo, and J.B. Horton, Atmospheric Corrosion Behavior of Aluminum-Zinc Alloy-Coated Steel, in *Atmospheric Factors Affecting the Corrosion of Engineering Metals*. STP 646, S.K. Coburn, Ed., American Society for Testing and Materials, 1978, p 165-184
45. S. Marut'yan, I. Boiko, V. Bobrova, and I. Legkova, Influence of Manganese on the Corrosion Resistance of Hot-Aluminized Steel, *Prot. Met.*, Vol 18 (No. 2), 1982, p 181-182
46. B. Allen and R. Heidersbach, "The Effectiveness of Cadmium Coatings as Hycrogen Barriers and Corrosion Resistant Coatings," Paper 230, presented at Corrosion/83, National Association of Corrosion Engineers, April 1983
47. D. Aylor, "Anticorrosion Barriers: Chemistry and Applications," Paper presented at the Philadelphia Symposium, American Chemical Society, Aug 1984
48. J. Mazia, In Search of the Golden Fleece—Aluminum in Focus, *Met. Finish.*, Vol 80 (No. 3), March 1982, p 75-80
49. M. El-Sherbing and F. Salem, Surface Protection by Ion Plated Coatings, *Anti-Corros.*, Nov 1981, p 15-18
50. V. McLoughlin, The Replacement of Cadmium for the Coating of Fasteners in Aerospace Applications, *Trans. IMF*, Vol 57, Part 3, Autumn 1974, p 102-104
51. H.S. Preiser, Jacketing and Coating, in *Handbook of Corrosion Protection for Steel Structures in Marine Environments*, American Iron and Steel Institute, 1981
52. S. Rodgers and R. Drisko, Painting Navy Ships, in *Steel Structures Painting Manual*, Vol 1, 2nd ed., *Good Painting Practice*, Steel Structures Painting Council, 1982
53. J. Smart, Marine Coatings, in *Marine Corrosion*, AIChE Today Series, American Institute of Chemical Engineers, 1985
54. "Recommended Practice: Corrosion Control on Steel, Fixed Offshore Platforms Associated with Petroleum Production," NACE RP-01-76, 1983 Revision, National Association of Corrosion Engineers
55. H. England and R. Heidersbach, "The Effects of Water Depth on Cathodic Protection of Steel in Seawater," Paper 4154, presented at the Offshore Technology Conference, Houston, TX, May 1981
56. D. Boening, "Offshore Cathodic Protection Experience and Economic Reassessment," Paper 2702, presented at the Offshore Technology Conference, Houston, TX, May 1976
57. E. Levings, J. Finnegan, W. McKie, and R. Strommen, "The Murchison Platform Cathodic Protection System," Paper 4565, presented at the Offshore Technology Conference, Houston TX, May 1983
58. J. Smart, Corrosion Failure of Offshore Steel Platforms, *Mater. Perform.*, May 1980, p 41
59. K. Fischer, P. Mehdizadeh, P. Solheim, and A. Hansen, "Hot Risers in the North Sea: A Parametric Study of CP and Corrosion Characteristics of Hot Steel in Cold Seawater," Paper 4566, presented at the Offshore Technology Conference, Houston, TX, May 1983
60. J. Fu, J. Chow, and S. Chan, "A New Versatile Method for Designing Cathodic Protection Systems for Marine Structures," Paper 4567, presented at the Offshore Technology Conference, Houston, TX, May 1983
61. R. Heidersbach, J. Fu, and R. Erbar, Computers in Corrosion Control, National Association of Corrosion Engineers, 1986
62. R. Strommen, "Computer Modeling of Offshore Cathodic Protection Systems Utilized in CP Monitoring," Paper 4367, presented at the Offshore Technology Conference, Houston, TX, May 1982
63. M. Haroun, "Cathodic Protection Modeling of Nodes in Offshore Structures," MS thesis, Oklahoma State University, 1986
64. J. Fu, *Corrosion*, Vol 38 (No. 5), May 1982, p 295
65. J. Fu and S. Chow, *Mater. Perform.*, Vol 21 (No. 10), Oct 1982, p 9

SELECTED REFERENCES

Corrosion and Biological Corrosion

- J.A. Beavers, G.H. Koch, and W.E. Berry, Corrosion of Metals in Marine Environments, MCIC Report 86-50, Metals and Ceramics Information Center, Battelle Columbus Laboratories, 1986
- S.C. Dexter, *Handbook of Oceanographic Engineering Materials*, Wiley-Interscience, 1979
- S.C. Dexter, Ed., *Biologically Induced Corrosion: Proceedings of the International Conference*, National Association of Corrosion Engineers, 1986
- F.W. Fink and W.K. Boyd, "Corrosion of Metals in Marine Environments," MCIC Report 78-37, Metals and Ceramics Information Center, Battelle Columbus Laboratories, 1978
- M.G. Fontana and N.D. Greene, *Corrosion Engineering*, 2nd ed., McGraw-Hill, 1978
- F.L. La Que, *Marine Corrosion*, Wiley-Interscience, 1975
- *Microbial Corrosion Proceedings*, The National Physical Laboratory and the Metals Society of London, March 1983

- H.H. Uhlig and R.W. Revie, *Corrosion and Corrosion Control*, 3rd ed., Wiley-Interscience, 1985

Seawater Properties

- C.G. Gorshkov, *Atlas of the Oceans—Pacific Ocean*, Ministry of Defense of the USSR, Military Sea Transport (in Russian; See also the *Atlas of the Mediterranean Sea*)
- J.P. Riley and G. Skirrow, Ed., *Chemical Oceanography*, Vol 1 and 2, 2nd ed., Academic Press
- H. Sverdrup, M. Johnson, and R. Fleming, *The Oceans*, Prentice-Hall, 1942

Biological Film Formation

- R.C.W. Berkeley, J.M. Lynch, J. Melling, P.R. Rutter, and B. Vincent, Ed., *Microbial Adhesion to Surfaces*, Ellis Horwood, 1980
- J.D. Costlow and R.C. Tipper, Ed., *Marine Biodeterioration: Proceedings of the Symposium*, Naval Institute Press, 1984
- K.C. Marshall, *Interfaces in Microbial Ecology*, Harvard University Press, 1976

Marine Atmospheres

- A.H. Ailor, *Atmospheric Corrosion*, John Wiley & Sons, 1982
- S.W. Dean and E.C. Rhea, Ed., *Atmospheric Corrosion*, STP 767, American Society for Testing and Materials, 1982
- M. Schumacher, Ed., *Seawater Corrosion Handbook*, Noyes Data Corporation, 1979

Corrosion in the Nuclear Power Industry

Chairman: Joseph C. Danko, American Welding Institute

THIS ARTICLE will review the major corrosion problems in light water reactors, the research on the corrosion mechanism(s), and the development of engineering solutions and their implementation. To understand the occurrence of corrosion problems, a brief historical perspective of the corrosion design basis of commercial light water reactors, boiling water, and pressurized water reactors is necessary. Although corrosion was considered in the plant designs, it was not viewed as a serious problem. This was based on the results of laboratory experiments and in-reactor tests that did not indicate any major corrosion problems with the materials selected for the plant construction. However, the laboratory tests did not necessarily reproduce the reactor operating conditions and the early in-reactor test did not fully represent the commercial reactor conditions in all cases, and, finally, the test times were indeed of short duration relative to the plant design lifetime of 40 years. Thus, the design basis for the materials selection was determined on the favorable but limited test data that were available, and corrosion limitations on component integrity were therefore not anticipated.

The initial operation of the early commercial light water reactors encountered few corrosion problems. These were quickly repaired and did not have a major impact on plant availability. As more plants entered service, and more operating time was accumulated on existing plants, more corrosion-related incidents appeared in the piping and other components. Eventually, the corrosion of the plant materials did impact on plant availability, economics, reliability, and, in some cases, plant safety.

The major corrosion-related problems addressed by the utility industry and the nuclear steam system suppliers include the intergranular stress-corrosion cracking of welded austenitic stainless steel pipes in boiling water reactors and the steam generator corrosion in the pressurized water reactors. These have had the greatest impact on plant availability and economics to the utility industry. Other corrosion problems reported in this article cover the influence of corrosion on radiation fields, stress-corrosion cracking in steam turbine materials, erosion-corrosion in steam lines, corrosion of nuclear fuel elements, and the effect of nuclear lubricants on corrosion. Corrosion problems have also occurred in the condenser systems; this problem is discussed in the article "Corrosion in Fossil Fuel Power Plants" in this Volume.

The final section of this article reviews ongoing developments in the disposal of high-level nuclear waste (processed, or spent, fuels) in deep underground repositories. Material selection for waste containment packages, which must keep radioactive isotopes contained for time periods up to 1000 years, and the primary geologic media considered for repository siting are discussed. As will be shown, the properties of geologic media, such as granite, salt deposits, and sub-seabed sediments, are critical because of their influence on the corrosion behavior (particularly stress-corrosion cracking properties) of container materials.

Corrosion in Boiling Water Reactors

Barry M. Gordon
Principal Engineer, Corrosion Performance
General Electric Company

Gerald M. Gordon
Manager, Plant Materials Technology
General Electric Company

Corrosion/materials problems are a major cause of lost availability for both the pressurized water reactor (PWR) and the boiling water reactor (BWR). Despite the differences in operating temperature, pressure, and coolant environment, the current primary materials problem for both of these nuclear power plant systems is stress-corrosion cracking (SCC). For the PWR, SCC of steam generator tubes occurs in both the primary and secondary circuits. In the case of the domestic BWR, the cracking of the heat-affected zones (HAZ) of welded type 304 and 316 stainless steel piping is responsible for the greatest materials impact on availability.

This Section will review only corrosion/materials problems in the BWR; the primary concern of SCC of BWR piping and the SCC of other structural components will be discussed. This Section will also present a discussion on another form of environmental cracking that can occur in the BWR and PWR—corrosion fatigue. Specifically, the BWR components discussed in this Section include piping, control rod collet retainer tubes, jet pump beams, core internals, and feedwater nozzles. However, before presenting case histories of these concerns and their subsequent resolution, it is important to review the basic mechanisms and the conditions necessary for producing these two forms of environmental cracking.

Basics of SCC

The visible manifestations of SCC are cracks that create the impression of inherent brittleness in the material, because the cracks propagate with little or no attendant macroscopic plastic deformation. A metal that suffers from SCC is usually characterized by its typical mechanical values (yield strength and tensile strength), and with the exception of the cracked region, the metal appears normal. Many alloys are most likely susceptible to SCC in at least one environment. However, SCC does not occur in all environments, nor does an environment that induces SCC in one alloy necessarily induce SCC in another alloy.

Perhaps the most critical factor to remember concerning SCC is that the three factors necessary for producing SCC must be simultaneously present. The elimination of any one of these factors or the reduction of one of these factors below some threshold level eliminates SCC. The three necessary conditions for SCC are:

- Susceptible material
- Tensile stress (applied and residual)
- Corrosive environment (an environment that can provide the chemical driving force for corrosion reaction)

This concept is illustrated in Fig. 1, in which the small region consisting of the overlapping areas of three circles represents SCC. If one circle is missing or reduced sufficiently, SCC susceptibility is reduced. The value of this representation is that it is now evident that there are three obvious methods for combating SCC:

- Reducing the tensile stress below some threshold value by new designs or processing techniques
- Eliminating or reducing the severity of the corrosive environment
- Changing the susceptible material by using an alternate material, alloying additions, cladding, or heat treatment

As noted in the article "Environmentally Induced Cracking" in this Volume, SCC can proceed through a material in two modes—intergranular and transgranular. The modes are sometimes

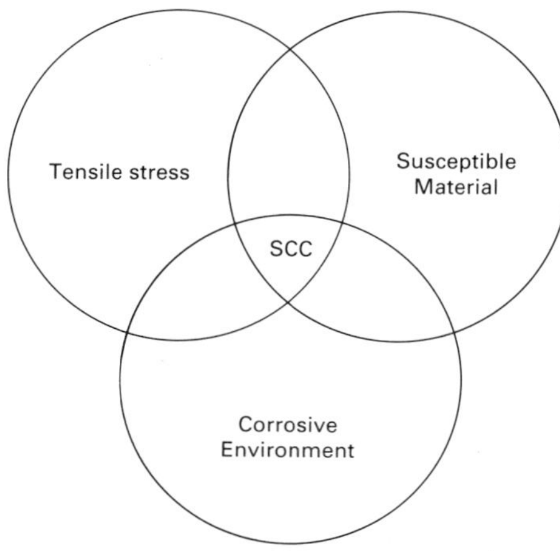

Fig. 1 The three necessary factors for producing SCC

mixed, or the mode switches from one morphology to the other. Intergranular SCC and transgranular SCC often occur in the same alloy, depending on the environment or the microstructure.

Cracking usually proceeds perpendicularly to the tensile stress. Another key characteristic of SCC is that the crack is usually tight; that is, the crack has a very large length-to-width ratio, indicating the absence of macroscopic plastic deformation. Cracks also vary in degree of branching or formation of satellite cracks. In numerous cases, the cracks are characterized by virtually no branches, but in some cases, they exhibit multibranched river delta patterns. This dependency is also a function of the environment, crack tip stress intensity, and microstructure, as detailed in the article "Modes of Fracture" in Volume 12 of the 9th Edition of *Metals Handbook.*

Increasing the stress level decreases the time for crack initiation, that is, the time to propagate to a specified shallow depth, such as 25 μm (1 mil). In the pure water specified for normal operation of a BWR, the threshold stress is believed to be greater than the at-temperature yield stress. In other alloy-environmental systems, it has been observed to be as low as 10% of the yield stress. It is critical to remember that any threshold value must be used with considerable caution because environmental conditions may change during operation. For example, a reactor component may be stressed below the SCC threshold value until a chloride or resin (sulfate) intrusion incident occurs, at which time SCC may initiate.

The stresses must be tensile and not compressive and must be of sufficient magnitude. Sources of tensile stresses include applied, residual, thermal, welding, and even corrosion product. Numerous cases of SCC have been observed in which no externally applied stress has been identified. As-welded material, for example, contains weld residual stresses approaching the yield point and often exceeding the yield point of the annealed material.

Basics of Corrosion Fatigue

Corrosion fatigue is fatigue aggravated by corrosion reactions. Because fatigue failures usually occur at applied stress levels below or at the operating temperature yield stress after numerous cyclic applications, the presence of an injurious environment reduces the number of cycles necessary to failure and reduces the stress level at which failure occurs. (It is important to note that stress raisers at initiation sites or the crack tip increase the local stress over the yield stress.) This phenomenon is illustrated in Fig. 2. In the case in which an alloy exhibits a fatigue endurance limit below which fatigue failures would not occur (body-centered cubic materials), the presence of a corrosive environment eliminates or reduces this threshold.

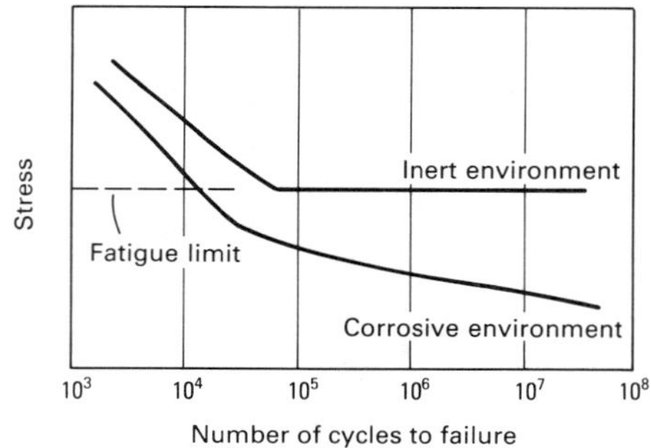

Fig. 2 Elimination of the fatigue limit due to the presence of a corrosion environment

Thus, corrosion fatigue is the reduction of fatigue resistance due to the presence of a corrosive medium. Corrosion fatigue is defined in terms of both environmental and mechanical properties. It may even be considered a special case of SCC. However, the mode of fracture and the preventive measures differ to a degree such that it is considered advisable to examine it separately.

Fatigue in the absence of environmental effects is characterized by fatigue resistance values, such as cyclic crack growth rates, that are nearly independent of the stress cycle frequency. This factor allows fatigue testing to be performed at accelerated rates. However, corrosion fatigue resistance is markedly affected by the stress cycle frequency for the following reasons. At extremely high cyclic frequencies, the material would not have the opportunity to react with the environment during each cycle, but at extremely low frequencies, passivation may take place during each step of the cycle. Thus, in the process of evaluating corrosion fatigue resistance, it is critical to conduct the test under conditions approaching those encountered in the field.

Corrosion fatigue cracks are typically transgranular, although they can be intergranular, especially for sensitized stainless steel in the BWR environment. The cracks are often branched, and several cracks are usually observed in the metal surface adjacent to the major crack. Fatigue cracks are also transgranular (with the exception of lead and tin), but rarely is there evidence of more than one crack. Corrosion pits may serve as nucleation sites for corrosion fatigue, but they are not a necessary precursor to cracking. Aqueous environments producing corrosion fatigue are numerous and not specific; this is contrary to the situation of SCC, for which only certain environment-metal combinations usually result in damage.

Intergranular SCC in BWR Piping

Although the intergranular SCC of BWR piping is not considered a safety problem, these cracking incidents affect plant availability, operating costs, and man-Rem exposure for inspection, repair, and so on. Figure 3 shows the major piping systems and other components affected by intergranular SCC.

Field Experience. The first observed BWR field intergranular SCC incident occurred in an early plant in the weld HAZ of a 150-mm (6-in.) type 304 stainless steel recirculation bypass line in late 1965 after 68 months of plant operation. This section of piping was replaced by type 304L stainless steel. Subsequently, in 1966 and 1967, there were a number of additional piping indications found by ultrasonic examination or by leakage. The first significant outbreak of intergranular SCC of welded type 304 stainless steel occurred in the fall of 1974 and early 1975. Sixty-four incidents of cracking were identified during this period, and all of them occurred in small-diameter lines (<250 mm, or 10 in.). As will be explained in the discussion "Mechanism of Intergranular SCC in BWR Piping" in this article, research revealed that this cracking was a result of the simultaneous interaction of weld-sensitized type 304 stainless steel, high tensile stress due to the combination of operating stresses and weld residual stresses, and high-temperature oxygenated BWR water (Ref 1).

During 1978, incidents of intergranular SCC were first noticed in a large-diameter (610 mm, or 24 in.) piping in an overseas BWR. This incident established additional concern for the main recirculation piping (Fig. 3) because replacement of these lines would be difficult, costly, and would require additional leak-before-break and degraded pipe behavior analysis. In 1982, extensive intergranular SCC of 710-mm (28-in.) recirculation piping was found in a domestic BWR. Additional inspections revealed cracking in large-diameter piping of other domestic BWRs.

Mechanism of Intergranular SCC in BWR Piping

As noted in the previous discussion on "Field Experience" in this article, the three necessary conditions for intergranular SCC of austenitic stainless steels in the BWR are:

- Sensitized microstructure (chromium depletion at the austenite grain boundaries)
- Tensile stress (applied plus residual) over the yield stress at temperature
- Oxygenated high-temperature water

Metallurgical Factors. When many non-stabilized austenitic stainless steels are heated in the temperature range of approximately 550 to 850 °C (1000 to 1550 °F), they undergo a solid state reaction and become sensitized and susceptible to intergranular corrosion. If a tensile stress is applied (or residually present) to a component characterized by this type of heat treatment, intergranular SCC can occur (if the environment is specifically corrosive to this particular metal system).

It is now accepted that this phenomenon in the BWR is related to local chromium depletion resulting from the precipitation of carbides at the austenite grain boundaries. To understand this phenomenon in terms of microstructure, it is instructive to examine the equilibrium relationships and carbon solubility in an 18Cr-8Ni alloy (Fig. 4). This simplified phase diagram indicates that austenite containing less than approximately 0.03% C should remain in solid solution over the

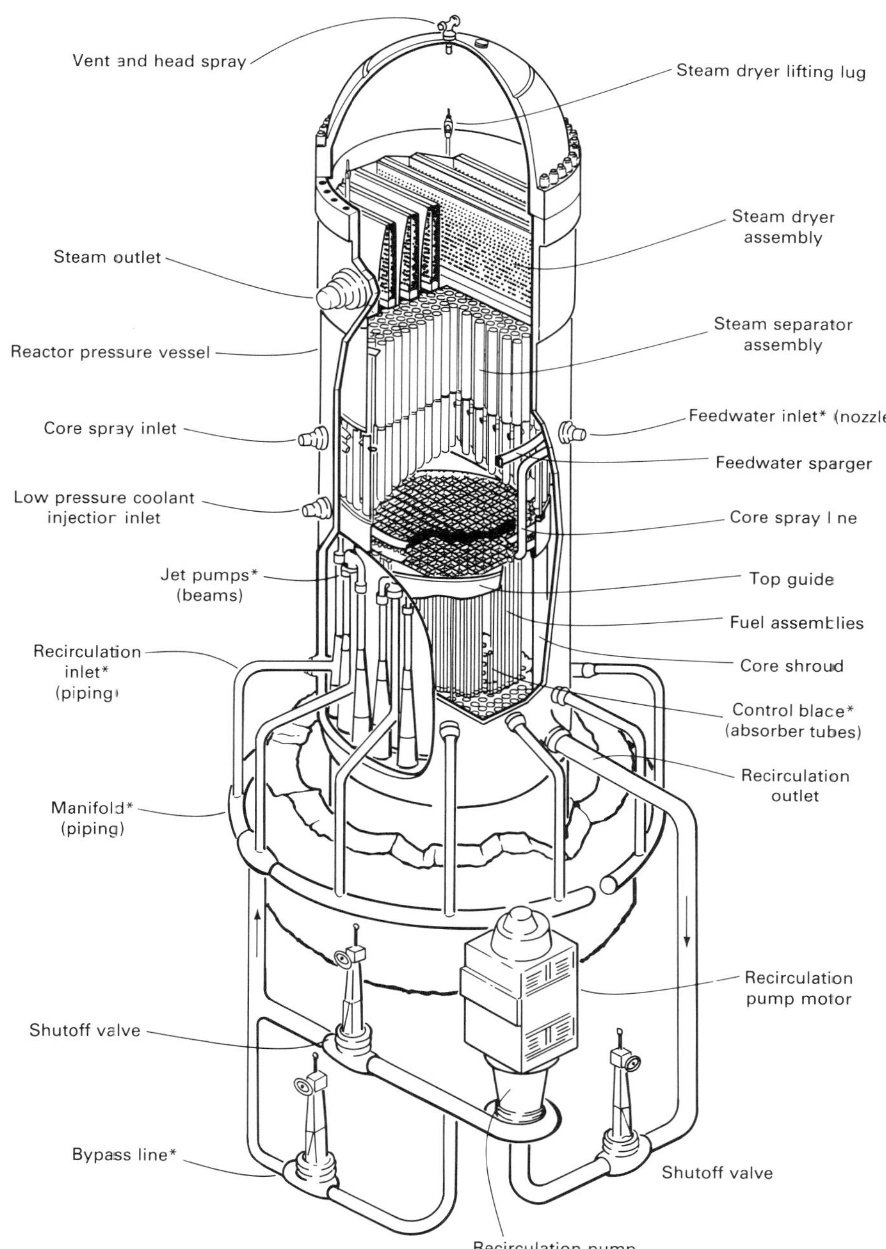

Fig. 3 Some of the key components of the BWR. Asterisks indicate areas of corrosion problems discussed in the Section "Corrosion in Boiling Water Reactors" in this article.

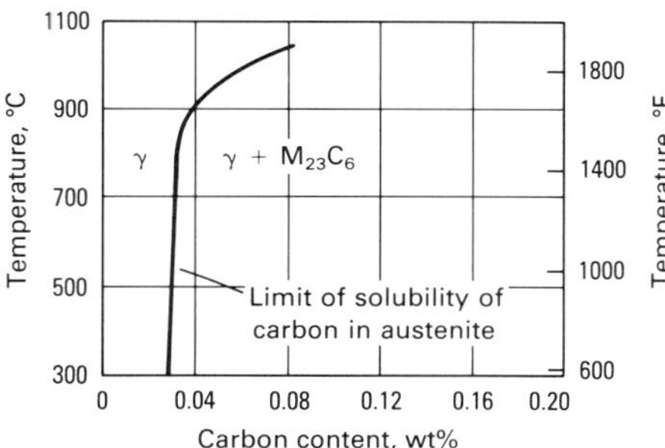

Fig. 4 Solid solubility of carbon in an Fe-18Cr-8Ni alloy. Source: Ref 2

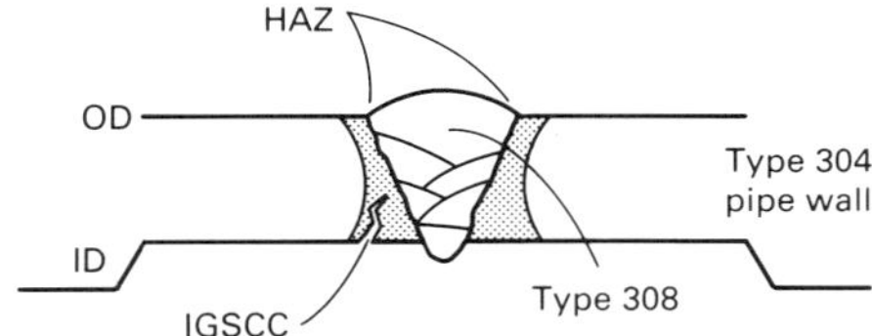

Fig. 5 Location of intergranular SCC in HAZ of type 304 pipe. OD, outside diameter; ID, inside diameter

temperature range shown. Austenite containing carbon in excess of 0.03% should precipitate $M_{23}C_6$ upon cooling below the solubility line. However, at relatively rapid rates of cooling, this reaction is partially suppressed, and the room temperature microstructure is characterized by austenite supersaturated with carbon.

If this supersaturated austenite is reheated to elevated temperatures within the $\gamma + M_{23}C_6$ field, further precipitation of the chromium-rich $M_{23}C_6$ (six carbon atoms tie up 23 chromium atoms) will take place preferentially at the austenite grain boundaries. Certain time-temperature combinations will be sufficient to precipitate this chromium-rich carbide, but will be insufficient to rediffuse chromium back into the austenite near the carbide. The net result is the formation of envelopes of chromium-depleted austenite around the carbide, which in certain environments will not resist intergranular attack or intergranular SCC. This chromium depletion results in a sensitized microstructure. This attack, when caused by welding, has been known as weld decay. However, more modern practice is to use the term sensitized to describe chromium depletion, irrespective of whether it has resulted from slow cooling, postweld heat treatment (a process designed to reduce weld residual stress in low-alloy steel pressure vessels), elevated-temperature service, or welding.

It is important, however, to distinguish among the various heat treatments because the stress-corrosion response differs with different heat treatments. For example, furnace-sensitized type 304 stainless steel is somewhat more susceptible to intergranular SCC than weld-sensitized material in an oxygenated pure water environment because of the higher integrated time at temperature. The sensitized regions adjacent to a weld are known as the HAZ (Fig. 5). Figure 6 shows an actual intergranular crack in the HAZ of a 400-mm (16-in.) pipe. The crack does not penetrate the weld metal itself. The use of this intergranular SCC immunity of the weld metal as a means of mitigating cracking will be discussed below in the discussion of "Corrosion-Resistant Cladding" in this article.

The chromium-depleted zone adjacent to the grain boundary (Fig. 7) is basically no longer a stainless steel but rather some type of low-alloy steel, because the chromium content may be considerably below the 10 to 12% required to create a stainless steel. The precipitation of $Cr_{23}C_6$ at the grain boundaries and the creation of a chromium-depleted zone have resulted in the formation of a metallurgical galvanic cell between a small anodic area of low-alloy steel (chromium-depleted zone) and a large cathodic area of stainless steel. This galvanic cell with unfavorable area anode/cathode ratios may result in more rapid attack of the impoverished area even in a low-conductivity oxygenated water environment. Additional information on sensitization of stainless steels and HAZ corrosion can be found in the articles "Corrosion of Weldments" and "Corrosion of Stainless Steels" in this Volume.

Tensile Stress Factors. There are primarily three sources of stress in the BWR (and PWR): fabricational stresses, primary stresses, and secondary stresses. Fabricational stresses consist of stresses introduced during fit-up and assembly in the shop or in the field, those introduced by machining or forming operations (such as surface grinding or cold straightening), and those intro-

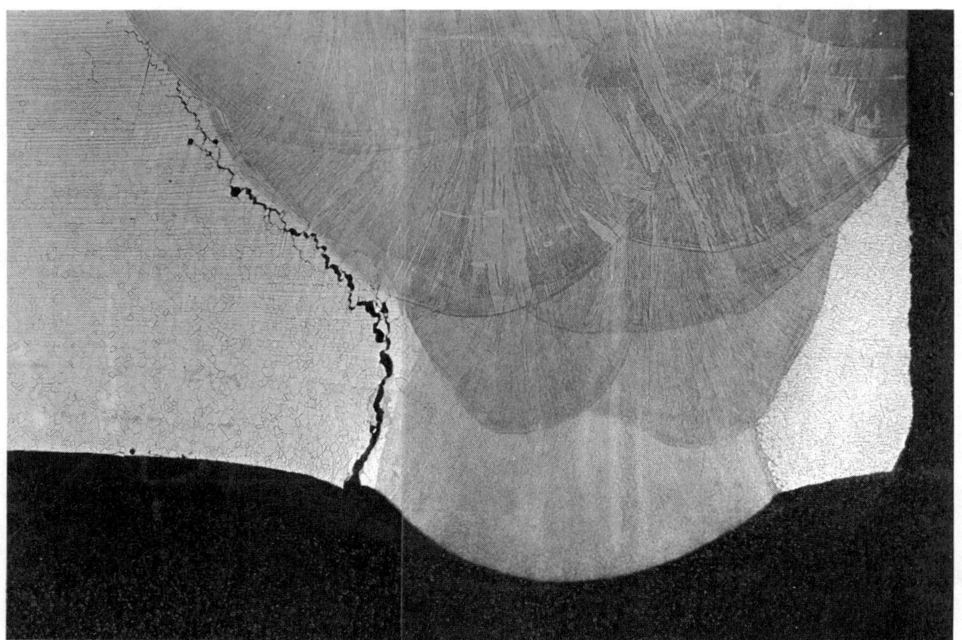

Fig. 6 Pipe test results showing intergranular SCC in a 400-mm (16-in.) type 304 stainless steel pipe HAZ. 13×

duced by other operations (such as welding). For example, abusive grinding can introduce surface tensile stresses of approximately the magnitude of the yield point or higher. Welding residual stresses near the yield point can also be present in pipes. Primary stresses from the operational forces on the equipment may be as high as the yield stress locally at constraints in an assembly. Secondary stresses—for example, from thermal expansion—may also locally reach the yield point. Finally, cyclic stresses from vibrations or from changes in operating mode can also add to the sum. Such varying stresses may be of great importance in initiating cracks or in restarting stopped cracks because they provide continuing plastic strain.

The effect of stress level on the time to failure of sensitized austenitic stainless steels has been studied in detail, and a well-defined dependence of stress on time to cracking has been found. Figure 8 shows this relationship for furnace-sensitized (24 h at 650 °C, or 1200 °F) type 304 stainless steel in 0.2-ppm oxygen (O_2) BWR-type water. As expected, the lower the tensile stress, the longer the time to cracking. Also, it seems apparent that the stress dependency curve is asymptotic to the at-temperature yield strength of the sensitized material.

Environmental Factors. The recirculating coolant in a BWR is high-purity neutral-pH water containing typically 0.2 mg/L (ppm) radiolytically produced dissolved oxygen with less-than-stoichiometric amounts of dissolved hydrogen (~0.02 mg/L). This amount of oxygen is sufficient to provide the chemical or electrochemical driving force for intergranular SCC. During the corrosion reaction, the dissolved oxygen is reduced to the hydroxyl ion, and the chromium-depleted zone of the stainless steel is oxidized.

Increasing the dissolved oxygen content to approximately 3 mg/L will increase the rate of cracking. Above 3 mg/L, the rate of oxygen arrival is limited to the limiting diffusion rate. The presence in the oxygenated environment of such anions as chloride, sulfate, and carbonate will significantly increase the propensity for intergranular SCC. Anions such as nitrate and fluoride have only a minor detrimental effect on type 304 stainless steel cracking propensities at operating temperatures when they are present in low concentrations (Ref 3-5).

Quantitative Model for Intergranular SCC in the BWR. A quantitative model for intergranular SCC (and corrosion fatigue) in the BWR has been developed (Ref 6). The model is based on the slip dissolution/film rupture model and relates crack advance to the oxidation reactions that are occurring at the crack tip where a thermodynamically stable oxide (or protective film) is ruptured by an increase in the strain in the underlying matrix. The amount of subsequent crack propagation is related via Faraday's law to the oxidation charge density associated with both dissolution and oxide growth on the bare surface. Such an environmentally controlled mode of crack advance is sustained by a further increment in matrix strain, which reruptures the oxide after a certain time period.

To validate the slip dissolution/film rupture hypothesis and to use it as a quantitative intergranular SCC prediction tool, several critical factors were evaluated:

- Metallurgical condition at the crack tip
- Crack tip environment, including electrochemical potential, pH, and anion content
- Oxidation rates on bare surfaces of alloy-environment systems symptomatic of those at the strain crack tip
- Crack tip strain rate

The model successfully predicts to within a factor of two the environmentally controlled crack propagation rates for stainless steel and pressure vessel steels in water at 288 °C (550 °F) over a range of system variables (Ref 6). As a point of comparison, it should be noted that the validation criteria of past models have been limited to orders-of-magnitude predictions for a specific material/environment/stress condition. This model is now being used in the BWR industry as a crack propagation prediction tool.

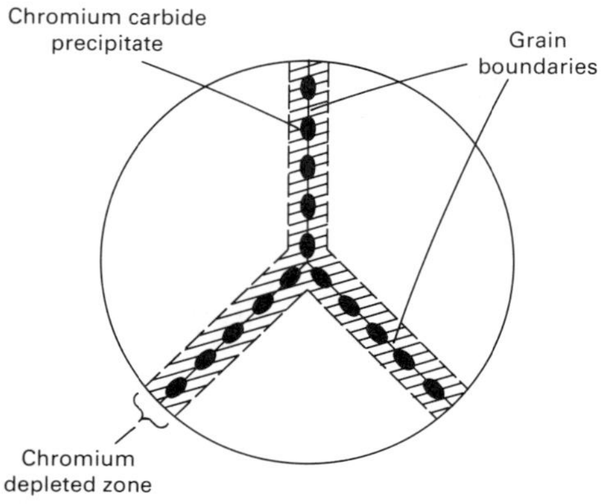

Chromium carbide precipitate
Grain boundaries
Chromium depleted zone

Fig. 7 Diagrammatic representation of a grain boundary in sensitized type 304 stainless steel

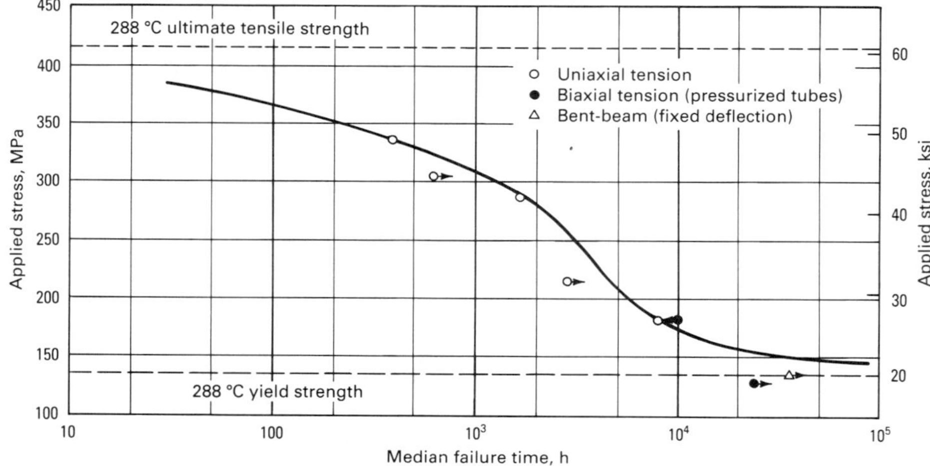

Fig. 8 Stress dependence of intergranular SCC of sensitized type 304 stainless steel in 288-°C (550-°F) water with 0.2 mg/L (ppm) O_2

Mitigation of Intergranular SCC in BWR Piping

The following discussion presents several intergranular SCC mitigation techniques for BWR piping. The mitigation methods are presented under general categories of material solutions, stress solutions, and environmental solutions. A few of the remedies address two of the three necessary intergranular SCC factors.

Materials Solutions

The material solutions to the piping intergranular SCC concern in the United States consist primarily of replacing the susceptible type 304 and 316 stainless steels with more sensitization-resistant materials, such as type 316 and 304 Nuclear Grades, redissolving the chromium carbides by solution heat treatment, and cladding with crack-resistant weld metal. Other materials, such as type-347 Nuclear Grade, have been used successfully overseas.

Nuclear Grade Stainless Steels. The replacement of current piping materials with materials more resistant to sensitization is a straightforward approach to mitigating intergranular SCC. Because it is well documented that decreasing the carbon content and increasing the molybdenum content of stainless steel would reduce the kinetics of sensitization (Ref 7-10), these two compositional modfications are synergistically beneficial for increasing the resistance of type 304 stainless steel to sensitization. However, type 316 Nuclear Grade and type 304 (with no added molybdenum) Nuclear Grade stainless steels take this theme a step further. Instead of the nominal 0.03% C maximum of the L-grade stainless steels, the Nuclear Grades are characterized by a maximum carbon content of 0.020%. The second important composition characteristic of type 304NG and type 316NG is the specification of 0.060 to 0.100% N. This modification is designed to recover the decrease in alloy strength due to the reduction of the carbon content. Another successful approach that has been used in the Federal Republic of Germany is the use of low-carbon niobium-stabilized type 347 stainless steel.

For the Nuclear Grade materials, full-size pipe tests have shown that factors of improvement over type 304 stainless steel performance can be expected to be at least 50 to 100 times in normal BWR operation (Ref 11). The necessary factors for improvement for the current 40-year service life is about 20. Therefore, the replacement of type 304 stainless steel piping with type 316, type 304, or type 347 Nuclear Grade will provide substantial resistance to intergranular SCC.

It is important to note that creviced or cold-worked type 316 or type 304 Nuclear Grade can suffer intergranular SCC. When crevices are present, the ionic impurity content in the water can greatly aggravate intergranular SCC. Among the specific impurities, sulfate ions are particularly deleterious (Ref 4). It is therefore extremely important to maintain good-quality water chemistry in the BWR to minimize cracking propensities.

Solution Heat Treatment. Immunity against intergranular SCC of type 304 stainless steel can be provided by eliminating weld-sensitized regions. This can be accomplished by solution heat treatment to redissolve the chromium carbides and eliminate chromium depletion around previously sensitized grain boundaries. Moreover, solution heat treatment will eliminate detrimental cold work and weld residual stress in the pipe.

Following a butt-welding operation, the entire pipe segment is solution annealed at 1040 to 1150 °C (1900 to 2100 °F) for 15 min per 25 mm (1.0 in.) of thickness but not less than 15 min or more than 1 h, regardless of thickness. The pipe segment is then quenched in circulating water to a temperature below 205 °C (400 °F). Solution heat treatment is generally limited to those weld joints made in the shop where heat treatment facilities are available, because of dimensional tolerance consideration, size constraints of the vendor facilities (furnace and quench tank) and cooling rate requirements (dead end legs).

Corrosion-Resistant Cladding. The corrosion-resistant cladding remedy achieves its resistance to intergranular SCC by using the intergranular SCC resistance inherent in duplex austenitic-ferritic weld metals (Ref 11). Although the carbide precipitation observed in the HAZ inside surface is also present in the weld metal, the nature of the duplex structure of the weld metal provides resistance to intergranular SCC in the BWR. In fact, intergranular SCC propagating from the weld HAZ is generally blunted when it reaches the weld metal if sufficient ferrite is present. As shown in Fig 6, the cracking will actually curve away from the weld metal.

Field experience has indicated that in the as-welded condition very little ferrite is required to prevent intergranular SCC. These results and numerous laboratory data generated on welded and furnace-sensitized type 308 and type 308L weld metal prompted the conclusion that a minimum amount of ferrite (8%) must be present to provide a high degree of resistance to intergranular SCC in BWR environments. As with type 304 stainless steel, reducing the carbon level is also beneficial.

In the corrosion-resistant cladding technique, type 308L weld metal is applied to the inside surface of the pipe at the pipe weld ends before making the final field weld. This duplex weld metal covers the region that will become sensitized during the final weld process, thus providing intergranular SCC resistance by maintaining low carbon and a sufficient ferrite level in the region that would normally be sensitized. Additional information on corrosion-resistant cladding can be found in the article "Corrosion of Clad Metals" in this Volume.

Weld Overlay Repair. Before any discussion of weld overlay, it is important to note that the weld overlay repair technique is not a fully qualified long-term mitigation technique for intergranular SCC. Currently, the Nuclear Regulatory Commission (NRC) has allowed up to three fuel cycles of operation (~54 months) on a cycle-by-cycle basis with weld overlay. However, laboratory environmental qualification programs and effective ultrasonic test procedures have established a basis for its extended lifetime beyond 22 fuel cycles (~396 months) of operation (Ref 12).

The weld overlay is similar to corrosion-resistant cladding in that it uses layers of intergranular SCC resistant duplex weld metal. For cases where Alloy 182 is to be overlayed, Alloy 82 weld metal is used. The most significant difference is that the layer of weld metal is placed on the outside surface of the pipe while the pipe is being cooled internally with water and is used to prevent an existing crack from penetrating through the wall. The weld overlay is also applied as a structural reinforcement to restore the original piping safety margins (Fig. 9). An equally important effect of the weld overlay is that it produces a favorable (compressive) residual stress pattern that can retard or arrest crack growth.

The weld overlay technique has the potential for being the most cost-effective method as compared to other repair techniques (pipe replacement, solution heat treatment, corrosion-resistant cladding), which require draining of the system. For additional information, see the article "Weld Overlays" in Volume 6 of the 9th Edition of *Metals Handbook*.

Tensile Stress Solutions

The tensile stress solutions primarily affect the weld residual stress profile by placing the inner surface weld residual stress in compression. These solutions, which are discussed below, include heat sink welding, induction heating stress improvement, and last pass heat sink welding.

Heat Sink Welding. If a pipe can be welded without producing a sensitized structure and high residual tensile stresses in the weld HAZ, the resultant component will be resistant to intergranular SCC in the BWR environment. The heat sink welding program has developed procedures that reduce the sensitization produced on the inside surface of welded pipe and, more important, change the state of surface residual welding stresses from tension to compression. This approach can be used in shop or field applications.

Heat sink welding involves water cooling the inside surface of the pipe during all weld passes subsequent to the root pass or first two layers. Water cooling can be applied by using flowing or turbulent water, by spray cooling through a sparger placed inside the pipe, or, in a vertical run, by still water.

Laboratory type 304 stainless steel butt welds have been produced to evaluate the inside surface heat sink welding techniques (Ref 13). Residual stresses were measured with strain gages. It was found that in a variety of pipe sizes the inside surface tensile residual stress is reduced substantially or changed from tension to compression as a result of this approach. Heat sink welding, as mentioned above, has a secondary benefit in that it reduces the time at temperature for sensitization due to the presence of the cooling water heat sink.

The induction heating stress improvement technique changes the normally high tensile stress present on the pipe inside surface of weld HAZs to a benign compressive stress (Ref 14, 15). This process involves induction heating of the outer pipe surface of completed girth welds to approximately 400 °C (750 °F) while simultaneously cooling the inside surface, preferably with flowing water (Fig. 10). Thermal expansion caused by the induction heating plastically yields the outside surface in compression, while the cool inside surface plastically yields in tension. After cooldown, contraction of the pipe outside

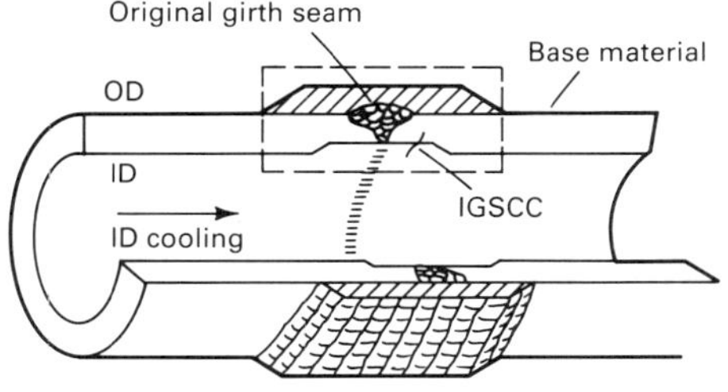

Fig. 9 Weld overlay intergranular SCC mitigation technique

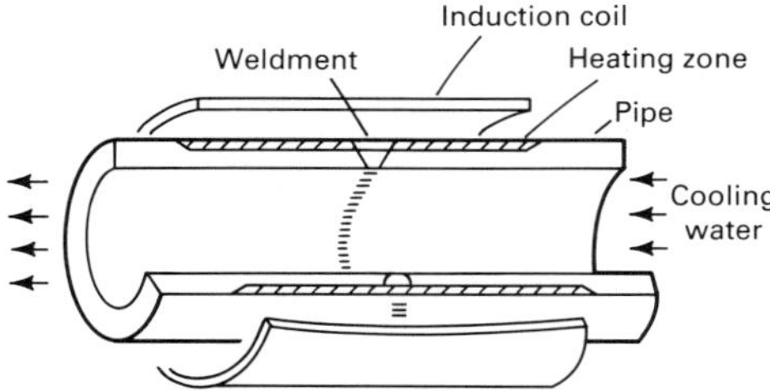

Fig. 10 Illustration of heating and cooling process for induction heating stress improvement

surface causes the stress state to reverse, leaving the inner surface in compression and the outside surface in tension (Fig. 11).

Qualification of the effectiveness of induction heating stress improvement has been accomplished by establishing the following (Ref 12, 15):

- Induction heating stress improvement treatment reliably reduces normally high tensile inside surface residual stresses to a zero compressive state
- Full-size environmental pipe testing and residual stress tests have demonstrated that these beneficial residual stresses result in a large improvement in intergranular SCC resistance
- Metallurgical investigations have shown that an induction heating stress improvement treatment produces no adverse effects. No increase in sensitization is found, and no significant variation in mechanical properties occurs
- A minimum life of 12 fuel cycles (~216 months) has been measured on induction heated stress improved precracked pipes in the laboratory if the initial cracking does not exceed about 20% of wall thickness

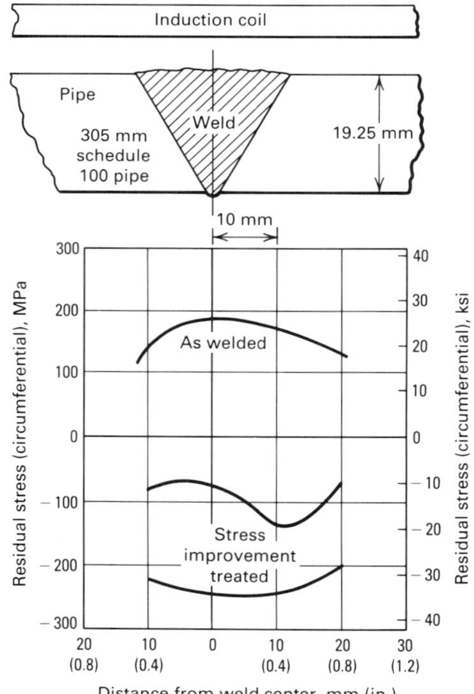

Fig. 11 Residual stress comparison for induction heating stress improvement

Last Pass Heat Sink Welding. As discussed in the previous section, induction heating stress improvement is an intergranular SCC mitigation technique that favorably alters the weld residual stress pattern. The welding torch is the heat source that initially produces the undesirable residual stress that induction heating stress improvement counterbalances. Analysis was able to establish that this residual stress state could be made compressive by the introduction of inside surface water cooling during the welding operation. Cooling during the entire welding process (heat sink welding) or just during the last pass could be effective in reversing the residual stresses analogous to the induction heating stress improvement process.

Qualification of the last pass heat sink welding process consisted of magnesium chloride ($MgCl_2$) residual stress tests, which verified that the last pass heat sink welding process does produce compressive residual stresses uniformly around the pipe circumference (Ref 16). Stress-relief strain gage measurements quantified the compressive axial stress and revealed that the stresses were compressive up to about 50% through-wall.

Finally, intergranular SCC improvement was evaluated by using pipe tests performed at applied stresses above yield in a high-oxygen (8 mg/L) high-temperature (288 °C, or 550 °F) water environment. As shown in Fig. 12, at the end of the program period, the pipes had been on test for over 5500 h, demonstrating a factor of more than 5.5 and more than 6.5 improvement, respectively, at the two test stresses (193.7 and 211 MPa, or 28.1 and 30.6 ksi, respectively). This factor of improvement approaches that determined for induction heating stress improvement.

Environmental Solution (Hydrogen Water Chemistry)

Numerous laboratory studies have indicated that susceptibility to intergranular SCC in stainless steels diminishes with decreasing oxygen content of the water below 0.2 mg/L. At 0.015 mg/L oxygen, intergranular SCC initiation is extremely difficult in the laboratory in sensitized type 304 stainless steel if good water purity (<0.3 μS/cm) is maintained (Ref 17).

The candidate oxygen-reducing additives to the BWR environment consisted of ammonia, hydrazine, ammonia-hydrazine, hydrazine morpholine, and hydrogen. Hydrogen was selected as the most suitable candidate, based on laboratory studies as well as an overall systems and economic analysis. Hydrogen also produces no changes in pH, results in minimum system impact, and is not corrosive or toxic.

Initial, very short term (~10 hours and 4 days) in-reactor hydrogen injection tests were performed as early as 1979 and 1981, respectively (Ref 18). After a somewhat longer (6 weeks) hydrogen demonstration test at the Commonwealth Edison Dresden-2 plant in 1982 (Ref 19), this plant became the first facility to initiate full-time operation on hydrogen injection. The test clearly demonstrated that the addition of hydrogen gas does indeed reduce the dissolved oxygen content below the intergranular SCC threshold level as measured by electrochemical potential techniques plus oxygen analysis and verified by in-reactor constant extension rate tests (Ref 20).

Potentials obtained during the normal Dresden-2 environment were characterized by an electrochemical potential for type 304 stainless steel of ~0 mV_{SHE} (standard hydrogen electrode), while the hydrogenated Dresden-2 environment reduced the type 304 stainless steel potential to ≤ -280 mV_{SHE}. This drop in potential at Dresden-2 and the data obtained at other BWRs and in the laboratory are illustrated in Fig. 13.

Crack growth data versus time and environment on precracked furnace-sensitized type 304 stainless steel were also obtained at Dresden-2 by using the reversing direct current (dc) electrical potential monitoring technique (Ref 21). The specimen was precracked in the laboratory in the nominal 0.2-mg/L O_2 and then transported to Dresden-2 for testing. The stress intensity factor (K_I) level for the specimen resembles that of cracks previously identified in a sensitized Dresden-2 piping-to-nozzle transition (safe end) at 27.5 MPa$\sqrt{m}$ (25 ksi$\sqrt{in.}$). The results of this test revealed that no significant crack extension occurred (Fig. 14). This result verifies the result of the mid-cycle and end-cycle in-service inspection of known pipe cracks at Dresden-2 that also did not grow during this 18-month hydrogen water chemistry exposure.

The laboratory hydrogen water chemistry materials program was characterized by an extensive test matrix (Ref 23). The testing techniques utilized in the program included full-size pipe tests; fracture mechanics studies; electrochemical investigations; constant extension rate tests; straining electrode tests; constant load tests; bent beam tests; fatigue testing; cyclic crack growth studies; general, galvanic, and crevice corrosion investigations; and corrosion oxide analysis. All of these laboratory studies clearly demonstrated that hydrogen water chemistry, which is current-

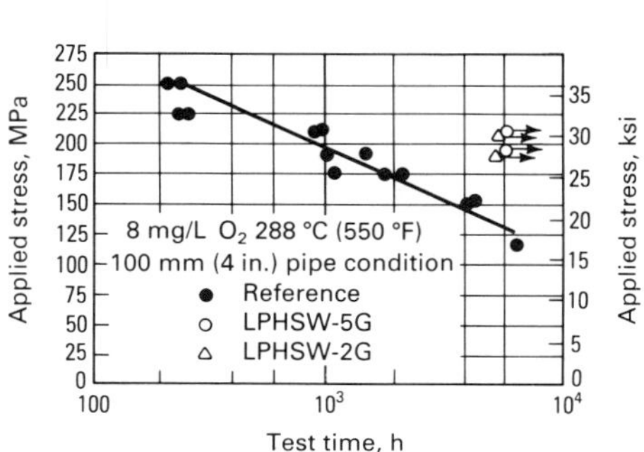

Fig. 12 Comparison of last pass heat sink welding (LPHSW) pipe tests with reference pipe tests

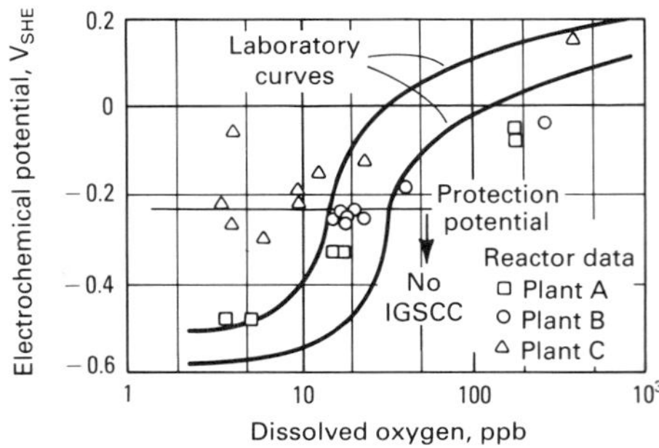

Fig. 13 Relationship between dissolved oxygen and electrochemical potential for type 304 stainless steel

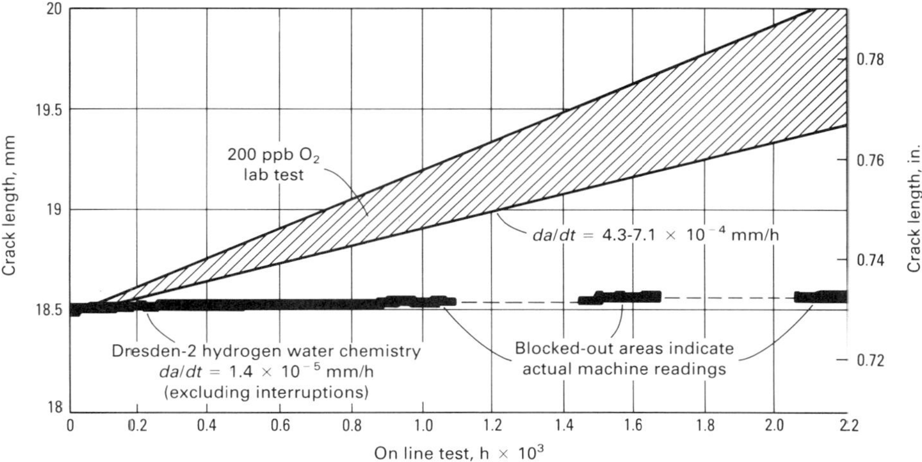

Fig. 14 Electrical potential monitoring results on furnace-sensitized type 304 stainless steel in Dresden-2 BWR. Crack growth is reduced by a factor of approximately 50. Source: Ref 22

ly defined as an electrochemical potential of type 304 stainless steel of <280 mV_{SHE} and water conductivity of <0.3 μS/cm, mitigates many forms of environmental cracking.

Summation of Intergranular SCC Mitigation

The above discussion has indicated the following conclusions concerning the mitigation of intergranular SCC in BWR piping systems and its impact on future BWR performance.

- The causes of the intergranular SCC of BWR piping (material, tensile stress, and environment) are understood and can be prevented
- Several countermeasures/mitigation techniques have been qualified and are available. To achieve a greater margin against intergranular SCC, more than one remedy should be applied to a plant, such as the use of type 316NG stainless steel and hydrogen water chemistry
- The contribution of intergranular SCC in piping to BWR plant unavailability can be expected to decrease in the future if one or more of the qualified countermeasures are implemented by the BWR utilities

SCC in Nitrided Stainless Steel

Positive core reactivity control is maintained in the domestic BWR by the use of movable control rods interspersed throughout the core. The rods are moved vertically by hydraulically actuated, locking-piston drive mechanisms. The drive mechanisms are bottom-entry upward-scramming drives that are mounted on a flanged housing on the reactor vessel bottom head (Fig. 15). The control rod drive mechanism is a double-acting hydraulic piston that uses condensate water as the operating fluid. An index tube and a piston, coupled to the control rod, are locked at fixed increments by a collet mechanism. One of the key components of this device is the collet retainer tube (CRT).

The CRT is a short tube welded to the upper end of the drive. Its three primary functions are to carry the hydraulic unlocking pressure to the collet piston, to provide an outer cylinder, with a suitable wear surface for the metal collet piston rings, and to provide mechanical support for the guide cap, a member that incorporates the cam surface for holding the collet fingers open and provides the upper rod guide or bushing.

The CRT (Fig. 16) is fabricated from wrought type 304 stainless steel. The inner surface, where contact is made with the collet piston seal ring, is hard surfaced by nitriding. Under normal conditions, the stresses on the CRT are quite small. However, during rapid (2.5 s) control rod insertions (scrams), the tube experiences high thermal stresses when the outer surface is initially subjected to a sudden flow of 288-°C (550-°F) water. The normal CRT operating range is 65 to 150 °C (150 to 300 °F). The water flow pattern during a scram is such that the inside of the CRT is cold, while the outside has alternating zones of hot and cold water.

Field Experience With the CRT. During a routine maintenance inspection of disassembled control rod drives at a domestic BWR in June 1975, fine cracks were observed by liquid-penetrant examination in a limited number of CRTs (Fig. 17). Subsequent inspection of drives in other operating BWRs also disclosed occasional occurrences of CRT cracking. Inspection of several test drives having simulated service histories for an excess of the 300 hot scram design life also indicated evidence of cracking. The cracks, which were determined to be intergranular, were generally circumferential and appeared with greatest frequency below and between the cooling ports in the area of the change in wall thickness (Fig. 18).

Mechanism of Cracking of the CRT. A detailed analytical and experimental investigation was performed to determine the cause of cracking and to assess the margin to failure (Ref 14). As was the case with stainless steel piping, the cracking was intergranular SCC produced by the simultaneous combination of sensitization, as discussed above, due to the nitriding thermal cycle, tensile stress, and highly oxygenated water. Specifically, it was determined that there were four sources of thermal stresses near the location of the cracks (Ref 24):

- Thermal shock stresses due to the sudden increase in temperature
- Bending stresses due to the through-wall temperature gradient
- Discontinuity stresses due to the difference in heat-up rates between the thick and thin sections
- Bending stresses due to circumferential temperature gradients

The environment of the control rod drive is typically supplied by the condensate storage tank. This cooling water was often characterized by a conductivity of ~1 μS/cm (25 °C, or 77 °F), and the typical dissolved oxygen content is 5 mg/L. This high-oxygen/high-conductivity environment will accelerate intergranular SCC.

Mitigation of Intergranular SCC in the CRT. It has been shown that the CRT cracking was due to intergranular SCC caused by high thermal stresses with material sensitized by nitriding plus exposure to a high-oxygen water environment. Therefore, the problem can be mitigated by attacking each one of these three necessary factors.

To eliminate the high thermal stresses, the design of the CRT was changed by installing a two-piece CRT, which eliminated the section change and thus reduced the discontinuity stress. To reduce stresses further, the new design (Fig. 18a) has 12 cooling water exit holes rather than the original 3. The effect is to reduce the circumferential variation in temperature below the flow holes during a scram cycle.

The CRT material was changed from wrought type 304 stainless steel to a centrifugally cast CF-3 with ferrite control, a proven BWR material with known resistance to intergranular SCC. The newer-design CRT is also hardfaced with Colmonoy 6, a nickel-base hardfacing material applied by the flame spray process rather than by nitriding.

The environment was changed by a recommendation to BWR operating plant utilities to use high-purity deaerated water (conductivity: ≤0.1 μS/cm, measured at 25 °C, or 77 °F; dissolved oxygen: <5 mg/L) for control rod drive cooling water. This reduction in dissolved oxygen can increase the time to crack initiation of the older CRT by a factor of 100. The recommendation identified approaches to achieving water quality limits, water sampling, and monitoring additions.

The main conclusions of this case study can be summarized as follows:

- The intergranular SCC observed in the CRT was due to the simultaneous interaction of sensitized material, high thermal stresses, and highly oxygenated water
- The problem has been mitigated by eliminating sensitization by a change of material to CF-3, by redesigning to reduce the discontinuity stresses and circumferential stresses, and by reducing the oxygen content and conductivity of the environment
- The intergranular SCC of BWR collet retainer tubes is understood, and a solution is now available

SCC of Alloy X-750 Jet Pump Beams

Jet pumps located within the reactor vessel are used in the BWR recirculation system (Fig. 3). The jet pumps, which have no moving parts, provide a continuous circulation path for a major portion of the core coolant flow. The water from the recirculation system flows up the riser pipe, reverses direction through the 180° bend nozzle, and draws surrounding water in the downcomer

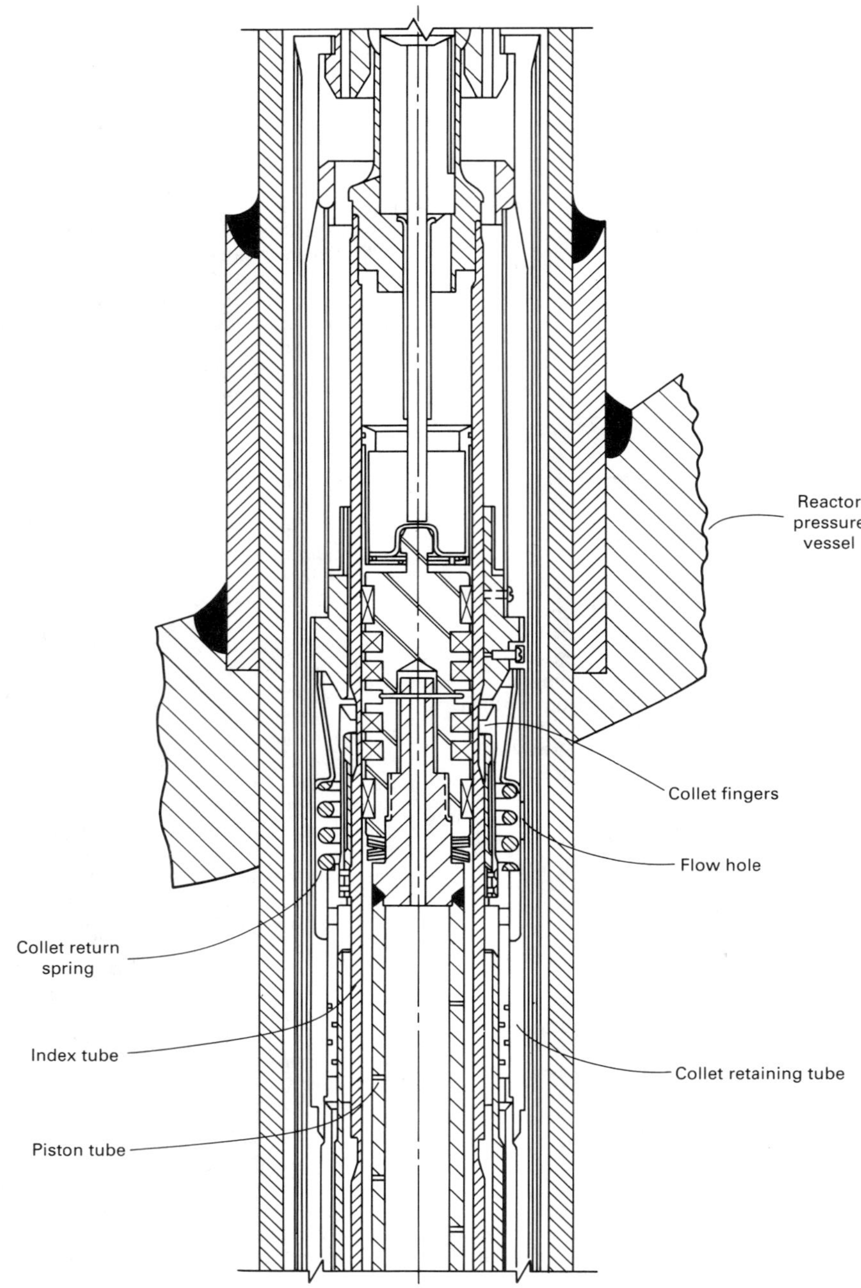

Fig. 15 Control rod drive schematic

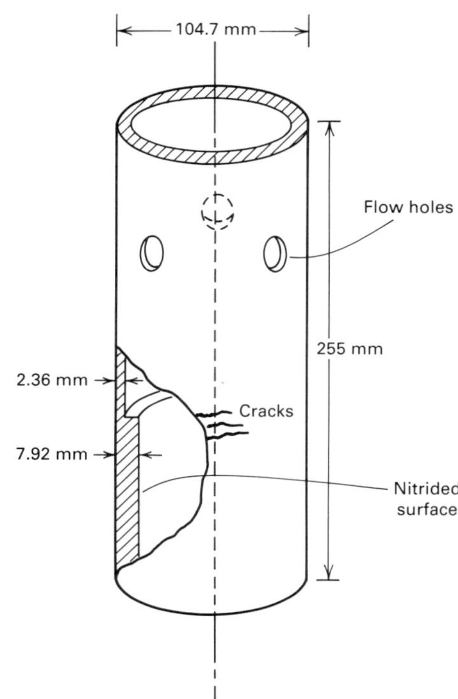

Fig. 16 Geometry of the collet retainer tube

Fig. 17 Intergranular SCC of collet retainer tube revealed by liquid-penetrant examination

region into the jet pump throat, where the two flows mix and then diffuse in the diffuser (Fig. 19).

Field Experience With Jet Pump Beams. The inlet mixer section of each jet pump, from the entrance to the 180° bend to the diffuser slip joints, is removable for inspection. The inlet mixer is held in place by a beam-bolt assembly located in the riser transition piece and is centered in the diffuser by a three-point contact of the restrainer bracket. The beam ends are positioned in pockets in the transition piece, and the beam load is transferred to the elbow through a bolt located in the center of the beam.

In mid-1979 and early 1980, a jet pump hold-down beam fabricated from Alloy X-750 containing 70% Ni (min), 14 to 17% Cr, and 5 to 9% Fe failed during plant operation at one overseas and one domestic BWR plant (Fig. 20). Metallurgical examinations of the failed beams revealed that intergranular SCC under sustained loading was responsible for the failure. Figure 21 shows a detailed illustration of the beam-bolt assembly and the beam top view.

Mechanism of Cracking of the Jet Pump Beam. The Alloy X-750 jet pump beams had received the following heat treatment: heating to 885 °C (1625 °F), holding at this temperature for 24 h (equalizing), followed by aging at 705 °C (1300 °F) for 20 h. This heat treatment is known as equalized and aged.

A stress analysis revealed that the maximum stress occurs at the beam mid-plane top surface. The results of this analysis are consistent with examination of actual beams with cracks in various stages of propagation. The older beam model, using a preload of 111.2 kN (25 kips), resulted in a maximum tensile stress of 593 MPa (86 ksi), but the newer-style beam with a larger cross section, using a preload of 133.5 kN (30 kips), resulted in a maximum tensile stress of 496 MPa (72 ksi).

As was the case for BWR piping, the 0.2-mg/L oxygenated 288-°C (550-°F) water in the BWR was sufficient to provide the chemical driving force for intergranular SCC of the jet pump beams. To quantify the effects of the environment on Alloy X-750, creviced and uncreviced tensile specimens were tested at constant load in air-saturated 8-mg/L oxygenated water and typical BWR 0.2-mg/L oxygenated water at 288 °C (550 °F) at various applied stress ratios (applied stress/yield stress).

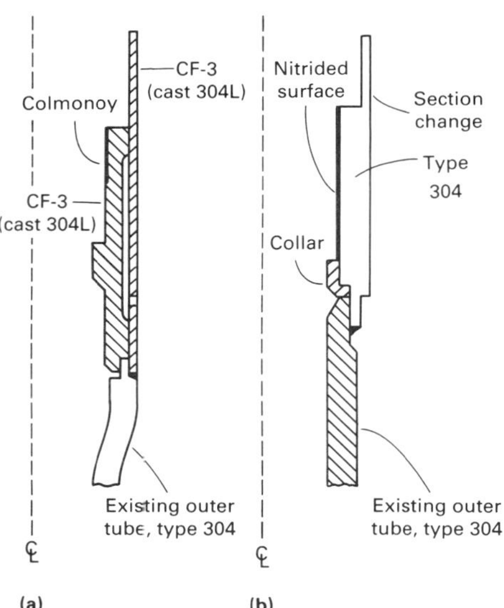

Fig. 18 Collet retainer tube improvements. (a) Improved design. (b) Old design

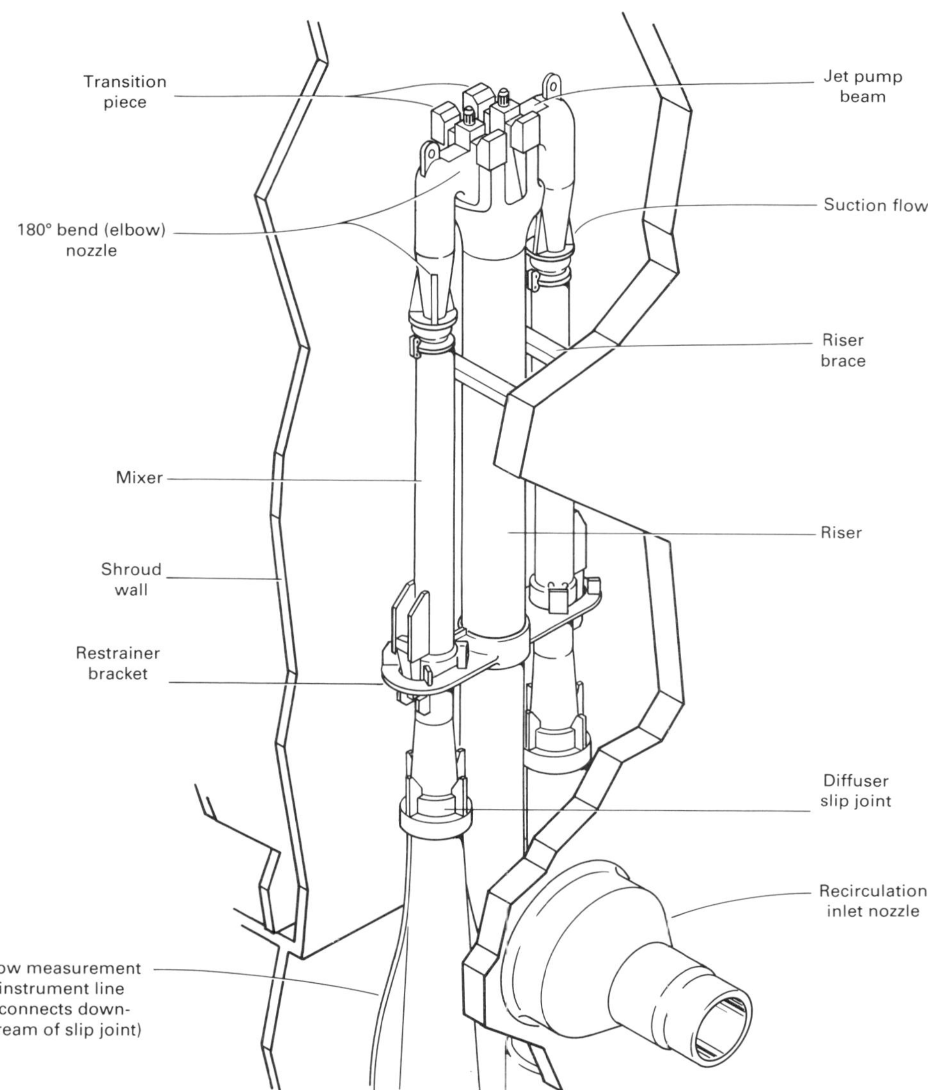

Fig. 19 Jet pump assembly

Mitigation of Intergranular SCC in Jet Pump Beams. A two-pronged attack was used to solve the intergranular SCC problem. Both the material and stress factors of the intergranular SCC three intersecting circles shown in Fig. 1 are addressed.

Extensive research and testing revealed that high-temperature annealing (1095 °C, or 2000 °F, for 1 h) can reduce the susceptibility of Alloy X-750 to intergranular SCC in oxygenated water. It is believed that this occurs because of the creation of new grain boundaries that are free of precipitates and impurities. By eliminating the equalizing step (885 °C, or 1625 °F, for 20 h) and retaining the direct aging step at 705 °C (1300 °F) for 20 h, the material attains high strength and is much more resistant to intergranular initiation (Ref 25). This heat treatment is known as high-temperature annealing and aging. It can also be applied to previously heat-treated beams and does not affect performance in hydrogen water chemistry environments (Ref 26).

To address the tensile stress component, evaluations revealed that the preload could be reduced on the newer-type jet pump beam from 133.5 to 111.2 kN (30 to 25 kips). This reduction in stress alone would result in a significant increase in lifetime of the equalized and aged heat-treated Alloy X-750 from a calculated average 13 years to 19 to 40 years or more for a ≤2.5% probability of crack initiation. The synergistic combination of lower stress and the high-temperature annealing and aged heat treatment will decrease the probability of cracking even further—to more than 40 years for a ≤2.5% probability of crack initiation.

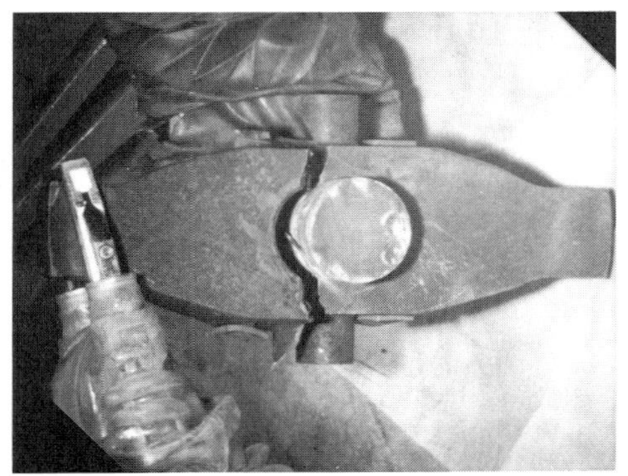

Fig. 20 Failed jet pump beam due to intergranular SCC. Approximately 1/3×

Conclusions. The primary conclusions of this case study are as follows:

- The intergranular SCC observed in the Alloy X-750 jet pump beam was a result of simultaneous interaction of a susceptibly heat-treated material, heat treatment, sufficient preload tensile stress, and high-temperature oxygenated water
- The problem has been mitigated by changing the heat treatment of Alloy X-750 beams and reducing the preload
- It is anticipated that the cracking of jet pump beams will not be a concern during future BWR operation, because all BWR utilities have addressed this problem by beam replacement, reheat treatment, or preload reduction

Irradiation-Assisted SCC

Irradiation-assisted SCC is an intergranular SCC phenomenon that can occur in BWR internals. Irradiation-assisted SCC does not require the presence of a sensitized microstructure or high tensile stresses.

The intergranular SCC described in the discussion on "Mechanism of Intergranular SCC in BWR Piping" in this article was the result of the simultaneous interaction of sensitized stainless steel (that is, chromium depletion at the grain boundaries), high tensile stresses (weld residual, pressure, and thermal), and oxygenated high-temperature water. The mechanism of irradiation-assisted SCC appears to involve the simultaneous interaction of highly irradiated nonsensitized material with diffusion of impurities (sulfur, silicon, and phosphorus) to the grain boundaries, low stress (fabrication, irradiation creep) and high-temperature water with short-

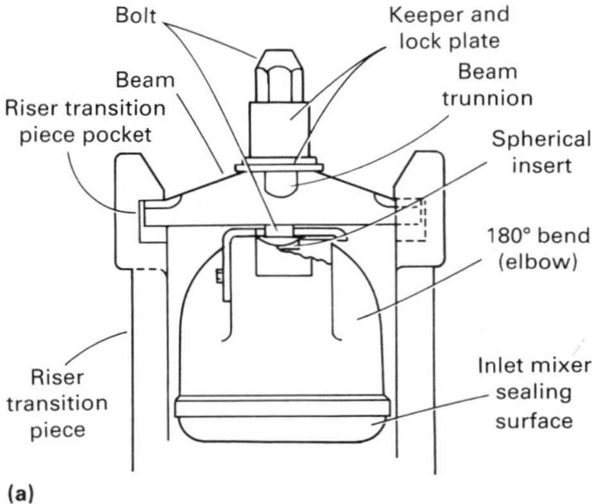

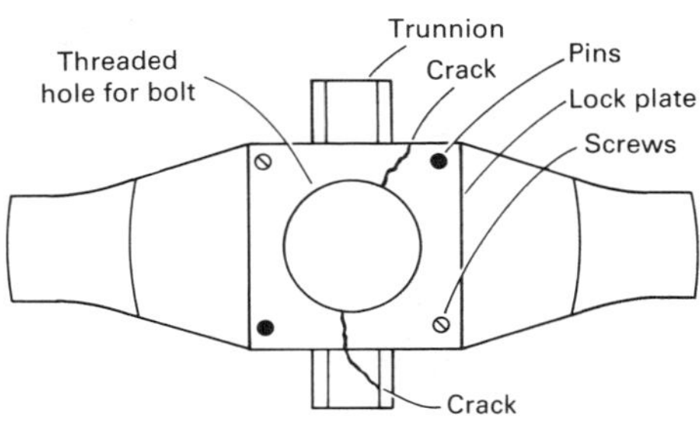

Fig. 21 Jet pump beam/bolt assembly (a) and beam top view (b)

lived oxidizing species (hydrogen peroxide, H_2O_2), gamma, and neutron flux. It is extremely important to note that irradiation-assisted SCC does not require the chromium depletion sensitization or high tensile stresses that are implicit in the failure of nonirradiated stainless steel components.

Field Experience With Irradiation-Assisted SCC. During the early history of the BWR (1960s), type 304 stainless steel was used as a fuel cladding material. Because this highly stressed material suffered extensive cracking, the type 304 stainless steel was replaced with Zircaloy-2. It is believed that this instance of cracking was the first indication that annealed type 304 stainless steel could suffer intergranular SCC in the BWR core environment. Because this type of intergranular cracking was produced in nonsensitized, highly stressed, highly irradiated components, this form of intergranular SCC is referred to as irradiation-assisted SCC. (As noted above, high tensile stresses are not necessary for irradiation-assisted SCC, but the presence of high tensile stress certainly exacerbates the phenomenon.)

Recently, core components such as type 304 stainless steel neutron source holders, control blade absorber tubes (Fig. 22), and nuclear instrument tube holders have cracked in the BWR environment. Table 1 presents a summary of significant field irradiation-assisted SCC experience. To date, the cracking has been limited to readily replaceable components. It is also important to note that PWRs have also experienced this type of corrosion phenomenon (Ref 27, 28). The results of the field irradiation-assisted SCC experience suggest the following:

- Dynamic strain or high stresses produced irradiation-assisted SCC in fuel cladding and absorber tubing

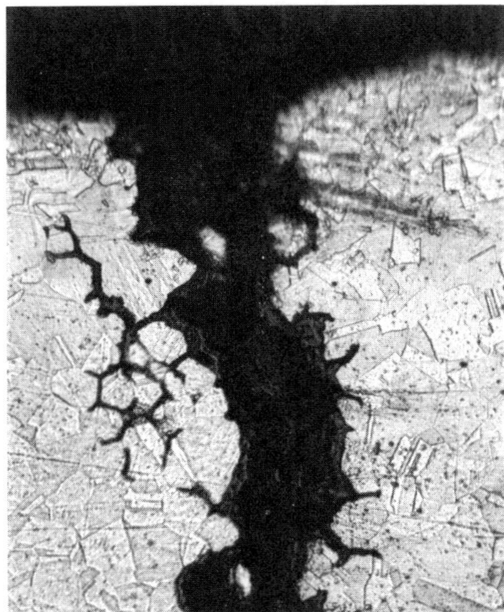

Fig. 22 Irradiation-assisted SCC of a control blade absorber tube. 500×

- More recent results with lower-stressed neutron source holders and instrument tubes indicate that the stress threshold for irradiation-assisted SCC may be lower than initially believed
- No field irradiation-assisted SCC has been observed at a fluence below about $\leq 5 \times 10^{20}$ NVT (neutrons per unit volume multiplied by time) (>1 MeV)
- Cracks may occur at lower stresses for higher fluences

Mechanism of Irradiation-Assisted SCC. The mechanism of irradiation-assisted SCC appears to involve highly irradiated annealed austenitic stainless steel with diffusion of impurities (sulfur, silicon, phosphorus) to the grain boundaries, low stress (fabrication, irradiation creep), and a highly oxidizing environment associated with the gamma and neutron flux (Ref 27, 28). Partial chromium depletion is also possible. To simulate these highly oxidizing conditions in the core, tests were performed on irradiated annealed type 304 stainless steels at different levels of fluence. The irradiation-assisted SCC phenomenon was reproduced, and the results indicate that the observed field irradiation-assisted SCC fluence threshold is consistent with the laboratory tests.

Mitigation of Irradiation-Assisted SCC. Fortunately, the same environmental solution for the intergranular SCC of piping (hydrogen water chemistry) appears promising for this phenomenon. Table 2 presents the results of SCC tests on highly irradiated commercial-purity annealed type 304 stainless steel in a high-oxygen (32 mg/L) environment simulating the highly oxidizing regions in the core and in hydrogen water chemistry. The difference in SCC response is dramatic. The highly oxidizing environment produced a fracture surface with 99% intergranular SCC, while the more reducing environment characteristic of hydrogen water chemistry produced ductile failure. The reason for this significant disparity may be explained by the electrochemical potential of type 304 stainless steel, which is significantly more oxidizing in the core region

Table 1 Summary of significant field irradiation-assisted SCC experience

Stainless steel component	Fluence (neutrons/cm^2)	Source of stress
Fuel cladding	5×10^{20} to 2×10^{21}	Fabrication; fuel cladding interaction
Neutron source holders	10^{21} to 10^{22}	Welding; beryllium swelling after initial crevice attack
Control rod absorber tubes..........	5×10^{20} to 3×10^{21}	B_4C swelling
In-core instrument tubes..........	10^{21} to 10^{22}	Fabrication

Table 2 Hot cell SCC test results on irradiated type 304 stainless steel

Environment	Fluence	Fracture
32 ppm O_2	2×10^{21} NVT (>1 MeV)	99% Intergranular SCC
Hydrogen water chemistry......	2×10^{21} NVT (>1 MeV)	Ductile

Note: NVT = neutrons per unit volume × time.

(~250 mV_{SHE}) than in the recirculation piping system (≈30 mV_{SHE}). Hydrogen water chemistry obtained by lowering the electrochemical potential appears to be a promising method of mitigation for irradiation-assisted SCC if sufficient reduction of electrochemical potential can be achieved in the core.

The second promising technique for mitigating irradiation-assisted SCC is the use of high-purity materials containing controlled amounts of such impurities as sulfur, phosphorus, and silicon (Ref 27). This mitigation technique is based on early laboratory testing in a highly oxidizing medium, such as boiling nitric acid (HNO_3) with Cr^{6+} ions, and on the highly successful in-reactor performance of high-purity type 348 stainless steel fuel cladding at the La Crosse BWR. In contrast, commercial-purity type 348 stainless steel materials with the nominal commercial levels of sulfur, phosphorus, and silicon installed at the La Crosse BWR suffered irradiation-assisted SCC. Other support for the implementation of high-purity materials is based on in-reactor studies (Ref 29) and on laboratory studies on nonirradiated stainless steel where the presence of these specific impurities increased the intergranular SCC susceptibility (Ref 30). Work is continuing in this area.

Conclusions. The results of in-reactor and laboratory investigations suggest the following conclusions concerning irradiation-assisted SCC in the BWR:

- The irradiation-assisted SCC of components in the BWR core was the result of the simultaneous interaction of irradiation-enhanced impurity segregation, tensile stress, and highly oxidizing environment
- Preliminary studies suggest that hydrogen water chemistry and high-purity materials are promising remedies for mitigating irradiation-assisted SCC

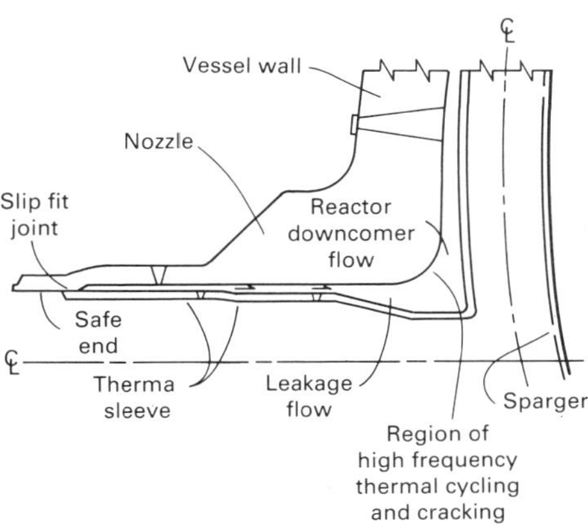

Fig. 23 Cross section of feedwater nozzle with cracking location

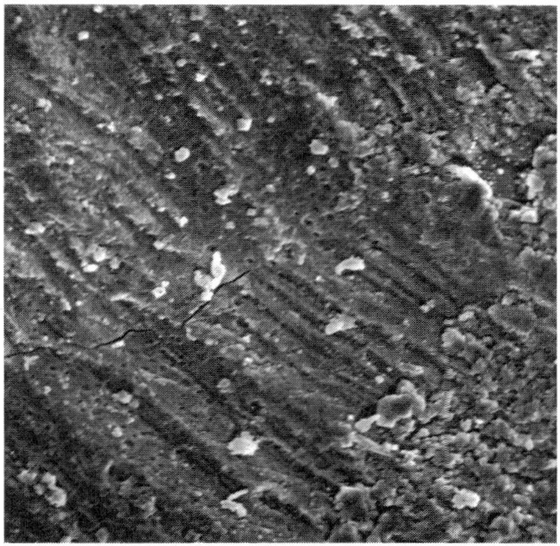

Fig. 24 Example of corrosion fatigue striations in a feedwater nozzle. 1500×

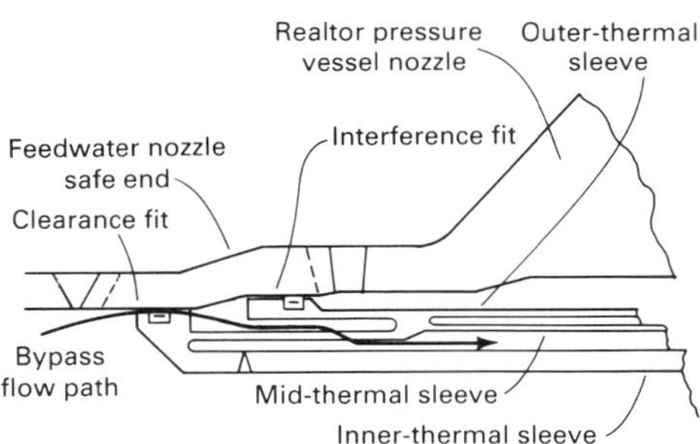

Fig. 25 Improved feedwater thermal sleeve design to eliminate corrosion fatigue

Case History: Corrosion Fatigue in Feedwater Nozzles

Fatigue evaluation of nuclear power plant equipment subjected to cyclic loading conditions is of primary concern to equipment designers and stress analysts. Because the number of loadings resulting in significant stresses that occur over the life of light water reactors seldom exceed several thousand, and in numerous cases do not even exceed several hundred, fatigue usage is generally classified as low cycle. However, in this particular case history on feedwater nozzle cracking, both low- and high-cycle fatigue were involved. Figure 23 shows a cross section of a feedwater nozzle that suffered from cracking of the cladding of the feedwater nozzle bed radii, a problem common to a number of BWRs.

Field Experience. Metallurgical examination of a boat sample (a wedge-shaped specimen cut out of a component) removed from the cladding of the feedwater nozzle bend radii revealed that the primary cause of the nozzle crack was thermal-induced corrosion fatigue (Fig. 24). It was further determined that there are two corrosion fatigue mechanisms present: a high-cycle corrosion fatigue mechanism, which initiates the cracks, and a low-cycle corrosion fatigue mechanism, which causes the cracks to propagate.

Mechanisms of Corrosion Fatigue in Feedwater Nozzles. The high-cycle mechanism was found to be primarily caused by leakage flow passing between the thermal sleeve and safe end. This leakage flow, which is at feedwater temperature, mixes in a turbulent manner with hot downcomer flow in the annulus between the nozzle and thermal sleeve. The mixing fluid impinges on the nozzle wall, causing thermal cycling of the metal surface. It has been determined by test and by field measurements at two BWRs that the metal temperature cycling, with leakage present, has a magnitude of up to 50% of the difference in temperature between the feedwater and the downcomer water. The cycling occurs with frequencies between 0.1 and 1 Hz and thus can initiate cracking rapidly with little if any environmental contribution. The exact time to crack initiation depends on several factors, including the duration of operation with low feedwater temperature.

The cracks initiated by the high-frequency cycling described above will arrest at a depth of approximately 6 mm (0.25 in.) from the metal surface. This arrest results from the fact that the high-cycle thermal input induces thermal stresses with steep gradients and shallow depths. Cracks that arrest, as described above, present no problem from an engineering standpoint, because they will not degrade safety or availability and would not have to be repaired under the rules of Section XI of the American Society of Mechanical Engineers Code.

There is another mechanism present in the feedwater nozzles that causes the cracks to continue to grow. This mechanism is the combined pressure and thermal cycles imposed by start-up/shutdown and feedwater on-off transients. These transients, although relatively few in number, produce large stress cycles in the nozzle and in time could drive the cracks to significant depths. This low-cycle fatigue crack propagation is environmentally accelerated under BWR conditions. The deepest cracks observed were up to 18% of wall and relatively short.

Fortunately, the cracking is readily detectable by dye-penetrant examination and/or ultrasonic examination from the outside of the nozzle. The deep cracks require repair and thus can result in a significant impact on plant availability and operating cost.

Leakage between the safe end and thermal sleeve also has an aggravating effect on the crack growth rate because it increases the heat transfer coefficient between the feedwater and the nozzle. The increased heat transfer coefficient increases the stresses in the nozzle during thermal transients, and because crack growth is dependent on stress to the fourth power, a significant effect results.

Mitigation of Cracking. Several solutions were derived to mitigate this problem from both a design and a corrosion/metallurgical viewpoint. The solution consisted of three parts: a revised sparger thermal sleeve design, use of unclad feedwater nozzles, and a revised system of configuration/operating procedures that mitigates the conditions tending to produce crack initiation and growth. Together these three elements constitute a solution to the problem with margin for unexpected conditions. The implementation of hydrogen water chemistry would also mitigate this problem because test results have indicated a decrease in corrosion fatigue crack propagation rate in hydrogen water chemistry (Ref 26).

A sparger thermal sleeve design has been developed that meets the following objectives:

- It can be installed and removed without cutting feedwater piping
- It protects the feedwater nozzle against the high-frequency thermal cycles that initiate nozzle cracks through the use of redundant metal O-ring seals
- It uses materials and processes that make the part immune to intergranular SCC

The resulting design is schematically shown in Fig. 25. The thermal cycling profiles for the previous and the newer-design feedwater spargers are shown in Fig. 26. The marked reduction in thermal cycling is evident.

The main conclusions concerning the corrosion fatigue cracking of BWR feedwater nozzles are:

- The corrosion fatigue of BWR feedwater nozzles was a result of the interaction of high-cycle fatigue crack initiation due to leakage flow passing between the thermal sleeve and safe end and low-cycle corrosion fatigue crack growth due to start-up/shutdown and feedwater on-off transients
- The cracking problem was mitigated by a redesign of the feedwater thermal sleeve to eliminate leakage and the removal of the nozzle cladding. Hydrogen water chemistry will also provide an additional margin against this phenomenon

Steam Generator Failure or Degradation

Stanley J. Green
Electric Power Research Institute

Steam generators in PWR power plants transfer heat from a primary coolant system (pressurized water) to a secondary coolant system. Primary coolant water is heated in the core and passes through the steam generator, where it transfers heat to the secondary coolant water to make steam. The steam then drives a turbine that turns an electric generator. Steam is condensed and returns to the steam generator as feedwater.

Two types of PWR steam generators are in use: recirculating steam generators (RSGs) and once-through steam generators (OTSGs). Most of the units are vertical, and this Section will be limited to vertical units. Some of these steam generators have operated with a minimum of problems while

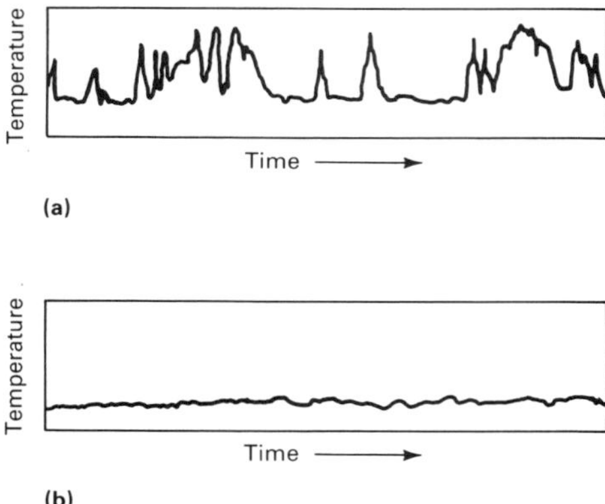

Fig. 26 Temperature variations with (a) and without (b) bypass leakage

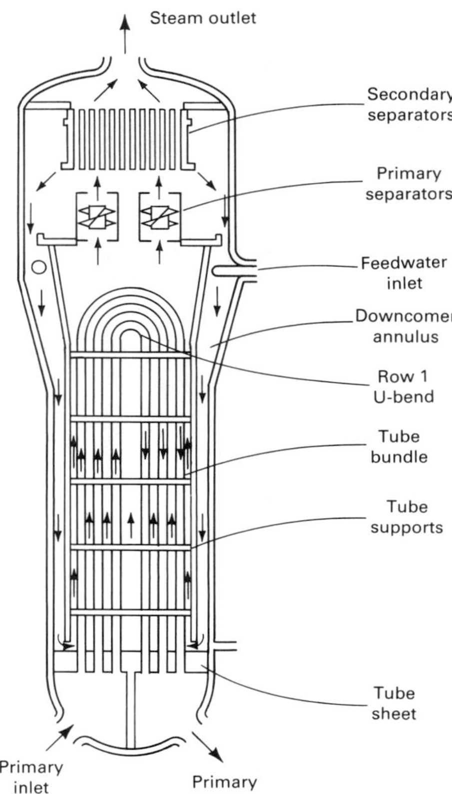

Fig. 27 Recirculating steam generator

other steam generator designs have experienced a variety of corrosion-induced and mechanically induced problems. The discussion will focus more on those designs that have experienced problems and where effort has been expended to correct them. The corrosion problems include denting, wastage, intergranular attack, SCC and pitting on the outside surfaces of the tubes, and SCC from the inner surfaces of the tube. The mechanical concerns have included water hammer, thermal stratification in feedwater pipes, fretting and wear of the tubes caused by excessive tube vibration, and erosion-corrosion. These problems have caused unscheduled outages and expensive repairs. Where most extensively affected, some steam generators have been replaced after 8 to 12 years of operation, which is far short of the expected plant operating period of 40 years.

The scope of this Section will be limited to the corrosion-related issues. A brief summary of other degradation phenomena is given at the end of this Section, and previous summaries of these issues are available in Ref 31 to 34. Also, specific preventive and corrective actions are discussed in Ref 35.

Steam Generator Design

A typical vertical recirculating steam generator (feedring-type) is shown in Fig. 27. (Other designs employ preheat features, where the cold feed is introduced into the lower part of the bundle.) Water is fed into the downcomer, where it is mixed with two to five volumes of recirculating water from the moisture separators. The downcomer water flows to the bottom of the steam generator, across the tubesheet, and then upward through the tube bundle where steam is generated.

The thermodynamic quality of the water-steam mixture at the top of the bundle is about 17 to 33% when it enters the steam separators, which corresponds to a circulation ratio in the range of 6:1 to 3:1. The pressure on the secondary side is about 4135 to 7240 kPa (600 to 1050 psia). The primary coolant flows through U-tubes at a pressure of about 9655 to 15515 kPa (1400 to 2250 psia). It enters the steam generator at about 310 to 325 °C (590 to 620 °F) and leaves at about 255 to 290 °C (495 to 550 °F). At the primary inlet, the temperature difference across the tube wall is about 35 to 50 °C (65 to 90 °F), corresponding to a heat flux of 315 460 to 441 640 W/m^2 (100 000 to 140 000 Btu/h·ft^2). At the primary outlet or cold side, the temperature difference between the primary and secondary sides is about 10 to 15 °C (20 to 25 °F), corresponding to a heat flux of about 94 640 W/m^2 (30 000 Btu/h·ft^2). The Westinghouse, Combustion Engineering, Kraftwerk Union (KWU), Framatome, and Mitsubishi designs have comparable operating parameters, while the Babcock and Wilcox (B&W)/Atomic Energy of Canada Limited (AECL) design operates at lower temperatures and pressures.

Knowledge of the materials, water chemistry, and tube support arrangements used in RSGs is required to understand the problems that have occurred. The tubes have been made primarily of Alloy 600, a nickel, chromium, iron alloy. In KWU and later B&W/AECL designs, the tubes have been made of Alloy 800, an iron-base superalloy. The mill-annealing conditions vary among the manufacturers, while B&W/AECL have used stress-relieved (605 °C, or 1125 °F, for 8 h) tubing for their Alloy 600 tubed steam generators. Some of the more recent designs using Alloy 600 tubing have thermally treated the tubing (705 °C, or 1300 °F, for 15 h) to improve resistance to SCC.

The tube support structures for most of the early units were made of carbon steel, while later units have switched to type 405, 409, and 410 stainless steels for additional corrosion resistance. Type 347 stainless steel has always been used for KWU steam generator tube support structures.

Tube support structures of early units used tube support plates with drilled holes (Fig. 28a and b), plates with broached holes (Fig. 28c), and lattice bars or egg crates (Fig. 28e). When the drilled-hole support plates were found to promote accumulation of corrodents, they were changed to a quatrefoil design hole with lands (Fig. 28d) or to a lattice support structure (Fig. 28e).

In some of the early designs, the tubes were only partly expanded into the lower end tubesheet, leaving a crevice between the outside diameter of the tube and the inside diameter of the hole in the tubesheet, that is, in the upper part of the tubesheet; in other designs, the tubes were expanded into the tubesheet along its full length. The tubes have been expanded into the tubesheet by mechanical, hydraulic, and explosive expansion methods. Generally, the tubes have been expanded for the full length of the tubesheet in the later designs. Most of the early RSGs employed coordinated sodium phosphate water treatment for the secondary side, which is the conventional water treatment method for fossil-fired boilers with similar steam pressures.

An OTSG is shown in Fig. 29. Water enters a feed annulus above the ninth tube support plate level. There it is mixed with steam aspirated from the tube bundle area and preheated to saturation. The saturated water flows down the annulus, across the lower tubesheet, and upward into the tube bundle, where it becomes steam. It reaches 100% quality (on the average) in the ninth and tenth support plate region, and achieves about 20 to 35 °C (40 to 60 °F) of superheat at about 6370 kPa (924 psia) at the top of the unit. The superheated steam flows radially outward and then down the annulus to the steam outlet connection. The primary coolant flow is from top to bottom. It enters at about 315 to 325 °C (600 to 620 °F) and leaves at about 290 to 295 °C (555 to 560 °F). The temperature difference between the primary and secondary sides at the bottom of the steam generator is similar to that on the cold leg of a recirculating steam generator.

Once-through steam generator materials are similar to those used in RSGs. Tubes are Alloy 600 in a mill-annealed plus a stress-relieved condition, which sensitizes the tubing (causes chromium-depleted grain boundaries). Tube support plates are carbon steel. However, the holes are not drilled round holes as in the early RSGs, but trefoil holes with three lands supporting the tubes (Fig. 28c). Tubesheets are low-alloy steel. Once-through steam generators have always used all-volatile water treatment water chemistry.

Prevention of Steam Generator Corrosion Problems

As noted above, these steam generators have experienced a variety of corrosion-induced and mechanically induced problems. The types of damage and the number of units affected are presented in Table 3, which indicates that essentially no units have operated trouble-free for more than 5 years. The industry has made a very substantial effort to prevent and minimize these steam generator problems. This includes a major program by the Steam Generator Owners Group, managed by the Electric Power Research Institute (EPRI), and substantial programs carried out by steam generator suppliers and utilities.

The approaches used to correct or prevent corrosion for both new and operating units are based on addressing the causes. The corrective measures can be divided into the following three categories:

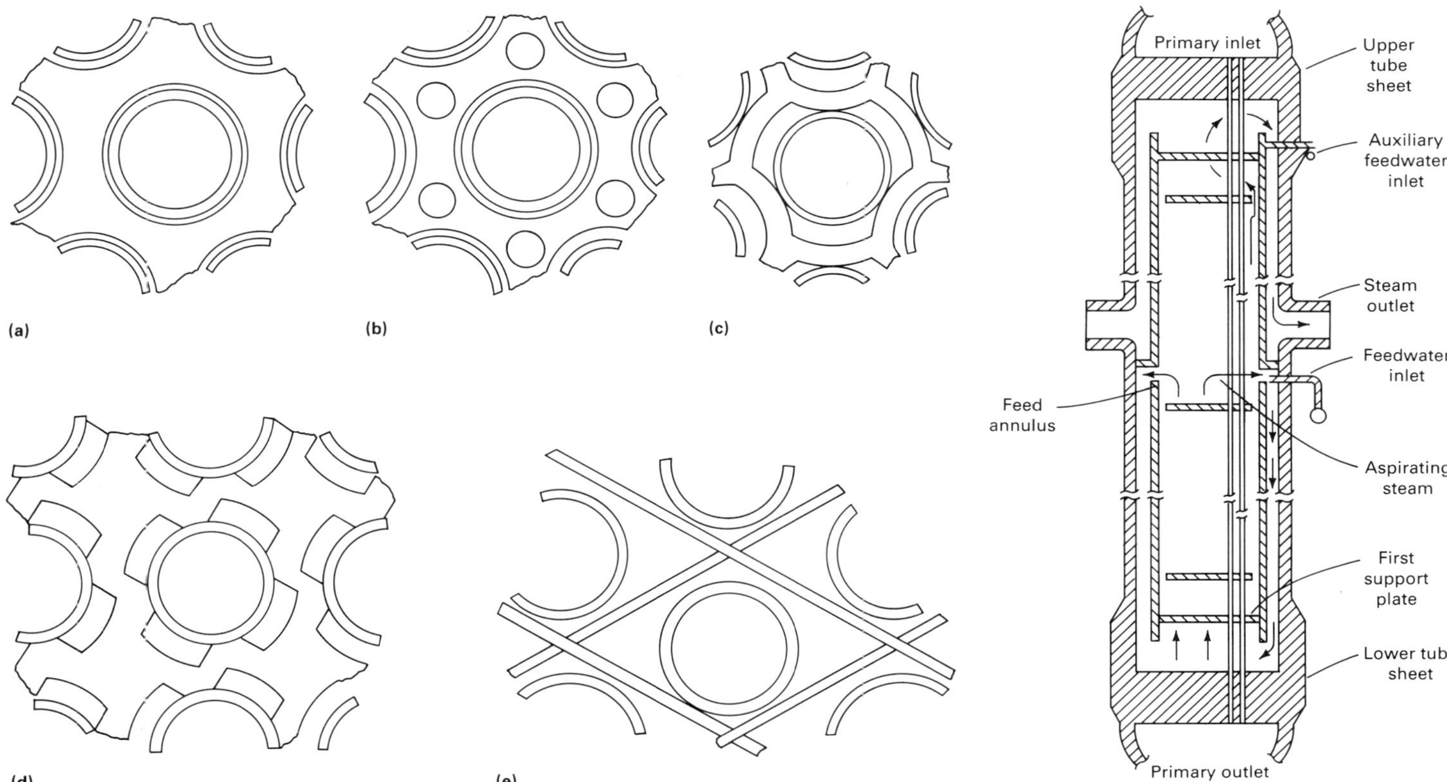

Fig. 28 Tube support device designs. (a) Drilled, without flow holes. (b) Drilled, with flow holes. (c) Broach-trefoil. (d) Broach-quatrefoil. (e) Egg crate

Fig. 29 Once-through steam generator

- Modifying the environment
- Modifying the materials
- Modifying the stresses

The corrosion problems and the types of corrective action developed for each issue are listed in Table 4. Each of the problems and corrective actions is discussed below.

Tube Wastage

Causes. Tube wastage, or thinning, was one of the first corrosion problems that occurred in recirculating steam generators operating with sodium phosphate as a secondary water treatment. Phosphate wastage was first observed in commercial PWR plants when phosphate treatment was changed to a low sodium-to-phosphate molar ratio control, in which the molar ratio of Na/PO_4 was maintained at about 2.0. This change was in response to a series of caustic SCC events attributed to operation with uncontrolled high Na/PO_4 ratios (above 2.8), from which free caustic could result. The incidence of caustic SCC dropped markedly, but the general corrosion, now known as phosphate wastage, began to be observed within approximately 1 year after the change. Some phosphate wastage observations are discussed in Ref 36; a detailed report of such events at two specific nuclear power installations is provided in Ref 37.

The phosphate wastage in PWR steam generators of the Westinghouse design occurred for the most part at the interfaces between hot leg tubes and the tops of sludge piles that accumulated on top of the tubesheet. Where the sludge pile was deep, the zone of wastage extended about 25 mm (1 in.) into the pile; in most cases, it did not penetrate appreciably into the tube/tubesheet crevice. Wastage was first noted in steam generators of the Combustion Engineering design. It was most extensive in the vicinity of antivibration straps, which were relatively wide and oriented in such a way as to define a region of steam blanketing. Corrosion was concentrated at the boundaries of this region, where a liquid/vapor interface presumably fluctuated over a short length of tubing.

As recognition grew that this new phenomenon was widespread and of a generic nature, major laboratory investigations were launched by the two United States vendors of RSGs. It was demonstrated in pot boilers that the location of attack was related to the concentration of aggressive species at steam/water interfaces. Emphasis was placed on investigating the chemistry of sodium phosphate solutions in high-temperature water, in addition to corrosion studies and model boiler tests. The laboratory work showed that the corrosivity of concentrated sodium phosphate solutions was related to both concentration and the Na/PO_4 molar ratio.

The laboratory results led to recommendations for Na/PO_4 molar ratio control in the relatively narrow band of 2.3 to 2.6. The lower limit was related to the rapidly increasing rate of wastage at lower ratios, and the higher limit was selected to avoid free caustic and concomitant caustic SCC. A number of plant operators were successful in controlling their steam generator chemistries within this restricted range of compositions, but eddy-current inspections indicated that the rate of attack was only slowed, not stopped. It appeared that attack could not be stopped unless the operators consistently operated within a narrow control band at the high end of the allowed range. Furthermore, a constantly increasing sludge burden, augmented by precipitated phosphate compounds, made such control increasingly difficult.

Corrective Actions. The above difficulties led the vendors of the affected plants to recommend

Table 3 Units affected by steam generator problems

Total units: 121 (65 units having >5 years of operation and 56 units having <5 years of operation)

Problem	Units affected (as of 1984)
Denting	
Denting at support plates	42
Denting at tubesheet	18
Tubing corrosion	
Wastage	31
Pitting	11
Outside-diameter initiated SCC	31
Inside-diameter initiated SCC	33
Outside-diameter initiated intergranular attack	28
Inside-diameter initiated intergranular attack	2
Mechanical damage	
Fretting	30
High-cycle fatigue	5
Erosion	2
Water hammer	10
Moisture carryover	6
No problems	37
No problems (for 65 units >5 years of operation)	1

Table 4 Corrective actions for steam generator corrosion problems

Problem	Modify environment Operating	Modify environment New	Modify materials Operating(a)	Modify materials New	Modify stress Operating	Modify stress New
Tube wastage	X	...	...	...	...	...
Denting	X	X	...	X	...	...
Inside-diameter SCC	...	...	X	X	X	X
Outside-diameter intergranular attack	X	X	X	X	...	X
Pitting	X	X	X	...	...	...
Corrosion fatigue	X	X	...	...	...	...

(a) By sleeving

that the operators adopt an all-volatile water treatment based on the use of ammonia and hydrazine. This recommendation was almost universally followed in the United States except for two plants. Steam generators manufactured by KWU continued to use phosphate water treatment. These steam generators, tubed with Alloy 800 material, have experienced a slower rate of wastage corrosion. It is not clear whether the slower rate is a function of the alloy, the low concentration of phosphate used, the relatively low Na/PO_4 ratio, or the relatively good impurity control history at several of these units. It is probable that all of these factors are involved. However, KWU has switched to all-volatile water treatment for their new units.

Thus, the cause for wastage was determined to be an aggressive environment, and the corrective action involved a change in the environment for operating units from a phosphate chemistry to all-volatile water treatment. However, as discussed below, this change led to other corrosion problems, namely denting.

Denting

Causes. Denting was discovered in 1975 when eddy-current probes were prevented from passing through tube/tube support plate intersections by tube diameter restrictions. By 1977, denting had become a widespread problem and had resulted in the formation of the Steam Generator Owners Group as a concerted effort to address the problem (Ref 38).

Denting is a term used to describe the localized tube diameter reduction that occurs when the hole of a carbon steel tube support plate corrodes to the point at which the corrosion products deform the steam generator tubing (Fig. 30). The cause of denting is best explained by Potter-Mann-type linear accelerated corrosion, in which a nonprotective oxide layer is formed as the corrosion progresses. These corrosion products, which have a bulk volume considerably larger than the volume of metal corroded, fill the original tube/tube support plate crevice and deform the tube and the tube support plate simultaneously.

Sample intersections of tubes and support plates removed from dented steam generators have shown local chloride concentrations of over 4000 ppm in the dented region. The high local chloride concentration is caused by local thermal-hydraulic conditions within the crevice between the tube and the tube support plate. The source of chloride is generally condenser leakage, particularly at plants cooled by seawater.

Tube deformation has been reported in some steam generators with carbon steel egg crate supports, implying that concentration of chemicals and carbon steel corrosion may also be occurring with this design. Similarly, distorted eddy-current signals (dings) have been observed at the ninth and tenth support plates of OTSGs that use a trefoil design support plate. This is a most likely region for corrosion because it is the region where dryout and most of the deposition of any chemicals present in the feedwater occur. It is the region corresponding to approximately 95 to 100% quality on the average. However, it is possible that these dings may be caused by tube vibration and impact on the tube support plates.

Denting has also been observed at the top of the tubesheet crevice in several RSGs, but it has progressed at a slower rate. Whether this is due to different geometry, the presence of sludge, or different materials (different grades of low-alloy or carbon steel are used in the tubesheet and the support plate) is not known at this time. Denting has not been reported in the KWU steam generators where the support structures are fabricated of type 347 stainless steel or in the B&W/AECL units that have used lattice bars and broached hole plates with both carbon steel and type 410 stainless steel.

Several major consequences can result from the uncontrolled progression of denting: tube cracking and leaking at U-bends, tube cracking and leaking of a highly concentrated acidic and oxidizing environment at the constricted crevices, tube support ligament failure, and gross deformation of tube support plates. Laboratory tests proved that denting is caused by the presence of a high concentration of chlorides in the support plate crevice and the presence of an oxidizing environment outside the crevice (Ref 39-41). This aggressive environment is brought about by the introduction of acid-forming impurities via the feed into the steam generator and subsequent concentration at the tube/tube support crevice. Concentration factors of greater than 20 000 times within the crevice have been observed in laboratory tests.

Corrective Actions. Based on this knowledge of the causes of denting, the corrective actions for operating plants have been to modify the environment and make it less aggressive. This has been accomplished by establishing secondary water chemistry guidelines based on laboratory and field data that recommend reduced levels of impurities in the steam generator (Ref 42) and by recommending methods for achieving these greatly lowered impurity levels. These methods include reducing condenser in-leakage (Ref 43, 44), reducing air in-leakage (Ref 45), producing purer makeup water, and using condensate polishers to purify the water. Another method employed to modify the environment is the addition of boric acid, which inhibits the acid chloride attack (Ref 46, 47). The application of these methods has led to greatly reduced denting rates. The PWR Secondary Water Chemistry Guidelines were prepared by a committee of industry experts under

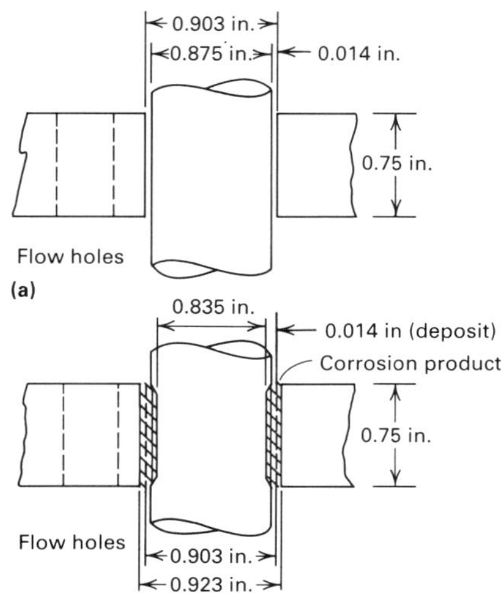

Fig. 30 Denting at tube/tube support plate intersection. (a) Normal tube/tube support plate intersection. (b) Dented intersection. Denting results in a reduction in tube diameter and thinning and cracking of support plate ligaments. 1 in. = 25.4 mm

the sponsorship of the Steam Generator Owners Group (Ref 42). The guidelines combine the results of laboratory corrosion studies and evaluation of good PWR operating pratice. The PWR Secondary Water Chemistry Guidelines consist of chapters covering:

- Management Responsibilities
- Recirculating Steam Generators
- Once-Through Steam Generators
- Analytical Methods
- Data Management and Surveillance

These guidelines have been endorsed by the Steam Generator Owners Group and have been generally adopted by the utilities. They have also been endorsed by the NRC and the Institute for Nuclear Power Operations.

For new plants, one vendor minimized the potential of formation of the aggressive environment by switching from drilled to broached support plates having flat lands. With this design, the concentration of chemicals within the tube/tube support crevice is greatly reduced. For new plants, the corrosion was reduced by the use of more corrosion-resistant materials. Types 405 and 409 stainless steels, which are more corrosion resistant, have replaced the use of carbon steel for United States manufacturers of RSGs (Fig. 31). (Type 410 stainless steel replaced carbon steel in the B&W/AECL units.) The differences in corrosion between carbon steel and type 405 stainless steel is less at higher salt concentrations (Ref 48, 49).

In summary, denting, brought about by the corrosion of carbon steel, is caused primarily by concentration in the tube/tube support crevices of impurities in the secondary coolant. The corrective actions include modifying the environment by reducing, inhibiting, or neutralizing the impurities in the steam generator; by using a tube support design that reduces the concentrating mechanism; and by using more corrosion-resistant materials.

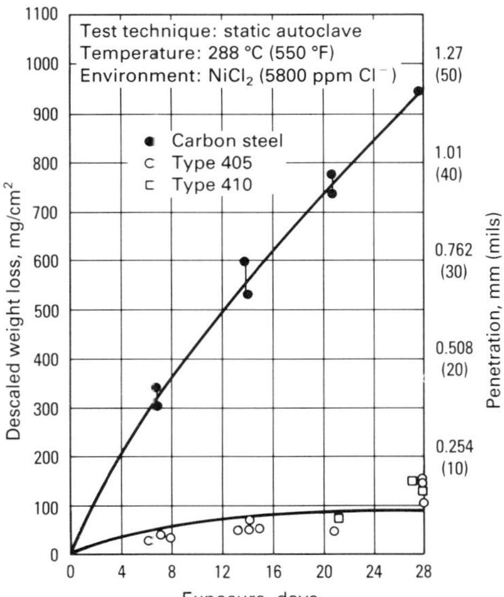

Fig. 31 Corrosion of carbon steel and 12% Cr stainless steels in nickel chloride ($NiCl_2$) solutions

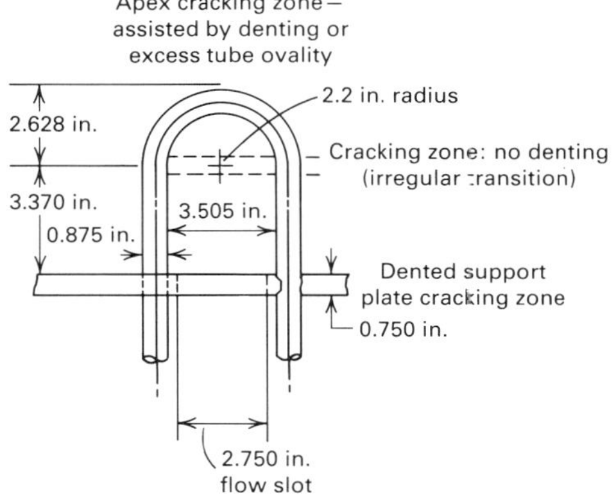

Fig. 32 Location of cracking in steam generator inner row U-bends. 1 in. = 25.4 mm

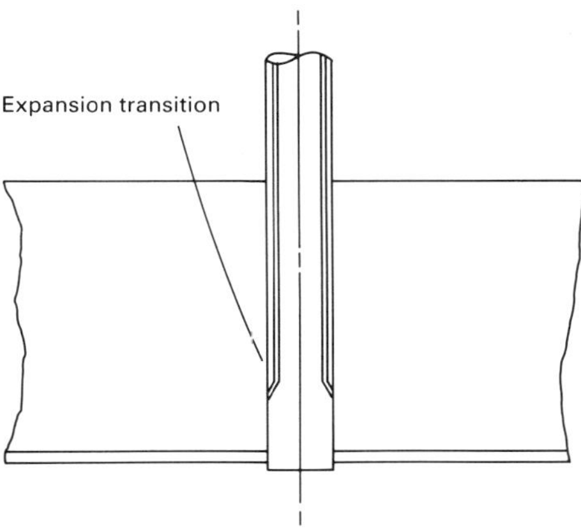

Fig. 33 Details of tubesheet expansion transition region. (a) Partial tubesheet expansion transition region. (b) Full tubesheet explosive, hydraulic, or roller expansion transition region

Table 5 Plants affected by primary-side cracking

Type of primary-side attack(a)	Number of plants affected (as of 1985)
U-bend tangent cracks	16
Roll transition cracks	16
Roll expansion cracks	10
Explosive expansion transition cracks	1
Total number of plants affected	**29**

(a) Excluding those caused by denting

Inside Surface (Primary-Side) SCC

Causes. Primary-side SCC of Alloy 600 steam generator tubing evolved from a laboratory prediction during the 1950s and 1960s to a major degradation mechanism during the 1970s and 1980s in operating steam generators. As early as 1957, cracking of high-nickel alloys in high-purity water at 350 °C (660 °F) was reported (Ref 50). During the following years, numerous laboratory tests were performed in different environments in an attempt to duplicate and explain these observations. This work was summarized in 1975 (Ref 51), in 1981 (Ref 52), and in 1983 (Ref 53). In 1971, the laboratory cracking phenomenon in high-purity water became an in-service degradation mechanism, with the first confirmed primary-side cracking of hot leg roll transition regions at the tubesheet and suspected primary cracking in U-bends (Ref 54). Leakage at U-bends was experienced in Obrigheim steam generators, manufactured by KWU, after only 2 years of operation.

In recent years, cracking of Alloy 600 tubes from the primary side has become a problem of increasing importance. As of early 1985, this type of cracking has been experienced by a number of plants (Table 5) of some steam generator designs.

Cracking in the U-bend has occurred mainly on the inner row at the apex and at the tangent points (Fig. 32). Cracks have also occurred in the tubesheet at transition expansion and roll expansion regions (Fig. 33).

Primary-side cracking is a form of intergranular SCC. This type of intergranular SCC, in common with other forms of SCC, occurs when certain environmental, tensile stress, and material susceptibility factors are sufficiently severe.

Environmental factors influence the initiation of primary-side SCC; of these, temperature must be considered a major factor. The first roll transitions experiencing SCC were located on the hot leg side. Because the hot leg at 320 °C (610 °F) is 30 to 40 °C (55 to 70 °F) hotter than the cold leg at 280 °C (535 °F), it would appear that temperature has a significant influence on SCC initiation, indicating a thermally activated process. The reaction rate can be characterized by an Arrhenius factor, exp ($-Q/kT$), where Q is an apparent activation energy, k is the Boltzmann constant, and T is absolute temperature. Thus, the influence of temperature can be graphically represented by an Arrhenius plot of reciprocal temperature versus the logarithm of crack growth rate (Ref 55) or time to failure (Ref 56). At a typical temperature difference of 30 °C (55 °F) between hot and cold legs, this energy of activation could account for a factor of four to five increase in the time to the onset of cracking.

Other environmental factors that might influence crack initiation are hydrogen gas and chemical contaminants. Various investigations have shown that dissolved hydrogen gas can make an environment more aggressive (Ref 55, 57, 58). It has also been determined that the addition of lithium hydroxide and boric acid along with hydrogen (simulating PWR primary water) appears to produce results intermediate between pure water (least aggressive) and pure water with hydrogen added (most aggressive) (Ref 55). Primary water is less aggressive than pure water at the same hydrogen concentration.

Tensile stresses also have a major impact on the initiation of SCC. The main source of stress is residual stress from tube manufacture and installation. Pressure and thermal stresses also play significant secondary roles. To date, only the most highly strained regions of steam generator tubing (that is, row-one U-bends, roll transition regions, expanded regions, and dented areas) have exhibited SCC. This has also been demonstrated by the results of numerous laboratory tests, which show that plastically deformed samples, that is, split tube U-bends, accelerate crack initiation (Ref 50, 55, 57-61). Stresses in excess of yield can lead to rapid cracking in susceptible material. A threshold stress, below which cracking will not occur, of roughly 0.8 yield for high-temperature water (Ref 55) and 0.5 yield for concentrated sodium hydroxide (Ref 61) has been determined.

Material susceptibility, in combination with the factors mentioned above, can cause SCC. However, it is important to emphasize that there is not a single product called "mill-annealed" Alloy 600 tubing. Each tubing manufacturer employs a different process to produce "mill-annealed" tubing, and the resistance to SCC varies greatly with the process. Some of the mill-annealed tubing has not experienced any SCC over extended periods of operation, while SCC has occurred in other tubing after 1 to 2 years of service.

This microstructural aspect of cracking involves mostly the final mill-annealing temperature and whether precipitation of grain-boundary carbides occurs during the annealing treatment. The most susceptible microstructures are those produced by low mill-annealing temperatures (Ref 57, 62) that develop fine grain size (ASTM 9 to 11) and a copious quantity of intragranular carbide. The grain boundaries of these materials usually contain little, if any, carbide phase (Ref 57, 58, 63).

The beneficial influence of grain-boundary chromium carbides on primary-side SCC resistance has received extensive evaluation over the past several years. As far back as 1973, results were published indicating that Alloy 600, given a heat treatment to precipitate grain-boundary chromium carbide, developed improved SCC resistance in high-purity water (Ref 59). Recent work has demonstrated that grain-boundary carbides improve primary-side SCC resistance with or without grain-boundary chromium depletion (Ref 57, 58, 60, 61). Annealing temperatures in the range of 980 to 1010 °C (1800 to 1850 °F) will avoid undue grain growth and provide enough dissolved carbon so that carbide precipitation will occur during cooling and subsequent thermal treatment for 15 h at 705 °C (1300 °F). The 15-h treatment provides enough time for grain-boundary chromium content to recover (Ref 58).

Corrective Actions. Preventive measures against SCC are discussed in Ref 64 and are summarized below.

Reduction of temperature in the inner row U-bends is one possible corrective action. By orificing the inlet, the flow to these tubes is reduced, which would cause the primary coolant temperature to reach the cold-side temperature at the apex and thus reduce the time to crack by factors of about two to three.

Stress Relief. For operating plants and for plants already built but not operating, reduction of tensile stresses by stress-relief heat treatment and peening (discussed below) of the inside-diameter surface are the most practical approaches. Stress relief includes the use of resistance heaters, induction heaters, and global heat treatment.

Resistance heaters can provide *in situ* stress relief of U-bends according to laboratory tests; during the stress-relieving process, the heater is pulled up to and around the bend (Ref 65). It is believed that temperatures in a fairly broad range, even for a short time, would significantly reduce stress and improve resistance to attack.

Induction heaters can be used to reduce stresses sufficiently in roll transitions to increase resistance significantly to primary-side attack (Ref 65, 66). This type of stress relief would have the advantage of decreasing both residual stresses at inner and outer surfaces in the roll transition area without causing significant sensitization. Induction heating equipment has been developed for use in brazing sleeves inside tubes and is thus available for *in situ* treatment. One question that still needs to be resolved is the extent of high-temperature corrosion that would occur from any residual salts that would be present in the gap created between the tube outside diameter and the tubesheet in the area that reaches high temperature. According to preliminary evaluation, the gap is acceptable because the crevice depth is relatively shallow (for example, 6 mm, or 1/4 in.) and comparable to tube support plate crevices already present in the generator. Another question that needs to be resolved is whether unacceptable tube axial stresses could be developed by the stress relief.

Global heat treatment as a means of stress relief involves heating the entire tubesheet area to a temperature of about 610 °C (1130 °F) in order to relieve residual stresses and to obtain microstructural improvement (carbide precipitation at the grain boundaries). It would be performed by using heaters placed under the tubesheet and along the outside of the shell up to about one shell diameter above the tubesheet. Separately controlled heaters would be located along the divider plate and around the channel head in order to ensure that large thermal expansion stresses do not develop at divider plate junctions. This procedure would require only a few weeks of critical path time and would minimize the amount of high-radiation exposure work required in the channel head.

Before application, plant operators should resolve specific questions, such as the effect of residual salts, the factor of improvement expected from the stress relief, and whether the carbide precipitation obtained during the stress relief could cause Alloy 600 to become susceptible to attack by sulfur species such as polythionic acid. This would require increased care in water chemistry control. In the evaluation of various stress-relieving methods, plant operators need to assess the benefits and risk factors connected with the different corrective measures.

Shot peening to produce residual surface compressive stresses is a well-known approach for providing resistance to stress corrosion. Shot peening has been investigated for use inside tubes in the tubesheet area of steam generators (Ref 67) and has been performed in a number of plants. Shot are blown up the tube so that the shot impinge on a conical deflector located in the area to be peened. It is possible that the same approach could be used for U-bends. Before applying this method to the U-bend and/or tubesheet region it is necessary to ensure that a compromise can be reached between the induced inside-diameter compressive stresses and the resultant outside-diameter tensile stresses. Thus, the total tube SCC resistance, inside and outside diameter, would be optimized.

Rotopeening has been performed for both radioactive and nonradioactive plants to reduce the stresses in the expansion transition and expanded areas. It is performed by using beads bonded to fabric flappers that are rotated in a flapper wheel arrangement such that the beads impact the tube inside-diameter surface.

The compressive stresses developed on the tube inside diameter by the peening must be balanced by tensile stresses in the remaining wall thickness. Excessive inside-diameter peening could lead to significant tensile outside-diameter surface stresses. These stresses, combined with applied pressure and thermal stresses, might then aggravate attack at the outer surface. This possibility needs to be evaluated in relation to water chemistry and possible sludge pile accumulation.

If cracks are already present at the time peening is performed, the effects of the stresses induced by the peening on crack growth need to be assessed. It is suspected that peening will serve to prevent initiation of new cracks, but will not prevent growth of existing cracks through the wall thickness. If this is correct, then peening will be most useful in cases in which most tubes have either no cracks or only very small ones.

Sleeving is another repair method for modifying the tubing material in operating plants. Sleeves up 1120 mm (44 in.) long have been installed in the region between the lower face of the tubesheet and the first support plate as a corrective measure against pitting and intergranular attack at the outer surface. In some cases, even longer sleeves have been installed. The sleeves bridge the damaged area and are attached to sound material beyond either end of the damage. The ends of the sleeves are expanded hydraulically or explosively and are in most cases sealed by rolling, welding, or brazing (examples of sleeves are shown in Fig. 34). A discussion of tube sleeving development for OSTGs is presented in Ref 68.

For more recent plants, the corrective action has been to use tubing made of thermally treated Alloy 600. Extensive tests have shown this material to be greatly superior to mill-annealed Alloy 600 (Ref 57, 58, 61). The thermal treatment involves a final mill-annealing temperature of 980 to 1010 °C (1800 to 1850 °F), followed by a 705-°C (1300-°F) treatment for 15 h to produce a semicontinuous grain-boundary precipitation (Ref 58). More recent steam generators are being fabricated by using thermally treated Alloy 690. It has been reported that this alloy is superior to thermally treated Alloy 600 from the standpoint of resistance to SCC.

Intergranular Attack and SCC at the Outside Surface

Causes. Intergranular corrosion in one or another of its various forms has been experienced in many steam generators operating with seawater- or freshwater-cooled locations. The rate of propagation has been shown to vary widely, depending on the form of the attack. Progression rates in some cases have been sufficiently rapid to require mid-cycle inspections and unscheduled outages to plug or repair leaking tubes and to cause significant economic loss to the PWR operator. Fortunately, most of the corrosion has been confined to crevice locations so that leaks have been small and without risk of a large rupture. Intergranular corrosion has been found to take various forms, such as intergranular SCC, intergranular attack, and intergranular penetration.

Intergranular SCC in Alloy 600 steam generator tubing is illustrated in Fig. 35. In the case of intergranular SCC, the corrosion morphology consists of single or multiple major cracks with minor-to-moderate amounts of branching. Cracks propagate intergranularly in essentially all cases in Alloy 600 tubing. Experience suggests that intergranular SCC requires stresses greater than 0.5 yield in order to propagate rapidly. At lower levels, propagation rates may approach 0, or the corrosion may take another intergranular form.

Intergranular attack, a second form of intergranular corrosion, has been described as general intergranular attack or volumetric intergranular attack (Fig. 36). Its morphology is characterized by a uniform or relatively uniform attack of all grain boundaries over the surface of the tubing. In the purest case, stress does not contribute to the morphology of intergranular attack, which distinguishes this phenomenon from intergranular SCC. However, the close relationship between intergranular attack and intergranular SCC is apparent in Fig. 36, in which a stress-assisted finger of corrosion penetrated from the layer of intergranular attack into the tube material.

Intergranular penetration, the third form of intergranular corrosion, has been variously described as a mixture or a hybrid of the other two forms.

Within the steam generator, intergranular corrosion has been found in a variety of locations. The most important of these locations are the tube/tubesheet crevice (an annular gap remaining after steam generator manufacture), as shown in

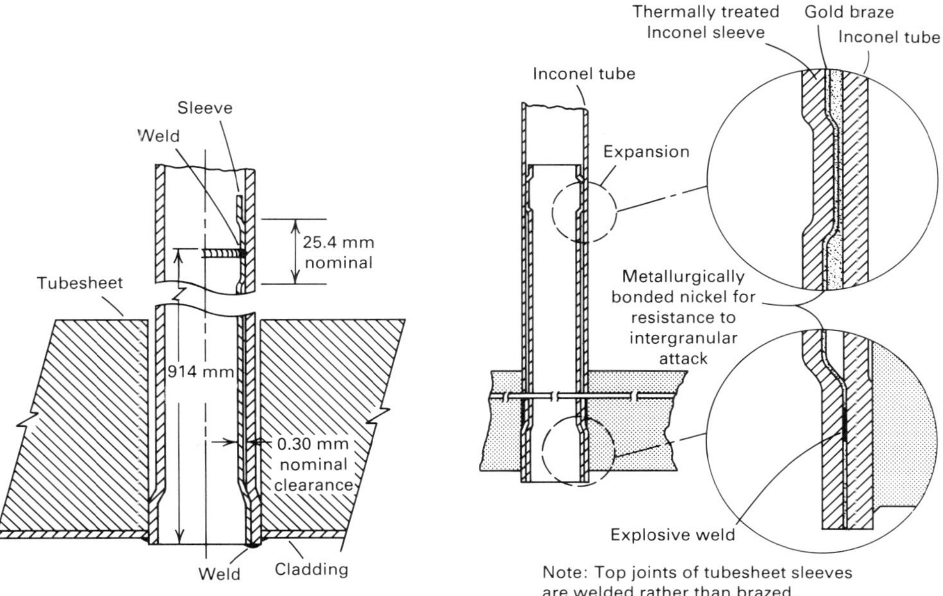

Hydraulically expanded
Hard mechanical roll
Brazed
Tubesheet
Hydraulically expanded
Hydraulically expanded
Hard mechanical roll
Hard mechanical roll
Hybrid expansion joint sleeve
Brazed joint sleeve

Notes: 1. Height above tubesheet selected based on defect locations and length of sleeve permitted by channel head to tubesheet clearance

2. This geometry is based on written descriptions and thus is only an approximate rendering of the actual design

(c)

Fig. 34 Example of sleeve designs for protecting tubing material. (a) Combustion Engineering welded sleeve. (b) B&W regular length sleeve. (c) Westinghouse-type sleeve

0.008in. (0.2mm)

Fig. 35 Intergranular SCC in Alloy 600 C-rings. Source: Ref 69

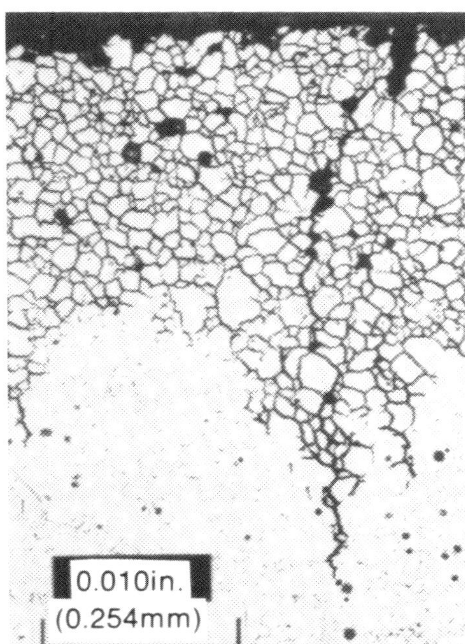

Fig. 36 Intergranular attack of Alloy 600 steam generator tubing. Etched sample. Source: Ref 70

Fig. 37, and the tube/tube support plate crevice. Intergranular corrosion at this latter location has been found at several Japanese sites, but has not occurred widely in the United States. Corrosion has also been observed in the sludge region above the tubesheet. All of the steam generators experiencing secondary-side intergranular corrosion are tubed with mill-annealed Alloy 600. In laboratory tests, however, intergranular corrosion has been produced in mill-annealed Alloys 600, 690, and 800. Based on these findings, there appears to be little difference in performance among these alloys in the mill-annealed condition (Ref 33, 34).

Tests have shown how aggressive chemicals can concentrate in these crevices and in the sludge under heat transfer conditions (Ref 71, 72); alternate wetting and drying is a particularly effective concentration mechanism. Five classes of environmental contaminants have been postulated to explain the occurrence of intergranular corrosion (Ref 33). These include high concentrations of sodium hydroxide (NaOH) and/or potassium hydroxide (KOH), the products from the reaction of sulfate ions with hydrazine or hydrogen (reactive sulfur-bearing species are postulated), the products of thermal decomposition of ion exchange resins (sulfates and organic residuals), highly concentrated salt solutions at neutral or

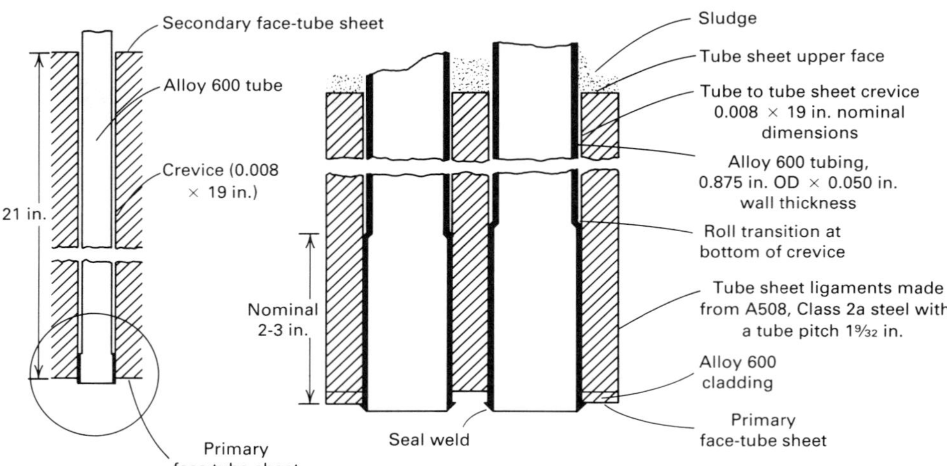

Fig. 37 Typical steam generator tube showing the tubesheet annular crevice. 1 in. = 25.4 mm

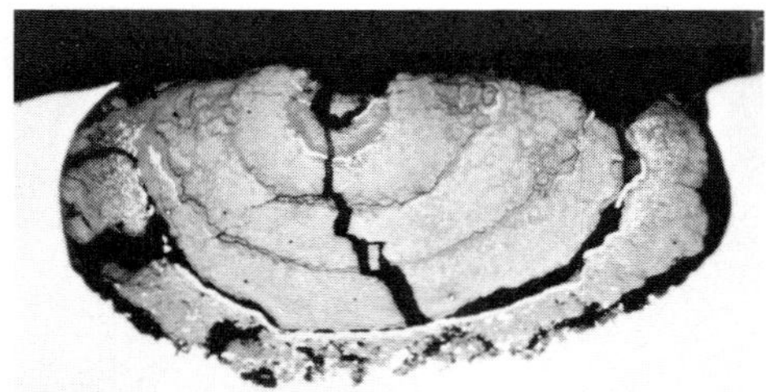

Fig. 38 Pitting of a steam generator tube. Source: Ref 73

nearly neutral pH (these salt solutions are the natural consequences of condenser leakage concentrated to high levels of salt by the boiling processes in the steam generator), and alkaline carbonates and/or their reaction or hydrolysis products (believed to affect the nature of the passive film on the alloy surface).

Corrective Actions. It has been shown with reasonable certainty that increased service stress, dynamic strain, and a high residual stress level can all be major factors in accelerating initiation or propagation of intergranular SCC. A similar statement for intergranular attack cannot be made, because experimental results have shown that intergranular attack can develop even with compressive stress (Ref 58).

Stress-relief treatment of tubing in new steam generators is accomplished by installing thermally treated Alloy 600 tubing that has been aged at 705 °C (1300 °F) for 15 h. Also, one manufacturer uses a full bundle stress relief after steam generator fabrication to provide stress relief. No practical procedure for the application of thermal stress relief to the secondary side of in-service tubing has yet been proposed.

For new plants, modification of the materials by thermal treatment to improve the microstructure can reduce the susceptibility to caustic-induced intergranular SCC. For operating plants, sleeving can be applied to mitigate the effect of intergranular corrosion. The modified material or sleeve should be resistant to faulted secondary environments; for example, sleeves have used a duplex sleeve with a nickel or nickel alloy outside layer (which is considered immune to caustic attack).

Modification of the crevice environment appears to be the most straightforward method of preventing or arresting intergranular attack and may apply equally to the intergranular SCC. Modification can include several factors, such as lowering the temperature, adding a pH neutralizer, removing the corrodent by flushing or soaking, and changing the concentration and/or ratio of bulk water contaminants. Laboratory studies have confirmed the benefit of several of these corrective measures; some of the modifications have been applied to operating steam generators.

The operating temperature of one unit was lowered for several years, which effectively reduced the rate of progression of the intergranular attack. The hot leg temperature was lowered to approximately the normal cold leg temperature. A necessary side effect of significant temperature reductions is loss of power. Based on the experience of one utility, it is judged that the temperature must be reduced to 300 °C (575 °F) or lower to have a major impact on corrosion rate.

Boric acid has been added during off-line crevice-flushing operations to reduce the pH level. The pH neutralizer was used on the basis of limited laboratory data regarding its effectiveness. Tubes subsequently removed from one of the steam generators were found to have boric acid present over the full length of the tube surface within the tubesheet crevice. No estimate can be made as to effectiveness. Equilibrium calculations suggest that high levels of boric acid (equivalent to an Na/B ratio approximately equal to 1) will be required to reduce the pH significantly when the pH is held in the alkaline range by NaOH or KOH (very strong bases). On-line addition of boric acid has been employed at several plants.

Flushing or soaking off-line maintenance procedures have been used with varying results at several plants. The flushing procedures are based on laboratory tests and involve a depressurization that causes boiling within the tube/tubesheet crevice and ejection of concentrated solutions from the crevice. Optimization of crevice-flushing procedures for each unit and periodic repetitive application of the optimized procedures are recommended. It should be recognized that flushing or adding a pH neutralizer may be difficult if denting has occurred at the top of the tubesheet or if the annulus is fouled with corrosion products, thus blocking access to the crevice.

The concentration of contaminants accessible to the crevice may be controlled by eliminating or reducing the entry of contaminants to the steam generator and by controlling the concentrating capability of the sludge pile above the crevice. Reducing the entry of contaminants is best accomplished by preventing condenser leaks, routing drains to the condenser (if condensate is subsequently treated), and properly treating makeup water sources. Use of full-flow condensate polishers for control of ionic species has not been shown to be effective in controlling the species that are probably responsible for intergranular attack and intergranular SCC, as was shown for control of chloride responsible for denting. Polishers are potential sources of sulfates (SO_4) and sodium (Na^+) if operated or regenerated improperly, and they do not effectively remove silica or organics. However, some plants have used these condensate polishers very effectively to minimize impurity entry into the steam generators.

Control of the sludge pile, which is an effective concentrating mechanism, requires three courses of action:

- Effective, periodic sludge lancing
- Minimization of particulate transport by preventing air entry and/or providing for feedwater filtration, such as by powdered resin condensate polishers
- Preventing the entry of chemical species that tend to promote agglomeration

Chemical cleaning has been used to remove the sludge on the tubesheet. One group of utilities has attempted to modify the aggressive environment in the tube/tubesheet crevices by fully expanding the tubes in the tubesheet, thus eliminating the crevice.

Pitting

Causes. Extensive pitting on the outer surfaces of tubes has been observed in two units. The pitting occurred primarily on the cold leg between the tubesheet and the first support plate in regions where sludge or tube scale was present. Figure 38 shows a typical steam generator pit. The laminar appearance is caused by the presence of metallic copper layers. In addition, the corrosion deposit is enriched in chromium and depleted in nickel and iron compared to the base metal. Laboratory tests have shown that pits can be formed in the presence of high chloride ion concentration from either seawater or copper chloride (Ref 74). Thus, it is concluded that the pits are caused by chloride, low pH, and an oxidant such as cupric chloride ($CuCl_2$) or oxygen. Temperatures above about 150 °C (300 °F) are required to form pits such as those observed in the operating units. It is further believed that sludge and scale act as a medium in which the bulk impurities are concentrated to higher levels by the boiling action.

Corrective Actions. For existing plants, the tubing is repaired and retained in service by the use of sleeves. Sleeves may be of one alloy or a bimetallic alloy in which the outer alloy is selected for resistance to pitting corrosion and the inner alloy is selected for primary-side corrosion resistance. The principal corrective action is to modify the environment to make it less corrosive. One approach is to reduce the sludge and scale by

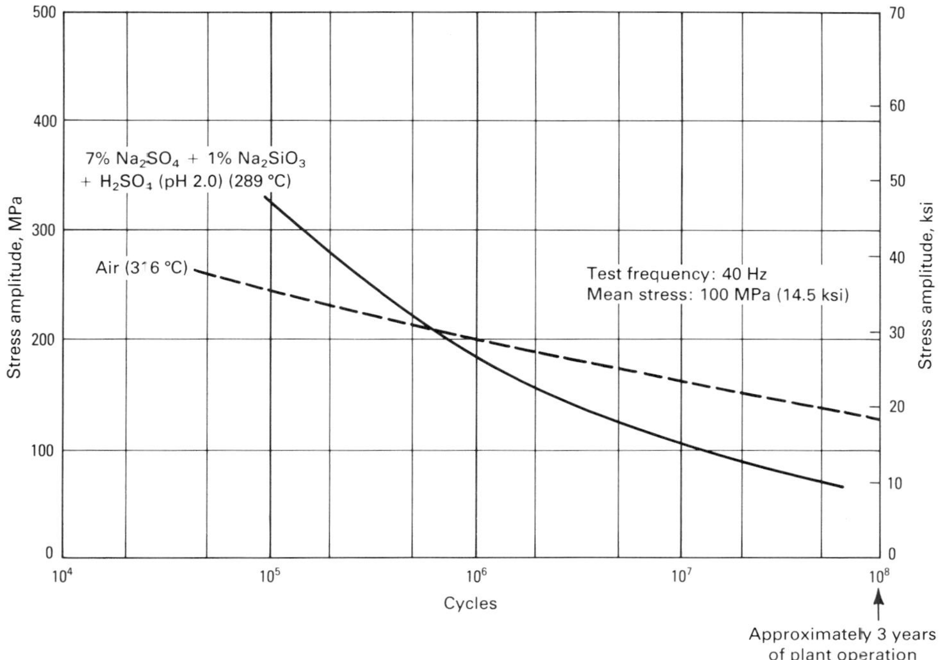

Fig. 39 Corrosion fatigue effects of Alloy 600

minimizing the entry of solids (higher pH and reduced air in-leakage), by sludge lancing, and by chemical cleaning. All of these methods are now being used. Another approach is to minimize the entry of soluble contaminants (principally chlorides and oxidants) by following the Secondary Water Chemistry Guidelines (Ref 42), by quickly repairing leaky condensers, and by deaerating auxiliary feedwater used for hot standby.

Corrosion Fatigue

Causes. Cracking of some tubes from the outer surface has occurred in the upper regions of several OTSGs. It is believed that these cracks are caused by corrosion fatigue resulting from small-amplitude vibration combined with the transport of impurities into the upper regions of the OTSG units, particularly in the open lane. Laboratory tests have shown a decrease in Alloy 600 fatigue strength in the presence of chemicals that are judged to be present in these upper regions (Ref 75).

In these tests, a substantial decrease in fatigue resistance of Alloy 600 was observed in acid sulfate/silicate solution. The environment was selected to be consistent with that postulated to exist in the region of interest at the top of an OTSG. Figure 39 shows the response of Alloy 600 in an aggressive (pH 2) sodium sulfate/sodium silicate/sulfuric acid environment. The fatigue strength at 10^7 cycles at 289 °C (552 °F) was found to be approximately 56% of the value measured in air at about the same temperature. Aerated acid sulfate/silicate and alkaline sulfate/silicate environments had a less deleterious effect.

Corrective Action. The corrective action is to modify the environment to make it less aggressive. This can be done by minimizing the entry of these impurities into the steam generator. Also, the use of mechanical flow-diverting lane blockers has been tested and shown to be a useful steam generator modification; they reduce the flow of liquid droplets to the upper regions via the open lane (Ref 76). In new plants, the open lane has been eliminated.

Other Phenomena

A variety of additional factors have contributed to lower-than-desired steam generator availability or have caused the utility operators to incur repair costs. These phenomena will not be discussed in this section. Some of the mechanical problems and proposed or actual corrective actions are summarized briefly in Table 6. References 77 and 78 provide information that is particularly useful to those utility operators that are assessing whether to repair or replace their steam generators.

Table 6 Corrective actions for steam generator mechanical problems

Problem	Corrective action
Erosion-corrosion of feedwater nozzles	Higher pH coolant; replace with higher chromium tubing
Loose parts fretting and wear	Inspection; keep loose parts out
Wear at antivibration bar locations	Replace with wider bars, improved materials; reduce tube-to-bar gap
Wear in preheat units	Improved flow distribution baffle; reduced tube to support plate clearance Split feed flow
Water hammer	Modify feed design and feed rings to avoid drainage of feed ring
Increased pressure drop in once-through steam generator	Flow surge to dislodge sludge; chemical cleaning

Corrosion of Zircaloy-Clad LWR Fuel Rods

A.J. Machiels
Electric Power Research Institute

Almost all power reactors in the world are cooled by either normal or heavy water and are termed light water reactors (LWRs) or heavy water reactors (HWRs), respectively. The heat-generating fuel elements, or fuel rods, are made of stacks of uranium dioxide (UO_2) pellets clad or sheathed in hermetically sealed tubes. Fuel rods are grouped together in assemblies whose designs differ markedly according to the type of reactor (Ref 79).

A primary fuel assembly reliability consideration is the ability to maintain optimum power generation over the design lifetime of the assembly without releasing the radioactive by-products of the nuclear fission process into the primary coolant (water). Therefore, the cladding or sheathing tubes separating the coolant water from the nuclear fuel and its by-products must resist the corrosive attacks of both environments while meeting all mechanical loads present or anticipated during power and fuel-handling operations.

Although the first LWRs in the United States used stainless steel as a cladding material, the economics of LWR-generated electricity favor the use of zirconium-base alloys (Zircaloys) wherever possible in the high-flux regions of the reactor core. The reason for this is the low neutron absorption cross section of zirconium. As a result, tubes and other assembly structural components are for the most part made of zirconium alloys.

The primary zirconium alloys are known as Zircaloy-2 and Zircaloy-4. They are mostly zirconium (~98 wt%) with low amounts of alloying elements (tin, iron, chromium, nickel, and oxygen), which confer to the alloys the desirable mechanical and corrosion properties. Zirconium-niobium alloys are also used in Canada and Russia.

Zircaloy-2 and Zircaloy-4 were selected during the early days of the U.S. Nuclear Navy program on the basis of relatively few data (Ref 80), and it is now generally accepted that the formulation of these alloys is not optimized with regard to their corrosion resistance to water, which is the topic of this Section. The discussion that follows is restricted to the water-side corrosion properties of Zircaloy under normal operating conditions in LWRs. Oxidation during high-temperature transients leading to core damage is beyond the scope of this Section, but was recently reviewed (Ref 81). A brief introduction to corrosion under isothermal conditions in high-temperature water is first presented. The applications to pressurized PWRs and BWRs are then discussed. Also, given the large number of reviews already devoted to the topic, this discussion is limited to the more recent developments that have, or will have, an impact on the corrosion performance of the Zircaloys. For additional information on the corrosion performance of Zircaloy alloys, see the article "Corrosion of Zirconium and Hafnium" in this Volume.

Corrosion in High-Temperature Water

Light water reactors normally operate with a coolant temperature between 250 and 350 °C (480 and 660 °F) at a pressure of either approximately 7 MPa (1000 psi) (BWRs) or 15 MPa (2200 psi) (PWRs). Therefore, out-of-reactor corrosion studies are usually conducted in high-pressure autoclaves, in which the water is either in the liquid state at temperatures up to approximately 360 °C (680 °F) or in the gaseous state (steam) at temperatures of approximately 400 °C (750 °F) and higher. Under these conditions, the attack of Zircaloy by water is generally uniform. With steam at high temperature (typically 500 °C, or 930 °F) and high pressure (typically 10 MPa, or 1500 psi), this uniform attack can be accompanied by a more localized form called nodular corrosion.

The reaction between Zircaloy and water can be written as:

$$Zr + 2H_2O \rightarrow ZrO_2 + 2(1 - x)H_2 + 4 \times H \quad \text{(Eq 1)}$$

where the first and second left-hand terms represent zirconium (of which Zircaloy is mostly made) and water, respectively; the right-hand terms represent zirconium oxide, hydrogen that is released into the corroding water, and hydrogen that is picked up by the Zircaloy; x is the pickup fraction. The pickup of hydrogen by the Zircaloy substrate leads to changes in its mechanical properties and, in particular, to a loss of ductility.

The formation of zirconium oxide results in a loss of metallic Zircaloy. Because the Pilling-Bedworth ratio is equal to 1.56 at ambient temperature, a given loss in metal thickness results in a larger gain in oxide thickness (the Pilling-Bedworth ratio is explained in the article "Fundamentals of Corrosion in Gases" in this Volume). In Fig. 40, the positions of the Zircaloy/ZrO_2 and ZrO_2/H_2O interfaces are shown relative to the position of the original Zircaloy/H_2O interface.

The kinetics of the uniform corrosion reaction can be conveniently divided into two main periods, referred to as pretransition and posttransition. In the first of the periods, a black coherent oxide film is formed, and the corrosion rate diminishes with time according to a rate law given by the simplified form: (weight gain) = constant x (time)^n, with the exponent n in the range of 0.25 to 0.5. A transition follows to a period where the rate law is closer to linear ($n \simeq$ 1). This transition is eventually accompanied by a conversion of the black oxide to a gray or white oxide. The corrosion kinetics of Zircaloy in high-temperature water actually follow a periodic behavior, and the apparently linear dependence of the posttransition rate is obtained by averaging the time-dependent rates over several periods (Ref 82). A progressive acceleration of the posttransition rate, especially after sufficiently long reaction times, has also been shown (Ref 83).

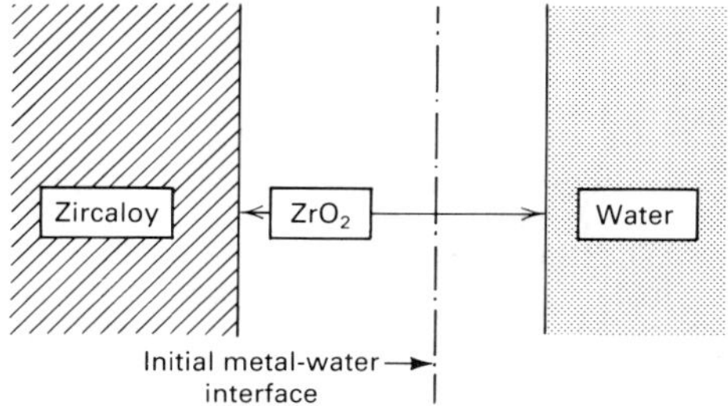

Fig. 40 Zircaloy/zirconium oxide/water interface

Given that zirconium is a most reactive metal, it should readily corrode according to Eq 1. However, because of the protective nature of at least part of the oxide film, Zircaloy belongs to the group of extremely corrosion-resistant alloys. The film acts as a barrier through which the reactive species (oxygen ions, electrons, etc.) must diffuse to sustain the corrosion reaction (Ref 84). During pretransition, reaction rates are determined by diffusion through a surface film of increasing thickness. At transition, which typically occurs when the oxide layer is 2 to 3 μm thick, part of the film undergoes a change in morphology and becomes nonprotective (Ref 85). During posttransition, the thickness of the protective part of the oxide film successively increases and abruptly decreases, yielding cyclically time-dependent kinetics. As discussed in the previous paragraph, linearization over a sufficiently long period of time can be readily performed. This is equivalent to assuming the existence of a protective barrier of constant thickness. Under typical high-temperature water autoclave conditions, this equivalent thickness is about 1 μm (Ref 86).

The out-of-reactor corrosion properties of Zircaloys are, in general, strongly dependent on the material microstructure and on temperature. In some cases, these properties also depend on the water chemistry conditions.

Zircaloy Material. The corrosion resistance of Zircaloys depends on a number of factors that include the concentration and distribution of alloying elements and impurity elements. These factors in turn depend on the thermomechanical processing history, which is an essential part of the manufacturing sequence. Other manufacturing variables that are of significance in determining corrosion resistance are related to the surface treatment of the final component, such as grinding, grit blasting, or pickling.

The fabrication variables or the manufacturing variables involved in the production of zirconium sponge are not considered in this section. However, corrosion properties are very sensitive to a number of key manufacturing parameters, such as quenching rates and annealing times and temperatures (Ref 87). Moreover, some variability in these parameters always accompanies the large-scale production of Zircaloy components, even under nominally identical conditions. Therefore, the corrosion performance of a given population of Zircaloy components is generally characterized by a range of corrosion behaviors, the extent of which can be very significant.

Temperature. Corrosion studies in autoclaves under isothermal conditions have led to a number of mathematical expressions, such as (Ref 88):

$$\text{Pretransition } \Delta W^3 = 5.07 \cdot 10^{13} \cdot \exp\left(\frac{-32\,289}{RT}\right)t \quad \text{(Eq 2a)}$$

$$\text{Posttransition } \Delta W = \Delta W_t + 2.21 \cdot 10^9 \cdot \exp\left(\frac{-28\,200}{RT}\right)(t - t_t) \quad \text{(Eq 2b)}$$

$$\text{Transition } \Delta W_t = 3.21 \cdot 10^8 \cdot \exp\left(\frac{-10\,763}{RT - 1.17 \cdot 10^2 \cdot T}\right) \quad \text{(Eq 2c)}$$

where ΔW is the specimen weight gain (in mg/dm^2); ΔW_t is the specimen weight gain at transition (in mg/dm^2); t is the total exposure time (in days); t_t is the time to transition (in days); R is the gas constant, 1.987 cal/(mol·K); and T is the absolute temperature (in degrees Kelvin). In the temperature range of 280 to 400 °C (535 to 750 °F), the pretransition rates increase by a factor of two for each 45- to 50-°C (80- to 90-°F) temperature increment; more important, the posttransition corrosion rates increase by a factor of two for each 16- to 20-°C (29- to 36-°F) temperature increment.

Water Chemistry Conditions. The protective nature of the diffusion barrier can be drastically altered by the presence of some species in the corroding water. Most autoclave experiments use pure water only. However, because of its relevance to LWR technology, lithium hydroxide (LiOH) is sometimes added to the water. At sufficiently high concentrations, the presence of LiOH can lead to rapid corrosion of the Zircaloys (Ref 89).

Corrosion in LWRs

Early experiments on zirconium-alloy corrosion in reactors showed that significant differences in corrosion behavior could occur (Ref 90, 91). Compared to a high-temperature water autoclave, the reactor environment is characterized by the presence of an intense radiation field, the existence of large temperature gradients across the Zircaloy-water contact layers, and the presence of impurities and soluble chemical additives that are usually not present in high-temperature water autoclaves. Subsequent investigations have shown that each of these factors, separately or in combination, can be important.

Water Chemistry. Boiling water reactors normally operate with high-purity water. However, radiolytic decomposition of the water eventually results in the production of stoichiometric amounts of hydrogen and oxygen. Partitioning of those two gases occurs between the steam phase, which is continuously extracted from the system, and the liquid phase, which is recirculated. Typical steady-state concentrations of hydrogen and oxygen in the recirculation loops are, respectively, 20 and 200 ppb. Pressurized water reactors operate with a hydrogen overpressure sufficient to inhibit the formation of radiolytic oxygen in the coolant, a basic additive (LiOH) to control the release of corrosion products into the coolant and their deposition on fuel surfaces, and a chemical shim (orthoboric acid, H_3BO_3) to control the nuclear reactivity. Under these conditions, typical hydrogen and oxygen concentrations are maintained at >2000 ppb and less than 5 ppb, respectively.

In both systems, the structural components of the reactor system are the source of corrosion products generally characterized by low solubilities. As they are transported throughout the system by the coolant, they tend to form deposits on both in-core and out-of-core surfaces. These deposits, along with the coolant-borne impurities, are referred to as crud. Crud deposits consist mostly of iron oxides with high porosities (65 to 85%). Magnetite (Fe_3O_4), in which the iron is partially replaced by the other constituents (such as nickel or chromium) of the alloys exposed to the coolant, and hematite (Fe_2O_3) are the most common crud constituents in PWRs and BWRs, respectively. Formation of crud deposits on fuel element surfaces is largely dictated by solubility and heat transfer considerations. Under conditions that favor negative solubility temperature coefficient (that is, solubility decreases when temperature increases), or the formation of concentration cells created by boiling heat transfer, or both, crud deposition readily occurs.

Therefore, the important water chemistry parameters include:

- Concentrations of dissolved oxygen and hydrogen
- Concentrations of chemical additives (PWR technology only)
- Concentrations and precipitation characteristics of the coolant-borne impurities

The effects of these parameters can be very important in the presence of a reactor radiation field, or a large enough temperature gradient, or both.

Effect of Radiation. Radiation can affect the corrosion behavior of Zircaloys by modifying the aggressivity of water through the formation of highly reactive, oxygenated radicals; by modifying the diffusive properties of the protective oxide layer through the formation of point defects; and by modifying the microstructure of the Zircaloy substrate by changing the concentration and distribution of some elements in the Zircaloy material close to the Zircaloy/zirconium oxide interface. Effects due to the second item appear to be relatively minor. Effects due to the last item are potentially important, but little is known in this area. Therefore, these radiation effects will not be considered any further in this section.

As already discussed, the BWR and PWR environments lead to significantly different oxygen concentrations in the coolant. In BWRs, where the oxygen content of the coolant water is classified as high, irradiation clearly enhances the oxidation rate. The formation of a uniform oxide layer proceeds at a rate greater than that measured under similar temperature conditions in an autoclave and is generally accompanied by the appearance of locally thicker patches of zirconium oxide having the form of nodules or pustules (Fig. 41). Initially, the nodules appear as white patches on the black pretransition oxide surface (Fig. 42); with exposure, they grow in diameter as well as in thickness, and they eventually cover the entire exposed surface.

In PWRs, where the oxygen content of the coolant water is very low, only the formation of a uniform film is observed. There remains some disagreement as to whether or not irradiation enhances oxidation under those PWR conditions (Ref 92). Estimates of the magnitude of acceleration generally vary between none and a factor of four. Up to oxide thicknesses of 5 μm, no acceleration is detected; however, for oxide thicknesses greater than 15 to 25 μm, some acceleration is usually observed.

Effect of Heat Flux. The transport of heat from the nuclear fuel pellets to the coolant produces large surface heat fluxes. Both the magnitude of the heat flux and the mode of heat transfer between the fuel element outer surface and the coolant have significant effects on the corrosion performance of the Zircaloy cladding. Temperature gradients in the cladding itself influence the distribution of the hydrogen picked up by the alloy during the corrosion process (Eq 1), and temperature gradients across the cladding/coolant interface increase the temperature of the protective oxide layer.

Acceleration of the corrosion process by the hydrogen produced by the corrosion reaction is a possibility, but the existence and magnitude of this effect under operating conditions applicable to nuclear fuel elements remain to be established. The thermal acceleration obtained by increasing the interface temperature is, in principle, more straightforward to evaluate. With reference to Fig. 43, which schematically represents the temperature profile across the Zircaloy/coolant interface at some axial elevation of a heat-generating fuel element, it can be seen that the average temperature of the protective oxide layer, T_I, located at the metal/oxide interface is given by:

$$T_I = T_B + \Delta T_{CO} + \Delta T_{CR} + \Delta T_{OX} \quad \text{(Eq 3)}$$

where T_B is the bulk coolant temperature and ΔT_{CO}, ΔT_{CR}, and ΔT_{OX} denote the temperature differences across the crud/coolant interface, the crud layer of thickness ρ, and the oxide layer of thickness δ, respectively.

Because the oxide thickness, δ, varies between 0 at the beginning of power generation to its maximum value, δ_{MAX}, at the end of power generation, ΔT_{OX} is a function of time and is given by:

$$\Delta T_{OX}(t) = q''(t) \cdot \frac{\delta(t)}{k_{OX}} \quad \text{(Eq 4)}$$

where t denotes time, q″ is the surface heat flux, and k_{OX} is the thermal conductivity of the oxide. Assuming steady-state reactor operating conditions resulting in constant T_B, ΔT_{CO}, and ΔT_{CR}, that is:

$$T_B + \Delta T_{CO} + \Delta T_{CR} = T_O \equiv \text{constant} \quad \text{(Eq 5)}$$

where T_O is the temperature of the outer surface of the oxide layer. Eq 3 can be rewritten as:

$$T_I(t) = T_O + q'' \cdot \frac{\delta(t)}{k_{OX}} \quad \text{(Eq 6)}$$

Transforming Eq 2b from weight gain to oxide thickness and differentiating with regard to time leads to:

$$\frac{d\delta}{dt} = K \exp\left(\frac{-Q}{RT}\right) \quad \text{(Eq 7)}$$

where K and Q are two constants that can be readily evaluated.

Substituting Eq 6 into Eq 7 yields after some manipulation:

$$\frac{d\delta(t)}{dt} \simeq \left(\frac{d\delta}{dt}\right)_{T = T_O} \cdot A(t) \quad \text{(Eq 8a)}$$

where $A(t)$ is a time-dependent thermal acceleration factor equal to:

$$A(t) = \exp\left[\beta_O \cdot q'' \cdot \delta(t)\right] \quad \text{(Eq 8b)}$$

with

$$\beta_O = \frac{Q}{k_{OX} R T_O^2} \quad \text{(Eq 8c)}$$

Therefore, in the presence of a heat flux, oxidation rates are obtained by multiplying the oxidation rate calculated at a temperature T_O by a time-dependent thermal acceleration factor that depends exponentially on the oxide layer thickness and on the surface heat flux.

$T_O < T_{SAT}$. When T_B is significantly lower than the saturation temperature, T_{SAT}, corresponding to the system pressure, as it is for the major part of a PWR core, T_O, is given by:

$$T_O = T_B + q'' \cdot \left(\frac{1}{\alpha} + \frac{\delta}{k_{CR}}\right) \quad \text{(Eq 9)}$$

where α is the convective heat transfer coefficient; α is a strong function of the local flow

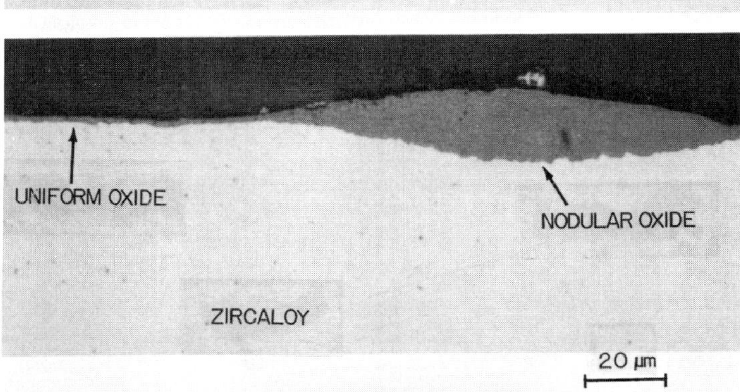

Fig. 41 Photomicrograph showing the uniform and nodular oxides

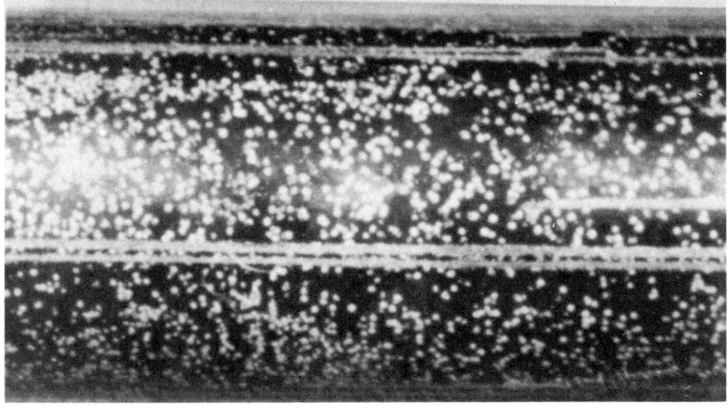

Fig. 42 White oxide nodules on a black, pretransition oxide

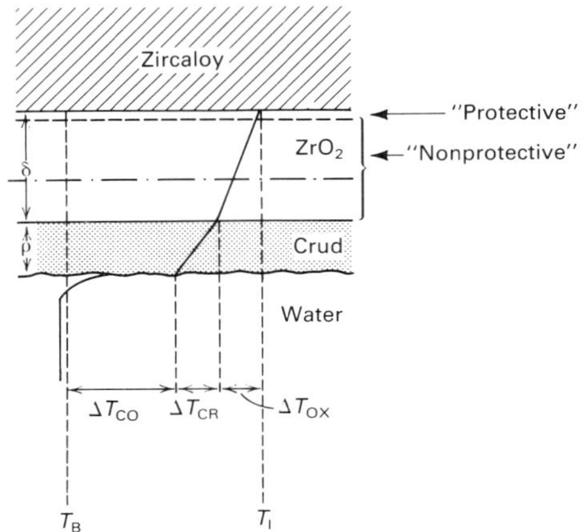

Fig. 43 Temperature profile across the water, crud, and oxide films

turbulence and assumes typical values of 3 to 6 $W/cm^2 \cdot K$ (Ref 88). The thermal conductivity of the crud is denoted by k_{CR} and assumes typical values of approximately 1 W/m·K (Ref 93). Equation 9 predicts that large heat fluxes produce large temperature differences; as a result, T_O can be significantly higher than T_B, especially when substantial amounts of crud deposits are present. Under present recommended operating practices for PWRs, crud accumulation is nominal ($\leq$10 μm); therefore, its impact tends to be small.

$T_O \simeq T_{SAT}$. When T_B is not significantly lower than T_{SAT}, as it is in BWRs and in parts of some PWRs, boiling heat transfer dominates. Except when dense crud deposits are present, the maximum value for T_O ($T_{O,MAX}$) is given by:

$$T_{O,MAX} = T_{SAT} + \Delta T_1 \qquad \text{(Eq 10)}$$

where ΔT_1 represents the amount of superheat needed to sustain the boiling process. The evaluation of ΔT_1 is complicated by the fact that boiling heat transfer promotes the concentration of coolant impurities and additives, especially when the change of phase occurs at fixed locations along the heat transfer surface. These conditions may exist with noncrudded rods when steam bubbles originate from the same sites (Ref 94), as well as with crudded rods when wick boiling is present (Ref 93). For example, a simple evaluation of the maximum concentration factor, F_{MAX}, that can be obtained under wick-boiling conditions is given by (Ref 95):

$$F_{MAX} = \frac{C_{MAX}}{C_B} = \exp\left(\frac{q'' \cdot \rho}{D \cdot p \cdot L_{VAP} \cdot d}\right) \qquad \text{(Eq 11)}$$

where C_B is the bulk coolant concentration of a given species, and C_{MAX} is its maximum concentration in a crud of thickness ρ. The diffusivity of the species in water is given by D, p is the crud porosity, L_{VAP} is the latent heat of evaporation, and d is the density of water. Soluble species may lead to an elevation of the boiling point when C_{MAX} is large enough. When this is the case, ΔT_1 is given by:

$$\Delta T_1 = \Delta T_{BPE} + \Delta T_2 \qquad \text{(Eq 12)}$$

where ΔT_{BPE} represents the boiling point elevation, and ΔT_2 is the superheat calculated from boiling heat transfer considerations. The value ΔT_2 is typically small (a few degrees Celsius). The value ΔT_{BPE} is expected to be negligible in BWRs, but not necessarily so in PWRs, because of the additives present in the coolant.

Species with a low-to-moderate solubility may precipitate for C_{MAX} large enough. A few impurities, most notably copper, lead to the formation of dense cruds, through which heat can be transferred by conduction only. The value $T_{O,MAX}$ is then given by:

$$T_{O,MAX} = T_{SAT} + \Delta T_1 + q'' \cdot \frac{\rho}{k_{CR}} \qquad \text{(Eq 13)}$$

Large amounts of dense cruds (~100 μm) can result in cladding temperatures that are approximately 50 °C (90 °F) higher than those obtained in the presence of porous cruds for typical heat fluxes in BWRs.

Boiling also influences the radiolytic conditions existing in the outer layers of the heat-generating fuel elements. A quantitative description of the effect of boiling on Zircaloy corrosion is proposed in Ref 96.

Application to BWRs. In BWRs, the temperature is practically constant along the length of a fuel element, and $T_{O,MAX}$ is approximately equal to the system saturation temperature, typically about 290 °C (555 °F). Under these conditions, uniform corrosion rates are low, even when an enhancement factor is factored in to take into account the oxygenated conditions of the coolant. The formation of a uniform film is accompanied by the nucleation and growth of oxide nodules. The nodular oxide thickness is mainly a function of burnup with a tendency to saturation (Ref 97).

Zircaloy corrosion in BWRs has not been a problem except in reactors characterized by high concentrations of soluble copper in the reactor water. Failures have been observed, in particular with fuel claddings having high nodular corrosion susceptibility (Ref 98, 99). Those claddings develop nodular oxide layers that contain sizable cracks through which heat transfer appears to be occurring by wick boiling. As discussed above, although wick boiling itself leads to negligible superheats, it can produce large concentration effects. When copper concentrations (C_{MAX}, Eq 11) are large enough, copper compounds precipitate and progressively plug the network of cracks necessary to sustain the wick-boiling process. This condition eventually leads to a significant temperature increase of the cladding because $T_{O,MAX}$ is now given by Eq 13 rather than Eq 10. Thermal acceleration of the corrosion reaction can eventually lead to fuel failures. As can be seen from Eq 11, C_{MAX} is linearly dependent on C_B, but depends exponentially on the heat flux, q", and the nodular corrosion properties of the cladding through ρ/p, where ρ and p now represent the thickness and porosity of the cracked nodular oxide. The latter depend on the corrosion properties of the Zircaloy material.

The recent implementation of adding hydrogen to the feedwater of several operating BWRs to prevent intergranular SCC in the recirculation system piping has created a water chemistry in the BWR core that is outside the experience base of either BWRs or PWRs. The effects of lowering the concentration of oxygen in the reactor water should be beneficial as far as oxidation rates are concerned. The effects on hydriding rates are more difficult to assess because the pickup fraction could be directly influenced by the hydrogen dissolved in the water. Overall, the effects of hydrogen addition on oxidation and hydriding are expected to be small (Ref 100). After 1 cycle of hydrogen water chemistry at Dresden-2, the fuel examination results show that cladding oxidation is well below any performance concerns. Also, for the Zircaloy components that were examined, it does not appear that, within the limits of experimental uncertainties, hydriding rates are directly influenced by the presence of hydrogen in the coolant. Additional examinations after 2 and 3 cycles of hydrogen water chemistry are planned (Ref 101).

Application to PWRs. Because boiling heat transfer also occurs locally in PWRs—especially in the more recently designed ones—$T_{O,MAX}$ is given by Eq 10, in which T_{SAT} is typically 340 °C (645 °F). The evaluation of ΔT_1 is complicated by the presence of LiOH and boric acid in the reactor coolant. Because boiling promotes the concentration of these additives on the outer surfaces of the fuel elements, some local elevation of the boiling point is likely. Lithium hydroxide has a weak effect on the boiling point elevation of water, but the impact of boric acid may be more significant (Ref 93). Uncertainties associated with the accurate evaluation of the interface temperature have hampered efforts to quantify the effects of other phenomena, such as those resulting from the presence of a radiation field.

Zircaloy corrosion in PWRs has become an area of prime concern from a fuel performance point of view. In particular, fuel failures and plant derating due to corrosion have occurred in plants that combine high coolant temperature and high heat flux. Therefore, the full economic benefits associated with high thermal efficiency, high burnup, higher-pH coolant (by adding greater amounts of LiOH), and higher fuel utilization (by fuel rod design changes) may not be obtainable in many existing plants wihout some improvement in the corrosion properties of the current zirconium alloys.

Influence of Corrosion on Radiation Fields

Robert A. Shaw
Electric Power Research Institute

Radiation fields exist in nuclear power plants primarily because of the deposition of radioisotopes on the surfaces of pipes and other components. These radiation fields can significantly influence the operation and maintenance of nuclear power plants. Consequently, the control of these radiation fields wields an influence over the cost of the generation of nuclear power.

The radiation exposure experienced by the personnel who work in nuclear power plants is a key measure of the effectiveness of radiation field control measures. Figure 44 traces the history of the median of plant exposures for BWRs and PWRs. It indicates a regular, continuing increase until the early 1980s and some reduction after 1983. Significant contributors to the exposure peaks in the early 1980s have been the materials problems experienced in these plants. Pipe cracking and the consequent inspection and replace-

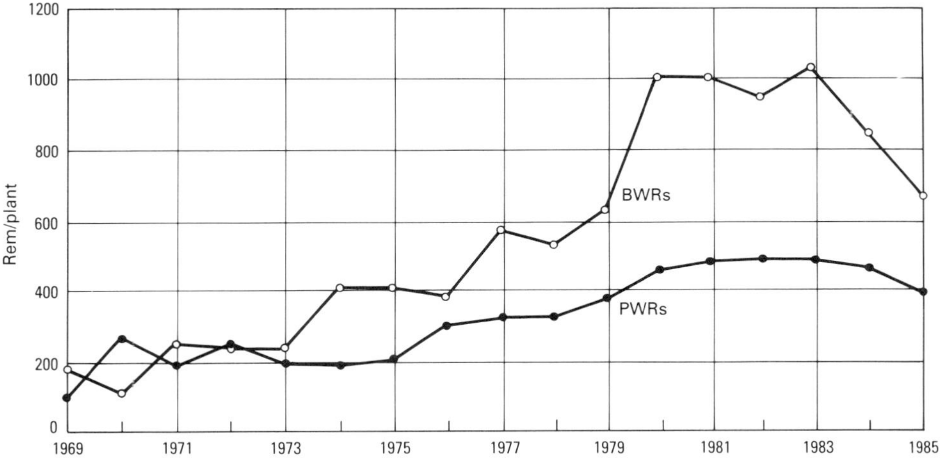

Fig. 44 Median of U.S. nuclear power plant radiation exposures. Rem, roentgen equivalent man

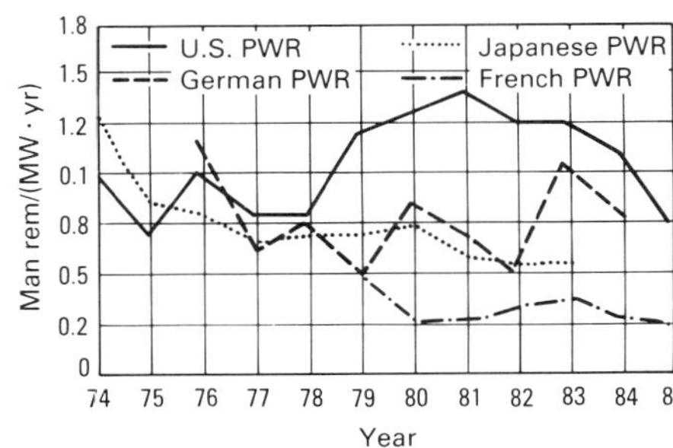

Fig. 45 Occupational radiation exposure per unit electricity generated by PWR nuclear power plants

ment of such pipes caused larger-than-normal radiation exposures to be experienced in BWRs. For PWRs, steam generator materials degradation caused significant increase in inspection and, in some cases, replacement of steam generators, causing a similar increase in exposure at roughly the same time.

Concurrent with these problems, which caused increases in exposure, has been the application of a number of techniques for reducing radiation exposure in plants. These have included techniques for reducing the radiation fields such as water chemistry control techniques, materials selection emphasizing replacement of cobalt alloys, and decontamination to remove radioisotopes from surfaces. In addition to these techniques for reducing radiation fields, radiation exposures have been reduced through the use of remote equipment, shielding, and extensive planning and training for high exposure tasks.

A second measure of expenditure of radiation exposure is presented in Fig. 45 and 46. These diagrams show the man rem/MW·yr, an expression of the radiation exposure that has been required in order to generate electrical energy at nuclear power plants. These two figures compare the experience of the United States with PWRs and BWRs, respectively, to that of other countries. It shows the experience in other countries to be superior to that of the U.S. and stresses the need for effective radiation control programs at U.S. plants. For additional information on techniques to control radiation fields in nuclear reactors, see Ref 102 to 106.

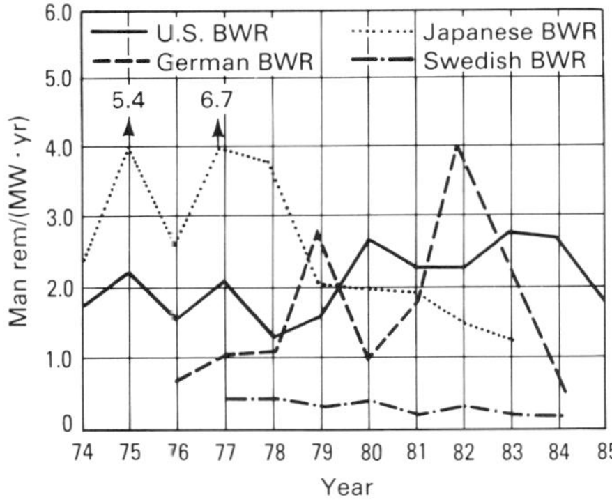

Fig. 46 Occupational radiation exposure per unit electricity generated by BWR nuclear power plants

Radiation Sources

There are two primary sources of radioisotopes generated in nuclear power plants. The first is within the fuel itself where the fissioning process creates fission products and their decay products, which are radioactive. In modern nuclear power plants, fission products are effectively constrained to remain within the Zircaloy cladding present on each fuel rod. Occasionally, in some plants, a few rods will experience pin hole or other types of penetrations through the Zircaloy, permitting the release of fission products into the reactor coolant. Two of the major constituents of fission product radioisotopes—radioactive iodine and radioactive cesium—are useful to illustrate the transport processses, which are typical for fission product.

Iodine, when released to the coolant, will predominantly shift toward the volatile gaseous species. As such, it will be processed and monitored with the gases that are removed from the reactor coolant. Iodine generally makes an insignificant contribution to the total exposure experienced at nuclear power plants, although there have been occasional situations in which iodine, which has dissolved in the coolant at the time of shutdown, has delayed shutdown operations pending its removal from the coolant system by ion exchange. Cesium, on the other hand, is readily soluble in reactor water, and therefore is removed from this water by the purification system. These purification systems generally use an ion exchange process, which removes the dissolved cesium from the coolant. Subsequently, the cesium ends up in the radioactive waste, where it is processed and packaged on site and shipped off site for disposal. For each of these particular isotopes, and similarly for most fission product isotopes, due to their chemical behavior and/or their low concentration in the coolant water, they contribute relatively insignificantly to the radiation fields that are present from deposition on the surfaces of the components of the reactor coolant system.

The other primary source of radiation fields is from the corrosion products generated at the surfaces of the iron-nickel alloys present as the primary constituent of the pressure boundary. The general corrosion that takes place on these surfaces releases dissolved metallic constituents, such as iron, nickel, manganese, chromium, and cobalt, into the cooling water. Some of these metallic ions are deposited either by particle deposition or by precipitation on hot fuel element surfaces in the core. There they are exposed to neutrons that are generated as a result of the fission process generating the power within the cores. These neutrons will cause the stable radioisotopes of these various metallic constituents to be transmuted into radioactive species. Two of the primary radioactive constituents generated here are ^{60}Co and ^{58}Co.

Cobalt-60 is generated as a result of neutron absorption on naturally occurring ^{59}Co. Although cobalt is present generally as an impurity in the iron-nickel alloys used in reactor power plants, its high susceptibility to neutron adsorption and the high energy of the radioactive emissions from ^{60}Co cause it to predominate over other radioisotopes formed from other elements that are present in higher concentrations. Cobalt-58 is formed as a result of a neutron knocking a proton from the nucleus of ^{58}Ni, a naturally occurring isotope of nickel.

Following the generation of radioisotopes from the deposits present on the surface of the fuel, various processes such as adsorption, dissolution, and erosion can cause these isotopes to be released from the fuel surface, returning them back to the reactor coolant. Transport with the reactor coolant will allow some of these to be deposited later and then incorporated into the growing corrosion films on the surfaces in the reactor coolant system.

To understand the various operational and design parameters that influence the buildup of radiation fields, it is necessary to discuss separately the BWR and the PWR. Such influencing features as materials present in the systems, the operational chemistry used in the systems, and the filtration and removal processes used differ sufficiently between these two to warrant separate discussions of these systems.

Boiling Water Reactors

In BWRs, the pressure in the reactor coolant system is maintained sufficiently low to allow boiling to take place. Consequently, the steam generated is fed directly to the turbine and is then condensed in the condenser. This condensate,

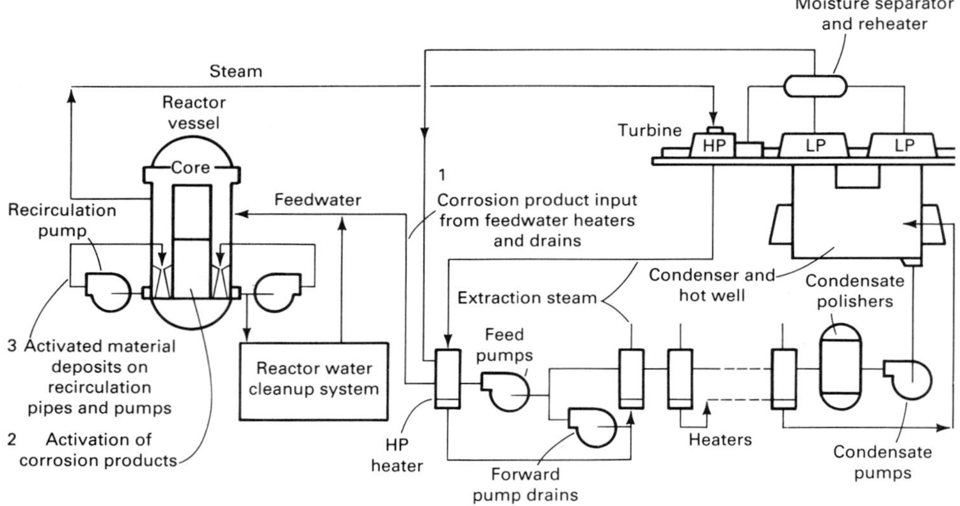

Fig. 47 Diagram of BWR circuit showing origin of radioactive contamination. HP, high pressure. LP, low pressure. Numbered areas indicate sequence of events.

Table 7 Estimates of principal sources of ^{59}Co for a BWR

System/component/alloy	^{59}Co input, g/yr
Forward-pumped heater drains	
Main steam valves + HP Turbine	95
Other valves	22
Nickel in carbon steel	2
Stainless steel	11
	130
Feedwater	
Valves	18
Stainless steel	5
Condensate treatment effluent	6
	29
Primary system	
Control blade pins/rollers	29
Jet pumps	23
Recirculating system valves	18
Stainless steel + nickel-chromium-iron alloy corrosion	7
Control rod drives	8
	85
Total	**244**
Off-line valve maintenance	30–90

which is collected in the hot well, is then purified through a condensate polisher. Such a condensate polisher can be either a deep bed ion exchange resin or a powdered resin. The former is more effective in removing dissolved material, whereas the latter is more effective in removing particulate material. This polished condensate water is then delivered through the feedwater heater section, where steam from the various turbine sections is used to preheat the feedwater prior to its return to the reactor vessel. Water in the reactor vessel that is not vaporized as steam is recirculated through a recirculation pipe and pump system back to the bottom of the reactor core, where it is fed upward through the core for cooling the fuel elements.

In considering the generation of corrosion products within a BWR in Fig. 47, a reasonable starting point is in the hot well. Here in the condensate, the primary species will be Fe_2O_3, which is produced primarily from steam impingement on the walls of the hot well. In addition, dissolved forms of other metallic species can be present, particularly from the condenser tubes, which may contain zinc, copper, aluminum, and other elements different from those found in the stainless steel used throughout the rest of the system.

The chemistry of the water exerts significant influence over the form and the amount of the corrosion products present. In the BWR, there are generally no additives. The objective is to keep the water as pure and clean as possible. The result is a neutral water chemistry with regard to pH and an intention of keeping the conductivity as low as possible. The latter is a measure of the impurities present in the system. In addition, as the water passes through the core, radiolysis (the dissociation of molecules by radiation) occurs, causing a small portion of the water molecules to be decomposed into its constituent elements—hydrogen and oxygen. The presence of these gases plus small amounts of air in-leakage, which occurs mostly in the region of the condenser and hot well, create an oxidizing condition in the cooling water.

The condensate polisher serves to remove impurities that are present in the condensate in the hot well. It generally performs its function quite effectively. The accepted measure of its effectiveness in the BWR is the conductivity of the outlet of the polisher.

As the feedwater proceeds up along the feedwater train, some of the impurities will deposit on surfaces and corrosion and other processes will cause materials to be leached from the system surfaces to the water. In addition, in some plants, the condensate in some of the feedwater heater drains is forward pumped into the system. That is, the drains are pumped directly into the feedwater lines at an appropriate point for the purpose of increasing thermal efficiency of the nuclear power plant. In so doing, this forward-pumped water has bypassed the condensate polisher, meaning that its impurities have not been removed by such a system. The result of all these processes is that at the final feedwater location, where the water is pumped into the reactor vessel, there is a concentration of impurities. The primary constituent will still be Fe_2O_3, but the presence of cobalt, nickel, copper, and zinc can influence the subsequent processes that determine radiation field buildup on pipe surfaces.

Hematite is the primary constituent of the deposit on the fuel surfaces. It also acts as the absorption medium for other constituents, including copper, zinc, and cobalt. For the BWR, ^{60}Co is the primary constituent of radiation fields, which generally contributes more than 95% of the radiation fields coming from plant surfaces. In addition to being present as an impurity in stainless steel, which makes up almost all the surface area outside of the Zircaloy core cladding, cobalt is present in small surface area cobalt-base alloy materials used in valves, pumps, and control rods. Table 7 shows the distribution of cobalt sources within a BWR.

Cobalt-60 has a relatively long half-life of 5.27 years. This means that any ^{60}Co generated in the system that is incorporated in the growing corrosion film on system surfaces will influence radiation fields in the system for quite a few years unless it is removed from such surfaces. The currently accepted model for corrosion product incorporation in a growing corrosion film on stainless steel surfaces is illustrated in Fig. 48. This illustration shows the particles of Fe_2O_3 and ^{60}Co being deposited on the surface and slowly incorporated in the growing thin film on the stainless steel. Steps 3 and 4 then show the development of the two layers and the corresponding ionic diffusion processes, finally resulting in an outer layer of Fe_2O_3 with an inner layer of Fe_3O_4. The Fe_3O_4 is generated in the oxygen-depleted region near the surface of the stainless steel.

The ^{60}Co and other impurity ions that are present in very low concentrations could be incorporated into this film in a variety of ways, including adsorption, crystallization, precipitation, and substitution. However, their presence is in such low concentration that chemical analyses

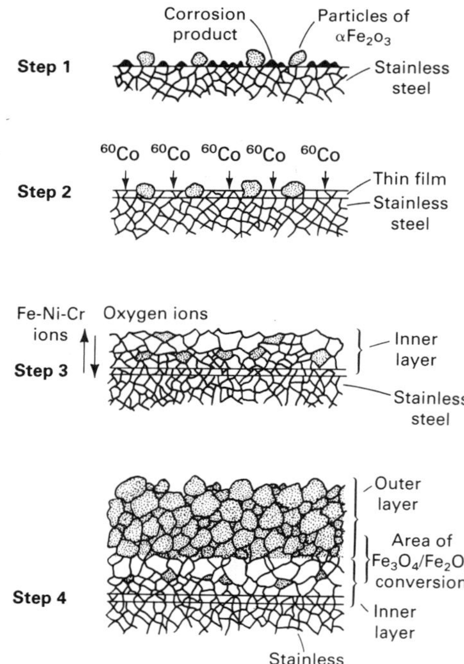

Fig. 48 Crud growth model for BWRs

have thus far been unable to determine the particular form of the ^{60}Co in this film.

It is important to note that the ^{60}Co is not in a volatile form and therefore stays in the water in the region near the reactor core. This means that the turbine, hot well, condenser, and feedwater systems are not significantly contaminated by this radioisotope or any other for that matter. Radiation fields in a BWR are predominantly, and almost exclusively, from the reactor pressure vessel, recirculation pipes and pumps, and the surfaces of the reactor water cleanup system.

Pressurized Water Reactors

In the PWR, higher pressure is sustained in the system; this prevents gross boiling so that the system is maintained liquid. The water is cooled by its passage through steam generator tubes. These tubes are made of Alloy 600 in U.S. PWRs. Secondary-side coolant on the outside of these steam generator tubes boils, creating steam that is used to turn the turbines in the system. In the primary circuit, after the water has passed through the steam generator tubes and through the pump, it is driven back through the core, where it is heated again. The interconnecting piping and the cladding of the pressure vessel are all of stainless steel. Therefore, in the PWR, the major surface areas are the zirconium alloy of the fuel cladding and the Alloy 600 of the steam generator tubes, with a secondary surface area of stainless steel. This system is illustrated in Fig. 49.

The water chemistry of the system is distinct from that in the BWR. In the PWR, boric acid is added as a neutronic control, and LiOH is added to create a basic pH condition in the primary loop coolant. In addition, a hydrogen overpressure is maintained on the coolant circuit, which creates a dissolved hydrogen concentration sufficient to recombine with the oxygenating species formed by radiolysis within the core. This creates a reducing condition in the PWR primary coolant.

The primary constituents released from the corrosion of the surfaces present in the PWR are iron and nickel. This fact, combined with the reducing condition present in the primary coolant, results in a deposit on the fuel that is nickel ferrite, where nickel replaces one of these iron atoms in Fe_3O_4 in a fraction of the lattice positions. Nickel will commonly replace one of the iron atoms in Fe_3O_4 in 30 to 60% of the molecules. Similar to the BWR, cobalt occurs within this deposit on the fuel. As a result of the presence of both cobalt and nickel on the fuel, ^{58}Co and ^{60}Co are both significant sources of radiation in the PWR system. Cobalt-58, however, has a much shorter half-life of 71 days and consequently is not retained from one refueling outage to the next as is ^{60}Co, but will be newly generated from fuel cycle to cycle.

The transport processes that conduct the cobalt radioisotopes to the out-of-core surfaces are similar to those for the BWR except that in the PWR, dissolution and precipitation probably play a much more significant role. These radioisotopes, however, will be transported to the out-of-core Alloy 600 and stainless steel surfaces, where they are incorporated in the corrosion product film as it develops on those surfaces.

The most significant region in the PWR for radiation control is within the channel heads of the steam generators. This is the region where maintenance and nondestructive examination of steam generator systems takes place. Consequently, this is also the region where most of the radiation exposures associated with primary circuit maintenance occur.

Radiation Control Techniques

There are a number of techniques that have been developed over the last few years that permit the operators of nuclear power plants to effect a reduction in the rate of buildup of radiation fields within their plants. These fall into four categories: water chemistry control, materials selection, surface treatment, and decontamination.

Water Chemistry Control. In BWRs, as was previously mentioned, the control of the amount of iron in the coolant is a key to reducing the deposition on the fuel and subsequent radiation field buildup. Control of the iron in the coolant is determined by the operation of the condensate polisher. The electrical conductivity of the water is used as the measure of determining how effective the condensate polisher is performing and is an indicator of any change in the performance of the polisher. Similarly, the reactor water cleanup system provides a kidney-type purification circuit for the water in the reactor vessel.

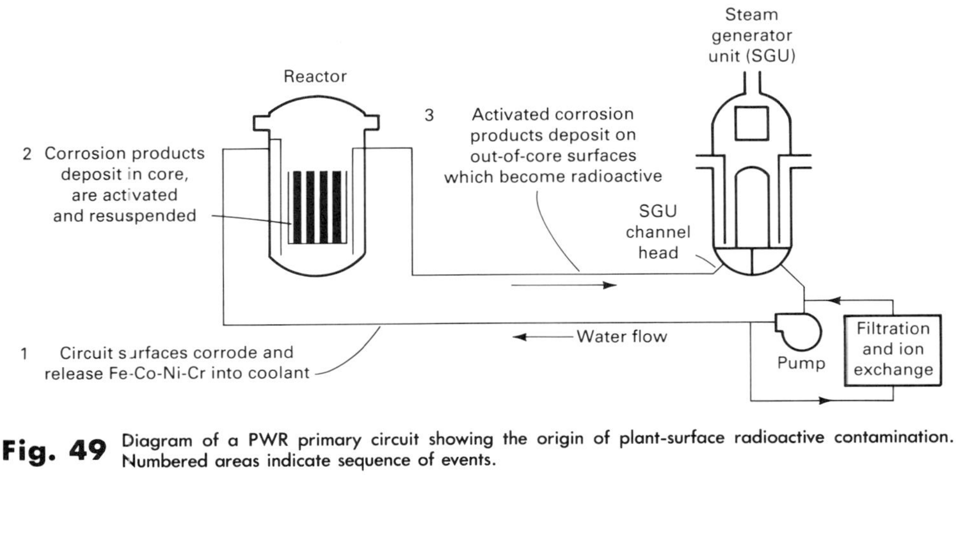

Fig. 49 Diagram of a PWR primary circuit showing the origin of plant-surface radioactive contamination. Numbered areas indicate sequence of events.

Dissolved oxygen also plays a role in BWRs. It has been determined that extremely low concentrations of oxygen, below of the order of 20 ppb, forms a corrosive that is very easily released from the steel surface. Consequently, it has been found desirable in BWRs to maintain the oxygen in concentration in the feedwater lines of 20 to 200 ppb.

Recent information indicates that certain metallic constituents can significantly influence the sites available for cobalt in the growing crystal lattices. In particular, the presence of zinc in the water has been found to interfere with the absorption of cobalt ions into the growing film. Consequently, small concentrations of zinc of the order of 5 to 10 ppb can reduce the deposition of ^{60}Co on pipe surfaces and therefore reduce the radiation field buildup in BWRs.

In PWRs where precipitation and dissolution from fuel surfaces have been shown to be a significant influencing factor in radiation field buildup, the pH of the coolant can be controlled in a manner to reduce the amount of precipitate on the fuel surfaces. LiOH that is added to the coolant is controlled so that as the boric acid is reduced through the fuel cycle, the LiOH is similarly reduced to maintain constant pH throughout the cycle. When this pH is high enough, in the region of 7.1 to 7.4 at operating temperatures, the thermodynamics of the solubility of nickel ferrite are such that nickel ferrite will not precipitate on the fuel at this pH. This reduction in the fuel deposition concurrently reduces the generation of cobalt isotopes and the subsequent buildup of radiation fields in the steam generator and on other surfaces.

Materials Selection. High-cobalt alloys are present in most nuclear reactor systems, usually in valves, pumps, and control rods. New materials containing low amounts of cobalt or no cobalt are being developed that are designed to serve the function of the high-cobalt alloys. These materials are currently in testing stages to show that their performance is sufficient for what is needed within the power plants. Some changes have already taken place, such as the replacement of pins and rollers in BWR control blades, which were previously of high-cobalt alloys and are now made of low-cobalt alloys. Such replacements offer significant opportunity for radiation field reduction in the future from ^{60}Co.

Surface Treatment. As both BWRs and PWRs are experiencing component replacements (recirculation pipes on the former and steam generators in the latter), the condition of the surfaces of these new components is also a consideration. Electropolishing has been shown to reduce the effective surface area up to a factor of five on pipe on steam generator surfaces. This is accomplished by trimming off the asperities present on the surfaces, thus creating a more uniform surface finish.

Decontamination solvents have been developed that are quite effective in removing the radioisotopes in the oxide layers that are present in previously operated nuclear power plants. These are used particularly when major maintenance and repair work is conducted on a piece of equipment that has very high radiation fields associated with it. Such decontamination can afford a utility an opportunity to make use of recently developed procedures, such as chemistry control or electropolishing, to reduce the subsequent radiation field buildups on the component.

SCC in Steam Turbine Materials

Floyd Gelhaus
Electric Power Research Institute

Stress-corrosion cracking is the fracture of a metal that results from the joint action of tensile stresses and a corrosive environment. Crack initiation and propagation are also dependent on metallurgical parameters (Ref 107). Cracking has occurred in both reheat and nonreheat turbines, in quenched-and-tempered 3.5NiCrMoV, and normalized-and-tempered 2.5NiCrMoV steels with a wide range of grain sizes. A typical intergranular crack is shown in Fig. 50. A single compilation of laboratory and field data, in conjunction with an annotated bibliography of more than 200 references, was published in 1984 (Ref 108).

Parameters of Influence

Stress. Until the introduction of the 500+ Mg (550 ton) monoblock rotor by Japanese steel manufacturers in the late 1970s, the large low-pressure steam turbine rotors that are utilized in nuclear power plants were fabricated by one of two techniques. Brown Boveri Company in Switzerland pioneered the welded rotor, in which several large disks are joined at the low-stress outer diameter with a sequence of gas tungsten arc and shielded arc welding processes. Other manufacturers, including the General Electric Company (Fig. 51) and the Westinghouse Electric Company (Fig. 52), have used the technique of thermally shrinking disks onto a stepped-diameter shaft, keying each disk/shaft interface to prevent independent rotation (Fig. 53 and 54). In addition to creating a zone of high local stress, the shaft keyways are flow/no-flow locations that promote the capture of chemical species. The welded rotor has no keyways. Since the catastrophic rupture of a shrunk-on disk at the Hinkley Point A Nuclear Station in England in September 1969 (Ref 109, 110), considerable attention has been given to understanding disk cracking and this potential failure mode.

Cracking has occurred not only in keyways but also on the bore surfaces, hub faces, web faces, and in the rim attachment area of shrunk-on disks (Fig. 55). Also, cracks have been found on the web faces and in the rim blade-attachment area of the integral disks of monoblock rotors. Monoblock and welded rotor configurations eliminate the keyway as a crack-starter location, but cracks persist in the rim blade-attachment area (Fig. 56).

Material. Most modern U.S. turbine rotors use 3.5NiCrMoV steel, conforming to the requirements of ASTM A471 (Class 1 through 9). Yield strength variation with tempering temperature is shown in Fig. 57, and the A471 temperature boundary for all classes (>839 K, 565 °C, or 1050 °F) is specifically noted. Chemical compositions for A471 Class 6 and A470 Class 5 are shown in Table 8.

Early correlations of crack growth rate data revealed a dependence on yield strength and an inverse dependence on temperature. This equation is generally expressed as:

$$\ln \dot{a} = C_1 - \frac{C_2}{T} + C_3\, s_y \qquad \text{(Eq 14)}$$

where $\dot{a}$ is the crack growth rate, T is the disk operating temperature, s_y is the room-temperature yield strength, and C_1, C_2, and C_3 are fitting constants. A recent analysis identified two new key variables: composition and tempering temperature (Ref 111). The latter variable does not exhibit a strong effect for materials meeting A471 tempering temperature/yield strength standards, and for these, the correlation with yield strength is sufficient. However, for NiCrMoV and other low-alloy steels, tempering temperature is a key variable, with a stronger but separate effect on crack growth rate as compared to yield strength.

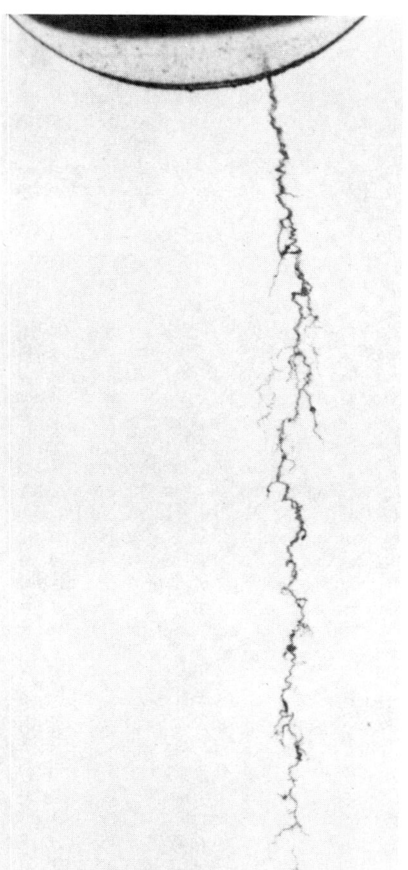
(a)

(b)

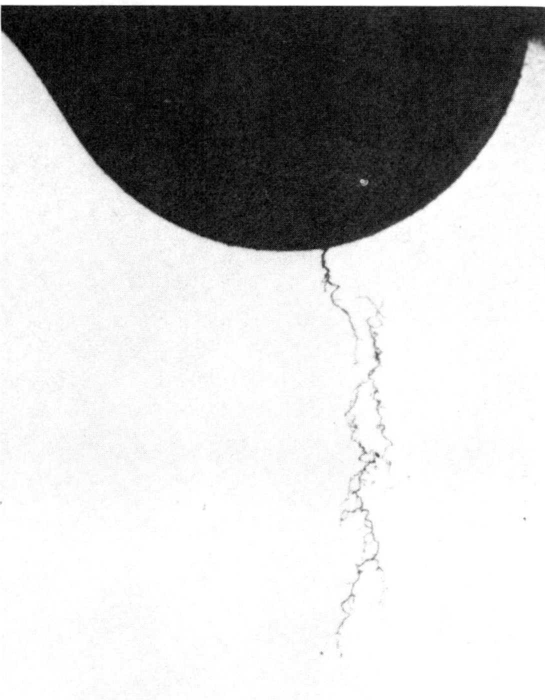
(c)

Fig. 50 Typical keyway intergranular stress-corrosion crack. Sections are orientated normal to the length of the keyway at different distances from the outlet face. (a) 28 mm (1.1 in.). (b) 10 mm (0.4 in.). (c) 75 mm (3 in.). Source: EPRI Report NP-3341

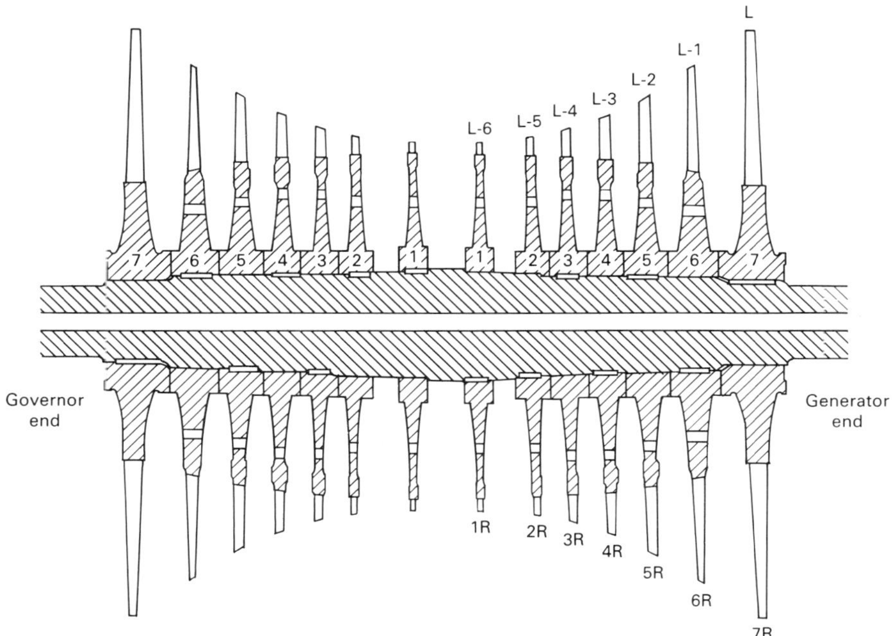

Fig. 51 Schematic of General Electric low-pressure turbine rotor. All disks are shrunk on, with one blade row per disk. Source: EPRI Report NP-2429-LD, Vol 6

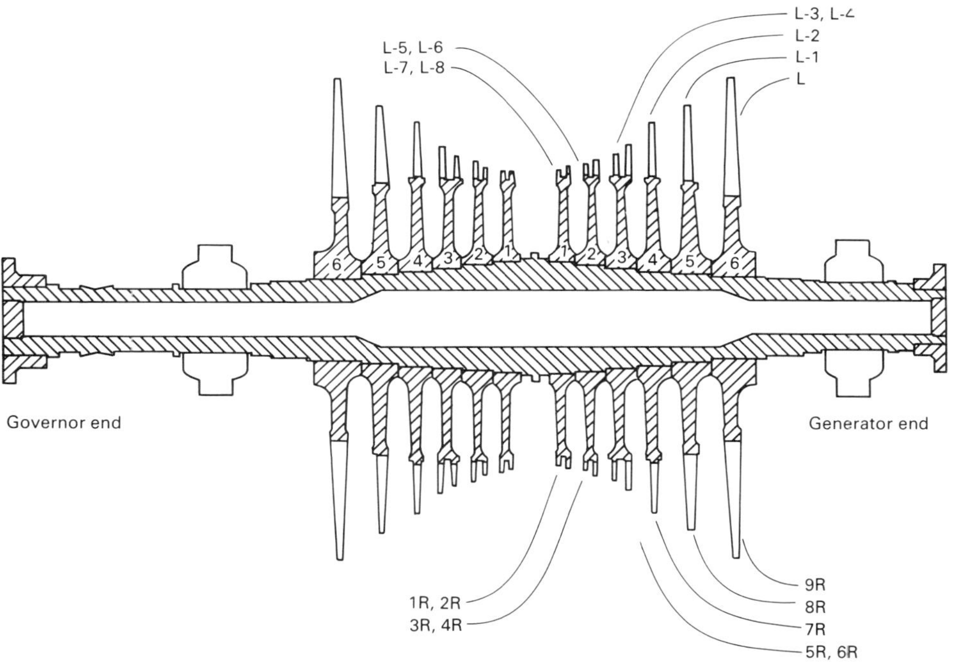

Fig. 52 Schematic of Westinghouse low-pressure turbine rotor. All disks are shrunk on, and the number of blade rows per disk is as indicated. Source: EPRI Report NP-2429-LD, Vol 6

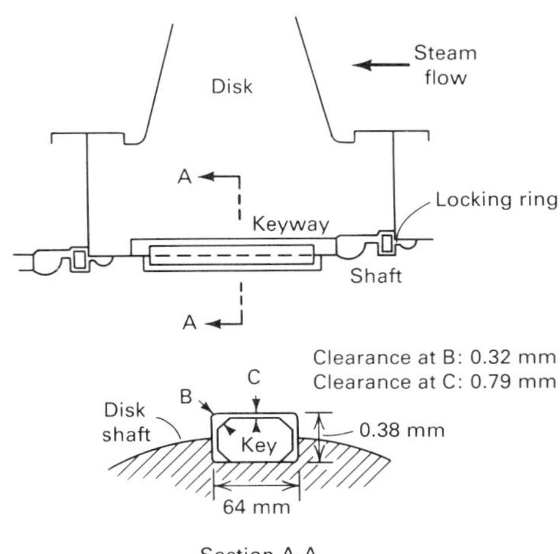

Fig. 53 Schematic of keyway design reportedly used by General Electric in shrunk-on disks of low-pressure turbine rotors. Source: EPRI Report NP-2429-LD, Vol 6

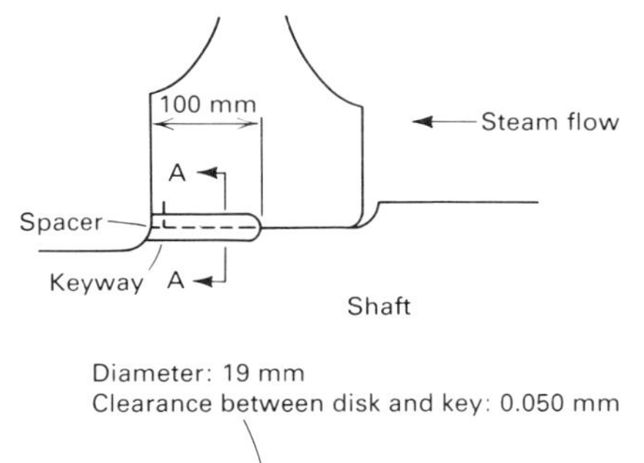

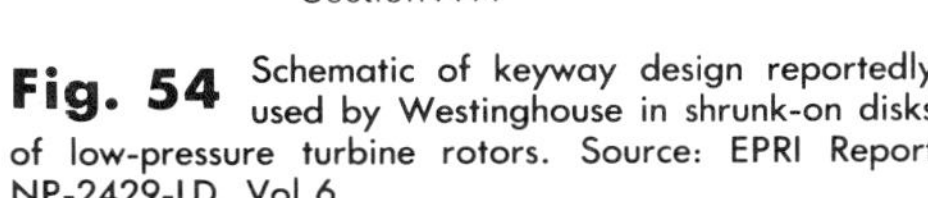

Fig. 54 Schematic of keyway design reportedly used by Westinghouse in shrunk-on disks of low-pressure turbine rotors. Source: EPRI Report NP-2429-LD, Vol 6

Regarding composition, manganese was indicated by this analysis to be a strongly correlated variable, and Eq 14 was modified to include that effect by adding a C_4Mn term, using the weight percent of manganese. This manganese dependence should be interpreted as an unresolved manganese/nickel/vanadium effect, because nickel and vanadium were highly correlated with manganese. Sulfur concentration was also identified as a key variable, but no model containing a sulfur term was calibrated to the data. For the two-, three-, and four-variable model, Eq 14 is:

$$\ln \dot{a} = -8.8 - \frac{4040}{T} + 0.0231\, s_y \quad \text{(Eq 15a)}$$

$$= -4.74 - \frac{9270}{T} + 0.3337\, s_y + 4.5\text{Mn} \quad \text{(Eq 15b)}$$

$$= -4.74 - \frac{9270}{T} + 0.0337\, s_y + 4.53\text{Mn} \quad \text{(Eq 15c)}$$

$$= 7.04 - \frac{9270}{T} + 0.0337\, s_y + 4.53\text{Mn} \quad \text{(Eq 15d)}$$

where $\dot{a}$ is given in inches per hour, T is in degrees Rankine, s_y is in kips per square inch, manganese (Mn) is in weight percent, and T_t, the tempering temperature, is in degrees Kelvin. Three of the Eq 15b constants were forced not to change when including the T_t term; Eq 15c permits a variety of low-alloy steels (single or multiheat) to be analyzed on a common basis.

Laboratory experiments to determine the effect of the segregation of phosphorus to the grain boundaries have shown little effect of this impurity on SCC at high stress levels (Ref 112, 113). All speci-

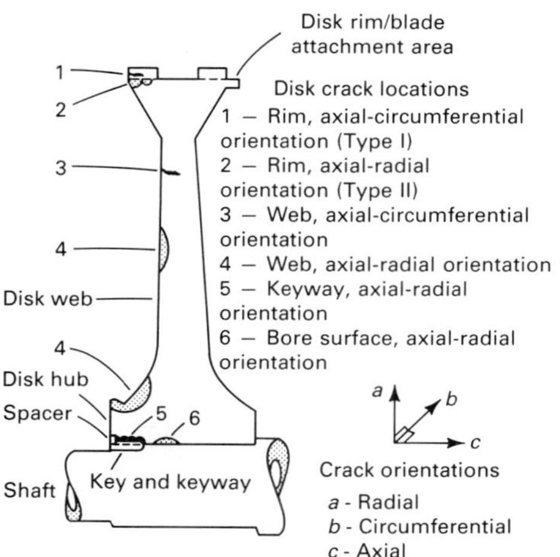

Fig. 55 Schematic showing typical locations and orientations of cracks in various U.S. low-pressure rotor disks. This illustration is of a shrunk-on disk with a semicircular keyway and an axial-entry fir-tree rim attachment configuration. Disks of this type have experienced cracking at all of the locations illustrated except at the central web in the axial-radial orientation and on the web below the rim (location 3). Cracks at the latter two locations have been found in integral disks. Source: EPRI Report NP-2429-LD, Vol 1

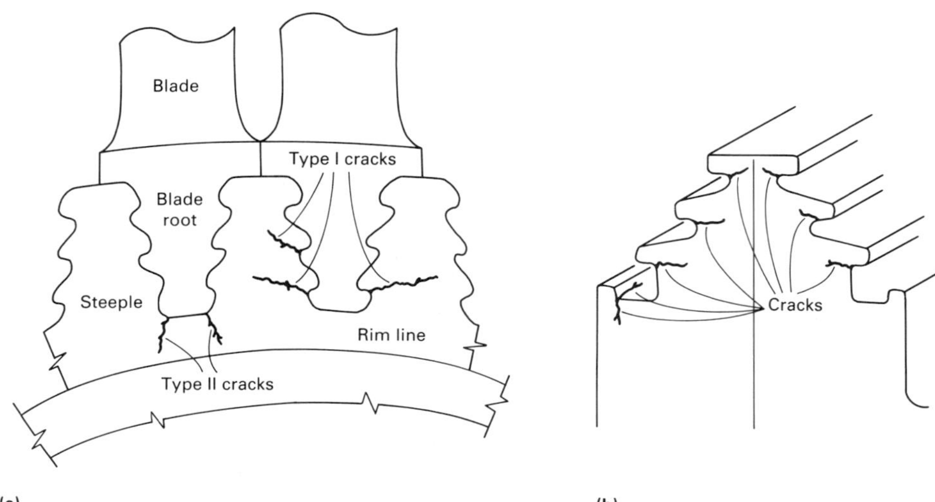

Fig. 56 Schematics of rim attachment configurations used in U.S. low-pressure rotors in which cracking has been found. Locations and orientations of rim attachment cracking experienced in various rotors are illustrated. (a) Axial-entry fir-tree. (b) Notch-entry dovetail. Source: EPRI Report NP-2429-LD, Vol 1

mens in both experiments failed in less than 500 h. When the test solution is neutral, segregated phosphorus has no effect on the resistance of an alloy to SCC. The response to a highly caustic solution (9 M NaOH) was, however, dependent on the applied potential, changing from a strong increase in the cracking susceptibility at −400 mV (Hg/HgO) to no effect at −800 mV (Hg/HgO).

Environment. Potential is a thermodynamic parameter that is a measure of the energy of a chemical reaction. It is a measure of the oxidizing power of a solution. For a corrosion reaction, potential is a measure of the energy necessary to cause a metal atom to be transferred from the metal lattice into solution as a metal ion. Both temperature and pH can either raise or lower the potential of a corrosion reaction, depending on the species involved and on other factors.

With no applied voltage, the rate of metal atom transfer would reach equilibrium, and those chemical processes would establish the free corrosion potential. To investigate the range of possible chemical reactions, a variable voltage is applied, and a polarization diagram is generated.

Figure 58(a) shows an anodic polarization diagram for a metal-solution combination that forms a protective corrosion layer (passivation film) over a range of potentials (E_{pr}). Figure 58(b) includes a second area of anodic activity ($E_{crit(2)}$), a phenomenon that can be associated with a redox reaction or with the dissolution of a second element, such as chromium, in low-alloy steels.

The importance of pH is shown by the curves in Fig. 59(a). The additional protection offered by neutral-to-higher pH in this example generally holds true for all the metals-environment combinations in steam turbines (type 403 stainless steel is used for turbine blades); therefore, the working fluid pH is held neutral for BWRs and between 8.9 and 9.6 for PWRs.

Figure 59(b) shows the influence of temperature at the neutral pH 7. The corrosion potential becomes more negative with increasing temperature.

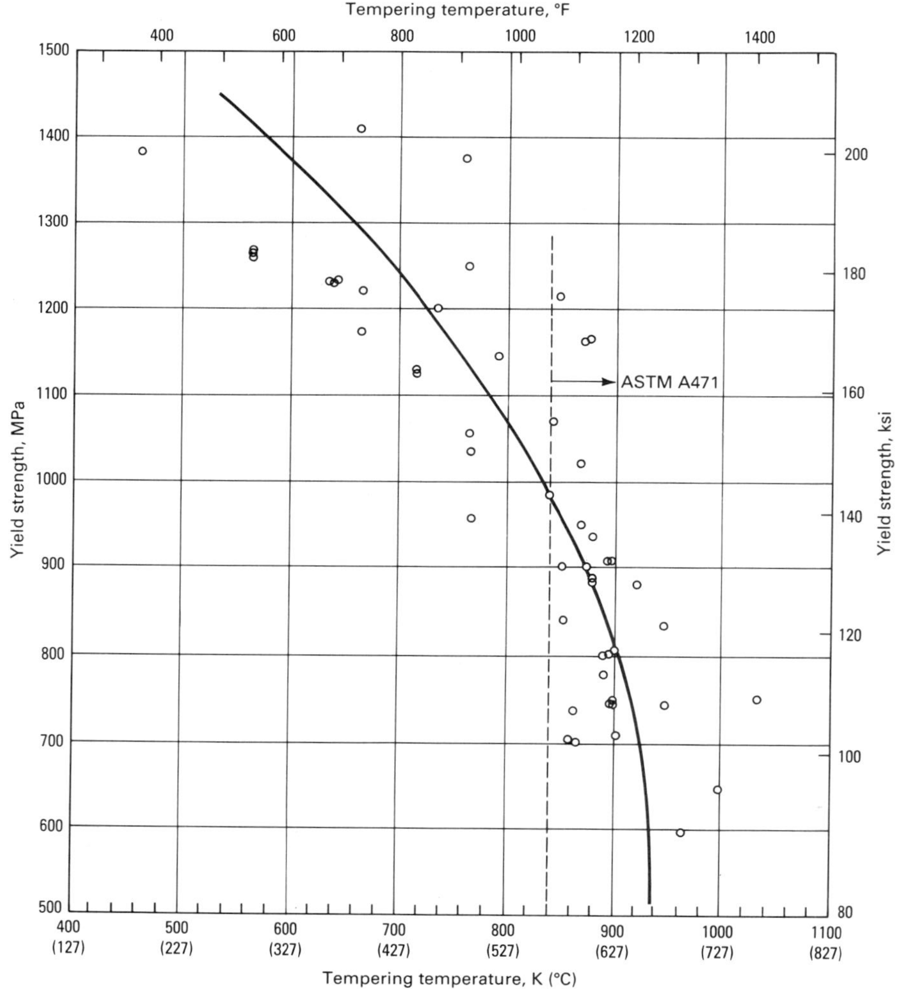

Fig. 57 Yield strength versus tempering temperature for low-alloy steels. Source: EPRI Report NP-4056

Table 8 Chemical specifications for representative steels used in modern U.S. low-pressure rotors with integral and shrunk-on disks

Element, wt%	ASTM A470 Class 5 steel for integral rotor disks	ASTM A471 Class 6 steel for shrunk-on rotor disks
Carbon (max)	0.28	0.28
Manganese	0.20–0.60	0.70 max
Phosphorus (max)	0.015	0.015
Sulfur (max)	0.018	0.015
Silicon (max)(a)	0.10	0.10
Nickel	3.25–4.00	2.00–4.00
Chromium	1.25–2.00	0.75–2.00
Molybdenum	0.25–0.60(b)	0.20–0.70
Vanadium	0.05–0.15	0.05 min

(a) Modern U.S. turbine rotor disks are made from vacuum-deoxidized steels for which the silicon limit is 0.10%. For nonvacuum-deoxidized steels, A470, Class 5 allows a silicon range of 0.15–0.30%, and A471, Class 6 allows a silicon range of 0.15–0.35%. (b) If desirable because of operating temperatures, a minimum molybdenum content of 0.40% may be specified by the purchaser. Source: EPRI Report NP-3634, Vol 2

The lower temperature allows the protective layer to remain intact over a much larger range of potential. It is important to note that, although type 403 stainless steel does suffer pitting (for example, at pH 10 in a solution of NaCl), the alloy tempered at 650 °C (1200 °F) is quite resistant to SCC.

The potential is most strongly influenced by the kind and concentration of species in solution, causing marked shifts in the corrosion rate (that is, the current density). Figure 59(c) shows how the polarization curve shifts with species kind, and Fig. 59(d) and (e) indicate the influence of concentration for the same species. The different behaviors exhibited by these latter curve sets are evidence of the complexity of the corrosion processes and point out that the behavior of an alloy in service is difficult to state in general terms, because the service environment is often complex and ill-defined. However, the general trends from laboratory tests help define what service conditions are required to get the observed response. For intergranular SCC, polarization curves for bulk alloys are helpful, but the real need is for data showing how grain-boundary compositions behave.

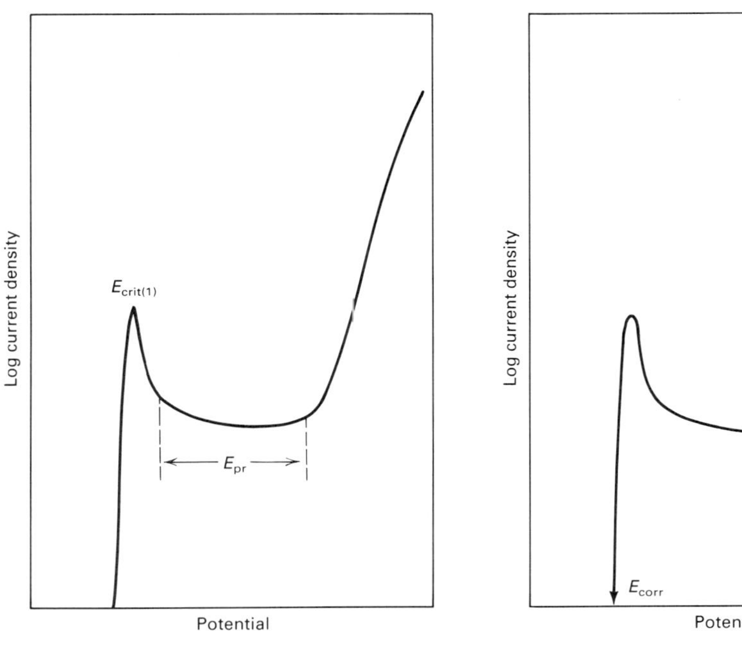

Fig. 58 Schematic of anodic polarization curves for (a) a metal that forms a protective layer and (b) an alloy that includes a second passivation peak where E_{corr} is the corrosion potential, E_{crit} is the critical potential of passivation, E_{ox} is the oxygen evolution potential, and E_{pr} is the passive potential range.

Laboratory Tests

Considerable data were obtained by U.K. experimenters following the failure at Hinkley Point, but most of this was generated using 3CrMo steel. Later work in the U.S. focused more on the 3.5NiCrMoV alloy, but much of these data were taken using strong caustic environments (35% NaOH). Table 9 summarizes the results of these earlier experiments. Analyses of these data show that several environments—for example, >1% NaOH, aerated/oxygenated water—produced higher crack growth rates than were estimated from the field data; models that are derived to help a utility predict the lifetime of turbine components (Eq 15) have to be based on laboratory data that approximate field conditions.

In-Plant Conditions

The majority of U.S. PWR power plants control steam generator water pH using phosphates (Na_3PO_4) or all-volatile treatment, that is, control by ammonia (NH_3) or by amines such as morpholine (C_4H_9NO). The major advantage for phosphate treatment is its capacity to buffer against both acidic and basic upsets, a protection only weakly afforded by the amines. The volatile amines introduce no solids into the system, whereas the salts added in the phosphate method can accumulate and concentrate. No chemical additions are made to control feedwater pH in BWR power plants.

Oxygen levels in power plant feedwater are controlled by adding scavenging chemicals (hydrazine, N_2H_4) or by mechanical means. Only two U.S plants currently use a feedwater deaerator, with most using only air ejectors on the deaerating section of the condensers. In BWRs, the reactor water oxygen level can reach several parts per million during start-up. Chemical upsets can be initiated by a condenser tube failure, and the analyses of chemicals that can be introduced is shown in Table 10. Air in-leakage rates of 2800 L/min can be experienced with severe condenser leaks, while a rate one-tenth that level is common. With 500 L/min in-leakage, the oxygen concentration in the condensate can increase to 30 ppm. Powder resin filters and deep bed demineralizers are two common full-flow condensate polishing systems used in U.S. plants. However, if not properly operated/regenerated, these systems can become sources of ionic impurities.

Table 11 shows typical power plant data, and such data are used to infer the concentration in the steam reaching the turbine systems. Because the solubility of salts in steam decreases as temperature and pressure are decreased, solubility limits are exceeded at some stage with the expansion of the steam in the turbine. The thermodynamic data required to predict the exact conditions under which various salts will deposit are not available. However, scrapings taken from turbine surfaces indicate that condensation to form concentrated solutions in the crevice areas (keyways, blade attachments) does occur and that SCC is a typical result. Field data on crack growth rate are comparable with lab data using deaerated water, steam, carbonated water, and 1% NaOH + 0.1% NaCl (Ref 53).

Data Analysis Results

Equation 15 gives the relationship between crack growth rate and key material/environmental parameters. Those relationships were developed using both laboratory and field data. Failure rate analysis of the field data show that rim crack growth rates are higher than the rates for cracks located at the keyway but, for an equivalent number of operating hours, that the keyway is the preferential cracking site. Figure 60 shows these data; rim, face, and bore data are combined because they are similar, and the three-parameter Weibull fit does indicate that this incidence of cracking increases slightly more rapidly than incidences of keyway cracking. A comparison of the two U.S. manufacturers of turbines used in nuclear plants is shown in Fig. 61. Although the cumulative fractions of rotors experiencing keyway cracking is equivalent for 3- to 9-year-old turbine systems, the data show a more rapid increase in failure rate with plant age in General Electric turbines. Figure 62 results from adding data from NEI Parsons (United Kingdom) and from AEG (South Africa), and the two hypothetical points assume that a crack in the oldest known Brown Boveri & Co. Ltd. (Switzerland) and in the Kraftwerk Union (Germany) rotors was discovered in mid-1984. The predicted failure rates for Brown Boveri & Co. Ltd. and Kraftwerk Union are lower than the observed rates for the other manufacturers.

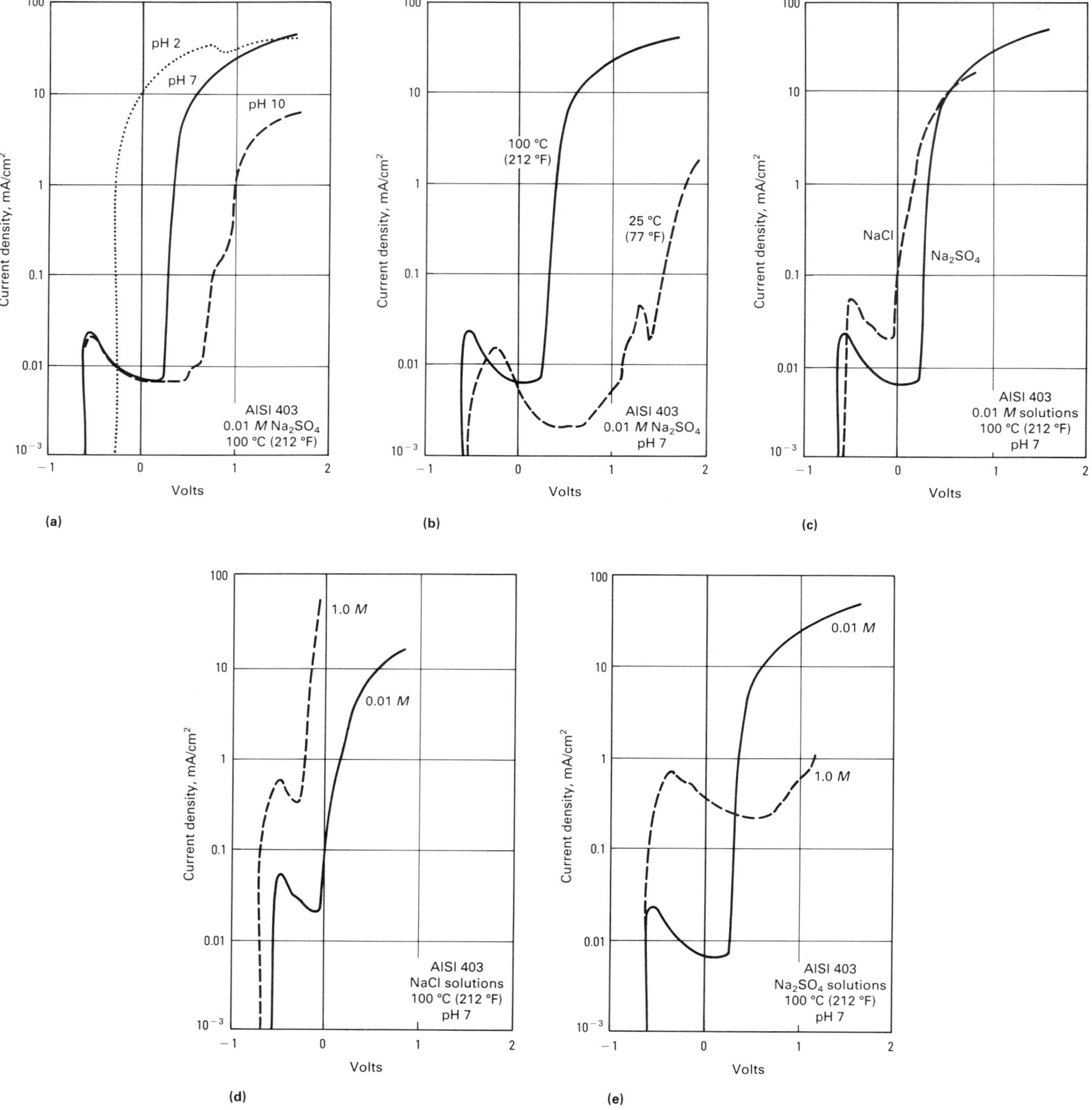

Fig. 59 Anodic polarization curves for type 403 stainless steel showing the effects of solution pH (a), temperature (b), different neutral pH solutions of the same molar concentration (c), sodium chloride (NaCl) solution concentration (d), and sodium sulfate (Na_2SO_4) solution concentration (e). In each of these figures, the current density is proportional to the corrosion rate.

Life Prediction Methodology

The key decision facing the turbine owner who has some reason to believe that disk cracks exist is whether to run, repair, or replace. Nondestructive examination indications may give evidence of a crack, but even the best techniques leave a large uncertainty in crack size and sometimes also in crack orientation. Nonetheless, a decision has to be made regarding the growth of this flaw. Some obvious questions include:

- Is it safe to continue to run with the crack growing?
- Is repair necessary while a new disk is on order (a whole stage of blades is often removed, and a pressure drop plate inserted, and the turbine run at a much reduced output)?
- Is immediate replacement called for because the suspected flaw is near critical size?

Life estimation calculations are made to provide insight for this complex decision. These predictions require input concerning the crack growth

Table 9 Collation of crack growth rates from precracked wedge-opening load specimens

Steel and condition	Environment	Temperature °C	Temperature °F	Yield strength(a) MPa	Yield strength(a) ksi	Potential, mV	Stress intensity MPa $\sqrt{m}$	Stress intensity ksi $\sqrt{in.}$	Crack growth rate mm/yr	Crack growth rate in./yr
3CrMo (AOH), embrittled	35% NaOH	115	240	700	102	−400(b)	18–39	16–35	50–278	1.9–10.9
	35% NaOH	115	240	700	102	−600(b)	13–18	12–16	63–83	2.5–3.3
	35% NaOH	115	240	700	102	−800(b)	8–36	7–33	57–631	2.2–24.8
	35% NaOH	115	240	700	102	−900(b)	18	16	71	2.8
3CrMo (AOH), de-embrittled	35% NaOH	85–115	185–240	700	102	−800(b)	13–38	12–34	13–85	0.5–3.4
	35% NaOH	85–115	185–240	700	102	−600(b)	19	17	52	2.05
	35% NaOH	85–115	185–240	700	102	−800(b)	7–37	6–34	47–662	1.8–2.6
	35% NaOH	85–115	185–240	700	102	−900(b)	20	18	4–39	0.16–1.5
3CrMo (AOH), as-received	35% NaOH	85–115	185–240	700	102	−800(b)	10–20	9–18	6.3–328	0.25–4.8
3CrMo (AOH), de-embrittled	35% NaOH	85–115	185–240	710	103	−800(b)	10–20	9–18	6.3–136	0.25–5.4
	35% NaOH	100	212	700	102	−800(b)	20	18	35–91	1.4–3.6
	10% NaOH	100	212	700	102	−800(b)	20	18	8–14	0.3–0.5
	4% NaOH	100	212	700	102	−800(b)	20	18	0.6–2.5	0.02–0.1
2NiCrMoV	35% NaOH	115	239	786	114	−250(b)	10–66	9–60	0.4–5	0.016–0.2
	28% NaOH(c)	110	230	615–1115	89–162	−250(b)	30–60	27–55	1–1135	0.04–44.7
3.5NiCrMoV (VCD)	25% NaOH	90	195	821	119	. . .	16–126	14.5–115	0.3–4.7	0.01–0.18
	25% NaOH	90	195	653	95	. . .	32–126	29–115	0.1–0.3	0.004–0.01
3CrMo (BE)	25% NaOH	90	195	681	99	. . .	32–48	29–44	1.2–12.6	0.05–0.5
3.5NiCrMoV	10% NaOH(d)	157	315	1124	163	−670 to −780(e)	22–88	20–80	4–1182	0.16–46.5
	10% NaOH(d)	157	315	876	127	−670 to −780(e)	22–88	20–80	3–334	0.01–13
	10% NaOH(d)	157	315	820	119	−670 to −780(e)	22–88	20–80	4–123	0.16–4.8
	10% NaOH(d)	157	315	731	106	−670 to −780(e)	22–66	20–60	7–244	0.3–9.6
	10% NaOH(d)	157	315	717	104	−670 to −780(e)	22–88	20–80	3.6–78	0.14–3.1
3.5 NiCrMoV	10% NaOH(c)	157	315	1124	163	30 to −210(e)	22–88	20–80	18–259	0.7–10.2
	10% NaOH(c)	157	315	876	127	30 to −210(e)	22–88	20–80	1–32	0.04–1.3
	10% NaOH(c)	157	315	820	119	30 to −210(e)	22–88	20–80	3.6–78	0.14–3.1
	10% NaOH(c)	157	315	731	106	30 to −210(e)	22–88	20–80	2–66	0.08–2.6
	10% NaOH(c)	157	315	717	104	30 to −210(e)	22–88	20–88	3.6–78	0.14–3.1
2Cr1Mo	10% NaOH(d)	157	315	706	102	30 to −210(e)	66	60	7.4	0.3
3Cr0.5Mo	8% NaOH	115	240	505–540	73–78	−354(e)	20	18	0.3–82	0.01–3.2
3NiCrMoV	1% NaOH + 0.1% NaCl(d)	157	315	1124	163	−700 to −760(e)	33–66	30–60	7–78	0.28–3.1
	1% NaOH + 0.1% NaCl(d)	157	315	876	127	−700 to −760(e)	33–66	30–60	2–4.4	0.08–0.17
	1% NaOH + 0.1% NaCl(d)	157	315	820	119	−700 to −760(e)	33–66	30–60	0.9–5.7	0.04–0.22
	1% NaOH + 0.1% NaCl(d)	157	315	731	106	−700 to −760(e)	33–66	30–60	3	0.12
3.5NiCrMoV	1% NaOH + 0.1% NaCl(f)	157	315	1124	163	120–170(e)	33–66	30–60	2.2–53	0.09–2.1
	1% NaOH + 0.1% NaCl(f)	157	315	876	127	120–170(e)	33–66	30–60	0.5	0.02
	1% NaOH + 0.1% NaCl(f)	157	315	821	119	120–170(e)	33–66	30–60	. . .	. . .
	1% NaOH + 0.1% NaCl(f)	157	315	731	106	120–170(e)	33–66	30–60	0.9–3.1	0.04–0.12
	1% NaOH + 0.1% NaCl(f)	157	315	717	104	120–170(e)	33–66	30–60	0.5–5.8	0.02–0.23
3.5NiCrMoV	Pure water(d)	157	315	1124	163	. . .	22–88	20–80	0.7–23	0.03–0.9
				876	127	. . .	22–66	20–80	0.4–4.5	0.016–0.18
				821	119	. . .	22–88	20–60	1.3–3.6	0.05–0.14
				731	106	. . .	22–66	20–60	0.4–4.2	0.016–0.165
				717	104	. . .	22–66	20–60	0.5–1.1	0.02–0.04
2Cr1Mo	Pure water(d)	157	315	706	102	. . .	66	60	4.2	0.165
3.5NiCrMoV	Pure water(f)	157	315	1124	163	30 to −200(e)	33–66	30–60	76–197	3–7.8
	Pure water(f)	157	315	876	127	30 to −200(e)	33–66	30–60	2–16	0.08–0.25
	Pure water(f)	157	315	820	119	30 to −200(e)	33–66	30–60	3.5–8	0.14–0.3
	Pure water(f)	157	315	731	106	30 to −200(e)	33–66	30–60	1.6–4.6	0.06–0.18
	Pure water(f)	157	315	717	104	30 to −200(e)	33–66	30–60	1.6–20	0.06–0.8
3.5NiCrMoV	Pure water(c)	157	315	1124	163	235 to 290(e)	25	23	3.2	0.13
	Pure water(c)	157	315	876	127	235 to 290(e)	33–63	30–57	11	0.4
	Pure water(c)	157	315	821	119	235 to 290(e)	32–62	29–56	. . .	. . .
	Pure water(c)	157	315	731	106	235 to 290(e)	32–66	29–60	0.9	0.035
	Pure water(c)	157	315	717	104	235 to 290(e)	32–65	29–59	4.5	0.18
3.5NiCrMoV	Pure water(g)	157	315	1124	163	130 to 250(e)	33–66	30–60	13.8–70	0.54–2.8
	Pure water(g)	157	315	876	127	130 to 250(e)	33–66	30–60	3–7.2	0.12–0.28
2Cr1Mo	Pure water	157	315	821	119	. . .	33–66	30–60	7.2	0.28
26NiCrMoV127	Pure water(d)	100	212	760	110	. . .	10–110	9–100	0.2–0.6	0.008–0.02
	Pure water(d)	100	212	1220	177	. . .	10–115	9–105	0.6–3248	0.02–128
	Pure water	100	212	635	92	. . .	30–60	27–55	0.06–19000	0.002–748
3CrMo (AOH)	Water	90	195	722–758	105–110	. . .	66	60	0.35–1.4	0.014–0.06
3.5NiCrMoV	Water	90	195	910	132	. . .	66	60	0.04–0.05	0.0016–0.002
3CrMo (AOH)	Steam	90	195	722–758	105–110	. . .	11–112	10–102	0–0.9	0–0.035
3.5 NiCrMoV	Steam	90	195	910	132	. . .	66	60	0.9–1.9	0.035–0.07

(continued)

AOH, acid open hearth; BE, basic electric; VCD, vacuum carbon dioxide.
(a) Yield strength values were intentionally varied by heat treatment. (b) Potential with respect to Hg/HgO. (c) Oxygenated. (d) Deaerated. (e) Potential with respect to standard hydrogen electrode (SHE). (f) Aerated. Source: Ref 114

Table 9 (continued)

Steel and condition	Environment	Temperature °C	Temperature °F	Yield strength(a) MPa	Yield strength(a) ksi	Potential, mV	Stress intensity MPa $\sqrt{m}$	Stress intensity ksi $\sqrt{in.}$	Crack growth rate mm/yr	Crack growth rate in./yr
3CrMo (AOH)	Steam	90	195	722–758	105–110	. . .	66	60	0.9–1.9	0.035–0.07
3CrMo (BE)	Steam	90	195	745–772	108–112	. . .	66	60	1.2–1.7	0.05–0.067
3.5NiCrMoV	Steam	90	195	910	132	. . .	66	60	0.3–1.3	0.012–0.05
2NiCrMoV	Steam	90	195	786–848	114–123	. . .	66	60	1	0.04
4.5NiCrMoV	Steam	90	195	896–938	130–136	. . .	66	60	1.9–3.5	0.07–0.14
3CrMoV	Steam	90	195	862–924	125–134	. . .	5–110	4.5–100	1.9–3.5	0.07–0.14
3.5NiCrMoV	Steam	120	250	896	130	. . .	66	60	3.1	0.12
3CrMo	Steam	120	250	772	112	. . .	66	60	13	0.5
3CrMo	Steam	120	250	772	112	. . .	5–110	4.5–100	1.6–3.2	0.06–0.13

AOH, acid open hearth; BE, basic electric; VCD, vacuum carbon dioxide.
(a) Yield strength values were intentionally varied by heat treatment. (b) Potential with respect to Hg/HgO. (c) Oxygenated. (d) Deaerated. (e) Potential with respect to standard hydrogen electrode (SHE). (f) Aerated. Source: Ref 114

Table 10 Typical cooling water analyses

Element	Fresh river water(a)	Fresh lake water(b)	Brackish water(c)	Seawater(d)
Calcium (ppm)	58	32	44	400
Magnesium (ppm)	15	11	78	1272
Sodium (ppm)	13	3.2	603	10 561
Potassium (ppm)	(e)		20	380
Lead (ppm)	(e)	0.004	(e)	0.21
Chloride (ppm)	4.8	2.1	1053	18 980
Bicarbonate (ppm)	217	149	68	142
Total alkalinity (ppm as calcium carbonate, $CaCO_3$)	178	(e)	56	(e)
Fluoride (ppm)	(e)	0.25	0.08	3.5
Bromide (ppm)	(e)	(e)	3.5	65
Sulfate (ppm)	45	7	220	2649
Nitrate (ppm)	(e)	1.6	1.2	10^{-3} to 7×10^{-1}
Phosphate (ppm)	(e)	0.6	(e)	10^{-3} to 10^{-1}
Silica (ppm)	14	5	8.6	0.01 to 7.0
Carbon dioxide (ppm)	(e)	3.8	2.9	6
Oxygen (ppm)	(e)	(e)	6.2	5
pH	(e)	(e)	6.2	5

(a) Mississippi River. (b) Lake Michigan. (c) Estuary on U.S. East Coast. (d) Typical ocean water. (e) Not determined. Source: EPRI Report NP 2429, Vol 4

rate (which requires material and environmental information), the stress levels as a function of operational conditions (changing thermal and load boundary conditions) within the geometry containing the crack, applicable stress concentration factors, the initial flaw size, shape, and orientation, material fracture toughness, and stress intensity expressions.

Fortunately, powerful numerical tools are available for these fracture mechanics life predictions (Ref 115) and can conveniently be used in conjunction with standard finite element temperature and stress distribution codes that calculate those conditions for the uncracked geometry. Simplifying assumptions/techniques are reasonable for most keyway cracking investigations. However, the blade attachment area requires a three-dimensional analysis. The nondestructive inspection techniques for this disk perimeter region are much less reliable than those for the bore and keyway. Also, vibration can be present at levels sufficient to induce corrosion fatigue, introducing the need for a new set of material-environment data, most of which are not currently available. The step-by-step procedures for a lifetime prediction analysis are given in Ref 114.

Solution Options

Remedies to eliminate and/or reduce SCC have been suggested and applied that focus on each of the three variables involved: stress, environment, and material. Both British and American manufacturers have offered designs that move the keyway from the disk/shaft interface to a lower-stress location along the common surface of adjacent disk faces (Fig. 63). Neither the new monoblock design nor the welded-rotor design has a keyway.

In producing the ingots required for the monoblock rotors, many significant metallurgical achievements have been demonstrated. Very low sulfur and phosphorus levels (for example, 30 ppm) are typical when using a combination of ladle refining and electric arc furnaces. Multiple heat treatments develop fine grain size after vacuum degassing has lowered both hydrogen (1 ppm) and oxygen levels (25 ppm). Although clean steel (Ref 116) SCC data are sparse, empirical data and analyses indicate better SCC resistance. This is particularly important, because blade at-

Table 11 High-pressure drain (HPD), moisture separator drain (MSD), and condensate (Cond.) chemistry during a power escalation

Nuclear PWR with recirculating steam generator

Power level, %	Na⁺(a) HPD(b)	Na⁺(a) MSD(c)	Na⁺(a) Cond.(d)	Cl(a) HPD(b)	Cl(a) MSD(c)	Cl(a) Cond.(d)	Mg^{2+}(a) HPD(b)	Mg^{2+}(a) MSD(c)	Mg^{2+}(a) Cond.(d)	Ca^{2+}(a) HPD(b)	Ca^{2+}(a) MSD(c)	Ca^{2+}(a) Cond.(d)
20	<10	. . .	<10	0.8	. . .	1.6	0.27	. . .	0.27	0.95	. . .	1.20
25	. . .	<10	0.8	. . .	1.6	0.27	. . .	. . .	0.27	0.95	. . .	1.20
32	. . .	. . .	<10	. . .	. . .	2.3	. . .	. . .	0.18	. . .	. . .	0.55
35	<10	<10	. . .	1.86	1.6	. . .	0.30	0.19	. . .	0.96	0.60	. . .
40	. . .	16	. . .	. . .	1.6	. . .	. . .	0.20	. . .	. . .	0.60	. . .
45	<10	. . .	<10	2.31	. . .	0.61	0.32	. . .	0.09	0.91	0.91	0.28
58	<10	. . .	13	1.30	. . .	0.90	0.25	. . .	0.33	0.80	. . .	0.28
64	. . .	. . .	16	. . .	. . .	0.64	. . .	. . .	0.25	. . .	. . .	0.96
68	30	30	. . .	1.30	0.86	. . .	0.25	0.40	. . .	0.80	0.86	. . .
80	<10	<10	. . .	0.96	0.80	. . .	0.13	0.33	. . .	0.50	0.80	. . .
82	. . .	. . .	15	. . .	. . .	0.76	. . .	. . .	0.25	. . .	. . .	0.96
86	<10	<10	<18	0.96	0.80	0.76	0.13	0.33	0.65	0.50	0.80	2.14
90	<10	<10	<10	. . .	0.80	0.62	. . .	0.33	. . .	. . .	0.80	. . .
93	<10	. . .	10	. . .	. . .	0.39	. . .	. . .	0.22	. . .	. . .	0.41
95	<10	. . .	<10	. . .	. . .	0.59	. . .	. . .	0.12	. . .	. . .	0.95

Concentration, ppb

(a) 10-ppb detection limit. (b) High-pressure drains. (c) Moisture separator drains. (d) Condenser condensate. Source: EPRI Report NP 2429, Vol 4 tachments have not undergone a significant design change and remain as areas of SCC concern.

Significant advances have been made by U.S. plant operators toward better water chemistry control. Guidelines exist for both PWR and BWR configurations (Ref 117), and sophisticated online measuring instruments are available to track a wide variety of anions and cations. The magnitude of allowable condenser leak rates is now much lower, and considerable care is being given to the problem of resin/chemical carryover following regeneration of water cleanup systems. However, the rate of crack extension for the current generation of turbine materials is not 0 in even ultrapure water. Analyses of data and case histories have indicated that the elimination of keyways, lower attachment area stresses, and cleaner steels and environment have all contributed to the decrease observed in SCC events in steam turbines since 1985.

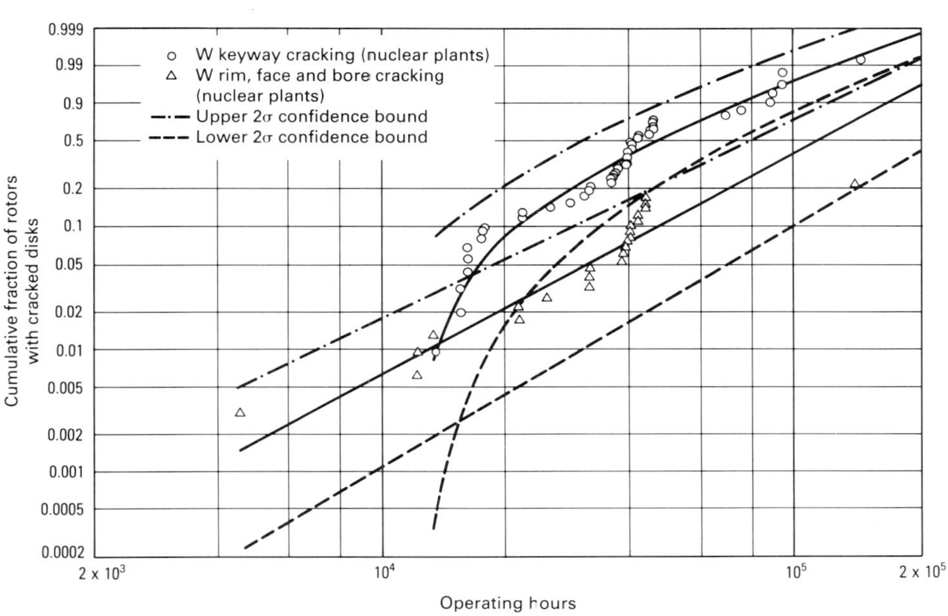

Fig. 60 Field data from Westinghouse (W) shrunk-on disks showing a comparison of keyway and disk cracking at other locations. See also Fig. 61 and 62. Source: Ref 111

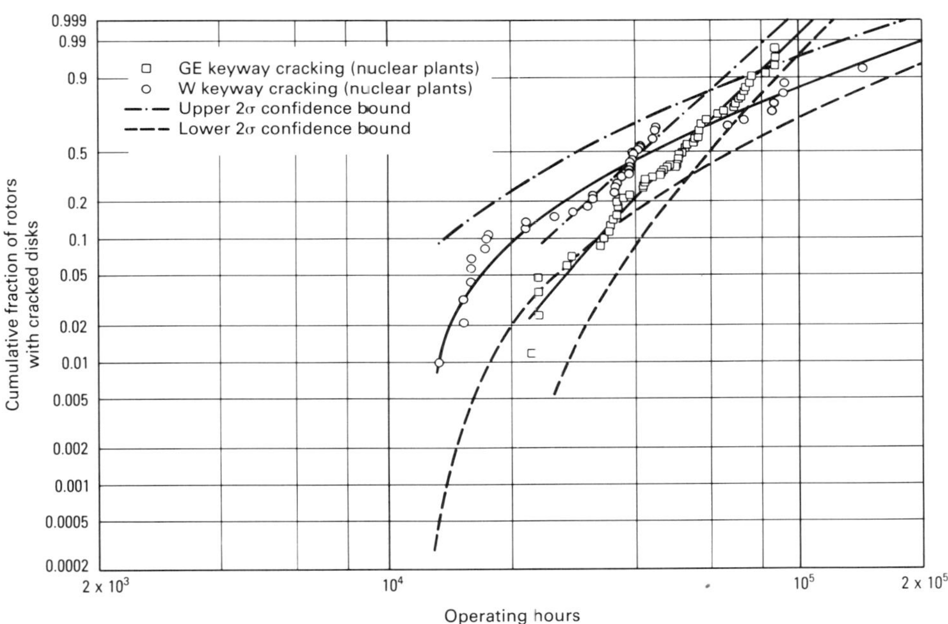

Fig. 61 Comparison of shrunk-on disk keyway cracking for General Electric (GE) and Westinghouse (W) steam turbine rotors. See also Fig. 62. Source: Ref 111

Effect of Nuclear Lubricants on Corrosion

Carl J. Czajkowski
Brookhaven National Laboratory

Lubricants are commonly used to prevent seizure/galling in threaded connections. They must allow easy disassembly of the connection and should aid in the control of the relationship between torque and clamping force.

Various failed components at nuclear power stations (turbine disks and bolts) have had the root cause of failure attributed to lubricant/coolant interactions in the reactor environment. The most commonly used lubricants for nuclear plant applications are solid-film lubricants. One of the primary reasons for using a solid-film lubricant versus an oil in power plant applications is the tendency for oil lubricants to dry out and oxidize at higher temperatures; also, oils cannot withstand extreme pressures at low speeds and tend to permit seizing (Ref 118).

One additional use of solid-film lubricants is corrosion protection. This can be accomplished through the use of a baked-on lubricant coating applied to the component.

Lubricant-Related Failures

Turbine Disk Investigations. In-service failure of steam turbines at nuclear power stations not only cause the loss of hundreds of thousands of dollars a day to a utility due to extended shutdowns but may also cause damage to the safety systems of the plant. One instance of a catastrophic accident was the Yankee-Rowe failure. This unit first started commercial operation in June 1961. The failure occurred during start-up of the unit, which had achieved 1800 rpm with 2% steam. The turbine disk, which was made of either ASTM A294 Class 3 or A471 Class 5 in the quenched-and-tempered condition, fractured and exploded into hundreds of pieces.

Examination of failed sections (Fig. 64) of the turbine disk disclosed numerous axial cracks running perpendicular to the circumferential direction of the hub. These thumbnail cracks (outlined in chalk in Fig. 64) were intergranular and had a tight, black adherent oxide associated with them. The bore of this turbine had been coated with a molybdenum disulfide (MoS_2) lubricant to facilitate disk assembly. This investigation concluded that the cause of the failure was a MoS_2/water/metal interaction resulting in intergranular SCC.

A second instance of turbine disk cracking attributed to MoS_2 was keyway cracking of a turbine disk (Fig. 65) at the Cooper Nuclear Station (Ref 119). The cracking of this quenched-and-tempered Fe-0.2C-0.4Mo-1.7Cr-3.5Ni steel was also predominantly intergranular with some evidence of pitting in the keyway area. Energy-dispersive spectroscopy (EDS) of the fracture surface and surrounding areas disclosed the presence of MoS_2 on the keyway surfaces of the disk. This investigation also attributed the intergranular SCC to a lubricant/moisture interaction. It was apparent from the two investigations that the potential for a stress-corrosion situation can result from the possible dissociation of MoS_2 in a moist environment.

Bolting Investigations. In 1983, the NRC reported 44 distinct instances of bolting degradation at nuclear power stations between October 1964 through March 1982 (Ref 120). The materials investigated included plain carbon steels, alloy

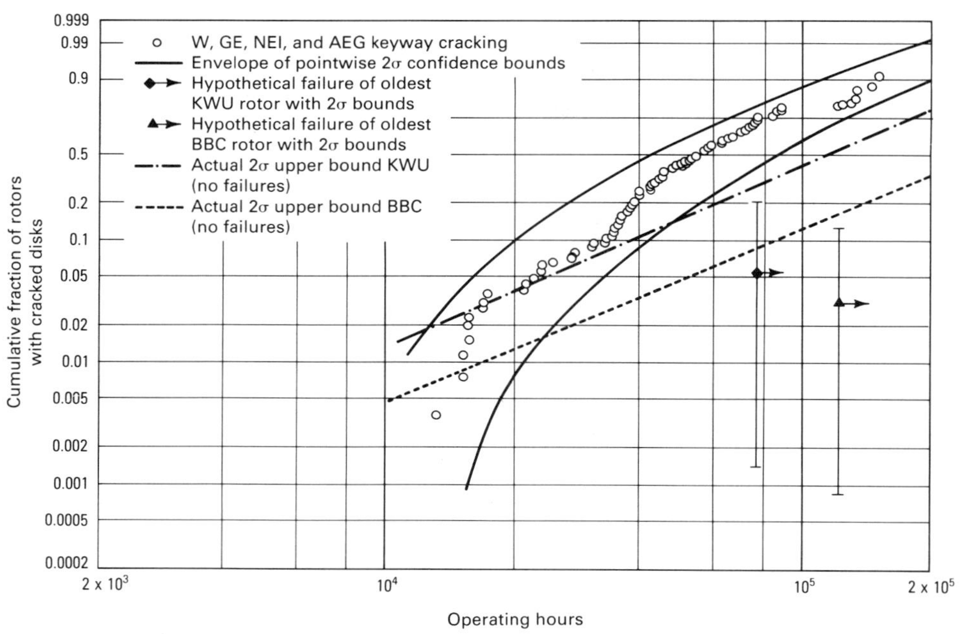

Fig. 62 Data from nondomestic manufacturers added to Fig. 61. Manufacturers include Brown Boveri & Co. (BBC), Switzerland; Kraftwerk Union (KWU), Germany; NEI Parsons (United Kingdom); and AEG (South Africa). Note that both the BBC and KWU curves result from hypothetical failure. Source: Ref 111

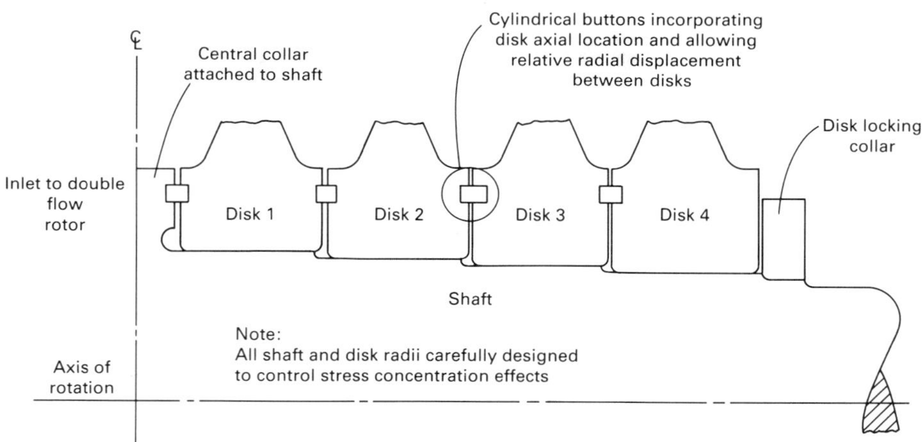

Fig. 63 Schematic of rehabilitated British rotor. Keyways have been moved from the shaft/disk interface to a lower-stress disk face position. Source: EPRI Report NP-2429, Vol 1

steels, and stainless steels. Nineteen of these were attributed to an SCC phenomenon, thirteen to boric acid wastage corrosion, three to fatigue, one to erosion-corrosion, and the remaining eight to material related causes. It was evident that the largest single cause of failure was SCC; the second most common cause of deterioration was corrosion of bolts adjacent to leaking gaskets.

Stress Corrosion of Steam Generator Manway Studs. Steam leaking through a gasket can react with the high-temperature lubricants used to assemble the bolted joint. This interaction occurred in at least two cases at PWRs. In both cases, failures of ASTM A540-B24 fasteners were attributed to a MoS_2/moisture reaction, which resulted in SCC (Ref 121, 122). The cracking was predominantly transgranular in nature with numerous secondary intergranular cracks.

Chemical Analysis of Various Lubricants

Eleven lubricants were evaluated at Brookhaven National Laboratory. These lubricants were types commonly used in construction and postoperational lubrication at nuclear power plants. The lubricants analyzed are listed in Table 12. Each of these lubricants was either sprayed or smeared onto carbon blocks and then placed into a scanning electron microscope and subjected to elemental EDS analysis. Energy dispersive spectroscopy will discern only elements with atomic numbers greater than 11 (sodium); therefore, certain light elements will not be detected. For detailed information on EDS, see the articles "Scanning Electron Microscopy" and "Electron Probe X-Ray Microanalysis" in Volume 10 of the 9th Edition of *Metals Handbook*. The scans for most of the lubricants displayed the expected result with three notable exceptions.

Figure 66(a) shows an EDS scan of a commercially available MoS_2 (lubricant 2) spray. The scan detected both antimony and titanium in addition to the molybdenum and sulfur. The titanium is sometimes used for coloring. The significance of the antimony addition is unknown. The EDS scan of lubricant 4, which is another commercial MoS_2 spray, is shown in Fig. 66(b). This scan showed only a sulfur peak with no molybdenum present. Lubricant 5, which is a graphite in isopropanol lubricant, was subjected to EDS, and the resultant scan is shown in Fig. 66(c). This lubricant showed a high sulfur peak with a trace of silicon also present. The presence of sulfur would not be normally anticipated in a graphite plus alcohol lubricant and might be detrimental to highly alloyed steels under certain conditions.

In addition to the 11 lubricants discussed above, a baked-on bonded solid-film lubricant was also evaluated. This lubricant was not subjected to EDS analysis because of significant outgasing of the material when it is subjected to an electron beam. The reported chemical specifications of the manufacturer of this lubricant were as follows:

Constituent	Amount (max)
Fluoride	29 ppm
Chlorine	200 ppm
Sulfur	98 ppm
Lead	2 ppm
Mercury	1 ppm
Arsenic	0.05%
Zinc	1 ppm

Note: The carrier material for this lubricant is proprietary. Similar lubricants, however, use silicone-base carriers.

It can be clearly seen from the EDS scans shown in Fig. 66 that there is a significant variability in the chemical composition of allegedly similar lubricants. The appearance of possibly detrimental elements in lubricants 2 (antimony) and 5 (sulfur) and the total absence of molybdenum in lubricant 4 show that independent chemical analysis of lubricants before their use in a nuclear power plant is advisable.

Testing of Nuclear Lubricants

Steaming Tests on MoS_2. Prior work on turbine disks had indicated that in moist conditions MoS_2 could hydrolyze and form detrimental sulfides that could promote SCC failure in low-alloy steels (Ref 119). This work involved frictional tests in flowing nitrogen/water mixtures. Because rubbing forces are at work only during the torquing of a bolt and would not normally occur during its service life, some steaming tests were performed to determine if steam alone could interact with MoS_2 in order to form potentially detrimental compounds.

Three tests were performed to determine if a gaseous sulfide could be produced by steaming MoS_2. Distilled water was poured into an autoclave, after which a glass dish with chemically pure MoS_2 was suspended above the water; the autoclave head was tightened and the water heated to 100 °C (212 °F). The autoclave head was vented through a plastic tube with a metering valve into a test tube of water and then into a test

Fig. 64 Fractured section of the generator end disk from the Yankee-Rowe power plant. Thumbnail axial cracks are outlined in chalk.

Fig. 65 Photograph of keyway crack after liquid-penetrant examination (top view)

tube of cadmium acetate ($Cd(C_2H_3O_2)_2$). The chemical reaction for this test is:

$$H_2S + Cd(C_2H_3O_2)_2 \rightarrow CdS \downarrow \text{(precipitate)}$$

After the experiment was completed, the amount of cadmium sulfide (CdS) produced was then weighed and calculated as sulfur in the water. All three tests showed that steaming MoS_2 will produce hydrogen sulfide (H_2S).

There has been a tremendous amount of controversy over the use of MoS_2 as a lubricant on nuclear power plant components. An investigation performed to explain the Hinkley Point A disaster found that MoS_2 reduces the crack initiation time of turbine steels by a factor of three in a steam environment (Ref 123). Other tests showed that the notched tensile strength of a disk steel can be reduced by a factor of 3.5 compared to a test in steam alone (Ref 119). Investigations have shown that MoS_2 accelerates the corrosive tendencies of metallic materials in moist environments (Ref 124, 125).

Early work showed that MoS_2 hydrolyzes in a moist environment to produce H_2S and that this hydrolytic reaction is reversible (Ref 126). A report for the Royal Aircraft Establishment (U.K.) cites accelerated corrosion of steel in contact with MoS_2 under conditions of high humidity (Ref 127). The products of this reaction were MoO_2 and sulfuric acid (H_2SO_4). Sulfuric acid is very detrimental to low-alloy steels. At least three other investigations have associated nuclear component failures with the use of MoS_2 (Ref 119, 128, 129).

Research conducted by AECL has shown that the following materials crack under BWR conditions when contaminated by MoS_2:

- Sensitized type 304 stainless
- Sensitized Inconel 600
- Inconel 718
- 17-4 PH in the H1025 and H1100 conditions
- PH 13-8 Mo
- Custom 455
- AM 355 SCT 1000

The AECL work is supported by a report that documents stress-corrosion tests on a number of materials with 12 different lubricants (Ref 130). One conclusion was that MoS_2 is the most aggressive lubricant. It readily cracked 17-4 PH (1100) and cold-worked type 304. After longer exposures, even annealed type 304 stainless steel cracked with this lubricant. Other studies have also shown that MoS_2 can produce H_2S under PWR conditions (Ref 131).

Table 12 Various solid lubricants analyzed by Brookhaven National Laboratory with the use of EDS

Lubricant number	Type
1	Chemically pure MoS_2
2	Commercial MoS_2 spray
3	Commercial MoS_2 spray
4	Commercial MoS_2 spray
5	Graphite in isopropanol
6	Graphite in ammonia
7	Nickel + graphite lubricant
8	Copper + graphite lubricant
9	Antiseizing lubricant (copper-aluminum)
10	Nickel-base never-seizing lubricant
11	Never-seizing lubricant (copper-zinc-aluminum)

Note: The three commercially available MoS_2 sprays (lubricant numbers 2, 3, and 4) were obtained from three different manufacturers.

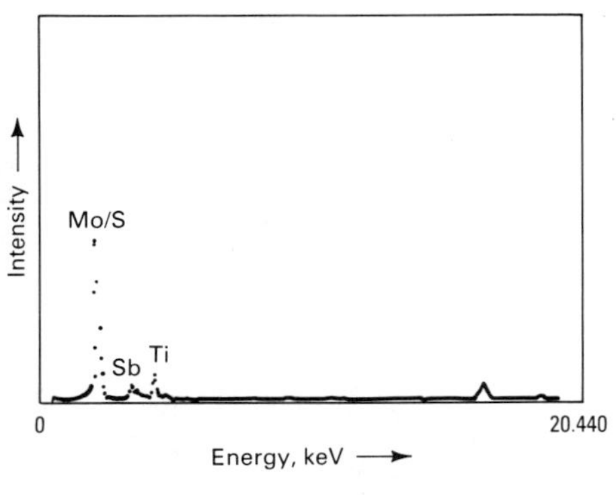

(a)

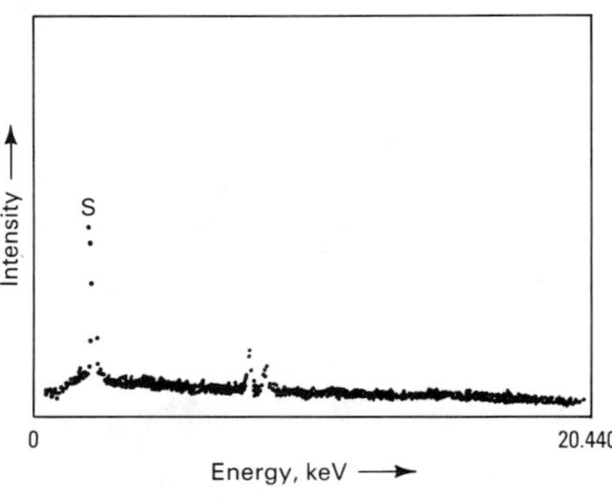

(b)

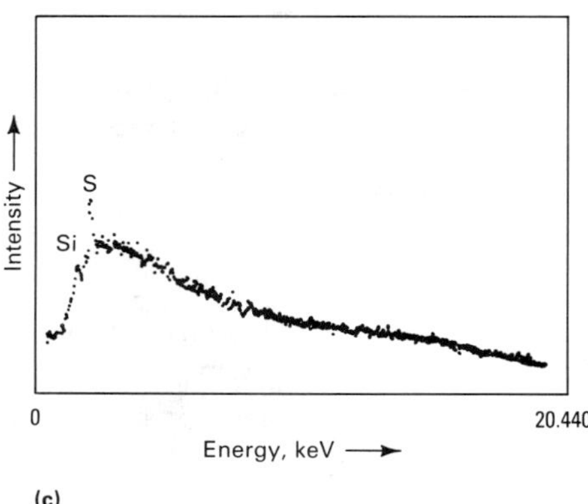

(c)

Fig. 66 EDS analysis of three solid lubricants. (a) Spectrum of Lubricant 2. (b) Spectrum of Lubricant 4. (c) Spectrum of Lubricant 5. Lubricant numbers are identified in Table 12.

The chemical instability of MoS_2 in aqueous media has been addressed in an EPRI report (Ref 132). The report states that MoS_2 is not stable at pH values greater than 4.5. The stable compounds formed are $HMoO_4^-$ or MoO_4^{2-}, with aqueous H_2S, HS^-, or S^{2-} also formed. These compounds could easily promote a sulfide SCC phenomenon in critical bolting applications in which moisture is present with MoS_2.

Coefficient of Friction Testing. Stress-corrosion cracking requires the following three conditions to be present simultaneously: susceptible material, corrosive environment, and applied tensile stress. If any of the these conditions is missing, SCC will not occur. In the case of bolting, the material and the environment may not be able to be altered, in which case the applied tensile stress (preload) must be controlled (if design conditions allow) to keep the applied tensile stress below some critical level. Because the preload depends on the coefficient of friction for a given lubricant on a given surface or substrate, a factor of two to three (some lubricants have ranges given) in this frictional coefficient may put the applied stress into the critical range of SCC susceptibility.

Both static and kinetic coefficient of friction measurements were made at Brookhaven National Laboratory. Some of the lubricants were tested both wet (as sprayed or brushed) and dry (if the lubricants were of the air-drying type). The technique used for these measurements was the sled and bed technique. The tests performed are similar to those outlined in ASTM D 1894 (Ref 133). The results of these tests are given in Table 13. The results show that the coefficients of friction, even for similar types of lubricants, vary widely. For example, the static values for lubricants 1 to 4 (all MoS_2-base lubricants), ranged from 0.026 to 0.273, depending on manufacturer and condition (wet or dry).

This wide variation of measured coefficients of frictions for similar solid lubricants shows that generalizations of this value for same type (for example, MoS_2-base, graphite-base, copper-base, or nickel-base) should not be made and that the coefficient of friction should be determined (or obtained) for the specific lubricant and substrate used.

Environmental Notch Tensile Testing. To ascertain if bolting materials (specifically A193-B7 and A540-B24 Class 2) would have an environmental interaction with any of the lubricants and steam, constant extension rate tests were performed at 100 and 280 °C (212 and 535 °F). Because ordinary tensile tests on smooth specimens will not indicate whether a material is prone to brittle fracture in the presence of a stress concentration (similar to the threads on bolts), notched tensile specimens were used. The specimens had a 50% diam notch (3.2 mm, or 0.125 in.) machined into them (Fig. 67).

The fracture face of each of the specimens was examined by scanning electron microscopy (SEM) after the constant extension rate test. Prior to examination by SEM, the specimens were electrolytically cleaned to remove the oxide film.

This investigation involved a total of 62 notched tensile specimens made from ASTM A193-B7 and A540-B24 Class 2 materials; 18 of these test specimens were coated with a bonded solid-film lubricant (Table 14).

All of the tests were conducted in a stainless steel autoclave. Approximately 1000 mL of demineralized water was added to the autoclave. The load was applied to the specimen when it was in the space above the water. The temperature of the solution was raised to test temperature—either 100 or 280 °C (212 or 535 °F)—through the use of resistance heating coils. After the required temperature was achieved and stabilized (approximately 30 min), the extension testing began in the steam environment. When the lubricant tests were conducted, the lubricants were either sprayed or brushed onto the notched area of the specimen. All specimens were tested at a strain rate of 5 to 9 × 10^{-7} s^{-1}. Fracture surfaces of two test specimens are shown in Fig. 68. The notched constant extension rate tests of the bolting materials showed that both A540-B24 Class 2 and

Table 13 Coefficient of friction for various solid lubricants

The tests performed are similar to those outlined in ASTM D 1894 (Ref 133). Carbon steel blocks were used as the substrate materials.

Lubricant number(a) and condition(b)	Coefficient of friction Static	Kinetic
1, dry	0.182	0.164
1, wetted with isopropanol	0.273	0.142
2, dry	0.266	0.223
2, wet	0.250	0.149
3, dry	0.223	...
3, wet	0.156	...
4, dry	0.026	0.150
4, wet	...	0.130
5, dry	0.130	0.160
5, wet	0.185	0.190
6, dry	0.351	0.170
6, wet	...	0.149
7, dry	0.055	0.126
7, wet	0.093	0.171
8, dry	0.042	0.296
9, wet	0.049	0.268

(a) Lubricant numbers are identified in Table 12. (b) Dry condition, air-drying type lubricant; wet condition, as sprayed or brushed

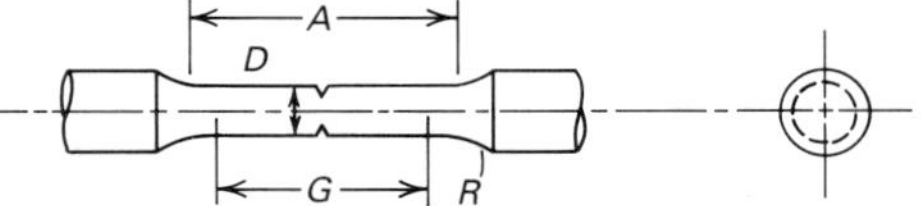

Nominal diameter, 6.4 mm

G—Gage length, 25.4 ± 0.125 mm
D—Diameter, 6.4 ± 0.125 mm
R—Radius of fillet (min), 4.8 mm
A—Length of reduced section (min), 31.8 mm

Fig. 67 Typical notched tensile specimen used in constant extension rate tests

A193-B7 materials are susceptible to a SCC failure in steam at 280 °C (535 °F).

Although it is not conclusive, the use of MoS_2 or a copper plus graphite lubricant seems to promote this susceptibility to SCC. The use of a solid bonded film lubricant does not significantly improve the bolting materials performance with either steam or MoS_2.

General Corrosion Adjacent to Leaking Gaskets and Corrosion Rate Testing. Pressurized water reactor gaskets around pumps have sometimes allowed seepage of small amounts of primary coolant. When this happens, the water flashes to steam, leaving behind a sludge or paste consisting of the dissolved boric acid (H_3BO_3) and LiOH from the primary coolant. This corrosive environment then attacks the bolt material. The most notable instance of this type of failure occurred at the Fort Calhoun Plant in May 1980, when certain bolts were degraded on reactor coolant pumps due to gasket leakage (Ref 121). The worst case of corrosion occurred when a bolt with an original diameter of 89 mm (3.5 in.) was reduced to a diameter of 28 mm (1.1 in.) by the H_3BO_3 over a period of approximately 7 years at about 150 to 200 °C (~300 to 400 °F).

Previously published data were used in the ensuing investigation on an AISI 4135 steel in H_3BO_3 and H_3BO_3 + KOH solutions at 20 and 60 °C (70 and 140 °F). In addition, data on A193-B7 and AISI 4130 material at higher temperatures were plotted to determine if extrapolations were

Table 14 Results of notched tensile constant extension rate tests

Specimen number	Temperature		Test duration, h	Total extension		Extension rate, (s^{-1})	Tensile strength		Coating(c)
	°C	°F		mm	in.		MPa	ksi	
1(a)	100	212	5:48	0.254	0.010	4.8×10^{-7}	1349	197	Bare
2(a)	100	212	...	0.711	0.280	...	1910	277	Bare
3(a)	100	212	6:40	0.381	0.015	6.25×10^{-7}	1574	228	Bare
4(a)	100	212	6:48	0.381	0.015	6.1×10^{-7}	1900	276	Bare
5(a)	100	212	7:01	0.572	0.0225	8.9×10^{-7}	1883	273	Bare
6(a)	100	212	6:22	0.572	0.0225	9.8×10^{-7}	1658	240	Bare
7(a)	280	535	7:28	0.660	0.0260	9.6×10^{-7}	1388	201	Bare
8(a)	280	535	4:56	0.622	0.0245	1.38×10^{-7}	1349	196	Bare
9(a)	280	535	5:03	0.508	0.020	1.1×10^{-6}	1310	190	Bare
10(a)	280	535	5:46	0.470	0.0185	8.9×10^{-7}	1495	217	Bare
11(a)	100	212	7:08	0.508	0.020	7.0×10^{-7}	1428	207	Lube 1
12(a)	100	212	6:31	0.584	0.023	9.8×10^{-7}	1827	265	Lube 1
13(a)	280	535	...	0.572	0.0225	...	1068	155	Lube 1
14(a)	280	535	6:43	0.445	0.0175	7.2×10^{-7}	1181	171	Lube 1
15(a)	280	535	7:40	0.787	0.0310	1.1×10^{-6}	1670	242	Lube 1
16(a)	280	535	7:12	0.686	0.0270	1.0×10^{-6}	1855	269	Lube 1
17(a)	280	535	7:13	0.927	0.0365	1.4×10^{-6}	1670	242	Lube 1
18(a)	280	535	15:26	0.787	0.031	5.58×10^{-7}	992	144	Lube 1
19(a)	280	535	17:47	0.572	0.0225	3.5×10^{-7}	826	120	Lube 1
20(a)	100	212	7:14	0.572	0.0225	8.6×10^{-7}	1405	204	Lube 2
21(a)	280	535	6:46	0.660	0.0260	1.06×10^{-6}	1574	228	Lube 2
22(a)	100	212	5:54	0.673	0.0265	1.2×10^{-6}	1602	232	Lube 3
23(a)	100	212	7:25	0.711	0.0280	1.0×10^{-6}	1743	253	Lube 3
24(a)	100	212	6:50	0.660	0.0260	1.0×10^{-6}	1658	240	Lube 3
25(a)	280	535	7:19	0.635	0.025	9.0×10^{-7}	1574	228	Lube 3
26(a)	280	535	7:41	0.762	0.030	1.1×10^{-6}	1647	239	Lube 3
27(a)	280	535	5:53	0.572	0.0225	1.06×10^{-6}	1518	220	Lube 4
28(a)	280	535	7:35	0.673	0.0265	9.7×10^{-7}	1687	245	Lube 5
29(a)	280	535	8:12	0.762	0.030	1.02×10^{-6}	1738	252	Lube 7
30(a)	280	535	...	...	...	...	...	...	...
31(a)	100	212	6:24	0.635	0.0250	1.08×10^{-6}	1574	228	Lube 8
32(a)	280	535	5:40	0.597	0.0235	1.15×10^{-6}	1011	147	Lube 8
33(a)	280	535	8:05	0.762	0.030	1.0×10^{-6}	1687	245	Lube 9
34(a)	280	535	7:00	0.673	0.0265	1.05×10^{-6}	1687	245	Lube 11
35(b)	100	212	6:07	0.826	0.0325	1.47×10^{-6}	1591	231	Bare
36(b)	280	535	6:19	0.330	0.0130	5.0×10^{-7}	787	114	Bare
37(b)	280	535	7:03	0.445	0.0175	6.9×10^{-7}	1602	232	Bare
38(b)	280	535	4:49	0.038	0.0015	8.64×10^{-8}	1152	167	Bare
39(b)	280	535	5:33	0.546	0.0215	1.07×10^{-6}	1518	220	Bare
40(b)	100	212	5:33	0.673	0.0265	1.3×10^{-6}	1597	232	Lube 1
41(b)	280	535	4:28	0.508	0.020	1.2×10^{-6}	1180	465	Lube 1
42(b)	280	535	4:14	0.419	0.0165	1.08×10^{-6}	1237	179	Lube 1
43(b)	280	535	5:04	0.483	0.0190	1.04×10^{-6}	1366	198	Lube 1
44(b)	280	535	7:00	0.762	0.030	1.2×10^{-6}	1771	257	Lube 8
45(a)	280	535	7:20	0.368	0.0145	5.5×10^{-7}	1687	245	Bare
46(a)	280	535	14:32	0.457	0.0180	3.4×10^{-7}	1181	171	Bare
47(a)	280	535	14:43	0.381	0.015	2.8×10^{-7}	1546	224	Bare
48(a)	280	535	14:01	0.826	0.0325	6.4×10^{-7}	1799	261	Bare
49(a)	280	535	13:38	0.533	0.0210	4.3×10^{-7}	1687	245	Bare
50(a)	280	535	5:06	0.457	0.0180	9.8×10^{-7}	1405	204	Lube 1
51(a)	280	535	24:00	0.813	0.0320	3.7×10^{-7}	2024	293	Lube 1
52(a)	280	535	8:08	0.559	0.0220	7.5×10^{-7}	1181	172	Lube 1
53(a)	280	535	14:36	0.445	0.0175	3.3×10^{-7}	1574	228	Lube 1
54(a)	280	535	7:09	0.191	0.0075	2.9×10^{-7}	1771	257	Lube 8
55(a)	280	535	7:45	0.559	0.0220	7.87×10^{-7}	1799	261	Bare
56(a)	280	535	14:27	0.330	0.0130	2.5×10^{-7}	1405	204	Bare
57(b)	100	212	12:53	0.572	0.0225	4.9×10^{-7}	1490	216	Bare
58(b)	100	212	12:01	0.305	0.0120	2.8×10^{-7}	1602	232	Bare
59(b)	280	535	13:30	0.406	0.0160	3.3×10^{-7}	1490	216	Bare
60(b)	100	212	12:51	0.318	0.0125	2.7×10^{-7}	1546	224	Lube 1
61(b)	100	212	15:13	...	...		...	...	Lube 1
62(b)	280	535	12:50	0.635	0.025	5.4×10^{-7}	1518	220	Lube 1

(a) Specimen made from quenched-and-tempered ASTM A540-B24 Class 2 nickel-chromium-molybdenum (modified 4340) steel. (b) Specimen made from quenched-and-tempered ASTM A193-B7 chromium-molybdenum steel. (c) Lubricant numbers are identified in Table 12.

valid (Ref 121). The temperature dependence of the corrosion rate is shown in Fig. 69. As the temperature increases to the boiling point of water, the corrosion rate increases at much faster rates. This increase continues until the boiling point is reached and then starts to diminish, probably because of the loss of water in solution. The decrease in corrosion rate continues through at least a temperature up to 178 °C (352 °F) in H_3BO_3 + LiOH. As higher temperatures are attained, water of hydration would also start to evaporate, which could lead to a deposit of lithium salt of metaboric acid ($LiBO_2$) ($LiBO_2$ has a melting point as high as 845 °C, or 1555 °F, in nonhydrated form). Corrosion products may have a similar stifling effect in the case of aqueous H_3BO_3 by itself, thus accounting for the drop in corrosion rate after H_2O is lost.

Nine specimens coated with the bonded proprietary solid-film lubricant described earlier were subjected to the same tests as the bare metal specimens. Five specimens were tested at 100 °C (212 °F) (3 for 132.5 h and 2 for 77 h). Two specimens were also tested at 178 °C (352 °F) for 120 h, and two specimens at 315 °C (600 °F) for 130 h. The working solution used for the testing was 4000 ppm boron (as H_3BO_3) + H_2O + LiOH titrated to a pH of 7.3. The data generated from these tests are plotted on the graph of the original test data in Fig. 69.

These corrosion rate experiments, using previously coated solid bonded film lubricant specimens, show a marked decrease in metal loss at

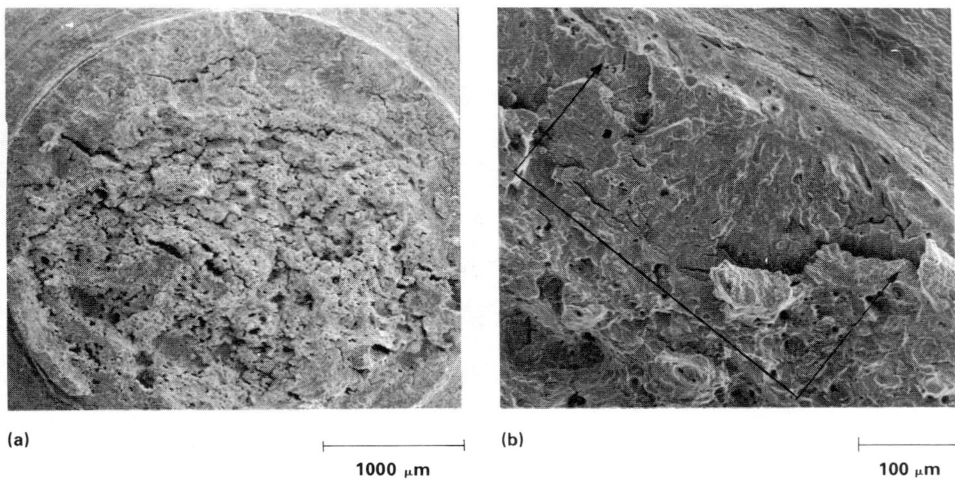

Fig. 68 Fracture surfaces of two failed notched tensile specimens. (a) Low-magnification SEM fractograph of the ductile + transgranular fracture face from specimen 42 in Table 14 (ASTM A193-B-7). (b) Higher-magnification SEM fractograph of the transgranular area seen on specimen 13 in Table 14 (ASTM A540-B24 Class 2)

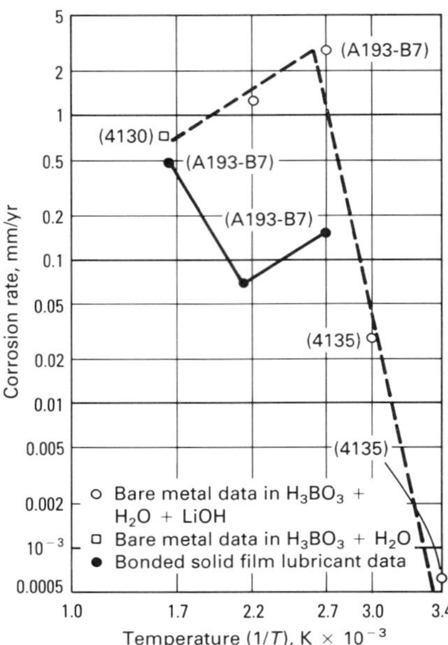

Fig. 69 Graph of the data from the corrosion rate versus temperature experiments for nonlubricated and lubricated specimens. All points represent the average of data collected.

100 and 178 °C (212 and 350 °F) compared to previously reported bare metal data. This lubricant protection disappears at 315 °C (600 °F), where the results on previously coated specimens differed little from results on bare metal specimens.

Observations on Copper Bearing Lubricants

The copper bearing lubricants have also been involved in at least one report as a crack initiator (Ref 130). The particular lubricant used (a copper plus graphite lubricant) produced cracks in 17-4 PH specimens. This result is complementary to the Brookhaven National Laboratory results, in which significantly lower notched tensile strength values were associated with the use of this lubricant. The effect of copper on low-alloy steels in as little as 1 ppm chloride solutions has been observed (Ref 134). In this case, the copper is speculated to have shifted the potential into the cracking range in a constant extension rate test at 268 °C (514 °F). This potential shift might well be applicable to the lubricants.

Erosion-Corrosion in Wet Steam Flow

Norris S. Hirota
Electric Power Research Institute

Erosion-corrosion is an accelerated form of corrosion caused by the relative motion between a corrosive medium and a metal surface. The erosion attack is directed at the protective layers and not on the metal itself. The metal corrosion is accelerated by the continuous erosive destruction of the protective layers. Wet steam erosion-corrosion of nuclear plant piping represents a potential industry-wide problem that can lead to costly outages and repairs as well as concerns for plant reliability and safety.

Carbon steel lines that carry wet steam, such as high-pressure turbine extraction and turbine exhaust piping in both PWRs and BWRs, are especially susceptible to erosion-corrosion. Pipe wall degradation rates as high as 1.0 to 1.5 mm/yr (40 to 60 mils/yr) have occurred and in some plants have resulted in pipe ruptures. Laboratory test results and extensive field tests have indicated that erosion-corrosion in wet steam (two-phase) lines is similar to that which occurs in single-phase (water) lines. The appearance of damage is usually characterized by patterns of grooves, waves, rounded holes, and valleys in some consistent direction (Fig. 70). Regions of high wear rates are those where liquid film was found (Ref 136). In wet steam lines constructed of carbon steel, the erosion-corrosion mechanism is complex and involves the electrochemical corrosion processes of oxidation and reduction, the convective mass transport of soluble compounds caused by the dissolution of the protective surface film, and the mechanical wear effects of abrasion, droplet impingement, and/or solid-particle erosion.

Theory

Experimental work using both single- and two-phase flow, as well as extensive field testing and experiences in nuclear plants, strongly indicates that wet steam erosion-corrosion is similar to that which occurs in single-phase situations. Pipe regions of high wear are typically those where the metal surfaces contact liquid water only. However, prediction of actual erosion-corrosion rates in operating plants using single-phase test data and empirical equations has achieved only limited success.

There are a multitude of possible iron and iron-alloy corrosion processes. The following represents only one of the self-consistent set of reactions in single-phase situations. In oxygen-free water, iron, being thermodynamically unstable, dissolves to form Fe^{2+} ions with the formation of Fe(II)-hydroxide. Under static conditions, the fluid layer near the metal becomes saturated with this hydroxide, and the corrosion process essentially stops. However, with fluid flow, complete Fe(II)-hydroxide saturation of the fluid layer near the wall becomes less likely because of the mass transport between the boundary layer and the main flow field.

This set of reactions is written as:

$$Fe \rightarrow Fe^{2+} + 2e^- \quad \text{(Eq 16a)}$$

$$2H_2O \rightarrow 2H^+ + 2OH^- \quad \text{(Eq 16b)}$$

$$2H^+ + 2e^- \rightarrow H_2 \quad \text{(Eq 16c)}$$

$$Fe + 2H_2O \rightarrow Fe(OH)_2 + H_2 \quad \text{(Eq 16d)}$$

The Fe(II)-hydroxide can be converted into Fe_3O_4 by the Schikorr reaction (Ref 137):

$$3Fe(OH)_2 + Fe_3O_4 + 2H^+ + 2e^- + 2H_2O \quad \text{(Eq 17a)}$$

$$2H^+ + 2e^- \rightarrow H_2 \quad \text{(Eq 17b)}$$

$$3Fe(OH)_2 \rightarrow Fe_3O_4 + H_2 + 2H_2O \quad \text{(Eq 17c)}$$

The reaction rate depends on temperature; at about 180 °C (355 °F), the Fe_3O_4 formation rate is high enough to protect against erosion-corrosion. The oxidation reaction rate of steel and the dissolution rate of the oxide are equal. The dissolution is produced by the reduction of Fe_3O_4 by hydrogen. In oxygenated water, the overall reaction is:

$$2Fe + 2H_2O + O_2 \rightarrow 2Fe^{2+} + 4OH^- \rightarrow 2Fe(OH)_2 \quad \text{(Eq 18)}$$

The Fe(II)-hydroxide, being unstable in oxygenated solutions, is oxidized to the ferric salt (Ref 138):

$$2Fe(OH)_2 + H_2O + \tfrac{1}{2}O_2 \rightarrow 2Fe(OH)_3 \quad \text{(Eq 19)}$$

Magnetite can then be formed through (Ref 137):

$$2Fe(OH)_3 + Fe(OH)_2 \rightarrow Fe_3O_4 + 4H_2O \quad \text{(Eq 20)}$$

The rate of erosion-corrosion varies according to:

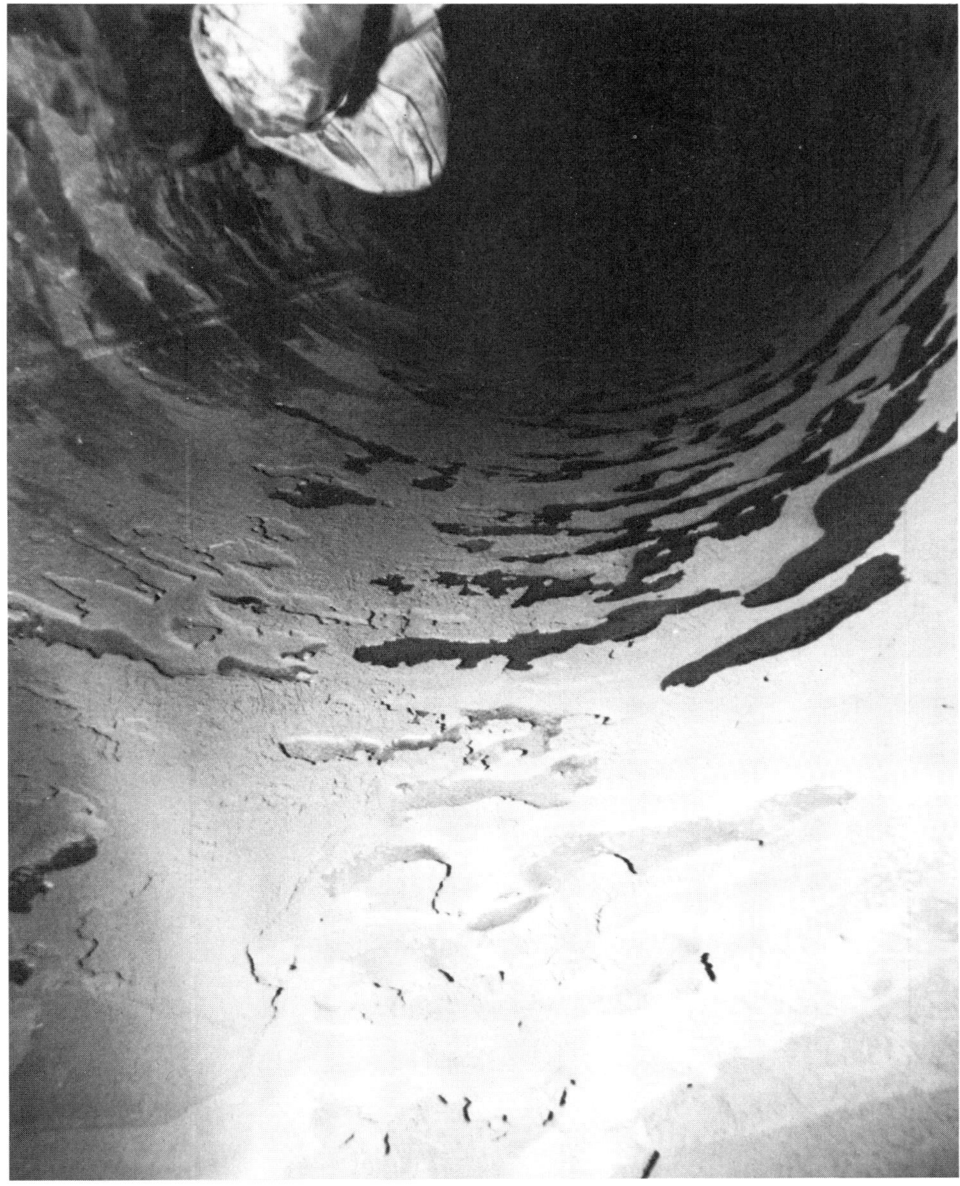

Fig. 70 Erosion-corrosion damage in wet steam piping. Source: Ref 135

$$\frac{dm}{dt} = k(C_s - C_b) \quad \text{(Eq 21)}$$

where k is the mass transfer coefficient, C_s is the concentration of soluble iron at surface, C_b is the concentration of iron in flow, and dm/dt is the rate of metal loss.

The iron concentration corresponding to the equilibrium oxide solubility can be substituted for C_s. This oxide solubility varies with the potential of the oxide surface; at negative potentials where Fe_3O_4 is stable, the solubility increases as the potential becomes more negative. In addition, the potential depends on the corrosion rate; as the corrosion rate increases, the potential becomes more negative. The solubility of Fe_3O_4 has been shown to depend on the square of the mass transfer coefficient, k. Thus, the erosion-corrosion rate is dependent on the cube of the mass transfer coefficient (k^3). The mass transfer coefficient varies with the local hydrodynamic conditions.

Parameters of Influence

Moisture. The lack of erosion-corrosion damage in dry steam lines indicates the importance of moisture. In several PWR and BWR steam systems, moisture percentages can reach 15%, and under off-design conditions, even higher percentages occur. Although test data showing the effects of varying moisture levels on the rate of erosion-corrosion are not available, inspection reports from operating power plants indicate that piping systems carrying steam of high moisture content suffer higher rates of erosion-corrosion. These findings are consistent with Keller's equation, which will be discussed later (see Eq 22 and the corresponding discussion).

Material Composition. The most widely used material for U.S. nuclear plant wet steam piping is carbon steel, which has shown a susceptibility to erosion-corrosion. Under certain conditions of alloying (with elements such as chromium, copper, and molybdenum), erosion-corrosion resistance can be significantly improved (Table 15). However, in some current turbine exhaust piping, carbon steel with trace amounts of copper is used, but field experience indicates that this material is also susceptible to erosion-corrosion damage.

In comparison to ordinary carbon steel, erosion-corrosion rates can be reduced by three times with carbon-molybdenum steel and by more than ten times with chromium-molybdenum steels. Field experience showed that 1.25Cr-0.5Mo and 2.25Cr-1Mo steels are virtually immune to erosion-corrosion. Additional tests conducted under PWR conditions indicate that 12% Cr steels have excellent resistance, 2.25Cr-1Mo steel performed better than steel containing copper, and steel with less than 1% Cr did not have adequate resistance. Figure 71 shows material wear rates for 11 steels and 2 carbon steel specimens for three typical pH-oxygen combinations in 180-°C (355-°F) PWR water flowing at 20 m/s (65 ft/s).

pH. Water chemistry and pH play an important role in changing erosion-corrosion rates. Pressurized water reactor turbine piping has experienced erosion-corrosion damage when condensate pH falls below 9.3; similar damage in BWRs is experienced where pH is in the neutral range (7.0). Significant reduction in erosion-corrosion is achieved in PWRs when pH is maintained above 9.3, as shown in Fig. 72. In fact, the erosion-corrosion rate drops significantly at pH above 9.2, independent of the oxygen content.

Both NH_3 and C_4H_9NO have been used successfully to maintain pH above 9.3. Morpholine provides a better distribution ratio, which allows more of it to be retained in the water phase at high temperatures. The resultant higher pH helps prevent localized damage. Plants with all steel materials should maintain pH between 9.3 and 9.6. Plants with copper alloys, however, should maintain condensate/feedwater pH between 8.8 and 9.2 to avoid excessive copper pickup if ammonia is used to control pH. Condensate polishers, if used, also determine what optimum pH level to maintain, depending on the choice of operating mode, that is, hydrogen or ammonia cycle (Ref 135, 139).

Temperature. Erosion-corrosion rates are highly temperature dependent, as shown in Fig. 73. These curves represent single-phase flow laboratory test results and indicate a well-defined maximum erosion-corrosion rate that is a function of hydrodynamic conditions. At higher temperatures, the protective layer formation changes from the intermediate stage of the primary Fe(II)-hydroxide formation to the direct Fe_3O_4 formation. Under two-phase conditions in ammoniated water, the maximum erosion-corrosion rate occurs at approximately 180 °C (355 °F). At temperatures less than 180 °C (355 °F), lower erosion-corrosion rates are attributable to a slower rate of chemical reaction.

Oxygen. Figures 74 and 75 illustrate the influence of oxygen on erosion-corrosion behavior. Under BWR conditions (neutral water), oxygen is beneficial (based on iron release data for carbon steel). The iron release rate may decrease by 100 times when raising the oxygen concentration from 1 to 200 ppb over the temperature range of

Table 15 Influence of chemical composition on wet steam erosion-corrosion test results

Specimen number	Chemical composition, wt %: C	Mn	Si	P	S	Cr	Mo	Al	Cu	Ni	Weight loss, mg
2.65	0.14	0.58	0.22	0.009	0.027	0.26	0.020	0.010	0.16	0.06	43.1
2.34	0.10	0.55	0.04	0.006	0.026	0.05	0.010	0.010	0.19	0.07	56.6
2.15	0.09	0.49	0.10	0.005	0.028	0.05	0.010	0.010	0.20	0.08	57.5
3.13	0.15	1.13	0.32	0.024	0.023	0.12	0.030	0.010	0.19	0.08	59.9
3.19	0.13	1.10	0.48	0.017	0.025	0.08	0.050	0.010	0.22	0.12	60.7
3.14	0.16	1.03	0.33	0.022	0.029	0.08	0.020	0.010	0.11	0.17	61.2
2.35	0.10	0.65	0.18	0.021	0.024	0.08	0.020	0.010	0.14	0.08	61.6
5.6	0.20	1.25	0.55	0.004	0.011	0.22	0.040	0.045	0.05	0.06	62.0
1.26	0.17	0.72	0.25	0.006	0.014	0.07	0.290	0.005	0.04	0.03	63.1
1.9	0.14	0.61	0.18	0.004	0.021	0.04	0.260	0.010	0.07	0.04	63.4
1.17–3	0.15	0.61	0.16	0.009	0.025	0.04	0.270	0.005	0.10	0.02	64.0
2.49–2	0.22	0.56	0.18	0.009	0.023	0.06	0.040	0.005	0.17	0.06	65.5
3.9	0.13	0.99	0.37	0.016	0.022	0.04	0.020	0.010	0.18	0.08	65.9
3.15	0.10	0.98	0.21	0.008	0.018	0.05	0.040	0.005	0.10	0.10	67.7
3.10	0.12	0.98	0.37	0.012	0.010	0.04	0.020	0.010	0.03	0.03	68.6
5.3	0.15	1.30	0.34	0.012	0.017	0.10	0.400	0.005	0.03	0.08	70.0
1.15	0.14	0.57	0.23	0.009	0.010	0.11	0.290	0.010	0.07	0.06	70.3
3.12	0.12	0.98	0.37	0.012	0.016	0.04	0.010	0.010	0.03	0.03	70.1
5.9	0.12	1.47	0.47	0.015	0.016	0.10	0.005	0.020	0.04	0.02	71.0
3.11	0.12	1.01	0.39	0.011	0.013	0.04	0.010	0.010	0.03	0.03	71.2
2.90	0.14	0.57	0.26	0.015	0.032	0.04	0.010	0.010	0.11	0.06	72.0
1.16	0.14	0.62	0.22	0.004	0.007	0.04	0.500	0.010	0.04	0.03	72.4
1.5	0.16	0.59	0.24	0.004	0.022	0.04	0.260	0.010	0.10	0.06	72.7
1.6	0.18	0.68	0.21	0.006	0.020	0.05	0.260	0.010	0.10	0.06	72.7
2.89–2	0.14	0.58	0.25	0.015	0.031	0.04	0.010	0.010	0.11	0.05	73.0
2.45	0.18	0.52	0.34	0.008	0.017	0.04	0.050	0.005	0.10	0.04	75.0
2.59	0.18	0.94	0.18	0.012	0.010	0.04	0.010	0.010	0.02	0.03	75.0
1.11	0.16	0.70	0.19	0.006	0.019	0.02	0.280	0.010	0.04	0.02	75.6
5.5	0.21	1.31	0.42	0.016	0.021	0.24	0.010	0.055	0.08	0.07	76.0
5.7	0.19	1.09	0.57	0.010	0.009	0.14	0.040	0.030	0.12	0.22	76.0
1.22	0.15	0.73	0.25	0.013	0.014	0.02	0.290	0.010	0.03	0.03	76.3
2.58	0.11	0.54	0.07	0.009	0.017	0.03	0.010	0.040	0.05	0.04	76.3
1.12	0.16	0.69	0.18	0.007	0.020	0.02	0.280	0.019	0.04	0.02	76.6
3.17	0.15	1.22	0.38	0.017	0.013	0.02	0.010	0.010	0.03	0.04	77.8
2.11	0.08	0.50	0.18	0.012	0.012	0.06	0.010	0.020	0.04	0.04	78.6
2.21	0.09	0.56	0.18	0.010	0.014	0.02	0.010	0.020	0.03	0.02	78.6
1.13	0.16	0.73	0.18	0.006	0.018	0.02	0.270	0.010	0.04	0.02	79.0
1.10	0.16	0.70	0.20	0.005	0.015	0.02	0.280	0.010	0.04	0.02	79.2
2.66	0.10	0.48	0.25	0.009	0.026	0.03	0.010	0.005	0.10	0.04	79.6
1.14	0.16	0.72	0.20	0.006	0.020	0.02	0.270	0.010	0.04	0.02	80.5
3.3	0.15	1.18	0.33	0.008	0.018	0.05	0.020	0.010	0.18	0.07	82.9
2.20	0.10	0.51	0.14	0.008	0.016	0.03	0.010	0.010	0.04	0.02	83.3
2.19	0.09	0.51	0.18	0.009	0.021	0.02	0.010	0.010	0.04	0.02	85.2
3.7	0.16	1.09	0.26	0.020	0.032	0.03	0.010	0.010	0.12	0.06	87.9
2.85	0.14	1.05	0.01	0.014	0.019	0.02	0.003	0.003	0.05	0.02	88.6
3.6	0.16	1.11	0.42	0.007	0.022	0.08	0.020	0.030	0.10	0.06	88.9
2.16	0.10	0.49	0.20	0.011	0.022	0.02	0.010	0.020	0.03	0.02	89.5
5.10	0.18	0.99	0.28	0.011	0.009	0.02	0.005	0.030	0.01	0.02	90.0
2.17	0.10	0.48	0.20	0.007	0.018	0.02	0.010	0.010	0.02	0.02	90.4
2.40–1	0.37	0.63	0.32	0.004	0.026	0.02	0.010	0.010	0.07	0.04	91.0
2.18	0.10	0.49	0.16	0.012	0.024	0.02	0.010	0.020	0.03	0.02	92.7
2.88	0.15	1.15	0.01	0.027	0.034	0.01	0.003	0.003	0.04	0.02	95.2
2.72	0.09	0.55	0.10	0.028	0.018	0.01	0.005	0.020	0.03	0.05	96.9
2.84	0.18	1.09	0.01	0.011	0.025	0.02	0.003	0.003	0.06	0.02	98.0
2.73	0.17	0.64	0.26	0.015	0.014	0.01	0.005	0.020	0.02	0.01	98.3
2.89–1	0.17	0.64	0.28	0.017	0.016	0.01	0.003	0.020	0.02	0.02	103.1
2.104	0.18	1.11	0.32	0.014	0.024	0.01	0.003	0.050	0.02	0.02	107.0
2.83	0.09	0.97	0.23	0.009	0.014	0.01	0.005	0.003	0.01	0.01	116.0
2.109	0.10	0.48	0.20	0.033	0.018	0.01	0.005	0.055	0.01	0.04	Not tested

Source: Ref 135

38 to 204 °C (100 to 3400 °F) in neutral water at 1.8 m/s (6 ft/s). This has been the experience in wet steam lines in BWR plants (Ref 135). Moreover, neutral water with low oxygen produced the maximum material wear. Consequently, some BWR plants maintain oxygen in the feedwater to the reactor above 15 to 20 ppb.

Under PWR conditions, the use of oxygen as a conditioning agent has met with limited success. Care must be taken, because steam generator tube materials, such as Incoloy 800 and Inconel 600, are sensitive to oxygen, especially in crevices. More success has been achieved by raising the pH of steam condensates by replacing ammonia with less volatile amines (Ref 140, 141). Although oxygen feed is beneficial in single-phase flow, it may not be as effective in two-phase situations; a higher oxygen concentration is required to offset the unfavorable partition coefficient of oxygen between steam and water. Moreover, chloride and sulfate contamination, together with higher levels of oxygen, could cause rapid pitting.

Flow Path Geometry and Velocity. Poor flow path geometry can significantly accelerate the rate of erosion-corrosion attack. Typical examples of poor piping layout are shown in Fig. 76 and 77. In most of these cases, severe moisture impingement and very high velocities were involved. Areas of severe attack were located near or at system discontinuities, such as elbows, branch connections, and weld areas where backing rings were used. In the laboratory, tests indicated that the erosion-corrosion rate increased exponentially with velocity, as illustrated in Fig. 78. Although these curves were generated for neutral water, a similar tendency is noticed in PWR secondary circuits with pH levels of 9 using NH_3 and an oxygen content of less than 5 ppb. However, the absolute magnitudes of the specific material wear rates are significantly lower because of the higher pH.

Locations of Damage. Nuclear plant turbine piping and associated components are subjected

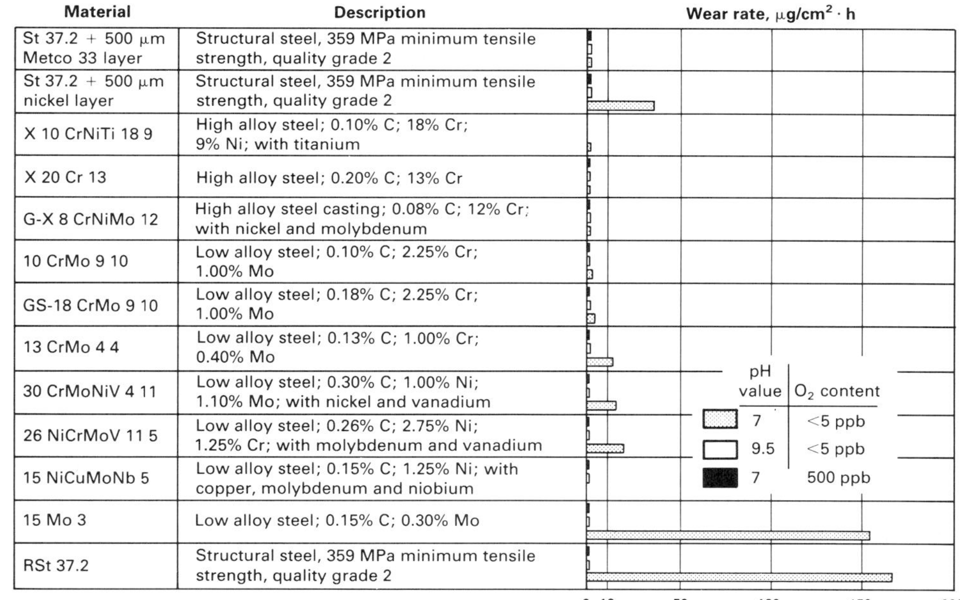

Fig. 71 Material wear due to erosion-corrosion of various metals. Pressure: 40 bar (4000 kPa, or 580 psi), Temperature: 180 °C (355 °F), Velocity: 20 m/s (65 ft/s). Source: Ref 137

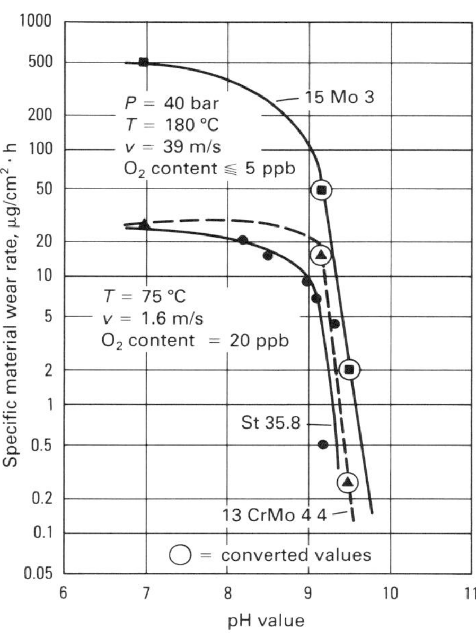

Fig. 72 Effect of pH value on material wear due to erosion-corrosion. *P*, pressure; *T*, temperature; *v*, velocity. For material composition, see Fig. 71. Source: Ref 137

to high-velocity steam of varying moisture content. Figures 79 and 80 show typical piping arrangements of affected systems. Some common piping systems that have suffered erosion-corrosion damage include high-pressure turbine exhaust piping, high-pressure turbine extraction piping, low-pressure turbine extraction piping, low-pressure turbine inlet piping (some plants), and miscellaneous wet steam lines (for example, auxiliary steam lines). Based on field experience, the following areas are the likely candidates for erosion-corrosion damage (Ref 135):

- Branch connections at the sides and bottoms of pipe runs are expected to contain a higher amount of moisture and consequently have a higher erosion-corrosion rate than branch connections located on the tops of pipe runs
- Where extraction steam piping is directed down into a feedwater heater with two inlet connections, the second connection will contain a higher moisture content and consequently have a higher erosion-corrosion rate unless the piping design incorporates a method to remove the moisture (Fig. 77)
- Short distances between changes in direction and other discontinuities do not allow the turbulence to dissipate. These areas are expected to have higher rates of erosion-corrosion. Usually, a length equivalent to 10 diameters of straight pipe is required for flow to stabilize. In systems in which all fittings are not being inspected, one area that should be inspected is where two changes in direction are separated by the least distance
- Branch connections of 90° generally have a higher erosion-corrosion rate than bends and elbows do and should also be of higher priority for monitoring than lateral connections
- Elbows with turning vanes are expected to have significantly lower erosion-corrosion rates than those without. However, areas just downstream of turning vanes have a higher erosion-corrosion rate because of higher turbulence intensities
- Areas where repairs involve internal placement of weld metal are particularly susceptible to severe erosion-corrosion. Areas around the repairs should be inspected, especially the leading and trailing edges with respect to flow
- As the radii of directional changes are reduced, the rate of erosion-corrosion is expected to increase. Short-radius elbows are more susceptible to erosion-corrosion than long-radius elbows, and long-radius elbows are more susceptible than 5-diameter bends. The likelihood that bends are fabricated with wall thicknesses closer to minimum than elbows should also be a consideration

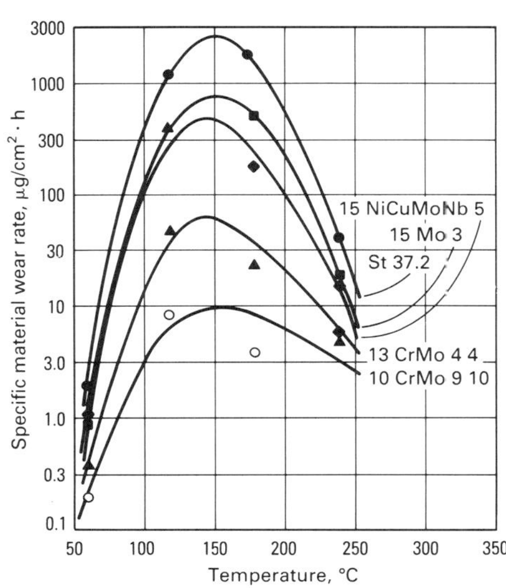

Fig. 73 Temperature effect on material wear due to erosion-corrosion. *P*: 40 bar, *v*: 35 m/s, pH: 7, O_2 content: <40 ppb, period of exposure: 200 h. See Fig. 71 for material composition. Source: Ref 137

A form of erosion-corrosion called tiger-striping has been identified in a few plants (Ref 135). Tiger-striping is characterized by a striping of uniform degradation on the pipe inside diameter. The damaged areas are not limited to areas of flow discontinuities and have been documented in numerous straight pipe sections. Why only some systems are subject to this unusual phenomenon has not been determined. Although occurring less frequently than the more general form of erosion-corrosion, tiger-striping is no less

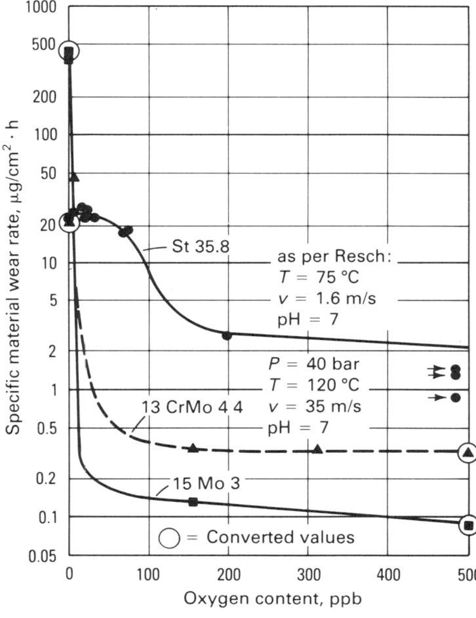

Fig. 74 Effect of oxygen content on material wear due to erosion-corrosion. See Fig. 71 for material composition. Source: Ref 137

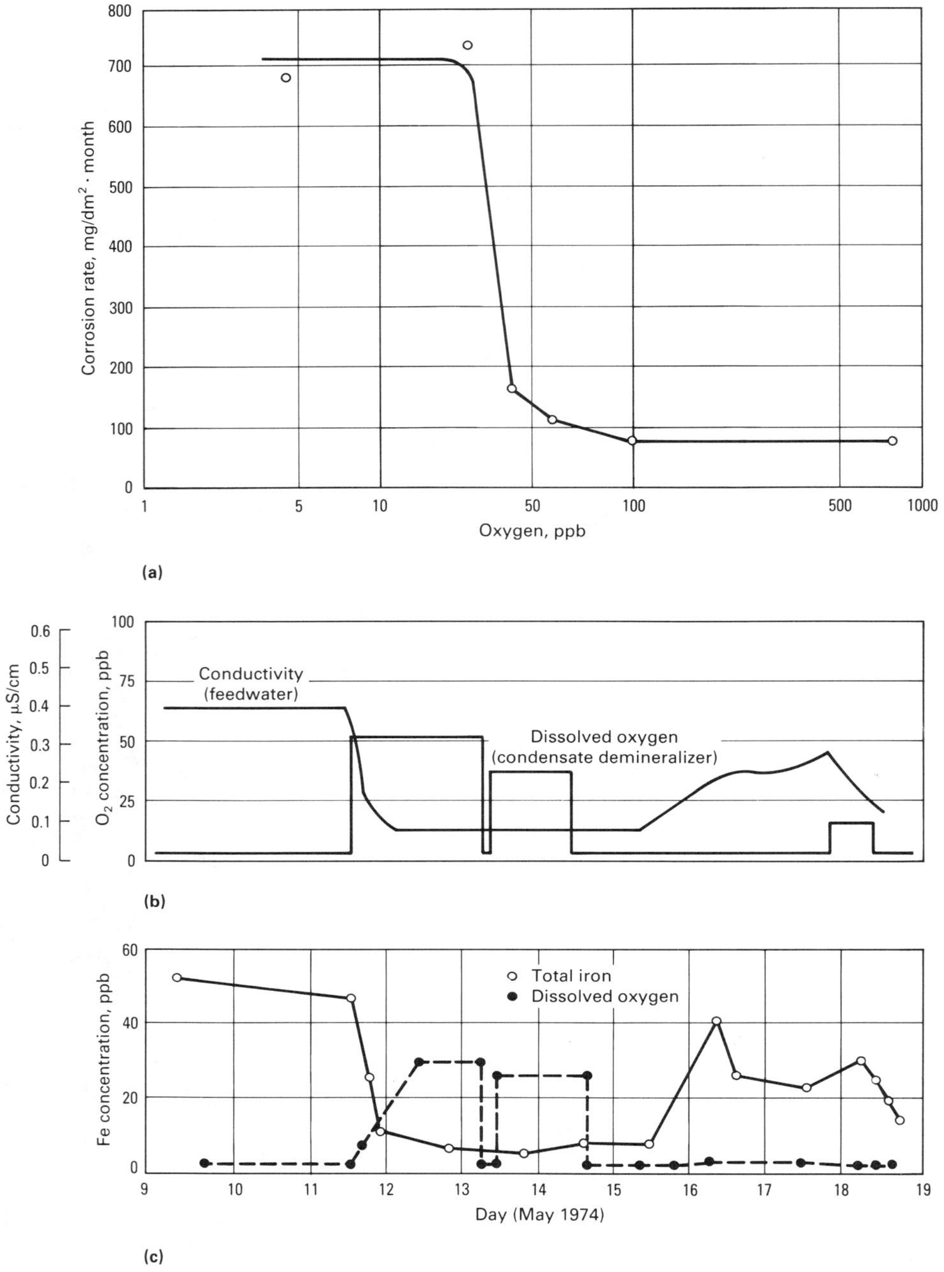

Fig. 75 Influence of oxygen on the corrosion behavior of carbon steel. (a) Effect of oxygen on the corrosion rate of carbon steel in neutral water at 100 °C (212 °F). (b) and (c) Time variation of total dissolved oxygen and total iron concentration in the feedwater with addition of oxygen gas to the feedwater system. The conductivity of the feedwater and dissolved oxygen concentration at the condensate demineralizer outlet are also shown. Source: Ref 135

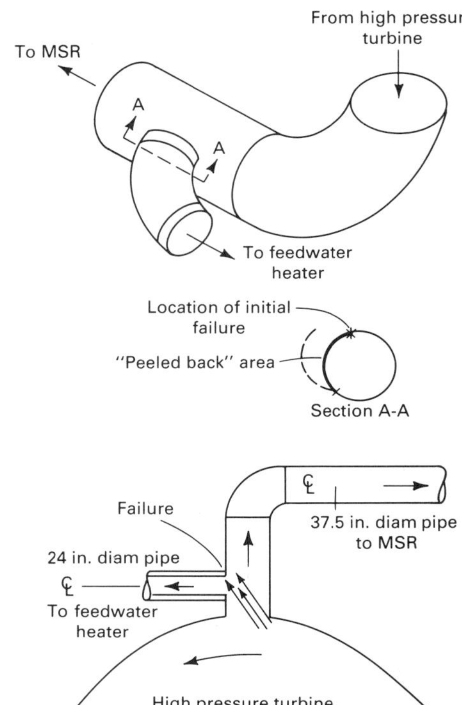

Fig. 76 Typical turbine steam extraction piping geometry problems. Source: Ref 135

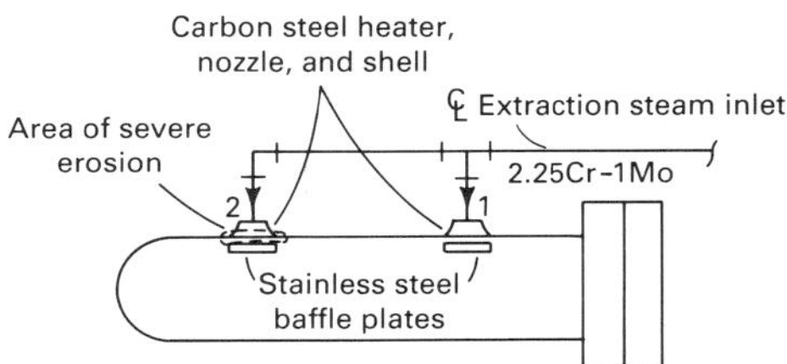

Fig. 77 Problems at extraction piping/feedwater heater interface. Moisture separation caused by piping geometry allows more liquid to impinge on the second inlet nozzle. Source: Ref 135

severe from the standpoint of wall thickness degradation, and site inspection procedures should be able to detect its existence.

Inspection Program

Approach. Failure to detect and repair erosion-corrosion degradation can result in a disruptive event that would possibly injure plant personnel and cause extensive plant equipment damage requiring costly repairs and plant downtime (Fig. 81). An analysis program can provide a means of ranking systems based on their susceptibility to erosion-corrosion and identifying the components within those systems that are most likely to incur severe damage. This prioritization criteria can be used to limit the number of plant inspection sites, providing a cost-effective basis on which to estimate the severity of the problem with a high confidence level.

An erosion-corrosion wear rate prediction model could serve as the basis for this methodology. By first calculating expected wear rates for potentially high erosion-corrosion susceptible areas and then determining the allowable wear for components in those areas, a wear period or remaining life can be established and used to identify high-priority inspection sites; a short wear period or remaining life identifies a high priority. The wear rate model provides only the first estimate of important inspection sites. Thereafter, the frequency and sites of future inspections should be based on the measured wear rates (Ref 135).

Keller's Equation. Keller proposed an empirical wear rate model in 1974 for power plant equipment. This model is based on velocity, temperature, configuration, and the amount of moisture. The model does not account for varying water chemistry (pH and oxygen). It was developed for two-phase flow conditions (temperatures between 50 and 150 °C, or 120 and 300 °F, and materials of low-carbon steel to 13% Cr steels) and for a wear rate range of 0.4 to 5 mm/yr (0.016 to 0.2 in./yr) (Ref 136).

The material loss prediction model can be used as the criterion for selecting and prioritizing components subjected to erosion-corrosion degradation for an inspection program. Keller's equation states:

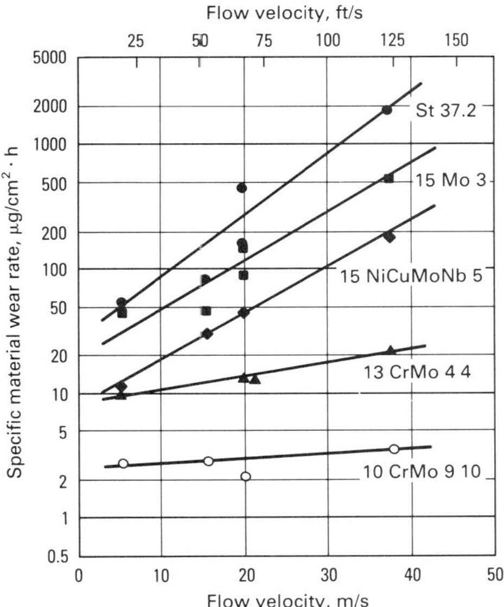

Fig. 78 Effect of flow velocity on material wear due to single-phase erosion-corrosion. *P*: 40 bar (4000 kPa, or 580 psi), *T*: 180 °C (355 °F), pH: 7, oxygen content: <5 ppb, period of exposure: 200 h. See Fig. 71 for material composition. Source: Ref 137

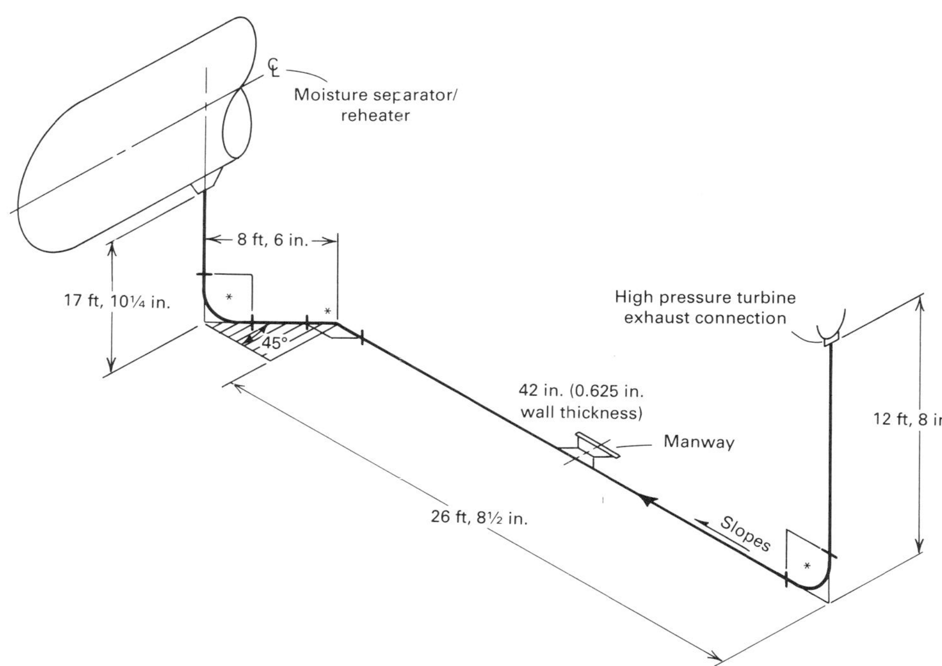

Fig. 79 Typical turbine exhaust piping arrangement. Turning vane elbows are marked with asterisks. Source: Ref 135

$$s = f(T)\, f(x)\, c\, K_c - K_s \qquad \text{(Eq 22)}$$

where s is the maximum local depth of material loss in mm/10 000 h; $f(T)$ is a dimensionless variable denoting the influence of temperature on erosion-corrosion damage ($f(T)$ versus T is shown in Fig. 82); $f(x)$ is a dimensionless variable denoting the influence of steam wetness on erosion-corrosion (for subcooled water, it has been suggested that this has a value of unity, but for two-phase mixtures, it has the form $f(x) = (1-x)K_x$, where x is the steam quality, and $0 < K_x < 1.0$; a value of $K_x = 0.5$ is considered to be most appropriate); K_c is a variable accounting for the effect of local geometry on the fluid flow (Fig. 83); c is the fluid velocity (in meters per second); and K_s is a threshold value that the product of $f(T)$, $f(x)$, c, and K_c must exceed before erosion-corrosion is observed. A value of 1.0 mm/10 000 h (0.04 in./10 000 h) has been given by Keller.

Keller also developed a set of influence factors, K_c, that adjust the predicted rate of erosion-corrosion as a function of steam path configuration. The predicted wear rates for components with configurations other than straight pipe can be easily calculated using the s value determined previously for the straight pipe. Although the parameters of Keller's equation are, to a large extent, representative of and significant to the influences of erosion-corrosion, this method of calculating wear rates is not precise in predicting actual rates or material life. Instead, the intent of this model in establishing the inspection program is to identify high-probability erosion-corrosion areas. Once established, actual wear rate data from site inspections should be used (Ref 135).

Inspection Techniques. Inspection should be conducted visually where practical and supplemented by ultrasonic testing to establish an accurate wall thickness. This method can be used very easily for systems that have personnel access. Where direct visual inspection is not possible, remote visual techniques such as fiber optics or a remotely operated pipe crawler equipped with a video camera would be recommended. With manual ultrasonic testing techniques, it is important that an inspection grid is established and that the grid size is not too large. Wall thickness measurements at the intersections of the grid and continuous scanning of the area bounded by the grid are required to identify areas of minimum wall thickness. Manual ultrasonic inspection, however, lacks the requisite degree of reproducibility. Automated ultrasonic inspection tools, usually computer enhanced, can produce a three-dimensional topographical representation of the pipe wall interior and can provide high reproducibility and improved records keeping.

Ultrasonic inspection requires pipe insulation removal and plant shutdown. Radiography can provide for the inspection of large areas of pipe wall interiors without costly insulation removal and ultrasonic test site preparation through a series of tangential shots of the pipe wall-insulation sectors. Because the pipe wall has substantially greater density than the insulation or fluid in the pipe does, contrast changes appear in the image at the inner and outer surfaces. The separation between these lines, which is a measurement of wall thickness, can be continuously recorded.

Prevention Considerations

In-Line Separators. In-line or preseparators can remove entrained moisture in steam as it leaves the high-pressure turbine. These devices have been used in French PWRs and in a few domestic nuclear plants. In addition to providing erosion-corrosion protection in wet steam piping, the separators can improve plant cycle efficiency through more effective moisture removal. Studies show definite economic advantages in retrofitting these devices in most domestic PWRs and BWRs (Ref 142). Typically, the in-line separators are placed in the turbine cross-around piping at the high-pressure turbine exhaust, where erosion-corrosion rates have been high.

Alternate Materials. Chromium-molybdenum steels have shown satisfactory service in power plant piping systems subjected to flashing liquids. In wet steam environments, 1.25 Cr-0.5Mo (P11 grade) and 2.25Cr-1Mo (P22 grade) alloy steels have shown good resistance to erosion-corrosion in both operating systems and laboratory tests (Ref 135). Of the two materials, P22 grade is more readily available in standard schedule pipe and fitting sizes and in plate for fabrication of the larger turbine exhaust piping. Grades P11 and P22, however, require special considerations for welding, including preheat and postweld heat treatment. At turbine exhaust and extraction temperatures, P11 and P22 have mechanical properties similar to those of carbon steel; therefore, thermal stresses and nozzle loadings should be similar.

Type 304 and 316 austenitic stainless steels have excellent resistance to erosion-corrosion. Type 304L has been successfully used as a carbon steel replacement in several power plants. Because of their improved resistance to intergranular stress corrosion after fabrication, the low-carbon grades are considered more reliable than the straight grades. Both low-carbon and straight grades of austenitic stainless steels are as readily available in standard schedule pipe, fittings, and plate as P22 alloy steel grades. Welding of austenitic stainless steels to existing carbon steel connections (that is, turbine casings, feedwater heater shells, and moisture separator reheater shells) is critical and requires appropriate procedures. Preheat and postweld heat treatment, however, are usually not required.

Austenitic stainless steel is susceptible to chloride SCC, and consideration must be given to using nonmetallic thermal insulation under wet conditions, because chloride and fluoride ions

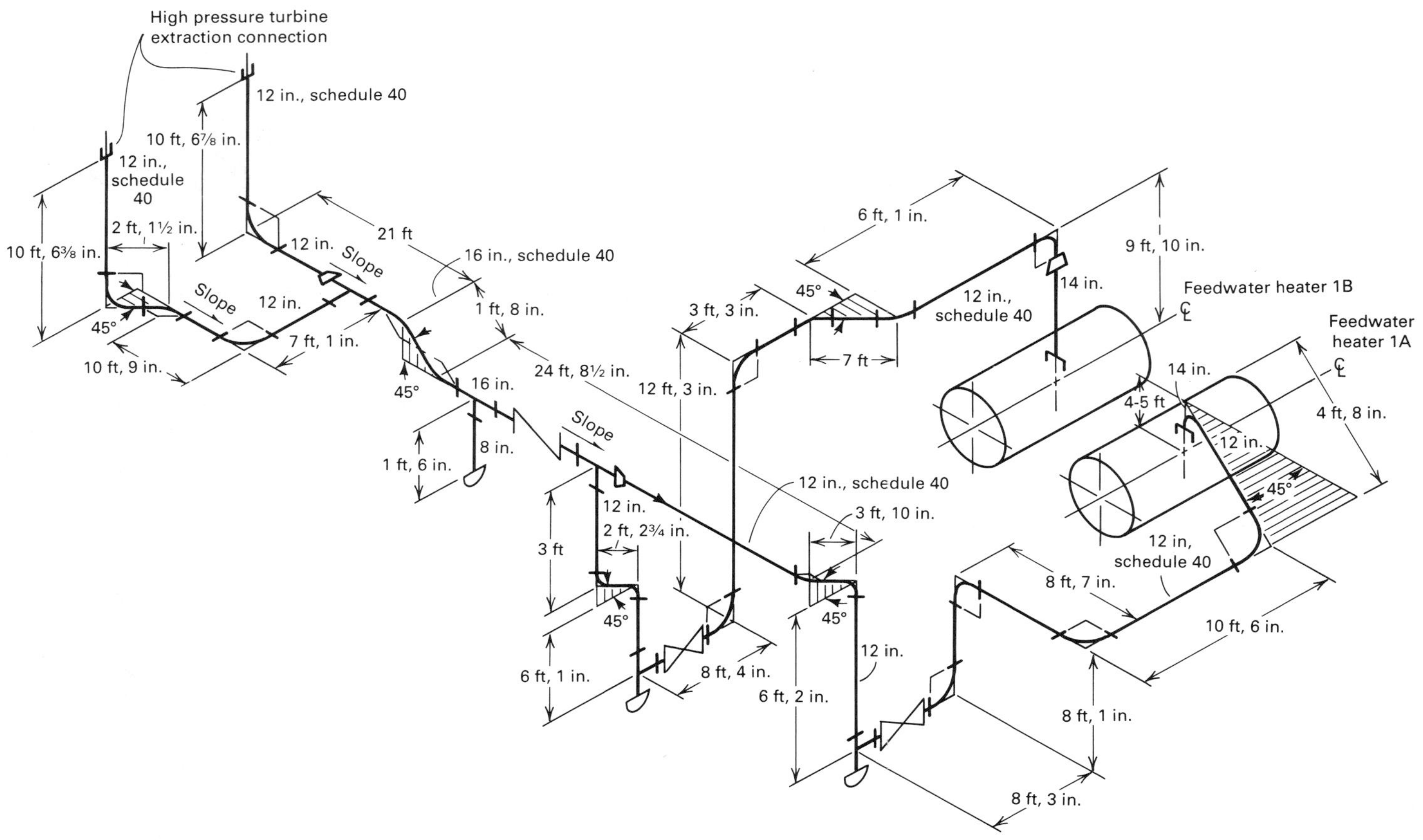

Fig. 80 Typical turbine extraction piping arrangement. Source: Ref 135

may leach onto the piping. Straight and low-carbon grades of austenitic steels will have the same dimensions and weight as the carbon steel piping. The thermal expansion coefficient, however, is about 1.4 times that of carbon steel. New piping analysis with possible layout changes and support modifications to control thermal stresses and nozzle loadings will be required (Ref 135).

Weld Repair and Cladding. Localized defects are good candidates for weld repair with the same filler metal as the base material or an alloy material more resistant to attack, such as 1.25Cr-0.5Mo. Experience in power plants, however, indicates that in many cases the areas adjacent to the weld repair were even more vigorously attacked, leaving the weld/clad itself almost untouched. Erosion-corrosion damage was typically apparent on both sides of the weld in the HAZ, which arises inherently during the welding pro-

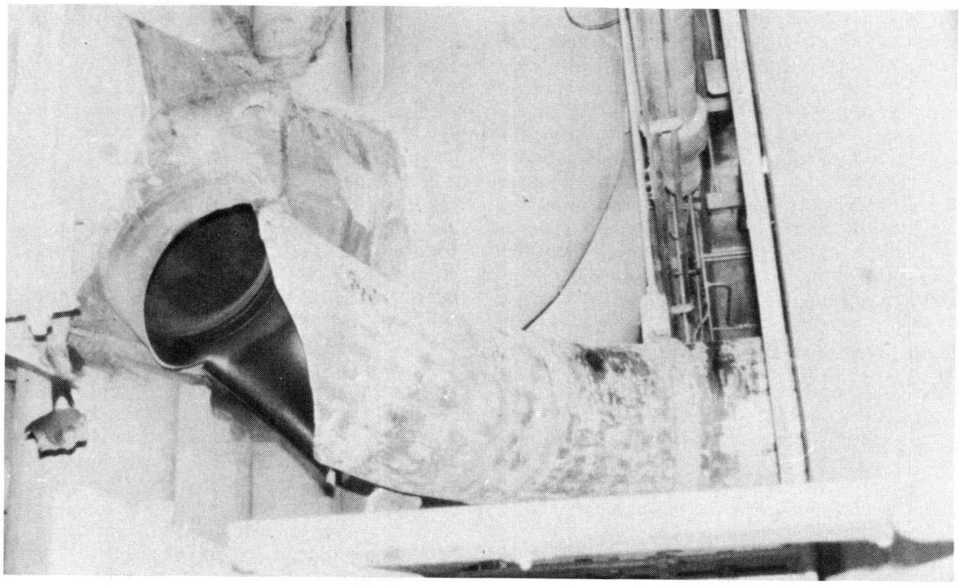

Fig. 81 Failure at high-pressure extraction piping elbow

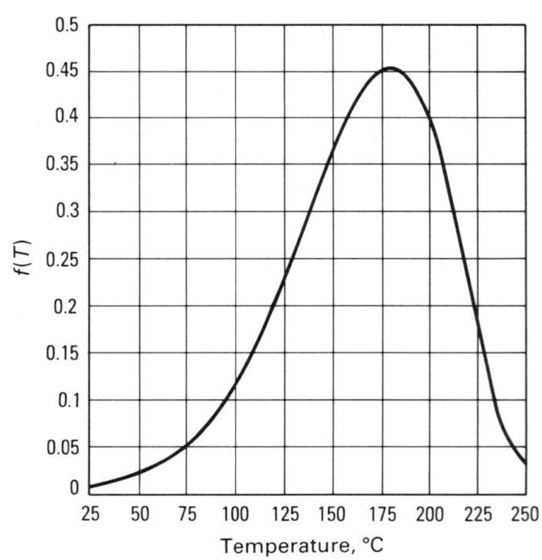

Fig. 82 Temperature influence function for Keller's equation (see Eq 22 in text for explanation). Source: Ref 135

Flow pattern			Reference velocity	K_c
Primary flow stagnation points	At pipes		Velocity of initial flow (upstream of stagnation obstacle)	1
	At blades			1
	At plates			1
	In pipe junctions			1
				0.8
Secondary flow stagnation points	$R_{mean}/D \cdot 0.5$	In elbow pipes	Flow velocity	0.7
	$R_{mean}/D \cdot 1.5$			0.4
	$R_{mean}/D \cdot 2.5$			0.3
	Behind pipe joints			0.2
Stagnation points due to vortex formation	Behind sharp edged admission pipes		Flow velocity	0.2
	At and behind barriers			0.2
No stagnation points	In straight pipes		Flow velocity	0.04
	In untight horizontal turbine joints		Velocity calculated from pressure drop	0.08
Complicated flow through turbine part	In turbine gland seal		Velocity calculated from pressure drop	0.08
	At and above turbine blades and at drainage collecting rings		Average circumferential blade velocity	0.3

Fig. 83 Flow path influence function for Keller's equation (see Eq 22). Source: Ref 135

cess. An alternate method of repair, especially for small piping, is to overlay metal on the outside surface of the pipe. Cladding extremely large areas, although possible, is not considered feasible because of the extremely high cost (Ref 135).

Corrosion of Containment Materials for Radioactive-Waste Isolation

J.W. Braithwaite
Sandia National Laboratories

The safe disposal of radioactive waste requires that the waste be isolated from the environment until radioactive decay has reduced its toxicity to innocuous levels. Many different types of radioactive waste are produced during commercial and defense nuclear fuel cycles. However, one type of waste, denoted high-level waste (HLW), contains the highest concentration of radiotoxic and heat-generating species. Because of this factor, the most stringent standards for disposing of radioactive wastes are being placed worldwide on HLW, and the majority of the radioactive waste management effort is being directed toward the HLW problem.

All of the countries currently studying the options for disposing of HLW have selected deep geologic formations to be the primary barrier for accomplishing this isolation. However, for conservatism, engineered components will also be used. The principal engineered component in this multibarrier approach is called the waste package and includes the waste itself, possibly a stabilizing matrix for the waste (together termed the wasteform), and a metallic container that encloses the wasteform.

The waste container may be a single vessel, but because of the many requirements being placed on the waste package, it will probably consist of two or more concentric vessels, each having a specific function. For example, the container that directly holds the wasteform may be designed to facilitate safe waste handling and emplacement operations. This container may be overpacked with a corrosion-resistant outer layer. Additional vessels may also be present to attenuate radiation and/or to resist external loads following emplacement. This section will specifically address the long-term corrosion behavior of HLW container materials. The information presented is also applicable to the selection of materials for shorter-term components (for example, to allow waste retrieval) and for use in the disposal of other, less toxic radioactive wastes.

Two basic types of HLW exist that will require disposal. The majority of the HLW in the United States will simply be spent nuclear fuel from commercial power reactors (Ref 143). Assemblies of spent fuel will either be placed intact into a container or first consolidated and then loaded into a container. The second type of waste results from the reprocessing of either defense-related or commercial spent fuel. Both sources of this reprocessed waste will be stabilized in a matrix material, probably a borosilicate glass, and then cast into a canister.

The programs of all the nations involved with HLW isolation recognize the advantages of having the container remain intact for the relatively long period of 300 to 1000 years. The major heat-producing species in HLW are two isotopes of cesium and strontium, both with 30-year half-lives. A 300- to 1000-year lifetime container allows the concentration of these two fission products, and thus the thermal output of the waste, to be reduced by a factor of at least 1000 before any dissolution and/or release of radioactivity can begin. The lower temperature of the waste is beneficial because dissolution of both wasteforms is thermally activated (Ref 144, 145). Also, after this time period, the radiotoxicity of the remaining fission products is less than that for a long-lived isotope of americium and less than four times that for the entire inventory at extended time periods (>10 000 years) (Ref 146, 147). Finally, a 300+ year container could assist any short-term waste retrieval operations, should that option need to be exercised.

In the U.S., the Nuclear Regulatory Commission (NRC) requires that the waste be reliably contained for a period of 300 to 1000 years (Ref 148). Other countries have adopted this requirement at least as a goal in their programs. The ultimate application for the container would be for it to function as an absolute barrier for the entire time that significant radiotoxicity is present. Canada and Sweden are investigating this possibility by determining the potential for containing the waste for periods of at least 100 000 years (Ref 147, 149).

Because the container must function both as a transportation vessel and as a corrosion barrier, almost all of the materials being considered for its construction are metallic. Given the environments and the candidate metals, the forms of corrosion that are applicable to waste containers include:

- General, or uniform, corrosion
- Localized attack (crevice, pitting, and intergranular attack)
- Stress-corrosion cracking
- Hydrogen-assisted cracking

The extremely long lifetime requirement being placed on the container forces the designer to consider only materials that have very predictable and reliable behavior. The rates of the latter three mechanisms are all sufficiently high or unpredictable that any design specifying a material that would fail by these types of nonuniform attack would be impractical. Thus, the challenge is to identify materials that will undergo only general corrosion in a given disposal environment. Materials that are acceptable in one environment may not be usable in another (for example, crystalline rock versus bedded salt).

Two basic approaches have evolved for potentially satisfying the lifetime requirement being placed on the waste container: use of a corrosion-resistant material or use of a corrosion-allowance material. For a reasonably designed container to survive for 300 to 1000 years, its corrosion rate must be relatively small. As will be shown, a number of materials have been identified with corrosion rates that are practically nil (micrometers or less per year). Very little corrosion allowance is required in the design of the waste package constructed with these "corrosion-resistant" materials. These metals usually owe their durability to very adherent oxide films, and as such, the major difficulty in their characterization is to demonstrate that the other, more insidious corrosion mechanisms will not be active during the long containment time. Examples of corrosion-resistant materials are titanium alloys, nickel-base alloys, and, in some applications, possibly stainless steels.

The alternate approach is to use "corrosion-allowance" materials that, in general, have more predictable corrosion characteristics. Although these materials have much higher general corrosion rates, their susceptibility to nonuniform corrosion is usually very limited due, in part, to a lack of protective oxide containing product layers. The corrosion rate of these active materials is often controlled by the rate of the cathodic process (oxygen reduction and/or hydrogen ion discharge). As will be shown later, the groundwater flow rates (oxidant supply) are very limited, and their pH is usually near neutral. Given these types of conditions, the general corrosion rate may be relatively low and can be measured reliably. Thus, an allowance for the material wastage can be incorporated into the design. Examples of these types of materials include mild steel, cast iron, and copper.

Given the above considerations, the materials characterization programs have two primary objectives:

- To quantify the general corrosion rate
- To demonstrate that no accelerated attack will occur. This requirement can be satisfied either by showing that the material is not susceptible to these forms of attack or that their time to initiation (the incubation period) is longer than the required lifetime of the container (for example, sensitization of stainless steel)

The impact of these two objectives on the corrosion programs will be evident in the discussion of "Corrosion Results From Site-Specific Studies" provided later in this section.

In addition to corrosion characteristics, cost and mechanical properties must also be considered in the selection of container materials. The mechanical properties can be very important when satisfying waste package requirements related to waste-handling and retrieval operations, but this subject is not relevant in this section. Container cost can be a factor. Unit material costs for the corrosion-allowance materials tend to be much lower than those for corrosion-resistant alloys; however, because wall thicknesses must be greater, this cost difference is reduced. Overall fabrication costs are very design dependent, and cannot be generalized. To compare costs properly for different container options, total disposal costs should be considered because the use of a corrosion-resistant material may allow a simple, lightweight design to be used with substantially fewer components, less fabrication, less emplacement-hole mining, and smaller handling equipment.

The remainder of the background information needed to evaluate properly the results of ongoing and completed corrosion testing is given in the next two parts of this section. The first, "Considerations for Waste Isolation," describes the unique problems associated with characterizing waste container behavior, and the second, "Environment," summarizes the expected container environments.

Considerations for Waste Isolation

The requirement that a waste container survive intact for periods of several hundred years in elevated-temperature irradiated geologic environments has created a difficult problem for materials and design engineers to solve. The unique aspect of this problem is associated with making very reliable predictions about the corrosion behavior or container materials for these extended periods of time. Many of the alloy systems being considered have been in existence for less than 100 years. Thus, a data base concerning long-term behavior is practically nonexistent. Because most countries, including the United States, do not plan to have their first repository operational until the late 1990s, sufficient time (10 to 15 years) exists to allow most of the needed long-term simulations and accelerated testing to be completed. In the U.S., an additional 50 years is available for testing before the repository is actually sealed (the "retrieval period"). Certainly, the important corrosion processes should be identifiable within these time frames.

The general problems associated with interpreting results from long-term and accelerated testing have been addressed by many investigators (Ref 150, 151, 152). For example, K. Nuttall discusses many of the considerations that should be given to identifying long-term failure processes and noted that unless the effect of a particular failure mechanism is large, it is difficult to characterize by using short-term tests (Ref 152). The real concern is that these unknown changes may lead to delayed, catastrophic stress-assisted failures (for example, SCC, hydrogen cracking, hydriding). The materials engineer should remember, however, that a proper waste package design will minimize container stress and thus greatly decrease the potential for these types of stress-related corrosion processes. A detailed methodology for predicting long-term behavior using accelerated testing is proposed in Ref 153.

The specific problems associated with predicting material behavior over extended time periods relate to identifying potential changes in both the environment and the metallurgical characteristics of the alloy. To understand these types of problems better, some examples are given below.

The important environmental parameters affecting corrosion include container temperature, groundwater chemistry, groundwater flow rate, hydrostatic and lithostatic pressure (influences water phase and container stress), and radiation flux. The predicted history for these parameters must be quantified before accurate and reliable assessments of the performance of the container can be made. The difficulty in easily predicting the effect of these environments on corrosion was demonstrated by some results from initial site studies performed in Canada. The groundwater in the crystalline-rock disposal environment in the Canadian shield was expected to be benign (no halide ion, neutral pH, low ion strength) (Ref 147). During a research-drilling program, however, groundwater containing 5.6 g/L of chloride ion was encountered. The existence of isolated pockets of saline groundwater in the deep rock formations supports the need for careful site characterization studies.

Metallurgical changes could be important because of their effect on the structures and properties of passive films, time to inducement of delayed mechanical failure, and time to sensitization of the microstructure. Again, the state of stress in the container could be a major factor that determines whether these metallurgically induced changes will actually lead to a failure. These potential effects are recognized and are being addressed by many investigators (Ref 151, 152, 154, 155).

Theoretically, radiation can affect the corrosion behavior of the container by affecting both the container environment and its metallurgical properties. However, the general conclusion reached by most investigators is that the types and dose rates of radiation emitted from decaying wastes are not sufficient to degrade the properties of either the container material or its passivating oxide layer and that the important effect of radiation is the change produced in the external environment due to groundwater radiolysis (Ref 152, 156). The effect of radiation on the environment is covered in the latter part of the next discussion. However, radiation has been shown to produce small changes in the properties of some metals that could lead eventually to increased corrosion rates (Ref 157). Such potential effects of radiation should be identified during long-term testing.

Environment

Numerous deep-geologic formations are being considered by the various countries as the media for isolating radioactive waste with the characteristics of many of these formations being quite similar. To aid in these discussions, this similarity allows the formations to be grouped into the following three categories:

- Crystalline rock (igneous and metamorphic)
- Salt deposits (bedded and domed)
- Sedimentary deposits (clay and seabed sediments)

The properties of the geologic formation are very important because of their influence on many of the factors that determine the corrosion behavior of container materials. These factors include the groundwater chemistry, the groundwater flow rate, the hydrostatic pressure, the lithostatic pressure, the phase of the water contacting the container (liquid and/or vapor), the temperature of the container, and the types and concentrations of radiolysis products. The container temperature and radiation output are influenced by the design and loading of the waste package (size, thermal output, radiation output), the rate and density of waste package emplacement, and the thermal properties of the formation.

A brief summary of the types of environments being used in the selection of waste container materials follows. Because radiation could have

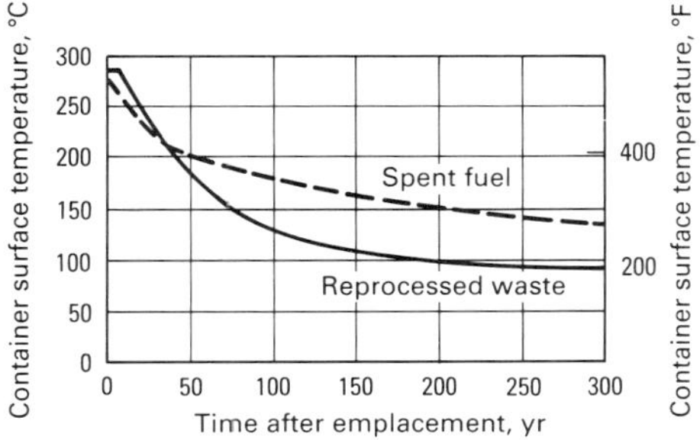

Fig. 84 The predicted temperature history for waste packages emplaced in a tuff formation at a 50-kW/acre areal loading. Power loading for a spent-fuel package is 3.3 kW and for a reprocessed-waste package is 2.2 kW. Source: Ref 158

Table 16 Typical peak container temperatures for selected repository locations

Host rock	Spent fuel °C	Spent fuel °F	Reprocessed HLW °C	Reprocessed HLW °F	Ref
Crystalline rock					
Basalt, U.S.	210	410	250	480	150
Tuff, U.S.	260	500	280	535	158
Granite, Sweden	80	175	80	175	149
Salt					
United States	180	355	235	455	159
West Germany	...	...	200	390	160
Sedimentary deposits					
Subseabed, U.S.	...	...	200	390	161
Subseabed, U.K.	...	...	100	212	162
Clay, U.K.	...	...	150	300	160

an important and potentially unique impact on the environment of the container through the production of radiolysis products, this discussion of environment is separated into two parts: general characteristics and radiation effects.

General Characteristics

Temperature. Because heat is a significant by-product of HLW decay, the temperature of all waste containers will initially increase and then decrease as the activity of the waste decays. The predicted temperature history for 3-kW waste packages emplaced in a consolidated volcanic ash (tuff) formation in the U.S. is shown in Fig. 84. Typical maximum container temperatures for a number of other repository locations are given in Table 16. Although these maximum temperatures are quite different, the general shape of the temperature curve shown in Fig. 84 should be similar for the other repositories. The variability in maximum temperature is due primarily to design philosophy. The temperature at a given location can be lowered by longer waste aging before emplacement, lower package loading, and lower overall repository loading. The lower temperatures will, in general, enhance the performance of the entire waste package and decrease the impact of emplacing waste on the geologic formation itself. However, a penalty is incurred to reduce temperatures because higher handling and emplacement costs, along with a larger usable area, are also required.

Water Chemistry. A summary of the expected composition of the groundwaters associated with several of the candidate formations is given in Table 17. The E_h, or oxidizing potential, of the groundwater in most cases will be determined by the presence of radiolysis products (see the discussion of "Radiation Effects" in this article). However, if radiation shielding is used in the package, conditions can range from slightly oxidizing (unsaturated tuff) to slightly reducing (basalt, granite, sub-seabed). The groundwaters associated with the crystalline-rock formations should all be relatively benign to most materials because of their low ionic strengths, near neutral pH, and low concentrations of halide ions. The corrosivity of these waters could be increased if significant groundwater vaporization occurs during the early times following emplacement when high container temperatures exist. Any brine contacting the container in the salt repositories will be quite corrosive because of the high concentration of halides and the potential for low pH due to magnesium salt hydrolysis (Ref 154). The situation is similar for the chloride-containing seabed environment. The corrosivity of groundwaters in a clay environment should fall between that of seawater and the crystalline-rock waters.

Water Availability. Because water movement is the primary mechanism for transporting radioactive species away from the package, an important criterion for selecting geologic formations is a lack of available water. The candidate formations have a low permeability and/or a small quantity of flowing water, resulting in a very low flow rate of groundwater past the waste containers. For crystalline-rock formations, groundwater will flow, because of hydraulic gradients, to the package either through the rock matrix or fractures and fissures. The water flow rate through the granite formation in Sweden and the tuff formation in the U.S. has been estimated to be of the order of 0.1 L/m^2/yr (Ref 149, 158).

Brine can be transported in the relatively dry (0.05 to 2% H_2O) salt formations by thermally induced migration of brine inclusions. This process has been estimated to transport only 8 L of brine to the package in the first 1000 years (Ref 151). Additionally, more significant quantities of brine could contact the package as a result of accidental flooding. Sub-seabed sediments will be saturated with essentially stagnant seawater (water velocity: <1 mm/yr, or 0.04 in./yr). Containers placed in clay will be exposed to humid air contaminated with volatile compounds released from the clay during heating or possibly to water due to intrusion from overlying aquifers.

Water Phase. Almost all of the repository sites being considered are located below the water table. During construction and water emplacement, these repositories will be dewatered and become dry. Immediately following closure, the container will be exposed to hot, relatively dry conditions, thus promoting oxidation. As water slowly resaturates the formation, inundated conditions may be produced and the hydrostatic pressure may allow liquid water to be present on the container at temperatures above 100 °C (212 °F). For repositories located in fractured, unsaturated rock, liquid water above 100 °C (212 °F) should never exist. For reference, the International Atomic Energy Agency qualitatively described many of the general features of the candidate sites that are relevant to this issue (Ref 169).

Radiation Effects

The dose rate of gamma radiation at the surface of the container will be determined by the concentration of the various radioactive isotopes within the package (a function of age, type, length of time the fuel is in the reactor) and the attenuation provided by the container. Currently, the U.S. plans to accept the most active waste fordisposal (5 years out of reactor). This waste will produce maximum dose rates at the container surface of 10^3 to 10^5 rads/h (Ref 159). Most of the corrosion testing conducted in the presence of radiation uses these maximum dose rate levels.

Table 17 Groundwater compositions (in parts per million) for selected repository locations at 25 °C (75 °F)

Ion	Basalt (Ref 163)	Tuff (Ref 158)	Granite (Ref 164, 165)	Salt (high-magnesium brines) (Ref 154, 166, 167)	Seawater (Ref 154)	Clay (Ref 168)
Na^+	250	51	0–106	6 500–42 000	10 600	63
K^+	1.9	5	...	10 500–30 000	380	7.4
Mg^{2+}	0.4	2	0–6	35 000–85 000	1 270	3.6
Ca^{2+}	1.3	14	10–40	600–14 700	400	21
Sr^{2+}	...	0.05	...	5	13	...
Fe^{2+}	...	0.04	0.02–5	...	...	189
NH_4^+	...	...	0.05–0.2	...	...	...
Cl^-	148	7.5	4–36	190 000–270 000	19 000	36
SO_4^{2-}	108	22	0.5–24	160–13 000	880	...
I^-	...	...	...	10	0.05	...
CO_3^{2-}	97	120	90–275	700	146	188
Br^-	...	...	...	400–2 400	65	...
BO_3^{3-}	...	...	...	1 200	...	...
NO_3^-	...	5.6	0.01–0.05	...	...	6
HS^-	...	...	0–0.5	...	...	...
F^-	37	2.2	0–2	...	...	817
$H_3SiO_4^-$	103	61	0–19	...	...	8
pH	9–10	7.1	7–9	6.5	8.1	7.4

Many researchers have addressed the effect of radiation on the environment of the waste container. The information available on this subject has been collected and analyzed as of 1981 (Ref 157). Based on this information, some general conclusions can be made, as follows. First, only gamma radiation from the HLW can affect the environment, because the other forms of radiation will not penetrate the container. The gamma radiation will cause radiolysis products to be formed in the groundwater or vapor environment around the container. Second, the net effect of the radiolysis products will be to make the environment more oxidizing. In general, the higher the solute concentration in the groundwater, the higher the concentration of radiolysis products. Lastly, the major radiolysis products associated with the three rock types can be summarized:

I Crystalline rock

Groundwater: H, OH, e^-_{aq}, H_2, H_2O_2, HO_2, O_2^-
Vapor: CO, O_3, HNO_3, NH_3, NO_x

II Salt deposits

Brine: species from Category I plus oxychlorides, oxybromides, Cl_2^-, Cl_2, Br_2, O_2
Vapor: species from Category I plus HCl, HBr

III Sedimentary deposits

Subseabed: Same as Category I with the addition of the chloride containing species from Category II
Clay: Based on groundwater chemistry, radiolysis products for clay should also be close to those for the hard-rock formations.

Radiolysis of the low ionic strength groundwaters associated with crystalline rock is very similar to that for pure water, with the stable species being molecular hydrogen (H_2) and hydrogen peroxide (H_2O_2). A detailed investigation of brine radiolysis and some semi-quantitative predictions of product yield are discussed in Ref 170 and 171.

Recently, a study was conducted to determine the effect prior irradiation of rock salt may have on the chemistry of the brine that may contact the container (Ref 172). It was concluded that the pH and total base of a saline solution made from irradiated rock salt can be significantly different from a solution made from nonirradiated salt. Either alkaline or acidic brine can result, depending on the shielding of the waste container and the quantity of brine present during the irradiation period.

Because metals corrode by oxidation, radiolysis may adversely affect corrosion rates. The strong oxidants produced in brines should make salt the most aggressive of the geologic media being considered. However, the presence of strongly adherent oxide films on certain metals can significantly reduce the impact of radiolysis products. For two localized corrosion mechanisms applicable to austenitic stainless steel (crevice and stress cracking), radiolysis has even been shown to have a beneficial effect by repairing localized defects in the oxide layer and thus reducing their susceptibility to these forms of corrosion (Ref 156, 157).

Simulating the effect of radiolysis products on the groundwater chemistry is a difficult task because most of the available information on radiolysis products is not quantitative and because many of the radiolysis products themselves, especially the free radicals, have very short lifetimes. For this reason, the corrosion tests completed to date that have included radiolysis products have been done under irradiated conditions. These tests require specialized facilities and procedures, factors that have dramatically limited the number of corrosion studies that include radiation effects. Attempts to produce simulated environments are continuing with limited success to date. For example, some of the electrochemical characteristics of type 316 stainless steel in low ionic strength solutions have been reproduced by simply adding H_2O_2 to the solution (Ref 173).

Corrosion Results From Site-Specific Studies

The first comprehensive programs undertaken to identify acceptable container materials started by selecting several classes of potentially suitable alloys and evaluating their performance in the laboratory and/or the field. In general, relatively simple corrosion tests were used, often under overtest conditions. This initial selection of candidate alloys was based on such criteria as corrosion behavior (general rate and susceptibility to localized/stress-assisted attack), material costs, and material availability (Ref 161). Based on the results of these screening tests, a few very promising materials were identified that are now being subjected to detailed characterization studies, the eventual objective of which is to qualify definitively a material for constructing licensable waste containers. Currently, none of the programs has completed the detailed-study phase. A list of the major candidate alloys, along with their nominal compositions, is given in Table 18.

This discussion will present corrosion results from both the screening tests and the detailed studies. These results are grouped by the three categories of rock types defined in the previous section. Because of space limitation, a comprehensive review is not possible and only selected information is given, the majority of which addresses general corrosion. However, a significant portion of the actual investigations is or was concerned with characterizing nonuniform corrosion behavior. The detailed results are available in the References. Two factors should be remembered when considering this information. First, proper container design could eliminate most tensile stresses, thus reducing the potential for stress-assisted types of failures, such as SCC, hydriding, and hydrogen-assisted cracking. Second, one testing parameter that was highly variable and is noted with all the results is the duration of the exposure. The short length of many tests should preclude the use of the associated results in any extrapolations of corrosion rates or behavior to long time periods.

For an excellent general reference, an in-depth review of this subject through 1981 was performed by the Brookhaven National Laboratory (Ref 156). Also, the Pacific Northwest Laboratory compiled all of the relevant results of corrosion testing in the U.S. up to 1983 (Ref 174). Following this section, the author's overall observations and conclusions based on this site-specific information are presented.

Crystalline Rock

Several countries are currently investigating the suitability of isolating high-level waste in a

Table 18 Nominal chemical composition of waste container candidate alloys

Alloy	Element, %								
	C	Mn	Si	Cr	Ni	Mo	Fe	Cu	Other
Ductile cast iron	3–4	0.5	2–3	. . .	0–1	. . .	bal	. . .	0.02–0.07Mg
Gray cast iron	3.75 (max)	0.9 (max)	2.4 (max)	1.25 (max)	5 (max)	1 (max)	bal	. . .	. . .
1018 carbon steel	0.18	0.75	0.25	. . .	. . .	. . .	bal	. . .	. . .
4130 alloy steel	0.3 (max)	0.5	0.3	1.0	. . .	0.2	bal	. . .	. . .
Corten A steel	0.1	0.4	0.5	1.0	. . .	. . .	bal	. . .	0.02Ti
2.25Cr-1Mo	0.2	0.8	0.3	2.2	. . .	1	bal	. . .	. . .
Naval brass	. . .	. . .	. . .	. . .	. . .	. . .	. . .	60	39Zn, 1Sn
90–10 cupronickel	. . .	. . .	. . .	. . .	10	. . .	1.3	88.7	. . .
Type 304	0.08 (max)	2.0 (max)	1.0 (max)	19	10	. . .	bal	. . .	. . .
Type 316	0.08 (max)	2.0 (max)	1.0 (max)	17	12	2.5	bal	. . .	. . .
Type 416	15 (max)	. . .	1.0 (max)	13	. . .	. . .	bal	. . .	. . .
Ebrite 26-1	0.01 (max)	0.4 (max)	0.4 (max)	26	0.5 (max)	1	bal	0.2 (max)	. . .
Monel 400	0.2	1.0	0.2	. . .	66.5	. . .	1.2	31.5	. . .
Incoloy 825	0.03	0.5	0.2	21.5	42	3.0	30	2.2	0.9Ti
Inconel 600	0.08	0.5	0.008	15.5	76	. . .	8	. . .	. . .
Inconel 625	0.05	0.2	0.2	21.5	61	9.0	2.5	. . .	4Nb, 0.2Ti
Hastelloy C-276	0.02 (max)	1 (max)	0.08 (max)	15	bal	16	5	. . .	4W
Hastelloy C-4	0.015 (max)	1 (max)	0.08 (max)	16	bal	16	3	. . .	. . .
Zircaloy 2	0.12	. . .	. . .	0.1	0.05	. . .	. . .	. . .	1.5Sn, 98.2Zr
Ti-Gr2	0.1 (max)	. . .	. . .	. . .	. . .	. . .	. . .	. . .	bal Ti
Ti-Gr7	0.1 (max)	. . .	. . .	. . .	. . .	. . .	. . .	. . .	0.2 Pd, bal Ti
Ti-Gr12	0.08 (max)	. . .	. . .	. . .	0.8	0.3	0.3 (max)	. . .	bal Ti

crystalline (metamorphic or igneous) rock formation. Formations included in this list are granite (Sweden and Europe), granitelike (Canada), basalt (U.S.), and tuff (U.S.). Because some of the characteristics for each type of formation are different, results will be presented separately below.

Granite. In the well-developed Swedish program, the emphasis on the selection of materials and ultimate design of the container was driven by a Swedish law that required the identification of demonstrably safe disposal methods (Ref 149, 165, 175). Container fabrication and handling costs were not considered to be important factors. Two metallic containers resulted from their assessment, both self-shielded to eliminate potential radiation effects. The first container, for reprocessed waste, consists of a 100-mm (4-in.) thick lead lining surrounded by a 6-mm (0.25-in.) thick unalloyed titanium canister. The second container, for spent fuel, is a 200-mm (8-in.) thick copper canister. Without radiation, the expected environment within the granite repository is slightly reducing. The general corrosion rate of the titanium was measured to be less than 0.25 μm/yr (0.01 mil/yr), and the susceptibility to localized attack judged to be very small. Although the risk of delayed failure due to hydrogen/stress effects was also believed to be small, such a failure mode could not be fully excluded. Furthermore, lead could cause solid-metal embrittlement of the titanium. Because of the potential hydrogen effects, their conservative approach is to not depend on the titanium, but to rely solely on the lead component for corrosion protection.

The maximum corrosion rate of both the lead and the copper canister under these conditions will be constrained by the rate of oxidant transport to the surface of the canister. The potential rate of oxidant (oxygen and sulfide) supply was determined assuming expected water flow rates (see the discussion on "Environment" in this article). Penetration rates were then calculated as follows. If all the attack is concentrated into one hemispherical pit in the lead, the liner would perforate in 4500 years. For the copper canister, even if pits 25 times deeper than the average penetration occurred, only 60 mm (2.4 in.) of the copper would be penetrated after a million years. Their conservative conclusions were that the titanium/lead container should last for thousands of years and the copper container for hundreds of thousands of years.

In Europe, emphasis has been placed on studying corrosion-resistant alloys (nickel-chromium base and Ti-Gr7), along with several low-alloy steels (Ref 164, 176). The general corrosion rate of the resistant alloys is usually less than 1 μm/yr (0.04 mil/yr). Under certain possible conditions, the corrosion behavior of the Ti-Gr7 alloy has been shown to be superior to that of Hastelloy C-4. In irradiated granitic solutions, Hastelloy C-4 was found to be susceptible to pitting. In high-temperature solutions, Hastelloy C-4 was susceptible to both crevice corrosion and SCC. The titanium alloy showed no susceptibility to these forms of attack.

Corrosion rates measured for the carbon steels ranged from 3 to 55 μm/yr (0.12 to 2.2 mils/yr), with one study showing that the rate reaches a maximum around 80 °C (175 °F) (Ref 177). It was concluded in Ref 178 that (1) the conditions that would lead to localized corrosion of carbon steels are quite specific and unlikely to be present in typical granitic groundwaters and (2) that hydrogen embrittlement and hydrogen blistering of low-alloys steels is possible in granitic environments with a high rate of hydrogen production. However, the first conclusion was qualified by the observation that localized corrosion should still be studied because the possibility exists that conditions in the future could produce the correct combination of passivating ions (for example, bicarbonate) and aggressive ions (chloride) to cause localized corrosion. If the hydrogen effects are found to be viable processes, then radiation shielding may be required.

The Canadian container program along with some recent results are described in Ref 147, 152, and 179. The alloy families being considered include austenitic stainless steels, copper, nickel-base superalloys, and titanium alloys. Results have shown that the susceptibility of Ti-Gr2 to crevice corrosion at temperatures above 100 °C (212 °F) may be inhibited by the presence of gamma radiation and that two nickel-chromium alloys, Hastelloy C-276 and Inconel 625, should not undergo localized corrosion at temperatures below 100 °C (212 °F) even in 20% NaCl solutions.

Basalt. The early screening studies performed with aerated basalt groundwater showed that low-alloy steels and irons could be viable container materials (Ref 180). The majority of the subsequent activity has focused on cast irons, steels, and titanium-base alloys, with a new comprehensive study devoted to copper-base alloys just beginning. The titanium alloys were selected as a potential alternative because of their excellent performance in more aggressive brine solutions. The corrosion-allowance materials (iron, steel, copper) are being studied because, without radiation, the basalt groundwater should be slightly reducing (similar to the Swedish granite environment). Thus, for shielded packages, the performance of these materials could be very good.

The evaluation program has consisted of three major testing elements: general corrosion, irradiation corrosion, and environmental cracking. Some results from the first two elements are shown in Table 19 and Fig. 85 and 86. Oxidizing conditions increase the corrosion rate over anoxic conditions by a factor of 9 to 10 (Table 19). It has been estimated (Ref 163) that irradiation enhances general corrosion by a factor of two to three for all the alloys being studied (Fig. 85 and 86). However, the maximum steady-state corrosion rate observed was only 11 μm/yr (0.43 mil/yr) under temperature and irradiation conditions that are conservatively severe. Based on these and other results, it has been concluded that cast iron or cast steel appear to be suitable materials for long-term barriers in basalt. If effective radiation shielding is used, a corrosion allowance of approximately 12 mm (0.47 in.) would be required. In the presence of radiation, an allowance of 25 mm (1 in.) might be adequate.

Corrosion rates for Ti-Gr2 and Ti-Gr12 in both oxygenated and irradiated basalt environments are very low—less than 2 μm/yr (0.08 mil/yr) (Ref 151, 182). Additionally, the titanium alloys have not been shown to be susceptible to any form of environmentally enhanced cracking (Ref 163). The long-term corrosion rates of many copper-base alloys are also sufficiently low, <20 μm/yr (0.78 mil/yr), at 200 °C (390 °F) that their use now appears feasible. More detailed copper evaluations are being conducted (Ref 183).

Tuff. The potential location of a waste repository in tuff is unique among all the sites because it is located in the unsaturated zone. This factor, along with the fractured character of the tuff, should preclude the possibility of the waste container becoming inundated at temperatures above 100 °C (212 °F), but could also keep any groundwater at the repository horizon aerated. Three test environments are applicable: hot, "dry" air with controlled humidity (below saturation) at temperatures from 95 to 300 °C (205 to 570 °F); moist air with humidity at or near saturation; and inundated conditions (Ref 158). Inundated conditions represent an overtest, but may exit tempo-

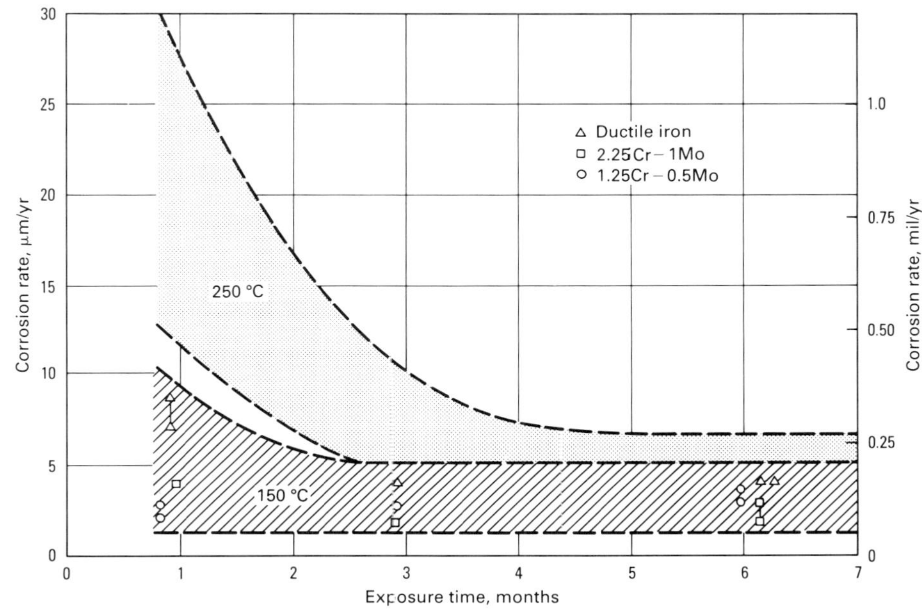

Fig. 85 General corrosion rate of cast ferrous alloys in basalt groundwater at 150 and 250 °C (300 and 480 °F). Source: Ref 182

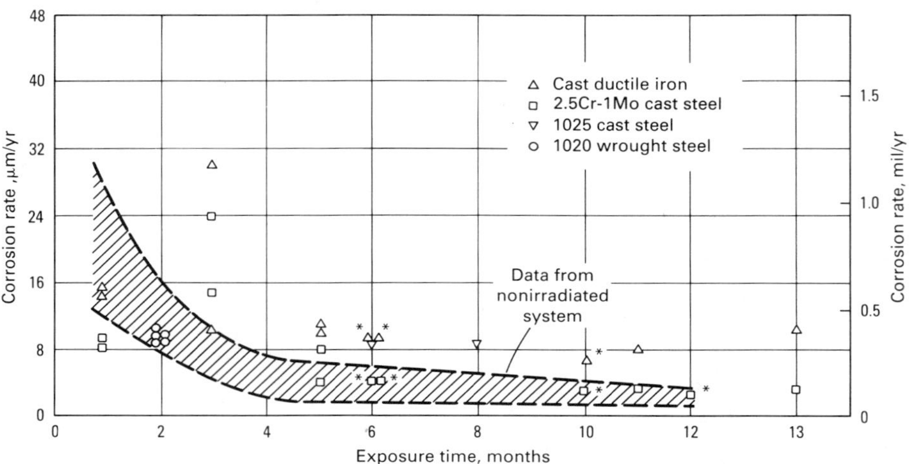

Fig. 86 General corrosion rate of cast ferrous alloys in synthetic basalt groundwater at 250 °C (480 °F) and a gamma irradiation rate of 3 × 10^5 rad/h. Source: Ref 163

Table 19 General corrosion rates of carbon steels under various conditions in basalt groundwater

Test duration: 2 weeks

AISI designation	Average corrosion rate, µm/yr (mils/yr) — Deoxygenated (<0.1 ppm O_2) 150 °C (300 °F)	Deoxygenated (<0.1 ppm O_2) 250 °C (480 °F)	Oxygenated (~8 ppm O_2) 150 °C (300 °F)
1006	13 (0.51)	12 (0.47)	118 (4.64)
1020	13 (0.51)	12 (0.47)	101 (3.97)
1025	11 (0.43)	13 (0.51)	105 (4.13)

Source: Ref 181

Table 20 General corrosion rates of candidate steels in 100-°C (212-°F) tuff-conditioned saturated steam and water

Test duration: 6 weeks

Alloy	Corrosion rate — Steam µm/yr	Steam mils/yr	Water µm/yr	Water mils/yr
1020 carbon steel	48(a,b)	1.89	28(a,b)	1.10
A36 carbon steel	49(a,b)	1.93	36(a,b)	1.42
A366 carbon steel	52(a,b)	2.05	28(a,b)	1.10
2.25Cr-1Mo alloy steel	12(a,b)	0.47	30(a,b)	1.18
9Cr-1Mo alloy steel	1.4	0.05	3.5	0.14
409 stainless steel	nil		nil	
416 stainless steel	0.2	0.008	0.3	0.012
304L stainless steel	0.1	0.004	0.1	0.004
316L stainless steel	nil		0.5	0.019
317L stainless steel	0.1	0.004	0.5	0.019

(a) Crevice corrosion. (b) Located surface attack. Source: Ref 158

rarily if unexpected episodic intrusions of water occur. The latter two environments were used in the material screening studies.

Because the groundwater composition is relatively benign (Table 17) and the temperatures associated with groundwater contact are low, several ferrous-base alloys were selected for the initial screening study. Results from testing with and without radiation are given in Tables 20 and 21, respectively. The general corrosion rates of the stainless steel alloys are all very low (<1 µm/yr, or 0.04 mil/yr). No susceptibility to localized attack or SCC was noted for the stainless steels in these screening tests.

For the low-alloy steels, the corrosion rates are much higher, and localized attack was observed on all the samples. The best nonstainless material tested was the 9Cr-1Mo alloy steel, but this material stress corrosion cracked during testing. Radiation did not significantly affect the corrosion behavior of any metal studied with the exception of the 9Cr-1Mo steel. This alloy showed some evidence of pitting attack that was not observed without radiation. A few years later, more detailed corrosion testing of the low-alloy ferrous materials was completed, with one of the objectives being to quantify the localized corrosion rates (Ref 184). Results are shown in Table 22. Although the general corrosion rates are quite different from those reported in Table 20, the importance of pitting and crevice corrosion for most of the alloys is evident.

The screening tests resulted in the selection of type 304L stainless steel as the reference material. Because it was recognized that a material with better resistance to localized attack and SCC may be required, several alternate stainless-type alloys are also being subjected to the detailed-testing phase (Ref 158). These alternate alloys include type 316L, type 321, and Incoloy 825, each offering the potential for some improvement in performance over type 304L, but each being more costly. Many of the detailed studies have been completed, and many are ongoing or planned. Subjects already addressed include the effect of stress on the behavior of type 304L (Ref 185), the potential for low-temperature sensitization of type 304L (Ref 155), and the investigation of corrosion processes in irradiated environments (Ref 173, 186). Activity is continuing on the identification and experimental detection of the critical sensitization parameters. In addition, a metallurgical examination was completed of a type 304L container from a spent-fuel field test in a similar environment with the result that after almost 3 years of exposure no observable cracking or general corrosion was identifiable (Ref 187).

Recently, the list of alternative materials being studied in detail was expanded to include several copper-base alloys: oxygen-free copper, aluminum bronze, and cupro-nickel (Ref 186, 188). These additions were made because of the recognized difficulty in reliably characterizing passive, metallurgically complex alloys such as type 304L. The initial feasibility study for these copper-base alloys has been completed, and detailed studies, including testing in irradiated environments, are ongoing. Corrosion rates measured after 6-month exposures for the copper alloys under nonirradiated conditions at 100 °C (212 °F) were all less than 3 µm/yr (0.12 mil/yr).

Salt Deposits

Waste isolation in bedded and domed salt deposits is being considered by the U.S. and several European countries, most notably West Germany. The first studies to screen potential materials were associated with Project Salt Vault and involved both laboratory and field testing in a bedded salt formation in Kansas. These tests were conducted in relatively dry salt (up to 2% H_2O). Some of the laboratory results are given in Table 23. The conclusion from this testing was that austenitic stainless steels are resistant to corrosion as long as the salt remains dry. However, the field tests confirmed that containers may not remain dry because SCC occurred in many of the stainless steel containers and instruments (Ref 189).

A large amount of corrosion data has been collected in the geothermal energy programs. This information is qualitatively relevant to waste isolation in salt because of the similarity in temperature and solution compositions. The missing environmental parameter is radiation. The majority of this source of information is summarized in Ref 174.

The results of the initial, comprehensive screening program using severe, high-temperature, inundated conditions were reported in Ref 161. These "overtest" conditions were used to allow the best performing materials to be identified. A partial list of the corrosion results from this study is given in Table 24. As shown, the majority of the 20 materials tested had fairly low general corrosion rates, and only a few experienced localized attack (pitting or crevice corrosion). The effect of radiation was found to increase with increases in the solute concentration in the water and the gamma dose rate. This study also showed that the magnesium-containing brines are the most corrosive of the various brines being investigated. This characteristic is probably due to magnesium salts that hydrolyze as the temperature of the brine is increased. The hydrolysis, in turn, significantly lowers brine pH. For consistency, all of the corrosion results given in this discussion were taken from testing that used magnesium-containing brines. Based on corrosion behavior, material cost, and material availability, several promising alloys were identified, with Ti-Gr12 being selected as the primary candidate.

During the same time period, a complementary screening study was being performed at the Pacific Northwest Laboratory (Ref 180). In this activity, some of the above tests were repeated (see bracketed values in Table 24), and a few more low-cost iron-base alloys evaluated. The available information on the corrosion of iron and steel has been compiled and is given in Fig. 87. A conclusion reached by the U.S. group investigating commercial HLW disposal (Salt Repository Project) was that some of the relatively inexpensive iron-base alloys may be usable in the high-temperature salt environment.

A few years later in West Germany, a more in-depth long-term screening study was performed. In this study, localized corrosion, along with the effects of gamma irradiation and brine chemistry, was quantitatively evaluated. Some of the results are given in Table 25. As can be seen, the effect of gamma irradiation is very significant for most of the alloys. These tests concluded that Ti-Gr7 has excellent corrosion resistance, even in the presence of gamma irradiation. The general corrosion rate should be less than 1 μm/yr (0.04 mil/yr) with no susceptibility to localized or stress-assisted attack. The only other alloys that are viewed as viable candidates are Hastelloy C-4 and the unalloyed steels. However, for these latter materials, radiation shielding must be included because of the deleterious effects of radiolysis products.

Most of the detailed characterization studies completed to date have been done in the U.S. These studies have focused on corrosion-resistant titanium alloys and inexpensive corrosion-allowance iron-base alloys. Sandia National Laboratories has been the dominant investigator of titanium behavior (Ref 154). General corrosion rates for titanium alloys are shown in Tables 26 and 27. The majority of the titanium studies have been directed at determining the susceptibility to nonuniform types of behavior, including localized corrosion, SCC, and hydrogen embrittlement. Other factors being considered are the effects of radiation, composition, microstructure, and processing history. The Sandia studies have shown that a container made from Ti-Gr12 should not fail by general corrosion during the initial 1000-year period with total penetration less than 1 mm (40 mils). Although a significant amount of research has been performed without identifying any susceptibility, the unresolved concerns are with SCC and hydrogen embrittlement. The results of other investigations addressing delayed failure in titanium alloys are given in Ref 190 and 191.

The Pacific Northwest Laboratory has been the prime investigator of iron-base alloy corrosion (Ref 166, 180, 192, 193). The effects of water content, time, and irradiation on the corrosion rate of several of the candidate alloys are shown in Fig. 88 to 90. The important conclusions reached from these studies included the following:

- In the absence of radiation, the corrosion rate of low-alloy ferrous materials is sufficiently low even under inundated conditions that 25 mm (1 in.) of material allowance would be adequate

Table 21 General corrosion rates for candidate steels in irradiated tuff environments at two temperatures

Test duration: 8 weeks

Alloy	Environment	Corrosion rate: 3×10^5 rad/h, μm/yr	mils/yr	6×10^5 rad/h, μm/yr	mils/yr
1025 carbon steel	Groundwater	36	1.42	47	1.85
	Groundwater and tuff	47	1.85	38	1.50
9Cr-1Mo alloy steel	Groundwater	15	0.59	13	0.51
	Groundwater and tuff	21	0.83	34	1.34
304L stainless steel (annealed)	Groundwater	0.3	0.012	0.4	0.016
	Groundwater and tuff	0.3	0.012	0.3	0.012
304L stainless steel (sensitized)	Groundwater	0.3	0.012	0.5	0.019
	Groundwater and tuff	0.3	0.012	0.5	0.019

Source: Ref 158

Table 22 General and localized corrosion rates of steels in tuff groundwater at different temperatures

Test duration: 9 weeks

Temperature, °C	°F	General corrosion rate, μm/yr	mils/yr	Pitting corrosion rate, μm/yr	mils/yr	Pitting factor	Crevice corrosion rate, μm/yr	mils/yr	Crevice factor
1020 carbon steel									
50	120	401	15.8	380	14.9	0.95	413	16.2	1.03
70	160	505	19.9	1018	40.1	2.02	359	14.1	0.71
80	175	531	20.9	465	18.3	0.88	472	18.6	0.89
90	195	414	16.3	1046	41.2	2.52	563	22.2	1.36
100	212	320	12.6	1018	40.1	3.18	635	25	1.98
Gray cast iron									
50	120	359	14.1	203	8.0	0.57	98.2	3.87	0.42
70	160	422	16.6	392	15.4	0.94	138	5.43	0.43
80	175	357	14.1	79.6	3.13	0.22	305	12	0.85
90	195	323	12.7	27.9	1.10	0.09	14.6	0.57	0.05
100	212	318	12.5	0	0	0	232	9.13	0.69
2.25Cr-1Mo alloy steel									
50	120	316	12.4	649	25.6	2.07	946	37.2	2.97
70	160	469	18.5	1448	57.0	3.07	1100	43.3	2.33
80	175	376	14.8	1089	42.9	2.90	1415	55.7	3.63
90	195	370	14.6	868	34.2	1.83	800	31.5	2.17
100	212	278	10.9	1352	53.2	4.87	781	30.7	4.50
9Cr-1Mo alloy steel									
50	120	21.2	0.83	42.3	1.67	1.83	202	7.95	14.6
70	160	21.2	0.83	152	5.98	7.47	450	17.7	22.5
80	175	14.1	0.56	246	9.68	17.4	319	12.6	23.1
90	195	8.3	0.33	0	0	0	84.7	3.33	10.1
100	212	6.8	0.27	0	0	0	33.0	1.30	8.1

Source: Ref 184

Table 23 General corrosion rates for various metals under conditions simulating relatively dry salt mine storage

Test duration: 4 weeks in autoclaves with specimens packed in the salt. Bracketed values are maximum pitting penetration.

Alloy	Corrosion rate, µm/yr (mils/yr): 99.5% NaCl + 0.5% H_2O, 200 °C (390 °F)	99.5% NaCl + 0.5% H_2O, 300 °C (570 °F)	99.75% NaCl + 0.25% $MgCl_2$, 200 °C (390 °F)	99.75% NaCl + 0.25% $MgCl_2$, 300 °C (570 °F)
Type 304L	5 (0.19) [0.010 mm, or 0.4 mil]	13 (0.51) [0.056 mm, or 2.2 mils]	18 (0.71) [0.005 mm, or 0.19 mil]	20 (0.79)
Hastelloy C	3 (0.12)	3 (0.12)	3 (0.12)	3 (0.12)
ASTM A108 carbon steel	33 (1.30) [0.046 mm, or 1.8 mils]	38 (1.49) [0.066 mm, or 2.6 mils]	69 (2.72)	89 (3.50)
Nickel	3 (0.12)	3 (0.12)	8 (0.31)	15 (0.59)
Commercial-purity aluminum	5 (0.19) [0.053 mm, or 2.1 mils]	Weight gain [0.084 mm, or 3.3 mils]	Weight gain	Weight gain [0.091 mm, or 3.6 mils]
Ti-Gr2	3 (0.12)	3 (0.12)	5 (0.19)	3 (0.12)

Source: Ref 174

Table 24 General corrosion rates of U.S. candidate alloys in 250-°C (480-°F) chloride solutions

Test duration: 4 weeks (deoxygenated), 2 weeks (oxygenated), 8 weeks (bracketed values)

Alloy	Corrosion rate, µm/yr (mils/yr): Deoxygenated (O_2:30 ppb), Brine	Deoxygenated, Seawater	Oxygenated, Brine (O_2:600 ppm)	Oxygenated, Seawater (O_2:1750 ppm)
1018 carbon steel	1700 (66.9)	400 (15.7)	7000 (275)	11 000 (433)
Corten A steel	900 (35.4)	200 (7.87)	...	...
2.25Cr-1Mo steel	1000 (39.4)(a)	200 (7.87)	...	...
Lead	500 (19.7)	300 (11.8)	1200 (47.2)	1000 (39.4)
Copper	70 (2.75) [415 µm/yr, or 16.4 mils/yr]	50 (1.97)	1200 (47.2)	5000 (197)
Naval brass	1000 (39.4)	1000 (39.4)	...	...
90–10 cupronickel	140 (5.51)	70 (2.75)	400 (15.7)	700 (27.5)
Type 304L	18 (0.71) [0.3 µm/yr, or 0.012 mil/yr]	6 (0.24)	...	...
Type 316L	15 (0.59)	5 (0.197)	...	...
Nitronic 50 stainless steel	8 (0.31)	3 (0.12)	...	...
20Cb3 stainless steel	7 (0.275)	5 (0.197)	100 (3.94)(a)	...
Ebrite 26-1 stainless steel	16 (0.63)	5 (0.197)	240 (9.45)	...
Monel 400	30 (1.18)	100 (3.94)	...	...
Incoloy	6 (0.24)	4 (0.157)	...	...
Inconel 600	9 (0.35) [1 µm/yr, or 0.04 mil/yr]	5 (0.197)	...	100 (3.94)
Inconel 625	5 (0.197)	12 (0.47)(b)	...	...
Hastelloy C-276	7 (0.275) [0.4 µm/yr, or 0.016 mil/yr]	2 (0.08)	60 (2.36)(b)	200 (7.87)(b)
Zircaloy 2	1 (0.04)	...	...	...
Ti-Gr2	14 (0.55)	12 (0.47)	...	...
Ti-Gr12	3 (0.12) [0.2 µm/yr, or 0.008 mil/yr]	1 (0.04)	0.4 (0.015)	0.6 (0.02)

(a) Crevice corrosion. (b) Pitting corrosion. Source: Ref 161 and 180 (bracketed values)

for a 1000-year time period. However, at a dose rate of 10^5 rads/h, the corrosion rate increases by a factor of six. If the container is thick enough (approximately 80 to 100 mm, or 3 to 4 in.) to reduce radiation levels to 10^3 rads/h, then only a small enhancement in rate would be expected

- The corrosion rates of a variety of low-alloy ferrous materials are all similar
- Some evidence was found that cast steel undergoes a limited amount of environmental embrittlement. However, no susceptibility to SCC was found
- A preliminary corrosion rate model was developed that includes the effects of temperature, brine flow rate, brine volume, container area, and radiation dose rate. Also, a less detailed corrosion model has been developed for a wider range of materials than is reported in Ref 194

Sedimentary Deposits

Two types of sedimentary deposits are being evaluated as potential hosts for a waste repository: sub-seabed sediments (U.S. and Europe) and clay formations (Europe). In general, the corrosion research portion of the programs investigating these deposits is not developed to the same level as that given in the previous discussions on "Salt Deposits" and "Crystalline Rock" in this section. Only the screening phase to select viable candidate materials has been completed.

Sub-Seabed Sediments. The U.S. program has been conducted at Sandia National Laboratories in parallel with the salt investigations reported above. The corrosion behavior of candidate alloys in heated seawater and sediments was shown in Tables 24 and 26. Because seawater is a diluted brine, the conclusions presented in the previous discussion on "Salt Deposits" for both the screening study and the detailed analyses of titanium and iron-base alloy corrosion are largely applicable. Supporting the detailed Sandia findings are the results discussed in Ref 15. This study investigated the corrosion behavior of Ti-Gr2 in hot (100 to 130 °C, or 212 to 265 °F) Baltic seawater and found no evidence of pitting corrosion and a general corrosion rate typically less than 0.1 µm/yr (0.004 mil/yr). Finally, some archeological data exist that show metals can survive for very long time periods (greater than 1000 years) in deep seawater environments (Ref 195, 196). For example, a 10- to 20-mm (0.4- to 0.8-in.) thick copper container could provide 1000 years of containment. These data must be used only as indicators of performance because heat and radiation were not present during these extended exposures.

Similar to their materials selection for the salt formations, the Europeans have chosen titanium alloys and low-alloy steels as their prime candidate materials (Ref 162). Some results from this European activity are shown in Fig. 91 and 92. For comparison, the corrosion rates given in Fig. 92 for steel in deaerated seawater may be 1000 times higher if the seawater is aerated. In the heated seabed environment, Ti-Gr2, Ti-Gr7, and Ti-Gr12 should have excellent resistance to general attack (maximum rate of 12 µm/yr, or 0.47 mil/yr). All the titanium alloys may crevice corrode if the temperatures are too high (130 °C, or

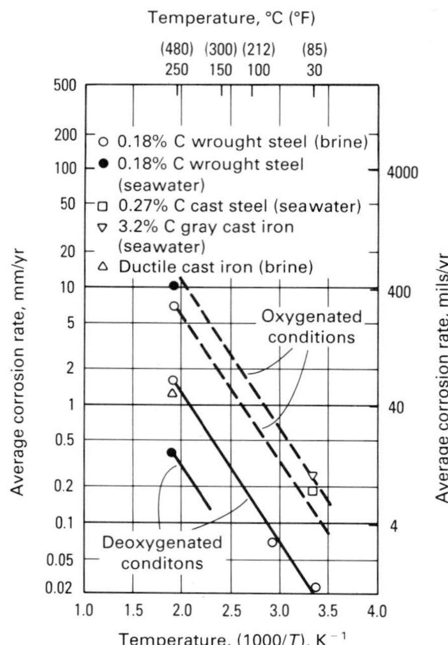

Fig. 87 The effect of temperature on the general corrosion rate of cast iron and steel in chloride solutions. Test duration: 2 to 8 weeks. Source: Ref 174

Table 25 General corrosion rates of West German candidate alloys in brine

Test duration: 70–115 weeks (without radiation), 21 to 86 weeks (with radiation)

Alloy	Temperature		Corrosion rate, μm/yr (mils/yr)	
	°C	°F	Without gamma irradiation	With gamma irradiation (10^5 rad/h)
Ti-Gr7	90	195	0.17 (0.007)	0.7 (0.03)
	200	390	0.14 (0.006)	...
Hastelloy C4	90	195	0.02 (0.0008)(a)	2.3 (0.09)(b)
	200	390	0.21 (0.008)	...
Fine-grain steel	90	195	35 (1.38)	464 (13.3)(b)
	200	390	590 (23.2)	...
Cast steel	90	195	33 (1.3)	666 (26.2)(b)
	200	390	508 (20)	...
Cast silicon-iron	90	195	3.8 (0.15)(b,c)	56 (2.20)(b)
	170	340	10.8 (0.43)(b,c)	...
Cast spheroidal iron	90	195	46 (1.81)(b,c)	165 (6.50)(b)
	170	340	91 (3.58)(b,c)	...
Ni-resist D2	90	195	2.5 (0.10)	157 (6.18)(b)
	170	340	15.1 (0.59)(b,c)	...
Ni-resist D4	90	195	2.1 (0.08)	77 (3.03)(b)
	170	340	5.1 (0.20)(b,c)	...

(a) Crevice corroded. (b) Pitted. (c) Intergranularly attacked. Source: Ref 167

Table 26 General corrosion rates of titanium alloys at 250 °C (480 °F) and two dissolved oxygen concentrations

Test duration: 4 weeks

Alloy	Corrosion rate, μm/yr (mils/yr) — Brine		Seawater	
	30 ppb O_2	450 ppm O_2	30 ppb O_2	500 ppm O_2
Ti-Gr2	14 (0.05)	3200 (126)	12 (0.47)	16 (0.63)
Ti-Gr7	2.4 (0.094)	0.4 (0.016)	1.1 (0.043)	0.6 (0.024)
Ti-Gr12	3.2 (0.126)	1.8 (0.070)	1.1 (0.043)	0.6 (0.024)

Source: Ref 154

Table 27 General corrosion rates of titanium alloys at three temperatures in deoxygenated brine

Test duration: 4 weeks; oxygen concentration: 30 ppb

Alloy	Corrosion rate, μm/yr (mils/yr) — 70 °C (160 °F)	150 °C (300 °F)	250 °C (480 °F)
Ti-Gr2	0.06 (0.0024)	2.6 (0.102)	14 (0.55)
Ti-Gr7	0.09 (0.0035)	0.03 (0.0012)	2.4 (0.094)
Ti-Gr12	0.07 (0.0028)	0.9 (0.035)	3.2 (0.126)

Source: Ref 154

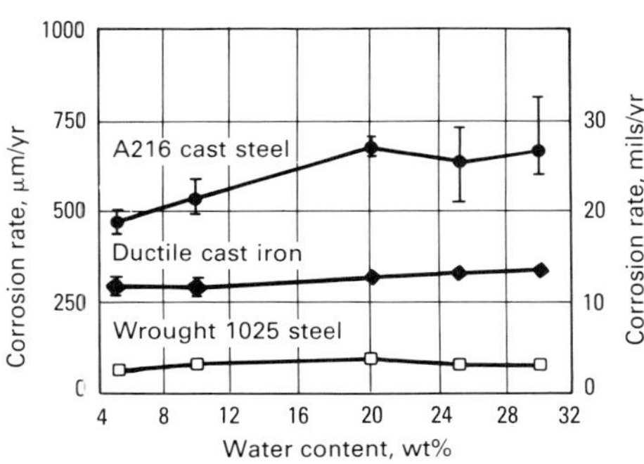

Fig. 90 The effect of low levels of magnesium-containing brine in NaCl on the general corrosion rate of cast irons and steel at 150 °C (300 °F). Test duration: 13 weeks. Source: Ref 166

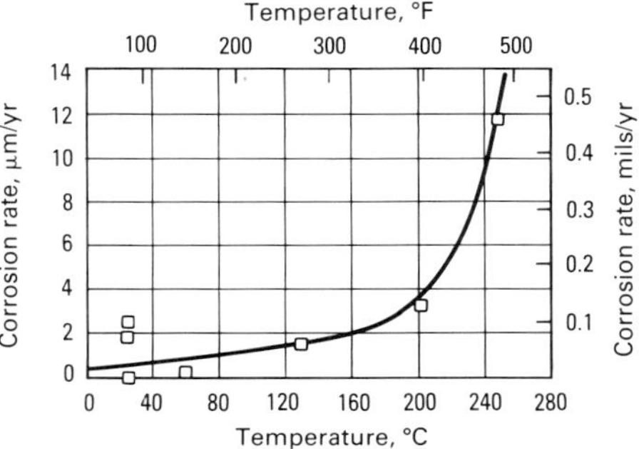

Fig. 91 Effect of temperature on the general corrosion rate of Ti-Gr2 in seawater. Source: Ref 162

265 °F, for Ti-Gr2; 170 °C, or 340 °F, for Ti-Gr7), but should not be susceptible to SCC. For the low-alloy steels under expected deaerated conditions, the corrosion rates will be small (8 μm/yr at 90 °C, or 0.31 mil/yr at 195 °F). However, if oxygen and/or radiolysis products are present or if the temperature exceeds about 150 °C (300 °F), then the rates may appreciably increase. For a steel container to be acceptable, a radiation shield

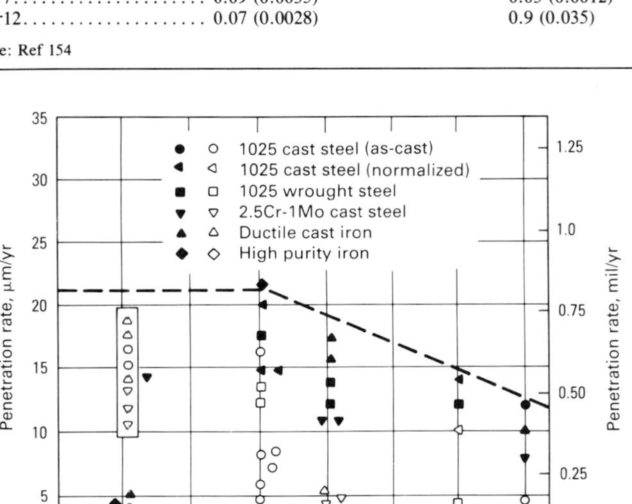

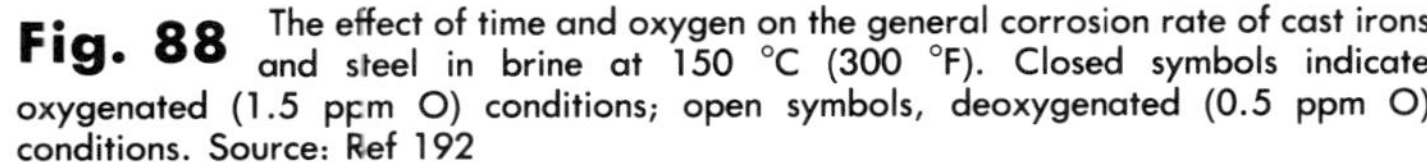

Fig. 88 The effect of time and oxygen on the general corrosion rate of cast irons and steel in brine at 150 °C (300 °F). Closed symbols indicate oxygenated (1.5 ppm O) conditions; open symbols, deoxygenated (0.5 ppm O) conditions. Source: Ref 192

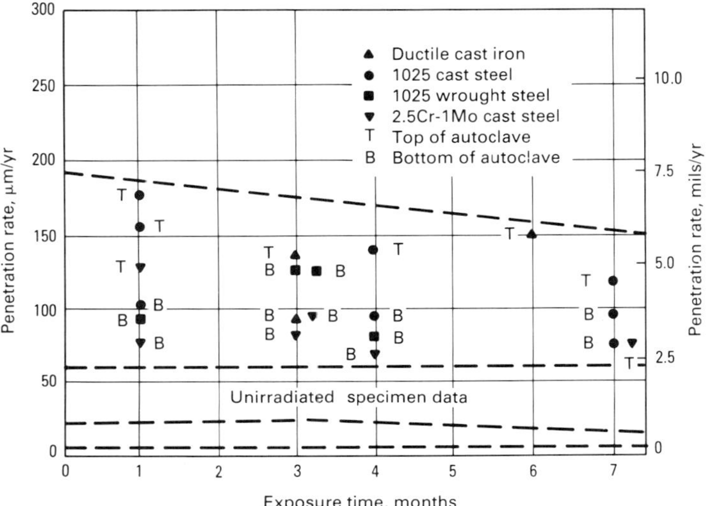

Fig. 89 The effect of time and radiation on the general corrosion rate of cast irons and steel in brine at 150 °C (300 °F). Dose rate: 10^5 rads/h. Source: Ref 192

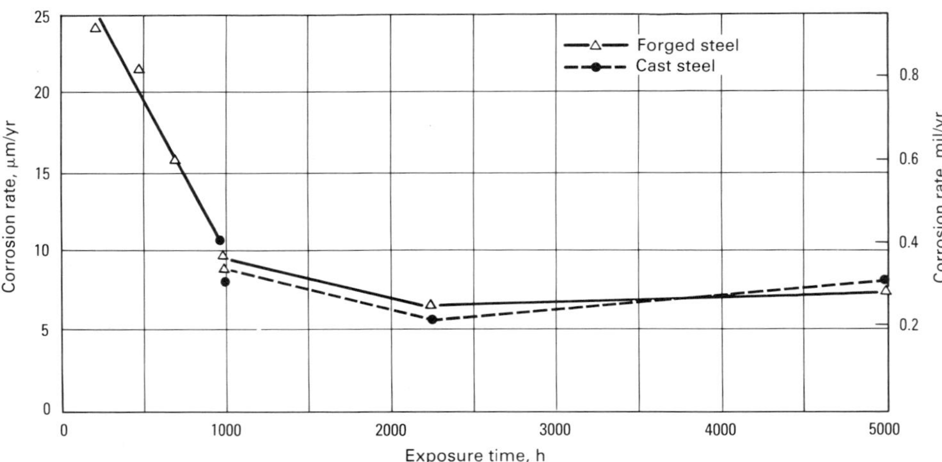

Fig. 92 Effect of time on the general corrosion rate of low-alloy steels in deaerated seawater at 90 °C (195 °F). Source: Ref 162

may have to be included as part of the waste package.

Clay Formations. The initial screening of materials for use in a clay formation was reported in Ref 168. The study identified sulfur and sulfur compounds as the corrosive agents released from heated clay that could ultimately contact the container. Carbon steel and chromized steel performed poorly in these environments. The general corrosion rates of the stainless steels, nickel-base alloys, and titanium alloys were all relatively low. The only materials that were not susceptible to localized attack were the titanium alloys and Hastelloy C-4. A more detailed screening study of metal behavior in clay was recently completed (Ref 164). Some of the results from this latter study are shown in Tables 28 and 29. The corrosion rates of the preferred resistant materials, Ti-Gr7 and Hastelloy C-4, are less than 1 μm/yr (0.04 mil/yr) under all experimental conditions. The corrosion rate of the carbon steel reference material depends on the temperature and the oxygen content of the aqueous phase and is thus highly variable, ranging from 2 to 300 μm/yr (0.08 to 11.8 mils/yr). Corrosion tests that include the effect of radiation have not been performed.

General Observations

Based on information presented and referenced in the discussion on "Corrosion Results From Site-Specific Studies" in this section, several common elements were observed that are, in general, applicable to the identification of container materials. These include:

- Most of the waste management programs have included low-alloy steels/irons and titanium-base alloys in their list of prime candidate container materials. The most notable exception to this general selection is Sweden, with a proposed very thick copper container for spent fuel. The nonuniform corrosion behavior of metallurgically complex alloys, such as the stainless steels and the nickel-base superalloys, is recognized to be very difficult to characterize reliably. This factor is limiting the acceptability of these types of alloys
- The corrosion rates of irons and steels may be low enough to allow thick containers to be usable in many environments. Titanium alloys have very low general corrosion rates (<1 μm/yr, or 0.04 mil/yr) and only one identified potential problem: hydrogen-assisted failure. However, with proper container design, a favorable stress state should exist that would eliminate concern for this and any other stress-assisted type of nonuniform failure
- The presence of gamma radiation has been shown to have an important, mainly detrimental, effect on the corrosion processes, especially for corrosion-allowance materials. To alleviate this potential problem, many designs will have to be self-shielded if corrosion-allowance materials are to be used
- The corrosivity of the geologic environments being studied decreases in the order salt > sub-seabed > clay > crystalline rock
- Theoretically, the corrosion characteristics of the candidate materials together with the total disposal costs (container, packaging, mining, emplacement) can be used to determine if the container should be constructed from a corrosion-resistant (for example, titanium) or a corrosion-allowance (for example, steel) material. This determination cannot currently be made primarily because the characterization process has not been completed and the exact requirements that must be satisfied to license a material for this application have not been specified. In this latter area, for example, what constitutes a failure? What failure frequency is acceptable? The author's personal speculation is that if a corrosion-resistant material can be satisfactorily qualified, then it will probably be cost effective to use in the corrosive environments (for example, bedded salt). Alternately, a relatively benign environment (such as basalt) could justifiably use a corrosion-allowance material. As such, the selection of the most cost-effective material will be possible only after more licensing details become available in each country.

Table 28 General corrosion rates of untreated candidate alloys in a gaseous clay-containing environment at two temperatures

Test duration: 17 weeks

Alloy	Corrosion rate, μm/yr (mils/yr) 25 °C (75 °F)	50 °C (120 °F)
Hastelloy C	0.23 (0.0090)	0.27 (0.0106)
Ti-Gr7	0.07 (0.0028)	0.20 (0.0078)
Carbon steel	28 (1.10)	107 (4.21)

Source: Ref 164

Table 29 General corrosion rates of candidate alloys in interstitial clay water at two temperatures

Test duration: 9 weeks

Alloy	Corrosion rate, μm/yr (mils/yr) 25 °C (75 °F)	75 °C (160 °F)
Ductile iron	105 (4.13)	2760 (109)
Type-304	0.22 (0.0086)	0.35 (0.0138)
Type-316	0.07 (0.0028)	0.1 (0.004)
Type-410	0.09 (0.0035)	1.1 (0.043)
Copper	44 (1.73)	92 (3.62)
Ti-Gr2	0.1 (0.004)	0.12 (0.0047)

Source: Ref 164

REFERENCES

1. H.H. Klepfer, "Investigation of Cause of Cracking in Austenitic Stainless Steel Piping," NEDO-21000-1, 75NED35, General Electric Company, July 1975
2. C. Husen and C.H. Samans, *Chem. Eng.*, 27 Jan 1969
3. R.B. Davis and M.E. Indig, "The Effect of Aqueous Impurities on the Stress Corrosion Cracking of Austenitic Stainless Steel in High Temperature Water," Paper 128, presented at Corrosion/83, Anaheim, CA, National Association of Corrosion Engineers, April 1983
4. W.J. Shack *et al.*, "Environmental Assisted Cracking in Light Water Reactors Semiannual Reports," NUREG/CR-4667, Vol 1, ANL-86-31, Argonne National Laboratory, June 1986
5. L. Ljungberg and E. Hallden, "BWR Water Chemistry-Impurity Studies—Literature Survey on Effects on Stress Corrosion Cracking," Report KM 84-92, ASEA-AT-OM, March 1984
6. F.P. Ford *et al.*, "Environmentally Controlled Cracking of Stainless and Low-Alloy Steels in Light-Water Reactor Environments," Final Report Nov 1981-July 1985, Electric Power Research Institute, to be published
7. C.S. Tedmon, *J. Elect. Soc.*, Vol 118, 1971
8. C.S. Tedmon and D.A. Vermilyea, Report 70-C-232, General Electric Corporate Research & Development, July 1970
9. C. Strawstrom and M. Hillert, *J. Iron Steel Inst.*, Vol 207, 1969, p 77
10. V. Cihal, *Corrosion-Traitements-Protection-Finition*, Vol 18, 1970
11. J. Alexander, "Alternative Alloys for BWR Pipe Applications," EPRI NP-2671-LD, Electric Power Research Institute, Oct 1982
12. A.E. Pickett, "Assessment of Remedies for Degraded Piping," Paper presented at 1986 Seminar on Countermeasures for Pipe Cracking in BWRs, Electric Power Research Institute, Nov 1986
13. N.R. Hughes, "Evaluation of Near-Term BWR Piping Remedies," EPRI NP-1222,

Electric Power Research Institute, May 1980

14. Fukushima Daiichi Nuclear Power Station Unit 3, "Primary Loop Recirculation System Work Report for Pipe Branches Replacement," Paper presented at BWR Operating Plant Technical Conference No. 8, Monterey, CA, General Electric Company, Feb 1983
15. H.P. Offer, "Induction Heating Stress Improvement," EPRI NP-3375, Electric Power Research Institute, Nov 1983
16. R.M. Horn, "Last Pass Heat Sink Welding," EPRI NP-3479-LD, Electric Power Research Institute, March 1984
17. M.E. Indig and A.R. McIlree, *Corrosion*, Vol 35, 1979, p 288-295
18. J. Magdalinski and R. Ivars, "Oxygen Suppression in Oskarshamn-2," Winter Meeting, Washington, DC, American Nuclear Society, Nov 1982
19. E.L. Burley, "Oxygen Suppression in Boiling Water Reactors—Phase 2 Final Report," DOE/ET/34203-47, NEDC-23856-7, General Electric Company, Oct 1982
20. B.M. Gordon, Mitigation of Stress Corrosion Cracking Through Suppression of Radiolytic Oxygen, in *Proceedings of the International Symposium on Environmental Degradation of Materials in Nuclear Power Systems—Water Reactors*, National Association of Corrosion Engineers, 1984, p 893-932
21. T.A. Prater, W.R. Catlin, and L.F. Coffin, "Surface Crack Growth in High Temperature Water," 83CRD277, General Electric Corporate Research & Development, Dec 1983
22. B.M. Gordon, "BWR Material Life Extension Through Hydrogen Water Chemistry," Paper presented at the Operability of Nuclear Power Systems in Normal and Adverse Environments, Albuquerque, NM, American Nuclear Society, Sept-Oct 1986
23. B.M. Gordon, "Laboratory Studies of Materials Performance in Hydrogen Water Chemistry," Paper presented at EPRI Seminar on BWR Corrosion, Chemistry and Radiation Control, Palo Alto, CA, Electric Power Research Institute, Oct 1984
24. S. Ranganath *et al.*, Failure Analysis, Testing and Product Improvement of a Control Rod Drive Component From a Boiling Water Reactor, in *Failure Prevention and Reliability*, American Society of Mechanical Engineers, 1977
25. R.A. Carnahan *et al.*, "Improvements in Jet Pump Hold-Down Beam Service Life," NEDE-24362-1, General Electric Company, Dec 1981
26. B.M. Gordon *et al.*, "Hydrogen Water Chemistry for BWR's—Materials Behavior—Interim Report, 1 April 1983-1 July 1986," Electric Power Research Institute, to be published
27. A.J. Jacobs and G.P. Wozadlo, "Irradiation-Assisted Stress Corrosion Cracking as a Factor in Nuclear Power Plant Aging," Paper presented at the International Conference on Nuclear Power Plant Aging, Availability Factor and Reliability Analysis, San Diego, CA, American Society for Metals, July 1985
28. J.S. Armijo, *Corrosion*, Vol 21, 1965, p 235
29. F. Garzarolli *et al.*, Deformability of Austenitic Stainless Steel and Ni-Base Alloys in the Core of a Boiling and Pressurized Water Reactor, in *Proceedings of the Second International Symposium on Environmental Degradation of Materials in Nuclear Power Systems—Water Reactors*, American Society for Metals, 1986, p 131-138
30. R.N. Duncan, "Stainless Steel Failure Investigation Program," Final Summary Report GEAP-5530, General Electric Company, 1968
31. S.J. Green and J.P.N. Paine, "Steam Generator Materials—Experience and Prognosis," Paper presented at the International Symposium on Environmental Degradation of Materials in Nuclear Power Systems-Water Reactors, Myrtle Beach, SC, National Association of Corrosion Engineers, Aug 1983
32. J.P.N. Paine and S.J. Green, Materials Performance in Nuclear Steam Generators, *Nucl. Technol.*, Vol 55, 1981, p 10-29
33. J.P.N. Paine, "Operating Experience and Intergranular Corrosion of Inconel Alloy 600 Steam Generator Tubing," Paper presented at Corrosion/82, National Association of Corrosion Engineers, March 1982
34. "Mechanisms for Formation and Disruption of Surface Oxides," Final Report, Research Project S302-16, Electric Power Research Institute, to be published
35. P. Hernalsteen and R. Houben, "Preventive and Corrective Actions for Doel 2 Steam Generators," Paper presented at The Specialist Meeting on Steam Generators, Stockholm, Sweden, NEA/CSNI-UNIPEDE, Oct 1984
36. J.R. Weeks, *Corrosion Problems in Energy Conversion and Generation*, The Electrochemical Society, 1974, p 322
37. H. Baschek and E. Sandona, "The Steam Generator Failure History of the Nuclear Power Plants Beznau 1 and Beznau 2," Paper presented at Educational Seminar, Colloquium on Steam Generator Tube Failures, Southwest Research Institute, 1974
38. L.S. Martel, "Steam Generator Owners Group Water Chemistry Program," Paper presented at the American Power Conference, Chicago, IL, Illinois Institute of Technology, April 1978
39. "PWR Steam-Side Chemistry Follow Program," EPRI NP-2541, Final Report, Research Project 699-1, Electric Power Research Institute, Aug 1982
40. "Causes of Denting: Laboratory Test Results," Vol 2, EPRI NP-3275, Final Report, Research Project S157-1, Electric Power Research Institute, May 1984
41. "Determination and Verification of Required Water Chemistry Limits," Vol 1, EPRI NP-3274, Final Report, Research Project S111-1, Electric Power Research Institute, March 1984
42. "PWR Secondary Water Chemistry Guidelines," EPRI NP-2704, Special Report, Electric Power Research Institute, Oct 1982
43. "Assessment of Condenser Leakage Problems," EPRI NP-1467, Final Report, Research Project TPS79-729, Electric Power Research Institute, Aug 1980
44. "Prevention of Condenser Failures—The State of the Art," EPRI 2282-SR, Special Report, Electric Power Research Institute, March 1982
45. "Condenser Inleakage Monitoring System Development," EPRI NP-2597, Final Report, Research Project S182-1, Electric Power Research Institute, Sept 1982
46. "Laboratory Studies Related to Steam Generator Tube Denting," Vol 1 and 2, EPRI NP-3023, Final Report, Research Project S112-1, Electric Power Research Institute, Sept 1983
47. "Neutralization of Crevice Acids," EPRI NP-3054, Final Report, Research Project 623-2, Electric Power Research Institute, May 1983
48. A.R. Vaia, G. Economy, M.J. Wootten, and R.G. Aspden, Denting of Steam Generator Tubes in PWR Plants, *Mater. Perform.*, Vol 19, Feb 1980, p 9
49. C.E. Shoemaker, "Selecting Support Structure Alloys for Nuclear Steam Generators," Paper presented at the Second International Symposium on Environmental Degradation of Materials in Nuclear Power Systems-Water Reactors, Monterey, CA, National Association of Corrosion Engineers, Sept 1985
50. H. Coriou, L. Grall, Y. Le Gall, and S. Vettier, *Corrosion Fissurante Sous Contrainte De L'Inconel Dans L'Eau A Haute Temperature*, North Holland, 1959, p 161
51. D. van Rooyen, Review of the Stress Corrosion Cracking of Inconel 600, *Corrosion*, Vol 31, 1975, p 327
52. "Stress Corrosion Cracking of Alloy 600," EPRI NP-2114-SR, Special Report, Electric Power Research Institute, Nov 1981
53. "Intergranular Stress Corrosion Cracking of Ni-Cr-Fe Alloy 600 Tubes in PWR Primary Water—Review and Assessment for Model Development," EPRI NP-3057, Final Report, Research Project S138-8, Electric Power Research Institute, May 1983
54. H.J. Schenk, "Investigation of Tube Failures in Inconel 600 Steam Generator Tubing at KWO Obrigheim," *Mater. Perform.*, Vol 15 (No. 3), March 1976, p 25-33
55. R. Bandy and D. van Rooyen, "Quantitative Examination of Stress Corrosion Cracking of Alloy 600 in High Temperature Water—Work During 1983," Paper presented at EPRI Workshop on Primary Side SCC and Secondary Side SCC and IGC of PWR Steam Generator Tubing, Clearwater Beach, FL, Electric Power Research Institute, Nov 1982
56. A. Stein, "Stress Corrosion Cracking of Alloy 600 in Primary Water," Paper presented at EPRI Workshop on Primary Side SCC and Secondary Side SCC and IGC of PWR Steam Generator Tubing, Clearwater Beach, FL, Electric Power Research Institute, Nov 1982
57. G.P. Airey, The Stress Corrosion Cracking (SCC) Performance of Inconel Alloy 600 in Pure and Primary Water Environments, in *Proceedings of the International Symposium on Environmental Degradation of Materials in Nuclear Power Systems-Water Reactors*, National Association of Corrosion Engineers, 1984
58. "Optimization of Metallurgical Variables to Improve Corrosion Resistance of Inconel Alloy 600," EPRI NP-3051, Final Report, Research Project 1708-1, Electric Power Research Institute, July 1983
59. J. Blanchet, H. Coriou, L. Grall, C. Mahieu,

C. Otter, and G. Turleur, Influence de la Contraine, des Traitments Thermiques et des Couplages sur la Fissuration Intergranulaire des Alliages Inconel 600 et X-750, *J. Nucl. Mater.*, Vol 55, 1975, p 187; see also *International Conference on SCC and Hydrogen Embrittlement of Iron Base Alloys*, R.W. Staehle, Ed., National Association of Corrosion Engineers, 1977

60. H. Domain, R.H. Emanuelson, L. Katz, L.W. Sarver, and G.J. Theus, Effect of Microstructure on Stress Corrosion Cracking of Alloy 600 in High Purity Water, *Corrosion*, Vol 33, 1977, p 26
61. "Stress Corrosion Cracking of Alloy 600 and Alloy 690 in All-Volatile Treated Water at Elevated Temperatures," EPRI NP-3061, Final Report, Research Project S192-2, Electric Power Research Institute, May 1983
62. A.A. Stein, A. DeLeon, and A.R. McIlree, "Prediction of Intergranular Stress Corrosion Cracking of Alloy 600 Steam Generator Tubing in Primary Water and Influence of Mill Annealing Temperature on the Susceptibility of Alloy 600 Steam Generator Tubing to Primary Water SCC," Paper presented at the Second International Degradation of Materials in Nuclear Power Systems-Water Reactors, Monterey, CA, National Association of Corrosion Engineers, Sept 1985
63. J. Engstrom and K. Norring, "Primary and Secondary Cracking at Ringhals 2," Paper presented at EPRI Workshop on Primary Side SCC and Secondary Side SCC and IGC of PWR Steam Generator Tubing, Clearwater Beach, FL, Electric Power Research Institute, Nov 1982
64. G. Frederick and P. Hernalsteen, "Comparative Evaluation of Preventive Measures Against PS SCC of MA I 600 Steam Generator Tubes," Paper presented at Post Structural Mechanics in Reactor Technology Conference Seminar No. 3, Ispra, Italy, Aug 1985
65. "In Situ Heat Treatment and Polythionic Acid Testing of Inconel 600 Row 1 Steam Generator U-Bends," EPRI NP-3056, Final Report, Research Project S191-3, Electric Power Research Institute, April 1983
66. "Stress Relief to Prevent Stress Corrosion in the Transition Region of Expanded Alloy 600 Steam Generator Tubing," EPRI NP-3055, Final Report, Research Project S192-3, Electric Power Research Institute, May 1983
67. G. Frederick and P. Hernalsteen, "Generic Preventive Actions for Mitigating MA Inconel 600 Susceptibility to Pure Water Stress Corrosion Cracking," Paper presented at The Specialist Meeting on Steam Generators, Stockholm, Sweden, NEA/CSNI-UNIPEDE, Oct 1984
68. B.L. Dow, Jr. and L.H. Bohn, "Steam Generator Tube Sleeving Development for Once-Through Steam Generators," Paper presented at Structural Mechanics in Reactor Technology Conference, Brussels, Belgium, Aug 1985
69. "Optimization of Metallurgical Variables to Improve Corrosion Resistance of Inconel Alloy 600," EPRI NP-3051, Final Report, Research Project 1708-1, Electric Power Research Institute, July 1983, p 3-35
70. "Examination of Three Steam Generator Tubes From the Point Beach Unit 1 Nuclear Power Plant," EPRI NP-2958-LD, Final Report, Research Project S138-1, Electric Power Research Institute, March 1983, p 3-8, 4-21, 4-95
71. "Steam Generator Sludge Pile Model Boiler Testing," EPRI NP-1941, Final Report, Research Project S119-1, Electric Power Research Institute, July 1981
72. "Tube-to-Tubesheet Joint Test," Vol 1 and 2, EPRI NP-3013, Final Report, Research Project S119-2, Electric Power Research Institute, March 1983
73. "Evaluation of Steam Generator Tube R12C66 From Indian Point 3," EPRI NP-3029, Final Report, Research Project S138-6, Electric Power Research Institute, May 1983
74. J.F. Sykes and M.J. Angwin, "The Causes of Major Pitting of Alloy 600 Steam Generator Tubing in Pressurized Water Reactors," Paper presented at the Second International Symposium on Environmental Degradation of Materials in Nuclear Power Systems-Water Reactors, Monterey, CA, National Association of Corrosion Engineers, Sept 1985
75. "Fatigue Performance of Ni-Cr-Fe Alloy 600 Under Typical PWR Steam Generator Conditions," EPRI NP-2957, Final Report, Research Project S110-1, Electric Power Research Institute, March 1983
76. "Model Tests of a Once-Through Steam Generator for Lane Blocker Assessment and THEDA Code Verification," EPRI NP-3042, Final Report, Research Project S186-1, Electric Power Research Institute, June 1983
77. L. Hunyadi, "Ringhals—Steam Generator Repair/Replacement Options," Paper presented at Post Structural Mechanics in Reactor Technology Conference Seminar No. 3, Ispra, Italy, Aug 1985
78. R. Vollmer, "U.S. Criteria for Repair and Replacement of Steam Generators and BWR Piping," Paper presented at Post Structural Mechanics in Reactor Technology Conference Seminar No. 3, Ispra, Italy, Aug 1985
79. J.T.A. Roberts, *Structural Materials in Nuclear Power Systems*, Plenum Press, 1981
80. H.G. Rickover, L.D. Geiger, and R. Lustman, "History of the Development of Zirconium Alloys for Use in Nuclear Reactors," TID-26740, United States ERDA Report, 1975
81. F.J. Erbacher and S. Leistikow, Zircaloy Fuel Cladding in a LOCA—A Review, in *Zirconium in the Nuclear Industry: Seventh Annual Symposium*, STP 939, American Society for Testing and Materials, 1987
82. J.S. Bryner, The Cyclic Nature of Corrosion of Zircaloy-4 in 633K Water, *J. Nucl. Mater.*, Vol 82, 1979, p 84
83. H.R. Peters, Improved Characterization of Aqueous Corrosion Kinetics of Zircaloy-4, in *Zirconium in the Nuclear Industry*, STP-824, American Society for Testing and Materials, 1984
84. H. Kaesche, *Metallic Corrosion: Principles of Physical Chemistry and Current Problems*, National Association of Corrosion Engineers, 1985
85. B. Cox, Processes Occurring During the Breakdown of Oxide Film on Zirconium Alloys, *J. Nucl. Mater.*, Vol 29, 1969, p 50
86. F. Garzarolli, I. Löh, H. Stehle, E. Steinberg, and M. Edeling, "KWU-Results on Waterside Corrosion of Zircaloy in PWRs and BWRs," Paper presented at IAEA Specialists' Meeting, Leningrad (USSR), June 1983
87. C.M. Eucken, P.T. Finden, R.A. Graham, and C.T. Wang, "Influence of Manufacturing Variables on Zircaloy Corrosion," Paper presented at EPRI Workshop on Zircaloy Corrosion, Charlotte, NC, Electric Power Research Institute, Aug 1986
88. F. Garzarolli, W. Jung, H. Schoenfeld, A.M. Garde, G.W. Parry, and P.G. Smerde, "Waterside Corrosion of Zircaloy Fuel Rods," EPRI NP-2789, Electric Power Research Institute, Dec 1982
89. S.G. McDonald, G.P. Sabol, and K.D. Sheppard, Effect of Lithium Hydroxide on the Corrosion Behavior of Zircaloy-4, in *Zirconium in the Nuclear Industry*, STP 824, American Society for Testing and Materials, 1984
90. B. Cox, Effects of Irradiation on the Oxidation of Zirconium Alloys in High-Temperature Aqueous Environments, *J. Nucl. Mater.*, Vol 28, 1968, p 1
91. F.H. Mergerth, C.P. Ruiz, and U.E. Wolff, "Zircaloy-Clad UO_2 Fuel Rod Evaluation Program," Report GEAP-10371, General Electric Company, June 1971
92. B. Cox, "Assessment of PWR Waterside Corrosion Models and Data," EPRI NP-4287, Electric Power Research Institute, Oct 1985
93. C. Pan and B.G. Jones, "Wick Boiling in Porous Deposits With Chimneys," EPRI RP-1250-08, Final Report, Electric Power Research Institute, to be published
94. Y. Asakura, M. Kikuchi, S. Uchida, and H. Yusa, Deposition of Iron-Oxide on Heated Surfaces in Boiling Water, *Nucl. Sci. Eng.*, Vol 67, 1978, p 1
95. G.M.W. Mann and R. Castle, "Salt Concentration in Heated Crevices and Simulated Scale," EPRI NP-3050, Electric Power Research Institute, Oct 1983
96. D.G. Franklin and C.Y. Li, Effects of Heat Flux and Irradiation-Induced Changes in Water Chemistry on Zircaloy Nodular Corrosion, in *Zirconium in the Nuclear Industry: Seventh Annual Symposium*, STP 939, American Society for Testing and Materials, 1987
97. F. Garzarolli and H. Stehle, "Behavior of Core Structural Materials in Light Water Cooled Power Reactors," Paper presented at IAEA International Symposium on Improvements in Water Reactor Fuel Technology, Stockholm, Sweden, Sept 1986
98. A. Garlick, R. Summerling, G.L. Shires, Crud-Induced Overheating Defects in Water Reactor Fuel Pins, *J. Br. Nucl. Energy Soc.*, Vol 16, 1977, p 77
99. M.O. Marlowe, J.S. Armijo, B. Cheng, and R.B. Adamson, Nuclear Fuel Cladding Localized Corrosion, in *Topical Meeting on Light Water Reactor Fuel Performance*, American Nuclear Society, April 1985
100. B. Cox, "Effect of Hydrogen Injection on Hydrogen Uptake by BWR Fuel Cladding," EPRI NP-3146, Electric Power Research Institute, June 1983
101. A.J. Machiels, Effects of Hydrogen Water

Chemistry on Fuel Performance, *EPRI J.*, April/May 1986, p 54-55
102. G.C.W. Comley, The Significance of Corrosion Products in Water Reactor Coolant Circuits, *Prog. Nucl. Energy*, Vol 16, 1985, p 41-72
103. C.J. Wood, Reduced Out-of-Core Radiation Eases Maintenance Activities at Nuclear Power Plants, *Power*, Vol 131, 1987, p 29-32
104. C.J. Wood, "Manual of Recent Techniques for LWR Radiation-Field Control," EPRI NP-4505-SR, Electric Power Research Institute, March 1986
105. Y. Solomon, An Overview of Water Chemistry for Pressurized Water Reactors, in *Proceedings of Water Chemistry of Nuclear Reactor Systems*, British Nuclear Energy Society, 1978, p 101-112
106. P. Cohen, *Water Coolant Technology of Power Reactors*, Gordon & Breach, 1969
107. Stress Corrosion Cracking, in *Failure Analysis and Prevention*, Vol 11, 9th ed., *Metals Handbook*, American Society for Metals, 1986, p 203-224
108. E. Eason, "Stress Corrosion Cracking in Steam Turbine Discs: Survey of Data Collection, Reduction, and Modeling Activities," EPRI NP-3691, Electric Power Research Institute, 1984
109. D. Kalderon, Steam Turbine Failure at Hinkley Point 'A', *Proc. IME*, Vol 186 (No. 31), 1972, p 341-377
110. J.L. Gray, Investigation Into the Consequences of the Failure of a Turbine-Generator at Hinkley Point 'A' Power Station, *Proc. IME*, Vol 186 (No. 32), 1972, p 379-390
111. E. Eason, "Stress Corrosion Cracking in Steam Turbine Discs: Analysis of Field and Laboratory Data," EPRI NP-4056, Electric Power Research Institute, 1985
112. C.L. Briant, "Effects of Impurity Segregation, Alloy Composition, and Microstructure on the Stress Corrosion Cracking and Temper Embrittlement of Rotor Steels," EPRI NP-444OM, Electric Power Research Institute, Feb 1986
113. H.B. Gayley, "Relationship Between Turbine Rotor and Disk Metallurgical Characteristics and Stress Corrosion Cracking Behavior," EPRI NP-4695M, Electric Power Research Institute, Sept 1986
114. P.K. Nair *et al.*, "Steam Turbine Disk Lifetime Prediction Manual," EPRI NP-4936, Electric Power Research Institute, Dec 1986
115. P. Besuner, "BIFIF—Fracture Mechanics Code for Structures," EPRI NP-1830, Electric Power Research Institute, 1981
116. R.I. Jaffee *et al.*, "Production and Properties of a Super Clean 3.5NiCrMoV LP Rotor Forging," Paper presented at the Tenth International Forging Conference, Sheffield, England, Institute of Metals, Sept 1985
117. "PWR Secondary Water Chemistry Guidelines," Revision 1, June 1984; "BWR Water Chemistry Guidelines," EPRI NP-3589-SR-LD, Electric Power Research Institute, April 1985
118. R. Holinski, "Metallurgical Changes During High Temperature Screw Lubrication," Paper presented at ASME/ASLE Lubrication Conference, Washington, DC, Oct 1982
119. C. Czajkowski and J. Weeks, Examination of Cracked Turbine Discs From Nuclear Power Plants, *Mater. Perform.*, Vol 22 (No. 3), March 1983, p 21-25
120. W. Koo, Report NUREG-0943, United States Nuclear Regulatory Commission, Jan 1983
121. C. Czajkowski, Corrosion and Stress Corrosion Cracking of Bolting Materials in Light Water Reactors, in *Proceedings of the International Symposium on Environmental Degradation of Materials in Nuclear Power Systems-Water Reactors*, National Association of Corrosion Engineers, 1983, p 192-208
122. L.H. Burck and W.J. Foley, Report IE-123, United States Nuclear Regulatory Commission, April 1981
123. D.V. Thornton, P.B. Mould, and E.C. Patrick, Conference on Grain Boundaries, The Institution of Metallurgists, 1976
124. S.F. Calhoun, Rock Island Arsenal Report 62-2752, U.S. Government Report, Aug 1962
125. C. Perna, Picatinny Arsenal Report DC3-1, U.S. Government Report, Jan 1961
126. A.J. Haltner and C.S. Oliver, Proceedings of American Chemical Society, Petroleum Division, Symposium on Chemistry of Friction and Wear, Vol 3 (No. 4), 1958, p A77-84
127. E. Kay, *Wear*, Vol 12, 1968, p 165-171
128. O. Jonas, Paper 55, presented at Corrosion/84, National Association of Corrosion Engineers, 1984
129. Report YACE-67, Westinghouse Electric Corporation, Nov 1958
130. M.C. Rowland and T.C. Rose, Report APED-4422, General Electric Company, Dec 1963
131. J.F. Hall, "Bolting Degradation or Failure in Nuclear Power Plants," Paper presented at EPRI seminar, Electric Power Research Institute, Nov 1983
132. C.M. Chen, K. Aral, and G.J. Theus, EPRI NP-3137, Vol 1, Project 1167-2, Final Report, Electric Power Research Institute, June 1983
133. "Standard Test Method for Static and Kinetic Coefficients of Friction of Plastic Film and Sheeting," D 1894, *Annual Book of ASTM Standards*, American Society for Testing and Materials
134. C.J. Czajkowski, Evaluation of the Transgranular Cracking Phenomenon on the Indian Point No. 3 Steam Generator Vessels, *Int. J. Pres. Ves. Piping*, Vol 26, 1986, p 97-110
135. G.A. Delp, J.D. Robison, and M.T. Sedlack, "Erosion/Corrosion in Nuclear Plant Steam Piping: Causes and Inspection Program Guidelines," EPRI NP-3944, Electric Power Research Institute, April 1985
136. L.E. Sanchez-Caldera, "The Mechanism of Corrosion-Erosion in Steam Extraction Lines of Power Stations, Ph.D. thesis, Massachusetts Institute of Technology, June 1984
137. H.G. Heitmann and W. Kastner, *VGB-Kraftwerkstechnik*, Vol 63 (No. 3), March 1982, p 180-187
138. M.G. Fontana and N.D. Greene, *Corrosion Engineering*, 2nd ed., McGraw-Hill, 1978
139. M.A. Sadler and M.R. Darvill, "Condensate Polishers for Brackish Water-Cooled PWR's," EPRI NP-4550, Electric Power Research Institute, July 1986
140. J.M. Riddle, W. Lechnick, and R. Nolan, "Survey of Domestic and Foreign PWR Experience With Morpholine in Chemistry Control by All-Volatile Treatment," EPRI NP-4671, Electric Power Research Institute, July 1986
141. J.M. Riddle, G.D. Burns, and L.J. Cain, "Chemistry Control With Morpholine at Beaver Valley Power Station," EPRI NP-4623, Electric Power Research Institute, June 1986
142. R.E. Anderson, K.L. Draper, R.A. Kadlec, and R.A. Stoudt, "Evaluation of a Moisture Removal Device for Turbine Steam Piping," EPRI NP-3927, Electric Power Research Institute, April 1985
143. "Generic Requirements for Mined Geologic Disposal Systems; OGR/B-2," Appendix B, United States Department of Energy, Sept 1984
144. J.H. Westsik, J.W. Shade, and G.L. McVay, Temperature Dependence for Hydrothermal Reactions of Waste Glasses and Ceramics, in *Science Underlying Radioactive Waste Management*, Vol II, Plenum Press, 1980, p 247
145. L.H. Johnson *et al.*, "The Dissolution of Unirradiated Uranium Dixode Fuel Under Hydrothermal Oxidizing Conditions," AECL-TR128, Atomic Energy of Canada, April 1981, p 10
146. J. Hamstra, Radiotoxic Hazard Measure for Buried Solid Radioactive Waste, *Nucl. Safety*, Vol 16 (No. 2), 1975, p 180
147. D. Cameron *et al.*, The Development of Durable, Man-Made Containment Systems for Fuel Isolation, *Can. Metall. Quart.*, Vol 22 (No. 1), 1983, p 89, 91, 94
148. "10 CFR 60: Disposal of High-Level Radioactive Wastes in Geologic Repositories; Licensing Procedures," Part 60.113(a)(1)(ii)(A), United States Nuclear Regulatory Commission, 1983
149. E. Mattsson, Corrosion Resistance of Canisters of Final Disposal of Spent Nuclear Fuel, in *Scientific Basis for Nuclear Waste Management*, Vol 1, Plenum Press, 1979, p 271-281
150. H.C. Claiborne *et al.*, "Repository Environmental Parameters Relevant to Assessing the Performance of High-Level Waste Packages," ORNL/TM-9522, Oak Ridge National Laboratory, May 1985, p 31, 32, 51-64, 101
151. P. Soo, "Review of DOE Waste-Package Program, Subtask 1.1," BNL-NUREG-51494: Vol 2, Brookhaven National Laboratory, April 1983, p 1-5, 1-43, 2-31 to 33
152. K. Nuttall, Some Aspects of the Prediction of Long-term Performance of Fuel Disposal Containers, *Can. Metall. Quart.*, Vol 22 (No. 3), 1983, p 404-406
153. R.E. Thomas and R.W. Cote, "Methodology for Predicting the Life of Waste Package Materials and Components Using Multifactor Accelerated Life Tests," ONWI-501, Battelle Memorial Institute, Sept 1983
154. M.A. Molecke *et al.*, Materials for High-Level Waste Canisters/Overpacks in Salt Formations, *Nucl. Technol.*, Vol 63, Dec 1983, p 476-506
155. M.J. Fox and R.D. McCright, "An Overview of Low Temperature Sensitization," UCRL-15619, Lawrence Livermore National Laboratory, Dec 1983, p 1-25

156. R. Dayal *et al.*, "Nuclear Waste Management Technical Support in the Development of Nuclear Waste Form Criteria for the NRC, Task 1: Waste Package Overview," BNL-NUREG-51458, Brookhaven National Laboratory, Feb 1982, p 122-193, 268-273
157. R.S. Glass, "Effects of Radiation on the Chemical Environment Surrounding Waste Canisters in Proposed Repository Sites and Possible Effects on the Corrosion Process," SAND81-1677, Sandia National Laboratories, Dec 1981, p 63-80
158. R.D. McCright *et al.*, "Selection of Candidate Canister Materials for High-Level Nuclear Waste Containment in a Tuff Repository," UCRL-89988, Lawrence Livermore National Laboratory, Nov 1983, p 1-42
159. S.J. Basham and J.A. Carr, Waste Package Designs for Disposal of High-Level Waste in Salt Formations, *Radioactive Waste Mgmt.*, Vol 2, 1983, p 412-414
160. "Geological Disposal of Heat Generating Radioactive Waste: Container Design Study," DoE Report RW/85.020, U.K. Department of the Environment, Feb 1985, p 17
161. J.W. Braithwaite and M.A. Molecke, Nuclear Waste Canister Corrosion Studies Pertinent to Geologic Isolation, *Nucl. Chem. Waste Mgmt.*, Vol 1, 1980, p 39-45
162. G.P. Marsh, "Influence of Temperature and Pressure on the Behaviour of High Level Waste and Canister Materials Under Marine Disposal Conditions, Part 2," Report S.P. 1.07.C2.85.51, Commission of European Communities, Dec 1985, p 1, 15-27
163. J.L. Nelson *et al.*, Irradiation-Corrosion Evaluation of Metals for Nuclear Waste Package Applications in Grande Ronde Basalt Groundwater, *Mater. Res. Soc.*, Vol 26, 1984, p 121-128
164. B. Haijtink, Corrosion Behaviour of Container Materials for Geologic Disposal of High-Level Waste, in *Nuclear Science and Technology*, Report EUR 10398, Commission of the European Communities, 1986, p 15-18, 74-85
165. *Final Storage of Spent Nuclear Fuel-KBS-3; III Barriers*, Swedish Nuclear Fuel Supply Company, Division KBS, 1983, p 10.8-10.15
166. R.E. Westerman and S.G. Pitman, Corrosion of Candidate Iron-Base Waste Package Structural Barrier Materials in Moist Salt Environments, *Mater. Res. Soc.*, Vol 44, 1985, p 282-285
167. E. Smailos *et al.*, Corrosion Behaviour of Container Materials for the Disposal of High-Level Wastes in Rock Salt Formations, in *Nuclear Science and Technology*, Report EUR 10400, Commission of the European Communities, 1986, p 15-24
168. F. Casteels *et al.*, Corrosion of Materials in a Clay Environment, in *Scientific Basis for Nuclear Waste Management*, Vol 2, Plenum Press, 1980, p 385-393
169. "Deep Underground Disposal of Radioactive Wastes: Near-Field Effects," Technical Report 251, International Atomic Energy Agency, 1985, p 6-58
170. G.H. Jenks, "Review of Information on the Radiation Chemistry of Materials Around Waste Canisters in Salt and Assessment of the Need for Additional Experimentation," ORNL-5607, Oak Ridge National Laboratory, March 1980
171. G.H. Jenks, "Radiolysis and Hydrolysis in Salt-Mine Brines," ORNL-TM 3717, Oak Ridge National Laboratory, March 1972
172. S.V. Panno and P. Soo, Potential Effects of Gamma Irradiation on the Chemistry and Alkalinity of Brine in High-Level Nuclear Waste Repositories in Rock Salt," *Nucl. Technol.*, Vol 67, Nov 1984, p 268, 280
173. R.S. Glass *et al.*, "Gamma Radiation Effects on Corrosion: I. Electrochemical Mechanisms," UCRL-92311, Lawrence Livermore National Laboratory, Feb 1985, p 9
174. M.D. Merz, "State-of-the-Art Report on Corrosion Data Pertaining to Metallic Barriers for Nuclear Waste Repositories," PNL-4474, Pacific Northwest Laboratory, Oct 1982, p 1.1-6.8
175. L.B. Nilsson and T. Papp, A Concept for Safe Final Disposal of Spent Nuclear Fuel, *Radioactive Waste Mgmt.*, Vol 3, 1983, p 93-106
176. M. Helie and G. Plante, HLW Container Corrosion in Geological Disposal Conditions, *Mater. Res. Soc.*, Vol 50, 1985, p 445-452
177. J.P. Simpson *et al.*, Corrosion Rate of Unalloyed Steels and Cast Irons in Reducing Granitic Groundwaters and Chloride Solutions, *Mater. Res. Soc.*, Vol 50, 1985, p 429-436
178. G.P. Marsh *et al.*, Evaluation of the Localized Corrosion of Carbon Steel Overpacks for Nuclear Waste Disposal in Granite Environments, *Mater. Res. Soc.*, Vol 50, 1985, p 421-428
179. R.B. Lyon and L.H. Johnson, A Review of Progress in the Canadian Nuclear Fuel Waste Management Program, *Mater. Res. Soc.*, Vol 50, 1985, p 59
180. R.E. Westerman *et al.*, "Investigation of Metallic, Ceramic, and Polymeric Materials for Engineered Barrier Applications in Nuclear Waste Packages," PNL-3484, Pacific Northwest Laboratory, 1980
181. R.P. Anantatmula *et al.*, Corrosion Behavior of Low-Carbon Steels in Grande Ronde Basalt Groundwater in the Presence of Basalt-Bentonite Packing, *Mater. Res. Soc.*, Vol 26, 1984, p 113-120
182. R.E. Westerman *et al.*, "General Corrosion, Irradiation-Corrosion and Environmental-Mechanical Evaluation of Nuclear Waste Package Structural Barrier Materials," PNL-4364, Pacific Northwest Laboratory, Sept 1982
183. D.R. Duncan *et al.*, "Feasibility Assessment of Copper-Base Waste Package Container Materials in a Repository in Basalt," Report SD-BWI-TA-023, Rockwell Hanford Operations, Sept 1986
184. R.D. McCright and H. Weiss, Corrosion Behavior of Carbon Steels Under Tuff Repository Environmental Conditions, *Mater. Res. Soc.*, Vol 44, 1985, p 287-294
185. M.C. Juhas *et al.*, "Behavior of Stressed and Unstressed 304L Specimens in Tuff Repository Environmental Conditions," UCRL-91804, Lawrence Livermore National Laboratory, Nov 1984
186. R.S. Glass *et al.*, "Corrosion Processes of Austenitic Stainless Steels and Copper-Based Materials in Gamma-Irradiated Aqueous Environments," UCRL-92941, Lawrence Livermore National Laboratory, Sept 1985
187. H. Weiss *et al.*, "Metallurgical Analysis of a 304L Stainless Steel Canister for the Spent Fuel Test—Climax," UCID-20436, Lawrence Livermore National Laboratory, April 1985
188. R.D. McCright, "FY 1985 Status Report on Feasibility Assessment of Copper Base Waste Package Container Materials in a Tuff Repository," UCID-20509, Lawrence Livermore National Laboratory, Sept 1985
189. T.M. Kegley and F.M. Empson, "Examination of Modified Pillar and Simulated Waste Container Test Heaters," ORNL-TM 2422, Oak Ridge National Laboratory, Jan 1969
190. T.F. Archbold and D.H. Polonis, "Assessment of Delayed Failure Models in Titanium and Titanium Alloys," PNL-4127, Pacific Northwest Laboratory, Dec 1981
191. N.R. Moody and S.L. Robinson, "Composition and Microstructural Effects on Hydrogen Embrittlement of ASTM Grade 12 Titanium," SAND85-8247, Sandia National Laboratories, May 1986
192. R.E. Westerman *et al.*, "Evaluation of Iron-Base Materials for Waste Package Containers in a Salt Repository," *Mater. Res. Soc.*, Vol 26, 1984, p 430-435
193. R.E. Westerman *et al.*, "Corrosion and Environmental-Mechanical Characterization of Iron-Base Nuclear Waste Package Structural Barrier Materials," PNL-5426, March 1986
194. D.H. Lester *et al.*, "Waste Package Performance Evaluation," ONWI-302, Battelle Memorial Institute, March 1983, p 5.15 to 5.21
195. A.B. Johnson and B. Francis, "Durability of Metals for Archeological Objects, Metal Meteorites and Native Metals," PNL-3198, Pacific Northwest Laboratory, 1980
196. R.F. Tylecote, "Durable Materials for Seawater: The Archeological Evidence," BNFL-314(R), British Nuclear Fuel Limited, 1977

Corrosion in Fossil Fuel Power Plants

Chairman: Barry C. Syrett, Electric Power Research Institute

Corrosion of Fossil Fuel Power Systems

R.I. Jaffee
Electric Power Research Institute

THE ELECTRIC POWER INDUSTRY uses three types of fossil-fired plants. The most common plant is the pulverized coal-fired steam power plant, which may be used either as a baseload plant (where the plant runs continuously at capacity except for scheduled outages for maintenance) or for intermediate loads between the steady baseload and higher loads needed daily. Gas turbines are used for peak loads that occur for an hour or two each day. Combined cycles using both gas and steam turbines are generally intended for baseload service, although they must also be capable of sustaining intermediate-load service.

The fuels used in fossil-fired plants are natural gas, petroleum, and coal. Natural gas is generally extremely pure and does not constitute a corrosion threat, unless firing is substoichiometric for reducing NO_x emissions. Petroleum fuels can be corrosive to boilers if they contain vanadium and alkali metals; these produce liquid vanadium oxides or alkali sulfates, both of which are highly corrosive to metals in the combustion chamber or hot-gas passages. Coal can contain such impurities as sulfur, chlorides, and alkali metals, which are extremely corrosive. Many coals are virtually unusable except under deaerated service conditions.

Steam Power Plants

Figure 1 shows an illustration of a fossil-fired steam power plant. Three fluid flow loops circulate through the system: fuel-air, water-steam, and condenser cooling. In the fuel-air loop, the fossil fuel is burned in air, transfers its heat to a series of heat exchangers, is cleaned of particulate matter, is scrubbed of sulfur oxides, and exits through the stack. In the water-steam loop, clean feedwater is converted into superheated steam in a boiler, which expands through a series of turbines, converting its heat into mechanical energy, and is condensed, conditioned, pumped, and heated as feedwater. In the condenser-cooling loop, cold water is passed through the condenser and can be recirculated if a cooling tower is used or can be exhausted back to the source of the cooling water. Each fluid loop possesses its unique corrosion problems.

The fossil fuel is burned in a very large chamber constructed of water walls consisting of vertical or spiral steel tubes about 60 mm (2.4 in.) in diameter that are welded together in a web about 20 mm (0.8 in.) wide. The feedwater ascends the water walls and is heated by the combusted fuel. In subcritical boilers, the generated steam is separated from the water such that the water is returned to the bottom of the water wall through downcomers, but the saturated steam is superheated in tubular heat exchangers suspended in the gas stream. In supercritical boilers, the pressure is above the critical point, and the liquid becomes superheated vapor without undergoing a phase change. The feedwater is conditioned to be slightly alkaline, but the fluid in the boiler may become acidic or caustic, depending on the presence of corrosion deposits and flow interruptions. Under acidic conditions, the steel boiler tubes may be hydrogen embrittled; under caustic conditions, the tubes may be caustic gouged. In supercritical boilers, overheating and excessive internal scaling in water walls may occur if boiling undergoes departure from nucleate boiling conditions at the region near the critical pressure.

The superheaters and reheaters are subject to steam oxidation on their inner surfaces and to hot corrosion on their outer surfaces. Steam oxidation results from attack by superheated steam, which acts similarly to oxygen at the same temperatures. For low-alloy steels, steam oxidation is of concern above 540 °C (1000 °F). This is about the temperature at which creep strengths limit low-alloy steels, and it becomes necessary to use alloys with higher chromium contents for higher allowable stress and for better steam oxidation resistance.

Fire-side corrosion in superheaters and reheaters is a typical problem. In coal-fired boilers, it exhibits a maximum rate at 700 to 750 °C (1290 to 1380 °F), where the corrodent is liquid, and decreases to a minimum at higher temperatures, where the corrodent does not condense. The liquid ash is generally an alkali sulfate or a complex alkali iron trisulfate. At higher temperatures, corrosion is predominantly the oxidation of uncooled parts, such as hangers.

Another type of boiler, the fluidized-bed boiler, was once thought to be free of fire-side corrosion because it operated under dry conditions at about 850 °C (1560 °F). The coal is burned in a fluidized bed composed of limestone or dolomite. The calcium sulfate ($CaSO_4$) product of desulfurization is in equilibrium with the calcium oxide (CaO) absorbant. At low oxygen potentials, such as those that occur in the condensed bed itself, the sulfur potential may rise high enough to sulfidize many otherwise very corrosion-resistant alloys.

Following the coal-air loop in a conventional power plant further, the flue gas passes through

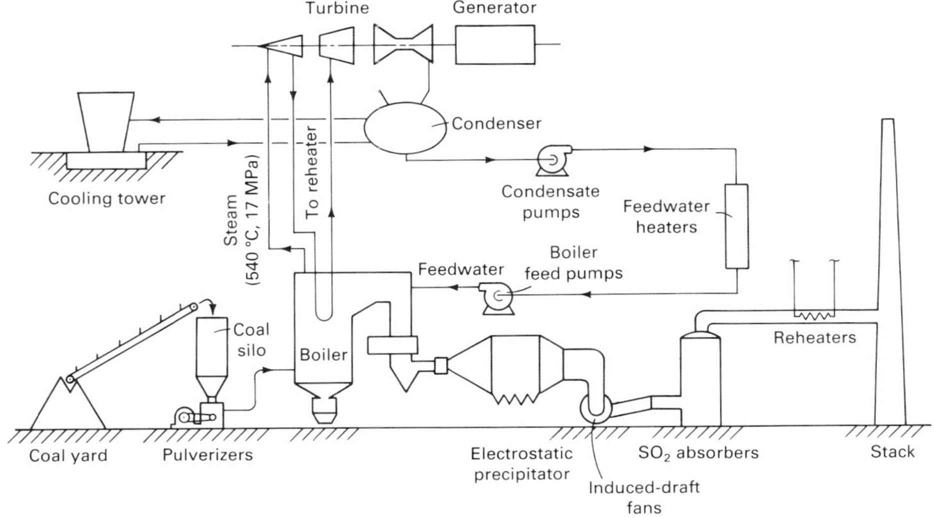

Fig. 1 Schematic of a coal-fired power plant

heat exchangers for preheating combustion air. It is important to maintain the flue gas temperature above its dew point in order to avoid the deposition of sulfuric acid (H_2SO_4). After passage through the precipitator while above its dew point, the flue gas is scrubbed of its sulfur dioxide (SO_2) content in a flue gas desulfurization scrubber. There are many points of corrosion concern in the scrubber, but the most serious areas are the inlet duct, where the SO_2-laden gas is hottest and a wet/dry interface exists, and the outlet duct, where the scrubbed gas, still containing sulfur trioxide (SO_3), begins to condense on the walls of the duct.

Once the steam has been compressed and superheated, it leaves the boiler through heavy-wall pipes and enters the high-pressure turbine. There it expands and returns to the boiler for reheating before entering the intermediate-pressure turbine for a second expansion. After one or two expansions and reheatings, the steam enters the low-pressure turbine.

As long as the steam is dry, there is little corrosion in the high- or intermediate-pressure turbines except when condensation of solid sodium hydroxide (NaOH) occurs. However, when expansion of the steam in the low-pressure turbine reaches the point of initial condensation (at the so-called Wilson Line), high-concentration chloride and sulfate salt solutions may deposit if the steam is contaminated. The distribution of salts between steam vapor and steam condensate is such that the condensate may be 10^6 times more concentrated than the vapor. Thus, the permissible impurity levels in the feedwater are measured in parts per billion in order to protect the low-pressure turbine at the Wilson Line, which generally occurs at the next to the last (L − 1, or last minus one) row of turbine blades. The Wilson Line shifts to a higher temperature point in the low-pressure turbine at reduced load such that in load-cycling plants there is alternate wetting and drying of the salt deposit as the load increases and decreases during cyclic operation. This is a serious condition for corrosion of blades and disks in the L − 1 row.

The salt solution is often acidic as a result of evaporation of ammonia (NH_3) from the water-conditioning process. During shutdowns, oxygen and carbon dioxide (CO_2) may dissolve in the acidic salt, aggravating the corrosive condition at the L − 1 row of a low-pressure turbine. The commonly used 12% Cr turbine blade alloy becomes pitted under these conditions and may lose up to 90% of its fatigue strength. Corrective measures include using blades designed to be strong enough to operate with pitted surfaces, cleaning and maintaining the steam to avoid corrosive salt deposition, using more corrosion-resistant low-pressure turbine blade materials (such as titanium alloys), or protecting 12% Cr steel blades with corrosion-resistant coatings.

Corrosion problems may occur in the condenser-cooling loop, especially if the cooling water is sulfide-contaminated seawater or brackish water acting on copper-base tubes or tubesheets. Also, pitting or crevice corrosion may occur under deposits or barnacles or between tubes and tubesheets. However, the primary concern with condensers is leakage of seawater or contaminated cooling water into the water-steam loop, which operates below atmospheric pressure; this leakage would result in drastic corrosion effects on boiler and turbine components.

Gas Turbines

In a gas turbine, inlet air is compressed in a compressor, reacted with fuel in a combustion chamber, and directed at stationary airfoil vanes and through rotor blades or buckets constituting the turbine stage. Thus, the entire fuel and air input passes through the gas turbine without an intermediate heat exchanger. Any corrosive impurities present in either fuel or air will affect the high-temperature components, primarily the combustor, nozzle diaphragm, and turbine. The principal threats are oxidation and hot corrosion. These have been largely met by using alloys of increased chromium content, particularly Ni-20Cr and Co-30Cr alloys. However, as the turbine inlet temperature increased to achieve higher thermal efficiency, it became necessary to strengthen the alloys, which resulted in lower chromium contents and greater vulnerability to oxidation and hot corrosion. The use of bypass air cooling of the hot-section parts essentially reduced the metal temperature to a point at which the high-temperature strength was sufficient, while the gas temperature increased progressively. The high-temperature high-velocity gas stream causes evaporation of volatile chromium trioxide (CrO_3) from otherwise protective chromic oxide (Cr_2O_3) scales. The alloys subjected to the highest turbine temperatures are protected by coatings containing over 5% Al, which form protective aluminum oxide (Al_2O_3) scales.

If the fuel or inlet air contains alkali metals, sulfur, or vanadium as impurities, hot corrosion may occur. This is combated by limiting these impurities in the fuel. Additives such as magnesium oxide (MgO) also help. Air filtration is used to reduce ingestion of airborn impurities. Vanes and blades are washed periodically to remove accumulated salts. The coatings that form Al_2O_3 protective scales are sometimes improved by platinum metal additions or sublayers. Also, coatings with small additions of yttrium promote adherence of the Al_2O_3 scale, which otherwise might spall off.

Combined Cycle Plants

In a combined cycle power plant, the gas turbine is used as a high-temperature topping cycle whose exhaust gas enters a waste heat boiler, which raises steam to operate the steam turbine and generator. To a large extent, corrosion problems in the combined cycle are simply the sum of the corrosion problems in the gas turbine and the steam boiler and turbine. Control of impurities in the inlet air and fuels is essential. Corrosion from the use of gasified coal, which may have caused severe problems in the gas turbine and steam generator, has largely been eliminated by scrubbing the gasified coal in a water-quenching operation. There are severe corrosion problems in the radiant cooler used to generate process steam from the raw gasified coal, which contains hydrogen sulfide (H_2S) and other corrosive agents before the scrubbing operation. These problems can be handled by limiting the temperature of the radiant cooler to be commensurate with the heat-exchanger material used, which is generally coated or clad steel.

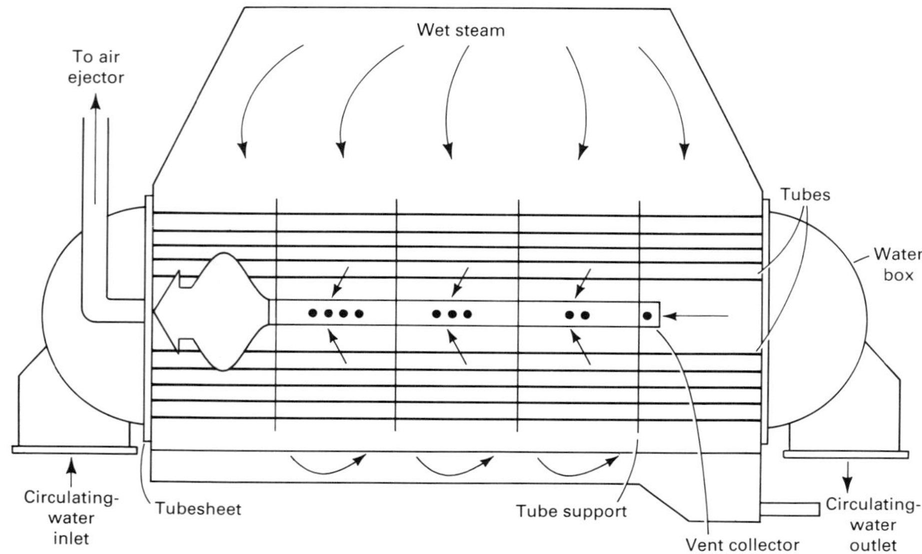

Fig. 2 Schematic of a typical condenser in an electric power plant

Corrosion of Condensers

Barry C. Syrett
Electric Power Research Institute
Roland L. Coit
Consultant

A steam surface condenser is a shell and tube heat exchanger that is positioned immediately downstream of the low-pressure steam turbine. Heat is transferred from steam on the outside of the condenser tubes (the shell side or steam side) to water on the inside (the tube side or water side). A schematic of a typical electric power plant condenser is shown in Fig. 2.

The condenser is a particularly critical component in a power plant because its failure can affect many other components in the steam-water cycle. The root cause of many of the corrosion problems in fossil fuel boilers, nuclear steam generators, low-pressure steam turbines, and feedwater heaters has been traced to condensers that have leaked and allowed contamination of the steam condensate with raw cooling water and air. Most tube leaks are caused by corrosion, but some failures are purely mechanical, such as those caused by steam impingement (erosion), tube-to-tubesheet joint leaks, mechanical ruptures from foreign object impact, and tube vibration resulting in fretting wear and fatigue. Purely mechanical failures will not be discussed further.

Corrosion mechanisms that have led to failures or serious problems in power plant condensers are summarized in Table 1. Table 1 reflects known service problems to date, rather than susceptibilities that might be inferred solely from laboratory tests; each form of failure occurs only under specific environmental and metallurgical conditions. Information on tubesheet materials requires special explanation. Because tubesheets are very thick (>25 mm, or 1 in.), corrosion rates can be 15 to 50 times higher than in condenser tubes and still be considered acceptable in most cases. Furthermore, even if corrosion is a serious problem in a tubesheet, inspections are usually scheduled frequently enough, and the tubesheet is thick enough, that suitable repairs can be made or corrosion protection procedures can be instituted long before the tubesheet is penetrated. Thus, although Table 1 indicates that copper alloy tubesheets have suffered significant (often severe) galvanic corrosion under certain conditions, leakage of cooling water through the tubesheet from the water side to the steam side has rarely occurred. Each of the corrosion mechanisms responsible for failures in condensers will be reviewed below. Some of the methods of preventing these failures will also be summarized.

Erosion-Corrosion

Erosion-corrosion is a relatively common water-side phenomenon that is a problem only in copper alloy condenser tubes. It occurs in areas where the turbulence intensity at the metal surface is high enough to cause mechanical or electrochemical disruption of the protective oxide film. In these turbulent regions, pitlike features develop. Turbulence increases with increasing velocity and is greatly influenced by geometry. For example, turbulence intensity is much higher at tube inlets than it is several feet down the tubes; this results in the phenomenon of inlet-end erosion-corrosion. Tube inserts have been used to circumvent this problem. A tube insert is a tightly fitting internal sleeve, typically 150 to 300 mm (6 to 12 in.) long, made from a material resistant to erosion-corrosion that shields the susceptible tube ends. However, unless there is a smooth transition between the end of the insert and the tube, the insert can itself create turbulent conditions and promote erosion-corrosion further down the tube (Fig. 3). Recent experiments have demonstrated that inlet-end erosion-corrosion can also be prevented by installing a cathodic protection system in the water box region.

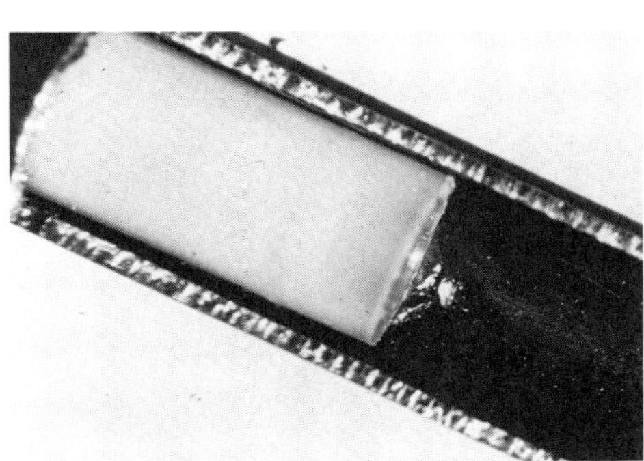

Fig. 3 Erosion-corrosion occurring immediately downstream of a nylon insert in an aluminum brass condenser tube cooled by seawater

Erosion-corrosion may also occur when marine life or debris in a tube creates a partial blockage, resulting in locally high velocities through the restricted opening. In these cases, the best solution is to keep the tubes clean, using one or more of the following methods:

- Install or improve intake screens
- Install on-line sponge ball cleaning
- Periodically reverse flow (backwash)
- Manually clean with brushes, balls, scrapers, and so on (off line)
- Prevent biofouling by chlorination or thermal shock

Alternatively, some copper alloys benefit from periodic dosing of the water with ferrous ions (Fe^{2+}), which are usually added as ferrous sulfate ($FeSO_4$) solution. The Fe^{2+} ions deposit as a protective lepidocrocite [FeO(OH)] layer on the copper alloy surface.

Sulfide Attack

This form of attack affects only copper alloys and occurs when the cooling water, most often brackish water or seawater, is polluted with sulfides, polysulfides, or elemental sulfur. As little as 10 mg/m^3 (10 ppb) of sulfide in the cooling water can have a detrimental effect, and concentrations far greater than this are often measured in polluted harbors and estuaries. Sulfide attack manifests itself in many ways. It can greatly increase general corrosion rates, and it can induce or accelerate dealloying, pitting, erosion-corrosion, intergranular attack, and galvanic corrosion. Penetration rates in polluted waters can be extraordinarily high, sometimes as high as 20 mm/yr (800 mils/yr). No copper alloy is resistant to sulfide attack, and the relative performance of copper alloys in polluted or brackish waters seems to depend on the precise environmental conditions.

If the incoming water contains sulfide and there is no obvious method of eliminating the source, the most successful method of reducing or preventing the problem is to dose the water periodically or continuously with $FeSO_4$ or some other source of Fe^{2+} ions. However, sulfide attack can also occur in condensers cooled with nominally unpolluted water if marine organisms trapped within the condenser during downtime are allowed to die and putrefy to produce sulfides. This can probably be prevented, or at least reduced, by turning on the pumps for an hour or two each day to flush out the condenser with fresh seawater. In addition, sulfate-reducing bacteria can produce sulfides under debris and deposits where the oxygen content is low. Thus, the risk of sulfide attack is greatly reduced if the copper alloy tubes are regularly cleaned.

Dealloying

Another water-side problem in brass-tubed condensers is dealloying. Dezincification is an example of dealloying that has been observed in utility condensers. In dezincification, zinc is selectively removed from brass alloys to leave a copper-rich surface layer.

Dealloying is rarely the cause of condenser tube failure, but when it does occur, it is normally restricted to localized areas, such as beneath deposits or at hot spots. This results in plug-type dealloying. Clearly, maintaining clean tubes will reduce the incidence of this type of failure. A much less localized form of dealloying, termed layer-type dealloying, is rare in tubes, but has occasionally occurred in brass tubesheets, particularly in conjunction with galvanic corrosion induced by titanium or stainless steel condenser tubes. Under such circumstances, a cathodic protection system installed in the water box will control both galvanic corrosion and dealloying.

Crevice Corrosion and Pitting

Some stainless steels and copper alloys are susceptible to water-side pitting and crevice corrosion. Brass and austenitic stainless steel condenser tubes, in particular, are known to have failed by pitting and crevice corrosion. There is limited evidence that copper alloys have ade-

Table 1 Corrosion mechanisms that have caused problems in power plant condensers under certain conditions

Alloy	Corrosion mechanism: Erosion-corrosion	Sulfide attack	Dealloying	Crevice corrosion/pitting	Galvanic corrosion	Environmental cracking	Condensate corrosion
Copper alloys							
Muntz metal (tubesheets)	N	(W)	(W)	N	W	N	N
Aluminum bronze (tubesheets)	N	(W)	N	N	(W)	N	N
Aluminum bronze	W	W	(W)	(W)(a)	(W)	(S)	(S)
90Cu-10Ni	W	W	(W)	(W)(a)	(W?)	N	(S)
70Cu-30Ni	W	W	N	(W)(a)	(W?)	N	(S)
Aluminum brass	W	W	(W)	W(a)	W	W/S	S
Admiralty brass	W	W	(W)	W(a)	W	W/S	S
Stainless steels							
AISI type 304	N	N	N	W	N(b)	N	N
AL6X	N	N	N	(W)	N(b)	N	N
AL29-4C	N	N	N	N	N(b)	(W)	N
Sea-Cure	N	N	N	(W)(c)	N(b)	(W)	N
Titanium alloys							
Commercial-purity titanium	N	N	N	N	N(b)	(N)	N

W, water-side problem; S, steam-side problem; N, not a problem; ( ), slight sensitivity to problem; ?, problems have occurred in similar alloys. (a) Perhaps a problem only when sulfide is present. (b) Can induce galvanic corrosion of adjacent copper alloys, iron, and carbon steels when used in seawater or other highly conductive waters. (c) A problem in heats containing only 25.5% Cr and 3% Mo

Fig. 4 Example of pitting in AISI type 316 stainless steel in seawater service

quate resistance to these forms of corrosion if the cooling water is completely free of sulfide. Certainly, susceptibility seems to be greatly increased when sulfide is present.

Pitting and crevice corrosion of stainless steels are more dependent on the chloride content of the cooling water than on the sulfide content, although laboratory data have demonstrated that the detrimental effects of chloride are accentuated in the presence of sulfide. Some alloys, such as AISI type 304 and 316 stainless steels, which generally perform well in freshwaters or slightly brackish waters, suffer rapid pitting and crevice corrosion in seawater (Fig. 4). The newer, more highly alloyed stainless steels, including AL6X (UNS NO8366), AL29-4C (Fe-29Cr-4Mo-0.35Si-0.02C-0.02N-0.24Ti), and Sea-Cure (Fe-27.5Cr-3.4Mo-1.7Ni-0.4Mn-0.4Si-0.02C-0.5Ti+Nb), generally perform well even in seawater. However, a few failures have been reported for AL6X and for some of the early heats of Sea-Cure.

Again, tube cleanliness is a critical issue because debris and deposits promote the formation of concentration cells (the precursor to crevice corrosion) and because they favor the production of sulfides. The tube-cleaning techniques summarized earlier in this section are therefore equally useful in preventing crevice corrosion and pitting in copper alloys and stainless steels.

Galvanic Corrosion

Galvanic corrosion is not a problem in poorly conducting waters, such as those normally found on the steam side of condensers. However, it can be a water-side problem in condensers cooled with seawater or with medium- or high-conductivity fresh and brackish waters.

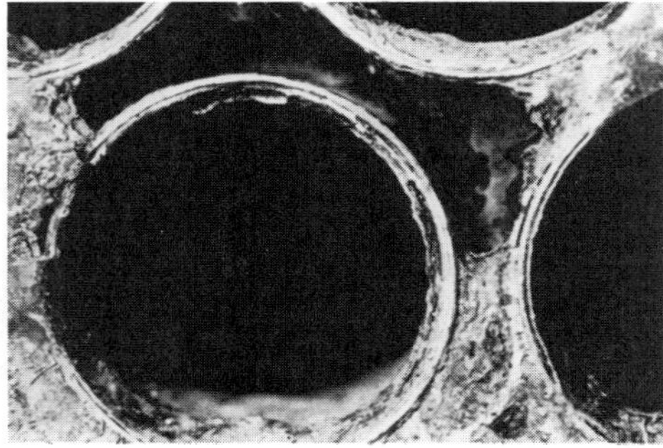

Fig. 5 Galvanic corrosion of a Muntz metal tubesheet, fitted with AL6X stainless steel tubes, after 1 year of service

In seawater, tube materials, such as copper-nickel alloys, stainless steels, and titanium, are more noble than tubesheet materials, such as Muntz metal and aluminum bronze. Consequently, the tubesheet may suffer galvanic attack when fitted with these more noble tubes (Fig. 5). Laboratory tests have demonstrated that the rate of galvanic corrosion of a Muntz metal tubesheet fitted with titanium or stainless steel tubes can exceed 5 mm/yr (200 mils/yr) in seawater. Similarly, if stainless steel inserts are installed in copper alloy tubes to prevent inlet-end erosion-corrosion, rapid galvanic corrosion can be promoted in the tube close to the insert/tube interface.

Other galvanic couples can exist in a condenser, but in each case, the recommended method of alleviating the problem is to install a cathodic protection system in the water box. Cathodic protection current requirements can be reduced by coating the tubesheet and water box with a nonconducting material.

Environmental Cracking

Stress-corrosion cracking (SCC) and hydrogen embrittlement cracking are forms of environmental cracking that can affect condensers. Hydrogen embrittlement cracking was identified as a problem on the water side of ferritic stainless steel tubes in a couple of condensers fitted with cathodic protection systems. It is believed that hydrogen was generated on the surface of the tubes by the passage of too high a cathodic protection current and that this hydrogen promoted slow crack growth and failures at the ends of the tubes. The tube ends were particularly susceptible to hydrogen embrittlement cracking because roller expansion during fabrication introduced higher-than-normal residual stresses in these zones.

Entry of hydrogen into cathodically protected titanium tubes is also possible if too high a cathodic protection current is delivered. In such cases, the absorbed hydrogen can react with the metal to form a brittle titanium hydride phase which could conceivably crack and lead to premature failure. However, so far no titanium condenser tube failures have been reported. One electric utility that grossly overprotected the waterbox region of a titanium-tubed condenser discovered that the ends of the tubes were severely hydrided, but even here the affected tubes did not leak.

Apart from the rather unusual failures in ferritic stainless steels, environmental cracking is a problem only in copper alloys, specifically the brasses. Here, SCC (not hydrogen embrittlement cracking) is the mechanism of failure. Most SCC failures initiate on the steam side of the tubes, and all occur when the steam condensate contains high concentrations of NH_3 and oxygen. The NH_3 is derived from the chemicals added for boiler feedwater chemistry control, and oxygen originates from air that leaks into the system through imperfectly maintained turbine glands, expansion joints, valve packing glands, and so on. The NH_3 and oxygen concentrations are particularly high in the air removal section, and it is here that SCC occurs most frequently.

Steam-side SCC can be controlled by ensuring that the condensate on the tubes has a low oxygen concentration. The maintenance of an airtight system requires continuing attention to all seals, glands, and joints that are subjected to internal pressures less than atmospheric during start-up, normal operation, or shutdown. Helium tracer and similar techniques allow leaks to be detected with moderate ease.

Water-side SCC of brasses has occurred less frequently than steam-side attack, and in most cases, the species responsible for the failure were not positively identified. However, NH_3 and its derivatives (nitrates and nitrites) are often suspected of promoting SCC. Possible sources of these species are farm fertilizers (runoff) and decaying organisms in polluted water. Water-side failures frequently initiate beneath surface depos-

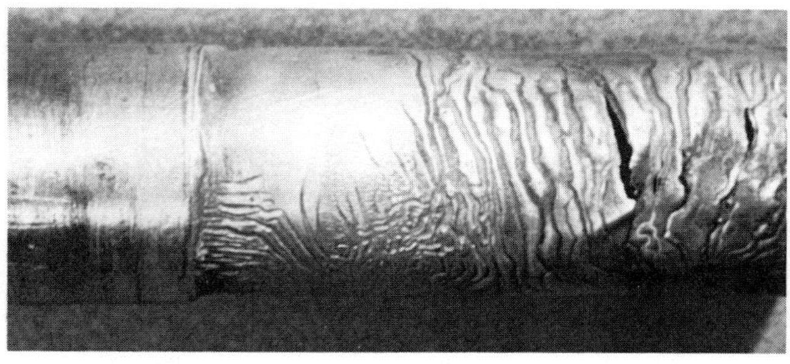

(a)

(b)

Fig. 6 Examples of NH_3 attack on admiralty brass. (a) The unattacked tube end (left) was protected by the tubesheet. (b) Condensate grooving that occurred at one side of a support plate its, probably because the deleterious species can concentrate beneath the deposit to levels that favor SCC. Thus, once again, tube cleanliness is important, and cleaning techniques can be used to minimize water-side SCC.

No matter which side of the tube is susceptible to environmental cracking, the incidence of cracking can be controlled by reducing or eliminating residual tensile stresses. Roller expansion of the tubes during installation will always introduce some residual stresses, but care should be taken to avoid expansion beyond the back of the tubesheet, an event that can lead to particularly high residual stresses. In addition, fully stress-relieved tubes should be used, and during installation, they should not be bent or mechanically abused.

Condensate Corrosion

Copper alloy condenser tubes, particularly brass condenser tubes, are susceptible to condensate corrosion in steam condensate that contains high concentrations of NH_3 and oxygen. Consequently, condensate corrosion, like SCC, is most prevalent in the air removal section. Condensate corrosion, also known as NH_3 attack, is a form of corrosion that is localized not by microstructural features in the metal but by the localization of the corrosive environment (Fig. 6). For example, the slight tilt routinely given to condenser tubes may promote flow of some of the steam condensate toward one side of a tube support plate. There, the flow from a large number of tubes can collect and run down the plate surface. Such localized flow can create deep circumferential grooves, termed condensate grooving, in the tubes immediately next to the support plate (Fig. 6b) if the condensate contains high levels of NH_3 and oxygen. Condensate corrosion can be controlled by reducing the oxygen concentration in the condensate, as discussed previously for steam-side SCC, or by selecting more resistant alloys, such as copper-nickel alloys, or completely resistant alloys, such as stainless steels and titanium.

Corrosion Prevention

Many corrosion modes operating in condensers can be prevented if only two maintenance procedures are followed. First, if air is eliminated from the steam, condensate corrosion and SCC of copper alloy tubes can be prevented. Second, if condenser tubes are kept clean and free of deposits, debris, and biofouling on the water side, sulfide attack, dealloying, erosion-corrosion, crevice corrosion, pitting, and SCC can be prevented or minimized.

Corrosion of Deaerators and Feedwater Heaters

Robert J. Bell
Heat Exchanger Systems, Inc.

Deaerators (direct contact deaerating feedwater heaters are used in fossil and a few nuclear power plants primarily to remove dissolved gases (mostly oxygen and nitrogen) from condensate/feedwater and to raise the condensate temperature by exchange with extraction steam by mechanical deaeration. Another function is to provide deaerator storage capacity and proper suction conditions for the boiler feed pump. Closed feedwater heaters (Fig. 7) are used in power plants to increase the overall cycle efficiency of the plant by delivering to the boiler or steam generator water at higher temperatures, thereby reducing the heat required to produce steam. This is accomplished by heating the feedwater (condensate) using extraction steam from the turbine.

Deaerators and feedwater heaters are susceptible to various forms of corrosion. This section will discuss corrosion in these applications as well as the measures that can be taken to minimize it.

Closed Feedwater Heaters

Corrosion problems in closed feedwater heaters are usually manifested as tube failures. Although tube failures may be a symptom of another problem, such as a baffle failure, the predominant causes of tube failures are destructive vibration, impingement erosion, cavitation-type erosion, and corrosion. Corrosion is a lesser cause than the others. The tube alloys most frequently used in closed feedwater heaters and their associated corrosion concerns are summarized in Table 2 and will be described in the rest of this section.

Stress-corrosion cracking occurs when an alloy under a tensile stress (applied and/or residual from manufacture or welding) is exposed to a specific corrosive environment. The chloride ion (Cl^-) induced SCC of austenitic stainless steels, for example, AISI type 304, in power plant systems is well documented. Failures due to SCC in feedwater heaters have not been significant in number, and gross deviation from normal feedwater chemistry is the universal cause. The potential for concentration of corrosively aggressive chlorides occurs mainly on the shell side in the desuperheating zone in the event of a small leak. This factor, combined with controlled feedwater conductivity, pH, and dissolved oxygen content, possibly explains why there have been few reported failures of stainless steel tubes due to SCC.

The copper alloys are susceptible to NH_3-induced SCC. When many plants switched to all-volatile water treatment (a source of NH_3), the incidence of SCC in copper alloys correspondingly increased. The failures typically occur in poorly vented areas where oxygen and NH_3 can concentrate. Stress-corrosion cracking can be minimized by reducing the residual stresses in the material, for example, by stress relieving tube U-bends. Because of the susceptibility of copper alloys to other corrosion mechanisms and because of the overall need to reduce copper within the system, the application of the copper alloys has been greatly diminished.

There have been a few reported failures in Monel 400 tubes, primarily at the U-bend because of SCC. Such failures can be prevented by limiting residual stress.

Exfoliation is a form of corrosion in which the corrosive penetration runs primarily on a plane that is parallel to the tube surface (Fig. 8). Generally, attack is along boundaries of grains that are elongated in the drawing direction. The expansive force of insoluble corrosion products tends to force these grains apart. The outside surfaces of exfoliated tubes exhibit heavy scaling with a leafy appearance; hence the term exfoliation. Both 70Cu-30Ni and 80Cu-20Ni alloys are susceptible to this form of attack; the degree of susceptibility increases with nickel content.

In the power generation industry, exfoliation was first encountered in units that were converted from baseload operation to peaking/cycling service (Ref 1). The peaking/cycling service results in greater and more frequent exposure to oxygen. This problem is reduced by steam or nitrogen blanketing when the unit is out of service.

Erosion-Corrosion. Many closed feedwater heater tubes are protected from accelerated corrosion by the formation of a corrosion product film, which in turn acts as a diffusion barrier that can limit the net diffusion rate of the corrodent

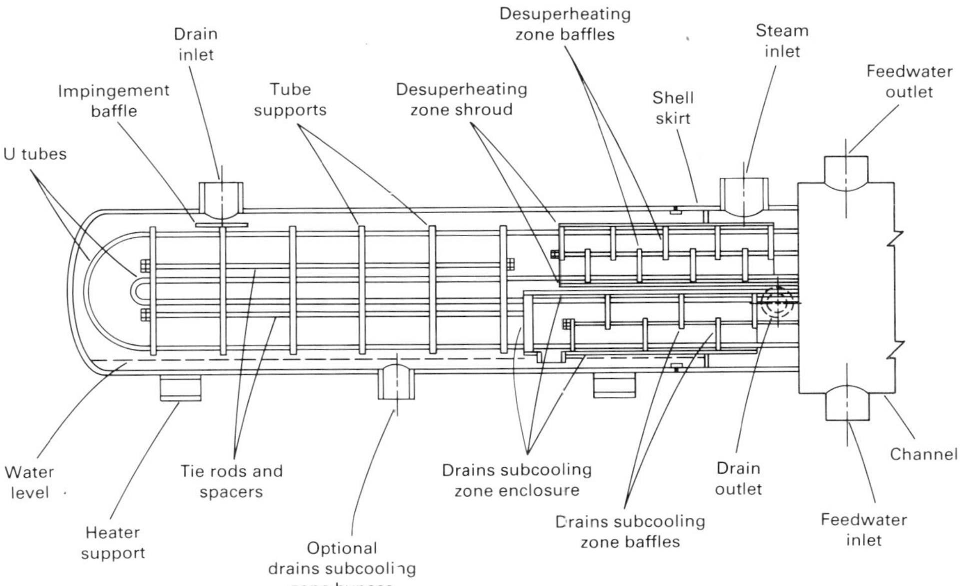

Fig. 7 Schematic of a 3-zone feedwater heater. Zones: desuperheating, condensing, and subcooling

Table 2 Tube alloys used in closed feedwater heaters and associated corrosion mechanisms

	Corrosion mechanism and alloy susceptibility(a)							
Alloy	SCC	Exfoliation	Erosion-corrosion	Crevice corrosion	Pitting	Condensate corrosion	General corrosion	Snake skin
Admiralty brass	S	I	I	I	I	S	S	I
90Cu-10Ni	I	I	I	I	I	I	S	I
80Cu-20Ni	M	S	I	I	I	S	S	I
70Cu-30Ni	M	S	I	I	I	I	S	S
Carbon steel	I	I	S	I	I	I	S	S
Austenitic stainless steels	S	I	I	S	S	I	I	I
Ferritic stainless steels (for example, AISI type 439)	S(b)	I	I	I	I	I	I	I
Monel 400	M	I	I	I	I	I	I	S

(a) S, susceptible; I, immune; M, marginally susceptible in the annealed state and susceptible in the drawn stress-relieved state. (b) AISI type 439 stainless steel has greater immunity than AISI type 304; AISI type 446 is virtually immune.

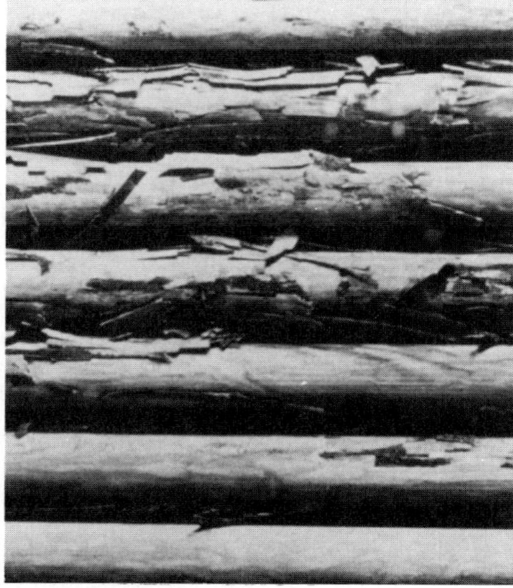

Fig. 8 Exfoliation of tubes in a closed feedwater heater

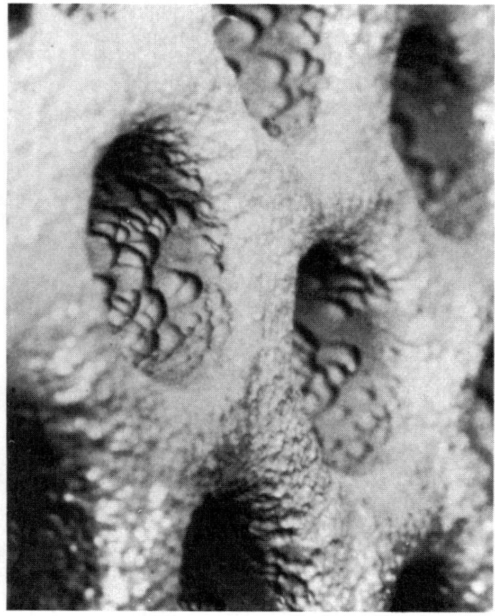

Fig. 9 Erosion-corrosion of carbon steel tubes in a closed feedwater heater. Courtesy of the Electric Power Research Institute

(for example, dissolved oxygen) to the underlying base metal. By limiting the net diffusion rate of corrodent through the film, the corrosion rate is also limited. Turbulence at tube entrances can produce excessive shear forces on the protective corrosion product film; these forces reduce the section thickness of the film and consequently accelerate corrodent flux and the corrosion process. This process is erosion-corrosion.

Although other tube alloys used in closed feedwater heaters are theoretically susceptible to erosion-corrosion, only carbon steels have exhibited widespread failures (Fig. 9). The primary causes are excessive water velocities, either by design or by abnormal operation of the feedwater heaters, and channel geometries that cause local zones of turbulence at the tubesheet. Low feedwater pH, for example, less than 8.8, accelerates this mechanism because of the impact of pH on the carbon steel corrosion product film.

Erosion-corrosion is also predominant at temperatures between 150 and 230 °C (300 and 450 °F). At these temperatures, the protective corrosion product film is soft and comparatively friable (Ref 2).

Pitting of feedwater heater tubes is generally limited to austenitic stainless steel alloys and carbon steels. Pitting of austenitic stainless steel alloys is caused by the contamination of feedwater by chlorides. Pitting typically occurs when chlorides are concentrated under deposits, within crevices, and through evaporation, for example, in the desuperheating zone. The few reported incidents of chloride-induced pitting in austenitic stainless steel (type 304) are generally associated with excessive seawater intrusion permitted by condenser tube leaks (Ref 2). There are no reported similar occurrences with ferritic stainless steels, such as AISI type 439.

Failure to control feedwater oxygen content properly (<7 ppb) has resulted in a few reported incidents of pitting in carbon steel tubing (Ref 3). One incident of carbon steel tube pitting was also attributed to excessive saltwater intrusion due to operating with a condenser tube leak (Ref 2).

General corrosion occurs in all closed feedwater heater tube alloys. Corrosion rate, however, varies substantially among the typical feedwater heater tube alloys. Passive materials, such as the austenitic and ferritic stainless steels, demonstrate negligible uniform corrosion rates and are therefore considered essentially immune to this form of corrosive attack. In general, susceptibility to general corrosion decreases as alloy selection changes from carbon steel to brasses, copper-nickels, Monel 400, and stainless steels (immune). The nature of the corrosion product film is the primary controlling variable.

In terms of the corrosion product film, two characteristics are important: the corrodent flux through it and the uniformity of diffusion resistance at all points. General corrosion typically occurs where diffusion resistance is essentially the same at all points.

The variables affecting general corrosion, and therefore the corrosion product film, are pH, temperature, fluid velocity, and concentration of corrodent. The magnitude of influence exerted by each variable significantly increases or diminishes as a function of the exposed tube material.

Two very critical parameters are introduced into the general corrosion reaction when cycling service begins. First, greater amounts of oxygen and CO_2 are present during low loads and during outages. Because the corrosion rate of carbon steel is directly proportional to corrodent concentration, greater corrosion is likely with cycling service. Second, cycling service creates thermal expansion-contraction problems that arise from the difference in the thermal expansion coefficients of the carbon steel tube and its corrosion product film. Therefore, cycling service leads to fracturing of the protective corrosion product film. The metal directly beneath these fracture sites has little, if any, remaining film to provide a protective diffusion barrier. It follows, then, that fossil plant cycling can accelerate the corrosion of carbon steel tubes.

Snake skins are a result of the redeposition of copper corrosion products on Monel 400 or 70Cu-30Ni high-pressure feedwater heater tubes. The deposition results in a thin flaky film that shrinks and subsequently sheds from the tubes when dried. The film appearance is similar to a skin shed by a snake (Fig. 10). It can cause a significant reduction in heat transfer. Low-pressure closed feedwater heaters employing admiralty brass tubes are usually the source of the copper.

Deaerators

Deaerators are subject to general corrosion, erosion-corrosion, SCC, and corrosion fatigue. The corrodent for the latter mechanism is dissolved oxygen—even at concentrations of 7 ppb or less. This corrosion process usually begins as oxygen-induced pits, which in turn act as stress risers, that ultimately promote the initiation of (typically) transgranular cracks. This process requires low-frequency stress cycles that produce greater contact time between the metal and corrodent.

Corrosion of Steam/Water-side Boilers

R.B. Dooley
Electric Power Research Institute

The cost to electric utilities from corrosion and deposition on steam- and water-side boilers is high—approximately $3.5 billion per year. One-half of the outages forced by plant failures have also been estimated to be attributable in some way to water-side corrosion. Contaminant depo-

Fig. 10 Snake skin formed by the redeposition of copper corrosion products on copper alloy reheater tubes. Courtesy of the Electric Power Research Institute

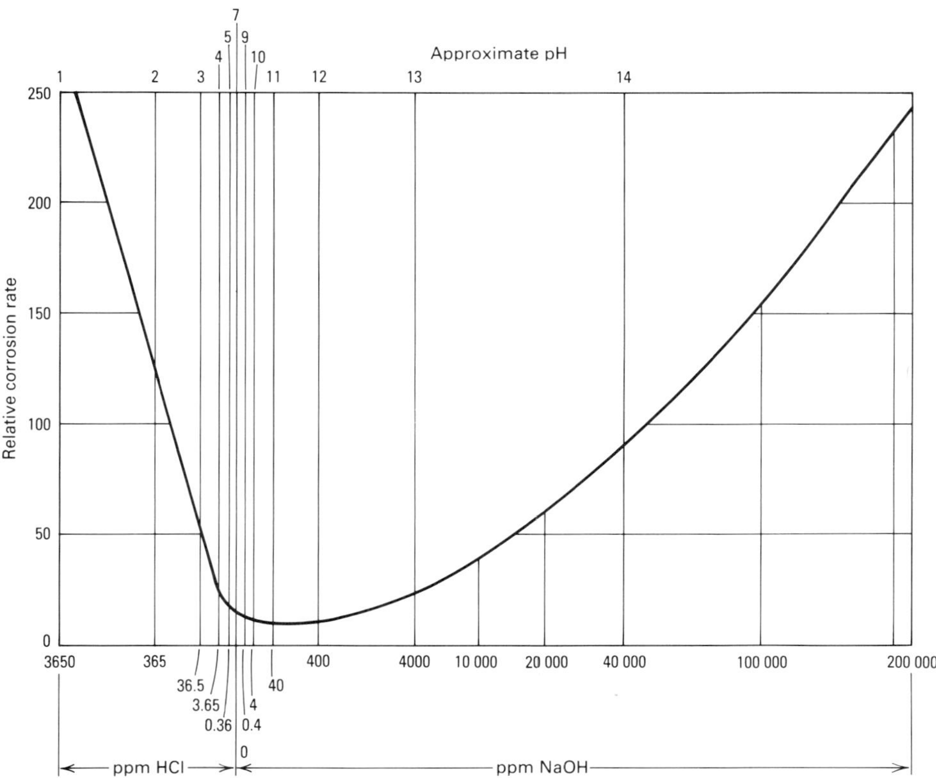

Fig. 11 Effect of pH on the corrosion rate of steel in water at 310 °C (590 °F). Upsets in water chemistry that increase or decrease the pH of boiler water or of wall deposits can result in corrosion of the water wall tubes. Source: Ref 4

sition in most cases reduces equipment efficiency and induces corrosion by a variety of mechanisms and is implicated in a variety of boiler tube failure mechanisms that are most common in water walls and economizers.

Water side deposits often begin as accumulations of corrosion products transported to the boiler from other parts of the system. The corrosion product deposit is porous, in contrast to the protective magnetite (Fe_3O_4) film. This porous deposit serves as a trap for corrosive impurities, such as caustic, chlorides, and acid sulfates.

Boiler tube failures initiating in steam-containing tubing have also almost exclusively been the result of entrainment of contaminants within the steam. Chlorides, sulfates, and caustic are the most common contaminants. However, the growth of Fe_3O_4 on the inside tube surface can also be a secondary contributor to tube failure. If its growth rate is excessive, this will act as a thermal barrier and cause the tube wall temperature to rise, sometimes above the point at which excessive creep damage will result in an overheating failure.

Water Walls and Economizers

These components are usually manufactured of plain carbon steel (ASTM A210, ASME SA-192); in a few exceptions, the tubes are a low-chromium ferritic material (ASME SA-213, grade T-11). In most cases, the tube metal temperatures are less than 400 °C (750 °F) for subcritical boilers.

Normal Protective Fe_3O_4 Growth. It has been pointed out many times that the only thing that protects a boiler is a thin film of Fe_3O_4 on the water-side tube surface. The ability to use inexpensive carbon or low-alloy steels in contact with water at high temperatures and pressures is due to the reaction between iron and oxygen-free, neutral, or slightly alkaline water:

$$3Fe + 4H_2O \rightarrow Fe_3O_4 + 4H_2 \quad \text{(Eq 1)}$$

by which a film of Fe_3O_4 is formed on the water-side surface of the tube; the kinetics of the reaction are parabolic, and the protective oxide consists of two layers described as the inner and outer. In practice, the outer layer is seldom formed, because as the iron diffuses outward the Fe_3O_4 formed at the outer interface usually becomes entrained in the boiler water flow and then deposits, together with feedwater corrosion products in another region of the boiler, which may be of higher heat flux.

The growth of Fe_3O_4 in economizer tubing occurs in a similar fashion to that in the water wall. However, a common observation is that the water-side surfaces are more uneven and contain more pits than those in the water walls.

Breakdown of the Normal Protection Mechanism. The corrosion resistance of water wall/economizer tubing depends on keeping the protective Fe_3O_4 in place, on the pH level of the water, and on the amount of contaminants. Figure 11 shows the effect of pH on the rate of corrosion of steel by water. Magnetite is unstable and soluble at pH values below 5 and above 12.

The significant categories of water-side corrosion failure mechanisms are caustic corrosion, hydrogen damage, and pitting (localized corrosion). A significant factor in these mechanisms is the amount of corrosion product deposited on the wall tube. Caustic corrosion and hydrogen damage result from the breakdown of the protective Fe_3O_4 layer by the concentration of corrosive chemicals within a wall deposit. As indicated below, different failure mechanisms will be experienced, depending on the contaminants present.

Caustic corrosion is sometimes referred to as caustic attack, caustic gouging, or ductile gouging. Caustic corrosion develops from the deposition of feedwater corrosion products in which NaOH can concentrate to high pH levels. At high pH levels, the protective Fe_3O_4 layer of the tube steel becomes soluble, and rapid corrosion occurs (Fig. 11).

Caustic corrosion is caused by the selective deposition of corrosion products and NaOH at locations of high heat flux. As porous deposits accumulate in high heat input areas, NaOH concentrates through a process known as wick boiling. The caustic levels can concentrate from less than 100 ppm NaOH in the bulk water to over 200 000 ppm adjacent to the tube surface. The electrochemical nature of the attack is shown in Fig. 12. The hydroxide ions (OH^-) are concentrated within the deposit layer so that the hydrogen ion (H^+) concentration is highest in the boiler water.

The susceptibility of high-pressure boilers to such corrosion damage may be reduced by minimizing the entry of deposit-forming materials and by performing periodic removal of the water-side deposits by chemical cleaning. Rigorous monitoring and control of the water chemistry is necessary to prevent high caustic levels.

Hydrogen damage develops from the generation of hydrogen during rapid corrosion of the internal surface of the tube. The atomic hydrogen migrates through the tube steel, where it can react with iron carbide (Fe_3C) to form methane

(CH_4). The larger CH_4 molecules become trapped at the grain boundaries and cause a network of discontinuous internal cracks to be produced. These cracks grow, and some will link up to cause a throughwall fracture.

Hydrogen damage is caused by operation of the boiler with low-pH water chemistry and the concentration of contaminants within the deposits on the internal tube wall. The electrochemical nature of hydrogen damage is shown in Fig. 13. Under acidic attack, H^+ ions are concentrated within the deposit so that the cathode site is localized and adjacent to the anode.

Monitoring and control of boiler water chemistry are important in preventing internal tube deposits and hydrogen damage. The most common method of preventing increased corrosion and hydrogen damage is to develop operating guidelines for boiler water and particularly for the action to be taken when the boiler water is outside the guidelines (see the section "Boiler Water and Steam Chemistry" in this article). For example, chemical cleaning should be immediately considered when the boiler water pH has been below 7 for more than 1 h.

Pitting. Boiler tube failures caused by pitting or localized corrosion result from oxygen attack or acid conditions on the internal surfaces of the boiler tube. The localized corrosion produces perforations of the tube wall when a small area on the tube becomes anodic to the rest of the tube surface and preferentially corrodes. The anodic condition can develop from exposure of the tube to water with high acid or oxygen concentrations or at crevices.

The oxygen in the boiler water reacts with and rapidly removes the hydrogen produced at the cathodes (Fig. 13), thus accelerating the cathodic reaction. The oxygen will also oxidize the Fe^{2+} ion. Hematite iron oxide (Fe_2O_3) will form as the corrosion product and cover the craterlike perforation in the tube wall.

Pitting failure can occur anywhere in the boiler, particularly in economizers, superheaters, reheaters, and the nonheated portions of water wall tubes. For full protection against oxygen pitting during shutdown, it is necessary to keep the boiler full with hydrazine-treated water and blanketed or capped with nitrogen. Oxygen pitting attack of economizer tubing can be prevented by proper operation of deaerators and their heaters; by elimination of air in-leakage paths in low-pressure feedwater heaters, extraction piping, and condensate piping; and by injection of an oxygen scavenger chemical.

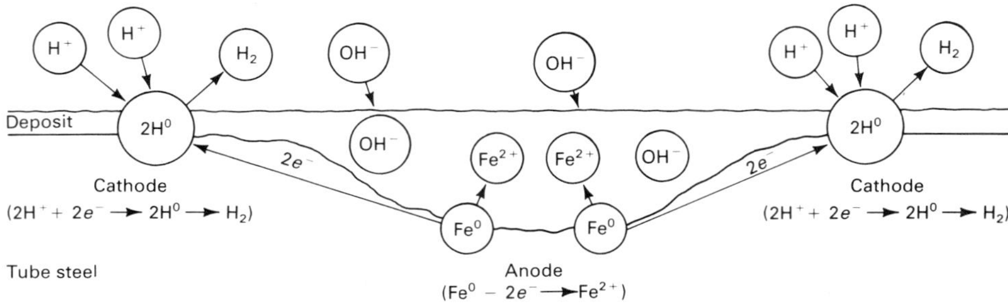

Fig. 12 Schematic of the mechanism of caustic corrosion. A caustic upset in boiler water conditions can result in concentration of OH^- ions in the deposit and generation of hydrogen gas into the boiler water. Source: Ref 5

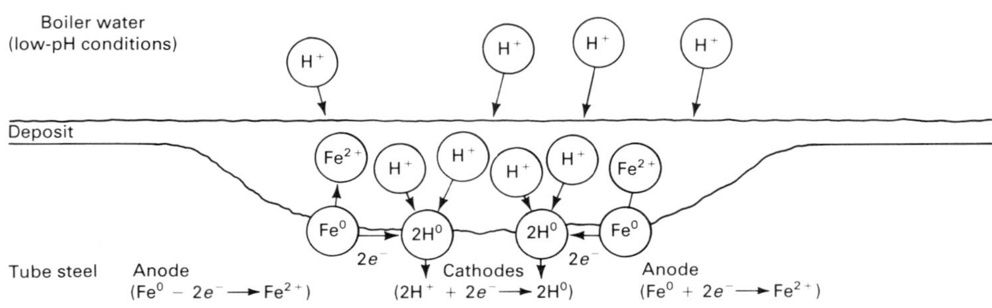

Fig. 13 Schematic of the corrosion mechanism of hydrogen damage. Acid (low-pH) boiler water conditions can result in concentration of H^+ ions in the deposit and generation of hydrogen atoms into the tube material. Source: Ref 5

Superheaters and Reheaters

Unlike water wall and economizer tubes, the tubes in the superheater and reheater are designed for a finite life, which is not based on failure but on a conservative creep criterion. For steam temperatures of 540 °C (1000 °F), tube metal temperatures can exceed 600 °C (1110 °F), especially in the last stages of the superheat and reheat sections. Tube materials can vary from carbon steels to low-chromium ferritic (ASME SA-213, grades T-11, T-22) to austenitic stainless steels (ASME SA-213, grades T304, T321, and T347).

Normal Protective Oxide Growth. Chromium-containing steels exposed to high-pressure power station steam at metal temperatures from 550 to 650 °C (1020 to 1200 °F) initially form a spinel-type oxide consisting of two layers whose relative thicknesses depend on the chromium content of the steel. In superheaters and reheaters, however, the outer layer is essentially pure Fe_3O_4, and the inner layer is an iron-chromium spinel-type oxide containing the steel alloying elements.

Breakdown of the normal protective oxide in superheaters and reheaters occurs similarly to that in water walls and economizers. If the normal protective growth continues as described above, no problems will exist. There are two ways in which this protection can break down; the first is excessive growth and exfoliation, and the second is SCC.

Excessive Growth and Exfoliation. Although excessive internal growth of oxide can elevate the tube temperature, a much more serious problem is solid-particle erosion of the turbine components when the oxide exfoliates and travels into the turbine. For ferritic tube materials, corrosion initially occurs parabolically with time, but at a later stage, it can become linear, depending on the temperature.

The duplex scales produced during the parabolic period are approximately of equal thickness and are parallel sided. Any deviation of the corrosion rate from parabolic is associated with a multilayer scale. Exfoliation occurs only in ferritic materials when this multilayer growth occurs. Exfoliation of the scale is related to the stresses that are induced in the scale during temperature cycles by differences in thermal expansion between the scale and the tube.

Austenitic stainless steels generally corrode more slowly than ferritic stainless steels under the same steam conditions; as a result, the scales are somewhat thinner. This is because of the higher chromium content in austenitic stainless steels.

The initiation of scale exfoliation is not as easily defined for austenitic stainless steels, but may correspond either to the breakdown of the inner layer into a laminated structure or to some critical level of defects (voids) at the oxide/oxide interface. The cause of exfoliation is again due to the difference in thermal expansion between the scale and the tube. Replacement of existing superheater or reheater tubes with chromate-treated, chromized, or stainless steels (for the original ferritic tubes) is an effective method of reducing scale exfoliation.

Stress-corrosion cracking failures in a boiler usually occur in the austenitic stainless steels used for superheater and reheater tubing. However, SCC failures can occur in some ferritic reheater tubing when high levels of caustic are introduced from the desuperheating or attemperator spray water station.

Conditions for SCC initiation and propagation arise from the contamination of boiler water or steam, the introduction of high tensile stresses from service conditions, or the production of high residual tensile stresses during fabrication and assembly. Common contaminants are chlorides and caustic, which result in transgranular cracks, and sulfur from chemical cleaning, which results in intergranular cracks.

Boiler Water and Steam Chemistry

The cycle chemistry is of paramount performance for all of the corrosion mechanisms that occur in water walls, economizers, reheaters and superheaters, operation, and control. Through proper understanding of these mechanisms, as well as the others around the cycle, it is possible to define water and steam quality limits in order to eliminate or lessen corrosion. The reliability and availability of the equipment can be improved by following water and steam quality guidelines for all types of operation (baseload, cycling, and peaking), by adopting target and action levels, and by taking the appropriate action.

Corrosion of Steam Turbines

Otakar Jonas, Consultant

The steam turbine is the simplest and most efficient engine for converting large amounts of heat energy into mechanical work. As the steam is allowed to expand, it acquires high velocity and exerts force on the turbine blades. Turbines range in size from a few kilowatts for one-stage units to 1300 MW for multiple-stage multiple-component units comprising high-pressure, intermediate-pressure, and up to three low-pressure turbines. For mechanical drives, single- and double-stage turbines are generally used. Most larger modern turbines are multiple-stage axial-flow units. Steam pressure and temperature conditions are governed by the boiler and range from less than 1.4 MPa (200 psi) saturated to more than 35 MPa (5000 psi) at 650 °C (1200 °F) superheated and supercritical (Ref 6).

Turbines are typically built for a 25- to 40-year life. Recently, a small 8-MW turbine built in 1908 was inspected and found to be in excellent condition. During this long life, corrosion and other material damage can accumulate and lead to premature failures. Corrosion usually results from a combination of water chemistry, design, and material selection problems.

From 1971 to 1980, steam turbines (excluding controls) contributed 6.7% of forced outages in United States utility units (Ref 7), which cost the utilities 192 575 GW·h in more than 9000 outages. Outages due to corrosion were a major part of this total. A survey of about 500 utility turbines larger than 100 MW showed that corrosion failures occur in 4 to 5% of operating turbines each year (Ref 8, 9). Considering that it takes an average of 4 years for corrosion to result in a failure, about 20% of operational large turbines are under attack at any given time. A survey of industrial turbines conducted by the American Society of Mechanical Engineers (ASME) found a similar degree of corrosion and deposit problems (Ref 10).

It has been estimated that corrosion losses in utility steam systems amounted to about $1.5 billion of the $70 billion annual cost of corrosion in the United States in 1978 (Ref 11); the losses today are about $3.5 billion. The cost of corrosion of fossil turbine blades in the United States is about $300 million annually (Ref 12). Adding the estimated costs of disk, bolt, bellows, and piping corrosion, the total annual cost of utility turbine corrosion in the United States is about $600 million. The cost of replacement power can be as much as two orders of magnitude higher than that of replacement parts.

Major Corrosion Problems in Steam Turbines

Corrosion fatigue, SCC, pitting, and erosion-corrosion are the primary corrosion mechanisms in steam turbines. Figure 14 and Table 3 show the distribution of corrosion within the turbine. Pitting and corrosion fatigue of turbine blades and SCC of disks are currently the two costliest problems. Much research has been devoted to these two problems (Ref 12-29), and progress is being made both in operation (better steam chemistry) and design (lower stresses, no crevices, limit on maximum strength to reduce susceptibility to SCC).

Statistics reveal interesting correlations. Blade failures are most frequent (in fossil utility turbines) in the L − 1 row, which is immediately before the saturation line (Ref 13-16). In industrial turbines, blade failures are also most frequent in the first row after the stage in which first moisture occurs (Ref 17). Blade pitting and cracking are more frequent in once-through boiler units (these units use all-volatile water treatment and condensate polishers) than in drum boiler units (Ref 18). Statistics from West Germany, the United Kingdom, and Japan indicate a very low incidence of blade failures. A recent survey of 494 utility units in the United States reported that 167 (34%) of the units experienced blade failures between 1970 and 1981. In 1980, 1981, and 1982 the failure rate was about 4.7% (Ref 12).

Stress corrosion of low-pressure turbine disks is specifically distributed for each type of turbine and each disk location, indicating a correlation between cracking and surface temperature and steam condition. This correlation is often related to the impurity concentration by evaporation of moisture. The tendency toward cracking increases with yield strength, stress, and operating temperature (Ref 8, 19-23).

There are pronounced thermodynamic effects that cause the concentration of impurities at surfaces and a turbine component failure. They are most frequent where the metal surface temperature is slightly above the saturation temperature of steam. For the low-pressure fossil utility turbines, this usually occurs on the L − 1 blades and on various surfaces of the last two disks. It is one of the reasons the L − 1 blades in fossil utility turbines have higher failure rates than any other blade row. Another reason for the high L − 1 blade failure rate could be the effects of transonic flow, vibratory stresses, and changing temperatures resulting from an interaction of the shock wave with the Wilson line (periodic destruction of the Wilson line) (Ref 30).

Erosion-corrosion of carbon steel wet-steam piping is of primary concern in industrial and nuclear turbine units. It is most pronounced in carbon steel pipes with high-velocity turbulent flow and low-pH moisture containing high concentrations of CO_2 or other acid-forming anions (Ref 31-36).

Corrosion of other turbine parts is due to one or more of the same causes described above (Table 3). There is a strong SCC low-cycle fatigue interaction in many turbine materials (well recognized in piping) that may be important in station-

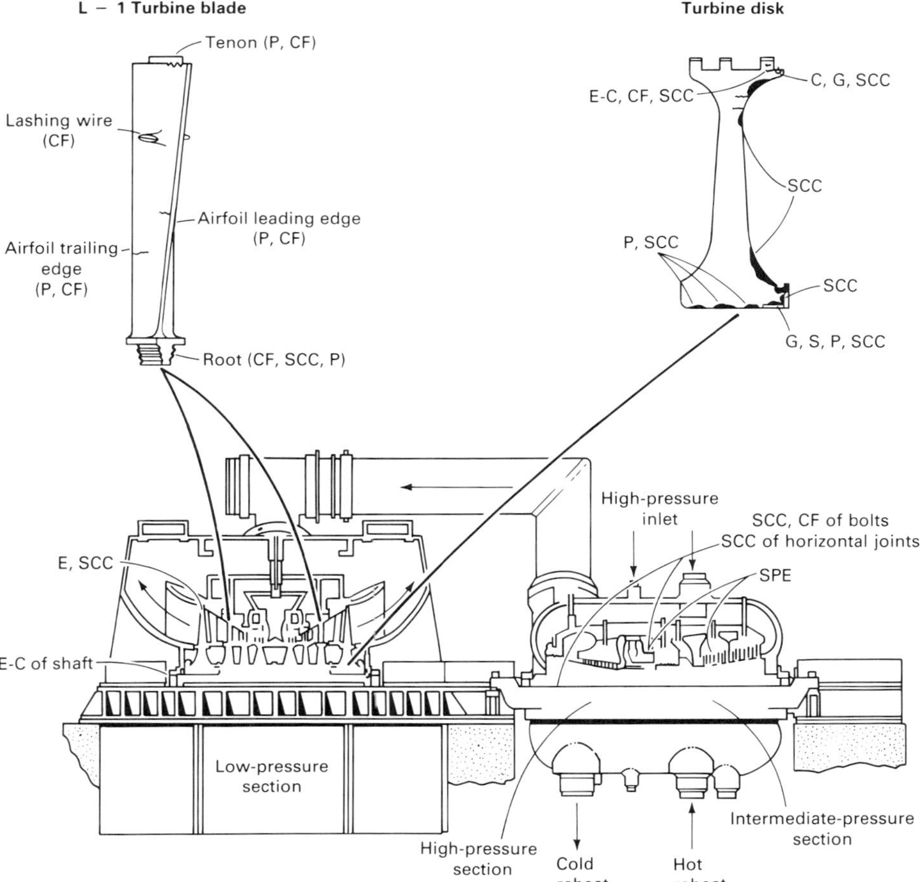

Fig. 14 Schematics showing locations of corrosion in steam turbine components. P, pitting; CF, corrosion fatigue; SCC, stress-corrosion cracking; C, crevice corrosion; G, galvanic corrosion; E, erosion; E-C, erosion-corrosion; SPE, solid-particle erosion

Table 3 Corrosion mechanisms in steam turbine components

Component	Material	Corrosion mechanisms(a)
Rotor	Forged Cr-Mo-V or Ni-Cr-Mo-V low-alloy steel	P, SCC, CF, E
Shell	Cast carbon or Cr-Mo-V low-alloy steel	SCC, E-C
Disks, bucket wheels	Forged Cr-Mo-V, Ni-Cr-Mo-V or Ni-Cr-Mo low-alloy steel	P, SCC, CF, E-C
Dovetail pins	Cr-Mo low-alloy steel	SCC
Blades, buckets	Stainless steels (12Cr or 17-4 PH), copper alloys	P, CF, SCC, E
Bucket tie wires	12Cr stainless steels (ferritic and martensitic)	SCC, P, CF
Shrouds, bucket covers	Stainless steels (12Cr or 17-4 PH)	P, SCC
Stationary blades	AISI type 304 stainless steel	SCC, SCC-LCF
Expansion bellows	AISI types 321 or 304 stainless steels, Inconel 600	SCC, SCC-LCF
Erosion shields	Weld-deposited or soldered Stellite type 6B; hardened blade materials (see above)	SCC, E
Bolts	Incoloy 901, Refractalloy 25, Pyromet 860	SCC, SCC-LCF
Wet-steam piping	Carbon steel	E-C
Valve bushings and stems	13Cr-Mo and other stainless steels	P, OX

(a) P, pitting; SCC, stress-corrosion cracking; CF, corrosion fatigue; E, erosion; E-C, erosion-corrosion; LCF, low-cycle fatigue; OX, oxidation in steam. General corrosion is experienced by all carbon and low-alloy steel components. Solid-particle erosion is experienced in high-pressure and intermediate-pressure inlets (nozzle blocks, stationary and rotating blades, and valves). It is caused by exfoliation of steam-grown oxides in superheater and reheater tubes and in steam pipes.

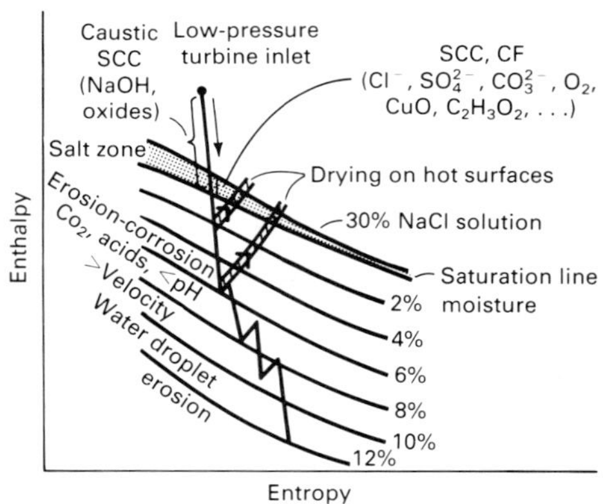

Fig. 15 Mollier diagram showing the low-pressure turbine steam expansion line, thermodynamic regions of impurity concentration (with impurities given in parentheses), and corrosion mechanisms. Source Ref 6 and 9

ary blade, bolt, expansion bellows, and turbine cylinder failures. The highest frequency of turbine bolting failures has been in high- and intermediate-pressure cylinders and nozzle blocks. Stress-corrosion cracking of expansion bellows (intermediate- and low-pressure pipes) used to be a problem in fossil units with high concentrations of NaOH (Ref 37). To find the true causes of corrosion, it is essential to analyze the local temperature, pressure, chemistry, moisture velocity, and stress and material conditions.

Turbine Materials

There is little worldwide variation in materials for blades, disks, rotors, and turbine cylinders, and only a few major changes have been introduced in the last decade. Titanium alloy blades are slowly being introduced for the last low-pressure stages. Also, improved melting practices, control of inclusions and trace elements, and weld repair techniques are being used for disks and rotors. In addition to the materials listed in Table 3, carbon steel is used for low-pressure casings (shells and cylinders), chromium-molybdenum steels and austenitic stainless steels for high-pressure piping, and miscellaneous other materials for other components.

Considering the typical design life of 25 to 40 years and the relatively high stresses, turbine materials perform remarkably well. Turbine steels are susceptible to SCC and corrosion fatigue in numerous environments, such as caustic, chlorides, hydrogen, carbonate-bicarbonate, carbonate-CO_2, acids, and, at higher stresses and strength levels, pure water and steam.

No turbine can permanently tolerate concentrated caustic, sodium chloride (NaCl), or acids. Under such conditions, disk and rotor materials would crack by SCC in a few hundred hours at stresses as low as 10% of the yield strength. Chromium steel (12% Cr) blades may tolerate NaOH, but would pit and crack by corrosion fatigue at very low vibratory stresses (as low as 6.9 MPa, or 1 ksi) in many other corrodents. With high concentrations of impurities for long periods of time, there would always be a weak link in the turbine or within the power cycle that would fail prematurely. Although titanium could be used for blades and would tolerate most of the impurities (except hydroxides), corrosion-resistant materials for the large forgings of rotors and disks are difficult to find because of the cost, forgeability, banding of alloying elements and impurities, and other problems. Plating and surface coatings appear to be only marginal temporary solutions (Ref 38).

In addition to the susceptibility to localized corrosion in concentrated impurities, the low-alloy and carbon steels used extensively in steam turbines have rather narrow ranges of passivity (Ref 39). This makes these steels vulnerable to pH and oxygen excursions and consequent pitting and other forms of localized corrosion during operation and layup. In chloride solutions, the pH region for passivation becomes even narrower. The material properties needed for designing against corrosion, failure analysis, and evaluation of residual life include fracture toughness (K_{Ic}), stress-corrosion threshold stress (σ_{SCC}), threshold stress intensity (K_{ISCC}), crack propagation rate ($\dot{a}_{SCC}$), corrosion fatigue limit, corrosion fatigue threshold stress intensity (ΔK_{th}), corrosion fatigue crack propagation rate (da/dN), pitting rate, and a pit depth limit.

Environment

Corrosive impurities are transported into the turbine steam from the preboiler cycle (in the feedwater) and the boiler as a total (mechanical plus vaporous) carryover (Ref 6, 8, 9). Their major sources include condenser leaks, air inleakage, makeup water, and improperly operated condensate polishers. The turbine environment is controlled through a control of impurity ingress and various feedwater, boiler water, and steam chemistry limits (Ref 6, 8, 40-45). The corrosiveness of the steam turbine environment is caused by one or more of the following:

- Concentration of impurities from low part per billion levels in steam to percent levels on surfaces and the formation of concentrated aqueous solutions (concentration by deposition or evaporation of moisture)
- Insufficient pH control (in both acid and alkaline regions)
- High-velocity, high-turbulence, and low-pH moisture

The situation is illustrated in Fig. 15, which shows a Mollier diagram with a typical turbine steam expansion line, the thermodynamic regions of impurity concentrations (NaOH and salts), and the resulting corrosion. The conditions on hot turbine surfaces (in relation to the steam saturation temperature) can shift from the wet-steam region into the salt zone and above. This is why SCC of disks often occurs in the wet-steam regions (Ref 19, 23, 29), and it emphasizes the need to consider local surface temperatures in design and failure analysis. The surfaces may be hot because of heat transfer through the metal (Ref 9, 22, 29) or because of the stagnation temperature effect (zero flow velocity at the surface and conversion of kinetic energy of steam into heat) (Ref 46). Depending upon their vapor pressures, impurities can be present as a dry salt or as an aqueous solution. In the wet-steam region, they are either diluted by moisture or could concentrate by evaporation on hot surfaces.

The steam impurities that are of most concern include chlorides, sulfates, fluorides, carbonates, hydroxides, organic and inorganic acids, oxygen, and CO_2. Their behavior in turbine steam and deposits is well documented (Ref 6, 9, 14, 30, 46-53). There are strong synergistic effects and interactions with metal oxides.

In addition to corrosion during operation, turbines can corrode during:

- Manufacture (machining fluids and lubricants)
- Storage (airborne impurities and preservatives)
- Erection (airborne impurities, preservatives, and cleaning fluids)
- Chemical cleaning (storage of acid in condenser hotwells)
- Nondestructive testing (cleaning and testing fluids)
- Layup (deposits plus wet air)

Many of the above substances may contain high concentrations of sulfur and chlorine, which could form acids upon decomposition. Decomposition of typical organics—for example, carbon tetrachloride (CCl_4)—occurs at about 150 °C (300 °F). Therefore, the composition of all of the above substances should be controlled (maximum of 50 to 100 ppm S and 50 to 100 Cl has been recommended), and most of them should be removed before operation. Molybdenum disulfide (MoS_2) is being increasingly implicated as a corrodent in power system applications (Ref 54-58).

Layup corrosion increases rapidly when the relative humidity of the air reaches about 60%. When deposits are present, layup corrosion can be in the form of pitting or SCC.

Design

Because of the long design life, steam turbines undergo a limited prototype testing in which the long-term effects of material degradation, such as corrosion, creep, and low-cycle fatigue, cannot be accurately simulated. When development was slow, relatively long-term experience was transferred into new products. Because of the rapid development of new turbine types, larger sizes, and new power cycles and water treatment practices during the last 25 years, the experience was short and limited, and some problems were developed that still need to be corrected and considered in new designs and redesigns. Design disciplines that affect turbine corrosion can be separated into four parts:

- Mechanical design (stresses, stress concentrations, and stress intensity, K_I)
- Heat transfer (surface temperatures and heated crevices)
- Flow (moisture velocity, location of the salt zone, stagnation temperature, and interaction of shock wave with Wilson line)
- Physical shape (crevices, obstacles to flow, and surface finish)

The various aspects of design against environment-sensitive fracture are well documented (Ref 9, 12, 13, 19-21, 24, 29, 30, 59-68), but only a few of these references deal with the complexity of the problem (Ref 9, 13, 14, 20, 60, 65-68). Probabilistic approaches are discussed in Ref 67 and 69. Problems with such approaches include the lack of a large number of statistical data on corrosion properties of materials, service stresses, and environments. The mechanical design concepts for avoiding turbine corrosion should include evaluation of safety factors against σ_{SCC}, K_{ISCC}, $\dot{a}_{SCC}$, (da/dN), ΔK_{th}, corrosion fatigue limit, pitting rate, and a pit depth limit.

A more conservative approach should be taken for high multiaxial tensile stresses, strain (mechanical or thermal) fluctuations, other mechanical interactions, and large component sizes. True residual stresses, both microscopic and macroscopic, should also be considered. Implementation of the above concepts requires that corrosion testing generate quantitative data that are useful to the designer.

Hot Corrosion in Coal- and Oil-Fired Boilers

Ian G. Wright
Battelle Columbus Division

Corrosion from the firing of coal or oil is essentially related to specific impurities in the fuels, which can lead to the formation of nonprotective scales or can disrupt normally protective oxide scales. The relevant impurities in coal are sulfur (about 0.5 to 5.2% in United States coals), sodium (0.01 to 0.7%), potassium (0.2 to 0.7%), and chlorine (0.01 to 0.28%). In oil, the important impurities are sodium, which from United States refineries may range up to 300 ppm by weight; vanadium (up to 150 ppm by weight); and sulphur (0.6 to 3.6%). During combustion, these impurities can be melted or vaporized and will deposit upon contact with surfaces at temperatures lower than the condensation temperatures of the specific species. This provides a mechanism for the accumulation of deposits of fly ash on cooled surfaces downstream of the burners (Ref 70, 71).

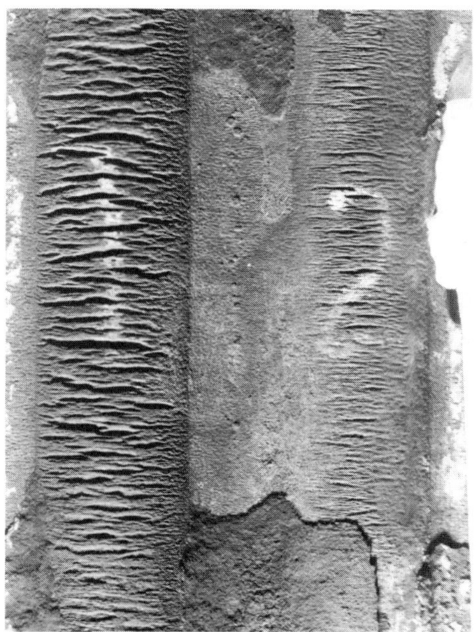

Fig. 16 Example of grooving in water wall tubes. Note the thick, adherent scale remaining in some areas.

Fig. 17 Cross section of circumferential cracks on the fire-side surface of a water wall tube. Source: Ref 72

Water wall fire-side corrosion is generally found in regions around the burners. The thick, hard external scales formed may be quite smooth, but often exhibit cracks or grooves that can resemble an alligator hide. Cracking or grooving is usually circumferential, and it is more common in supercritical than subcritical boilers. It occurs in the areas of the water walls that receive the highest heat flux and is apparently a result of superimposed thermal stress. Figure 16 illustrates the appearance of grooving. Figure 17 shows the grooves to be sharp-pointed cracks filled with corrosion product (mostly iron oxide) but with a central core of iron sulfide. Figure 18 shows a cross section of an ASME SA-213, grade T-11 water wall tube that has formed a thick, smooth scale. The outer scale is a mixture of iron sulfide and iron oxide, with the chromium from the alloy dispersed in the inner layer. The major causes of corrosion of the water wall tubes are, first, the reducing (substoichiometric) conditions caused by impingement of incompletely combusted coal particles and flames and, second, molten salt or slag-related attack.

Gas Phase Corrosion

Reducing atmosphere corrosion may be a result of direct reaction of the water wall tubes with a substoichiometric gaseous environment containing sulfur or with deposited, partially combusted char containing iron pyrites, which gives rise to a very localized corrosive environment. Reducing conditions have two main effects on corrosion. First, they tend to lower the melting temperature of any deposited slag, which increases its ability to dissolve the normal oxide scales on the tubes, and second, the stable gaseous sulfur compounds under these conditions include H_2S (Ref 73-76), which is considerably more corrosive than the SO_2 that predominates under oxidizing conditions. Under reducing conditions, iron sulfide is the expected corrosion product on iron, rather than the oxides (Fe_3O_4 and Fe_2O_3). Sulfide scales allow significantly higher rates of transport of iron cations than oxides do and so are less protective. However, Cr_2O_3, which is a very protective oxide, would be expected to form on chromium under reducing conditions; therefore, alloys capable of forming Cr_2O_3 scales, such as AISI type 310 stainless steel, Inconel 671, or chromized coatings, would be expected to exhibit good corrosion resistance in this application.

Molten Salt Corrosion

Slag-related attack takes several forms. Local disruption of the normal oxide film on the wall tubes by intrusion of molten slag can lead to accelerated oxidation or (if sulfur species are present in the slag) to oxidation-sulfidation. In coal-fired boilers, alkali sulfates deposited on the water walls may react with SO_2 or SO_3 to form pyrosulfates, such as potassium pyrosulfate ($K_2S_2O_7$) and sodium pyrosulfate ($Na_2S_2O_7$), or possibly complex alkali-iron trisulfates. The latter compounds are formed in thicker deposits after long periods of time at about 480 °C (900 °F) (Ref 77). The K_2SO_4-$K_2S_2O_7$ system forms a molten salt mixture at 407 °C (764 °F) when the SO_3 concentration is 150 ppm. The corresponding sodium system can become liquid at 400 °C (750 °F), but it requires about 2500 ppm SO_3 for this to

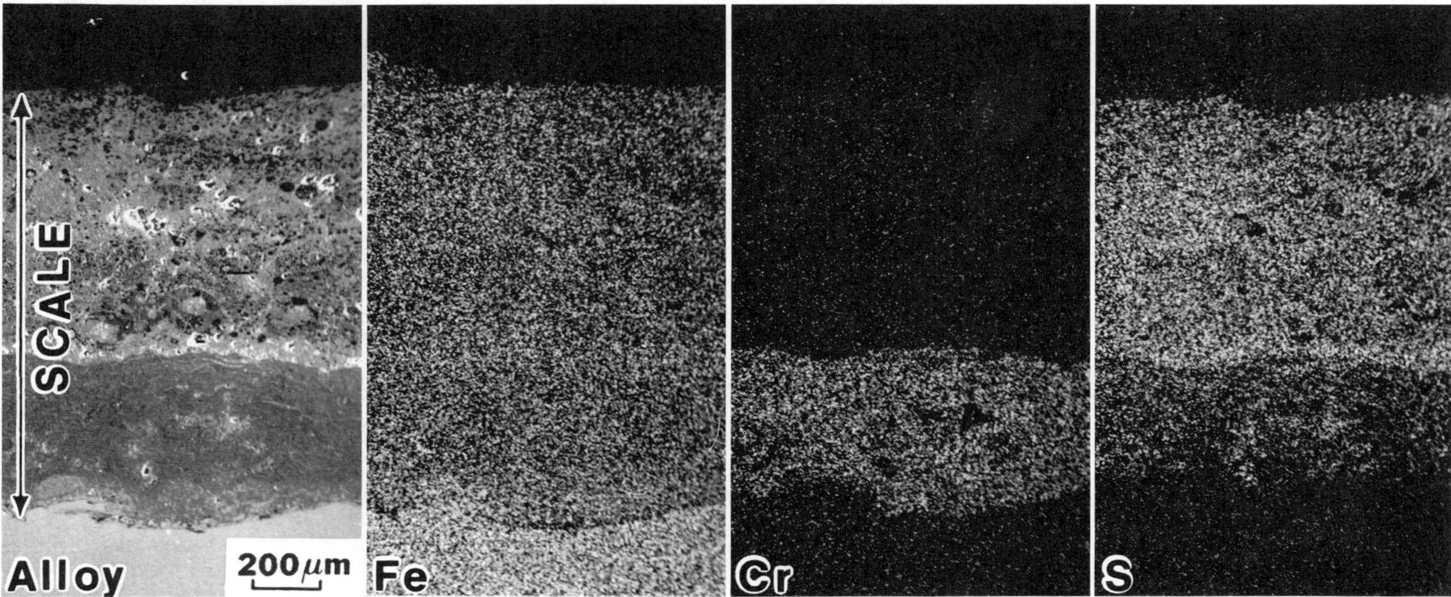

Fig. 18 Cross section of the fire-side face of a water wall tube subjected to reducing (substoichiometric) combustion conditions. From left to right: optical micrograph of deposit and x-ray elemental dot maps for iron, chromium, and sulfur, respectively

occur; such levels of sulfur oxides are likely only under deposits (Ref 70). Thus, molten salt attack on the tube metal by $K_2S_2O_7$ is more likely and occurs according to the reaction:

$$K_2S_2O_7 + 3Fe \rightarrow FeS + Fe_2O_3 + K_2SO_4 \quad \text{(Eq 2)}$$

By such a mechanism, $K_2S_2O_7$ can react aggressively with any protective iron oxide scales on the tubes and lead to accelerated wastage through fluxing of the oxides and attack of the substrate metal. Differential scanning calorimetry of samples of deposits taken from water wall tubes typically indicates melting points in the range of 335 to 410 °C (635 to 770 °F) (Ref 78).

The role of chlorine in fire-side corrosion is difficult to define. However, it seems likely that under reducing conditions HCl or NaCl can render the oxide scales less protective by causing them to blister or crack or by reactions with the oxides or base metal to form volatile products (Ref 71).

Molten salt related corrosion of water walls is seldom encountered in oil-fired boilers. This is presumably related to the absence of chlorine in the oil and to the lower ash content of oil compared to coal (about 0.5% in oil, and up to 35% in U.S. coals); therefore, the deposits formed on the water walls of oil-fired boilers are very thin and do not provide the necessary conditions for the formation of pyrosulfates. In addition, the tube metal temperature in the water wall region of a furnace is usually below 400 °C (750 °F), which is lower than the melting points of any compounds possible between sodium and vanadium oxides. Therefore, any compounds formed are unlikely to melt and pose a corrosion threat under normal operating conditions.

Prevention of Corrosion

Solutions to fire-side corrosion of furnace water walls are available from changes in operating procedures and changes in tube materials. Where the corrosion results from the presence of reducing conditions near the water walls, operational actions include adjusting the air and fuel distribution to individual burners and among burners in order to promote better mixing and more uniform combustion conditions, as well as resetting to design specification the coal fineness delivered to the burners from the milling plant. Flame impingement can be rectified by changing the characteristics of the offending burners through adjustment of secondary air registers to control air flow and degree of swirl. The most accurate way of making such burner adjustments is by monitoring the composition of the gas close to the affected water walls to ensure that oxidizing conditions (or acceptably low carbon monoxide, CO, levels: <1%) are achieved.

Another method of countering reducing conditions near the water walls is to introduce a flow of air along the walls through openings in the membrane between water wall tubes. This is often referred to as air blanketing or curtain air, and it should also be implemented and adjusted in conjunction with local monitoring of the gas composition.

Furnace wall corrosion can also be lessened by reducing the levels of the chemical species in the coal that are responsible for corrosion. Approximately one-half the sulfur and alkali metal content of coal can be removed by standard coal-washing procedures. However, washing generally does not remove the chlorine-containing species from coal; therefore, the net effect is that the chlorine content of washed coal is increased. An alternative strategy is to blend the coal to reduce the average content of corrosive species. Both of these strategies may involve some increase in fuel costs, which must be weighed against the increased tube lifetimes. The same approaches for control of reducing environments near the water walls described for coal-fired boilers can be applied to oil-fired boilers.

Materials solutions to water wall fire-side corrosion problems involve either direct replacement with tubes of a more corrosion-resistant alloy or the application of a corrosion-resistant alloy as a cladding on the affected tube. Replacement of the tubes with the same material and the use of regular wall thickness monitoring where wastage rates are only slightly greater than allowable have been recommended to give the desired tube life (Ref 79). For more severe wastage, little difference in performance has been found for alloys containing less than 9% Cr. In this case, the materials choice is probably between thicker-walled carbon steel tubes and tubes with a coextruded outer layer of a high-chromium alloy, such as AISI type 310 stainless steel or 50Cr-50Ni. The technique of cladding a tube fabricated from the alloy typically chosen for the application (based on strength considerations) with an outer layer of a corrosion resistant alloy by coextrusion can provide a cost-effective solution to this problem. The Central Electricity Generating Board has found coextruded tubes of type 310 on low-carbon steel to be economically viable for base-loaded units and has used such tubes in boilers since 1974 (Ref 80).

Increased corrosion resistance can also be gained by enriching the surfaces to be protected with such elements as chromium or aluminum. This can be achieved by diffusion treatments or by the spraying of metal overlay coatings. Although pack aluminizing can be applied commercially to tubes up to about 6 m (20 ft) long (Ref 81), there are as yet no reports of utility boiler experience with such materials. Flame or plasma spraying has been used to apply high-chromium, high-aluminum, iron-chromium-aluminum alloy compositions to water wall tubes. Although such alloys have shown good corrosion resistance in laboratory tests, problems have been experienced in practice as a result of a lack of reproducibility of coating application techniques.

Hot Corrosion in Boilers Burning Municipal Solid Waste

H.H. Krause
Battelle Columbus Division

The corrosion problems experienced in boilers fueled with municipal refuse are different from those encountered with fossil fuels in that chlorine rather than sulfur is primarily responsible for the attack. The average chlorine content of municipal solid waste is 0.5%, of which about one-half is present as polyvinylchloride (PVC) plastic. The other half is inorganic, principally NaCl. The chlorine in the plastic is converted to hydrochloric acid (HCl) in the combustion process. The inorganic chlorides are vaporized in the flame and ultimately condense in the boiler deposits or pass through the boiler with the flue gases. Zinc, lead, and tin in the refuse also play a role in the corrosion process by reacting with the HCl to form metal chlorides and/or eutectic mixtures with melting points low enough to cause molten salt attack at wall tube metal temperatures.

Gas Phase Corrosion

Hydrochloric acid alone has little effect on carbon steel at temperatures below 260 °C (500 °F) (Ref 82). However, in the presence of excess air in the boiler, the reaction to form ferrous chloride ($FeCl_2$) on the steel surface occurs more readily. This corrosion product is stable at water wall tube temperatures, and the curve for metal wastage versus time is parabolic. However, at a metal temperature of about 400 °C (750 °F), which may occur in the superheater, the $FeCl_2$ is further chlorinated to the readily volatile ferric chloride ($FeCl_3$) (Ref 83). If the gas temperature in the area exceeds 815 °C (1500 °F), the $FeCl_3$ will evaporate rapidly, and breakaway corrosion can then occur. These effects are shown in Fig. 19, which presents data from corrosion probe exposures conducted in a municipal incinerator.

The corrosion probe results also demonstrated that $FeCl_2$ was formed as a corrosion product at temperatures of 150 to 260 °C (300 to 500 °F); in this range, HCl is not particularly corrosive (Ref 85). This result can be explained only by attack from elemental chlorine rather than HCl. In one study, corrosion of carbon steel by chlorine reached the breakaway condition at 205 °C (400 °F) (Ref 82); therefore, this agent must be responsible for any low temperature attack.

A reducing atmosphere can also contribute to wall tube corrosion. This effect was first observed in European incinerators, in which excessive corrosion of side walls and furnace corners was found to be associated with reducing conditions in addition to chloride deposits. A recent study investigated the combined effect of CO and HCl on corrosion (Ref 86). Corrosion rates were linear with time in a simulated flue gas containing 400 ppm HCl, 10% CO, 10% H_2O, 0.5% SO_2, and the balance nitrogen. At 400 °C (750 °F), the slope of the metal loss versus time curve was 1.14 mm/yr (45 mils/yr). Without HCl in the gas mixture, a parabolic rate curve was obtained that leveled off at a maximum metal loss of only 127 μm (5 mils). The synergistic action of HCl and CO in promoting corrosion was attributed to disruption of the oxide layer on the metal surface, followed by spalling of the oxide, thus exposing the metal to further oxidation.

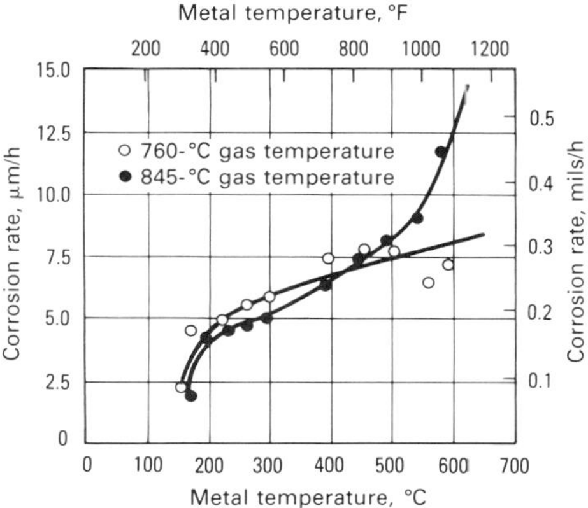

Fig. 19 Corrosion rates at two temperatures of carbon steel in a boiler burning municipal refuse. Source: Ref 84

Molten Salt Corrosion

Incinerator boiler deposits usually contain several percent of zinc and lead, with a lesser amount of tin. In combination with chlorine, these elements can form low-melting compounds or eutectic mixtures that would be molten at tube metal temperatures. Thus, for example, stannous chloride ($SnCl_2$) melts at 246 °C (475 °F) and zinc chloride ($ZnCl_2$) at 283 °C (541 °F). The eutectic mixture of 50.2% lead chloride ($PbCl_2$) and 49.8% $FeCl_3$ melts at 175 °C (347 °F). Examination of the layer of scale at the corroded metal surface of boiler tubes and probe specimens by the electron microprobe and by energy-dispersive x-ray analysis has shown these metals to be associated with chlorine. Consequently, molten salt attack by these metal chlorides can occur and will be more severe than gas phase corrosion.

A recent investigation of an incinerator wall tube that was corroding at a rate of 2 mm/yr (80 mils/yr) showed that zinc and sodium were both associated with chlorine in the deposit (Ref 87). The presence of NaCl was confirmed by x-ray diffraction. However, the high corrosion rate could not be accounted for in terms of attack by NaCl or HCl. Consequently, laboratory tests were conducted to demonstrate that the corrosion could be caused by the eutectic mixture of 84% $ZnCl_2$ and 16% NaCl, which has a melting point of 262 °C (504 °F). After a 336-h exposure to this mixture at 315 °C (600 °F), carbon steel had a corrosion rate of 23 mm/yr (910 mils/yr), indicating that such molten salt attack was the likely mechanism in the incinerator. There is as yet no evidence for participation of $SnCl_2$ in the incinerator corrosion reactions. However, its low melting point and the possibility of forming a eutectic mixture with NaCl that melts at 199 °C (390 °F) make it a likely contributor to molten salt corrosion.

Prevention of Corrosion

Most of the methods for preventing incinerator wall tube corrosion exact some penalty in boiler efficiency. The practice of studding the tubes and covering them with silicon carbide refractory has been widely used in European incinerators, but this remedy reduces heat transfer. Increasing overfire air or blanketing the walls with air to prevent reducing conditions in the flue gas has been effective, but either approach will reduce boiler efficiency. Lowering tube metal temperatures by operating at lower steam pressure also has a cost in efficiency.

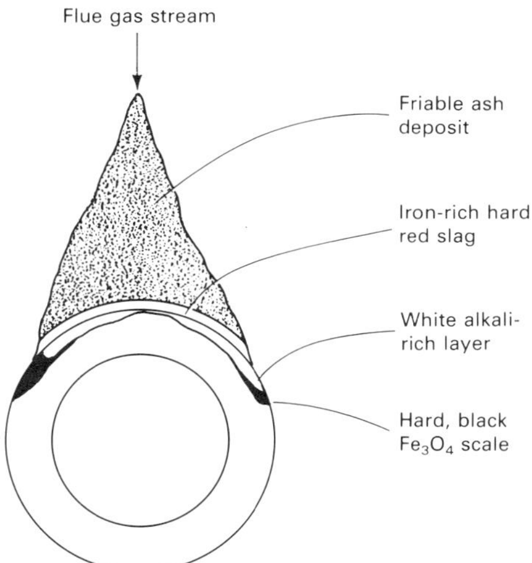

Fig. 20 Deposit layers on a corroding superheater or reheater tube

However, upgrading the boiler tube material to a corrosion-resistant alloy does not involve an efficiency penalty. Although capital costs will be greater, the extended tube life resulting from the use of more resistant alloys can offset the initial expense and can be a cost-effective solution to the problem. Extensive corrosion probe studies in municipal incinerators showed that in the temperature range of 150 to 315 °C (300 to 600 °F) a number of alloys provided good performance in resisting high-temperature corrosion (Ref 88). In decreasing order, the better alloys were Incoloy 825; AISI types 446, 310, 316L, 304, and 321 stainless steels; and Inconel alloys 600 and 601. However, when subjected to moist deposits, simulating boiler downtime conditions, all of the austenitic stainless steels underwent chloride SCC. The type 446 stainless steel, Inconel 600, and Inconel 601 suffered pitting. Consequently, unless the boilers were to be maintained at a temperature above the HCl dew point during downtime, only Incoloy 825 was recommended.

The laboratory tests discussed in Ref 87 showed that Inconel alloys 600, 625, and 690, Incoloy 800, and type 304 and 309 stainless steels were resistant to the $ZnCl_2$-NaCl mixture at 315 °C (600 °F). In addition, an overlay of Inconel 625 applied to the carbon steel tubes in the incinerator offers promise. Recent field tests conducted at a municipal waste incineration boiler indicated that alloy 556 (iron-nickel-cobalt-chromium) has excellent resistance in such applications (Ref 89). The test rack was exposed to flue gases at 800 °C (1475 °F) for 950 h. Maximum depths of attack, that is, metal loss and maximum internal penetration, were 0.21 mm (8.2 mils) for alloy 556, 0.34 mm (13.4 mils) for AISI type 309 stainless steel, 0.69 mm (27 mils) for Hastelloy X, and more than 1.1 mm (45 mils) for Inconel 690. The Inconel 690 coupon was perforated. Sulfidation was reported to be the major mode of attack. Consequently, Inconel 625, Incoloy 825, alloy 556, and possibly Incoloy 800 are the preferred materials solutions to wall tube corrosion in municipal incinerators.

Corrosion of Superheaters and High-Temperature Air Heaters

John Stringer
Electric Power Research Institute

Corrosion of superheaters and air heaters is inevitable in coal-fired boilers and in fluidized-bed combustors. This section will discuss factors that cause corrosion as well as measures that can be taken to minimize corrosion in these applications.

Superheater Corrosion in Coal-Fired Boilers

The fire-side corrosion of superheaters in pulverized coal-fired utility boilers is one of the principal problems that have limited main steam temperatures to 540 °C (1000 °F) for the last 30 years. Allowing for the temperature decrease through the tube walls and the typical non-uniformities in the temperature distribution across a superheater bank, this corresponds to an outer maximum metal surface temperature of approximately 620 to 650 °C (1150 to 1200 °F). The metals and alloys used for superheater tubing are primarily determined by the ASME Boiler Code, which is concerned only with the strength of the materials.

For typical steam pressures and tube wall thicknesses, Table 4 shows the maximum-use temperature for several boiler alloys for which code standards exist. These are the wall midsection temperatures, which are typically 25 °C (45 °F) lower than the outer surface temperature. A further limitation is imposed by the oxidation resistance of the steels. The ASME code does not mention oxidation, and some allowance can of course be made by increasing the tube wall thickness. However, for typical conditions, Table 4 also lists the maximum metal surface temperature that can be tolerated from an oxidation wastage point of view.

Table 4 Temperature limits of superheater tube materials covered in ASME Boiler Codes

Material	Maximum-use temperature: Oxidation/graphitization criteria, metal surface(a), °C	°F	Strength criteria, metal midsection, °C	°F
SA-106 carbon steel	400–500	750–930	425	795
Ferritic alloy steels				
0.5Cr–0.5Mo	550	1020	510	950
1.25Cr–0.5Mo	565	1050	560	1040
2.25Cr–1Mo	580	1075	595	1105
9Cr–1Mo	650	1200	650	1200
Austenitic stainless steel				
Type 304H	760	1400	815	1500

(a) In the fired section, tube surface temperatures are typically 20 to 30 °C (35 to 55 °F) higher than the tube midwall temperature. In a typical U.S. utility boiler, the maximum metal surface temperature is approximately 625 °C (1155 °F).

The fire-side corrosion discussed in this section is the process by which the wastage rate is increased above these design rates. The enhanced corrosion is generally associated with the formation of a deposit on the tube walls. Although the presence of a deposit does not necessarily indicate accelerated corrosion, accelerated corrosion is never experienced in the absence of such a deposit.

Nature of the Deposit. Figure 20 shows an illustration of the deposit formed on superheater tubes. The deposit may be several inches thick, with the outer layers formed of loosely sintered fly ash. The outer part of this deposit will be essentially at the gas temperature. The white inner layer is rich in alkali sulfate. The black layer is largely Fe_3O_4. Sulfur prints indicate that sulfide is present at the metal surface.

It has been shown that deposition of alkali sulfate, as well as corrosion, is often a more or less linear function of the chlorine content of the coal. It is now generally believed that chlorine plays no direct role in the corrosion process; instead, it acts as a release agent for the alkali metals in the coal.

It is generally believed that the salt deposit must be molten for accelerated corrosion to occur. The melting point of sodium sulfate (Na_2SO_4) is 884 °C (1623 °F); that of potassium sulfate (K_2SO_4) is 1069 °C (1956 °F). The minimum melting point of a mixture of the two is 823 °C (1513 °F). Because the molten phase must be close to the metal, there is no possibility that temperatures as high as this could be attained. In the deposit on tubes cooled to room temperature, a complex sulfate having the general formula $(Na,K)_3Fe(SO_4)_3$ has often been identified. This phase has a minimum melting point of 554 °C (1029 °F) at a sodium:potassium ratio of 2:3. However, it is relatively unstable and at 540 °C (1000 °F) will decompose to the solid constituent oxides unless the local SO_3 partial pressure exceeds 25 Pa (0.004 psi). This is approximately five times the normal SO_3 partial pressure within a boiler, but it has been shown that the local partial pressure may be significantly higher near the corroding surface.

Corrosion beneath the deposit is at a maximum toward the edges of the deposit at the 5 and 7 o'clock positions. The metal temperature will be highest under the deposit at these locations because beneath the thick part of the deposit the metal surface will be insulated from the hot gas. Generally, corrosion consists of pits with relatively smooth metal/scale interfaces. There is little evidence of internal penetration or internal sulfidation. The pits eventually overlap to produce an apparently general corrosion.

It is difficult to obtain much information on the influence of different factors on the corrosion from in-service data because the corrosion is very specific to a particular boiler burning a particular fuel under a definite set of conditions. In addition, different materials are usually not exposed simultaneously. On the basis of laboratory tests, it appears that the accelerated attack is present only for metal temperatures in the range of 560 to 700 °C (1040 to 1290 °F), with a maximum at approximately 670 °C (1240 °F). It is believed that the minimum temperature corresponds to the solidus temperature of the salt deposit and that the maximum is the dissociation temperature of the low-melting complexes. Figure 21 shows the relative rates of attack for a range of typical boiler tube materials, including some coating and cladding alloys.

For a given metal temperature, the corrosion rate increases with gas temperature. Corrosion

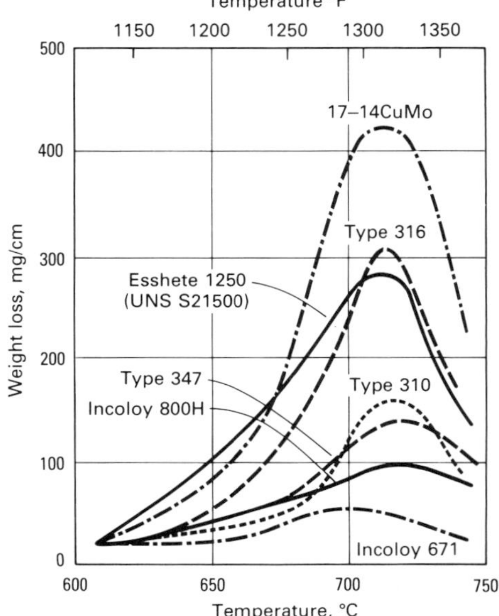

Fig. 21 Corrosion rates of alloys in a laboratory test using synthetic ash (37.5 mol% Na_2SO_4, 37.5 mol% K_2SO_4, and 25 mol% Fe_2O_3) in a synthetic flue gas (80% nitrogen, 15% CO_2, 4% oxygen, and 1% SO_2 saturated with water). Exposure time: 50 h

rate increases by a factor of three at a metal temperature of 650 °C (1200 °F) as the gas temperature rises from 800 to 1400 °C (1470 to 2550 °F).

Corrosion Prevention. Empirical procedures are available for predicting the probable corrosivity of a given coal. A known corrosive coal may be blended with another to produce a less corrosive ash, and it may also be possible to use additives, such as limestone. If it is known at the design stage that a corrosive coal must be burned, the superheater can be positioned further back along the gas path where the gas temperature is lower. Such positioning also reduces the amount of corrosive material reaching the superheater, because some corrodent is deposited on upstream tubing.

If control of coal chemistry or design changes are not feasible, corrosion-resistant materials must be used. For a boiler that has exhibited corrosion, it may be sufficient to wrap strips of AISI type 310 stainless steel around the leading tubes in a platen; this is termed bandaging. If bandaging is insufficient, it may be necessary to replace the finishing superheater tubing with a coextruded material, in which the inner layer is a strong coded alloy and the outer layer is a corrosion-resistant material, such as type 310 stainless steel or Incoloy 671 (Ni-50%Cr). Over the years, there have been many efforts to develop coating systems that could be applied in the field by such techniques as plasma spraying. However, the results have generally been disappointing, although for a particular situation this may still be a cost-effective approach.

Corrosion of Superheaters and High-Temperature Air Heaters in Fluidized-Bed Combustors

A recent development as an alternative to the pulverized coal-fired boiler is the fluidized-bed combustor. In this unit, the coal is burned in a fluidized bed of particles; very good combustion efficiency is achieved at relatively low temperatures (typically 850 to 900 °C, or 1560 to 1650 °F). This results in low NO_x emissions. In addition, sulfur can be captured in the bed by adding an absorbant, such as CaO.

Because of the low combustion temperature, it was anticipated that there would be no ash fusion, little or no ash deposition on the in-bed heat-exchanger surface, and limited alkali release. It was therefore assumed that in-bed fire-side corrosion of superheaters by a mechanism similar to that described in the preceding section would be unlikely. This appears to be true. However, a different mechanism of corrosion does present a potential problem; this will be described in the section "Corrosion Mechanism" in this article. In addition, the in-bed environment is potentially erosive, and the combination of erosion and corrosion can result in accelerated wastage.

The In-Bed Environment. The bed consists of particles with an average size typically close to 0.8 mm (30 mils). The major constituent is $CaSO_4$, and there are relatively minor amounts of coal ash, still smaller amounts of unreacted CaO, and typically less than 2% C. At metal temperatures above approximately 500 °C (930 °F), reddish-brown deposits form that consist mostly of $CaSO_4$, with about 15% coal ash related material. The deposit is extremely dense; its porosity is below the limits of detection of most conventional techniques. It is very well bonded to the oxide on the metal, and when it detaches, for example, upon thermal cycling, it generally takes the metal oxide with it.

The oxygen potential in the bed varies from approximately 10^{-1} to 10^{-14} atm. The oxygen potential fluctuates relatively rapidly (a frequency of the order of 0.5 to 2 Hz). In some regions, the oxygen potential is generally low, with peaks to high values; in other regions of the bed, the potential is generally high, with peaks to low values. If local equilibrium were established, this low oxygen potential, combined with the $CaO/CaSO_4$ equilibrium, would generate a sulfur partial pressure of the order of 10^{-5} atm.

The Corrosion Mechanism. Materials within the bed can suffer a form of corrosion in which the normally protective oxide (usually Cr_2O_3) is disrupted; this allows the rapid growth of the less protective oxides of the base metal. Within the metal itself, chromium-rich sulfides are present. In extreme cases, with nickel-base alloys at temperatures above about 650 °C (1200 °F), liquid nickel sulfides may appear, and the degradation of the alloy becomes catastrophic. It has been shown that corrosion is more severe in regions of the bed where the oxygen potential is generally low, with peaks to higher values; regions near the coal feed ports are particularly corrosive.

Materials for In-Bed Superheaters and Air Heaters. The conditions for in-bed superheaters are similar to those in a conventional boiler in terms of the internal temperature and pressure and the lifetime generally required for commercial success. The maximum metal temperature is of the order of 650 °C (1200 °F), and the available materials are those defined by the ASME Boiler Code. The conditions for an air heater generating hot air to be expanded through a gas turbine are somewhat different. The internal pressures are significantly lower than those in utility boiler tubes (of the order of 1 to 1.5 MPa, or 145 to 218 psi, as opposed to 17 to 25 MPa, or 2465 to 3625 psi), but the temperatures are considerably higher. The maximum metal temperature is essentially equal to the bed temperature.

Nickel-base alloys are generally very sensitive to this form of corrosion, and their use should be avoided within fluidized beds unless low local oxygen activities can absolutely be eliminated. Stainless steels, such as type 304 and type 347, are relatively resistant. Type 310 is very resistant to the corrosion, but its mechanical properties at elevated temperatures make it unsuitable for boiler tubes, although it could be used as a cladding. Incoloy 800 has a higher nickel content and good high-temperature properties. Its behavior in the bed is intermediate; that is, in locations where the oxygen activity is always high, it has excellent corrosion resistance, but it can suffer very rapid attack in regions of low oxygen activity.

Therefore, the materials of choice for the superheater are the austenitic stainless steels. The principal question is whether these steels may suffer breakdown corrosion in several thousand hours in this type of environment. Long-term investigations of this question are in progress or have recently been completed. If it is demonstrated that breakdown does take place or if the simultaneous action of erosion leads to accelerated attack, coextruded tubes may be necessary.

Selection of materials for air heaters is considerably more difficult because many of the alloys with adequate strength at the higher temperatures have high nickel contents. In some cases, the corrosion-sensitive alloy Incoloy 800H has been selected, with efforts to locate the hotter tubes in regions of the bed with high oxygen activities. This presents a long-term risk because it is difficult to be sure that the conditions within the bed will remain constant, and fluctuations in the distribution of coal or air in the bed can profoundly affect local oxygen activities. Again, coextruded tubes may be necessary.

Corrosion of Combustion Turbines

R. Viswanathan
Electric Power Research Institute

Combustion turbine components are susceptible to aqueous corrosion and to oxidation-sulfidation corrosion occurring at elevated temperatures. Aqueous corrosion problems are encountered in the compressor section, but oxidation and sulfidation problems are encountered by blades and vanes in the hot-gas path of the turbine section.

Corrosion of Compressor Blades and Disks

Compressor components are generally made of ferritic stainless steels or low-alloy steels. At many utilities, the humidity in the atmosphere is sufficiently high that water condenses on these components during operation. In addition, water often collects on the compressor components of peaking units that are idle; some utilities heat their turbines when they are not used to avoid this problem. General corrosion and pitting can occur as a result of the accumulation of water on compressor blades and disks. The corrosion product fouling of compressor blades associated with corrosion reduces the efficiency of the compressor and therefore increases the heat rate of the turbine. If corrosion is extensive, parts may need to be replaced. Extensive corrosion is often encountered in the low-pressure compressor section of turbines, in which the disks are made of low-alloy steel. Corrosion problems are compounded when aggressive environments are present, such as salt air or air containing contaminants from other units at the plant.

Periodic cleaning of the compressor has been known to restore the efficiency lost because of corrosion. Utilities use many methods of cleaning, including nutshelling (abrasive blasting using crushed nutshells) as well as washing with water, water plus detergent, or water plus organic solvents. Cleaning often cannot be performed without interrupting the operation of the turbine; this presents problems for the operators of turbines used in intermediate- or baseload service. Nutshelling may also lead to clogging of the fuel nozzle and does not provide any cleaning to the turbine section.

Many coatings have been used with mixed results in an effort to reduce the extent of corrosion in compressors. The most disappointing has been the nickel-cadmium family of coatings. These coatings are subject to pitting in acidic environments. They are cathodic to the blades and therefore can accelerate the rate of localized blade corrosion at coating defects. In addition to pitting, bare spots in the coating are often present where the electrodes were placed during electrodeposition of the coating, and this further accelerates corrosion. In some turbines, nickel-cadmium coatings on the blades of the latter stages have been lost, reportedly by vaporization of the cadmium.

More recently, sacrificial coatings (that is, anodic to the blades) have been used in the compressor section. These aluminum-base coatings have met with moderate success; many utilities have reported reduced corrosion. In some cases, however, the rate of corrosion (or at least the rate of fouling) was essentially unchanged, presumably because of the porosity of the coatings. This problem is reportedly being addressed by some coating suppliers through use of a subsequent sealer overcoating. In the cases in which coated blades become fouled, the coatings still have been found to be beneficial because the coating facilitates cleaning of the blades. In these cases, the recovery of efficiency by cleaning was greater for coated blades than that for uncoated blades.

Hot Corrosion of Turbine Blades and Vanes

Blades and vanes located in the hot-gas path in the turbine section are subject to a combined oxidation-sulfidation phenomenon that is commonly referred to as hot corrosion. These components are generally made from nickel- or cobalt-base superalloys. Three basic types of hot corrosion attack have been recognized (Fig. 22). In the temperature range of 650 to 705 °C (1200 to 1300 °F), layer-type corrosion characterized by an uneven scale/metal interface and the absence of subscale sulfides is observed. At temperatures above 760 °C (1400 °F), nonlayer-type corrosion (type I) is observed. Type I corrosion is characterized by a smooth scale/metal interface and a continuous, uniform precipitate-depleted zone containing discrete sulfide particles beneath the scale. The transition from one type to the other, which occurs in the range of 705 to 760 °C (1300 to 1400 °F), is characterized by an uneven scale/metal interface containing intermittent pockets of subscale precipitate-depleted zones and sulfides. The layer-type and the transitional corrosion together are variously referred to as type II hot corrosion, low temperature hot corrosion, and low-power corrosion.

Figure 23 shows the essential features of the corrosion products associated with the three types of corrosion. In the nonlayer-type high-temperature form of hot corrosion (Fig. 23c), discrete chromium and titanium sulfide particles are present in a region of the matrix depleted in these elements, adjacent to the base metal. The surface scales consist of protective Cr_2O_3 with some titanium oxides. With decreasing temperature, the chromium and titanium sulfides are increasingly agglomerated into large interconnecting sulfide networks, and the surface scales contain predominantly the oxides of nickel and cobalt (Fig. 23b). Complete layer-type corrosion (Fig. 23a) is characterized by the chromium and titanium sulfides forming a continuous layer. The surface scale in this case contains only the unprotective oxides of nickel and cobalt.

In addition to the corrosion features described above, grain-boundary spikes (sharp-pointed cracks) are present in the zone of transition-type corrosion. The spikes usually contain sulfides alone or sulfides followed by oxide penetration. They occur over a narrow region of the blade in a manner suggesting that spike formation is dependent on stress.

The high temperature form of hot corrosion involves the formation on the hot-gas path parts of condensed salts that are often molten at the turbine operating temperature. The major components of such salts are Na_2SO_4 and/or K_2SO_4, which are apparently formed in the combustion process from sulfur from the fuel and sodium from the fuel or the ingested air. Because potassium salts act very similarly to sodium salts, specifications limiting alkali content in fuel or air are usually taken to be the sum total of sodium plus potassium.

Very small amounts of sulfur and sodium or of potassium in the fuel and air can produce sufficient Na_2SO_4 in the turbine to cause extensive corrosion problems because of the concentrating effect of the turbine pressure ratio. For example, a threshold level has been suggested for sodium in air of 0.008 ppm by weight; hot corrosion will not occur below this level. Therefore, nonlayer-type hot corrosion is possible even when premium fuels are used. This has been especially true in aircraft-derivative turbines, which have turbine blades made from B-1900 (UNS N13010). Alloy B-1900 has performed well with ultraclean aircraft fuels, but has experienced numerous corrosion problems in land-based service. Other fuel (or air) impurities, such as vanadium, phosphorus, lead, and chlorides, may form with Na_2SO_4 mixed salts having reduced melting temperatures and thus broaden the range of conditions over which attack by molten salts can occur. Agents such as unburned carbon can also promote deleterious interactions in the salt deposits.

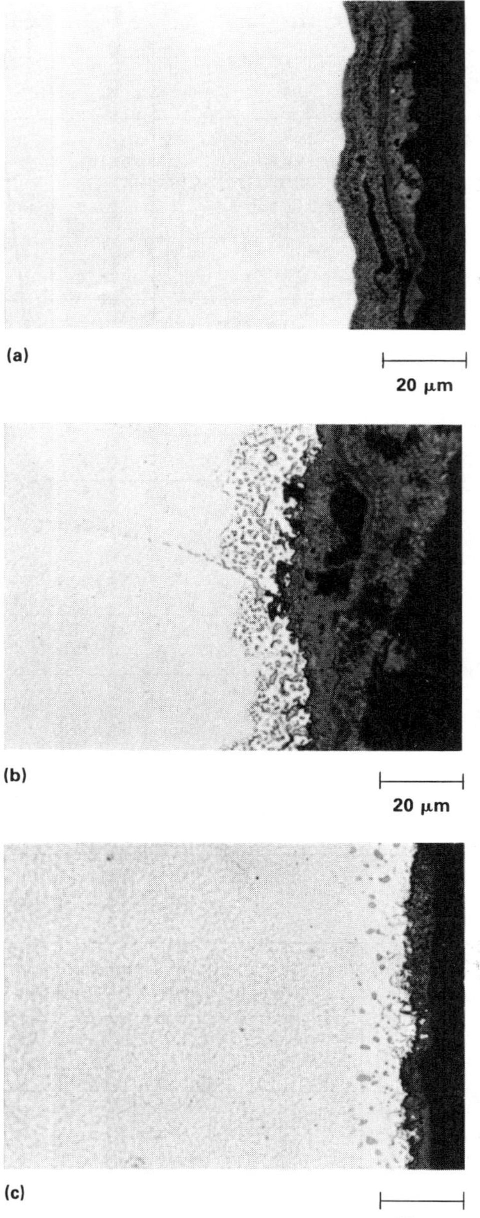

Fig. 22 Three forms of hot corrosion in Udimet 710 turbine blades. (a) Layer type. (b) Transition type. (c) Nonlayer type

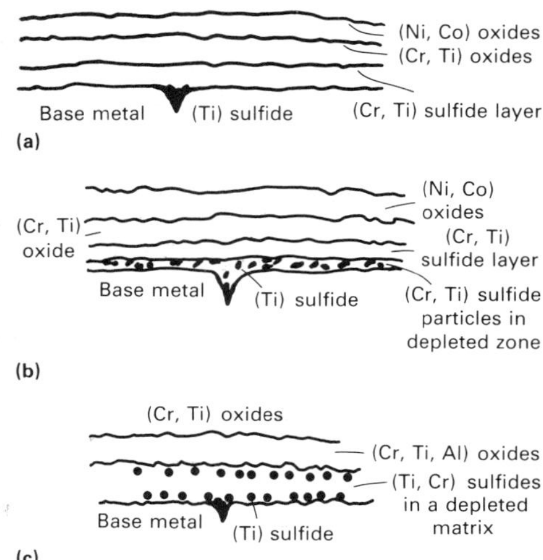

Fig. 23 Schematic of corrosion products formed in the three types of hot corrosion of turbine blades. (a) Layer type. (b) Transition type. (c) Nonlayer type

Research over the past 15 years has led to better definition of the relationships among temperature, pressure, salt concentration, and salt vapor-liquid equilibria so that the location and rate of salt deposition in an engine can be predicted. In addition, it has been demonstrated that a high chromium content is required in an alloy for good resistance to type I hot corrosion. The trend toward lower chromium levels with increasing alloy strength has therefore rendered most superalloys inherently susceptible to this type of corrosion. The effects of other alloying additions, such as tungsten, molybdenum, and tantalum, have been documented; their effects on rendering an alloy more or less susceptible to hot corrosion are known. The near standardization of such alloys as IN-738 and IN-939 for first-stage blades and buckets, as well as FSX-414 (Co-0.25C-29.5Cr-10.5Ni-7W-2maxFe-1maxMn-1maxSi-0.012B) for first-stage vanes and nozzles, implies that these are the accepted best compromises between high-temperature strength and hot corrosion resistance. It has also been possible to devise coatings with alloying levels adjusted to resist type I hot corrosion. The use of such coatings is essential for the protection of most modern superalloys intended for duty as first-stage blades or buckets.

The low-temperature form of hot corrosion produces severe pitting and results from the formation of low-melting eutectic mixtures of essentially Na_2SO_4 and cobalt sulfate ($CoSO_4$), a corrosion product resulting from the reaction of the blade surface with SO_3 in the combustion gas. The melting point of the Na_2SO_4-$CoSO_4$ eutectic is 545 °C (1013 °F). Unlike type I hot corrosion, a partial pressure of SO_3 in the gas is critical for the reactions to occur in low-temperature hot corrosion. Knowledge of the relationships between SO_3 partial pressure and temperature inside a turbine allows some prediction of where layer-type hot corrosion can occur. Because first-stage blade metal temperatures in heavy-duty engines range from about 650 to 855 °C (1200 to 1575 °F), all three types of hot corrosion can occur when sulfur and sodium are present in sufficient quantities.

To avoid hot corrosion in stationary combustion turbines, fuel specifications for sulfur, sodium, potassium, and vanadium are typically set at approximately 1% S, 0.2 to 0.6 ppm Na + K, and 0.5 ppm V. Impurity content limitations can be varied if blade coatings are used, and corrosion inhibitors, such as magnesium, can be added to the fuel. Where the ambient air at the site is contaminated, as in industrial or coastal locations, air filtration is also often practiced.

Problems have been experienced with occasional batches of fuel containing higher-than-specified levels of impurities. A problem that has had to be addressed is the difficulty in accurately measuring low levels of elements such as sodium in fuel oil. Compliance with stringent specifications requires careful supervision and the use of such techniques as centrifuging the oil, which result in increased costs. Impurities from other sources, such as the plum stones (which contain sodium and potassium) used in a carboblast cleaning technique at low engine power, have led to cracking of aluminide coatings and corrosion of blades and vanes. Entrapment of plum stone fragments in these components places the corrosive species in contact with the alloy surfaces.

Air filtration is not a panacea. It is expensive and requires proper maintenance and monitoring to prevent the periodic release into the engine of material captured on the filters. For example, there are reported instances of collected contaminants being washed off of high-efficiency filters and into engines by sudden heavy storms.

Overall, with stringent control of fuel specifications and good air filtration, essentially no unexpected corrosion-related problems are encountered. The life limitation is then the creep strength or thermal fatigue strength of the first-stage blades or vanes. Where such controls cannot be exerted, alloys with some inherent corrosion resistance are used, together with a coating. The alloy used and the type and thickness of the coating are generally the least costly options that correspond to the planned engine maintenance schedules.

Historically, the development of corrosion-resistant coatings was aimed at combating high-temperature hot corrosion. The earliest coatings were the diffusion aluminides. It was found that chromium-modified aluminides offered little additional protection against high-temperature hot corrosion compared to the basic aluminides, but that the platinum-aluminides offered superior protection compared to the basic aluminides. The chromium-modified aluminides have since been found to be particularly beneficial against low-temperature hot corrosion, giving results equivalent to those of the platinum-aluminides; both modified aluminides performed better than the basic aluminides.

Although these diffusion aluminides have been successful in reducing hot corrosion, the chemistry of these coatings is not readily modified for further improvement in corrosion resistance. Thus, increased attention has been given to the development of overlay coatings, which offer significant compositional flexibility.

The actual compositions of these coatings depend on their intended use. Because Al_2O_3 is used for protection against high-temperature hot corrosion, coatings that exhibit the greatest high-temperature protection are generally high in aluminum (11%) and low in chromium (<23%). Low-temperature hot corrosion, on the other hand, depends primarily on Cr_2O_3 for protection; therefore, coatings exhibiting the greatest low-temperature corrosion protection are high in chromium (>30%) and low in aluminum. Other elements, such as silicon, hafnium, tantalum, and platinum, are added to these coatings in an attempt to improve resistance to corrosion and spalling. High-chromium MCrAlY coatings have been developed to offer superior low-temperature protection without sacrificing high-temperature protection because industrial gas turbines sometimes operate under varying load conditions that could result in exposures to both low- and high-temperature conditions. Overlay coatings have been applied by such techniques as electron beam physical vapor deposition, plasma spray, and sputtering.

Apart from overall material wastage due to hot corrosion, an additional concern has been the degradation of mechanical properties, particularly creep and fatique resistance. Figure 24, based on accurate laboratory tests, depicts for several alloys the variation of the ratio of time to rupture in a salt environment to the time to rupture in air with applied stress at 705 °C (1300 °F). The pronounced degradation of the stress rupture life (sometimes by a factor of as much as 10^5) due to the corrosive action of the salt mixture is evident. The plots tend to converge at low stresses; this indicates that a threshold stress level may exist for each alloy, below which rupture life may become insensitive to corrosion. Additional studies are needed to understand the mechanism of and implications of the environmentally induced mechanical property degradation in the context of the field performance of components.

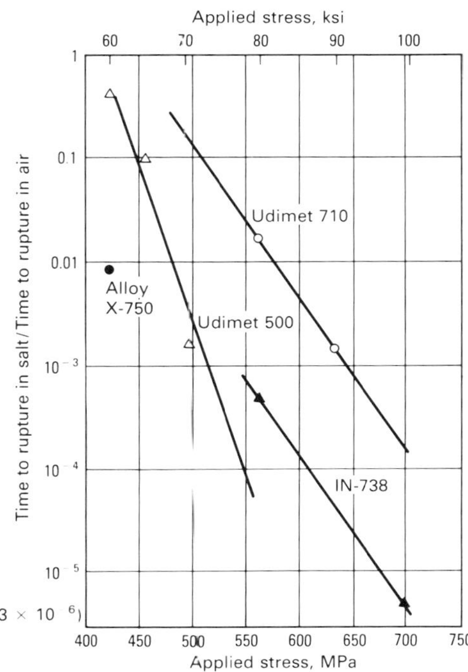

Fig. 24 Degradation in rupture life for various superalloys due to hot corrosion at 705 °C (1300 °F)

Components Susceptible to Dew-Point Corrosion

W.M. Cox, D. Gearey, and G.C. Wood
Corrosion and Protection Centre
Industrial Services
University of Manchester
Institute of Science and Technology

Dew-point corrosion is the attack in the low-temperature section of combustion equipment resulting from acidic flue gas vapors that condense and cause corrosion damage to the plant materials. It occurs when gas is cooled below the saturation temperature pertinent to the concentration of condensable species contained by the gas. Waste flue gas produced by the combustion of fossil fuels may contain several components, such as SO_3, HCl, and H_2O, and therefore may display several dew-point temperatures at which the various species begin to condense. Corrosion rate peaks related to individual condensation processes may appear, but the formation of protective corrosion products and the deposition of soot and ash can moderate the corrosive effects of deposited acids. In conventional boiler plants, the risk areas normally include the posteconomizer flue gas handling section, that is, air heaters, ducting and precipitators, induced-draft fans, and chimney stacks.

Dew-point corrosion problems in nominally dry flue gas handling systems will be discussed in this section. Problems in wet FGD systems are covered in the section "Corrosion of Flue Gas Desulfurization Systems" in this article. This section will emphasize the large coal- and oil-fired power generation equipment, but similar problems are found in many other combustion systems. The underlying mechanisms of attack are similar, but their detailed natures are affected by the operation variables. Dew-point corrosion and the secondary factors affecting it are thoroughly discussed in Ref 90 and 91.

Most fossil-fired power plants are constructed of carbon steel, and past work has largely been conducted on this material. In specialized applications, low-alloy steels, stainless steels, nickel-base alloys, and organic and inorganic coatings are also used. Much recent dew-point corrosion work is reported in Ref 90. The following discussion draws heavily on the information in Ref 90 and on a recent 6-year collaborative investigation sponsored by the United Kingdom Department of Trade and Industry (Ref 91).

Areas Susceptible to Attack

A conventional coal-fired power generation boiler is illustrated in Fig. 25; the locations of items prone to dew-point corrosion are indicated. The most susceptible areas are discussed below.

Penthouse Casing and Hanger Bars. The penthouse casing encloses the tube header pipework above the boiler furnace roof. Flue gas leakage from the furnace can cause corrosion of the casing and furnace support steelwork. The hanger bars holding the furnace tube bundles are particularly susceptible to attack near the seals retaining the bars passing from the casing to the external environment (Fig. 26). Bars can corrode such that insufficient cross section remains to support the load, and failure allows collapse of the furnace roof and superheater tube bundles.

Air Heater Cold Ends. Air heaters are commonly either of honeycomb matrix (Ljungström or Rothemühle) or shell and tube construction.

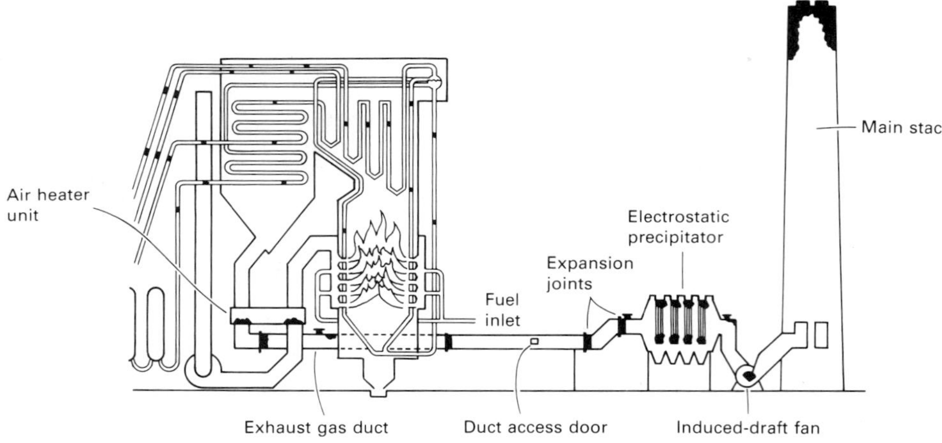

Fig. 25 Schematic of a fossil-fired power generation boiler showing areas susceptible to dew-point corrosion (black areas)

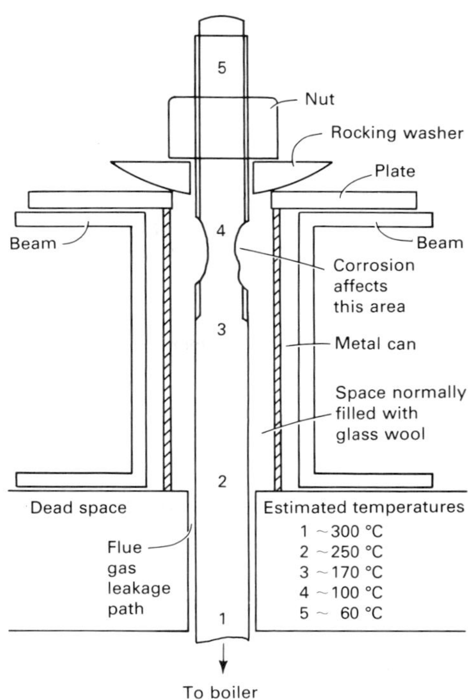

Fig. 26 Schematic of a hanger bar packing box showing areas prone to corrosion. Source: Ref 90

For both, damage is usually most severe at the cold end.

A Ljungström rotating matrix air heater is shown schematically in Fig. 27. Rothemühle air heaters are similar except that the matrix is stationary and the flue gas/inlet air is supplied by an arrangement of rotating hoods. In the Ljungström type, the ductwork is stationary, and the heat-exchanger matrix rotates in either a horizontal or a vertical plane.

Corrosion damage is sustained by the thin steel heat-exchanger elements and by support steelwork and air seal materials. It is partly caused by cold air leaking from the inlet to the outlet duct (bypassing the air heater matrix), but damage is often worsened by mechanical interaction between the rotating and fixed components as well as by displaced heat-exchanger elements that have fallen from their baskets. The highest dew-point corrosion rates are often associated with operation of a cold-end sootblower, which removes fouling deposits by steam or air jets. Increased dew-point attack seems linked to moisture droplets entrained in the blowing medium impinging directly on the lower edges of the air heater elements, causing dissolution of aggressive salts, removal of protective bonded deposits, mechanical abrasion or erosion of the element surface, and fatigue cracking of the element plates. The local injection of considerable moisture may also tend to increase the acid dew-point temperature of the flue gas in a region where the metal temperature may already approach this dew point.

In tube-type air heaters, corrosion is again normally found at the cold end. The same temperature and sootblowing parameters described for matrix air heaters also apply to tube-type heaters, but bypass air leakage is not a problem until tubes have been perforated. However, tube air heaters are often prone to poor gas distribution across the heat-exchanger surface. This leads to localized cool spots where dew-point corrosion takes place.

Ductwork, Expansion Joints, Inspection or Sampling Ports, and Access Doors. Damage at the locations shown in Fig. 25 is often caused by constant low operating temperatures, which are normally related to air entry at fabrication faults, leaking expansion joints and door seals, or careless replacement of sampling port covers. Such leaks can cause an appreciable reduction in duct metal temperature due to stratification effects within the gas stream. Attack rates of 5 mm/yr (200 mils/yr) have been reported in precipitator outlet manifold ductwork. Such damage substantially affects ductwork integrity and frequently leads to increased attack rates on plant components downstream, particularly on the electrostatic precipitator housing and fittings and on the chimney stack.

Electrostatic Precipitators and Filter Bag Houses. Fly ash precipitators are not normally used on oil- or gas-fired equipment, but are usually present on coal-fired systems or furnace exhaust streams. Filter bags are sometimes used instead; however, both methods of dust collection are located in large insulated housings, and their internal components are expensive to maintain and replace.

Precipitator housings are normally constructed of low-carbon steel and concrete. Corrosion attack of the housing is often associated with high levels of air in-leakage upstream due to inadequate maintenance or poor housekeeping. Occasionally, poor internal gas flow distribution may allow cool zones to develop, especially under low-flow conditions. Damage appears as casing perforation and sulfation or spalling of concrete, corrosion of rebars, and loss of structural integrity.

Damage to precipitator components is very expensive and necessitates plant shutdown for repair or replacement. Low-carbon steel collector plates and emitter wires may sustain attack during in-service periods of low flue gas temperature, which result from excessive air entry upstream of the precipitators or from inadequate sealing of the dust discharge doors and plate suspension masts (Fig. 28). Off-line attack resulting from the sweating of acid-saturated pulverized fuel ash may also be important. Precipitator plate attack is more marked because the thin-gage plates are attacked simultaneously from both sides; that is, the plate thinning rate is doubled.

Filter bag installations are normally less expensive than precipitators, but require more maintenance. Dew-point corrosion is frequently displayed on the housings because repeated access to replace holed or detached bags tends to cause wear to access door seals. Low-temperature operation or frequent start-up routines may subject the bag house, bag support, and rapping equipment to extended periods of corrosion attack; this causes blinding of bags and premature holing of the bag material due to chafing during operation or rapping cycles.

Induced-Draft Fan Seals. Corrosion damage to the induced-draft fan housing, control vanes, and impeller seals is again due to air in-leakage. Severe damage is normally associated with neglect of seal condition, and attack of the fan housing itself can rapidly become severe if neglected. Fan housing washing drains should not be left open when the fan is in service. Impeller damage is less common because this component contacts the bulk flue gas.

Chimney Stacks. The stack is often the most vulnerable unit in the gas-handling system because it experiences the results of poorly controlled operation of the furnace and other components of the gas train. Damage is normally due to continuous condensation during operation and buildup of the agglomerated deposits, which insulate the stack inner surface. Stack temperature monitoring is often ignored or neglected. Stacks frequently suffer severe attack before the damage is evident, and inconvenient emergency repairs, necessitating unscheduled unit shutdown, must be made. Damage can occur to liners and/or to the body of the structure, whether it is low-carbon steel, nickel alloy, or concrete (Fig. 29). In prefabricated sectional steel stacks, corrosion occurs preferentially on the inner surface near the external flanges or ladder attachments because of local cooling effects.

Mitigation of Dew-Point Corrosion

In boiler technology, the "acid dew point" usually means the H_2SO_4 dew-point temperature, because this is the highest temperature at which

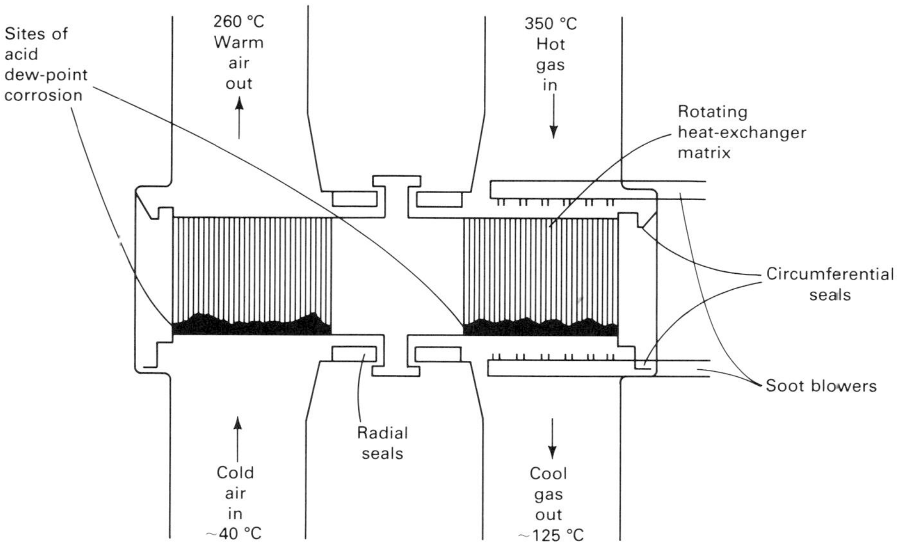

Fig. 27 Schematic of a Ljungström air heater showing location of the air seal and cold end basket corrosion sites

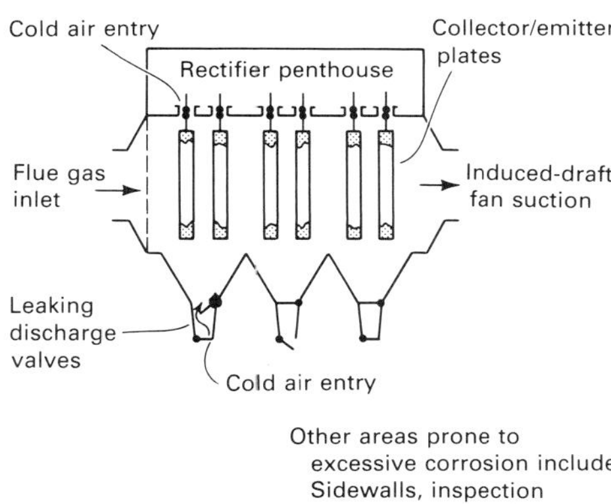

Fig. 28 Schematic of an electrostatic precipitator showing areas prone to excessive corrosion (shaded)

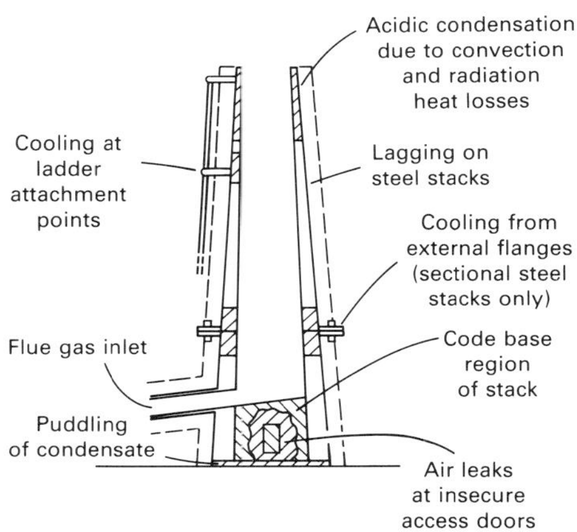

Fig. 29 Schematic of a chimney stack showing sites prone to corrosion (shaded)

acid condensation occurs. Earlier studies showed that, in dust-free flue gases, low-carbon steel corrosion reaches a maximum at about 25 to 50 °C (45 to 90 °F) below the acid dew-point temperature. More recently (Ref 90) it has been seen that at much lower temperatures (below 60 °C, or 140 °F) the corrosion rate of low-carbon steel increases rapidly to at least twice the high-temperature maximum, and to at least three times the high-temperature maximum if HCl is present (Fig. 30).

During the period from 1942 to 1955, it was found difficult to define a precise temperature below which rapid corrosion would start and continue; therefore, a threshold figure was sought above which the plant could be operated confidently. A dew-point probe and meter was developed to identify the dew-point temperature, and maintenance of the flue gas temperature at 20 to 30 °C (36 to 54 °F) higher than the dew-point temperature was recommended to avoid condensation and corrosion.

Increasing fuel costs, higher overall efficiencies, and environmental considerations have focused renewed effort on this subject and the reexamination of earlier philosophies. The drive toward less expensive low-grade (more sulfur-rich fuels), together with improved heat extraction processes, has produced a move toward lower back-end temperatures with better combustion control, materials selection, and use of additives to minimize dew-point corrosion. A further refinement arises from the development of continuous in-plant monitors of dew-point corrosion that can identify harmful operating regions directly.

It is difficult to give blanket advice on lessening dew-point corrosion because much depends on the precise plant configuration and service environment. However, general comments on materials selection, plant operation, use of neutralizing additives, maintenance, good housekeeping, and lagging (insulation) are offered below.

Materials Selection

Most power boiler exhaust components are made of carbon steel. The material can last indefinitely in conventional plants if metal temperature is maintained above the acid dew-point temperature when the plant is in service and if reasonable care is taken to prevent excessive off-line attack during shutdown periods. In marginal conditions, cast iron may extend service life, but normally, this is simply due to its greater cross section.

A few low-alloy steels have shown superior performance to low-carbon steel in particular applications. Notably, Cor-Ten, a high-strength low-alloy weathering steel, has been used with fair success for cold-end air heater elements. The improved corrosion performance seems related to the formation of a thin protective layer of corrosion product.

Expensive alloy materials have been used where plain-carbon steel corrosion is severe, but they have not always been successful. The good but sometimes unreliable performance of such materials is due to the presence of a preformed protective oxide. Failure is associated with deposit formation, which causes differential aeration and concentration cells, producing pitting attack.

Coating use is hazardous because of difficulties in surface preparation, quality of application, physical durability, adhesion under cycling temperature, and dust abrasion. However, in certain practical application tests, good performance has been reported. For concrete stack lining in power stations, an expensive fluoroelastomer has given outstanding service. Less expensive modified coal tar epoxy and glass flake polyester coatings have performed well, and an isocyanate-cured epoxy material has also been successful in less troublesome plant conditions. For air heater elements, double-dipped enamel coatings are likely to be most effective. Major difficulties in ensuring defect-free adherent coatings have occurred with arc-sprayed aluminum, chromia, and alumina materials, but arc-sprayed aluminum has been reported to be generally satisfactory for high-temperature gas turbine stack applications. However, in these stacks, severe thermal conditions can lead to exfoliation and acidic rust flaking.

Plant Operation

Although materials selection can be pertinent in certain plant components, careful plant operation, good housekeeping, and comprehensive maintenance represent a major and more general route to technically efficient, fuel economical, and cost-effective operation. Classical dew-point conditions do not necessarily yield unacceptably high corrosion rates in practice. Control of the secondary factors alone can lead to significant savings by minimizing replacement costs in key areas. Improved on-line monitoring capabilities allow the onset of corrosion to be detected and avoided, often by minor operational refinement, and the possibility of lowering back-end temperatures without risk to the plant can be explored confidently. Precise operational control is important. The inherent variability of excess oxygen, furnace temperature, combustion method, and fuel ashing characteristics can affect low-temperature corrosion more than changes in fuel sulfur content.

Maintenance, Good Housekeeping, and Lagging

A good housekeeping list can be established and adequate planned maintenance procedures introduced for any particular installation because practical dew-point factors are now better understood. Poorly designed or maintained thermal insulation, badly sealed inspection doors, leaking expansion joints, and so on, are unacceptable if

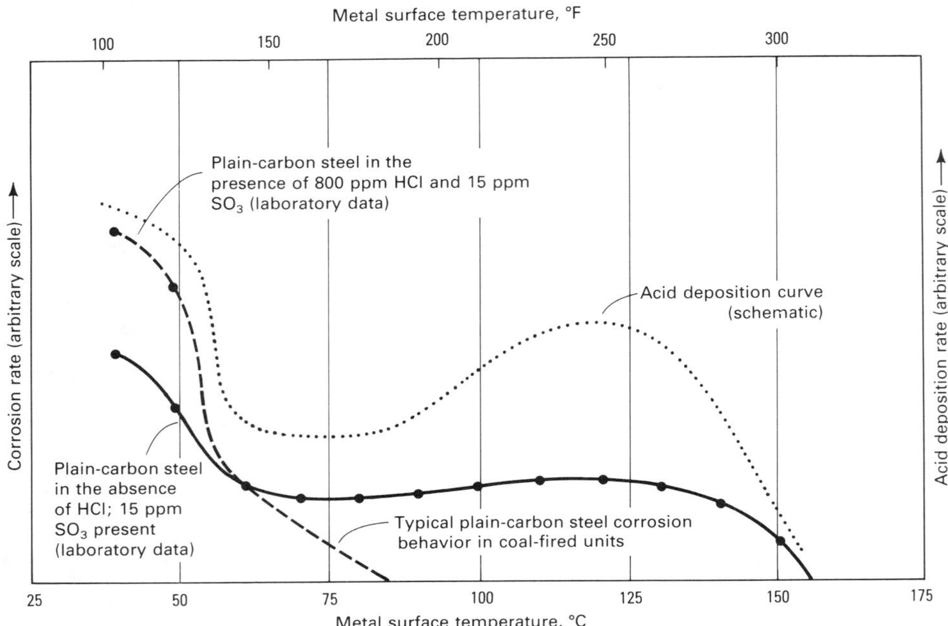

Fig. 30 Modified dew-point curve showing the corrosion rate of a freshly prepared plain-carbon steel surface exposed to dust-free flue gas in a laboratory test rig. The increase in corrosion rate at lower temperatures in the presence of HCl is shown (Ref 91). Modification of classic dew-point corrosion behavior in coal-fired power plants is also shown.

future installations are to operate at optimum levels.

Additives

Neutralizing additives, usually calcium or magnesium oxide/hydroxide, tend to have an *ad hoc* usage, especially in oil-fired situations. Their benefit in preventing acid smuts is proven, but their ability to reduce corrosion rates consistently is less certain. Modern electrochemical monitoring techniques can provide a continuous means of assessing additive performance, although in many cases fluctuations in fuel composition and operational control may mask the contribution of the additive. Direct injection of NH_3 has been employed to reduce flue gas acidity, but the resultant formation of sticky bisulfite deposits can lead to severe fouling.

Guidance for Specific Sections of the Plant

Penthouse Casing and Hanger Bars. The casing and hanger bars should be maintained during operation at a temperature in excess of the acid dew-point temperature. This is achieved by external insulation of all items. Hanger bar packing boxes should be routinely checked to ensure that excessive clearance has not developed between the packing and the bar, allowing flue gas leakage. This is especially important above the superheater pendant tubes, which tend to require more maintenance than other tube bundles.

Air Heater Cold Ends. In matrix-type airheaters, air leakage should be controlled, possibly by allowing only a predetermined acceptable leakage rate, and condensate injection should be prevented, especially that due to leaking sootblower steam isolation valves. Compressed air is preferred as the blower medium, and slight preheating to dry the air should be considered. Adequate condensate drainage facilities must be maintained for all types of blowers. Air heaters should not be water washed on line unless absolutely necessary, but thorough water washing off line is recommended. In many cases, the need to do this may point to inadequate operational control, for example, during low-load conditions. For lower outlet flue gas temperature units, the use of Cor-Ten steel or double-dipped enameled elements should be considered for severe service. Adequate specification and coating quality control must be ensured when using enameled elements. Good gas distribution across the air heater minimizes cold spots. The use of fuel additives should be considered for combating unavoidable corrosion attack.

Special care should be taken in shell and tube air heaters to ensure good temperature distribution across the unit. Buildup of gas-side deposits, which may allow sub-dew-point metal temperatures to occur beneath, must not be permitted. Leakage from perforated cold-end tubes must be minimized.

Ductwork, Expansion Joints, Sampling Ports, and Access Doors. All air in-leakage must be prevented. Door seals should be replaced routinely, and a gastight fit ensured. Thermal insulation should be maintained in good condition, leaking expansion joints replaced, and sampling port covers properly refitted after use.

Electrostatic Precipitators and Filter Bag Houses. Adequate maintenance can prevent cold air entry, especially at discharge doors and electrode support masts, and access door seals. External lagging must be in good condition to maintain housing temperature above the acid dew-point. Air entry upstream of the precipitator or bag house should be minimized. If possible, the unit should be kept warm during hot-standby conditions. Operation periods in low gas flow or start-up conditions should be minimized.

Induced-draft fan seals should be maintained in good condition on a routine basis. Fan housing washing drains should not be left uncovered.

Stacks. Adequate temperature monitoring should be ensured at top and bottom. Stack operating temperature should be maintained within design limits at all times. For low-carbon steel stacks, adequate insulation and lining condition must be ensured. Stacks should not be run wet or cold unless specifically designed to do so. Routine inspection programs for monitoring stack lining and structural integrity should be initiated.

Corrosion of Flue Gas Desulfurization Systems

G.H. Koch and H.S. Rosenberg
Battelle Columbus Division

The control of SO_2 emissions is a primary concern in coal-fired power plants. This can be done either by removing sulfur from the coal before combustion or by FGD in scrubbers. A variety of methods of both types are being developed, and some are already in commercial operation.

Among the FGD processes, lime and limestone wet scrubbing are the most developed. A schematic of a typical system of this type is shown in Fig. 31. Hot, dry flue gas at about 150 °C (300 °F) contacts a spray of an aqueous lime or limestone slurry in the absorber. The SO_2 in the flue gas dissolves in the slurry to form sulfites, which may be oxidized to sulfates by oxygen in the flue gas. The pH of the absorber slurry typically ranges from 5 to 6.5. Scrubbed and cooled gas at 50 to 55 °C (120 to 130 °F) passes through mist eliminators and outlet ducting to the stack. The scrubber gas is sometimes reheated to 60 to 95 °C (140 to 200 °F) to increase its buoyancy by mixing with unscrubbed gas (bypass reheat) or by using special reheaters. When bypass reheat is used, the SO_2 and SO_3 in the gas readily react with the water in the scrubbed gas to form H_2SO_4, which can create an extremely oxidizing condensate on the duct wall. Supplemental information on corrosion of and materials for FGD systems is available in the article "Corrosion of Emission-Control Equipment" in this Volume.

Materials for FGD Systems

A variety of construction materials have been used in operating FGD systems. These include metals, organic linings and plastics, and ceramic and inorganic materials. Metals ranging from carbon steel to nickel-base alloys have been used in most components of FGD systems.

Unlined carbon steel or high-strength low-alloy steels, such as Cor-Ten, can be used where alkaline (pH >7) conditions are maintained—for example, in lime/limestone storage silos or in the piping that carries the fresh slurry to the absorber. Rubber-lined carbon steel is frequently used for protection against abrasion. Under more acidic conditions, the steels must be protected by a lining. Where unlined metal is exposed to acidic environments, corrosion-resistant alloys are required. Stainless steels—for example, type 316L, type 317L, and alloy 904L—have been used in such FGD system components as prescrubbers, absorbers, spray nozzles, reheaters, and damp-

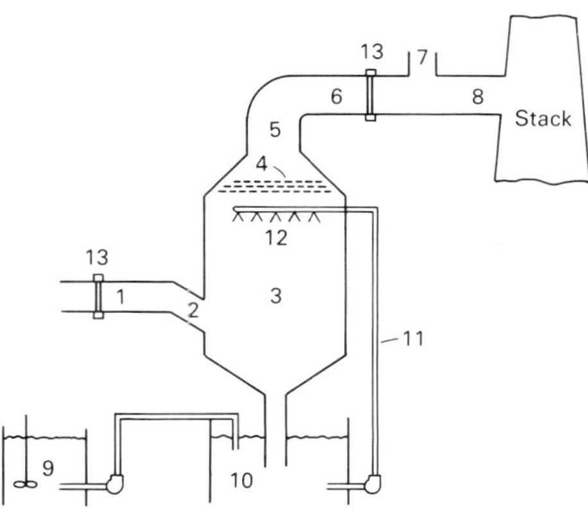

Fig. 31 Schematic of a wet FGD system. 1, inlet ductwork; 2, absorber inlet; 3, absorber body; 4, mist eliminators; 5, absorber outlet; 6, outlet ductwork; 7, reheaters and associated ductwork; 8, stack breeching; 9, alkali tanks; 10, recycle tanks; 11, slurry piping; 12, slurry nozzles; 13, dampers

ers. In critical locations with more aggressive environments, such as outlet ducts, outlet dampers, and reheaters, nickel-base alloys (Inconel 625, and Hastelloy alloys C-276 and G3) and titanium (grades 2 and 7) have been used. When the alloys fail, it is usually by general corrosion, pitting, crevice corrosion, erosion-corrosion, or SCC.

Organic linings have been extensively used in various FGD components, including prescrubbers, absorbers, tanks, outlet ducts, and stacks. These linings include epoxies, vinyl-esters, polyesters, fluoroelastomers, and rubber. The use of organic linings is attractive because it can provide the lowest-initial-cost material for many of these components.

The linings are generally applied in liquid form by brushing, rolling, or spraying. To provide an effective barrier, the organic linings used in FGD systems are often applied in several layers to achieve a thickness greater than that of typical coatings. Moreover, the linings often contain a flake-type filler, which decreases the permeability of the lining and provides reinforcement. In such components as venturi prescrubbers, in which high abrasion resistance is required, organic linings reinforced with glass cloth or inert matting have been used. These linings may also contain special abrasion-resistant fillers, such as Al_2O_3, to improve wear resistance. Although rubber is an organic lining, it is often placed in a separate category because it is applied in sheet rather than liquid form. Failure of organic linings can occur by blistering, debonding, and wear from abrasive slurries. Excursions to high temperatures can damage or destroy organic linings.

Nonmetallic inorganic materials are used in prescrubbers, spray nozzles, slurry pumps, outlet ducts, and stacks. The materials used for these components include prefired bricks and shapes, hydraulically bonded concretes and mortars, and chemically bonded concretes and mortars, all of which are used as lining materials where temperature resistance, chemical resistance, or abrasion resistance are required. Prefired shapes are also used for spray nozzles and pump components. Acid-resistant bricks are commonly used as construction materials for stack linings. Alumina bricks are more abrasion resistant than acid-resistant bricks and will also withstand hot H_2SO_4; therefore, they have been used to line venturi throats. Silicon carbide shapes have high abrasion and chemical resistance and are used for spray nozzles and pump components as well as for lining venturi throats. Design aspects that could cause failure include mechanical stresses from shrinkage, insufficient thermal expansion allowances, vibration, and thermal or mechanical shock.

Hydraulic-setting cement-bonded concretes (hydraulic concretes) are used as linings for prescrubbers, outlet ducts, and stacks. These concretes contain calcium aluminate ($CaAl_2O_4$), which can withstand temperatures of 260 °C (500 °F) without strength degradation. However, in H_2SO_4 solutions (pH <4), the cement is attacked because of dissolution of the $CaAl_2O_4$. This dissolution of the bond results in loosening of the aggregate and erosive wear of the lining.

Chemically bonded mortars are commonly used to bond acid-resistant brick stack linings. Chemically bonded concrete mixes can also be used as linings. They generally contain siliceous aggregates; a sodium silicate (Na_2SiO_3), potassium silicate (K_2SiO_3), or colloidal silica bond phase; and a silicofluoride, phosphate, or organic bond gelling agent. Because chemically bonded concretes can withstand hot acidic environments, they offer an abrasion-resistant alternative to organic linings. However, like hydraulic cements, the chemically bonded cements have a finite permeability; therefore, any condensed acid will ultimately reach the substrate to which they are applied, unless the substrate is protected by an impermeable membrane. More information on chemically bonded concrete linings is available in the article "Chemical-Setting Ceramic Linings" in this Volume.

Borosilicate glass block is used in outlet ducts and stacks. The blocks are usually bonded with urethane asphalt. The combination of these two materials results in a lining that is extremely resistant to permeation by aggressive condensate. Moreover, the thermal insulating properties of the glass block and the flexibility of the membrane material contribute to the good performance of the lining. However, the borosilicate glass can be easily damaged, and failures can occur due to abrasion, puncturing, and cracking. Also, scale that forms on the lining cannot be easily removed. Specific materials of construction that are currently used in major FGD components (prescrubbers, absorbers, outlet ducts, and stacks) will be reviewed in the following sections.

Materials for Specific FGD Components

Prescrubbers. The flue gas from the air preheater (about 150 °C, or 300 °F) is cooled in a prescrubber to the adiabatic saturation temperature of about 50 °C (120 °F). There are basically three types of prescrubbers:

- A quench or saturation duct with water sprays, which cools hot gas already cleaned of particulates by an electrostatic precipitator
- A quencher zone in the bottom of the absorber or in a separate vessel
- A rod, constricted throat, plumb bob, or flooded disk venturi system that cools the gas and removes particulates

Scrubbing liquor is usually used in the prescrubber so that it operates at a pH of about 4 to 6. In some cases, makeup water or reclaimed water is used in quench ducts, and the operating pH is then 1 to 3. This low-pH solution can be extremely aggressive toward materials of construction, particularly when the SO_2 scrubber operates in a closed-loop mode, which can increase the chloride concentration to 40 000 to 50 000 ppm.

Quench ducts and quench zone systems are usually constructed of either concrete-lined carbon steel or an alloy ranging from type 317LM stainless steel to nickel-base alloys, such as Hastelloy C-276. In cases where the pH is low (<3), the concrete deteriorates and spalls off the carbon steel substrate, exposing the carbon steel to the aggressive solution. In these environments, stainless steel and low nickel-base alloys fail by pitting and crevice corrosion, and only highly resistant alloys, such as Hastelloy C-276 and titanium grade 7 (titanium-palladium), can withstand this extremely aggressive environment.

Absorbers. The flue gas enters the absorber at either about 50 °C (120 °F) or about 150 °C (300 °F), depending on whether or not a prescrubber is used. The conditions in absorbers are similar to those in prescrubbers in which lime or limestone scrubbing is used, that is, pH of 5 to 6.5. The chloride concentration of the scrubbing liquor depends on how the prescrubber and absorber loops are tied together and whether an open- or closed-loop water system is used. Depending on these conditions, the chloride concentration can range from a few hundred to 50 000 ppm.

Absorbers incorporate a variety of designs and many different materials of construction. Generally, absorbers use a combination of materials, the primary categories being stainless steel, rubber-lined carbon steel, organic-lined carbon steel, and ceramic-lined carbon steel.

Molybdenum-containing stainless steels, such as types 316L, 316LM, 317L, and 317LM, and alloy 904L have been used for construction of several FGD absorbers, as well as such components as trays, plates, supports, and fasteners. High chloride concentrations of the scrubber slurries can result in pitting and crevice corrosion, and abrasion by these slurries can sometimes cause wear failures.

Natural rubber and synthetic rubber (neoprene and chlorobutyl) are widely used as lining materials for absorber walls. Generally, rubber linings perform well in absorber environments. Failures as a result of improper application or curing occur as debonding or blister formation. In addition, the absence of industry standards for composition and formulation of rubber products also plays an important role in rubber lining failures.

Several absorbers are in operation with glass flake filled polyester or epoxy linings. The organic linings have generally performed satisfactorily in the relatively mild absorber environments.

Outlet ducts carry the flue gas from the absorber to the stack and are thus exposed to saturated (wet) flue gas at about 50 °C (120 °F) and/or reheated flue gas, depending on the existence and location of a reheater. Part of the outlet duct may also be exposed to hot flue gas during periods of scrubber shutdown if the FGD system has a bypass duct that joins the outlet duct ahead of the stack. The outlet duct has been a major problem area, particularly in units with duct sections that handle both hot gas and wet gas. Failure of an outlet duct may require complete boiler shutdown and loss of generating capacity for long periods of time because of the lack of standby components or bypass capability.

The severe attack of metallic and nonmetallic materials of construction has been attributed to low-pH conditions (<3) and to the presence of chlorides. The low pH is attributed to the presence of unscrubbed SO_2 and HCl in the flue gas, which can reach high concentrations when bypass reheat is used, and to the absence of the buffering effect of lime or limestone. The chlorides, which come from the coal as well as the makeup water, can reach very high concentrations when FGD systems are operated in a closed-loop mode. Recent laboratory studies on several stainless steels, nickel-base alloys, and titanium-base alloys have indicated that trace elements originating from the coal and the makeup water, such as iron, aluminum, chromium, copper, phosphorus, and fluorine, have a significant effect on the corrosion performance of alloys.

Operating experience in wet SO_2 scrubber outlet ducts has indicated that many types of organic and inorganic linings provide good service in wet-gas and reheated-gas environments. However, wet-gas and hot-gas conditions caused by bypass for reheat or for scrubber maintenance typically result in less than 2 years of service for many linings. Although lining application conditions and procedures are considered to be very important to the success of a lining, environmental conditions in the duct, such as temperature distribution and condensate chemistry, have a considerable effect on the life of a lining.

Glass flake polyester, vinyl-ester, fluoroelastomer, and multifunctional epoxy are the most commonly used lining materials in outlet ducts. The performance of the linings is generally excellent in wet outlet duct environments, but may be poor in wet-dry environments that result from bypass.

Inorganic cementitious linings are generally not used in wet environments, because of their high permeability, but their use in hot environments is increasing because of the temperature problems with organic linings. Hydraulically bonded concrete has failed in several wet environments, but can give reasonable life under dry conditions. The use of an organic membrane between the carbon steel substrate and the cementitious lining has proved effective in keeping the condensate away from the steel. However, corrosion will occur when holidays are present in the organic membrane. The ability of concretes to absorb the condensates that form on the surface is being used to develop a method of protecting the metallic substrate by means of electrochemical potential control.

Finally, borosilicate glass blocks with urethane asphalt mortar and membrane have been successfully applied in several outlet ducts. As indicated earlier, the primary disadvantage of this lining is its low mechanical strength, which renders it susceptible to abrasion by fly ash or scrubber slurry carryover and damage due to impact from tools and maintenance equipment. When heavy scale forms on the surface, its weight may tear off part of the glass blocks. Also, it is extremely difficult to clean a scaled surface. To overcome this shortcoming, cementitious coatings are often applied over the glass block surface, particularly in floor applications.

Until recently, alloys were not commonly used in outlet ducts. Type 316L and 317LM stainless steels, alloy 904L, and Hastelloy G have been used in a few ducts, but all have experienced pitting and crevice corrosion. Recent application of high nickel-base alloys and titanium have proved successful. Particularly, the application of a 1.5-mm (1/16-in.) thick sheet of Hastelloy C-276 onto a carbon steel substrate has provided a cost-effective alternative to full-thickness alloy walls and nonmetallic linings.

Stacks. The flue gas from an FGD system exits to the atmosphere through a chimney or stack. The stack consists of an outer shell constructed of concrete or carbon steel and one or more flues constructed of brick, metal, or fiberglass-reinforced plastic (FRP) that carry the flue gas (Fig. 32). A stack is a critical component in that, as in the case of outlet ducts, failure may require complete shutdown of the boiler, which results in loss of generating capacity for extended periods of time. The performance of a stack lining depends strongly on whether the flue gas is wet or reheated and whether the stack is also used for hot bypassed gas.

Stack flues that are exposed only to wet gas are located downstream from FGD systems without any type of reheat and without FGD bypass ducts. The flue gas is typically at 50 to 55 °C (120 to 130 °F), and the flue is not exposed to high temperatures. Under these relatively mild operating conditions, several materials of construction can be used. For example, Inconel 625, FRP, inorganic-lined carbon steel (borosilicate glass block), and organic-lined carbon steel have been installed in stacks and have generally performed well. Furthermore, acid-resistant brick is being widely used as a material of construction for stacks.

Although acid-resistant brick linings have exhibited good performance in all categories of operating conditions, a leaning phenomenon has recently been reported for several free-standing brick flues. This phenomenon has required some utilities to perform considerable maintenance work near the top of the stack to eliminate rain cap rubbing and/or to provide adequate clearance to accommodate the wind load deflections of the concrete shell. In one case, the flue was completely rebuilt.

The leaning phenomenon appears to correlate with mixing wet gas and hot gas in the stack, that is, separate breechings for scrubbed gas and bypassed gas. Less severe leans have occurred when wet gas and hot gas have been mixed in the outlet duct near the breeching in the flue. Preliminary results from the analysis of core samples from several leaning flues indicate that differential moisture expansion of the bricks is a primary cause of the lean phenomenon.

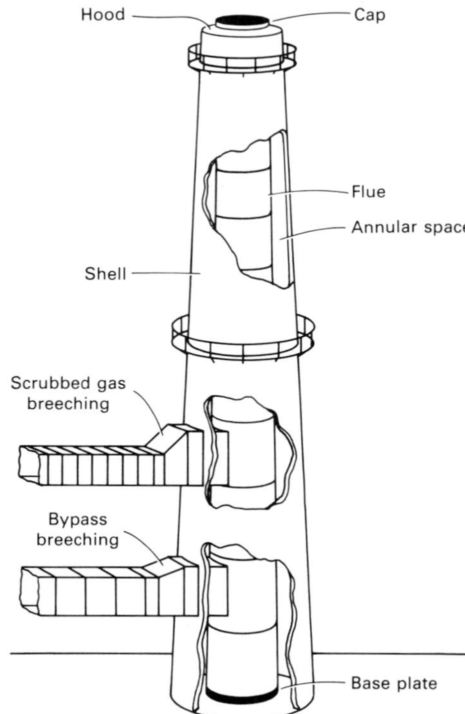

Fig. 32 Schematic of a typical stack

Corrosion of Generators

R. Viswanathan
Electric Power Research Institute

The most significant corrosion problem in generators is SCC of retaining rings. Retaining rings are massive steel rings that are shrunk onto the generator rotor to retain the circumferential conductor coils wound around the two ends of the rotor. The rings currently in use are made of an austenitic steel containing 18% Mn and 5% Cr. Retaining rings are highly stressed, critical components of a generator, and several catastrophic incidents involving failures of these rings have occurred worldwide.

Forms of Corrosion

Stress-corrosion cracking is the single major cause of failure of retaining rings. The stress required can arise from the residual stresses from cold fabrication and machining, shrink-fit assembly stresses, or operating stresses. The aggressive environments causing cracking may include lubricating oils and fluids used during machining, solvents and paints used for cleaning and dye penetrant inspection, condensed moisture, salt water and industrial air environments encountered during storage and shipping, and moist hydrogen or water used for generator cooling. Residual stresses from fabrication alone are sometimes sufficient to cause cracking of the rings, as evidenced by instances of cracking in storage.

For machines with hydrogen-cooled rotors, the periods of highest risk are frequently during storage and assembly, that is, before the machine is closed. During this time, retaining rings are exposed to the ambient plant environment, and their temperature may easily drop below the dew point. Humidity may enter the machine in a number of ways during operation—for example, imperfectly dried hydrogen; from leaks in water-fed heat exchangers; in the oil of the hydrogen seal circuits; or, in the case of direct water cooling of the rotor, from simple leakage. In general, however, the increased temperature of rings in operation is enough to avoid condensation, which will more readily accumulate on the cold surfaces of the water circuit. As a further precaution, the dew point should be correctly monitored and maintained at an acceptable level.

The generator rotor must be kept dry during manufacture, transportation, storage, and periodic shutdown to protect the electrical insulation and to prevent general rusting of the rotor. In addition, water condensation should be prevent-

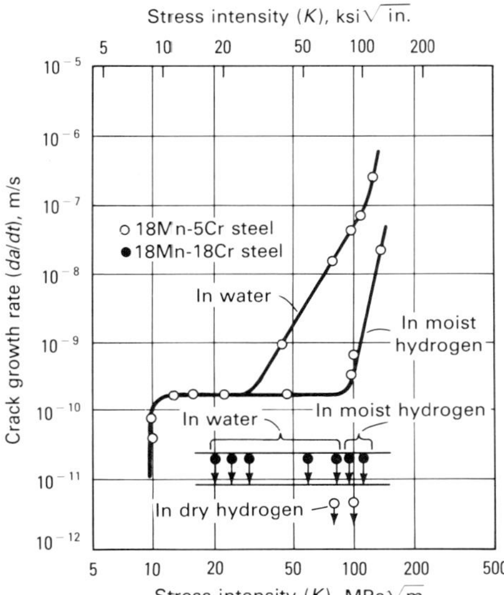

Fig. 33 Crack growth rate versus stress intensity for two generator retaining ring steels in three environments. In dry hydrogen, neither material showed any crack growth at K up to 100 MPa$\sqrt{m}$ (91 ksi$\sqrt{in.}$). The 18Mn-5Cr alloy, however, exhibits crack growth in moist hydrogen and water; the 18Mn-18Cr steel does not.

ed from forming on the retaining ring surface. Several methods are possible, but application of each is determined by considering such factors as the climate of the site location and the shipping route. For example, special packing is applied to keep the rotor dry when it is shipped to tropical countries. The packing is opened immediately before the start of rotor assembly.

The rate of crack growth da/dt as a function of stress intensity K in dry hydrogen, moist hydrogen, and water environments is illustrated in Fig. 33. No crack growth occurs in dry H_2 even up to K values of 100 MPa$\sqrt{m}$ (91 ksi$\sqrt{in.}$), which indicates that the threshold stress intensity for crack propagation in dry hydrogen K_H exceeds this value. On the other hand, in moist hydrogen, rapid crack growth occurs at all values of $K > 10$ MPa$\sqrt{m}$ (9.1 ksi$\sqrt{in.}$). The same behavior is observed in distilled water. It is clear from these data that the threshold value of K for SCC propagation in aqueous environments (K_{ISCC}) is as low as 10 MPa$\sqrt{m}$ (9.1 ksi$\sqrt{in.}$). Other investigators have reported slightly higher K_{ISCC} values in the range of 15 to 25 MPa$\sqrt{m}$ (13.7 to 22.8 ksi$\sqrt{in.}$). Once this value is exceeded, the crack propagates rapidly (within a matter of hours or days) to reach the critical crack size a_{crit}. Therefore, in the presence of corrosive environments, failure is no longer governed by K_{Ic} but by K_{ISCC}.

Based on K_{ISCC} = 10 MPa$\sqrt{m}$ (9.1 ksi$\sqrt{in.}$), a_{crit} for various values of stress can be estimated with the help of fracture mechanics evaluations. It can be shown that over the range of working stresses of practical interest a_{crit} is much below the sensitivity of detection by current nondestructive testing techniques. In fact, at a stress level of 690 MPa (100 ksi), even pits only as deep as 0.14 mm (5.4 mils) appear to be capable of causing rapid crack propagation. This means that any crack that is detectable during an inspection or any pitting that is present can potentially compromise ring integrity and therefore indicate the need for immediate corrective action. For this reason, the concept of a tolerable defect size does not exist, and inspection procedures have placed more emphasis on crack detection than on crack sizing. Furthermore, visual inspections that can readily detect rusting and pitting have a prominent role among inspection techniques.

Despite the extreme susceptibility of 18Mn-5Cr steel to SCC crack growth, retaining rings made of this steel have served the industry well. The reasons for this are twofold. First, initiation of pits and/or cracks is also a function of stress, and it can take many years to occur. Second, the propagation of cracks is a discontinuous process because exposure to moist environments occurs only intermittently.

In view of the extreme susceptibility to SCC of 18Mn-5Cr steel rings, manufacturers have recently developed an alternative steel containing 18% Mn and 18% Cr. Extensive laboratory tests have shown the 18Mn-18Cr steel to be very resistant to SCC. Rings made of the new alloy are currently being used in new and replacement generator rotors by most manufacturers. With the introduction of the new material, it is anticipated that ring corrosion problems in generators will cease to be of concern in the future.

Corrosion of Ash-Handling Systems

L.D. Fox
Tennessee Valley Authority

Ash handling is a major problem for utilities and industrial owners using coal as a primary fuel. The firing concept used, that is, cyclone, pulverized-coal, or fluidized-bed firing, determines the type and characteristics of the ash. Waste ash (fly ash and bottom ash) is generated in large volumes and must be disposed of in an environmentally acceptable manner. Fly ash comprises small dustlike particles (100 μm or less in diameter); bottom ash consists of much larger slag particles removed at furnace bottoms.

Conventional ash-handling systems collect, convey, and dispose of waste ash by methods that depend on site-specific considerations, regulatory (local, state, and federal) requirements, and economic considerations. Generally, the ash will be handled wet or dry, but the handling system may include combinations of wet and dry conveying. Fly ash is collected dry and is generally conveyed dry to storage or disposal. Wet conveying and storage of fly ash has become environmentally more difficult, and regulations now favor dry conveying systems. Bottom ash is collected in a water bath and is almost always handled wet. Most ash-handling systems convey bottom ash and fly ash separately to storage or disposal, but conveyance in the same pipeline is also acceptable. Typical operating problems associated with conventional fly ash/bottom ash handling systems may be grouped into two categories:

- Corrosion and erosion (material deterioration)
- Scaling and plugging (material buildup)

The above problem areas must be addressed by designers of ash-handling systems and plant operators, who should consider such factors as equipment selection, material applications, maintenance policy, and system concept. The handling system concepts for fly ash/bottom ash are dry/dry, dry/wet, wet/dry, or wet/wet. The subject of ash handling is discussed in detail in Ref 92. Corrosion and corrosion-related problems are more prevalent in wet systems than in dry systems. Corrosion-related problems by system concept are reviewed below.

Fly Ash Systems

Dry Fly Ash Systems. The pneumatic conveying of coal ash particles can produce plugging of conveying lines, sticking of the ash to chutes and hoppers, and erosion of internal parts of ash-handling equipment. Corrosion of internal components is normally small to nonexistent. Only the introduction of unwanted moisture will create an environment that will produce corrosion that is severe enough to cause system outages or to require frequent maintenance.

Wet Fly Ash Systems. The transport of fly ash with water as the conveying medium is normally achieved by using a slurry of 10 to 15% ash by weight. The introduction of conveying water sets up an array of water chemistry situations that can lead to scaling and/or corrosion. The tendency for scaling and corrosion depends in part on the chemical characteristics of the ash sluice water and in part on the composition of coal ash.

Table 5 shows the variations in coal ash composition with coal rank. The soluble ash species of such elements as iron, calcium, sodium, magnesium, potassium, and a variety of trace elements will produce a wide range of pH levels in the ash slurry. For example, a dramatic change in coal supply can result in changes of 1 to 2 pH units. The dominant alkaline constituents are Fe_2O_3, CaO, MgO, Na_2O, and K_2O; the acid constituents are SiO_2, Al_2O_3, and TiO_2 (Ref 93). This interaction of water and ash is the cause of most corrosion problems.

Scale formation on equipment internals and pipeline walls is always a frequent maintenance problem. A characterization of typical scale indicates that compounds of calcium, magnesium, sodium, and silica go into solution, concentrate, and precipitate out on internal surfaces. The reduction in pipeline diameter by heavy scale buildup, pump and valve scaling (reduction in internal clearance), and plugging of small control lines will require dismantling and descaling, or replacement. Most corrosion or erosion-related failures are complicated by scale formation.

Wet Bottom Ash Systems

Hot ash (clinkers or agglomerated slag) deposits of varying sizes are quenched and collected for conveying in the bottom ash system. As noted above, bottom ash systems are generally conveyed wet from collection to final disposal. However, some systems incorporate dewatering equipment and haul the ash by mobile equipment or belt conveyors to suitable disposal sites. The pumping and pipeline conveying of bottom ash slurries (normally 15 to 20% ash by weight) generally results in a highly abrasive but only moderately corrosive condition. Agglomerated granular (fused) bottom ash, with its irregular shape and large size, gives the slurry its abrasive nature. As compared to fly ash, it is inert and insoluble; therefore, chemical interaction be-

Table 5 Variations in coal ash composition by coal type

Coal type	Composition, %									
	SiO_2	Al_2O_3	Fe_2O_3	TiO_2	CaO	MgO	Na_2O	K_2O	SO_3	Ash
Anthracite	48–68	25–44	2–10	1–2	0.2–4	0.2–1	. . .	. . .	0.1–1	4–19
Bituminous	7–68	4–39	2–44	0.5–4	0.7–36	0.1–4	0.2–3	0.2–4	0.1–32	3–32
Subbituminous	17–58	4–35	3–19	0.6–2	2.2–52	0.5–8	. . .	. . .	3–16	3–16
Lignite	6–40	4–26	1–34	0–8	12.4–52	2.8–14	0.2–28	0.1–1.3	8.3–32	4–19

Source: Ref 92

tween the bottom ash and conveying water is less worrisome.

Current environmental regulations have resulted in the development of the zero discharge (closed-loop) system. The reuse of conveying water in either bottom ash or fly ash systems will concentrate soluble salts and will usually result in a more corrosive water chemistry. Therefore, most failures result from erosion, scaling, or plugging, but corrosion may become a problem if soluble salts are allowed to accumulate too long in closed-loop systems.

Mitigating the Problems

Operating problems due to corrosion, erosion, scaling, and plugging can be minimized by consideration of the following:

- Piping systems should have gradual bends, turns, and transitions and should be without sharp angles to reduce erosion and/or plugging
- Materials of construction (metals, nonmetals, and coatings) that are resistant to acid attack should be selected
- All storage vessels should have constant agitation to keep solids in suspension; this facilitates pumping
- Pipe diameters and line velocity must be such that suspended solids do not cause plugging
- Piping systems should be flushed during shutdown to remove solid before solids cementation reactions can occur
- Abrasive-resistant liners (replaceable metal, plastic, ceramic, and elastomers) should be selected for system components (pump impellers and casings, agitator blades, and pipe liners) for better control of erosion and corrosion of wetted parts

It is not possible to recommend a single approach or a material of construction that will work for every ash-handling application. Corrosion problems are reduced and equipment lifetime is extended by good case-by-case material selection, the existence of adequate quality control, operational simplicity, and the use of inspection/maintenance practices.

The Future

Present and pending regulations (mainly federal) tend to restrict the wet-dry options for utility and industry owners. Systems of the future will be dry processes for fly ash and will require suitable reuse or environmentally safe disposal of the ash. Wet systems for both bottom ash and combined fly ash/bottom ash will drift to the zero discharge concept and recycle their conveying water. Corrosion considerations will take on a new dimension, and the existing experience base will be used to solve future corrosion problems.

REFERENCES

1. G.C. Woerson and E.A. Tice, Corrosion of Feedwater Heater Tubing Alloys in Peaking Service, *Trans. ASME*, July 1985
2. G.E. Moller and B.C. Syrett, "Corrosion Related Failures in Feedwater Heaters," EPRI CS/NP-3743, Electric Power Research Institute, Oct 1984
3. A.R. Jacobstein *et al.*, "Failure Cause Analysis Feedwater Heaters," EPRI CS-1776, Electric Power Research Institute, April 1981
4. H.A. Grabowski and H.A. Klein, Corrosion and Hydrogen Damage in High Pressure Boilers, in *Second Annual Educational Forum on Corrosion*, National Association of Corrosion Engineers, 1964
5. "Manual for Investigation and Correction of Boiler Tube Failures," EPRI CS-3945, Electric Power Research Institute, April 1985
6. O. Jonas, Steam, in *Kirk-Othmer Encyclopedia of Chemical Technology*, Vol 21, 3rd ed., John Wiley & Sons, 1983
7. F.C. Olds, Power Eng., Vol 87 (No. 2), 1983, p. 42
8. A.F. Aschoff, Y.H. Lee, D.M. Sopocy, and O. Jonas, "Interim Consensus Guidelines on Fossil Plant Cycle Chemistry," EPRI CS-4629, Electric Power Research Institute, June 1986
9. O. Jonas, "Understanding Steam Turbine Corrosion," Paper 55, presented at Corrosion/84, National Association of Corrosion Engineers, 1984
10. Results of 1982 Survey, Committee on Water in Thermal Power Systems, Industrial Subcommittee, Steam Purity Task Group, American Society of Mechanical Engineers
11. L.H. Bennett *et al.*, "Economic Effects of Metallic Corrosion in the U.S.—A Report to the Congress by the National Bureau of Standards," United States Department of Commerce, March 1978
12. R.P. Dewey *et al.*, Paper 83-JPGC-Pwr-20, Presented at Joint Power Generation Conference, American Society of Mechanical Engineers, 1983
13. R.I. Jaffee, Ed., *Corrosion Fatigue of Steam Turbine Blade Materials*, Pergamon Press, 1983
14. O. Jonas and A. Pebler, Characterization of Operational Environment for Steam Turbine-Blading Alloys, EPRI CS-2931, Electric Power Research Institute, 1984
15. Research and Development Status Report, Coal Combustion Systems Div., *EPRI J.*, Vol 5 (No. 3), 1980, p 44
16. F.J. Heymann *et al.*, Steam Turbine Blades: Considerations in Design and a Survey of Blade Failures, EPRI CS-1967, Electric Power Research Institute, Aug 1981
17. A. Whitehead and G.F. Wolfe, "Steam Purity for Industrial Turbines," Paper presented at the Eighth ASME Industrial Power Conference, Houston, TX, American Society of Mechanical Engineers, Oct 1980
18. B.W. Bussert, R.M. Curran, and G.C. Gould, Paper 78-JPGC-Pwr-9, Joint Power Generation Conference, American Society of Mechanical Engineers, 1978
19. F.F. Lyle, Jr. and H.C. Burghard, Jr., Paper 216, presented at Corrosion/82, National Association of Corrosion Engineers, 1982
20. F.F. Lyle, Jr. and H.C. Burghard, Jr., "Steam Turbine Disc Cracking Experience," EPRI NP-2429-LD, Electric Power Research Institute, June 1982
21. Workshop on Stress Corrosion Cracking of Turbine Rotors and Discs, Electric Power Research Institute and Central Electricity Generation Board, Nov 1979
22. J.M. Hodge and I.L. Mogford, *Proc. Inst. Mech. Eng.*, Vol 193, 1979, p 93
23. W. Engelke *et al.*, "Design, Operating and Inspection Considerations to Control Steam Corrosion of LP Turbine Discs," Paper presented at American Power Conference, Chicago, IL, April 1983
24. Research Projects RP-311, RP-502, RP-700, RP-912, RP-969, RP-1068, RP-1124, RP-1264, RP-1398, RP-1886, RP-1929, RP-2712, TPS82-633, RFP-2297, Electric Power Research Institute
25. "Annotated Bibliography of Public Literature on Stress Corrosion of Turbine Discs," EPRI TPS82-683, Electric Power Research Institute, to be published
26. H. Haas, Major Damage Caused by Turbine or Generator Rotor Failures in the Range of the Tripping Speed, *Der Maschinenschaden*, Vol 50, 1977, p 6
27. A. Atrens *et al.*, BBC Experience With Low-Pressure Steam-Turbine Blades, in *Corrosion Fatigue of Steam Turbine Blade Materials*, Pergamon Press, 1983
28. O. Jonas, *J. Test. Eval.*, Vol 6 (No. 1), 1978, p 40
29. J.L. Gray, *Proc. Inst. Mech. Eng.*, Vol 186 (No. 32), 1972, p 379
30. O. Jonas, W.T. Lindsay, Jr., and N.A. Evans, in *Turbine Steam Purity—1979 Update: Water and Steam*, J. Straub and K. Scheffler, Ed., Pergamon Press, 1980
31. H.G. Heitmann and P. Schub, "Initial Experience Gained With a High pH Value in the Secondary System of Pressurized Water Reactors," Paper presented at Water Chemistry III, British Nuclear Engineering Society, 1983
32. H.G. Heitmann and W. Kastner, *VGB-Krafwerkstechnik*, Vol 57 (No.6), 1982, p 211
33. J.P. Cerdan *et al.*, "Erosion Corrosion in Wet Steam: Impacts of Variables and Possible Remedies," Paper presented at Water

Chemistry and Corrosion in Steam Water Loops of Nuclear Power Stations Symposium, March 1980
34. J. Marceau, "Erosion Corrosion by Wet Steam: Design Choices Sizing, Materials, Manufacturing," Paper presented at Water Chemistry and Corrosion in Steam-Water Loops of Nuclear Power Stations Symposium, March 1980
35. G.J. Bignold, "Erosion Corrosion in Nuclear Steam Generators," Paper presented at Water Chemistry II, British Nuclear Engineering Society, 1980
36. P. Berge *et al.*, "Effects of Chemistry on Corrosion-Erosion of Steels in Water and Wet Steam," Paper presented at Water Chemistry II, British Nuclear Engineering Society, 1980
37. L.D. Kramer *et al.*, *Mater. Perform.*, Vol 14, 1975, p 15
38. J. Mancuso *et al.*, "Development of Low-Pressure Turbine Coatings Resistant to Steam-Borne Corrodents," EPRI CS-3139, Electric Power Research Institute, June 1983
39. R. Garnsey, *Combustion*, Vol 52 (No. 2), 1980, p 39
40. O. Jonas, *Combustion*, Vol 50 (No. 6)
41. "Boiler Water Limits and Steam Purity Recommendations," American Boiler Manufacturers Association, 1982
42. "Chemical Control of Boiler Feedwater, Boiler Water, and Saturated Steam for Drum-Type and Once-Through Boilers," Generation Operation Memorandum 72, Issue 4, Central Electricity Generation Board
43. New Guidelines for Feedwater and Boiler Water for Power Stations, *VGB-Kraftwerks, Technik Mitteilugen*, Vol 52, 1972, p 167
44. "Water Quality of the Feedwater and the Boiler Water for Recirculating Boilers," JIS B8223, "Feedwater Quality for Once-Through Boilers," JIS B8224, Japanese Institute of Standards
45. D.E. Simon II, in *Proceedings of the 36th International Water Conference*, Engineering Society of Western Pennsylvania, 1976, p 65
46. O. Jonas, M. Roidt, and A.S. Manocha, "Dynamic Deposition and Solubility of NaCl in Superheated Steam," Paper 19, presented at the 44th International Water Conference, Pittsburgh, PA, Engineering Society of Western Pennsylvania, Oct 1983
47. W.E. Allmon *et al.*, "Deposition of Corrosion Salts From Steam," EPRI NP-3002, Electric Power Research Institute, April 1983
48. W.T. Lindsay, Jr., *Power Eng.*, May 1979, p 68
49. H.G. Heitmann, *Mitl. Ver. Grosskesselbetr.*, Vol 90, 1964, p 171
50. F.F. Straus, "Steam Turbine Blade Deposits," Bulletin 59, University of Illinois, June 1946
51. M.A. Styrikovich, O.I. Martynova, and Z.S. Belova, *Therm. Eng.*, Vol 12 (No. 9), 1965, p 115
52. O.I. Martynova, "Transport and Concentration Processes of Steam and Water Impurities in Steam Generating Systems," Moscow Power Institute, 1962, p 547-562
53. O. Jonas, "Water, Steam, and Turbine Deposit Chemistry in Phosphate Treated Drum Boiler Units," EPRI RP 1886-9, Electric Power Research Institute, to be published
54. C.J. Czajkowski, "Corrosion and SCC of Bolting Materials in Light Water Reactors," Paper 10, presented at the International Symposium on Environmental Degradation of Materials in Nuclear Power Systems-Water Reactors, Myrtle Beach, SC, Aug 1983
55. D.J. Turner, "SCC of LP Turbines: The Generation of Potentially Hazardous Environments From Molybdenum Compounds," RD/L/N 204/74, Central Electricity Research Laboratory, 1974
56. J.F. Newman, "The SC of Turbine Disc Steels in Dilute Molybdate Solutions and Stagnant Water," RD/L/N 215/74. Central Electricity Research Laboratory, 1974
57. C.J. Czajkowski and J.R. Weeks, "Examination of Turbine Discs From Nuclear Power Plants," Paper 220, presented at Corrosion/82, National Association of Corrosion Engineers, 1982
58. *Elect. Week*, 10 April 1978, p 9
59. M.O. Speidel and A. Atrens, Ed., *Corrosion in Power Generating Equipment*, Plenum Press, 1984
60. M.O. Speidel, in *International Conference on Materials*, Vol 1, Aug 1979
61. "Characterization of Environmentally Assisted Cracking for Design, State of the Art," NMAB-386, National Research Council, 1982
62. International Conference on Advances in Life Prediction Methods, Albany, NY, American Society of Mechanical Engineers, April 1983
63. R. Pigott *et al.*, "Increasing Availability of LP Steam Turbines by Design and Materials Selection," Paper presented at the American Power Conference, Chicago, IL, April 1982
64. Major Damage to Steam-Turbosets, *Der Maschinschaden*, Vol 50 (No. 6), 1977, p 193
65. International Symposium on Environmental Degradation of Materials in Nuclear Power Systems-Water Reactors, Myrtle Beach, SC, Aug 1983
66. Carbon Steel Workshop, Electric Power Research Institute, May 1979
67. W.G. Clark, Jr. *et al.*, Paper 81-JPGC-Pwr-31, presented at Joint Power Generation Conference, American Society of Mechanical Engineers, 1981
68. O. Jonas, Guidelines Help Designers Protect Against Localized Corrosion, *Power*, Vol 130 (No. 8), Aug 1986, p 35-38
69. J.M. Bloom and J.C. Ekvall, Ed., *Probabilistic Fracture Mechanics and Fatigue Methods: Applications for Structural Design and Maintenance*, STP 798, American Society for Testing and Materials, 1983
70. W.T. Reid, *External Corrosion and Deposits—Boilers and Gas Turbines*, Elsevier, 1971
71. E. Raask, *Mineral Impurities in Coal Combustion*, Hemisphere, 1985
72. EPRI Research Project RP 1890-4, Electric Power Research Institute, 1985
73. I.P. Ivanova and L.A. Svistunova, Corrosion of 12Kh1MF Steel and Various Corrosion Resistant Coatings in a Medium of Flue Gases During Combustion of Anthracite, *Teploenergetika*, Vol 18 (No. 1), 1971, p 60-63
74. I.P. Ivanova, V.P. Kaminskii, and A.G. Belyaeva, High Temperature Corrosion of Waterwall Tubes in Supercritical Boilers Burning Anthracite Fines, *Teploenergetika*, Vol 19 (No. 1), 1972, p 16-18
75. I.P. Ivanova and Yu.L. Marshak, High-Temperature Corrosion of Screens During Combustion of Anthracite Culm," *Teploenergetika*, Vol 22 (No. 2), 1975, p 15-18
76. S.F. Chou, P.L. Daniel, A.J. Blazewicz, and R.F. Dudek, Hydrogen Sulfide Corrosion in Low-NO_x Combustion Systems, in *High-Temperature Corrosion in Energy Systems*, M.F. Rothman, Ed., American Institute of Mining, Metallurgical, and Petroleum Engineers, 1985
77. R.C. Corey, H.A. Grabowski, and B.J. Cross, *Trans. ASME*, Vol 71 (No. 8), 1949, p 951-963
78. D.N. French, *Metallurgical Failures in Fossil-Fired Boilers*, John Wiley & Sons, 1983
79. D.R. Holmes and D.B. Meadowcroft, "Fireside Corrosion and Problems of Tube Life Prediction," Paper presented at Symposium on Thermal Utilities Boiler Reliability, Hamilton, Ontario, McMaster University, May 1983
80. T. Flatley, E.P. Latham, and C.W. Morris, *Mater. Perform.*, Vol 20 (No. 5), 1981, p 12-17
81. W.A. McGill and M.J. Weinbaum, *Mater. Perform.*, Vol 17 (No. 1), 1978, p 16-20
82. M.H. Brown, W.B. DeLong, and J.R. Auld, Corrosion by Chlorine and by Hydrogen Chloride at High Temperatures, *Ind. Eng. Chem.*, Vol 39, 1947, p 839-844
83. Y. Ihara, H. Ohgame, and K. Sakiyama, The Corrosion Behavior of Iron in Hydrogen Chloride Gas and Gas Mixtures of Hydrogen Chloride and Oxygen at High Temperatures, *Corros. Sci.*, Vol 21 (No. 12), 1981, p 805-817
84. D.A. Vaughan, P.D. Miller, and W.K. Boyd, Fireside Corrosion in Municipal Incinerators Versus PVC Content of the Refuse, in *Proceedings of the 1974 National Incinerator Conference* (Miami, FL), May 1974, p 179-190
85. H.H. Krause, High Temperature Corrosion Problems in Waste Incineration Systems," *J. Mater. Energy Syst.*, Vol 7 (No. 4), 1986, p 322-332
86. S. Brooks and D.B. Meadowcroft, The Influence of Chlorine on the Corrosion of Mild and Low Alloy Steels in Substoichiometric Combustion Gases, in *Corrosion Resistant Materials for Coal Conversion Systems*, D.B. Meadowcroft and M.I. Manning, Ed., Applied Science, 1983, p 105-120
87. P.L. Daniel, J.D. Blue, and J.L. Barna, Furnace-Wall Corrosion in Refuse-Fired Boilers, in *Proceedings of the 1986 National Waste Processing Conference* (Denver, CO), 1986, p 221-228
88. H.H. Krause, D.A. Vaughan, and P.D. Miller, Corrosion and Deposits From Combustion of Solid Waste, *J. Eng. Power, (Trans. ASME)*, Vol 95 (No. 1), 1973, p 45-52
89. G.Y. Lai, "High Temperature Corrosion in Various Waste Incineration Environments, Paper presented at CORROSION/86, National Association of Corrosion Engineers, March 1986
90. D.R. Holmes, Ed., *Dewpoint Corrosion*, Ellis Horwood Ltd., 1985
91. "Final Report of the Dewpoint Investigation," U.K. Department of Trade and Industry, June 1986
92. "Coal Ash Disposal Manual," EPRI CS-2049, Electric Power Research Institute, Oct 1981
93. *Steam: Its Generation and Use*, 39th ed., The Babcock & Wilcox Company, 1978, p 15-4

SELECTED REFERENCES

- J.A. Beavers, A.K. Agrawal, and W.E. Berry, "Corrosion-Related Failures in Power Plant Condensers," EPRI NP-1468, Electric Power Research Institute, August 1980
- J.A. Beavers and G.H. Koch, Review of Corrosion Related Failure in Flue Gas Desulfurization Systems, *Mater. Perform.*, Vol 21, Oct 1982, p 13
- *Combustion: Fossil Fuel Power Systems*, 3rd ed., Combustion Engineering Corporation, 1981
- R.B. Dooley and H.J. Westwood, "Analysis and Preventions of Boiler Tube Failures," Report 83-237G, Canadian Electrical Association, Nov 1983
- D.N. French, *Metallurgical Failures in Fossil-Fired Boilers*, John Wiley & Sons, 1983
- "Interim Consensus Guidelines on Fossil Plant Cycle Chemistry," EPRI CS-4629, Electric Power Research Institute, June 1986
- G.H. Koch, J.A. Beavers, N.G. Thompson, and W.E. Berry, "Literature Review of FGD Construction Materials," EPRI CS-2533, Electric Power Research Institute, Aug 1982
- G.H. Koch and B.C. Syrett, Progress in EPRI Research on Materials for Flue Gas Desulfurization Systems, in *Dew Point Corrosion*, D.R. Holmes, Ed., Ellis Horwood Ltd., 1985
- "Manual For Inspection and Correction of Boiler Tube Failures," EPRI CS-3945, Electric Power Research Institute, April 1985
- W.C. Martin, D.A. Froelich, C.V. Weilert, and P.N. Dyer, "Acid Deposition on Ductwork," EPRI CS-3240, Electric Power Research Institute, Nov 1983
- D.B. Meadowcroft and M.I. Manning, Ed., *Corrosion Resistant Materials for Coal Conversion Systems*, Applied Science, 1983
- "Proceedings of Ninth Symposium on Flue Gas Desulfurization," EPRI CS-4390, Electric Power Research Institute, Jan 1986
- W.T. Reid, *External Corrosion and Deposits—Boilers and Gas Turbines*, Elsevier, 1971
- H.S. Rosenburg, C.W. Kistler, L.O. Nilsson, L.J. Nowacki, J.A. Beavers, G.H. Koch, and H.H. Krause, "Construction of Materials for Wet Scrubber: Update Volume 1 and 2," EPRI CS-3350, Electric Power Research Institute, July 1984
- M.F. Rothman, Ed., *High-Temperature Corrosion in Energy Systems*, The Metallurgical Society, 1984
- *Steam: Its Generation and Use*, 39th ed., The Babcock & Wilcox Company, 1978
- B.C. Syrett, Ed., "Seminar Proceedings: Prevention of Condenser Failures—The State of the Art," EPRI CS-4329-SR, Electric Power Research Institute, August 1980
- B.C. Syrett, G.H. Koch, Ed., and F. Mansfeld, Review of Corrosion in SO_2 Scrubbers, in *Proceedings of the Fourth Asian-Pacific Corrosion Control Conference* (Tokyo, Japan), May 1985

Corrosion in the Automotive Industry

CORROSION has in recent years become a major concern of automakers around the world. The use of deicing chemicals has increased tenfold in North America since the mid-1950s (Ref 1) and is the major cause of corrosion of automobile body panels in the "salt belt" (Fig. 1). Other environmental factors, including air pollution in industrial areas, exposure to marine atmospheres in coastal regions, acid (low-pH) precipitation, and dust control procedures on rural roads, also contribute to the increased corrosion of automobile body panels and other components.

The auto industry has responded to the corrosion problem in a number of ways, including the use of more corrosion-resistant materials, improved paint systems, and better design practices. This article will discuss the use of precoated sheet steels, paint systems and methods of application, and the use of new, more corrosion-resistant materials to combat corrosion.

Forms of Corrosion

The automotive environment can result in several forms of corrosion. Chief among these are uniform or general corrosion, crevice corrosion, galvanic corrosion, poultice corrrosion, and pitting corrosion. Other forms include inside-out corrosion, outside-in corrosion, scab corrosion (similar to outside-in corrosion, but it usually occurs at joints and crevices where dirt and moisture are trapped), saponification, cathodic delamination, and filiform corrosion. These are further discussed in Ref 1.

Uniform corrosion occurs over the entire exposed surface of a component in the form of a general thinning of the material. Because of the uniform nature of the attack, this is the least damaging of the forms of automotive corrosion. More information on this form of corrosion is available in the article "General Corrosion" in this Volume.

Crevice corrosion (Fig. 2) is a severe form of localized attack that is normally associated with small pockets of stagnant electrolyte that can form at holes or joints or under fasteners. The attack in this form of corrosion is rapid, and it can result in early failure. The section "Crevice Corrosion" of the article "Localized Corrosion" in this Volume gives detailed information on this form of attack.

Galvanic corrosion results from the contact of two dissimilar metals in the presence of an electrolyte. The more active metal or alloy becomes the anode of the couple and may be subject to rapid attack. This type of corrosion has limited the automotive applications for aluminum alloys, because aluminum will corrode preferentially in contact with steel. For this reason, aluminum must be insulated from direct contact with steel through the use of nonconductive or barrier-type spacers or sealers (Ref 1). Figure 3 illustrates schematically the mechanism of galvanic corrosion. More information on this form of corrosion is available in the section "Galvanic Corrosion" of the article "Localized Corrosion" in this Volume.

Poultice corrosion is a form of crevice corrosion that occurs under deposits of road debris, such as mud that may be deposited on the underside of fenders and at other locations. The deposits hold corrosive substances such as road salt in contact with the body material and retard or prevent runoff. Electrolyte composition gradients (Fig. 4) are thought to be the most common cause of this form of corrosion (Ref 2).

Pitting corrosion (Fig. 5) is similar to crevice corrosion in that it is a localized attack. It occurs most often in areas of low pH that are depleted in oxygen but have a relatively high concentration of chlorides (Ref 2). Once pitting is initiated, the mechanism is similar to that of crevice corrosion, with the pit itself acting as the crevice (Ref 1). Pits can initiate at metal inhomogeneities, breaks in a protective coating, surface deposits, or other defects. More information on pitting corrosion is available in the section "Pitting" of the article "Localized Corrosion" in this Volume.

All of these forms of corrosion are the result of reaction with the environment and all require the presence of water. These and other forms of corrosion that are prevalent in the automotive environment are discussed further in Ref 1.

Precoated Steels

Beginning in the late 1950s, automakers responded to the challenge of increased corrosion of body panels through the use of precoated sheet steels, especially galvanized (zinc-coated) steels. The use of zinc-coated steels has steadily increased ever since, reaching an estimated total of 156 kg (344 lb) in the typical U.S. automobile in 1986 (Table 1). This represents an increase of 13.5% over the 137 kg (303 lb) of zinc-coated steel per car used in 1985 (Ref 3).

The auto industry has also employed other types of coatings, including zinc-aluminum and zinc-iron alloys, aluminum, tin, and long terne (lead-tin alloy). Composite coatings are being developed that offer better corrosion resistance in laboratory tests than Zincrometal, which is steel coated with a zinc-rich organic primer.

Coatings have traditionally been designated according to standard American Society for Testing and Materials coating weight designations, but the desire for more consistency in formability and weldability has placed an increased emphasis on tighter coating weight limits. Automakers are specifying more and more coated steels in terms of grams of coating per square meter (g/m^2) (Ref 1) or ounces per square foot (oz/ft^2). Some coatings are still described by coating thickness; for example, Zincrometal generally has a coating that is 15 μm thick.

Zinc-Coated Steels

Zinc-coated (galvanized) steels are the most common coated steels used as body panels in the automotive industry (Fig. 6). They can be obtained with widely varying coating weights (20 to

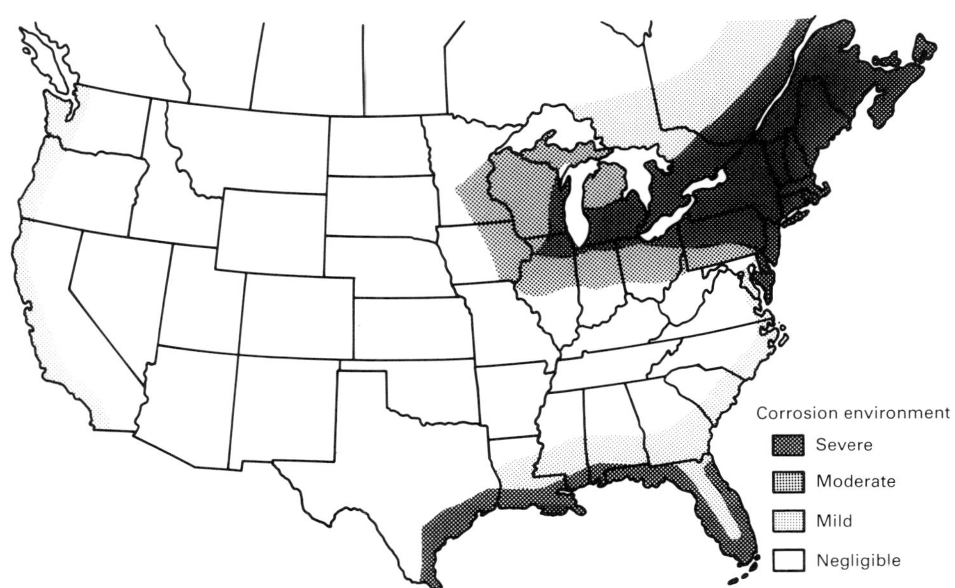

Fig. 1 Map showing the severity of automotive corrosion environments in the United States and southern Canada. Source: Ref 1

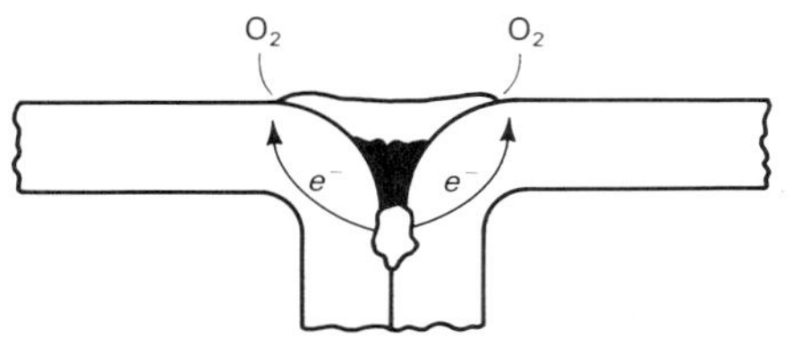

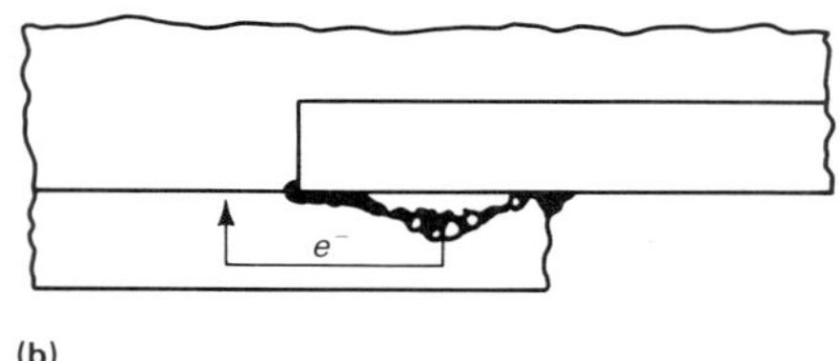

Fig. 2 Schematics showing the mechanism of crevice corrosion at a joint. Crevice corrosion is common at weldments or sheet metal joints (a) and rough surfaces where electrolyte can be trapped, and it can occur at tightly sealed lap joints, as in (b). Source: Ref 2

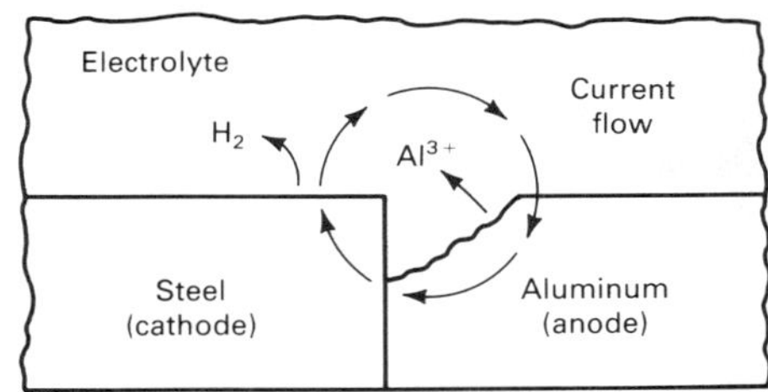

Fig. 3 Schematic showing the mechanism of galvanic corrosion, which occurs when two dissimilar metals are placed in contact with one another in the presence of an electrolyte. The more electrochemically active of the two metals will act as an anode and will corrode preferentially, while the less active metal is protected. Source: Ref 2

160 g/m^2) and with one side coated, both sides coated equally, or different coating weights on each side.

One-side galvanized steel is produced by either hot dipping or electrogalvanizing. The zinc-coated surface faces inward in the body panel to prevent inside-out corrosion. One-side galvanized steel is commonly used for hoods, doors, and quarter panels (Ref 4).

One-and-a-half side galvanized steel uses pure zinc on interior body surfaces and a zinc-iron alloy on the outer surface. This enhances cosmetic corrosion resistance on exterior surfaces of body panels in the same types of applications as one-side galvanized steel.

Two-side galvanized steel uses pure zinc coatings on both sides of the steel sheets in such applications as floor pans, wheel houses, and interior and exterior structural surfaces (for example, door inner panels and structural components).

Hot-dip galvanized steel has seen the widest application by automakers. It provides effective corrosion protection and is more economical than steels coated by other processes for a given coating weight (Ref 5).

Hot dip galvanized steel is produced from cold-rolled steel by heat treating or normalizing on the coating line to obtain the desired mechanical properties, followed by hot dipping. The coated strip can then be postannealed to improve formability and is finally tempered to obtain the required flatness and surface texture. Formability and weldability vary with coating thickness and with the number of coating defects (Ref 5). Two-side hot dip galvanized steel is currently being used in most auto body panels by one U.S. automaker. Paintability is being enhanced by blowing a fine zinc powder onto the surface of the steel as it emerges from the molten zinc bath (Ref 6).

Hot dip galvanizing is also being used to improve the corrosion resistance of various automotive components after fabrication (Ref 7). This type of process is beneficial in combating undervehicle corrosion; suspension parts, other undervehicle components, and (on an experimental basis) even entire floor pans have been galvanized after fabrication. Parts can be readily formed and fabricated without fear of damaging a prefabrication coating (Ref 7).

Another two-side hot dip product, galvannealed steel, uses a zinc-iron alloy coating on both sides of the sheet. It is produced by hot dipping and then annealing or wiping the steel sheet to form the zinc-iron alloy. Galvannealed steel offers better weldability and paintability than hot dip galvanized steel, and accelerated corrosion tests have shown it to possess corrosion resistance equal to or better than that of conventional hot dip galvanized steel (Ref 4). Galvannealed steel is being used in both body and underbody applications. Steelmakers have recently focused their development efforts on improving the formability, mechanical properties, fabricability, and paintability of hot-dip galvanized steels for the automotive industry (Ref 8).

Electrogalvanized steels have had limited application in the automotive industry mainly because of their higher cost relative to hot dip galvanized steels. Production costs are largely related to the amount of electricity required to produce coatings of sufficient thickness. Despite this, at least one U.S. automaker plans to employ a differentially coated, two-side electrodeposited zinc alloy (Zn-15Fe) coated steel for body panels (Ref 4).

Although initial reports based on laboratory salt spray corrosion tests indicated that electrodeposited zinc and zinc alloy coatings offered two to forty times the corrosion resistance of hot dip coatings of equivalent thickness, a comprehensive undervehicle testing program revealed that this is not the case (Ref 9). These tests compared a variety of electrogalvanized zinc and zinc alloy coatings with heavier hot dip zinc and zinc alloy coatings, and they showed that the heavier hot dip coatings were superior in resistance to all forms of corrosion investigated (Table 2). The performance of the hot dip galvanized steels was rated as excellent. The poor performance of the electrogalvanized steels was attributed to the rougher surface of the electrodeposited coatings, which resulted in higher electrochemical corrosion currents. Current work is focusing on characterizing the surface roughness and texture of electrodeposited coatings and on correlating surface characteristics and corrosion resistance in undervehicle tests (Ref 9).

The surface roughness of electrogalvanized steels is advantageous for paint adhesion, however, and electrocoated steels also offer better formability and better weldability than hot dip steels (Ref 1). Electrogalvanized surfaces generally contain fewer defects than hot-dip steels, so it is easier to achieve a Class A paint finish on electrocoated steels. For these reasons, use of electrocoated steels (both pure zinc and zinc alloy coatings) in the automotive industry is on the rise; this trend can be expected to continue if research into more efficient electrodeposition processes proves beneficial and if the cost of these materials is lowered.

Zincrometal is also used extensively for outer body panels in automobiles. First introduced in 1972, Zincrometal is a coil-coated product consisting of a mixed-oxide underlayer containing metallic zinc particles and a zinc-rich organic (epoxy) topcoat. It is weldable, formable, paintable, and compatible with commonly used adhesives. Zincrometal is used primarily in one-side applications to protect against inside-out corrosion. The corrosion resistance of Zincrometal is not as good as that of hot dip galvanized steels (Ref 4), and its use is expected to decline as more electrogalvanized steels and other types of coatings are employed.

Zinc-alloy coated steels have also been developed. Coatings include zinc-iron (15 to 80%

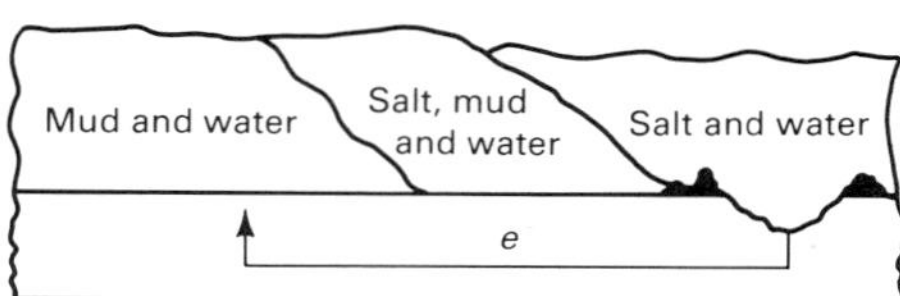

Fig. 4 Schematic showing the mechanism of poultice corrosion. The most common cause of this type of corrosion is thought to be electrolyte composition gradients. In the example shown, clumps of mud and water have collected, and the varying concentrations of salt and water within the clump encourage corrosion. Source: Ref 2

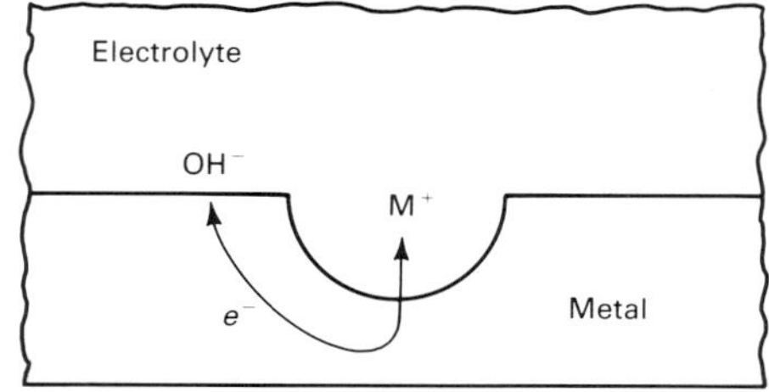

Fig. 5 Schematic showing the mechanism of pitting corrosion. Like crevice corrosion, pitting occurs at localized areas depleted of oxygen, low in pH, and high in chlorides. Source: Ref 2

Table 1 Use of zinc-coated steel for a typical 1986 model U.S. car

Type	Amount of steel kg	Amount of steel lb	Amount of zinc kg	Amount of zinc lb
One-side galvanized	33.5	74	0.55	1.21
Two-side galvanized	93	205	3.05	6.72
Zincrometal	29.5	65	0.19	0.41
Net total	156	344	3.8	8.34

Source: Ref 3

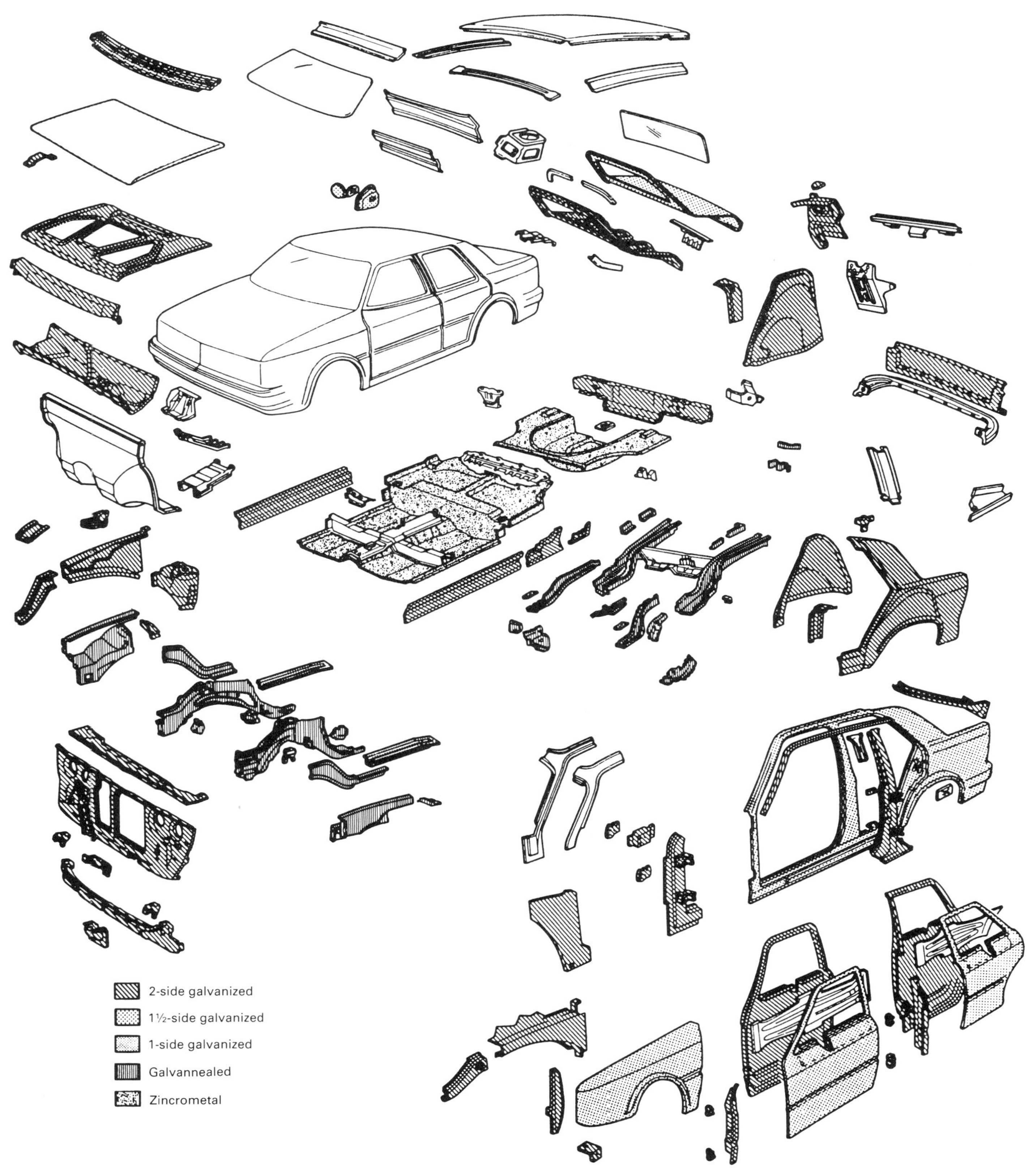

Fig. 6 Use of zinc-coated steels in a 1987 model by one U.S. automaker. Source: Ref 4

Table 2 Corrosion of unpainted coated steel test coupons after 2 years of undervehicle exposure

Material	Coating weight per side g/m²	Coating weight per side oz/ft²	Steel thickness mm	Steel thickness in.	Surface area showing base metal attack, % Vehicle 1	Surface area showing base metal attack, % Vehicle 2	Average pit depth Vehicle 1 μm	Average pit depth Vehicle 1 mils	Average pit depth Vehicle 2 μm	Average pit depth Vehicle 2 mils
Hot dip										
Galvanized 1	120/150	0.39/0.49	0.71	0.028	0.6	14	0	0	15	0.6
Galvanized 2	100/120	0.33/0.39	0.90	0.035	0.3	27.3	11	0.43	56	2.2
Galvanized 3	55/90	0.18/0.30	0.45	0.018	0.5	5.0	0	0	15	0.6
Galvannealed 1	80/120	0.26/0.39	1.42	0.055	0	1.0	0	0	22	0.87
Galvannealed 2	75/85	0.25/0.28	0.89	0.035	0.3	32.8	11	0.43	86	3.4
One-side galvannealed	66	0.22	0.66	0.025	25	56.5	48	1.9	67	2.6
One-side electrodeposited										
Zn	90	0.30	0.88	0.034	61	86	64	2.5	120	4.7
Zn-15Ni-0.4Co	37	0.12	0.70	0.0275	46	67.5	75	3	81	3.2
Zn-16Ni	20	0.065	0.68	0.027	85	93.5	83	3.3	100	4
Zn-16Ni	40	0.13	0.68	0.027	38	79.3	73	2.9	128	5
Zn-16Al	25	0.08	0.68	0.027	59	84.3	64	2.5	97	3.8
Zn-22Al	40	0.13	0.68	0.027	54	76.5	64	2.5	90	3.5
Zinc-rich primer										
One-side Zincrometal	40	0.13	0.92	0.036	10.8	17.3	53	2.1	73	2.9
Uncoated										
Cold-rolled steel	...	...	0.51	0.020	100	100	>250	>10	>250	>10

(a) Vehicle 1, 660 days, 51 000 km (31 700 miles); vehicle 2, 660 days, 53 500 km (33 250 miles). Source: Ref 9

Fe) and zinc-nickel (10 to 14% Ni) alloys. These coatings have been developed for the most part by Japanese steelmakers and are applied by electrodeposition. Zinc-iron coatings offer excellent corrosion resistance and weldability. Zinc-nickel coatings are more corrosion resistant than pure zinc coatings, but problems include brittleness from residual stresses and the fact that the coating is not completely sacrificial, as is a pure zinc coating. This can lead to accelerated corrosion of the steel substrate if the coating is damaged (Ref 10). Multilayer coatings that take advantage of the properties of each layer have been developed in Europe. An example of this is Zincrox, a zinc-chromium-chromium oxide coating (Ref 10). The CrO_x top layer of this coating acts as a barrier to perforation and provides excellent paint adhesion and weldability (Ref 10).

Another relatively new development in zinc alloy coatings is Galfan, a Zn-5Al-mischmetal alloy coating applied by hot dipping. Applications in the United States are limited, but European automakers have used Galfan in such applications as brake servo housings, headlight reflectors and frames, and universal joint shrouds (Ref 11). Galfan is also being considered for oil pans, fuel tanks, and heavily formed body panels. Table 3 compares the undervehicle corrosion resistance of Galfan with that of hot dip galvanized steel. More information on zinc coatings is available in the articles "Corrosion of Zinc," "Electroplated Coatings," and "Hot Dip Coatings" in this Volume.

Other Coated Steels

Aluminum-coated (aluminized) steels containing 8 to 12% Si in the coating are used by automakers for applications involving high-temperature corrosion resistance, such as in exhaust systems, heat shields, and underhood components. The Al-45Zn coated steels are used in similar applications (Ref 1).

Long terne coated (lead-tin alloy, usually 3 to 8% Sn) steels offer corrosion protection in gas tanks, fuel lines, and brake lines and do not contaminate gasoline (Ref 1). Terne is cathodic to the steel substrate and therefore does not offer the sacrificial corrosion protection of galvanized coatings. Terne-coated steels are also sometimes given a thin coating of electrodeposited nickel as an intermediate layer; this material is used in applications similar to those of regular terne (Ref 1).

Organic composite coated steels have been developed mainly by Japanese steelmakers in cooperation with automakers in that country, although development is underway in other countries as well. These coil-coated products generally employ an electroplated zinc alloy base layer and a chemical conversion coating under a thin organic topcoat containing a high percentage of metal powder (Ref 12-14). The thinness of the organic topcoat allows for good formability without the risk of damaging the coating.

Figure 7 (Ref 12) compares the corrosion resistance of one of these organic composite coated sheet steels to cold-rolled steel and to Zincrometal. Another of these products uses an organic-silicate composite topcoat only about 1 μm thick and has corrosion resistance and weldability superior to that of Zincrometal (Ref 13). A bake-hardenable version of this material has also been developed (Ref 13). Researchers at a third Japanese steel company have developed a bake-hardenable organic composite coated sheet steel with a 0.8- to 1.5-μm thick organic topcoat. The material possesses corrosion resistance, formability, and weldability equivalent to that of Zincrometal-KII, which uses a 7-μm thick top coat (Ref 14). Production of these composite coated materials is increasing in anticipation of increased demand from Japanese automakers.

A similar material has been developed in the United States. This material has an electrodeposited zinc alloy base coat, a mixed intermediate layer of chromium oxide and zinc dust, and an organic topcoat for barrier protection (Ref 15). Figure 8 is a micrograph showing the cross section of the composite coated steel. In salt spray tests comparing this material to electrodeposited zinc-nickel and Zincrometal, zinc-nickel failed after 216 h, Zincrometal at 480 h, and the composite coating at 960 h (Ref 15). This material was developed to have weldability, formability, and adhesive compatibility similar to that of Zincrometal, and developmental work is continuing.

Paint Systems

The primary function of automotive paint systems is to provide a protective barrier against corrosive substances in the outside environment. This is accomplished through the use of a paint system comprising a conversion coating, one or more coats of primer, and a colored topcoat. A typical system might employ a conversion coating, an electrodeposited primer 30 μm thick, a base color coat 15 μm thick, and a clear topcoat

Table 3 Corrosion of hot-dip galvanized and Galfan coated steel specimens in undervehicle testing

Exposed in Buffalo and Detroit for the entire winter of 1981–1982

Material	Specimen orientation	Coating weight Exposed side g/m²	Coating weight Exposed side oz/ft²	Coating weight Protected side g/m²	Coating weight Protected side oz/ft²	Change in coating weight, %	Pitting	Surface appearance
Buffalo								
Galfan	Horizontal	216	0.71	223	0.73	3	Slight	Smooth, uniform gray
Galvanized	Horizontal	174	0.57	272	0.89	36	Severe(a)	Rough, dirty brown
Detroit								
Galfan	Vertical	216	0.71	238	0.78	9	Slight	Smooth, uniform gray
Galvanized	Vertical	207	0.68	253	0.83	18	Severe(a)	Rough, dirty brown

(a) Galvanized samples averaged 53 times as many pits per unit area than Galfan samples. Source: Ref 11

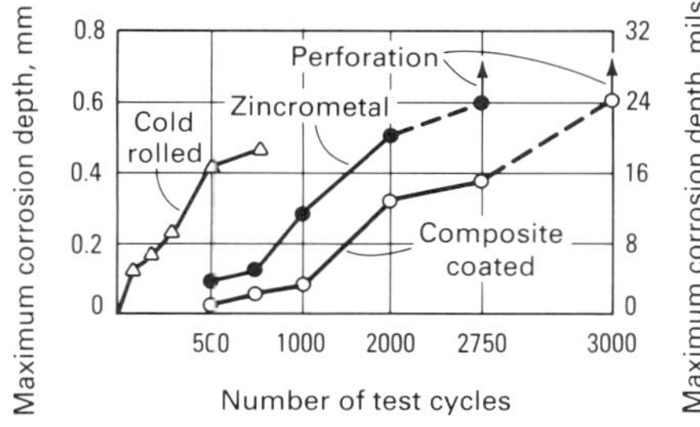

Fig. 7 Corrosion of heavily worked samples of a composite coated steel, Zincrometal, and cold-rolled steel in a laboratory cyclic test. Test consisted of 28-min cycles of dipping in 5% saline solution at 40 °C (100 °F), humidifying at 50 °C (120 °F), and drying at 60 °C (140 °F). Source: Ref 12

40 μm thick. Because corrosion usually initiates at coating defects, emphasis is placed on obtaining the most defect-free coating system possible. Detailed information on the types of paint formulations available, the application of paints, and corrosion of painted metals is available in the article "Organic Coatings" in this Volume, and Table 4 outlines the development in the last 25 years of primer and paint systems at one U.S. automaker.

Surface Preparation

Metal Cleaning. Surface preparation for painting begins with cleaning of the metal. Cleaning usually involves the use of an alkaline cleaner and one or more water rinses to remove dirt and contaminants—for example, oil left from the stamping operation—that can limit adhesion of the subsequent coatings. More information on chemical cleaning of metals is available in the article "Cleaning for Surface Conversion" in this Volume.

A phosphate conversion coating is then applied to the clean metal by either spraying or immersion. A typical process sequence involves application of the phosphate solution to the metal, a cold water rinse, and a chromic acid rinse to seal the conversion coating and to enhance corrosion resistance. More information on the application and corrosion resistance of phosphate conversion coatings is available in the article "Phosphate Conversion Coatings" in this Volume; chromate conversion coatings are discussed in the article "Chromate Conversion Coatings" in this Volume.

The conversion coating enhances the corrosion resistance of the metal surface, but more importantly, the crystalline nature of the coating provides excellent paint adhesion for subsequent application of primer and topcoats. For this reason, a conversion coating with small, dense crystals and minimal porosity is desirable (Ref 4).

Primers

Primers are applied to body panels to enhance corrosion resistance, to give a better surface appearance to the finished panel, and to provide an adherent surface for subsequent organic coatings. Parts are primed immediately after the conversion coating process. Currently, the most frequently used primer materials are the high-film build epoxy resins, usually with corrosion-inhibiting pigments and a small amount of solvent added to aid in flow during the priming operation (see the article "Organic Coatings" in this Volume for more information on paint formulations).

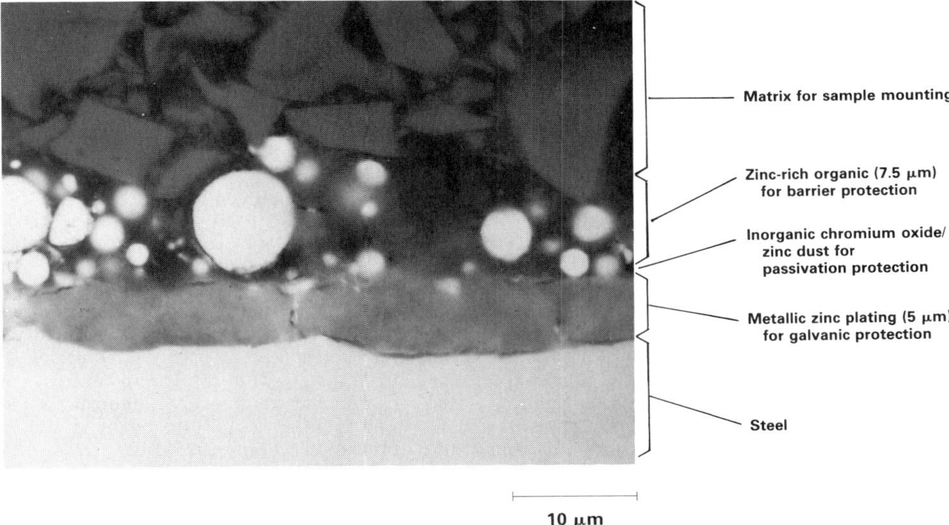

Fig. 8 SEM micrograph of cross section through a composite coated sheet steel. Source: Ref 15

Most primers are thermosetting compounds that require curing after application. Either single or multiple primer coats can be applied by spraying, dipping, flowing, or the cathodic electrocoat process (Ref 16).

Spray application of primer can be accomplished either manually or automatically, depending on the complexity of the parts being processed. Both electrostatic and airless spraying are employed. The spraying process gives uniform coating thicknesses and flexibility (coating thickness and areas to be covered are easily varied) (Ref 16).

In dip coating, the parts to be primed are immersed in a large tank containing the primer. Care must be taken when dipping parts containing inner panels or other features that could result in the formation of air bubbles, which prevent paint coverage in the area of the bubble. Adequate drainage must be provided to prevent collection of excess primer in crevices and shelf areas of the workpiece. Both of these potential problems can be minimized by good part design (for example, adding vent holes to parts with inner and outer panels, such as hoods).

In flow coating the primer is dispensed through large nozzles; excess primer is then allowed to drip off of the part and is collected for reuse. Flow coating usually gives better coverage of multipiece components than spraying, because the primer is allowed to flow around and through the part.

Electrocoating. The state-of-the-art in primer application is electrodeposition. Cathodic electrodeposition has become the dominant method of primer application for automakers around the world. In 1982, 33.4 million of the approximately 37 million cars and trucks built throughout the world used electrodeposited primers; of these, 25.4 million were coated by the cathodic electrodeposition process (Ref 17).

In the cathodic electrodeposition process, the workpiece is negatively charged and is immersed in a large tank containing the primer, which is positively charged. The primer is electrically attracted to the metal, resulting in a uniform coating thickness, excellent adhesion, a smooth surface, and excellent corrosion resistance. The electrodeposition process lends itself to automation, saves paint, is free from dripping and sagging, and is extremely reliable. Film thicknesses of 13 to 18 μm can easily be deposited, and high-film build primer formulations can allow deposition of thicknesses up to 35 μm (Ref 17).

Topcoats

The colored organic topcoat in automotive applications provides an additional barrier against the outside environment as well as a pleasing appearance. A wide range of topcoat formulations have been used in the automotive industry; nonaqueous acrylic dispersion enamels, high-solids solution enamels, thermoplastic acrylic lac-

Table 4 Evolution of primer and paint systems at one U.S. automaker

Primer or paint	Year of introduction	Curing temperature °C	°F	Minimum thickness mm (mils)
Primers				
Water-reducible dip primer	1960	165	325	0.015 (0.6)
High-solids spray primer	1981	165	325	0.019 (0.75) interior, 0.03 (1.2) exterior
Cathodic electrodeposited primer	1983	180	350	0.013 (0.5) interior, 0.03 (1.2) exterior
Color coats				
Conventional high-efficiency acrylic enamels	1979	120	250	0.043 (1.7)
Basecoat/clear coat acrylic enamels	1981	120	250	0.02 (0.8) basecoat, 0.03 (1.2) clear coat

Source: Ref 4

quers, and high-solids basecoat/clear coat enamels are currently used in North America (Ref 1). Topcoats are usually applied by spraying to obtain the best possible finish and high gloss. Electrostatic spraying techniques help to maintain a uniform film thickness and appearance.

The topcoat system currently used by one U.S. automaker uses a urethane-acrylic enamel color coat followed by a clear coat. The clear acrylic topcoat is applied to the wet color coat, and it protects the color coat from ultraviolet radiation from the sun, preventing color changes in the color coat pigment and resin. The clear topcoat can be polished to remove dirt and defects; polishing also gives a high gloss to the topcoat system (Ref 4).

Trends in topcoat formulation and application are being determined more by stricter environmental regulations and the desire to reduce production costs than by need for improved paint performance (Ref 1). Air quality legislation mandating reduced solvent emissions has resulted in the development of waterborne base coats and high-solids paints; automated cathodic electrodeposition of primers and electrostatic spraying of topcoats has increased production efficiency and reduced costs (Ref 1).

Corrosion in Other Automotive Systems

This article thus far has dealt mainly with corrosion and corrosion protection for automotive body materials. This section will discuss other areas of the automobile that are subject to corrosion. Underhood and underbody corrosion affects such vital components as fuel systems, cooling systems, electrical systems, and exhaust systems. Although perhaps not of cosmetic concern, corrosion in these areas can affect the safe operation of the vehicle.

Fuel Systems. As mentioned earlier in this article, fuel tanks and lines are generally fabricated in the United States from long terne (lead-tin) coated steels. Terne-coated steel has good overall corrosion resistance in this application, but is subject to pinhole-type corrosion if water is trapped in the fuel tank (Ref 18). Electrogalvanized steel is also used by some overseas manufacturers, and it has good resistance to pinhole-type corrosion. Over time, however, the inner surfaces of an electrogalvanized fuel tank may form white zinc corrosion products (white rust), with subsequent attack of the base metal (Ref 18). For other components of the fuel system, such as fuel lines, hot dip Zn-5Al-mischmetal (Galfan) coated steel is being employed (Ref 11).

Cooling Systems. Corrosion of automotive cooling systems is accelerated by dissimilar-metal couples, exhaust gas leakage, high operating temperatures, aeration, poor quality water, and coolant flow (Ref 19). A wide variety of materials are used in the typical automotive cooling system. Wrought brass or aluminum is used for radiators and heater cores, stamped steel for various small components and housings, and aluminum for parts such as coolant pumps (Ref 19).

Radiator tubes can be attacked by general corrosion, and brass tubes are subject to dezincification on both internal and external surfaces. Also, because brass tubes are fabricated by soldering, solder flux may cause stress-corrosion cracking (SCC) (Ref 20). A Cu-35Zn-0.3Al-0.2Sn-0.02P alloy has been developed that gives resistance to dezincification equal to that of arsenical brass in laboratory tests, and it has higher resistance to SCC than other brasses tested (Ref 20).

Components such as the coolant pump are subject to cavitation damage from collapsing vapor bubbles in the coolant; excessive coolant flow can also cause impingement damage to radiator tubes. Because of the variety of materials used in the cooling system, galvanic corrosion is of concern when dissimilar-metal parts are placed in electrical contact by the conducting coolant. Both of these problems can be alleviated somewhat by good design practices (Ref 19).

Proper maintenance is also important in minimizing cooling system corrosion, but surveys show that proper maintenance practices and manufacturers' recommendations are often neglected in the United States (Ref 21). Use of antifreeze at the proper concentration—U.S. automakers and antifreeze suppliers usually recommend a mixture of 50 to 70% antifreeze with water—is important, and coolant formulation can have an effect on corrosion if proper inhibitors are not used (Ref 21). No single inhibitor can protect all of the metals in an automotive cooling system, so many coolant formulations use a combination of inhibitors (Ref 19).

Electrical systems are subject to a wide variety of corrosion problems caused by the severity of the automotive environment. Corrosion problems in conventional electrical systems can be minimized at the design stage by sealing components whenever possible. A wide variety of organic and inorganic compounds are used for this purpose, including paints and primers, silicone compounds, varnishes, numerous plastics, and oils and greases (Ref 22).

The use of electronic components in automobiles has increased dramatically in the past decade. These systems were first used for engine control, but they now perform a wide variety of control and monitoring functions (Ref 23). Electronic components are subject to a variety of corrosion problems, including corrosion-induced leakage and shorts on printed circuit boards, metal migration problems, and corrosion in plastic-packaged devices (see the article "Corrosion in the Electronics Industry" in this Volume). Proper design, manufacturing, and quality assurance procedures can help to minimize these types of problems in automotive electronic systems.

Automotive exhaust systems are subject to general and localized external corrosion and to internal corrosion caused by exhaust gas condensates (Ref 24). Exhaust gas temperatures near the exhaust manifold have been measured at up to 870 °C (1600 °F), with corresponding metal temperatures as high as 595 °C (1100 °F) (Ref 24); therefore, exhaust system components are subject to high-temperature oxidizing conditions.

The materials used to combat corrosion in exhaust systems include aluminized steel and type 409 stainless steel. The use of type 409 stainless steel is increasing because of its relatively low cost and good resistance to corrosion and high-temperature oxidation.

The use of catalytic converters, which began in the United States in the 1975 model year, has presented some special problems for automakers. Because of the oxidizing, catalyzed reaction that takes place in the exhaust gas stream in the converter, exhaust gas temperatures and the amount of corrosive substances, such as sulfuric acid, in the exhaust gas stream are increased (Ref 24). The converter itself essentially consists of a noble metal coated ceramic substrate, often housed in a type 409 stainless steel canister. The location of the converter between the exhaust manifold and the exhaust pipe has prompted the use of more corrosion-resistant materials downstream from the converter.

Other Automotive Systems. A variety of coated steels are used for automotive suspension components, including hot-dip galvanized and galvannealed steels, electrogalvanized steel, aluminized steel, and cathodic electrocoated or epoxy powder coated steels. Some suspension components (for example, rear cross members) and structural components (such as bumper reinforcements and body side rails) are being fabricated from high-strength low-alloy steels to reduce weight, but formability problems with these materials have thus far limited their use to structural-type applications (Ref 25). High-strength steels, although thinner in section than carbon steels, are protected in much the same way by a variety of coating materials. More information on the types of microalloyed steels used, the strength ranges employed, and the automotive applications for such materials is available in Ref 26 to 32.

Clad metals are used in the auto industry for their decorative appearance and corrosion resistance, for example, in automotive trim applications (Ref 33). Figure 9 shows a stainless steel clad aluminum alloy used for trim applications. The aluminum inner surface, being more electrochemically active than steel, acts as a sacrificial anode to prevent corrosion of the adjacent steel body panel. The outer stainless steel surface provides corrosion resistance, abrasion and dent resistance, and a decorative appearance. This material is also used for auto and truck bumpers to reduce weight while maintaining corrosion resistance (see the article "Corrosion of Clad Metals" in this Volume).

Design Considerations

As mentioned earlier in this article, design can play an important role in determining the corrosion resistance of an automotive assembly or component. In fact, the configuration of a part or assembly is often the determining factor in the type and severity of corrosion that occurs in service (Ref 1). Design factors that can influence corrosion resistance are reviewed in Ref 34 and 35, and additional information on this subject is available in the article "Design Details to Minimize Corrosion" in this Volume.

Corrosion Testing

The automotive industry employs many common corrosion test methods; for example, laboratory salt spray and electrochemical tests are often used in the development and evaluation of new materials (see the article "Laboratory Testing" in this Volume). Tests that are specific to the industry also are used, such as proving ground vehicle testing, mobile testing using underbody test racks, and field surveys. Recent emphasis has been on the development of laboratory tests that can closely approximate the results of much longer and more costly proving ground or undervehicle tests. This section will focus on both field and laboratory testing methods that are specific to the automotive industry.

Mobile testing using test racks mounted under or on the vehicle is commonly used to evaluate as-received precoated steels and primed and

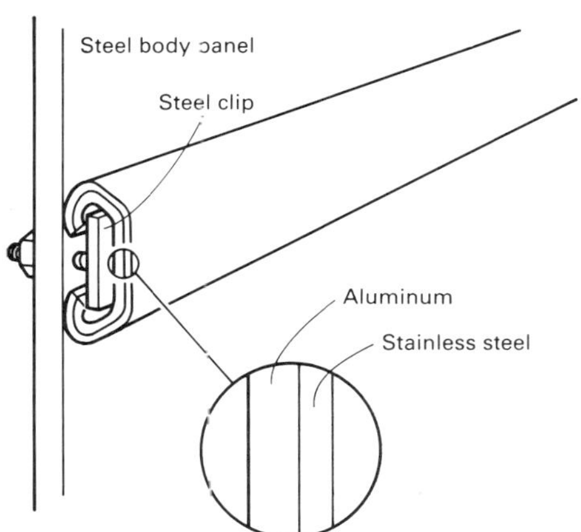

Fig. 9 Stainless steel clad aluminum alloy used in automotive trim applications. The use of steel clips to attach the trim strip prevents galvanic corrosion of the aluminum in contact with the steel body panel.

painted specimens. Of the commonly used test methods, mobile testing probably comes the closest to simulating actual service conditions (Ref 1), but it requires test periods of up to several years. The measurements taken usually include percent surface area of base metal attacked and depth and density of pitting. The test methods used and the results of two of these tests are documented in Ref 36 and 37, and Ref 38 correlates the results of an undervehicle testing program with those obtained from a laboratory test method.

Proving Ground Testing. In this method, prototype or production vehicles are corrosion tested on the proving grounds of the company. Test cycles that produce accelerated attack are used, with test times varying from 10 weeks to 10 months (Ref 1). Proving ground testing can simulate service conditions for some types of corrosion, but other forms, such as perforation corrosion, are more difficult to accelerate.

Field surveys, when properly conducted, offer a direct method of comparing the performance of various types of coatings and other materials. These surveys often involve the destructive inspection of actual vehicles with well-defined service histories in corrosive environments (Ref 1). The results of one recent survey of 5- and 6-year-old vehicles are documented in Ref 39.

Laboratory Tests. Cyclic laboratory tests have been developed in recent years to simulate in a relatively short time the effects of many years of actual operation. One such test uses three environment chambers to subject the test material to extreme climatic conditions, and it uses other equipment, such as a gravel blower, to simulate actual road use as closely as possible (Ref 40). This test is intended to simulate 6 years of actual service over a 14-week test period. Other such tests have been developed, but correlation to actual service tests has not been good. More development is required to make these tests reliable indicators of the corrosion resistance of materials for automotive use.

REFERENCES

1. "Cracking Down on Corrosion: Cooperative Efforts Toward Vehicle Durability," American Iron and Steel Institute, 1985
2. J.C. Bittence, Waging War on Rust, Part I: Understanding Rust, *Mach. Des.*, 7 Oct 1976, p 108-113; Part II: Resisting Rust, 11 Nov 1976, p 146-152
3. "US Automotive Market for Zinc Coatings 1984-1986," Zinc Institute Inc.
4. D.J. Bologna, Corrosion Resistant Materials and Body Paint Systems for Automotive Applications, SAE Paper 862015, in *Proceedings of the Automotive Corrosion and Prevention Conference*, P-188, Society of Automotive Engineers, 1986, p 69-80
5. M. Chilaud, S. Mathieu, P. Pichant, and G. Quinchon, "Hot Dip Galvanized Steel—Product of the Future," SAE Paper 860271, Society of Automotive Engineers, 1986
6. D.F. Baxter, Jr., Developments in Coated Steels, *Met. Prog.*, May 1986, p 31-35
7. Auto Makers Take the Plunge Into Hot Dip Galvanizing, *Zinc*, No. 1, Zinc Institute Inc., 1985, p 5
8. H. Kunitake, The Challenge of the Automotive Industry, in *Proceedings of the 18th Annual Meetings of the IISI*, International Iron and Steel Institute, 1984
9. R.J. Neville and K.M. DeSouza, Electrogalvanized or Hot Dip Galvanized—Results of Five Years of Undervehicle Corrosion Testing, SAE Paper 862010, in *Proceedings of the Automotive Corrosion and Prevention Conference*, P-188, Society of Automotive Engineers, 1986, p 31-40
10. M. Memmi *et al.*, A Qualitative and Quantitative Evaluation of Zn + Cr-CrO_x Multilayer Coating Compared to Other Coated Steel Sheets, SAE Paper 862028, in *Proceedings of the Automotive Corrosion and Prevention Conference*, P-188, Society of Automotive Engineers, 1986, p 175-185
11. R.F. Lynch and F.E. Goodwin, "Galfan Coated Steel for Automotive Applications," SAE Paper 860658, Society of Automotive Engineers, 1986
12. Y. Shindou *et al.*, Properties of Organic Composite-Coated Steel Sheet for Automobile Body Panels, SAE Paper 862016, in *Proceedings of the Automotive Corrosion and Prevention Conference*, P-188, Society of Automotive Engineers, 1986, p 81-90
13. M. Yamashita, T. Kubota, and T. Adaniya, Organic-Silicate Composite Coated Steel Sheet for Automobile Body Panel, SAE Paper 862017, in *Proceedings of the Automotive Corrosion and Prevention Conference*, P-188, Society of Automotive Engineers, 1986, p 91-97
14. T. Mohri *et al.*, Newly Developed Organic Composite-Coated Steel Sheet With Bake Hardenability, SAE Paper 862030, in *Proceedings of the Automotive Corrosion and Prevention Conference*, P-188, Society of Automotive Engineers, 1986, p 199-208
15. T.E. Dorsett, Development of a Composite Coating for Pre-Coated Automotive Sheet Metal, SAE Paper 862027, in *Proceedings of the Automotive Corrosion and Prevention Conference*, P-188, Society of Automotive Engineers, 1986, p 163-173
16. W.E. Tudor, A Primer—Automotive Finishing and Corrosion Protection, SAE Paper 780914, in *Designing for Automotive Corrosion Prevention*, P-78, Society for Automotive Engineers, 1978, p 36-42
17. F.M. Loop, High Film Build Cathodic Electrodeposition Provides Improved Corrosion Protection, SAE Paper 831813, in *Proceedings of the 2nd Automotive Corrosion Prevention Conference*, P-136, Society of Automotive Engineers, 1983, p 35-44
18. D.J. Bologna and H.T. Page, Corrosion Considerations in Design of Automotive Fuel Systems, SAE Paper 780920, in *Designing for Automotive Corrosion Prevention*, P-78, Society of Automotive Engineers, 1978, p 65-70
19. E. Beynon, N.R. Cooper, and H.J. Hannigan, Cooling System Corrosion in Relation to Design and Materials, SAE Paper 780919, in *Designing for Automotive Corrosion Prevention*, P-78, Society of Automotive Engineers, 1978, p 56-64
20. J. Miyake, M. Tsuji, and S. Kawauchi, Corrosion Prevention for Automobile Radiator Tubes, SAE Paper 862021, in *Proceedings of the Automotive Corrosion and Prevention Conference*, P-188, Society of Automotive Engineers, 1986, p 117-121
21. N.R. Cooper, H.J. Hannigan, and J.C. McCourt, A One Thousand Car Assessment of the U.S. Car Population Cooling Systems, SAE Paper 831821, in *Proceedings of the 2nd Automotive Corrosion Prevention Conference*, P-136, Society of Automotive Engineers, 1983, p 121-130
22. M.M. Jones and E.E. Welker, Electrical Component Corrosion Prevention, SAE Paper 780924, in *Designing for Automotive Corrosion Prevention*, P-78, Society of Automotive Engineers, 1978, p 107-118
23. J.P. Cook and G.E. Servais, Corrosion Failures in Semiconductor Devices and Electronic Systems, SAE Paper 831830, in *Proceedings of the 2nd Automotive Corrosion Prevention Conference*, P-136, Society of Automotive Engineers, 1983, p 187-197
24. W.R. Patterson, Materials, Design, and Corrosion Effects on Exhaust System Life, SAE Paper 780921, in *Designing for Automotive Corrosion Prevention*, P-78, Society of Automotive Engineers, 1978, p 71-106
25. S. Dinda, C. Belleau, and D.K. Kelley, High Strength Low Alloy Steels in Automotive Structures, in *HSLA Steels: Technology and Applications*, American Society for Metals, 1984, p 475
26. D.A. Wilkinson and D.D. Rogers, A New HSLA Steel for an Automotive Steering Coupling Component, in *HSLA Steels: Technology and Applications*, American Society for Metals, 1984, p 459
27. R.G. Davies, Forming Problems Encountered in Application of High Strength Steels to Automotive Components, in *HSLA Steels: Technology and Applications*, American Society for Metals, 1984, p 467
28. H.E. Chandler, High Strength Sheet Forms Like Mild Steel, *Met. Prog.*, Nov 1985, p 63-66
29. M. Takahashi *et al.*, Criteria of High Strength Steels for Applying to Automobile Frame Components, in *HSLA Steels: Technology and Applications*, American Society for Metals, 1984, p 493
30. K. Tamura and M. Shiokawa, Application of Higher Strength Steel Sheets and Its Process in Nissan Motor Company, in *HSLA Steels: Technology and Applications*, American Society for Metals, 1984, p 503
31. G.T. Halmos, Roll Forming HSLA Steels, in *HSLA Steels: Technology and Applications*,

American Society for Metals, 1984, p 515
32. J.C. Kopchick, Automotive Application of Ultra-High Strength Steel Sheet, in *HSLA Steels: Technology and Applications*, American Society for Metals, 1984, p 523
33. R. Baboian, Causes and Effects of Corrosion Relating to Exterior Trim on Automobiles, SAE Paper 831835, in *Proceedings of the 2nd Automotive Corrosion Prevention Conference*, P-136, Society of Automotive Engineers, 1983, p 223-227
34. "Prevention of Corrosion of Metals," Handbook Supplement HSJ 447, Society of Automotive Engineers, 1981
35. L.C. Rowe, "The Application of Corrosion Principles to Engineering Design," Paper 770292, presented at the SAE Automotive Engineering Congress, Society of Automotive Engineers, Feb 1977
36. R.D. McDonald and R.R. Ramsingh, Eighteen Months of Underbody Automotive Materials Testing, *Mater. Perform.*, Vol 24 (No. 4), April 1985, p 48-53
37. R.D. McDonald and R.R. Ramsingh, "Corrosion Testing by the Under-Car Method, December 1978 to May 1980," Report MRP/PMRL 81-46(TR), Canadian Centre for Mineral and Energy Technology, Physical Metallurgy Research Laboratories, July 1981
38. R.D. McDonald, "Corrosion of Automotive Steels in Deicing Salt Environments: Comparison of a Laboratory Method With Undercar Testing," Report MRP/PMRL 83-71(J), Canadian Centre for Mineral and Energy Technology, Physical Metallurgy Research Laboratories, March 1983
39. A.W. Bryant and W.C. Oldenburg, 1985 Body Corrosion Survey—5 and 6 Year Old Vehicles, SAE Paper 862025, in *Proceedings of the Automotive Corrosion and Prevention Conference*, P-188, Society of Automotive Engineers, 1986, p 143-154
40. R. Dietz, A Three-Chamber Corrosion Test Method for Passenger Cars, SAE Paper 831814, in *Proceedings of the 2nd Automotive Corrosion Prevention Conference*, P-136, Society of Automotive Engineers, 1983, p 47-56

Corrosion in the Aircraft Industry

CORROSION CONTROL is of the utmost concern in the aircraft industry because of its potential impact on human safety and on extremely expensive aircraft. In the military, winning the war on corrosion is essential to military preparedness and national security. Military aircraft are very expensive; some, such as the B-1B strategic bomber, cost over $200 million each. To achieve high levels of payload and performance in military aircraft, materials are chosen more for their mechanical properties than for their inherent corrosion resistance. They depend on coatings and routine maintenance to preserve their integrity. Military aircraft are flown throughout the world and are therefore exposed to the most severe corrosive environments on earth. The contracting agency imposes numerous specifications on suppliers to control corrosion on military aircraft.

An excellent text on corrosion is MIL-HDBK-729, *Corrosion and Corrosion Prevention—Metals*. The military services sponsor a Tri-Service Corrosion Conference every 2 or 3 years to provide a forum for corrosion control and experiences with various weapon systems. In addition, the Corrosion Information Analyses Center serves the Department of Defense (DOD) as part of the Metals and Ceramics Information Center. Each of the military branches have corrosion control centers and laboratories serving them.

Corrosion control of commercial aircraft is also of paramount importance for similar reasons. Flight safety is essential to the airline industry. Again, commercial airlines are exposed to highly corrosive environments all over the world. Commercial aircraft represent investments of up to $100 million per unit for some of the widebody aircraft. Full-scale inspections to determine structural integrity often exceed $2.5 million. Even more severe economic consequences could result from lawsuits or the loss of a manufacturer's reputation in the event of a corrosion-related mishap. Consequently, although only a short reference to the need for corrosion control to ensure airworthiness and safety is found in Federal Aviation Administration regulations (Federal Airworthiness Regulation 25.609), commercial aircraft manufacturers take as much care in this area as the military.

Commercial airlines, which are responsible for corrosion control once the aircraft is delivered, are provided with detailed documents from the aircraft manufacturers, such as corrosion control handbooks and customer service documents, to ensure proper servicing and inspection. Nevertheless, in spite of extensive corrosion control efforts, problems are experienced. This article will present typical examples of corrosion problems in the aircraft industry to provide additional insight into the causes and corrective actions.

Corrosion of Airframes

Michael L. Bauccio
The Boeing Company

The potential for corrosion of aircraft structures is a major consideration in the design of the aircraft. Corrosion can be related to various types of material deterioration. This is because the corrosion process can be defined as the degradation of a material or materials by a reaction with the environment. Usually, the reaction is electrochemical, and the material or materials are metallic (Ref 1).

Corrosion phenomena often occur on the surfaces of aircraft structures (Ref 2). As shown in Fig. 1, the physical effects of corrosion can be categorized as follows:

- Local pitting or crevice corrosion (one-dimensional local irregularity)
- General surface roughening due to uniform, filiform, or fretting corrosion (two-dimensional discontinuity)
- Intergranular and transgranular cracking, which is observed in stress-corrosion cracking (SCC) (two-dimensional discontinuity)
- Degradative transformation of materials that occurs on a larger scale, which is referred to in Fig. 1 as a three-dimensional bulk reduction of material. This is observed in galvanic and exfoliation corrosion

The detrimental effects of aircraft corrosion are not due to corrosion alone, but to the interaction of corrosion with fatigue, wear (including erosion and fretting), erosion, and stress resulting in premature fracture. The most significant aspects concerning aircraft corrosion that will be presented in this section are that:

- Airframe corrosion problems vary in severity
- Cosmetic corrosion, which simply mars the appearance of airframe surfaces, can develop into a widespread form of material deterioration
- Catastrophic mechanical failures can develop from corrosion that is permitted to spread and cause a significant reduction in structural strength
- Environmental and mechanical factors combine to produce the types of aircraft corrosion that usually lead to catastrophic in-flight failures. The tropical marine environment is one of the most severe that modern aircraft are exposed to. An example of wing strut deterioration that was produced by this highly corrosive environment is shown in Fig. 2. The corroded wing strut was

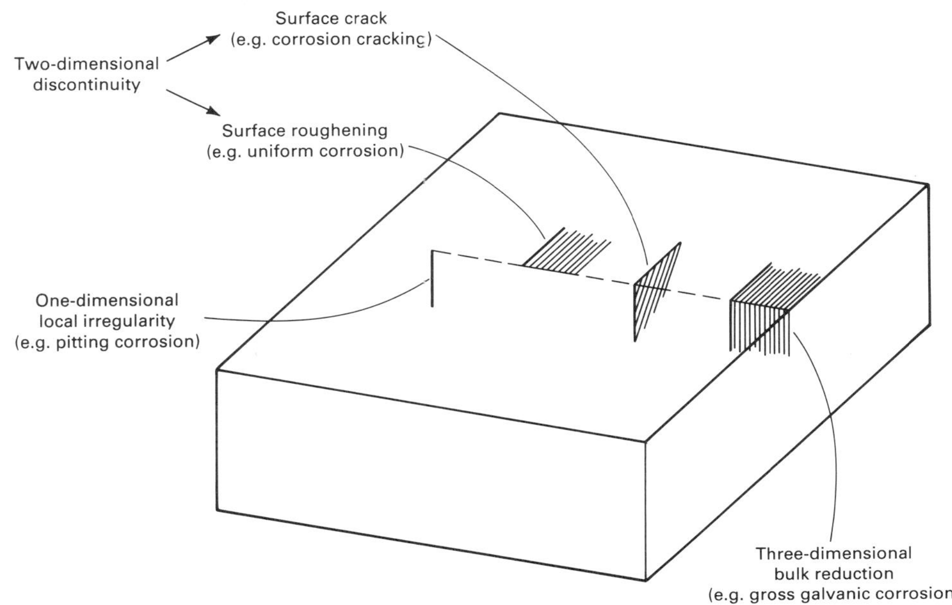

Fig. 1 Schematic showing the physical effects of corrosion on metallic aircraft materials. See text for details.

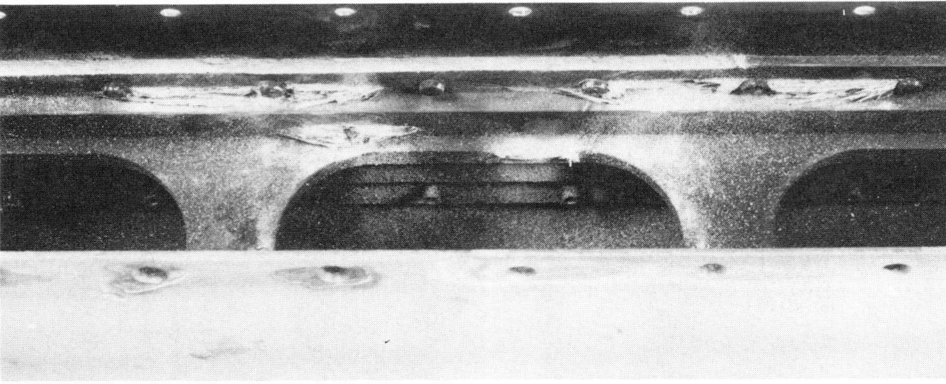

(a)

(b)

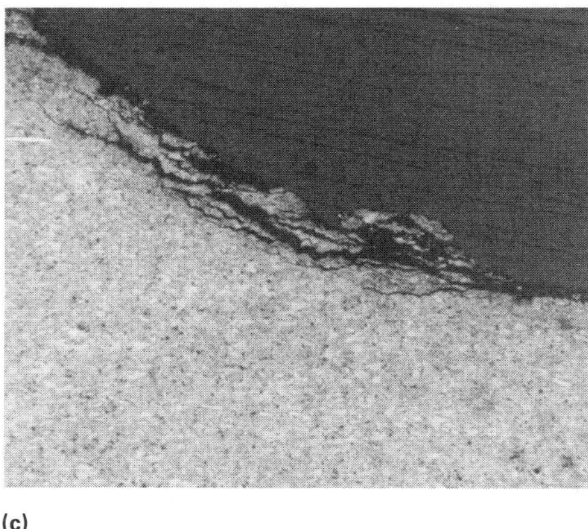

(c)

Fig. 2 Corroded aluminum alloy 2024-T351 wing strut (a) with attached leading edge. (b) Close-up of corrosion on the wing strut with the leading edge skin removed. (c) Corrosion on the wing strut. 25×

fabricated from aluminum alloy 2024-T351 (QQ-A-200/3) extruded bar

- Some of the most critical types of aircraft corrosion that can result in fracture of components or airframe members are SCC, corrosion fatigue, and hydrogen embrittlement (Ref 3). Selected examples of these aircraft fracture failures are illustrated in Fig. 3 to 8

New Material and Process Solutions From Old Corrosion Problems

Of the numerous cases of aircraft corrosion that the author has reviewed, many have appeared in various forms in the past 30 years. However, those corrosion problems that have challenged aircraft materials and process engineers have usually led to positive developments, which have resulted in greater control of aircraft corrosion.

Some very recent corrosion cases have also led to significant new corrosion-preventive measures for aircraft applications. For example, in 1985, a C-5B military cargo aircraft was grounded so that about 11 000 aluminum nuts could be replaced by the specified cadmium-plated steel nuts (Ref 6). An additional 40 000 aluminum nuts are to be replaced on four other C-5B aircraft. The aluminum nuts, which were installed despite the engineering drawing requirement for cadmium-plated steel nuts, were discovered during an inspection by the manufacturer. The positive side of this story is that the C-5B manufacturer has developed and patented a magnesium chromate ($MgCrO_4$) sealant that can be used for corrosion prevention in structures that have mating surfaces of aluminum and steel (Ref 6). Aluminum nuts can be used for reduced weight, and steel is used for greater strength on the C-5B, which contains about 4 000 000 structural fasteners.

Other corrosion-related incidents that have been investigated by aerospace materials engineers have been reported many years ago and are still being studied today so that improved solutions can be obtained. One such case is concerned with the hydrogen embrittlement and SCC of high-strength steel aircraft components. A monograph on hydrogen stress cracking and hydrogen embrittlement of low-alloy aircraft steels was published in 1956 (Ref 7). The failure of 4140 steel nacelle eyebolts was reported in 1971 (Ref 7). This bolt failure was due to a baking time that was too short to provide the steel bolts with

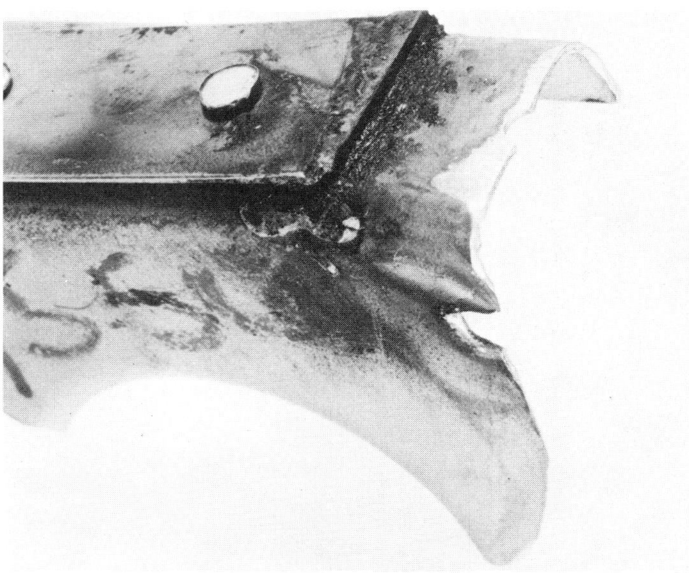

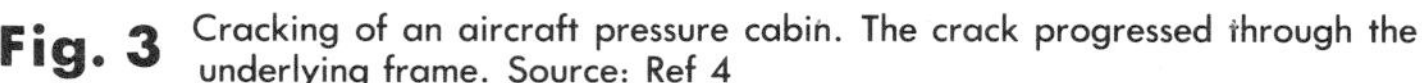

Fig. 3 Cracking of an aircraft pressure cabin. The crack progressed through the underlying frame. Source: Ref 4

Fig. 4 Service failure of an aircraft pivot bracket. Arrow points to fracture origin. Source: Ref 4

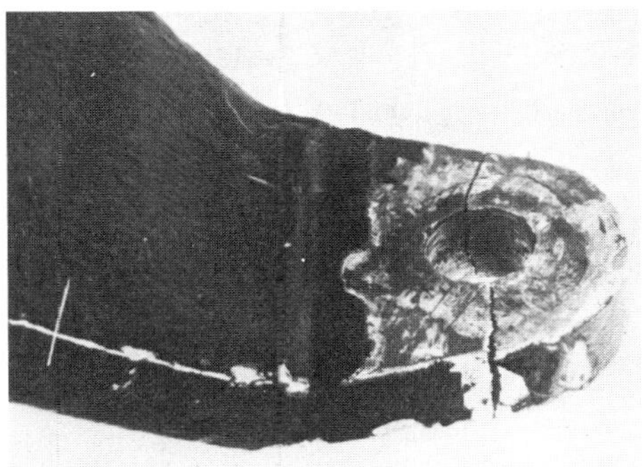

Fig. 5 Service failure of a helicopter rotor drive yoke. Source: Ref 4

proper hydrogen embrittlement relief, which would have enhanced the mechanical strength of these fasteners. In-service failures due to hydrogen embrittlement and SCC have also occurred in chromium-plated landing gears and in cadmium-plated steel fasteners (Ref 7).

Fig. 6 Service failure of an aircraft main-plane spar boom. Arrows point to fracture origins. Source: Ref 4

Approximately 14 years after the above case histories were published, a report confirmed that baking time is the only significant variable in determining the final hydrogen concentration in cadmium-plated high-strength steels (Ref 8). The importance of baking time for hydrogen embrittlement relief was recognized by several aerospace companies in the early 1960s. The process specifications that had been written by these companies were revised to require 23 h of baking at 190 °C (375 °F), instead of the 3-h baking time that was found in military specifications.

One approach that can be used to prevent hydrogen embrittlement is to avoid the processing of steel components in plating or pickling baths. Another technique is to apply ion vapor deposited (IVD) aluminum coatings. These coatings have recently been reported as an effective method for protection against hydrogen-induced embrittlement, as well as SCC, and can withstand service temperatures up to 495 °C (925 °F) (Ref 9).

Ion vapor deposited coatings must meet the requirements of military specification MIL-C-83488 (Ref 10). Aluminum IVD coatings have been applied to the following aircraft structures with favorable results in terms of improved corrosion resistance (Ref 9):

- Steel and titanium fasteners installed in aluminum aircraft structures
- A fatigue-critical aluminum alloy wing skin
- A DC-10 aft engine hanger
- A high-strength steel landing gear

Exfoliation corrosion, which has a long history in connection with airframe deterioration, was observed many years ago where cadmium-plated fasteners were installed in high-strength aluminum alloys. Exfoliation corrosion occurs in metallic materials that have a directionally oriented grain structure. Riveted aircraft structures have a high vulnerability to exfoliation corrosion because the rivet holes provide an unobstructed pathway for corrosive electrolytes to reach metallic airframe materials, especially aluminum, which is the material selected for many aircraft parts (Ref 11).

Exfoliation corrosion initiates between bimetallic couples and progresses along grain boundaries as an intergranular crack. This intergranular crack widens into a crack plane and enlarges into multiple crack planes. Corrosive oxides press outward against the adjacent metal, thus producing a pattern of delamination.

An illustration of this type of failure, originating at a fastener hole, is shown in Fig. 9. The best alternative available for preventing exfoliation corrosion is to select a corrosion-resistant alloy and heat treatment. Chromate-inhibited elastomeric sealants can also be applied to protect airframe fasteners against corrosion (Ref 12). These fasteners must be wet-installed with the chromated sealant material in order to obtain the highest degree of corrosion protection.

Another example of exfoliation corrosion is illustrated in Fig. 10. The failed airframe structure shown was removed from an aircraft that operated primarily in a marine environment. The structure is a tail plane attachment fitting made of an aluminum alloy (2024-T4) that meets federal specification QQ-A-250/4. The arrow in Fig. 10(a) points to the corrosion. This corrosion problem was primarily caused by inadequate sealing of the bolt hole during installation of the cadmium-plated steel bolt; this allowed seawater to attack the aluminum alloy.

Additional Corrosion Cases

There are many more examples of how the scope of aircraft structural corrosion spans long periods of time between the origination of a specific problem and the development of an improved material or process, or both, for mitigating the corrosion problem. Aircraft corrosion cases are often solved within much shorter periods than those problems described above. However, the same immediate, or quick-fix, solutions will be reevaluated and modified over the years by aircraft manufacturers for:

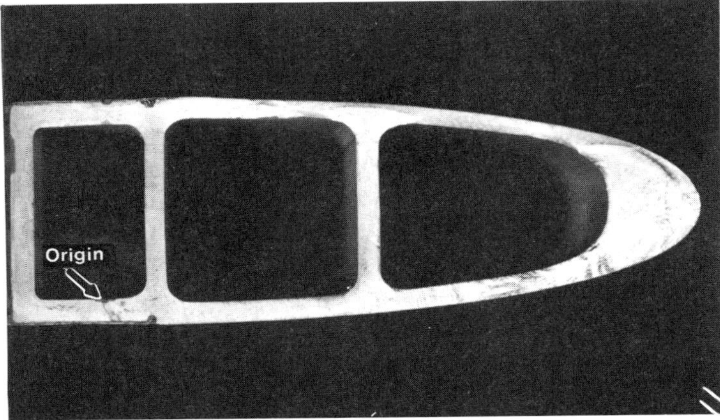

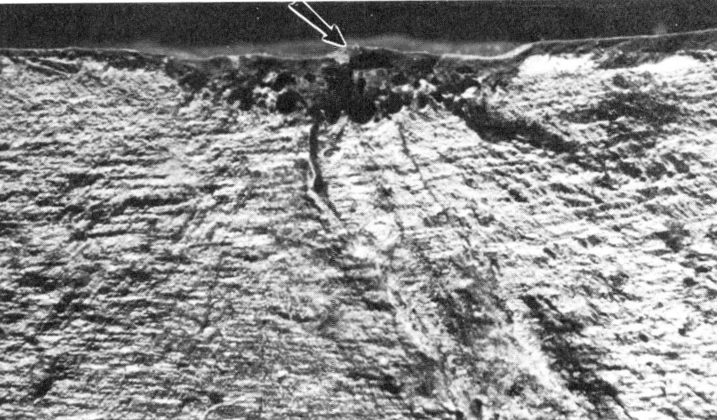

Fig. 7 Service failure of a helicopter rotor blade extrusion. Source: Ref 4

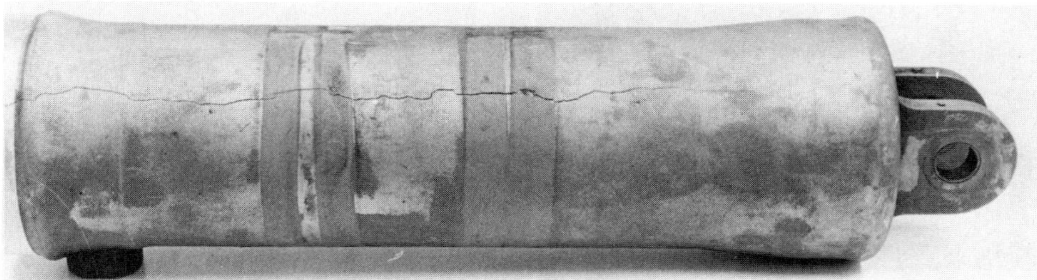

Fig. 8 Fracture of an aluminum alloy 2014-T6 main landing gear actuator from a C-141 military cargo plane. Exfoliation induced by differential aeration was indicated as the cause of this failure, based on the appearance of the fracture near the origin. Source: Ref 5

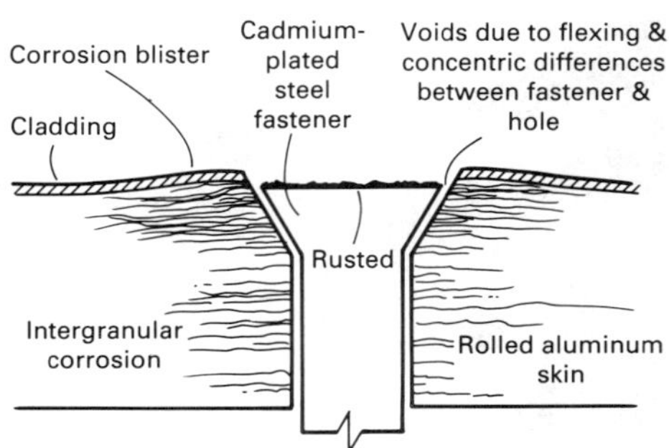

Fig. 9 Schematic of exfoliation in an aluminum aircraft panel. Source: Ref 12

- Increased reductions in the gross weight and cost of aircraft systems
- Materials and processes that will meet the increasingly stringent environmental and occupational safety regulations of state, local, and federal government agencies

Corrosion-preventive technology also requires continued study and modification. This is for the purpose of ensuring that corrosion-resistant materials and corrosion control techniques are effectively applied to the diverse variety of airframe components. For example, thick organic coatings usually cannot be used on mating parts that have close tolerances. These parts are often protected by inorganic coatings that are relatively thin (about 1 to 35 μm, or 0.04 to 1.4 mils), such as the wear-resistant chromium platings or anodized coatings on aluminum structures, the black oxide or phosphatized coatings on steel parts, and the dichromate and anodic (Dow 17) treatments used on magnesium alloy structures.

Some materials problems, such as microbial deterioration of integral wing fuel tanks, have plagued the aircraft industry since the beginning of the jet age (Ref 13). Such ongoing problems illustrate the need for continuing research and development of new materials and protection systems for corrosion control.

Aircraft materials that are resistant to corrosive deterioration should be selected during the design phase. This objective can be facilitated by following the general techniques and design rules for corrosion prevention and control presented in Table 1. When selecting aluminum alloys for SCC resistance, aircraft designers should consult Table 2. The process of selecting corrosion-resistant steels for aircraft systems can be facilitated by reference to Table 3, which provides the resistance to general corrosion and SCC of several classes of steel.

Aircraft Corrosion-Related Failures

The best way for professional engineers to grasp the extent of aircraft-related corrosion problems and solutions is to become familiar with many of the corrosion-related cases that have been documented. The following discussions in this section provide a comprehensive overview of many aircraft corrosion problems that have been described in the technical literature. For most of these corrosion cases, the successfully applied corrosion-preventive treatments, which were developed to avert future corrosion problems, are described. These airframe corrosion cases were selected because they can be of great assistance to design and corrosion engineers in alleviating possible airframe corrosion problems in the future.

Galvanic Corrosion. When dissimilar and unprotected structural materials with different electrochemical potentials are assembled and exposed to a corrosive environment, galvanic corrosion (dissimilar-metal corrosion) occurs. The more active (less noble) material of the galvanic couple becomes the anode and therefore undergoes dissolution and corrosion (Ref 16).

One of the best ways of identifying galvanic corrosion is to examine the severity of corrosion damage at the junction between the dissimilar metals. Galvanic corrosion usually predominates at the point of contact between the two materials. Several examples of galvanic corrosion, which have been described in the literature on aircraft structural corrosion, are provided below (Ref 16).

Cadmium-Plated Steel Fasteners. These fasteners were in contact with a 7075-T6 aluminum skin (Fig. 11). Severe corrosion damage occurred on the aluminum skin around the periphery of the fastener heads.

Aluminum Alloy Skin. Galvanic corrosion in this aluminum alloy skin was caused by an electrically conductive adhesive compound. The corrosion occurred along the edges of an anti-icing boot that was bonded to the wing leading edges.

7075-T6 Kingpin Lug. Galvanic corrosion on this lug was caused by direct contact between this part and steel bushings. The bushings were used to protect the lug mounting holes from mechanical damage. This failure could have been prevented by the proper application (and continual inspection) of a protective coating on the aluminum alloy, such as a hard anodized coating.

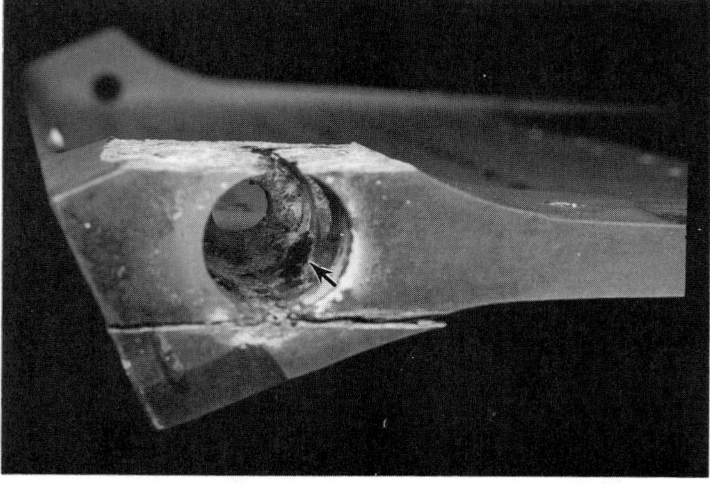

(a)

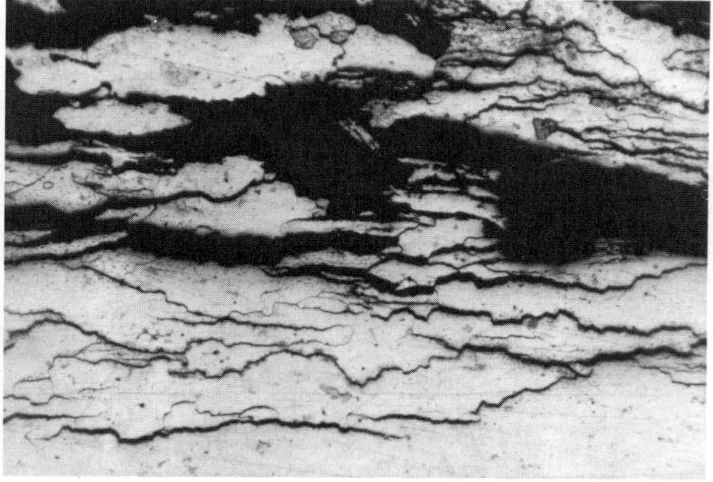

(b)

Fig. 10 Example of exfoliation corrosion. (a) Failed aluminum alloy 2024-T4 tail plane fitting. Arrow points to corrosion that was produced by direct contact between a cadmium-plated steel bolt and the aluminum fitting. (b) Exfoliation in the tail plane fitting. 55×

Table 1 General techniques and design considerations for minimizing corrosion

Item	Suggestions
Eliminate areas where trapped moisture is held in contact with metal	Avoid such features at the design stage by careful attention to design or structure details. Provide properly located drain holes. Minimum hole size should be 3.2 mm (1/8 in.) to prevent plugging.
Choose nonabsorbent, nonwicking materials	Determine water absorption qualities of materials to be used. Use epoxy and vinyl tapes and coatings, wax, or latex for protective barriers. Avoid, if possible, use of wood, paper, cardboard, open cell foams, and sponge rubbers.
Protect all faying surfaces	Use proper sealing materials (tapes, films, sealing compounds) on all faying surfaces. Use primers. Lengthen continuous liquid path to prevent formation of an electrolytic cell.
Use compatible metals	For magnesium-aluminum couples, 5000- and 6000-series aluminum alloys are the most compatible. For magnesium-steel couples, use tin or cadmium plated steel. For bimetallic couples use metals or alloys in the same group per MIL-STD-889, or as close as possible. Use tapes or primers on faying surfaces to prevent metallic or electrical contact.
Select proper finishing systems	Choose chemical treatments, paints, plating on basis of service requirements. Service test system before setting up production run. Use past experience in similar applications as guide to choice.

Source: Ref 14

Table 2 Relative resistances of aluminum alloys to SCC

Alloy and temper	Product form: Rolled plate	Rod and bar	Extruded shapes	Forgings
2014-T6	Poor	Poor	Poor	Poor
2024-T3, T4	Poor	Poor	Poor	...
2024-T6	...	Good	...	Poor
2024-T8	Good	Excellent	Good	Intermediate
2124-T851	Good	...	...	...
2219-T3, T37	Poor	...	Poor	...
2219-T6, T8	Excellent	Excellent	Excellent	Excellent
6061-T6	Excellent	Excellent	Excellent	Excellent
7049-T73	Excellent	...	Good	Good
7149-T73	...	...	Good	Good
7049-T76	...	...	Intermediate	...
7*x*75-T736	...	...	...	Good
7050-T736	Good	...	Good	Good
7050-T76	Intermediate	...	Intermediate	...
7*x*75-T6	Poor	Poor	Poor	Poor
7*x*75-T73	Excellent	Excellent	Excellent	Excellent
7*x*75-T76	Intermediate	...	Intermediate	...

Source: Ref 15

Table 3 Resistance of some stainless steels to general corrosion and SCC

Alloy	General corrosion resistance	SCC resistance
Austenitic grades		
Type 301	Excellent	Excellent
Type 316	Excellent	Moderate
Type 347	High	Excellent
A286	High	Excellent
Type 321	High	Excellent
Type 304 (ELC)	High	Excellent
Type 302	High	Excellent
Type 304	High	Excellent
Type 310	Excellent	Excellent
Martensitic grades		
440C	Low to moderate	Susceptibility varies significantly with composition, heat treatment, and product form.
420 410 416	Low to moderate; will develop superficial rust film with atmospheric exposure	
Precipitation-hardenable grades		
21-6-9	Moderate	Susceptibility varies significantly with composition, heat treatment, and product form.
PH13-8Mo	Moderate	
PH15-7Mo	Moderate	
PH14-8Mo	Moderate	
17-4PH	Moderate	
15-5PH	Moderate	
AM355	Moderate	
AM350	Moderate	

Source: Ref 15

Cast Magnesium Flap Control Lever Arm. The lever arm was manufactured from a magnesium casting alloy. A zinc-chromate primer and an epoxy topcoat had been applied to the lever arm structure for corrosion protection. Galvanic corrosion occurred in the sockets of the levers, where steel balls were located. The movement of these steel balls in the socket caused mechanical deterioration of the protective coating. The highly anodic magnesium, therefore, was in direct contact with the steel balls, causing the galvanic corrosion. Galvanic corrosion also was found around a press-fitted aluminum bushing in the lever arm. Current aircraft design requirements (in most cases) prohibit the use of magnesium castings, especially for components that will be in contact with moving parts. Application of the sealants and organic coating systems according to the guidelines in Ref 17 might have averted this corrosion problem.

In summary, galvanic corrosion can be prevented by avoiding dissimilar-metal contact. Whenever joints between dissimilar metals must be made, metals should be selected whose galvanic potentials in seawater (or salt water) are closest in order to minimize corrosion rates. Generally, potential differences of less than 0.25 V will be satisfactory without protection.

Where possible, a high ratio of anodic-to-cathodic area is desirable in designing to prevent galvanic corrosion. Therefore, the use of a fastener that is slightly cathodic to the base metal is often desirable. Furthermore, materials that naturally form protective oxides should be selected. These materials, including stainless steel and aluminum, often do not become active in galvanic cells. Other design approaches include eliminating the corrosive environment and providing an organic or inorganic barrier coating between the dissimilar metals. This involves the installation of steel fasteners into aluminum with a wet chromate primer or the use of an anodized aluminum structure in contact with a steel part. More information on galvanic corrosion is available in the section "Galvanic Corrosion" of the article "General Corrosion" in this Volume.

Uniform corrosion, or general corrosion, is a type of material deterioration that develops evenly over large areas of aircraft structures. Uniform corrosion is produced by many closely spaced anodic and cathodic sites on unprotected or partially protected surfaces (Ref 16).

Because large areas are affected in uniform corrosion, this type of damage can be recognized and remedied relatively easily as compared to other types of corrosion. The recognition of uniform corrosion is also enhanced by the unique characteristics of the corrosion products that are produced by various engineering materials (Ref 16). Typical problems and corrosion-preventive treatments pertaining to uniform corrosion that have been described in the technical literature include the following (Ref 16).

Adjuster and Eye of a Steel Track Rod. Rust occurred on these components after the cadmium plating on the threads became damaged during assembly or maintenance. This problem may have been prevented by the application of a protective coating, such as an organic paint primer, over the cadmium plating.

Aluminum Alloy Stringer. General corrosion was found on the surface of this component. Deterioration occurred in areas where the chromate pretreatment and primer coating had flaked off of the stringer.

Magnesium Alloy Skin of a Helicopter. This case is similar to the previous example. The uniform corrosion on this component also developed from the failure of the chromate conversion coating that had been applied to the magnesium skin.

Aluminum-Honeycomb Structure on a Fighter Aircraft. Water penetrated the inner surface of the aluminum skin through small holes. A foam rubber filler between the skin and the honeycomb core possibly contributed to the uniform corrosion that was produced in this case.

Structural Steel Fasteners. The surfaces of the nuts had the greatest degree of corrosion damage. In some areas, the corrosion penetrated very close to the bolt surface.

In summary, uniform corrosion can be prevented by selecting proper protective coatings, by ensuring proper surface preparation and coating application, and by touching up deteriorated surfaces as soon as possible. Sacrificial coatings, such as alcladding on aluminum, spread corrosion over large areas and thus enable aluminum sheet metal products to retain their mechanical strength. Information on the mechanism of uniform corrosion can be obtained in the article "General Corrosion" in this Volume.

Pitting corrosion produces deterioration of airframe structures by forming cavities and oxidation products in small (localized) areas of the affected components. The severity of pitting corrosion is determined by:

- The susceptibility of the airframe material to pitting attack. Unprotected, active metals, such as magnesium, are most susceptible
- The severity of the environment. Marine environments often cause pitting in airframe structures (Ref 16). This is because the chloride (Cl^-) ions in seawater promote the destruction of protective oxide films on the metallic materials that are used in airframes

Damage that was produced by pitting corrosion in airframe components has been documented for many cases that provide good examples of this type of aircraft degradation. Several of these incidents will be discussed for the following airframe structures.

Helicopter Structural Cleat (Ref 16). This magnesium alloy casting had been given the corrosion-preventive treatments of chromating and sealing (Ref 16). Pitting deterioration was enhanced by local damage in the protective coating. The alleged cause of this damage was impact from tools that were used for installing fasteners.

Stringer End Cap (Ref 16). This airframe structure was also made of magnesium, and it was chromated and primed. Moisture exposure contributed to the severe degree of pitting corrosion that occurred in this part.

Spring Wire. This hard-drawn high-carbon steel structure was observed to have been damaged by fatigue cracking in an area that had deteriorated by pitting corrosion (Fig. 12). Cyclic loading that is exerted on aircraft components during takeoff and landing promotes this type of fatigue fracturing in airframes that deteriorate because of corrosion. This particular case can also be classified as an example of corrosion fatigue (see the discussion "Corrosion Fatigue" in this section).

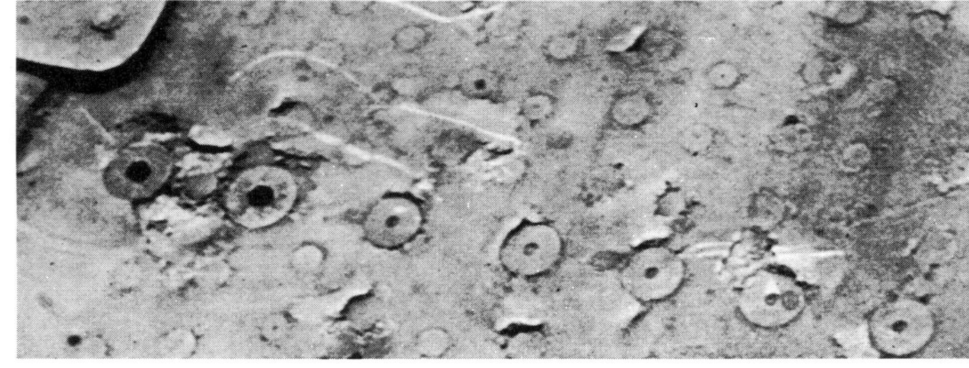

(a)

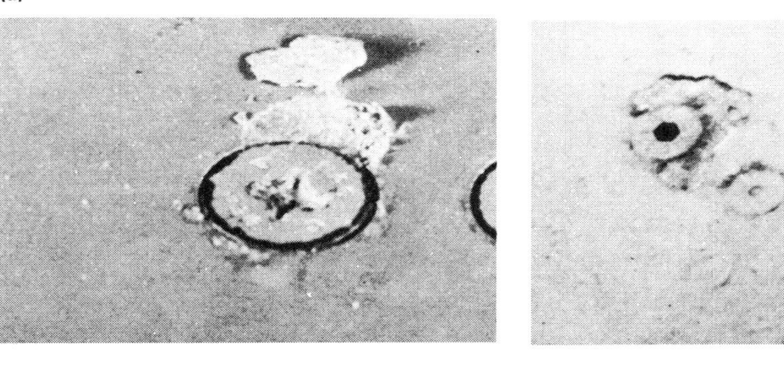

(b) (c)

Fig. 11 Galvanic corrosion of aluminum alloy 7075-T6 aircraft skin around cadmium-plated steel fasteners. (a) General view. (b) and (c) Close-ups of individual fasteners. Source: Ref 16

Fig. 12 Fatigue cracking of a hard-drawn carbon steel spring. Courtesy of Aeronautical Research Laboratories, Australia

Hydraulic Cylinders. An inner landing gear door actuator cylinder that failed on an F-101 aircraft is shown in Fig. 13. This part was fabricated from aluminum alloy 2024-T4. A close-up view of this failed component is shown in Fig. 13(b), which clearly indicates surface pitting corrosion. This pitting contributed to the fracture that is illustrated in Fig. 13(c). From Fig. 13(c), it was determined that exfoliation corrosion was involved in the initiation of the failure process.

Cargo Aircraft Brake Lining Carriers. Damage produced by pitting corrosion was observed in the brake lining carrier structure on the C-5A cargo aircraft (Ref 18). This brake lining carrier serves as a heat sink (for energy dissipation) during aircraft landings. Figure 14(a) shows the entire part, and Fig. 14(b) provides a magnified view of the component. Pitting deterioration was responsible for inducing filiform corrosion in the brake lining carrier, which consisted of the purest available beryllium material—approximately 99% Be. After this case was detected, a chromate conversion coating that conforms to MIL-C-5541C was applied to this part to minimize the probability of similar corrosion problems in the future. However, in 1985, the corrosion-prone beryllium was replaced by a carbon (graphite) composite material in the manufacturing of the later-model C-5B cargo aircraft (Ref 18). The carbon-composite material was selected by the C-5B manufacturer because it provided reduced weight (the new brake lining carriers saved up to 180 kg, or 400 lb, per aircraft in comparison to the beryllium structures), lower cost, and longer life, with better corrosion resistance and durability than the beryllium component.

Tip Tank Latch Knob (Ref 16). Pitting corrosion and fatigue occurred on this AISI 4340 steel structure. Several failures were recorded for this component. Fractures occurred below the head of the latch knob at the curved surfaces (radii). The fracture surfaces are shown in Fig. 15 at magnifications of 2.5 and 5×. Pitting corrosion was determined to be the primary cause of the fatigue failures that occurred with this latch knob. A change of the 4340 steel material to 17-4PH stainless steel was recommended as the long-term solution to this problem. This is because 17-4PH steel has significantly greater corrosion resistance than 4340 steel. However, periodic inspections also were required for increased protection against pitting, crevice, and edge corrosion in a marine environment.

Wing Flap Hinge Bearings (Ref 16). These structures consisted of chromium-plated type 440C martensitic stainless steel. The inner diameter of the failed bearings had numerous corrosion pits. Fracture occurred intergranularly, either by SCC or by hydrogen embrittlement. An electroless nickel plating was used as a substitute for the chromium plating in order to alleviate this pitting corrosion problem.

In summary, pitting corrosion in aircraft systems is often caused by:

- Local breakdown in a protective film on an alloy. This is usually accelerated when the material comes into contact with chloride-containing solutions
- Alloys susceptible to pitting because of localized impurities in the alloys. These impurities are either anodic or cathodic to the base metal
- Deposits of heavy metals (from water) on aircraft surfaces
- Localized damage (holidays) in applied protective coatings
- The accumulation of deposits (dirt, dust, grease) on bare aluminum, stainless steel, or steel surfaces

Whenever pitting corrosion occurs, a review of its causes is necessary to determine if a change in material (or materials), design, or protective coatings will most effectively deter future pitting corrosion problems. Because pitting can cause perforation of aircraft parts (especially thin, sheetmetal structures) or fatigue failures, care must be taken to:

- Neutralize pitting corrosion whenever it is detected
- Make repairs and provide local strengthening by using doublers (small patches of boron fiber reinforced plastic)
- Provide proper coating protection and regularly scheduled maintenance inspections

More information on pitting corrosion is available in the section "Pitting" of the article "Localized Corrosion" in this Volume.

Crevice corrosion occurs on aircraft structures when a corrosive fluid, such as salt spray, enters crevices that are located in individual parts or in between different components of a structural assembly. An anodic region usually develops at the bottom of the crevice, producing corrosive attack in the structure. Differences in the concentrations of dissolved salts or dissolved oxygen in the corrosive fluid will produce concentration

(a)

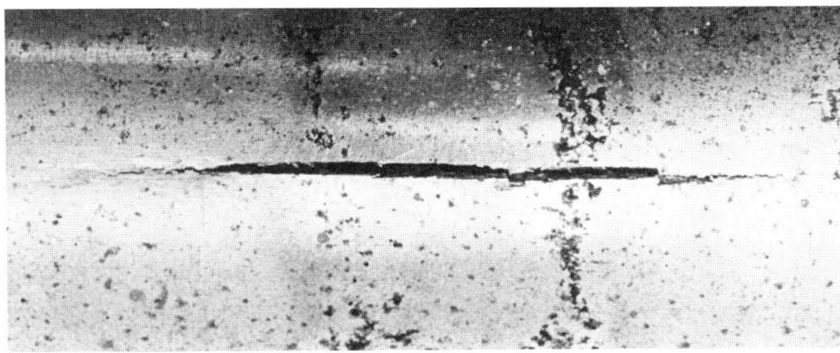

(b)

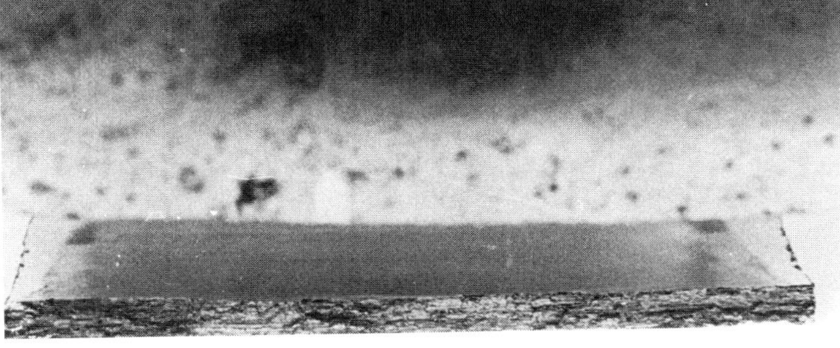

(c)

Fig. 13 Aluminum alloy 2024-T4 landing gear door actuator (a) that failed because of pitting corrosion. Arrow points to crack. (b) Close-up of crack in landing gear door actuator in (a). (c) SEM of the fracture surface of the door actuator. Source: Ref 5

(a)

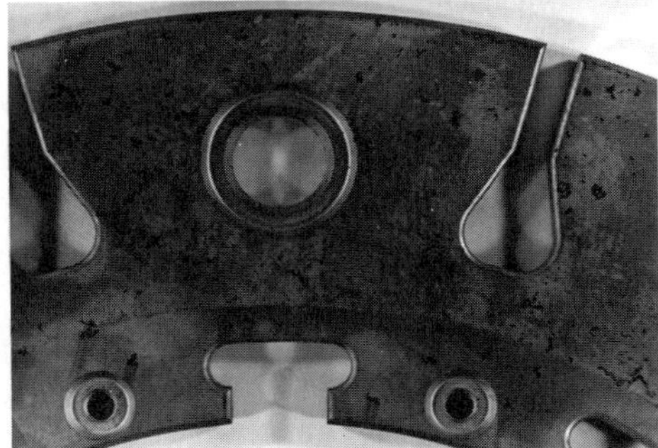

(b)

Fig. 14 Pitted beryllium brake lining carrier (a) from a C-5A transport plane. (b) Close-up showing pitting and filiform corrosion. Courtesy of the National Association of Corrosion Engineers

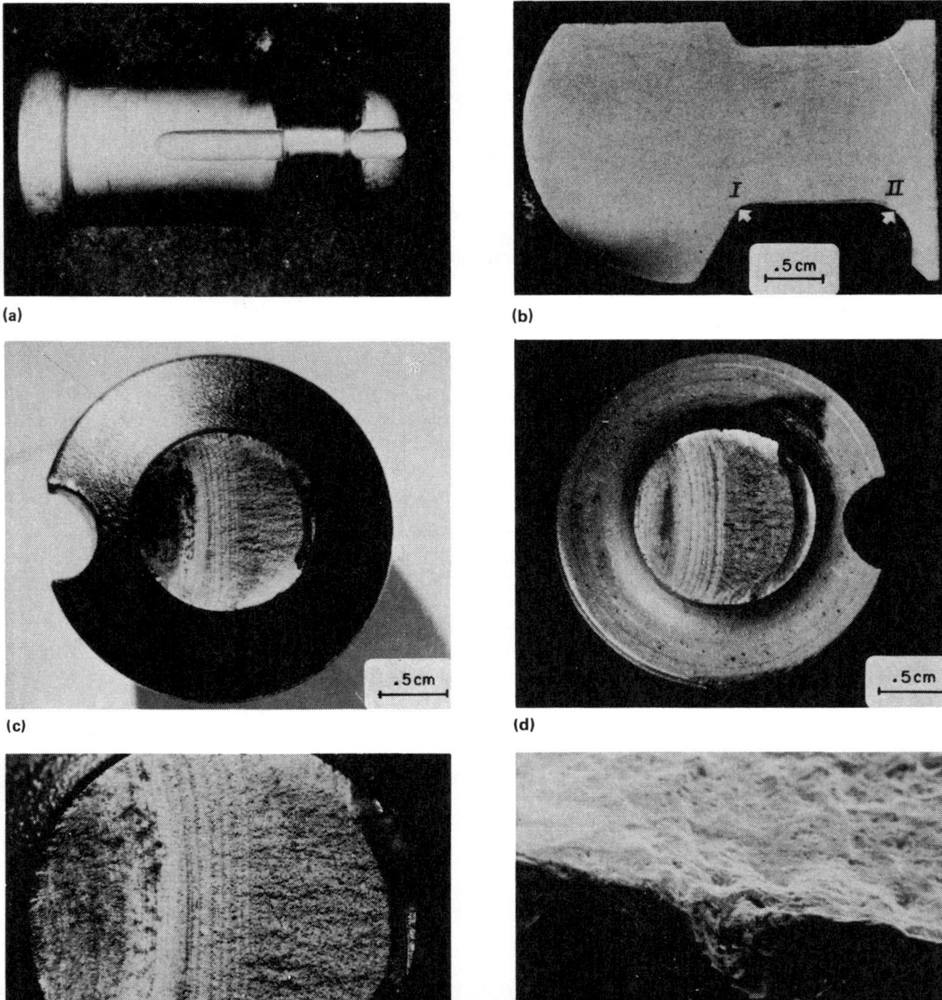

Fig. 15 Pitting corrosion and fatigue failure of a 4340 steel tip tank latch knob. (a) General view of the latch knob. (b) Cross section through the knob showing crack initiation sites. (c) Fracture surface of crack that initiated at site I. (d) Fracture surface of crack that initiated at site II. (e) Fracture surface at site I. (f) Pitting at a fatigue crack initiation site. Source: Ref 16

cells and thus promote the development of anodic sites in structural airframe crevices (Ref 16). Typical cases of crevice corrosion have been observed in the following aircraft components (Ref 16).

Magnesium Panel. Crevice corrosion perforated this structure, which was in a dismantled joint area from a transport aircraft structure. Inadequate sealing was determined to be the cause of this corrosion problem.

Magnesium Alloy Skin Joint. This example is similar to the previous one. The cause of this problem was attributed to unsatisfactory sealing practices.

Point of Contact Between a Magnesium Alloy Floor Panel and Aluminum Alloy Frames and Angle Plates. Both galvanic corrosion and crevice corrosion occurred in this case. Both of these forms of corrosion could have been prevented or minimized by a better protective coating, a more effective fluid removal (drainage) system, and better utilization of sealants (Ref 16). Crevice corrosion also occurs in many alloy systems other than magnesium, including stainless steels and aluminum alloys.

In summary, two methods are commonly used to prevent crevice corrosion. The first is proper design. This includes provisions for drainage, sealants that are applied to faying surfaces, or beads of sealant that cover crevices formed by mating surfaces. The second method consists of spraying aluminum airframes with water-displacing (penetrating) oils that seep into crevices and prevent water ingestion. The section "Crevice Corrosion" of the article "Localized Corrosion" in this Volume contains more information on this form of attack.

Stress-Corrosion Cracking. The majority of the aluminum airframe structures that have been documented as failing by SCC have been manufactured of alloys that contain aluminum, copper, zinc, and magnesium (Ref 19). These are the 2000- and 7000-series aluminum alloys. Stress-corrosion cracking failures have also been observed in high-strength steels, such as:

- A martensitic steel spindle sleeve on the main rotor hub assembly of a military helicopter
- Large (38-mm, or 1.5-in., diam) high-strength steel bolts. When these parts failed, the head of the bolt completely separated from the shaft (Ref 19)
- A 300M steel landing gear drag strut on a military helicopter (Ref 20)

Stress-corrosion cracking is difficult to recognize in aluminum forgings because there is often no visual indication of surface corrosion products. The cracks can also be very long and deep (Ref 16). Stress-corrosion cracks usually occur in the end grain of the forged component at the parting plane of the forging. The cracking that occurs in airframe stress-corrosion failures is usually intergranular.

Many airframe SCC failures have involved structures that were manufactured from aluminum alloys, especially 7079-T6 and 7075-T6. These include the following airframe components that were fabricated from aluminum and that have been observed to fail by SCC (Ref 16) (the specific aluminum alloy is provided in parentheses):

- A main landing gear locking cylinder (7079-T6)
- A main landing gear H-link structure (7079-T6). This damaged component is illustrated in Fig. 16. The stress corrosion was induced by the precipitation of magnesium aluminide (Mg_2Al_3), which caused the grain boundaries in the aluminum forging to deteriorate anodically
- The front and rear spars of a vertical fin (7079-T6). As shown in Fig. 17, many of the cracks in this failure propagated from fastener holes. These spars had received corrosion-preventive surface treatment. However, some of this protection was inadvertently removed during the installation of bolts. Bare metal, therefore, was exposed to a high-humidity environment and sustained high tensile stresses that were produced by the installation of fasteners into these structures. This problem was remedied by the use of 7075-T73 aluminum forgings for the front and rear spars. The latter material provides greater resistance to SCC than 7079-T6
- The bearing housing of a vertical stabilizer beam (7079-T6)
- A main landing gear bogie, which has the appearance of a beam or strut-type structure (7075-T6)
- The hydraulic cylinders that serve as actuators for a main landing gear door (7075-T6). Views of the fracture surface, including the appearance of intergranular fracture, are shown in Fig. 18. This problem could have been alleviated by the application of a better corrosion-protective treatment in order to minimize the degree of pitting corrosion that occurred during the storage of these cylinders. The use of the more SCC-resistant aluminum alloy 7075-T73 also would have helped to prevent this failure
- The fork and strut components of a nose landing gear (7075-T6)
- A fuselage frame structure, in which SCC occurred in between fastener holes (7075-T6)
- A nose landing gear strut (7075-T6)

Stress-corrosion cracking has also been observed in an aircraft aluminum-copper main landing gear forging and in four main landing gear retraction cylinders that were made of 4340 steel (Ref 16). Several views of this main landing gear forging failure are shown in Fig. 19. Deep corrosion pits were observed in the area of stress-corrosion crack initiation (Fig. 19c and d). This main landing gear forging was manufactured from an alloy that contains aluminum, copper, silicon, and manganese. This alloy is similar to aluminum alloy 2017. Aircraft structures that are fabricated from the 2017-T4 and 2017-T451 alloys can fail by SCC when sustained tensile stresses are exerted in the transverse direction relative to the grain structure.

A representative example of SCC in an aircraft tubing structure is shown in Fig. 20. This part was fabricated from AM350 precipitation-hardened stainless steel. An intergranular type of cracking can be seen in the photomicrograph in Fig. 20(b). Fracture was intergranular (Fig. 20c).

Good examples of SCC in airframe fasteners have been presented in the literature. Most of these fractured parts were H-11 steel bolts manufactured according to AMS 6487. The usual corrosion-protective treatment for these bolts was either:

- A coating of fluoborate cadmium that complies with NAS 672 (for fasteners that have minimum tensile strengths of 1517 MPa, or 220 ksi)

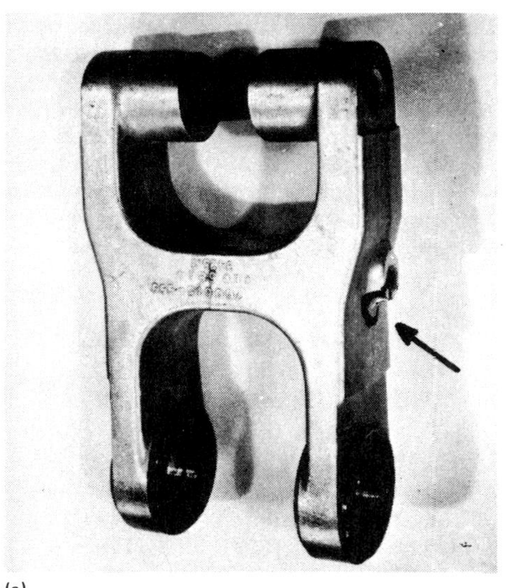
(a)

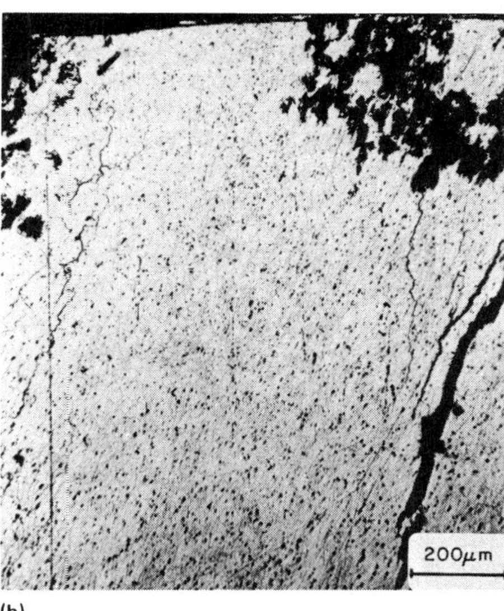

(b)

Fig. 16 SCC in an aluminum alloy 7079-T6 main landing gear H-link. (a) Overall view of H-link. (b) Pitting and intergranular corrosion that initiated SCC. Source: Ref 16

- A vacuum-deposited cadmium finish, as specified by MIL-C-8837 (for fasteners that have minimum tensile strengths of 1793 MPa, or 260 ksi)

As shown in Fig. 21, the initiation of SCC in a failed aircraft alloy steel bolt occurs at a surface corrosion pit. Propagation of this corrosion-induced crack gradually continues until the bolt fractures, because the stress-corrosion crack significantly reduces the amount of mechanical stress the bolt can sustain.

Two SCC case histories involving H-11 alloy steel fasteners have resulted in beneficial airframe design improvements. These incidents, and the design solutions, are described below.

In the first case history, cadmium-plated H-11 fasteners became pitted and eventually failed by SCC on a widebody transport (cargo) aircraft. These fasteners were installed in the engine-to-pylon attachment structures. The cadmium-plated steel fasteners had been installed into a titanium airframe component. This coupling of dissimilar airframe materials produced galvanic corrosion, severe pitting, and the eventual failure of these fasteners by SCC (Ref 21).

As a result of this incident, all of the fasteners that were considered to be critical to the engine-to-pylon attachments were replaced by cold-worked and aged A286 stainless steel and cold-worked and aged Inconel alloy 718. Generally, neither cadmium- nor silver-plated parts should be used in contact with titanium under compressive loading. This condition causes the cadmium or silver to migrate into the intergranular structure of the titanium and induce cracking. This migration of cadmium or silver into the intergranular structure of titanium is commonly referred to as poisoning. Although suitable for installation into steel structures, cadmium- and silver-plated parts are to be avoided in titanium assemblies that will be subjected to high compressive loads, heat, or both.

In the second case history, SCC failures of 1517 MPa (220 ksi) H-11 fasteners occurred in the wing attachment area on several general-purpose aircraft. Alloy steel fasteners installed in this wing attachment structure also fractured. These alloy steel fasteners had maximum tensile strengths of 1103 and 1240 MPa (160 and 180 ksi). The design solution to this problem was to have all of the noncorrosion-resistant steel fasteners, including the alloy steel ones, replaced by 1240 MPa (180 ksi) and 1517 MPa (220 ksi) Inconel 718 fasteners (Ref 21).

There are a few important guidelines that can be followed in order to minimize the probability of fastener failure due to SCC or other types of corrosive deterioration (Ref 22). First, fasteners should not produce adverse effects on the structures that they are joining. These fasteners also should not be installed into certain materials that would make these fasteners susceptible to corrosion. Second, fasteners should be selected that are slightly cathodic to the parts they will join. This is especially advisable in aggressive environmental conditions, such as in tropical-marine areas. Materials that are susceptible to hydrogen embrittlement, such as high-strength steels, generally should not be used for fastening applications in aircraft structures. Finally, critical joints should not be connected with fastening materials that are susceptible to corrosive deterioration in the environment in which the actual aircraft system will be operating.

Aircraft service safety and airworthiness can be significantly improved by selecting airframe materials that are resistant to SCC and brittle fracture (Ref 23). Aircraft design engineers can minimize airframe failures due to SCC by ensuring that:

- Materials that are selected for aircraft applications have low rates of SCC (Ref 23). The T6-tempered 2000-series aluminum alloys (except for 2024 aluminum) should be avoided. The 7000-series aluminum alloys should be in the T-73 or T-76 tempers. Similarly, stress-corrosion-resistant tempers of steel alloys should be used
- Cracks in selected airframe materials are detected prior to a significant reduction in the static strength of the airframe part (Ref 23)
- Tensile and residual stresses, which may become significant during fabrication and assembly, are minimized. This can be done by select-

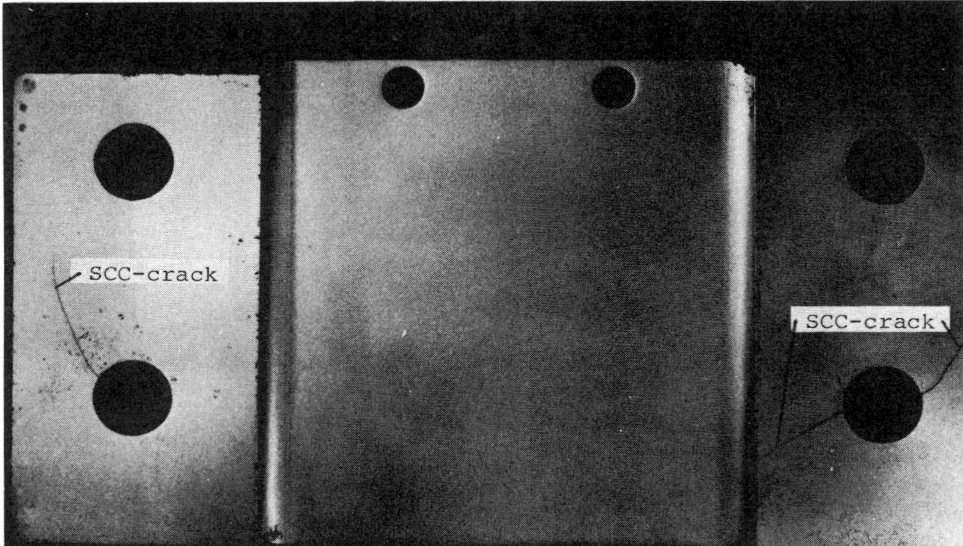

(a)

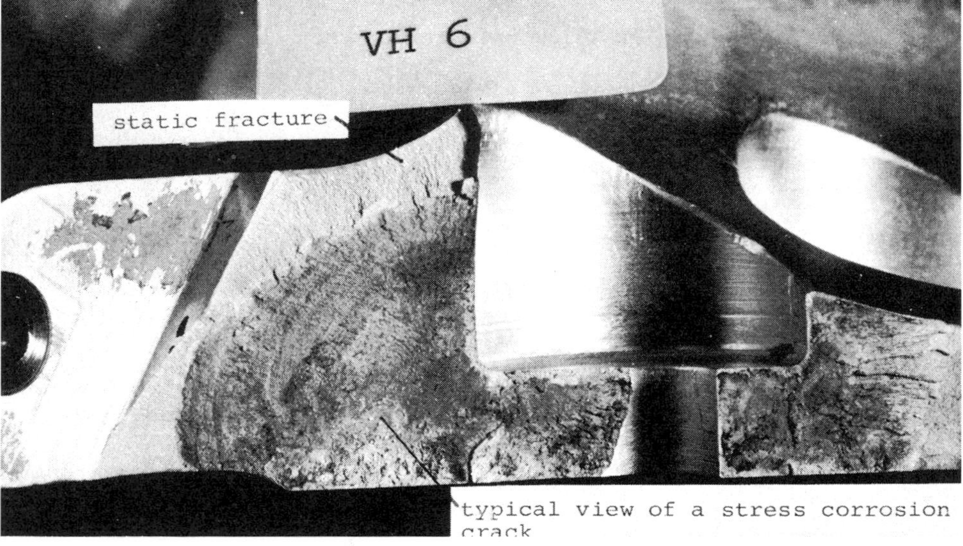

(b)

Fig. 17 SCC in aluminum alloy 7079-T6 spars of a vertical fin. (a) Cracks in the mating surface of the rear spar. (b) Fracture surface of a statically broken front spar. Source: Ref 16

ing stress-relieved tempers for specific alloys (Ref 24) and by minimizing assembly stresses by shimming, avoiding interference fits, and shot peening
- Compressive stresses are properly applied on machined surfaces by shot peening
- Organic coatings are used, whenever possible (Ref 24)
- Inorganic corrosion-protective methods are used, such as cladding on aluminum or electroplating on steel, whenever possible (Ref 24)

More information on the mechanism of SCC is available in the section "Stress-Corrosion Cracking" in the article "Environmentally Induced Cracking" in this Volume.

Intergranular Corrosion. A severe case of intergranular corrosion was described in the beginning of this section (see Fig. 9 and 10). This discussion will provide some additional corrosion problems and solutions involving intergranular corrosion and the more damaging form of intergranular corrosion—exfoliation. The purpose of this discussion is to provide a more detailed explanation of how these forms of corrosion are interrelated.

In intergranular corrosion and in exfoliation corrosion, the grain boundaries of the corroded metal become anodic. The bulk material in between the grain boundaries is not affected and therefore is cathodic. Corrosion products and, occasionally, cracking are produced on the surface of materials that corrode intergranularly. This form of corrosion sometimes penetrates underneath the metal surface, making it difficult to detect the damage without the aid of a microscope (Ref 16). Intergranular corrosion can occur either alone, in conjunction with pitting corrosion, or with exfoliation corrosion.

When intergranular corrosion occurs along grain boundaries that are parallel to the plane of the material, such as in a flat plate, the corrosion proceeds in the short-transverse direction along these elongated grain boundaries. The combination of the corrosion and the entrapped corrosion products causes the plate to delaminate.

When intergranular corrosion occurs, a network of fine surface cracks can develop. This fine cracking pattern has been observed in an aircraft hydraulic valve made of a forged aluminum alloy that deteriorated by intergranular corrosion (Ref 16).

An example of exfoliation corrosion of an aluminum alloy stabilizer bracket from a light aircraft is shown in Fig. 22. This deterioration started as intergranular corrosion but gradually became more severe and propagated as exfoliation corrosion. The horizontal surface of the stabilizer bracket had been exposed to atmospheric moisture and contaminants, which collected at the interface between the bracket and a nylon bushing. No corrosion was found on bracket surfaces that were protected by a chemical conversion coating. This problem could have been prevented by effective sealing of the bracket-to-bushing interface, along with regular inspections (Ref 16). Some typical, documented examples of intergranular corrosion and exfoliation corrosion have also occurred on the following structures.

Wing Box Lower Panel of a Fighter Aircraft (Ref 16). This panel was made of aluminum alloy 7075-T6. Most of the corrosion occurred around fastener holes. Extensive intergranular cracking was observed. The report on this case also indicated that pitting occurred in the bores and countersinks of the fastener holes. Filiform corrosion was also detected in the fastener hole areas. This problem was solved by applying a conversion coating to the fastener hole bores and countersinks. Next, the fasteners were wet assembled using a strontium chromate primer and an acrylic topcoat. Another part of the solution in this case was the development of a new aluminum alloy, 7475-T761 (Ref 16). This material has a high level of resistance to exfoliation corrosion and therefore has been considered by aircraft designers as a favorable replacement for the 7075-T6 alloy.

Main Rotor Blade of a Helicopter (Fig. 23). Intergranular and exfoliation corrosion predominated in the area between the leading edge spar and the surface skin of the blade. Extensive corrosion accumulated at the leading edge, causing the skin of the blade to lift off of the spar. The leading edge spar was manufactured from aluminum alloy 2024. During a metallurgical examination, copper aluminide was found in the grain boundaries of this material. It was therefore determined that the 2024 aluminum structure was improperly heat treated. More information on this form of attack is available in the section "Intergranular Corrosion" of the article "Metallurgically Influenced Corrosion" in this Volume.

Filiform Corrosion. One of the more insidious types of electrochemical degradation in aircraft structures is filiform corrosion. This form of corrosion usually occurs beneath the cladding of clad aluminum alloys, under organic coatings applied to airframe surfaces, and at fastener holes (Ref 1, 16). Filiform corrosion has been described as a type of anodic undermining, with reference to its occur-

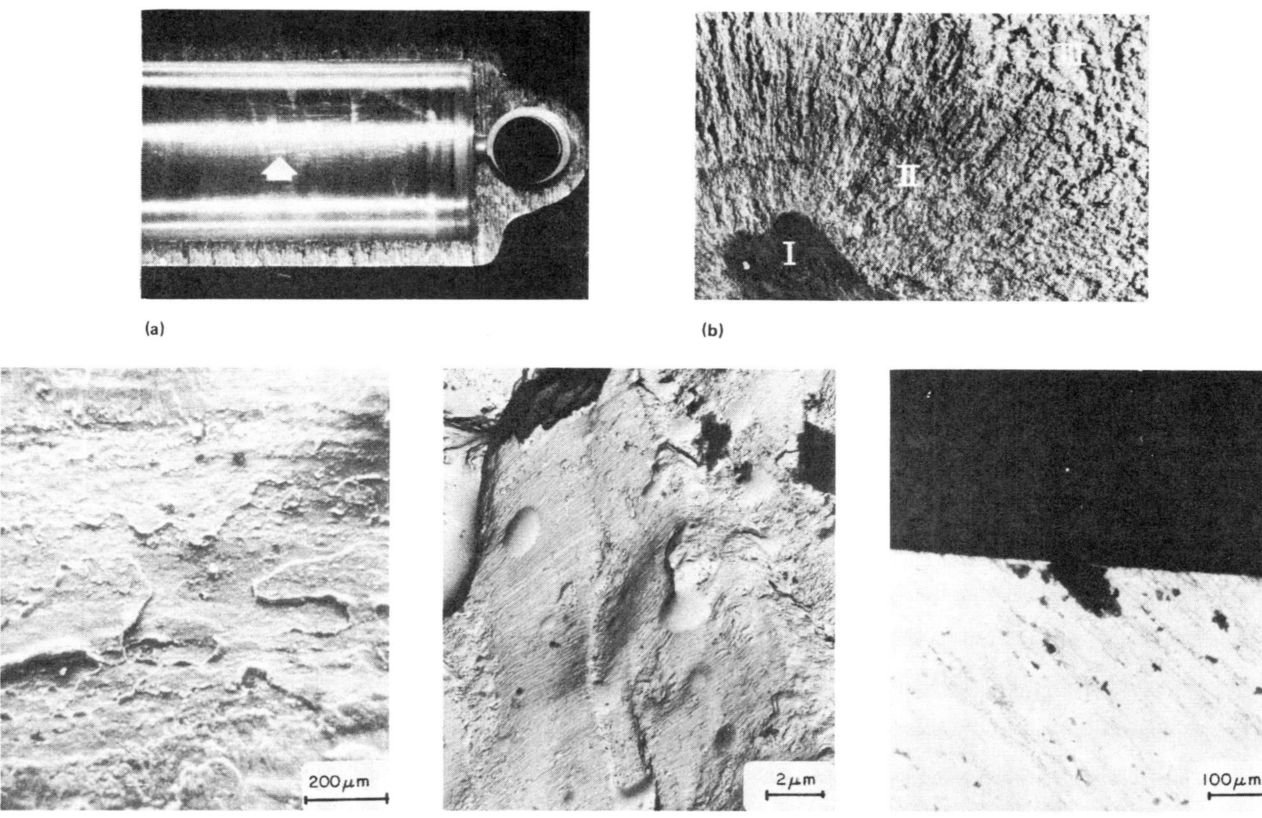

Fig. 18 SCC of an aluminum alloy 7075-T6 hydraulic cylinder. (a) Section through cylinder showing cracks on the inside surface (arrow). (b) Fracture surface showing three regions of cracking. (c) Appearance of intergranular fracture in region I. (d) Fatigue striations observed in region II. (e) Cross section through the inside surface showing corrosion pit. Source: Ref 16

rence on metallic structures that are protected by organic primers and topcoats (Ref 1).

Filiform corrosion usually begins as a shallow corrosion pit, which continues its attack on the base metal by spreading laterally along the surface of the structure. This appears as an irregular pattern of thin filaments of corrosion products on the affected metallic material. A schematic illustration of filiform corrosion is presented in Fig. 24.

One theory that is applicable to aircraft systems proposes the following series of events that promote filiform corrosion on coated metallic airframe parts (Ref 25):

- In-flight, minor relative motion (vibration) produces very thin discontinuities, which also have been called hair cracks, in the airframe paint coating. Salts that primarily consist of sodium and calcium are present in the areas where these small cracks originate
- These salts (inside the hair cracks) absorb moisture from the atmosphere by osmosis. High humidity accelerates the propagation of filiform corrosion (Ref 16). Chlorine also can activate this type of corrosion (Ref 25)
- Moisture absorption at the airframe structure produces an aqueous electrolyte, which causes anodic and cathodic reactions to take place
- The anodic reaction produces pitting that does not penetrate very deeply into the aluminum airframe material
- Aluminum hydroxide ($Al(OH)_3$) is precipitated because of the formation of hydroxyl (OH^-) groups
- The head and body of the corrosive filament develop and separate from the areas where the original cracks were formed
- The corrosive filamentary head continues to deposit corrosion products because water is replenished in the head by osmosis

Because filiform corrosion propagates on structural areas that are either clad or coated with organic paint, this type of material deterioration can spread extensively before it is detected by aircraft maintenance personnel. In typical cases, filiform corrosion develops around fastener holes on airframe sheet structures. Paint blistering around the rivet holes is a characteristic feature of this type of corrosion. Some of the materials that are known to have been affected by filiform corrosion are magnesium, aluminum, steel, and chromium-plated nickel. Documented case histories of filiform corrosion of various components include the following.

Fuselage Skins. On a Boeing 707 aircraft operated by a major commercial airline, filiform corrosion occurred on fuselage skins along rows of fasteners (Ref 26). Paint blistering is produced on airframe aluminum sheetmetal structures when this form of attack occurs (Ref 26).

Areas Around Steel Fasteners. Filiform corrosion was observed in the areas around steel fasteners, which originally were affected by intergranular corrosion. The corroding fasteners were installed in the lower wing skins of the Boeing 707. Figure 25(a) illustrates this deterioration prior to paint removal, while Fig. 25(b) shows the filiform corrosion damage after the paint coating was stripped from the airframe surface.

Lower Wing Skin. When a Boeing 747 aircraft was first placed into service, filiform corrosion was detected on the lower wing skins of one of these aircraft (Ref 26). This corrosion developed from intergranular corrosion around titanium fasteners that were inserted into the airframe structure.

Pylon Tank. Filiform corrosion caused the perforation of one area of an aluminum alloy 6061-T6 pylon tank (Fig. 26). Pitting and intergranular corrosion were also detected on the pylon tank during the investigation of this problem (Fig. 26c). The aircraft that operated with this tank had been flying in the hot and humid Mediterranean environment. The proper application and maintenance of epoxy or polyurethane paint finishes would have minimized the amount of deterioration on this structure.

Other Components. A structural engineer for a commercial airline reported that filiform corrosion was detected on horizontal and vertical stabilizers, trailing edge flaps, and an aluminum metal sprayed surface on fiberglass wing-body fairing panels (Ref 26).

For the areas on the Boeing 707 that were affected by filiform corrosion, a corrosion-inhibiting compound called LPS-3 was found to be a successful treatment for the prevention and control of corrosive (filiform) growth. This compound is a greasy hydrocarbon-base preservative that minimizes oxidation by displacing water

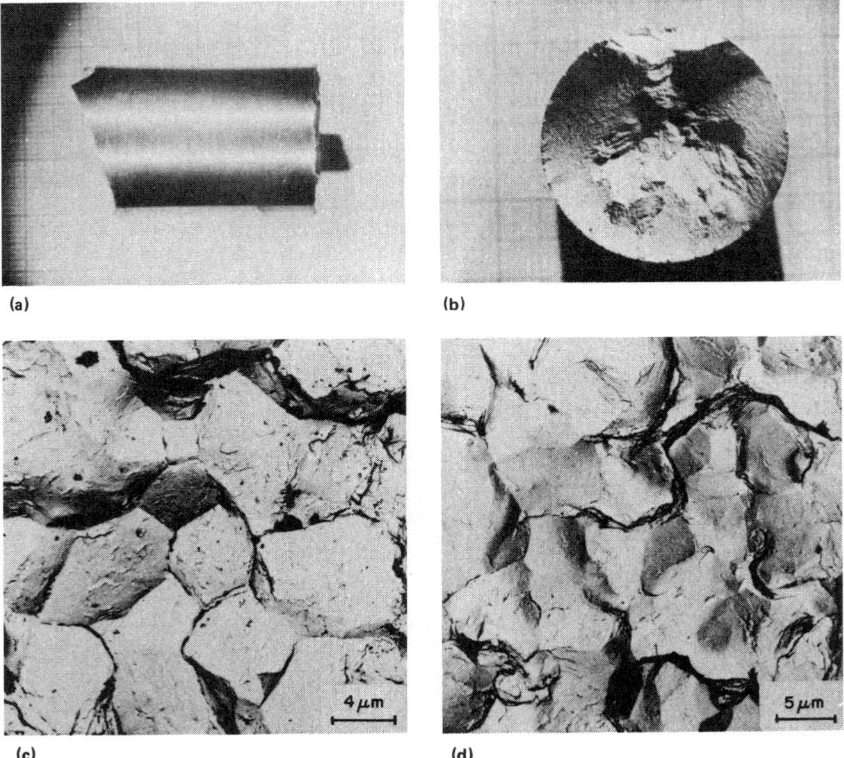

Fig. 19 SCC in a 4340 main landing gear pivot pin. (a) Central portion of the broken pin. (b) One of two fracture surfaces on the piece shown in (a). (c) TEM of the fracture surface. (d) Fracture surface of another specimen that failed during a sustained load test

away from the head of the corrosive filament (Ref 26).

The following sequential treatment has been determined to be effective in alleviating filiform corrosion of airframes:

- Glass bead blasting to remove observable corrosion products
- Chemical conversion coatings
- Boeing Material Specification (BMS) 10-79 strontium chromate epoxy primer
- BMS 10-60II flexible polyurethane enamel

The commercial airline engineer that reported the above airframe treatment for filiform corrosion stated that this treatment provides 5 years of service life with only minor maintenance (Ref 26).

A polysulfide-calcium-strontium rubber primer covered with polyurethane paint, which has been used for corrosion protection on military aircraft, has also been found to be effective in deterring filiform corrosion (Ref 26). In one case, for example, filiform corrosion was observed on some aluminum alloy 2024 aircraft ailerons and flaps (Ref 16). This largely superficial corrosion initiated at the surfaces around fastener holes. The recommended solution was to replace the nitroalkyd resin paint coating with a finishing system that consisted of an epoxy primer and a polyurethane topcoating. The section "Filiform Corrosion" of the article "Localized Corrosion" in this Volume provides more information on filiform corrosion.

Fretting corrosion occurs on airframe structural surfaces that move against each other in an environment that is conducive to corrosion. Often, the relative movement is barely discernible, but results in significant deterioration. This type of corrosion is very significant in operational aircraft systems because airframes are subject to high vibration levels and because airframe components must also endure various types of mechanical stress other than vibration, resulting in small relative movement between parts (Ref 27).

Some of the airframe parts that have been documented as being susceptible to fretting corrosion are (Ref 27):

- Hinge point bearings on ailerons, elevators, rudders, and flap structures
- Control pulley bearings
- Universal joint bearings
- Propeller and propeller control bearings, housings, and shafts
- Instrument bearings
- Spline connections
- Pin joints
- Clamps
- Riveted lap joints

Fretting actually appears as a combination of wear and corrosive deterioration. Fretting corrosion occurs in the presence of oxygen, which oxidizes the small wear particles that form in structural contact areas. These small particles develop because of the oscillatory motion that exists in contacting metallic materials. The hard oxides then become trapped in the point of contact and further promote intense wear on the two rubbing surfaces (Ref 28).

Concerning the development of aluminum helicopter rotor blades, it has been noted that fretting produces a significant reduction in the fatigue strength of the root retention area of the blade (Ref 28). With little or no fretting, the maximum allowable alternating stress, at 10^7 cycles, is about 21 MPa (3000 psi). Fretting causes the maximum allowable stress to drop to between 8 and 10 MPa (1200 and 1500 psi) in joints where there will be high stress concentrations (Ref 28). Representative descriptions of documented cases of fretting corrosion in aircraft structures include the following.

Flying Control Hinge Pins. Fretting deterioration on these cylindrical structures appeared as broad bands of accumulated corrosion products. Severe (deep) pitting corrosion was also observed in this case (Ref 16). This damage possibly occurred as a result of both wear (due to large and small amplitude displacements) and fretting corrosion (resulting from oxidation).

Propeller Shaft Ball Bearings. Fretting corrosion was observed in the inner and outer races of these propeller aircraft bearings. The bearings were manufactured from 51100 bearing steel, and they were heat treated to produce a fine tempered martensite (hardness 63 HRC). The locations on the outer bearing race at which fretting corrosion occurred appeared to be slightly brighter than the unaffected regions of the structure. The fretted areas were also damaged by pitting corrosion (Ref 16).

Lower Boom of the Wing Forward Spar. This structure was fabricated from an aluminum alloy 2014 forging (Ref 23). A fatigue crack started at the edge of a bolt hole in this bomber aircraft component. This crack extended to only 0.5% of the total cross-sectional area of this part. The initiation of this fatigue crack, which eventually caused the in-flight failure of the aircraft, was attributed to fretting (Ref 23).

Engine Fuel Tanks. Failures in these components occurred because of fretting that caused poorly soldered lap joints to separate (Ref 29). Mechanically locked joints were the recommended solution for alleviation of this problem. The redesign of these fuel tanks also included use of a thicker terne (lead-tin) plate on tank inner surfaces (Ref 29).

One of the most widely used techniques for averting fretting corrosion is to apply a continuous film of lubrication between the contacting surfaces (Ref 28). A second method consists of increasing the load (pressure) at the point of surface contact. This will reduce the amount of slip (relative, oscillatory motion) at the contact point.

Yet another method involves selecting materials with a high resistance to fretting corrosion. Metallic combinations that have high, medium, and low resistances to fretting corrosion are presented in Table 4. These results are from a study on fretting wear and corrosion that was conducted by the Massachusetts Institute of Technology. Fretting tests were performed under dry conditions both in air and in a vacuum. In this investigation, low fretting corrosion resistance was attributed to the high hardness of metal oxides that were formed when certain pairs of metals were moved against each other. For example, the low resistance to fretting corrosion of the aluminum-to-aluminum combination was probably a result of the high hardness of aluminum oxide, which is formed during fretting and has a Moh's hardness of 9 (the maximum Moh's hardness value is 10) (Ref 28). Preloading assembled joints and the use of sealants can reduce this problem in aluminum-to-aluminum joints. Anodizing of mating parts is also effective because a continuous, hard aluminum oxide is produced on the mating surfaces.

(a)

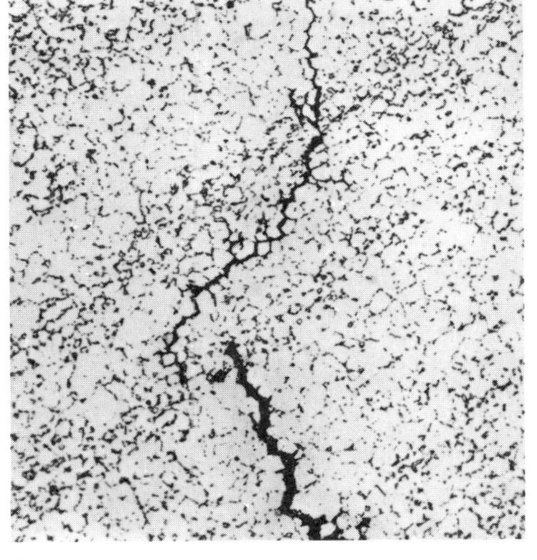

(b)

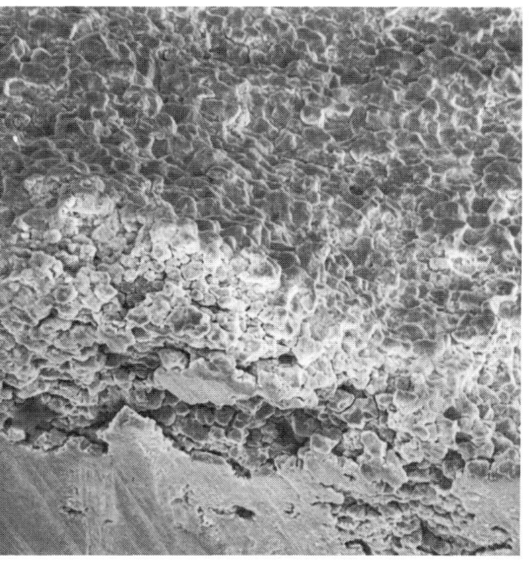

(c)

Fig. 20 Intergranular SCC in an aircraft pipe structure fabricated from AM-350 precipitation-hardened stainless steel. (a) Overall view of pipe. (b) Micrograph showing intergranular nature of cracking. (c) SEM of crack shown in (b). 400×

Corrosion Fatigue. Aircraft structures are subjected to in-flight mechanical stresses as well as to corrosive environments. The mechanical stresses are cyclic in nature; that is, periods of high stress, or loading, are followed by relatively lower stresses, or unloading (Ref 1). This cyclic stress results in fatigue loading of airframe structures.

When airframe structures are adversely affected by the combined effects of corrosion and fatigue, the accelerated deterioration that occurs is referred to as corrosion fatigue. Airframe structures fracture in a shorter time because of fatigue that occurs in combination with corrosion (Ref 16). Further, corrosion reduces the fatigue (endurance) limit of airframe components. Corrosion fatigue is most severe in those airframe parts that are most vulnerable to both fatigue and corrosion, such as fastener holes (Ref 30).

A typical example of the corrosion fatigue failure of a military helicopter rotor assembly is discussed in Ref 16, and more information on this failure is also provided in Example 16 in the section "Appendix: Case Histories and Failures" in this article. This rotor assembly consisted of several components—a horizontal hinge pin, a nut, and a locking washer—that were manufactured from 4340 and 4130 steel (Fig. 27a to c). Fracture was observed on the horizontal hinge pin. A flat beach mark on the hinge pin indicated the location of crack initiation in the structure. Figures 27(d) and (e) show the respective fracture surface and beach mark. One side of the threaded area of the hinge pin had a dent (Fig. 27f). Corrosion pits were discovered close to the fracture initiation site. Cracks that ran parallel to the primary fracture surface appeared to be emanating from small pits in the hinge pin (Fig. 27g). An embedded thread in the hinge pin was formed by contact of the 4130 low-alloy steel nut with the 4340 pin (Fig. 27h).

The failed hinge pin had been cadmium plated. Pitting corrosion in the hinge pin promoted the deterioration of the cadmium plating and the exposure of the 4340 steel base material.

This corrosion fatigue failure led to a complete inspection and overhaul of all components in this fleet of helicopters. The inspected (undamaged) hinge pins were stripped of the cadmium plating, shot peened, and replated to a minimum thickness of 12.5 μm (0.5 mil) (Ref 16).

Other typical corrosion fatigue failures have been documented, including the following (Ref 16):

- A main landing gear wheel, made of magnesium alloy AZ91C
- A main landing gear wheel manufactured from magnesium alloy QE22A. Figure 28 illustrates the deterioration that occurred in this case. The fracture of this part started on the outer surface and was probably caused by a defective corrosion-protective coating system. Regular inspection and repair of the paint system that was applied to this structure could have prevented this damage
- Rivets fabricated from aluminum alloy 5052. These rivets were used for the installation of a helicopter filter housing. The damaged rivets are shown in Fig. 29. Figure 29 also presents views of the fracture surface and the grain structure of the rivet material

In summary, special corrosion prevention and control treatments must be provided to aircraft that operate in environments conducive to corrosion fatigue. Fatigue resistance can be enhanced in critical areas by designing parts with generous corner radii and by shot peening. Adequate corrosion protection must be used and then followed up by continuous in-service inspections. Pitting must be removed from critical parts because corrosion fatigue often nucleates in corrosion pits. More information on corrosion fatigue is available in the section "Corrosion Fatigue" of the article "Mechanically Assisted Degradation" in this Volume.

Microbiological Corrosion. This type of airframe corrosion is also called biological corrosion (Ref 31). Microbiological corrosion was discovered in the mid-1950s in aircraft integral fuel (wing) tanks (Ref 13). The microbes used kerosene fuels and salt water as growth media (Ref 16, 31). Certain types of synthetic rubber have also been found to promote microbial growth (Ref 14, 32). This support of microbial growth by synthetic rubber was discovered in cases of microbiological corrosion on the military C-130 transport aircraft (Ref 32).

Most of the observed airframe corrosion has been produced by a group of microbes called fungi. The greatest number of cases of aircraft integral fuel tank corrosion have been attributed to the fungus *Cladosporium Resinae* (Ref 16). This microorganism was recently identified by some Argentinian investigators in an integral fuel tank corrosion incident (Ref 33). The fuel tank material was aluminum alloy 2024. The fungus was most active (and, therefore, most corrosive) at the fuel/water interface inside of the 2024 aluminum fuel tank.

Microbiological corrosion can also be caused by bacteria and yeasts (Ref 16). The bacterium *Pseudomonas aeruginosa* has been observed in

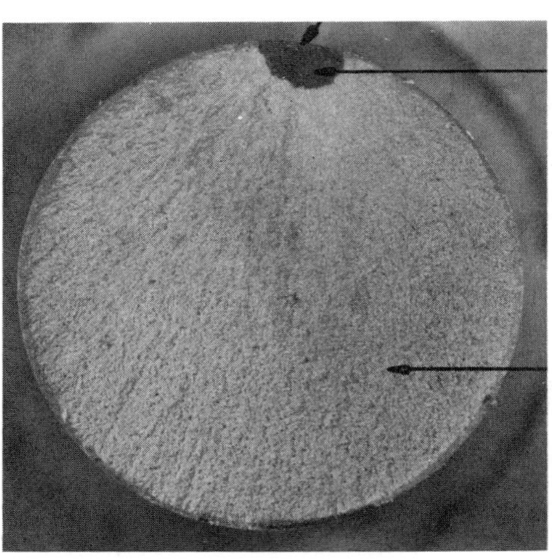

(a)

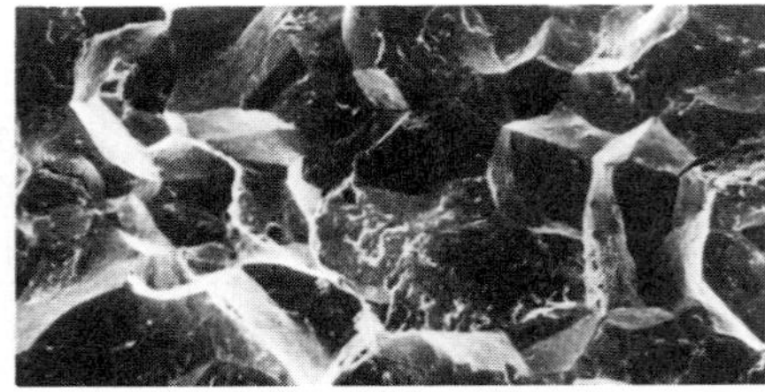

(b)

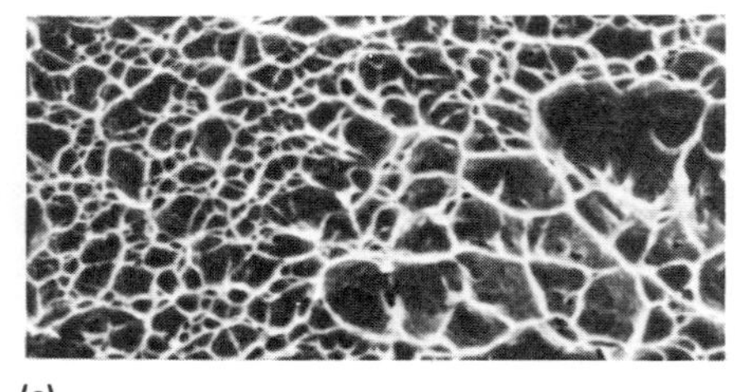

(c)

Fig. 21 SCC of an alloy steel aircraft bolt. (a) Fracture surface showing origin at a corrosion pit on the bolt surface. (b) SEM showing brittle, intergranular SCC near the origin. (c) SEM of the region of fast fracture showing ductile nature of fracture in this area. Source: Ref 21

aircraft fuel tank corrosion cases, but much less frequently than *Cladosporium resinae*.

Aircraft that operate in tropical environments appear to be most vulnerable to microbiological corrosion (Ref 16). Pitting corrosion is often present in microbiological corrosion cases, with pit depths ranging from 1.5 to 3.2 mm (0.06 to 0.125 in.) (Ref 13). One report on microbial corrosion, involving commercial transportation aircraft, indicated that the structural fastener areas had the highest level of deterioration, based on the severity and frequency of corrosive attack (Ref 13).

Microbiological corrosion problems have been observed both in military and commercial aircraft (Ref 34). One case that involved a DC-9 wing fuel tank led to the conclusion that microbiological corrosion can develop on airframes protected by the polyurethane coating that conforms to military specification MIL-C-27725 (Ref 34). One researcher confirmed that microorganisms can penetrate polyurethane coatings in a report on C-130 integral wing tank corrosion problems (Ref 32). The remedy that was selected for prevention and control of C-130 microbiological corrosion was to remove the synthetic rubber lining and to substitute it with a polyurethane coating that contains a biocidal green dye (Ref 32).

A commercial airline system that started a fuel quality improvement program in the early 1970s has reported that microbiological corrosion problems in its fleet of aircraft have been almost completely eliminated. Therefore, it can be concluded that two actions are necessary for the prevention and control of airframe microbiological corrosion problems. These are the selection of the proper structural materials, including biocidal protective coatings, and the implementation of stringent controls on the quality of fuel that is used in aircraft systems. More information on attack by microorganisms is available in the section "General Biological Corrosion" of the article "General Corrosion" and in the section "Biological Corrosion" of the article "Localized Corrosion" in this Volume.

Corrosion in Aircraft Bilges. Corrosion often occurs in aircraft bilge areas, which are sinks (or sumps) into which waste fluids and solids collect. These materials—especially oil, water, and dissimilar metallic drill chips—can set up electrochemical (galvanic) cells in the bilge (Ref 35). Because these corrosion reactions occur in areas that normally are not accessible to aircraft maintenance personnel, the potential for extensive airframe deterioration due to corrosion is significant (Ref 36).

One of the best ways to control bilge area corrosion in aircraft systems is to provide access through stairways or rampways so that mechanics can conveniently inspect and maintain these areas (Ref 36). Additional measures that can be taken, especially during the aircraft design phase, are to (Ref 36):

- Consider using aluminum alloys that have greater resistance to general corrosion and SCC, such as 7075-T73 and 7475-T761
- Use an external organic primer and topcoat finishing system. This system can, for example, consist of the MIL-P-23377 epoxy-polyamide primer (one coat) and the MIL-C-83286 polyurethane topcoat (two coats)
- Seal all faying surfaces of the airframe structure. A sealant meeting MIL-S-81733, which contains corrosion-inhibitive chromates, is often used

Hydrogen-induced failures of aircraft equipment result from a chemical and mechanical interaction with hydrogen. The hydrogen that enters metallic airframe structures causes mechanically induced deterioration by applying very high pressures within small cavities in the affected airframe structure.

Another term that is used for hydrogen-induced fracture is hydrogen embrittlement. Hydrogen embrittlement can cause dangerous and sometimes catastrophic failures in some steel parts that are used in aircraft. Two possible processing sources of hydrogen embrittlement are pickling solutions, which are used for scale and rust removal, and electroplating solutions (Ref 16). Hydrogen embrit-

(a)

(b)

Fig. 22 Exfoliation corrosion of an aluminum alloy stabilizer bracket. (a) Heavy surface corrosion on the stabilizer bracket. (b) Cross section through the bracket showing corroded surface grains and corrosion at grain boundaries of elongated grains. Source: Ref 16

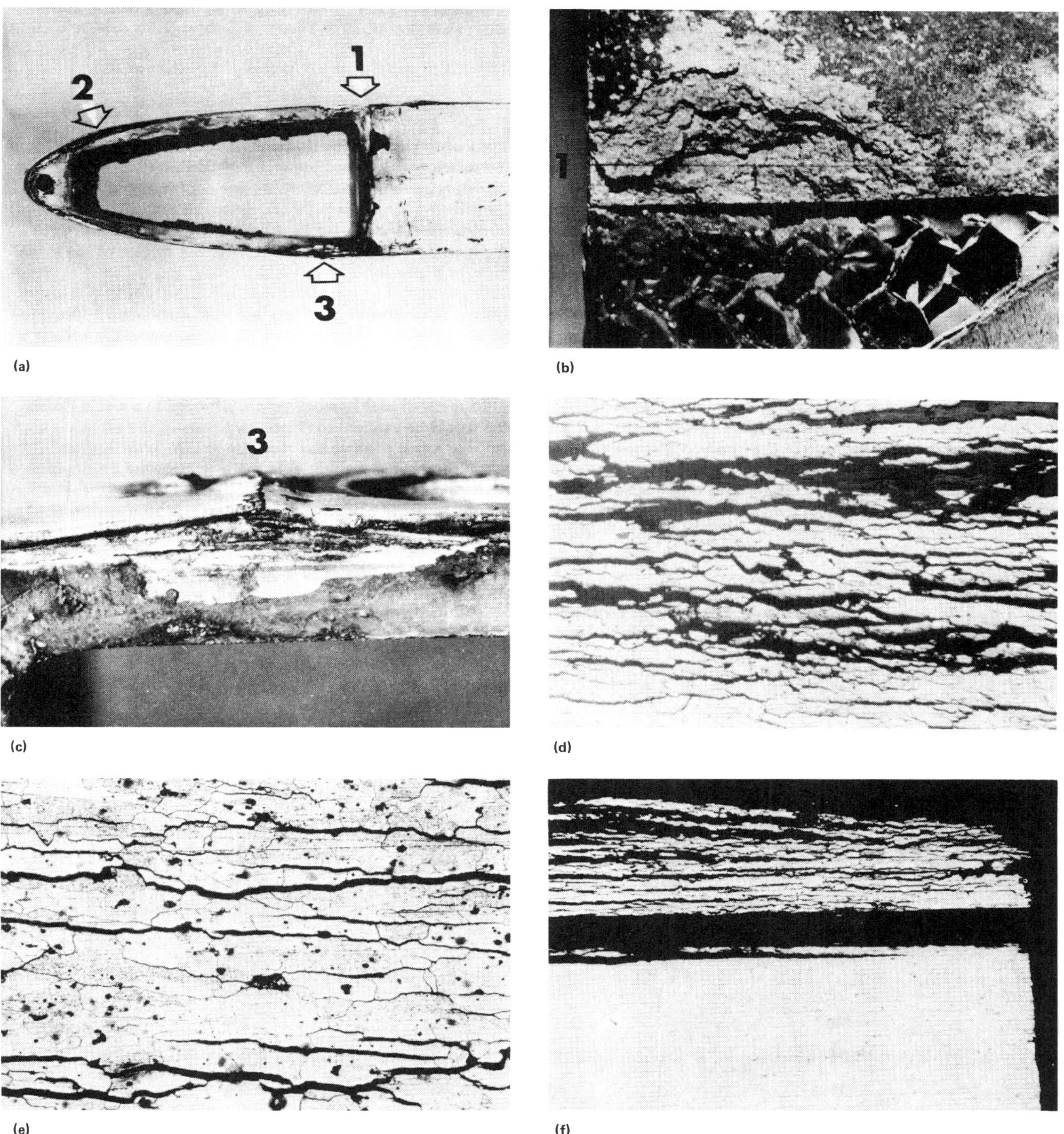

Fig. 23 Corrosion of an aluminum alloy 2024 helicopter rotor blade. (a) Leading edge at the blade tip showing three areas of severe corrosion. (b) Corrosion in the aluminum alloy skin at area 1. (c) Rupture of the surface skin at area 3 due to buildup of corrosion products in the underlying spar. (d) and (e) Intergranular corrosion in the spar. (f) Exfoliation in the surface skin. Source: Ref 16

tlement may also result in areas where steel acts as the cathode during galvanic (dissimilar metal) corrosion.

Hydrogen-induced cracking is one of the more severe forms of deterioration because it can result in delayed fracture under sustained stress. Hydrogen embrittlement has been observed (and has caused the failure of) some of the high-strength steel alloys used in aircraft landing gears (Fig. 30). Representative cases involving this type of corrosion were discussed in the introduction to this section. Additional examples of hydrogen-induced airframe corrosion are discussed below for various aircraft structures.

Landing Gear. Several C-141 landing gear cylinders failed because of hydrogen embrittlement (Ref 37). Hydrogen entered these 4340 steel cylinders during electroplating operations. Subsequently, 300M steel cylinders, coated on the interior surfaces with a hard chromium plating, have been used as a result of a redesign and retrofitting of the C-141 landing gear cylinders (Ref 37).

Propeller Retaining Bolt (Fig. 31). The composition of this steel part—which failed by hydrogen embrittlement—was Fe-0.5C-2.5Ni-0.75Cr. This steel bolt had been heat treated to a strength of 1379 MPa (200 ksi), and a bright electroplated cadmium coating had been applied.

Nose Gear Strut. Hydrogen embrittlement was the primary cause of the failure of this steel

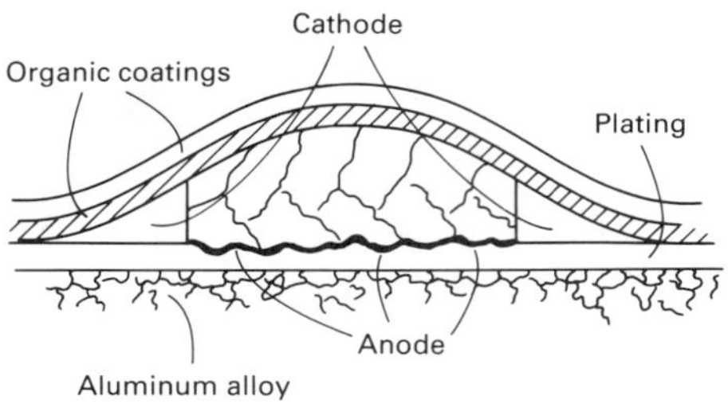

Fig. 24 Schematic of the development of filiform corrosion on an aluminum alloy. Source: Ref 25

aircraft component (Ref 38). The strut separated during the taxiing of a small bomber. The severely corrosive Southeast Asian environment probably abetted the ultimate failure of this structure.

Flap Control Return Spring (Ref 16). The internal crack in this carbon steel part indicated that the failure resulted from hydrogen embrittlement. This flap control return spring was cadmium plated, but there was no evidence that the part had been baked to remove hydrogen. Precautions were subsequently taken to ensure that proper baking is performed.

Main Landing Gear Pivot Pins (Ref 16). The intergranular fracture of these 4340 steel components was caused by hydrogen penetration into the metallic structures. Sustained loads were a major contributing factor in this failure. The pivot pins were chromium plated, and it was determined that the hydrogen came from the plating process. Therefore, the baking procedure, which is performed after electroplating, was modified to enhance the degree of hydrogen removal.

Main Landing Gear Drag Link Bolt (Ref 16). This 4340 steel part had been cadmium plated. Fracture occurred in the threaded region of the bolt.

Hydrogen tolerance for steels is inversely proportional to their strength levels. Hydrogen embrittlement can be avoided by using lower-strength steels, by limiting immersion times in acid or plating baths, and by in-process (intermediate) baking. Plated parts are usually baked at 190 °C (375 °F) to provide hydrogen embrittlement relief. A safer baking duration for high-strength steels (>1100 MPa, or 160 ksi) is 23 h. Longer baking times may be required for parts that have large cross sections, bright cadmium treatments, or very high strength. More information on hydrogen-induced failures is available in the section "Hydrogen Damage" of the article "Environmentally Induced Cracking" in this Volume.

Corrosion of Bonded Airframe Structures. Adhesive-bonded airframe structures can deteriorate by delamination and electrochemical corrosion of one or both of the metallic structural face sheets that eventually separate from the bondline. The long-term environmental susceptibility of bonded structures to this type of corrosion is known to be primarily caused by (Ref 39):

- Water or moisture intrusion into the adhesive bondline. Perforated honeycomb core sections of these bonded parts promote the collection of moisture (Ref 40). This leads to deterioration of the honeycomb core due to freezing-thawing cycles, corrosion, or both (Ref 41)
- Deficient surface preparation of face sheet and core materials that are used in the complete adhesive-bonded part

Improper surface preparation was the documented cause of the failure of a C-141A wing trailing edge upper surface panel. This was caused by a separation of the outer face sheet because of poor adhesion to a corner section of the wing upper surface. Corrosion of the aluminum face sheet started after the face sheet separated from the bonded structure. The disbonded aluminum face sheet had been prepared by using the sodium dichromate-sulfuric acid etch process, which is also called the Forest Products Laboratory (FPL) etch. The failure of the C-141A outer wing panel was attributed to inadequate process control while the FPL etching process was being accomplished (Ref 39).

Another reported adhesive bonding failure occurred on the C-141A petal door inner skin assembly. This delamination and corrosion also occurred because of poor surface preparation and moisture penetration between the bonded structural panels. Extensive corrosion was noted near the hinge fitting of this skin assembly (Ref 39).

Delamination and corrosion of adhesive-bonded airframe components have also been reported on the C-141 main landing gear doors (Ref 41) and on the more highly stressed aircraft control surfaces, such as spoilers, on one fleet of commercial airliners (Ref 42). These latter parts consisted of alclad aluminum alloy 7075 bonded to aluminum honeycomb with an adhesive that cured at 120 °C (250 °F).

A significant contribution to the engineering effort to mitigate these delamination and corrosion problems has been the development of the phosphoric acid anodizing process (Ref 39-42). This solution to adhesive structural disbonding resulted from 10 years of research at the Boeing and Douglas aircraft companies. Phosphoric acid (H_3PO_4) anodizing enhances the durability of adhesive-bonded structures by improving the resistance of the bondline to hydration and by reducing the extent of corrosion of exposed alclad aluminum surfaces (Ref 42).

In addition to H_3PO_4 anodizing, the following techniques have helped to reduce delamination and corrosion of bonded airframe structures:

- Curing the film adhesive at 175 °C (350 °F) instead of at 120 °C (250 °F) (Ref 42)
- Coating the surfaces with a corrosion-inhibiting primer that cures at 120 °C (250 °F) (Ref 42)
- Better-quality adhesive bonding through the use of adhesives that have greater long-term resistance to plastic deformation. This increased toughness is obtained by the crosslinking of these two-part 100% solids paste adhesives during room-temperature curing (Ref 40)

Because corrosion of alclad aluminum spreads in the plane of the sheet, corrosion of alclad airframe parts results in the delamination of bonded honeycomb panels. Therefore, alclad sheet is often avoided for honeycomb facesheets.

Another effective corrosion control approach is to apply only nonperforated honeycomb core. This prevents the moisture penetration that can occur in cellular honeycomb parts. Sealing is performed on the edges of honeycomb panels to protect the core from moisture ingestion.

Erosion-corrosion is corrosion that is accelerated by the impingement of moving fluids, which may contain solid particles, onto the surfaces of

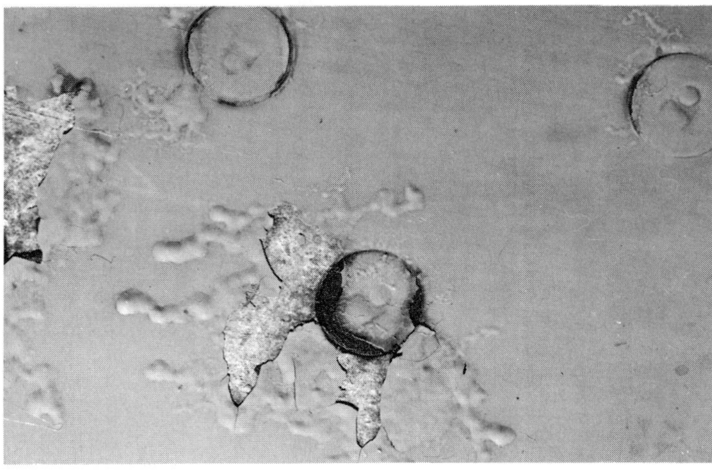

(a)

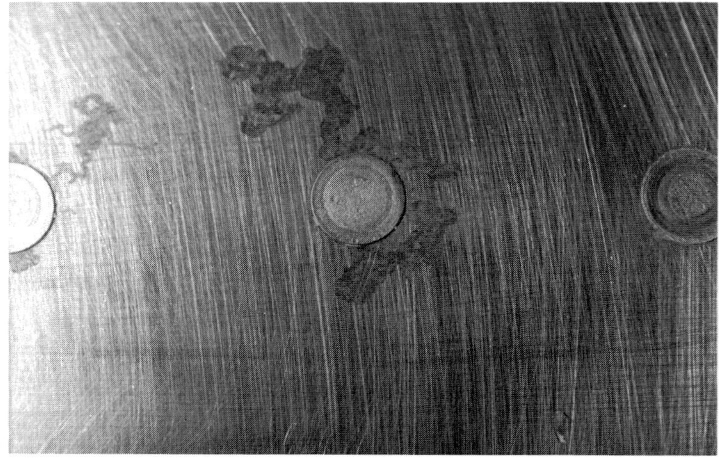

(b)

Fig. 25 Filiform corrosion of an aluminum aircraft skin around steel fasteners. (a) Before paint removal showing paint cracking and blistering. (b) After paint removal. Courtesy of Qantas Airways and the Society for the Advancement of Material and Process Engineering

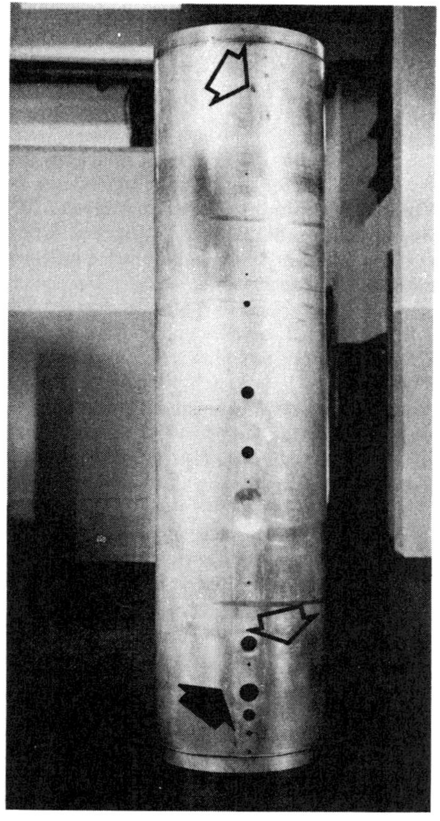
(a)

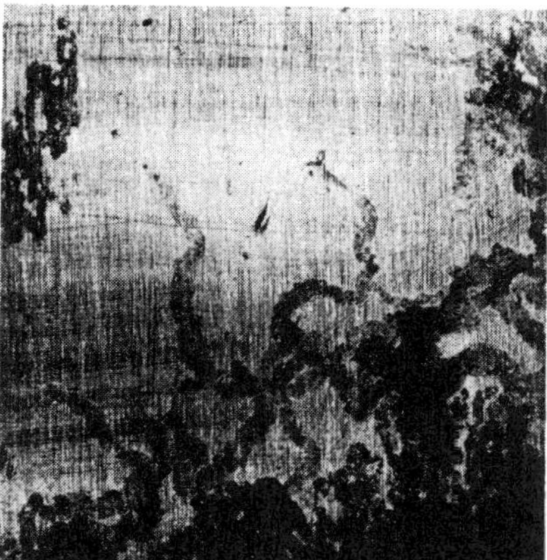
(b)

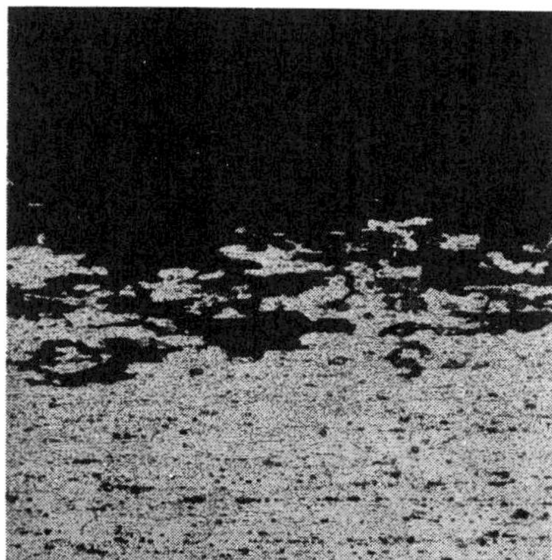
(c)

Fig. 26 Filiform corrosion of a fighter aircraft pylon tank. (a) Overall view of the tank showing uniform corrosion (open arrows) and penetration (solid arrows). (b) Indications of filiform corrosion. (c) Pitting and intergranular corrosion. Source: Ref 16

materials. Erosion-corrosion in aircraft is most severe when the fluid impinges upon the aircraft surface at a high velocity and when the moving fluid contains abrasive solid matter, especially sand. High temperatures also accelerate the deterioration process when erosion-corrosion occurs.

One reported case in which erosion-corrosion was recorded in aircraft structures involved damage to the undercarriage parts of a transport aircraft. This large air vehicle had been operating on a desert landing strip. The erosion-corrosion damage on this structure, therefore, was accelerated by extremely high temperatures and impact by sand (Ref 16).

Prevention and Control of Airframe Corrosion

The cost escalation attributed to corrosion problems in the United States has been significant—from $5.5 billion in 1947 to $167 billion in 1985 (Table 5). Therefore, it is important that the manufacturers and users of aircraft systems implement an effective corrosion prevention and control program.

There are many techniques that are available for preventing and controlling airframe corrosion. Some of these corrosion control methods have been used for many years, such as the application of greases to bearings in aircraft control mechanisms, wheels, rudder posts, and other components (Ref 44). Other relatively new approaches have been utilized, including the application of advanced composite materials to secondary airframe structures (Ref 45). In cases in which carbon composite materials have been applied, the proper corrosion prevention and control methods had to be developed and practiced for preventing galvanic corrosion between the carbon composites and other less noble materials, especially aluminum.

In addition to noncritical (secondary) structures, airframe manufacturers have also applied reinforced plastic and carbon composites to primary structures, such as the fuselage, wings, and empennage, which have a greater degree of safety-of-flight importance than secondary structures. On the C-130, for example, the wing contains about 80% advanced composite materials (Ref 45). Advanced composites are also used in 71% of the structure of one business aircraft (Ref 45). Advanced composite materials are being more widely used in aircraft systems, primarily because of their strength at increasing aircraft velocities (Ref 46).

Corrosion-preventive technology must be properly and regularly applied to airframe components to ensure that the airworthiness and flight safety of operational aircraft are not jeopardized. The corrosion cases that have been presented in this section demonstrate that the objectives of airframe corrosion prevention and control can be attained by implementing new and improved engineering design solutions.

Table 4 Resistance to fretting corrosion of various material couples under dry conditions

Couple	Resistance to fretting
Steel on steel	Low
Nickel on steel	Low
Aluminum on steel	Low
Antimony plate on steel	Low
Tin on steel	Low
Aluminum on aluminum	Low
Zinc-plated steel on aluminum	Low
Iron-plated steel on aluminum	Low
Cadmium on steel	Medium
Zinc on steel	Medium
Copper alloys on steel	Medium
Zinc on aluminum	Medium
Copper plate on aluminum	Medium
Nickel plate on aluminum	Medium
Iron plate on aluminum	Medium
Silver plate on aluminum	Medium
Lead on steel	High
Silver plate on steel	High
Silver plate on aluminum plate	High
Steel with conversion coating on steel	High

Source: Ref 28

The essential elements of an airframe corrosion prevention and control program are proper material selection, an adequate finish specification, a thorough plan for effective maintenance, inspection, and repair (Ref 47). During the material selection phase, considerable trouble can be avoided if aluminum alloys and tempers are selected based on SCC resistance, exfoliation corrosion resistance, fracture toughness, and mechanical strength properties. Tempered steel having a yield strength below 1379 MPa (200 ksi) should be selected whenever possible in order to avoid hydrogen embrittlement and SCC problems. The corrosion prevention and control specification must ensure that these materials receive satisfactory hydrogen embrittlement relief during processing. Precipitation-hardening steels that have been heat treated for maximum resistance to SCC should be used.

The task of preparing an adequate corrosion prevention and control plan is facilitated by feedback from aircraft operators and maintenance technicians. The engineers and other specialists who provide feedback to corrosion engineers and designers should be thoroughly familiar with the operational profile of the aircraft system for which the specific corrosion prevention and control plan is being written (Ref 48). Feedback will not only assist airframe designers in writing a plan and specifications for corrosion prevention and control but will also enable these professionals to understand how and where corrosion problems tend to occur in particular airframe structures and to determine the most useful (and most cost-effective) applications for airframe corrosion-preventive technology.

Determining the most useful applications for corrosion-preventive technology will help to minimize repeated occurrences of airframe corrosion problems. An example of this is a significant metallurgical contribution to the C-130 airframe corrosion prevention and control program—the use of aluminum alloy 7075-T73 (instead of the less corrosion resistant 7075-T6 aluminum alloy) in the upper-center wing and other components of the C-130 (Ref 32). The main structural (aluminum) parts of the C-130 are provided with additional corrosion

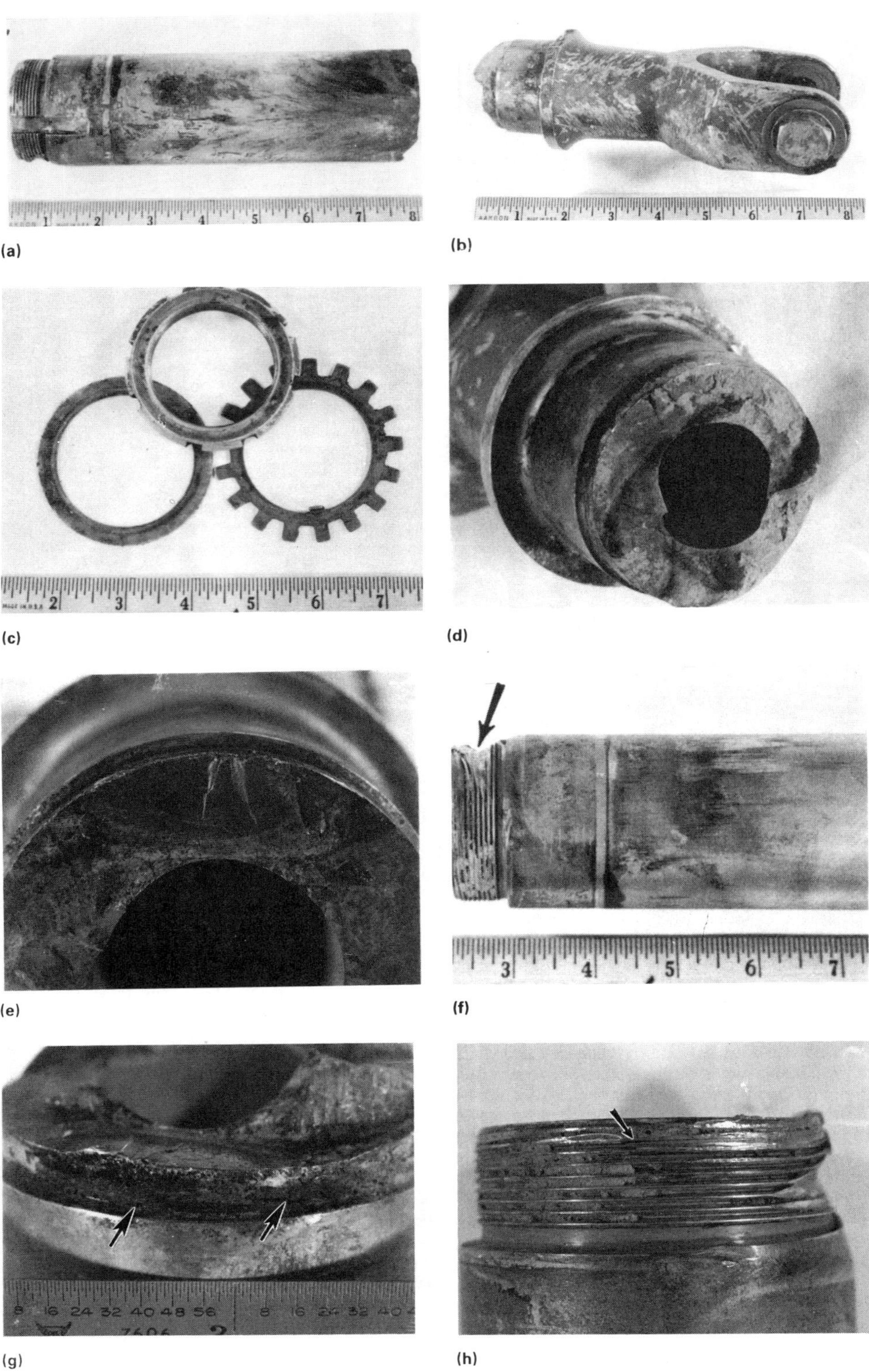

Fig. 27 Corrosion fatigue of a 4340 steel helicopter rotor assembly. (a) Horizontal hinge pin. (b) Nut. (c) Locking washers. (d) and (e) Views of the hinge pin fracture surface. A beach mark is visible in (e). (f) Dent on one of the threaded surfaces of the horizontal hinge pin. (g) Corrosion pits and cracks that emanated from pits in the hinge pin. (h) Embedded thread from the nut in the hinge pin threads. Source: Ref 16

protection by anodizing and then coating them with a polysulfide primer (Ref 32).

An adequate finish specification must be prepared and enforced in the early stages of aircraft design. During the design period, qualified corrosion engineers should review the initial drawings; this would prevent potential corrosion problems from occurring.

Several documents that provide comprehensive requirements for corrosion prevention and control have been published. These standards were written for military airframe systems, but they can also be applied to corrosion protection on commercial airframes with little or no modification. The following United States DOD specifications can be used as a guide in selecting or preparing an adequate finish specification:

- Military Specification MIL-F-7179: "Finishes and Coatings: General Specification for Protection of Aerospace Weapons Systems, Structures, and Parts"
- Military Standard MIL-STD-1568: "Materials and Processes for Corrosion Control"
- Standardization Document SD-24: "General Specification for Design and Construction of Weapon Systems"
- Military Specification MIL-S-5002: "Surface Treatments and Inorganic Coatings for Metal Surfaces"
- Aeronautical Design Standard-13: "Air Vehicle Materials and Processes"

The corrosion prevention and control guidelines included in the above specifications have been widely used by government and industry. One of the most significant of these guidelines is the use of compatible materials in dissimilar-metal couples, as summarized in Table 6.

In addition to having a written plan, it is important to have designated corrosion and design engineers that are appointed to corrosion-prevention teams or advisory boards. These groups have the responsibility of establishing materials and process requirements and design improvements for ensuring effective corrosion protection on specific aircraft systems. These requirements are generated from documented cases in which corrosion has been observed on airframe components, especially those parts that have deteriorated on earlier models of the same or similar aircraft systems.

Airframe maintenance and repair should be performed to alleviate airframe corrosion. Examples of these actions include:

- The application of corrosion-preventive (water-displacing) compounds to airframe surfaces (Ref 49)
- Use of corrosion-inhibitive elastomeric sealants (Ref 50)
- Repair of stress-corrosion cracks on internal aluminum wing structures by applying bonded doublers consisting of boron fiber reinforced plastic (Ref 51)

Another important step in the corrosion control effort is the drafting of a preventive maintenance and inspection plan (Ref 47). This plan provides an outline of corrosion maintenance, inspection, and repair requirements that include the locations and frequencies of inspections and the corrosion-contributing factors that should be considered by professionals who inspect aircraft structures for deterioration. Some of the factors that can induce the corrosion of airframes include the environmental conditions to which the airframe is (or will be) exposed and corrosive spillage (usually alkaline or acidic chemicals) from galleys and toilets.

The inspection and maintenance plan must be defined and administered as early as possible. This is very significant because early corrosion control actions are crucial for the avoidance of costly aircraft damage or failures.

(a)

(b)

Fig. 28 Corrosion fatigue failure of a magnesium alloy QE-22A main landing gear wheel. (a) General view of wheel. Crack is shown by the dashed line. (b) Fracture surface. Origin is indicated by the arrow. Source: Ref 16

Corrosion of Powerplants

Michael L. Bauccio
The Boeing Company

Aircraft powerplants have the important function of providing thrust, or propulsion, to enable aircraft to take off and remain in flight. Powerplant systems also provide reverse thrust to enable aircraft to land safely. The development and proper utilization of corrosion-preventive technology has enabled aircraft powerplants to fulfill these significant functions.

In aircraft powerplant systems, including engine and hydraulic power units, several types of corrosion have been observed. Among these types of powerplant corrosion (but not the only kinds) are:

- High-temperature corrosion. This has also been referred to as sulfidation, or hot corrosion. High-temperature corrosion of powerplant components is also manifested as hot-salt-induced SCC and as very rapid oxidation (fire). Both of the latter forms of high-temperature corrosion have been observed primarily in titanium structures
- Cold corrosion. This type of powerplant corrosion originally referred to all types of corrosion other than hot corrosion, which primarily includes pitting and fretting corrosion of aircraft powerplant components. The cold form of pitting and fretting corrosion develops at considerably lower temperatures than those attained under hot corrosion conditions
- Chemical corrosion
- Erosion corrosion

The objective of this section is to focus on the above generic classes of aircraft powerplant corrosion. The problems and corrosion-preventive solutions that have proved to be significant developments for minimizing future occurrences of these corrosion problems will also be discussed. The thesis of this section is that powerplant corrosion problems can be minimized by regularly scheduled maintenance inspections and by effective applications of corrosion-preventive treatments to powerplant components.

Most of the literature on the corrosion of aircraft powerplants is concerned with high-temperature corrosion, which occurs above approximately 315 °C (600 °F). This is because aircraft powerplants operate at extremely high temperatures. Some aircraft powerplant sections—such as the turbine rotor inlet—reach temperatures of about 1240 °C (2260 °F) (Ref 52). These high-temperature operating cycles have produced hot corrosion in powerplant components that did not have adequate corrosion protection. Some reciprocating parts, such as exhaust valves, are protected against high-temperature corrosion by coatings of high-temperature alloys (Ref 2). These metallic coatings are welded onto the valve faces and then machined to the required angle of either 30 or 45° (Ref 53).

Corrosion of powerplants has also been found in many cases at much lower temperatures than those attained in hot corrosion. One of the most cold corrosion prone areas of the aircraft—the engine air inlet, or frontal, area and the cooling air vents—has experienced corrosive deterioration at relatively low temperatures. The frontal areas are vulnerable to erosion that is caused by airborne particulate matter, such as dirt, sand, and gravel. This abrasive action removes protective treatments and coatings from the frontal engine structures. This is eventually followed by corrosive deterioration in and around the areas that had been mechanically damaged by airborne solids (Ref 54). Certain engine parts, including reciprocating engine cylinder fins and accessory mounting bases, have small, unpainted areas that are susceptible to corrosion by airborne salts and moisture (Ref 54). The most effective technique for preventing and controlling corrosion in engine intake areas is to clean, examine, and refurbish these areas regularly, especially when the aircraft system must operate under aggressive conditions, such as in a tropical marine environment (Ref 55).

The materials that are usually affected by the highly corrosive powerplant environment are metal alloys that are designed for long-term service at high temperatures. These alloys consist primarily of nickel, cobalt, and titanium. Some of the older nickel-base superalloy materials, such as Udimet alloy 700, have a low resistance to fatigue, which compromises the high-temperature static strength of these alloys (Ref 56). Therefore, powerplant corrosion can be minimized by the increased use of materials that have the best combination of fatigue strength and corrosion resistance at elevated temperatures.

Case Histories

Many typical case histories, or lessons learned, have been described in the literature on powerplant corrosion. Discussions of the actual forms of deterioration, as well as the corrosion-protective measures that were implemented to control or avert future problems, are presented below. These corrosion problems have led to significant improvements in the materials and processes

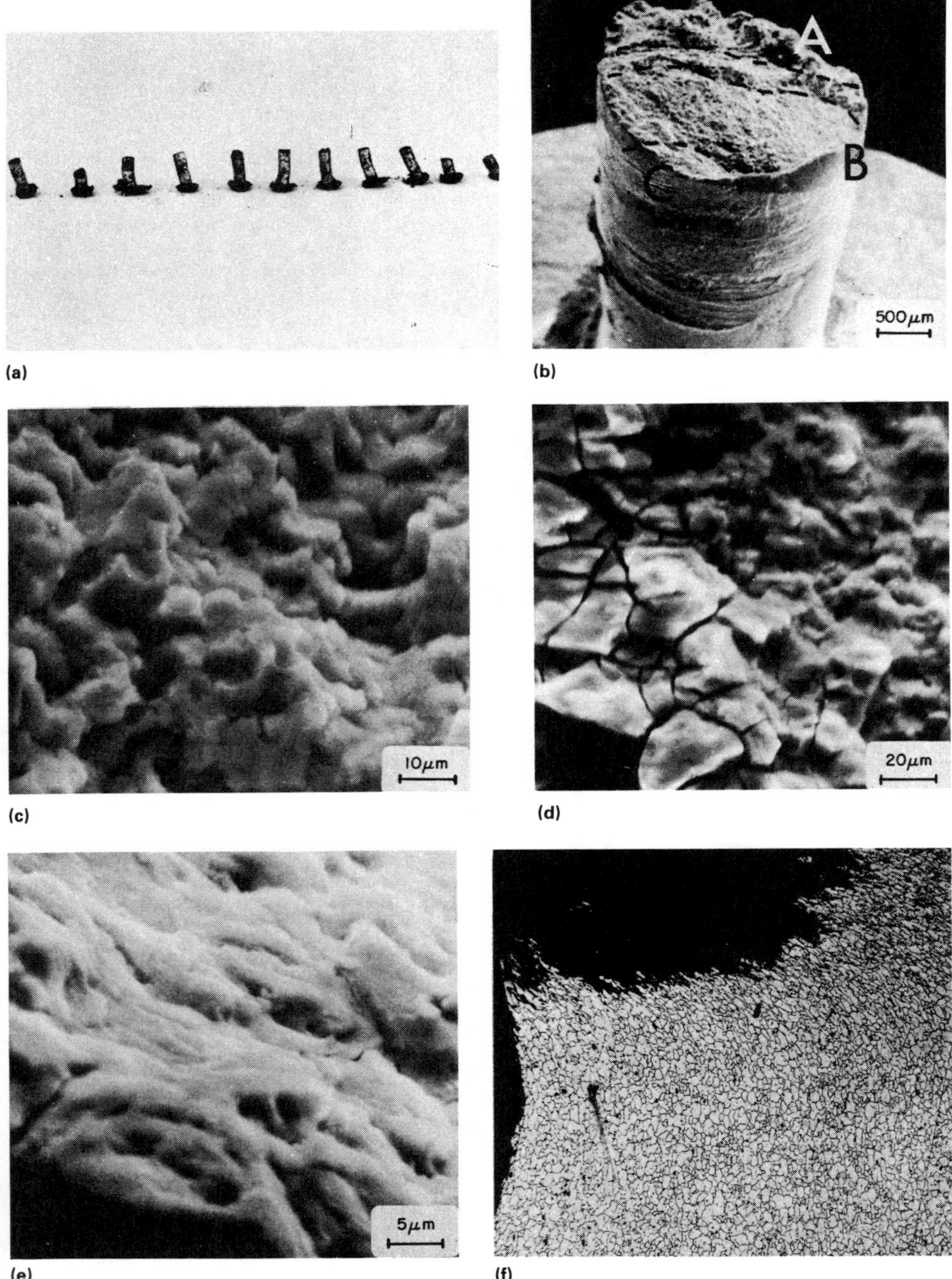

Fig. 29 Corrosion fatigue of aluminum alloy 5052 rivets. (a) Rivets removed from the structure for examination. (b) Fracture surface of a rivet showing three zones of fracture. 20×. (c) Intergranular fracture observed in zone A. 1000×. (d) Mud crack pattern in zone A. 500×. (e) Fracture surface in zone B. 2000×. (f) Grain structure of the rivets. The deformation is caused by tearing in an overload region. 500×. Source: Ref 16

used for the protection of aircraft powerplant components and for the assurance of aircraft flight safety.

High-Temperature Corrosion

There are three predominant types of high-temperature corrosion that have caused mechanical failures of aircraft powerplant components. These are sulfidation, hot-salt SCC, and fires.

Sulfidation has occurred in gas turbine parts that were made of nickel and cobalt alloys. Hot corrosion was observed in the 1960s on gas turbine blades that were manufactured from Nimonic 105 (Ni-20Co-15Cr-5Mo-4.5Al-1.4Ti) and GMR-235 (Ni-15.5Cr-10Fe-5.25Mo-3Al-2Ti). This corrosive deterioration was called black plague (Ref 57). Some of the cobalt-base superalloys, which are also used in gas turbine powerplants, have also required advanced protective coatings for the mitigation of hot corrosion (Ref 58). The nature of these coatings will be discussed later in this section.

Hot-salt SCC has been found primarily in titanium-alloy powerplant parts, such as the compressor blades of military supersonic trainer engines (Ref 59). A case history concerning hot-salt SCC in a titanium powerplant component is given in the discussion "Hot-Salt SCC" in this section.

Fires have occurred in powerplant components that are fabricated from titanium and other materials. This phenomenon is an extremely rapid form of oxidation that can cause extensive damage within a brief period in aircraft powerplants. In the case of titanium fires, this is a very critical problem because titanium oxidation occurs so quickly that it produces combustion. The deterioration that is caused by combustion spreads throughout the powerplant because titanium is widely used in aircraft powerplants because of its high strength-to-weight ratio, ability to maintain adequate strength at moderately high temperatures, and good corrosion resistance at ambient temperatures.

Hot Corrosion (Sulfidation). An example of hot corrosion in the rotating blades of an aircraft powerplant is shown in Fig. 32. This part was manufactured from the nickel-base superalloy IN-738. The creep failure of these blades—which also are called buckets—was initiated by sulfidation. Magnified cross sections of these blades at 75 and 130×, respectively, are shown in Fig. 32(b) and (c).

Hot corrosion problems have also been found in gas turbine stator (stationary) vanes (Ref 59). In gas turbine blades and vanes that were made of the IN-713 nickel-base alloy, hot corrosion has produced deterioration that appeared as cracking, swelling, or spalling. Oxides have been observed below the surface of components that are subjected to high-temperature corrosive environments. The subsurface oxides consist of nickel and large quantities of chromium, aluminum, titanium, and niobium. Nickel-rich metallic material has been found to lie adjacent to these oxides (Ref 59).

The dynamic nature of the environment in aircraft powerplants promotes this type of high-temperature oxidation (sulfidation) of the main burner components and of downstream parts, such as turbine blades, vanes, and seals. Very high thermal and mechanical strains are produced on these parts because of powerplant operation at elevated temperatures (Ref 60). The corrosive ash that is usually present in this dynamic-oxidation environment is sodium sulfate (Na_2SO_4), which reacts with powerplant materials and causes a depletion of the alloying elements required to form a protective oxide film on the surfaces of these parts. The formation of a protective film can also be inhibited by consumption of the alloying elements by means of either acidic or basic oxide fluxing reactions (Ref 52, 60).

Hot corrosion protection has been obtained primarily by the use of advanced-technology coatings and materials that are specifically designed for long-term endurance in dynamic corrosion environments. One such coating is a proprietary mixture of cobalt, chromium, aluminum, and yttrium (CoCrAlY). Aluminum is the most significant element in this coating because it reacts with oxygen to form a thin layer of corrosion-resistant aluminum oxide on the surface of the powerplant parts (Ref 61). Coatings of this type have been applied by electron beam plasma vapor deposition techniques.

Hot corrosion problems in aircraft powerplants have also been alleviated by the application of pack cementation type coatings (Ref 58). These coatings consist of either of the following elements:

Fig. 30 Hydrogen embrittlement failure of a high-strength steel aircraft landing gear component

- Aluminum only, or a combination of aluminum and other elements, such as chromium and silicon
- Chromium alone, or chromium added to other elements, such as tantalum and aluminum

Another type of advanced coating that has been investigated for protection of powerplant components against hot corrosion is the thermal-barrier coating, which consists of various combinations of ceramic materials. These coatings are plasma sprayed, and they are composed of a specific ceramic material that is applied over a layer of an oxidation-resistant metal. Accelerated hot corrosion tests on ceramic thermal-barrier coatings have determined that the failure mechanism of these coatings involves delamination, which occurs prior to surface cracking or spalling (Ref 62). Despite the failure of some thermal-barrier coatings, the aluminum-ceramic coatings have been successfully applied in the last 20 years to various gas turbine components, including blades and vanes for corrosion protection to temperatures of about 650 °C (1200 °F) (Ref 63).

Other materials have been developed and used for improving the resistance of powerplant components to hot corrosion, such as oxide dispersion-strengthened alloys (Ref 64), which include the following materials:

- Inconel alloy MA-754, which was developed for use in gas turbine vanes
- Incoloy alloy MA-956, which was developed for powerplant combustion chamber burner cans
- Alloy MA-6000E, which was designed and produced for application in gas turbine blades

In addition to the above oxide dispersion-strengthened materials, the following materials have been designed for mitigating high-temperature corrosion in powerplant systems:

- Nickel-, cobalt-, and iron-base superalloys that contain platinum and platinum-group metals. Alloying with platinum group metals provides a great improvement in overall hot corrosion resistance (Ref 65)
- Directionally solidified alloys. These materials include columnar-grained and single-crystal forms (Ref 56). In addition to having a significantly higher resistance to hot corrosion, these materials (such as the directionally solidified MAR-M200 nickel-base alloy) have up to 100 times the high-temperature fatigue life of the polycrystalline forms of the same alloy (Ref 56)

Hot-salt SCC has been produced in the laboratory in titanium alloys that are susceptible to embrittlement and fracture. The conditions required to induce this type of failure are a combination of the following (Ref 66-68):

- High stress
- High temperature (approximately 345 °C, or 650 °F)
- Exposure to a high concentration of chloride salts, which is anticipated in a marine environment

Hot-salt SCC usually produces intergranular-type corrosion in titanium alloys. In cases in which hot-salt SCC occurs because of the exposure of titanium powerplant parts to solid chloride salts, the minimum temperature at which hot-salt SCC develops is approximately 290 °C (550 °F) (Ref 68).

Hot-salt SCC of titanium powerplant components has been reported as the cause of powerplant deterioration in one incident that occurred during ground testing. In this case, a compressor rotor assembly in a turbine engine failed during an overspeed acceleration test. Two compressor disks, manufactured from titanium alloy Ti-7Al-4Mo, fractured in several locations. Cracking occurred at the bolt holes, where dissimilar-metal (A-286) bolts were in contact with the titanium disk. This failure is illustrated in Fig. 33. The two failed disks had large silver deposits in the tie bolt holes, which was indicated by surface attack at the crack initiation sites. Silver chloride was formed during the test cycle by a chemical reac-

(a)

(b)

Fig. 31 Hydrogen embrittlement of a high-strength steel propeller retaining bolt. (a) Overall view of bolt. (b) Cross section on the bolt. 215×. Courtesy of Aeronautical Research Laboratories, Australia

Table 5 Cost escalation for corrosion in the U.S. from 1947 to 1985

Year	Cost, billions of dollars
1947	5.5
1965	>6
1967	>10
1975	70
1982	126
1985	167

Source: Ref 43

Table 6 Metals and alloys compatible in dissimilar-metal couples

Group number	Metallurgical category	emf, V	Anodic index(a), V	Compatible couples(b)
1	Gold, solid and plated; gold-platinum alloys; wrought platinum	+0.15	0	
2	Rhodium plated on silver-plated copper	+0.05	0.10	
3	Silver, solid or plated; high-silver alloys	0	0.15	
4	Nickel, solid or plated; monel metal, high-nickel-copper alloys	−0.15	0.30	
5	Copper, solid or plated; low brasses or bronzes; silver solder; German silvery high copper-nickel alloys; nickel-chromium alloys; austenitic corrosion-resistant steels	−0.20	0.35	
6	Commercial yellow brasses and bronzes	−0.25	0.40	
7	High brasses and bronzes; naval brass; Muntz metal	−0.30	0.45	
8	18% Cr type corrosion-resistant steels	−0.35	0.50	
9	Chromium plated; tin plated; 12% Cr type corrosion-resistant steels	−0.45	0.60	
10	Tin-plate; terneplate; tin-lead solder	−0.50	0.65	
11	Lead, solid or plated; high-lead alloys	−0.55	0.70	
12	Aluminum, wrought alloys of the 2000 series	−0.60	0.75	
13	Iron, wrought, gray or malleable; plain carbon and low-alloy steels, armco iron	−0.70	0.85	
14	Aluminum, wrought alloys other than 2000 series aluminum, cast alloys of the silicon type	−0.75	0.90	
15	Aluminum, cast alloys other than silicon type; cadmium, plated and chromated	−0.80	0.95	
16	Hot-dip zinc plate; galvanized steel	−1.05	1.20	
17	Zinc, wrought; zinc-base die-casting alloys; zinc plated	−1.10	1.25	
18	Magnesium and magnesium-base alloys, cast or wrought	−1.60	1.75	

(a) Anodic index is the absolute value of the potential difference between the most noble (cathodic) metals listed and the metal or alloy in question. For example, the emf of gold (group 1) is +0.15 V, and the emf of wrought 2000-series aluminum alloys (group 12) is −0.60 V. Thus, the anodic index of wrought 2000-series aluminum alloys is 0.75 V. (b) "Compatible" means the potential difference of the metals in question is not more than 0.25 V. An open circle indicates the most cathodic members of a series; a closed circle indicates an anodic member. Arrows indicate the anodic direction. Source: Ref 17

tion between silver and chlorine, which were present in the high-temperature environment. Silver had been used as an antifretting coating on the tie bolts. As a consequence of this failure, the silver coating was replaced by a molybdenum disulfide dry-film lubricant (Ref 67, 69).

Rapid Oxidation. Corrosion problems produced by rapid oxidation (combustion) in aircraft powerplants have predominantly involved titanium. In some isolated cases of rapid oxidation, the damaged parts were fabricated from steel. The total number of known occurrences of titanium combustion in powerplants has been reported to be 144. In 85 of these incidents, the combustion did not penetrate the case, or skin, of the powerplant. Details of the deterioration that occurred in the 144 documented cases of titanium combustion are available for only two of these incidents (Ref 70).

The ignition temperature of titanium in air is approximately 1625 °C (2960 °F). This is significantly higher than the maximum service temperature for titanium. Therefore, either a mechanical or aerodynamic upset of the operation of the powerplant must occur in order to raise the temperature of titanium components to the ignition point. Several events can cause mechanical rubbing or jamming in powerplant structures, including (Ref 70):

- Interference by loose solid particles with rotating and static structures. These particles can bend blades, causing the blades and vanes to rub against each other continuously.
- Rotor imbalance, which can cause severe rubbing. This develops from the failure of rotating disks, spools, or blades
- Radial or axial displacement of the rotor, which can be caused by a failed bearing
- Displacement of the rotating powerplant blades into stationary parts, which produces compressor stall

The most significant of these causes of powerplant titanium combustion are trapped blades and radial displacement of the rotor (Ref 71). Only a small number of titanium combustion incidents have occurred because of aerodynamic heating, stall, or surge (Ref 71).

Powerplant case penetration has been observed in 57% of the titanium combustion incidents involving high-bypass ratio turbofan powerplants. Case penetration is illustrated in Fig. 34. In most of the case penetration incidents, small holes were observed in various locations on the case. Some incidents of case penetration have been characterized by complete deterioration of the case that covers the compressor section of the powerplant (Ref 71).

Titanium fires that are caused by blade tip rubbing can be prevented by (Ref 71):

- Increasing the clearances (tolerances) between the powerplant blade tips and case interior surfaces, although this is in opposition to the current trend toward tighter tolerances for improved performance
- Applying blade-tip materials that are resistant to wear. Abrasion-resistant materials have been investigated for this purpose. A plasma-sprayed Co-30WC tip treatment compound provided the best results in one study (Ref 72)
- Protecting the entrance edges of bleed air manifolds with steel grommets (Ref 70)

As previously mentioned, steel components have also failed by rapid oxidation in aircraft powerplants. These powerplant corrosion problems are described as breech chamber failures.

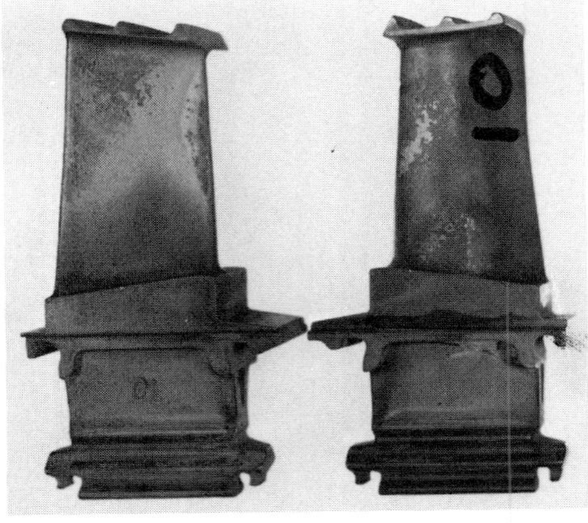

(a)

(b)

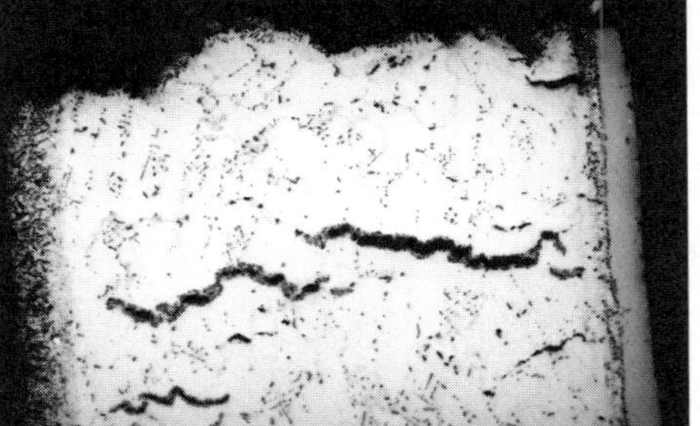

(c)

Fig. 32 Aircraft powerplant turbine blades (a) that were damaged by hot corrosion. (b) Higher-magnification view of corrosion. 35×. (c) Corroded turbine blade. 55×

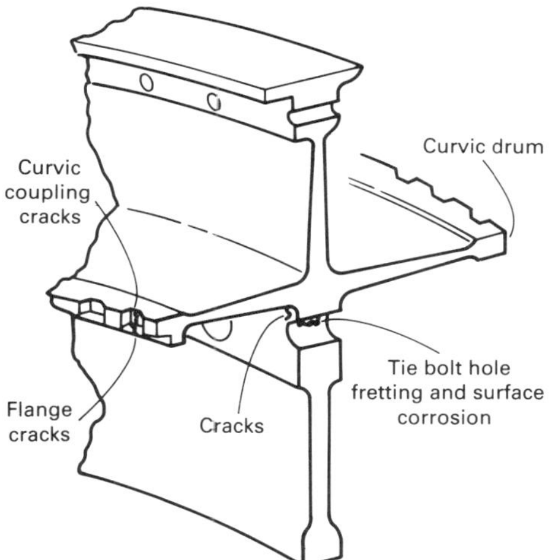

Fig. 33 Cracking of a Ti-7Al-4Mo aircraft powerplant compressor disk

Fig. 34 Case penetration produced by titanium combustion in a high-bypass turbofan aircraft powerplant. Courtesy of C.W. Elrod, Wright-Patterson Air Force Base

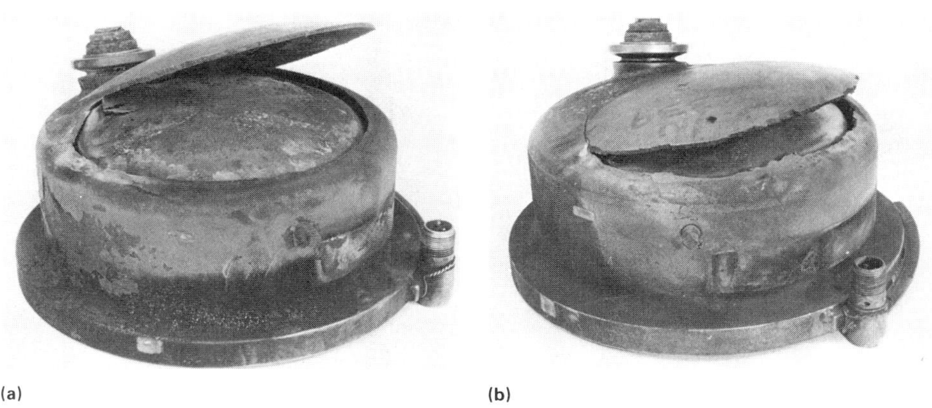

Fig. 35 Typical failures of 4340 steel aircraft powerplant breech chambers. Source: Ref 5

Breech chambers are used in various types of aircraft engines. These chambers are part of the cartridge, or pneumatic, starters (Ref 5). Combustion products from the breech chambers are exhausted toward the turbine blades to start the engine rapidly. Typical breech chamber failures are shown in Fig 35. These components were fabricated from 4340 steel heat treated to between 40 and 45 HRC. As indicated in Fig. 35, the breech chamber dome bursts at a point that is opposite to the hot gas discharge nozzle. These failures occur where stress concentrations are the highest, which is where the dome contacts the breech chamber body (Ref 5).

A fracture at the hot gas nozzle of the breech chamber is shown in Fig. 36. At the base of the nozzle (beyond the location of the fracture), a pattern of trenches was observed on the external chamber surface (Fig. 36b). The most probable explanation for these breech chamber failures is a defective coating system. There was no finish on the internal surfaces of these failed chambers. The domes were coated with electroless nickel, which caused galvanic corrosion on the more active (anodic) steel chambers (Ref 5). Future breech chamber failures can be avoided by coating the entire chamber with one of the high-temperature resistant coatings mentioned in the discussion "Hot Corrosion (Sulfidation)" in this section. The electroless nickel finish, applied directly to the steel chamber dome, should not be used, because of the possibility of galvanic corrosion at the nickel and steel contacting surfaces.

Cold Corrosion

Fretting corrosion is a very important design consideration for titanium parts in aircraft powerplants. This is because fretting corrosion can significantly reduce the fatigue strength of titanum alloy Ti-6Al-4V (Ref 73). Fatigue strength reductions of greater than 50% have been documented (Ref 67).

Fretting-induced deterioration has occurred in aircraft powerplant control bearings (Ref 67). The components that are most susceptible to significant fretting deterioration are the fan roller bearings on aircraft turbofan engines (Ref 74).

Fretting is also a significant factor in the design of powerplant compressor blades fabricated from titanium. Fretting corrosion can occur on the root and midspan shroud of these blades unless adequate design precautions are taken to minimize the potential for fretting to occur (Ref 67).

In addition to powerplant bearings and compressor blades, the following components of aircraft engines are known to be susceptible to fretting corrosion (Ref 74):

- Airfoil roots in rotors and stators
- Stator vanes and shroud interfaces
- Splines and rotor assembly stackup interfaces
- Piston ring secondary seals
- Torque pins for seal mountings
- Fasteners, including threads, the surfaces of bolted joints, and disconnect rings
- Metal static seals
- Engine mounts and thrust links
- Slip joints, which are used in burners and tubing assemblies
- Disk spacers
- Planetary gear shafts (Fig. 37). These parts have incurred fretting damage primarily in the gear bore. A smaller amount of fretting deterioration has also been observed in the bearing bore
- Clutches
- Joints and interfaces between titanium, aluminum, and austenitic stainless steel parts

The most significant fretting problems in aircraft powerplants have occurred in seals, splines, blade mountings, and blade dampers (Ref 74).

The damage produced by fretting corrosion usually appears as pitting, polishing, or both. Galling is produced in powerplant structures when fretting continues over relatively long peri-

(a) (b)

Fig. 36 Fracture in the hot gas nozzle (a) of a powerplant breech chamber. (b) Trench pattern at the base of the breech chamber hot gas nozzle shown in (a). Source: Ref 5

Fig. 37 Fretting corrosion of a planetary gear shaft from an aircraft powerplant. Courtesy of the Advisory Group for Aerospace Research and Development

ods of time (Ref 74). Both galling and fretting have been observed on oil-film dampers that are coupled with thrust washers. Additional powerplant components that have been subject to fretting are discussed below.

Over-Running Clutches. These components are used in helicopter drive trains (Ref 74). Figure 38 shows the fretting damage that has been observed in the powerplant components. This problem is caused by vibration (and rubbing) of the balls in the inner and outer races of bearings that support the rotor shaft. These inner and outer bearing races rotate at the same speed when the overrunning clutches are engaged (Ref 74).

Splines. In one case involving fretting on a compressor-turbine shaft spline, the spline failed and resulted in an engine fire, which caused the aircraft to crash. A spline structure surface that suffered fretting corrosion damage is shown in Fig. 39.

Fretting of splines is inhibited by maintaining a lubricating oil film on the surfaces of these parts. Solid film lubrication has proved to be an effective antifretting treatment for splines. The use of beryllium-copper material instead of SAE 4140 steel has also enhanced the fretting resistance of powerplant spline structures (Ref 74). Applications of vacuum coatings, such as by sputtering or ion plating, have also proved to be effective in alleviating spline fretting corrosion problems (Ref 74). Components that have been subject to fretting corrosion are discussed below.

Powerplant Rotor Blade Root Mounting Surfaces. An illustration of this fretting corrosion problem is provided in Fig. 40. Fatigue cracks can propagate as a result of this fretting damage, leading to eventual failure of these blades. Plasma-sprayed coatings, which have good adherence and friction-reducing properties, can be applied to blade surfaces in order to minimize the deterioration and subsequent mechanical failures caused by fretting.

Helicopter Reciprocating Engine Connecting Rod. This component failed in flight by fatigue that was caused by fretting corrosion on a small area on the bore of the connecting rod (Ref 55). Fretting corrosion developed in this connecting rod because the shell bearing, inside of the large-end bore of the rod, moved in a rapid rotational-oscillatory manner. The design solution to this problem was to increase the circumference of the shell bearings and to apply greater torquing force onto the bolts. This sufficiently reduced the rotational-oscillatory movement that had caused this problem, and further fretting corrosion was prevented (Ref 55).

Low-Pressure Compressor Casing. This part was made from magnesium alloy EZ33. Fretting corrosion occurred on this component because of mechanical rubbing action by a row of steel stator vanes on an annular ring. Because this fretting damage resulted from contact between dissimilar

Fig. 38 Fretting corrosion of the bearing race of an aircraft powerplant over-running clutch. Courtesy of the Advisory Group for Aerospace Research and Development

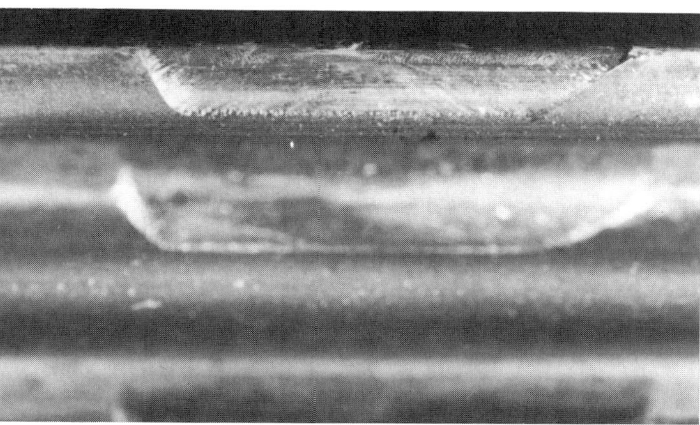

Fig. 39 Fretting corrosion on an aircraft powerplant spline structure. Courtesy of the Advisory Group for Aerospace Research and Development

metals (steel against magnesium), this incident is an example of fretting corrosion that occurred in a dissimilar-metal couple (Ref 55). The softer metal in the couple (EZ33 magnesium) was damaged more severely than the harder metal (steel). This problem can be alleviated by changing the EZ33 magnesium casing material to a harder, more wear-resistant alloy or by applying a wear-resistant (hard-facing) coating to the areas of the EZ33 magnesium casing that will come into contact with the steel stator vanes.

Stress-corrosion cracking has been reported as the dominant type of deterioration in the following failures of aircraft powerplant components.

Engine Exhaust Tailcone Assembly (Ref 5). A large section (about one-third) of this component fell away from an A-7D aircraft during flight. Circumferential resistance-welded stiffeners (Fig. 41) were used for stabilizing the tailcone structure on the outer surface of the assembly. The material used in the manufacture of this powerplant assembly was corrosion-resistant steel. In addition to SCC, intergranular degradation was observed where the tailcone was welded to the circumferential stiffener. During the analysis of this failure, the A-286 alloy was found to be prone to intergranular corrosion due to the formation of precipitates in the grain boundaries of the A-286 material.

High-Pressure Powerplant Compressor Vane (Ref 39). This failure occurred on a military fighter aircraft. The SCC originated at a point on the vane where pitting corrosion had developed. Several actions have been taken to prevent future powerplant vane failures, including changing the vane material, coating the vanes (nickel plating was one of the evaluated coatings), and periodically washing the compressor with water.

Pitting Corrosion. Cases involving pitting corrosion in powerplant components have developed over a relatively long period of time as compared to other types of ambient-temperature corrosion. The pitting corrosion of rotor and stator blades made of low-alloy steels and martensitic stainless steels is a significant powerplant deterioration problem that has produced fatigue failures and compressor stalls (Ref 75). The severity of these pitting problems was reduced by using better coatings and by applying improved corrosion control procedures, such as washing of the unprotected parts with water and inhibitors (Ref 76). Several types of coatings have been applied for the alleviation of pitting corrosion in these steel parts, including (Ref 77, 78):

- Paint
- Overlay coatings
- Metal-ceramic barrier coatings
- Nickel-cadmium diffusion coatings

A proposed solution to this problem was to change the blade material from steel to a nickel-base superalloy, IN-718 (Ref 77). The cost of this material change, however, would be significantly greater than the alternative of applying a specific coating to these parts. Powerplant blades and vanes can be coated with aluminum for pitting corrosion protection. Aluminum coatings can be applied by using the following techniques:

- The slurry method (Ref 79)
- Pack diffusion (Ref 79)
- Hot dipping and diffusion treatments (Ref 79)
- Ion vapor deposition (Ref 80)

Pitting corrosion has occurred in powerplant compressor stator vanes despite the use of a ceramic barrier coating material (Ref 81). Severe pitting was noted in this case. Some of the pits developed at critical locations, where the strength of the stator vane could be adversely affected. Inadequate process controls during the application of the ceramic slurry (bisque) to the stator vanes contributed to the rapid acceleration of the pitting corrosion that occurred in this case (Ref 81). Additional incidents involving pitting corrosion in powerplant components are discussed below.

AM-355 Stainless Steel Compressor Blades (Ref 81). Severe pitting corrosion was observed in the dovetail section, which is where the blade meets the compressor wheel. This problem was solved by replacing the pitted blades and by applying corrosion inhibitors to the newly installed parts. Pitting corrosion was prevented by the application of two different corrosion-preventive compounds (Ref 81).

Thrust Reverser Doors (Ref 55). Pitting corrosion caused the in-flight detachment of the thrust reverser doors of a commercial aircraft. One of the thrust reverser door driver links fractured because of fatigue. The fatigue crack started where pitting corrosion developed on the surface of the driver link, which was fabricated from type 422 stainless steel. The driver link had been coated with a

Fig. 40 Fretting corrosion on the root surface of an aircraft powerplant compressor blade. Courtesy of the Advisory Group for Aerospace Research and Development

(a)

(b)

Fig. 41 Circumferential stiffener (a) from a tailcone assembly that fell off an aircraft during flight. (b) Resistance weld on the stiffener in (a). Source: Ref 5

dry-film lubricant consisting of molybdenum disulfide. Pitting damage in the stainless steel part was caused by the accumulation of fuel combustion by-products and moisture in localized areas where the coating was defective. Preventive measures that can be taken to eliminate this problem and similar problems consist of: (1) reducing the high stress level in the thrust reverser door drive link (this is a longer-term solution because it would require a redesign of the drive link); (2) recoating the links with a more protective material, which will prevent the future development of pitting corrosion; and (3) establishing a continuous inspection program to ensure that corrosion is identified early, if it occurs.

Engine Exhaust Valve (Ref 55). This fatigue failure occurred during the flight of a single-engine aircraft. The fatigue-induced fracture started at a local area on the valve stem that was severely damage by pitting corrosion. Pitting reduced the section diameter of the exhaust valve (near the location of the fracture) by about 0.25 mm (0.01 in.). This pitting deterioration was produced by acidic compounds in old engine oil that had not been drained and replaced with new oil. Timely draining and replacement of engine oil is required to maintain the required concentration of corrosion inhibitors in the oil for preservation of the engine components.

Erosion-Corrosion and Cavitation. Erosion-corrosion occurs when solid particles collide at high velocities with powerplant components. This action removes the protective coatings and oxide films, resulting in corrosion of the unprotected metallic parts.

Cavitation produces deterioration of powerplant components by the high-velocity impingement of water droplets on these parts. Titanium is one of the best materials to use in turbine blades for resistance to cavitation (Ref 67).

Severe erosion problems have developed in magnesium splitter vanes (Ref 75). The Naval Air Development Center has studied the use of explosively welded aluminum claddings as a possible solution to this problem. Aluminum alloy substitutions have also been considered as a solution to erosion-corrosion problems in magnesium air inlet housings (Ref 76). Erosion attack has been especially noted in the air stream where protective coatings had worn away (Ref 76).

Hot desert climates can cause very severe erosion-corrosion problems in aircraft that operate in this type of environment (Ref 55). Dust and sand, with high salt concentrations, can erode surface finishes on powerplant components. Corrosion is accelerated by the extremely high temperatures that are reached in desert areas.

Erosion-corrosion has a high probability of occurring on various powerplant parts, including the following:

- Impellers
- Compressor blades

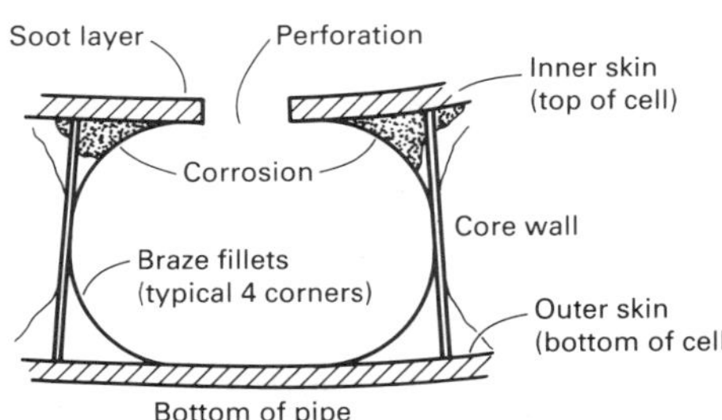

Fig. 42 Schematic showing an individual cell of a honeycomb sandwich engine tailpipe that failed because of corrosion

(a)

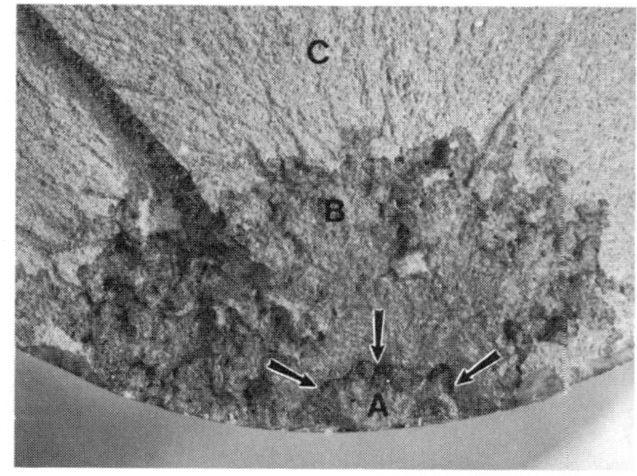

(b)

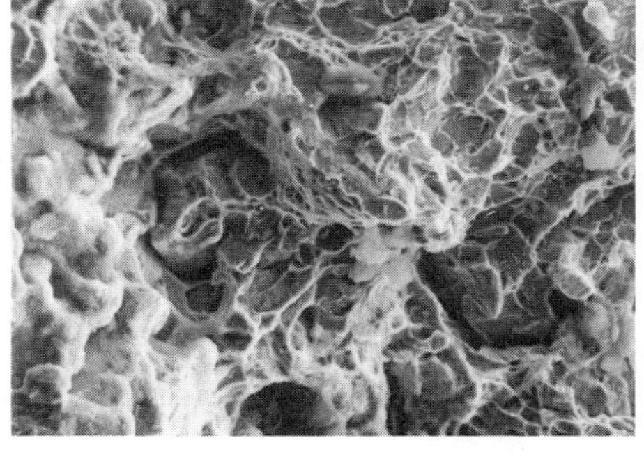
(c)

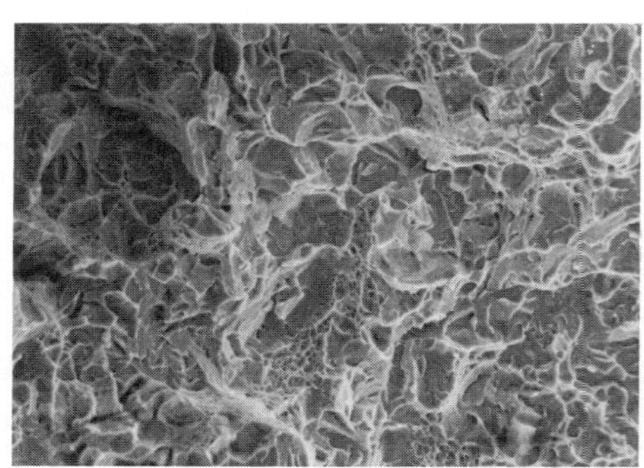
(d)

Fig. 43 Parallel lines of corrosion (a) on the shank of a PH13-8Mo stainless steel aircraft attachment bolt. (b) Close-up of fracture surface of bolt showing corroded area. Arrows point to one possible crack arrest line. (c) SEM fractograph of area in B in Fig. 2. Note corrosion product (left) and ductile dimples in the center. 265×. (d) SEM fractograph of area C in (b). Area of fast fracture shows cleavage and dimples. 265×

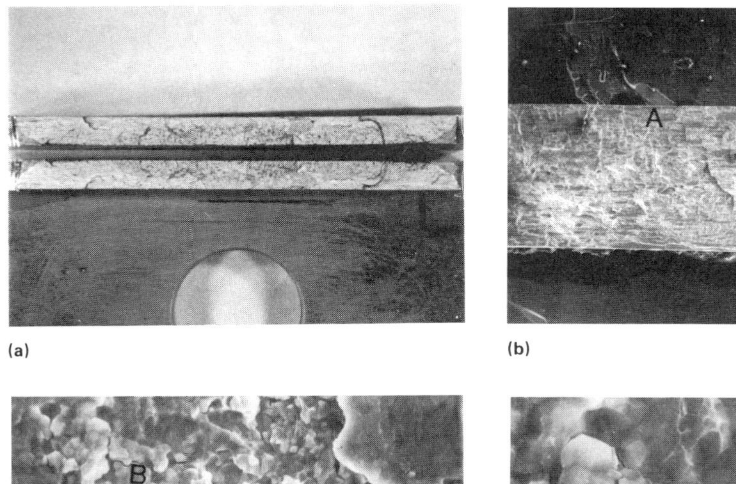

(a) (b) (c) (d)

Fig. 44 Opened crack (a) in aluminum alloy 7075-T651 ejection seat swivel fixture that failed by SCC. Note crack propagation markings that suggest the crack initiated on the inside wall of the fixture and woody appearance of the fracture. (b) Higher-magnification view of fracture surface from (a). Note woody appearance, which indicates a precrack mechanism. The inner wall of the fixture is at the top. 5×. (c) SEM fractograph of area A from (b). Note slight intergranular appearance of structure. 425×. (d) SEM fractograph of area B from (c) showing intergranular facets on the fracture surface. 1060×

- Turbine blades
- Turbine nozzles
- Turbine guide vanes

These powerplant components are vulnerable to erosion-corrosion that is caused by solid particles that impact these parts in the air intake area or in the hot gas stream. Powerplant components can be protected from erosion-corrosion by coatings that consist of hard, erosion-resistant materials. Many new erosion-resistant coatings are currently being evaluated (Ref 55).

Miscellaneous Types of Powerplant Corrosion

This discussion will cover various case histories regarding powerplant corrosion. The types of deterioration that caused these corrosion incidents are general corrosion; chemical corrosion, which is produced by chemical spills; intergranular corrosion; and corrosion fatigue.

General corrosion was found on aluminum-brazed titanium tailpipe extensions that were attached to the engine of a commercial aircraft (Ref 82). This deterioration was observed at the conclusion of a 3-year in-service evaluation of these tailpipe extensions. Among the factors that contributed to this corrosion problem were operation of the aircraft in rainy and humid conditions and a hydrophilic and sooty coating on the inner surfaces of the tailpipes.

The skins, or outer surfaces, of the tailpipe extension surrounded a honeycomb core. The skins were perforated for sound attenuation. The perforation resulted in atmospheric exposure and subsequent corrosion of the aluminum-brazed titanium component.

The hydrophilic (sooty) coating and the general corrosion are illustrated in Fig. 42. As a result of this case, it was concluded that aluminum-brazed titanium is not a suitable material for jet engine tailpipes that have perforated skins and honeycomb cores.

Chemical corrosion has been documented for components of the F-14 aircraft radar antenna hydraulic drive system (Ref 83). Corrosion on the internal surfaces of these parts was produced by contamination of the hydraulic fluid by water and halogenated solvents, primarily trichlorotrifluoroethane. The concentrations of these liquid contaminants exceeded the limits of the hydraulic oil specification (MIL-H-5606). These limits were 25 ppm for halogenated solvents and 150 ppm for water (Ref 83). The parts that were damaged consisted of 4130, nitralloy, and tool steels.

This corrosion did not lead to a failure of the hydraulic drive system. However, the long-term effect would have been a reduction in the reliability of the system. This problem was eliminated by ensuring that periodic draining and refilling with fresh hydraulic fluid was performed. This use of vacuum degassing procedures and desiccant filters also helped to alleviate this deterioration problem (Ref 83).

The emergency power unit of the F-16 military aircraft contains mixed hydrazine fuel consisting of 70% hydrazine and 30% water (Ref 84). Although this fuel can cause serious corrosion problems, no such cases have been published. Hydrazine-induced deterioration has been prevented on the F-16 emergency power unit by the development and application of an epoxy-type sealant that is resistant to hydrazine (Ref 84).

Intergranular corrosion was found to be the cause of a failure of a light aircraft engine exhaust pipe (Ref 55). The exhaust pipe separated from the aircraft engine during flight. The fracturing of the austenitic stainless steel pipe had occurred near its flange support. Intergranular corrosion cracks produced stress concentration sites. These stress raisers initiated fatigue due to vibration of the exhaust pipe and a muffler that was attached to the pipe. The muffler was cantilevered and unsupported—a design that produced high cyclic stresses, which was a major cause of the exhaust pipe failure.

The severely corrosive environment produced by exposure of the pipe to combustion by-products and condensates at high and low temperatures also played a major role in initiating the fracture of the exhaust pipe. It is possible that the exposure of the austenitic stainless steel to hot exhaust gases caused the sensitization of the steel. When sensitization occurs, the grain boundaries become susceptible to intergranular corrosion. A greater degree of microstructural stability can be achieved in austenitic steels by reducing the carbon content of these materials (see the article "Corrosion of Stainless Steels" in this Volume). This intergranular corrosion problem was solved by adding a structural support for the muffler in order to minimize the level of cyclic stress that this component would have to endure.

Corrosion fatigue has been reported as the cause of failure of an aircraft powerplant rocker-arm journal bearing. The bearing surface was examined, and deep pits and score marks were found (Ref 55). Fatigue cracks were observed to propagate radially from cavities along the surface of this part. This journal bearing was determined to be cast aluminum alloy 242. It was determined that there was gross porosity in the bearing casting. The material was therefore unsatisfactory for fatigue-critical applications. Proper quality control during the metallurgical processing of the casting to prevent porosity would have helped to produce a stronger, more fatigue-resistant part. General corrosion abetted the failure process.

Corrosion fatigue has also been reported to be the cause of steel engine shaft failures. These failures can originate at bolt holes or at the roots of gear teeth. The original corrosion damage then propagates by fatigue (Ref 85).

Case Histories and Failures

Ron Williams
Air Force Wright Aeronautical Laboratories

The Air Force Wright Aeronautical Laboratories' Materials Laboratory Structural Failure Analysis Group is responsible for performing metallurgical failure analyses on a variety of aircraft structural components. This includes airframes, propulsion and missile systems, and ground support equipment. Some of these analyses concern failures that occurred by corrosion mechanisms. This number has averaged approximately 15% of the total analyzed failures for the last several years. Most of these incidences are

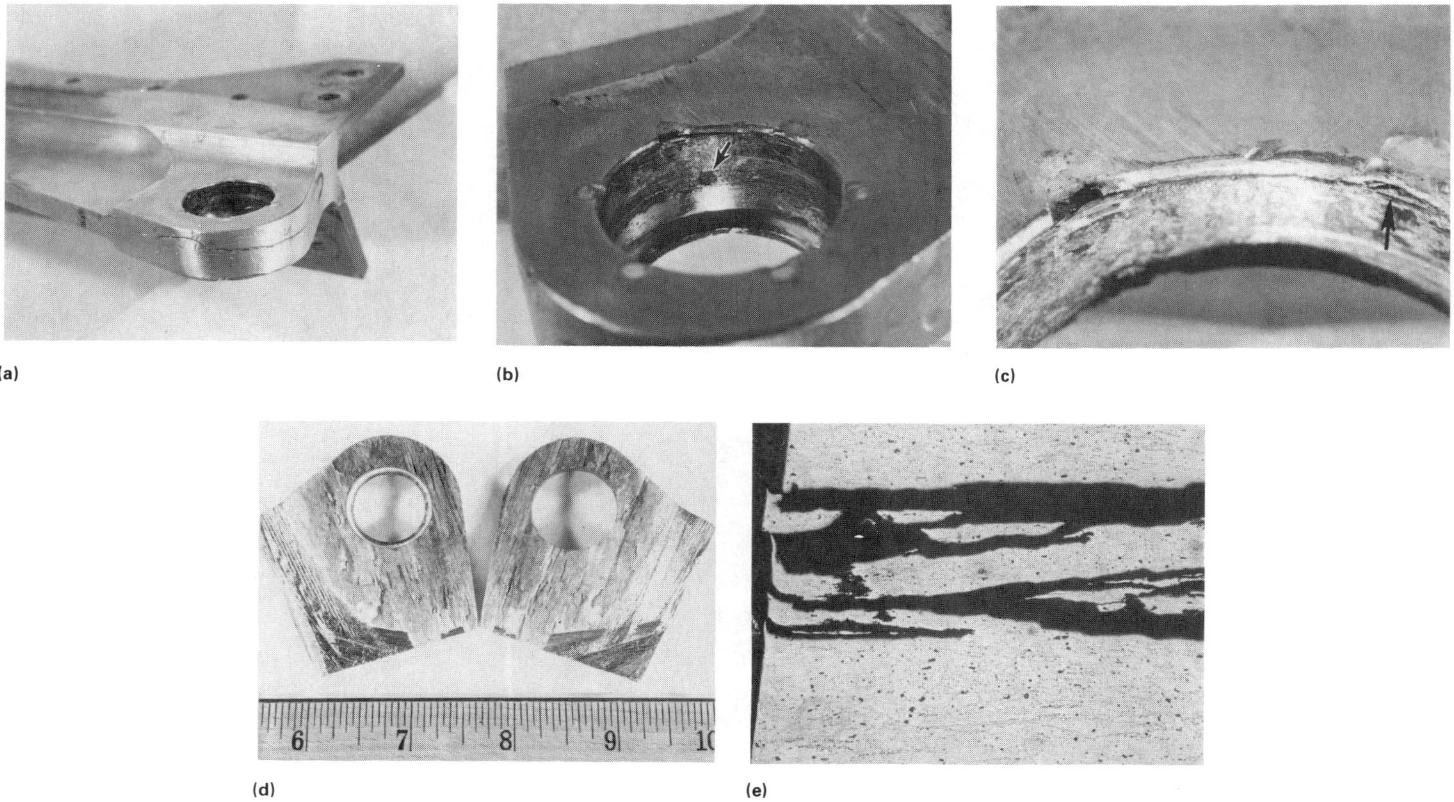

Fig. 45 End of aluminum alloy 7178-T6 aircraft wing bracket (a) showing cracking. (b) View of bracket showing symmetrical indentations on the top surface. Arrow shows a pit on the inner wall. 1×. (c) Close-up of indentations showing deformed surface (arrow) indicating directional movement of the bushing. 4×. (d) Failed fracture surfaces of bracket showing the woody fracture appearance characteristic of exfoliation. (e) Cross section of bracket showing delamination caused by exfoliation. 105×

Fig. 46 SEM of fracture surface (a) from a failed 17-7PH stainless steel aircraft controller diaphragm showing intergranular fracture indicative of SCC. 170×. (b) SEM fractograph of area adjacent to that shown in (a) showing intergranular fracture, secondary cracking, and little or no evidence of ductility; this suggested a brittle fracture mechanism. 170×. (c) Microstructure of diaphragm taken perpendicular to the fracture face. The carbide network at the grain boundaries is evident. 210×

either manufacturing/quality control related or involve materials and/or heat treatments used in older systems. The following discussions represent a number of case histories of failures associated with corrosion and the recommended corrective actions to preclude repetitive occurrences.

Examples of Aircraft Structural Corrosion Failures

Example 1: Aircraft Attachment Bolt Failure. During a routine inspection on an aircraft assembly line, an airframe attachment bolt was found to be broken. The bolt was one of 12 that attach the lower outboard longeron to the wing carry through structure. Failure occurred on the right-hand forward bolt in this longeron splice attachment. The bolt was fabricated from PH13-8Mo stainless steel heat treated to have an ultimate tensile strength of 1517 to 1655 MPa (220 to 240 ksi). A water-soluble coolant was used in drilling the bolt hole where this fastener was inserted.

Investigation. Figure 43(a) shows a bolt shank with corrosion evident on surfaces in contact with the splice joint, while Fig. 43(b) shows the fracture and initiation site of the bolt failure. Surface pitting on the bolt shank and subsequent corrosion cracking are shown. Scanning electron microscopy (SEM) examination of the fracture revealed the topography to be intergranular at the initiation site and to a depth of 8.4 mm (0.33 in.) (Fig. 43c). The remainder of the fracture showed a mixed topography of cleavage and ductile dimples (Fig. 43d).

Examination of corrosion products on the fracture by Auger emission spectroscopy and secondary imaging spectroscopy showed the presence of elements typically found in tap water. Chemical analysis of the bolt material showed the composition to be within specification limits for PH13-8Mo stainless steel. Rockwell C hardness measurements taken on the bolt produced values ranging from 47 to 48 HRC, which would corre-

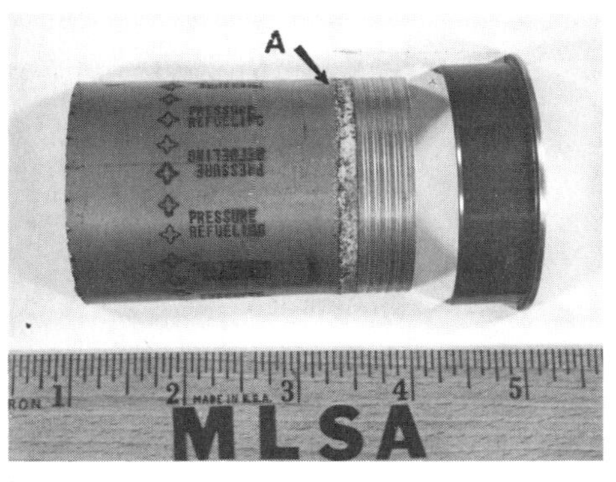

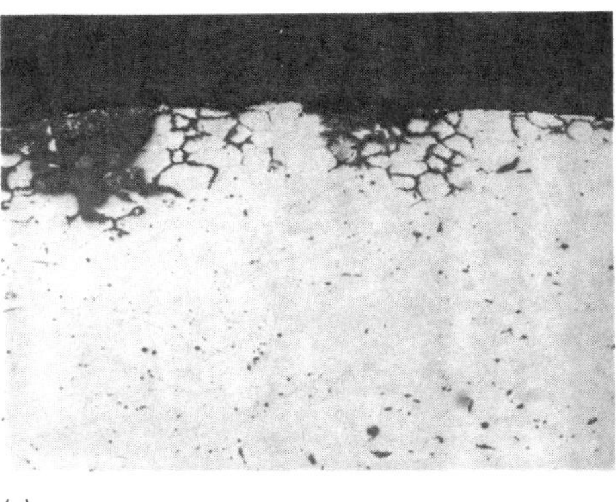

(a) (b) (c)

Fig. 47 Corrosion (a) of aluminum alloy 6061-T6 aircraft fuel line (arrow). (b) Close-up of corrosion on fuel line. Note pitting and corrosion products. (c) Intergranular corrosion of the fuel line at area A from (a)

spond to the specified ultimate tensile strength of 1517 to 1655 MPa (220 to 240 ksi).

Conclusion. It was concluded from the study that failure of the attachment bolt was caused by stress corrosion. The source of the corrosive media was the water-soluble coolant used in boring the bolt holes.

Recommendations consisted of inspecting for corrosion all the bolts that were installed using the water-soluble coolant at the spliced joint areas, rinsing all machined bolt holes with a noncorrosive agent, and installing new PH13-8Mo stainless steel bolts with a polysulfide wet sealant.

Example 2: Failure by SCC of an Ejection Seat Swivel. A routine examination on a seat ejection system found that the catapult attachment swivel contained cracks on opposite sides of the part. This swivel, or bath tub, does not experience any extreme loads prior to activation of the catapult system. Some loads could be absorbed, however, when the aircraft is subjected to G loads. The bath tub is fabricated from aluminum alloy 7075-T651 plate.

Investigation. Visual examination of the part revealed that cracks were positioned near the base of the bath tub configuration and extended through the wall thickness. One of the cracks was opened (Fig. 44a); this indicated that the fractures initiated on the inner walls of the fixture.

Electron optical examination of the fracture at low magnifications revealed a woody appearing topography (Fig. 44b). Further electron optical examination of the fracture at 800 and 2000× (Fig. 44c and d) showed that the cracking pattern initiated and progressed by an intergranular failure mechanism. This fracture topography indicated that cracking was due to stress corrosion.

Examination of the microstructure near the fracture revealed that the crack was progressing parallel to the transverse grain flow direction and further suggested SCC. Chemical analysis and hardness tests conducted on the submitted material showed it to be within specification requirements for 7075-T651 aluminum base material.

Conclusion and Recommendation. It was concluded that failure of the catapult attachment swivel fixture occurred by SCC, and it was recommended that the 7075 aluminum ejection seat fixture be supplied in the T-73 temper to minimize susceptibility to SCC.

Example 3: Cracking of an Aircraft Wing Bracket. During an inspection cycle, cracking was detected in a wing fillet flap bracket. The cracking was located on the end of the bracket, as shown in Fig. 45(a). The configuration of the end of the bracket suggested that a bushing and rod were integral working components to the bracket hardware.

Investigation. Visual examination of the bushing seat area showed the presence of surface corrosion pits (Fig. 45b). Also shown are six symmetrical indentations that were produced during staking to prevent shifting of the bushing. Further visual examination revealed deformation adjacent to the indentations (Fig. 45c), indicating that the bushing had deformed the material.

Optical examination of the opened fracture showed a woody, delaminated, fibrous-textured fractured surface (Fig. 45d). These fracture characteristics indicated that failure progressed by exfoliation corrosion. A cross section taken through the fracture surface revealed the presence of delamination due to exfoliation, as shown in Fig. 45(e).

A chemical analysis of the material revealed it to be within specification for the required aluminum alloy 7178 except for a slightly lower than required zinc percentage. A hardness survey taken on the hardware found the values to range from 85 to 87 HRB and suggested that the material was in the T-6 condition.

Conclusions. From this analysis, it was concluded that failure of the wing fillet flap bracket

(a)

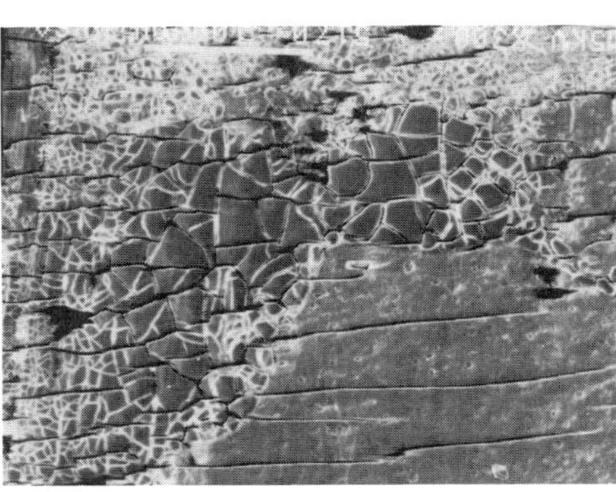

(b)

(c)

(d)

Fig. 48 Aluminum alloy 7075-T6 aircraft wing panel (a) showing unusual surface appearance. (b) SEM of the panel surface showing cracked anodized coating. 160×. (c) SEM showing the anodized coating flaking away and corrosion deposit under the coating. 85×. (d) Cross section of corrosion site on panel showing depth of intergranular attack. 265×

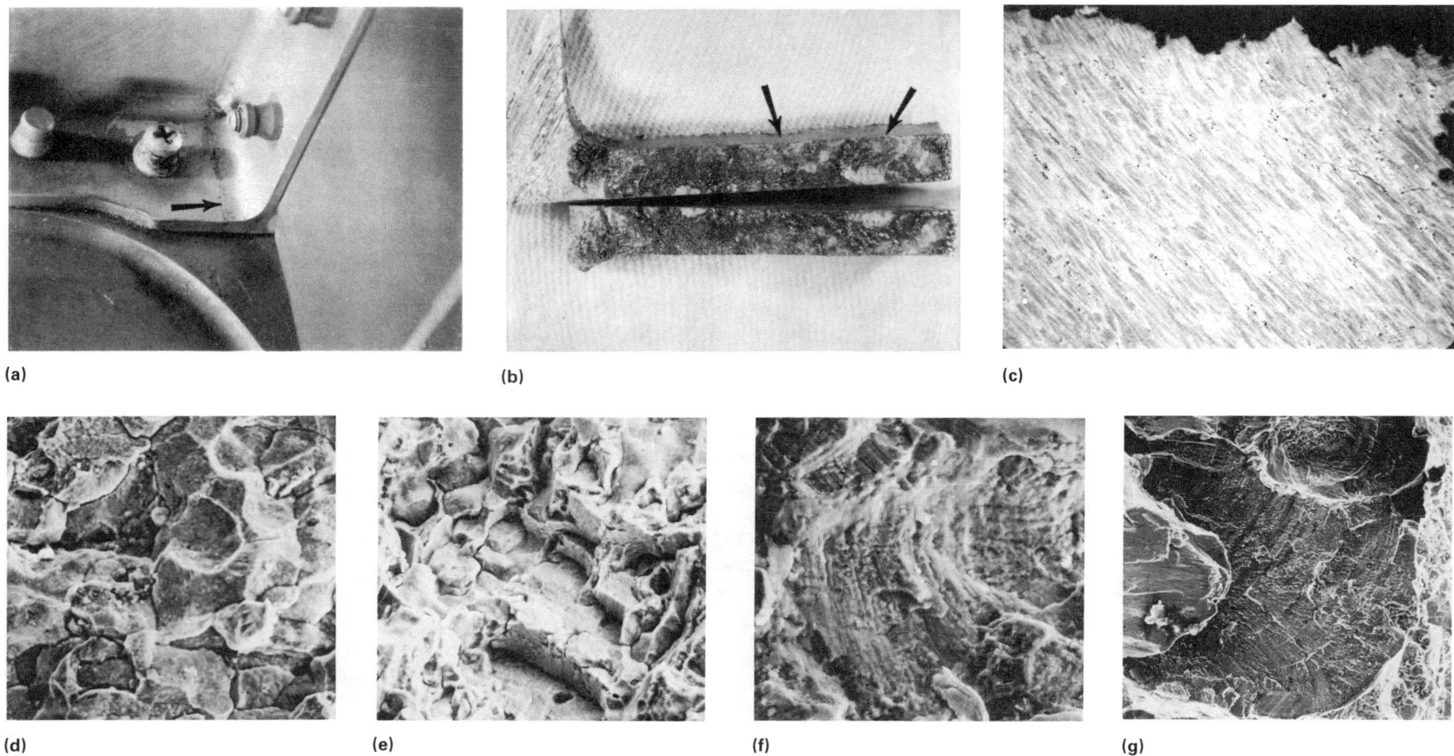

Fig. 49 Aluminum alloy 7079-T6 aircraft wing spar (a) showing crack (arrow). (b) Fracture surfaces of opened spar crack. Note clamshell marks at termination of the crack (left). Suspected multiple initiation sites are located between arrows. 1.5×. (c) Section of flange with surface at right. Grain flow in this area was at an angle to applied stress, which resulted in end grain exposure. 105×. (d) SEM fractograph taken between the arrows in (b). Note intergranular fracture pattern indicative of SCC. 95×. (e) SEM taken near the termination of the fracture showing the crack still progressing by SCC. 190×. (f) SEM showing fatigue striations near the crack origin. 235×. (g) Fatigue striations near the termination of cracking. 30×

was due to surface corrosion pits on the extrusion bracket hole wall surface. Crack progression occurred by exfoliation corrosion and was aided by a contributing stress introduced by movement of the bushing.

Recommendations. It was recommended that a material substitution be made of the 7178-T6 material because of its susceptibility to exfoliation corrosion. Candidate replacements included aluminum alloys 7175, 7050, or 7049.

Example 4: Failure of an Aircraft Controller Diaphragm. The diaphragm from a side controller was found during a preflight inspection to be broken. The controller diaphragm was fabricated from 17-7PH stainless steel in the RH 950 heat treatment condition. Failure occurred by cracking of the base of the flangelike diaphragm. The crack traveled 360° around the diaphragm.

Investigation. Examination of the submitted piece by SEM revealed that the failure occurred by a brittle intergranular mechanism and indicated a failure mode of selective grain-boundary separation (Fig. 46a and b). The diaphragms are heat treated in batches of 25. An improper heat treatment could have resulted in the formation of grain-boundary precipitates. These grain-boundary precipitates would include chromium carbides. A loss of chromium from the matrix adjacent to the grain boundaries increases the susceptibility of this material to grain-boundary attack by oxidizing media. This action could result in intergranular corrosion and/or SCC if the part was placed under load in an oxidizing medium.

Metallographic examination of a section taken perpendicular to the fracture revealed a martensitic structure with grain-boundary attack on the outer surfaces. The grain-boundary attack most probably occurred during heat treatment. Figure 46(c) shows a microstructure that was etched to reveal the presence of carbide precipitates. The grain-boundary network of carbides strongly suggested that the material was exposed to an adverse heat treatment with chromium depletion of the matrix adjacent to the grain boundaries.

Conclusions. It was concluded that failure of the diaphragm was due to a combination of sensitization caused by improper heat treatment and subsequent SCC.

Recommendations. It was recommended that the remaining 24 sensor diaphragms from the affected batch be removed from service. In addition, it was recommended that a sample from each heat treat batch be submitted to the Strauss test (ASTM A 262, practice E) to determine susceptibility to intergranular corrosion. It was also recommended that a stress analysis be performed on the system to determine whether or not a different heat treatment (which would offer lower strength but higher toughness) could be used for this part.

Example 5: Fuel Line Corrosion. Inspections revealed fuel line corrosion beneath ferrules (Fig. 47). The cause of the corrosion was traced to the fuel line marking process, which involved electrolytic labeling of ferruled aluminum alloy 6061-T6 tubes. Although subsequent rinsing of the fuel lines washed off most of the electrolyte, some was trapped between the 6061-T6 tubing and the ferrule. This condition made corrosion of the fuel lines inevitable.

Investigation. Microstructural analysis revealed extensive intergranular corrosion of the 6061-T6 tubing beneath the ferrule (Fig. 47c). This attack caused grains to become dislodged, giving the appearance of pitting. Corrosion penetrated approximately 0.13 mm (0.005 in.) into the tubing.

In an attempt to determine if the corrosion products were active, two specimens from the corroded fuel lines with corrosion products were mounted and soaked in distilled water at room temperature for 2 and 4 days. The 2-day exposure resulted in a localized intergranular corrosion on the inside diameter of the tubing, while the 4-day exposure resulted in extensive intergranular corrosion of the tube cross section from the inside diameter to the outside diameter.

Corrosion products from beneath the ferrules were placed on a piece of uncorroded 6061-T6 tubing in an attempt to substantiate further whether or not the corrosion products were active. Electrical tape loosely applied to the specimen held the products in place while the test specimens were submersed in distilled water for 5 days. Subsequent inspection of the specimen revealed that corrosion did not occur during the 5 days.

Emission spectroscopy of the corrosion products showed that small amounts of aluminum (4%), sodium (3%), cobalt (2%), chromium (0.35%), boron (0.25%), and iron (0.05%) were present. The remaining 90% of the material analyzed was nonmetallic.

Conclusions. It was concluded that the marking electrolyte used for labeling was trapped

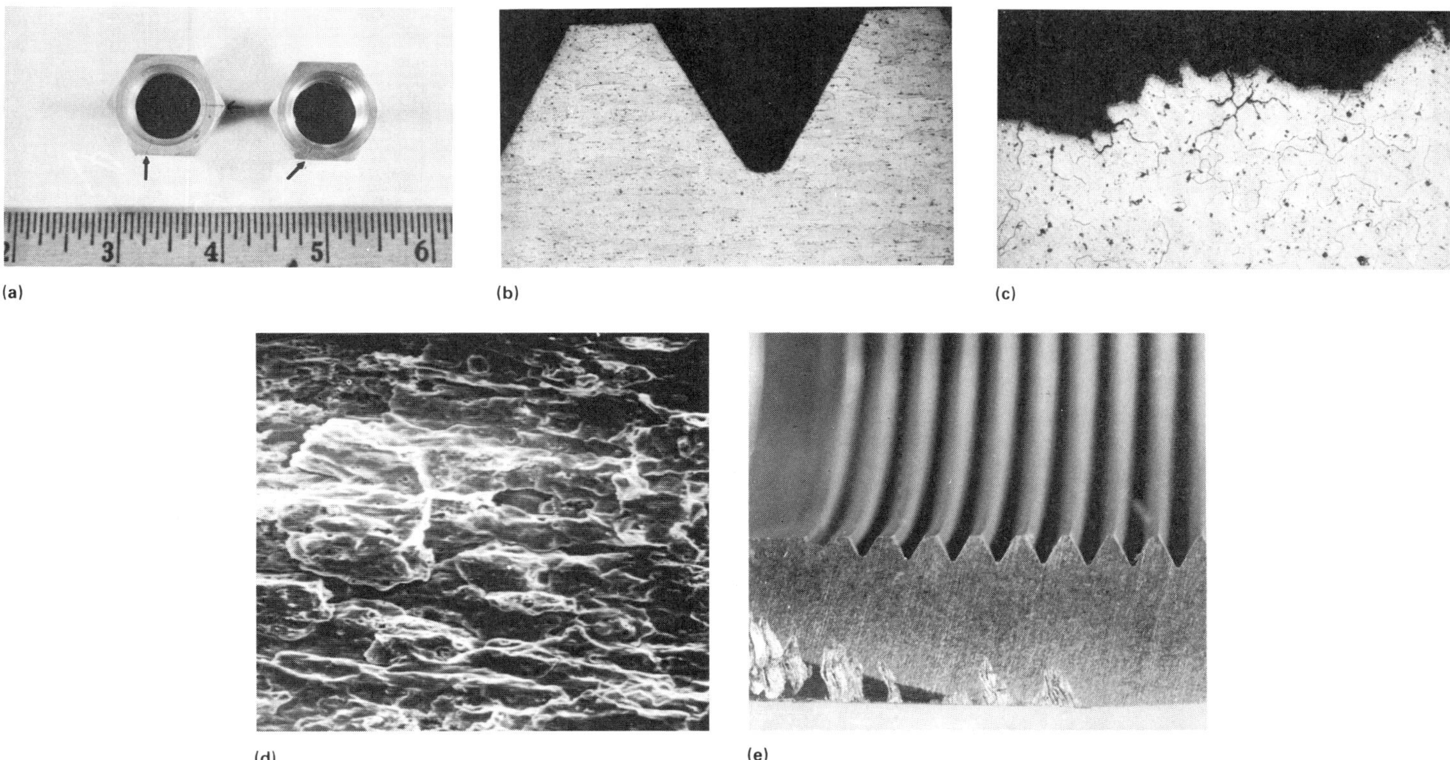

Fig. 50 Top view (a) of cracked aluminum alloy 2024-T351 pitostatic connectors. Arrows indicate cracks. (b) Cross section of one connector showing elongated grains that were cut to form connector threads. 25×. (c) Cross section showing intergranular cracking with multiple branching in one connector. 105×. (d) SEM fractograph showing intergranular cracking and separation of elongated grains. 130×. (e) Cross section of connector threads showing incomplete thread form resulting from improper tapping

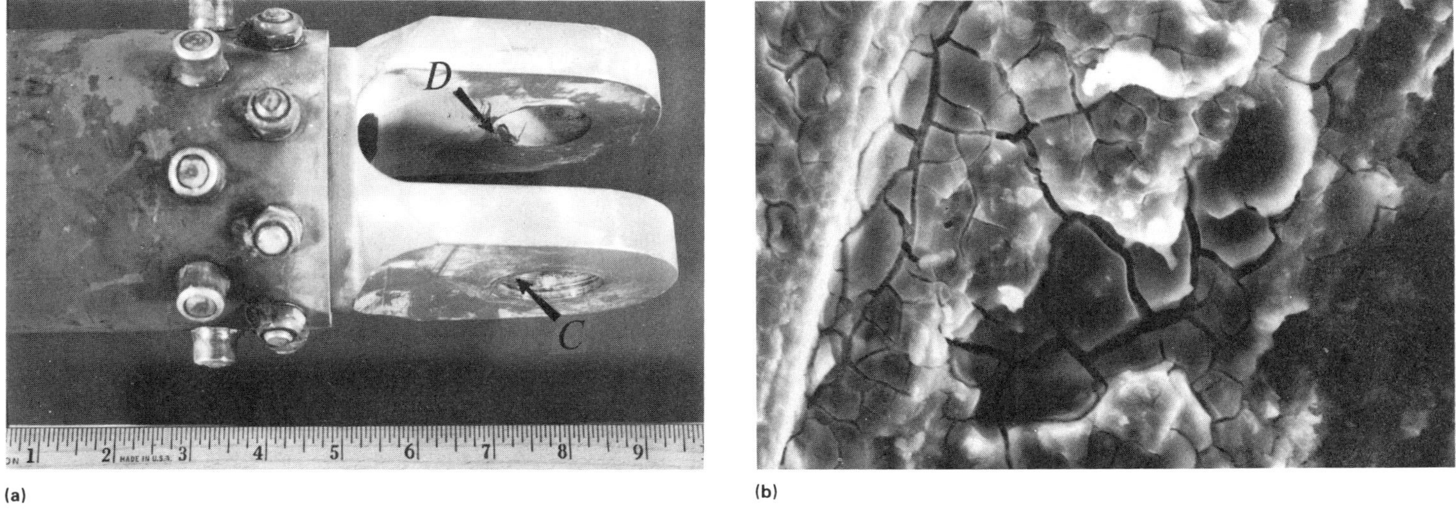

Fig. 51 Cracked aluminum alloy 7075-T6 aircraft pylon strut (a) with arrows indicating cracks. (b) SEM of crack C from (a) showing the mud crack pattern indicative of a corrosion mechanism. 820×

between the 6061-T6 tubing and the ferrule. This fostered intergranular corrosion. Experiments indicated that the corrosion products were inactive.

Recommendations. It was recommended that another marking process be used that does not involve corrosive materials. The prevention of electrolyte from being trapped between the tubing and ferrules by using a MIL-S-8802 sealant was recommended.

Example 6: Corrosion of Aluminum Alloy 7075-T6 Wing Panel. New aircraft wing panels extruded from 7075-T6 aluminum were reported to be discolored, exhibiting an unusual pattern of circular black interrupted lines (Fig. 48a). The black marks were coherent with the metal and could not be removed by scouring or light sanding. The panels, subsequent to profiling and machining, were required to be penetrant inspected, shot-peened, H_2SO_4 anodized, and coated with MIL-C-27725 integral fuel tank coating on the rib side.

During processing, the extrusions are machined on the flat side, oiled, deburred, hot formed, cleaned, penetrant inspected, covered with oil, and then shotpeened. They are then recoated with oil, shipped to a second vendor, handwiped with a solvent, alkaline cleaned, acid desmutted, sulfuric acid anodized, and hot water sealed.

(a)

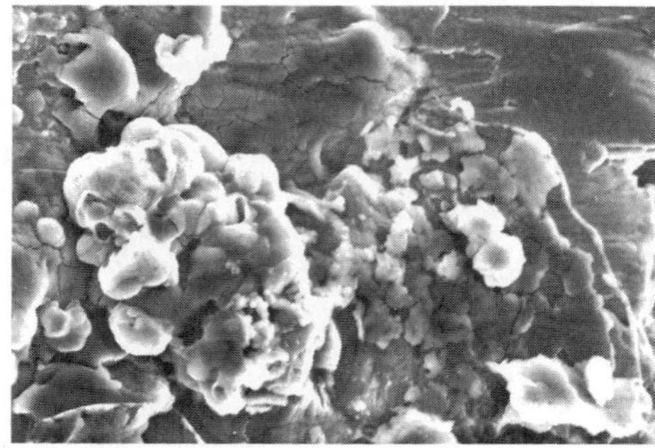

(b)

Fig. 52 Cadmium-plated 4340 steel ballast gas elbow assembly (a) with arrow showing hole where corrosion was found. (b) SEM of corrosion products on inside hole surface. 430×

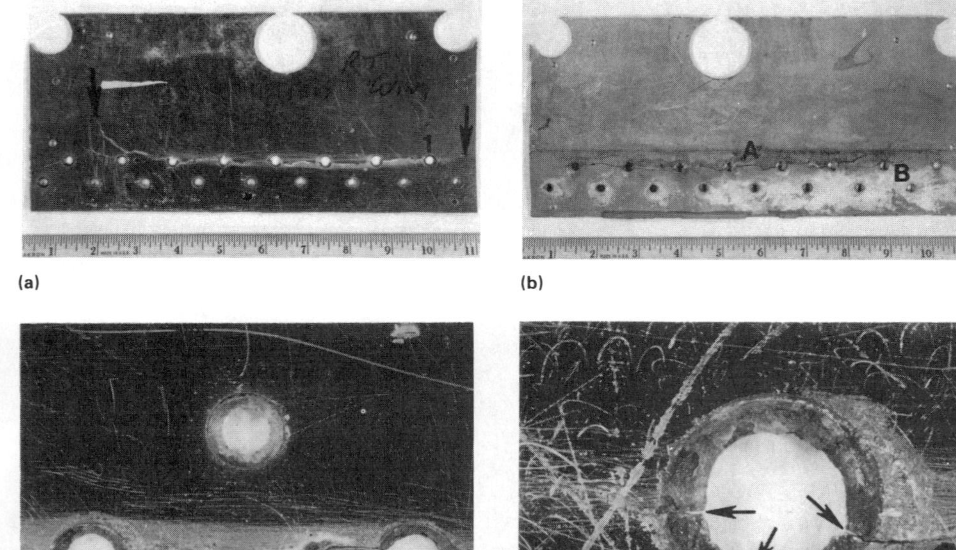

Fig. 53 Overall view (a) of cracked magnesium alloy AZ31B aircraft wing leading edge panel. Arrows show the length of the crack. (b) Other side of panel shown in (a). A denotes the primary crack; B shows a second, smaller crack. (c) Close-up of fastener holes through which the crack progressed. Note that the two bottom holes are not beveled. (d) Close-up of fastener hole showing cracking at three sites (arrows)

Investigation. The panels were studied using the scanning electron microscope and microprobe analysis. Both conventional energy-dispersive and Auger analyzers were employed. Figures 48(b) and (c) illustrate the contention that the anodic coating was applied over an improperly cleaned and contaminated surface. It was evident that the expanding corrosion product had cracked and in some places had flaked away the anodized coating. The corrodent had penetrated the base aluminum in the form of subsurface intergranular attack (Fig. 48d). The depth of attack was measured to be 0.035 mm (0.0014 in.).

Microprobe analysis of the corrosion product did not reveal any clues concerning the reason or origin of the corrodent. A high sulfur concentration was found associated with the corrosion product and on surface areas away from the products. It was suspected that the origin of the sulfur was the hydrocarbon oil. When the anodized layer was stripped from the panels using a phosphoric-chromic acid solution, the evidence of sulfur disappeared. This same stripping procedure did not remove the black corrosion product.

Energy-dispersive analysis of the corrosion product revealed the presence of iron, calcium, phosphorus, and chromium in excess. No chlorides were detected. Auger spectroscopy revealed the presence of large amounts of carbon and nitrogen. The MIL-C-27725 coating was removed from a portion of the rib side by using a paint stripper. No corrosion or discoloration of the aluminum was observed.

Conclusions. It was concluded that the corrosion of the anodized panel most probably resulted from improper and insufficient cleaning prior to anodizing. The preservation oils used during the various steps of manufacture and their incomplete removal prior to anodizing were highly suspect.

Recommendations. It was recommended that a vapor degreaser be used during cleaning prior to anodizing. A hot inhibited alkaline cleaner was also recommended during cleaning prior to anodizing. The panels should be dichromate sealed after anodizing. The use of deionized water was also recommended during the dichromate sealing operation. In addition, the use of an epoxy primer prior to shipment of the panels was endorsed. Most importantly, surveillance of the anodizing process itself was emphasized, including continual monitoring of bath acid concentration, solution cleanliness, temperature control, and voltage/amperage control.

Example 7: Cracked Aircraft Wing Spar. A crack (Fig. 49) was found in an aircraft main wing spar flange fabricated from aluminum alloy 7079-T6 during a routine nondestructive x-ray inspection after the craft had logged 300 h.

Investigation. Visual examination of the crack edge shown in Fig. 49(a) revealed that the installation of the fasteners produced a fit up stress, as indicated by the approximate 0.75-mm (0.03-in.) springback of the flange after the crack propagated through the hardware. Further inspection of the opened fracture (Fig. 49b) showed that the crack had been present for some time because a heavy buildup of corrosion products was seen on the fractured surface. The fracture initiated at multiple origins between the arrows shown in Fig. 49(b). Metallographic examination of the flange in the area of fracture initiation showed the presence of end grain exposure (Fig. 49c), which would promote SCC.

Electron optical examination of the fracture shown in Fig. 49(b) produced the scanning electron fractographs shown in Fig. 49(d) to (g). Figures 49(d) and (e) show an intergranular topography, while the fractographs in Fig. 49(f) and (g) reveal fatigue striations. This clearly shows the flange was cracking by a mixed mode of stress corrosion and fatigue.

Chemical analysis of the flange showed that the material met compositional requirements for 7079 aluminum base material. Hardness measurement of 85 HRB showed the material was in the T-6 heat treat condition.

Conclusions. It was concluded that the cracking of the flange occurred by a combination of stress corrosion and fatigue. The cracking was accelerated because of an inadvertent fit up stress during installation. The age of the crack could not be established. However, a reevaluation of prior x-ray inspections in this area would result in some close estimate of the age of the crack. End grain exposure further promoted SCC.

Example 8: SCC of Pitostatic System Connectors. Pitostatic system connectors were being found cracked on several aircraft. The cracks were not restricted to any particular group of aircraft. Two of the cracked connectors were submitted for failure analysis. Both were reportedly made of 2024-T351 aluminum. The connectors had cut pipelike threads that are sealed with teflon-type tape when installed.

Investigation. Longitudinal cracks were located near the opening of the female ends of each connector (Fig. 50a). Both connectors had the same size female end but different size male ends.

(a)

(b)

(c)

(d)

Fig. 54 Overall view (a) of external tank pressure/vent valve. (b) Partially disassembled valve. A, solenoid switch; B, dual-position valve; C, valve housing. (c) Segment of air line showing residue on check valve poppet (arrow). (d) Close-up view of check valve poppet

The connector with the large diameter and longer male end had two cracks, while the connector with the small diameter and shorter male end had only one crack. This size difference was believed to have had no bearing on the cracking.

The connector with the large male end was sectioned, and part of the fracture was metallographically examined. The connector exhibited an elongated recrystallized grain structure with cut threads (Fig. 50b). A cross section through the fracture showed intergranular cracking and branching of the crack (Fig. 50c), characteristic of SCC.

Corrosion deposits were chemically removed from one section of the fracture surface, and the surface was examined in the scanning electron microscope. The fracture surface exhibited intergranular cracking of elongated grains (Fig. 50d).

A section of the connector with the large male end and some thin transparent film found on the threads of the connector were chemically analyzed. The connector was determined to be either 2014 or 2017 aluminum alloy, and the film was determined to be fluorinated hydrocarbon teflon-type tape.

Hardness checks on both connectors showed the large male end connector to be 75 HRB and the small male end connector to be 77 HRB. Electrical conductivity checks on both connectors showed the large male end connector to have a conductivity of 31% IACS (International Annealed Copper Standard) and the small male end connector to have a conductivity of 27.5% IACS.

The threads of all connector components were incompletely formed with a bottom tap and therefore produced a tapered or pipe-type thread. The large male end connector had only one to two threads cut full depth (Fig. 50e).

Conclusions. It was concluded that the pitostatic system connectors failed by SCC. The corrodent involved could not be conclusively determined. The stress was caused by forcing the improperly threaded female nut over its fully threaded male counterpart to effect a seal. The pipelike, incomplete threads produced high hoop stresses when torqued down over a fully formed thread. The one connector tested for chemical composition was not made of 2024 aluminum alloy as reported but of 2017 aluminum. Hardness and conductivity data on both connectors were compatible with a T351 condition for a 2024 alloy.

Recommendations. It was recommended that the pitostatic system connector manufacturing process be revised to produce full-depth threads rather than pseudo pipe threads. It was also recommended that the wall thickness be increased to increase the hoop stress bearing area if pipe threads were to be used. A determination of proper torque values for tightening the connectors was also suggested.

Example 9: Cracking of Aircraft Pylon Strut. A pylon strut was submitted for failure analysis. Cracks were found in two locations on the ears of the strut (Fig. 51a). Because the part was still intact, the cracks had to be forced open so that the fractures could be examined.

Investigation. Chemical analysis and hardness measurements indicated that the strut was 7075-T6 aluminum. Scanning electron microscopy of the opened cracks showed that the crack surfaces were covered with a mud crack pattern suggestive of SCC (Fig. 51b). The T6 temper is susceptible to SCC.

Conclusions. It was concluded that cracking of the strut could have been aggravated by the hard landing experienced by the aircraft. The strut, however, contained stress-corrosion cracks which were present before the landing.

Recommendation. It was recommended that an inspection for SCC be made of all pylon struts with a similar service life.

Example 10: Corrosion of a Ballast Gas Elbow Assembly. A cadmium-plated 4340 steel ballast elbow assembly (Fig. 52) was submitted to the laboratory for failure analysis. It was requested that a determination be made regarding the element or radical present in an oxidation product found inside the elbow assembly.

Investigation. It was determined by energy-dispersive x-ray analysis in the SEM that iron was the predominant species, presumably in an oxide form. No cadmium was present inside the hole where oxidation occurred. However, there was cadmium on the outer surface. The inside surface had the appearance of typical corrosion products (Fig. 52b). Hardness measurements indicated that the 4340 steel was heat treated to a strength of approximately 862 MPa (125 ksi).

Conclusions. From these analyses, it was concluded that the oxide detected on the ballast elbow was iron oxide. The possibility that the corrosion products would eventually create a blockage of the affected hole was great considering the small hole diameter (4.2 mm, or 1/6 in.). It was recommended that a quick fix to stop the corrosion would be to apply a corrosion inhibitor inside the hole. This, however, would cause the possibility of inhibitor buildup and the eventual clogging of the hole.

Recommendations. A change in the manufacturing process to include a cadmium plating on the hole inside surface was recommended. This was to be accomplished in accordance with MIL specification QQ-P-416, Type II, Class 1.

A material change to 300-series stainless steel was also recommended. Its strength is slightly lower than that of the 4340 material, but its corrosion resistance is superior, and no surface treatment would be required.

Example 11: Failure of Aircraft Wing Leading Edge Panel. Cracks were found on the wing leading edge of a test aircraft. The cracks were located on the inboard side of the No. 2 and No. 3 engines. Crack lengths were approximately 230 mm (9 in.) long on the left side and approximately 130 mm (5 in.) long on the right side. The cracks ran parallel to the leading edge. The 230-mm (9-in.) crack was received for examination.

Investigation. Visual examination of the submitted panel revealed two cracks. One crack ran through six adjacent fastener holes (Fig. 53a and b). A close-up of the fastener holes shown in Fig. 53(c) and (d) revealed that sections of the beveled edges of the holes were missing; corrosion was evident. Further visual examination of the fastener holes after separation of the crack showed that the fracture faces were corroded.

Optical examination of either side of the middle group of fastener holes showed that the area of suspected crack initiation had suffered excessive corrosion. Although a slight beach mark appearance was seen adjacent to the holes, a definite cause for the cracking could not be established. Examination of the holes on the end of the crack showed fracture characteristics typical of fatigue and/or corrosion fatigue.

A chemical analysis of the plate material showed the composition to be an AZ31B magnesium alloy.

Conclusion and Recommendation. It was concluded that crack propagation of the fracture in the wing panel occurred by a combination of corrosion and high-cycle fatigue in the end fastener holes. Conclusive proof of the cause of failure initiation in the middle fastener holes was masked by excessive corrosion. It was recom-

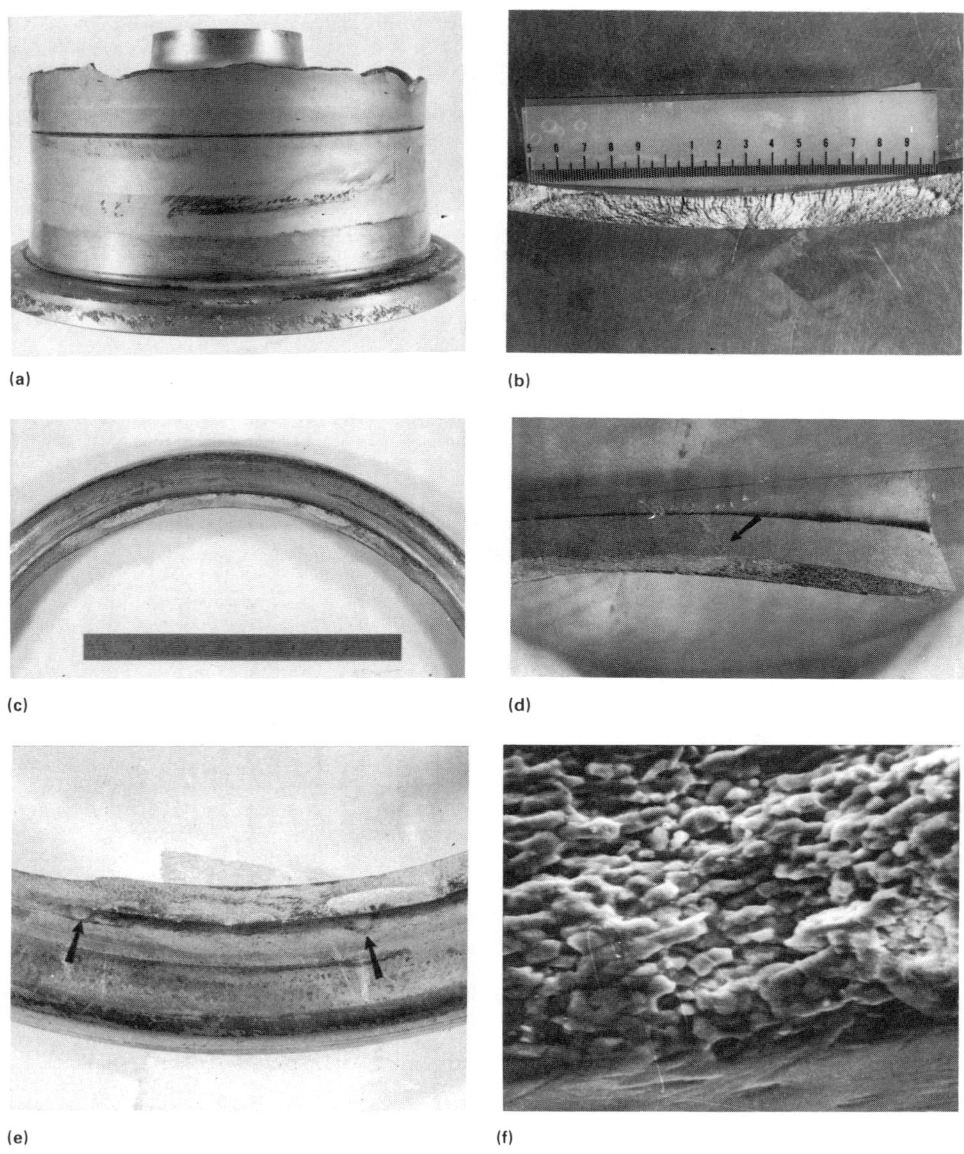

Fig. 55 Aluminum alloy 2014-T6 aircraft nose wheel (a) that failed at the flange. (b) Close-up of tube well on wheel 31. (c) Appearance of flange failure on wheel 67. The topography is typical of other flange failures. (d) Close-up of wheel 31; note indentation (arrow). (e) Close-up of wheel shown in (a); note surface blemishes (arrow). (f) SEM of typical fracture initiation site showing an angular, blocky structure indicative of a corrosion-related failure mechanism. 780×

mended that future panels be manufactured of 2024 aluminum.

Example 12: Failure of an External Tank Pressure/Vent Valve. The pressure/vent valve shown in Fig. 54(a) was submitted for laboratory analysis. The external tank pressure/vent valve regulates the external tank fuel feed system, which transfers fuel under pressure to the internal tanks of the aircraft. It is stated in the technical order that when the solenoid of the pressure/vent valve (part A, Fig. 54b) is energized, service air at 448 kPa (65 psig) shifts the dual-position valve (part B, Fig. 54b).

It was reported that the dual-position valve was found to be sticking at the intermediate positions. In addition, it was found that service air check valves located on the incoming lines contained poppets that were being stuck in a closed or partially closed position because of suspected corrosion product (Fig. 54c and d). These two factors prevented full pressurization of the external tank and caused a subsequent degradation in fuel flow to the internal tanks.

Investigation. Residue taken from the check valve poppet and from the dual-position valve was chemically analyzed. Only small samples (103 mg) of residue could be obtained from the parts for analysis. For this reason, each of the samples could be analyzed only for a single element. Chloride was selected for the analysis and was present in both samples. The residue found inside the check valve had a reddish color, indicating that the chloride-containing compound reacted with the anodized, dichromate sealed check valve housing.

It was noted that the dry-film lubricant applied to the dual-position valve was of different consistency and was heavier than that found on an identical valve assembly also submitted to the laboratory. In addition, the coating of the failed valve was flaking off. Although it did not cause the initial malfunction of the pressure/vent valve assembly, subsequent application of the lubricant without the complete removal of previous coats was a potential for longer-term binding problems.

Conclusions. It was suspected that moisture entering the service air lines left a chloride-containing compound upon evaporation within the air check valves and pressure/vent assembly. This compound subsequently reacted with the check valve housing to lock the check valve poppets in a closed or partially closed position, decreasing the actual pressure being supplied to the pressure/vent valve.

Recommendations. It was recommended that an inspection be conducted to ensure that the service air check valves are operating properly prior to removal and servicing of the pressure/vent valve assembly. It was also recommended that the dry-film lubricant be checked to ensure that it meets specifications for the pressure/vent valve assembly.

Example 13: Corrosion Fatigue of Aircraft Nose Wheels. Four nose wheel failures were received by the laboratory for determination of the cause of failure. The wheels were fabricated from 2014-T6 aluminum material and were cold worked at the flange.

Investigation. Visual examination showed that the failure started in the tube well area on the wheel with serial number 31. The failure initiated in the flange fillet on wheels with serial numbers 67, 217, and 250. Figure 55(a) shows a typical example of these failures. Further visual examination of the wheel fractures indicated that failure progressed because of fatigue (Fig. 55b and c). There was a superficial indentation adjacent to the origin on wheel 31 (Fig. 55d), and there were superficial periodic blemishes on the fillet of nose wheels 67, 217, and 250 (Fig. 55e). The indentation on wheel 31 could have contributed to the cracking found in the tube well; however, the blemishes at the fillet of wheels 67, 217, and 250 were merely superficial and were not thought to be deleterious.

Scanning electron microscopy examination of the fractures showed that failure initiated by SCC or a corrosion pit on all failures examined. Figure 55(f) shows a typical example. The failures then progressed by fatigue.

A chemical analysis conducted on the submitted wheels showed that the wheels met the composition requirements for 2014 aluminum base material. A hardness survey indicated the wheels were in the T-6 tempered condition.

The wheels were examined by dye penetrant to determine if the remaining sections contained additional flaws. No additional flaws were seen on the wheels that had failed in the flange area. There was, however, one flaw area in the flange of the wheel that failed in the tube well. This flaw resembled a corrosion pit.

Conclusions. It was concluded that failure of nose wheels 67, 217, and 250 was caused by cracking due to SCC or pitting. The failures progressed by fatigue. Because failure occurred in the same general area on all three wheels, these locations are suspect as being underdesigned.

Recommendations. It was recommended that consideration be given to redesign of the nose wheel and that additional service data be accumulated in order to understand the contributing factors that result in failure of the wheel.

Example 14: Failure of Nose Gear Door Bolts. Nose gear door securing bolts were reported to be failing; three separate incidents were cited. One of these bolts, measuring 25 × 32 mm (1 × 1¹/₄ in.), was submitted for analysis. The bolt was a cadmium-plated, countersunk head type with a common screwdriver slot.

Investigation. Figure 56(a) shows the fracture face of the submitted bolt with the nut still attached. The fracture originated at a thread root and propagated across the cross section. An arrest pattern characteristic of fatigue that terminates at the smooth, featureless final overload area can be seen. The topography of the fracture was excessively rough and more granular than would be expected from pure mechanical fatigue; this indicated an allied corrosion mechanism.

Cracks other than the one leading to failure were observed in the bolt. The large crack at the bolt head shown in Fig. 56(b) was near separation. Metallographic examination of the bolt cross section showed many cracks typical of stress-corrosion damage. The scanning electron micrograph (Fig. 56c) of the bolt surface clearly showed extensive intergranular SCC.

Conclusion. The bolt failed by a combination of SCC and fatigue. The dual interaction of stress corrosion and mechanical fatigue is exemplified by the scanning electron photograph of the fracture surface (Fig. 56d), which shows characteristics of both cyclic fatigue and stress corrosion.

Recommendations. It was recommended that aerospace-quality fasteners meeting NAS 7104, NAS 7204, or NAS 7504 be used to replace the currently used fasteners.

Example 15: Failure of Tool Steel Pylon Attachment Stud. The failed pylon attachment stud illustrated in Fig. 57(a) was reportedly found during a routine walk-around inspection. Half of the stud was found lying on the apron under the aircraft.

Investigation. The stud exhibited gross localized corrosion pitting at several different areas on its surface. Light general rust was also evident. Severe pitting near the fracture location is illustrated in Fig. 57(b). The extent of the corrosion damage into the stud is shown on the fracture surface (Fig. 57c). The relatively clean and recent area of the fracture face is not entirely overload failure mechanism, but contains evidence of intergranular SCC as well as ductile dimples. Figure 57(d) is a scanning electron micrograph showing intergranular SCC on the fracture face.

The bolt is specified to be H-11 tool steel heat treated to 46 to 49 HRC (ultimate tensile strength: 1517 to 1655 MPa, or 220 to 240 ksi). Hardness measurements of 47 HRC confirmed the heat treatment to be correct. The protective coating was found to be an inorganic water-base aluminide coating having a coating thickness of 7.5 to 13 μm (0.3 to 0.5 mil). It was noted that the coating was of a nonuniform mottled nature.

Conclusions. It was concluded that the failure of the pylon attachment stud was caused by general corrosion followed by SCC. The stud was not adequately protected against corrosion by the coating.

Recommendations. It was recommended that the coating be applied to a thickness of 38 to 75 μm (1.5 to 3 mils) to provide long-time corrosion resistance. The coating must be either burnished or cured at 540 °C (1000 °F) to provide cathodic protection to the steel. Other coatings, such as cadmium or aluminum, were also recommended if a thinner coating is needed.

(a)

(b)

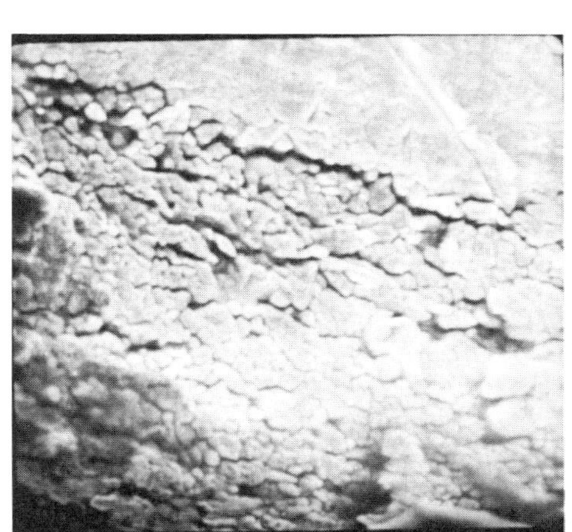
(c)

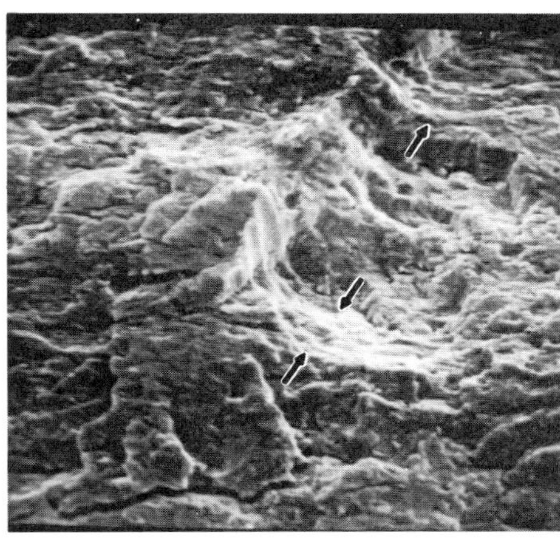
(d)

Fig. 56 Fracture surface (a) of failed cadmium-plated 1040 steel nose gear door bolt. The crack propagation pattern (arrow indicates the origin) and topography suggest both fatigue and corrosion. (b) Head of bolt showing cracking (arrow) that would lead to separation in a short time. 3×. (c) SEM of bolt surface showing extensive intergranular SCC. 650×. (d) SEM fractograph showing fatigue striations (arrow) interspersed with secondary cracking and evidence of SCC. 650×

Example 16: Corrosion of a Laser Mirror. A failed laser mirror and another complete mirror of the same construction were submitted to the laboratory for analysis. The laser mirror consisted of three layers of material brazed together to form channels through which the cooling water flows. Portions of the top and middle layers from an area of the mirror that was visually determined to be the most damaged were included in the parts forwarded for analysis (Fig. 58a).

Investigation. Samples were analyzed with light optical and scanning electron microscopy. Portions of the mirror were sent out for analysis of the base material and visually identified corrosion products.

A portion of the middle layer from area A was noted to exhibit the highest corrosion attack, with many visible areas where the corrosion had completely perforated the base material (Fig. 58b). The corrosion product, which appeared under the SEM as distinct granular particles (Fig. 58c), was shown by chemical analysis to contain molybdenum and copper with a trace of gold. The base material was analyzed as molybdenum with negligible alloying additions.

The primary mode of corrosion attack on the base material appeared to be intergranular (Fig. 58d), although uniform corrosion was also evident as a general thinning of the material cross section. The corrosion product was shown to adhere loosely to the base material, and it was found that a gentle wash with acetone could dislodge the particles. This finding gave rise to another possible source for material degradation—that of erosion-corrosion, which may in part attribute to the worn, rounded appearance of the modules under the SEM (Fig. 58c).

Scanning electron microscopy examination of the portion of the sample with nodules on both sides (middle layer) showed that the nodules on one side failed in a ductile/tensile mode (Fig. 58e), while those on the opposite side appeared to have failed in a compressive/shearing mode (Fig. 58f). These findings are consistent with the concept of micro-bowing expected to be present in this type of failure, that is, the water pressure forcing apart the two surfaces in the upper layer (tensile) while imparting compressive stresses to the layers below.

Portions of the mirror analyzed were seen to have areas in which the brazing alloy was corroded but the base material was relatively unaffected. This finding suggested the possibility of a galvanic couple between the brazing alloy and the base metal.

The sample mirror was radiographed and ultrasonically inspected. The radiograph did not show any areas of corrosion, with the ultrasonic c-scan able to image the rectangular pattern of pin-type spacers.

(a)

(b)

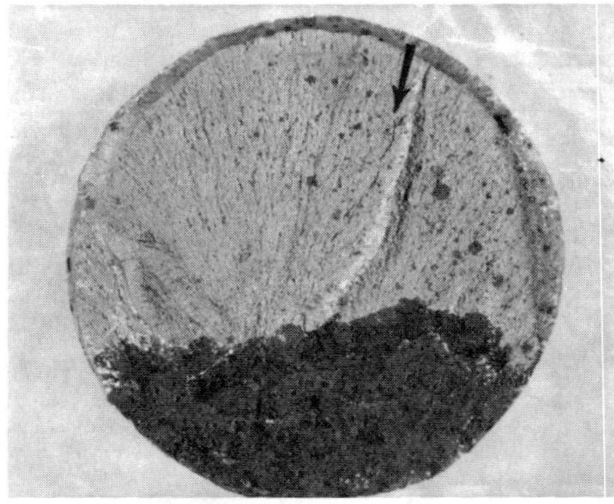
(c)

(d)

Fig. 57 H-11 tool steel pylon attachment stud (a) that failed by corrosion. (b) Gross pitting corrosion on the stud surface near the fracture site. Note the irregular, mottled appearance of the coating. (c) Fracture surface of the stud showing extent of corrosion within the stud (dark area). The arrow points to one major crack caused by stress corrosion. (d) SEM showing stress-corrosion damage on the fracture face. 315×

Conclusions. It was concluded that the corrosion attack noted sufficiently weakened the base material and the brazed joints, allowing catastrophic failure of the mirror due to the pressure of the cooling water.

Recommendations. It was recommended that the mirrors be cleaned of all corrosion products present as a result of past service conditions. The mirrors should be proof tested to determine if the residual structural integrity is sufficient for future operational requirements. It was recommended that the water system consisting of deionized water and formaldehyde be replaced with water having a low oxygen content and a cathodic inhibitor (oxygen scavenger).

Example 17: Failure of a Helicopter Rotor. Several rotor blade components were received for laboratory analysis. These included the horizontal hinge pin and the associated nut and locking washer (Fig. 27a, b, and c in the section "Corrosion of Airframes" in this article).

Investigation. Visual examination of the submitted parts revealed that the hinge pin, fabricated from 4340 steel, was broken and that the fracture face showed a flat beach mark pattern indicative of a preexisting crack (Fig. 27d and e). Also noteworthy on the hinge pin was a ding or dent in the threaded area that occurred on only one side (Fig. 27f).

Optical examination of the hinge pin showed that the beach mark or preexisting crack had progressed to approximately one-fourth the total cylinder cross-sectional area. Further examination of the area at the fracture initiation site revealed corrosion pits and secondary cracking on the outer circumference of the pin adjacent to the primary fracture face. The cracks ran parallel to the primary fracture and propagated from small pits (Fig. 27g).

Close scrutiny of the threaded area of the pin revealed an embedded thread that did not appear to come from the pin (Fig. 27h). A chemical analysis was conducted on the embedded thread and on an associated attachment to determine the origin of the thread. This analysis showed that the thread and nut were 4140 steel.

Scanning electron fractographic examination of the fracture initiation site strongly suggested that the fracture progressed by fatigue. Hardness measurements taken on the pin produced a value of 41 HRC, which was within the 39 to 43 HRC drawing requirements.

Conclusions. It was concluded that the failure of the horizontal hinge pin initiated at areas of localized corrosion pits. The pits in turn initiated fatigue cracks, resulting in a failure mode of corrosion fatigue. The embedded thread was the same material as the associated nut and probably came from the attachment nut.

Recommendations. It was recommended that all of the horizontal hinge pins be inspected to determine the current existence of cracks, corrosion pits, and general corrosion. Those pins that are determined to be satisfactory for further use should be stripped of cadmium, shot peened, and coated with cadmium to a minimum thickness of 0.0127 mm (0.0005 in.).

Example 18: Corrosion Failure of Wing Flap Hinge Bearings. Three wing flap hinge bearings were received by the laboratory for analysis (Fig. 59a). The bearings were fabricated from chromium-plated type 440C martensitic stainless steel.

Investigation. Visual analysis of the hardware showed that two of the three bearings were cracked. One of these two bearings, designated bearing 1 in Fig. 59(a), contained two separate cracks. One of these cracks took the path shown in Fig. 59(b), while the second crack progressed straight across the bearing cross section. Bearing 2 in Fig. 59(a) exhibited the crack path seen in Fig. 59(c). Further visual examination of the bearing crack fractures showed that the cracks were flat and slightly fibrous (Fig. 59d and e). The fracture face of bearing 2 differed slightly from the two fractures seen on bearing 1 in that there was a chip missing on the outer edge of the bearing.

Optical examination of the bearings revealed numerous corrosion pits on the inner diameter of the failed bearings and on bearing 3, which had not failed (Fig. 59f). Metallographic examination showed that the pitting progressed in an intergranular corrosive pattern (Fig. 59g). The microstructure was acceptable for type 440C martensitic stainless steel. Electron optical examination of the fractures by transmission electron microscopy produced fractographs that depicted an intergranular failure mechanism (Fig. 59h and i). Chemical analyses and hardness measurements of bearings 1 and 2 showed the material to be within compositional requirements for type 440C stainless steel and heat treated to 54 to 55 HRC.

Conclusion. The intergranular fracture pattern seen in the electron fractographs, coupled with the corrosion pits observed on the inner diameter of the bearings, strongly suggested that the failure initiated by pitting and progressed by SCC or hydrogen embrittlement from the plating operation.

Recommendations. It was recommended that the extent of the flap hinge bearing cracking problem be determined by using nondestructive inspection because it is possible to crack hardened type 440C during the chromium plating process. An inspection for pitting on the bearing inner diameter was also recommended. It was suggested that electroless nickel be used as a coating for the entire bearing. Electroless nickel, when heated to 345 °C (650 °F) for 1 h after plating, has wear resistance equivalent to that of chromium. A review of the chromium plating and baking sequence was also recommended to ensure that a source of hydrogen is not introduced during the plating operation.

REFERENCES

1. M.L. Bauccio, Case Studies of Material Deterioration and Failure, in *Materials Selection for Military Equipment (A Handbook of Critical Properties and Applications)*, National Association of Corrosion Engineers, 1987, p 1
2. L.M. Bland, Aircraft Structural Life Monitoring and the Problem of Corrosion, in *Aircraft Structural Fatigue*, Structures Report 363, Materials Report AR-000-724, Proceedings of a symposium held in Melbourne, Oct 1976, Aeronautical Research Laboratories, 1977, p 227-253
3. V.S. Agarwala, D.A. Berman, and G. Kohlhaas, Causes and Prevention of Structural Materials Failures in Naval Environments, *Mater. Perform.*, Vol 24 (No. 6), 1985, p 9-16

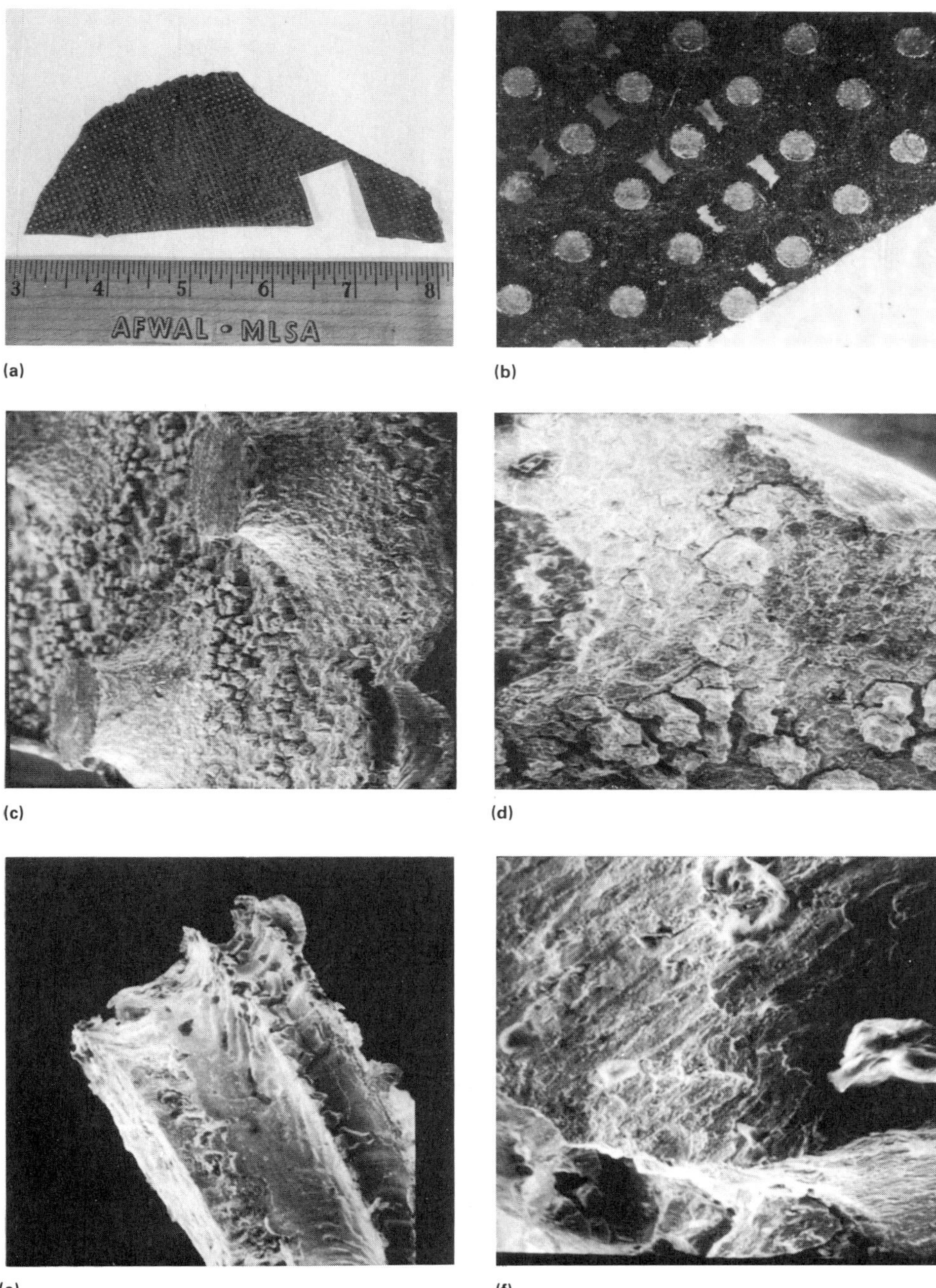

Fig. 58 Section of the most damaged area (a) of a failed laser mirror. (b) Section of area shown in (a) showing corrosion product and deterioration of the base metal. 5×. (c) Corrosion product and nodules in the cooling water channels. 30×. (d) SEM showing corrosion product and intergranular attack on the base metal. 40×. (e) Nodule on upper surface of middle layer of laser mirror showing failure in a ductile/tensile mode. (f) Nodule from lower surface of middle layer of the mirror showing failure due to compressive/shearing stresses

4. W.T. Kirkby, Examples of Aircraft Failure, in *Fracture Mechanics of Aircraft Structures*, AGARDograph 176, H. Liebowitz, Ed., Advisory Group for Aerospace Research and Development, 1974, p 8-13
5. W.R. Coleman, R.J. Block, and R.D. Daniels, Corrosion Problems in Aircraft Components—Case Studies of Failures, in *Proceedings of the 1980 Tri-Service Corrosion Conference*, AFWAL-TR-4019, Vol II, Air Force Wright Aeronautical Laboratories, 1981, p 241-270
6. Lockheed Grounds C-5B to Replace Fasteners, *Aviation Week Space Technol.*, Vol 123 (No. 13), 1985, p 26
7. W. Beck, E.J. Jankowsky, and P. Fischer, "Hydrogen Stress Cracking of High Strength Steels," Report NADC-MA-7140, Naval Air Development Center, 1971
8. D.A. Berman, The Effect of Baking and Stress on the Hydrogen Content of Cadmium Plated High Strength Steels, *Mater. Perform.*, Vol 24 (No. 11), 1985, p 36-41
9. D.E. Muehlberger and J.J. Reilly, "Improved Equipment Productivity Increases Applications for Ion Vapor Deposition of Aluminum," Report 830691, Society of Automotive Engineers, 1983
10. "Coating, Aluminum, Ion Vapor Deposited," MIL-C-83488B, U.S. Air Force Materials Laboratory, Dec 1978
11. D.P. Lahiri and A.V. Reddy, Exfoliation Corrosion in Aircraft Structural Member Made of Al-Cu-Mg Alloy, *Trans. Indian Inst. Met.*, Vol 35 (No. 5), 1982, p 456-460
12. R.N. Miller, Inhibitive Sealing Compounds and Coating Systems Solve Aircraft Corrosion Problems, *SAMPE J.*, April/May 1970, p 54-58
13. M.H. Trimble, The Need for Improved Materials in Integral Aircraft Fuel Tanks, in *Materials and Processes in Service Performance*, Society for the Advancement of Material and Process Engineering, 1977, p 3-8
14. N. Kackley and M. Levy, "Combatting Corrosion in Army Aircraft," Paper presented at Corrosion/87, San Francisco, CA, National Association of Corrosion Engineers, March 1987
15. "Materials and Processes for Corrosion Prevention and Control in Aerospace Weapons Systems," MIL-STD-1568, Revision A, Air Force Wright Aeronautical Laboratories, Oct 1979
16. *Aircraft Corrosion: Causes and Case Histories*, Vol 1, AGARD Corrosion Handbook, AGARD-AG-278, Advisory Group for Aerospace Research and Development, 1985
17. "Finishes for Ground Electronic Equipment," MIL-F-14072, U.S. Army Electronics Research and Development Command, June 1986
18. M.L. Bauccio, "Properties and Applications of Beryllium and Beryllium-Copper Aerospace Materials," Paper 109, presented at Corrosion/84, New Orleans, LA, National Association of Corrosion Engineers, 1984
19. J. Fielding, Stress Corrosion Cracking and the Aircraft Industry, *J. Inst. Met.*, Vol 101 (No. 9), 1973, p 238-240
20. M.L. Bauccio, Corrosion Problems in the U.S. Army BLACK HAWK Helicopter, in *Proceedings of the 1980 Tri-Service Corrosion Conference*, AFWAL-TR-81-4019, Vol I, Air Force Wright Aeronautical Laboratories, 1981
21. T.A. Roach, Aerospace High Performance Fasteners Resist Stress Corrosion Cracking, *Mater. Perform.*, Vol 23 (No. 9), 1984, p 42-45
22. V.R. Pludek, *Design and Corrosion Control*, Macmillan, 1977, p 68
23. W. Barrois, Service Failures and Laboratory Tests, in *Fracture Mechanics of Aircraft Structures*, AGARDograph 176, H. Liebowitz, Ed., Advisory Group for Aerospace Research and Development, 1974, p 325-345
24. P.M. Toor, Inhibition of Stress Corrosion Cracking in the Design of Aircraft Structures, in *Materials and Processes—Continuing Innovations*, 28th National SAMPE Symposium and Exhibition, Society for the Advancement of Material and Process Engineering, 1983, p 973-981
25. H. Lajain, Corrosion Protection Schemes for Aircraft Structures: Some Examples for the Corrosion Behavior of Al-Alloys, in *Aircraft Corrosion*, AGARD-CP-315, Proceedings of the 52nd Meeting of the AGARD Structures and Materials Panel, Advisory Group for

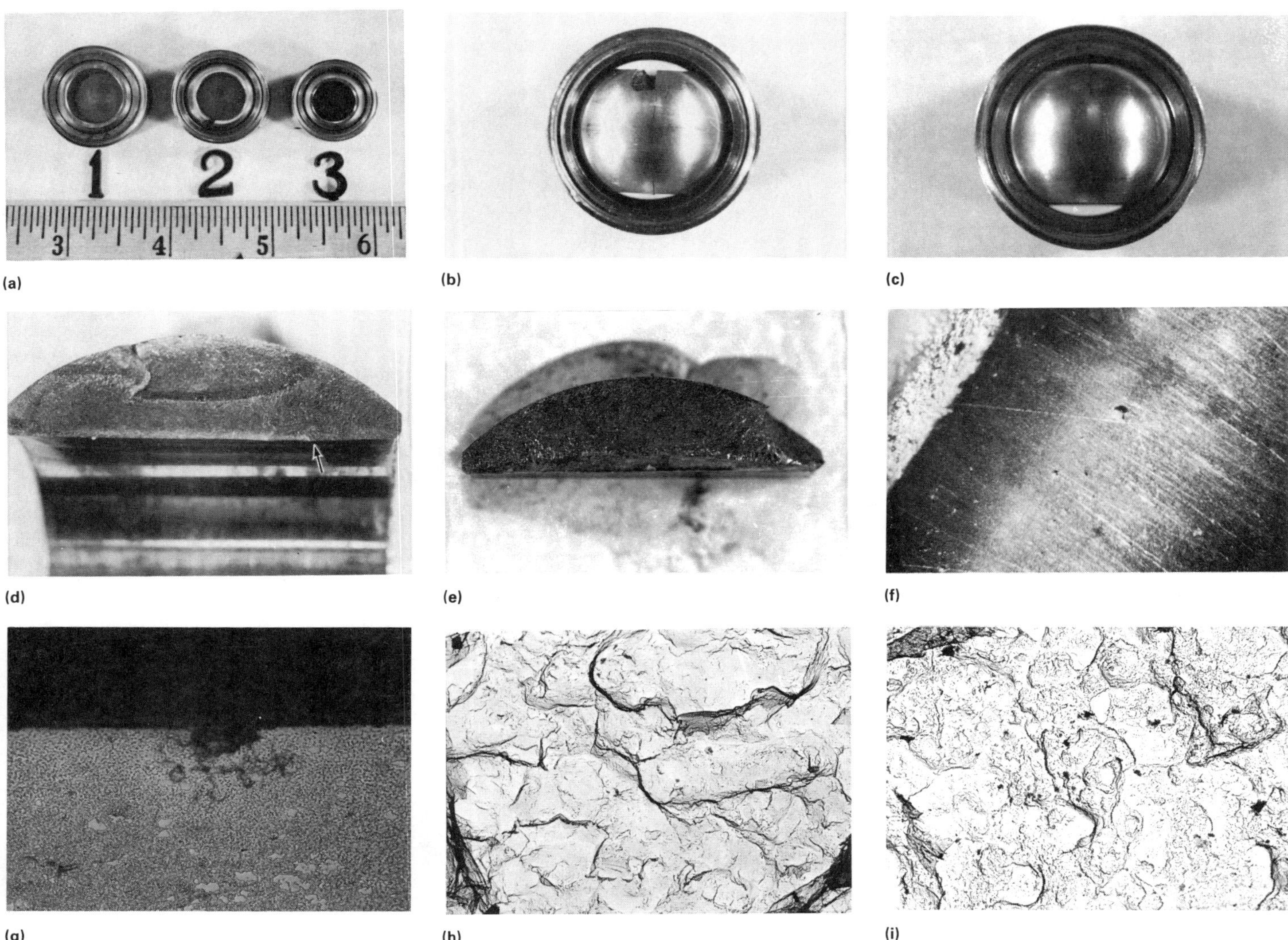

Fig. 59 Cracked type 440C stainless steel (a) aircraft wing flap hinge bearings. (b) Crack configuration of bearing 1 from (a). (c) Crack configuration of bearing 2 from (a). (d) Fracture surface of second crack in bearing 1. Arrow shows the probable fracture origin. 2.5×. (e) Fracture surface of bearing 2. The crack front was abraded, so the area of origin would not be clearly defined. 2.5×. (f) Corrosion pits on the inside surface of one of the bearings. All three bearings submitted had this type of damage. 10×. (g) Cross section of corrosion pit on one of the bearings; note the intergranular attack. 265×. (h) SEM fractograph of bearing 1 showing intergranular mechanism of fracture. 2690×. (i) SEM fractograph of bearing 2 showing intergranular fracture mechanism. Corrosion products are evident on the intergranular facets. 2690×

Aerospace Research and Development, 1981, p 14-1 to 14-16

26. W.M. Ryan, Filiform Corrosion on Painted Aluminum Alloy Surfaces, in *The Enigma of the Eighties: Environment, Economics, Energy*, Vol 24, Book 1 of 2, Proceedings of the 24th National SAMPE Symposium and Exhibition, Society for the Advancement of Material and Process Engineering, 1979, p 638-648
27. R.B. Waterhouse, *Fretting Corrosion*, Pergamon Press, 1972, p 1-5 and 36-43
28. Aeronautical Information Report 47, Society of Automotive Engineers, 1956
29. M. Levy and P.A.M. Farrell, The CPC Mission, U.S. Army Corrosion/Deterioration Problems, U.S. Army ManTech J., Vol 10 (No. 4), 1985, p 5-14
30. R.J.H. Wanhill, J.J. DeLuccia, and L.B. Vogelesang, Environmental Fatigue of Aluminum Alloy Structural Joints, in *Seventh International Light Metals Conference* (Leoben, Vienna), June 1981, p 92-93
31. *Military Standardization Handbook, Corrosion and Corrosion Prevention*, Metals, MIL-HDBK-729, Department of Defense, U.S. Army Materials Technology Laboratory, 1983, p 67
32. R.N. Miller, The Evolution of the Corrosion Free Airplane, *Mater. Perform.*, Vol 25 (No. 3), 1986, p 57-59
33. B. Rosales and M. Del Carmen, "The Predominance of Microbial Growth Versus Metallurgical Characteristics of the Corrosion of 2024 Al Alloy Through Electrochemical Data," Paper 124, presented at Corrosion/86, National Association of Corrosion Engineers, 1986
34. *Wing Tank Microbial Growth and Corrosion*, Proceedings of the Boeing/Airline Regional Conference, Boeing Commercial Airplane Company, 1980
35. "Corrosion Control for Aircraft," Advisory Circular 43-4, U.S. Department of Transportation, Federal Aviation Administration, 1973, p 10
36. "Lessons Learned Bulletin: Corrosion," Directorate of Systems Support, Air Force Acquisition Logistics Center, 1985, p 83
37. L.D. Griffin and D. Latterman, C-141A Service Experience—Materials and Processes, *SAMPE J.*, Vol 14 (No. 2), 1978, p 9-16
38. J.S. Leak, Corrosion—A Study of Recent Air Force Experience, in *Proceedings of the NACE 26th Conference*, National Association of Corrosion Engineers, 1970, p 497-503
39. L.D. Griffin and D. Latterman, C-141A Service Experience—Materials and Processes, *SAMPE J.*, Vol 14 (No. 2), 1978, p 9-16
40. V. Wigotsky, Metal Bonding Slowly Winning Over Reluctant Designers, *Aerospace Am.*, Vol 22 (No. 9), 1984, p 82-86
41. R.E. Horton, Demonstration of an Improved Method for Repair of Bonded Aircraft Structure, in *The Enigma of the Eighties: Environment, Economics, Energy*, Vol 24, Book 1 of 2, Proceedings of the 24th National SAMPE Symposium and Exhibition, Society for the

Advancement of Material and Process Engineering, 1979, p 659-668

42. M.H. Kuperman, Structural Adhesive Bond Repair of Aircraft Flight Control Surfaces, in *The Enigma of the Eighties: Environment, Economics, Energy*, Vol 24, Book 2 of 2, Proceedings of the 24th National SAMPE Symposium and Exhibition, Society for the Advancement of Material and Process Engineering, 1979, p 1126-1139
43. H.J. Singletary, "Aircraft Service Life Extension Through Corrosion Control," Paper presented at Corrosion/87, San Francisco, CA, National Association of Corrosion Engineers, March 1987
44. "Corrosion Preventive Characteristics of Aircraft Greases," Aeronautical Information Report 40, Society of Automotive Engineers, 1955
45. J.M. Margolis, Advanced Composites for Primary Exterior Aircraft Structures, *Plast. Des. Forum*, Vol 10 (No. 5), 1985, p 78-84
46. M.A. Steinberg, Materials for Aerospace, *Sci. Am.*, Vol 255 (No. 4), 1986, p 67-72
47. Corrosion Conference, Boeing/Airplane Regional Conferences, Boeing Commercial Airplane Company, 1982, p IV.9-IV.10
48. C.R. Pye, Recorder's Report—Session II, in *Aircraft Corrosion*, Conference Proceedings 315, Advisory Group for Aerospace Research and Development, 1981, p R2-1-R2-2
49. A.E. Hohman, *Design and Manufacturing Practices to Minimize Corrosion in Aircraft*, R-714, Advisory Group for Aerospace Research and Development, March 1984, p 7.1-7.5
50. J.J. DeLuccia and S.F. Saletros, Naval Air Systems Command Corrosion Control Program, in *Proceedings of the 1980 Tri-Service Conference on Corrosion*, Vol II, F.H. Meyer, Ed., AFML-TR-81-4019, Air Force Materials Laboratory, Nov 1980
51. A.A. Baker and M.M. Hutchison, Fibre Composite Reinforcement of Cracked Aircraft Structures, in *Aircraft Structural Fatigue*, Structures Report 363, Materials Report 104, Proceedings of a symposium held in Melbourne, Oct 1976, Aeronautical Research Laboratories, 1977, p 419-457
52. M.L. Bauccio, "Hot Corrosion: A Review of Chemical Mechanisms and Protective Coatings," Paper 109, presented at Corrosion/82, National Association of Corrosion Engineers, 1982
53. *Airframe and Powerplant Mechanics Powerplant Handbook*, U.S. Department of Transportation, Federal Aviation Administration, 1976, p 20
54. "Corrosion Control for Aircraft," Advisory Circular 43-4, U.S. Department of Transportation, Federal Aviation Administration, 1973, p 14
55. *Aircraft Corrosion: Causes and Case Histories*, Vol 1, AGARD Corrosion Handbook, AGARD-AG-278, Advisory Group for Aerospace Research and Development, 1985
56. M. Gell, G.R. Leverant, and C.H. Wells, The Fatigue Strength of Nickel-Base Superalloys, in *Achievement of High Fatigue Resistance in Metals and Alloys*, STP 467, American Society for Testing and Materials, 1970, p 113-153
57. P.R. Belcher, R.J. Bird, and R.W. Wilson, "Black Plague" Corrosion of Aircraft Turbine Blades, in *Hot Corrosion Problems Associated With Gas Turbines*, STP 421, American Society for Testing and Materials, 1967, p 123-145
58. H. Grimm and M.P. Malik, Corrosion of Aircraft Structures: Significant Problems, Control Methods, and Possible Improvements, in *Proceedings of the Sixth European Congress on Metallic Corrosion (Eurocorr '77)*, Society of the Chemical Industry, 1977, p 595-605
59. P.A. Bergman, Corrosion Problems in Aircraft Gas Turbines, in *Proceedings of the Air Force Materials Laboratory Fiftieth Anniversary Technical Conference on Corrosion of Military and Aerospace Equipment*, AFML-TR-67-329, Air Force Materials Laboratory, 1967, p 1413-1431
60. R.R. Dils and P.S. Follansbee, Dynamic Oxidation and Corrosion in Power Generating Units, *Corrosion*, Vol 33 (No. 11), 1977, p 385-402
61. New Coating Combats "Hot Corrosion" in Gas Turbine Buckets, *Anti-Corros. Methods Mater.*, Vol 27 (No. 12), 1980, p 7
62. R.A. Miller and C.E. Lowell, "Failure Mechanisms of Thermal Barrier Coatings Exposed to Elevated Temperatures," Technical Memorandum 82905, National Aeronautics and Space Administration, 1982
63. D.F. Baxter, Jr., Automotive Technology: Developments in Coated Steels, *Met. Prog.*, Vol 129 (No. 6), 1986, p 31-35
64. J.H. Weber, High Temperature Oxide Dispersion Strengthened Alloys, in *The 1980's—Payoff Decade for Advanced Materials*, Vol 25, Proceedings of the National SAMPE Symposium and Exhibition, Society for the Advancement of Material and Process Engineering, 1980, p 752-764
65. D.R. Coupland, C.W. Corti, and G.L. Selman, The PGM Concept: Enhanced Corrosion Resistant Superalloys for Industrial and Aerospace Applications, in *Proceedings of the Conference on Behavior of High Temperature Alloys in Aggressive Environments*, Petten, Oct 1979, p 525-536
66. H.R. Gray, Hot-Salt Stress-Corrosion of Titanium Alloys as Related to Turbine Engine Operation, in *Proceedings of the Second International Conference on Titanium Science and Technology*, Vol 4, 1973, p 2627-2638
67. J.R. Myers, H.B. Bomberger, and F.H. Froes, Corrosion Behavior and Use of Titanium and Its Alloys, *J. Met.*, Oct 1984
68. S.P. Rideout, M.R. Louthan, Jr., and C.L. Selby, Basic Mechanisms of Stress Corrosion Cracking of Titanium, in *Stress-Corrosion Cracking of Titanium*, STP 397, American Society for Testing and Materials, 1966, p 137-151
69. R.E. Duttweiler, R.R. Wagner, and K.C. Antony, An Investigation of Stress-Corrosion Failures in Titanium Compressor Components, in *Stress-Corrosion Cracking of Titanium*, STP 397, American Society for Testing and Materials, 1966, p 152-178
70. T.R. Strobridge, J.C. Moulder, and A.F. Clark, "Titanium Combustion in Turbine Engines," Reports FAA-RD-79-51 and NBSIR 79-1616, U.S. Department of Transportation, Federal Aviation Administration, July 1979
71. C.W. Elrod, The Combustion of Titanium in Gas Turbine Engines, in *Proceedings of the 1980 Tri-Service Conference on Corrosion*, AFWAL-TR-81-4019, Vol II, Air Force Wright Aeronautical Laboratories, 1981, p 55-92
72. B.A. Manty, V.G. Anderson and H.M. Hodgens, *Blade Tip Treatment—Titanium Alloy Compressor Blades*, AFWAL-TR-80-4149, Air Force Wright Aeronautical Laboratories, 1980, p 119
73. R.B. Waterhouse, *Fretting Corrosion*, Pergamon Press, 1972, p 36
74. R.L. Johnson and R.C. Bill, Fretting in Aircraft Turbine Engines, in *Fretting in Aircraft Systems*, AGARD-CP-161, Advisory Group for Aerospace Research and Development, 1975
75. R.G. Mahorter, Cold Corrosion in Aircraft Engines, in *Proceedings of the 1976 Tri-Service Conference on Corrosion*, S.J. Ketcham, Ed., Naval Air Development Center, 1976, p 257-274
76. R.G. Mahorter, Current and Anticipated Materials Problems of Aircraft Engines in Marine Atmospheres, in *Proceedings of the 1974 Conference on Gas Turbine Materials in the Marine Environment*, MCIC-75-27, Metals and Ceramics Information Center, 1975, p 1-10
77. E.J. Hammersley, Corrosion in Airframes, Power Plants and Associated Aircraft Equipment, in *The Theory, Significance, and Prevention of Corrosion in Aircraft*, AGARD-LS-84, Advisory Group for Aerospace Research and Development, 1976, p 4-11
78. M.P. Malik, Corrosion of Aircraft Structures—Significant Problems, Control Methods and Possible Improvements, in *Proceedings of the 6th European Congress on Metallic Corrosion (Eurocorr '77)*, Society of the Chemical Industry, 1977, p 602-603
79. R. Baboian *et al.*, Aluminum Coating of Steel, in *Surface Cleaning, Finishing and Coating*, Vol 5, 9th ed., Metals Handbook, American Society for Metals, 1982, p 339
80. D.E. Muehlberger, Presentation (seminar) given on the technology of ion-vapor-deposition of aluminum, Bell Helicopter TEXTRON, Inc., May 1986
81. J.E. Newhart, "Cold" Corrosion Evaluations Resulting from Full-Scale, Controlled Environment, Turbine Engine Testing, in *Proceedings of the 1972 Tri-Service Conference on Corrosion*, MCIC-73-19, M.M. Jacobson and A. Gallaccio, Ed., Metals and Ceramics Information Center, 1973, p 237-238
82. S.D. Elrod, "Service Evaluation of Aluminum-Brazed Titanium (ABTi) Jet Engine Tailpipe Extensions," NASA Contractor Report 3617, National Aeronautics and Space Administration, Scientific and Technical Information Branch, 1982
83. L.C. Lipp, "Halogenated Solvent-Induced Corrosion in Hydraulic Systems," Preprint 78-AM-4A-2, American Society of Lubrication Engineers, 1978
84. H. Weltman, Corrosion Characteristics and Control of Mixed Hydrazine Fuel, in *Proceedings of the 1980 Tri-Service Conference on Corrosion*, AFWAL-TR-4019, Air Force Wright Aeronautical Laboratories, 1981, p 343-358
85. B. Cohen, Corrosion Fatigue in the Aerospace Industry, in *Corrosion Fatigue: Chemistry, Mechanisms and Microstructure*, National Association of Corrosion Engineers, 1972, p 65-83

Corrosion in the Aerospace Industry

CORROSION CONTROL is an essential part of the design of aerospace hardware, because a corrosion failure could result in the loss of a satellite, a booster, or the space shuttle orbiter. Of the three, satellites experience the most benign environments; they are usually handled in clean rooms with controlled humidity from the time they are built until they are launched. In low earth orbit, the corrosion effects of oxygen and moisture no longer exist, although some surface erosion from atomic oxygen can destroy optical (thermal control) surfaces. At the other extreme, a booster, such as the solid rocket motor case of the space shuttle system, becomes immersed in seawater after launch and must be pulled from the ocean and thoroughly cleaned to avoid saltwater corrosion. A solid rocket case is expected to be reused up to 19 additional times.

The space shuttle orbiter was originally designed to be maintenance free for 10 years. From a corrosion standpoint, this is a rather severe requirement for the highly corrosive Cape Canaveral environment. Because of unexpected delays in launching schedules, there is a high probability that shuttle orbiters will see service in excess of 20 years. Corrosion and structural inspection plans have recently been initiated to ensure the safety of the airframe as well as the systems, which are subjected to highly corrosive fluids. In these systems, corrosion failure from perforation could result in explosions or fire, or corrosion deposits in valves could cause malfunctions, possibly resulting in the loss of the spacecraft. Therefore, control of corrosion is considered to be vital for the safety and success of aerospace hardware.

Applicable specifications imposed by the National Aeronautics and Space Administration (NASA) on subcontractors to control or restrict material selection and to control corrosion are:

- SE-R-0006: NASA-JSC Requirements for Materials and Processes
- MSFC-SPEC-250(1): Protective Finishes for Space Vehicle Structures and Associated Flight Equipment, General Specification for
- MSFC-SPEC-522A: Design Criteria for Controlling Stress Corrosion Cracking
- NHB-8060.1: Flammability, Odor, and Offgassing Requirements and Test Procedures for Materials in Environments That Support Combustion
- JSC-30233: General Specification—Space Station Requirements for Materials and Processes

The following sections will detail the steps taken by engineering, design, and material and process groups to avoid spacecraft corrosion. Case histories of failures and their solutions will also be reviewed.

Corrosion of Manned Spacecraft

L.J. Korb
Rockwell International
Space Transportation & Systems Division

The prevention or control of the corrosion of manned spacecraft presents a wide variety of challenges to the corrosion engineer for at least four reasons. First, a manned spacecraft, such as the space shuttle orbiter, has structures and systems operating at temperatures ranging from −253 °C (−423 °F) for liquid hydrogen tanks to approximately 1455 °C (2650 °F) for critical metallic pressure ports of the nose cap. Several hundred different alloys are required, including the light metals, such as beryllium, magnesium, aluminum, and titanium; steels, such as low-alloy, tool, corrosion resistant, precipitation hardenable, and maraging; nickel and nickel-base alloys, including pure nickel, Monel alloys, Inconel alloys, and other nickel- and cobalt-base superalloys; the refractory metals, principally niobium and molybdenum; the copper-base alloys, including pure coppers, beryllium coppers, bronzes, and brasses; the precious metals; and a variety of plating alloys. Quite often, intimate knowledge of the corrosion behavior of these alloys in the intended environment is simply lacking.

Second, because weight is so critical for manned spacecraft, the alloys were often chosen for their inherent high strength-to-density ratios rather than for their corrosion resistance. Early studies indicated that the savings per pound of weight in the orbiter vehicle structure or systems amounted to well over $30 000 during the mission life. Therefore, emphasis was directed toward corrosion control coatings for airframe structures rather than the selection of corrosion-resistant alloys. In fluid systems, however, material compatibility played an important role.

Third, the various parts of a spacecraft are subject to a wide variety of environments in addition to the temperature regimes discussed above. The seacoast exposure at Cape Kennedy, for example, is very severe because of the heat, high humidity, salt air, and the daily condensation of dew onto the structures. Fluid systems, on the other hand, are exposed to very aggressive chemicals: hot gaseous oxygen, nitrogen tetroxide, high-pressure hydrogen, hydrazine, and ammonia.

Finally, corrosion protection often must be subordinated to other critical functions, such as the spacecraft electrical grounding requirements. Often, where a corrosion engineer may wish to isolate a galvanic couple electrically by using an insulator, the requirements for grounding to ensure lightning protection and control of electromagnetic interference must overrule. Corrosion-protective measures must then be taken for the area of the grounded connection.

Early manned spacecraft, such as the Mercury, Gemini, Apollo, and Sky Lab programs, were each designed for a single launch. The Mercury, Gemini, and Apollo capsules parachuted into the ocean and could not be refurbished economically. The Skylab, having no reentry capability, broke up and burned at the end of its useful life. The major emphasis in corrosion protection for these vehicles was to ensure freedom from general corrosion and stress corrosion for the 3 to 5 years preceding launch. In the case of the Apollo, an in-depth age life and corrosion reevaluation was made to ensure a 10-year operational life from its manufacture to its use on Skylab missions.

The space shuttle orbiter was originally designed to avoid corrosion for a 10-year life, and it was anticipated that it would fulfill its 100-mission launch capability during this period. Because of unforeseen launch problems, some orbiters have experienced less than 10 missions while approaching their 10-year design life. Reevaluations will be made to extend this life to 20 years and to maintain the corrosion protection systems as necessary. Because of its multiple-launch capability and its long design life, the space shuttle orbiter is unquestionably the most challenging of the manned spacecraft from a corrosion control standpoint. Its corrosion protection and the problems experienced will be discussed in detail. However, because other manned space vehicles have experienced some unusual corrosion problems, many of which involved unknown incompatibilities at that time, this article will also cover representative case histories of corrosion of several manned spacecraft.

Operational Mission of the Space Shuttle Orbiter

The space shuttle system is designed to provide an economical system for transporting personnel and supplies into low earth orbit. The system (Fig. 1) consists of four major elements: the solid rocket boosters, the space shuttle main engines, the external tank, and the space shuttle orbiter. Elements other than the orbiter itself are discussed in the section "Corrosion of Space Boosters and Space Satellites" in this article. The system is capable of launching up to 29 500 kg (65 000 lb) of payload into orbit and returning up to 14 500 kg (32 000 lb) of payload from orbit (Ref 1).

Nearly 89% of the thrust for launch is provided by the solid rocket boosters. The remaining 11% is delivered by the main engines, which burn hydrogen with oxygen supplied by the external

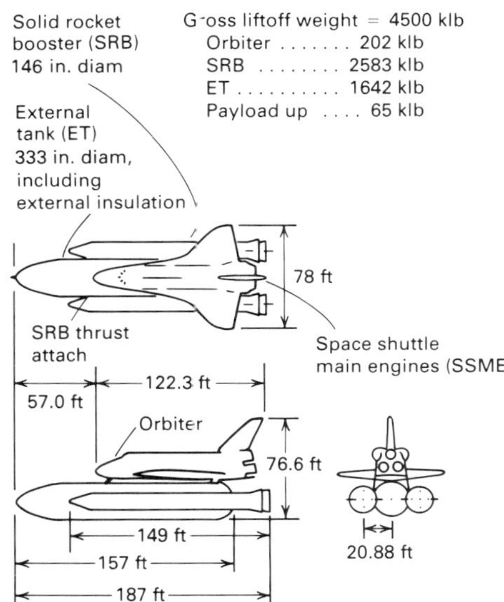

Fig. 1 Space shuttle system. The orbiter weight includes the 165-klb orbiter (including engines) plus 37-klb crew, consumables, and propellants. The external tank diameter includes external insulation. klb = 1000 lb

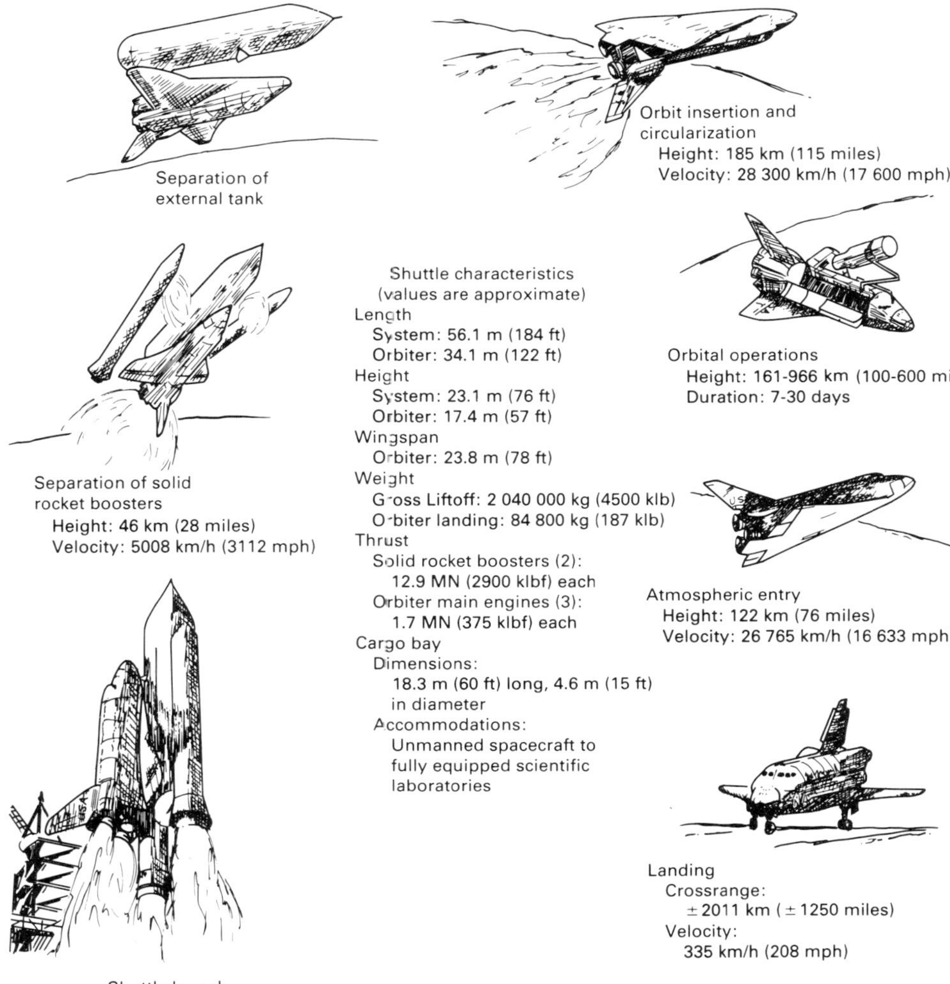

Fig. 2 Profile of shuttle mission. Each shuttle orbiter can fly a minimum of 100 missions and carry as much as 29 500 kg (65 000 lb) of cargo and up to seven crew members into orbit. It can return 14 500 kg (32 000 lb) of cargo to earth. klb = 1000 lb; klbf = 1000 lbf

tank. The orbiter, launched vertically in a piggyback position on the external tank, houses the astronauts, the main engines, and the payloads.

The orbiter must function as both a spacecraft and an aircraft. During entry from orbit, it must be protected from temperatures exceeding 1260 °C (2300 °F) on its lower fuselage and in excess of 1455 °C (2650 °F) along the leading edges and nose cap. At an altitude of approximately 46 km (150 000 ft), the orbiter will slow to about eight times the speed of sound and will pass through its maximum heating. At 15 km (50 000 ft), the orbiter will be maneuvered aerodynamically to land as a glider. A typical mission profile is shown in Fig. 2.

The requirement to achieve a minimum-weight orbiter (68 000 kg, or 150 000 lb, dry weight) has necessitated use of the most efficient structural materials and processes. The requirement for 100-mission reuse has extended advancements in thermal protection materials well beyond the state-of-the-art existing at the inception of the design.

Corrosion Control Program: Space Shuttle Orbiter

The key to a successful corrosion control program for the space shuttle orbiter was to develop sound technical and management programs. Although the major structural parts of the orbiter, such as the wings, tail, fuselage, and cabin, were manufactured by only a few companies, it was estimated that more than 20 000 suppliers were responsible for providing systems and parts for the vehicle. It was necessary to review and control all orbiter parts to provide the high levels of reliability required.

The material and process management program consisted of the following key elements:

- A material and process group in engineering
- A drawing review system requiring sign-off by a materials engineer
- A tracking system for all materials
- An orbiter materials and process control specification
- A corrosion-control and finishes specification
- A stress-corrosion control plan

Each material application was reviewed by a qualified material and process engineer who had sign-off authority on drawing, engineering orders, and material rework dispositions. A material tracking system was set up at the inception of the program to prevent 12 material-related hazards from occurring on the orbiter. These include controls for atmospheric corrosion and stress corrosion, fluid and propellent incompatibilities, age life, flammability, toxicity, offgassing, and condensation of volatile condensible matters. Materials and finishes were identified, evaluated, and, when accepted, entered into a computer. Material identification even included those materials used in minute quantities, such as the ink used to stamp part numbers or the nearly invisible cetyl alcohol lubricants on fasteners. A master directory (index) of the behavior of each material in the 12 hazardous categories was maintained and used as a reference.

In each hazardous category, a series of encoded, acceptable engineering approaches for each "buyoff" was listed to assist the engineer. For example, a part may be made from a material having a stress-corrosion threshold of 50% of its tensile yield strength, yet be acceptable because:

- It is adequately coated
- It experiences no significant tensile stresses in the critical stress-corrosion direction (including residual and installation stresses)
- It is in a benign environment, such as the cabin

In a few cases, the complexity of the part, such as a motor, precluded a separate evaluation of each material, and the entire configuration was qualification tested in its intended-use environment to avoid these hazards.

A Material and Process Control Specification (MC999-0096) was placed on all major subcontractors. A similar specification controlled parts that were designed and manufactured in-house. These specifications included controls for fluid systems compatibility, stress corrosion, atmospheric corrosion, and galvanic corrosion. The controls imposed are summarized below.

Control of Fluid Systems Compatibility. A fluid systems compatibility analysis is required that covers all fluids and materials used in the system, such as testing, processing, inspection, and operation, along with known or expected trace contaminants. Fluid system compatibility refers to interaction problems involved with materials and the liquid or gaseous subsystems. The problems experienced generally fall into the following categories:

- *Autoignition:* Spontaneous ignition of the material or the fluid
- *Impact ignition:* Ignition brought about by shock or impact within the fluid
- *Catalytic reaction:* Reactions such as the catalytic decomposition of the fluid
- *Material degradation:* This includes such phenomena as chemical attack, corrosion, galvanic corrosion, stress corrosion, hydrogen embrittlement, and crack growth acceleration with metals and includes embrittlement, abnormal swelling, leaching of plasticizers, ultraviolet degradation, and so on, with nonmetallic materials
- *Fluid degradation:* Reactions in which the physical or chemical characteristics of the fluid are altered
- *Potential ignition:* Ignition due to proximity to electrical ignition sources

Materials selection was required to minimize the compatibility problems with the fluid systems. Material-fluid combinations that result in autoignition, impact ignition, or another catastrophic mode of failure were not permitted.

The use of electrical and electronic components exposed to nondielectric fluid systems was avoided. Buyer approval was required prior to using electronic components in hazardous fluid systems.

Metallic materials listed in Appendix I of MC999-0096 are rated for compatibility with gaseous oxygen (GOX), liquid oxygen (LOX), nitrogen tetroxide (N_2O_4), hydrazine (N_2H_4), monomethyl hydrazine (MMH), and low-pressure ($\leq$3.1 MPa, or 450 psi) and high-pressure ($>$3.1 MPa, or 450 psi) hydrogen. Nonmetallic materials listed in Appendix II of MC999-0096 are rated for compatibility with low- and high-pressure GOX, LOX, N_2O_4, N_2H_4, MMH, liquid hydrogen, and hydraulic fluid.

Materials that are compatible and noncompatible with titanium are listed in MF0004-018.

Lubricants for static service with special fluids (application to elastomers, metals, and threads) are:

Fluid	Lubricant
Ammonia	Krytox 240 AC; Braycoat 3L-38RP; Braycoat 815Z oil
Deionized water	Krytox 240 AC; Braycoat 3L-38RP; Braycoat 815Z oil
Freon 21	DC F-6-1101
FC 40	DC F-6-1101

Lubricants for dynamic service with special fluids must be resolved on an individual basis. Use of the above lists did not absolve the seller of full responsibility for verifying compatibility under the particular design conditions used by his fluid system.

Control of Stress Corrosion. The subcontractor was required to prepare a stress-corrosion plan utilizing MSFC Specification 522A as a guideline for controlling stress corrosion and to take the actions necessary to prevent such failures. Wherever possible, the supplier was required to select materials that are either not susceptible to stress corrosion or have a high resistance to stress corrosion in the anticipated life cycle environment. Where susceptible materials were used, the supplier was required, at the minimum, to take the following actions to reduce stress-corrosion problems to the extent feasible:

- Select less susceptible alloys, tempers, or clad products
- Reduce sustained stress levels on the part below stress-corrosion threshold levels, especially in the more susceptible short-transverse grain direction
- Protect the part from the detrimental environment by hermetically sealing or coating the part or by inhibiting the environment (closed system)
- Avoid or reduce residual stresses in parts or assemblies by stress relieving, by avoiding interference fits, or by shimming assemblies
- Avoid galvanic couples, which may tend to accelerate stress corrosion
- Provide for regular inspection of parts to determine surface flaws and cracking during the life cycle of the part
- Improve the surface quality by reducing surface roughness and/or increasing surface compressive stresses
- Avoid the use of titanium in contact with silver, silver-plated material, or silver-plated fasteners, such as silver-plated A-286 nuts

Control of Galvanic Corrosion. Dissimilar metals were not to be used in intimate contact unless they were suitably protected against gal-

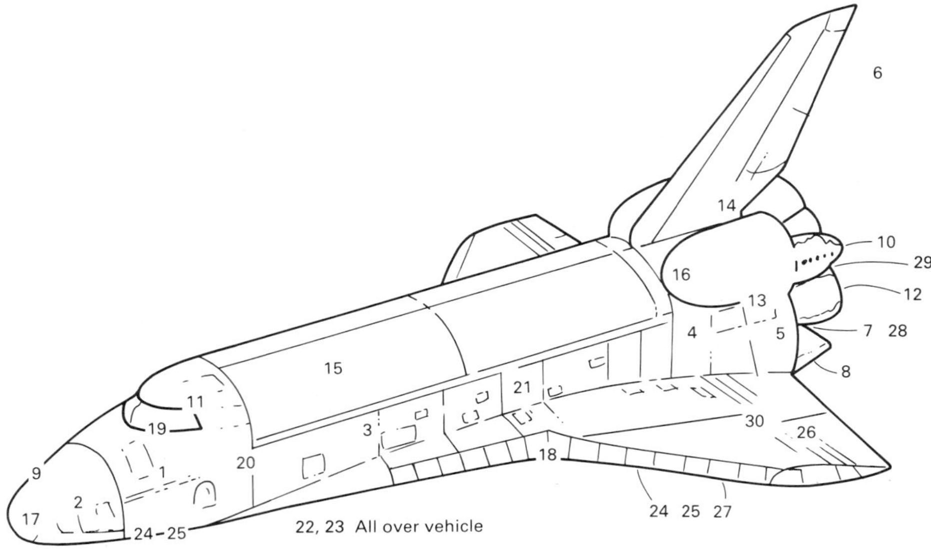

1 Cabin: aluminum alloy 2219
2 Forward fuselage: aluminum alloy 2024
3 Mid fuselage: aluminum alloys 2124, 7075
4 Aft fuselage: aluminum alloy 2124
5 Aft heatshield: aluminum alloy 2124
6 Tail: aluminum alloys 2124 and 2024
7 Engine mounted heat shield: Inconel alloy 625 sandwich
8 Body flap: aluminum alloy 2024
9 ACS engine nozzles: niobium alloy C103
10 OMS engine nozzles: niobium alloy FS-85
11 Window frames: S-65 Beryllium
12 Thrust structure: Ti-6Al-4V (diffusion bonded), Ti-6Al-4V and boron epoxy
13 APU: René 41 turbine, Hastelloy alloy B, Hastelloy alloy X, René 95
14 Tail conical seal: Inconel alloy 718 sandwich
15 Cargo bay doors: Graphite/epoxy
16 OMS pod structure: Graphite/epoxy
17 Nose cone: RCC
18 Wing leading edges: RCC
19 Windows: Quartz (outer), aluminosilicate (inner)
20 Pressure vessels: Kevlar-Ti-8Al-4V, Kevlar-Inconel alloy 718, Ti-6Al-4V, aluminum alloy 2219
21 Mid fuselage support tubes: Boron-aluminum, diffusion bonded Ti-6Al-4V
22 Plumbing systems: Type 304 and 21-6-9 stainless steels, Ti-3Al-2.5V, Inconel alloy 718
23 High-strength fasteners: A-286, MP35N, Inconel alloy 718
24 Landing gear: 300M steel, aluminum alloy 7075
25 Brakes: Beryllium, carbon-carbon
26 Flipper doors/rub panels: Ti-6Al-4V, Inconel alloy 625
27 ET door: Beryllium S-65
28 High-temperature fasteners: Udimet 500
29 Main engine nozzle: Inconel alloy 718, A-286 tubes
30 Elevon cove seal: Niobium alloy C103

Fig. 3 Typical materials of construction for the space shuttle orbiter vanic corrosion. Because of the seriousness of galvanic corrosion, every effort was made to avoid the use of dissimilar metals, to exclude moisture or other electrolytes from the system, and to protect metal surfaces in the contact area. Metals were considered compatible if they were in the same grouping as specified in MSFC-SPEC-250, Class II, or if the difference in solution potential was 0.25 V or less.

Control of Atmospheric Corrosion. All parts, assemblies, and equipment, including spares, were finished to provide protection from corrosion in accordance with the requirements of MSFC-SPEC-250, Class II, as a minimum. All organic finishes and anodized aluminum that contact titanium were limited to surfaces not normally exposed to propellants. A finish specification delineating the finishes used on each specific material in any particular application and corrosion control procedure was prepared by the orbiter contractor. The finish specification and related procedures from the subcontractor were required to provide for in-process corrosion control. Specific requirements were also given for:

- Surface preparation for adhesive bonding
- Finish systems for interior and exterior surfaces (including those surfaces to which the thermal protection system was to be bonded)
- Fastener installations
- Joints and faying surface sealing
- Protection for parts to be shipped for vehicle final assembly

Designing to Control Corrosion of the Space Shuttle Orbiter

From a corrosion control standpoint, it is convenient to separate the space shuttle orbiter into four categories: primary structure, fluid systems, mechanical systems, and avionics systems. Each of these areas has its own unique problems.

Primary Structure

Weight and cost both dictated that the primary structure of the orbiter be made from aluminum. The majority of this structure was made from heat-treated alloys of the 2000 and 7000 series (Fig. 3). However, some 5000- and 6000-series alloys were also used.

Aluminum Airframe. Prior to alloy selection, two surveys were conducted. The first was to identify steps to be taken to avoid stress-corrosion problems with aluminum alloys, and the second to identify a corrosion protection system that could survive the unique spacecraft environment.

The stress-corrosion survey, conducted in the early 1970s, indicated that virtually all stress-corrosion failures in service occurred in the 2000-series alloys in the T3, T4, and T6*XX* tempers and in the 7000-series alloys in the T6*XX* tempers and perpendicular to the short-transverse direction. Only alloys 2024, 2124, and 2219 have high SCC resistance in the T6*XX* tempers. In forgings, stress corrosion occurred in end grain runout along the forging parting lines. In extrusions and plate, failures occurred where parts were severely formed, where interference fits had occurred, or where assembly stresses were high. Theoretically, stress corrosion will not occur until all three of the essential elements are present: a susceptible microstructure, a corrosive environment, and surface tensile stresses. If any one of these was eliminated, stress corrosion should not occur. The approach used on the orbiter, however, was to eliminate or minimize all three of these conditions to the maximum practical extent.

First, only aluminum alloys were permitted with a minimum 170-MPa (25-ksi) stress-corrosion threshold (according to the supplier's standard test methods: 30 days in salt spray) in all directions. This eliminated the T3, T4, and T6*XX* tempers of nearly all the 2000- and 7000-series alloys, whose stress-corrosion thresholds could be as low as 48 MPa (7 ksi) or less in the short-transverse direction. The preferred microstructure of the −T73 and −T76 tempers was chosen for 7000-series alloys, and although some weight penalty was incurred, program reliability was well served. Because most of the aluminum structure is designed for compressive loading (buckling, crippling) or shear, a very small weight penalty actually resulted. For the 2000-series alloys, the T8*XX* tempers were used predominantly; however, in a few cases, the T6 or T62 tempers were used for alloys 2124 and 2219.

Second, every effort was made to reduce residual stress levels. Mill products were ordered in stress-relieved tempers (for example, T651 or T851) wherever possible to reduce machining distortion and susceptibility to stress corrosion. Interference fits were limited to stress levels below 67% of the stress-corrosion thresholds. Tables were prepared to allow the materials and process engineer to ascertain stress levels resulting from interference fit pins and bushings into various size lugs. Residual stresses in assembly were minimized by shimming. Forming by bending, which put the short-transverse direction into tension, was not permitted.

Finally, the corrosion-protective paint system (described below) was applied to all aluminum parts. As of 1986, no incidents of aluminum stress-corrosion failures have been reported on the orbiter since its inception in 1972.

The second survey conducted in the early 1970s involved the selection of a paint system that would meet the unique requirements of the shuttle. The system had to provide protection to aluminum from corrosion for a minimum of 10 years of seacoast exposure, without touchup, because it must also serve as the base to which the thermal protection system (TPS) tiles of the shuttle are bonded. Unlike commercial aircraft, the external surfaces could not be washed, repainted, or protected from water intrusion and crevice corrosion by using water-displacing chemicals.

Table 1 Metals and alloys compatible in dissimilar-metal couples

Group number	Metallurgical category	emf, V	Anodic index(a), V	Compatible couples(b)															
1	Gold, solid and plated; gold-platinum alloys; wrought platinum	+0.15	0	○															
2	Rhodium plated on silver-plated copper	+0.05	0.10	●	○														
3	Silver, solid or plated; high-silver alloys	0	0.15	●	●	○													
4	Nickel, solid or plated; monel metal, high-nickel-copper alloys	−0.15	0.30		●	●	○												
5	Copper, solid or plated; low brasses or bronzes; silver solder; German silvery high copper-nickel alloys; nickel-chromium alloys; austenitic corrosion-resistant steels	−0.20	0.35		●	●	●	○											
6	Commercial yellow brasses and bronzes	−0.25	0.40			●	●	●	○										
7	High brasses and bronzes; naval brass; Muntz metal	−0.30	0.45				●	●	●	○									
8	18% Cr type corrosion-resistant steels	−0.35	0.50				●	●	●	●	○								
9	Chromium plated; tin plated; 12% Cr type corrosion-resistant steels	−0.45	0.60					●	●	●	●	○							
10	Tin-plate; terneplate; tin-lead solder	−0.50	0.65						●	●	●	●	○						
11	Lead, solid or plated; high-lead alloys	−0.55	0.70							●	●	●	●	○					
12	Aluminum, wrought alloys of the 2000 series	−0.60	0.75								●	●	●	●	○				
13	Iron, wrought, gray or malleable; plain carbon and low-alloy steels; armco iron	−0.70	0.85									●	●	●	●	○			
14	Aluminum, wrought alloys other than 2000 series aluminum, cast alloys of the silicon type	−0.75	0.90										●	●	●	●	○		
15	Aluminum, cast alloys other than silicon type; cadmium, plated and chromated	−0.80	0.95											●	●	●	●	○	
16	Hot-dip-zinc plate; galvanized steel	−1.05	1.20															●	○
17	Zinc, wrought; zinc-base die-casting alloys; zinc plated	−1.10	1.25																●
18	Magnesium and magnesium-base alloys, cast or wrought	−1.60	1.75	●															

(a) Anodic index is the absolute value of the potential difference between the most noble (cathodic) metals listed and the metal or alloy in question. For example, the emf of gold (group 1) is +0.15 V, and the emf of wrought 2000-series aluminum alloys (group 12) is −0.60 V. Thus, the anodic index of wrought 2000-series aluminum alloys is 0.75 V. (b) "Compatible" means the potential difference of the metals in question, which are connected by lines, is not more than 0.25 V. An open circle indicates the most cathodic members of a series; a closed circle indicates an anodic member. Arrows indicate the anodic direction.

Table 2 Fluids used on the space shuttle orbiter

See also Fig. 4.

Index number	Fluid/gas	Location	System	Quantity	Explosive limit	Threshold limit value, ppm	Remarks
1.	Ammonia (NH_3)	Aft fuselage	ECLSS	44.27 kg (97.6 lb)	16%	25	Two tanks
2.	Breathing oxygen (GOX)	Mid fuselage	ECLSS	32.21 kg (71 lb)	NA(a)	(b)	One tank
3.	Freon-21	Mid and aft fuselage	ECLSS	272.16 kg (600 lb)	NA	1000	System
4.	Freon-1301	Crew compartment	ECLSS (fire extinguisher)	5.17 kg (11.4 lb)	NA	1000	Three tanks
5.	Fluorinert, FC-40	Mid fuselage	EPS	17.5 kg (39 lb)	NA	...	Fuel cell coolant loops
6.	Helium	Forward RCS module	Forward RCS	3.81 kg (8.4 lb)	NA	(c)	Two tanks
		OMS/RCS modules	OMS	44 kg (97 lb)	NA	(c)	Two tanks
		Aft fuselage	MPS	529.52 L (18.6 ft^3)	NA	(c)	Four tanks
		Mid fuselage	MPS	1843.4 L (65.1 ft^3)	NA	(c)	Six tanks
		Aft fuselage	Aft RCS	7.62 kg (16.8 lb)	NA	(c)	Four tanks
7.	Hydrazine (N_2H_4)	Aft fuselage	APU	131.99 kg (291 lb)	4.7%	1	Three tanks
8.	Hydraulic fluid	Forward, mid, and aft fuselage	HYD	344 L (90.9 gal)	204 °C (400 °F)	(d)	Three systems
		Landing gear struts	LDG	13.6 kg (30 lb)	110 °C (230 °F)	(d)	Nose and main gear
9.	Liquid hydrogen (LH_2)	Aft fuselage	MPS	142.43 kg (314 lb)	4%	(c)	Feedlines and main engine
		Mid fuselage	EPS	83.47 kg (186 lb)	4%	(c)	Two tanks
10.	Liquid oxygen (LOX)	Aft fuselage	MPS	2219 kg (4892 lb)	NA	(b)	Feedlines and main engine
		Mid fuselage	EPS, LSS	708.5 kg (2162 lb)	NA	(b)	Two tanks
11.	Lubricating oil	Aft fuselage	APU	8.16 kg (18 lb)	245 °C (475 °F)	(d)	Three systems
12.	Monomethyl hydrazine (MMH)	Forward RCS module	Forward RCS	480.75 L (127 gal)	3.0%	0.2	One tank
		Aft RCS modules	Aft RCS	961.5 L (254 gal)	3.0%	0.2	Two tanks
		OMS modules	OMS	4845.31 L (1280 gal)	3.0%	0.2	Two tanks
13.	Nitrogen	Mid fuselage	ECLSS	103.42 kg (228 lb)	NA	(c)	Four tanks
14.	Nitrogen tetroxide (N_2O_4)	Forward RCS module	Forward RCS	465.6 L (123 gal)	NA	5	One tank
		Aft RCS module	Aft RCS	935 L (247 gal)	NA	5	Two tanks
		OMS	OMS	4845.31 L (1280 gal)	NA	5	Two tanks
15.	Water (deionized)	Crew module	ECLSS	29.5 kg (65 lb)	NA	None	Two cooling loops
		Aft fuselage	ECLSS	194.14 kg (428 lb)	NA	None	Three tanks
16.	Water (potable)	Lower crew module	LSS	289.4 kg (638 lb)	NA	None	Four tanks
17.	Water (waste)	Lower crew module	LSS	72.7 kg (159.5 lb)	NA	None	One tank

(a) NA, not applicable. (b) No threshold limit value; upper limit is 6 h at 1 atm of pressure, lower limit is 19%. (c) Simple asphyxiant, no threshold limit value. (d) No threshold limit value; inhalation of vapors not encountered in normal use.

The paint system had to endure temperatures of 175 °C (350 °F) during entry and landing because heat from reentry soaked back into the structure. It had to be capable of surviving space vacuum and low temperatures (−155 °C, or −250 °F) without degradation. Minimum offgassing was desirable to avoid giving off toxic fumes inside the cabin (crew hazard) or the condensation of volatile material on windows or optical (thermal) control surfaces. The need for exceptional corrosion resistance was further mandated by the floating bilge; that is, the orbiter is stacked and launched in a vertical attitude, operates in zero gravity, and reenters and lands horizontally. It was not possible to ensure that all water drains out of the structure in all attitudes.

The paint system chosen was a chromate-inhibited epoxy polyamine primer. This system was tough, abrasion resistant, and durable. Surfaces to be painted were either anodized according to MIL-A-8625 type II, class 1, or chemically filmed according to MIL-C-5541, class 1A. Each coat of paint was 0.015 to 0.023 mm (0.6 to 0.9 mil) thick. A single coat of the chromated epoxy polyamine primer demonstrated 1500 h of salt spray protection without corrosion, even in areas scratched through to bare aluminum.

The surface to which the external TPS was bonded had a single coat of the chromated epoxy polyamine paint. It achieved additional corrosion protection from a room-temperature vulcanized (RTV) adhesive layer 0.13 to 0.23 mm (5 to 9 mils) thick used to bond the tiles. In the cargo bay area, the single coat of epoxy polyamine primer was overcoated with one coat of polyurethane (MIL-C-83286 or MIL-C-81773) to achieve the required optical properties—absorptivity (α), emissivity (ϵ), and the proper α/ϵ ratio for heat control. The interior of the cabin required the use of nonglare coatings and selected colors. Again, a single undercoat of the chromated epoxy polyamine primer was coated with polyurethane. In this case, the polyurethane not only provided a durable color but also acted as a barrier to unacceptable offgassing products of the primer.

Parts were painted as details, drilled and assembled, and then repainted upon assembly to coat the fasteners. Although it was desirable from a corrosion standpoint to install all rivets wet, practical manufacturing considerations did not permit it. Automatic riveting machines, which were used to install nearly 90 000 rivets in the wing, could not use wet rivet installations.

The weight reduction demands of the program resulted in the elimination of the use of two coats of paint on interior surfaces. More than 500 kg (1100 lb) of paint were used to cover 8175 m^2 (88 000 ft^2) of surface. By substituting an anodize coating for one coat of primer, significant weight savings were realized. Consequently, the finish system for *Discovery* and *Atlantis* followed the general scheme:

Area	Coating
Exterior TPS surface	Anodize + 1 coat chromated epoxy polyamine primer + 1 coat RTV adhesive
Exterior non-TPS surfaces	Anodize + 1 coat chromated epoxy polyamine primer
Interior surface	Anodize + either 1 coat chromated epoxy polyamine primer or 1 coat polyurethane
Crew compartment	Chemical film or anodize + 1 coat chromated epoxy polyamine primer + 1 coat polyurethane or anodize only

The forward fuselage was fabricated as a sheet metal skin-stringer design in aluminum alloy 2024-T6. Suspended inside the forward fuselage was an all-welded aluminum pressurized cabin made from aluminum alloy 2219 using the T6 and T8 tempers. It was approximately conical in shape, about 5 m (17 ft) long and tapering from 5

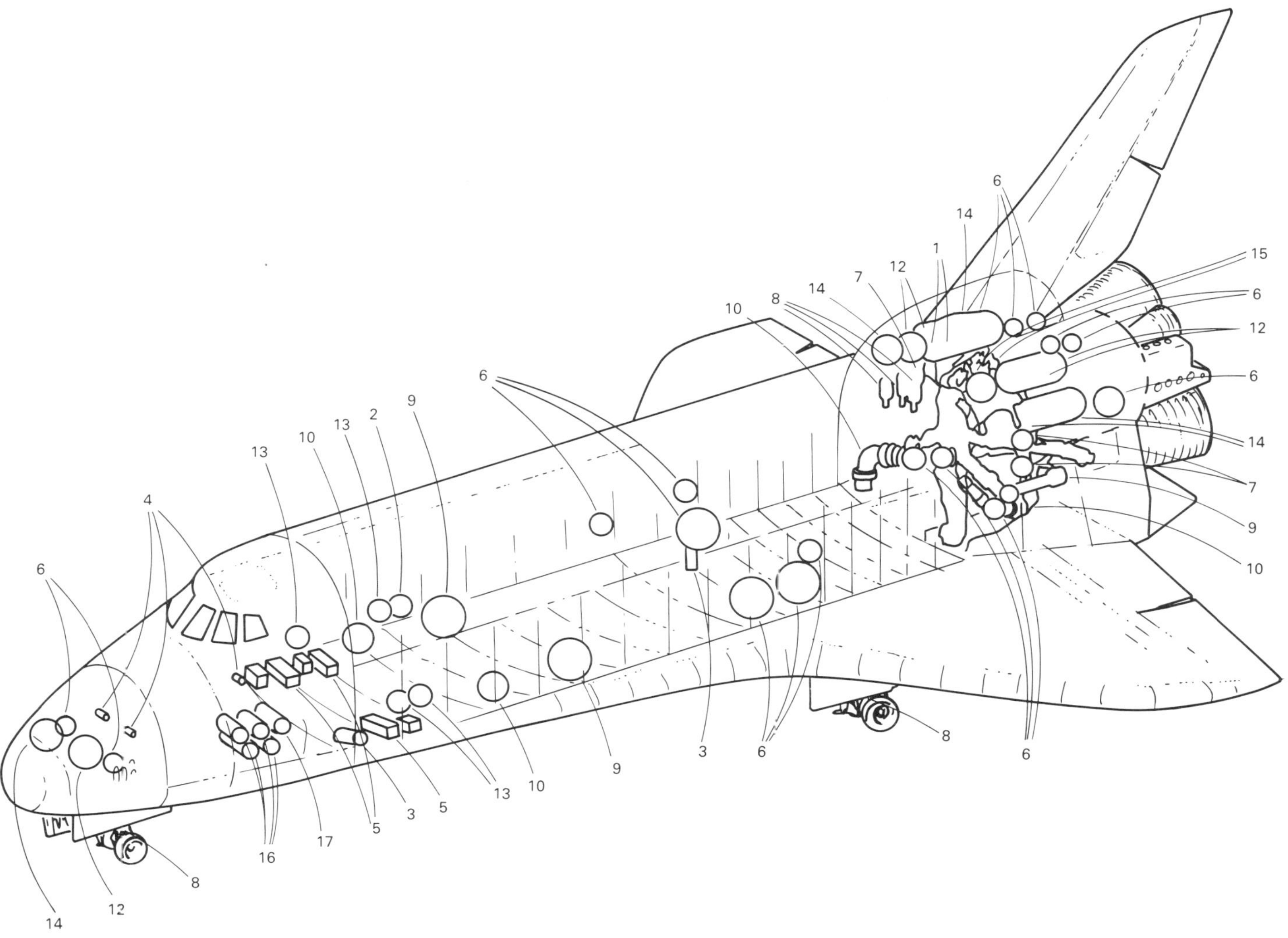

Fig. 4 Locations of liquids and gases in the space shuttle orbiter. Numbers correspond to the fluid index numbers used in Table 2.

to 2.4 m (17 to 8 ft) in diameter at its forward end. The mid and aft fuselage structures were machined from aluminum alloy 2124-T851 plate. Major frames were of aluminum alloy 7075 in the T76 or T73 tempers.

Aluminum honeycomb sandwich was extensively used in the wing and body flap areas. Corrosion protection systems must prevent corrosion of the thin (0.025 to 0.075 mm, or 1 to 3 mil) honeycomb core (usually aluminum alloy 5056-H39) and delamination of the skins. Face sheet skins could not be alclad, because corrosion would proceed in the plane of the sheet, resulting in delamination of the bond line. To prevent corrosion, all aluminum cores were protected with conversion coatings and were nonperforated; the face sheets used corrosion-resistant adhesive primers, and the sandwich assemblies were sealed at the edges to prevent water entry.

Structural Joints and Fasteners. The corrosion engineers favored the assembly of structural joints with RTV faying surface sealants; however, the electrical bonding requirements or grounding of each panel eliminated this approach. Electrical bonding requires a maximum dc resistance of 2.5 mΩ across joints requiring lightning protection or radio frequency grounding for electrical or electronic equipment. For a typical faying surface joint, local removal of the paint or anodize on the detail (down to bare metal) is required in the area of the fastener. Bare aluminum surfaces are subsequently coated with a chemical film (MIL-C-5541, class 1A). The joint is then bolted with stainless steel fasteners using stainless steel washers under the bolthead and nut to protect aluminum surfaces during application of torque. Fasteners are installed with the shank portion wetted with chromated epoxy polyamine primer. Joints are subsequently touched up with a chemical film and a coat of chromated epoxy polyamine primer around the washers. Faying surfaces are sealed with a continuous fillet of RTV 577, which is a white, thixotropic silicone rubber material.

There are approximately 30 different kinds of electrical grounding joints available to suit various designs on the orbiter; the grounding techniques used include jumpers, spot welds, staples, metallized tape, and the method described above. Even an adhesively bonded edge member or a T-section bonded to a honeycomb face sheet must be provided with a ground to the face sheet itself.

Dissimilar-metal joints are permitted on the orbiter without additional galvanic protection if they fall within the range shown in Table 1. Table 1 should be used only as a guideline, and such factors as cathode-to-anode area ratios, corrosive environments, and other detrimental factors must be evaluated.

The fasteners chosen for the spacecraft are all basically corrosion resistant, but care must be taken with dissimilar-metal combinations. Bolts are typically made of alloys A-286 (965 to 1380 MPa, or 140 to 200 ksi), Inconel alloy 718 (1240 MPa, or 180 ksi), and MP35N (1655 MPa, or 240 ksi). For applications up to 870 °C (1600 °F), Udimet 500 is used (1035 MPa, or 150 ksi). Of these materials, only A-286 has shown any corrosion in service. The

Table 3 Major shuttle orbiter pressure vessels

Fluid/gas	System	Alloy	Number per vehicle	Size, diameter or length × diameter, mm	in.	Factor of safety	Operating pressure, MPa	psi	Proof pressure, MPa	psi	Burst pressure, MPa	psi
Ammonia (NH_3)	ECLSS	Ti-6Al-4V	2	437	17.2	4.0	2.2	320	4.3	625	8.9	1290
Breathing oxygen	ECLSS	Kevlar/Inconel alloy 718	1	660	26	1.5	22.8	3300	29.2	4240	34.1	4950
Freon-21	ECLSS	Aluminum 6061-T6, AM350 bellows contain the fluid	2	686 × 330 (accumulator)	27 × 13	2.0	1.6	230	2.4	345	3.2	460
Helium	Forward RCS	Kevlar/Ti-6Al-4V(a)	2	475	18.7	1.5	27.6	4000	36.6	5310	41.4	6000
	Aft RCS	Kevlar/Ti-6Al-4V(a)	4	475	18.7	1.5	27.6	4000	36.6	5310	41.4	6000
	OMS	Kevlar/Ti-6Al-4V(a)	2	1024	40.3	1.5	33.6	4875	44.9	6510	50.5	7325
	MPS	Kevlar/Ti-6Al-4V(a)	7	663	26.1	1.5	31.0	4500	42.7	6190	46.5	6750
	MPS	Ti-6Al-4V(a)	2	249	9.8	4.0	5.9	850	11.7	1700	23.4	3400
Hydrazine (neat)	APU	Ti-6Al-4V(a)	3	711	28	3.35	2.4	355	6.7	970	7.4	1070
Hydraulic fluid	HYD	Chromium-plated 4130 steel cylinder, 2024-	3	635 × 86 (accumulator)	25 × 3.4	12.0	1.7	250	20.7	3000	41.4	6000
		T851 aluminum piston		635 × 86 (accumulator)	25 × 3.4	4.0	20.7	3000	41.4	6000	82.7	12 000
Liquid hydrogen (LH_2)	EPS	Aluminum 2219-T6	2–4	1214	47.8	1.5	2.2	320	2.4	350	3.3	480
Liquid oxygen (LOX)	EPS	Inconel 718(b)	2–4	973	38.3	1.5	7.1	1035	8.6	1240	109	1575
Monomethyl hydrazine (MMH)	Forward RCS	Ti-6Al-4V(b)	1	991	39	1.5	2.4	350	3.2	465	3.6	525
	Aft RCS	Ti-6Al-4V(b)	2	991	39	1.5	2.4	350	3.2	465	3.6	525
	OMS	Ti-6Al-4V(b)	2	2438 × 1245	96 × 49	1.5	2.2	315	2.4	345	3.2	470
Nitrogen	ECLSS	Kevlar/Ti-6Al-4V(a)	4	660	26	1.5	22.8	3300	28.8	4175	34.1	4950
	OMS	Ti-6Al-4V(a)	2	84	3.3	2.5	3.1	450	6.2	900	7.4	1080
	OMS	Ti-6Al-4V(a)	2	132	5.2	40	20.7	3000	41.4	6000	82.7	12 000
Nitrogen tetroxide (N_2O_4)	Forward RCS	Ti-6Al-4V(b)	1	991	39	1.5	2.4	350	3.2	465	3.6	525
	Aft RCS	Ti-6Al-4V(b)	2	991	39	1.5	2.4	350	3.2	465	3.6	525
	OMS	Ti-6Al-4V(b)	2	2438 × 1245	96 × 49	1.5	2.2	315	2.4	345	3.2	470
Water, deionized	ECLSS	6061-T6 aluminum shell	3	737 × 396	29 × 15.6	2.0	0.3	37	0.4	58	0.5	74
	ECLSS	Inconel alloy 718 bellows, and Inconel alloy 625 end fittings contact water	2	208 × 152 (accumulator)	8.2 × 6	2.0	0.6	90	0.9	135	1.2	180
Water, potable and waste	Crew module	Same as above	4	902 × 394	35.5 × 15.5	2.0	0.1	20	...	...	0.3	40
	Crew module	Same as above	1	902 × 394	35.5 × 15.5	2.0	0.1	20	...	...	0.3	40
Water, cooling	APU	Ti-6Al-4V(a)	1	432	17	4.0	3.0	435	4.5	655	12.1	1755
	APU	Ti-6Al-4V(a)	1	244	9.6	8.0	0.7	100	4.1	600	5.5	800

(a) Annealed. (b) Solution treated and aged. Note: The pressure vessels listed in Table 3 may differ slightly from Table 2, because Table 3 includes major accumulators, pressure vessels integral to a particular hardware system, and extra pressure vessels needed for some missions.

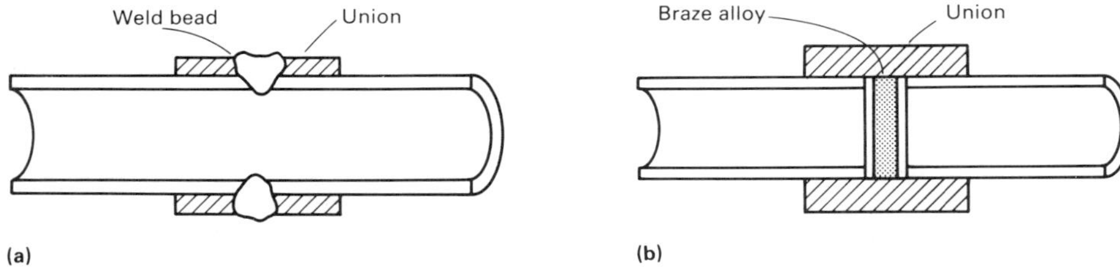

Fig. 5 Stainless steel and Inconel alloy fluid system permanent joints. (a) Automatic weld, inert gas, tungsten arc. (b) Braze joint

corrosion is only superficial and of no real concern. It is removed only for cosmetic reasons. None of these alloys is susceptible to hydrogen embrittlement in orbiter vehicle service.

Nuts are made from A-286 and Inconel alloy 718 and are lubricated with a thin (0.005 to 0.01 mm, or 0.2 to 0.4 mil) silver plate. Bolts and nuts are always installed with washers. The aluminum, therefore, never contacts the silver plate. Because the hole, as previously mentioned, is coated with wet chromated epoxy polyamine primer, no moisture can penetrate between the stainless or nickel shank and the aluminum hole. Stainless steel washers are separated from the aluminum surface with a dry coat of primer (where electrical grounding is not required) or a chemical film coating plus a touchup of the primer around the washer (where electrical grounding is required). No problems with galvanic corrosion are experienced in these installations.

Inserts, made from A-286, are silver plated and must also be installed into aluminum wetted on their exterior with the chromated epoxy polyamine primer. Where nuts are used in contact with titanium, only molybdenum disulfide-type dry film lubricants are used. Experience has shown that silver in contact with titanium at approximately 265 °C (500 °F) or above can bring about stress-corrosion cracking (SCC) of titanium (Ref 2-4). Titanium contact with silver is also prohibited by MIL-S-5002.

Titanium pin and collar fasteners are used for shear applications. To save weight, the orbiter uses aluminum alloy 2024 collars rather than A-286. Again, the holes are wet coated with the chromated epoxy polyamine primer by applying primer to the fastener shank away from the threads. Both ends of the fastener are subsequently touched up with the primer. Where such fasteners are used on the graphite cargo bay doors, it is necessary to use RTV rubber as a corrosion barrier, thus completely encapsulating the aluminum collar to prevent corrosion.

Rivets used on the spacecraft are made from aluminum alloy 2219-T62. These rivets provide good shear strength while avoiding the need to "ice box" rivets after solution treating, as is required with aluminum alloy 2024 rivets. Further, aluminum alloy 2024 rivets would undergo aging at entry temperatures of the orbiter (175 °C, or 350 °F), resulting in an increased susceptibility to corrosion as grain-boundary precipitation initiated. Aluminum alloy 2219-T81 rivets, although

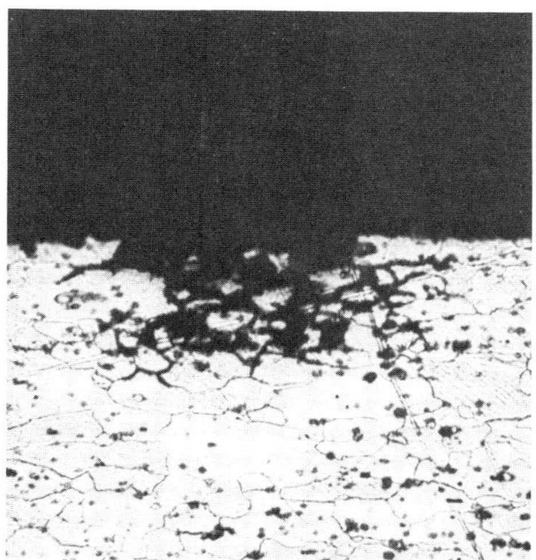

Fig. 6 Pitting corrosion of an aluminum alloy 2014-T5 sheet. Pitting occurred during the manufacturing cycle. Note the intergranular nature of the pit. 150×.

(a)

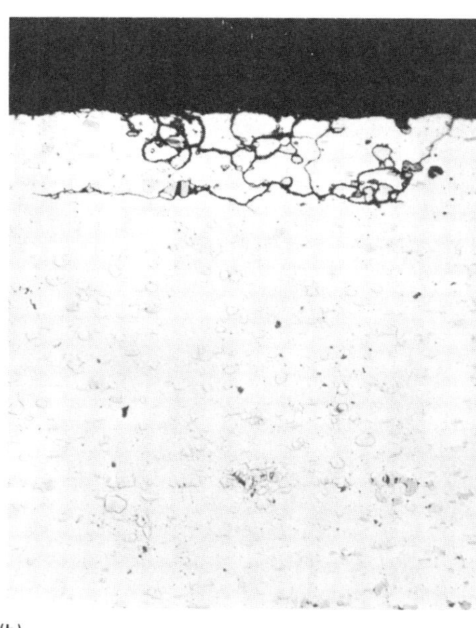

(b)

Fig. 7 Heat-treated aluminum pitted from 6 months of in-plant exposure. (a) Aluminum alloy 2024 exposed in the metal-processing area. (b) Aluminum alloy 2014 exposed in the manufacturing area. Note the intergranular nature of the pits. Both 285×

also commercially available, lack sufficient ductility to prevent cracking of driven heads. (The widely used aerospace aluminum alloy 7050-T73 rivets were unavailable when the shuttle orbiter was being built.)

Other Structural Alloys. No corrosion protection (except for passivation treatments after fabrication) is considered necessary for stainless steels. Stainless steels, particularly the precipitation-hardenable grades, will often display light surface corrosion products after extensive exposure. A light abrasion will remove the corrosion. No effort is made to passivate stainless parts as installed, because chemical spillage is considered more detrimental to the structure than any enhanced corrosion protection gained from passivation.

Nickel alloys such as Inconel alloy 718 and Inconel alloy 625 are used for elevated-temperature service with no corrosion protection. Inconel alloy 718 brazed honeycomb panels are used for the conical seals on the vertical stabilizer and for outboard elevon rub panels and flipper door panels. Inconel alloy 625, made as a resistance-welded sandwich, is used to temperatures of 870 °C (1600 °F). The surface of the sandwich is coated with a wear-resistant high-emittance chromium oxide coating.

Titanium also requires no further corrosion protection. Titanium, principally as Ti-6Al-4V, is widely used as forging, bar, and plate products throughout the spacecraft. Many other high-strength titanium alloys are also used. The major structural members of the aft thrust structure are made of Ti-6Al-4V. These transmit the thrust of the liquid rocket engines to the orbiter structure. Titanium honeycomb sandwiches, made by the liquid interface diffusion (LID) bonding process, are used as inboard elevon and flipper door panels. Although the honeycomb has a perforated core, no corrosion is experienced.

Because of prior experience in which processing and testing solutions had resulted in SCC of titanium alloys, a control specification, MF0004-018, has been imposed. This specification defines a list of fluids that are suitable for titanium and the specific conditions under which their contact is appropriate.

Beryllium alloy S-65 (99% Be min) is used structurally for the external tank door and for windshield retainers. Beams providing structural support in the windshield area use either S-65 or CIP HIP-1 (Ref 5), which is also used nonstructurally for the navigational base and the star tracker boom, as well as for heat sinks. Beryllium must be protected in service. The beryllium is anodized according to a Rockwell internal specification and is painted with one coat of chromated epoxy polyamine paint or chemically filmed in a manner similar to that used for aluminum (MIL-C-5541). Two coats of the chromated epoxy polyamine paint are then applied. The anodized coating (0.05 mm, or 0.2 mil, minimum) reveals no corrosion when tested in 168 h of salt spray according to ASTM B 117.

Steels must be protected in service from the seacoast environment. Steels are often painted with chromated epoxy polyamine paint or plated with nickel or chromium, depending on the service. Cadmium plating is not used except under rare circumstances, because it can easily sublime in space and redeposit on cooler adjacent surfaces. To avoid problems with SCC and hydrogen embrittlement, steel alloys are restricted to 1380 MPa (200 ksi) or less in tensile applications, and precipitation-hardenable steels are restricted to the H1000 or higher-temperature tempers. Steels with tensile strengths as high as 2070 MPa (300 ksi) can be used for applications that involve bearing, compressive, or shear loads; such applications include ball or roller bearings, valve seats, and springs. A more complete description of the corrosion protection of steel alloy parts that are in moving contact with each other can be found in the discussion "Mechanical Systems" in this article.

Niobium is used for low-stress applications in the orbiter airframe structure. Tubes and nozzle parts fabricated from niobium alloy C-103 (Nb-10Hf-1Ti-0.5Zr) are used in the reinforced carbon-carbon nose cap for the shuttle entry air data system (SEADS) program involving measurements of aerodynamic pressures. These parts have a VH109 silicide coating to prevent high-temperature oxidation. The coating was chosen because of its performance at the design temperatures (1455 °C, or 2650 °F) (see the discussion "Case Histories" in this section). Niobium alloy C-103 parts are used as closeout members in elevon seals to shield the hot plasma from the interior structure and mechanisms. These parts are coated with an R512E silicide coating. They were designed for service at maximum temperatures of 1370 °C (2500 °F). Silicide coating systems are ceramic in nature and may be chipped by impact on edges or surfaces. Tests were

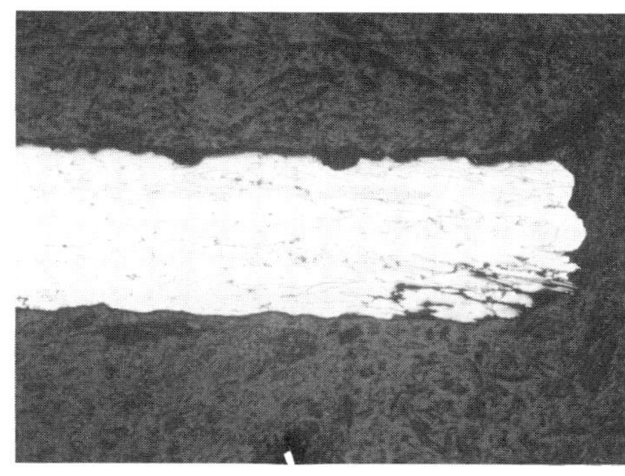

Fig. 8 Pitting corrosion of an aluminum alloy 7075-T6 aluminum radial shear beam from the Apollo program. The beam is machined and chemically milled from a 64-mm (2.5-in.) thick plate to a final nominal thickness of 0.45 mm (0.018 in.). Pitting occurred from improper protection either during manufacturing or in service. 30×

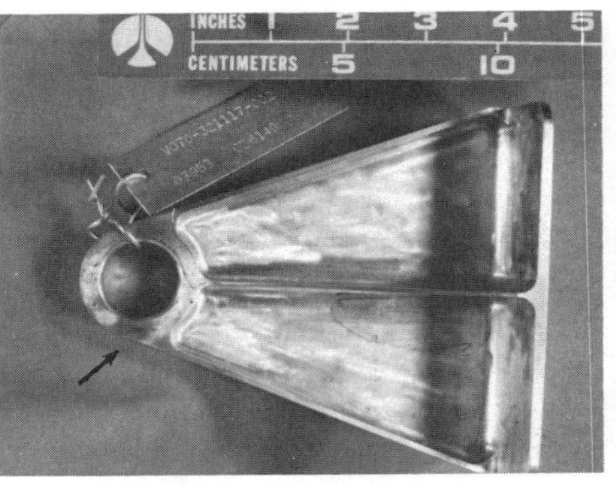

(a)

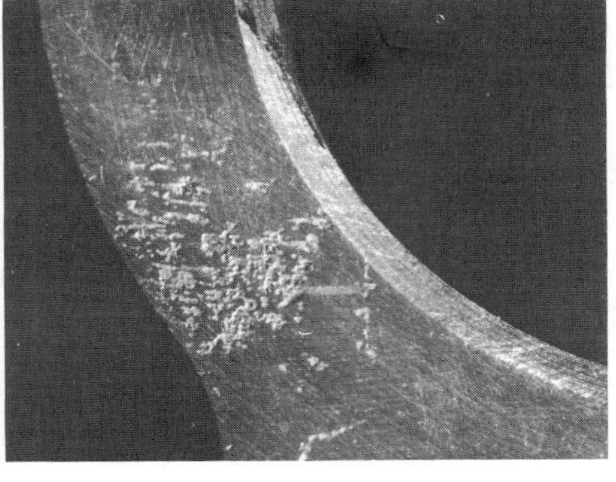

(b)

Fig. 9 Pitting corrosion of an aluminum alloy 2024-T62 structural fitting used on the space shuttle orbiter. (a) Fitting that was pitted from lack of interim protection after machining. 0.25×. (b) Enlargement of pitted surface. ~4×

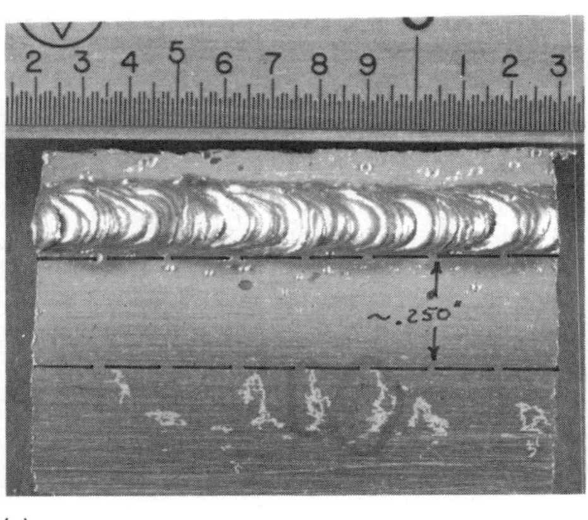

(a)

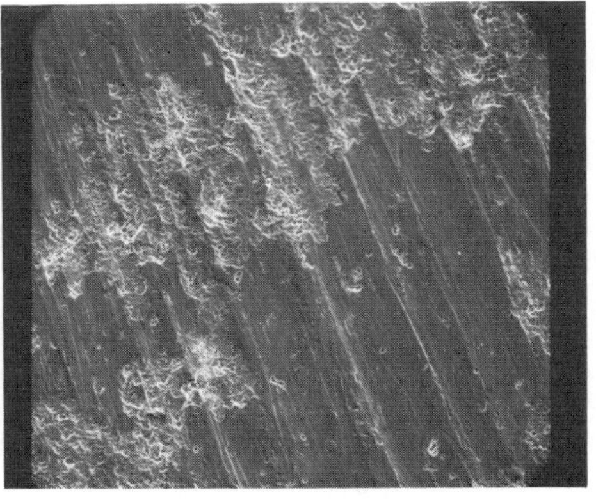

(b)

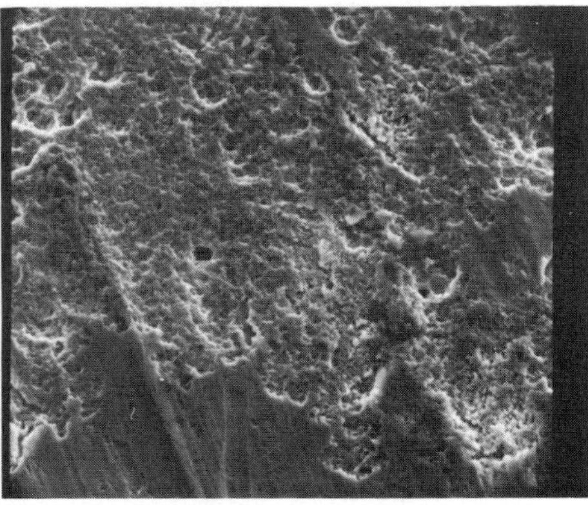

(c)

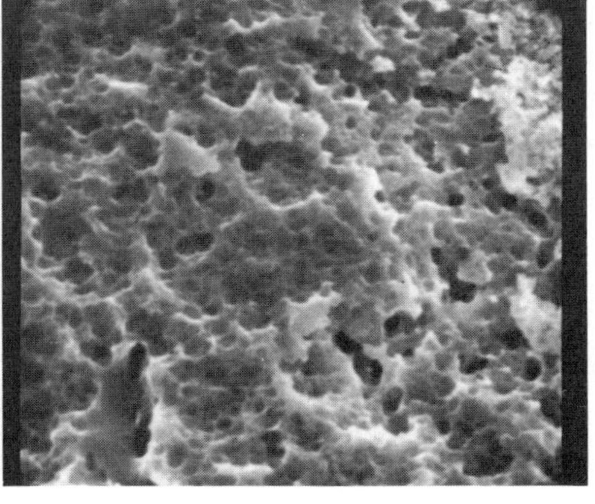

(d)

Fig. 10 Filiform corrosion in an aluminum alloy 2219-T6 hydrogen tank used on the space shuttle orbiter. Attack was due to atmospheric humidity. Small spherical beads in (a) are splatter from electron beam welding. (a) Root side of weld showing filiform corrosion beyond the HAZ. (b) to (d) Enlargements of the corrosion attack at 60×, 235×, and 1180×, respectively

conducted to verify that parts with coating damage down to bare metal could still function for a limited number of flights (see the discussion "Case Histories" in this section).

Composite materials are widely used on the orbiter and present no corrosion problems except for graphite epoxy structures. Although graphite is compatible with titanium, corrosion-resistant steels (A-286 and 300-series stainless steels), nickel, and cobalt-base alloys, the galvanic potential between graphite and aluminum or graphite and steel requires special design considerations. Suitable galvanic isolation is accomplished by using a layer of titanium foil, Tedlar, Kapton, or type 120 glass fabric with suitable resin plus two coats of chromate epoxy polyamine primer. All edges of the joints between the graphite and the aluminum or steel are sealed with RTV silicone to preclude moisture intrusion.

More than 300 boron/aluminum composite tubes with diffusion-bonded Ti-6Al-4V clevises are used on the orbiter, principally in the mid fuselage to stabilize frames or as pressure vessel supports. The aluminum portion is painted with chromated epoxy polyamine primer. These present no special corrosion design problems. Also, there are no corrosion problems with boron epoxy-bonded reinforcements on titanium thrust structure tubes in the aft thrust structure.

Fluid Systems

The space shuttle orbiter fluid systems must provide for the storage, transfer, and regulation of 17 different fluids, as shown in Table 2. The fluid systems can be grouped into major functional areas:

- Environmental control and life support system (ECLSS)
- Electrical power system (EPS)
- Reaction control system (RCS)
- Orbital maneuvering system (OMS)
- Main propulsion system (MPS)
- Auxiliary power unit (APU)
- Hydraulic system (HYD)

The locations of the pressure vessels used to contain these fluids are shown in Fig. 4. Design information covering these pressure vessels is given in Table 3.

Plumbing Lines. The philosophy followed in corrosion control for fluid systems and pressure vessels was to select materials that were compatible with the fluids without protective coatings and to control the fluid chemistry, not only as purchased but also during vehicle loading and operations. Stainless steel lines are used for all systems except hydraulic fluids and hot gaseous oxygen. Hydraulic fluids are contained in Ti-3Al-2.5V lines. Inconel alloy 718 is used for hot gaseous oxygen lines.

Because many of the fluids are hazardous (toxic, explosive), metallurgical joints were used in all permanent connections of stainless and Inconel alloy lines. Type 304L stainless steel was used for service with helium, hydrazine, liquid hydrogen, liquid oxygen, monomethyl hydrazine, nitrogen, and nitrogen tetroxide. Type 304L was selected to avoid potential sensitization. Most of the permanent joints were automatically welded by orbiting arc equipment, using an external sleeve of the same alloy to supply reinforcement to the weld bead, as shown in Fig. 5(a). The bead geometry on the inside of the tube is smooth and free from crevices that may initiate chemical attack. The stainless steel tubing, in many cases, is in the one-eighth hard condition, and the sleeve also provides added strength to compensate for the localized annealing at the welds and heat-affected zone (HAZ); this results in the full efficiency of the one-eighth hard material.

Alloy 21-6-9 stainless steel was used in lines carrying ammonia, breathing oxygen, freons, oxygen, hot gaseous hydrogen (280 °C, or 540 °F), nitrogen, and waters. The stainless steel was usually brazed with gold alloy Nicoro 80 (81.5Au-16.5Cu-2Ni) in a configuration shown in Fig. 5(b). This configuration also eliminates an internal crevice in the lines, because the brazing alloy flows along the capillary between the tube and

(a)

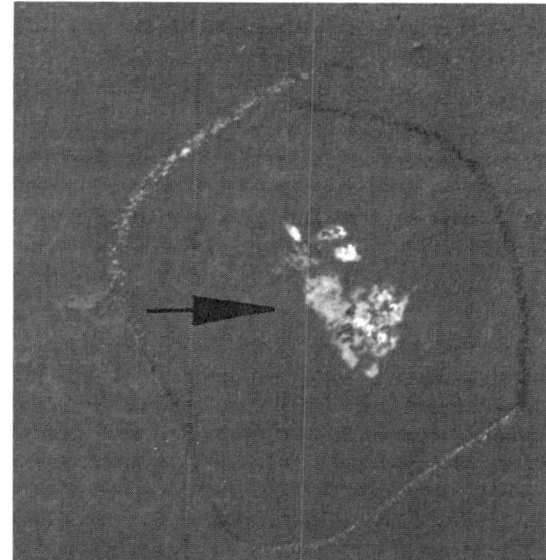

(b)

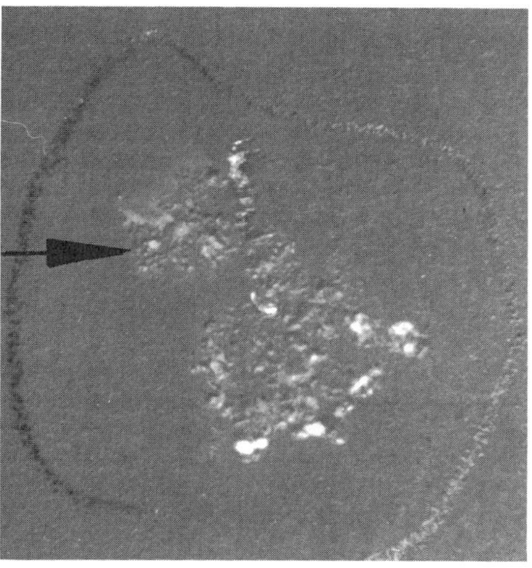

(c)

Fig. 11 Galvanic corrosion of aluminum alloy 2024-T81 sheet of orbiter front wing spar from the *Columbia* orbiter. (a) Corrosion appeared as aluminum oxide deposits (arrow) along the edge of an Inconel alloy 601 foil covered insulation blanket. (b) Open pits (arrow) in an area that was in contact with the foil insulation blanket. (c) Oxide buildup under paint in a similar contact area causes formation of "bubbles" (arrow)

sleeve, seals it, and forms a fillet along the periphery of the sleeve. There were two concerns with this braze combination. First, would the braze alloy create a galvanic-corrosion problem? Testing of the electrode potentials and exposures with actual fluids showed no galvanic effect that could be detected. Second, would the copper in the braze alloy cause liquid-metal embrittlement? The literature indicated the 21-6-9 alloy to be particularly susceptible to copper embrittlement. Joints were made and sectioned to reveal the microstructure of the brazed joint.

Intrusion of the brazing alloy into grain boundaries did occur, but never exceeded 0.05 mm (2 mils) even after four induction brazing cycles. The intrusion did not affect the static or fatigue strength of the joint; in fact, it appeared to give a superior attachment to the substrate.

In many cases, stainless steel lines must connect pressure vessels made from titanium alloys such as Ti-6Al-4V. Typically, a bimetallic joint is used in which a steel tube is joined to a titanium tube or fitting, which is then welded to the pressure vessel. The steel-to-titanium tube joints are made by coextrusion or, in the case of the APU water tank, by swaging. Aluminum-stainless steel joints are required for attachment of plumbing to the aluminum alloy 2219-T6 tank. These joints are made by friction welding. No galvanic-corrosion problems have been experienced with the fluids involved.

For hot gaseous oxygen lines operating to 31 MPa (4500 psi) and 300 °C (570 °F), it was more efficient to use Inconel alloy 718 to obtain the required strength at temperature. Inconel joints were welded in the same manner as the type 304L stainless steel joints described above. The Inconel alloy joints also had a type 304L stainless steel sleeve, which provided sufficient reinforcement to ensure that nearly the full heat-treat properties of the Inconel alloy 718 tubing were realized (ultimate tensile strength: 1240 MPa, or 180 ksi; tensile yield strength: 1035 MPa, or 150 ksi). Initial attempts were made to braze this alloy, but because of its tenacious high-temperature oxide film, repeat braze-debraze cycles (even over nickel-plated ends) could not be achieved.

The environmental control and life support system was designed to provide:

- Atmospheric control to the pressurized crew cabin (oxygen, nitrogen, carbon dioxide, water vapor, odor)
- Pressure control to the crew cabin
- Thermal control to the crew cabin and avionics boxes
- Potable water and waste management control

The temperature of the crew cabin is maintained between 16 and 32 °C (61 and 90 °F). Oxygen partial pressure is maintained at 22 000 N/m^2 (3.2 psi), and nitrogen is added to achieve pressures of 70 500 N/m^2 (10.2 psi) for space operations and 101 000 N/m^2 (14.7 psi) for launching conditions. Relative humidity is controlled to prevent condensation of moisture. Therefore, the cabin atmosphere is a benign environment from a corrosion standpoint.

Aluminum cold plates are used to remove heat from the electronic boxes in the mid and aft fuselages. The cold plates are fluxless brazed with aluminum alloy 6951 face sheets and aluminum alloy 6061 cores. Freon 21 (dichloromonofluoromethane) is the coolant. Stainless steel cold plates are used in the crew cabin to carry heat from avionics boxes within the cabin. These coldplates are brazed from AISI type 304L sheet using Amdry 930 (Ni-22.5Mn-5Cu-7Si). Deionized water is used in the stainless steel cold plates. A water loop transfers the excess heat from the cabin and cabin avionics equipment to the freon cooling loop by way of the cabin heat exchanger. The freon cooling loop delivers this heat, together with the heat from the fuel cells, payloads, and mid and aft avionics equipment, to a large aluminum radiator (111 m^2, or 1195 ft^2) where the heat is radiated into space. When the cargo bay doors are closed during ascent or immediately before reentry, an active thermal rejection system, the flash evaporator, is employed. Heat is rejected by the boiling of water in this system.

Stainless steel was chosen for water lines because of the prior difficulties encountered on the Apollo program with aluminum lines and cold plates. Despite the use of inhibitors such as triethanolaminephosphate (TEAP) and sodium mercaptobenzothiozole (NABT) on Apollo aluminum systems, serious corrosion problems were encountered. Solutions had to be continuously circulated to avoid solution concentration gradi-

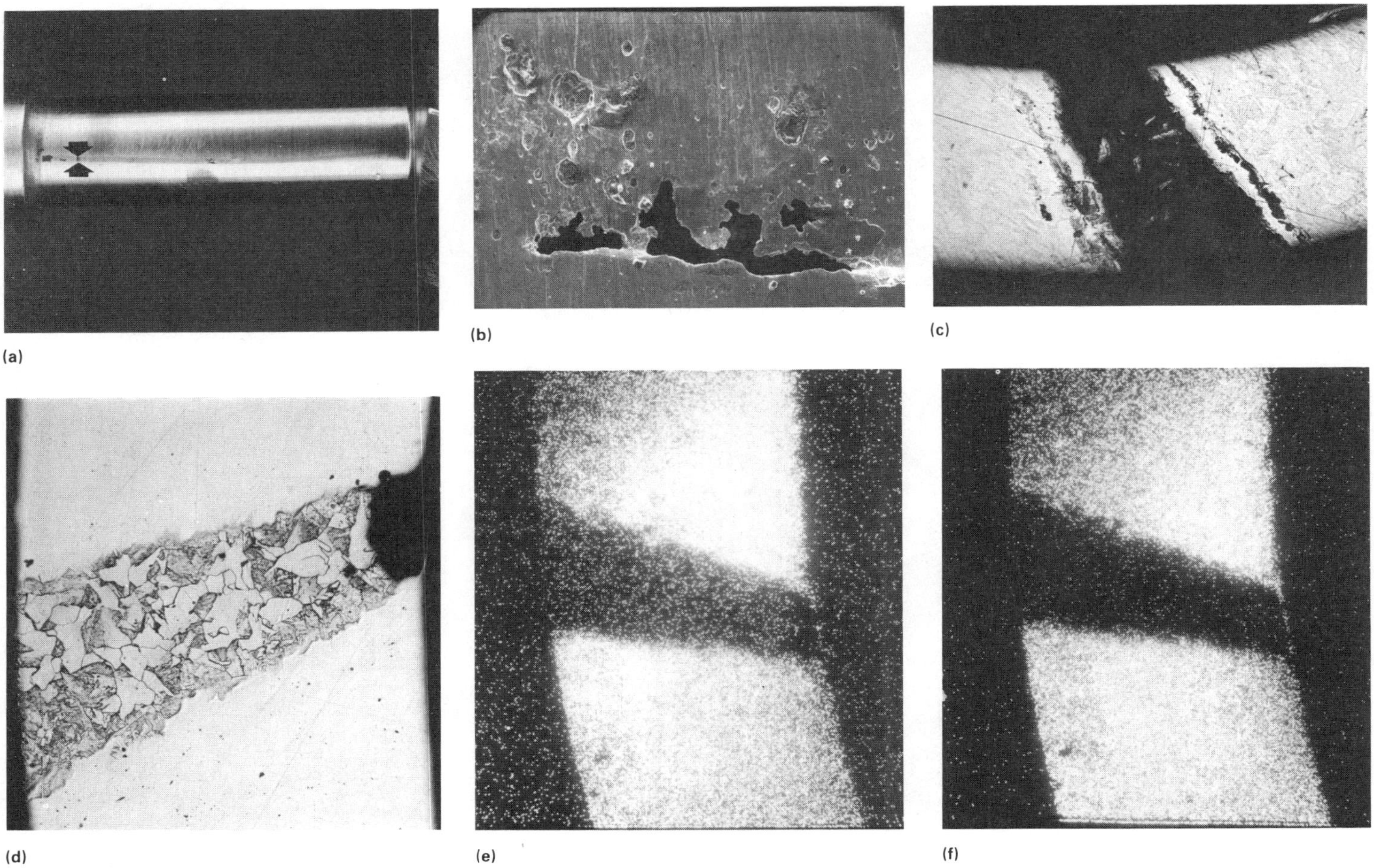

(a) (b) (c)

(d) (e) (f)

Fig. 12 Galvanic corrosion of forged type 304L stainless steel ECLSS outlet water valve used on the shuttle orbiter. (a) Tubular section of water valve with embedded longitudinal inclusion (between arrows). 2.5×. (b) Enlargement of inclusion revealing perforation. 125×. (c) Section through perforated area. 45×. (d) Section through inclusion showing beginning of galvanic attack. Nital etch (3%) used to bring out structure of carbon steel. 85×. (e) Electron dot map of nickel content through the inclusion. 55×. (f) Electron dot map of chromium content through the inclusion. 55×

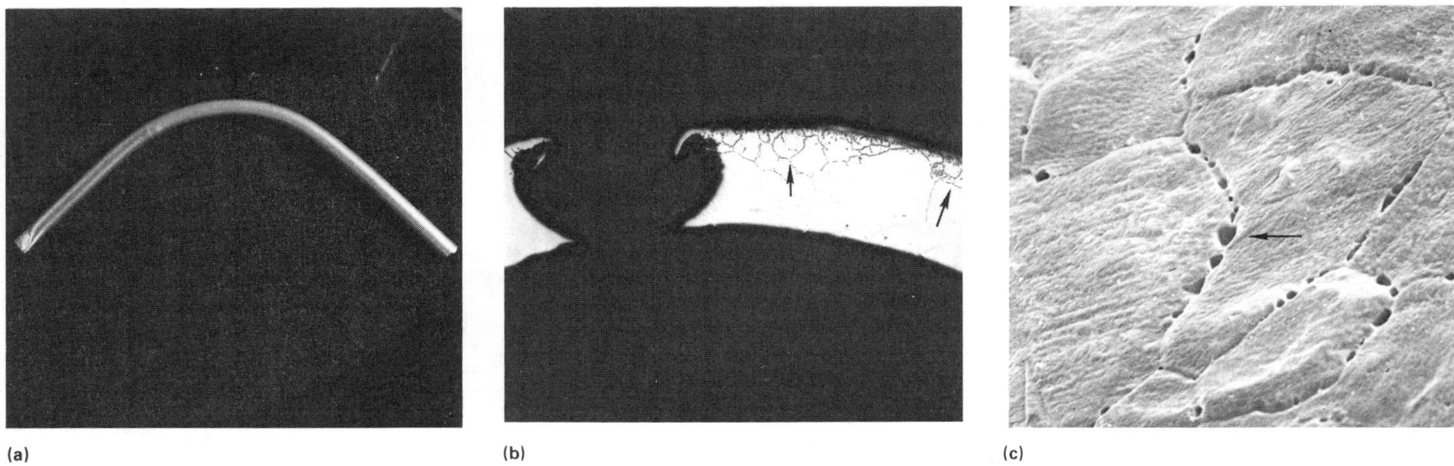

(a) (b) (c)

Fig. 13 Intergranular corrosion of a type 304L stainless steel tube in a shuttle orbiter ammonia boiler. (a) Test performed to show tube ductility. 1×. (b) Cross section through the thin-wall (0.2 mm, or 8 mils) tube revealing sensitization on outside diameter due to carbonaceous deposit formed during brazing. 75×. (c) Surface SEM showing grain-boundary carbides are being removed from outside diameter during corrosion. 980×

ents leading to localized corrosion (pitting) of the aluminum lines. Pits as deep as 0.7 mm (28 mils) were found in aluminum lines after limited service.

To preclude problems with corrosion in water systems on the space shuttle orbiter, it was decided that stainless steel would be used for all water (cooling and potable) lines. Each material in the water path was identified, evaluated and accepted or changed, and tested in a control loop at NASA (Houston) for up to 2 years of service. Orbiter water chemistry was controlled to limit

(a)

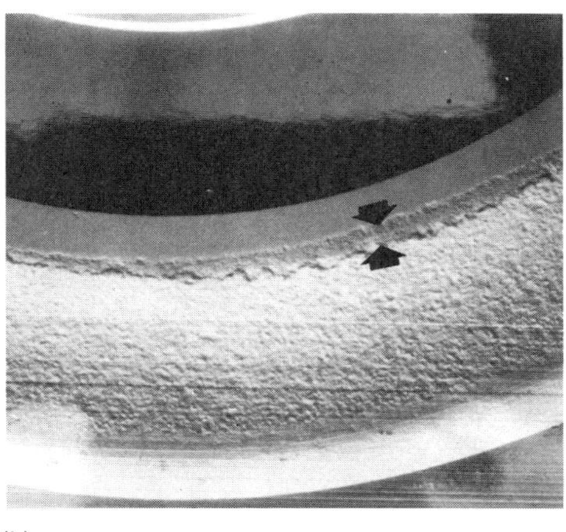

(b)

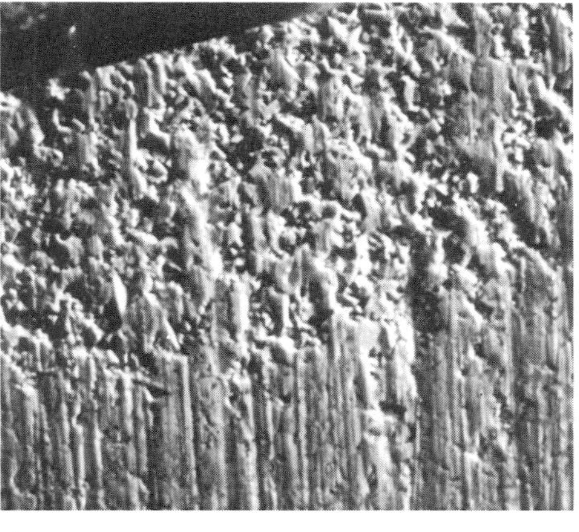

(c)

Fig. 14 Selective leaching of a tungsten carbide valve seat in a shuttle orbiter APU gas generator valve module. Leaching of the cobalt binder was caused by excessive exposure to water during ultrasonic cleaning and hot water rinsing. (a) Valve seat showing narrow sealing surface 0.1 to 0.15 mm (4.5 to 6 mils) wide. 8×. (b) Loss of sealing surface due to selective leaching and poppet impact. 45×. (c) SEM cross section showing depth of leaching. 1600×

oxygen content to 0.5 ppm (to avoid localized oxygen concentration cells), conductivity was limited to $3.3 \times 10^{-6}\ \Omega^{-1} \cdot cm^{-1}$, and the pH was controlled within the 6.0 to 8.0 range. After more than 5 years of service, no corrosion problems have been experienced with stainless steel.

Waste water is handled by 21-6-9 stainless steel lines and is stored in aluminum alloy 6061-T6 tanks. Waste water tends to be extremely corrosive to aluminum. However, coatings such as Tufram, a tetrafluoroethylene-impregnated anodize, have been used effectively to protect aluminum tanks from urine.

The ammonia (NH_3) boiler provides for heat rejection at altitudes below approximately 30.5 km (100 000 ft); at these altitudes, the cargo bay doors are closed, and the boiling of the flash evaporator can no longer provide sufficient cooling to the freon. The ammonia boiler is a shell and tube heat exchanger with a single pass on the ammonia side and two passes for each Freon-21 coolant loop. The ammonia flows through the bank of 77 small-diameter stainless steel tubes, and the Freon-21 flows over the exterior of the tubes. Because these tubes have such thin walls (0.2 mm, or 8 mils), perforation by corrosion is a major concern. Corrosion has been experienced as isolated areas of intergranular attack due to misprocessing of type 304L stainless steel tubing. Very small amounts of carbon residue left on a thin-wall tube such as this will result in sensitization during its brazing cycle. A change has recently been made to a stabilized grade (type 347 stainless steel) to avoid these problems.

Ammonia is stored in a titanium Ti-6Al-4V pressure vessel. No corrosion problems have been experienced in this application. Nitrogen is stored at 22.8 MPa (3300 psi) in Ti-6Al-4V pressure vessels that are filament overwrapped with Kevlar 49 aramid. It presents no corrosion problems.

Gaseous breathing oxygen is stored in a pressure vessel made from Inconel alloy 718 that is also overwrapped with Kevlar 49 aramid filament. It also operates at 22.8 MPa (3300 psi). Extreme care must be taken when designing either liquid or gaseous oxygen systems because of the dangers of ignition with metals and organic materials (Ref 6). Titanium and magnesium are not used in oxygen systems for this reason and are considered to be highly reactive under impact. Materials that are considered satisfactory for use must pass an impact test consisting of 20 consecutive impacts at an energy level of 98 J (72 ft·lb) when tested in the Army Ballistic Missile Agency (ABMA) Impact Tester according to NHB 8060.1, Test 13. Further, when gaseous oxygen pressures exceed 6.9 MPa (1000 psi), materials must also pass dynamic qualification (pneumatic impact) according to NHB 8060.1, Test 14.

Although extreme care is taken in material selection for oxygen systems, this is not enough to ensure an ignition-free system (see the discussion "Case Histories" in this section). Because an oxygen ignition is a catastrophic event that results in an explosion and significant molten metal, it is not always possible to reconstruct the precise cause of a failure. Metallic materials that pass the above tests can still ignite if design or manufacturing operations result in:

- Energetic particle impact, caused by particles accelerated to sonic velocities
- Contamination
- Pneumatic shock from rapid valve opening across large pressure differences
- Fretting or galling, generating particles and localized high temperatures
- Frictional heating, such as by flowing past a feather edge
- Gaseous heating, such as by adiabatic compression and Helmholz resonance (resonance in blind columns) (Ref 7, 8)

Of the various engineering metals tested, Inconel alloy 718 and Monel alloy 400 offer superior resistance to ignition problems and are preferred in valves in which dynamic problems can occur. When properly designed and manufactured, stainless steel has been successfully used in oxygen and LOX valves. Aluminum, stainless steel, and Inconel have been employed without incident in numerous oxygen pressure vessels.

The electrical power system uses fuel cells to generate electricity by combining gaseous hydrogen and oxygen. The electrolyte in the fuel cells—potassium hydroxide—is contained between

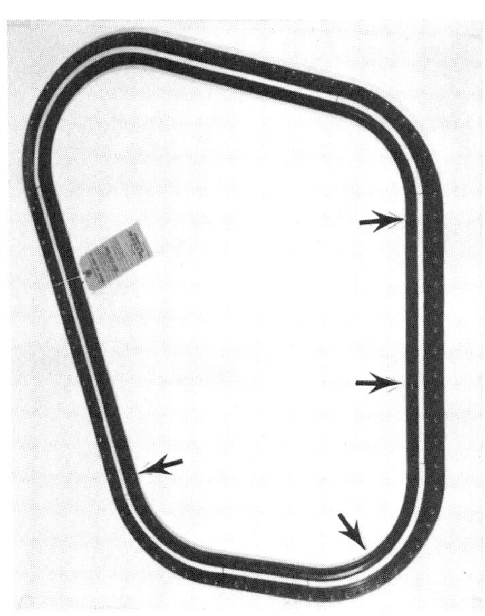

(a)

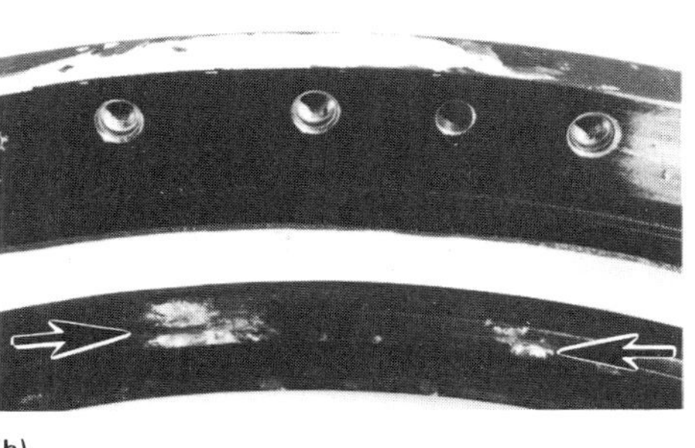

(b)

Fig. 15 Crevice corrosion of an anodized aluminum alloy 2024-T851 window frame from the space shuttle *Challenger*. Corrosion occurred along both thermal and environmental sealing grooves. (a) Window frame showing locations of corrosion (arrows). (b) Enlargement of (a) showing corrosion in Viton seal area (arrows). Rain water carrying dissolved salt deposits from the window was the corrosive medium.

gold-plated magnesium electrodes. For corrosion protection, the magnesium substrate is plated with zinc, copper, and nickel beneath the gold.

The hydrogen and oxygen are stored as supercritical gases in double-wall pressure vessels. The hydrogen is contained in an aluminum alloy 2219-T6 inner shell that has been electron beam welded; the oxygen is stored in an inner pressure vessel of electron beam welded Inconel alloy 718. Both tanks are suspended within pressure-tight external shells made of aluminum alloy 2219-T6. A series of 12 S-glass/epoxy straps suspend the inner tank from the forged girth ring of the outer shell. The annulus between the vessels contains multilayered reflective insulation. A pressure level of 1×10^{-5} torr is maintained by a vacuum ion pump. Aluminum alloy 2219-T6 is very susceptible to intergranular pitting corrosion, and care must be taken during fabrication to prevent dirt particles and/or moisture from coming in contact with it (see the discussion "Case Histories" in this section).

The reaction control system uses rocket engines burning monomethyl hydrazine (MMH) with N_2O_4 to achieve the desired attitude control of the orbiter while in orbit. The 38 reaction control engines are each capable of providing 3870 N (870 lbf) of thrust. Six vernier RCS engines allow fine tuning of the orbiter attitude. They develop 111 N (25 lbf) of thrust each.

Both N_2O_4 and MMH are stored in pressure vessels made of Ti-6Al-4V. Several precautions must be taken in the case of N_2O_4. First, ingestion of moisture will result in the formation of nitric acid, which can be corrosive to some elements of the system.

Second, N_2O_4 spills can be dangerous because of toxicity and can be destructive to spacecraft hardware. Nitrogen tetroxide will readily corrode nickel, strip off protective paints, and dissolve nylon. In addition, because it is an aggressive oxidizer, N_2O_4 can react violently and ignite organic materials.

Third, N_2O_4 will cause SCC of titanium in the absence of a trace of nitric oxide (NO) (Ref 9-12) (see the discussion "Case Histories" in this section). Current specifications call for 1.5 to 3% NO; but repeated loading causes volatile loss of NO, and storage may cause stratification. Although only a trace of NO (perhaps as little as 0.2%) will prevent SCC, care must be taken to ensure at least 0.6% NO to be safe.

Fourth, impact ignition of titanium can occur in N_2O_4. Tests have shown that threaded fasteners can suffer localized melting and that even an impact of inert material, such as glass or sand, on the titanium surface can result in localized melting in N_2O_4. Unlike the oxygen reaction, the reaction is quickly quenched and occurs as low as the 54- to 68-J (40- to 50-ft·lb) level on the ABMA Impact Tester. Therefore, only aluminum fasteners threaded into titanium N_2O_4 tanks are permitted on the space shuttle orbiter; this ensures that designs are free of contamination and potential impacts.

Finally, one of the major problems with N_2O_4 is not the problem of spacecraft corrosion but the deposition of corrosion products (picked up during storage) into valves, preventing valve closure and restricting flow. Nitrogen tetroxide dissolves small amounts of iron (a few parts per million) from storage tanks. The solubility of the iron is a function of temperature, water content, and NO content. Proper conditioning of N_2O_4 prior to loading will precipitate out complex iron nitrate compounds; this will prevent problems caused by

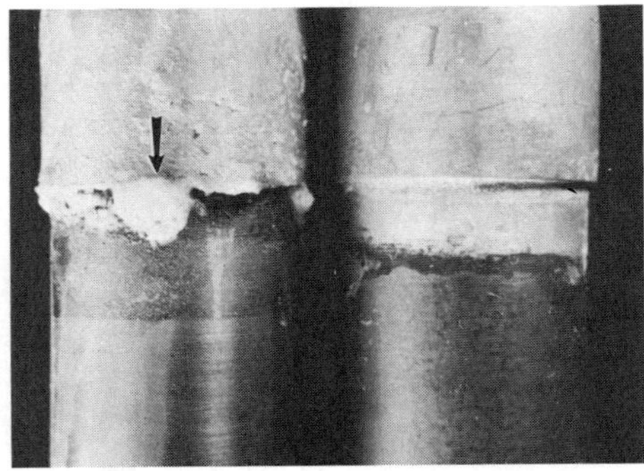

(a)

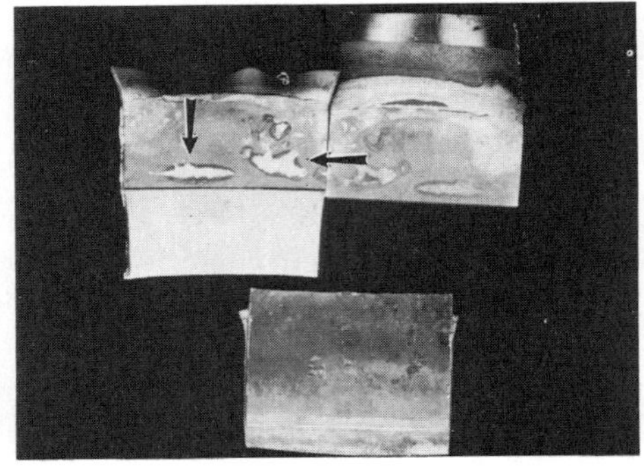

(b)

Fig. 16 Crevice corrosion of an Apollo aluminum-stainless steel brazed joint caused by bleedout of the brazing alloy. Upper portion is aluminum alloy 6061-T6, lower portion is tin-plated type 304L stainless steel. Brazing alloy was 718 aluminum. (a) Foaming of aluminum hydroxide corrosion products (arrows). Entrapped flux exposed to air caused corrosion. 1×. (b) Cross section through pockets of entrapped flux (arrows)

corrosion product deposition (see the discussion "Case Histories" in this section).

Monomethyl hydrazine itself can cause problems. Although it is not as unstable or reactive as neat (pure) hydrazine, which will be covered in detail in the discussion "Auxiliary Power Unit" in this section, it must be handled in essentially the same way as pure hydrazine.

Rocket chambers used in the reaction control system are made from niobium alloy C-103. These are coated with an R512A silicide coating that prevents oxidation in high-temperature service (up to 1315 °C, or 2400 °F). No oxidation problems with reaction control engine chambers have been experienced in service. The primary chambers are film cooled with hydrazine. Burnthrough has occurred in laboratory testing under off-limit conditions (engine instability); however, current design modifications prevent further structural damage to the spacecraft by using automatic sensing devices that shut off fuel and oxidizer valves if penetration of chamber walls occurs.

The vernier engines have also shown localized burnthrough during extensive laboratory testing when hot oxidizer impinges on the silicide coating. This condition is aggravated by the thousands of thermal cycles (thermal fatigue) required by this engine, the inability of the coating to accommodate minor machining offsets or discrepancies without fracture, the lack of a fuel-cooling film along the inside of this chamber design, and a doublet-type injector that limits the mixing of the fuel and oxidizer (see the discussion "Case Histories" in this section).

Helium pressure vessels (28 MPa, or 4000 psi) made of annealed Ti-6Al-4V liners overwrapped with Kevlar 49 aramid filament provide the pressure necessary to feed the propellants. Some corrosion problems have been encountered in helium systems. The orbiter has experienced problems with hydrazine vapors migrating into the fine orifices of helium valves. Reactions with surface contaminants have resulted in plugging of orifices due to complex hydrazine deposits (see the discussion "Case Histories" in this section). Helium systems must always be designed to be compatible with the fuels and oxidizers with which they are used, because back migration (diffusion) of these substances will occur.

The orbital maneuvering system provides the propulsion to insert the shuttle orbiter into earth orbit, to change orbit, to rendezvous, and to deorbit. As with the reaction control system, it also uses the storable propellants N_2O_4 and MMH, as well as a helium pressurant system. The two OMS engines are capable of providing 267 000 N (6000 lbf) of thrust. The propellant tanks have the capacity to provide a change in velocity of 300 m/s (1000 ft/s) when carrying a full payload of 29 500 kg (65 000 lb).

The nozzles are made of niobium alloy FS-85 coated with an R512E silicide coating that is used for the RCS rocker chambers. These nozzles are used for service to 1350 °C (2480 °F) (Ref 13). Injectors are made from diffusion-bonded 300-series stainless steel platelets. Platelets have injector hole patterns etched in them by a photographic etching process.

The pressure vessels for the fuel and oxidizers, operating at 2.2 MPa (315 psi), are made from annealed Ti-6Al-4V. The helium pressure vessels (33 MPa, or 4800 psi) are made from annealed Ti-6Al-4V overwrapped with Kevlar 49 aramid filament.

The fuel and oxidizer present the same types of problems as those experienced in the RCS, except for the nozzle extensions. High-temperature oxidation of the nozzle extensions has occurred in areas where mechanical deformation (buckling) caused cracking and spalling of the silicide coating and where thermocouple attach brackets were broken off in service (see the discussion "Case Histories" in this section).

The main propulsion system provides the vacuum-jacketed lines for carrying liquid hydrogen and liquid oxygen from the external tank to the shuttle main engines. These lines are also used to carry high-pressure high-temperature gaseous oxygen and hydrogen back to the external tank as a pressurant to expel the liquid hydrogen and liquid oxygen.

The compatibility problems with gaseous oxygen have been covered in the discussion "Environmental Control and Life Support System" in this section. With gaseous hydrogen, the compatibility issue concerns the embrittlement of metals by hydrogen under certain specific conditions. The extent of embrittlement is a function of hydrogen pressure, strain level, and, probably, time of exposure. The embrittlement is thought to occur from rupture of the protective oxide film of the metal, followed by some mechanism by which atomic (nascent) hydrogen enters the metal. In mild cases, this embrittlement takes the form of a reduction in notched strength in hydrogen compared with specimens exposed to helium or air.

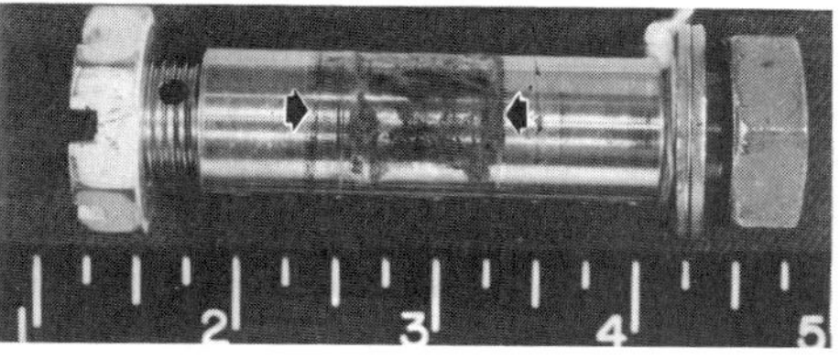

(a)

(c)

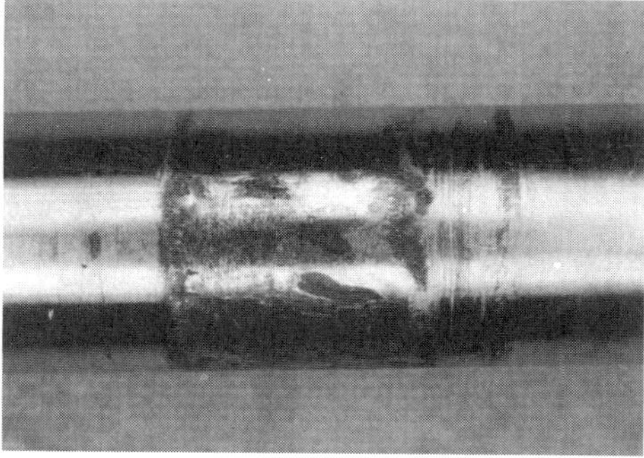
(b)

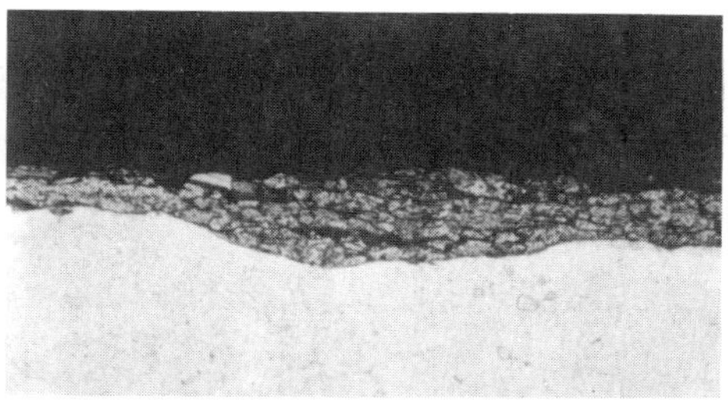
(d)

Fig. 17 Fretting corrosion of a steel spacer used to mount the rudder speed brake on the shuttle orbiter. The spacer is made of 17-4PH H1150M stainless steel. (a) Spacer on bolt shows contact area with an Inconel alloy 718 spherical bearing. Fretted area is between arrows. (b) Enlargement of fretting corrosion. 1×. (c) Mating Inconel alloy 718 bearing showing a similar pattern but only superficial marring of surface. 1.5×. (d) Cross section through fretting corrosion. 175×

In more severe cases, the ductility or tensile strength of the alloy changes significantly.

In the case of titanium, embrittling hydrides are formed; within a few minutes, these hydrides can result in the destruction of the entire cross section of a part (see the discussion "Case Histories" in this section). For this reason, exposure of titanium to hydrogen is totally avoided.

With other metals, it is important to verify compatibility at the maximum service pressures. Most metals do not show significant property changes below 3.45 MPa (500 psi). At hydrogen pressures of 13.8 MPa (2000 psi), materials such as Inconel alloy 718 show measurable loss of ductility, and at 69 MPa (10 000 psi), even austenitic stainless steel is affected.

No accepted criterion exists for the use of metals in gaseous hydrogen. At the Rockwell Space Transportation System Division, for example, an alloy is used in gaseous hydrogen if its sharp notched ($K_t \sim 17$) to unnotched strength ratio at maximum design pressure does not fall below 1.0 and if a factor of safety of four is maintained in the system. At the Rockwell Rocketdyne division, materials are strain limited, and materials susceptible to high-pressure gaseous hydrogen are copper plated (0.1 to 0.25 mm, or 4 to 10 mils) to ensure that no adverse reactions occur. Welds in nickel alloys are overlayed on the root side by two layers of Inconel alloy 903.

Leakage of hydrogen is a major concern in design because hydrogen can form an explosive mixture with air at concentrations between 4 and 96% hydrogen. The aft areas of the orbiter are extensively purged with helium before launch to avoid this problem if a leak occurs.

The vacuum-jacketed lines carrying LOX and LH_2 are of two sizes—305 mm (12 in.) and 430 mm (17 in.) in diameter—and are made of welded Inconel alloy 718. Inconel alloy 718 is ideal for this service because it is compatible with both liquid and gaseous hydrogen (at these pressures) and maintains high ductility below −253 °C (−423 °F). The vacuum-jacketed lines must accommodate expansion and contraction and are designed to do so by articulation and angulation at joints. Bellows of Inconel alloy 718 accommodate angulation while containing the cryogens. The articulation is provided by gimbal rings and internally supported ball-strut tie-rod assemblies. The hardened balls, ranging in size from 32 to 57 mm ($1^1/_4$ to $2^1/_4$ in.) in diameter, are made of a tungsten carbide (Stoody 2) alloy.

Inconel alloy 718 (see the discussion "Plumbing Lines" in this section) is used for the high-pressure high-temperature oxygen lines (31 MPa, or 4500 psi, at 280 °C, or 540 °F) that deliver pressurant to the external tank. Helium pressure vessels are made from annealed Ti-6Al-4V and annealed Ti-6Al-4V overwrapped with Kevlar 49 filament.

The auxiliary power unit provides hydraulic pressure for the actuation of a number of flight control systems, including the rudder, speed brake, body flap, elevons, landing gear, and brakes. Power is obtained by vaporizing and decomposing neat (pure) hydrazine in a catalyst bed and passing it through a two-stage turbine, which in turn drives a hydraulic pump. Each of the three APUs develops 100 kW (135 hp), and the pump delivers 18 MPa (3000 psi) to the hydraulic system.

The compatibility of hydrazine with materials in a system must consider several points. First, hydrazine can be unstable and will decompose into N_2, H_2, and NH_3, causing a rapid pressure rise and, if not properly vented, an explosion. Neat (pure) hydrazine tends to be more reactive (unstable) than monomethyl hydrazine. Decomposition is catalyzed by metal surfaces and/or contaminants left on surfaces, even after cleaning. The data base in aerospace is large and highly inaccurate regarding metal-hydrazine compatibility for several reasons:

- Many of the early investigators rated compatibility as satisfactory based on appearance of the metal specimen exposed, not the fluid reaction
- No uniform cleaning method or decomposition criteria existed when data were generated
- No uniform controls were used in testing

The current approach at the Space Transportation and System Division is to use materials that have a history of satisfactory hydrazine service at the temperatures expected. Where a new material is used, pressures rise tests of the new material versus time are made, along with controls of compatible materials cleaned in the same manner. Literature data, because of their unreliability, are not used except to indicate which materials should be tested.

Second, a metal should also be checked with hydrazine to determine its autoignition temperature (AIT). This is done to ensure the AIT is safely below operating temperature. No standards are available for autoignition testing. Frequently, the fluid is allowed to drip onto a heated plate of the test metal, and the temperature of the test metal is slowly raised until ignition occurs.

Third, hydrazine, with even short exposure to air (a few seconds), will react with CO_2 to form carbazic acid—a very viscous, sticky compound which can clog lines, orifices, or valves. It can also aggressively attack certain metals, such as cobalt and nickel. Hydrazine systems must always be kept under an inert gaseous blanket whenever opened.

Fourth, hydrazine (and carbazic acid) will dissolve or selectively leach certain metals. This may later result in malfunction of a valve by the flow-decay phenomenon, in which hydrazine salts of these metals end up precipitating at valve seat areas (see the discussion "Case Histories" in this section).

Fifth, every effort must be taken to avoid hydrazine spills, because hydrazine is very toxic (Table 2) and highly flammable in air if spread out over a large area (large surface-to-volume ratio). Ignition has occurred when hydrazine was spilled on orbiter thermal protection system tiles made from sintered pure SiO_2 filaments.

The metals used in the APU in contact with hydrazine include 300-series stainless steels, precipitation-hardening steels, tungsten carbide (for valve seats), and Hastelloy alloy B. After the hydrazine has been decomposed by the catalyst bed, its major compatibility problem is with the formation of nitrides by the hot ammonia gas formed. The catalyst bed exceeds 925 °C (1700 °F), and the turbine operates at 595 °C (1100 °F). Alloys are chosen both for high-temperature properties and resistance to nitriding. The turbine wheel and blades are made from René 41. The wheel has a circumferential Inconel alloy 625 shroud welded with Hastelloy alloy W wire to the blade tips. The injector section and catalyst bed housing are from Hastelloy alloy B.

The hydraulic system uses predominately type 300 series stainless steel valves and compo-

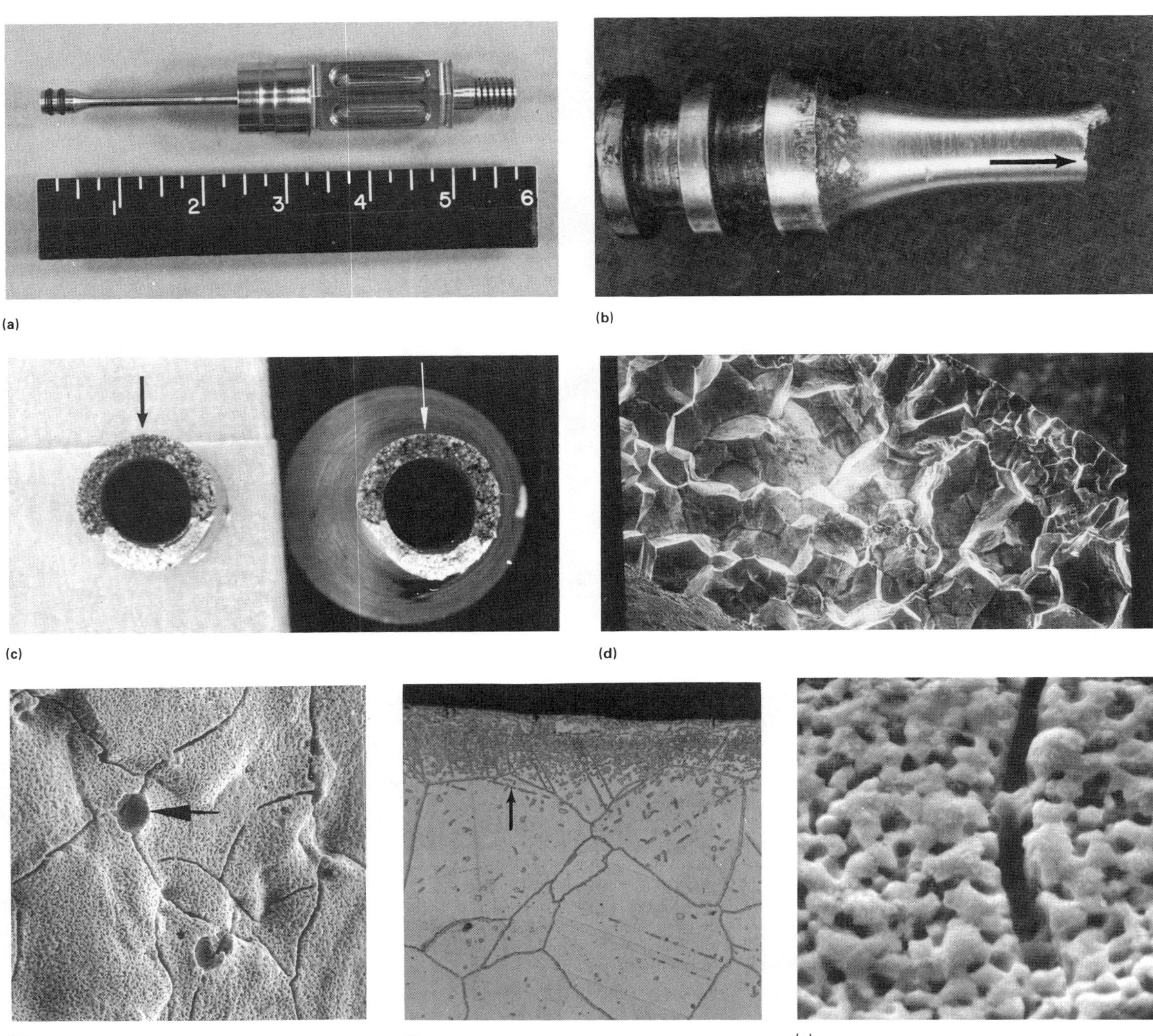

(a) (b) (c) (d) (e) (f) (g)

Fig. 18 Stress-corrosion cracking of a Hastelloy alloy B orbiter APU injector tube in a hydrazine environment. (a) Typical injector tube and catalyst bed. (b) Failed tube of APU #1. Arrow shows area of SCC. (c) Fracture faces of failed APU #2 injector. Dark areas (arrows) have been stress corroded. Bright areas were ductile fracture from tensile load used to separate parts. 6×. (d) Enlarged fracture face of (c) showing intergranular character. 940×. (e) Surface SEM of tube inside diameter showing etching of carbides in grain boundaries (arrow) and surface. 370×. (f) Sensitized surface layer on inside diameter due to carbide contamination (from electrical discharge machining process) entering braze cycle. 285×. (g) Enlargement of surface corrosion. 3450×

nents attached to Ti-3Al-2.5V lines. The titanium lines were chosen because they saved approximately 270 kg (600 lb) over stainless steel. Permanent joints are externally "swaged" with Permaswage fittings that also incorporate RTV 630 rubber seal rings as a backup to the metal-to-metal seal. In the presence of hydraulic oil, the rubber expands, ensuring a tight joint capable of sealing hydraulic oil. Hydraulic oil per MIL-H-83282A is used throughout the spacecraft except in the landing gear struts, where oil per MIL-H-5606C is used. Control of the chlorine content of hydraulic oil is the most important approach to ensuring corrosion-free systems. (Chlorine concentrations above 100 ppm can cause corrosion problems.) No major corrosion problems have been experienced in the shuttle orbiter hydraulic systems.

Mechanical Systems

Mechanical systems include primarily mechanical devices, pyrotechnical devices, and landing gear systems. These systems mainly employ high-strength steels, Inconel alloys, and some aluminum alloys.

Mechanical Devices. Steels, because of their high strengths and hardnesses, are often preferred for mechanical devices. Higher loading can be directed into a smaller space volume than with aluminum or titanium, which are less dense. Typical applications involve 4130 and 4340 low-alloy steels for mechanical devices and 4340 or alloy steel carburized grades for gears. Low-alloy

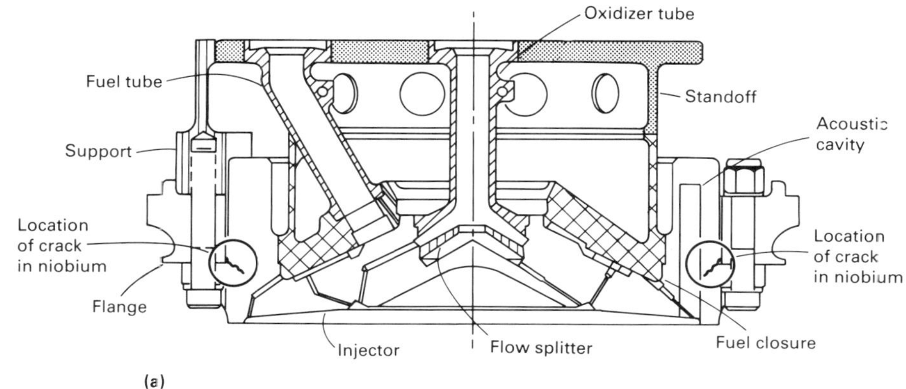

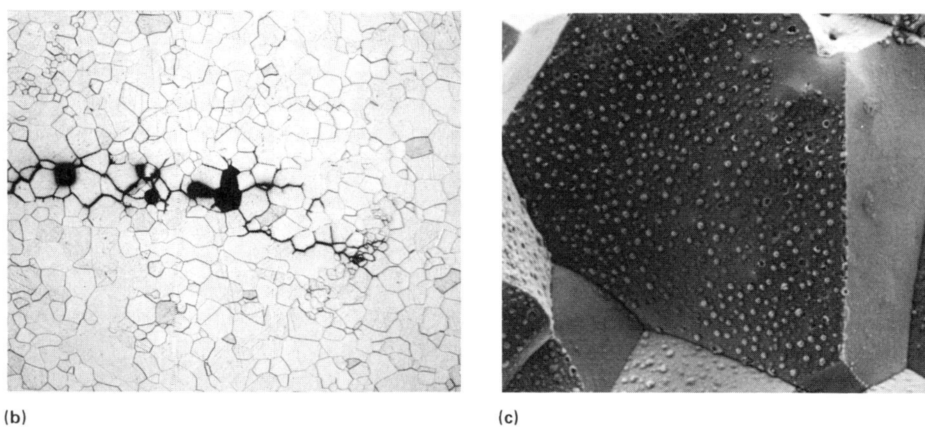

Fig. 19 Fluoride hot salt SCC of niobium alloy C 103 injector used on the orbiter RCS chambers. (a) Schematic of C 103 injector and titanium bolting ring showing failure area. (b) Cross section through failure showing intergranular attack. 60×. (c) Fracture face showing grain boundaries and microscopic eruptions of niobium oxide. 1035×

steel parts must be protected from corrosion. Depending on the design and service, the following types of systems are used:

- Paints, such as chromated epoxy polyamines
- Phosphate coating (DOD-P-16232) with oils
- Cadmium-titanium or cadmium plating plus an overcoating
- Electroless nickel
- Chromium, electroplated
- Black oxide
- Nickel/tin, electroplated
- Electroless nickel plus electroplated chromium
- Vapor-deposited aluminum

In addition, lubrication coatings such as Braycoat grease (space compatible), molybdenum disulfide in an organic matrix, or Vitrolube NPI-1220 (a high-temperature vitreous-base lubricant) are used. Where sufficient corrosion resistance cannot be achieved with steels, corrosion-resistant alloys are used, such as Inconel alloy 718, titanium Ti-6Al-4V or Ti-6Al-2Sn-4Zn-6Mo, and precipitation-hardenable steels (such as 17-4PH, 15-5PH, AM350, AM355, and Custom 455).

For bearings, type 440C stainless steel and 52100 bearing steel are used, protected with Braycoat grease. High-strength springs are made of type 301 stainless steel in the condition B spring temper, Elgiloy, and 17-7PH in the CH900 condition. High-strength Belleville washers, using 6150 low-alloy steel, exhibit the best resistance to hydrogen embrittlement if they are coated with vapor-deposited aluminum (see the discussion "Case Histories" in this section). Some use is also made of maraging steels and 9Ni-4Co-0.3C steel for such applications as hydraulic cylinders.

Pyrotechnic devices are used to separate the external tank from the orbiter, to release umbilicals, and to open emergency escape hatches or panels. Pyrotechnic devices operate in a number of different ways. In some designs, the explosive device is within the bolt or nut, forcing its separation. Other designs may cut a panel with a linear-shaped charge or blow a panel joint apart far enough to break bolts at a prenotched section. Explosively actuated guillotines have been used on the Apollo to cut cables or shrouds and have been used in some shuttle applications. Pyrotechnic devices have been used to deploy parachutes or to release landing gear. On the shuttle orbiter, most pyrotechnic devices are made of Inconel alloy 718. This includes the frangible nuts attaching the external tank to the orbiter in the aft section, the explosive bolts that mate the external tank with the orbiter in the forward section, and the crew escape and emergency egress hatches.

Guillotine blades have been made of Inconel alloy 718, A-286, and corrosion-protected tool steels. No corrosion problems have occurred in these areas.

The orbiter landing gear system is a conventional aircraft tricycle configuration with steerable nose gear and the main left and right landing gear. The major parts of the system include the shock strut assembly, wheels and tires, axles, brakes and antiskid controls, and nose wheel steering and damping controls. Initiation of the system hydraulically releases uplock hooks that permit the landing gear to free fall into the extended position. Springs and hydraulic actuators assist in the free fall. Pyrotechnic actuators may also unlock the uplock hooks if the hydraulic system malfunctions. When fully extended, the gear is locked by spring-loaded bungees.

Minimum weight was a prime consideration in the design of the landing gear. This resulted in the need to use high-strength materials, highly efficient braking materials, and minimum thicknesses for tires and wheels. Tires were designed for 2 landings, brakes for 5 landings, and wheels for 100 missions, although increased loading has reduced wheel life somewhat. The steel used for parts carrying high loads was primarily 300M at a 1895-MPa (275-ksi) strength level. The use of high-strength steels in this manner represented a deviation from limitations placed by the Material and Process Group on steel strength levels throughout the spacecraft (see the discussion "Primary Structure" in this section). At this high strength level, steel is notch and impact sensitive, has limited toughness, is prone to stress-corrosion problems, and is highly susceptible to hydrogen embrittlement. Its selection, however, was based on a history of satisfactory use in landing gear applications for both civilian and military aircraft over a 20-year period. Nevertheless, serious embrittlement problems were encountered during development and in early orbiter service.

The 300M steel was corrosion protected on functional surfaces by either cadmium alone or by chromium plating. Nonfunctional surfaces were coated with a cadmium-titanium plating plus a chromated epoxy polyamine paint primer and a polyurethane topcoat. Normally, cadmium plating would be avoided because of its tendency for sublimation and redeposition in the space environment, but the environment of the landing gear in the orbiter wheel wells precluded space exposure and sublimation problems. No corrosion problems have been identified with the 300M finish system in service; however, several misprocessing problems with chromium plating and plating of the cadmium-titanium finish have resulted in hydrogen embrittlement failures (see the discussion "Case Histories" in this section). The cadmium-titanium plating is considered to be a low-embrittlement plating process and is currently covered by MIL-STD-1500.

The current brake designs use thermal grade beryllium heat sink disks (both rotors and stators) with reinforced carbon-carbon linings. They are chemically filmed in a manner similar to that described for aluminum in MIL-C-5541 and are unprotected on their mating (working) surfaces. Temperatures during braking can be very high locally because of uneven pressure distributions and have occasionally resulted in beryllium carbide formation (>1095 °C, or 2000 °F) during high-energy stops in early brake designs. Carbide formation results in embrittlement, cracking, and potential failure of the braking system. Unfailed parts showing this condition are scrapped. Galvanic corrosion between the carbon and beryllium would appear to be a concern, but such a problem has never occurred. Newer brake designs will use structural carbon-carbon to replace

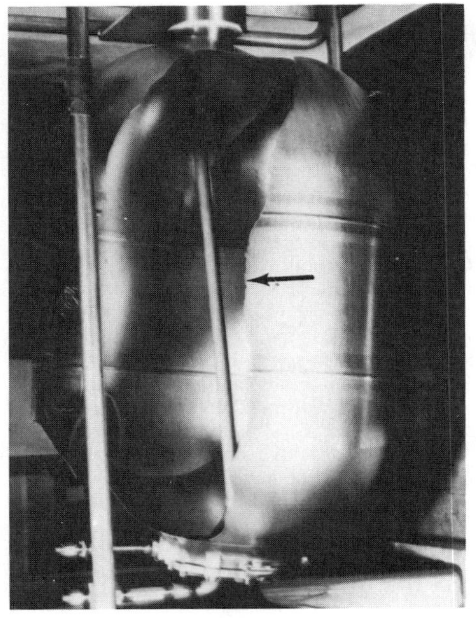

(a)

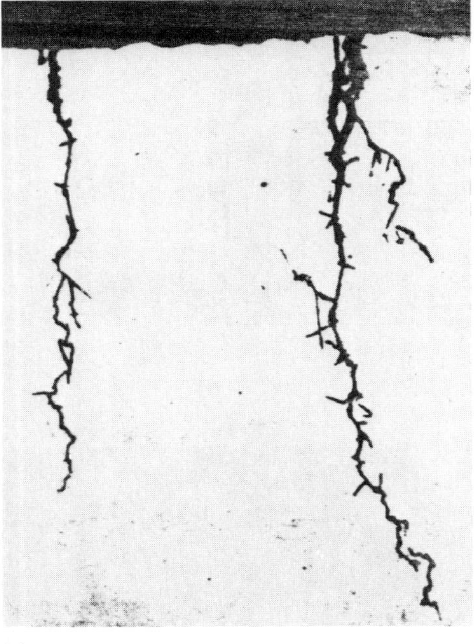

(b)

(c)

(d)

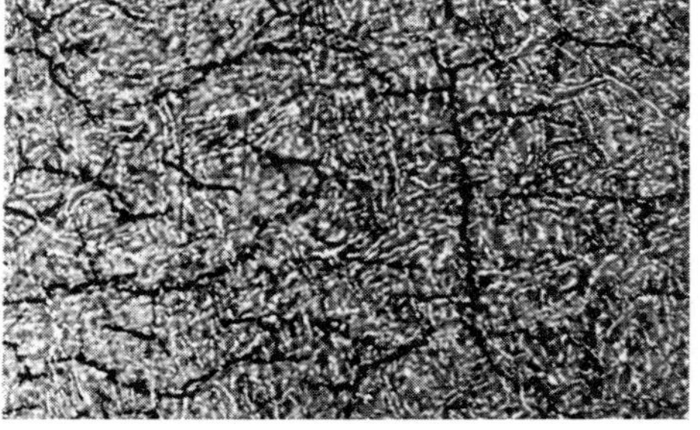

(e)

Fig. 20 Stress-corrosion failure of an Apollo Ti-6Al-4V RCS pressure vessel due to nitrogen tetroxide. (a) Failed vessel after exposure to pressurized N_2O_4 for 34 h. (b) Cross section through typical stress-corrosion cracks. 250×. (c) Correlation between the number of cracks per square inch and stress level. (d) Cracking in cylindrical section where hoop stress predominates. (e) Cracking in biaxial area where stresses are approximately equal. (d) and (e) Both 35×

the beryllium; this will result in longer life (lower lifetime costs) and higher energy absorption capability, but higher axle and wheel temperatures are expected.

Axles were made of 300M steel. The nose landing gear axle was later changed to Inconel alloy 718 for improved fatigue life. The cadmium-titanium coating on the 300M main landing gear axles will not exceed 230 °C (450 °F) during normal operation and is safe from liquid cadmium embrittlement. In a rare, high energy level emergency stop, this temperature could be exceeded, resulting in a potential for cadmium embrittlement. This would occur as heat from the brakes soaks back into the axle well after the vehicle has stopped. Under these conditions, the axle would be replaced.

Wheels are currently made of die-forged halves of aluminum alloy 7049-T73 that are bolted together with high-strength MP35N fasteners. The main wheels are chemically filmed according to MIL-C-5541, followed by a chromated epoxy polyamine primer (MIL-P-23377) and a topcoat of polyurethane according to MIL-C-83286 for corrosion protection. The nose wheel substitutes anodizing for the chemical film treatment; otherwise, it is identical. No corrosion problems have been experienced.

Aluminum alloy 7075-T73 is used for hydraulic actuators. Exterior surfaces are protected by chromate conversion coating plus chromated epoxy polyamine primer and polyurethane topcoat. Interior surfaces are immersed in hydraulic fluid and are not coated. No corrosion problems have occurred with the aluminum acuators.

Avionics

All orbiter avionics systems are located within the spacecraft structure or pressurized cabin; therefore, from a corrosion standpoint, they experience a controlled and relatively benign atmosphere that is free of rain and salt spray. Electronic equipment located within the orbiter cabin or cargo bay areas, because of the controlled humidity, cannot experience condensation of moisture, but equipment within the aft equipment bays can experience potential condensation. Based on the types of equipment, corrosion protection can be further examined in three areas:

- Black boxes
- Electronic circuits
- Electrical connectors

Black boxes are typically made of the relatively corrosion-resistant 6000-series aluminum alloys, either welded or dip brazed and resin sealed. The boxes are painted or anodized according to MIL-A-8625 on the exterior. They are chemically filmed according to MIL-C-5541 on the interior and on areas that contact coldplates to ensure maximum heat transfer and electrical grounding. The paint system is not fixed; it may be the chromated epoxy polyamine primer used for the spacecraft or a system chosen by the manufacturer; however, it must meet the requirements for exposure to humidity and salt spray of Federal Test Method Standard 141, Methods 6201 and 6061, respectively. Some boxes are sealed with a gasket; others are environmentally sealed and filled with an inert gas. Air-cooled black boxes located in the crew cabin are unsealed.

All electrically active surfaces of electrical and electronic circuits, including solder joints, printed circuits, and wire terminations, are con-

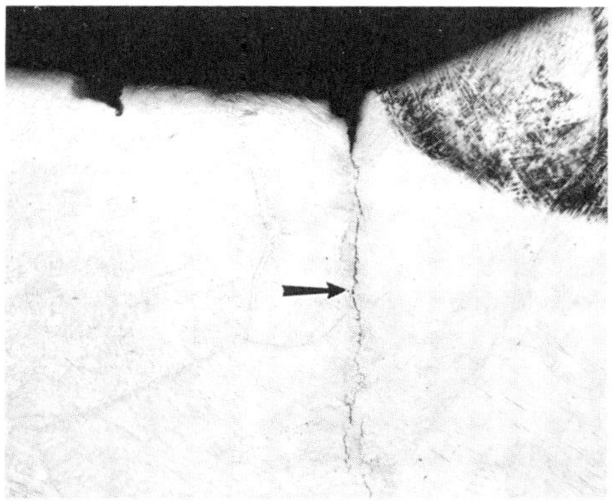

(a)

(b)

(c)

(d)

Fig. 21 Stress-corrosion cracking of a solution-treated and aged Ti-6Al-4V Apollo SPS fuel pressure vessel during a system checkout test. Fluid test medium was methanol. (a) Cross section adjacent to weld in cracked vessel. 65×. (b) Another crack near the same weld. 65×. (c) and (d) TEM fractographs of fracture surface showing no particular stress-corrosion features. Both 2500×

formally coated. The coatings used include Dow Corning 3140 and 3145, Columbia Technology Hysol and Furane Plastics polyurethanes, and General Electric RTV 560 and RTV 566. Printed circuit conformal coating thicknesses range from 50 to 250 μm (2 to 10 mils). Coatings on other surfaces range from 100 to 375 μm (4 to 15 mils).

Electrical connector metallic parts are made of various aluminum alloys or 300-series stainless steels. All aluminum parts are nickel plated. Plating thickness is not always specified, but connectors must pass salt spray corrosion tests per MIL-STD-202, method 101, test condition B.

Connector pin and socket contacts are electrolytically gold plated over a copper and nickel strike. A gold plate thickness of 1.25 μm (50 μin.) is considered to be borderline between porous and nonporous plating. All orbiter connector contacts include a minimum gold plating thickness of 2.5 μm (100 μin.). When mated, the connectors are environmentally sealed by peripheral gaskets.

No major problems have occurred with corrosion of avionics devices. Nitrogen tetroxide spills in the OMS pod areas and forward RCS areas have required in-depth evaluation of the suitability of connectors and pin contacts; however, no abnormal functions were noted even though connector bodies turned green.

Case Histories

Despite extensive efforts to anticipate and avoid corrosion problems with manned spacecraft, such problems will inevitably occur. Corrosion problems, as a group, have clearly become the most critical and costly problems with metals used on space vehicles. Because many corrosion problems can result in unexpected or catastrophic failures, especially those of SCC, hydrogen embrittlement, and metal ignition (in oxidizers), their impact is usually severe. Extensive efforts must be made to identify the precise cause of failure and to provide suitable inspections and rework for existing hardware; the lives of the astronauts cannot be jeopardized nor can the risk be taken of severe damage to a $2 billion orbiter. As a consequence, some corrosion problems have resulted in rather expensive launch delays, and the inevitable questions are asked: "Why did this occur? Why was this not foreseen?"

There are no simple answers. Field experience is unparalled in revealing differences in behavior from laboratory test results or the deficiencies in the hardware designs. The reasons for corrosion problems discussed in this section can be broadly categorized as follows:

- Lack of adequate protection of parts during the manufacturing cycle
- Failure to remove processing chemicals or fluids completely
- Failure to provide an adequate hydrogen embrittlement relief
- Use of improper or contaminated fluids during manufacturing or testing
- Failure to control fluid chemistry within a spacecraft fluid system
- Inadequate corrosion protection
- Unknown reactions or unforeseen problems

In this section, a variety of different corrosion problems that have occurred will be presented. These embrace many metal systems, such as aluminum, stainless steel, low-alloy and precipitation-hardenable steels, nickel, titanium, and niobium, as well as several different types of corrosion attack. The descriptions that follow will identify the specific causes of the corrosion, its characteristics, and how it was corrected.

Pitting Attack

Pitting attack of metals is caused by local breakdown in a protective oxide film due to local surface inclusions or defects, local changes in microstructure (such as by precipitates that strengthen the metal), or local corrosion cells brought about by deposition of more cathodic materials on a surface. When a pit occurs, the bottom of the pit is anodic and often continues to grow until perforation occurs.

Pitting attack of the 2000- and 7000-series aluminum alloys often occurs during the manufacturing cycle if parts are not properly protected with oil or chemical film (MIL-C-5541). Many parts are transported to machining sources, chemical milling plants, and heat-treating companies for processing, often in open-top trucks in California. Occasionally, parts are not unloaded over the weekend and are left exposed to the elements. In other cases, parts may be stored outside, perhaps protected by plastic or a tarp, or even stored inside next to doors; these methods of storage allow condensation to form on the parts. No outdoor storage is satisfactory. Dust that collects on the parts absorbs moisture from the air, resulting in significant pitting—in some cases within 24 h. Moisture condensation under plastic in contact with the part results in severe attack.

Aluminum Spacecraft Structural Parts. Typical pitting attack (Fig. 6) occurred on an aluminum alloy 2014-T6 test part during the Apollo program. The maximum depth of attack was 0.1 mm (4 mils), about the same as the diameter. The pitting is intergranular. If the quench rate is sufficiently rapid during the heat treating of aluminum, no intergranular corrosion occurs within a pit (see the article "Heat Treating of Aluminum Alloys" in Volume 4 of the 9th Edition of *Metals Handbook*). Because pits such as that shown in Fig. 6 are only slightly larger in diameter than a human hair, they may be missed unless close surface examination of aluminum is made at a magnification of 5 to 10×. This has made it difficult to convince in-house manufacturing personnel or subcontractors of the seriousness of protecting surfaces, and it is only when a gross area of pitting becomes obvious that arguments cease. Another Rockwell facility in Tulsa insisted that the pitting problem was characteristic of the California environment and even ran 6-month exposure tests of unprotected aluminum to prove it. Results are shown in Fig. 7. The coupon shown in Fig. 7(a) was heat-treated aluminum alloy 2024 exposed in the metal-processing area, and the one shown in Fig. 7(b) was heat-treated aluminum alloy 2014 exposed in the manufacturing area.

Typical pitting on improperly protected spacecraft parts is shown in Fig. 8. The pits shown are

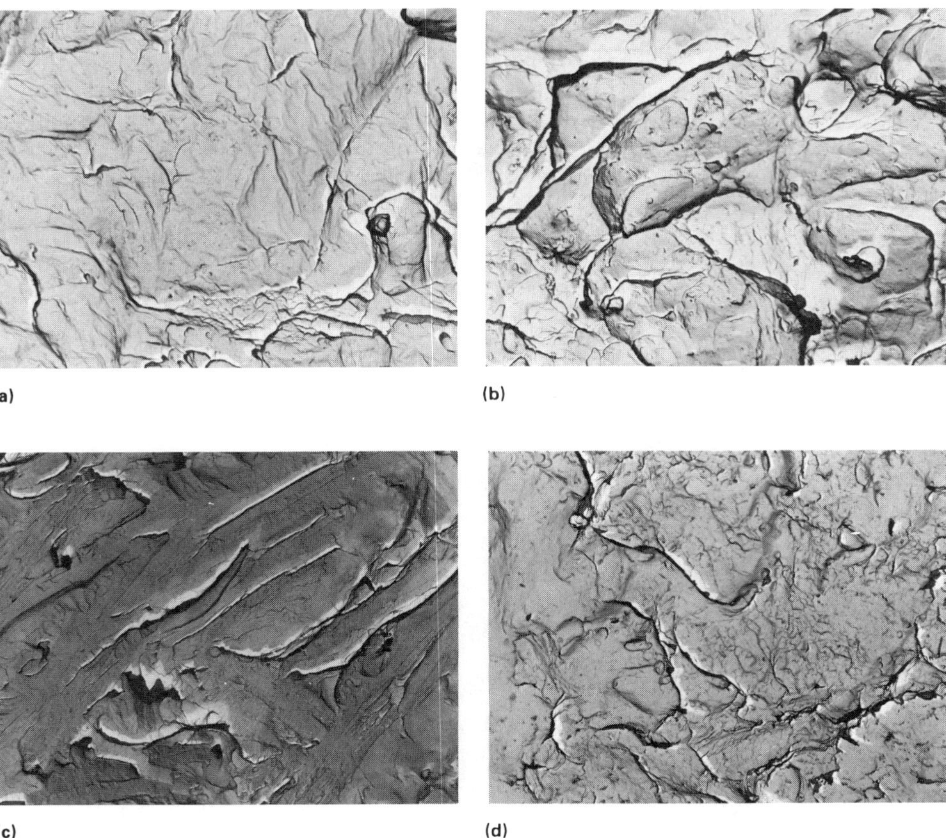

Fig. 22 Stress-corrosion failure in an Apollo Ti-6Al-4V pressure vessel development test. (a) and (b) TEMs of fracture face of stress-corrosion crack in the vessel in a 24-h distilled water exposure after contamination of titanium with soap prior to heat treatment (aging). Fine hairlike wrinkles are characteristic of stress corrosion. (c) and (d) Stress-corrosion failure of first Apollo SPS development pressure vessel of Ti-6Al-4V. Cause unknown

typical of those found in a 7075-T6 radial shear beam of the Apollo Service Module after a fatigue test failure. Although the fatigue test represented an overtest of the structure, pitting and extensive end-grain attack from the chemical milling process contributed to the failure. Figure 9 shows typical pitting corrosion of an aluminum alloy 2024-T62 fitting resulting from inadequate protection during manufacturing.

Pitting of structures may lead to premature fatigue failure in critically loaded or cycled parts; however, in many cases, it is not critical. Pitting of honeycomb sandwich face sheets may undermine the integrity of the sandwich by moisture ingestion. Pitting of gear teeth or springs often leads to failure in mechanical systems. Pitting in fluid-containing hardware (pressure vessels, tubes) may lead to perforation and leakage. By the time pitting is discovered on structural parts, the part may have already cost $25 000 because of extensive machining, or it may be part of a welded assembly worth over $1 million dollars; therefore, every effort is made to save the part.

Pitting corrosion in aluminum must be deactivated, or pits will continue to grow as the aluminum oxide that is formed continues to absorb moisture from the air. Pits can be deactivated in place by using a proprietary deoxidizer, followed by a deionized water rinse, wipe, and pH test of the pits. Local masking is used. Pits are then brush chemically filmed according to MIL-C-5541 and painted. Where stress analysis permits, pits can be sanded flush or removed with a dental drill or hole drill. In some cases, bonded doublers are added to restore strength.

Filiform Corrosion

Filiform corrosion is a special form of corrosion that occurs underneath a protective film. It is a moving oxygen concentration cell. Corrosion takes the form of threadlike or filamentary trails and proceeds along the metal surface rather than penetrating through the thickness. Significant filiform corrosion can occur in a matter of hours or days. It develops in the presence of relative humidities as low as 60%. A key condition for the development of filiform corrosion is that the film is semipermeable, permitting oxygen as well as humidity to pass through it. Filiform corrosion, therefore, is essentially a form of crevice corrosion in which one member forming the crevice (the protective film) is semipermeable.

In filiform corrosion, an anodic head, typically 0.08 to 0.13 mm (3 to 5 mils) wide, advances and dissolves the metal in its path. The pH of the head is highly acidic (often as low as 1), and the tail or trail increases in pH, often exceeding a pH of 8. As the anodic head advances, the cathodic region behind it fills with corrosion products. Filiform corrosion can be recognized by the following characteristics:

- Relatively shallow corrosion of the metal surface
- Meandering, filamentary pathways
- Occurrence below a paint or other protective film

To prevent filiform corrosion, coatings with lower permeability to water and oxygen are required. The following is an example of failure of a pressure vessel by filiform corrosion.

Liquid Hydrogen Pressure Vessel. During modification of the *Columbia* vehicle, an aluminum alloy 2219-T6 liquid hydrogen tank used for supplying hydrogen to the fuel cells was removed from the vehicle and stored inside the plant for a 3-month period; the tank was inadvertently exposed to the atmosphere. When this was discovered, a moisture test was conducted of the air inside the vessel. It indicated higher than acceptable levels. The exterior of the pressure vessel was covered by a vacuum jacket (see the discussion "Electrical Power System" in this section) and protected from corrosion. Boroscopic examination of the interior of the tank revealed a corroded zone completely encircling the vessel on both sides of the electron beam girth weld. The zone was approximately 4.7 mm ($^3/_{16}$ in.) wide and centered 9.5 mm ($^3/_8$ in.) from the edge of the girth weld (Fig. 10).

It was not possible to verify the depth of the attack or to deactivate the corrosion in place. Therefore, it was necessary to remove the tank from service and examine it metallographically. Although the attack turned out to be superficial, the fatigue life of the vessel could not have been evaluated nondestructively.

At the time of examination, it was believed that the corrosion zones on both sides of the welds were brought about by a preferential anodic phase in the weld HAZs or a preferential anodic zone caused by differences in permeability and thickness of the protective oxide coating. If moisture had condensed in the tank, corrosion would have taken place at the tank bottom. Because the corrosion path encircled the weld, it was suspected that corrosion occurred in the water vapor phase.

Surprisingly, metallographic examination revealed that the corroded areas were entirely free of corrosion products (Fig. 10). This indicated that the corrosion had occurred at an earlier stage of manufacture and was passivated and removed by chemical cleaning. Although searches of manufacturing records could not verify it, the most probable cause of corrosion was believed to be filiform corrosion occurring under a tape used to attach a protective paper or film over the surfaces prepared for welding. No other explanation adequately accounted for the lack of corrosion products, the shallowness of attack (0.01 mm, or 0.4 mil), and the filamentary network observed.

Galvanic Corrosion

Galvanic corrosion results from or is accelerated by dissimilar-metal contact. The more electronegative metal becomes the anode and is corroded. The more electropositive metal becomes the cathode and is not attacked. The severity of galvanic corrosion depends on the flow of current; therefore, corrosion rates tend to be accelerated when larger differences in potential exist between the two metals. Corrosion damage to the anode becomes more severe as the cathode-to-anode ratio increases. Corrosion rates also increase as solution conductivities increase. To prevent galvanic corrosion, parts can be isolated from each other electrically,

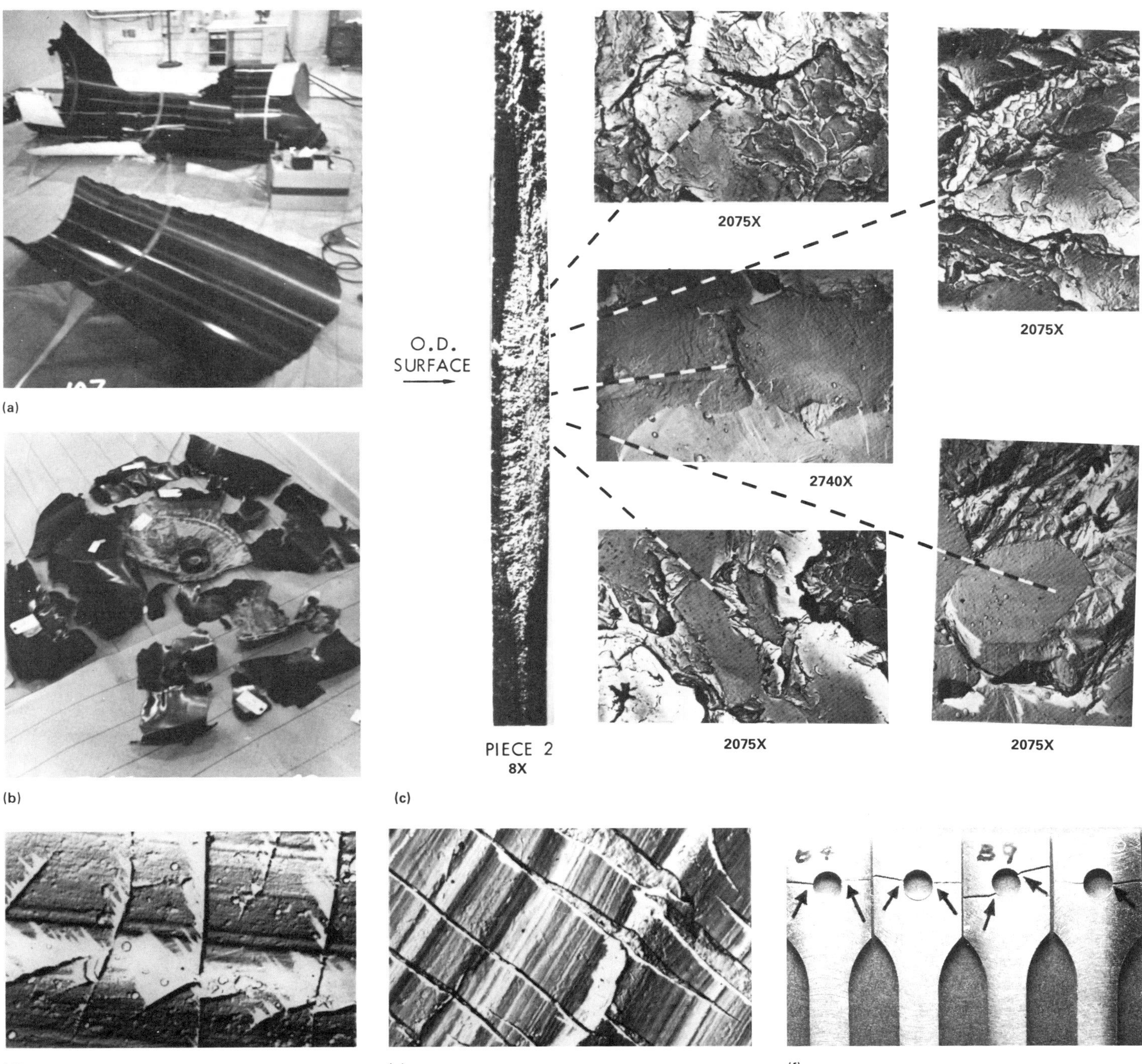

Fig. 23 Stress-corrosion failure of a solution-treated and aged Ti-6Al-4V Apollo SPS fuel pressure vessel during a sustained pressure test in methanol. (a) Explosive failure of the tank occurred, resulting in severe ripping of the cylinder section. (b) Dome section fragmented by explosion. (c) Fracture origin and TEM fractographs of fracture face showing quasi-cleavage failure. (d) Surface of outside diameter showing crazing of oxide film under load and machining lines. 2500×. (e) Similar view of inside diameter surface. 2500×. (f) Titanium notched stress-corrosion specimens cracked at pin-loading areas in methanol. Pins were austenitic stainless steel.

can be coated, or the anode can be cathodically protected by impressed current or other sacrificial, more anodic materials or coatings. If paint is used, the most effective member to coat is the cathode because the reduction in the cathode-to-anode ratio is extremely effective. If the anode alone were coated, corrosion would rapidly occur at coating defects accelerated by the very large cathode-to-anode area ratios.

When galvanic couples occur in spacecraft structures that exceed the requirements given in Table 1, the accepted design practice is to use two coats of paint as a moisture barrier and for electrical isolation. Designs should prevent entrapment of water and should provide for sealing of crevices. In fluid systems, detrimental galvanic couples are avoided. In the first 6 years of service of the space shuttle orbiter, four major galvanic problems arose: three with structural components and one in a fluid system.

Forward Wing Spar. Attached to the forward wing spar on the space shuttle orbiter are A-286 stainless steel fittings that support the reinforced carbon-carbon leading edges of the wing. Leading edge temperatures exceed 1260 °C (2300 °F), and the wing spar made from an adhesive-bonded aluminum sandwich structure

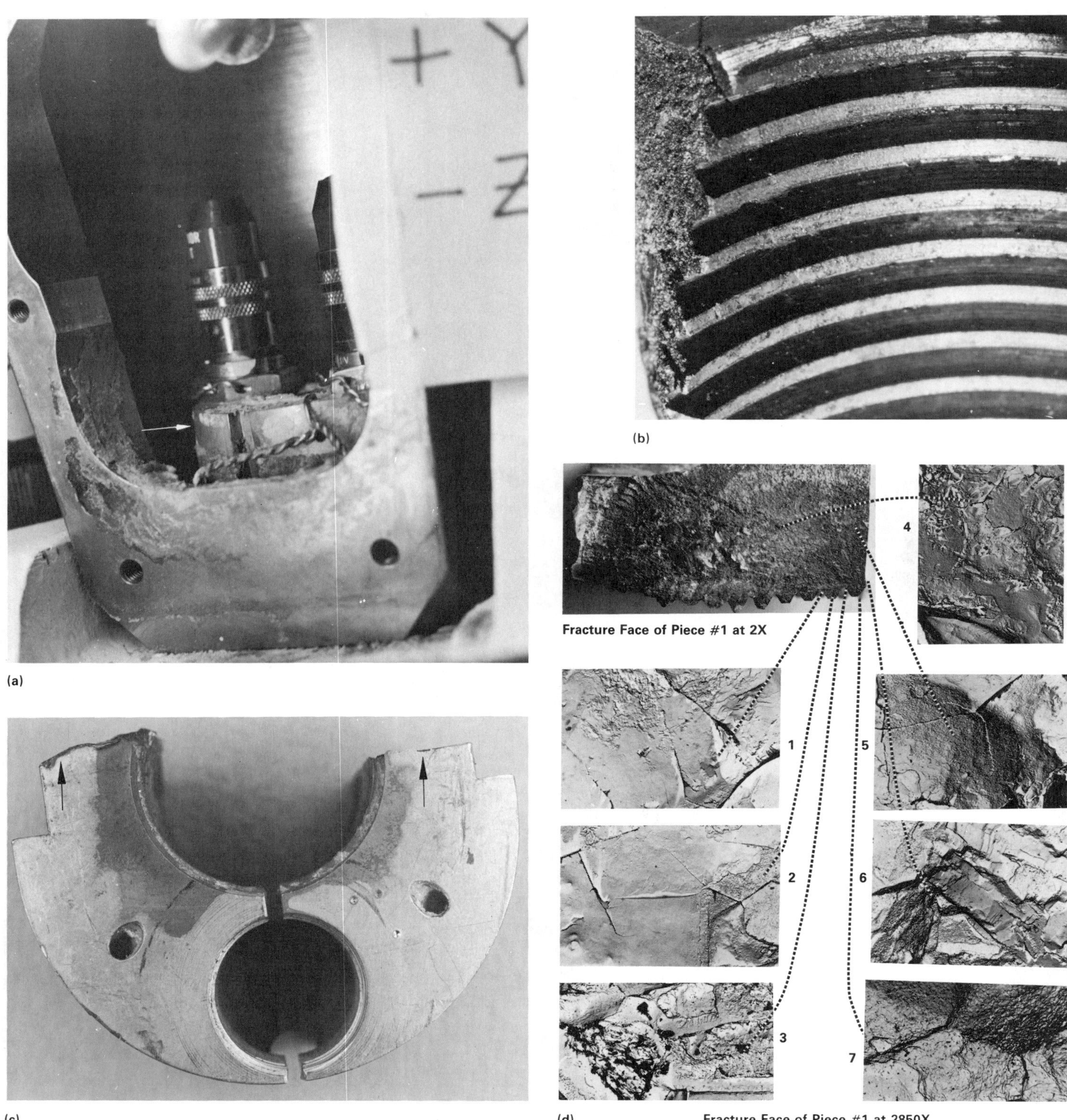

Fig. 24 Stress-corrosion failure of an Apollo launch escape tower leg frangible nut. The nut was made of 4340 low-alloy steel heat treated to 1520 MPa (220 ksi). The corroding medium was ammonium fluoborate leached from the ablative on the solid rocket engine nozzle extensions of launch escape tower. Note salt accumulation. (a) Frangible nut (arrow) located within launch escape tower leg. (b) Fracture face of failed nut in (a) showing extensive rusting. (c) Top view of nut. The fracture face is at top. (d) TEM fractographs. Views 6 and 7 are origin area. Note intergranular attack and corrosion of grain faces.

must be insulated to keep its temperature below 175 °C (350 °F). The spar face sheets are made of aluminum alloy 2024-T81 and have been chemically milled down to 0.35 mm (0.014 in.) in some areas. The insulation blanket material is a Dynaflex (alumina silica chromia fibers) felt with a density of 130 to 390 kg/m^3 (8 to 24 lb/ft^3). It was packed in an embossed, resistance-welded Inconel alloy 601 foil that is 0.1 mm (0.004 in.) thick.

During inspection of the wing of the *Columbia* vehicle after one of its early flights, blisters were

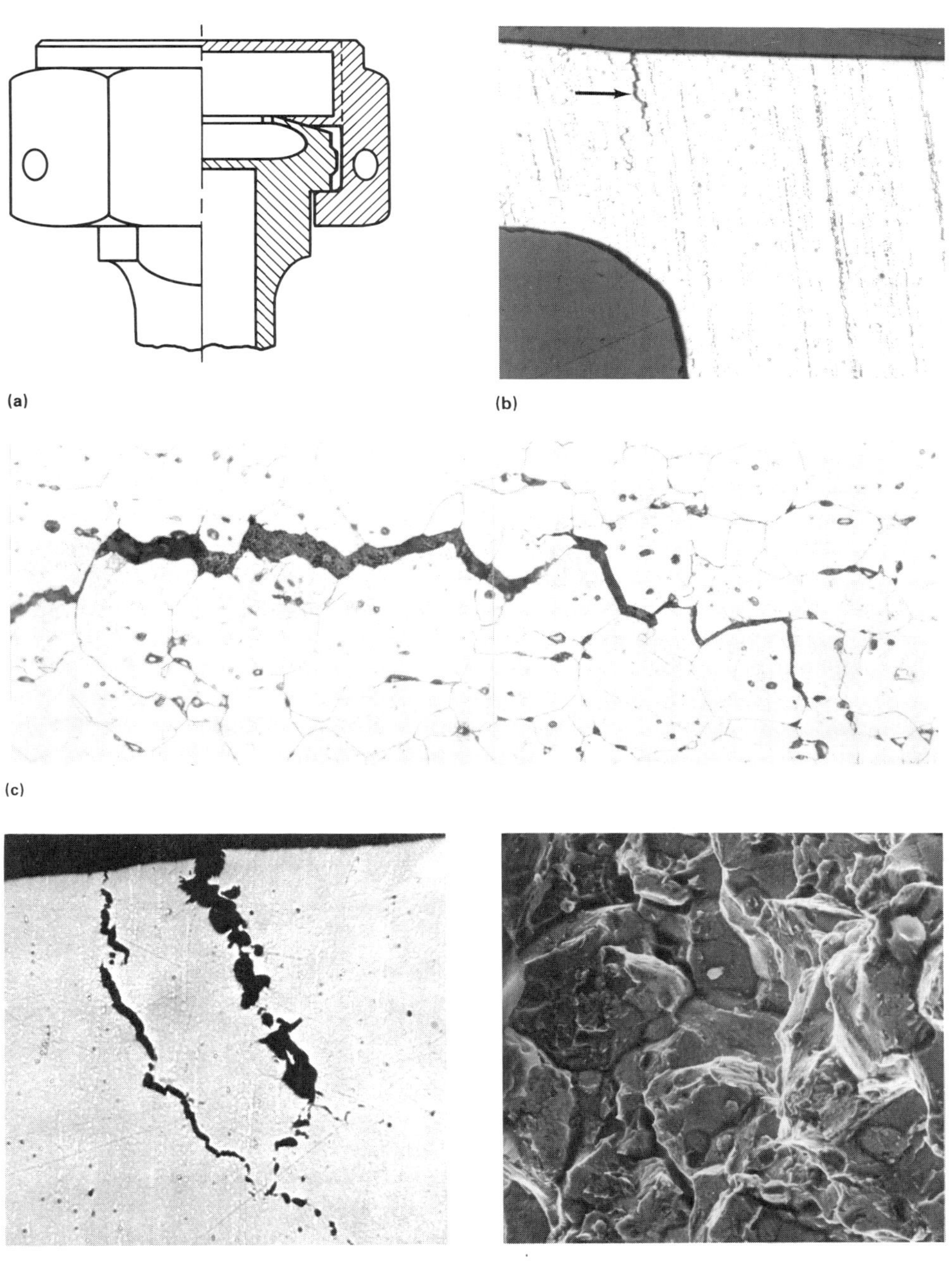

Fig. 25 Stress-corrosion cracking of 17-7PH (TH1060) fluid fittings on the Apollo and shuttle programs. (a) Illustration of assembled bending beam type fluid fitting showing the exterior (left) and cross section (right). (b) Cross section of stress-corrosion crack in location of maximum bending stress. Etchant: Kroll's reagent. 60×. (c) to (e) Stress-corrosion cracking in same alloy, temper, and design occurring 15 years later on shuttle orbiter. (c) Close-up showing intergranular cracking. 320×. (d) Similar location to cross section in (b) showing stress-corrosion crack. 105×. (e) SEM of fracture surface showing intergranular facets. 700×

seen in the chromated epoxy polyamine paint used to protect the spar (Fig. 11). The blisters occurred mostly along the edge pattern of the Inconel blankets. Removal of the paint revealed highly localized pitting with depths ranging from 12 to 350 μm (0.5 to 14 mils), perforating the face sheets in the thinnest areas.

The interface between the Inconel foil package and the wing spar permitted capillary moisture entrapment, especially while the orbiter was vertical on the launch pad. Although it was protected with two coats of the chromated epoxy polyamine paint, the galvanic current was able to perforate the paint. Because the foil blankets were unpainted, the cathode-to-anode ratio at paint flaws was extremely large.

Accelerated salt spray testing showed that corrosion could appear within 300 h with two coats of paint, but aluminum was fully protected from the galvanic corrosion with a 75-μm (3-mil) coating of RTV 560 or three coats of paint. The pits on the orbiter were deactivated. Pit depths were measured. More than 700 local areas exceeded acceptable pit depth criteria and required the bonding of approximately 200 doublers to prevent fatigue or perforation into the honeycomb core. The spar was refinished with three coats of paint; the edges of the foil blankets were coated with 75 μm (3 mils) of RTV 560.

Elevons. Similar galvanic corrosion of aluminum alloy 2024-T81 occurred in the elevons under flipper doors and rub panels where Inconel alloy 601 foil insulation blankets rested against painted aluminum honeycomb face sheet. Corrosion was deactivated; however, in this case, a nylon-type surface insulation was used to replace the foil blankets, eliminating the galvanic couple.

Orbital Maneuvering System Pod Structure. Galvanic attack similar to that discussed above occurred in the OMS pod of the *Columbia*, where goldized Kapton multilayer reflective insulation was in contact with painted aluminum. In this case, a faulty environmental seal permitted water entry, but no provisions were available to drain the water. The pits were sanded, ground out, or deactivated. The area was chemically filmed and painted with two coats of chromated epoxy polyamine paint and then recoated with 75 to 125 μm (3 to 5 mils) of RTV 560. Organic-coated aluminized Kapton insulation was also substituted for the goldized Kapton insulation.

Environmental Control and Life Support System Outlet Water Valve. After the sixth flight of the orbiter, a potable water valve was discovered to be leaking. The leak emanated from a zone of corroded material that traversed the length of the valve forging. Although the valve body was manufactured from a type 304L stainless steel forging, the material within the corroded zone was found to be low-carbon steel. It was learned that the carbon steel had been accidentally introduced in the ingot in the final stages of solidification. Subsequently, the ingot was rolled into bar and then forged. Because any of more than 50 other valves from that heat could have the same defect, a complete review of all parts was made. The defect could be detected easily by the copper sulfate test according to Method 102 of MIL-STD-753A. It was recommended that this test be performed followed by repassivation.

Figure 12(a) shows the tube section with a longitudinal streak through it (appears similar to a crayon mark). Figure 12(b) shows the corroded area that leaked. Attack is seen completely through the tube wall in Fig. 12(c). The microstructure was then etched with 3% nital to bring out the carbon steel (Fig. 12d). Figures 12(e) and (f) show x-ray dot maps of nickel and chromium concentrations, respectively, through the inclusion shown in Fig. 12(d). Figures 12(b) to (d) show that the carbon steel is anodic to the stainless steel in water and that corrosion was accelerated by this galvanic couple. Although this problem could not have been anticipated, the supplier would have recognized the problem if he had passivated and inspected the stainless steel properly.

Intergranular Corrosion

Intergranular corrosion results where the microstructure of an alloy provides a preferred corrosion path along grain boundaries. Often, the material adjacent to the grain boundary is depleted of a particular element that precipitates out as an intermetallic compound at the grain boundary. In the case of aluminum alloys of the 2000 series, the precipitation of $CuAl_2$ at grain boundaries occurs. The $CuAl_2$ is more noble than the adjacent aluminum, and intergranular corrosion of

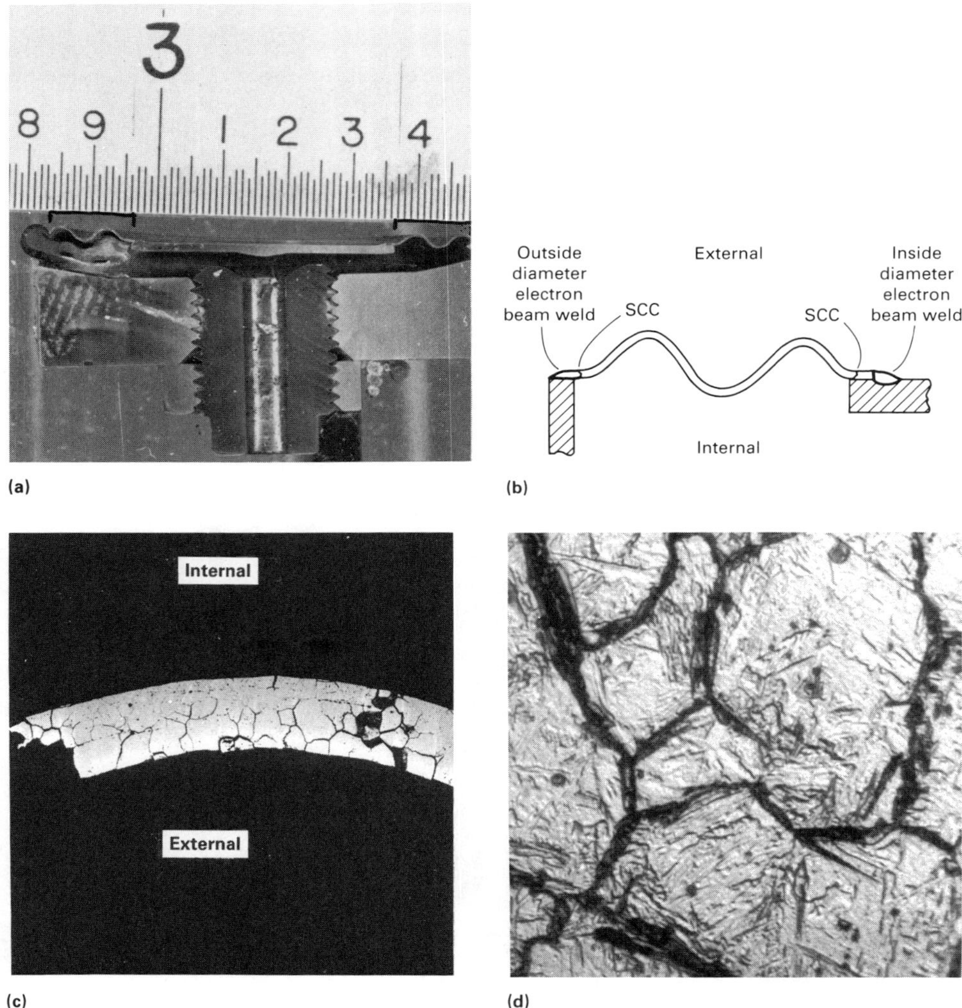

Fig. 26 Stress-corrosion cracking of the 17-7PH stainless steel diaphragm of a potable water transducer used on the shuttle orbiter. (a) Cross section of transducer and corrugated diaphragm. (b) Sketch of diaphragm showing electron beam welds and cracks at points of highest bending. (c) Intergranular corrosion at the failure location. The part was sensitized along its entire length. 130×. (d) Cross section showing martensitic grain boundaries with grain-boundary carbides surrounding austenitic grains. 140×

the adjacent aluminum takes place. Artificial aging of aluminum causes precipitates to occur throughout the grains as well as in the grain boundaries, thus minimizing localized corrosion. Stainless steels, when heated into or cooled through the 425- to 870-°C (800- to 1600-°F) range, will precipitate chromium carbides preferentially at grain boundaries, resulting in anodic paths adjacent to these grain boundaries. Lowering carbon levels to below 0.03% (to avoid a continuous carbide grain-boundary network) or adding niobium or titanium (to precipitate carbides) will permit the chromium in the grain boundaries to keep the steel from becoming sensitized.

Intergranular corrosion is insidious in that little actual corrosion is required before the structural integrity of a part or the perforation of a fluid container occurs. A knowledge of metallurgy is important in preventing intergranular corrosion attack because microstructure is sensitive to composition, temperatures, cooling rates, and surface contamination.

The ammonia boiler, as previously described, is a heat exchanger using Freon-21 and ammonia. The Freon-21 picks up heat from the orbiter electronic boxes, the crew cabin, and the payloads, and the heat is dissipated by boiling ammonia when the orbiter radiators cannot be deployed, such as during the final stages of reentry. Accidental contamination of these tube surfaces prior to brazing of the heat exchanger resulted in the sensitization of some of the thin-wall (0.2 mm, or 8 mil) type 304L stainless steel tubes. The sensitized tubes were attacked and perforated by the fluid (Fig. 13).

The corrective action was to change the tubing to type 347 stainless steel to prevent chromium carbide precipitation at grain boundaries if parts were accidentally contaminated prior to brazing. In addition, procedures to prevent contamination in the future were upgraded. Particular emphasis was placed on removal of carbonaceous drawing compounds from tube surfaces before brazing.

Selective Leaching

Selective leaching (also called parting, dealloying, or demetallification) occurs when the corrosion process removes one or more elements from the alloy matrix. Specific categories of selective leaching often carry the name of the dissolved element in their title, such as dezincification, dealuminification, denickelification, or decobaltification. In the case of gray cast iron, selective leaching is called graphitic corrosion. The process can occur in single-phase alloys, for example, with brasses having high zinc content such as a 70Cu-30Zn alloy. It can also occur in multiphase alloys.

In the selective leaching process, typically one of two mechanisms occurs: alloy dissolution and replating of the cathodic element or selective dissolution of an anodic alloy constituent. In any case, the matrix that is left is spongy and porous and has very little strength or integrity. In some cases, a large, localized area of metal (a plug) is attacked.

Selective leaching can be prevented by proper alloy selection, that is, matching the alloy system to a particular environment. Changing ratios of alloying elements and adding inhibiting elements are often satisfactory approaches. In aerospace, selective leaching is a rare problem. Where it occurs, a proper alloy selection or changing the processing medium is the key to prevention.

Gas Generator Valve Module (GGVM) Valve Seats. The APU of the shuttle orbiter contains a GGVM that utilizes four tungsten carbide valve seats to regulate hydrazine flow to the catalyst bed. The 5-mm (0.2-in.) diam valve seats have sealing lands that are only 115 to 150 μm (0.0045 to 0.0060 in.) wide. The seats are manufactured from sintered KZ-96 tungsten carbide containing 5% Co as a binder.

In mid-1985, valve leakage problems during acceptance testing were traced to a breakdown of the valve seat sealing lands. The problem became widespread, with a high rejection rate during acceptance testing procedures (ATP).

It was determined that revised cleaning procedures that had been implemented recently to correct a GGVM contamination problem early in the program were subjecting the seats to long periods of exposure to deionized water and ultrasonic cleaning. Tests proved that such exposure produced leaching of the cobalt binder from the sintered tungsten carbide alloy. Subsequent impacting of the valve seats by the poppets (during ATP) resulted in breakdown of the seats and the noted leakage failures (Fig. 14).

As a by-product of the testing, it was found that a small amount of leaching of the cobalt also occurs from exposure to decomposition by-products of hydrazine, the system fluid. This leaching action is several orders of magnitude less than that from deionized water. Tests were underway in the time period from 1985 to 1986 to identify alternative tungsten carbide alloys that might be acceptable substitutes for KZ-96. Alloys under investigation included those with different binders and those with finer grain sizes. The corrective action taken during the manufacturing cycle was to substitute isopropyl alcohol for deionized water in all operations.

Crevice Corrosion

Crevice corrosion results from a concentration cell formed between the electrolyte within the

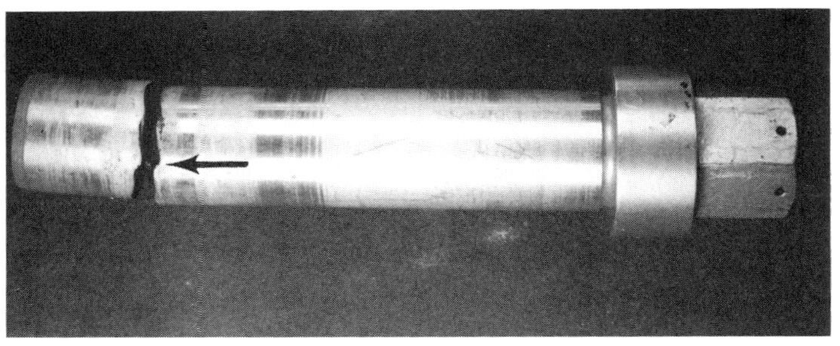

(a)

(b)

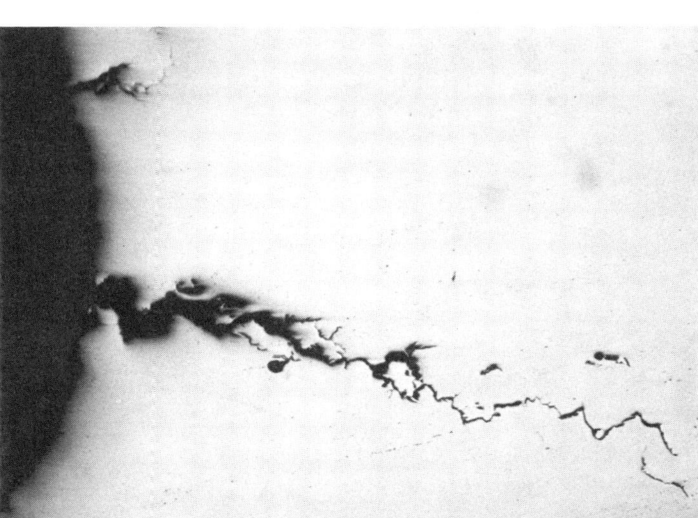

(c)

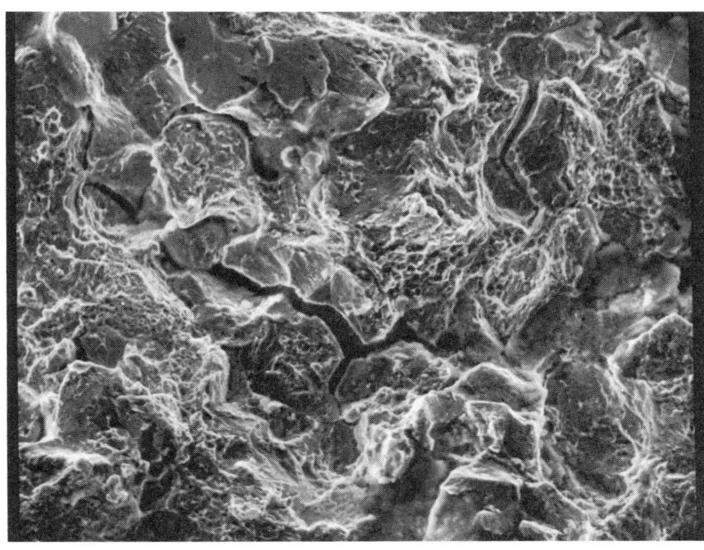

(d)

Fig. 27 Stress-corrosion failure of a 300M steel shear pin used for a shuttle ferry flight. The 300M alloy is heat treated to 1895 MPa (275 ksi). (a) Shear pin showing failure at the runout area of internal threads. (b) Fracture face showing zone where SCC occurred (between arrows). (c) Cross section through cracked area showing characteristic branched SCC. 110×. (d) SEM view of intergranular corrosion. Note pitting of grain faces. 525×

crevice, which is oxygen starved, and the electrolyte outside the crevice, where oxygen is more plentiful. The material within the crevice acts as the anode, and the exterior material becomes the cathode. This is similar to pitting, in which the base of the pit becomes the anode. The author finds it convenient to view a crevice as a plane of corrosion, that is, a two-dimensional pit. The resistance of materials to crevice corrosion varies widely. Those metals whose protective oxides films result from oxygen adsorbed to the surfaces and are in dynamic equilibrium with the outside environment, such as stainless steels, appear to suffer most when shut off from oxygen by crevices.

In good corrosion design practice, crevices are avoided whenever possible. Crevices not only trap water or the chemical being processed but also become sumps for the other contaminants in the system, often resulting in major corrosion problems. Spacecraft structural faying surfaces, for example, either include a faying surface sealant or the edges are fillet sealed (see the discussion "Structural Joints and Fasteners" in this section). Sometimes, however, a crevice cannot be avoided, as shown in the cases below. Therefore, additional care is required to avoid detrimental corrosion.

Anodized Aluminum Window Frames. The exterior window frames on the space shuttle orbiter are made of aluminum alloy 2124-T851 that has been anodized according to a Rockwell specification by using a sulfuric acid anodize, a black dye, and a sodium dichromate seal. The window frames are protected from heat during spacecraft entry by pure silica tiles. Two grooves within each window frame contain the fibrafax thermal seal (to prevent hot gas plasma flow) and the Viton pressure (environmental) seal. The side windows, when the orbiter is stacked for launch on the pad, provide for possible water entrapment in a small portion of the periphery of the windows. Rain can wash down window surfaces, picking up salt deposits from seacoast exposure. The aluminum window frame peak temperature is approximately 55 °C (130 °F).

After the eighth flight of the space shuttle orbiter, two side windows were removed from the *Challenger* vehicle for examination. The window surfaces had appeared to have a hazy opacity even after polishing. The opacity was due to microscopic erosion of unknown origin. Examination of the window frame showed several localized areas of corrosion through the anodized coating in and adjacent to the seal grooves. The areas away from the crevice were uncorroded (Fig. 15). The recommended corrective action for new window frames was to add a coat of chromated epoxy polyamine primer, followed by a coat of polyurethane over the black anodize.

Aluminum Brazed Joints. The nature of the brazing process is to provide a gap between faying surfaces that will act as a capillary for braze alloy flow. Selection of the braze processes used for the plumbing systems lines on the Apollo and the space shuttle orbiter depended on the braze alloy filling the capillary gap to ensure the inside of the tube would not have a crevice (see the discussion "Plumbing Lines" in this section). These stainless steel brazes were made under an inert gas shield, and no flux was required.

In brazing aluminum, however, it is necessary to use a flux to remove its tenacious oxide film and to ensure wetting of the faying surfaces. The brazing fluxes usually consist of mixtures of alkali and alkaline earth chlorides and fluorides, sometimes containing aluminum fluoride or cryolite ($3NaF \cdot AlF_3$). Braze fluxes will often become entrapped between the faying surfaces, and if the joint is completely sealed by a fillet, no problems arise. However, when incomplete fillets occur, the brazing fluxes, which are generally hygroscopic, will bleed out and cause corrosion.

Cleaning of the part and verifying that surfaces are free of fluorides and chlorides do not always solve the problem, because weeks later the part can be covered with flocculent aluminum hydroxide corrosion products. Brazing of aluminum is generally avoided on the shuttle orbiter wherever possible. Aluminum cold plates are made by

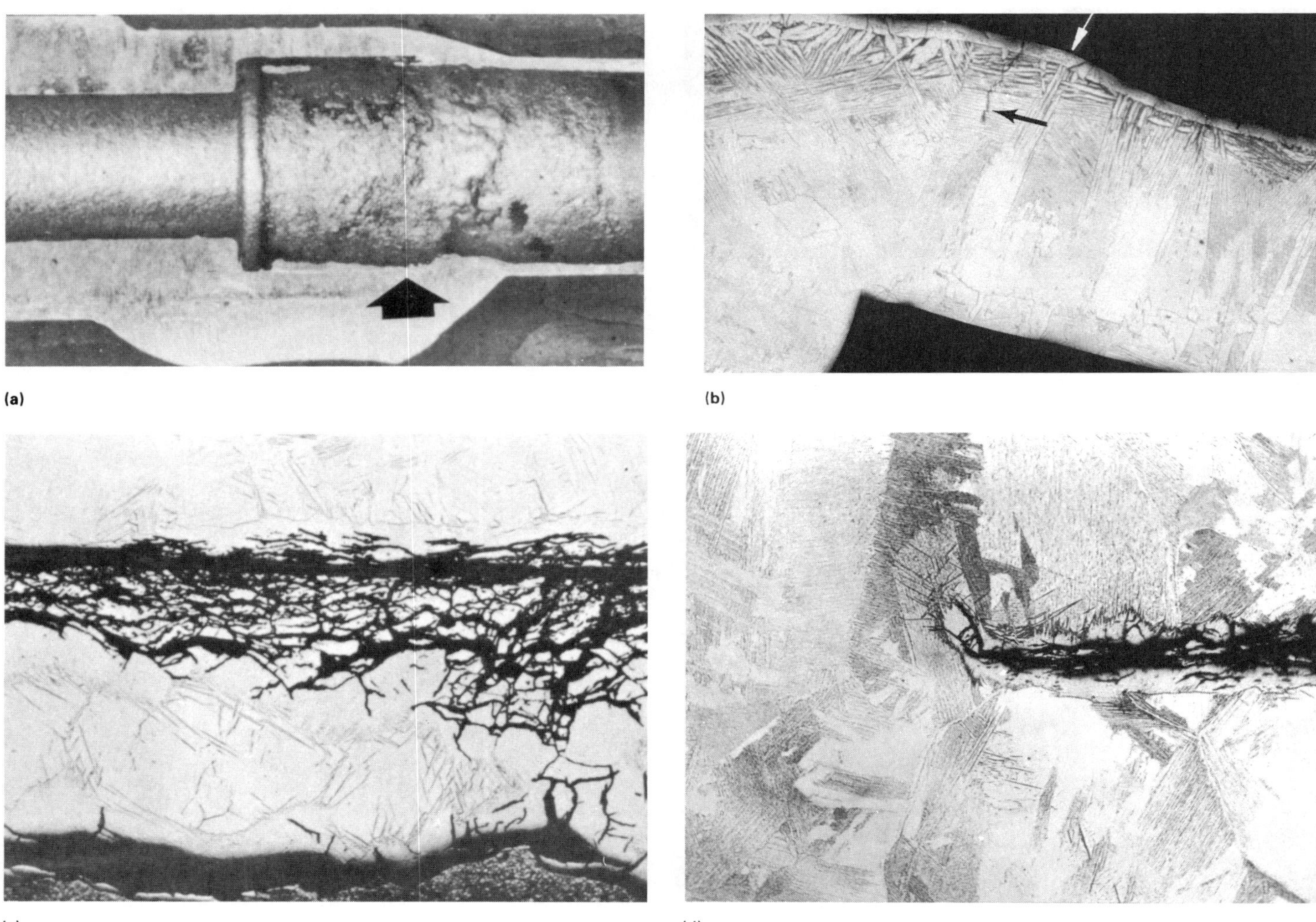

(a) (b) (c) (d)

Fig. 28 Hydrogen embrittlement failure of A55 titanium tubing in a cryogenic hydrogen storage system on the Apollo program. (a) Tube with hydrided joint area (arrow). (b) Hydride layer on the inside diameter of a welded joint showing cracks. 105×. (c) Massive hydride layer in faying surface between welded tube and sleeve. 175×. (d) Hydrides discovered in tube joined to pressure vessel not in a direct hydrogen flow area. 85×

fluxless brazing in inert gas under temperatures and pressure sufficient to ensure braze alloy flow. In electronic boxes in which braze designs have been used, the joints are vacuum impregnated with a resin seal to avoid bleedout of entrapped fluxes. This appears to be quite satisfactory.

On the Apollo program, aluminum-to-stainless steel tube brazes were used in discrete applications. The stainless tube was tin plated. Problems experienced with braze bleedout are shown in Fig. 16.

Fretting Corrosion

Fretting is the abrasive wear of two touching surfaces subject to cyclic relative motions of extremely small amplitude. Fretting corrosion is an increased degree of deterioration that occurs because of repeated corrosion or oxidation of the freshly abraded surface and the accumulation of abrasive corrosion products between these surfaces. Although fretting is often limited to small, localized patches of wear, it can eventually provide a path for leakage (for example, valve seats) or an initiation site for fatigue. Fretting corrosion can be controlled by lubrication of the faying surfaces, restricting the degree of movement, or by the selection of materials and combinations that are less susceptible to fretting (see Table 4 in the article "Corrosion of Airframes" in this Volume).

Rudder Speed Brake Power Drive Unit Spacer. The mounting bolts of the rudder speed brake power drive unit of the space shuttle orbiter are made of A-286 stainless steel heat treated to 965 MPa (140 ksi). Bolts are either 15.9 or 22.2 mm (5/8 or 7/8 in.) in diameter. The bolts are sleeved with a spacer that passes through a spherical bearing. The spacer is made of 17-4 PH steel (H1150M), and the bearing is Inconel alloy 718 heat treated to 1240 MPa (180 ksi). The surface finish on the spacer is 16 RHR (root height reading). When the power drive unit was removed from the *Enterprise* vehicle, fretting corrosion was discovered on the exterior of the spacer and the interior of the ball. The fretting corrosion of the 17-4PH was quite severe, as shown in Fig. 17. The corrective action consisted of changing the spacer to Inconel alloy 718 and applying a dry film (molybdenum disulfide) coating.

Stress-Corrosion Cracking

Stress corrosion requires the simultaneous occurrence of three conditions: a susceptible material or microstructure, a corrosive environment, and surface tensile stresses. Control of stress corrosion is achieved by avoiding any one of these conditions (see the discussion "Control of Stress Corrosion" in this section). Each metal family has its own unique environments in which it displays susceptibility to stress corrosion.

Stress-corrosion cracking is one of the most insidious forms of corrosion because it often comes on without any warning and results in major, sometimes catastrophic, structural failures. In fluid systems, the presence or absence of a trace element can make the difference between no reaction and a major failure. Prior laboratory testing may have missed a potential problem that occurs in service, as illustrated in some of the failures below. It is mandatory that the cause of a stress-corrosion failure be demonstrated in the laboratory after a suspected stress corrodent has been identified and that testing be conducted as closely as possible to the chemical and metallur-

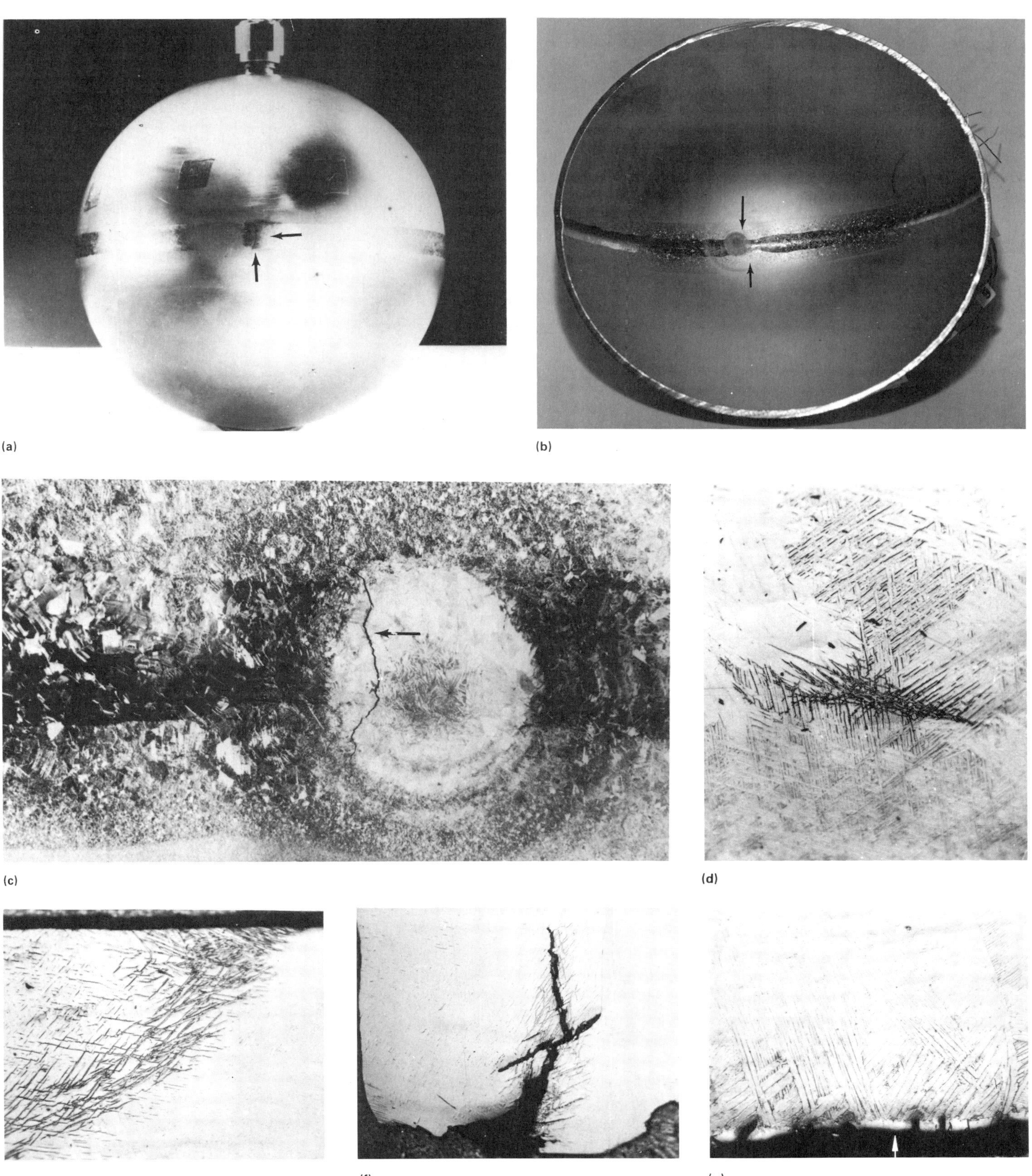

Fig. 29 Hydrogen embrittlement failure of a titanium pressure vessel for nitrogen storage in the Apollo fuel cells. (a) Pressure vessel made of Ti-5Al-2.5Sn showing weld repair (arrows). (b) Internal view showing crack in repair area. (c) Close-up view of crack in the repair. (d) and (e) Hydrides on the vessel outside diameter. (d) 325×. (e) 215×. (f) and (g) Hydrides on the pressure vessel inside diameter. Hydrogen came from a leaking furnace valve during weld stress relieving. (f) 165×. (g) 215×

(a)

(b)

(c)

Fig. 30 Hydrogen embrittlement failure of a 4335V modified steel solid rocket motor case used in the Apollo launch escape tower. Steel is heat treated to 1445 MPa (210 ksi). The pressure test medium was inhibited distilled water. (a) Failed case with failure origin removed from longitudinal rupture along weld. (b) Notched toughness specimen from HAZ tested in air. K_{Ic} = 90 MPa$\sqrt{m}$ (82 ksi$\sqrt{in.}$). Specimen has electrically discharge machined notch and was fatigue cracked prior to toughness testing. (c) Notched toughness specimen from weld made in a similar fashion. K = 83.5 MPa$\sqrt{m}$ (76 ksi$\sqrt{in.}$) in air for 120 min. No flaw growth. K_i = 83.5 MPa$\sqrt{m}$ (76 ksi$\sqrt{in.}$) in inhibited water. The specimen failed in 53 min.

gical conditions experienced by the failed hardware.

The injector tube of the space shuttle orbiter auxiliary power unit carries liquid hydrazine to a catalyst bed, where it is heated and decomposes into nitrogen, hydrogen, and a trace of ammonia. The hot decomposed gases drive a turbine wheel to generate secondary power for spacecraft systems. Shortly after touchdown from the ninth launch of the orbiter, two of three APUs detonated (Ref 14). Extensive investigation determined that the cause of the detonations was the decomposition of hydrazine. While the orbiter was in orbit, hydrazine had leaked through cracks in the Hastelloy alloy B injector tube walls of two different APUs. In the space vacuum, evaporation withdrew heat from the leaking fluid, resulting in the formation of hydrazine snow balls. During reentry, the snow balls melted and ignited—somewhere below 12 km (40 000 ft) in altitude—and the heat resulted in decomposition and detonation of the hydrazine.

The fractures on each tube were intergranular, started on the inside diameter, occurred in almost identical locations, and extended 220 to 240° around the periphery (Fig. 18). The cracks were determined to be caused by stress corrosion, as evidenced from their appearance and by extensive testing that eliminated all other failure mechanisms.

It was demonstrated by laboratory testing that ammonia or ammonium hydroxide was the only potential fluid that could cause stress corrosion of Hastelloy alloy B on the spacecraft. The ammonium vapors resulted from decomposition of hydrazine in the catalyst bed, probably as a result of hydrazine leaking into the injector tubes through the valve seat or from the equilibrium of the gas generator environment within the tube after shutdown. Moisture, resulting in the formation of ammonium hydroxide, was available from atmosphere migrating back into the exhaust duct. Misalignment and cocking of the injector tubes during installation resulted in high stresses in the failure area (up to yield stress). A sensitized microstructure of precipitated carbides along inside diameter grain boundaries provided a preferred path for stress corrosion to proceed. A carbide network found on the inside diameter of the injector tubes was the result of carbon deposited during the electrical discharge machining process to machine the injector tube bore. Subsequent injector tube brazing and cool-down cycles permitted grain-boundary diffusion and precipitation.

The solution was to eliminate the preload stresses on the injector tube by instrumenting the installation and to eliminate the sensitized carbide network by reaming the tube inside diameter. The braze cycle was raised to 1185 °C (2165 °F) to ensure uniform diffusion of any carbides, followed by rapid cooling to ensure that any carbides inherent to the basic alloy (0.05 max) would not precipitate at grain boundaries. Later designs of injector tubes were also internally coated with a thin chromized layer.

The Reaction Control System injector of the Space Shuttle orbiter is made from uncoated niobium alloy C 103. This injector is bolted to a titanium mounting ring section. A niobium alloy C 103 chamber is then welded to the injector face. Cracking was observed in the sharp radius of the injector adjacent to the bolting ring. Examination of the fracture face showed that the attack was intergranular and had fine eruptions (popcorn balls) of niobium oxide on its surface.

A review of the manufacturing sequence isolated the problem to those steps performed between the bolting of the titanium flange section to the injector flange and the final operation in which the engine was baked at 315 °C (600 °F) for 20 h to remove resin from the insulation encasing the engine. The most probable cause of cracking was thought to be stress corrosion occurring from entrapment of an etchant (50% HNO_3-50% HF) used to remove traces of iron and copper from the niobium prior to welding of the chamber to the injector. Repeated laboratory attempts to duplicate failure were unsuccessful until a method was devised to entrap the etchant between two pieces of niobium while tensile stresses were applied during sustained heating in the 290- to 315-°C (550- to 600-°F) range. The failure was duplicated and attributed to hot fluoride salt SCC (Fig. 19) (Ref 15). Performing the acid etching and rinsing prior to bolting the injector assembly prevented future occurrences.

Reaction Control System Oxidizer Pressure Vessels. Nitrogen tetroxide, a storable hypergolic oxidizer, was used in the service propulsion system (SPS) and in the reaction control system on the Apollo program. The SPS provided the propulsion for orbit insertion, lunar flight, return from the moon, and deorbit for entry. The RCS provided for vehicle attitude control through roll, pitch, and yaw engines. Titanium alloy Ti-6Al-4V was chosen for pressure vessels in both systems as a result of laboratory tests on corrosion, stress corrosion, and impact ignition with N_2O_4.

Qualification testing of the SPS pressure vessel for 46 days of exposure under a membrane stress of 690 MPa (100 ksi) was completed without problems in mid-1964. In January 1965, an RCS pressure vessel of the same alloy, protected from N_2O_4 by a teflon positive expulsion bladder, cracked in six adjacent locations (Ref 9-12). The cracks were parallel to each other and perpendicular to the maximum stress. The fracture surface

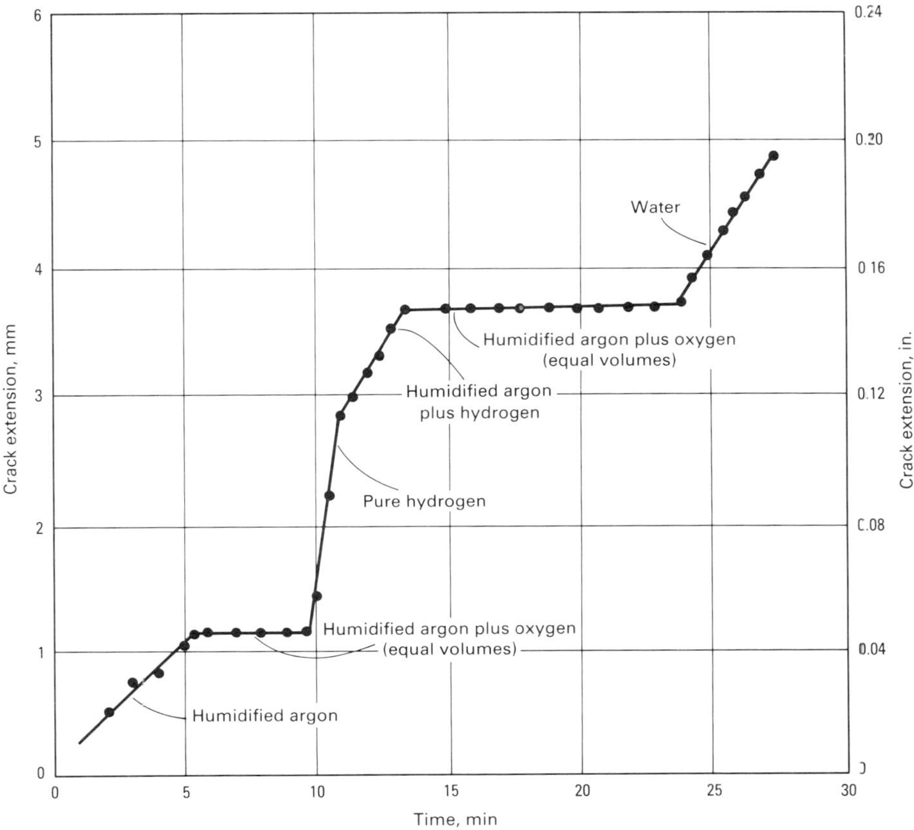

Fig. 31 Subcritical crack growth in H-11 steel (yield strength: 1585 MPa, or 230 ksi) in water, water vapor, hydrogen, and oxygen environments

had a red stain and was flat and brittle in appearance. Because no N_2O_4 was supposedly in contact with the vessel, it was believed that misprocessing may have caused the fractures. Subsequent testing of ten pressure vessels without bladders in June of that year was conducted to ascertain if misprocessing could be bracketed to certain manufactured pressure vessel lots over a period of time. Within 34 h after testing, one of the pressure vessels blew up (Fig. 20) and within a few days most of the others had failed.

The cause of the failure was immediately suspected to be stress corrosion, but the aggressive fluid that contacted these surfaces was not immediately determined. Cracks occurred very close together on the inside of the pressure vessel, and the number of cracks per unit area was proportional to the local stress. Pressure vessels of the same design, when tested with N_2O_4 by Rockwell on the West Coast, did not fail. The RCS tank contractor, Bell Aerosystems Company, was soon able to demonstrate coupon failures in N_2O_4, but Rockwell could not, although over 300 specimens were cleaned and tested under more than 40 variables, including various contaminants.

Chemical testing of the propellants used at Bell and Rockwell revealed no differences or out-of-specification conditions. Existing chemical techniques were not capable of accurately quantifying all species present in the N_2O_4, especially compounds of nitrogen; however, cooling of the propellants revealed a color difference in N_2O_4 between the supplies at Bell and Rockwell. The N_2O_4 at Rockwell, when cooled to −18 °C (0 °F), turned green because of the presence of nitric oxide (NO), but the N_2O_4 at Bell was yellow. The green color resulted from a mixture of N_2O_4 (yellow) and dissolved NO as N_2O_3 (blue). Rockwell supplies of N_2O_4 had been purchased earlier than those of Bell Aerospace.

Investigation revealed that a change in the military specification MIL-P-26539 had been made during this time period to improve the specific impulse of N_2O_4 by oxygenating the trace quantities of residual NO. This simple change had a devastating effect on its stress-corrosion behavior with titanium. Testing indicated as little as 0.2% NO was probably sufficient to inhibit the SCC of titanium. Specifications were changed thereafter to require a minimum NO content of 0.6%; present grades contain 1.5 to 3% NO. Addition of NO to existing supplies solved the problem.

Early in the investigation, it was believed that no stress corrosion could occur in solutions as nonconductive as N_2O_4 (specific conductivity at 25 °C, or 80 °F, is $3.1 \times 10^{-13}\ \Omega^{-} \cdot cm^{-1}$). However, because of the low conductivity of N_2O_4, only closely spaced local cathodes and anodes could carry corrosion currents (Fig. 20). This resulted in a large number of cracks (up to 70 cracks/in.) rather than a single crack. This investigation also illustrated how past stress-corrosion test results or pressure vessel qualification can be voided by minor chemical changes in the corroding medium. This is further illustrated in the following discussion.

Service Propulsion System Fuel Tanks. The storable hypergolic fuel used for the service propulsion system on the Apollo Service Module was a blend of 50% hydrazine and 50% unsymmetrical dimethyl hydrazine. It was contained in two titanium Ti-6Al-4V pressure vessels approximately 1.2 m (4 ft) in diameter and 3 m (10 ft) long. Because hydrazine compounds are toxic and dangerous to handle, a "referee" fluid with similar density and flow characteristics was used in system checkout testing. Methanol was chosen as a safe fluid based on subcontracted studies conducted for the program. Methanol had been successfully used as a fluid in a tri-flush cleaning process for propellant systems at that point in time. Among the other advantages, it had a low explosive potential, was miscible with both fuel and water, and would leave surfaces residue free.

During an acceptance test of the Apollo Spacecraft 101 service module prior to delivery, an SPS fuel pressure vessel (SN054) containing methanol developed cracks adjacent to the welds (Fig. 21). The test was stopped. This acceptance test had been run 38 times on similar pressure vessels without problems. Failure analysis could not reveal the cause of the cracking. The fractures had branching cracks characteristic of stress corrosion yet fracture faces exhibited a somewhat featureless quasi-cleavage appearance without typical stress-corrosion features. The adjacent material was ductile and within chemistry.

Misprocessing of pressure vessels during manufacturing was suspected, because the supplier had previously demonstrated that pickup of cleaning agents and contaminants on a Ti-6Al-4V pressure vessel prior to a heat-treat aging cycle would result in delayed stress-corrosion failure. These included such contaminants as finger prints, chlorinated kitchen cleansers, and liquid hand soap. A similar crack adjacent to the weld occurred on the first development pressure vessel for this program was thought to be caused by such contamination (Fig. 22).

The Spacecraft 017 service module was then put into test. An additional test was initiated to ensure that no marginal SPS pressure vessels would pass through system checkout. This additional test consisted of 25 pressure cycles, followed by a 24-h pressure hold. After only a few hours into the hold cycle, the replaced SPS pressure vessel failed catastrophically (Fig. 23).

The investigation conducted after this failure disclosed that titanium will undergo SCC in methanol (Ref 9, 16-20). Attack is promoted by crazing of the protective oxide film. It was learned that minor changes in the testing procedures could inhibit or accelerate the reaction. For example, the addition of 1% H_2O inhibited the reaction completely. It could be restarted by a 5 ppm addition of chloride. Initial stress-corrosion testing in the laboratory was performed with available aluminum test fixtures, with 300-series stainless steel, and then with titanium. No failures occurred in aluminum fixtures (apparently it provided cathodic protection). Failures in stainless steel fixtures occurred at the specimen pin areas, not at the sharp notches used to initiate stress cracking. Failures occurred in only a few hours with stainless steel because it apparently accelerated the reaction by galvanic coupling. Specimens tested to the same stress levels in titanium fixtures often took several times as long to fail.

The obvious solution to the problem was to replace the methanol with a suitable alternate fluid. Isopropyl alcohol was chosen after considerable testing. This incident further resulted in the imposition of a control specification (MF0004-

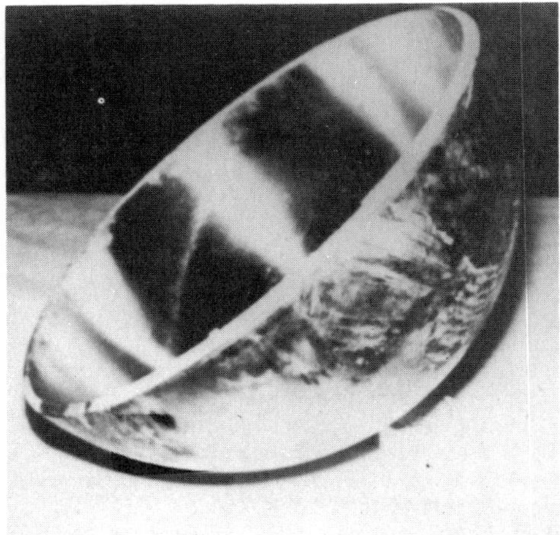

(a)

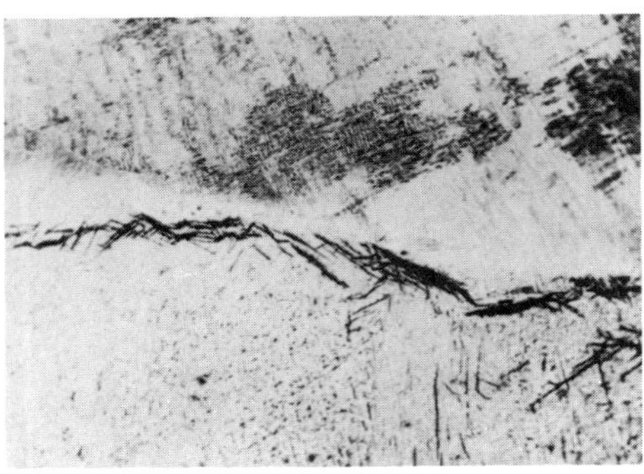

(b)

(c)

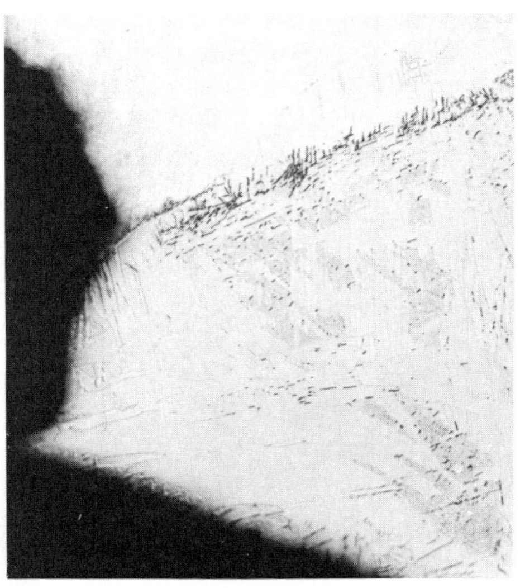

(d)

Fig. 32 Hydrogen embrittlement failure of a Ti-6Al-4V helium pressure vessel used on the Saturn IV B. Similar hydriding occurred in Apollo SPS pressure vessels. (a) Failed pressure vessel due to brittle hydride formation along weld bead made with commercially pure titanium. (b) Hydride at edge of pure metal weld bead in Ti-6Al-4V vessel shown in (a). 620×. (c) Section through a Ti-6Al-4V Apollo SPS pressure vessel welded with pure titanium on the root side. Coupon was stress cycled 200 times, then tested in tension to failure (lower arrow indicates direction of test loading). Hydride needles adjacent to the weld bead are shown. 45×. (d) Path of hydride needles changes direction and becomes parallel to load across weld. 70×

018) for all fluids that contact titanium for existing and future space designs.

This failure further illustrates the importance of trace chemicals in accelerating or inhibiting a failure. It also points out that test fixture materials can affect test results. Although it is preferable to use fixtures of the same alloy, it may not always be critical in the screening of possible stress-corroding fluids, because it is often desirable to accelerate stress-corrosion testing during the screening phase. This acceleration can be carried out by significantly increasing the stress (using a notch), by raising the temperature (usually), or by making the specimen the anode in a fluid. It is notable that failures involving stainless steel fixtures did not occur when 1% H_2O was added to methanol, nor did failures occur in isopropyl alcohol, benzene, Freon TF, Freon MF, and distilled water and MMH, even if HCl was added or bubbled through the solutions.

Tower Leg Frangible Nut. The launch escape tower was designed to pull the Apollo command module free of the Saturn V launch system in the event of an abort to ensure astronaut safety. Under normal launch conditions, the tower is jettisoned. It is separated from the command module by firing two pyrotechnic charges at opposite sides of each of the four attach nuts (frangible nuts). The detonation separates each of the frangible nuts into two pieces. The tower attachment to the command module lay inside the tower leg well. The frangible nut was made of 4340 steel heat treated to 1520 MPa (220 ksi). A few days before a scheduled launch, one of the nuts was reported to have fractured (Fig. 24).

The part and fracture face were rusted and encrusted with salt. The access door to the tower leg well had been left open in a rain storm. Laboratory efforts to duplicate the stress-corrosion failure in salt (NaCl) water were unsuccessful after 30 days of exposure. The encrusted salt on the fractured nut was analyzed during this testing period; it was ammonium fluoborate, not salt from the sea coast. Stress-corrosion testing in this solution produced an immediate failure. Subsequent investigation disclosed that the ammonium fluoborate was a salt added to the ablative material on the launch escape motor nozzle skirts. Apparently, this salt had been leached out during a storm and ran down the tower leg in through the open access door. Obviously, this is not a scenario that would have been anticipated during design. The corrective action was to ensure that the doors were sealed in the rain and that the ablative was painted.

Aluminum Structural Elements. In October 1967, a structural failure occurred on a Lunar Excursion Module (LM) test article (Ref 17). A crack occurred in a web splice plate made of aluminum alloy 7075-T651. Investigation revealed that shims had been omitted from the assembly and that high installation stresses resulted in a stress-corrosion failure in the atmospheric environment. The low stress-corrosion threshold of the alloy and temper (<55 MPa, or 8 ksi) was contributory. In December 1967, numerous aluminum alloy 7075-T6 tubes on the same test article were also found to be cracked. The ends of the tubes had been swaged down to mate with a tubular fitting. A tolerance buildup problem resulted in high sustained tensile stresses and the eventual SCC of 20 tubes. This prompted a thorough review of all Apollo and LM hardware to evaluate whether other parts had similar problems. More than 130 LM parts were found to have been cracked as a result of stress corrosion. The corrective action taken involved shimming, relieving fabrication stresses, shot peening, and the application of protective coatings.

Fluid Fittings. Separable mechanical fittings of the B-nut type used in aerospace hardware as AN, MS, and MC fittings in the 1950s and 1960s were often very troublesome in terms of leakage and SCC. Because of the toxicity and the potentially explosive nature of aerospace fuels and oxidizers, it was decided early in the Apollo program to use a bending beam type conical seal as the standard separable fitting (Fig. 25a).

This seal, which was made of precipitation-hardenable stainless steel, had both the conical face and the mating surface polished to an 8 RHR finish. As the seal was tightened, the face was spring loaded against its mating part, acting in some respects as a Belleville spring. Fittings were made from either 17-4PH or 17-7PH alloys. Rockwell used 17-4PH fittings and experienced no problems. An Apollo RCS engine subcontractor, however, reported circumferential cracking of the face seal in 1965 after 10 months of service in N_2O_4. The failure analyses conducted at Rockwell erroneously reported that the failure was due to overstressing, perhaps aggravated by cycling of the fretted areas on the face. Interestingly, however, the failure was intergranular in nature and had the characteristics of stress-corrosion branching networks (Fig. 25c to e). The subcontractor switched to 17-4PH fittings.

Fifteen years later, this time in the space shuttle orbiter program, the subcontractor for the gas generator valve module of the auxiliary power unit reported cracking of the same fitting design made from 17-7PH stainless steel. This fitting also failed in a circumferential manner, and cracking was intergranular in nature. Calculations indicated that high tensile stresses (1240 MPa, or 180 ksi) were applied to the face when the nut was tightened well above the 515-MPa (75-ksi) stress-corrosion threshold for this alloy as supplied in the TH1060 condition. In this case, chlorides from a machining (cutting) lubricant were accidentally introduced to the inside diameter of this assembly. The chlorides brought about SCC. Replacement of these fittings with 17-4PH steel (H1075) was made in 1980.

Potable Water Pressure Transducer. A 17-7PH stainless steel diaphragm of a potable water

(a)

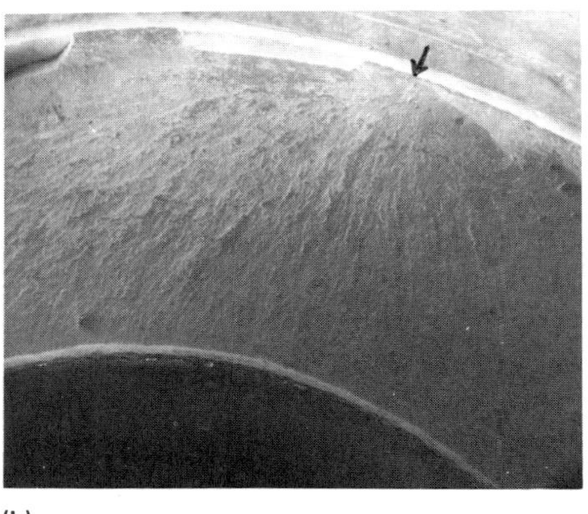
(b)

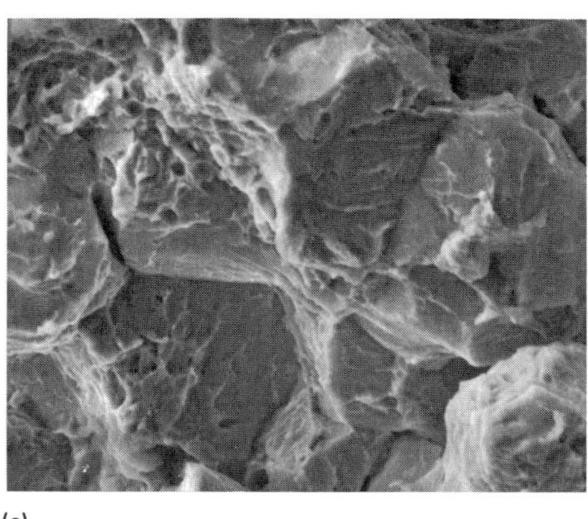
(c)

Fig. 33 Hydrogen embrittlement failure of a 300M steel orbiter nose landing gear steering collar pin. The pin was heat treated to a 1895-MPa (275-ksi) strength level. The part was plated with chromium and titanium-cadmium. (a) Pin showing location of failure. Actual size. (b) Failure origin (arrow). 9×. (c) Brittle intergranular fracture face characteristic of hydrogen embrittlement. Parts did not receive a hydrogen embrittlement relief bake due to processing error. 1380×

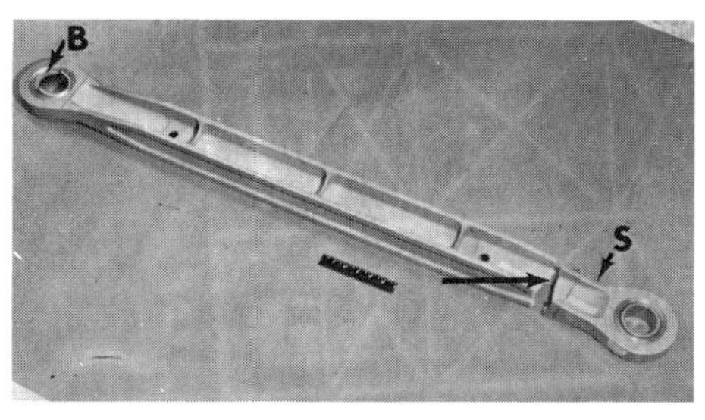

(a)

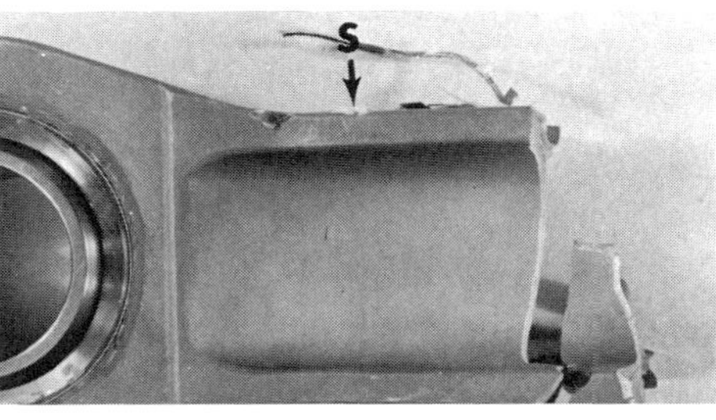

(b)

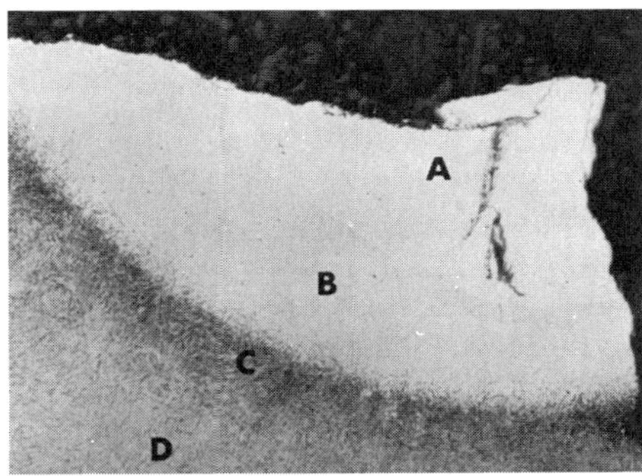

(c)

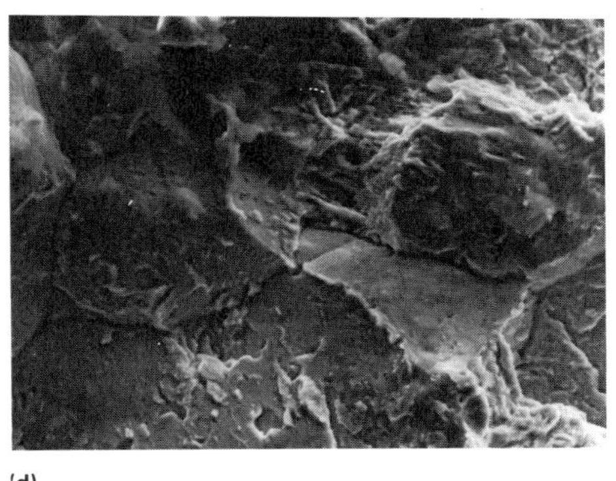
(d)

Fig. 34 Hydrogen embrittlement of an orbiter landing gear lower drag brace made of 300M steel. Steel was heat treated to 1895 MPa (275 ksi). The part was plated with titanium-cadmium. Wear surfaces were chromium plated. (a) Drag brace showing tensile failure that occurred at 50% of ultimate tensile strength during a qualification test. (b) Close-up of failure. (c) A section through the initiation site showing an arc burn. A, area melted by arc burn; B, untempered martensite; C, overtempered martensite; D, tempered martensite of the base material. 145×. (d) Fracture face away from initiation area showing intergranular failure characteristic of hydrogen embrittlement. 870×

pressure transducer (Fig. 26) failed during the eighth flight of the space shuttle orbiter, causing the transducer to malfunction. The pressure transducer had completed three prior flights without failure and experienced only a nominal pressure and temperature of 215 to 240 kPa and 25 °C (31 to 35 psi and 75 °F). The 17-7PH steel was in the TH1050 condition. The failure occurred near an electron beam weld and was intergranular; therefore, sensitization was suspected. Hardness was 34 HRC rather than the 38 to 44 HRC expected. A review of the specification and heat-treat records of the supplier indicated that the part was processed properly.

Metallography revealed carbide precipitation in grain boundaries, with primary austenite in the grains and a martensitic network in the grain boundaries (Fig. 26d). Because the alloy is a martensitic alloy, this microstructure was highly irregular. Contamination could explain grain-boundary carbides but could not explain why grains failed to transform. A leading raw material supplier of 17-7PH was contacted and verified that this alloy, if improperly balanced in chemistry (even though within specification limits), may exhibit this unusual behavior.

Although the entire diaphragm was sensitized and had anodic (martensitic) grain boundaries, failure occurred as a circumferential crack in the area of highest bending stress, a stress calculated to be approximately 690 MPa (100 ksi), which is well above the stress-corrosion threshold of the alloy. The failure was therefore judged to be SCC.

Ferry Lock Pin. During ferry flight of the space shuttle orbiter back to Cape Kennedy, the thrust vector control servoactuator system is locked into place with shear pins. Originally, 17-4PH stainless steel (AMS 5643) was used. Later, the design was changed to 300M steel, which was heat treated to 1895 MPa (275 ksi) minimum ultimate tensile strength and 1585 MPa (230 ksi) minimum tensile yield strength. The part is internally threaded for about one-third of its length and silver plated in accordance with AMS 2412 after a nickel and copper strike. The total plating thickness does not exceed 10 μm (0.4 mil), and the part is baked for hydrogen embrittlement relief for 23 h at 190 °C (375 °F).

The pin failed during torquing prior to ferrying the shuttle orbiter back to the Cape after the fourth shuttle flight (Fig. 27). Examination by scanning electron microscopy (SEM) (Fig. 27d) revealed that the final failure was ductile. A visible rusted area on the bolt showed intergranular attack, and the metallographic cross section showed a branching crack network (Fig. 27c). Failure appeared to start on the interior threaded area where maximum tensile stress occurred. The SCC was probably accelerated by the silver plating through galvanic action. The inability to plate a threaded hole uniformly to this depth probably contributed to the corrosion problem. Even though this part was not associated with the launch vehicle configuration, this failure points out how other interfacing parts could result in expensive failures if careful engineering, manufacturing, and quality control steps are not taken.

Hydrogen Embrittlement

The term hydrogen embrittlement, as used in this discussion, refers to the pickup of hydrogen by metals that results in either the immediate loss of ductility or the delayed failure by subcritical

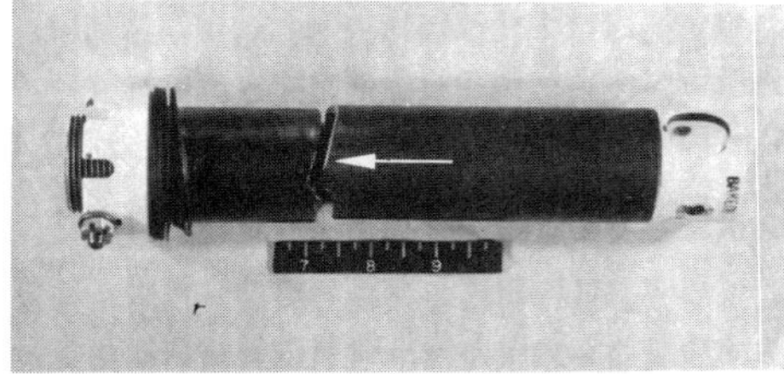

(a)

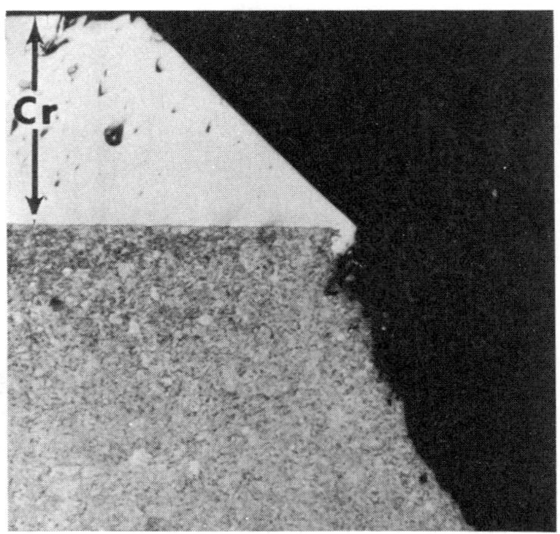

(b)

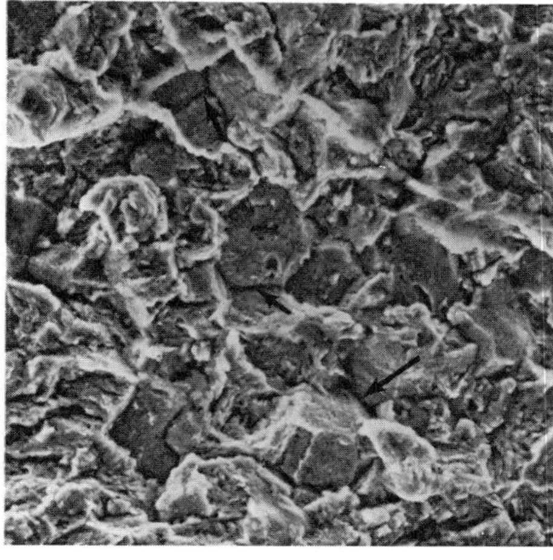

(c)

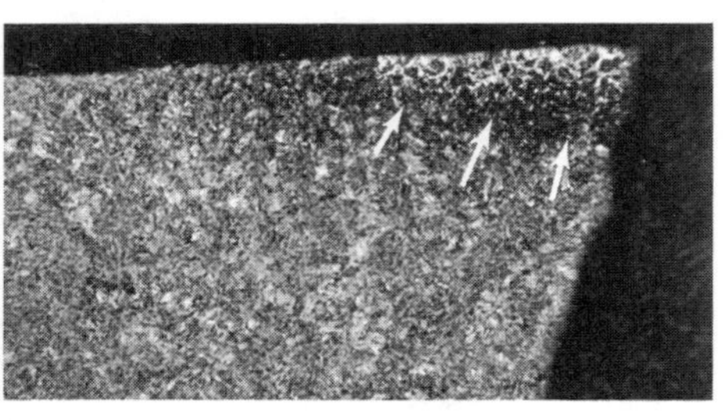

(d)

Fig. 35 Hydrogen embrittlement of an orbiter nose landing gear trunnion pin. Pin was made from 300M steel heat treated to 1895 MPa (275 ksi). Wear surfaces were chromium plated, and nonwear surfaces were plated with titanium-cadmium. (a) Failed trunnion pin showing fracture (arrow). Pin is loaded in shear and bending. 1/5×. (b) Fracture surface and very thick chromium plating. 85×. (c) Fracture face showing intergranular failure propagated by hydrogen. 875×. (d) Localized grinding burn and untempered martensite (arrows) where cracking initiated. 95×

crack growth under sustained stresses. Hydrogen may enter the metal from ionic species in processing solutions, from the gaseous phase at moderate or high pressure or temperature where the protective oxide is broken or permeable, or from the corrosion process itself, where the metal acts as the cathode.

To prevent hydrogen embrittlement on manned spacecraft, several steps are taken. For example, titanium is no longer used in hydrogen systems (Ref 21) (see the following discussion). Steel parts are restricted in strength to below 1380 MPa (200 ksi) for tensile-loading applications. Steel parts with tensile strengths greater than 965 MPa (140 ksi), which are processed in potentially embrittling fluids, must receive in-process baking as well as a final 23-h bake at 190 °C (375 °F). Certain highly embrittling platings are avoided. Finally, by designing to avoid fluid entrapment where possible and by proper coating selection, parts can be protected from the corrosion process. Nevertheless, hydrogen embrittlement never seems to be eliminated totally, because hydrogen always seems to find new ways to manifest itself.

Some of the examples below point out that hydrogen failures often occur where preexisting defects are present. The hydrogen assists subcritical crack growth until the crack becomes unstable. It is believed many of these failures would not have occurred in either the absence of the defect or the absence of hydrogen. Testing a failed part to determine the existing hydrogen content is almost always useless because the destructive hydrogen has long since left the failed part.

Liquid Hydrogen Pressure Vessel. The supercritical cryogenic hydrogen pressure vessels used in the Apollo service module were made from Ti-5Al-2.5Sn ELI. The ELI grade was chosen to ensure adequate ductility to −253 °C (−423 °F). The pressure vessel was designed for 1.7-MPa (245-psi) service and was of a dewar design described in the discussion "Electrical Power System" in this section. As a part of the development testing of the cryogenic system, the hydrogen tank and lines were being flow tested at fuel cell purge rates; that is, the hydrogen was heated and allowed to pass through the welded titanium plumbing joints to an exhaust. During this period, which was approximately 1 h long, the titanium alloy tubes failed, literally flaking and crumbling (Fig. 28). The walls were almost completely consumed (Ref 22).

Investigation disclosed that hydride embrittlement occurred at the weld and that hydride layers were formed in discrete areas along the pure titanium tube. It was postulated that this purging flow of hydrogen, traveling at supersonic speeds, heated the inside of the tube and that turbulence in the weld area, perhaps associated with defects, stripped off the protective oxide.

Destructive evaluation was made of an existing pressure vessel at some 30 different welded connections. A few of these showed hydriding that was superficial in nature except for one tubular section outside the flow stream (Fig. 28d) where bending may have occurred. Corrective action consisted of changing all plumbing to type 304L stainless steel. Future spacecraft cryogenic systems, such as the space shuttle orbiter, used aluminum alloy 2219-T6 for pressure vessels.

Fuel Cell Nitrogen Pressure Vessel. The nitrogen pressure vessel in Apollo electrical power system fuel cells (Fig. 29) provides operating pressure for the KOH electrolyte. The pressure vessel, made of Ti-5Al-2.5Sn, would normally rupture above 59.3 MPa (8600 psi). Minimum system design pressure was 39.8 MPa (5780 psi). During proof pressure testing, one vessel failed (leaked) at 10.4 MPa (1500 psi). Investigation revealed that the vessel had been repair welded locally in a dry box with insufficient gas coverage to prevent oxygen embrittlement of the weld. Metallographic examination also revealed concentrated hydride needles lining the edges of the repair bead and localized surface hydriding. This could not be explained by the welding process.

Further investigation revealed the vessel was stress relieved after welding at approximately 595 °C (1100 °F) in a controlled-atmosphere furnace. Unfortunately, the furnace also had the capability of providing a hydrogen-reducing atmosphere. The investigation disclosed that a hydrogen valve near this part in the furnace was found to be leaking. This leakage was eventually very costly to the supplier because all of the Apollo parts were reevaluated and/or scrapped and the supplier had to investigate and reverify the quality of all other titanium parts processed in that furnace over the past few years. Because the supplier also manufactured jet engines, intensive efforts were made to ensure the complete safety of the delivered hardware.

Launch Escape Tower Solid Rocket Motor Case. The launch escape motor was a solid rocket motor designed to pull the Apollo command module free of the Saturn V stack in the event of an aborted launch. The motor case was 660 mm (26 in.) in diameter and 3.8 m (12.5 ft) long. The motor case was made of 4335V modified low-alloy steel heat treated to 1445 MPa (210 ksi). It was welded longitudinally in the cylindrical section and circumferentially where the forged heads joined the cylinders. Sixty cases were initially ordered, and all passed the hydrostatic proof pressure test. Approximately 2 years later, six additional rocket motor cases were ordered. Two of these burst during the hydrostatic test.

An investigation revealed that two factors contributed to the failure. First, the manufacturer ran out of weld wire and substituted a higher-carbon wire (0.40% C versus 0.35% C) in weld repair areas. Second, because of government safety regulations, the manufacturer changed from proof testing with oil to testing with inhibited (0.2% sodium dichromate) water. Inhibited water *per se* does not prevent hydrogen embrittlement. As a result, the weld repair areas were highly susceptible to hydrogen embrittlement. Rupture of the existing surface oxide during proof testing permitted hydrogen embrittlement of the case, and the case rupture occurred in the area with the lowest tolerance to hydrogen, that is, the higher strength weld (Fig. 30). This susceptibility was confirmed in the laboratory with fracture mechanics test specimens.

The literature had reported premature failure of high-strength steel pressure vessels in inhibited water in the early 1960s. The mechanism can be explained by Fig. 31, which shows that hydro-

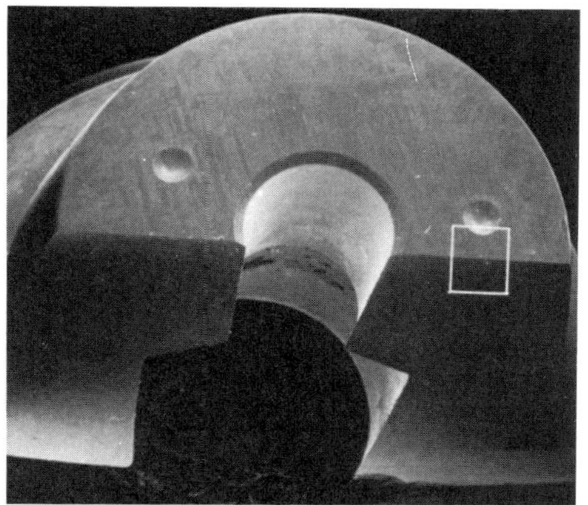

(a)

(b)

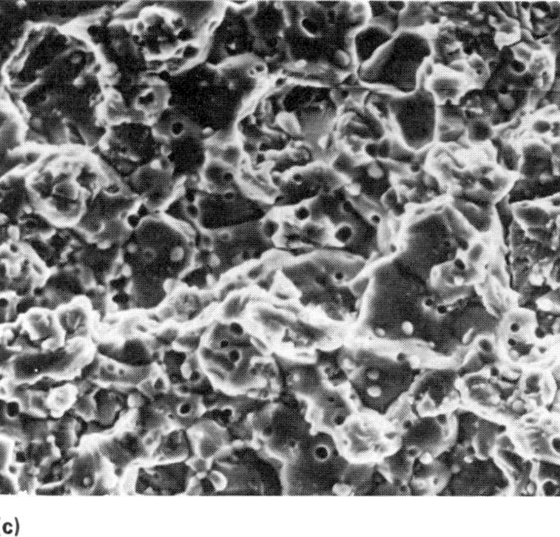

(c)

Fig. 36 Hydrogen embrittlement of a type 440A stainless steel valve seat from an orbiter solenoid latching valve. Seat is hardened to 52 HRC. (a) Sectioned valve seat showing area of cracking (inside box). 9×. (b) Cracking (arrows) appears to originate in hardness indentation. 75×. (c) Fracture surface of crack showing intergranular nature of the failure. Failure was caused by hydrogen in etching solution and residual stress at the hardness indentation, not by previous high-pressure hydrogen exposure. 1440×

gen-assisted crack growth is inhibited in the presence of oxygen but continues in an oxygen-free environment (Ref 23). Figure 31 also explains why hydrogen embrittlement seldom occurs when a high-strength steel is corroded in air, even though hydrogen is generated, but occurs in water tests of pressure vessels where the availability of oxygen is limited.

Pressure Vessel Welds. During the Apollo program, pressure vessels made of Ti-6Al-4V, an α-β alloy, were welded with either Ti-6Al-4V wire or commercially pure titanium wire. The commercially pure wire was softer and more easily handled. One aerospace supplier reported that an explosion occurred in a Ti-6Al-4V helium pressure vessel and that the explosion resulted in the failure of other helium pressure vessels in the vicinity. The failed pressure vessels used multipass welding and commercially pure wire to join thick weld joints (>25 mm, or 1 in.).

Examination of the failed weld showed excessive hydride needle formation following the contour of the molten weld metal, essentially perpendicular to the transverse loading across the bead. Failure occurred because of the brittle hydride needles (Fig. 32). It was postulated that the high solubility for hydrogen of Ti-6Al-4V alloy compared to that of commercially pure titanium was the source of the hydrogen and that repeated weld cycles allowed diffusion of hydrogen to pure titanium followed by precipitation at the pure metal/Ti-6Al-4V interface. Theories of strain aging were also advanced, suggesting that hydrogen migrated to the highest strain areas.

All of the solution-treated and aged Ti-6Al-4V pressure vessels on the Apollo command and service modules were welded with Ti-6Al-4V wire except for the final pass on the SPS fuel and oxidizer tanks. In this case, the root side of weld was back chipped, and a filler pass was made with commercially pure wire.

Specimens from SPS qualification tanks were examined and, in some cases, cycle tested. Hydride formation, although limited in nature, did follow the inside diameter bead contour. Because the pass was wide and shallow, the hydride path changed from perpendicular to the weld to a plane parallel to the surface. This undoubtedly explained why no problem had occurred with these pressure vessels. Existing SPS vessels were considered safe based on proof pressure, x-ray, and penetrant inspection.

In future programs, commercially pure titanium wire was not permitted for welding α-β titanium alloys. This experience pointed out that the migration and precipitation of hydrogen must be considered in α-rich titanium structural members that undergo thermal cycles.

Landing Gear Parts. The space shuttle orbiter landing gear is made principally from 300M steel heat treated to 1895 MPa (275 ksi). At this strength level, steel is highly susceptible to SCC and hydrogen embrittlement. Because of the critical orbiter weight requirements (see the discussion "Orbiter Landing Gear System" in this section), this alloy and temper were used despite objections from material and process engineers. Over a short period of time, three separate hydrogen embrittlement failures took place. Each pointed out processing deficiencies that had to be corrected immediately.

The first failure occurred the day before the initial rollout of the shuttle orbiter in September 1976. While the orbiter was resting on the landing gear in Palmdale, CA, a technician noticed the head of a bolt on the ground near the gear. A quick inspection revealed that it came from a 300M pin that retains the steering actuator. The pin, nut, and washer were still on the spacecraft. The pin failed by delayed fracture under the low installation preload used with a shear fastener (170 MPa, or 25 ksi, tension) (Fig. 33).

Metallurgical analyses revealed that cracks occurred in three locations at the radius of the head of the pin. An embrittled microstructure (rock candy in appearance) was located in these areas with no evidence of corrosion. A significant portion of the fracture face was characteristic of hydrogen embrittlement, although final failure was ductile under more rapid fracture. Investigation revealed this pin was part of a lot that had been reworked to correct a plating error. The pin had both chromium plating and titanium-cadmium plating on the part. Chromium was plated on the pin shank and head. Cadmium-titanium was plated on the head radius and internal hole. A review of the records at the vendor failed to disclose any evidence of a hydrogen embrittlement relief baking step.

Corrective action consisted of a reinspection and rebaking of all reworked parts. Future reworks required full manufacturing planning, not just a material review disposition.

The second part that failed was the lower drag brace of the main landing gear. It was also made of 300M steel at the same heat-treat level, had chromium-plated wear surfaces, and had cadmium-titanium plating on other surfaces for corrosion control. The brace failed under a 2-h sustained load of 950 MPa (138 ksi), or 50% of its ultimate tensile strength, during static load qualifications testing. Failure analysis again determined that crack propagation was by hydrogen. The initiation sites were arc burns on the part caused by accidental contact with a hand-held electrode used to ensure more uniform plating (Fig. 34). All parts that had been plated with a hand-held electrode had to be restripped and inspected. For these reworked parts and future parts requiring a hand-held electrode, the electrode was adequately protected by wrapping with nylon or other approved organic web fabrics. This incident was also an indication that baking in this heavy section was marginal for the removal of hydrogen. A complete review of baking procedures, times, and temperatures was made to ensure that no deficiencies existed.

The third part to fail was the trunnion pin of the nose landing gear. The pin failed again under static load and was made from 300M steel heat treated to the same levels as the other parts. This part was chromium plated on the shank area and had a titanium-cadmium plating applied to threads and to other surfaces requiring corrosion protection. The thread plating ranged in thickness from 5 to 7.5 μm (0.2 to 0.3 mil), while other corrosion protection plating was 12.5 to 17.5 μm (0.5 to 0.7 mil) thick.

Failure analysis disclosed the same grain-boundary fracture characteristic of hydrogen embrittlement—again with no corrosion on fracture faces. The part had a few local areas of untempered martensite from grinding burns (Fig. 35), and some areas had chromium plating as thick as 0.3 mm (12 mils). The drawings called for a nominal thickness of 0.06 mm (2.5 mil). Some cadmium plating solution had entered the fracture surface and had both plated and deposited salts. It was believed that overheating due to

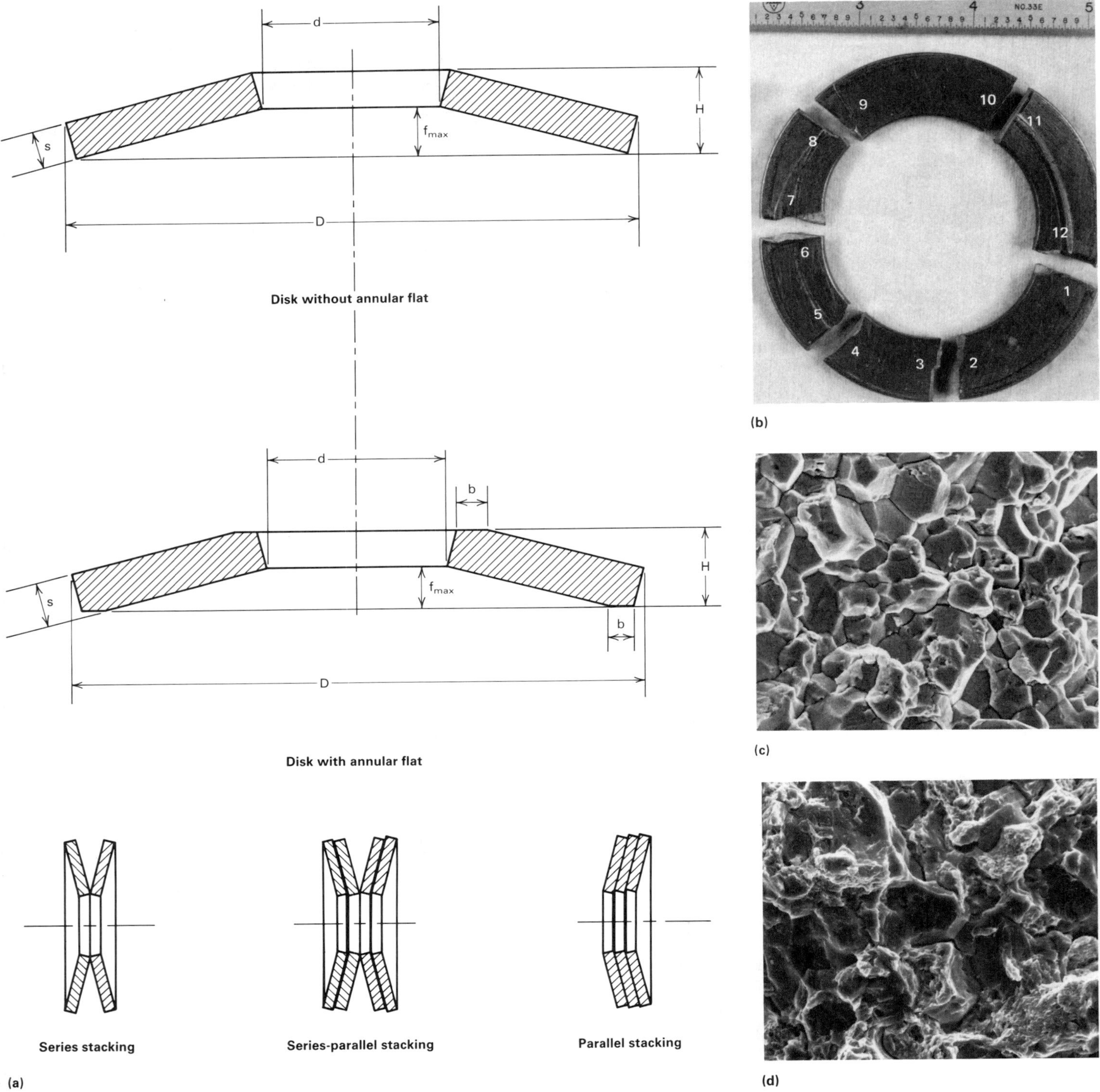

Fig. 37 Hydrogen embrittlement of alloy steel Belleville springs for the space shuttle orbiter program. (a) Illustration of spring design and stacking arrangements. (b) Belleville spring that failed in service. (c) Fracture face of a cadmium-plated Vascomax 300 maraging steel spring that failed from hydrogen embrittlement in saltwater immersion. 1080×. (d) Fracture face of cadmium-plated 6150 alloy steel Belleville spring that failed by hydrogen embrittlement due to inadequate baking. 1440×

grinding resulted in untempered martensite, which cracked either before or upon immersion in the plating bath. Residual hydrogen left in the part after plating migrated to the crack areas when the parts were under sustained load, resulting in slow crack growth leading to failure. Fracture mechanics analyses were able to show that the initial flaw size would grow by subcritical crack growth to the final size (2.5 mm deep × 2.4 mm long, or 0.100 × 0.095 in.) that failed under load.

In this case, special controls had to be placed on grinding, including lubricants, pressures, speeds, and feeds. It is significant that all failures occurred during a 4- to 5-month period and that, since then (9 additional years), no subsequent failures have been noted.

High-Pressure Hydrogen Valve Seat. Early shuttle orbiter flights required high-pressure hydrogen valves for the fuel cell system. A solenoid valve, used successfully in the Apollo program for high-pressure helium in the reaction control system, was evaluated as a candidate valve for use with the high-pressure hydrogen. Testing consisted of exposing the valve to a 16.5-MPa (2400-psi) pressure for 24 h, followed by 200 actuation cycles at 2.4 MPa (350 psi). After hydrogen testing, the valve was disassembled for metallurgical examination. Hardness tests were performed and cross sections were made. A valve seat made of type 440A stainless steel was found to be cracked (Fig. 36).

The crack originated from a hardness indentation, and concern was expressed regarding whether the hydrogen gaseous exposure or the metallographic examination caused the failure. Using other available valve seats, unexposed to gaseous hydrogen, it could be demonstrated that the combination of the hardness indentation and the acid etchant (a mixture of HNO_3, HCl, and water) was the cause of the failure. Again, hydrogen was a culprit, but this time it was not from high-pressure hydrogen gas but from metallographic preparation procedures. The valve was qualified for use.

A Belleville spring is a convex-concave washer that stores energy when flattened. It is widely used in aerospace in bungee applications. When a Belleville spring is compressed, very high tensile forces are put on its periphery. Because these springs are made of high-strength steel alloys, stress corrosion and hydrogen embrittlement become real concerns. Depending on the application, the springs are stacked in series, parallel, or series-parallel stackings (Fig. 37).

The Belleville springs used on the space shuttle orbiter may be made from Vascomax 300 maraging steel or 6150 steel. The springs have been plated with cadmium. Cadmium plating is permitted because the springs are totally contained and will not be exposed to the space vacuum (see the discussion "Structural Joints and Fasteners" in this section). Testing has shown that cadmium-plated maraging steel springs will withstand 30 days of salt fog without failure, even with breaches in the cadmium plating, but the springs fail in a 30-day saltwater exposure because of hydrogen embrittlement as a result of cadmium cathodically protecting the steel.

On several occasions, Belleville springs made of cadmium-plated 6150 steel have failed within minutes after loading. In these cases, hydrogen embrittlement from the plating process is suspected. On one occasion, 40 springs were replaced in a bungee. When reloaded, 17 new springs failed within a short period of time. Repeated baking at 190 °C (375 °F) for 23 h has not completely solved the problem. The literature indicates that cadmium-plated springs should be baked at 260 °C (500 °F) for 1 h or at 230 °C (450 °F) for 4 h. The current approach used on the orbiter has been to use vacuum plating, thus avoiding any exposure to hydrogen pickup during electroplating.

Fig. 38 Hydrogen embrittlement of a low-alloy steel Apollo test part plated with nickel-tin and tested under sustained load. Nickel-tin plating is 5 to 10 μm (0.2 to 0.4 mil) thick. No hydrogen embrittlement relief baking was used in the test part. The part was tested at 50% of its ultimate strength (1380 MPa, or 200 ksi) and failed in less than 7 h, beginning at external threads.

Nickel-Tin-Plated Steel Parts. Three weeks before the first manned flight of the Apollo, a 4340 steel parachute fitting failed. Metallurgical examination of the fracture face revealed a rock candy intergranular fracture typical of hydrogen embrittlement. Investigation disclosed that the parachute system subcontractor specified a 3-h hydrogen embrittlement relief bake at 190 °C (375 °F) instead of the 23 h required by the Apollo contractor after application of the plated nickel-tin coating.

The nickel-tin coating had originally been developed to replace cadmium plating on fasteners because cadmium plating sublimes in the vacuum of space. The total coating is 5 to 10 μm (0.2 to 0.4 mil) thick and is excellent for close-tolerance threads. Investigation of the records at the plating shop, however, revealed that the plater performed no hydrogen relief bakeout, because the military specification for tin plating at that time did not require it. The plater ignored the drawing callouts.

Over 1000 different spacecraft part designs were analyzed to determine which parts needed to be inspected and/or replaced. Extensive testing was performed to find the threshold stress levels of parts that had not been baked. Efforts were concentrated on evaluating the highest strength, most highly loaded threaded parts first, because these have the least tolerance for hydrogen and the highest probability of failure. One such configuration is shown in Fig. 38. This part, made of low-alloy steel heat treated to 1380 MPa (200 ksi) tensile strength and nickel-tin plated, failed within 7 h at a stress of 69 MPa (100 ksi). Through inspections of critical parts, torque level verification, and associated testing, the safety of flight parts was ensured. No failure had been found on any flight safety critical parts.

One may question why there were no previous part failures by hydrogen embrittlement given that the same plating procedures had been used for 3 to 4 years. A plausible explanation is that hydrogen will find its way out of steel over a period of time by diffusion through plated coatings. The nickel-tin coating was far more permeable to hydrogen than cadmium platings. Therefore, parts that did not encounter sustained tensile loads shortly after plating were eventually relieved of hydrogen. As it happened, a single supplier had plated nearly 9000 spacecraft parts by the time this problem surfaced. Many of these parts were already installed into the Apollo vehicles under production. Fortunately, the problem

(a)

(b)

Fig. 39 Oxygen ignition of a type 316 stainless steel check valve that occurred during testing a shuttle orbiter LOX flow control valve. (a) The check valve (right) ignited during test at 31 MPa (4500 psi) oxygen pressure. Also shown is the LOX flow control valve (left). (b) Close-up of failed check valve

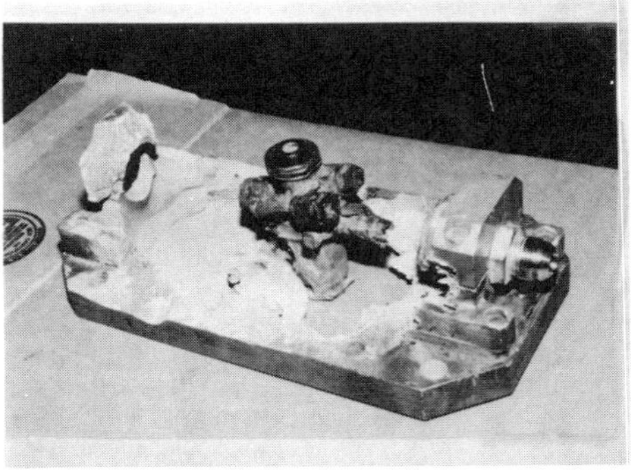

(a)

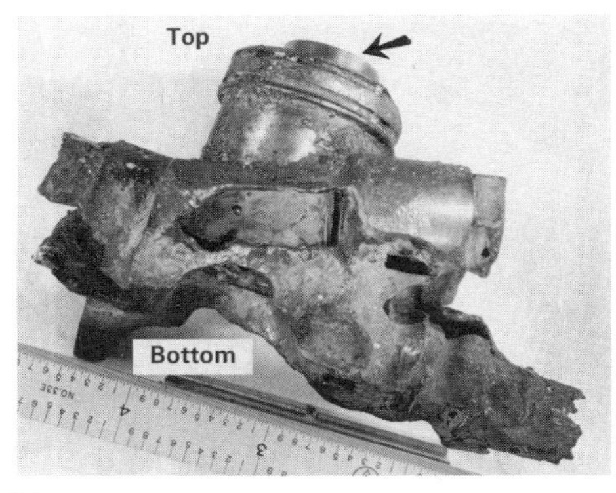

(b)

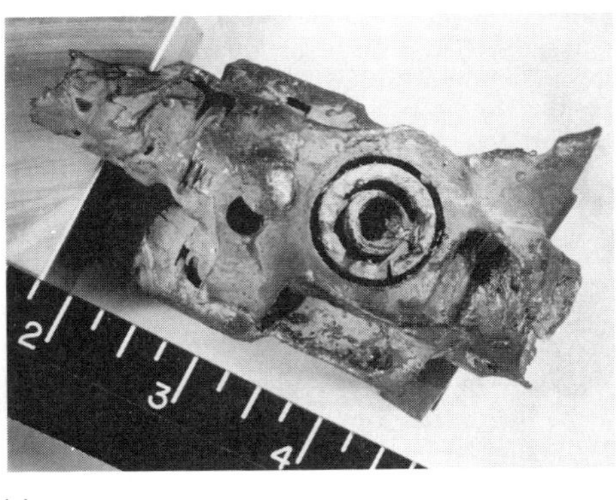

(c)

Fig. 40 Damage to an Inconel alloy 718 shuttle LOX flow control valve and aluminum test fixture due to ignition of a silicone O-ring seal in an austenitic stainless steel test fitting. (a) Test article and fixture after ignition. (b) Side view of valve showing localized melting. Arrow indicates area where solenoid screws on. (c) Bottom view of valve

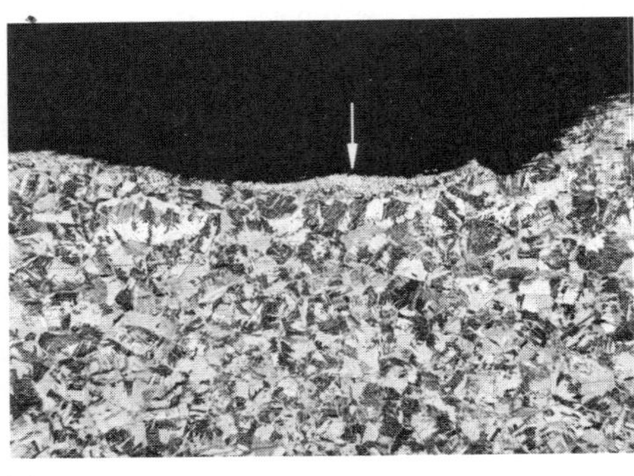

(a)

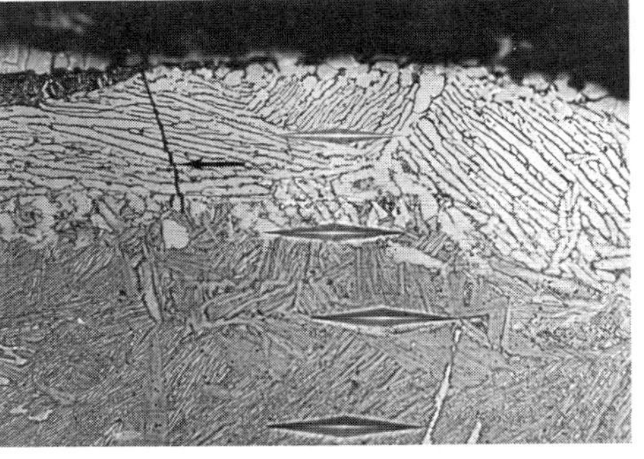

(b)

Fig. 41 Oxygen embrittlement of an extruded Ti-6Al-4V launch escape tower tube for the Apollo spacecraft. (a) Cross section of tube inside diameter showing pitting and α case (arrow). 30×. (b) Cracking of the α case (arrow). 260×. The glassy coating used to protect the part during processing was not continuous, resulting in high-temperature oxidation in air.

could be resolved without vehicle disassembly or scrapping of parts.

Oxygen Ignition

Oxygen is widely used in spacecraft operations as either liquid oxygen or gaseous oxygen. Liquid oxygen will react under certain impact conditions with nearly any metal. Metals such as titanium and magnesium are relatively easy to ignite, while aluminum and stainless steels require considerably more energy to ignite and are used in LOX tubing, valves, and pressure vessel designs. Inconel alloy 718 is one of the most resistant metals to LOX ignition and is widely used in orbiter LOX or GOX applications. In gaseous oxygen, both mechanical impact and pneumatic impact can cause ignition (see the discussion "Main Propulsion System" in this section). Organic materials also readily react with oxygen under impact conditions. When ignition occurs, the part is often so badly melted that no direct identification of the cause can be determined, and the causes of failure can only be inferred from detailed analyses of design and operating or test parameters.

Extravehicular Mobility Unit. A fire destroyed an extravehicular test unit and space suit at the NASA Johnson Spacecraft Center in Houston. Although the location of the ignition was pinpointed, the cause of the fire could never be positively identified. Particle impact, as well as design and manufacturing defects, could not be ruled out. Aluminum and 300-series stainless steel were used in the design. The aluminum was severely burned.

The oxygen flow control valve controls the flow of hot gaseous oxygen at 280 °C (540 °F) and at 31 MPa (4500 psi) to the external tank of the space shuttle system. Two explosions occurred during testing of this valve. In both cases, the valve was a victim of the explosion, not the cause. The first, in January 1977, took place 7 min into a flow test of the valve at a test facility. A facility check valve, which was acting as a shutoff valve against 31-MPa (4500-psi) oxygen, failed at ambient temperatures. The valve was made of type 316 stainless steel with a Stellite ball on the valve stem. The ensuing ignition damaged the stainless steel flow control valve (Fig. 39). The cause was suspected to be contamination in the system.

Concern by NASA for the safety of the orbiter LOX flow control valves resulted in the decision to make these valves from Inconel alloy 718; this decision was based on tests conducted at NASA White Sands using high-velocity particle impacts. During the acceptance testing of a valve for the *Atlantis* spacecraft in June 1984, an ignition occurred in an adjacent 300-series stainless steel fitting. The ignition melted the stainless steel, aluminum base plate, and part of the Inconel alloy 718 valve (Fig. 40). This failure was attributed to the ignition of a loose silicone rubber O-ring seal after approximately 600 s of flow at 195 °C (380 °F) and 27.6 MPa (4000 psi).

High-Temperature Gaseous Reactions

High-temperature gaseous reactions occur during mill processing, heat treating, and surface hardening of metals. During heat treating, detrimental reactions with metals take the form of carburizing or decarburizing in steel, intergranular oxidation in nickel-base superalloys, and formation of an α case on titanium alloys. Surface attack on aluminum alloys is cosmetic in nature and not particularly detrimental to its properties. To prevent these reactions, furnaces with inert, controlled, or vacuum atmospheres can be used, the part can be coated or protected, or the detrimental surfaces can be machined off, grit blasted, pickled, and so on.

During surface hardening, carburizing or nitriding atmospheres are used to achieve the desired surface hardnesses. These are normally well controlled by specifications and quality control sampling to prevent detrimental surfaces from being accepted.

High-temperature detrimental gas reactions become a major concern when they are unanticipated. There may be insufficient allowance on raw material to remove unacceptable layers. Reactions with hardware have occurred where parts, in final dimensions, have little or no allowance for property losses or embrittled surfaces. The examples presented below describe such typical problems.

Launch Escape Tower Tubular Members. The function of the launch escape tower in the Apollo program was to pull the command module free of the Saturn V launch system in the event of an abort. The launch escape tower, a titanium tubular truss structure about 3 m (10 ft) high, was attached to a solid rocket motor case on the upper end and the command module on the lower end through the tower leg bolts. The titanium tubing specified was Ti-6Al-4V with an 89-mm (3.5-in.) outside diameter and a 3.2-mm (0.125-in.) wall. The titanium was produced by a hot extrusion process in which the billet is coated with a glass layer, which not only lubricates the extrusion but also protects it from oxidation.

Excessive surface roughness (85 to 190 RHR) and pitting on the inside of a lot of tubing prompted a destructive microsectioning to determine the cause of the problem (Fig. 41). The inside surface was found to contain a brittle α case and localized

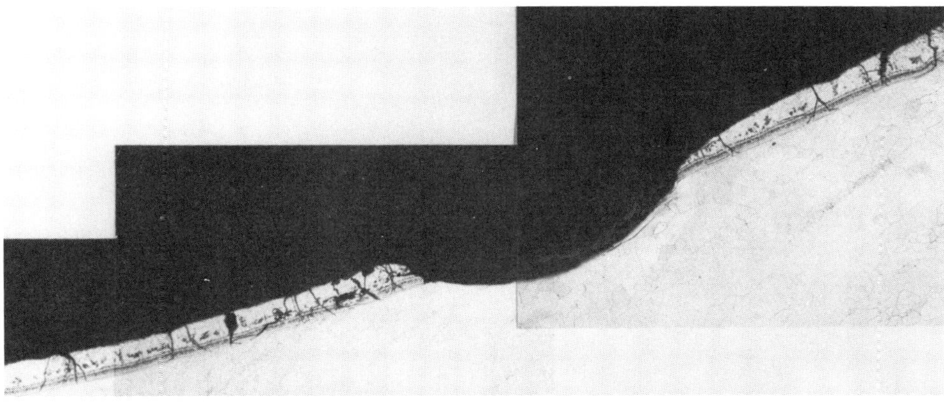

(a)

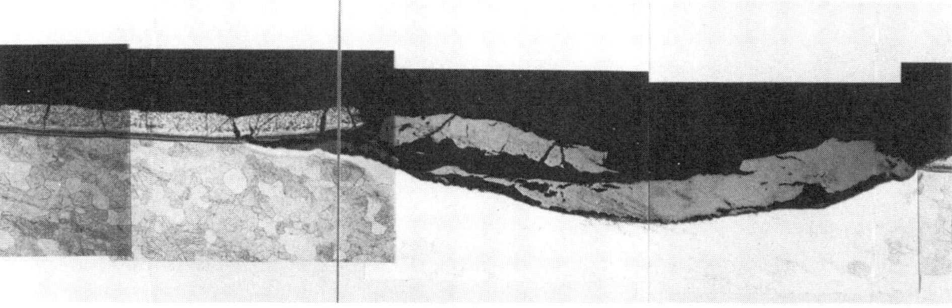

(b)

Fig. 42 Oxidation of a niobium alloy C 103 RCS engine chamber after cyclic temperature testing. The thermal expansion mismatch between the protective silicide coating and the niobium substrate caused the coating to spall. (a) Coating failure site and oxidation of the C 103 substrate after ~81 700 s and 267 000 cycles at 650 to 815 °C (1200 to 1500 °F). (b) As oxidation of the substrate progresses the adjacent coating fails. Here the coating is being lifted from the surface.

cracks indicative of a high-temperature reaction with oxygen (>705 °C, or 1300 °F). Because there was no practical way to rework the inside of the tubing at that time, the lot was scrapped. Future lots provided for adequate material removal on the tube inside diameter.

Reaction Control System Vernier Engine Chambers. The reaction control system on the space shuttle orbiter provides the rocket propulsion to change the attitude of the orbiter with regard to the sun or earth. The RCS vernier engine chamber, made of niobium alloy C 103, must function to 1315 °C (2400 °F). The chambers are protected with an R512A silicide coating. Localized failure of the coating was observed in a vernier RCS engine that had undergone an extensive number of firing cycles. In one case, failure occurred at a slight (75 μm, or 3 mil) mismatch between two machining cuts. This resulted in an offset in the coating, accelerating localized failure. To provide the greatest coating cyclic life capability, action was taken to ensure blending of all machine cuts, to use a dual coating thickness, and to ensure a minimum total coating thickness of 0.1 mm (4 mils).

The silicide coating used on the RCS engine chambers can also fail when exposed to a cyclic, low-temperature (650 to 815 °C, or 1200 to 1500 °F) oxidizing environment for extended periods. The low-temperature failure is caused by the thermal expansion mismatch between the silicide coating and the niobium alloy C 103 substrate. Cracks in the brittle coating fill with oxides, eventually causing spalling of the coating. Once the coating spalls, oxygen can reach the niobium alloy substrate. Oxidation of the niobium alloy at these temperatures is relatively slow, but as the substrate oxidizes, the adjacent coating is undermined. This results in more spalling, which enlarges the coating failure site (Fig. 42).

Orbital Maneuvering System Nozzle Extension. The orbital maneuvering system provides the rocket propulsion for orbit insertion, translation, rendezvous, and deorbit of the space shuttle orbiter. The conical OMS nozzle extension is approximately 1.3 m (50 in.) long and 1.2 m (46 in.) in diameter. It is a welded sheet metal structure made of niobium alloy FS-85, typically 1.5 mm (0.060 in.) thick, and has an R512E silicide coating to protect it from oxidation at temperatures to 1360 °C (2480 °F).

A nozzle was removed from the *Challenger* vehicle when cracks adjacent to the weld bead were found (Fig. 43). The fracture face showed brittle quasi-cleavage with some grain-boundary fractures rather than ductile dimpling. A hardness traverse indicated the fracture to be brittle. Because welding was performed in a controlled chamber to maximum oxygen levels of 2 ppm, no weld contamination during manufacturing was suspected. Inspection of the nozzle showed the coating had been breached and spalled in eyebrow-shaped areas as a result of the nozzle flexing under pressure. A redesigned, stiffer nozzle was already available for replacement; the failed lightweight nozzle represented an earlier design.

Auxiliary Power Unit Gas Generator Catalyst Bed. The auxiliary power unit of the space shuttle orbiter uses an iridium catalyst bed to decompose neat N_2H_4. The decomposed gases are then directed into a high-temperature turbine,

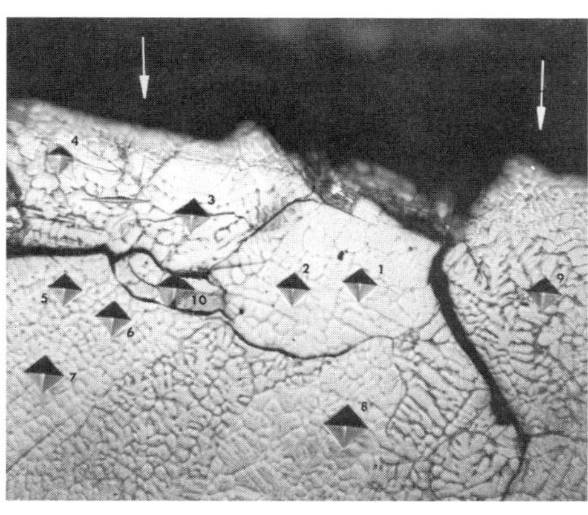

(a)

(b)

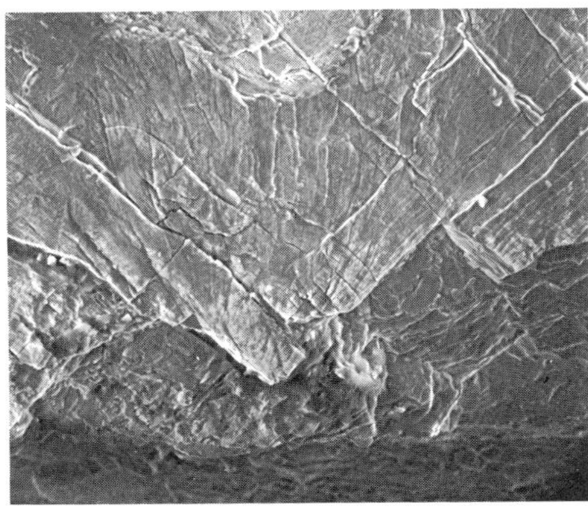

(c)

Fig. 43 High-temperature oxidation embrittlement of a niobium orbiter OMS rocket nozzle extension due to mechanical damage to the silicide coating. (a) Fracture of FS-85 alloy showing that hardness increases near the failed edge. Hardness ranges from 67 HRC at location 4 to 46 HRC at location 8. 145×. (b) Fracture face showing brittle combined quasi-cleavage and intergranular failure. 90×. (c) Enlargement of fracture face showing quasi-cleavage failure with no apparent ductility. 295×

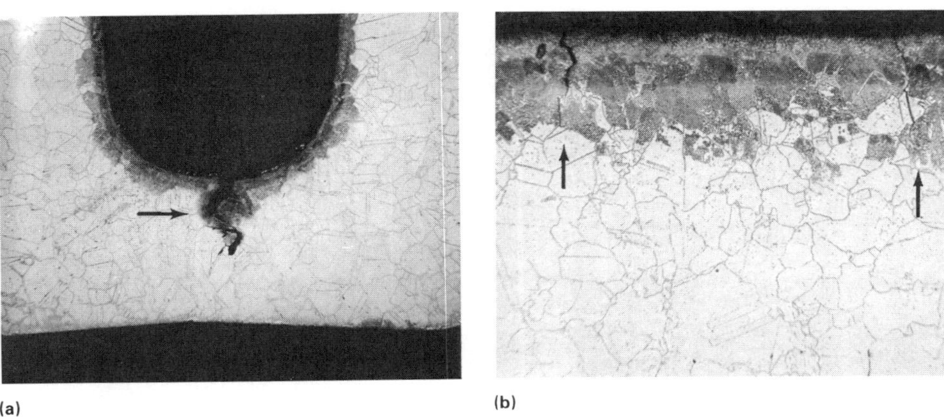

(a) (b)

Fig. 44 High-temperature nitriding of an orbiter APU gas generator catalyst bed housing. The housing is made of Hastelloy alloy B and is in service to 925 °C (1700 °F) in the presence of ammonia formed by hydrazine decomposition. (a) Nitriding and crack originating from embrittled area. 25×. (b) Another area of housing showing nitriding and cracking. 40×.

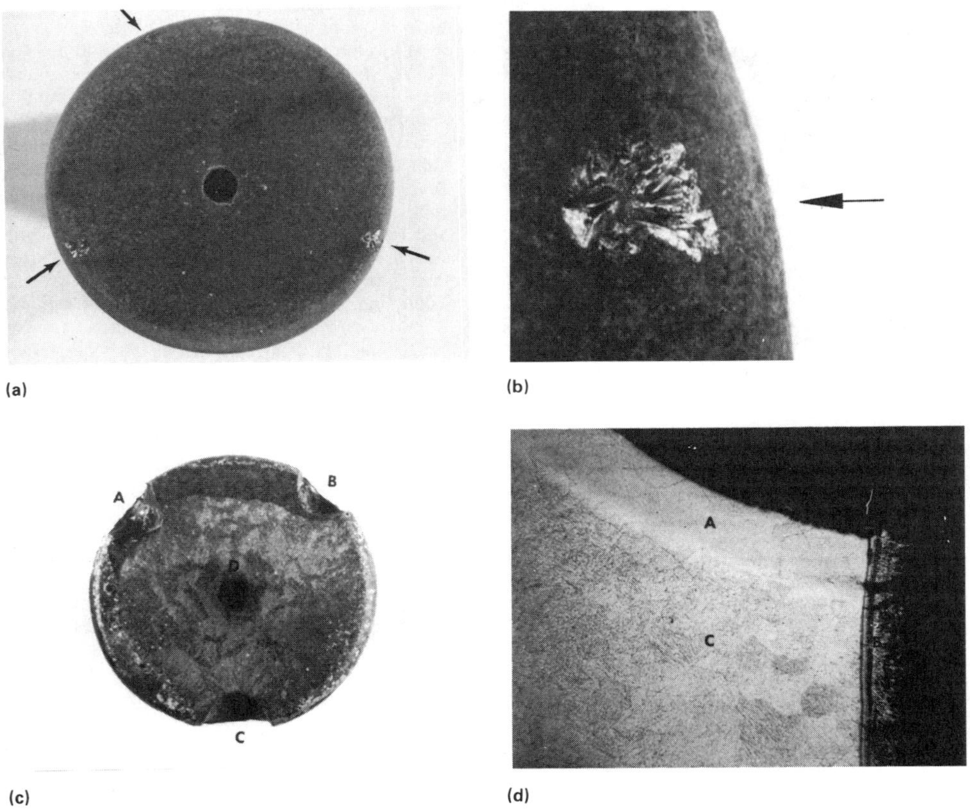

(a) (b) (c) (d)

Fig. 45 High-temperature oxidation of a C 103 niobium alloy nose cap plug in the orbiter SEADS program. The plug is coated with a silicide coating, which was purposely damaged for testing to verify flight safety. (a) Three perforations (arrows) in the coating result in spalled areas of 1 × 1.5 mm (0.040 × 0.060 in.) each. 2×. (b) Enlargement showing base metal damage 1/4 × 1/2 × 3/4 mm deep (0.010 × 0.020 × 0.030 in. deep). 15×. (c) Oxidation of niobium during one simulated entry cycle at a 1485-°C (2700-°F) peak temperature using 0.57-kPa (12-psi) oxygen partial pressure. Craters A, B, and C grew in damaged areas to 6.3 mm diameter × 2.5 mm deep (1/4 × 0.10 in.). 1.5×. (d) Embrittlement of crater surface in section A is 0.2 mm (7 mils) deep. Section C is undamaged base metal. 50×

which in turn drives hydraulic pumps for orbiter hydraulic pressure. The gases resulting from the decomposition are nitrogen and hydrogen with a small amount of ammonia. The latter gas will, at the temperatures of the catalyst (>930 °C, or >1700 °F), cause nitriding of the Hastelloy alloy B used to house the catalyst bed. Hastelloy alloy B was chosen for this application because of its relatively high resistance to nitriding. Nevertheless, nitriding does occur, occasionally resulting in cracking of parts (Fig. 44). Fortunately, the cracked parts are under low loads, failure is not critical, and no high-temperature coating is required.

The shuttle entry air data system nose cap is a modified reinforced carbon-carbon composite nose cap in which 14 pressure ports were added to provide air pressure distribution data throughout the flight entry profile. Holes drilled in the composite nose cap permitted insertion of a niobium alloy C-103 plug that was connected on the internal side to thin-wall niobium tubing. These tubes terminate at pressure sensors. Concern was for the survival of the niobium plugs, because loss of a plug could result in ingestion of a stream of extremely hot plasma that could potentially damage the spacecraft. Although the niobium was coated with a VH109 silicide coating (a proprietary mixture of chromium, titanium, silicon, and hafnium), there was concern that the silicide coatings might experience inadvertent and undetected damage during the manufacturing or launch cycle.

A test program was run to determine whether a niobium port with damage through the coating to bare metal would survive a reentry temperature at the maximum oxygen partial pressure expected during the heating peak, thus ensuring fail-safe mission behavior (Fig. 45). Flaws were made in the coating of a size readily visible by inspection using a chisel point indenter. Each defect resulted in a coating spall area approximately 1 × 1.5 mm (0.040 × 0.060 in.). Penetration into the bare metal was approximately 0.25 × 0.5 × 0.75 mm (0.010 × 0.020 × 0.030 in.) deep.

A maximum reentry cycle is expected to reach 1430 °C (2600 °F). The part shown in Fig. 45 represents an overtest; it was heated to 1485 °C (2700 °F). Normally, reaction of niobium in air results in a canary yellow voluminous oxide, but flow characteristics caused the oxide to be blown free of the surface. Craters grew at damaged coating areas during this cycle to sizes of approximately 6.3 mm in diameter × 2.5 mm deep (0.25 × 0.10 in.), and some subsurface embrittlement was experienced. Niobium, used in the highest-temperature location on the space shuttle orbiter, showed that it is capable of surviving at least one reentry regime even if it has local coating damage.

Liquid-Metal Cracking

Liquid-metal cracking is cracking in a base metal that occurs in the presence of a liquid metal. Often, parts that crack are under stress. Intrusion of the liquid metal into grain boundaries often occurs, but in some cases, intrusion is difficult to detect. The term liquid-metal embrittlement is often used because liquid-metal cracking often causes embrittlement.

Liquid-metal cracking does not always lead to catastrophic failure of parts, as shown in the brazing example below, but it often totally destroys the structural integrity of the part. The major structural materials on a manned space vehicle are primarily made of alloys of aluminum, steel, stainless steel, titanium, and nickel. These are embrittled by liquid metals shown below:

Alloy	Liquid metals causing embrittlement
Aluminum	Mercury, indium, tin, zinc
Steel	Tin, cadmium, zinc, lead, copper, lithium
Stainless steel	Cadmium, aluminum, lead, copper
Titanium	Cadmium, mercury
Nickel	Zinc, cadmium, mercury

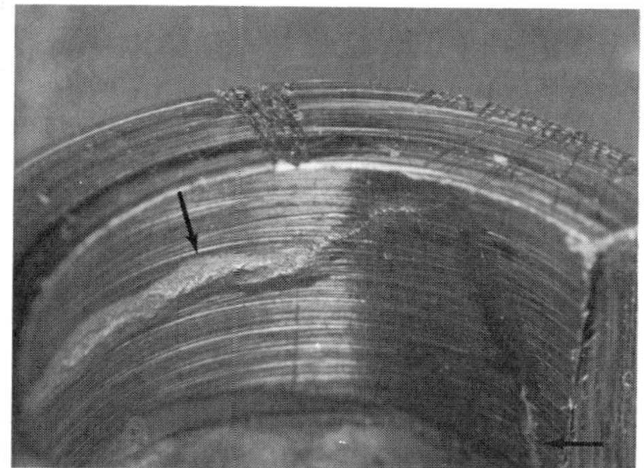

(a)

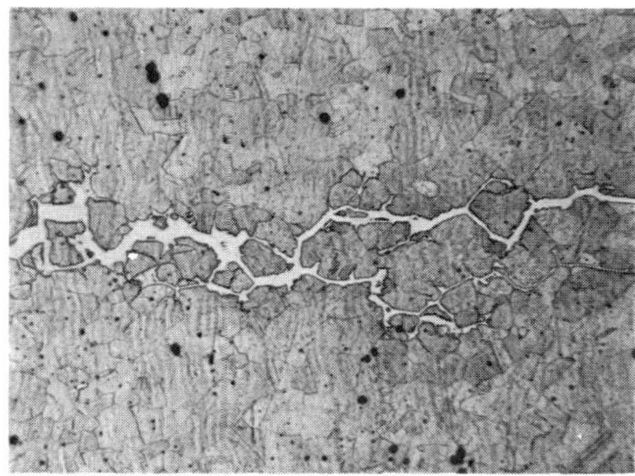

(b)

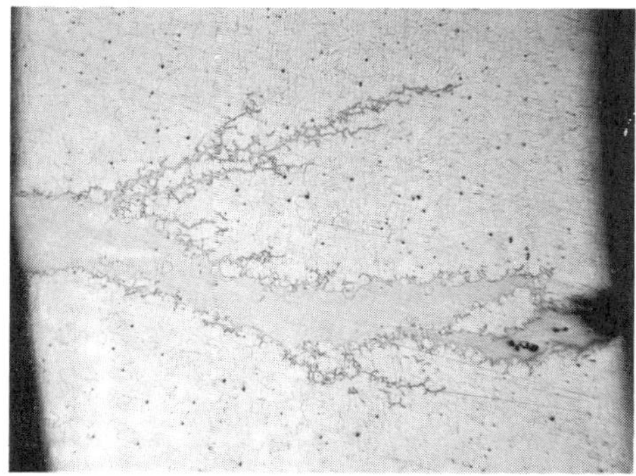

(c)

Fig. 46 Liquid-metal cracking of brazed plumbing lines and unions in the shuttle orbiter due to overheating. Brazing alloy is Nicoro 80 (81.5Au-16.5Cu-2Ni). (a) Braze-filled crack (arrows) on Inconel alloy 718 tube end. 5×. (b) and (c) Intrusion of brazing alloy into 21-6-9 stainless steel. (b) 130×. (c) 45×

Most of these metals have a very low melting point. The lowest, mercury, is often used in switches, instruments (thermometers, manometers), and vapor arc lamps. Mercury is prohibited from use on the space shuttle orbiter to avoid contamination from spillage. Cadmium, due to its sublimation in space, is again highly restricted and is never permitted to exceed 120 °C (250 °F). This is well below the liquid-metal embrittlement temperatures of the plated steel. Zinc, except as an alloy constituent, is generally not used on manned spacecraft.

Many of the embrittling elements find their way into spacecraft usage in three families. First, these elements are used in low-temperature sol-

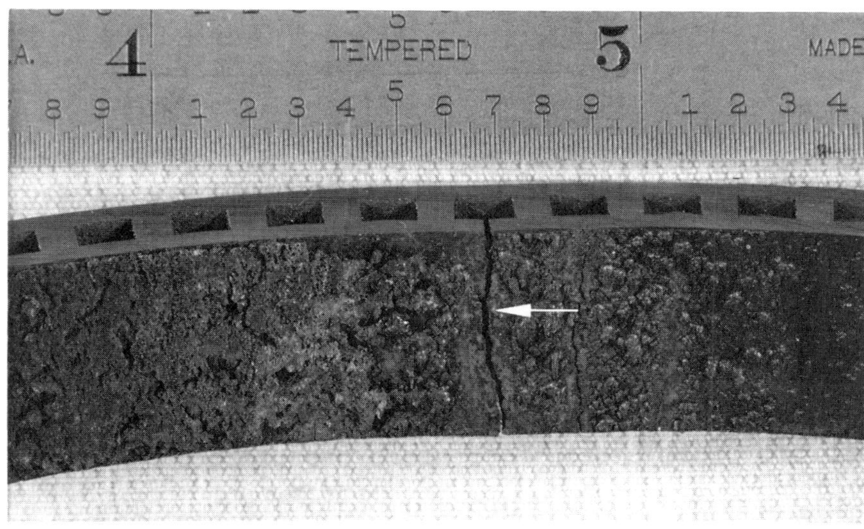

(a)

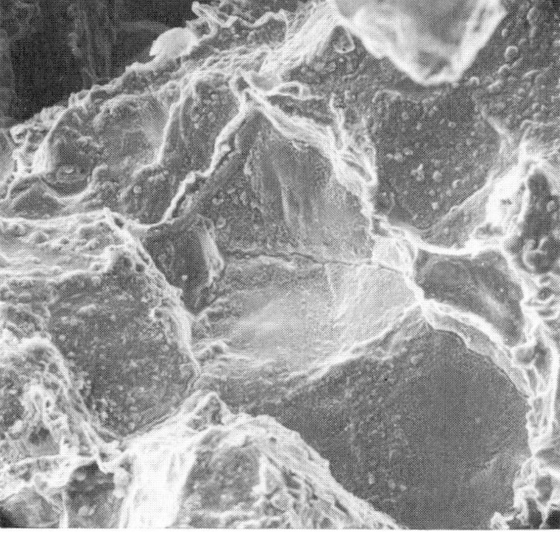

(c)

(b)

(d)

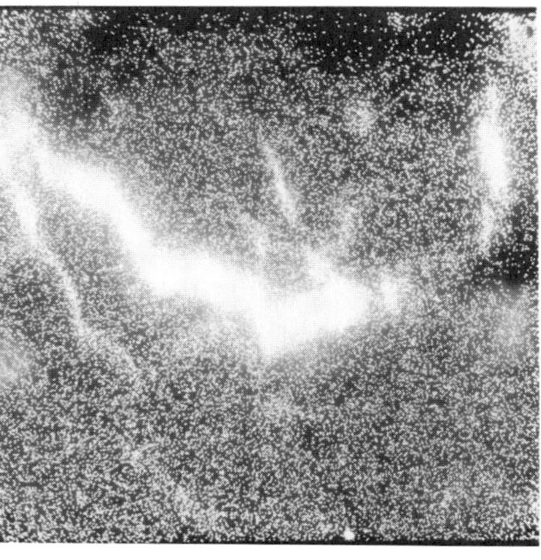

(e)

Fig. 47 Liquid-metal cracking of an orbiter OMS rocket chamber during testing. The chamber is regeneratively cooled and made from type 304L stainless steel. Copper wiring broke loose and melted inside the chamber during instability testing, resulting in liquid copper embrittlement. (a) Section of chamber showing a crack (arrow) in the fuel channel. (b) Crack cross section. 70×. (c) Fracture face of crack showing intergranular appearance. 355×. (d) Enlargement of copper-filled crack in (b). 125×. (e) Electron beam dot map of copper concentration in crack in (d). 125×

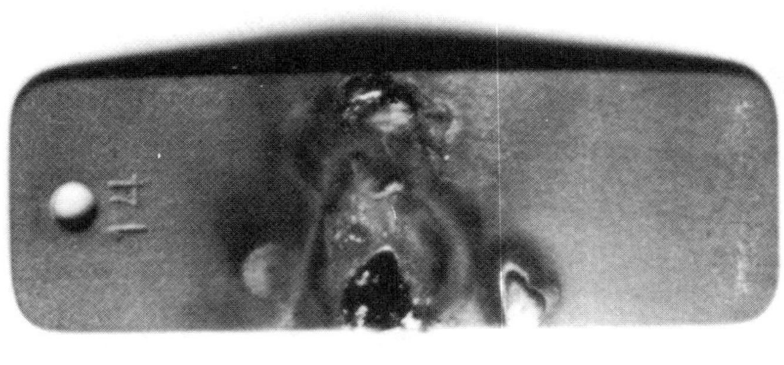

(a)

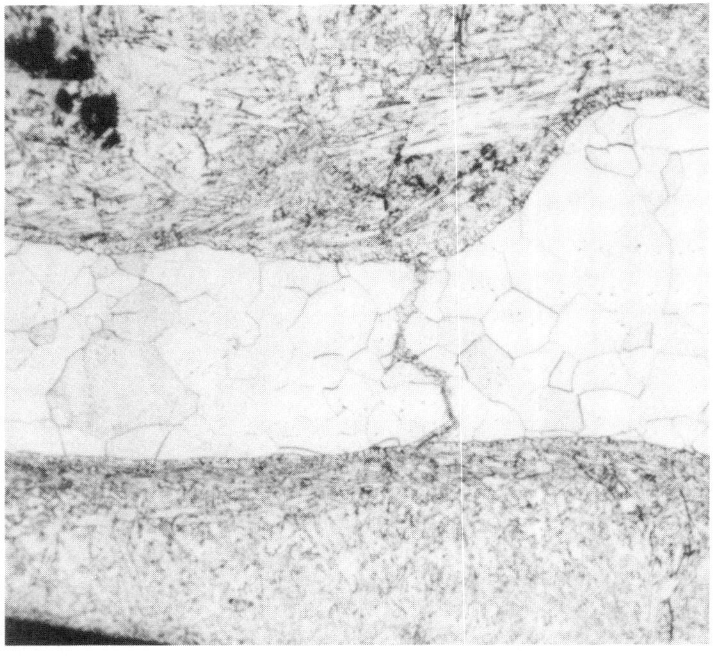

(b)

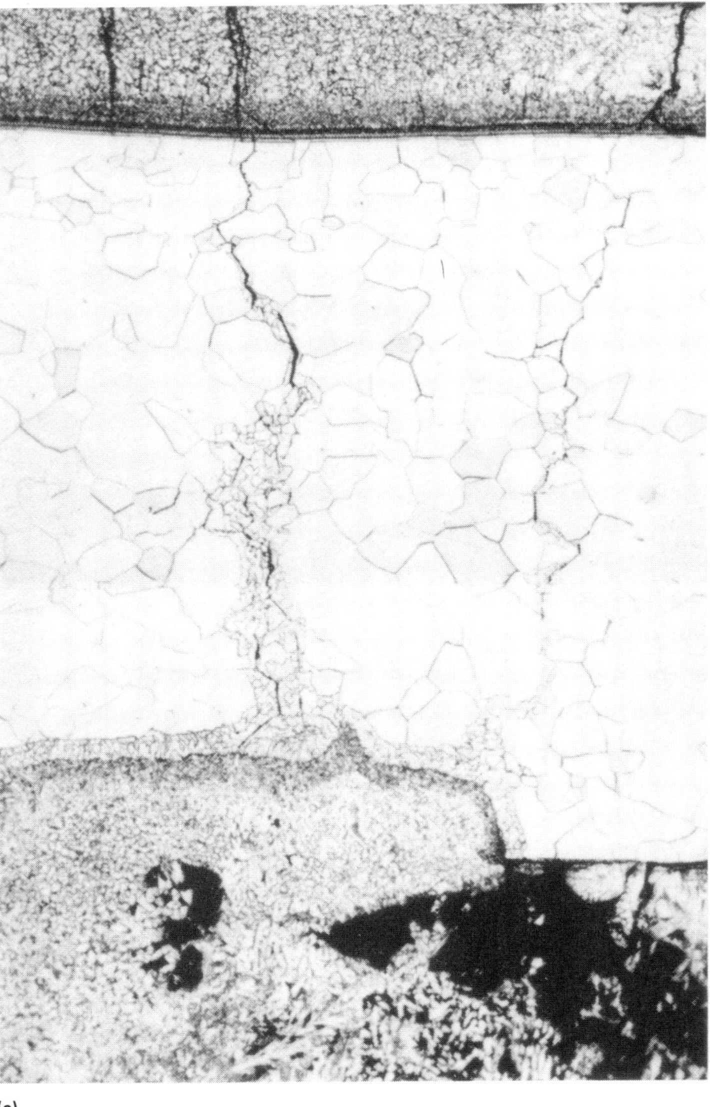

(c)

Fig. 48 Liquid-metal embrittlement of a niobium alloy C 103 test coupon intended to simulate embrittlement of a C 103 shuttle RCS nozzle. The coupon, with a silicide coating, was stressed, and fine chromel and alumel wires were melted on its surface. The resulting coating damage is shown in (a). (b) Cross section showing damage caused by liquid alumel. 100×. (c) Cross section showing damage from liquid chromel. 100×

ders for electrical and avionics applications. These include lead-tin for electrically soldered connections and some indium-base solders for glass-to-metal seals. Second, these embrittling elements are used as silver solders for higher-strength applications. Most of these solders melt in the range of 605 to 800 °C (1125 to 1475 °F) and may contain copper, cadmium, zinc, lithium, and tin. These have caused liquid-metal cracking, especially with incompatible metals under stress. Particular attention is given to these applications. They can be quite troublesome where torch brazing techniques are used, because temperature control depends upon the skill of the operator. Third, many brazing alloys used on the spacecraft for stainless steel plumbing systems contain copper. Many years of brazing experience with automated equipment has proved this to be acceptable (see the following discussion).

Brazed Plumbing Joints. Radiographic inspection of three brazed manifold tube joints for the shuttle orbiter revealed that the 21-6-9 stainless steel tubing had cracked. The tubing was 13 mm (1/2 in.) in diameter and was brazed by automatic equipment with Nicoro 80 braze alloy (81.5Au-16.5Cu-2Ni). Metallurgical examination showed that all brazed joint cracks were completely filled with the braze alloy. A review of all shuttle orbiter brazing up to this time indicated that eight of the 1165 joints (~0.5%) had similar braze-filled cracks. The cracks were both longitudinal and cicumferential; they occurred in type 304L stainless steel, Inconel alloy 718, and 21-6-9 stainless steel in line sizes of 9.5, 13, and 19 mm (3/8, 1/2, and 3/4 in.) (Fig. 46). All cracks were completely under the braze union.

During the Apollo program, 30 tube ends of 5500 braze joints in type 304L stainless steel showed the same indication. Extensive pressure testing of braze-filled cracked tube ends indicated that these tubes would burst in the tubular wall section rather than in the braze-filled crack under the union. These joints were also found to be satisfactory in thermal shock and vibration tests. Based upon this experience, tubes with braze-filled cracked ends, which remain totally under the braze union, are acceptable.

The cause of the problem is thought to be liquid-metal attack. The tube end may contain residual tensile stresses during its induction brazing or may contain small cracks that grow in contact with the braze metal. The fresh surface of the crack is readily wet by the braze alloy. Although the initial induction brazing heats the joint to approximately 980 °C (1800 °F), reheating a braze joint with the same induction brazing parameters increases the temperature because the braze alloy now conducts heat from the sleeve to the tube end more readily than in the initial braze, in which heat is transferred across the capillary gap to the tube by radiation. It is believed that temperatures in tube ends may exceed 1095 °C (2000 °F) during reheat cycles used to reflow braze or debraze.

Copper is known to be capable of causing liquid-metal embrittlement of stainless steels

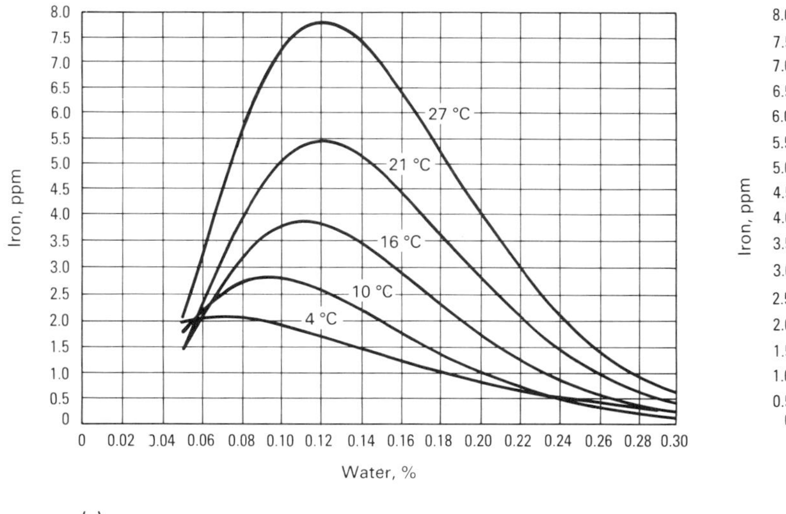

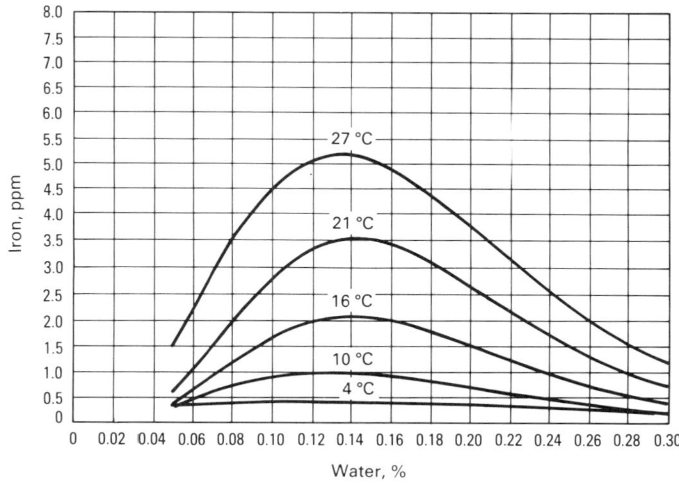

Fig. 49 Solubility of iron in N_2O_4 as a function of water, temperature, and NO content. (a) Solubility curves at 2.5% NO. (b) Solubility curves at 1.25% NO

Fig. 50 Failure of an orbiter RCS AC motor relief valve from flow-decay products in N_2O_4 testing. (a) Front side of poppet. The round spots are deposits which caused the valve to malfunction. (b) Back side of poppet showing similar surface deposits

above 1095 °C (2000 °F). It is suspected that the high copper content of the braze alloy initiates the cracking when it occurs.

The OMS rocket engine combustion chamber is a regeneratively cooled, axial flow type chamber in which the fuel passes through passages in the chamber wall. The fuel acts as a coolant and at the same time is preheated before entering the firing chamber. The chamber is made from type 304L stainless steel. Its exterior is straddle milled to make longitudinal grooves, which are then closed into channels by electroless nickel plating.

The testing program for the space shuttle orbiter OMS engine combustion chamber requires the evaluation of engine combustion instability. This is simulated by igniting a small bomb in the combustion chamber during a firing sequence. (The niobium nozzle extension, normally bolted to the engine in service, is not attached during this test.) Immediately after one such instability test, a portion of the chamber was found to be cracked (Fig. 47).

The cracked chamber and some bomb residue, scabbed onto the inner surface, were examined. The scab material contained copper, copper oxide, and calcium carbonate. The copper apparently came from copper wiring associated with the bomb. The source of the calcium could not be determined. Copper was found in crack areas and on crack faces and had penetrated the metal along grain boundaries.

Reaction Control System Chamber. The reaction control system is used to control the roll, pitch, and yaw movements of the space shuttle orbiter and employs an engine using a silicide-coated niobium alloy C 103 chamber. The hypergolic propellants N_2O_4 and hydrazine react within the chamber to provide engine thrust. Normally, chambers are internally film cooled by the flow of the hydrazine fuel along chamber walls and operate well below 1315 °C (2400 °F). Under conditions of engine combustion instability, loss of the protective film may result in wall temperatures exceeding 1535 °C (2800 °F), and chamber burnthrough may take place. To prevent any detrimental action to the surrounding spacecraft structure in the event of an instability and burnthrough, a wire sensing system on the chamber exterior will trigger shutoff of fuel and oxidizer valves.

During a test of the wire sensing system, a reaction control chamber was purposely modified to induce instability, and significant cracking of the chamber resulted. Cracking was associated with the melting of chromel-alumel thermocouple wires at temperatures slightly above 1425 °C (2600 °F). A test was set up in the laboratory on stressed coupons to simulate the suspected liquid-metal embrittlement failure.

The test specimen measured 25 × 75 × 1.4 mm (1 × 3 × 0.056 in.), was coated with a silicide coating (75Si-2Cr-5Ti), and was subjected to elastic bending stresses. Fine thermocouple wires were then placed on the niobium surface. Figure 48(a) shows the extent of surface attack that occurred instantaneously as wires were melted. Figures 48(b) and (c) show cross sections of liquid-metal embrittlement caused by melting the alumel (95Ni, Al, Si, Mn) and the chromel (90Ni-10Cr) thermocouples, respectively. This example illustrates that even protective ceramic coatings can be breached during liquid-metal attack.

Precipitation of Corrosion Products

To function properly, a valve for a manned spacecraft must be free of corrosion products. Some spacecraft valves, in a full-open position, only move 0.25 mm (10 mils) off the valve seat. Some have sealing lands only 125 μm (5 mils) wide. One of the cases of valve failures is the clogging that occurs from flow-decay products, that is, precipitation of dissolved metals as complex salts.

Reaction Control System Quick Disconnect. After servicing of the forward reaction control system on the space shuttle orbiter in preparation for the second launch, a quick disconnect failed, spilling nitrogen tetroxide on the spacecraft *Columbia*. The N_2O_4 attacked the epoxy amine primer and resulted in the loosening or loss of about 100 tiles. It also entered the spacecraft, exposing structure to corrosion and electrical wiring to possible damage. Examination of

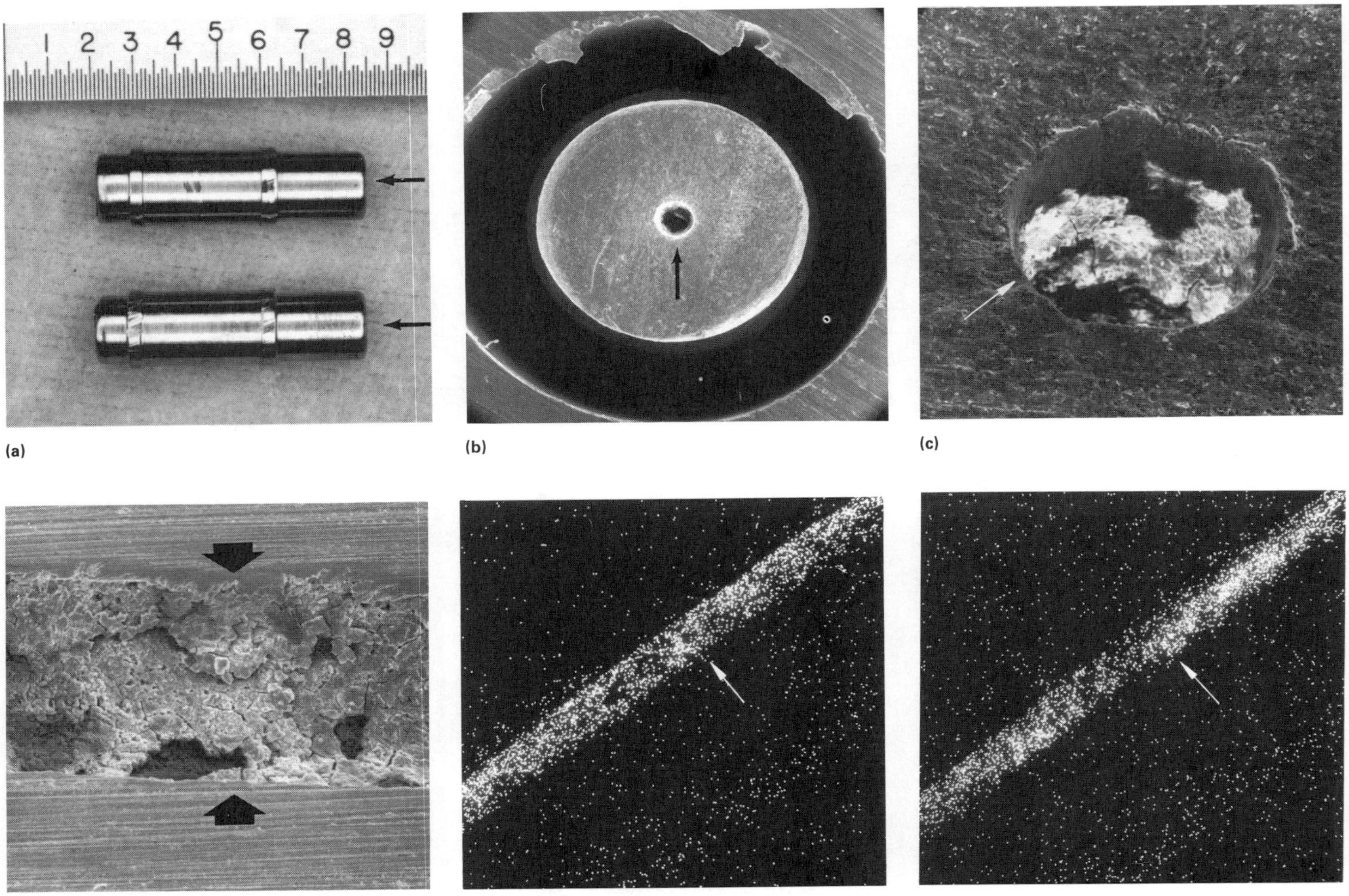

(a) (b) (c)

(d) (e) (f)

Fig. 51 Failure of an orbiter RCS helium regulator valve due to contamination of the sensor orifice by hydrazine salts. (a) Sensor orifice assembly made from austenitic stainless steel. (b) Orifice shown with screen removed. Orifice is 0.18 mm (7 mils) in diameter. 35×. (c) Deposits of hydrazine salts (arrow). 300×. (d) Cross section through orifice showing salt deposits. (e) and (f) Electron dot maps of orifice cross section. The diagonal streak is the orifice. (e) Phosphorus. (f) Sulfur. Sulfur and phosphorus came from soap contaminants. These reacted with hydrazine vapors to block orifice. (d), (e), and (f) 145×

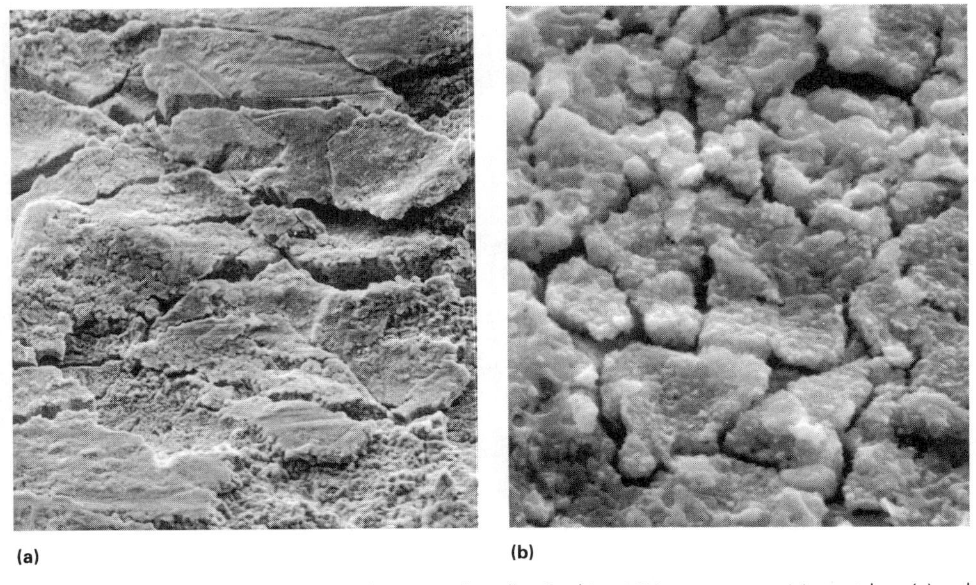

(a) (b)

Fig. 52 Encrustation of hydrazine salts in Hastelloy alloy B orbiter APU gas generator injector tubes. (a) and (b) Deposits occurring in sections of tubes above the 375 °C (700 °F) temperature range. Both 415×

the quick disconnect revealed deposition of complex iron nitrate compounds on its moving surfaces, causing jamming of the disconnect in an open position.

Nitrogen tetroxide is commonly handled and stored in steel containers. In the process, iron dissolves into the N_2O_4, and because it is present at concentrations of only a few parts per million, the integrity of the storage lines and containers are not jeopardized. Under certain conditions of temperature and pressure, however, the iron will precipitate out as complex salts or form gels that cause valves to malfunction. Figures 49(a) and (b) show the solubility of iron in N_2O_4 as a function of water content, temperature, and NO content. At 20 °C (70 °F), the solubility of iron is 5.5 ppm when the NO content is 2.5% and water is 0.12%. If the temperature were to cool to 10 °C (50 °F) after the spacecraft was loaded with this iron-saturated N_2O_4, only 2.6 ppm would remain soluble. The rest could precipitate out. A quick glance at the curves will show that under these conditions, a pickup of moisture by N_2O_4 (or loss) drastically decreases solubility, as does lowering the temperature and losing NO, a somewhat more volatile constituent.

The failure of the quick disconnect delayed the launch of the second flight by more than a month for repair of the thermal protection system and evaluation and/or repair of other spacecraft problems created by the spillage. These problems were minimized in future launches by the use of chillers and filters to process N_2O_4 prior to its loading. Furthermore, samples are being taken at the spacecraft interface to verify that iron levels are below 2 ppm.

The AC motor relief valves are used to control the flow of propellants (fuel and oxidizer) to the RCS engines. Flow decay products from nitrogen tetroxide caused an AC motor relief valve to fail to open at the specified pressure in an orbiter reaction control system test program. The flow decay products are shown on the poppet in Fig. 50.

Reaction Control System Helium Regulator Sensor Tubes. In the reaction control system of the space shuttle orbiter, helium is used to expel propellants to the RCS rocket engines and vernier rocket engines. The helium regulator valve, which controls this function, uses a sensor that has a fine orifice (0.18 mm, or 7 mils, in diameter) through which helium passes. The pressure drop in this flow channel is a measure of the helium demand of the system. Because the hole is only about twice the diameter of a human hair, it is essential that no contamination, flow decay, or corrosion products restrict its flow.

One would expect a helium system to be totally free of corrosion or precipitation products. However, it must be recognized that molecular diffusion permits fuels and propellants to backstream into helium components. Figure 51 shows a result of contamination in the sensor passage as a combination of entrapped cleaning solution and reaction of this product with hydrazine. The tube passage, on one of the shuttle flights, resulted in the sluggish response of the valve to the system demand. Laboratory testing indicates that repeated wetting and drying with hydrazine vapors alone can cause plugging of this passage with decomposition products. Experience has shown that the roughness of the walls of the hole probably contributed to iron dissolution and subsequent precipitation of complex iron hydrazine salts. Fortunately, the valve can be checked out adequately prior to flight. Studies were underway in 1986 to determine what design or processing changes could be made to minimize this problem.

The APU injector tube of the space shuttle orbiter is made of Hastelloy alloy B. It is the conduit for liquid hydrazine to the catalyst bed in which the hydrazine decomposes to nitrogen, hydrogen, and some NH_3 gases. As the hydrazine approaches the catalyst bed, it is heated by the catalyst bed heater to about 650 °C (1200 °F). Metals dissolved in hydrazine, especially iron and molybdenum complex salts, precipitate on these walls when hydrazine vaporizes, as shown in Fig. 52. This precipitation is insufficient to restrict flow in this application. If precipitation of this nature were to occur on moving valve parts, it could result in valve leakage, sluggish response, or jamming.

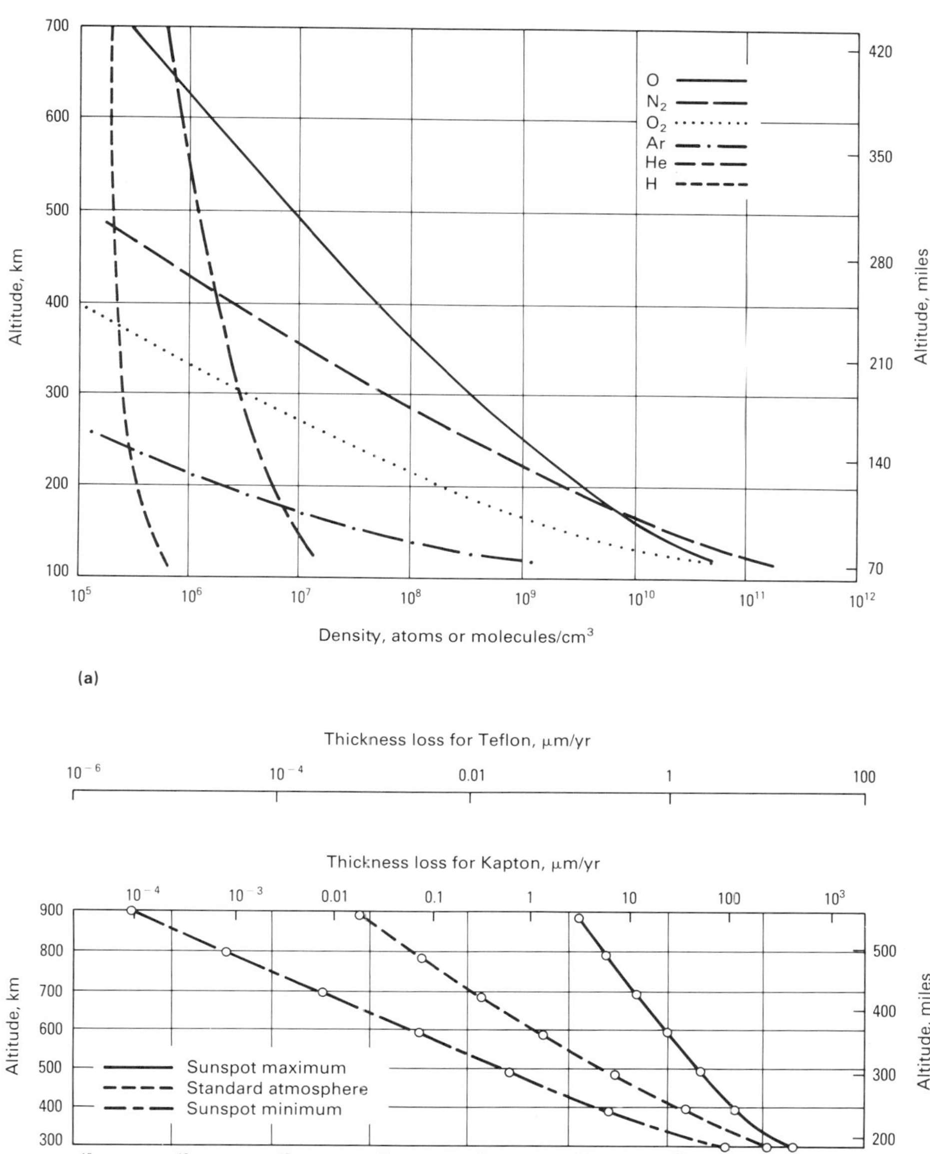

Fig. 53 Atmospheric composition (a) as a function of altitude. Source: Ref 24. (b) Effect of solar (sunspot) activity and atomic oxygen flux on thickness loss of Kapton- and Teflon-covered solar inertial facing surfaces (two sides exposed). Source: Ref 25

Atomic Oxygen in Low Earth Orbit

Atmospheres in low earth orbit principally consist of atoms and/or molecules of oxygen, nitrogen, argon, helium, and hydrogen (Fig. 53a). Although the densities of these constituents are low at typical spacecraft altitudes of 300 to 500 km (185 to 310 miles), the orbital velocity of the spacecraft results in flux densities of 10^{13} to 10^{15} atoms/cm^2 · s. The interactions with the spacecraft produce drag effects that can limit orbital life in lower earth orbits as well as cause changes in the optical (thermal control) and physical properties of coatings and polymers by reflection, adsorption, or chemical reactions.

In the altitude range of approximately 200 to 650 km (125 to 400 miles), the atmosphere consists primarily of atomic oxygen. The interaction of materials with atomic oxygen, which has impacting energy levels of 5 eV, can produce significant surface changes in some materials. The effect of atomic oxygen depends on the fluence (or total flux of atoms per unit area), the impact angle, and the material impacted. The greatest damage occurs to surfaces moving in the direction of flight. The flux level of atomic oxygen depends on solar activity and is related to the 11-year solar (sun spot) cycles, as shown in Fig. 53(b).

Atomic oxygen degrades organic materials by breaking atomic bonds, by oxidation, and by causing loss of volatile species. This degradation takes the form of surface erosion. In the case of organic polymers, losses of 0.2 to 0.35 mm (8 to 14 mils) could be expected on a space station

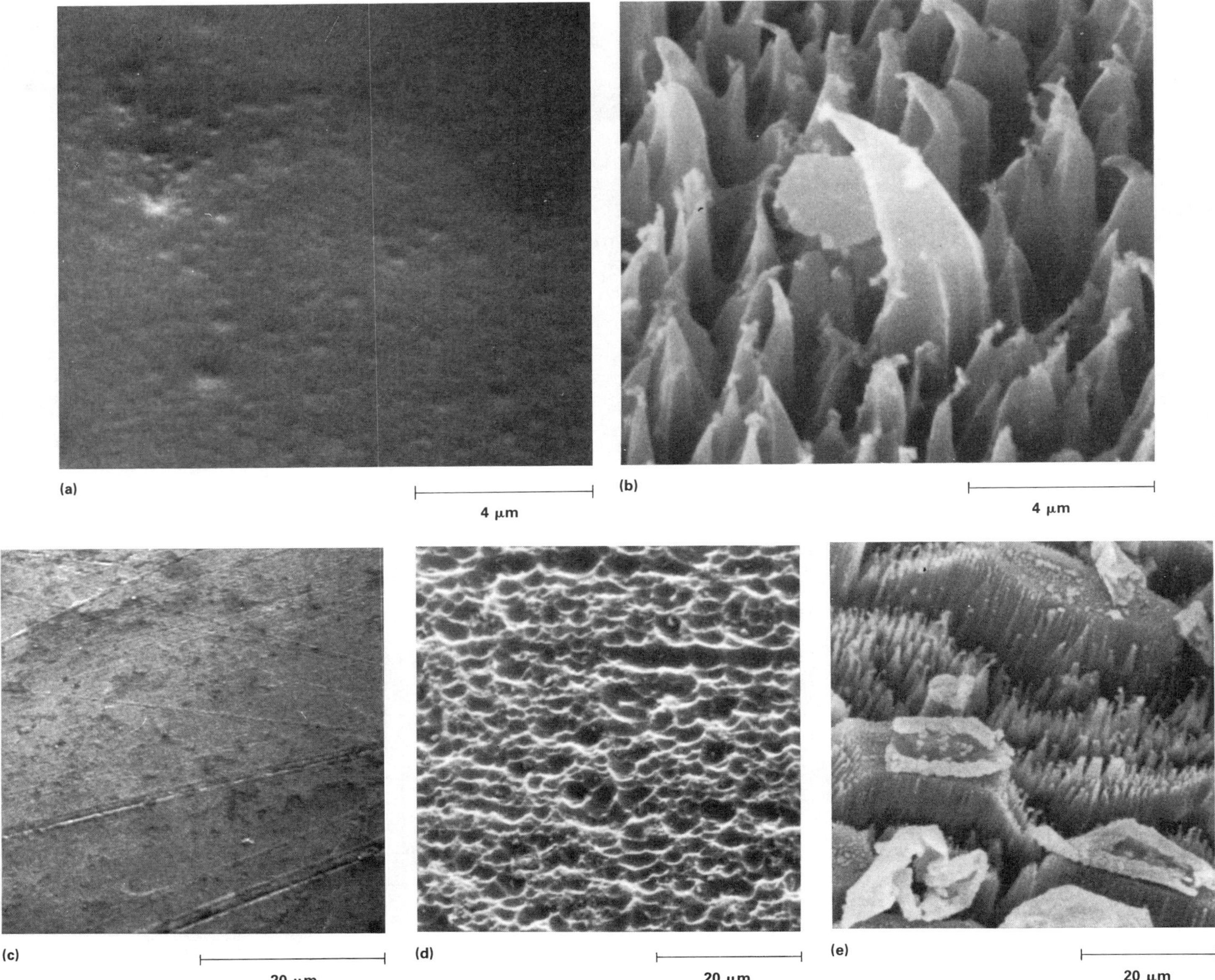

Fig. 54 Atomic oxygen degradation of materials in low earth orbit. (a) and (b) Polyurethane paint before (a) and after (b) 40 h of low earth orbit exposure in the velocity direction on the eighth shuttle flight. Both 14 000×. (c) and (d) Sample of Kapton H (polyimide) film before (c) and after (d) approximately 4 years of exposure on a satellite in low earth orbit. (e) Surface of a metallized Teflon film after about 4 years of exposure on a satellite in low earth orbit. The surface had been coated with 100 nm of vapor-deposited Inconel alloy over 150 nm of vapor-deposited silver. Both metals appear to have been completely eroded away. Courtesy of Materials and Process Laboratories, NASA George C. Marshall Space Flight Center

every solar cycle (11 years), depending on the orientation. Silicone polymers and perfluorinated polymers are approximately 50 times more resistant to surface loss than organic polymers. Only two metals show significant reaction—silver and osmium. Silver oxidizes readily, and expected surface losses, based on Shuttle SPS-8 tests (Ref 26), could exceed 0.2 mm/yr (8 mils/yr). Osmium reacts to form OsO_4, a volatile gas; this reaction also results in significant losses. All other metals are nonreactive except for copper, which forms a superficial protective oxide film and undergoes essentially no changes in properties or dimensions.

To prevent detrimental damage due to atomic oxygen, proper materials selection and the use of coatings are often acceptable approaches. Some metals appear resistant in thicknesses as thin as 15 nm.

Atomic Oxygen Degradation of Polymeric Insulation Films. The first shuttle flight resulted in surface roughening of the Kapton (polyimide) thermal control blankets used in the cargo bay areas. The surfaces of the film, which were normally glossy and amber in color, changed to a flat, translucent yellow appearance that resulted from microscopic roughening of these surfaces. Areas shielded by other parts showed no changes. From an analysis of the data and environments, it was concluded that atomic oxygen was the responsible active species causing this surface reaction/erosion. Subsequently, panels of test materials were exposed on racks in the shuttle cargo bay on the fifth and eighth flights to study the extent of degradation and to determine material reaction efficiencies.

Samples of polyurethane paint before and after 40 h of exposure in the velocity direction on the eighth orbiter flight are shown in Fig. 54(a) and (b). Samples of Kapton H polyimide film before and after 4 years of exposure on a satellite in low earth orbit are shown in Fig. 54(c) and (d). Figure 54(e) shows the surface of a 0.05-mm (2-mil) thick sheet of FEP teflon coated with 100 nm of Inconel over 150 nm of silver and exposed to atomic oxygen and ultraviolet radiation in low earth orbit for 4 years. The metal coating appears to be completely stripped from the surface.

Corrosion of Space Boosters and Space Satellites

Douglas B. Franklin
George C. Marshall Space Flight Center
National Aeronautics and Space Administration

The selection of corrosion prevention systems for space boosters and satellites requires the consideration of many different criteria related to the expected operating environments and various functional requirements. Both boosters and satellites may be exposed to a seacoast environment for prolonged periods as well as to various propellants and operating fluids. Satellites are also exposed to the high vacuum and solar radiation of space, and their external coating must provide controlled values for solar absorption and emittance to ensure proper temperature control. Boosters, on the other hand, often have severe cryogenic and elevated temperature exposures.

The need for lightweight structures, coupled with the low factors of safety used in component design, result in the use of materials selected more for strength than corrosion resistance. This creates a high potential for stress-corrosion failure and makes the effects of surface corrosion very significant. When the high reliability requirements for spacecraft are added to these considerations, it places a very demanding performance requirement on the corrosion protection system used. No reuse was planned in earlier space booster programs, but for the space shuttle, economy demands the reuse of most of the major components. Perhaps the most demanding environmental exposure is the solid rocket booster, which must survive periodic immersion in seawater for several days as well as damage to protective coatings resulting from towing and recovery operations (Fig. 55). The boosters must be capable of up to 20 reuses.

Fig. 55 Ocean recovery of a shuttle solid rocket booster

Aluminum Alloys

The primary structural materials used for many of the propulsion components in the space shuttle transportation system (Fig. 56) are high-strength aluminum alloys. This includes aluminum alloy 2219 for welded structures (such as the propellant tankage) and aluminum alloys 7075 and 2024 for structures where mechanical joining methods can be used. The protective system used on aluminum surfaces where exterior exposure is the primary concern consists of a chemical conversion coating (MIL-C-5541) to promote paint adhesion, followed by a chromate-inhibited epoxy primer 0.025 mm (1 mil) thick (Ref 27) and an epoxy topcoat 0.025 to 0.045 mm (1.0 to 1.8 mils) thick (Ref 28).

This system is used on the upper and lower skirt structures of the solid rocket boosters that are recovered from the ocean and reused. With some modification, it also is used on the external tank (Fig. 56) and provides a good base for bonding thermal insulation to provide protection from aerodynamic heating and heat from rocket exhaust plumes. To obtain good paint adhesion, it has been found to be extremely important to maintain surface cleanliness between each processing step, particularly after application of the chemical conversion coating. In addition, strict compliance with the recommendations of the paint manufacturer for drying times and application procedures is required. To ensure adequate quality control, the coatings are applied and tested to the requirements of MIL-F-18264.

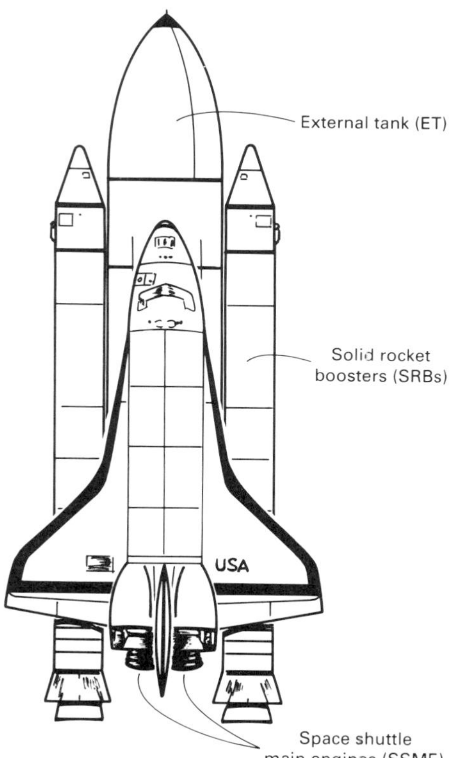

Fig. 56 Major propulsion components of the space shuttle transportation system

Where immersion in seawater and component reuse without refurbishment are required, all faying surfaces must be completely sealed (wet lay-up) with a polysulfide sealant meeting the requirements of MIL-S-8802. It is very important that there are no open gaps or voids not completely filled by the sealant, because seawater can be forced into these areas during recovery and will cause a serious corrosion problem. Fasteners should also be installed with wet sealant, and fastener heads must be completely oversealed. These procedures cannot be followed where electrical bonding is required. For these areas, jumper cables are used with the contact surfaces bare. The connection is then completely oversealed with polysulfide sealant.

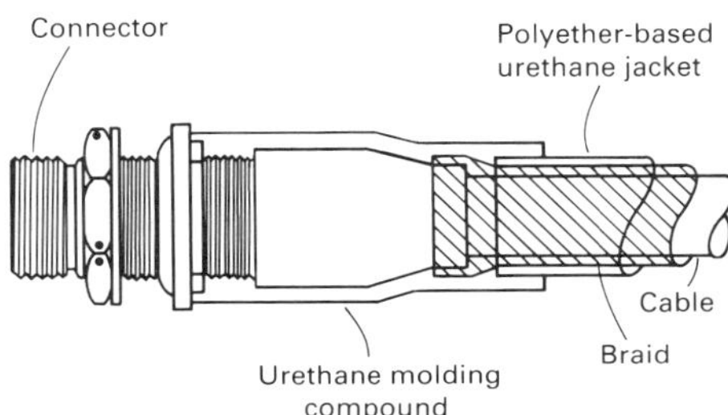

Fig. 57 Schematic showing electrical cable with watertight jacket

Another area requiring special consideration for protection is under the polyurethane foam insulation used on the external tank exterior to prevent excessive cryogenic propellant boil-off (liquid hydrogen and oxygen) during launch preparation and flight. Because of the need for good cryogenic adhesive properties, the paint system previously described cannot be used. It was also found that the bonding primer originally selected to promote adhesion between the tank surface and the spray foam would not adequately protect the aluminum alloy 2219 surface during extended storage, shipping, and launch preparations, particularly where multiple exposures to cryogenic temperatures are involved. After laboratory tests, the material finally selected for this purpose consists of an epoxy bonding primer with an increased amount of strontium chromate pigment (Ref 29). This primer provides improved corrosion protection under the foam insulation without any significant degradation of cryogenic adhesive properties.

Steel Alloys

One major area in which aluminum alloys are not used is the motor case of the solid rocket booster. This application uses D6AC low-alloy steel heat treated to an ultimate tensile strength of 1340 to 1550 MPa (195 to 225 ksi). The paint system selected for corrosion protection is a sacrificial zinc-rich epoxy-polyamide primer 0.038 to 0.063 mm (1.5 to 2.5 mils) thick (Ref 30) and an epoxy-polyamide topcoat 0.038 to 0.063 mm (1.5 to 2.5 mils) thick (Ref 31). The steel surfaces are gritblasted to white metal (Steel Structures Painting Council Specification SSPC-SP6) prior to painting. Although this system provides good protection to steel surfaces for prolonged periods of time, the paint is removed (by gritblasting) and reapplied after each flight because of magnetic-particle inspection requirements imposed on the motor case surfaces.

There are a few areas in which paint cannot be used to provide corrosion protection. These are the areas where the solid rocket booster motor case segments are joined and where the skirt segments join the motor case segments. The material used to provide protection here is a heavy-duty calcium-base grease with special corrosion inhibitors for use in seawater (Ref 32). The grease is carefully applied to all bare areas during assembly and is removed and reapplied after each flight. The grease protects the surfaces during storage and preflight operations and provides excellent protection for several days of ocean exposure during recovery of the segments and until joint refurbishing operations can be initiated, which can be several weeks later. The grease can also be applied diluted with trichloroethane or other solvents to provide protection to bare motor case segments during initial shipment by rail from the manufacturer as well as after the paint is stripped from used motor cases during refurbishment.

Graphite/Epoxy Motor Case

A graphite/epoxy motor case that provides significant weight savings for the SRBs is under development. Because of structural considerations and design requirements, D6AC steel adapter rings are used to join the motor case segments. To prevent galvanic corrosion on the

Table 4 Stress-corrosion failures in space boosters

Alloy	Material form	Failure occurrence	Component name	Program
Aluminum alloy 7079-T6	Forging	Prelaunch	LOX dome	Saturn IB
AM-355 stainless steel	Bar	Prelaunch	Flared tubing sleeve	Saturn I
17-7PH stainless steel	Sheet	Prelaunch	Wave spring	Saturn IB
Aluminum alloy 7079-T6	Forging	Manufacture	Prevalve control support link	Saturn V
Aluminum alloy 7075-T6	Plate	Test	Splice angle	Saturn V
Aluminum alloy 7075-T6	Bar	Assembly	Prevalve control piston cylinder	Saturn IB
Aluminum alloy 2024-T4	Bar	Test	Oxidizer check valve body	Saturn IB
17-7PH stainless steel	Sheet	Test	Actuator spring	Saturn V
17-7PH stainless steel	Sheet	Test	Prevalve Belleville spring	Saturn V
Aluminum alloy 7178-T6	Forging	Storage	Upper E-beam	Saturn IB
7079-T652	Forging	Storage	Rear spar	Saturn IB
7079-T6	Forging	Test	Hold-down fitting	Saturn IB
7079-T6	Forging	Test	Actuator body	Saturn V

(a)

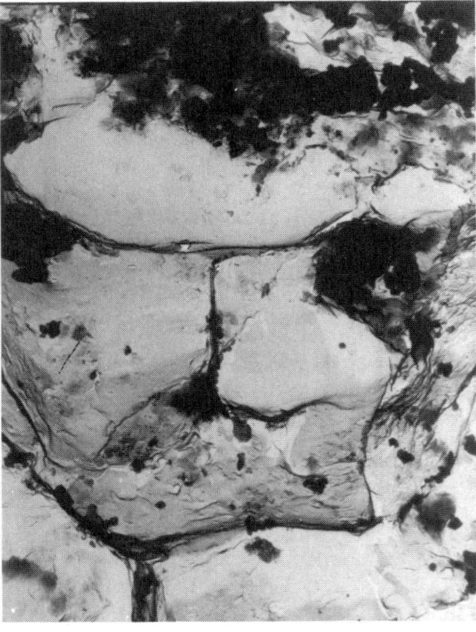

(b)

Fig. 58 Stress-corrosion failure of 7079-T6 aluminum forging. (a) Failed housing. (b) Fractograph of failure

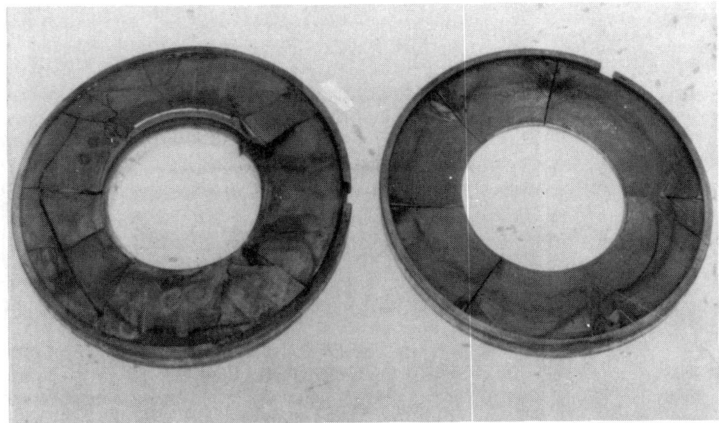

(a)

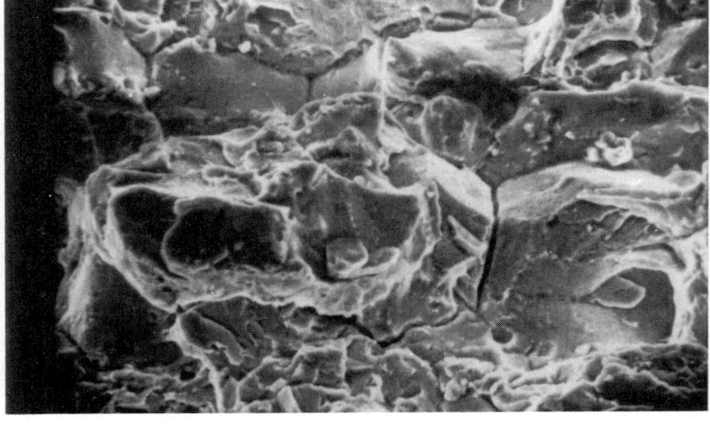

(b)

Fig. 59 Stress-corrosion failure of 17-7PH RH950 stainless steel valve actuator Belleville spring. (a) Failed spring washers. (b) SEM of fracture surface. 2000×

Table 5 Alloys with high resistance to SCC

Alloy	Condition
Ferrous alloys	
Carbon steel (1000 series)	Below 1240 MPa (180 ksi) ultimate tensile strength
Low-alloy steel (4130, 4340, D6AC, etc.)	Below 1240 MPa (180 ksi) ultimate tensile strength
Music wire (ASTM 228)	Cold drawn
HY-80 steel	Quenched and tempered
HY-130 steel	Quenched and tempered
HY-140 steel	Quenched and tempered
1095 spring steel	Quenched and tempered
300-series stainless steels (unsensitized)	All
21-6-9 stainless steel	All
20Cb stainless steel	All
20Cb-3 stainless steel	All
A-286 stainless steel	All
AM350 stainless steel	SCT1000 and above
AM355 stainless steel	SCT1000 and above
Almar 362 stainless steel	H1000 and above
Custom 455 stainless steel	H1000 and above
15-5PH stainless steel	H1000 and above
PH14-8 Mo stainless steel	CH900 and SRH950 and above
PH15-7 Mo stainless steel	CH900
17-7PH stainless steel	CH900
Nitronic 33	All
Wrought aluminum alloys	
1000 series	All
2011	T8
2024 rod, bar	T8
2219	T6, T8
3000 series	All
5000 series	(a)
6000 series	All
7049	T73
7149	T73
7050	T73
7075	T73
7475	T73
Cast aluminum alloys	
355.0, C355	T6
356.0, A356.0	All
357.0	All
B358.0	All
359.0	All
380.0, A380.0	As cast
514.0	As cast
518.0	As cast
535.0	As cast
A712.0, C712.0	As cast
Nickel-base alloys	
Hastelloy alloy C	All
Hastelloy alloy X	All
Incoloy alloy 800	All
Incoloy alloy 901	All
Incoloy alloy 903	All
Inconel alloy 600	Annealed
Inconel alloy 625	Annealed
Inconel alloy 718	All
Inconel alloy X-750	All
Monel alloy K-500	All
Ni-Span-C alloy 902	All
René 41	All
Unitemp 212	All
Waspaloy	All
Copper alloys(b)	
C11000	37
C17000	AT, HT (c)
C17200	AT, HT (c)
C19400	37
C19500	90
C23000	40
C42200	37
C44300	10
C51000	37
C52100	37
C61900	40
C68800	40
C70600	50
C72500	50, Annealed
Miscellaneous alloys	
Beryllium, S-200C	Annealed
Haynes alloy 25	All
Haynes alloy 188	All
MP35N	All
Ti-3Al-2.5V	All
Ti-6Al-4V	All
Ti-13V-11Cr-3Al	All
Magnesium, M1A	All
Magnesium, LA141	Stabilized
Magnesium, LAZ933	All

(a) High-magnesium alloys 5456, 5083, and 5086 should be used only in controlled tempers (H111, H112, H116, H117, H323, H343) for resistance to SCC and exfoliation. Alloys with more than 3% Mg are not recommended for applications above 66 °C (150 °F). (b) For copper alloys, numbers under "Condition" indicate the percentage of cold work. (c) AT, annealed and precipitation hardened; HT, work hardened and precipitation hardened

Table 6 Alloys with moderate resistance to SCC

Alloy	Condition
Ferrous alloys	
Carbon steel (1000 series)	1240 to 1380 MPa (180–200 ksi) ultimate tensile strength
Low-alloy steel (4130, 4340, D6AC, etc.)	1240 to 1380 MPa (180–200 ksi) ultimate tensile strength
Nitronic 32	All
Nitronic 60	All
Types 403, 410, 416, 431 stainless steels	(a)
PH13-8Mo stainless steel	All
15-5PH stainless steel	Below H1000
17-4PH stainless steel	All
Wrought aluminum alloys	
2024 rod, bar, extrusion	T6, T62
2024 plate, extrusions	T8
2124 plate	T8
2048 plate	T8
4032	T6
7001	T75, T76
7049	T76
7050	T736, T76
7075	T76
7175	T736, T76
7475	T76
7178	T76
Cast aluminum alloys	
319.0, A319.0	As cast
333.0, A333.0	As cast
Magnesium alloys	
AZ31B	All
AK60A	All

(a) Tempering between 370 and 595 °C (700 and 1100 °F) should be avoided because resistance to SCC and corrosion is lowered.

D6AC rings, the exterior surfaces of the graphite/epoxy are coated with the epoxy-polyamide topcoat discussed previously, and the machined joint area is sealed with a nonconductive sealant. This minimizes the cathode surface area and reduces the galvanic current. The D6AC rings are protected as described in the previous paragraphs.

Electrical Cables

One area of the solid rocket booster that has been very difficult to protect is the electrical cables and connectors. Because seawater contacting the ends of the stranded electrical wiring will quickly permeate through the wire and prohibit any reuse capability, electrical cables are enclosed in a watertight jacket of a polyether-based urethane plastic. This requires bonding of the jacket to the connector at each end of the cable, as shown in Fig. 57. In addition, the stainless steel connectors have a watertight O-ring seal to prevent the intrusion of seawater. To provide additional protection against inadvertent leakage, the connector pins are coated with a film of heavy-duty calcium grease (Ref 32) prior to assembly. The female connector sockets are designed so that the grease film is wiped off during insertion of the male pins and necessary electrical continuity through the connector is maintained.

Cathodic Protection

Carbon fiber reinforced phenolic ablators are used to line the solid rocket motor nozzles. These materials appeared to aggravate corrosion in the solid rocket booster aft skirt during immersion in the ocean. Sacrificial zinc anodes (MIL-A-18001) were therefore added to provide additional protection. This includes zinc anodes for several individual aluminum components, the use of zinc for several nonstructural components, and the use of flame sprayed zinc on several aluminum components. In addition, zinc anodes are attached at various locations by divers prior to towing the solid rocket boosters back to land for refurbishment. These anodes have reduced significantly the galvanic attack of the aluminum surfaces in the aft skirt of the solid rocket booster.

Other Alloys

Other alloys used for space booster systems include type 304 stainless steel, type 321 stainless steel (for welded components), Inconel alloy 718, titanium alloys Ti-6Al-4V and Ti-3Al-2.5V, and MP35N nickel-cobalt alloy. Although these alloys are inherently corrosion resistant, special treatments are usually required to ensure that exposed surfaces are passivated to reduce possible pitting problems. Surfaces exposed to seawater during solid rocket booster recovery are flushed with potable water (200 ppm chloride max), washed with a nonionic detergent (0.5% solution at 57 °C, or 135 °F), flushed again with potable water, and rinsed with deionized water until residual surface chlorides are below 50 ppm. Components are refurbished as necessary after each flight to ensure that their integrity is not compromised.

Control of Stress Corrosion

One other area that is carefully considered in design is the control of stress corrosion, particularly because of the widespread use of high-

Table 7 Alloys with low resistance to SCC

Alloy	Condition
Ferrous alloys	
Carbon steel (1000 series)	Above 1380 MPa (200 ksi) ultimate tensile strength
Low-alloy steel (4130, 4340, D6AC, etc.)	Above 1380 MPa (200 ksi) ultimate tensile strength
H-11 steel	Above 1380 MPa (200 ksi) ultimate tensile strength
440C stainless steel	All
18Ni maraging steel, 200 grade	Aged at 480 °C (900 °F)
18Ni maraging steel, 250 grade	Aged at 480 °C (900 °F)
18Ni maraging steel, 300 grade	Aged at 480 °C (900 °F)
18Ni maraging steel, 350 grade	Aged at 480 °C (900 °F)
AM350 stainless steel	Below SCT1000
AM355 stainless steel	Below SCT1000
Custom 455 stainless steel	Below H1000
PH15-7 Mo stainless steel	All except CH900
17-7PH stainless steel	All except CH900
Wrought aluminum alloys	
2011	T3, T4
2014	All
2017	All
2024	T3, T4
2024 forging	T6, T62, T8
2024 plate	T62
7001	T6
7039	All
7075	T6
7175	T6
7079	T6
7178	T6
7475	T6
Cast aluminum alloys	
295.0	T6
B295.0	T6
520.0	T4
707.0	T6
D712.0	As cast
Copper alloys(a)	
C26000	50
C35300	50
C44300	40
C67200	50, annealed
C68700	10, 40
C76200	25, 50, annealed
C76600	38
C77000	38, 50, annealed
C78200	50
Magnesium alloys	
AZ61A	All
AZ80A	All

(a) For copper alloys, number under "Condition" indicates percentage of cold work.

Table 8 Relative resistance to hydrogen embrittlement of various alloys in high-pressure hydrogen at room temperature

Alloy	Stress concentration factor, K_t	Pressure MPa	Pressure ksi	Ratio H_2/He(a)
250 maraging steel	8	69	10	0.12
Type 410 stainless steel	8	69	10	0.22
1042 steel (quenched and tempered)	8	69	10	0.22
17-7PH (TH1050)	8	69	10	0.23
HP9-4-20 alloy steel	8	69	10	0.24
H-11 high-strength steel	8	69	10	0.25
Inconel alloy X-750	6.3	48	7	0.26
René 41	8	69	10	0.27
ED nickel	8	69	10	0.31
4140 steel	8	69	10	0.40
Inconel alloy 718	8	69	10	0.46
MP35N	6.3	69	10	0.50
Type 440C stainless steel	8	69	10	0.50
Ti-6Al-4V (solution treated and aged)	8	69	10	0.58
Monel alloy 400	6.3	48	7	0.65
D-979 stainless steel	6.3	48	7	0.69
Nickel 270	8	69	10	0.70
CG27 stainless steel	6.3	48	7	0.72
ASTM A515, grade 70	8	69	10	0.73
HY-100 steel	8	69	10	0.73
ASTM A372, type IV	8	69	10	0.74
1042 steel (normalized)	8	69	10	0.75
Inconel alloy 625	8	34	5	0.76
ASTM A517, grade F	8	69	10	0.77
ASTM A533, type B	8	69	10	0.78
Waspaloy	6.3	48	7	0.78
Ti-6Al-4V (annealed)	8	69	10	0.79
1020 steel	8	69	10	0.79
HY-80 steel	8	69	10	0.80
Inconel alloy 706	6.3	48	7	0.80
Ti-5Al-2.5Sn	8	69	10	0.81
ARMCO iron	8	69	10	0.86
P/M Inconel alloy 718	6.3	48	7	0.86
Type 304 stainless steel	8	69	10	0.87
Type 321 stainless steel	8	34	5	0.87
Hastelloy alloy X	8	34	5	0.87
Type 305 stainless steel	8	69	10	0.89
Astroloy	8	34	5	0.90
Type 347 stainless steel	8	34	5	0.91
Haynes alloy 188	6.3	48	7	0.92
Type 304N stainless steel	6.3	103	15	0.93
Type 310 stainless steel	8	69	10	0.93
Beryllium-copper	8	69	10	0.93
RA330	6.3	48	7	0.95
A-286	8	69	10	0.97
21-6-9 stainless steel	6.3	48	7	0.97
Aluminum alloy 7075-T73	8	69	10	0.98
Incoloy alloy 802	6.3	48	7	0.99
Aluminum alloy 6061-T6	8	69	10	1.00
Copper (C10100)	8	69	10	1.00
Type 316 stainless steel	8	69	10	1.00
Incoloy alloy 903	8	34	5	1.00

(a) Ratio of notched strength in hydrogen to notched strength in helium

strength alloys that generally have poor resistance to stress corrosion. Several stress-corrosion failures have occurred in earlier programs, many of which resulted in significant program impact.

Table 4 lists these failures. As shown in Table 4, most of the failures have occurred in high-strength aluminum alloys and in the precipitation-hardening stainless steels in the seacoast environment. There have also been instances in which unique environments were not adequately considered. For example, stress-corrosion failure of a beryllium copper spring occurred because it was not recognized that a small amount of hydrazine could leak past an O-ring, thus exposing the spring to ammonia, a decomposition product of hydrazine.

Figure 58 shows an actual stress-corrosion failure that occurred in an aluminum alloy 7079-T6 forging used as a main housing in a hydraulic actuator. The failure occurred at the forging parting plane and was caused by excessive moisture in the hydraulic oil, residual stress from heat treating, and the interference fit of a small check valve that was press fit into the forging at the parting line. The failure was found during acceptance testing of the part prior to installation.

An example of stress-corrosion failure in 17-7PH RH950 stainless steel Belleville washers is shown in Fig. 59. The washers were used as a Belleville spring (a stack of 54 washers) in a valve actuator. Failure resulted from exposure to a humid atmosphere (water was trapped in the actuator housing) and the installation stresses caused by the preload on the spring stack.

The best method for controlling stress corrosion is to select, where possible, materials that are highly resistant to stress corrosion. To this end, guidelines (MSFC-SPEC-522) have been prepared to aid the designer in the selection of materials for use in space booster and satellite systems. Tables 5 to 7 list alloys grouped to show their comparative stress-corrosion resistance when exposed to a seacoast environment. The materials listed in Table 5 are considered resistant to stress corrosion in a seacoast atmosphere and can be used without restrictions.

The materials listed in Tables 6 and 7 should not be used unless specific evaluation and justification are made for each specific application. This requires a careful consideration of various

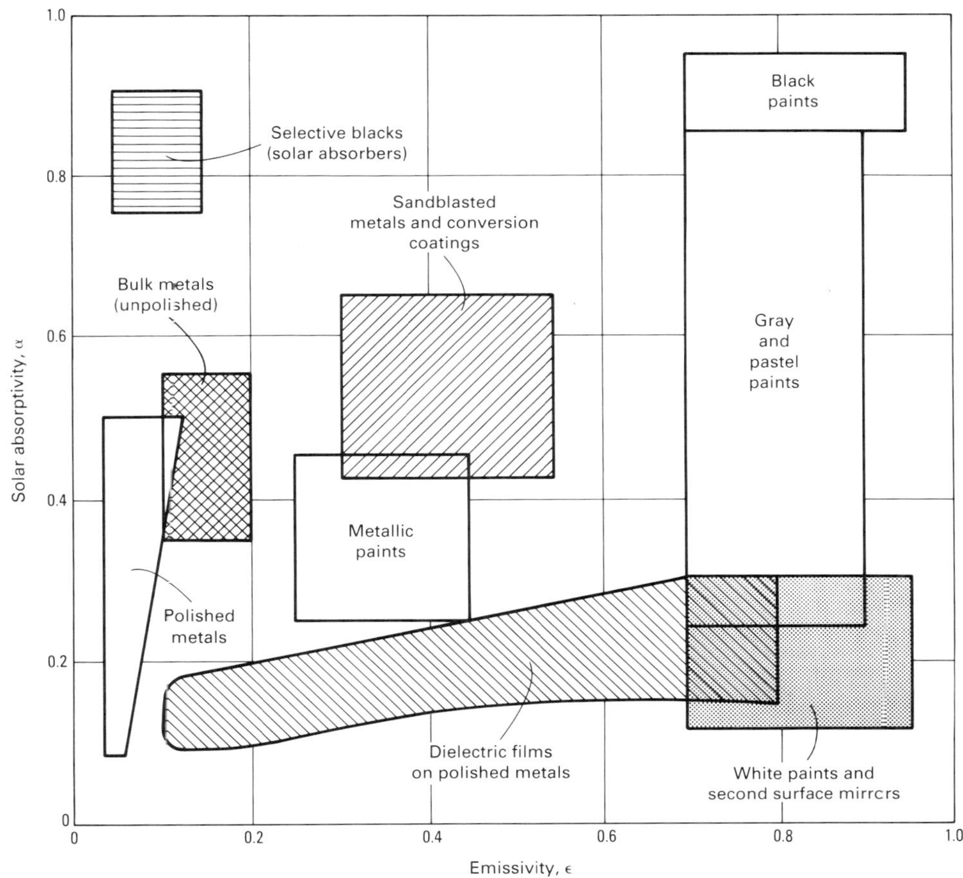

Fig. 60 Chart showing the range in optical properties for several types of coatings and surfaces

factors in addition to the susceptibility of the material to stress corrosion. This includes exposure conditions, protective treatments used, and sustained tensile stresses that may be present. It is extremely important to consider all sources of stress; not only operational stresses but also residual stresses and stresses that may be induced into the part during assembly must be considered. Residual stresses can be the result of machining, forming, and heat-treating processes, while assembly stresses can result from improper fit-up tolerances, overtorquing, press fits, high-interference fasteners, and welding. A more detailed discussion of these factors can be found in MSFC-SPEC-552.

Special Environmental Effects

In addition to ordinary atmospheric environmental effects, corrosion control procedures for space boosters and satellites must take into account other special environmental factors. Many of these are related to propellant compatibility.

Oxygen. Most organic materials are not compatible with oxygen systems and can ignite under service conditions. Because of a lack of compatibility with liquid oxygen and stringent cleanliness requirements, organic coatings are not used on aluminum propellant tank (external tank) interiors. Corrosion on these surfaces (aluminum alloy 2219-T87) is prevented by use of a chemical conversion coating (MIL-C-5541). In addition, rigorous drying procedures are required after conversion coating and tank cleaning (usually done in one continuous process), and the humidity inside the tanks is controlled to below 60% relative humidity during storage and purged to −12 °C (10 °F) dew point before shipping to the launch site.

The primary metallic materials not considered acceptable in oxygen systems are tin, magnesium, and titanium alloys. Their use may result in catastrophic impact ignition. Tin is particularly reactive, and it has been found that copper alloys with a tin content as low as 2% can react when impacted in liquid oxygen. All materials used in oxygen systems are selected to meet the requirements of NASA Handbook NHB 8060.1 at the temperature and pressure that will be encountered during use.

Hydrogen. Propellant compatibility is also of concern in hydrogen systems, particularly at the high pressures—up to 48 MPa (7000 psi)—found in the space shuttle main engine (SSME). Table 8 shows the effect of exposure to high-pressure hydrogen on the notched strength ratio of several metal alloys. One technique that has been used to protect alloys susceptible to hydrogen embrittlement is to copper plate the exposed surfaces. A 0.13-mm (5-mil) thick coating of electroplated copper has been used for protecting Inconel alloy 718 in several SSME components. Gold plating that is 0.13 mm (5 mils) thick has also been used for this purpose on Waspaloy alloy turbine disks. For these coatings to be effective, very careful procedures are required to ensure that coating adhesion and integrity are of the highest quality.

Hydrazine and Nitrogen Tetroxide. Several other propellants are encountered in space booster and satellite systems that present unique compatibility problems. The two most common are hydrazine and nitrogen tetroxide. The major concern with hydrazine systems is not corrosive attack but the decomposition of hydrazine, which can result in failure due to a large volumetric change. For nitrogen tetroxide systems, most metallic materials are resistant when the N_2O_4 is dry. However, because moisture can easily contaminate such systems (resulting in the formation of nitric acid), the primary materials of construction are those that also have high resistance to nitric acid, such as the aluminum alloys, stainless steels, and titanium alloys.

Corrosion Control for Satellites

The criteria for selection of coatings for satellite systems are usually related to their thermal control properties and resistance to the effects of the space environment. Properties can vary widely depending upon the specific requirements needed. Figure 60 illustrates the variation in types of coatings and surface treatments that may be used.

Corrosion protection properties are usually of secondary importance. Consequently, the environmental exposure conditions, particularly during manufacture and storage, must be carefully controlled to prevent corrosion and to prevent the deterioration of critical surfaces from contamination. This means stringent controls on packaging of individual components, humidity control during component assembly, and environmental control during storage of completed assemblies. Because most systems are assembled in a clean room to prevent surface contamination, keeping the relative humidity below 70% should prevent most corrosion problems during the assembly process. For storage purposes, particularly long-term storage, environmental conditions should be regulated so that the maximum relative humidity is below 60% and preferably below 50%. Storage in uncontrolled environments is not permitted.

REFERENCES

1. Press Information, Space Shuttle Transportation System, Rockwell International, 1982
2. "The Effects of Silver on the Properties of Titanium," DMIC Technical Note, Defense Metals Information Center, Battelle Memorial Institute, 1965
3. "The Stress-Corrosion and Accelerated Crack-Propagation Behavior of Titanium and Titanium Alloys," DMIC Technical Note, Defense Metals Information Center, Battelle Memorial Institute, 1966
4. W.K. Boyd and F.W. Fink, "The Phenomenon of Hot-Salt Stress-Corrosion Cracking of Titanium Alloys," NASA CR-117, Battelle Memorial Institute
5. L.B. Norwood, "Application of Beryllium on the Space Shuttle Orbiter," Paper presented at the 15th National SAMPE Conference, Cincinnati, OH, Society for the Advancement of Materials and Process Engineering, 1983
6. "Reactivity of Metals With Liquid and Gas-

eous Oxygen," DMIC Technical Note, Defense Metals Information Center, Battelle Memorial Institute, 1963

7. R.L. Johnston, "Multidisciplinary Approach to the Design of High Pressure Oxygen Systems," Paper presented at Multidisciplinary Analysis and Optimization Symposium, 1984
8. A.C. Bond, H.O. Pohl, N.H. Chaffee, W.W. Guy, C.S. Allton, R.L. Johnston, W.L. Castner, and J.S. Stradling, *Design Guide for High Pressure Oxygen Systems*, Reference Publication 113, National Aeronautics and Space Administration, 1983
9. C.B. Brownfield, "The Stress Corrosion of Titanium in Nitrogen Tetroxide, Methyl Alcohol and Other Fluids," SID 67-213, North American Aviation Inc., Space and Information Systems Division, 1967
10. "Stress Corrosion of Ti-6Al-4V in Liquid Nitrogen Tetroxide," DMIC Technical Note, Defense Metals Information Center, Battelle Memorial Institute, 1966
11. G.F. Kappelt and E.J. King, "Observations on the Stress Corrosion of 6Al-4V Titanium Alloy in Nitrogen Tetroxide," Paper presented at the AFML 50th Anniversary, Corrosion of Military and Aerospace Equipment Technical Conference, Denver, CO, Air Force Materials Laboratory, 1967
12. R.E. Johnson, G.F. Kappelt, and L.J. Korb, "A Case History of Titanium Stress Corrosion in Nitrogen Tetroxide," Paper presented at the National Metals Conference, Chicago, IL, American Society for Metals, 1966
13. V. Frick, "Materials Selection Criteria for the OMS Rocket Engine," Paper presented at the Golden Gate Welding Metals Conference, Interfaces in Industry, San Francisco, CA, 1975
14. L.J. Korb, D.C. Augustine, W.L. Castner, and C.D. Brownfield, "A Metallurgical Analysis of the Failure of Two Auxiliary Power Units of the Spacecraft Columbia," Paper presented at the SME Space Tech Conference and Exposition, Anaheim, CA, 1985
15. F.K. Lampson, The Marquardt Corporation, private communication, 1986
16. L.J. Korb and R.E. Johnson, "Stress Corrosion of Titanium Tanks in Methanol," Paper presented at the AFML Technical Conference on Corrosion of Military and Aerospace Equipment, Denver, CO, Air Force Materials Laboratory, 1967
17. R.E. Johnson, "Apollo Experience Report—The Problems of Stress Corrosion Cracking," NASA TN D-7111, NASA Technical Note, National Aeronautics and Space Administration, 1973
18. W.B. Lisagor, "Some Factors Affecting the Stress Corrosion Cracking of Ti-6Al-4V in Methanol," NASA TN D-5557, Langley Research Center, 1969
19. R.L. Johnston, R.E. Johnson, G.M. Ecord, and W.L. Castner, "Stress Corrosion Cracking of Ti-6Al-4V in Methanol," NASA TN D-3868, National Aeronautics and Space Administration, 1967
20. F. Mansfeld, The Effect of Water on Passivity and Pitting of Titanium in Solutions of Methanol and Hydrogen Chloride, *J. Electrochem. Soc.*, 1971
21. R.R. Boyer and W.F. Spurr, Characteristics of Sustained-Load Cracking and Hydrogen Effects in Ti-6Al-4V, *Metall. Trans. A*, Vol 9A, Jan 1978
22. "Reaction of Titanium With Gaseous Hydrogen at Ambient Temperatures," DMIC Technical Note, Defense Metals Information Center, Battelle Memorial Institute, 1965
23. H.H. Johnson and A.M. Willner, Moisture and Stable Crack Growth in High Strength Steel, *Appl. Mater. Res.*, Vol 4, 1965, p 34
24. L.J. Leger, J.T. Visentine, and J.A. Schliesing, "A Consideration of Atomic Oxygen Interactions With Space Station," Paper presented at the AIAA 23rd Aerospace Sciences Meeting, Reno, NV, American Institute of Aeronautics and Astronautics, Jan 1985
25. L.J. Leger, J.T. Visentine, and J.F. Kuminecz, "Low Earth Orbit Atomic Oxygen Effects on Surfaces," Paper presented at the AIAA 22nd Aerospace Sciences Meeting, Reno, NV, American Institute of Aeronautics and Astronautics, Jan 1984
26. H.J. Rockoff, "Materials System Requirements and Challenges on the Operational Space Station," Paper presented at the SME Space Tech Conference and Exposition, Anaheim, CA, Sept 1985
27. Product Bulletin 463-6-3, Sikkens Aerospace Finishes Divison, Akzo Coatings America, Inc.
28. Product Bulletin 400 Series, Sikkens Aerospace Finishes Division, Akzo Coatings America, Inc.
29. Product Bulletin 515-346, Desoto, Inc.
30. Technical Data Form 1042, Rust-Oleum Corporation
31. Technical Data Form 1047, Rust-Oleum Corporation
32. Product Bulletin Code 9760, Conoco, Inc.

Corrosion in the Electronics Industry

Jack D. Guttenplan, Rockwell International Corporation

THE EFFECTS OF CORROSION on material degradation and component performance in the electronics industry have long been recognized. Recent studies have shown that corrosion is becoming an even more significant factor in the reliability of electrical and electronic equipment. This has occurred because of the following trends:

- Designs requiring higher component density and faster signal processing, resulting in smaller components with closer spacings and thinner metallic sections
- New requirements for low-resistance, electrically stable grounding paths and electrical bonds to protect against stray electromagnetic radiation
- The exposure of electronics to more severe environments

The effects of environment and corrosion on the reliability of electronic equipment became very evident during the Vietnam War. Military electronics were not designed to resist the effects of the hot, humid environment and the monsoon rains to which they were exposed, and high failure rates resulted. Other examples of severe environments to which military electronic equipment are exposed include avionic systems in ship-based naval aircraft (Ref 1) and guidance systems on nuclear submarines (Ref 2).

The problem of corrosion of electronic equipment, however, is not unique to the military. It exists throughout commercial and military applications worldwide and can occur in various severity levels of indoor and outdoor environments.

Frequently, minute amounts of contaminants or corrosion products can cause serious degradation or complete failure. Thin films of oxidation or corrosion products, invisible to the eye, can result in noise or resistance buildup in electrical contacts and subsequent system failure. Extremely low levels of moisture and corrosive contaminants have been known to cause corrosion problems with printed circuit boards, encapsulated integrated circuits, nichrome-film resistors, electrical connectors, other discrete devices, and a wide range of plated components.

Studies have been initiated to determine both the nature of field environments and to develop laboratory test methods applicable to various classes of operating environments. These studies have shown that parts per billion levels of selected pollutants are sufficient under proper conditions of temperature and humidity to accelerate corrosion reactions in electronic equipment (Ref 3). Corrosion can occur during manufacturing, storage, shipping, and service. Moisture and such corrosive agents as chlorides, fluorides, hydrogen sulfide (H_2S), sulfur dioxide (SO_2), nitrogen compounds such as ammonia (NH_3), and other airborne contaminants are the major culprits (Ref 4). The sources of these corrosives have been the subject of intensive investigation. Some sources that have been identified include solder flux residues, residual electroplating or other processing chemicals, sulfur from storage container materials, vaporized contaminants from adhesives, reactive substances in plastic materials and glass, environmental acid deposition, and a wide range of reactive airborne contaminants.

Most of the same principles and mechanisms apply in the corrosion of electronic equipment as in, for example, the corrosion of a large structure. However, the interaction among electrical, metallurgical, and environmental conditions, together with severe dimensional constraints, can lead to a unique set of corrosion problems for electronic systems. These problems will be discussed briefly in the following sections.

Corrosion Problems and Preventive Methods

Moisture Intrusion Into Black Boxes. Many manufacturers of electronic equipment have resorted to sealed black boxes to prevent corrosion. However, moisture intrusion remains a major factor in electronic equipment failures. The optimal way to exclude moisture is to make the box hermetic (airtight) by fusion. Even if the box is hermetic, materials inside the box must be baked out to eliminate outgassing of moisture, and the box should be evacuated and pressurized with a dry inert gas. This procedure will provide confidence of low relative humidity throughout the service life.

However, the requirement for internal maintenance/testing and the increasing number of through-wall electrical input/output connections make fusion impractical in many cases. Designers must select other approaches for sealing boxes. Elastomers (rubber compounds) in the form of gaskets and O-rings are commonly used to seal lids and other points of entry. However, water vapor will eventually permeate through any elastomeric sealant and thus increase internal humidity. Designers need to predict the long-term entry of water vapor through these seals. Desiccants can postpone an increase in relative humidity, but the rate of moisture influx must still be estimated.

A recent study provided data on the moisture vapor transmission rates for typical elastomeric sealants and illustrations on the use of these data in calculating the rates of moisture influx into sealed boxes (Ref 5, 6). It was concluded that a butyl rubber compound was the best sealant against water vapor permeation, with ethylene propylene rubber a less desirable alternative. In an example using butyl rubber gaskets to seal a box, the calculations showed that, starting at 0% relative humidity inside and 90% relative humidity on the exterior, the time to reach 50% relative humidity at 40 °C (100 °F) inside the box was 7 years. With a proper desiccant in the box, the time was extended to 20 years. Using the same example, it can be calculated that the time for a silicone rubber gasket (without desiccant) would be 3 weeks and that the time for an ethylene propylene rubber gasket would be 1.9 years (see Ref 5 for the particular rubber compounds used).

An additional conclusion of this study was that in order to have high confidence that corrosion will not be a problem inside a sealed black box the relative humidity should be maintained at 40% or less at room temperature (Ref 6). Also, to prevent condensation, care must be taken to keep the internal temperature of the box from dropping below the dew point. This study did not address gross leakage of moisture due to improperly designed or seated seals and defective materials or the incompatibility of sealant materials with environmental fluids.

One of the standard methods of cooling an electronic enclosure is the use of forced ventilation, with air generally drawn from the atmosphere surrounding the equipment enclosure. In sites with aggressive atmospheres, this type of cooling will greatly accelerate corrosion because the circulating contaminated air comes in intimate contact with sensitive electronics. Unless the outside environment is benign, the introduction of outside air into an electronic equipment cabinet should be eliminated or limited to the lowest rate possible (Ref 7).

For equipment that is neither sealed nor pressurized, corrosion protection should be designed for worst-case conditions on the assumption that moisture will get in. Corrosion can be minimized by:

- Providing low point drains so that water cannot collect
- Encapsulating components so that moisture cannot reach them
- Conformal coating printed circuit boards for protection against moisture and contaminants
- Mounting printed circuit boards vertically and well above the bottom of the housing to prevent moisture and debris from collecting on the boards
- Locating edge connectors on the vertical sides, not the bottom, of the printed circuit board so that moisture and debris will not collect in the mated connectors and degrade the contacts
- Locating feed-through connectors on the sides of the box, not on the bottom, for similar reasons

- Designing cabling to lead down and away from connectors and providing drip loops where possible
- Avoiding hygroscopic materials that will hold or wick up moisture
- Avoiding materials that emit corrosive vapors
- Using a volatile corrosion inhibitor that is carefully selected to protect those metals of concern in a particular application

Dissimilar-Metal Corrosion. Electronic design is unique in the wide variety of metals used because of particular physical and electrical properties. Some of the more common metals and their uses in an electronic system are given in Table 1. These metals are combined to form a myriad of dissimilar-metal couples in electronic equipment. In the presence of moisture (an electrolyte), destructive galvanic corrosion can take place (see the section "Galvanic Corrosion" of the article "General Corrosion" in this Volume).

The principles of galvanic corrosion are discussed elsewhere in this Volume, and will not be dealt with in this article. To minimize the effects of galvanic corrosion in electronic equipment, the joining of dissimilar metals as defined in MIL-STD-889 should be avoided wherever possible. Where similar metals cannot be used because of design requirements, one or more of the following steps should be taken:

- Design the couple so that the area of the more noble metal (cathode) is appreciably smaller than the area of the more active metal (anode). Decrease the cathode area by painting or coating
- Plate the cathode (and/or anode) with a compatible metal
- Interpose a compatible metallic washer or gasket between the dissimilar metals
- Interpose a nonabsorbing, insulative washer, gasket, or coupling between the joined metals
- Paint the faying surfaces prior to joining
- Seal the interfaces to preclude entrance of moisture
- Place the electronics in a hermetically sealed enclosure and pressurize with an inert gas. If impossible to hermetically seal, use elastomeric seals and maintain relative humidity below 40% at room temperature (see the section "Moisture Intrusion Into Black Boxes" in this article)
- Where electrical bonding is required, follow the guidelines discussed in the section "Electromagnetic Interference (EMI)" of this article.

The prolific use of noble (more cathodic) metal platings on anodic substrates as corrosion barriers for electronic components is a related problem. Where the plating is excessively thin (porous), is cracked due to flexure or differential thermal expansion, or suffers mechanical damage, then the anodic base metal is exposed. In the presence of moisture, accelerated galvanic corrosion will occur. This corrosion is particularly severe because of the unfavorable area relationship (large cathode-to-anode area ratio). Care must be taken to ensure that minimum thicknesses of plating necessary to eliminate porosity are specified on engineering drawings and are enforced by quality control testing. Reflowing of tin and solder coatings helps to eliminate porosity. Where flexure of the plating is anticipated, for example, on component leads, a low-stress ductile plating should be used. In this example, a solder plate would be preferred to the more brittle electroless nickel plate.

Table 1 Metals and alloys commonly used in electronic systems

Metal	Uses
Gold	Electrical connector contacts, printed circuit board edge connectors, leaf-type relays, miniature coaxial connectors, semiconductor leads, and microminiature and hybrid circuits
Silver	Protective coating on relay contacts, wave guide interiors, wire, high-frequency cavities, EMI/EMP shields, and EMI gaskets
Magnesium alloys	Radar antenna dishes and lightweight structures, such as chassis, supports, and frames
Iron, steel, and ferrous alloys	Component leads, magnetic shields, magnetic coatings on memory disks, transformers, brackets, racks, hermetic electrical connector shells, and fastener hardware
Aluminum alloys	Equipment housings, chassis, mounting racks, supports, frames, electrical connector shells, and printed circuit board heat sinks
Copper and copper alloys	Wire, printed circuit board circuitry and heat sinks, component leads, terminals, bus bars, nuts and bolts, and radio frequency gaskets
Cadmium plating	Sacrificial protective coating on ferrous fastener hardware and on electrical connectors
Nickel plating	Barrier-type layer between copper and gold in electrical contacts, for corrosion protection on electrical connectors, printed circuit board heat sinks, electrical bonds in EMI applications, and for compatibility in dissimilar-metal junctions
Tin plating	For corrosion protection, solderability, and compatibility between dissimilar metals, on electrical connectors, radio frequency shields, filters, small enclosures, component leads, and automatic switching devices
Solder and solder plating	For joining, solderability, and corrosion protection
Beryllium	Inertial guidance instruments

Source: Ref 8

Electromagnetic Interference (EMI). Even the most reliable electronic circuits and components are susceptible to malfunction due to interference from natural and man-made electromagnetic emissions. Filtering, shielding, and grounding are three ways to minimize these effects, known as electromagnetic interference, and to keep the equipment itself from being a source of interference to other electronics (Ref 9).

Shielding and grounding involve surrounding the electronics with a conductive shield or envelope grounded back to the main structural section or airframe. This requirement has resulted in an increasing number of electrically bonded interfaces and grounding paths in which low direct current (dc) resistance must be maintained. Extensive use of the light (active) metals for housings and chassis in electronic systems, together with the inability to use insulating-type coatings at electrically bonded interfaces, has compounded the difficulties of corrosion control. Normal corrosion rates, accelerated by galvanic action where dissimilar metals must be bonded, can lead to an increase in electrical resistance due to oxide and corrosion product formation. This will result in the eventual loss of EMI protection.

Protective treatments for electrically bonded interfaces must satisfy the conflicting requirements of corrosion control while maintaining good electrical continuity. Several methods are helpful in meeting these requirements, including plated metals, chemical conversion films, metal-to-metal contacts sealed with an organic moisture barrier, water-displacing ultrathin-film corrosion-preventive compounds as specified in MIL-C-81309, and combinations of the above.

The use of MIL-C-5541, Class 3, chemical conversion films for the protection of aluminum in electrically bonded interfaces has been investigated for aerospace electronic equipment (Ref 10). These Class 3 films combine good corrosion resistance (that is, they must pass 168 h of exposure to 5% salt spray test per ASTM B 117) with low electrical resistance, as specified in MIL-C-81706.

The effects of temperature, mechanical action, and aging on Class 3 chemical film properties have also been evaluated (Ref 10). All three effects are important because they could influence the reliability of EMI protection. High temperatures are known to degrade the corrosion resistance of chemical films. Scratches or abrasion could expose base metal and degrade corrosion protection. Finally, long storage and service life requirements of 10 years or more for military hardware, much of which is without inspection, make the effects of aging an important consideration. Electrical resistance must remain stable to meet the requirements for radio frequency junctions specified in MIL-B-5087, "Bonding, Electrical, and Lightning Protection for Aerospace Systems." The requirement is a maximum dc resistance of 0.0025 Ω/junction. Three principal conclusions were reached in this investigation.

First, Class 3 chemical films of MIL-C-5541 will withstand temperatures to 65 °C (150 °F) for extended periods of time without loss of properties. The maximum exposure without serious degradation is a few hours at 95 °C (200 °F). Thermal treatments or processing that exceed this exposure cannot be tolerated.

Second, scratches up to 1.6 mm (1/16 in.) wide and mechanical abrasion where the chemical film is not completely removed pose no threat. Apparently, hexavalent chromium (Cr^{6+}) leaching slowly out of the film acts as a corrosion inhibitor to protect bare aluminum.

Finally, preliminary aging results indicate that the properties of chemical films remain stable at both low (20% relative humidity) and high (cycling temperature humidity per MIL-STD-202, Method 106) humidities after 1 year and 2 years of exposure. The low-humidity test represented exposure inside a sealed black box, but 10 days of exposure to the high humidity is a standard accelerated corrosion test for electronic systems.

These results indicate that MIL-C-5541, Class 3, chemical films are an excellent choice for the protection of aluminum in electrical bonding applications. Galvanic effects are nonexistent, as they are with noble metal platings; the chemical films are easily repaired; and the properties remain stable. Where aluminum is joined to more noble metals (for example, nickel, stainless steel, or silver), sealing of the interface with an organic sealant will be required.

A recent trend in commercial electronic systems is the movement away from metal and toward plastic enclosures (Ref 11). Metals are excellent shielding materials, while plastics are transparent to electromagnetic signals. Two approaches are being used to impart metal shielding properties to plastics. One is to incorporate conductive fillers in molded plastics, and the other approach is to apply metallized surface coatings to the interior surfaces of plastic enclosures. Both techniques have some major problems that are currently being addressed.

System-Generated Electromagnetic Pulse. A problem related to EMI protection is protection against system-generated electromagnetic pulse. This is a new requirement for military hardware. Secondary electrons emitted from an irradiated metal and accelerated out through the double layer at the metal surface at high velocity are capable of harming electronic circuits or components that they might contact. To prevent this, a gas fill should be used inside of a sealed black box that will slow down or absorb electrons, and a low atomic number (low-Z) coating must be applied to the inside surfaces of the box to reduce the electron emission efficiency of those surfaces. The low-Z coating is defined as a coating containing less than 1% by mass of elements of atomic number (Z) greater than 9. The thickness and the coverage requirements for the low-Z coating will vary, depending on the atmosphere/lack of atmosphere within the box. In many cases, a clear epoxy polyamide coating 0.025 to 0.1 mm (1 to 4 mils) thick has been used for this purpose. The low-Z coating can substitute for a corrosion-preventive coating on interior surfaces.

Flux Residues. The main functions of a soldering flux are to remove oxides, tarnish films, and other impurities from the surfaces of metals being soldered and to exclude atmosphere from the surfaces during soldering to prevent the formation of new oxides. A soldering flux should also lower the surface tension of the molten solder, allowing the solder to flow readily and adhere to more of the base metal surfaces (Ref 12).

In the case of liquid-type soldering fluxes, many organic and inorganic materials provide excellent fluxing action, but most of these materials leave extremely corrosive residues after cooling (Ref 13). Corrosive flux residues that attack and consume the solder alloy and base metals can weaken or embrittle the soldered connections. They can also increase the electrical resistance of the connections or open them entirely. On the other hand, conductive corrosive residues can pick up moisture from the atmosphere and can lower insulation resistance and form conductive paths, causing current leakage or short circuits. Nonconductive corrosion products, especially those carried by fumes, can cause damage by building up on electrical contact surfaces.

The fluxes used for electronic soldering can be divided into two categories: the solvent-soluble, or rosin types and the water-soluble types. In general, the water-soluble fluxes and their residues are much more active than the rosin fluxes and residues, but mild water-soluble fluxes and highly activated rosin fluxes are available. Inorganic zinc chloride-hydrochloric acid ($ZnCl_2$-HCl) type fluxes are included in the water-soluble category, but their highly corrosive nature and the difficulty in removing the residues preclude their use in electronic soldering.

Water white rosin, a widely used soldering flux material, leaves essentially noncorrosive, nonconductive, and nonhygroscopic residues. Because rosin is a very weak organic acid, its practical ability to clean a metal oxide surface is limited. To increase the ability of rosin fluxes to clean oxide surfaces, activating agents are added to produce rosin mildly activated (RMA) and rosin activated (RA) fluxes. For RMA fluxes, the activator may be any of a number of amines, organic acids, amides, or halogen-containing materials (Ref 14). Because RMA fluxes must be noncorrosive, as specified in military specification MIL-F-14256, only a small, limited amount of these materials may be added. The activating agents for RA fluxes are usually amine-neutralized HCl or occasionally a halogen-substituted organic material. These proprietary materials are water soluble and are the primary active constituents of the organic chloride type fluxes. When heated, these activators decompose and liberate HCl. This liberated acid, unlike rosin, not only reduces the metal oxide but readily attacks the cleaned metal surface as well as the surface of the solder.

The organic chloride type fluxes are the most active fluxes used in soldering electronic assemblies. The active components are usually amine hydrochlorides (or hydrobromides) combined with water-soluble organic acids. Glutamic acid hydrochloride is an example. Fast and efficient removal of the residues is necessary to prevent initiation of corrosion.

Water-soluble chloride-free fluxes have gained some popularity. The active components of these fluxes are water-soluble organic acids. The residues are similar to those of the organic chloride type fluxes, but they eliminate the possibility of chloride ion (Cl^-) entering the corrosion cycle. If left on the board too long, they may discolor the solder surface. A water rinse is usually sufficient for flux removal. If polymerization occurs during soldering, alkaline cleaners may facilitate removal.

In a study performed for the U.S. Army Electronics Command, it was determined that the single largest factor causing corrosion of printed circuit board assemblies was flux residues (Ref 15). To prevent this form of corrosive attack, the lowest acid content flux possible should be used for soldering (Ref 8). In addition, cleaning processes should be employed that are tailored to the type of flux used and will completely remove all residues.

Component Lead Materials/Finishes. High reliability in the operation of moisture-sensitive microelectronic devices is often achieved by packaging these devices in hermetically sealed containers. This generally limits the electrical feed-through leads to low coefficient of expansion materials such as Kovar, Dumet, and Alloy 42 because of the need to form a seal with glass. These low-expansion materials, however, rate very poorly for solderability. They are finished, therefore, with a metal that is selected for its solderability and compatibility with internal wire bonding and die-bonding operations. The commonly used lead materials are difficult to plate, and many solderable finishes cannot be plated directly onto these materials. Gold is predominantly used because it can be plated directly on Kovar, has good solderability, is corrosion resistant, and is compatible with bonding operations. Tin and tin alloy electroplated finishes are also used in many cases for solderability and corrosion control.

Although gold itself is quite inert, severe galvanic action can occur on gold-plated Kovar if the coating is not pore free (Ref 16). A potential difference of approximately 0.6 V can be generated by the dissimilar-metal couple of gold to Kovar (Ref 17), and the area relationships are poor (large cathode-to-anode area ratio), resulting in accelerated attack of the Kovar base material.

Tin or tin-lead solder coatings, on the other hand, are preferable for corrosion control because they are close to Kovar in the electromotive series and because the potential difference is low (0.02 V) (Ref 17). Quite often, gold-plated component leads are pretinned, that is, immersed in the solder pot to give a solder coating. Studies have shown that gold coatings as thick as 7.5 μm (300 μin.) completely dissolve in molten solder to give a solder-Kovar bond. On most metals, 2.5 μm (100 μin.) of gold is sufficient for solderability. Greater thicknesses are of little value because the gold simply dissolves in the molten solder. The pretinning procedure effectively eliminates the galvanic-corrosion problem.

Other corrosion-sensitive areas for component leads are the lead-glass interface at the lead egress from the container and lead bends, which are susceptible to stress-corrosion cracking. When the glass is fused during the sealing process, a meniscus is formed, resulting in a thin coating of glass extending out on the lead material. The lead is then gold plated up to this meniscus. The thin coating of glass can be cracked or broken away during plating and subsequent handling, packaging, lead bending, and soldering, leaving bare lead material. This site is very susceptible to corrosive attack because of the galvanic couple formed between the more noble material finish and the base metal.

Stress-corrosion failures of leads have been observed for both transistor cans and integrated circuit packages when stress and moisture were combined (Ref 16, 18). Tensile stresses arise from lead bending during installation or even by differential thermal expansion in a rigid mounting because of the low expansion of Kovar alloy and the high expansion of the rigid circuit board. The practice of pretinning and organic conformal coating has been an effective solution to both of the above corrosion-sensitive areas.

Special care must be taken in the pretinning of leads on nonhermetic devices, particularly nichrome-film resistors. Thermal stress from the solder-dipping process can damage epoxy-to-metal end seals and allow the entry of fluxes, chlorinated solvents, and moisture. The nichrome film is susceptible to corrosion in the presence of moisture and a chloride contaminant; corrosion (electrolysis) can become rapid if a bias is imposed. The temperature of the solder, the residence time in the solder, and the proximity of molten solder to the body of the resistor are critical parameters that must be regulated. A solder temperature of no higher than 260 °C (500 °F), a residence time of 5 s, and immersion to no closer than 0.64 mm (0.025 in.) from the body are recommended. The use of heat-sinking lead holders, the elimination of fluxes more aggressive than RMA fluxes, and the minimal use of trichloroethane and water rinses were also recommended (Ref 19).

Organic Outgassing Products. Many organic materials commonly used in electronic equipment can cause corrosion damage by the outgassing of vapors, particularly within nonbreathing enclosures. Potentially corrosive materials include adhesives, resins, plastics, elastomers, sealants, and organic finishes (Ref 18). Reference 20 describes corrosive vapors (hydrogen chloride) given off by polyvinyl chloride insulation on wiring. Other examples of materials that might cause damage inside of closed compartments include:

- Acetic acid from acid condensation cured room-temperature vulcanized (RTV) silicone sealants
- Sulfur compounds from polysulfide sealants and paper or cardboard used for packaging
- Phenolic constituents from molded phenolic resins
- Ammoniacal vapors from molded resins
- Esters or organic acid vapors from plasticized resins
- Organic acids (generally acetic) from raw wood used in packing cases
- Sulfurous or acidic vapors from cushioning materials used for packaging
- Amine catalysts from epoxy materials

Damage from outgassing can be minimized by the following techniques (Ref 21):

- Organic materials should be thoroughly cured before assembly, especially acid-activated plastics
- Polyvinyl chloride should not be used in closed compartments
- Where practical, the use of neoprene, phenolic, polysulfide, and vinyl should be avoided in closed areas
- Where possible, materials should be given a bake to outgas moisture
- Cadmium should not be used in closed compartments

This last point refers to the fact that cadmium is very susceptible to attack by corrosive outgassing products in confined spaces. The use of cadmium is banned in this application by federal and military specifications and standards, such as QQ-P-416, MIL-S-5002, and MIL-STD-1250.

Metal Whiskers. Certain metals, including tin, zinc, cadmium, and silver, are subject to the growth of metallic filaments known as whiskers. Given sufficient time and the proper conditions, these filaments can grow long enough to short out adjacent circuitry and cause failures. High voltage quickly burns off the whiskers, but the low voltages characteristic of much electronic equipment cannot (Ref 22).

Whiskers are of particular interest in the case of tin because of the wide use of tin plate on miniature electronic equipment. The exact mechanism of whisker growth is not known. It appears to be related to metallurgical imperfections in these metals (Ref 18). In the case of tin, the most prevalent hypothesis states that whiskers grow because of strains set up in the tin plate during the plating process, especially if various organics required for the process are co-deposited (Ref 23). The growth of whiskers is slightly enhanced by elevated temperature and humidity. Whisker growth is greatly accelerated in the presence of high stresses. Other empirical data on tin whisker formation include (Ref 23):

- Bright tin in which organics are co-deposited has a propensity to produce whiskers, as opposed to a matte tin process
- Use of high current densities during plating has a tendency to induce stresses in tin plate that produce whiskers
- Thin tin-plated coatings (<1.3 μm, or 50 μin.) have more of a propensity toward whisker growth than tin plating of 2.5 μm (100 μin.) or more
- Baking after tin plating has helped to produce surfaces that are free of whiskers

Whisker growth can be minimized by the use of thicker coatings, preferably fused or hot dipped, or by stress relief if the coating is electrodeposited. Apparently, thin barrier coatings are not effective in eliminating whisker growth. Consideration should be given to adequate spacing, positive barriers of insulating material, and maintaining a low level of humidity in the equipment. Finally, the co-deposition of small amounts of lead, in the case of tin, has been found to eliminate whisker growth (Ref 18). This is further clarified in Ref 24. The addition of 1.5% Pb to the tin coating (or reflowing) has been shown to eliminate whisker formations on tin-plated contacts.

Silver migration is a form of electrolysis in which silver ions dissolve from one conductor in a circuit and, under the impetus of a direct current and in the presence of moisture, migrate across an insulator to a second conductor. The rate of migration is affected by the magnitude of the potential difference, the humidity, the type of substrate, and the contamination present on the substrate. It is postulated that the silver is oxidized and that the resulting silver cations are carried in an electrolyte film to the negative conductor, where they are reduced back to metallic silver. Upon drying, silver will be found on the insulator, giving low insulation resistance or even conductivity.

Silver migration is minimized by using conformal coatings, by maintaining humidity at a level that precludes condensation, by using nonhygroscopic insulation, by removing contaminants, and by maintaining the spacing between conductors at different voltage potentials as wide as possible. Where practicable, gold, platinum, or tin-lead coatings are recommended for use instead of silver (Ref 18, 21).

Brittle Intermetallics. Gold has an extremely high rate of dissolution in molten 60-40 tin-lead solder, leading to the formation of the intermetallic compound $AuSn_4$ (Ref 25). Being extremely brittle, $AuSn_4$ can give rise to severe embrittlement in a solder joint. Several investigators in this field have studied the effect of gold coatings on joint strength on the basis that all coatings are converted to $AuSn_4$. The critical composition of the intermetallic compounds in a solder joint that causes embrittlement appears to be about 0.6% Au.

When the solder bath becomes saturated with gold, the suspended intermetallic material gives rise to grittiness in the appearance of the joint. However, embrittlement is the main problem, particularly if a device is to be exposed to thermal cycling. In the solder pot, a level of as low as 0.02% and as high as 0.2% Au has been reported to be the critical amount of gold contamination (Ref 26).

As mentioned previously, gold-plated leads are usually pretinned, thus dissolving the gold and giving a solder-to-base metal bond. Multiple dips are often specified to ensure complete removal of the gold. If it is necessary to solder over gold, care should be taken to minimize the formation of the brittle gold-tin intermetallic by one or more of the following methods (Ref 21):

- Use of extremely pure (99.99% +) gold
- Use of thin gold plate
- Use of minimum soldering time at minimum temperature

The soldered joints should be protected with a conformal coating. In addition, the gold content of the solder pot should be maintained below 0.2%.

Gold-Aluminum. The failure of gold-aluminum thermocompression bonds in integrated circuit devices due to electrical opens or intermittencies was initially attributed to a phenomenon known as purple plague (Ref 27). However, this turned out to be somewhat of a misnomer. Purple plague is defined as a brittle gold-aluminum compound formed in the presence of silicon (Ref 21). The gold-aluminum equilibrium diagram shows that five intermetallic compounds can be formed. Only one of these, $AuAl_2$, in the aluminum-rich phase, appears purple in the crystalline form.

An in-depth study of the cause of these thermocompression bond failures found no purple $AuAl_2$ formation (Ref 28). The study concluded that the failures occurred in the gold-rich phase because of the formation of Au_4Al_2 and Au_2Al. The failure mechanism postulated was the formation and transformation of these brittle intermetallics accompanied by the formation of strained interfaces. This was caused by the mismatch of atomic lattices and changes in volume, inducing stresses into the thermocompression bond. The strained interfaces thus formed are susceptible to fracture in the hard, brittle intermetallic phases. Methods of preventing this type of failure include the use of a third metal as a diffusion barrier and/or thermal treatments that will minimize or eliminate formation of intermetallics.

Electrical Contacts. One of the more severe corrosion problems unique to electronic systems is the degradation of electrical contacts. The reliability of modern systems is increasingly governed more by the reliability of interconnections than by active solid-state devices. This is due in part to the low connector contact forces and low voltages used in modern circuits and therefore to the sensitivity to very small amounts of corrosion products and other contaminants.

Contact corrosion is dependent on the materials with which the contact is plated. Nonnoble contact finishes, with tin as the major example, are susceptible to oxide formation that can result in a gradual increase in contact resistance. The dominant failure mechanism for tin-plated contacts is fretting corrosion (Ref 29, 30), a repetitive oxidation of contact points due to small-amplitude contact movement. The potential for fretting corrosion can be minimized by providing mechanically stable contacts, usually through high contact force, or by the use of a contact lubricant specifically formulated to be effective against fretting corrosion.

Noble metal finishes, of which gold is the major choice, do not react with the environment, but because of economic and other reasons, the plating is generally thin and has pores in which galvanic corrosion can occur. Corrosion of and migration of corrosion products from the copper alloy spring materials used in electrical contacts can lead to contact degradation in certain environments, particularly those containing chlorides and sulfides. In moisture-laden rooms, in which the relative humidity can reach 90%, a galvanic reaction takes place between the gold and copper at the pores; corrosion products form and gradually creep over the gold surface, resulting in poor or intermittent contact. Even where moisture is controlled at lower levels, high concentrations of sulfides in the environment can give the same effect. At one time, silver was used as an underplating for gold. In this case, the presence of sulfides caused rapid degradation of the contacts (Ref 2). At higher temperatures, solid-state diffusion of copper to the surface of the thin gold layer and subsequent reaction with the environment can cause increases in contact resistance.

Nickel barrier plating of the contact spring is effective in minimizing these aspects of copper corrosion. Housing design can also be effective in restricting the access of the environment to the contact interface or contact spring (see the section "Moisture Intrusion Into Black Boxes" in this article). A chromate conversion coating was developed and applied successfully to the cleaning and refurbishment of electrical contacts on submarines (Ref 2).

Design Considerations

The following are guidelines that should be considered for corrosion prevention and control in the design of electronic equipment. Some of these guidelines can be used in addition to the methods of corrosion prevention and control described previously in this article.

General. Methods for minimizing or preventing corrosion should be considered at the beginning of the design stage. Corrosion engineers, as well as electronics and radiation-hardening engineers, should be consulted, and designs should be based on the assumption that moisture will be present in the intended application.

Black Box Design. Hermetic (fusion) sealing should be used whenever possible. Butyl rubber O-rings and gaskets should be used for environmental sealing if butyl rubber is compatible with environmental fluids. Black boxes should be pressurized with dry inert gas, and a warning signal to indicate loss of pressure should be used. Electrical bonding or grounding interfaces and EMI gaskets should be sealed against moisture or fluid intrusion. Low point drains should be designed into nonhermetic and nonpressurized equipment. The introduction of outside air into electronic enclosures for cooling should be minimized or eliminated, and printed circuit boards should be mounted vertically and conformal coated in accordance with MIL-I-46058. Edge connectors should be located on vertical sides of printed circuit boards, drip loops should be used on electrical cables, and feed-through connectors should be located on the sides of the housing.

Materials Selection. Metals should be used in their most corrosion-resistant form (heat treatment, surface treatment, passivation, and so on) with the least amount of residual stressing possible. Hygroscopic materials should be avoided, as should RTV materials that contain acetic acid and cushioning materials that can deteriorate (revert), such as ester-type polyurethanes. Silver, cadmium, or bright tin plating should not be used on the inside of containers. To avoid whiskers, tin plate should be matte rather than deposited with organic brighteners. Formation of tin whiskers can also be avoided by fusing or alloying the tin plate. Silver or silver plating on circuit conductors or electrical contacts should be avoided. Funginert materials should be used, as should low-outgassing materials or bake-out materials, prior to sealing of the black box. Guidelines in MIL-STD-1250 and MIL-STD-454 should be followed.

Dissimilar Metals. If it can be avoided, dissimilar metals as defined in MIL-STD-889 should not be used in contact with each other. If such use cannot be avoided, plate with a metal (or metals) that will reduce the potential difference of the couple to 0.25 V or less, interpose a compatible metallic washer or gasket, interpose an insulative washer, gasket or coupling, or paint the faying surface. All interfaces should be sealed, anode-to-cathode area ratios should be maintained as large as possible, the cathodic (more noble) material should be painted, and dissimilar-metal inserts or fasteners should be installed by the wet primer method. Magnesium should never be mated to a metal more noble than aluminum, and it must be remembered that graphite, graphite composites, graphite lubricants, and graphite-, silver-, or copper-filled conductive paint or EMI gaskets act as a noble metal in dissimilar-metal contact. These materials should not be coupled with aluminum or any structural metal.

Finishes. All aluminum, unless it is plated, should be coated with a chemical film in accordance with MIL-C-5541 or anodized in accordance with MIL-A-8625. If conductivity is required, aluminum can be coated with a MIL-C-5541, Class 3, film or plated. Nickel barrier plating should be used under gold plating. Where a more noble metal plating is used on an active base metal, any pore or defect in the plating will be a site for galvanic corrosion with a very poor area ratio; proper thickness, handling, and processing should be ensured and a barrier plating used if necessary. Adequate dimensional tolerances must be allowed for protective coatings, and paint should be applied wherever possible. A complete paint system consisting of a conversion coating, primer, and topcoat in accordance with MIL-S-5002, MIL-F-7179, and MIL-F-18264 should be used. When coating the cathodic member of a dissimilar-metal couple, a paint resistant to alkaline attack (not an alkyd-base paint) should be used. High-strength materials, if plated, should be baked to minimize susceptibility to hydrogen embrittlement, and all stainless steel components should be passivated in accordance with federal specification QQ-P-35.

Joining. Low-carbon (for example, type 304L and type 316L) or stabilized (type 321 and type 347) austenitic stainless steels should be used in welded structures to minimize susceptibility to sensitization and intergranular attack. Joints of dip-brazed aluminum structures should be thoroughly cleaned and then vacuum impregnated in accordance with MIL-I-6869, Class 2 and MIL-STD-276; all light metal castings should also be vacuum impregnated. Machining into a dip-brazed joint should be avoided. Solder fluxes with the lowest possible acid content should be used, and joints should be rigorously cleaned and tested for ionic contamination after soldering. In the thermocompression bonding of gold wire to aluminum metallization on semiconductor devices, a barrier plate or thermal treatment should be used to prevent the formation of brittle gold-aluminum intermetallics.

Maintainability. It should be remembered that metals can corrode during fabrication and storage and before assembly; protection is required. Cooling systems that remove moisture and particulate matter should be used for completed assemblies, and filters and traps should be cleaned or replaced regularly. For closed containers, desiccant systems with visual indicators should be employed. The temperature inside a sealed black box should never be allowed to drop below the dew point.

REFERENCES

1. A.F. Carrato, I.S. Shaffer, and R.H. Richwine, "Corrosion in Naval Aircraft Electronic Systems," Paper presented at Corrosion/78, New Orleans, LA, National Association of Corrosion Engineers Conference, Oct 1978
2. J.D. Guttenplan and L.N. Hashimoto, Corrosion Control for Electrical Contacts in Submarine Based Electronic Equipment, *Mater. Perform.*, Vol 18 (No. 12), Dec 1978, p 49-55
3. W.H. Abbott, "Field Versus Laboratory Experience in the Evaluation of Electronic Components and Materials," Paper presented at Corrosion/83, Anaheim, CA, National Association of Corrosion Engineers, April 1983
4. R. Baboian, Corrosion—A National Problem, *ASTM Stand. News*, March 1986, p 37
5. J.M. Kolyer, "Rate of Moisture Permeation in Elastomer-Sealed Electronic Boxes," Paper presented at the 30th National SAMPE Symposium, Anaheim, CA, Society for the Advancement of Material and Process Engineering, March 1985
6. J.D. Guttenplan and J.M. Kolyer, "Prevention of Moisture-Induced Corrosion in Electronic Enclosures by Sealing with Elastomers," in preparation
7. W.J. Curren, J.R. Martin, D.W. Noon, and P.E. Gilson, "Environmental Impact on Electronic Control Systems: How to Evaluate—How to Cope," Paper presented at Corrosion/86, National Association of Corrosion Engineers, March 1986
8. *Design Guidelines for Prevention and Control of Avionic Corrosion*, NAVMAT P 4855-2, Department of the Navy, June 1983
9. R.H. Sparling, "Corrosion Prevention at the Drawing Board," Paper presented at the NACE Western Regional Conference, Anaheim, CA, National Association of Corrosion Engineers, Sept 1963
10. M.E. Farmer and J.D. Guttenplan, Electrical Resistance Stability of Chemical Films on Aluminum, *Mater. Perform.*, April 1984, p 9-13
11. T. Dixon, EMI Threatens Electronic Systems Operation, *Elect. Pack. Prod.*, Dec 1983, p 86-91
12. *Soldering Manual*, American Welding Society, 1959
13. W.R. Studnick and C.C. Foune, Do Residues From Today's Fluxes Corrode Copper on PW Boards?, *Circuits Mfg.*, April 1973, p 34-39
14. P.G. Casperson, The Nature of Residues Presented to the Cleaning Machine, *Elect. Pack. Prod.*, Jan 1980, p 73-79
15. E.R. Brands and F.Y. Hayashi, "Final Report, Fungus and Corrosion Induced Failure Mechanisms," Report C73-973.8/201, Prepared for U.S. Army Electronics Command, Rockwell International Corporation, April 1974
16. D.A. Colling and T.J. Dowling, Corrosion Protection of Lead Materials, *Elect. Pack. Prod.*, June 1970

17. "Electromotive Series for Aerospace Metals, Alloys and Materials," Rocketdyne Report MPR 73-863, Rockwell International Corporation, Sept 1973
18. R.G. Baker, "Corrosion and Plating Problems in Electronic Equipment for the Communications Industry," Paper presented at Corrosion/73, Anaheim, CA, National Association of Corrosion Engineers, March 1973
19. J.D. Guttenplan and J.S. Rollins, *Rockwell Int. Corros. Informer*, Vol 13 (No. 2), Aug 1986
20. "Wire, Cables and Harnesses, Insulation, Polyvinyl Chloride (PVC)," ALERT LC-A-80-01, U.S. Government Printing Office, April 1980
21. "Corrosion Prevention and Deterioration Control in Electronic Components and Assemblies," MIL-STD-1250(MI), U.S. Government Printing Office, March 1967
22. F.A. Lowenheim, Ed., *Guide to the Selection and Use of Electroplated and Related Finishes*, STP 785, American Society for Testing and Materials, 1982, p 48
23. N. Lycoudes, "Tin Plated Surfaces and the Whisker Growth Phenomenon," Reliability Report RIC-1695, Motorola Inc., May 1976
24. J.T. Menke, "The Anatomy of an Electronic Black Box," Paper presented at Corrosion/83, Anaheim, CA, National Association of Corrosion Engineers, April 1983
25. C.A. MacKay, Causes and Effects of Solder Contamination: Part II, *Electri-Onics*, April 1983, p 41-44
26. H.H. Manko, *Solders and Soldering*, 2nd ed., McGraw-Hill, 1979, p 76
27. W.C. Shumay, Microjoining for Electronics, *Adv. Mater. Process.*, Nov 1986, p 38-42
28. L.E. Colteryahn, Metallurgical Mechanisms for Au/Al TC Bond Failure, in *Proceedings of the Second Physics of Failure (CQAP) Colloquium* (Anaheim, CA), Rockwell International Corporation, June 1965
29. R.S. Mooczkowski, "Corrosion and Electrical Contact Interfaces," Paper presented at Corrosion/85, Boston, MA, National Association of Corrosion Engineers, March 1985
30. M. Antler, Survey of Contact Fretting in Electrical Connectors, *IEEE Trans. Components, Hybrids, Mfg. Technol.*, Vol CHMT-8 (No. 1), March 1985

SELECTED REFERENCES

General

- U.R. Evans, *The Corrosion and Oxidation of Metals*, Edward Arnold, 1960, supplemental volumes, 1968 and 1976
- M.G. Fontana and N.D. Greene, *Corrosion Engineering*, 2nd ed., McGraw-Hill, 1978
- L.L. Shreir, *Corrosion*, Vol I and II, 2nd ed., Newnes Butterworth, 1976
- N.D. Tomashov, *Theory of Corrosion and Protection of Metals*, Macmillan, 1966
- H.H. Uhlig, *Corrosion Handbook*, John Wiley & Sons, 1948

Special Technical Publications

- *A Handbook of Protective Coatings for Military and Aerospace Equipment*, TPC Publication 10, National Association of Corrosion Engineers, 1983
- F.A. Lowenheim, Ed., *Guide to the Selection and Use of Electroplated and Related Finishes*, STP 785, American Society for Testing and Materials, 1982

Military Publication/Standards/Specifications

- "Bonding, Electrical, and Lightning Protection for Aerospace Systems," MIL-B-5087B(ASG), Interim Amend. 3, U.S. Government Printing Office, Dec 1984
- *Corrosion and Corrosion Prevention—Metals*, MIL-HDBK-729, Military Standardization Handbook, U.S. Government Printing Office, Nov 1983
- "Corrosion Prevention and Deterioration Control in Electronic Components and Assemblies," MIL-STD-1250(MI), U.S. Government Printing Office, March 1967
- *Design Guidelines for Prevention and Control of Avionic Corrosion*, NAVMAT P 4855-2, Department of the Navy, June 1983
- "Dissimilar Metals," MIL-STD-889D, Notice 1, U.S. Government Printing Office, Nov 1979
- "Finishes, Coatings, and Sealants for the Protection of Aerospace Weapons Systems," MIL-F-7179F, U.S. Government Printing Office, Sept 1984
- "Finishes: Organic, Weapons System Application and Control of," MIL-F-18264D, Amend. 1, U.S. Government Printing Office, April 1971
- "Materials and Processes for Corrosion Prevention and Control in Aerospace Weapons Systems," MIL-STD-1568A(USAF), U.S. Government Printing Office, Oct 1979
- *Prevention and Control of Corrosion and Fungus in Communication, Electronic, Meteorological, and Avionic Equipment*, Technical Manual, T.O. 1-1-689, Change 1, U.S. Government Printing Office, July 1, 1975
- *Protective Finishes for Metal and Wood Surfaces*, MIL-HDBK-132A, Military Standardization Handbook, U.S. Government Printing Office, May 1984
- "Standard General Requirements for Electronic Equipment," MIL-STD-454J, Notice 3, U.S. Government Printing Office, June 1985
- "Surfaces Treatments and Inorganic Coatings for Metal Surfaces of Weapons Systems," MIL-S-5002, Amend. 1, U.S. Government Printing Office, Aug 1978

Case Histories and Failures of Electronics and Communications Equipment

Eddie White, George Slenski, and Bill Dobbs
Electronic Failure Analysis Group,
Air Force Wright Aeronautical Laboratories

THE ELECTRONIC FAILURE ANALYSIS GROUP of the Air Force Wright Aeronautical Laboratories' Materials Laboratory has investigated a large number of electronic and electrical failures. It has been established that about 83% of these failures are caused by materials and manufacturing process defects. Also, it has been verified that about 20% of the failures are caused by corrosion problems.

Several different testing and analysis techniques will be presented to identify ways of investigating corrosion problems in electronic failures. These brief discussions are intended only as introductions to the techniques; more information is available in the literature on the technique. Following this, a number of case histories of electronic failures associated with corrosion problems will be presented with recommended corrective actions. All of the corrective actions interrupt the electrochemical circuit required for corrosion to occur (anode, cathode, and electrolyte to conduct electricity between the anode and the cathode). Corrective actions consist of removing the electrolyte, insulating the anode, or removing the cathode. This usually results in a significant cost savings for the customer and a marked improvement in the reliability of the equipment.

Analysis Techniques

The penetration of moisture and contaminants into electronic systems has many detrimental effects, including corrosion. In most electronic systems, dimensions have been minimized for faster signal processing and higher density. This means that most metallizations are thin, or small in cross-sectional area, and that the individual metallizations are close together. In such systems, trace amounts of moisture and contamination may cause system failure. If the aluminum metallized surface of an integrated circuit is contaminated and if moisture is present, a slight amount of corrosion may result in an open circuit. Corrosion of 1 pg of aluminum is sufficient to open an integrated circuit conductor. This extreme sensitivity requires special caution when dealing with corrosion in electronic systems.

Environmental Testing

Failure modes in electronic components and systems can be identified and related to field failures with environmental testing techniques. The four techniques commonly used are thermal and humidity cycling, humidity testing, salt fog tests, and combined environments reliabilty testing.

The thermal and humidity cycling test, sometimes called the moisture resistance test, is performed for the purpose of evaluating in an accelerated manner the resistance of component parts and constituent materials to the deteriorative effects of high humidity and heat conditions. Most degradation results directly or indirectly from the absorption of moisture vapor or films by vulnerable insulating materials and from the surface wetting of metals and insulation. These phenomena produce many deleterious effects, including corrosion, physical distortion and decomposition of organic materials, leaching out and spending of constituents of materials, and detrimental changes in electrical properties.

This test derives its effectiveness from its use of temperature cycling, which provides the alternate periods of condensation and drying essential to the development of the corrosion processes and produces a breathing action of moisture into partially sealed containers. Increased effectiveness is also obtained by the use of a high temperature, which intensifies the effect of humidity. Provision is made for the application of a polarizing voltage across insulation to investigate the possibility of electrolysis, which can promote eventual dielectric breakdown. Results obtained with this test are reproducible and have been confirmed by investigations of field failures. This test has proved to be reliable for indicating those parts that are not suitable for use in high-temperature and humid environments. Figure 1 shows a typical 24-h temperature and humidity cycle.

The humidity test requires that specimens be placed in a chamber and exposed to a relative humidity of 90 to 95% and a temperature of 40 °C (100 °F) for a selected period of time. This test is performed to evaluate the properties of materials used in components as they are influenced by the absorption and diffusion of moisture. This is an accelerated test that is accomplished by the continuous exposure of the specimen to high relative humidity at an elevated temperature. These conditions impose a vapor pressure on the material under test that constitutes the force behind the moisture migration and penetration. Hygroscopic materials are sensitive to moisture and deteriorate rapidly under humid conditions. Absorption of moisture by many materials results in swelling, which destroys their functional utility and causes loss of physical strength as well as changes in other important mechanical properties. Insulating materials that absorb moisture may suffer degradation of their electrical properties. This test is useful in determining the moisture absorption of insulating materials.

The salt fog test requires that the specimen be placed in a chamber in which air is circulated through a 5% sodium chloride (NaCl) solution maintained at 30 °C (85 °F). The relative humidity of the air should be 95 to 98% when released from the salt solution. The salt fog test, in which specimens are subjected to a fine mist of salt solution, has several useful purposes; however, this test has been erroneously considered by many as an all-purpose accelerated corrosion test that, when passed, will guarantee samples to be satisfactory under any corrosive condition. Experience has shown that there is seldom a direct relationship between resistance to salt fog corrosion and resistance to corrosion in other media, even in so-called marine environments. However, some idea of the relative service life and behavior of different samples of a similar material can be gained by means of the salt fog test, provided accumulated data from correlated field service tests and laboratory salt fog tests show that such a relationship exists. Such correlation tests are also necessary to show the degree of acceleration, if any, produced by the laboratory test.

The salt fog test is generally considered unreliable for comparing the general corrosion resistance of different kinds of metals or for predicting their comparative service life. The salt fog test has received its widest acceptance as a test for evaluating the uniformity of protective coatings. In this connection, the test is useful for evaluating different lots of the same product. When used

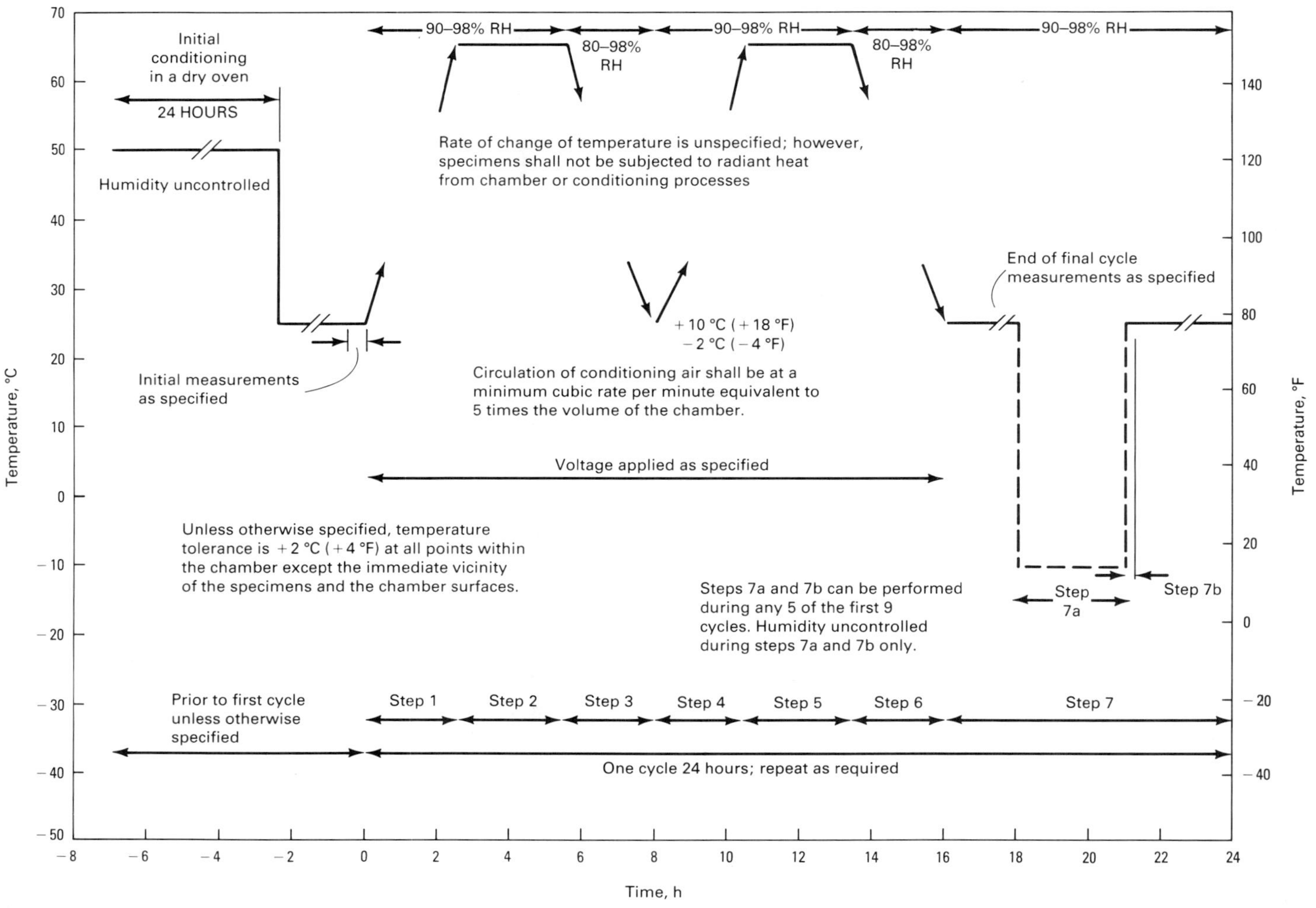

Fig. 1 Graphic representation of thermal and humidity cycling test

to check the porosity of metallic coatings, the test is more dependable when applied to coatings which are cathodic, rather than anodic, toward the base metal.

Combined Environments Reliability Testing (CERT). The purpose of this test is to identify failures that temperature, humidity, vibration, and altitude can induce in electronic equipment either individually or in any combination. Studies have shown that thermal effects, vibration, humidity, and, in certain cases, altitude have the greatest effect on the life of avionics in the operational environment. It has been suggested that these four factors account for 88% of all the environmentally induced failures in the field. Temperature, humidity, vibration, and altitude can interact to produce failures in electronic equipment mounted inside of an aircraft in synergistic ways that are not apparent when stressed independently.

The system under test is placed inside a chamber in which the four stresses may be applied. A temperature and humidity cycle similar to that shown in Fig. 1 is combined with vibration and altitude cycles to provide the final stress cycle that the equipment experiences. The equipment is electrically tested during environmental stress testing so that failures may be identified. This testing procedure is a very effective tool for identifying system deficiencies; the correction of these deficiencies enhances the reliability of avionic equipment.

Chemical Analysis Techniques

Infrared, visible, and ultraviolet spectroscopies are used for the qualitative and quantitative analysis of organic compounds and alloys. These methods are based on the fact that within any molecule the atoms vibrate with a few definite, sharply defined frequencies characteristic of that molecule. When a sample is placed in a beam of infrared radiation, it adsorbs energy at those frequencies characteristic of the molecule and transmits all other frequencies. The resulting pattern, known as an infrared spectrum, is thus a fingerprint that is characteristic of the compound. The intensity of an absorption band is a known, reproducible function of the amount of material in the sample beam. Therefore, measurement of the absorption intensity yields quantitative values of the material present in a sample. However, the instrument slit width plays an important part in quantitative analysis, and better techniques are available. More information on infrared spectroscopy is available in the article "Infrared Spectroscopy" in Volume 10 of the 9th Edition of *Metals Handbook*.

Ultraviolet and visible spectroscopies provide useful information that complements the infrared analysis. An ultraviolet or visible light of variable wavelength is selectively passed through a sample solution. When absorbance occurs in the ultraviolet or visible range, it indicates that electrons in the sample molecules are undergoing transition from the normal bonding molecular orbitals to higher-energy antibonding orbitals. The absorbances at these wavelengths indicate the presence of conjugated double-bond systems. The intensity of absorbance is proportional to the concentration of the molecules. The article "Ultraviolet/Visible Absorption Spectroscopy" in Volume 10 of the 9th Edition of *Metals Handbook* contains detailed information on the theory and use of this technique.

Raman Spectroscopy. Raman scattering relates to the change of frequency observed when a monochromatic light beam is scattered by polyatomic molecules. The Raman spectrum characterizes the scattering medium and is independent of the incident radiation frequency. It corresponds to the frequencies of the atom oscillations in polyatomic structures. The principal utility of Raman spectroscopy is in dealing with

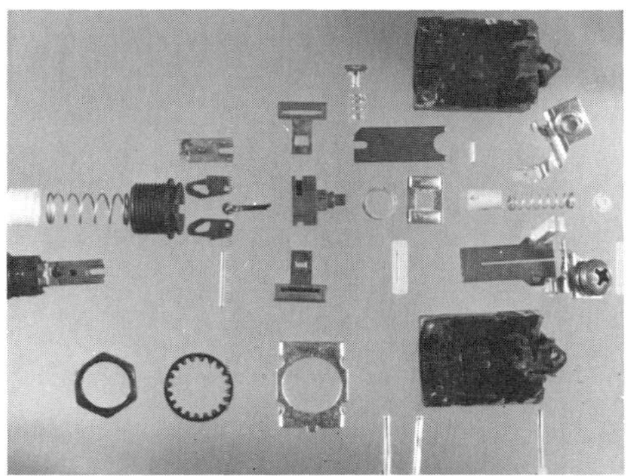

Fig. 2 Exploded view of an aircraft circuit breaker

highly symmetrical molecules or chromophores (functional groups that produce color) within a molecule. Raman spectra can be used to characterize and identify chemical species (inorganic, organic, and biological molecules) and precisely determine their structures in all phases of matter. Thus, vibrational Raman spectra can be regarded as unique fingerprints and often contain information on the local molecular environment. Detailed information on Raman spectroscopy can be obtained in the article "Raman Spectroscopy" in Volume 10 of the 9th Edition of *Metals Handbook*.

X-Ray Diffraction. X-rays are electromagnetic radiation and, like visible light, can be diffracted. When a crystal is placed in a collimated x-ray beam, it behaves as a three-dimensional grating that produces a diffraction pattern. The wavelength, λ, of the incident radiation is related to the interplanar grating spacing, d, of the crystal by the equation known as Bragg's law:

$$n\lambda = 2d \sin \theta \qquad \text{(Eq 1)}$$

where n is an integer (the order of diffraction), and θ is the angle of incidence. Equation 1 founded the science of x-ray diffraction and determines the crystal parameter, d, when λ is known and θ, or 2θ, is measured. This technique can be used to identify elements by their crystal structure and their relative degree of crystallinity (see the article "X-Ray Powder Diffraction" in Volume 10 of the 9th Edition of *Metals Handbook*).

Atomic Absorption Spectroscopy (AAS) is known as a very specific technique for the quantitative analysis of organic and inorganic materials for metals in nanogram to picogram levels. In this technique, a sample solution is atomized in a flame or furnace in which the ground-state metal atoms are excited to a higher-energy state in the presence of specific wavelengths of light. By measuring the amount of light absorbed, a quantitative determination of the amount of the element present in the sample can be made.

With the use of a furnace, atomic absorption spectroscopy (FAAS) has enhanced the capability for determining hard-to-analyze elements, such as boron, lead, aluminum, chromium, bismuth, and molybdenum, at trace levels. Although flame AAS is a convenient and fast technique, FAAS is extremely efficient for the measurement of low concentrations. A small sample volume is required for complete analysis, and solid samples can be directly analyzed with no sample preparation. More information on AAS is available in the article "Atomic Absorption Spectrometry" in Volume 10 of the 9th Edition of *Metals Handbook*.

Emission spectroscopy is the technique used for the simultaneous analysis of all metallic elements in a sample. Metal atoms of ground-state energy are excited by thermal energy (in the form of an arc or a spark) to higher-energy states. These atoms then undergo transitions back to the ground state, accompanied by the emission of energy in the form of light and at discrete wavelengths. Each metal atom can undergo many different transitions, the sum of which is the characteristic emission spectrum for that element.

The spectra of all metals in the sample are superimposed, resolved into the component wavelengths by the grating and optics of the instrument, and recorded as lines on photographic plates. After processing, the plates are read on a densitometer, and an element is identified from the wavelength of its emission lines. This technique is most useful for the qualitative and semiquantitative analysis of metals. Emission spectroscopy is more versatile than atomic absorption in that several elements can be identified simultaneously. Quantitative analysis with this method is very time consuming, but once the elements are identified, quantitative analysis can be performed with atomic absorption (flame or furnace) and inductively coupled plasma emission spectroscopic methods. More information on emission spectroscopy is available in the article "Optical Emission Spectroscopy" in Volume 10 of the 9th Edition of *Metals Handbook*.

Gas and Liquid Chromatography. Gas chromatography (GC) is basically a technique for quantitatively separating components in a mixture of organic compounds. The components are partitioned between two phases, an inert moving gas, and a stationary liquid film. The liquid film is supported in a long column through which the vaporized sample and gas move. The separation depends on the affinity each component has for the stationary phase. The compounds with the least affinity will emerge from the column first, while those with more affinity will emerge after some time interval. Upon leaving the column, the component passes through a detector, and its presence is indicated by a peak on a recorder. The identity of the peak can be determined by comparison with standards, and the concentration of the component is derived from the peak area. In order to analyze a sample by GC, it must have some vapor pressure below about 35 °C (95 °F). The article "Gas Chromatography/Mass Spectrometry" in Volume 10 of the 9th Edition of *Metals Handbook* contains detailed information on this technique.

Modern high-performance liquid chromatography (HPLC) has become an extremely powerful technique for separating mixtures into their components. With HPLC, compounds of poor thermal stability and low volatility that cannot be separated by gas chromatography can often be separated for either qualitative or quantitative analysis. The components of a sample move through the column at different rates, and the amount of information that can be derived depends on the type of detector used. The standard technique of retention time comparison with known standards is practiced. In this technique, fractions of a sample can be collected in solution and can be analyzed by any other analytical technique. Other detectors can also be interfaced with this technique to diversify its applications. More information on HPLC can be obtained in the article "Liquid Chromatography" in Volume 10 of the 9th Edition of *Metals Handbook*.

Mass spectroscopy is used to analyze samples according to their mass. This technique is ideal for isotope and molecular weight indentification. The sample is vaporized, and the atoms or molecules are then ionized by a directed electron beam. The ionized beam is accelerated through a magnetic field, then detected by an ion collector. The resulting information from this process is a plot of the ion mass to charge ratio versus the ion relative abundance. Identification of the constituents of a sample that can be vaporized is readily aided by mass spectroscopy (see the article

(a)

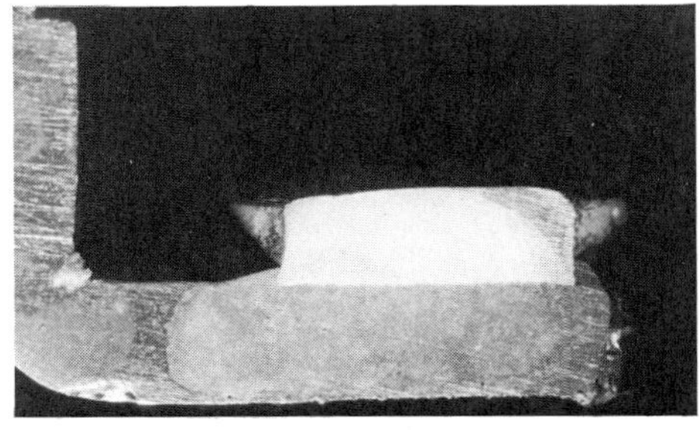

(b)

Fig. 3 Cross sections of circuit breaker contacts after 20-day humidity test. The tungsten/silver contact (a) is severely corroded, while the cadmium/silver contact (b) shows little corrosion.

Fig. 4 Failed circuit breaker contact after 48-h salt fog test

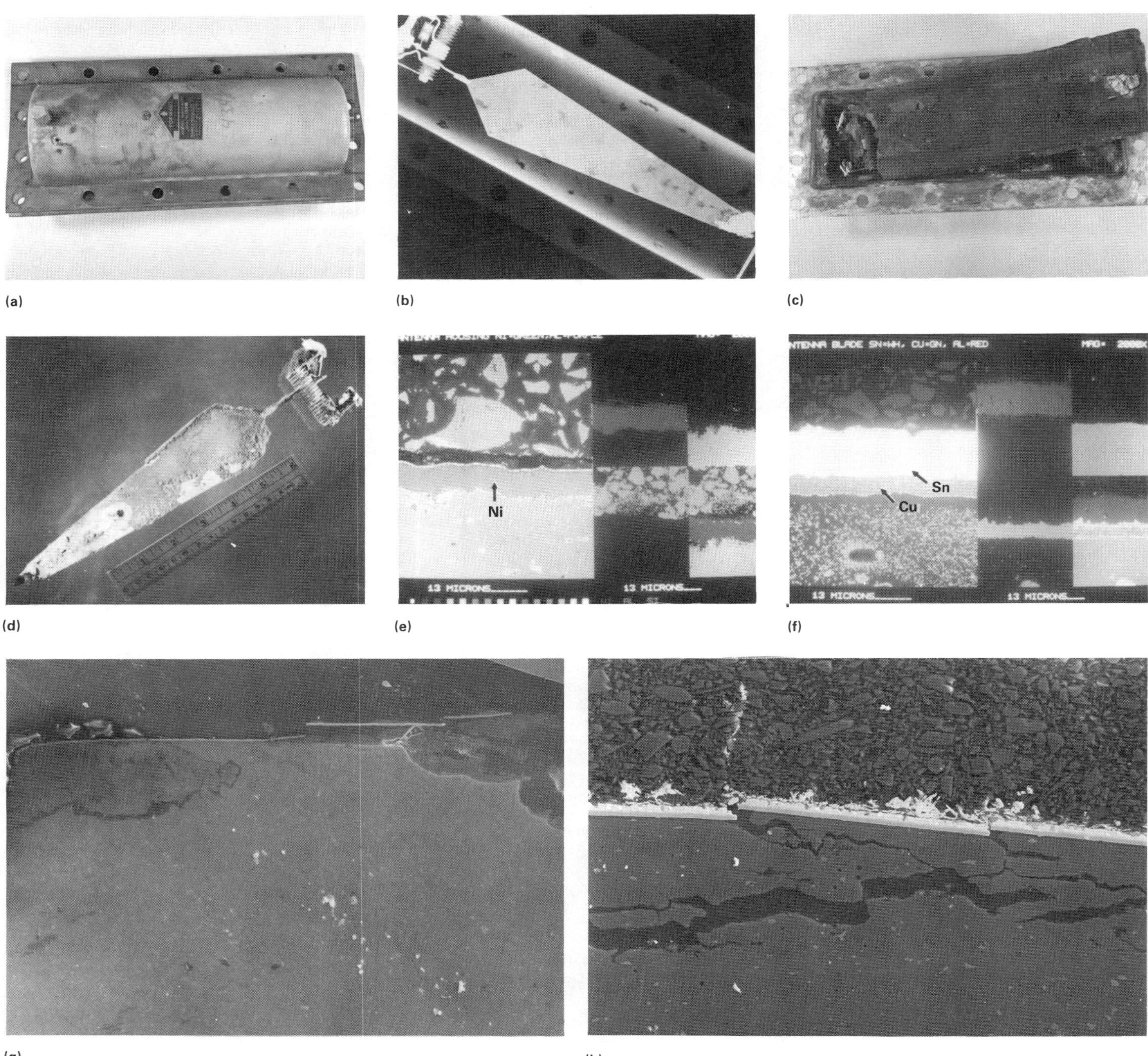

Fig. 5 Aircraft antenna marker beacon (a) that failed by corrosion. (b) X-ray radiograph of beacon showing areas of corrosion. (c) Polyamide foam inside the beacon. (d) Corroded aluminum antenna blade removed from housing. (e) X-ray map showing thickness of nickel plate (arrow) on aluminum beacon housing. (f) X-ray map showing the thickness of copper/tin plate (arrows) on aluminum antenna blade. (g) SEM micrograph of nickel plate on beacon housing showing areas of lifted plating. 50×. (h) SEM micrograph of copper/tin plate on aluminum antenna blade showing areas of lifted plating. 185×

"Spark Source Mass Spectrometry" in Volume 10 of the 9th Edition of *Metals Handbook*).

Wet chemistry and microelemental analysis techniques are used to investigate the chemistry of materials in the following ways:

- Quantitative elemental (composition) analysis
- Qualitative identification of material type
- Qualitative detection of component moieties
- Check on quantitative instrumental methods
- Isolation and characterization of inclusions and phases
- Determination of oxidation state

Modern instrumental techniques for quantitative elemental analysis are capable of determining large numbers of sample constituents in very short periods of time. Even so, wet chemical analysis remains important for several reasons. First, small samples of almost any shape can be used. Second, chemical results average over the entire sample, while most spectrographic techniques analyze small areas of the sample surface. Third, some spectrographic techniques are sensitive to the thermal and mechanical treatment of the sample, while wet chemistry is free of these problems. Fourth, and perhaps most significant, wet chemistry provides the calibration for nearly all spectrographic techniques. More information on wet chemical analysis is available in the article

(a)

(b)

Fig. 6 Component side of failed printed circuit board (a) showing flux contamination (arrows). (b) Close-up of one of the dual in-line packages shown in (a). Leads are shorted out by flux contamination (arrow).

(a)

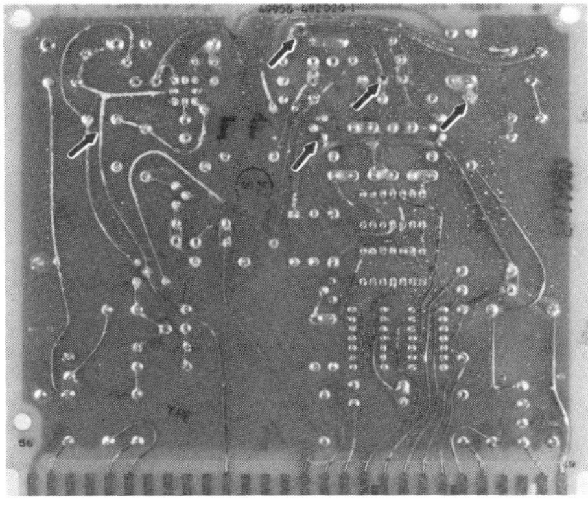
(b)

Fig. 7 Component side of failed printed circuit board (a) from a ground-base radar system showing corroded conductors. (b) Back side of board shown in (a); note corrosion on conductors (arrows) and white areas of mealing on the board.

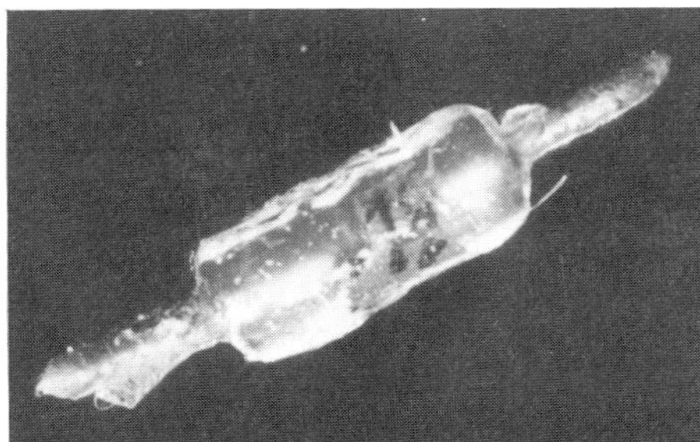

Fig. 8 Corroded diode removed from a failed printed circuit board

"Classical Wet Analytical Chemistry" in Volume 10 of the 9th Edition of *Metals Handbook*.

Electron Optics

Scanning Electron Microscopy (SEM). The scanning electron microscope, like the transmission electron microscope, uses electronics to form a magnified image of the specimen. Unlike the transmission electron microscope, however, the electron beam of the scanning electron microscope does not penetrate the specimen. The electron beam incident on the specimen is rastered (similar to the television scan) across the object. The detector records either the secondary or the backscattered electrons emitted from the top side of the sample, near where the electron beam strikes it. The sample does not have to be thin, but it must be reasonably conductive so that it does not charge and deflect the electron beam. Nonconductive specimens are coated with a thin layer of gold, chromium, or carbon before SEM examination. Integrated circuits can be examined in the scanning electron microscope while bias is applied to the integrated circuit. A great deal can be learned about a device from this technique. Detailed information on the theory and applications of SEM is available in the article "Scanning Electron Microscopy" in Volume 10 of the 9th Edition of *Metals Handbook*.

Transmission Electron Microscopy (TEM). The transmission electron microscope is a device that forms magnified images by means of electrons. The electrons are usually accelerated to 50 keV or above. The microscope magnifies in several stages by means of electromagnetic or electrostatic lenses. Transmission electron microscopes at higher energies (several MeV) have resolutions of a few angstroms. The specimen must be cut into a very thin sample so that the electron beam can penetrate it. After the beam penetrates the sample, a photograph of the sample is taken and kept as a permanent record. The microscope can be adjusted so that an electron diffraction pattern from the sample is obtained. This provides information about the crystallographic structure of the sample. The transmission electron microscope usually provides the highest resolution image that can be obtained from a specimen. The article "Analytical Transmission Electron Microscopy" in Volume 10 of the 9th Edition of *Metals Handbook* provides detailed information on all aspects of TEM.

Characteristic x-ray analysis is frequently done in the scanning electron microscope, and it identifies the elemental constituents in the sample. The electron beam in the microscope is directed on the specimen, and the incident electrons knock electrons free from the bound atomic states of the atoms. As other free electrons drop into these empty bound states, they emit photons or x-rays with energies that are characteristic of that particular element. The emitted x-rays are compared to standards for the identification of the element. This technique permits the elemental analysis of microminiature samples and is discussed in detail in the article "Electron Probe X-Ray Microanalysis" in Volume 10 of the 9th Edition of *Metals Handbook*.

Package Moisture Content Analysis

The moisture content of sealed packages can be analyzed by mass spectroscopy. The package is maintained at a temperature above the dew point when punctured. The gases from the package are directed into the mass spectrometer, which measures the relative amount of moisture in the gases. This type of analysis is important in assessing the potential for corrosion inside hermetically sealed packages. The article "Gas Analysis by Mass Spectrometry" in Volume 10 of the 9th Edition of *Metals Handbook* contains information on this technique.

Surface Analysis

Auger spectroscopy is used to study the sample surface to a depth of about 0.5 to 2 nm (5 to 20 Å). When the sample surface is bombarded with electrons in the energy range of 20 to 2500 eV, Auger electrons are emitted from the surface with characteristic energies that are determined by the elements constituting the sample. When the emitted Auger electron energies are identified, the elements in the sample are also determined. Depth profiling can be accomplished by sputtering the sample surface at a controlled rate with an ionized gas. More information on this technique is available in the article "Auger Electron Spectroscopy" in Volume 10 of the 9th Edition of *Metals Handbook*.

Electron spectroscopy for chemical analysis (ESCA), also known as x-ray photoelectron spectroscopy (XPS), uses the photoelectric effect to study the electronic structure of surfaces as well as the chemical identity of the surface components. With this technique, the excitation source is the x-ray; therefore, disturbance of the sample surface is minimized. The ejected electrons possess energies that are characteristic of given elements, but these energies are also dependent on the chemical environment of the element so that some information is obtained about the surface chemistry. The depth of surface examined may be as deep as 10 nm (100 Å) for organic materials. Detailed information on this technique is available in the article "X-Ray Photoelectron Spectroscopy" in Volume 10 of the 9th Edition of *Metals Handbook*.

(a)

(b)

Fig. 9 Cross section (a) of printed circuit board contaminated by chlorine from a highly reactive water-soluble soldering flux. Left to right: void, conformal coating, solder, copper, and the epoxy/fiberglass board matrix. 130×. (b) X-ray dot map showing chlorine contamination (arrow) in the area shown in (a). The light dots indicate the presence of chlorine.

(a)

(b)

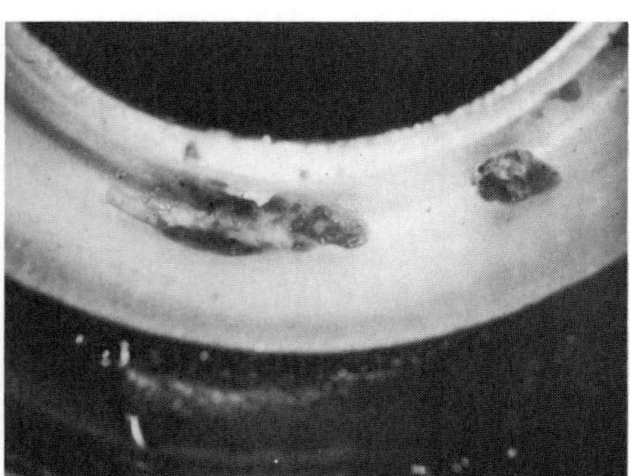

(c)

(d)

Fig. 10 Triode amplifier tubes (a) showing corrosion. (b) Exterior corrosion on the triode of a tube. (c) Interior corrosion on a triode (the glass tube envelope has been removed). (d) Cross section of triode showing entrapped corrodent (arrow)

Secondary ion mass spectroscopy (SIMS) bombards the sample surface with selected ions, usually argon, at designated energies, generally less than 1 keV. As a result of this ion bombardment, secondary ions are sputtered off the sample surface. The secondary ions are collected and analyzed with a mass spectrometer. The technique usually provides detection sensitivity of 1 ppb or better with minimal sample surface damage and can be used for depth profiling (see the article "Secondary Ion Mass Spectroscopy" in Volume 10 of the 9th Edition of *Metals Handbook*).

Ion scattering spectroscopy (ISS) requires that the excitation beam be an ion source, similar to SIMS, and that the sample surface atoms scatter the incident ions through binary collisions. The scattered primary beam is then energy analyzed. This measured intensity is plotted versus the ratio of the scattered energy divided by the initial energy. Scattering peaks in the plotted curve correspond to elements in the sample surface. These peaks are calibrated against known elements for identification. The article "Low-Energy Ion-Scattering Spectroscopy" in Volume 10 of the 9th Edition of *Metals Handbook* contains more information on this technique.

Examples of Electronic and Electrical Corrosion

The following are a few representative examples of corrosion-induced failures of electronic devices. The analysis techniques discussed above were used to identify the corrosion mechanism.

Circuit Breakers. Numerous aircraft circuit breakers (Fig. 2) were identified as failed because their contact resistance when closed was considered too high. A number of circuit breakers from several different manufacturers were tested in the laboratory to determine the cause of the high resistance. The circuit break contacts were identified as either tungsten/silver or cadmium/silver mixtures. Nine circuit breakers were exposed to a 10-day humidity test (40 °C, or 100 °F, and 95% relative humidity). A destructive physical analysis of the parts found a small amount of corrosion on the tungsten/silver contacts. Because of this, a 20-day humidity test (49 °C, or 120 °F, and 95% relative humidity) was conducted on two contacts that had been cross sectioned. After the 20-day test, the tungsten/silver contacts were severely corroded, while the cadmium/silver contacts exhibited very little corrosion (Fig. 3).

A 48-h salt fog test (35 °C, or 95 °F, 5% NaCl, pH 6.5 to 7.2) was conducted on seven of the circuit breakers. Two of the circuit breakers failed. This was caused by salt condensation on the contacts (Fig. 4).

(a)

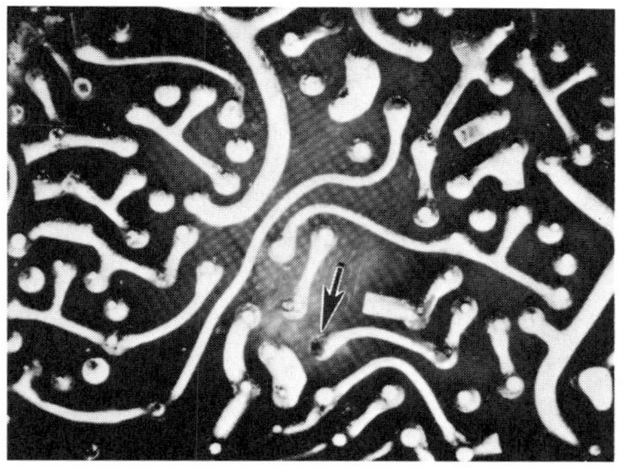
(b)

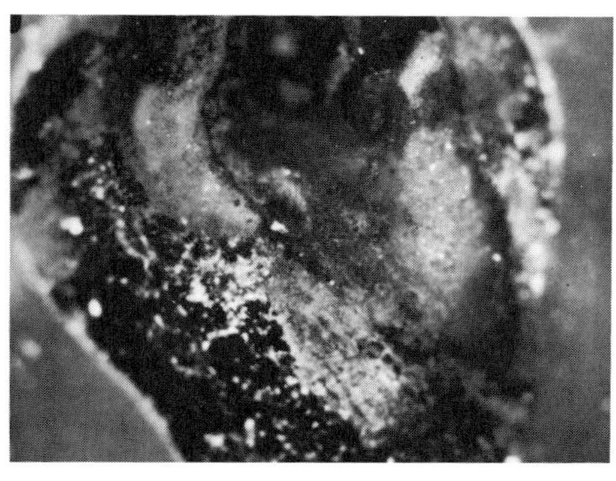
(c)

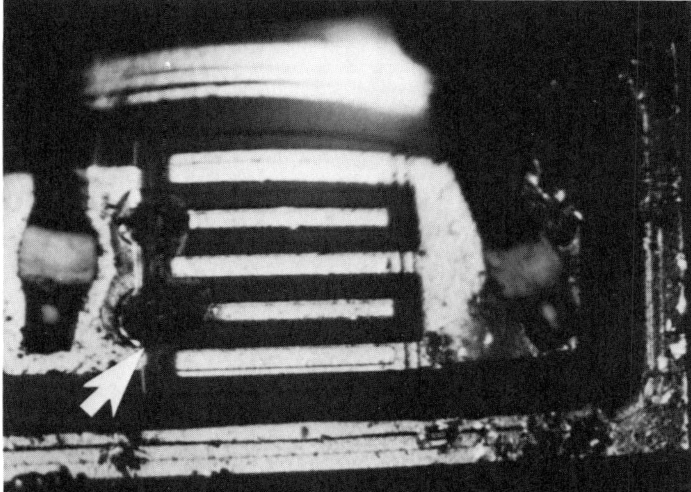
(d)

(e)

Fig. 11 Fuse that failed from corrosion after several years of storage. (a) Fuse electronics removed from package showing white vibration-dampening foam between printed circuit boards. (b) Corrosion (arrow) caused by chloride contamination on a printed circuit board. (c) Enlargement of area shown in (b). (d) Failed transistor (arrow marks site of open aluminum conductors) with high surface chloride contamination. (e) SEM micrograph of area marked by arrow in (d)

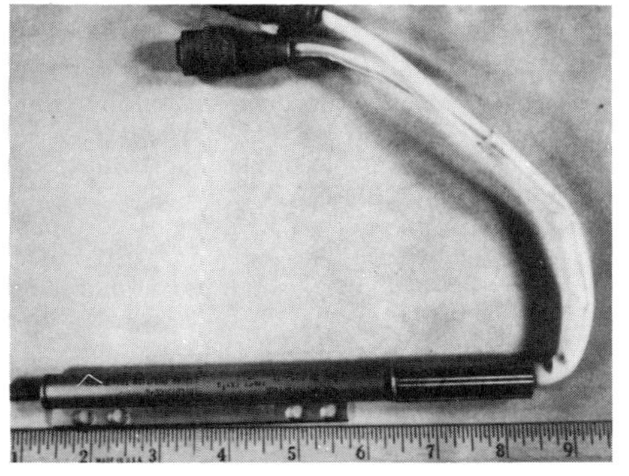
(a)

(b)

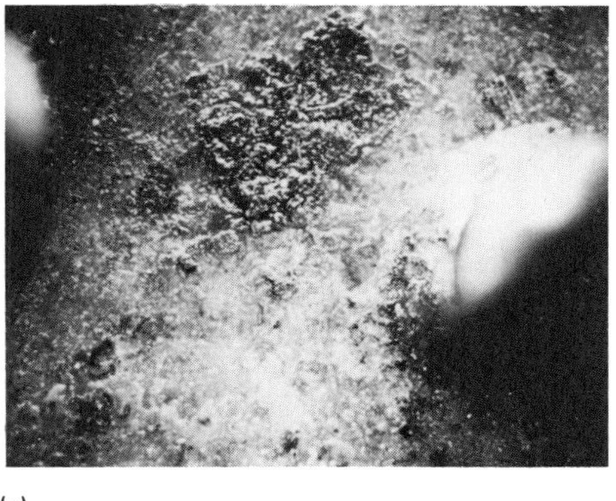
(c)

Fig. 12 Linear feedback potentiometer (a) used for aircraft nosewheel steering. (b) Corrosion product (arrow) on potentiometer exterior that shorted out electrical leads. (c) Enlargement of site marked by arrow in (b)

The results of the analysis indicate that the tungsten/silver contacts corrode much more severely than the cadmium/silver contacts. Braze fluxes used to join the contacts to the copper arm may contribute to corrosion. The contacts in all circuit breakers are improved by the mechanical wiping action of opening and closing the contacts several times. This should be done at periodic intervals. If the conduction of large currents is not required, the cadmium/silver contacts show superior resistance to corrosion.

Antenna Marker Beacon. A corroded antenna marker beacon (Fig. 5) was removed from an aircraft so that the source of corrosion could be determined. The beacon was x-rayed, and corrosion sites were visible in the radiograph (Fig. 5b). The antenna was opened and found to contain a polyamide foam (Fig. 5c). The antenna blade was removed from the foam (Fig. 5d) and observed to be corroded also. Both the housing and blade were found to be aluminum. The housing was

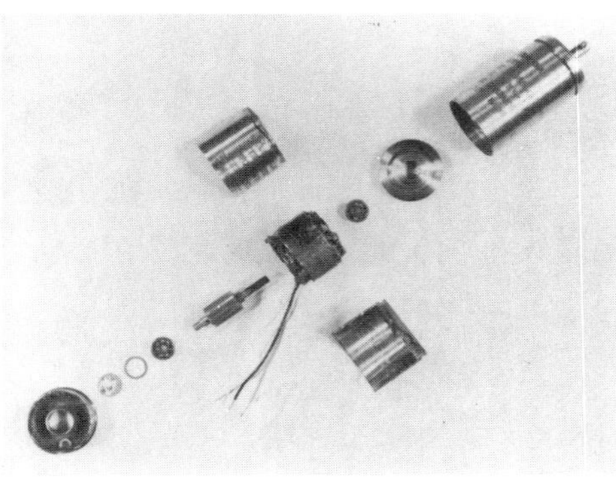

(a)

(b)

(c)

Fig. 13 Exploded view (a) of stepper motor. (b) Cracked polyimide insulation on field coils. (c) Arrow shows site of open field coils.

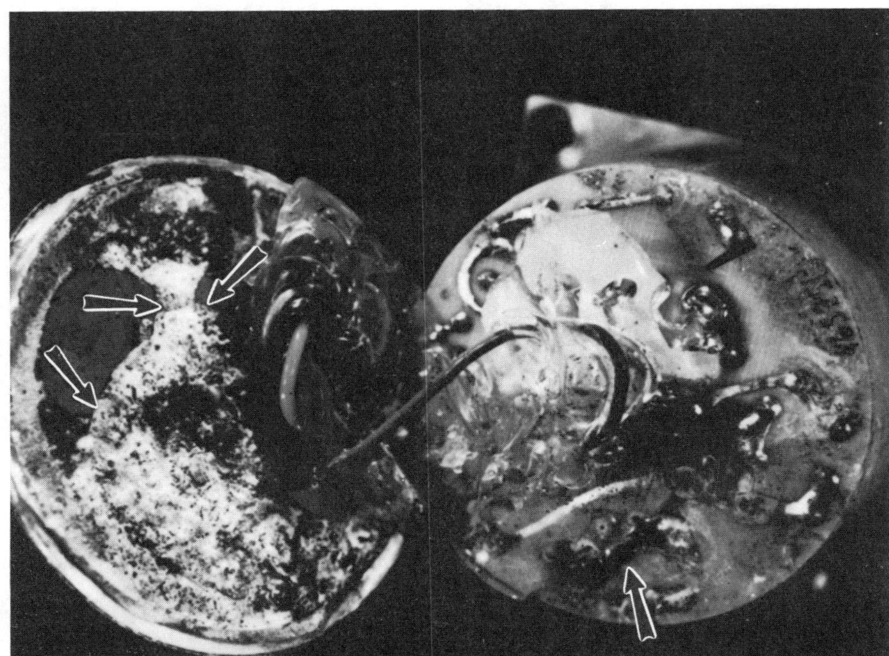

Fig. 14 Aircraft accelerometer that failed after 150 cycles in the CERT test. Arrows show regions of corrosion on the opened accelerometer.

found to be plated with electroless nickel (Fig. 5e), and the blade was plated with a copper strike followed with a tin plate (Fig. 5f).

It is believed that both the nickel plate and the copper/tin plate were porous. This permitted the penetration of moisture to the plating and aluminum interface. The presence of moisture and the anodic relationship of aluminum to nickel or tin resulted in a galvanic cell that caused pitting corrosion in the aluminum (Fig. 5g and h). These plating systems were used to maintain a high electrical surface conductance on the aluminum components. The corrosion resulted in nonconductive surfaces that affected the electrical performance of the antenna. It was recommended that, instead of plating the aluminum, a chromate conversion coating be used.

Printed wiring boards are subject to failure if a highly reactive soldering flux is used or if the flux is not completely removed after the soldering operation.

Flux Contamination on Dual In-line Packages. A printed circuit board (Fig. 6a) was removed from an aircraft because of short circuits on the board. Blue-green corrosion products appeared around and under dual in-line packages on the board (Fig. 6b). Copper chloride was detected in the corrosion residue by characteristic x-ray analysis in the scanning electron microscope. Chlorides are a common contamination resulting from active solder fluxes and poor cleaning processes. It was recommended that a rosin-type solder flux be used and that a board cleanliness test be performed after cleaning.

Corroded Conductor Traces on a Ground-Base Radar System. A ground-base radar system was removed from a 9-month storage for testing. All electrical systems failed because of short circuits in many of the multilayer printed circuit boards. Upon examination, the circuit boards appeared to have corroded conductor traces. Contamination and corrosion products had migrated to regions between conductor traces so that the traces were shorted out (Fig. 7). It was established by characteristic x-ray analysis that the contamination resulted from solder flux. A different cleaning procedure was recommended.

Flux on Board Components. Components from a printed circuit board have also been found to be contaminated with solder fluxes. Figure 8 shows a corroded diode taken from a board. Inadequate removal of flux residues was determined to be the cause of this problem.

Water-Soluble Flux. The use of a highly reactive solder flux may sometimes have unexpected consequences. Of course, the reactive flux makes the soldering procedure easier. Components or boards that have poor solderability may sometimes be used, or some other device defects may be overcome; however, the reactive flux usually contaminates some parts of the electronic system so that it eventually causes more problems than it may solve.

A reactive water-soluble flux was used on a number of printed circuit boards because some of the component leads exhibited poor solderability. The flux had a high chlorine content. During the soldering operation, some of these chlorides were absorbed in the epoxy/fiberglass board surface. After soldering, the boards were cleaned and conformally coated, which trapped the flux near the surface of the board. The board cross section and an x-ray dot map of this area are shown in Fig. 9. This chlorine may eventually affect the copper conductors in the board because the conformal coating will absorb moisture, which could result in the formation of HCl. Without the reactive water-soluble flux, this chloride surface contamination does not occur. If the components to be soldered are handled properly, the soldering operation can be accomplished with less reactive fluxes. This is the best procedure.

Vacuum Tubes. Triode amplifier tubes were failing because of severe corrosion of the exterior and interior of the tubes (Fig. 10). It was determined that the contaminant was a chloride and that its source was most likely a highly corrosive chloride flux. Figure 10(b) shows localized corrosion on the outside of the tube, and Fig. 10(c) shows the corrosion on the inside of the tube. Figure 10(d) shows a tube cross section with the entrapped green corrodent clearly visible. It was recommended that all fluxes be thoroughly cleaned from the tube before sealing.

Fuses. After storage for several years, some fuses were tested. The fuse failure rate was

(a) (b)

Fig. 15 Disk recorder head that failed by pitting corrosion. (a) Arrow marks metallic head. (b) Close-up view of head marked by arrow in (a)

exceedingly high. The cause of this high failure rate required identification. Figure 11(a) shows the fuse electronics when removed from the package. The white vibration-dampening foam is visible between the circular printed circuit boards. Upon analysis, the foam was identified as polyvinyl chloride (PVC). The circular boards were examined, and it was found that they were heavily contaminated with chlorides. One of the boards was placed in a humidity cabinet for 24 h. The results of this test are shown in Fig. 11(b) and (c). Several failed plastic-encapsulated transistors from the board were opened (Fig. 11d and e). The arrow in Fig. 11(d) marks the region in which the aluminum metallization was completely melted. The transistor surface was heavily contaminated with chlorides. It was found that these chloride contaminants were originating from the vibration-dampening PVC foam. The chloride penetrated the transistor by wicking up the metal leads extending through the plastic package. The PVC foam was replaced with a polyacrylic elastomer foam, which does not emit chloride contamination.

Steering Potentiometer. A linear feedback potentiometer is often used in the steering system of an aircraft nose wheel. A potentiometer of this type is shown in Fig. 12(a). Several failures of this device have been observed. The cause of these failures was investigated. It was found that the ends of the potentiometer were exposed to the environment. Sufficient moisture and contamination were collected on the potentiometer in this area to cause corrosion. Figures 12(b) and (c) show corrosion products collecting on the glass header insulation around the incoming electrical wires. These corrosion products were shorting out the electrical leads on the potentiometer. The problem was corrected by providing this area of the potentiometer with sufficient protection to prevent the entrance of the moisture and the contamination.

Stepper motors (Fig. 13a) are often submerged inside a fuel tank for cooling purposes. Of course, this means that the motor components are then exposed to the fuel environment. In all cases, there is a small amount of water in the fuel. With sufficient time, this water will hydrolyze polyimide wire insulation.

The stepper motor shown in Fig. 13 had polyimide insulation on the field coils. After sufficient exposure to the water in the fuel, some of the coil insulation cracked (Fig. 13b).

This defect exposes the copper of the field coils to the fuel. If some contamination in the fuel is available, copper ions will migrate into the fuel. This produces open circuits in the field coils, as shown in Fig. 13(c). This type of failure can be avoided by hermetically sealing the field coils from the fuel or by using polysulfide insulation.

An aircraft accelerometer was environmentally sealed instead of hermetically sealed. After extended use, the accelerometer failed. Several accelerometers were tested in the CERT chamber to establish the failure mode. The stresses used in the CERT chamber were selected to approximate the actual aircraft environment and included temperature cycling, humidity cycling, and altitude cycling. The accelerometer was electrically operational during a part of the cycle and failed after 150 test cycles. The package was opened (Fig. 14), and it was easily seen that corrosion residues were shorting out various parts of the circuits. This failure can be eliminated by hermetically sealing the accelerometer package.

Disk recorder heads (Fig. 15) were failing at a very high rate. This was occurring in an area where the recorder operator was smoking cigarettes. Some of the particulate matter in the smoke was being trapped between the disk and the head. This contamination, along with the moisture available from the air, was causing pitting corrosion in the head (Fig. 15b). It was recommended that the operator avoid smoking in the recording area and that the heads be cleaned with alcohol rather than a halogenated cleaning solvent.

Electrical connectors are subject to corrosion resulting from a variety of factors, including improper processing and poor design. Two such failures will be discussed.

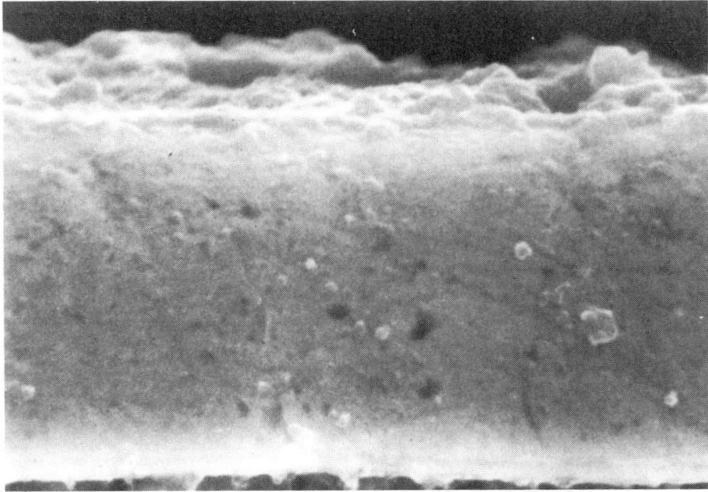

(a) (b)

Fig. 16 Cross sections of good (a) and corroded (b) connectors from a radar modulator. The silver plating on (a) was 6.7 μm thick; the plating on (b) was only 5 μm thick. Both 8200×

Fig. 17 Corroded low-voltage electrical connector on avionic equipment

Radar Modulator. The cause of failure of a radar modulator was identified as a corroded connector. Two connectors, one corroded and the other not, were joined by a wire insulated with a silicone rubber that had a chloride content under 0.16%. The surfaces of both connectors were silver plated. The connectors were sectioned (Fig. 16), and the base material was identified as leaded brass. The silver plating on the uncorroded connector (Fig. 16a) was 6.7 μm thick. The silver plate on the corroded connector (Fig. 16b) was 5 μm thick. Both plating thicknesses are too thin for adequate protection. The plating process should meet Federal Specification QQ-S-365C, which requires a nickel strike and a silver plating thickness of 13 μm. The corrosion was associated with a thin silver plating, and it is believed a silver plating of specified thickness will eliminate the corrosion problem.

Low-Voltage Connectors. Corrosion of electrical connectors (Fig. 17) causes a large number of electrical failures. These problems can be minimized by:

- Installing the connectors in a horizontal position
- Placing a loop in the wire so that water will not flow down the wire into the connector
- Using inhibitors on the connector pins and receptacle interior mating areas (MIL-C-81309, Type III)
- Using the proper inhibitors on the external connector surfaces (MIL-C-85054)

It is recommended that an aluminum connector with cadmium plate be used. The possibility of corrosion can be further minimized by using a chromate conversion coating over the cadmium plate.

A microwave detector (Fig. 18a) failed because of corrosion. The ferrite core surface and cavity of the detector are shown in Fig. 18(b) and (c). The detector case was constructed of nickel-plated aluminum. The cavity lid was removed and is shown in Fig. 18(d). Corrosion is visible along the edge. This corrosion is attributed to delamination of the nickel coating and the exposure of a dissimilar-metal couple to high levels of moisture. The nickel coating delamination is the result of a poor plating process. This is usually caused by inadequate cleaning of the aluminum surface prior to plating. The pitting corrosion of the aluminum case and the poor lid-to-case seal allowed moisture to enter the internal case cavity, which resulted in the corrosion shown in Fig. 18(b) and (c).

(a)

(b)

(c)

(d)

Fig. 18 Microwave detector (a) that failed by corrosion. (b) Corroded ferrite core of detector. (c) Corrosion in detector cavity. (d) Corrosion on the nickel-plated aluminum lid

Table 1 Faying surface resistance of plates exposed to high humidity

Sample number	Average surface resistance at 1 A, μΩ: Initial	After 24 h	After 48 h	After 96 h	After 192 h	Final 240 h
1	140	245	256	201	226	183
2	130	282	263	220	208	214
3	134	207	230	219	230	243
4	146	322	205	202	198	202
5	144	374	333	346	356	326

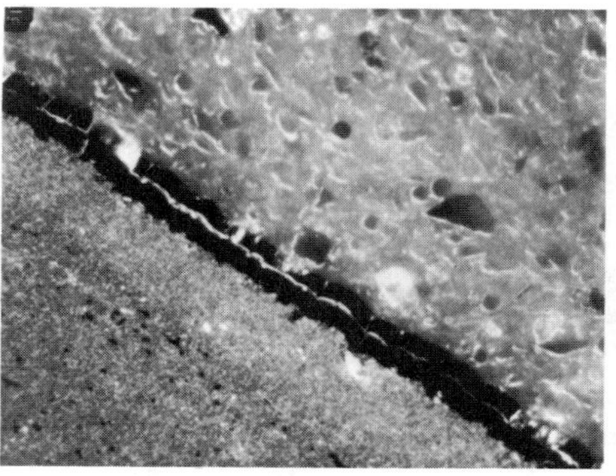

Fig. 19 Cross section of nickel/boron plating on aluminum panel for electromagnetic interference protection

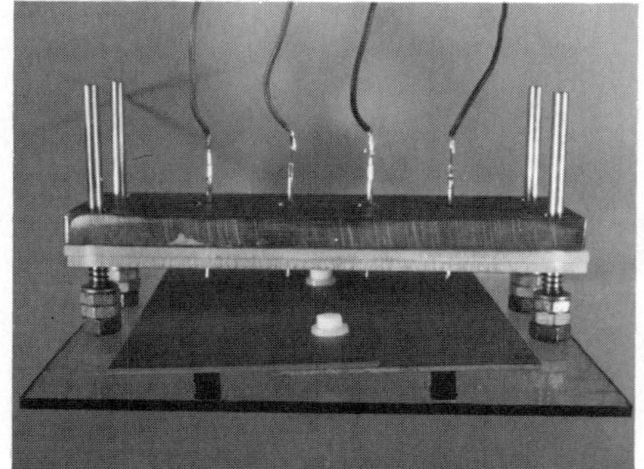

Fig. 20 Test fixture used to measure surface electrical resistance of nickel/boron-plated aluminum test panels

Fig. 21 Typical test panel (sample 1) after humidity testing

An alternate plating for the aluminum should be considered. Ion vapor deposited aluminum should be used instead of nickel. A better lid-to-case seal should be obtained through the use of an appropriate gasket material.

Nickel/Boron-Plated Panels. For many applications, materials are required to meet conflicting requirements. An example of this is when metal panels are required to have corrosion resistance and good electromagnetic interference (EMI) protection. Such panels can be used to package electronic equipment.

Twelve nickel/boron-coated aluminum plates were given electrical and environmental tests. Samples 1, 2, and 3 were aluminum alloy 6062; samples 4 and 5 were aluminum alloy 2024. The nickel/boron coating was 90% Ni with 10% B diffused into the nickel. A plate was sectioned, and it was found that the coating was made in two layers (Fig. 19). Each layer was 16 μm (0.625 mils) thick. The faying surfaces were prepared for testing by overlapping the pieces and drilling holes through the panels. The two plates were bolted together with nylon screws, and a torque wrench was used to apply 0.1 J (1 in. · lb) of torque to each bolt.

The EMI protection requirement is that all faying surfaces maintain a resistance that does not exceed 2.5 mΩ after environmental exposure. A four-point probe test fixture (Fig. 20) was designed to measure the faying surface conductivity. The outer probes carried 1 A of current, and the inner probes were used to measure the voltage drop. For all measurements, the current was reversed, and the resistance values were averaged.

Five of the prepared specimens were submitted to humidity testing (95% relative humidity at 49 °C, or 120 °F, for 10 days), and one was kept as a reference. The specimens were periodically removed for visual inspection and, after retorquing, were measured for surface resistance. The surface resistance measurements for the specimens are shown in Table 1. The condition of sample 1, which was typical of all the samples, after the 10-day humidity exposure is shown in Fig. 21.

The five samples were disassembled and cleaned with distilled water in preparation for a salt fog test. The chamber was set up in accordance with ASTM B 117. A 5% salt solution of 35 °C (95 °F) and a condensation rate of 1 to 2 mils/h/80 cm inside the chamber were used to create the salt fog atmosphere. The specimen plates were bolted together again, as during the humidity test, and exposed to the salt fog for 336 h, or 14 days. The samples were removed periodically for visual inspection and electrical testing. The samples were washed with distilled water, dried for 2 h, and retorqued before being electrically tested. After testing, the samples were then returned to the chamber.

The results of the electrical testing are shown in Table 2. All five samples exhibited signs of corrosion after 144 h. The number 4 and 5 samples exhibited severe corrosion (Fig. 22a). The surface resistances of these samples were higher than the other samples, but still below the maximum of 2.5 mΩ. After 336 h in the salt fog, samples 4 and 5 again exhibited severe corrosion (Fig. 22b). Only the surface resistance of sample 4 was above 2.5 mΩ after 192 h. After electrical testing, the samples were disassembled. The over-

Table 2 Faying surface resistance of plates exposed to salt fog

Sample number	Average surface resistance at 1 A, μΩ: Initial	After 24 h	After 48 h	After 96 h	After 192 h	Final 240 h
1	198	155	143	203	274	277
2	212	136	134	211	255	260
3	273	162	152	158	155	161
4	220	176	200	799	15 369	20 727
5	198	285	298	332	400	632

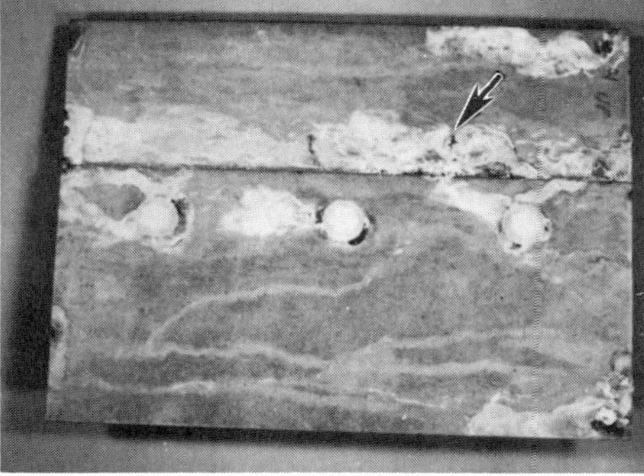

(a)

(b)

Fig. 22 Severely corroded test panel (sample 4) after 144 h of salt fog testing (a) and after 336 h of salt fog testing (b)

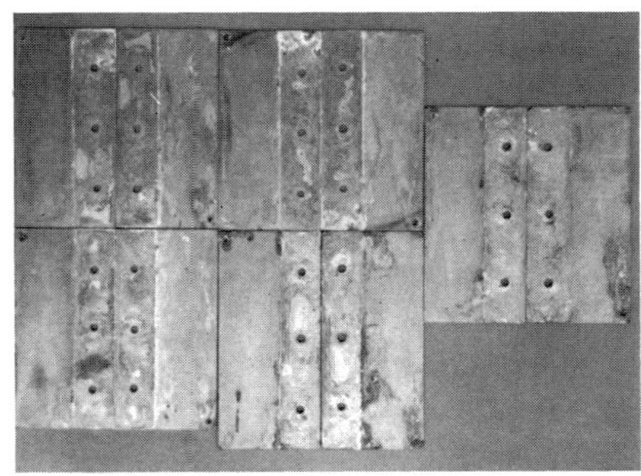

(a)

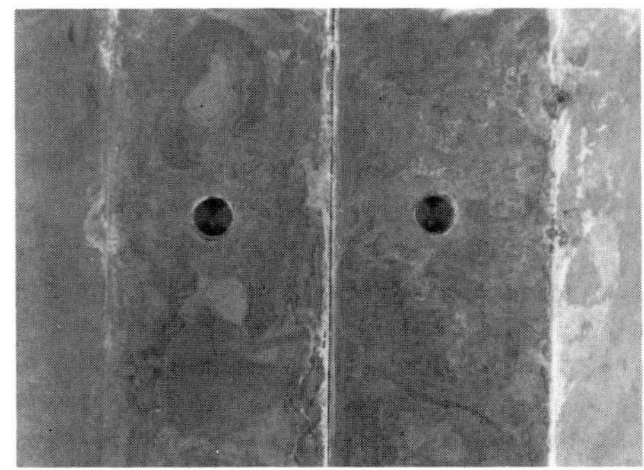

(b)

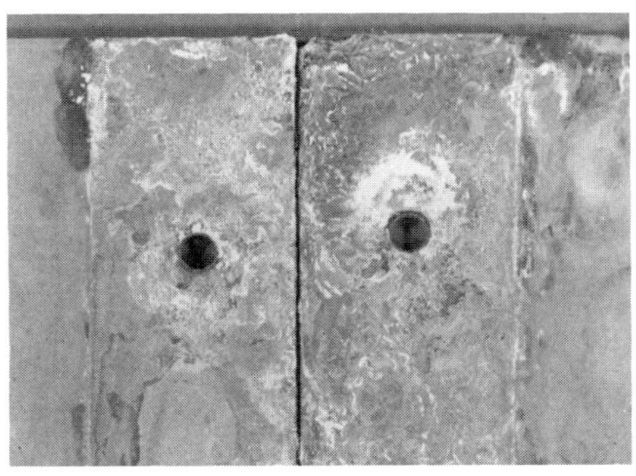

(c)

Fig. 23 Surfaces (a) of all samples after 336 h of salt fog testing. (b) Surface of the least corroded sample after 336 h. (c) Surface of the most corroded sample after 336 h

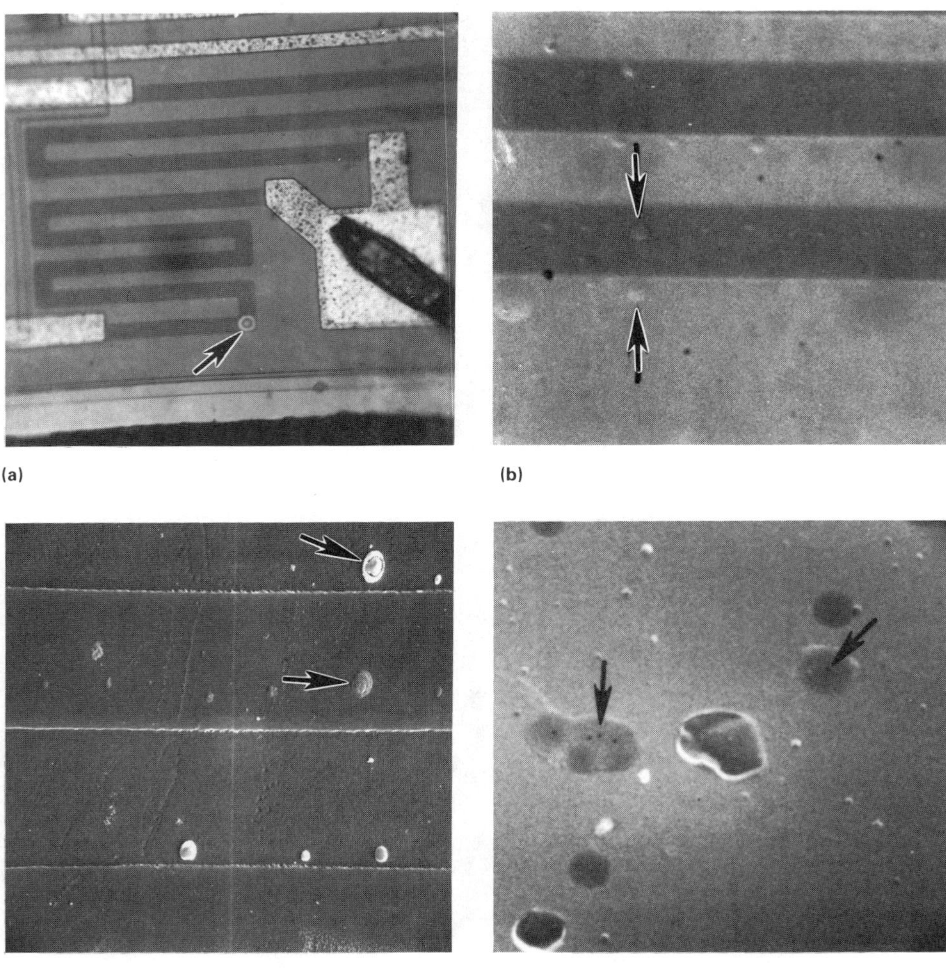

Fig. 24 Thin-film chromium resistor (a) in an integrated circuit showing location of failure (arrow). (b) Close-up of contaminated thin-film chromium resistor. Arrows mark contamination under glass film. (c) SEM micrograph of area shown in (b) with protective glass removed. Arrows mark the same contamination particles as those shown in (b). (d) Holes in thin-film chromium resistor (arrows)

Fig. 25 Thin-film tantalum nitride resistor chips (a) in a failed photodetector. Arrows locate the resistor chips. (b) Open tantalum nitride resistor (arrow) between aluminum contacts. 170×. (c) Erosion along the edges of thin-film tantalum nitride resistors. 170×

lapping surfaces of all samples exhibited corrosion (Fig. 23a); Fig. 23(b) and (c) compare the samples with the least and most corrosion.

The nickel/boron-coated aluminum plates passed the humidity testing, but they did not perform as well in the salt fog testing. Corrosion was most severe when a break occurred in the nickel/boron coating and allowed the salt solution to penetrate into the coating/aluminum interface. The anodic relationship of aluminum to nickel resulted in pitting corrosion of the aluminum and in the formation of aluminum oxide. Aluminum oxide is an excellent insulator and will significantly increase the surface resistance of the panel. The plating should be free of surface imperfections and protected from scratches that could break the coating and result in corrosion.

Integrated circuits are highly susceptible to corrosion-induced failure because of their geometries and high densities.

Thin-film chromium resistors in integrated circuits (Fig. 24) were found to fail frequently at the site marked by the arrow in Fig. 24. The package moisture content was analyzed and found to be as high as a few percent in a volume of about 0.5 cm^3 (0.03 in.^{3}). The thin-film resistors were found to be contaminated. Figure 24(b) shows a resistor with the protective glass coat on the integrated circuit. The arrows mark two particles of contamination. Figure 24(c) shows the same two particles of contamination shown in Fig. 24(b), but with the glass coat etched off. The contamination and moisture caused corrosion to occur in the chromium. Figure 24(d) shows holes, marked by the arrows, in the chromium. These holes may be mobile with high current densities and may cluster together. Regardless, the corrosion enlarges the size of the hole until the cross-sectional area of the chromium becomes too small to carry the required current. At this point, the chromium vaporizes and condenses on the glass coating, as indicated by the arrow in Fig. 24(a). This problem is eliminated when the sources of contamination are removed from the resistor and the package moisture content is reduced below a few tenths of a percent.

Thin-Film Resistor in a Photodetector. A photodetector was found to fail because the thin-film resistors failed. Figure 25(a) shows the location of the resistor chips on the substrate. The resistors were found to be tantalum nitride with aluminum metallization contact overlays; one of the resistors was completely open (Fig. 25b). There was no protective glass coating on the resistor chips. An Auger spectrum revealed that the resistors were contaminated with phosphorus. The package moisture content was found to be as high as 6%. The high moisture content and the contamination caused the corrosion of the resistor. The disappearance of the metal along the resistor edges is evident in Fig. 25(c).

Hybrid microcircuits are subject to corrosion-induced failures if packages are not properly sealed or are otherwise contaminated. Two such failures will be discussed.

A hybrid digital-to-analog (D/A) converter failed because of corrosion caused by high moisture levels and surface contamination. Figure 26(a) shows the voltage reference integrated circuit of the converter. The arrows mark damage

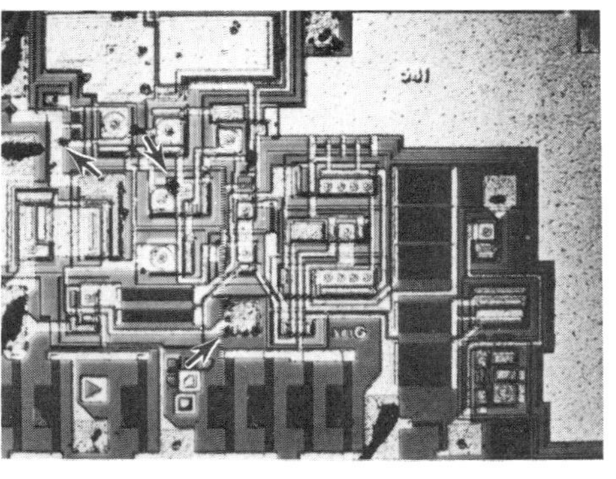

(a)

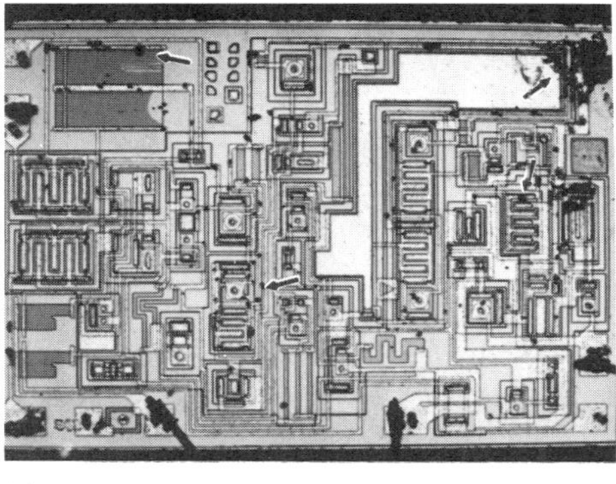

(b)

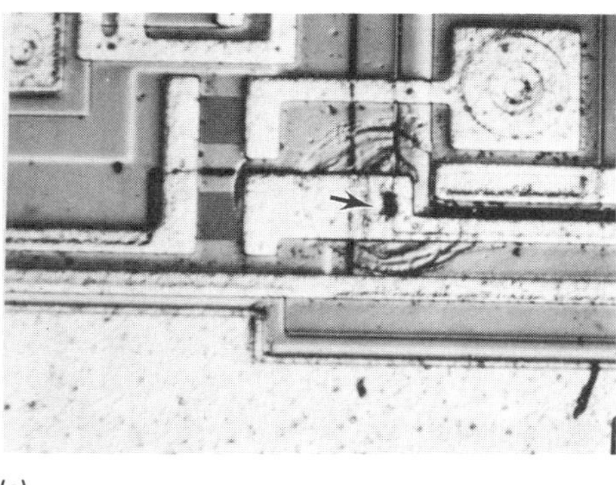

(c)

Fig. 26 Failed voltage reference integrated circuit (a) in hybrid D/A converter. Arrows mark corrosion sites. (b) Failed second operational amplifier integrated circuit from hybrid D/A converter. Arrows show corrosion sites. (c) Enlargement of one of the damage sites shown in (b)

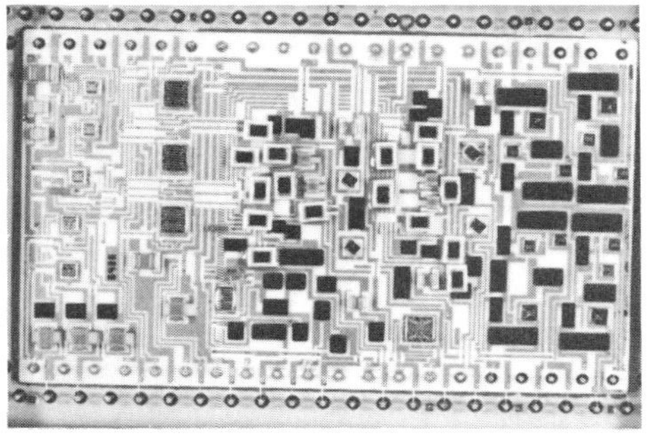

(a)

(b)

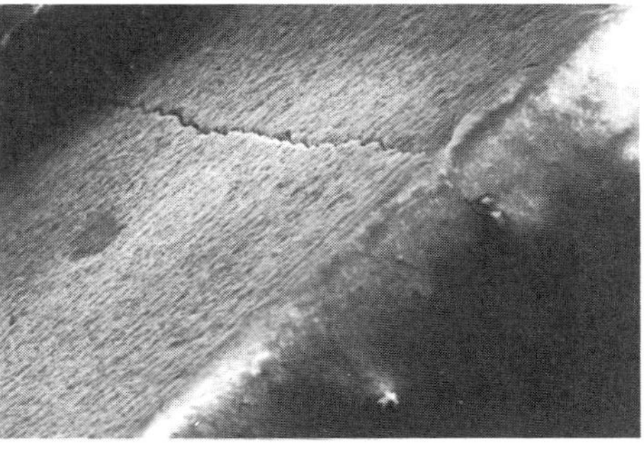

(c)

Fig. 27 Hybrid microcircuit (a) from an amplifier. (b) Bridging between conductors on the microcircuit shown in (a). (c) SEM micrograph of cracked ceramic chip conductor from microcircuit. 80×

sites on the aluminum metallization. Figure 26(b) shows the second operational amplifier. The arrows mark the damage sites. The wire bond damage can be seen in the upper right. Figure 26(c) shows an enlargement of the damage site on the operational amplifier. The surface contamination was identified by characteristic x-rays to be chlorine, sodium, and potassium. Corrosion will be minimized if the contamination is eliminated and if the moisture content is reduced.

Amplifier hybrid microcircuits (Fig. 27a) were analyzed to determine the cause of failure. The gases in the package were analyzed and found to contain high moisture levels as well as NH_3. This environment and some contamination caused the bridging between conductors shown in Fig. 27(b). Many of the chip capacitors were cracked because of poor handling techniques (Fig. 27c). When moisture penetrated this crack into the capacitor interior, significant changes occurred in the value of the capacitance. These failure modes can be minimized by ensuring that the package is a hermetic seal, that the gas contained in the package is dry, that the chip capacitors are not broken, and that the amount of epoxy used inside the package is small.

Klystron electron tubes were failing because of corroded output connectors (Fig. 28). There were some white and green corrosion residues on the connector. With characteristic x-ray analysis,

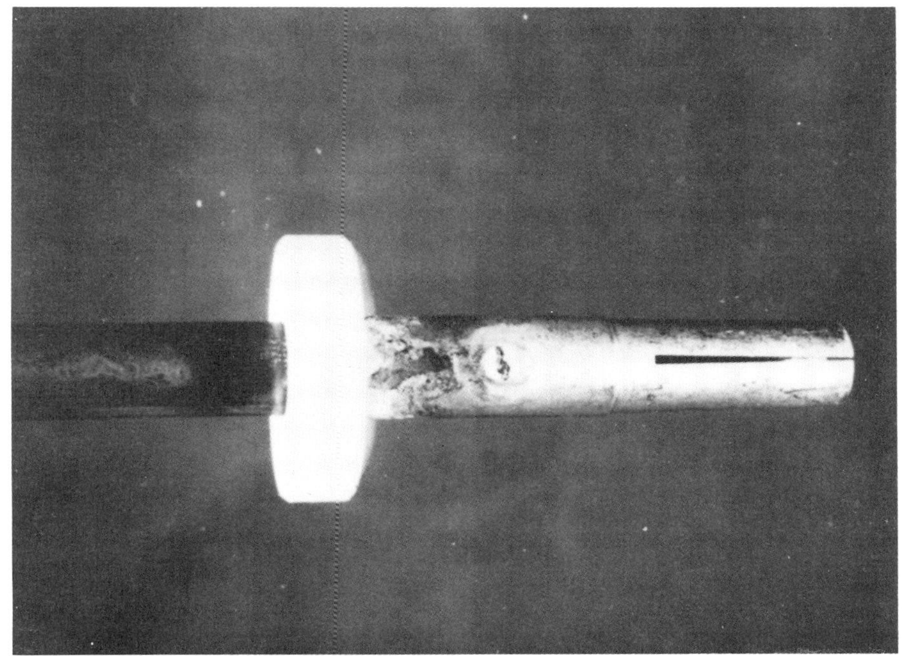

Fig. 28 Corroded output connector from a failed klystron electron tube

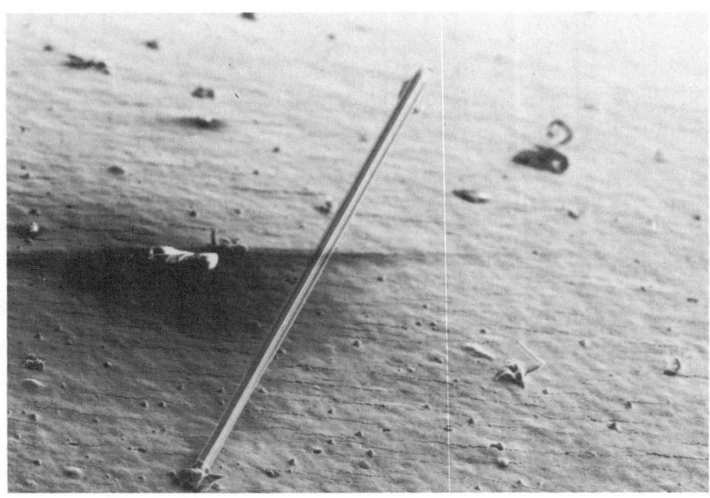

(a)

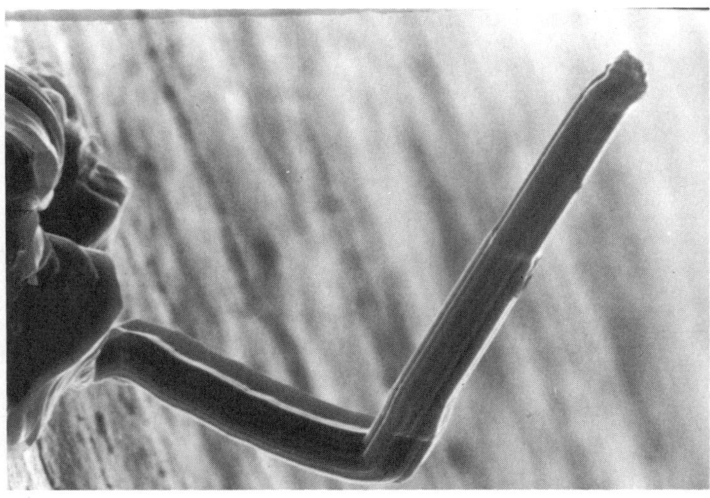

(b)

Fig. 29 SEM micrographs of tin whiskers on the interior surface of a hybrid device lid. (a) 245×. (b) 3280×

the white residue was identified as a form of tin chloride. The green corrosion was found to contain copper, zinc, and chlorine. The connector was found to be copper plated with silver, followed with a thin plate of gold. The solder used to join the connector pieces was tin, and it appeared that the solder flux was responsible for the chloride contamination. Both the gold plating and the silver plating were found to be thin and porous. It was recommended that the solder flux be cleaned from the connector and that the plating thicknesses be increased so that porosity is eliminated.

Tin Whiskers. Certain hybrid packages were found to have many cases of intermittent operation and failures that could not be duplicated upon retesting. In an investigation of these devices, it was found that whiskerlike growths were extending from the interior surface of the hybrid package lid. The lid was a Kovar metal that had been tin plated and then soldered with a tin/lead solder to the substrate. The whiskerlike growths (Fig. 29) were suspected of shorting out parts of the circuits during certain times, then moving to a different position.

The actual cause of the whisker growth is unknown, but it is suspected to be an electrochemical phenomenon similar to concentration cell corrosion. It is known that the whiskers do not occur if the tin plate is fused or if tin/lead solder is plated instead of tin.

Corrosion in Telephone Cable Plants

George Schick, Bell Communications Research, Inc.

THE TELEPHONE CABLE PLANT consists of many different metals and alloys that may be directly exposed to the environment or may become partially exposed because mechanical, electrical, or biological effects, such as dig-ins, lightning, or termites, respectively, have damaged their coating or jacketing. This article will describe the components of the telephone plant where corrosion can take place and will discuss how the environment interacts with such components. This article will also describe various causes of corrosion and their prevention in the telephone plant. Finally, corrosion case histories and their solutions will be described in the aerial, underground, and buried plants.

Metallic Components of the Telephone Cable Plant

The telephone cable plant consists of many materials. In this section, the various metals that most often pose a corrosion problem and the typical locations in the plant where these problems occur will be identified.

Although only small amounts of new lead-sheathed cable are placed in the telephone plant, it still represents a substantial percentage of in-place cables, especially in the underground plant. The amount of polyethylene-jacketed cables is increasing. These cables contain aluminum or aluminum and steel shields. The multipair cables contain an enormous amount of copper. Some plastic-jacketed cables crossing rivers, lakes, or seas are mechanically protected with bare, jute and tar covered, or neoprene-jacketed galvanized steel armor wires.

Where the ends of two cables are joined or where part of a large cable branches off, the connections (splices) are enclosed in closures usually made of galvanized cast iron held together with stainless steel fasteners (aerial, underground, and buried plant), tin-coated steel (buried plant), aluminum (aerial plant), lead (aerial, underground, and buried plant), stainless steel (underground plant), or plastic (aerial, underground, and buried plant). Coaxial cables in the underground plant have tin-coated brass terminals. In the buried plant, metallic closures are protected with field-applicable coverings; in the underground plant, they are usually left bare. Carrier systems, which require signal amplification with repeaters, include apparatus cases of galvanized steel and galvanized cast iron held together with a stainless steel band clamp; repeater housings made of plastics are joined with stainless steel fasteners and may have tin-coated brass inserts (valves) for gas pressurization.

Support hardware in manholes, guy wires, rods, and anchors and most aerial strands are composed of galvanized steel. The wires used for lashing cables to aerial strands are made of galvanized steel or stainless steel. Some of the aerial strands have aluminum coatings; until about 30 years ago, telephone companies used Monel support hardware in very corrosive manholes.

Grounding systems use a variety of metals. Copper, copper-clad steel, galvanized steel, and stainless steel are used in rod or wire form to contact the earth, and copper (bare or insulated) wires, copper, and tin-coated copper ribbons or braids serve as conductors between the plant and the components that contact the earth. Connectors and clamps, depending on the area and type of exposure, are of copper, brass (tin coated or bare), bronze (underground and buried plants), steel, galvanized steel, stainless steel, and aluminum (for exposure only to the atmosphere). Protector devices contain brass components that are exposed to the atmosphere because their plastic or aluminum housings require occasional reentry; thus, they are not sealed against moist and polluted air.

Access to the underground plant is provided with precast or cast-in-place concrete manholes. Because wet concrete is a conductive medium, the steel reinforcing bars of these manholes are indirectly exposed to the environment.

Interaction Between Telephone Cable Plants and the Environment

In this section, only the chemical effects of the environment and its pollutants or additives will be considered in general terms. Under non-extreme conditions (absence of stray dc currents, specific extremely corrosive chemicals, and strong bacteriological colonies), directly exposed metals and alloys often fare better, because their selection criteria include corrosion resistance. Coated or jacketed metals and alloys rely on separation from the environment for corrosion protection. Man-made corrodents, design effects, and special environmental effects, such as stray currents, galvanic couples, and bacterial corrosion, respectively, will be discussed in the section "Causes of Corrosion" in this article.

Aerial Plant. The three most distinguishable environments to which the aerial plant can be exposed are industrial, marine, and rural. There are ever-increasing areas where these environments overlap. For example, some industrial complexes are located very close to the seashore, creating marine-industrial atmospheres. In suburban areas, pollution by automobile exhaust systems increases the corrosivity of the otherwise rural atmosphere, and the acid rain and chemicals used for fertilizers in rural areas can create corrosive conditions.

Industrial Atmospheres. The primary pollutants in the industrial atmosphere are carbon dioxide (CO_2) and sulfur dioxide (SO_2), but it can also contain nitrogen dioxide (NO_2), hydrogen sulfide (H_2S), chlorine (Cl_2), and ammonium compounds. These pollutants create acidic compounds that readily attack zinc and steel. Sulfur dioxide and NO_2 are also oxidizing agents, and together with O_2 in the air, they maintain the passive film on aluminum. Thus, aluminum or aluminum-coated steel may corrode less than galvanized steel in some industrial environments. Chemical plants may create extremely aggressive environments in their vicinity. For example, a plant that produces Cl_2 gas from salt brine electrolysis may occasionally release Cl_2 into the atmosphere. Because Cl_2 in contact with moisture forms hydrochloric acid (HCl), the telephone plant at such a location can experience corrosion failures of galvanized steel, discontinuity of aluminum shields at the non-hermetically sealed splice closures of polyethylene-jacketed cables, and advanced corrosion of brass binding posts, which may in turn lead to current leakage and noise.

Marine atmospheres can deposit airborne salt solution droplets, which generally attack galvanized steel. At seacoastal areas where fog frequently causes condensation of a thin moisture layer on the aerial plant, the corrosive conditions increase because the chloride salt deposit is reactivated by the moisture, and the moisture is saturated with air, which acts as a cathodic depolarizer, thus increasing the rate of corrosion. In such areas, even 400-series stainless steel lashing wire may corrode. Chloride salts can also corrode aluminum. This corrosion is usually localized, and it causes pitting. Marine atmospheres can affect copper and copper-base alloys, but their rates of corrosion are low compared to that of galvanized steel or aluminum.

Rural atmospheres, where the only corrosive medium is rain or moisture condensate containing CO_2, are the least aggressive to the aerial plant. Corrosion, if any, usually appears in the form of dull haze on galvanized steel, aluminum, lead, and copper or brass surfaces. Areas affected by acid rain may exhibit a similarity to industrial atmosphere.

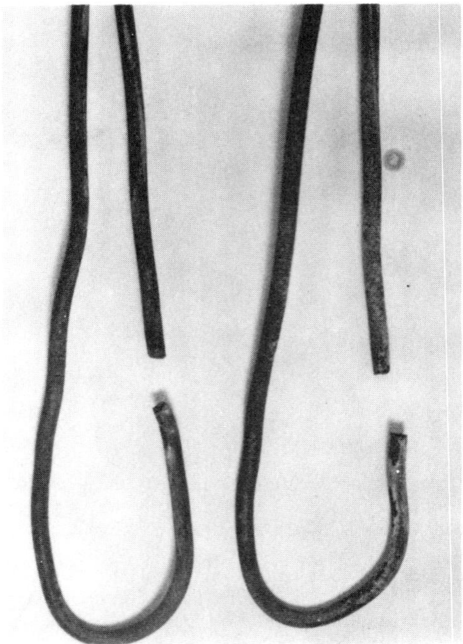

Fig. 1 Broken copper tail wires of an aerial clamp

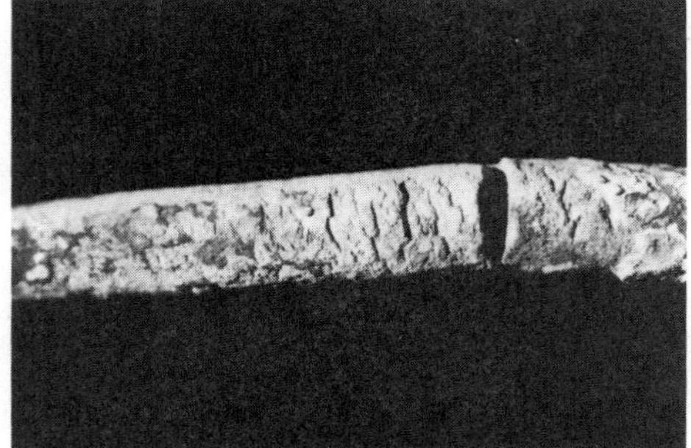

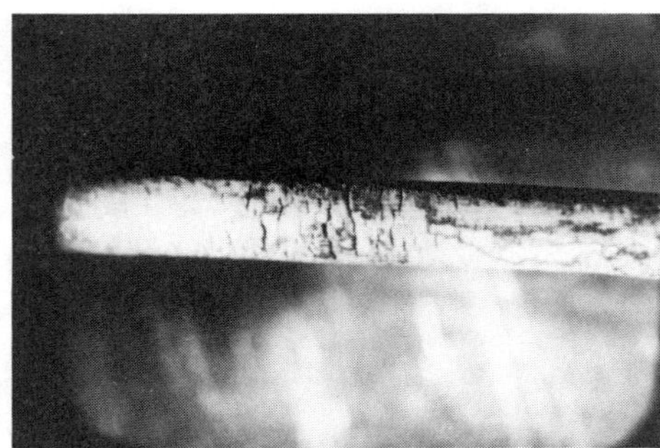

Fig. 2 Tail wires near failures. 3×

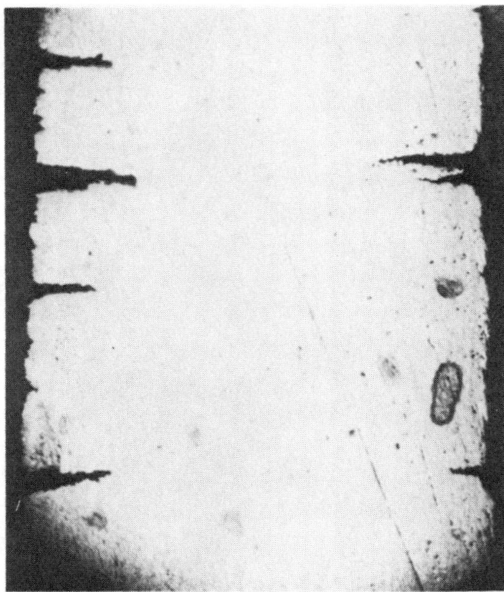

Fig. 3 Photomicrograph indicating many closely spaced cracks near a failure. 14×

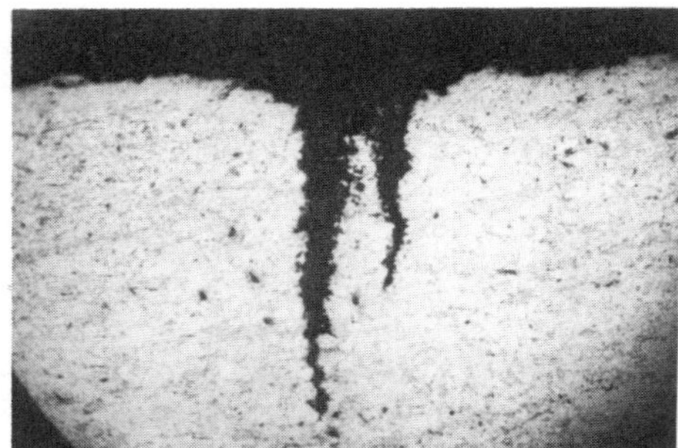

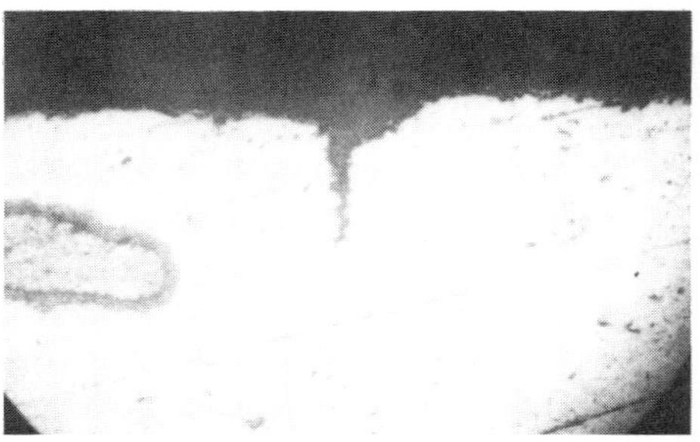

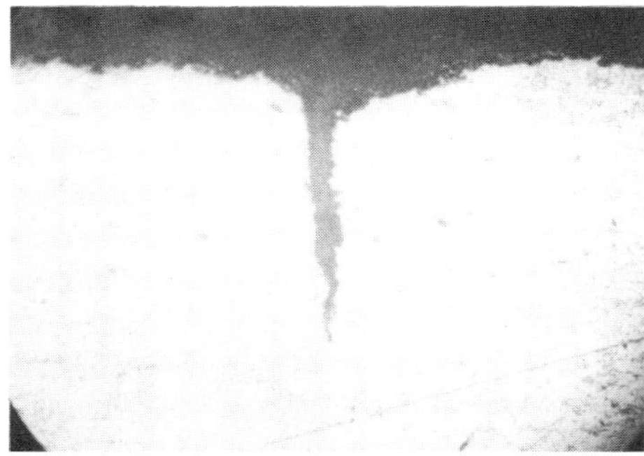

Fig. 4 Photomicrographs of transgranular cracks on copper tail wires. The cracks are initiated at the bottoms of the pits. 60×

The underground plant contains cables and hardware placed in conduits and manholes. Although new conduits are made of plastics, such as polyvinyl chloride (PVC) and polyethylene, the majority of those in the field are made of clay (old conduits), concrete, and fiber-cement (transite), all of which conduct electric current when permeated with moisture. The manholes are usually precast or cast-in-place concrete, both of which are reinforced with a network of steel bars. Older manholes are made of brick without steel reinforcing bars. Concrete, when permeated by moisture, also becomes a conductive medium. Because fresh concrete is alkaline, it passivates the reinforcing bar surfaces, thus rendering them more corrosion resistant than before passivation.

A large percentage of manholes are fully, partially, or seasonally flooded, and the waters contain soluble salts of the native soil. Some of these salts may be corrosive. Conduits may also contain mud and silt that can cause localized corrosion of lead cable sheaths. Manholes located close to the sea may be flooded with seawater, which is corrosive to all bare metallic components. Conduits and manholes may also contain man-made pollutants, such as deicing salts, industrial effluents, and fertilizers. Fertilizers containing ammonium compounds can cause stress-corrosion cracking (SCC) of stressed brass components. It is interesting to note that in the past, when cable ducts were made of wood, many lead cable sheaths failed by corrosion. Investigations indicated that acetic acid (CH_3COOH) leached from the wood caused these failures.

The buried plant is directly exposed to the earth, although some of the cables may be pulled into buried plastic conduits. Some of the older lead-sheathed cables are exposed to a wide variety of salts in various concentrations and to soils of widely varying pH. The soil pH in the United States ranges from 2.3 (acidic) to 9.5 (alkaline). Local conditions may cause even wider variations; for example, cinders may be strongly acidic. The buried plant can also be exposed to man-made corrodents, such as deicing salts, industrial effluents, and fertilizers. In addition to lead-sheathed cables, bare galvanized steel mounting posts of pedestal terminals and ground rods in the buried plant are exposed to the soils.

Causes of Corrosion

In this section, the corrosion-accelerating effects of electric currents, dissimilar metals, variations in the environment, combination of stresses and corrodents, and bacterial effects will be discussed. The various built-in and externally applied corrosion control measures will be discussed in the section "Corrosion Prevention" in this article.

Stray Currents and Interference Currents. Stray currents and cathodic protection interference currents create the most aggressive corrosive environment for the underground and buried telephone plants and for some components of the aerial plant, such as ground rods, guy rods, and anchors. Stray currents originate from grounded dc-powered systems, such as dc rail transportation (trains and subways), mining operations, welding plants, and high-voltage dc power transmission. The direct current from stray current

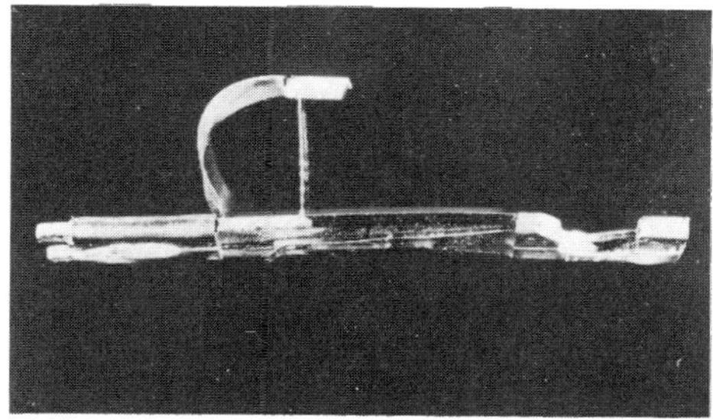

(a)

(b)

Fig. 5 Nickel-silver aerial plant fuses. (a) New exposed fuse. (b) Fuse that failed because of SCC

Fig. 6 Field failures of type 410 stainless steel bolts

sources usually travels through the environment and enters the telephone plant where its resistance to ground is low, for example, central offices, manholes, or bare lead cable sheaths. The current enters the environment from the telephone plant near its source, seeking a path of least resistance back to its origin. The location where the current leaves the telephone plant to enter the surrounding environment is where the plant corrodes. Inside multipair cables (not gas pressurized or filled) where water can enter the cable core and where small defects in the polyethylene wire insulation expose bare copper, the central office battery can provide driving voltage for corrosive current flow between wires that have a 48-V potential difference.

Interference currents originate at impressed-current cathodic protection installations of other (foreign) underground structures. The telephone plant can pick up such currents by crossing or closely paralleling the foreign structure, for example, gas pipelines. A potential gradient field exists around the impressed-current anode (anode field), with the potential being the highest (most positive) at the anode and with the current flowing away from the anode. The current flowing through the earth toward the protected structure creates a cathode field around it and renders the structure negative. Where the telephone plant passes through the anode field, the interference current enters the plant near the anode, flows on the cable in two opposite directions, and leaves (causes corrosion) the plant at two different locations some distance away, depending on the extent of the anode field and the resistance of the plant to the ground. When the telephone plant passes through a cathode field, it collects stray currents from extensive areas, and the current enters the environment (causes corrosion) at an area near the connecting point of the foreign plant to the negative output of the rectifier.

Galvanic Effects. Corrosive galvanic couples are unavoidable in the telephone cable plant. Although bare metallic components are necessary to provide electric safety through grounding, metals at various positions on the galvanic scale are used because of strength, conductivity, formability, and cost requirements. Galvanic corrosion may also take place between the telephone plant and other grounded, foreign metallic structures, such as copper or copper-clad ground rods.

In the aerial plant, galvanic corrosion is not a primary concern. This is because the corrosive environments exist in the form of thin moisture layers; thus, the electrolyte represents a relatively high-resistance path. Galvanic corrosion is expected in industrial and seacoastal environments where lead-sheathed cable is lashed to galvanized steel strand with galvanized steel or stainless steel wire. Corrosion caused by the bimetallic coupling is restricted to the vicinity of the contact area.

In the underground plant, the electrolyte is contaminated water, mud, and silt in manholes and conduits. The plant components that corrode the most, as a result of bimetallic coupling, are galvanized steel support hardware and galvanized steel or galvanized cast iron closures. The corrosion-accelerating cathodes are steel manhole reinforcing bars (passivated by the alkalinity of the concrete), lead cable sheath, copper-base alloy (brass, bronze) closures and hardware, stainless steel closures and fasteners, and tinned or bare copper bonding conductors. Copper pipes or grounding conductors in foreign plants bonded to the telephone plant may also act as corrosion-accelerating cathodes.

In the buried plant, lead cable sheath or galvanized steel pedestal mounting posts can galvanically corrode through direct or indirect contact to copper ground rods, grounding conductors, or power cable concentric neutrals. Although, strictly speaking, galvanized steel guy rods and anchors are part of the aerial plant, they are directly

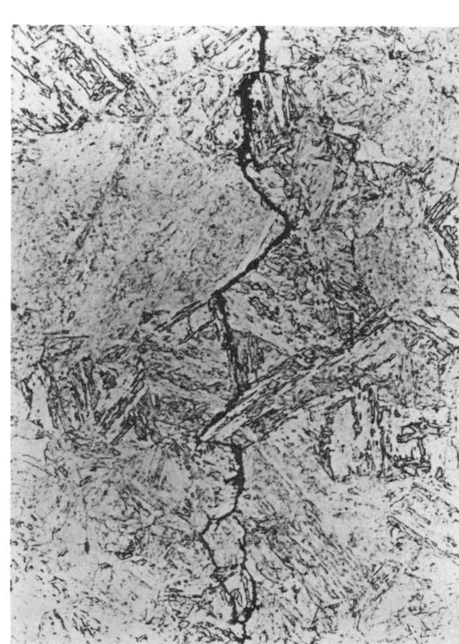

Fig. 7 Section of type 410 stainless steel bolt. The bolt failed after 3 months of service in a flooded manhole. 180×

buried in the soil. Copper and copper-clad ground rods of the power company cable plant may be indirectly connected to the rods and anchors through bond wires and cable shields. Such bimetallic coupling can cause galvanic corrosion of the galvanized steel.

Concentration Cell Effects. Bare lead sheath cables in conduits that contain water usually corrode by differential aeration cells where the cables are in contact with the conduit wall. The presence of mud or silt in the conduit further

Fig. 8 Hydrogen-stress cracking of type 410 stainless steel bolts. (a) Quenched from 1010 to 65 °C (1850 to 150 °F) in oil, then tempered at 535 °C (1000 °F) for 1 h. Bolt exposed in 5% H_2SO_4 solution and 1 mL/L Rodine inhibitor as cathode with platinum anode. Applied current: 180 mA/2 bolts. (b) Quenched from 1010 to 65 °C (1850 to 150 °F) in oil, then tempered at 425 °C (800 °F) for 1 h. Bolt exposed in 5% H_2SO_4 and 1 mL/L Rodine inhibitor as cathode with platinum anode. Applied current: 180 mA/2 bolts. (c) Bolts as-received exposed in 5% H_2SO_4 and 1 mL/L Rodine inhibitor as cathode with platinum anode. Applied current: 180 mA/2 bolts

enhances this effect. Directly buried cables and other buried structures, such as fuel storage tanks, compact the soil below them more than at their tops or sides. Because compacted soil contains less dissolved oxygen than loose soil, differential aeration cells develop under these conditions. Concentration cell corrosion can also take place where a bare lead cable sheath passes through soils of various composition or where ground rods or guy rods pass through soil layers of different compositions.

Stress-Corrosion Cracking. Although many telephone plant components have residual manufacturing stresses or applied stresses, they are seldom exposed in environments that induce SCC. The most notable exceptions to this rule are the stressed copper-base alloys. Low concentration of ammonia or ammonium compounds present in industrial pollutants or in fertilizers have caused SCC of copper-base alloys. Most of these failures involved yellow brass (65Cu-35Zn) or cartridge brass (70Cu-30Zn).

Hydrogen-stress cracking usually takes place where hydrogen enters the stressed metal. In the telephone cable plant, this phenomenon was found on some stainless steel fasteners in contact with galvanized steel or galvanized cast iron. The galvanic coupling generated the hydrogen on the stainless steel cathodes.

Corrosion Fatigue. The aerial cable plant is exposed to winds and, as a result, to cyclic movement. This movement exerts cyclic stresses. The copper tail wire of aerial clamps exposed to cyclic stresses in a marine atmosphere can experience corrosion fatigue failure.

Bacterial Corrosion. The most frequently encountered bacterial corrosion in the telephone cable plant is caused by sulfate-reducing bacteria. Such bacteria can cause corrosion of lead cable sheath in anaerobic soils (dense clay-type soils) in the pH range of 5.5 to 8.5. The same bacteria can also thrive in stagnant manhole water and corrode the galvanized steel support hardware. The bacteria reduces inorganic sulfates to sulfides when organic matter is present according to the following reactions:

At the anode $4Fe \rightarrow 4Fe^{2+} + 8e^-$

At the cathode $8H_2O \rightarrow 8H$ (adsorbed on the metal) $+ 8OH^- - 8e^-$

$$8H \text{ (adsorbed)} + Na_2SO_4 \xrightarrow{\text{bacteria}} 4H_2O + Na_2S$$

$$Na_2S + 2H_2CO_3 \rightarrow 2NaHCO_3 + H_2S$$

The overall reaction is $4Fe + 2H_2O + Na_2SO_4 + 2H_2CO_3 \rightarrow 3Fe(OH)_2 + FeS + 2NaHCO_3$.

Corrosion Prevention

In general, telephone cable plant corrosion-preventive techniques fall into two major categories: those incorporated into the design and those applied in the field when conditions require them.

Corrosion Prevention Incorporated in the Design. Most new cables placed in service since the 1950s are polyethylene jacketed. The polyethylene outer jacket reduces the stray and interference current pickup and discharge areas and protects the cables from chemicals, such as organic acids and nitrates, as well as galvanic and concentration cell effects that attack the lead cable sheath. In the aerial plant, self-supporting cable encloses both the cable and support strand in polyethylene, and a continuous polyethylene web holds the two together, describing a figure 8 cross section. This construction protects both cable and strand from the environment and precludes the need for lashing wire, thus eliminating a galvanic couple between galvanized steel and stainless steel. Many splice closures and apparatus cases are also made of noncorroding plastics with stainless steel fasteners and generally corrosion-resistant tin-coated copper-base alloy inserts, for example, pressure valves. As a second line of defense, the aluminum and steel cable shields, under the polyethylene outer jackets, are covered with flooding compounds that can seal small defects on the polyethylene outer jacket and prevent moisture entry. Further protection of polyethylene-insulated wire pairs in the cable core is accomplished with a gel-type filling compound that occupies the space between the wires and prevents moisture entry. Similarly, filling compounds are also used in some splice closures to prevent the corrosion of splices.

Metallic coatings are also used for corrosion protection of the telephone cable plant. Practically all support hardware is made of galvanized steel. The thickness of galvanizing depends on the severity of the exposure. For example, support strands have 205 to 285 g/m² (0.65 to 0.90 oz/ft²) of zinc (class A galvanizing) at mild, rural areas and 470 to 850 g/m² (1.5 to 2.7 oz/ft²) of zinc (class C galvanizing) at industrial and seacoastal areas. Because aluminum coating is several times more resistant in industrial atmospheres than zinc of the same thickness, aerial support strands

Fig. 9 Section of type 410 stainless steel bolt. Quenched from 1010 °C (1850 °F), then tempered at 425 °C (800 °F) for 1 h. Bolt (cathode) broke under a torque of 70 N · m (600 in. · lb) after being hydrogen charged in 5% H_2SO_4 solution. 185×

(a)

(b)

Fig. 10 Type 301 half-hard stainless steel band clamp illustrating a characteristic brittle hydrogen-stress crack. (a) Loop. (b) Hinge

are also used with aluminum coating in industrial environments (Ref 1).

Brass housings for terminals or transducers, brass valves, most of the copper bonding ribbons, and some of the copper-base alloy connectors used in manholes are tin coated. The coating is applied primarily to reduce the galvanic potential differences between the copper or copper-base alloys and the more active metals (galvanized steel, lead) in the manhole.

When the strength, forming, conductivity, or cost requirements override galvanic-corrosion considerations, mitigation measures must be applied to reduce the corrosive effect of bimetallic couples. One such method is to design the component made of the more noble metal (cathode) small as compared to the more active metal. Stainless steel bolts and nuts holding together galvanized cast iron splice cases or tin-coated brass valves in galvanized cast iron splice cases exemplify this method.

The steel reinforcing bars in precast or cast-in-place concrete manholes are passivated by the alkalinity of the concrete. Their potential can be as noble as −0.2 V versus a standard copper/copper (cupric) sulfate reference electrode. The potential of galvanized steel support hardware in the same manhole may be as active as −1.1 V versus the same reference. The surface area of the reinforcing bars can be considerably larger than that of the support hardware. These conditions can lead to galvanic corrosion of the galvanized steel in flooded manholes. At areas where the manhole reinforcing bars need not be part of the ground, an option is given to break the bimetallic couple. The reinforcing bar system is contacted through a welded steel stud in each manhole section, and short sections of tin-coated bonding ribbons are attached to them inside the manhole. The support hardware is mounted on the manhole wall through nonconductive bushings to avoid accidental direct contact with the reinforcing bars. This system can provide or eliminate contact between the reinforcing bars and the rest of the plant.

Materials selection and heat treatment can reduce susceptibility to SCC. Using annealed red brass (85Cu-15Zn) instead of stressed cartridge brass (70Cu-30Zn) and using commercial bronze (90Cu-10Zn) instead of yellow brass (65Cu-35Zn) where applied stress cannot be eliminated are good examples of these methods.

Corrosion Prevention for Existing Plants. Because the telephone cable plant corrodes the most under the influence of stray and interference currents, the most important preventive measures are to reduce such currents entering the plant and, if that is not possible, to provide a metallic path for these currents to return to their origin. Central offices usually represent low resistance to ground and thus serve as current pickup areas. Insulating joints that cut into the cable shields just before crossing through the wall of a cable entrance facility can stop such current pickup. Insulating joints can also be cut in cables where they enter other large buildings (low-resistance grounds) or, because the shield of an aerial cable can carry stray currents, on the riser poles connecting aerial to underground or aerial to buried plants. Strain insulators are often placed in the guy wires to prevent stray, interference, or galvanic currents from corroding guy anchors and rods.

In stray or interference current areas, a potential survey over the plant can establish the location(s) where such currents corrode the telephone plant. Current leaving the plant through the environment will render its potential more positive (less negative) than in the absence of such currents. Low-resistance drainage or interference bonds connecting the telephone plant to the system where the stray or interference current originates serve as a conducting path to avoid corrosion. Where the stray current originates at a dc transport system, the magnitude and direction of the current changes with time and may flow toward the telephone plant, causing corrosion at another location. In order to prevent this undesirable current flow, a reverse current switch is placed in the drainage bond, thus allowing the current to flow away from the telephone plant but blocking the current flow in the opposite direction.

Cathodic protection is used as a corrosion-reducing measure in stray and nonstray current areas. At stray or interference current locations, galvanic or impressed-current systems are used when bonds and insulating joints do not solve the problem. Care must be taken to avoid overprotecting the lead-sheathed cable. Lead that is polarized to a more negative potential than −1.5 V may be cathodically overprotected. The cathodic reaction generates a strongly alkaline environment at the lead surface. Lead, an amphoteric element, dissolves in strong alkalies; thus, cathodic overprotection leads to lead corrosion.

Most coated steel fuel tanks are cathodically protected. Small tanks that do not have the ground paved over them are protected with galvanic anodes. Large tanks, and where the surface is paved over the tank, are protected with impressed-current systems. In flooded manholes, the cathodic protection applied consists primarily of magnesium anodes.

The central office batteries are grounded at the positive terminal to prevent the corrosion of lead cable sheath and conductors at wire insulation defects in polyethylene-jacketed cable if and when water enters the cable core. Negative grounding would corrode lead cable sheath the same way as stray currents would. Inside the cable, one member of the wire pair is grounded together with the bare aluminum cable shield. The other conductor of the pair is connected to the −48 V terminal of the central office battery. If water enters the cable core, the corrosion currents will flow from a very large surface area of the shield (as compared to the size of an insulation defect on the wire), and the very low corrosion current density will cause negligible corrosion of the shield. Other field-applied corrosion control measures are wrapping galvanized cast iron splice closures or tin-coated steel sleeves with hot- or cold-applied tapes for buried plant use and protecting lead sheath cable segments with irradiated cross-linked heat-shrinkable polyethylene sleeves.

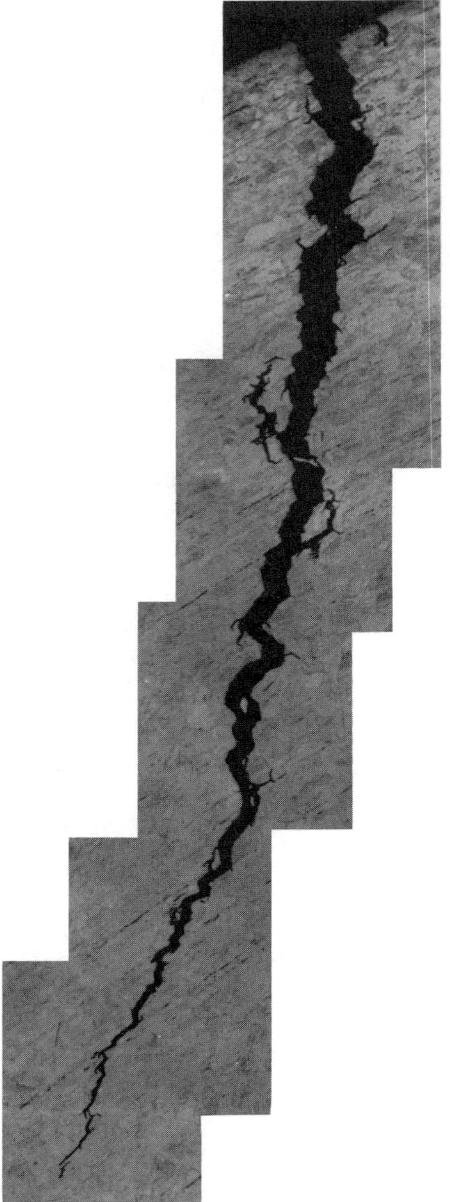

Fig. 11 Photomicrograph of a type 301 half-hard stainless steel band clamp loop field failure by hydrogen-stress cracking. 250×

Case Histories

Example 1: Corrosion Fatigue of the Tail Wire of an Aerial Clamp. The small cable (drop wire) providing service for individual subscribers from the aerial plant is held in place by a clamp made of a tin-coated brass body (attached to the cable) and a copper tail wire loop (attached to a galvanized steel hook or to a porcelain insulator). The tail wire is 2.6-mm (0.102-in.) diam annealed copper, and the clamp assembly must withstand a 2470-N (555-lb) load without breaking or slipping.

A number of these clamps, located a few hundred feet from the ocean, have failed (Fig. 1). The sharply broken wire indicated no weakening by abrasion. The combination of the corrosive action of saltwater spray and cyclic stresses arising from the "dancing" of aerial cables can cause failure by corrosion fatigue. The copper tail wire failures indicated characteristics that are generally associated with corrosion fatigue. The broken wires showed multiple cracks near the failure (Fig. 2 and 3). The cracks are transgranular and originate at the bases of pits (Fig. 4). These pits can act as stress points; therefore, the stress and number of stress cycles that cause corrosion fatigue failure of a metal are lower than the fatigue limit of the same metal without the corrosive environment.

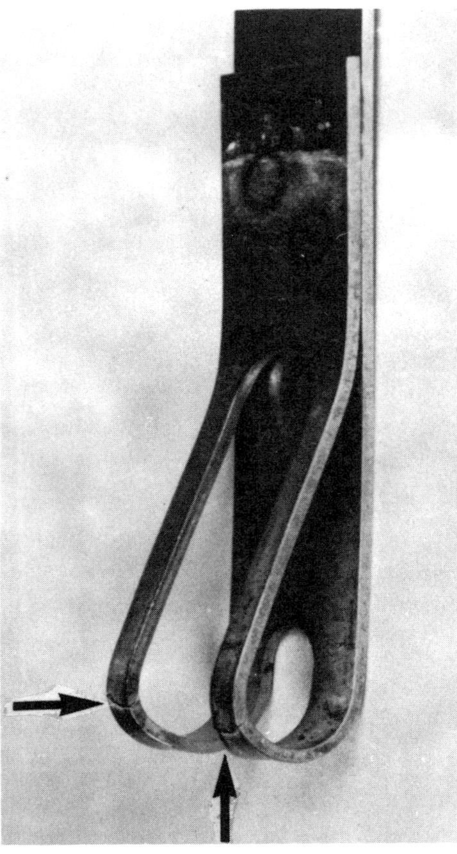

Fig. 12 Type 301 half-hard clamp loops that were hydrogen charged in diluted NaCl solution with platinum anode in the presence of MoS_2 lubricant for 5 days

Based on the circumstances of exposure and the characteristics of the cracks, it was diagnosed that the copper tail wire failures were due to corrosion fatigue. The solution to this problem was to change the tail wire material for direct seashore exposure from annealed copper to annealed Monel. The annealed Monel has a corrosion fatigue limit in saltwater that is 2.8 times higher than that of annealed copper (Ref 2).

Example 2: SCC of Aerial Plant Fuses. Several fuses, made of nickel silver (57 to 61% Cu, 11 to 13% Ni, bal Zn), exposed in central offices where the air contained industrial atmospheric contaminants that were particularly high in ammonium and nitrate ions failed by SCC (Fig. 5). Test solutions of 1 *N* ammonium nitrate (NH_4NO_3) and a 1:1 mixture of 1 *N* sodium nitrate ($NaNO_3$) and 1 *N* calcium nitrate ($Ca(NO_3)_2$) were prepared (NH_4^+, Na^+, and Ca^{2+} were the major cations in the pollutants). In addition, stressed fuses made of nickel silver and of cupro-nickel (80Cu-20Ni) were exposed to 1 drop of corrosive solution at the stressed area. All specimens were exposed for 42 days in a container in which the relative humidity was maintained at 76%.

All nickel silver specimens failed after 2 days of exposure to NH_4NO_3 solution. However, 17% of them failed and 67% showed crack initiation but no failure after 42 days of exposure to $NaNO_3$ + $Ca(NO_3)_2$ solution. None of the cupro-nickel specimens failed, but among those exposed to NH_4NO_3, 17% displayed crack initiation and 83% showed partial dealloying after 42 days. Based on the test results, the fuse material was changed from nickel silver to cupro-nickel, and this material change solved the SCC problem.

Example 3: Hydrogen-Stress Cracking of Type 410 Stainless Steel Splice Case Bolts. Type 410 stainless steel bolts were used to hold together galvanized gray cast iron splice case halves. Before installation, the bolts were treated with molybdenum disulfide (MoS_2) antiseize compound. The torque applied to these bolts can exceed 70 N · m (600 in. · lb). In flooded manholes, the galvanic-corrosion reaction between zinc (galvanizing) and stainless steel can generate hydrogen at the stainless steel cathode. The rate at which absorbed hydrogen on the cathode combines to form H_2 gas is affected by the catalytic properties of the metal surface. Good catalysts, such as iron, increase the rate of H_2 formation; poor catalysts do not. Catalyst poisons, such as sulfides, retard the formation of H_2, increase the concentration of atomic hydrogen on the cathode surface, and therefore facilitate the entrance of hydrogen atoms into the metal lattice, causing loss of ductility. In stressed high-strength alloys, such as type 410 stainless steel, they induce hydrogen-stress cracking (Ref 3). For the type 410 splice case bolts, the MoS_2 antiseize compound provided the catalyst poison.

Several failures of splice case bolts were discovered in flooded manholes after they were in service for 3 to 4 months (Fig. 6 and 7). Laboratory experiments were conducted to determine if the failure mode was hydrogen-stress cracking, if sulfides accelerate the failure, if heat treatment can improve the resistance against this failure mode, and if the type 305 austenitic stainless steel would serve as a replacement material.

The experiments were conducted in sulfuric acid (H_2SO_4) solution that contained 1 mL/L Rodine inhibitor. The test specimens were tightened in a high-strength steel fixture with 60 N · m (540 in. · lb) of torque. Three threads of each bolt were exposed to the solution, and the remainder of the assembly was covered with paraffin wax. The bolts (cathodes) were connected to the negative terminal of a battery through a current-limiting resistor against a platinum anode and cathodically charged with hydrogen. The time to failure was measured. All type 410 stainless steel bolts were quenched from 1010 to 65 °C (1850 to 150 °F) in oil, then tempered.

For the first series of experiments, all bolts were tempered at 370 °C (700 °F) for 1 h, and the test solution was 0.33 *N* H_2SO_4. The average failure time of these bolts was 22 ± 2 h. The experiment was repeated with 1 g/L sodium monosulfide (Na_2S) added to the solution. The bolts failed in 6.3 ± 2.5 h, indicating that the sulfide acts as a catalyst poison. For the next series of experiments, the solution was 5% H_2SO_4, and the bolts were tempered for 1 h at

370, 425, and 535 °C (700, 800, and 1000 °F) (Fig. 8 and 9). They failed after 16 ± 3 h, 20 ± 4 h, and 23 ± 4 h, respectively. These results indicate that tempering has only a limited effect on the failure time and does not solve the problem.

Finally, type 305 stainless steel bolts were tested, and they did not fail after 219 h under test conditions similar to those used to establish the effect of tempering. Based on these results, the solution to the hydrogen-stress cracking problem consists of changing the bolt from type 410 to 305 stainless steel, eliminating use of MoS_2, and limiting the torque to 60 N · m (540 in. · lb).

Example 4: Hydrogen-Stress Cracking of Type 301 Stainless Steel Clamp. When they have been severely cold worked, even austenitic stainless steels can hydrogen stress crack (Ref 4). Several type 301 stainless steel clamps, which were used to hold cylindrical galvanized steel covers to galvanized cast iron bases, failed in flooded manholes after 1 to 6 months of service. Like the type 410 stainless steel splice case bolts, they were treated with antiseize compound containing MoS_2. These clamps were made of half-hard type 301 stainless steel. Further stresses were introduced by forming small-diameter loops at the two ends of the clamp and by applying 30 N · m (260 in. · lb) of torque for clamping. Based on the conditions (the clamp is the cathode of a galvanic cell with zinc) and the brittle nature of the cracks, the failures were diagnosed as hydrogen-stress cracking (Fig. 10 and 11).

Laboratory experiments were conducted to substantiate the above diagnosis and to evaluate the effect of annealing and the hydrogen-stress cracking behavior of type 316 stainless steel. Half-hard type 301, annealed type 301, and annealed type 316 stainless steel clamp ends were placed in 0.1 *N* sodium chloride (NaCl) solution with or without MoS_2 treatment. Half of each specimen type was coupled to galvanized steel panels for 106 days, and the other half was connected as cathode to a 3.2-V potential source against platinum anodes for 5 days (Fig. 12).

One of the half-hard type 301 specimens that was MoS_2 treated and charged with hydrogen from an outside power source cracked after 5 days. All the other half-hard 301 specimens showed cracks after hydrogen charging regardless of the presence of MoS_2 or the method of exposure (coupled to a galvanized steel panel or connected to an outside potential source). When tensile tested, they all showed brittle fracture. The hydrogen charging reduced both the breaking load and the ductility, and MoS_2 treatment further reduced both. Although the annealed type 301 specimens exhibited ductile failure, they also indicated reduction of breaking load and ductility after hydrogen charging. Annealed hydrogen-charged type 316 specimens showed ductile failure, and had no adverse effect on hydrogen charging. The problem was solved by changing the clamp material from type 301 to type 316 stainless steel and by eliminating the MoS_2 antiseize compound.

Table 1 Corrosion of galvanized steel mounting posts

Environment: 11 200-Ω · cm soil; length of test: 1159 days

Number of mounting posts connected to the power cable	Average corrosion rate: Posts coupled to carbon black filled polyethylene-jacketed cable, μm/yr	mils/yr	Posts connected to cable with bare concentric neutral wires, μm/yr	mils/yr
1	254	10.00	41.1	1.62
2	193	7.58	23.6	0.93
3	169	6.65	21.3	0.84
4	159	6.25	23.1	0.91

Example 5: Corrosion Failures in Buried Plants. Corrosion failures of concentric power cable neutrals necessitated the development of corrosion control measures for the neutral wires. One measure is the application of a carbon black filled outer polyethylene jacket extruded over the concentric neutral wires. This jacket separates the corrosive environment from the neutral wires while ensuring continuous grounding of the plant. Telephone and power cables are often buried jointly in the same trench. For electric safety, the power and telephone plants are bonded together. Because the carbon black (approximately 35% by weight) filled polyethylene can act as a large cathode, it accelerates the corrosion rate of relatively small bare metallic components of the telephone cable plant, for example, galvanized steel mounting posts. This galvanic corrosion, although normally innocuous, can be particularly aggressive in porous and oxidizing soils in which oxygen or other oxidizing agents act as cathodic depolarizers (Ref 5).

Laboratory studies in 0.1 *N* NaCl solution predicted a penetration (corrosion) rate of 0.75 to 1.6 mm/yr (30 to 65 mils/yr) in galvanized steel when coupled to carbon black filled polyethylene-jacketed power cable at an anode/cathode area ratio of 0.015, and 50 to 150 μm/yr (2 to 6 mils/yr) when the same ratio was 1.0 (Ref 6). Similar experiments with 14-gage copper wire anode at a 0.0003 anode/cathode area ratio produced the failure of copper in 8 months. Field experiments conducted for 1159 days in 11 200-Ω · cm resistivity soil with galvanized steel pedestal mounting posts showed that the posts connected to carbon black filled polyethylene-jacketed power cable corroded at a rate 7.3 times higher than that of similar posts connected to bare concentric neutral-type power cable (Table 1). Mounting an additional galvanized steel post on the pedestal terminal and using solid, uninsulated No. 6 American Wire Gage copper conductor for bonding can solve this galvanic-corrosion problem.

REFERENCES

1. L.L. Shreir, *Corrosion*, Vol 2, John Wiley & Sons, 1963, p 14.24
2. H.H. Uhlig, *Corrosion and Corrosion Control*, 2nd ed., John Wiley & Sons, 1971, p 151
3. H.H. Uhlig, *Corrosion and Corrosion Control*, 2nd ed., John Wiley & Sons, 1971, p 48-50
4. H.H. Uhlig, *Corrosion and Corrosion Control*, 2nd ed., John Wiley & Sons, 1971, p 316
5. G. Schick, The Effect of Oxygen, Salt Concentration, and pH on Galvanic Cells Between a Carbon-Black Filled Polyethylene Cathode and Various Metal Anodes, *J. Test. Eval.*, Vol 8 (No. 3), May 1980, p 143-154
6. G. Schick, Galvanic Corrosion of Metals Coupled to Carbon-Black Filled Polyethylene, in *Corrosion and Corrosion Protection*, Vol 8, R.P. Frankenthal and F. Mansfield, Ed., Proceedings of an International Symposium Honoring Professor H.H. Uhlig on his 75th Birthday, The Electrochemical Society, 1981, p 259-265

Corrosion in the Chemical Processing Industry*

Chairman: Thomas F. Degnan, Consultant

THIS ARTICLE covers two groups of subjects: one concerns significant problems in the control of corrosion and protection of chemical-processing equipment, and the other concerns the handling and corrosive effects of the most commercially important acids and bases produced and used by the chemical-processing industries.

Environmental protection is of primary importance to the processing industries. Unexpected corrosion failures, which could cause the release of chemicals, cannot be tolerated. Methods of preventing and detecting corrosion hidden under thermal insulation are described in the Section "Corrosion Under Thermal Insulation" in this article. The causes of unexpected corrosion failures from process and environmental variables are covered in the Section "Effects of Process and Environmental Variables" in this article. Fouling of equipment and piping can lead to premature failures, release of chemicals, and reduced efficiency. Three methods of cleaning are described in the Section "Chemical Cleaning of Process Equipment" in this article.

The Sections on corrosion by commercially important acids and bases briefly discuss the various grades, methods of manufacture, the most common materials of construction, and their effect on a broad range of metals and nonmetals. The latest developments in corrosion-resistant alloys, plastics, and electrochemical techniques, such as anodic polarization, applicable to each chemical are also covered in this article.

Many of the materials and techniques used to combat corrosion in the chemical-processing industries were developed by processing companies in response to corrosion problems in their plants. One large chemical company, for example, recognized more than 50 years ago the basic importance of both unit operations and materials in the design and construction of the large-scale plants emerging from their research and established a technical division with two groups at its experimental station. The materials group developed silicon bronze (Copper Development Association, CDA, C65500), Alloy Casting Institute (ACI) CN-7M stainless steel casting alloy (Alloy 20), 20Cb-3 stainless steel (Ref 1), dispersion-strengthened nickel-base corrosion- and wear-resistant alloys, and other significant advancements in new metals. Several of the larger chemical companies were also instrumental in the commercialization of titanium (Ref 1). Corrosion engineers employed by a large German chemical company were responsible for the development of Hastelloy C-276.

Corrosion monitoring using instantaneous corrosion measurement is based on an electrochemical theory formulated by the employees of a major chemical industry firm (Ref 2). Their work on polarization also led to the design of the titanium-palladium alloy, with its broader range of passivity in reducing environments. The many chemically resistant plastics and elastomers used in chemical-processing industry plants today were also developed by the industry.

The chemical-processing industry has undergone profound changes in the past decade. Energy costs, environmental considerations, and employee and public health concerns have brought about changes in plant construction. Plants are now more energy efficient and less polluting. Also, fewer large plants are being built. High-value, low-volume specialty chemicals and biotechnical products are favored over the commodity products. The resulting products are often of high purity. Electronic grades of acids are not uncommon. High-quality products often require more corrosion-resistant materials for equipment and containers for shipping and storage. Titanium alloys are routinely considered for chemical applications, and zirconium, niobium, and tantalum are also being used more often (Ref 3). More expensive materials are more affordable in small equipment, for which design and fabrication costs represent a higher percentage of the installed cost than the cost of the materials used in fabrication. Purity levels can also be improved by electrochemical means, such as the anodic or cathodic protection of steel storage tanks and stainless steel coolers and liners.

Effects of Process and Environmental Variables

Thomas F. Degnan
Consultant

Two chemical-processing plants making the same product and using the same or a similar process will sometimes have different experiences with corrosion. At one plant, a steel pipeline may last for many years in a given service, yet an identical pipeline many fail within weeks or months in the same service at another plant. A major piece of equipment may suddenly fail after 15 or 20 years of service as a result of less than 1 ppm of metal ion contamination in a new source of raw material.

In designing a chemical-processing plant, considerable attention is paid to fluid flows, sizings of lines, pumps and processing equipment, and temperatures and pressures. Materials are selected on the basis of past experience, corrosion tests, the literature, and the recommendations of material suppliers. The nominal compositions of process streams and raw materials are known with some accuracy. Occasionally, however, variables come into play that cause corrosion failures. Table 1 lists variables that can affect the performance of metals in corrosive environments.

Plant Environment

One factor that is occasionally overlooked is the surrounding environment. For example, the instrument tubing at a new facility, built at an existing chemical plant in West Virginia, consisted of a combination of copper (C12200) tubing and yellow brass (C27000) fittings. However, within a few weeks, the fittings had failed by stress-corrosion cracking (SCC). The new building was located near, and downwind of, another facility that occasionally vented small amounts of nitric oxide (NO) to the atmosphere (Ref 4).

Tests later showed that brass samples would fail by SCC if exposed to the source of NO emissions. An expensive shutdown was necessary to replace all of the instrument tubing and fittings with stainless steel.

Similar failures of brass heat exchangers have occurred as a result of their proximity to sources of ammonia (NH_3). In one case, cooling tower water became contaminated with a few parts per million of NH_3 when an ammonia-manufacturing plant was built nearby. In another case, the admiralty brass (C44400) heat-exchanger tubing in a large air compressor stress cracked in cold-worked areas after it was installed about 30 m (100 ft) from a building in which laboratory reagents, including ammonium hydroxide (NH_4OH) and nitric acid (HNO_3), were bottled.

Cooling Water

The quality of water used to cool shell and tube heat exchangers can vary from season to season and year to year, particularly if the water is taken from a river that discharges into a bay or sea. In one case, for example, a plant had 64 identical AISI type 304 stainless steel condensers cooled on the shell side, with river water that varied in chloride content from less than 50 ppm to as high as 2000 ppm. The stainless steel tubes failed by external pitting at baffles. It was discovered that

Table 1 Effects of process variables on corrosion of metals in aggressive environments

Process variable	Carbon steels	Austenitic stainless steels	Nickel, nickel-copper alloys	Nickel-copper, nickel-chromium-molybdenum alloys	Titanium	Zirconium	Copper, copper alloys
Velocity							
Increasing	>(a)	0 to >	0 to >	0 to >	0(b)	0	>
Decreasing	<L(c)(d)	<L	0 to <	0 to <	0	0	<
Aeration	>	0 to <	>	V(e)	0 to <	0	>
Galvanic couple	>	0	0	0	HE(f)	HE	0
Impurities							
Nonchloride oxidants	V	<	>	V	0	0	>
$FeCl_3$, $CuCl_2$	>	>L, SCC(g)	>	0 to >	0	>E(h)	>
Neutral chlorides	>	>L, SCC	0	0 to >	0, unless H(j)	0	>
Ammonia	0	0	>	0	0	0	>SCC
Sodium, potassium	0, SCC if H	0, SCC if H	0	0	0, unless H	0	0, unless H
Fluorides	0	0	0	0	>	>	0
Nitrites	V	V	>	V	0(k)	0	>SCC
Nitrates	<, SCC	0 to <	0	0	0	0	0
HCl in chlorine	>	>	>	>	>(m)	>	>
Mercury	0	0	0, SCC if H	0	0, unless H	0	>SCC
Sulfur compounds	V	0, SCC if S(n)	>	0, SCC if S	0	0	>

(a) >, generally increases attack. (b) 0, generally has no effect on metal in question. (c) <, generally decreases attack. (d) L, local attack or crevice corrosion is possible. (e) V, may increase or decrease attack, depending on specific alloy and environment. (f) HE, hydrogen embrittlement is possible. (g) SCC, stress-corrosion cracking is possible. (h) E, embrittlement. (j) H, hot (from 80 to 150 °C, or 175 to 300 °F, depending on variable and alloy). (k) Except in fuming HNO_3. (m) Except for dilute solutions. (n) S, sensitized by heat treatment

an almost perfect correlation could be obtained by plotting the average monthly chloride content against the number of leaking condensers retubed that month.

In a similar case, a series of stainless steel heat exchangers cooled with recirculating treated water had been in service for more than 10 years (Ref 5). During an emergency, untreated river water was used to cool the units for about 48 h. Several weeks later, five of the coolers failed by massive chloride SCC. There was a large buildup of dried mud in the units, which caused reduced heat transfer and higher wall temperatures.

Steam

The corrosive properties of the steam that will be used in a plant are seldom considered, particularly if the steam will be supplied by a public utility or power plant that has good boiler water treatment that will prevent condensate corrosion. However, the impurities in steam can cause corrosion problems.

At one site, for example, high-pressure steam was purchased from a central power station; pressure was reduced to 2.76 MPa (400 psi) as it entered the plant. Some years later, plant operators decided to cool the superheated steam by spraying with deionized water at a point just beyond the depressurizing valve. There were no problems until 15 years later, when it became necessary to replace and reroute part of the steam line inside the plant. Within a few months, cracks and resulting leaks appeared adjacent to and across the welded joints in the new piping. The plant operators blamed the welding contractor for poor welding.

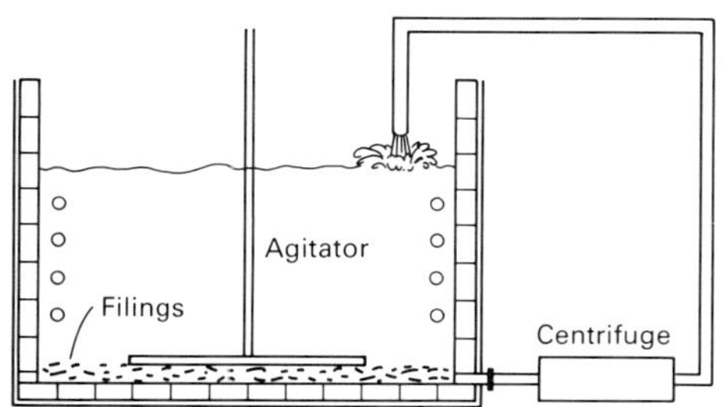

Fig. 1 Equipment used for iron reduction and removal of spent filings

However, metallographic examination showed that the cracking was the result of caustic embrittlement. The deionized water contained 10 ppm of sodium, which formed concentrated sodium hydroxide (NaOH) as the water evaporated. The older piping did not crack, because it had been stress relieved by the superheated steam over a period of years before cooling of the steam was initiated.

A similar case was reported in which a stainless steel bellows in a 2.76-MPa (400-psi) steam line failed by caustic cracking (Ref 4). The failure was attributed to carryover of boiler compounds.

Aeration

The design of the plant and, in particular, equipment selection can influence the amount of air introduced into a process stream, which in turn may profoundly affect corrosion. This can account for different corrosion experiences in plants having identical chemistries and products.

Solid-liquid separation (Fig. 1) can be accomplished in several ways. Some methods, such as centrifuging, can cause more aeration than others—pressure filtration, for example—with occasional disastrous results.

In iron reductions, iron filings and dilute hydrochloric acid (HCl) are agitated with an organic nitro compound, which is reduced to an amine by the hydrogen evolved. In the design of one such unit, for example, a Monel 400 centrifuge, coils, and agitator were selected based on laboratory corrosion tests and general experience with units using filter presses having no return stream. Aeration caused by the centrifuge, together with a return pipe that terminated in the vapor space of the vessel, caused conversion of ferrous chloride ($FeCl_2$) to ferric chloride ($FeCl_3$), with resulting rapid corrosion of all Monel 400 parts. The temporary solution involved blanketing the equipment with nitrogen, extending the return pipe below the liquid level, and adding a corrosion inhibitor.

In a caustic chlorine plant, steel piping that transferred saturated alkaline brine at ambient temperature failed by crevice corrosion and pitting shortly after start-up, although steel piping had been successfully used at other plants. Failure was attributed to aeration caused by centrifuges removing precipitated salt from caustic liquors and returning the salt to the dissolver.

Start-Up and Shutdown Conditions

Start-ups and shutdowns are variables that can easily be overlooked in the design stage or conducted improperly by operations. Such a case was reported in a high-sulfur coal gasification plant operating at 430 to 595 °C (800 to 1100 °F) (Ref 6). A designer selected type 310 stainless steel rather than type 304 because he felt that the higher chromium content of the type 310 would result in improved high-temperature resistance. The unit failed, however, by massive intergranular polythionic acid SCC that occurred during the first shutdown.

This type of corrosion occurs in sensitized stainless steels in an acidic sulfide environment in the presence of air. It can be mitigated by following the guidelines in National Association of Corrosion Engineers (NACE) standard RP-01-70 (Ref 7). Perhaps a better approach is to use a carbide-stabilized grade of stainless steel, such as type 347 or type 321.

In another example, silicate cement was selected for use with unalloyed tantalum repair plugs in a new glass-lined vessel and its accessories for a bromination process, because bromine would attack the organic cement that is normally used with the plugs. The vessel was obtained plug-free, but the agitator and baffle both required tantalum plugs. As part of the start-up procedure, plant operators boiled out the equipment with water. This dissolved the silicate cement. When the vessel was placed in operation, the bromine attacked the steel under the plugs. The agitator dissolved, leaving only the stub revolving above the liquid level.

In a third case, type 304 stainless steel was correctly selected for a 99% sulfuric acid (H_2SO_4) pipeline. During the first shutdown, the plant operators decided to blow the acid out of the line. They used compressed air, which was unwise, because compressed air is commonly saturated with water. Within a few days, leaks were discovered adjacent to some welds where dilute acid had accumulated. Dry nitrogen is now used to blow out the line, which has been in service for 20 years and as many shutdowns.

Seasonal Temperature Changes

Ambient temperature is an important variable that can vary widely. Metal parts exposed to the sun may reach 60 °C (140 °F) in the summer. In one case, for example, a steel pipeline carrying 98% H_2SO_4 was supplied with properly designed steam tracing that maintained a maximum temperature of 50 °C (125 °F) during the winter (Ref 6). During the summer, the line failed by corrosion. The steam had not been disconnected, and the temperature of the line exceeded 70 °C (150 °F).

Heat exchangers (Fig. 2) are usually designed conservatively for summer water temperatures. During the winter, the colder cooling water can cause a lower exit process temperature than desired. To correct this problem, plant operators will frequently decrease the water flow. All too often, the water flow is throttled at the water inlet. This can result in a fluctuating liquid level, with buildup of solids, pitting, and/or SCC of tubes. The correct arrangement, with the water control valve on the discharge line, is shown in Fig. 2.

Variable Process Flow Rates

Plants are frequently operated at flow rates other than design capacity. They may be turned down to considerably less than capacity or pushed above it by eliminating bottlenecks. Either condition can cause corrosion problems.

In an H_2SO_4 plant, for example, the last 150 mm (6 in.) of the steel waste heat boiler tubes were safe-ended with alloy 20Cb-3 stainless steel to protect against dew-point corrosion. (Safe ending is the welding of a length of corrosion-resistant tubing to the end of a less resistant material.) Unfortunately, the steel tubes corroded upstream of the alloy. The process was not operated within reasonable design limits. Corrosion resistance in this case would have required 1.8 m (6 ft) rather than 150 mm (6 in.) of safe-ending because the tubes were operating at 66 °C (119 °F) below the design temperature and well below the dew point.

In another case, a five-pass Monel 400 heat exchanger was used to vaporize anhydrous hydrofluoric acid (HF). Severe corrosion occurred throughout the fifth and part of the fourth pass, with no corrosion in the first three passes. Insufficient flow caused nonvolatile acid impurities, present in the HF at part per million levels, to be concentrated in the last one-plus passes.

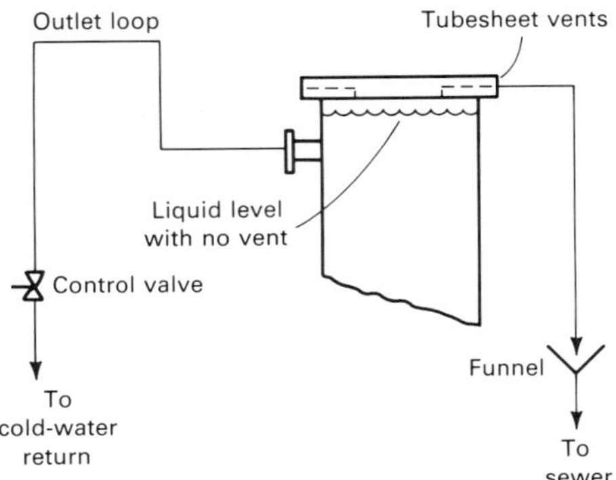

Fig. 2 Correct piping arrangement for venting fixed tubesheet heat exchangers. Source: Ref 8

In a third example, plant operators decided to install a higher-capacity pump to reduce the time required to transfer a charge of 98% H_2SO_4 through a steel line from 1 h to 15 min. This also increased the velocity from less than 0.6 m/s (2 ft/s) to more than 1.5 m/s (5 ft/s). The line failed within a week.

In another instance, a Monel 400 shell and tube heat exchanger had given satisfactory service in an organic acid oil for more than 5 years. The exchanger was a two-pass unit with a divider plate in the bottom head. The bottom head also contained the inlet and outlet nozzles. The divider plate was gasketed to a groove in the tubesheet. The velocity through the unit was increased by increasing the pump pressure. Because it was not designed for the higher pressure, the divider plate deflected, allowing the gasket to blow out. The product bypassed the divider at high velocity, which caused the unit to fail by erosion-corrosion.

Impurities

A principal cause of unexpected corrosion failures is the presence of small amounts of impurities in the chemical environment. Producers of H_2SO_4 sometimes provide facilities to recover spent acid as a service to their customers, who would otherwise have difficulty in disposing of the acid in an environmentally acceptable manner. The acid, contaminated with organic matter, is burned, and the hot gases travel through a waste heat boiler.

The operators of one such plant, for example, discovered that the steel tubes of their waste heat boiler had short service lives. Examination of the tubes showed that they were being attacked by a molten slag from the spent acid. Analysis showed that the slag had a high phosphate and lead content, a low melting point, and was very corrosive to steel when molten.

In another plant, zinc chloride ($ZnCl_2$) is made by dissolving zinc recovered from galvanizing operations and from other sources in HCl, treating the various chemicals, and concentrating the salt solution by evaporation. The nickel coils in a concentrating tank pitted and had short service lives. After a 30-day test, a zirconium coil appeared to be unaffected; therefore, a zirconium coil was installed in the tank. It failed after only 6 months of service. Failure was traced to a galvanizing company that had used fluorides in its flux. Zirconium is not resistant to fluorides.

At a New Jersey plant, a titanium-clad autoclave was used for the air-HNO_3 procedure in the manufacture of an organic acid. After several years of successful operation, the liquid-phase surfaces were found to be pitted and cratered, although they were unattacked when inspected a few weeks earlier. The presence of H_2SO_4 was suspected. Investigation revealed that a trailer of mixed H_2SO_4 and HNO_3 had been mistakenly unloaded into an HNO_3 tank.

In another example, a Monel 400 petrochemical plant extraction unit handling 50 to 65% H_2SO_4 and alcohol at 30 to 40 °C (85 to 100 °F) had given satisfactory service for 17 years, with a 5-year life expectancy for heat-exchanger tubes (Ref 9). Unexpectedly, tubes began to fail within 5 weeks. Replacement piping failed in as little as 3 weeks, predominately at welds. Laboratory analyses revealed that the plant solutions contained copper in quantities greater than could be explained by the simple dissolution of Monel 400. Tests showed that copper was being leached from the surfaces in a manner similar to zinc in a dezincification process.

Investigation uncovered that there had been a change in the feedstock supply. The new supply was contaminated with a few parts per million of copper ions. This contamination initiated the corrosion process, which fed upon itself by leaching more copper ions from the surface of the Monel 400.

In a similar case, a Monel 400 isomerization reactor was built based on laboratory corrosion tests and plant tests, with corrosion coupons in different locations in a glass-lined isomerizer (Ref 6). The process involved heating terpenes in the presence of H_2SO_4, followed by decantation, with the acid layer returned to the reactor. Precautions were taken to provide a nonaerated system by using a nitrogen blanket.

Shortly after the Monel 400 reactor was installed, the H_2SO_4 was found to be discolored, and excessive corrosion was discovered. Successive isomerizations using the recycled H_2SO_4 had caused a buildup of cupric ion (Cu^{2+}). Sulfuric acid with 0.5 ppm or more Cu^{2+} ion will cause autocatalytic destruction of Monel 400, as was found in the earlier case. The copper can come from an outside source, excessive corrosion, or buildup of corrosion products from recycling. Once started, this reaction is almost impossible to stop, because the Monel 400 surfaces become a ready source of more copper.

At another plant, 20Cb-3 stainless steel piping had been used for many years to handle denitrated 80% H_2SO_4 at 50 °C (120 °F). Unexpectedly, a number of failures occurred. Potentiostatic studies, using diluted reagent grade acid, showed that 20Cb-3 stainless steel should be passive under these conditions. The plant operators, some distance from the laboratory, continued to complain about the shortened service life of the piping, and plant acid was eventually shipped to the laboratory. The potentiostatic plot showed borderline active-passive behavior. Further studies showed that an extremely small amount of HNO_3 present in the denitrated acid caused the unstability and that the absence of all HNO_3 or the presence of more than a critical amount promoted passivity. The efficient, but not perfect, operation of the denitrating system caused the corrosion.

Water

One of the most important process variables, from the corrosion viewpoint, is the amount of water in a process stream. The water may be present as water vapor in a gas stream or may be soluble. Water is most likely to contribute to corrosion when it is present as a separate liquid phase, particularly in the presence of hydrogen chloride ($HCl \cdot H_2O$), which has a great affinity for water (Fig. 3).

When the water is present as vapor, it is likely to condense at cold spots, such as pipe supports welded directly to process piping or in nozzles in the top or vapor phase of reactors, even though the lines and nozzles are insulated. Hydrogen chloride raises the dew point. Chlorinations are a particular problem for several reasons. Commer-

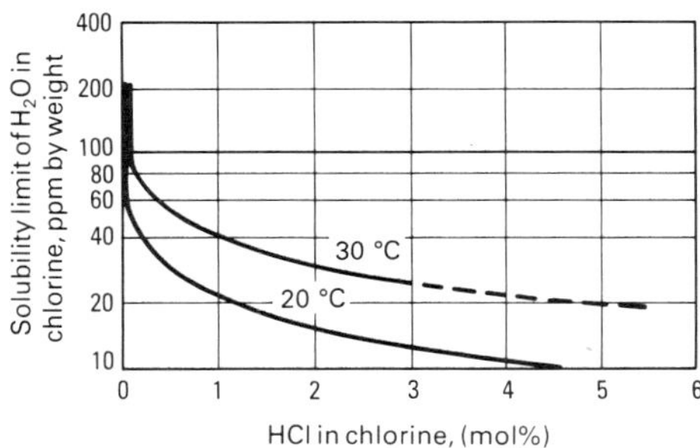

Fig. 3 Effect of HCl on solubility of water in chlorine. Source: Ref 10

cial chlorine contains variable amounts of water, up to 50 ppm or more, because the solubility limit of water in HCl is about 200 ppm at room temperature. The $HCl{\cdot}H_2O$ formed during chlorinations reduces the solubility of water in chlorine to well below 50 ppm at room temperature (Fig. 3). The condensate of water-chlorine-$HCl{\cdot}H_2O$ is corrosive to all metals except tantalum.

For example, stainless steel reactors are used in the manufacture of fluorinated hydrocarbon refrigerants from chlorinated hydrocarbons by reacting them with HF in the presence of chlorine and a catalyst. After many years of service, the cover of a reactor exhibited chloride SCC just below two nozzles. Higher-than-normal water content in one or more of the reactants had caused condensation in the blind nozzles. The condensate evaporated when it ran onto the cover, causing concentration of chloride salts and subsequent chloride SCC. Steam tracing of the nozzles (Fig. 4) prevented condensation and corrected the problem.

In the same plant, a column was installed to recover excess chlorine from a supposedly dry process stream containing fluorocarbon, $HCl{\cdot}H_2O$, traces of HF, and chlorine. The brine-cooled condenser had 20Cb-3 stainless steel tubes. A few weeks after start-up, the tubes failed from process-side corrosion. Attack occurred preferentially on the coldest side of the tubes facing the brine flow. The unit was temporarily retubed with heavy-wall steel tubes. After a few weeks, these tubes leaked. The brine was drained from the shell, and almost immediately, the steel tubes caught on fire. The hydrogen generated by corrosion of the outside of the tubes reacted exothermically with the chlorine present, raising the temperature of the tubes to the ignition temperature of steel in chlorine (about 250 °C, or 480 °F). The process of removing chlorine by distillation was abandoned. No metallic heat-exchanger tube is available that will resist aqueous HCl saturated with chlorine in the presence of acid fluorides.

The trace amounts of water in the stream fed to the column were not detectable by the analytical methods used at the plant; the water was present in small but variable amounts. The presence of large amounts of $HCl{\cdot}H_2O$ raised the dew point enough to cause the water to condense on the coldest parts of the tube walls.

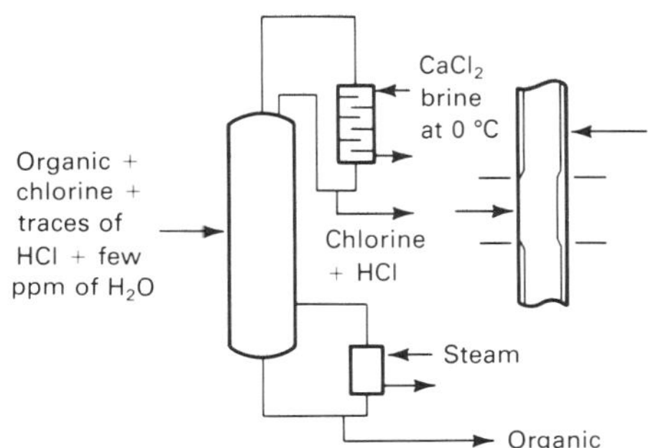

Fig. 4 Arrangement of chlorine recovery column and corrosion pattern on condenser tubes. Source: Ref 11

Chemical Cleaning of Process Equipment

L. Jones
ERT, A Resource Engineering Company
J.D. Haff and B.J. Moniz
E.I. Du Pont de Nemours & Company, Inc.

Process equipment and piping must be cleaned to prevent contamination of a process or product, to improve the operation of a process, to reduce the opportunity for premature failure, and to prepare equipment for inspection. However, equipment should be cleaned only for good reason. In addition to the cost of unnecessary cleaning, problems may be introduced. For example, most chemical cleaning processes cause some metal loss. In other cases, washing before cleaning may cause accelerated corrosion, such as during the preparation of a concentrated H_2SO_4 storage tank for inspection. Other potential problems are:

- Difficulties associated with pumping hot corrosives through temporary connections
- Difficulties associated with a crowded work space, for example, during a turnaround
- The need to dispose of wastes
- The possibility of generating toxic or flammable by-products during cleaning

There are four types of equipment cleaning: preoperational, chemical, mechanical, and on-line. These must be evaluated for each job in order to select the most cost-effective. To make a sound evaluation, the deposit to be removed must be thoroughly characterized.

Fouling of Equipment

Deposits that cause fouling accumulate in equipment and piping and impede heat transfer or fluid flow or cause product contamination. Deposits may be organic, inorganic, or a mixture of the two. Scales are crystalline deposits that precipitate in a system (Table 2). There are four principal sources of deposits: water-side, fire-side, process-side, and preoperational.

Water-side deposits are of many types. Hardness (calcium and magnesium) based deposits and iron oxide are the most common water-side deposits and often affect boilers and cooling systems. Process and oil leaks can foul boilers and cooling systems. Biofouling, mud, and debris are often found in cooling systems. Treatment chemicals, if not properly controlled, can add to deposits and scale. Silica can form hard, adherent deposits in boilers, steam turbines, and cooling systems. Corrosion products can add to deposits.

Table 2 Summary of common types of scale-forming minerals

Scale	Chemical formula
General	
Sodium iron silicate	$NaFe(SiO_3)_2$
Barium sulfate	$BaSO_4$
Sodium aluminum silicate	$NaAl\ Si_2O_6{\cdot}H_2O$
Aragonite (rhombic crystals)	$CaCO_3$
Calcium carbonate—(hexagonal crystal)	$CaCO_3$
Calcium sulfate	$CaSO_4$
Magnesium carbonate and hydroxide	$3MgCO_3{\cdot}Mg(OH)_2{\cdot}3H_2O$
Calcium phosphate	$Ca_{10}(OH)_2(PO_4)_6$
Iron oxide	Alpha FeO (OH)
Iron oxide—magnetite	Fe_3O_4
Iron oxide—red	Fe_2O_3
Iron chrome spinels	$CrFe_2O_4$
Iron sulfide	FeS
Magnesium hydroxide	$Mg(OH)_2$
Magnesium oxide	MgO
Manganese dioxide	MnO_2
Aluminum silicate	$Al_2O_3{\cdot}4SiO_2{\cdot}4H_2O$
Sodium aluminum silicate	$Na_8Al_6Si_6O_{24}{\cdot}SO_4$
Calcium sodium silicate	$4CaO{\cdot}Na_2O{\cdot}6SiO_2{\cdot}H_2O$
Magnesium silicate	$Mg_3Si_2O_7{\cdot}2H_2O$
Silica	SiO_2
Sodium aluminum silicate	$Na_8Al_6Si_6O_{24}{\cdot}Cl_2$
Magnesium iron aluminum silicate	$(Mg,Fe)_3(Si,Al)_4O_{10}(OH)_2{\cdot}4H_2O$
Calcium silicate	$5CaO{\cdot}5SiO_2{\cdot}H_2O$
Copper or copper alloy equipment	
Copper iron sulfide	CuFeS
Copper sulfide	CuS and Cu_2S
Basic copper chloride	$CuCl_2{\cdot}3Cu(OH)_2$
Copper oxide	Cu_2O
Chalcopyrite	$CuFeS_2$
Beta zinc sulfide	ZnS
Green basic carbonate	$CuCO_3Cu(OH)_2$

Fire-side deposits can be extremely corrosive. Slags from burning oil and wastes can corrode boiler equipment if they become moist. Fly ash deposits can accumulate in coal-fired boilers. Gas-fired boilers are generally clean. Some compounds that are burned in incinerators or waste heat boilers can seriously corrode or erode boiler tubes.

Process-Side Deposits. There are many types of process-side deposits. Organic residues, tars, and coke are common in the petroleum and petrochemical industries. Iron oxide and sulfides are often present in these organic deposits. Sulfate deposits are common in H_2SO_4 plants. Iron-, copper-, and nickel-containing deposits often occur in HF plants.

Organic deposits may develop through the polymerization of leaking gases or from the decomposition of process constituents. In some cases, organics help to bond inorganic deposits, such as iron oxides or sulfides.

Some process-side deposits are pyrophoric when exposed to air or oxygen. The most common is iron sulfide, which is likely to be found in natural-gas- and petroleum-refining processes or when aqueous solutions of hydrogen sulfide (H_2S) are dried in the absence of air.

Preoperational deposits are formed during the fabrication and erection of process equipment and piping. In addition to mill scale residues, metal surfaces become coated with dirt, oil, grease, weld spatter, pipe-threading compound, protective shop coatings, and corrosion products.

Highly alloyed materials, such as stainless steels, nickel-base alloys, reactive metals, or high-temperature alloys, may become contaminated with iron from tooling; zinc, cadmium, and aluminum from scaffolding; and zinc, sulfur, and chlorine in certain marking materials. These elements can cause corrosion or embrittlement.

Critical Equipment Areas

Requirements for cleaning will vary with the type of equipment. The operating characteristics and design must be assessed before selecting a cleaning method.

Columns. The two critical areas for deposit formation in a column are at the trays, where vapor passes through a valve, sieve, flapper, or riser, and in the flash zone, where vapor condenses. Operating history sometimes indicates which areas require cleaning; for example, the vapor line is suspect if the column vapor rate becomes limiting. Inspection is necessary to determine the extent and location of fouling.

Glass-lined vessels require special attention when their water jackets are chemically cleaned. The recommendations of the manufacturer must be followed. The most commonly recommended cleaning solution is dilute alkaline sodium hypochlorite (NaClO) (Ref 12). If strong acids are used, atomic (nascent) hydrogen formed by corrosion diffuses into the shell and recombines as hydrogen molecules at the glass/metal interface, which causes spalling of the glass.

Oxygen, chlorine, and fluorine piping systems must be free of organic contaminants. Organic materials, particularly hydrocarbon greases and oils, react violently with these chemicals. Preoperational cleaning is mandatory in such cases (Ref 13). After cleaning, the lines should be blown dry, using oil-free nitrogen or air.

Identification of Deposits

To select an effective cleaning procedure, the deposit must be characterized, or identified. The sample should represent the deposit in the most critical fouling area. For exchangers and boilers, this is the highest heat transfer section. Expediency should not dictate the location of the sample. A cleaning procedure should not be based on a sample of loose deposit from a noncritical area, because the sample at this location may not be representative. Table 3 lists some common components of boiler deposits.

When removed by scraping, the samples should be as intact as possible. They should be removed to the base metal, taking care not to introduce any metallic chips from the blade or substrate. Thickness, density, porosity, type (homogeneous or layered), and color should be noted.

When only a limited amount of deposit is available, replication tape is a useful method of removing it. Polyvinyl chloride (PVC) or other chloride-containing tapes should not be used on stainless steels, which are susceptible to chloride pitting and stress cracking.

Many analytical techniques are used to characterize deposit samples. Typical methods include x-ray diffraction, optical emission spectroscopy, and x-ray spectrometry. Most chemical cleaning contractors, water treatment supplies, and analytical laboratories have the facilities to characterize deposits.

Preoperational Cleaning

Unlike process- or water-side deposits, the types of deposits in original equipment are easily categorized (Ref 14). Preoperational cleaning should include consideration of the degree of cleanliness required and the material of construction. Areas where preoperational cleaning is used include:

- Process equipment start-up, boilers, and steam-generating and compression systems
- Lubricating oil systems before oil flushing
- Critical services, such as oxygen, chlorine, or fluorine piping
- Water treatment and inhibition programs

Boilers are cleaned to remove oils, grease, and mill scale. When boilers are coated with heavy protective greases, two-stage cleaning (for organic and inorganic deposit removal) should be used. A degreasing step using alkaline boilout solutions or emulsions is first used. Common second-step solvents include chelants, organic acids, or HF.

Columns contain similar contaminants. They are cleaned by fill and soak, cascade, or foam methods, using solvents similar to those used for boilers. The design of the column may eliminate certain methods, such as cascade cleaning for a packed column.

Shell and Tube Heat Exchangers. The most serious fouling is found on the interior (tube side) or exterior (shell side) of the tubes. Other locations are on the shell side at baffles or drain nozzles. Figure 5 shows a fouled heat-exchanger tube bundle.

The head should be removed for inspection if tube-side fouling of the tubes is suspected. The

Fig. 5 Fouled heat-exchanger tube bundle

Table 3 Components of boiler deposits

Mineral	Formula	Nature of deposit	Usual location and form
Acmite	$Na_2O{\cdot}Fe_2O_3{\cdot}4SiO_2$	Hard, adherent	Tube scale under hydroxyapatite or serpentine
Alpha quartz	SiO_2	Hard, adherent	Turbine blades, mud drum, tube scale
Amphibole	$MgO{\cdot}SiO_2$	Adherent binder	Tube scale and sludge
Analcite	$Na_2O{\cdot}Al_2O_3{\cdot}4SiO_2{\cdot}2H_2O$	Hard, adherent	Tube scale under hydroxyapatite or serpentine
Anhydrite	$CaSO_4$	Hard, adherent	Tube scale, generating tubes
Aragonite	$CaCO_3$	Hard, adherent	Tube scale, feed lines, sludge
Brucite	$Mg(OH)_2$	Flocculent	Sludge in mud drum and water wall headers
Copper	Cu	Electroplated layer	Boiler tubes and turbine blades
Cuprite	Cu_2O	Adherent layer	Turbine blades, boiler deposits
Gypsum	$CaSO_4{\cdot}2H_2O$	Hard, adherent	Tube scale, generating tubes
Hematite	Fe_2O_3	Binder	Throughout boiler
Hydroxyapatite	$Ca_{10}(PO_4)_6(OH)_2$	Flocculent	Mud drum, water walls, sludge
Magnesium phosphate	$Mg_3(PO_4)_2$	Adherent binder	Tubes, mud drum, water walls
Magnetite	Fe_3O_4	Protective film	All internal surfaces
Noselite	$4Na_2O{\cdot}3Al_2O_3{\cdot}6SiO_2{\cdot}SO_4$	Hard, adherent	Tube scale
Pectolite	$Na_2O{\cdot}4CaO{\cdot}6SiO_2{\cdot}H_2O$	Hard, adherent	Tube scale
Serpentine	$3MgO{\cdot}2SiO_2{\cdot}H_2O$	Flocculent	Sludge
Sodalite	$3Na_2O{\cdot}3Al_2O_3{\cdot}6SiO_2{\cdot}2NaCl$	Hard, adherent	Tube scale
Xonotlite	$5CaO{\cdot}5SiO_2{\cdot}H_2O$	Hard, adherent	Tube scale

shell is more difficult to inspect, unless the tube bundle is removable, but limited information may be gained through nozzles.

Heat-exchanger tubes may be cleaned mechanically or chemically. Mechanical cleaning may damage tubes. Individual tubes should not be steam blown, because this may damage rolled tube joints. Tubes should not be hammered with any metallic tool, and scraping or rodding should be done with care because any scoring or gouging can lead to premature failure. High-pressure and ultrahigh-pressure water cleaning are preferred.

Chemical cleaning methods use circulation, fill and soak, or foam. However, severely blocked tubes may resist the entry of the cleaning agent or may retain it beyond the neutralization step of the cleaning process, leading to corrosion during shutdown or in service.

Heat-exchanger shells are normally chemically cleaned using the circulation or the fill and soak method. If the tube bundle is removable, mechanical cleaning with high-pressure or ultrahigh-pressure water is a good technique.

Boilers. Typical water-side deposits found in boilers are listed in Table 3. Deposits vary depending on raw water composition, feedwater treatment, and operating pressure.

The heaviest deposition occurs in tubes with the highest heat input, an area that may be physically impossible to inspect. A tube section can be taken from the area where deposition is known to be heaviest in order to characterize the deposit. Although various yardsticks (in grams of scale per square foot) have been proposed for determining the need to clean, each case should be individually evaluated. Factors to be considered are the degree of fouling, the type of service, the reliability required, the operating history, and future operation.

Chemical cleaning of the water side is generally more effective than mechanical cleaning, particularly in designs with heavily swaged tubes and tight bends. Preoperational cleaning of boilers must be conducted to allow the steel surface to develop a protective film of magnetite (Fe_3O_4) when the boiler is put into service and to remove mill scale.

Furnaces. The external fouling of furnace tubes depends on the nature of the fuel. Oil-burning furnaces usually have significantly more deposit formation and corrosion problems than coal-burning types, while natural-gas-burning furnaces have very few problems.

Slag accumulates when metallic salts and oxides are vaporized and condense in various parts of the furnace. Because its melting point is relatively low, the slag forms a sticky corrosive deposit of various salts, primarily sodium and vanadium. These slags should be mechanically removed by chipping or dry sandblasting. Wet cleaning methods may cause acid formation. For internally coked tubes, steam-air decoking or mechanical cleaning is preferred.

Pumps and Compressors. Cooling water jackets are often chemically cleaned to remove iron oxide, water-formed scale, and possible oil infiltration. All loose material is first removed by opening the clean-out plates and flushing. A two-stage chemical cleaning process is then used, first to dissolve any organic deposits and then to remove inorganic scales. The acidic cleaner selected for inorganic scales should be compatible with the materials of construction.

Piping may contain various contaminants, including dirt, loose paint, sand and grit, varnish, grease and oils, weld spatter, mill scale, and rust. Piping should first be inspected and all construction debris removed. Dirt, loose paint, sand, and grit are removed by flushing with clean water or blowing with dry compressed air or steam. Varnish, grease, and oils are removed by steam blasting with detergent or hot water containing an alkaline degreasing agent. Mechanical cleaning may be required, depending on the amount of weld spatter, mill scale, and rust. The piping may then be chemically cleaned if necessary, using organic acids and chelants, followed by neutralizing and passivating.

Moisture removal may be required for such specific applications as compressor or refrigeration piping. When all traces of moisture must be removed, the system can be filled with alcohol, evacuated to evaporate the alcohol, then flushed with an inert gas.

Piping carrying oxygen, chlorine, and fluorine requires stringent cleaning to remove organic contamination. No organic-containing residues can be permitted (Ref 13).

Chemical Cleaning

Chemical cleaning is the use of chemicals to dissolve or loosen deposits from process equipment and piping. It offers several advantages over mechanical cleaning, including more uniform removal, no need to dismantle equipment, lower overall cost (generally), and longer intervals between cleanings. In some cases, chemical cleaning is the only practical method.

The primary disadvantages of chemical cleaning are the possibility of excessive equipment corrosion and solvent disposal. Chemical cleaning solvents must be assessed in a corrosion test program before their field acceptance.

Chemical cleaning is performed by a contractor who specializes in this work. Some cleaning procedures are protected by patents.

Chemical Cleaning Methods

There are six major chemical cleaning methods: circulation, fill and soak, cascade, foam, vapor phase organic, and steam-injected cleaning. A seventh variation is discussed in the section "On-Line Cleaning" in this article.

Circulation, the most common method, is applied to columns, heat exchangers, cooling water jackets, and so on, where the total volume required to fill the equipment is not excessive. The equipment is arranged such that it can be filled with the cleaning solution and circulated by a pump to maintain flow through the system. Movement of solution through the equipment greatly assists the cleaning action. As cleaning progresses, temperature and concentration are measured in order to monitor the progress. The cleaner may be replenished (sweetened) occasionally to maintain efficiency. Corrosion coupons or on-line monitoring determines the effect of the cleaning chemicals on the equipment materials.

With circulation cleaning, the rate of flow through the equipment is critical. Large-diameter connections are preferred, and a high-capacity pump may be necessary to produce the required circulation. After cleaning, the equipment is drained, neutralized, flushed, and passivated.

Fill and soak cleaning involves filling the equipment with the cleaner and draining it after a set period of time. This may be repeated several times. The equipment is then water flushed to remove loose insolubles and residual chemicals.

Fill and soak cleaning offers limited circulation. The poor access of fresh cleaning solution to the metal, together with the inability to maintain solution temperature, may cause the cleaning action to cease.

The method is limited to relatively small equipment containing light amounts of highly soluble fouling and to equipment in which circulation cannot be properly controlled. Because good agitation is achieved only during the flushing stage, flushing should be as thorough as possible. Circulation and fill and soak cleaning are sometimes used alternately.

Cascade cleaning, a modification of the circulation method, is usually applied to columns with trays. The column is partially filled, and the liquid is continuously drawn from the reservoir and pumped to the highest point. The liquid then cascades down through the column, cleaning surfaces as it passes over them. The liquid draw-off point must be suitably located to avoid recirculation of loosed foulants. High-capacity pumps and large-diameter piping are required to achieve the necessary transfer of liquid to produce a flow pattern that will contact all fouled surfaces within the column.

The cascade method is primarily used in large columns and is suitable for most types, except for packed columns. Cleaning is not effective in inaccessible areas, such as the underside of trays, due to poor contact with the cleaning solution. Contact may be improved by injecting air or nitrogen at the base of the column. If steam is used to heat the chemicals, the location of the steam injection point should not lead to localized overheating. High temperature can also increase corrosion in the vapor space.

Foam cleaning uses a static foam generator that employs air or nitrogen to produce a foamed solvent (Ref 15). Foam stabilizers are required to prolong foam life and increase the effectiveness of the cleaning chemicals. Foam cleaning is used on equipment that cannot support full or partial filling with liquid. Foam cleaning results in significantly less liquid volume for disposal compared with other methods.

Vapor phase organic cleaning is used in equipment that is difficult to clean with liquids. For example, vaporized organic solvents are used to remove organic deposits from columns. The organic solvent is vaporized, injected into the top of the column, condensed, collected in a circulation tank, and revaporized.

The principal concerns are the handling and disposal of the solvent and its flammability (when applicable). The recirculating tank should be purged and blanketed with nitrogen, fitted with an adequate venting and condensing system, and grounded to prevent accumulation of an electrical charge.

Steam-injected cleaning involves the injection of a concentrated mixture of cleaning chemicals into a stream of fast-moving steam. The steam is injected at one end of the system and condensed at the other. The steam atomizes the chemicals, increasing their effectiveness, and ensures good contact with the metal surface.

Steam-injected cleaning is very effective for critical piping systems. As with foam cleaning, the method produces a relatively low amount of liquid for disposal.

Chemical Cleaning Solutions

A wide variety of standard chemical cleaning solutions are available (Table 4). Many proprietary solutions are based on these chemicals. Some are patented or involve patented equipment. Chemical cleaning contractors are the best source of information on standard or patented techniques.

Most chemical cleaning contractors calculate the concentration of chemicals in weight percent, but some use volume percent. The user must be aware of this. For example, a 10 wt% solution of HCl is equivalent to 25 vol% of the normal 30% concentrated HCl.

Chemical cleaning solutions include mineral acids, organic acids, bases, complexing agents, oxidizing agents, reducing agents, and organic solvents. Inhibitors and surfactants are added to reduce corrosion and to improve cleaning efficiency. Following the cleaning cycle, a passivating agent can be introduced to prevent further corrosion or to remove trace ion contamination.

Mineral acids are strong scale dissolvers. They include HCl, hydrochloric/ammonium bifluoride (HCl/NH_4HF_2), sulfamic acid (NH_2SO_3H), HNO_3, phosphoric acid (H_3PO_4), and H_2SO_4.

Organic acids are much weaker. They are often used in combination with other chemicals to complex scales. An advantage of organic acids is that they can be disposed of by incineration. They include formic (HCOOH), hydroxyacetic-formic, acetic (CH_3COOH), and citric acid.

Bases are principally used to remove grease or organic deposits. They include alkaline boilout solutions and emulsions.

Complexing agents are chemicals that combine with metallic ions to form complex ions, which are ions having two or more radicals capable of independent existence. Ferricyanide $[Fe(CN_6)]^{4-}$ is an example of a complex ion. Complexing agents are of two types: chelants and sequestrants. Chelants complex the metallic ion into a ring structure that is difficult to ionize, and sequestrants complex the metallic ion into a structure that is water soluble.

Oxidizing agents are used to oxidize compounds present in deposits to make them suitable for dissolution. They include chromic acid (H_2CrO_4), potassium permanganate ($KMnO_4$), and sodium nitrite ($NaNO_2$).

Reducing agents are used to reduce compounds in deposits to a form that makes them suitable for dissolution and to prevent the formation of hazardous by-products. They include sodium hydrosulfite ($NaHSO_2$) and oxalic acid.

Inhibitors are specific compounds that are added to cleaning chemicals to diminish their corrosive effect on metals. Most inhibitors are proprietary, and recommendations for their use are available from the supplier.

Surfactants are added to chemical cleaning solutions to improve their wetting characteristics. They are also used to improve the performance of inhibitors, emulsify oils, improve the characteristics of foaming solvents, and act as detergents in acid and alkali solutions. As with inhibitors, most surfactants are proprietary products.

Hydrochloric acid is the least expensive and most widely used solvent for water-side deposits on steels. The concentration and temperature vary from 5 to 15% and 50 to 80 °C (120 to 175 °F). The acid must be replenished if the concentration falls below 4%.

Hydrochloric acid must always be used with a filming inhibitor to minimize corrosion. Inhibited HCl is usually suitable for cleaning carbon or alloy steel, cast iron, brasses, bronzes, copper-nickel alloys, and Monel 400.

Hydrochloric acid, even inhibited, is not recommended for cleaning stainless steels, Incoloy 800, Inconel 600, aluminum, or galvanized steel. Its applicability for cleaning titanium or zirconium depends on the oxidizing contaminants present in the acid.

Hydrochloric Acid/Ammonium Bifluoride. The addition of about 1% NH_4HF_2 releases HF, which not only improves the potency of the HCl but, more important, also aids in dissolving silicate scales. A filming inhibitor should be added. Materials compatibility is similar to that of HCl except that titanium, zirconium, or tantalum should never be exposed to an acid containing NH_4HF_2.

Hydrofluoric acid can be used as a preoperational cleaner to remove mill scale, at 1 to 2% and 80 °C (175 °F). It is not widely used in North America, because of unfamiliarity and perceived handling difficulties.

Sulfuric acid is used on a limited scale, from 5 to 10% at temperatures up to 80 °C (180 °F). Although inexpensive, it has several disadvantages, including the extreme care required during handling and the precipitation of calcium sulfate ($CaSO_4$) with deposits containing calcium salts.

When inhibited, H_2SO_4 may be used on carbon steel, austenitic stainless steels, copper-nickel alloys, admiralty brass, aluminum bronze, and Monel 400. Sulfuric acid should not be used on aluminum or galvanized steel.

Nitric acid is primarily used for cleaning stainless steel, titanium, or zirconium. It is an extremely strong oxidizer and is rarely used, because of handling difficulties. Conventional inhibitors, detergents, and other additives are not stable in HNO_3. It should not be used on carbon steel, copper alloys, or Monel 400. Piping for handling HNO_3 should be lined with stainless steel or polytetrafluoroethylene (PTFE).

Phosphoric acid is sometimes used at 2% and 50 °C (120 °F) for 4 to 6 h to pickle and passivate steel piping. It is not as effective as HCl in removing iron oxide scale, but is preferred for cleaning stainless steels.

Phosphoric acid was originally used for removing mill scale from new boilers because it also helped passivate the surface. It is also used to brighten aluminum.

Sulfamic acid is used from 7 to 10% up to 60 °C (140 °F) to remove calcium and other carbonate scales and also iron oxides. It is not as effective as HCl on iron oxides. If NH_2SO_3H is heated above 75 to 80 °C (170 to 180 °F), it begins to hydrolyze to H_2SO_4, greatly reducing its effectiveness. If mixed with sodium chloride (NaCl), HCl is slowly released, and the mixture is more effective in removing iron oxide deposits.

Although relatively expensive, NH_2SO_3H is reasonably safe to handle and may be transported solid and diluted on-site. This greatly facilitates handling and makes it preferred for in-house cleaning.

Inhibited NH_2SO_3H can be used to clean carbon steel, copper, admiralty brass, cast iron, and Monel 400. Without the chloride addition, it is effective on most stainless steels. The addition of

Table 4 Scales and solvents

Scale component	Solvent(a)	Testing conditions
Iron oxide	5 to 15% HCl	65–80 °C (150–175 °F)
Fe_3O_4 (magnetite or mill scale)	2% hydroxyacetic/1 formic	65–80 °C (150–175 °F) circulating
Fe_2O_3 (red iron oxide or red rust)	Monoammoniated citric acid	85–105 °C (185–220 °F) circulating
	Ammonium EDTA	75–150 °C (170–300 °F) circulating
	EDTA organic acid mixtures	40–65 °C (100–150 °F) circulating
Copper, oxides	Copper complexor in HCl	65 °C (150 °F)
	Ammoniacal bromate	50–85 °C (120–185 °F)
	Monoammoniated citric acid	60–85 °C (140–185 °F), pH 9 to 11
	Ammonium persulfate	Below 40 °C (100 °F)
	Ammonium EDTA	65–85 °C (150–185 °F), pH 9 to 11
Calcium carbonate	5 to 15% HCl	Preferably not above 65 °C (150 °F)
	7 to 10% sulfamic acid	Do not exceed 60 °C (140 °F)
	Sodium EDTA	Circulate at 60–150 °C (150–300 °F)
Calcium sulfate	Sodium EDTA	Circulate at 60–150 °C (150–300 °F)
	1% NaOH-5% HCl	Circulate at 50–65 °C (120–150 °F)
	EDTA organic acid mixtures	40–65 °C (100–150 °F) circulating
Hydroxyapatite of phosphate compounds ($Ca_{10}(OH)_2{\cdot}(PO_4)_6$)	5 to 10% HCl	Preferably above 65 °C (150 °F)
	Sodium EDTA	Undesirable to add fluoric
		Circulate at 65–150 °C (150–300 °F)
	Sulfamic acid 7 to 10%	Do not exceed 60 °C (140 °F)
Silicate compounds, for example, acmite ($NaFe(SiO_3)_2$) and analcite ($NaAlSi_2O_6{\cdot}H_2O$)	Prolonged treatment with 0.5 to 1% soda ash at 345 kPa (50 psi), follow with HCl containing fluoride	Alkaline preboil at 345–690 kPa (50–100 psi) for 12 to 16 h
Pedtolite ($4Ca{\cdot}Na_2O{\cdot}6SiO_2{\cdot}H_2O$)	HCl containing ammonium bifluoride	65–80 °C (150–175 °F)
Serpentine ($Mg_3Si_2O_7{\cdot}2H_2O$)		
Sulfides ferrous: troilite (FeS) and pyrrhotite (FeS)	HCl, inhibited	Heat slowly to avoid sudden release of H_2S toxic gas
Disulfides: FeS_2, marcasite and FeS_2, pyrite	Chromic acid, followed by HCl	Boiling 7 to 10% chromic acid, followed by inhibited HCl
Organic residues Organo lignins Algae Some polymeric residues	Potassium permanganate, followed by HCl containing oxalic acid	Circulate at 100 °C (210 °F), add 1 to 2% $KMnO_4$ solution. Oxalic acid added to HCl controls release of chlorine toxic gas

(a) The chemicals listed are to be considered possible solvents only. There are many alternative solvents for each deposit listed.

ferric ion (Fe^{3+}) is used to protect 400-series stainless steels.

Formic acid is generally used as a mixture with citric acid or HCl because alone it is unable to remove iron oxide deposits. Formic acid can be used on stainless steels, is relatively inexpensive, and can be disposed of by incineration.

Hydroxyacetic-formic acid is used as a 2:1% hydroxyacetic-formic mixture at 80 to 105 °C (180 to 220 °F) to remove iron oxide and calcium deposits. When added, NH_4HF_2 will also remove silica-containing deposits. The acid must be inhibited for use on carbon steel.

Hydroxyacetic-formic acid is especially advantageous in nondrainable sections of reactors or boilers, because it may be decomposed to harmless by-products by heating to 150 to 175 °C (300 to 350 °F). The primary use of this cleaner is to remove Fe_3O_4 in critical boilers and in superheater sections of other boilers. It can be used on stainless steels and is applicable in mixed-metal systems.

Acetic acid is used to clean calcium carbonate scales, but it is ineffective in removing iron oxide deposits. Weaker than formic acid, it may be preferred where extremely long contact times are necessary.

Citric acid at 3% and 65 °C (150 °F) is used to clean iron oxide deposits from aluminum or titanium. Monoammoniated citric acid is effective in removing iron oxide deposits at pH 3.5. By adjusting the pH to 9 and adding an oxidant, such as $NaNO_2$, copper can be removed with the same solution. This technique is patented. Citric acid is also used to scavenge residual iron oxide from boilers after HCl cleaning in order to facilitate passivation treatment.

Alkaline boilout solutions are used to emulsify and disperse various oils, greases, and organic contaminants from new steel or stainless steel equipment, such as drum boilers and retubed heat exchangers. Examples are combinations of NaOH, trisodium phosphate (Na_3PO_4, or TSP), sodium carbonate (Na_2CO_3, or soda ash), and sodium silicate (Na_2SiO_3). These are added in concentrations of 0.5 to 1% and boiled or circulated for many hours, often under pressure. A low-sudsing dishwasher-type detergent is usually added. Proprietary compositions are available from suppliers.

Some operators avoid using NaOH because of the possibility of SCC (caustic embrittlement) of steel. A controlled amount of sodium nitrate ($NaNO_3$) may be added to the NaOH to inhibit cracking.

Alkaline emulsions are used to remove coatings of oily material from inorganic deposits that prevent their dissolution by cleaning chemicals. For light-bodied oil, a Na_3PO_4-surfactant combination may be circulated for a few hours at 95 °C (200 °F) to remove it. The equipment is then flushed with water.

Chelants will complex such deposits as iron oxide or copper. They include ethylenediaminetetraacetic acid (EDTA), citric acid, gluconic acid, and nitroacetic acid.

Chelating agents must be pH adjusted by using acids or alkalis, depending on the type of chelating agent and the scale constituents to be complexed. For example, under alkaline conditions, the sodium salt of EDTA will complex calcium-containing deposits, but it will not remove iron oxide.

The most common chelating agent is the ammonium salt of EDTA, which is used to remove Fe_3O_4 scale from boilers at 75 to 150 °C (170 to 300 °F). Following this, the solution is oxidized using air to remove copper. This solution must be inhibited to prevent corrosion of the steel.

Some chelating processes are patented, such as the Alkaline Copper Removal Process (Ref 16), which uses triammonium EDTA at a pH of 9.2 for removing copper and iron oxides from a boiler while it is intermittently fired. Another patented chelating agent is the Citrosolv Process (Ref 17), which uses inhibited ammoniated citric acid to dissolve iron oxides and copper in utility steam generators. Chelating agents can often be incinerated, minimizing disposal problems.

Copper complexers, such as thiourea and hexhydropyrimidine-2-thione, are added to HCl when both iron and copper oxide deposits are present. The complexers help dissolve the copper and prevent it from plating out on steel in the system.

Chromic acid is used at 7 to 10%, primarily to remove iron pyrite and certain carbonaceous deposits that are insoluble in HCl. The iron pyrite is oxidized to the sulfate without the evolution of H_2S. With carbonaceous deposits, H_2CrO_4 degrades the tar that holds the deposits together.

Chromic acid is a strong oxidizing agent and should not be used on copper, brass, bronze, aluminum, zinc, or cast iron. It is acceptable on carbon steel and stainless steel. Steel piping should be used for handling it because nonmetallic hose materials, other than PTFE, are rapidly degraded.

Caustic permanganate is widely employed as a pretreatment in refinery-cleaning operations to remove carbonaceous deposits coupled with pyrophoric sulfides. A solution of 1% NaOH and 0.7% $KMnO_4$ is used at 65 to 80 °C (150 to 175 °F). This converts iron sulfide to iron oxide and sulfur, both of which can be removed with acid cleaning. Little or no H_2S is generated. If HCl is used for the acid-cleaning step, oxalic acid should be added to reduce any chlorine gas that might be evolved upon reaction with the insoluble manganese formed during the first stage of cleaning.

Organic solvents used to dissolve organic deposits include kerosene, xylene, toluene, xylene bottoms, heavy aromatic naphtha, 1,1,1-trichloroethane, N-methyl-2-pyrrolidone, alcohols, glycols, and ortho-dichlorobenzene (Ref 18).

Some solvents are strictly regulated by Environmental Protection Agency (EPA), Occupational Safety and Health Administration (OSHA), and state and local regulations. The use of any solvents should be carefully reviewed. With few exceptions, rubber hoses or rubber-lined equipment should not be used to handle organic solvents.

Organic emulsions are solvent-based cleaners for heavy organic deposits, such as tar, asphalts, and polymers. These solutions typically consist of an aromatic solvent, a chlorinated hydrocarbon, and emulsifying agents. The mixture should be capable of being heated to 65 to 70 °C (150 to 160 °F). A thorough rinsing step is necessary to remove the sludge left behind when the cleaning agent is drained. The possibility of hydrolysis of the chlorinated hydrocarbons prevents use of these mixtures on stainless steels.

Passivating solutions are used to prevent carbon steel from rusting after acid cleaning, before it is returned to service. The term passivation is also applied to procedures that are used to remove surface iron contamination from stainless steel and titanium equipment.

Most passivating solutions for carbon steel are alkaline and contain an oxidizing agent, even if it is only air. They are applied after rinsing and neutralization.

For the passivation of stainless steel and titanium, mild iron contamination may be removed by using a mixture of 1% each of citric acid and HNO_3. For more persistent contamination, strong HNO_3 solutions must be used (Ref 19).

Corrosion inhibitors are added to chemical cleaning agents to permit effective cleaning without excessive corrosion of the equipment. Most inhibitor formulations are organic compounds and function in three ways (Table 5):

- *Cathodic inhibitors* impede the cathodic half of the corrosion reaction, for example, $2H^+ + 2e^- \rightarrow H_2$
- *Anodic inhibitors* limit the anodic half of the corrosion reaction, for example, $Fe \rightarrow Fe^{2+} + 2e^-$
- *Adsorption inhibitors* form a physical barrier on the metal surface and prevent corrosion

Inhibitors must be capable of dispersing or dissolving in the chemical cleaning solution, both fresh and contaminated. They must also be compatible with the equipment materials of construction, for example, low-chloride inhibitors should be used on stainless steels.

The concentration of the inhibitor used is related to the type of solvent, its temperature, the equipment materials of construction, the surface-to-volume ratio in the equipment, and the degree of turbulence. Typical film-forming inhibitor concentrations for mineral acids on steel are of the order of less than 0.2%.

Most common inhibitors are proprietary formulations, and technical data may be obtained from suppliers. Inhibitors often contain dyes, which confirm whether or not the inhibitor has been added.

Chemical Cleaning Procedures

In addition to selecting the optimum solvent, other factors essential to successful chemical cleaning are the development of a safe and effective procedure and the prevention of excessive equipment corrosion. Selection of the chemical cleaning contractor is important because the quality of individual crews can vary. Also, the contractor must have the ability and experience to use specific procedures, some of which may require unique steps or patented procedures. The proposed procedure should be written and included in a scope of work that is mutually agreeable to the owner and the contractor. When the job is complete, the key elements must be documented.

Planning Stage. The proposed cleaning procedure should be reviewed with the contractor or personnel who will conduct the cleaning. The

Table 5 Classification of selected corrosion inhibitors

Cathodic	Anodic	General adsorption
Alkylamines	Benzotriazole (H)	Acetylenic alcohols
Benzotriazole (L)	Mannich bases	Benzotriazole
Diphenylamine	Pyridines	Dibenzylsulfoxide
Thioureas (L)	Quinolines	Diphenylamine
	Thiols	Furfuraldehyde
	Thioureas (H)	

L, at low concentration; H, at high concentration

procedure should outline the specific duties of the contractor and the owner and should include the following:

- Listing of all chemicals used, including primary solvent and corrosion inhibitor, surfactants, emulsifiers, neutralizers, passivators, and antifoams
- Solution contact times, temperatures, and pressures
- Method of corrosion control and monitoring
- Special contractor equipment required, such as filter elements, high-volume pumps, and heat exchangers
- Contractor safety procedures and owner safety requirements, for example, methods of treating hazardous by-products, such as H_2S and hydrogen cyanide (HCN), which may be generated during cleaning
- Disposal requirements for waste material, including adherence to federal, state, and local regulations
- Lay-up requirements after cleaning, if equipment is not to be used immediately

When mutually agreed upon, the procedure is developed into a scope of work.

Cleaning Stage. The equipment is readied for cleaning. After isolation, all restrictions, such as orifice plates, should be removed. Tubes of a shell and tube heat exchanger should be cleaned of obstructions before chemical cleaning. Failure to free the tubes could result in solvent remaining in tubes and possible damage at start-up. In general, the chemical cleaning operation consists of the following steps:

- Leak testing of contractor fittings
- Hooking up of cleaning system
- Establishment of circulation rate and temperature
- Blending in of chemicals
- Circulation and monitoring, which include recording of solvent strength and deposit removal. If the solvent strength remains constant for two or more consecutive readings, cleaning is usually considered complete
- Draining system of solvent
- Flushing with water
- Neutralizing, if an acidic solvent was used
- Passivation, if required
- Inspection and evaluation of cleaning

Documentation of Cleaning. A simple format will include relevant information on the equipment temperature, materials of construction, chemicals, method of circulation, time of circulation, and evaluation of equipment. The format may be modified for different pieces of equipment.

Corrosion Monitoring in Chemical Cleaning. Chemical cleaning introduces the real possibility of equipment damage from corrosion. Various precautions may be taken to eliminate damage or to reduce the corrosion rate to acceptable levels, such as reducing the cleaning temperature or contact time. Corrosion monitoring during chemical cleaning may consist of on-line electrochemical monitoring, corrosion coupons, or a bypass spool piece containing a sample of the deposit.

On-line electrochemical monitoring provides an instant record of the corrosion rate during cleaning. However, localized corrosion effects are not indicated.

Corrosion coupons should be representative of the materials of construction of the equipment being cleaned. They should be insulated from one another to avoid galvanic effect. If necessary, special coupons, such as U-bends, should be included to check for SCC. The corrosion rates of coupons are generally higher than the actual corrosion rate of the equipment during the cleaning cycle. However, excessively high coupon rates suggest that cleaning has been performed outside of the desired range of temperature, chemical concentration, or circulation rate. Experience dictates acceptable corrosion rates obtained from corrosion coupons. Localized corrosion observed on coupons is unacceptable. A bypass spool piece containing a sample of the deposit is the ideal coupon because it will indicate the effectiveness of the cleaning as well as the condition of the metal after cleaning.

Corrosion should be monitored during every chemical operation. Post-inspection is extremely important to check for damage and to gage the effectiveness of cleaning. Certain scales may increase the corrosiveness of the chemical cleaning solution as they dissolve.

Corrosion Rate. Chemical cleaning contractors measure the general corrosion rate in pounds per square foot per day ($lb/ft^2/d$). This unit is converted to mils per year (mils/yr) by using a factor that varies with the density of the metal under consideration. To obtain mils per year, multiply pounds per square foot per day by 70.3×10^3/density (grams per cubic centimeter). For steel, 0.01 $lb/ft^2/d$ = 91 mils/yr.

A significantly higher corrosion rate is tolerable during chemical cleaning than would be allowed in service. For example, 400 mils/yr obtained from coupons may be acceptable for carbon steel. With relatively thick components, such as pump casings, even higher rates may be tolerable. Less than 200 mils/yr should be the goal. Localized corrosion—for example, pitting or SCC—is not tolerable.

Localized Corrosion. The principal forms of localized corrosion that occur during chemical cleaning are pitting, SCC, and under-deposit corrosion (crevice corrosion).

Pitting occurs when filming inhibitors break down or there is insufficient inhibitor addition. Pitting will also occur with HCl and incompatible metals, such as aluminum or stainless steels. Oxidizing conditions, such as free Fe^{3+} ion, and the presence of the chloride ion (Cl^-) will also encourage pitting in metals resistant to general corrosion, especially the active-passive metals mentioned. Pitting is especially harmful on thin-wall equipment, for example, heat-exchanger tubes.

Stress-corrosion cracking is likely to occur in specific chemical alloy combinations when the threshold temperature and concentration requirements for cracking are exceeded. This usually occurs in vapor spaces, splash zones, or crevices where the cleaning agent can concentrate. If the chemical cannot be washed out, SCC may occur later, for example, during service. The two most common situations to guard against are the hydroxyl ion (OH^-), for example, from caustic, with carbon steel and the Cl^- ion, for example, from HCl, with austenitic stainless steels. In some cases, an inhibitor can be added to minimize cracking. Hydrogen-assisted cracking may also occur in hardenable low-alloy and stainless steels in specific heat-treated conditions during acid cleaning.

Under-deposit corrosion occurs when cleaning is incomplete and the deposits left behind harbor or concentrate corrosives. The corrosion is difficult to see because it is hidden by the deposits.

Temperature, Pressure, and Flow Rate. Of these factors that affect corrosion rate, temperature is the most significant. The higher the temperature of the solvent, the more effective the cleaning generally is. However, corrosion rate increases significantly with temperature and limits the temperature that can be used (Fig. 6). Increasing temperature also increases the amount of inhibitor required.

Each cleaning operation should include a review of the pressure (hydrostatic and gas) that may be experienced. Acceptable limits should not be exceeded. For example, in one operation, impervious graphite heat-exchanger tube failures were significantly reduced when the recirculating solvent pressure was decreased from 690 to 207 kPa (100 to 30 psig).

It is extremely difficult to relate flow rate or solution velocity directly to the corrosion rate. This is because the key to increased corrosion rate is turbulence. Experience is the best indicator of velocity or flow limitations during chemical cleaning.

Ferric ion corrosion is a form of general or localized corrosion that occurs on carbon steel, nickel, Monel 400, copper, brass, and zirconium during chemical cleaning. The effect is worst with HCl (because of the presence of the Cl^- ion), but can also occur in other mineral acids, citric acid, EDTA chelant formulations, and HCOOH.

Ferric iron corrosion occurs when there are heavy deposits of iron oxide coupled with aeration during cleaning. Dissolution during cleaning, together with aeration, generates the Fe^{3+} ion, which corrodes steel as follows:

$$2Fe^{3+} + Fe \rightarrow 3Fe^{2+} \quad \text{(Eq 1)}$$

Other metals can also be corroded because of the oxidizing nature of the Fe^{3+} ion, especially if the Cl^- ion is present.

Ferric ion corrosion may be prevented with specific inhibitors, for example, stannous chloride ($SnCl_2$). It prevents the formation of Fe^{3+} ion, as follows:

$$Sn^{2+} \rightarrow Sn^{4+} + 2e^-$$

$$Sn^{2+} + 2Fe^{3+} \rightarrow Sn^{4+} + 2Fe^{2+} \quad \text{(Eq 2)}$$

Proprietary Fe^{3+} ion corrosion inhibitors are also available.

Even if an Fe^{3+} ion inhibitor is added, the total iron content (ferrous and ferric) must be monitored regularly (at least every 30 min) during cleaning. Measurement of the Fe^{3+} ion content may be misleading. When the iron content reaches 2 to 3%, the solution should be removed.

Sulfide Corrosion. Sulfides are present in many refinery deposits, and H_2S may be released when sulfides are contacted by acid cleaning solutions. Sulfides and H_2S are extremely corrosive and reduce the effectiveness of corrosion inhibitors. Hydrogen sulfide will also crack hardenable low-alloy steels and blister low-strength carbon steels.

Corrosion problems with sulfides and H_2S may be reduced by lowering the acid concentration and temperature. Inhibitors are effective in preventing cracking or blistering.

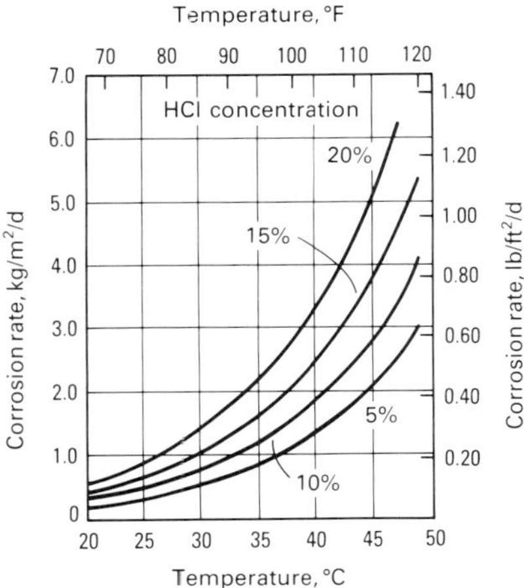

Fig. 6 Effect of temperature on corrosion of low-carbon steel in uninhibited HCl

Sulfide stringers present in free-machining steels have increased susceptibility toward corrosion during acid cleaning. Selenium-containing free-machining steels or low-sulfur regular grades are less susceptible to attack.

Effectiveness of Corrosion Inhibitor. A useful qualitative test to check whether a filming inhibitor has been added to the system before circulation is the steel wool test. In this test, a small wad of steel wool is completely immersed in a beaker containing the inhibited acid at the strength and temperature to be used during cleaning. If the inhibitor addition is effective, there should be no hydrogen evolution from the beaker for at least 5 min and the wad of steel wool should remain at the bottom of the beaker.

As noted above, by-products formed during cleaning can reduce the effectiveness of the corrosion inhibitor. Also, increasing temperature and concentration of cleaning chemicals usually necessitates a greater dose of inhibitor.

Mechanical Cleaning

Mechanical cleaning is extensively used to clean many types of equipment. There are three broad categories: hydraulic, abrasive, and thermal.

The hydraulic category includes water cleaning, high-pressure water blasting, ultrahigh-pressure water blasting, and ultrasonic cleaning. The abrasive category includes rodding, drilling, sandblasting, pigging and scraping, turbining, and explosive removal. The thermal category includes steam cleaning.

Hydraulic cleaning is carried out at pressures from 2070 kPa to 221 MPa (300 to 32 000 psig).

Water cleaning is often used to flush mud and debris from pipelines at pressures to 2070 kPa (300 psig). Slug flow flushing with air injection will obtain better scouring and will conserve water. Water, coupled with a pipeline pig, is often effective in removing scale and corrosion products from lines.

High-pressure water cleaning, the most common form of mechanical cleaning, is carried out at pressures of 6.9 to 69 MPa (1000 to 10 000 psig). It is used to remove such deposits as light rusting, minerals, and polymers. There are three types of high-pressure water cleaning: tube lancing, line moling, and shotgun jetting.

In tube lancing, a rigid or flexible lance is used to clean the inside of tube bundles, and the operation is repeated for each tube. In line moling, a short self-propelled jet nozzle (mole) and a high-pressure hose are used to clean the inside of piping. In shotgun jetting, a relatively short hand-held gun (shotgun) is used to clean surfaces that cannot be cleaned practically by the other means. Shotgun jetting is not effective in cleaning tubes.

High-pressure water cleaning reduces the risk of corrosion damage to equipment, and disposal of the water is less of a problem than with chemical cleaning. Also, the risk of generating hazardous or toxic by-products is eliminated.

The jetted stream is provided by an operator holding a water lance. The lowest pressure with appropriate volume of water is selected, consistent with achieving the required degree of cleaning. This is dictated by experience and risk to personnel.

High-pressure water cleaning must be done by trained personnel using appropriate safety devices and regularly inspected equipment. The area around the job should be suitably barricaded and sign posted.

Line friction leads to a significant loss of pump pressure. To counteract friction loss when pumping through a long hose, organic polymers have been developed that, when added to water in concentrations as low as 0.05 to 0.1%, reduce friction loss as much as 90%.

Ultrahigh-pressure water blasting is conducted at 221 MPa (32 000 psig), using a flow rate of 1 to 4 gal/min. The low flow rate means there is a very low back thrust, allowing the equipment to be operated in congested work spaces with a higher degree of safety. The ultrahigh pressure and low volume of delivery are achieved by using intensifiers rather than the standard plunger pumps. Nozzle configurations are tailored to the application. The nozzle is usually held 19 to 100 mm (3/4 to 4 in.) away from the workpiece. The tooling is specially designed for specific applications, primarily scale, rust, and polymer removal; tube bundle cleaning; and such specialized items as cleaning turbines, diaphragms, and rotors. Cleaning is not effective around more than one 90° turn. The low volume of water minimizes disposal problems.

The technique is particularly advantageous where high-pressure water blasting is ineffective and sandblasting is not desirable, for example, adjacent to pumps, valves, and compressors. Ultrahigh-pressure water blasting has been cost-effective in preparing surfaces for inspection, such as the wet fluorescent magnetic-particle inspection of deaerator welds. In these cases, an inhibitor is added to the water to prevent flash rusting.

Ultrahigh-pressure water blasting is a proprietary process and is limited to a few vendors who tailor their tools to specific jobs. It is more expensive than high-pressure water blasting, but has the advantages of quality and speed.

Abrasive cleaning is implemented where hydraulic cleaning may not be practical.

Rodding is used for lightly plugged heat-exchanger tubes where scale has not built up. Care should be taken to prevent scoring of the tubes.

Drilling may be used on tightly plugged heat-exchanger tubes. The drill is hollow, and a fluid connection is used to flush out freed material, eliminating the need to pull the drill out to clear the point and remove loose material.

Sandblasting involves cleaning the internal surface of process lines and equipment by using an abrasive blasting tool. These are available for different pipe sizes. A standard sandblast nozzle is used that impinges the sand against the conical carbide tip of the nozzle, spraying the sand out radially against the walls of the tube. A circular end plate removes most of the sand as the tool is pulled back out of the pipe.

Piping and Scraping. Pipeline pigs and scrapers are sometimes used, alone or with special chemicals, to clean pipelines. This is a combination of mechanical and chemical cleaning.

The pig is a flexible bullet-shaped foam cylinder that is propelled through the pipeline with water. A typical system consists of several pigs, launching and trapping facilities, and a water source. Pigging has been used to clean lines from 25 to 914 mm (1 to 36 in.) in diameter.

Turbining is a tube- or pipe-cleaning method that uses air, steam, or water to drive a motor that turns cutters, brushes, or knockers in order to remove deposits. Many different heads are available for different deposits and equipment, including models for curved tubes.

Overzealous turbining may damage pipes because the cutter types will cut into the base metal. Also, turbining is relatively slow and costly compared to other mechanical and chemical methods.

Explosive removal of pipe deposits requires experience and has been used on hard and brittle deposits in piping. Primacord is manufactured in various degrees of explosive power; selection of the proper strength depends on pipe size and type of deposit.

Thermal cleaning is the third major mechanical cleaning method.

Steam cleaning can be used to remove mill scale and debris from lines and equipment. When mixed with oxygen, steam cleaning can remove coke and polymers from equipment.

On-Line Cleaning

The cleaning methods discussed so far require that the equipment be removed from service and opened to some degree for the cleaning. In some cases, on-line cleaning is effective and more convenient. On-line cleaning can be chemical or mechanical, infrequent or regularly scheduled. Effective on-line cleaning can save time and labor and can prevent a shutdown. Examples of on-line cleaning include:

- Side-stream filtration
- Air bumping of heat exchangers
- Washing soft deposits from steam turbines
- Reversing flow in heat exchangers
- Passing brushes or sponge balls through exchanger tubes
- Depressing pH or adding sequestrants and dispersants in a cooling water system
- Increasing chelant or dispersant feed to a boiler

Not all cleaning jobs are conducive to on-line cleaning. Hard or insoluble deposits may not be removed effectively. In boilers, if the deposit is heavy, on-line cleaning may cause the deposit to slough and plug tubes—possibly leading to tube failures. Also, some on-line cleaning costs can be high, for example, the initial cost of installing brushes or sponge balls to clean heat exchangers. Each case must be evaluated individually to determine if on-line cleaning will be cost-effective.

Corrosion Under Thermal Insulation

William G. Ashbaugh
Cortest Engineering Services

The proliferation in recent years of corrosion failures of both steel and stainless steel under insulation has caused this problem to be of great concern to the operators of petroleum-, gas-, and chemical-processing plants. Piping and vessels are insulated to conserve energy by keeping cold processes cold and hot processes hot. Once a vessel is covered with insulation and operating satisfactorily, concern for the condition of the metal under insulation usually diminishes. Thus, corrosion of steel and SCC of stainless steel begins and develops insidiously, often with serious and costly consequences many years later. To deal with corrosion under insulation effectively and economically, a systems approach must be developed that considers the metal surface, temperatures, water, insulation, and design.

Corrosion of steel under insulation did not receive a great deal of attention by the corrosion engineering community until the late 1970s and early 1980s. Although considerable attention was being given to the SCC of stainless steel under insulation (see below), steel vessels and piping were rarely mentioned. This situation existed because of the slow corrosion rate of the steel equipment, together with the fact that many large hydrocarbon-processing plants were built 10 to 20 years earlier.

The most significant turn of events, which fortunately revealed the corrosion problem to many, was the energy shortage and the resultant emphasis on energy conservation. The effort to conserve energy soon led to replacement of much of the 10- to 20-year-old insulation with more efficient systems. As the old insulations were removed, localized, often severe corrosion damage was found.

In 1978, the Materials Technology Institute (MTI) of the Chemical Process Industry initiated a project at Battelle Columbus Laboratories to define the problem of corrosion of steel equipment under thermal insulation and identify nondestructive techniques potentially capable of detecting such corrosion (Ref 20). This work on nondestructive testing for corrosion under insulation has been continued by MTI through a second and third phase, leading to a new electromagnetic technique. Another example of the recent interest in this corrosion problem was the Joint Conference on Corrosion Under Insulation held in San Antonio, TX, in October of 1983 (Ref 21).

Stress-corrosion cracking of stainless steel under thermal insulation has been observed and recognized for many years (Ref 22, 23). The collection of and concentration of chlorides on austenitic stainless steels can lead to rapid failure of expensive equipment. The study of this phenomenon has attracted much attention, and many approaches have been attempted in order to alleviate the problem. The roles of the insulation materials, atmospheres, moisture entry, and operating temperature have been investigated (Ref 21). Many of these concepts can also be applied to the general corrosion of steel under insulation. This section is divided into two parts: one discusses steel, and the other discusses stainless steel, because the nature of the problem, aside from the presence of insulation, is different.

Corrosion of Steel Under Insulation

The Problem. Steel does not corrode simply because it is covered with insulation. Steel corrodes when it contacts water and a free supply of oxygen. The primary role of insulation in this type of corrosion is to produce an annular space in which water can collect on the metal surface and remain, with full access to oxygen (air).

The most active sites are those where the steel passes through the insulation and on horizontal metal shapes—for example, insulation support rings where water can collect. Figures 7 to 9 show some of the types of damage that occur near insulation entry points.

The other major corrosion problem develops in situations where there are cycling temperatures that vary from below the dew point to above ambient. In this case, the classic wet/dry cycle occurs when the cold metal develops water condensation that is then baked off during the hot/dry cycle. The transition from cold/wet to hot/dry includes an interim period of damp/warm conditions with attendant high corrosion rates. Figures 10 and 11 show examples. In both cases, if the lines had operated constantly either cold or hot, no significant corrosion would have developed.

The Mechanism. The corrosion rate of steel in water is largely controlled by two factors: temperature and availability of oxygen. In the absence of oxygen, steel corrosion is negligible.

In an open system, the oxygen content decreases with increasing temperature to the point where corrosion decreases even though the temperature continues to increase (Fig. 12). A num-

Fig. 7 Heavy corrosion of a steel support bracket where it passed through cellular glass insulation on a cold-temperature column. Moisture condensed on the column shell where the support was warmed by the extended attachment.

Fig. 8 Corrosion around the ladder clip of a gas-processing column. The column was painted with an inadequate protective coating (0.05 mm, or 2 mils, of epoxy) that could not prevent corrosion near the moisture entry points.

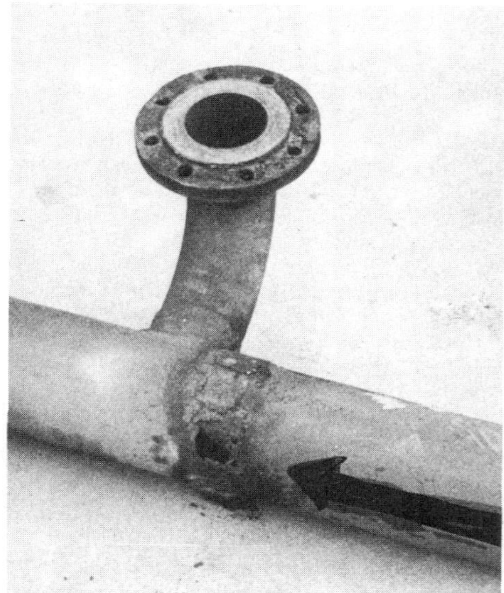

Fig. 9 Steel pipe corroded on the bottom side where a poorly sealed insulation joint allowed water to enter the insulation/pipe annulus. The moisture ran around the pipe and collected on the bottom.

Fig. 10 Heavy general corrosion of a steel line insulated with cellular glass and aluminum outer jacketing. The line is at ambient temperature except when it is used to vent hydrocarbons. Temperature then decreases, causing moisture condensation, followed by warming to ambient temperature and slow corrosion. The line was replaced after 11 years of service.

Fig. 11 Severe corrosion of a steel line in cyclic hot and cold service. The line was originally in fuel gas service, painted with inorganic zinc, and left uninsulated. It was changed to hydrocarbon/steam cyclic service and insulated for safety reasons. Wetting of the line under the end of the insulation and failure of the inorganic zinc paint caused corrosion in less than 3 years.

ber of recent case histories of corrosion under insulation have been reviewed, and the estimated corrosion rates have also been plotted in Fig. 12. These field data confirm that corrosion of steel under insulation increases steadily with increasing temperature, matching the curve for a closed system.

The problem of steel corrosion under insulation can be classified as equivalent to corrosion in a closed hot-water system. The thermal insulation does not corrode the steel, but, more correctly, forms an annular space where moisture collects. The insulation also forms a barrier to the escape of water or water vapors, which hastens the normal corrosion rate of wet steel.

In an attempt to determine which insulation contributes most and which contributes least to corrosion of steel, tests were conducted that compared 12 common insulation materials (Ref 24). After a 1-year exposure to the elements, dousing with water, and daily heating with steam, it was concluded that, although the insulation materials that absorbed water developed higher corrosion rates than those that did not, the difference between the two was not great. Therefore, efforts to select a type of insulation to prevent corrosion would not be practical. Field observation and reports from the literature confirm that corrosion of steel occurs under any and all types of insulation.

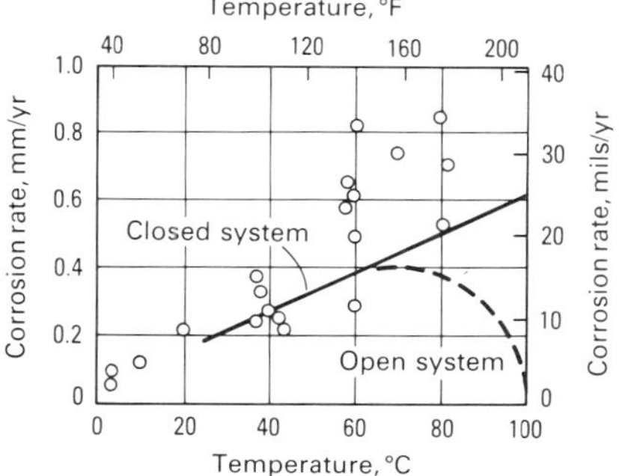

Fig. 12 Effect of temperature on corrosion of steel in water. Data points are from actual plant measurements of corrosion under insulation.

Prevention. One attempt to prevent corrosion damage involves adding a corrosion inhibitor to the insulation. Inhibitors such as Na_2SiO_3 are added for this purpose. The practical effects of such an approach are questionable. In one set of laboratory tests, inhibited insulation material was submerged in water, then set out to dry. After several cycles of wetting and drying, the insulation was found to have no benefit over uninhibited insulation of the same brand.

A corrosion inhibitor for insulation must be water soluble so that it can be dissolved and carried to the steel surface. If the insulation is waterproofed on the outside, water will enter through cracks and openings in the insulation system. This water is in the annulus and contacts only the inner surface of the insulation. The water-soluble inhibitor will soon be extracted and displaced by the water running down the vessel walls. Therefore, relying upon a consumable inhibitor for long-term protection may not be dependable.

If water entry to the insulation/steel annulus is the problem, perhaps the solution lies in more effective waterproofing. Advancements in various metal jacketings, sealants, and mastics have improved the protection of insulation materials from water, weather, and man. However, these waterproofing systems probably do not prevent water entry, because, in order to seal and prevent the annulus from breathing with temperature changes, the seal would have to be equivalent to a pressure vessel.

Waterproofing systems are designed to keep the insulation dry. They are not capable of preventing water vapor and air from contacting the annular space. In fact, a waterproofing system keeps the water vapor in the annulus and effectively produces the hot-water closed system, with attendant high corrosion rates.

Therefore, selection of insulation, inhibiting the insulation, and waterproofing of the insulation are not effective deterrents to corrosion. Instead, a high-quality protective coating, correctly applied to the steel before insulation, can offer long-term protection. Although a specific coating system cannot be recommended, the requirements that must be met by the coating can be specified.

In the insulation industry, a primer is sometimes used, particularly under spray-on foam insulation materials. It should be understood that the purpose of the primer is to present a clean surface for bonding of the insulation. The primer-type paints are not intended to and will not prevent corrosion by hot water. The key to specifying a paint system is to remember that it will be exposed to hot water vapors, a very severe environment for paints.

The use of inorganic zinc paint as a primer, coupled with the hot water, will not cause accelerated corrosion of the steel. Zinc and the silicate binder are both dissolved by hot water. Even if conditions favored the zinc becoming cathodic to steel, the protective coating binder would dissolve and the coating would break down. The piping shown in Fig. 11 had inorganic zinc paint, which dissolved when the pipe was insulated and put into hot/cold cyclic service.

The advisability of using inorganic zinc under other protective coatings is not as clear. Some tests seem to indicate good performance. Others show that inhibited primers with topcoats achieve maximum performance in hot-water systems. Therefore, the main criterion for a protective coating under insulation is that it resist hot water and water vapors. This severe service will also require high-quality surface preparation.

Inspection. The final subject regarding corrosion of steel under insulation is monitoring or inspection. Due to the widespread occurrence of corrosion under insulation, the need for rapid, quantitative nondestructive testing inspection of the steel has developed. The most common procedure is to cut plugs or windows in the insulation, exposing the steel for ultrasonic thickness testing.

Spot thickness testing is relatively accurate, but covers only a very small portion of the vessel or piping. Corrosion under insulation tends to be localized; therefore, spot checking can miss potential trouble areas. Several private inspection companies are working on devices to measure corrosion of steel through insulation by eddy-current techniques. Although these seem to have merit, they are greatly restricted by the usually complicated geometry of a chemical plant piping system.

In 1979, MTI commissioned a study to evaluate this corrosion problem and review the state-of-the-art for inspection through insulation (Ref 20). This project was followed by a second phase to evaluate use of x-rays and fluorescent image enhancers plus continuous television monitoring (Ref 25). The results showed that piping could be inspected at 1.5 m/s (5 ft/s) and revealed significant corrosion. The practical problems of using the massive inspection equipment in the field makes its usefulness in the plant uncertain.

A third MTI study was authorized to develop a special-frequency electromagnetic analyzer. The work is in progress and is reportedly meeting with success. Field trials are planned to evaluate the practical feasibility of this method in the plant.

However, the established procedure of cutting windows and measuring directly is still the most widely used. With an enhanced understanding of the mechanism, the inspectors can make better choices about where to spot check in order to locate corrosion. Specific areas that should be on the inspection checklist include:

- All surfaces exposed to frequent hot/cold temperature cycling
- Cold-temperature equipment where nozzles, clips, and brackets extend through the insulation
- The hot-to-cold interface area of cold-temperature distillation columns with a hot base temperature
- Horizontal piping, particularly at joints or piping branches and on the bottom of the pipe
- Wherever the insulation weather barrier has been mechanically damaged or removed
- Wherever the insulation has changed shape or started to swell, indicating a possible rust build-up

Once the steel equipment has been covered with a good protective coating, the inspection at windows can be made less frequently and is necessary only to monitor the condition of the paint. Thickness readings are not necessary unless the paint film has failed.

SCC of Stainless Steel Under Insulation

The Problem. During and after World War II, many new petrochemical plants were built in the United States. Many of these plants contained equipment built with the new austenitic stainless steels. These alloys, such as types 304, 316, and 347, were widely used to combat process corrosion and to maintain product purity. As in all chemical processes, a large percentage of this equipment was insulated for thermal efficiency.

During this period, corrosion and materials engineers discovered some of the limitations of these stainless "wonder metals." In addition to crevice corrosion and weld decay, the engineers encountered SCC. In 1956, a paper was published describing a lab test for evaluating insulation materials as SCC initiators (Ref 26).

As the insulation systems and their weather barrier coatings aged, more incidents of SCC under insulation began to occur in the chemical-processing industry (Fig. 13). In the following years, much work was done to investigate the problem and the counter measures (Ref 23, 26).

The Mechanism. Soon after the widespread use of the 18-8 austenitic stainless steels, beginning in the late 1930s, it became clear that the Cl^- ion in water could be very damaging (Ref 27). In addition to causing localized corrosion, such as pitting and crevice corrosion, rapid failure was seen in the form of a fine network of transgranular cracking (Fig. 14). This pattern of cracking is very destructive and is found on the surface (Fig. 13) as well as in cross section (Fig. 14).

The theoretical processes of chloride SCC of the 18-8 stainless steels are under investigation. However, it has been established that four conditions are necessary for SCC to develop:

- An 18-8 austenitic stainless steel
- The presence of residual or applied surface tensile stresses
- The presence of chlorides; bromide (Br^-) and fluoride (F^-) ions may also be involved
- The presence of an electrolyte (water)

When these conditions are present, the occurrence of SCC is highly probable.

The stainless steel alloys susceptible to SCC are generally classified as the 18-8's. This includes the molybdenum-containing grades (types 316 and 317), the carbon-stabilized grades (types 318, 321, and 347), and the low-carbon grades (types 304L and 316L). Many variations of the basic 18-8 have been developed in order to combat SCC. These variations are higher in nickel, chromium, and molybdenum, and there are also grades with lower nickel and higher chromium (duplex stainless steels). All of these alloys are highly resistant to SCC and therefore not part of the problem of SCC under insulation.

For SCC to develop, sufficient tensile stress must be present in the material. If the tensile stress is eliminated or greatly reduced, cracking will not occur. The threshold stress required to develop cracking depends somewhat on the severity of the cracking medium. Most mill products, such as sheet, plate, pipe, and tubing, contain enough residual tensile stresses from processing to develop cracks without external stresses. When the austenitic stainless steels are cold formed and welded, additional stresses are imposed. As the total stress increases, SCC becomes more severe.

The Cl^- ion is damaging to the passive protective layer on the 18-8 stainless steels. Once the passive layer is penetrated, localized corrosion cells become active. Under the proper set of circumstances, SCC can lead to failure in only a few days or weeks. Sodium chloride, because of its high solubility and widespread presence, is the most common culprit (Ref 28). This neutral salt is the most common, but not the most aggressive. Chloride salts of the weak bases and light metals, such as LiCl, $MgCl_2$, and $AlCl_3$, can rapidly crack the 18-8 stainless steels under the right conditions of temperature and moisture content. The practical aspects of SCC of stainless steels are discussed in Ref 27.

The concentration of chlorides necessary to initiate SCC is difficult to ascertain. Researchers have developed cracking in solutions with remarkably low levels of chlorides ($<$10 ppm). The situation of chlorides under insulation is unique

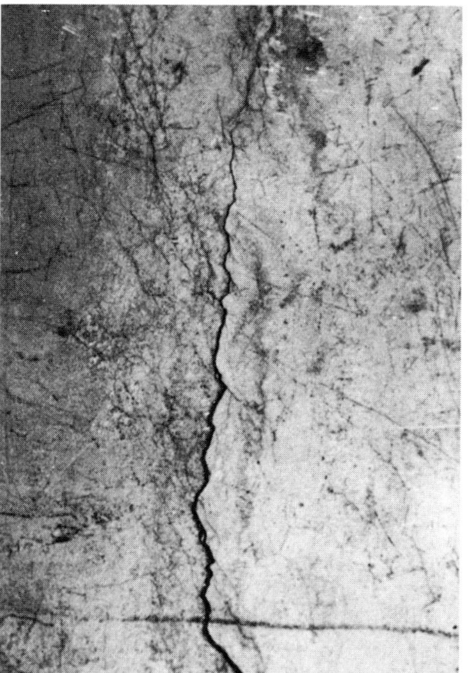

Fig. 13 Typical chloride SCC of stainless steel under insulation. The major crack is accompanied by a branched network of very fine cracks. Approximately actual size

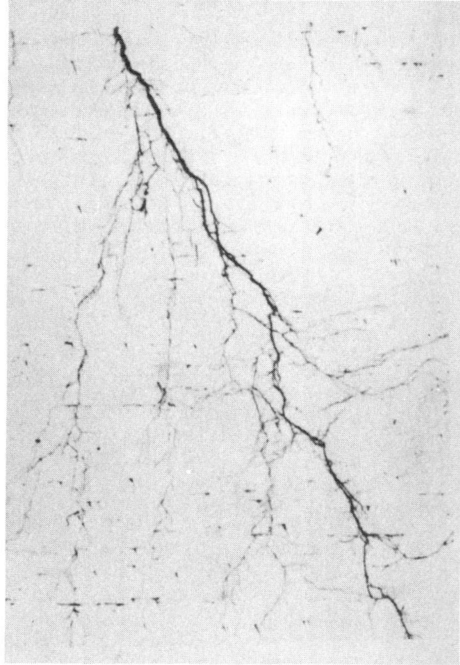

Fig. 14 Fine transgranular cracks typical of those found in 18-8 stainless steels. 80×

Fig. 15 Type 316 stainless steel pipe flange lap showing external SCC

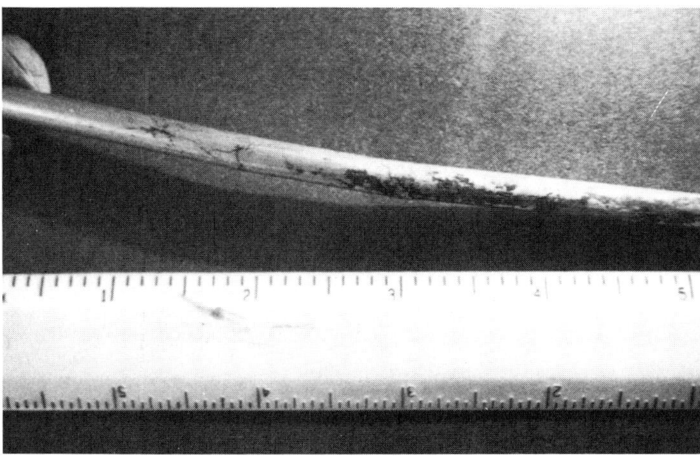

Fig. 16 Type 304 stainless steel instrument tube showing external SCC under an asbestos tape wrapping used to protect plant personnel

and ultimately depends on the concentration of chlorides deposited on the external surface of the metal. The concentrating mechanism is discussed in some detail in the following Section of this article. Therefore, the amount of chlorides present becomes a debatable issue. If any chlorides are detected, there will probably be some localized sites of high concentration.

The most important condition affecting chloride concentration is the temperature of the metal surface. Temperature has a dual effect: first, elevated temperatures will cause water evaporation, which in turn concentrates the chloride salts, and, second, as the temperature increases, the rate of the corrosion reaction increases.

Finally, chlorides have been discussed because they are the most common and aggressive of the halogen family. Bromides and fluorides may also cause SCC, but are less common and probably less aggressive (Ref 27).

Water is the fourth necessary ingredient in SCC. Because SCC is an electrochemical reaction, it requires an electrolyte. As water penetrates the insulation system, it plays a key role at the metal surface, depending on the equipment operating conditions.

Examination of the phenomenon of corrosion of steel under insulation provides a better appreciation of the widespread intrusion of water (Ref 29-31). In effect, water must be expected to enter the metal/insulation annulus at joints or breaks in the insulation and its protective coating. The water then condenses or wets the metal surface, or if it is too hot, the water is vaporized (Ref 29).

This water vapor (steam) penetrates the entire insulation system and settles into places where it can recondense. Because the outer surface of the insulation is designed to keep water out, it also serves to keep water in. The thermal insulation does not have to be in poor condition or constantly water soaked. A common practice in chemical plants is to turn on the fire protection water systems on a regular basis. This deluges the equipment with chloride-bearing water. Some seacoast locations use seawater for the fire protection water. Hot food-processing equipment is regularly washed with tap water, which contains chlorides. All insulation system water barriers eventually develop defects. As the vessel/insulation system breathes, moist air contacts the metal surface. From the insulation standpoint, the outer covering acts as a weather barrier to protect the physical integrity of the insulation material. The outer coverings are not intended, nor can they be expected, to maintain an air- and watertight system.

Other Problems. External SCC has been found in other situations similar to those that occur under thermal insulation. One of the more common locations is under slip-on flanges, which create crevices that often fill with rust from corrosion of the flanges. This rust pocket is an ideal place for the collection and holding of water and chlorides. Figure 15 shows a cracked type 316 stainless steel lap nozzle.

Another example is the common instrument tube. Because these tubes often carry steam or hot process materials, it is necessary to wrap them for the protection of personnel. Figure 16 shows widespread external SCC of stainless steel tubing under asbestos tape wrapping.

In yet another application, polyethylene tape was wrapped around a painted stainless steel line at each pipe support location. The hot line caused the tape to loosen, taking some of the paint with it. The loose tape also created pockets where water and chlorides collected, resulting in external SCC.

In a final example, a pipe fabricator affixed the company logo sticker onto the stainless steel pipe. The pipe was subsequently painted according to specifications, but without first removing the label. The label eventually peeled off, which exposed the bare pipe to a severe chloride-filled atmosphere. This situation resulted in external SCC (Fig. 17).

Prevention of External SCC. As an understanding of external SCC and its mechanisms developed, selection of a preventive method became much easier. Table 6 summarizes the possible methods of prevention that apply to each causative agent. As can be seen, application of a suitable protective coating system is generally the most economical method, although other methods are included and may be practical under certain circumstances.

The critical step then becomes the implementation of whichever preventive method has been selected. Assuming a protective coating has been chosen, it is necessary to convince project or operating managers of the benefits of painting stainless steel. Although painting stainless steel may at first seem unusual, experienced field personnel will usually understand the need for preventive measures.

Application of any protective coating requires a good specification and inspection. The use of manufacturer application guidelines and the knowledge of an experienced inspector are indispensable in producing an acceptable protective coating.

Fortunately, protective coatings work very well on stainless steel because the metallic substrate does not oxidize or rust. Therefore, coating adhesion failures are very rare. The main objective is to achieve a continuous coating at a reasonable cost that will resist the hot-water environment encountered under insulation.

Fig. 17 External SCC under a logo sticker on a type 316 stainless steel pipe. The label was painted over, then came loose, allowing water to contact the bare steel. The label adhesive may also have been high in chlorides.

Corrosion by Sulfuric Acid

S.K. Brubaker
E.I. Du Pont de Nemours & Company, Inc.

Sulfuric acid is the largest volume inorganic acid currently in use and is generally considered to be the most important industrial chemical. Sulfuric acid is made by the contact process, in which elemental sulfur or sulfur-containing waste is burned to form sulfur dioxide (SO_2). Sulfur dioxide is converted to sulfur trioxide (SO_3) by contact with a vanadium catalyst. The SO_3 is absorbed in oleum (fuming H_2SO_4) and H_2SO_4 in a series of towers. Because of heightened environmental concerns, sources of sulfur, such as smelter or power plant stack gases, are now being converted into H_2SO_4.

The corrosiveness of H_2SO_4 depends on many factors, particularly temperature and concentration. Strong, hot conditions present the greatest problems, and few materials except platinum, tantalum, fluorocarbon plastic, and brick-lined steel will resist 60 to 98% H_2SO_4 at 120 °C (250 °F). However, other variables also influence the resistance of materials to H_2SO_4. The presence of oxidizing or reducing contaminants, velocity effects, solids in suspension, and galvanic effects can alter the serviceability of a particular material of construction.

It is unwise to select materials of construction for equipment that will handle H_2SO_4 solely on the basis of published corrosion data unless the conditions involved are adequately and specifically covered by the reference data. Seemingly minor differences in impurities or environmental conditions may significantly affect actual service corrosion rates. Impurities such as halides generally increase corrosion. Aeration or the presence of oxidizing agents generally accelerates corrosion of nonferrous materials and reduces corrosion of stainless alloys, but the extent of these effects depends on specific conditions. Hot-wall effects are frequently overlooked, and heating coils made of the same material as the containing vessel can corrode rapidly while the condition of the vessel itself remains satisfactory.

It is, therefore, advisable to consider all general corrosion data only as an indicator of relative resistance and as a guide by which the limiting conditions of materials may be further reviewed. Final selection of materials for specific equipment depends, of course, on such factors as allowable corrosion rate, desired mechanical and physical properties, fabrication requirements, availability, and cost (Ref 32).

Corrosion Mechanisms

The corrosion of metals in H_2SO_4 is complex, and an understanding of electrochemical theory is useful. The electrochemical behavior of most metals falls into three categories: active, passive, or active-passive (Fig. 18).

For active behavior (Case I), the corrosion potential is in the active region, and a wide range of corrosion rates are possible. For passive behavior (Case III), the anodic and cathodic current curves intersect in the most stable passive region; these alloys generally passivate spontaneously and exhibit low corrosion rates. For active-passive behavior (Case II), the cathodic curve intersects the anodic curve at three potentials: one in the active region and two in the passive region. Because the middle region is not stable, the intersections indicate the possibility of unpredictable and erratic corrosion rates, depending on the environment. Slight changes in the environment may cause a shift to active behavior. In fact, a regular oscillation is often observed between active and passive behavior. Thus, Case II corrosion is difficult to study and presents the most risk for materials selection. Environmental conditions that appear identical from one lab to another, from one plant to another, and from one day to another can produce widely varying results. Stainless steels and nickel-base alloys frequently exhibit Case II behavior when being pushed to their limits in H_2SO_4 service (Ref 33-35).

Carbon Steel

General Resistance. Steel has long been used in handling concentrated H_2SO_4 at ambient temperatures under static and low-velocity conditions (<0.9 m/s, or 3 ft/s). A soft sulfate film forms that is highly protective unless physically disturbed. The actual corrosion rate of steel depends on temperature, acid concentration, iron content, and flow, because these parameters determine the dissolution rate of the protective sulfate film. Figure 19 shows the corrosion resistance of steel as a function of temperature for nonflowing conditions at concentrations above 65%. In the concentration range of 65 to 100%, steel is applicable for ambient-temperature storage tanks and piping at low velocities. Within the 100 to 101% range, steel exhibits potentially catastrophic increases in corrosion rates as the temperature exceeds 25 °C (75 °F) and is not recommended, except where low temperatures and velocities can be ensured. Steel is used in

Table 6 Guidelines for selecting external SCC preventive

Causative agent	Preventive method	Comments	Evaluation
Austenitic stainless steel	Change to SCC-resistant alloy	Stainless steel alloys with >30% Ni and the duplex stainless alloys are alternative choices, but cost considerably more and may not be readily available.	Extra cost compared to other preventive methods makes this an unwise choice.
Tensile stress	Thermal treatment (anneal or stress relieve)	Annealing at 1065 °C (1950 °F), followed by water quenching will distort and scale equipment severely. Stress relieving at 955 °C (1750 °F) and slow cooling will sensitize the grain structure and cause some warpage and scaling. Note: A stress-relieved vessel or pipe will be subjected to tensile stresses in assembly and under operating conditions. May override the thermal treatment	Generally not practical for piping and vessels; may be used for small individual components.
	Shot peen	Shot peening converts the surface stresses to compressive stress and is a proven SCC preventive method. It is a delicate process requiring specific skills and experience. May be costly or difficult to apply in the field	Should be considered, but may be more costly and difficult to obtain than other prevention methods.
Chlorides	Remove or eliminate Cl^- ion	Because of their widespread occurrence, highly soluble chloride salts are difficult to avoid or keep off of equipment.	Not practical
	Apply barrier coating to stainless steel	Use of a protective coating on the stainless steel surface can prevent Cl^- contact with the alloy.	This is a practical and proven preventive method.
		Wrap stainless steel with aluminum foil, which serves as both a barrier coating and cathodic protection anode.	Being used with success; extended life of the aluminum has not been determined.
Water	Improve waterproofing to prevent water entry	No type of coating, cementing, or wrapping of insulation can keep air and water from entering the insulation system, except for constructing an external pressure shell. Note: The application and maintenance of a weather barrier is important to good insulation performance and should have a high maintenance priority.	Not practical to expect a wrap or coating to keep water out
	Apply barrier coating to stainless steel	A carefully selected protective coating can provide long-term protection for stainless steel equipment.	This is a practical and proven preventive method.
		Use of aluminum foil wrap as above	Limited use, but with success
		Note: Use of inorganic zinc primer or paint system is not safe due to the possibility of liquid-metal embrittlement upon subsequent welding or exposure to extreme heat.	

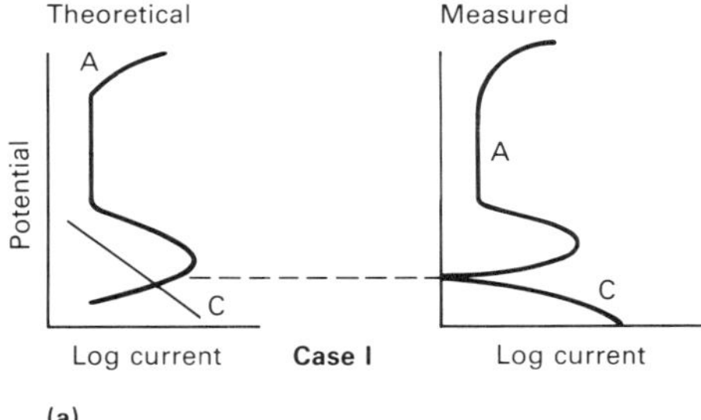

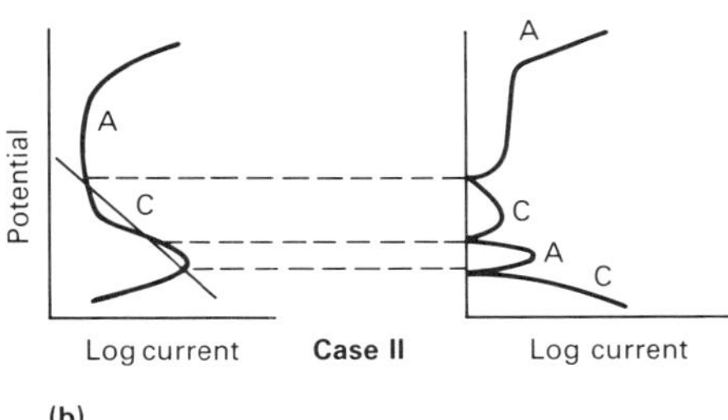

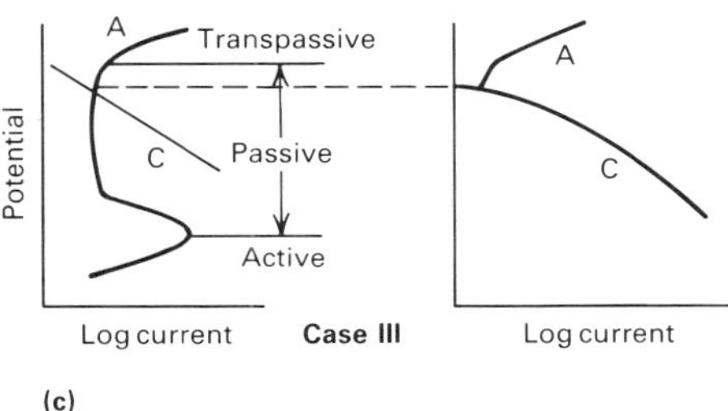

Fig. 18 Comparison of theoretical and measured anodic polarization curves for (a) active (Case I), (b) active-passive (Case II), and (c) passive (Case III) behavior. A, anodic current; C, cathodic current. Source: Ref 33

oleum service above 101% acid concentration at ambient and moderate temperatures.

Localized attack can occur even at flow velocities within the prescribed limits. Discontinuities such as short-radiused elbows, excessive penetration of welds, and pipe mismatch may cause sufficient downstream turbulence to disturb the protective sulfate film, resulting in high corrosion rates. Weldments must be thoroughly inspected to ensure that they contain no slag, surface porosity, laps, excessive penetration, or other welding defects that might initiate accelerated corrosion. In addition, steel vessels and piping should be free of mill scale, or serious pitting or nonuniform corrosion may occur.

Hydrogen grooving is another form of localized attack that occurs on vertical or inclined surfaces exposed to the liquid phase. During the corrosion of steel by H_2SO_4, atomic hydrogen is evolved. If produced in sufficient quantities, the hydrogen combines to form small bubbles that stream along preferred paths on vertical and inclined surfaces, disrupting the soft protective iron sulfate film. Channels and deep grooves may eventually be formed. Grooving is commonly observed in the tops of horizontal manways on the side of storage tanks and on the top 180° of horizontal pipe runs (Fig. 20). In piping, stagnant conditions promote grooving; therefore, a minimum velocity of 0.3 m/s (1 ft/s) is often recommended.

Grooving, combined with erosion-corrosion, also occurs on the sidewalls of tanks (Fig. 21). Location of the liquid inlet in the roof near the shell has, in at least two cases, resulted in combined erosion-corrosion and hydrogen grooving, which caused catastrophic rupture of large storage tanks (Ref 36).

Influence of Copper. The addition of copper to steels for improved resistance to H_2SO_4 is a controversial subject. Steels containing 0.02% Cu were reported to have rates 16 times higher than steels containing 0.10% Cu in 42 wt% H_2SO_4 (Ref 37). The associated electrochemical study showed that the effects of additional copper dissolved in steel reduced the exchange current of the cathodic reaction, thereby decreasing the corrosion rate. Another study confirmed the beneficial effect of 0.1 to 0.5% Cu in H_2SO_4 solutions up to about 55% (Ref 38). However, in concentrated H_2SO_4 solutions above 60%, the corrosion mechanism appears to differ, and the beneficial effect was not observed. It is generally believed that the advantages, if any, of the use of copper-bearing steels for handling strong H_2SO_4 are not sufficient to warrant the additional expense.

Anodic protection has been used to lower corrosion rates on steel tanks storing 93 to 99% H_2SO_4. The intent is to prolong tank life and to minimize iron pickup. Typically, corrosion rates can be lowered by 50 to 80%. In this process, the steel is electrochemically driven from the active to the passive region by the application of an applied potential between a cathode and the steel. If proper passivity is achieved, the corrosion rate decreases to less than 0.1 mm/yr (3.9 mils/yr) (Ref 39). The use of stainless steel and alloy 20Cb-3 in tanks being protected must be carefully analyzed because the passive protection potential for steel may overlap active regions for these alloys.

With improvements in the H_2SO_4 production process, acid with lower iron content is being produced. This acid is more corrosive to steel than the high-iron acid. Thus, the use of anodic protection will become increasingly important.

Mechanism of Corrosion Protection. Carbon steel corrodes in the active state at all concentrations up to 100% H_2SO_4. At concentrations below 50%, the iron sulfate corrosion product readily goes into solution, and corrosion rates are high. At higher concentrations, the initial corrosion rate is high, but is quickly reduced by the accumulation of iron sulfate corrosion product. It has been shown that the dissolution and diffusion of the ferrous sulfate away from the surface is the rate-limiting step (Ref 40). The effects of velocity, concentration, and temperature on the corrosion process in piping and storage tanks have been modeled (Ref 41, 42). At concentrations from 65 to 100% and at ambient temperatures, the sulfate layer diffuses into solution sufficiently slowly for the use of steel as a material of construction. Corrosion rates are typically 0.15 to 1.0 mm/yr (5.9 to 39 mils/yr). In the oleum range above 101% H_2SO_4 concentration (4% free SO_3), steel corrodes in the passive region because of the oxidizing effect of SO_3. Corrosion rates are typically 0.1 mm/yr (3.9 mils/yr) at ambient temperature. In the 100 to 101% concentration range, the steel goes through an active-passive transition state, and measured corrosion rates are erratic and excessive above 25 °C (75 °F).

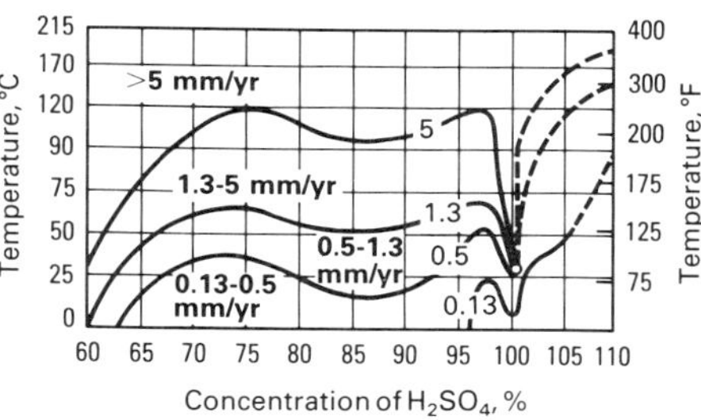

Fig. 19 Corrosion of steel by H_2SO_4 as a function of temperature and acid concentration. Source: Ref 35

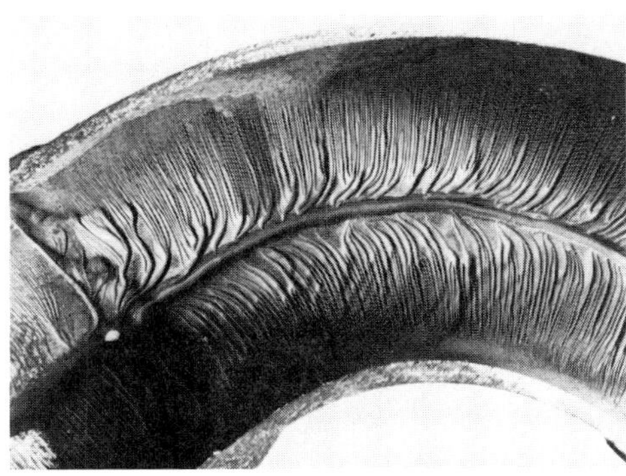

Fig. 20 Hydrogen grooving of a 75-mm (3-in.) diam steel elbow. The elbow was sectioned; the top half is shown.

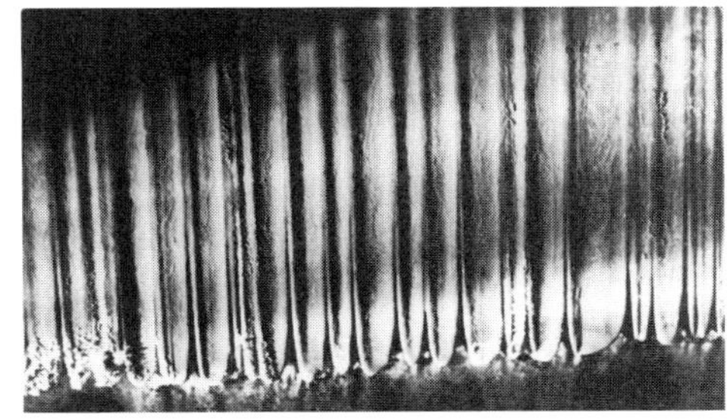

Fig. 21 Hydrogen grooving on the sidewall of an H_2SO_4 storage tank

Cast Irons

Gray cast iron has been used for piping and coolers for H_2SO_4 since acid plants were first built. Cast iron is at least as resistant to corrosion as steel is in the 65 to 100% acid concentration range at ambient temperature. The higher carbon and silicon content of cast irons leads to conditions favoring superior resistance at higher velocities and elevated temperatures, at least in the concentration range of 90 to 100%. Gray cast iron is less sensitive to velocity than steel and is frequently used up to 1.7 m/s (5.6 ft/s) in larger-diameter piping. The superior resistance of cast iron may be due to interference of the graphite flake network with the reaction between the acid and the metallic matrix (Ref 43). Graphite flakes may act as cathodic areas and shift the corrosion potential to a more favorable region (Ref 44).

Gray cast iron normally should not be used in the oleum range because of a tendency toward violent cracking. Free SO_3 is thought to attack the graphite flakes. Corrosion products form in the voids and strain the structure by volumetric expansion. Cracking of gray cast iron has also occurred in concentrated H_2SO_4 service, in which the cast iron was attached to stainless steel coolers protected by anodic protection. It is theorized that the anodic current causes the cast iron to behave as though it is in oleum service. This problem has been solved by placing a 3-m (10-ft) ductile iron pipe spool between the cooler and the gray cast iron pipe. Ductile cast iron is not subject to the same phenomena as gray cast iron, because the graphite is in the form of isolated nodules and the metal matrix is substantially stronger.

Gray cast iron is brittle and frequently ruptures catastrophically. Care must be taken to support cast iron pipe properly and to replace it before it thins excessively because of corrosion. For replacement or new piping, ductile cast iron, stainless steels with and without anodic protection, and plastic-lined pipe are now favored.

Ductile cast iron is nearly as corrosion resistant as gray cast iron in concentrated H_2SO_4 service. Laboratory testing comparing the materials has been inconclusive, probably because chemical compositions vary considerably. Actual plant experiences with piping tend to indicate up to 50% higher corrosion rates for ductile cast iron versus gray cast iron, but this is usually compensated for by the additional corrosion allowance used with ductile cast iron.

A modified ductile cast iron was recently introduced that reportedly has twice the corrosion resistance of conventional ductile cast iron. This alloy has about 3.5% Si versus the 1.8 to 2.8% for standard ductile iron.

High-Silicon Cast Iron. Another material that has long been used for handling H_2SO_4 is high-silicon iron. Iron with 14.5% Si has exceptional resistance to H_2SO_4 in all concentrations to 100% up to the atmospheric boiling points. Corrosion rates are normally less than 0.12 mm/yr (5 mils/yr), as shown in Fig. 22.

The resistance of high-silicon iron derives from the formation of a strong silicon-rich abrasion-resistant film. Even the most severely abrasive slurries can be handled.

The high-silicon irons are available only in the cast form. In addition, they have low tensile strength and virtually no ductility. High-silicon irons are susceptible to thermal and mechanical shock. Sharp hammer blows and rapid temperature fluctuations have caused the material to fail. High-silicon irons are rapidly attacked in oleum or other services containing free SO_3. Only SO_3, SO_2, and fluorine contaminants are known to alter drastically the corrosion resistance of high-silicon iron to H_2SO_4.

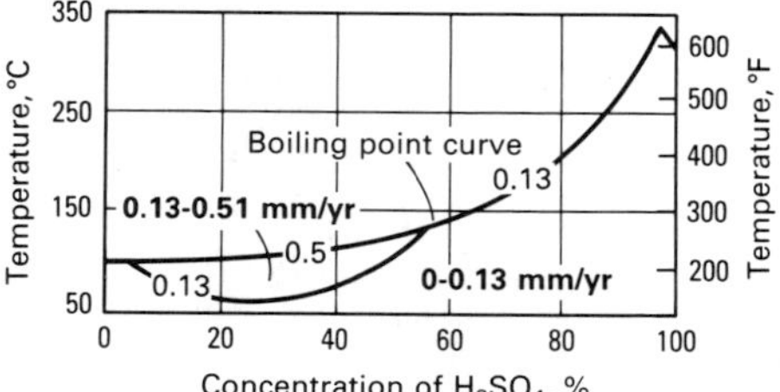

Fig. 22 Corrosion of high-silicon iron in H_2SO_4 as a function of temperature and acid concentration

Austenitic Stainless Steels

The resistance of austenitic stainless steels to H_2SO_4 is complex due to the active-passive nature of the alloys. An excellent summary that includes corrosion rate data is provided in Ref 45.

Mechanism of Protection. Stainless steels depend on electrochemical passivity—Case III behavior—for resistance to corrosion in H_2SO_4 solutions. Stable passivity is achieved at ambient temperatures in the very low and very high concentrations and in oleum.

Corrosion Resistance. At ambient temperatures, austenitic stainless steels, for example, type 304, exhibit stable Case III passivity in H_2SO_4 above 93% concentration and are frequently used for piping and tankage where product purity is desirable. The corrosion rates are essentially nil as compared with 0.15 to 1.0 mm/yr (5.9 to 39 mils/yr) for steel. Molybdenum stretches the passive region, making type 316 and 317 acceptable above 90% concentration at ambient temperature. The upper temperature limit for stable passivity for types 304 and 316 in 93% H_2SO_4 is believed to be around 40 °C (105 °F). For 98.5% H_2SO_4, the upper stable passive limit is believed to be above 70 °C (160 °F) (Ref 46). As concentration increases above 99%, corrosivity decreases rapidly, allowing the use of stainless steels above 100 °C (210 °F).

In dilute acid, only the molybdenum grades, such as types 316 or 317, are useful, although type 304 may be used when only a trace of acid is present. Figure 23 shows corrosion data for these alloys in as-mixed and refluxed (aerated) H_2SO_4 solutions. Stainless steels have poor resistance to deaerated dilute solutions. Type 310 stainless steel with 25% Cr and no intentionally added molybdenum is more resistant than the molybdenum-bearing grades when oxidizing agents are present. This is attributed to the higher chromium content of the type 310.

Effect of Velocity. If a stainless steel is solidly in Case III passive behavior, velocity appears to have little effect. Laboratory tests of type 304L with velocities to 6 m/s (20 ft/s) in 93% H_2SO_4 at ambient temperature have shown Case III passive behavior (Ref 46). However, once the alloy drops to Case II (active-passive) behavior, usually because of increasing temperature, velocity has a major effect. Under abrasive conditions, cast stainless steels have shown Case II active-passive behavior in 96% H_2SO_4 even at ambient temperatures (Ref 48).

Effect of Aeration and Oxidants. Highly aerated solutions are much more suitable for these alloys than air-free ones. Similarly, the presence of oxidizing impurities stabilizes the passive film, and the resistance to H_2SO_4 of austenitic stainless steels improves markedly. Cations that are easily reducible, such as Fe^{3+}, Cu^{2+}, stannic (Sn^{4+}) and cerric (Ce^{4+}) ions, are oxidizing agents that can inhibit the attack of stainless steels in H_2SO_4 solutions. It was found that 0.19 g/L of Fe^{3+} ion was sufficient to cause passivity and low corrosion rates in boiling 10% H_2SO_4, but 0.115 g/L did not give inhibition (Ref 49).

Other oxidizing agents, such as H_2CrO_4 and HNO_3, were shown to be effective in reducing corrosion rates (Ref 50). Nitric acid concentrations as low as 1.5% were found to inhibit the corrosion of stainless steel over a wide range of H_2SO_4 concentrations at ambient and elevated temperatures (Ref 51). Oxidants in sufficient quantities were shown to reduce the corrosivity of H_2SO_4 on stainless steel by shifting the corrosion potential from an active to a passive state (Ref 52).

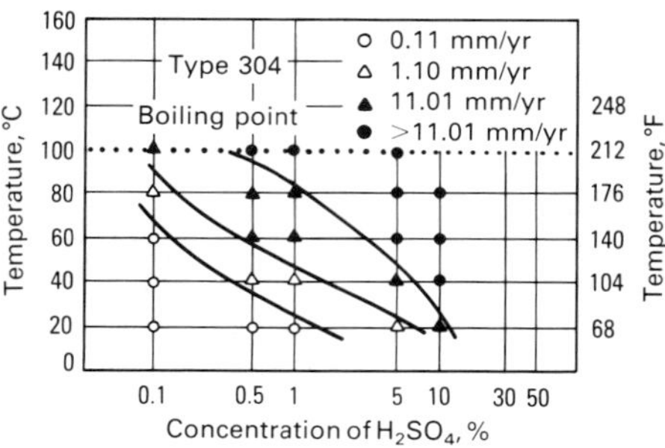

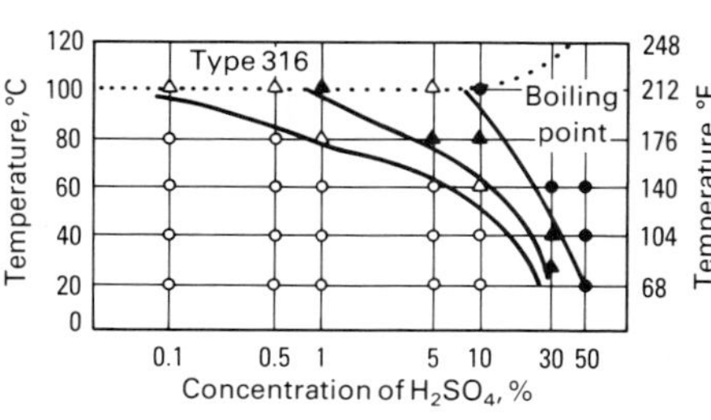

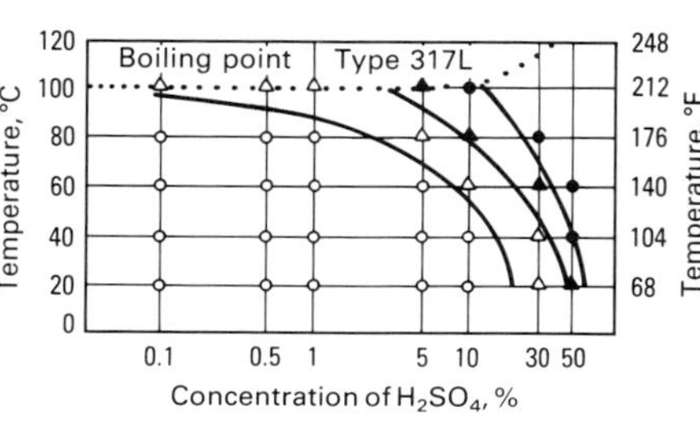

Fig. 23 Isocorrosion diagrams for (a) type 304, (b) type 316, and (c) type 317L stainless steels in aerated H_2SO_4 up to 50% concentration. Source: Ref 47

Effect of SO_3. In strong H_2SO_4 (above 97% concentration) and in oleum, the increased SO_3 content has a strong oxidizing effect, and corrosion rates are reduced dramatically. Figure 24 shows an isocorrosion diagram generated for type 304L stainless steel in an absorption tower environment (Ref 53, 54). Most stainless steels and nickel-base alloys likely have similar reductions in corrosion rates with increasing concentration. However, the molybdenum-containing alloys are distinctly inferior unless the chromium content is high, as is the case with E-Brite stainless steel (Table 7).

Extreme care must be taken when using stainless steels in the 98 to 100% concentration at high temperatures; velocity conditions, reductions in acid concentration, or changes in oxidant levels may initiate high corrosion rates. For contrast, compare the corrosion data generated in flowing

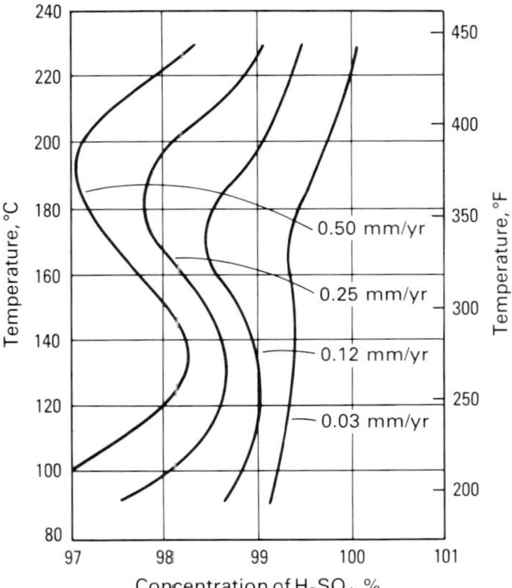

Fig. 24 Isocorrosion diagram of type 304 stainless steel in highly concentrated H_2SO_4. Source: Ref 54

98.7% H_2SO_4 at 100 °C (210 °F) shown in Table 8 with the data in Table 7. In the oleum range, stainless steels are free from the concerns about minor concentration variations and corrosion resistance is extended well in excess of 100 °C (210 °F).

Anodic protection is a practical method of extending the useful temperature and concentration range for stainless steels in H_2SO_4. With anodic protection, a stainless steel component (anode) is held in the passive (Case III) condition by an impressed current from a cathode.

Fortunately, H_2SO_4 is a good conductor of electricity and therefore has good throwing power. Complex shell sides of stainless steel shell and tube coolers handling concentrated H_2SO_4 can be easily protected using Hastelloy C-276 cathodes that extend the full length of the tube bundle. Piping is more difficult to protect because a cathode must throw the current a greater distance from the cathodes, which are typically located every 4.5 to 6 m (15 to 20 ft). Stainless steel may be protected in 93% H_2SO_4 up to 70 °C (160 °F) and in 98% H_2SO_4 up to 120 ° (250 °F). Corrosion rates can be reduced to 0.01 to 0.1 mm/yr (0.4 to 4 mils/yr). Approximately 400 anodically protected coolers have been installed in acid recirculation loops in H_2SO_4 plants throughout the world.

Table 7 Corrosion rates of various metals in 99% H_2SO_4 at 100–120 °C (212–250 °F)

Alloy	Corrosion rate mm/yr	mils/yr
Steel	>2.4	94.5
Cast iron	0.12	4.7
Ductile iron	0.25	9.8
Type 304L	0.02	0.8
Type 316L	0.06	2.4
Alloy 904L	0.19	7.5
Alloy 20Cb-3	0.08	3.1
Alloy C-276	0.33	13.0
Alloy B-2	2.3	90.6
A-611	0.04	1.6
E-Brite 26-1	<0.01	0.4

Silicon Stainless Steels. Austenitic stainless steels containing 5 to 6% Si have interesting corrosion characteristics in concentrated H_2SO_4 service. Cast and wrought versions are available. The cast version has a typical composition of Fe-5Si-21Cr-16Ni-0.02C (Ref 56). The wrought version of this alloy, A-611, has a typical composition of Fe-5.3Si-18Cr-18Ni-0.02C. The A-611 alloy has useful corrosion resistance in 99% H_2SO_4 up to 120 °C (250 °F) without anodic protection. Corrosion protection is obtained by the formation of a tenacious silicon-rich film formed on the surface during the initial days of corrosion. Corrosion resistance of the cast alloy is similar. Piping, distributors, and pump tanks handling hot 98 to 99% H_2SO_4 have been made from the A-611 alloy.

Cast Stainless Steels

Cast stainless steels have essentially the same corrosion resistance to H_2SO_4 as their wrought counterparts. Because the cast versions contain second-phase ferrite for castability, care must be taken that proper heat treatment is performed for maximum corrosion resistance. Preferential corrosion as the result of the duplex structure has been shown to be a problem. However, properly cast and heat-treated duplex materials perform well (Ref 45).

One cast alloy that does not have a close wrought counterpart is ACI CD-4MCu. Corrosion resistance lies between 20Cb-3 stainless steel and type 316. Figure 25 shows the isocorrosion diagram for this alloy. Corrosion resistance of CD-4MCu in H_2SO_4 extends over the entire concentration range at ambient temperatures, and the alloy is suitable above 100 °C (210 °F) in oleum service.

Table 8 Corrosion test in flowing 98.7% H_2SO_4 at 100 °C (212 °F)

Alloy	Corrosion rate mm/yr	mils/yr
Type 304	0.5	19.7
Type 316	3.44	135.0
Alloy 904L	2.3	90.6

Source: Ref 55

Higher Austenitic Stainless Steels

Like the austenitic stainless steels, the corrosion resistance of the higher austenitic stainless steels is also complex. However, the range of passivity and corrosion resistance is extended because of the higher alloy content. Like the stainless steels, Case I (active) and Case II (active-passive cyclic behavior) are the modes of corrosion. Resistance is achieved in the Case III (passive) state. A summary that includes corrosion data is provided in Ref 45.

Iron-base nickel-chromium-molybdenum alloys. This class of alloys contains approximately 25% Ni, 20% Cr, and 4.5% Mo; copper, titanium, and niobium are sometimes added as stabilizing elements. Alloys in this category that do not contain copper are generally more corrosion resistant than type 316 stainless steel and include Hastelloy M-532 and alloy JS-700 (Fe-25Cr-20Ni-3Mn-3Mo). Figure 26 shows an isocorrosion diagram of Hastelloy M-532. The temperature and concentration range of this alloy has been extended beyond that for type 316 as the result of the increased alloy content.

The copper-bearing alloys in this class are more resistant in H_2SO_4 than the copper-free alloys. Figure 27 shows an isocorrosion diagram for a 25Ni-20Cr-4.5Mo-1.5Cu alloy such as alloy 904L. The copper additions make the alloy suitable for the entire concentration range at ambient temperature.

The 20-Type alloys are usually the first considered when an H_2SO_4 environment is too corrosive for the use of steel, 300-series stainless steels, or cast iron. This group contains both wrought (20Cb-3) and cast alloys (ACI CN-7M) that are roughly equivalent in resistance to H_2SO_4.

Cast ACI CN-7M. Figure 28 shows an isocorrosion diagram for cast ACI CN-7M. This alloy is generally suitable to 80 °C (175 °F) at concentrations to 50%. For higher concentrations, good corrosion resistance is expected to 65 °C (150 °F).

Wrought Alloy 20Cb-3. The wrought counterpart to cast ACI CN-7M was developed in 1947.

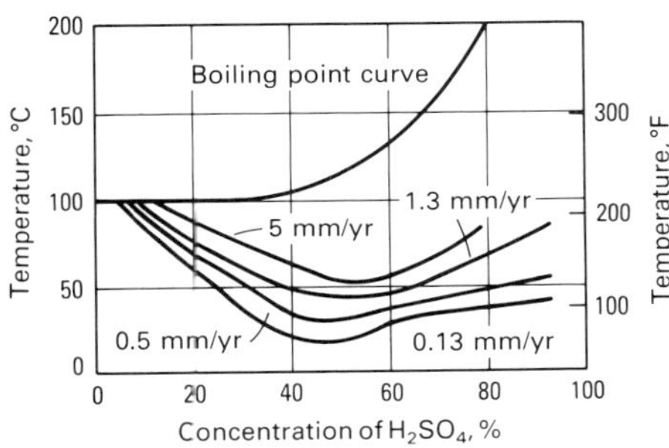

Fig. 25 Isocorrosion diagram for ACI CD-4MCu in H_2SO_4. Source: Ref 45

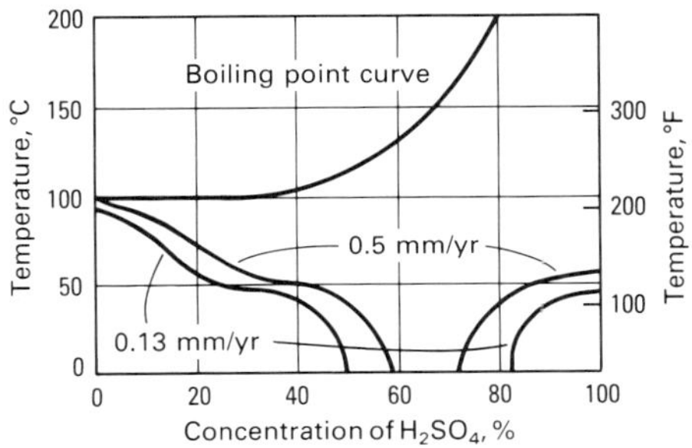

Fig. 26 Isocorrosion diagram for Hastelloy M-532 in H_2SO_4. Source: Ref 45

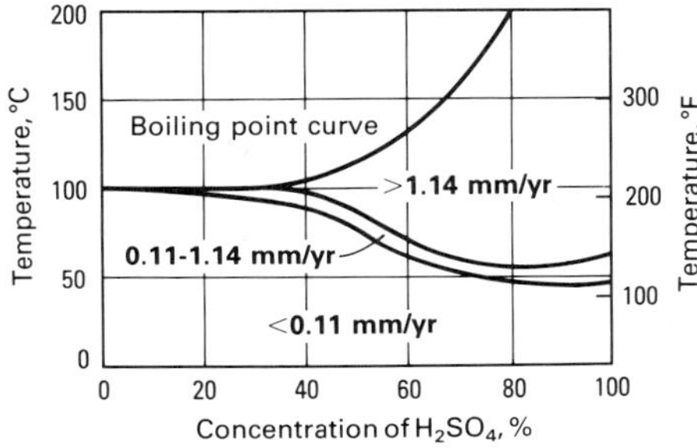

Fig. 27 Isocorrosion diagram for alloy 904L in H_2SO_4. Source: Ref 45

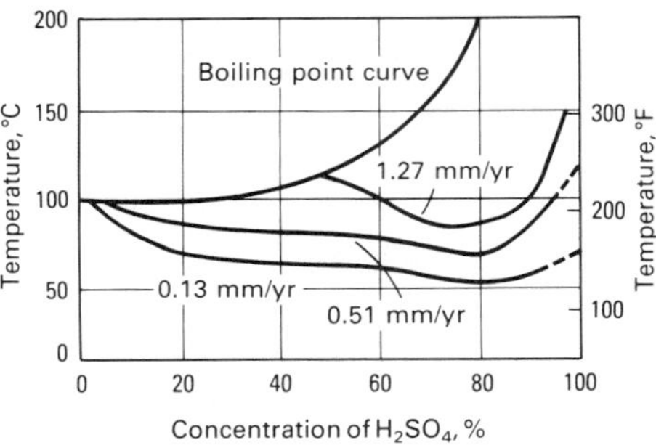

Fig. 28 Isocorrosion diagram for ACI CN-7M in H_2SO_4. Source: Ref 45

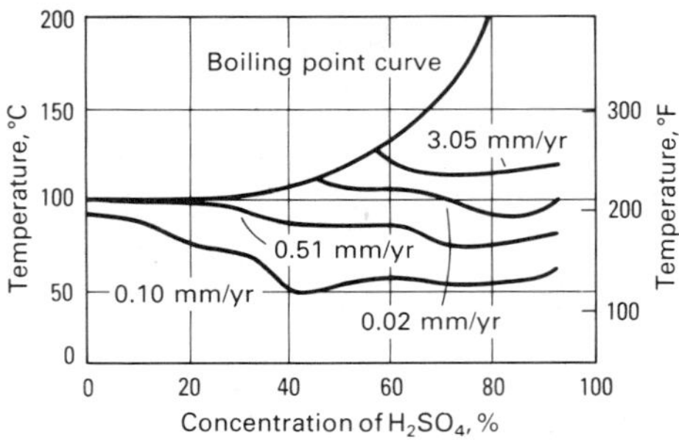

Fig. 29 Isocorrosion diagram for alloy 20Cb-3 in H_2SO_4. Source: Ref 45

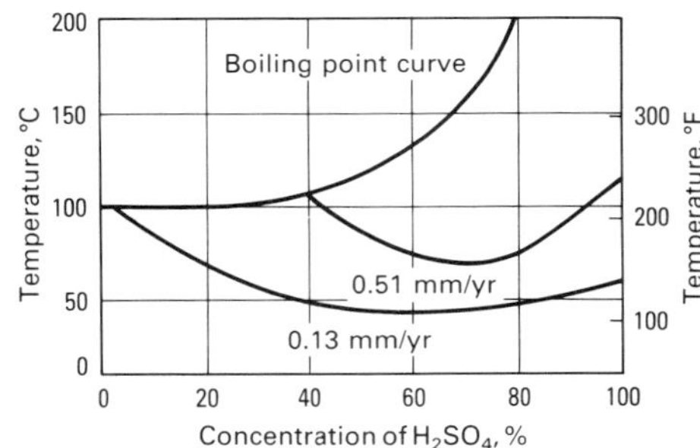

Fig. 30 Isocorrosion diagram for Incoloy 825 in H_2SO_4. Source: Ref 45

In 1948, niobium was added to this alloy for stabilization against sensitization and intergranular attack. In 1963, the nickel content was raised to 33 to 35% in order to give greater resistance to chloride SCC and to improve resistance to boiling H_2SO_4 under heat transfer conditions. Minor changes were subsequently made to impart greater resistance to intergranular corrosion. This alloy is now known as 20Cb-3. A typical composition is Fe-34Ni-20Cr-3.3Cu-2.5 Mo. Corrosion resistance of 20Cb-3 is similar to that of CN-7M. Figure 29 shows an isocorrosion diagram for 20Cb-3.

Nickel-Base Alloys

The nickel-base austenitic alloys have superior resistance to corrosion in H_2SO_4 up to 95% concentration because of their high alloy content. Frequently, low corrosion rates occur in both the active and passive corrosion states. Thus, reliable corrosion behavior is achieved over a wide range of concentrations, temperatures, and impurity levels. Extensive discussion, data, and references are given in Ref 45, 57, 58.

Nickel-base iron-chromium-molybdenum-copper alloys. Incoloy 825 has excellent resistance to H_2SO_4 up to 40% concentration and in concentrated acid (Fig. 30). The higher nickel content of this alloy versus the 20-Type alloys offers slightly improved corrosion resistance at high temperatures in low and high concentrations of H_2SO_4.

Hastelloy G and G-3 are modifications of the now obsolete Hastelloy F. A typical composition is 45Ni-22Cr-6.5Mo-2Cu. Hastelloy G-3 is a modified version of Hastelloy G with the same general corrosion resistance. It was modified to resist formation of grain-boundary precipitates during prolonged heating, such as stress relief. These alloys are promoted for their resistance to halide-contaminated H_2SO_4 environments. Figures 31 and 32 show isocorrosion diagrams for Hastelloy G.

Nickel-copper alloys. Monel 400 is used for handling H_2SO_4 under reducing conditions. Thus, this alloy offers an alternative to stainless steels and other alloys exhibiting active-passive behavior when H_2SO_4 solutions are not oxidizing, such as deaerated dilute acid. Monel 400 exhibits reasonably low corrosion rates in air-free H_2SO_4 up to 85% concentration at 30 °C (85 °F) and up to 60% concentration at 95 °C (205 °F).

Nickel-molybdenum-chromium-iron alloys. These nickel-base alloys, containing about 16% Mo, 16% Cr, and 3 to 5% Fe, are available in wrought (Hastelloy C-276 and C-4) and cast form (ASTM A494, grades CW-12MW and CW-7M). In general, the corrosion resistance is excellent (Fig. 33 and 34). At room temperature, the corrosion rate for all concentrations is less than 0.1 mm/yr (3.9 mils/yr). Because of the higher chromium content, these alloys are more resistant to H_2SO_4 containing oxidizing contaminants, for example, Fe^{3+} or Cu^{2+} ions, than to such materials as Monel 400 or Hastelloy B-2. Another material in this group is Inconel 625 (Fig. 35).

Nickel-Base Molybdenum Alloys. These nickel-base alloys, containing about 30% Mo, 3 to 6% Fe, and 1% Cr, are available in the cast and wrought forms. The corrosion resistance of these alloys is excellent in pure H_2SO_4 over a wide range of temperatures and concentrations. However, oxidizing contaminants, such as Fe^{3+} ions, increase corrosion rates considerably and have caused premature failure. Chlorides also increase corrosion rates. Hastelloy B-2 is the wrought material in this alloy group (Fig. 36). The cast version with similar corrosion resistance is ASTM A494, grade N-12MV. Cast alloy ASTM A494, grade N-7M, also falls into this general grouping.

Nickel-base chromium-iron-cobalt-silicon alloys are proprietary alloys developed specifically for hot concentrated H_2SO_4 pump and valve applications. Alloy 55 is one such cast alloy (Fig. 37). Figure 38 shows the effect of temperature in 98% H_2SO_4 as determined by laboratory tests. Alloy 66 is a ductile cast alloy that can also be made in the wrought form. Both forms have excellent corrosion resistance in the 0 to 60% and 80 to 90% ranges, but performance is erratic in the 60 to 80% range (Fig. 39).

Nickel-base chromium-molybdenum-copper alloys are proprietary alloys designed to resist H_2SO_4 concentrations to 98% at temperatures to 100 °C (212 °F). Illium 98 is a weldable, machinable cast alloy. Illium B, also a cast alloy, is a version of 98 that has been modified for enhanced corrosion resistance; however, illium B is not easily welded (Fig. 40).

Zirconium

The most prevalent application of unalloyed zirconium in the chemical industry has been in hot H_2SO_4. Zirconium has excellent resistance to H_2SO_4 up to 50% concentration at temperatures to boiling and above. From 50 to 65% concentration, resistance is generally excellent at elevated temperatures, but the passive film is a less effective barrier. Experience has shown that welding and oxidizing species in more than 50% H_2SO_4 can encourage selective attack. In concentrated H_2SO_4 above 70%, the corrosion rate of zirconium increases rapidly with increasing concentration. Because welds are known to be slightly less resistant than the parent metal, zirconium is seldom used above 60 to 65% concentration in the welded condition unless it has been heat treated to maximize corrosion resistance (Fig. 41).

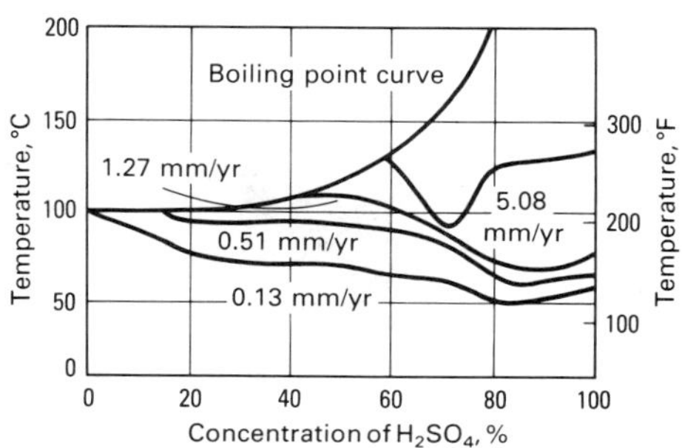

Fig. 31 Isocorrosion diagram for Hastelloy G in H_2SO_4. Source: Ref 45

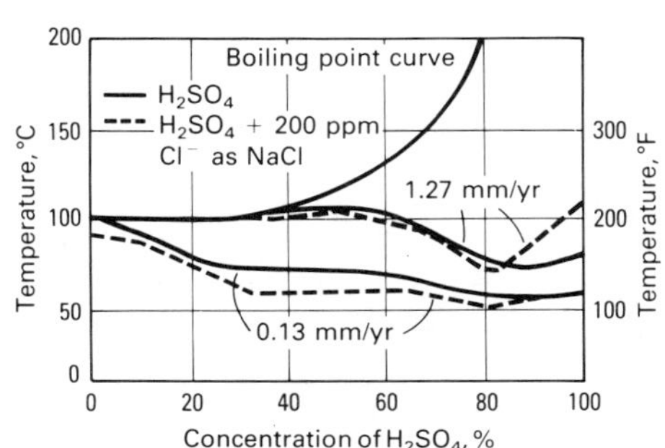

Fig. 32 Isocorrosion diagram for Hastelloy G in H_2SO_4 solutions contaminated with Cl^- ion. Source: Ref 45

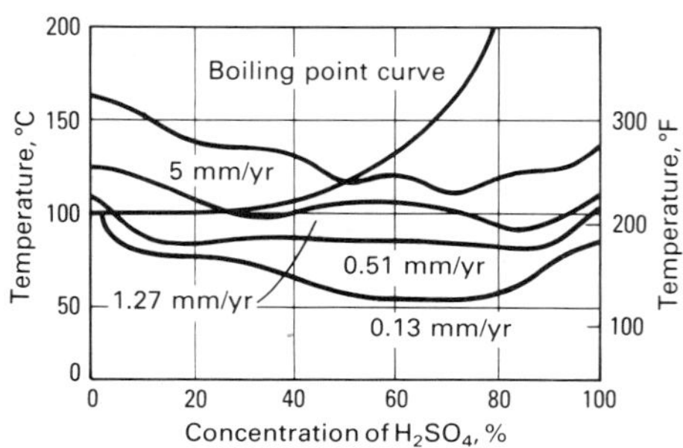

Fig. 33 Isocorrosion diagram for Hastelloy C-276 in H_2SO_4. Source: Ref 45

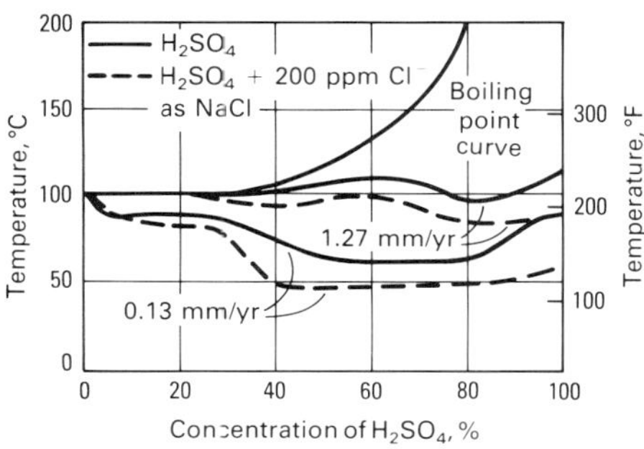

Fig. 34 Isocorrosion diagram for Hastelloy C-276 in H_2SO_4 and in H_2SO_4 contaminated with Cl^- ion. Source: Ref 45

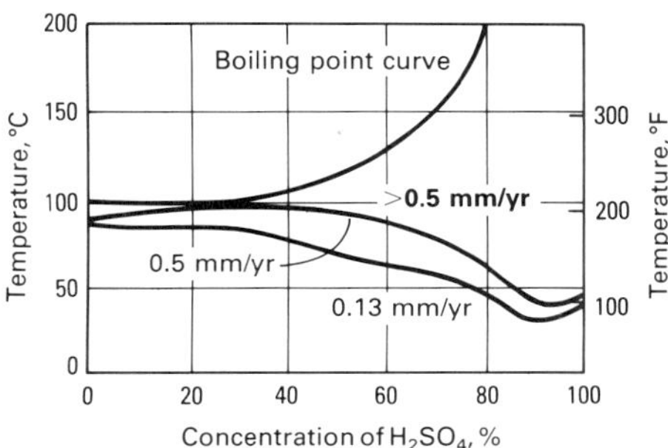

Fig. 35 Isocorrosion diagram for Inconel 625 in H_2SO_4. Source: Ref 58

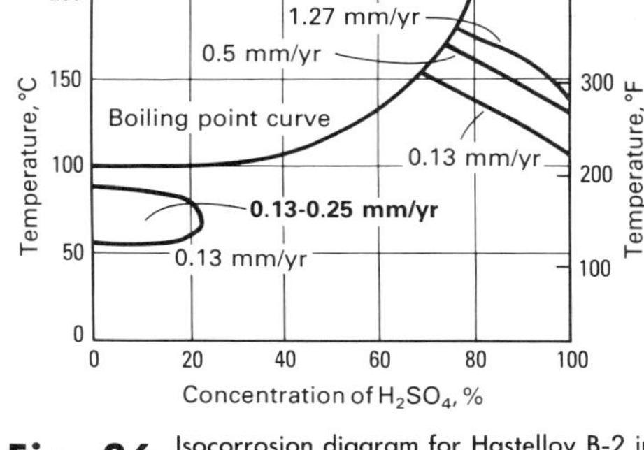

Fig. 36 Isocorrosion diagram for Hastelloy B-2 in H_2SO_4. Source: Ref 45

The corrosion resistance of zirconium in H_2SO_4 depends on the formation of a passive film. The protective film consists of predominantly cubic zirconium dioxide (ZrO_2), which is highly ordered and corrosion resistant. Electrochemical measurements show the corrosion potential of zirconium in H_2SO_4 to be located solidly in the passive region when it is resisting corrosion. However, as temperature and/or concentration increases, the transpassive (breakdown) potential decreases, and there is less tolerance for oxidizing agents, such as Fe^{3+}, Cu^{2+}, and nitrate (NO_3^-) ions. At high temperatures in concentrations exceeding 50%, oxidizing agents in sufficient quantities delete the passive region, causing zirconium to corrode actively.

Other Metals and Alloys

Unalloyed tantalum has resistance to H_2SO_4 over the entire range of concentration and temperature except for very strong and exceptionally hot conditions. Figure 42 shows the corrosion rates of tantalum in 98% acid and oleum. A common use is for reboilers and concentrators of H_2SO_4 in the 70% concentration range.

Unalloyed titanium is rapidly attacked by all concentrations of H_2SO_4 except very dilute solutions. Impurities in the form of oxidizing agents may act as inhibitors; for example, titanium resists 0 to 50% concentrations of H_2SO_4 saturated with chlorine (Ref 61, 62). Titanium has also been used in the H_2SO_4 leaching of nickel ores. Resistance is attributed to the presence of heavy-metal oxidizing agents, such as Fe^{3+} and Cu^{2+}, ions (Ref 63, 64).

Lead resists H_2SO_4 but its protective sulfate film is increasingly solubilized above 90% concentration. The film is easily damaged by erosion or abrasion even at low velocities, with the rate at attack increasing rapidly with concentration. Lead is useful as a pan material to catch acid drippings. It is also used as a membrane behind brick in H_2SO_4 concentrating units and scrubbing units.

Copper and copper-base alloys are not widely used in relatively pure H_2SO_4 because of the drastic effect of oxidizing conditions. The resultant Cu^{2+} ions also cause autocatalytic corrosion. Brass is not often used, because of the potential for dezincification. Bronzes have the greatest applicability, with the normal copper-tin-bronzes showing acceptable service below 60% concentration at 79 °C (174 °F). The silicon-bronzes extend the suitable range to about 70% concentration. The aluminum-bronzes have resistance approaching that of Monel 400; however, dealuminification can be a problem, particularly with alloys containing more than 8% Al (Ref 18).

Nonmetals

Nonmetallic materials of construction have wide application in H_2SO_4. Most of these materials have good corrosion resistance to the pure acid, particularly in dilute concentrations, and are primarily restricted by their mechanical properties at temperature. They are relatively unaffected by most inorganic contaminants, except for such strongly oxidizing agents as HNO_3, peroxides, and dichromates.

Brick linings have been used for the most severe H_2SO_4 conditions. Under most conditions, acid-resistant fire clay brick can be used. For extreme conditions, silica brick is used. High-alumina (Al_2O_3) refractories are inferior at high temperatures and concentrations. At concentrations below 70% lead, plastic or elastomeric corrosion-resistant membranes are required behind the brick. The brick serves as a thermal and abrasion barrier. At higher concentrations, brick can be used directly on steel, although tetrafluoroethylene (TFE) or mortar barriers are often used. For moderate concentrations and temperatures, plastic mortars, such as furans and phenolics, are used. For high temperatures and concentrations, the silicate cements are required. Sodium silicate cement has been the most common type even though it slowly converts to sodium sulfate (Na_2SO_4), which hydrates and degrades. Potassium silicate (K_2SiO_3) is now preferred because its sulfate salt does not hydrate. Mortars based on pure colloidal silica are also available for extreme conditions.

Polyvinyl chloride has excellent resistance to H_2SO_4 to approximately 93% concentration at ambient temperature. Chlorinated PVC resists slightly higher concentrations and temperatures. These materials have been used for pipelines, but are not popular for longer runs, because they have high coefficients of expansion and require essentially continuous support. Many companies prohibit the use of solid plastic piping for safety considerations because of the potential for breakage.

Lined Pipe. Polyvinylidene chloride (PVDC), polypropylene (PP), polyvinylidene fluoride (PVDF), and PTFE all are used for linings in pipe for handling H_2SO_4. The range of usefulness is shown in Table 9. Although PP is shown as resistant to 98% H_2SO_4 at ambient temperatures, there is considerable controversy on this point because high concentrations can cause charring and SCC. To avoid this, the PP must either be kept in compression or a copolymer must be used.

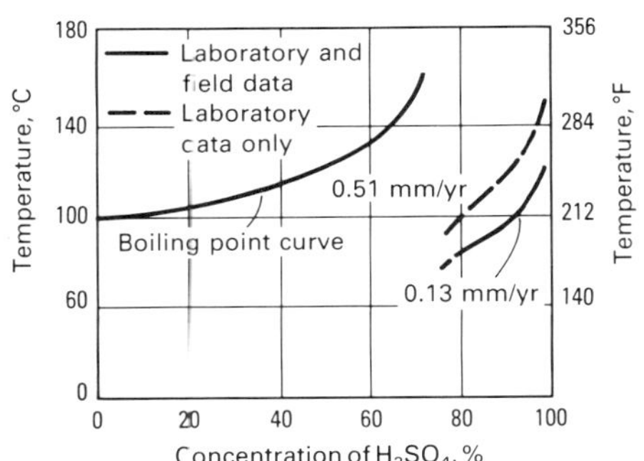

Fig. 37 Isocorrosion diagram for alloy 55 in H_2SO_4. Source: Ref 45

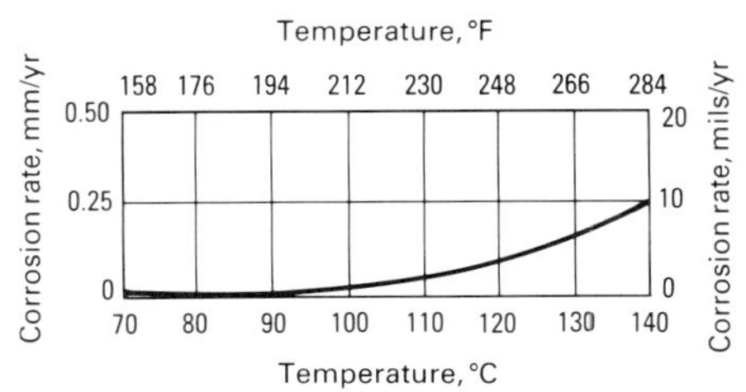

Fig. 38 Corrosion of alloy 55 in 98% H_2SO_4. Source: Ref 45

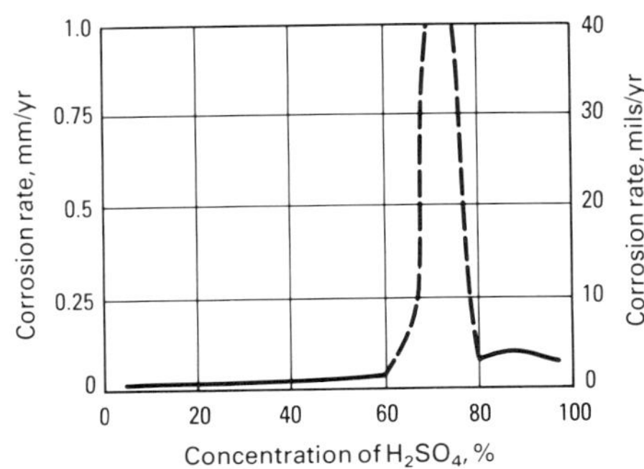

Fig. 39 Corrosion rates of alloy 66 in H_2SO_4 solutions at 100 °C (212 °F). Source: Ref 45

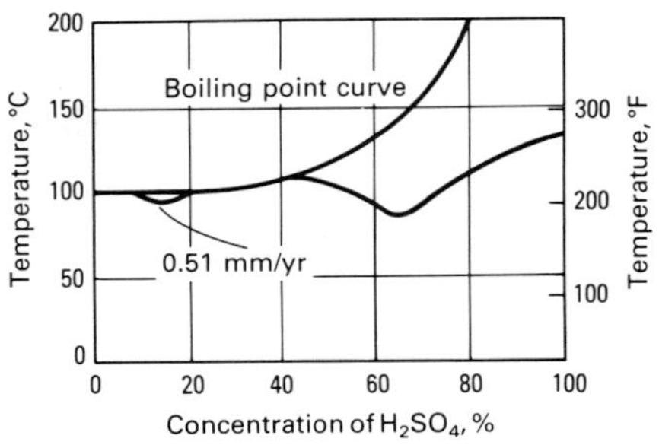

Fig. 40 Isocorrosion diagram for Illium B in H_2SO_4. Source: Ref 45

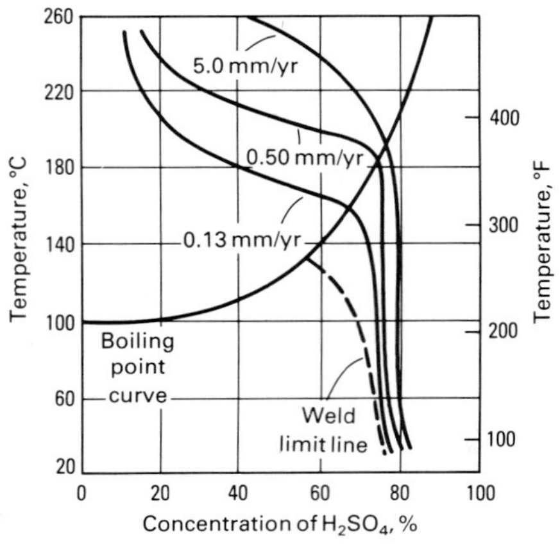

Fig. 41 Corrosion of zirconium by H_2SO_4 as a function of temperature and acid concentration. Source: Ref 59, 60

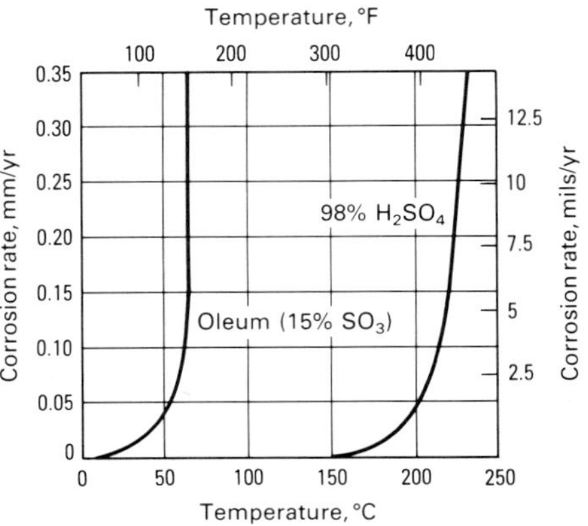

Fig. 42 Corrosion of tantalum in 98% H_2SO_4 and in oleum

Polyethylene possesses a high degree of resistance to attack by H_2SO_4 up to 98% concentration at ambient temperature. The high-density resins show virtually no attack in 98% acid at temperatures to 50 °C (120 °F). The use of polyethylene equipment for handling H_2SO_4 must be restricted to relatively low temperatures and low loads unless continuously supported. Polyethylene is widely used in laboratories for beakers, bottles, and so on, which do not require heating over direct flames. Tanks constructed of molded, seamless, high-density, cross-linked polyethylene have been used to store relatively small volumes (3800 L, or 1000 gal.) of concentrated H_2SO_4 up to 96%.

Fluoroplastics. Fully fluorinated plastics, such as PTFE, fluorinated ethylene propylene (FEP), and perfluoroalkoxy (PFA), are unattacked by H_2SO_4 and oleum at all concentrations. Other fluoroplastics, including PVDF, ethylene-chlorotrifluoroethylene (ECTFE), and ethylene-tetrafluoroethylene (ETFE), are resistant to acid up to 98% concentration, but have less resistance to oleum and SO_3. Table 10 lists temperature limitations and polymer designations.

Fluoroplastics are being increasingly used for H_2SO_4 applications, such as linings in pumps and valves. Sheet linings of FEP, ECTFE, and PVDF permit fabrication of large components, such as tanks. Spaghetti-type FEP heat exchangers have been used to cool concentrated H_2SO_4 in contact plants for more than two decades. In Europe, the heat recovered is often used for town heating. These units cannot be used to cool oleum, because the reaction of free SO_3 and water vapor permeating the FEP tubes causes excessive temperatures and destruction of the resin. Recently, PFA construction and design modifications have improved exchanger temperature and pressure capabilities. Dispersion of powder linings of PVDF, ECTFE, and PFA, with or without fiber reinforcement, are applied to a broad range of steel equipment (Ref 65). Equipment size is limited only by available ovens.

Table 9 Suitability of plastic-lined piping systems in H_2SO_4

Material	Maximum temperature(a) °C	°F	Concentration of H_2SO_4, %
PVDC	52	126	<16
	24	75	30
	NR		>60
PP	93	200	<60
	80	175	93
	65	150	96
	50	120	98
PVDF	120	250	<16
	105	220	30–60
	80	175	85–93
	65	150	94–98
TFE	260	500	0–100

(a) Temperature may be further limited by mechanical conditions under vacuum. NR, not rated

Polyester Resins. Polyester and vinyl-ester resins, normally used to form a fiberglass-reinforced laminate, show virtually no attack in dilute H_2SO_4 at 90 °C (200 °F) and are used in pickling operations where 15 to 20% acid may be encountered at temperatures to 90 °C (195 °F). At 50% concentration, the polyesters can be safely used to 65 °C (150 °F) and can even find application in special cases where the temperature approaches 90 °C (200 °F). At 75% acid, the temperature must be restricted to less than 65 °C (150 °F). At 90%, most polyester and vinyl-ester resins are rapidly degraded even at ambient temperatures.

Phenol-formaldehyde and furfural alcohol resins are blended with reinforcing glass or other filler to make corrosion-resistant composite pipe and equipment. Some phenol-formaldehyde composites resist H_2SO_4 at concentrations to 50% at 150 °C (300 °F), to 70% at 100 °C (212 °F), and to 96% at ambient temperature. Furfural alcohol resin composites resist H_2SO_4 at concentrations to 50% at 90 °C (195 °F) and to 70% at ambient temperature.

Carbon and Impervious Graphite. Impervious graphite has excellent corrosion resistance to all but highly oxidizing concentrations of H_2SO_4. Impervious graphite coolers are used in 93% H_2SO_4 (Ref 66). Care must be taken that the impregnant is resistant to the solution. Polyester, furan, and PTFE impregnants are available. Carbon is somewhat superior to graphite in the higher concentrations and is generally preferred for hot acid at concentrations above 60%. Carbon is less conductive and lacks strength and is therefore unsuitable for cooler construction. Both carbon and graphite are, of course, brittle (Ref 67, 68).

Table 10 Suitability of nonmetal linings in H_2SO_4

Material	Maximum temperature °C	°F	H_2SO_4 concentration, %
PTFE	260	500	0–100
PFA	260	500	0–100
FEP	205	400	0–100
ECTFE	150	300	<98
ETFE	150	300	<98
PVDF	120	250	<16
	105	220	30–60
	80	175	85–93
	65	150	94–98

Coatings and Linings. Heat-cured phenolic linings on steel are often used to prevent iron contamination in 93% H_2SO_4 and have reasonable life in 98% H_2SO_4 at ambient temperature. These thin-film linings should not be used for concentrations below 70%, because of rapid corrosion of exposed steel substrate at pinholes.

Butyl rubber and Neoprene exhibit good resistance to 50% H_2SO_4 at modest temperatures and will resist 75% acid under ambient conditions. A sulfonated chlorinated polyethylene elastomer has been successfully used in plant applications involving dilute H_2SO_4 as well as hoses handling 93% H_2SO_4 acid.

Glass and glass-lined equipment is widely used in H_2SO_4 service. For severe H_2SO_4 applications, glass-lined steel is a popular material of construction.

Corrosion by Nitric Acid

Ronald D. Crooks
Hercules, Inc.

Nitric acid is typically produced by the air oxidation of NH_3. This catalyzed reaction takes place at very high temperatures. The gaseous oxidation product is condensed to an aqueous liquid of about 65% concentration. During the high-temperature oxidation, corrosion of the plant materials is of secondary concern. The elevated operating temperatures dictate that the high-temperature properties of the materials are the primary design consideration. Corrosion considerations prevail during and after condensation and at lower temperatures.

The concentration of HNO_3 up to 99% requires secondary processing to remove excess water. This involves mixing 65% HNO_3 with another substance having a greater affinity for water (such as H_2SO_4), then separating the mixed acids by distillation and condensation processes.

The basic materials of construction used for equipment to carry out the process described above are reasonably standardized and are listed below:

Equipment	Material
Ammonia converter...	Austenitic stainless steel or nickel-base alloy
Acid coolers and condensers.....	Type 304L (a very low-carbon, 0.030 max, version of type 304 stainless steel
65% acid storage	Type 304L
Acid concentrator	Type 304L
Distillation column ...	Corrosion-resistant cast iron or glass-lined steel
Concentrated acid condensers.........	Aluminum; glass/PTFE/Alloy 610
Concentrated acid storage	Aluminum alloy 3003, glass-lined steel

Additional information is available in Ref 69 to 81.

Commercially produced HNO_3 is available in concentrations from 52 to 99%. Nitric acid over 86% is described as fuming. Nitric acid up to 95% is stored and shipped in type 304 stainless steel. Concentrated acid above 95% is handled in Aluminum Association (AA) aluminum alloys 1100 or 3003. Figure 43 shows the reason for this; the corrosion rate of type 304 stainless steel increases rapidly above 95% concentration, while that of aluminum 3003 remains essentially constant to 100%. A new stainless steel containing 4% Si—alloy A-610—shows excellent resistance to concentrated HNO_3 (Fig. 48); unfortunately, this advantage does not extend to lower concentrations.

Nitric acid is a strong oxidizing agent and attacks most metals, such as iron, by oxidizing the metal to the oxide. A secondary effect of oxidation is the generation of hydrogen at the metal/acid interface, which can cause hydrogen embrittlement of some materials, for example, high-strength steels. Metals and alloys that are able to form adherent oxide films, such as austenitic stainless steels and aluminum alloys, are protected by their oxide films from corrosion by HNO_3.

Metals

Austenitic Stainless Steels. The basic corrosion data for the austenitic (300-series) stainless steels have been reduced to the isocorrosion diagram format (Fig. 44). This diagram shows the effect of temperature and HNO_3 concentration on the corrosion of type 304 stainless steel. Increasing either or both raises the corrosion rate; nevertheless, there is a large useful area extending from 0 to 90% concentration and up to the boiling point below 50% concentration in which the predicted corrosion rate is less than 0.13 mm/yr (5 mils/yr).

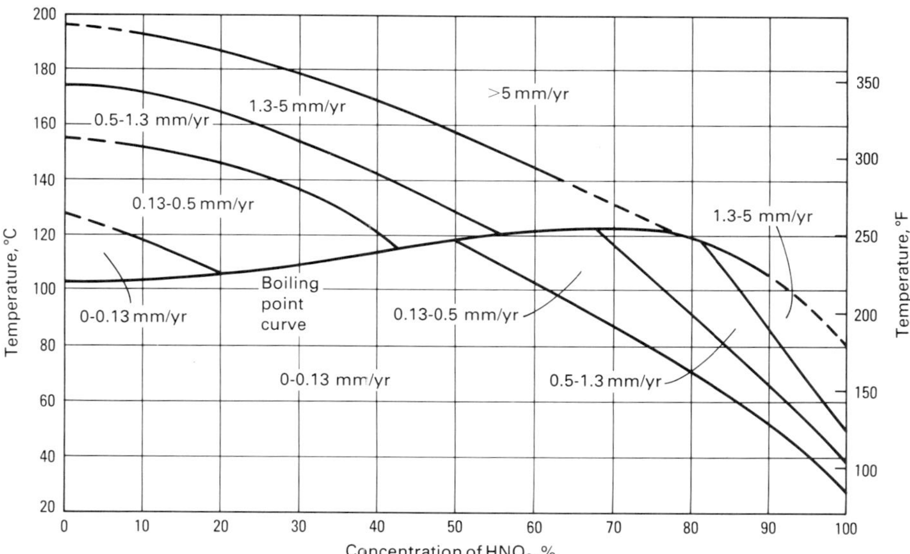

Fig. 44 Isocorrosion diagram for annealed type 304 stainless steel in HNO_3. Source: Ref 43

Experience has shown that, although all austenitic stainless steels behave in this fashion, types 304 and 304L, when welded, are clearly superior to the others. Therefore, they are the most popular grades for HNO_3 service.

Figure 44 also applies in general to the cast equivalents of wrought stainless steels. However, the generally higher carbon content of cast alloys and the propensity of castings to have high-carbon surfaces often leads to selective intergranular corrosion in strong HNO_3.

Although austenitic stainless steels are commonly used in HNO_3, they are not without problems. One of the most prevalent is selective corrosion associated with chromium carbides precipitated around grain boundaries in the weld HAZ. Because few pieces of industrial equipment are made without welding, this is a serious shortcoming. There are three methods of avoiding this problem:

- Use low or extra-low carbon content (Fig. 45)
- Add carbide stabilizers to the alloy, as in types 321 and 347
- Solution anneal after welding to reduce the chromium carbide gradient

Of these alternatives, solution annealing is frequently not practical, and the choice of stabilized alloys is not always an option. Therefore, when welding is planned, use of low-carbon stainless steels is the most popular alternative.

Sensitization of stainless steel refers to the precipitation of chromium carbides and the resultant depletion of the matrix of chromium as a result of heating from 480 to 760 °C (900 to 1400 °F). Figure 46 shows that the effect of such heating on corrosion rate in 65% HNO_3 is detrimental in all cases. Sigma phase, which may also form during prolonged heating of austenitic stainless steels, is preferentially and rapidly attacked by 65% HNO_3. Solution heat treating the alloy will restore corrosion resistance.

The corrosion of austenitic stainless steels in HNO_3 is accompanied by the formation of

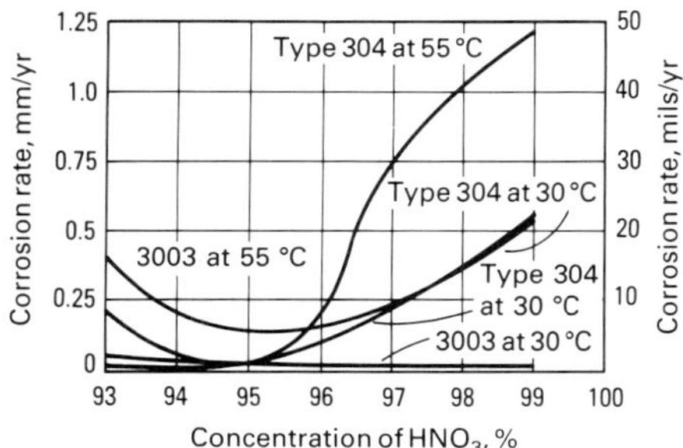

Fig. 43 Comparison of corrosion of aluminum alloy 3003 and type 304 stainless steel in HNO_3

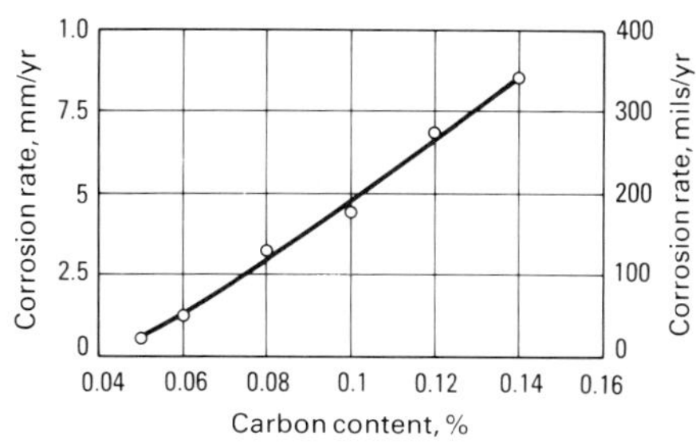

Fig. 45 Effect of carbon content on corrosion rate of type 304 stainless steel in boiling 65% HNO_3. Source: Ref 82

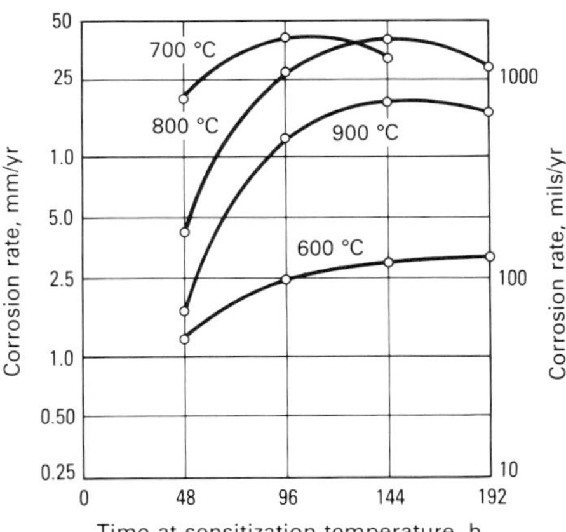

Fig. 46 Effect of sensitization time and temperature on the corrosion of type 304 stainless steel in boiling 65% HNO_3. Source: Ref 78

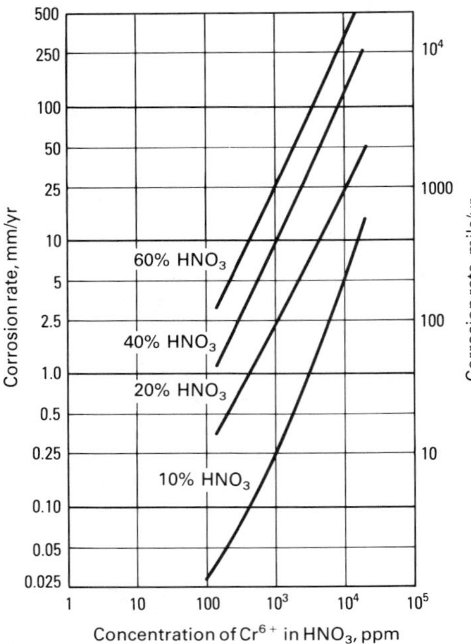

Fig. 47 Effect of hexavalent Cr^{6+} contamination on the corrosion rate of type 304 stainless steel in HNO_3. Test duration: 40 h. Source: Ref 79

hexavalent chromium (Cr^{6+}), a complex chromium compound that increases the corrosivity of HNO_3 solutions. The effect of Cr^{6+} buildup is shown in Fig. 47 to be clearly detrimental. In general, the presence of chlorides and fluorides in HNO_3 solutions tends to increase the corrosion rate of stainless steels.

Selective corrosion along grain boundaries is common in austenitic stainless steels exposed to HNO_3, especially strong acid. There is some evidence to support the view that this cannot be entirely prevented; however, maintaining low carbon and avoiding sensitization will help.

For very concentrated acid (>95%), the addition of silicon to iron and to austenitic stainless steels is beneficial. Cast iron with 14% Si is very resistant to acid over 50% concentration. Recently, two new stainless alloys of 4 and 6% Si—alloys A-610 and A-611—have been produced (Ref 83). These new alloys have remarkable resistance to HNO_3 above 95% (Fig. 48). At lower concentrations, they offer no advantage over type 304.

Aluminum, as previously mentioned, shows an advantage over type 304 at acid concentrations exceeding 95% (Fig. 43). However, if acid

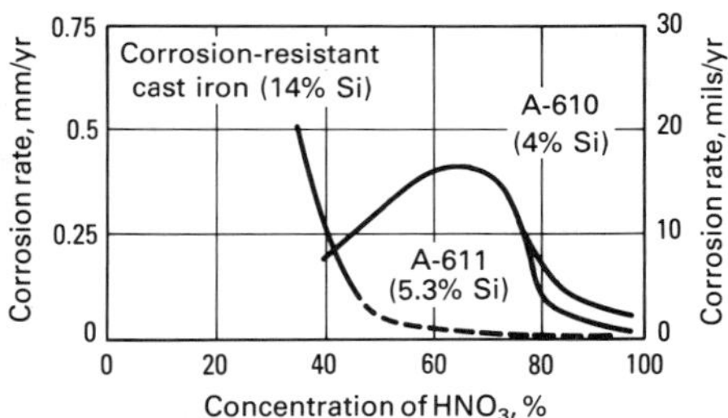

Fig. 48 Effect of silicon content on the corrosion of iron and iron-chromium-nickel alloys in boiling HNO_3. Source: Ref 84

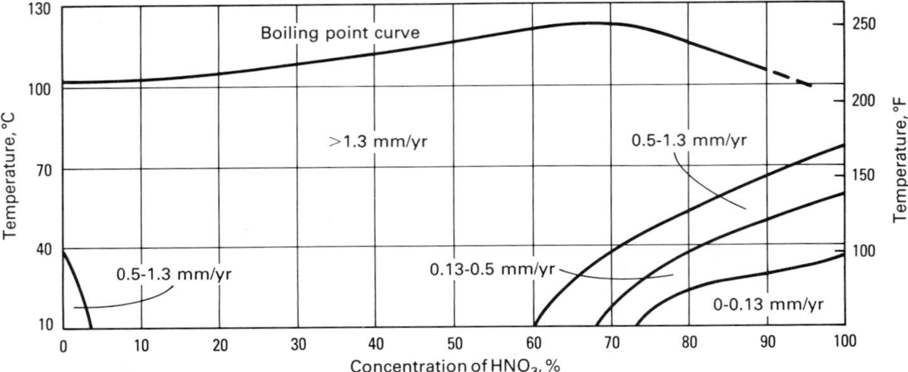

Fig. 49 Isocorrosion diagram for aluminum alloy 1100 in HNO_3. Source: Ref 43

concentration falls below 80% or if temperature rises above 40 °C (100 °F), much higher corrosion rates can be expected. This relationship is clearly shown in the isocorrosion diagram for aluminum (Fig. 49).

The preferred aluminum alloys for HNO_3 service are alloys 1100 and 3003 (Ref 85). If higher strength is required, alloys 5052 or 5454 can be used.

Aluminum alloys may also suffer degradation by improper welding or by using an incorrect filler metal, which leads to selective weld corrosion. To avoid these problems, the filler metals listed below should be used for the indicated alloy:

Aluminum alloy	Filler metal
1100	1100
3003	1100
5052	5356
5454	5356

Higher-alloy materials, such as 20Cb-3 stainless steel, Hastelloy C-276, Chlorimet 3, Illium 98, and Incoloy 825, are adequately resistant to HNO_3, but are not often used, because the more economical austenitic stainless steels are just as good. Some exceptions where the higher alloys are useful include very high velocities, mixed acids (especially HF), or the presence of other contaminants in the HNO_3 tending to increase the corrosivity, for example, Cl^- and F^- ions.

Titanium and titanium-palladium alloys resist concentrated HNO_3 from 65 to 90% and dilute acid of less than 10% (Fig. 50). At concentrations above 90%, however, titanium is subject to SCC. Titanium is never used in red fuming HNO_3, because a pyrophoric reaction can occur

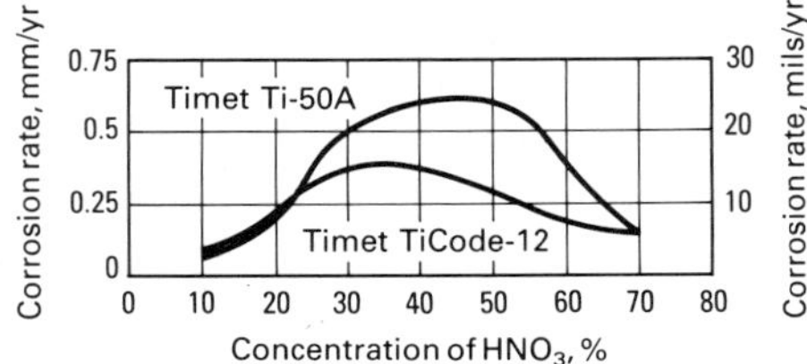

Fig. 50 Corrosion of titanium in boiling HNO_3. Source: Ref 80

if water content is less than 1.34% and nitrous oxide (N_2O) exceeds 6%.

Zirconium is even more resistant to HNO_3 than titanium is. Zirconium has a corrosion rate of less than 0.025 mm/yr (1 mil/yr) at all concentrations up to the boiling point. This corrosion resistance extends to 230 °C (450 °F) at concentrations under 65%. The response of zirconium to concentrations exceeding 65% and above the boiling point is uncertain. Unfortunately, zirconium is subject to SCC in HNO_3 at high concentrations and therefore must be used with caution in stronger acids (Ref 86, 87).

Nonmetals

Glass and ceramic materials, unlike most metals, do not corrode in acid, but they are changed by exposure to it. Elements such as iron and silica may leach out, resulting in decreases in strength and density. Glass-coated steel equipment is fully resistant to HNO_3 up to 70% concentration and to 125 °C (260 °F). Exposure to conditions beyond these limits will reduce the life of glass-coated equipment (Ref 88).

Nitric acid attacks most nonmetals, but some can be used to a limited degree in immersion service (Table 11). The temperature limits given in Table 11 apply to tank linings and coatings as well as composite structures embodying these resins.

Coatings

Frequently, there are requirements for coatings resistant to splash and spillage of HNO_3, which is frequently washed down. None will withstand concentrated acid, but for 50% concentration or less, epoxies, epoxy-phenolics, chlorinated rubber, and aliphatic urethanes have given good service.

Table 11 Maximum-use temperatures for plastics in HNO_3 service

Material	Concentration of HNO_3	Temperature °C	Temperature °F
PTFE	70	232	450
Polyethylene	30	60	140
PVDC	50	52	125
PVDF	50	52	125
Polyester	50	43	110
Vinyl-esters	20	49	120
Furan	5	66	150

Corrosion by Organic Acids

G.B. Elder
Union Carbide Corporation

Organic acids constitute a group of the most important chemicals currently in use in industry. The acids are produced more as precursors for other chemicals than for end use as organic acids. Acetic acid is the best known member of the group and is produced in the largest volume, but the other organic acids are also important for the preparation of compounds used in daily life—compounds from aspirin to plastics and fibers.

The subject of corrosion by organic acids is complicated not only by the numerous acids to be considered but also because the acids typically are not handled as a pure chemical but as process mixtures with inorganic acids, organic solvents, salts, and mixtures of several organic acids. They are even used as solvents for other chemical reactions.

This section will present some of the corrosion characteristics shared by the organic acids, will detail information on corrosion by formic acid (HCOOH), acetic acid (CH_3COOH), and propionic acid (CH_3CH_2COOH), and will provide selected data on longer-chain organic acids.

Corrosion Characteristics

Organic acids are weak acids, but provide sufficient protons to act as true acids toward most metals. Because most organic acids are neither oxidizing nor reducing to metals, they are handled successfully in such metals as copper, which do not directly displace hydrogen from acids. In organic acids, the corrosion behavior of the 400-series or type 304 stainless steels, which are protected by oxide films, is mixed. This makes contaminants extremely important, because they tend to shift the oxidizing capacity of the acid mixture. Air, Fe^{3+} ions, peracids, or peroxides will cause rapid attack of copper, and the presence of chlorides can have disastrous effects on stainless steels. Additional information on the role of contaminants in CH_3COOH corrosion is provided in Ref 89.

Corrosion testing can be difficult in organic acid media. Electrochemical measurements are most successful in dilute aqueous solutions of the acids because the conductance is very low in high concentrations or in solutions in nonaqueous solvents. The addition of sodium or chloride salts is reported to allow electrochemical measurements (Ref 90). Reported electrochemical data obtained in strong CH_3COOH, acetic acid anhydride [$(CH_3CO)_2O$)], and HCOOH solutions showed active-passive behavior that is consistent with field experience (Ref 91).

Data obtained by immersion tests in the laboratory often show erroneous results unless the atmosphere is carefully controlled. Without atmospheric control, the solution will be saturated with air at the beginning of the test, but will lose air as the temperature is increased until, at boiling conditions, almost all of the air will be removed. This situation can lead to results that vary widely, depending on the length of the test. Short tests of metals that exhibit active-passive behavior can also be misleading, because the metal may remain passive for a short time but may corrode actively after a longer exposure.

Truly anhydrous organic acids are usually much more corrosive to stainless steels than organic acids containing even traces of water. Corrosion rates reported for glacial (concentrated) acids often reflect this effect, because the acids may be truly anhydrous or may contain small amounts of water.

Formic Acid

Formic acid is the most highly ionized of the common organic acids and therefore the most corrosive. It reacts readily with many oxidizing and reducing compounds and is somewhat unstable as the concentration approaches 100%, decomposing to carbon monoxide and water. The following discussion is divided into metal and alloy groups used to handle HCOOH.

Steel is attacked quite rapidly by HCOOH at all concentrations and temperatures and is normally not considered for HCOOH service. Alloy steels that have been heat treated to high strengths and placed under stress fail by hydrogen cracking when exposed to inorganic acids. Stressed 18% Ni maraging steel cracks when exposed at ambient temperatures to 10% aqueous HCOOH, but does not crack in 91% HCOOH, or in either glacial acid or 10% CH_3COOH (Ref 92).

Aluminum shows fair resistance to HCOOH at any concentration at ambient temperatures as long as there is no contamination of the acid (Fig. 51). However, contamination with a wide variety of materials—for example, heavy-metal salts—can cause severe corrosion of aluminum. Aluminum has been used to ship concentrated HCOOH of about 95 to 99% concentration.

Typical rates of attack on aluminum alloy 5086 in HCOOH at 45 °C (115 °F) are shown in Fig. 52, which illustrates that the rate of attack increases rapidly with decreasing concentration at the maximum probable temperatures encountered during shipment. In addition, the acid used in the tests became turbid with aluminum salts, destroying the acid purity, and aluminum could not be used for shipment of this grade of HCOOH. Less meticulous grades of 95 to 99% acid could be shipped satisfactorily.

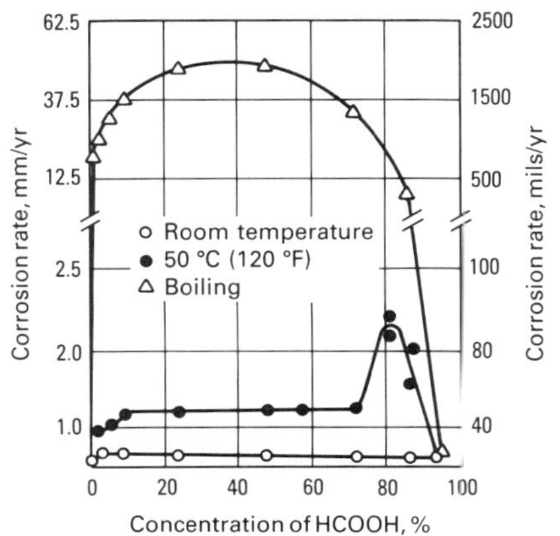

Fig. 51 Corrosion of aluminum alloy 1100-H14 in aqueous reagent grade HCOOH solutions. Source: Ref 93

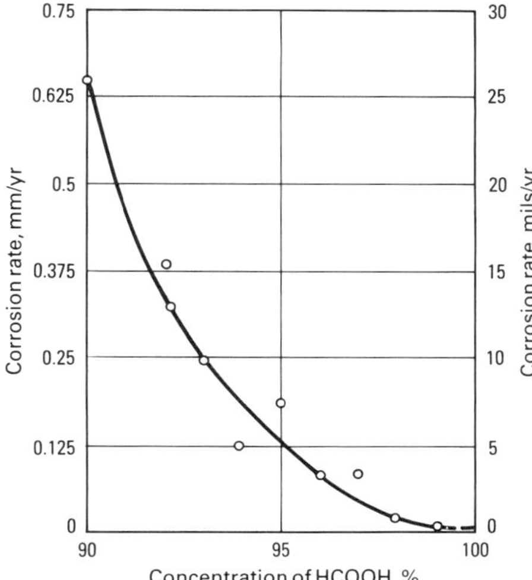

Fig. 52 Corrosion of aluminum alloy 5086 in HCOOH at 45 °C (115 °F). Source: Ref 93

Copper and its alloys, except yellow brasses, which dezincify, respond in approximately the same manner as aluminum to exposure to HCOOH. Table 12 shows typical corrosion rates for copper (C10300) and 90Cu-10Ni copper-nickel (C70600) in various concentrations of HCOOH. Aluminum bronze is reported to be a superior copper alloy for handling HCOOH (Ref 94).

The resistance of copper and its alloys to HCOOH depends on the presence or absence of oxygen or other oxidizing agents. If free air or other oxidants are present, high corrosion rates will be encountered; if the acid is free of air and other oxidants, copper will provide usable resistance to HCOOH at all concentrations to the atmospheric boiling point and even at higher temperatures. Copper and its alloys are probably the most widely used materials for handling HCOOH. The anomalies in the data shown in Table 12, such as the higher rate of attack in 50 and 70% HCOOH, are probably caused by incomplete deaeration during laboratory tests, although some increase in the rate of corrosion in intermediate acid strengths is to be expected because of maximum dissociation in these concentrations.

Stainless steels. The 400-series stainless steels are usually not resistant to HCOOH, except for very dilute, cool solutions, and they are seldom used in HCOOH service. Type 304 stainless steel has excellent resistance to HCOOH at all concentrations at ambient temperatures, and it is the preferred material of construction for storage of the acid. However, type 304 stainless steel is resistant to only 1 or 2% HCOOH at the atmospheric boiling temperature, and corrosion tests are advisable whenever type 304 stainless steel is considered for handling HCOOH at elevated temperatures. Table 13 shows typical rates of attack on various stainless steels in several concentrations of HCOOH at the atmospheric boiling temperature.

Type 316 stainless steel shows excellent resistance to HCOOH in all concentrations at ambient temperatures and is resistant to at least 5% HCOOH acid at the atmospheric boiling temperature. However, type 316 stainless steel can be seriously attacked by intermediate strengths of

HCOOH at higher temperatures, and corrosion tests are advisable.

The 20-type alloys, such as 20Cb-3, are more resistant to HCOOH than type 316 stainless steel is, and their use should be considered in higher concentrations at higher temperatures. Other alloys with chromium and nickel contents higher than those in type 316 stainless steel also show superior resistance to mixtures of HCOOH and CH_3COOH and would be expected to perform better in HCOOH itself. Duplex alloys are also reported to be superior to type 316 stainless steel (Ref 94). Weld overlays of 20-type alloys have also been used to alleviate the crevice corrosion of type 316 stainless steel—for example, under gaskets.

A low-carbon niobium-bearing variant of type 446 stainless steel—alloy S44627—appears to have exceptional corrosion resistance to HCOOH in preliminary laboratory studies and plant usage. This alloy should be considered for handling HCOOH.

Titanium has shown outstanding resistance to HCOOH in laboratory tests and in field usage. However, titanium can be attacked at truly catastrophic rates by anhydrous HCOOH. Additional work is needed to determine the precise conditions under which titanium is attacked in HCOOH, and until these parameters have been established, titanium should be tested very carefully before use in HCOOH approaching 100% concentration at elevated temperatures.

Other Alloys. Several of the high-nickel alloys, such as Hastelloy B, Hastelloy C-276, and Hastelloy C-4, have shown outstanding resistance to HCOOH in process equipment and are reported to exhibit very good resistance even at temperatures above the atmospheric boiling point. Figure 53 shows corrosion data for Hastelloy C. Lead is substantially nonresistant to HCOOH and other organic acids. The refractory metals tantalum, niobium, and zirconium resist HCOOH, as does silver.

Nonmetals. Formic acid is an excellent solvent and is destructive to most plastic materials and coatings. Therefore, plastics are not normally considered for handling HCOOH. Exceptions are the polyolefins, which are sometimes used to handle small quantities of acid; fluorocarbons, which are resistant; and some rubber linings, which can be used if discoloration of the acid is not critical. All coatings are normally unsatisfactory for strong HCOOH service.

Acetic Acid

Acetic acid, as well as its derivatives, is produced in large quantities. It is the most important organic acid and is frequently encountered as a contaminant in other organic chemical processes.

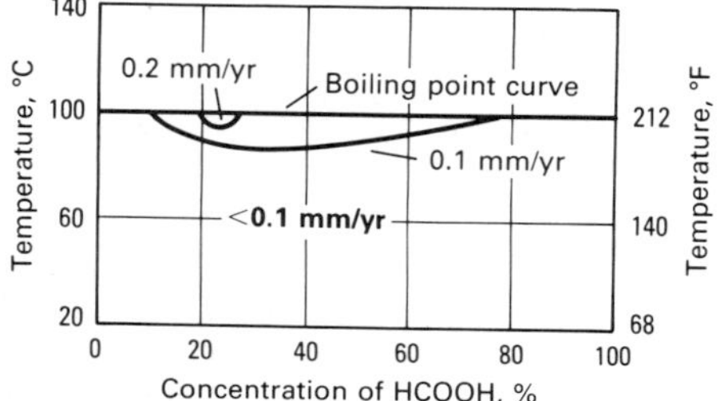

Fig. 53 Corrosion of Hastelloy C in HCOOH

Table 12 Corrosion of copper and copper-nickel by HCOOH

Laboratory tests in deaerated acid at atmospheric boiling temperature; test duration: 96 h

Acid concentration, %	Corrosion rate: Copper C10300, mm/yr	Copper C10300, mils/yr	Copper-nickel C70600, mm/yr	Copper-nickel C70600, mils/yr
1.0	0.02	0.8	0.02	0.9
5.0	0.02	0.7	0.02	0.9
10.0	0.02	0.6	0.02	0.7
20.0	0.20	7.8	0.40	15.7
40.0	0.14	5.5	0.34	13.3
50.0	0.26	10.2	0.54	21.1
60.0	0.05	2.0	0.03	1.3
70.0	0.76	30.0	0.76	30.0
80.0	0.20	7.8	0.13	5.0
90.0	0.22	8.7	0.19	7.6

Acetic acid is even used as a solvent for some organic reactions. Consequently, knowledge of its corrosivity is essential. Additional information on corrosion by CH_3COOH is provided in Ref 91 and 95.

Acetic acid is classified as a weak acid, but the effective acidity in aqueous streams increases rapidly with concentration (Fig. 54). The following sections discuss the resistance of various materials to CH_3COOH.

Steel is attacked quite rapidly by all concentrations of CH_3COOH, even at room temperature. Glacial CH_3COOH at room temperature is less aggressive than aqueous solutions of the acid, but still gives a rate of attack of 0.8 to 1.3 mm/yr (30 to 50 mils/yr). Therefore, steel is normally unacceptable for use in CH_3COOH service.

Aluminum exhibits good resistance to nearly all concentrations of CH_3COOH at room temperature and has been used extensively for storage and shipment. It is fairly resistant to 97 to 99% CH_3COOH to the boiling point, but is attacked very rapidly in concentrations near 100% or containing excess $(CH_3CO)_2O$. Aluminum again becomes resistant to pure $(CH_3CO)_2O$, although it causes contamination of the anhydride due to formation of a white crystalline solid, aluminum triacetate $(Al(C_2H_3O_2)_3)$, which precipitates in the liquid.

An excellent summary of the use of aluminum in CH_3COOH and $(CH_3CO)_2O$ is available in Ref 96. Figures 55 and 56 show the resistance of aluminum in CH_3COOH and $(CH_3CO)_2O$. The data are for aluminum alloy 1100, but similar rates would be expected for such alloys as 3003, 6063, and 5086.

The corrosion resistance of aluminum in CH_3COOH is strongly affected by contaminants. Aluminum can corrode in almost any concentration of CH_3COOH at any temperature if the acid is contaminated with the proper species.

Copper and its alloys, except those with high zinc content (>15% Zn), show good resistance to all concentrations of CH_3COOH up to and even above the atmospheric boiling temperature in the absence of oxygen or other oxidants. Copper was used almost exclusively to handle CH_3COOH until the advent of the stainless steels, but today, type 316 stainless steel and higher alloys are often used.

The absence of oxidizing agents is essential for copper to resist attack by CH_3COOH and other organic acids. Copper is nearly immune to attack by pure, uncontaminated CH_3COOH, yet slight contamination with air through storage under an air atmosphere or by entry of air through a pump seal can increase the rate of attack in a copper column to hundreds of mils per year. One set of laboratory tests at room temperature in 50% CH_3COOH showed corrosion rates of 1.8 mm/yr (71.5 mils/yr) when the solution was sparged with oxygen, but only 0.08 mm/yr (3.1 mils/yr) when the solution was nitrogen sparged.

The addition of nickel to copper moderates the effect of oxidants. Tests in boiling 50% CH_3COOH sparged with air for 120 h gave rates of 7.9 mm/yr (310 mils/yr) for copper, 4.8 mm/yr (188 mils/yr) for copper alloy C71500 (copper-nickel, 30%), and 2.1 mm/yr (84 mils/yr) for copper containing 67% Ni. Similar reductions with increasing nickel content were noted when Fe^{3+} ion was added to the solution; however, rates still remained quite high.

Stainless Steels. The chromium stainless steels of the 400 series occasionally exhibit low corrosion rates in laboratory tests in dilute CH_3COOH. However, because field experience with these materials indicates high corrosion rates and pitting attack, they are rarely used for CH_3COOH production equipment. Type 304 stainless steel is the lowest grade commonly used. Exceptions include the high-purity ferritic stainless steels, which show good resistance.

Type 304 stainless steel finds wide application in dilute CH_3COOH and in the shipment and storage of concentrated CH_3COOH. Data show that glacial CH_3COOH can be handled in type 304 stainless steel to a temperature of about 80 °C

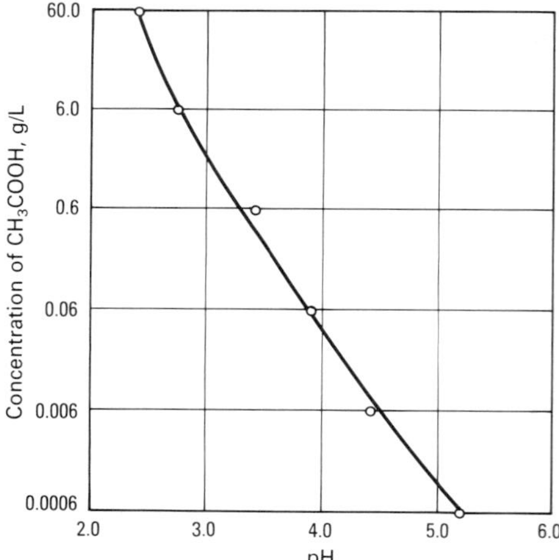

Fig. 54 Concentration versus pH of aqueous CH_3COOH solutions

Table 13 Corrosion of stainless steel by HCOOH

96-h laboratory tests at atmospheric boiling temperature

Acid concentration, %	Corrosion rate									
	Type 304(a)		Type 316(a)		Type 316(b)		20Cb-3(c)		UNS S44627(c)(d)	
	mm/yr	mils/yr	mm/yr	mils/yr	mm/yr	mils/yr	mm/yr	mils/yr	mm/yr	mils/yr
1.0	0.17	6.8	0.09	3.5	...	...	...	...	...	...
5.0	0.79	31.1	0.04	1.5	...	...	...	...	<0.03	<1.0
10.0	1.35	53.0	0.27	10.5	...	...	...	...	...	...
20.0	1.93	75.9	0.28	10.9	...	...	...	...	...	...
40.0	3.45	136	0.20	7.8	...	...	...	...	...	...
50.0	4.26	168	0.51	20.0	0.38	15.0	0.03	1.0	...	...
60.0	3.45	136	0.47	18.5	...	...	...	...	...	...
70.0	4.04	159	0.50	19.5	0.33	13.0	...	...	<0.03	<1.0
80.0	4.29	169	0.47	18.7	...	...	...	...	...	...
90.0	3.28	129	0.42	16.5	0.15	6.0	0.10	4.0	...	...
100.0	...	...	...	...	0.10	4.0	...	...	...	...

(a) Oxygen not controlled. (b) Deaerated. (c) 48-h exposure. (d) Low-carbon molybdenum-bearing version of type 446

(175 °F) and that type 304 has been satisfactory for lower concentrations to the boiling point of the acid (Ref 92). At temperatures above 60 °C (140 °F), use of the low-carbon type 304L is advisable for welded construction in order to prevent intergranular attack of heat-affected zones (HAZs).

Type 316 stainless steel is the alloy most commonly used in CH_3COOH processing equipment. It will resist glacial acid to temperatures above the atmospheric boiling point. As with type 304 stainless steel, the low-carbon grade (type 316L) is required for the higher-temperature application.

Acetic anhydride was produced as a coproduct in the old acetaldehyde oxidation process for CH_3COOH and is found in other acid streams. When CH_3COOH is truly anhydrous or contains small quantities of $(CH_3CO)_2O$, the rate of attack on type 316 stainless steel increases dramatically. Experience has shown that the introduction of a few tenths of a percent of water will reduce the corrosion. Figure 57 illustrates the reduction in corrosion rate as $(CH_3CO)_2O$ is added.

Contamination with chloride can cause pitting, rapid SCC, and accelerated corrosion of type 316 stainless steel. Up to 20 ppm of chloride can be tolerated (Ref 91), but higher concentrations are likely to cause rapid equipment failure.

Transferring heat through a metal wall, as in heat exchangers, can drastically alter the corrosion characteristics of the metal. A method to test metals under heat transfer conditions and data on stainless steels and some nickel alloys in HCOOH and CH_3COOH under heat transfer conditions are given in Ref 97. Higher alloys, such as 20Cb-3 and Incoloy 825, show better resistance to CH_3COOH than type 316 stainless steel does.

Titanium resists all concentrations of CH_3COOH up to the atmospheric boiling point. Electrochemical studies in CH_3COOH solutions suggest that it is possible to attack titanium in anhydrous CH_3COOH, but titanium has been used very successfully (Ref 98). The high-strength titanium alloys should not be used, because of their susceptibility to SCC.

Other alloys. Hastelloy alloys C-276 and B resist CH_3COOH solutions at all concentrations and normal temperatures. These materials are sometimes used where the acid is used in conjunction with inorganic acids and salts that limit the use of stainless steels or copper alloys. Hastelloy B is used under reducing conditions, such as with combinations of CH_3COOH and H_2SO_4, while Hastelloy C-276 is commonly used in highly oxidizing CH_3COOH solutions.

Silver has been frequently used in Europe to handle CH_3COOH, and it is quite resistant to all concentrations at normal temperatures. Because of cost, silver has been used very little in the United States.

Lead has very limited resistance to CH_3COOH. It has been used to store glacial CH_3COOH where temperature, degree of aeration, and velocity are low, but dilute CH_3COOH, even at room temperature, attacks lead at rates exceeding 1.3 mm/yr (50 mils/yr). These rates increase rapidly with increasing aeration and velocity.

Wood tanks have been extensively used for the storage of dilute CH_3COOH, but stainless steels are now normally used. Many wooden tanks are still in use after many years of service.

Rubber linings, either hard or semihard, have been used successfully in CH_3COOH storage tanks and for those applications where discoloration of the acid is not objectionable. Glass linings have also been used to handle CH_3COOH. No attack on alkaline borosilicate glass has been reported below about 150 to 175 °C (300 to 350 °F) (Ref 99).

Propionic Acid

CH_3CH_2COOH has corrosion characteristics very similar to, but somewhat less aggressive than, those of CH_3COOH. The following sections will discuss various materials for use in CH_3CH_2COOH.

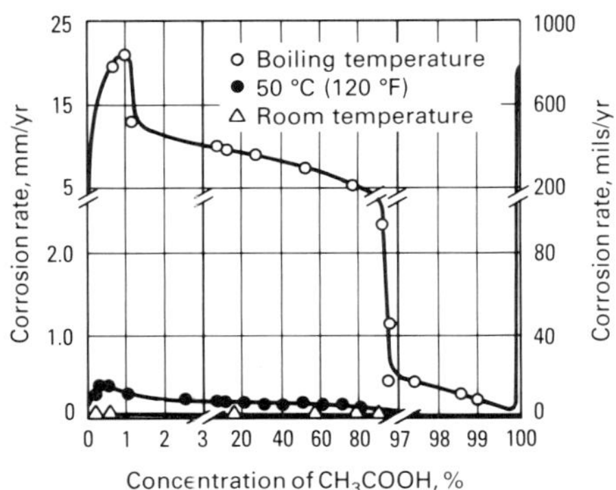

Fig 55 Effect of acid concentration and temperature on the corrosion of aluminum alloy 1100-H14 in CH_3COOH

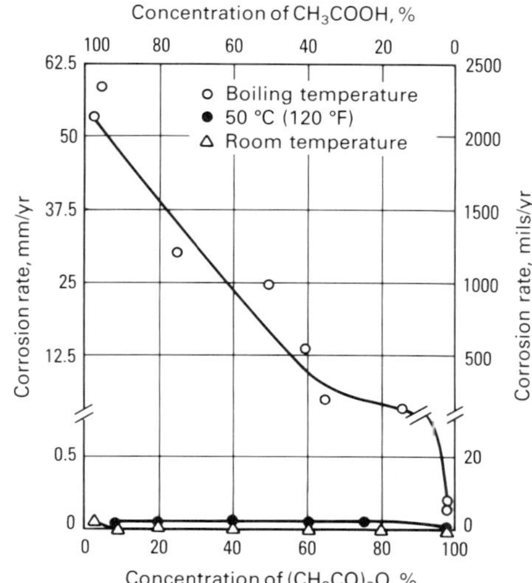

Fig. 56 Metal loss of aluminum alloy 1100-H14 in CH_3COOH-$(CH_3CO)_2O$ solutions at atmospheric boiling temperature

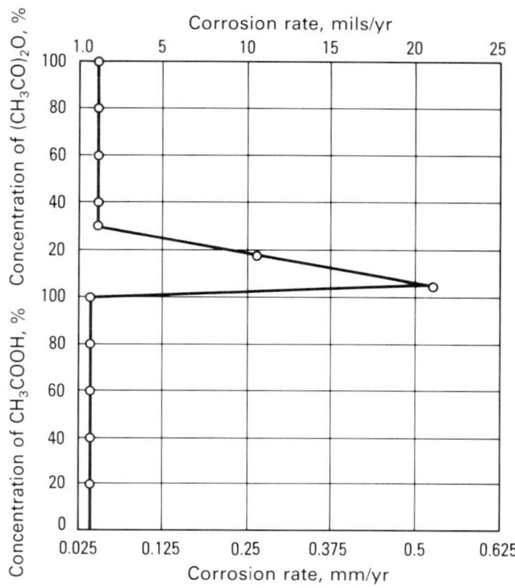

Fig. 57 Corrosion of type 316 stainless steel in CH_3COOH and in CH_3COOH-$(CH_3CO)_2O$ mixtures

Steel is attacked at rates of about 0.6 mm/yr (25 mils/yr) in pure CH_3CH_2COOH at room temperature and at much higher rates in aqueous solutions of the acid. Therefore, steel has very limited use in handling CH_3CH_2COOH.

Aluminum. The corrosion characteristics of aluminum in CH_3CH_2COOH are shown in Fig. 58. The rates of attack are very similar to those in CH_3COOH. Again, the rates can be significantly affected by contamination.

Copper and copper alloys are excellent for handling all concentrations of CH_3CH_2COOH. The data shown in Fig. 59 indicate attack on copper in boiling 100% CH_3CH_2COOH, but this is believed to be an anomaly caused by incomplete deaeration of the solution. As with HCOOH and CH_3COOH, copper and its alloys are satisfactory only if the solutions are completely deaerated and do not contain other oxidizing agents.

Stainless Steels. The 400-series stainless steels are not used in CH_3CH_2COOH, because of their propensity for pitting. They should be thoroughly tested in the exact environment for which their use is proposed. Type 304 stainless steel shows good resistance to CH_3CH_2COOH at room temperature and to aqueous solutions up to about 50% concentration at boiling temperature.

The atmosphere over the laboratory test solutions (Fig. 59) was not controlled, and the tests were of short duration (24 to 75 h), which caused two misleading results. The rate of corrosion of copper was found to increase rapidly above 65% concentration, which is true with air present in the solution, but copper would have very acceptable rates of corrosion in all concentrations of acid in the absence of air or other oxidants. Type 304 stainless steel was shown to have decreasing corrosion rates between 80 and 100% concentration; field experience indicates that type 304 stainless steel exhibits borderline passivity in this range of concentrations at the boiling temperature and should not be considered for such service.

Type 316 stainless steel is the preferred material for handling hot concentrated CH_3CH_2COOH solutions. The low-carbon grade (type 316L) should be used to avoid possible intergranular attack, unless corrosion tests in the exact environment show that intergranular attack is not a problem.

Other Metals. Hastelloy alloys B and C show excellent resistance to CH_3CH_2COOH solutions under reducing and oxidizing conditions, respectively. Other nickel alloys exhibit good resistance to low concentrations of CH_3CH_2COOH, but are not normally as good as type 316L stainless steel for the higher concentrations at high temperatures.

Other Organic Acids

It is impossible to cover each of the remaining hundreds of organic acids. However, Table 14 lists the corrosion rates of various metals in several of the longer-chain aliphatic acids and aromatic acids and some dicarboxylic acids. Table 15 provides data on the corrosion of some less familiar alloys in various organic acids. It is possible to make some generalized statements concerning materials for use in the higher molecular weight acids.

Steel is usable at ambient temperatures in the higher molecular weight acids, and it is used conventionally to store many of the acids and their corresponding anhydrides.

Aluminum shows good resistance to the acids at room temperature and is widely used for their handling. Some of the higher molecular weight acids cause severe attack of aluminum at highly elevated temperatures; therefore, the use of aluminum must be considered for the specific acid and temperature desired.

Copper and copper alloys exhibit good resistance to all of the higher molecular weight acids and can be used quite widely to handle the acids, even at elevated temperatures in the absence of oxidants.

Stainless Steels. Type 304 stainless steel has excellent resistance to the higher molecular weight organic acids at room temperature and at lower concentrations at high temperatures. With the concentrated acids, type 304 is sometimes severely corroded. Type 316 stainless steel is then required and is usable in almost all of the acids, even at elevated temperatures.

Other Alloys. The nickel-molybdenum and nickel-molybdenum-chromium alloys show excellent resistance to the higher molecular weight acids, but the expense of these alloys is rarely justified unless other contaminants, such as inorganic acids, are also present. The nickel alloys, particularly nickel-copper, have been used to process the higher molecular weight acids at elevated temperatures. They can be particularly useful when contamination of the acids prohibits the use of the type 316 stainless steel.

Corrosion by Hydrogen Chloride and Hydrochloric Acid

Thomas F. Degnan
Consultant

Hydrochloric acid is an important mineral acid with many uses, including acid pickling of steel, acid treatment of oil wells, chemical cleaning, and chemical processing. It is made by absorbing hydrogen chloride in water. Most acid is the by-product of chlorinations. Pure acid is produced by burning chlorine and hydrogen. Hydrochloric acid is available in technical, recovered, food-processing, and reagent grades. Reagent grade is normally 37.1% (Ref 100).

Hydrochloric acid is a corrosive, hazardous liquid that reacts with most metals to form explosive hydrogen gas and causes severe burns and irritation of eyes and mucous membranes. Safe handling procedures for HCl are described in Ref 101. Additional safety information is available from the manufacturer.

Concentrated HCl is transported and stored in rubber-lined tanks, although custom-fabricated polyester reinforced thermoset plastic (RTP) storage tanks have been used. Pipelines are usually plastic-lined steel. Processes involving aqueous acid are commonly carried out in glass-lined steel equipment. Nonmetallic materials are normally preferred because of the corrosive action of this strongest of acids on most metals.

Candidate metals and alloys for handling HCl are shown in Fig. 60. Suitable materials are judged to be those with a corrosion rate under 0.5 mm/yr (20 mils/yr) when exposed to uncontaminated HCl. In practice, contamination is not uncommon and can be catastrophic. Selection of a candidate metal should be based on extensive corrosion testing or, preferably, field experience, using the grade of acid that will be available.

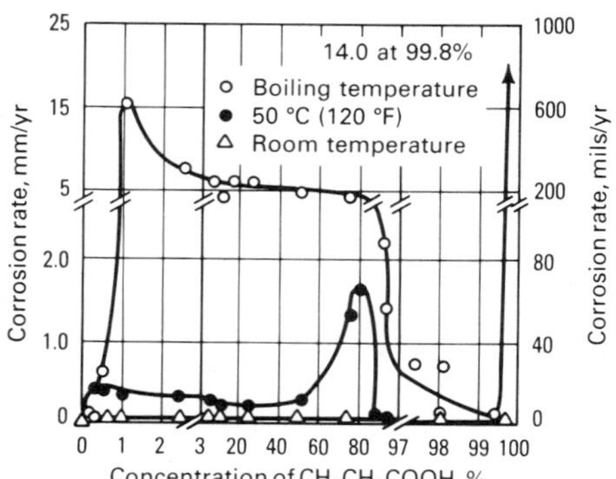

Fig. 58 Corrosion of aluminum alloy 1100-H14 in CH_3CH_2COOH solutions at various temperatures

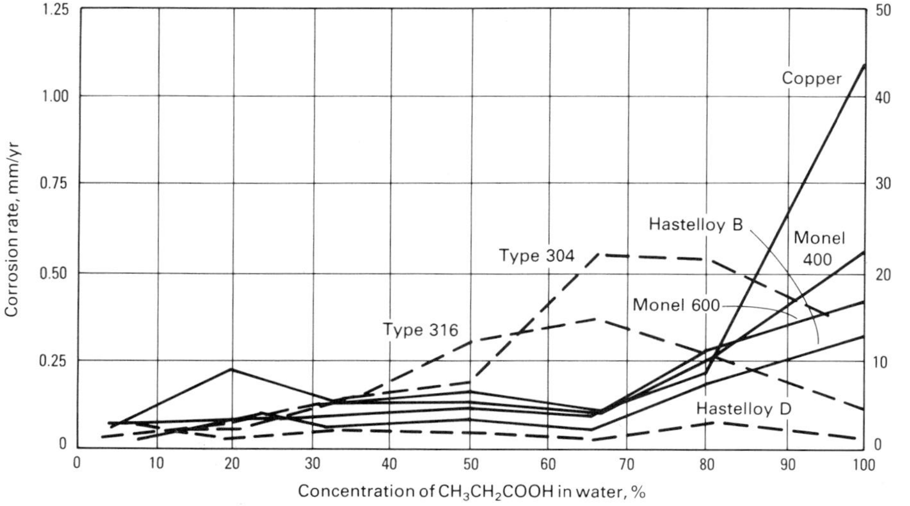

Fig. 59 Corrosion of metals in CH_3CH_2COOH at boiling temperature

Table 14 Corrosion of metals in refined organic acids

Acid	Corrosion rate: Steel mm/yr	Steel mils/yr	Copper mm/yr	Copper mils/yr	Silicon-bronze mm/yr	Silicon-bronze mils/yr	Type 304 stainless steel mm/yr	Type 304 stainless steel mils/yr	Type 316 stainless steel mm/yr	Type 316 stainless steel mils/yr
50% acrylic in an ether at 88 °C (190 °F)	...	...	...	...	...	...	<0.02	<1	<0.02	<1
90% benzoic at 138 °C (280 °F)	...	...	...	...	...	...	0.38	15	0.13	5
Butyric										
Room temperature	0.15	6	0.05	2	0.05	2	<0.02	<1	<0.02	<1
115 °C (240 °F)	...	...	...	...	...	...	0.08	3	0.08	3
Boiling (163 °C, or 325 °F)	...	...	...	...	...	...	1.42	56	0.12	5
Crotonic (crude product), 92 °C (200 °F)	...	...	...	...	...	...	<0.02	1	<0.02	1
2-ethylbutyric										
Room temperature	0.18	7	0.02	1	0.02	1	<0.02	<1	<0.02	<1
150 °C (300 °F)	0.86	34	0.41	16	0.23	9	0.53	21	<0.02	<1
2-ethylhexoic										
Room temperature	0.02	1	<0.02	<1	<0.02	<1	<0.00	<1	<0.02	<1
190 °C (375 °F)	1.27	50	<0.02	<1	<0.02	<1	0.20	8	<0.02	<1
Heptanedionic (pimelic), 225 °C (435 °F)	...	...	...	...	...	...	0.94	37	0.18	7
Hexadienoic (sorbic) as water slurry, 88 °C (190 °F)	...	...	...	...	...	...	<0.02	<1	<0.02	<1
Iso-octanoic										
Room temperature	<0.02	<1	<0.02	<1	<0.02	<1	<0.02	<1	<0.02	<1
190 °C (375 °F)	0.89	35	<0.02	<1	<0.02	<1	0.20	8	<0.02	<1
Iso-decanic										
Room temperature	<0.02	<1	<0.02	<1	<0.02	<1	<0.02	<1	<0.02	<1
190 °C (375 °F)	0.84	33	<0.02	<1	<0.02	<1	0.20	8	<0.02	<1
2-methylpentanoic										
Room temperature	0.02	1	0.08	3	0.10	4	<0.02	<1	<0.02	<1
150 °C (300 °F)	0.53	21	0.30	12	0.08	3	<0.02	<1	<0.02	<1
Pentanedioic (gluloric), 210 °C (410 °F)	...	...	...	...	...	...	0.68	27	0.20	<1
Pentanoic (valeric)										
Room temperature	0.05	2	0.05	2	0.05	2	<0.02	<1	<0.02	<1
114 °C (237 °F)	1.37	54	0.68	27	0.13	5	<0.02	<1	<0.02	<1

Effect of Impurities

Fluorides. Acid recovered from the manufacture of fluorocarbons may contain trace amounts of HF. It has been reported that such acids may contain more than 0.5% HF (Ref 103). Commercial suppliers remove most of the fluoride from HCl by selective absorption, and it is unlikely that an unsuspecting customer would receive acid containing as much as 0.5%. However, glass-lined steel and the refractory metals, such as zirconium, niobium, tantalum, and titanium (but not molybdenum), have very low tolerance levels for fluorides. Zirconium is reported to tolerate less than 10 ppm (Ref 104). Tantalum may tolerate 10 ppm or more. The limits are undefined except for a few specific cases, and it is best to consider all of these metals and glass-lined steel to be essentially nonresistant of fluorides and to know the source and specification limits of the acid if these materials are used.

Ferric Salts. The presence of Fe^{3+} ions in HCl has a profound effect on the corrosion of many metals and alloys otherwise resistant to HCl. Nickel-base alloys, including Hastelloy B-2, the copper alloys, and unalloyed zirconium, are affected. Although the acid specification may be low in iron, acid can easily become contaminated during shipment and handling.

Cupric Salts. Cupric ions have an accelerating effect on the corrosion of many metals that is similar to that of Fe^{3+} ions. Like Fe^{3+} ions, Cu^{2+} ions can cause pitting and SCC of zirconium. The presence of Cu^{2+} ions can also lead to the autocatalytic acceleration of nickel-copper and copper-nickel alloys. It is unlikely that commercial acid would contain Cu^{2+} salts. This is more of a problem of in-process contamination by exposure of copper-containing metals or introduction as an

Table 15 Corrosion of miscellaneous alloys by organic acids

48-h tests at atmospheric boiling temperature; atmosphere not controlled

Test medium	Corrosion rate: Type 329 mm/yr	Type 329 mils/yr	Tantalum mm/yr	Tantalum mils/yr	Titanium mm/yr	Titanium mils/yr	Zirconium mm/yr	Zirconium mils/yr	Crucible 223 mm/yr	Crucible 223 mils/yr	UNS S44627(a) mm/yr	UNS S44627(a) mils/yr	MP35N mm/yr	MP35N mils/yr
Glacial CH_3COOH	<0.02	<1	nil		<0.02	<1	nil		0.18	7	<0.02	<1	0.05	2
99% CH_3COOH, 1% $(CH_3CO)_2O$	...	...	nil		<0.02	<1	<0.02	<1	23	900	...	...	...	...
90% CH_3COOH, 10% $(CH_3CO)_2O$	...	...	nil		<0.02	<1	<0.02	<1	...	...	...	...	...	...
50% CH_3COOH, 50% $(CH_3CO)_2O$	0.58	23	nil		0.18	7	<0.02	<1	...	...	0.20	8	<0.02	<1
90% CH_3COOH, 10% HCOOH	0.71	28	nil		<0.02	<1	<0.02	<1	0.13	5	...	...	...	...
70% HCOOH	1.27	50	<0.02	<1	...	...	<0.02	<1	...	...	<0.02	<1	0.13	5
20% HCOOH	...	...	<0.02	<1	...	...	<0.02	<1	4.75	187	<0.02	<1	...	...
2-ethylbutyric acid	...	...	nil		<0.02	<1	<0.02	<1	0.02	1	<0.02	<1	...	...
10% aqueous oxalic acid	0.36	14	nil		...	...	<0.02	<1	0.58	23	0.36	14	0.10	4

(a) Low-carbon molybdenum-containing variant of type 446

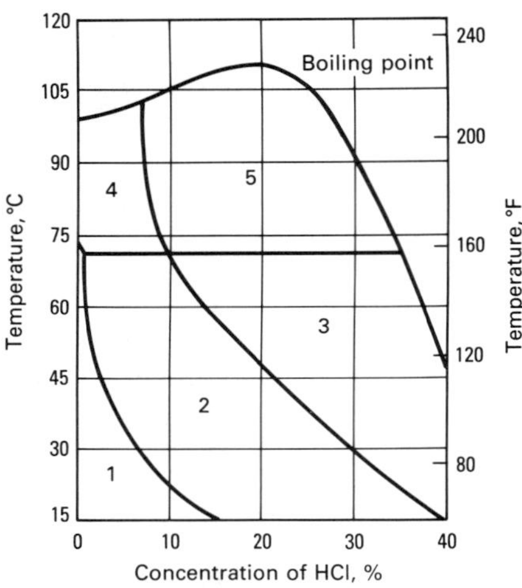

Fig. 60 Alloys for HCl service. Source: Ref 102

Metals with reported corrosion rates of <0.5 mm/yr (<20 mils/yr) in HCl

Zone 1	Zone 2	All zones (including 5)
ACI CN-7M(a)(c)(f)	Silicon bronze(b)(f)	Platinum
Monel 400(b)(c)(f)	Silicon cast iron(c)(g)	Tantalum
Copper(b)(c)(f)	**Zone 3**	Silver(c)(f)
Nickel 200(b)(c)(f)	Silicon cast iron(c)(g)	Zirconium(c)(f)
Silicon bronze(b)(c)(f)	**Zone 4**	Hastelloy B-2(c)(f)
Silicon cast iron(c)(g)	Tungsten	Molybdenum(c)(f)
Tungsten	Titanium, grade 7(e)	
Titanium, grade 7		
Titanium, grade 2(d)		

(a) <2% HCl at 25 °C (75 °F). (b) No aeration. (c) No $FeCl_3$ or $CuCl_2$ contamination. (d) <10% HCl at 25 °C (75 °F). (e) <5% HCl at boiling temperature. (f) No free chlorine. (g) Contains chromium, molybdenum, and nickel

impurity in a chemical or raw material used in the process.

Aeration, although less damaging than the presence of oxidizing metal salts, accelerates the corrosion of many metals. Figure 61 shows the effect of aeration on the corrosion rate of Nickel 200 and Monel 400.

Chlorine contamination accelerates the corrosion of all metals except unalloyed tantalum and noble metals; however, unalloyed titanium can be protected by the presence of chlorine in dilute HCl (Ref 105). Chlorine may be present in acid recovered from a chlorination process, but would be removed before sale.

Organics. Hydrochloric acid can become contaminated with organic solvents, such as carbon tetrachloride (CCl_4) or chlorobenzene (C_6H_5Cl), when recovered as a by-product of a chlorination process. Even a few parts per million of organic contaminants can, over a period of time, destroy rubber linings, elastomer membranes behind brick linings, and certain plastics and elastomers. It is unlikely that such acid would be shipped, although organic contamination is often encountered in plant processes.

Corrosion of Metals in HCl

Most corrosion data and graphical information published by metal suppliers and others are based on tests conducted in reagent grade HCl. In some cases, as in Fig. 60, limitations will be noted regarding the presence of aeration or impurities. As mentioned above, these data should be used only as a guide in selecting metals for further testing and evaluation.

Carbon and alloy steels are unsuitable for exposure to HCl except during acid cleaning.

Austenitic Stainless Steels. The commonly used austenitic stainless steels, such as types 304 and 316, are nonresistant to HCl at any concentration and temperature. At ambient temperatures and above, corrosion rates are high. Nickel, molybdenum, and, to a lesser extent, copper impart some resistance to dilute acid, but pitting, local attack, and SCC may result (Ref 106). Subambient temperatures will slow the corrosion rate, but will invite SCC. Type 316 stainless steel has been known to crack in 5% HCl at 0 °C (32 °F) (Ref 107). At high corrosion rates (>0.25 mm/yr, or 10 mils/yr), SCC is unlikely to occur. However, the corrosion products, particularly $FeCl_3$, will cause cracking. Chlorides can penetrate and destroy the passivity (oxide film) that is responsible for the corrosion resistance of stainless steels, and the corrosion engineer should resist every attempt to use stainless steels in environments containing chlorides.

The standard ferritic stainless steels, such as types 410 and 430, should not be considered, because their corrosion resistance to HCl is lower than that of carbon steel. An exception is 29-4-2 stainless steel, which reportedly resists up to 1.5% HCl to the boiling point and remains passive (Ref 108). However, it is not suitable at higher concentrations, and the alloy is susceptible to SCC, although its resistance is reported to be high.

Some stainless steels—such as 20Cb-3, with its high nickel content (32 to 38%), 2 to 3% Mo, and 3 to 4% Cu—resist dilute HCl at ambient temperatures. However, 20Cb-3 is susceptible to pitting and crevice attack in acid chlorides and should be used with caution.

Nickel and Nickel Alloys. Nickel 200 and Monel 400 have good resistance (<0.25 mm/yr, or 10 mils/yr) to dilute (<10%) HCl in the absence of air or oxidizing agents at ambient temperatures. Monel 400 has been used at concentrations below 20% at ambient temperatures under air-free conditions and under 10% concentration aerated, but penetration rates generally exceed 0.25 mm/yr (10 mils/yr) and may approach 1 mm/yr (40 mils/yr).

Increasing the temperature affects the corrosion rate of Nickel 200 more than that of Monel 400 in 5% HCl (Fig. 62). When HCl is formed by hydrolysis of chlorinated hydrocarbons, acid concentrations are often less than 0.5%. Under these

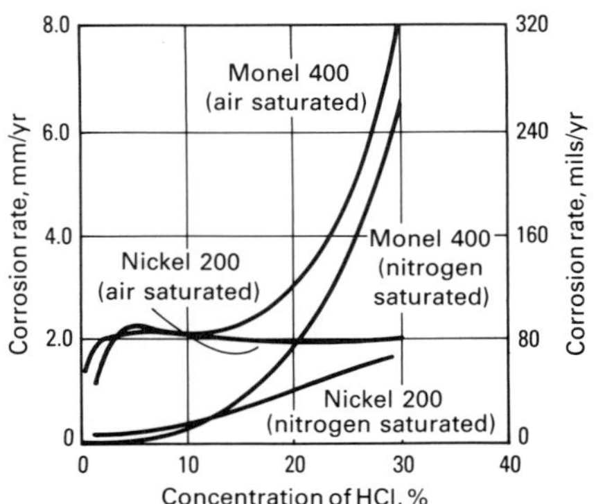

Fig. 61 Corrosion of Nickel 200 and Monel 400 in HCl solutions at 30 °C (85 °F). Source: Ref 104

Table 16 Laboratory corrosion tests of Inconel 625 in HCl solutions at 66 °C (150 °F)

Acid concentration, wt%	Corrosion rate mm/yr	Corrosion rate mils/yr
5	1.8	71
10	2.1	81
15	1.7	65
20	16.5	650
25	0.96	38
30	0.86	34
37 (concentrated)	0.38	15

Source: Ref 104

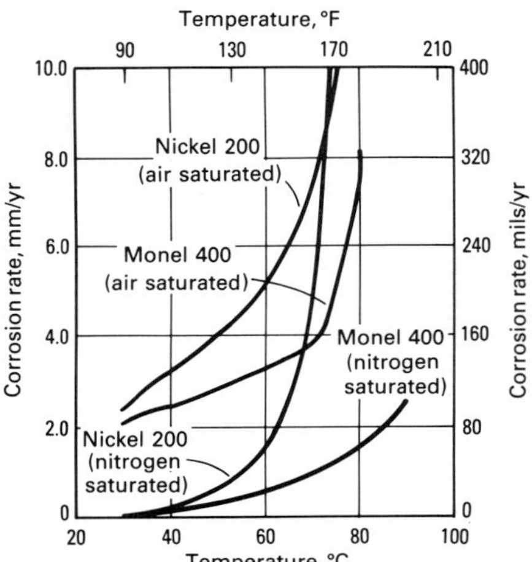

Fig. 62 Effect of temperature on the corrosion of Nickel 200 and Monel 400 in 5% HCl. Source: Ref 104

Table 17 Laboratory corrosion tests of Incoloy 825 in HCl

Acid concentration, wt%	Temperature °C	Temperature °F	Corrosion rate mm/yr	Corrosion rate mils/yr
5	Room		0.13	5
10	Room		0.18	7
15	Room		0.18	7
28	50	120	0.9	36

Source: Ref 104

conditions, Nickel 200 and Monel 400 have found application at temperatures below 200 °C (390 °F) (Ref 104).

Inconel 600, although it has useful resistance to cold dilute HCl, exhibits corrosion resistance which is inferior to that of Nickel 200 and Monel 400. However, this alloy is useful in handling wet halogenated solvents, because it has little, if any, catalytic effect on hydrolysis.

Inconel 825, Hastelloy G, and Inconel 625 contain appreciable amounts of chromium and increasing amounts of molybdenum, and they have useful resistance to all concentrations of HCl below 40 °C (100 °F). Inconel 625 has good resistance to concentrated reagent grade acid at ambient temperatures. Corrosion data on Inconel 825 and 625 in various concentrations are given in Tables 16 and 17.

These alloys have good resistance to dilute (<5%) acid at higher temperatures and are less affected by aeration than Nickel 200 and Monel 400. They are all very resistant to chloride SCC. Resistance to pitting and crevice corrosion in acid chloride solutions improves with increasing chromium and molybdenum contents. Thus, Inconel 625 (22Cr-9Mo) is more resistant than Hastelloy G (22Cr-6.5Mo), which in turn is more resistant than Inconel 825 (21Cr-3Mo). Figure 63 shows an isocorrosion diagram of Hastelloy G in HCl.

The most corrosion resistant of the nickel-base alloys to HCl are Hastelloy B-2 (Ni-28Mo) and Hastelloy C-276 (Ni-16Cr-15.5Mo). Hastelloy B-2 is one of the few metals with a corrosion rate under 0.5 mm/yr (20 mils/yr) in all concentrations and temperatures up to the atmospheric boiling point in nonaerated acid in the absence of oxidizing agents. Solution heat treatment after welding is required for maximum corrosion resistance.

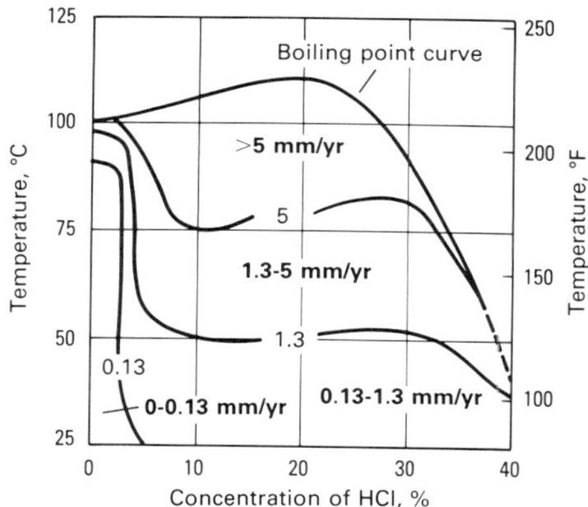

Fig. 63 Isocorrosion diagram for Hastelloy G in HCl

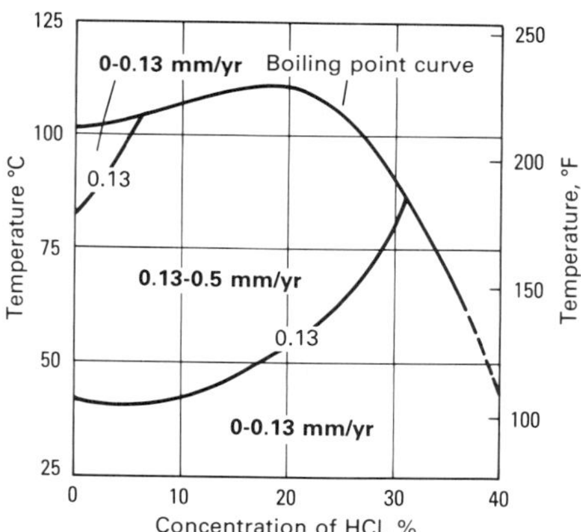

Fig. 64 Isocorrosion diagram for Hastelloy B-2 in HCl

Figures 64 and 65 show isocorrosion diagrams for unaerated and oxygen-purged HCl.

Hastelloy C-276 has excellent corrosion resistance (<0.13 mm/yr, or 5 mils/yr) in all concentrations of HCl at room temperatures and good resistance (<0.5 mm/yr, or 20 mils/yr) to all concentrations up to 50 °C (120 °F). At concentrations under 10%, its resistance often exceeds that of Hastelloy B-2. Figure 66 shows an isocorrosion diagram for Hastelloy C-276 in unaerated HCl. Oxygen and strong oxidizing agents accelerate corrosion, although markedly less than for Hastelloy B-2.

Copper and copper alloys have limited utility in HCl service because they are so sensitive to velocity, aeration, and oxidizing impurities. However, copper is one of the few common metals above hydrogen in the electromotive force (emf) series, and under reducing conditions, it can have low corrosion rates. An example is the use of silicon-bronze (CDA C65500) agitators and hardware in the manufacture of $ZnCl_2$ by dissolving zinc in HCl (Ref 107), with the evolution of hydrogen.

Typical corrosion rates for silicon-bronze in HCl under nonaerated nonoxidizing conditions are 0.08 to 0.1 mm/yr (3 to 4 mils/yr) in up to 20% acid and 0.5 mm/yr (20 mils/yr) in concentrated acid at 25 °C (75 °F). At 70 °C (160 °F), rates are approximately 1 mm/yr (40 mils/yr) in up to 20% acid and over 6.4 mm/yr (250 mils/yr) in concentrated (35 to 37%) acid.

Corrosion-Resistant Cast Iron. A high-silicon iron alloyed with small amounts of molybdenum, chromium, and copper has good resistance to all concentrations of HCl to temperatures as high as 95 °C (200 °F). It is one of the few metals commonly used for pumps and valves in handling commercial grades of acid (Ref 106). Fluoride impurities are damaging.

Zirconium. The corrosion resistance of zirconium to HCl exceeds that of most metals, except tantalum and such noble metals as gold and platinum. Aeration does not have an appreciable effect. Corrosion rates are less than 0.13 mm/yr (5 mils/yr) at all concentrations to the atmospheric boiling point and above (Ref 109). However, chlorine and relatively small amounts of Cu^{2+} and Fe^{3+} ions accelerate corrosion and can cause pitting and embrittlement. Figure 67 shows an isocorrosion diagram of zirconium in HCl.

It is important to avoid galvanic effects when connecting zirconium to other metals immersed in an electrolyte, because zirconium, like all the reactive metals, is sensitive to hydrogen embrittlement when it is the cathode of an electrochemical cell. Applications of zirconium equipment in HCl include pumps, valves, piping, and heat exchangers (Ref 109).

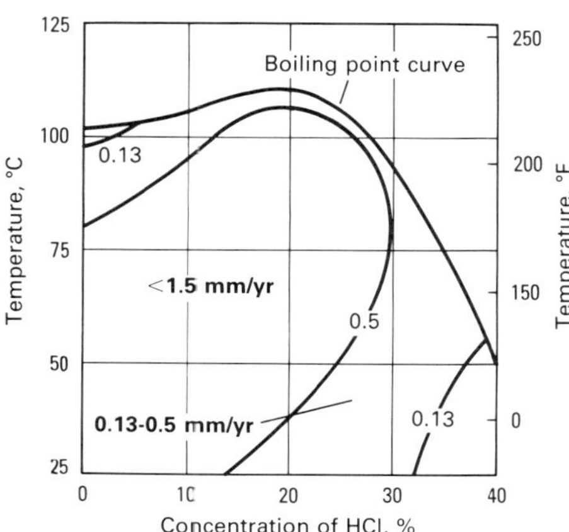

Fig. 65 Isocorrosion diagram for Hastelloy B-2 in oxygen-purged HCl

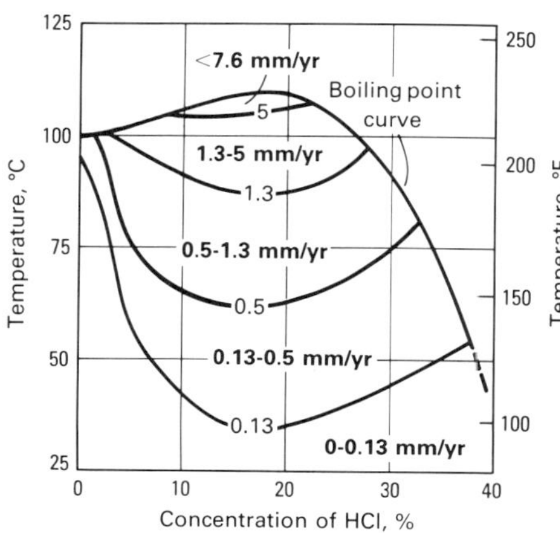

Fig. 66 Isocorrosion diagram for Hastelloy C-276 in HCl

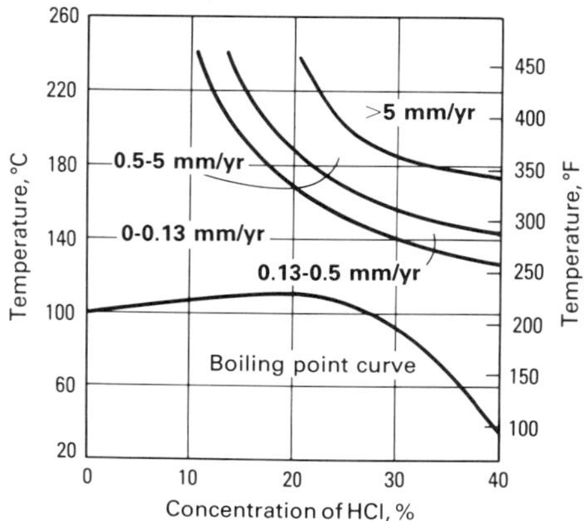

Fig. 67 Isocorrosion diagram for zirconium in HCl solutions. Source: Ref 109

Titanium and Titanium Alloys. Although titanium has limited resistance to HCl, it is, unlike other metals, passivated rather than corroded by the presence of dissolved oxygen, Fe^{3+} and Cu^{2+} ions, nitrates, chromates, chlorine, and other oxidizing impurities (Ref 105). In commercial applications involving hot, dilute acid, enough impurities are often present to provide a high degree of protection. For example, the corrosion rate of unalloyed titanium in boiling 4% HCl was lowered from 21.4 mm/yr (843 mils/yr) to 0.01 mm/yr (0.4 mils/yr) by the addition of 0.2% $FeCl_3$ (Ref 105).

An isocorrosion diagram for unalloyed titanium and two titanium alloys is shown in Fig. 68. No titanium alloy has resistance to concentrated grades of HCl.

Tantalum and its alloys are the most resistant to HCl. Tantalum resists concentrations below 25% up to 190 °C (375 °F) and concentrations to 37% at temperatures to 150 °C (300 °F) (Ref 110). It was found that tantalum is embrittled by concentrations of 25% or higher at 190 °C (375 °F) (Ref 110). Corrosion rates at that temperature were less than 0.025 mm/yr (<1 mil/yr) at 25% concentration or less, 0.01 mm/yr (3.9 mils/yr) at 30%, and 0.29 mm/yr (11.6 mils/yr) at 37%. Embrittlement was most pronounced in 37% acid. It also was found that corrosion and embrittlement could be avoided by coupling tantalum with platinum. This discovery formed the basis for the later development of the titanium-palladium alloy.

Tantalum, like zirconium and titanium, is embrittled by the absorption of atomic hydrogen. This occurs if the metal is corroded in a nonoxidizing chemical or if it becomes the cathode of an electrolytic cell. For example, it is possible to embrittle tantalum plugs in a glass-lined vessel, if the vessel is equipped with an agitator of a metal lower in the emf series.

Noble Metals. Silver forms a protective chloride film in HCl and is resistant as long as the film is not dissolved or disturbed. At ambient temperatures, in the absence of oxidizing agents, silver resists all concentrations, with a corrosion rate of less than 0.025 mm/yr (<1 mil/yr) at concentrations of 20% or lower. At the boiling point, corrosion rates vary from 0.025 mm/yr (1 mil/yr) at 5% to 0.5 mm/yr (20 mils/yr) at 20%. Increasing the temperatures, aeration, and velocities will increase corrosion rates (Ref 106).

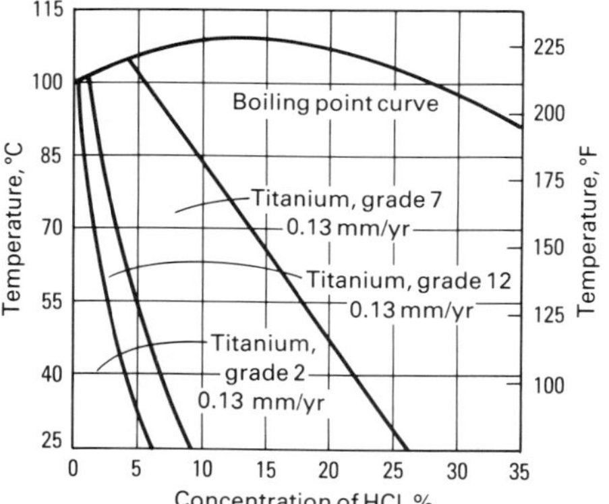

Fig. 68 Isocorrosion diagram for titanium in naturally aerated HCl solutions. Source: Ref 104

Refined gold is extremely resistant to all concentrations of HCl to and above the atmospheric boiling point. Oxidizing agents such as $FeCl_3$ and HNO_3 cause attack (Ref 106).

Refined platinum is almost as resistant as gold. It is slightly attacked by 36% acid at 100 °C (212 °F), the point at which gold shows no appreciable attack (Ref 106).

Nonmetallic Materials

Elastomers used in HCl service include natural rubber, neoprene, and others.

Natural rubber is used to line steel tanks, tank cars, and piping for handling commercial, uncontaminated, concentrated HCl (Ref 100). Natural rubber forms a protective rubber hydrochloride surface that slows further penetration, and a life expectancy of 20 years at ambient temperature is not uncommon. The lining must be properly compounded for the service, must be of sufficient thickness (a minimum of 6.4 mm, or 0.25 in.), and must be properly applied and inspected. Soft rubber linings are recommended to 40 to 45 °C (100 to 110 °F), depending on the manufacturer (Ref 111, 112). A three-ply construction (soft-hard-soft) is often used to take advantage of the impermeability of the harder rubber and the resistance to damage, thermal shock, and abrasion of the soft rubber. Three-ply construction is recommended to temperatures of 55 °C (130 °F) (Ref 111). For higher temperatures, semihard rubber is recommended to 70 °C (150 °F), and hard rubber or flexible ebonite to a maximum temperature of 95 °C (200 °F) (Ref 113). The latter must be protected against exposure to sunlight and thermal and mechanical shock.

Natural rubber is severely affected by many organic contaminants, some of which may be present in by-product acid. Because the effect is cumulative, even trace amounts will eventually cause failure.

Neoprene. Although relatively unaffected by ambient-temperature HCl, neoprene is not recommended for linings, because of its permeability to concentrated acid (Ref 111, 112).

Other Elastomers. A number of elastomers have resistance to HCl that is as good as or better than that of rubber, but they are seldom used as linings for this service because of economics and difficulty of application. However, they should be considered for exposure to sunlight, organics, higher temperatures, or flexing (the hard rubber hydrochloride surface cracks when flexed). These include butyl rubber, nitrite butyl rubber (NBR), ethylene propylene diene monomer (EPDM), and chlorosulfonated polyethylene elastomer (Ref 113, 114). Fluoroelastomers have good resistance to hot acid, but because of high cost, are limited to small parts, such as gaskets and O-rings.

Thermoplastics. A large number of thermoplastics are suitable for handling all concentrations of HCl. Some of the most commercially important are described below.

Polypropylene. Polypropylene-lined steel pipe has largely replaced rubber-lined pipe for commercial HCl service, primarily because of economics and ease of field fit-up. Lined pipe is recommended for all concentrations of HCl to temperatures of 110 °C (225 °F) (Ref 115). Polypropylene is also available in small-to-moderate size tanks and various molded shapes.

Polyethylene is unaffected by ambient-temperature acid, but its lack of stiffness, high thermal coefficient of expansion, and rapid falloff of strength with increasing temperature limit its use to underground or continuously supported pipelines. High-density cross-linked polyethylene is stronger and stiffer and is available in tanks up to 3785 L (1000 gal).

Polyvinyl Chloride. Both plasticized and unplasticized PVC have good resistance to HCl at ambient temperature. Unplasticized PVC piping systems are practical for complex systems that can be properly supported.

Polyvinylidine chloride has been used for many years as a lining for HCl pipelines. It is recommended for all concentrations to 80 °C (175 °F) (Ref 115). It has better resistance than rubber or polypropylene to certain hydrocarbons that may be present in process streams.

Polyvinylidine Fluoride. Pipe lined with PVDF is recommended for all concentrations of HCl to 135 °C (275 °F) (Ref 115). Polyvinylidine fluoride is also available in lined valves and pumps. It resists a wide variety of other chemicals and solvents that may be present in process streams or in recovered acid. Linings, frequently reinforced with graphite, glasscloth, or veil, have been applied with multiple thermally fused layers with varying degrees of success. Each case should be approached with caution.

Fluoroplastics. In addition to PVDF, a number of other fluorocarbon plastics resist all concentrations of HCl at temperatures as high as 260 °C (500 °F) (Ref 115). The fully fluorinated PTFE and PFA have the highest temperature resistance and are the most expensive. Fluorinated ethylene propylene is limited to about 200 °C (400 °F), but is more easily molded than PTFE. Ethylene-chlorotrifluoroethylene and ETFE are limited to 150 °C (300 °F), but have superior abrasion and permeability resistance (Ref 116). Most fluoroplastics are available as lined pipe, valves, and pumps (Ref 116).

Other Thermoplastics. Other plastics, such as acrylonitrile-butadiene-styrene (ABS), polysulfone, polyphenylene sulfide, and polyphenylsulfone, are resistant to HCl. Nonresistant thermoplastics include such polyimides as nylon and methacrylates.

Reinforced thermoset plastics for HCl service include polyester, epoxy, phenolic, and furan resins.

Polyester Resins. There have been a number of failures of glass-reinforced polyester resin storage tanks in concentrated HCl service. A life expectancy of 5 to 10 years has been reported (Ref 117). However, there are custom-built glass-reinforced plastic storage tanks that have given satisfactory service in concentrated HCl for 10 to 20 years and longer (Ref 107).

Standard, filament wound storage tanks are not suitable for concentrated acid. Vessels should be fabricated in accordance with the standards detailed in Ref 118. The inner resin-rich layer should be thicker than 0.8 mm (1/32 in.) with two layers of veil (Ref 107). Bubbles and porosity are not acceptable, because they will cause blistering, which will lead to degradation. The novalac epoxy vinyl-ester resins resist blistering better than other resins, although none are recommended for storage of concentrated acid (Ref 117).

In Europe, glass-reinforced plastic tanks for storing concentrated acid are frequently lined with PVC sheet. The novalac resins have superior resistance to some chlorinated hydrocarbons (Ref 119). Custom-built storage tanks using this resin have been in service for over 10 years in concentrated acid containing CCl_4 in excess of

Table 18 Recommended maximum temperature limits for fiberglass-reinforced plastic piping in HCl

Acid concentration, wt%	Maximum recommended temperature: Epoxy resin °C	°F	Vinyl-ester resin °C	°F
1	93	200	93	200
10.5	66	150	93	200
20	24	75	93	200
36.5 (conc)	NR		66	150(a)

NR, Not recommended. (a) Maximum temperature tested; could be suitable at higher temperature. Source: Ref 120

solubility (Ref 107). Pumps and piping systems of polyester glass-reinforced plastic are also used for moving concentrated acid.

The epoxy resins are generally less resistant to HCl than the resistant grades of polyesters. Manufacturers of glass-reinforced plastic pipe limit recommendations for their epoxy pipe to concentrations of 10 or 20% at ambient temperatures, while the polyester pipe is recommended for all concentrations at higher than ambient temperatures. Recommendations by a leading manufacturer are given in Table 18. It should be noted, however, that a noted pump manufacturer makes a silica-filled epoxy pump that is widely used for pumping concentrated HCl.

Phenolic Resins. Vessels, ducts, and piping systems fabricated of suitably filled phenolic resins will resist all concentrations of acid to temperatures as high as 175 to 200 °C (350 to 390 °F) (Ref 121). The phenolics are also resistant to many solvents. Because phenolic resins are weak in tension, equipment and pipe are often armored with epoxy-impregnated glass cloth or are filament wound. Piping systems must be carefully designed and installed with adequate support to minimize tensile stresses. Phenolic resins are often used as cements for brick linings.

Table 19 Suggested upper temperature limits for continuous service in dry HCl gas

Material	Temperature °C	°F
Platinum	1200	2190
Gold	870	1600
Nickel 201	510	950
Inconel 600	480	900
Hastelloy B	450	840
Hastelloy C	450	840
Type 316 stainless steel	430	805
Type 310Cb stainless steel	430	805
Type 304 stainless steel	400	750
Carbon steel	260	500
Monel 400	230	445
Silver	230	445
Cast iron	200	390
Copper C11000	90	195

Source: Ref 126

Furan resins are also used to construct vessel duct and piping systems that are resistant to all concentrations of wet HCl (Ref 121). The solvent and alkali resistance of furan resins exceeds that of phenolics, and they are easier to handle during fabrication. Thus, furan glass-reinforced plastic vessels are available, and furan-glass membranes are used in brick-lined equipment. Temperature resistance is somewhat lower than that of the phenolics. Furans are the most common of the plastic cements used for brick linings.

Other nonmetallic materials for HCl service include impervious graphite, carbon, glass, glass-lined steel, and wood.

Impervious graphite is the most commonly used material of construction for HCl absorption and refining equipment. Resin-impregnated graphite equipment resists all concentrations of HCl to temperatures of 165 to 185 °C (330 to 365 °F), depending on impregnating resin and manufacturer. Polytetrafluoroethylene-impregnated graphite is reported to be resistant to a maximum temperature of 230 °C (450 °F) (Ref 122). By using graphite impregnated with pure carbon, it is possible to increase the maximum-use temperature in air to 400 °C (750 °F) (Ref 122). Graphite has excellent thermal conductivity and chemical resistance (except to strong oxidants and concentrated H_2SO_4). However, it is extremely weak in tension and is seldom used for piping. Heat exchangers are available in block and crossbore construction to obtain more rugged units than shell and tube designs.

Carbon has poor heat transfer but excellent abrasion resistance. It is available as bricks for brick-lined construction. Like graphite, carbon resists all concentrations of HCl to 400 °C (750 °F) in air.

Glass. Complete small chemical plants are available in glassware. Piping systems can be obtained armored with glass fabric. Piping systems must be properly supported, because tensile stresses can result in delayed fracture.

Glass-Lined Steel. The corrosion resistance of a proprietary glass lining in HCl is shown in Fig. 69. The minimum thickness of high voltage tested glass linings is about 1 mm (40 mils). The calculated life expectancy may occasionally be somewhat greater than actual experience because of localized chains of bubbles. However, these data provide a reasonably realistic design basis.

Wood is an economical material for handling dilute acids. Tanks and large-diameter piping cost appreciably less than rubber-lined steel or thermosetting plastic construction. Although unsuitable for concentrated HCl, wood has excellent resistance up to 10% concentration. The National Wood Tank Institute (NWTI) recommends wood in 2% HCl to temperatures of 60 °C (140 °F) and 10% HCl to 70 °C (160 °F) (Ref 124). Cypress and fir were unaffected after 31 days of exposure in both 5 and 10% acid at room temperature and showed slight fiber disintegration and slight embrittlement and softening after 8 h in boiling 10% HCl (Ref 125).

Hydrogen Chloride Gas

The recommended upper temperature limits for various metals in HCl gas are shown in Table 19. These guidelines were developed from short-term tests. Because corrosion rates generally

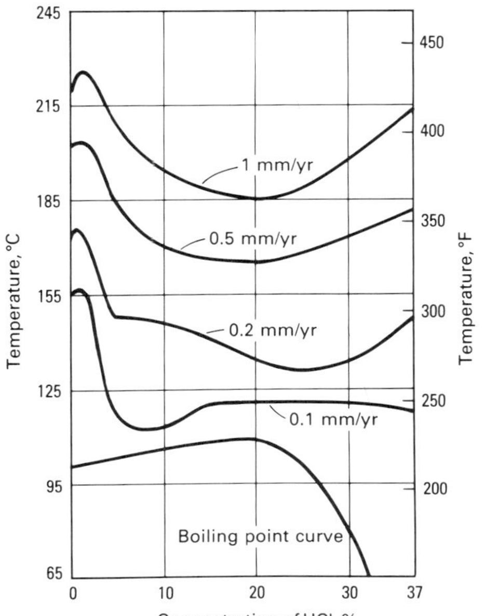

Fig. 69 Isocorrosion diagram for Glasteel 5000 in HCl. Source: Ref 123

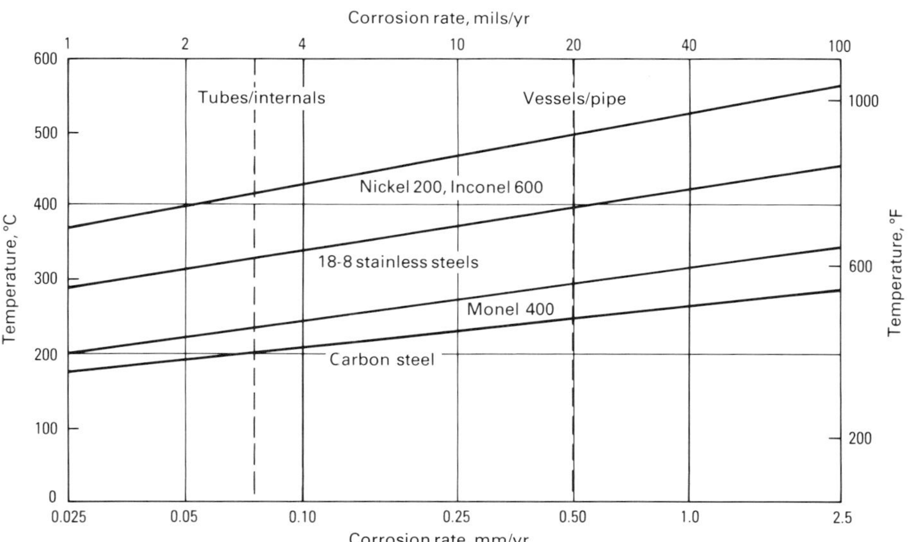

Fig. 70 Maximum temperature limits for various alloys in HCl service. Source: Ref 127

decrease with time as films and scales form, these limits have proved to be conservative. Temperature versus design corrosion rates for steel and several alloys are shown in Fig. 70.

The addition of water vapor to HCl gas has little effect on upper temperature limits. The addition of 0.2% water vapor by weight has been found to have no effect (Ref 126). However, water vapor has a severe effect at lower temperatures.

In a recent study (Ref 128), the suggested lower limit for carbon steel in a 60% H_2O, 40% HCl environment was 200 °C (390 °F) or slightly lower. The lower temperature limits for nickel alloys were 115 °C (240 °F) for Hastelloy C-276, 140 °C (285 °F) for Inconel 625, and 260 °C (500 °F) for Nickel 200, Hastelloy B-2, and Inconel 600. Grade 7 (0.2% Pd) titanium had satisfactory resistance up to 120 °C (250 °F). None of the materials tested, including Nickel 200, titanium, zirconium, and cobalt-base alloys, was suitable at or slightly below the dew point. Minimum temperatures for metals ranged from 30 to 100 °C (85 to 210 °F) above the dew point, depending on the hygroscopicity of the corrosion products.

Corrosion by Hydrogen Fluoride and Hydrofluoric Acid

Thomas F. Degnan
Consultant

Anhydrous hydrogen fluoride (AHF) and aqueous HF are of great industrial importance. Anhydrous hydrogen fluoride is the foundation of the multibillion dollar fluorocarbon industry, which encompasses essentially all refrigerants, a fire-extinguishing agent, ultrasonic cleaning fluids, fluorocarbon plastics, and fluorocarbon elastomers. A popular process for alkylation of petroleum to enhance yields of gasoline depends on the use of AHF. Aqueous HF is used in large quantities to pickle stainless steels, to acid treat wells, and to etch glass.

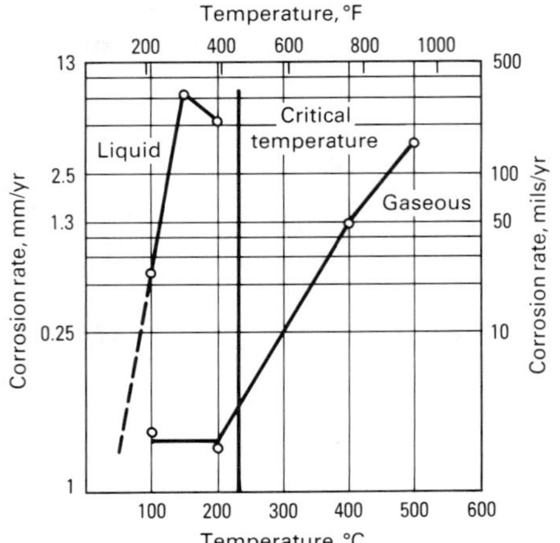

Fig. 71 Corrosion of steel in liquid and gaseous AHF. Source: Ref 135

Aqueous HF and AHF are hazardous chemicals. Fluoride salts, although added to potable waters to prevent tooth decay, are toxic in higher concentrations. Painful, persistent burns result from contact with aqueous HF or AHF, and inhalation of high concentrations of the vapors causes lung damage. The Occupational Safety and Health Administration has ruled that an employee's exposure to HF vapor in any 8-h work shift of a 40-h work week shall not exceed a time-weighted average of 3 ppm HF vapor by volume (Ref 129).

Essential information for the safe handling and use of HF is provided in Ref 130. Recommended respiratory protection guidelines are given in Ref 129. The latest product bulletins issued by the manufacturers contain current information.

Anhydrous hydrogen fluoride is manufactured by the reaction of H_2SO_4 and calcium fluoride (CaF_2) in horizontal kilnlike reactors, usually constructed of steel with a high-nickel alloy liner at the feed and discharge ends. The popular Buch process premixes the CaF_2 and H_2SO_4 in a high-nickel alloy mixer that reduces corrosion of the steel reactor. The reaction is endothermic and therefore requires a source of heat.

Information on suitable materials of construction for handling HF of various strengths and at various temperatures is given in Ref 131 and in the manufacturer's literature. However, many factors can influence the performance of these materials that are not shown in charts, and adequate experience or testing under the proposed conditions is necessary to avoid serious corrosion problems. The duration of the test is important because most metals form protective films or scales that decrease initially high rates to the low rates found in commercial use (Ref 132). Therefore, the validity of short-term rates should be discounted unless the validity of the short-term test has been confirmed by longer tests.

Table 20 lists data on the corrosion resistance of various metals in liquid AHF and AHF-hydrocarbon mixtures at temperatures of 15 to 90 °C (60 to 190 °F). Data on the corrosion resistance of various metals in selected concentrations of aqueous HF are shown in Tables 21 and 22.

Carbon and Low-Alloy Steels

Carbon steels have useful corrosion resistance from 64 to 100% HF. Resistance to aqueous acid is limited to ambient temperatures. Liquid AHF is commonly handled in carbon steel at temperatures to 65 °C (150 °F). Gaseous AHF is less corrosive and can be handled at temperatures to 300 °C (570 °F). Figure 71 shows a graph of corrosion rate versus temperature for liquid and gaseous AHF.

Although Interstate Commerce Commission (ICC) regulations permitted transportation of aqueous HF of 60% or greater concentration in steel containers that were passivated with 58% acid, one study found that even passivated drums developed excessive hydrogen pressure from corrosion with 60% acid (Ref 136). The results of this testing, shown in Table 23, indicate that the minimum concentration should be 64%. In practice, the closest commercial grade is 70% HF, and the maximum recommended temperature for storage and handling is 30 °C (90 °F).

Steel valves for AHF service often use high-nickel alloy trim. Closely fitting steel parts adhere to each other as a result of the nature of the corrosion products.

Alloy steels often have higher corrosion rates than carbon steel. In one case, a section of low-temperature pipe (ASTM A333, grade 9) con-

Table 20 Corrosion of metals and alloys in AHF

Exposure for 6 to 40 days

	Corrosion rate at temperature, °C (°F)											
	15–25 (60–80)		25–40 (80–100)		40–95 (100–200)		55 (130)		70 (160)		80–90 (180–190)	
Metal	mm/yr	mils/yr	mm/yr	mils/yr	mm/yr	mils/yr	mm/yr	mils/yr	mm/yr	mils/yr	mm/yr	mils/yr
Carbon steel	0.07	2.8	0.16	6.2	...	...	0.35	14	...	...	2.3	89
Low-alloy steel	0.15	6	0.14	5.9	...	...	...	...	...	...	2	78
Austenitic stainless steel	0.16	6.2	0.12	4.8	...	...	...	...	...	...	0.06	2.4
Monel 400	0.08	3.2	0.02	0.9	0.12	4.7	...	...	...	...	...	...
Copper	0.33	12.9	...	...	...	...	...	...	...	...	...	...
Nickel 200	0.06	2.5	...	...	...	...	...	...	0.12	4.6	...	...
70–30 copper-nickel	...	...	0.05	2	...	...	0.008	0.3	...	...	0.25	10
80–20 copper-nickel	...	...	0.13	5.2	...	...	...	...	...	...	...	...
Red brass	0.76	30	0.4	16	...	...	...	...	...	...	1.3	50
Admiralty brass	0.25	10	0.33	12.8	...	...	0.01	0.4	...	...	0.5	20
Aluminum-bronze	0.37	14.4	...	...	...	...	...	...	...	...	...	...
Phosphorus-bronze	0.5	20	0.48	18.8	...	...	...	...	...	...	1.5	60
Inconel 600	...	...	0.067	2.6	...	...	...	...	...	...	...	...
Duriron	...	...	1.1	45	...	...	...	...	...	...	...	...
Aluminum	0.52	20.4	...	...	...	...	...	...	...	...	24.8	976
Magnesium	0.13	5.2	0.43	17.1	...	...	nil		...	...	nil	

Source: Ref 133

Table 21 Corrosion of selected metals and alloys in boiling 48% HF

Exposure for 40 days; vapor phase purged with nitrogen

Metal	Corrosion rate: Liquid phase, mm/yr	Liquid phase, mils/yr	Vapor phase, mm/yr	Vapor phase, mils/yr
Platinum	nil		nil	
Silver	nil		nil	
Copper	0.045	1.8	0.058	2.28
90–10 copper-nickel	0.067	2.64	0.018	0.72
80–20 copper-nickel	0.12	4.68	0.018	0.72
70–30 copper-nickel	0.08	3.25	0.02	0.84
Monel 400	0.46	18	0.85	33.6
Lead	5	200	3.8	150

Source: Ref 134

Table 23 Corrosion of steel in HF at ambient temperature

Exposure for 25 days

Concentration of HF, %	Corrosion rate, mm/yr	Corrosion rate, mils/yr
58	3	120
60	2.53	99.6
61	2.1	81.6
62	1.5	58.8
63	0.24	9.6
64	0.05	2.0
65	0.055	2.2
67.5	0.048	1.9
69.9	0.064	2.5

Source: Ref 136

taining nickel and copper was installed in a carbon steel piping system handling a mixture of organics and AHF at an above-ambient temperature. The section failed by uniform thinning, while the carbon steel fittings on either end were relatively unaffected (Ref 135).

It has been found that dilute acids (1.4 to 1.7 *N* HF, H_2SO_4, perchloric acid ($HClO_4$), and HCl) used for chemically cleaning heat exchangers corroded iron and carbon steel at about equivalent rates but that HF caused exceptionally high corrosion rates on steels containing 1.25 to 18% Cr (ASTM A335 grades and type 430 stainless steels) (Ref 137). Data comparing the corrosion rates in dilute HF, H_2SO_4, and HCl are given in Table 24.

Hydrogen Embrittlement and Blistering of Steels. Both HF and AHF cause hydrogen embrittlement of hardened carbon and alloy steels as well as hydrogen blistering and stepwise cracking of plate and pipelines.

When nonoxidizing acids, including HF, corrode steel, atomic hydrogen is formed. The atomic hydrogen may either combine to form gaseous molecular hydrogen or may be absorbed into the steel. Sulfides and arsenic, present in small quantities in HF, inhibit, or poison, the formation of molecular hydrogen and thus promote the entry of atomic hydrogen into the steel, where it may recombine at inclusion sites or laminations to form blisters or may migrate to dislocations in hardened steel to cause hydrogen embrittlement. Even purified (electronic) grade HF, however, exhibits similar but diminished hydrogen effects, as do other nonoxidizing acids to a lesser degree.

Hydrogen sulfide is the best known and most potent causative agent for hydrogen blistering and embrittlement; suitable materials and hardness levels to prevent embrittlement are covered in NACE standard MR-01-75 (Ref 138). Hydrofluoric acid is the second most potent agent for hydrogen blistering and embrittlement, with the added problem that some nickel-copper alloys, notably Monel 400 and Monel K500, acceptable in sulfide service, can fail by SCC in HF (Ref 139, 140).

To prevent hydrogen embrittlement in welded structures, the requirements of NACE Standard RP-04-72 should be followed (Ref 141). Contrary to a comment in the standard, shielded metal arc welds have cracked, and welds produced by all welding methods should be hardness tested. As pointed out in the standard, stress-relief heat treatment, although desirable, will not necessarily reduce hardness, particularly that of high-manganese welds, and does not avoid the necessity for testing. Techniques such as the uphill welding of vertical seams and the use of temper beads are useful in controlling weld hardness. The hardness of the HAZ can be reduced by using pre- and postweld heating and lower carbon parent metal. Alloy steels and carbon steels with tensile strengths exceeding 483 MPa (70 ksi) should be avoided in welded structures.

Alloy steel fasteners have been a source of many failures in AHF service, particularly the use of ASTM A193, grade B7, a chromium-molybdenum steel. When a high-strength fastener must be used, grade B7M, the same steel tempered to a lower hardness of 201 to 235 HB, is a better choice. However, these bolts will crack if stressed beyond their yield point in an HF environment, and the hardness range has been a difficult one for manufacturers to meet and still comply with the minimum tensile requirements.

Hydrogen blistering and stepwise cracking of steel plates have been troublesome with HF as well as with sulfides in sour crudes (Ref 132, 133, 142). Figure 72 shows a photograph of the cross section of a blistered plate from an AHF storage sphere.

One study reported that blistering occurred in a semikilled steel containing many inclusions, but not in a cleaner fully killed steel. On the other hand, another researcher found no blistering in 43 semikilled ASTM A285, grade C, plates, nor in 54 fully silicon killed (ASTM A515, grade 70) plates in AHF tank cars and storage spheres. However, blistering was found in 55 of 442 plates of aluminum-silicon killed ASTM A516, grade 70, steel. Fine-grain practice (killed) steels are often preferred for ambient and lower-temperature service because their improved notch toughness, particularly when normalized or quenched and tempered, minimizes the possibility of brittle fracture.

Blistering occurs more readily in fine-grain practice steels because the added aluminum forms clusters of Al_2O_3 particles that segregate in the center of the plate and form sites for hydrogen accumulation. Type II manganese sulfide inclusions, which predominate in killed steels, flatten out and also serve as sites for hydrogen accumulation.

A promising solution to the problem of obtaining a clean steel with good notch toughness at a reasonable cost has been the availability in recent years of low-sulfur (0.005 to 0.010% maximum) steels with inclusion shape control. These steels are vacuum poured and then treated in the ladle with calcium or a rare earth, which removes the Al_2O_3 particles as calcium aluminate and spheroidizes the manganese sulfide particles. These steels were developed to prevent laminar tearing during welding by improving transverse strength and ductility.

Austenitic Stainless Steels

Austenitic stainless steels have good resistance to liquid AHF at somewhat elevated temperatures (Ref 133). Unpublished data show corrosion rates of less than 0.13 mm/yr (<5 mils/yr) for type 304 stainless steel at 100 °C (210 °F), but high

Table 22 Corrosion of selected metals and alloys in commercial HF solutions

Exposure for 35 days at 60 °C (140 °F); vapor phase purged with nitrogen

Metal	50% Liquid mm/yr	50% Liquid mils/yr	50% Vapor mm/yr	50% Vapor mils/yr	65% Liquid mm/yr	65% Liquid mils/yr	65% Vapor mm/yr	65% Vapor mils/yr	70% Liquid mm/yr	70% Liquid mils/yr	70% Vapor mm/yr	70% Vapor mils/yr
Platinum	nil		nil		nil		nil		...	...	...	...
Silver	0.009	0.36	nil		0.018	0.72	0.0005	0.02	0.018	0.72	0.00025	0.01
Monel 400	0.46	18.12	0.12	4.68	0.13	4.8	0.058	2.28	0.14	5.4	0.05	2.03
Magnesium	0.21	8.4	0.03	1.2	0.058	2.28	0.058	2.28	...	...	...	...
Hastelloy C	0.74	29.3	0.6	24	0.19	7.55	0.24	9.6	...	...	...	...
Illium R	0.22	8.65	0.08	3.23	0.2	8	0.02	0.84	...	...	...	...

Source: Ref 134

Table 24 Corrosion rates of iron, carbon steel, and alloy steels in dilute acids

Exposure for 3 h at 45 °C (115 °F)

Metal	1.7 *N* HF		1.5 *N* H_2SO_4		1.4 *N* HCl	
	mm/yr	mils/yr	mm/yr	mils/yr	mm/yr	mils/yr
Armco iron	14	551	22	866	13	512
ASTM A106 carbon steel	26	1.02 in./yr	20	787	23	906
ASTM A335						
Grade P11 (1.5Cr–0.25Mo)	290	11.4 in./yr	41	1.6 in./yr	4	157
Grade P22 (2Cr–1Mo)	640	25.2 in./yr	330	13 in./yr	10	394
Grade P5 (5Cr–0.5Mo)	>290	>11.4 in./yr	94	3.7 in./yr	3	118
Grade P9 (10Cr–1Mo)	330	13 in./yr	83	3.27 in./yr	6	236

Source: Ref 137

corrosion rates at 150 °C (300 °F) (Ref 135). Type 304 stainless steel has good resistance to AHF gas to 200 °C (390 °F). Cast ACI CF-8M stainless steel is used for AHF pumps. However, both type 304 stainless steel and carbon steel have been known to fail by direct impingement of AHF streams.

Austenitic stainless steels have limited resistance to dilute HF. Type 304 stainless steel has poor resistance to any significant concentration, but type 316 has useful resistance at ambient temperatures and concentrations below 10%.

Cold-worked type 303 stainless steel failed rapidly when used as a fastener material in an HF plant. Cold-worked type 304 stainless steel fasteners (ASTM A193, grade B8, class 2) had few failures. The failed fasteners were strongly magnetic. It is believed that α-martensite is the susceptible phase. Alpha-martensite is formed when such alloys as type 301, type 303, and, to a lesser extent, type 304 stainless steels are cold worked. Figure 73 shows the effect of cold work on the magnetic permeability of common austenitic stainless steels. Type 316 is metallurgically stable, and type 304 only marginally so.

Annealed austenitic stainless steels are resistant but not immune to SCC by HF. In one study, types 304 and 316 both failed from transgranular cracking in impure 12% HF at 70 °C (160 °F); type 304 failed in hot 40 to 50% HF (Ref 133). However, these conditions are more severe than are reasonable for the use of stainless steels in HF. This study also reported that cold-worked 18-8 stainless steel did not suffer SCC in a 7-day test in AHF at 70 °C (150 °F). Another study showed that the threshold Cl^- ion concentration necessary to cause intergranular SCC of sensitized type 304 stainless steel was greatly reduced by the presence of fluorides (Ref 143).

Fig. 72 Blister section from a sphere plate that was in HF service. Source: Ref 142

High-Alloy Stainless Steels and Superalloys

Higher alloys such as 20Cb-3 stainless steel and Incoloy 825 have good resistance to all concentrations of HF at ambient temperatures and to 0 to 10% concentrations at 70 °C (160 °F) (Ref 131). Corrosion data are shown in Table 25. The preferred material for pumps and valves for 70% HF at ambient temperatures and for valve trim for AHF is ACI CN-7M casting alloy.

Nickel and Nickel Alloys

Nickel 200 is less resistant than Monel 400 to aqueous HF. Also, oxygen has a greater accelerating effect on corrosion of Nickel 200. In aqueous HF, use of Nickel 200 is limited to completely air-free systems below 80 °C (175 °F). Although there are reports of SCC of Nickel 200 in aqueous HF, they appear to be related to such impurities as cupric fluoride (CuF_2) (Ref 144). Nickel 200 is one of the most resistant alloys to hot AHF vapor, but it may be embrittled by sulfur compound impurities.

Monel 400 is used extensively in HF alkylation units and in the manufacture and handling of HF. It has excellent resistance to liquid HF over the entire concentration range in the absence of oxygen to at least 150 °C (300 °F).

Monel 400 is subject to SCC when exposed to wet vapors of HF in the presence of oxygen. Cracking is intergranular. Figure 74 shows a photomicrograph of SCC of a cold-bent Monel 400 instrument tubing that was inadvertently exposed to moist vapors and air during a shutdown. Stress-corrosion cracking of both Monel 400 and Monel K500 in moist vapor phase conditions has been widely reported in the literature (Ref 140).

The mechanism of cracking was unknown until it was found that aqueous HF solutions containing appreciable concentrations of cupric chloride ($CuCl_2$) would cause rapid cracking of stressed Monel 400 (Ref 144). As shown in Fig. 75, the least resistant nickel-copper composition corresponds to that of Monel 400 (33% Cu). Copper-nickel alloys with 30% or less nickel are resistant.

The reason that SCC is usually limited to the vapor phase rather than liquid is the enrichment of a thin layer of aqueous HF in the vapor with copper fluoride corrosion products. The presence of oxygen accelerates corrosion and forms CuF_2 from CuF. The much greater dilution of corrosion products prevents reaching a critical concentration of CuF_2 in the liquid phase.

Recommendations for prevention of SCC include immersion of all parts or purging of the vapor spaces with nitrogen. Thermal stress relief, although beneficial, has not always been successful in preventing SCC. As shown in Fig. 76, SCC can occur at relatively low stress levels.

Inconel 600 is resistant to dilute aqueous HF at ambient temperatures and to AHF. It has been used in valves and other equipment in place of Monel 400 to avoid possible SCC. It is the most widely used alloy for hot HF vapors, combining excellent chemical resistance and comparatively good metallurgical stability. Figure 77 shows the corrosion rate for Inconel 600 in dilute HF at 75 °C (165 °F).

High-Performance Nickel Alloys. The Hastelloy alloys (C-276, B-2, G, and G-3) and Inconel 625 all have good-to-excellent resistance to aqueous and anhydrous HF and to high-temperature HF vapors. The presence of air or oxidizing agents can cause fairly high corrosion rates in hot aqueous acids. Table 26 gives corrosion data for the original (now obsolete) compositions of alloys of this type.

In one study, a series of corrosion tests was conducted in the liquid phase of 6.8 *M* (13.6%) HF-1.35 *M* zirconium fluoride (ZrF_4) and in the vapor phase of 6.8 *M* (13.6%) HF at 80 °C (175 °F) using 75-mm (3-in.) test vessels of Inconel 625 and Hastelloy C-4 (Ref 146). Welded, stressed U-bend samples of both materials were exposed. Previous screening tests showed that these alloys were the most promising of a group that included Hastelloy C-276 and G, Inconel 690, and Incoloy 825. Corrosion rates in the HF-ZrF_4 solution at 80 °C (165 °F) were 1.1 mm/yr (44 mils/yr) for Hastelloy C-4 and 2.4 mm/yr (96 mils/yr) for Inconel 625.

After evaluation of the results, Hastelloy C-4 with a carbon content of 0.006% maximum (versus the normal limit of 0.01%) was recommended. This was based on the observation of attack of the HAZs (weld decay) adjacent to a butt weld in the test vessel. Analysis of the vessel wall showed 0.011% C. Welded specimens with 0.001 to 0.006% C showed no such attack, nor did a replacement vessel made of pipe with 0.002% C.

Inconel 625 U-bend samples, which were exposed to the vapor phase, showed severe end grain attack and embrittlement within the first week of exposure, while Hastelloy C-4, although showing appreciable general corrosion, did not.

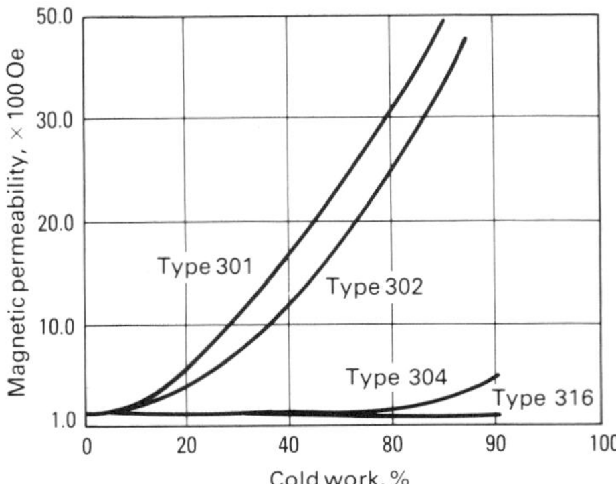

Fig. 73 Effect of cold work on the magnetic permeability of austenitic stainless steels

Table 25 Corrosion of nickel stainless steels and chromium-nickel-molybdenum-iron alloys in aqueous HF
Laboratory tests with no aeration or agitation (except boiling tests)

Concentration HF, %	Temperature °C	Temperature °F	Test duration, days	Corrosion rate: Type 304 mm/yr	Type 304 mils/yr	Type 316 mm/yr	Type 316 mils/yr	Type 309Cb mm/yr	Type 309Cb mils/yr	Alloy 20Cb-3 mm/yr	Alloy 20Cb-3 mils/yr	Incoloy 825 mm/yr	Incoloy 825 mils/yr
0.05	60	140	10	0.3	12	...	...	0.25	10	...	...	...	...
0.1	60	140	10	0.64	25	...	...	0.69	27	...	...	...	...
0.15	60	140	10	1.2	47	...	...	1.1	44	...	...	...	...
0.2	60	140	10	1.6	62	...	...	1.4	54	...	...	...	...
10	16	60	30	0.01	0.4	<0.002	<0.1	...	...	...	...	...	...
20	102	215	3	...	...	...	...	...	...	...	...	1.04	41
38	110	230	2	...	...	51	2000	...	...	...	...	...	...
38	Boiling		4	...	...	...	...	...	...	...	...	0.25	10
48	Boiling		4	...	...	...	...	...	...	...	...	0.23	9
50	60	140	35	...	...	...	...	...	...	...	...	0.05	2
65	60	140	35	...	...	...	...	...	...	...	...	0.13	5
70	60	140	35	...	...	...	...	...	...	...	...	0.13	5
70	21	70	42	...	...	1.24	49	...	...	0.38	15	0.35	14
90	4	40	0.2	0.9	35	...	...	...	...	...	...	...	...
90	21	70	1	0.76	30	...	...	...	...	...	...	...	...
90	21	70	1	0.28	11(a)	...	...	...	...	...	...	...	...
98	34–44	95–110	3.5	...	...	0.05	2	...	...	...	...	...	...

(a) Velocity: 0.14 to 0.43 m/s (0.4 to 1.4 ft/s)

Both alloys showed preferential weld attack. Neither alloy exhibited SCC. It should be noted that these high-performance alloys have good resistance to hot H_2SO_4-HF mixtures and are used in the manufacture of HF.

Other Metals and Alloys

Molybdenum has good resistance to both aqueous and anhydrous HF. The corrosion rates of molybdenum in 25 to 50% HF at 100 °C (210 °F) are 0.4 to 0.5 mm/yr (16 to 20 mils/yr) in aerated acid. Rates in the absence of air are so small as to be negligible (Ref 147).

Precious metals, including gold, silver, and platinum, are all more or less unaffected by AHF or by aqueous HF of any concentration at ambient temperatures and, in most cases, to the boiling point or higher (Ref 148, 149). Silver can be affected by the presence of sulfides and oxygen (Ref 149). Platinum and gold are unaffected at high temperatures.

Magnesium has good resistance to HF because the protective film formed is insoluble in aqueous acid.

Reactive and Refractory Metals. Tantalum, niobium, titanium, and zirconium are all nonresistant to HF and AHF in even trace amounts. Metals and welding products that contain appreciable amounts of these elements will suffer accelerated corrosion by HF under certain circumstances. The presence of several percent niobium in Monel 400 welds has been known to cause preferential weld attack. Titanium additions have been found to be less harmful (Ref 152).

Nonmetallic Materials

The uses and limitations of plastics, elastomers, carbon, and graphite are discussed in Ref 150.

Glass, materials reinforced with glass fibers, and materials rich in silica and silicon are nonresistant to HF.

Plastics. Polyethylene, PP, PVDF, and carbon-filled phenolics are limited to a maximum concentration of about 70% HF. Because the concentration of HF in the vapor phase above 70% HF is enriched, a limit of 60% is conservative. Polyvinyl chloride is limited to about 50% concentration. Silica-filled phenolics should be avoided. Furan resins are inferior to phenolics and may crack.

Butyl and chlorobutyl elastomers are used to line tank cars for 70% HF, but are attacked in the vapor phase of 70% acid at temperatures above 30 °C (90 °F). Chlorosulfonated polyethylene and EPDM elastomers also have useful resistance to aqueous HF at concentrations below 70%. Natural rubber and neoprene are useful up to 50% acid.

Fig. 74 SCC of a Monel 400 line in HF. 70×. Source: Ref 135

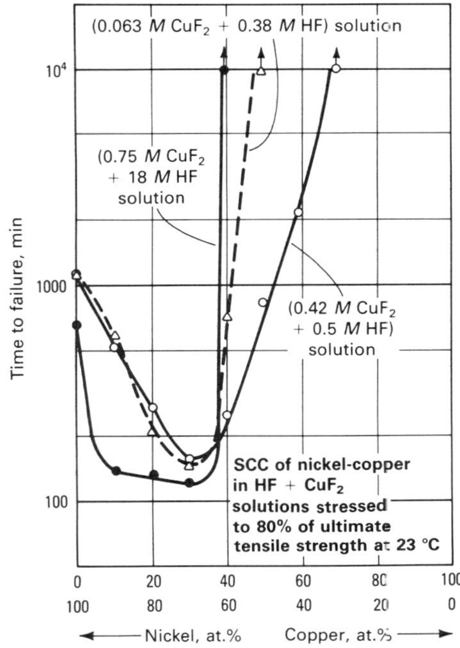

Fig. 75 Time to failure of stressed nickel-copper alloys in HF + CuF_2 solutions. Source: Ref 144

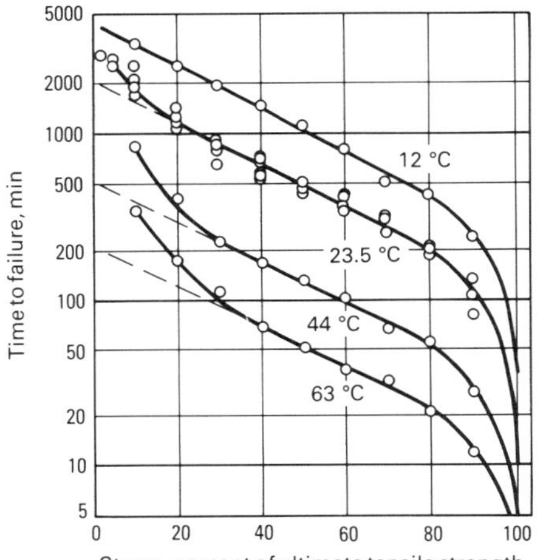

Fig. 76 Effect of stress and temperature on SCC resistance of Monel 400 in HF + CuF_2 solution. Source: Ref 144

Table 26 Corrosion of Hastelloy alloys in aqueous HF
Laboratory test duration of 3.6 to 5 days

Concentration HF, %	Temperature °C (°F)	Corrosion rate: Hastelloy B mm/yr	Hastelloy B mils/yr	Hastelloy C mm/yr	Hastelloy C mils/yr	Hastelloy D mm/yr	Hastelloy D mils/yr	Hastelloy F mm/yr	Hastelloy F mils/yr
5	Room	0.1	4	0.025	1	0.025	1	0.05	2
25	Room	0.13	5	0.13	5	0.15	6	0.3	12
40	Room	0.07	2.6	0.074	2.9	0.025	1	...	...
40	54 (130)	0.025	1	0.025	10	0.07	2.6	...	...
45	Room	0.076	3	0.15	6	0.1	4	0.38	15
50	Boiling (95 °C, or 205 °F)	...	...	4.6	180	...	...	...	...
60	Room	0.04	1.6	0.09	3.6	0.06	2.4	...	...
65	Boiling (70 °C, or 160 °F)	...	...	0.43	17	...	...	...	...
98	34–44 (95–110)	0.1	4	0.025	1	...	...	...	...

Source: Ref 152

It should be noted that the resistance of elastomers to HF is strongly influenced by compounding. Silica and magnesia materials must be avoided. The supplier must be informed of the intended service. Anhydrous HF is resisted by many fluoroelastomers and fluoroplastics, including TFE, FEP, PFA, and monochlorotrifluoroethylene (CTFE).

High-Temperature Gas

A series of short-term corrosion tests was conducted on the rate of attack of hot, gaseous AHF on various metals (Ref 151). Results are shown in Table 27. Nickel 200, Monel 400, and Inconel 600 all have useful resistance to gaseous AHF at temperatures to 600 °C (1110 °F). Austenitic stainless steels are inferior to carbon steel in this regard. Although corrosion rates are exaggerated and distorted by the short 3- to 15-h test periods, other data confirm the general ranking of the materials. More realistic high-temperature data for nickel-containing alloys are shown in Table 28. Although the tests were run for only 36 h, the results agree with experience.

Inconel 600 has been successfully used in the hydrofluorination of metal oxides at temperatures of 370 to 590 °C (700 to 1095 °F). Nickel 200 and the copper-nickel alloys are susceptible to intergranular embrittlement by sulfur compounds at temperatures above about 300 °C (570 °F) (Ref 152). Trace amounts of these compounds are present in commercial AHF. Inconel 600 resists embrittlement and has been successfully used at these temperatures in gaseous HF environments (Ref 152). It should be noted that many nickel alloys, including Hastelloy C-276 and Inconel 625, undergo metallurgical changes, such as aging or long-range ordering at temperatures above 400 to 500 °C (750 to 930 °F). Hardness and susceptibility to hydrogen embrittlement also increase. Inconel 600 is relatively unaffected.

Corrosion by Chlorine

E.L. Liening
The Dow Chemical Company

This Section considers corrosion of metals by dry chlorine in both the liquid and gaseous forms, moist chlorine, and chlorine-water. Materials (including nonmetals) used in chlorine-manufacturing environments are reviewed in Ref 153. Reference 154 is a good source for references on corrosion by chlorine published before 1976.

Dry Chlorine

Dry chlorine is not corrosive to steels, stainless steels, or nickel alloys at ambient temperatures. It is commonly shipped and handled in carbon steel equipment, with higher-alloy materials such as nickel, Monel 400, and Hastelloy C usually used for critical parts. Valves are frequently steel with Monel 400 or Hastelloy C trim and stem. Steel is usable up to about 150 °C (300 °F) and possibly higher under certain conditions. Stainless steels are usable up to about 300 °C (570 °F), and nickel is commonly used up to about 500 °C (930 °F). Inconel 600 or low-carbon nickel is often substituted for nickel at temperatures where graphitization may occur. Moisture will greatly accelerate attack on any of these materials, with the additional danger of SCC of stainless steels.

Iron and Steel. Eight-hour tests showed corrosion rates for steel to be below 0.0025 mm/yr (0.1 mils/yr) at up to 250 °C (475 °F), but ignition occurred at 250 °C (485 °F) (Ref 155). A slightly lower ignition temperature was reported for 16- and 20-h tests (Ref 156). Ignition of steel wool (grade 00) occurred at temperatures as low as 185 °C (365 °F) (Ref 155).

A maximum-use temperature for steel is given in Ref 156 and 157 as 205 °C (400 °F), but the discussion in Ref 157 suggests that a prudent maximum temperature may be nearer 150 °C (300 °F) because of exotherms from reaction with grease-contaminated equipment. For grease-free and properly cleaned equipment, however, a maximum-use temperature of 200 °C (390 °F) may be acceptable.

Tests with various irons and steels in flowing chlorine indicated that ignition occurs at lower temperatures for steels with higher alloy contents, especially for carbon (Ref 158, 159). Ignition at temperatures as low as 220 °C (430 °F) was found for iron containing 0.3% C and 6% other alloy content. This probably accounts for the lower use temperature for cast iron versus steel given in Ref 156. Iron alloys with silicon contents in the 10 to 15% range, however, reportedly resisted attack by dry chlorine (Ref 160, 161), but their poor impact properties limit practical applications in such critical service.

Below 250 °C (480 °F), the presence of oxygen and moisture had little effect on iron chlorination rates (Ref 162). Because iron is attacked less rapidly by hydrogen chloride than chlorine, the presence of hydrogen chloride in dry chlorine should have little effect on iron chlorination rates. In practice, however, the use of steel is avoided where moisture may be present.

The data in Ref 156, presented in Table 29, show corrosion rates and suggested upper temperature limits for use of steel and many other alloys in dry chlorine, based on 2- to 20-h tests. The temperature limits for many of the alloys correspond to higher corrosion rates than might usually be tolerated for expensive alloys. The rates shown, however, are in most cases higher than those that would occur in prolonged exposures. This is because, for most of the alloys, passivation occurs by the formation of a metal chloride film, after which corrosion decreases rapidly. In many cases, corrosion at higher temperatures is roughly proportional to the vapor pressure of the particular metal chlorides formed.

Suggested upper design limits for steel and some other alloys in dry chlorine are shown in Fig. 78, based on corrosion rates of approximately 0.08 mm/yr (3 mils/yr) for tubes and internals, and 0.5 mm/yr (20 mils/yr) for vessels and pipes. Similar design guidelines are discussed in Ref 127, 163, and 164.

Aluminum. Care must be taken when evaluating results for aluminum because the protective aluminum oxide film may delay the onset of corrosion. In one study, there was a 5-h delay before reaction in dry chlorine at 500 °C (930 °F) (Ref 158). Aluminum was reported to be usable up to about 120 °C (250 °F) (Ref 156), and moisture in dry chlorine at room temperature increased attack (Ref 165). This is supported by the work documented in Ref 166 and is attributable to condensation. At higher temperatures (130 to 630 °C, or 265 to 1165 °F), the presence of water was reported to decrease attack on aluminum (Ref 167), as discussed in the section "Moist Chlorine" in this article.

Copper. A maximum-use temperature of 205 °C (400 °F) was suggested for copper in dry chlorine (Ref 156).

Ignition of copper was observed at temperatures as low as 260 to 300 °C (500 to 570 °F) at high velocities (Ref 168). Ignition occurred at 290 to 310 °C (555 to 590 °F) at a velocity of 250 mL/min, and no ignition occurred at 40 mL/min (Ref 162). The same researchers reported that below 200 °C (390 °F) the presence of oxygen and water

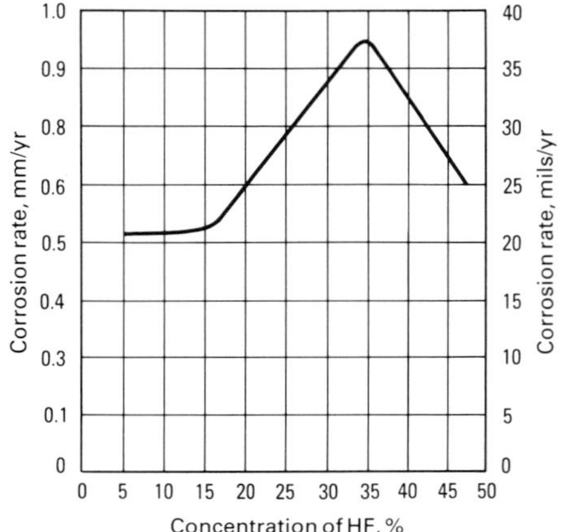

Fig. 77 Corrosion of Inconel 600 in HF at 75 °C (165 °F). Source: Ref 145

Table 27 Corrosion of metals and alloys by gaseous AHF at elevated temperatures

Exposure for approximately 4 h

Metal	Corrosion rate at temperature, °C (°F): 500 (930) mm/yr	500 (930) mils/yr	550 (1020) mm/yr	550 (1020) mils/yr	600 (1110) mm/yr	600 (1110) mils/yr
Nickel 200	0.9	36	...	...	0.9	36
Monel 400	1.2	48	1.2	48	1.8	72
Copper	1.5	60	...	...	1.2	48
Inconel 600	1.5	60	...	...	1.5	60
Aluminum alloy 1100	4.9	192	...	...	14.6	576
Magnesium G	13.8	542	...	...	...	...
1020 steel	15.5	612	14.6	576	7.6	300
Type 430 stainless steel	1.5	60	9.1	360	11.6	456
Type 304 stainless steel	...	...	...	...	13.4	528
Type 347 stainless steel	183	7200	457	18 000	177	6960
Type 309Cb stainless steel	5.8	228	42.7	1680	168	6600
Type 310 stainless steel	12.2	480	100.6	3960	305	12 000

Source: Ref 151

vapor accelerates attack on copper, while above 250 °C (480 °F) they reduce the rate of attack and move the ignition temperature to near 350 °C (660 °F) (Ref 162, 169).

Stainless Steels. Austenitic stainless steels were shown to be significantly more resistant to dry chlorine than steel, aluminum, or copper (Ref 156, 158). Type 304 and type 316 stainless steels may be used up to 300 °C (570 °F). Moisture was reported to accelerate attack below 370 °C (700 °F) but to exert little effect or even decrease attack above that temperature (Ref 156, 167). The presence of moisture also increases the possibility of SCC.

Nickel and Nickel Alloys. Nickel 200 and nickel-base alloys show excellent resistance to dry chlorine, as indicated in Table 30. The good performance of nickel was supported by data in Ref 171 for tests in chlorine at a pressure of 13.2 kPa (0.13 atm) and in Ref 158 in flowing chlorine. However, the reports indicated higher corrosion rates (2.4 and 2 mm/yr, or 95 and 81 mils/yr) for nickel at 525 to 540 °C (975 to 1000 °F) than the study described in Ref 156. A temperature of 500 °C (930 °F) seems a wise upper limit for routine use of nickel in dry chlorine.

Water vapor was found to increase attack on nickel below 550 °C (1020 °F), but had little effect above that temperature (Ref 167). Between 425 and 760 °C (795 and 1400 °F), graphitization of Nickel 200 may occur, and a low-carbon version of nickel with a maximum of 0.02% C is normally used. When sulfur compounds are present, Inconel 600 is often substituted for nickel to avoid intergranular attack.

The data given in Table 29 indicate that Inconel 600 and Hastelloy B perform nearly as well as nickel in dry chlorine, and Hastelloy C somewhat less well. Chromel A and Monel 400 perform much better than stainless steels, but not as well as the other alloys mentioned above. Cast ACI alloy CW-12M (Ni-18Cr-18Mo) was reported to corrode in dry chlorine at 0 to 60 °C (32 to 140 °F) at about the same rate as Hastelloy C (Ref 62, 172).

Monel 400 is commonly used as trim on valves, but it should be used with care in refrigerated systems for the reasons discussed in the section "Refrigerated Liquid Chlorine" in this article. Water vapor in chlorine at temperatures below the dew point is corrosive to a wide variety of nickel, nickel-copper, nickel-chromium-iron, and nickel-chromium-molybdenum alloys (Ref 173).

Other Alloys. Limited information on the performance of various other alloys in dry chlorine is also available. Magnesium was found to perform as well as Chromel A, and actually better than Monel 400 (Ref 156). A maximum-use temperature of 455 °C (850 °F) was suggested, but use of magnesium in chlorine is not widespread.

Lead was found to be resistant in dry chlorine up to 275 °C (525 °F) (Ref 174). Average corrosion rates for lead in dry flowing chlorine were reported as 0.06 mm/yr (2.4 mils/yr) at 200 °C (390 °F), 0.13 mm/yr (5.1 mils/yr) at 250 °C (480 °F), 0.14 mm/yr (5.5 mils/yr) at 275 °C (525 °F), 1.5 mm/yr (59 mils/yr) at 295 °C (565 °F), and 2.5 mm/yr (98 mils/yr) at 310 °C (590 °F) (Ref 159, 167).

Unalloyed zirconium corrodes at less than 0.13 mm/yr (5 mils/yr) in dry chlorine near room temperature, but is not resistant in wet chlorine (Ref 62, 172, 175-177). Reactor grade zirconium tubing was reported to stress corrosion crack in 0.01 mg/cm^3 of chlorine gas at 360 to 400 °C (680 to 750 °F) (Ref 178).

Titanium was found to ignite in dry chlorine at temperatures as low as −18 °C (0 °F) (Ref 62, 172, 179, 180). However, small amounts of moisture in chlorine can passivate titanium (see the section "Moist Chlorine" in this article).

Niobium suffered no attack in dry chlorine up to approximately 200 °C (390 °F) (Ref 181). Tantalum performs well in dry chlorine at up to 250 °C (480 °F) (Ref 182, 183). Attack on tantalum was shown to begin at 250 °C (480 °F), to be violent after 35 min at 450 °C (840 °F), and to be instantaneous at 500 °C (930 °F) (Ref 184). This is consistent with the data given in Ref 167. Pitting of tantalum was reported in a mixture of dry chlorine and anhydrous methanol at 65 °C (150 °F), presumably caused by the presence of halogenated HCOOH contamination (Ref 178).

Refrigerated Liquid Chlorine

Refrigerated liquid chlorine can also be handled in steel, but special care should be taken at

Table 28 Corrosion of metals and alloys in HF gas

36-h laboratory test in which 3.2 kg/h (7 lb/h) HF at 27.6 kPa (4 psig) at 500 to 600 °C (930 to 1100 °F) was passed through a furnace for hydrofluorination of metal oxides

Alloy	Corrosion rate mm/yr	Corrosion rate mils/yr
Hastelloy C	0.008	0.3
Inconel 600	0.018	0.7
Hastelloy B	0.05	2
Nickel 200	0.23	9
Nickel 201	0.36	14
Monel 400	0.33	13
Monel K-500	0.4	16
70–30 copper-nickel	0.4	16

Source: Ref 152

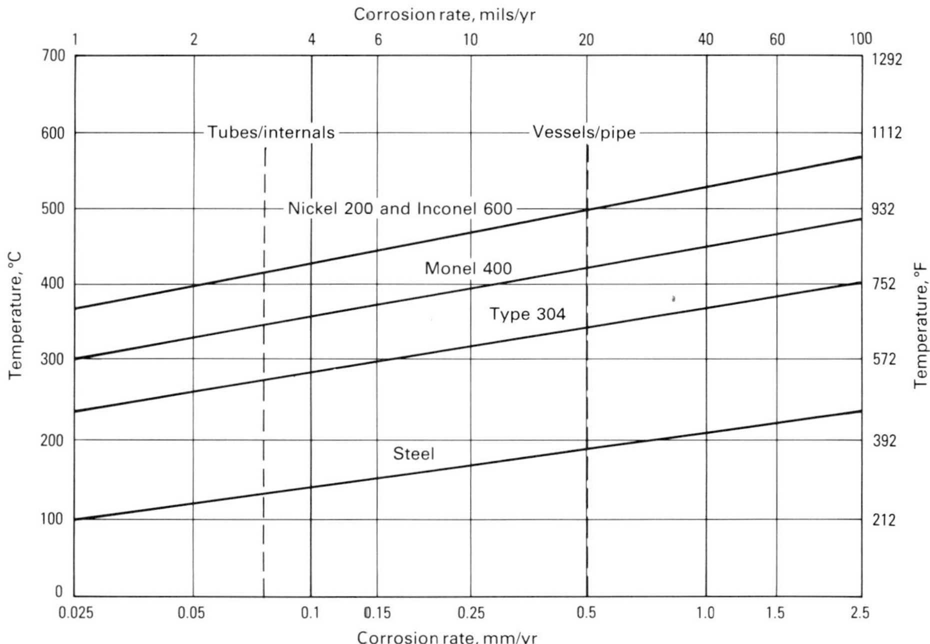

Fig. 78 Design guidelines for use in dry chlorine. Source: Ref 127

Table 29 Approximate temperatures at which the indicated corrosion rates occur in dry chlorine

Alloy	Temperature for corrosion rate, mm/yr (mils/yr)											
	0.76 (30)		1.5 (60)		3 (120)		15 (600)		30 (1200)		Maximum(a)	
	°C	°F	°C	°F	°C	°F	°C	°F	°C	°F	°C	°F
Nickel 201	510	950	540	1000	590	1100	650	1200	680	1250	540	1000
Inconel 600	510	950	540	1000	565	1050	650	1200	680	1250	540	1000
Hastelloy B	510	950	540	1000	590	1100	650	1200	...	...	540	1000
Hastelloy C	480	900	540	1000	560	1050	650	1200	...	...	510	950
Magnesium	450	850	480	900	510	950	540	1000	565	1050	450	850
Ni-20Cr-1Si	425	800	480	900	540	1000	620	1150	...	...	450	850
Monel 400	400	750	450	850	480	900	540	1000	540	1000	420	800
Type 316 stainless steel	310	600	345	650	400	750	450	850	480	900	340	650
Type 304 stainless steel	290	550	315	600	340	650	400	750	450	850	310	600
Platinum	480	900	510	950	540	1000	560	1050	560	1050	260	500
Hastelloy D	205	400	230	450	290	550	...	...	...	...	205	400
Deoxidized copper	180	350	230	450	260	500	260	500	290	550	205	400
Carbon steel	120	250	180	350	205	400	230	450	230	450	205	400
Cast iron	90	200	120	250	180	350	230	450	230	450	180	350
Aluminum alloy 1100	120	250	150	300	150	300	180	350	180	350	120	250
Gold	120	250	150	300	180	350	200	400	200	400	...	...
Silver	40	100	65	150	120	250	230	450	260	500	...	...

(a) Suggested upper temperature limit for continuous service. Source: Ref 156

potential leak sites, such as at valves and nonwelded fittings. Because the chlorine is refrigerated, pipelines and associated equipment become encased in ice formed from the moisture in the air. Chlorine from even small leaks is then trapped beneath the ice, forming wet chlorine gas that is corrosive even at low temperatures. Therefore, Alloy 20 materials are often used for valves and other fittings in refrigerated liquid chlorine equipment.

Similarly, ordinarily nonwetted parts, such as bonnet bolts in valves, are typically replaced by more resistant alloys. Nickel-copper alloys, such as Monel 400, have been used, but they are sensitive to oxidizing corrosives. Figure 79 shows an example of corrosion of Monel 400 bonnet bolts. Chlorine had leaked along the valve stem and was trapped beneath the ice, where it became wet and very corrosive. Nickel-chromium-molybdenum alloys, such as Hastelloy C-276, are used in order to avoid this type of attack on critical components such as bonnet bolts.

High-Temperature Mixed Gases

Several investigations were conducted on alloys in high-temperature mixed gases of chlorine, oxygen, and argon. Various alloys were tested in an atmosphere of argon, 20% O_2, and 2% Cl_2 at 900 °C (1650 °F) for 8 h (Ref 185). Alloy 214 performed the best, with metal loss of less than 0.025 mm/yr (1 mil/yr). Alloy R-41 was the next best performer, followed by Inconel 601 and 600, type 310 stainless steel, and Inconel 625. Hastelloy S was less resistant than the above group, followed by Incoloy 800H, Hastelloy X, Hastelloy C-276, and alloy 188. Longer-term tests in a similar environment yielded the following in order of performance: alloys 214, 601, R-41, X, C-276, and 625.

Corrosion in argon-oxygen-chlorine mixtures was investigated on a more basic level (Ref 186). Work was performed in environments containing argon, 20% O_2, and 0.25% Cl_2 at 900 to 1000 °C (1650 to 1830 °F) and in air and 2% Cl_2 at the same temperatures (Ref 187, 188). A report on the corrosion at 260 to 425 °C (500 to 795 °F) of a variety of engineering alloys considered for use in a waste heat recovery system for a chlorinated hydrocarbon incinerator is given in Ref 189.

Table 30 Corrosion of cast alloys by water-saturated chlorine gas

Exposure for 6 weeks at room temperature

Alloy	Wrought version	Corrosion rate mm/yr	Corrosion rate mils/yr
ACI CF-8M	Type 316	0.79	31
ACI CN-7M	...	1.04	41
ACI CD-4MCu	Duplex stainless steel	1.24	49
ACI CW-12M	Hastelloy C	0.056	2.2

Source: Ref 170

Moist Chlorine

Many industrial chlorine environments contain substantial water, particularly those in chlorine manufacture prior to the drying operation. Wet chlorine gas is extremely corrosive at temperatures below the dew point because the condensate is a very acidic and oxidizing mixture. One of the most commonly used metals for wet chlorine service is titanium, especially for wet chlorine compressors. Titanium is perfectly passive if there is enough water in the chlorine, but it ignites if there is not enough. With sufficient water, titanium is resistant up to at least 175 °C (345 °F) and probably higher. Wet chlorine gas can be handled at room temperatures with stainless steels similar to Alloy 20 if condensation is not too great and if corrosion rates up to 1.3 mm/yr (50 mils/yr) are tolerable. Hastelloy C-276 is suitable to somewhat higher temperatures, and niobium or tantalum is required at even higher temperatures.

Nonmetals are frequently used in wet chlorine gas because of its corrosivity to most metals. Reinforced thermosetting plastics are widely used for handling wet chlorine gas, especially the gases coming from chlorine cells. Nonasbestos reinforcement is expected to predominate as asbestos becomes unavailable because of its toxicity. Glass and ceramics are also used, but thermal and mechanical shock problems must be dealt with. Reference 153, a summary of replies to a questionnaire on materials for handling industrial chlorine mixtures, discusses materials used for wet chlorine gases, particularly nonmetals. The materials used in equipment for the production of chlorine are reviewed in Ref 190.

Iron and Steel. The response of iron-base alloys to moisture in chlorine depends on the temperature range and particular class of alloys. At practical-use temperatures, small amounts of

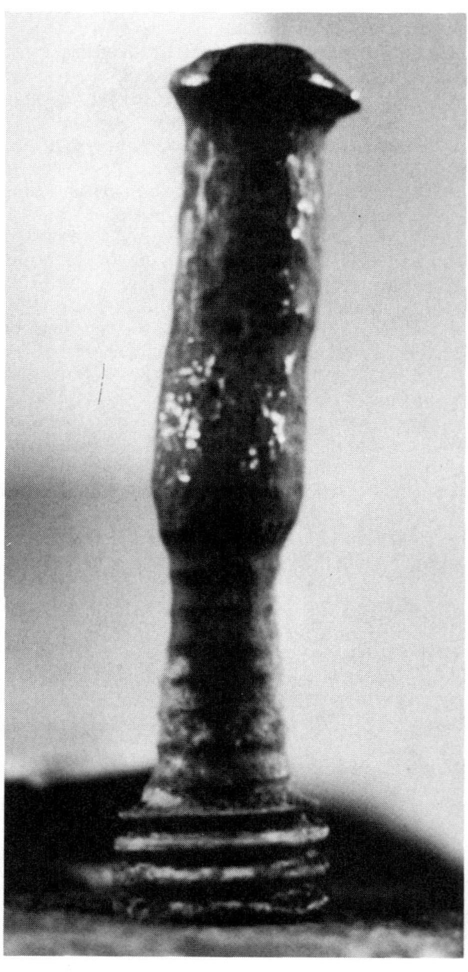

Fig. 79 Monel 400 bonnet bolt extensively corroded by chlorine gas trapped beneath ice covering a valve in liquid chlorine service

moisture and oxygen were reported to have little effect on iron chlorination rates; these contaminants were found to reduce chlorination rates above 300 °C (570 °F) (Ref 158, 169, 191). In practice, steel is generally avoided in any service where chlorine gas may collect water vapor.

Stainless Steels. Moisture has a significant effect on the corrosion of type 304 stainless steel in chlorine, as indicated in Ref 156, which documents tests conducted on type 304 stainless steel in chlorine containing 0.4% H_2O. Rates of approximately 30.5 mm/yr (1200 mils/yr) were found at 40 °C (105 °F), versus an estimated 0.3 mm/yr (12 mils/yr) at 100 °C (210 °F) in dry chlorine. Corrosion in wet chlorine decreased with increasing temperature until about 370 °C (700 °F), at which point the corrosion was about 4.6 mm/yr (180 mils/yr) and the effect of the moisture disappeared. The detrimental effect of moisture at low temperatures is believed to exist for chromium and austenitic stainless steels in general.

Several alloys were tested in water-saturated chlorine gas for 6 weeks at room temperature (Ref 170). The results (Table 30), show substantially higher rates than in dry chlorine. The data support the view that moisture in room-temperature chlorine gas increases corrosion rates on stainless steels, including cast stainless steels.

Aluminum. Aluminum alloy 1100 is more readily attacked by wet than dry chlorine at room temperature, especially if condensation is present. At temperatures above 130 °C (265 °F), however, the presence of moisture greatly reduces corrosion on aluminum. Data from Ref 167 (Table 31) show that larger amounts of water more effectively reduce corrosion on aluminum. Aluminum appears to be usable to 130 °C (265 °F) at 0.06% H_2O, 200 °C (390 °F) at 1.5% H_2O, and 545 °C (1015 °F) at 30% H_2O.

Copper and copper alloys were found to suffer accelerated attack in chlorine gas saturated with water vapor (Ref 192). Copper alloys do not have adequate corrosion resistance for practical use in moist chlorine at temperatures above 200 °C (390 °F). Water vapor at room temperature was shown to accelerate attack on copper and copper alloys (Ref 193). The same effect was reported on copper at temperatures below 200 °C (390 °F) (Ref 162). Above 250 °C (480 °F), however, water vapor and oxygen in chlorine was reported to reduce attack of copper and to move its ignition temperature from about 300 °C (570 °F) to about 350 °C (660 °F) (Ref 162, 169).

Nickel and nickel alloys are adversely affected by the presence of moisture in chlorine at temperatures up to their maximum-use temperatures in dry chlorine. Water vapor at 1.5% was found to double the reaction rate between chlorine and nickel, while 30% H_2O increased the rate from 2 to 20 times (Ref 162, 167). Above 550 °C (1020 °F), moisture was reported to have little effect, but the rates at that temperature even in dry chlorine make nickel marginal.

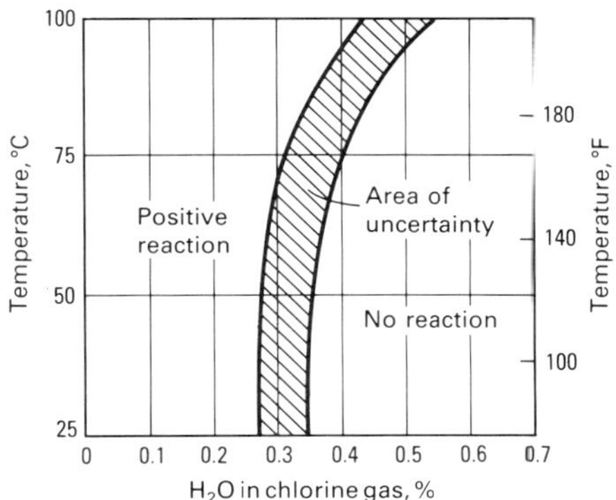

Fig. 80 Estimated water required to passivate unalloyed titanium in chlorine gas. Source: Ref 80

Hastelloy C-276 corrodes 2 to 1000 times faster, and Chlorimet 3 100 to 1000 times faster, in wet chlorine than in dry. Numerous other nickel, nickel-copper, nickel-chromium-iron, and nickel-chromium-molybdenum alloys are also reported to suffer greatly accelerated attack by wet chlorine at such temperatures.

Titanium is well known for its resistance to corrosion by wet chlorine and is widely used in various chlorine-manufacturing equipment, such as compressors that are exposed to wet chlorine. However, a minimum quantity of water is required to maintain the passivity of titanium. The amount of water required at temperatures between 25 and 175 °C (75 and 350 °F) was found to depend on chlorine pressure, temperature, flow rate, purity, and degree of surface abrasion of the titanium (Ref 179). Figure 80 shows a general guideline for water content needed to maintain passivity of commercial-purity titanium in chlorine at temperatures to 105 °C (220 °F).

Crude cell chlorine tested under static conditions required about 0.5% H_2O at 125 °C (255 °F) and 1.2% H_2O at 175 °C (350 °F) (Ref 179). Less water was required at flow rates above 0.15 m/s (0.5 ft/s). Pure (99.5%) chlorine requires relatively more water: about 0.93% at room temperature and 1.5% at 200 °C (390 °F) under static conditions.

Table 31 Effect of water in chlorine gas on corrosion of aluminum alloy 1100

Temperature		Corrosion rate for amount of water in chlorine, wt%					
		0.06		1.5		30	
°C	°F	mm/yr	mils/yr	mm/yr	mils/yr	mm/yr	mils/yr
130	265	<0.5	<20	<0.5	<20	. . .	. . .
140	285	. . .	. . .	. . .	. . .	<0.5	<20
170	340	3.6	140	. . .	. . .	0.53	21
200	390	. . .	. . .	<0.5	<20	. . .	. . .
290	555	. . .	. . .	1.04	41	<0.5	<20
320	610	. . .	. . .	1.04	41	. . .	. . .
350	660	. . .	. . .	5.3 m/yr	210 in./yr	. . .	. . .
400	750	. . .	. . .	21 m/yr	830 in./yr	. . .	. . .
545	1015	. . .	. . .	. . .	. . .	<0.5	<20
615	1140	. . .	. . .	. . .	. . .	5.3	210
630	1165	. . .	. . .	. . .	. . .	7.9 m/yr	311 in./yr

Source: Ref 167

Table 32 Corrosion in chlorine-saturated water

Temperature: 25 °C (75 °F)

Alloy	Test duration, days	Corrosion rate mm/yr	mils/yr
ACI CF-8M	42	0.013	0.50(a)
Type 316	56	0.008	0.30
ACI CN-7M	42	0.05	1.8(a)
20Cb-3	56	0.008	0.30
ACI CD-4MCu	42	0.06	2.5(a)
ACI CW-12M	42	0.023	0.90
Hastelloy C-276	56	0.0025	0.10
Monel 400	56	24	948
Titanium, grade 2	56	0.0005	0.02

(a) Crevice corrosion. Source: Ref 170

Table 33 Corrosion of alloys in chlorine-ice

147-day tests at −20 °C (−4 °F)

Alloy	Corrosion rate μm/yr	mils/yr
Steel	38	1.5
ACI CF-8	0.25	0.01
ACI CF-8M	0.25	0.01
ACI CN-7M	<0.25	<0.01
ACI CD-4MCu	0.25	0.01
Alloy 255	<0.25	<0.01
N-12M	76	3.01
ACI CW-12M	3.8	0.15
M-35	29	1.15

Source: Ref 170

Other Alloys. Relatively limited information is available for other materials. Unalloyed zirconium was found to corrode in wet chlorine at a rate of 2 mm/yr (80 mils/yr) at 15 °C (60 °F) (Ref 62) and 4.9 mm/yr (192 mils/yr) at 25 °C (75 °F) (Ref 176). It also corroded at over 1.3 mm/yr (50 mils/yr) in room-temperature chlorine containing 0.3% H_2O (Ref 177).

Niobium was found to be resistant to wet chlorine at up to 100 °C (210 °F) and tantalum at up to 150 °C (300 °F) (Ref 161, 183, 194, 195). Tantalum reportedly performed well in chlorine plus 1.5% H_2O at up to 375 °C (705 °F) and in chlorine plus 30% H_2O at up to 400 °C (750 °F) (Ref 184). However, other studies indicated higher rates under these conditions (Ref 167). Another source reported corrosion of tantalum in wet chlorine at temperatures above 350 °C (660 °F), but the amount of water was not specified (Ref 194).

Chlorine-Water

Chlorine dissolved in water forms a mixture of HCl and hypochlorous acid (HClO). The latter is very oxidizing, which makes the mixture extremely corrosive. Relatively little information is available on corrosion in water containing substantial levels of chlorine, especially near saturation. Aluminum was reported to be unsuitable in HClO and to be attacked with extensive pitting in chlorine-water environments (Ref 166). Zirconium was found to corrode at less than 0.025 mm/yr (1 mil/yr) in chlorine-saturated water (Ref 175).

Corrosion of several alloys in chlorine-saturated water at 25 °C (75 °F) was investigated (Ref 170), and the results, presented in Table 32, show that rates are generally low for the chromium-containing alloys tested and for titanium. Only Monel 400 showed a very high rate (24 mm/yr, or 948 mils/yr), which is to be expected because of its sensitivity to oxidants. No large difference in performance between cast alloys and their wrought equivalents was found, except for crevice attack on ACI CF-8M, CN-7M, and CD-4MCu.

Related tests were performed in chlorine ice at −20 °C (−4 °F) (Ref 170). The results, given in Table 33, show that corrosion rates for chromium-containing alloys are below 0.0025 mm/yr (0.1 mils/yr), except for ACI CW-12M at 0.0038 mm/yr (0.15 mils/yr). Higher, but not unacceptable, rates were found for steel, N-12M, and M-35. These alloys are sensitive to oxidants in acidic environments, which explains their poorer performance.

Table 34 Corrosion of metals and alloys in caustic concentrations from 5 to 14%

Metal or alloy	Corrosion rate in NaOH: 5–10%(a) mm/yr	5–10%(a) mils/yr	10%(b) mm/yr	10%(b) mils/yr	14%(c) mm/yr	14%(c) mils/yr
Titanium	0.001	0.04	nil		...	...
Zirconium	0.005	0.2(d)	0.0018	0.07	...	...
Nickel	0.005	0.2	8×10^{-5}	0.003	0.0005	0.02
Monel	0.008	0.3	nil		0.0013	0.05
Inconel	0.0013	0.05	nil		0.0008	0.03
Low-carbon	0.1	4(d)	0.015	0.6(e)	0.21	8.2
Cast iron	...	...	...	...	0.21	8.2
Ni-Resist I	...	...	...	...	0.074	2.9

(a) 124-day tests at 21 °C (70 °F) (Ref 196). (b) 207-day exposure to effluent from an electrolytic chlorine cell containing 15% NaCl at 80 °C (180 °F) (Ref 62). (c) 90-day exposure at 90 °C (190 °F) in first effect of a multi-effect evaporator (Ref 196). (d) Slight pitting attack. (e) Slight attack under spacer

Corrosion by Alkalies and Hypochlorite

James K. Nelson
PPG Industries, Inc.

Caustic soda (NaOH), caustic potash (potassium hydroxide, KOH), and soda ash (Na_2CO_3) are true alkaline chemicals. They will be reviewed in this Section and will provide a basis for general discussion of various alkaline exposures. Technically, hypochlorites are alkaline oxidizing salts, but they exhibit behavior that is much different from alkalies. Some of the topics that will be addressed include:

- Applications for caustic and caustic-containing solutions in which carbon steel provides acceptable service with and without special treatments
- Use of austenitic stainless steels in alkaline solutions, addressing caustic embrittlement and SCC for alkaline chloride solutions
- Decomposition behavior of hypochlorites and its relationship to metallic corrosion and nonmetallic degradation

Sodium Hydroxide

Caustic soda is the most widely used and available alkaline chemical. Most NaOH is produced as a coproduct of chlorine through the use of electrolytic cells; the cells are of the diaphragm, mercury, or membrane type. Some NaOH is marketed as produced in the cells; most is evaporated and sold as 50 and 73% solutions or as anhydrous beads. Most caustic end uses require solutions of relatively low concentrations. The following materials discussions will present information on a spectrum of materials within a wide range of caustic temperatures and concentrations.

A number of reference tables provide resistance data for many metals; these data are included in Tables 34 to 41. Table 38 relates specifically to corrosivity relationships of diaphragm and mercury cell caustic.

No discussion of materials of construction for NaOH would be complete without stressing the need to consider safety in every application. Caustic solutions, especially when hot, are extremely damaging to the human body. Exposure can cause immediate and severe burns; eyes are an especially sensitive area. Frequently, the materials selection decisions must account for potential personnel exposure and dictate a more costly but completely reliable material.

Steel and Iron. The possibilities of product contamination or SCC (often called caustic embrittlement) are primary restrictions to the use of iron and steel in caustic service. These problems limit applications, yet low-carbon steel remains the most frequently used material. It is effective in caustic solutions of up to 50% concentration and at temperatures to 90 °C (190 °F).

The addition of OH^- ions to an aqueous medium decreases the corrosion rate of steel as well as that of other metals. Hydroxyl ions in solution decrease corrosion rates by acting as an anodic inhibitor, which increases anodic polarization by helping to form a protective film and keep it repaired. Figure 81 shows an alternative description of this effect. A pH of 12 is obtained by adding about 0.5 g/L NaOH. Benefits are increased by adding even larger quantities of caustic, up to 54% (Table 42).

Effective use is made of bare steel tanks storing 50% caustic at temperatures from ambient to 65 °C (150 °F). Iron contamination is frequently a concern because many caustic applications must limit the presence of iron to only a few parts per million.

When tanks are first filled, there is a sharp increase in the iron content of the caustic. If the tank contents remain relatively undisturbed for several weeks, the sampled iron content will decrease to levels of 4 ppm or less. This reduction is attributed to settling of precipitated ferric hydroxide ($Fe(OH)_3$), which then coats the steel with a gelatinous protective film. The preceding reactions would be an initial corrosion of the steel to form ferrous hydroxide ($Fe(OH)_2$), followed by oxidation to the less soluble $Fe(OH)_3$. As the caustic level is lowered, the protective hydroxide film further oxidizes to more voluminous ferric oxide (Fe_2O_3), which may be sloughed off or is redissolved as the tank level is raised.

It is easy to understand why bare steel is not acceptable if a low iron contaminant level is a requirement. In these cases, lined equipment is often used. If iron contamination is not a problem, the next limiting factor is SCC.

This problem has been discussed widely by producers and users debating practical and theoretical application limits. Figure 82 shows two of the more prominent studies on the stress-cracking range.

The upper area, which was developed in the laboratory (Ref 203), showed no cracking below 100 °C (212 °F) over the 62-day stressed coupon test. The lower curve is based on information related to field service (Ref 202). Cracking was reported at temperatures as low as 50 °C (120 °F). In general, field experience is preferred to lab testing; however, in this case, the accuracy and completeness of field information assembled seemed to lead to an overly conservative result. A suggested practical limit is 65 °C (150 °F) at 50% concentration, allowing higher temperatures for dilute, nonconcentrating solutions.

Steel construction is suitable in NaOH concentrations of 10 to 20% at 90 °C (190 °F), where no potential exists for localized solution concentration or severely stressed conditions. Corrosion rates would be in the range of 0.05 to 0.4 mm/yr

Table 35 Corrosion of metals and alloys in NaOH at various concentrations

Laboratory tests at room temperature

Material	Corrosion rate in NaOH: 5% mm/yr	5% mils/yr	15% mm/yr	15% mils/yr	25% mm/yr	25% mils/yr	35% mm/yr	35% mils/yr	45% mm/yr	45% mils/yr
Aluminum-bronze	0.09	3.8	0.018	0.7	0.008	0.3	0.01	0.4	0.005	0.2
Deoxidized copper	0.086	3.4	0.008	0.3	0.008	0.3	0.025	1	0.005	0.2
Monel	<0.0025	<0.1	<0.0025	<0.1	<0.0025	<0.1	<0.0025	<0.1	<0.0025	<0.1
Nickel	<0.0025	<0.1	<0.0025	<0.1	<0.0025	<0.1	<0.0025	<0.1	<0.0025	<0.1
Type 304 stainless steel	0.0025	0.1	<0.0025	<0.1	<0.0025	<0.1	<0.0025	<0.1	<0.0025	<0.1

Source: Ref 197

Table 36 Corrosion of metals and alloys in 30–50% NaOH

16-day exposure in a single-effect evaporator concentrating NaOH from 30 to 50%. Average temperature: 80 °C (180 °F)

Material	Corrosion rate mm/yr	mils/yr
Nickel	0.0025	0.1
Monel	0.005	0.2
Copper-nickel-zinc (75-20-5)	0.013	0.5
Copper	0.06	2.3
Low-carbon steel	0.09	3.7
Cast iron	0.18	7
Chromium steel (14Cr)	0.84	33

Source: Ref 196

(2 to 15 mils/yr), depending on agitation, potential concentrating effects, high local surface (metal) temperature, and stress levels. Stress conditions can result from fabrication and joining procedures, such as forming, welding, or bending, and may occur from an externally applied load.

As the magnitude of stress increases, the time for stress cracking to occur will usually decrease when above the SCC threshold. It is generally accepted that stresses approaching the yield point are required, yet relatively small stresses have in some cases produced failure. Thermal stress relieving minimizes this phenomenon and should be implemented whenever there is a question regarding suitability of steel for the particular application.

Stress-relieving steel equipment used in cell liquor (a 10% NaOH, 15% NaCl solution) applications at 90 °C (190 °F) significantly reduces intergranular attack at weld areas. It should be noted that the presence or absence of salt has no apparent effect on caustic SCC of steel.

Other means of avoiding SCC are commonly implemented. Isolating the steel from caustic solutions with an organic lining, for example, is very effective. Stressed areas rarely, if ever, crack after the lining ultimately fails even though they may be exposed for months to hot, strong caustic. A possible explanation is that the steel undergoes a low-temperature stress relief during the life of the lining. There is evidence that the application of cathodic protection also prevents SCC and can even halt crack propagation. Practical applications of cathodic protection to prevent steel from cracking in caustic solutions are not known. The corrosion behavior of alloy steels, high-strength low-alloy construction steels, and abrasion-resistant steels does not differ significantly from that of carbon steel.

Cast iron also shows corrosion performance similar to steel and has proven quite useful in certain applications up to 70% concentration and 90 °C (190 °F). Stress-corrosion cracking is not a problem for cast irons. It is interesting to note that for many years cast iron pots were used to evaporate caustic to anhydrous NaOH at temperatures over 370 °C (700 °F). Corrosion rates in these cases were generally high (Table 41). The normal service life was several years, owing to the relatively thick-wall castings used. Nickel alloys were not applicable, because various color and clarity agents containing sulfur were routinely added.

Austenitic cast irons (Ni-Resist types) show superior resistance to caustic solutions up to 70% concentration and approaching boiling temperatures; corrosion rates are generally less than 0.25 mm/yr (10 mils/yr) (Ref 196). Nickel-containing cast irons generally benefit by additional nickel alloying as long as sulfur or sulfur compounds are not present (Table 43).

Stainless Steels. Austenitic and ferritic stainless steels are often used to fill an economic slot in what could be termed intermediate ranges of caustic service. Austenitic stainless steels, primarily types 304 and 316, are very resistant to caustic in concentrations up to 50% and temperatures to about 95 °C (200 °F) (Tables 35, 37-39, and 44). Usage in these ranges has increased in recent years for economic reasons even though some potential problems must be addressed.

Figure 83 shows corrosion rates and a stress-corrosion boundary for austenitic stainless steels exposed to NaOH solutions. No distinction is made between types 304 and 316, because their behaviors are so similar. Of significance is the SCC zone, based on known failures, which indicates potential cracking problems above about 105 °C (220 °F). It is generally accepted that this caustic SCC mechanism occurs regardless of austenitic alloy type (304 versus 316). Cracking is affected by stress level and temperature. To provide a measure of insurance, the suggested maximum service temperature is 95 °C (200 °F). Microscopic stress cracking has been reported in laboratory testing at concentrations as low as 10% with temperatures of 100 °C (210 °F) (Ref 204).

Although the austenitic stainless steels crack readily in neutral and acid chlorides above 60 °C (140 °F), the effect of chlorides in an alkaline solution seems to be nil (Ref 205). As long as the solution remains alkaline, the mode of stress cracking is that of caustic embrittlement and continues to occur only in the range indicated in Fig. 83. A solution of 0.5 g/L NaOH with a pH of 12 is sufficiently alkaline. The concern for chloride contamination in caustic when using stainless steel equipment is not well founded.

More important concerns are those extraneous to the caustic solution. Failures resulting from such factors as external exposure, faulty insulation, contaminated test water, and improper cleaning and storage have produced more problems than handling caustic.

Because chloride pitting and/or chloride stress cracking are often primary concerns, alloy selection should consider relative performance in these areas. Although types 304 and 316 perform comparably in caustic and in standard stress-cracking tests, type 316 shows improved overall performance because of its pitting resistance (the role of dissolved oxygen may be significant in this effect) (Ref 206). In addition, the low-carbon grades perform marginally better because of their resistance to sensitization (Ref 205). This suggests that type 316L (molybdenum-containing low-carbon type 316) should be used unless significant controls are to be placed on the total exposure of the equipment. Painting the exteriors of stainless equipment is also recommended, especially if insulation will be applied (see the Section "Corrosion Under Thermal Insulation" in this article).

Applications of stainless steel for caustic services include piping, valves, pumps, and equipment. Transfer piping applications are quite common. Problems rarely occur whenever 10 to 20% solutions are involved, because clean-out and freezing considerations are minimal.

Another possible hazard involves introduction of mercury as a contaminant in mercury cell caustic. This can contribute to cracking or pitting of austenitic stainless with mercury concentrations as low as a few parts per million (see the article "Liquid-Metal Embrittlement" in this Volume). Also, there are cases in which stainless steels cannot be used, because of concern in returning Cr^{6+} ions to a mercury cell operation.

Cast stainless pumps and valves have performed very well in caustic applications. The nature of cast surfaces minimizes SCC problems, and castings are usually acceptable in situations considerably beyond the capabilities of wrought products. Corrosion rates are similar to those of the wrought products.

One other practical application of stainless steel in caustic service is in evaporation processes, in which high-purity alloy 26-1 (a low-carbon version of type 446 stainless steel contain-

Table 37 Corrosion of metals and alloys in 50% NaOH at various temperatures

Material	Corrosion rate at temperature, °C (°F): 40 (100)(a)		60 (135)(b)		55–75 (130–165)(c)		150 (300)(d)	
	mm/yr	mils/yr	mm/yr	mils/yr	mm/yr	mils/yr	mm/yr	mils/yr
Titanium	0.00025	0.01	0.013	0.5	...	...	...	...
Zirconium	0.0023	0.09	0.002	0.08	...	...	...	...
Nickel	0.00023	0.009	0.0005	0.02	0.0005	0.02	0.013	0.5
Inconel	0.0002	0.008	0.0005	0.02	0.0008	0.03	...	...
Monel	0.0005	0.02	0.0005	0.02	0.0008	0.03	0.013	0.5
70-30 copper-nickel	...	...	...	...	0.0013	0.05	...	...
Deoxidized copper	...	...	...	...	...	...	0.14	5.5
Aluminum bronze	...	...	0.025	1	...	...	0.08	3
Type 304 stainless steel	...	...	...	...	0.0025	0.1	1.2	47
Low-carbon steel	0.018	0.7	0.13	5	0.2	8	...	...
Ni-Resist I	...	...	...	...	0.05	2	...	...
Cast iron	...	...	...	...	0.27	10.5	...	...

(a) 162-day exposure (Ref 62). (b) 135-day exposure (Ref 62). (c) 30-day exposure (Ref 196). (d) Laboratory test (Ref 197)

Table 38 Comparison of corrosiveness of NaOH manufactured in diaphragm cells and mercury cells

	Diaphragm cell corrosion rate				Mercury cell corrosion rate			
	50% NaOH at 35–90 °C (95–190 °F)		73% NaOH at 100–125 °C (212–260 °F)		50% NaOH at 40–80 °C (100–180 °F)		73% NaOH at 115 °C (240 °F)	
Material	mm/yr	mils/yr	mm/yr	mils/yr	mm/yr	mils/yr	mm/yr	mils/yr
Nickel 200	<0.0025	<0.1	<0.005	<0.2	<0.0025	<0.1	0.008	0.3
Inconel 600	<0.0025	<0.1	0.005	0.2	<0.0025	0.1	0.005	0.2
Monel 400	<0.0025	<0.1	0.01	0.4	0.0025	<0.1	0.013	0.5
Incoloy 800	<0.0025	<0.1	0.04	1.6	<0.0025	<0.1	0.008	0.3
20Cb-3	<0.0025	<0.1	0.02	0.8	<0.0025	<0.1	0.01	0.4
Cast alloy 20	<0.0025	<0.1	0.09	3.5	<0.0025	<0.1	0.01	0.4
Ni-Resist type 3	0.0064	0.25	0.094	3.7	0.0025	0.1	0.03	1.2
Type 316 stainless steel	0.017	0.65	0.24	9.3	0.0025	0.1	0.25	10
Type 304 stainless steel	0.005	0.2	0.4	15.8	0.019	0.75	0.38	15
Type 430 stainless steel	0.01	0.4	>0.97	>38	>0.14	>5.4	1.5	60
Low-carbon steel	0.12	4.7	>0.87	>34.2	0.09	3.4	1.8	71
Cast iron	0.11	4.2	1.06	41.7	0.08	3.3	2.1	82
Ductile cast iron	0.1	3.9	1.7	65.7	0.06	2.4	2.6	103

Source: Ref 198, originally from NACE Task Group T-5A round-robin testing

ing niobium and molybdenum), heat-exchanger tubing has been used. Some high-temperature applications have proven effective where corrosion rates on nickel are excessive. This is probably due to the presence of hypochlorite or chlorate contaminants. The debate between nickel and ferrous alloys continues, yet both materials have shown satisfactory performance. The 26-1 alloy is useful up to 175 °C (350 °F), depending on the caustic, chloride, and contaminant concentrations (Ref 207). When 26-1 is attacked, the normal failure mode seems to be intergranular. At this time, there are no specific criteria to provide recommendations regarding the use of ferritic stainless steel. Decisions should be governed by economics, taking into account the impact of heat transfer coefficients as well as the cost of tubing.

A comprehensive summary of criteria for selection of stainless steel for process equipment is given in Ref 205. There probably are good applications for the new ferritic and duplex steels; however, little testing has been completed.

Nickel and Nickel Alloys. Nickel 200 or 201 and nickel-base Monel 400 and Inconel 600 have outstanding resistance to caustic solutions over a wide range of concentrations and temperatures. Nickel has a maximum corrosion rate of 0.025 mm/yr (1 mil/yr) for all concentrations up to 73% boiling solutions; corrosion rates are summarized in Tables 34 to 41. Some of the high nickel-chromium alloys and high-nickel austenitic cast irons perform very well, showing low corrosion rates in 73% NaOH up to 125 °C (255 °F) (Table 38).

Nickel or nickel-base alloys are extensively used in more severe applications. The very low corrosion rates also ensure low metal ion contamination. Nickel has the lowest corrosion rates, even in molten anhydrous NaOH up to 540 °C (1000 °F) and is essentially immune to caustic SCC.

Even with its high level of resistance, nickel is sometimes used in conjunction with cathodic protection. This technique is implemented in nickel concentrators and storage tanks when metal ion contamination must be minimized. Current densities as low as 0.11 A/m^2 (0.01 A/ft^2) have been effective.

When exposure in concentrated solutions above 315 °C (600 °F) is anticipated, low-carbon Nickel 201 (0.02% C maximum) should be selected. This material was specifically designed to avoid potential intergranular corrosion, which may occur in Nickel 200 (0.15% C maximum). In areas of high stress, this type of attack has resulted in cracking after intergranular corrosion, but is mechanically driven.

A common application of nickel and Monel 400 is in fabricating clad vessels and equipment. In these cases, special welding procedures are required to maintain iron contamination levels in welds within acceptable ranges. Typically, the nickel-clad thickness represents 20% or less of the base steel thickness. Nickel welds containing up to 25% Fe showed little change in corrosion rate in 73% caustic at 120 °C (250 °F) (Ref 208). Current welding practice has improved, and maximum iron levels of about 5% can be specified and maintained.

Repair welding of nickel and nickel alloys in caustic service is even more critical. It is absolutely necessary to remove corrosion products and any foreign material from the vicinity of the weld area. Chemical cleaning with a pickling acid is usually very effective; however, careful mechanical cleaning with a fine wheel or disk grinder is acceptable. If proper cleaning is not carried out, the weld will often become embrittled and crack.

The mercury contamination discussed earlier can also affect nickel and nickel alloys. Even though the mercury cell caustic is no more corrosive than diaphragm cell production under normal conditions (Table 38), a mercury-contaminated product could create cracking or pitting problems. The chances of this, however, have proven to be remote. The effect of oxidizable sulfur compounds can also be detrimental at higher temperatures and is probably encountered more frequently. The effects of sulfur, added for color shading, are given in Table 41 for nickel in caustic fusion.

Perhaps the contaminants most discussed in recent years among producers of caustic chemicals have been hypochlorite and/or chlorate in caustic evaporation services. Increased corrosion rates on nickel evaporator tubes have been attributed to increased sodium chlorate ($NaClO_3$) levels in the caustic (Ref 209). The chlorate levels studied were high (300 to 700 ppm), yet laboratory tests showed corrosion rates significantly lower than actual experience. Certainly, velocity and abrasion are principal factors in increasing actual service rates. Most of the studies focused on chlorate levels. Recent laboratory work showed significant modification to corrosion behavior resulting from hypochlorite additions as low as 1 to 2 ppm (Ref 210). Further studies are required before definitive recommendations can be made. It is recognized that chlorate and hypochlorite have a negative effect on nickel and nickel alloys in caustic service (Ref 207, 211, 212).

Monel 400 and Inconel 600, although excellent in caustic service, have higher corrosion rates than the Nickel 200 and 201 metals. Inconel 600 has shown a slight sensitivity to SCC above 190

Table 39 Corrosion of metals and alloys in 70–73% NaOH at various temperatures

	Corrosion rate at temperature, °C (°F)							
	110 (230)(a)		90–115 (190–240)(b)		120 (240)(c)		130 (265)(d)	
Material	mm/yr	mils/yr	mm/yr	mils/yr	mm/yr	mils/yr	mm/yr	mils/yr
Nickel	0.0025	0.1	0.0025	0.1	0.005	0.2	0.025	1
Monel	0.0025	0.1	0.028	1.1	0.013	0.5	0.023	0.9
Inconel	0.0025	0.1	0.008	0.3	0.005	0.2	0.025	1
Zirconium	0.02	0.8	...	...	...	...	0.05	2
Titanium	0.05	2	...	...	...	...	0.18	7
Aluminum-bronze	0.023	0.9	...	...	0.15	6.1	...	...
Type 304 stainless steel	...	...	0.69	27	0.26	10.2	...	...
Steel	0.99	39	1.45	57	...	...	>2	>80(e)

(a) 126-day exposure (Ref 62). (b) 90-day exposure (Ref 196). (c) 180-day exposure (Ref 197). (d) 200-day exposure (Ref 62). (e) Duplicate specimens consumed during test

Table 40 Corrosion of cast irons by molten NaOH at 510 °C (950 °F)

14-day test in anhydrous NaOH containing 0.5% NaCl, 0.5% Na_2CO_3, and 0.03% Na_2SO_4

Material	Corrosion rate mm/yr	Corrosion rate mils/yr	Pit depth mm	Pit depth mils
Gray iron	2.5–3.4	97–135	0.13	5
Ductile iron	5.3	207	...	...
White iron	3.8	151	0.5	20
3% nickel-iron	1.8	71	...	...
Austenitic, type 1	15.9	628	1.5	60
Austenitic, type 2	24.2	954	1.8	70
Ductile austenitic, type 2	11.8	466	1.5	60
Austenitic, type 3	2.2	87	none	
Austenitic, type 4	13.6	534	1.0	40
Wrought nickel	0.23	9	...	...

Source: Ref 199

Table 42 Corrosion of steel in NaOH solutions (22-day laboratory test at room temperature)

Concentration of NaOH, g/L	Corrosion rate mm/yr	Corrosion rate mils/yr
0	0.05	2
0.001	0.05	2
0.01	0.05	2
0.1	0.05	2
1.0	0.018	0.7
10	nil	
100	0.0025	0.1
540	nil	

Source: Ref 201

°C (375 °F). However, it replaces nickel for direct fire production of anhydrous NaOH with sulfur-containing fuels. Monel 400 has a generally high corrosion rate at concentrations above 75% and temperatures above atmospheric boiling. Additional information on nickel and nickel alloys in various alkaline services is available in Ref 196.

Other Metals. A few other metals have useful resistance to NaOH, although use is generally limited to very special circumstances, frequently involving dual service. Silver has good resistance and at one time was used in evaporating to anhydrous NaOH. Zirconium and titanium are very good in moderate concentrations and temperatures (Tables 34, 37, 39, and 41). Zirconium is considered useful to 73% concentrations at temperatures to 140 °C (280 °F). Titanium, when attacked, absorbs hydrogen liberated in the corrosion process, limiting service to temperatures of about 100 °C (210 °F) and to concentrations generally below 50%. Iron-base nickel-chromium alloys (including 20Cb-3) show appreciable resistance that generally surpasses that of the austenitic stainless steels. Copper and some copper-base alloys (aluminum-bronze and copper-nickel) show useful resistance (Table 38). Higher nickel in the copper-nickel alloys indicates better resistance; alloy 70-30 (CDA 71500) has been used where copper contamination can be tolerated.

Certain metals should not be used with caustic because of excessive corrosion rates. These include aluminum, magnesium, zinc, tin, chromium, and zinc-containing brasses or bronzes.

Nonmetals. Many nonmetallic materials have excellent resistance to NaOH within their normal temperature ranges. Temperature limits are often much more restrictive than for resistant metals. Among the resistant nonmetals are types of organic linings and elastomers, some thermoplastics, and some reinforced plastics.

Linings were mentioned earlier as a means of using the economics of steel construction while achieving the corrosion resistance and product protection provided by organic, elastomeric, and even metallic linings. Many steel tanks in caustic service are lined not so much for corrosion protection as for minimizing iron contamination. It is recognized that as caustic concentrations increase, the recommended temperature limit also increases for most linings. The lower the concentration of the caustic, the more readily it permeates and attacks the linings. The common linings for these applications are spray-applied neoprene latex and phenolic-modified epoxy.

The neoprene latex material has found extensive application in caustic service for more than 30 years. Applied at thicknesses of 0.3 mm (12 mils) for tank cars and 0.5 mm (20 mils) for storage tanks, it is recommended for continuous service up to 115 °C (240 °F) for concentrations between 50 and 73%. It has been used at lower concentrations, but this application is questionable at high temperatures.

The modified epoxy-phenolic linings have also been used for many years. Typically applied at thicknesses of 0.25 to 0.38 mm (10 to 15 mils), these linings are often used for multiservice applications because they have a wider range of chemical resistance. They have been recommended for caustic applications up to 73% at 130 °C (270 °F); low-concentration high-temperature service is more effective, depending on the individual formulation.

Spray-applied amine-epoxy linings have been successfully used in equipment and railcars handling solid caustic beads. Probably the most severe exposure results from hot-water washing of residues.

Several types of elastomeric sheet linings possess excellent caustic resistance. For economic reasons, their use is generally limited to small tanks or equipment. Recommended materials include neoprene, chlorosulfonated polyethylene, EPDM rubber, and certain formulations of natural rubber. They are usually used for applications up to 50% caustic, with temperatures limited by individual elastomer type. Neoprene is recommended to as high as 110 °C (230 °F), and chlorosulfonated polyethylene to as high as 120 °C (250 °F). Abrasive services are also effectively served by the elastomeric linings.

Fabricated rubber parts (seals, o-rings, diaphragms) are also made of these elastomers and are useful in caustic service. The perfluoroelastomers are very resistant, but fluoroelastomers should not be used in strong or hot solutions.

A large number of thermoplastics have excellent resistance to caustic solutions up to their normal temperature limits. Included are PVC, chlorinated polyvinyl chloride (CPVC), polyethylene, PP, and the fluorocarbons (FEP, TFE, and PFA). Polyvinylidene fluoride is limited to 70% concentrations and temperatures of 120 °C (250 °F). Stress-cracking problems have been noted in some high-density polyethylenes; generally, cross-linked polyethylene has performed better.

Thermoplastic materials are used in a wide variety of caustic-handling applications, including valves, pipe, small vessels, pumps, gaskets, and lined pipe. The use of lined pipe (PP is common) is significant because it offers a plastic application with a high level of safety.

Fiberglass-reinforced plastics are used in various caustic applications, especially in piping and

Table 41 Corrosion of metals and alloys in caustic fusion process

Material	Laboratory test(a) mm/yr	Laboratory test(a) mils/yr	Plant test(b) mm/yr	Plant test(b) mils/yr
Silver	0.13	5.3	...	...
Nickel	1.3–1.8	52–72	6.6	260
Zirconium	...	...	2.8	110
Cast iron	3.3	130	5.3	210
Ni-Resist type 3	3.3	130	...	...
Ni-Resist type 2	3.8	150	...	...
Monel	6.6	260	9.7	380
Low-carbon steel	...	...	12.7	500
Inconel	...	...	49	1930
Type 302 stainless steel	45.7	1800	...	...
Chromium steel (18.5% Cr)	68.6	2700	...	...

(a) 75% NaOH concentrated to anhydrous; maximum temperature: 480 °C (900 °F). (b) 73% NaOH concentrated to anhydrous; maximum temperature: 540 °C (1000 °F). Batch shaded with sulfur and Na_2NO_3; exposure time: 2.5 days. Source: Ref 198

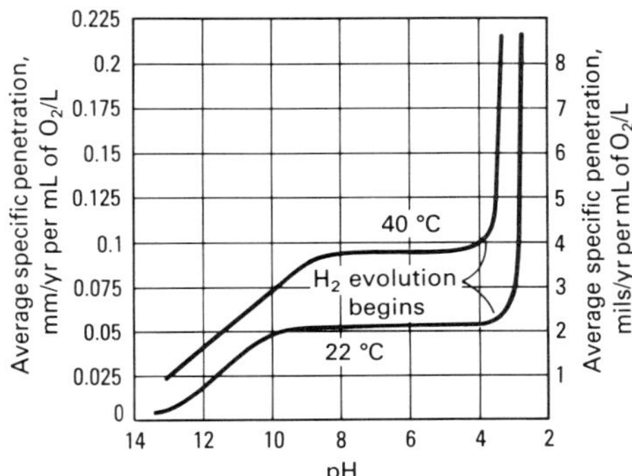

Fig. 81 Effect of pH on corrosion of low-carbon steel in aerated water containing NaOH and HCl. Source: Ref 200

Table 43 Effect of nickel additions on corrosion of cast iron in boiling 50–65% NaOH

81-day test under 660 mm (26 in.) of mercury

Nickel, %	Corrosion rate mm/yr	mils/yr
0	1.9	73
0	2.3	91
0	2.2	86
3.5	1.2	47
5	1.24	49
15	0.8	30
20	0.08	3.3
20 (plus 2% Cr)	0.15	6
30	0.01	0.4

Source: Ref 196

equipment. Often, the lower-concentration applications are more aggressive to fiberglass-reinforced plastic. Because glass is known to be attacked in caustic service, effective designs usually incorporate a special interior surface treatment. Synthetic surfacing veils (polyester and acrylic are common ones) should be specified for caustic applications. The selection and protection of a highly resistant inner surface, often called the corrosion barrier, is critical. Applications in high-concentration high-temperature service are usually not suited to fiberglass-reinforced plastic.

Epoxy resins generally have better resistance to caustic and are useful up to 95 °C (200 °F). Vinyl-esters follow, with potential application up to 80 °C (175 °F). Polyesters are generally comparable to the vinyl-esters; some are slightly better (bisphenol-fumarate), and some worse (isothalics and blends). The notable exception is chlorinated (chlorendic anhydride) polyester, which should not be used in NaOH. Any potential fiberglass-reinforced plastic applications for caustic service should be carefully reviewed with resin suppliers and manufacturers.

Glass and ceramics have very limited application in NaOH, especially if high temperatures and concentrations are involved. Borosilicate glass is especially sensitive to caustic attack in hot solutions (95 °C, or 200 °F) as low as 5% concentration. Alkali-resistant glass on steel may be used to pH 14 at 65 °C (150 °F) or pH 13 at 80 °C (175 °F).

Potassium Hydroxide

Potassium hydroxide is also known as caustic potash. It has a higher boiling point than NaOH

Table 45 Corrosion of metals and alloys in KOH solutions

Material	Corrosion rate 13% KOH(a) mm/yr	mils/yr	50% KOH(b) mm/yr	mils/yr
Titanium	0.023	0.9	0.01	0.4
Zirconium	0.005	0.2	0.0015	0.06
Nickel	nil		0.00008	0.003
Monel	nil		0.00005	0.002
Inconel	nil		nil	
Low-carbon steel	0.013	0.5(c)	0.0013	0.05

(a) 207-day test at 30 °C (85 °F); 13% KCl added to solution. (b) 207-day test at 25 °C (80 °F). (c) Slight attack under spacer. Source: Ref 62

at equal concentrations. Consequently, it is rarely evaporated above a 90% strength. It is very similar to NaOH in corrosion behavior as well as safety requirements.

Materials. Corrosion and degradation of metals and nonmetals is the same in KOH as in NaOH. Indications are that KOH is slightly more aggressive, but this generally relates to its higher boiling point and the increased likelihood of higher-temperature exposure. Table 45 shows the corrosion rates of some metals in KOH. These data do not indicate any higher rates, with the possible exception of titanium. Still, all of the rates are very low. It has been reported that nickel alloys are less sensitive to stress cracking in KOH than in NaOH (Ref 196).

Most literature for such products as plastics and linings indicates the same levels of resistance for KOH and NaOH. One exception is that fiberglass-reinforced plastic shows a generally lower resistance for KOH, regardless of the resin type.

Soda Ash

Sodium carbonate is an alkaline salt. It is commonly called soda ash. This chemical was extensively used in alkaline processes to supply sodium oxide (Na_2O) equivalence. Over the years, some uses have been replaced with caustic. At one time, Na_2CO_3 was primarily synthetically produced using an ammonia-soda (Solvay) process. Current production is largely of the natural ash, which is mined and refined.

When in solution, Na_2CO_3 creates less alkalinity than the hydroxides (Fig. 84). A 0.1% solution creates a pH of 11; a fully saturated solution is 35%, which has a pH of 12.5. The related bicarbonates (sodium bicarbonate, $NaHCO_3$) are less alkaline, but they will convert to carbonates at

Fig. 82 Temperature and concentration of NaOH required to cause cracking of steel. Source: Ref 202, 203

high temperatures and increase the pH of the solution. The safety requirements for Na_2CO_3 can be considered less demanding because of its lower alkalinity.

Metals. Soda ash solutions can be handled easily in anything that is suitable for caustic. Similar to other alkaline solutions, the presence of carbonate ions (CO_3^{2-}) usually provides an inhibiting effect on metals. Carbon steel has good resistance and is used extensively in Na_2CO_3 solutions. There is a smaller possibility of alkaline stress cracking, and steel is routinely used up to boiling temperatures.

When nil corrosion rates are deemed necessary, the stainless steels and nickel alloys perform very effectively. Stainless steel has an SCC range similar to that in caustic, but at considerably elevated temperatures. Aluminum is attacked depending on concentration and temperature, but can be inhibited with silicates in dilute solutions. Similarly, zinc and zinc brasses are attacked, but also show beneficial results from various inhibitors.

Nonmetals. Because of the relative resistance of iron and steel, there is less need to consider nonmetals. Most linings perform well. Thin-film linings provide reasonable contamination protection when required. Most plastics have good resistance to Na_2CO_3; the same reactions that were present with caustic still occur, but to a much lesser degree. Although weaker solutions of NaOH or KOH are more aggressive at high

Table 44 Corrosion of stainless steel in NaOH solutions

Type	Concentration of NaOH, %	Temperature °C	°F	Test duration, days	Corrosion rate mm/yr	mils/yr
302	20	50–60	120–140	134	<0.0025	<0.1
304	22	50–60	120–140	133	<0.0025	<0.1
309	20	50–60	120–140	134	<0.0025	<0.1
310	20	50–60	120–140	134	<0.0025	<0.1
410	20	50–60	120–140	134	0.0025	0.1
430	20	50–60	120–140	134	0.0025	0.1
304	72(a)	120–125	245–255	119	0.09	3.7
316	72(a)	120–125	245–255	119	0.08	3.1
329	72(a)	120–125	245–255	119	0.0025	0.1
21Cr-4Ni-0.5Cu	72(a)	120–125	245–255	119	0.15	6
410	72(a)	120–125	245–255	119	0.8	32
302	73(b)	100–120	210–245	88	0.97	38
304	73(b)	100–120	212–245	88	1.1	45

(a) Solution moderately agitated. (b) No aeration. Source: Ref. 199

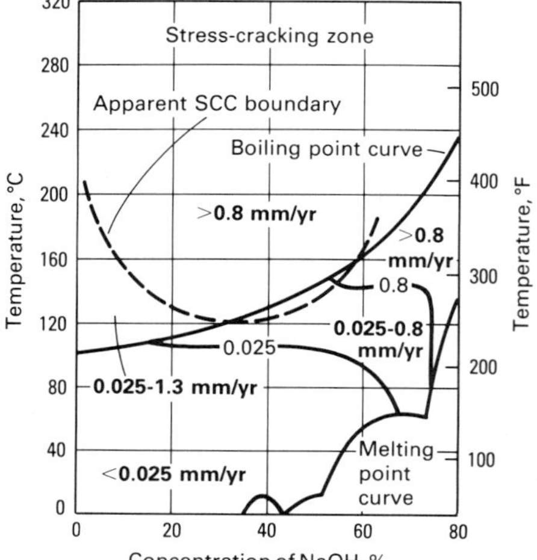

Fig. 83 Isocorrosion diagram for type 304 and 316 stainless steels in NaOH

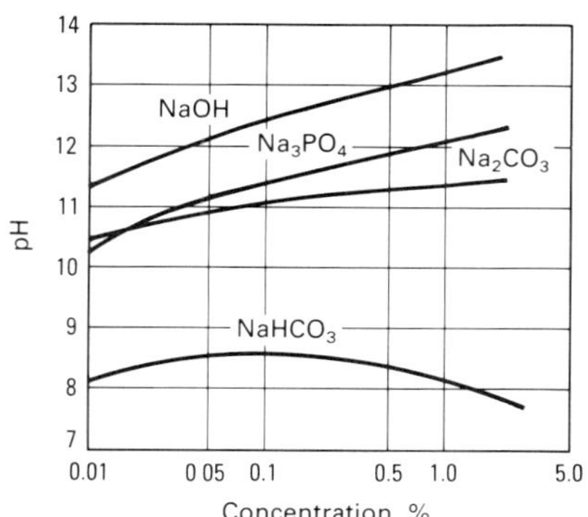

Fig. 84 Concentration versus pH of various alkaline solutions

temperatures, this is not the case with Na_2CO_3 solutions, because they are so much less alkaline.

Other Alkalies

Many of the corrosion rates of materials in alkaline service are attributed to the pH of the solution. Figure 84 compares these values with concentration for some industrial alkalies.

Essentially, all of the hydroxides behave similarly and can be handled in materials recommended previously for comparable concentrations and temperatures of caustic. The alkaline salts relate to Na_2CO_3 and can be treated similarly with few exceptions; salts would include borates, silicates, and phosphates. Sulfides are also considered alkaline salts, but are generally more corrosive to metals. Reactions with nonmetallic materials are similar to the other alkaline salts. Careful checking is warranted when selecting a lining for alkali service because performance variations with temperature and concentration can be significant.

Hypochlorites

These alkaline oxidizing salts are among the most corrosive salts. Sodium hypochlorite is produced as a liquid by chlorinating 20% NaOH. Resulting soda bleach solutions contain about 15% NaOCl with some residual NaOH to aid stability. These industrial-strength bleaches are diluted to concentrations of about 5% for household use. To enhance stability further, the NaOH used should not contain heavy metals, and the NaOCl should be stored in dark or polyethylene bottles at temperatures below 30 °C (85 °F). Sodium hypochlorite does not exist as a stable solid.

The solid hypochlorites include lithium, calcium, strontium, and barium, although the only major applications are for calcium hypochlorite ($Ca(OCl)_2$). The product is made by various processes, usually resulting in a dihydrate salt, $Ca(OCl)_2{\cdot}2H_2O$, containing 65% available chlorine. Solid $Ca(OCl)_2$ will decompose rapidly at high temperatures (175 °C, or 350 °F), releasing chlorine. It also reacts vigorously with many oxidizable organic compounds. Calcium hypochlorite solutions are encountered during production and in other chemical processes for bleaching, sanitizing, or deodorizing. The solid product is often dissolved in water to form dilute solutions used for bleaching operations and water purification.

Hypochlorite solutions are found in industrial applications, such as bleaching in the pulp and paper industry, odor control, and waste treatment/disposal. Solutions are often produced when neutralizing vent gas and other waste streams. Hypochlorite solutions are subject to decomposition, which is influenced by concentration, ionic strength, pH, temperature, light, and impurities. Trace metal contamination of cobalt, nickel, and copper catalyzes decomposition; iron and manganese provide the same effect, but to a lesser degree. The resulting decomposition reaction creates an aggressive environment. This is why many metals and plastics do not provide adequate resistance to hypochlorite solutions.

Sodium Hypochlorite

Metals. The hypochlorite ion (OCl^-) is similar to wet chlorine gas in its effects on materials. Not many metals show good resistance even at low temperatures and concentrations. Because hypochlorite solutions are unstable at neutral and lower pHs, they normally contain excess alkali, which modifies the aggressiveness somewhat. Corrosion rates of HOCl (OCl^- ion present at low pH) are compared with those found in batch manufacture of NaOCl in alkaline solution in Table 46.

Titanium is the only metal that provides consistently good performance in NaOCl. Although special titanium alloys are available, the commercial grade 2 is usually suitable for a full range of concentrations and temperatures. Although titanium has been known to suffer crevice corrosion in hot wet chlorine, it apparently has not shown the same susceptibility in hypochlorite solutions. A wide variety of titanium products are used, including piping, pumps, valves, heat exchangers, fans, and vessels.

Tantalum is also very resistant to hypochlorite, but its use is limited because of its relatively high cost. Nickel-base iron-chromium-molybdenum alloys, wrought and cast (Hastelloy C, Chlorimet 3, and so on), show good resistance in various solutions. The high iron-chromium-nickel alloys, such as 20Cb-3, and silicon cast irons have also been used, especially as pump materials, in the less aggressive environments. Many of these materials have shown unpredictable performance in seemingly similar applications.

Sodium hypochlorite is usually maintained at low temperatures to prevent decomposition. The effect of higher temperature, even in the dilute solutions, is reflected by corrosion rates (Table 47). The presence of heavy-metal ions, especially in the case of higher solution concentrations, often causes instability with decomposition. This potential may rule against metals that might otherwise be acceptable, because the occurrence of decomposition often triggers a corrosion reaction.

The relative rankings in a more recent study (Table 47, exposure 2) offer an interesting insight for what may be new metals for hypochlorite service. Some of the newer super stainless steel materials indicate useful resistance in certain NaOCl solutions (Ref 207). Successful performance of a high-alloy duplex stainless (Ferralium 255) in a NaOCl scrubber was recently reported (Ref 214). When the stainless steels are attacked in NaOCl, the mode is usually pitting and/or crevice corrosion.

Nickel and some nickel alloys can be protected by alkaline inhibitors, such as Na_2SiO_3 and Na_3PO_4. The corrosion rates are significantly reduced, and the mode of corrosion becomes more uniform (Table 48). However, nickel finds few successful applications in NaOCl service, contrary to its dominance in NaOH.

Aluminum and aluminum alloys are also inhibited by silicates. Hypochlorites usually promptly destroy their protective oxide film and cause rapid attack.

Copper is normally attacked by localized pitting, with corrosion rates exceeding 0.5 mm/yr (20 mils/yr).

Nonmetals. The instability of NaOCl has a significant impact on the resistance of nonmetals. Generally, those materials that are resistant to wet chlorine gas are resistant to hypochlorite at any normal concentration and temperature. Unlike the metals, there are many plastics, rubber and elastomers, glass, and ceramics that fall into this category. Because NaOCl is normally maintained at low or moderate temperatures, most resistant nonmetals are serviceable to their usual temperature limit.

A notable exception is occasionally found when service is at or near the NaOCl decomposition temperature (60 °C, or 140 °F). Some nonmetals are severely affected by the decomposition reaction, which is initiated by temperature or catalyzed by heavy-metal ions.

Coated or lined steel is common in NaOCl service, and a wide variety of resins are used. The fluorocarbon resins are essentially unaffected. Polyvinyl chloride and CPVC also show excellent resistance; steel vessels lined with unplasticized PVC sheet were used for making soda bleach for 20 years. Loose PVC liners have been used to extend the service lives of deteriorated lined tanks. Polyvinylidene chloride has provided service in some specialized bleaching solutions. Polypropylene is also used extensively in lined piping and pumps, although there have been reported incidents of stress cracking.

Many of the thermoplastics are used as solid products. In addition to many of those identified as lining materials, polyethylene (cross-linked and high-density) provides useful service as tanks and small vessels. The variations in formulation of thermoplastics should reinforce the requirement to compare specific products in any proposed solution if there is any question regarding resistance.

A number of fiberglass-reinforced plastics resist NaOCl at normal temperatures and concentrations. Many resins provide useful service. Generically, these would include bisphenol and chlorendic-type polyesters, vinyl-esters, and, to a limited extent, epoxy. It would seem that the use of fiberglass-reinforced plastic will be grow-

Table 46 Corrosion of metals in HOCl and NaOCl

	Corrosion rate			
	17% HOCl(a)		16% NaOCl(b)	
Material	mm/yr	mils/yr	mm/yr	mils/yr
Titanium	<0.0025	<0.1	<0.0025	<0.1
Zirconium	0.05	2(c)	<0.0025	<0.1
Hastelloy C	0.23	9	0.0025	0.1
Chlorimet 3	1.0	40	0.008	0.3
Durachlor	...	...	0.02	0.8

(a) 203-day test in 17% HOCl with some free chlorine and chlorine monoxide (Cl_2O) at 10 °C (50 °F). (b) Repeated exposures to batch manufacture of NaOCl, starting with 18–20% NaOH chlorinated to 16% NaOCl. Test duration: 170 days. Temperature: 20 °C (70 °F). (c) Severe attack under spacer. Source: Ref 62

Table 47 Corrosion of metals and alloys in NaOCl solutions at elevated temperatures

Material	Corrosion rate at 65–95 °C (150–200 °F)(a)		Relative corrosion rate at 50 °C (120 °F)(b)	
	mm/yr	mils/yr	mm/yr	mils/yr
Titanium	0.0025	0.1	0.000025	0.001
Remanit 2800	...	...	0.000025	0.001
E-Brite	...	...	0.00025	0.01
2Mo-0.4Cu stainless steel	...	...	0.00025	0.01
Type 317 stainless steel	...	...	0.00025	0.01
Inconel 600	...	...	0.005	0.02
Zirconium	0.1	4	...	...
Durichlor	0.18	7	...	...
Duriron	0.3	12	...	...
Incoloy 825	...	...	0.005	0.2
Hastelloy C	1.2	46	0.005	0.2
Type 316 stainless steel	>2.5	100 (consumed)	0.008	0.3
Type 304 stainless steel	...	...	0.025	1.0
Low-carbon steel	>5	>200 (consumed)	0.025	1.0

(a) 72-day test in 1.5–4% NaOCl with 12–15% NaCl and 1% NaOH (Ref 62). (b) NaOCl solution (pH 9) with 500 ppm active chlorine and 1.2% NaCl (Ref 213)

ing; applications include piping, equipment, tanks, pumps, and many special fabrications.

A number of factors control the usefulness of fiberglass-reinforced plastic in hypochlorite solutions (Ref 216). Foremost are factors that affect the stability of the solution and the degree of chlorination or, conversely, the level of excess alkalinity. It has already been indicated that temperature, pH, and concentration contribute to decomposition. Because heavy metals also trigger decomposition, any potential source of metal ions should be evaluated. This should include thixotropic additives, fillers, ultraviolet inhibitors, metallic contaminations, and curing systems.

The most common fiberglass-reinforced plastic curing system uses methyl ethyl ketone-peroxide (MEKP) with a cobalt-containing promoter—an obvious source of cobalt ions. The resin manufacturers provide mixed recommendations regarding alternative curing systems, some indicating that increased fabrication difficulty more than offsets the benefit of eliminating the cobalt. The quality of fabrication is a most important factor in successful applications of fiberglass-reinforced plastic and should not be compromised. Most resin suppliers recommend synthetic veils. Synthetic materials should certainly be used for strongly alkaline solutions. Postcuring is also recommended and provides a more resistant fabrication. Full curing of all secondary joints is most important. Finally, thixotropic agents should not be added to resin systems, especially when used in surfaces exposed to hypochlorite.

Glass-flake reinforced spray linings are also used in hypochlorite services. The problems encountered indicate that these applications should be limited to low-temperature services or applied where the environment is not overly aggressive to the substrate.

Elastomeric lining is common in a wide variety of hypochlorite applications—a result of typically low temperatures. Probably the most frequently used lining material is chlorobutyl, because of its good resistance coupled with moderate cost. Other sheet lining materials with good resistance include EPDM and chlorosulfonated polyethylene. Natural rubber also finds some useful low-temperature applications. Other elastomeric applications might use the fluoroelastomers, which are good in any hypochlorite service, or neoprene, which has limited range.

Glass is unaffected by hypochlorite within a moderate range of alkalinity and temperature. Glass-tubed heat exchangers have been used to cool soda bleach during manufacture. Vitrified clay pipe and other ceramic materials show excellent resistance to hypochlorite. Under certain conditions, concrete is resistant and has been used for manufacturing and storage tanks.

Calcium Hypochlorite

Although produced as a solid, the reactions of $Ca(OCl)_2$ in solution are very similar to those of NaOCl. In general, the recommended temperature levels for $Ca(OCl)_2$ are slightly higher, probably owing to the higher decomposition temperature. Like NaOCl, the calcium product is unstable at lower pH, but it can be concentrated to a higher degree. The corrosion rates for some metals are shown in Table 49.

Plastics and elastomerics can usually be used in $Ca(OCl)_2$ to slightly higher concentrations and temperatures than in NaOCl. Virtually the same applications are served by these materials in both products.

The solid $Ca(OCl)_2$ product is typically packed in polyethylene or polyethylene-lined containers. Epoxy-phenolic lining performs effectively for trucks and railcars. It is critical to keep the product dry and away from organic fluids. Aluminum is sometimes used in handling solid $Ca(OCl)_2$ because any corrosion residue does not discolor the product.

Table 48 Effect of inhibitors on corrosion in NaOCl

16-h beaker test at 40 °C (105 °F); no agitation

Solution composition, g/L			Corrosion rate					
Available chlorine	Na_2SiO_3	Na_3PO_4	Monel 400		Nickel 200		Inconel 600	
			mm/yr	mils/yr	mm/yr	mils/yr	mm/yr	mils/yr
6.5	...	...	2.9	113	1.3	52	0.3	12
6.5	0.5	...	0.46	18	0.25	10	0.08	3
6.5	...	0.5	0.2	8	0.5	20	0.08	3
6.5	2.0	...	0.05	2	0.025	1	0.025	1
6.5	...	2.0	0.08	3	0.23	9	0.025	1
3.3	...	...	1.0	40	0.8	30	0.13	5
3.3	0.5	...	0.025	1	0.1	4	0.025	1
3.3	...	0.5	0.1	4	0.15	6	0.025	1
0.1	...	...	0.1	4	0.1	4	0.05	2
0.1	0.5	...	0.008	0.3	0.013	0.5	0.018	0.7
0.1	...	0.5	0.033	1.3	0.015	0.6	0.018	0.7

Source: Ref 215

Table 49 Corrosion of metals and alloys in $Ca(OCl)_2$

204-day test in 18–20% $Ca(OCl)_2$ at 20–24 °C (70–75 °F)

Material	Corrosion rate		Pitting
	mm/yr	mils/yr	
Titanium	nil		none
Zirconium	0.025	1(a)	none
Hastelloy C	<0.0025	<0.1	none
Chlorimet 3	0.025	1	none
Type 316 stainless steel	0.25	10(a)	severe

(a) Severe attack under spacer

Corrosion by Ammonia

A.S. Krisher
ASK Associates

Anhydrous ammonia, a major commercial chemical, is used in the manufacture of fertilizers, HNO_3, acrylonitrile, and other products. Except for a sensitivity to SCC, carbon steel is fully acceptable in NH_3 service. Stress-corrosion cracking of carbon steel NH_3 storage vessels was first observed in the early 1950s. In most cases, the developing cracks have been detected by inspection before leakage or rupture. However, there have been a few catastrophic failures. For example, in France in 1968, a tanker ruptured, killing 5 people. A second case was in South Africa, where a large tank failed in 1973 with 22 fatalities.

Ammonia is stored under three conditions. It can be stored by cooling it to a low enough temperature, (−34 °C, or −29 °F) to maintain it in the liquid state at atmospheric pressure. This method is frequently described as cryogenic storage. A second approach is to contain the ammonia under sufficient pressure (about 2070 kPa, or 300 psig) to maintain the ammonia in the liquid phase at ambient temperature. Cylindrical pressure vessels are often used for fairly small quantities. Spherical pressure vessels are used for larger quantities. The third condition involves some degree of refrigeration combined with pressurization. This is termed semirefrigerated storage.

Most cases of SCC have occurred in ambient-temperature pressurized storage vessels, for the most part in spheres. A few problems have been observed in semirefrigerated storage. There have been no documented cases of SCC in cryogenic storage vessels. When SCC does occur, cracks are primarily transgranular and progress at a relatively slow rate compared to other SCC phenomena.

Laboratory Studies

One investigation using statically loaded tuning fork type specimens and tensile bars showed that NH_3 SCC is accelerated by cold work, welding, applied stresses, and the use of higher-strength steels. It was found that air contamination promotes SCC and that water in amounts greater than 0.1% inhibits cracking (Ref 217).

Other experiments using slow strain rate tests and a low-alloy steel also found that air contamination promoted SCC and that water at a level greater than 0.09% was an effective inhibitor. Electrochemical studies showed that the SCC involves an anodic chemical process (Ref 218, 219).

Field tests were conducted using specimens stressed by residual stresses from welding (Ref 220). Results indicated that high-strength steels fail more rapidly than low-carbon steel and that hard welds (welds that are harder than the base material) tend to accelerate cracking. Thermal stress relieving was also found to be effective in preventing SCC.

Another investigation using low-alloy steels and slow strain rate test methods produced SCC at temperatures as low as 0 °C (32 °F). Again, air contamination and low water content promoted SCC (Ref 221).

Results of an industry-sponsored technical investigation that used both slow strain rate tests and fracture mechanics type specimens are documented in Ref 222. It was found that oxygen levels greater than 5 ppm are required for SCC. Levels as low as 1 ppm caused cracking if carbon dioxide was also present and water was absent. This work also suggested that hydrazine ($NH_2 \cdot NH_2$), ammonium carbonate [$(NH_4)_2CO_3$], and ammonium bicarbonate (NH_4HCO_3) might be inhibitors. The fracture mechanics test methods were not successful, possibly because of the slow rate of cracking.

Other tests using low-alloy steel and the slow strain rate test confirmed again that oxygen as a contaminant is damaging, with indications that levels as low a 0.01 ppm might be sufficient to cause the cracking, at least in low-alloy steel (Fig. 85). Nitrogen also appeared to be a cracking accelerator in combination with oxygen. The lower limit of the required water content to inhibit cracking was found to be about 0.08 wt% (Fig. 86). This work showed $NH_2 \cdot NH_2$ to be an effective inhibitor at 0.025 wt% for a contamination level of 200 ppm O_2.

This body of laboratory work (seven studies over a period of 19 years by six different investigators using three different methods in four different countries) is impressive in its consistency. All of the studies showed that the primary causes of the cracking are high stresses and air contamination. Nitrogen and carbon dioxide were suggested by separate investigators as promoting SCC. Cracking is accelerated by the use of high-strength steels, the presence of hard welds, and air contamination. The cracking mechanism can be inhibited by water above about 0.1%. Thermal stress relief, if done properly, reduces stress below the critical level.

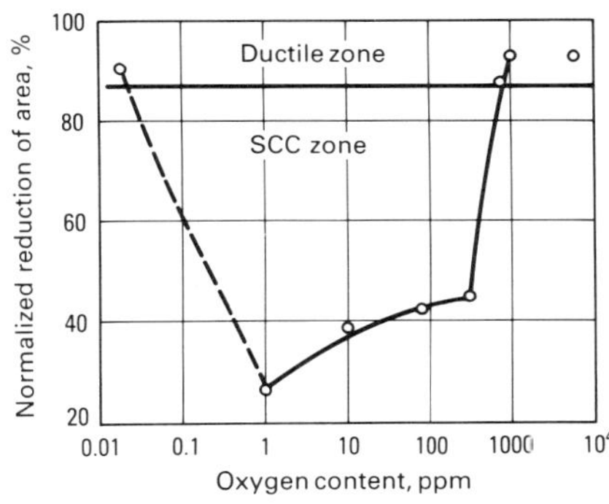

Fig. 85 Effect of oxygen content on apparent ductility observed in slow strain rate tests of low-alloy steel in liquid NH_3. Source: Ref 223

Field Experiences

Some reports suggest that water is not always an effective inhibitor, especially when water is added after SCC is detected. The significance of these reports is clouded by a lack of evidence that adequate control systems were used to ensure that a sufficient level of water was maintained. The research studies discussed previously do not address the effectiveness of water addition in slowing the growth of pre-existing cracks.

There is also a problem area with the vapor phase of NH_3 tanks. Water is considerably less volatile than NH_3, resulting in a lower water content in the vapor phase than in the liquid. If NH_3 vapor condenses on the wall of the vessel, the water content will probably be inadequate for inhibition, and SCC in the vapor phase is possible.

There are also reports of recracking of vessels that cracked, were repaired, and then were stress relieved. It is extremely difficult to repair vessels that have suffered SCC. There are many cracks in the equipment, including some of submicroscopic size. It is extremely difficult to prevent these cracks from propagating later. Stress relief of a vessel that has suffered SCC is also very likely to be unsuccessful. The very small cracks are contaminated to some degree. When this metal is subjected to the stress-relief thermal cycle, such

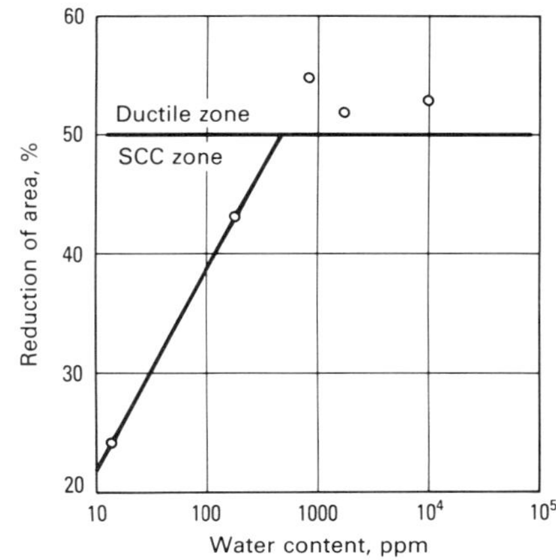

Fig. 86 Effect of water content on apparent ductility observed in slow strain rate tests of low-alloy steel in liquid NH_3. Oxygen content was 200 ppm, added as air. Source: Ref 223

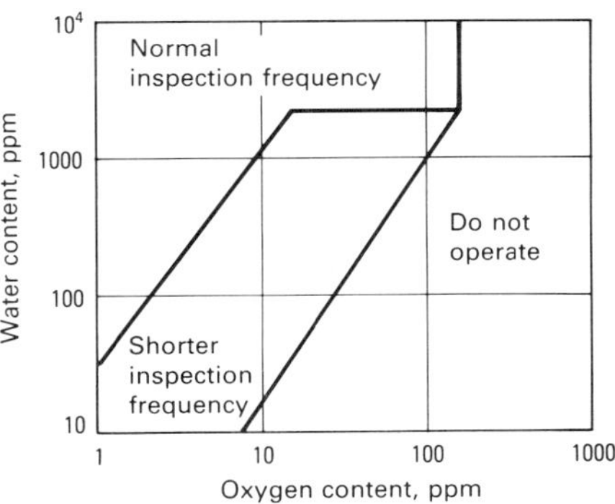

Fig. 87 Guidelines for changes in inspection frequency when oxygen or water content is outside preferred range. Source: Ref 224

phenomena as nitriding and carburizing may occur and promote further cracking.

Practical Operating Guidance

It is apparent that SCC of carbon and low-alloy steel NH_3 storage vessels can be a problem if proper procedures in design, fabrication, operation, inspection, and maintenance are neglected. If the degree of such neglect is large enough, catastrophic failure is possible. However, it is also apparent that application of proper procedures will ensure satisfactory long-term storage. Reference 224 discusses such practices. General recommendations for design, fabrication, operation, and inspection and maintenance practices are presented below.

Design and Fabrication. Normal design methods used for vessels to contain hazardous fluids should be followed, including all requirements of governing codes and agencies. Design should also be reviewed using fracture mechanics concepts to assess the risk of brittle fracture. Fabrication should be carefully inspected by a properly qualified engineer representing the end user.

A low-strength (specified tensile strength not exceeding 483 MPa, or 70 ksi) grade of carbon steel should be used. The hardness of welds should be specified to be 225 HB maximum, and the weld hardness should be checked in the field.

Postweld heat treatment (stress relief) at 595 °C (1100 °F) minimum should be specified for all pressure vessels. The lower temperature/longer time alternatives for such treatment allowed in some codes are less effective in reducing residual stress levels.

Operating practices should minimize air contamination. Water content should be maintained at 0.2% minimum if water is not objectionable to the user of the NH_3. Water (and, if feasible, oxygen) content should be checked by routine sampling and analysis.

Inspection and Maintenance. All tanks in NH_3 storage service should be carefully inspected on a routine basis. A new tank should be carefully inspected by the wet fluorescent magnetic-particle method after 1 to 2 years service. If no cracks are found, a somewhat longer inspection interval may be appropriate. If cracks are found, their severity should be assessed and appropriate actions taken. These actions may involve simply recording location and size of cracks, grinding out the cracked areas, or grinding out and rewelding the cracked areas. Figure 87 shows guidelines regarding modifications of inspection frequency as a function of oxygen and water content.

Any existing tank that has not been so inspected and has been in service longer than 2 years should be inspected at the first opportunity. Stress relief after repairs is not recommended. As noted previously, it is unlikely to be beneficial and may be harmful.

ACKNOWLEDGMENT

The author wishes to thank R.E. Tatnall, J.L. Cooney, W.I. Pollock, B.J. Moniz, and E.L. Perry for their assistance in the preparation of this article.

REFERENCES

1. W.D. Manly, Materials Industry: Present Trends/Future Outlook, in *Advances in Materials Technology for Process Industries' Needs*, National Association of Corrosion Engineers, 1985, p 10-13
2. M. Stern and A.L. Geary, *J. Electrochem. Soc.*, Vol 104, 1957, p 56
3. D.A. Rickard, Process Technology Industries—Where They are Today and Where They Will be Tomorrow, in *Advances in Materials Technology for Process Industries' Needs*, National Association of Corrosion Engineers, 1985, p 1-5
4. G.B. Elder, *Met. Prog.*, Vol 111 (No. 4), 1977, p 44
5. O.W. Siebert, *Mater. Perform.*, Vol 17 (No. 4), 1978, p 33
6. O.W. Siebert, *Mater. Perform.*, Vol 22 (No. 10), 1980, p 9
7. "Protection of Austenitic Stainless Steel from Polythionic Acid Stress Corrosion Cracking During Shutdown of Refinery Equipment," RP-01-70, National Association of Corrosion Engineers, 1970
8. C.P. Dillon *et al.*, *Guidelines For Control of Stress-Corrosion Cracking of Nickel Bearing Stainless Steels and Nickel-Bearing Alloys*, Manual No. 1, Materials Technology Institute, 1979, p 159
9. R.J. Twigg, *Can. Chem. Process*, Vol 65 (No. 2), 1981, p 27
10. J.A.A. Ketelaar, *Electrochem. Technol.*, Vol 5 (No. 3-4), 1967, p 143
11. T.F. Degnan, *Mater. Perform.*, Vol 22 (No. 10), 1983, p 45
12. S. Lederman, "Glass Lined Equipment—Typical and Atypical Glass Failures Encountered in the Field," Paper presented at Corrosion/82, National Association of Corrosion Engineers, 1982
13. "Cleaning Equipment for Oxygen Service," Pamphlet G-41, Compressed Gas Association, Inc.
14. A.W. Fynsk and R.T. Harris, Preoperational Cleaning of a Steam Generator System at a New Chemical Plant, in *Proceedings of the 41st International Water Congress*, 1980
15. L.C. Slabodrick, private communication
16. Patent assigned to Dow Chemical Company
17. Patent assigned to Pfizer Chemicals Div.
18. *Industrial Cleaning Manual*, TPC-8, National Association of Corrosion Engineers, 1982
19. "Cleaning and Descaling Stainless Steels," Publication SS103-582, 10M-GP, American Iron and Steel Institute
20. G.J. Falkenback, J.R. Fox, and R.P. Meister, "Investigation of Nondestructive Testing Techniques for Detecting Corrosion of Steel Under Insulation," Technical Report 4, Project 12, Phase I, Materials Technology Institute, July 1981
21. W.I. Pollock and J.M. Barnhart, Ed., Corrosion of Metals Under Insulation, in *STP 880*, American Society for Testing and Materials, 1984
22. A.W. Dana and W.B. DeLong, *Corrosion*, Vol 12, July 1956, p 309t
23. W.G. Ashbaugh, *Mater. Protect.*, May 1965
24. T.F. Laundrie and W.G. Ashbaugh, A Study of Corrosion of Steel Under a Variety of Thermal Insulation Materials, in *STP 880*, W.I. Pollock and J.M. Barnhart, Ed., American Society for Testing and Materials, 1980
25. O.A. Ullrich, "Investigation of an Approach to Detection of Corrosion Under Insulation," Technical Report 7, Project 12, Phase II, Materials Technology Institute, March 1982
26. W.G. Ashbaugh, External Stress Corrosion Cracking of Stainless Steel Under Thermal Insulation—20 Years Later, in *STP 880*, W.I. Pollock and J.M. Barnhart, Ed., American Society for Testing and Materials, 1980
27. C.P. Dillon and D.R. McIntyre, "Guidelines for Preventing Stress Corrosion Cracking in the Chemical Process Industries," Publication 15, Materials Technology Institute, March 1985
28. D.R. McIntyre, Factors Affecting the Stress Corrosion Cracking of Austenitic Stainless Steels Under Thermal Insulation, in *STP 880*, American Society for Testing and Materials, 1980
29. P. Lazar, Factors Affecting Corrosion of Carbon Steel Under Insulation, in *STP 880*, American Society for Testing and Materials, 1980
30. T. Sandberg, Experience With Corrosion Beneath Thermal Insulation in a Petrochemical Plant, in *STP 880*, American Society for Testing and Materials, 1980
31. V.C. Long and P.G. Crawley, Recent Experiences With Corrosion Beneath Thermal Insulation in a Chemical Plant, in *STP 880*, American Society for Testing and Materials, 1980
32. "Materials of Construction for Handling Sulfuric Acid," Technical Committee Report 5A151, National Association of Corrosion Engineers, 1985
33. O.W. Siebert, *Mater. Perform.*, Vol 20 (No. 2), Feb 1981, p 38
34. H.S. Tong, "Corrosion and Electrochemical Behavior of Fe-Cr-Ni Alloys in Concentrated Sulfuric Acid Solutions," Paper presented at Symposium on Progress in Electrochemical Corrosion Testing, American Society for Testing and Materials, 20-25 May 1979
35. MG. Fontana, *Ind. Eng. Chem.*, Vol 43, August 1951, p 65a
36. M. Tiivel and F. McGlynn, "Avoiding Problems in Sulfuric Acid Storage," Paper presented at AIChE Meeting, New Orleans, LA, American Institute of Chemical Engineers, April 1986
37. E. Williams and M.E. Komp, *Corrosion*, Vol 21 (No. 1), Jan 1965, p 9-14
38. H. Endo and S. Morioka, "Dissolution Phenomenon of Copper-Containing Steels in Aqueous Sulfuric Acid Solutions of Various Concentrations," Paper presented at the Japanese Metal Association 3rd Symposium, April 1938
39. D. Fyfe *et al.*, *Chem. Eng. Prog.*, March 1977
40. B.T. Ellison and W.R. Schmeal, *Elec. Soc.*, Vol 125, 1978, p 524
41. S.W. Dean, Jr. and G.D. Grab, "Corrosion of Carbon Steel by Concentrated Sulfuric Acid," Paper 147, presented at Corrosion/84, National Association of Corrosion Engineers, 1984
42. S.W. Dean and G.D. Grab, "Corrosion of Carbon Steel Tanks in Concentrated Sulfuric Acid Service," Paper 298, presented at Corrosion/85, National Association of Corrosion Engineers, 1985
43. M.G. Fontana and N.D. Greene, *Corrosion Engineering*, McGraw Hill, 1967
44. E. Maahn, *Br. Corros. J.*, Vol 1, Nov 1966
45. "The Corrosion Resistance of Nickel-Containing Alloys in Sulfuric Acid and Related Compounds," Corrosion Engineering Bulletin 1, The International Nickel Company, Inc., 1983
46. J.E. Strutt, "Corrosion Resistance of Stainless Steels in 93% and 98.5% Sulfuric Acid," Materials Technology Institute, Sept 1985
47. H. Abo, M. Ueda, and S. Noguchi, *Boshoku Gijutso*, Vol 23, 1974, p 341-346 (in Japanese)
48. P.F. Wieser *et al.*, *Mater. Protec.*, Vol 12 (No. 7), July 1973, p 34-38
49. M.A. Streicher, *Corrosion*, Vol 14 (No. 2), Feb 1958, p 59t-70t
50. G.C. Kiefer and W.G. Renshaw, *Corrosion*, Vol 8, Aug 1950, p 235
51. J.R. Auld, Effect of Heat Treatment and Welding on Corrosion Resistance of Austenitic Stainless Steels, in *Proceedings of the Second International Conference on Metallic Corrosion*, National Association of Corrosion Engineers, 1967
52. T.N. Anderson *et al.*, *Metall. Trans.*, Vol IIA, Aug 1980
53. D.R. McAlister *et al.*, U.S. patent 4,576,813, granted March 1986, Heat Recovery From Concentrated Sulfuric Acid
54. D.R. McAlister *et al.*, "A Major Breakthrough in Sulfuric Acid," Paper presented at AIChE 1986 Annual Meeting, New Orleans, LA, American Institute of Chemical Engineers, April 1986
55. M. Renner *et al.*, "Corrosion Resistance of Stainless Steels and Nickel Alloys in Concentrated Sulfuric Acid," Paper 189, presented at Corrosion/86, National Association of Corrosion Engineers, 1986
56. D.J. Chronister and T.C. Spence, Influence of Higher Silicon Levels on the Corrosion Resistance of Modified CF-Type Cast Stainless Steels, in *Proceedings of the NACE Corrosion/85 Symposium on Corrosion in Sulfuric Acid*, National Association of Corrosion Engineers, 1985, p 75
57. N. Sridhar, "Mechanisms of Corrosion in Concentrated Sulfuric Acid," Paper presented at Sulfur '85 International Conference, London, 10-13 Nov 1985
58. J.R. Crum and M.E. Adkins, Correlation of Alloy 625 Electrochemical Behavior with the Sulfuric Acid Isocorrosion Chart, in

Proceedings of the NACE Corrosion/85 Symposium on Corrosion in Sulfuric Acid, National Association of Corrosion Engineers, 1985, p 23

59. R.T. Webster and T.L. Yau, Zirconium in Sulfuric Acid Applications, in *Proceedings of the NACE Corrosion/85 Symposium on Corrosion in Sulfuric Acid*, National Association of Corrosion Engineers, 1985, p 69
60. M.A. Maguire and T.L. Yau, "Corrosion-Electrochemical Properties of Zirconium on Mineral Acids," Paper 265, presented at Corrosion/86, National Association of Corrosion Engineers, 1986
61. L.W. Gleekman, *Corrosion*, Vol 14, Sept 1958
62. P.J. Gegner and W.L. Wilson, *Corrosion*, Vol 15 (No. 7), 1959
63. J.P. Cotton, *Chem. Eng. Prog.*, Vol 66 (No. 10), Oct 1970
64. N.G. Feige, "The Industrial Applications of Titanium in the Chemical Industry," Paper presented at the Symposium on Titanium-Zirconium for the Chemical Process Industries, New Orleans, LA, Nov 1975
65. R.E. Tatnall and D.J. Kratzer, The Use of Fluoroplastics in Sulfuric Acid Service, in *Proceedings of the NACE Corrosion/85 Symposium on Corrosion in Sulfuric Acid*, National Association of Corrosion Engineers, 1985, p 85
66. J.R. Schley, *Chem. Eng.*, 18 Feb 1974
67. J.R. Schley, *Chem. Eng.*, 18 March 1974
68. E. Sheilds and W.J. Dessert, *Pollut. Eng.*, Dec 1981
69. J.A. Beavers, R.R. White, and W.E. Berry, "Corrosion Studies in Fuel Element Reprocessing Environments Containing Nitric Acid," ORNL/Sub 7327/13, Oak Ridge National Laboratory, April 1982
70. C. Chakrabarty, M.M. Singh, and C.U. Agarwal, *J. Electrochem. Soc.*, Vol 31, 1982, p 165-169
71. G.L. Delaney and T.F. Lemke, *Corros. Australasia*, Vol 6, 1981, p 4-6
72. C. Hahin. R.M. Stoss, B.H. Nelson, and P.J. Reucroft, *Corrosion*, Vol 32 (No. 6), June 1976
73. M.J. Johnson, J.R. Kearns, and H.E. Deverall, "The Corrosion of the New Ferritic Stainless Steels in Nitric Acid," Paper 144, presented at Corrosion/84, National Association of Corrosion Engineers, 2-6 April 1984
74. M. Kobayashi, M. Aoki, M. Ohkubo, and M. Miki, "Development of Nitric Acid Resistant Stainless Steels," Paper 145, presented at Corrosion/84, National Association of Corrosion Engineers, 2-6 April 1984
75. A. Natarajan, T.S. Lakshmanon, K.V. Nagarajan, and B.K. Sarkar, *Trans. Indian Inst. Met.*, Vol 35 (No. 1), Feb 1983
76. S. Sadigh, "Inspection Guidelines for Aluminum Tank Cars in 98% HNO_3 Service," Hercules Engineering Document, Hercules, Inc., 13 Aug 1980
77. S. Sadigh, "Prevention of Corrosion by Hot Concentrated HNO_3 in Condensers and Dehydration Columns," Proprietary Hercules Report, Hercules, Inc., 30 Dec 1981
78. J.E. Slater and R.W. Staehle, "A Study of the Mechanism of Stress Corrosion Cracking in the Iron-Nickel-Chromium Alloy System in Chloride Environments," Contract No. AT(11-1)2069, United States Atomic Energy Commission, 30 Sept 1970
79. M.W. Wilding and B.E. Paige, "Idaho National Engineering Laboratory Survey on Corrosion of Metals and Alloys in Solutions Containing Nitric Acid," Report N77-32302, National Technical Information Service, Dec 1976
80. "Corrosion Resistance of Titanium," Timet Corporation
81. 26th Biennial Materials of Construction Report (Handling Nitric Acid), *Chem. Eng.*, 11 Nov 1974, p 129
82. R.S. Stewart, *Met. Prog.*, Vol 52, Dec 1947
83. K.F. Krysiak, "Weldability and Corrosion Performance of a 4% Silicon Stainless Steel for 99% HNO_3 Service," Proprietary Hercules Report, Hercules, Inc., 29 May 1984
84. "Product Information on Stainless Steel for the Nitric Acid Industry," TOK31/08.82, Vereinigte Edel Stahlwerke
85. D.D.N. Singh, R.S. Chaudhary, and C.V. Agarwal, *J. Electrochem. Soc.*, Sept 1982, p 1869
86. J.A. Beavers, J.C. Griess, and W.K. Boyd, *Corrosion*, Vol 36 (No. 5), May 1981
87. T.L. Yau, *Corrosion*, Vol 5, May 1983
88. J.P. Bennett, "Corrosion Resistance of Selected Ceramic Materials to Nitric Acid," Report 8851, Sup. Docs. 128.23.8851, United States Department of the Interior Bureau of Mines
89. C.P. Dillon, *Mater. Protect.*, Vol 21 (No. 9), 1965, p 4
90. A.E. Tsinman *et al.*, *Elektrokhimiya*, Vol II (No. 1), 1975, p 127
91. "Corrosion Resistance of Nickel-Containing Alloys in Organic Acids and Related Compounds," The International Nickel Company, Inc., 1979, p 5
92. G.B. Elder, Corrosion by Organic Acid, in *Process Industries Corrosion*, National Association of Corrosion Engineers, 1975, p 247
93. NACE Technical Committee, *Mater. Perform.*, Vol 13 (No. 7), 1974, p 13
94. B. Larsson and U. Lundell, "Special Stainless Steels for the Process Industries," A.B. Sandvik Steel, 1983
95. "Corrosion by Acetic Acid," Technical Committee Report 5A157, National Association of Corrosion Engineers
96. A.B. McKee and W.W. Binger, *Corrosion*, Vol 13, 1957, p 786t
97. N.D. Groves *et al.*, *Corrosion*, Vol 17 (No. 4), 1961, p 173t
98. A.E. Tsinman *et al.*, *Zashch. Met.*, Vol 8, 1972, p 567
99. R.F. Miller *et al.*, *Corrosion*, Vol 10 (No. 1), 1954, p 7
100. "Storage and Handling Hydrochloric Acid," Technical Bulletin, E.I. Du Pont de Nemours & Company, 1972
101. "Manufacturing Chemists Association Chemical Safety Data," Sheet SD-39, Manufacturing Chemists Association, 1970
102. *Corrosion Data Survey—Part I, Metals*, National Association of Corrosion Engineers, 1974
103. L.W. Gleekman, *Chemical Process Industries Symposium*, National Association of Corrosion Engineers, 1975, p 225
104. "Resistance to Corrosion," 4th ed., Inco Alloys International, 1985, p 21-24
105. "Corrosion Resistance of Titanium," Titanium Metals Corporation of America
106. "Engineering Guide to Du Pont Elastomers," E.I. Du Pont de Nemours & Company
107. E.I. Du Pont de Nemours & Company, private communication
108. "Properties of a High Purity 29% Cr-4% Mo-2% Ni Ferritic Alloy for Aggressive Environments," Allegheny Ludlum Steel Corporation, 1982, p 9
109. "Zircadyne Corrosion Properties," Teledyne Wah Chang Albany, 1981
110. C.R. Bishop and M. Stern, *Corrosion*, Vol 17 (No. 8), 1961, p 379t
111. "Uniroyal Rubber Linings," Uniroyal, Inc., 1968
112. "B.F. Goodrich Tank Linings," B.F. Goodrich, Inc., 1974
113. NACE Technical Committee, *Corrosion*, Vol 17 (No. 9), 1961, p 453t
114. "The General Chemical Resistance of Various Elastomers," The Los Angeles Rubber Group Inc., 1970
115. "Chemical Resistance Guide for Systems Using Dow Plastic-Lined Piping Products," The Dow Chemical Company, 1978
116. R.E. Tatnall and D.J. Kratzer, Paper 307, presented at Corrosion/85, National Association of Corrosion Engineers, 1985
117. J.E. Niesse, *Mater. Perform.*, Vol 21 (No. 1), 1982, p 25
118. T.G. Priest and O.W. Seibert, *Mater. Perform.*, Vol 20 (No. 10), 1981, p 38
119. W.W. McClellan, T.F. Anderson, and R.F. Stavinoha, 30th Anniversary Technical Conference, Section 6A, Reinforced Plastics/Composites Institute, The Society of the Plastics Industry, Inc., 1975
120. "Chemical Resistance Red Thread, Green Thread, Polythread, Chemline Piping Systems for the Chemical Process Industries," Product Bulletin 9002, A.O. Smith-Inland Inc.
121. 21st Biennial Report—Materials of Construction—Section 1, *Chem. Eng.*, Vol 71 (No. 24), 9 Nov 1964, p 170
122. "Graphitar, A Challenge to Corrosion," Carbone-Lorraine Industries Corporation
123. "Pfaudler 5000 Glassteel Technical Data," The Pfaudler Company Inc., 1979
124. "Wood Tanks for Corrosive Applications," Technical Bulletin 758, National Wood Tank Institute, 1975
125. A.P. Pfeil, *Des. Eng.*, June 1961, p 114
126. M.H. Brown, W.B. DeLong, and J.R. Auld, *Ind. Eng. Chem.*, Vol 39 (No. 7), 1947, p 839-844
127. C.M. Schillmoller, *Chem. Eng.*, Vol 87 (No. 5), 10 March 1980, p 161
128. J.P. Carter, B.S. Covino, Jr., T.J. Driscoll, W.D. Riley, and M. Rosen, *Corrosion*, Vol 40 (No. 5), 1984, p 205
129. "Hydrofluoric Acid," Technical Bulletin, E.I. Du Pont de Nemours & Company, Inc., 1978
130. "Manufacturing Chemists Association Chemical Safety Data Sheet," SD-25, Manufacturing Chemists Association
131. "Materials for Receiving, Handling, and Storing Hydrofluoric Acid," Technical Committee Report 5A171, National Association of Corrosion Engineers, 1983
132. M.E. Holmberg and F.A. Prange, *Ind. Eng. Chem.*, Vol 37, 1945, p 1030
133. "Hydrofluoric Acid Alkylation," Phillips Petroleum Company, 1946

134. H.A. Pray, F.W. Fink, B.E. Friedl, and W.J. Brain, Report 268, Battelle Memorial Institute, 1953
135. E.I Du Pont de Nemours & Company, private communication
136. G.C. Whitaker, *Corrosion*, Vol 6 (No. 9), 1950, p 283
137. G. Trabanelli, A. Frifnani, G. Brunoro, C. Monticelli, and F. Zucchi, *Mater. Perform.*, Vol 24 (No. 6), 1985, p 33
138. "Sulfide Stress Cracking Resistant Metallic Materials for Oil Field Equipment," MR-01-75, National Association of Corrosion Engineers, 1984
139. F.A. Prange, *Corrosion*, Vol 8 (No. 10), 1952, p 355
140. H.R. Copson and C.F. Cheng, *Corrosion*, Vol 12 (No. 12), 1956, p 647
141. "Methods and Controls to Prevent In-Service Cracking of Carbon Steel (P-1) Welds in Corrosive Petroleum Refining Environments," RP-04-72, National Association of Corrosion Engineers
142. R.L. Schuyler III, *Mater. Perform.*, Vol 18 (No. 8), 1979, p 9
143. M. Takemoto, T. Shonohara, M. Shirai, and T. Shinogaya, *Mater. Perform.*, Vol 24 (No. 6), 1985, p 26
144. L. Graf and W. Wittich, *Werkst. Korros.*, Vol 17, 1966, p 385
145. "Huntington Alloys Creating Change in the Chemical Processing and Related Industries," Inco Alloys International, 1979
146. B.E. Paige, *Mater. Perform.*, Vol 17 (No. 5), 1978, p 15
147. "Technical Notes," Climax Molybdenum Company, 1959
148. H.H. Uhlig, *Corrosion Handbook*, John Wiley & Sons, 1948, p 115, 303, 315, 769
149. F.L. LaQue and H.R. Copson, *Corrosion Resistance of Metals and Alloys*, 2nd ed., A.C.S. Monograph 158, Reinhold, 1963
150. T.F. Degnan, Materials for Handling Hydrofluoric, Nitric, and Sulfuric Acids, in *Process Industries Corrosion*, National Association of Corrosion Engineers, 1975, p 233
151. W.R. Myers and W. B. DeLong, *Chem. Eng. Prog.*, Vol 44, 1948, p 359
152. "Corrosion Engineering Bulletin CEB-5," The International Nickel Company Inc., 1975
153. "Summary of Replies to Questionnaire on Handling of Chlorine Mixtures," Technical Committee Report 5A259, National Association of Corrosion Engineers
154. Bibliography of Corrosion by Chlorine, TPC-4, National Association of Corrosion Engineers, 1976
155. G. Heinemann, F.G. Garrison, and P.A. Haber, *Ind. Eng. Chem.*, Vol 38 (No. 5), 1946, p 497
156. M.H. Brown, W.B. DeLong, and J.R. Auld, *Ind. Eng. Chem.*, Vol 39 (No. 7), 1947, p 839
157. W.Z. Friend and B.B. Knapp, *Trans. AIChE*, Section A, 25 Feb 1943, p 731
158. K.L. Tseitlin and J.A. Strunkin, *J. Appl. Chem. (USSR)*, Vol 31 (No. 12), 1958, p 1832
159. K.L. Tseitlin, *J. Appl. Chem. (USSR)*, Vol 28 (No. 5), 1955, p 467
160. W.A. Luce and R.B. Seymour, *Chem. Eng.*, Vol 57 (No. 10), 1950, p 217
161. S.D. Kirkpatrick and J.R. Callahan, *Chem. Eng.*, Vol 57 (No. 11), 1950, p 107
162. P.L. Daniel and R.A. Rapp, *Hologen Corrosion of Metals*, Vol 5, *Advances in Corrosion Science and Technology*, M.G. Fontana and R.W. Staehle, Ed., Plenum Press, 1976, p 55
163. G.N. Kirby, *Chem. Eng.*, Vol 87 (No. 23), 1980, p 86
164. N.C. Horowitz, *Chem. Eng.*, Vol 88 (No. 7), 1981, p 105
165. E. Rabald, *Corrosion Guide*, American Elsevier, 1968
166. P. Juniere and M. Sigwalt, *Aluminum—Its Application in the Chemical and Food Industries*, Crosby Lockwood & Son Ltd., 1964
167. K.L. Tseitlin and J.A. Strunkin, *J. Appl. Chem. (USSR)*, Vol 29 (No. 11), 1956, p 1793
168. K.L. Tseitlin, *J. Appl. Chem. (USSR)*, Vol 27 (No. 9), 1954, p 889
169. K.L. Tseitlin, *J. Appl. Chem. (USSR)*, Vol 29 (No. 2), 1956, p 253
170. E.L. Liening, Report ME-4242, The Dow Chemical Company, April 1980
171. B.J. Downey, J.C. Bernel, and P.J. Zimmer, *Corrosion*, Vol 25 (No. 12), 1969, p 502
172. R.S. Sheppard, D.R. Hise, P.J. Gegner, and W.L. Wilson, *Corrosion*, Vol 18 (No. 6), 1962, p 211t
173. "Huntington Alloys—Resistance to Corrosion," Publication 25M(11-70)S-37, Huntington Alloys, Inc., 1970
174. *Lead for Corrosion Resistant Applications*, Lead Industries Association, 1974
175. "Zirconium and Hafnium," Publication 10M-101570, Amax
176. L.B. Golden, I.R. Lane, Jr., and W.L. Acherman, *Ind. Eng. Chem.*, Vol 44 (No. 8), 1952, p 1930
177. "Zircadyne Corrosion Data," TWCA-8101Zr19, Teledyne Wah Chang, 1981
178. E. Rabald, *Werkst. Korros.*, Vol 12 (No. 11), 1961, p 695
179. E.E. Millaway and M.H. Kleinman, *Corrosion*, Vol 23 (No. 4), 1967, p 88
180. G.E. Hutchinson and P.H. Permar, *Corrosion*, Vol 5 (No. 10), 1949, p 319
181. "Columbium," Publication 313-PD1, KBI Division of Cabot Corporation, 1985
182. "Tantalum," Publication 312-PD1, KBI Division of Cabot Corporation, 1985
183. "Corrosion Resistant Materials," Bulletin 104 PD1, Kawecki Berylco Industries, Inc., 1977
184. M. Schussler, *Corrosion Data Survey on Tantalum*, Fansteel Inc., 1972
185. S. Baranow, G.Y. Lai, and M.F. Rothman, "Materials Performance in High Temperature, Halogen-Bearing Environments," Paper 16, presented at Corrosion/84, National Association of Corrosion Engineers, 1984
186. M.J. McNallan, J.M. Oh, and W.W. Liang, "High-Temperature Corrosion of Metals in Argon-Oxygen-Chlorine Mixtures," DOE/ER-12093-T1, Gas Research Institute, 1982
187. M.J. McNallan, M.H. Rhee, S. Thongtem, and T. Hansler, "The Effect of Temperature on the High-Temperature Corrosion of Superalloys in Argon-20% Oxygen-0.25% Chlorine," Paper 11, presented at Corrosion/85, National Association of Corrosion Engineers, 1985
188. P. Elliot, A.A. Ansari, R. Prescott, and M.F. Rothman, "Behavior of Selected Commercial-Base Alloys During High Temperature Oxychlorination," Paper 13, presented at Corrosion/85, National Association of Corrosion Engineers, 1985
189. W.C. Fort III and W.R. Dicks, *Mater. Perform.*, Vol 25 (No. 3), 1986, p 9
190. W.H. Shearon, Jr., F. Chrencik, and C.L. Dickinson, *Ind. Eng. Chem.*, Vol 40 (No. 11), 1948, p 2002
191. E.E. Millaway and L.C. Covington, "Resistance of Titanium to Gaseous and Liquid Fluorine," Titanium Metals Corporation of America, 1959
192. V. Pershke and L. Pecherkin, *Khimistroi*, Vol 6, 1934, p 140
193. The American Brass Company, *Chem. Eng.*, Vol 58 (No. 1), 1951, p 108
194. "Corrosion Resistance of Tantalum and Niobium Metals," Bulletin 3000, NRC Inc.
195. W.E. Bratt, L.R. Scribner, and C.G. Chisholm, *Chem. Eng.*, Vol 54 (No. 2), 1947, p 219
196. "Corrosion Resistance of Nickel and Nickel-containing Alloys in Caustic Soda and other Alkalies," Corrosion Engineering Bulletin CEB-2, The International Nickel Company, Inc., 1973
197. Ampco Metal Div., Ampco-Pittsburgh, unpublished research, 1951
198. P.J. Gegner, "Corrosion Resistance of Materials in Alkalies and Hypochlorites," Paper 27, Process Industries Corrosion Short Course, National Association of Corrosion Engineers, 1974
199. F.L. LaQue and H.R. Copson, *Corrosion Resistance of Metals and Alloys*, Reinhold, 1963
200. W. Whitman, R. Russell, and V. Alteri, *Ind. Eng. Chem.*, Vol 16, 1924, p 665
201. E. Heyn and D. Bauer, *Mitt. Kgl. Material Prufungsamt Prussia*, Vol 26, 1908
202. H.W. Schmidt, P.J. Gegner, G. Heinemann, C.F. Pogacar, and E.H. Wyche, *Corrosion*, Vol 7, 1951, p 295
203. A.A. Berk and W.F. Waldeck, *Chem. Eng.*, Vol 57 (No. 6), 1950, p 235
204. C.W. Funk and G.B. Barton, "Caustic Stress Corrosion Cracking," Paper 54, presented at Corrosion/77, National Association of Corrosion Engineers, 1977
205. E.C. Hoxie, "Some Considerations in the Selection of Stainless Steel for Pressure Vessels and Piping," The International Nickel Company, Inc., 1975
206. M. Kowaka and T. Kudo, *Sumitoma Search*, No. 18, Nov 1977
207. I.A. Franson, Internal Laboratory Reports, Allegheny Ludlum Steel Corporation, 1982
208. P.J. Gegner, *Corrosion*, Vol 12, June 1956
209. B.M. Barkel, "Accelerated Corrosion of Nickel Tubes in Caustic Evaporation Service," Paper 13, presented at National Association of Corrosion Engineers, 1979
210. B.A. Maloney, Internal Laboratory Reports, PPG Industries, Inc., 1985
211. J.R. Crum and W.G. Lipscomb, "Correlation Between Laboratory Tests and Field Experience for Nickel 200 and 26-1 Stainless Steel in Caustic Service," Paper 23, presented at Corrosion/83, National Association of Corrosion Engineers, 1983
212. M. Yasuda, F. Takeya, and F. Hine, *Corrosion*, Vol 39 (No. 10), Oct 1983
213. S.R. Seagle, *Pulp Paper*, Vol 53 (No. 10), Sept 1979

214. *Cabot Dig.*, Vol 36 (No. 5), Sept 1985
215. "Resistance of Nickel and High Nickel Alloys to Corrosion by Hydrochloric Acid, Hydrogen Chloride and Chlorine," Corrosion Engineering Bulletin CEB-3, The International Nickel Company, Inc., 1972
216. J.C. Miller, D.M. Longenecker, and G.G. Greth, "Factors Affecting Performance of Reinforced Plastics in Sodium Hypochlorite Environments," Paper presented at the 25th Annual Technical Conference, The Society of the Plastic Industry, 1970
217. A.W. Loginow and E.H. Phelps, *Corrosion*, Vol 18 (No. 8), 1962, p 299-309
218. D.C. Deegan and B.E. Wilde, *Corrosion*, Vol 29 (No. 8), 1973, p 310-315
219. D.C. Deegan, B.E. Wilde, and R.W. Staehle, *Corrosion*, Vol 32 (No. 4), 1976, p 139-142
220. T. Kawamoto, T. Kenjo, and Y. Imasaka, *IHI Eng. Rev.*, Vol 10 (No. 4), 1977, p 17-25
221. F.F. Lyle and R.T. Hill, "SCC Susceptibility of High-Strength Steels in Liquid Ammonia at Low Temperatures," Paper 225, presented at Corrosion/78, National Association of Corrosion Engineers, 1978
222. K. Farrow, J. Hutchings, and G. Sanderson, *Br. Corros. J.*, Vol 16 (No. 1), 1981, p 11-19
223. B.E. Wilde, *Corrosion*, Vol 37 (No. 3), 1981, p 131-141
224. J.M.B. Gotch *et al.*, "Code of Practice for the Storage of Anhydrous Ammonia Under Pressure in the United Kingdom," Chemical Industries Association Ltd., 1980

Corrosion in the Pulp and Paper Industry

Chairman: Andrew Garner, Pulp and Paper Research Institute of Canada

IN THE PAST DECADE, the understanding of corrosion phenomena in the pulp and paper industry has progressed from one based largely on practical experience to the point at which an appreciation of the corrosion mechanism is considered an essential first step to solving a problem. Recent cracking problems with kraft continuous digesters provide a good case in point. In the early 1960s, when the continuous digester process was first adopted, the industry had limited familiarity with caustic cracking. The occasional report of a cracked batch digester was considered to be an isolated incident, and its significance to the new continuous technology was not appreciated. This is evident from the fact that beginning in the mid-1960s, many continuous digesters were built with nonstress-relieved upper cooking zones. It is now known that these vessels have caustic levels and temperatures so close to the caustic cracking range that full stress relief is essential. Recognition of this was possible only because considerable investigation and research provided a mechanistic understanding of the problem and proposed solutions.

A parallel progression of events can be traced for paper machine corrosion phenomena. Process changes over the past three decades, including system closure and new brightening practices, have created a series of problems resulting in the failure of existing materials of construction. Components with high failure rates include bronze fourdrinier wires, CA-15 cast stainless steel suction press rolls, cast iron vacuum pump impellers, and type 304 stainless steel process piping. For most of these components, there has been a sustained effort to understand the failure mechanism, and in some of the later cases, a detailed understanding has been achieved.

For example it is now understood that pitting of type 304 stainless steel occurs when the pit site is sufficiently acidified by anions such as chloride or even sulfate; that pitting can be accelerated by thiosulfate, which is thought to deliver sulfur to the pit; and that pitting can be inhibited by bisulfite, which buffers the pit pH. The relative proportions of these anions determine whether or not pitting occurs. Therefore, the corrosiveness of white water can be predicted and controlled by process changes. The current understanding of these and many other corrosion problems in the pulp and paper industry are covered in this article.

Paper Machine Corrosion

Donald A. Wensley
MacMillan Bloedel Research

Corrosion problems in paper machines are generally most severe in the wet end and in ancillary equipment handling white water (Fig. 1). Metal surfaces in the wet end are exposed to the white water environment by immersion, splashing, vapor mist, and the formation of crevices beneath pulp pads or other deposits. White water environments cause rapid corrosion of bare carbon steel, appreciable corrosion of copper alloys and cast irons, and can even cause localized corrosion of stainless steels. Most wetted paper machine surfaces are constructed from austenitic stainless steels; types 304L or 316L are the predominant choices. The critical components in a paper machine wet end are discussed below.

Paper Machine Components

Stock Piping. Paper machine stock piping is typically thin-gage (schedule 10) type 304L or 316L austenitic stainless steel. Corrosion is generally not a problem in stock lines if the correct stainless grades and weld filler metals have been chosen for the white water environment and if surfaces are kept clean. Surfaces that have become roughened or have weld projections that have not been removed provide sites for stock hang-up. Pitting or microbiological attack is likely to occur under deposits. Fatigue cracking problems in stock piping are associated with poor welding practice.

Headboxes. The headbox (Fig. 2) has the critical function of transferring flow formerly contained in a round pipe to a flat flow of uniform consistency across the width of the paper machine. Clean, smooth surfaces inside the headbox are essential. Aside from the corrosion aspect, the release of pulp pads resulting from stock hang-up may cause product quality problems or even sheet breakage. Pickling, passivation, and buffing with abrasives are commonly used to provide a smooth surface finish. Electropolishing improves the finish for surfaces of 0.4 μm (15 μin.) and above.

It is especially important to implement proper cleaning techniques during fabrication. Stainless steels are naturally passive in most papermaking environments; however, the presence of embedded iron, iron oxides, heat tint, weld spatter, slag, and other surface contaminants impairs the corrosion resistance of the underlying metal. The misuse of carbon steel wire brushes for weld cleanup may cause serious pitting problems because iron particles will become embedded in the stainless steel surface. Iron contamination is best removed by passivating with a nitric acid solution, which dissolves the iron and leaves the stainless steel intact. Pickling solutions, which typically contain fluoride in addition to nitric acid, are much more aggressive. Pickling solutions also attack the stainless steel in addition to removing iron and weld heat tint; therefore, pickling should be done only as a final cleanup step on

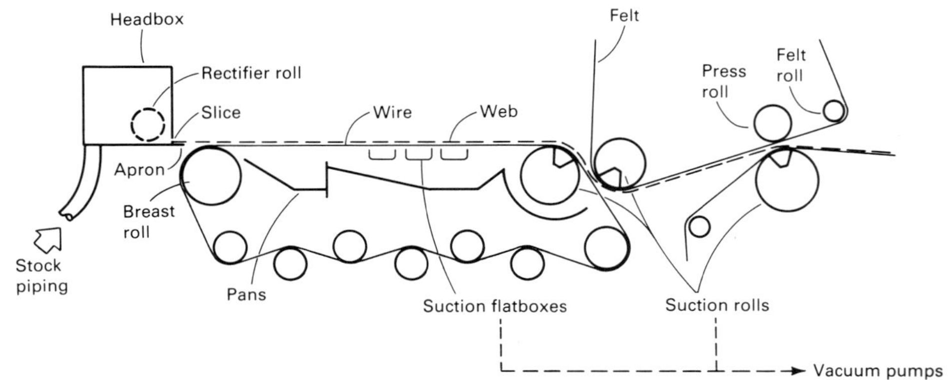

Fig. 1 Schematic diagram of the wet end of a fourdrinier paper machine

Fig. 2 View inside a paper machine headbox showing rectifier rolls and overhead shower nozzles. Material of construction is type 304L stainless steel.

Fig. 3 Severe pitting corrosion of a type 316L stainless steel slice lip

equipment for which excellent surface finish is not required.

Paper machine headboxes are often constructed from highly corrosion-resistant alloys as additional insurance against the possibility of corrosion roughening of internal surfaces. The existence of crevices inside the headbox, the presence of an air/liquid interface, the potential for internal mechanical damage (from rectifier rolls), and even the possibility of stray current damage make the headbox environment more severe than that inside stock piping. A headbox can be constructed from type 317L stainless steel if type 316L stainless steel is used for other components. Overmatched fillers (fillers with higher alloying contents, particularly molybdenum contents, than the base metal) have been used to provide additional corrosion resistance for weldments. Some headboxes have even been constructed of UNS N10276, a nickel-base alloy that is usually used in much more severe corrosion applications. The reasoning behind this material selection is that immunity to corrosion means freedom from maintenance.

Headbox apron and slice lips are very prone to pitting corrosion attack (Fig. 3). Pits initiate on surfaces outside of the liquid flow and then tend to grow in size, eventually reaching the edge of the lip. Corrosion extending to lip edges has a detrimental effect on sheet formation. The use of specialty stainless steels, such as UNS S31254 (254SMO), which contains 6% Mo, has been successful in combating pitting of apron and slice lips.

Paper Machine Wire. The wire is so named because a phosphor bronze wire mesh was originally used as the moving screen onto which the dilute fiber suspension was distributed from the headbox. Bronze wire life was limited by corrosion attack. Corrosion inhibitor additions were found to be beneficial. Wires coated with tin, nickel, or plastic also gave superior service. Synthetic plastic mesh fabrics are now commonly used for paper machine wires. Although this eliminates the wire corrosion problem, the use of plastic wires appears to have contributed to the increased corrosion of bronze couch rolls. This is attributed to the removal of an effective source of cathodic protection (that is, the metal wires anodically sacrificed themselves and protected the rolls).

Wet End Structures. White water draining through the wire is caught in stainless steel collecting trays and pans. Corrosion is likely to occur wherever pulp deposits can accumulate on metal surfaces. Structural members are generally protected with stainless steel cladding. Where this cladding is incomplete or has been removed, white water can gain access to the underlying structural steel. The white water pits or chests beneath the wet end of a paper machine can be of concrete or tile-lined construction.

Vacuum Pumps. Liquid ring vacuum pumps typically have ductile cast iron rotors (Fig. 4) and gray cast iron casings. From a corrosion standpoint, it is desirable to use fresh seal water. Paper machine white water is considered unsuitable for seal water use because of its high temperature and low pH. Even freshwater may be corrosive if the flow rate is so low that the temperature increases inside the pump or if the freshwater becomes contaminated by excessive white water carryover. A discharge seal water temperature of 50 °C (122 °F) and/or a pH less than 4.5 can cause accelerated cast iron corrosion. High discharge seal water conductivity (in excess of 200 μmho) may indicate excessive white water carryover. Accelerated impeller corrosion can occur at rotor tip speeds above 27 m/s (80 ft/s). The presence of sand or grit in the water will also shorten liquid ring vacuum pump life because of erosion.

As cast irons corrode in seal water or white water a residue of graphite is left behind (Fig. 5). Pearlitic ductile iron forms a more protective graphite surface layer than ferritic ductile iron because the three-dimensional carbonaceous residue left behind from pearlite dissolution acts as a binder for the graphite (Fig. 6). Rapid liquid ring vacuum pump corrosion can be expected if environmental, velocity, or erosion conditions are such that the protective graphite film is removed. In cases of severe corrosion, liquid ring vacuum pumps have been built with stainless-lined bodies and stainless steel rotors.

Suction Rolls. The corrosion-related failures of suction roll shells represent the most serious materials and corrosion problem in modern paper machines. A variety of alloys are used for suction roll shells, including bronzes and various grades of stainless steel (martensitic, austenitic, duplex, precipitation hardening). Failures are due to corrosion thinning, pitting, corrosion fatigue, and stress-corrosion cracking (SCC). More detailed information can be found in the section "Suction Roll Corrosion" in this article.

Roll Journals. Paper machine roll journals are subject to corrosion fatigue cracking (Fig. 7). Cracking develops at changes in section size—most often where journals meet roll heads. Depending on the location of the roll, the environment causing corrosion fatigue may be the humid paper machine atmosphere or may involve splashing of white water and/or the formation of damp pads of pulp on the journals. Felt roll journals may also be exposed to felt cleaning chemicals, some of which are acidic. Journals are constructed from a variety of materials, including medium- or high-carbon steels, low-alloy steels such as AISI 4140 and AISI 4340, cast irons, and various grades of stainless steel. No material is immune to corrosion fatigue; indeed, there is no fatigue limit below which indefinitely long operation could be guaranteed.

Improved corrosion fatigue performance can be realized by protecting journals from the environment, by redesigning to increase their diameter and to remove step changes in section size, by polishing to remove surface flaws, and by selecting a material with superior corrosion fatigue resistance. Wet end roll heads and journals often have stainless steel shields to protect them from exposure to white water. Coatings, including paint and thermally applied metal spray, have also been used to protect journals. The disadvantage of such physical barriers is that they may hinder inspection of surfaces for cracking. Du-

Fig. 4 Corroded ductile iron rotor of a liquid ring vacuum pump

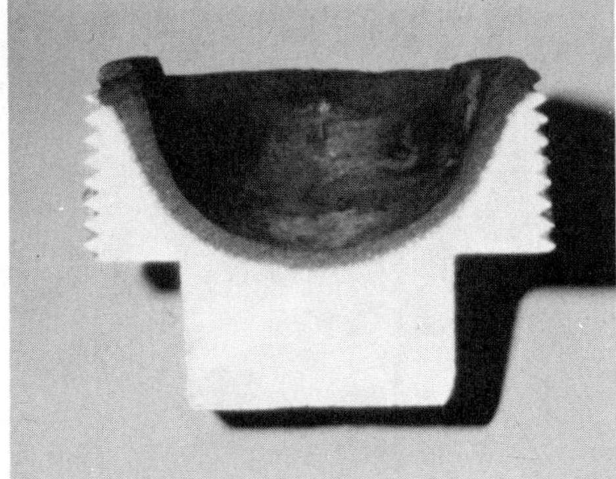

Fig. 5 Cross section of a corroded gray cast iron pipe plug removed from white water service clearly showing the graphite residue left behind as the iron matrix dissolved

plex stainless steels and precipitation-hardening steels (such as 17-4PH, condition H1100) appear to offer much better corrosion fatigue resistance than the more conventional journal materials. Integral roll heads and journals made from ductile cast iron also appear to offer superior service if there is a generous radius between the head and journal.

White Water

The wet end of a paper machine produces a uniform paper web from an aqueous suspension of pulp fibers. Stock comes to the paper machine at about 3% consistency and is further diluted to about 0.5 to 1.0% at the fan pump before the headbox. Most of the white water is removed by drainage through the paper machine wire. The white waters from different sources along the wet end are kept separate and the rich white waters (containing more fibers) are recycled for incoming stock dilution. Lean white waters can be used elsewhere on the machine as shower water or as makeup water.

Closure of papermaking systems involves the recycle of white water, either directly or after treatment. The degree of closure can be expressed as the amount of water consumed per air-dried ton (adt) of paper produced. Closed mills may discharge less than 1000 gal/adt per day. An open mill, on the other hand, may discharge over 10 000 gal/adt. Closure results in an increase in the concentration of dissolved inorganic and organic solids, a decrease in pH, and an increase in temperature.

Closure is considered detrimental from a corrosion standpoint; however, there is no clear relationship between increasing closure and paper machine corrosion. Although accelerated corrosion may be expected for construction materials that undergo general corrosion attack (carbon steels, cast irons, copper alloys), no effect on stainless steel corrosion may be noticed until a critical concentration of aggressive ions and/or temperature is attained. Increasing closure may only serve to decrease the margin of safety for corrosion-free service with a given stainless steel. Beyond a certain degree of closure, spontaneous pitting of the less corrosion-resistant stainless steels (types 304 and 321) may occur. Greater degrees of closure can be tolerated in those mills in which white waters were not particularly corrosive to begin with because the corrosivity of a given white water ultimately depends on its composition.

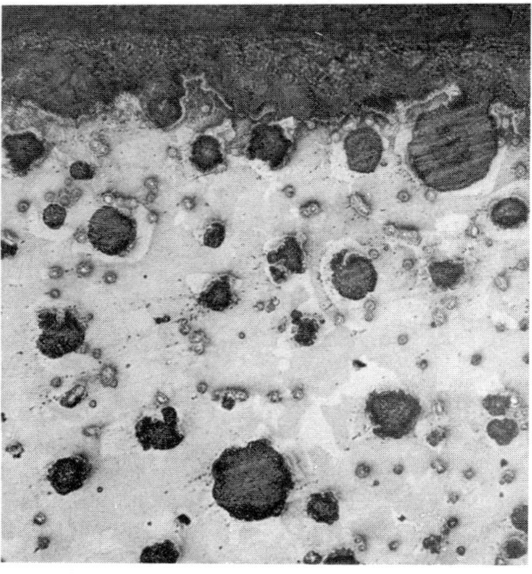

Fig. 6 Micrograph of a cross section of a pearlitic ductile iron coupon exposed for 1872 h in vacuum pump discharge seal water

Composition. Paper machine white waters can vary widely in composition from mill to mill, within a given mill, or even from day to day on a given paper machine. The corrosivity of white water depends primarily on the pH, temperature, and the concentrations of aggressive inorganic anions, such as chloride and thiosulfate. The concentrations of dissolved inorganic compounds found in white waters will depend on whether the wood was seaborne, the type of pulping process used, whether bleaching or brightening was carried out, and the types of chemicals used in the wet end of the machine. The chemicals used include such additives as fillers, sizes, retention aids, defoamers, slimicides, and dyes. Mechanical pulps are often brightened with sodium hydrosulfite, a chemical that decomposes rapidly in the pulp stream to form other sulfur-containing ions that persist in the white water. An average white water composition—typical of a West Coast newsprint mill using a furnish of semibleached kraft and hydrosulfite-brightened thermomechanical pulp and groundwood—is:

Temperature	118 °F (48 °C)
pH (initial)	4.6
pH (on standing)	3.0
Sodium chloride (NaCl)	311 ppm
Sodium sulfate	855 ppm
Sodium thiosulfate	34 ppm
Sodium sulfite	75 ppm
Conductivity	1749 μmho

Temperature. It is widely understood that increased temperature is detrimental from a corrosion standpoint. In general, most chemical reaction rates (including corrosion reactions) double for each 10-°C (18-°F) increase in temperature. This is likely the case for general corrosion of carbon steel, cast iron, and copper alloys; however, to initiate pitting corrosion of stainless steels, it is necessary to exceed a certain critical temperature. This critical temperature will be lower at higher chloride concentrations. Closure will contribute to an increase in white water temperature; however, increased temperatures may offer considerable energy savings. High temperatures also help to improve drainage in the wet end of a paper machine, thus permitting an increase in machine speed.

pH. Although there are both acid and alkaline papermaking processes, most papermaking uses the acid process. Typical white water pHs are in the range 4 to 6. The pH is controlled on the acid

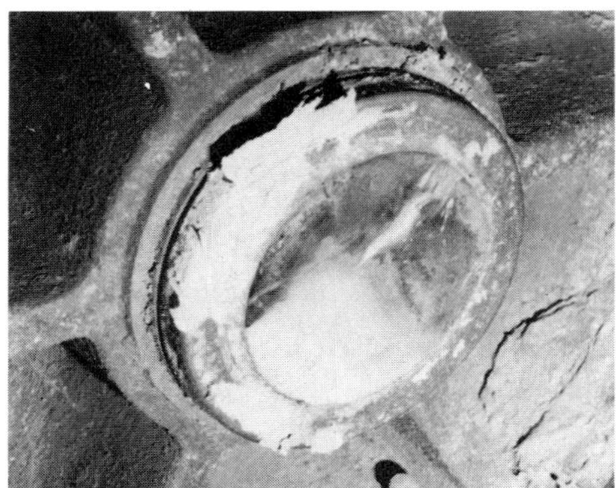

Fig. 7 View of the fracture surface of a felt roll journa that failed by corrosion fatigue. Material of construction is AISI 4140 low-alloy steel quenched and tempered to 300 HB.

side by alum additions and/or souring with sulfur dioxide, sulfite, or sulfuric acid.

Below about pH 4, the general corrosion rates of many alloys are accelerated. Hydrogen ions become so plentiful that their reduction replaces the reduction of dissolved oxygen molecules as the predominant cathodic half reaction in the corrosion process. Corrosion processes are no longer diffusion rate limited. Decreased pH also lowers the stability of passive films on stainless steels; this makes the steel more susceptible to pitting initiation and increases the probability that a pit, once initiated, will continue to grow.

The pH of a white water may decrease upon standing; pH may decrease by 0.5 to 2.0 units within approximately 1 day of sampling. The reasons for this decrease are not clear, but it is known that the continued oxidation of dissolved hydrosulfite and bisulfite generates free hydrogen ions. One would expect that white waters kept in storage (for example, during shutdowns) may become more corrosive. It has been observed that the instantaneous corrosion rate of carbon steel (a measure of white water corrosivity) often correlates well with the magnitude of the pH decrease subsequently measured in white water samples. The white waters showing greater pH decreases are those that originally gave higher carbon steel corrosion rates.

Chlorides. Coastal mills that rely on the transportation of logs in seawater may have chloride concentrations ten times higher than those found in inland mills. Chloride ions are very stable in white waters and will tend to build up in concentration as closure proceeds. Chlorides can cause pitting and crevice corrosion of austenitic stainless steels. Chlorides are also agents for the SCC of stainless steels, but usually at temperatures above 60 °C (140 °F). Stress-corrosion cracking is therefore not a serious problem in most white waters. Chlorides accelerate the general corrosion of carbon steels; however, they do not appear to produce significant acceleration of the corrosion of copper-base alloys.

Sulfates. Sulfate ions (SO_4^{2-}) appear in white water as a residual chemical from the kraft pulping process, from seawater, from alum or sulfuric acid additions, and from the oxidation of more reduced sulfur-containing ions, such as sulfite (SO_3^{2-} or HSO_3^-), thiosulfate ($S_2O_3^{2-}$), and hydrosulfite ($S_2O_4^{2-}$). Sulfate is the thermodynamically stable form of sulfur in white waters.

Sulfate is not an aggressive anion; indeed, it can act as an inhibitor for chloride-related pitting if it is present in (molar) excess over chloride ions. On the other hand, sulfate ions make a significant contribution to white water conductivity and therefore indirectly tend to increase the rates of electrochemical corrosion reactions by facilitating charge transfer in the white water electrolyte. Sulfate ions also provide the food supply for sulfate-reducing bacteria, which in turn cause microbiological corrosion of paper machine equipment. Further, in white waters containing $S_2O_3^{2-}$, an excess of SO_4^{2-} ion is required for thiosulfate pitting to become established.

Thiosulfates. Anaerobic decomposition of warm hydrosulfite solutions produces both thiosulfate and bisulfite ions. Additions of hydrosulfite in excess of that required to meet target paper brightness levels serve only to introduce even more thiosulfate into the white water system. Indeed, it appears that the excess hydrosulfite is very rapidly, and stoichiometrically, converted to thiosulfate. The kinetics of oxidation to sulfate are sufficiently slow that appreciable thiosulfate concentrations can build up.

Thiosulfate concentrations in white water can be reduced by ensuring that the concentration in the original source is as low as possible. Storage of hydrosulfite solutions at 2 to 9 °C (36 to 48 °F) and/or the addition of an alkaline stabilizer to maintain a pH of 10 or more will minimize thiosulfate formation. Thiosulfate ions have been found to be particularly aggressive pitting agents for stainless steels (see the discussion "Thiosulfate Pitting" in this section). Pitting can occur at much lower concentrations (for example, 10 ppm $Na_2S_2O_3$) than that required for chloride pitting (>200 ppm NaCl). Thiosulfates also increase general corrosion rates of copper-base alloys.

Sulfites. Sulfite additions are often made to control the pH of pulp stock. Bisulfite (the stable form of sulfite in white water) is also produced during the decomposition of hydrosulfite brightening solutions. Sulfite is not considered to be aggressive; in fact, sulfites are often used as oxygen-scavenging corrosion inhibitors in other aqueous systems, such as boiler feedwater. Sulfites are readily oxidized to sulfate in the presence of dissolved oxygen. Sulfites may have inhibitive properties in paper machine white water. On the other hand, SO_2 evolution can occur under conditions of high sulfite concentration and low pH, leading to severe atmospheric corrosion problems.

Conductivity. White water conductivity depends on the concentrations of all dissolved ionic species, both inorganic and organic. For carbon steel, cast iron, and copper-base alloys, higher conductivity is an indication of higher corrosivity. This is not the case for stainless steels as long as they are passive and are not undergoing localized corrosion attack.

The corrosivity of carbon steel due to paper machine white water has been monitored by taking instantaneous linear polarization corrosion rate measurements together with simultaneous white water samples for analysis. Linear regression of the carbon steel corrosion rate with individual white water properties reveals that conductivity provides the best regression coefficient, r^2:

Property	r^2
Conductivity	84
ΔpH (on standing)	74
Na_2SO_4	72
NaCl	66
pH (on standing)	64
pH (initial)	18
Temperature	2

Corrosion Mechanisms

Compared with other pulp and paper mill process streams, paper machine white waters are not particularly aggressive. The conventional austenitic stainless steels used for the construction of wetted paper machine components are nonetheless subject to various forms of corrosion attack. The three major corrosion problems affecting stainless steels in paper machine white water systems are:

- Chloride pitting and crevice corrosion
- Thiosulfate pitting
- Microbiological attack

In recent years, there have been reports of rapid attack of type 304 stainless steel equipment in mills where this alloy had served well for decades. The recent unsuitability of type 304 stainless steel can be attributed to closure, hydrosulfite brightening, or both.

Chloride Pitting and Crevice Corrosion. Stainless steels rely on the stable formation of a passive surface film for immunity to corrosion in paper machine white waters. Oxidizing conditions must prevail for the passive film to form and be maintained. Dissolved oxygen from the air is sufficient to maintain stable passivity. Stainless steels can also withstand slightly reducing conditions in white waters without suffering serious attack.

There is a certain safe range of oxidizing conditions within which the stainless steel corrosion potential can vary (these oxidizing conditions can be measured electrochemically as a range of oxidizing potentials). In the presence of aggressive anions such as chlorides, however, this safe potential range is narrowed. If a certain critical corrosion potential (called the breakdown potential) is exceeded, chloride ions can attack the stainless steel surface. The attack manifests itself as pits because passive film breakdown occurs only in isolated locations, such as weak spots in the film due to defects in the underlying metal. Pitting attack is favored by the following conditions:

- Higher chloride concentration
- Higher temperature
- Highly oxidizing conditions
- Low pH
- Stagnant or low-velocity conditions
- Lower molybdenum content in the stainless steel

A new chloride pit tends to repassivate unless the favorable conditions for initiation are maintained until the pit has become established. The metal within the pit becomes anodic with respect to the surrounding metal outside, which becomes cathodic because of ready access to dissolved oxygen for the reduction half of the net corrosion reaction. Anodic dissolution and subsequent hydrolysis of metal ions inside the pit result in the generation of free hydrogen ions, which in turn promote the diffusion of additional anions into the pit to maintain charge neutrality. Because chloride anions are much more mobile than sulfate anions, chlorides preferentially migrate into the pit. Eventually, the solution inside a growing pit may come to resemble a solution of hydrochloric acid more than paper machine white water, and pit growth may continue almost independently of external conditions.

Crevice corrosion initiates beneath deposits or in other areas shielded from direct contact with the white water environment (Fig. 8). Crevice corrosion initiates more readily than pitting attack, which it closely resembles, because the conditions for an oxygen concentration cell already exist. The metal surface inside a crevice has difficulty maintaining passivity because of reduced access to dissolved oxygen; on the other hand, the surrounding metal has ready access to dissolved oxygen. Passive film breakdown within the crevice is thus facilitated by the natural tendency of the crevice to become anodic and the surrounding metal to become cathodic. Once crevice corrosion is initiated, its growth mechanism is the same as that for pitting corrosion.

The primary corrosion concern with white water system closure is that the critical chloride concentration and/or temperature for stainless steel breakdown will be exceeded. Under such conditions, pitting corrosion may occur spontaneously. This is a particular concern with type 304L stainless steel. Stainless steels are chosen for service in chloride pitting environments on the basis of molybdenum content. Because higher molybdenum alloys are also more expensive, it is common to select the alloy with the minimum molybdenum content required for resistance to pitting. A hierarchy of austenitic alloys, representing increasing resistance to chloride pitting and crevice corrosion, can be listed in increasing order of minimum molybdenum content:

Alloy	Molybdenum, %
Type 304L	0
Type 316L	2
SIS 2353	2.5
Type 317L	3
UNS N08904	4
UNS S31254	6
UNS N10276	15

Although type 316L stainless steel has thus far appeared to be resistant under conditions of high closure, it is often produced to have a molybdenum content barely above the minimum of 2% (rather than on the high end of the specified 2 to 3% range). A greater margin of safety for chloride pitting resistance can be ensured by instituting a supplementary requirement that the molybdenum content be no less than 2.5% or by purchasing according to Swedish specification SIS 2353.

Thiosulfate Pitting. Thiosulfate is an aggressive pitting agent, especially for stainless steels that do not contain molybdenum. Thiosulfate pitting, unlike chloride pitting, occurs below, rather than above, a certain critical potential—the thiosulfate reduction potential. Reduction of thiosulfate in the presence of hydrogen ions produces an adsorbed sulfur monolayer on the metal surface. The adsorbed sulfur activates the anodic dissolution of the metal and hinders repassivation. Excess hydrogen ions must be present for acidification of the pit; further, there must also be a larger amount of inert ions (sulfate and chloride) that can be transported into the pit to meet charge transfer requirements. The worst case for thiosulfate pitting occurs within the molar concentration ratio:

Fig. 8 Crevice corrosion of a type 316L checking plate located adjacent to a headbox apron. Corrosion developed under pulp pads that formed despite the highly polished surface.

$$\frac{Na_2SO_4 + 1/2NaCl}{Na_2S_2O_3} = 10 \text{ to } 20$$

Above the range represented by the ratio, there is insufficient thiosulfate to reach the pit nucleus. Below this range, there is too much thiosulfate reduction, which prevents acidification of the pit.

Thiosulfate corrosion particularly affects those grades of stainless steels that do not contain molybdenum. Once formed, pits are very stable and are not subject to spontaneous repassivation. Scratches encourage the initiation of pits. Few, large pits tend to form rather than many, small pits, as in chloride pitting. Sensitized type 304 stainless steel (weld heat-affected zones, HAZs) is particularly susceptible to thiosulfate pitting. Type 316L stainless steel is the minimum grade that should be used for white water service where high thiosulfate levels may exist. It is recommended that thiosulfate levels be controlled below 5 and 10 ppm for equipment made of type 304 and 316 stainless steel, respectively.

Microbiological Corrosion. Paper machine white waters contain nutrients that can sustain bacterial growth. Microbiological growth thrives in near-neutral pH environments. White water temperatures are also usually within the favorable range of 40 to 50 °C (105 to 120 °F). Although higher temperatures may prevent the growth of some forms of bacteria, increased temperatures can increase the metabolism of those bacteria that can adapt to heat. The result is the formation of slimes.

Stock and white water flow systems are designed to minimize slime accumulations. Surfaces are polished and weld projections removed to prevent hang-ups. Wherever slime deposits can build up, however, microbiological corrosion can occur. Once a deposit has grown to a sufficient thickness to exclude oxygen, a colony of sulfate-reducing bacteria (*Desulfovibrio desulfuricans*) can become established. Enzymes produced by these anaerobic bacteria catalyze the reduction of sulfates to form free sulfide ions. Chemically reducing conditions quickly develop, resulting in the depassivation of the stainless steel surface beneath the deposit. Active corrosion in the form of pitting then proceeds.

Corrosion due to *Desulfovibrio desulfuricans* is manifested by the presence of large, shallow pits covered with a black crust (Fig. 9). Perforations through stainless steel equipment are usually small because the entry of oxygen at a leak will stop the activity of sulfate-reducing bacteria (Fig. 10).

Corrosion Control in Pulp Bleach Plants

Andrew Garner
Pulp and Paper Research Institute of Canada

Pulp mill bleach plants have traditionally used austenitic stainless steels because of their combination of good corrosion resistance and weldability. Type 317L (18Cr-14Ni-3.5Mo) has been the typical bleach plant alloy for oxidizing acid chloride environments. However, bleach plants have become more corrosive over the past 20 years as mills have closed wash water systems and reduced effluent volumes. In modern closed bleach plants, type 317L is no longer adequate for long-term service (Ref 1), and many mills have turned to higher-alloy stainless steels, nickel-base alloys, and titanium for better corrosion resistance. Metals are chosen over nonmetals for moving equipment, such as washers. Metals are stronger, tougher, have better fatigue properties, and, if they have sufficient corrosion resistance, require virtually no maintenance. However, the more corrosion-resistant alloys are more costly, and the challenge is to choose an alloy with just enough resistance to avoid corrosion problems.

A wide selection of alloys is available for bleach plant applications. The list includes three families of stainless steels (austenitic, ferritic, and duplex), whose differing merits are outlined in Table 1. Table 2 lists most of the commercially available candidate alloys and their chemical compositions. Table 3 outlines the influence of each alloy component on bleach plant corrosion resistance.

Fig. 9 Section of type 304L stainless steel plate removed from a tapered header used to deliver stock to a headbox showing severe microbiological corrosion

Fig. 10 External view of the type 304L stainless steel tapered header in Fig. 9, showing leakage occurring at small perforations

Table 1 Characteristics of three families of stainless steels for bleach plant service

Family	Examples	Characteristics	Comments
Austenitics	316L 317L 904L 254SMO AL-6XN	Tough, ductile, readily welded without loss of corrosion resistance; corrosion resistance related to alloy content	Bleach plant steels traditionally chosen from this group
	Nitronic 50	As above, with better pitting resistance than type 317L	Manganese-substituted austenitic, may be better value than type 317L; not common at present
Ferritics			
Low-interstitial type	29-4-2 29-4	Not as tough as austenitics, particularly after welding thicker sections. Special precautions needed for welding to avoid N_2 pickup. Corrosion resistance, related to alloy content, can be very good.	Higher alloys have remarkable corrosion resistance. Thin-section (<3 mm, or 0.120 in.) may find applications as corrugated deck or tubing; not common at present. (C+N) ≤ 0.025%
Ti- or Ti+Nb stabilized type	29-4C NYBY MONIT SEA-CURE	0.02% C max versions	Less expensive 0.02% C version of low-interstitial ferritics, for thin-section weldments (≤1.14 mm, or 0.045 in.)
Duplex	2205 Ferralium 255	Tough, ductile, should be weldable without significant loss in corrosion resistance	Combines toughness and corrosion resistance of austenitics and ferritics. Can be made highly resistant to SCC

Austenitic stainless steels, including the AISI 300 series and enriched variations of these steels, are tough and easy to weld. Their corrosion resistance ranges from fair to excellent, depending on the alloy content (Table 1).

Ferritic stainless steels in the 400 series are not used in bleach plants, because of their poor corrosion resistance, particularly after welding. However, there is a new generation of extralow carbon and nitrogen (low interstitial) grades that retain postweld corrosion resistance. As with austenitic stainless steels, the corrosion resistance of these ferritic stainless steels ranges from fair to excellent, depending on alloy content. However, steels such as 29-4 have not been used in the bleach plant, because of problems with embrittling precipitation in thicker sections (3 mm, or 0.12 in., and over) and because of the special precautions required to avoid nitrogen contamination during welding.

In contrast to the ferritic grades, the properties of austenitic and duplex stainless steels are enhanced by nitrogen. As a result, duplex stainless steels such as 2205 and Ferralium 255 have recently been developed with improved corrosion resistance.

Laboratory Assessment of Candidate Alloys

The relative corrosion resistance of stainless steels and nickel-base alloys can be assessed with the ferric chloride ($FeCl_3$) test, which was recently shown to be an appropriate assessment of bleach plant performance (Ref 1). The relative resistance to pitting corrosion of a range of commercial stainless steels is shown in Fig. 11. A critical temperature has been measured for each alloy below which no pitting will occur in $FeCl_3$ (Ref 2, 3). The higher the cited pitting temperature, the more resistant the steel is to pitting. Steels such as type 316L and 317L, for example, have comparatively poor pitting resistance. The 904L-type alloys with about 4.5% Mo provide somewhat better pitting resistance, and the 6% Mo steels, such as 254SMO, are remarkably resistant. Based on these results, one might predict that the duplex steel, Ferralium 255, and the manganese-substituted austenitic, Nitronic 50, should outperform type 317L in the bleach plant.

Generally, similar conclusions can be drawn from crevice corrosion tests in $FeCl_3$ (Ref 4, 5). Such data are presented in Fig. 12 and 13. The critical temperatures for crevice corrosion are lower than those for pitting, indicating that crevice corrosion is more readily initiated. If equipment is not designed to avoid crevice corrosion, then this will be the mode of failure. An example of this form of attack is shown in Fig. 14, which shows a type 317L corrugated deck from a C-stage washer that failed because of crevice corrosion.

Field Testing of Candidate Alloys

Field testing of candidate alloys has been performed by a number of researchers (Ref 1, 6-9). An early study measured pitting attack on a range of welded alloys in comparatively benign Nordic bleach plants (Ref 6). Pitting and crevice attack were later tested in more closed (more corrosive) Canadian bleach plants. This program focused on a few representative alloys, examined the selection of welding electrodes, and compared gaseous and liquid exposure in C- and D-stage washers (Ref 1). Figure 15 shows the corrosion products covering stainless steel test coupons that were exposed to a D-stage washer environment.

Two other exposure programs of note were carried out by the Technical Association of the Pulp and Paper Industry (TAPPI) Corrosion and Materials Engineering Committee in U.S. and Canadian mills. Unwelded alloys were tested in the first program (Ref 7), while the second program tested welded alloys (Ref 8, 9). The alloys tested included almost all available choices for bleach plant applications. However, alloy development has been such an active field in recent years, other promising steels, such as Ferralium 255, AL-6XN, and 29-4C, have been commercialized since the comprehensive TAPPI exposure programs.

Data from all these test rack programs can be interpreted as follows:

- The premium bleach plant alloy of the future: It will probably be chosen from 254SMO, Hastelloy alloy G-3, Sanicro 28, or 20Mo-6; VDM Cronifer 19/25 HMO, AL-6XN, and 29-4C (thin section) were not tested, but they should also be competitive
- Attractive alloys close to the type 317L cost level: Nitronic 50 and 1.4439 (317LMN) are promising alternatives to type 317L. Ferralium

Table 2 Typical chemical analyses of commercially available bleach plant alloys

Alloy	Composition, wt% Cr	Ni	Mn	C	N	Si	P	S	Mo	Fe
Austenitic stainless steels										
316L	16	13	1.6	0.03	...	0.1	0.021	0.012	2.8	bal
317L	18	14	1.9	0.02	...	0.5	0.029	0.009	3.2	bal
Eastern SS 317LM	18	14	1.2	0.02	0.07	0.7	0.029	0.008	4.0	bal
Uddeholm 34L	17	15	1.7	0.03	...	0.6	0.033	0.010	4.3	bal
Uddeholm 34LN	18	14	1.4	0.03	0.15	0.6	...	...	4.7	bal
1.4439	18	14	1.5	0.03	0.13	0.5	...	...	4.3	bal
Nitronic 50	21	14	6.2	0.05	0.22	0.3	0.022	0.011	2.2	bal(a)
Carpenter 20Cb-3	20	33	0.3	0.04	...	0.4	0.015	0.005	2.4	bal(b)
Uddeholm 904L	20	25	1.8	0.02	...	0.4	0.025	0.004	4.2	bal(c)
Sandvik 2RK65	20	25	1.8	0.02	...	0.5	0.020	0.005	4.5	bal(d)
Jessop JS700	21	25	1.7	0.03	...	0.5	...	...	4.5	bal(d,e)
Haynes H20M	22	26	0.8	0.03	...	0.6	0.013	0.010	4.2	bal(f)
VDM Cronifer 19/25 LC	20	25	1.4	0.02	...	0.4	0.018	0.003	4.8	bal(g)
AL-6X	20	24	1.7	0.02	...	0.3	0.021	0.001	6.6	bal(h)
AL-6XN	20	24	1.7	0.02	0.2	0.3	0.021	0.001	6.0	bal
Avesta 254SMO	20	18	0.5	0.02	0.21	0.5	0.015	0.002	6.1	bal(j)
VDM Cronifer 19/25 HMO	21	25	1.3	0.02	0.14	0.3	0.018	0.010	5.9	bal(k)
Ferritic stainless steels										
29-4	29	0	0.1	0.003	0.01	0.1	...	0.010	4.0	bal
29-4-2	29	2	0.1	0.003	0.01	0.1	...	0.010	4.0	bal
29-4C	29	...	...	0.02	0.02	0.35	...	0.002	4.0	bal(m)
NYBY MONIT	25	4	0.3	0.02	0.01	0.2	...	...	4.0	bal(n)
SEA-CURE	27.5	1.7	0.4	0.020	0.025	0.4	0.020	0.010	3.4	bal(p)
Duplex stainless steels										
2205	22	5.5	1.5	0.03	0.14	0.5	...	...	3.0	bal
Ferralium 255	25	5	0.7	0.03	0.16	0.3	0.021	0.002	...	bal(q)
Nickel-base alloys										
Carpenter 20Mo-6	24	bal	0.3	0.02	...	0.3	0.021	0.004	5.6	31(r)
Incoloy 825	22	bal	0.3	0.02	...	0.2	...	0.002	2.9	28.5(s)
Hastelloy alloy G	22	bal	1.3	0.01	...	0.4	0.014	0.002	6.4	19.8(t)
Hastelloy alloy G-3	22	bal	0.7	0.01	...	0.3	0.011	0.002	7.1	19.9(u)
Inconel 625	22	bal	0.1	0.01	...	0.2	0.010	0.007	9.5	3.9(v)
Hastelloy alloy C-276	16	bal	0.5	0.01	...	0.1	0.011	0.002	15.6	5.6(w)
Hastelloy alloy C-22	22	bal	0.5	0.01	...	0.1	0.011	0.002	13.0	3.0(x)

(a) 0.18 V. (b) 3.4Cu, 0.83(Nb + Ta). (c) 1.5Cu. (d) Nominal composition. (e) 0.2Nb. (f) 0.36Ti. (g) 1.6Cu. (h) 0.05Al, 0.07Ce. (j) 0.7Cu. (k) 1.7Cu. (m) Ti = 6×(C+N). (n) 0.3Cu-0.6Ti. (p) 0.5Ti+Nb. (q) 1.6Cu. (r) 3.3Cu-0.05(Nb+Ta). (s) 2.2Cu-0.06Al-0.88Ti. (t) 2.18(Nb+Ta)-2.07Co-0.79W. (u) 0.5(Nb+Ta)-3.2Co-0.8W. (v) 0.12Al-0.24Ti-3.61(Nb+Ta). (w) 1.13Co-0.2V-3.49W. (x) 1Co, 0.2V, 2.5W

Table 3 Effect of alloy components on the corrosion resistance of stainless steels in bleach plant applications

Element	Effect	Comments
Beneficial alloy additions		
Chromium	Enhances resistance to initiation of pitting and crevice corrosion	Steel must have more than 11% Cr to exhibit stainless property
Nickel	Enhances resistance to propagation of pitting and crevice corrosion	Higher levels of nickel enable partly corroded component to remain functional; little or no nickel in ferritic grades
Molybdenum	Enhances resistance to initiation and propagation of pitting and crevice corrosion	Over three times more effective than chromium against pitting and crevice attack, but has solubility limit of about 7% in stainless steels
Nitrogen	Enhances pitting resistance, particularly in combination with molybdenum	Used in austenitic and duplex grades only; increases strength of the steel
Detrimental residual elements		
Carbon	More than 0.03% can cause sensitization, making heat-affected zones of welds less corrosion resistant.	Oxidized out of steel during refining, down to limit set by simultaneous, costly, chromium oxidation
Phosphorus	Can cause hot cracking, that is, cracks formed in weld metal upon cooling. Hot cracks are sites for crevice corrosion, which looks like pitting attack.	Can only be controlled by use of low-phosphorus charge materials. Less than 0.015% P is respectable.
Sulfur	As with phosphorus, can cause hot cracking	Can be controlled to very low levels (<0.005% S) by good steel-making practice. Less than 0.015% S is respectable. Less than 0.005% S is excellent.

Note: Silicon, manganese, and copper are added for steelmaking reasons or sulfuric acid resistance (copper).

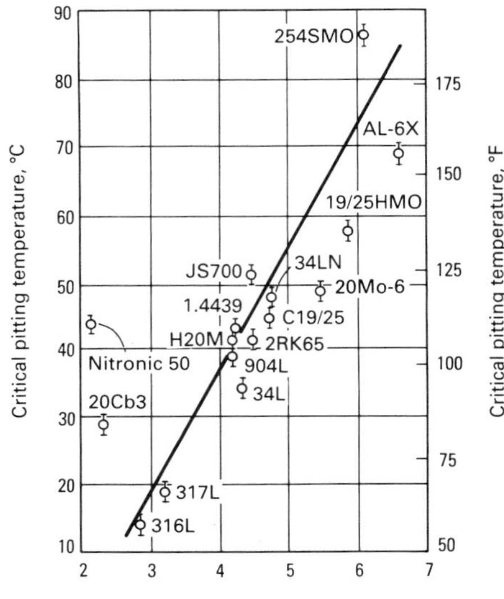

Fig. 11 Effect of molybdenum content on the $FeCl_3$ critical pitting temperature of commercial stainless steels. The more resistant steels have higher critical pitting temperatures

255 was not tested but should also be competitive

- Alloys with less competitive price or performance: Titanium, Hastelloy alloy C-276, Inconel alloy 625, and 29-4-2 all performed exceptionally well, but are expensive; 904L and related alloys appear to perform slightly below the level required for a premium alloy

Bleach Plant Environments

Residual oxidants, such as chlorine (Cl_2) and chlorine dioxide (ClO_2), are the primary cause of corrosion in the bleach plant. The corrosive influence of Cl_2 has been demonstrated with coupon testing at higher Cl_2 levels (Fig. 16) (Ref 10); other work has indirectly identified 25 ppm Cl_2 or ClO_2 as the level above which corrosion reactions are driven by residual oxidants (Ref 11). It seems probable that the 25 ppm of Cl_2 determined by iodine titration is close to zero actual Cl_2, because the titration is also sensitive to traces of oxidizing organics present in C-stage filtrates.

Any steps that can be taken to lower residual oxidants to below 25 ppm will lower corrosion rates. Options include automatic chlorine control, chlorine dioxide sensor/controls, SO_2 antichlor, and NaOH additions (ClO_2 is not very corrosive at pH 7, and NaOH additions to a pH 4 filtrate will transform ClO_2 to ClO_2^- after a few hours).

Recycling of filtrates can compound corrosion problems. For example, residual ClO_2 can be recycled with a D2-stage filtrate to the D1-stage washer showers; therefore, an increase from 50 to 150 ppm ClO_2 has been measured in a D1-stage filtrate when D2-stage SO_2 additions were cut off during high ClO_2 usage. Recycling of D1- or D2-stage filtrate to the C-stage washer should be avoided completely, because the more acidic C-stage filtrate (pH 2) regenerates ClO_2 from chlorite ions, thus rendering the shower water highly corrosive to stainless steels. Many C-stage

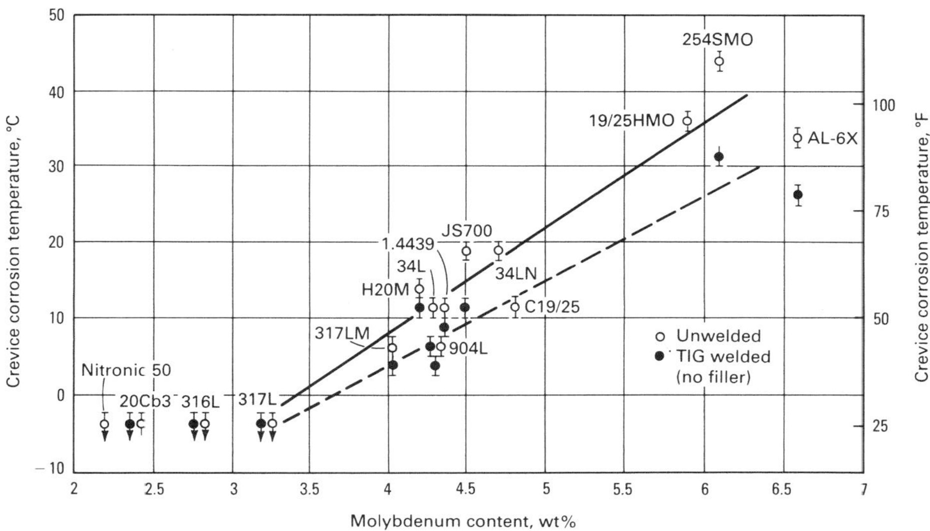

Fig. 12 Effect of molybdenum content on the crevice corrosion temperature of commercial stainless steels. The more resistant steels have higher crevice corrosion temperatures in the $FeCl_3$ test.

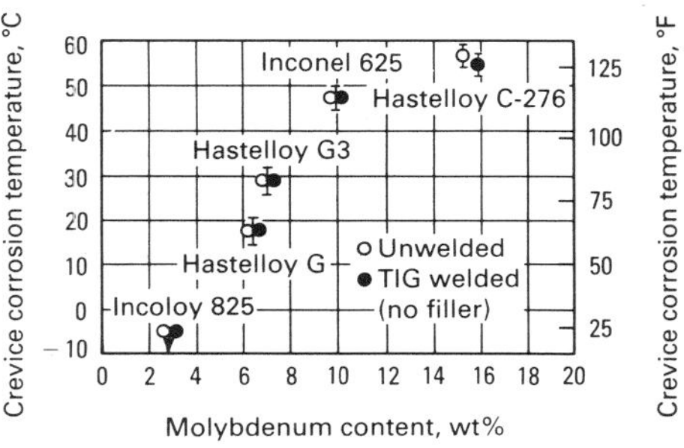

Fig. 13 Effect of molybdenum content on the crevice corrosion temperature of nickel-base alloys. Note the superior performance of Inconel alloy 625 and Hastelloy alloy C-276. Compare with Fig. 12.

washers were lost because of this practice when recycling was first carried out.

Effect of Temperature, Chlorides, and pH

Temperatures and chloride ion concentrations are raised after bleach plant closure (Ref 1). The influence of these two changes on corrosion, although significant, is usually far overshadowed by the corrosive effect of residual oxidants. Typical environmental conditions for bleach plant washers are given in Table 4.

Recycling can lower pH in all acidic washing stages. Lower pH will create a corrosion problem only during the last stage of bleach plant closure, namely when C-stage filtrate is used for C-stage tower dilution. With this practice, lower pH can give the pulp the viscosity protection required for the consequent higher-temperature chlorination. However, when pH decreases to 1.5 or 1.2 as compared to the normal pH 2, stainless steel corrosion rates increase dramatically. It is probable that a critical lower limit pH exists for any given stainless below which corrosion rates are high, but such limits have not yet been identified with any degree of certainty. What can be assumed is that more highly alloyed materials should have lower limiting pHs and that nickel and molybdenum should be the most influential alloying elements.

Vapor-Phase Corrosion

The corrosion of nonwetted components, such as shower pipes and metal vats above the stock-level, is a major problem in C-stage washers. This attack is caused by excess chlorination, in which gaseous chlorine from the filtrate makes small droplets of condensation highly corrosive. Chlorination stage gas-phase attack is probably the most aggressive in all the bleach plant. Methods of avoiding this problem include improving chlorine control; using titanium or nonmetallic-coated shower pipes; and, for stainless steel vats, cladding with 6% Mo stainless or nickel-base alloys to just below the liquid level or lining the whole vat with nonmetallic coating (good design and workmanship is essential).

The vapor space above D-stage washers is not very corrosive unless SO_2 is used in excess, is poorly controlled, or is badly mixed. When both SO_2 and ClO_2 are present in the vapor space, any condensation will contain hydrochloric acid (HCl) and sulfuric acid (H_2SO_4), mixtures of which are very corrosive to stainless steel.

D-Stage Washer Corrosion

Sodium hydroxide or SO_2 additions are made to bleached pulp before the pulp machine in order to improve drainage and to limit brightness reversion. Such additions are often made before the D2 washer so that the washer may also be protected from residual-related corrosion. In some cases, SO_2 additions are made before the D1-stage washer for corrosion protection, and in Nordic countries, SO_2 is even added before some C-stage washers. Sulfur dioxide additions are an effective (if costly) way of limiting corrosion; they work because SO_2 reacts irreversibly with the residual oxidant to form nonoxidizing reaction products, for example, with ClO_2 (Ref 1):

$$2ClO_2 + 5SO_2 + 6H_2O \rightarrow 2HCl + 5H_2SO_4$$

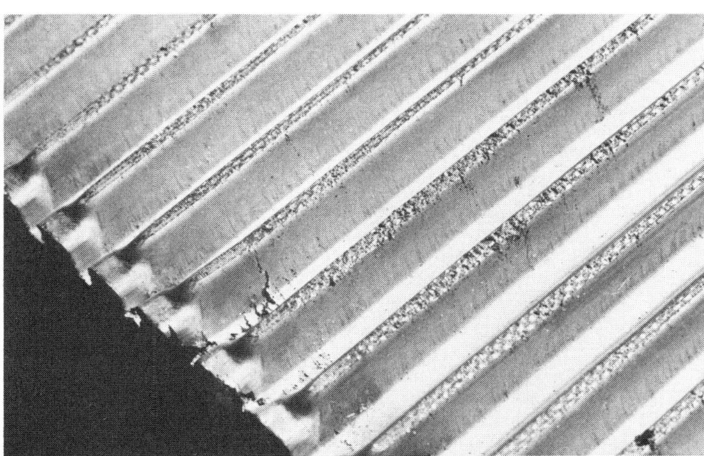

Fig. 14 Failure of corrugated type 317L washer-deck commonly associated with crevice corrosion

Fig. 15 Stainless steel coupons of type 316L, 317L, and 904L on a rack exposed below the incoming stock of a D-stage washer. Profuse ferrous oxide corrosion products cover the coupons.

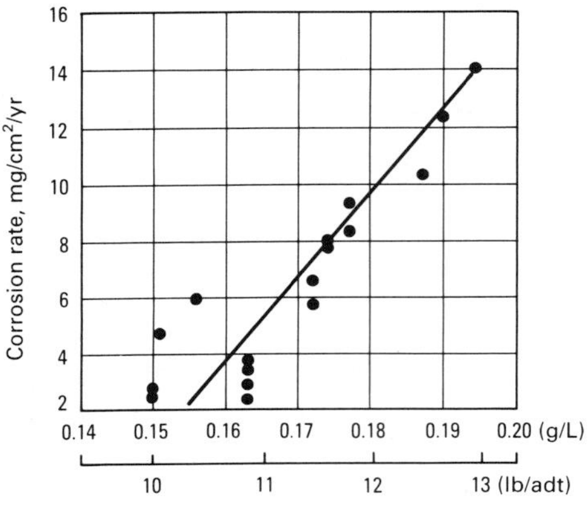

Fig. 16 Effect of residual chlorine on the corrosion rate of test coupons in the C-stage washer

The amount of acid formed by this reaction has a negligible effect on washer corrosion, and sufficient SO_2 should be used to maintain a trace of residual SO_2 at all times. Sulfur dioxide control can be automatic or manual; in the latter case, a target maximum pH of 3 is often used for additions before the D2-stage washer.

If SO_2 additions are discontinued—for example, because of SO_2 shortages—rust nodules or "barnacles" may appear within a few weeks on the washer (Fig. 15). Residual ClO_2 (present because SO_2 was discontinued) has caused the iron oxide [FeO(OH)] corrosion deposits to form (Ref 1). The deposits are insoluble above pH 3.5, and their presence exacerbates the corrosion problem, because additional under-deposit formation occurs. Deposits can be avoided by increasing the pH to above 5.5 (target pH 7) with NaOH and by holding residual ClO_2 to less than 25 ppm.

When a D-stage washer drum has been in operation for a number of years it is not advisable to change to C-stage service if there is any evidence of rust deposit. The deposits may plug corrosion pits, and in C-stage washing at pH 2, the deposits will be dissolved quickly, causing the washer to leak.

Hypochlorite, Oxygen, and Peroxide

Because bleaching with these three oxidants is usually carried out under alkaline conditions, stainless steels are much less subject to pitting and crevice corrosion. Hypochlorite washers are commonly made of type 316L to resist crevice attack. Oxygen reactors are often made of high-nickel stainless steels such as 20Cb3 to guard against chloride SCC, which can occur at higher temperatures in pressure vessels. Some lower alloys have been used in this latter application, apparently without problems, although type 304 stainless steel has failed by chloride SCC. However, chloride pitting corrosion can also occur in oxygen reactors, and alloys containing higher molybdenum levels may be necessary. Alkaline peroxide is used to brighten mechanical pulps in newsprint production without any corrosion consequences. Peroxide has also been used in place of ClO_2 in chemical-pulp bleach plants and appears to present no more corrosion problems than ClO_2.

Table 4 Typical environmental conditions for bleach plant washers

Environment	Chlorination	Chlorine dioxide	Hypochlorite
Oxidant	30 ppm Cl_2	30 ppm ClO_2	30 ppm NaOCl
pH	2	4	9
Temperature, °C (°F)	45 (115)	65 (150)	40 (105)
Chloride ion (Cl^-), ppm	1500	1000	2000
Electrochemical potentials, mV_{SCE}			
Redox potential	1000	500	...
Washer potential	750	500	...

Note: SCE, saturated calomel electrode

Corrosion of Welds

Stainless steel washers fail occasionally because of weld-related corrosion. The principal causes of weld-related corrosion are detailed in Table 5 and outlined in Ref 12.

Welding without filler metal creates a preferential attack site on austenitic stainless steel and should be avoided in washer construction. It is important not to select a filler metal that gives a deposit that is less corrosion resistant than the base metal. For type 316L, the American Welding Society standard filler is adequate (Ref 13). However, for type 317L and the more highly alloyed materials, recent field and laboratory tests have shown that a filler metal with a composition similar to that of the base metal can have much lower pitting resistance (Ref 1, 14).

The optimal weld metal for all 4.5 to 6% Mo austenitic stainless steels is Inconel alloy 112 or an equivalent (for example, Inconel alloy 625 or Avesta alloy P12). Inconel alloy 112 is a good choice because:

- It is metallurgically compatible with all austenitic stainless steels
- As-welded Inconel alloy 112 is highly resistant to pitting and crevice corrosion
- There is no significant galvanic effect between Inconel alloy 112 and austenitic stainless steels in bleach plant liquors
- If microfissures or hot cracks occur in the weld metal, they will not be preferentially attacked by crevice corrosion. This is a particular problem for most ferrite-free stainless steel fillers

Microfissuring or hot cracking is a phenomenon associated with thermal stresses during welding. These stresses usually cause small cracks to form in the weld metal or HAZ of a stainless steel or nickel-base alloy weldment. Higher nickel content alloys, which have a greater coefficient of thermal expansion, are more susceptible to hot cracking. Cracking is more likely to occur because of higher phosphorus (>0.015% P) and sulfur (>0.015% S) in the alloy or contamination of the weld area. It is most commonly seen in the HAZ in the previous pass of a multiple-pass weld. Hot cracking rarely has a detrimental effect on the mechanical properties or structural integrity of a fabrication. However, it can be very detrimental to the corrosion properties of a weldment. Microfissures form crevice corrosion sites that are readily attacked.

Recent laboratory tests have shown that other fillers can be used for 904L, such as Sanicro 27.31.4L CuR, Smitweld NiCro 31/27, and Thermanit 30/40 E. These fillers would be appropriate for microfissure-free shop construction. However, contaminated weldments are best repaired with Inconel alloy 112 or an equivalent filler.

Another common problem with stainless steel weldments—sensitization in the HAZ—is avoided in bleach plants by the use of low-carbon steels (0.03% C max for austenitics). Similarly, fusion-line attack (sometimes called knife-line attack) due to precipitation of carbides at the fusion line in niobium- and titanium-stabilized austenitic steels is rarely seen, because these steels have been made obsolescent by new steelmaking technology.

Table 5 Principal causes of corrosion of austenitic stainless steel weldments

Attack site and mode	Reason for attack	When is it a problem?	How to avoid
Weld metal			
Pitting	Welding with no filler	All molybdenum-containing austenitics(a)	Use appropriate filler.
	Welding with underalloyed filler	3 to 6% Mo austenitics(a)	Use appropriate filler, for example, IN112.
Crevice corrosion	Microfissures in weld metal create sites for crevice corrosion (looks like pitting)	In ferrite-free stainless steel weld metal, for example, fillers commonly recommended for 904L(a)	Use IN112 electrode for 3 to 6% Mo austenitics.
	Lack of penetration	In one-side or stitched butt-weld joints	Ensure full penetration, and do not use stitchwelds on process side.
	Entrapped welding flux	Shielded metal arc welded joints	Use electrode with good flux detachment.
Heat-affected zone	Precipitation of carbides during welding	When steel has over 0.03% C	Use steel with 0.03% C max.
Fusion line	Unmixed zone formed at fusion line	With high-alloy steels close to their corrosion limits	Use lower heat input on final pass.
	Precipitation of carbides at fusion line (knife-line attack	In niobium- or titanium-stabilized steels	A very rare problem. Niobium- and titanium-stabilized steels not common

(a) Particularly after high heat input welding

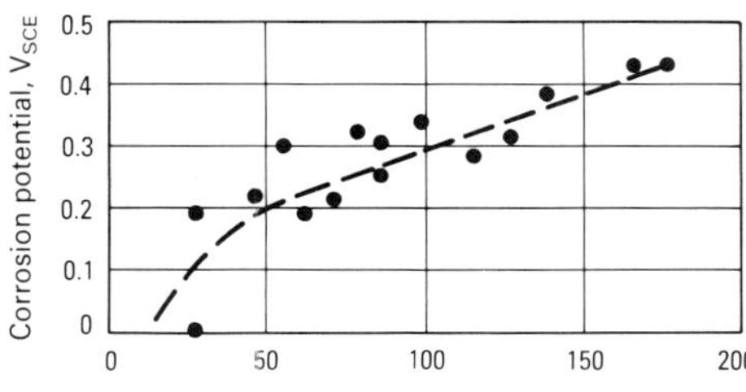

Fig. 17 Effect of residual Cl_2 concentration in the washer vat on the free-corrosion potential of a type 317L stainless steel bleach plant washer. Source: Ref 15

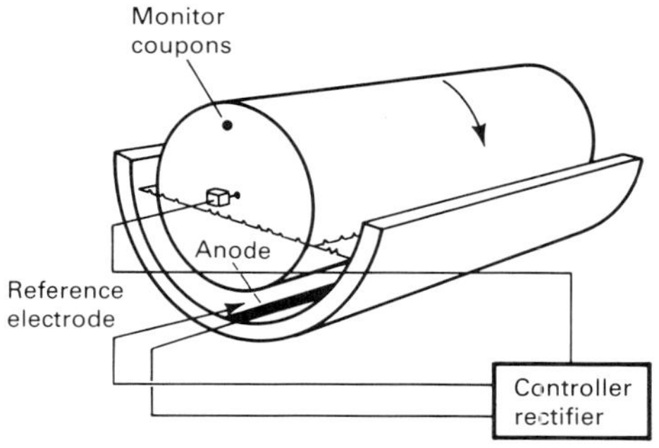

Fig. 18 Components of a washer electrochemical protection system

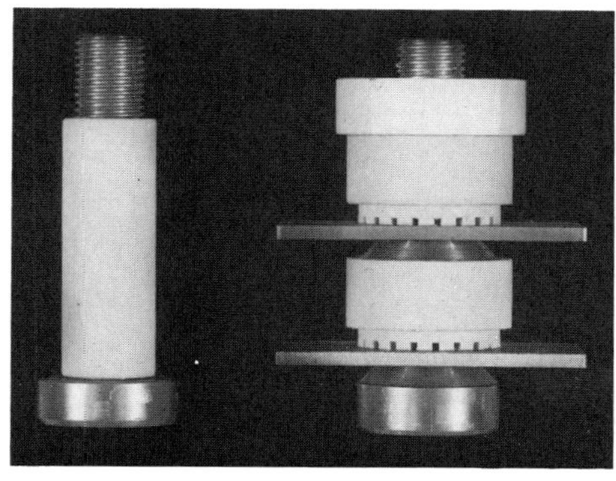

Fig. 19 Crevice corrosion monitor assembly

However, attack at the fusion line is possible when overalloyed fillers such as Inconel alloy 112 are used with high heat input welding. Such welding can create zones consisting of melted base metal that is not mixed with weld filler—called unmixed zones—at the fusion line. Cases of unmixed zone corrosion have occasionally been observed in the bleach plant. In practice, this can be minimized by the use of lower heat input on the final weld passes.

Electrochemical Protection

The discussion thus far has centered on the selection of material or the control of the environment to minimize corrosion. A third approach has recently become available with the development of electrochemical protection for bleach plant washers (Ref 10, 15, 16).

General Description. The life of a stainless steel washer can be greatly extended if the washer is cathodically polarized from the oxidizing potentials imposed by residual oxidants such as chlorine (Fig. 17) to a more negative, passive potential by the use of a rectifier and a platinized anode mounted in the washer vat (Fig. 18). Electrical contact to the washer is made through a rotating mercury contactor; the washer potential is measured with a reference electrode and is automatically controlled with a feedback-controlled rectifier to a potential set point or window that minimizes corrosive attack.

Principle of Protection. A detailed description of the principle of electrochemical protection is provided in Ref 15. However, the essential feature of the technique is the use of the electron as an antichlor for corrosion protection. Electrons are fed from the rectifier to the washer in the form of electric current. At the washer surface, they react with chlorine (Ref 2):

$$Cl_2 + 2e^- \rightarrow 2Cl^-$$

Therefore, like SO_2, electrons react with the corrosive oxidant (chlorine) to form relatively harmless chloride ions. For corrosion protection, the electron has a clear advantage over SO_2: It can be delivered to the cathodic reaction site. For this reason, comparatively few electrons are needed. The required current is low, and running costs are negligible.

Because of the comparatively low cost of electrochemical protection and the ease of retrofitting to existing washers, commercialization of the technique has found wide acceptance. The first successful operation of one of these systems was in Nova Scotia, Canada, in 1978. By the end of 1986, there were about 90 installations worldwide.

Monitoring Technique. The corrosion rate of each protected washer is monitored with coupons, using a technique designed for comparison of protected and unprotected coupons (Ref 17). A mounting bolt is welded to the end face of the rotating washer as shown in Fig. 18, and two coupons, together with segmented crevicing disks, are mounted so that one is in electrical contact with the washer and the other is isolated, with all other mounting details being identical (Fig. 19). The degree of protection is assessed by comparing the corrosion of protected and unprotected coupons after a 60-day exposure period (Fig. 20).

Results to Date. Both weight loss and depth of attack measurements have been made on coupons to assess the comparative severity of pitting or crevice corrosion (Ref 10). The ratio of unprotected and protected coupon weight loss is used as a measure of protection. Table 6 lists protection ratios measured for washers that have been protected for up to 5 years (6 washers types/212 test coupons). These results show that protection lowers corrosion rates. On the average, unprotected coupons lose six times more weight than protected coupons.

Improved performance with protection can be compared with improved performance after alloy upgrading in the absence of protection. The results of an extensive bleach plant exposure program, which involved 880 coupons, 40 test racks, 10 mills, and no electrochemical protection (Ref 1), were reexamined to obtain the average weight loss ratios (using type 317L as a base case) for alloys 904L and 254SMO. These data are given in Table 7; greater ratios indicate better performance. Type 317L coupons lose 5.5 times as much

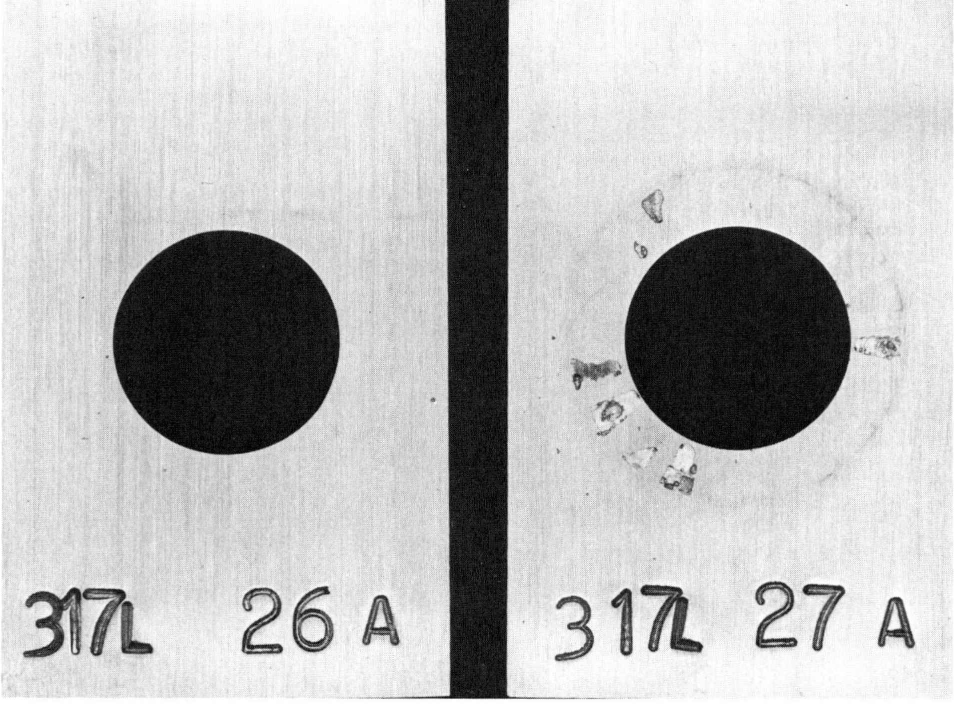

Fig. 20 Type 317L stainless steel monitor coupons after a 60-day exposure attached to a C-stage washer. The coupon on the left was in electrical contact with the washer and was therefore protected. The coupon on the right was isolated and unprotected. Substantial crevice corrosion occurred on the right-hand coupon at 7 of the 20 possible crevicing sites.

Table 6 Type 317L test coupon data from the first six electrochemical protection systems

Washer	Protection period, years	Average protection ratio(a)
C-stage	5.4	4.9
D2-stage	4.0	2.6
D2-stage	3.3	12.3
D1-stage	2.9	8.2
D2-stage	2.9	1.9
D1-stage	1.8	6.2
General average		**6.0**

(a) Unprotected ÷ protected type 317L coupon weight loss (larger ratios indicate better performance).

Table 7 Test rack weight loss data for unprotected steels

Steel	Average weight loss ratio(a)
Type 317L	1
904L	2.5
254SMO	5.5

(a) This ratio is obtained, for example, by determining the average weight loss for type 317L coupons ÷ average weight loss for 904L coupons (larger ratios indicate better performance).

weight as 254SMO coupons. Therefore, the improved performance achieved by the electrochemical protection of type 317L appears to be very close to that gained by upgrading type 317L to unprotected 254SMO.

Economics. The installation and operating costs of electrochemical protection systems are small compared to the resulting cost savings. Capital cost savings are such that if the life of a washer is extended from 5 to 10 years by protection, the protection system will have a payback period of about 1 year.

Even this substantial saving can be overshadowed by savings in chemical costs in mills that add NaOH or SO_2 to pulp before washing. Experience has shown that, in general, SO_2 additions need not be made ahead of a protected washer to protect it from corrosion. If pulp souring is still required, then this can be done immediately after the washer, where much less SO_2 will be needed. Additional savings can be realized by eliminating SO_2 use in a closed bleach plant, because SO_2 free recycled filtrate used for tower dilution consumes much less unreacted ClO_2. Protection systems have been installed on washers made from types 316L, 317L, and 904L, and 254SMO stainless steels. A protected 254SMO washer probably represents the state-of-the-art for corrosion control in the most severe washer environments.

Corrosion by Sulfite Pulping Liquors

C.B. Thompson and Andrew Garner
Pulp and Paper Research Institute of Canada

Sulfite pulping, one of the oldest methods of chemical pulping, dates back to the 1860s. For many years, sulfite pulping was the primary chemical method for making paper, although since the 1950s it has been overtaken in importance by the kraft process to such an extent that the sulfite industry was perceived by many as dying out. In recent years, however, a growing appreciation of the versatility of the sulfite process and the appearance of combined sulfite and mechanical pulping methods (semichemical or chemi-mechanical) have combined to ensure the future of sulfite pulping. These changes in the industry are thoroughly documented in Ref 18).

This section will discuss the various methods of sulfite pulping, the principal corrosion mechanisms occurring in sulfite environments, and the major corrosion problem areas found in sulfite mills. Information on corrosion in the disk refiners used in the later stages of semichemical and chemi-mechanical pulping can be found in the section "Corrosion of Mechanical Pulping Equipment" in this article.

Sulfite Pulping

Sulfite Chemistry. Sulfite pulping liquors are prepared by dissolving SO_2 in a solution of calcium, sodium, magnesium, or ammonium hydroxide. The pH of the resultant cooking liquor depends on the base and the amount of SO_2 dissolved, and it can range from 1 to above 13. The exact value is chosen according to the particular mill process and the desired pulp characteristics. Fresh sulfite liquor is essentially a mixture of sulfite and bisulfite ions in an aqueous solution of SO_2. The ratios of these three components can vary widely, according to the liquor pH. The relative concentrations of bisulfite and sulfite ions at different pH levels are shown in Fig. 21; bisulfite dominates at pH values less than 6, while sulfite ions dominate at pH values above 7. The concentration of aqueous SO_2 also increases at lower pHs (Ref 19).

Traditional Sulfite Pulping. For many years, most sulfite mills used a calcium hydroxide base cooking liquor, mainly because of the economy of the calcium carbonate feedstock. The calcium-base process is restricted to very low pH (typically 1.5) and low cooking temperatures (140 to 150 °C, or 295 to 300 °F) because of scaling and solubility problems. Furthermore, the process is difficult to adapt to high-yield pulping. These reasons, as well as increasingly stringent antipollution legislation, have led to the expanded use of soluble sodium, magnesium, and ammonium bases in recent years.

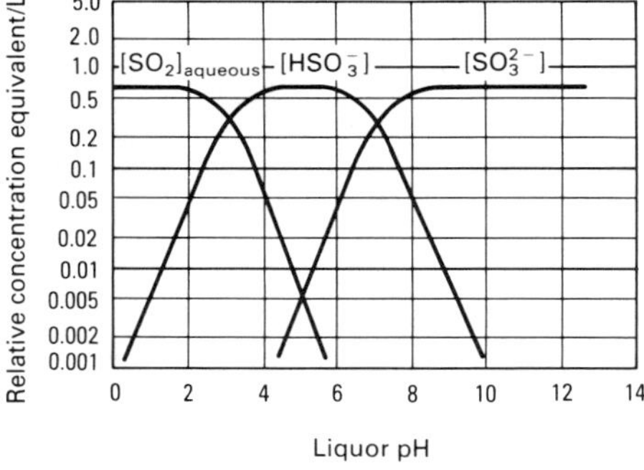

Fig. 21 Relative concentrations of bisulfite (HSO_3^-), sulfite (SO_3^{2-}), and aqueous sulfur dioxide (SO_2) as a function of liquor pH at 130 °C (265 °F). Aqueous sulfur dioxide and bisulfite dominate at acidic pHs, sulfite dominates at alkaline pHs. Source: Ref 19

A typical sulfite operation is shown in Fig. 22. Sulfur (either in powder or liquid form) is burned to produce SO_2. The gas is cooled rapidly to minimize the formation of sulfur trioxide (SO_3) and then passed into absorption towers, where it is dissolved in the hydroxide base to form the raw pulping liquor. Pulping is carried out in either batch or continuous digesters; when the cook is complete the chips are released into a blow tank. The chips are then taken for washing, screening, and bleaching as required. The spent cooking liquor (red liquor) can be evaporated and burned in a furnace for steam generation and/or recovery of cooking chemicals, or it can be further treated to obtain chemical by-products. A typical operation might also include equipment for SO_2 recovery from the cook and an acid accumulator.

Recent developments in sulfite mills have centered around the need to decrease pollution, to increase pulping efficiency (that is, increase the proportion of usable fiber obtained per cook), and to adapt to market demands for specific grades of pulp. These requirements have resulted in a variety of processes, such as high-yield sulfite pulping and chemi-thermomechanical pulping, in which sulfite chemical treatment of the chips is combined with additional treatment of the fiber in a disk refiner. The interrelationship of these processes with pure chemical pulping and pure mechanical pulping is shown in Fig. 23. An important practical point to note is that high-yield and very high-yield sulfite pulping require the use of a digester, but chemi-mechanical and chemi-thermomechanical pulping use smaller impregnation vessels or steaming tubes for pretreatment of the chips.

Corrosion Mechanisms in Sulfite Liquors

Stainless steels are used for most process equipment in the sulfite mill. As in any industrial plant, the steels in sulfite mills can experience many different types of corrosion. The most serious ones are usually connected in some way to the presence or absence of SO_2. Chloride-induced localized corrosion and SCC may also be problems, particularly in coastal mills.

The presence of SO_2 in sulfite liquors can affect corrosion of stainless steels in three ways. It can:

- Maintain passivity against acidic cooking liquors
- Form H_2SO_4 by decomposition of the liquor
- Form H_2SO_4 by oxidation to SO_3

These effects are discussed in more detail below.

Maintaining Passivity. Sulfur dioxide in solution helps to maintain the passivity of stainless steels such as types 316L and 317L that are commonly used in sulfite mills (Ref 20, 21). Therefore, a potential corrosion problem exists in any situation in which the concentration of dissolved SO_2 becomes very low. This may happen, for example, in vacuum evaporators or in batch digesters during a cook. However, practical experience has shown that this type of corrosion is usually not of concern with batch digesters. Unless conditions are especially severe (that is, high temperature, low pH, high chlorides), austenitic stainless steels containing more than 2.7% Mo can withstand short periods without any dissolved SO_2 present. Opening the digesters to the

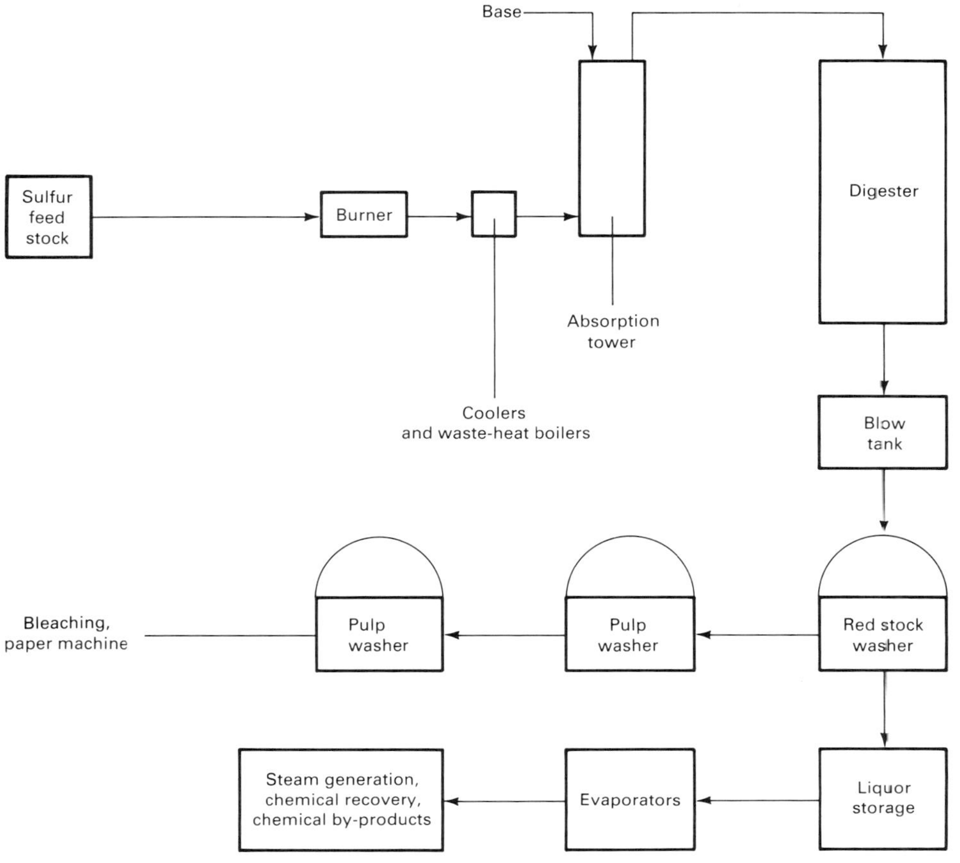

Fig. 22 Typical layout for a traditional low-yield sulfite mill operation

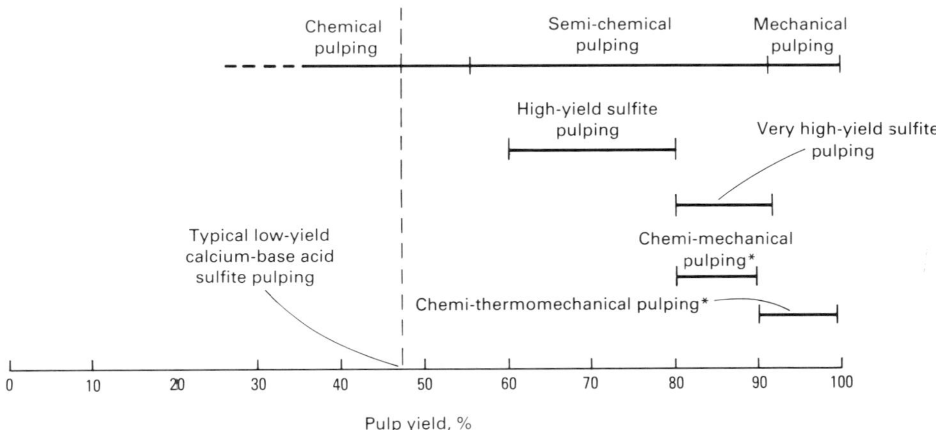

Fig. 23 Interrelationship between chemical, semichemical, and mechanical pulping processes. The processes marked with an asterisk use impregnation vessels or steaming tubes rather than digesters. Source: Ref 18

atmosphere between cooks also helps to maintain passivity.

Sulfuric Acid Formation by Liquor Decomposition. Acid bisulfite liquors can spontaneously decompose to form H_2SO_4 (Ref 22):

$$3SO_2 + 2H_2O \rightarrow 2H_2SO_4 + S$$

$$3NaHSO_3 \rightarrow NaHSO_4 + Na_2SO_4 + S + H_2O$$

The decomposition reactions are accelerated by increasing temperature, the presence of thiosulfate, and low liquor pH (<4). Thiosulfate forms spontaneously in acid bisulfite and sulfite liquors and is therefore always present to some extent. The presence of trace concentrations of cobalt, molybdenum, and selenium can also catalyze the decomposition reactions, even in cases in which the liquor pH is greater than 4. Liquor decomposition is dangerous for corrosion due to the formation of H_2SO_4, but it is also very serious for pulp production because it can result in a charred cook. Consequently, thiosulfate levels are regularly monitored in acid bisulfite and bisulfite mills. However, despite these precautions, liquor decomposition can still occur in isolated areas, such as under pulp deposits or in stagnant areas of a digester (Ref 23). Liquor degradation due to thiosulfate is usually of more concern in mills producing bleached pulp (that is, long cooking times) than in the case of short-cook high-yield mills.

Sulfuric Acid Formation by Oxidation. Free SO_2 is formed at acid pHs and raised temperatures because of the dissociation of bisulfite ions in the liquor:

$$HSO_3^- \rightleftarrows SO_2 + OH^-$$

$$H^+ + HSO_3^- \rightleftarrows H_2O + SO_2$$

Therefore, any vapor space above sulfite liquors may contain gaseous SO_2. If oxygen is present in the system, corrosion can then occur from the direct formation of H_2SO_4 on the exposed steel surfaces. This occurs by the oxidation of SO_2 to SO_3, followed by the condensation of H_2SO_4, particularly on cooler parts of process equipment (Ref 24):

$$2SO_2 + O_2 \rightarrow 2SO_3$$

$$SO_3 + H_2O \rightarrow H_2SO_4$$

Corrosion Problem Areas

A general guide to corrosion in sulfite mills is provided in Ref 23. In addition, the formation of SO_3 in sulfite pulping operations has been thoroughly documented in Ref 25.

Liquor Preparation. In the majority of cases, sulfur is burned to obtain SO_2. The gas is then cooled before being passed into absorption towers, where it is dissolved in the chosen base. Sulfur feedstocks may be either liquid or powder. In the case of powder sulfur, precautions must be taken to prevent weathering of the pile in order to minimize the *in-situ* formation of thiosulfate and H_2SO_4. Similarly, moisture must be avoided in liquid sulfur storage tanks to reduce the possibility of H_2SO_4 attack at the waterline.

The most serious potential corrosion problem in the liquor preparation area is the formation of SO_3. This can occur by inefficient sulfur burning or by the entry of air into the burner-SO_2 absorption system. Precautions include cooling the SO_2 rapidly after its formation and ensuring that all metallic parts in contact with SO_2 in the absorber tower are continuously sprayed by the absorbing solution. In addition, the amount of iron in contact with the SO_2 gas is minimized because the metal catalyzes formation of the trioxide.

Digesters. Both batch and continuous digesters are used in sulfite mills. Continuous digesters are generally preferred for neutral sulfite semichemical and chemi-mechanical pulping applications. In North America, batch digesters for acid sulfite pulping are normally made of carbon steel plate lined with acid-resistant carbon brick. Metallic fittings within the digester are made from a minimum of type 316L stainless steel with a molybdenum content of 2.7% or greater. Corrosion is not usually a problem, although the bricks degrade and must be replaced after 6 to 10 years. Some batch digesters are stainless steel lined; this practice is common in Scandinavia. Localized corrosion and SCC have been found in these digesters, often associated with high (>1000 ppm) chloride levels in the liquor at coastal mills. As a result, these digesters are usually constructed of type 317L stainless steel (Ref 26). More highly alloyed stainless steels, such as type 317LMN and 904L, and nickel-base alloys have also been used in particularly corrosive conditions.

Digesters for use in neutral, alkaline, and chemi-mechanical sulfite pulping can be made from a range of materials, including carbon steel and austenitic stainless steels. In one case, accelerated corrosion in the vapor phase of a type 317L stainless steel continuous high-yield bisulfite digester was recently a problem because of SO_3 formation (Ref 27). Formation of SO_3 was minimized by raising the liquor pH from 4 to 6, by using a plug screw feeder to reduce the entry of oxygen with the chips, by feeding chips in below the liquor level, and by preheating the water in the adjacent blow tank to lower its dissolved oxygen content. At the same time, the vapor space of the digester was clad in 904L stainless steel to provide increased corrosion resistance.

Liquor Recovery Systems. Evaporators are commonly made from type 316L or 317L stainless steel, although type 304L or carbon steel has been used for the final effects in sodium-base systems. Pitting and crevice corrosion under scale deposits are the most commonly reported problems. Liquor decomposition to form H_2SO_4 may also occur. In general terms, magnesium- and calcium-base systems are considered to be most corrosive towards evaporators (Ref 26).

Recovery boilers have been largely constructed from carbon steel, although increasing use has been made of composite stainless steel-carbon steel water tubing to combat corrosion. Flame- and plasma-sprayed stainless steel have also been used for the same purpose. Sulfite mill recovery boiler exhaust gases contain significantly more sulfur dioxide than those from kraft recovery boilers; they may also contain greater amounts of SO_3, hydrogen sulfide (H_2S), or HCl (Ref 28). Type 316L and 317L stainless steels and fiber-reinforced plastics have been used for scrubber bodies and internals, but have sometimes been replaced by higher-alloyed stainless steels or nickel-base alloys because of severe pitting and crevice corrosion. Attack can be especially severe in the inlet quenching/chloride removal zone, where even titanium and nickel-base alloys have suffered extensive general dissolution. Special care must be taken in these areas to avoid scale buildup and the presence of wet-dry zones, where chloride concentration can occur.

Corrosion in the Recovery Boiler*

Recovery boilers are used in the wood pulp industry to recover pulping chemicals and to raise steam by burning the organic residues present in spent pulping liquors. By far, most of the recovery boilers in North America burn kraft process liquor (black liquor). When evaporated to about 65% solids and immediately before firing in the boiler, a heavy black liquor contains 50% organic solids and approximately 6% total sulfur, mostly in the form of Na_2SO_4 and $Na_2S_2O_3$. Some coastal or closed cycle mills operate with high sodium chloride (NaCl) content (for example, 12% NaCl) black liquor.

One typical boiler configuration is shown in Fig. 24. It differs from an electric utility boiler in that the underlying firing zone is a bed of molten salt (the smelt bed) that operates under reducing conditions and is contained in a water-walled furnace bottom. In addition, the combustion zone is extraordinarily tall so that chemical reaction products can condense and run down the water wall before reaching the superheater.

In and above the smelt bed, chemical reactions take place that reduce sulfate to sulfide and form sodium carbonate (Na_2CO_3). The operating philosophy of the boiler places safety and a continuous supply of regenerated liquor (green liquor, made by dissolving smelt in water) ahead of steam supply and energy efficiency. Steam pressures are generally in the range 4135 to 10 340 kPa (600 to 1500 psi), and a typical boiler might generate 135 000 kg (300 000 lb) of steam per hour, burning 725 to 900 Mg/day (800 to 1000 tons/day) of solids.

Corrosion Mechanisms

There are numerous mechanisms of corrosion or chemical degradation of the recovery boiler. Whether or not corrosion can take place and the rate at which corrosion proceeds depend on the composition and metallurgical condition of the material of construction, on the environment to which the materials are exposed (for example, liquid or gaseous phase, concentrations of chemicals), and on the temperature of the material and environment.

Composition of some alloys and coatings are listed in Table 8. The corrosion products formed on the surface of the metal depend on the metal composition and on environmental factors. These

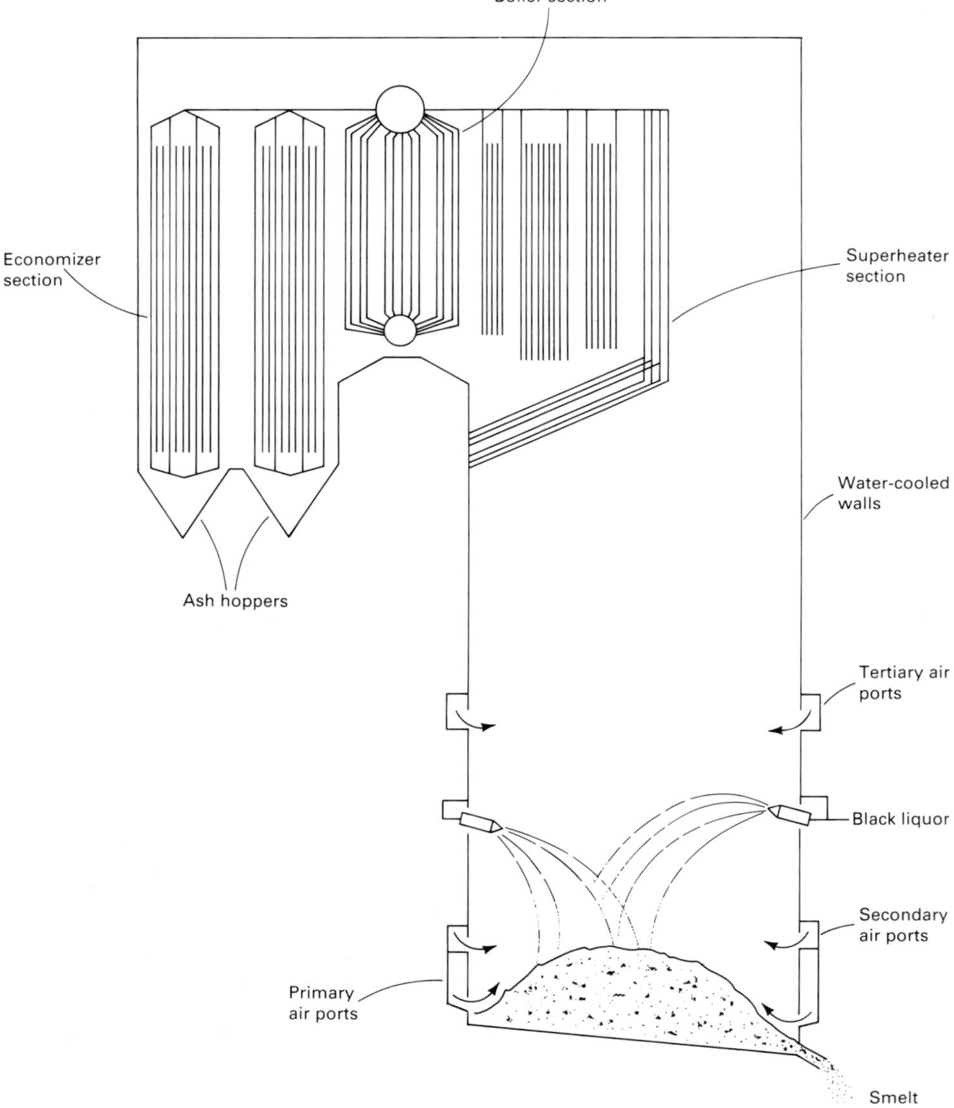

Fig. 24 Schematic of typical recovery boiler used in the wood pulp industry

*Adapted with permission from *Recovery Boiler Manual*, J.H. Jansen, Ed., Vol 3, American Paper Institute, 1985

corrosion products often protect the underlying metal from exposure to the corrosive environment and therefore reduce or prevent further corrosion.

The environmental characteristics determine whether or not corrosion may be expected. Metals can be exposed to corrosive chemicals in solid, liquid, or gaseous phase, with the latter two being most harmful.

The temperature of the metal in contact with the corrosion environment will determine the rate at which corrosion reactions will take place. Sometimes, a corrosion reaction proceeds slowly because of the temperature. Sometimes, corrosion reactions take place only in a specific temperature range. Table 9 shows the typical locations where the different types of corrosion have been found in recovery boilers.

Metal wastage due to corrosion can occasionally be accelerated by and confused with erosion and overheating, although erosion and overheating fall outside the strict definition of corrosion. However, the combination of metal properties, composition of corrosive chemicals in contact with the metal, thermal cycling, and the temperature will determine the rate at which corrosion will take place.

Corrosion in the Furnace

Corrosion, in combination with erosion, can severely damage tube metal, pin studs, stack studs, flat studs, and protective coatings in the furnace hearth zone. The corrosion rate of carbon steel in recovery boilers rapidly increases above 315 °C (600 °F). Operating pressures above 6900 kPa (1000 psig) could result in external tube metal temperatures at or above this value.

Corrosion by Flue Gas in the Absence of Sulfur Compounds. When the bare metal surface of carbon steel is exposed to flue gases containing oxygen at elevated temperatures, corrosion may occur as follows:

$$3Fe + 2O_2 \rightarrow Fe_3O_4 \quad \text{(Eq 1)}$$

$$4Fe + 3O_2 \rightarrow 2Fe_2O_3 \quad \text{(Eq 2)}$$

$$2Fe + O_2 \rightarrow 2FeO \quad \text{(Eq 3)}$$

The corrosion products formed, magnetite (Fe_3O_4), ferric oxide or hematite (Fe_2O_3), or ferrous oxide (FeO), depend on oxygen concentrations and temperatures. Of these products, only Fe_3O_4 is formed as a dense surface layer that prevents further attack of the underlying metal. The other iron oxides, Fe_2O_3 and FeO (formed at temperatures above 565 °C, or 1050 °F), do not form a dense protective layer, and corrosion proceeds in the underlying metal at the same rate.

When the steel is alloyed with such metals as chromium, nickel, or molybdenum, the formation of Fe_3O_4 will prevail; in such cases, only minor amounts of Fe_2O_3 are formed, and no FeO is formed. These alloys are, therefore, more corrosion resistant. At higher alloy contents, other metal oxides will also be formed that protect against further oxygen attack. The corrosion resistance of higher-alloyed steels is primarily due to the formation of a protective layer of chromium oxide (Cr_2O_3).

Corrosion by the Flue Gas in the Presence of Sulfur Compounds. In the presence of sulfur-bearing compounds, such as in kraft recovery boilers, the dominating corrosion reaction of carbon steel is (sulfidation reaction):

$$Fe + S \rightarrow FeS \quad \text{(Eq 4)}$$

Iron sulfide is formed as a porous layer on the metal surface and does not protect the underlying metal. Because of the chemical structure of iron sulfide, the underlying iron can migrate through the sulfide layer and react with flue gas. Thus, corrosion will proceed, and iron sulfide will continue to build up on the outer surface. At high sulfur concentrations, an outer layer of iron disulfide (FeS_2) can also be formed:

$$Fe + 2S \rightarrow FeS_2 \quad \text{(Eq 5)}$$

Elemental sulfur can be formed as a result of the following reactions in the boilers:

Table 9 Corrosion mechanisms and typical locations in recovery boilers

Mechanisms	Locations
Sulfidation reaction	Water wall tubes, studs, and air ports in furnace hearth zone. Both furnace-side and back-side surface of tubes
Molten hydroxides	Around primary air ports on back-side surface of tubes
Sulfidation-oxidation reaction	Lower superheater section, or further downstream, when liquor/smelt carryover is excessive
Sulfur trioxide adsorption to form liquid pyrosulfates	Superheater and boiler bank
Sodium chloride enrichment and deformation	Superheater and boiler bank
Sulfur trioxide adsorption to form liquid bisulfates	Economizer and back-end rows of boiler bank
Acid dew-point corrosion	Cold-side of economizer, direct contact evaporator, precipitator, ID fan, and all flue gas ducting connecting these sections

Table 8 Metal alloy and coating compositions for recovery boiler tubing

Alloy/coating	Composition, wt%(a)								
	C	Mn	P	S	Si	Cr	Ni	Mo	Fe	Other
Tubing alloys										
A192	0.06–0.18	0.27–0.63	0.048	0.058	0.25	. . .	. . .	. . .	bal	. . .
A210, grade A-1	0.27	0.93	0.048	0.058	0.1 (min)	. . .	. . .	. . .	bal	. . .
A210, grade C	0.35	0.29–1.06	0.048	0.058	0.1 (min)	. . .	. . .	. . .	bal	. . .
T-1	0.10–0.20	0.30–0.80	0.045	0.045	0.10–0.50	. . .	. . .	0.44–0.65	bal	. . .
T-9	0.15	0.30–0.60	0.030	0.030	0.25–1.00	8.00–10.00	. . .	0.90–1.10	bal	. . .
T-11	0.15	0.30–0.60	0.030	0.030	0.50–1.00	1.00–1.50	. . .	0.44–0.65	bal	. . .
T-22	0.15	0.30–0.60	0.030	0.030	0.50	1.90–2.60	. . .	0.87–1.13	bal	. . .
Type 304H	0.04–0.10	2.00	0.040	0.030	0.75	18.0–20.0	8.00–11.00	. . .	bal	. . .
Type 310	0.15	2.00	0.040	0.030	0.75	24.0–26.0	19.0–22.0	. . .		
Type 327H	0.04–0.10	2.00	0.040	0.030	0.75	17.0–20.0	9.00–13.00	. . .	bal	Ti(b)
Type 347H	0.04–0.10	2.00	0.040	0.030	0.75	17.0–20.0	9.00–13.00	. . .	bal	Nb+Ta(c)
Incoloy alloy 800H	0.05–0.10	1.5	. . .	0.015	1.0	19.0–23.0	30.0–35.0	. . .	39.5 (min)	0.75 Cu; 0.15–0.60 Al, Ti
Inconel alloy 690	0.05	0.50	. . .	0.015	0.50	27–31	58.0 (min)	. . .	7–11	0.50 Cu
Composite boiler tubing										
Inside tube										
A210, grade A-1	0.27	0.93	0.048	0.058	0.1 (min)	. . .	. . .	. . .	bal	. . .
Outside tube										
Type 304	0.08	2.00	0.040	0.030	0.75	18.0–20.0	8.00–11.00	. . .	bal	. . .
Coatings										
Metco alloy 444	. . .	. . .	. . .	. . .	. . .	9.0	bal	5.5	5.0	7.0 Al
Metco alloy 465	. . .	. . .	. . .	. . .	. . .	27.5	. . .	2.0	bal	6.0 Al
Hastelloy alloy C-276	0.02	1.00	0.040	0.030	0.08	14.5–16.5	bal	15.0–17.0	4.00–7.00	3.00–4.50 W; 2.50 Co

(a) Maximum wt% unless otherwise specified. (b) Type 321H shall have a titanium content of not less than four times the carbon content and not more than 0.60%. (c) Type 347H shall have a niobium plus tantalum content of not less than eight times the carbon content and not more than 1.0%.

$$Na_2S + 2CO_2 \rightarrow Na_2CO_3 + CO + S \quad \text{(Eq 6)}$$

$$2Na_2S + 3SO_2 \rightarrow 2Na_2SO_3 + 3S \quad \text{(Eq 7)}$$

$$2H_2S + O_2 \rightarrow 2H_2O + 2S \quad \text{(Eq 8)}$$

$$Fe + S \rightarrow FeS \quad \text{(Eq 9)}$$

Of these sulfur-producing reactions, the reaction of sodium sulfide (Na_2S) with carbon dioxide ($2CO_2$) as shown in Eq 6 appears to predominate. Some of this elemental sulfur then reacts with iron to form iron sulfides.

Laboratory experiments have shown that iron monosulfide (FeS) will be formed at low reaction rates when oxygen is not present. The corrosion rate increases considerably when oxygen is also present, especially at temperatures above 150 °C (300 °F). At a hydrogen sulfide to oxygen ratio of approximately 1 (in the gas phase), the formation of FeS (Eq 4) is at a maximum rate. When the oxygen concentration is increased further, the corrosion reaction will increasingly lead to the formation of Fe_3O_4 (Eq 1) and FeS_2 (Eq 5), which develop protective layers against continuing corrosion of underlying metal.

Oxygen and gaseous sulfur concentrations can vary widely in time and place in the recovery boiler furnace, and little is known about actual concentration profiles and gradients. They are dependent on such variables as the sodium and sulfur content in the black liquor, bed temperature, bed size, primary air flow, and primary air pressure.

Water wall tubes in the furnace hearth zone are usually covered with a layer of solidified smelt. This solid smelt layer decreases the corrosion rate because the tube metal is not exposed directly to molten smelt or corrosive gases and because the tube surface temperatures are lowered due to the insulating properties of the solidified smelt. One method of corrosion protection in the furnace hearth zone is to increase the thickness of the smelt layer by attaching pin studs to the tubes. The studs cool the smelt and serve as anchors to hold the solidified smelt layer onto the tubes. Corrosion from the smelt itself is negligible at normal tube metal temperatures (below 345 °C, or 650 °F). Stud wastage will occur and require repair before wastage of the underlying tube metal occurs.

In nonwelded wall recovery units, furnace gases and smelt may penetrate to the outside or back-side surface of the tubes because of improper tube or stud alignment and/or flat (fin) stud wastage. This can lead to severe cold-side corrosion, as described in the following discussion "Corrosion by Molten Hydroxides" in this section.

When liquor or smelt is carried into the superheater section, where tube metal temperatures are often above 425 to 510 °C (800 to 950 °F), corrosion may take place as the direct result of the smelt deposits on the metal (see the discussion "Corrosion in the Superheater" in this section).

Some research work indicates that the solidified smelt layer is not always directly attached to the tube metal, but is separated from it by a thin layer of irregularly spaced porous material, probably dried black liquor. This provides a space between tube metal and smelt layer through which corrosive gases may attack the tube metal. The formation of this porous layer spacing will largely depend on operating variables.

Corrosion by Molten Hydroxides. Highly volatile hydroxides of sodium and/or potassium are released from the furnace bed and may diffuse to the cold side of the furnace tubes, especially close to the primary air ports. Because temperatures are much lower here, condensation/sublimation of the hydroxides may occur.

Liquid hydroxide causes corrosion of the tube metal. Sodium hydroxide has a melting point of 315 °C (600 °F) and may exist in the liquid state in high-pressure boilers. In combination with carbonates, sulfates, and potassium, deposit melting points have been found as low as 250 °C (480 °F). Laboratory experiments show even lower melting points (as low as 170 °C, or 340 °F), when substantial amounts of potassium hydroxide are also present. The condensation/sublimation deposits on lower-temperature metal surfaces are characterized by low sulfide content.

In practice, molten hydroxide corrosion does not appear to take place in boilers with operating pressures below 6000 kPa (870 psig), because of lower saturation temperatures. The potassium content, which greatly affects the deposit melting point, can vary from mill to mill, depending on wood species and the source of chemical make-up.

Molten hydroxides are much more corrosive toward stainless steel than carbon steel. There have been numerous reports of tube-side corrosion on the water walls of composite tube recovery boilers. Once cladding is wasted to expose underlying carbon steel, subsequent attack is much slower. This form of attack of stainless steel occurs at ports in the water walls, particularly the primary air ports.

Corrosion in the Superheater

Tube metal temperatures are highest in the superheater section. Superheated steam temperatures at the superheater outlet can be as high as 510 °C (950 °F). Outside tube metal temperatures make the superheater metals vulnerable to high rates of corrosion. At these temperatures, carbon steel is not corrosion resistant, even if the corrosive environment consists only of oxygen with no smelt or sulfur-bearing compounds present. Superheater tube materials in the high-temperature section consisting of structural steels, such as T-11 and T-22 (see Table 8 for compositions), and stainless steels have a higher corrosion resistance than carbon steel. Even these higher-alloy steels may also be vulnerable to corrosion.

Corrosion by Flue Gas Constituents. Under normal conditions, ash and/or black liquor carryover is deposited on the superheater tubes and dislodged by sootblower action. The porous structure of the ash deposits will allow bare tube metal to be exposed to the flue gases. The exposed surfaces are subject to corrosion by the flue gases, which contain the following:

Compound	Concentration
Oxygen (O_2)	0–5%
Hydrogen sulfide (H_2S)	0–50 ppm
Sulfur dioxide (SO_2)	0–500 ppm
Sulfur trioxide (SO_3)	0–50 ppm
Carbon monoxide (CO)	0–5%
Water vapor (H_2O)	30–35%
Carbon dioxide (CO_2)	10–15%
Nitrogen (N_2)	50–55%

Oxygen. Unprotected metals at high superheater temperatures will react with excess oxygen in the flue gas, as described in the discussion "Corrosion by Flue Gas in the Absence of Sulfur Compounds" in this section.

Hydrogen Sulfide. Hydrogen sulfide and other total reduced sulfur (TRS) compounds are generated in the furnace hearth zone and, with proper combustion, will be oxidized to SO_2 by the time the flue gas reaches the superheater section. It is possible that the TRS concentration leaving the combustion zone is in the range of 0 to 1000 ppm, while O_2 concentrations are normally 2 to 5%. In this case, the H_2S/O_2 ratio will be far below 1, which is the critical ratio for iron attack by dissociated H_2S. Hydrogen sulfide dissociates as follows:

$$2H_2S + O_2 \rightarrow 2H_2O + 2S \quad \text{(Eq 8)}$$

$$Fe + S \rightarrow FeS \quad \text{(Eq 4)}$$

When TRS generation in the furnace is excessive and air supply to the tertiary and/or secondary combustion zones is insufficient to oxidize TRS, then the H_2S/O_2 ratio may increase to its critical ratio of 1, and iron attack by sulfur will be at a maximum rate. Therefore, excessive TRS and the use of low excess air may lead to this type of superheater corrosion. However, generation of TRS in the furnace is minimized by proper air distribution. Maintenance of proper air distribution usually avoids this type of corrosion. Further, proper air distribution is essential for compliance with air emission standards.

Sulfur dioxide is formed in the combustion zone by the oxidation of sulfur bearing compounds. Sulfur dioxide causes negligible corrosion in the superheater.

Sulfur trioxide can be formed through the oxidation of SO_2 at high temperatures in the gas phase:

$$2SO_2 + O_2 \rightleftarrows 2SO_3 \quad \text{(Eq 10)}$$

Equation 10 is an equilibrium reaction, and under normal furnace conditions, only a small amount of the SO_2 present will react to form SO_3.

Another mechanism for formation of SO_3 is the catalytic action of metal surfaces along with the flue gases flow. Oxides of iron and oxides of vanadium are known catalysts. Excessive firing of oil with a high vanadium content (fuel oil may contain up to 500 ppm V and 3% S) may lead to deposits of vanadium pentoxide (V_2O_5), which enhances the formation of SO_3.

Sulfur trioxide in the gas phase does not impose a corrosion problem. However, when it combines with water vapor and condenses, corrosive H_2SO_4 and/or sulfurous acid (H_2SO_3) are formed. This type of corrosion is known as dew-point corrosion and occurs when flue gases are cooled to the temperature at which condensation takes place (see the discussion "Corrosion in the Economizer" in this section). Another harmful effect of SO_3 is adsorption on Na_2SO_4 deposits in the superheater and boiler bank areas; this lowers the pH of these deposits and causes sticky ash formation.

Carbon Monoxide. Incomplete combustion leads to CO in the flue gas. Carbon monoxide alone does not cause corrosion, but it slightly increases the corrosion process by H_2S gases due to its reducing characteristics. Maintaining sufficient and proper air delivery, especially secondary/tertiary air, will minimize or eliminate the presence of CO in the flue gases.

Water Vapor. The reaction of iron metal with water vapor:

$$3Fe + 4H_2O \rightleftarrows Fe_3O_4 + 4H_2 \quad \text{(Eq 11)}$$

is favored by thermodynamics to proceed to the right. The reaction rate, however, is slow, and corrosion inside or outside of the tubes is not usually due to this reaction.

Carbon dioxide alone seldom causes corrosion. In combination with Na_2S, however, it releases sulfur compounds from the smelt (Eq 6), leading to the corrosion reaction of sulfur with iron.

Nitrogen is inert and does not participate in any of the combustion or corrosion processes in recovery boilers.

Formation of Superheater Ash Deposits

Deposits will normally be present on superheater tubes. The source, structure, and composition of these deposits can vary. Recent research has established that two basically different mechanisms cause the formation of superheater deposits:

- Carryover of black liquor particles
- Vaporization/condensation of sodium compounds

These two mechanisms act alone or in combination, depending on liquor composition, bed temperature, combustion air delivery system, and liquor spray pattern. The structure and composition of the deposits vary according to aging factors and the mechanism by which they were formed. The sodium to sulfur ratio in the flue gas is an important parameter of corrosion/deposition chemistry. This ratio is usually 1.4 or higher in kraft recovery flue gases.

Deposits From Black Liquor Carryover. Black liquor particles can be entrained in the combustion gases. The organics will be burned off, leaving predominantly Na_2CO_3, Na_2SO_4, Na_2S, and some NaOH. These deposits are found at lower elevations of the radiant superheater surfaces close to the furnace, especially at the leading side of the tubes. The carryover deposits may sinter and fuse, thus forming a hard, thick layer on the tube surface. This type of deposit formation may be found on units that are significantly overloaded.

Deposits From Vaporization/Condensation of Sodium Compounds. Vaporization of sodium and sodium compounds begins in the furnace bed. These compounds condense on the cooler sections of the superheater and the boiler bank tubes, forming a thin white deposit consisting mainly of sodium sulfate and some sodium carbonate. A minor amount of sodium chloride may also be present. Condensation of these compounds may occur on bare tubes or in the porous spaces between the previously laid down carryover deposits. This results in a mixture of carryover and condensation deposits.

Deposits formed by the vaporization/condensation process are likely to be found at higher, cooler elevations away from the furnace and especially on the trailing side of the tubes where flue gas velocities are lower. Also, at lower elevations closer to the furnace, some condensation products may be found on the trailing side, where they are shielded from the gas stream.

Table 10 Typical tube deposit analysis

Deposit	Superheater tube leading side; carryover	Boiler bank tube trailing side; vaporization/condensation
Na_2CO_3,%	37.6	0
Na_2SO_4,%	60.5	97.6
Other, %	1.9	2.4
pH saturated solution	12.0	4.4

The type of process by which the deposits are laid down determines their chemical characteristics and composition. Carryover deposits will be richer in sodium carbonate. The pH of a saturated solution of this material is usually well above 7 (alkaline). Vaporization/condensation deposits usually do not contain much sodium carbonate, but consist mainly of sodium sulfate. The pH of a saturated solution is slightly below 7 (acid) or can be much lower because of SO_3 adsorption on the surface. A typical deposit analysis is shown in Table 10.

Deposit Enrichment With Chlorides. The two mechanisms of deposit formation—liquor carryover and vaporization/condensation—occur independently and may lead to a mixture of deposits. The porous spaces in the carryover deposits may be filled with deposits from condensation. These deposits are commonly found close to the tube metal where the temperature is lowest.

Sodium chloride (NaCl), potassium chloride (KCl), and potassium hydroxide (KOH) are among the most volatile chemicals present in the recovery furnace. They may also be found in high concentrations, together with deposits from condensation (Na_2SO_4), closest to the tube metal.

Research work indicates that Cl/Na and K/(K + Na) ratios increase when progressing from the outside of the deposit layer to the inside toward the tube metal surface. The increased amount of chlorides and potassium in the deposits is of particular concern with closed chemical recovery systems (that is, those with bleach plant effluent recovery) and in mills processing seawater-borne logs. In other mills, chloride levels in the liquor cycle can also be excessive. Sodium chloride concentrations up to 40 g/L in the white liquor have been reported (the corrosive influence of chlorides is explained in the discussion "Effects of Chloride Compounds" in this section). A comparison of kraft recovery superheater deposit compositions from different types of mills is given in Table 11.

Corrosion Mechanisms in Superheater Ash Deposits

The adsorption of SO_3 on ash deposits and the resulting decrease in pH impose a serious corrosion problem in that an acidic environment is created. This type of corrosion may occur especially in sulfite liquor systems with high sulfidities, but may also be found in kraft liquor systems. It has been found that Na_2SO_4 and SO_3, in combination with water vapor, may form sodium pyrosulfate ($Na_2S_2O_7$) and sodium bisulfate ($NaHSO_4$) according to:

$$Na_2SO_4 + H_2O + SO_3 \rightarrow Na_2S_2O_7 + H_2O \quad \text{(Eq 12)}$$

Table 11 Examples of kraft recovery superheater deposit compositions

Deposit, %	Mill A(a)	Mill B(b)	Mill C(c)	Mill D(d)
NaCl	2.2	16.0	22.5	3.7
Cl/Na (molar)	4.9	25.5	26.8	4.4
K/(K+Na) (molar)	1.7	5.7	7.5	11.4

(a) Inland pulp mill with bleach plant (open cycle). (b) Pulp mill with bleach plant (closed cycle). (c) West Coast pulp mill using seawater-born logs. (d) Pulp mill using hardwood chips

$$Na_2S_2O_7 + H_2O \rightarrow 2NaHSO_4 \quad \text{(Eq 13)}$$

The adsorption of SO_3 and the reaction with Na_2SO_4 occur only in deposits depleted of Na_2CO_3. Sodium trioxide will be neutralized when carbonate is present, such as in the deposits from black liquor carryover (that is, on the superheater and boiler bank tubes closest to the furnace), according to:

$$Na_2CO_3 + SO_3 \rightarrow Na_2SO_4 + CO_2 \quad \text{(Eq 14)}$$

This reaction proceeds rapidly at temperatures above 400 °C (750 °F).

When carbonate is no longer present in the deposits and SO_3 is still present in the flue gas, $Na_2S_2O_7$ may be formed. At lower temperatures, $NaHSO_4$ may also be formed.

The corrosive attack from $Na_2S_2O_7$ is thought to occur only in the liquid state. The liquid state of pyrosulfate may occur in many areas of the recovery boiler because its melting point is about 400 °C (750 °F). Solid $Na_2S_2O_7$ is not particularly corrosive. In recovery systems with high potassium content, sodium and potassium pyrosulfates ($Na_2S_2O_7/K_2S_2O_7$ eutectic) may be liquid at temperatures as low as 280 °C (540 °F).

Sodium bisulfate is stable at temperatures below 280 °C (540 °F) and will therefore not be found in the higher-temperature sections in the superheater. However, $NaHSO_4$ can form on lower-temperature surfaces of the superheater, boiler bank, and economizer. Fouling and corrosion of these metal surfaces may result because of the presence of molten $NaHSO_4$. This corrosion mechanism may be found in kraft recovery boilers having low sodium to sulfur ratios and high SO_3 concentrations in the flue gas and in a narrow temperature range from approximately 185 to 280 °C (370 to 540 °F).

In kraft pulp mills having normal white liquor sulfidities (25 to 30%) and normal recovery furnace operation, the concentration of SO_3 in the flue gas will be low, and it all will likely be neutralized by Na_2CO_3. Some of the following conditions, however, may increase SO_2 and therefore SO_3 formation and lead to acidic deposits or sticky ash:

- High white liquor sulfidity and low sodium to sulfur ratio in the liquor
- Low char bed temperature
- Excessive or uncontrolled spent acid additions
- Excessive firing of oil with high sulfur and vanadium content (catalytic formation of SO_3)
- High potassium content in the liquor

Any one of these conditions or a combination may be conducive to the formation of acidic sulfates and may require special attention and correction.

Effects of Chloride Compounds. The role of chloride compounds in superheater corrosion is not clear. Little research has been done on these corrosion mechanisms, and it is thought that chloride compounds may either directly attack tube material or decrease the melting point of deposits, thus causing fluxing of normal protective metal oxides.

Sodium chloride vapor, when condensed in the tube surface deposits, may release HCl in accordance with the following reaction:

$$4NaCl + 2SO_2 + O_2 + 2H_2O \rightarrow 4HCl + 2Na_2SO_4 \quad \text{(Eq 15)}$$

Subsequent reactions of HCl fumes with tube material (either metal or metal oxides) may involve the formation of volatile iron chlorides and/or metal oxychlorides.

From laboratory experiments, it is known that the deformation and melting points of a mixture of Na_2S, Na_2CO_3, and Na_2SO_4 can be greatly affected by the NaCl content. The lowest first deformation point of a mixture of Na_2S, Na_2CO_3, Na_2SO_4 and NaCl has been found below 595 °C (1100 °F). The deformation and melting points can be lowered by as much as 220 °C (400 °F) with NaCl concentrations above 10%. Potassium may decrease these melting points even further.

In closed cycle mills with bleach plant effluent recovery or in mills using seawater-borne logs, the chloride input to the recovery boiler can be high. Because of the condensation/vaporization mechanism, the NaCl content of superheater deposits has been shown to be above 15%, and deposit deformation and melting may occur around 595 °C (1100 °F). Erosion-corrosion of tube metal may occur when 70% or more of the total deposit is in the liquid phase, thus making the entire deposit flux.

In kraft pulp mills without bleach plant effluent recovery or seawater-borne logs, the deposit melting point is not significantly affected, and subsequent fluxing of deposits does not usually occur. However, some conditions may cause high chloride levels or other conditions conducive to deposit fluxing, such as:

- High chloride input to the system due to a high chloride content in the spent acid from ClO_2 generation
- Excessive or uncontrolled spent acid additions producing temporarily high inorganics input to the liquor
- Local hot spots in the superheater due to flue gas channeling or improper design features

These conditions should be avoided or minimized.

Effects of Miscellaneous Chemistry. Superheater metal wastage may also occur as a result of the reaction and/or decomposition of sodium-sulfur compounds at elevated temperatures. The Na_2S and $Na_2S_2O_3$ contents in smelt or ash deposits between the furnace hearth zone and the superheater will generally decrease from their concentrations in the smelt to deplete the superheater tube deposits unless excessive liquor/smelt carryover takes place. Usually, a small amount of Na_2S can be found in the deposits at the superheater bends.

In laboratory experiments, it has been shown that sodium-sulfur compounds may react or decompose at elevated temperatures, resulting in the formation of free sulfur. These experiments (at 370 °C, or 700 °F) indicated the following reactions:

$$Na_2SO_4 + Na_2S \rightarrow Na_2SO_3 + S + Na_2O^* \quad \text{(Eq 16)}$$

$$Na_2S + O_2 \rightarrow Na_2S_2O_3 + Na_2O^* \quad \text{(Eq 17)}$$

$$Na_2S_2O_3 \rightarrow Na_2SO_3 + S \quad \text{(Eq 18)}$$

$$Na_2S + 2CO_2 \rightarrow Na_2CO_3 + CO + S \quad \text{(Eq 6)}$$

$$2Na_2S + 3SO_2 \rightarrow 2Na_2SO_3 + 3S \quad \text{(Eq 7)}$$

Equations 6, 7, and 16 to 18 also occur in the furnace bed and around the water wall tubing where tube metal temperatures are much lower.

The formation of free sulfur at the superheater lower bends may result in metal attack:

$$Fe + S \rightarrow FeS \quad \text{(Eq 4)}$$

when bare tube metal is not protected.

*Sodium oxide (Na_2O) is very reactive and combines quickly with CO_2 or H_2O to form Na_2CO_3 or NaOH, respectively.

Corrosion in the Boiler Generating Bank

Tube metal temperatures in the boiler bank area are lower than in the superheater. This makes boiler bank tubing less vulnerable to corrosion.

Certain conditions, however, may cause elevated temperatures in the boiler bank riser tubes. This may lead to accelerated corrosion. In localized areas, conditions that cause overheating include:

- Internal scale
- Flue gas channeling
- Mechanical blockage of water flow
- Low water level
- Improper water distribution/circulation

At moderately elevated temperatures, accelerated corrosion may result when the tube metal is subjected to a corrosive environment.

The corrosion mechanisms in the boiler bank are the same as for the superheater, that is, corrosion by flue gas constituents or by ash deposits. Lower temperatures in the boiler bank result in lower corrosion rates. In addition, there is less chance for the formation of chloride-containing deposits.

Corrosion at the Drum Tube Seats. Tube metal wastage may occur at the drum tube seats. Corrosion is due to the combined action of chemicals and water and may be accelerated in the high stress areas resulting from the tube rolling technique or vibration-induced stresses. The water present may be from tube seat leakage, water-washing remains, soot blowing, or absorption by deposits around the tube seat.

Erosion by Sootblower Steam/Water Mixtures. Boiler bank tubes, superheater tubes, and economizer tubes may be subject to steam/water mixture sootblower action, resulting in tube metal wastage. The damage usually appears as localized external wall thinning, and the affected area usually has a smooth, polished appearance. Proper sootblower spacing and care in piping design (which ensures that only dry steam is used for sootblowing) and proper maintenance of steam traps should be exercised. Leaking poppet valves cause steam to condense in the lance tube when the sootblower is idle. This water runs from the nozzle and may cause corrosion of wall tubes and casing. Leaking poppet valves and defective steam traps should be promptly repaired.

Corrosion in the Economizer. External corrosion of economizer tubes may occur when economizer flue gas exit temperatures and/or feedwater temperatures are low. The flue gas exit temperature and feedwater temperature entering the economizer determine the tube metal temperature. The tube metal temperature may drop below the acid dew point. The dew-point temperature is highly affected by SO_3 concentrations in the flue gas. An increase in SO_3 in the flue gas elevates the dew-point temperature. Therefore, condensation is more likely to occur.

It is generally good practice to maintain the flue gas temperature at the economizer outlet at or above 160 °C (325 °F) and to maintain the feedwater temperature entering the economizer at or above 135 °C (275 °F) in order to prevent condensation. If condensation occurs, the product is dilute H_2SO_4:

$$SO_3 + H_2O \rightarrow H_2SO_4 \quad \text{(Eq 19)}$$

The H_2SO_4 attacks metals and protective iron oxide coatings:

$$Fe + H_2SO_4 \rightarrow FeSO_4 + H_2 \quad \text{(Eq 20)}$$

$$Fe + Fe_3O_4 + 4H_2SO_4 \rightarrow 4FeSO_4 + 4H_2O \quad \text{(Eq 21)}$$

Therefore, corrosion may take place, especially at localized areas where ambient air can leak through doors or holes in the casing. In this type of corrosion, the rate increases as the temperature decreases, not because the corrosion mechanism itself proceeds faster at lower temperatures, but because more of the corrosive H_2SO_4 is formed.

Economizers are particularly susceptible to this type of corrosion because operating gas temperatures are near the acid dew point and any ambient air leakage or other air infiltration, such as at vents, will tend to lower gas temperatures in localized areas. It should be noted that economizers are particularly susceptible to high leakage/infiltration due to high-draft conditions.

Acid dew-point corrosion can also occur in other downstream components of the system, such as electrostatic precipitators and flue gas scrubbers. Corrosion in the economizer can also be caused by the formation of molten bisulfates below 280 °C (540 °F), as explained in the discussion "Adsorption of SO_3" in this section.

Suction Roll Corrosion

Max D. Moskal
Stone Container Corporation

Suction rolls are used to remove water from paper at the wet end of the paper machine. One way to accomplish this is by passing the paper web through a roll nip, one roll of which is the suction roll. The suction roll is drilled to an open

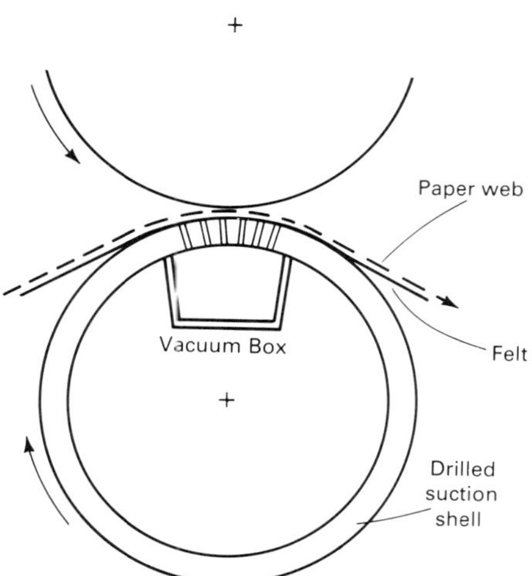

Fig. 25 Cross section of suction roll configuration

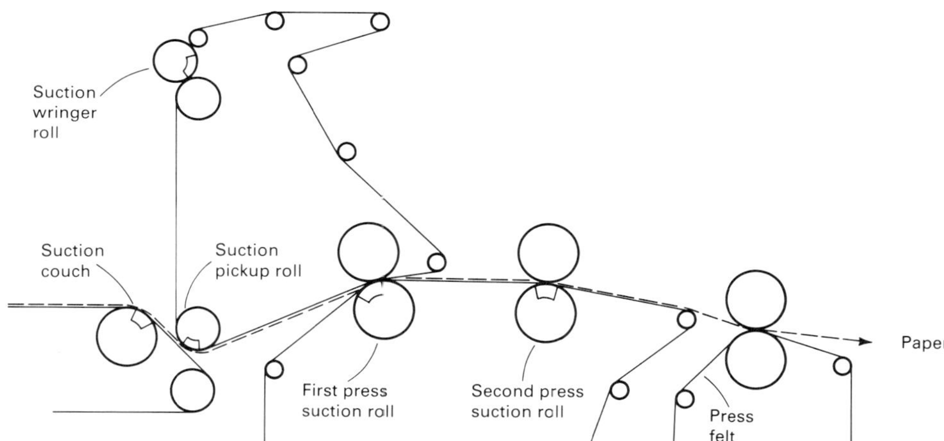

Fig. 26 Schematic diagram of wet end of paper machine showing types of suction rolls

area of about 20%, and a vacuum is applied to the inside of the roll. A typical cross section through a suction roll is shown in Fig. 25. A rubber cover is often used on the suction press roll or mating roll to give a desired nip pressure relationship. Many machine designs have been used in the wet end of paper machines throughout the years. A typical paper machine may use three or more suction rolls designated as couch, wringer, pickup, or press rolls (Fig. 26).

The critical component in the suction roll is the drilled shell, which is subject to corrosion and fatigue cracking. During the past 30 years, changes have evolved in design and machine environment that have increased the cracking and corrosion tendency of the suction shell. For example, demands for increased machine production and speed have been resolved by increasing roll nip pressures. Wider machines have also been designed. These new designs have required heavier and stronger shells to meet the increased stress that is imposed. The corrosive environment in the paper machine wet end has also become more severe as water systems have been closed, water temperatures increased, and chemicals used to clean paper machine felts. Numerous changes have also occurred in a search for materials and designs that better resist shell corrosion and corrosion-assisted cracking.

This section will describe the history of suction roll and materials development, the operating environment, corrosion fatigue of suction rolls, shell manufacturing methods, and corrective measures.

History of Suction Roll Alloys

Prior to about 1950, suction shells were cast from bronze having a composition of approximately 85% Cu, 5% Pb, 5% Zn, and 5% Sn. Bronze has many advantages, such as castability, machineability, corrosion resistance, and relatively low cost. However, bronze lacked the strength needed to resist the higher press loads that were introduced at that time, and it lacked the stiffness necessary to maintain shape and roundness on larger and wider paper machines. In 1964, it was noted that suction roll nip pressures had increased from the original levels of 26 kN/m (148 lb/in.) up to 79 kN/m (451 lb/in.) (Ref 29). Designs with nip pressures to 131 kN/m (750 lb/in.) were proposed in the 1960s and were actually achieved in the 1970s. Also, paper machine widths increased from typically 3.8 m (150 in.) in the 1950s to over 6.3 m (248 in.) in the 1970s, with maximum machine widths achieved in excess of 10 m (394 in.).

The first stainless steel materials to be widely used were forged type 410 martensitic alloy and centrifugally cast CF-3M and CF-8M austenitic alloys. The austenitic alloys were deemed especially desirable for low-stress, high-corrosion applications, such as the couch position. From 1955 to 1965, the influence of corrosion on fatigue was not widely recognized. In one case (Ref 29), shell designs were based on the fatigue strength of these alloys in air (Fig. 27).

In 1957, the first centrifugally cast CA-15 martensitic stainless steel shells were cast. Also during this early period, shells were cast from alloy A-63 stainless steel and aluminum bronze. Tables 12 and 13 list the nominal compositions of the alloys used for suction shells.

Between 1969 and 1977, a host of new centrifugally cast stainless steel alloys was introduced for suction shells. These were DSS69, C169, alloy 70 (martensitic), and alloys A170, A171, A271, and A-75 (duplex/ferritic-austenitic). Also during this period, rolled and welded stainless steel shells were produced in Sweden from type 316 and 3RE60 stainless steels. Continuous-cast bronzes GC-CuSn5ZnPb and GC-CuAl9,5Ni also emerged during this period.

In 1971, the reduced fatigue strength of bronzes and stainless steels in synthesized white water was reported (Ref 30). The importance of residual stresses on fatigue cracking was also more widely recognized at this time.

Between 1978 and 1986, additional new stainless alloys were introduced. These included centrifugally cast precipitation-hardening duplex alloys VK-A378 and KCR-A682 and duplex alloy A-86. In 1978, the argon-oxygen decarburization process was introduced in the United States for suction shells. This permitted more economical melting and casting of stainless alloys with a very low carbon content. Centrifugally cast CF-3M has since been produced instead of CF-8M. Several forged alloys were also introduced during this period: PM-4-1300 and PM-4-1300M (martensitic), PM-3-1811MN (austenitic), PM-3-1804M, and PM-2-2505 (duplex).

During the 1980s, emphasis has been placed on stainless steel development with greater corrosion and fatigue resistance and low residual stresses.

Paper Machine Environment

A detailed discussion of the paper machine corrosion environment is given in the section "Paper Machine Corrosion" in this article. However, considerations specific to suction roll corrosion will be covered in this discussion.

Suction rolls are subjected to a variety of corrosive environments. The severity of the environment depends primarily on the type of paper produced on the machine, the source of water, and the degree of closure of the mill. The corrosivity of the environment depends primarily on pH, temperature, dissolved solids content, chlorides, and the presence of sulfur compounds. The area around the suction rolls is subjected to splash from stock and white water, and deposits of paper fiber are present in crevices. Bacteriological growth can also occur, especially in areas that are difficult to treat with biocide chemicals (commonly used biocides are listed in the article "Control of Environmental Variables in Water Recirculating Systems" in this Volume). Suction rolls are also subject to corrosion from chemicals and waters used in showers around the rolls and from felt treatment chemicals.

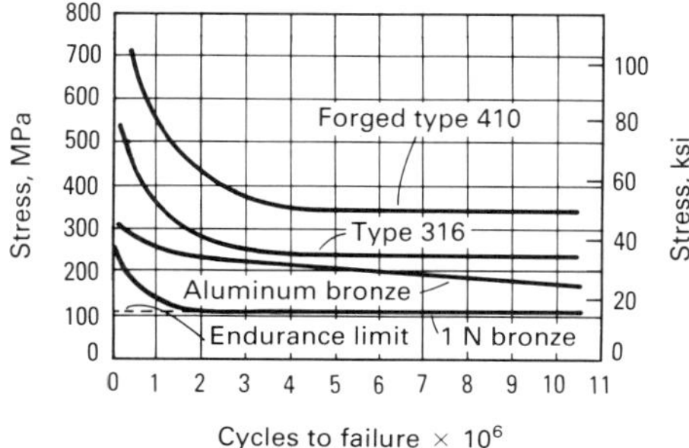

Fig. 27 Typical fatigue curves of suction shell alloys as tested in air. These data were used to establish early designs of suction rolls. Courtesy of the Technical Section, Canadian Pulp and Paper Association

Table 12 Composition of copper-base suction shell alloys

Material	Composition, %							
	Cu	Sn	Pb	Zn	Al	Ni	Fe	Mn
IN bronze(a)	85.4	4.7	4.6	4.6	...	0.6	...	...
GC-CuSn5ZnPb(b)	84–86	4–6	4–6	4–6	...	...	...	...
GC-CuA19,5Ni(b)	81.5–83.5	...	...	...	9.0–9.5	4.0–4.5	3.0–3.5	0.5–1.0

(a) Typical composition. (b) Composition range

Table 13 Nominal composition of stainless steel suction shell alloys
Materials are categorized according to structure.

Material	Composition, %(a)						
	C	Cr	Ni	Mo	Cu	Mn	Si
Austenitic							
CF3M	0.02	17.7	13.8	2.3	...	1.3	0.8
CF8M(b)	0.05	17.7	13.8	2.3	...	1.3	0.8
PM-3-1811MN	0.015	16.5	13.5	2.1	...	1.6	0.5
Martensitic							
C-169, CA-15	0.07	12.4	0.6	0.5	...	0.5	0.6
PM-4-1300	0.12	12.5	0.4	0.4	...	0.7	0.5
DSS-69(c,d)	0.04	12.4	4.0	0.7	...	0.7	0.6
A-70(d)	0.03	11.9	4.0	1.5	...	0.8	0.5
Duplex							
A-63(d)	0.05	21.8	9.4	2.7	...	0.8	1.3
A-170(d)	0.07	23.3	10.7	2.1	...	0.7	1.5
A-171	0.07	22.2	8.3	1.2	...	0.8	1.1
A-271	0.06	24.6	4.3	0.7	...	0.7	1.3
3RE60(e)	0.02	18.5	4.7	2.8	...	1.5	1.7
A-75	0.02	26.0	6.8	...	...	0.8	0.5
A-86	0.02	26.0	6.0	...	2.8	0.8	0.7
VK-A378(f), A682(g)	0.06	20.0	5.0	2.0	4.5	0.6	0.8
PM-3-1804M	0.06	17.9	4.0	2.0	...	0.6	0.6
PM-2-2505	0.07	26.0	4.0	0.8	...	1.2	1.3

(a) Remainder of compositions is iron. (b) Alloy has been discontinued for suction shells in North America. (c) Details of composition not published, but similar to ASTM A-296, alloy CA-6NM (discontinued specification). (d) Obsolete alloy. (e) Contains nitrogen. (f) Contains nitrogen and tungsten. (g) Alloy A-682 has a composition similar to that of VK-A378, only reduced chromium and nickel.

Fig. 28 Corrosion damage to a CA-15 stainless steel shell caused by muriatic acid felt cleaning

Laboratory studies have shown that $Na_2S_2O_3$ in stock and white water systems results in severe pitting corrosion of the less resistant martensitic stainless steels, and thiosulfate is believed to be important in corrosion and cracking of suction roll shells. Sodium thiosulfate, residual sodium hydrosulfite ($NaHSO_2$), and the sulfide ion are also highly corrosive to bronze.

Biological corrosion has also been observed within the drilled holes of suction shells. The hole area is highly favorable to the growth of microbes because of the presence of dissolved organic materials, dissolved inorganic salts, and a favorable temperature range of 40 to 50 °C (105 to 120 °F). Anaerobic sulfate-reducing bacteria thrive in crevice locations and under salt and fiber deposits. Therefore, because of the concentration of ions and the localized pH reduction, the chemical environment beneath deposits can become far more aggressive than the bulk stock or white water chemistry. Additional information on biological corrosion can be found in the articles "Effects of Environmental Variables on Aqueous Corrosion," "General Corrosion," "Localized Corrosion," and "Evaluation of Microbiological Corrosion" in this Volume.

The showers and chemicals used on the paper machine also influence the environment around suction rolls. Showers are used to clean machine felts, to remove deposits from drilled holes in suction rolls, and to lubricate roll seal strips. A wide variety of shower cleaning chemicals are used for felts and rolls, including acids, aliphatic and aromatic solvents, or animal and vegetable oils. Lubricating and needle showers are often supplied with white water at temperature ranges of 40 to 50 °C (105 to 120 °F); this white water is more corrosive to the suction roll than freshwater. Attack from chemicals has also contributed to failure of elastomeric covers on suction rolls (Ref 31). Figure 28 shows damage to a CA-15 martensitic stainless steel suction shell from muriatic acid used for felt cleaning.

Off-line chemical cleaning has been used to remove deposits from drilled holes in suction rolls. There have been numerous cases of severe damage to both shells and rubber covers when corrosive cleaning chemicals were used (Ref 32).

Corrosion Fatigue Testing

In the 1960s, several investigators recognized the need to obtain corrosion fatigue data for suction shell materials using a synthesized paper machine white water environment. Corrosion fatigue testing of suction materials was first described by R. Thompson, who used a conventional R.R. Moore rotating-beam machine equipped with a corrosion chamber around the specimen (Ref 33). Liquid was continuously dropped onto the stressed portion of the specimen. Corrosion fatigue tests were conducted using a reverse bending plate with a single hole drilled in the stressed regions (Ref 34). Crack growth rate tests were employed to evaluate the corrosion fatigue characteristics of suction roll materials (Ref 35, 36). These investigators used compact-tension specimens subjected to cyclic loading in an electrohydraulic fatigue machine.

Corrosion fatigue test data have been published largely by shell producers and paper machine equipment manufacturers. A variety of test data have been shown for all of the alloys used for suction rolls. Because different test methods and a variety of environments have been used for tests, direct comparison of corrosion fatigue strength data is very difficult.

Rotating-bending fatigue test data for several alloys in white water at pH 3.5 and varying chloride and sulfate contents are shown in Fig. 29. Rotating-bending fatigue test data by a different investigator for alloys in a highly aggressive white water are shown in Fig. 30.

In 1985, the TAPPI Corrosion and Materials Engineering Committee summarized survey results on the correlation among corrosion fatigue testing methods. The survey included data provided by ten investigators on the type of testing used, specimen configuration, test conditions, and environment (Ref 37). The TAPPI committee also published documents describing standard methods and test environments for suction shell alloys (Ref 38, 39).

Recent investigations in near-threshold fatigue crack growth testing have indicated that mean stresses could have a significant effect on the threshold for crack growth in duplex stainless steel alloys (Ref 36). These data correlate with the concept that high residual tensile stresses in suction rolls decrease fatigue life in service.

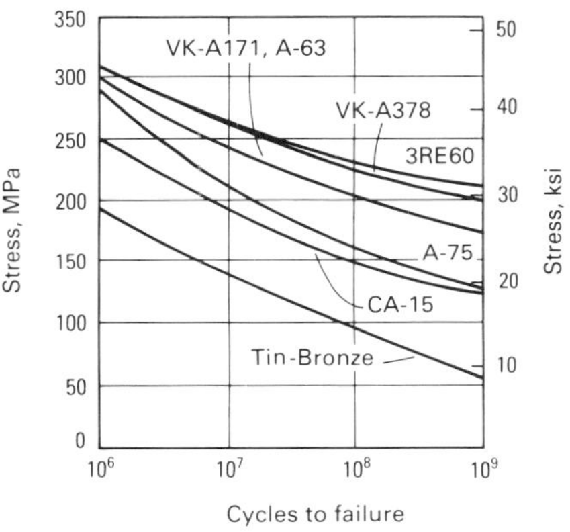

Fig. 29 Fatigue curves in white water based on the data of material suppliers. 1500- to 1750-rpm rotating-bending test, pH = 3.5, Cl^- = 20-400 ppm, SO_4^{2-} = 250-1000 ppm. Courtesy of the Technical Association of the Pulp and Paper Industry

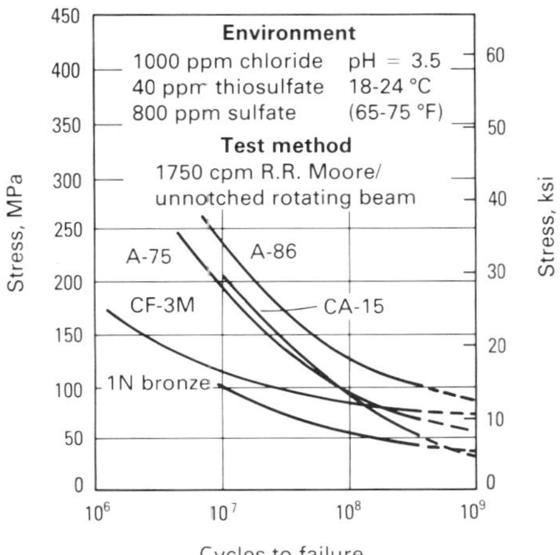

Fig. 30 Fatigue curves for suction shell alloys in an aggressive environment. Courtesy of Sandusky Foundry & Machine Company

Roll Failures

Corrosion and cracking failures have been reported for couch, press, and felt wringer rolls. There has been no widespread problem with the lightly loaded suction pickup rolls. Failure by cracking is frequently unexpected, and is usually the most cause for concern. When cracks progress a significant distance around the shell, the suction roll begins to vibrate and must be immediately removed from the paper machine Deterioration of shells by general corrosion and pitting is another important means of failure. Severe roughening of the outside diameter surface of bronze couch rolls is known to reduce the life of fourdrinier wires. An example of outside diameter surface corrosion in bronze is shown in Fig. 31. General corrosion in drilled holes results in hole enlargement, decreasing the land area between holes to a point at which corrosion fatigue cracking can occur (Fig. 32).

Most suction rolls that are reported to have failed have been removed from service because of cracking. Cracked shells of bronze and martensitic stainless alloys almost always exhibit a visible condition of surface corrosion or pitting in the drilled holes. Higher-alloyed duplex and austenitic stainless steels seldom show visible corrosion, although fatigue failures in these alloys are known to be corrosion assisted.

Failure Experiences. Two separate surveys covering roll failures have been published since 1979 (Ref 32, 40). The surveys indicated that most failures occurred in bronze and martensitic stainless steel shells. However, it was also noted that there are far more bronze and CA-15 shells in service as compared to other alloys. Cracking was also more prevalent in the higher-speed and more corrosive newsprint machines than in fine paper, pulp, or board machines.

Corrosion Fatigue Failures. Fatigue cracking is first observed in shells as single cracks in the land area between two drilled holes or as cracks that begin at a hole and have not yet progressed to a neighboring hole. Cracking then proceeds across the land regions connecting several holes, eventually joining and proceeding circumferentially around the shell (Fig. 33). Cracking usually manifests on the roll inside diameter surface near the mid-span, where operating stresses are the highest. Martensitic stainless steels show many fatigue initiation sites through the cross section of the fractured surface (Fig. 34). Bronze and duplex stainless steels seldom show well-defined fatigue initiation sites.

Longitudinal cracking is often observed near the mid-span of austenitic and duplex stainless steel shells, and this has been attributed to overloading or residual stresses from heat treatment. Longitudinal cracks eventually turn to the circumferential direction and proceed around the roll to ultimate failure.

Rolls may develop only a single crack that progresses to failure, but more often, they will develop tens or hundreds of cracks before final failure occurs. Bronze and martensitic steel shells usually exhibit many small cracks before a larger crack develops. Duplex stainless steels develop fewer cracks or a single large crack.

Microexamination of the cracked region frequently shows cracking associated with corrosion pitting (Fig. 35). Cracking has also been attributed to manufacturing defects, poor drilling quality (Fig. 36), excessive porosity (Fig. 37), welding repairs, and electrical discharge machining used to remove broken drill bits. Intergranular cracking has been observed in austenitic stainless steel that was sensitized during heat treatment and in bronze shells that were exposed to mercury and ammonium contamination.

Rubber Cover Failures. A recent Canadian survey showed that the cost of premature rubber cover failures on suction rolls was about as large as the cost of shell failures (Ref 40). In the survey, 38 mills were questioned about rubber cover failures on suction and nonsuction rolls for the year 1983. A total of 70 cover failures were experienced in the 38 mills surveyed.

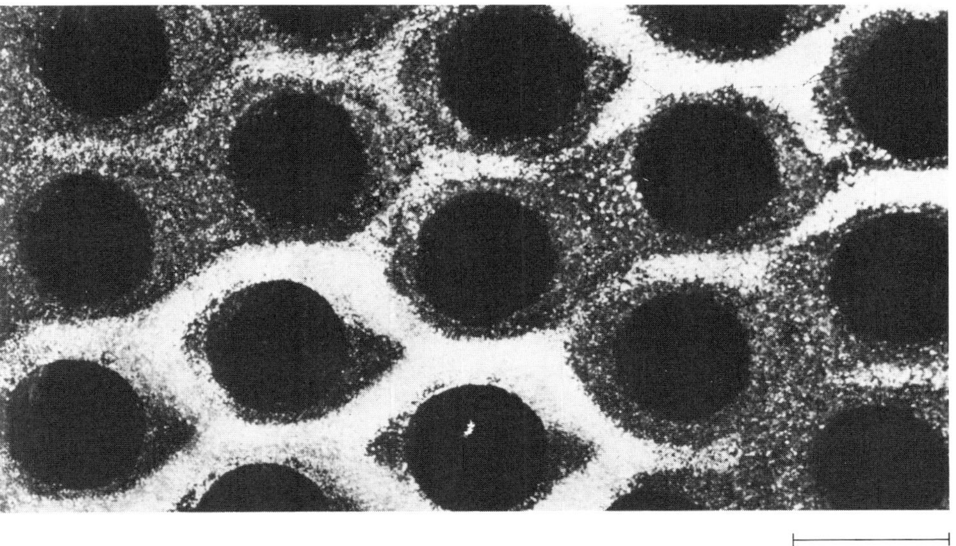

Fig. 31 Thiosulfate corrosion created surface roughening (dark areas) on this uncovered bronze roll. Courtesy of the Technical Section, Canadian Pulp and Paper Association

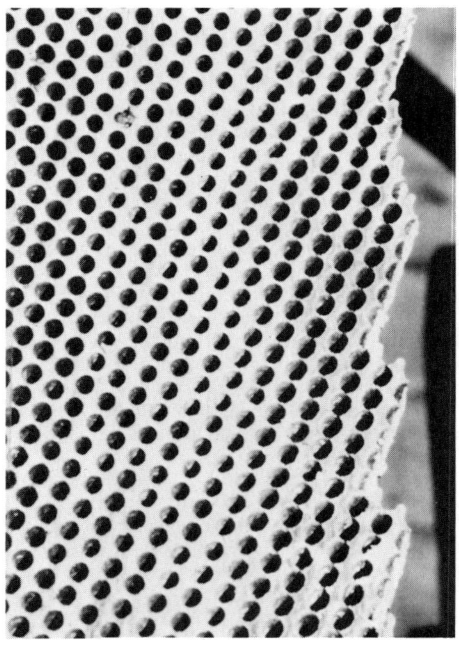

(a)

(b)

Fig. 32 General corrosion in drilled holes. (a) Inside surface of bronze shell adjacent to fracture. The hole enlargement varied periodically along the length of the shell and corresponded to shower nozzle spacing. (b) Close-up view of hole corrosion. Courtesy of MacMillan-Bloedel, Ltd.

Fig. 33 Typical cracking between drilled holes on the outside surface in a stainless steel alloy shell. Courtesy of The Institute of Paper Chemistry

A significant cause of rubber cover failures is deterioration from chemicals used on the suction rolls, both on-line and off-line. Organic solvents have been found to be highly detrimental to rubber covers, especially when used in a continuous mode as opposed to batch treatment. High-pressure hydraulic cleaning can produce mechanical damage and detachment of rubber covers.

Shell Manufacturing Methods

Most suction roll shells are produced by the centrifugal casting method, followed by continuous-casting (bronze), rolled and welded (stainless steel), and forging processes (stainless steel). Each method of manufacturing claims certain advantages. After production of the rough shell, it may be heat treated (depending on the alloy), machined to near-final shell dimensions, and drilled. The shell is then finished on the inside diameter surface and head seats machined. The outside diameter surface is finished by turning, belt sanding, or grinding. A smooth outside surface finish is not required if the shell is to be subsequently rubber covered.

Castings. The process of centrifugal casting is the oldest and most versatile method of shell production. It is applicable for both bronze and stainless steel alloys. Metal is poured into a spinning mold that is horizontally positioned, and the spinning motion is continued until the metal shell has completely solidified. In this process, the last metal to freeze is at the inside diameter surface of the casting. Nonmetallic and intermetallic impurities concentrate on this inner surface, which is later machined away.

Continuous casting is a relatively new method of producing cast shells. It is applicable only to bronze alloys. In this method, the molten metal is poured through holes in a distributing vessel into a vertical collar-type mold. Metal flow, speed of extraction from the mold, and cooling are coordinated so that the casting process is continuous. An advantage of continuous casting is that the uniformity of metal properties through the wall can be readily controlled.

Weldments. Welded stainless steel shells have been produced in Sweden since the early 1970s. Most production has been limited to 3RE60 duplex stainless steel alloy, which is hot formed from plate into semicylinders. The semicylinders are subsequently welded longitudinally to form short cylinders, and several short cylinders are welded at the ends to form the suction shell.

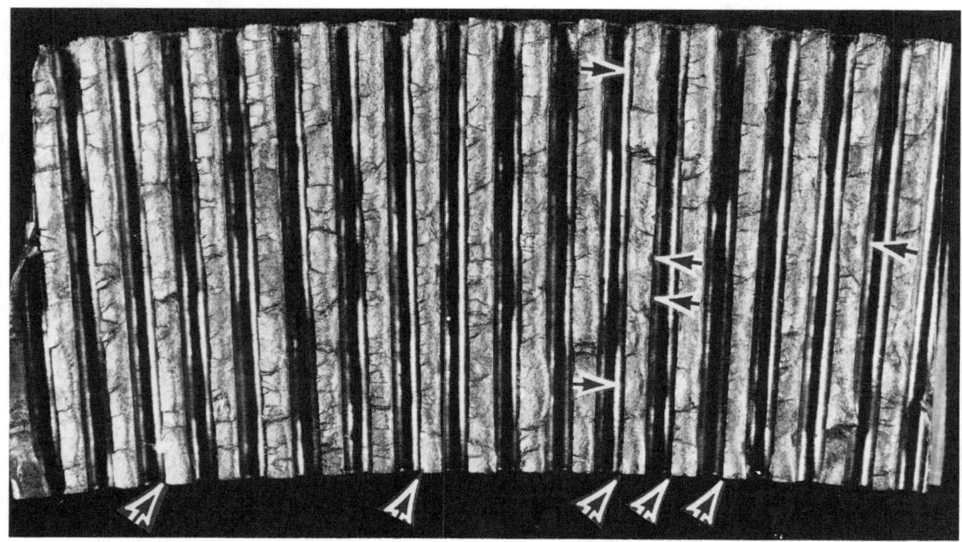

Fig. 34 Fracture surface of CA-15 stainless steel suction shell showing several fatigue nucleation sites (arrows). See also Fig. 35.

Fig. 35 Photomicrograph of specimen in Fig. 34 showing a fatigue crack initiating at the base of a corrosion pit. Etched with Marbles reagent. 75×

Fig. 36 Fatigue cracking of CA-15 stainless steel suction roll due to poor drilling quality

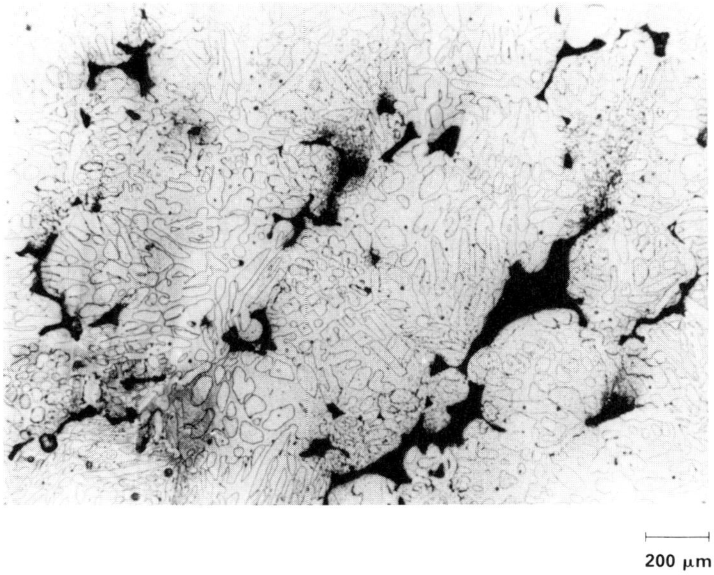

200 μm

Fig. 37 Casting porosity in duplex stainless steel shell

Forgings. Forged suction shells are produced by first upsetting and punching a hollow cylinder from a solid cylinder, then enlarging and forging the shell on a pipe-forming mandrel. The process minimizes the porosity and segregation that may be found in castings, but can be subject to inclusions and other internal flaws.

Heat treatment of suction shells has proved to be a critical operation affecting the corrosion and the residual stress properties of shells. Current practice in heat treatment is limited to stainless steel suction shell alloys. Although aluminum bronze is heat treatable, shells from this material are currently furnished in the as-cast condition.

Each stainless steel alloy requires a specific heat treatment to provide the best combination of corrosion resistance and low residual stresses. Slow cooling from the final heat treatment produces lower residual stresses, but (depending on the alloy) may produce undesirable intermetallic phases and a low corrosion fatigue resistance (Ref 41, 42). Austenitic alloys may be given a single-solution anneal treatment, but martensitic alloys are quenched and tempered. Duplex alloys may be subjected to a single-solution treatment or, in the case of copper-bearing duplex alloys, may be given one or two additional subcritical aging treatments.

Drilling. Suction shells are drilled using multiple-spindle machines. Couch rolls are frequently drilled with larger hole sizes of up to 6.3 mm (0.25 in.), but press roll drilling is typically 4.0 mm (0.16 in.). The drilling process is critical in that hole spacing must be maintained to within close dimensional tolerances.

Gun drilling was first developed in Europe for drilling stainless steels and is particularly applicable when drilling long, straight holes in thick-wall stainless steel shells. Twist drilling of these alloys sometimes results in converging of drilled holes at the inside diameter surface (Fig. 36). Gun drilling also has the advantage of producing a smooth surface finish within the drilled hole. This reduces the tendency toward the initiation of corrosion and deposit formation. Gun-drilled holes have been measured to have a surface finish of $R_a = 0.5$ μm (20 μin.) as compared to $R_a = 5.0$ μm (200 μin.) for twist-drilled holes, where R_a is the arithmetic average roughness. A process of reaming after twist drilling has also been used to produce a good surface finish in the drilled holes.

Shell Quality. Several types of manufacturing defects have been attributed to the failure of suction shells, as discussed above. In an analysis of 52 suction roll failures, about 50% of the failures were attributed to defects in manufacture or design, 35% were due to improper material selection for the environment, and 15% could not be attributed to any cause (Ref 43).

Corrective Measures

Three factors influence suction shell resistance to corrosion fatigue: stress, corrosion environment, and the corrosion fatigue strength of the alloy. A favorable change in any of these factors will produce an increase in shell life.

Bronze shells are occasionally removed from service because of general or pitting corrosion. The factors that influence these failures are the corrosion environment and the resistance of the alloy to the environment.

Good judgment in roll design, material selection, and control of the paper machine environment all contribute to the long life of suction rolls. Good quality control during the manufacture of the shell and timely in-service cleaning and roll inspection also contribute to a favorable service life.

Operating Stresses. Both the applied stress and internal residual stresses inherent in the shell material contribute to the stress in operation of the suction roll. The responsibility for the level of applied stress rests chiefly with the roll designer who selects the limit of design stress for the material selected. Each shell material will have a limit of stress that will give long life in a particular environment. The magnitude of applied stress is a function of roll configuration and applied loads. Drilled hole patterns must also be favorably designed to reduce applied stress. The most important factor influencing the stress in the roll is shell thickness; of course, stress decreases as the shell thickness increases. A stress calculation procedure has been developed by TAPPI that can be used as a guide for determining stresses in suction rolls (Ref 44).

The residual stress that is inherent in the alloy and the heat treatment are also important factors contributing to roll life. Shell materials with known high residual stresses should be designed more conservatively or, preferably, not used at all. This is covered in the discussion "Material Selection" in this section.

Gun-drilled or reamed holes for improved surface finish will reduce corrosion and improve the cleanliness of drilled holes in service. Gun drilling or reaming is usually not performed on bronze shells, but would be helpful in the maintenance of the shell in service.

Corrosion Environment. The papermaker often has little control of some factors influencing the corrosion environment, yet there are many conditions that are within his control. For example, thiosulfate ion contamination and residual hydrosulfite ($NaHSO_2$) in newsprint paper machine white water can occur in $NaHSO_2$ brightening; practices for reducing this problem are addressed in Ref 40. Freshwater showers will provide more effective cleaning and reduce corrosive effects. Good maintenance of the roll is also an important factor in minimizing corrosion. Drilled holes must be kept clean of deposits by the use of needle showers and periodic off-machine cleaning. A careful program of biological control should be maintained. The papermaker should also avoid the use of cleaning chemicals (such as muriatic acid) that are known to be damaging to shell alloys and roll covers. The use of a corrosion inhibitor such as sodium mercaptobenzotriazole should be considered to reduce the corrosion of bronze rolls. The use of inhibitors for couch rolls is described in Ref 45. An example of severe pickup roll corrosion due to thiosulfate that was corrected with inhibitors is discussed in Ref 46.

Material Selection. When a new or replacement roll is contemplated, the question of material selection is of primary importance. Past operating experience with existing suction rolls should be used as a guide in selecting the new shell material. Reference 47 describes conditions that contribute to roll failure and will aid in new shell material selection.

Duplex stainless steel has the highest resistance to fatigue cracking, followed by martensitic stainless steel and bronze. Austenitic stainless steel has good corrosion resistance to most environments, but has relatively low fatigue resistance and moderate-to-high residual stresses (Ref 41, 42). Duplex stainless steels have the highest fatigue strength and generally have moderate-to-low residual stresses; bronze has low residual stress, as does CA-15 martensitic stainless steel.

Selection of a new alloy should involve an evaluation of the condition of the existing roll and the factors contributing to its deterioration. The changes in the corrosion environment that have occurred on the paper machine should also be evaluated. One is tempted to select an identical alloy if the previous material served for 15 or 20 years. However, it is more likely that changes in the machine environment have occurred at a recent time and that the life of a new roll of the same alloy would be substantially reduced. This factor accounts for the recent popularity of the stronger corrosion-resistant duplex stainless steel alloys.

Suction shell material cost is also a consideration. In North America and Europe, bronze is the lowest cost shell alloy, followed by martensitic stainless steel, austenitic stainless, and the duplex alloys. In Japan, bronze is seldom used because of availability and cost.

Manufacturing Quality. There is little published information relating to suction shell manufacturing quality. Shells should be inspected using liquid penetrant prior to drilling to detect such flaws as porosity, slag, and cracks. Skillful control of melting and casting procedures can minimize these defects. Weld repairs should be used with discretion, and only before shell heat treatment. Type 1N bronze should not be weld repaired, because of hot shortness, which develops in the alloy. Careful attention should be given to drilling and machining quality and dimensional control. Industrywide standards giving acceptance criteria for casting flaws and drilling quality have not yet been developed.

In-Service Inspection. Periodic inspection of the suction roll is a critical step that will help to extend its life and minimize unexpected shutdowns. Rolls should be removed from the machine, disassembled, and thoroughly cleaned and inspected at least annually. Inspection should include an evaluation of drilled hole cleanliness, fatigue cracking of the shell, and rubber cover deterioration or detachment. Water-washable liquid penetrant has been successfully used to detect cracks on the inside and outside surfaces. A careful visual examination should also be made for corrosion and cracks.

When cracking is observed during an annual inspection, immediate consideration must be given to replacement because of the long lead time required for shell manufacture. Consideration should also be given to inspecting the roll more often than once per year to observe the rate of crack growth. Thus, a careful record must be made whenever cracks are observed.

Corrosion by Kraft Pulping Liquors

R.A. Yeske
The Institute of Paper Chemistry

The kraft process is the predominant pulping process used in North America to extract fibers from wood for use in the manufacture of paper, tissue, and board. The term kraft is derived from the German word for strong, which reflects the high strength of paper products derived from kraft pulp. This high strength, together with effective methods of recovering pulping chemicals, explains the popularity of kraft pulping. In 1984, more than three-fourths of the pulp produced in the United States—approximately 39 000 000 metric tons, or 43 000 000 short tons—was manufactured by the kraft process (Ref 48).

In the kraft process, hot alkaline sulfide liquor is used to dissolve the lignin from wood chips and to separate individual wood fibers for use in papermaking (Ref 49). Wood chips are exposed to cooking liquors for several hours at elevated temperature and pressure in a process called digestion. Digestion may occur by repetitive batch processes in small batch digesters, or the process may occur continuously in larger continuous digesters. The contents of the digester are then discharged under pressure into a receiver called a blow tank. Finally, the fibers are separated from the spent liquor in a series of washing stages.

Pulping chemicals are recovered from the spent liquor in a series of process steps shown schematically in Fig. 38. First, the spent liquor (known as black liquor) extracted from the pulp is concentrated to 60% solids content in multiple-effect evaporators. A soapy by-product called tall oil is usually extracted from the liquor at some stage of the concentration process. The heavy black liquor is burned as fuel in a chemical recovery boiler. The steam produced by the combustion of the organic constituents of black liquor is used for process use and for power generation. The inorganic constituents in the spent liquor fall to the bottom of the recovery boiler, where they are extracted as a molten salt called smelt.

The smelt recovered from the boiler is converted into cooking liquor in a process known as recausticizing. The molten smelt, which consists primarily of Na_2CO_3 and Na_2S, is dissolved in water to make green liquor. The green liquor is first clarified to remove insoluble dregs and then causticized by exposure to slaked lime. In causticizing, Na_2CO_3 is converted to NaOH by the reaction:

$$Na_2CO_3 + Ca(OH)_2 \rightarrow 2NaOH + CaCO_3 \quad \text{(Eq 22)}$$

The $CaCO_3$ precipitate is removed from the liquor by clarification or filtration, leaving a white liquor with approximately 100 g/L of NaOH, 30 g/L of Na_2S, and lesser amounts of residual carbonates and sulfoxy compounds. The white liquor is stored for reuse as cooking liquor. Meanwhile, the $CaCO_3$ precipitate (lime mud) is washed and converted to lime in a lime-burning kiln. This lime is later slaked and reused in the causticizing processes.

Although some details vary from mill to mill, the process equipment used in kraft pulping is more or less the same throughout the industry. The principal items include pressure vessels for batch or continuous digestion, rotary drum pulp washers, multiple-effect liquor evaporators, storage tanks for various liquors, chemical recovery boilers, electrostatic precipitators, rotary lime kilns, slakers, green and white liquor clarifiers, and liquor storage tanks. Ancillary equipment includes pumps, valves, piping, heat exchangers, control instrumentation, and equipment for tall oil processing. To some extent, diffusion washers are replacing rotary drum washers and pressure filters are replacing clarifiers in the modern pulp mill.

Much of the equipment used in the kraft pulp mill is fabricated from plain carbon steel, although carbon steel has limited resistance to corrosion and cracking when exposed to kraft process liquors. In critical locations (such as evaporator tubes in high-temperature evaporator effects), stainless steel is routinely used because of rapid attack of plain carbon steel. In other locations, stainless steel can be used to avoid the inconvenience of periodic replacement of carbon steel equipment. In general, the high temperatures and alkalinity of kraft liquors prevent the use of polymeric materials of construction in pulp mill equipment. However, brick and tile linings, as well as sprayed-on concrete linings, are often found in recausticizing equipment, particularly where abrasion is a concern. Periodically, pulp mill equipment is acid cleaned by recirculating inhibited acid—usually HCl—to remove carbonate deposits that build up on screens and heat-exchanger surfaces.

Corrosion in Pulp Mill Equipment

Batch digesters were initially thought to be immune to the severe corrosion affecting digesters used in the sulfite pulping process. In 1930, a catastrophic digester failure killed several workers and prompted the industry to measure the wall thickness of kraft digesters (Ref 50). The digesters were inspected, and several were retired because of corrosion-induced wastage of the vessel wall. Thereafter, the industry was vigilant in monitoring the thickness of batch digester walls and retiring vessels that had corroded beyond safe limits.

Corrosion damage appears in batch digesters in several forms, including uniform wastage, large gouges, pitting, and occasional cracking (Ref 51). In some cases, the gouges are clearly associated with erosion caused by steam impingement during direct steam heating of the digester, by recirculating liquors, or by impingement of high-velocity pulp slurries discharged at the end of a cook. Figure 39 shows an example of erosion-corrosion of a blow target plate used to break the momentum of the pulp mass discharged from the digester under pressure. Grooves are often found beneath blind nozzles and other areas where liquor-saturated pulp can lodge, allowing liquor to run down the walls between cooks. Corrosion is often more severe in the bottom cone area, where flow rates are higher during indirect heating and during blows. In some digesters, corrosion is most severe in the splash zone, where cooking liquor splashes on the hot sidewalls during digester charging. Welds are often sites of preferential attack in batch digesters, including accelerated corrosion and occasional episodes of SCC.

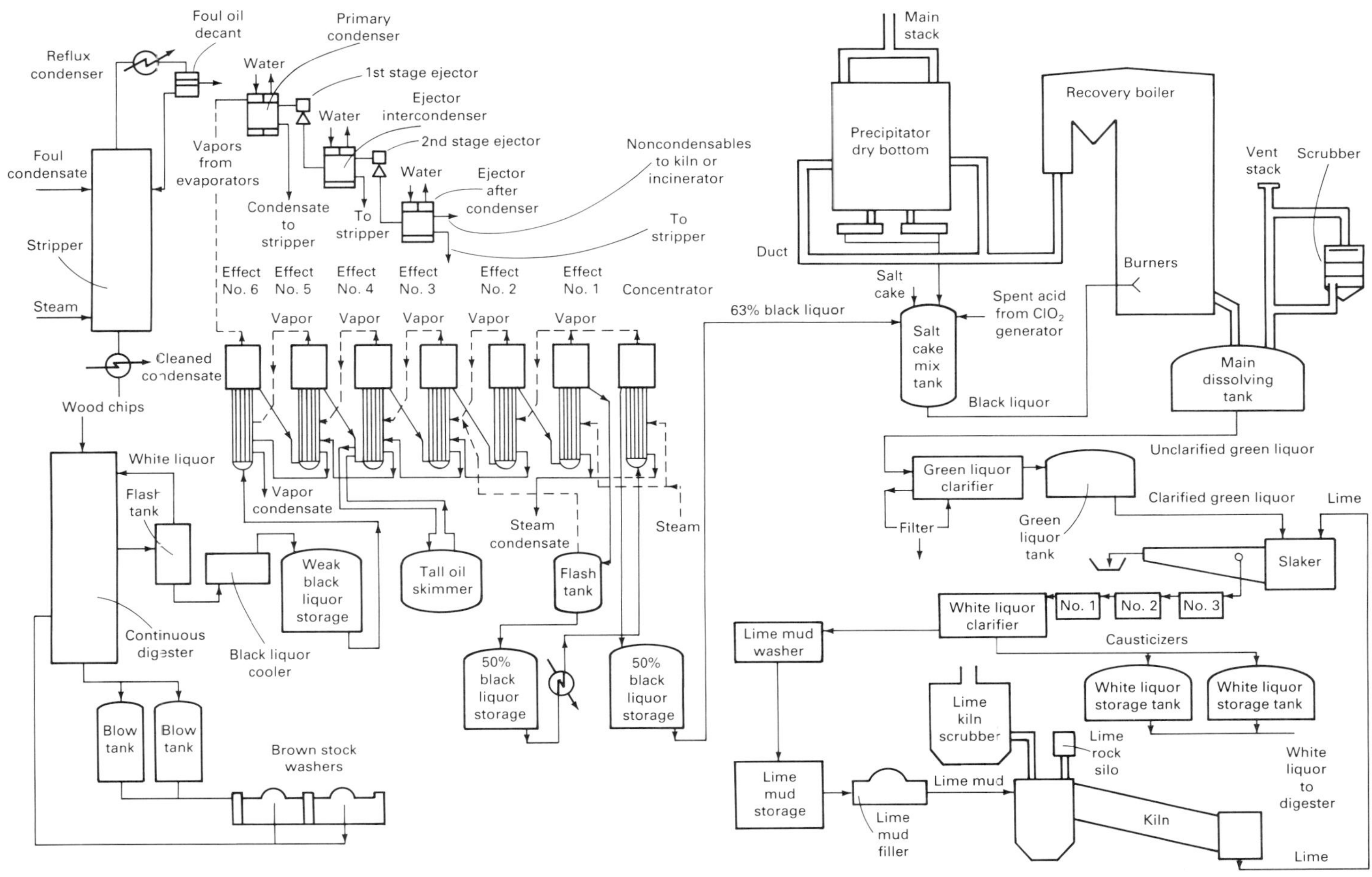

Fig. 38 Schematic representation of kraft pulping and chemical recovery

In the late 1940s, the industry became alarmed over an abrupt decrease in the service lives of batch digester vessels because of a corrosion-induced loss of wall thickness. At one mill with 20 digesters in operation, the service life of digesters installed after 1947 was only 3 to 5 years, while the lifetime for digesters installed before 1947 was 12 to 14 years (Ref 52). Several events apparently combined to increase the corrosion rate. The modern low-odor recovery boilers introduced about this time were more effective in retaining sulfur in the process, with resultant increases in the sulfidity of the cooking liquor. Batch digester production was increased by using more frequent batch cooks of greater severity (higher temperatures and more aggressive cooking liquors). Furthermore, semikilled steel was used to fabricate digesters, replacing the rimmed Bessemer steels used in earlier generations of digesters. Each of these factors played a role in the apparent increase in digester wall corrosion.

Considerable progress was made in understanding and controlling batch digester corrosion as a result of the intensive investigations completed in the early 1950s. The effect of liquor composition was thoroughly investigated through surveys and laboratory corrosion studies. General wastage was greatly accelerated by thiosulfates and low concentrations of polysulfides in the cooking liquor (Ref 53, 54) and by the introduction of air into the digester during charging (Ref 55). Linear regression equations were developed to relate the corrosion rate to concentrations of various minor and major constituents in white liquor (Ref 55, 56). Laboratory and field studies showed that most digester corrosion occurred at the start of the batch cook (Ref 57, 58), and corrosion rates were described in the arcane unit, mils per two-thousand cooks (MTTC). Digester walls were found to be passivated during later stages of the cook and reactivated when raw liquor was introduced into the vessel for the next cook.

Particularly troublesome was a phenomenon called hot plate boiling, in which liquor that was charged into the vessel splashed off the chip pile onto sidewalls that were still hot from the previous cook (Ref 59). The evaporative concentration of dissolved chemicals and the damage to the passive film from the thermal shock combined to reactivate the surface of the vessel for further corrosion during the early stages of the next cook.

The influence of the composition and the microstructure of digester steel on susceptibility to corrosion remains controversial. One researcher noted that steels that were deoxygenated with silicon corroded more rapidly during contact with cooking liquor than steels with lower silicon and higher oxygen levels (Ref 60). This observation explains the preferential attack of high silicon welds that is frequently noted. Several laboratory studies confirmed the effect of silicon on the corrosion rate of steel (Ref 61-63), while other studies dismissed the effect as insignificant (Ref 64). A similar controversy regarding the benefits of normalizing heat treatments also developed (Ref 55, 65). The issue of silicon content has never been fully resolved, but many digesters have been subsequently fabricated with special digester steel (for example, ASTM A285 Grade C, containing no more than 0.03 wt% Si) and low-silicon weld filler metal (for example, E6010) in final passes on structural welds.

Digester lifetimes were extended by a variety of measures involving mill operations, the use of linings, and the specification of low-silicon steels. Chip- and liquor-charging practices were modified to eliminate hot plate boiling. This was accomplished by introducing cooking liquor from the bottom of the digester, by using special liquor injection nozzles, and by simultaneous charging of chips and liquor. Graphite brick linings were once extensively used, but stainless weld overlays and thermal-sprayed coatings are now preferred.

The durability of stainless weld overlays has often been inadequate in spite of a TAPPI standard for acceptable overlay practice (Ref 66). Pitting, cracking, and interpass corrosion are the most common complaints. Excessive dilution of the filler metal and production of partially martensitic overlays have been implicated in the

Fig. 39 Erosion-corrosion of a blow target plate by pulp discharged at high velocities from a batch digester

poor performance of overlays. The choices of filler metal composition, heat input, and travel speed affect the uniformity of the overlay and the dilution of the filler metal. The weld overlay composition recommended by TAPPI for batch digester protection is as follows (Ref 66):

Alloying element	Composition, wt%
Carbon	0.15 max
Manganese	2.50 max
Phosphorous	0.045 max
Sulfur	0.030 max
Silicon	1.00 max
Chromium	18.0 min
Nickel	8.0 min
Molybdenum, when desired	Cr:Mo ≥ 8:1

The Schaeffler diagram predicts that overlays of the specified composition will be devoid of the ferrite required to prevent hot cracking. Furthermore, the high allowable carbon concentration may result in sensitization of the overlay by the heat of subsequent weld passes. Consequently, mill operators often specify a higher-chromium equivalent and lower carbon in the as-deposited overlay.

Blow Target Plates. Erosion-corrosion of the target plate in blow tanks is a continuing maintenance problem, but some advantage can be obtained by using more highly alloyed material. For example, Hastelloy alloy C-276 has endured for 4 years in target plate service where 18-8 austenitic stainless steels were lasting for only a few months.

Piping and Ancillary Equipment. Carbon steel piping is subject to corrosion and erosion-corrosion in the kraft pulp mill, but is often chosen for economy. Elbows in blow lines and other areas of high abrasion are particularly susceptible to attack. Type 304L stainless steels is effective in controlling corrosion damage in piping. Pump impellers, pump casings, and valves made from cast CF-8M are usually resistant to attack by kraft liquors.

Continuous Digesters. In contrast to the cyclic environments encountered in batch digesters, wetted surfaces in continuous digesters experience the same environment for months at a time, although the exact environment depends on the elevation in the digester. Wood chips are added at the top of the vessel, and the pulp mass works its way downward while the chips are successively saturated with liquor, cooked, washed, cooled, and discharged from the bottom of the vessel. Cooking, extraction, and wash liquors are circulated through the pulp through a series of internal screens and downcomer pipes.

Although the rate of uniform wastage in continuous digesters is usually quite low—of the order of 0.13 to 0.25 mm/yr (5 to 10 mils/yr)—higher rates of attack have been observed locally. In particular, erosion-corrosion can be a problem where fresh cooking liquor impinges on steel surfaces near the top of the digester. Pitting of welds and preferential corrosion of weld filler metal have also been observed in continuous digesters, but this attack may be a result of acid cleaning practices, rather than attack by cooking liquor.

A severe corrosion problem in continuous digesters literally burst upon the scene in 1980 with the catastrophic failure of a large continuous digester vessel during routine operation (Ref 67). The digester ruptured near the top of the vessel because of extensive SCC of shell girth welds. This cracking was restricted to the welds in the impregnation zone at the top of the digester. Caustic SCC was implicated because of the branched, intergranular appearance of the cracks and the presence of sodium hydroxide in the hot cooking liquor.

Subsequent inspections of similar digesters revealed that more than half of the 140 continuous digesters operating in North America exhibited cracking in structural welds (Ref 68). Cracking has been found in girth and vertical shell welds, attachment welds inside the digester, and nozzle welds. Both longitudinal and transverse cracking have been observed, but the deepest cracking has been found in longitudinal cracks in the weld HAZ. Examples of cracking in actual digester welds are shown in Fig. 40.

Most of the cracking has been found in the impregnation zone, where the caustic concentration in the cooking liquor is highest. Furthermore, the residual stresses in impregnation zone welds were often not relieved by postweld heat treatment, because the shell is thinner in this zone and stress relief is not required under provisions of the ASME Boiler and Pressure Vessel Code.

An extensive survey of digester cracking statistics failed to reveal differences in digester design or operation that would account for differences in cracking susceptibility (Ref 68). As shown in Fig. 41, postweld heat treatment significantly reduced, but did not eliminate, susceptibility to severe cracking.

Cracking similar to digester cracking was reproduced in the laboratory by using accelerated slow strain rate and fracture mechanics tests (Ref 69). These tests indicated that the caustic concentration in cooking liquors at the impregnation zone was sufficient for caustic cracking of pressure vessel weldments. Furthermore, caustic cracking occurred only when the potential of the digester steel was within a 100-mV range centered close to digester potentials. Figure 42 shows the dependence of cracking susceptibility on potential as determined in slow strain rate tests performed on welded specimens in simulated impregnation zone liquor (Ref 70). Potential measurements made on an operating digester indicated that the digester rest potential remained above the cracking range, except for a few days following an upset in operating routine (Ref 71).

Two measures have been successful in controlling digester weld cracking: high alloy barrier coatings placed over susceptible welds and anodic protection. Other remedial measures, such as shot peening, temper-bead weld repair, *in situ* stress relief, and unsealed thermal spray coatings, have not been uniformly successful in preventing recracking.

Weld overlays and thermal sprayed coatings have both been successfully used to control cracking. Weld overlays—primarily Inconel alloy 82 and type 309L stainless steel—have been applied in bands over structural welds to isolate them from contact with cooking liquor. The overlay bands are themselves resistant to corrosion damage, but there have been cases of cracking of the carbon steel substrate in the HAZ at the edge of the overlay band. Experience with thermal sprayed coatings has been limited, but it appears that flame-sprayed and plasma-sprayed stainless steel coatings protect underlying welds if the coating is sealed with a silicone-modified furan after application (Ref 72). Additional information is available in the article "Thermal Spray Coatings" in this Volume.

Anodic protection has been effective in controlling both corrosion and caustic cracking in a number of continuous digesters (Ref 73). Protection is achieved by passing a controlled direct current (dc) through an electrolytic cell consisting of the digester wall, the cooking liquor, and a special cathode installed inside the digester. Currents as high as 1000 A (at 12 V dc) may be required to passivate the vessel initially, but only a few hundred watts of electrical power is required to maintain anodic protection. The digester potential is maintained approximately 100 mV above the upper limit of the potential range required for cracking. Anodic protection has suppressed further cracking in several digesters previously susceptible to severe and chronic cracking. Additional information is available in the article "Anodic Protection" in this Volume.

Liquor Reheater Tubing. Corrosion and cracking have been encountered in stainless steel tubes in shell and tube heat exchangers used to heat recirculating liquors. An example of a cracked liquor reheater tube is shown in Fig. 43. In most cases, severe transgranular cracking occurs when liquor leaks onto the steam side of the tubes, where it is concentrated and heated by the steam. Caustic cracking and chloride SCC have both been implicated. Differential thermal expansion between the stainless steel tube bundle and the carbon steel shell apparently provides the stress required for cracking. Cracking has been eliminated in more than 100 kraft mills by the use of a duplex stainless steel (3RE60, see Table 13 for composition) in liquor heater tubes; Inconel heater tubes have also been successfully used (Ref 74, 75).

Corrosion of stainless steel liquor heater tubing is more puzzling, because stainless steel is generally resistant to attack by kraft liquors at liquor heater temperatures. Corrosion of liquor heater tubes may actually result from improper acid cleaning—for example, by using excessive temperatures or uninhibited HCl.

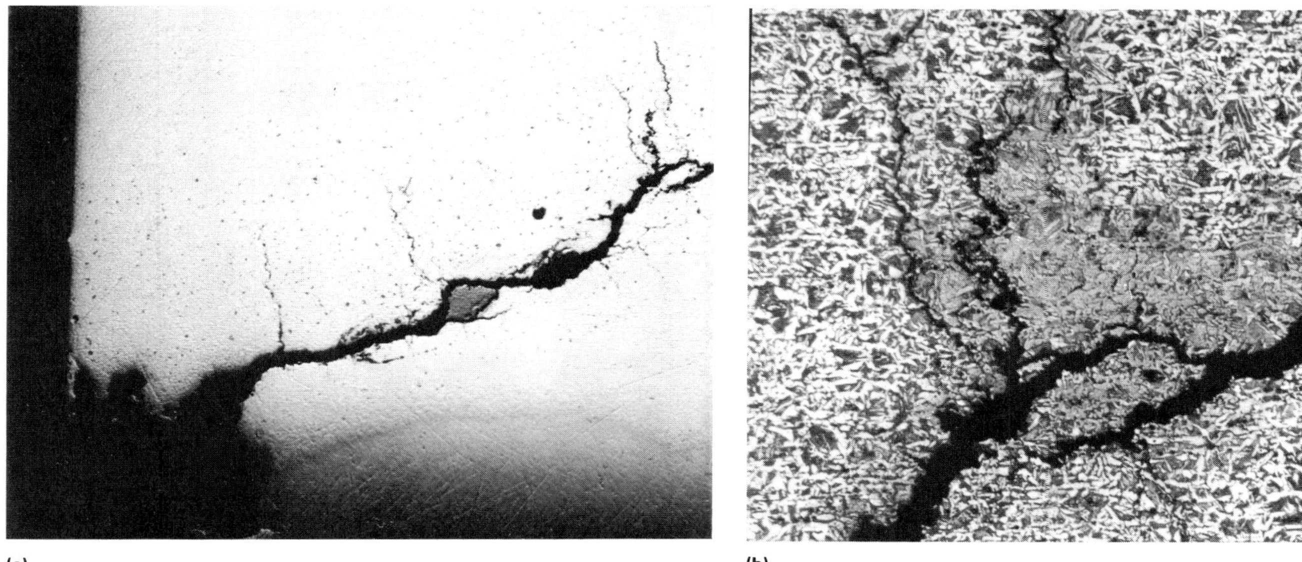

Fig. 40 Examples of digester weld cracking. (a) Macrograph showing cracking in a sample taken from a continuous digester weld. 10×. (b) Photomicrograph showing branched, intergranular nature of cracking in an actual continuous digester weld. 40×

Corrosion in Recausticizing Equipment

The recausticizing plant provides a variety of corrosive media, including white and green liquor, lime mud slurries, lime mud wash waters, and lime kiln scrubber waters. White and green liquors attack storage tanks, clarifiers, and piping fabricated from plain carbon steel, but have little effect on stainless steel equipment (in the absence of abrasion). Erosion-corrosion is frequently seen on equipment in contact with abrasive lime mud slurries, such as clarifier rakes, lime mud slurry pump casings and impellers (Fig. 44), and lime mud discharge lines. Corrosion also occurs on lime kiln scrubber components.

Plain carbon steel storage tanks and clarifiers in white and green liquor service suffer severe corrosive attack, particularly in zones where liquor levels fluctuate frequently. Figure 45 shows the variation in corrosion rate at different elevations along the sidewall of a white liquor clarifier (Ref 76). In the zone immediately above the outlet bustle pipe, the corrosion rate approaches 0.75 mm/yr (30 mils/yr). Areas that are seldom immersed experience a much lower rate of corrosion, as do surfaces near the bottom, which are protected by a layer of lime mud sediment. In multitray clarifiers, the trays experience a similar form of attack. Little corrosion is seen on the top of the tray where lime mud settles, but severe attack occurs on the underside of the tray where air bubbles entrained in the liquor remain in contact with the tray. White liquor clarifiers and storage tanks often need extensive repair or replacement after only 10 years of service, and cases of severe corrosion in less than 2 years have been reported.

The corrosivity of white liquor is strongly influenced by the liquor composition. In simulated white liquors containing only the important pulping chemicals—NaOH and Na_2S—the rate of corrosion of carbon steel is less than 0.25 mm/yr (10 mils/yr) (Ref 77). However, the presence of thiosulfates and low concentrations of polysulfide in white liquor can increase the corrosion rate to more than 1.3 mm/yr (50 mils/yr), as shown for thiosulfate in Table 14. The effect of thiosulfate and polysulfide is greater in liquors containing high concentrations of NaOH and Na_2S. The other species typically found in white liquor—sulfates, sulfites, chlorides, and carbonates—are apparently innocuous (Ref 78). Because the hydrosulfide in white liquor is rapidly converted to thiosulfate by air contact, the high rate of corrosion at air/liquor interfaces is probably related to the thiosulfate effect shown in Table 14.

Liquor velocity relative to steel surfaces can also have a large effect on the rate of corrosion (Ref 79). As shown in Table 15, the rate of corrosion of carbon steel by white liquor is increased fivefold by a change from stagnant to laminar flow conditions. Under turbulent flow

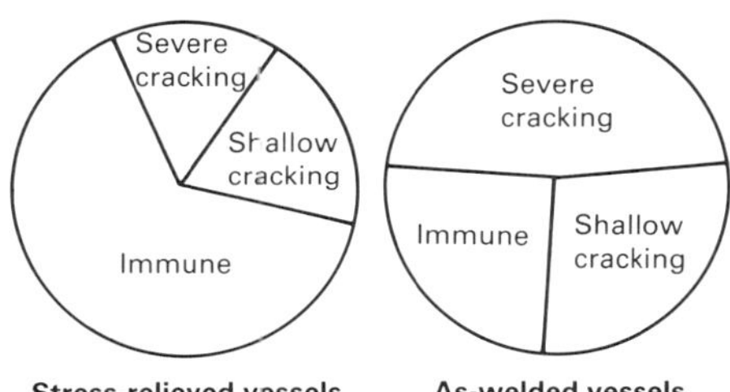

Fig. 41 Effect of stress relief on SCC susceptibility of continuous digesters

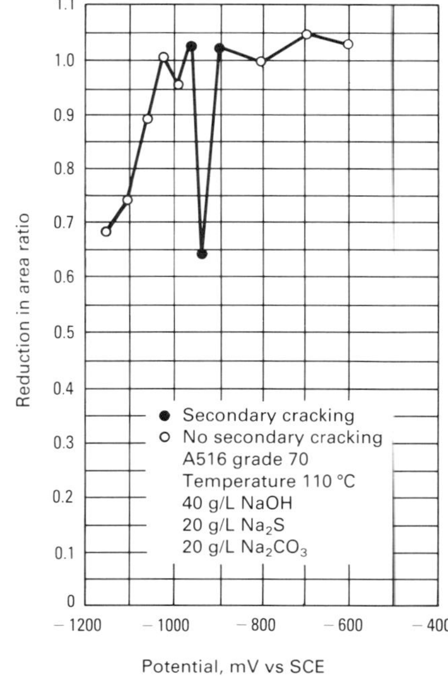

Fig. 42 Plot showing the effect of potential on cracking severity in controlled-potential slow strain rate testing of digester steels exposed to a simulated impregnation zone liquor. SCE, saturated calomel electrode

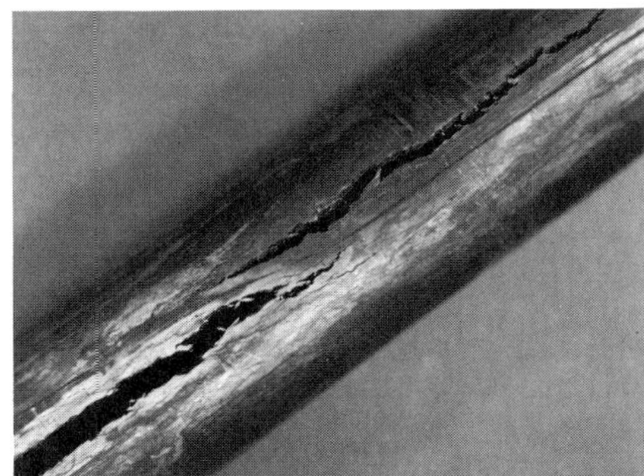

Fig. 43 Transgranular SCC of type 304L stainless liquor reheater tubing

Fig. 44 Erosion-corrosion of a lime mud pump impeller

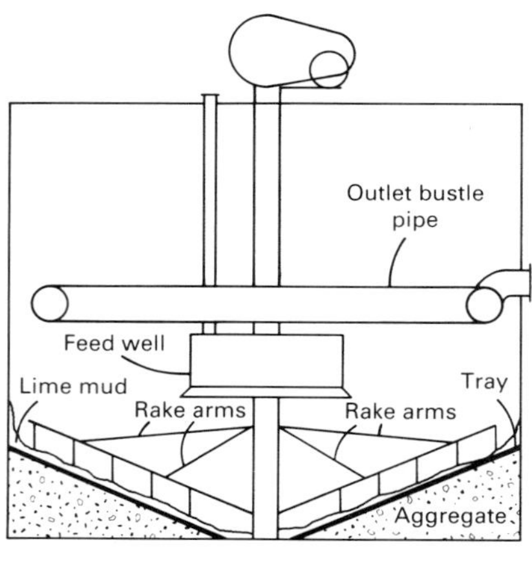

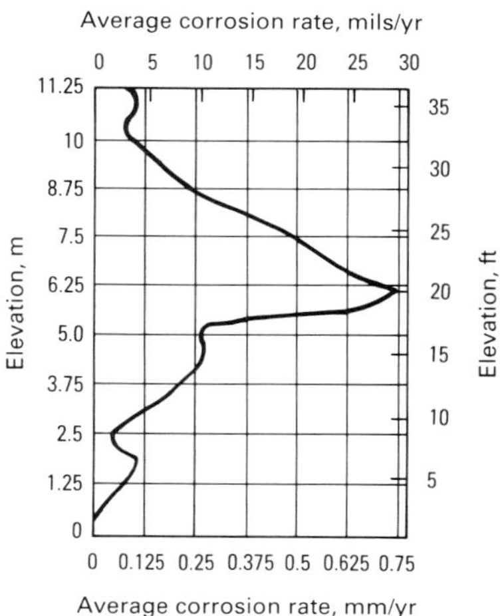

Fig. 45 Variation of corrosion rate with elevation at the wall of a white liquor clarifier

conditions, corrosion rates exceeding 3.8 mm/yr (150 mils/yr) are possible.

Several corrosion control measures have been effective in recausticizing applications, including stainless steel fabrication, specification of thick-wall carbon steel at critical sites, controlling thiosulfate and polysulfide levels, the use of barrier linings, and anodic protection.

Type 304L stainless steel is not affected by white or green liquor, and stainless steel recausticizing equipment will probably last for the life of the mill. The higher cost of stainless steel can be mitigated to some extent by eliminating the corrosion allowance on wall thickness. There is little justification for using more costly type 316L stainless steel in recausticizing applications; the high pH of the liquors inhibits chloride attack and precludes the need for molybdenum in the stainless steel. In vessels with corrosion at the liquid level line as shown in Fig. 45, stainless steel belly bands can be installed to prolong tank lifetime. Usually, formed stainless steel plates are seal welded to the inside wall. Careful installation is required to prevent liquor from leaking between the stainless cladding and the wall and accelerating wall corrosion.

Experience with alkali-resistant concrete and brickwork linings in recausticizing vessels has been checkered. In some cases, the lining prevents further corrosion damage. In many cases, liquor penetrates behind the barrier linings, allowing renewed attack of the wall. Installation of a protective polymeric membrane between the vessel wall and the lining may eliminate damage if the lining is breached.

Thiosulfate levels can be reduced by minimizing air contact with liquor in order to lower corrosion rates in white and green liquor systems. Thiosulfates are formed wherever dissolved sulfides in the liquors come in contact with air. Likely sites for thiosulfate formation include shatter jets (where smelt is dispersed as it falls into the dissolving tank liquor), splitter box weirs, storage tank and clarifier liquid surfaces, and leaking pump seals. Steam shatter jets are preferred over compressed air jets for reducing thiosulfate levels. In some cases, thiosulfates are generated when weak wash liquor is used to scrub lime kiln flue gases before being used as dissolving tank makeup liquor. Scrubber waters should be completely oxidized to convert thiosulfates to innocuous sulfates before use as dissolving tank makeup liquor.

Natural polysulfide levels are negligible in most white liquors. However, polysulfide-accelerated attack can occur if emulsified sulfur is added (for sulfur makeup) in the recausticizing plant, because elemental sulfur is rapidly converted to polysulfide. Makeup sulfur should instead be added to black liquor to minimize corrosion damage.

Recently, anodic protection has been shown to be an effective and economical corrosion control

Table 14 Effect of thiosulfate concentration on white liquor corrosivity

Duration of exposure: 8 weeks

Concentration, g/L			Corrosion rate	
NaOH	Na_2S	$Na_2S_2O_3$	mm/yr	mils/yr
80	20	0	0.18	7
80	20	2.5	0.18	7
80	20	10	0.28	11
80	20	25	0.71	28
80	20	50	1.35	53
80	40	2.5	0.15	6
80	40	5	0.51	20
80	40	10	0.53	21
80	40	25	1.09	43
80	40	50	2.21	87

Source: Ref 78

Table 15 Effect of liquor velocity on white liquor corrosivity of simulated liquor, 100 g/L NaOH + 33 g/L Na_2S

Velocity		Corrosion rate	
m/s	ft/s	mm/yr	mils/yr
0.0	0.0	0.13	5
0.14	0.46	1.65	65
0.25	0.82	2.03	80
0.30	0.98	2.51	99
0.43	1.41	3.05	120
0.50	1.64	3.63	143
0.86	2.82	3.50	138
1.32	4.33	3.81	150
2.62	8.60	3.76	148

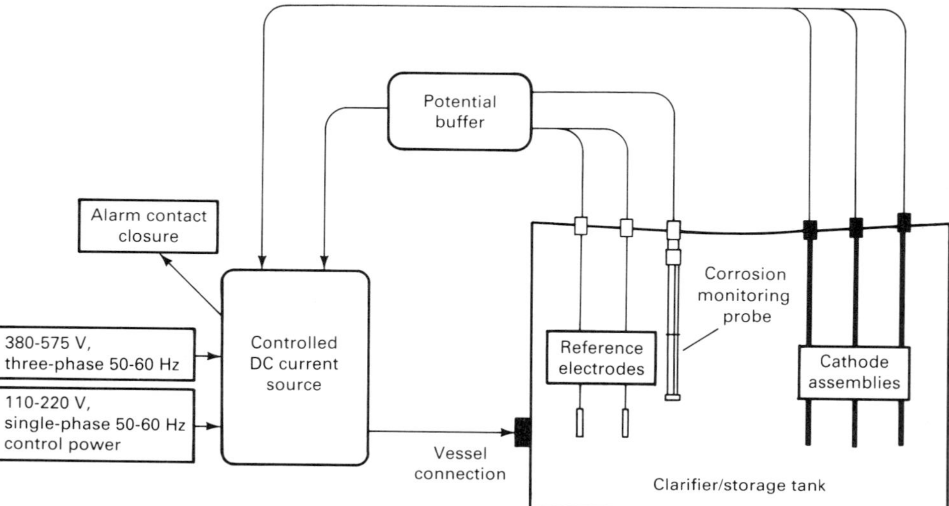

Fig. 46 Schematic diagram of an anodic protection system for recausticizing storage tanks and unit clarifiers

Fig. 47 Pitting of a carbon steel root section in a black liquor storage tank

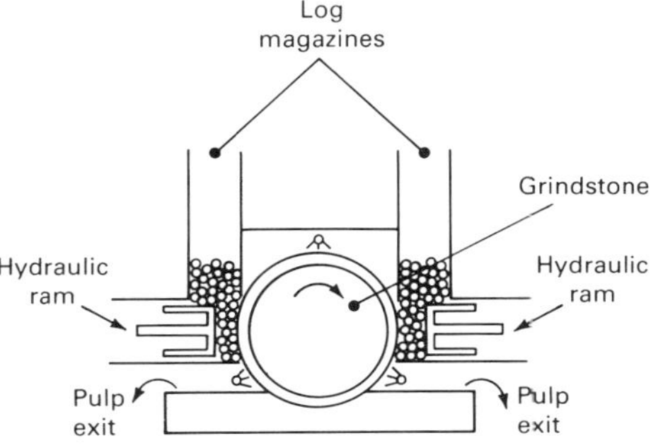

Fig. 48 Schematic sectional view of a stone groundwood machine. Cut, debarked logs are fed into magazines, from which they are pressed against a rotating grindstone to be pulped.

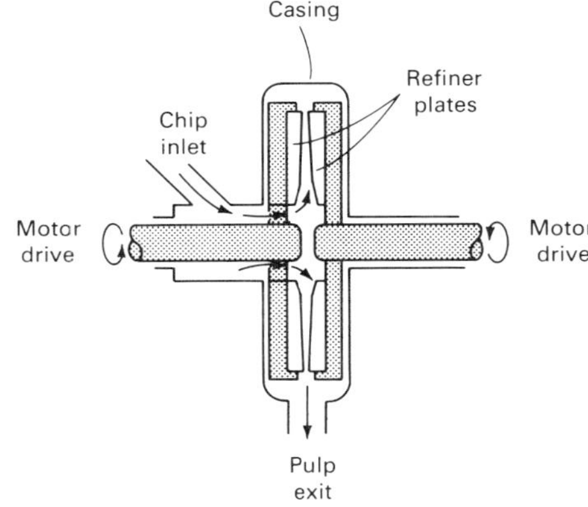

Fig. 49 Schematic sectional view of a disk refiner. Pulping takes place between rotating rings of refiner plates. In this example, the refiner plates are mounted on opposing rotor assemblies (shaded) that counterrotate at high speed. Such a unit is called a double-disk refiner. Single-disk refiners, in which only one set of refiner plates is rotated, are also common.

measure for use in white liquor storage tanks and clarifiers (Ref 76). In a prototype installation on an operating clarifier with severe corrosion damage, corrosion rates were reduced by nearly an order of magnitude by holding the rest potential of the vessel in the passive range. The anodic protection system is shown schematically in Fig. 46. Although relatively large currents were required to achieve initial protection, little power is needed to maintain protective passivation.

Erosion-corrosion is a severe problem wherever abrasive lime mud is found. Stainless steel fabrication is appropriate wherever abrasive conditions prevail, as in rakes and rake arms in white and green liquor clarifiers. In severely abrasive conditions, such as lime mud pump impellers, hard-facing overlays such as cobalt-base Stellite alloys may be effective, but periodic replacement of components is usually necessary.

Corrosion in Black Liquor Processing Equipment

Black liquor varies considerably, depending on the pulping process used, the pulping yield, and the stage of chemical recovery. Weak black liquor, the filtrate obtained from brown stock washing, contains approximately 20% solids consisting of organic compounds extracted from the wood (principally lignin compounds), inorganic salts not consumed in the pulping process, organic acids, and various hemicelluloses and sugars. The pH of weak black liquor is of the order of 12, and the residual active alkali (NaOH and Na_2S) can range from 4 to 30 g/L. The solids content of heavy black liquor is increased by evaporation into the 60 to 75% range, and heavy black liquor at room temperature is viscous and semisolid. Black liquor oxidation can be used to convert reduced sulfur compounds in black liquor to thiosulfates and sulfates in order to reduce odorous emissions from the recovery boiler and to conserve sulfur in the process.

Corrosion of brown stock washing equipment is rarely a concern in the kraft pulp mill. Brown stock washing is usually carried out on rotary drum vacuum washers, although displacement washing is becoming more common. Washer drums are usually fabricated from carbon steel, while backing wires and face wires are usually stainless steel. The moderately high pH of weak wash filtrate, together with the low concentration of NaOH and the inhibitive effect of the dissolved organic species, generally prevent severe corrosion of brown stock washing equipment in spite of the dissimilar-metal construction. However, corrosion can become a problem if the pH of the weak wash filtrate is depressed by the use of acid bleach plant filtrate as shower water or by the addition of neutral sulfite semichemical pulping liquors. When acidic filtrates are used, it is common practice to raise the pH above 10 to reduce corrosion and to prevent odorous emission of hydrogen sulfide gas.

The details of black liquor evaporation equipment vary from mill to mill. Multiple-effect evaporators are generally used to increase the black liquor solids to the 55% level, while concentra-

Fig. 50 General view of a refiner plate. The large bars at the bottom are called breaker bars. The bars become progressively finer towards the top edge of the plate.

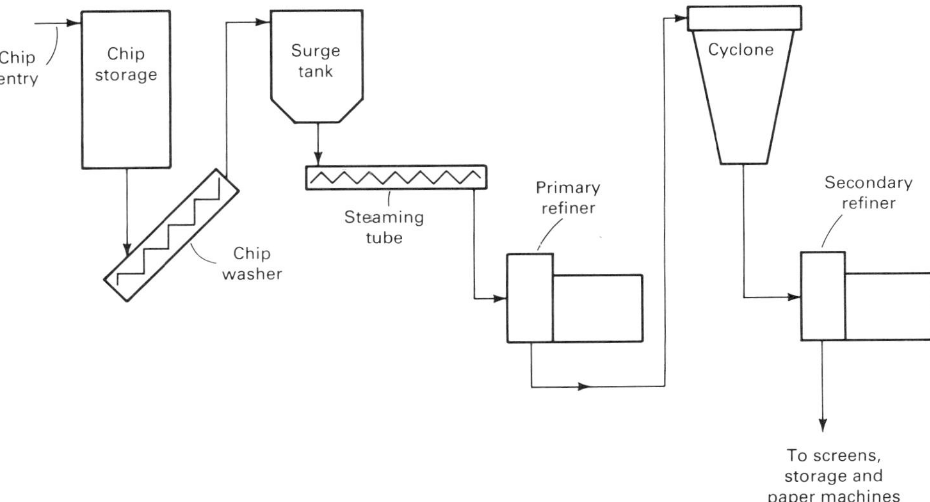

Fig. 51 Schematic of a typical thermomechanical pulping system layout

Table 16 Typical refiner plate alloy types and compositions

Alloy type	Composition, wt% C	Cr	Ni	Si	Mo	Mn	P	S
Martensitic white cast iron	3.3	2.4	4.0	1.2	0.3	0.6	0.03	0.02
Martensitic high-chromium	2.7	23.2	0.4	0.5	2.0	0.7	0.02	0.02
white cast irons	3.0	27.7	1.1	0.6	0.1	0.5	0.03	0.03
Martensitic cast	0.8	18.0	1.4	0.9	0.5	0.6	0.03	0.02
stainless steels	1.6	16.3	1.5	1.1	0.9	0.6	0.01	0.02

tors or cascade evaporators complete the evaporation. The shell and tube heat exchanger (rising film, long tube vertical evaporators) is the most common evaporator design, but other configurations with plates and falling films are becoming more popular. Plain carbon steel and type 304L and 316L stainless steels are common materials of construction. Stainless steel tubes are used in the higher-temperature evaporator effects (>95 °C, or 200 °F); both stainless and carbon steel tubes are used at lower temperatures. Black liquor storage tanks and oxidation equipment are usually made of plain carbon steel. Stainless steel piping is commonly used for black liquor transport.

In general, black liquor is much less corrosive toward carbon steel than the recausticizing liquors, because the concentrations of NaOH and Na_2S in black liquor have been reduced by the pulping reactions. The residual alkali in black liquor maintains the pH near 12, which stabilizes a protective film on the steel surfaces. In spite of this film, carbon steel storage tanks wetted by black liquor may experience pitting and crevice corrosion, particularly at the liquor level line and underneath sediments and scales. Pitting attack, such as that shown in Fig. 47, is often encountered on roofs and other surfaces that are exposed to vapors emanating from the black liquor. This pitting is attributed to volatile organic acids that condense on cooler surfaces but are not neutralized by direct contact with the alkaline black liquor.

Remedial measures for corrosion in black liquor storage tanks are limited. Generally, the moderate rate of attack does not justify the use of stainless steel for tanks and vessels. Although the high temperatures and organic sulfur compounds in black liquor make fiber-reinforced plastic generally unsuitable for immersion service, some mills have installed fiber-reinforced polymeric roofs on black liquor storage tanks to minimize pitting attack. Thick-wall carbon steel construction and the use of stainless belly bands may also be beneficial when attack is restricted to one area, such as the liquor level line.

Stainless steels are generally resistant to corrosion by black liquors at temperatures normally encountered in evaporation and concentration equipment. Over the years, experience has shown which evaporator components must be stainless to prevent excessive corrosion; therefore, severe evaporator corrosion is now relatively infrequent. Tubes, vapor deflectors, and liquor boxes are generally stainless steel or stainless-clad carbon steel, particularly in the high-temperature effects. When localized corrosion does occur on carbon steel surfaces in modern evaporators, it can usually be traced to unanticipated contact between hot liquor and the steel due to foaming, excessive throughput, or other operating misadventures. In these cases, stainless steel weld overlays or thermal spray coating applied to the carbon steel surface usually forestalls continued corrosion.

Although black liquor oxidation produces high levels of thiosulfate in the black liquor, the thiosulfates are much less injurious to carbon steel in black liquor service as compared to the recausticizing situation. Oxidized black liquors have low residual alkali levels because of consumption of NaOH in the digestion process and complete oxidation of Na_2S. In the absence of significant concentrations of NaOH and Na_2S, thiosulfates are relatively innocuous toward carbon steel. Consequently, it is desirable to oxidize weak black liquor, rather than strong black liquor, so that more of the black liquor process equipment benefits from the reduced corrosion that accompanies black liquor oxidation. Carbon steel is generally adequate for oxidation equipment and oxidized black liquor storage.

Corrosion of Mechanical Pulping Equipment

C.B. Thompson
Pulp and Paper Research Institute of Canada

D.A. Wensley
MacMillan Bloedel Research

In mechanical pulping, wood fibers are separated from each other by rolling, rubbing, or teasing, either against themselves or against a harder, specially designed surface. Chemical additions or the application of heat and pressure can be used to increase the efficiency of the process, but the predominant pulping forces are mechanical. Mechanical pulping can be divided into two main areas: groundwood and disk refining. In overall terms, mechanical pulping forms an important part of the pulp and paper industry and is notable for its technological and economic expansion in recent years (Ref 80).

Mechanical Pulping Systems

Groundwood is one of the oldest commercial pulping processes and, until recently, was the predominant means of mechanical pulping. The most common version is stone groundwood, in which pulp is formed by pressing debarked logs against a rotating grindstone, as shown in Fig. 48. Water sprays are directed onto the stone to prevent the fibers from burning, and the pulp is removed in the form of a slurry. Other variations include pressurized groundwood, in which pulping occurs at elevated temperature, and chemigroundwood, in which a chemical pretreatment is used.

Corrosion is usually not of concern in these units, largely because most pulping is done by the stone groundwood method, in which water sprays are the only potentially corrosive environment. As a result, attack is limited to general wastage of exposed carbon steel.

Disk refining has overtaken groundwood in importance since the 1970s, primarily because of its ability to produce higher-strength pulps and to utilize chips or sawmill residues as feedstock.

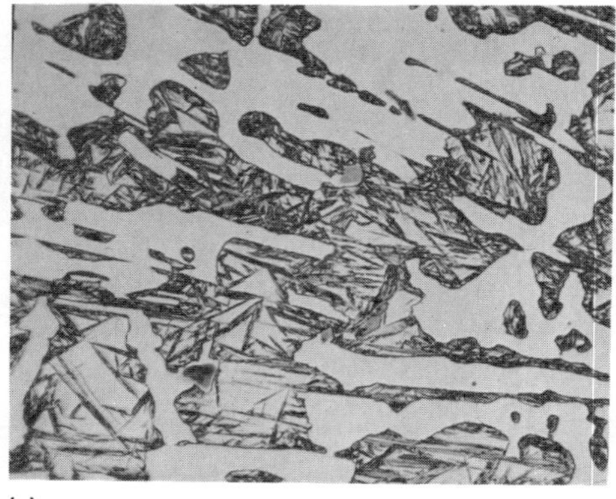
(a)

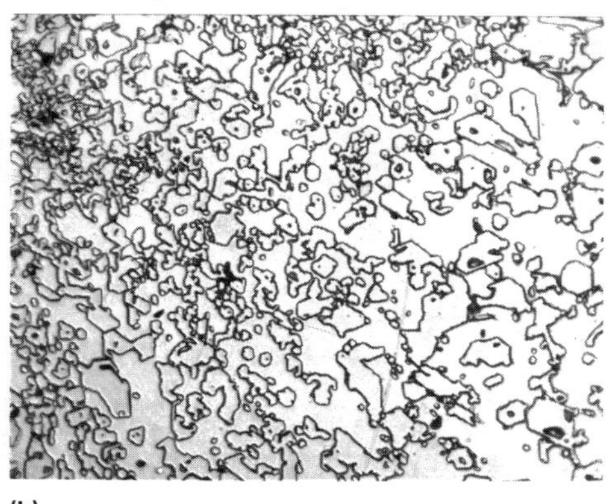
(b)

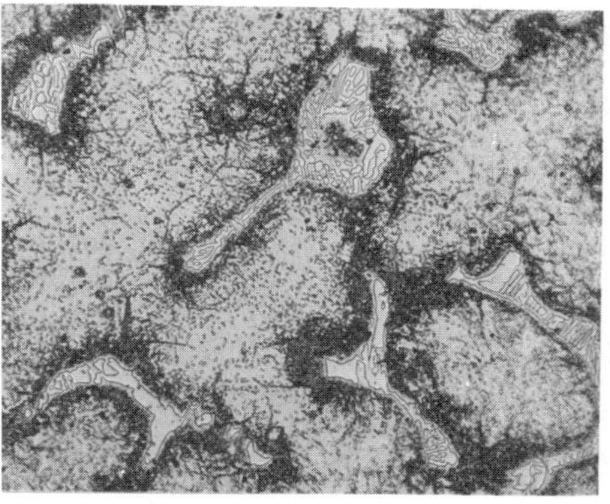
(c)

Fig. 52 Typical refiner plate alloy microstructures. (a) Low-chromium white cast iron; martensite in a eutectic matrix of martensite and iron carbide. Etched with 2% nital. (b) High-chromium white cast iron; iron-chromium primary carbides in a martensitic matrix. Depending on the heat treatment, a large amount of retained austenite may be present in the structure. Etched with 2% nital. (c) Cast stainless steel; iron-chromium primary carbides in a martensitic matrix containing secondary carbides. Etched with mixed acids. All 230×

Fig. 53 Severe bar rounding (arrows) on a cast stainless steel refiner plate. Approximately $^{3}/_{5}\times$

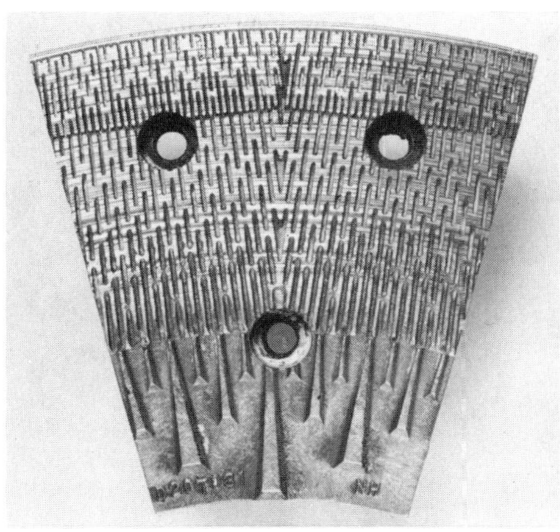

Fig. 54 Clashed surface on a white cast iron refiner plate. The circumferential serrations are typical of this type of wear.

Fig. 55 Heavy cavitation attack on the intermediate refining zone bars of a high-chromium white cast iron refiner plate

Modern disk refiners are typically large and powerful; disk diameters up to 1.8 m (70 in.) are common, with installed motor powers of 13 400 kW (18 000 HP) (Ref 80). A schematic sectional view of a refiner is shown in Fig. 49.

In operation, refiner pulp is produced by passing wood chips between two or more closely spaced serrated disks, at least one of which is being rotated at high speed by an electric motor. Each disk is made by bolting refiner plates onto a backing plate. A typical refiner plate is illustrated in Fig. 50, showing the characteristic series of raised bars. The space between two opposing sets of refiner plates is called the refining zone. As wood chips pass outward through the refining zone, they are gradually reduced to pulp by interaction with refiner plate bars. Although many plate bar patterns exist, it is common to have large bars near the inlet to cause rapid breakdown of the chips. As the pulp progresses through the outer refining zone, the bars become finer in order to impart maximum refining energy to the stock. In most cases, it is usual to have two or more refiners operating in series to obtain optimum refining efficiency.

A large number of refiner pulping processes have been developed, primarily as a result of the need to produce tailor-made pulps and to decrease energy requirements. The simplest process is refiner mechanical pulping, in which the chips are fed directly to the refiner with no prior use of heat or chemical dissolution. For newsprint production, this process has now been largely superceded by thermomechanical pulping, in which the chips are softened in a steaming tube before entry into the primary refiner, and by chemi-thermomechanical pulping, in which both heat and chemical additives are used in order to accelerate the pulping. In both thermomechanical and chemi-thermomechanical systems, chip presteaming and primary refining take place at temperatures above 100 °C (212 °F); in the secondary refiner, the pulp is usually discharged to atmosphere. A schematic diagram of a thermomechanical pulping system is shown in Fig. 51.

In addition to the pure mechanical pulping described above, refiners are also used in a variety of chemi-mechanical pulping methods, in which chips are typically pretreated in a digester before entering the refiners. The principal difference between the two types of pulping is the pulp yield. Chemi-mechanical pulps are in the range 55 to 90% yield; the yield for pure mechanical pulping exceeds 90%.

Corrosion is of more concern in disk refining than in stone groundwood due to the higher process temperatures and the presence of pulping chemicals. In addition, significant wear can occur because of the action of high-velocity jets of steam and water within the refiner and the presence of sand and grit entrained in the stock. Presently, the major problem areas for corrosion and wear are the refiner plates and the steaming tubes; these are discussed below in more detail.

Fig. 56 View inside a horizontal chemi-thermomechanical pulping steaming tube with the screw removed for inspection. The vessel is of UNS NO8904 construction and has three longitudinal rub bars of the same composition stitch-welded to the shell. See also Fig. 59.

Corrosion and Wear of Refiner Plates

Most refiner plates are cast, although wrought plate sets are available for low-consistency units.

Fig. 57 Cross section of a piece cut from a type 316 stainless steel steaming vessel that failed by chloride SCC. Cracking occurred under deposits on the lee-side of a 13-mm (0.5-in.) square rub bar and progressed through the 6-mm (0.25-in.) thick shell.

Fig. 58 Photograph of the inside surface of a failed AISI 316 steaming tube showing chloride SCC that developed beneath deposits (the cracks have been opened by subsequent deformation). The lower part of the surface was swept clean by chip movement and did not crack.

75 μm

Fig. 59 Micrograph of transgranular SCC observed in a UNS NO8904 rub bar removed from the steaming tube shown in Fig. 56

Plates are produced by either refiner manufacturers or specialist refiner plate foundries. Alloys are often of proprietary composition, although most fall into the general classes of white cast iron or cast stainless steel (Ref 81, 82). Some typical compositions are shown in Table 16. Refiner plates cast in white cast iron generally have a large volume fraction of primary carbides set in a martensitic matrix, which sometimes contains large amounts of retained austenite. Cast stainless steel refiner plates usually possess a lower concentration of carbides set in a martensitic matrix. Some typical microstructures are shown in Fig. 52.

Corrosion and Wear. Premature wastage of refiner plates by corrosion and wear adversely affects such critical pulp quality factors as burst, tear, and shives (Ref 83, 84). Worn plates can also decrease paper machine runnability; this can result in significant costs due to machine stoppages or lower speeds (Ref 85) and the need for chemical reinforcing pulps. The types of damage caused by the corrosion and wear of refiner plates are detailed below. It should be noted, however, that little systematic study has been reported in this area, and there is no consensus of opinion as to the actual mechanisms by which such wastage takes place.

Bar rounding is usually found in the outer refining zone on the bar leading edges. Severe bar rounding leads to loss of profile and a reduction in refining efficiency; a typical example is shown in Fig. 53. Bar rounding appears to take place by a combination of corrosion and low-stress abrasion due to sand and grit in the stock. Examination of rounded surfaces at high magnification shows numerous short abrasion furrows, often with associated lips or platelets of deformed material. Typically, the primary carbides are left in relief on the surface.

Clashing. During operation, it is common for opposing sets of refiner plates to touch each other. This is called clashing, and it can be due to loss of stockfeed or a temporary disruption of the pulp pad between the plates. Because of the high power applied during refining, clashing produces severe plate damage, commonly in the form of deep, circumferential grooves and serrations. A typical example is shown in Fig. 54. Some investigators have attributed clashing to electrochemical dissolution rather than mechanical contact, primarily because of the remarkably smooth surfaces that are often found on clashed plates (Ref 86). However, thermal cracking (Ref 87) and evidence of surface melting and smearing (Ref 82) have been observed on clashed plates, and it is probable that in most cases metal wastage due to clashing is primarily mechanical in nature. To reduce the frequency of clashing, various types of proximity sensors have been developed to monitor the gap between opposing plate sets (Ref 88).

Cavitation and Liquid Droplet Impingement Erosion. Heavy pitting at the base of breaker bars and on the edges and top surfaces of intermediate and outer zone refining bars is commonly attributed to cavitation erosion (Ref 89, 90). A typical example is shown in Fig. 55. In some cases, similar attack can be produced by liquid droplet impingement erosion, which is caused by water jets or condensing steam (Ref 91). Metal wastage due to these mechanisms is

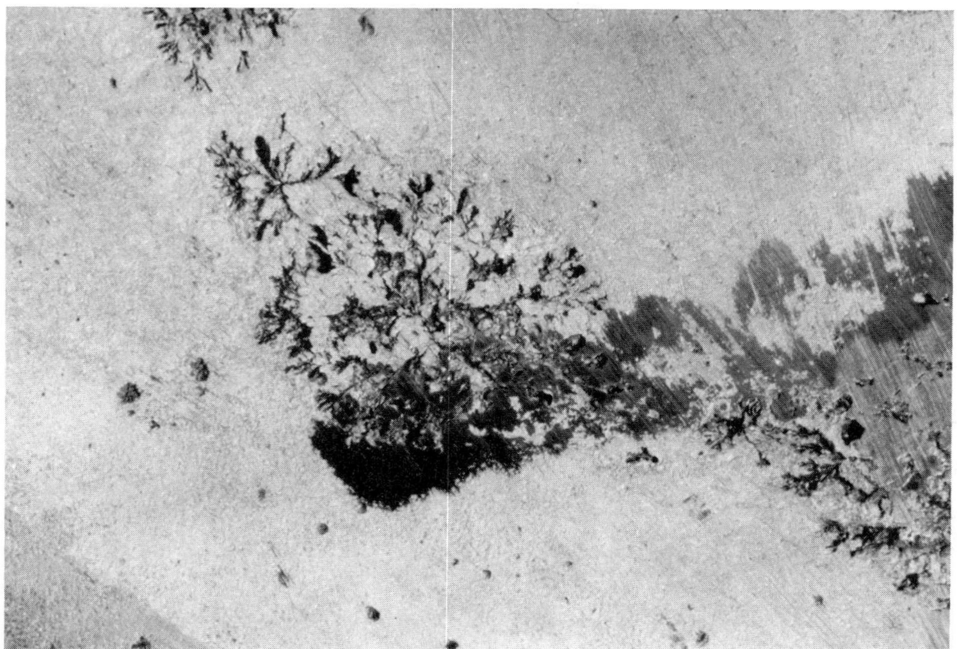

Fig. 60 Crows-foot SCC observed on the inside top surface of the shell of a UNS NO8904 steaming tube. Approximately 1×

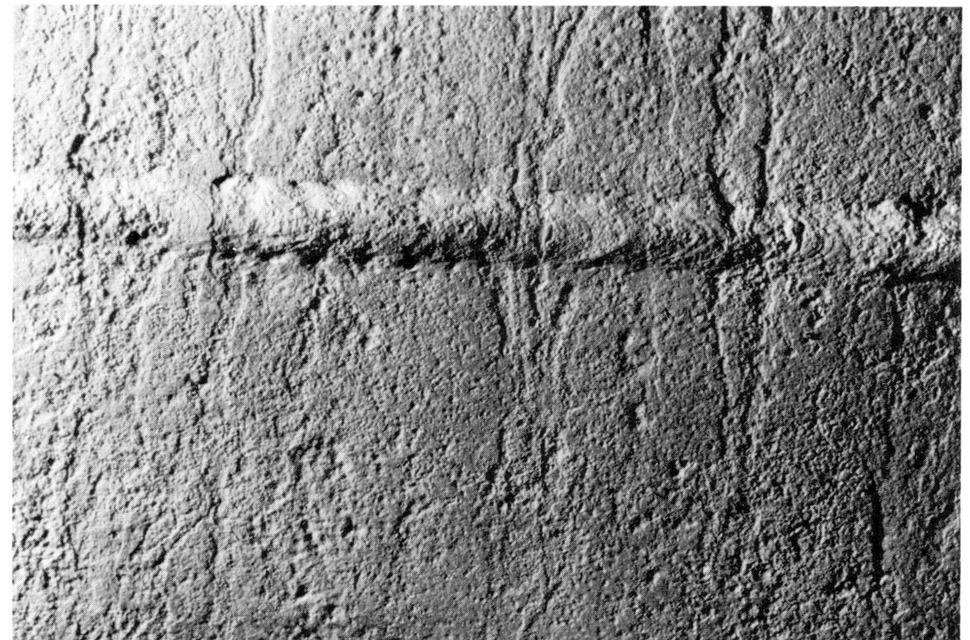

Fig. 61 Acidic condensate corrosion runs observed across a longitudinal seam weld on the wall of a UNS NO8904 steaming tube (see Fig. 56 left side wall)

generally regarded as secondary in importance to bar rounding and clashing; plates are changed only if there is a danger of bar segments being undermined and breaking free during operation. In some cases, however, cavitation has caused such rapid loss of bar profile that plates have had to be changed after as little as 300 h of service. Such intense attack is probably due more to the design and operational characteristics of the particular refiner than to any deficiency in the cavitation resistance of the refiner plate.

Corrosion in Pressurized Equipment

Corrosion of pressurized equipment constructed from stainless alloys, such as thermomechanical or chemi-thermomechanical pulping steaming vessels and bisulfite chemi-mechanical pulping digesters, can cause serious materials reliability problems if either chlorides or acidic sulfur compounds are present. Problems include chloride SCC, pitting, and crevice corrosion, as well as general wastage due to sulfuric acid condensation.

Thermomechanical and Chemi-Thermomechanical Pulping

Thermomechanical and chemi-thermomechanical steaming vessels may be of horizontal, vertical, or inclined configuration. Steaming is typically carried out at 130 to 150 °C (260 to 300 °F) and at 200 to 350 kPa (30 to 50 psig). Horizontal steaming tubes (Fig. 56) experience corrosion problems in areas not swept clean by the moving chip bed. Deposits can form on the vessel walls and on the lee sides of longitudinal rub bars. The rub bars are used to guide the chip mass along the vessel and to protect the vessel from accidental damage by the screw.

Corrosion by Chlorides. Chips from seaborne logs may introduce appreciable amounts of chloride. The liquid film surrounding damp wood chips may contain 2500 ppm NaCl and may also have an acid pH (4 to 6) due to the release of wood acids. Chlorides can concentrate beneath deposits, perhaps by evaporation of areas intermittently wetted with chloride-containing water. Chemi-thermomechanical pulping steaming tube deposits have been analyzed to contain as much as 38 000 ppm NaCl.

Chloride SCC has caused the rupture of a type 316 stainless steel steaming tube. Severe cracking was associated with the residual tensile stress fields around rub bar stitch welds (Fig. 57). Extensive cracking was also found beneath vapor space deposits in the same vessel—well removed from any welds (Fig. 58). Chloride SCC has also been observed in horizontal steaming vessels and screws constructed from UNS NO8904 stainless steel (Fig. 59 and 60).

Pitting or crevice corrosion has also been observed in steaming vessels, often in association with general attack due to condensation of acidic sulfur compounds. Chloride pitting attack tends to be particularly widespread beneath deposits that form on the screw shaft cladding and flights.

Corrosion by Acidic Condensates. Additions of Na_2SO_3 can be made prior to the steaming tube. Sulfur dioxide evolution occurs from sulfite solutions under the acidic conditions that exist naturally with damp wood chips. Sulfur dioxide is converted into the insidious sulfur trioxide in the presence of oxygen. Sulfur trioxide may condense or dissolve in condensed water films as sulfuric acid.

Stainless steels are resistant to H_2SO_4 only in either very dilute or very concentrated solutions. Condensates of intermediate concentration are therefore liable to be quite corrosive. Indeed, condensate corrosion runs have been observed on a steaming tube wall (Fig. 61). Condensate drips that form directly overhead produce circular areas of corrosion. The progress of acidic

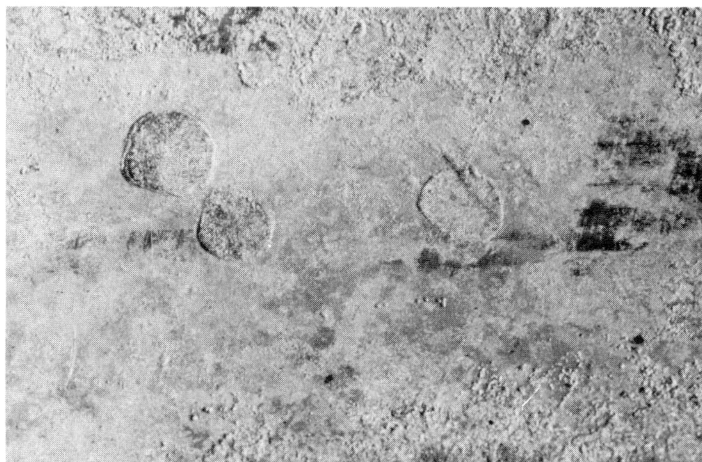

Fig. 62 Circular areas of corrosion observed directly overhead in a UNS NO8904 steaming tube due to formation of acidic condensate drips. Approximately 3/4×

Fig. 63 Photograph 16 months later of the same area as in Fig. 62 showing the progress of corrosion from acidic condensation. Approximately 3/4×

condensate corrosion in a UNS NO8904 steaming tube has been monitored by photographs taken 16 months apart (Fig. 62 and 63).

Bisulfite Chemi-Mechanical Pulping Digesters

Bisulfite chemi-mechanical pulping involves cooking wood chips in continuous digesters with $NaHSO_3$ liquors at 160 °C (320 °F) and at 620 kPa (90 psig). Depending on the amount of liquid in the digester, cooking may be in either the liquid or vapor phase. The conditions that promote the formation of vapor-phase condensates containing high concentrations of H_2SO_4 are the presence of air and low pH (Ref 92). If no effort is made to exclude air, the raw liquor pH must be 7.6 to prevent H_2SO_4 formation. At the optimum process pH of 6, conversely, extremely corrosive condensates containing 34.9% SO_4^{2-} and having a contact pH of 0 can be formed. Corrosion-free operation at pH 6 has been obtained by strict exclusion of air by the following means:

- Presteaming of the chips
- Plug-screw feeding into the digester, below the liquid level
- Preheating the cold blow water used to fill the digester prior to start-up
- Replacing the air in the vessel with steam prior to start-up

REFERENCES

1. A. Garner, *Pulp Paper Can.*, Vol 82 (No. 12), 1981, p T414
2. R.J. Brigham and E.W. Tozer, *Corrosion*, Vol 29, 1973, p 33
3. A. Garner, *Corrosion*, Vol 35, 1979, p 108
4. R.J. Brigham, *Corrosion*, Vol 30, 1974, p 396
5. A. Garner, *Corrosion*, Vol 37, 1981, p 178
6. B. Wallen, in *Pulp and Paper Industry Corrosion Problems*, Vol 2, National Association of Corrosion Engineers, 1977, p 43
7. A.H. Tuthill, J.D. Rushton, J.J. Geisler, R.H. Heasley, and L.L. Edwards, *TAPPI J.*, Vol 62 (No. 11), 1979, p 49
8. A.H. Tuthill, in *Pulp and Paper Industry Corrosion Problems*, Vol 4, Swedish Corrosion Institute, 1983
9. J. Hill, in *Pulp and Paper Industry Corrosion Problems*, Vol 4, Swedish Corrosion Institute, 1983
10. L.H. Laliberté and A. Garner, *TAPPI J.*, Vol 64 (No. 1), 1981, p 47
11. S. Henrikson and V. Kuchera, in *Pulp and Paper Industry Corrosion Problems*, Vol 3, National Association of Corrosion Engineers, 1983, p 137
12. A. Garner, How Stainless Steel Welds Corrode, *Met. Prog.*, Vol 127 (No. 5), April 1985, p 31
13. American Welding Society Specification AWSA5.9-69
14. A. Garner, *Weld. J.*, Vol 62 (No. 1), 1983, p 27
15. A. Garner, *Mater Perform.*, Vol 21 (No. 5), 1982, p 43
16. A. Garner and L.H. Laliberté, U.S. Patent 4,285,787, 1981
17. A. Garner, U.S. Patent 4,285,232, 1981
18. O.V. Ingruber, M.J. Kocurek, and A. Wong, Ed., *Pulp and Paper Manufacture*, Vol 4, *Sulfite Science and Technology*, Joint Textbook Committee of the Paper Industry of the United States and Canada, 1985
19. S.A. Rydholm, *Pulping Processes*, 1st ed., John Wiley & Sons, 1965, p 469
20. P.-E. Ahlers, The Corrosion of Stainless Steels in Passive and Active States Under the Conditions of Sulfite Cooking, in *Pulp and Paper Industry Corrosion Problems*, Vol 3, National Association of Corrosion Engineers, 1982, p 66
21. H.-J. Rocha, Sulphurous Acid as Corrosion Inhibitor in Sulphite Mills, *DEW-Technische Berichte*, Vol 10 (No. 2), 1970, p 124
22. O.V. Ingruber, Sulfite Pulping Cooking Liquor and the Four Bases, in *Pulp and Paper Manufacture*, Vol 4, *Sulfite Science and Technology*, Joint Textbook Committee of the Paper Industry of the United States and Canada, 1985
23. D. McGovern, A Review of Corrosion in the Sulfite Pulping Industry, in *Pulp and Paper Industry Corrosion Problems*, Vol 3, National Association of Corrosion Engineers, 1982, p 60
24. S.K. Murarka and W.T.A. Dwars, Potential Formation of Acidic Condensates in Vapour Phase Sulphite Pulping, in *Proceedings of the 71st Technical Section Annual Meeting*, Canadian Pulp and Paper Association, 1985, p A109
25. G.J.C. Potter and R.B. Kesler, The Formation of Sulphur Trioxide and Calcium Sulphate in the Sulphite Process, *TAPPI J.*, Vol 41 (No. 2), 1958, p 183A
26. K.-E. Jonsson, Stainless Steels in the Pulp and Paper Industry, in *Handbook of Stainless Steels*, D. Peckner and I.M. Bernstein, Ed., McGraw-Hill, 1977, p 43-47
27. S.K. Murarka, W.T.A. Dwars, and J.F. Langlois, Potential Formation of Corrosive Acidic Condensate and Its Monitoring in a Continuous Vapour Phase Bisulphite Chemi-Mechanical Pulping (BCMP) Process, in *Pulp and Paper Industry Corrosion Problems*, Vol 5, Canadian Pulp and Paper Association, 1986, p 187
28. B. Wallen, M. Liljas, and J. Olsson, Performance of a High Molybdenum Stainless Steel in Gas Cleaning Systems With Particular Reference to the Pulp and Paper Industry, *Mater. Perform.*, Vol 21 (No. 6), 1982, p 40
29. J.B. Stirling, Modern Suction Roll Design and Construction, in *CPPA Technical Section Proceedings*, Canadian Pulp and Paper Association, 1964, p D74-D79
30. R.M. Vadas, The Suction Roll—Its Problems and Its Future, *Pulp Paper Can.*, Vol 72 (No. 11), 1971, p 66-70
31. J.P. McNamee, Influence of Suction Roll Covers on Shell Corrosion, in *Pulp and Paper Industry Corrosion Problems*, Vol 1, National Association of Corrosion Engineers, 1974, p 94-98
32. F.S. Wooster, M.D. Moskal, and J.P. McNamee, *TAPPI J.*, Vol 62 (No. 9), 1979, p 71-74
33. R. Thompson, The Suction Roll—Its Problems and Its Future. Part II, *Pulp Paper Can.*, Vol 72 (No. 11), 1971, p 70-76
34. C.B. Dahl and E.P. Neubauer, Modern Practices Can Present Hazards to Stainless Steel Suction Rolls, *Paper Trade J.*, Vol 155 (No. 34), 1971, p 46, 48, 50
35. C. Kelley, J. Vestola, V. Sailas, and R. Pelloux, Corrosion-Fatigue Crack Propagation of High-Strength Stainless Steels Used in Suction Rolls, *TAPPI J.*, Vol 58 (No. 11), 1975, p 80-85
36. R.A. Yeske, "Prediction of Suction Roll Performance From Laboratory Testing," Paper 147, presented at Corrosion/86, National Association of Corrosion Engineers, 1986
37. D.F. Bowers and W.S. Butterfield, Corrosion Fatigue Data for Paper Machine Suction Roll Alloys—Its Significance, in *Proceedings of the TAPPI Engineering Conference*, Technical Association of the Pulp and Paper Industry, 1986
38. "Plate Bending Fatigue Testing in a Corrosive Environment," TIS 0402-08, Technical Association of the Pulp and Paper Industry, 1984
39. "Rotating Bar Bending Fatigue Testing in a Corrosive Environment," TIS 0402-09, Technical Association of the Pulp and Paper Industry, 1984
40. A. Garner, Suction Roll Failures in Canada, *Pulp Paper Can.*, Vol 86 (No. 3), 1985, p 81-89
41. M.D. Moskal and G.W. Eggeman, The Effect of Residual Stresses on Corrosion Fatigue of Suction Rolls, in *Pulp and Paper Industry Corrosion Problems*, Vol 2, National Association of Corrosion Engineers, 1977, p 94-99
42. C.B. Dahl and C.W. Rainger, Stainless Steel Suction Roll Performance Design and Materials, in *Pulp and Paper Industry Corrosion Problems*, Vol 2, National Association of Corrosion Engineers, 1977, p 105-110
43. V. Sailas, Critical Components of the Paper Machine: Damage Caused by Metallic Corrosion, in *Pulp and Paper Industry Corrosion Problems*, Vol 2, National Association of Corrosion Engineers, 1977, p 100-104
44. "Guide for Evaluation of Paper Machine Suction Roll Shells," TIS 0402-10, Technical Association of the Pulp and Paper Industry, 1985
45. C.C. Nathan and A.J. Piluso, Wet End Corrosion Problems in Paper Mills, in *Pulp and Paper Industry Corrosion Problems*, Vol 2, National Association of Corrosion Engineers, 1977, p 126
46. W.S. Butterfield, private communication, 1986
47. "Suction Roll Shell Service Report," TIS 014-56, Technical Association of the Pulp and Paper Industry, 1982
48. *Lockwood's Directory of the Paper and Allied Trades*, Vance Publishing, 1985
49. R.G. MacDonald and J.N. Franklin, Ed., *The Pulping of Wood*, McGraw-Hill, 1969
50. W.D. Halsey, *Paper Trade J.*, Vol 90, 1930, p 140-143
51. *Inspection of Digesters*, Monograph 12, Technical Association of the Pulp and Paper Industry, 1954
52. T.C. Johnson, *TAPPI J.*, Vol 33 (No. 10), 1950, p 481-484
53. L. Ruus and L. Stockman, *TAPPI J.*, Vol 38 (No. 3), 1955, p 156A-161A
54. W. Mueller, *Corrosion*, Vol 17 (No. 12), 1961, p 557
55. L. Stockman and L. Ruus, *Svensk Papperstid.*, Vol 57 (No. 22), 1954, p 831-838
56. C.B. Christiansen and J.B. Lathrop, *Pulp Paper Mag. Can.*, Vol 55, 1954, p 113
57. W. Mueller, *Can. J. Technol.*, Vol 34, 1956, p 162-181
58. J. Hassler, *TAPPI J.*, Vol 38 (No. 5), 1955, p 265-274
59. F. Flynn, F. Richter, and F. Snyder, *TAPPI J.*, Vol 36 (No. 10), 1953, p 433-444

60. C. von Essen, *Svensk Papperstid.*, Vol 52, 1949, p 549
61. B. Roald, *Norsk Skogind.*, Vol 11, 1957, p 446-450
62. E. Hopper, *TAPPI J.*, Vol 36 (No. 8), 1953, p 345-352
63. L. Ruus and L. Stockman, *Svensk Papperstid.*, Vol 56 (No. 22), 1953, p 857-865
64. N. Shoumatoff and H.O. Teeple, *TAPPI J.*, Vol 38 (No. 4), 1955, p 202-207
65. R. Huseby and M. Scheil, *TAPPI J.*, Vol 34 (No. 5), 1951, p 202-209
66. "Stainless Steel Weld Overlay in Sulfate and Soda Pulp Digester Vessels," TIS 405-5, Technical Association of the Pulp and Paper Industry, 1981
67. K. Smith, *Pulp Paper*, Vol 55 (No. 10), 1981, p 66-69
68. R. Yeske, *Pulp Paper*, Vol 57 (No. 10), 1983, p 75
69. D. Singbeil and A. Garner, in *Proceedings of the Fifth International Symposium on Corrosion in the Pulp and Paper Industry*, National Association of Corrosion Engineers, 1986, p 267-271
70. D. Singbeil and A. Garner, "Second Semi-Annual Progress Report on the Research Program to Investigate Cracking of Continuous Digesters," Pulp and Paper Research Institute of Canada, 1984
71. R. Yeske and C. Guzi, *TAPPI J.*, Vol 69 (No. 5), 1986, p 104-108
72. C. Guzi, in *Proceedings of the Fourth International Symposium on Corrosion in the Pulp and Paper Industry*, National Association of Corrosion Engineers, 1983, p 37
73. I. Munro, in *Proceedings of the 1985 TAPPI Engineering Conference*, Technical Association of the Pulp and Paper Industry, 1985, p 37
74. T. Lemke and D. Graver, *TAPPI J.*, Vol 59 (No. 2), 1976, p 134-135
75. D. Jedlica and E. Montrone, in *Proceedings of the 1977 TAPPI Engineering Conference*, Technical Association of the Pulp and Paper Industry, 1977, p 327-329
76. R. Yeske, in *Proceedings of the Fifth International Symposium on Corrosion in the Pulp and Paper Industry*, National Association of Corrosion Engineers, 1986, p 219
77. R. Yeske, Paper 245, presented at Corrosion/86, National Association of Corrosion Engineers, 1986
78. D. Crowe, The Institute of Paper Chemistry, private communication
79. R. Yeske, *Kraft Recovery Operations Seminar*, Technical Association of the Pulp and Paper Industry, 1985, p 211-222.
80. G.A. Smook, *Handbook for Pulp and Paper Technologists*, Canadian Pulp and Paper Association, 1982, p 44-57
81. C.B. Thompson and A. Garner, The Metallurgy and Wear of Refiner Plate Alloys, in *1985 TAPPI Engineering Conference*, Technical Association of the Pulp and Paper Industry, 1985, p 223-231
82. P. Clayton and D. Christensen, Metallurgical Aspects of Refiner Plates, in *Pulp and Paper Industry Corrosion Problems*, Vol 5, Canadian Pulp and Paper Association, 1986, p 175-179
83. W.G. Mihelich *et al.*, Single-Stage Chip Refining—Some Major Operating Parameters and Their Effects on Pulp Quality, *Pulp Paper Mag. Can.*, Vol 73 (No. 5), 1972, p 78-82
84. E.W. Nystrom and R.H. Okell, Sawdust Refining at Crofton, *Pulp Paper Mag. Can.*, Vol 70 (No. 4), 1969, p 83-87
85. F.S. Giffin, Effect of TMP Primary Refiner Plate Rounding on Pulp Quality and Paper Machine Operation, in *Proceedings of the 68th CPPA Technical Section Annual Meeting*, Canadian Pulp and Paper Association, 1982, p B157-B162
86. D.J. Rideout, R.M. Hopkins, and J. Molgaard, Corrosion and Wear of Plates in the Production of Refiner Mechanical Pulp, *Pulp Paper Can.*, Vol 83 (No. 5), 1982, p 54-58
87. C.B. Thompson and A. Garner, Wear Characteristics of Refiner Plates, in *Proceedings of the 72nd CPPA Technical Section Annual Meeting*, Canadian Pulp and Paper Association, 1986, p A117-A122
88. J.S. Jack, C. Mills, and P.H. Baas, A Device for Detection and Prevention of Plate Clash in Disc Refiners, *Pulp Paper Can.*, Vol 82 (No. 9), 1981, p T311-314
89. P.O. Kettunen, Wear and Corrosion Resistant Materials for Refiner Blades, in *Proceedings of the Fourth International Symposium on Corrosion in the Pulp and Paper Industry*, National Association of Corrosion Engineers, 1983, p 98-102
90. R.N. Beaudry, Improving Refiner Plate Costs, in *Proceedings of the 1984 TAPPI Pulping Conference*, Technical Association of the Pulp and Paper Industry, 1981, p 275-279
91. W.C. Frazier, W.H. Arnold, T.C. Williams, and R.S. Charlton, Performance, Design and Metallurgy of Refiner Plates for TMP, *Pulp Paper Can.*, Vol 82 (No. 3), 1981, p 43-55
92. S.K. Murarka, W.T.A. Dwars, and J.F. Langlois, Potential Formation of Corrosive Acidic Condensate and its Monitoring in a Continuous Vapour Phase Bisulfite Chemi-Mechanical Pulping (BCMP) Process, in *Pulp and Paper Industry Corrosion Problems*, Vol 5, Canadian Pulp and Paper Association, 1986, p 187

SELECTED REFERENCES

Paper Machine Corrosion

- R.W. Barton, Thiosulfate Generated by Excess Application of Sodium Hydrosulfite, *Pulp Paper*, Vol 59 (No. 6), 1985, p 108
- D.C. Bennett and C.J. Federowicz, Prediction of Localized Corrosion of Stainless Steels in White Water, *Mater. Perform.*, Vol 21 (No. 4), 1982, p 39
- B.M. Blakey and P.H. Thorpe, Failure and Redesign of Press Roll Ends in Paperboard Machines, *APPITA*, Vol 33, 1979, p 45
- D.F. Bowers, Changes in Water Properties and Corrosivity with Closure, *TAPPI J.*, Vol 66 (No. 9), 1983, p 103
- W.S. Butterfield and P.E. Glogowski, Determination and Corrosivity of Thiosulfate Ion in Paper Mill Systems, in *TAPPI Engineering Conference*, Technical Association of the Pulp and Paper Industry, 1984, p 61
- E. Danielsson, Corrosion and Corrosion Control in the Paper Machine Wet End, in *Pulp and Paper Industry Corrosion Problems*, Vol 3, National Association of Corrosion Engineers, 1980, p 191
- J.L. Ewald and D.P. Hundley, Surface Finish Requirements of the Headbox, *TAPPI J.*, Vol 63 (No. 11), 1980, p 121
- A. Garner, Thiosulfate Corrosion in Paper Machine White Water, *Corrosion*, Vol 41, 1985, p 587
- J.P. Gerhauser, Corrosion of Fourdrinier Wires, *TAPPI J.*, Vol 43 (No. 4), 1960, p 207A
- V.V. Gorelov and A.K. Talybly, Corrosion Protection of the Equipment in Closed and Reduced Water Recycling Systems, in *Pulp and Paper Industry Corrosion Problems*, Vol 4, Swedish Corrosion Institute, 1983, p 113
- M. Kurkela, N. Suutala and J. Kemppainen, On the Selection of Stainless Steels in Bleach Plants and White Water Systems, in *Pulp and Paper Industry Corrosion Problems*, Vol 5, Canadian Pulp and Paper Association, 1986, p 127
- W.A. Mueller and J.M. Muhonen, Pitting Corrosion of Stainless Steels in Six Paper Machine Headboxes: Mechanism and Prevention, *TAPPI J.*, Vol 55 (No. 4), 1972, p 589
- J.M. Muhonen, Corrosion of Stainless Steels in Whitewater, in *Pulp and Paper Industry Corrosion Problems*, Vol 1, National Association of Corrosion Engineers, 1974, p 75
- C.C. Nathan and A.J. Piluso, Wet End Corrosion Problems in Paper Mills, in *Pulp and Paper Industry Corrosion Problems*, Vol 2, National Association of Corrosion Engineers, 1977, p 126
- R.C. Newman, Pitting of Stainless Alloys in Sulfate Solutions Containing Thiosulfate Ions, *Corrosion*, Vol 41, 1985, p 450
- A.J. Piluso and C.C. Nathan, Chemical Treatment to Control Corrosion in the Wet-End Operations of Pulp and Paper Mills, in *Pulp and Paper Industry Corrosion Problems*, Vol 1, National Association of Corrosion Engineers, 1974, p 14
- J.D. Rushton and P.A. Kelly, White Water Corrosion Data from an Integrated Newsprint Operation, in *TAPPI Engineering Conference*, Technical Association of the Pulp and Paper Industry, 1984, p 41
- M.K. Smith and D.A. Wensley, Effects of White Water System Closure on Paper Properties and Machine Corrosion, in *67th Annual Meeting, Technical Section*, Canadian Pulp and Paper Association, 1981, p A159
- G. Sund and S. Strom, The Consequences of System Closure for Corrosion in Swedish Pulp and Paper Mills, in *Pulp and Paper Industry Corrosion Problems*, Vol 5, Canadian Pulp and Paper Association, 1986, p 51
- P.H. Thorpe, Corrosion in Paper Machines—An Overview, in *Pulp and Paper Industry Corrosion Problems*, Vol 3, National Association of Corrosion Engineers, 1980, p 184
- P.H. Thorpe, Microbiological Corrosion of Stainless Steel in Paper Machines and its Causes, in *Pulp and Paper Industry Corrosion Problems*, Vol 5, Canadian Pulp and Paper Association, 1986, p 169

Recovery Boiler Corrosion

- P.E. Ahlers, Chloride in Kraft Mill Recovery Systems, in *Pulp and Paper Industry Corrosion Problems*, Vol 2, National Association of Corrosion Engineers, 1977, p 23
- R. Backman, M. Hupa, and P. Hyoty, Corrosion Related to Acidic Sulphates in Sulphate and Sodium Sulphite Recovery Boilers, *TAPPI J.*, Vol 67 (No. 12), 1984, p 60
- R. Backman, M. Hupa, and E. Uppstu, Fouling and Corrosion Mechanisms in Recovery Boiler Superheater Area, in *Pulp and Paper*

Industry Corrosion Problems, Vol 5, Canadian Pulp and Paper Association, 1986, p 243

- F. Bruno, Primary Air Register Corrosion in Kraft Recovery Boilers, in *Pulp and Paper Industry Corrosion Problems*, Vol 4, Swedish Corrosion Institute, 1983, p 68
- R. Cawein and C. Nin, Detection and Analysis of Generating Tube Thinning in Recovery Boilers, in *TAPPI Engineering Conference Proceedings*, Technical Association of the Pulp and Paper Industry, 1982, p 539
- D.G. Chakrapani and H. Czyzewski, Corrosion Fatigue in Steam Generating and Processing Systems of the Pulp & Paper Industry, in *Pulp and Paper Industry Corrosion Problems*, Vol 3, National Association of Corrosion Engineers, 1980, p 255
- J.A. Dickinson, M.E. Murphy, and W.C. Wolfe, Kraft Recovery Boiler Furnace Corrosion Protection, in *Tappi Engineering Conference Proceedings*, Technical Association of the Pulp and Paper Industry, 1981, p 607
- G. Hough, Ed., *Chemical Recovery in the Alkaline Pulping Process*, Technical Association of the Pulp and Paper Industry, 1985
- A. Jaakkola and T. Roos, Operational Experiences of Austenitic Superheater Tube Bends in Recovery Boilers, in *Pulp and Paper Industry Corrosion Problems*, Vol 4, Swedish Corrosion Institute, 1983, p 82
- K. Kuukkanen, S. Nikkanen, and P. Hyoty, "Operational Experiences in the Rauma-Repola High Sulfidity Recovery Boiler," Paper presented at the TAPPI International Sulfite Pulping Conference, Technical Association of the Pulp and Paper Industry, 1982
- M.K. Minty, An Emission Control System for Recovery Boilers and Materials to Handle the Corrosive Environment, in *Pulp and Paper Industry Corrosion Problems*, Vol 4, Swedish Corrosion Institute, 1983, p 91
- O. Moberg, Recovery Boiler Corrosion, in *Pulp and Paper Industry Corrosion Problems*, Vol 1, National Association of Corrosion Engineers, 1974, p 125
- C.R. Morin and J.E. Slater, Corrosion Problems in Power Generation Equipment, Individual Case Histories, in *Pulp and Paper Industry Corrosion Problems*, Vol 3, National Association of Corrosion Engineers, 1980, p 234
- K.W. Morris, A.L. Plumley, and W.R. Roczniak, The Effect of Chlorides on Recovery Unit Superheater Wastage: An R & D Progress Report, in *Pulp and Paper Industry Corrosion Problems*, Vol 3, National Association of Corrosion Engineers, 1980, p 47
- T. Odelstam, Performance of Composite Furnace Tubes in Recovery Boilers, in *Pulp and Paper Industry Corrosion Problems*, Vol 4, Swedish Corrosion Institute, 1983, p 64
- A.L. Plumley and W.R. Roczniak, Recovery Unit Waterwall Protection: a C-E Status Report, *TAPPI J.*, Vol 58 (No. 9), 1975, p 118
- A.L. Plumley, W.R. Roczniak, and B.E. Lefebvre, "Recovery Unit Superheater Wastage and Control—Progress Report II," Paper presented at Black Liquor Recovery Boiler Symposium, Helsinki, Finland, 1982
- P. Pollak and R. Oesterholm, Corrosion Resistant, Safe and Economical Composite Tubes for Black Liquor Recovery Boilers, in *Pulp and Paper Industry Corrosion Problems*, Vol 2, National Association of Corrosion Engineers, 1977, p 147
- D.W. Reeve, D.C. Pryke, J.A. Lukes, D.A. Donovan, G. Valiquette, and E.M. Yemchuk, Chemical Recovery in the Closed Cycle Mill. Part 1: Superheater Corrosion, *Pulp Paper Mag. Can.*, Vol 84 (No. 1), 1983, p 58
- D.W. Reeve, D.C. Pryke, and H.N. Tran, Corrosion in the Closed Cycle Mill, in *Pulp and Paper Industry Corrosion Problems*, Vol 4, Swedish Corrosion Institute, 1983, p 85
- W.R. Reeve, H.N. Tran, and D. Barham, Superheater Fireside Deposits and Corrosion in Kraft Recovery Boilers, *TAPPI J.*, Vol 64 (No. 5), 1981, p 110
- W.B.A. Sharp, Composite Furnace Tubes for Recovery Boilers—a Problem Solved, *TAPPI J.*, Vol 64 (No. 7), 1981, p 113
- O. Stelling and A. Vegeby, Corrosion on Tubes in Black Liquor Recovery Boilers, *Pulp Paper Mag. Can.*, Vol 70 (No. 15), 1969, p T236
- R.G. Tallent and A.L. Plumley, Recent Research on External Corrosion of Waterwall Tubes in Kraft Recovery Furnaces, *TAPPI J.*, Vol 52 (No. 10), 1969, p 1955
- H.N. Tran, D.W. Reeve, and D. Barham, Formation of Kraft Recovery Boiler Superheater Fireside Deposits, *Pulp Paper Mag. Can.*, Vol 84 (No. 1), 1983, p 36
- H.N. Tran, D. Barham, D.W. Reeve, P.H. Davis, and C.E. Guzi, Acidic Sulphate Corrosion in Kraft Recovery Boilers, in *Pulp and Paper Industry Corrosion Problems*, Vol 5, Canadian Pulp and Paper Association, 1986, p 201
- D.A. Wensley, Corrosion and Cracking of Composite Boiler Tubes, in *1986 Kraft Recovery Operations Seminar*, Technical Association of the Pulp and Paper Industry, 1986, p 231

Corrosion in the Brewery Industry

Edgar W. Dreyman, PCA Engineering, Inc.

BREWERIES are unique in many aspects in the food-processing industry. The product and its production, as well as cleaning procedures, storage, and bottling, all use great quantities of water. This requires a large amount of tankage and extensive hot water facilities; furthermore, the product is an acidic liquid that is aggressive to low-carbon steel. This is further complicated by the fact that the presence of iron ions in the product drastically affects its shelf life.

Wet, damp, and high-humidity conditions all contribute to plant corrosion and premature equipment failure if not properly treated. These factors make the typical brewery a challenge to the corrosion engineer.

Each brewery, whether old or modern, has a cross section of unique equipment ranging from standard power plant equipment, such as boilers, condensers, and oil-handling units, to highly sophisticated laboratory and pilot plant equipment for handling yeasts, malts, and hops essences. All types of materials are employed, from wood to glass, with many metals and special alloys used in large quantities. A brewery materials engineer must be familiar with the materials of construction for all of this equipment and must also be aware of diverse corrosion control techniques involving coatings, metals, cathodic protection, plastics, and even the use of inhibitors.

Corrosion Control Methods

As in any other industry, there are always several ways to solve a given materials problem in the brewery. The task of the materials (corrosion) engineer is to sort out the available choices and to arrive at the most cost-effective solution to the problem.

Traditionally, brewmasters, who at one time exercised complete control of all functions in a brewery, employed set conditions for plant equipment, such as wood with wax linings (tanks with slotted or perforated false bottoms used for filtering clear liquid from the grain mash) for fermentation and storage, copper for kettles and Lauter tubs, and low-carbon steel for pasteurization. This was due in part to the training programs available in Germany and later in the United States. A series of brewing academies or schools provided most of the training for brewmasters and also acted as consultants to most breweries. Their influence dictated what type of equipment was considered suitable for brewing beer.

This has changed dramatically in recent years, and many plants now have production superintendents in addition to or instead of brewmasters. With the consolidation of smaller plants into giant, multiplant brewing empires, the selection of equipment has become a staff engineering function. This group generally works out of the home offices and dictates the policy for all plants, because a major concern is that products made at any location taste alike. Some of the materials of construction used will be discussed below.

Wood. Many smaller, older plants still use wood fermentors and storage tanks. These tanks must be internally lined with an odorless wax or pitch that imparts no odor or taste to the product. Corrosion of steel hoops and rods with turnbuckles had been a problem that required good coating systems, such as the wax or pitch used to line the barrel or food-grade enamels. The key to using wood was to keep it moist so that it remained swelled tight.

In trying to achieve longer life of the linings, several breweries switched to epoxy linings. These were completely impervious to moisture; consequently, cracking of the lining occurred as the wood staves separated from each other. Some internally pitched, wooden beer barrels are still in use, but both of these areas have yielded to the use of metals. Wood was also susceptible to dry out, which imparted an unacceptable flavor to the product.

Steel. Because iron affects the shelf life of beer, steel was not a major material of construction for tankage in contact with beer until superior linings or coatings became available to the industry. Initially, pitches or wax-type coatings were employed. These coatings had to be applied hot, and this was difficult to accomplish in the cold cellars. Some of these coatings had poor abrasion resistance and would crack and chip; therefore, they gave way to the tasteless, pure, food and drug approved epoxies.

This opened a new field for factory, epoxy-lined fermentation and storage tanks. Tank size was limited only by railroad rights-of-way; therefore, 3.7-m (12-ft) diam x 16.5-m (54-ft) long tanks became common in the industry.

Concurrently, there was an increase in the use of glass-lined steel tanks and finally thin-wall stainless steel tanks. Because the cost of nickel has decreased, new construction is virtually all stainless steel tankage.

Stainless Steel. Most of the brewery equipment currently being installed is fabricated from AISI type 304 stainless steel and includes kettles, tanks, tubs, plate coolers, and even some pasteurizers. Stainless steel hot water tanks have been a problem; there has been some major stress-corrosion cracking of heavy-wall vessels holding up to 500 barrels. The problem has been one of chloride ion (Cl^-) concentration due to evaporation under hot conditions and cracking of shaped plate sections at the bottom of the tanks. Remedial measures have consisted of treated water plus complete drainage and flushing.

Stress-corrosion failures in stainless steel thin-wall piping have been common (Fig. 1). Stress-corrosion cracking has taken place where low-chloride water leaking from valve stems saturated insulation and concentrated at temperatures as low as 70 °C (160 °F). Remedial measures have consisted of replacing the piping, coating the exposed stainless steel with a chloride-free coating, and maintaining better valve maintenance.

Stress-corrosion cracking has also been observed where heavy scaling has occurred because of hard water conditions, with chloride concentration in the hardness salt deposits. Remedial

Fig. 1 Stress-corrosion cracking of thin-wall stainless steel hot water piping

action has involved acid treating the water to remove carbonates.

Another problem area has been contact knobs in plate-type coolers. Where deposits have collected, cracking has occurred because of vibration during operation. Remedial action has consisted of using special cleaning compounds to keep surfaces clean.

Special Alloys. Type 444 ferritic stainless steel has been successfully used where SCC problems have occurred because of deposit formation. Monel has replaced type 316 stainless steel for beechwood-chip baskets in one proprietary application in which cracking had occurred around holes in the screening.

Copper has been the traditional metal in breweries for centuries, but with the advent of new alkaline cleaners, some corrosion problems have occurred. Large-diameter brewing kettles are joined by solder joints; these joints have been preferentially corroded and have required silver soldering repair. The use of in-place cleaning with jet sprays has resulted in metal thinning of the copper plates at points of impingement. Rotary cleaners appear to be superior to stationary jets.

Kettle floor thinning has also been observed; cleaning compounds and mechanical abrasion from brushing have been responsible for this. Stainless steels are much more resistant to this type of action.

On percolator refits using stainless steel units with stainless straps, cases of galvanic attack have been reported on bolting and strapping. Copper or brass fittings should not be used in contact with large areas of stainless steel.

Beer is a very good conductor of electricity (low resistance of 200 Ω/cm) due to high CO_2 loading. Even hot water will promote galvanic effects. Thus, precipitation of scales on kettle surfaces can be protective, but when scales are removed with cleaners such as sulfamic acid, the surfaces can become very active. Experience has shown that type 304 or 316 stainless steel will perform well in this environment.

Coatings. Breweries are large consumers of quality coatings, not only for tankage but also for structural steel, flooring, and other working areas. The coatings used range from high-heat silicones for stacks to special superresistant grouts for floor pavers. With a special emphasis on sanitation, many coatings containing antibacterial and mold-control agents are employed throughout the industry.

Bottling and canning operations have very severe conditions for most coatings. Broken bottles can spill product on the coating, and bottle washers use highly alkaline cleaners and label removers at high temperatures. High-build epoxies with inhibitive pigment epoxy primers perform well in many cases.

The water-tolerant polyamide epoxies perform better in cellars and other damp areas. Special low-temperature curing epoxies have been most successful in cellars. Considerable evaluation has been done on water-base epoxies because odors can be readily picked up by the beer or other ingredients, such as hops or malts. Because quick maintenance repairs are critical in large, high-volume breweries, many products that are user friendly must be employed (for example, products that cure quickly on damp surfaces with low odor).

Again, with the large amount of water vapor present in many areas, the use of polyurethane topcoats over epoxies is finding greater acceptance. The polyurethanes do not chalk, and they hold their gloss with good color retention. This is especially true of bright greens, blues, or reds.

Floor Materials. Considerable glazed tile is used in breweries, and special epoxies with good adhesion to very smooth surfaces have been employed to coat glazed ceramic tile in order to prevent crazing (cracking). Epoxy grouts (American Tile Council Specifications) have solved the problem of grout deterioration around kettles, syrup tanks, and other areas susceptible to acidic and microbiological attack. Finally, the problems associated with floors have been addressed by many coating producers with mixed results.

Beer, as well as corn syrup, will attack concrete; therefore, many areas have tile pavers installed. However, as previously mentioned, proper grouts must be used. Where the concrete must be coated, epoxies with embedded grit (skidproofing) are being used in many areas. Chlorinated rubber has been used on cellar floors (Fig. 2), but it generally requires more maintenance because it does not have the abrasion resistance of the epoxies.

Proper surface preparation is most important in brewery floor work. On new floors, acid etching or a light sandblast will suffice. On old floors, removal of a minimum of 50 mm (2 in.) of concrete is mandatory; more extensive concrete removal is sometimes necessary, depending on the degree of contamination. Bacterial contamination deep in the pores of the concrete is a common occurrence. If floors are not properly sealed, corrosion of concrete rebars and structural steel can result, with eventual cracking and spalling of the concrete (Fig. 3).

Glass Linings. At one point in the evolution of the brewing industry, glass-lined tanks were introduced for fermentors, storage, and filling tanks. Glass is an excellent material for preserving the purity of the product and is resistant to the mild acids and alkaline cleaners used to remove scales and deposits. Glass linings are applied by spraying a borosilicate glass frit on the sandblasted surface of the tank, and then firing the entire tank in a furnace until a smooth glass surface is achieved.

Glass linings were installed at thicknesses of 0.4 to 0.8 mm (15 to 30 mils) and gave good service, but were prone to maintenance problems. For example, the lining could be fractured by workmen dropping tools or hose fittings on the glass. Initially, factory pinhole repairs were made by gold filling, similar to a dental inlay. This was followed by repairs with sprayed tin; even large repairs, such as the knuckle areas on tank heads, were repaired in this manner. Tank perforations were repaired with tantalum plugs. As epoxies became available, glass repairs were performed with these materials.

In tanks with multiple pinholing, the glass lining was removed by gritblasting, and the tanks were recoated with epoxies. Cathodic protection was also employed in some cases; this will be discussed later in this article.

Because of the large capital investment required to produce glass linings and the major reduction in the number of breweries, glass linings are no longer produced commercially. Despite this, many tanks are still in use and require maintenance.

Plastics. From a corrosion control point of view, plastic materials are very useful; therefore, they have found application in breweries. Water treatment tanks, acid storage, roofing, and gutters are applications for plastics that are common to most industrial activity and as such are used by breweries. Fiberglass and polyvinyl chloride are among the plastics that have been employed. Small polypropylene tanks for yeast culture and other specialty service have some record of use.

Inhibitors. Where large volumes of water are being used, constant attention must be paid to reducing the water demand, not only because water is becoming expensive and scarce but also because wastewater must be handled in an ecologically sound manner. Brewery waste is high in bacteria and, if untreated, has a very high biological oxygen demand.

Consequently, many closed cooling systems are being employed. This necessitates cooling water treatment, and even before the current environmental regulations, breweries could not use additives containing chromates. This means that zinc/phosphate-type inhibitors and proprietary, pure, food and drug approved systems must be employed.

Even boiler water treatment must be monitored because live steam is used for heating water

Fig. 2 Chlorinated rubber floor coating in a cellar

Fig. 3 Cracking and spalling of a concrete floor

that goes into the product and would carry chemicals over into the product. Beer production is sensitive to most chemicals in terms of the organoleptic properties of the product and the fragility of yeast cultures.

Cathodic protection is an electrochemical method of corrosion control that impresses a direct current onto the structure to be protected. This in turn overrides the many small dc corrosion cells existing on that structure and stops corrosion (see the article "Cathodic Protection" in this Volume).

This method of corrosion control can overcome the effects of galvanic cells due to dissimilar metals and of concentration cells due to differential oxygen concentration, temperatures, or stresses. Combined with good coatings systems, cathodic protection is a most cost-effective way of controlling the corrosion of immersed and underground structures.

Breweries employ cathodic protection on waste storage tanks, fermentor tanks, beer storage tanks, hot water tanks, and underground pipelines for oil, water, sewerage, and, in some cases, product lines. Water treatment facilities also employ this type of corrosion control.

Equipment Problems

Air Conditioning Equipment. Humidity control in cellars will reduce paint deterioration and mold growth and will promote working safety with relatively dry floors. Lithium chloride units used for moisture control have had some deterioration problems that have necessitated the use of exotic alloys and superior coating systems.

In dusty areas, such as malt plants, the poultice corrosion of aluminum fins on cooling units has been a problem (Fig. 4). However, as with many maintenance problems, proper cleaning on a scheduled basis can prevent the failure of such components.

Cooling systems require water treatment, and water tower pans should be properly coated with such products as coal tar epoxies. On occasion, sacrificial zinc anodes have been used for cathodic protection of bare areas in coatings to prevent corrosion. As in all industrial situations, cooling towers must not be placed where they become scrubbers for stack gases. Absorption refrigeration systems use sacrificial anodes in waterboxes

Fig. 4 Corrosion on air conditioner cooling fins

and require periodic eddy current inspection of the tubes to minimize corrosion failures.

Compressors. Many older breweries still use CO_2 and ammonia (NH_3) compressors, which may have intercoolers. Closed systems, if used, require water treatment and the use of sacrificial zinc anodes in cooler heads.

The brine systems associated with older breweries are gradually being phased out, but at one time, they were a major corrosion problem requiring proper inhibition. As mentioned, because chromate inhibitors are not permitted, mixed phosphate inhibitors are used for corrosion control.

Pasteurizers. A major part of a brewery is the bottling and canning operation. One of the most expensive pieces of equipment in the bottleshop is the pasteurizer. The old units were basket pasteurizers that dipped baskets of bottles into water at various temperatures to kill any residual yeast in the product. If this is not done, fermentation can occur in the bottle or can, and gushing takes place when the container is opened.

Modern pasteurizers are continuous-belt operations in which water at different temperatures is sprayed over the bottles or cans. This highly oxygenated hot water is quite corrosive, and the use of inhibitors is limited by the potential for staining of the package.

The normal pasteurizer is of stainless steel construction in the spray areas, with low-carbon steel tanks under each temperature zone. It is in these tanks that problems can be caused by broken beer bottles, which introduce trash and debris onto the tank floors.

The lower-temperature zones of a pasteurizer encourage bacterial growth, and oxygen cells can form under the biological deposits. This corrosion is further enhanced by copper heating coils in the low-resistivity water and by concentration cells produced by hot wall effects in the various compartments. Coatings have not solved this problem, because of mechanical damage from broken glass and the effects of high temperatures on the coating.

Cathodic protection has performed well. Originally, magnesium galvanic anodes were used, but the consumption rate was high; replacement was required every 2 years. High-purity zinc anodes have given 10-year service lives, but extreme care must be exercised so that water chemistry does not cause reversal problems where the steel becomes anodic to the zinc. This can be a problem in some very soft waters.

Ultimately, impressed-current systems have proved to be the most efficient corrosion control method for pasteurizers. The anode materials used include resin-impregnated carbon, high-silicon cast iron, or platinized niobium or titanium.

Current densities of about 160 to 270 mA/m^2 (15 to 25 mA/ft^2) of bare, immersed steel or copper are used. Anodes are mounted on epoxy-coated stand-off holders that hold the anode 150 to 300 mm (6 to 12 in.) from the floor, walls, or heating coils.

A dc rectifier provides direct current through four circuits, each of which is resistorized to reflect the variance in resistivity due to water temperature. Normally, this requires anodes operating at water temperatures of 10 °C (50 °F), at 20 to 40 °C (70 to 100 °F) for preheat, at 50 to 60 °C (120 to 140 °F) for heat, at 75 °C (165 °F) (the pasteurizing temperature), and at 25 °C (80 °F) for cooling and final rinse.

Anode lives are calculated from the following rates of consumption:

- *Carbon*: 0.45 kg/A·yr (1 lb/A·yr) consumption
- *High-silicon cast iron*: 0.34 kg/A·yr (0.75 lb/A·yr) consumption
- *Platinum*: 0.006 g/A·yr (2×10^{-4} oz/A·yr consumption

The platinum anodes can be much smaller because the allowable current density is much higher:

- *Carbon*: 10.8 A/m^2 (1 A/ft^2) anode surface
- *High-silicon cast iron*: 27 A/m^2 (2.5 A/ft^2) anode surface
- *Platinum*: 540 A/m^2 (50 A/ft^2) anode surface

These values have been compensated for use in freshwaters. In hard waters (high in carbonates), some scaling will occur on the cathodic surfaces, and overprotection is discouraged because it will cause stripping of coatings near anodes.

Polarized potentials should not exceed −1.3 V versus a copper/copper sulfate reference electrode if epoxies are being used. Alkyd or zinc-rich coatings are not recommended. Although air-cooled rectifiers perform well as a power source, oil-cooled, sealed units are recommended because of the extensive use of water during cleanup.

Finally, some plants have pasteurizers dedicated to can pasteurization only. When tin-plated steel cans were used, the seam solder would fall to the tank floors, and galvanic pitting of the floor would take place. Most plants now use deep-drawn aluminum cans with rolled tops; therefore, this is no longer a problem.

Beer Barrels. Barrels are generally not thought of as equipment, but they represent a large brewery investment and have been a source of corrosion problems in the past. Except for the steel hoops, wooden barrels are not a problem; however, the lack of suitable wood, the few available coopers, and the heavy weight of wood has led to the use of aluminum and stainless steel barrels.

Initially, aluminum appeared to be the best material for this purpose, but shortly after it was put into use, pitting appeared in the barrels. This ultimately led to perforation, and the rough, pitted surface became hard to clean. The cause of this corrosion was found to be the tin-plated, brass tap that was commonly used. This brass tube in close proximity to the aluminum became the cathode in beer and resulted in pitting of the aluminum. This required pitching the interior of the barrels, thus negating the cost savings for aluminum. Currently, the preferred material of construction for barrels is stainless steel.

Tanks. In addition to the use of coatings in tanks, some cladding with stainless steel has been employed.

Lauter Tubs. The metal cladding of Lauter tubs, especially the bottom of the tub where a perforated or slotted stainless or brass false bottom rests, has been successful. The stand-off legs of the drain plates would perforate the floor coating and cause galvanic cell corrosion action at these areas. Organic coatings were tried first, but the impact was too severe. Therefore, stainless pads or complete cladding with thin sheets of type 304 stainless steel was employed to correct the problem.

Concrete Tanks. Several breweries conceived the idea of constructing buildings in a honeycomb fashion, with concrete tanks being a part of the building support. This practice was employed in Europe and the United States.

Originally, these tanks were coated with wax, which would crack. Beer would be lost in the walls of the building, with the result being odor, corrosion of concrete rebars, and microbial contamination. Attempts were made to correct this problem with fiberglass (cloth) reinforced epoxy. This was very expensive, and failures in the lining occurred unless substantial layers of contaminated concrete were removed.

A firm in the United Kingdom worked out procedures to apply thin-sheet type 304 stainless steel to concrete without expansion buckling, and this technique was used successfully in Scotland and Holland. Similar techniques were then used in the United States with acceptable performance.

The main problem in using thin type 304 stainless sheet lies in preventing movement, which can cause wrinkling and, ultimately, cracking. Proper welding techniques and support spacing are necessary to achieve a clad tank without movement of the cladding.

Another problem area was the tank doors, which today are primarily stainless steel. Coated low-carbon steel and glass-coated doors chipped readily at the edges, which are in contact with the beer; heavy pitting in these areas resulted. The solid stainless door currently provides satisfactory service.

The fermentation tank environment is more severe than that in storage tanks. Fermentation tanks fabricated of stainless steel provide good service in most plants; however, many low-carbon steel tanks still exist in working breweries and present maintenance problems. Fermentation tanks have copper heating and cooling coils inside the tank to control the rate of fermentation. The area ratio of bare steel (breaks in the coating or glass) to bare copper is very unfavorable and results in very rapid attack of the steel in the acidic CO_2-saturated product.

Repairs made to fermentation tanks using epoxies have performed well, but care must be exercised in grinding out the pits to ensure absolute cleanliness. The resin must not only conform to pure food regulations but must also exhibit low shrinkage characteristics so that it does not pull away from the edge of the pit. The resin must not be feathered over the edge onto the good coating or glass surface; if this occurs, the thin coating may fail, and the product will get into the edges of the pit with the resin coming out as a plug.

Cathodic Protection

Another method used to prevent corrosion in fermentation tanks is the use of cathodic protection to monitor the size of pinholes in coatings and linings. Galvanic anodes would introduce metal ions into the product; therefore, they cannot be used. Consequently, impressed-current systems must be employed, with anode selection becoming most important.

High-silicon cast iron anodes cannot be used, because this would introduce iron ions into the beer. Platinum would appear to be the best choice because it is the most inert. Unfortunately, oxygen is liberated at the anodic surface and this would cause deterioration of the product. Carbon, therefore, is the most logical choice for anodes because the oxygen produced combines with the carbon and produces CO_2, which occurs naturally in the beer. A second caution is that oil-impregnated carbon must not be used, because any uncarbonized oil would contaminate the product and inhibit foam formation, which is critical for a good head on the beer.

Resin-impregnated carbon anodes 75 mm (3 in.) thick and 200 mm (8 in.) in diameter have been used in fermentors. These are mounted on stand-off insulators installed in the tank floor and penetrating the tank bottom so all wiring is on the outside of the tank. This hardware has a porcelain insulator with stainless fittings. These stand-off insulators isolate each anode and provide better throwing power of the current; therefore, fewer anodes are required.

Current demand for such a tank is quite low, using 21.5 to 54 mA/m^2 (2 to 5 mA/ft^2) of bare area (including bare coil area). Typical current for an 1800-barrel tank is 0.3 to 0.5 A at 3.5 to 5 Vdc.

In order to maintain a balanced direct current load to each tank, the doors are fitted with waterproof microswitches that transfer the load to a dummy resistor when the tank is open. This prevents any current flow to the tank when workers are inside. Generally, an entire cellar is equipped with anodes from one power supply with 6 to 12 circuits. Power supplies are air cooled and are installed outside of wet cellars.

The bare steel must not be overprotected with excessive current, because coating stripping can occur. This is especially critical when phenolic coatings are used that are sensitive to alkalies. The result of the cathodic reaction is to increase the pH at the tank metal surface, which can soften phenolic resins.

In addition, hydrogen is liberated at the cathode and can be absorbed into the steel. Excessive hydrogen can cause lining failure in glass-lined tanks; therefore, excess current must be avoided. Potentials of less than −1.00 V versus a silver/silver chloride reference electrode have provided steel protection without coating or lining breakdown.

Hot Water Tanks. Many breweries using low-carbon steel hot water tanks that are bare, coated, or lined with gunited (sprayed concrete) linings have employed cathodic protection. Anodes have been resin-impregnated carbon; 75-mm (3-in.) × 1.5-m (60-in.) hanging anodes have performed well, with service lives of 7 to 10 years. These systems have been most effective where flat copper heating coils are used.

One system has been in service for 30 years in a tank that was originally scheduled for replacement. It must be remembered that only the submerged parts of a tank are protected; therefore, water tanks must be coated on the top, or dome, of the tank and down the walls to the waterline. In addition, complete coating should be considered, because this will substantially reduce the size of any cathodic protection system to be employed.

The coating used must be compatible with impressed current. Inert coatings such as epoxies, chlorinated rubbers, or vinyls will function well. Alkyds and inorganic zinc-rich coatings or primers are not recommended.

Cold Water Tanks. The same techniques employed to protect hot water tanks also are used for cold water storage and fire protection tanks. Some fire protection storage tanks are used for dual service; these may contain 3.8 × 10^6 L (1 × 10^6 gal) of water, with 950 000 to 1.9 × 10^6 L (250 000 to 500 000 gal) of this being used for process water. This means that the water throughput in the upper third of the tank may be substantial. This can be highly oxygenated, aggressive, agitated water; consequently, current demand for a cathodic protection system may be very high in that area (323 to 540 mA/m^2, or 30 to 50 mA/ft^2) for proper corrosion control. These design factors must be considered for such systems.

Plant Structures

Buildings. Although most structures are brick and concrete, which require good coatings and sealers (silicone), corrugated siding also is employed. Tedlar (polyvinyl fluoride) coated siding has shown good service for the costs involved. Care should be exercised to employ proper fasteners as specified by the siding vendor.

Galvanized and aluminized siding and roofing panels are commonly employed and should be specified with sufficient zinc or aluminum thickness to ensure reasonable life. If metal panels are to be recoated, special pretreatment is generally required, and coating suppliers should be consulted for the proper materials to use as primers with their top coats.

Rebar Protection. The use of epoxy-coated concrete rebars for buildings and slab construction where water intrusion may be a problem is highly recommended to prevent concrete spalling. Where deicing salt contamination is a problem, as in heated driveways or parking garages, the small additional cost is well worthwhile.

Fig. 5 Corrosion failure of an underground cast iron pipe

Where concrete delamination is occurring in existing structures, cathodic protection of rebars has been successfully applied. Carbon strands or platinum wire anodes or conductive polymer anodes can be successfully used for rebar protection. Anodes for this purpose must be used with conductive coatings to distribute a low dc current density over the salt-impregnated areas.

Buried Tanks and Piping. Buried steel fuel and product storage tanks, such as solvent tanks for can coatings, can be readily protected with a combination of coatings, isolation from electrical contact with other structures, and cathodic protection. The Steel Tank Institute has specifications and designs for applying protection during tank manufacturing, with 30-year warranties against corrosion. Steel fuel storage tanks for large truck fleets, when these are a part of the brewery operation, can be readily protected. Fiberglass tanks are an alternative choice, but require more stringent installation procedures than steel.

Underground fire protection lines, gas lines, water lines, and waste lines are all readily protected with coatings and cathodic protection. If cast iron bell-and-spigot pipe (Fig. 5) is being employed, it must be bonded across the joints. Thermit bonds are preferred to brass wedges, which can loosen and fall out.

An alternative to cathodic protection is the use of polyethylene, plastic sleeve encasement. This technique was developed by the Cast Iron Pipe Research Institute, and it has performed well in the United Kingdom as well as in the United States.

Miscellaneous Structures. Depending on the brewery location, the type of subsoils, and the proximity to rivers and lakes, other structures, such as steel bearing piles or sheet piling, must be protected. When cosmetic requirements are not a consideration, such coatings as black coal tar epoxies or polyurethanes can be used in conjunction with cathodic protection.

Water and waste treatment plants require protection for the steel structures, such as settling basins, flocculators, and clarifiers, and for the concrete. Superior concrete protection must be employed if soils are high in sulfates, with coal tar epoxies giving good performance. The selection of corrosion control materials and systems must be predicated upon the expected life of the structure, the expected difficulty of future maintenance, and the overall cost to the owner.

SELECTED REFERENCES

- W.D. Rigg, "Accelerated Galvanic Corrosion in Beer Tanks in the Presence of Dissimilar Metals," Paper presented at the MBAA 47th Annual Convention, New York, Master Brewers Association of America, 1954
- H.B. Dwight, Calculation of Resistance to Ground, *Electr. Eng.*, Dec 1936, p 1319-1329
- Cathodic Protection Frequently Practical in Corrosion of Brewery Vessels, *Corrosion*, Vol 13, Jan 1957, p 134
- J.D. Redmond, Solving Brewery Stress Corrosion Cracking Problems, *MBAA Tech. Quart.*, Vol 21, Nov 1984
- Solving Design Problems for Cathodic Protection of Glass-Lined Domestic Water Heaters, *Corrosion*, Vol 16, Sept 1960, p 9-17
- E.W. Dreyman, Some Successful Applications of Cathodic Protection in Breweries, *Mater. Prot.*, Feb 1962, p 58-62
- C.W. Ambler and W.E. Allen, "Zinc Galvanic Anodes as a Supplement to Coatings in a Brewery Pasteurizer," Paper presented at the Northeast Regional Conference, National Association of Corrosion Engineers, Oct 1958

Corrosion in the Pharmaceutical Industry

Ralph J. Valentine, VAL-CORR

THE PREVENTION AND MITIGATION of corrosion in the pharmaceutical industry presents a demanding challenge to materials engineers. In most cases, they are working with processes that require equipment and piping systems to be fabricated from material having extremely low corrosion rates when exposed to a wide variety of corrosive media and operating conditions. The substances produced by corrosion reactions contaminate the product being manufactured. This contamination must be removed in one of the subsequent process steps so that the product can pass the quality control tests required for compliance with the stringent purity and quality demands established by the applicable government regulatory agencies.

Materials of Construction

The materials of construction found in pharmaceutical production facilities include:

- Metals and alloys, both solid and clad
- Thermosetting plastics and thermoplastics, both solid and as linings
- Ceramics, both solid and as linings
- Impregnated carbon

Stainless Steels

The austenitic stainless steels, especially the low-carbon and stabilized grades, have been the workhorse alloys in the pharmaceutical industry for many years. These alloys exhibit good corrosion resistance in many media, are readily fabricated, have excellent strength over a wide temperature range, offer good availability, and are relatively inexpensive. The surface condition of the austenitic stainless steels is critical where the pharmaceutical product must not be contaminated and where the stainless steel is required to resist an aggressive environment. The highly protective chromium oxide film that gives stainless steel its corrosion resistance is tenacious, durable, and self-healing in the presence of oxygen; however, this film can be damaged during equipment fabrication and postfabrication cleanup practices. Fortunately, serious problems can be minimized by following good procurement, handling, design, fabrication, and cleanup practices.

The austenitic stainless steels are widely used in oxidizing environments, high-purity water service, and in fine chemical and pharmaceutical production equipment and piping. They are not suitable for use in chloride-containing environments, particularly at high chloride concentrations and at high temperatures.

In recent years, stainless steel producers have developed the so-called super austenitic stainless steels having excellent resistance to general corrosion and pitting/crevice attack in chloride-containing environments. These alloys are highly resistant to intergranular corrosion and stress-corrosion cracking (SCC); this makes them useful in oxidizing chloride solutions, oxidizing acids, and brines. The high molybdenum (2.5 to 6.5%) content and increased chromium and nitrogen give the super austenitics good resistance to pitting and crevice corrosion. The relatively high nickel content (18 to 31%) and the high chromium and molybdenum levels give the alloys excellent SCC resistance. The presence of copper in the alloys improves resistance to sulfuric (H_2SO_4), phosphoric (H_3PO_4), and acetic acids.

The super stainless alloys are less costly than the nickel-base alloys and are readily available in a wide range of product forms, such as pipe and tubing, sheet, plate, and forgings, as well as a full range of welding consumables. In addition, the super austenitic stainless steels are more workable and have better weldability than high-alloy ferritic steels. When the super austenitics are specified, the welding specifications of alloy producers should be followed explicitly so that the full chemical and cracking resistance of the alloy is maintained.

The duplex stainless steels are, as the name implies, duplex in structure. At room temperature, their equilibrium structure is a mixture of austenite and ferrite phases. The compositions of these alloys are carefully controlled to maintain the proper balance of austenite to ferrite. Most of the physical properties of the duplex stainless steels are between those of the austenitic and ferritic stainless alloys. The thermal conductivity of the duplex stainless alloys is less than half that of carbon steel, but about 25% higher than that of the austenitic stainless steels.

The coefficient of expansion of carbon steel is similar to that of the stainless duplex alloys and is about 40% less than that of the austenitic stainless alloys. The duplexes have excellent toughness as well as high strength. Compared with the ferritic stainless steels, the ductile-to-brittle transition of the duplex alloys is more gradual and occurs at a lower temperature, thus allowing the production of a wide range of product forms.

The duplex alloys are not suitable for cryogenic service; the austenitic grades are preferred in this application. The duplex grades have good resistance to chloride SCC; however, the various alloys can show variability in pitting and crevice corrosion susceptibility because of the segregation of the ferrite and austenite phases. The high chromium content of the duplex stainless steels makes them strongly resistant to oxidation. However, prolonged exposure at elevated temperatures (above 345 to 370 °C, or 650 to 700 °F) can affect toughness and corrosion resistance in aqueous media and should be avoided.

All commonly used duplex stainless steels are included in a variety of specifications for sheet, strip, plate, and seamless and welded tubing and pipe issued by the American Society for Testing and Materials (ASTM). Many duplex stainless steels are also included in Section VIII, Division 1, of the American Society of Mechanical Engineers (ASME) Boiler and Pressure Vessel Code, in which they are identified by their UNS designations.

The best corrosion resistance and mechanical properties in welded duplex stainless steels are achieved when the welding practice encourages the formation of a 50-50 austenite-ferrite phase in both the weld metal and the heat-affected zone (HAZ). Welding practices are required that emphasize cleanliness, the avoidance of carbon contamination, and the use of dry inert gas shielding. Because of the sensitivity of ferrite to hydrogen embrittlement, shielding gases containing hydrogen should not be used. Readily available duplex alloys include alloy 2205 (UNS S31803), 44LN (UNS S31200), and Ferralium 255 (S32550).

High-Purity Ferritic Stainless Steels. The super ferritic stainless steels were introduced in the United States during the past decade. The oldest and best known is E-Brite 26-1, an alloy containing 26% Cr and 1% Mo. E-Brite was electron beam refined to reduce carbon and nitrogen to very low levels. This process is no longer in use because of manufacturing problems; E-Brite is now made by vacuum melting. Other super ferritic alloys with approximately the same chromium and molybdenum contents with stabilizing additions of titanium have been introduced to the industry. The 26-1 alloy generally has corrosion resistance equal to or better than that of AISI types 304 and 316 stainless steel, the workhorse alloys of the pharmaceutical industry. In addition, the 26-1 alloy is resistant to SCC, a major shortcoming of the austenitic stainless steels. Because of the resistance of the high-purity ferritic alloys to a wide variety of aggressive environments, many compositions have been developed, such as 18Cr-2Mo, 29Cr-4Mo, 29Cr-4Mo-2Ni, and 27Cr-3.5Mo-2Ni.

As-welded super ferritic stainless steels typically have poor weld zone ductility and are notch sensitive in the HAZ. If these alloys are heated to between 400 and 480 °C (750 and 900 °F) for prolonged times or are slowly cooled within this temperature range, notch toughness is further reduced, and the material becomes brittle. The use of the super ferritics at low temperatures is limited because of their high ductile-to-brittle transition temperatures. If the corrosion resistance of the super ferritic alloys is to be maintained, extreme care must be taken to avoid contamination with nitrogen or carbon during welding. Special welding procedures developed by the alloy producer should be used. Because of problems with welding and controlling the ductile-to-brittle transition temperatures, these alloys are generally used in sheet thicknesses under 3.2 mm (1/8 in.) and for tubing. More information on the corrosion of all types of stainless steels is available in the article "Corrosion of Stainless Steels" in this Volume.

Nickel and Nickel-Base Alloys

Commercially pure nickel (Nickel 200) is highly resistant to many corrosive media. It is most useful in reducing environments, and it can be used under oxidizing conditions that cause the development of a passive oxide film. Nickel has poor corrosion resistance in H_2SO_4, HCl, HNO_3, and H_3PO_4. Pure nickel has outstanding resistance to alkalies, the exception being ammonium hydroxide (NH_4OH), which rapidly corrodes nickel. Oxidizing acid chlorides such as ferric, cupric and mercuric are very corrosive. Nickel is used for containing very reactive chlorides, such as phosphorus oxychloride, phosphorus trichloride, nitrosyl chloride, benzyl chloride, and benzoyl chloride. Pure nickel resists anhydrous chlorine, anhydrous hydrogen chloride, phenol, and bromine.

Nickel-Copper Alloys. Monel alloy 400 is more resistant than nickel to corrosion under reducing conditions and more resistant than copper to corrosion under oxidizing conditions. As a solid-solution alloy, Monel 400 is free from the corrosion that can result from local galvanic action between the phases of multiphase alloys. Monel is generally resistant to SCC; exceptions are mercury and solutions of its salts, fluorosilicates, concentrated caustic soda (NaOH), and potassium hydroxide (KOH). Monel 400 is resistant to all common dry gases at room temperatures. It is not resistant to chlorine, bromine, nitric oxides, ammonia, sulfur dioxide, and hydrogen sulfide in the presence of moisture. Monel is useful in handling H_2SO_4 under air-free reducing conditions. Aeration causes a sharp increase in the corrosion rate. Monel 400 can handle aerated HCl at 10% concentration at room temperature; above room temperature, applications are usually limited to 3 to 4% HCl. In unaerated hydrofluoric acid (HF), Monel 400 is resistant to all concentrations up to the boiling point. Monel 400 has very poor corrosion resistance in HNO_3. Monel 400 has good resistance to NaOH up to about 50% concentration and to NH_4OH up to 3% concentration.

Nickel-Chromium Alloys. The high nickel content of these materials gives them considerable resistance to corrosion under reducing conditions and in strong alkaline environments. Also, because of the high nickel content, the alloys are virtually immune to chloride SCC. However, at high temperature and in contact with concentrated alkalies, they are subject to SCC. Mercury will also cause SCC at elevated temperatures.

Nickel-molybdenum alloys were developed to be resistant to HCl at all temperatures and concentrations. The nickel-molybdenum alloys have good corrosion resistance to other nonoxidizing environments, including boiling 60% H_2SO_4, pure H_3PO_4 at most concentrations and temperatures, wet hydrogen chloride gas, hydrogen chloride to 455 °C (850 °F), and wet halogenated organics. The presence of ferric or cupric salts or other oxidizing agents will cause rapid corrosion of these alloys.

Nickel-Chromium-Molybdenum Alloys. The addition of chromium to the nickel-molybdenum alloys increases the resistance to oxidizing environments, giving them good resistance to HNO_3 and H_3PO_4, as well as to most chloride salts. The alloys are very resistant to pitting and crevice corrosion because of the molybdenum content. The nickel-chromium-molybdenum alloys are the most versatile corrosion-resistant alloys available. They are some of the few alloys that are resistant to wet chlorine gas, hypochlorite, and chlorine dioxide solutions. They have good resistance to ferric and cupric chlorides and other oxidizing salts, alkalies, and acids.

Nickel-chromium-molybdenum-copper alloys were developed to resist H_2SO_4 and HNO_3 over a wide range of concentrations and temperatures. They are resistant to H_3PO_4 even when the acid contains fluorides or oxidizing compounds. The article "Corrosion of Nickel-Base Alloys" in this Volume contains more information on corrosion of nickel and nickel-base alloys.

Titanium

Titanium forms a tight, adherent oxide film that makes it resistant to many oxidizing reagents, including HNO_3 and chromic acid (H_2CrO_4). Titanium is attacked by reducing acids such as H_2SO_4 and H_3PO_4. It is useful in moist chlorine gas and hypochlorite at ambient and elevated temperatures; most organic acids; and water, seawater, and brine solutions at temperatures to the boiling point. At elevated temperatures, titanium is subject to pitting and crevice corrosion in a chloride environment. Titanium will not resist red fuming HNO_3 or HF in any concentration or temperature and will ignite at very low temperatures in dry chlorine gas. More information on the corrosion of titanium and its alloys is available in the article "Corrosion of Titanium and Titanium Alloys" in this Volume.

Zirconium

The two zirconium-base alloys used most frequently in the pharmaceutical and chemical-processing industries contain some hafnium, which is metallurgically and chemically similar to zirconium and does not reduce its corrosion resistance. The alloy UNS R60702 contains a minimum of 99.2% Zr + Hf, with a maximum hafnium content of 4.5%. The second alloy, UNS R60705, contains a minimum of 95.5% Zr + HF, with a maximum of 4.5% HF and 2 to 3% Nb. Both are approved for use in the construction of pressure vessels according to the ASME Boiler and Pressure Vessel Code, Section VIII.

Zirconium has excellent resistance to HCl at all concentrations up to temperatures of 120 °C (250 °F). It resists HNO_3 in all concentrations up to 90% and temperatures to 150°C (300 °F); however, SCC may occur above 70% concentration if high tensile stresses are present. Zirconium is corrosion resistant in H_3PO_4 in concentrations up to 55% at temperatures of 175 °C (350 °F). Above 55% concentration, the corrosion rate increases with temperature, but even in 85% H_3PO_4 at 60 °C (140 °F) the corrosion rate is still less than 0.13 mm/yr (5 mils/yr). If there are fluoride ion impurities present at any concentration of H_3PO_4, zirconium may be subject to rapid corrosion attack. Zirconium alloys are resistant to H_2SO_4 up to 75% concentration and at temperatures to boiling. Ferric, cupric, and nitrate ion impurities cause corrosion of zirconium in H_2SO_4 concentrations above 65%. Fluoride ion concentrations as low as 1 ppm in 50% H_2SO_4 will cause corrosion of zirconium alloys. Zirconium has no corrosion resistance to HF and is rapidly attacked at concentrations as low as 0.001%.

Zirconium is resistant to virtually all alkaline solutions, either fused or in solution to boiling temperature. Zirconium has excellent resistance to corrosion in most organics and organic acids. It should be noted that SCC of zirconium occurs in ferric and cupric chloride solutions, concentrated HNO_3, methanol-hydrochloric acid and methanol-iodine solutions, and liquid mercury or cesium. Detailed information on the corrosion of zirconium and its alloys is available in the article "Corrosion of Zirconium and Hafnium" in this Volume.

Impervious Graphite

Impervious graphite is made by impregnating, under pressure, raw graphite with phenolic, furan, or fluorocarbon resins. The resulting nonporous graphite is impermeable to gases and liquids and is highly corrosion resistant in acids and many solvents. Impervious graphite is dimensionally stable, does not fatigue, and will withstand thermal shock. The phenolic and furan impregnants leach out from the graphite when exposed to ammoniacal compounds, NaOH and KOH, wet halogens, hydrogen peroxide (H_2O_2), strong HNO_3, hypochlorites, and some solvents.

The introduction of the fluorocarbon impregnants to graphite has markedly increased corrosion resistance in solvents, acids, and alkalies. The maximum service temperature for the phenolic and furan impregnants is 170 °C (340 °F); for the Teflon impregnant, 205 °C (400 °F).

Fluoropolymers

All fluorocarbons have high molecular weights, high melting points, and excellent chemical resistance. They have found wide application in chemical and pharmaceutical plants as pipe liners, nozzle liners, gaskets, expansion joints, valve liners, diaphragms for valves and pumps, seals and seal components, and barrier linings for vessels.

Polytetrafluoroethylene (PTFE) has a service temperature of 245 to 260 °C (475 to 500 °F) and is immune to most corrosive environments. Among the materials that attack PTFE are molten alkali metals and free fluorine. This material is rapidly permeated by bromine and oxides of nitrogen, and low molecular weight amines tend to plasticize the polymer. It can also be used at cryogenic temperatures, giving it the widest temperature range of any polymer.

Perfluoroalkoxytetrafluoroethylene (PFA) is a copolymer of tetrafluoroethylene and a perfluorovinyl ether. This material has a service temperature of 195 to 230 °C (425 to 450 °F) and is similar in corrosion resistance to PTFE. It has a lower permeability to most chemicals than PTFE, which is always helpful for lining/barrier applications, and possesses higher tensile properties at elevated temperatures.

Fluorinated Ethylene Propylene (FEP) is a copolymer of tetrafluoroethylene and hexafluoropropylene. It is a fully fluorinated thermoplastic with a service temperature of 175 °C (350 °F). Like PTFE and PFA, this material is chemically inert, with a slightly lower permeability than that of PTFE. The uses for FEP are similar to those for PFA and PTFE.

Ethylene-Chlorotrifluoroethylene (ECTFE) is the result of a 1:1 alternating copolymer of ethylene and chlorotrifluoroethylene having a service temperature of 160 °C (320 °F). Because ECTFE is not completely fluorinated, there are sites along the polymer chain at which chemical attack may occur. This material is attacked by aromatic solvents above 120 °C (250 °F), chlorinated hydrocarbons, ethers, methanol, butanol, and ketones above 65 °C (150 °F), acetic acid above 95 °C (200 °F), and H_2SO_4 and aromatic amines above 65 °C (150 °F). It resists mineral acids up to 120 °C (250 °F), inorganic alkalies, inorganic salts, and oxidizing acids at room temperatures.

Polyvinylidene fluoride (PVDF) is a crystalline, high molecular weight, partially fluorinated polymer of vinylidenedifluoride having a service temperature of 135 °C (275 °F). It is attacked by hot alkalies, hot H_2SO_4, solvents, and warm organic acids. This material has good resistance to chlorine, bromine, and their compounds; however, it will exhibit cracking when subjected to stress in the presence of nascent halogens. Virgin unplasticized PVDF has been successfully used for high-purity water piping as a replacement for electropolished type 316L stainless steel.

Glass-Lined Steel

This material offers the corrosion resistance of glass combined with the strength of steel, making it useful for process equipment operating at elevated pressure and temperature. Glass-lined steel has excellent resistance to corrosion over a wide range of pH and environments. Most glass-lined steel applications will not adversely affect product purity, flavor, or color. Glass-lined steel has an extremely smooth surface that resists fouling and is easily cleaned, which makes it attractive for use in pharmaceutical and fine chemical manufacture.

Material and Corrosion Failures Encountered

The types of failures experienced in the pharmaceutical industry are similar in many ways to those seen in the chemical-processing industries (see the article "Corrosion in the Chemical Processing Industry" in this Volume). Three primary causes of failure in the manufacture of pharmaceuticals—embedded iron, failures of glass linings, and corrosion under thermal insulation—will be discussed in this section.

Embedded Iron

A common problem during the fabrication of stainless steel equipment is the embedding of iron in the stainless steel surface. The iron corrodes when exposed to moist air or when wetted, leaving rust streaks. Larger embedded iron particles can also initiate crevice corrosion attack in the stainless steel. Embedded iron cannot be tolerated in a fabrication destined for an application in the pharmaceutical industry in which the stainless steel is used to prevent product contamination.

Fabrication Practices. The following practices are recommended for minimizing embedded iron in fabrication. First, sheet, strip, and pipe are usually purchased in a surface finish known as AISI 2B (a bright, cold-rolled finish; see the article "Cleaning and Finishing of Stainless Steel" in Volume 5 of the 9th Edition of *Metals Handbook*). Plate is normally hot rolled, annealed, and pickled and is furnished with a 2B mill finish. If a plate having a better surface finish is required, it should be specified during the procurement stage. When cleanliness is very important, sheet and plate can be ordered with a protective adhesive paper that can be left in place during storage and fabrication. Pipe and tubing can be ordered with protective end covers, especially if the pipe and tubing is to be stored outdoors.

Second, sheet and plates should be stored indoors and upright in racks, not horizontally on the floor. The dragging of sheets and plates over each other and worker foot traffic are often primary causes of embedded iron and deep surface scratches.

Third, care should be exercised in handling the sheet and plate on layout tables, forming roll aprons, and benches. This will minimize iron contamination.

Lastly, equipment design plays an important role in iron contamination. Equipment and piping should be free draining. If internal attachments are needed, they should not interfere with free drainage. Bottom connections should be completely free draining. Vessel bottoms used as a work area during construction collect debris, and the foot traffic grinds the debris into the surface. It is suggested that the vessel bottom be flushed down and drained completely at the end of the workday to remove collected dirt and debris. If the vessel is a large flat-bottom unit, a slatted wood floor should be installed to reduce the grinding of contaminants into the vessel bottom by foot traffic.

Testing new fabrications for embedded iron is relatively easy. The surfaces should be washed with clean water, drained completely, and after a 24-h waiting period, inspected for rust streaks on the surface. Water testing of the fabrication as a minimum should definitely be part of the equipment purchase order. For items to be used in a pharmaceutical plant, a more sensitive test—the ferroxyl test (ASTM A 380) for free iron—should be called for in the purchase order. This test can be easily performed in the field as well as in the fabricator's shop.

Removal of Embedded Iron. Pickling is the most effective method of removing embedded iron. The surfaces must be cleaned of all surface oil, grease, and other organic materials so that the surface becomes wet by the pickling solution. The pickling solution is a mixture of HNO_3 and HF at 50 °C (120 °F). The solution removes the embedded iron and other metallic contaminants and leaves the surface clean and in its most corrosion-resistant condition. It should be noted that HNO_3 alone will remove only superficial iron contamination and will leave the deeply embedded particles. Small items are usually pickled by immersion. Piping and vessels that are too large for immersion can be pickled by circulating the pickling solution through them. It is recommended that a competent chemical cleaning contractor be employed for the pickling operation. If the ferroxyl test shows only spotty patches of iron contamination, then it is recommended that these be removed by the use of an HNO_3-HF pickling paste, rather than a complete pickling bath.

Another method of cleaning the stainless steel surfaces is the use of glass bead blasting. The beads should be clean and of a proper size to abrade the surface slightly and remove the contamination. Gritblasting and sandblasting are not recommended; they leave a rough profile that makes the stainless steel prone to crevice corrosion.

Organic contamination on stainless steel surfaces increases crevice corrosion. Contaminants include grease, oil, marking crayons, paint, and adhesive tape. Removal of organic contaminants is best accomplished by the use of a nonchlorinated solvent. It is important that nonchlorinated solvents be used. If a proprietary degreasing solvent is used, it should be tested to ensure that it does not contain chlorides. Residual chlorides remain in crevices and cause chloride SCC of austenitic stainless steels.

Weld Defects. Austenitic stainless steel surfaces can be affected by slag from coated welding electrodes, arc strikes, welding stop points, grinding marks, and weld spatter. These factors have initiated corrosion in aggressive environments that normally do not attack stainless steels. Arc strikes damage the protective oxide film of the stainless steel and create crevicelike imperfections in or near the HAZ. Weld stops create pinpoint defects in the weld metal. Arc strikes and weld stop points are actually more damaging than embedded iron, because they occur where the protective film has been weakened by the heat of welding.

Weld stop defects can be avoided by using runout tabs, by beginning the arc immediately ahead of the stop point, and by welding over each intermediate stop point. Arc strikes are more difficult to eliminate. Initially, the arc can be struck on a runout tab. It can also be struck on the weld metal when the filler metal will tolerate arc strikes. If the filler metal will not tolerate arc strikes, the arc must be struck alongside the filler metal in or adjacent to the HAZ.

Weld spatter creates a small weld in which the molten glob of metal touches and adheres to the surface. The protective oxide film is penetrated, and small crevices or pits are formed where the film has been weakened.

Heat tint formation also weakens the oxide film. This weakening is greater for some degrees of heating than for others, as indicated by the extent of color change. The necessity of removing heat tint is greatest where the environment is very aggressive and the stainless steel approaches the limit of its corrosion resistance. Pickling by immersion in the standard HNO_3-HF solution is the simplest and preferred method of heat tint removal when size permits. Glass bead blasting, using beads that are clean and of proper size in order to prevent overroughening of the surface, can also be employed to remove the heat tint.

Small slag particles from coated electrodes resist cleaning and tend to collect in slight undercuts or other irregularities. To remove slag from 300-series stainless steel, wire brushes fabricated from 300-series stainless steel should be used. For critical service, brushing should be followed by local pickling or glass bead blasting. Grinding is frequently used to remove slag, arc strikes, weld spatter, and other imperfections. Grinding wheels and continuous-belt grinders can overheat the surface and reduce corrosion resistance; therefore, they have limited usefulness. Abrasive disks and flapper wheels are not as harmful to the metal surface. Disks must be kept clean and replaced frequently. These procedures are good commercial fabrication practices and should be specified during the bidding and procurement stages in an effort to eliminate cost overruns and poor service performance.

Failures of Glass-Lined Steel Equipment

Most glass-lined equipment failures are not related to chemical deterioration of the lining, but rather to mechanical and thermal influences. The typical failures encountered in the use of this type of equipment in the pharmaceutical and fine chemical industries usually involve mechanical shock, corrosion, abrasion, thermal shock, and thermal stress.

Mechanical shock is the most common cause of glass failure in pharmaceutical production facilities. It accounts for approximately 70% of the failures in glass-lined steel process equipment and is frequently the result of human error.

The most obvious cause of glass failure due to mechanical shock is objects falling on either the exterior or interior of the vessel. Care must be observed at all times when working near glass-lined equipment because a shock to the outside of a vessel may cause damage to the glass lining.

Lifting lugs are normally supplied on glass-lining equipment and should be used for lifting the equipment and setting it in place. The lugs are specifically designed for this purpose and should be used in accordance with manufacturer's recommended procedures for handling and rigging. Shortcuts in rigging, such as using a nozzle as a lifting lug, can easily subject the glass lining to undue stress and cause damage to the lining. If mechanical shock has occurred or is suspected, the equipment interior should be inspected immediately and, if necessary, repaired.

Entering the interior of glass-lined steel equipment always creates a potential for mechanical damage. In addition to compliance with the routine safety precautions, the mechanic should wear clean, soft rubber soled shoes or sneakers to prevent scratching of the lining and should remove all loose objects from his pockets prior to entering the lined equipment. Even metal belt buckles should be removed to prevent accidental scratching of the glass on the entry manway. Tools can be lowered to the mechanic when he is safely inside.

If it becomes necessary to remove product or by-products from the walls of glass-lined steel equipment, care must be taken to avoid scratching of the glazed surface. Metal tools should never be used. Plastic or wood scrapers can be used; high-pressure water jets are preferable.

The introduction of nucleated glass linings by fabricators of glass-lined steel equipment has reduced the damage caused by impact and, to some extent, has lessened its occurrence. Nucleated or partially nucleated glass linings have higher tensile strength and fracture energy than conventional glassed steel linings. These properties have lessened the tendency toward spalling from mechanical shock, the releasing of internal stresses, or thermal shock. Nucleated glass linings cannot prevent cracking entirely, but they tend to limit the extent of the damage by restricting it to a small area, usually requiring only a tantalum plug for repair.

Corrosion Failures of Glass Linings. Glass-lined steel is not completely inert and is constantly undergoing local chemical reactions at the glass surface. Glass-lined steel can be used with corrosive materials because of the low rate of reaction; the slower the rate, the longer the useful life of the glass lining.

Acids. Except for HF, concentrated H_3PO_4, and phosphorous acid (H_3PO_3) above 85%, glass lined steel is resistant to corrosion by acids. Generally, the corrosion rate decreases with concentration, but accelerates with increasing temperature. Acid attack is more severe in the vapor phase than in the liquid phase, especially in more dilute solutions, because of the water vapor in the vapor phase. Usually, acid attack will result in a gradual loss of the fire-polished surface, but the lining will generally retain a dull, smooth finish.

Hydrofluoric acid will completely destroy glass-lined steel equipment. Even with concentrations as low as 20 ppm, fluorides in an acid environment corrode glass severely, especially in continuous reactions in which the fluorides are repeatedly replaced. Hydrofluoric acid reacts with silicon dioxide, the primary ingredient in glass, and destroys the silicon dioxide structure, forming silicon tetrafluoride and water vapor. The silicon tetrafluoride then hydrolyzes into silicon dioxide and hydrofluorine silicic acid, which are absorbed by the condensing water vapor. When this contaminated condensate reaches the heated vessel wall in the vapor space, it evaporates, depositing silicon dioxide and liberating the hydrofluorine silicic acid, which then disintegrates into silicon tetrafluoride and HF. The chain reaction keeps repeating itself with continuous replenishment of HF. The corrosion occurs both in the liquid phase and in the cooler areas of the vapor space in which the fluoride vapor can condense.

Glass is not attacked by fluorine and its compounds in an alkaline environment, nor is it attacked by anhydrous hydrogen fluoride gas. The prerequisite for HF attack in the vapor space is the formation of water; thus, the corrosion rate will be greatly reduced if the vapor area is heated. In the liquid phase, fluorides will severely etch the glass and produce a roughened surface with a complete loss of the fire-polished surface. In the vapor phase, the attack is more localized and concentrated; chipping and pinholes will be seen, but with considerably less loss of the fire-polished glass surface.

Reagents that contain fluoride impurities must be carefully analyzed to determine the fluoride level before they are used. Frequently, technical grade H_3PO_4 and its salts are often contaminated with fluoride, as are other mineral acids. It is important to realize this when using these chemicals in recovery operations.

Alkaline attack of glass linings is much more severe than acid attack. The attack takes place only in the liquid phase in the case of nonvolatile alkalies. The greater the concentration and pH of the alkali, the greater the amount of corrosion. Corrosion by alkalies is evidenced by pinholes, chipping, and a severe loss of the fire-polished surface.

Many glass-lined steel reactors have been lost prematurely in service by improper charging of reactants into the vessels. Caustic reactants charged into a vessel should always be fed directly into the liquid phase. If fed through a nozzle, the alkali will run down the side wall of the reactor in the vapor space and cause severe alkaline attack, especially if the reactor is being heated.

Water can cause severe corrosion of glass-lined steel, and the severity increases with water purity and temperature, becoming greatest above the boiling point. When water droplets condense on the relatively cool surface of the glass-lined equipment in the vapor space, they leach out alkali ions from the glass and form an alkaline solution that attacks the glass. A small amount of acid added to the water usually slows the corrosion caused by condensation in the vapor space. This addition of acid is frequently useful in steam distillations.

Abrasion Failures. Failure of glass linings by abrasion alone is not very common. It is evidenced by a loss of fire polish of the glazed surface and results in a rough, sandpaperlike finish. Abrasion in conjunction with acid corrosion results in severe failure. The abrasive action weakens the silica network mechanically, allowing acid corrosion to accelerate rapidly.

Thermal shock failure occurs because of abrupt changes in the temperature of the glass lining and results in relatively small but thick pieces of glass spalling off in rigid fractures. There are four operations in which sudden temperature variations can cause thermal shock:

- Sudden cooling of a glass-lined surface by subjecting a preheated surface to a cold liquid
- Sudden heating of a glass-lined steel wall by rapidly circulating a very hot fluid through the jacket of a cold vessel
- Sudden heating of the glass-lined steel surface by introducing a hot fluid into a cold vessel
- Sudden cooling of the vessel wall by rapid circulation of a cold fluid through the jacket of a preheated vessel

Thermal shock is strictly an operational problem and can be eliminated by adequate process controls. Unlike failure due to mechanical shock, thermal shock usually damages the glass lining so that repairs with tantalum plugs are not practical or possible. Reglassing of the vessel is then required.

Failure due to thermal stress is caused by differential heating or cooling that is not instantaneous. Thermal stress may occur on the vessel wall just below the area at the top jacket closure ring or in the area where the bottom jacket closure ring is welded to the vessel. In either case, the inside of the jacket can be heated or cooled, while the unjacketed area is not. At temperatures approaching 205 °C (400 °F), sufficient strain can be developed in the glass lining at these areas to cause the glass to crack.

Thermal stress can be reduced by careful control of heating or cooling operations. Also, insulation of the unjacketed areas will help to reduce the extreme temperature variations.

Water hammer has also been known to cause shock waves that add to the thermal stresses at the area of the jacket closure. If these stresses are allowed to persist for long periods of time, cracks can develop with or without chipping.

Overstressing of Nozzles. Glass is strong in compression, but a concentrated point load can damage a glass-lined nozzle. The convex radius at the top of the nozzle also makes it susceptible to damage due to overstressing. The two situations that can cause overstressing of the nozzle and lead to possible failure are overtorquing and overstressing by the attached piping.

Overtorquing of bolts or clamps used to secure nozzles on glass-lined vessels to piping can cause glass lining to spall in large segments that may include a considerable area of the nozzle. Overstressing of nozzles through external piping can break the glass lining on the nozzle face. The design of external piping systems should include expansion joints or bellows as close as possible to the nozzles to minimize eccentric loads and moments and to allow for thermal expansion of the piping and the vessel. Cold springing when connecting pipe to a nozzle cannot be tolerated. Supporting heavy gear drives on a nozzle can also create an unsafe condition with regard to the glass lining.

Failures of Repair Plugs. Tantalum is the most commonly used repair material for glass linings. Its corrosion resistance is very close to that of glass, except in fuming H_2SO_4 above 65 °C (150 °F), H_2SO_4 above 98%, free sulfur trioxide (SO_3) above 65 °C (150 °F), or nascent hydrogen.

A repair made with tantalum will protect a glass-lined vessel for its useful life if the repair plug is installed in accordance with the recommendations of the manufacturer. However, the repair plug must be inspected periodically to ensure that it is secure.

Galvanic corrosion can cause hydrogen embrittlement of tantalum repair plugs, because the tantalum is in contact with the steel substrate. If a second dissimilar metal is present in the vessel, such as a metallic dip tube fabricated from a material other than tantalum, a galvanic cell is produced. Tantalum, being more noble, will generally act as the cathode where hydrogen is liberated.

Failure Due to Hydrogen Damage. Acids on the exterior steel surfaces of glass-lined steel equipment will, in time, react with the steel, forming nascent hydrogen. This hydrogen diffuses through the steel behind the glass and causes the glass to spall because of pressure buildup.

To avoid this type of failure, all acid spills on the outside of a glass-lined steel vessel should be washed off immediately to avoid acid attack. The so-called acid-resistant coatings give some degree of protection, but none is completely effective against acid attack on the steels.

Jacket cleaning can create another source of nascent hydrogen. The cleaning solutions recommended by the manufacturer should be used for cleaning the inside jacket of a glass-lined vessel. Strong acid solutions, inhibited or otherwise, should never be used for descaling the jacket in order to avoid any possibility of nascent hydrogen formation.

Corrosion Beneath Thermal Insulation

Serious corrosion problems frequently occur under thermal insulation applied to vessels and piping components in pharmaceutical plants when the insulation becomes wet. Corrosion beneath insulation is an insidious problem that is also discussed in the article "Corrosion in the Chemical Processing Industry" in this Volume. The insulation usually conceals the corroding metal, and the situation can go undetected until metal failure occurs. The corrosion of metal components beneath insulation has led to very high maintenance costs and lost production time and has frequently required complete replacement of major components. In addition, operator and plant safety may be jeopardized.

Thermal insulation received from manufacturers and distributors is dry, or nearly so. Therefore, if the insulation remains dry, there is no corrosion problem. Possible solutions to the problem of metallic corrosion beneath wet insulation include keeping the insulation dry or protecting the metal.

Unfortunately, the application of this solution is not that simple. Insulation can become wet in storage and during field erection. Moisture or weather barriers are not always installed correctly, or they are not effective in preventing water entry. Weather barriers and protective coatings can become damaged and are often not maintained and repaired.

The problem is further complicated by the fact that the degree of corrosion by wet insulation appears to be dependent on the insulation material as well as the atmospheric contaminants and moisture entering from external sources. Water extracts from calcium silicate base insulation, fiberglass, cellular glass, and ceramic fiber are generally neutral to alkaline, with pH values in the range of 7 to 11. Cellular glass is free of soluble chloride, while calcium silicate, fiberglass, and some ceramic fibers contain chlorides. Mineral wool gives a neutral environment when wet, usually a pH value of 6 to 7 with a low chloride content (2 to 3 ppm). Water extracts from organic foams can be quite acidic, with pH values of 2 or 3. In addition, where halogenated fire retardants have been added to the foam, water extracts show high levels of free halide, depending on the degree of hydrolysis achieved.

Carbon and low-alloy steels are normally passive in alkaline environments and have minimal corrosion rates. However, chloride ions (Cl^-), either from the insulation material itself or from airborne or waterborne contaminants, tend to break down the passivity locally and initiate pitting corrosion. If penetration by acidic airborne or waterborne contaminants of sulfur or nitrogen oxides is possible or if water extracts from the insulation are acidic, such as from organic foams, then general corrosion occurs. Occasionally, airborne or waterborne contaminants, notably the nitrate anion (NO_3^-), cause external SCC of nonstress-relieved carbon and low-alloy steel systems, especially if a cyclic wetting and drying concentration mechanism is present. Generally, plant facilities operating continuously or intermittently between 65 and 205 °C (150 and 400 °F) are subject to corrosive attack.

The most significant corrosion problem that occurs when insulated austenitic stainless steels are subjected to moisture is external SCC. The problem occurs because chlorides tend to concentrate under the insulation at the surface of the metal when the insulation becomes wet. The moisture can leach soluble chlorides out of the insulation, or the entering moisture may already contain chloride from the environment. At the warm metal surface, the moisture is vaporized, leaving behind an increasing concentration of chlorides. During operation, when the equipment or piping is in the susceptible temperature range, chloride SCC can then occur rapidly. The four factors necessary for SCC are an austenitic stainless steel, Cl^- ions, tensile stress on the metal, and temperatures between 50 and 230 °C (120 and 450 °F).

Unfortunately, thermal stress relief, which is usually effective in preventing SCC of carbon and low-alloy steels, is normally not practical for austenitic stainless steels. However, there are numerous controllable factors in the design, construction, and maintenance of insulated equipment that have a marked effect on the amount of damage caused by corrosion under thermal insulation.

Equipment Design. The design of pressure vessels, tanks, and piping generally includes numerous details for support, reinforcement, and connection to other equipment. These details may include stiffening rings, insulation support rings, gussets, brackets, reinforcing pads, ladder brackets, flanges, and hangers. The design of equipment, including these details, is the responsibility of engineers/designers utilizing construction codes to ensure reliable designs for both insulated and uninsulated equipment. Unfortunately, consideration of the problem of insulating these details and of leaving adequate room for the insulation is completely lacking in these codes and in the instructions to the engineers/designers. As a result, the items are designed as though they will not be insulated.

Undesirable geometries and design features include:

- Flat horizontal surfaces (such as vacuum rings)
- Structural shapes that trap water (H-beams, channels)
- Shapes or configurations that are impossible to weatherproof properly (structural members, gussets)
- Shapes that lead moisture and contaminants into the insulation hanger rods (angle iron brackets)
- Inadequate spacing that causes interruption of the vapor or weather barrier (nozzle extensions, ladder brackets, deck or grating supports). The weather barrier on such designs is frequently broken because of inappropriate details for insulated equipment or the lack of space for the specified insulation thickness

The consequence of an incomplete moisture barrier is that more water and contaminants get into the insulation at each exposure cycle; this increases the time required for drying, cools the insulated equipment to temperatures at which corrosion is possible, and thus increases the damage. Some equipment details, such as gussets, brackets, and hangers, actually funnel water and contaminants into the insulation. Another consideration is the increased cost of insulating equipment that was not designed for insulation. In such cases, the insulation and jacketing must be cut and fitted by installers; thus, needless man-hours are spent insulating a complicated detail, with the result being an installation that is doomed to failure. The solution to such problems is to specify the type of insulation, thickness, and weather jacketing in the design stage.

The service or operating temperature of the equipment is very important in corrosion beneath thermal insulation. Higher temperatures make water more corrosive, and paints and caulking will fail prematurely.

Generally, corrosion associated with equipment operating below freezing temperatures is corrosion outside of, not under, the insulation. Equipment operating between freezing and the atmospheric dew point is subject to continuous corrosion, and damage can occur as quickly as it does under warm insulation. However, corrosion beneath warm insulation is more difficult to control, because of the drying out of water entering the insulation and the concentration of contaminants carried in with the water drying out repeatedly in the same location.

Insulation Materials. Corrosion is possible beneath all types of insulation. The insulation material is only a contributing factor. The insulation characteristics that are most influential in the corrosion of metal beneath insulation include water absorbency, chemical contributions to the water phase (not only from the insulation but also from external sources), and service temperature. Some of the more widely used insulation materials are discussed below.

Polyurethane foam is primarily used for cold and antisweat service. It does not absorb or wick water as long as the cell structure remains intact. It is permeable to water vapor in cold service when the required vapor barrier fails. Vapor diffuses through cell walls to the temperature zone in which it condenses and diffuses further to the point at which it freezes. The maximum service temperature is 80 °C (180 °F). If used in continuously cold service, it does not corrode unprotected metal surfaces. If in intermittent service to its maximum service temperature, it can cause corrosion of unprotected wet metal surfaces from released chlorides in fire retardants and blowing agents. The ultraviolet rays from the sun will decompose this insulation.

Polyisocyanurate foam is a fire-resistant organic foam having a low flame propagation rate. It does not absorb and wick water as long as the cell structure remains intact. It is permeable to water vapor in cold service when the required vapor barrier fails. The maximum service temperature is 120 °C (250 °F).

When polyisocyanurate is exposed to heat and moisture, the cell structure in the heated zone breaks down. The decomposition products may contain chlorides from the fire retardant and blowing agent and may aggressively corrode unprotected metal surfaces.

Flexible foamed elastomer does not readily absorb or wick water, and it has a maximum service temperature of 80 °C (180 °F). Although not corrosive by itself, it will support the corrosion of unprotected metal surfaces when water is present, especially when the water contains chlorides from an external source.

Cellular glass is a rigid glass foam whose blowing agent contains carbon dioxide and hydrogen sulfide. It does not absorb and wick water. The maximum service temperature is 480 °C (900 °F). When water is present and the cell structure is damaged, release of the foam blowing agent may cause corrosion on unprotected carbon steel surfaces.

Glass fiber for insulation is usually a pure glass fiber containing various types of binders. Fiberglass will absorb and wick water; however, it drains excess moisture better than other types of insulation. The maximum service temperature is 230 °C (450 °F), with special formulations to 455 °C (850 °F).

Mineral wool is a mineral or metal slag fiber that is basically an impure glass. It readily absorbs and wicks water. Maximum service temperatures range from 650 to 980 °C (1200 to 1800 °F), depending on type and manufacturer. The fact that mineral wool will wick and hold water makes it a contributory factor in the corrosion on unprotected wet metal surfaces.

Calcium silicate insulation is a cementitious mixture and readily absorbs and wicks water. Calcium silicate insulation can hold up to 400% of its own weight in water without dripping. The maximum service temperature is 650 °C (1200 °F). Although its pH is initially high (10 average), it is aggressive in supporting corrosion on unprotected wet metal surfaces because of its moisture retention, particularly when the moisture contains chlorides from an external source.

Protective coatings are extremely important in preventing the corrosion of metal surfaces beneath thermal insulation. In the past, the attitude has been that a single coat of primer would be adequate based on the assumption that the weatherproofing never allowed water to penetrate into the insulation system. This was not the case.

Basically, service under thermal insulation is virtually an immersion service. Once the weather- or vaporproofing barrier is broken, metal surfaces under isolation are wet longer than the surfaces of most uninsulated equipment. Under warm insulation, the coating is subject to higher temperatures than most coated, uninsulated equipment. The coatings beneath thermal insulation fail because of chemical degradation and the permeability of the coating. Highly permeable coatings allow corrosion to initiate behind the coating film, even in the absence of breaks or pinholes in the coating.

In selecting a protective coating for use beneath insulation, consideration should be given to its abrasion resistance, temperature resistance, chemical resistance, and its ability to resist hot water vapors. Organic coating materials are discussed in detail in the article "Organic Coatings and Linings" in this Volume.

In general, for carbon steel and stainless steel piping and equipment operating at -45 to 120 °C (-50 to 250 °F) under insulation, it is recommended that the surface be abrasive blast cleaned to National Association of Corrosion Engineers (NACE) No. 2 or Steel Structures Painting Council (SSPC) SP10 Near White metal standards having a surface profile of 50 to 75 μm (2 to 3 mils). The freshly cleaned surfaces should then be coated with 305 μm (12 mils) (total dry-film thickness) of an epoxy phenolic coating applied in two or more coats or with 305 μm (12 mils) (total dry-film thickness) of a high melting point, amine-cured, coal tar epoxy applied in one or two coats.

For carbon steel piping and equipment operating at temperatures of 120 to 260 °C (250 to 500 °F) with intermittent cycling service into the hot water range under insulation, it is recommended that the surface be abrasive blast cleaned to NACE No. 1 or SSPC SP10 Near White metal standards having a surface profile of 50 to 75 μm (2 to 3 mils). The freshly cleaned surfaces should then be coated with 150 μm (6 mils) (total dry-film thickness) of a copolymerized silicone resin applied in two or more coats.

Carbon steel piping and equipment operating continuously above 260 °C (500 °F) under insulation should have the surface abrasive blast cleaned to NACE No. 3 or SSPC SP6 Commercial Blast standards having a surface profile of 25 to 50 μm (1 to 2 mils). The freshly cleaned surfaces should then be coated with 100 μm (4 mils) (total dry-film thickness) of a black zinc-free modified silicone coating applied in two coats.

For stainless steel piping and equipment operating at temperatures from 20 to 370 °C (250 to 700 °F) under insulation, it is recommended that the surface be cleaned according to SSPC SP1 Solvent Cleaning standards. The freshly cleaned surfaces should be coated with 100 μm (4 mils) (dry-film thickness) of a black zinc-free modified silicone coating applied in two coats.

Inorganic and organic zinc-rich primers have given very poor performance under thermal insulation and should not be used. Possible reasons for this poor performance include:

- The likelihood of reversal of polarity of galvanic couples with increasing temperature
- Chemical salts carried in and deposited with the water that interfere with or destroy the effectiveness of the coating
- The environment beneath the insulation is not freely ventilated and may not have adequate oxygen or carbon dioxide for film forming reactions to occur

Weatherproofing and Vaporproofing. The outer covering of the insulation system is critical. It is the principal barrier to the water necessary for the corrosion of metals beneath thermal insulation. Also, it is the only part of the insulation system that can be quickly inspected and economically repaired.

The purpose of a weather barrier, which should be used on warm or hot equipment, is to keep liquid out but to permit the evaporation of any moisture that gets in. The purpose of a vapor barrier is to keep both liquid and vapor out of the insulation system. Vapor barriers should be used on all cold and dual-service insulation systems.

SELECTED REFERENCES

- G.E. Moller, Designing With Stainless Steels for Service in Stress Corrosion Environments, *Mater. Perform.*, Vol 16 (No. 5), p 32-44
- A. Garner, "Corrosion of High Alloy Austenitic Stainless Steel Weldments in Oxidizing Environments," Paper presented at Corrosion/82, National Association of Corrosion Engineers, 1982
- Technical Bulletins S-1, 871-ENG; S-1, 872-ENG; S-1, 874-ENG; S-1, 880-ENG; S-1, 884-ENG; S-1, 885-ENG; S-51-26-ENG; S-52-27-ENG; S-51-28 ENG; S-51-33-ENG; S-52-72-ENG; and S-54-18-ENG, Sandvik Steel AB
- Technical Bulletins H-2000A, H-2002A, and H-2010A, Haynes International, Inc.
- D.H. DeClerck and A.J. Patarcity, 32nd Report on Materials of Construction, *Chem. Eng.*, Vol 93 (No. 22), p 46-63
- Technical Bulletins T-5, T-7, T-15, T-37, T-40, T-42, and S-37, Inco Alloys International, Inc.
- W.I. Pollock and J.M. Barnhart, Ed., *Corrosion of Metals Under Thermal Insulation*, STP 880, American Society for Testing and Materials, 1985
- S. Lederman, "Glass Lined Equipment—Typical and Atypical Failures Encountered in the Field," Paper presented at Corrosion/82, National Association of Corrosion Engineers, 1982
- Technical Bulletins TWCA-8102ZR and TWCA 8103ZR, Teledyne Wah Chang Albany

Corrosion in Petroleum Production Operations

Chairman: James E. Donham, Consultant

THE PRODUCTION of oil and gas, its transportation and refining, and its subsequent use as fuel and raw materials for chemicals constitute a complex and demanding process. Various problems are encountered in this process, and corrosion is a major one. The costs of lost time, the replacement of materials of construction, and the constant personnel involvement in corrosion control are substantial and, if not controlled, can be catastrophic. The control of corrosion through the use of coatings, metallurgy, nonmetallic materials of construction, cathodic protection, inhibitors, and other methods has evolved into a science in its own right and has created industries devoted solely to corrosion control in specific areas.

This article will discuss the particular corrosion problems encountered, the methods of control used in petroleum production, and the storage and transportation of oil and gas to the refinery. Refinery corrosion is discussed in the article "Corrosion in Petroleum Refining and Petrochemical Operations" in this Volume.

Causes of Corrosion

Arthur K. Dunlop
Corrosion Control Consultant

The fundamentals of corrosion and the forms of corrosion are discussed in detail in the Sections so named in this Volume. Therefore, this article will concentrate on those aspects that tend to be unique to corrosion as encountered in oil and gas production.

Most unique, of course, are the environments encountered in actual production formations, which, in the absence of contamination, are devoid of oxygen. *In situ* corrosives are limited to carbon dioxide, hydrogen sulfide, polysulfides, organic acids, and elemental sulfur. Additional unique aspects are the extremes of temperature and, particularly, pressure encountered. In deep gas wells (6000 m, or 20 000 ft), temperatures approaching 230 °C (450 °F) have been measured, and partial pressures of CO_2 and H_2S of the order of 20.7 MPa (3000 psi) and 48 MPa (7000 psi), respectively, have been encountered.

Convenient access to the most important literature on H_2S corrosion (particularly with regard to sulfide stress cracking) and CO_2 corrosion is available in Ref 1 and 2. Supplemental information is provided in Ref 3 and 4.

To the initially oxygen-free geologic environment, a variety of oxygen-contaminated fluids may be introduced. Examples are drilling muds, which are used during drilling and maintenance of wells; dense brines; water and carbon dioxide injected for secondary oil recovery; and hydrochloric acid injected to aid formation permeability. Some of these fluids are inherently corrosive; others are potentially corrosive only when contaminated with oxygen.

Oxygen is also responsible for the external corrosion of offshore platforms and drilling rigs. In oil and gas production, highly stressed structural members are directly exposed to a corrosive seawater environment. This makes corrosion fatigue a particular concern.

Another unique aspect of oil and gas production operations, particularly in older fields, is the almost exclusive use of carbon and low-alloy steels. Only recently has any extensive use of corrosion-resistant alloys been justified.

Oxygen

Although it is not normally present at depths more than approximately 100 m (330 ft) below the surface, oxygen is nevertheless responsible for a great deal of the corrosion encountered in oil and gas production. However, oxygen-induced internal corrosion problems tend to be greater in oil production, where much of the processing and handling occurs at near-ambient pressure. This makes oxygen contamination through leaking pump seals, casing and process vents, open hatches, and open handling (as in mud pits during drilling, trucking, and so on) highly likely. Also, failure of oxygen removal processes (gas stripping and chemical scavenging) is a relatively common occurrence in waterflood systems (see the discussion "Corrosion in Secondary Recovery Operations" in this article).

A number of the properties of oxygen contribute to its uniqueness as a corrosive. Oxygen is a strong oxidant. This means that even trace concentrations can be harmful, and the corrosion potential of steel (almost 1.3 V) is high enough to overcome very substantial potential drops between anodic and cathodic sites. Also, the kinetics of oxygen reduction on a metal or conductive oxide surface are relatively fast. This, coupled with the low solubility of oxygen in water and brines, tends to produce conditions in which the mass transport of oxygen is the rate-limiting step in the corrosion of carbon and low-alloy steels in nonacidic environments.

Mass transport is important in a number of aspects of oxygen corrosion and corrosion control. On newly installed bare steel offshore structures, mass transport of oxygen governs current requirements for cathodic protection. Poor mass transport under deposits and in crevices promotes localization of attack. In the final analysis, limiting the mass transport of oxygen plays a critical role in much of the corrosion control in oxygenated systems.

The crucial role of mass transport can be illustrated as follows. At ambient conditions, water equilibrated with air will contain of the order of 7 to 8 ppm of oxygen. Under such conditions, mass transport limited rates of general corrosion of steel range from about 0.25 mm/yr (10 mils/yr) in a stagnant system to 15 mm/yr (600 mils/yr) in a highly turbulent one. However, by chemically scavenging the oxygen concentration down to the order of 7 to 8 ppb, the corresponding rates are reduced to less than about 0.01 mm/yr (0.4 mils/yr). Such rates are acceptable. However, under these conditions, magnetite forms as a stable protective corrosion product film and further lowers the corrosion rate by introducing a slower, anodically controlling step.

An even more fundamental role of magnetite should be acknowledged. This is the protection of the steel surface from reaction with water or the hydrogen ions contained in the water. Therefore, if an excess of a chelating agent such as ethylenediamine tetraàcetic acid (EDTA) dissolves a protective magnetite film, as would normally occur in regions of high turbulence, rapid corrosion ensues, despite the absence of oxygen.

Hydrogen Sulfide, Polysulfides, and Sulfur

Hydrogen sulfide, when dissolved in water, is a weak acid and is therefore corrosive because it is a source of hydrogen ions. In the absence of buffering ions, water equilibrated with 1 atm of H_2S has a pH of about 4. However, under high-pressure formation conditions, pH values as low as 3 have been calculated.

Hydrogen sulfide can also play other roles in corrosion in oil and gas production. It acts as a catalyst to promote absorption by steel of atomic hydrogen formed by the cathodic reduction of hydrogen ions. This accounts for its role in promoting sulfide stress cracking of high-strength steels (yield strength greater than approximately 690 MPa, or 100 ksi) (Ref 5).

Hydrogen sulfide also reacts with elemental sulfur. In a gas phase with a high H_2S partial pressure, sulfanes (free acid forms of a polysulfide) are formed so that elemental sulfur is rendered mobile and is produced along with the remaining gaseous constituents. However, as the pressure decreases traveling up the production tubing, the sulfanes dissociate and elemental sulfur precipitates. Various solvent treatments are used to avoid plugging by such sulfur.

In the aqueous phase, under acidic conditions, sulfanes are also largely dissociated into H_2S and elemental sulfur. However, enough strongly oxidizing species can remain either as polysulfide ions or as traces of sulfanes to play a significant role in corrosion reactions. Oxygen contamination of sour (H_2S-containing) systems can also result in polysulfide formation.

Iron sulfide corrosion products can be important in corrosion control. Because of the low solubility, rapid precipitation, and mechanical properties of such corrosion products, velocity effects are generally not encountered in sour systems. Satisfactory inhibition at velocities up to 30 m/s (100 ft/s) has been proved.

The great range of possible iron sulfide corrosion products and their possible effects on corrosion have been extensively studied (Ref 6-9), but little of immediate practical value has resulted. At lower temperatures and H_2S partial pressures, an adequately protective film often forms. The absence of chloride salts strongly promotes this condition, and the absence of oxygen is absolutely essential.

At the high temperatures (150 to 230 °C, or 300 to 450 °F) and H_2S partial pressures (thousands of pounds per square inch) encountered in deep sour gas wells, a so-called barnacle type of localized corrosion (Ref 10) can occur, resulting in corrosion rates of several hundred mils per year (Ref 11). This type of attack is strongly promoted by polysulfide-type species and requires the presence of some minimum chloride concentration. Although initially recognized in deep sour well environments, this same mechanism may operate at lower rates under much milder conditions.

In the barnacle mechanism (Fig. 1), corrosion can be sustained beneath thick but porous iron sulfide deposits (primarily pyrrhotite, $FeS_{1.15}$) because the FeS surface is an effective cathode. The anodic reaction beneath the FeS deposit is dependent on the presence of a thin layer of concentrated $FeCl_2$ at the Fe/FeS interface. This intervening $FeCl_2$ layer is acidic due to ferrous ion hydrolysis, thus preventing precipitation of FeS directly on the corroding steel surface and enabling the anodic reaction to be sustained by the cathodic reaction on the external FeS surface.

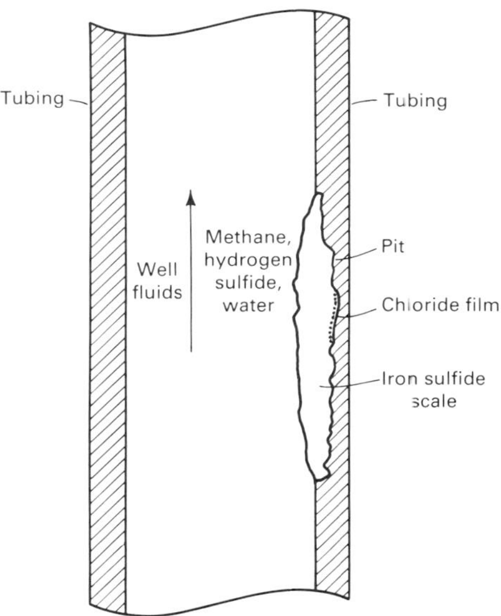

Fig. 1 Schematic showing the barnacle mechanism of sour pitting corrosion. Source: Ref 11

Carbon Dioxide

Carbon dioxide, like H_2S, is a weakly acidic gas and becomes corrosive when dissolved in water. However, CO_2 must first hydrate to carbonic acid (H_2CO_3) (a relatively slow reaction) before it is acidic. There are other marked differences between the two systems. Velocity effects are very important in the CO_2 system; corrosion rates can reach very high levels (thousands of mils per year), and the presence of salts is often unimportant.

Whether or not corrosion in a CO_2 system is inherently controlled or uncontrolled depends critically on the factors governing the deposition and retention of a protective iron carbonate (siderite) scale. On the other hand, there are the factors that determine the rate of corrosion on bare steel. These latter factors govern the importance of maintaining corrosion control.

Bare steel corrosion rates can be calculated from Eq 1, which was developed on the basis of electrochemical studies of the aqueous CO_2/carbon steel system (Ref 12):

$$\log R = A - \frac{2320}{t + 273} - \frac{5.55t}{1000} + 0.67 \log \overline{p} \quad \text{(Eq 1)}$$

where R is the corrosion rate, t is temperature (°C), A is a constant, and $\overline{p}$ is CO_2 partial pressure. When R is calculated in millimeters per year and $\overline{p}$ is in atmospheres, A = 7.96. When R is calculated in mils per year and $\overline{p}$ is in pounds per square inch, A = 8.78.

Corrosion rates calculated with Eq 1 reach 25 mm/yr (1000 mils/yr) at 65 °C (150 °F) and 1 MPa (150 psi) CO_2 pressure, and 250 mm/yr (10 000 mils/yr) at 82 °C (180 °F) and 16 MPa (2300 psi) CO_2 pressure. Obviously, such rates are unacceptable. An alternative, idealized condition occurs when a protective carbonate scale is present and when the corrosion rate is limited by the need to replenish the film lost due to solubility in the aqueous phase. Under such conditions, the rates calculated for a hypothetical sweet (CO_2-containing) gas well reached a maximum of about 0.15 mm/yr (6 mils/yr) as compared to calculated bare metal rates of 500 to 2000 mm/yr (20 000 to 80 000 mils/yr) (Ref 13).

Conditions favoring the formation of a protective iron carbonate scale are:

- Elevated-temperature (decreased scale solubility, decreased CO_2 solubility, and accelerated precipitation kinetics)
- Increased pH, as occurs in bicarbonate-containing waters (decreased solubility)
- Lack of turbulence

Turbulence is often the critical factor in pushing a sweet system into a corrosive regime. Excessive degrees of turbulence prevent either the formation or retention of a protective iron carbonate film.

The critical velocity equation has been used to estimate when excessive turbulence can be expected in a CO_2 system (Ref 14). There is no doubt that the velocity effect is real, but there is some question as to whether this exact form of the equation is the appropriate one:

$$\text{Critical velocity} = \frac{K}{\sqrt{\rho}} \quad \text{(Eq 2)}$$

where velocity is calculated in feet per second, K is a constant, and ρ is the density of the produced fluid (liquid + gas combined). When ρ is in kilograms per cubic meter, K = 7.6; for ρ in pounds per cubic foot, K = 100.

When both H_2S and CO_2 are present, simplified calculations indicate that iron sulfide may be the corrosion product scale when the H_2S/CO_2 ratio exceeds about 1/500 (Ref 13); sour system considerations would then be expected to apply. Even in a strictly CO_2 system, iron carbonate may not always be the corrosion product. Magnetite may form instead. Figure 2 shows the stability fields expected for siderite and magnetite as a function of the redox potential (expressed here in terms of hydrogen fugacity) of the system (Ref 13). In actual experience, corrosion product scales are often found to consist of mixtures or layers of siderite and magnetite.

Iron carbonate lacks conductivity and therefore does not provide an efficient cathode surface. Thus, the types of pitting mechanisms found in oxygenated and in H_2S-containing systems do not occur. Rather, generalized corrosion occurs at any regions not covered by the protective scale. The result is that on any bare metal the anodic and cathodic regions are so microscopically dispersed that salt—to provide conductivity—is not needed to achieve the corrosion rates predicted by Eq 1.

Strong Acids

Strong acids are often pumped into wells to stimulate production by increasing formation permeability. For limestone formations, 15 and 28% hydrochloric acids are commonly used. For sandstones, additions of 3% HF are necessary. In deep sour gas wells where HCl inhibitors lose effectiveness, 12% formic acid has been used.

Corrosion control is normally achieved by a combination of inhibition and limiting exposure time to 2 to 12 h. If corrosion-resistant alloys are present (austenitic and duplex stainless steels, and so on), concern for stress-corrosion cracking (SCC) and inhibitor ineffectiveness (respectively) may rule out the use of HCl.

Concentrated Brines

Dense halide brines of the cations of calcium, zinc, and, more rarely, magnesium are sometimes used to balance formation pressures during various production operations. All can be corrosive because of dissolved oxygen or entrained air. In addition, such brines may be corrosive because of the acidity generated by the hydrolysis of the metallic ions, as illustrated in Eq 3:

$$Zn^{2+} + H_2O = ZnOH^+ + H^+ \quad \text{(Eq 3)}$$

Corrosivity due to acidity is worst with dense zinc brines. More expensive calcium bromide

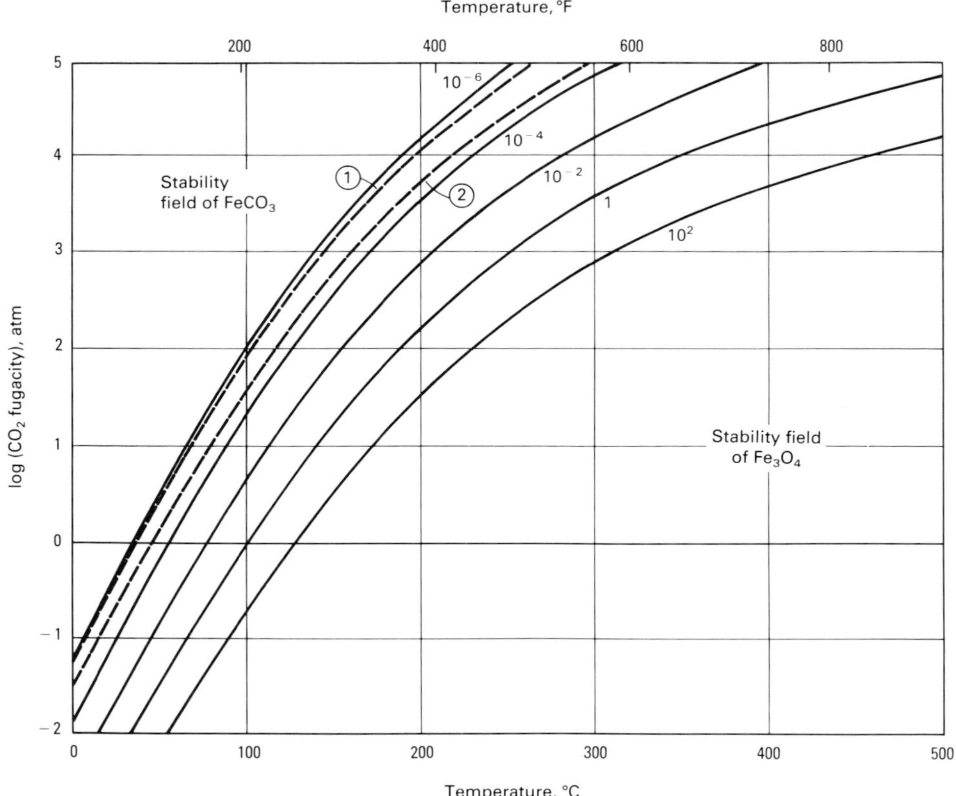

Fig. 2 Effect of temperature and hydrogen fugacity on the stability of $FeCO_3$ and Fe_3O_4 in contact with aqueous CO_2. Equilibrium calculations determine boundaries (indicated by the iso-hydrogen fugacity curves, with fugacities given in atmospheres) between $FeCO_3$ and Fe_3O_4 stability fields in the produced fluid, which contained CO_2, water, and traces of hydrogen. Curves 1 and 2 locate boundaries for locations 5180 m (17 000 ft) deep and at the wellhead, respectively, of a 170 000 m^3/day (6 000 000-ft^3/day) well corroding at a rate of 0.75 mm/yr (30 mils/yr). Source: Ref 13

brines are now often used at densities above about 1.7 g/cm^3 (14 lb/gal) (attainable with $CaBr_2$ brines) to avoid long-term exposure to $ZnCl_2$ brines.

Stray-Current Corrosion

If an extraneous direct current (dc) in the earth is traversed by a conductor, part of the current will transfer to the lower-resistance path thus provided. Direct currents are much more destructive than alternating currents (ac); an equivalent ac current causes only about 1% of the damage of a dc current (Ref 15). Regions of current arrival (where electrons depart) will become cathodic, and those regions where the current departs will become anodic. With corrodible metals such as carbon and low-alloy steels, corrosion in the anodic areas is the result. For example, 1 A · yr can corrode 9 kg (20 lb) of steel.

Cathodic protection systems are the most likely present-day sources of stray dc currents in production operations. More detailed discussions are available in Ref 4 and 16. The section "Stray-Current Corrosion" of the article "General Corrosion" in this Volume also contains information on the causes and mechanisms of stray-current corrosion.

Under-Deposit (Crevice) Corrosion

This is a form of localized corrosion found almost exclusively (if not exclusively) in oxygen-containing systems. Such corrosion is usually most intense in chloride-containing systems. It is essential to have some form of shielding of an area on a metal such that it is wet by an electrolyte solution but is not readily accessible to oxygen, the diffusing corrosive species.

This type of attack is usually associated with small volumes of stagnant solution caused by surface deposits (sand, sludge, corrosion products, bacterial growth), crevices in joints, and gasket surfaces. Crevice corrosion is discussed in Ref 17, and a quantitative treatment of crevice corrosion (particularly of stainless steels) is provided in Ref 18 to 20.

The mechanism of crevice corrosion hinges upon the environmental conditions resulting from the loss of hydroxide production with cessation of the cathodic reaction when the initial oxygen in the shielded region is exhausted. Thus, in the shielded region, the anodic corrosion reaction continues because the corrosion potential is maintained by the reduction of oxygen on the outside surface. However, chloride or other anions now migrate into the developing anodic region to maintain electroneutrality. Thus, a relatively concentrated, essentially ferrous chloride solution accumulates in the shielded region. As a result of the hydrolysis of the ferrous ions, the pH drops to a value of 2 or 3.

At this point, the crevice corrosion type of localized attack is fully established. The anodic reaction continues in the shielded region because in the low-pH environment the ferrous ions go readily into solution and have little tendency to precipitate as an oxide or hydroxide on the surface and thus stifle the anodic reaction. Outside, the cathodic reaction continues unperturbed. Because of the large ratio of cathodic-to-anodic surface area, high rates of localized corrosion can be maintained with very modest cathodic current densities. More information on the mechanisms of crevice corrosion is available in the article "Localized Corrosion" in this Volume.

Galvanic Corrosion

When two dissimilar metals are electrically coupled—both electronically by a metal bond and ionically through an electrolyte—the more active (electronegative) metal tends to become a sacrificial anode and supply cathodic protection to the more noble metal. Such situations are often encountered in heat exchangers in which carbon steel tubesheets are used with copper alloy tubes and at junctions between piping, fasteners, or corrosion-resistant sheeting with containers of another material.

Problems with galvanic corrosion are the most acute when the cathode-to-anode area ratio is large. Such situations are often encountered inadvertently. This has happened when the normal polarity difference between zinc and steel in a galvanized pipe reversed in a bicarbonate/chloride brine so that the steel pipe walls at pinholes in the galvanizing perforated rapidly while trying to protect the extensive adjacent galvanized area. Another situation is when plastic-coated steel is coupled to more noble metal. At any pinholes in the coating, a very adverse area ratio will exist, and rapid corrosion rates can result. The section "Galvanic Corrosion" of the article "General Corrosion" in this Volume contains more information on this form of attack.

Biological Effects

The most important biological effect on corrosion in oil and gas production is the generation of H_2S by sulfate-reducing bacteria (*Desulfovibrio Desulfuricans*). These are anaerobic bacteria that metabolize sulfate ions (using an organic carbon source) and produce hydrogen sulfide. They can thus introduce H_2S and all its corrosive ramifications into normally H_2S-free systems.

Colonies of sulfate-reducing bacteria can also form deposits that are conducive to under-deposit corrosion. Contrary to previous beliefs, any resultant corrosion appears to be due to a mechanical shielding action, rather than any depolarizing action resulting from the metabolic processes of the SRB. However, this is not to deny that the introduction of H_2S (whatever the source) into a crevice region could have an accelerating effect on corrosion, because H_2S is known to be an anodic stimulant. More information on biological corrosion is available in the articles "General Corrosion" and "Localized Corrosion" in this Volume.

Mechanical and Mechanical/Corrosive Effects

Cavitation. This metal removal—often grain by grain—is due to high-pressure shock wave impingement resulting from the rapid collapse of minute bubbles created under certain conditions in high-velocity fluid handling equipment. It is usually found on pump impellers operating with too low a suction pressure.

Erosion. Most commonly, this is direct metal removal by the cutting action of high-velocity abrasive particles. Erosion failures (washouts) are seen in drill pipe when leaks (loose connections or a corrosion fatigue crack) allow drilling mud to flow through the wall under a high-pressure gradient. Erosion of flow lines at bends and joints by produced sand is probably the other most common occurrence of erosion in oil and gas production.

Erosion-Corrosion. Strictly speaking, in erosion-corrosion, only the protective corrosion product film is removed by erosive forces; however, with the protective film absent, corrosion can occur at a greatly accelerated rate. Erosion-corrosion may play a role in CO_2 corrosion (Ref 21), and sand, under mild flow conditions, may also cause erosion corrosion. Erosion-corrosion has also been noted in heavy anchor chains where their use in an abrasive bottom mud allowed corrosion at contact regions to proceed at a rate of many hundreds of mils per year.

Corrosion fatigue results from subjecting a metal to alternating stresses in a corrosive environment. At points of greatest stress, the corrosion product film becomes damaged during cycling, thus allowing localized corrosion to take place. Eventually, this leads to crack initiation and crack growth by a combination of mechanical and corrosive action. Because of this combined action, damage per cycle is greater at low cycling rates, where corrosion can play a larger role. Also, in corrosion fatigue, a fatigue limit does not exist; rather than leveling out as in simple fatigue, the usable stress level continues to decrease with increasing cycles.

The greatest concern for corrosion fatigue arises in connection with highly stressed, submerged, offshore structures. Welded connections on drill ships and on drilling and production platforms are particularly susceptible to this form of structural impairment. More information on attack resulting from combined corrosion and mechanical effects is available in the article "Mechanically Assisted Degradation" in this Volume.

Corrosion Control Methods*

Arthur K. Dunlop
Corrosion Control Consultant
David H. Patrick
ARCO Resources Technology
Donald E. Drake
Mobil Corporation

Materials Selection

Traditionally, carbon and low-alloy steels were virtually the only metals used in the production of oil and gas. This resulted from the fact that very large quantities of metal are required in petroleum production, and until a few years ago, crude oil and gas were relatively low-value products. In addition, insurmountable corrosion problems were not encountered.

This situation changed when gas and oil prices increased dramatically and deeper wells were drilled that encountered corrosive environments of greatly increased severity. The final factor that made the current widespread use of corrosion-resistant alloys possible was the development of high-strength forms of these alloys. This allowed thinner pipe and vessel walls and greatly reduced the amount of material required.

The result of this situation is that essentially all of the high-tonnage uses of corrosion-resistant alloys in oil and gas production involve alloys in a high-strength form. Yield strengths typically span the range of 550 to 1250 MPa (80 to 180 ksi), but can reach nearly 1750 MPa (250 ksi) in wire lines.

Metallurgical Considerations

The great majority of corrosion-resistant alloys used in oil and gas production were originally developed for other applications not requiring high strength. Therefore, many of these have an austenitic microstructure and can be strengthened only by some form of cold working. This presents no problem for the production of tubulars used underground in the well, because they are joined by threaded connections. However, for other applications, such as welded flow lines and cast and forged valves, different techniques must be used.

In addition to the problems of environmental compatibility, there are a number of difficulties of a metallurgical nature that will be briefly mentioned. For high-strength materials, particularly at the upper end of the range, adequate ductility is often difficult to achieve. Also, heat treatments developed for other applications may not be optimal for petroleum protection environments.

Environmental Considerations

There are several environmental factors that are more or less unique to oil and gas production. One factor is the general absence of oxygen in produced fluids. The dominant naturally occurring corrosives are carbon dioxide and hydrogen sulfide. However, in wells with high hydrogen sulfide concentrations, elemental sulfur—a relatively strong oxidizing agent—can also be present.

Water, which must be present to make sulfur or the acid gases (CO_2 and H_2S) corrosive, can be assumed to be present in all productive geologic formations. Therefore, liquid water is present from the bottom of the well up to the point in the flow system at which it is removed. Corrosivity is aggravated by the presence of salt in formation waters. Such brines can range from a few percent sodium chloride up to saturated solutions containing as much as 300 000 ppm of total dissolved salts.

Temperature also affects corrosivity. Deep well temperatures up to 205 °C (400 °F) are not uncommon. Therefore, temperatures from these levels down to ambient must be considered.

Corrosion

From other applications, enough is generally known to select alloys that are resistant to general corrosion. However, resistance to localized corrosion (pitting and crevice corrosion) often requires some experimental study.

Environmental Stress Cracking. In corrosion-resistant alloy selection, environmental stress cracking normally requires the greatest portion of experimental effort. This follows from the use of high-strength alloys in environments in which there is relatively little experience with these alloys.

The problem is complicated by the fact that the lowest alloy content and the maximum reliable strength level are needed to achieve an economically viable choice. Because this involves design close to the limits of the material, it is necessary to define these limits as accurately as possible.

The term environmental stress cracking was selected because it includes two different mechanisms: cathodic and anodic. The cathodic mechanism is that of hydrogen embrittlement found in sulfide stress cracking of high-strength carbon and low-alloy steels in the presence of hydrogen sulfide. Here, H_2S promotes entry of cathodically evolved hydrogen atoms into the metal.

The anodic mechanism is that involved in the chloride SCC of austenitic stainless steels at temperatures generally above about 65 °C (150 °F). Nickel-base alloys are normally immune to this type of chloride cracking. However, the presence of H_2S can induce susceptibility in normally resistant alloys by acting to promote anodic dissolution. In the most resistant alloys, such as Hastelloy alloy C-276 (Table 1), elemental sulfur and a temperature of about 230 °C (450 °F) must also be present to cause cracking. Cracking will occur only in high-strength, heavily cold worked samples.

Effects in the Cathodic Mechanism. In corrosion-resistant alloys, environmental effects on cathodic stress cracking are generally similar to those encountered in the sulfide stress cracking of carbon and low-alloy steels. However, in extreme cases, the temperatures needed to avoid cracking susceptibility in the highest-strength materials can be as high as 120 °C (250 °F). An example is Hastelloy alloy C-276 at a hardness of 45 HRC. In contrast, the highest minimum temperature given in NACE standard MR-01-75 for high-strength low-alloy tubulars is 80 °C (175 °F).

Also, on corrosion-resistant alloys, chloride ions can have a significant deleterious effect in cathodic cracking. This results from deterioration of passivity in the presence of chloride. The resulting higher anodic corrosion rate can then support a higher cathodic rate and thus gives rise to enhanced cathodic charging of the corrosion-resistant alloy with atomic hydrogen.

Effects in the Anodic Mechanism. Determining regions of susceptibility to anodic cracking is much more complicated than defining regions of susceptibility to cathodic cracking. This is due to the multidimensional nature of the problem. Therefore, in anodic cracking, susceptibility is a strong function of such environmental factors as temperature, chloride concentration, and hydrogen sulfide partial pressure; in addition, pH, oxidants (elemental sulfur or polysulfides), and galvanic coupling can be involved. Further, alloy composition (corrosion resistance), strength level, and stress all influence cracking susceptibility.

The effect of H_2S requires some elaboration. In cathodic cracking, there is a well-established threshold for the initiation of the sulfide stress cracking phenomenon; this is not the case in anodic cracking. The effect in anodic cracking is more monotonic. Therefore, there can almost be a different critical value for every combination of alloy, strength, applied stress, and environmental

*The section on materials selection was written by A.K. Dunlop; the section on coatings was written by D.H. Patrick; and the sections on cathodic protection, inhibitors, nonmetallic materials, and environmental control were written by D.E. Drake.

Table 1 Compositions of corrosion-resistant alloys

Alloy	UNS designation	C	Cr	Fe	Ni	Mo	Other
		Composition, %(a)					
Hastelloy alloy C-276	N10276	0.02	14.5–16.5	4–7	bal	15–17	2.5Co, 1.0Mn, 4.5W, 0.35V
Hastelloy alloy G	N06007	0.05	21–23.5	18–21	bal	5.5–7.5	2.5Nb, 2.5Co, 2.5Cu, 2.0Mn, 1W, 1Si
Haynes No. 20 Mod	N08320	0.05	21–23	bal	25–27	4–6	2.5Mn, 1Si, Ti-4 × C min
Sanicro 28	N08028	0.03	26–28	bal	29.5–32.5	3–4	1.4Cu, 2.5Mn, 1Si
Inconel alloy 625	N06625	0.10	20–23	5	bal	8–10	0.4Al, 4.15Nb, 0.5Mn, 0.4Ti
Incoloy alloy 825	N08825	0.05	19.5–23.5	bal	38–46	2.5–3.5	0.2Al, 3Cu, 1Mn, 1.2Ti, 0.5Si
Monel alloy 400	N04400	0.3	...	2.5	63–70	...	bal Cu, 2Mn, 0.5Si
MP35N	R30035	0.025	19–21	1.0	33–37	9–10.5	bal Co, 0.15Mn, 1Ti
Type 316 stainless steel	S31600	0.08	16–18	bal	10–14	2–3	2Mn, 1Si
Type 410 stainless steel	S41000	0.15	11.5–13.5	bal	...	...	1Si, 1Mn
Ferralium 255	S32550	0.04	24–27	bal	4.5–6.5	2–4	2.5Cu, 1.5Mn, 0.25N, 1Si

(a) Maximum unless range is given or otherwise indicated.

condition. However, it is safe to assume that research in this area is leading to the development of some generalized patterns.

Selection of Corrosion-Resistant Alloys

Alloy selection, from a corrosion standpoint, can be considered to be a three-step process. First, resistance to general corrosion must be ensured. This is primarily a function of the chromium content of the alloy. Second, resistance to localized attack also must be ensured. This is primarily a function of molybdenum content. Finally, resistance to environmental stress cracking is sought at the highest feasible strength level. Nickel content plays a principal role in this instance, particularly in providing resistance to anodic cracking.

The close correlation between pitting resistance and resistance to anodic cracking should be noted. This apparently results from the ease of crack initiation under the low-pH high-chloride conditions found in pits. Therefore, higher molybdenum can also increase resistance to anodic cracking.

With the procedures given below, regions of alloy applicability can be shown schematically as a qualitative function of environmental severity. This has been attempted in Fig. 3, in which an aqueous, CO_2-containing environment (hence low pH) has been assumed and the effects of temperature, chloride, and H_2S concentration are illustrated. The effect of yield strength is not shown, but if environmental cracking is the limiting factor, reducing the yield strength should extend applicability to more severe environments.

The reader should be cautioned that a diagram such as Fig. 3 is really more of a guide to alloy qualification than to direct selection for a particular application. Therefore, it may aid in developing a more efficient approach to alloy testing.

Testing for Resistance to Environmental Stress Cracking

The most directly applicable results are obtained by exposing samples of commercially produced alloys to an environment simulating as closely as possible that expected in actual production operations. Fortunately, an understanding of the principles involved allows considerable simplification to be made without significantly altering the value of the results.

Two simplifications can be readily made in the environmental parameters, as follows. First, only the CO_2 and H_2S partial pressures are usually reproduced. The overburden of methane pressure to create the actual total pressure of the environment is dispensed with. This can substantially lower the pressure ratings of test vessels. Second, only the chloride content of the brine is typically reproduced, rather than trying to simulate the total ionic spectrum of the produced fluids. Often, no reliable analysis is available. Assumption of a saturated sodium chloride solution is then a relatively conservative approach.

A wide variety of test specimens are used, but for high-strength corrosion-resistant alloy tubulars, C-rings are particularly convenient. Double cantilever beam specimens are very attractive in that they can provide a quantitative measure of fracture toughness, which can then be used in mechanical design. Slow strain rate testing can also be very useful in assessing alloy limits.

Another aspect of testing is the imposed stress level. Exposure at 100% of yield strength is the only conservative approach. However, some investigators place unwarranted reliance on design values and test at some lesser fraction of the yield strength. Detailed information on testing for resistance to environmental cracking is available in the articles "Evaluation of Hydrogen Embrittlement" and "Evaluation of Stress-Corrosion Cracking" in this Volume.

Coatings

Internal protective coatings have been used to protect tubing, downhole equipment, wellhead components, Christmas trees (manifolds used to control the rate of production, receive the produced fluids under pressure, and direct the produced fluids to the gathering point), and various downstream flow lines and pressure vessels for more than 30 years. Because internal coatings are subject to damage, successful use is usually accompanied by chemical inhibition or cathodic protection as part of the entire protective program. Most of the coating use has been below 175 °C (345 °F).

Tubing. The benefits derived from coating tubing depend on the coating remaining intact. Because no coating can be applied and installed 100% holiday-free, inhibition programs are commonly employed to accommodate holidays and minor damage. The suitability of the service is dependent on specific testing and an effective quality control program.

Inhibitor and volume will not change even though an operator decides to use coated tubing rather than bare tubing. The use of coated tubing improves the protection in shielded areas that are inaccessible to inhibitors. The two greatest dan-

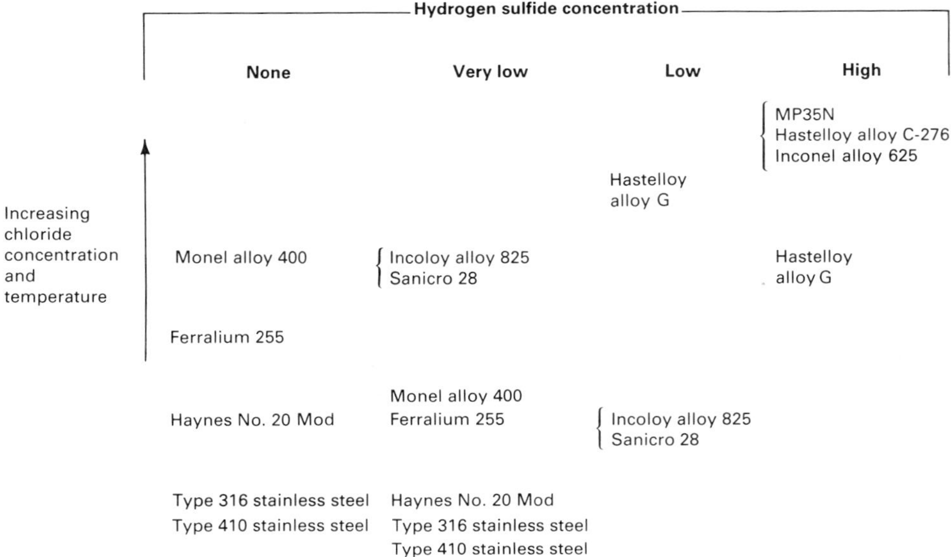

Fig. 3 Schematic showing corrosion-resistant alloy selection for production environments containing aqueous CO_2 and H_2S

gers to coated tubing are wireline damage and improper joint selection. Wireline damage can be minimized by adjusting running procedures to include wireline guides and to slow wireline speed (<0.5 m/s, or <100 ft/min). Inhibiting immediately after wireline work is good practice.

Proper joint selection involves choosing a joint that allows coating to be applied around the pin nose into the first few pin threads and from the first few coupling threads into the coupling body. The proper joint allows the coating to remain undamaged. Often, a corrosion barrier compression ring is used to accomplish this end. Metal-to-metal sealing joint designs are not joints that can be coated.

The coatings used for tubing protection are polyurethane, phenolformaldehyde, epoxized cresol novolac, and epoxy resins. Suitability for service is and should be based on laboratory testing using the specific environment proposed for the service.

The quality control parameters of concern are tubing surface finish/preparation, application techniques, coating thickness, holiday detection, joint condition, and inspection. Inspection is required to ensure the suitability of the other parameters. Quality control and surveillance are as much a part of a successful protective coating program as choosing the appropriate coating. The production coating must be applied in the same way the coating was applied to the test specimens.

Coated pipe and couplings must be carefully handled after coating, during shipping, in storage, and at the well site. The threads must be protected from impact with other pipe and objects.

Wireline work is necessary. These operations can be accomplished with a minimal amount of damage to the tubing if the wireline speeds are kept to 0.5 m/s (100 ft/min) or less, if all sharp edges are removed from the tools, if all tools are plastic coated or covered with a plastic sleeve, and if wheeled centralizers are used on all center hole jobs. Using the above precautions, many wireline trips can be made with little or no damage. The fact that coating should not be used because the wireline will cut the coating and cause accelerated corrosion in the wireline track is not true. When the wireline cuts uncoated tubing, it causes cold work. The wireline track then becomes anodic to the surrounding bare steel, and corrosion is accelerated. In coated tubing, the coating electrically insulates the cathodic areas so that the corrosion rate in the track is essentially the same as that for uncoated steel without wireline damage.

Wellheads, Christmas Trees, and Downhole Equipment. Exposed surfaces of wellhead equipment, Christmas trees, and downhole equipment must be coated or manufactured of corrosion-resistant materials. This starts with the tubing hanger, which is threaded onto the production tubing, and continues through the tubing bonnet (tubing adapter), the master valve(s), the tee or cross, and the wing and crown valve(s) and into the choke. Ring gasket grooves, valve seat pockets and other compression fitted parts must not be coated. These areas can be overlaid with corrosion-resistant alloys, and the coating can be applied over a transition area up to the corrosion-resistant alloy overlay. Valve internal cavities need not be coated.

Generally, tubing hangers are difficult to coat and are therefore made of corrosion-resistant alloys for corrosive service. Hangers without back pressure valve threads can be coated, but the cost of coating the carbon steel hanger may be equivalent to the cost of a corrosion-resistant alloy hanger.

Downhole equipment (nipples, polished bore receptacles, seal subs, tie backs, millout subs, packers, and so on) uses the same standards as tubing. Most downhole equipment is considered uncoatable because it was not designed for coating. Such equipment is usually coated as an assembled unit, rather than as individual pieces. If installation is below the packers, both internal and external surfaces should be coated.

Coated surface flow lines usually employ flanges or threads and couplings to join them. Joining by welding damages the coating, and field-applied repair coating is not recommended.

Vessels. The vessel must be designed as a coated vessel. Coating a vessel that was not designed for coating is rarely successful. Vessel design for coating involves welded hangers for anodes in the fluid zone, flanged access, removable internals, smooth internal surfaces, and good access to all internal surfaces to be coated. As a general rule, coating over bolted assemblies is not successful.

Cathodic Protection

Corrosion occurs when an anode and a cathode are electrically connected in the presence of an electrolyte, and electrical currents leave a metal and go into an electrolyte. If another power source can be used to oppose these corrosion currents sufficiently, the metal will be protected from corrosion. This technique is known as cathodic protection. The entire metal surface is converted into a cathode, while the corrosion currents are transferred to an auxiliary anode in which corrosion can proceed. Cathodic protection has been used in the oil field to protect pipelines, well casings, tanks and production vessels, and offshore platforms. More information on principles and applications of cathodic protection is available in the article "Cathodic Protection" in this Volume.

Types of Cathodic Protection Systems

There are two methods of applying cathodic protection: the sacrificial anode method and the impressed-current method. Because some metals are less noble (more electronegative) than the most common oil field material—steel—in the galvanic series, they will become the anode (site of corrosion attack) when coupled to steel in the presence of an electrolyte. The most common of these materials are magnesium, zinc, and aluminum; these are called sacrificial anodes. These types of anodes are used when:

- Current requirements are relatively low
- Electric power is not readily available
- Short system life dictates a low capital investment

The anode is usually electrically connected by a wire or steel strap to the structure to be protected. Magnesium and zinc are usually used in soils, while zinc can also be used in brine environments.

In the impressed-current method, an external energy source produces an electric current that is sent to the impressed-current anodes. The most common types of these materials include graphite, high-silicon cast iron, lead-silver alloy, platinum, and even scrap steel rails. These types of anodes are used when:

- Current requirements are high
- Electrolyte resistivity is high
- Fluctuation in current requirements will occur. These types of systems can be adjusted to compensate for varying current requirements
- Electrical power is readily available, although this is not now as severe a limitation as it was in past years

In a typical impressed-current system, alternating current from a power line flows into a rectifier where it is converted into direct current. The dc current then flows to the anode groundbed. Other means of supplying this electrical current include solar energy and thermoelectric generators. These methods are applicable at locations where conventional electric power is not economically available.

Solar energy has powered cathodic protection for well casings in Kansas (Ref 22) and Saudi Arabia (Ref 23); segments of a 480-km (300-mile) long, 0.8-m (32-in.) pipeline in Libya (Ref 24); and segments of a natural gas distribution system in Washington (Ref 25). In these systems, silicon semiconductor devices convert sunlight directly into dc electricity, which is then used for the anode groundbed and to charge batteries. These batteries provide current to the anode groundbeds during periods of little or no sunlight. The batteries must be checked periodically for proper electrolyte levels. Solar panels can be easily replaced or increased to attain a higher current output because they are fabricated in modules.

One oil company uses solar energy to protect about 800 well casings in western Kansas. The solar panels consist of individual silicon solar cells connected in series to form modules. These modules are then connected in parallel to form a panel that is rated at about 12 A and 4 V. Because casing in this field requires 1 to 2 A for cathodic protection, 2-V, 500-A · h lead-acid batteries are used. A rheostat controls the rate of current flow from the batteries, and the voltage regulator controls the battery charge rate. The batteries can provide electrical current to the anodes for 10 days even if the sun is completely blocked.

Another unconventional electrical current source is the thermoelectric generator. One type of thermoelectric generator is a system that uses a burner to heat an organic liquid in a vapor generator. This vapor then expands through a turbine wheel, thus producing power to a shaft to drive an alternator, where ac power is produced. The vapor then passes through a condenser where it is cooled and condensed back into a liquid to start the cycle again. Once the ac power is produced by the alternator, it is then sent to the rectifier to be converted to dc. Figure 4 shows an illustration of this type of thermoelectric generator. These systems can produce a maximum of 80 A at 21 V (Ref 26).

Application of Cathodic Protection to Oil Field Equipment

Pipelines. Cathodic protection of pipelines is very common in oil field operations. As discussed earlier in this article, sacrificial anodes or impressed current can be used for cathodic protection (Fig. 5). If the pipeline is well coated and not very long, the current requirements will probably be achieved with sacrificial anodes. If bare, a

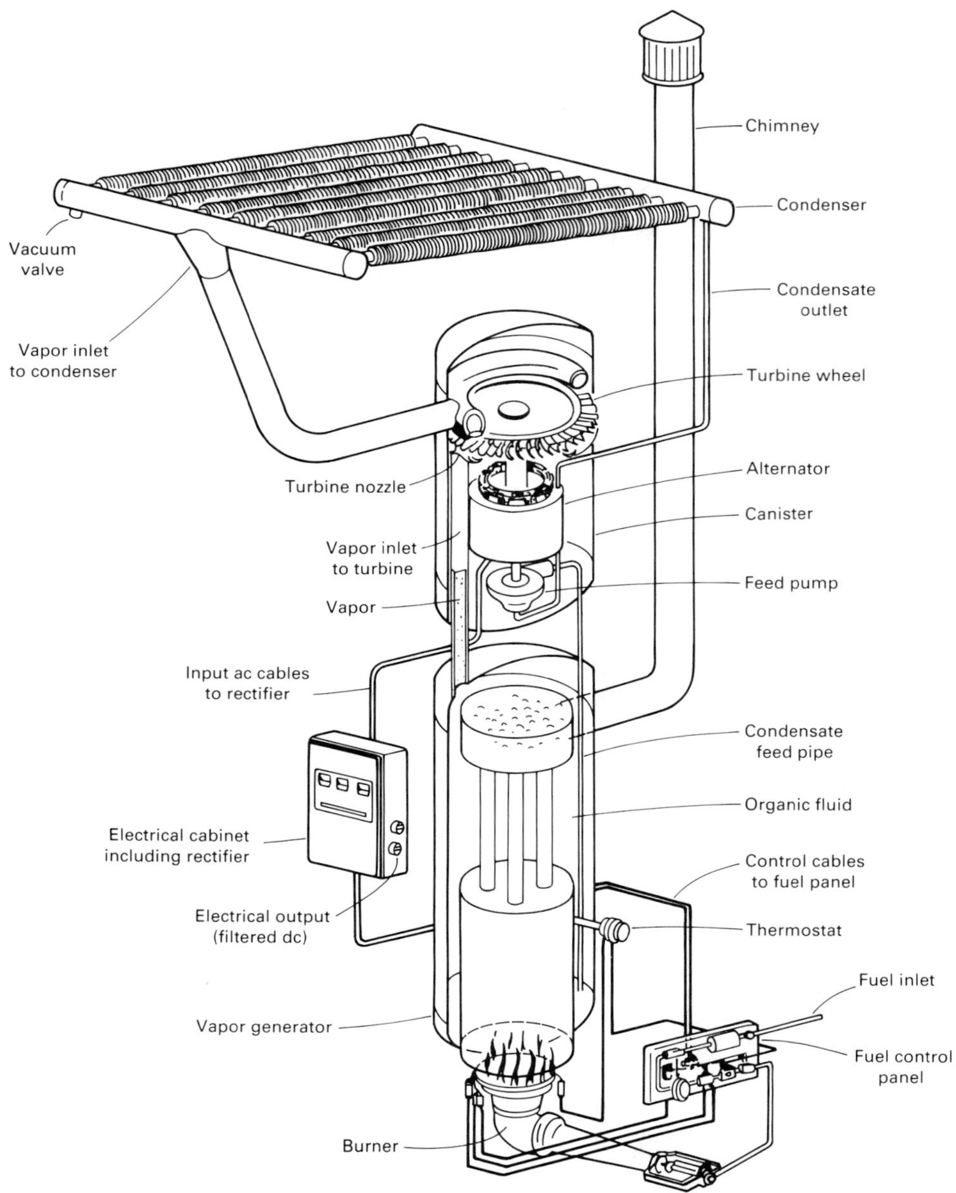

Fig. 4 Thermoelectric generator used to power cathodic protection systems in remote locations. Source: Ref 26

steel pipeline could require 1.1 mA/m^2 (0.1 mA/ft^2) in soil, while a very well coated pipeline could require only 0.003 mA/m^2 (0.0003 mA/ft^2) or less for cathodic protection (Ref 27).

The resistivity of soils (and therefore their corrosivity) will also vary with location. Differences in aeration, soil composition (sand or clay), and the presence of chemical spills are just a few of the factors that will affect the corrosivity of the soil. Sometimes, the resistivity of the surface soils is so high that conventional groundbeds (1.8 m, or 6 ft, deep) cannot be used. Conventional groundbeds are normally used when the soil resistivities near the surface of the ground are less than 5000 Ω · cm (Ref 28).

In high-resistivity soils, deep groundbeds can be used where the anodes are installed vertically in holes at depths of 15 m (50 ft) or more. Deep groundbeds (Fig. 6) are normally used to provide a better distribution of current than conventional groundbeds. These types of groundbeds also minimize right-of-way problems and are essentially unaffected by seasonal moisture variations. They are more expensive to install than conventional groundbeds, and it is usually impossible to repair any damage to the cable insulation. These systems use the same type of anodes as the conventional groundbeds. The major difference is that a perforated vent pipe can be installed to prevent chlorine gas from accumulating around the anodes. If these gases collect around the anodes, they form an insulating barrier that increases the resistance of the groundbed and eventually causes the groundbed to become ineffective.

Well Casings. The first step in externally protecting well casing is to cement through any corrosive zones. Cement acts as a coating and will significantly reduce, but not completely stop, corrosion of the casing. Therefore, cathodic protection is needed to supplement the cement (Fig. 7). The experience of one company with the cathodic protection of well casings showed an 88% success rate in preventing predicted casing failures (Ref 31). Although a single anode bed for a buried pipeline may protect as much as 80 km (50 miles), the maximum amount of casing that needs to be protected usually does not exceed 2.4 to 3.2 km (1.5 to 2 miles). One company coated nine casing strings in 3500-m (11 500-ft) wells with fusion-bonded epoxy in Florida (Ref 31). The coating was used to reduce current requirements and to improve current distribution. Uncoated casing strings were protected with 22 to 25 A, and even then there was incomplete corrosion control. Only 10 A were needed to protect the coated casing. Some of these casing strings were pulled because of other operational problems, and the coating was found to be in excellent condition.

Another method of reducing the current requirement is to place an insulating joint in the flow line at the wellhead. This joint prevents current from the flow line from flowing into the wellhead and down the casing. This current would leave the casing at low-resistivity zones (Ref 30).

Two methods are generally used to determine current requirements for well casing: casing potential profile and *E*-log *I*. The casing potential profile is measured by using a tool that consists of two spring-loaded probes approximately 7.6 m (25 ft) apart. This tool is pulled through the casing, and voltage readings are taken between the probes as they contact the casing every 15 or 30 m (50 or 100 ft). A plot of potential versus depth is made (Fig. 8). A slope upward and to the left indicates an anodic area, while a slope upward and to the right indicates a cathodic area. Current from a temporary groundbed is then applied to the casing for protection, and another potential profile is taken. Current is increased until the profile slopes are upward and to the right. Providing a profile slope to the right does not necessarily mean that all of the casing is protected, but it does mean that all gross corrosion areas have been eliminated.

Another technique for determining the required current is the *E*-log *I* curve. It is less expensive and does not require the disturbance of subsurface equipment. The flow line to the well, however, must be isolated from the well casing. The pipe-to-soil potential relative to the Cu-$CuSO_4$ reference electrode is measured, and a small amount of current from a temporary groundbed is applied. The current is then interrupted, and the pipe-to-soil potential is measured as quickly as possible. The current is then increased a small amount, and the process is repeated to obtain a curve similar to that shown in Fig. 9. Generally, the current required corresponds to the break in the curve.

Tanks and Production Vessels. Internal corrosion of water-handling tanks and vessels can be controlled by the use of cathodic protection. Even if a coating has been applied to the interior of a water storage tank, there will always be imperfections where corrosion can occur; therefore, cathodic protection is needed. Sacrificial anodes can be suspended from the top of the tank as shown in Fig. 10 to offer protection to the portion of the tank that is covered with water. Cathodic protection will not help in the vapor area of the tank. Coatings must be used to protect this area. Sacrificial anodes have also been placed on concrete blocks in tanks. These blocks insu-

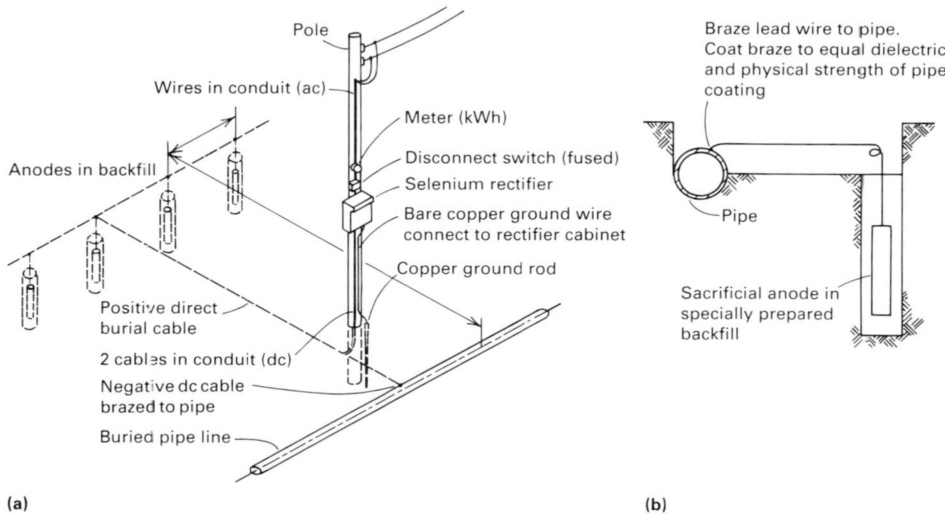

Fig. 5 Typical cathodic protection installations. (a) Impressed current. (b) Sacrificial anode

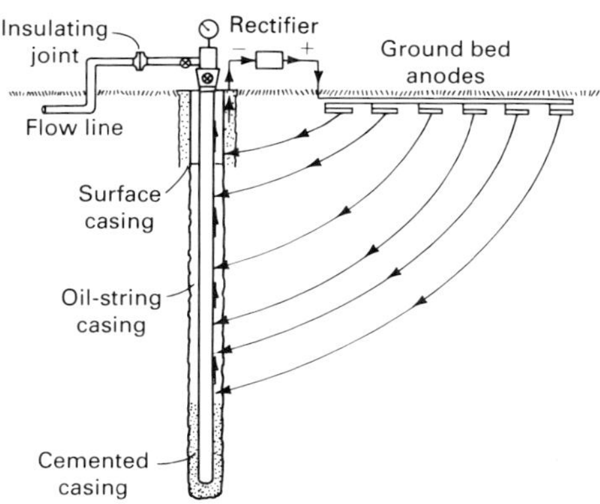

Fig. 7 Cathodic protection installation for a well casing. Source: Ref 30

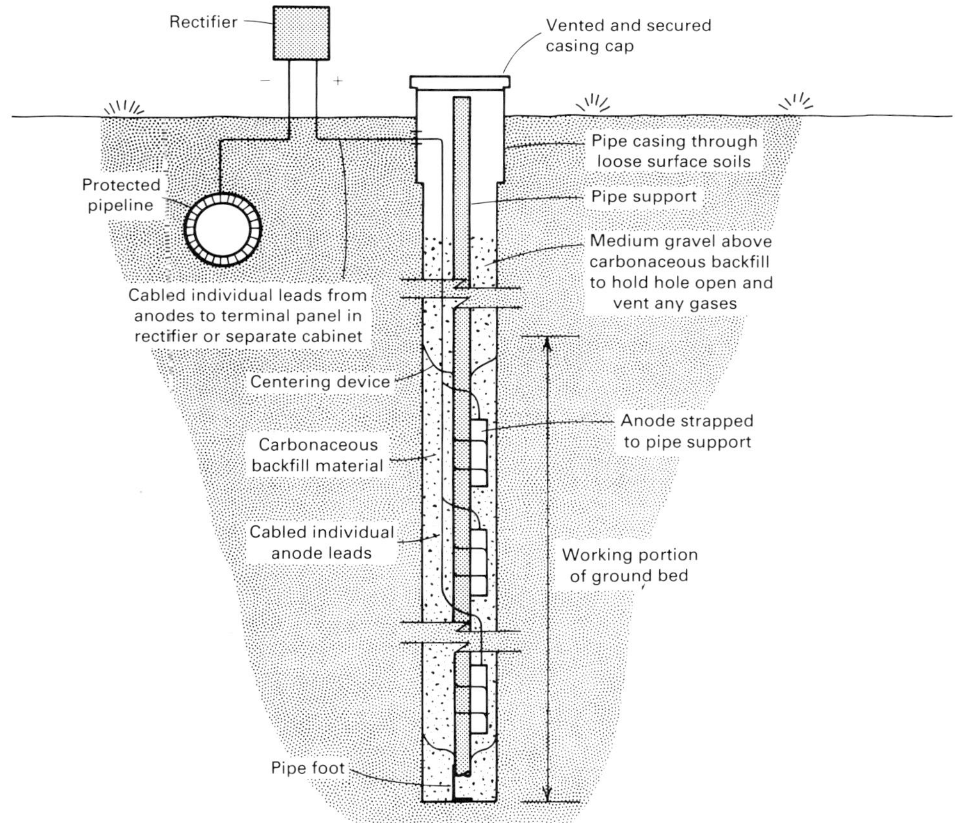

Fig. 6 Typical deep groundbed cathodic protection installation. Source: Ref 29

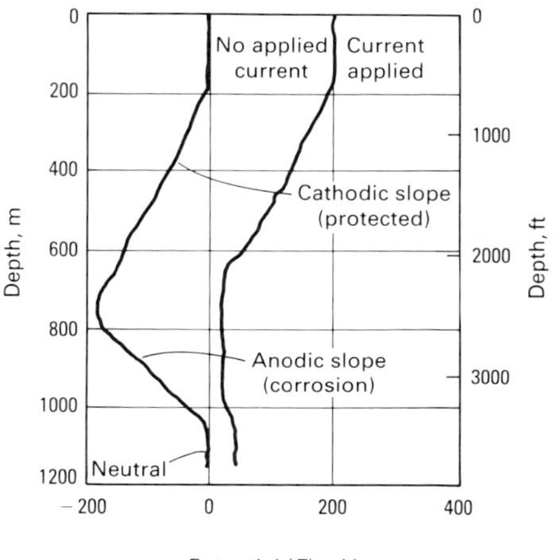

Fig. 8 Casing potential profile curve. Source: Ref 32

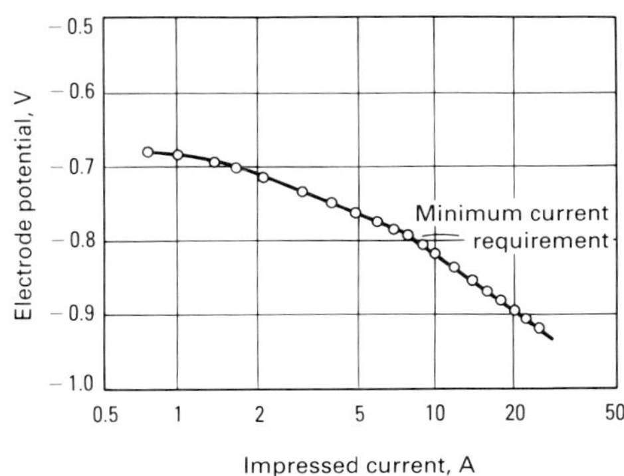

Fig. 9 Typical *E*-log *I* curve. The break in the curve indicates the minimum cathodic protection current requirement. Source: Ref 32

late the anode from the tank and allow decomposition products to fall away from the anode. In most cases, a lead wire is brought outside of the tank and welded to the tank. These anodes should also not be allowed to touch the sides of the tank and should be uniformly distributed within the tank to give uniform current distribution.

In vessels that have several sections separated by steel plates, the anodes might be shielded from protecting all of the vessel. In these cases, the only safe procedure is to install an anode in each compartment. Fire tubes in similar tanks without cathodic protection in one southern Texas field failed from corrosion in 3 months (Ref 34). When ac power is available, impressed-current systems using high-silicon cast iron, graphite, or platinized titanium anodes in through-the-wall mounts have also been used. Crude oil and some oil field chemicals have tended to stifle the flow of current from these anodes.

Offshore Platforms. The subsea zone of an offshore platform includes the area from the splash zone to and including the pilings below the mudline. Cathodic protection is the principal means of preventing corrosion in this zone, but

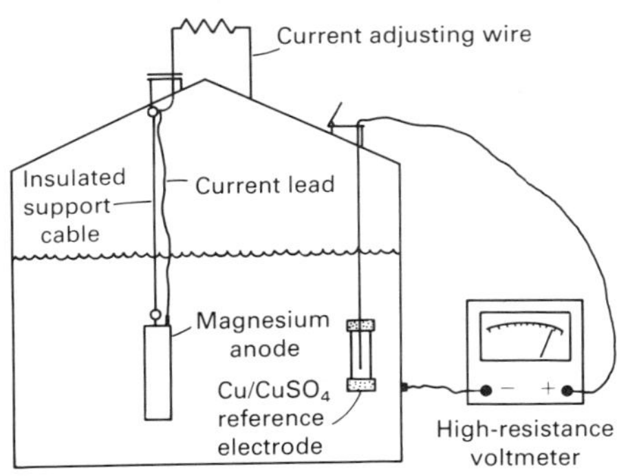

Fig. 10 Cathodic protection system for a water storage tank. Source: Ref 33

some companies also use coatings in conjunction with cathodic protection (Ref 35). The amount of electric current required to protect the bare steel varies with location. Typical current density values range from 54 to 65 mA/m^2 (5 to 6 mA/ft^2) in the Gulf of Mexico, 86 to 160 mA/m^2 (8 to 15 mA/ft^2) in the North Sea, and as high as 375 to 430 mA/m^2 (35 to 40 mA/ft^2) in the Cook Inlet (Ref 27). Current densities in the Cook Inlet are high because of the 8-knot tidal currents experienced there. In the mud zone, current densities of 10.8 to 32 mA/m^2 (1 to 3 mA/ft^2) are needed for protection, and an allowance of 3 A per well is customary for well casings (Ref 36). Environmental factors, such as oxygen content, water salinity, temperature, velocity, erosive effects, marine growth, and calcareous deposits, are largely responsible for the differences in current densities. It is very important that the current demand be conservatively estimated.

Partial protection of the steel in seawater usually means that the area of corrosion is reduced, while the unprotected areas continue to corrode at a high rate. Some companies report pits as deep as 13 to 16 mm (0.5 to 0.625 in.) and, in many cases, holes in platform members after less than 5 years on location without adequate cathodic protection. This result corresponds to a corrosion rate of 2.5 to 3.2 mm/yr (100 to 125 mils/yr) (Ref 37).

As in all applications of corrosion control on offshore platforms, the first step for cathodic protection in the subsea zone is design. Tubular members should be used whenever possible. Recessed corners in channels and I-beams are difficult to protect. Even crevices formed by placing channels back-to-back and noncontinuous welded joints cannot be protected. Bolted and riveted fittings should be avoided. Piping such as grout lines, discharge lines, water supply casings, and pipeline risers, if clustered around a platform leg, can cause shielding and interfere with the flow of cathodic protection current. If economically feasible, piping that is not necessary for platform operations should be removed. A minimum clear spacing of 1½ diameters of the smaller pipe should be provided, and coatings on the pipe can also be used to minimize shielding. Corrosion will be negligible on the internal surfaces of structural members that are sealed and have no contact with either the atmosphere or the seawater. During launch, some structural members are flooded for the life of the platform. To prevent any internal corrosion, the flooding valves should be closed to isolate the flooded chambers from contact with fresh seawater or oxygen in the atmosphere.

Sacrificial Anode Systems. The early offshore platforms installed in the late 1940s and early 1950s used 45- and 68-kg (100- and 150-lb) magnesium anodes supported from horizontal braces. A low-carbon steel wire rope connected the anode to the brace. These anodes had a 2-year design life, which was normally shortened to 1 year or less because of hurricane or rough weather losses. These swinging anodes tangled with subsea braces, shorted, and rubbed the conductor wires to failure. A variable resistor was also connected in series with the anode and the connection at the brace. This resistor was used to regulate the current output of the anode to achieve maximum efficiency. Unfortunately, this system could never be maintained. Magnesium anodes in seawater have a high current output and corrode rapidly; therefore, they must be replaced frequently. This type of system has been discontinued for offshore use.

Zinc anodes have been used since the early nineteenth century. However, impurities, such as iron, were responsible for erratic performance. Virtually all zinc anodes are now fabricated from high-purity zinc meeting the military specification MIL-A-18001-H. More efficient aluminum anodes have also been developed. A mercury-zinc-aluminum alloy anode provides 2½ times more current output than a zinc anode on a pound-for-pound basis. An anode weight of 330 kg (725 lb) is common for the initial system, while 150 kg (325 lb) is common for the replacement system. Control of impurities in the anode is essential for best performance.

The number of anodes needed depends on the size of the anode and its useful life. A design life of 20 years is common. The distribution of anodes is also important, because poor distribution and the use of too few anodes will result in underprotection, particularly at welded joints. Individual anodes should be mounted at least 0.3 m (12 in.) from the structure, or a dielectric shield should be used beneath them to improve the current distribution. Anodes should not be located in either the splash zone or on bottom bracings. The anode will not function properly if it is intermittently in and out of the seawater, and mercury-containing aluminum anodes will passivate and not function if covered by mud. Some of the earliest anodes used were prone to being knocked off of the platform during installation because the standoff posts were either too small or did not provide adequate area for contact welding of the anode to the member. Gussets and doubler plates can help obtain better anode attachment. A large fraction of premature cathodic protection failures have been traced to an inadequate number of anodes installed or excessive losses during pile driving. Some of these losses during pile driving are due to poor weld quality in attaching the anodes to the members.

Impressed-Current Systems. In an impressed-current system, the three essential components are the rectifier, the anodes, and the cable joining them together. The anode materials include graphite, high-silicon cast iron, lead-silver, and platinum wound on a niobium rod. Permanently mounted anodes, retrievable anodes, and remote anode sleds have all been successfully used. Because anodes in impressed-current systems generally produce considerably more current than sacrificial anodes, there may be only 6 or 8 impressed anodes on a structure that might do the same job as 50 to 70 sacrificial anodes. The location of these impressed-current anodes is very important in order to ensure that adequate current distribution is obtained to cover the entire surface. The connecting wiring is the critical part of an impressed-current system, especially in the splash zone, where the cable can be subjected to severe wave pounding if it is not housed in a protective conduit. Even these conduits can be torn away from the platform during a hurricane if they have been underdesigned. It is often necessary to protect the structural member near the anode with fiberglass coating or a wrap called a dielectric shield. This shield prevents excessive current consumption at this area.

Whether sacrificial or impressed anodes are used, cathodic protection currents will promote the formation of hydroxyl (OH^-) ions at cathodic areas (the entire platform, it is hoped) and cause a pH shift in the seawater near the platform. Also, the concentration of calcium and magnesium ions tends to increase in the film of seawater over the cathode. As a result of these changes, the solubility of calcium carbonate and magnesium hydroxide is exceeded, and a calcareous coating is deposited. These mineral deposits provide the primary corrosion control, and the cathodic protection current demand drops to a level sufficient to repair this coating when it is damaged. For example, if a current density of 540 mA/m^2 (50 mA/ft^2) is applied to a platform for the first 5 days on location, protection can be maintained with a current density of 32 mA/m^2 (3 mA/ft^2).

There appears to be less tendency for these mineral deposits to form in the deep ocean. In an experiment conducted by the U.S. Navy, sacrificial anodes were effective at providing cathodic protection to bare steel in seawater at depths of 1700 m (5600 ft) (Ref 38). However, the anodes were consumed more rapidly than if they were located near the surface. Because the pH is lower at great depths and the calcium carbonate concentration is below saturation, higher currents are required to achieve protection.

Inhibitors

Corrosion inhibitors are materials that, when present in a system in relatively small quantities, produce a reaction in metal loss due to corrosion attack. These inhibitors can interfere with the anodic or cathodic reaction, can form a protective barrier on the metal surface against corrosive agents, or can work by a combination of these actions. For oil field corrosion inhibitors, organic compounds containing nitrogen (amines) dominate because of their effectiveness and availability. These inhibitors usually contain three elements:

- One or more active inhibitor components
- A solvent base
- Certain additives, such as surfactants, dispersants, demulsifiers, and defoamers

Solvents are used to dilute inhibitors to control physical characteristics (such as viscosity and pour point), to aid in obtaining proper inhibitor concentration and placement during treating, to assist inhibition, and to maintain a reasonable cost per unit volume (Ref 39).

Physical characteristics of inhibitors must be considered when evaluating a potential application. These include:

- Physical form
- Solubility
- Emulsion forming tendencies
- Thermal stability
- Compatibility with other chemicals

Physical Form. Inhibitors may take either a solid or liquid form. Solid inhibitors have been made in the shape of a stick that will sink to the bottom of a well and then slowly dissolve and be produced back. These sticks are rarely used. Most corrosion inhibitors are liquid form and have densities that range from 840 to 1440 g/L (7 to 12 lb/gal) (Ref 40). These liquids must not freeze when exposed to the coldest of field conditions and must be stable with a minimum loss to the vapor state when exposed to the hottest of field conditions.

Solubility. The formation of an inhibitor film and its life are primarily governed by the solubility of that product in the system. There are three categories of solubility: soluble, insoluble, and dispersible.

A product is soluble in a fluid when it forms a clear mixture that does not separate. A product is insoluble in a fluid when it will separate after mixing to form an identifiable layer. Materials are dispersible in a fluid if they form a mixture that is not clear and separates slowly, if at all.

Different solubilities are required for different applications of corrosion inhibitors. An inhibitor to be added continuously to a waterflood should be water soluble or highly dispersible. Similarly, an inhibitor to be used for a squeeze treatment should be completely soluble in the carrying fluid to facilitate placement of the inhibitor without plugging the formation. On the other hand, where the only method of application is a periodic treatment, continuing protection requires some degree of insolubility of the inhibitor in the fluids to which it is exposed. In practical terms, this means that an inhibitor used in tubing displacement cannot be completely soluble in the well fluids. Also, a dispersion must be stable enough to remain intact until the inhibitor reaches the metal surface to be protected.

Emulsion-Forming Tendencies. Because of the chemical nature of most corrosion inhibitors, there is a positive tendency in water-oil systems to form emulsions. Some of these emulsions will break down quite readily, while others are extremely stable and practically impossible to break. When squeezed, an incorrectly selected inhibitor can form an emulsion in the formation that blocks or severely restricts further production.

The inclusion of a demulsifier in a corrosion inhibitor is no guarantee against the formation of stable emulsions. Produced fluids from each field must be tested to provide reasonable assurance that no stable emulsion will be formed upon application of a specific corrosion inhibitor.

Thermal Stability. Corrosion inhibitors generally have temperature limits above which they will lose their effectiveness and change their chemical composition. This temperature may be variable for any one inhibitor, depending on such conditions as pressure and presence of water. A typical example is that of an acid-amine salt. Under atmospheric conditions, this salt will yield water and form an amide at 70 to 90 °C (160 to 190 °F). However, this chemical can be used in oil wells in the presence of water at these temperatures with no apparent degradation. Of course, exposure to high temperatures at low pressures will result in the vaporization of the solvent systems in these inhibitors.

Compatibility With Other Chemicals. The compatibility of corrosion inhibitors with other chemicals is ordinarily not troublesome when the inhibitor and the other chemicals are present in parts per million concentrations. However, chemical users frequently want to mix various chemicals so that a single chemical pump can be used for injection. Many products are not compatible with corrosion inhibitors, because of variations in solvent systems, type of chemicals (cationic versus anionic), and so on.

Most oil field corrosion inhibitors are cationic to some extent; that is, they carry a positive electrical charge. Mixing a cationic inhibitor with an anionic chemical, such as a scale inhibitor or certain surfactants, will likely produce a reaction product that can have characteristics that are entirely different from those of either of its parent products. At best, the new material may function poorly; at worst, it may not function at all or may even form deposits in the system. When the two chemicals must be used, this problem can be prevented by using separate injection points that are not closely spaced. It should probably be standard practice never to mix any two different products. This would avoid any potential problems.

A final example of operating problems can be found where a conventional inhibitor is used in a gas stream upstream of a compressor. The nonvolatile components in the inhibitor could be left behind to foul the valves of the compressor.

Selection of Inhibitors

Many factors are involved in the selection of inhibitors, including the following:

- Identification of the problem to be solved
- Corrosives present
- Type of system (which influences the treatment method)
- Pressure and temperature
- Velocity
- Production composition

Although problems such as rod breaks and flow line leaks may initially be seen as purely corrosion failures, the actual cause of the problem could be oxygen, scale, or bacteria. Rod coupling failures could be caused by poor assembly or corrosion fatigue. Overstressing of rods greatly accelerates these failures. In such cases, mechanical measures could reduce or eliminate the need for chemical treatment. If attack is due to oxygen entry, installation of gas blankets or closing of the casing valve could greatly reduce corrosion.

The presence of corrosives such as H_2S and CO_2 greatly influences the choice of an inhibitor. Some inhibitors perform best in sweet fluids, while other inhibitors work best in sour fluids. Even the concentration of NaCl has a bearing on the choice of an inhibitor. With increasing NaCl content, some inhibitors will become insoluble and deposit.

The type of system also has an effect on the selection of an inhibitor. The correct inhibitor to use is determined by whether the system is a pumping oil well, a gas-lift well, a gas well, a waterflood system, or a flow line. For example, a weighted inhibitor is seldom recommended in dry gas wells, because water is required to release the inhibitor before it becomes effective (Ref 41). Also, when a gas-lift well is treated, the inhibitor is injected into the gas-lift lines. Therefore, the inhibitor must not have any tendency to form gunky deposits.

Both temperature and pressure have an influence on inhibitor selection. Bottom hole temperatures and pressures may get so high that inhibitors polymerize and form a sludge. Pressure influences the corrosivity of CO_2 and H_2S.

Velocity is yet another factor to consider. With pipelines, low velocity might be insufficient to displace water from low areas in the line. In the case of dry gas pipelines with low velocity, a water-soluble inhibitor should be selected and shuld be injected continuously. If the velocity is high enough to prevent any accumulations of water in low areas of a dry gas line, then an oil-soluble inhibitor should be batch treated.

The composition of the produced water also determines the choice of inhibitor. Criteria such as water/oil ratio, salinity of water, and acidity of the water and oil are vital to the correct selection of the inhibitor.

Tests are conducted in the laboratory using the standard wheel test. Although not infallible, this test attempts to duplicate field conditions as closely as possible. Parameters such as temperature, water-cut, batch or continuous treatment, and whether the system is sweet or sour are reproduced as closely as possible to field conditions. Usually, these tests involve weight loss coupons.

The rate at which an inhibitor forms a film is completely dependent on the product and its environment. However, it can generally be said that film formation is a function of time and is not instantaneous. The concentration of inhibitor required to develop an adequate film is also directly related to the characteristics of the product and system. Many factors affect the dosage and frequency of treatment, including the following:

- Severity of corrosion
- Total amount of fluid produced
- Percentage of water
- Nature of corrodent
- Chemical selected
- Fluid level in the casing annulus

Because no laboratory test can take into account all of the conditions imposed by the oil well, the dosage and frequency of treatment must be constantly reviewed.

There are two general rules to follow for dosages. First, for continuous injection, a dosage of 10 to 20 ppm based on total produced fluid is used as a starting point. Second, for batch treatment, weekly batch frequency is used with a starting dosage of 3.8 L (1 gal) per week for each 100 barrels of daily fluid production.

Also, if the corrosivity of the system is known, the following general criteria can be used to define more accurately a treatment rate for continuous injection (Ref 42):

Mild corrosion	10–15 ppm
Moderate corrosion	15–25 ppm
Severe corrosion	>25 ppm

Several major problems can occur with inhibitors, including foaming, emulsions, scale removal and plugging, and safety and handling. Corrosion of other metals can also be a problem.

The most appropriate action to take in avoiding difficulty from foaming is to determine where foam-forming conditions exist in the system. These will consist of places where the inhibitor-containing fluid is agitated with a gas, such as in a gas separator, a countercurrent stripper, or an aerator. The next step is to obtain a sample of the fluid and gas from the process step, add the inhibitor in question, adjust the temperature to that corresponding to the process step, and shake vigorously. If this test produces a stable foam, a potential problem exists.

There are three alternative remedies. First, an antifoaming agent can be added (this must be tested also); second, tests can be conducted to select an inhibitor that does not cause foaming; and third, the system can be shut down periodically and treated with a slug of persistent inhibitor. The last two remedies are the least palatable because the need for an inhibitor is at hand and there are few processes that can be shut down with sufficient frequency to maintain effective inhibition by slug treatment.

Emulsions are another problem that can occur when the wrong inhibitor is used. The use of other chemicals, heat, or both can usually break these emulsions. A great variety of chemicals are used for this purpose, but no one material has proved effective for all emulsions.

A system can be plugged as the result of an inhibitor loosening scale and suspending it in the fluid. This problem is best avoided by planning. The best preventive measure is to clean the system thoroughly, if possible, before inhibitor is applied. An alternative or supplementary method in systems that are very sensitive to suspended solids is to protect the sensitive parts with temporary filters.

As with most industrial chemicals handled in large volume on a regular basis, oil field corrosion inhibitors should be treated with respect from a safety standpoint. Although these products are not generally highly toxic (many acid corrosion inhibitor formulations are toxic), they can produce reactions because of the amines and aromatic solvents present. Reactions usually consist of skin burns from contact and dizziness from inhalation of the vapors. Repeated contact with amines will cause the development of sensitization to these products in some individuals. To avoid these problems, any contacted body areas should be washed, and contaminated clothing should be changed as soon as possible. Any accumulation of vapors should be eliminated from confined spaces.

Another possible adverse effect of inhibition is an increased rate of corrosion of a metal in the system other than the one for which the inhibitor was selected to protect. For example, some amines protect steel admirably, but will severely attack copper and brass. Nitrites may attack lead and lead alloys, such as solder. In some cases, the inhibitor may react in the system to produce a harmful product. An illustration of this is the reduction of nitrate inhibitors to form ammonia, which causes SCC of copper and brass. The only way to avoid these problems is to know the metallic components of a system and to be thoroughly familiar with the properties of the inhibitor to be used.

Application of Inhibitors

Choosing the proper inhibitor for treating a corrosion problem in the oil field is important; however, it is equally important to select the correct treating method. The best inhibitor available will not successfully control corrosion if it does not reach the trouble area. To be effective and economical, a corrosion inhibitor:

- Must be present at an initial concentration sufficient to promote complete coverage of all steel surfaces
- Must be replenished as necessary to repair washed-away portions of the protective inhibitor film

Batch treatments (Fig. 11) are commonly used in producing wells and, in some cases, in gas lines and crude flow lines. Inhibitor can be batched down the tubing-casing annulus, through the tubing, or between pigs (in the case of a pipeline). The various types of batch treatment are:

- Standard batch
- Extended batch
- Annular slug
- Tubing displacement
- Between pigs batch

Standard Batch. This method is used for producing wells that are not equipped with packers. The inhibitor is put into the annulus, and the well is placed on circulation to distribute the inhibitor throughout the system. Normally, the longer the well is circulated, the better the inhibitor film. The application of this treatment in low fluid level wells depends on the fluid level maintained in the annulus. The method would not be recommended in wells that pump off. It would be estimated that a fluid level of at least 46 m (150 ft) should be maintained. In placing the treatment in operation in these wells, it would be recommended that the initial treatment be immediately displaced into the tubing and that a second batch of inhibitor be placed in the annulus.

Extended Batch. This method is a variation of the standard batch treatment, but in this case, the inhibitor is left in the annulus. As the annular fluid level fluctuates, small amounts of inhibitor are carried in the oil into the tubing, thus giving the well periodic treatments weeks or months after the actual treatment. This type of treatment has lasted up to 6 months in some wells of Oklahoma (Ref 43). It must be remembered that this technique depends on a substantial fluid level in the annulus because the inhibitor is inventoried in the oil of the annular space.

Annular Slug. There is one technique for batch treating pumping wells that allows the well to continue full production while being treated. A water-dispersible or water-soluble inhibitor is mixed with water and placed in the annulus. This mixture will fall through the oil phase in the annulus. This technique will work if there is little or no water level in the annulus, but probably will not work if there is a substantial water level in the

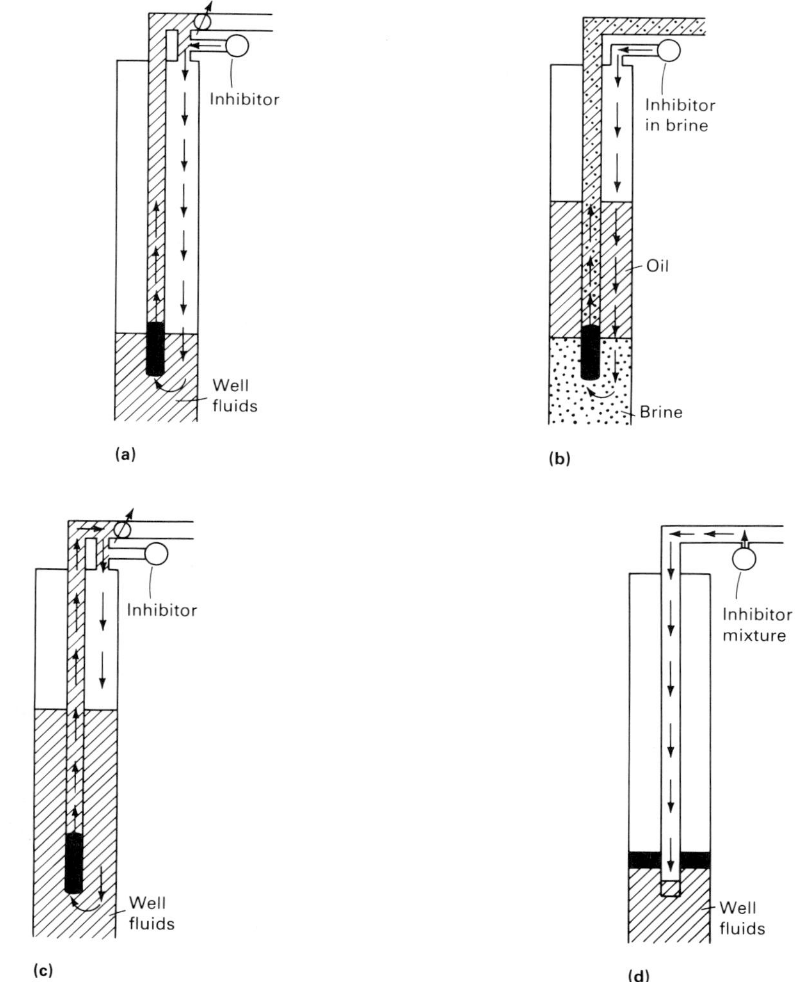

Fig. 11 Four types of batch inhibitor treating techniques. (a) Standard batch. (b) Annular slug. (c) Extended batch. (d) Tubing displacement

annulus. Frequency of treatment ranges from twice weekly to monthly.

Tubing Displacement. Wells that are set on packers or gas-lift wells are frequently treated by tubing displacement. The inhibitor is either dispersed or put in solution in water or hydrocarbon. The water may be fresh or produced. The hydrocarbon may be produced, or it may be a refined product, such as kerosene or diesel. The inhibitor is usually used at about 10% concentration in the water or hydrocarbon. The desired amount of this mixture is then introduced into the tubing. If the well is a dry gas well, the mixture will fall to the bottom if sufficient shut-in time is given (from several hours to overnight, depending on the depth of the well). If the tubing contains liquids, the mixture must be displaced to the bottom of the well by pumping liquid (usually produced fluids) in behind the mixture. The amount of displacing liquid is calculated by determining the volume of the tubing and subtracting the volume of inhibitor mixture. After the inhibitor has been displaced to the bottom, the well is usually shut in for 2 to 24 h. The well is then put into normal operation in the usual manner.

The tubing displacement technique is also known as a kiss squeeze. This type of treatment will last from a week to several months, depending on the system and the inhibitor, and is normally used on flowing oil wells.

Between Pigs Batch. This method is used to control corrosion in gas pipelines and is only used by itself in moderately corrosive systems. The volume of inhibitor mixture needed to give a 3-mil-thick coating can be calculated from an equation that takes into account the pipe diameter and length (Ref 44, 45).

Continuous treatment is used on producing wells, injection wells, pipelines, and flow lines. Continuous treatment simply involves introducing inhibitor on a continuous basis so that its concentration in the corrosive fluids is maintained at a level sufficient to prevent or reduce corrosion. This concentration may vary from a few parts per million to 50 ppm or more, depending on the severity of attack. There are many ways to continuously treat producing wells:

- Introduce inhibitor into the line that bypasses part of production into the annulus
- Inject inhibitor into the power oil of a subsurface hydraulic pump
- Inject inhibitor into a small string of tubing that is run downhole (small-bore treating string)

Squeeze Treatment. This is a combination batch-continuous method in which the inhibitor solution is placed into the formation. The inhibitor and diluent are displaced down the tubing and into formation by 25 to 75 drums of displacing fluid, which is usually clean crude, diesel fuel, or nitrogen. When the well is returned to production after a squeeze, the initial concentration of chemicals in the returned fluid is high and decreases very rapidly. The inhibitor continuously returns from the formation to repair any breaks in the inhibitor film. The second squeeze and successive treatments all give a longer treatment life than the first squeeze. Possibly, a portion of the chemical used in the first squeeze is trapped in the formation and cannot return to the well bore. This action is shown in Fig. 12. The advantages of squeeze treatment include:

- Can be used in tubingless or multiple completion wells
- Treating frequency is reduced and ranges from 6 to 18 months, depending on the inhibitor, the formation, placement technique, and the fluids being produced

The disadvantages of squeeze treatment are as follows:

- High cost
- Possible clay swelling
- Emulsion blocks that restrict production
- Injection pressure must be kept below the pressure necessary to fracture the formation

This method is used on gas-lift wells having a high-pressure, a high gas-oil ratio, and a high rate of water production. More information on the use of inhibitors in the oil patch is available in the article "Corrosion Inhibitors for Oil and Gas Production" in this Volume.

Nonmetallic Materials

In recent years, there has been an increased use of nonmetallic materials in oil field operations. These materials are being used because they do not corrode in the environments in which steel readily corrodes. They are also lightweight, suitable for rapid installation, and, in most cases, less expensive than steel. In a 1982 American Gas Association survey of 56 gas utility companies, it was found that nonmetallic pipe systems failed at only 13.2% the rate of metallic pipe systems when excavation damage is excluded (Ref 46). Nonmetallic pipe can be classified into three major categories: thermoplastic materials, fiber-reinforced materials, and cement-asbestos.

Thermoplastic materials can be repeatedly heated, softened, and reshaped without destruction. The most commonly used themoplastic pipe materials are (Ref 34):

- Polyvinyl chloride (PVC)
- Chlorinated polyvinyl chloride (CPVC)
- Polyethylene (PE)
- Polyacetal (PA)
- Acrylonitrile-butadiene-styrene (ABS)
- Cellulose acetate butyrate (CAB)

Glass fiber reinforced thermoset materials are chemically set and cannot be softened or reshaped by the application of heat. There are two major classes of these materials in oil field use:

- Fiberglass-reinforced epoxy (FRE)
- Fiberglass-reinforced polyester (FRP)

Cement-asbestos is the oldest nonmetallic material in use in the oil field. It is a combination of portland cement, asbestos fibers, and silica. It can be obtained with an epoxy lining, but most of this pipe currently in use is unlined (Ref 47).

Joining Methods

The methods used to join various types of nonmetallic pipe are shown in Table 2. The heat welding method uses a heating element to soften the ends of the joints, which are then pushed together and held until the joint cools. About 25% of all thermoplastic pipe joints are made by this method (Ref 34). Solvent welding can be used on some of the thermoplastic pipe materials and on both of the thermosetting materials. This method uses both a solvent and a glue to hold the joints together. Finally, threads can be used on all nonmetallic pipe materials.

Advantages and Disadvantages

The advantages of nonmetallic materials include the following (Ref 34):

- They are generally immune to corrosion in aqueous systems
- They are lightweight and are therefore easier to handle
- Nonmetallic pipe is quickly joined and installed
- No external protection, such as coatings or cathodic protection, is required

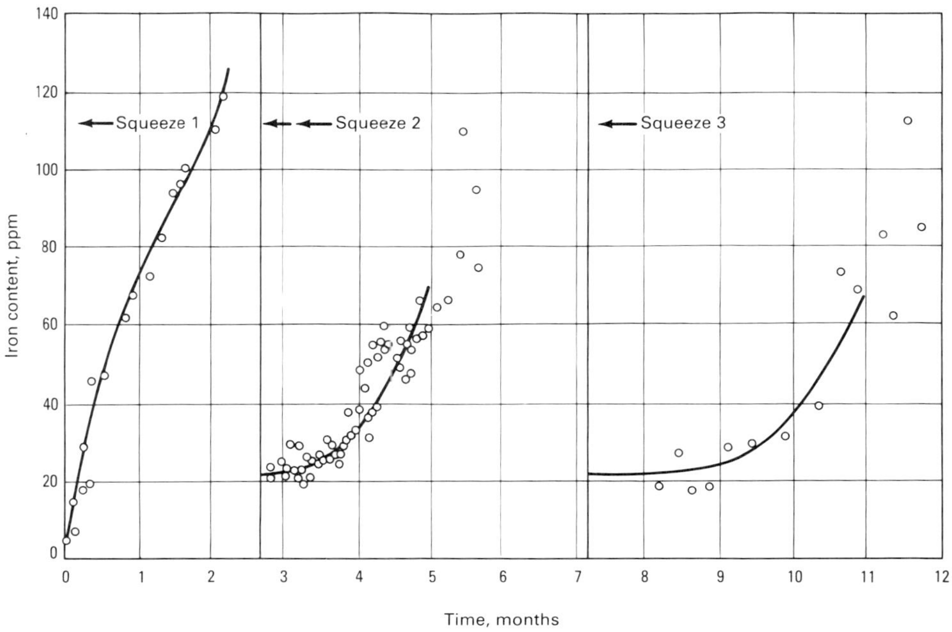

Fig. 12 Iron content of produced water after squeeze treatment. Iron content is one measure of inhibitor effectiveness.

Table 2 Joining methods for nonmetallic pipe materials

Material	Joining method: Heat	Solvent	Thread
Thermoplastic materials			
Polyvinyl chloride	...	X	X
Chlorinated PVC	...	X	X
Polyethylene	X	...	X
Polypropylene	X	...	X
Polyacetal	X	...	X
Acrylonitrile-butadiene-styrene	X	X	X
Cellulose acetate butyrate	X	X	X
Thermosetting materials			
Glass-reinforced epoxy	...	X	X
Glass-reinforced polyester	...	X	X
Cement asbestos	(Rubber ring seal)		X

Source: Ref 34

- The smooth internal surface of nonmetallic pipe results in lower fluid friction loss

Among the disadvantages of nonmetallic materials are (Ref 34):

- Nonmetallic pipe has a more limited working temperature and pressure. These limits are also more difficult to predict with assurance than the limits of steel pipe
- Careful handling is required in loading, unloading, and installation
- Nonmetallic pipe should be buried to protect it from sunlight, mechanical damage, freezing, and fire
- Nonmetallic pipe has very low resistance to vibration and pressure surges

Typical Applications

Thermoplastics have seen use in flow lines, gathering lines, saltwater disposal lines, liners for steel pipe in high-pressure operations, and fuel lines for gas engines. Polyvinyl chloride has a maximum temperature limit of 65 °C (150 °F) and a maximum operating hoop stress of 27.5 MPa (4000 psi). Polyethylene has a maximum operating temperature of 40 °C (100 °F) and a maximum operating hoop stress of 4.3 MPa (625 psi) (Ref 30).

Glass fiber reinforced thermoset materials have also seen use in flow lines, gathering lines, saltwater disposal lines, liners for steel pipe in high-pressure operations, and fuel lines for gas engines. They have also been used for tubing in disposal and injection wells. Neither FRE nor FRP should be used for a well production flow line or gas gathering system at pressures above 2.1 MPa (300 psi) and temperatures of 65 °C (150 °F). These materials should not be used in vacuum systems or where repetitive vacuum surges are likely to occur and in lines handling sand-laden fluid.

In addition, FRP has been used for stock tanks and barrels ranging in size from small chemical tanks of 1890 L (500 gallons) or less to 500 barrels or larger. Even sucker rods have had their bodies made of FRP.

Cement-asbestos materials have been used in low-pressure saltwater disposal lines. They have a maximum temperature rating of 95 °C (200 °F).

Environmental Control

Oxgyen dissolved in oil field water is one of the primary causes of corrosion. Dissolved oxygen is needed at 25 °C (75 °F) for an appreciable corrosion rate in neutral waters, while even in high-salinity brines at 150 °C (300 °F), the corrosion rate is low once the oxygen is removed (Ref 32). This type of corrosion is usually a localized form of attack, such as pitting, rather than a uniform attack. Oxygen also causes the growth of aerobic bacteria, algae, and slime, which can create plugging and enhance pitting. Also, mixing an oxygen-containing water with oil field waters containing dissolved iron or hydrogen sulfide can cause precipitation of iron oxides, iron hydroxides, or free sulfur, thus causing serious plugging problems. In one case, some injection wells of a waterflood in west Texas were filled with as much as 23 m (75 ft) of iron hydroxides after a few months of service (Ref 48). Even if there are other corrosive agents present, air-free operation is needed in order for film-forming corrosion inhibitors to work (Ref 49); the presence of dissolved oxygen will significantly reduce the effectiveness of corrosion inhibitors.

Both mechanical and chemical means have been used to remove dissolved oxygen from oil field waters. The mechanical means are countercurrent gas stripping and vacuum deaeration, while the chemical means include sodium sulfite, ammonium bisulfite, and sulfur dioxide. The choice of oxygen removal method depends on economics. Usually, the mechanical means are used when large quantities of oxygen are to be removed. Chemical removal is usually employed to remove small quantities of oxygen and even sometimes for the removal of residual oxygen after the mechanical means have been used.

Mechanical Methods

Gas stripping is performed in either a packed column or a perforated tray column. Perforated tray columns are preferred because they are not as easily fouled with suspended solids or bacterial slime as packed columns. Figure 13 illustrates a tray-type gas stripping column. Oxygenated water flows into the top of the column, while the stripping gas flows through the bottom inlet. As the gas bubbles up through the water, oxygen comes out of solution. The trays or packing in the column increases the contact area. These systems are designed to use not more than 0.06 m^3 (2 ft^3) of gas per barrel of water being stripped (Ref 30). The gas source should be free of both oxygen and H_2S. Either natural gas or exhaust gas from engines is commonly used. The principle of removal is to reduce the concentration of oxygen in the gas coming in with the water by dilution with the stripping gas.

In vacuum deaeration, a vacuum is created in a packed tower, and as the oxygenated water is passed over the packing, the low pressure causes the oxygen to bubble out of solution. The vacuum pump pulls the oxygen, water vapor, and other gases from the top of the tower. The tower usually consists of several different pressure stages, as shown in Fig. 14. In a packed column, each stage consists of a height of packing, which is sealed from the stage below by a layer of water in the bottom of the packing (Ref 33). A single-stage tower will economically remove oxygen only to a lower limit of 0.1 ppm because of the excessive vacuum pump horsepower required to achieve lower concentrations. Therefore, multistage columns are needed. Dissolved oxygen concentrations as low as 0.01 ppm have been achieved in three-stage towers (Ref 33).

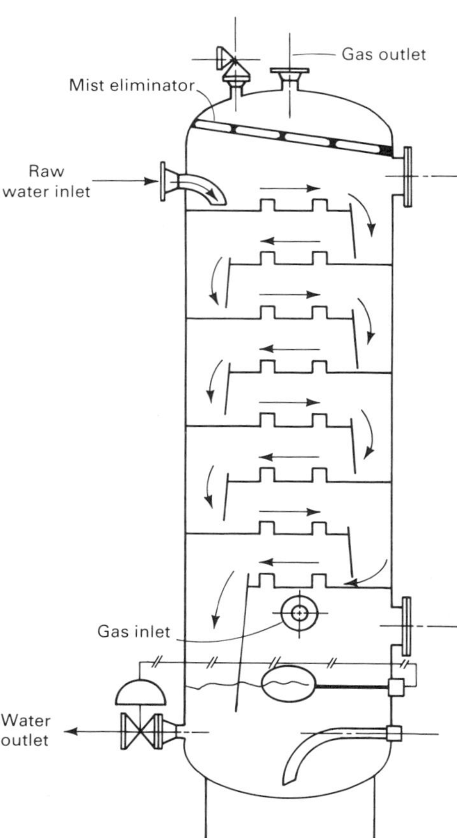

Fig. 13 Tray-type countercurrent gas stripping column. Source: Ref 33

Combination Vacuum Deaeration and Gas Stripping. Vacuum deaeration with the use of 0.003 m^3 (0.1 ft^3) of natural gas has been used to reduce the oxygen content of water from 5 to 0.05 ppm or less in a 40 000 barrel per day waterflood in west Texas. Single-stage vacuum deaeration reduced the oxygen content of the water to 0.17 ppm, while the gas stripping further reduced the oxygen content to 0.05 ppm. Corrosion rates in the water were reduced from 0.36 to 0.04 mm/yr (14 to 1.6 mils/yr) (Ref 50).

Chemical Methods

Sodium sulfite is used to scavenge oxygen from water and is available as a liquid or as a powder. It reacts with oxygen according to Eq 4:

$$2Na_2SO_3 + O_2 \rightarrow 2Na_2SO_4 \quad \text{(Eq 4)}$$

Approximately 8 ppm of Na_2SO_3 is required to react with 1 ppm O_2. A 10% excess is usually required for complete reaction, and a catalyst such as cobalt chloride (0.1 ppm) is needed to scavenge to acceptable levels within a few minutes. Because Na_2SO_3 solutions will react with atmospheric oxygen, an inert gas blanket is required on the storage tank.

Ammonium bisulfite is a liquid scavenger and reacts with oxygen according to Eq 5:

$$2NH_4HSO_3 + O_2 \rightarrow (NH_4)_2SO_4 + H_2SO_4 \quad \text{(Eq 5)}$$

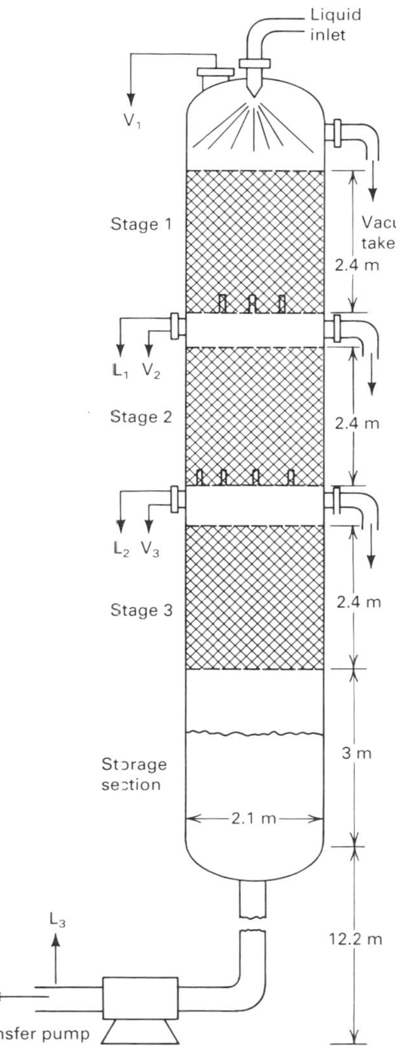

Fig. 14 Three-stage vacuum deaerator tower. L_1, L_2, L_3 and V_1, V_2, V_3 are liquid and vapor sample points, respectively. Pressure decreases as the liquid moves down the tower. Source: Ref 33

An 80% solution of NH_4HSO_3 requires a 10:1 ratio by weight for the reaction. A 10% excess is needed to complete the reaction. Ammonium bisulfite does not react with air and can be stored in open containers. A catalyst is not usually needed for oil field brines. Because the chemical is supplied as a solution with a pH of 4 to 4.5, it must be stored in a corrosion-resistant vessel. Type 304 stainless steel is commonly used (Ref 33).

Sulfur dioxide is a chemical scavenger that can be either supplied as a liquified gas under pressure in a cylinder or generated by burning sulfur. The reaction between sulfur dioxide and oxygen proceeds according to Eq 6:

$$SO_2 + H_2O + \tfrac{1}{2}O_2 \rightarrow H_2SO_4 \quad \text{(Eq 6)}$$

A quantity of 4 ppm by weight of SO_2 is required to remove 1 ppm of oxygen. A 10% excess and a catalyst such as cobalt chloride are needed to complete the reaction. Sulfur dioxide from cylinders is applied by using a bypass line that handles approximately 10% of the total fluids, as shown in Fig. 15. The scavenger is added to the bypass fluids. The materials used in this bypass line should be resistant to acid attack because of the low pH formed from the reaction. Use of SO_2 cylinders is most advantageous in small systems (less than 10 000 barrels per day) or where small concentrations of dissolved oxygen are encountered in larger systems (Ref 52).

When larger volumes of water are to be treated, it may be more economical to produce SO_2 gas by burning sulfur. This gas is then dissolved in a sidestream of the water to be treated, pumped through the packed column, and then back into the main line.

Precautions

Some precautions involving oxygen scavengers should be noted (Ref 33):

- Oxygen scavengers will react with chlorine and hypochlorite (ClO^-), which are added to injection water for bacterial control. Therefore, these chemicals should be added downstream of the point of scavenger injection to allow completion of the scavenger-oxygen reaction
- Any organic chemicals, such as biocides, scale inhibitors, and corrosion inhibitors, can possibly interfere with the scavenger-oxygen reaction and should be selected with care
- Oxygen scavengers cannot normally be used in sour systems. If H_2S is present, it may react with the cobalt chloride catalyst to form insoluble sulfides (Ref 33)

Oxygen Exclusion

It is usually more economical to exclude oxygen from oil field equipment than to remove it after it has entered the system. The most common means of excluding oxygen is through the use of gas blankets on water supply wells and water storage tanks. Maintenance of valve stems and pump packing is also important.

All tanks handling air-free water should be blanketed with an oxygen-free gas such as natural gas or nitrogen. Most tanks require only a few ounces of pressure (Ref 49). The regulator should be sized to supply gas at a rate adequate to maintain pressure when the fluid level drops.

Oil blankets should not be used in place of gas blankets. Oxygen may be 5 to 25 times as soluble in hydrocarbons as in oil field waters (Ref 49). Oil blankets will coat precipitates in the water, which can lead to well plugging problems. Some bacteria will even thrive at the oil/water interface (Ref 33).

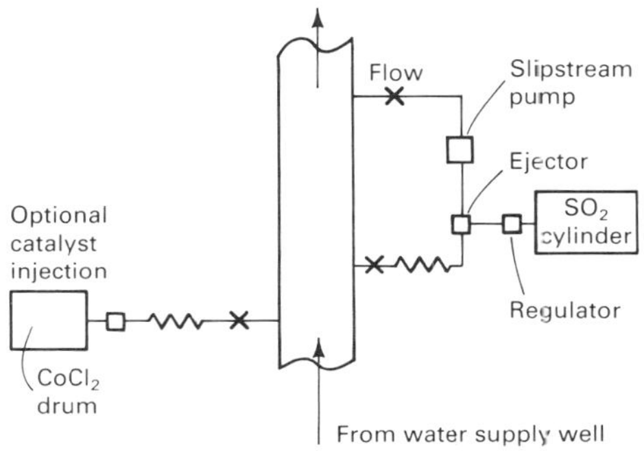

Fig. 15 Application of SO_2 through the use of a bypass line. Source: Ref 51

Even supply wells and producing wells may need to be gas blanketed to prevent oxygen entry. If these wells are operated cyclically without gas blankets, oxygenated air will be drawn into the annulus every time the well is turned on and the fluid level drops.

Oxygen can also enter a pump on its suction side if a net positive suction head is not maintained. If the seals start to leak, air can then be sucked into the pump (Ref 49).

Problems Encountered and Protective Measures*

H.E. Bush
Consultant
James E. Donham
Consultant
John D. Alkire
Amoco Corporation
S. Ibarra and T.M. Stastny
Amoco Corporation
M.C. Place, Jr.
Shell Oil Company

The problems encountered and protective measures to be discussed in this section are based on the state-of-the-art as practiced daily by corrosion and petroleum engineers and production personnel. These are by no means all of the methods employed for corrosion protection, but they represent the most commonly used processes.

Drilling Fluid Corrosion

Due to the nature of drilling conditions, corrosion is a problem in water-base drilling fluids. Important considerations are the causes of corrosion and the rate and forms of corrosion attack.

Causes of Corrosion

The major environmental causes of corrosion in drilling fluids are oxygen, carbon dioxide, hydrogen sulfide, ionic concentration, and low pH. Physical conditions causing corrosion include metal composition, metal properties, string design, stress, and temperature. Combined physicochemical corrosion accelerators include stress corrosion and erosion-corrosion. Microorganisms also introduce biological causes of corrosion in drilling environments.

Combined Effects. The forms of corrosion attack will provide characteristic patterns that can be identified and used in selecting preventive methods (Ref 53). The major forms of attack are crevice corrosion under deposits, corrosion fatigue, and SCC. Erosion-corrosion, uniform attack, and galvanic corrosion are also common problems. More than one form of attack can occur, and one form can transform into a second

*The section "Drilling Fluid Corrosion" was written by H.E. Bush; the sections "Primary Production," "Carbon Dioxide Injection," and "Storage of Tubular Goods" were written by J.E. Donham; the section "Corrosion in Secondary Recovery Operations" was written by J.D. Alkire; the section "Corrosion of Offshore Production Platforms" was written by S. Ibarra and T.M.Stastny; and the section "Corrosion of Gathering Systems, Tanks, and Pipelines" was written by M.C. Place, Jr.

form. For example, a pit can deepen and increase stress, initiating stress corrosion or fatigue (Table 3).

Failure Analysis

Analysis of used and failed equipment can provide a means of developing corrosion prevention methods. Identification of corrosion products (Ref 54) and corrosion forms (Ref 17, 55) can be developed into cause-effect mechanisms.

Monitoring

Drill pipe corrosion coupons are used to measure the rate, form, and cause of corrosion (Ref 56). Corrosion rates of 2.4 to 9.8 kg/m^2/yr (0.5 to 2 lb/ft^2/yr) free of pitting is an acceptable range. The lower rate should be used for deviated holes, deep drilling, and/or high-stress conditions. Special monitoring using linear polarization (Ref 57) or galvanic probes (Ref 58) will provide instant detection of corrosion and changes in rates that can be related to system conditions on a real time basis.

On-site chemical and physical analyses of drilling fluid properties are conducted on a frequent basis (Ref 56). Test procedures should include oxygen, CO_2, H_2S, and bacterial analysis for comprehensive monitoring of corrosion problems. The *Drilling Manual*, published by the International Association of Drilling Contractors, presents equipment inspection methods as well as information on the care, handling, and specifications of tool joints, drill pipe, casing, and tubing. Reference is made to this manual to provide comprehensive information on drilling and production equipment.

Oxygen Corrosion Control

Oxygen causes crevice corrosion under deposits and is considered to be the most serious corrosion accelerator in drilling environments. Oxygen enters the drilling fluid system externally from the atmosphere, usually by way of solids control and mud mixing equipment (Ref 59). The operation of this equipment to reduce air entrapment into the circulating system is an effective technique for limiting oxygen levels. Foaming problems are characteristic of some mud systems and can result in high oxygen levels on the high-pressure side of the pump. Defoaming the fluid or maintaining properties to release gas quickly is required to overcome this problem.

Oxygen scavengers, such as sodium sulfite or ammonium bisulfite, are used to remove oxygen from drilling fluid (see the discussion "Environmental Control" in this article). Treatment methods involve a continuous addition of chemical at the rate of 10 mg/L sulfite ion for each 1 mg/L of oxygen present in the fluid. A residual sulfite concentration of approximately 100 mg/L is maintained in the drilling fluid as a functional means of controlling oxygen in drilling systems. Oxygen scavenger catalysts are frequently required to overcome interfering side reactions that prevent the oxygen-sulfite reaction. Calcium in the fluid can combine with sodium sulfite and form calcium sulfite precipitate, thus preventing the sulfite ion from scavenging oxygen. Aldehydes and chlorine dioxide used as biocides in drilling fluids react with sulfite ions and may prevent oxygen removal. The addition of cobalt or nickel catalysts overcomes many of these problems by increasing oxygen-sulfite reaction rates.

Passivating compounds, such as sodium chromate or nitrite, are used to protect equipment during air, mist, or foam drilling operations. Treatment levels range from approximately 500 to 2000 mg/L of chromate or nitrite ion in fresh to slightly brackish fluid. Higher concentrations are required in high-brine solutions, and sodium nitrite is not recommended above approximately 25 000 mg/L of chloride ion concentration. A noteworthy disadvantage of using passivating agents is the tendency toward accelerated pitting attack if treatment levels are too low or if deposits exist under which the metal cannot be passivated. Zinc compounds are often combined with passivating agents to reduce pitting tendencies. Treatments for controlling deposits are recommended to mitigate under-deposit attack and are covered in the discussion "Scale and Deposit Control" in this section. Chromate or nitrite compounds are not compatible with sulfite-type oxygen scavengers.

Care should be exercised in the use and disposal of chromate compounds. These materials are classified as carcinogens, and personnel safety should be ensured. Injection of drilling fluid within deep formations has been used to dispose of excess or waste fluid.

A clear advantage is gained in the use of sodium chromate or nitrite chemicals when hydrogen sulfide is encountered in the well. These compounds oxidize and remove H_2S (see the discussion "Hydrogen Sulfide Corrosion Control" in this Section).

Atmospheric corrosion occurs on drilling equipment in urban, polluted, tropical, and marine environments. Protective coatings are commonly applied at the steel mill or storage yard and

Table 3 Drilling fluid corrosion control trouble shooting chart

Corrosion cause	Primary source	Identification	Major corrosion forms	Remedies
Oxygen	Atmosphere, mud conditioning, equipment, oxidizing agents, air drilling	Oxygen test, iron oxide by-products	Underdeposit corrosion, pitting	Avoid mechanical air entrapment in mud pits or defoamers; use oxygen scavengers for normal drilling; use passivating agents for air drilling operations.
Hydrogen sulfide	Formation, bacteria, chemical or thermal degradation	H_2S analysis, iron sulfide test	Underdeposit corrosion, uniform corrosion, sulfide stress cracking	Control pH $\geqq$ 9.5; use sulfide scavengers or filming inhibitors, reduce stress; change to oil mud in severe H_2S environments.
Carbon dioxide	Formation, bacteria, chemical or thermal degradation	CO_2 analysis, iron carbonate by-products	Underdeposit corrosion, pitting	Control pH $\geqq$ 9.5; use calcium hydroxide to combine with and precipitate CO_2 products; use filming inhibitors.
Dissolved salts	Formation, chemical additives	Water analysis	Underdeposit corrosion, uniform corrosion, chloride SCC	Control pH $\geqq$ 9.5; remove oxygen, H_2S, and/or CO_2; use filming inhibitors.
Bacteria	Makeup water, formation	Culture tests for sulfate-reducing bacteria and/or slime forms	Underdeposit corrosion, sulfide stress cracking	Control pH $\geqq$ 9.5; add biocides.
Temperature	Formation heat, friction	Test	Sulfide stress cracking, pitting	Select temperature-stable chemicals; use friction/torque reducers; cool mud; use oil mud.
Abrasion (erosion-corrosion)	Formation, directional or deviated hole	Observation	Erosion-corrosion	Use lubricants, torque reducers, and filming inhibitors; Control solids; use oil muds.
Metal composition				
Carbon Steels				
HRC $\leq$ 22	...	...	Underdeposit corrosion, uniform corrosion, pitting	Control pH $\geqq$ 9.5; control oxygen, H_2S, and CO_2; limit stress.
HRC > 22	...	...	Underdeposit corrosion, sulfide stress cracking, pitting	Control H_2S to very low levels.
Stainless alloys	...	...	Chloride SCC	Limit salts and temperature exposure; control deposits.
Aluminum alloys	...	...	Underdeposit corrosion, galvanic corrosion, pitting	Limit pH to 9.5–10.5; use passivation.

periodically between drilling operations. Many coating compositions are commercially available for both short- and long-term storage (2 or 3 years). The filming inhibitors that are typically used during drilling often provide good atmospheric protection for short periods between jobs. For long-term exposure, careful surface cleaning and selected atmospheric coating are recommended.

Hydrogen Sulfide Corrosion Control

Hydrogen sulfide causes two forms of corrosion surface attack—under-deposit (crevice) corrosion and sulfide stress cracking. Corrosion control methods include selecting resistant materials, removing the H_2S from the fluid, and reducing stress. Hydrogen sulfide enters the drilling fluid primarily from the formation, but it can also come from thermally degraded mud products, sulfate-reducing bacteria, and makeup water.

Alkaline pH control and sulfide scavengers are used to neutralize, precipitate, and/or oxidize H_2S. Film-forming amine-type inhibitors are recommended for coating the drill string. Caustic soda or calcium hydroxide treatments are used to neutralize the acid gas. Alkaline pH above 9.5 results in the production of sodium bisulfide or sodium sulfide products that are almost totally water soluble. This treatment provides both personnel safety and corrosion protection.

Compounds of iron oxide (Fe_3O_4), zinc carbonate, zinc oxide, and zinc chelates are used to precipitate sulfide ions from solution. Pretreatments of approximately 1 kg/barrel (2 lb/barrel) of one of the scavengers are commonly recommended as a precaution against a small influx of H_2S entering the mud system and causing damage. Tests are used to monitor scavenger concentrations and treatment requirements. Sodium chromate, zinc chromate, or sodium nitrite compounds are used to oxidize H_2S to sulfate or elemental sulfur. The oxidizing process is a fast and efficient method of removing H_2S from the system. There is no compatibility problem with the sulfide scavengers listed above. Formaldehyde and chlorine dioxide are compounds that are frequently used as drilling fluid biocides. These products react with hydrogen sulfide, offsetting its corrosive action; however, their biocidal properties are diminished or eliminated in the process.

Filming amine inhibitors provide protection from hydrogen sulfide surface attack and hydrogen embrittlement. Oil-soluble filming inhibitors applied directly on the drill pipe are recommended to offset corrosion fatigue and hydrogen embrittlement. Care should be taken with cationic filming inhibitors, which can damage mud properties by flocculating the anionic clays in drilling systems. Oil muds provide the most effective protection against all corrosion causes, including H_2S. The oil phase provides a nonconductive film covering exposed equipment and thus preventing the corrosion process.

Stress reduction by mechanical changes, such as rotary speed and less weight on the bit, is effective in reducing sulfide-induced embrittlement failures. Torque-reducing agents, particularly in high-angle drilling, are effective in lowering stress.

Material selection for drill pipe and casing can have a significant effect in controlling sulfide stress cracking. The brittle failures related to H_2S are linked to the hardness and yield strength of the steel. Steels with hardness levels below 22 HRC or with maximum yield strengths of 620 MPa (90 ksi) have few sulfide stress cracking problems. Cold work, such as tong or slip marks, increases the hardness of steel, and sulfide stress cracking then becomes a problem. The service stresses in drilling frequently demand materials of great strength, requiring hardness and strength levels that are susceptible to sulfide stress cracking. Because of such requirements, the primary means of avoiding sulfide stress cracking is by control of the drilling fluid. A full discussion of the metals used in sulfide environments is provided in NACE MR-01-75.

Higher temperatures (above 80 °C, or 175 °F) reduce sulfide stress cracking failures on high-strength steel. This factor can become advantageous in drilling and production operations if properly controlled. For example, an influx of H_2S while drilling may not cause damage if the fluid temperature is above 80 °C (175 °F) in the hole. If H_2S is detected, scavenging should always be completed before operations are begun that would lower the metal temperature, such as pulling the drill pipe from the hole.

Carbon Dioxide Corrosion Control

Carbon dioxide causes pitting primarily by under-deposit corrosion cell action. Corrosion control methods involve controlling the pH in the higher alkaline ranges. An effective technique is to treat the mud with calcium hydroxide to neutralize this acid-forming gas and to precipitate carbonates, thus lowering CO_2 levels. Film-forming inhibitors of the oil-soluble amine type applied by spraying the outside of the drill pipe and batch treatments for inside diameter filming are recommended to penetrate pits and deposits, stopping their corrosion action. Control of CO_2 is quite similar to H_2S corrosion control, and these two gases often enter the mud from the formation together.

Scale and Deposit Control

Mineral scale, corrosion by-products, and mud that form deposits on exposed metal are a major factor in setting up conditions that result in under-deposit pitting attack. The prevention and removal of these deposits with scale inhibitors is quite effective in offsetting this most serious drilling fluid corrosion problem. Inhibitors such as organic phosphonate, phosphate esters, and others of the acrylic, acrylamide, or maleic acid base structures have been effective. Products that exhibit threshold effect, temperature stability, and strong surface-active characteristics are useful. Treatments are variable because of environmental conditions, which differ greatly in drilling fluid compositions. As general rules apply, treatments of 15 to 75 mg/L are used on a daily basis for most mineral scale control. Treatments above this level are used to control deposits of metal corrosion by-products. Considerably higher treatment levels, up to 1000 mg/L, are used to provide corrosion protection. Care should be exercised in using the higher treatment levels, because these compounds may alter mud properties because of their dispersing characteristics.

Primary Production

There are two main types of producing oil wells: artificial lift wells and flowing wells. Artificial lift wells can be further divided according to the method used to pump the hydrocarbon to the surface. These include rod-pumped wells, wells that use downhole hydraulic pumps, and gas-lift wells. Approximately 90% of the artificial lift wells in the United States are rod pumped.

Artificial Lift Wells

Rod-pumped wells. In a rod-pumped well, the potential for corrosion damage is aggravated by the sucker rods alternately being stretched and compressed and by the abrasion of the rod couplings on the inside of the tubing. It is common for a well to have continuing sucker rod failures. Pulling and replacing the rods is a quick fix, but the problem will continue to exist until the root cause of the failure is identified and corrected. Identifying the problem is the most important step, because corrective action cannot be taken if the cause is not clear. Rod breaks should be inspected immediately after the rod string is pulled to determine if corrosion is occurring and to determine the steps that can be taken immediately to prevent a recurrence.

Corrosion in rod-pumped wells can be caused by several mechanisms, as discussed below. Galvanic corrosion is caused by dissimilar metals in contact or by the difference in metallurgy between two areas on a sucker rod. Most galvanic corrosion on rods is caused by differences in metal condition caused by hammer, wrench, or tong marks and the grooves left by rod-straightening machines. The impact area will be cathodic to the body of the rod, and corrosion will occur adjacent to the mark. Sucker rods have a soft decarburized layer or skin of low-carbon steel 0.13 to 0.2 mm (5 to 8 mils) thick. This layer can be broken by careless handling.

Bent rods are sometimes straightened and used again. This is poor practice, because a bent rod is permanently damaged and should be discarded. The rod-straightening machine will put spiral grooves around the rod, and corrosion will occur directly adjacent to the groove.

Any of these conditions will lead to pitting, and stress raisers will be set up. The cyclic stresses resulting from alternately stretching and compressing the rods during pumping operations will lead to rapid failure.

Stray current from surface equipment or leakage from a cathodic protection system will cause severe corrosion where the current leaves the rod string. It is usually seen on couplings or the part of the rod that is close to the coupling.

Damage from oxygen corrosion may take place when the rods are stored outdoors or when oxygen enters the wellbore through the annulus. Rusting of stored rods will often cause pitting, and rust deposits can set up concentration cells or under-deposit corrosion when the rods are run in the hole. Oxygen entry into the wellbore in wells that pump off or oxygen introduced during inhibitor treating operations will aggravate other forms of corrosion by depolarizing the cathodes on the metal surface during the corrosion reaction. Oxygen corrosion generally occurs in the lower part of the well: the casing, pump, tubing, and the lower part of the rod string. The effect lessens in the upper part of the well, because the oxygen is depleted by the corrosion reaction.

Carbon dioxide or sweet corrosion is caused by CO_2 from produced gases dissolving in water and forming carbonic acid. The carbonic acid ionizes to bicarbonate ions and hydrogen ions. A low pH results, and the bicarbonate and the carbonic acid will react directly with the steel rod and cause metal loss and pitting. The pits formed are usually round bottomed with sharp sides, and they may be connected in a line or will sometimes form a ring around the rod. Fatigue cracks will be initiated at the bottom of the pits.

Carbon dioxide corrosion is aggravated by the presence of oxygen and organic acids. Oxygen depolarizes the cathodes, and organic acids deplete the bicarbonate ion concentration, which dissolves protective carbonate scale. The formation of iron carbonate scale is the major limiting factor in CO_2 corrosion.

Many pumping wells are in the temperature range (<100 °C, or 212 °F) that is most conducive to CO_2 pitting. At these temperatures, the iron carbonate scale is formed mainly away from the surface, with some forming as a noncontinuous layer. Accelerated metal loss occurs in the gaps in the scale layer, and pits are formed.

Carbon dioxide corrosion may be sudden and catastrophic when breakthrough takes place in CO_2 floods. Wells that have been noncorrosive have failed within weeks after breakthrough.

Hydrogen sulfide ionizes in water to form HS^- and hydrogen ions. It is characterized by metal loss and pitting and can be quite severe. The iron sulfide formed generally does not form a protective layer and is usually cathodic to the metal surface. Even if a protective sulfide layer is formed, a break in this layer will result in pitting.

The presence of oxygen will increase the corrosion rate, and oxygen, in addition to depolarizing the cathodes, reacts with iron sulfide and forms elemental sulfur. This removes any protective sulfide layer and increases the corrosion rate. Elemental sulfur is corrosive to steel and further increases the corrosion rate.

Organic acids increase the corrosion rate by dissolving iron sulfide scale and leaving the metal bare. They also lower the pH and increase the driving force of the corrosion reaction. At pH less than 6, hydrogen sulfide reacts directly with the metal, and little or no iron sulfide is formed on the surface.

The pits formed during H_2S corrosion are generally small, round, and cone shaped. The acute angle at the bottom of the pit is a stress raiser, and it leads to cracking. Pits are usually not connected and are in a random pattern.

The amount of H_2S present has a direct effect on the time to failure of rods due to cracking. In some cases, corrosion pits are so small as to be undetectable before cracking occurs, or cracking may take place quickly at dents or nicks on the rods.

In addition to metal loss and pitting, sulfide stress cracking may occur in H_2S corrosion. The corrosion reaction generates hydrogen ions that combine to form atomic hydrogen. Atomic hydrogen penetrates the metal along grain boundaries and recombines to molecular hydrogen. The molecular hydrogen generates high pressures, and the metal cracks at grain boundaries. The microcrack acts as a stress raiser and quickly propagates.

Rod-on-tubing abrasion is common, and it aggravates corrosion reactions. Surface scale is removed, leaving bare metal. The adjacent areas covered with scale are cathodic to the bare metal and increase metal loss. Both the rod couplings and the tubing are damaged. Severe abrasion will lead to galling or removal of large portions of metal, which are literally torn away.

The flow velocities in pumping wells are generally not high enough to influence the corrosion rate, but localized areas of high velocity around rod protectors and restrictions due to scale build-up in the tubing could occur. High velocities can remove protective scale and inhibitor films, particularly if solids are present.

Elimination of Sucker Rod Corrosion. Once the cause of the corrosion has been found, corrective action is required so that the problem does not recur. The first step is to determine if the pumping program is correct. If the rod string and pumping procedure are not dynamically balanced, excessive tensile and compressive stresses are applied that will hasten fatigue failure and cracking due to corrosion.

The range of load, that is, the difference between the load of the upstroke and the downstroke, should be kept to a minimum. Long strokes at low speed will give the lowest load. The load is due to the weight of the rods and the fluid column on the upstroke and the weight of the rods on the downstroke.

The upstroke causes stretching, and the downstroke releases this stress, causing flexing of the rod. This cyclic stress induces fatigue failure; therefore, minimizing stress will reduce breaks caused by corrosion-induced cracking.

Proper rod string makeup will also reduce failures. Recommended torque loading should be followed when making up the string to be sure that the coupling is not in excessive stress or is not subject to play or movement resulting from too little torque during makeup. Hitting the rods or couplings with hammers and the use of pipe wrenches on the string should be avoided to eliminate marks that can lead to cracking.

Fluid pounding should be avoided. Fluid pounding is caused by the pump not filling completely on the upstroke and the plunger hitting the fluid on the downstroke. The sudden stop of movement causes a shock wave to propagate up the rod string. Fluid pounding can be the most damaging factor in rod failure. Rod guides can be installed to prevent rod-on-tubing wear.

Once mechanical deficiencies are corrected, an inhibition program should be initiated. Corrosion inhibitors can prevent or greatly reduce failures caused by pitting or fatigue and will ensure that the changes made in rod loading and handling will be effective.

Sucker rods are sometimes stored outdoors or in areas where internal storage is conducive to corrosion, such as coastal and industrial areas and in oil fields that produce hydrogen sulfide. Oxygen corrosion or rust is aggravated by the deposition of salt from marine environments, such as spray on offshore platforms and coastal areas. In warehouses and under sheds, the presence of sulfur dioxide, oxides of nitrogen, and H_2S will initiate corrosion attack and increase rusting.

The rod body and threads should be regularly inspected for corrosion damage. After inspection, the rods should be cleaned, and protective coatings should be applied.

Suitable coatings that will provide protection for a minimum of 2 years should be applied by the manufacturer over rods and couplings. An oil-soluble coating is preferred, and it should be maintained by reapplication during storage. Used sucker rods should be cleaned and coated before storage.

Sucker rods should be protected when pulled during workover operations. A batch of oil and inhibitor solution can be pumped into the tubing before pulling, or the rods can be coated after pulling.

Couplings should be dipped in or brushed with an oil-inhibitor mixture before makeup. Care should be taken so that the amount of inhibitor added is not excessive. Thus, proper makeup can be performed.

It is recommended that inhibitor be added to the tubing when the rods are run in the hole for initial filming. When the well is placed in production, one tubing volume of fluid should be circulated.

Once the well is in production, an inhibition and monitoring program should be initiated. An inhibitor is selected by testing for efficiency, usually with laboratory tests. These tests may include a wheel test, in which the inhibitor is added to bottles or high-pressure cells, rotated in a heated oven, and compared to an untreated control for percent protection and lack of pitting. Other tests include the stirred flask test and flow tests. All of these tests are designed to duplicate field conditions or to determine response to different corrodents under specified parameters.

There are several methods by which the cell can be treated. These include batch, continuous, and squeeze treatment, which are covered in the discussion "Inhibitors" in this Section and are discussed in the article "Corrosion Inhibitors for Oil and Gas Production" in this Volume. Other methods include tubing displacement after unseating the pump and the use of weighted inhibitors, sticks, or encapsulated inhibitors.

Downhole hydraulic pumps operate by pumping clean crude oil with a surface engine-driven pump down a string of tubing to operate a downhole hydraulic pump. The downhole pump lifts one barrel of fluid for each barrel of power fluid. The power oil is comingled with the produced fluid and separated on the surface. Problems can arise if the power oil carries water and solids or if CO_2, H_2S, or organic acids are present. The use of corrosion-resistant alloys and inhibitors can alleviate corrosion. Inhibitors are continuously added to the power fluid at the surface pump suction. Scale inhibitors and demulsifiers can also be added to the power fluid to prevent scale deposition and carryover of water into the power fluid.

In gas-lift wells, pressurized gas is injected into the annulus and through a gas-lift valve into the tubing. Fluid is displaced upward and out of the well by the gas. The process is repeated in batches or slugs in an intermittent system, or as a steady stream in a continuous-flow system. The velocities and turbulence encountered may increase corrosion initiated by H_2S and CO_2. Corrosion-resistant alloys can be used in gas-lift valves, and tubing can be protected by inhibitors.

Inhibitors are added into the lift gas at the surface and are carried with the gas stream into the tubing. Protection is provided above the lowest gas-lift valve. If corrosion occurs below the valve, batching or squeezing may be required for complete protection. The inhibitor selected is usually oil soluble and water dispersible, and it can be diluted with hydrocarbon to assist in carrying it downhole.

Flowing Wells

Corrosion problems in flowing wells are somewhat different from those encountered in artificial lift wells in that velocity becomes an important factor, and higher pressures lead to higher CO_2 partial pressures. Treating methods are more limited because of completion requirements.

Gas condensate wells may produce gas, hydrocarbons, formation water, acid gases (CO_2 and

H_2S), and organic acids. If the producing conditions allow liquid water to be produced or to condense on the tubing, corrosion is likely.

In wells producing formation water, corrosion may occur anywhere in the tubing string, wellhead, and flow line. Temperatures in the wellbore will affect the corrosion rate, and flow velocities also affect metal loss. The salinity of the water and acid gas content are factors in corrosion rates.

Wells that produce no formation water will corrode where the dew or condensation point of water is reached and free water condenses on the tubing. The water will dissolve CO_2 or H_2S and become corrosive.

Carbon dioxide corrosion is particularly damaging in condensed water. Dissolved CO_2 can lower the pH of water to less than 4.5 at CO_2 partial pressures of 69 kPa (10 psi) and temperatures of 75 °C (170 °F). Carbon dioxide corrosion can cause severe pitting when conditions of temperature and salinity form iron carbonate scale in a noncontinuous or spotty layer. Organic acids increase CO_2 corrosion rates by dissolving iron carbonate scale and by lowering bicarbonate content so that further iron carbonate scaling is prevented.

H_2S also dissolves in water, although the pH reduction is not as great as that found with CO_2. Metal loss and pitting, along with hydrogen embrittlement and sulfide stress cracking, may be present.

Oxygen is not present in the production stream, and it is not a problem unless it is introduced into the system by corrosion treatments. It then can have some effect on corrosion and will cause pitting of ferritic stainless steels.

Materials Selection. Most gas wells are completed with low-alloy steels for economic reasons. These steels will perform satisfactorily in most wells, and the application of coatings and the use of corrosion inhibitors permit their use in severe environments of high temperature, pressure, and CO_2 content.

Many Tuscaloosa Trend wells completed with carbon steel tubing are producing with no corrosion failures when coated and inhibited. Producing conditions range to 230 °C (450 °F) bottom hole temperature, 124 to 138 MPa (18 to 20 ksi) pressure, and CO_2 content of 5% or more. Hydrogen sulfide is also found in some wells at concentrations of 20 to 50 ppm.

Alloy Tubulars. Where conditions and economics warrant, corrosion-resistant alloys can be used. Steel with 9% Cr and 1% Mo has low corrosion rates up to 100 °C (212 °F). Higher corrosion rates and pitting become a problem at higher temperatures. The partial pressure of CO_2 is not a factor at temperatures below 240 °C (465 °F). Steel with 13% Cr is effective up to 150 °C (300 °F). Oxygen will cause severe pitting of 13% Cr steel; therefore, chemical injection systems must be kept oxygen free by an inert gas blanket on storage tanks.

If H_2S is present, 9% Cr and 13% Cr steels can be used at hardnesses below 22 HRC. Of the two, 13% Cr steel is more resistant to chloride cracking.

Coatings. Low-alloy steels can be coated for corrosion resistance. Coatings include baked-on phenolics, epoxies, and polyurethanes with fillers to give the required thickness, coating integrity, and corrosion resistance. Proper application is required for an intact coating that conforms to requirements. Tubing surface preparation, application methods, coating thickness, and holiday detection are part of the inspection and quality assurance process.

Joints and connections should be designed so that the continuity of the coating is unbroken. The first few threads inside the female connection and the pin nose must be coated. A compression ring can be installed to ensure joint integrity.

Special care must be taken when the wireline operations are carried out in the coated tubing. Coatings are easily damaged or scratched, and once the coating is broken, corrosion and disbonding of the remaining coating can take place.

Wireline guides and running speeds of less than 0.5 m/s (100 ft/min) will minimize damage. A corrosion inhibitor should be used directly after wireline operations. The wireline tools should not have sharp edges and should be plastic coated. Wireline centralizers should also be used.

Coatings are also subject to disbondment if pressures are released suddenly. Gases can penetrate the coating, and when a sudden pressure drop occurs, the gases will expand and lift the coating.

Inhibitors. Corrosion inhibitors are an effective means of corrosion control, and they are required in highly corrosive environments in which carbon steel is used. They are needed even if the tubing is coated, because a holiday-free coating does not exist. Combination coating-inhibitor procedures are particularly effective.

The most commonly used inhibitors function by forming a film on metal surfaces that stops the flow of corrosion current. Nearly all inhibitors are fatty amines or quaternary ammonium compounds. The nitrogen in the molecule possesses a strong cationic charge and is chemically absorbed onto anodic sites on the metal surface. Cross-bonding of the film and the attraction of a layer of oil aid in isolating the surface.

Inhibitors are selected for several characteristics. The major consideration is the lowest corrosion and pitting rate, followed by film persistence, nondamaging to producing formations when squeezed, and minimal system upsets due to emulsion stabilization.

Once a inhibitor is selected, a treating method is used that fulfills the system requirements. Several methods are commonly used to treat flowing wells; batch treating, continuous injection, and squeezing will be discussed below.

Batch treating involves the intermittent addition of relatively large quantities of inhibitor solution to the annulus or down the tubing of a gas condensate well. A batch treatment in a flowing well consists of dumping a solution of inhibitor in condensate or diesel fuel down the tubing, shutting the well down to allow the inhibitor solution to fall to the bottom, and repeating at a set interval. The disadvantage of this treatment is that the inhibitor may not go to bottom. The tubing may contain up to 50% of its volume of water and oil, and the bottom of the well below the static shut-in fluid level may not be treated.

A method of treatment that ensures that the batch will reach the bottom of the well is tubing displacement. A batch of inhibitor in oil, usually one-third of the tubing volume, is pumped in, and enough condensate or oil is pumped in to displace the batch to bottom. The well is shut in for a few hours and brought back on production. This ensures that the tubing is treated all the way to bottom. Tubing displacements may last from a few days to a month or so, depending on the severity of the corrosion problem, the produced fluids, the flow velocity, and the ability of the inhibitor to form a persistent film.

Nitrogen or another gas can be used to displace the inhibitor solution instead of liquid. This is of value if the well has a low bottom pressure, because filling the tubing may permanently stop production or "kill" the well. It is also of value where volumes of oil cannot be easily moved around. A variation of the nitrogen batch is to atomize the inhibitor solution into the nitrogen as it is pumped into the well. The inhibitor selected should have good film persistency.

Continuous injection consists of constant addition of small concentrations of inhibitor into a producing well. The chemical can be added into a chemical or capillary string or down the annulus of a packerless completion. Chemical injection valves in a side pocket mandrel can be installed so that the solution can be pumped continuously into the annulus of a well with a packer.

In the Tuscaloosa Trend, wells were originally completed with a Y-block and kill string. This string was used for chemical injection. Wells were then completed with a packer and a chemical string, and later with a packerless completion where the inhibitor was added down the annulus.

In deep, hot wells, the inhibitor is added diluted in condensate. This is necessary because the gas is undersaturated with hydrocarbon. At high pressures, gas acts as a liquid and may strip the solvent from the inhibitor. The amount of condensate is calculated to saturate the gas. Some wells have been treated with a water solution or dispersion of inhibitor instead of condensate.

Capillary strings are small-diameter, armored tubing that is strapped to the outside of the tubing as it is run into the well. A surface tank, pump, and filter are installed. The filter is necessary to prevent particles from plugging the small-diameter tubing. Inhibitors must be selected that do not polymerize, because this would also plug the capillary.

A recent method of treating a well with a packer consists of using a perforating gun to shoot holes in the tubing. The inhibitor is pumped down the annulus and through the holes. This method is said to be more economical than recompleting the well.

Continuous treating of deep, hot wells requires an inhibitor that will not break down or form a gunk or char. This is particularly important in wells treated with a capillary string or down the annulus where the inhibitor solution must remain for an extended period of time. A surface filtering system is also required for capillary string treating.

The deep, hot wells in the Tuscaloosa Trend require an inhibitor that will withstand temperatures to 230 °C (450 °F) without breaking down. Although the chemical strings in the wells that have not been converted to packerless completions are large in diameter (25 to 50 mm, or 1 to 2 in.), plugging problems may occur. Most of this is due to salt plugs, and the condensate has a natural fouling tendency. Therefore, any tendency to form an insoluble residue by the inhibitor adds to the problem. A high-pressure high-temperature stability test is run in lease fluids to ensure the stability of the inhibitor. The amount of inhibitor required will range from 10 to 100 ppm under most conditions. Extremely corrosive wells may require more.

Squeezing involves placing an inhibitor solution into the producing formation far enough back from the wellbore so that a continuous feedback of inhibitor is obtained. The squeeze is sized so that a predetermined life is obtained. Field crude, condensate or diesel oil are commonly used as diluents for squeeze treatment.

The inhibitor must have the proper solubility in the diluent, and it must not form a gunk or severe emulsion with produced water. Either condition could cause temporary or permanent loss of permeability and subsequent loss of production. A core test is sometimes conducted to select an inhibitor for a tight, or fairly low permeability, formation. Film persistence is not as important as continuous protection, because inhibitor will be present in the production stream at all times.

In some reservoirs, condensate is above critical temperature and therefore exists as a gas. This condition is known as a retrograde reservoir, because when the pressure is lowered, condensate comes out of solution with the gas, rather than the normal condition in which lowering the pressure vaporizes the condensate.

A dry reservoir that contains no liquid condensate should not be squeezed. Permanent loss of relative permeability will occur, and gas production rates and hydrocarbon recovery will be decreased.

One common objection to squeezing is that inhibitors are cationic and will oil-wet the formation. The wetting characteristics of a surface-active material are based more on its hydrophile-lipophile balance (HLB), which is a measure of the tendency of the inhibitor to water-wet or oil-wet a surface, than on reservoir properties.

The HLB is determined by the size and type of oil- or water-soluble parts of the molecule. Nonionic surfactants are used to oil-wet metals in lube oils and are used to water-wet materials in cleaners. Sulfonates are excellent water wetters, while other sulfonates are used as oil wetters. Cationics follow the same rules. In fact, polyamines and quaternary ammonium compounds are used in workover fluids to water-wet silicates.

It is most likely that some oil wetting of the formation occurs (the inhibitor goes into the oil in the reservoir). This is what causes a squeeze to work. Nevertheless, any change in the wettability of the formation is reversible. The formation immediately begins to return to its original state once the wetting agent is removed or begins to dissipate. Natural or simulated core tests can be conducted to ensure that no formation damage will occur from the inhibitor.

The loss of production that can result from a squeeze is due to the formation of a stable emulsion in the area immediately adjacent to the wellbore. This emulsion is nearly always a water-in-oil emulsion, which is very viscous. The high-viscosity emulsion will not flow through the pore throats. Emulsion blocking can be prevented by proper inhibitor selection and by adding demulsifiers to the squeeze.

A typical squeeze can be performed in the following manner. First the amount of inhibitor required for the projected life of the squeeze should be calculated:

$$V = \frac{42 \cdot P \cdot D \cdot 3 \text{ (ppm)}}{1\,000\,000} \qquad \text{(Eq 7)}$$

where V is the volume of inhibitor (gallons), P is the total daily production in barrels (including both oil and water), D is the expected squeeze life, and ppm is the amount of inhibitor feedback desired (this is multiplied by 3, because it is assumed that only one-third of the inhibitor will remain in place that will desorb and feed back). The remaining five steps are as follows:

- Dilute the inhibitor with crude, condensate, or diesel oil to 10%
- Pump a spearhead of 5 to 10 barrels of oil with 19 L (5 gal) of demulsifier
- Pump the main body of the squeeze treatment into the formation
- Overflush with one tubing volume plus one day's production volume of oil (19 to 38 L, or 5 to 10 gal, of demulsifier can be added to the overflush)
- Shut in the well for 12 to 24 h

This procedure can be modified to suit the requirements of a particular situation.

In many applications, the amount of overflush needed to place the inhibitor properly is too large, or filling the tubing may kill a low-pressure flowing well. The use of nitrogen instead of hydrocarbons overcomes these restrictions.

In a nitrogen squeeze, the inhibitor solution is displaced downhole and into the formation with an equivalent amount of nitrogen. This leaves the tubing empty and charges the formation so that it flows back readily.

This procedure can be modified so that the inhibitor solution is atomized into the nitrogen as it is injected. Both of these procedures have been used with excellent results. The wells could be returned to production in 4 h, and due to the charging effect, the increased production rates for a day or so compensated for the production loss during the squeeze.

Monitoring

Once the well, line, or vessel is treated, it is necessary to evaluate the effectiveness of the treatment program and to determine when to re-treat or change dosage levels. The methods used include iron counts, weight loss coupons, test nipples and spools, electrical resistance and linear polarization methods, and waiting for the tubing to fail. This last method is not popular with operators. However, the collection of failure data, such as tubing failures, is a valuable source of monitoring information.

Iron counts consist of taking a representative sample of produced water and testing for iron content. The sample must be representative. The iron counts can be plotted for easier understanding. Some computer programs will present iron counts on a computer plot in color. Care must be taken so that the iron from the formation is not assumed to be metal loss from the tubing. Some of the deep, hot Tuscaloosa Trend wells produce water with over 100 ppm of iron. A base count should be conducted on a downhole sample, if possible.

Corrosion coupons may be flat or cylindrical and may be installed in any accessible location. It must be remembered that coupons measure corrosion only where they are placed. Coupons show corrosion that has already taken place, and a single coupon will not show whether the corrosion was uniform or occurred all at once.

Different types of coupon holders or chucks are used, depending on the system, the pressure, the location, or other factors. Most coupons are run in a 25- or 50-mm (1- or 2-in.) threaded plug. Flat coupon holders hold two coupons, while cylindrical coupon chucks may contain eight or more. The multichuck coupons allow a coupon to be pulled at intervals to see if the corrosion rate is uniform or not.

High-pressure systems require a special coupon check and insertion device. The insertion tool fits into a special attachment on the pipe or vessel that has a high-pressure chamber with a valve on each end. The inner valve is closed, the retrieval tool inserted, and the inner valve opened. The tool is then run in and left. The procedure is reversed to remove the coupon.

The industry guide for preparing, installing, and interpreting coupons is NACE standard RP-07-75. The primary consideration is that all coupons be treated exactly alike. A method of preparation that does not alter the metallurgy of the coupon is required. Grinding and sanding of coupons should be controlled to avoid metallurgical changes and to provide a consistent and reproducible surface finish.

Coupons should be handled carefully and stored in noncorrosive envelopes until they are installed. Rust spots caused by improper handling, fingerprints, and so on, may initiate a pit that is not representative of the system being evaluated. Prior to installation, the weight, serial number, date installed, name of system, location of coupon, and orientation of the coupon and holder should be recorded. The coupons are left in the system for a predetermined number of days and then removed.

When the coupons are removed, the serial number, date removed, observations of any erosion or mechanical damage, and appearance should be recorded. A photograph of the coupon may be valuable in some cases. The coupons should then be placed in a moisture-proof envelope impregnated with a vapor phase inhibitor and taken immediately to the laboratory for cleaning and weighing. The coupons can be blotted (not wiped) dry prior to placing in the envelope.

The laboratory receives the coupon and inspects, cleans, and weighs it. A report is issued showing the thickness loss, any pitting observations, and any other observations of interest.

Electrical resistance probes function by reading the resistance to current flow of a thin loop of metal installed in the system. The loop of metal is part of an electrical bridge circuit. As the loop corrodes and loses cross-sectional area, electrical resistance increases, and the current flow decreases. This unbalances the bridge and reads out directly on a meter. It may also be recorded on a strip recorder. As the reading changes, the points are plotted on a graph. The slope of the line is translated into a corrosion rate. The slope will change when a well is treated or when any event occurs that changes corrosion rate.

The metal loop is fragile, and it can be broken if foreign objects, chunks of scale, and similar obstructions are present in the flow. The loop may become coated with paraffin and will not give a true reading. Pitting rates cannot be determined with a resistance probe.

Linear polarization instruments apply a voltage to a pair of electrodes, compare it to a reference voltage, and then read and record the current flow. The voltage is impressed in a forward direction until breakdown occurs, that is, until a small increase in voltage causes a large increase in current. The voltage is increased some distance past breakdown, and the polarity is reversed. The positive electrode then becomes the negative electrode.

The voltage is then increased in the opposite direction past the original voltage starting point. These voltages are recorded and plotted continuously on graph paper. The position and shape of the curves will show the corrosion rate, maximum corrosion rate, and the influence of oxygen.

The principal advantage of this instrument is that a corrosion rate can be determined immediately, and a pitting rate can be measured without waiting a month or so to pull a coupon or calculate a slope with a resistance probe. Probe locations are very important, because the probe must be immersed in water to give an accurate reading. Probes can become coated with paraffin and will show an erroneous corrosion rate. They should be located in the bottom of a line or in a bypass loop so that they are continously immersed in water. The probe may also short out if a piece of metal is across the probe or if corrosion by-products and other deposits collect on the probe.

A test loop can be installed in a system for better control. A test loop is simply a bypass with valves for controlling flow, and it may contain weight loss coupons as well as probes. The system can be monitored, and different inhibitors can be evaluated at the same time.

A caliper survey can be conducted to determine if pitting and general metal loss have been halted. The caliper log can be easily compared with a log run before the treatment is begun.

Other monitoring methods include hydrogen probes, galvanic probes, and electromagnetic logging devices. Chemical analysis of produced water for alloying metals, such as manganese and chromium, can be conducted.

Collection of Field Data. It is important to collect and chart failure rates. Some failures may occur over a period of time and may erroneously indicate that the treatment is not effective. However, through proper charting and comparing with previous failure rates, the effectiveness of proper treating will be shown.

Corrosion in Secondary Recovery Operations

Secondary recovery, or waterflooding, generally increases the corrosion problems in existing producing wells. It also creates a new set of problems because of the facilities required to reinject the produced water. This section will discuss the corrosion problems that are specific to the various types of environments or equipment used in secondary recovery, that is, producing wells, producing flow lines, separation facilities, tanks, injection pumps, injection lines, and injection wells. Although not specifically addressed, disposal wells (wells that are used for produced water disposal rather than reinjection into producing formations) are considered to be the same as injection wells. Corrosion mitigation methods and guidelines are then discussed for each type of environment.

Types of Corrosion Problems

Producing Wells. The corrosion mechanisms in secondary recovery are similar to those in primary production. The primary causes of corrosion are dissolved acid gases (H_2S and CO_2) in the produced fluids. Naturally occurring organic acids are often present and can aggravate H_2S and CO_2 problems. Corrosion will generally increase in secondary recovery because of the large increase in water production caused by waterflooding. The fraction of water produced, or water-cut, may increase to 90% or more. This increases the potential for corrosion, because more of the metal surfaces may be water-wet rather than oil-wet. The increased volume of water can increase pumping equipment stresses. Increased stress levels can cause more corrosion fatigue related failures.

Corrosion mechanisms may change during waterflooding. For example, a normally sweet field (that is, the produced fluids contain no H_2S) may begin to produce H_2S because of the growth of sulfate-reducing bacteria in the formation. This can cause unexpected corrosion related to the H_2S, pitting under sulfate-reducing bacteria deposits, or failures from sulfide stress cracking in high-strength materials.

Mineral scale problems, such as the deposition of $CaCO_3$, $CaSO_4$, or $BaSO_4$, may increase during waterflooding. This is usually the result of changes in the formation water brought about by injecting waters from sources other than the original reservoir. Although not strictly a corrosion problem, scale deposition can cause increased failures due to wear and under-deposit corrosion.

Producing Flow Lines. Corrosion mechanisms in producing flow lines are similar to the mechanisms downhole, but generally occur at a lower rate because temperatures and pressures are lower at the surface. Corrosion is often localized to the bottom of flow lines if flow rates are low enough to permit water stratification, which allows the bottom of the line to be continuously water-wet. Under-deposit corrosion and sulfate-reducing bacteria related pitting are often severe under sludge or scale deposits that accumulate in the low flow rate lines.

Oil/Water Separation Facilities. Corrosion in these facilities is normally related to attack by corrodents in produced fluids and deposit-related problems. Separation facilities are unique in that they often use heat to aid in oil/water separation. Heat transfer surfaces are usually subject to mineral scale deposition because of solubility changes caused by temperature increases. Scale deposition can result in severe under-deposit corrosion because metal surface temperatures increase due to the reduced heat transfer. Creep rupture failure can occur in direct fired heaters if deposition is severe enough to cause very high metal temperatures. Some separation equipment is open to the atmosphere, thus allowing oxygen contamination of the produced fluids and causing increased corrosion in equipment handling the water phase.

Tanks/Water Storage. Tanks are subject to corrosion by acid gases (CO_2, H_2S) carried over with the produced water. Under-deposit corrosion can be severe under accumulated sludge and debris in tank bottoms. These deposits are also prime areas for the growth of sulfate-reducing bacteria. Tank roofs often fail because of condensation. As water condenses on the roof, it will absorb acid gases from the tank fluids. This can cause severe pitting. Oxygen contamination often occurs in tanks. Obviously, open tanks are subject to contamination. Contamination can occur in normally closed tanks if hatches and vent systems are poorly maintained. Although oxygen can be somewhat corrosive by itself, its primary role in waterflood system corrosion is to significantly increase the rate of attack of other corrodents already in the system.

Injection pumps can fail by normal corrosion mechanisms as well as by cavitation and erosion. Pump intake piping design must take into account the presence of dissolved H_2S and CO_2 in the water. These gases can affect net positive suction head calculations. If sufficient net positive suction head is not provided, cavitation can occur. Erosion and erosion-corrosion can occur because of solids in the water. Solids normally consist of corrosion products, formation fines, and mineral scale particulates. Alloy materials such as type 304 and 316 stainless steels are often used for pump internal parts. These alloys can fail by chloride SCC in produced brines if temperatures are above 52 to 65 °C (125 to 150 °F). Pumps are subject to cyclic stresses. Corrosion fatigue failure can occur at sharp changes in cross section, grooves, and at pitted areas, all of which cause stress concentrations.

Injection Flow Lines and Wells. Corrosion mechanisms are generally the same for producing well flow lines and tubulars, that is H_2S, CO_2, and organic acids. Under-deposit problems in the bottoms of lines and under mineral scales can also occur, as can problems with sulfate-reducing bacteria. Oxygen contamination will greatly accelerate all but the sulfate-reducing bacteria mechanism. Sulfate-reducing bacteria corrosion can still occur even in aerated systems because localized areas under scales, sludges, or aerobic bacterial slimes can become anaerobic and thus support the growth of sulfate-reducing bacteria.

Injection wells and flow lines may require periodic acidizing to reduce pressure drops and to restore the injectivity lost because of the buildup of corrosion products and mineral scales. Severe corrosion can occur if acidizing fluids are not properly inhibited and flushed from the system.

Corrosion Mitigation Methods

Producing Wells. Corrosion control methods for secondary recovery are typically the same as those used for primary recovery. The particular method implemented will depend on the type of production method used (that is, beam lift, electric submersible pump, or gas lift), well design, and the economics of the individual situation.

Corrosion inhibitors are widely used to protect tubulars and other downhole equipment in all types of producing wells. The most common methods of sending the inhibitor downhole where it can protect the well equipment are referred to as squeeze treatment, batch treatment, and continuous treatment (see the discussions "Inhibitors" and "Primary Production" in this Section). References 60 and 61 contain detailed descriptions of the various methods as well as guidelines for selecting a particular method. Regardless of the method used, the inhibitor must be effective against the particular type of corrosion occurring, that is, H_2S, CO_2, or both. Laboratory tests should be performed if there are any questions regarding the effectiveness of the inhibitor for a given type of corrosion.

The type of corrosion inhibitor used (oil-soluble, oil-insoluble, water-dispersible, water-soluble, and so on) will depend on the treatment method. Batch treatment is a widely used method of treatment for beam lift wells. Corrosion inhibitor solutions are periodically injected into the casing-tubing annulus and flushed to the bottom of the well with produced fluids, diesel oil, or water. A water-dispersible inhibitor is normally used because of the high percentage of water in the well stream. However, increased water dispersibility can cause problems with oil/water separation because of the tendency for dispersion chemicals in the inhibitor to cause emulsions to form. Tests should be performed with actual well fluids to determine the emulsion tendency of the particular inhibitor being considered for use. Often, any one of several inhibitors may be able to provide the necessary corrosion protection; however, there will be vast differences in emulsion formation.

Continuous injection of inhibitor may be necessary for wells with high fluid levels in the annulus above the pump. Water-soluble inhibitors are normally specified for this type of treatment. Studies have shown that continuous treatment may not be as effective as periodic batch treatment under most conditions (Ref 60). Emulsion problems are sometimes worse with water-soluble inhibitors than with oil-soluble or water-dispersible inhibitors because of the increased use of surfactants in water-soluble inhibitors.

The frequency of treatment and the quantity of inhibitor used will generally have to be increased during secondary production. In general, it is more effective to increase the frequency of inhibition (assuming a batch treatment procedure is used) rather than the quantity, although both may need to be adjusted in some cases. Treatment should be adjusted on the basis of corrosion-monitoring results and well equipment life. Corrosion monitoring can be accomplished in a variety of ways. Corrosion coupons installed in flow lines near the wellhead are the most common. Downhole monitoring is more difficult. Preweighed, short (0.6 m, or 2 ft) sucker rods can be used as downhole corrosion coupons, as can short joints of production tubing. Information on the preparation, installation, and evaluation of corrosion coupon data is provided in NACE RP-07-75.

Downhole equipment should be carefully examined for signs of corrosion whenever it is removed from the well. The occurrence of sucker rod failures is a common measure of downhole inhibition effectiveness in rod pumped wells (Ref 61). The number of failures that can be tolerated will depend on the economics of each producing situation. A general guideline is one corrosion-related failure per well per year. It should be remembered that corrosion fatigue failures of sucker rods are a function of corrosion and stress. Therefore, heavily loaded rods will tolerate less corrosion before failure than rods with lower stress levels.

Corrosion inhibitors are less effective in sucker rod pumps because of wear. Corrosion-related failures are generally controlled by changing the pump metallurgy. Guidelines for selecting pump materials are provided in NACE MR-01-76. Galvanic corrosion problems can be quite severe in pumps and are best controlled by eliminating or reducing the extent of dissimilar metals in contact with each other in the pump. This also applies to coatings used for wear resistance, such as chromium and nickel plating. Rapid failure can often occur in underlying steel if these coatings become damaged by wear. If the wear resistance of chromium plating is required, it may be necessary to upgrade the base material to avoid galvanic corrosion problems.

Wear of sucker rod strings can be controlled through the use of centralizing rod guides. A variety of nonmetallic materials are molded or physically attached to the sucker rod to prevent it from contacting the tubing. Welded or metal guides should not be used. Sucker rod couplings are normally coated with a corrosion-resistant alloy by flame spraying or similar techniques. This will provide both wear and corrosion protection. Similar coatings can be applied to the rods; however, these have not been widely used because of the high cost involved.

Fiber-reinforced plastic sucker rods can be used to reduce corrosion fatigue failures; however, their primary benefit comes from production concerns rather than corrosion. Corrosion inhibition is still necessary when FRP rods are used to protect the steel end connections of the rod and steel well tubulars. In addition, steel rods are not entirely eliminated from the string when FRP rods are used. Internal tubular coatings are not widely used in rod-pumped wells, because they rapidly fail from rod wear. Fiber-reinforced plastic tubing is not widely used for the same reason.

Electric submersible pump wells are treated in much the same way as rod-pumped wells. Electric submersible pump wells pose an additional problem in that the pump fluid intake is above the motor housing. This means that inhibitors injected into the annulus do not reach the housing. A variety of methods have been used to reduce the corrosion of housings, including applying corrosion-resistant coatings and selecting corrosion-resistant alloys for the housing. Special inhibitor injection systems using small-diameter tubing to release inhibitors below the motor have also been employed. Corrosion of electric submersible pump internal parts is not typically a problem because of the widespread use of corrosion-resistant alloys. Internal tubular coatings can be used with electric submersible pump wells, because they are not subject to wear. Fiber-reinforced tubing has found application in a limited number of electric submersible pump wells.

Gas-lift wells are commonly treated by atomizing inhibitor solutions into the lift gas. This can provide protection to the tubulars only above the lowest operating gas-lift valve. Internal tubular coatings, FRP tubulars, and corrosion-resistant alloys can be used above or below the operating valve to provide corrosion protection.

Producing Flow Lines. Carryover from downhole corrosion inhibition is often sufficient to protect flow lines. In extremely corrosive conditions, additional inhibitor injection, either batch or continuous, may be required. Internal coatings can be used on flow lines; however, obtaining protection in the area of pipe joints can be difficult. A variety of methods have been developed to minimize damage to the coating even in welded lines. Fiber-reinforced plastic line pipe is becoming more widely used for flow lines, because it is inherently corrosion resistant. Polyethylene lines are also used in low-temperature low-pressure applications.

Oil/Water Separation Facilities. Supplemental inhibitor injection is often used to help protect these facilities. In addition, vessels such as separators are often internally coated. Organic coatings are normally used, but platings such as electroless nickel are also employed. Noble platings, such as electroless nickel, can cause severe galvanic corrosion of underlying steel if the coating is cracked or otherwise damaged. Internal cathodic protection with sacrificial anodes is also used in vessels. Internal baffles and other pieces can be fabricated from corrosion-resistant alloys. Corrosion-resistant alloy linings can also be used.

Heat transfer surfaces and vessel bottoms should be periodically cleaned of scale and debris. Scale inhibitors should be used if continuous scale deposition problems occur. Chromium-containing steels (2.25 to 12% Cr) can be used for heat transfer surfaces in direct fired heaters to reduce the possibility of creep rupture failures in applications subject to severe scale formation. However, these materials can be rapidly attacked in the presence of H_2S. High-nickel corrosion-resistant alloys can also be used to help prevent under-deposit corrosion problems.

Tanks/Water Storage. Internal coating is a common method of protecting tanks. Organic coatings are typically used. Steel tank life is often extended by the use of FRP linings, especially tank bottoms. Both chopped and mat systems are used. A variety of nonmetallic liners have also been used. Fiber-reinforced plastic tanks are becoming more popular in smaller sizes. Internal cathodic protection can also be used, normally in conjunction with internal coatings. Tanks should be periodically cleaned to remove the accumulated sludge and debris that hinder normal corrosion control methods and promote under-deposit and sulfate-reducing bacteria problems.

Tanks are usually the first source of oxygen contamination in the injection system. Open tanks and pits should be avoided. Various methods of excluding oxygen in open tanks have been attempted. Oil layers are ineffective. Several floating systems have been developed that are useful to some degree, but are not totally effective. It must be remembered that as little as 0.01 ppm oxygen is sufficient to cause major increases in corrosion rates. Oxygen also renders many corrosion inhibitors ineffective. Closed tanks can also allow oxygen entry. Poorly maintained hatch seals and venting systems are notorious as sources of contamination. The optimal method of excluding oxygen is to ensure that all openings to the tank are properly maintained and that a low-pressure inert gas blanket is used. Gas blanketing provides a slight positive pressure that will keep air from entering. Gas blankets can be part of the vapor recovery system, if used, or can be externally supplied from bottled gases, such as nitrogen.

Oxygen can enter the injection system in other ways. Often, additional water must be obtained to augment produced water volumes. Freshwater can be obtained from lakes, rivers, or wells drilled into aquifers. Seawater is used in offshore and coastal locations. All of these waters will have some amount of oxygen contamination. Severe corrosion can result if this contamination is not removed. Common removal methods include the use of chemicals or scavengers, such as sodium sulfite or ammonium bisulfite, and vacuum or gas stripping (see the discussion "Environmental Control" in this article).

Tanks are also excellent locations for the growth of sulfate-reducing bacteria. If tanks become contaminated with sulfate-reducing bacteria, they must be cleaned and sterilized with biocides. Cleaning is a necessity, because it is impossible for biocides to penetrate adequately the large amounts of sludge and debris on the tank bottom.

Injection Pumps. Corrosion-resistant alloys are widely used in injection pumps and ancillary equipment. The particular choice of materials used will depend on the nature of the fluids handled and the type of pump involved. Specific material recommendations are provided in NACE

RP-04-75. Caution should be exercised, because this specification does not address the temperature limitations of the materials. Chloride SCC can occur in 300-series stainless steels if they are used in saline waters above 52 to 65 °C (125 to 150 °F). Also, pitting of these materials can occur in aerated salt water if they are left stagnant in a pump. For example, it is common practice to have standby equipment piped into a system and to test the equipment periodically. Flushing the equipment with deaerated and inhibited freshwater is recommended to prevent pitting corrosion.

Flow Lines and Injection Wells. All potential corrosion mechanisms must be dealt with to obtain acceptable service lives of injection systems. This includes corrosion by dissolved acid gases, growth of sulfate-reducing bacteria, oxygen contamination, and scale/sludge deposition.

Corrosion inhibitors can be used to control flow line and injection well corrosion. Treatment is usually continuous, but batch treatment can also be used. Both oil-soluble/highly water-dispersible and water-soluble chemicals are used. Flow lines can be internally coated with organic coatings. Cement and other nonmetallic linings are also used. Fiber-reinforced plastic flow lines are widely used, even in high-pressure injection systems. The lack of an American Petroleum Institute (API) specification for high-pressure FRP line pipe and the lack of standardization in the FRP pipe industry have limited the use of these materials. Standardization of pressure rating methods and improved quality control of the products will greatly increase FRP use.

Injection well tubulars can be bare steel if corrosion inhibition is used. Internal coating is also widely used even with corrosion inhibition (see the discussion "Coatings" in this Section). Care must be taken when handling internally coated tubing to prevent coating damage. Special guides must be used when the tubing is installed to prevent damage to the pin nose. Makeup equipment must not deform the tubing enough to crack the coating. Standard API couplings are routinely internally coated in the standoff thread area. The recent advent of flush joint tubing connections using nonstandard couplings has helped to make internally coated tubing applications more reliable. The new connections help to seal the end of the tubing joints in the coupling. This has long been a problem area in internally coated tubing because it is easily damaged during handling and installation.

Corrosion-resistant alloy tubulars are used on some occasions, but their high cost is usually prohibitive. Fiber-reinforced tubing is used to some extent; however, again, the lack of standardization has been a limiting factor. Handling and makeup procedures are critical for successful fiber-reinforced tubing application. Many failures have resulted from overtorquing of FRP connections by crews used to handling steel tubulars. No reliable method has been developed for accurately predicting the long-term performance of FRP tubulars subject to both internal pressure and axial load.

Injection wells frequently require acidizing to restore injectivity. Typical acids used are 15% HCl and 12% HCl-3% HF. Severe corrosion can result if these acids are not properly inhibited. Corrosion inhibitors are available from acid service companies. Inhibitor concentration should be such that the corrosion rate of low-carbon steel is less than 245 g/m^2 (0.05 lb/ft^2) over the length of time the acid is to be in the well. It is good practice to ensure that the acid delivered to the job site actually contains the inhibitor and is the strength called for in the workover procedure. A simple test procedure for determining the presence or absence of inhibitor is given in API Bulletin D 15. This test is not designed to determine inhibitor effectiveness at well conditions nor to compare different inhibitors. Laboratory testing is necessary to establish inhibitor effectiveness.

Acid exposure can have a wide range of effects on the internal tubular coatings that may be present. Laboratory testing should be conducted if there is any doubt regarding the ability of the coating to withstand the acid exposure without damage. Fiber-reinforced plastic tubulars can also be damaged by exposure to mineral acids. Although tubing manufacturers do not prohibit acid exposure, they all recommend that temperatures and exposure times be kept to absolute minimums. The use of hydrofluoric acid in acidizing fluids is not recommended if FRP tubing is installed.

Carbon Dioxide Injection

Secondary recovery by waterflooding can greatly increase the amount of oil recovered over primary production, but may still leave up to 80% of oil in place in the reservoir. Tertiary recovery by injecting CO_2 will remove the oil not obtained by waterflooding. Carbon dioxide can be used at much lower pressures than other gases, such as nitrogen or methane, because it dissolves readily in some crudes and can cause up to a tenfold viscosity reduction in heavy crudes.

Oils with an API gravity of 25 or higher are candidates for miscible flooding. This process can recover oil from low-permeability reservoirs. Oils with gravities down to API 15 are recovered by an immiscible process based on oil swelling and viscosity reduction.

Carbon dioxide injection uses gas from fields that produce almost pure CO_2 from burning of lignite and recovered CO_2 from industrial combustion gases. These gases are purified and compressed, and in some cases, they are pipelined for hundreds of miles to the fields to be flooded. The Texas Permian Basin, North Dakota, the Texas Gulf Coast, and the California area have had CO_2 injection projects in operation for several years.

Because CO_2 is an acid gas, production problems are encountered when CO_2 is injected. Carbon dioxide ionizes in water to form carbonic acid and will react directly with carbon steel. The corrosion rates can be quite high, and pumps can fail in a matter of days after breakthrough of the CO_2 into producing wells. Some scaling problems may arise because carbonic acid may dissolve calcium carbonate from the formation. The calcium bicarbonate formed during this reaction may come out of solution in heaters and vessels as calcium carbonate when CO_2 is lost. Calcium sulfate ($CaSO_4$) will also dissolve and may cause scaling in surface equipment.

Emulsion treating characteristics may change when CO_2 dissolves in oil. Asphaltenes may cause problems by dissolving in CO_2 as it sweeps through the formation and then coming out of solution on the surface.

Elastomers must be selected with care, because they may swell or lose strength when exposed to CO_2. Leaking packers due to seal failure will cause pressure on the annulus of CO_2 injection wells and annular space corrosion.

Carbon Dioxide Production Facilities

Carbon dioxide source wells may produce from a few percent to almost pure CO_2. They may produce both liquid and vapor phase CO_2. The presence of water in the produced CO_2 will cause hydrate formation and corrosion. Hydrate formation can be controlled by glycol dehydration, but special measures must be taken to control corrosion.

A corrosion inhibitor can be added to the producing well to control corrosion. Continuous injection downhole of a water-soluble filming amine inhibitor should protect the tubing and wellhead. The use of type 316L and 304L stainless steels and FRP for completions and flow lines is an alternative to inhibitor use.

In a typical CO_2 production facility, the gas travels through a wellstream heater to a contactor in which water is removed. It is then scrubbed, compressed, and sent to the pipeline. Materials selection in the design of the system is the key to corrosion control in the processing plant. Corrosion-resistant alloys can be used in areas of high corrosion, and carbon steel is used where conditions allow its use. A maximum water content of 60% of saturation is obtained by dehydration so that corrosion of the pipeline is prevented. Dehydration also prevents hydrate formation when temperatures are low.

Injection systems

Water and CO_2 are injected alternately in some systems, such as the SACROC unit in the Kelly Snyder field in west Texas. This is known as the water and gas process.

The distribution system consists of parallel separate lines for water and CO_2 that are carbon steel coated externally and cathodically protected. Carbon dioxide in the line contains less than 50 ppm water, so internal corrosion is minimal. Valves in the system range from bare carbon steel to plastic coated with type 316 stainless steel trim. Fluorocarbon and nylon O-rings have performed satisfactorily, and Buna N rubber is used for stem sealing, although these materials swell somewhat.

Water lines are cement lined with sulfate-resistant cements and artificial pozzolans, as specified in API RP-10E. Most leaks have been due to the failure of asbestos gaskets. The use of grout instead of gaskets has been effective. Water-soluble inhibitors are added to protect voids in cement linings and plastic coatings.

Carbon dioxide injection systems have suffered corrosion problems when the mixing of water and CO_2 at each cycle of alternate CO_2/water injection occurs. Plastic coating and type 316 stainless steel trim, ceramic gate valves with electroless nickel-coated bodies, and electroless nickel-coated check valves were tried. The type 316 stainless steel and ceramic gates performed well, but the other methods failed.

Injection wells originally used type 410 stainless steel wellheads and valves. Severe pitting occurred under deposits laid down from suspended solids in the injection water.

The type 410 stainless steel was plastic coated, and the gates and seats were changed to type 316 stainless steel to correct the problem.

Failures occurred in the couplings of the plastic-coated tubing in the injection wells when the seal rings failed. This was corrected by changing the coating on the couplings from an epoxy-modified phenolic to a polyphenylene sulfide.

Production Systems

Most failures in the SACROC unit were due to rod breakage. Inhibitor programs were satisfactory in some wells, but many did not respond. Plastic-coated rods and spray metal coating with type 316 stainless steel helped to alleviate the problem. The use of fiberglass rods in the upper 70% of the string, along with stainless steel coated rods on the bottom, reduced rod breakage to an average of 1.1 per well per year.

Tubing leaks can be controlled with coatings and the use of 9Cr-1Mo and 13%Cr steel tubing where inhibitors fail to control. Flow line corrosion can be controlled by the use of fiberglass-epoxy lines. Other systems have experienced problems similar to those found in the SACROC unit and have successfully controlled corrosion with the previously described methods. Inhibitor selection by field testing with linear polarization techniques has resulted in improved protection of producing wells. The linear polarization technique is also used in routine monitoring, along with coupons, iron counts, and caliper surveys.

Corrosion of Oil and Gas Offshore Production Platforms

Offshore structures have been in service in various parts of the world for over 40 years. Early experience was in the Gulf of Mexico with water depths of less than 90 m (300 ft). Technology has advanced to the point at which the largest drilling and production platform stands in more than 305 m (1000 ft) of water.

A platform consists of three parts. The jacket is a welded tubular space frame that is designed as a template for pile driving and as lateral bracing for the piles. The piles anchor the platform permanently to the sea floor and carry both vertical and lateral loads. The superstructure is mounted on top of the jacket and consists of the deck and trusses necessary to support operational and other loads. Generally, platforms are carried from the fabrication yard to the site on a barge and are either lifted or launched off the barge into the water. After positioning the jacket, the main piles are driven through the legs of the jacket, one through each leg. Other piles, known as skirt piles, can be driven around the perimeter of the jacket as needed.

Current design and fabrication practices related to fixed steel offshore structures can be found in industry publications, professional journals, and the proceedings of technical conferences. The most basic American document on this subject is API RP-2A (Ref 62), which was first issued in October 1969 and has had many subsequent editions. Because American experience has been mainly in the Gulf of Mexico, API RP-2A generally represents that experience.

General Corrosion

Marine structures operate in a complex environment that can vary significantly according to site location and water depth. Figure 16 shows the four main platform corrosion zones: soil, seawater, splash zone, and marine atmospheric.

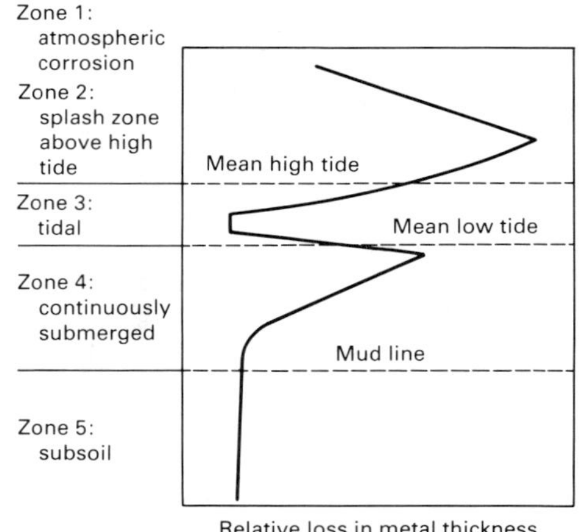

Fig. 16 Corrosion zones on fixed offshore structures. Source: Ref 63

Marine atmospheric corrosion problems occur on the portion of the jacket above the splash zone and on the superstructure. Exposed steel surfaces suffer corrosion from an environment of water condensation, rain, salt precipitation, sea mist, and oxygen. Corrosion rates can range from 0.05 to 0.64 mm/yr (2 to 25 mils/yr). Corrosion is particularly severe at crevices and sharp-edged areas, such as skip-welded plates and steel structural shapes. Attention to design and fabrication details can eliminate most of these problem areas. Atmospheric corrosion can be minimized by using coatings or by substituting nonferrous materials, such as copper alloys, nickel alloys, and FRP for steel components. Care must be taken not to create a galvanic-corrosion problem by coupling dissimilar metals. Table 4 gives a summary description of several marine zones and the characteristic behavior of the steel.

The splash zone is defined in NACE RP-01-76 (Ref 65) to be the area of the platform that is alternately in and out of the water because of tides, winds, and sea. It does not include surfaces that are only wetted during major storms. The splash zone of the platform can cover an interval of 1.5 to more than 12 m (5 to more than 40 ft), depending on location. Generally, the area of the platform that suffers the most severe steel corrosion is the splash zone, as shown in Fig. 17. Common methods of controlling corrosion in the splash zone include applying coatings, increasing jacket wall thickness by 6.4 to 19 mm (0.25 to 0.75 in.) in the splash zone to compensate for the higher corrosion rates, or applying a Monel alloy wrapper.

Corrosion of steel in seawater is a function of water salinity, temperature, oxygen content, velocity, resistivity, and chemistry. Table 5 summarizes the effects of these and other factors on the corrosion of steel in seawater. Several of the variables controlling corrosion are interrelated. As an example, Table 6 demonstrates the relationship between temperature and oxygen solubility in seawater. The lower the temperature, the higher the solubility. As temperature or oxygen levels increase, corrosion rates will increase. To control seawater corrosion, the steel jackets are normally cathodically protected. The types of cathodic protection systems used are sacrificial anode, impressed current, or a combination of the two (see the discussion "Cathodic Protection" in this article). Occasionally, cathodic protection will be used in combination with coatings. Not only does cathodic protection control corrosion but it also eliminates concern over a corrosion fatigue failure of the jacket. Corrosion fatigue is discussed in more detail below. Typical cathodic protection system design parameters are given in Tables 7 to 9.

The major platform components below the mudline are the jacket piles. In general, steel corrosion rates are low below the mudline. The exception is when the mud contains sulfate-reducing bacteria. Because the piles have electrical continuity with the jacket, the jacket cathodic protection system will normally protect the piles from corrosion in saline muds.

Fatigue

Corrosion is of particular concern for the platform tubular welded joints, called nodes. The nodes are areas of high stress due to their complex geometries (Ref 67-69). The points of maximum stress in the nodes occur at the toe of the welds joining the tubular members. Cyclic stresses result from environmental factors, such as waves, tides, and operating loads. Platforms are designed to handle both a maximum stress and fatigue. The maximum stress is usually based on 100-year storm conditions. Platform fatigue life is based on a environmental stress distribution analysis, along with analysis of the stress cycles (Ref 70). Fatigue design curves have been published by the American Welding Society, British Standards Institute, American Petroleum Institute, and Det Norske Veritas.

Corrosion can reduce the fatigue life of platform anodes. Galvanic corrosion of nonstress-relieved welds and pitting corrosion can result in stress raisers. Therefore, corrosion can lead to the initiation of cracks and can increase the growth rate of existing cracks, reducing fatigue life. The fatigue life of steel exposed to seawater is shorter than that of steel exposed to air because of corrosion fatigue. As Fig. 18 illustrates, steel immersed in seawater does not exhibit an endurance limit. Because there is no endurance limit, unprotected steel exposed to seawater is susceptible to fatigue failure even at low stress levels after long-term cyclic service. An API study discussed crack initiation in smooth, notched, and welded specimens and summarized a number of earlier investigations (Ref 72). Table 10 ranks according to importance the various seawater environmental variables influencing corrosion fatigue crack initiation of carbon steel. The corrosion fatigue effects are eliminated by the application of cathodic protection (Fig. 19).

Inspection

The purpose of periodic inspection is to ensure that the structure is fit for continued service. Through the inspection program, a company is protecting its personnel and assets. In some locations of the world, governmental bodies have established legislation or code agencies that determine minimum inspection requirements. Elsewhere, the operator decides on the minimum inspection needs for the platform.

Inspection is required even though platforms are designed and constructed to conservative codes (Ref 73). Inspection allows confirmation that the codes are adequate. It should be noted that the codes represent the best experience and knowledge at the time they are written. Often, the design parameters must be extrapolated for use in new frontier environments that were not

Table 4 Classification of typical marine environments

Marine zone	Description of environment	Characteristic corrosion behavior of steel
Atmosphere (above splash)	Minute particles of sea salt are carried by wind. Corrosivity varies with height above water, wind velocity and direction, dew cycle, rainfall, temperature, solar radiation, dust, season, and pollution. Even bird droppings are a factor.	Sheltered surfaces may deteriorate more rapidly than those boldly exposed. Top surfaces may be washed free of salt by rain. Coral dust combined with salt seems to be particularly corrosive to steel equipment. Corrosion usually decreases rapidly as one goes inland.
Splash	Wet, well-aerated surface; no fouling	Most aggressive zone for many metals, for example, steel. Protective coatings are more difficult to maintain than in other zones.
Tidal	Marine fouling is apt to be present to high watermark. Oil coating from polluted harbor water may be present. Usually, ample oxygen is available.	Steel at the tidal zone may act cathodically (well aerated) and receive some protection from the corrosion just below tidal zone in the case of a continuous steel pile. Isolated steel panels show relatively high attack in the tidal zone. Oil coating on surface may reduce attack.
Shallow water (near surface and near shore)	Seawater usually is saturated with oxygen. Pollution, sediment, fouling, velocity, and so on, all may play an active role.	Corrosion may be more rapid than in marine atmosphere. A calcareous scale forms at cathodic areas. Protective coatings and/or cathodic protection can be used for corrosion control. In most waters, a layer of hard shell and other biofouling restricts the available oxygen at the surface and thus reduces corrosion (increased stress on structure from the weight of fouling must be provided for).
Continental-shelf depths	No plant fouling, much less animal (shell) fouling with distance from shore. Some decrease in oxygen, especially in the Pacific, and lower temperature	...
Deep ocean	Oxygen varies, tending to be much lower than at the surface in the Pacific but not too different in the Atlantic. Temperature near 0 °C (32 °F). Velocity low; pH lower than at surface.	Steel corrosion is often lower. Anode consumption is greater to polarize the same area of steel as at the surface. There is less tendency for protective mineral scale formation.
Mud	Bacteria are often present, for example, sulfate-reducing types. Bottom sediments vary in origin, characteristics, and behavior.	Mud is usually corrosive, occasionally inert. Mud-to-bottom water corrosion cells seem possible. Partly embedded panels tend to be rapidly attacked in mud. Sulfides are a factor. Less current than in seawater is consumed to obtain cathodic polarization for the buried part of the structure.

Source: Ref 64

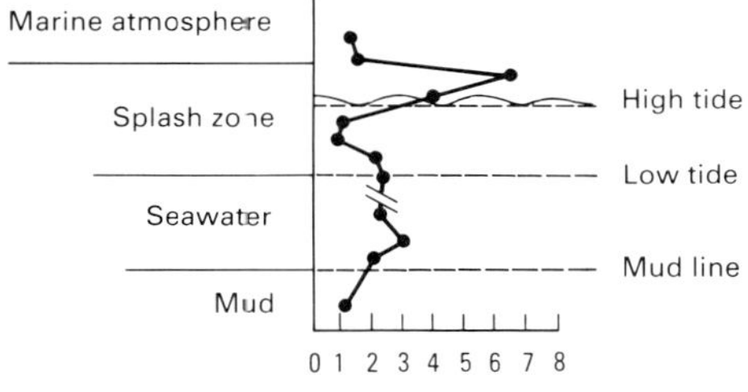

Fig. 17 Zones of corrosion for steel in seawater and the relative corrosion rate in each zone. Source: Ref 65

foreseen when the codes were written. Inspection results provide the information necessary for updating the codes to account for these new environments. Inspection of the platform jacket is designed to assess corrosion, fatigue cracks, joint and brace failure, impact damage, marine growth, scour, and debris accumulation. Inspection techniques include visual inspection by divers and remote operated vehicles, still and video photography, ultrasonic thickness measurements, cathodic potential surveys, magnetic-particle inspection, and vibration frequency attenuation.

Corrosion of Gathering Systems, Tanks, and Pipelines

Gathering systems are defined as all production facilities from the wellhead choke (or pumping T) to the sales point (oil and/or gas); subsystems include flow lines, separation, and dehydration. Gas processing will be reviewed briefly. Sulfur plants (conversion of H_2S to elemental sulfur), gas transmission lines and oil pipelines, gasoline plants, and water-handling facilities for disposal are beyond the scope of this discussion, although the same principles will apply. Internal and external corrosion alleviation systems will be reviewed, respectively.

Internal corrosion is dependent on the composition, temperature, pressure, and flow regime of the produced fluids. Although the general direction of the effect of each of these items is known, the magnitude of the effect is not precisely known, and the interrelationship of all these effects is not well understood. In short, corrosion alleviation is not an exact science (Ref 74).

A natural gas reservoir is a reservoir that, under initial conditions, is a single, gaseous hydrocarbon phase. If this gaseous phase contains hydrocarbons that are recoverable as liquids on the surface, the reservoir is a gas condensate reservoir. A gas well is a well that produces fluids from a gas or gas condensate reservoir. An oil well is a well that produces from a hydrocarbon reservoir that is either a two-phase system or a single liquid phase (Ref 75). Because produced fluids essentially determine the internal environment (weather conditions and process design can affect temperatures and pressures), gathering systems for gas wells and oil wells will be discussed separately.

From both a metallurgical and corrosion viewpoint, it is important to distinguish between sweet gas wells and sour gas wells. If the partial pressure of H_2S is greater than 0.34 kPa (0.05 psia), the gas stream is sour and materials that resist sulfide stress cracking must be used. The latest revision of NACE MR-01-75 lists materials that are recognized to have acceptable resistance to sulfide stress cracking.

The determination of whether H_2S or CO_2 corrosion mechanisms will predominate is not as simple. Early investigators believed that CO_2 had a synergistic effect on H_2S corrosion. Subsequent investigators indicate that it is the ratio of partial pressures of CO_2 to H_2S that controls the corrosion mechanism. According to studies by one researcher, unless the ratio of CO_2 to H_2S partial pressure is greater than 500, the corrosion mechanism is dominated by H_2S (Ref 76). More recent investigations suggest that at some temperatures when the ratio of CO_2 to H_2S partial pressure is plotted versus corrosion, there is a sharp discontinuity as the ratio is reduced when the controlling mechanism changes from CO_2 to H_2S, but not always at 500. At other temperatures, the data suggest that the change in corrosion mechanism from CO_2 to H_2S controlled as the ratio of partial pressure of CO_2 to H_2S is reduced is a continuous function, rather than the above step function (Ref 77).

Because CO_2 corrosion is usually more severe than H_2S corrosion at cool conditions and because most production facilities are relatively cool, CO_2 corrosion should be considered whenever the ratios of CO_2 to H_2S partial pressures are greater than 100. Oxygen is never present naturally in oil and gas reservoirs, and without exception, it is preferable to keep it out rather than to alleviate its corrosion effects.

Sweet Gas

With regard to CO_2 corrosion alleviation in flow lines, there are several choices. First, low-

Table 5 Factors that affect corrosion of carbon steel immersed in seawater

Factor	Effect on iron and steel
Chloride ion	Highly corrosive to ferrous metals. Carbon steel and common ferrous metals cannot be passivated. Sea salt is about 55% chloride.
Electrical conductivity	High conductivity makes it possible for anodes and cathodes to operate over long distances; therefore, corrosion possibilities are increased, and the total attack may be much greater than that for the same structure in freshwater.
Oxygen	Steel corrosion is cathodically controlled for the most part. Oxygen, by depolarizing the cathode, facilitates the attack; therefore, a high oxygen content increases corrosivity.
Velocity	Corrosion rate is increased, especially in turbulent flow. Moving seawater may destroy the rust barrier and provide more oxygen. Impingement attack tends to promote rapid penetration. Cavitation damage exposes fresh steel surfaces to further corrosion.
Temperature	Increased ambient temperature tends to accelerate attack. Heated seawater may deposit protective scale or lose its oxygen; either or both actions tend to reduce attack.
Biofouling	Hard-shell animal fouling tends to reduce attack by restricting access of oxygen. Bacteria can take part in the corrosion reaction in some cases.
Stress	Cyclic stress sometimes accelerates failure of a corroding steel member. Tensile stresses near yield also promote failure in special situations.
Pollution	Sulfides, which are normally present in polluted seawater, greatly accelerate attack on steel. However, the low oxygen content of polluted water could favor reduced corrosion.
Silt and suspended sediment	Erosion of the steel surface by suspended matter in the flowing seawater greatly increases the tendency toward corrosion.
Film formation	A coating of rust or rust and mineral scale (calcium and magnesium salts) will interfere with the diffusion of oxygen to the cathode surface, thus slowing the attack.

Source: Ref 64

Table 6 Solubilities of various gases in ocean water

Gas	Partial pressure in dry air, kPa	Partial pressure in dry air, atm	Solubility, mL/L·atm, 0 °C (32 °F)	Solubility, mL/L·atm, 24 °C (75 °F)	Equilibrium concentration in surface seawater, mL/L, 0 °C (32 °F)	Equilibrium concentration in surface seawater, mL/L, 24 °C (75 °F)
Helium	5.3×10^{-4}	5.2×10^{-6}	8.0	6.9	4.1×10^{-5}	3.4×10^{-5}
Nitrogen	79.1	0.781	18	12	14	9
Oxygen	21.2	0.209	42	26	8.8	5.5
Carbon dioxide	0.032	3.2×10^{-4}	1460	720	0.47	0.23

Source: Ref 66

alloy steel with a corrosion allowance can be used; a nomograph establishes the maximum corrosion rate for CO_2 (Ref 78). Velocity may also be important; several authors have suggested that there is a critical velocity above which CO_2 corrosion is very difficult to control (Ref 79). Again, more recent data suggest that at the same temperature, pressure, composition, and pH, the CO_2 corrosion-velocity relationship is a continuous relationship rather than a step function (Ref 77). If the flow lines are welded, the operator should be certain that the weldments are at least as corrosion resistant (and where H_2S is present, as cracking resistant) as the pipe body.

A second choice is to use corrosion-resistant materials, alloys, or coatings. With regard to CO_2, either type 316 stainless steel or duplex stainless steel will provide sufficient internal corrosion resistance. If H_2S is present, then NACE MR-01-75 must be followed. Type 316 stainless steel is subject to chloride SCC at elevated temperatures, and both alloys (type 316 stainless steel and duplex) may be subject to external pitting corrosion or crevice corrosion. Both alloys also require special care from the time they are installed to the time they are put into service. Oxygen and perhaps bacteria will result in pitting corrosion. Metallurgical solutions, if properly executed, result in permanent low-maintenance corrosion alleviation systems.

A third choice is to internally line low-alloy steel pipelines wth a corrosion-resistant material. These systems may have advantages over solid corrosion-resistant alloys. First, they may be less expensive, and second, the alleviation of external corrosion problems of low-alloy steel is well understood (years of history with large quantities of pipelines) and less sophisticated. There are two disadvantages. Special welding procedures are required, and when the metallurgical coating is not bonded, buckling of the liner may occur, particularly in bends. This buckling will inhibit the use of tools pumped through the flow line.

Two other internal coatings are also commonly used: organic polymers (plastic coatings) and cement linings. Both systems can be economic successes. However, both systems have difficulty in maintaining corrosion resistance at the joints, are difficult to install holiday-free, have limited life, and may suffer disbondment and failure when the pipe is improperly handled or distorted. The plastic coatings are permeable to the produced fluids, and eventually (3 to 5 years), the corrosive fluids will permeate the coating. The resulting corrosion products cause disbonding of the coating and complete loss of its corrosion resistance. Cement linings are thicker and less subject to produced fluid penetration than plastic coatings. They are heavy and may have limited resistance to acids.

A final alternative is to use nonmetallic pipe materials such as FRP or polymerized hydrocarbons. The advantage of these materials is their complete resistance to corrosion. Their disadvantages are low allowable temperatures, low fatigue resistance, low strength, low resistance to mechanical damage, and high combustibility. They also have problems with joint integrity. Finally, they may be vulnerable to attack from the produced fluids (CO_2 may dissolve the resins from fiberglass pipe, and unsaturated hydrocarbons may dissolve the polymerized hydrocarbon pipes).

An alternative to corrosion-resistant materials is to use corrosion inhibition. If corrosion inhibition is used to protect the gas well downhole tubulars, the same system will protect the flow lines if the flow lines are properly designed. The flow lines should be sized to ensure turbulent flow at a velocity that is not significantly higher than that present in the tubing. Turbulent flow will ensure inhibitor contact with the entire internal pipe surface, and limited velocities (flow lines are cooler than gas well tubulars) will ensure that flow line conditions are not significantly more corrosive than the gas well tubulars. Because the economic consequences of downhole tubular failures are usually much greater than flow line failures, designing the surface flow lines to utilize downhole inhibition systems effectively will result in a successful flow line inhibition system. If downhole inhibition is not used, then surface inhibition can be used to protect the flow lines. In low velocity/low liquid flow regimes, periodic inhibition with a technique that inhibits the entire internal surface can be successful. For higher velocities (turbulent flow) and/or high liquid content, continuous inhibition will probably be necessary.

Separation and dehydration facilities offer fewer alternatives. Because inhibitors usually stay in the liquid phase in multiphase systems and usually in the hydrocarbon phase if one is present, reliable inhibition of the vapor space is impossible. Fortunately, corrosion is usually not severe, because only water of condensation is present. For very severely corrosive environments, such as wet CO_2, type 316 stainless steel or type 316 stainless steel internal cladding is usually used. For normal gas production with CO_2, low-alloy steel, combined with a corrosion allowance and monitoring, is used.

For gas dehydration systems, low-alloy steel with corrosion allowance is normally sufficient when combined with pH control of the glycol. For gas streams high in CO_2 concentrations, some operators have found it necessary to internally clad the wet gas portions with type 316 stainless steel.

Sour Gas

Sour gas wells present a more difficult problem. First, all materials must be resistant to sulfide stress cracking. It is essential that no equipment suffer catastrophic cracking failures. General corrosion is less severe for H_2S than for CO_2 at the lower temperatures usually encountered in flow line systems, and H_2S corrosion is not velocity dependent. As long as sufficient liquid hydrocarbon is present, corrosion is usually minor. When hydrocarbon liquids exceed 100 barrels per 28 300 m^3 (1 000 000 ft^3) of gas, when the water content of the liquid phase is less than 10%, and when the flow is turbulent (to avoid

Table 7 Design criteria for offshore cathodic protection systems

Production area	Water resistivity, Ω·cm(a)	Environmental factors(b): Water temperature °C	°F	Turbulence factor (wave action)	Lateral water flow	Typical design current density(c) mA/m²	mA/ft²
Gulf of Mexico	20	22	70	Moderate	Moderate	54–65	5–6
U.S. West Coast	24	15	60	Moderate	Moderate	76–106	7–10
Cook Inlet	50	1	35	Low	High	380–430	35–40
North Sea(d)	26–33	0–12	32–55	High	Moderate	86–216	8–20
Persian Gulf	15	30	85	Moderate	Low	54–86	5–8
Indonesia	19	24	75	Moderate	Moderate	54–65	5–6

(a) Water resistivity is a function of chlorinity and temperature, and it decreases as both chlorinity and temperature increase. (b) Typical values and ratings based on average conditions, remote from river discharge. (c) In ordinary seawater, a current density less than the design value will suffice to hold the platform at protective potential once polarization has been accomplished and calcareous coatings are built up by the design current density. It should be noted, however, that depolarization can result from storm action. (d) Conditions in the North Sea can vary greatly from the northern to the southern area, from winter to summer, and during storm periods. Source: Ref 65

Table 8 Energy capabilities and consumption rates of sacrificial anode materials in seawater

Anode material	Energy capability(a) A·h/kg	A·h/lb	Consumption rate kg/A·yr	lb/A·yr	Anode to water(b) closed circuit potential, V versus Ag/AgCl reference electrode
Aluminum-zinc-mercury	2750–2840	1250–1290	3.1–3.2	6.8–7	−1 to −1.05
Aluminum-zinc-indium	2290–2600	1040–1180	3.4–3.8	7.4–8.4	−1.05 to −1.1
Aluminum-zinc-tin	925–2600	420–1180	7.4–20.8	16.3–45.9	−1 to −1.05
High-purity zinc	780–815	354–370	10.7–11.2	24.8–23.7	−1 to −1.05
Magnesium alloy H-1	1100	500	8.0	17.5	−1.4 to −1.6

(a) Data are ranges taken from field tests conducted by the Naval Research Laboratory at Key West, FL and from manufacturers' long-term field tests. (b) Measured potentials can vary because of temperature and salinity differences. Source: Ref 65

Table 9 Consumption rates of impressed-current anode materials

Material	Typical anode current density in saltwater service A/m²	A/ft²	Nominal consumption rate g/A·yr	lb/A·yr
Pb-6Sb-1Ag	160–220	15–20	15–86	0.03–0.2(a)
Pb-6Sb-2Ag	160–220	15–20	13–25	0.03–0.06(a)
Platinum (on titanium, niobium, or tantalum substrate)	540–3200	50–300	3.6–7.3	0.008–0.016(b)
Graphite	10–40	1–4	230–450	0.5–1.0
Fe-14.5Si-4.5Cr	10–40	1–4	230–450	0.5–1.0

(a) Very high consumption rates of lead-silver anodes have been experienced at depths in excess of 30 m (100 ft). (b) This figure can increase when current density is extremely high and/or in waters of low salinity. Source: Ref 65

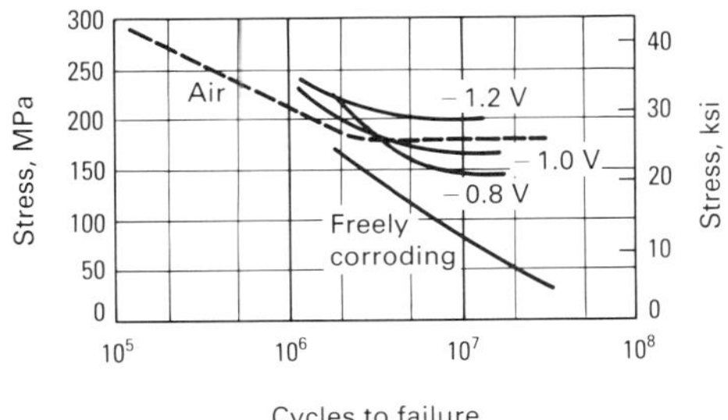

Fig. 18 Fatigue data for carbon steel in seawater as a function of specimen potential. Source: Ref 71

water phase in the bottom of the line), no serious corrosion problem should be expected. In cases in which the above conservative criteria are not met, inhibition may be necessary. The same rationale in regard to the use of inhibitors in flow lines with CO_2 applies to flow lines for H_2S. Inhibition systems that protect downhole tubulars will protect the flow lines if the entire internal wall is inhibited. Processing facilities, gas separators, and sweeting systems (sulfinol, amines, and so on) are usually constructed of low-alloy steel with a corrosion allowance and are regularly monitored.

Oil Wells

Oil wells, as previously defined, are wells that produce from a hydrocarbon reservoir that is either two-phase or a single liquid phase. Oil well flow line corrosion is much easier to handle than gas well flow line corrosion. First, sour oil wells are usually beyond the scope of NACE MR-01-75. Second, CO_2 corrosion is seldom, if ever, a problem. Most crude oils (and associated gas) either contain no CO_2 or sufficient H_2S for H_2S to be the controlling corrosion mechanism. Corrosion problems are often minor until water-cuts approach 30%. Normally, when water-cuts are high enough for saltwater (with or without H_2S) corrosion to be a problem in the flow lines, corrosion will be more severe on the downhole tubulars, and the inhibition system protecting the tubulars will protect the flow lines. Flow lines with stratified flow that allow free water to flow or stagnate in the bottom of the flow lines may suffer corrosion problems. In these cases, inhibitors in the oil phase from the well may not help. This problem is successfully handled by frequent pigging in order to clear the flow lines of water; to clean out sediment, which will foster crevice corrosion; and to distribute inhibitor over the entire internal surface.

Gas separation facilities and free water knock outs are usually made of low-alloy steel. Free water knock outs are often internally coated with an organic coating. Coatings, if properly applied to clean dry surfaces, may extend the vessel life a few years. Generally, coatings do not remain holiday-free for very long in saltwater service. The only other alternative to low-alloy steel with corrosion allowance and internal coating is a corrosion-resistant material. The low corrosivity, however, does not justify the cost of a corrosion-resistant alloy, either solid or internally clad. The combination of size and pressure usually eliminates the use of materials such as fiberglass.

Corrosion problems can be severe in saltwater-handling facilities, although they are usually mild to moderate. Corrosion alleviation systems are limited to cathodic protection, organic coatings, and, occasionally, nonmetallic vessels.

Crude oil is usually dehydrated by using gravity separation of the lighter oil from the unwanted water. Heat, chemicals, and electric fields are often used to accelerate the gravity separation. The separation vessel should not experience severe corrosion if the system is kept oxygen-free. Cathodic protection is only partially successful for the vessel, because only the water-wet portion is protected and the oil/water interface fluctuates. The heating coils often suffer more severe corrosion than the vessel body. As long as these coils are in the water portion of the vessel, well designed and maintained cathodic protection systems are successful in alleviating corrosion of the heating surfaces. When large tanks are used, galvanizing significantly prolongs the tank life in the absence of H_2S. Hot-dip galvanizing is considered to be the most effective treatment, but it is limited to bolted tanks. The bolted tank gaskets, in turn, limit the temperatures at which the vessel can be operated.

The gas separated from the oil is low pressure and can be safely handled with low-alloy steel with a corrosion allowance. The storage of dehydrated oil (usually about 1% H_2O) poses no internal corrosion problems, except in the tank bottom, where salt water will accumulate.

External Corrosion

External corrosion can also be a serious and costly problem. In wet or corrosive soil, low-alloy steel flow lines should be coated. There are a variety of successful coating systems. Usually, the external coating system consists of two parts: a mastic that coats and protects the pipe and a coating or wrapping that protects the mastic. If the flow lines are needed for long periods of time, then the external coating should be supplemented

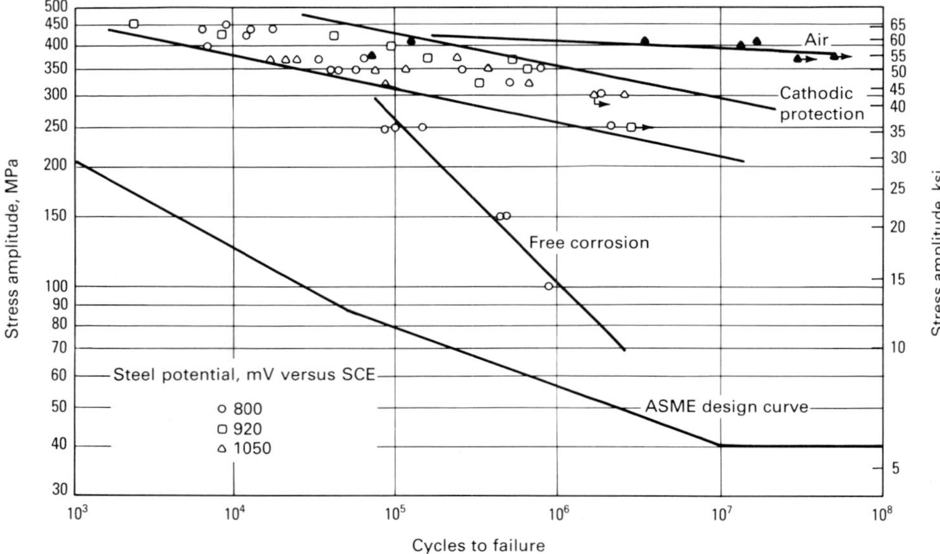

Fig. 19 Effect of cathodic protection on the fatigue performance of alloy steel in seawater. Tests performed on 6.4-mm (1/4-in.) diam specimens at a mean stress of 425 MPa (69 ksi).

Table 10 Summary of major variables influencing the corrosion fatigue crack initiation behavior of carbon steels in seawater

Variable	Effect
Cyclic frequency	Slower frequencies cause reduced fatigue resistance for unprotected steel.
Cathodic potential	Adequate cathodic protection restores fatigue resistance to levels observed in air.
Oxygen level	Fatigue resistance in deaerated seawater is similar to that in air.
Temperature	Although decreasing temperature results in increased oxygen levels, the overall effect of decreasing temperature is that fatigue resistance is improved to some extent in range of 13 to 45 °C (55 to 115 °F).
pH level	Over a broad range of values (4 to 10), there is little effect of pH on fatigue resistance. Low values (<4) decrease fatigue resistance, and high values (>10) improve fatigue resistance to levels similar to those observed in air.

Source: Ref 72

with cathodic protection. Surprisingly, it is probably more important to cathodically protect corrosion-resistant alloy flow lines than low-alloy steel flow lines. The corrosion-resistant alloys—type 316 stainless steel or duplex stainless steel—are subject to crevice corrosion in the presence of oxygen. Because these stainless steel lines are much more expensive than low-alloy steel lines, they cannot be allowed to fail by external corrosion. Therefore, external cathodic protection of these flow lines is essential when they are buried or submerged. To reduce the current level required for protection, the corrosion-resistant alloy, like the low-alloy steel pipeline, is usually externally coated.

The other components of the gathering system are the storage tanks, gun barrels, and surge tanks. As has been reviewed earlier, these vessels all internally accumulate salt water on the bottom. Similarly, these tank bottoms are all subject to external corrosion. Because the tank bottoms are relatively thin and may suffer internal corrosion, they should be protected from external corrosion. Therefore, in wet environments or where long service is needed, cathodic protection of tank bottoms should be considered. For piping and vessels above ground, painting is the accepted method of protecting against corrosion.

Monitoring

Monitoring is an essential part of any corrosion alleviation system. No corrosion alleviation system is completely reliable, and in many cases failure can be catastrophic, both from a personal safety perspective and from an environmental and/or economic perspective. There are a wide variety of inspection techniques. The thoroughness and frequency of the monitoring must be weighed against the consequences of failure, and the type of monitoring must be tailored to the particular system. Certainly, it is more catastrophic to have a high-pressure sour gas vessel failure than an atmospheric saltwater tank failure.

Flow lines can be monitored with calipers that are pumped through the line, x-rayed, or, where warranted, cut open and inspected. Vessels can be visually and/or ultrasonically inspected. When ultrasonic insection is used, reference points are usually permanently fixed to the vessel external wall so that the ultrasonic test is conducted at the same location each time.

Storage of Tubular Goods

Tubular goods used in oil-producing and drilling operations are sometimes stored outdoors or in areas where internal storage is conducive to corrosion. This is particularly true in coastal regions and industrial areas in which acid gases and pollutants are present as well as in oil fields that produce hydrogen sulfide.

Oxygen corrosion, or rust, is aggravated by the deposition of salt from marine environments, such as that encountered in wind-driven spray on offshore platforms, and airborne salt in coastal areas. Pipe yards situated close to the beach are particularly susceptible to severe atmospheric corrosion. In warehouses and under sheds, the presence of industrial pollutants such as SO_2, oxides of nitrogen, and other gases will initiate corrosion attack and increase rusting when they react with moisture in the air or on the pipe.

Even in areas of relatively low salt and pollution content, severe corrosion may occur if the relative humidity is high. The pipe will cool off during the night, and dew will fall, covering the pipe with a conductive layer of electrolyte. The rust that is already on the pipe is hygroscopic and will remain moist after the free water has evaporated. This leads to concentration cell attack and severe pitting. Pit depth and size are of particular importance, because failures may occur when the pipe is put into service under pressure.

Inspection

Before being put into service, tubular goods that have been stored for any length of time should be inspected. Particular attention should be given to the following areas:

- *External rusting:* The percent of surface area covered by rust should be recorded
- *Presence of mill scale, lacquer-type mill coatings, or other temporary coatings:* This is important, because areas not covered may corrode, while protected areas may set up concentration cells and accelerate localized corrosion
- *Internal corrosion:* The interior of the pipe should be inspected for rusting and pitting
- *Condition of threads:* Threads should be examined to determine if corrosion damage has occurred that could prevent proper makeup

Cleaning

After inspection, the pipe should be cleaned prior to applying any protective coatings. The pipe can be cleaned mechanically or with acids or rust dissolvers.

If the pipe is used, or has been stored in marine or industrial environments, it should be water blasted to remove any salt or acid deposits, weathered for a period of time to allow underdeposit salts to migrate to the surface, and water blasted again. One month is usually sufficient weathering time.

A water-soluble phosphate-base temporary rust inhibitor can be applied during the weathering period to minimize further rusting. The inside of the pipe should also be protected because water may collect on the bottom and cause pitting.

The pipe can now be physically cleaned. Wire brushing with an automated machine is a preferred method, but a rotary hand-held wire brush can be used if a machine is not available. If rusting and pitting are severe, the pipe should be sand blasted. The inside of the pipe can be cleaned with a mechanical rattler or a round brush.

An alternative cleaning method for lightly rusted pipe or where mechanical cleaning facilities are unavailable is the use of an acid-base rust remover/chelant. These are usually based on phosphoric acid and will contain inhibitors and passivators to prevent removal of metal. The pipe can be soaked in a trough of the inhibited acid cleaner until deposits are dissolved.

Application of Protective Coatings

Once the pipe is cleaned, temporary rust-preventive coatings can be applied to halt further rusting. They can be applied by the automatic machine that cleans the pipe or by hand sprayers, dipping, or brushing. The important factor is that the pipe is completely covered with the coating. Temporary coatings are manufactured in several types, as discussed below.

Lacquer coatings may consist of an oil-soluble resin in a volatile hydrocarbon solvent. These coatings may be brittle and may flake off and expose bare metal.

Slushing compounds usually consist of asphalt dissolved in a nonevaporating hydrocarbon so that a thick, oily layer is present on the surface. These compounds are resistant to mechanical damage, but may be difficult to apply and remove.

Polymeric coatings may contain acrylics, chlorinated vinyl chlorides, or other materials that will dry and polymerize from an aqueous solution. The coating is similar to that found when floor wax is applied. These coatings may contain metal passivators and rust converters that will be anodic to the metal surface and greatly alleviate further rusting. Polymeric coatings are easy to apply and are sometimes used without precleaning the surface, because the converters and passivators are designed to modify corrosion by-products to eliminate concentration cell corrosion.

Sulfonate-base coatings are formulated from petroleum sulfonates, waxes, and materials that form a flexible and nondrying coating. Pigments and fillers can be added for appearance and durability. The sulfonate acts as a rust inhibitor, and the other materials seal the surface to prevent moisture entry. These are relatively easy to apply and are resistant to mechanical damage.

Rust passivators are of the same type as the phosphoric acid base rust removers, or may be similar to the polymeric coatings. They are used for rust prevention under sheds or in warehouses and may not be adequate for outdoor storage in corrosive areas

The application method used will depend on the type of coating selected. A hydrocarbon-base coating can be brushed or sprayed, or the pipe can be dipped in a trough containing the coating. Dilution with naphtha, aromatic solvents, or diesel oil can be done, depending on the recommendations of the manufacturer. A hydrocarbon-base coating should not be applied to wet or damp pipe. Some coatings are reported to be able to displace water from a metal surface, but care should be taken that the surface is not too wet.

A water-base coating will probably be applied full strength. It can also be sprayed or brushed. The recommendations of the manufacturer should be followed to ensure proper coverage.

A suitable thread-protecting compound should be applied and thread protectors screwed onto the threads before the pipe is coated. It may be desirable to coat the inside of the pipe with an oily coating and to use pipe caps to prevent water entry during storage.

If the coating is air sprayed, the pipe should be coated one layer at a time, making certain that the spray is adjusted to provide coverage without excessive loss. The pipe should be rolled, and the other side covered. The use of wood spacers between layers of pipe will allow any moisture to drain and dry during subsequent outdoor storage. Once the pipe is coated, sufficient drying time should be allowed before the pipe is moved.

Continuing Maintenance

The pipe should be inspected at preselected intervals to ensure that the coating is performing satisfactorily. The program can be modified, or the pipe can be recoated when needed. Complete records should be maintained so that corrosion prevention is an on-going and effective process.

Industry Standards

Donald S. Burns
Spraymetal, Inc.
James W. Johnson
WKM Division of Cooper Industries

This section will discuss the applicable standards for metallic materials used in critical environmental conditions that may cause resistant materials to behave abnormally. Selection of materials for drilling, completion, production, and transportation of oil and gas is covered in several standards. In some cases, one standard may override another that addresses the same product. The following standards will be discussed: NACE MR-01-75, API 6A and 6D, ASME section IX and API 1104 (welding), API 14D, and ANSI standards.

Materials and Design Specifications

NACE MR-01-75 (1978 revision) and all later revisions cover the requirements for metallic materials used in the drilling and completion of oil and gas wells and the production of oil and gas that contain hydrogen sulfide (H_2S). Hydrogen sulfide is highly toxic; concentrations as low as 1000 ppm can cause death. The safe use of H_2S is discussed in NACE TM-01-77.

In addition to the necessary safety considerations, oil and gas that contain H_2S can cause materials that are normally strong and ductile to fail in a sudden and brittle manner at very low stress levels. The special metallurgical requirements for metallic materials in this environment are discussed in NACE MR-01-75.

In general, all carbon and low-alloy steels that are properly heat treated to a maximum hardness of 22 HRC are acceptable for use in H_2S service with two exceptions: free-machining steels and steels containing more than 1% Ni. Laboratory tests have indicated that these materials can fail at hardnesses below 22 HRC.

As their hardnesses exceed 22 HRC, carbon and low-alloy steels become progressively more susceptible to failure. Laboratory tests have indicated almost instantaneous failures at stresses considerably below yield for hard materials; therefore, all carbon and low-alloy steels can be made susceptible to sulfide stress cracking by improper thermal treatment and/or by mechanical damage (cold work) due to handling problems.

In addition to carbon and low-alloy steels, NACE MR-01-75 discusses stainless and other high-alloy steels. Some of these steels can be used at hardnesses above 22 HRC when additional strength is needed. These materials should be selected with care.

API 6A and 6D present design and material requirements for oil and gas production and pipeline equipment. The API standards usually contain references to ASTM, AISI, ASME, or other general specifications for acceptable materials for these products and always include a reference to NACE MR-01-75 when the products are to be used in H_2S service. The engineer must be aware that in certain cases the NACE standard may overrule API or other standards and that under these conditions design considerations must be addressed.

Recent concern about consistency in quality and product performance prompted API to rewrite API 6A. The new standard was issued in April 1986, and it requires manufacturers of API equipment to produce and maintain a documented control system that can be audited. This control system must also be capable of providing engineering, manufacturing, and quality guidelines for the consistent production of products meeting the requirements of the revised edition. Before approval, each manufacturer certified will be audited by an independent API-approved auditing team.

The new document also introduces four product specification levels, which can be used to specify the desired quality level. In addition to the quality levels, API has added various environmental conditions. These enable the end user to be very specific in terms of product end use, the expected quality level, and product performance.

ASME section IX and API 1104 discuss the welding of pressure vessels and line pipe, including the requirements for qualifying welding procedure, welders, and the quality level of the production weldments. These standards also identify the variables that dictate when different procedures are required to cover material thickness, material chemistry, joint design, welding position, and thermal treatments. If weldments are to be used in H_2S environments, such factors as additional hardness testing, the type of welding electrode, and the wire/flux combination used should be considered.

The American Petroleum Institute and the National Association of Corrosion Engineers publish documents that indicate whether weldments made with certain wire/flux combinations have experienced failure in H_2S environments under conditions in which failures were unexpected. Weldments with high manganese and silicon contents have been shown to be very susceptible to sulfide stress cracking at hardnesses at or below 22 HRC even after thermal treatments.

If temperature is an environmental consideration, the designer should be aware that the ability of materials and weldments to withstand sudden impact loading decreases significantly with temperature. Materials and weldments that have ductile characteristics at room temperature (20 °C, or 70 °F) may behave in a brittle manner when exposed to lower temperatures. In addition to the typical mechanical-property testing of materials and weldments, impact testing is required to ensure that the product is properly designed for the anticipated operating temperatures. Standards from ASTM, ASME, and others address these design requirements and generally specify materials that have Charpy V-notch impact test requirements added to the normal mechanical-property tests.

In the case of weldments, Charpy V-notch impact testing of the base metal, the heat-affected zone, and the weld metal is also required to ensure that the weldment will function satisfactorily at the design temperature. Welding must be performed with care to ensure that the completed weldment is produced in accordance with standard procedures.

API 14D establishes requirements for wellhead surface safety valves (SSV valves), underwater safety valves (USV valves), and their actuators (SSV/USV actuators). The requirements include design, material quality, performance testing, and functional testing.

Equipment manufactured to this specification is primarily intended for service in outer continental waters (OCS) contiguous to the United States. The equipment is identified with the API 14D monogram. It is also eligible for the OCS monogram when manufactured under a quality program conforming to ANSI/ASME-SPPE-1 specifications. The federal government has mandated that all such equipment installed in federal OCS waters have both monograms.

The ANSI/ASME-SPPE-1 standards establish requirements for quality programs, accreditation of quality programs, and reporting of malfunctions and failures. Manufacturers conforming to these standards are authorized by ASME to apply the OCS monogram to SSVs (valve and actuator).

REFERENCES

1. R.N. Tuttle and R.D. Kane, Ed., *H_2S Corrosion in Oil & Gas Production—A Compilation of Classic Papers*, National Association of Corrosion Engineers, 1981
2. L.E. Newton, Jr. and R.H. Hausler, Ed., *CO_2 Corrosion in Oil and Gas Production—Selected Papers, Abstracts and References*, National Association of Corrosion Engineers, 1984
3. C.C. Nathan, Ed., *Corrosion Inhibitors*, National Association of Corrosion Engineers, 1973
4. *Corrosion Control in Petroleum Production*, TPC Publication 5, National Association of Corrosion Engineers, 1979
5. "Sulfide Stress Cracking Resistant Metallic Material for Oil Field Equipment," NACE MR-01-75 (Latest Revision), Material Requirement, National Association of Corrosion Engineers
6. J.B. Sardisco and R.E. Pitts, Corrosion of Iron in an H_2S-CO_2-H_2O System—Composition and Protectiveness of the Sulfide Film as a Function of pH, *Corrosion*, Nov 1965, p 350-354
7. R.A. King and D.S. Wakerley, Corrosion of Mild Steel by Ferrous Sulfide, *Br. Corros. J.*, Vol 8, Jan 1973, p 41
8. R.A. King, J.D.A. Miller, and J.S. Smith, Corrosion of Mild Steel by Iron Sulfides, *Br. Corros. J.*, Vol 8, 1973, p 137
9. J.S. Smith and J.D.A. Miller, Nature of Sulfides and Their Corrosive Effect on Ferrous Metals: A Review, *Br. Corros. J.*, Vol 10 (No. 3), 1975, p 136
10. P.R. Rhodes, Corrosion Mechanism of Carbon Steel in Aqueous H_2S Solutions, Abstract 107, in *Extended Abstracts*, Vol 76-2, The Electrochemical Society, 1976, p 300
11. M.C. Place, Jr., "Corrosion Control—Deep Sour Gas Production," Paper presented at the 54th Annual Fall Technical Conference and Exhibition of the Society of Petroleum Engineers of AIME, Las Vegas, NV, Society of Petroleum Engineers, Sept 1979
12. C. deWaard and D.E. Milliams, "Prediction of Carbonic Acid Corrosion in Natural Gas Pipelines," Paper F1, Presented at the First International Conference on the Internal and External Protection of Pipes, BHRA Fluid Engineering, University of Durham, Sept, 1975
13. A.K. Dunlop, H.L. Hassell, and P.R. Rhodes, "Fundamental Considerations in Sweet Gas Well Corrosion," Paper 46, presented at Corrosion/83, Anaheim, CA., National Association of Corrosion Engineers, April 1983
14. D.R. Fincher, J.J. Marr and J.W. Ward, Paper 7, presented at Corrosion/75, Toronto, Canada, National Association of Corrosion Engineers, April 1975
15. S.P. Ewing, Corrosion by Stray Current, in *Corrosion Handbook*, H.H. Uhlig, Ed., John Wiley & Sons, 1948, p 601-606
16. H.H. Uhlig, *Corrosion and Corrosion Control*, John Wiley & Sons, 1963
17. M.G. Fontana and N.D. Greene, *Corrosion Engineering*, 1st ed., McGraw-Hill, 1967; 2nd ed., 1978
18. J.W. Oldfield and W.H. Sutton, Crevice Corrosion of Stainless Steels—I. A Mathematical Model, *Br. Corros. J.*, Vol 13 (No. 1), 1978, p 13-22
19. J.W. Oldfield and W.H. Sutton, Crevice Corrosion of Stainless Steels—II. Experimental Studies, *Br. Corros. J.*, Vol 13 (No. 13), 1978, p 104-111
20. J.W. Oldfield, Crevice Corrosion of Stainless Steels—The Importance of Crevice Geometry and Alloy Composition, *Métaux-Corros.-Ind.*, Vol 56 (No. 668), April 1981, p 137-147
21. R.H. Hansler, Ed., *Advances in CO_2 Corrosion*, Vol I, National Association of Corrosion Engineers, 1985
22. Solar Energy Tapped for Cathodic Protection of Casing, *Oil Gas J.*, Oct 1980, p 113
23. J. Leavenworth, Solar Powered Cathodic Protection for Saudi Arabian Oilfields, *Mater. Perform.*, Dec 1984, p 21
24. G.W. Curren, Sun Powers Libya Cathodic Protection System, *Oil Gas J.*, March, 1982, p 177
25. J. Evans, Gas Utility Uses Sun Power to Cathodic Protect Gas Mains, *Pipe Line Ind.*, Sept 1984, p 23
26. N.S. Christopher, Cathodic Protection Power Source Designed for Remote Locations, *Pipe Line Ind.*, Oct 1985, p 47
27. R.S. Treseder, Ed., *Corrosion Engineer's Reference Book*, National Association of Corrosion Engineers, 1980
28. M.T. Chapman, Control of External Casing Corrosion, *Mater. Prot. and Perform.*, Sept 1973, p 10
29. A.W. Peabody, *Control of Pipeline Corrosion*, National Association of Corrosion Engineers, 1976, p 105
30. T. Allen and A.P. Roberts, *Production Operations*, Vol 2, Oil & Gas Consultants International, Inc., 1982
31. W. F. Gast, "Has Cathodic Protection Been Effective in Controlling External Casing Corrosion for Sun Exploration & Production Co.? A 20 Year Review Tells the Story!," Paper 151, presented at Corrosion/85, Boston, MA, National Association of Corrosion Engineers, March 1985
32. A.G. Ostroff, Understanding and Controlling Corrosion, in *Corrosion Control Handbook*, Petroleum Engineering Publishing, 1975
33. C.C. Patton, *Oilfield Water Systems*, Campbell Petroleum Series, 1981
34. *Corrosion Control in Petroleum Production*, TPC Publication 5, National Association of Corrosion Engineers, 1979, p 60
35. F.W. Schremp, Corrosion Prevention for Offshore Platforms, *J. Petrol. Technol.*, April 1984, p 609
36. C.E. Hedborg, "Corrosion in the Offshore Environment" OTC Paper 1958, presented at Offshore Technology Conference, Houston, TX, May 1974
37. J. Davis, E.P. Doremus, and R. Pass, "Worldwide Design Considerations for Cathodic Protection of Offshore Facilities Including Those in Deep Water," OTC Paper 2306, presented at Offshore Technology Conference, Houston, TX, May 1975
38. M. Schumacher, Ed., *Seawater Corrosion Handbook*, Noyes Data Corporation, 1979, p 78
39. L. Coker, "Some of the Things You Always Wanted to Know About Corrosion Inhibitors But Didn't Ask," Paper presented at the Permian Basin Meeting, National Association of Corrosion Engineers, 1978
40. *Corrosion Control in Petroleum Production*, TPC Publication 5, National Association of Corrosion Engineers, 1979, p 47
41. *Corrosion Control in Petroleum Production*, TPC Publication 5, National Association of Corrosion Engineers, 1979, p 50
42. H.J. Endean, "Procedures for Evaluating Corrosion and Selecting Treating Methods for Oil Wells," Corrosion Control Course, The University of Oklahoma, 1977
43. *Corrosion Control in Petroleum Production*, TPC Publication 5, National Association of Corrosion Engineers, 1979, p 49
44. R.L. Steelman, Use of Corrosion Inhibitors in Offshore Gas Pipeline Protection, *Oil Gas J.*, Oct 1980, p 154
45. L.W. Gatlin and H.J. Endean, "Water Distribution and Corrosion in Wet Gas Transmission Systems," Paper 174, presented at Corrosion/75, National Association of Corrosion Engineers, 1975
46. P.D. Schrickel, Plastic Pipe Meets Gas Industry Needs, *Pipe Line Ind.*, Oct 1984, p 19
47. G.L. Davis, "Selection and Use of Nonmetallic Pipe," Corrosion Control Course, The University of Oklahoma, 1977
48. R.F. Weeter, Desorption of Oxygen From Water Using Natural Gas for Countercurrent Stripping, *J. Petrol. Technol.*, May 1965, p 515
49. H.G. Byars and B.R. Gallop, Injection Water + Oxygen = Corrosion and/or Well Plugging Solids, *Mater. Perform.*, Dec 1974
50. W.J. Frank, "Efficient Removal of Oxygen in a Waterflood by Vacuum Deaeration," SPE Paper 4064, Oct 1972
51. D.C. Scranton, Practical Applications of Oxygen Scavengers in the Oilfield—A Review,

Mater. Perform., Sept 1979, p 47
52. R.F. Weeter, Conditioning of Water by Removal of Corrosive Gases, *J. Petrol. Technol.*, Feb 1972, p 182
53. J.T.N. Atkinson and H. VanDroffelaar, chapter 6, in *Corrosion and Its Control: An Introduction to the Subject*, National Association of Corrosion Engineers, 1982
54. "Collection and Identification of Corrosion Products," NACE RP-01-73, National Association of Corrosion Engineers,
55. C.P. Dillon, Ed., *Forms of Corrosion Recognition and Prevention*, NACE Handbook 1, National Association of Corrosion Engineers, 1982
56. "Drill Pipe Corrosion Ring Coupon Test Procedure," API RP-13B, API Standard Procedures for Testing Drilling Fluids, Appendix A, American Petroleum Institute
57. *Modern Electrical Methods for Determining Corrosion Rates*, NACE Publication 3D170, National Association of Corrosion Engineers,
58. "Proposed Use of Galvanic Probe Corrosion Monitor In Oil and Gas Drilling and Production Operations," NACE Committee Report T-10-16, National Association of Corrosion Engineers
59. B.Q. Bradley, Oxygen: A Major Element in Drill Pipe Corrosion, *Mater. Prot.*, Dec 1967
60. W.J. Frank, Here's How to Deal With Corrosion Problems in Rod-Pumped Wells, *Oil Gas J.*, May 1976
61. Recommendations of Corrosion Control of Sucker Rods by Chemical Treatment, NACE Task Group Report, *Mater. Perform.*, May 1967
62. "Recommended Practice for Planning, Designing and Constructing Fixed Offshore Platforms," API RP-2A, American Petroleum Institute
63. W.J. Graff. *Introduction to Offshore Structures*, Gulf Publishing, 1981
64. M. Schumaker, Ed., *Seawater Corrosion Handbook*, Noyes Data Corporation, 1979,
65. "Corrosion Control of Steel, Fixed Offshore Platforms Associated With Petroleum Production," NACE RP-01-76, National Association of Corrosion Engineers
66. W.S. Broecker, *Chemical Oceanography*, Harcourt Brace Jovanovich, 1974
67. B. Tomkins, "Fatigue Design Rules for Steel Welded Joints in Offshore Structures," Paper OTC 4403, presented at the 14th Offshore Technology Conference, May 1982, Houston, TX
68. W.D. Kover and S. Dharmavasan, "Fatigue Fracture Mechanics Analysis of T and Y Joints," Paper OTC 4404, presented at the 14th Offshore Technology Conference, May 1982, Houston, TX
69. A. Mukhopadhyay, Y. Itoh, and J.C. Bouwkamp, "Fatigue Behavior of Tubular Joints in Offshore Structures," Paper OTC 2207, presented at the Third Offshore Technology Conference, May 1975, Houston,TX
70. R.M. Kenley, "Measurement of Fatigue Performance of Forties Bravo," Paper OTC 4402, presented at the 14th Offshore Technology Conference, May 1982, Houston, TX
71. Y. Minami and H. Takada, Corrosion Fatigue and Cathodic Protection of Mild Steel, *Boshoku Gijutsu*, Vol 7 (No. 6), 1958, p 336
72. C.E. Jaske *et al.*, Corrosion Fatigue of Structural Steels in Seawater for Offshore Application, in *Corrosion-Fatigue Technology*, STP 642, American Society for Testing and Materials, 1978
73. E.C. Faulds, "Structural Inspection and Maintenance in a North Sea Environment," Paper OTC 4360, presented at the 14th Offshore Technology Conference, May 1982, Houston, TX
74. M.C. Place, Jr., in *Corrosion Control Considerations for High Pressure Sour Gas Reservoirs*, Proceedings of Sulfur/84, Sulphur Development Institute of Canada, 1984, p 387
75. B.C. Craft and M.F. Hawkins, *Applied Petroleum Reservoir Engineering*, Prentice-Hall, 1959, p 5
76. A.K. Dunlop, "Fundamental Considerations in Sweet Gas Well Corrosion," Paper 46, presented at Corrosion/83, Anaheim, CA, National Association of Corrosion Engineers, April 1983
77. S.D. Kapnsta, private communication
78. C. DeWaard and D.E. Milliams, Carbonic Acid Corrosion of Steel, *Corrosion*, Vol 31 (No. 5), 1975
79. D.R. Fincher, J.J. Marr, and J.W. Ward, Inhibiting Gas-Condensate Wells Can Become Complicated Problem, *Oil Gas J.*, Vol 73 (No. 23), 1975, p 52

SELECTED REFERENCES

Primary Production

- J.B. Bradburn and S.K. Kalra, Corrosion Mitigation—A Critical Facet of Well Completion Design, *J. Petrol. Technol.*, Sept 1983
- "Care and Handling of Sucker Rods." API RP 11BR, American Petroleum Institute
- *Corrosion Control in Petroleum Production*, National Association of Corrosion Engineers, 1979
- *Corrosion of Oil and Gas Equipment*, National Association of Corrosion Engineers and the American Petroleum Institute, 1958
- J. E. Donham, "Recent Developments in Corrosion Inhibitors and Their Use," Paper presented at the Offshore Production Chemicals Conference, Norwegian Society of Chartered Engineers, June 1983
- A.K. Dunlop, H.L. Hassell, and P.R. Rhodes, "Fundamental Considerations in Sweet Gas Well Corrosion," Paper presented at Corrosion/83, Anaheim, CA, National Association of Corrosion Engineers, April 1983
- S. Evans, J.M. Phelan, and M.E. Williams, "Batch Treatment of Offshore Wells in the East Cameron and Vermilion Areas," Paper presented at the 17th Annual Offshore Technological Conference, Houston, TX, May 1985
- R.H. Hausler, and S.G. Weeks, "Low Cost Low Volume Continuous Corrosion Inhibitor Application to Gas Production Tubulars," Paper presented at Corrosion/86, Houston, TX, National Association of Corrosion Engineers, March 1986
- C.J. Houghton and R.V. Westermark, "North Sea Downhole Corrosion; Identifying the Problem, Implementing the Solutions," Paper presented at 1983 Offshore Technological Conference, Houston, TX, May 1983
- G.C. Huntoon, "Completion Practices in Deep Sour Tuscaloosa Wells," Paper presented at the 57th Annual Fall Technical Conference and Exhibition of the Society of Petroleum Engineers of AIME, New Orleans, LA, Society of Petroleum Engineers, Sept 1982
- T. Murata, E. Sato, and R. Matsuhashi, "Factors Controlling Corrosion of Steels in CO_2 Saturated Environments," Paper presented at Corrosion/83, Anaheim, CA, National Association of Corrosion Engineers, April 1983
- *Primer of Oil and Gas Production*, 3rd ed., American Petroleum Institute, 1978
- W.B. Steward, Sucker Rod Failures, *Oil Gas J.*, April 1973

CO_2 Injection

- J.C. Ader and M.H. Stern, Slaughter Estate Unit Tertiary Miscible Gas Pilot Reservoir Description, *J. Petrol. Technol.*, May 1984, p 837
- B.W. Bradley, "CO_2 EOR Requires Corrosion Control Program in Gas Gathering Systems," Paper presented at the Permian Basin Corrosion Symposium, Odessa, TX, National Association of Corrosion Engineers, Nov 1985
- R.L. Mathis and S.O. Spears, "Effect of CO_2 Flooding on Dolomite Reservoir Rock, Denver Unit, Wasson (San Andres) Field, Texas," Paper presented at the 59th Technical Conference and Exhibition of the Society of Petroleum Engineers of AIME, Houston, TX, Society of Petroleum Engineers, Sept 1984
- L.E. Newton, Jr., "SACROC CO_2 Project—Corrosion Problems and Solutions," Paper presented at Corrosion/84, New Orleans, LA, National Association of Corrosion Engineers, April 1984
- B.C. Price and F.L. Gregg, "CO_2/EOR, From Source to Resource," Paper presented at the 62nd Annual GPA Convention, San Francisco, CA, Gas Processors Association, March 1983
- W.B. Saner and J.T. Patton, CO_2 Recovery of Heavy Oil; Wilmington Field Test, *J. Petrol. Technol.*, July 1986, p 24

Corrosion in Petroleum Refining and Petrochemical Operations

J. Gutzeit, Amoco Corporation
R.D. Merrick, Exxon Research and Engineering Company
L.R. Scharfstein, Mobil Research and Development Company

CORROSION has always been an unavoidable part of petroleum refining and petrochemical operations. Although certain materials problems are caused by other factors, a predominant number are due to various aspects of corrosion. Corrosion problems increase operating and maintenance costs substantially. Scheduled and unscheduled shutdowns for repairing corrosion damage in piping and equipment can be extremely expensive, and anything that can be safely done to keep a process unit on stream for long periods of time will be of great benefit. A large proportion of corrosion problems are actually caused by shutdowns. When equipment is opened to the atmosphere for inspection and repair, metal surfaces covered with corrosion products will be exposed to air and moisture. This can lead to pitting corrosion and stress-corrosion cracking unless preventive measures are implemented. When equipment is washed with water during a shutdown, corrosion can be caused by pockets of water left to dry.

Most petroleum refining and petrochemical plant operations involve flammable hydrocarbon streams, highly toxic or explosive gases, and strong acids or caustics that are often at elevated temperatures and pressures. Among the many metals and alloys that are available, relatively few can be used for the construction of process equipment and piping (Ref 1). These include carbon steel; some cast irons; certain low-alloy steels and stainless steels; and, to a much lesser degree, aluminum, copper, nickel, titanium, and their alloys. This article will present the considerations and concerns involved in selecting materials for process equipment in refineries and petrochemical plants. In addition, specific information on mechanical properties, corrosion, stress-corrosion cracking (SCC), erosion, and corrosion control will be provided.

Materials Selection

The selection of materials of construction has a significant impact on the operability, economics, and reliability of refining units and petrochemical plants. For this reason, materials selection should be a cooperative effort between the materials engineer and plant operations and maintenance personnel. Reliability can often be equated to predictable materials performance under a wide range of exposure conditions. Ideally, a material should provide some type of warning before it fails; materials that fracture spontaneously and without bulging as a result of brittle facture or SCC should be avoided. Uniform corrosion of equipment can be readily detected by various inspection techniques. In contrast, isolated pitting is potentially much more serious because leakage can occur at highly localized areas that are difficult to detect. The effect of environment on the mechanical properties of a material can also be significant. Certain exposure conditions can convert a normally ductile material into a very brittle material that may fail without warning. A material must not only be suitable for normal process conditions but must also be able to handle transient conditions encountered during start-up, shutdown, emergencies, or extended standby. It is often during these time periods that equipment suffers serious deterioration or that failure occurs.

Of particular concern is what will happen to equipment during a fire. Unexpected exposure to elevated temperatures can not only affect mechanical properties but can also produce detrimental side effects. Although all possible precautions should be taken to minimize the probability of a fire, the engineer responsible for materials selection must recognize that a fire may occur and that the equipment is expected to retain its integrity in order to avoid fueling the fire. This limits the application of materials with low melting points or those that may become subject to damage by thermal shock when fire-fighting water is applied, particularly in the case of refinery piping and equipment used to handle highly flammable hydrocarbon streams. On the other hand, fire resistance need not be considered for cooling-water or instrument-air systems. Although petrochemical plants may include some processes that involve nonflammable or nonhazardous streams, most equipment must be resistant to fires. Lack of fire resistance rules out the use of plastic components in refineries and petrochemical plants despite their excellent resistance to many types of corrosives. In addition, plastic components tend to be damaged by steam-out during a shutdown; this is required in order to free components of hydrocarbon residues and vapor before inspection or maintenance operations. The final step in the materials selection process is a reliability review of the materials and the corrosion control techniques that were selected. There must be total assurance that a plant will provide reliable service under all conditions, including those that occur during start-ups, shutdowns, downtime, standby, and emergencies.

Principal Materials

Materials selection criteria for a number of ferrous and nonferrous alloys used in petroleum-refining and petrochemical applications are presented in this section. Additional information on selecting the proper metal or alloy is available in the article "Materials Selection" in this Volume.

Carbon and Low-Alloy Steels. Carbon steel is probably used for at least 80% of all components in refineries and petrochemical plants because it is inexpensive, readily available, and easily fabricated. Every effort is made to use carbon steel, even if process changes are required to obtain satisfactory service from carbon steel (Ref 2). For example, process temperatures can be decreased, hydrocarbon streams dried up, or additives injected in order to reduce potential corrosion problems with carbon steel (Ref 3). In refineries, fractionation towers, separator drums, heat-exchanger shells, storage tanks, most piping, and all structures are generally fabricated from carbon steel. Carbon molybdenum steels, primarily the C-0.5 Mo grade, can offer substantial savings over carbon steels at temperatures between 425 and 540 °C (800 and 1000 °F). Because C-0.5 Mo steel has better resistance than carbon steel to high-temperature hydrogen attack, it has been extensively used for reactor vessels, heat-exchanger shells, separator drums, and piping for processes involving hydrogen at temperatures above 260 °C (500 °F). Recently, however, questions have been raised regarding the effect of long-term hydrogen exposure on C-0.5 Mo steel. As a result, low-alloy steels are preferred for new construction.

Low-alloy steels for refinery service are the chromium-molybdenum steels containing less than 10% Cr. These steels have excellent resistance to certain types of high-temperature sulfidic corrosion as well as to high-temperature hydrogen attack. To improve resistance to hydrogen stress cracking, low-alloy steels normally require postweld heat treatment. For refinery reactor vessels, which operate at high temperatures and pressures, 2.25Cr-1Mo steel is widely used. For improved corrosion resistance, these are often overlayed with stainless steel. Other applications for low-alloy steels are furnace tubes, heat-exchanger shells, and piping and separator drums.

Additional information is provided in the articles "Corrosion of Carbon Steels" and "Corrosion of Alloy Steels" in this Volume.

Stainless steels are extensively used in petrochemical plants because of the highly corrosive nature of the catalysts and solvents that are often used. In refineries, stainless steels have been primarily limited to applications involving high-temperature sulfidic corrosion and other forms of high-temperature attack (Ref 4). Most stainless steels will pit in the presence of chlorides (Ref 5).

Martensitic stainless steels, such as type 410 (S41000), must be postweld heat treated after welding to avoid hydrogen stress cracking problems as a result of exposure to hydrogen sulfide containing environments. Typical applications include pump components, fasteners, valve trim, turbine blades, and tray valves and other tray components in fractionation towers. Low-carbon varieties of type 410 stainless steel (S41008) are preferred for furnace tubes and piping, often in combination with aluminizing. Ferritic stainless steels, such as type 405 (S40500), are not subject to hydrogen stress cracking and are therefore a better choice than type 410 (S41000) stainless steel for vessel linings that are attached by welding (Ref 6). Austenitic stainless steels, such as type 304 (S30400) or type 316 (S31600), have excellent corrosion resistance, but are subject to SCC by chlorides. If sensitized, they are also subject to SCC by polythionic acids (Ref 7, 8). Typical applications include linings and tray components in fractionation towers; piping; heat-exchanger tubes; reactor cladding; tubes and tube hangers in furnaces; various components for compressors, turbines, pumps, and valves; and reboiler tubes. Additional information is available in the article "Corrosion of Stainless Steels" in this Volume.

Cast irons, because of their brittleness and low strength, are normally not used for pressure-retaining components for handling flammable hydrocarbons. The main exceptions are pump and valve components, ejectors, jets, strainers, and fittings in which the high hardness of cast iron reduces the velocity effects of corrosion, such as impingement, erosion, and cavitation. High-silicon cast irons (with 14% Si) are extremely corrosion resistant because of a passive surface layer of silicon oxide that forms during exposure to many chemical environments (except hydrofluoric acid). Typical refinery and petrochemical plant applications include valve and pump components for corrosive service. High-nickel cast irons (with 13 to 36% Ni and up to 6% Cr) have excellent corrosion, wear, and high-temperature resistance because of the relatively high alloy content (Ref 9). Typical uses are valve components, pump components, dampers, diffusers, tray components, and compressor parts. Additional information is provided in the article "Corrosion of Cast Irons" in this Volume.

Copper and aluminum alloys are usually restricted to applications below 260 °C (500 °F) because of strength limitations. Admiralty metal (C44300) tubes have been extensively used in water-cooled condensers and coolers at most refineries, but have often performed poorly in overhead condensers, compressor aftercoolers, and other locations where high concentrations of hydrogen sulfide and ammonia are encountered in aqueous condensate. The usual failure modes are pitting, ammonia SCC, and dezincification. Aluminum alloys, at one time, were proposed for refinery use as a substitute for carbon steel and admiralty metal (C44300) heat-exchanger tubes in cooling-water service (Ref 10-12). Aluminum tubes were found to be highly resistant to aqueous sulfide corrosion in overhead condensers. Unfortunately, fouling and pitting corrosion on the water side have always been a problem, and except for certain limited applications, most refineries do not use aluminum tubes. The only other major refinery use of aluminum has been in vacuum towers, in which aluminum provides resistance to the naphthenic acid corrosion of tray components. Aluminum is also used, in the form of aluminized coatings, to protect low-alloy steels against high-temperature sulfidic corrosion. Additional information is available in the articles "Corrosion of Copper and Copper Alloys" and "Corrosion of Aluminum and Aluminum Alloys" in this Volume.

Nickel alloys are especially resistant to sulfuric acid, hydrochloric acid, hydrofluoric acid, and caustic solutions, all of which can cause corrosion problems in certain refinery and petrochemical operations (Ref 13). As the nickel content is increased above 30%, austenitic alloys become, for all practical purposes, immune to chloride SCC. Nickel also forms the basis for many high-temperature alloys, but nickel alloys can be attacked and embrittled by sulfur-bearing gases at elevated temperatures. Alloy 400 (N04400) is extensively used as a lining for carbon steel equipment to prevent corrosion by hydrochloric acid and chloride salts (Ref 14). For the same reason, Alloy 400 (N04400) tubes have been used in overhead condensers. Alloy 400 (N04400) is also used against corrosion by hydrofluoric acid. High-nickel alloys, including alloy 625 (N06625) and alloy 825 (N08825), are used to reduce the polythionic acid corrosion of flare-stack tips. Alloy B-2 (N10665) is particularly well suited to handling hydrochloric acid at all concentrations and temperatures (including the boiling point), but is attacked if oxidizing salts are present (Ref 15, 16). Alloy B-2 (N10665), alloy C-4 (N10002), and alloy C-276 (N10276) have excellent resistance to all concentrations of sulfuric acid up to at least 95 °C (200 °F). Although expensive, these alloys are used for specific applications to overcome unusually severe corrosion problems. Additional information is provided in the article "Corrosion of Nickel-Base Alloys" in this Volume.

Titanium is a relative newcomer to the refining industry, but it has been extensively used in certain petrochemical processes. Titanium is not a high-temperature metal; welding and cutting must be done under inert gas atmospheres to prevent embrittlement (Ref 17, 18). From a practical point of view, the use of titanium in refinery and petrochemical plant service is limited to temperatures below 260 °C (500 °F) (Ref 19, 20). If hydrogen is present, temperatures should not exceed 175 °C (350 °F) in order to prevent embrittlement due to hydride formation. Titanium is fully resistant to many process streams. Tubes made from titanium grade 2 (R50400) are extensively used in overhead coolers and condensers on a number of refinery units to prevent corrosion by aqueous chlorides, sulfides, and sulfur dioxide. These tubes can corrode, however, beneath acidic deposits. Titanium tubes are often required when seawater or brackish water is used for cooling. Where underdeposit corrosion of pure titanium is a problem, titanium grade 12 (R53400), alloyed with nickel and molybdenum, should be used. Anodizing and high-temperature air oxidizing of titanium grade 2 (R50400) have been shown to be beneficial from a corrosion point of view (Ref 21). Additional information is provided in the article "Corrosion of Titanium and Titanium Alloys" in this Volume.

Codes and Standard Specifications

Rules for the design, fabrication, and inspection of pressure vessels, piping, and tanks are provided by codes that have been developed by industry and/or regulatory agencies in various countries, as shown by the listing in Table 1. In the United States, the ANSI/ASME Boiler and Pressure Vessel Code, Section VIII, which covers unfired pressure vessels, is used by most industries and fabricators. In most states, it is mandatory that the code be followed, and with the heightened concern over industrial safety, the number of states that require code compliance is increasing. Therefore, the first step in selecting materials of construction is to know what the code covers and what it does not.

The ANSI/ASME Boiler and Pressure Vessel Code also provides a list of acceptable steels and allowable stress values. The detailed specifications for these steels are provided in Sections II A and II B, which are based on ASTM standard specifications (Table 2). The code also provides the method for calculating the required minimum thickness of various components based on design temperature and pressure. The need for heat treating during fabrication and inspection requirements is also defined based on the alloy selected and the pressure-wall thickness. For welded pressure vessels, Section IX of the code defines the requirements for qualifying the welding process to be used.

The code does not consider the effect of process environment on the materials selected. The code recognizes that corrosion can and does occur, and it provides rules for including corrosion allowances in the calculation of the required pressure-wall thickness; but suitable values for the corrosion allowance must be specified by the designer. It is also the responsibility of the designer to specify any special heat treatments, hardness limitations, or other details that may be required as a result of environmental factors. Similarly, the designer must determine accurately the full range of likely operating conditions, including upsets, that may be encountered so that the design criteria are met.

Mechanical Properties

Elevated-Temperature Properties. As mentioned in the preceding section, the applicable code will specify the allowable stress that is to be used for a particular steel in the design of a given piece of equipment. This allowable stress is based on the temperature to which the equipment will be exposed. Steels operating under normal plant conditions can be exposed to these temperatures for prolonged periods of time without adverse effects on their allowable strength if there is no corrosion. As working temperatures increase, the mechanical strength of most materials decreases. In actual practice, however, a material is more likely to fail at elevated temperatures by creep (elongation) or stress rupture than from a decrease in tensile or yield strength.

For example, Table 3 shows the short-term, elevated-temperature yield strengths of several carbon and low-alloy steels. As can be seen from

Table 1 Construction codes for refinery process equipment

Country	Issuing organization	Source document(s)
Pressure Vessels		
United States	American Society of Mechanical Engineers American National Standards Institute	Boiler and Pressure Vessel Code, Section VIII
Great Britain	British Standards Institution	BS 1515: Fusion Welded Pressure Vessel for use in the Chemical, Petroleum and Allied Industries BS 5500: Unfired Fusion Welded Pressure Vessels
Germany	Arbeitsgemeinschaft Druckbehalter (published by Carl Heymans Verlag KG)	A.D. Merkblatter
Italy	Associazione Nazionale per il Controllo della Combustione	ANCC Code
Netherlands	Dienst voor Stoomwezen	Regels Voor Toestellen (Rules for Pressure Vessels)
Sweden	Tryckkarlskomissionen (Swedish Pressure Vessel Commission)	Swedish Pressure Vessel Code
Piping		
United States	American Society of Mechanical Engineers American National Standards Institute	B31.3 Code for Pressure Piping
Great Britain	British Standards Institution	BS 3351: Piping Systems for Petroleum Refineries and Petrochemical Plants
Tanks		
United States	American Petroleum Institute	API 620: Recommended Rules for Design and Construction of Large Welded Low Pressure Storage Tanks API 650: Welded Steel Tanks for Oil Storage
Great Britain	British Standards Institution	BS 2654: Vertical Steel Welded Storage Tanks for the Petroleum Industry

the tabulated data, all three steels (carbon steel, C-0.5 Mo steel, and 2.25Cr-1Mo steel) have satisfactory yield-strength values up to 480 °C (900 °F). These values do not, however, adequately represent the long-term resistance of the steels to creep when stressed at elevated temperatures. Instead, creep resistance values are a more accurate measure of elevated-temperature mechanical strength. Creep resistance values are obtained from creep and stress rupture tests at elevated temperatures over a period of 10 000 h and are usually extrapolated to 100 000 h. Table 4 shows creep resistance values for the three steels discussed above. The deterioration of creep resistance of carbon steel at 480 and 540 °C (900 and 1000 °F) is readily apparent, as is the marked improvement afforded by use of 2.25Cr-1Mo steel. Table 5 lists suggested maximum service temperatures for five different steels and alloys based on creep or rupture data. In some applications, such as furnace tubes, code-allowable stresses need not be followed, and equipment may be operated at temperatures and stresses that can lead to creep failure. In order to predict failure with greater accuracy, equipment operating in the creep range should be periodically inspected as the design life is approached.

Hardness. The hardness of steels is not considered by the code as a specified property. Whether it is the result of forming or welding operations, hardness has, however, a distinct effect on the suitability of a steel for a particular environment. Although carbon steel normally has low hardness values, cooling from elevated temperatures, such as those encountered during welding, may result in localized hard zones. If hardness values exceed 200 HB, carbon steel may become subject to cracking in aqueous sulfide environments. For this reason, it is often necessary to set a maximum hardness limit for carbon steel used in pressure vessels. In some cases, it is desirable to impose a uniform hardness limitation on all pieces of fabricated equipment because the originally intended service application might be changed at some future date to one in which the component would be exposed to an aqueous sulfide environment. A high hardness value is also indicative of an increase in tensile strength and a corresponding decrease in ductility. Steels with high hardness values can be expected to behave in a brittle manner. Low-alloy steels often require heat treatment after welding to reduce hardness in the weld area and to reduce the stresses associated with welding.

Fatigue Strength. Certain components, such as compressors or pumps, require materials with good fatigue resistance properties. Fatigue resistance also needs to be considered when bolting and piping materials are selected. Fatigue resistance is the ability of a load-carrying component to resist fracture from cycles of repeatedly applied forces, such as vibrational or rotational stresses. One common rule of thumb is to limit the average fatigue stress to approximately one-half the ultimate tensile strength of the material involved. Because fatigue involves crack formation, crack propagation, and residual strength, several mechanical properties are involved in determining fatigue resistance. Obviously, higher strength will help a material resist crack formation and final fracture. Other factors to be considered are the cleanliness of the material and whether it hardens or softens under plastic strain. Clean material with a low inclusion content and fine grains will improve fatigue resistance because inclusions can be a source for crack initiation, while fine grains will assist in allowing the crystallographic planes to slip without cracking.

Low-Temperature Properties. Carbon steel begins to lose its toughness and ductility as service temperatures decrease below ambient. Because most equipment in refineries and petrochemical plants is made of carbon steel, insufficient low-temperature toughness could represent a potentially serious problem. Fortunately, few operations are carried out at low temperatures and most equipment made of carbon steel operates at temperatures ranging from ambient to approximately 425 °C (800 °F). Refinery and

Table 2 ASTM standard specifications for refinery steels

Carbon and alloy steel bolts and nuts covered by A193, A194, A320, A354, A449, A453, A540, A563

Material	Pipes and tubes	Plates	Castings	Forgings
Carbon steel	A53, A106, A120, A134, A135, A139, A178, A179, A192, A210, A211, A214, A226, A333, A334, A369, A381(a), A524, A587, A671, A672, A691	A283, A285, A299, A442, A455, A515, A516, A537, A570, A573(a)	A27(a), A216, A352	A105, A181, A234, A268, A350, A372, A420, A508, A541
C-0.5Mo steel	A161(a), A209, A250, A335, A369, A426, A672, A691	A204, A302, A517, A533	A217, A352, A487	A182, A234, A336, A508, A541
1Cr-0.5Mo steel	A213, A335, A369, A426, A691	A387, A517	...	A182, A234, A336
1.25Cr-0.5Mo steel	A199, A200(a), A213, A335, A369, A426, A691	A387, A389(a), A517	A217, A389(a)	A182, A234, A336, A541
2Cr-0.5Mo steel	A199, A200(a), A213, A369	...	...	...
2.25Cr-1Mo steel	A199, A213, A335, A369, A426, A691	A387, A542	A217, A487	A182, A234, A336, A541, A542
3Cr-1Mo steel	A199, A200(a), A213, A335, A369, A426, A691	A387	...	A182, A336
5Cr-0.5Mo steel	A199, A200(a), A213, A335, A369, A426, A691	A387	A217	A182, A234, A336
7Cr-0.5Mo steel	A199, A200(a), A213, A335, A369, A426	A387	...	A182, A234
9Cr-1Mo steel	A199, A200(a), A213, A335, A369, A426	A387	A217	A182, A234, A336
Ferritic, martensitic, and austenitic stainless steel	A213, A249, A268, A269, A271(a), A312, A358, A376, A409, A430, A451, A452, A511(a)	A167, A176(a), A240, A412, A457	A297(a), A351, A447(a)	A182, A336, A403, A473(a)

(a) These specifications are not approved by either the ANSI/ASME Boiler and Pressure Vessel Code or by the ANSI/ASME Code for Pressure Piping B31.3.

Table 3 Short-term elevated-temperature yield strengths

Test temperature		Carbon steel, 0.2% yield strength		C-0.5Mo steel, 0.2% yield strength		2.25Cr-1Mo steel, 0.2% yield strength	
°C	°F	MPa	ksi	MPa	ksi	MPa	ksi
25	80	248	36.0	276	40.0	272	39.5
150	300	208	30.2	241	34.9	247	35.8
260	500	192	27.8	212	30.7	238	34.5
370	700	175	25.4	190	27.6	234	34.0
480	900	148	21.5	175	25.4	193	28.0

Table 4 Creep resistance extrapolated to 100 000 h

Test temperature		Carbon steel, stress for creep rate of 1%		C-0.5Mo steel, stress for creep rate of 1%		2.25Cr-1Mo steel, stress for creep rate of 1%	
°C	°F	MPa	ksi	MPa	ksi	MPa	ksi
425	800	95	13.8	150	21.8	...	...
480	900	41	6.0	98	14.2	152	22.0
540	1000	18	2.6	43	6.2	55	8.0

petrochemical plant equipment and processes that may require special low-temperature toughness grades of steels include liquified-propane storage, ammonia storage, solvent dewaxing units, and liquified petroleum gas (LPG) processing. It is possible, by specifying certain additional requirements, to obtain carbon steels that are suitable for temperatures as low as −45 °C (−50 °F), depending on thickness. To resist brittle fracture at lowered temperatures, steels should be fully killed, fine grained, normalized, and should have received postweld heat treatment.

Typical American Society for Testing and Materials (ASTM) standard specifications for carbon steels with enhanced ability to perform at low temperatures are given in Table 6. Steels, alloyed with 2 to 9% Ni, and austenitic stainless steels can extend the range of available notch-tough steels to even lower temperatures. The simplest quality control test (although not always adequate) for ensuring proper notch toughness is the Charpy V-notch impact test carried out at the minimum design temperature, or lower. A minimum value of approximately 20.5 J (15 ft · lb) is the usual acceptance criterion. The Charpy test is designed to simulate failure of a pressure vessel, containing a fabrication- or service-induced cracklike defect, by rapid crack propagation (brittle failure) when stressed at low temperatures.

Embrittlement Phenomena. There are a number of environmental effects on the mechanical properties of low-alloy steels and stainless steels used for refinery and petrochemical plant construction that need to be considered. In almost all cases, the effect is one of embrittlement due to an increase in hardness or a reduction in the notch ductility of the material. Detailed information on embrittlement mechanisms and the resulting fracture appearance can be found in the article "Visual Examination and Light Microscopy" in Volume 12 of the 9th Edition of *Metals Handbook*.

Temper embrittlement causes a significant increase in the brittle-to-ductile transition temperature of low-alloy steels containing 1 to 3% Cr that are exposed to above 370 to 540 °C (700 to 1000 °F) for some period of time. Brittle failure at weld defects can occur when process equipment made from these steels is fully pressurized during start-up or shutdown. Therefore, pressure should be limited to 25% of design when temperatures are below 150 °C (300 °F) (Ref 22). Ideally, equipment made from steels that have become temper embrittled should be preheated to above 120 °C (250 °F) before pressurization following a shutdown. Temper embrittlement is caused by the segregation of residual steel elements to the grain boundaries, and this greatly reduces the intercrystalline strength. Limiting the acceptance levels of such elements as manganese, silicon, phosphorus, tin, antimony, and arsenic can improve the temper embrittlement resistance of 2.25Cr-0.5Mo, 2.25Cr-1Mo, and 3Cr-1Mo steels. Frequent nondestructive testing of major weld seams is recommended to determine if equipment has become embrittled.

885-°F (475-°C) Embrittlement. Another embrittling phenomenon, referred to as 885 Embrittlement, occurs with ferritic stainless steels containing 12% or more chromium after long-term exposure to temperatures between 400 and 540 °C (750 and 1000 °F). Heat treatment at about 620 °C (1150 °F), followed by rapid cooling, will restore ductility to embrittled low-alloy and ferritic stainless steels.

Sigma-phase embrittlement can occur in austenitic stainless steels as well as in straight-chromium stainless steels. Of the austenitic stainless steels, the most susceptible compositions contain approximately 25% Cr and 20% Ni. The straight-chromium steels that are most susceptible to σ-phase formation contain 17% or more chromium. Sigma-phase formation increases room-temperature tensile strength and hardness while decreasing ductility to the point of extreme brittleness. As a result, cracks are very likely to develop during cooling from operating temperatures. Sigma phase most commonly forms in equipment operating in a temperature range of 650 to 760 °C (1200 to 1400 °F). Because σ phase can be dissolved at temperatures above 980 °C (1800 °F), the original properties of stainless steels can be restored by a suitable heat treatment.

Creep embrittlement is the stress-dependent embrittlement of low-alloy steels operating in the creep range. The result is a reduction in the stress rupture ductility. Creep embrittlement is caused by the formation of precipitates within the grains and by elongated grain-boundary carbides. Detrimental effects can be eliminated by annealing the steel.

Fabricability

With very few exceptions, process equipment and piping are fabricated by welding wrought steels. The shells of pressure vessels are usually made from rolled plate, while nozzles are forgings. This requires that the steels have sufficient ductility for forming and are readily weldable. Weldability of steels is important not only for initial fabrication but also for future field repairs or modifications. Weld repairs and postweld heat treatments can affect the mechanical properties of components that have been normalized or quenched and tempered.

Welding may result in certain other problems. Hydrogen dissolved in liquid weld metal can cause cracking during solidification, as well as embrittlement of the weld. The risk is reduced by the use of low-hydrogen electrodes, careful drying of electrodes, and close control of pre- and postweld heat treatments. Stress relief or reheat cracking is intergranular cracking in the weld heat-affected zone (HAZ). The HAZ cracking occurs when weldments are heated during postweld heat treatment, or it occurs by subsequent exposure to elevated service temperatures. Low-alloy steels are especially susceptible to the above phenomena, but hydrogen cracking can occur with any of the ferritic steels if proper care is not taken.

Corrosion Resistance

The effects of the environment need to be considered when materials of construction are specified. General corrosion (uniform metal loss) is the easiest form of metal deterioration that can be considered in the design phase because additional metal can be provided in the form of a corrosion allowance. It is also the easiest form of corrosion that can be detected by nondestructive testing techniques. In the case of pitting corrosion, it is possible to provide a pitting allowance. Because metal loss due to general corrosion is

Table 5 Suggested maximum temperatures for continuous service based on creep or rupture data

Material	Maximum temperature based on creep rate		Maximum temperature based on rupture	
	°C	°F	°C	°F
Carbon steel	450	850	540	1000
C-0.5 Mo steel	510	950	595	1100
2.25Cr-1Mo steel	540	1000	650	1200
Type 304 stainless steel	595	1100	815	1500
Alloy C-276 nickel-base alloy	650	1200	1040	1900

Table 6 ASTM standard specifications for carbon steel with enhanced resistance to brittle fracture at lowered temperatures

Product form	Temperature: To −30 °C (−20 °F)	Temperature: To −45 °C (−50 °F)
Plate	A516, normalized (may require impact testing)	A516 normalized, stress relieved and Charpy impact tested
Pipe	A524	A333 grade 1 and grade 6
Tube	A210	A334 grade 1 and grade 6
Forgings	A727 and A350 grade LF1	A350 grade LF2
Fittings	A420 WPL6	A420 WPL6
Castings	A352 grade LCA	A352 grade LCB and grade LCC

often not significant under pitting conditions, this approach would represent a rather expensive method of protecting equipment. Instead, it would be more practical to avoid process conditions that produce pitting or to change to a material that will not pit. Stress-corrosion cracking is one of the most serious forms of metal deterioration because it can result in the complete fracture of equipment and considerable losses to an operating facility. Stress-corrosion cracks are very difficult to detect because they may occur during operation and because they are usually not uniformly distributed over the metal surfaces. Austenitic stainless steels are highly susceptible to SCC and consequently are usually avoided for the primary pressure boundary of components. However, they are used as protective, internal linings.

Corrosion

For practical purposes, corrosion in refineries and petrochemical plants can be classified into low-temperature corrosion and high-temperature corrosion (Ref 23, 24). Low-temperature corrosion is considered to occur below approximately 260 °C (500 °F) in the presence of water. Carbon steel can be used to handle most hydrocarbon streams in this temperature range, except where aqueous corrosion by inorganic contamination, such as hydrogen chloride or hydrogen sulfide, necessitates selective application of more resistant alloys. High-temperature corrosion is considered to take place above approximately 260 °C (500 °F). The presence of water is not necessary, because corrosion occurs by the direct reaction between metal and environment.

Low-Temperature Corrosion

Most corrosion problems in refineries are not caused by hydrocarbons that are processed but by various inorganic compounds, such as water, hydrogen sulfide, hydrochloric acid, hydrofluoric acid, sulfuric acid, and caustic (Ref 25). There are two principal sources of these compounds: feedstock contaminants and process chemicals, including solvents, neutralizers, and catalysts. Generally, the same applies to corrosion problems in petrochemical plants except that corrosion is also caused by organic acids, such as acetic acid, that may be used as solvents. In addition, corrosion problems are caused by the atmosphere, cooling water, boiler feedwater, steam condensate, and soil.

Low-Temperature Corrosion by Feed-Stock Contaminants

The major cause of low-temperature (and, for that matter, high-temperature) refinery corrosion is the presence of contaminants in crude oil as it is produced. Although some contaminants are removed during preliminary treating in the fields, most end up in refinery tankage, along with contaminants picked up in pipelines or marine tankers. In most cases, the actual corrosives are formed during initial refinery operations. For example, potentially corrosive hydrogen chloride evolves in crude preheat furnaces from relatively harmless calcium and magnesium chlorides entrained in crude oil (Ref 26). In petrochemical plants, certain corrosives may have been introduced from upstream refinery and other process operations. Other corrosives can form from corrosion products after exposure to air during shutdowns; polythionic acids fall into this category. The following discussion will focus on the most important crude oil contaminants that have caused corrosion problems.

Air. During shutdowns or turnarounds, most plant equipment is exposed to air. Air also can enter the suction side of pumps if seals or connections are not tight. In general, the air contamination of hydrocarbon streams has been more detrimental with regard to fouling than corrosion. However, air contamination has been cited as a cause of accelerated corrosion in vacuum transfer lines and vacuum towers of crude distillation units. Air contamination has supposedly increased the overhead corrosion of crude distillation towers, but this has been difficult to reconcile with the fact that oxygen in air reacts with hydrogen sulfide to form polysulfides, which tend to inhibit corrosion.

Water. Water is found in all crude oils and is difficult to remove completely. In addition, water originates with stripping steam for fractionation towers and is produced in hydrotreating operations. Water not only functions as an electrolyte but also hydrolyzes certain inorganic chlorides to hydrogen chloride, as noted above. Water is primarily responsible for various forms of corrosion in fractionation tower overhead systems. In general, whenever equipment can be kept dry through suitable process or equipment changes, corrosion problems will be minimized.

The combination of water and air can be especially detrimental. Moisture and air are drawn into storage tanks during normal breathing as a result of pumping and changes in temperature. Tank activity and corrosion are closely interrelated. Because crude and heavy oils form a protective oil film on the working areas of a tank shell, corrosion is generally limited to the top shell ring and the underside of the roof. Tank bottom corrosion occurs mostly with crude oil tanks and is caused by water and salt entrained in the crude oil. A layer of water usually settles out and can become highly corrosive. Alternate exposure to sour crude oils and salt water causes especially severe corrosion (Ref 27). Mill scale tends to accelerate tank bottom corrosion because cracks in the mill scale form anodic areas that pit, while the remaining mill scale acts as the cathode.

Light stocks do not form protective oil films, and corrosion occurs primarily at the middle shell rings because these are exposed to more wetting and drying cycles than other tank areas (Ref 28). Corrosion is in the form of pitting under globules of water that attach themselves to the tank wall. Pitting becomes so extensive that metal loss appears as more or less uniform corrosion. The rate of corrosion is proportional to the water and air content of light stocks. Contamination from chloride and hydrogen sulfide accelerates attack.

Hydrogen Sulfide. Sour crude oils and gases that contain hydrogen sulfide are handled by most refineries (Ref 29). Hydrogen sulfide is also present in some feed stocks handled by petrochemical plants. During processing at elevated temperatures, hydrogen sulfide is also formed by the decomposition of organic sulfur compounds that are present. Corrosion of steel by hydrogen sulfide forms the familiar black sulfide film seen in almost all refinery equipment (Ref 30). Hydrogen sulfide is the main constituent of refinery sour waters and can cause severe corrosion problems in overhead systems of certain fractionation towers, in hydrocracker and hydrotreater effluent streams, in the vapor recovery (light ends) section of fluid catalytic cracking units, in sour water stripping units, and in sulfur recovery units (Ref 31, 32). These will be discussed in greater detail in the section "Sour Water" in this article.

In general, carbon steel has fairly good resistance to aqueous sulfide corrosion because a protective iron sulfide film is formed (Ref 33). To avoid hydrogen stress cracking (sulfide cracking), hard welds (above 200 HB) must be avoided, if necessary, through suitable postweld heat treatment (Ref 34). Excessive localized corrosion in vessels has been resolved by selective lining with alloy 400 (N04400), but this alloy can be less resistant than carbon steel to aqueous sulfide corrosion at temperatures above 150 °C (300 °F). If significant amounts of chlorides are not present, lining vessels with type 405 (S40500) or type 304 (S30400) stainless steel can be considered. More recently, titanium grade 2 (R50400) tubes have been used as replacements for carbon steel tubes to control aqueous sulfide corrosion in heat exchangers at a number of units (Ref 35, 36).

Hydrogen Chloride. In refineries, corrosion by hydrogen chloride is primarily a problem in crude distillation units and, to a lesser degree, in reforming and hydrotreating units. In petrochemical plants, hydrogen chloride contamination can be present in certain feed stocks or can be formed by the hydrolysis of aluminum chloride catalyst.

In most production wells, chloride salts are found either dissolved in water that is emulsified in crude oil or as suspended solids. Salts also originate from brines injected for secondary recovery or from seawater ballast in marine tankers. Typically, the salts in crude oils consist of 75% sodium chloride, 15% magnesium chloride, and 10% calcium chloride (Ref 37). When crude oils are charged to crude distillation units and heated to temperatures above approximately 120 °C (250 °F), hydrogen chloride is evolved from magnesium and calcium chloride, while sodium chloride is essentially stable up to roughly 760 °C (1400 °F). Hydrogen chloride evolution takes place primarily in crude preheat furnaces. Dry hydrogen chloride, especially in the presence of large amounts of hydrocarbon vapor or liquid, is not corrosive to carbon steel (Ref 38).

When steam is added, however, to the bottom of the crude tower to facilitate fractionation, dilute hydrochloric acid forms in the top of the tower and in the overhead condensing system. Severe aqueous chloride corrosion of carbon steel components can occur at temperatures below the initial water dew point (Ref 39). Corrosion rate increases with a decrease in pH value of overhead condensate water. Corrosion is mostly in the form of droplet-impingement attack at elbows of the overhead vapor line and at inlets of overhead condensers. Corrosion also occurs on condenser tubes that are at the temperatures where most of the water condenses out. Often, droplets of dilute hydrochloric acid become entrapped under deposits that are present on tower trays, in condenser shells, and at baffles. The resultant underdeposit corrosion is highly localized and usually quite severe.

Overhead condensing systems of both the crude and vacuum towers of crude distillation units are generally made from carbon steel. Coolers and condensers in cooling-water service usually use admiralty metal (C44300) tubes to reduce corrosion and fouling on the water side. Where aqueous chloride corrosion is a problem on the process side, titanium grade 2 (R50400) tubes should be considered. The top of the crude tower can be lined with alloy 400 (N04400), and tray components made from alloy 400 (N04400) can be used for the upper five or so trays to combat aqueous chloride corrosion. Alloy 400 (N04400) tubes usually have not been cost effective in overhead coolers and condensers, but alloy 400 (N04400) has been successfully used for selective strip lining of those areas of the overhead system where excessive corrosion occurs despite the implementation of other corrosion control measures.

To minimize aqueous chloride corrosion in the overhead system of crude towers, it is best to keep the salt content of the crude oil charge as low as possible, preferably below 1 pound per thousand barrels (PTB), corresponding to roughly 4 ppm. This is done by proper tank-settling, desalting, or, if necessary, double desalting (Ref 40-42). Another way to reduce overhead corrosion would be to inject caustic (sodium hydroxide) into the crude oil downstream of the desalter. Up to 3 PTB (10 ppm) caustic can usually be tolerated from a process point of view, while higher concentrations increase fouling of crude preheat exchangers, boiler corrosion by sodium vanadate (when reduced crude is burned as boiler fuel), or coking in lines and heaters of coking units (Ref 43). Caustic should not be used when reduced crude is charged directly to catalytic cracking or hydrotreating units, because of possible catalyst deactivation.

Neutralizers are injected into the overhead vapor line of the crude tower to maintain the pH value of stripping steam condensate between 5 and 6 (Ref 44-47). A pH value above 7 can increase corrosion with sour crudes, as well as fouling and underdeposit corrosion by neutralizer chloride salts. Where fouling becomes a problem, water should be injected, either intermittently or continuously, to dissolve salt deposits in those areas of the overhead system that are not exposed to stripping-steam condensate (Ref 48). Filming-amine corrosion inhibitors can be injected into the overhead vapor line to provide additional insurance against excessive corrosion (Ref 49-51).

In downstream refining equipment, chlorides accelerate corrosion by penetrating protective surface films, increasing electrolyte conductivity, or complexing with steel surfaces (Ref 52). In reforming units, organic chlorides are often used to regenerate reformer catalyst. Hydrogen chloride is stripped off the catalyst if excessive moisture is present in the reformer feed; this causes increased corrosion, not only in reforming units but also in hydrotreating units that use excess hydrogen (make-gas) from the reformer. As in the case of crude distillation units, water washing and injection of neutralizers and/or filming-amine corrosion inhibitors can be used to control fouling and corrosion by chloride salts. Hydrogen make-gas can be passed through a water scrubber to remove hydrogen chloride. Selective alloying with alloy 825 (N08825), alloy 400 (N04400), or titanium grade 2 (R50400) can be required to control chloride attack in heat exchangers and separator drums.

Nitrogen Compounds. Organic nitrogen compounds, such as indole, carbuzole, pyridine, or quinoline, are present in many crude oils, but do not contribute to corrosion problems unless converted to ammonia or hydrogen cyanide (Ref 53). This occurs primarily in catalytic cracking, hydrotreating, and hydrocracking operations where ammonia and hydrogen cyanide, in combination with hydrogen sulfide and other constituents, become the major constituents of sour water that can be highly corrosive to carbon steel (Ref 54).

Ammonia is also produced in ammonia plants to become a raw material for the manufacture of urea and other nitrogen-base fertilizers. Ammonia in synthesis gas at temperatures between 450 and 500 °C (840 and 930 °F) causes nitriding of steel components. When synthesis gas is compressed to up to 34.5 MPa (5000 psig) prior to conversion, corrosive ammonium carbonate is formed, requiring various stainless steels for critical components. Condensed ammonia is also corrosive and can cause SCC of stressed carbon steel and low-alloy steel components (Ref 55).

Sour Water. The term sour water denotes various types of process water containing primarily hydrogen sulfide, ammonia, and hydrogen cyanide, often in combination with certain other organic and inorganic compounds, including phenols, mercaptans, chlorides, and fluorides. Sour waters are removed from refining units by settling in overhead reflux drums, separator drums, water coalescer drums, and other specialized equipment. Depending on their exact composition, sour waters can become highly corrosive. Sour water corrosion is of particular concern in the vapor recovery (light ends) section of catalytic cracking units and in reactor effluent and light ends sections of hydrotreating and hydrocracking units, in which high concentrations of ammonia can saturate process water with ammonium bisulfide and cause serious corrosion of carbon steel components. Ammonium bisulfide will also rapidly attack admiralty metal (C44300) tubes. Sour water corrosion is a major problem at some sour water stripping units, in which exceptionally high concentrations of ammonium bisulfide build up in the thin film of condensed water on overhead condenser tubes. The resultant corrosion can be so severe that even tubes made from austenitic stainless steels are attacked, and only titanium grade 2 (R50400) tubes have sufficient resistance to be used in this service.

Normally, all components in the vapor recovery (light ends) sections of catalytic cracking units are made of carbon steel. Exceptions to this rule include tower internals made of type 405 (S40500) or 410 (S41000) stainless steel and tubes in overhead condensers and compressor aftercoolers made from admiralty metal (C44300), alloy 400 (N04400), or titanium grade 2 (R50400). Corrosion problems of carbon steel components are often closely associated with hydrogen blistering. Admiralty metal (C44300) tubes in overhead condensers may typically last only 5 years, with leaks finally occurring as a result of ammonia SCC. Depending on the particular process conditions, admiralty metal (C44300) tubes can also corrode by severe localized attack. Admiralty metal (C44300) tubes in compressor aftercoolers often need to be replaced with titanium grade 2 (R50400) tubes. The biggest recurring problem has been corrosion and hydrogen blistering of carbon steel in coolers, separator drums, absorber/stripper towers, and, occasionally, overhead condensers at a number of locations. These will be discussed in greater detail in the section "Hydrogen Blistering" in this article.

Components in hydrotreating and hydrocracking units that operate at temperatures below about 260 °C (500 °F) are typically made from carbon steel. Where aqueous ammonium bisulfide corrosion becomes a problem, generous corrosion allowances may have to be provided for carbon steel (Ref 56, 57). Selective alloying with alloy 825 (N08825), alloy 400 (N04400) or titanium grade 2 (R50400) may be required for heat exchangers and separator drums to control excessive corrosion. On some units, corrosion is accompanied by hydrogen blistering. Hydrotreating and hydrocracking units that experience fouling problems due to ammonium sulfide or ammonium chloride deposition may require intermittent or continuous water injection to dissolve these salt deposits. It is of prime importance, however, that sufficient coalescer capacity be available or provided in order to ensure that the injected water is removed. Otherwise, serious corrosion can occur when the water ends up in downstream equipment.

All equipment and piping of reforming units that operate at below approximately 260 °C (500 °F) are usually made from carbon steel. Although admiralty metal (C44300) tubes are often used in water-cooled effluent cooler and condensers, chloride attack may necessitate selective alloying with alloy 400 (N04400) or titanium grade 2 (R50400). In some cases, carbon steel tubes are superior to admiralty metal (C44300) tubes, provided the cooling water is properly treated. Similar considerations apply to water-cooled coolers and condensers in the overhead systems of prefractionator, splitter, debutanizer, and other fractionation towers. Filming-amine corrosion inhibitors can be used to help control overhead corrosion.

The principal material of construction for sour water stripping units is carbon steel. There are several varieties of sour water strippers, but nonacidified condensing and noncondensing strippers are most commonly used (Ref 58, 59). The stripping medium is primarily steam. Stripper towers are generally made from carbon steel with type 316 (S31600) stainless steel, aluminum, or carbon steel internals, depending on corrosion experience. To control tower corrosion, a minimum top temperature of 80 °C (180 °F) is required. Below this temperature, hydrogen sulfide will concentrate in the upper part of the tower but will not be carried overhead. Feed charge pumps are usually made from cast iron or cast steel, including the impellers. Feed piping, bottoms piping, and the feed/bottoms heat exchanger can be made from carbon steel. Carbon steel has also been satisfactory for stripper reboilers that may be used instead of live stripping steam. Thermosyphon reboilers (with sour water in the tubes) are recommended over kettle reboilers because the latter are often prone to fouling and resultant underdeposit corrosion.

Most corrosion problems have been in overhead condensers of condensing sour water strippers (Ref 60). Although a variety of alloys have been used for overhead condenser tubes, only aluminum and titanium grade 2 (R50400) can be relied on to provide adequate resistance to the highly corrosive conditions encountered in many overhead systems. Carbon steel is usually satisfactory for the overhead vapor line, condenser shell, run down lines, accumulator drum, and reflux lines. All welds in these components should be postweld heat treated to avoid hydrogen stress cracking. Reflux pumps can be made of carbon steel or type 304 (S30400) stainless steel, but for optimum performance, alloy 20 (N08020) is recommended. Hydrogen blistering often accompanies corrosion in overhead condenser shells and reflux drums. Water-soluble filming-amine corrosion inhibitors can be injected into the overhead vapor line to help control both corrosion and hydrogen blistering. Few, if any, corrosion problems have been experienced with noncondensing sour water strippers.

Serious sour water corrosion of carbon steel components can occur in the overhead system of amine regenerators (strippers) of gas-treating or sulfur recovery units, especially if all of the water condensate is returned to the tower as reflux. Corrosion is usually accompanied by hydrogen blistering. Continuous or periodic blowdown of sour water to the sour water stripping unit should be employed to lower the concentrations of hydrogen sulfide, ammonia, and cyanide in the overhead water condensate. If this fails to control corrosion, carbon steel condenser tubes may have to be replaced with titanium grade 2 (R50400) tubes. In addition, corrosion can be minimized by operating the regenerator so that roughly 0.5% amine is taken overhead to act as a corrosion inhibitor.

Polythionic Acids. Combustion of H_2S in refinery flares can produce polythionic acids of the type $H_2S_xO_y$ (including sulfurous acid) and cause severe intergranular corrosion of flare tips made of stainless steels and high-nickel alloys (Ref 61). Corrosion can be minimized by using nickel alloys, such as alloy 825 (N08825) or alloy 625 (N06625). Polythionic acids also cause SCC during shutdown, as discussed in the section "SCC and Embrittlement" in this article.

Low-Temperature Corrosion by Process Chemicals

Severe corrosion problems can be caused by process chemicals, such as various alkylation catalysts, certain alkylation by-products, organic acid solvents used in certain petrochemical processes, hydrogen chloride stripped off reformer catalyst, and caustic and other neutralizers that, ironically, are added to control acid corrosion. Filming-amine corrosion inhibitors can be quite corrosive if injected undiluted (neat) into a hot vapor stream. Another group of process chemicals that are corrosive, or become corrosive, is solvents used in treating and gas-scrubbing operations.

Acetic Acid. Corrosion by acetic acid can be a problem in petrochemical process units for the manufacture of certain organic intermediates, such as terephthalic acid. Various types of austenitic stainless steels are used, as well as alloy C-4 (N06455), alloy C-276 (N10276), and titanium, to control corrosion by acetic acid in the presence of small amounts of hydrogen bromide or hydrogen chloride.

As a rule, even tenths of a percent of water in acetic acid can have a significant influence on corrosion. Type 304 (S30400) stainless steel usually has sufficient resistance to the lower concentrations of acetic acid up to the boiling point. Higher concentrations can also be handled by type 304 (S30400) stainless steel if the temperature is below about 90 °C (190 °F). Increasing the chromium and/or nickel content has little effect on resistance to acetic acid. Addition of molybdenum, however, markedly increases the resistance of austenitic stainless steels, and type 316 (S31600) and type 317 (S31700) stainless steels are used for the overwhelming majority of hot acetic acid applications. Corrosion by acetic acid increases with temperature. Bromide and chloride contamination causes pitting and SCC, while addition of oxidizing agents, including air, can reduce corrosion rates by several orders of magnitude.

Aluminum Chloride. Certain refining and petrochemical processes, such as butane isomerization, ethylbenzene production, and polybutene production, use aluminum chloride as a catalyst (Ref 62). Aluminum chloride is not corrosive if it is kept absolutely dry. If traces of water or water vapor are present in hydrocarbon streams, aluminum chloride hydrolyzes to hydrochloric acid, which can of course be highly corrosive. To control corrosion in the presence of aluminum chloride, feed is dried in calcium chloride dryers. During shutdowns, equipment should be opened for the shortest possible time. Upon closing, it should be dried with hot air, followed by inert gas blanketing. Equipment that is exposed to hydrochloric acid may require extensive lining with nickel alloys, such as alloy 400 (N04400), B-2 (N10665), C-4 (N06455), or C-276 (N10276) (Ref 63).

Organic Chlorides. Organic chlorides in crude oils will form various amounts of hydrogen chloride at the elevated temperatures of crude preheat furnaces, depending on the chlorides involved. Many crude oils contain small amounts of organic chlorides (5 to 50 ppm), but the major problem is contamination with organic chloride solvents during production. Although major producers are aware of the problem, some small operators may still use organic chloride solvents to remove wax deposits in oil field tankage and associated equipment and piping. Spent solvent is then simply added to the crude oil. Organic chloride solvents are also extensively used for metal-degreasing operations in and out of the refinery. Spent solvent is often discarded with slop oil, which is added to the crude oil and charged to the crude distillation unit.

Contaminated crude oils have been found to contain as much as 7000 ppm chlorinated hydrocarbons. Such crude oils not only cause severe corrosion in the overhead system of crude distillation towers but also affect reformer operations (Ref 64). Typical problems in the latter category include runaway cracking, rapid coke accumulation on the catalyst, and increased corrosion in fractionator overhead systems (Ref 65, 66). Obviously, every effort must be made to avoid charging contaminated crude oil. Organic chlorides cannot be removed by desalting. If contaminated crude oil must be run off, the usual approach is to blend it slowly into uncontaminated crude oil.

Hydrogen Fluoride. Some alkylation processes use concentrated hydrofluoric acid instead of sulfuric acid as the catalyst. In general, hydrofluoric acid is less corrosive than hydrochloric acid because it passivates most metals by the formation of protective fluoride films. If these films are destroyed by diluted acid, severe corrosion occurs. Therefore, as long as feed stocks are kept dry, carbon steel—with various corrosion allowances—can be used for vessels, piping, and valve bodies of hydrofluoric acid alkylation units. Alloy 400 (N04400) is used selectively at locations where excessive corrosion has been experienced. Hydrofluoric acid can cause hydrogen blistering of carbon steel equipment and hydrogen stress cracking of hardened bolts (Ref 67).

By following proper design practices and prescribed maintenance procedures and by diligently keeping feed stocks and equipment dry, there will be few corrosion problems. All carbon steel welds that contact hydrofluoric acid should be postweld heat treated (Ref 68). This applies especially to welds in various vessels. Vessels should be radiographed to check for slag inclusions in plates and welds; slag inclusions are attacked by hydrofluoric acid. Hydrofluoric acid has the capability of finding the smallest holes in welds or threads. During welding, each preceding pass must be properly cleaned. All threaded connections should be seal welded. Where leaks do show up after start-up, small holes can often be peened shut, or small bits of copper or lead can be peened into larger holes to seal a leak. Any subsequent repair welds should also be postweld heat treated.

Fractionation towers should have type 410 (S41000) stainless steel tray valves and bolting; alloy 400 (N04400) tray valves and bolting are preferred for the deisobutanizer tower. The acid rerun tower usually requires cladding with alloy 400 (N04400) and alloy 400 (N04400) tray components. To avoid SCC, alloy 400 (N04400) welds that contact hydrofluoric acid should be postweld heat treated. No asbestos gasketing should be used on trays. Soft iron gaskets are used on channel head-to-shell joints of heat exchangers. Spiral-wound alloy 400/Teflon gaskets are also used but are more expensive. Carbon steel U-tube bundles are preferred for all exchangers that contact hydrofluoric acid; alloy 400 (N04400) tubes have been found to offer few advantages (Ref 69). Tube ends and tubesheet holes should be carefully cleaned to ensure tube rolls that are tight against hydrofluoric acid. Seal welding of tubes may be required. Internal bolting should not be used in exchangers.

The piping is generally carbon steel with welded connections that have received postweld heat treatment. All taps should be self-draining and should have double block valves. Instrument connections should be made from the top. Valve bodies on gate and plug valves are usually carbon steel, with Teflon packing and Teflon seats. Relief valves should have alloy 400 (N04400) trim. Teflon tape sealing should be used on any threaded connections. Pumps in hydrofluoric acid service normally have carbon steel casings that are weld overlayed with alloy 400 (N04400). Impellers and sleeves should be alloy 400 (N04400); shafts should be alloy K500 (N05500).

Specific areas where corrosion is likely to occur include the bottom of the acid rerun tower, the feed inlet areas of the deisobutanizer and depropanizer towers, the overhead condensers of these towers, the reboiler of the propane stripper, and piping around the acid rerun tower (Ref 70). Trouble areas in vessels are often selectively strip lined with alloy 400 (N04400). Dimpling of tray valve caps during manufacture reduces their tendency to stick to trays because of corrosion products. Alloy 400 (N04400) piping is used to replace carbon steel piping, which corrodes at excessive rates; welds should be postweld heat treated.

Experience has shown that most corrosion problems in hydrofluoric acid alkylation units occur after shutdowns because pockets of water have been left in the equipment. This water is from the neutralization and washing operation required for personnel safety before the equipment can be opened for inspection. It is very important that equipment be thoroughly dried by draining all low spots and by circulating hydrocarbon before the introduction of hydrofluoric acid catalyst at start-up. Corrosion by hydrofluoric acid is occasionally accompanied by hydrogen blistering. Filming-amine corrosion inhibitors have been injected into the overhead systems of various towers, sometimes in conjunction with injection of dilute soda ash solutions. Because the primary goal of proper operations is to keep the unit as dry as possible, intentional addition of water in any form should be considered only as a last resort.

Sulfuric Acid. Certain alkylation units use essentially concentrated sulfuric acid as the catalyst; some of this sulfuric acid is entrained in reactor effluent and must be removed by neutralization with caustic and scrubbing with water. Acid removal may not be complete, however, and traces of acid—at various concentrations (in terms of water)—remain in the stream. Sulfuric acid can be highly corrosive to carbon steel, which is the principal material of construction for sulfuric acid alkylation units. Because the boiling point of sulfuric acid ranges from 165 to 315 °C (330 to 600 °F), depending on concentration, entrained acid usually ends up in the bottom of the first fractionation tower and reboiler following the reactor; this is where the entrained acid becomes concentrated.

Acid concentrations above 85% by weight are usually not corrosive to carbon steel if temperatures are below 40 °C (100 °F). Cold-worked metal (usually bends) should be stress relieved. Under ideal operating conditions, few, if any, corrosion and fouling problems occur (Ref 71, 72).

Carbon steel depends on a film of iron sulfate for corrosion resistance, and if its film is destroyed by high velocities and flow turbulence, corrosion can be quite severe. For this reason, flow velocities should be below 1.2 m/s (4 ft/s). Attack in the form of erosion-corrosion can occur at piping welds that have not received postweld heat treatment. This highly localized attack immediately downstream of piping welds has been attributed to a spheroidized structure; a normalizing postweld heat treatment at 870 °C (1600 °F) is required to minimize corrosion (Ref 73). Erosion-corrosion is also a problem at other locations of high turbulence or velocity (Ref 74). Alloy 20 (N08020) is more resistant than carbon steel to this type of corrosion. In extreme cases, however, even alloy 20 (N08020) will be damaged by erosion-corrosion, and the selective use of alloy B-2 (N10665) may be required.

Carbon steel valves usually require alloy 20 (N08020) internals or trim because even slight attack of carbon steel seating surfaces is sufficient to cause leakage (Ref 75). Pump internals and injection and mixing nozzles in concentrated or spent sulfuric acid service are often made of alloy 20 (N08020), alloy B-2 (N10665), and alloy C-4 (N10002) or C-276 (N10276). For hydrocarbon streams containing only traces of concentrated or dilute sulfuric acid, steel-body valves with type 316 (S31600) stainless steel trim can be used. In this service, steel pump casings that are weld overlayed with aluminum bronze have been successfully used. Pump impellers made from high-silicon cast iron are often used.

Piping for hydrocarbon/acid mixing lines ahead of the reactors may require alloy 20 (N08020), because water contamination of feed stocks can cause severe corrosion of carbon steel. Alloy 400 (N04400) has been found to be useful for reactor effluent lines around the caustic and wash-water injection points. Valve trays in fractionation towers require type 405 (S40500) or 410 (S41000) stainless steel tray valves and bolting. In general, organic coatings are not resistant to concentrated sulfuric acid. Teflon has excellent resistance to sulfuric acid and is extensively used for gaskets, pump valve packing, and mixing nozzles.

In addition to sulfuric acid, reactor effluent contains traces of alkyl and dialkyl sulfates from secondary alkylation reactions (Ref 76). These esters decompose in reboilers to form sulfur dioxide and polymeric compounds (the latter are notorious foulants). Sulfur dioxide combines readily with water in the upper part and overhead system of fractionation towers; the resultant sulfurous acid can cause severe corrosion in overhead condensers. In some units, carbon steel or admiralty metal (C44300) tubes in overhead condensers, particularly those of the deisobutanizer tower, may have to be replaced with alloy 400 (N04400) or titanium grade 2 (R50400) tubes. As a rule, however, titanium is not resistant to sulfuric acid corrosion. Neutralizers can be injected into the overhead vapor lines of various towers to maintain the pH value of aqueous condensate near 7. Filming-amine corrosion inhibitors can also be injected.

Caustic. Sodium hydroxide is widely used in refinery and petrochemical plant operations to neutralize acidic constituents. At ambient temperature and under dry conditions, caustic can be handled in carbon steel equipment. Carbon steel is also satisfactory for aqueous caustic solutions between 50 and 80 °C (120 and 180 °F), depending on concentration. For caustic service above these temperatures but below approximately 95 °C (200 °F), carbon steel can also be used if it has been postweld heat treated to avoid SCC at welds. Austenitic stainless steels, such as type 304 (S30400), can be used up to approximately 120 °C (250 °F), while nickel alloys are required at higher temperatures.

Severe caustic corrosion of the crude transfer line, which is immediately downstream of the caustic injection point, can occur in crude distillation units when 40% (by weight) caustic solution is injected into hot, desalted crude oil to neutralize any remaining hydrogen chloride. Predilution of the caustic with water to form a 3% (by weight) solution minimizes this problem. Better dispersion of the more diluted solution in the hot crude oil prevents puddles of molten caustic from collecting along the bottom of the transfer line. If caustic is injected too close to an elbow of the transfer line, impingement by droplets of caustic can cause severe attack and hole-through at the elbow.

There are some unusual situations in which caustic corrosion is encountered. For example, traces of caustic can become concentrated in boiler feedwater and cause corrosion (gouging) and SCC (caustic embrittlement). This occurs in boiler tubes that alternate between wet and dry conditions (steam blanketing) because of overfiring. In some petrochemical processes, caustic gouging is found under deposits in heat exchangers that remove heat by generation of steam. For example, vertical heat exchangers for cracked gas in ethylene units are especially vulnerable if deposits are allowed to accumulate on the bottom tubesheet. Boiler feedwater permeates these deposits and evaporates, and this causes the caustic to concentrate in any liquid that is left behind. The caustic content of such trapped liquid can reach several percent, which is more than enough to destroy the normally protective iron oxide (magnetite) film on boiler steel and thus cause severe corrosion.

Amines. Corrosion of carbon steel by amines in gas-treating and sulfur recovery units can usually be traced to faulty plant design, poor operating practices, and solution contamination (Ref 77). In general, corrosion is most severe in systems removing only carbon dioxide and is least severe in systems removing only hydrogen sulfide. Systems handling mixtures of the two fall between these two extremes if the gases contain at least 1 vol% hydrogen sulfide. Corrosion in amine plants using monoethanolamine is more severe than in those using diethanolamine, because the former is more prone to degradation.

Corrosion is not caused by the amine itself, but is caused by dissolved hydrogen sulfide or carbon dioxide and by amine degradation products (Ref 78). Corrosion is most severe at locations where acid gases are desorbed or removed from rich amine solution. Here, temperatures and flow turbulence are highest. This includes the regenerator (stripper) reboiler and lower portions of the regenerator itself (Ref 79). Corrosion can also be a significant problem on the rich-amine side of the lean/rich-amine exchanger, in amine solution pumps, and in reclaimers. Hydrogen blistering has been a problem in the bottom of the contactor (absorber) tower and in regenerator overhead condensers and reflux drums (Ref 80).

The common material of construction for amine units is carbon steel. To prevent alkaline SCC, welds of components in both lean and rich-amine service should be postweld heat treated regardless of service temperature (Ref 81). Postweld heat treatment also protects against hydrogen stress cracking. On the whole, there have been relatively few corrosion problems in most amine units.

Operating guidelines usually set limits on amine concentration (20%), acid-gas loading (0.35 mol per mole of amine), and reboiler steam temperature (150 °C, or 300 °F) (Ref 82, 83). These guidelines should be followed to minimize corrosion. Sidestream filtration is also extremely beneficial. Filming-amine corrosion inhibitors are often ineffective. Several proprietary oxidizing corrosion inhibitors based on sodium metavanadate are available. These have been successfully used in certain cases, but licensing costs tend to be high for any but the smaller units.

Regenerator towers usually should be lined with type 405 (S40500) stainless steel, and tower internals are often made of type 304 (S30400) stainless steel. Where applicable, type 304 (S30400) stainless steel is required for the rich-amine pressure of the let-down valve, as well as for piping downstream of the let-down valve, to control corrosion accelerated by high flow turbulence.

Corrosion in the regenerator reboiler is usually in the form of pitting and groove-type corrosion of tubes and is caused by localized overheating inside baffle holes (Ref 84). If thermosyphon reboilers are undersized, part of the tube bundle will become vapor blanketed, and the tubes will overheat. Subsequent exposure of the hot tubes to amine solution will cause severe turbulence and velocity-accelerated corrosion. Vapor blanketing also occurs if tubes are allowed to fill partially with steam condensate; this reduces the amount of tube surface available for heat transfer and increases the heat flux through the remainder of the tubes. Unless faulty reboiler operation can be corrected, carbon steel tubes may have to be replaced with type 304 (S30400) or 316 (S31600) stainless steel tubes. Alloy 400 (N04400) reboiler tubes have been successfully used in amine units that handle only carbon dioxide.

As a rule, carbon steel tubes are satisfactory for regenerator overhead condensers. As discussed in the section "Sour Water" in this article, high corrosion rates can occur at this location, and carbon steel tubes may have to be replaced with titanium grade 2 (R50400) tubes. Carbon steel tubes are used in reclaimers with proper neutralization of acidic constituents (Ref 85). Because the reclaimer can be taken out of service at any time, periodic retubing with carbon steel presents no problems. Cast iron pumps normally are used in low-pressure amine service. If corrosion problems occur, high-silicon cast iron impellers can be used. In high-pressure amine service type 316 (S31600) stainless steel pumps may be needed.

Phenol. Phenol (carbolic acid) is used in refineries to convert heavy, waxy distillates obtained by crude oil distillation into lubricating oils. As a rule, all components in the treating and raffinate recovery sections, except tubes in water-cooled heat exchangers, are made from carbon steel. If water is not present, few significant corrosion problems can be expected to occur in these sections. In the extract recovery section, however, severe corrosion can occur, especially where high flow turbulence is encountered. As a result, certain components require selective alloying with type 316 (S31600) stainless steel. Typically, stainless steel liners are required for the top of the dryer tower, the entire phenol flash tower, and various condenser shells and separator drums that handle phenolic water. Tubes and headers in the extract furnace should also be made of type 316 (S31600) stainless steel, with U-bends sleeved with alloy C-4 (N06455) on the outlet side to minimize velocity-accelerated corrosion.

High-Temperature Corrosion

High-temperature corrosion problems in refineries are of considerable importance (Ref 86). Equipment failures can have serious consequences because processes at high temperatures usually involve high pressures as well. With hydrocarbon streams, there is always the danger of fire when ruptures occur. On a more positive note, high-temperature refinery corrosion is primarily caused by various sulfur compounds originating with crude oil. Over the years, extensive research has been done to establish the mechanism of various forms of high-temperature sulfidic corrosion. Corrosion rate correlations are available; therefore, equipment life can be predicted with some degree of reliability.

Sulfidic Corrosion. Corrosion by various sulfur compounds at temperatures between 260 and 540 °C (500 and 1000 °F) is a common problem in many petroleum-refining processes and, occasionally, in petrochemical processes. Sulfur compounds originate with crude oils and include polysulfides, hydrogen sulfide, mercaptans, aliphatic sulfides, disulfides, and thiophenes (Ref 87). With the exception of thiophenes, sulfur compounds react with metal surfaces at elevated temperatures, forming metal sulfides, certain organic molecules, and hydrogen sulfide (Ref 88, 89). The relative corrosivity of sulfur compounds generally increases with temperature. Depending on the process particulars, corrosion is in the form of uniform thinning, localized attack, or erosion-corrosion. Corrosion control depends almost entirely on the formation of protective metal sulfide scales that exhibit parabolic growth behavior (Ref 90). In general, nickel and nickel-rich alloys are rapidly attacked by sulfur compounds at elevated temperatures, while chromium-containing steels provide excellent corrosion resistance (as does aluminum). The combination of hydrogen sulfide and hydrogen can be particularly corrosive, and as a rule, austenitic stainless steels are required for effective corrosion control.

Sulfidic Corrosion Without Hydrogen Present. This type of corrosion occurs primarily in various components of crude distillation units, catalytic cracking units, and hydrotreating and hydrocracking units upstream of the hydrogen injection line. Crude distillation units that process mostly sweet crude oils (less than 0.6% total sulfur, with essentially no hydrogen sulfide) experience relatively few corrosion problems. Preheat-exchanger tubes, furnace tubes, and transfer lines are generally made from carbon steel, as is corresponding equipment in the vacuum distillation section. The lower shell of distillation towers, where temperatures are above 230 °C (450 °F), is usually lined with stainless steel containing 12% Cr, such as type 405 (S40500). This prevents impingement attack under the highly turbulent flow conditions encountered, for example, near downcomers. For the same reason, trays are made of stainless steel containing 12% Cr. Even with low corrosion rates of carbon steel, certain tray components, such as tray valves, may fail in a short time because attack occurs from both sides of a relatively thin piece of metal.

Crude distillation units that process mostly sour crude oils require additional alloy protection against high-temperature sulfidic corrosion. The extent of alloying needed also depends on the design and the operating practices of a given unit. Typically, such units require low-alloy steels containing a minimum of 5% Cr for furnace tubes, headers and U-bends, and elbows and tees in transfer lines. In vacuum furnaces, tubes made from chromium steels containing 9% Cr are often used. Distillation towers are similar to those of units that process mostly sweet crude oils. Where corrosion problems persist, upgrading with steels containing a greater amount of chromium is indicated.

The high processing temperatures encountered in the reaction and catalyst regeneration section of catalytic cracking units require extensive use of refractory linings to protect all carbon steel components from oxidation and sulfidic corrosion. Refractory linings also provide protection against erosion by catalyst particles, particularly in cyclones, risers, standpipes, and slide valves. Stellite hardfacing is used on some components to protect against erosion. When there are no erosion problems and when protective linings are impractical, austenitic stainless steels, such as type 304 (S30400), can be used. Cyclone diplegs, air rings, and other internals in the catalyst regenerator are usually made from type 304 (S30400) stainless steel, as is piping for regenerator flue gas. Reactor feed piping is made from low-alloy steel, such as 5Cr-0.5Mo or 9Cr-1Mo, to control high-temperature sulfidic corrosion.

The main fractionation tower is usually made of carbon steel, with the lower part lined with stainless steel containing 12% Cr, such as type 405 (S40500) (Ref 91). Slurry piping between the bottom of the main fractionation tower and the reactor may receive an additional corrosion allowance as protection against excessive erosion. As a rule, there are few corrosion problems in the reaction, catalyst regeneration, and fractionation sections (Ref 92).

Hydrocracking and hydrotreating units usually require alloy protection against both high-temperature sulfidic corrosion and high-temperature hydrogen attack (Ref 93, 94). Low-alloy steels may be required for corrosion control ahead of the hydrogen injection line.

The so-called McConomy curves can be used to predict the relative corrosivity of crude oils and their various fractions (Ref 95). Although this method relates corrosivity to total sulfur content, and thus does not take into account the variable effects of different sulfur compounds, it can provide reliable corrosion trends if certain corrections are applied. Plant experience has shown that the McConomy curves, as originally published, tend to predict excessively high corrosion rates. The curves apply only to liquid hydrocarbon streams containing 0.6 wt% S (unless a correction factor for sulfur content is applied) and do not take into account the effects of vaporization and flow regime. The curves can be particularly useful, however, for predicting the effect of operational changes on known corrosion rates.

Over the years, it has been found that corrosion rates predicted by the original McConomy curves should be decreased by a factor of roughly 2.5, resulting in the modified curves shown in Fig. 1. The curves demonstrate the beneficial effects of alloying steel with chromium in order to reduce corrosion rates. Corrosion rates are roughly halved when the next higher grade of low-alloy steel (for example, 2.25 Cr-1Mo, 5Cr-0.5Mo, 7Cr-0.5Mo, or 9Cr-1Mo steel) is selected. Essentially, no corrosion occurs with stainless steels containing 12% or more chromium. Although few data are available, plant experience has shown that corrosion rates start to decrease as temperatures exceed 455 °C (850 °F). Two explanations frequently offered for this phenomenon are the possible decomposition of reactive sulfur compounds and the formation of a protective coke layer.

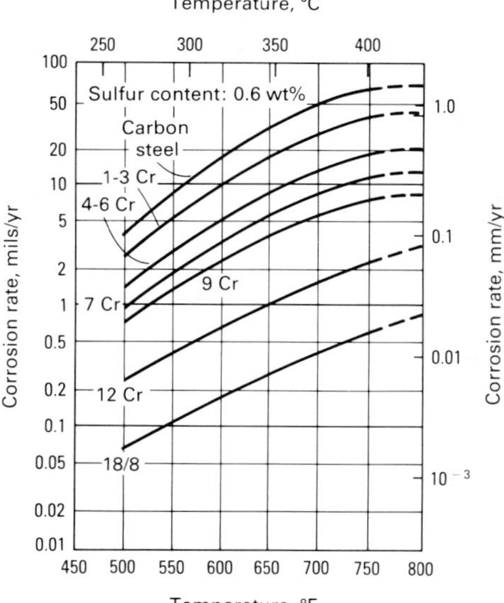

Fig. 1 Modified McConomy curves showing the effect of temperature on high-temperature sulfidic corrosion of various steels and stainless steels. Source: Ref 96

Fig. 2 High-temperature sulfidic corrosion of 150-mm (6-in.) diam carbon steel tube from radiant section of crude preheat furnace at crude distillation unit. Note accelerated attack on fire side

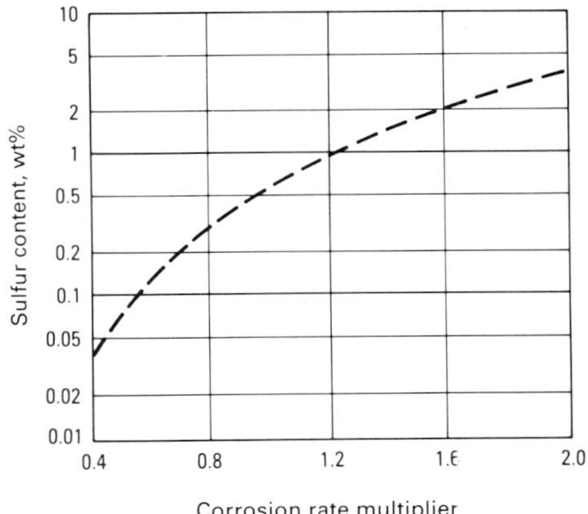

Fig. 3 Effect of sulfur content on corrosion rates predicted by modified McConomy curves in 290- to 400-°C (550- to 750-°F) temperature range. Source: Ref 96

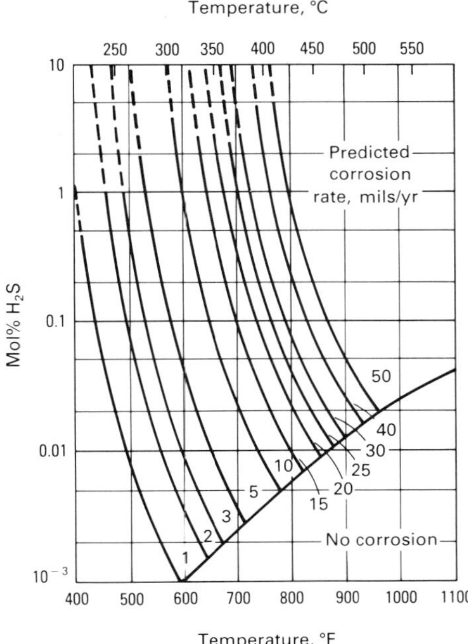

Fig. 4 Effect of temperature and hydrogen sulfide content on high-temperature H_2S/H_2 corrosion of carbon steel (naphtha desulfurizers). 1 mil/yr = 0.025 mm/yr. Source: Ref 96

Metal skin temperatures, rather than stream temperatures, should be used to predict corrosion rates when significant differences between the two arise. For example, metal temperatures of furnace tubes are typically 85 to 110 °C (150 to 200 °F) higher than the temperature of the hydrocarbon stream passing through the tubes. Furnace tubes normally corrode at a higher rate on the hot side (fire side) than on the cool side (wall side), as shown in Fig. 2. Convective-section tubes often show accelerated corrosion at contact areas with tube hangers because of locally increased temperatures. Similarly, replacement of bare convective-section tubes with finned or studded tubes can further increase tube metal temperatures by 85 to 110 °C (150 to 200 °F).

Correction factors for process streams with various total sulfur contents, averages of those proposed originally by McConomy, are shown in Fig. 3. As can be seen, doubling the sulfur content can increase corrosion rates by approximately 30%. To allow for the fact that the proportion of noncorrosive thiophenes is greater in high-boiling cuts (and residuum) than in the original crude charge, a corrosion factor ranging from 0.5 to 1 may have to be applied to the total sulfur content so that realistic corrosion rates can be obtained for such cuts. The degree of vaporization and the resultant two-phase flow regimes can have a significant effect on high-temperature sulfidic corrosion.

Sulfidic Corrosion With Hydrogen Present. The presence of hydrogen in, for example, hydrotreating and hydrocracking operations, increases the severity of high-temperature sulfidic corrosion. Hydrogen converts organic sulfur compounds in feed stocks to hydrogen sulfide; corrosion becomes a function of hydrogen sulfide concentration (or partial pressure).

Downstream of the hydrogen injection line, low-alloy steel piping usually requires aluminizing in order to minimize sulfidic corrosion. Alternatively, type 321 (S32100) stainless steel can be used. Tubes in the preheat furnace are aluminized low-alloy steel, aluminized 12% Cr stainless steel, or type 321 (S32100) stainless steel. Reactors are usually made of 2.25Cr-1Mo steel, either with a type 347 (S34700) stainless steel weld overlay or an internal refractory lining. Reactor internals are often type 321 (S32100) stainless steel (Ref 97).

Depending on the expected corrosion rates, reactor effluent piping operating above approximately 260 °C (500 °F) is made of type 321 (S32100) stainless steel, aluminized low-alloy steel, regular low-alloy steel, or carbon steel with suitable corrosion allowances. When selecting materials for this service, the recommendations of the American Petroleum Institute (API) should be followed to avoid problems with high-temperature hydrogen attack (Ref 98). The same considerations generally apply to separator drums and heat-exchanger vessels operating at temperatures above 260 °C (500 °F). Type 321 (S32100) stainless steel is usually required for heat-exchanger tubes at these temperatures.

A number of researchers have proposed various corrosion rate correlations for high-temperature sulfidic corrosion in the presence of hydrogen (Ref 99-105), but the most practical correlations seem to be the so-called Couper-Gorman curves. The Couper-Gorman curves are based on a survey conducted by National Association of Corrosion Engineers (NACE) Committee T-8 on Refining Industry Corrosion (Ref 106).

The Couper-Gorman curves differ from those previously published in that they reflect the influence of temperature on corrosion rates throughout a whole range of hydrogen sulfide concentrations. Total pressure was found not to be a significant variable between 1 and 18 MPa (150 and 2650 psig). It was also found that essentially no corrosion occurs at low hydrogen sulfide concentrations and temperatures above 315 °C (600 °F) because formation of iron sulfide becomes thermodynamically impossible. Curves are available for carbon steel, 5Cr-0.5Mo steel, 9Cr-1Mo steel, 12% Cr stainless steel, and 18Cr-8Ni austenitic stainless steel. For the low-alloy steels, two sets of curves apply, depending on whether the hydrocarbon stream is naphtha or gas oil. The curves again demonstrate the beneficial effects of alloying steel with chromium to reduce the corrosion rate.

Modified Couper-Gorman curves are shown in Fig. 4 to 11. To facilitate use of these curves, original segments of the curves were extended (dashed lines). In contrast to sulfidic corrosion in the absence of hydrogen, there is often no real improvement in corrosion resistance unless chromium content exceeds 5%. Therefore, the curves for 5Cr-0.5Mo steel also apply to carbon steel and low-alloy steels containing less than 5% Cr. Stainless steels containing at least 18% Cr are often required for essentially complete immunity to corrosion. Because the Couper-Gorman curves are primarily based on corrosion rate data for an all-vapor system, partial condensation can be expected to increase corrosion rates because of droplet impingement.

When selecting steels for resistance to high-temperature sulfidic corrosion in the presence of hydrogen, the possibility of high-temperature hydrogen attack should be considered. Conceivably, this problem arises when carbon steel and low-alloy steels containing less than 1% Cr are chosen for temperatures exceeding 260 °C (500 °F) and hydrogen partial pressures above 689 kPa (100 psia) and when corrosion rates are expected to be relatively low.

Naphthenic Acids. Naphthenic acids are organic acids that are present in many crude oils, especially those from California, Venezuela, Eastern Europe, and Russia. Although minor amounts of other organic acids can also be present, the main acids from naphthenic-base crudes are saturated ring structures with a single carboxyl group. Their general formula may be written as $R(CH_2)_nCOOH$, where R is usually a cyclopentane ring. The higher molecular weight acids can be bicyclic ($12 < n < 20$), tricyclic ($n > 20$), and even polycyclic (Ref 107, 108). Naphthenic acid content is generally expressed in terms of the neutralization number (total-acid number), which is determined by titration with potassium hydroxide, as described in ASTM D 664 (Ref 109).

Naphthenic acids are corrosive only at temperatures above approximately 230 °C (450 °F) in the

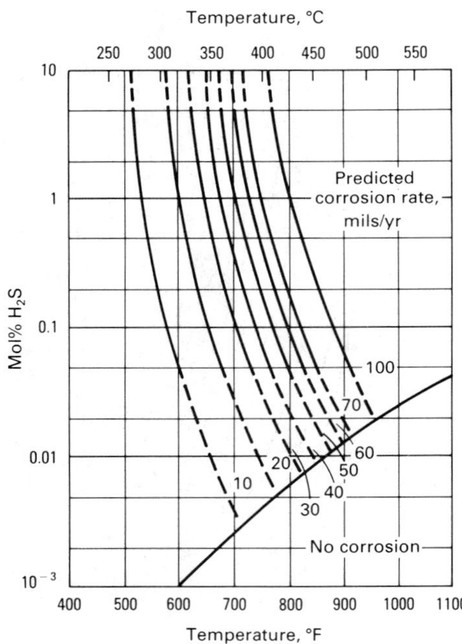

Fig. 5 Effect of temperature and hydrogen sulfide content on high-temperature H_2S/H_2 corrosion of carbon steel (gas oil desulfurizers). 1 mil/yr = 0.025 mm/yr. Source: Ref 96

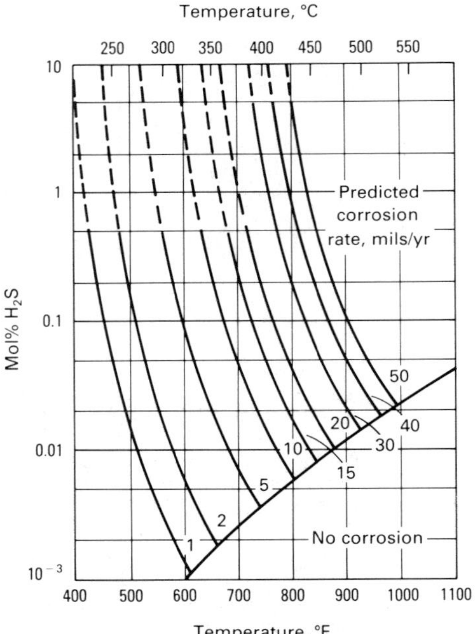

Fig. 6 Effect of temperature and hydrogen sulfide content on high-temperature H_2S/H_2 corrosion of 5Cr-0.5Mo steel (naphtha desulfurizers). 1 mil/yr = 0.025 mm/yr. Source: Ref 96

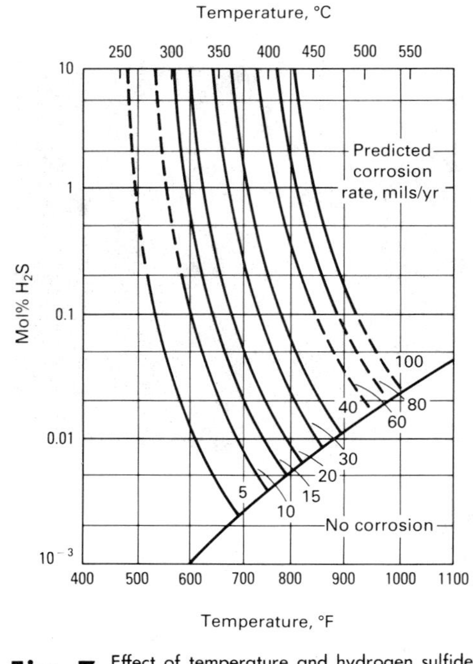

Fig. 7 Effect of temperature and hydrogen sulfide content on high-temperature H_2S/H_2 corrosion of 5Cr-0.5Mo steel (gas oil desulfurizers). 1 mil/yr = 0.025 mm/yr. Source: Ref 96

1 to 6 neutralization number range encountered with crude oil and various sidecuts. At any given temperature, corrosion rate is proportional to neutralization number. Corrosion rate triples with each 55-°C (100-°F) increase in temperature. In contrast to high-temperature sulfidic corrosion, no protective scale is formed, and low-alloy and stainless steels containing up to 12% Cr provide no benefits whatsoever over carbon steel (Ref 110). The presence of naphthenic acids may accelerate high-temperature sulfidic corrosion that occurs at furnace headers, elbows, and tees of crude distillation units because of unfavorable flow conditions.

Severe naphthenic acid corrosion has been experienced primarily in the vacuum towers of crude distillation units in the temperature zone of 290 to 345 °C (550 to 650 °F) and sometimes as low as 230 °C (450 °F) (Ref 111). Damage is in the form of pitting and localized (lake-type) attack of tray components and vessel walls. Attack is often

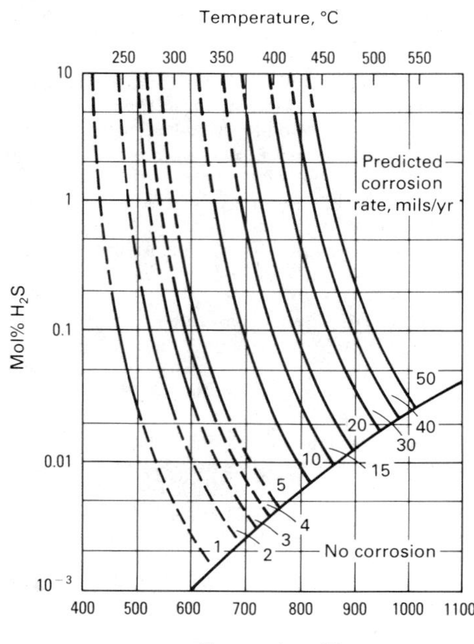

Fig. 8 Effect of temperature and hydrogen sulfide content on high-temperature H_2S/H_2 corrosion of 9Cr-1Mo steel (naphtha desulfurizers). 1 mil/yr = 0.025 mm/yr. Source: Ref 96

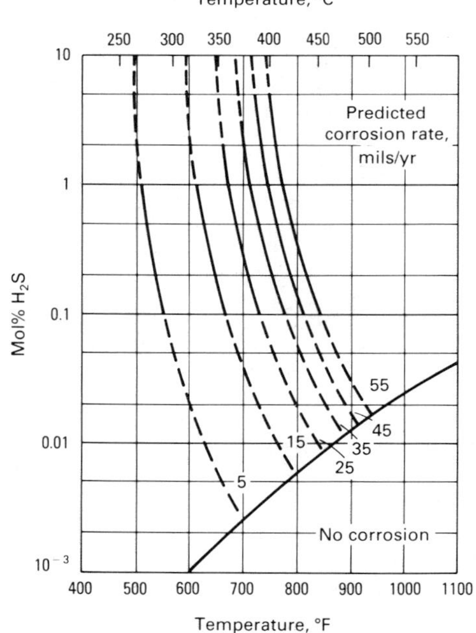

Fig. 9 Effect of temperature and hydrogen sulfide content on high-temperature H_2S/H_2 corrosion of 9Cr-1Mo steel (gas oil desulfurizers). 1 mil/yr = 0.025 mm/yr. Source: Ref 96

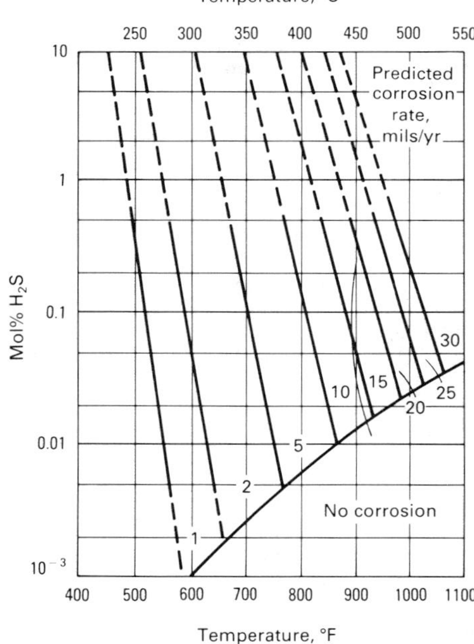

Fig. 10 Effect of temperature and hydrogen sulfide content on high-temperature H_2S/H_2 corrosion of 12% Cr stainless steel. 1 mil/yr = 0.025 mm/yr. Source: Ref 96

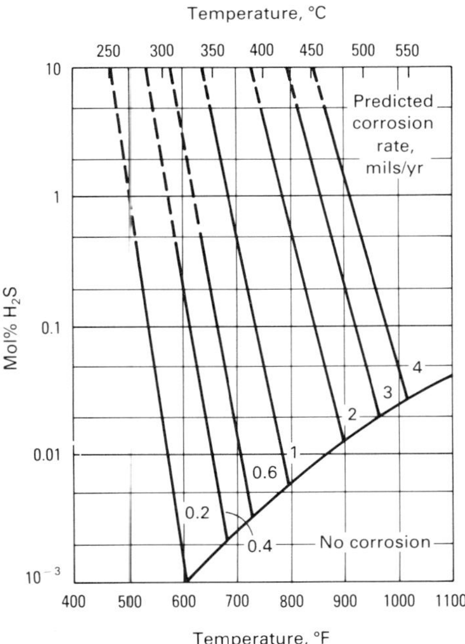

Fig. 11 Effect of temperature and hydrogen sulfide content on high-temperature H_2S/H_2 corrosion of 18Cr-8Ni austenitic stainless steel. 1 mil/yr = 0.025 mm/yr. Source: Ref 96

Fig. 12 Naphthenic acid corrosion on top of 150-mm (6-in.) bubble caps made from type 317 (S31700) stainless steel containing 2.95% Mo. Tray temperature was 305 °C (580 °F)

limited to the undersides of tray floors and to the inside and very top of the outside surfaces of bubble caps, as shown in Fig. 12. These areas are normally not covered by a layer of liquid, which suggests that the attack was caused by impinging droplets of the condensing acids. No corrosion damage is found at temperatures above 345 °C (650 °F), probably because a protective coke layer is formed.

Naphthenic acid corrosion is most easily controlled by blending crude oils having high neutralization numbers with other crude oils. Blending is designed to reduce the naphthenic acid content of the worst sidecut. In practice, blending means that the charge to the crude distillation unit has a neutralization number no higher than 0.5 to 1.0. However, this does not prevent corrosion of vacuum tower internals operating in the 290- to 345-°C (550- to 650-°F) range. These should be made from type 316 (S31600) or, preferably, type 317 (S31700) stainless steel containing at least 3.5% Mo. The vacuum tower lining in this temperature range should also be type 317 (S31700) stainless steel. Aluminum has excellent resistance to naphthenic acid corrosion in vacuum towers and can be used if its strength limitations and low resistance to velocity effects are kept in mind. Alloy 20 (N08020) and titanium grade 2 (R50400) are also resistant to naphthenic acid corrosion. In contrast, aluminized carbon steel tray components, such as bubble caps, have performed poorly.

Fuel Ash. Corrosion by fuel ash deposits can be one of the most serious operating problems with boiler and preheat furnaces. All fuels except natural gas contain certain inorganic contaminants that leave the furnace with products of combustion. These will deposit on heat-receiving surfaces, such as superheater tubes, and after melting can cause severe liquid-phase corrosion. Contaminants of this type include various combinations of vanadium, sulfur, and sodium compounds (Ref 112-114). Fuel ash corrosion is most likely to occur when residual fuel oil (Bunker C fuel) is burned.

In particular, vanadium pentoxide vapor (V_2O_5) reacts with sodium sulfate (Na_2SO_4) to form sodium vanadate ($Na_2O \cdot 6V_2O_5$). The latter compound reacts with steel, forming a molten slag that runs off and exposes fresh metal to attack. The cathodic part of the corrosion reaction is reduction of the pentoxide to the tetroxide (V_2O_4); therefore, the most common ingredient of superheater deposits is sodium vanadyl vanadate ($Na_2O \cdot V_2O_4 \cdot 5V_2O_5$). Table 7 lists the ash fusion temperatures of a number of fuel ash ingredients that can contribute to corrosion and fouling in boiler and preheat furnaces.

Corrosion increases sharply with increasing temperature and vanadium content of fuel. If the vanadium content in the fuel oil exceeds 150 ppm, the maximum tube wall temperature should be limited to 650 °C (1200 °F). Between 20 and 150 ppm V, maximum tube wall temperatures can be between 650 and 845 °C (1200 and 1550 °F), depending on sulfur content and the sodium-vanadium ratio of the fuel oil. With 5 to 20 ppm V, the maximum tube wall temperature can exceed 845 °C (1550 °F).

Table 7 Ash fusion temperatures of slag-forming compounds

Chemical compound	Chemical formula	Ash fusion temperature °C	°F
Vanadium pentoxide	V_2O_5	690	1274
Sodium sulfate	Na_2SO_4	890	1630
Nickel sulfate	$NiSO_4$	840	1545
Sodium metavanadate	$Na_2O{\cdot}V_2O_5$	630	1165
Sodium pyrovanadate	$2Na_2O{\cdot}V_2O_5$	655	1210
Sodium orthovanadate	$3Na_2O{\cdot}V_2O_5$	865	1590
Nickel orthovanadate	$3NiO{\cdot}V_2O_5$	900	1650
Sodium vanadyl vanadate	$Na_2O{\cdot}V_2O_4{\cdot}5V_2O_5$	625	1155
Sodium iron trisulfate	$2Na_3Fe[SO_4]_3$	620	1150

In general, most alloys are likely to suffer from fuel ash corrosion. However, alloys with high chromium and nickel contents provide the best resistance to this type of attack. Sodium vanadate corrosion can be reduced by firing boilers with low excess air (< 1%). This minimizes formation of sulfur trioxide in the firebox and produces high-melting slags containing vanadium tetroxide and trioxide rather than pentoxide. In the temperature range of 400 to 480 °C (750 to 900 °F), boiler tubes are corroded by alkali pyrosulfates such as sodium pyrosulfate and potassium pyrosulfate, when appreciable concentrations of sulfur trioxide are present.

Additives can be helpful in controlling corrosion, particularly in conjunction with firing with low excess air. The effectiveness of the additives varies. The most useful additives are based on organic magnesium compounds. Additives raise the melting point of fuel ash deposits and prevent the formation of sticky and highly corrosive films. Instead, a porous and fluffy deposit layer is formed with additives that can be readily removed by periodic cleaning. Magnesium-type additives offer additional benefits with regard to cold-end corrosion in boilers. Sulfuric acid condenses at temperatures between 150 and 175 °C (300 and 350 °F), depending on sulfur content of the fuel oil, and can cause serious corrosion problems. Additives neutralize any free acid by forming magnesium sulfate.

Oxidation. Carbon steels, low-alloy steels, and stainless steels react at elevated temperatures with oxygen in the surrounding air and become scaled. Nickel alloys can also become oxidized, especially if spalling of scale occurs. The oxidation of copper alloys is usually not a problem, because these are rarely used where operating temperatures exceed 260 °C (500 °F). Alloying with both chromium and nickel increases scaling resistance. Stainless steels or nickel alloys, except alloy 400 (N04400), are required to provide satisfactory oxidation resistance at temperatures above 705 °C (1300 °F). Thermal cycling, applied stresses, moisture, and sulfur-bearing gases will decrease scaling resistance. In refineries and petrochemical plants, high-temperature oxidation is primarly limited to the outside surfaces of furnace tubes, tube hangers, and other internal furnace components that are exposed to combustion gases containing excess air.

At elevated temperatures, steam decomposes at metal surfaces to hydrogen and oxygen and may cause steam oxidation of steel, which is somewhat more severe than air oxidation at the same temperature. Fluctuating steam temperatures tend to increase the rate of oxidation by causing scale to spall and thus expose fresh metal to further attack.

SCC and Embrittlement

Stress-corrosion cracking and environmental embrittlement are the most insidious forms of failure that can be experienced by process equipment, because they tend to strike without warning. Usually, there is no noticeable yielding or bulging of the component, there is no measurable metal loss, and through-thickness cracks can form in as little as 1 to 2 h after initial exposure to a crack-inducing environment. For example, cracking throughout an entire furnace coil occurred within 1 h after exposure to air and the resultant formation of polythionic acids. Towers and heat exchangers had to be scrapped because of hydrogen blistering, embrittlement, and stress cracking at welds. High-temperature hydrogen attack has resulted in the sudden rupture of pressure vessels. With consequences such as these, the possibility and/or probability of SCC and embrittlement occurring in a given environment cannot be taken too lightly.

Stress-Corrosion Cracking

Stress-corrosion cracking that results from exposure to five different environments is summarized below. These environments are:

- Chlorides
- Caustic
- Ammonia
- Amines
- Polythionic acid

Information on the mechanisms of SCC can be found in the article "Environmentally Induced Cracking" in this Volume. A detailed review of tests for determining susceptibility to SCC can be found in the article "Evaluation of Stress Corrosion Cracking" also in this Volume.

Chloride Cracking. Chlorides are perhaps the most common cause of SCC of austenitic stainless steels and nickel alloys. The literature abounds with studies of the mechanism of cracking, specific environments that accelerate cracking, and tests for predicting cracking tendency. In theory, one would need only a single chloride ion in water, with sufficient oxygen and residual stresses present, to cause cracking. In practice, however, the permissible limits on chloride ion content are higher.

The usual failure mode of chloride SCC in austenitic stainless steels is the transgranular, highly branched cracking illustrated in Fig. 13. Intergranular cracking is sometimes associated with transgranular cracking, but this is not at all common. If it occurs, it is usually because of a sensitized microstructure.

Based on laboratory tests in boiling 42% magnesium chloride solution, austenitic stainless steel and nickel alloys are subject to chloride SCC if their nickel content is less than about 45%. The heat treatment of an alloy was found to have no effect on its resistance to chloride SCC. In practice, however, stainless steel and nickel alloys containing greater than 30% Ni will be immune to chloride SCC in most refinery environments.

Factors that influence the rate and severity of cracking are chloride content, oxygen content, temperature, stress level, and pH value of an aqueous solution. It has been established that oxygen is required for chloride cracking to occur. Refinery and petrochemical plant experience confirms that stainless steel components, such as heat-exchanger tube bundles, usually do not crack until removed from operation and exposed to air during a shutdown. Increased oxygen content decreases the critical chloride content for cracking to occur, as shown in Fig. 14.

The severity of cracking increases with temperature. Cracking of austenitic stainless steel components rarely occurs at ambient temperatures. Stainless steel pump impellers in seawater service have shown no cracking problems despite the fact that both chloride and oxygen contents are high. Cracking has been found to occur, however, at tropical locations where exposure to direct sunlight can increase metal temperatures significantly above ambient. As a general rule, chloride SCC of process equipment occurs only at temperatures above about 65 °C (145 °F).

The stresses required to produce cracking can be assumed to be always present. Residual stresses from forming, bending, or joining operations are sufficient for cracks to form. Thermal stress-relief treatments at 870 °C (1600 °F) can effectively prevent cracking if done correctly and without the necessity of subsequent cold working (to correct distortion, for example).

In alkaline solutions, the likelihood of chloride SCC is greatly reduced. Consequently, austenitic stainless steels are frequently used for equipment exposed to amine solutions in gas-treating and sulfur recovery units. A survey of plant experiences has shown no reported instances of cracking despite the fact that chloride contents as high as 1000 ppm were measured in the circulating amine solution.

Most cracking problems occur when unexpected chloride concentrations are found in process streams or in the atmospheric environment. For example, chloride SCC was caused by seawater spray carried by prevailing winds. The spray soaked the insulation over type 304 stainless steel, chlorides were concentrated by evaporation, and cracking occurred at areas with residual weld stresses. Other frequent causes of cracking are water dripping on warm pipe and water leaching chlorides from insulation.

As discussed previously, chlorides are present in a number of refining units, including crude distillation, hydrocracking, hydrotreating, and reforming. Chlorides are also found in other units as contamination from upstream processing, or they are introduced with stripping stream, process water, or cooling water. The latter is a particular problem in petrochemical processes that use stainless steel heat exchangers to make steam as a means of recovering waste heat. Any chloride contamination of boiler feedwater can result in chlorides concentrating on heat-exchanger tubes and can cause pitting and SCC. As a rule, austenitic stainless steels are not recommended for components in which water is likely to evaporate or condense out.

When good resistance to aqueous sulfide corrosion is required, ferritic stainless steels or duplex stainless steels can be substituted for austenitic stainless steels (Ref 116, 117). Ferritic stainless steels, such as type 405 (S40500) or type 430 (S43000), are not susceptible to chloride SCC. The duplex stainless steels have a mixed ferritic-austenitic structure and are resistant to chloride SCC, but are not immune to highly aggressive chloride environments. For example,

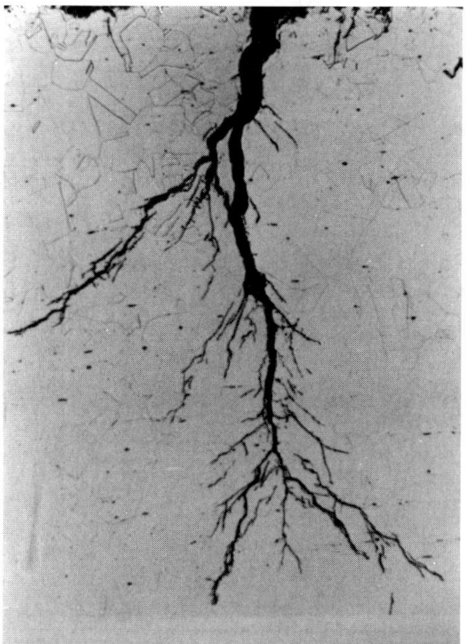

Fig. 13 Chloride SCC of type 304 (S30400) stainless steel tube by chloride-containing sour water. 70×

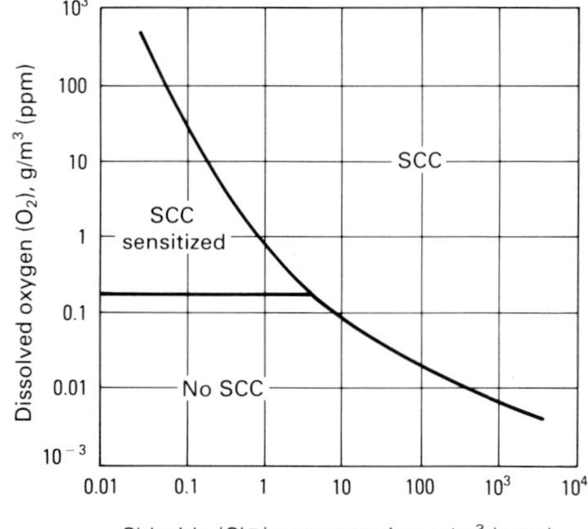

Fig. 14 Synergistic effect of chlorides and oxygen on the SCC of type 304 (S30400) stainless steel. The tests were conducted at 250 to 300 °C (480 to 570 °F) at a strain rate of $< 10^{-5} \cdot s^{-1}$. Source: Ref 115

cold-worked type 329 (S32900) stainless steel has cracked when chlorides were concentrated by vaporization of a process stream, as shown in Fig. 15. Some of the new proprietary duplex stainless steels, such as 3RE60 (S31500) and 2205 (S31803), have reportedly shown increased resistance toward chloride SCC.

There are no simple methods of preventing SCC when an austenitic stainless steel must be used in an environment known to contain chlorides. Chloride SCC in refineries and petrochemical plants often occurs under shutdown conditions when air and moisture enters equipment opened for inspection and repair. It has been found that the precautionary measures outlined in NACE RP-01-70 for the prevention of cracking by polythionic acids also help prevent cracking by chlorides (Ref 118). In particular, excluding air and moisture by nitrogen blanketing and rinsing equipment with an aqueous 0.5% sodium nitrate solution have been shown to inhibit chloride SCC. To prevent chloride SCC on the outside of insulated pipe, aluminum foil has been wrapped between the insulation and pipe to provide some measure of cathodic protection. One method of preventing the catastrophic failure of components by chloride SCC would be the use of austenitic stainless steel as an internal cladding. The highly branched mode of any cracking would effectively prevent the development of stress raisers. Carbon or low-alloy steel base metal would not be susceptible to cracking in chloride solutions, but some localized corrosion may occur. This type of construction would also provide resistance to cracking when chlorides are liable to contact the outside of the components, as in external insulation, for example.

Caustic Cracking. Stress-corrosion cracking of various steels and stainless steels by caustic (sodium hydroxide) is also fairly common in refinery and petrochemical plant operations.

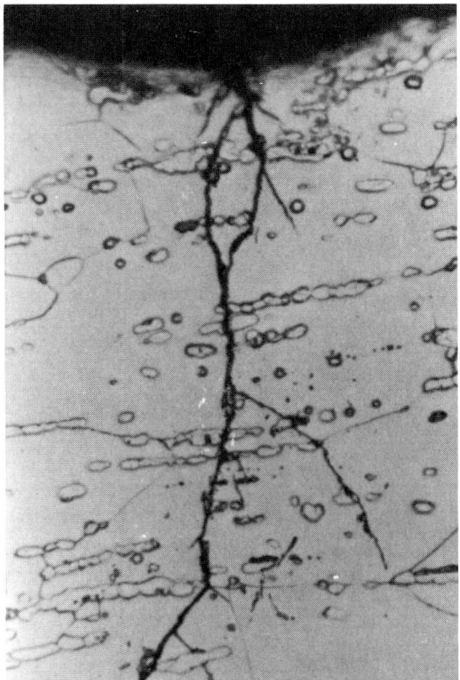

Fig. 15 Chloride SCC of type 329 (S32900) stainless steel by chloride salts that concentrated as water evaporated

Caustic is added in the form of 5 to 40% aqueous solution to certain process streams in order to neutralize residual acid catalysts, such as sulfuric acid, hydrofluoric acid, and hydrochloric acid. Caustic is also added to cooling water and boiler feedwater to counteract large decreases in pH value due to process leaks.

Although caustic attack is primarily in the form of localized corrosion (gouging) in some process streams (for example, crude oil), in others it may take the form of SCC. Traces of caustic can become concentrated in boiler feedwater and cause SCC (caustic embrittlement). This occurs in boiler tubes that alternate between wet and dry conditions (steam blanketing) because of overfiring. Locations such as cracked welds or leaky tube rolls can form steam pockets with cyclic overheating and quenching conditions. These frequently lead to caustic embrittlement.

Caustic SCC of carbon steel occurs at temperatures above 50 to 80 °C (120 to 180 °F), depending on caustic concentration. Welded carbon steel components that are exposed to caustic solutions above these temperatures should be postweld heat treated at 620 °C (1150 °F) for 1 h per 25 mm (1 in.) of metal thickness. Caustic SCC of austenitic stainless steels occurs between 105 and 205 °C (220 and 400 °F), depending on caustic concentration.

Cracking of austenitic stainless steels is often difficult to distinguish from cracking by chlorides, particularly because common grades of caustic also contain some sodium chloride. As a general rule, however, SCC by chlorides is usually, but not always, in the form of transgranular cracking, while caustic causes intergranular cracking, sometimes accompanied by transgranular cracking due to the presence of chlorides.

Caustic SCC of carbon steel is often initiated at discontinuities in areas of surface deformation as a result of cold-working or welding operations (Ref 119). Although caustic SCC occurs over a wide range of temperatures, there appears to be no correlation between temperature and time to failure. Because few failures have been reported at near-ambient temperatures, it appears that crack initiation times are inordinately long unless precracking, for example, in the form of weld defects, has occurred.

Caustic cracking of carbon steel has been found to occur over a narrow range of potentials near the active current peak of potential/log current curves. Typically, this potential range is centered about −700 mV versus the standard hydrogen electrode (SHE). The most negative (active) potential for inducing caustic cracking coincides with the potential for initiating passivation by magnetite (Fe_3O_4) formation. Cracking is promoted by small amounts of dissolved oxygen, sodium chloride, lead oxide, silica, silicates, sulfates, nitrates, permanganate, and chromates that cause the active corrosion potential to move slightly in the positive (noble) direction. In contrast, large amounts of these substances act as inhibitors by pushing the corrosion potential into the passivation range. Phosphates, acetates, carbonates, and tannins also act as inhibitors.

Ammonia Cracking. Ammonia has caused two types of SCC in refineries and petrochemical plants. The first is cracking of carbon steel in anhydrous ammonia service, and the second type is cracking of copper alloys, such as admiralty metal (C44300). In copper alloys, SCC can occur by ammonia contamination of process streams or by ammonia-base neutralizers that are added to control corrosion.

Carbon steel storage vessels, primarily spheres, have developed stress-corrosion cracks in anhydrous ammonia service at ambient temperature but elevated pressure. In most cases, cracking was detected by inspection before leakage or rupture, but there were at least two catastrophic failures (Ref 120). There have been few problems with semirefrigerated storage vessels and no documented cases of SCC in cryogenic storage vessels. The primary causes of cracking are high stresses, hard welds, and air contamination.

To minimize the likelihood of cracking, only low-strength steels, with a maximum tensile strength of 483 MPa (70 ksi), should be used in anhydrous ammonia service. Welds should be postweld heat treated at 595 °C (1100 °F) or higher, with a maximum allowable hardness of 225 HB. A water content of at least 0.2% should be maintained in the ammonia because water has been found to be an effective inhibitor of cracking. Air contamination increases the tendency toward cracking and should be minimized, if necessary, by the addition of hydrazine to the water. With a water content of 10 ppm, the oxygen content should be below 10 ppm for safe operation (Ref 121). The permissible oxygen content increases to 100 ppm with a water content of 0.1 percent. Regular inspection of all components in anhydrous ammonia service is recommended.

Cracking of admiralty metal (C44300) heat-exchanger tubes has been a recurring problem in a number of refining units and petrochemical process units. For example, ammonia is often used to neutralize acidic constituents, such as hydrogen chloride or sulfur dioxide, in overhead systems of crude distillation or alkylation units, respectively. Stripped sour water containing residual ammonia is used as desalter water at some crude distillation units. This practice causes ammonia contamination of the overhead system even if no ammonia is added intentionally. Ammonia is formed from nitrogen-containing feed stocks during catalytic cracking, hydrotreating, and hydrocracking operations. As a rule, cracking of admiralty metal (C44300) tubes occurs only during shutdowns when ammonia-containing deposits on the tube surfaces become exposed to air. To prevent cracking, tube bundles should be sprayed with a very dilute solution of sulfuric acid immediately after they are pulled from their shells in order to neutralize any residual ammonia. Cracking of admiralty metal (C44300) tubes has occasionally been attributed to traces of ammonia in cooling water.

Amine Cracking. Stress-corrosion cracking of carbon steel by aqueous amine solutions, which are used to remove hydrogen sulfide and carbon dioxide from refinery and petrochemical plant streams, has been a recurring problem for a number of years. In one case involving 20 wt% monoethanolamine solution, the affected equipment included two amine storage tanks, four absorber towers, a rich-amine flash drum, a lean-amine carbon treater, and various piping (Ref 122). Cracking was found primarily at welds exposed to amine solutions at temperatures ranging from 50 °C (125 °F) to below 95 °C (200 °F). Cracking was intergranular, with crack surfaces covered by a thin film of magnetite. No cracking was found in piping that had received postweld heat treatment and was operating at temperatures as high as 155 °C (310 °F). Consequently, most of the affected components were replaced with new ones that had received postweld heat treatment. After careful magnetic-particle inspection, the rest of the components were repaired; welds

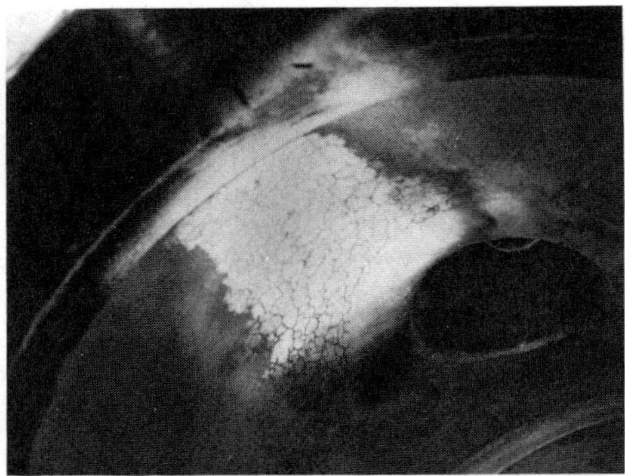

Fig. 16 Polythionic acid SCC of roll-bonded type 304 (S30400) stainless steel cladding. Note that cracking stops at the type 304 (S30400) weld overlay around the nozzle opening

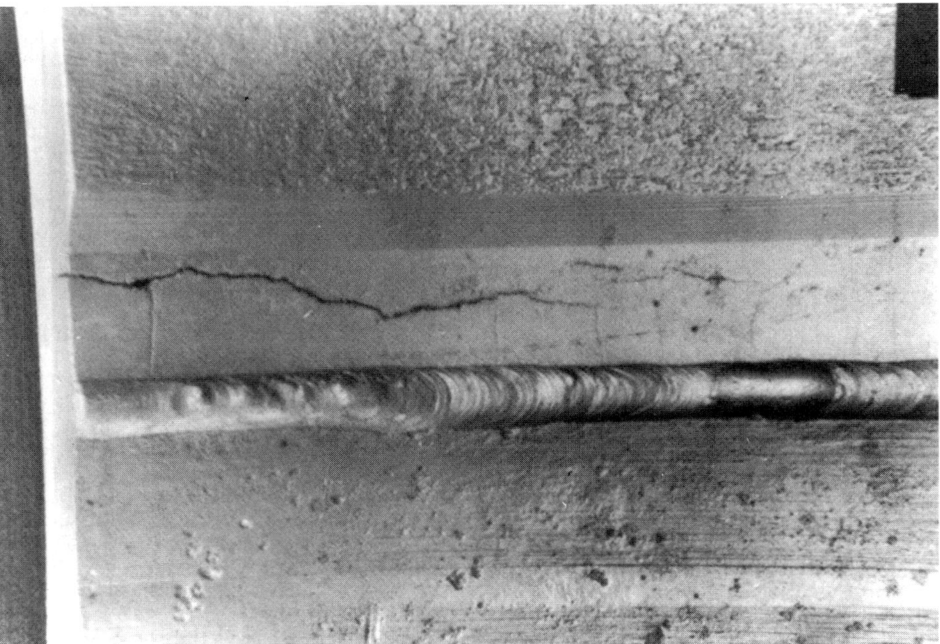

Fig. 17 Polythionic acid SCC of type 304 (S30400) furnace tube near weld to carbon steel tube. Cracking is both parallel and perpendicular to weld but not in the weld

received were postweld heat treated *in situ* or were stress relieved by shot peening.

In another, more recent case, a number of leaks were discovered at piping welds in lean-amine service at temperatures between 40 and 60 °C (100 and 140 °F). None of these welds had been postweld heat treated. Again, the affected components were replaced with new ones that had received a postweld heat treatment (Ref 123). Cracking of piping welds has also occurred in lean-amine piping of several gas-treating plants, but in all cases, temperatures were well above 95 °C (200 °F). For various reasons, these welds had not been postweld heat treated. Different types of amine solutions, including monoethanolamine, diethanolamine, and sulfinol (containing diisopropanolamine), were involved; this confirms that cracking is not limited to monoethanolamine solutions.

Amine SCC appears to be a form of alkaline SCC that is similar in many ways to caustic SCC. The failure mode is intergranular cracking in otherwise ductile material, usually without the formation of visible corrosion products. Cracks, which typically run parallel to the weld, are found in the weld metal, in the base metal, (~ 5 mm, or 0.2 in., away from the weld), and in the HAZ. Cracking is not related to weld hardness. To prevent amine SCC, postweld heat treatment at 620 °C (1150 °F) was recommended in the past for carbon steel welds exposed to amine solutions at temperatures exceeding 95 °C (200 °F). In light of the recently reported failures, welds of carbon steel components in amine service should be postweld heat treated regardless of service temperature.

Polythionic Acid Cracking. Stress-corrosion cracking of austenitic stainless steels by polythionic acids was first identified with the introduction of hydrotreating units. Austenitic stainless steels were required to provide resistance to high-temperature sulfidic corrosion in the presence of hydrogen. It was found that unstabilized austenitic stainless steel, such as type 304 (S30400), would crack adjacent to weldments during shutdowns. Typically, cracks were found to penetrate piping with a wall thickness of 12 mm (0.5 in.) in less than 8 h. Failures have been limited mostly to furnace tubes, heat-exchanger tubes, thermowells, and vessel linings (Ref 124). Similar cracking was also found in hydrocracking units and, more recently, in catalytic cracking units, in which austenitic stainless steels have found greater use because of an increase in catalyst regeneration temperatures (Ref 125).

Examples of SCC by polythionic acids are shown in Fig. 16 to 18. The cracking in roll-bonded cladding of type 304 (S30400) stainless steel (Fig. 16) is similar to mud cracking. Figure 17 shows the cracking that occurred in a type 304 (S30400) furnace tube near the weld to a carbon steel tube. These cracks are both parallel and perpendicular to the weld, reflecting different stresses in the weldment. The intergranular mode of crack propagation is shown in Fig. 18 and clearly distinguishes SCC by polythionic acids from chloride SCC (but not from caustic SCC).

Polythionic acid SCC occurs only in austenitic stainless steels and nickel-chromium-iron alloys that have become sensitized through thermal exposure (Ref 126, 127). Sensitization occurs when the carbon present in the alloy reacts with chromium to produce chromium carbides at the grain boundaries. As a result, the areas adjacent to the grain boundaries become depleted in chromium and are no longer fully resistant to certain corrosive environments.

Sensitization of type 304 (S30400) stainless steels normally occurs at temperatures between 400 and 815 °C (750 and 1500 °F), whenever the alloy is slowly cooled through this temperature range (such as during welding and heat treating), or during normal process operations. The higher the temperature, the shorter the time of exposure required for sensitization. Addition of stabilizing elements, such as titanium or niobium, or limiting the amount of carbon are two methods for reducing the effects of welding and heat treating on sensitization. However, they are not effective for long-term exposure to temperatures above 430 °C (800 °F). The resistance of titanium-stabilized type 321 (S32100) stainless steel to polythionic SCC can be significantly improved by a thermal stabilization at approximately 900 °C (1650 °F) and holding for 2 h, with no specific limits on the cooling rate. Thermal stabilization causes the precipitation of carbon as titanium carbide rather than chromium carbide and therefore decreases the amount of carbon available for chromium carbide formation upon subsequent high-temperature exposure. Also, any chromium depletion that does occur near the grain boundaries during this time period will be counteracted by chromium diffusion from within the alloy.

Laboratory studies and plant experiences have demonstrated that austenitic stainless steels are not sensitized when applied as a weld overlay

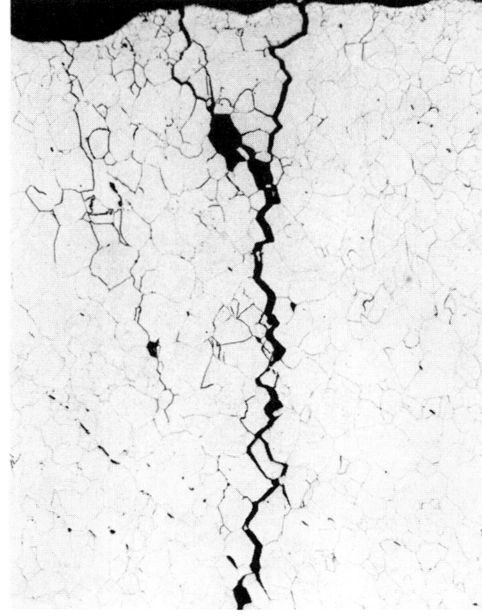

Fig. 18 Intergranular cracking typical of polythionic acid SCC in type 304 (S30400) stainless steel. 75×

over carbon or low-alloy steels. The lack of sensitization under these conditions was verified by testing stressed samples in a solution containing polythionic acids. As can be seen in Fig. 16, SCC of the roll-bonded cladding stops at the weld overlay around the nozzle.

Polythionic acids of the type $H_2S_xO_y$ (including sulfurous acid) are formed by the reaction of oxygen and water with the iron/chromium sulfide scale that covers the surfaces of austenitic stainless steel components as a result of high-temperature sulfidic corrosion. Because neither oxygen nor water is present during normal operation under conditions in which austenitic stainless steels would be used, SCC evidently occurs during shutdowns. Oxygen and water originate from steam or wash water used to free components of hydrocarbons during shutdown before inspection or simply from atmospheric exposure. In catalytic cracking units, oxygen and water can be present during normal operations at certain locations of the catalyst regeneration system because of steam purges and water sprays for preventing catalyst accumulation. The components involved include air rings, plenums, slide valves, cyclone components, and expansion joint bellows in the catalyst regenerator and associated lines.

In general, however, SCC by polythionic acids is considered to be a problem primarily during shutdown periods; suitable procedures to prevent cracking are outlined in NACE RP-01-70 (Ref 118). These procedures include nitrogen purging of components that were opened to the atmosphere, purging with dry air having a dew point below −15 °C (5 °F), or neutralizing any polythionic acids that are formed, by washing components with a 2% aqueous soda ash (sodium carbonate) solution.

Soda ash solution should also be used for hydrotesting prior to returning components to service. Residues of soda ash solution should be left on components during temporary storage to prevent SCC. The need for this can be illustrated by an experience with a U-tube heat-exchanger bundle fabricated of type 304 (S30400) stainless steel. After the bundle had been removed from a hydrotreating unit, the external surfaces of the tubes were washed with soda ash solution, which was allowed to dry. Before storing the bundle outdoors, instructions were given to cover it to prevent rainwater from washing off the soda ash residues. It was later discovered, however, that the U-bends had not been covered and that extensive SCC had taken place at these locations.

Hydrogen Damage

Corrosion of carbon and low-alloy steels by aqueous hydrogen sulfide solutions or sour waters can result in one or more types of hydrogen damage. These include loss of ductility on slow application of strain (hydrogen embrittlement), formation of blisters or internal voids (hydrogen blistering), and spontaneous cracking of high-strength or high-hardness steels (hydrogen stress cracking).

Atomic hydrogen (H) forms as part of the corrosion process and then evolves from cathodic areas of the metal as molecular hydrogen (H_2). When corrosion rates are high enough, desorption of molecular hydrogen from the surface becomes rate controlling. As the concentration of atomic hydrogen builds up at the surface, it diffuses into the steel and reduces its ductility. This embrittling effect is caused by hydrogen atoms collecting interstitially between metal atoms. The metal lattice becomes locally distorted, which restricts the mobility of dislocations and therefore the ability of the lattice to deform. Hydrogen atoms preferentially diffuse along grain boundaries and zones where the lattice has already been distorted by cold working or hardening. Atomic hydrogen can also combine to form molecular hydrogen at voids, such as manganese sulfide inclusion or laminations. Because of their larger size, hydrogen molecules cannot diffuse out of the steel, and blistering and fissuring are the result. Hydrogen stress cracking of embrittled metal is caused by static external stresses, transformation stresses (for example, as a result of welding), internal stresses, cold working, and hardening. As a rule, cracking does not occur in ductile steels or in steels that have received a proper postweld heat treatment.

Hydrogen damage occurs primarily when steel is exposed to aqueous hydrogen sulfide solutions having low pH values. Hydrogen sulfide, chemisorbed to the steel surface, partially poisons the reaction between hydrogen atoms that yields molecular hydrogen. Aqueous hydrogen sulfide solutions having high pH values can also cause hydrogen damage if cyanides are present. In the absence of cyanides, aqueous hydrogen sulfide solutions with pH values above 8 do not corrode steel, because a protective iron sulfide film forms on the surface.

Cyanides destroy this protective film and convert it into soluble ferrocyanide $[Fe(CN)_6{}^{-4}]$ complexes. As a result, the now unprotected steel can corrode very rapidly. For practical purposes, the corrosion rate depends primarily on the bisulfide ion (HS^-) concentration and, to a lesser extent, on the cyanide ion (CN^-) concentration. The more bisulfide ion that is present, the more cyanide that is required to destroy the protective iron sulfide film. It has been shown experimentally that corrosion of steel in aqueous ammonia/sulfide/cyanide solutions with pH values above 8 is always accompanied by hydrogen damage (Ref 54).

Hydrogen embrittlement is characterized by decreasing ductility with decreasing strain rate; this is contrary to metal behavior in most other types of embrittlement (Ref 128). For example, the ductility of carbon steel has been reported to drop from 42 to 7% when charged with hydrogen (Ref 129). This loss of ductility is only observed during slow strain rate testing and conventional tensile tests, but not during impact tests, such as the Charpy V-notch test. Failure, in the form of cracking, usually occurs some time after a load is applied to hydrogen-charged steel. Because this phenomenon is also known as static fatigue, the minimum load for failure to occur is known as the static fatigue limit.

Hydrogen embrittlement is temporary and can be reversed by heating the steel to drive out the hydrogen. The rate of recovery depends on time and temperature. Heating to 230 °C (450 °F) and holding for 1 h/25 mm (1 in.) of thickness has been found to be adequate to prevent cracking after welding. Although temperatures as high as 650 °C (1200 °F) for 2 h or as low as 105 °C (225 °F) for 1 day have reportedly been used to restore full ductility, even the heat of the sun on a summer day was found to be sufficient to restore ductility to a high-carbon cold-drawn steel wire that had been embrittled by exposure to wet hydrogen sulfide. As a rule, however, heating to temperatures above 315 °C (600 °F) for any length of time should be avoided to lessen the possibility of high-temperature hydrogen attack.

Titanium can also become embrittled by absorbed hydrogen as a result of corrosion or exposure to dry hydrogen gas (Ref 20). When hydrogen is absorbed by titanium in excess of about 150 ppm, a brittle titanium hydride phase will precipitate out, as shown in Fig. 19. Embrittlement due to titanium hydride precipitation is usually permanent and can be reversed only by vacuum annealing, which is difficult to perform. Absorption of hydrogen by titanium dramatically increases once the protective oxide film normally present on the metal is damaged through either mechanical abrasion or chemical reduction. Hydrogen intake is accelerated by the presence of surface contaminants, including iron smears, and occurs predominantly as temperatures exceed 70 °C (160 °F).

Hydriding can be minimized by anodizing or thermal oxidizing treatments to increase the thickness of the protective oxide film. If it is impractical to apply these treatments, acid pickling of titanium components—with 10 to 30 vol% nitric acid containing 1 to 3 vol% hydrofluoric acid at 49 to 52 °C (120 to 125 °F) for 1 to 5 min—can be performed to remove iron smears. Acid pickling is also recommended for cleaning titanium components after inspection and repairs during shutdowns, especially components exposed to concentrated acetic acid in certain petrochemical operations. To minimize hydrogen pickup during pickling, the volume ratio of nitric acid to hydrofluoric acid should be near 10. In some highly aggressive process environments, titanium components may have to be electrically insulated from more anodic components, such as aluminum, to prevent hydride formation as a result of hydrogen evolution on titanium surfaces. When process streams contain a significant volume of hydrogen (for example, reactor effluent from hydrotreating units), titanium should be used only at temperatures below 175 °C (350 °F).

Information on the mechanism of hydrogen embrittlement is available in the article "Environmentally Induced Cracking" in this Volume. Test procedures are reviewed in the article "Evaluation of Hydrogen Embrittlement" also in this Volume.

Hydrogen blistering has been a problem primarily in the vapor recovery (light ends) section of catalytic cracking units and, to a lesser degree, in the low-temperature areas of the reactor effluent section of hydrotreating and hydrocracking units (Ref 130-132). Hydrogen blistering has also been seen in the overhead systems for sour water stripper towers and amine regenerator (stripper) towers, as well as in the bottom of amine contactor (absorber) towers.

An example of hydrogen blistering in an absorber/stripper tower of a catalytic cracking unit is shown in Fig. 20. Hydrogen blistering often accompanies hydrogen embrittlement as a result of aqueous sulfide corrosion. Internal hydrogen blistering on a microscopic scale along grain boundaries (fissures) can lead to hydrogen-induced stepwise cracking. Cracking proceeds as metal ligaments between adjacent fissures crack because of applied stresses. As a rule, the severity of hydrogen blistering depends on the severity of corrosion, but even low corrosion rates can produce enough hydrogen to cause extensive damage. In some cases, hydrogen blistering is limited to dirty steel with highly oriented slag inclusions or laminations. Vapor/liquid interface areas in equipment often show

Fig. 19 Hydride formation in titanium grade 2 (R50400) after galvanic coupling to carbon steel in sour water at 110 °C (230 °F)

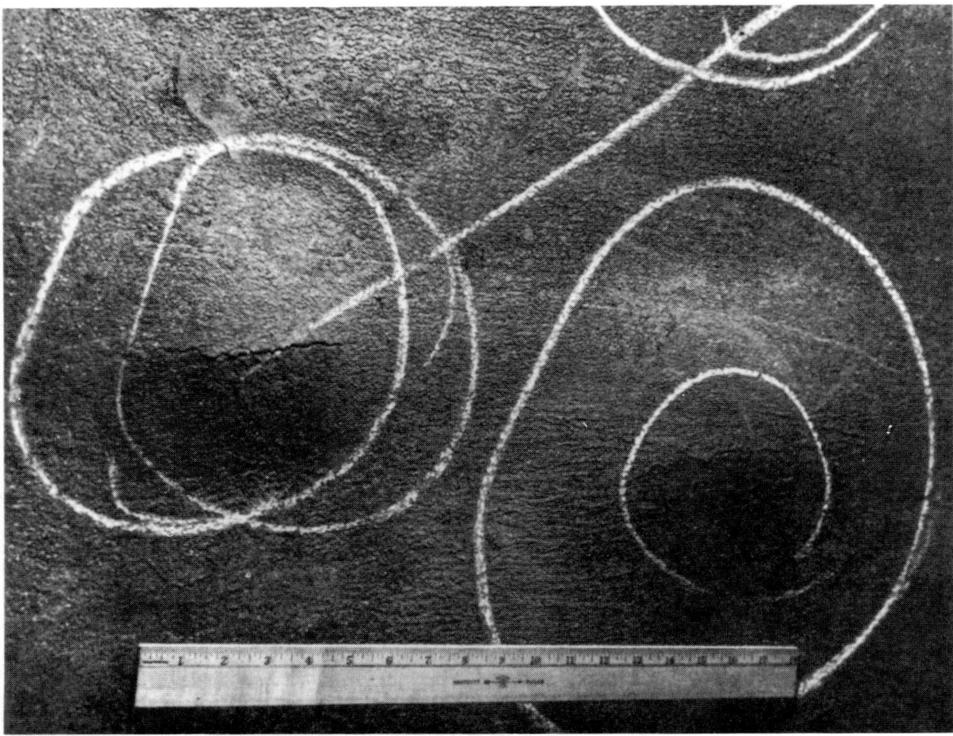

Fig. 20 Hydrogen blistering of a carbon steel shell of an absorber/stripper tower in the vapor recovery (light ends) section of a catalytic cracking unit. Note that the blisters have cracked open to the vessel interior

most of the damage, probably because ammonia, hydrogen sulfide, and hydrogen cyanide concentrate in the thin water films or in water droplets that collect at these areas.

The basic approach toward reducing corrosion and hydrogen blistering in the various vapor-compression stages of catalytic cracking units should be aimed at decreasing the concentration of cyanide and bisulfide ions in water condensate. Several methods for accomplishing this have been tried over the years (Ref 133, 134). Conversion of cyanide to harmless thiocyanate (SCN^-) by injection of air or polysulfide solutions at various locations has often produced undesirable side effects, such as accelerated corrosion and fouling at stagnant-flow areas. In contrast, water washing of the compressed wet-gas streams, in conjunction with corrosion inhibitor injection, has been found to be very effective when applied correctly and consistently (Ref 135). Water washing reduces the concentration of cyanides by improved contacting of vapors and dilution of water condensate. To prevent dissolved and suspended solids from fouling the compressor aftercooler, only water of fairly good quality, such as boiler feedwater or steam condensate, should be injected. To reduce the amount of freshwater used, stripping-stream condensate from the reflux drum can be used. As a rule, there is sufficient stripping-steam condensate to meet the wash-water requirements.

It is important that the waste sour water from the interstage and high-pressure separator drums be sent directly to waste disposal rather than first being recycled to the reflux drum. Waste water is often recycled for convenience so that its pressure can be reduced in the reflux drum prior to disposal. This alleviates the need for an external depressuring drum, but will build up the concentration of ammonia, hydrogen sulfide, and, especially, hydrogen cyanide in the wet gas leaving the reflux drum. Consequently, excessive concentrations of cyanides will be found in water condensing in the high-pressure stage. Water washing of the overhead of the debutanizer and depropanizer is indicated only if serious fouling problems occur. Normally, these streams are quite dry and should be kept that way to minimize corrosion and hydrogen blistering problems. With proper water washing of the compressed wet-gas stream, water washing of the overhead vapor streams of the debutanizer and depropanizer towers becomes unnecessary.

Corrosion inhibitors help control aqueous sulfide corrosion and hydrogen blistering even though cyanides may still be present. Hydrogen activity probes and chemical tests of water condensate are used to monitor the effectiveness of water washing and inhibitor injection. Where limited hydrogen blistering occurs in certain components of hydrotreating and hydrocracking units, it is usually sufficient to line affected areas with stainless steel or alloy 400 (N04400). This also applies to components of overhead systems for sour water stripper towers and amine regenerator (stripper) towers or the bottoms of amine contactor (absorber) towers.

Hydrogen Stress Cracking. Sour water containing hydrogen sulfide can cause spontaneous cracking of highly stressed high-strength steel components, such as bolting and compressor rotors (Ref 136). Cracking has also occurred in carbon steel components containing hard welds (Ref 137). Cracking is typically transgranular and will contain sulfide corrosion products, as shown in Fig. 21 and 22. Cracking of this type has become known as hydrogen stress cracking or sulfide cracking and should not be confused with hydrogen-induced stepwise cracking.

Hydrogen stress cracking was first identified in the production of sour crude oils when high-strength steels used for well-head and down-hole

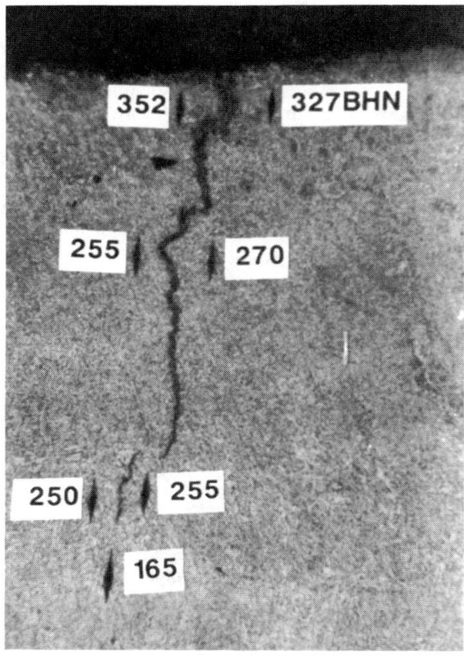

Fig. 21 Hydrogen stress cracking of a hard weld of a carbon steel vessel in sour water service. BHN = Brinell hardness. 40×

Fig. 22 Hydrogen stress cracking of hard HAZ next to weld in A516-70 pressure vessel steel after exposure to sour water. 35×

equipment cracked readily after contacting produced water that contained hydrogen sulfide. Hydrogen stress cracking was not experienced by refineries and petrochemical plants until the introduction of high-pressure processes that required high-strength bolting and other components in gas compressors. With the increased use of submerged arc welding for pressure vessel construction it was found that weld deposits significantly harder and stronger than the base metal could be produced. This led to transverse cracking in the weld deposit.

The mechanism of hydrogen stress cracking has been the subject of many investigations, most of which attempted to address the cracking seen in high-strength steels instead of the lower-strength steels used in refinery and petrochemical plant equipment. In general terms, hydrogen stress cracking occurs in the same corrosive environments that lead to hydrogen embrittlement. Hydrogen sulfide affects the corrosion rate and the relative amount of hydrogen absorption, but otherwise does not appear to be directly involved in the cracking mechanism. As a general rule of thumb, hydrogen stress cracking can be expected to occur in process streams containing in excess of 50 ppm hydrogen sulfide (although cracking has been found to occur at lower concentrations).

There is a direct relationship between hydrogen sulfide concentration and the allowable maximum hardness value of the HAZ on one hand and cracking threshold stress on the other. Typically, the allowable maximum hardness value decreases 30 HB, and the allowable threshold stress decreases by 50% for a tenfold increase in hydrogen sulfide concentration (Ref 138). Also, hydrogen stress cracking occurs primarily at ambient temperature. As in the case of hydrogen embrittlement and hydrogen blistering, hydrogen stress cracking of steel in refineries and petrochemical plants often requires the presence of cyanides.

The most effective way of preventing hydrogen stress cracking is to ensure that the steel is in the proper metallurgical condition. This means that weld hardness is limited to 200 HB (Ref 139). Because hard zones can also form in the HAZs of welds and shell plates from hot forming, the same hardness limitation should be applied in these areas. Guidelines for dealing with the hydrogen stress cracking that occurs in refineries and petrochemical plants are given in API 942 (Ref 140) and NACE RP-04-72 (Ref 141).

Postweld heat treatment of fabricated equipment will greatly reduce the occurrence of hydrogen stress cracking. The effect is twofold: First, there is the tempering effect of heating to 620 °C (1150 °F) on any hard microstructure, and second, the residual stresses from welding or forming are reduced. The residual stresses represent a much larger strain on the equipment than internal pressure stresses.

A large number of the ferrous alloys, including the stainless steels, as well as certain nonferrous alloys, are susceptible to hydrogen stress cracking. Cracking may be expected to occur with carbon and low-alloy steels when the tensile strength exceeds 620 MPa (90 ksi). Because there is a relationship between hardness and strength in steels, the above strength level approximates the 200 HB hardness limit. For other ferrous and nonferrous alloys used primarily in oil field equipment, limits on hardness and/or heat treatment have been established in NACE MR-01-75 (Ref 142). Although oil field environments can be more severe than those encountered during refining, the recommendations can be used as a general guide for material selection.

Hydrogen Attack

The term hydrogen attack (or, more specifically, high-temperature hydrogen attack) refers to the deterioration of the mechanical properties of steels in the presence of hydrogen gas at elevated temperatures and pressures. Although not a corrosion phenomenon in the usual sense, hydrogen attack is potentially a very serious problem with regard to the design and operation of refinery equipment in hydrogen service (Ref 143, 144). It is of particular concern in hydrotreating, reforming, and hydrocracking units at above roughly 260 °C (500 °F) and hydrogen partial pressures above 689 KPa (100 psia) (Ref 145). Under these conditions, molecular hydrogen (H_2) dissociates at the steel surface to atomic hydrogen (H), which readily diffuses into the steel. At grain boundaries, dislocations, inclusions, gross discontinuities, laminations, and other internal voids, atomic hydrogen will react with dissolved carbon and with metal carbides to form methane. The large size of its molecule precludes methane diffusion. As a result, internal methane pressures become high enough to blister the steel or to cause intergranular fissuring (Ref 131). If temperatures are high enough, dissolved carbon diffuses to the steel surface and combines with atomic hydrogen to evolve methane. Hydrogen attack now takes the form of overall decarburization rather than blistering or cracking.

The overall effect of hydrogen attack is the partial depletion of carbon in pearlite (decarburization) and the formation of fissures in the metal, as shown in Fig. 23. As attack proceeds, these effects become more pronounced, as shown in Fig. 24, in which partial depletion of carbon is evident in some of the grains while others are completely decarburized. Hydrogen attack is accompanied by loss of tensile strength and ductility. Consequently, unexpected failure of equipment without prior warning signs is the primary cause for concern.

Forms of Hydrogen Attack. Hydrogen attack can take several forms within the metal structure, depending on the severity of the attack, stress, and the presence of inclusions in the steel. The following discussion will illustrate these. General surface attack occurs when equipment, which is not under stress, is exposed to hydrogen at elevated temperatures and pressures. As a rule, decarburization is not uniform across the surface or through the thickness; instead, it takes place at various locations within the structure. The fissures that form are parallel to the metal surface. The fissures themselves are small and are not linked together, as may happen with more severe stages of attack.

Hydrogen attack often initiates at areas of high stress or stress concentration in the steel because atomic hydrogen preferentially diffuses to these areas. Isolated fingers of decarburized and fissured material are often found adjacent to weldments and are associated with the initial stages of hydrogen attack. It is also evident that the fissures tend to be parallel to the edge of the weld rather than the surface. This orientation of fissures is probably the result of residual stress

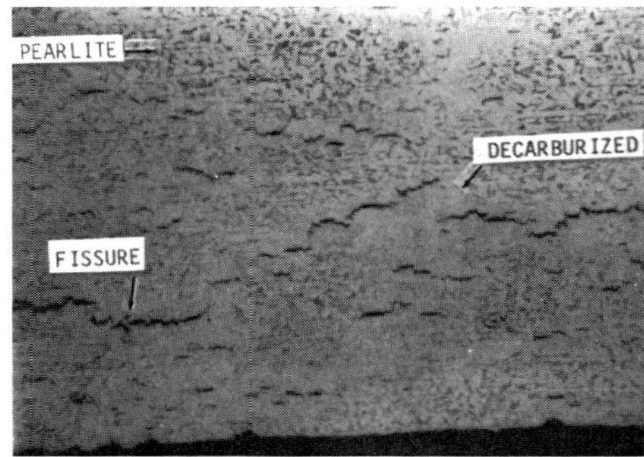

Fig. 23 High-temperature hydrogen attack of carbon steel in the form of decarburization and fissuring. 50×

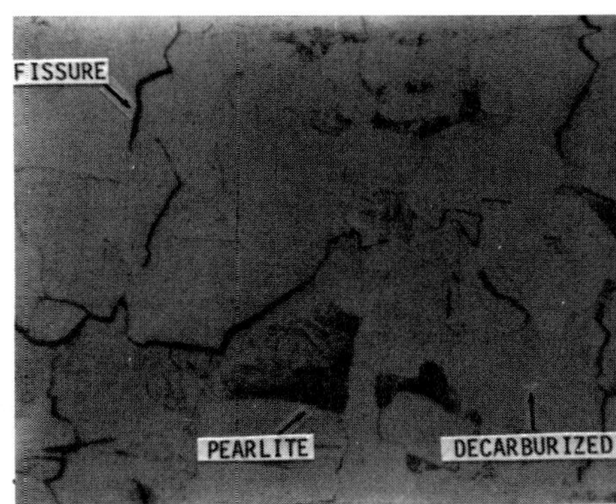

Fig. 24 Depletion of carbon in pearlite colonies and formation of grain-boundary fissures due to high-temperature hydrogen attack of carbon steel. 140×

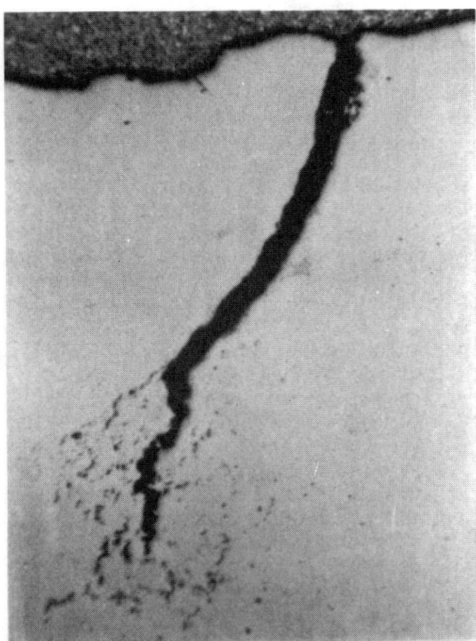

Fig. 25 High-temperature hydrogen attack, in the form of localized fissuring, at the tip of a fatigue crack that initiated at the toe of a fillet weld. 70×

adjacent to the weldment. Fissures in this direction can form through-thickness cracks.

The necessary stress for inducing localized hydrogen attack is not limited to weldments. Hydrogen attack has been found to be concentrated at the tip of a fatigue crack that initiated at the toe of a fillet weld and propagated along the HAZ of the weld. In this case, the hydrogen-containing process stream evidently entered the fatigue crack and caused fissuring around the crack tip, as shown in Fig. 25. Although no evidence of attack was found in adjacent portions of the piping system, the localized attack was the cause of a major failure.

Severe hydrogen attack can result in blisters and laminations, as shown in Fig. 26. This is an advanced stage of hydrogen attack, and it is accompanied by complete decarburization throughout the cross section of the steel. The laminar nature of the fissures is typically obtained when no local stresses are present, but the physical appearance of this blistering is quite similar to hydrogen blistering (described earlier).

Prevention of Hydrogen Attack. The only practical way to prevent hydrogen attack is to use only steels that, based on plant experience, have been found to be resistant to this type of deterioration. The following general rules are applicable to hydrogen attack:

- Carbide-forming alloying elements, such as chromium and molybdenum, increase the resistance of steel to hydrogen attack
- Increased carbon content decreases the resistance of steel to hydrogen attack
- Heat-affected zones are more susceptible to hydrogen attack than the base or weld metal

For most refinery and petrochemical plant applications, low-alloy chromium- and molybdenum-containing steels are used to prevent hydrogen attack. However, questions have recently been

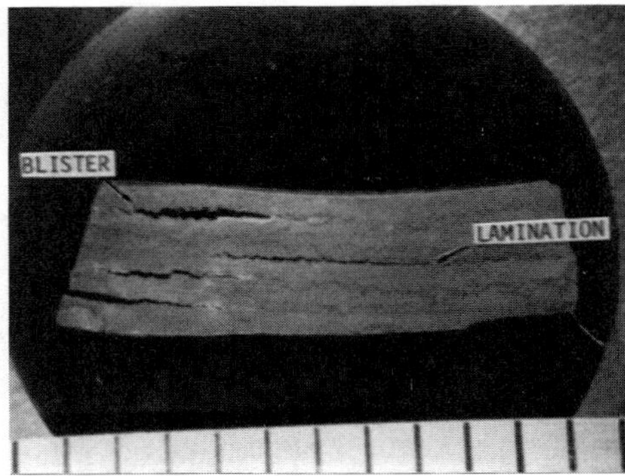

Fig. 26 High-temperature hydrogen attack in the form of blistering and laminar fissuring throughout the wall thickness of a carbon steel pipe

raised regarding the effect of long-term hydrogen exposure on C-0.5Mo steel (Ref 146). As a result, low-alloy steels are preferred over C-0.5Mo steel for new construction.

The conditions under which different steels can be used in high-temperature hydrogen service are listed in API 941 (Ref 147). The principal data are presented in the form of Nelson curves, as shown in Fig. 27. The curves are based on long-term refinery experience, rather than on laboratory studies. The curves are periodically revised by the API Subcommittee on Materials Engineering and Inspection, and the latest edition of API 941 should be consulted to ensure that the proper steel is selected for the operating conditions encountered.

In addition to hydrogen attack, hydrogen stress cracking can occur at carbon and low-alloy steel welds that were in hydrogen service above approximately 260 °C (500 °F). Cracking is intergranular and typically follows lines of high, localized stress and/or hardness. Cracking is caused by dissolved hydrogen and is prevented by postweld heat treatment. Proper hydrogen outgasing procedures should be followed when equipment is depressurized and cooled prior to shutdown.

Stainless steels with chromium contents above 12% and, in particular, the austenitic stainless steels are immune to hydrogen attack. It should be noted, however, that atomic hydrogen will diffuse through these steels; as a result, they will not provide protection against hydrogen attack if applied as a loose lining or an integral cladding over a nonresistant base steel.

Corrosion Fatigue

Corrosion, in conjunction with cyclic stressing, can bring about a significant reduction in the fatigue life of a metal. Failure under these circumstances is described as corrosion fatigue. Rotating equipment, valves, and some piping runs in refineries and petrochemical plants may be subject to corrosion fatigue. In particular, pump shafts and various springs are the two most likely candidates for corrosion fatigue. The types of springs involved include those of scraper-blade devices in a wax production unit, internal springs in relief valves, and compressor valve springs.

Prevention of Corrosion Fatigue. A number of corrective procedures are available for preventing corrosion fatigue. These include increasing the fatigue resistance and corrosion resistance of the metal involved, reducing the number of stress cycles or the stress per cycle, and removing or inhibiting the corrosive agent in the environment. Fatigue life can often be increased through heat treatments or alloy changes, which make the metal stronger and tougher. Corrosion resistance can be improved by applying protective coatings or by a material change. A design change can eliminate vibration or (in a spring) reduce the stress per cycle. Finally, adding a corrosion inhibitor or removing a source of pitting, such as chlorides, can often increase the corrosion fatigue life of the failing part. Additional information on corrosion fatigue is available in the article "Mechanically Assisted Degradation."

Liquid-Metal Embrittlement

Although liquid-metal embrittlement has been recognized for at least 50 years, it has received far less attention than the more commonly encountered hydrogen embrittlement or stress-cor-

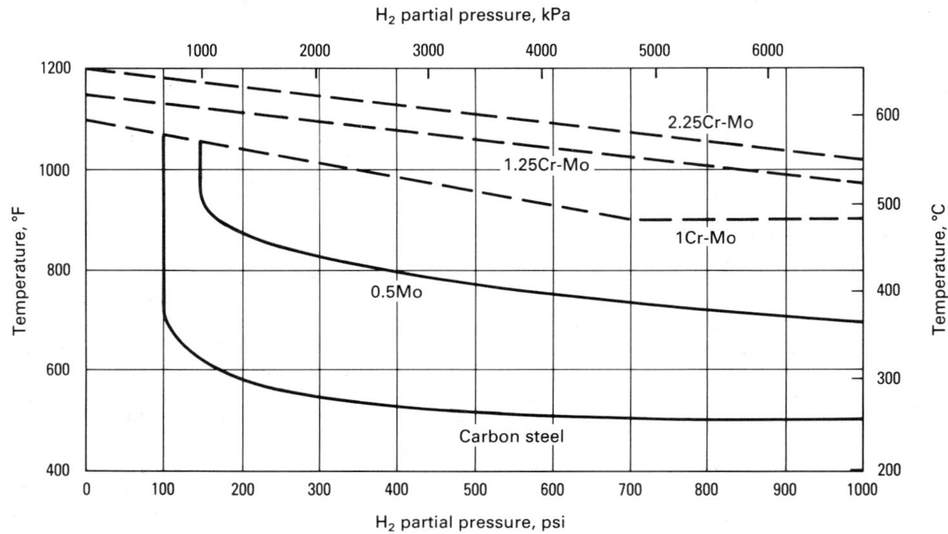

Fig. 27 Operating limits for various steels in high-temperature high-pressure hydrogen service (Nelson curves) to avoid decarburization and fissuring. Source: Ref 146

rosion cracking. This is due in part to the fact that the probability of liquid-metal contact occurring in refineries and petrochemical plants is normally rather small. In situations in which liquid-metal embrittlement has occurred, it has been mainly due to the zinc embrittlement of austenitic stainless steels. Isolated failures have been attributed to welding in the presence of residues of zinc-rich paint or to the heat treating of welded pipe components that carried splatter of zinc-rich paint. However, most of the reported failures due to zinc embrittlement have involved welding or fire exposure of austenitic stainless steel in contact with galvanized steel components.

For example, in one case, severe and extensive cracking in the weld HAZ of process piping made from austenitic stainless steel occurred in a petrochemical plant during the final stages of construction. Much of the piping had become splattered with zinc-rich paint. Although the welders had been instructed to clean affected piping prior to welding, no cleaning and only limited grinding were performed. After welding, dye penetrant inspection revealed many thin, branched cracks in the HAZ of welds, as shown in Fig. 28.

In many cases, through-wall cracks cause leaks during hydrotesting. Typically, zinc embrittlement cracks contain zinc-rich precipitates on fracture surfaces and at the very end of the crack tip. Cracking is invariably intergranular in nature (Ref 148).

Several different models for the zinc embrittlement of austenitic stainless steel have been proposed. The most accepted model involves the reduction in atomic bond strength at a surface imperfection or crack tip by chemisorbed zinc metal. Zinc embrittlement is a relatively slow process that is controlled by the rate of zinc diffusion along austenitic grain boundaries. Zinc combines with nickel, and this results in nickel-depleted zones adjacent to the grain boundaries. The resulting transformation of face-centered cubic austenite (γ) to body-centered cubic ferrite (α) in this region is thought to produce not only a suitable diffusion path for zinc but also the necessary stresses for initiating intergranular cracking. Externally applied stresses accelerate cracking by opening prior cracks to liquid metal.

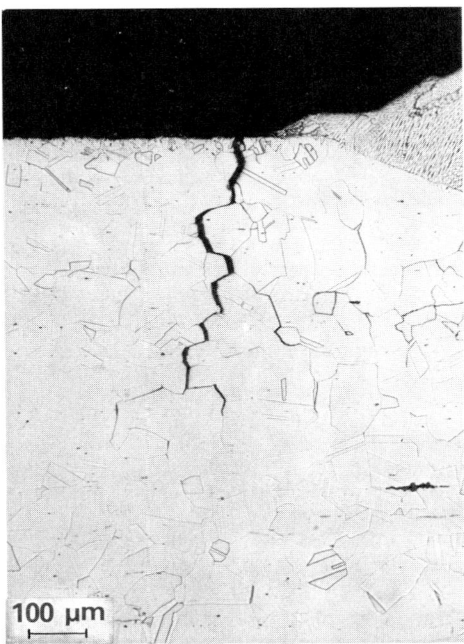

Fig. 28 Intergranular cracking in HAZ of stringer-bead weld on type 304 (S30400) stainless steel pipe due to zinc embrittlement. Weld area had been covered with zinc-rich paint.

Although the melting point of zinc is 420 °C (788 °F), no zinc embrittlement has been observed at temperatures below 570 °C (1380 °F), probably because of phase transformation and diffusion limitations. There is no evidence that an upper temperature limit exists. In the case of zinc-rich paints, only those having metallic zinc powder as a principal component can cause zinc embrittlement of austenitic stainless steels. Paints containing zinc oxide or zinc chromates are known not to cause embrittlement.

Prevention of Zinc Embrittlement. Obviously, the best approach to the prevention of zinc embrittlement is to avoid or minimize zinc contamination of austenitic stainless steel components in the first place. In practice, this means using no galvanized structural steel, such as railings, ladders, walkways, or corrugated sheet metal, at locations where molten zinc is likely to drop on stainless steel components if a fire occurs. If zinc-rich paints will be used on structural steel components, shop priming is to be preferred. Field application of zinc-rich paints should be done after all welding of stainless steel components has been completed and after insulation has been applied. Otherwise, stainless steel components should be temporarily covered with plastic sheathing to prevent deposition of overspray and splatter.

If stainless steel components have become contaminated despite these precautionary measures, proper cleaning procedures must be implemented. Visible paint overspray should be removed by sandblasting, wire brushing, or grinding. These operations should be followed by acid pickling and water rinsing. Acid pickling will remove any traces of zinc that may have been smeared into the stainless steel surface by mechanical cleaning operations. Suitable acid pickling solutions include 5 to 10% nitric acid, phosphoric acid, or sulfuric acid. Hydrochloric acid should not be used in order to avoid potential pitting or SCC problems. After removal of all traces of acid by water rinsing, final cleaning with naphtha solvent should be performed immediately before welding. Additional information on liquid-metal embrittlement and a related phenomenon—solid-metal embrittlement—can be found in the article "Environmentally Induced Cracking" in this Volume.

Erosion-Corrosion

Various materials of construction for refinery and petrochemical plant service may exhibit accelerated metal loss under unusual fluid-flow conditions. Attack is caused by a combination of flow velocity (mechanical factors) and corrosion (electrochemical factors) known as erosion-corrosion. Affected metal surfaces will often contain grooves or wavelike marks that indicate a pattern of directional attack. Soft metals, such as copper and aluminum alloys, are often especially prone to erosion-corrosion, as are metals such as stainless steels, which depend on thin oxide films for corrosion protection. Most cases of erosion-corrosion can be mitigated by proper design and/or material changes. For example, by eliminating sharp bends, erosion-corrosion problems can be significantly reduced in process piping. Increasing the pipe diameter of vapor lines will reduce flow velocities and therefore the erosion-corrosion by impinging droplets of liquid. Piping immediately downstream of pressure let-down valves often must be upgraded to prevent accelerated attack due to high flow turbulence.

Cavitation. Cavitation damage is a fairly common form of erosion-corrosion of pump impellers or hydraulic turbine internals. Cavitation is caused by collapsing gas bubbles at high-pressure locations; adjacent metal surfaces are damaged by the resultant hydraulic shock waves. Cavitation damage is usually in the form of loosely spaced pits that produce a roughened surface area. Subsurface metal shows evidence of mechanical deformation. As a general rule, cast alloys are likely to suffer more damage than wrought versions of the same alloy. Ductile materials, such as wrought austenitic stainless steels, have the best resistance to cavitation. Damage can be reduced by design changes, material changes, and the use of corrosion inhibitors. Smooth finishes on pump impellers will reduce damage. Some coatings can be beneficial. Design changes with the objective of reducing pressure gradients in the flowing liquid are most effective.

Mixed-Phase Flow. Accelerated corrosion due to mixed vapor/liquid streams is primarily found in crude and vacuum furnace headers and transfer lines of crude distillation units, in overhead vapor lines and condenser inlets on various fractionation towers, and in reactor effluent coolers of hydrocracking and hydrotreating units.

In general, increases in vapor load and mass velocity increase the severity of high-temperature sulfidic corrosion by crude oils and atmospheric residuum (reduced crude) (Ref 149). Corrosion is least severe with flow regimes in which the metal surface is completely wetted with a substantial liquid hydrocarbon layer. Corrosion is most severe with the spray flow that results from vapor velocities above 60 m/s (200 ft/s) and vapor loads above 60%.

Under these conditions, corrosion rates of certain components, such as furnace headers, furnace-tube return bends, and piping elbows, could increase by as much as two orders of magnitude. This phenomenon is caused by droplet impingement, which destroys the protective sulfide scale normally found on steel components, as shown in Fig. 29. Such impingement damage is usually not seen in straight piping, except immediately downstream of circumferential welds. Damage is usually in the form of sharp-edged laketype corrosion that, because of its appearance, is often confused with naphthenic acid corrosion. As a rule, 5Cr-0.5Mo steel components have sufficient resistance to all but severe cases of droplet impingement in transfer lines. Higher alloys should be used for furnace tubes and associated components, such as headers and return bends.

Corrosion damage at elbows of overhead vapor lines is often caused by droplet impingement as a result of excessively high vapor velocities. Typical impingement-type corrosion of tubes and baffles just below the vapor inlet of overhead condensers is shown in Fig. 30. As a general rule, overhead vapor velocities should be kept below 7.5 m/s (25 ft/s) to minimize impingement-type corrosion. In addition, horizontal impingement baffles can be mounted just above the top tube row of overhead condensers.

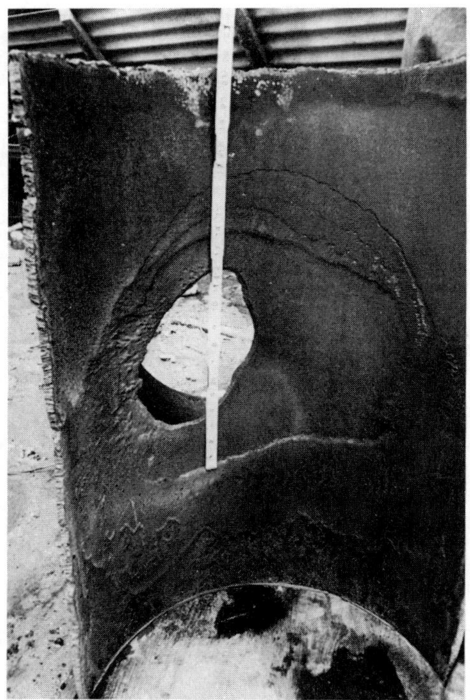

Fig. 29 Accelerated high-temperature sulfidic corrosion in 500-mm (20-in.) diam pipe of vacuum furnace outlet header due to droplet impingement at high vapor velocities

Fig. 30 Accelerated aqueous chloride corrosion below inlet nozzle of crude tower overhead condenser due to droplet impingement. Note partial loss of carbon steel baffles and localized corrosion along top of admiralty metal (C44300) tubes.

Air-cooled reactor effluent coolers of hydrocracking and hydrotreating units are also prone to impingement-type corrosion. Poor flow distribution through large banks of parallel air coolers can result in excessive flow velocities in some coolers, usually those in the center. The resulting low flow velocities in the outer coolers can cause deposition of ammonium sulfide and/or chloride in these coolers; this blocks the tubes and further increases velocities in the remaining air coolers (Ref 57). This problem is aggravated by low, nighttime air temperatures, which increase deposition problems. Installation of protective sleeves (ferrules) at the inlet tube end has helped to reduce attack in some cases; in others, it has only moved the area of attack to an area immediately downstream of the sleeves. Careful attention to proper flow distribution through redesign of the inlet headers is often the only way of controlling air cooler corrosion.

Entrained Catalyst Particles. Accelerated corrosion due to entrained catalyst particles can occur in the reaction and catalyst regeneration sections of catalytic cracking units. Refractory linings are required to provide protection against oxidation and high-temperature sulfidic corrosion, as well as erosion by catalyst particles, particularly in cyclones, risers, standpipes, and slide valves. Stellite hardfacing is used on some components to protect against erosion. When there are no erosion problems and when protective linings are impractical, austenitic stainless steels such as type 304 (S30400) can be used. Cyclone diplegs, air rings, and other internals in the catalyst regenerator are usually made of type 304 (S30400) stainless steel, as is piping for regenerator flue gas. The main fractionation tower is usually made of carbon steel, with the lower part lined with a ferritic or martensitic stainless steel containing 12% Cr such as type 405 (S40500) or 410 (S41000). Slurry piping between the bottom of the main fractionation tower and the reactor may receive an additional corrosion allowance as protection against excessive erosion.

Corrosion Control

A component in refinery or petrochemical service will require replacement when:

- Corrosion or other deterioration has made it unfit for further service
- It no longer performs satisfactorily, although it may still be operational
- It has become completely nonoperational

Based on the results of a failure analysis, certain corrective measures can be implemented. These include, for example, the use of alternative materials of construction, changes in equipment design and process conditions, the use of corrosion inhibitors, and the application of protective coatings and linings.

Materials Selection. The most common method for preventing repeated failures is selection of improved metals or alloys. If a given piece of equipment fails every other year or so, substitution of a more expensive metal that provides better performance can invariably be justified. In contrast, there is no justification for replacing a carbon-steel tank with one made from a more corrosion-resistant alloy if the tank corroded out after 35 years of service (unless, of course, all of the corrosion occurred during the last 6-month period).

Design Changes. Proper design of equipment is usually as important as proper selection of material. Design changes should involve consideration of changes in strength requirements, changes in material, and the need for additional allowances for corrosion, depending on past experience. Certain rules should be followed to reduce corrosion-related problems:

- Welded joints can be substituted for flanged joints to minimize crevice corrosion
- Equipment can be modified to permit easy cleaning and draining
- Piping for certain equipment, such as pumps, can be modified to allow bypassing or replacement during service
- Mechanical and vibrational stresses can be minimized by providing additional supports to avoid stress-related failures
- Sharp bends in piping can be realigned to prevent erosion-corrosion
- Heat exchangers can be modified to reduce temperature gradients that cause high stresses and also accelerate corrosion problems

Further information on design considerations is available in the articles "Design Details to Minimize Corrosion" and "Corrosion of Weldments" in this Volume.

Process Changes. Process changes that can be considered for reducing corrosion and other failures include the following:

- Temperature can be decreased to decrease corrosion rates
- Concentrations of critical corrosive species can be adjusted
- Flow velocity can be reduced to prevent erosion-corrosion
- Oxygen (air) can be removed by the use of scavenging chemicals
- Water entry can be controlled by installation of calcium chloride drying equipment, settling drums, or demister screens

Changing the concentration of corrosive contaminants in hydrocarbon streams is usually effective. Certain catalysts, such as hydrofluoric acid, are virtually inert when highly concentrated, but when diluted by water, they become extremely corrosive. Reducing the amount of acid entrained in hydrocarbon streams invariably reduces corrosion problems.

Changes in the flow regime of mixed vapor/liquid streams as a result of velocity changes can have a pronounced effect on corrosion. Very high velocities produce erosion-corrosion problems. Similarly, stagnant conditions must be avoided with metals that form passive oxide films, such as stainless steels.

Deaeration finds widespread application in the treatment of boiler feedwater. The corrosion of certain nickel alloys by hydrofluoric acid can be controlled by excluding air. However, metals that depend on passive oxide films for corrosion resistance may actively corrode in certain environments if air is excluded.

Corrosion Inhibitors. Filming-amine corrosion inhibitors are added in small concentrations to various process streams to help control corrosion. Filming-amine inhibitors become ineffective at temperatures above approximately 175 °C (350 °F)—not because they decompose, but because the rate of desorption begins to exceed the rate of adsorption. The inhibitors are primarily used to protect overhead condensing equipment. So-called water-dispersible types contain surfactants to allow addition to streams that contain excess water. In general, filming-amine inhibitors are continuously injected at a rate that is just sufficient to maintain protection. Higher-than-normal dosages are required for several days to establish a protective film after inhibitor injection has been interrupted (for example, because of pump failure) or after a shutdown. Most filming-amine inhibitors will tend to solubilize prior corrosion products; this can lead to fouling problems in downstream equipment. In amine units, these inhibitors usually cause foaming problems. Detailed information is available in the articles "Corrosion Inhibitors for Oil and Gas Production" and "Corrosion Inhibitors for Crude Oil Refineries" in this Volume.

Protective Coatings. Although most refineries and petrochemical plants do not rely on organic coatings for corrosion control, because a fire would immediately destroy these, extensive use is made of metallic cladding and refractory linings. Cladding is usually performed at the mill by rolling a relatively thin sheet of a corrosion-resistant metal or alloy to a thicker base metal at elevated temperatures. The finished product will exhibit a fully bonded, metallurgical joint. Clad plates can be formed and fabricated into vessels or piping by welding the joints of the base metal and overlaying the welds. Thin sheet strips of corrosion-resistant alloy can be attached by spot welding in the field to protect an unclad vessel. This process is known as strip lining. For long-term service, however, cladding tends to be more reliable than strip lining. Strip lining can become a source of problems with process streams that contain hydrogen at elevated temperatures and pressures or that evolve hydrogen because of corrosion. In either case, pockets of hydrogen can build up behind the lining and cause it to fail. Cladding by explosive bonding is primarily used for new construction to attach titanium or nickel-alloy-overlays.

Overlays of corrosion-resistant alloys can also be attached in the form of adjacent weld beads. For example, reactor vessels for hydrocracking and hydrotreating units are protected against high-temperature sulfidic corrosion with a weld overlay of type 347 (S34700) stainless steel.

Spray metallized coatings, as a rule, have not been extensively used in refinery and petrochemical plant applications, because such coatings have not been very reliable. In contrast, diffusion coatings, particularly aluminized coatings, have been highly successful as an alternative approach to controlling high-temperature sulfidic corrosion of carbon steel and low-alloy steel components. Aluminizing not only provides resistance to high-temperature sulfidic corrosion but also reduces scaling, carburization, coking, and erosion problems (Ref 150). Aluminized components are not necessarily less costly than those made of higher-alloy steels or stainless steels and a detailed cost analysis is usually required. Aluminizing is a proprietary process in which steel components are packed into a retort and exposed to aluminum vapor at a temperature above about 925 °C (1700 °F). The aluminum diffuses into the steel, and this forms a true metallurgical alloy containing over 50% Al at the surface.

Other methods for forming a corrosion-resistant iron-aluminum layer on steel surfaces include flame spraying and hot dipping, either with or without subsequent diffusion heat treatments. As a rule, these methods have been less effective than aluminizing, primarily because of nonuniform thickness, porosity, and holidays (Ref 151).

The aluminizing of carbon steel or low-alloy steel components typically increases their resistance to high-temperature sulfidation by two orders of magnitude (Ref 152). However, the diffused iron-aluminum layer is only 0.08 to 0.4 mm (3 to 15 mils) thick, depending on the base alloy, and can be readily damaged by mechanical (tube drilling for coke removal) or chemical means (acid or caustic cleaning). Aluminized components also require special fabrication techniques, including welding and roller expanding. Distortion and shrinkage during aluminizing can be a problem with some components.

A variety of processes for applying protective coatings are discussed in the Section "Corrosion Protection Methods" in this Volume. Methods for applying weld claddings are discussed in the articles "Weld Overlays," "Explosion Welding," and "Solid-State Welding" in Volume 6 of the 9th Edition of *Metals Handbook*.

Refractory Linings. Refractory materials are used in refineries and petrochemical plants as linings to protect steel from thermal degradation, erosion damage, and both low- and high-temperature corrosion. Unlike thin organic linings and coatings, refractory linings are applied in thicknesses of 25 to 100 mm (1 to 4 in.) or more. Metal wire reinforcement is used to help hold the lining in place.

Portland cement/sand is the lining that is most commonly used to protect steel against mild corrosives. This lining (frequently called gunite or shotcrete, after the application technique) has been used in refineries for more than 50 years. The primary application is sour water service with a pH value between 4 and 8. The material is low cost, and the application is straightforward. The mixture is applied with a pneumatic gun that shoots the premixed cement, sand, and water onto the vessel surface. A chicken-wire metal reinforcement is usually provided at the mid-thickness of the 25- to 40-mm (1- to 1.5-in.) thick lining. As with any other concrete, the lining must be cured before it is placed into service. During ambient-temperature curing, shrinkage cracks may occur, but they are not detrimental to the serviceability of the lining. Acidic water entering the cracks is neutralized by the alkalinity of the concrete and is trapped in the crack. However, continuing acid attack will result in gradual loss of thickness (much like corrosion), and aggregate, which is carried away in the process fluid, may cause erosion or plugging in downstream equipment, such as filters, pumps, and valves.

For highly acidic solutions or for higher temperatures, more resistant cements and aggregate are used, such as Lummite-Haydite cement. Special acid-resistant cements based on silicate, furane, phenolic, and sulfur compounds are used for highly acidic services, but at the cost of reduced resistance to alkaline solutions. In addition to gunning, these cements can be applied as thinner coatings by troweling.

For hot, highly acidic solutions, acid-proof brick linings can be used. The essential components of such a lining are a membrane lining against the steel shell, topped with brick-and-mortar construction. A membrane is placed between the bricks and the steel shell because brick-and-mortar construction is inherently porous and subject to capillary leakage. Membrane materials need not be completely immune to attack by the corrosive, because any swelling of the membrane will tend to seal off the bottoms of the capillary channels. Furthermore, the brick lining does provide thermal insulation, which reduces the temperature and therefore the aggressiveness of the corrosive. Typically, one course of 115-mm (4½-in.) thick brick raises the temperature limit of a membrane material by about 30 °C (55 °F), and two courses by about 50 °C (90 °F).

The membranes range from thermoplastic resins and elastomers to sheet or bonded lead linings, depending on the corrosive handled and the need to accommodate relative movement between the shell and the brick lining. Ceramic brick is satisfactory for all common operating conditions and corrosives, except strong alkalies, hydrofluoric acid, and fluorides, which require carbon brick. As a rule, acid-proof brick linings have limited applications in refineries and petrochemical plants, the principal uses being storage tanks for sulfuric acid. They are seldom used for pressure vessels, especially those that operate at elevated temperatures and pressures.

Heat-resistant linings can reduce the cost of pressure vessels that will operate at temperatures above 345 °C (650 °F). Allowable stresses, given in the ANSI/ASME Boiler and Pressure Vessel Code, Section VIII, for the design of carbon steel vessels, decrease with increasing temperatures. Because the maximum permissible temperature is 510 °C (950 °F), alloy steels must be used above this temperature. A heat-resistant lining, by reducing the metal temperature, will reduce the required wall thickness of carbon steel vessels, such as reactors. Alternatively, heat-resistant linings allow the use of carbon steel vessels at higher temperatures, such as reactors in fluid catalytic cracking units. For either of these applications, the integrity of the lining is important because loss of insulation can result in areas of the vessel reaching temperatures above the acceptable mechanical design limit (hot spots). When hot spots are found, temperatures can be reduced by air blowing or steam sprays. If this is not effective, the unit must be immediately taken out of service for repair of the lining. Properly applied linings will be effective for extended time periods and require repairs only during normally scheduled shutdowns.

The heat-resistant and insulating linings used in pressure vessels are refractory materials that are applied as monolithic linings by pneumatic gunning, casting, or hand packing. The most common types are the hydraulic-setting castable refractories. Their ease of application, generally good performance, and variety of strength and insulating characteristics make them versatile high-temperature materials. The quality of the applied castable material depends on the product quality as produced by the manufacturer, the experience of the applicator, and the manner in which heat is first applied to the material after it is installed. Erosion-resistant linings are relatively thin (40 mm, or 1 1/2 in.) layers of dense refractories that are supported by V-anchors or hexmesh welded directly to the vessel shell. To improve the serviceability of V-anchored linings, metal fibers are often added to the refractory mix before placement.

Catalytic cracking units, with their associated cyclones, transfer lines, and slide valves, represent the single largest application of refractories in refineries (Ref 153). The regenerator vessels, and sometimes the reactors, are lined with 100 to 150 mm (4 to 6 in.) of refractory concrete, which must withstand oxidizing (regenerator) or reducing (reactor) conditions at 540 to 760 °C (1000 to 1400 °F). The moving catalyst bed produces mild erosion, and mechanical or thermal spalling is frequently encountered. Cyclones and catalyst transfer lines are usually subjected to extreme erosion from fluidized catalyst at temperatures between 315 and 760 °C (600 and 1400 °F). Transfer lines are subjected to heavy loadings of catalyst at velocities of 7.5 to 15 m/s (25 to 50 ft/s). Slide valves in transfer lines and standpipes are subject to severe erosion, but must maintain their original thickness so they can control catalyst flow. Care must be exercised in designing the slide valve because of thermal expansion differentials. Plug valves have more tolerance for thermal expansion differentials and are frequently lined with refractory.

Because operating conditions in reformer reactors include hydrogen-rich vapors at temperatures near 500 °C (950 °F), special care is required in designing refractory liners because of the increased thermal conductivity of hydrogen-saturated refractories. Reformers in hydrogen and ammonia plants operate at temperatures of 1100 to 1370 °C (2000 to 2500 °F) and therefore require refractories of very low silica content. Furnaces are typically lined with refractory castables or insulating fire brick. Thermal shock and temperature cycling are the primary agents of attack. Mechanical movement, nut blasting, and similar forces also contribute to deterioration. Floors are usually made of dense refractory concrete or fire brick in order to resist foot traffic and mechanical impact during turnarounds. Stacks and breechings for most types of refinery units have similar service requirements: strength at high temperatures and resistance to corrosion, erosion, and spalling. Temperatures range from 205 to 815 °C (400 to 1500 °F), and the flue gas may contain catalyst, sulfur oxides, hydrogen sulfide, or carbon monoxide. Water or steam may be injected into the gas stream to control temperature.

If conditions are mild, with temperatures below 540 °C (1000 °F) and little or no erosion, insulating or semi-insulating refractory concrete on independent anchors makes a serviceable lining. More erosive conditions may require the use of dense high-strength refractory castable.

High-sulfur fuels generate sulfuric acid in the cooler zones of stacks and breechings where temperatures drop below the acid dew point, ranging from 150 to 177 °C (300 to 350 °F). Acid attacks many conventional refractory castables containing calcium aluminate cement. The alkali-silicate and some calcium aluminate base materials are moderately acid resistant and will withstand condensing flue gas vapors. For many furnace stack linings, densely applied semi-insulating refractory concrete is satisfactory.

Incinerators operate at temperatures as high as 1540 °C (2800 °F) and are particularly susceptible to slag and fly-ash attack because they often burn waste products. Problems can be expected with fluids containing alkalies and transition metals. Thermal shock can be a problem if water is injected to control temperatures or if the burner unit is allowed to cycle on and off several times a day, with periods of cooling in between. Phosphate-bonded alumina type castables are probably the best materials for this type of service. Additional information on refractory linings can be found in the article "Chemical-Setting Ceramic Linings" in this Volume.

REFERENCES

1. B.B. Morton, Metallurgical Methods for Combatting Corrosion and Abrasion in the Petroleum Industry, *J. Inst. Petrol.*, Vol 34 (No. 289), 1948, p 1-68
2. E.L. Hildebrand, Materials Selection for Petroleum Refineries and Petrochemical Plants, *Mater. Prot. Perform.*, Vol 11 (No. 7), 1972, p 19-22
3. A.J. Freedman, G.F. Tisinai, and E.S. Troscinski, Selection of Alloys for Refinery Processing Equipment, *Corrosion*, Vol 16 (No. 1), 1960, p 19t-25t
4. *The Role of Stainless Steels in Petroleum Refining*, American Iron and Steel Institute, 1977
5. Selection of Steel for High-Temperature Service in Petroleum Refinery Applications, in *Properties and Selection of Metals*, Vol 1, 8th ed., *Metals Handbook*, American Society for Metals, 1961, p 585-603
6. G.E. Moller, I.A. Franson, and T.J. Nichol, Experience With Ferritic Stainless Steel in Petroleum Refinery Heat Exchangers, *Mater. Perform.*, Vol 20 (No. 4), 1981, p 41-50
7. A.J. Brophy, Stress Corrosion Cracking of Austenitic Stainless Steels in Refinery Environments, *Mater. Perform.*, Vol 13 (No. 5), 1974, p 9-15
8. A.S. Couper and H.F. McConomy, Stress Corrosion Cracking of Austenitic Stainless Steels in Refineries, *Proc. API*, Vol 46 (III), 1966, p 321-326
9. T.P. May, J.F. Mason, Jr., and W.K. Abbott, Austenitic Nickel Cast Irons in the Petroleum Industry, *Mater. Prot.*, Vol 1 (No. 8), 1962, p 40-55
10. E.D. Verink, Jr., and F.B. Murphy, "Solving Refinery Corrosion Problems With Aluminum," Paper presented at the NACE 16th Annual Conference, Dallas, TX, National Association of Corrosion Engineers, March 1960
11. E.E. Kerns and W.E. Baker, Use of Aluminum in Petroleum Refinery Equipment, *Proc. API*, Vol 31 (III), 1951, p 89-98
12. R.L. Hilderbrand, Aluminum Exchanger and Condenser Tubes in Petroleum Service, *Proc. API*, Vol 40 (III), 1960, p 118-130
13. J. Kolts, J.B.C. Wu, and A.I. Asphahani, Highly Alloyed Austenitic Materials for Corrosion Service, *Met. Prog.*, Vol 125 (No. 10), 1983, p 25-36
14. J.F. Mason, Jr., The Selection of Materials for Some Petroleum Refinery Applications, *Corrosion*, Vol 12 (No. 5), 1956, p 199t-206t
15. *Corrosion Resistance of Hastelloy Alloys*, The Cabot Corporation, 1978
16. A.I. Asphahani, Corrosion Resistance of High Performance Alloys, *Mater. Perform.*, Vol 19 (No. 12), 1980, p 33-43
17. I.A. Franson and L.C. Covington, Application of Titanium to Oil Refinery Environments, *Proc. API*, Vol 56 (III), 1977, p 26-36
18. D.M. McCue, "Design Considerations for Titanium Heat Exchangers," Paper 60, presented at Corrosion/81, Houston, TX, National Association of Corrosion Engineers, 1981
19. J.A. McMaster, Selection of Titanium for Petroleum Refinery Components, *Mater. Perform.*, Vol 18 (No. 4), 1979, p 28-34
20. R.L. Jacobs and J.A. McMaster, Titanium Tubing: Economical Solution to Heat Exchanger Corrosion, *Mater. Prot. Perform.*, Vol 11 (No. 7), 1972, p 33-38
21. R.W. Schutz and L.C. Covington, Effect of Oxide Films on the Corrosion Resistance of Titanium, *Corrosion*, Vol 37 (No. 10), 1981, p 585-591
22. C.D. Clauser, L.G. Emmer, A.W. Pense, and R.D. Stout, A Phenomenological Study of the Susceptibility to Temper Embrittlement of 2.25%Cr-1%Mo, *Proc. API*, Vol 52 (III), 1972, p 790
23. Z.A. Foroulis, Corrosion and Corrosion Inhibition in the Petroleum Industry, *Werkst. Korros.*, Vol 33 (No. 2), 1982, p 121-131
24. L. Engel, Korrosion in Mineralolraffinerien, *Erdol Kohle*, Vol 27 (No. 6), 1974, p 301-306
25. Conditions Causing Deterioration or Failure, in *Guide for Inspection of Refinery Equipment*, 2nd ed., American Petroleum Institute, 1973
26. G.J. Samuelson, Hydrogen-Chloride Evolution from Crude Oils as a Function of Salt Concentration, *Proc. API*, Vol 34 (III), 1954, p 50-54
27. C.G. Munger, Deep Pitting Corrosion in Sour Crude Oil Tankers, *Mater. Perform.*, Vol 15 (No. 3), 1976, p 17-23
28. W.J. Neill and Z.A. Foroulis, Internal Corrosion in Floating Roof Gasoline Storage Tanks, *Mater. Perform.*, Vol 15 (No. 9), 1976, p 37-39
29. C.M. Hudgins, Jr., A Review of Sulfide Corrosion Problems in the Petroleum Industry, *Mater. Prot.*, Vol 8 (No. 1), 1969, p 41-47
30. S.P. Ewing, Electrochemical Studies of the Hydrogen Sulfide Corrosion Mechanism, *Corrosion*, Vol 11 (No. 11), 1955, p 497t-501t
31. R.L. Piehl, How to Cope With Corrosion in Hydrocracker Effluent Coolers, *Oil Gas J.*, Vol 66, 8 July 1968, p 60-63
32. C.B. Hutchison and W.B. Hughes, Oxidation Reduction Potential as a Control Criterion in Inhibition of Refinery Sulfide Corrosion, *Corrosion*, Vol 17 (No. 11), 1961, p 514t-518t
33. F.H. Meyer, O.L. Riggs, R.L. McGlasson, and J.D. Sudbury, Corrosion Products of Mild Steel in Hydrogen Sulfide Environments, *Corrosion*, Vol 14 (No. 2), 1958, p 109t-115t

34. "Controlling Weld Hardness of Carbon Steel Refining Equipment to Prevent Environmental Cracking," Publication 942, 2nd ed., American Petroleum Institute, 1982
35. D.B. Bird, Titanium Exchanger Tubing Works Well in Refinery Service, *Hydrocarbon Process.*, Vol 55 (No. 5), 1976, p 105-107
36. W.J. Neill, Experience With Titanium Tubing in Oil Refinery Heat Exchangers, *Mater. Perform.*, Vol 19 (No. 9), 1980, p 57-63
37. A.S. Couper, Corrosion Control in Crude Oil Distillation Overhead Condensers, *Proc. API*, Vol 44 (III), 1964, p 172-178
38. M.H Brown, W.B. DeLong, and J.R. Auld, Corrosion by Chlorine and by Hydrogen Chloride at High Temperatures, *Ind. Eng. Chem.*, Vol 39 (No. 7), 1947, p 839-844
39. K. Reiser, The Control of Corrosion in Refinery Distillation Units, *J. Inst. Petrol.*, Vol 53 (No. 527), 1967, p 352-366
40. D.L. Kronenberger and D.A. Pattison, Troubleshooting the Refinery Desalter Operation, *Mater. Perform.*, Vol 25 (No. 7), 1986, p 9-17
41. L.E. Fisher, G.C. Hall, and R.W. Stenzel, Crude Oil Desalting to Reduce Refinery Corrosion Problems, *Mater. Prot.*, Vol 1 (No. 5), 1962, p 8-11 and 14-17
42. L.C. Waterman, Crude Desalting: Why and How, *Hydrocarbon Process.*, Vol 44 (No. 2), 1965, p 133-138
43. M.J. Humphries and G. Sorell, Corrosion Control in Crude Oil Distillation Units, *Mater. Perform.*, Vol 15 (No. 2), 1976, p 13-21
44. J.A. Biehl and E.A. Schnake, Corrosion in Crude-Oil Processing—Low pH vs. High pH, *Proc. API*, Vol 37 (III), 1957, p 129-134
45. J.A. Biehl and E.A. Schnake, Processing Crude Oil at Low pH, *Proc. API*, Vol 39 (III), 1959, p 214-221
46. C.J. Scherrer, C.R. Baumann, and G.J. Jarno, "Crude Units: Focusing on Corrosion of Initial Condensation Equipment," Paper 66, presented at Corrosion/81, National Association of Corrosion Engineers, 1981
47. R.D. Merrick and T. Auerback, Crude Unit Overhead Corrosion Control, *Mater. Perform.*, Vol 22 (No. 9), 1983, p 15-21
48. R.H. Carlton, Continuous Injection of Overhead Receiver Water Controls Refinery Condenser and Exchanger Corrosion by Ammonium Chloride, *Mater. Prot.*, Vol 2 (No. 1), 1963, p 15-20
49. R.J. Hafsten and K.R. Walston, Use of Neutralizers and Inhibitors to Combat Corrosion in Hydrocarbon Streams, *Proc. API*, Vol 35 (III), 1955, p 80-91
50. D.L. Burns, R.L. Hildebrand, and P.D. Thomas, Corrosion Inhibitors in Refinery Process Streams, *Proc. API*, Vol 40 (III), 1960, p 155-162
51. C. Baumann and C. Scherrer, Evaluation of Inhibitors for Crude Topping Units, *Mater. Perform.*, Vol 18 (No. 11), 1979, p 51-57
52. R.T. Foley, The Role of the Chloride Ion in Iron Corrosion, *Corrosion*, Vol 26 (No. 2), 1970, p 58-70
53. J. Scherzer and D.P. McArthur, Tests Show Effects of Nitrogen Compounds on Commercial Fluid Cat Cracking Catalysts, *Oil Gas J.*, Vol 84, 27 Oct 1986, p 76-82
54. J. Gutzeit, Corrosion of Steel by Sulfides and Cyanides in Refinery Condensate Water, *Mater. Prot.*, Vol 7 (No. 12), 1968, p 17-23
55. G. Kobrin and E.S. Kopecki, Choosing Alloys for Ammonia Services, *Chem. Eng.*, Vol 85, 18 Dec 1978, p 115-128
56. R.L. Piehl, Survey of Corrosion in Hydrocracker Effluent Air Coolers, *Mater. Perform.*, Vol 15 (No. 1), 1976, p 15-20
57. E.G. Ehmke, Corrosion Correlations With Ammonia and Hydrogen Sulfide in Air Coolers, *Mater. Perform.*, Vol 14 (No. 7), 1975, p 20-28
58. "1972 Survey of Materials Experience and Corrosion Problems in Sour Water Strippers," Publication 944, American Petroleum Institute, 1974
59. "Survey of Construction Materials and Corrosion in Sour Water Strippers—1978," Publication 950, American Petroleum Institute, 1983
60. R.L. Hildebrand, Sour Water Strippers—A Review of Construction Materials, *Mater. Perform.*, Vol 13 (No. 5), 1974, p 16-19
61. J.E. Cantwell and R.E. Bryant, Failure of Refinery Piping by Liquid Metal Attack and Materials Experience With Smokeless Flares, *Proc. API*, Vol 53 (III), 1973, p 412-430
62. R.S. Treseder and A. Wachter, Corrosion in Petroleum Processes Employing Aluminum Chloride, *Corrosion*, Vol 5 (No. 11), 1949, p 383-391
63. J.F. Mason, Jr., and C.M. Schillmoller, Minimum Corrosion for Butane Isomerization Units, *Corrosion*, Vol 15 (No. 4), 1959, p 185t-188t
64. K. Brooks, Organic Chloride Contamination Rears Its Ugly Head Again, *Oil Gas J.*, Vol 60, 26 Nov 1962, p 74-75
65. D.H. Stormont, Chlorides in Crude Oil Plague Refiners, *Oil Gas J.*, Vol 67, 14 April 1969, p 94-96
66. E.B. Backensto and A.N. Yurick, Chloride Corrosion and Fouling in Catalytic Reformers With Naphtha Pretreaters, *Corrosion*, Vol 17 (No. 3), 1961, p 133t-136t
67. R.J. Schuyler III, Hydrogen Blistering of Steel in Anhydrous Hydrofluoric Acid, *Mater. Perform.*, Vol 18 (No. 8), 1979, p 9-16
68. K. Forry and C. Schrage, Trouble-Shooting HF Alkylation, *Hydrocarbon Process.*, Vol 45 (No. 1), 1966, p 107-114
69. D.P. Thornton, Jr., Corrosion-Free HF Alkylation, *Chem. Eng.*, Vol 77, 13 July 1970, p 108-112
70. National Petroleum Refiners Association Questions and Answers—Alkylation, *Oil Gas J.*, Vol 71, 7 May 1973, p 56-61; Vol 75, 23 May 1977, p 73-75; Vol 78, 14 July 1980, p 162-165
71. N.P. Lieberman, Basic Decision Key to Alky Problems, *Oil Gas J.*, Vol 78, 23 June 1980, p 141-144
72. National Petroleum Refiners Association Questions and Answers—Alkylation, *Oil Gas J.*, Vol 66, 26 Feb 1968, p 96-105; Vol 67, 27 Jan 1969, p 162-167; Vol 75, 23 May 1977, p 66-73
73. G.A. Nelson, Prevention of Localized Corrosion in Sulfuric Acid Handling Equipment, *Corrosion*, Vol 14 (No. 3), 1958, p 145t-149t
74. H.W. Van der Hoeven, Stromungsgeschwindigkeit als besonderer Faktor bei der Schwefelsaure-Korrosion, *Werkst. Korros.*, Vol 6 (No. 2), 1955, p 57-62
75. V.J. Groth and R.J. Hafsten, Corrosion of Refinery Equipment by Sulfuric Acid and Sulfuric Acid Sludges, *Corrosion*, Vol 10 (No. 11), 1954, p 368-390
76. R.B. Martin, *The Processing of Alkylation Unit Feedstocks and Reactor Effluent Streams*, Petrolite Corporation
77. API Survey Shows Few Amine Corrosion Problems, *Petrol. Refiner*, Vol 37 (No. 11), 1958, p 281-283
78. R.V. Comeaux, The Mechanism of MEA Corrosion, *Proc. API*, Vol 42 (III), 1962, p 481-489
79. F.S. Lang and J.F. Mason, Jr., Corrosion in Amine Gas Treating Solutions, *Corrosion*, Vol 14 (No. 2), 1958, p 105t-108t
80. A.J.R. Rees, Problems With Pressure Vessels in Sour Gas Service (Case Histories), *Mater. Perform.*, Vol 16 (No. 7), 1977, p 29-33
81. G.L. Garwood, What to do About Amine Stress Corrosion, *Oil Gas J.*, Vol 52, 27 July 1953, p 334-340
82. K.F. Butwell, How to Maintain Effective MEA Solutions, *Hydrocarbon Process.*, Vol 47 (No. 4), 1968, p 111-113
83. J.C. Dingman, D.L. Allen, and T.F. Moore, Minimize Corrosion in MEA Units, *Hydrocarbon Process.*, Vol 45 (No. 9), 1966, p 285-290
84. D. Ballard, How to Operate an Amine Plant, *Hydrocarbon Process.*, Vol 45 (No. 4), 1966, p 137-144
85. G.D. Hall and L.D. Polderman, Design and Operating Tips for Ethanolamine Gas Scrubbing Systems, *Chem. Eng. Prog.*, Vol 56 (No. 10), 1960, p 52-58
86. E.N. Skinner, J.F. Mason, and J.J. Moran, High Temperature Corrosion in Refinery and Petrochemical Service, *Corrosion*, Vol 16 (No. 12), 1960, p 593t-600t
87. Z.A. Foroulis, High Temperature Degradation of Structural Materials in Environments Encountered in the Petroleum and Petrochemical Industries: Some Mechanistic Observations, *Anti-Corros.*, Vol 32 (No. 11), 1985, p 4-9
88. A.S. Couper and A. Dravnieks, High Temperature Corrosion by Catalytically Formed Hydrogen Sulfide, *Corrosion*, Vol 18 (No. 8), 1962, p 291t-298t
89. A.S. Couper, High Temperature Mercaptan Corrosion of Steels, *Corrosion*, Vol 19 (No. 11), 1963, p 396t-401t
90. K.N. Strafford, The Sulfidation of Metals and Alloys, *Metall. Rev.*, Vol 138, 1969
91. F.A. Hendershot and H.L. Valentine, Materials for Catalytic Cracking Equipment (Survey), *Mater. Prot.*, Vol 6 (No. 10), 1967, p 43-47
92. N. Schofer, Corrosion Problems in a Fluid Catalytic Cracking and Fractionating Unit, *Corrosion*, Vol 5 (No. 6), 1949, p 182-188
93. S.L. Estefan, Design Guide to Metallurgy and Corrosion in Hydrogen Processes, *Hydrocarbon Process.*, Vol 49 (No. 12), 1970, p 85-92
94. L.T. Overstreet and R.A. White, Materials Specifications and Fabrication for Hydrocracking Process Equipment, *Mater. Prot.*, Vol 4 (No. 6), 1965, p 64-71
95. H.F. McConomy, High-Temperature Sulfidic

Corrosion in Hydrogen-Free Environment, *Proc. API*, Vol 43 (III), 1963, p 78-96
96. J. Gutzeit, High Temperature Sulfidic Corrosion of Steels, in *Process Industries Corrosion—The Theory and Practice*, National Association of Corrosion Engineers, 1986
97. D.W. McDowell, Jr., Refinery Reactor Design to Prevent High Temperature Corrosion, *Mater. Prot.*, Vol 5 (No. 11), 1966, p 45-48
98. "Steels for Hydrogen Service at Elevated Temperatures and Pressures in Petroleum Refineries and Petrochemical Plants," Publication 941, 3rd ed., American Petroleum Institute, 1983
99. E.B. Backensto, R.D. Drew, and C.C. Stapleford, High Temperature Hydrogen Sulfide Corrosion, *Corrosion*, Vol 12 (No. 1), 1956, p 6t-16t
100. G. Sorell and W.B. Hoyt, Collection and Correlation of High Temperature Hydrogen Sulfide Corrosion Data, *Corrosion*, Vol 12 (No. 5), 1956, p 213t-234t
101. C. Phillips, Jr., High Temperature Sulfide Corrosion in Catalytic Reforming of Light Naphthas, *Corrosion*, Vol 13 (No. 1), 1957, p 37t-42t
102. G. Sorell, Compilation and Correlation of High Temperature Catalytic Reformer Corrosion Data, *Corrosion*, Vol 14 (No. 1), 1958, p 15t-26t
103. W.H. Sharp and E.W. Haycock, Sulfide Scaling Under Hydrorefining Conditions, *Proc. API*, Vol 39 (III), 1959, p 74-91
104. J.D. McCoy and F.B. Hamel, New Corrosion Data for Hydrodesulfurizing Units, *Hydrocarbon Process.*, Vol 49 (No. 6), 1970, p 116-120
105. J.D. McCoy and F.B. Hamel, Effect of Hydrodesulfurizing Process Variables on Corrosion Rates, *Mater. Prot. Perform.*, Vol 10 (No. 4), 1971, p 17-22
106. A.S. Couper and J.W. Gorman, Computer Correlations to Estimate High Temperature H2S Corrosion in Refinery Streams, *Mater. Prot. Perform.*, Vol 10 (No. 1), 1971, p 31-37
107. J.J. Heller, Corrosion of Refinery Equipment by Naphthenic Acid, *Mater. Prot.*, Vol 2 (No. 9), 1963, p 90-96
108. B. Danilov, The Control of Corrosion in Refinery Vacuum Plants, *Anti-Corros.*, Vol 22 (No. 8), 1975, p 3-6
109. "Standard Test Method for Neutralization Number by Potentiometric Titration," D 664, *Annual Book of ASTM Standards*, American Society of Testing and Materials
110. W.A. Derungs, Naphthenic Acid Corrosion—An Old Enemy of the Petroleum Industry, *Corrosion*, Vol 12 (No. 12), 1956, p 617t-622t
111. J. Gutzeit, Naphthenic Acid Corrosion in Oil Refineries, *Mater. Perform.*, Vol 16 (No. 10), 1977, p 24-35
112. W.T. Reid, *External Corrosion and Deposits—Boilers and Gas Turbines*, American Elsevier, 1971
113. A.L. Plumley, J. Jonakin, and R.E. Vuia, "A Review Study of Fire-Side Corrosion in Utility and Industrial Boilers," Paper presented at Corrosion Seminar, Hamilton, ON, McMaster University and Engineering Institute of Canada, May 1966
114. G.W. Cunningham and A. de S. Brasunas, The Effects of Contamination by Vanadium and Sodium Compounds on the Air-Corrosion of Stainless Steel, *Corrosion*, Vol 12 (No. 8), 1956, p 389t-405t
115. D.R. McIntyre and C.P. Dillon, *Guidelines for Preventing Stress Corrosion Cracking in the Chemical Process Industries*, Publication 15, Materials Technology Institute of the Chemical Process Industries, 1985
116. R.F. Steigerwald, New Molybdenum Stainless Steels for Corrosion Resistance: A Review of Recent Developments, *Mater. Perform.*, Vol 13 (No. 9), 1974, p 9-15
117. S. Bernhardsson, P. Norberg, H. Eriksson, and O. Forsell, "Duplex and High Nickel Stainless Steels for Refineries and the Petrochemical Industry," Paper 165, presented at Corrosion/85, Houston, TX, National Association of Corrosion Engineers, 1985
118. "Protection of Austenitic Stainless Steel From Polythionic Acid Stress Corrosion Cracking During Shutdown of Refinery Equipment," NACE RP-01-70 (1985 Revision), National Association of Corrosion Engineers, 1985
119. C.S. Carter and M.V. Hyatt, Review of Stress Corrosion Cracking in Low-Alloy Steels With Yield Strengths Below 150 KSI, in *Stress Corrosion Cracking and Hydrogen Embrittlement of Iron Base Alloys*, National Association of Corrosion Engineers, 1977
120. A.S. Krisher, Material Requirements for Anhydrous Ammonia, in *Process Industries Corrosion—The Theory and Practice*, National Association of Corrosion Engineers, 1986
121. J.M.B. Gotch, *Code of Practice for the Storage of Anhydrous Ammonia Under Pressure in the United Kingdom*, Chemical Industries Association, Ltd., 1980
122. P.G. Hughes, Stress Corrosion Cracking in a MEA Unit, in *Proceedings of the 1982 U.K. National Corrosion Conference*, Institution of Corrosion Science and Technology, 1982, p 87
123. J. Gutzeit and J.M. Johnson, Stress-Corrosion Cracking of Carbon Steel Welds in Amine Service, *Mater. Perform.*, Vol 25 (No. 7), 1986, p 18-26
124. H. Nishida, K. Nakamura, and H. Takahashi, Intergranular Stress Corrosion Cracking of Sensitized 321 SS Tube Exposed to Polythionic Acid, *Mater. Perform.*, Vol 23 (No. 4), 1984, p 38-41
125. J.E. Cantwell, Embrittlement and Intergranular Stress Corrosion Cracking of Stainless Steels After Elevated Temperature Exposure in Refinery Process Units, *Proc. API*, Vol 63 (III), 1984, p 32-37
126. C.H. Samans, Stress Corrosion Cracking Susceptibility of Stainless Steels and Nickel-Base Alloys in Polythionic Acids and Acid Copper Sulfate Solution, *Corrosion*, Vol 20 (No. 8), 1964, p 256t-262t
127. R.L. Piehl, Stress Corrosion Cracking by Sulfur Acids, *Proc. API*, Vol 44 (III), 1964, p 189-197
128. H.C. Rodgers, Hydrogen Embrittlement in Engineering Materials, *Mater. Prot.*, Vol 1 (No. 4), 1962, p 26-33
129. G.A. Nelson and R.T. Effinger, Blistering and Embrittlement of Pressure Vessel Steels by Hydrogen, *Welding J.*, Vol 34 (No. 1), 1955, p 12S-21S
130. W.A. Bonner, H.D. Burnham, J.J. Conradi, and T. Skei, Prevention of Hydrogen Attack on Steel in Refinery Equipment, *Proc. API*, Vol 33 (III), 1953, p 255-272
131. T. Skei, A. Wachter, W.A. Bonner, and H.D. Burnham, Hydrogen Blistering of Steel in Hydrogen Sulfide Solutions, *Corrosion*, Vol 9 (No. 5), 1953, p 163-172
132. R.T. Effinger, M.L. Renquist, A. Wachter, and J.G. Wilson, Hydrogen Attack of Steel in Refinery Equipment, *Proc. API*, Vol 31 (III), 1951, p 107-133
133. W.A. Bonner and H.D. Burnham, Air Injection for Prevention of Hydrogen Penetration of Steel, *Corrosion*, Vol 11 (No. 10), 1955, p 447t-453t
134. E.F. Ehmke, "Use Ammonium Polysulfide to Stop Corrosion and Hydrogen Blistering," Paper 59, presented at Corrosion/81, Houston, TX, National Association of Corrosion Engineers, 1981
135. B.W. Neumaier and C.M. Schillmoller, Deterrence of Hydrogen Blistering at a Fluid Catalytic Cracking Unit, *Proc. API*, Vol 35 (III), 1955, p 92-109
136. G.B. Kohut and W.J. McGuire, Sulfide Stress Cracking Causes Failure of Compressor Components in Refinery Service, *Mater. Prot.*, Vol 7 (No. 6), 1968, p 17-21
137. E.L. Hildebrand, Aqueous Phase H_2S Cracking of Hard Carbon Steel Weldments—A Case History, *Proc. API*, Vol 50 (III), 1970, p 593-613
138. T.G. Gooch, Hardness and Stress Corrosion Cracking of Ferritic Steel, *Weld. Inst. Res. Bull.*, Vol 23 (No. 8), 1982, p 241-246
139. D.J. Kotecki and D.G. Howden, Wet Sulfide Cracking of Submerged Arc Weldments, *Proc. API*, Vol 52 (III), 1972, p 631-653
140. "Controlling Weld Hardness of Carbon Steel Refinery Equipment to Prevent Environmental Cracking," Recommended Practice 942, 2nd ed., American Petroleum Institute, 1983
141. "Methods and Controls to Prevent In-Service Cracking of Carbon Steel (P-1) Welds in Corrosive Petroleum Refinery Environments," NACE RP-04-72 (1976 Revision), National Association of Corrosion Engineers, 1976
142. "Sulfide Stress Cracking Resistant Metallic Materials for Oil Field Equipment," NACE MR-01-75 (1980 Revision), National Association of Corrosion Engineers, 1980
143. G. Sorell and M.J. Humphries, High Temperature Hydrogen Damage in Petroleum Refinery Equipment, *Mater. Perform.*, Vol 17 (No. 8), 1978, p 33-41
144. A.R. Ciuffreda and W.R. Rowland, Hydrogen Attack of Steel in Reformer Service, *Proc. API*, Vol 37 (III), 1957, p 116-128
145. R.D. Merrick and A.R. Ciuffreda, Hydrogen Attack of Carbon-0.5 Molybdenum Steels, *Proc. API*, Vol 61 (III), 1982, p 101-114
146. R. Chiba, K. Ohnishi, K. Ishii, and K. Maeda, Effect of Heat Treatment on Hydrogen Attack Resistance of C-0.5Mo Steels for Pressure Vessels, Heat Exchangers, and Piping, *Corrosion*, Vol 41 (No. 7), 1985, p 415-426
147. "Steels for Hydrogen Service at Elevated Temperatures and Pressures in Petroleum Refineries and Petrochemical Plants," Publication 941, 3rd ed., American Petroleum

Institute, 1983
148. J.M. Johnson and J. Gutzeit, Embrittlement of Stainless Steel Welds by Contamination With Zinc-Rich Paint, *Proc. API*, Vol 63 (III), 1984, p 65-72
149. G.R. Port, Hydrogen Sulfide Corrosion in a Distilling Unit, *Proc. API*, Vol 41 (III), 1961, p 98-103
150. W.A. McGill and M.J. Weinbaum, Aluminum-Diffused Steel Lasts Longer, *Oil Gas J.*, Vol 70, 9 Oct 1972, p 66-69
151. C.A. Robertson and H.L. Meyers, Application and Use of Aluminum Coatings in Oil Refinery Processes, *Mater. Prot.*, Vol 6 (No. 9), 1967, p 23-26
152. W.A. McGill and M.J. Weinbaum, The Selection, Application and Fabrication of Alonized Systems in the Refinery Environment, *Proc. API*, Vol 54 (III), 1975, p 125-159
153. M.S. Crowley, Refractories, in *Process Industries Corrosion—The Theory and Practice*, National Association of Corrosion Engineers, 1986

Corrosion of Pipelines

C.G. Siegfried, Ebasco Services Inc.

PIPELINES play an extremely important role throughout the world as a means of transporting gases and liquids over long distances from their sources to the ultimate consumers. The general public is not aware of the number of pipelines that are continually in service as a primary means of transportation. A buried operating pipeline is rather unobtrusive and rarely makes its presence known except at valves, pumping or compressor stations, or terminals. Because pipelines are hidden from view, they are not as noticeable as drilling rigs, refineries, or gas-processing plants.

At present, there are approximately 460 000 km (285 000 miles) of common carrier pipelines transporting about 46% of all the crude oil and refined products moved in the United States. In 1984, there were more than 1.6×10^6 km (10^6 miles) of natural gas pipelines in service in the United States; about 25% of this was in interstate service. At that time, there were approximately 280 000 km (174 000 miles) of pipeline carrying liquids in interstate service in the United States. Construction continues to add more miles of pipelines. In 1986, 9800 km (6090 miles) of natural gas pipeline were constructed, along with 5740 km (3567 miles) of crude oil pipelines and 2660 km (1652 miles) of pipeline for refined products. With this vast network of constantly functional pipelines in use moving natural resources and end products to locations where they are utilized, it becomes apparent that keeping them in service through the prevention of corrosion is technologically and economically advantageous.

Corrosion control of pipelines throughout the world is accomplished in the vast majority of cases through the use of cathodic protection combined with a suitable dielectric coating. The cathodic protection system applies protective current to the outer surface of the pipeline steel where it is exposed to the adjacent soils at imperfections in the coating system. The coating system serves to reduce greatly the total amount of protective current required during the operating life of the pipeline. Corrosion control considerations should begin in the design phase of a pipeline and should be continued through the construction phase and the entire economic life of the pipeline.

This article will address only transmission and distribution pipelines and pipe-type electrical system cables; information on the corrosion of underwater pipelines is available in the article "Marine Corrosion" in this Volume. Corrosion of buried telephone cables is discussed in the article "Corrosion in Telephone Cable Plants" in this Volume. The analysis of pipeline failures is addressed in the article "Failures of Pipelines" in Volume 11 of the 9th Edition of *Metals Handbook*.

Causes of Pipeline Corrosion

Dissimilar Soils. A buried pipeline, even one of a relatively short length, will almost inevitably encounter soils that have varying compositions. There can be variations of a physical nature (for example, differences in coarseness and grain size) as well as variations in type (for example, rock, loam, and clays). Additional variations can be of a chemical nature, such as pH and chemical constituents.

When a pipeline traverses dissimilar soils, the pipeline steel in a particular soil electrolyte will often assume a galvanic potential that is somewhat different from the potential of portions of the same pipeline traversing dissimilar soils elsewhere along the pipeline route. Such galvanic potential differences between different areas of a single pipeline can occur on a macroscale (that is, over many miles in the route of the pipeline) or on a microscale (within inches of each other or even over shorter distances).

Differential Aeration. A pipeline traversing soils that have varying levels of oxygen concentration will be subject to corrosion cell activity where the portion of the pipeline steel in the area of lowest oxygen concentration is anodic to other areas of the pipeline where there is a greater concentration of oxygen. This form of corrosion activity is also referred to as a concentration cell.

Dissimilar Metals. Pipelines having dissimilar materials of construction (for example, carbon steel pipe with brass cocks or valves) that are in contact with a common electrolyte with no electrical isolation between the two metals can be subject to intense corrosion (see the section "Galvanic Corrosion" of the article "General Corrosion" in this Volume). The metal that is highest in the galvanic series (Table 1) will be anodic to metals lower in the series. An example of dissimilar-metal corrosion is the use of magnesium alloys, aluminum alloys (in seawater), and zinc as galvanic (sacrificial) anodes for the protection of carbon steel structures such as pipelines.

New and Old Steel Pipe. As indicated in Table 1, the potential of steel located in a neutral electrolyte as measured against a copper-copper sulfate reference electrode can vary within a range of −0.2 to −0.8 V. The potential of a particular piece or section of steel is largely determined by the condition of the steel surface that contacts the electrolyte. Bright, new steel or old steel that has been mechanically restored to a shiny pristine surface condition will exhibit a potential between −0.5 and −0.8 V. Old, rusty steel will have a potential from −0.2 to −0.5 V. With bright, new steel pipe that is electrically connected to older or superficially rusted steel pipe in an electrolyte, galvanic potential differences as great as 0.5 V can exist in which the new pipe is anodic to the older, rusty pipe.

Table 1 Practical galvanic series of metals and alloys in neutral soils and water

Metal or alloy	Potential, V(a)
Most active	
Commercially pure magnesium	−1.75
Magnesium alloy (Mg-6Al-3Zn-0.15Mn)	−1.16
Zinc	−1.1
Aluminum alloy (5% Zn)	−1.05
Commercially pure aluminum	−0.8
Low-carbon steel (clean and shiny)	−0.5 to −0.8
Low-carbon steel (rusted)	−0.2 to −0.5
Cast iron (not graphitized)	−0.5
Lead	−0.5
Low-carbon steel in concrete	−0.2
Copper, brass, bronze	−0.2
High-silicon cast iron	−0.2
Mill scale on steel	−0.2
Carbon, graphite, coke	+0.3
Most noble	

(a) Typical potential normally observed in neutral soils and water, measured with respect to copper sulfate reference electrode. Source: A. W. Peabody, *Control of Pipeline Corrosion*, National Association of Corrosion Engineers, 1967

Bacteriological Corrosion. Microorganisms existing in a pipeline trench can affect the control of corrosion either directly or indirectly. Anaerobic bacteria, which thrive in the absence of oxygen, are sulfate-reducing organisms that consume hydrogen and cause a loss of polarization at the steel pipe surface. This loss of polarization can make the attainment of successful cathodic protection much more difficult.

Other bacteria that oxidize sulfur can exist in aerated environments. These bacteria (*Thiobacillus Thioxidans*) consume oxygen and oxidize sulfides into sulfates, such as sulfuric acid (H_2SO_4). By their metabolic processes, these bacteria can create concentrations of H_2SO_4 as high as 10%. Such an environment can be particularly hazardous to pipeline steel. Microbiological corrosion activity is discussed at length in the section "Localized Biological Corrosion" of the article "Localized Corrosion" in this Volume.

Interference-current effects are also referred to as stray-current effects. They usually occur when the direct currents associated with a foreign metallic system (one not directly associated with the pipeline of concern) use the pipeline steel as a preferential conductor in returning to their source. When this occurs, the currents will couple to the pipeline steel from the soil and flow longitudinally on the steel to a location or locations where they discharge from the steel to the adjacent soils in order to complete their circuit.

The hazard to the affected pipeline is primarly to the steel in the locations where current discharges to the earth; in so doing, it is in essence creating anodic areas with the attendant loss of pipe metal. Current pickup locations can also be hazardous to the corrosion control system of the affected pipeline if the current density being collected is of sufficient magnitude to incur hydrogen overvoltage conditions. The formation of gaseous hydrogen at dielectric coating imperfections can cause coating damage and can create hydrogen cracking effects on certain higher-strength steels. There are generally two types of interference-current activity: steady state and transient.

Steady State. In this type of interference-current activity, the magnitude of the interference current is essentially a constant. Such currents are usually incurred by the operation of other nearby impressed-current cathodic protection systems. Because all of the systems involved (the affected pipeline and the other cathodic protection system) are fixed with regard to location, the stray-current flow patterns are constant, and analysis through field testing is relatively straightforward.

Transient. In this type, the source of the stray current is the operation of a traction system (electrified railroad, subway, mining equipment) that uses direct current (dc) series motors for propulsion and the rail system as the return leg to their dc power source (for example, a substation). Such traction equipment will often have the positive feed to the vehicle motor through an overhead catenary wire or third rail. The negative return leg consisting of the surface-mounted rails can allow the return currents to couple at least partially to buried metallic shunt conductors, such as pipelines and buried metallic sheath cables.

A rail system with poorly bonded rails can represent a high-resistance intended current path. This current path can multiply the resultant stray-current magnitude picked up on adjacent pipelines, cables, and other subsurface metallic structures.

Again, the major hazard caused by a dc traction system to a buried pipeline is the area or areas where the stray current discharges from the pipeline to the adjacent soil to return to its source. In addition, the magnitude of these stray traction currents can be hundreds or even thousands of amperes.

Assessment of these stray traction currents can be very difficult because their magnitude varies as a function of the accelerating and loading demands on all the individual dc motors in the system. Because the stray-current sources are usually moving vehicles, their locations, and therefore the locations at which the stray-current couples initially to earth, can be a constantly shifting pattern. More information on this phenomenon is available in the section "Stray-Current Corrosion" of the article "General Corrosion" in this Volume.

Corrosion Control and Prevention

Materials Selection. The selection of materials for pipeline construction is limited when all of the aspects of safety, structural integrity, operating life, and economic considerations are taken into account and acted upon.

Carbon steel is the almost exclusive choice of pipeline designers. This is true for pipeline systems that are used to gather or collect the natural gas, crude oil, or water; it is also true for those pipelines that are used to transport substances over distances of hundreds of feet to hundreds of miles. It is also the case for piping systems that are used to distribute natural gas, water, water-refined liquids, and so on, to the end user.

Cast iron is extensively used in water and natural gas distribution systems. In recent years, nonmetallic materials have found application in natural gas distribution systems as carrier vehicles and as liners for restoring failed metallic piping to service without the need for trenching and replacement.

Carbon steels and, in certain types of service or environmental conditions, alloy steels are by far the most commonly used pipeline materials of construction. Table 2 lists compositions of some standard pipe steels.

Protective (Dielectric) Coatings. The function and desired characteristics of a dielectric-type pipeline coating are covered in NACE RP-01-69. This specification states that the function of such coatings is to control corrosion by isolating the external surface of the underground or submerged piping from the environment, to reduce cathodic protection requirements, and to improve (protective) current distribution. Coatings must be properly selected and applied, and the coated piping must be carefully installed to fulfill these functions. Different types of coatings can accomplish the desired functions. The desired characteristics of the coatings are:

- Effective electrical insulation. For preventing the electrolytic discharge of current from the steel surface of the pipe, the coating must have the characteristics of an effective dielectric material
- Effective moisture barrier. The permeation of a coating material by soil moisture would significantly reduce its dielectric properties
- Application considerations. The coating must be capable of being readily applied to the pipe, and it must not of itself or through required application procedures adversely affect the properties of the pipe. In addition, the coating system must be capable of application to the pipe with a minimum of defects
- Good adhesion to the pipe surface is required to prevent soil moisture migration between the pipe steel surface and the coating inner surface
- Ability to resist holidays with time. With a pipeline in place and backfilled, soil stress and soil contaminants—two primary considerations—can cause coating degradation. Soil stress, which can be brought about by seasonal variations in soil moisture content, can create significant forces on the coating that can wrinkle, tear, or thin areas of the coating. Contaminants in the soils adjacent to the coating can also seriously affect its long-term performance. Such contaminants can be chemical, such as solvents from leaks in a product pipeline. Microorganisms in the soil can also have a marked effect, as described in the section "Bacteriological Corrosion" in this article
- Ability to resist damage during handling, storage, and installation
- Ability to maintain substantially constant electrical resistivity over time. Because the corrosion control systems of most pipelines consist of the combination of coating and cathodic protection, it is desirable that the dielectric properties of the coating deteriorate at the minimum rate possible. Significant deterioration of the coating with time would otherwise result in a constant requirement for additional cathodic protection current, which would be both a technical and economic burden on the pipeline operator
- Resistance to disbonding. With the cathodic protection current applied and collecting on

Table 2 Typical compositions of ASTM and API pipe steels

Steel	Composition, %												
	C	Mn	Si	P	S	Cr	Mo	Nb	V	Ti	Al	B	Ni
A106, Grade A	0.25 max	0.27–0.93	0.10 min	0.048 max	0.058 max	...	...	...	...	...	...	...	...
A106, Grade B	0.30 max	0.29–1.06	0.10 min	0.048 max	0.058 max	...	...	...	...	...	...	...	...
A335, Grade P2	0.10–0.20	0.30–0.61	0.10–0.30	0.045 max	0.045 max	0.50–0.81	0.44–0.65	...	...	...	...	...	...
A335, Grade P5	0.15 max	0.30–0.60	0.50 max	0.030 max	0.030 max	4–6	0.45–0.65	...	...	...	...	...	...
A335, Grade P7	0.15 max	0.30–0.60	0.50–1	0.030 max	0.030 max	6–8	0.44–0.65	...	...	...	...	...	...
A335, Grade P11	0.15 max	0.30–0.60	0.50–1	0.030 max	0.030 max	1–1.50	0.44–0.65	...	...	...	...	...	...
A335, Grade P22	0.15 max	0.30–0.60	0.50 max	0.030 max	0.030 max	1.90–2.60	0.87–1.13	...	...	...	...	...	...
A381, Class Y52	0.26 max	1.40 max	...	0.040 max	0.050 max	...	...	...	...	...	...	...	...
API 5L-X46	0.30 max	1.35 max	...	0.04 max	0.05 max	...	...	...	...	...	...	...	...
API 5L-X60	0.26 max	1.35 max	...	0.04 max	0.05 max	...	...	0.05 min	0.02 min	0.03 min(a)	...	...	...
API 5L, Grade X52	0.21	0.90	0.26	0.015 max	0.015 max	...	...	...	0.09	...	0.030	...	...
API 5A, Grade K-55	0.45	1.30	0.26	0.015 max	0.015 max	...	...	...	...	...	0.007		...
API 5AX, Grade N-80	0.28	1.48	0.26	0.015 max	0.015 max	0.20	0.10	...	...	...	0.007	...	...
API 5AX, Grade P-110	0.28	1.48	0.26	0.015 max	0.015 max	0.22	0.23	...	...	...	0.007	...	...
API 5AC, Grade C-90	0.29	0.50	0.26	0.015 max	0.015 max	1.08	0.33	...	0.03	...	...	0.0015 min	...
API 5L, Grade A	0.17	0.50	...	0.020	0.020	...	...	...	...	...	...	...	
API 5L, Grade X60	0.05	1.11	0.017	0.007	0.006	...	...	0.045	...	...	0.045	...	...

(a) Niobium, vanacium, and titanium are used at manufacturer's option.

those portions of the pipeline steel exposed to the soil at coating holidays, there is the possibility of generation of gaseous hydrogen at the steel surface. The gaseous hydrogen can lift the coating edge adjacent to the holiday, thus worsening a bad situation. Therefore, resistance to electrically induced disbondment is an important coating property
- Resistance to chemical degradation
- Ease of repair
- Retention of physical characteristics

Types of Coatings. The pipe coating systems currently in use include the following generic types. Detailed information on paint and coating formulation and application for corrosion protection is available in the article "Organic Coatings and Linings" in this Volume.

Bituminous enamels are formulated from coal tar pitches or petroleum asphalts and have been widely used as protective coatings for more than 65 years. Coal tar and asphalt enamels are available in summer or winter grades. These enamels are the corrosion coating; they are combined with various combinations of fiberglass and/or felt to obtain mechanical strength for handling. The enamel coatings have been the workhorse coatings of the industry, and when properly selected and applied, they can provide efficient long-term corrosion protection.

Enamel systems can be designed for installation and use within an operating temperature range of 1 to 82 °C (30 to 180 °F). When temperatures fall below 4.4 °C (40 °F), added precautions should be taken to prevent cracking and disbonding of the coating during field installation. Enamels are affected by ultraviolet rays and should be protected by kraft paper or whitewash. Enamels are also affected by hydrocarbons, and the use of a barrier coat is recommended when known contamination exists. Bituminous enamel coatings are available for all sizes of pipe. In recent years, the use of enamels has declined for the following reasons:

- Reduced number of suppliers
- Restrictive environmental and health standards from the Occupational Safety and Health Administration, the Environmental Protection Agency, and the Food and Drug Administration
- Increased acceptance of plastic coating materials
- Alternative use of coating raw materials as fuels

Asphalt mastic pipe coating is a dense mixture of sand, crushed limestone, and fiber bound together with a select air-blown asphalt. These materials are proportioned to secure a maximum density of approximately 2.1 g/cm^3 (132 lb/ft^3). This mastic material is available with various types of asphalt. Selection is based on operating temperature and climatic conditions to obtain maximum flexibility and operating characteristics. This coating is a thick (12.7 to 16 mm, or ½ to ⅝ in.) extruded mastic that results in a seamless corrosion coating.

Extruded asphalt mastic pipe coating has been in use for more than 50 years. It is the thickest of the corrosion coatings and is cost effective for offshore installations.

Asphalt mastic systems can be designed for installation and use within an operating temperature range of 4.4 to 88 °C (40 to 190 °F). Precautionary measures should be taken when handling asphalt mastics in freezing temperatures. Whitewash is used to protect it from ultraviolet rays, and this should be maintained when in storage. This system is not intended for use above ground or in hydrocarbon-contaminated soils. This coating is available on 11.4- to 122-cm (4½- to 48-in.) outside diameter pipe.

Liquid Epoxies and Phenolics. Many different liquid systems are available that cure by heat and/or chemical reaction. Some are solvent types, and others are 100% solids. These systems are primarily used on larger-diameter pipe when conventional systems may not be available or when they may offer better resistance to operation temperatures in the 95-°C (200-°F) range.

Generally, epoxies have an amine or a polyamide curing agent and require a near-white blast-cleaned surface (NACE No. 2 or SSPC SP10). Coal tar epoxies have coal tar pitch added to the epoxy resin. A coal tar epoxy cured with a low molecular weight amine is especially resistant to an alkaline environment, such as that which occurs on a cathodically protected structure. Some coal tar epoxies become brittle when exposed to sunlight.

Extruded plastic coatings fall into two categories based on the method of extrusion, with additional variations resulting from the selection of adhesive. The two methods of extrusion are the crosshead or circular die and the side extrusion or T-shaped die. The four types of adhesives are asphalt-rubber blend, polyethylene copolymer, butyl rubber adhesive, and polyolefin rubber blend.

To date, of the polyolefins available, polyethylene has found the widest use, with polypropylene being used on a limited basis for its higher operating temperature. It is a recognized fact that the perfect pipe coating system does not exist. Each type or variation of adhesive and method of extrusion offers different characteristics based on the degree of importance to the user of certain measurable properties.

Fusion-bonded epoxy coatings are heat-activated, chemically cured coating systems. The epoxy coating is furnished in powdered form and, with the exception of the welded field joints, is plant applied to preheated pipe, special sections, connections, and fittings using fluid bed, air spray, or electrostatic spray methods.

Fusion-bonded epoxy coatings were introduced in 1959 and were first used as an exterior pipe coating in 1961. These coatings are applied to preheated pipe surfaces at 218 to 244 °C (425 to 475 °F). Some systems may require a primer system, and some require postheating for complete cure. A NACE No. 2 (SSPC SP10) near-white blast-cleaned surface is required. The coating is applied to a minimum thickness of 0.3 mm (12 mils); in some applications, coating thicknesses range to 0.64 mm (25 mils), with the restriction not to bend pipe coated with a film thickness greater than 0.4 mm (16 mils). The epoxy coatings exhibit good mechanical and physical properties and are the most resistant to hydrocarbons, acids, and alkalies.

A primary advantage of the fusion-bonded pipe coatings is that they cannot hide apparent steel defects; therefore, the steel surface can be inspected after it is coated. The number of holidays that occur is a function of the surface condition and the thickness of the coating specified. Increasing the thickness minimizes this problem, and the excellent resistance to the electrically induced disbondment of these coatings has resulted in their frequent use as pipeline coatings.

Mill-applied tape systems have been in use for more than 30 years on pipelines. For normal constructon conditions, prefabricated cold-applied tapes are applied as a three-layer system consisting of a primer, corrosion-preventive tape (inner layer), and a mechanically protective tape (outer layer). Tape systems are available on 5- to 305-cm (2- to 120-in.) outside diameter pipe.

The function of the primer is to provide a bonding medium between the pipe surface and the adhesive or sealant on the inner layer. The inner layer tape consists of a plastic backing and an adhesive. This layer is the corrosion-protective coating; therefore, it must provide a high electrical resistivity, low moisture absorption and permeability, and an effective bond to the primed steel surface. The minimum thickness is usually 3.7 mm (145 mils), with the total system being a minimum of 10 mm (400 mils) thick. The outer layer tape consists of a plastic film and an adhesive composed of the same types of materials used in the inner tape or materials that are compatible with the inner layer tape. The purpose of the outer layer tape is to provide mechanical protection to the inner layer tape and to be resistant to the elements during outdoor storage. The outer layer tape is usually a minimum of 0.64 mm (25 mils) in thickness.

The cold-applied multilayer tape systems are designed for plant coating operations and result in a uniform, reproducible, holiday-free coating over the entire length of any size pipe. The multiple-layer system allows the coating thickness to be custom designed to meet specific environmental conditions. These systems have been engineered to withstand normal handling, outdoor weathering, storage, and shipping conditions.

Waxes. Wax coatings have been in use for more than 48 years and are still employed on a limited basis. Microcrystalline wax coatings are usually used with a protective overwrap. The wax serves to waterproof the pipe, and the wrapper protects the wax coating from contact with the soil and affords some mechanical protection. The most prevalent use of wax coating is the over-the-ditch application with a combination machine that cleans, coats, wraps, and lowers into the ditch in one operation. The lack of objectionable or toxic fumes or smoke should make this system more acceptable.

Polyurethane Thermal Insulation. Efficient pipeline insulation has grown increasingly important as a means of operating hot and cold service pipelines. This is a system for controlling heat transfer in above- or belowground and marine pipelines. Polyurethane insulation is generally used in conjunction with a corrosion coating, but if the proper moisture vapor barrier is used over the polyurethane foam, effective corrosion protection is attained.

Concrete. Mortar linings and coatings have the longest history of use in protecting steel or wrought iron from corrosion. When steel is encased in concrete, a protective iron oxide film forms. As long as the alkalinity is maintained and the concrete is impermeable to chlorides and oxygen, corrosion protection is obtained.

The use of concrete as a corrosion coating is currently limited to internal lining. External application is usually employed over a corrosion-resistant coating for armor protection and negative buoyancy in marine environments.

Metallic (Galvanic) Coatings. Pipe coated with a galvanic coating, such as zinc (galvanizing) or cadmium, should not be utilized in direct burial service. Such metallic coatings are intended for the mitigation of atmospheric-type corrosion activity on the substrate steel.

Cathodic Protection. Some of the mechanisms or circumstances that can create anodic and cathodic areas on a steel pipeline or can make areas on the steel pipeline anodic to adjacent buried metallic structures were discussed in the section "Causes of Pipeline Corrosion" in this article. At the anodic areas on a pipeline, the electrochemical phenomenon known as corrosion causes metallic ions to be lost to the electrolyte in an irreversible manner. At the anodic areas, the conventional current discharges from the pipeline steel flow through the electrolyte to cathodic areas or structures.

The magnitude of the corrosion cell current is proportional to the driving potential between the anodic and cathodic areas and is inversely proportional to the resistance of the total path along which it flows. The total path resistance consists of the anode/electrolyte (soil) interface, the electrolyte itself, the electrolyte/cathodic interface, and the longitudinal resistance of the metallic paths traversed.

Two principal methods are used to eliminate, or at least significantly reduce, the corrosion current flow. The first approach involves increasing the resistance of the circuit over which the corrosion cell potential is impressed. Dielectric coatings attempt to accomplish this, but because holidays in a coating are inevitable in the case of a soil-buried pipeline, the corrosion circuit or circuits retain a finite resistance path.

The second approach consists of eliminating the open-circuit potential differences between anodic and cathodic areas. When there is no driving potential across the corrosion cell circuits, no corroding current can flow. Cathodic protection can be used to accomplish this method of control. Cathodic protection is obtained by causing a direct current to flow from a source external to the pipeline, through the electrolyte, and to the pipeline steel surface in contact with the electrolyte.

If all of the steel surface is collecting current from the adjacent electrolyte, no current can be discharging from the previously anodic areas of the pipeline; therefore, corrosion activity ceases on the pipeline steel. In essence, if the application of the protective current causes all exposed steel surfaces to become cathodes, cathodic protection has been properly established.

To establish a cathodic protection system for a pipeline, the protective current is discharged into the electrolyte from its source. This source consists of objects, usually metallic, that are properly located in the electrolyte and commonly referred to as anodes or a ground bed. These devices are correctly referred to as anodes because, in discharging the protective current to the electrolyte, they are consumed through the irreversible loss of metallic ions.

There are two primary methods of establishing a cathodic protection system for a pipeline. One is to use anodes made of metals or alloys that, by virtue of their place in the galvanic series of metals, will be anodic to the steel when connected to the pipeline electrically and located in the same electrolyte. Such anodes are known as galvanic anodes or sacrificial anodes. Common anode metals for land or marine pipelines are high-purity zinc and certain magnesium alloys. For marine pipelines, a number of aluminum alloys are also used.

In the second method of establishing cathodic protection for pipelines, the current for the anode system is provided by an external source. The most common source used for the protective current is a transformer-rectifier system in an appropriate housing. The input of the source is commercial alternating current (ac) that is transformed down to the required level and then rectified to a dc voltage of that level. The dc voltage output, when applied between the pipeline steel and the anode system through insulated conductors, creates a protective current flow. Such powered cathodic protection systems are commonly referred to as impressed-current systems.

The materials used for the anodes in impressed-current systems are usually selected for their low consumption rates, that is, relatively low amounts of metal consumed per ampere of current discharged per year. Such materials include graphite, high-silicon cast iron, a thin layer of platinum overlaying a titanium or niobium substrate, certain ferrites, certain conductive polymers, and carbon steel pipe or rails. Steel does not have a low consumption rate, but is usually used when availability of the pipe or rails has economic benefit.

Alternative power sources for impressed-current systems include engine-driven generators, photovoltaic arrays, fuel cells, thermoelectric generators, and wind-powered generators. As a general rule, galvanic anodes are used where relatively low amounts of protective current are required from an individual protective system. Impressed-current installations can provide hundreds of amperes each, but for pipeline applications, typical dc output ratings for individual installations would range from 10 to 60 A at 10 to 70 V. More information on equipment and applications for cathodic protection is available in the article "Cathodic Protection" in this Volume.

Design Considerations. Proper corrosion control of a pipeline is most effectively and economically begun during the design stage of the pipeline. Factors affecting the efficacy of the corrosion control program for the pipeline are discussed below.

The terrain, soils, and waters in which the pipeline will be located should be considered in order to determine the most effective and most durable coating system for the pipeline. These factors also have a direct bearing on the type and amount of cathodic protection that will be required for corrosion control. Harsher and more aggressively corrosive soils can require protective current densities to be raised, thus increasing total current requirements for a given pipeline system.

Adjacent or crossing pipeline systems to the proposed pipeline right-of-way must also be considered. The greater the number of nearby or crossing foreign pipeline systems relative to the proposed pipeline, the greater the chance for interference-current activity. This includes both steady-state interference current and transient current exchange between pipelines.

Urban Environments. The potential for damage to the pipeline coating or the pipeline itself from construction activity will usually be much greater in an urban environment. Therefore, the total current requirements for cathodic protection should be generously rated during design to provide for future increases beyond that which would be considered for normal coating degradation.

Adjacent overhead electric circuits can cause induced or conductive couplings between the electric circuit and the pipeline steel. Such couplings, if of sufficient transient or steady-state magnitude, can result in ac potentials on the pipeline steel that are injurious to personnel, pipeline equipment, or the pipeline dielectric coating. Although not strictly a corrosion consideration, such analysis is usually performed by corrosion/cathodic protection personnel because they are best equipped to make such analyses.

Electrical isolation must be established for those portions of the pipeline system intended for cathodic protection. Electrical isolation of the pipeline segments that will be placed under cathodic protection is required for two reasons. First, it minimizes the total amount of protective current needed to establish overall protection, and second, it enhances the ability of personnel to obtain and interpret field survey data relative to the levels of protection being attained.

The best time to analyze the system for the determination of electrical isolation requirements is during the design phase. This allows required isolation devices to be available for installation during pipeline construction and avoids retrofit requirements. Electrical isolation devices include flange insulation kits, insulating couplings, insulating unions, and other devices that block the flow of longitudinal current.

Test stations should be established for conducting electrical field surveys intended to determine the performance of the cathodic protection system. Test stations are devices that are similar to electric junction boxes in which insulated electric wires connected to the pipeline steel are terminated. These flush or aboveground stations permit cathodic protection personnel to analyze electrically the operation of the cathodic protection system.

With regard to electrical isolation equipment, there is economic justification for having such test stations and their pipe-connected test wires installed during pipeline construction before the pipe in the trench is backfilled. The number and type of test stations will usually be determined primarily by the type of pipeline system (transmission, distribution, and so on) and the area in which the pipeline will be located.

Construction Techniques. Corrosion control of a pipeline system is normally accomplished through the synergistic effects of cathodic protection and dielectric coatings. Corrosion cell activity is halted by the application of current components in proper densities at pipe steel exposed to the electrolyte. The coating system reduces, by many orders of magnitude in most cases, the total current output required to achieve full protection. Throughout a pipeline service life of, for example, 25 years, any factor that operates to minimize total current requirements can provide significant benefit in terms of operating costs.

Therefore, the pipeline system operator should use a high-quality coating that is well suited to the conditions and location in which the pipeline will operate. In addition, significant care and attention should be paid to ensure that the integrity of the coating is as close to perfect as possible.

Therefore, the primary corrosion considerations in terms of construction are careful handling and transporting of the pipe from the time the coating is applied to the point at which the pipe is backfilled in place. Extreme care in handling, loading, unloading, burying, and backfilling the pipe are paramount concerns. Most pipeline companies follow the practice of locating holidays in the coating by using a high-voltage detector just before or while the pipe is being lowered into the trench. Any holidays detected in this manner are repaired with techniques appropriate to the type of coating used. The purpose of this method is to optimize coating quality immediately before the coating becomes inaccessible.

Corrosion of Specific Types of Pipelines

Transmission Pipelines. Mitigation of corrosion in onshore transmission pipelines is primarily accomplished by the combination of cathodic protection and dielectric coating systems. The design of such cathodic protection systems is reasonably straightforward, because the corrosion engineer can often predict the distance at which protective current application from a remote anode bed will effectively protect the pipeline in both directions from a current drain point attached to the pipeline steel. The distance of effective full protection can be estimated if the pipeline diameter, steel type, wall thickness, soil characteristics, and general coating quality are known. This is so because a pipeline has an attenuation characteristic to current pickup from the electrolyte and longitudinal flow that is analogous to a leaky electric transmission line.

It is interesting to note that it is possible to protect the pipeline to a greater distance from a given anode location on a larger-diameter pipeline than on a pipeline with a smaller diameter. This is because the resistance of the longitudinal pipe steel current path is lower per unit length in proportion to the pipe diameter.

In establishing anode beds for transmission pipelines, an attempt is usually made to place the anodes at remote earth relative to the pipeline. Remote earth in most soils can be a minimum distance of 60 to 150 m (200 to 500 ft) from the pipeline. The higher the resistivity of the indigenous soil, the farther the anode bed must be from the nearest point on the pipeline to be at remote earth.

A remote earth anode bed placement is desirable because it results in the maximum spread of protective current in each direction from an individual anode bed. On larger-diameter (75 to 90 cm, or 30 to 36 in.) pipelines, corrosion can be controlled on 33 to 66 km (30 to 40 miles) of pipeline, assuming a reasonably satisfactory dielectric coating is in place. Remote anode beds are established in primarily two types of installations: Surface anode beds and deep anode beds.

Surface anode beds are typically installations in which a multiple number of cylindrical anodes 7.5 cm (3 in.) in diameter and 152 cm (60 in.) long are installed vertically to a depth of 4.5 m (15 ft) at 6-m (20-ft) centers. The individual anodes are connected in parallel to the positive-polarity output terminal of a suitably rated transformer-rectifier unit by an insulated anode header cable.

Deep anode beds are often used where right-of-way, urban buildup in the area, or soil resistivity conditions dictate their necessity. A deep anode is a vertical installation in which the individual anodes such as those described above are installed in a vertical array in a 20-cm (8-in.) hole drilled to a depth of 30 to 90 m (100 to 300 ft). The intent with a deep anode is usually to attain a remote earth installation without the necessity of departing a long distance from the pipeline right-of-way or attaining a lower electrical resistance (greater operating efficiency) anode installation by placing the anodes in lower resistivity, subsurface soil horizons.

Where localized protection of discrete areas on a transmission pipeline is required, either properly sized galvanic anode arrays or distributed impressed-current anode systems are usually used. Distributed impressed-current anode systems consist of one or more cylindrical anodes installed relatively close to the pipeline (7 to 30 m, or 25 to 100 ft) such that a modest current discharge from each anode will couple to the exposed pipeline steel in its vicinity.

Distribution Pipelines. Cathodic protection of distribution pipelines is accomplished with the same techniques and systems as those cited for transmission pipelines, that is, galvanic anodes or impressed-current systems. However, the installation, surveillance, and maintenance of distribution pipeline cathodic protection systems can be much more complex than with a transmission pipeline, which is often a simple buried, coated steel cylinder running between points A and B.

Distribution pipeline systems are often installed in a grid configuration conforming to the roadways fronting the individual structures that the system services. Whether rural or urban, they are usually in place in commercial areas, often under paved roadways and in close proximity to other buried piping and electrical and telephone systems.

In addition, distribution pipeline systems serve multiple delivery points, all of which require the installation and maintenance of electric isolation devices, which are often not in readily accessible locations, such as building basements. Failure or inadvertent short circuiting of just one such electric insulation device, out of a total of perhaps hundreds, can render a well-designed cathodic protection system ineffective.

For this reason, many distribution pipeline operators sectionalize their distribution pipeline systems into discrete areas of electrically continuous piping. This protects against the possibility of one short-circuited isolation device disabling a large portion of the cathodic protection system.

By virtue of the close proximity of a distribution piping system to adjacent buried metallic structures that may not be cathodically protected (for example, cast iron water or sewer systems), inadvertent contacts to such systems can also disable an otherwise effective cathodic protection system. In heavily trafficked urban areas, such contacts can be extremely difficult to locate and to clear. In addition, assessment of the continuing effectiveness of a distribution pipeline cathodic protection system through electrical test techniques is also difficult because the system is usually located under paved surfaces.

Despite all of the difficulties cited above, distribution pipeline systems are protected against corrosion activity. Both galvanic anode and impressed-current systems are used by system operators. The most prevalent type of design is the distributed anode type, often with galvanic anodes installed at specific intervals that are a function of the diameter of the pipeline. Impressed-current systems are also used, but with their typically higher operating voltages and protective current magnitudes, they are more susceptible to interference-current effects from adjacent buried metallic piping systems.

Electrical System Pipe-Type Cables. Pipe-type cables are basically pipelines that are used as buried electrical transmission circuits—usually in urban environments in which aerial rights-of-way are not available. The buried pipe contains three insulated phase conductors and a dielectric medium, such as insulating oil or an inert gas.

The carrier pipe will typically be 20 cm (8 in.) in diameter and have an appropriate dielectric coating similar to conventional pipelines. Unlike conventional pipelines, the pipe-type cable carrier pipe must have connections available to the electric safety grounding system at the electric utility facilities at each end of the carrier pipe. This is required to permit ground fault current to flow from a faulted phase conductor(s), thus causing safety relaying devices to operate equipment so as to clear the fault, that is, deenergize the circuit.

The buried carrier pipe is subject to all the sources of corrosion cell activity of any other buried pipeline and therefore will require the application of cathodic protection. It would be possible to cathodically protect the carrier pipe if it were to be solidly connected to the grounding system at both ends. Large amounts of total current would probably be required, however, with the additional possibility of interference-current activity also being a consideration in urban areas.

Two principle systems are used that satisfy the requirement for corrosion control of the carrier pipe while maintaining an electric safety grounding system connection for ground fault protection. They will be discussed below.

Dropping Resistor. This system has both ends of the carrier pipe tied to station ground through large-amperage resistor elements of typically 3 mΩ. Impressed-current rectifiers are mounted at each resistor and operated so as to circulate 100 A through the resistor from the station ground side to the carrier pipe. This current flow creates a voltage drop across the resistor such that the carrier pipe is held at a potential of −0.3 V relative to the potential of the station grounding network.

Polarization Cells. In this system, the carrier pipe is connected to the station grounds through high-amperage non-ohmic resistance cells. Cathodic protection is applied to the carrier pipe as desired. To the low-level dc potentials associated with cathodic protection, the cell appears as an open circuit, and protective current is confined to the carrier pipe.

In the event of a ground fault of one or more phase conductors in the carrier pipe, the increase in potential of the carrier pipe steel causes the polarization cell to go into a low-resistance conductive mode that results in coupling of the carrier pipe to the station ground. This action is analogous to a fast-response lightning arrester with amperage sufficiently great to conduct the anticipated fault current magnitude. When the ground fault is cleared, the cell is restored to an open-circuit mode.

SELECTED REFERENCES

- J.R. Meyers and M.A. Aimone, *Corrosion Control For Underground Steel Pipelines: A Treatise on Cathodic Protection*, James R. Myers and Associates
- J.H. Morgan, *Cathodic Protection*, Macmillan, 1960
- A.W. Peabody, Control of Pipeline Corrosion, National Association of Corrosion Engineers, 1967
- R.N. Sloan, "Mill Applied Coatings For Underground Pipelines," Paper presented at the Northeast Regional Meeting, New York, National Association of Corrosion Engineers, Oct 1979
- H.H. Uhlig, *The Corrosion Handbook*, John Wiley & Sons, 1948

Corrosion in the Mineral Industry

G.A. Minick, A.R. Wilfley & Sons, Inc., Centrifugal Pump Division
D.L. Olson, Center for Welding Research, Colorado School of Mines

THE MINING, mineral-processing, and extractive metallurgy industries are concerned with a wide range of corrosive media and must consider material selection as perhaps the most important general approach to corrosion resistance. It is common to find manganese steel in crushing and grinding service, variations of cast (hard) iron grinding balls and mill liners, hardenable and carburized grades of low-alloy steel gears and geared transmissions mining machinery, steel and carbide tools and drills, high nickel and cobalt hardfacing alloys, stainless steel leaching tanks and pumps, and high nickel-chromium and titanium alloys for the most severe environments. This wide selection of materials and their service lives affect operation cost and productivity. With the increasing mechanization in mines, corrosion problems and material performance become increasingly important to the mining industry.

Corrosion associated with the mining industry can be characterized as electrochemical attack enhanced by abrasion. All ingredients for corrosion attack are available, including highly conductive water, grinding media, dissimilar materials, oxygen, large pH range, and the presence of many well-known corrosive species in solution.

Mine atmospheres and mine waters are unique in that they vary widely from mine to mine. For example, temperatures have been found to range from approximately 5 to 30 °C (40 to 90 °F) in coal mines and above 40 °C (100 °F) in metal mines. Refrigeration and air conditioning have become necessary for improved working conditions. Humidity levels between 90 and 100% are common. Mine water also varies in mineral content, pH, and corrosivity.

Tables 1 and 2 list the contents of some Canadian mine waters. These constituents have been found in mines in the United States and in other countries, although the amounts can vary. Values of pH range from 2.8 (very acidic) to 12.3 (basic). High values are often the result of the lime content of the cement added to backfill. Chlorine ion (Cl^-) values show a wide range (from 5 to 25 000 ppm), and sulfate ion (SO_4^{2-}) values range from 57 to 5100 ppm. Chloride and sulfate are considered to be the most aggressive ions present in mine waters and account for the high corrosivity of most mines.

Important constituents of dissolved colloidal or suspended matter that contribute to the corrosive environment can be classified as dissolved gases, mineral constituents, organic matter, and microbiological organisms (Ref 3). Dissolved oxygen enhances the corrosion process, especially in waters with low pH, but oxygen content can vary widely in mine waters. Oxygen concentration also decreases with increasing temperature. One example of the effect of oxygen on corrosion rate is that aeration with oxygen has been found to increase grinding ball wear by 13 to 16% (Ref 4).

Carbon dioxide (CO_2) in mine water is associated with carbonate ion (CO_3^{2-}) content. Calcium bicarbonate ($Ca(HCO_3)_2$) has been reported to be the predominant source, and magnesium carbonate ($MgCO_3$) has been found to be a less active secondary source (Ref 3). Carbon dioxide has little effect on corrosion rate except when pH levels are below 7.

Mineral content of mine water begins with the breakdown of iron sulfide minerals, principally pyrite and marcasite, that are generally associated with most mineralization species. Oxidation of pyrite produces sulfuric acid (H_2SO_4). Consequently, mine water with pH as low as 2 is produced. The acid water accelerates the breakdown of minerals, increasing the concentration of silicon (Si^{4+}), aluminum (Al^{3+}), calcium (Ca^{2+}), magnesium (Mg^{2+}), and manganese ions in mine waters.

Anaerobic and aerobic microorganisms are well-known corrosion-producing agents. Principle acid-producing species are *Thiobacillus thiooxydans*, an aerobic species (Ref 3, 5). Their presence oxidizes sulfur or sulfur compounds, producing H_2SO_4 and contributing to the acidity of mine water.

Another species of aerobic bacteria, *Ferrobacillus ferro-oxydans*, is associated with the *Thiobacillus* type. When both types of bacteria are present, their synergistic effect has been reported to increase H_2SO_4 production by four times as compared to the production rate when no bacteria are present (Ref 6). Acid mine waters produced by these microorganisms can approach very acidic conditions. Anaerobic microorganisms such as *Sporovibrio desulfuricans*, are a sulfate-reducing type and are responsible for rapid corrosion of iron and steel structures (Ref 7). They are found in most soils. As mentioned earlier, they also reduce sulfates and sulfides by

Table 1 Analyses of some Canadian mine waters

Values have been converted from grams per liter to parts per million.

Mine number(a)	pH	Ionic species present, ppm: Cl^-	SO_4^{2-}	Ca^{2+}	Mg^{2+}	HCO_3^-	Cu^{2+}	Dissolved solids, ppm(b)
1a	5.4	20	1080	210	80	<100	2	1740
b	5.9	60	1380	390	55	<100	1	2290
c	12.3	40	240	520	0.7	<100	<0.1	2610
2a	6.3	5	57	14	3	nil	0.3	94
b	2.8	170	5100	230	580	<50	23	8740
3	8.3	9	120	49	4	<100	<0.5	320
4	7.0	10 500	1000	3450	490	200	<0.1	24 700
5	3.2	840	540	470	180	nil	<0.2	2640
6	3.3	<5	1830	31	290	nil	<0.2	3010

(a) 1a, b, and c and 2a and b are different samples from the same mine. (b) Dried at 110 °C (230 °F). Source: Ref 1

Table 2 Analyses of three Ontario mine waters

Values in parts per million unless otherwise stated

Source of mine water	Levack Mine (sulfide)	Helen Mine (iron)	Leitch Mine (gold)
Carbon dioxide,			
calculated	...	10	4
pH	3.4	7.2	7.6
Calcium carbonate			
(total)	0.0	83.1	77.7
Hardness (total)	1175	1079	2054
Calcium carbonate	0.0	996	1977
Calcium	405	245	744
Magnesium	42	113	48
Iron			
Total	0.46	0.06	1.4
Dissolved	0.36	0.0	0.11
Aluminum	1.4	0.3	0.12
Manganese			
Total	3.5	0.29	2.0
Dissolved	3.0	0.00	1.2
Copper	1.58	0.005	0.05
Zinc	0.54	1.7	0.05
Sodium	119	25.0	6870
Potassium	16.6	8.4	33
Ammonia			
Total	23.8	...	...
Dissolved	0.56	...	...
Carbonate	0.0	0.0	0.0
Bicarbonate	0.0	101	94.7
Sulfate	685	830	126
Chloride	629	125	12 078
Fluoride	0.4	0.18	0.37
Phosphate	0.01	0.0	0.37
Nitrate	0.3	37	19
Silica	20	3.1	5.6
Sodium, %	16	4.8	88

Source: Ref 2

Table 3 Selection of materials for mining and mill application

Application	Environment	Materials
Crushing and grinding	Heavy pressure, shock-impact loading	Austenitic manganese steels (ASTM A128), AISI 4300 series, ASTM A579 alloy steel, AISI 8600 series
Mill liners, grates, and abrasion-resistant plates	Severe gouging, crushing impact and wear, wet (pH 5–8)	Austenitic manganese steels, martensitic chromium-molybdenum white cast iron, martensitic high-chromium white cast iron, martensitic nickel-chromium white iron, martensitic medium-carbon chromium-molybdenum steel, austenitic 6Mn-1Mo steel, pearlitic high-carbon steel, pearlitic white cast iron
Grinding balls (Ref 8, 9)	Severe gouging, crushing impact and wear, wet (pH 6–8)	Pearlitic white cast iron, martensitic white cast iron, forged (0.8% C) steel, AISI 4155, Ni-Hard type 1, Ni-Hard type 4
Grinding rods (Ref 8, 9)	Severe gouging, crushing impact and wear, wet (pH 6–8)	Heat-treated alloy steel, AISI 52100, hot-rolled AISI 1095 modified with 1.2% Mn, hot-rolled AISI 1095 with 0.4% Mn
Gearing for mining machinery	Wet, lubricated (pH 5–8) wear, light duty	Carburized AISI 1015, 1020, 1022, 1117, 1118, heat-treated AISI 4340, 8645
	Wet, lubricated (pH 5–8) wear, moderate duty	Carburized AISI 8628, 4620, 4615, or equivalent
	Wet, lubricated (pH 5–8) wear, heavy duty	Carburized AISI 4820, 4320, 2320, or equivalent; nitrided AISI 4340, 4140, 4350, and 2.5% Cr steel
Load-haul-dump equipment	Wet (pH 5–8) wear, impact	AISI 1020, cast carbon steel, cast austenitic manganese steels (ASTM A128), cast ASTM A579 steel, ASTM A514 steel
Percussion drilling tools	Wear, impact, gouging (pH 6–8)	Carburized AISI 4320, 8620, and 9315; quenched-and-tempered 4140
Hardfacing	High-impact, wet (pH 6–8)	Austenitic manganese steels
	Unlubricated metal to metal rolling or sliding	Self-hardened, air-hardened steels
	Highly abrasive conditions, wet (pH 6–8)	High-carbon high-chromium white cast iron, high-chromium white cast iron
	Sliding abrasion on cutting edge of drilling tools, wet (pH 6–8)	Special tungsten- and boron-containing weld deposits
	Abrasion at high temperature and/or corrosion	High-nickel or -cobalt weld deposits
Pumps	pH (0–13), abradants	AISI type 304 or 316 stainless steel, Ni-Hard types 1 and 4, 27% Cr white cast irons
	pH (0–13), abradants	Low-carbon, high-manganese steels
	pH (0–13), abradants	Low-alloy cast iron
	pH (0–13), abradants	ACI CD-4MCu
	pH (0–13), abradants	ACI CF-3M (low carbon for as-welded corrosion resistance)
	pH (0–13), abradants	ACI CN-7M (niobium or titanium for as-welded corrosion resistance)
	pH (0–13), abradants	ASTM A484 steel (low-carbon for as-welded corrosion resistance)
	High-temperature environments	ASTM A743 or A744 nickel-base alloy, grade CZ-100
Flotation cells	Corrosive, pH (0–5)	Ni-Hard type 1, type 316 stainless steel
Paddles	Corrosive, pH (0–5)	Fiberglass-reinforced plastic and rubber
Spirals	Abrasive	Ni-Hard type 1
Classifiers blades	...	Ni-Hard type 4
Ore chutes	Impact, gouging, abrasion, pH 6–8	Ni-Hard, nickel-containing manganese steel
Scrapers	Impact loading, gouging, abrasion	Cast ASTM A579 steel, Ni-Hard cast iron, hard cast irons
Wire rope	Corrosive-abrasive, pH 2–12	Kevlar, steel wire rope
Piping	Corrosive-abrasive	Type 316 stainless steel, CN-7M, Ni-Hard cast irons, rubber covered fiberglass-reinforced plastic
Scrubbers	Off-gas products	High-grade nickel alloys
Chain conveyors	Corrosive-abrasive	Plated (nickel, cadmium, or zinc) steels

using available hydrogen from either organic compounds or aqueous electrolyte at the cathodic interface of metal, producing hydrogen sulfide (H_2S).

High humidity, high ambient temperature, dusts, fumes, breakdown of minerals that form acid mine waters, and microbiological microorganisms all contribute to corrosion and lead to degradation of mining equipment. Examples of the corrosion of principle mining equipment will be discussed below. Table 3 lists some of the various engineering materials that are used for mining and milling equipment.

Wire Rope

Mine shaft depths of 1830 m (6000 ft) are not uncommon; gold mines in South Africa approach depths of 2285 m (7500 ft) (Ref 10). The hoisting equipment used in these mines, especially ropes, on which lives of personnel depend, is subjected to the corrosive environment of the mines. Although wear is also a factor, corrosion is perhaps the most serious aspect of mine safety. Corrosion is difficult to evaluate and is a more serious cause of degradation than abrasion (Ref 11). If corrosion is evident, the remaining strength cannot be calculated with safety, nor is there any reasonable way to determine whether or not the rope is safe except by the judgement of the inspector.

Statistical analysis from the results of rope tests on mine-hoist wire ropes has shown that 66% of the ropes exhibited the greatest strength loss in the half of the ropes nearest the conveyance (Ref 12). This is the portion of the rope that is in contact with the shaft environment during most of its service life.

Replacement of hoist rope is a routine procedure in most mines and is suggested every 18 to 36 months, depending on the mine environment and use (Ref 1, 13). Some regulatory agencies will not allow the use of a shaft rope on which marked corrosion is evident (Ref 11).

Adequate service life of hoist rope is economically desirable. Therefore, the composition of present-day hoist rope has been extensively studied. Carbon steel strand wire has competition from such substitute materials as stainless steel (Ref 14) and synthetic fibers (Ref 15). Austenitic stainless steel rope (15.5 to 18.5% Cr and 11 to 13% Ni) is available and has endurance strength of 72 to 83% of that of carbon steel wire. Much can be said for synthetic fiber rope construction. For example, Aramid fiber rope has exceptional strength-to-weight ratios, outstanding tension-tension-fatigue performance (Ref 15), and excellent corrosion resistance.

Roof Bolts

Roof bolts are extensively used for roof support in underground mines. More than 120 million roof bolts are used per year in the United States mining industry (Ref 11). Roof bolts made of low-carbon steel in a number of design variations are subject to corrosion attack in the mine environment. In sulfide mines, the roof bolts have been reported to fail within 1 year by breaking at a distance of approximately 355 mm (14 in.) inside the drill hole (Ref 1). This roof bolt failure has been related to stress-corrosion cracking. Roof falls are associated with such roof bolt failures.

Pump and Piping Systems

Corrosion in pump and piping systems is well known in the mining and mineral-processing industries. The first indication of pump corrosion is that the pump no longer meets the flow demands of capacity and head requirements. Also, the external surfaces are corroded and encrusted with corrosion product.

Because recognition of corrosion type is so important in diagnosing corrosion problems and their prevention and because published information on case histories is scarce, corrosion types will be discussed and illustrated in the following sections in this article.

Uniform Corrosion. The most common form of pump corrosion is characterized by uniform attack on the entire exposed surface. Figure 1 shows uniform corrosion on a stainless steel pump impeller that was exposed to 50% phosphoric acid (H_3PO_4) and 10% gypsum pumping fluid

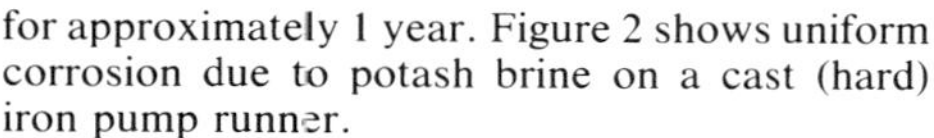

Fig. 1 Uniform corrosion of ACI CD-4MCu cast stainless steel pump impeller after 1 year in an environment containing 50% H_3PO_4 and 10% gypsum. Courtesy of A.R. Wilfley & Sons, Inc., Pump Division

(a)

(b)

Fig. 2 Uniform corrosion of an abrasion-resistant iron pump runner that contacted potash brine slurry. (a) End view. (b) Side view. Courtesy of A.R. Wilfley & Sons, Inc., Pump Division

for approximately 1 year. Figure 2 shows uniform corrosion due to potash brine on a cast (hard) iron pump runner.

Pitting corrosion on an Alloy Casting Institute (ACI) CF-8M stainless steel casting pump case is illustrated in Fig. 3. This pump case, which was exposed to low pH and high Cl^- concentration, failed after approximately 3 years of service.

Erosion-corrosion of an ACI CN-7M stainless steel cast impeller after exposure to hot concentrated H_2SO_4 with solids present is shown in Fig. 4. Erosion-corrosion is evident in Fig. 5, which shows that the erosion-corrosion damage increased on the portion of the impeller that had the greatest fluid velocity. The impeller, cast from an abrasion-resistant white iron, was used to pump fluids containing 30% solid (iron ore tailings) at a pH of 11.2.

Crevice Corrosion. Figure 6 shows crevice corrosion on an ACI CF-8M cast stainless steel pump case that was gasket sealed on the discharge flange. Attack is evident in the region where the gasket was placed. It is important to design mining and milling equipment for easy drainage and cleaning in order to prevent the buildup of stagnant water that will produce concentration cells and lead to crevice corrosion and pitting. Figure 7 illustrates both poor and improved design for avoidance of localized attack. More information on proper design is available in the article "Design Details to Minimize Corrosion" in this Volume.

Intergranular corrosion is common to stainless steel pump castings. Figure 8 shows the excessively attacked grain boundaries. A quick test to identify this type of corrosion is to peen the pump casting with a small hammer. Loss of acoustical properties is evidence of intergranular grain-boundary attack, especially in sensitized stainless steel castings. A form of intergranular corrosion associated with weld deposits that is commonly called weld decay is shown in Fig. 9, which illustrates a field-weld repair of an ACI CN-7M stainless steel impeller that was not postweld heat treated (solution annealed and quenched) to restore corrosion resistance. This weld-repaired casting was exposed to a phosphoric anhydride (P_2O_5) solution at 80 °C (175 °F). It is evident that the weld decay occurred in the heat-affected zone of the weld deposit.

Dealloying. Figure 10 shows selective leaching (dealloying) of a high-nickel cast iron. The physical dimension of the pump component remained constant, while porous, selectively leached regions grew into the casting. This porous layer consists of residual graphite contained in the cast iron and corrosion product. This pump component was exposed to fluosilicic acid (H_2SiF_6) and failed after 12 days of service.

Galvanic corrosion between two stainless steels is illustrated in Fig. 11. These AISI type 304 stainless steel stud bolts held together an Alloy 20 (ACI CN-7M) pump housing. The bolts became anodic to the housing in 45% H_2SO_4 and subsequently failed. Table 4 lists various combinations of pump and valve trim materials and indicates the combination that may be susceptible to galvanic attack.

Cavitation is a familiar term within the industry. The causes are also familiar; principally, there is a lack of net positive suction head, which is the suction pressure that should be available to the pump for correct performance. Cavitation usually manifests itself by another familiar pump characteristic—noise. The buildup and subsequent collapse of bubbles on the impeller create the familiar popcorn noise.

The effects of cavitation (the violent collapse of bubbles) are illustrated in Fig. 12 and 13. Plastic and metal impellers have been found to be susceptible to cavitation damage. A contributing cause of cavitation has been found to be plugged filters on the intake (suction) side of the pump, coupled with the marginal net positive suction head available to the pump. Table 5 lists engi-

Fig. 3 Pitting corrosion of an ACI CF-8M stainless steel pump case used to pump a nickel plating solution with a high concentration of Cl^- and a high operating temperature. This damage occurred during 3 years of service. Courtesy of A.R. Wilfley & Sons, Inc., Pump Division

Fig. 4 Erosion-corrosion of ACI CN-7M stainless steel pump components that pumped hot H_2SO_4 with some solids present. Note the grooves, gullies, waves, and valleys common to erosion-corrosion damage. Courtesy of A.R. Wilfley & Sons, Inc., Pump Division

Fig. 5 Erosion-corrosion of an abrasion-resistant iron pump runner used to pump 30% iron tailings in a fluid with a pH of 11.2. This runner had a service life of approximately 3 months. Note that most of the damage is on the outer peripheral area of the runner where fluid velocity is the highest. Courtesy of A.R. Wilfley & Sons, Inc., Pump Division

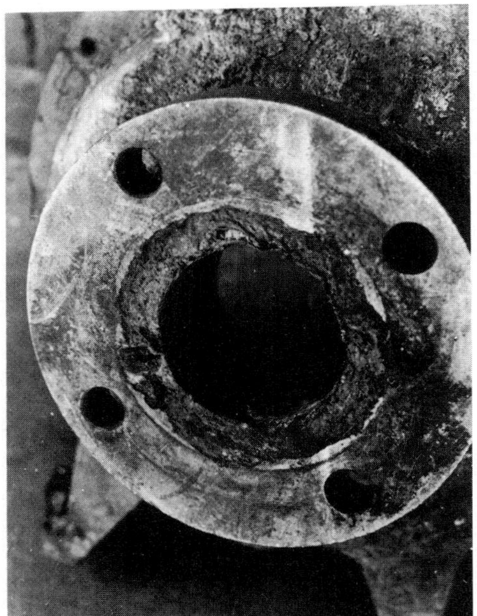

Fig. 6 Crevice corrosion at the intake flange of an ACI CF-8M stainless steel pump case. Notice that the corrosion damage occurred under the gasket. Courtesy of A.R. Wilfley & Sons, Inc., Pump Division

neering materials based on their resistance to cavitation damage.

Figure 12 shows the craterlike damage caused by cavitation on a plastic impeller. Figure 13 shows cavitation damage on an ACI CN-7M stainless steel impeller that pumped hot ammonium nitrate (NH_4NO_3); the pumping installation had a total lack of a net positive suction head.

Erosion-Corrosion. During mill processing, fluids are pumped containing particulates that are usually carried in a corrosive medium. Pumping this slurry promotes erosion-corrosion in piping, tanks, and pumps. Erosion-corrosion is a function of the fluid velocity and the nature of the particulates and fluid. The following procedures can be used to reduce erosion-corrosion or to increase the service lives of piping and pumping systems:

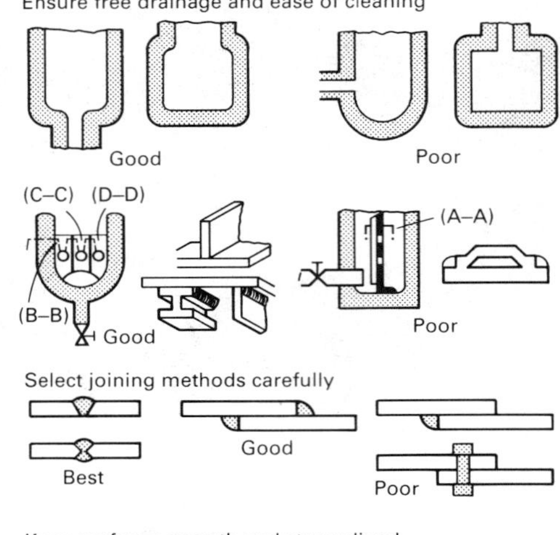

Fig. 7 Poor and improved engineering design to avoid crevice corrosion. Source: Ref 16

- Increase the thickness of pipes
- Use larger inside diameter pipes to reduce fluid velocity for the transport of a specific fluid volume
- Streamline bends in piping to ensure laminar flow
- Use nonmetallic ferrules inserted in the inlet ends of pipes
- Design for easy replacement of parts that experience severe erosion-corrosion
- Use coatings that produce an erosion-corrosion resistant barrier, such as rubber coatings (Ref 18)

Material selection is an important consideration for erosion-corrosion resistance. Alloy hardness has also been shown to be a factor in erosion-corrosion resistance. Generally, soft alloys are more susceptible to erosion-corrosion than their harder counterparts, but the relative hardness properties of the alloy can be misleading, because the hardening mechanism affects resistance to erosion-corrosion (Ref 8). For example solid-solution hardening has been found to offer greater resistance than that provided by conventional heat treatment. One example of this is the cast precipitation-hardening alloy ACI CD-4MCu, which outperforms Alloy 20 (CN-7M) and austenitic stainless steels in many applications.

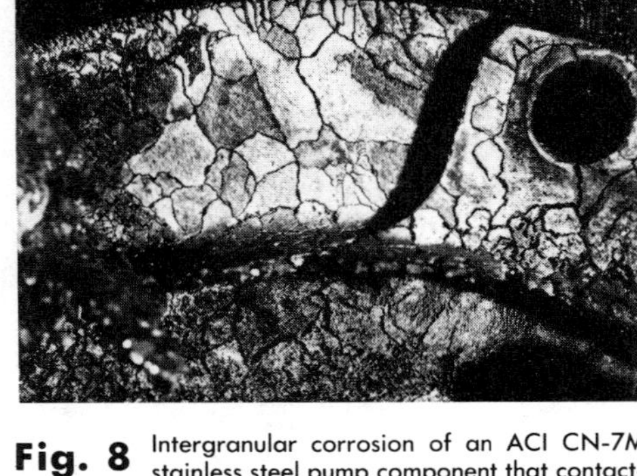

Fig. 8 Intergranular corrosion of an ACI CN-7M stainless steel pump component that contacted HCl-Cl_2 gas fumes. Note the grain-boundary attack. Courtesy of A.R. Wilfley & Sons, Inc., Pump Division

Economics often enters into material selection. Cast iron is relatively more economical and frequently exhibits better erosion-corrosion resistance than cast steel. High-silicon cast iron (14.5% Si) has been found to be an economical selection.

Active-passive materials, such as stainless steels and titanium, owe their corrosion resistance to their developing a protective passive oxide film. This protective film, however, can be continuously damaged by erosive-abrasive processes. Selection of passive alloys should be based only on experience and/or laboratory test results.

Joints must be reliable. Welded pipe, such as carbon steel or stainless steel, is free of flanges but is costly to install. Nonwelded joints are susceptible to crevice corrosion; therefore, stainless steel, in particular, will not attain its expected service life.

Tanks

Most tanks are made from low-carbon steel for economic considerations. The most common corrosion protection for these tanks is the use of coatings and linings. Cathodic protection can also

(a)

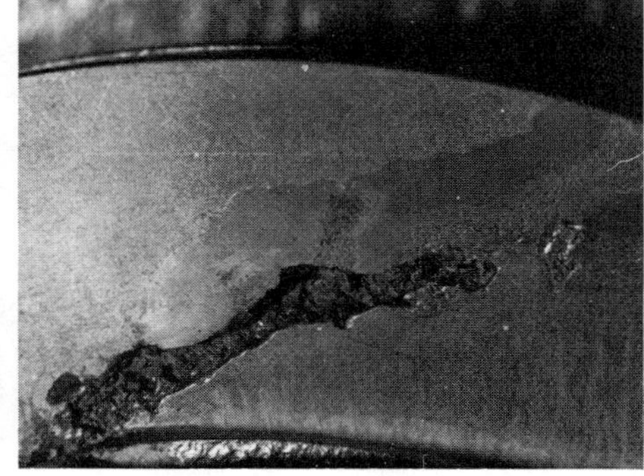

(b)

Fig. 9 Weld decay of an ACI CN-7M stainless steel pump impeller that was field weld repaired with no postweld heat treatment. The pump service was P_2O_5 solution at 80 °C (175 °F). (a) Overall view of impeller. (b) Closeup view of the weld repair and the associated weld decay, which occurred adjacent to the weld deposit. Courtesy of A.R. Wilfley & Sons, Inc., Pump Division

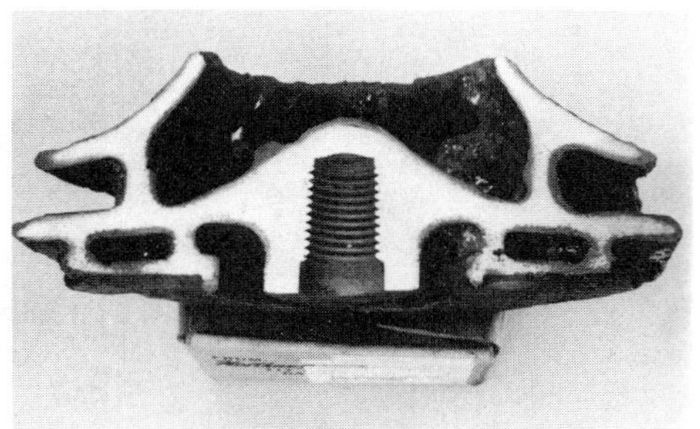

Fig. 10 Selective leaching of a cast iron pump impeller after 12 days of service in H_2SiF_6. Section through the impeller shows the selectively leached layer, which contains graphite and corrosion product. Courtesy of A.R. Wilfley & Sons, Inc., Pump Division

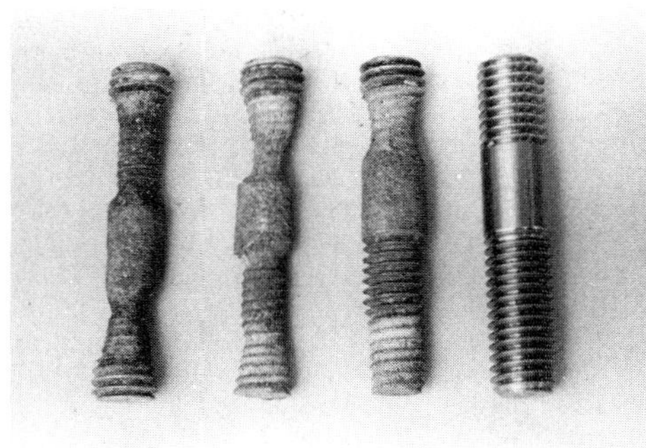

Fig. 11 Galvanic corrosion of AISI type 304 stainless steel stud bolts that fastened two Alloy 20 (ACI CN-7M) pump components. The pump was pumping 45% H_2SO_4 at 95 °C (200 °F). The stud bolts were anodic to the Alloy 20 pump housings. Courtesy of A.R. Wilfley & Sons, Inc., Pump Division

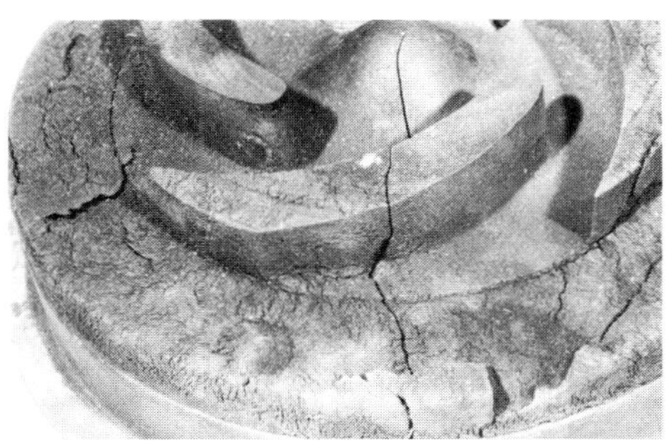

Fig. 12 Cavitation damage of phenolic plastic pump impeller. Note the craterlike depression on the damaged surface caused by the collapse of bubbles on the impeller surface. Courtesy of A.R. Wilfley & Sons, Inc., Pump Division

Fig. 13 Cavitation damage of an ACI CN-7M stainless steel pump impeller that pumped NH_4NO_3 solution at 140 °C (280 °F). Courtesy of A.R. Wilfley & Sons, Inc., Pump Division

be used and is commonly employed in conjunction with a coating (Ref 18). Coating materials can be classified as cement, epoxy, epoxy-phenolic, neoprene, latex, sprayed polyresin coating, polyesters and vinyl esters (heavy coatings), and baked phenolic. Steel tanks are also lined with natural rubber, synthetic elastomers, rubber-backed polypropylene, and glass. Glass lining would require an oven bake.

Table 4 Galvanic compatibility of materials used for pump components

Body material	Trim: Brass or bronze	Trim: Nickel-copper alloy	Trim: Type 316
Cast iron	Protected	Protected	Protected
Austenitic nickel cast iron	Protected	Protected	Protected
M or G bronze			
70–30 copper nickel	May vary(a)	Protected	Protected
Nickel-copper alloy	Unsatisfactory	Neutral	May vary(b)
Alloy 20	Unsatisfactory	Neutral	May vary(b)

(a) Bronze trim commonly used. Trim may become anodic to body if velocity and turbulence keep stable protective film from forming on seat. (b) Type 316 is so close to nickel-copper alloy in potential that it does not receive enough cathodic protection to protect it from pitting under low-velocity and crevice conditions. Source: Ref 17

Table 5 Rating of materials for cavitation resistance

Most resistant

Stellites
17Cr-7Ni stainless steel welding rod
18–8 stainless steel welding rod
Bronze welding rod (Cu-10Al–1.5Fe)
25Cr-20Ni weld
Eutectic-Xyron 2–24 weld
Ampco bronze casting
18–8 cast stainless steel
Nickel-aluminum bronze, cast
13% Cr cast iron
Manganese bronze, cast
18–8 stainless steel spray metallizing
Cast steel
Bronze
Rubber
Cast iron
Aluminum

Least resistant

Source: Ref 17

Table 6 Fatigue and corrosion fatigue strengths of various alloys at 10^7 cycles

Metal	Fatigue strength, MPa	Fatigue strength, ksi	In Levack water, MPa	In Levack water, ksi	In Helen water, MPa	In Helen water, ksi	In Leitch water, MPa	In Leitch water, ksi
T1 tool steel	414	60	131	19	145	21	152	22
Abrasion-resistant steel	307	44.5	152	22	145	21	124	18
Low-carbon steel	214	31	152	22	159	23	138	20
Stelcoloy-G steel	269	39	138	20	114	16.5	124	18
Aluminum alloy 6061-T6	107	15.5	69	10	107	15.5	55	8

(Corrosion fatigue strength columns: In Levack water, In Helen water, In Leitch water.) Source: Ref 2

Stainless steel and titanium alloys have also been used for tanks. Their use depends on their specific corrosion resistance to the solution. Selection of an alloy type becomes a question of economics.

Reactor Vessels

A variety of materials are used, depending on the corrosivity of the media being contained. With neutral or alkaline pH, carbon steels are often used. With increasing corrosivity, consideration is first given the austenitic stainless steels, then iron-nickel-chromium superalloys, and finally nickel-base superalloys. For special environments, copper, copper-nickel, and nickel-copper (Monel-type) alloys are used (Ref 19). Titanium is known to have excellent corrosion resistance in some of the most aggressive solutions.

Cyclic Loading Machinery

The mining and mineral industry use large numbers of rotary and cyclic loaded equipment. This equipment is subject to fatigue and corrosion fatigue. Table 6 illustrates the significant reduction of fatigue strength of various materials that were tested in mine water and compared to fatigue strength in air. This problem can be reduced by designing heavier sections into the part to reduce load and by applying protective coatings (Ref 18).

REFERENCES

1. G.R. Hoey and W. Dingley, Corrosion Control in Canadian Sulfide Ore Mines and Mills, *Can. Min. Metall. Bull.*, Vol 64, May 1971, p 1-8
2. G.J. Biefer, Corrosion Fatigue of Structural Metals in Mine Shaft Waters, *Can. Min. Metall. Bull.*, Vol 58, June 1967, p 675-681
3. N.S. Rawat, Corrosivity of Underground Mine Atmospheres and Mine Waters: A Review and Preliminary Study, *Br. Corros. J.*, Vol 11 (No. 2), 1976, p 86-91
4. I. Iwasaki, K.A. Natarajan, S.C. Riemer, and J.N. Orlich, Corrosion and Abrasive Wear in Ore Grinding, in *Wear of Materials 1985*, American Society of Mechanical Engineers, 1985, p 509-518
5. T.P. Beckwith, Jr., The Bacterial Corrosion of Iron and Steel, *J. Am. Water Works Assoc.*, Vol 33 (No. 1), June 1941, p 147-165
6. B. Intorre, E. Kaup, J. Hardman, P. Lanik, H. Feiler, S. Zostak, and W.E. Rinne, Complete Water Reuse Industrial Opportunity, in *Proceedings of the National Conference*, American Institute of Chemical Engineers, 1973, p 88
7. F.N. Speller, *Corrosion: Causes and Prevention*, McGraw-Hill, 1951, p 208
8. S.L. Pohlman and R.V. Olson, "Corrosion and Material Problem in the Copper Production Industry," Paper 229, presented at Corrosion/84, National Association of Corrosion Engineers, 1984
9. K. Adam, K.A. Natarajan, S.C. Riemer, and I. Iwasaki, Electrochemical Aspects of

Grinding Media—Mineral Interaction in Sulfide Ore Grinding, *Corrosion*, Vol 42 (No. 8), 1980, p 440-446
10. S.A. Bryson, Repair Work and Fabrication in Gold Mining Environments, *FWP J.*, Vol 24 (No. 2), 1984, p 35-48
11. J.M. Karhnak, "Corrosion and Wear Problems Associated With the Mining and Mineral Processing Industry," Paper 230, presented at Corrosion/84, National Association of Corrosion Engineers, 1984
12. R.L. Jentgen, R.C. Rice, and G.L. Anderson, Preliminary Statistical Analysis of Data From Ontario Special Rope Tests on Mine-Hoist Wire Ropes, *Can. Min. Metall. Bull.*, Vol 77 (No. 11), 1984, p 50-54
13. H. Precek and J. Zeigler, Ropes for Use at Great Depths in Mining, *Wire Ind.*, Vol 52 (No. 8), 1985, p 486-487
14. H. Hartmann, Hauling Ropes for Shaft Installations Under Extreme Corrosive Conditions, *Wire Ind.*, Vol 46 (No. 3), 1979, p 179
15. N. O'Hear, Developments in Aramid Fibre Ropes, *Wire Ind.*, Vol 49 (No. 11), 1982, p 845-850
16. R.F. Steigerwald, Corrosion Principles for the Mining Engineer, in *Symposium Materials for Mining Industry*, AMAX Molybdenum, Inc. 1974
17. M.G. Fontana and N.D. Greene, *Corrosion Engineering*, McGraw-Hill, 1967
18. L.D. Eccleston, Protective Coatings in the Mining Industry, *Can. Min. Metall. Bull.*, Vol 72 (No. 3), 1979, p 170-173
19. A.I. Asphahani and P. Crook, "Corrosion and Wear of High Performance Alloys in the Mining Industry," Paper 228, presented at Corrosion/84, National Association of Corrosion Engineers, 1984

Corrosion in Structures

John E. Slater, Invetech, Inc.

THE PREVENTION of metallic corrosion in structures, particularly the consideration of its effects in initial design and fabrication or later during in-service inspection and retrofit, is vital in ensuring the expected longevity of the structure. Indeed, in many cases, structures may have to last well beyond their originally anticipated lifetimes—for example, the extended lives of structures in Europe that contain metal, some of which date back many hundreds of years.

In considering the ramifications of corrosion and its prevention in different types of structures, it is useful to group such structures into various categories. Thus, the modern high rise—whether an office building, an apartment building, a condominium structure, or a special-purpose building such as a hospital—poses particular problems related to the corrosion of major structural components (which may be either totally metallic or may contain metallic material) and creates the necessity for corrosion considerations in any connector assembly used to tie the curtain-wall system on the building back to the structural frame or to the backup wall system. In low-rise structures, similar concerns exist, but may be less critical when the structure is only a few stories high.

Under such circumstances, the possibility of danger to people and property resulting from failure of, for example, curtain-wall tie systems may be less. Nevertheless, failure of metallic materials within the structure may lead to unsightliness or to possible lack of weathertight behavior. Both may require extensive reworking of the structure to reestablish adequate building performance.

Parking structures are frequently of conventionally reinforced concrete or prestressed/posttensioned construction. Where such structures are in the snow belt areas of the country or where chloride may intrude because of the proximity of marine environments, the reinforcement may be at risk from corrosion. Stadiums are another example in which either steel-frame or concrete-frame approaches can be used. In these cases, certain approaches may be needed to ensure structural integrity, particularly in view of the safety of the many thousands of people who may visit the structure and fill it to capacity.

Bridges have received much attention with regard to corrosion. This has been particularly prevalent in the snow belt states, where deicing salt application has led to significant premature deterioration of decks and supporting structures. However, supporting structures have also suffered damage because of saline water intrusion, particularly in the splash zone of reinforced concrete structures. Also, the use of weathering steels can be a problem in such structures, especially where design or construction practice does not adequately account for the particular limitations in the use of this type of steel.

Other specialized structures in which specific corrosion conditions may be a concern include sewage plants, which may have isolated situations related to contaminated water and the use of treatment chemicals. Other buildings or structures can be used to house or support chemical plants and/or utility plants. In the latter case, specific aspects of corrosion prevention in nuclear plants—especially containment buildings—have been the subject of critical review. Because many nuclear containment vessels are designed with posttensioned concrete, the longevity of any corrosion protection system applied to the posttensioning tendons is critical. This article will not address specific interactions with manufactured industrial chemicals, because this area is more reasonably the province of general or specific knowledge related to the particular chemical to which the building material will be exposed.

Metal/Environment Interactions

The most common metallic material used in structures is low- or medium-carbon steel. Specialized or strengthened materials, including heat-treated steels, stainless steels, and nonferrous metals, are sometimes employed; use of these materials is the exception rather than the rule (they will be discussed when appropriate in this article). The specific interactions that will be considered are the reactions of steel with the atmosphere in all of its forms—including polluted atmospheres, interior atmospheres, and internal atmospheres (for example, in a cavity wall)—in which the major criteria determining the performance of unprotected metal include the corrosivity of the atmosphere, the temperature, and the time of wetness of the metal.

A major research effort over the past 20 years has been expended in determining the reaction of steel with cementitious materials, which include mortar, concrete, and their variants. The chemical that typically controls corrosion behavior under these circumstances is the calcium hydroxide ($Ca(OH)_2$) introduced into the cementitious material by the portland cement or, in the case of mortars, lime. The reaction of $Ca(OH)_2$ with atmospheric carbon dioxide (CO_2) is an important degradation mechanism that will radically change the behavior of steel that is in contact with or embedded in the cementitious material. The other major factor that influences the corrosion of steel in this environment is the introduction of chloride ion (Cl^-) into the cementitious material. Chloride may play a decisive role in causing the cementitious material to change from a protective to a nonprotective environment with regard to embedded steel.

This article will discuss the generic situation of metallic materials—specifically steel—in reacting with the environments found in structures. These environments will be discussed in specific terms, particularly as related to atmospheric conditions and cementitious environments. The utility of different corrosion protection methods will be described in relationship to particular structures and to the different environments, either atmospheric or cementitious, encountered by the steel. Finally, examples of problems that have arisen in the corrosion performance of metallic materials in these environments will be delineated, with particular attention paid to the different problems as they relate to different structures in which the metallic material may exist.

General Considerations in the Corrosion of Structures

Corrosion of Steel in the Atmosphere. Atmospheric corrosion is discussed at length elsewhere in this Volume (see the articles "General Corrosion" and "Corrosion of Carbon Steels") and in Volume 1 of the 9th Edition of *Metals Handbook*. However, a brief discussion of factors affecting such behavior in structures is necessary. Steel corrosion in the atmosphere is typically a function of temperature, humidity, and the presence in the atmosphere of components that increase the corrosivity of the environment. The temperature/humidity ratio is particularly important, because the interaction among these factors leads to a function known as the time of wetness, during which a film of liquid is present on the surface of the steel. Corrosion occurs by a typical aqueous corrosion mechanism during the time that this film is present on the steel surface. The thinner the film, the easier the diffusion of oxygen through the film that drives the corrosion reaction.

The presence of agents in the atmosphere that can dissolve in the liquid film and promote or inhibit its production by changing the dew point can markedly influence the corrosion behavior of the steel. Thus, the presence of pollutants in the atmosphere—particularly sulfur dioxide (SO_2) and related compounds, which can lead to so-called acid rain—will influence the corrosion behavior of the steel by acting in several ways, such as:

- An increase in the conductivity, and therefore the corrosivity, of the liquid film
- Changes in the relative humidity at which the film may form

- Influence of the dissolved constituent on the formation and protectiveness of any corrosion product films that may exist on the steel

Therefore, the corrosion behavior of steel used in structures can be expected to be a function of the geographical location, which in turn will influence the temperature, humidity, and degree of pollution that exists. This behavior is indicated in Table 1.

Corrosion in marine environments is a specialized subset of atmospheric corrosion. This is because of the influence of wind-blown or particulate sea salt that may contact exposed steel. Sea salt is particularly aggressive to steel, possibly because of the concentration of magnesium chloride ($MgCl_2$) it contains. Therefore, chloride from the major constituent of sea salt, sodium chloride (NaCl), is deleterious to the corrosion behavior of steel because of its effect on the conductivity of the liquid film and its destruction of protective corrosion product films, but $MgCl_2$ acts to acidify the liquid film and, by its deliquescent action, to increase the time of wetness. This is the primary reason why the corrosion rates of structures in marine environments are considerably higher than in any other type of location. Of course, the possibility of splash zone action on the corrosion of structures is beyond the scope of this article (see the article "Marine Corrosion" in this Volume). The dramatic increase in the corrosion rates of steel in marine locations is indicated in Table 1.

Specialized environments exist in which specific pollutants are present as a function of the particular use or location of the structure. Therefore, chemical plant and refinery buildings and structures may be particularly vulnerable to certain types of atmospheric pollutants produced by the processes occurring within the plant and may require particular protective measures. Again, this is beyond the scope of this article, and the reaction of steel to such environments is best determined by noting the effect of the pollutant on the corrosion of steel in such publications as Ref 2.

The atmosphere within closed structures will usually be very different from that on the external surfaces. In general, the presence of pollutants may be significantly reduced, and the control of temperature and humidity may mean that the overall environmental corrosivity is significantly lower than that on the outside surfaces. However, there can be action that affects this circumstance. For example, the well-known barrier effect of internal walls, which is related to their containment of the internal environment, may be breached at certain locations, with the result that air may escape from the interior of the building into the cavity. If structural or other steel is present in this cavity, the interaction of the steel with air, which may be deliberately humidified, in contact with a possibly cold surface can lead to condensation and subsequent corrosion problems that would otherwise be unanticipated.

Such problems have been noted in the past—not only from the viewpoint of structural steel framing but also from the possibility of corrosion of elements in the system that tie the external cladding back to the frame of the building. This will be discussed in more detail later in this article. Therefore, concern is necessary not only for the influence of the external environment on the corrosion of steel but also for the effects of the internal environment, particularly the lack of containment.

Fireproofing is often mandated by fire codes for structural steelwork within buildings. In the past, fireproofing was used that contained corrosive constituents. Under circumstances in which moisture levels at the structural steel may increase (because of the exfiltration of humid air or the infiltration of water), the corrosive constituents in the fireproofing can lead to unacceptable corrosion of the steel. This potential problem is addressed in ASTM E 937, which is a standardized test for determining the corrosivity of such materials under standardized conditions (Ref 3). The use of coatings that meet conditions related to this test is required by many codes and should be mandated by the specifying architect or engineer.

Whenever steel is used in buildings, there is always the possibility that galvanic contact may occur. Therefore, galvanized conduit may contact bare steel. Copper-base alloys are used as wiring—for example, in lightning conduction rods. In general, the relative positions in the galvanic series of metals in seawater will normally dictate whether or not there is a potentially significant problem with galvanic corrosion when structural elements are in contact. The environment surrounding the potential galvanic couple will have a significant effect; more severely corrosive environments allow more galvanic activity to occur. Therefore, unanticipated contact between different metals and the impact of such contact on corrosion behavior must be an ever-present concern.

Although the use of the lower-strength steels generally indicates that environmental/mechanical interactions leading to unexpected brittle failure are very unlikely to occur in the normal environments used in structures, the use of higher-strength steels (particularly those of the quenched-and-tempered variety) may increase the likelihood of such stress-corrosion or hydrogen embrittlement failures, especially in more aggressive environments such as marine conditions. Compounding the problem is the welding that may be performed—for example, on bridge structures. Therefore, quenched-and-tempered steels of the ASTM A709, grade 100, type, which have yield strengths in the 690-MPa (100-ksi) range, must be welded carefully so as not to create hard regions in heat-affected zones that may be susceptible to hydrogen embrittlement in corrosive environments.

Similar considerations pertain to the area of embedded posttensioning steels. These steels are typically of the cold-drawn type (not quenched and tempered), corresponding to such standards as A416, grade 270 (1860 MPa, or 270 ksi, minimum ultimate tensile strength for seven-wire strand), and A421 (1655 MPa, or 240 ksi, minimum ultimate tensile strength for 6.4-mm, or 1/4-in., diam wire). Although laboratory test data indicate that cold-drawn steels are considerably less susceptible to failure by hydrogen embrittlement than their quenched-and-tempered counterparts, care must be taken to minimize the aque-

Table 1 Corrosion rates of carbon steel calibrating specimens at various locations

Location	Type of environment	Corrosion rate(a) µm/yr	Corrosion rate(a) mils/yr
Norman Wells, NWT, Canada	Polar	0.76	0.03
Phoenix, AZ	Rural arid	4.6	0.18
Esquimalt, Vancouver Island BC, Canada	Rural marine	13	0.5
Detroit, MI	Industrial	14.5	0.57
Fort Amidor Pier, Panama, CZ	Marine	14.5	0.57
Morenci, MI	Urban	19.5	0.77
Potter County, PA	Rural	20	0.8
Waterbury, CT	Industrial	22.8	0.89
State College, PA	Rural	23	0.9
Montreal, Que. Canada	Urban	23	0.9
Durham, NH	Rural	28	1.1
Middletown, OH	Semi-industrial	28	1.1
Pittsburgh, PA	Industrial	30	1.2
Columbus, OH	Industrial	33	1.3
Trail, BC, Canada	Industrial	33	1.3
Cleveland, OH	Industrial	38	1.5
Bethlehem, PA	Industrial	38	1.5
London, Battersea, England	Industrial	46	1.8
Monroeville, PA	Semi-industrial	48	1.9
Newark, NJ	Industrial	51	2.0
Manila, Philippine Islands	Tropical marine	51	2.0
Limon Bay, Panama, CZ	Tropical marine	61	2.4
Bayonne, NJ	Industrial	79	3.1
East Chicago, IN	Industrial	84	3.3
Brazos River, TX	Industrial marine	94	3.7
Cape Canaveral, FL (18-m, or 60-ft, elevation, 55 m, or 60 yd, from ocean)	Marine	132	5.2
Kure Beach, NC (250 m, or 800 ft, from ocean)	Marine	147	5.8
Cape Canaveral, FL (9-m, or 30-ft, elevation, 55 m, or 60 yd, from ocean)	Marine	165	6.5
Daytona Beach, FL	Marine	295	11.6
Cape Canaveral, FL (ground level, 55 m, or 60 yd, from ocean)	Marine	442	17.4
Point Reyes, CA	Marine	500	19.7
Kure Beach, NC (25 m, or 80 ft, from ocean)	Marine	533	21.0
Galeta Point Beach, Panama, CZ	Marine	686	27.0
Cape Canaveral, FL (beach)	Marine	1070	42.0

Source: Ref 1

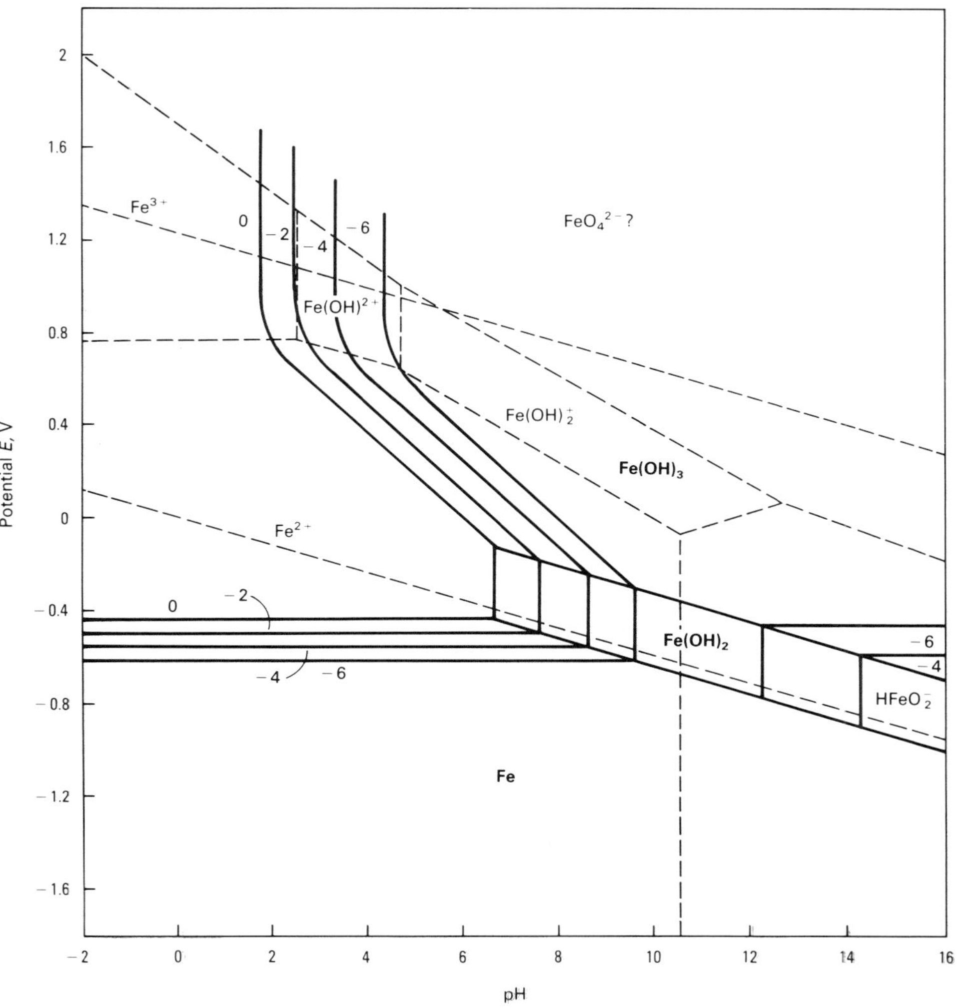

Fig. 1 Pourbaix (potential-pH) diagram for the system iron-water at 25 °C (75 °F). Source: Ref 4

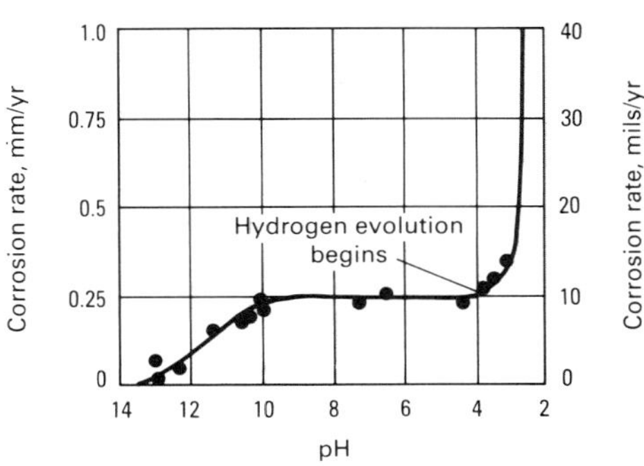

Fig. 2 Effect of pH on the corrosion rate of iron in aerated soft water at room temperature

ous and/or atmospheric corrosion of such steels before and after concrete placement.

With regard to this aspect, the atmospheric corrosion of other metallic materials commonly used in structures is of less concern. Aluminum can be used for a variety of building components, including windows and door frames. If the aluminum is protected from direct contact with uncured mortar or concrete and if a commonly used anodized coating is specified, little atmospheric corrosion is noted except under severe marine conditions. Zinc is typically used only as a protective coating for steel, and as such, the corrosion behavior will be covered in a later section in this article.

Steel in Cementitious Materials. The behavior of steel in contact with cementitious materials (concrete and mortar) is generally governed by the properties of the portland cement that is a constituent of the mortar or concrete. Portland cement is an alkaline material; the alkalinity results from the presence of $Ca(OH)_2$ and other soluble alkali salts. The pH of a saturated $Ca(OH)_2$ solution is approximately 12.5. Reference to the Pourbaix (potential-pH) diagram for iron (Fig. 1) shows that, under these circumstances and in the presence of moisture and oxygen, the steel will be in a passivated condition because of the formation of a thin film of oxide—generally considered to be γ FeOOH.

The effect of pH on the corrosion rate of steel in aerated water is shown in Fig. 2. If the oxygen is depleted of if the cementitious material is allowed to dry significantly, then the passive film may well be disrupted. However, under these conditions, the corrosion rate is expected to be extremely low because of the high resistivity and low driving force for the reaction. As will be discussed later in this article, there are various pollutants to the cementitious environment that can modify this behavior, leading to significant corrosion of steel in contact with cementitious materials.

Concrete is a hard, dense material that consists of cement paste surrounding aggregate particles. The major difference between mortar and concrete is in the size of the aggregate. Concrete contains a high proportion of large aggregate, such as gravel and crushed rock. Steel that is embedded in concrete, generally as reinforcement, can be of two main types. In conventional reinforced concrete, reinforcing bars are used to support the tensile loading, because of the relative weakness of unreinforced concrete in tension versus its significant strength in compression. Such reinforcement will be sized and placed according to requirements determined by the structural engineer. Bar sizes are usually stated as number X, where X represents the numerator in a fraction of eighths.

Other types of reinforcement in concrete are deliberately stressed in tension to place the concrete into residual compression. For prestressed concrete, this is typically accomplished by placing the stressed steel into the concrete form and pouring the concrete around the steel. Once the concrete has cured, then the external tension on the steel is released. The bond between the steel and the concrete, however, now places the surrounding concrete into compression, leading to the well-known concept of prefabricated, prestressed concrete.

Another method of accomplishing this same effect is posttensioning, in which the unstressed steel is placed in the concrete form before the concrete is poured. The steel is typically prevented from developing a bond with the concrete during the pouring process. After pouring and curing of the concrete, the steel now embedded in (but separated from) the concrete is tensioned by a variety of mechanisms, and the tensioned steel is then anchored to the ends of the concrete structure. In this way, the concrete around the steel is placed into tension.

Other steel components that may be embedded in concrete include such items as water lines (which may typically be galvanized steel) and the forms used as containment during the concrete pouring process. In many cases, these forms can be removed; in others, the forms will remain in place. As such, the forms are exposed to the atmosphere on their outer surface and to the concrete environment on their inner surface. The problems that have occurred during the use of such forms, particularly where additives to the concrete have caused a corrosive situation to occur, will be discussed later in this article.

In other cases, unanticipated contact may occur between different materials and steel embedded in concrete. Aluminum can be used for various articles embedded in or located on concrete—for example, balustrades and lampposts. If contact can occur between embedded steel and such aluminum in environments that increase the corrosivity of the steel, as will be determined later in this article, accelerated corrosion of the aluminum can occur, with corresponding galvanic protection of the steel. Although it is conceivable that, for example, copper water pipes may contact embedded steel reinforcement, there are no recognized instances of problems resulting from such contact.

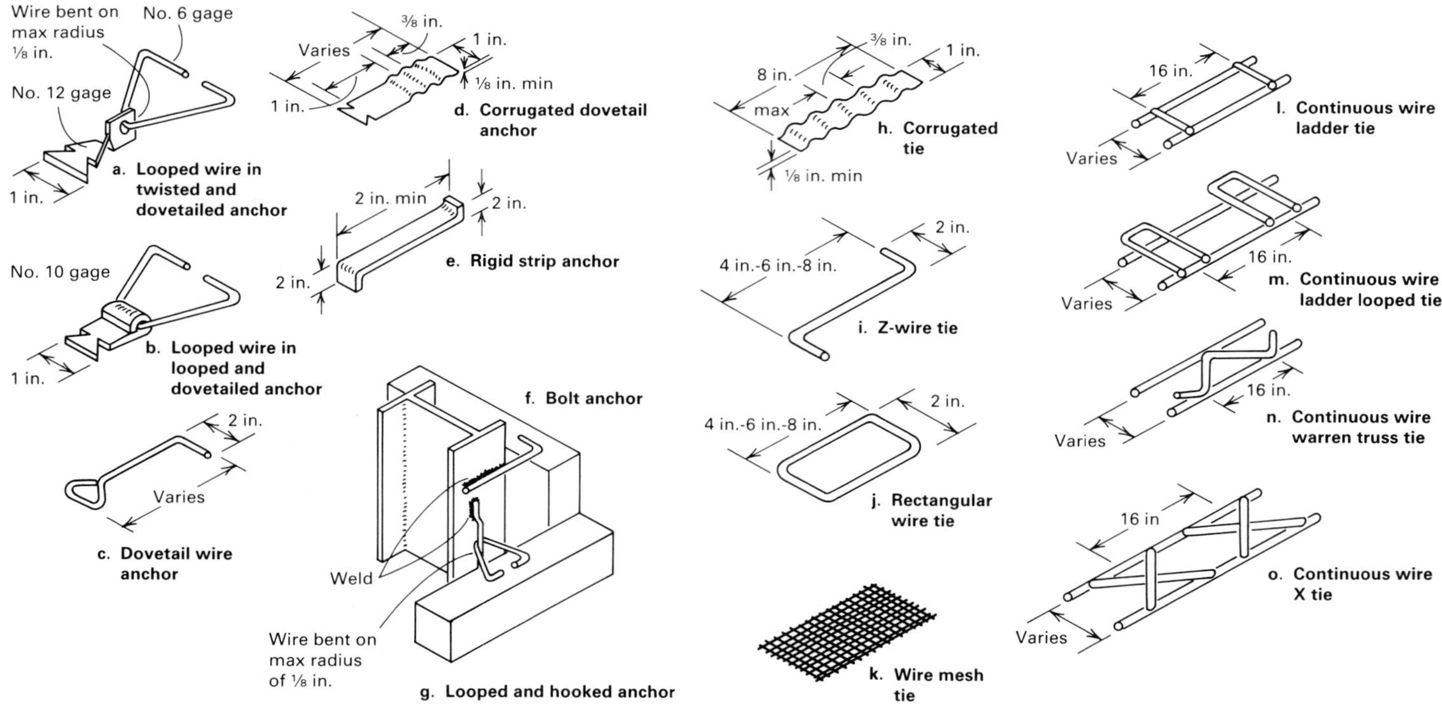

Fig. 3 Typical metallic anchors (a to g) and ties (h to o) used in masonry walls

The use of mortar, which is a cementitious material, is a necessary factor in the construction of brick masonry, and steel is used in such masonry in a variety of applications. The use of steel joint reinforcement to enhance the strength of the masonry by placing a mesh or ladder-type system within the bed joint is a common occurrence. Such a system can also be used to connect two wythes of masonry. The use of ties or anchors to anchor masonry used as a curtain-wall system back to the building frame or to a supporting structure is also necessary. Typical reinforcements, ties, and anchors are shown in Fig. 3. In these cases, although bare steel would suffice in uncontaminated high-pH mortar, it is common practice to use a more corrosion-resistant material for reasons that will be discussed later. Typically, galvanizing is recommended and used—either hot-dip or electroplated.

Other metallic components that may contact mortar include window frames and doorways. These may be steel or aluminum, if metallic. Such steel components can be protected against atmospheric corrosion with organic coatings, but in other cases, metallic coatings, such as galvanizing, can be used. Aluminum components can corrode in the uncured or green mortar environment, but in the absence of further corrosion enhancers, the corrosion rate normally drops to a very low level once curing has occurred and the mortar dries.

The behavior of steel in concrete and mortar as it is influenced by high pH can be drastically altered by two primary factors. The first is carbonation, or the reaction of the $Ca(OH)_2$ with CO_2 from the air. The reaction to form calcium carbonate ($CaCO_3$) lowers the pH of the cementitious material; under these conditions, the steel is no longer passive, but can become active and corrode significantly. This process of carbonation is a function of the porosity of the concrete or mortar, which is a function of such factors as the water/cement ratio, placement adequacy, and vibration/consolidation. In general, however, mortar is much more permeable than concrete and carbonates much more rapidly. Under these circumstances, the mortar in masonry structures must be considered to undergo carbonation at a relatively early point in its life. To prevent the corrosion of embedded steel due to carbonation, it is common practice to use protective coatings on steels (as has already been discussed, galvanizing is common) or to use inherently corrosion-resistant materials, such as stainless steels.

The second major contributor to the corrosion of embedded steel in cementitious materials is the influence of chloride. Chloride in sufficient quantities prevents the formation of the initial protective oxide film on steel (if the chloride is present in the material mix during the curing process) or, if added later, breaks down the passive film and allows corrosion to proceed by a mechanism similar to the pitting of stainless steel passive films by chloride.

Possible chloride sources in the mixing process are the aggregate, chloride introduced with the mixing water (a problem frequently encountered in the Middle East, where brackish water is often the only water available), and chloride-containing admixtures deliberately added to the mix as a set accelerant or so-called antifreeze—most commonly calcium chloride ($CaCl_2$). Chloride can also find its way into the cementitious material after placement and curing, because of the application of chloride to the outer surface. Sources of such chloride are marine environments. Deicing salt placed onto concrete bridge decks, parking structures, and so on, may also find its way onto buildings as a result of splash from roadways. Another possible source for chloride in mortar is the absorption of hydrochloric acid washing solutions applied to the masonry surface.

The quantity of chloride necessary to cause corrosion of bare steel in uncarbonated cementitious systems depends initially on the stage at which the chloride is introduced. For bare steel in initially chloride-free concrete that is subjected to chloride infiltration from the external surface, corrosion has been found to initiate on the steel when the chloride level reaches between 0.02 and 0.04% by weight of concrete at the steel interface. When the chloride is present in the mix, the tricalcium aluminate of the concrete can insolubilize a certain portion of the chloride during the curing process. Recognition of this fact initially led to the allowance of additions of 2% $CaCl_2 \cdot H_2O$ by weight of cement to concrete to accelerate set. More recent work, together with the release of bound chloride as concrete ages, has led to a recognition that maximum allowable chlorides in concrete should be a function of both the nature of the steel and the expected service environment of the concrete. The recommendations of American Concrete Institute Committee 201, which are being revised at this time, are given in Table 2.

A compounding factor in the chloride-induced corrosion of reinforcement is the action of macrocells. Because of the electrically continuous and extensive nature of the typical reinforcement in concrete, differences in chloride concentration and oxygen level at different locations on the steel can be anticipated. Under these conditions, separated anodic (high-chloride, low-oxygen) and cathodic (low-chloride, high-oxygen) areas can easily develop along the reinforcement, particularly where chloride enters the concrete from an exterior surface. This separate, macrocell action frequently leads to severe corrosion at the anodic areas. Macrocell action has been worsened unintentionally during concrete repair.

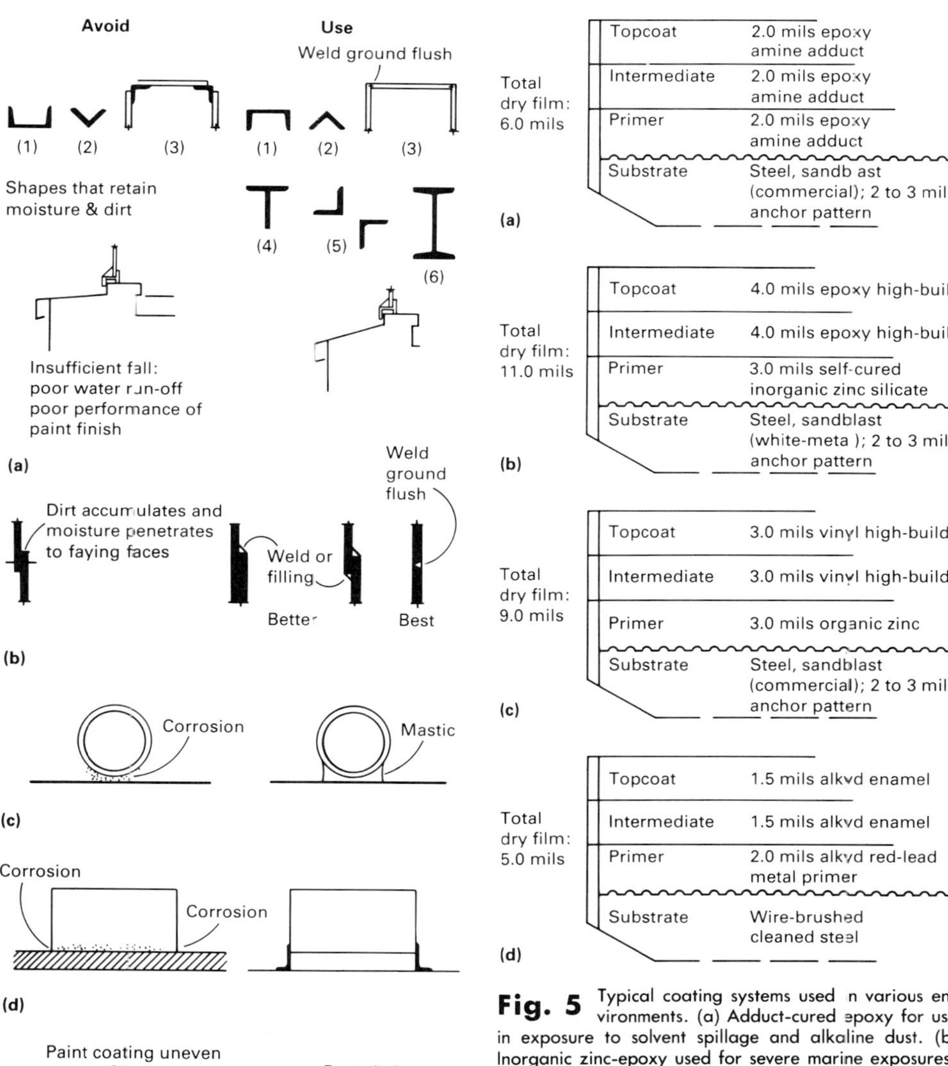

Fig. 4 Design and fabrication details to be considered in corrosion prevention. (a) Constructional members, sills, etc. (b) Joints. (c) Crevices. (d) Air circulation. (e) Corners, edges, and surfaces

Fig. 5 Typical coating systems used in various environments. (a) Adduct-cured epoxy for use in exposure to solvent spillage and alkaline dust. (b) Inorganic zinc-epoxy used for severe marine exposures. (c) Inorganic zinc/vinyl coating system for use in mild industrial environments. (d) Alkyd-base coating for mild inland atmospheric exposure. Source: Ref 5

Therefore, the replacement of deteriorated, spalled, chloride-contaminated concrete adjacent to reinforcing steel with fresh, chloride-free concrete can introduce a potent cathode into the system, with the result being accelerated corrosion of the surrounding reinforcement. This can lead to the well-known spall around a spall deterioration phenomenon.

The effect of corrosion induced by such a mechanism is somewhat dependent on the nature of the steel. For reinforcing steel, the major problem arising from an increased corrosion rate is the production of a voluminous oxide, which occupies a greater volume than the steel from which it was produced. Under these circumstances, significant tensile stresses can be introduced into the surrounding cementitious material, leading to cracking of the brittle cementitious material. Calculation of the Pilling-Bedworth ratio (see the article "Fundamentals of Corrosion in Gases" in this Volume) allows the magnitude of the volume expansion to be assessed, but this is typically found to be a factor of between 2 and 10, depending on the exact nature of the corrosion product.

The effect on high-strength prestressing steel may not be the same. Such steel typically has an ultimate tensile strength exceeding 1380 MPa (200 ksi) that is produced by cold forming. As discussed previously, steels of this hardness may be susceptible to stress-corrosion cracking or, more probably, hydrogen embrittlement. In this context, then, corrosion processes that may produce hydrogen can be very deleterious to the structure containing the posttensioning or prestressing. This is due to the possibility of hydrogen embrittlement leading to brittle fracture of the steel and therefore to the loss of compression of the surrounding concrete. Examples of this problem will be discussed later in this article.

Table 2 Recommended maximum water-soluble chloride levels in concrete mix (prior to service) for different reinforced concrete exposures

Type of exposure	Maximum recommended chloride level, wt% of concrete
Prestressed concrete	0.06
Conventionally reinforced concrete in a moist environment and exposed to chloride	0.10
Conventionally reinforced concrete in a moist environment but not exposed to chloride (includes locations where the concrete will be occasionally wetted, such as parking garages, waterfront structures, and areas with potential moisture condensation)	0.15
Aboveground building construction where the concrete will stay dry	No limit

Protection Methods

Atmospheric Corrosion. One of the most important aspects of corrosion protection is corrosion prevention through the consideration of the fundamental aspects of good design practice. Such considerations include:

- Avoidance of upturned angles, channels, and so on, that can collect moisture
- Avoidance of pockets within welded structures
- Grinding welds flush
- Elimination of crevices that can lead to accelerated corrosion

Examples of details to avoid, as well as more corrosion-resistant details, are shown in Fig. 4. Although some of these discrepancies can be mitigated by other corrosion protection methods, the selection of weathering steels makes adherence to good design and fabrication detailing mandatory.

Various methods are available for protecting steel against corrosion in so-called atmospheric conditions. The protection system is generally a barrier coating or a metal alloying of the steel that effectively introduces a barrier coating by a normal corrosion process.

The barrier coatings used to protect steel from atmospheric corrosion are of three main types: organic coatings, inorganic coatings, and metallic coatings. The selection of one of these coatings is a function of the expected environmental severity in which the structure is to perform, the corrosivity of that environment, the expected lifetime of the structure, and the possibility of further maintenance coating. In other words, coating selection is a combination of both technical and economic considerations.

Organic coatings are perhaps the most frequently used coatings for protecting steel from atmospheric corrosion. This is true with regard to bulk structural steel. However, such components as fasteners rarely have organic coatings, because of the difficulties related to the fit of fastener systems incorporating such coatings and the poor durability of such coatings under abrasion and tightening conditions. A review of the different types of organic coatings that are available is included in the article "Organic Coatings and Linings" in this Volume. Nevertheless, it is important to highlight in this article certain as-

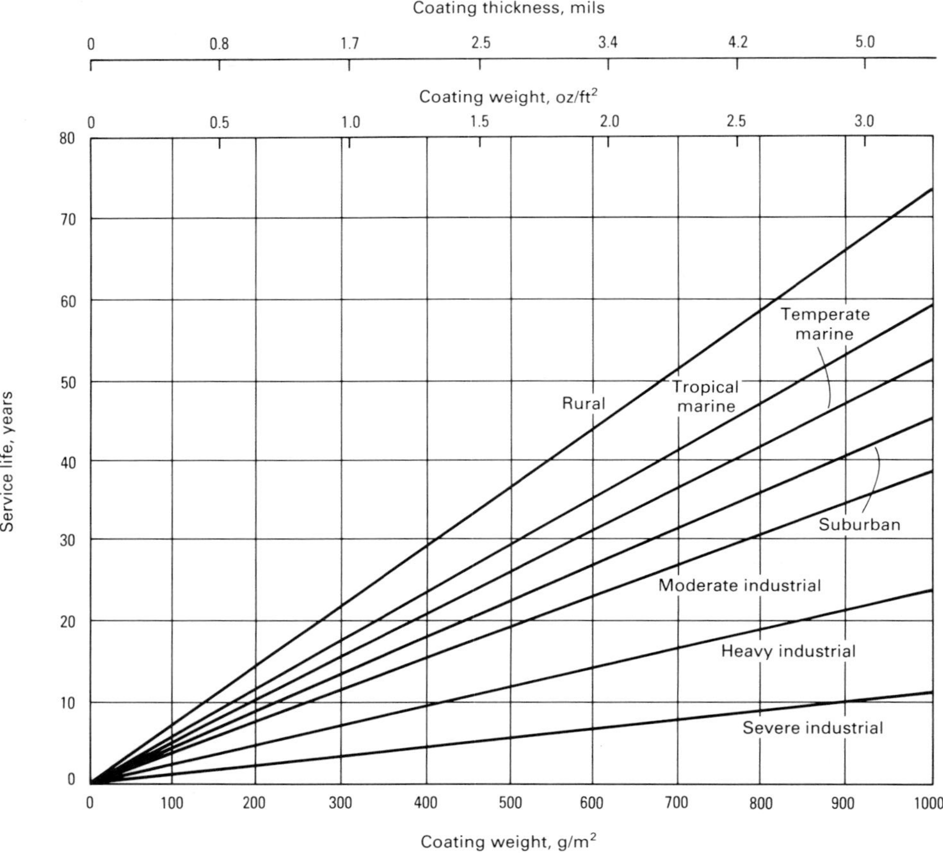

Fig. 6 Effect of zinc coating weight on service life of galvanized steel sheet in various environments. Service life is measured in years to the first appearance of significant rusting.

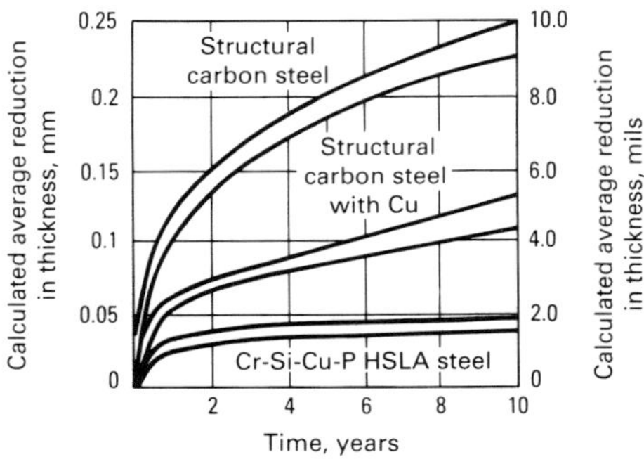

Fig. 7 Corrosion of three types of steels in an industrial atmosphere. Shaded areas indicate range for individual specimens

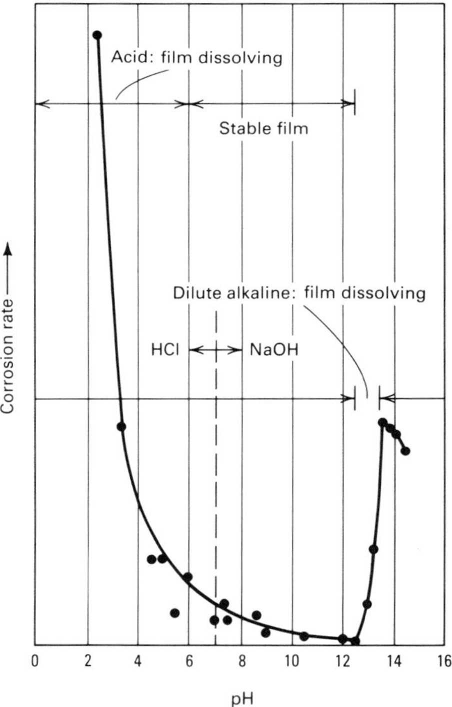

Fig. 8 Corrosion of zinc in aqueous solutions as a function of pH

pects of coatings procedures that are of vital importance in the longevity of the coating.

Although some coatings have recently become available that are less susceptible to the problems of incomplete surface preparation, it is generally recognized that the quality of the surface preparation is a major factor in the service life of the coating. Perhaps the most extensive evaluation of paint behavior as a function of substrate preparation has been conducted by the Steel Structures Painting Council, which has issued a series of standards on different degrees of cleaning prior to the application of paints.

The degree of cleanliness to be attained depends on the sensitivity of the chosen paint system to the level of surface cleanliness and on the required longevity of the system. In this regard, it is important to recognize that different methods of surface preparation are available, including centrifugal blast cleaning, abrasive air-blast cleaning, water-blast cleaning, and hand and power tool cleaning. Hand tool and power tool cleaning are typically employed only where small areas must be cleaned of preexisting scale and paint. It is also important to recognize that contaminants on steel surfaces that are not visible to the naked eye may influence the longevity of the subsequent coating. Therefore, inorganic salts present on the surface may dissolve in water that permeates through the coating, leading to significant corrosion of the substrate. Standard blast-cleaning techniques typically do not remove such contaminants, but water-blast cleaning does. However, this again becomes a matter of economics and the sensitivity of the coating system to such surface contamination.

The selection of a coating system is influenced by consideration of the likely corrosive environment and the consequences of coating failure. Figure 5 shows some typical preparation and coating formulations. Therefore, the need for increased durability of the coating increases with the corrosivity of the environment. In this regard, much steel is enclosed within building walls, and such steel is frequently only lightly shop primed. Indeed, it has been questioned whether such coating is in fact necessary, based on the environment in which the steel is to perform.

One of the problems in making decisions of this type is that the severity of the environment may not be adequately documented. Therefore, if there is exfiltration from the building, then condensation on the steel members can occur, resulting in an extended time of wetness and a corrosion rate that is significantly greater than might otherwise be anticipated. Under these circumstances, the thin shop primer may be much less than adequate, and a more rigorous corrosion prevention system should be considered. Adverse effects on such steel may include the pickup of corrosive constituents (particularly chloride) from other building materials—for example, from mortar during passage of water through masonry walls. If flashing systems are inadequate, then such water may reach structural steel and cause significant corrosion damage. This is another factor that needs to be considered whenever the so-called benign environment within, for example, the walls of a building is considered.

Inorganic Coatings. Other forms of coating that are midway between the use of a metallic coating and an organic coating are the inorganic zinc-rich coatings. The ultimate life expectancy of this type of material in severe weathering service has not yet been established, but over 20 years of experience has shown that complete protection of the steel substrate is still being provided. Protection in this case is largely the result of the good corrosion resistance of zinc itself. Indeed, it has been pointed out that the cured inorganic zinc-silicate films can be thought of as a cross between hot-dip galvanizing and a fused ceramic. Because of the absence of organics, inorganic coatings are considered to have the best solvent resistance of any type of protective coating. This makes them very useful for struc-

Fig. 9 Heavy buildup of corrosion scale on weathering steel structural members in conditions of poor air circulation, high humidity, and no wetting/drying

Fig. 10 Corrosion scale buildup on weathering steel structural members, which were in a sheltered area on a building exterior where wetting and drying did not occur

tures in which vapors are present, such as chemical plants, petroleum refineries, and production facilities.

The metallic coating system most commonly used for steel in structures is zinc coating. These coatings can be applied by hot-dip galvanizing or by electroplating for smaller components, such as bolts. Electroplating generally lays down a considerably thinner zinc layer than hot-dip galvanizing. Therefore, the service life of this layer can be considered to be shorter. The results of numerous investigations of the service lives of galvanized coatings show that the life of the coating is a function of the coating thickness and the environment in which it will operate. This is illustrated schematically in Fig. 6. Therefore, the thickness of galvanizing to be placed on a piece of structural steel will depend on an assessment of the nature of the environment and the required longevity. In this regard, the severity of the environment within a cavity wall where a zinc coating material may frequently be used has recently become a significant concern; this issue will be discussed later in this article. Additional information on zinc coatings is available in the articles "Hot Dip Coatings" and "Corrosion of Zinc" in this Volume.

For small components, other protective systems are available. These include cadmium plating, which is considered to be more corrosion-resistant than zinc in marine environments. Cadmium plating is currently viewed with caution because of the toxicity of the plating solutions and the difficulty of disposal. Nevertheless, the general corrosion resistance of cadmium plate makes the process worthy of consideration where the small steel components used in structures must be protected.

Weathering Steels. The final method of protection against atmospheric corrosion is the use of weathering steels. In this class of materials (typified by ASTM specifications A 588 for buildings and A 709 for bridges), small amounts of alloying elements—typically nickel, chromium, and copper—are added to the steel (Ref 6, 7). Under certain fairly specific situations, these alloying elements are incorporated into the oxide layer that forms on the steel, leading to the formation of a dense, more protective oxide. This oxide then serves as a barrier to further penetration of moisture and effectively acts almost as a "self-painted" coating, lessening the need for any other coating protection. An example of the improvement in corrosion behavior of such steels in an industrial environment is shown in Fig. 7.

Although such behavior is valuable, it must be recognized that it will be observed only under certain circumstances. Interestingly, the improvement is generally best in an industrially polluted environment; less improvement is noted in environments containing chloride—for example, marine environments. Also, the film or patina must undergo repeated wetting and drying to develop. Any use of this type of steel in a structure in which this type of exposure is not available will not allow development of the protection. In this context, it is vital to adhere to good corrosion-resistant design techniques when using weathering steels to ensure that the steel performs adequately. Any design or fabrication detail that incorporates crevices, improperly bolt-

Fig. 11 Heavy corrosion scale buildup on structural members of weathering steel at a pocket where water could collect and stand

Fig. 12 Corrosion-induced spalling of overlying concrete on reinforced columns. See also Fig. 13

Fig. 13 Severe corrosion on reinforcing steel from the column shown in Fig. 12

Fig. 14 Unconventional masonry tie constructed from flattened C-channel and bent to enter the brick core

Fig. 15 Severe corrosion and cracking (arrows) on wall tie

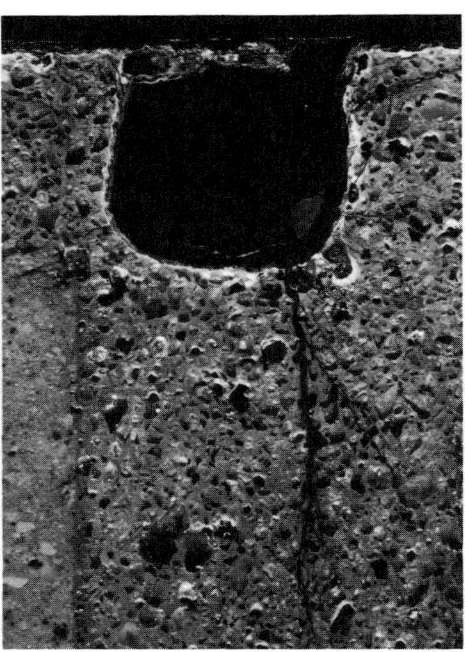

Fig. 16 Corrosion and resulting masonry cracking on anchor embedded in high-band mortar. See also Fig. 17

ed connections, and so on, will lead to behavior in which weathering steels exhibit no advantage over other steels. Examples of problems related to the use of weathering steels will be discussed later in this article.

The ultimate situation in using alloying elements to promote a barrier layer on the surface of steel involves the introduction of sufficient chromium into the steel to allow the formation of stable chromic oxide. The stainless steels contain a minimum of 12% Cr and are highly resistant to normal atmospheric conditions. More highly alloyed steels may be required for resistance to severe marine environments. However, the high cost of these materials precludes their use except in critical situations. Such critical situations may, for example, arise in connection bolts, ties, and other components that must perform for the expected service life of the building and are usually not located where inspection, maintenance, or replacement can be easily carried out. Under these circumstances the use of stainless steel becomes a highly advantageous proposition.

Cementitious Systems. As discussed previously in this article, cementitious systems normally provide a protective environment to steel. Indeed, the longevity of reinforced steel structures supports this situation. The major factors that cause steel embedded in cementitious systems to corrode and lead to significant problems are the influences of carbonation and chloride infiltration. Under both of these circumstances, the normally protective oxide film on steel breaks down, and corrosion will proceed at a rate that is sufficient to cause loss of cross section, buildup of voluminous corrosion product with resultant spalling of overlying material, or, in specific cases, hydrogen embrittlement.

One of the principal methods of preventing this problem in construction has been to place over the steel a concrete or cementitious material of sufficient quality and thickness to reduce dramatically any infiltration of chloride or CO_2 to the steel surface. Because of the diffusion-controlled nature of this infiltration, the depth and permeability of the material have a significant effect on the time at which sufficient chloride or CO_2 reaches the steel surface to allow corrosion to initiate and propagate. The obvious first step in preventing such corrosion is to ensure adequate concrete depth (typically taken as 50 mm, or 2 in., of clear cover over reinforcing steel) and to use as low a water-to-cement ratio as feasible to promote impermeability within the concrete.

Although this approach may be considered adequate for reducing carbonation problems within the expected service lives of structures, it may not be sufficient to mitigate chloride-induced corrosion. The reasons are twofold. First, the chloride may have been added to the mix; therefore, it would already be present at the level of the reinforcing steel, irrespective of the type, quality, or depth of concrete cover used. Such problems can be prevented at the specification stage; the use of any chloride-containing admixtures for any purpose whatsoever should be prohibited. Also, research has shown that prestressing steel may be even more deleteriously affected by chloride-induced corrosion than conventional reinforcing steel; limits are currently being considered on the amount of chloride permissible within the concrete from any source. Therefore, chloride-bearing aggregate could have a considerable influence on this situation.

Second, in the case of marine environments, and particularly in the application of deicing salt, it becomes difficult, if not impossible, to ensure that the depth of cover and the concrete quality will protect the steel against chloride-induced corrosion attack over the expected service life of the structure. In this case, alternative measures become appropriate. Such measures involve the application of a barrier coating to the steel or to the cementitious surface or the application of cathodic protection to bare steel, generally at a point in the service life of the structure at which corrosion damage has already occurred.

Various coatings for steel in reinforced concrete structures have been evaluated. These can generally be grouped into either organic coatings and inorganic or metallic coatings. Extensive work has demonstrated that the alkaline nature of concrete and the intrusion of chloride into the concrete are best resisted by an epoxy coating on the steel. Moreover, the reinforcing steel must be carefully prepared prior to application of the coating, and the application method is critical—typically, an electrostatic powder spray, followed by baking to fuse the coating. Bars coated in accordance with this technique have been successfully used in bridge decks since 1973. As necessary, repair of such coatings in the field at bar ends and holidays, and so on, can be undertaken using repair materials.

Epoxy coatings have been viewed with some disfavor on the basis of bond strength, that is, the possible tendency of the coating to prevent adequate bond development by lessening adhesion at the deformation/concrete interface. However, it would appear that recommended coating thicknesses, which are typically 0.18 ± 0.05 mm (7 ± 2 mils), mitigate this problem and do not lead to a significant reduction in bond strength.

Additional concerns when using epoxy coatings include the possibility that small holidays in the coating may be subject to accelerated attack, particularly if epoxy-coated steel is used only in the top mat of the structure, and that bare steel is used in, for example, the bottom mat. This is a traditional construction method when deicing salt is applied to the top surface; the bare steel at the bottom is rarely subjected to any chloride levels, because of the large distance (up to 250 mm, or 10 in.) from the upper surface of the concrete to the level of the lower mat. Although it is not clear that this concern has been completely eliminated, testing has indicated that it may not be a major problem. In any case, such a problem could of course be largely eliminated through the use of epoxy-coated reinforcing steel throughout the entire structure.

Metallic coatings for steel in cementitious materials are typically limited to zinc and nickel. As previously discussed, small items, such as threaded fasteners, can be cadmium plated. Cadmium plate is apparently slightly more resistant to chloride attack in cementitious materials than zinc

Fig. 17 Corrosion and resulting masonry cracking on an anchor embedded in high-band mortar. See also Fig. 16

Fig. 18 Corrosion of posttensioning anchorage. Note severe corrosion at the two wedge halves

because of the apparent formation of a basic cadmium chloride. Furthermore, metallic cadmium does not appear to form expansive corrosion products. However, the problems with continued cadmium coating related to environmental concerns have already been addressed.

By far the most common coating for steel in contact with cementitious materials is zinc—either applied by hot-dip galvanizing or plating. The factors that determine the rate at which zinc and zinc-coated steel corrode in cementitious materials are directly linked to the formation and preservation of protective films. A stable film, and therefore low corrosion rates, occurs on the zinc in the pH range of about 6 to 12.5.

Figure 8 shows the influence of pH on the corrosion rate of zinc. It is interesting that the pH of concrete or mortar in the noncarbonated state is typically about 12.5, which is the minimum on the pH versus corrosion rate curve for zinc. This observation is supported by laboratory studies evaluating the corrosion rate of zinc. Therefore, for a noncarbonated mortar, the corrosion rate of zinc in contact with this mortar is effectively zero. Once carbonation occurs, and the pH falls below 10, then the corrosion rate increases—typically to a level of approximately 0.5 to 0.8 μm/yr (0.02 to 0.03 mils/yr). Therefore, the typical use of hot-dip galvanizing as a protective measure for steel exposed to carbonated concrete or, more particularly mortar, is well founded. A typical 64-μm (2.5-mil) galvanized coating would be expected to last over 80 years under these circumstances.

Zinc does not perform as well when chloride is present in the cementitious material either as a result of the original additions or the infiltration from the exterior. The use of zinc coatings under these circumstances has been the subject of much debate. Indeed, the results of laboratory testing—including experiments related to the use of simulated environments such as $Ca(OH)_2$ solution with and without added quantities of chloride as well as laboratory-prepared specimens of zinc-coated steel in contact with chloride-contaminated concrete—may be in disagreement with studies of large-scale structures that have been fabricated with galvanized steel. Some of these structures are bridge decks; those that are most commonly used to promote galvanized reinforcement are structures in the Caribbean, where no cracking of the structure was noted even after chloride up to a level of approximately 0.26% by weight of the concrete was present at the rebar interface. In this case, however, approximately one-tenth of the original galvanized coating had been removed, and further corrosion could be anticipated. In other locations where galvanized steel has been used in bridge decks, no reports are currently known in the open literature regarding the performance of such steel as compared to bare steel in the same type of environment.

Studies on reinforced concrete specimens containing galvanized steel, exposed to either marine environments or to artificial ponding in NaCl solution, have led to a variety of conclusions related to the effectiveness of the galvanizing. Therefore, this range is all the way from a supposedly accelerated cracking of concrete containing galvanized steel to a retardation of such cracking as compared with bare steel. However, it should be emphasized that in most of these studies cracking of the overlying concrete has occurred, indicating that galvanizing is at best a palliative coating for steel in chloride-contaminated cementitious materials.

As far as is known, there is no direct evidence regarding the ability of galvanized steel to withstand a certain level of chloride before significant corrosion can occur. In studies using electrical resistance probes, it was found that for chloride added to mortar during the mixing phase the corrosion rate of zinc was of the order of 0.5 μm/yr (0.02 mils/yr) for a chloride content of approximately 0.15% by weight of the mortar. However, in the presence of carbonation, this corrosion rate increased to 10 μm/yr (0.4 mils/yr). Therefore, for a standard galvanized coating approximately 64 μm (2.5 mils) thick, the coating would survive for only 6.5 years under this latter condition. This is notwithstanding possible problems related to the production of nonprotective and expansive corrosion products from the zinc layer.

Zinc hydroxychloride has been found to form during the corrosion of zinc in chloride-contaminated cementitious materials. This component is significantly expansive compared to the zinc from which it was produced and can therefore cause cracking of overlying cementitious materials even before corrosion of substrate steel has occurred.

Therefore, in general, although zinc-coated steel is somewhat more resistant to chloride-induced corrosion than bare steel, the zinc cannot be relied on to provide protection indefi-

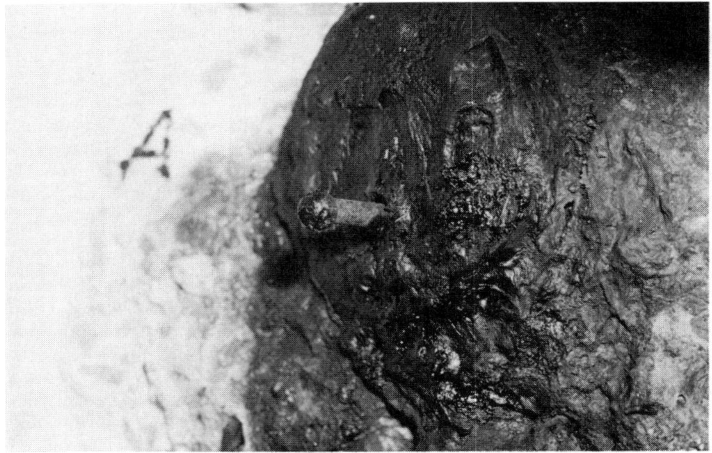

Fig. 19 Posttensioning anchorage with broken wire extending from anchor plate

Fig. 20 Pitting corrosion adjacent to fracture on failed posttensioning wire

nitely. Indeed, the products of corrosion of the zinc itself may lead to cracking of the overlying material.

In addition to applying barrier coatings on the embedded steel to resist corrosion, a similar barrier surface may be applied to the concrete or masonry surface to resist the penetration of constituents contributing to corrosion. These generally take the form of physical barriers, including:

- Organic membranes applied to the concrete surface (often used on parking structures)
- Low-permeability cementitious-base overlays and repairs (incorporating latex)
- Polymer impregnation of the concrete (rarely used)
- Waterproofing agents, for example, drying oils, stearates, or silanes

In using one of these systems, care must be taken not to trap the corrosive environment within the structure. Silanes applied to surfaces are effective in preventing this action; they allow water vapor transmission out of the structure but no liquid penetration into the structure. This prevents oxygen transport to the steel while allowing the structure to dry.

The final method of protection for steel in cementitious materials that may be subject to corrosion is cathodic protection. Cathodic protection is a well-established means of preventing corrosion in a variety of environments, particularly for pipelines buried in soil or immersed in seawater. It has also been successfully used to ensure the protection of steel pipelines that are buried or encased in concrete.

It was first applied to the problem of deterioration of conventionally reinforced concrete structures in California. During early experiments, it was necessary to use a total surface anode over the entire structure to be protected—in this case, the top surface of the bridge deck—because of the low throwing power of the anode and general high electrical resistance of concrete. Using this technique, large decks could be protected with only 10 W of power. The surface anode method did, however, require the embedment of current supply anodes in a relatively soft overlay composed of coke-asphalt mix, which itself could be subject to fairly rapid deterioration by traffic. This technique has been used by others with some success, and it continues to be improved.

One of the more important developments in the cathodic protection of structures is the use of different anode systems that allow horizontal surfaces other than upward facing to be protected. Therefore, the development of anode systems incorporating conductive polymers, conductive paints, and conductive concrete (containing typically coke or graphite) has made possible the cathodic protection of the underside of vertical surfaces of reinforced concrete structures. An example of such an application will be discussed later in this article.

Impressed-current systems using platinized wires in slots in the concrete surface, surrounded by conductive grout, have been used. Experience with this system has been clouded because of the deterioration of the grout and surrounding concrete by the low pH developed at the anode and by the poor throwing power of the anode. Studies have also been conducted on galvanic (sacrificial anode) systems, which are to be distinguished from the impressed-current systems previously discussed. In this type of system, for example, zinc wires are placed in slots above reinforcing steel, and the sacrificial action of the zinc is then transferred to the steel directly below.

Some additional observations can be made with regard to the cathodic protection of reinforced steel in concrete. First, care must be exercised whenever cathodic protection systems are considered for application to high-strength steels, such as those used in posttensioning and prestressing systems. This is because of the previously discussed possible susceptibility of these steels to hydrogen embrittlement, which may result from the overprotection of the steel and subsequent hydrogen evolution at the steel/concrete or mortar interface.

Furthermore, unbonded systems are expected to be difficult to protect because of the possible insulating effects of the sheath that is present around the steel. Indeed, even if such unbonded systems could be protected, the poor throwing power due to the lack of contact between the posttensioning steel and the surrounding concrete is probably an insurmountable problem.

Finally, considerable attention is currently being given to the appropriate criteria for cathodic protection of steel in cementitious materials. Although a transfer of the usual protection criteria from, for example, pipe in soil systems has been made, there is some doubt as to the reasonability or, in some cases, the adequacy of such protection criteria. This is due to the different nature of the environment between concrete and soil, particularly as it relates to pH. Therefore, rather than relying on a −0.85-V copper sulfate electrode as a standard criteria for protection, there is more interest in using a potential shift mechanism and in using E-log i plots for establishing the adequacy of the protection. More information on cathodic protection is available in the article "Cathodic Protection" in this Volume.

Case Histories

This section will cite specific examples of failures and problems that have occurred within structures to illustrate the general principles discussed previously, to point out those factors that bear most heavily on the development of a particular problem or failure, and to determine how protective or preventive measures can be implemented to repair the structure or to prevent the occurrence of failure in the future. The case history data to be discussed have been primarily taken from investigations of buildings or structures conducted by the author. Other examples of failures are available in the literature.

Failures Involving Corrosion of Structural Steel

In general, failures related to the corrosion of conventional structural steel (that which has been painted or otherwise protected) are, in the author's experience, rare. Cases do exist in which excessive humidity or chloride-laden water has contacted the metal. The use of weathering steels, however, has caused significant problems, particularly where the necessary design features and environment have not been carefully considered during selection of the material. Two examples of problems involving weathering steels will be discussed.

Example 1: Weathering Steel Corrosion in a Stadium. A large sports stadium situated about 300 m (1000 ft) from the ocean was built with weathering steel in the major structural members. The steel was used not only for the exposed portions of the structure but also beneath the stands. Significant corrosion and rust flaking were noted on this steel at several locations:

- Underneath the stands, where air circulation was poor and no standard wetting/drying could be anticipated (Fig. 9)
- At sheltered locations on the exterior, again where standard wetting/drying was not possible (Fig. 10)
- At joint details, where the important concepts of removal of crevices and pockets to retain water had not been practiced in the design of joints for the structure (Fig. 11)

The major problem associated with this structure was the amount of corrosion occurring on the structural steel beneath the stands. Although the loads in these structures were determined to be low, it was apparent that corrosion protection would be necessary. A series of tests was conducted to determine the optimum coating protocol, including surface preparation, type of coating, thickness, and number of coats.

This example serves to point out the important factors to be considered by the designer when weathering steel is selected. In particular, the protective patina can be developed adequately only with exposure to weather. Furthermore, marine environments are not conducive to the formation of such a patina, which apparently develops best in an industrial environment with relatively high sulfur levels in the atmosphere.

Example 2: Corrosion of Weathering Steel in a Hotel Parking Garage. A hotel parking garage in the Northeast was constructed with weathering steel in the columns and beams, along with conventional reinforced concrete slabs placed as the decks. The hotel and garage were situated in an area that experienced considerable amounts of snow and freezing temperatures. Deicing salt was commonly applied to roadways adjacent to the structure and was also probably applied to the reinforced concrete slabs themselves. Severe deterioration was noted in the weathering steel beams and columns, particularly those adjacent to leakage points of water, and in expansion joints. This corrosion was caused by contact with the deicing salt laden water, effectively destroying any patina that may have been expected to develop on the steel and leading to the production of voluminous, nonprotective oxides.

The failure in this case was caused by a lack of appreciation of the environment to which the weathering steel was to be exposed. Possible solutions to the problem include the use of coatings on the steel to protect it from further contact with chloride-laden water and the correction of water paths that lead to contact with the deicing salt. The sheltered location of most of the steel, however, would not allow effective patina development. Prohibiting the use of chloride deicing salt on the garage decks would probably reduce the problem somewhat, although pickup of deicing salt from roadways and track-in into the garage is a perpetually deleterious condition that is unavoidable.

Fig. 21 Metallographic cross section through the fracture initiation region of posttensioning wire. Note secondary cracks. Etched with 2% nital. 55×

Corrosion of Conventional Reinforcement

Corrosion of steel in conventionally reinforced concrete in deicing salt application areas and in marine areas is a well-known phenomenon on such structures as bridge decks and bridge support structures. This corrosion phenomenon has received sufficient illustration in the literature and will not be repeated here. An example of a similar problem that has occurred in a building not subjected to deicing salt or ambient marine environments will be used as an illustrative example.

Example 3: Corrosion of Reinforcing Steel in Building Columns. Figure 12 shows an example of a supporting column in a dormitory building on a Midwestern campus. The columns are of conventionally reinforced concrete, with a spiral of reinforcement that closely approaches the surface of the concrete column. The concrete showed cracking and spalling within a few years after its installation. Laboratory examination indicated that the concrete cover was low, that the concrete over the steel was carbonated close to the outer surface, and that the concrete contained a significant amount of chloride, apparently added during construction, that was due to either the presence of chloride-bearing aggregate or the deliberate addition of admixtures such as $CaCl_2$. Corrosion of embedded steel was severe in places (Fig. 13).

Because of the likelihood that carbonation might progress into the concrete and, together with the chloride, affect the more deeply embedded steel, the following repair plan was devised. First, loose and spalled concrete was chipped out of the columns. Second, the columns were shotcreted to a depth of approximately 25 mm (1 in.) above the topmost reinforcement. Finally, an impressed-current cathodic protection system was placed on the columns, using conductive polymer anodes with integral lead wires. This system was positioned above the surface of the concrete with stands, and the entire system was finally covered with an additional layer of concrete. This minimal concrete removal, along with the subsequent buildup and installation of a cathodic protection system, was far more cost effective than the alternatives, which included complete demolition of the column or removal of sufficient overlying concrete cover to necessitate shoring of the column to support the building above.

Corrosion of Ties and Anchors

The corrosion of steel ties and anchors used to attach masonry walls to the building frame, to the backup (usually concrete block) wall, or, more recently, to steel studs has been a source of significant concern. Cases are known in which such ties or anchors have corroded, either in the mortar or in the airspace. Two examples will be given.

Example 4: Corrosion of Nonstandard Wall Ties. In a hospital in the Midwest, cracking of masonry walls at attachment points to a steel stud backup was observed. When the interior wallboard was removed, the tie between the brick wall and the steel stud was found to be in the form of a light-weight C-channel that had been flattened and bent to form a 90° angle (Fig. 14). The flattened portion of the anchor was inserted into the masonry bed joint and down into the core of the brick. The remaining, intact portion of the channel was attached to the backup steel stud by a self-tapping screw. No effective corrosion protection had been applied to the angles; they exhibited remnants of shop primer at certain locations, but the efficacy of any coating had been destroyed during the bending and flattening process.

Severe corrosion of these angles had occurred in the airspace adjacent to the brick masonry. This was apparently caused by water running down the interior wall and becoming trapped. Also, the water would become trapped in the crevices formed by the flattened channel.

One effect of this corrosion was the interaction with the low cyclic stresses imposed on the tie due to movements between the stud and the wall, leading to cracking of the tie. This was particularly prevalent where corrosion had reduced the tie thickness to a fraction of its original dimension (Fig. 15).

This problem could have been prevented through the use of several alternative techniques, including:

- The use of more conventional ties incorporating corrosion-resistant coatings such as galvanizing
- The use of heavy organic coatings after bending to prevent corrosion. If this type of tie was to be mandated, then heavy organic coatings applied after bending would have been appropriate

The solution to this problem could involve the use of alternative supplementary anchors or the dismantling of portions of the wall containing such ties and replacement with masonry containing more appropriate anchors to ensure continued support of the masonry wall.

In addition to the corrosion of steel in the wall cavity itself, which can generally be lessened by the use of galvanizing, there is also concern regarding the corrosion that may occur in the portion of the tie within the mortar. This is the case where significant chloride is present in the mortar, particularly where chloride and carbonation can interact to produce a significantly corrosive environment. In these cases, the commonly used galvanized thicknesses may not be sufficient to protect the steel over the expected service life of the building.

Example 5: Cracking of Masonry Caused by Corrosion of Ties and Anchors. In many structures that have incorporated high-bond masonry mortar additives, the release of Cl^- due to alkaline hydrolysis has significantly corroded uncoated and coated (zinc and cadmium) steel. This has been the case for laid-in-place buildings (utilizing conventional ties) and for panelized buildings (in which embedment of connection devices in the mortar is used to affix the panels to the building frame). In this situation, connection devices may corrode within the mortar, with subsequent cracking of overlying masonry due to the buildup of corrosion product.

Examples of masonry exhibiting such cracking, which radiates from corroded embedded ties and anchors, are shown in Fig. 16 and 17. Under these circumstances, the integrity of the anchor becomes extremely suspect. No effective method is known for alleviating this problem once it has occurred. Alternative approaches have involved the use of heavily organic coated anchors (for example, epoxy coatings) or the use of austenitic, molybdenum-containing stainless steels, which should resist the onslaught of the Cl^-.

Corrosion of Posttensioning and Prestressing Structures

Unlike other cases of corrosion of steel in structures, the corrosion of posttensioning structures can be troublesome from two viewpoints. First, the possibility exists that substantial corrosion can lead to a loss of cross section and therefore failure of the gripping mechanism for the posttensioning strand or of the posttensioning steel itself. Second, corrosion products on the surface of the material may release sufficient hydrogen to cause hydrogen embrittlement. Two examples will be given.

Example 6: Corrosion of Posttensioning Anchorages. A posttensioned garage in the snow belt area of the United States exhibited significant corrosion of conventional reinforcement, as noted by the presence of cracking and spalling of overlying concrete. On one occasion, a posttensioning tendon failed. This led to a large-scale investigation of posttensioning members, particularly the anchorages. The anchorages were found to be significantly corroded because of their location adjacent to leaking expansion joints and because they were surrounded by poor, badly consolidated concrete. This had allowed deicing salt to penetrate to the level of the anchorages, leading to some severe corrosion. This was particularly true on the gripping wedges, as illustrated in Fig. 18.

Several of the anchorages were sufficiently corroded such that alternative anchorages had to be installed. In others, the corrosion was slowed by the injection of water-displacing grease into the anchorage through a grease fitting. This case history shows how poor-quality concrete can significantly affect the performance of metals embedded within it, particularly when the metals are of vital importance to the longevity and safety of the structure.

Example 7: Hydrogen Embrittlement of Posttensioning Wires. Single 6.4-mm (0.25-in) posttensioning wires failed in a parking garage in the southern portion of the United States. A typical anchorage with a broken buttonhead wire is shown in Fig. 19. Several such wires were removed, and the lengths were examined for signs of corrosion. Localized shallow pitting was common, as illustrated in Fig. 20. Chloride was detected in some of these pits.

Scanning electron microscopy and metallography of the ends of the fractures revealed an initial crack that was probably caused by hydrogen embrittlement (Fig. 21). Apparently, this hydrogen embrittlement had occurred because of the presence of the corrosion on the external surface. No chloride was detected on the fracture surface, and none was detected in the overlying concrete. This suggested that the corrosion observed may have initiated before placement of the tendons within the concrete. There is no known method of mitigating this type of problem once it has occurred, although it can be prevented through the use of judicious and careful corrosion-preventive techniques during the storage of the tendons before placement.

REFERENCES

1. S.K. Coburn *et al.*, Corrosiveness of Various Atmospheric Test Sites as Measured by Specimens of Steel and Iron, in *Metal Corrosion in the Atmosphere*, STP 435, American Society for Testing and Materials, 1968, p 360
2. D.L. Graver, Ed., *Corrosion Data Survey—Metals Section*, National Association of Corrosion Engineers, 1985
3. "Standard Test Method for Corrosion of Steel by Sprayed Fire-Resistant Material Applied to Structural Members," E 937, *Annual Book of ASTM Standards*, American Society for Testing and Materials
4. M. Pourbaix, *Atlas of Electrochemical Equilibria in Aqueous Solutions*, Pergamon Press, 1966
5. R. Zidell, Coatings for Steel, in *Paint Handbook*, G.E. Weismantel, Ed., McGraw-Hill, 1981
6. "Standard Specification for High-Strength Low-Alloy Structural Steel With 50 ksi (345 MPa) Minimum Yield Point to 4 in. (100 mm) Thick," A 588, *Annual Book of ASTM Standards*, American Society for Testing and Materials
7. "Standard Specification for Structural Steel for Bridges," A 709, *Annual Book of ASTM Standards*, American Society for Testing and Materials

Corrosion of Metal-Processing Equipment

METAL-PROCESSING EQUIPMENT is exposed to numerous corrosive environments and corrosion mechanisms. Heat-treating equipment is subject to high-temperature oxidation, carburization, and sulfidation. Corrosion by molten salts and molten metals is also of concern for heat-treating furnaces and accessories (see the section "Corrosion of Heat-Treating Furnace Accessories" in this article). Equipment for plating, pickling, and anodizing is exposed to acid and alkali solutions at temperatures up to or higher than 100 °C (212 °F) (see the section "Corrosion of Plating, Anodizing, and Pickling Equipment" in this article). Information on materials for and prevention of corrosion in these applications is also available in Volumes 4 and 5 of the 9th Edition of *Metals Handbook*.

Corrosion of Heat-Treating Furnace Accessories

G.Y. Lai and C.R. Patriarca
Haynes International, Inc.

Heat-treating furnace accessories include a wide variety of components, such as trays, baskets, pots, blowers, thermowells, belts, hangers, bellows, and dampers. Typical heat treatments include annealing, normalizing, hardening, carburizing, nitriding, carbonitriding, brazing, galvanizing, and sintering.

The medium or environment used for heat treating varies from process to process. The high-temperature corrosion of furnace components depends heavily on the environment (or atmosphere) involved in the operation. Typical environments are air, combustion atmospheres, carburizing and nitriding atmospheres, molten salts, and protective atmospheres (such as endothermic atmospheres, nitrogen, argon, hydrogen, and vacuum). Protective atmospheres are used to prevent metallic parts to be heat treated from forming heavy oxide scales during heat treatment. The environment can often be contaminated by impurities, which can greatly accelerate corrosion. These contaminants (such as sulfur, vanadium, and sodium) generally come from fuels used for combustion, from fluxes used for specific operations, and from drawing compounds, lubricants, and other substances that are left on the parts to be heat treated.

The modes of high-temperature corrosion that are most frequently responsible for the degradation of furnace accessories are oxidation, carburization, sulfidation, molten-salt corrosion, and molten-metal corrosion. Each mode of corrosion, along with the corrosion behavior of important engineering alloys, will be discussed in detail in this section. The compositions of the alloys under discussion are given in Table 1.

Oxidation

Oxidation is probably the predominant mode of high-temperature corrosion encountered in the heat-treating industry. The oxidation discussed in this section involves air or combustion atmospheres with little or no contaminants, such as sulfur, chlorine, alkali metals, and salt.

Carbon steel and alloy steels generally have adequate oxidation resistance for reasonable service lives for furnace accessories at temperatures to 540 °C (1000 °F) (Ref 1). At intermediate temperatures of 540 to 870 °C (1000 to 1600 °F) heat-resistant stainless steels, such as AISI types 304, 316, 309, and 446, generally exhibit good oxidation resistance (Ref 1). Very few oxidation data have been reported in this temperature range. As the temperature increases above 870 °C (1600 °F), many stainless steels begin to suffer rapid oxidation. Better heat-resistant materials, such as the nickel-base high-performance alloys, are needed for furnace components in order to combat oxidation at these high temperatures.

Numerous oxidation tests on commercial alloys have been performed at 980 °C (1800 °F) or higher. For example, in one investigation, cyclic oxidation tests were conducted in air, with each cycle consisting of exposing the samples at 980 °C (1800 °F) for 15 min, followed by a 5-min air cooling (Ref 2). The performance ranking, in order of decreasing performance, was found to be as follows: Inconel alloy 600, Incoloy alloy 800, type 310 stainless steel, type 309 stainless steel, type 347 stainless steel, and type 304 stainless steel. Similar cyclic oxidation tests performed in air at 1150, 1205, and 1260 °C (2100, 2200, and 2300 °F), cycling to room temperature by air cooling after every 50 h at temperature, showed Inconel alloy 601 to be the best performer, followed by Inconel alloy 600 and Incoloy alloy 800 (Ref 3).

In another study, ferritic stainless steels such as E-Brite and type 446 were shown to be significantly better than type 310 and Incoloy alloy 800H in terms of cyclic oxidation resistance in air (Ref 4). These test results showed weight change data of 2.2 mg/cm^2 for E-Brite, 10.0 mg/cm^2 for type 446 stainless steel, −83.2 mg/cm^2 for alloy 800, and −90.3 mg/cm^2 for type 310 stainless steel after exposure of the samples for a total of 1000 h with 15 min at 980 °C (1800 °F) and 5 min at room temperature. A separate test was also conducted. This test involved exposure of the samples at 980 °C (1800 °F) for 1000 h in air with interruptions after 1, 20, 40, 60, 80, 100, 220, 364, and 512 h for cooling to room temperature. The weight change results of these four alloys were −12.9, 9.2, 1.7, and 3.0 mg/cm^2 for E-Brite, type 446 stainless steel, type 310 stainless steel, and alloy 800, respectively (Ref 4).

An oxidation data base for a wide variety of commercial alloys, including stainless steels, iron-nickel-chromium alloys, nickel-chromium-iron alloys, and high-performance alloys, was recently generated (Ref 5). Tests were conducted in air at 980, 1095, 1150, and 1205 °C (1800, 2000, 2100, and 2200 °F) for 1008 h. The samples were cooled to room temperature once a week (each 168 h) for visual inspection. The results are summarized in Table 2.

Type 304 stainless steel and type 316 stainless steel both exhibited severe oxidation attack at 980 °C (1800 °F), while type 446 showed relatively mild attack. Many higher alloys, such as Incoloy alloy 800 and the nickel- and cobalt-base alloys, showed little attack. At 1095 °C (2000 °F), type 446 stainless steel suffered severe oxidation. Iron-nickel-chromium alloys, such as Incoloy alloy 800H and RA330, also suffered significant oxidation. Many nickel-base alloys, however, still exhibited little oxidation. At 1150 °C (2100 °F), most alloys suffered unacceptable oxidation, with the exception of only a few nickel-base alloys. At 1205 °C (2200 °F), all alloys except Haynes alloy 214 suffered severe attack. Alloy 214 showed negligible oxidation at all the test temperatures. This alloy is different from all of the other alloys tested in that it forms an aluminum oxide (Al_2O_3) scale when heated to elevated temperatures. Other alloys tested form chromium oxide (Cr_2O_3) scales when heated to elevated temperatures.

The alloy performance rankings (Ref 6) generated from the field in the furnace atmosphere produced by the combustion of natural gas were found to correspond closely to the air oxidation data presented in Table 2. The alumina-forming alloy 214 was found to be the best performer (Ref 6).

Carburization

Materials problems due to carburization are quite common in heat-treating components associated with carburizing furnaces. The environment in the carburizing furnace typically has a carbon activity that is significantly higher than that in the alloy of the furnace component. Therefore, carbon is transferred from the environment to the alloy. This results in the carburization of the alloy, and the carburized alloy becomes embrittled.

Nickel-base alloys are generally considered to be more resistant to carburization than stainless steels. The results of 25-h carburization tests performed at 1095 °C (2000 °F) in a gas mixture consisting of 2%

Table 1 Nominal chemical compositions of high-temperature alloys

Alloy	Composition, %									
	C	Fe	Ni	Co	Cr	Mo	W	Si	Mn	Other
AISI type 304 stainless steel	0.08(a)	bal	8	...	18	...	...	1.0(a)	2.0(a)	...
AISI type 309 stainless steel	0.20(a)	bal	12	...	23	...	...	1.0(a)	2.0(a)	...
253MA	0.08	bal	11	...	21	...	...	1.7	0.8(a)	0.17N, 0.05Ce
AISI type 310 stainless steel	0.25(a)	bal	20	...	25	...	...	1.5(a)	2.0(a)	...
AISI type 316 stainless steel	0.08(a)	bal	10	...	17	2.5	...	1.0(a)	2.0(a)	...
AISI type 446 stainless steel	0.20(a)	bal	...	...	25	...	...	1.0(a)	1.5(a)	0.25N
E-Brite	0.002	bal	0.15	...	26	1.0	...	0.2	0.1	...
Incoloy alloy 800H	0.08	bal	33	...	21	...	...	1.0(a)	1.5(a)	0.38Al, 0.38Ti
RA330	0.05	bal	35	...	19	...	...	1.3	1.5	...
Multimet	0.10	bal	20	20	21	3	2.5	1.0(a)	1.5(a)	1.0Nb+Ta, 0.5Cu, 0.15N
Haynes alloy 556	0.10	bal	20	18	22	3	2.5	0.4	1.0	0.2Al, 0.8Ta, 0.02La, 0.2N, 0.02Zr
Incoloy alloy 825	0.05(a)	29	bal	...	22	3	...	0.5(a)	1.0(a)	2Cu, 1Ti
Inconel alloy 600	0.08(a)	8	bal	...	16	...	...	0.5(a)	1.0(a)	0.35Al(a), 0.3Ti(a), 0.5Cu(a)
Haynes alloy 214	0.04	2.5	bal	...	16	...	...	...	...	4.5Al, Y
Inconel alloy 601	0.10(a)	14.1	bal	...	23	...	...	0.5(a)	1.0(a)	1.35Al, 1Cu(a)
Inconel alloy 617	0.07	1.5	bal	12.5	22	9	...	0.5	0.5	1.2Al, 0.3Ti, 0.2Cu
Hastelloy alloy S	0.02	3(a)	bal	2.0(a)	15.5	14.5	1.0(a)	0.4	0.5	0.2Al, 0.02La, 0.009B
Hastelloy alloy X	0.10	18.5	bal	1.5	22	9	0.6	1.0(a)	1.0(a)	...
Inconel alloy 625	0.10(a)	5(a)	bal	...	21.5	9	...	0.5(a)	0.5(a)	0.4Al(a), 0.4Ti(a), 3.5Nb+Ta
Haynes alloy 230	0.10	3(a)	bal	3(a)	22	2	14	0.4	0.5	0.3Al, 0.005B, 0.03La
RA333	0.05	18	bal	3	25	3	3	1.25	1.5	...
Hastelloy alloy N	0.06	5(a)	bal	...	7	16.5	0.5(a)	1.0(a)	0.8(a)	0.35Cu(a)
Haynes alloy 188	0.10	3(a)	22	bal	22	...	14	0.35	1.25(a)	0.04La
Haynes alloy 25	0.10	3(a)	10	bal	20	...	15	1.0(a)	1.5	...
Alloy 6B	1.2	3(a)	3(a)	bal	30	1.5(a)	4.5	2.0(a)	2.0(a)	...

(a) Maximum

methane (CH_4) and 98% hydrogen revealed the weight gain data of 2.78, 5.33, 18.35, and 18.91 mg/cm^2 for Inconel alloy 600, Incoloy alloy 800, type 310 stainless steel and type 309 stainless steel, respectively (Ref 7). Extensive carburization tests were recently performed to investigate 22 commercial alloys, including stainless steels, iron-chromium-nickel alloys, nickel-chromium-iron alloys, and nickel- and cobalt-base alloys (Ref 8). Tests were performed for 215 h at 870 and 925 °C (1600 and 1700 °F) and for 55 h at 980 °C (1800 °F) in a gas mixture consisting of 5 vol% hydrogen, 5 vol% CH_4, 5 vol% carbon monoxide (CO), and the balance argon. The results failed to reveal any correlation between carburization resistance and the alloy base. Nevertheless, it was found that the alumina-forming Haynes alloy 214 was the most resistant to carburization among all of the alloys tested.

These findings were confirmed in 24-h tests performed at 1095 °C (2000 °F) in the same gas mixture. The carburization data are summarized in Table 3. In this study, alloy 214 (an alumina former) was found to be significantly better than the chromia formers tested. Among the chromia formers, however, there is some question regarding the significance of the differences within the carbon absorption range of 9.9 to 14.4 mg/cm^2. Perhaps less severe environments are required to separate the capabilities of these alloys. Field testing will be an excellent way of determining alloy performance ranking. However, few data are available. Field tests were recently conducted in a heat-treating furnace used for carburizing, carbonitriding, and neutral hardening operations (Ref 9). Both RA333 and Inconel alloy 601 were found to exhibit better carburization resistance than any of the alloys tested, which included RA330, Incoloy alloy 800, and alloy DS.

Metal dusting is another frequently encountered mode of corrosion that is associated with carburizing furnaces. Metal dusting tends to occur in a region where the carbonaceous gas atmosphere becomes stagnant. The alloy normally suffers rapid metal wastage. The corrosion products (or wastage) generally consist of carbon soots, metal, metal carbides, and metal oxides. The attack is normally initiated from the metal surface that is in contact

Table 2 Results of 1008-h cyclic oxidation test in flowing air at temperatures indicated

Specimens were cycled to room temperature once a week.

Alloy	Oxidation rate at temperature															
	980 °C (1800 °F)				1095 °C (2000 °F)				1150 °C (2100 °F)				1205 °C (2200 °F)			
	Metal loss		Average metal affected(a)		Metal loss		Average metal affected		Metal loss		Average metal affected		Metal loss		Average metal affected	
	mm	mils	mm	mils	mm	mils	mm	mils	mm	mils	mm	mils	mm	mils	mm	mils
Haynes alloy 214	0.0025	0.1	0.005	0.2	0.0025	0.1	0.0025	0.1	0.005	0.2	0.0075	0.3	0.005	0.2	0.018	0.7
Haynes alloy 230	0.0075	0.3	0.018	0.7	0.013	0.5	0.033	1.3	0.058	2.3	0.086	3.4	0.11	4.5	0.2	7.9
Hastelloy alloy S	0.005	0.2	0.013	0.5	0.01	0.4	0.033	1.3	0.025	1.0	0.043	1.7	>0.81	>31.7(b)	>0.81	>31.7
Haynes alloy 188	0.005	0.2	0.015	0.6	0.01	0.4	0.033	1.3	0.18	7.2	0.2	8.0	>0.55	>21.7	>0.55	>21.7
Inconel alloy 600	0.0075	0.3	0.023	0.9	0.028	1.1	0.041	1.6	0.043	1.7	0.074	2.9	0.13	5.1	0.21	8.4
Inconel alloy 617	0.0075	0.3	0.033	1.3	0.015	0.6	0.046	1.8	0.028	1.1	0.086	3.4	0.27	10.6	0.32	12.5
AISI type 310	0.01	0.4	0.028	1.1	0.025	1.0	0.058	2.3	0.075	3.0	0.11	4.4	0.2	8.0	0.26	10.3
RA333	0.0075	0.3	0.025	1.0	0.025	1.0	0.058	2.3	0.05	2.0	0.1	4.0	0.18	7.1	0.45	17.7
Haynes alloy 556	0.01	0.4	0.028	1.1	0.025	1.0	0.067	2.6	0.24	9.3	0.29	11.6	>3.8	>150.0	>3.8	>150.0
Inconel alloy 601	0.013	0.5	0.033	1.3	0.03	1.2	0.067	2.6	0.061	2.4	0.135	5.3	0.11	4.4	0.19	7.5
Hastelloy alloy X	0.0075	0.3	0.023	0.9	0.038	1.5	0.069	2.7	0.11	4.5	0.147	5.8	>0.9	>35.4	>0.9	>35.4
Inconel alloy 625	0.0075	0.3	0.018	0.7	0.084	3.3	0.12	4.8	0.41	16.0	0.46	18.2	>1.21	>47.6	>1.21	>47.6
RA330	0.01	0.4	0.11	4.3	0.02	0.8	0.17	6.7	0.041	1.6	0.22	8.7	0.096	3.8	0.21	8.3
Incoloy alloy 800H	0.023	0.9	0.046	1.8	0.14	5.4	0.19	7.4	0.19	7.5	0.23	8.9	0.29	11.3	0.35	13.6
Haynes alloy 25	0.01	0.4	0.018	0.7	0.23	9.2	0.26	10.2	0.43	16.8	0.49	19.2	>0.96	>37.9	>0.96	>37.9
Multimet	0.01	0.4	0.033	1.3	0.226	8.9	0.29	11.6	>1.2	>47.2	>1.2	>47.2	>3.72	>146.4	>3.72	>146.4
AISI type 446	0.033	1.3	0.058	2.3	0.33	13.1	0.37	14.5	>0.55	>21.7	>0.55	>21.7	>0.59	>23.3	>0.59	>23.3
AISI type 304	0.14	5.5	0.21	8.1	>0.69	>27.1	>0.69	>27.1	>0.6	>23.6	>0.6	>23.6	>1.7	>68.0	>1.73	>68.0
AISI type 316	0.315	12.4	0.36	14.3	>1.7	>68.4	>1.7	>68.4	>2.7	>105.0	>2.7	>105.0	>3.57	>140.4	>3.57	>140.4

(a) Average metal affected = metal loss + internal penetration. (b) All figures shown as greater than stated value represent extrapolation of tests in which samples were consumed in less than 1008 h. Source: Ref 5

Table 3 Results of 24-h carburization tests performed at 1095 °C (2000 °F) in Ar-5H$_2$-5CO-5CH$_4$

Alloy	Carbon absorption, mg/cm^2
Haynes alloy 214	3.4
Inconel alloy 600	9.9
Inconel alloy 625	9.9
Haynes alloy 230	10.3
Hastelloy alloy X	10.6
Hastelloy alloy S	10.6
AISI type 304	10.6
Inconel alloy 617	11.5
AISI type 316	12.0
RA333	12.4
Incoloy alloy 800H	12.6
RA330	12.7
Haynes alloy 25	14.4

with the furnace refractory. The furnace components that suffer metal dusting include thermowells, probes, and anchors. Figure 1 illustrates the metal dusting attack on Multimet alloy. The component was perforated as a result of metal dusting. Metal dusting problems have also been reported in petrochemical processing (Ref 10).

Metal dusting has been encountered with straight chromium steels, austenitic stainless steels, and nickel- and cobalt-base alloys. All of these alloys are chromia formers; that is, they form Cr_2O_3 scales when heated to elevated temperatures. No work has been reported on the alloy systems that form a much more stable oxide scale, such as Al_2O_3. The Al_2O_3 scale was found to be much more resistant to carburization attack than the Cr_2O_3 scale (Ref 8). Because metal dusting is a form of carburization, it would appear that alumina formers, such as Haynes alloy 214, would also be more resistant to metal dusting.

Sulfidation

Furnace environments can sometimes be contaminated with sulfur. Sulfur can come from fuels, fluxes used for specific operations, and cutting oil left on the parts to be heat treated, among other sources. Sulfur in the furnace environment could greatly reduce the service lives of components through sulfidation attack.

It is well known that nickel-base alloys are highly susceptible to catastrophic sulfidation due to the formation of nickel-rich sulfides, which melt at about 650 °C (1200 °F). Figure 2 illustrates catastrophic failure of a nickel-chromium-iron alloy tube due to sulfidation attack in a heat-treating furnace. The liquid-appearing nickel-rich sulfide phases are clearly visible.

The sulfidation of metals and alloys has been the subject of numerous investigations. However, the investigations involving commercial alloys examined only a limited number of alloys in each case. This frequently does not provide designers or engineers with a sufficient number of alloys to make an informed materials selection. A comprehensive sulfidation study was recently undertaken to determine the relative alloy rankings of base alloys (Ref 11). Tests were performed at 760, 870, and 980 °C (1400, 1600, and 1800 °F) for 215 h in a gas mixture consisting of 5% hydrogen, 5% CO, 1% carbon dioxide (CO_2), 0.15% hydrogen sulfide (H_2S), 0.1% H_2O, and the balance argon. The cobalt-base alloys were found to be the best performers, followed by iron-base alloys, and then nickel-base alloys, which,

(a)

(b)

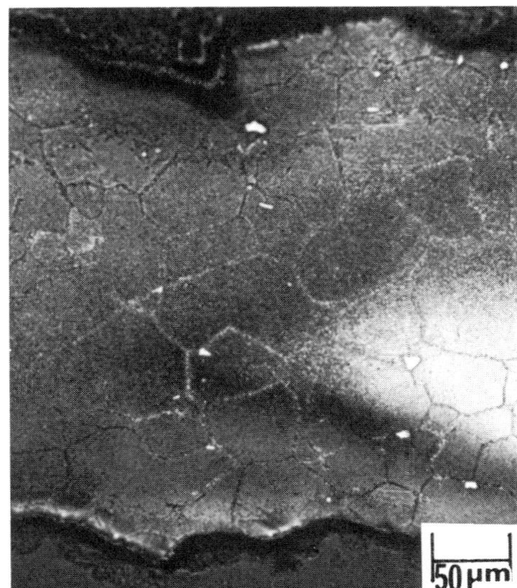

(c)

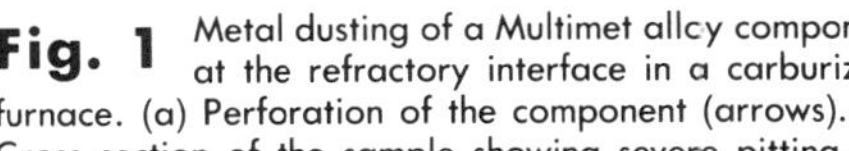

Fig. 1 Metal dusting of a Multimet alloy component at the refractory interface in a carburizing furnace. (a) Perforation of the component (arrows). (b) Cross section of the sample showing severe pitting. (c) Severe carburization beneath the pitted area

as a group, were generally the worst performers. Among iron-base alloys, the iron-nickel-cobalt-chromium alloy 556 was better than iron-nickel-chromium alloys such as Incoloy alloy 800H and type 310 stainless steel. The test results of representative alloys from each alloy base group are summarized in Table 4.

(a)

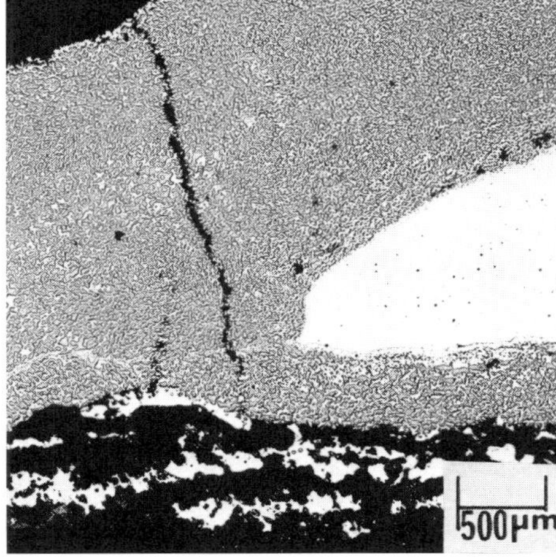

(b)

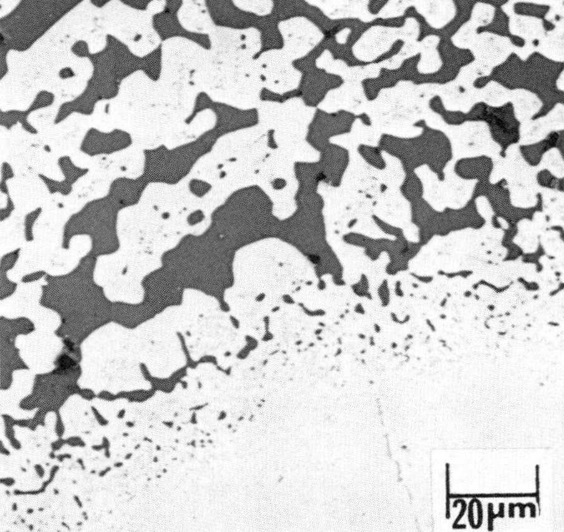

(c)

Fig. 2 Catastrophic sulfidation of an Inconel 601 furnace tube. The furnace atmosphere was contaminated with sulfur; the component failed after less than 1 month at 925 °C (1700 °F). (a) General view. (b) Cross section of the perforated area showing liquid-appearing nickel-rich sulfides. (c) Higher-magnification view of nickel-rich sulfides

Molten-Salt Corrosion

Molten salts are widely used in the heat-treating industry for tempering, annealing, hardening, reheating, carburizing, and other operations. The salts that are commonly used include nitrates, carbonates, cyanides, chlorides, and caustics, depending on the operation. For example, a mix-

Table 4 Results of 215-h sulfidation tests conducted in Ar-5H$_2$-5CO-1CO$_2$-0.15H$_2$S-0.1H$_2$O

Alloy	Average metal affected at temperature(a) 760 °C (1400 °F) mm	mils	870 °C (1600 °F) mm	mils	980 °C (1800 °F) mm	mils
Alloy 6B	0.038	1.5	0.064	2.5	0.11	4.2
Haynes alloy 25	0.046	1.8	0.036	1.4	0.046	1.8
Haynes alloy 188	0.084	3.3	0.074	2.9	0.048	1.9
Haynes alloy 556	0.097	3.8	0.297	11.7	0.05	2.0
AISI type 310	0.23	9.1	0.34	13.5	0.19	7.4
Incoloy alloy 800H	0.28	11.2	0.49	19.2	0.59	23.2
Haynes alloy 214	0.42	16.7	>0.45	>17.7	>0.45	>17.7
Inconel alloy 600	0.55	>21.7	>0.55	>21.7	>0.55	>21.7
Hastelloy alloy X	0.749	>29.5	>0.55	>21.7	>0.55	>21.7
Inconel alloy 601	0.749	>29.5	>0.55	>21.7	>0.55	>21.7

(a) Average metal affected = metal loss + internal penetration. Source: Ref 10

Table 5 Corrosion rates in molten NaNO$_3$-KNO$_3$ at 675 and 705 °C (1250 and 1300 °F)

Alloy	Corrosion rate 675 °C (1250 °F), 14-day test mm/yr	mils/yr	675 °C (1250 °F), 80-day test mm/yr	mils/yr	705 °C (1300 °F), 30-day test mm/yr	mils/yr
Haynes alloy 214	...	...	0.4	16	0.53	21
Inconel alloy 600	0.3	12	0.25	10	0.99	39
Hastelloy alloy N	0.33	13	0.23	9	1.22	48
Inconel alloy 601	0.48	19	0.48	19	1.24	49
Inconel alloy 617	0.36	14	...	...	...	...
Hastelloy alloy S	0.4	16	...	...	...	...
Inconel alloy 690	0.56	22	...	...	...	...
RA333	0.69	27	...	...	...	...
Inconel alloy 625	0.74	29	...	...	...	...
Hastelloy alloy X	1.04	41	...	...	...	...
Incoloy alloy 800	1.07	42	1.85	73	6.58	259
Haynes alloy 556	1.75	69	...	...	...	...
AISI type 310	2.0	79	...	...	...	...
AISI type 316	2.1	81	...	...	...	...
AISI type 317	2.11	83	...	...	...	...
AISI type 446	2.38	94	...	...	...	...
AISI type 304	2.67	105	...	...	...	...
RA330	2.77	109	...	...	...	...
253MA	2.97	117	...	...	...	...
Nickel 200	8.18	322	...	...	...	...

Source: Ref 11

ture of nitrates and nitrites is normally used for tempering and quenching. An alkali chloride-carbonate mixture is used for annealing ferrous and nonferrous metals. Neutral salt baths containing mixed chlorides are used for hardening steel parts.

Carbon steels, alloy steels, stainless steels, and iron-nickel-chromium alloys have been used for various furnace parts, such as electrodes, thermocouple protection tubes, and pots for salt baths. However, few corrosion data have been reported involving heat-treating salts.

One investigation recently provided corrosion data in molten sodium-potassium nitrate (NaNO$_3$-KNO$_3$) salts. The results of this study are given in Table 5. The nickel-chromium-iron-aluminum-yttrium alloy (alloy 214), nickel-chromium-iron alloys (Inconel alloys 600 and 601), and nickel-chromium-molybdenum alloys (Hastelloy alloys N and S) performed significantly better than stainless steels and iron-nickel-chromium alloys such as Incoloy alloy 800H and RA330. The data generated from 1-month tests in a neutral salt bath containing a mixture of barium, potassium, and sodium chlorides (BaCl$_2$, KCl, and NaCl) at 845 °C (1550 °F) were recently reported (Ref 6). The results are summarized in Table 6. The cobalt-base Haynes alloy 188 was found to be the best performer; the nickel-chromium-iron Inconel alloy 600 was the worst. Somewhat similar results were obtained in laboratory tests conducted for 100 h at 845 °C (1550 °F) in NaCl. These data are given in Table 7. Haynes alloy 188 was found to be the best, and Inconel alloy 600 the worst.

In some cases, the salt vapors could cause high-temperature corrosion attack that is significantly worse than that caused by contact with the molten salt. For example, it was found that the corrosion of a nickel-base alloy salt pot containing molten BaCl$_2$-KCl-NaCl mixture at 1010 °C (1850 °F) was significantly different between the air side (outside of the pot) and the molten salt side (inside) (Ref 14). The outside of the pot (that is, the air side contaminated with salt vapors) suffered three times as much attack as the inside of the pot, which was in contact with molten salt. No corrosion data were reported with respect to this type of accelerated oxidation attack involving molten salt vapors.

Molten-Metal Corrosion

Some heat-treating operations involve molten metals. Lead is used as a heat-treating medium. Cast iron and carbon steels have been used for components in contact with molten lead at temperatures to 480 °C (900 °F). In a 1242-h test in molten lead in an open crucible at 600 °C (1110 °F), Inconel alloy 600 was not appreciably attacked either at or below the liquid-metal surface (Ref 15). At 675 °C (1250 °F) for 1281 h in molten lead, Inconel alloy 600 suffered severe corrosion attack (Ref 15). Few corrosion data in molten lead have been reported in the literature.

The molten-zinc bath is used for galvanizing processes. Few corrosion data are available in the literature to allow engineers to make an informed materials selection for furnace components in contact with molten zinc. Carbon steel is generally used for the furnace components. Iron-nickel-cobalt-chromium alloys such as Haynes alloy 556 also have been reported for use as baskets.

Corrosion of Plating, Anodizing, and Pickling Equipment

Earl C. Groshart
Boeing Aerospace Company

An electroplating, anodizing, or pickling shop environment is highly corrosive to metals. The metals used in the construction of equipment to be employed in this environment must be inert to the environment or must be protected from it.

Tanks

The common materials for constructing tanks for plating, anodizing, and pickling are carbon steel and stainless steel, usually one of the austenitic (18Cr-8Ni) varieties. Concrete, plastic, and fiberglass, either alone or with reinforcing, as well as wood, also are used. Of these, only the metals offer any serious corrosion problems. The nonmetals, however, must be carefully selected to prevent contamination of the processing solutions. Table 8 recommends tank materials for the various solutions found in the finishing shop.

Metal Tank Design to Minimize Corrosion. Tank closure welds should be smooth or should be ground to provide a smooth surface. This not only facilitates the application of liners but also eliminates rough welds (inside the tank or outside) in which solution can be entrapped. Entrapped solution can result in the formation of a corrosion cell. Outside reinforcing should be welded with continuous welds. Tack or intermittent welds should not be used, because they leave open and unprotected overlapping areas in which solutions can be trapped and crevice corrosion can occur. If intermittent welds cannot be avoided, the overlaps should be sealed as shown in Fig. 3. Whenever possible, structural supports should be positioned vertically; again, this is done to help prevent trapping of the solutions. Corrosion occurs around outlets and flanges that are not designed to prevent the solution from contacting the metal. This is also true of overflows built into the tank; if the overflow is deep and/or narrow, it becomes difficult to line and protect it from solutions. The use of O-rings in flanges is a means of keeping the flanges free of solution, but O-rings are expensive and therefore not widely used. More information on designing to prevent or minimize corrosion is available in

Table 6 Results of 30-day field tests performed in a neutral salt bath containing $BaCl_2$, KCl, and NaCl at 845 °C (1550 °F)

Alloy	Average metal affected(a), mm	mils
Haynes alloy 188	0.69	27
Multimet	0.75	30
Hastelloy alloy X	0.97	38
Hastelloy alloy S	0.1	40
Haynes alloy 556	0.11	44
Haynes alloy 214	1.8	71
AISI type 304	1.9	75
AISI type 310	2.0	79
Inconel alloy 600	2.4	96

(a) Average metal affected = metal loss + internal penetration. Source: Ref 6

Table 7 Results of 100-h tests performed in NaCl at 845 °C (1550 °F)

Alloy	Average metal affected(a), mm	mils
Haynes alloy 188	0.05	2.0
Haynes alloy 556	0.066	2.6
Haynes alloy 214	0.079	3.1
AISI type 304	0.081	3.2
AISI type 446	0.081	3.2
AISI type 316	0.081	3.2
Hastelloy alloy X	0.097	3.8
AISI type 310	0.107	4.2
Incoloy alloy 800H	0.110	4.3
Inconel alloy 625	0.112	4.4
RA330	0.117	4.6
Inconel alloy 617	0.122	4.8
Haynes alloy 230	0.14	5.5
Hastelloy alloy S	0.168	6.6
RA333	0.19	7.5
Inconel alloy 600	0.196	7.7

(a) Average metal affected = metal loss + internal penetration. Source: Ref 13

the article "Design Details to Minimize Corrosion" in this Volume.

Corrosion Protection for Steel Tanks. The outside of a carbon steel tank must be protected, regardless of what is contained in the tank. For example, a steel tank containing an alkaline cleaner will not need protection on the inside, where the alkaline cleaner will keep the steel passivated and at a very low corrosion rate; however, the outside surfaces, which are exposed only to water and the corrosive environment of the shop atmosphere, will corrode severely. Protection of the outside surfaces of corrosion-resistant steel tanks is generally not necessary, except where crevices are possible and oxygen cells can be set up. However, because the tank material is corrosion resistant, overlapping joints are frequently ignored; this provides an initiation site for corrosion.

Protective coating for the outside of a steel tank should include a good two- or three-coat paint system. Epoxies, vinyls, or polyurethanes can be used. A system that consists of an epoxy primer applied over a clean, sandblasted surface, followed by two coats of a two-part urethane topcoat, will provide adequate protection in a finishing shop environment. Vinyl systems are immune to all solutions except very strong solvents and are excellent coatings for tank surfaces. Vinyl systems are also adequate for the inside surfaces of rinse and holding tanks. Vinyls should be applied only to clean, sandblasted (white metal) surfaces and should be a complete system; that is, primer and topcoat should both come from the same manufacturer and should be matched to each other.

Coal tar and epoxy-modified coal tar systems are excellent tank coatings. These coatings can be used on the inside and outside surfaces of cold tanks, such as rinse tanks. Epoxy-modified coatings tend to soften in contact with hot cleaning, pickling, or plating baths. The epoxy-modified coatings, however, are excellent for tanking understructures, such as the I-beams that keep the tanks off the floor and for the floor itself. All of these coatings can be applied to corrosion-resistant steel tanks, if required. The coal tars and the epoxy-modified coal tars are especially useful as coatings on the bottoms of tanks and between the tank and floor supports, where they are used to prevent crevices in overlapping joints.

The insides of carbon steel tanks used in the metal-finishing industry can also be protected by nonmetal liners. Materials for this application are listed in Table 8. Liners that are applied directly to the steel walls (hot melts, Table 8) should offer sufficient protection for the steel. The drop-in liners—whether thay are made outside of the tank by molding (or other methods) or are made inside of the tank by welding or adhesive bonding—leave the tank wall unprotected and create an ideal crevice corrosion area. Water and solutions will leak between the liner and the tank regardless of how carefully the flanges are made and sealed to the tank. The inside surfaces of these tanks should ideally be protected in the same manner as the outside. However, a heavy coat (or two light coats) of primer on these steel surfaces will offer sufficient protection so that wetting of the inside of the tank does not become an emergency.

Other Plating, Anodizing, and Pickling Equipment

In addition to tanks, other equipment, such as rectifiers, heaters, pumps, racks, bus bars, and wiring, is subject to corrosion in the environment of the finishing shop.

Wiring. In general, installation of electrical wiring is controlled by an electrical code that does not allow bare wires. The wire terminations, however, are subject to corrosion, especially the dc termination. Only copper wire is recommended. Each termination should be made bare and then overcoated with a good grease. Because this

Table 8 Relative corrosion resistance of tank materials, coatings, and linings for metal-finishing shops

Material	Finishing shop solution/relative corrosion resistance(a): Acid plating baths	Acid pickling baths: HNO_3	Acid pickling baths: HCl	Acid pickling baths: H_2SO_4	Alkaline cleaners/caustics	Alkaline plating baths	Anodizing baths: CrO_3	Anodizing baths: H_2SO_4	Cyanide plating baths: General	Cyanide plating baths: For Cd and Zn	HF and HBF_4 plating baths	Electrocleaning solutions	1,1,1 Trichloroethane
Metals													
Carbon steel	NR	NR	NR	NR	VG	S	NR	NR	S	G	NR	G	S
Stainless steel	NR	NR	NR	NR	G	G	NR	NR	G	G	NR	G	VG
Liners													
Natural rubber	VG	NR	G	G	G	G	NR	G	G	G	NR	G	NR
Vinyl chloride	G	S	G	G	G	G	S	G	G	G	G	G	NR
Neoprene	VG	NR	NR	G	G	G	NR	G	G	G	G	G	NR
Chlorosulfonated polyethylene	G	NR	G	G	G	G	S	G	G	G	NR	G	NR
Butyl/chlorobutyl rubber	VG	G	G	G	G	G	NR	G	G	G	G	G	NR
Fluorocarbons	VG	G	G	G	G	G	G	G	G	G	G	G	G
Coatings													
Asphaltic coal tar/epoxy	G	NR	G	G	S	G	S	G	G	G	G	NR	NR
Furan	S	NR	S	S	G	G	NR	G	G	G	S	G	G
Epoxy	S	NR	S	NR	G	G	NR	G	G	G	G	G	NR
Polyester	G	G	G	G	NR	G	G	G	G	G	G	NR	NR
Vinyl ester	G	G	G	G	G	G	G	G	G	G	G	G	S
Urethane	G	G	G	G	G	G	NR	G	G	G	NR	G	NR
Hot melts													
Polyethylene	G	G	G	G	G	G	G	G	G	G	G	G	NR
Polypropylene	G	G	G	G	G	G	G	G	G	G	G	G	NR
Polyvinyl chloride	G	G	G	G	G	G	G	G	G	G	G	G	NR

(a) VG, very good; G, good; S, satisfactory; NR, not recommended

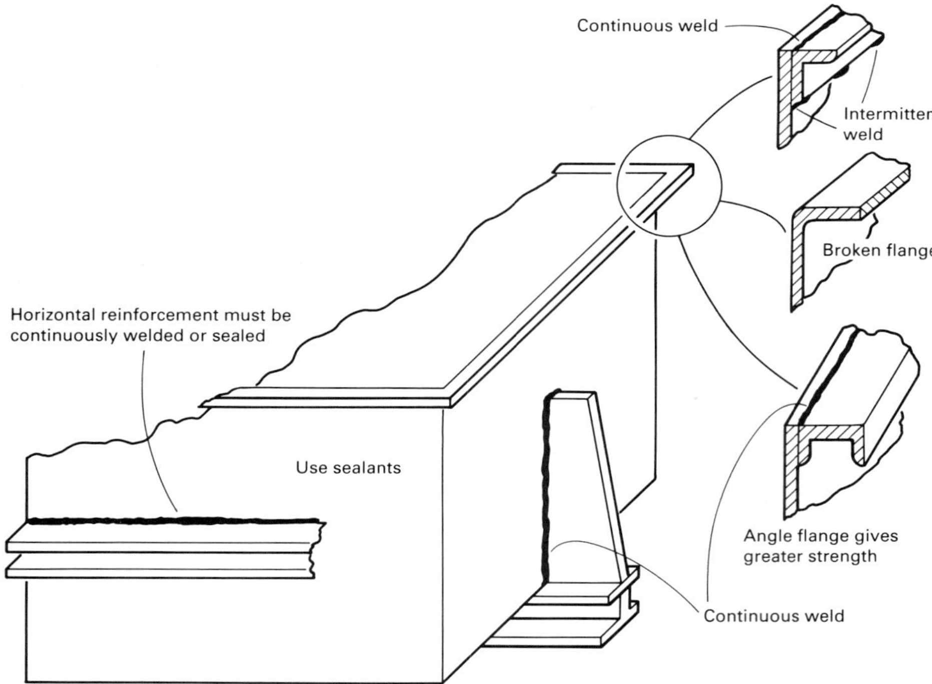

Fig. 3 Tank design details for preventing or minimizing corrosion

will not prevent corrosion of these electrical connections, they should be inspected regularly. When a corrosion product is seen at a joint, the corrosion product should be removed, the joint cleaned and reconnected, and the protective grease replaced. For dc connections to bus bars, a lead or lead-tin coating applied to the terminal lugs, the bar, washers, and bolts offers excellent protection.

Bus Bars. Copper bus bars require constant cleaning not only to remove corrosion products but also to ensure good connections to the rack splines. Tin, lead, and silver coatings on the bars provide temporary protection, but do not eliminate the need for cleaning. The anode bar should be cleaned once or twice a week, and the cathode bar should be cleaned daily.

Racks. A properly designed rack will have bare areas only in the area of the hook, which is required for contact with the bus bar, and in the area of the tip, which is required to make contact with the part. Both of these areas are subject to corrosion from the shop atmosphere during storage and from the solutions during use. These areas require constant maintenance. Aluminum and titanium anodizing racks are subject to anodization of the rack tips. Steel or stainless steel is often used as a rack material for carrying the load of the parts being treated—for example, a heavy crank shaft in a chromium plating bath. Both of these materials require complete masking, even where they make connection with the part. Otherwise, corrosion will not only destroy the rack and weaken the joint but will also contaminate the plating solution. The copper current-carrying member of the rack also requires masking.

Anode splines are generally copper above the plating solution and can be coated with any rack coating, except at the hook where electrical connection is made. Anode splines submerged in the solutions, except those made of the metal that is to dissolve in the bath, must be inert or masked. Rack coatings can be used for this purpose. Inert anodes (that is, those that carry current but do not dissolve), such as titanium in a nickel bath or lead in a chromium bath, are corrosion resistant and require little care. Iron anodes used in an alkaline tin bath, however, are subject to rusting when not in the bath. These units require protection when not in use or require cleaning before use.

Pumps. Conventional metal pumps will corrode in acid solutions. The current trend is to make pumps out of plastic (especially pumps that will be used for small tanks). The body of the pump is usually a grade of phenolic, and the impellers are hard rubber. In this case, the only metal part exposed to the solution is the impeller shaft, which can be fabricated from stainless steel or one of the acid-resistant nickel or cobalt alloys. Metal pumps require constant care to prevent or minimize corrosion.

Heaters. Hot water or steam tank heaters should be treated as part of the boiler system in order to prevent corrosion from the inside. Iron can be used in alkaline solutions, copper can be used in water and neutral solutions, and stainless steel can be used in mildly acid solutions. Stabilized lead can be used in chromic and fluoboric acid solutions. Carbon and ceramic heat exchangers can be used in all solutions except those containing fluoborates and fluorides. The same is true of immersion ceramic or quartz heaters if the framework is coated.

REFERENCES

1. *Selection of Stainless Steels*, American Society for Metals, 1968
2. E.N. Skinner, J.F. Mason, and J.J. Moran, *Corrosion*, Vol 16, p 593
3. INCONEL alloy 601 brochure, INCO Alloys International, Inc.
4. F.K. Kies and C.D. Schwartz, *J. Test. Eval.*, Vol 2 (No. 2), March 1974, p 118
5. M.F. Rothman, Cabot Corporation, private communication, 1985
6. D.E. Fluck, R.B. Herchenroeder, G.Y. Lai, and M.F. Rothman, *Met. Prog.*, Sept 1985, p 35
7. INCO Alloys International, Inc., unpublished research
8. G.Y. Lai, in *High Temperature Corrosion in Energy Systems*, M.F. Rothman, Ed., Symposium Proceedings, The Metallurgical Society, 1985, p 551
9. G.R. Rundell, Paper 377, presented at Corrosion/86, Houston, TX, National Association of Corrosion Engineers, March 1986
10. G.L. Swales, in *Behavior of High Temperature Alloys in Aggressive Environments*, I. Kirnan *et al.*, Ed., Proceedings of the Petten International Conference, The Metals Society, 1980, p 45
11. G.Y. Lai, in *High Temperature Corrosion in Energy Systems*, M.F. Rothman, Ed., Symposium Proceedings, The Metallurgical Society, 1985, p 227
12. J.W. Slusser, J.B. Titcomb, M.T. Heffelfinger, and D.R. Dunbobbin, *J. Met.*, July 1985, p 24
13. M.F. Rothman and G.Y. Lai, *Ind. Heat.*, Aug 1986, p 29
14. D.E. Fluck, Cabot Corporation, private communication, 1985
15. INCO Alloys International, Inc., unpublished research

Corrosion in Batteries and Fuel-Cell Power Sources

The ASM Committee on Corrosion of Electrochemical Power Sources*
Chairman: Wendy R. Cieslak, Exploratory Batteries Division, Sandia National Laboratories

BATTERIES AND FUEL CELLS, as electrochemical power sources, provide energy through controlled redox reactions. Because these devices contain electrochemically active components, they place metals in contact with environments in which the metals may corrode. The shelf lives of batteries, particularly those that operate at ambient temperatures depend on very slow rates of corrosion of the electrode materials at open circuit. The means of reducing this corrosion must also be evaluated for its influence on performance.

A second major corrosion consideration in electrochemical power sources involves the hardware. Again, shelf lives and service lives depend on very good corrosion resistance of the containment materials and inactive components, such as separators. In those systems in which electrolyte purity is important, even small amounts of corrosion that have not lessened structural integrity can degrade performance.

There is a wide variety of batteries and fuel cells, and new systems are constantly under development (Ref 1-5). Therefore, to illustrate the types of corrosion phenomena that occur, this article will discuss the following systems: lead-acid batteries, alkaline batteries (in terms of the sintered nickel electrode only), lithium ambient-temperature batteries, aluminum/air batteries, sodium/sulfur batteries, phosphoric acid (H_3PO_4) fuel cells, and molten carbonate fuel cells.

Lead-Acid Batteries

The principal use of lead in its primary and recycled states is in the manufacture of lead-acid storage batteries. A lead-acid battery consists of groups of positive and negative plates immersed in a concentrated electrolyte solution of sulfuric acid (H_2SO_4) (Ref 6-10). In its operating form, positive plates contain a web of lead alloy grids coated with lead dioxide (PbO_2). Negative plates contain a web of lead alloy grids coated with pure lead, usually known as sponge lead because of its porous nature. The following reactions occur during charging, and the reverse of these occur upon discharge:

$$\text{Positive: } PbSO_4 + 2H_2O \rightleftharpoons PbO_2 + 3H^+ + HSO_4^- + 2e^- \quad \text{(Eq 1)}$$

$$\text{Negative: } PbSO_4 + H^+ + 2e^- \rightleftharpoons Pb + HSO_4^- \quad \text{(Eq 2)}$$

During initial charging (first time), on the positive grid, a lead sulfate/lead oxide paste and a very small amount of the lead of the grid surface are converted to PbO_2; consequently, good bonding is obtained between the paste and the grid. On the negative grid, the paste is merely converted to a pure lead sponge. The major corrosion problem in a lead-acid battery occurs during subsequent cyclic discharging and recharging. Overcharging can lead to further conversion of lead to PbO_2 on the positive grid, according to the reaction:

$$Pb + 2O^{2-} \rightleftharpoons PbO_2 + 4e^- \quad \text{(Eq 3)}$$

In general, the rate of this process is higher at lower acid concentrations (for example, 2.17 *N*, or 1.065 specific gravity) than at higher concentrations (Ref 11). It is necessary to design the grids such that corrosion of the positive grid is uniform rather than intergranular. Grid corrosion is discussed in Ref 12 to 19.

Extremely fine grain sizes (3 to 5 μm) promote a uniform corrosion morphology (Ref 12). However, corrosion mechanisms in wrought and cast grids may differ considerably. For fine-grain cast grids, the equiaxed microstructure is susceptible to corrosion creep, or grid growth, in which a combination of corrosion products buildup and mechanical creep shorts the positive and negative grids (Ref 13, 14). Control of wrought alloy microstructures to produce thin, elongated grains can yield superior performance, because the small grain thickness (1 to 2 μm) provides intergranular corrosion resistance and the large grain length (1 to 2 mm, or 0.04 to 0.08 in.) in the grid growth direction provides creep resistance.

Various alloys are used to make grids, including pure lead, lead-antimony alloys with antimony levels from 0.5 to 6%, and, more recently, lead-calcium alloys (Ref 15). Pure lead has limited application as a grid material because it is extremely soft and is prone to severe corrosion creep. However, to take advantage of the good corrosion resistance of the pure metal (better than the commonly used alloys), batteries for certain standby power applications have been designed to allow for creep.

Antimony increases the strength of lead. However, because antimony decreases the hydrogen overvoltage of lead, the battery starts gassing (hydrogen evolution on the negative grid) at normal charging voltages, leading to loss of water from the electrolyte. To reduce this water loss, the trend has been to lower the content of antimony in the grids to 1% or less. With decreased antimony, the alloy becomes softer and the grain size increases, resulting in lower strength and more severe intergranular corrosion. Grain refiners, such as tellurium, are being tried in the industry to combat the intergranular corrosion problem. Nonetheless, modern designs are moving away from antimony alloys, because recent work suggests that even batteries with the low-antimony alloys exhibit up to six times greater water loss than batteries with wrought lead-calcium-tin grids (Ref 16).

Lead-calcium alloys, like pure lead, have extremely high hydrogen overvoltages, a desirable feature for batteries. Most of the automotive industry currently uses batteries with lead-calcium grids, known as maintenance-free batteries, because water loss and corrosion creep problems are minimized for these grids. In general, calcium increases the corrosion rate of lead, especially when the calcium content exceeds 0.075%. Aluminum additions of 50 to 200 ppm help reduce intergranular corrosion by promoting a uniform distribution of hardening precipitates, rather than the grain-boundary precipitation that occurs in aluminum-free lead-calcium alloys (Ref 17). Lead-calcium-tin alloys that have been developed to improve battery performance offer even better resistance to corrosion and grid growth.

Each grid is coated with a paste of lead oxide (PbO), lead sulfate ($PbSO_4$), water, H_2SO_4, binder, and other materials that affect the performance and life objectives of this complex mixture in its final state. If the sulfate content of the paste mixture is higher, the initial grid corrosion rate is lower because of the effect of sulfate on the acid concentration during charging. If the paste layer is thicker, the corrosion rate is higher because the concentration of acid at the grid/paste interface is lower. Battery manufacturers would prefer to use a low acid concentration for the initial fill because

*Alan P. Brown, Chemical Technology Division, Argonne National Laboratory; Vani K. Dantam, Delco-Remy Division, General Motors Corporation; Barry D. Lichter and Sandeep R. Shah, Department of Mechanical and Materials Engineering, Vanderbilt University; William W. Paden, College of Engineering, Architecture, and Technology, Oklahoma State University; Sam F. Pensabene, Battery Business Department, General Electric Company; Philip N. Ross, Jr., Materials and Molecular Research Division, Lawrence Berkeley Laboratory

low specific gravity acid is more conductive and the batteries are easier to charge. However, as mentioned above, low specific gravity (low concentration) acids are more corrosive. Thus, a balance is established, depending on paste composition, paste layer thickness, and grid alloy corrosion resistance. Lower sulfate levels, thicker paste layers, and grids with lower corrosion resistance need a higher concentration of acid to lessen the chances of grid corrosion.

Trace impurities in the acid, paste, and grids exert considerable influence on grid corrosion. Elements such as copper, nickel, selenium, tellurium, and chromium strongly reduce the hydrogen overvoltage of the battery and thus cause gassing. Gas bubbles that nucleate at the paste/grid interface cause the paste pellets to break loose, thus increasing corrosion rates by exposing a fresh lead surface (causing another problem since the loose pellets can fall to the bottom of the cell, accumulate, and eventually short electrodes). Iron has a similar effect, but to a lesser extent. Elements such as phosphorus and zinc slow corrosion rates by changing the morphologies and structures of corrosion products, but they also adversely affect battery performance.

Two more factors influence grid corrosion by affecting the acid concentration. The first is the acid-to-material ratio, defined as the theoretical number of ampere-hours required to dissociate the H_2SO_4 completely divided by the theoretical number of ampere-hours generated when all PbO_2 for positive plates is converted to $PbSO_4$. A higher ratio indicates a higher reserve of acid and therefore a smaller chance of corrosion due to low specific gravity acid. The second is acid stratification, a severe problem in stationary batteries. Constant vibration in automotive batteries and motive power batteries keeps the acid mixed. In stationary batteries, however, the acid stratifies with time, and the low specific gravity acid in the upper part of the battery is more corrosive.

Lastly, corrosion and the associated corrosion creep increase with temperature. For example, batteries in Florida, Arizona, or southern California, fail earlier because of grid corrosion than batteries in colder climates. The warmer climates also allow the grid alloys to soften because of overaging, further worsening the creep problem. Also, the trend in automotive design has been toward higher underhood temperatures in order to increase fuel economy and engine efficiency. If this trend continues, the battery may have to be moved to a cooler region of the vehicle.

Temperature must also be monitored during initial charging because the heat generated increases the temperature of the battery. Conventional practice involves charging the battery in steps, with some delay in between each step to reduce the temperature of the battery and minimize corrosion during initial charging.

A fine-grain (<3 μm) grid alloy microstructure, superior grid alloy mechanical properties (ultimate tensile strength: >45 MPa, or 6500 psi, at room temperature), and the absence of detrimental trace elements normally translate into satisfactory corrosion behavior in a properly designed battery. Maintenance-free batteries in passenger cars typically last from 3 to 7 years. Nonetheless, corrosion resistance of lead-acid battery grids is one of the attractive areas of research for increasing the life of the batteries. The development of aluminum-containing lead-calcium-tin alloys has already made it easier to manufacture high-quality maintenance-free long-lasting lead-acid batteries. With further improvements, the life of the lead-acid battery may double within the next decade.

Sintered Nickel Electrode in Alkaline Batteries

The $Ni(OH)_2/NiOOH$ electrode is used in several rechargeable alkaline battery systems, such as nickel-iron, nickel-zinc, nickel-cadmium, nickel-hydrogen, and nickel-metal hydride. Numerous variations of this electrode exist as a result of the differing requirements of the various battery systems and the optimization of each system for specific applications. Several battery systems and nickel electrode variations are discussed in Ref 2 and 20. This section will discuss the corrosion of a typical sintered electrode.

The electrode consists of a porous nickel matrix that holds the active materials. The matrix is made by sintering nickel powder onto a substrate, such as a nickel-plated perforated steel strip. The electrochemical reaction during charging of the electrode is represented by:

$$Ni(OH)_2 + OH^- \rightleftharpoons NiOOH + H_2O + e^- \quad \text{(Eq 4)}$$

Upon discharge of the electrode, the reverse reaction occurs. During overcharge, the following electrochemical reaction occurs at the electrode:

$$4OH^- \rightleftharpoons 2H_2O + O_2 + 4e^- \quad \text{(Eq 5)}$$

The nickel in the sintered matrix may corrode if sufficient continuous anodic polarization is present under an unfavorable set of conditions, such as elevated temperature, high concentration of alkali hydroxide electrolyte, or high carbonate impurity levels. Such corrosion can be represented by the following:

$$2Ni + O_2 + 2H_2O \rightleftharpoons 2Ni(OH)_2 \quad \text{(Eq 6)}$$

Equation 6 shows that nickel, oxygen, and water are consumed to produce nickel hydroxide ($Ni(OH)_2$). Consumption of water increases the electrolyte concentration to produce a condition even more favorable for corrosion. Excessive corrosion results in loss of electrical continuity between the active material surface and the current collection terminal. However, practical battery systems may cease useful operation long before loss of continuity occurs. Cessation may be associated with corrosion phenomena or may be related to independent causes.

For example, in sealed nickel-cadmium cells designed with a minimum amount of electrolyte, the system may cease operation in the following ways:

- Consumption of the nickel metal matrix due to the production of $Ni(OH)_2$, resulting in an overload of that structure. This weakening of the matrix can lead to excessive swelling and subsequent shorting
- Loss of electrolyte fluidity, or dry out, resulting directly from the consumption of water by the corrosion reaction (Eq 6)
- Loss of electrolyte fluidity due to the production of $Ni(OH)_2$ and consumption of oxygen by the corrosion reaction (Eq 6). This case is different from the one immediately above, as described below

At the negative electrode, while charging a ventable, sealed nickel-cadmium cell, the favored reaction is the reduction of cadmium hydroxide ($Cd(OH)_2$) to cadmium:

$$Cd(OH)_2 + 2e^- \rightleftharpoons Cd + 2OH^- \quad \text{(Eq 7)}$$

Under normal conditions of overcharge, the preferred reaction at the negative electrode is instead the reverse of Eq 5. As previously noted, the preferred reaction at the positive electrode during continuous overcharge is described by Eq 5. This reaction cannot begin, however, until all of the available $Ni(OH)_2$ is oxidized, including the additional amount produced by the corrosion reaction (Eq 6). Because no oxygen is available until after this additional amount of $Ni(OH)_2$ is oxidized, the preferred reaction at the negative electrode is Eq 7, provided a sufficient amount of uncharged $Cd(OH)_2$ is present in the negative electrode. Furthermore, the amount of oxygen consumed by corrosion reaction (Eq 6) is not available for the reduction reaction represented by the reverse of Eq 5. This condition again favors Eq 7. Sufficient corrosion leads to charging all of the available uncharged $Cd(OH)_2$. When this occurs, the negative electrode will evolve hydrogen instead of reducing oxygen, as described by the following reaction:

$$2H_2O + 2e^- \rightleftharpoons H_2 + 2OH^- \quad \text{(Eq 8)}$$

The net effect of Eq 8 leads to excessive internal gas pressures, causing the gases from the electrolysis of water to be lost from the system through a resealable vent.

The extent to which these corrosion phenomena influence the useful life of a battery system cannot generally be quantified because of the variety of sintered nickel electrodes, battery system designs, and applications. However, tailoring the sintered nickel electrode design and the battery system design to the specific application permits minimization of the corrosion reactions and extends the useful life of the battery.

Lithium Ambient-Temperature Batteries (LAMBs)

Lithium batteries, primarily by virtue of their very high energy density, are being increasingly used for military and commercial applications. Lithium ambient-temperature batteries may be either of the solid cathode (for example, Li/CuO, Li/V_2O_5, $Li/Bi_2Pb_2O_5$, Li/MnO_2, and Li/CF_x) or liquid cathode type (for example, Li/SO_2 and $Li/SOCl_2$) (Ref 21). A corrosion process common to all these systems, although particularly severe for the liquid cathode types, is the formation of a protective passive layer on the lithium anode (Ref 22, 23). Reference 22 reviews the properties of the lithium anode film, which enables the lithium, although thermodynamically unstable, to resist corrosion in nonaqueous electrolytes. However, because the film is also responsible for temporary voltage depressions on load (commonly known as voltage delay), optimization of both storage life and operating response depends on the properties of this layer. These properties are very sensitive to electrolyte composition and impurities, and this is an area of active research (Ref 22-27). Corrosion of cell hardware by the electrolyte is also a concern, and the literature includes references to corrosion problems in the Li/SO_2 (Ref 28-36), $Li/SOCl_2$ and other oxyhalides (Ref 37-41), Li/V_2O_5 (Ref 42-44), and Li/I_2 (Ref 45) systems, as discussed below.

Lithium/Sulfur Dioxide (Li/SO_2) Batteries. The electrolyte in Li/SO_2 batteries contains LiBr as the electrolyte salt and SO_2 as the cathode reactant dissolved in acetonitrile, sometimes mixed with propylene carbonate (Ref 21, 28, 29). Hardware corrosion problems in Li/SO_2 cells include glass corrosion, tantalum corrosion in welded regions, corrosion of lithium in contact with nickel, and stress-corrosion cracking (SCC) of the battery can. Figure 1 shows each of these sites on a simplified battery schematic. Glass corrosion is primarily the result of attack by lithium metal to produce lithium oxide (Li_2O) (Ref 28, 30, 31):

$$4Li + SiO_2 \rightleftharpoons 2Li_2O + Si \quad \text{(Eq 9)}$$

The corrosion product Li_2O further corrodes the glass:

$$Li_2O + SiO_2 \rightleftharpoons Li_2SiO_3 \quad \text{(Eq 10)}$$

$$2Li_2O + SiO_2 \rightleftharpoons Li_2SiO_4 \quad \text{(Eq 11)}$$

The conductive reaction products contain both lithium metal and oxide and progress across the glass from the negatively charged header to the center pin. Failure occurs either by the conductive product bridging the battery terminals or by thinning of the glass until the stress is sufficient to crack it. Low-silica glasses are more resistant to lithium metal corrosion, and a low-silica glass (TA-23) has been developed that offers satisfactory corrosion resistance (5 years predicted life) (Ref 32). This corrosion resistance has been further improved by at least a factor of ten in new nonsilicate glasses (Ref 33). The mechanism of glass corrosion is common to all lithium cells in which a liquid electrolyte containing lithium ions (Li^+) wets the glass, and the phenomenon has been noted in Li/V_2O_5, $Li/SOCl_2$, and other systems (Ref 29).

Corrosion of tantalum in Li/SO_2 cells has caused failure of the internal contact of the aluminum cathode tab to the tantalum foil strip, which was connected to a tantalum center pin (Ref 28, 30, 31). Both the strip and the pin exhibited intergranular corrosion, with the worst attack occurring in the weld heat-affected zone (HAZ). It was suggested that corrosion was caused by bromine, which was produced by electrolyte decomposition during storage. This problem was solved by changing to a molybdenum pin, which does not appear to be susceptible to corrosion in the battery electrolyte. Furthermore, the weak link—the foil between the aluminum tab and the pin—was removed.

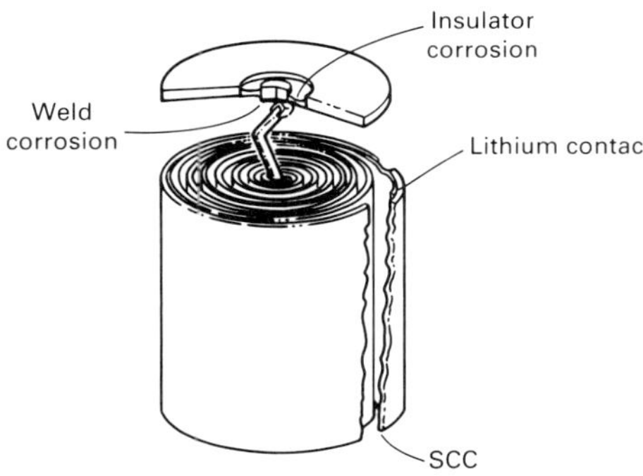

Fig. 1 Sites of corrosion in Li/SO_2 batteries. Courtesy of S.C. Levy

Corrosion of lithium in contact with nickel was manifested by open circuiting of lithium anodes at the point of contact with either the nickel anode strip or the nickel-plated steel can (Ref 28, 30). A study of galvanic corrosion of lithium in the battery electrolyte showed that nickel generated the least galvanic corrosion of the possible contact materials. Therefore, the material was not changed, but the geometry was altered by embedding a nickel grid in the lithium electrode so that contact was made throughout the anode rather than at a single point.

Stress-corrosion cracking of Li/SO_2 battery cans was first observed in nickel-plated low-carbon steel cans (Ref 28, 34, 35). Failures at the pressure relief vents and along the rim of the base were traced to intergranular SCC initiating on the inside surfaces. Cells at 70 °C (160 °F) failed in this manner in less than 2 months. The problem was alleviated by redesigning the base radius to lower the stresses, by proper annealing after forming, and by altering the microstructure to produce elongated grains, which cause SCC to follow a less favorable path.

Proper design to reduce stresses is also important for stainless steel cans, because AISI type 316 stainless steel has been shown to be susceptible to SCC in the Li/SO_2 electrolyte (Ref 36). Stress-corrosion cracking probably also contributed to the fracture of welds attaching headers to battery cans. Mixing of the nickel plating with the carbon steel cans in the weld fusion zone resulted in solidification to austenite, with subsequent transformation to martensite. Stress-corrosion cracking followed the prior-austenite grain boundaries. The weld was redesigned to lower the stresses and correct a weld solidification cracking problem, which should help alleviate SCC.

Lithium/Oxyhalide Cells. The electrolytes in lithium/oxyhalide ($SOCl_2$, SO_2Cl_2, and $POCl_3$) cells are very corrosive to organic materials. Elastomers are unstable, and Teflon is one of the only plastics that is compatible with these electrolytes, although it does react with lithium (Ref 37, 38). Separators are often made of borosilicate glasses, although these materials also react with lithium. Cell hardware must be made of metals that do not form acidic halides or electronically conductive salt films (Ref 38). Nickel and stainless steels have been shown to resist corrosion in solutions of $SOCl_2$ and SO_2Cl_2, while copper, tin, iron, steel, zinc, magnesium, aluminum, and titanium corrode actively (Ref 37). However, the resistant metals corrode rapidly upon contact with humidity after exposure to oxyhalide electrolytes. The mechanism of corrosion protection of type 316L stainless steel and nickel in 1.5 *M* $LiAlCl_4/SOCl_2$ electrolyte is the *in situ* formation of a protective metal chloride film (Ref 39, 40). Other metals, including platinum, tantalum, molybdenum, and tungsten, appear to be inert to $SOCl_2$ (Ref 40, 41). The passive and the inert metals can both be used in the construction of neutral electrolyte $Li/SOCl_2$ batteries.

The passive metals may be susceptible to environmental cracking in oxyhalide electrolytes. Postmortem analyses of nickel strips that were part of the anode circuit internal to $Li/SOCl_2$ batteries have revealed intergranular failures. Slow strain rate testing of nickel coupled to lithium in the battery electrolyte has demonstrated that intergranular environmental cracking can occur; however, the severity and possible mechanisms have not yet been reported. The phenomenon does not appear to be a practical problem unless the nickel strip is made very small in cross section (quantitative data are not available).

Lithium/Vanadium Pentoxide (Li/V_2O_5) Cells. Corrosion of the type 304 or 316 stainless steel positive grid was the primary cause of failures in Li/V_2O_5 batteries in the early 1970s (Ref 42, 43). The corrosion was thought to be caused by products of the reactions between water and the V_2O_5 or the $LiAsF_6$ + $LiBF_4$/methyl formate electrolyte. However, corrosion severity was found to be directly related to chloride contamination (Ref 42). Chloride levels are currently maintained at less than 100 ppm to avoid corrosion failures. Although certain inhibitors for chloride corrosion, particularly nitrate, looked promising, none was found to work under all conditions (Ref 44).

Lithium/iodine (Li/I_2) batteries, unlike the other lithium batteries discussed, are entirely solid-state. Potential corrosion of the stainless steel cans for Li/I_2 batteries was investigated when a new design placed the iodine-polyvinylpyridine (PVP) depolarizer material in contact with the type 304L stainless steel cans (Ref 45). No significant corrosion occurred in the absence of moisture; however, cells that were fabricated in a more humid environment than usual (1.5% relative humidity versus 0.5%) experienced fine pitting corrosion. The corrosive was suggested to be hydro iodic acid (formed by reaction of water with iodine), and because the corrosion was self-limiting, the acid was thought to be consumed in the corrosion reaction.

The corrosion problems in LAMBs (with the exception of glass corrosion) tend to be specific to each particular system and, as such, must be considered individually for each system. In general, alternate materials or designs have alleviated the problems, enabling the development of batteries capable of long (at least 5 to 15 years) shelf lives.

Aluminum/Air Batteries

Aluminum/air batteries have been proposed as an alternative energy source for transportation (Ref 46). The cell consists of an air cathode, an aluminum anode, and an electrolyte. The cathodic reduction of the oxygen in air occurs at a metal or carbon electrode of various designs. The electrolytes used include aqueous neutral chloride solutions and alkaline solutions, such as sodium hydroxide and potassium hydroxide (Ref 46). Parasitic corrosion of the aluminum anode is the major corrosion problem in this system.

Parasitic corrosion can be inhibited by alloying additions to the aluminum (Ref 46 to 56) and/or anion additions to the electrolyte (Ref 46, 51, 57). Alloy additions have various effects. Alloying elements that increase corrosion resistance include indium, thallium, phosphorus (Ref 46), and zinc (Ref 53), manganese and iron (Ref 55). Alloying elements that decrease corrosion resistance include copper (Ref 53), gallium and mercury (Ref 46). Magnesium has little effect on the corrosion resistance of aluminum (Ref 55). Because multialloy systems are more practical than binary alloys from a refining standpoint, investigation of multiaddition alloys are more relevant to the future commercialization of these battery systems (Ref 49, 50, 52, 53, 55, 56).

The inhibition of corrosion by the addition of anions to neutral and alkaline electrolytes has also been investigated. Anions that decrease corrosion rates include phosphate (PO_4^{3-}), chromate

(CrO_4^{2-}), molybdate (MoO_4^{2-}) (Ref 51), tartrate ($C_4H_4O_6^{2-}$), and borate (BO_3^{3-}) (Ref 57). Anions that exacerbate corrosion include nitrite (NO_2^-), permanganate (MnO_4^-), and silicate (SiO_3^{2-}) (Ref 51). However, in most cases, the influence of these variables on battery performance has not been well documented.

Sodium/Sulfur Batteries

The sodium/sulfur battery, a high-temperature system that operates between 325 and 350 °C (615 and 660 °F), is still under development. The individual cells feature a liquid sodium negative electrode and a liquid positive electrode of sulfur in the charged state or sodium polysulfides in the discharged state. Solid β″-alumina (Al_2O_3), usually in the form of a tube, or conductive glass fibers are used as the electrolyte. In tubular-cell designs (Fig. 2), either the sodium or the sulfur electrode can be on the inside of the tube. In either design, the major materials problem is with the positive current collector; sulfur and sodium polysulfides at 350 °C (660 °F) are extremely corrosive. With the central-sodium cells, the positive current collector is also the cell container so that, in addition to being corrosion resistant and a good electronic conductor, the positive current collector material must also be inexpensive, lightweight, and easy to fabricate into a sealed can. An excellent description of the sodium/sulfur cell and a review of the materials problems are provided in Ref 59.

Of the large number of metals and alloys that have been tested for compatibility with the sulfur electrode environments, few have shown both good corrosion resistance and high electrical conductivity. Carbon and graphite are corrosion resistant and have been used as positive current collectors in the central-sulfur cells; their relatively low electronic conductivity, however, limits cell length if good cell performance is to be maintained at high charge/discharge rates (Ref 60-62). Molybdenum (Ref 60, 63), high-chromium stainless steels, and nickel-base superalloys (Ref 64-70) have been used as cell containers/positive current collectors in central-sodium cells. None of these materials is satisfactory for production cells because of high cost and high density. Molybdenum is excellent for corrosion protection, but is difficult to fabricate and weld. Although molybdenum has been used as a liner for stainless steel cans (Ref 68), recent tests have shown its performance as a liner to be unacceptable (Ref 59). The corrosion resistance of the high-chromium alloys is marginal, with corrosion rates markedly increasing when the polysulfides are present (Ref 71, 72). The extensive formation of partially soluble corrosion products, such as Na_3FeS_3, NiS_2, and $NaCrS_2$ (Ref 71, 72), will reduce the amount of sulfur available for cell operation (thus reducing cell capacity) and possibly also affect cell chemistry and performance (Ref 73, 74). Also, post-test analysis of cycled cells has shown that $NaCrS_2$ has a tendency to accumulate at the electrolyte/polysulfide interface (Ref 75).

The most successful approach to the materials problem has been the use of protective coatings on a stainless steel or low-carbon steel substrate. Of the coating materials examined, chromium is by far the most studied and the most widely used in cell construction and testing (Ref 63, 76-79). The corrosion rates for these chromium coatings, however, are such that cell lifetimes much in excess of 5 years cannot be expected. This is adequate for an electric-vehicle battery, but is unsatisfactory for load-leveling applications for electric utilities. Other metal coatings, such as molybdenum (Ref 68, 80); alloy coatings based on the Co-Cr-Al-Y system (Ref 81); and electronically conducting refractory coatings, such as various carbides, nitrides, and oxides (Ref 80, 82-85), are being researched as possible solutions to the materials problem.

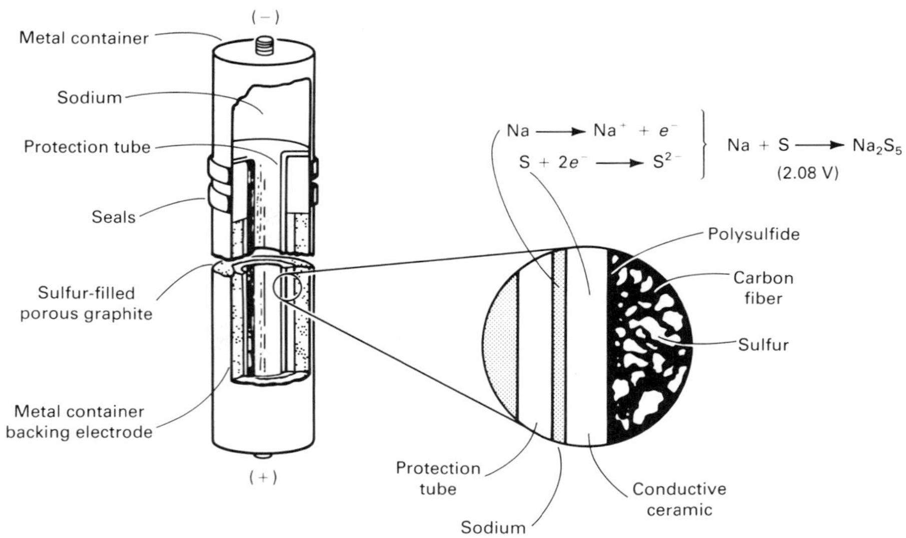

Fig. 2 Schematic of a sodium-sulfur cell. Source: Ref 58

Fuel Cells

This discussion will consider two types of fuel-cell technology whose state of development has advanced to the point at which reliability and service life are important issues in the evolution to a commercial product. Corrosive degradation of the components is now known to be a key factor in both the reliability and the service lives of these power plants. Because the two types of fuel cells use different families of materials to form the basic components, the two technologies will be discussed separately. The discussion will concern only corrosive degradation of the electrochemical energy conversion section of the power plant, omitting the fuel conditioning, thermal management (heat exchangers), and power-conditioning sections.

Phosphoric Acid Fuel Cell (PAFC)

The basic repeating elements in the electrochemical energy conversion section of the PAFC are shown schematically in Fig. 3. The repeating elements forming a single electrochemical cell are stacked to form a multicell module connected electrically in series. With the exception of the catalyst in the electrodes, the materials in the repeating elements are nonmetallic and are primarily allotropic forms of elemental carbon.

The basic structural component in the PAFC stack is the bipolar plate, which channels the gas flows across the backsides of the anode and cathode and also carries the electrical current from one cell to the next. The bipolar plate is a graphite-carbon composite fabricated by heat treating a graphite powder impregnated with hydrocarbon resin (Ref 86). Pyrolysis of the resin generally produces a glassy carbon second phase as opposed to the graphitic second phase (Ref 87) characteristic of synthetic graphite produced from coke-pitch (Ref 88) processing. The electrodes have a bi-layer composite structure, with a porous gas-supplying substrate layer fabricated from graphite fibers supporting the catalyst layer, which is a high-area carbon black impregnated with microcrystallites (2 to 10 nm) of platinum (Ref 89). A fluorocarbon polymer, such as polytetrafluoroethylene (PTFE), is used as a binder and wet-proofing agent in the electrode structure. The electrolyte, very concentrated H_3PO_4 (95 to 99%), is retained in a matrix of silicon carbide powder. The operating conditions for PAFCs vary, depending on the application and the size of the unit, from 478 K at 300 to 500 kPa (44 to 73 psi) to 433 K at ambient pressure. The pressurized operating conditions are more corrosive because of the higher temperature and the higher water activity.

At the anode, hydrogen is electrochemically converted to solvated protons by the reaction:

$$H_2S + 2S \rightleftharpoons 2(HS)^+ + 2e^- \qquad \text{(Eq 12)}$$

where S = H_2O,H_3PO_4. Oxygen is consumed at the cathode by reaction with solvated protons to form water by the reaction:

$$2(HS)^+ + \tfrac{1}{2}O_2 + 2e^- \rightleftharpoons H_2O + 2S \qquad \text{(Eq 13)}$$

The primary corrosion reactions are the anodic oxidation of the various allotropic forms of carbon and the dissolution of platinum (probably to form complexed platinum ion, Pt^{2+}, in solution). The following generic cell reaction is representative of the carbon oxidation:

$$C + 2H_2O + 4S \rightleftharpoons CO_2 + 4(HS)^+ + 4e^- \qquad \text{(Eq 14)}$$

Graphite is the most corrosion-resistant allotropic form of carbon. It is easily shown that the galvanic corrosion of graphite in oxygen-saturated acid is limited by the kinetics of the cathodic half-reaction, which is oxygen reduction on the graphite surface. However, if

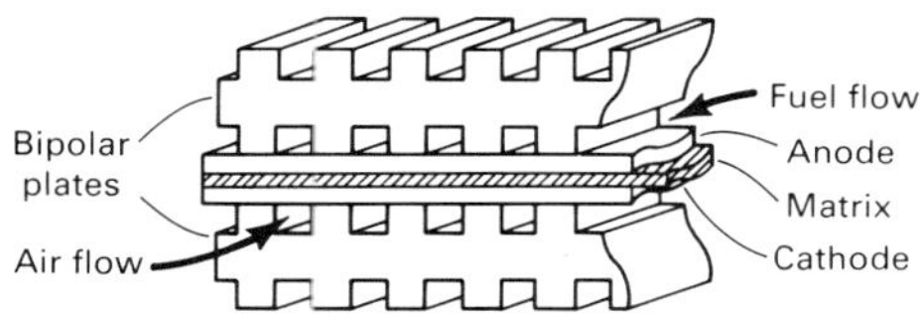

Fig. 3 Schematic of the repeating components of the PAFC power generation section

the surface of the graphite is both electronically and ionically coupled to an oxygen reduction catalyst, such as platinum, then the kinetics of the anodic half-reaction are controlling. This latter condition is clearly seen to be the case in the fuel-cell cathode because all components are in electrical contact, and at some points of physical contact, the components are wetted with electrolyte and share ionic pathways with the platinum crystallites. Extensive postmortem analyses have documented these corrosion-susceptible regions. Material loss, as a fraction of the material initially present, is greatest in the catalyst layer, followed by the losses in the substrate and then loss from the bipolar plate. This ranking is what would be expected based on relative surface area and extent of electrolyte wetting of the different components, assuming that the intrinsic corrosion resistance of each component is the same.

Reasonably extensive comparative studies of the kinetics of anodic oxidation of the various allotropic forms of carbon have been conducted. For example, a fairly extensive study of the corrosion of candidate high-area carbon blacks for use in the catalyst found that a correlation exists between corrosion rate (on a unit area basis) and *c*-axis layer plane spacing determined by x-ray powder diffraction analysis of heat-treated furnace blacks (Ref 90). Furnace blacks graphitized by heat treatment at 3000 K appear to be the material of choice for this component. Corrosion rates for graphite fiber materials were also compared with those of graphitized furnace blacks; on a unit area basis, the rates were found to be equivalent (same order of magnitude).

In addition, extensive studies of the corrosion of various types of carbon composite materials used in the bipolar plate have been reported in which the progressive development of corrosion resistance in the resin phase was followed by using thermal treatment (Ref 91). Lastly, the only fundamental study of the kinetic parameters of the corrosion reaction using single-phase materials, such as glassy carbon and single-crystal graphite, is discussed in Ref 92. This study confirmed three suggestions from previous work: that graphite is the most corrosion-resistant form of carbon (with the possible exception of diamond and diamondlike forms, which are electrically nonconductive and generally considered to be unsuitable), that the basal plane of graphite has much higher corrosion resistance than the edge-plane, and that the corrosion rate is first-order in water activity.

Platinum corrosion (dissolution to solvated ions) is also a serious problem in PAFCs. In one investigation, the solubility of bulk platinum was measured in concentrated H_3PO_4 at 443 and 463 K (Ref 93). The potential dependence of the solubility was found to be $RT/2F$, consistent with the Nernst equation for Pt/Pt^{2+} (as opposed to Pt/Pt^{4+}). Using these data, another researcher showed that platinum dissolution becomes rapid for cell voltages above 0.8 V and will result in loss of platinum from the cathode and precipitation at the anode (Ref 94). Platinum corrosion is a phenomenon for which there appears to be no remedy other than operating the fuel cell in such a way as to avoid cathode potentials above about 0.8 V for extended periods of time.

Molten Carbonate Fuel Cell (MCFC)

The MCFC component assembly follows the bipolar principle of the PAFC stack, but it uses completely different materials. Figure 4 shows a schematic diagram of the basic repeating elements in an MCFC stack. The electrolyte is a molten lithium carbonate (Li_2CO_3), potassium carbonate (K_2CO_3), and/or sodium carbonate (Na_2CO_3) mixture, with the most commonly chosen composition being $62Li_2CO_3$-$38K_2CO_3$ (usually referred to as a tile because the carbonate salts and the aluminate powder are hot pressed to form a solid piece) at 923 K. The ionic current in the cell is carried by the carbonate ion (CO_3^{2-}); therefore, the half-cell reactions at the anode and cathode are different from their PAFC counterparts. In this case, the cathode reaction is the reduction of oxygen by reaction with carbon dioxide (CO_2) to form CO_3^{2-}:

$$\tfrac{1}{2}O_2 + CO_2 + 2e^- \rightleftharpoons CO_3^{2-} \quad \text{(Eq 15)}$$

At the anode, hydrogen reacts with the CO_3^{2-} ion to form CO_2 and water:

$$H_2 + CO_3^{2-} \rightleftharpoons H_2O + CO_2 + 2e^- \quad \text{(Eq 16)}$$

To maintain invariance in electrolyte composition with time, it is necessary to recycle CO_2 from the anode gas to the cathode gas streams continuously. The current collector material is typically a high-chromium stainless steel, and the electrodes are nickel-base materials (metallic porous nickel at the anode and porous nickel oxide, NiO, at the cathode).

The components subject to corrosion and the nature of the corrosion processes are reasonably well understood from postmortem analyses of tested cells. With type 316 stainless steel as the current collector, moderate-to-severe attack of the anode current collector was observed after a few thousand hours of operation, with much less attack observed at the cathode (Ref 95). The region of the current collector that was most severely attacked was the point of contact with the anode material (and thus in contact with the carbonate melt) and areas near the point of contact, probably owing to creep of carbonate onto the current collector.

Early cells showed severe attack in the region of the wet-seal, a component that seals the anode from the external ambient gas (Ref 95). The wet-seal corrosion problem appears to have been solved by use of an aluminum coating on the collector housing that forms a dense insulating layer of Al_2O_3 that breaks the ionic path between the housing surface and the electrolyte and prevents galvanic corrosion (Ref 96).

A more serious and as yet unsolved corrosion process is the finding of significant solubility of NiO in the carbonate melts in use. Early theoretical analysis of metal corrosion in molten alkali carbonates presumed that NiO dissolution occurs by the reaction (Ref 97):

$$NiO + CO_2 \rightleftharpoons Ni^{2+} + CO_3^{2-} \quad \text{(Eq 17)}$$

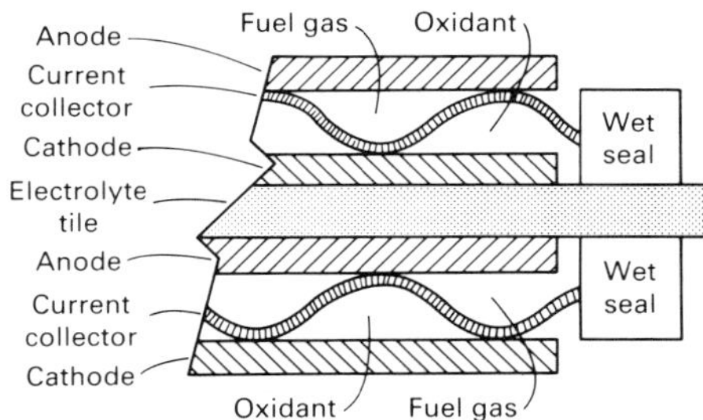

Fig. 4 Schematic of the repeating components of the MCFC power generation section

This analysis also predicted that NiO solubility would increase in proportion to CO_2 pressure. In the early period of MCFC technology development, it was believed that the CO_2 partial pressure at the cathode would be within a range in which NiO dissolution would not be significant. A more recent theoretical analysis of NiO dissolution (Ref 98) and experimental results (Ref 99) have shown the importance of other dissolution reactions for NiO, particularly those involving the formation of soluble anions according to:

$$NiO + \tfrac{1}{2}Li_2CO_3 + \tfrac{1}{2}O_2 \rightleftharpoons NiO_2^- + \tfrac{1}{2}CO_2 + Li^+ \quad \text{(Eq 18)}$$

$$NiO + Li_2CO_3 = NiO_2^{2-} + 2Li^+ + CO_2 \quad \text{(Eq 19)}$$

Equation 6 predicts an oxygen partial pressure dependence for NiO solubility that would enhance solubility with pressurization of the reactant gases. Recent measurements of NiO solubility and transport in the electrolyte tile have been reported (Ref 100). Nickel oxide solubility is now recognized as a serious technical problem for MCFC technology, and alternative cathode materials are being sought (Ref 5).

General corrosion studies of simple metals and selected alloys in molten carbonate melts date back to the earliest time of interest in MCFC technology. A relatively recent bibliography on molten salt corrosion covers several types of molten salts in addition to carbonates and provides a useful index to the literature before about 1976 (Ref 101). Reference 102 contains a section on carbonates and includes references from the Russian literature. Work that addresses corrosion phenomena specific to MCFC technology is much less plentiful. The theoretical overview of the requirements for corrosion resistance in MCFC applications provided in Ref 103 has served as a guide for much of the recent work.

ACKNOWLEDGMENT

Preparation of the lithium battery review performed at Sandia National Laboratories was supported by the United States Department of Energy under Contract No. DE-ACO4-76DP00789. Preparation of the sodium/sulfur review was supported by the United States Department of Energy, Office of Energy Storages, under Contract No. W-31-109-Eng-38. Funding for the preparation of the fuel cell review was provided by the Assistant Secretary for Fossil Energy, Office of Coal Utilization, Advanced Conversion and Energy Systems Division of the United States Department of Energy under Contract No. DE-ACO3-76SF00098. Philip N. Ross, Jr. acknowledges valuable input from Dave Shores, Charles Baumgartner, and John Appleby.

REFERENCES

1. D. Linden, Ed., *Handbook of Batteries and Fuel Cells*, McGraw-Hill, 1984
2. K. Othmer and M. Grayson, Ed., Batteries and Electric Cells, in *The Encyclopedia of Chemical Technology*, Vol 3, 3rd ed., John Wiley & Sons, 1978, p 503-663
3. C.A. Vincent *et al.*, *Modern Batteries, An Introduction to Electrochemical Power Sources*, Edward Arnold, 1984
4. G.W. Heise and N.C. Cahoon, Ed., *The Primary Battery*, Vol I, John Wiley & Sons, 1971; Vol II, 1976
5. A.J. Appleby *et al.*, An Overview of Fuel Cell Technology; Past, Present, and Future, *Energy*, Vol 11, 1986, p 1-231
6. J. Burbank, A.C. Simon, and E. Willihnganz, The Lead-Acid Cell, in *Advances in Electrochemistry and Electrochemical Engineering*, Vol 8, P. Delahay and C.W. Tobias, Ed., John Wiley & Sons, 1971
7. H. Bodie, *Lead-Acid Batteries*, R.J. Brodd and K.V. Kordesch, Trans., John Wiley & Sons, 1977
8. A.T. Kuhn, Ed., *Electrochemistry of Lead*, Academic Press, 1979
9. B.D. McNicholl and D.A. Rand, Ed., *Power Sources for Electric Vehicles*, Elsevier, 1984
10. D. Pavlov, Ed., *Advances in Lead-Acid Batteries*, Vol 84-14, The Electrochemical Society, 1984
11. J.J. Lander, Anodic Corrosion of Lead in Sulfuric Acid Solutions, *J. Electrochem. Soc.*, Vol 98, 1951, p 213-219; Vol 103, 1956, p 1-8
12. V.K. Dantam, "Effect of Tin on the Corrosion Properties of Wrought Lead-Calcium Alloys," Report V223022, General Motors Corporation, Delco Remy Division, April 1984
13. J.J. Lander, Effect of Corrosion and Growth on the Life of Positive Grids in the Lead-Acid Cell, *J. Electrochem. Soc.*, Vol 99, 1952, p 467-473
14. A.G. Cannone, D.O. Feder, and R.V. Biagetti, Positive Grid Design Principles, *Bell Syst. Tech. J.*, Sept 1970, p 1279-1303
15. M. Torralba, Present Trends in Lead Alloys for the Manufacture of Battery Grids—A Review, *J. Power Sources*, Vol 1, 1976-1977, p 301-310
16. V.K. Dantam, "Testing Low Antimony Alloy Grids in a F-II Design Battery," Report V223128, General Motors Corporation, Delco Remy Division, July 1986
17. V.K. Dantam, "Effect of Aluminum on Microstructural Properties of Wrought Lead-Calcium-Tin Alloys," Report V223049, General Motors Corporation, Delco Remy Division, June 1984
18. D. Marshall and W. Tiedeman, Microstructural Aspects of Grid Corrosion in the PbO_2 Electrode, *J. Electrochem. Soc.*, Vol 123, 1976, p 1849-1855
19. R.K. Galgali, P.V. Vasudeva Rao, and H.V.K. Udupa, Studies on Corrosion of Lead-Acid Battery Grids With Addition of Cobalt and Silver Compounds, *J. Electrochem. Soc.*, Vol 126, 1979, p 725-728
20. P. Oliva *et al.*, Review of the Structure and Electrochemistry of Nickel Hydroxides and Oxy-Hydroxides, *J. Power Sources*, Vol 8, 1982, p 229-255
21. J.P. Gabano, Lithium Battery Systems; An Overview, in *Lithium Batteries*, J.P. Gabano, Ed., Academic Press, 1983, p 1-12
22. A.N. Dey, Lithium Anode Film and Organic and Inorganic Electrolyte Batteries, *Thin Solid Films*, Vol 43, 1977, p 131-171
23. E. Peled, Lithium Stability and Film Formation in Organic and Inorganic Electrolytes for Lithium Batteries Systems, *Thin Solid Films*, Vol 43, 1977, p 43-72
24. R.V. Moshtev, Y. Geronova, and B. Puresheva, The Primary Passive Film on Li in $SOCl_2$ Electrolyte Solutions, *J. Electrochem. Soc.*, Vol 128, 1981, p 1851-1857
25. R.V. Moshtev and B. Puresheva, AC Impedance Study of the Lithium Electrode in Propylene Carbonate Solutions, *J. Electroanal. Chem.*, Vol 180, 1984, p 609-617
26. N.A. Fleischer, J.R. Thomas, and R.J. Ekern, Reduction of Voltage Delay in the $Li/SOCl_2$ Systems via Suitable Choice of Electrolyte Salts, *J. Electrochem. Soc.*, Vol 131, 1984, p 1733-1738
27. J.W. Boyd, The Effect of Polyvinyl Chloride and Fe on Film Growth and Voltage Delay in $SOCl_2$ Electrolytes, *J. Electrochem. Soc.*, Vol 134, 1987, p 18-24
28. P. Bro and S.C. Levy, Lithium Sulfur Dioxide Batteries, in *Lithium Battery Technology*, H.V. Venkatasetty, Ed., John Wiley & Sons, 1984, p 79-126
29. C.R. Walk, Lithium-Sulfur Dioxide Cells, in *Lithium Battery Technology*, H.V. Venkatasetty, Ed., John Wiley & Sons, 1984, p 281-302
30. S.C. Levy, Modified Li/SO_2 Cells for Long-Life Applications, in *Proceedings of the 29th Power Sources Symposium*, The Electrochemical Society, June 1980, p 96-109
31. S.C. Levy, Corrosion Reactions in Lithium-Sulfur Dioxide Cells, in *Corrosion in Batteries and Fuel Cells and Corrosion in Solar Energy Systems*, C.J. Johnson and S.L. Pohlman, Ed., The Electrochemical Society, 1983, p 9-16
32. U.S. Patent Application Serial No. 786561, filed 11 Oct 85
33. Powdered Preforms Boost Interest in Corrosion-Resistant Glass, *Glass Ind.*, Vol 65 (No. 9), Sept 1984, p 26
34. A. Attewell, "Metallurgical Evaluation of Cracking in Bases From Lithium-Sulfur Dioxide Cells," RAE (F) MT4/4/E1141, Royal Aircraft Establishment, 1982
35. C.C. Wang, Duracell International, Inc., private communication, 1983
36. W.R. Cieslak, D.R. McIntyre, and S.M. Wilhelm, Stress-Corrosion Cracking of Carbon Steel and Stainless Steel in Li/SO_2 Electrolytes, Abstract 56, in *Extended Abstracts*, Vol 85-2, The Electrochemical Society, 1985
37. J.J. Auborn, K.W. French, and A. Heller, Corrosion and Compatibility of Materials in Inorganic Oxyhalides, in *Corrosion Problems in Energy Conversion and Generation*, C.S. Tedman, Jr., Ed., The Electrochemical Society, 1979, p 56-61
38. C.R. Schlaikjer, Lithium-Oxyhalide Cells, in *Lithium Batteries*, J.P. Gabano, Ed., Academic Press, 1983, p 303-370
39. F.M. Delnick and W.R. Cieslak, Abstract 145, in *Extended Abstracts*, Vol 84-2, The Electrochemical Society, 1984
40. W.R. Cieslak, F.M. Delnick, D.E. Peebles, and J.W. Rogers, Jr., Passive Film Formation on Metals in Thionyl-Chloride Electrolytes for Lithium Batteries, in *Proceedings of the Symposium on Surfaces, Inhibition, and Passivation*, Fall Meeting, San Diego, CA, The Electrochemical Society, to be published
41. F.M. Delnick and W.R. Cieslak, "The Oxidation of $SOCl_2$ Electrolyte on Mo, Ta, W, and Pt Electrodes," SAND85-0015, Sandia National Laboratories, 1985
42. W.B. Ebner and W.C. Merz, Corrosion—A Major Failure Mode of Li/V_2O_5 Cells, in *Proceedings of the 29th Power Sources Symposium*, The Electrochemical Society, June 1980, p 214-219
43. C.R. Walk, Lithium-Vanadium Pentoxide Cells, in *Lithium Batteries*, J.P. Gabano, Ed., Academic Press, 1983, p 265-280
44. W.B. Ebner, Honeywell Power Sources Center, private communication, 1986
45. W.R. Brown, C.F. Holmes, and R.C. Stinebring, Corrosion Resistance of Lithium/Iodine Batteries Fabricated in an Extremely Dry Environment, in *Corrosion in Batteries and Fuel Cells and Corrosion in Solar Energy Systems*, C.J. Johnson and S.L. Pohlman, Ed., The Electrochemical Society, 1983, p 27-31
46. A. Despic, and D. Drazic, "Electrochemical Properties of Alloys of Aluminum with Gallium and Phosphorus," Final Report, Laboratory for Electrochemical Energy Conversion, Institute of Electrochemistry ICTM, University of Belgrade, June 1982
47. L. Bockstie *et al.*, Control of Aluminum Corrosion in Caustic Solutions, *J. Electrochem. Soc.*, Vol 110, April 1963, p 267-271
48. J.D. Talati, Aldehydes as Corrosion Inhibitors for Al-Mg Alloys in KOH, *Werkst. Korros.*, Vol 29 (No. 7), 1978, p 461-468
49. D.J. Levy *et al.*, Performance of a Rapidly Refuelable Aluminum-Air Battery, in *Proceedings of the 18th Intersociety Energy Conversion Engineering Conference*, American Institute of Chemical Engineers, 1983, p 1635-1640
50. M. Katon *et al.*, Aluminum-Air Battery, *J. Electrochem. Soc. Jpn.*, Vol 38, 1970, p 753-757
51. K.M. El-Sobki *et al.*, Corrosion Behavior of Aluminum in Neutral and Alkaline Chloride Solutions Containing Some Anions, *Corros. Prev. Control*, Vol 28, Dec 1981, p 7-12
52. H.P. Godard, An Insight Into the Corrosion Behavior of Aluminum, *Mater. Perform.*, Vol 20, July 1981, p 9-15
53. K. Tohma *et al.*, Compound Effects of Additional Zinc, Copper and Manganese on Electrochemical Properties and Corrosion Resistance of Aluminum, *Keikinzoku (J. Jpn. Inst. Light Met.)*, Vol 33 (No. 9), Sept 1983, p 39-56
54. N.D. Koshe *et al.*, Electrochemical Dissolution of Aluminum in a Slotted Cell With Flowing Electrolyte Kinetics of Processes in KOH Solutions Containing Mercury, *Soviet Electrochem.*, Vol 15 (No. 5), 1979, p 404-408

55. M. Zamin, The Role of Manganese in Corrosion Behavior of Aluminum-Manganese Alloys, *Corrosion*, Vol 37, 1981, p 627-632
56. B. Jovanovic *et al.*, "Electrochemical Properties of Certain New Ternary and Quaternary Alloys Based on Aluminum," UCRLTrans-11535, Report at the 20th Congress of Serbian Chemists, Jan 1977
57. D.M. Drazic *et al.*, The Effect of Anions on the Electrochemical Behavior of Aluminum, *Electrochim. Acta*, Vol 28 (No. 5), 1983, p 751-755
58. M.L. McClanahan, Electrolyte Durability From a User's Viewpoint, in *Proceedings of the 5th DOE/EPRI Beta Battery Workshop*, EPRI EM-3631 SR, Electric Power Research Institute, 1984, p 6-25
59. J.L. Sudworth and A.R. Tilley, *The Sodium-Sulfur Battery*, Chapman and Hall, 1985, p 199-226
60. T.L. Markin, A.R. Junkison, R.J. Bones, and D.A. Teagle, Current Collectors for Sodium-Sulfur Batteries, *J. Power Sources*, Vol 7, 1979, p 757-767
61. R.W. Minck, The Performance of Shaped Graphite Electrodes in Sodium Sulfur Cells, in *Proceedings of the Symposium on Advanced Battery Research*, ANL-76-8, Argonne National Laboratory, 1976, p B/199-B/209
62. J.L. Sudworth, A.R. Tilley, and J.M. Bird, A Potentially Low Cost Sodium/Sulfur Cell for Load Levelling, Abstract 88, in *Extended Abstracts*, Vol 77-2, Society Meeting, The Electrochemical Society, 1977, p 156-157
63. R.R. Dubin, Evaluation of Container Materials for the Sodium-Sulfur Battery, *Mater. Perform.*, Vol 20, 1981, p 13-18
64. R. Bauer, W. Haar, H. Kleinschmager, G. Weddigen, and W. Fischer, Some Studies on Sodium/Sulfur Cells, *J. Power Sources*, Vol 1, 1975, p 109-126
65. R.J. Bones, R.J. Brook, and T.L. Markin, Polarization and Corrosion Experiments in Sodium-Sulfur Cells, *J. Power Sources*, Vol 5, 1975, p 539-557
66. W. Fischer, W. Haar, B. Hartmann, H. Meinhold, and G. Weddigen, Recent Advances in Na/S Cell Development—A Review, *J. Power Sources*, Vol 3, 1978, p 299-309
67. B. Hartmann, Casing Materials for Sodium/Sulfur Cells, *J. Power Sources*, Vol 3, 1978, p 227-235
68. S. Hattori, M. Yamaura, S. Kimura, and S. Iwabuchi, "A New Design for the High-Performance Sodium-Sulfur Battery," Paper 770281, in *Proceedings of the SAE Congress*, Society of Automotive Engineers, 1977
69. K.R. Kinsman and D-G. Oei, Corrosion of Ni-Base Superalloys in Sodium Tetrasulfide, Abstract 57, in *Extended Abstracts*, Vol 77-2, The Electrochemical Society, 1977, p 156-157
70. S.P. Mitoff, "Development of Sodium Sulfur Batteries for Utility Application," EPRI EM-683, Electric Power Research Institute, 1978
71. A.P. Brown and J.E. Battles, "Materials Corrosion in Sulfur and Sodium Polysulfides," Paper presented at the 6th DOE/EPRI Beta Battery Workshop, Snowbird, UT, Electric Power Research Institute, 1985
72. A.P. Brown and J.E. Battles, The Corrosion of Metals and Alloys by Sodium Polysulfide Melts at 350 °C, Abstract 88, in *Extended Abstracts*, Vol 86-2, The Electrochemical Society, 1986
73. R.M. Dell, J.L. Sudworth, and I.W. Jones, Sodium/Sulfur Battery Development in the United Kingdom, in *Proceedings of the 11th Intersociety Energy Conversion Engineering Conference*, Vol 1, American Society of Mechanical Engineers, 1976, p 503-509
74. G. May and I.W. Jones, Materials Selection for Sodium Sulfur Batteries, *Metall. Mater. Technol.*, Vol 8, 1976, p 427-431
75. J.A. Smaga and J.E. Battles, "Post-Test Examination of FACC Sodium/Sulfur Cells," Paper presented at the 6th DOE/EPRI Beta Battery Workshop, Snowbird, UT, Electric Power Research Institute, 1985
76. J.P. Dumas and A. Vicker, Protective Chromium Layer, in *Proceedings of the 5th DOE/EPRI Beta Battery Workshop*, EPRI EM-3631-SR, Electric Power Research Institute, 1984, p 6/37-6/44
77. D.S. Park and D. Chatterji, Characterization and Corrosion Behavior of Duplex-Chromized Steel for the Sulfur Container in Na-S Cells, *Thin Solid Films*, Vol 83, 1981, p 429-435
78. A. Wicker, "Corrosion of Chromium-Coated Steel in Sodium Polysulfide Environments," EPRI EM-2947, Electric Power Research Institute, 1983
79. A. Wicker, G. Desplanches, and H. Saisse, Behavior of Chromium-Coated Steels in Sodium Polysulfide Environments, *Thin Solid Films*, Vol 83, 1981, p 437-447
80. A.R. Tilley and M.L. Wright, Development of a Corrosion Resistant Current Collector for the Sodium Sulfur Cell, in *Proceedings of the 16th Intersociety Energy Conversion Engineering Conference*, Vol 1, American Society of Mechanical Engineers, 1981, p 841-845
81. R. Knodler, Brown, Boveri, & Cie, private communication, 1985
82. K.R. Kinsman and W.L. Winterbottom, The Use of Coatings in High Temperature Battery Systems, *Thin Solid Films*. Vol 83, 1981, p 417-428
83. M. Mikkor, Graphite Aluminum- and Silicon Carbide-Coated Current Collectors for Sodium-Sulfur Cells, *J. Electrochem. Soc.*, Vol 132, 1985, p 991-998
84. G.R. Miller, Corrosion of Oxides, in *Proceedings of the 5th DOE/EPRI Beta Battery Workshop*, EPRI EM-3631-SR, Electric Power Research Institute, 1984, p 6/47-6/58
85. H.S. Wroblowa, R.P. Tischer, G.M. Crosbie, G.J. Tennenhouse, and V. Markovac, Stability of Materials in Sulfur-Polysulfide Melts, in *Proceedings of the 5th DOE/EPRI Beta Battery Workshop*, EPRI EM-3631-SR, Electric Power Research Institute, 1984, p 6/73-6/94
86. A.J. Appleby, Carbon Components in the PAFC-An Overview, in *Proceedings of the Workshop on the Electrochemistry of Carbon*, Vol 84-5, S. Sarangapani, J. Akridge, and B. Schumm, Ed., The Electrochemical Society, 1984, p 251-273
87. W.A. Nystrom, Raw Material and Processing Effects on the Electrochemical Wear of Carbon Composite Materials, in *Proceedings of the Workshop on the Electrochemistry of Carbon*, Vol 84-5, S. Sarangapani, J. Akridge, and B. Schumm, Ed., The Electrochemical Society, 1984, p 363-387
88. P.A. Thrower, Microstructure of Carbon Materials, in *Proceedings of the Workshop on the Electrochemistry of Carbon*, Vol 84-5, S. Sarangapani, J. Akridge, and B. Schumm, Ed., The Electrochemical Society, 1984, p 40-60
89. H. Kunz and G. Gruver, The Catalytic Activity of Platinum Supported on Carbon for Electrochemical Oxygen Reduction in Phosphoric Acid, *J. Electrochem. Soc.*, Vol 122, 1975, p 1279
90. P. Stonehart and J. MacDonald, "Stability of Acid Fuel Cell Cathode Materials," EPRI EM-1664, Electric Power Research Institute, 1981
91. L. Christner, H. Dhar, M. Farooque, and A. Kush, "Corrosion of Graphite Composites in PAFCs," Paper 80, presented at *Corrosion/86*, National Association of Corrosion Engineers, 1986
92. P.N. Ross, Corrosion of Graphite and Graphite-Carbon Composite Materials in Phosphoric Acid at Elevated Temperature and Pressure, Abstract 40, in *Extended Abstracts*, Vol 85-2, The Electrochemical Society, 1985
93. P. Bindra, S. Clouser, and E. Yeager, Platinum Dissolution in Concentrated Phosphoric Acid, *J. Electrochem. Soc.*, Vol 126, 1979, p 1631
94. P.N. Ross, "Deactivation and Poisoning of Fuel Cell Catalysts," LBL-19766, Lawrence Berkeley Laboratory, 1985
95. N.S. Choudhury, "Development of MCFCs for Power Generation," Final Report to the United States Department of Energy, Contract No. DE-ACO3-77ET11319, General Electric Company, 1980
96. R. Swaroop, J. Sim, and K. Kinoshita, Corrosion Protection of Molten Carbonate Fuel Cell Gas Seals, *J. Electrochem. Soc.*, Vol 125, 1978, p 1799
97. M. Ingram and G. Janz, The Thermodynamics of Corrosion in Molten Carbonates. Applications of E/T CO_2 Diagrams, *Electrochim. Acta*, Vol 10, 1965, p 783
98. D. Shores, C. Iacovangelo, and R. Wilson, Abstract 344, in *Extended Abstracts*, Vol 82-2, The Electrochemical Society, 1982
99. M. Orfield and D. Shores, Abstract 50, in *Extended Abstracts*, Vol 84-2, The Electrochemical Society, 1984
100. C. Baumgartner, Solubility and Transport of NiO Cathodes in Molten Carbonate Fuel Cells, *J. Am. Ceram. Soc.*, Vol 69, 1986, p 162
101. G. Janz and R. Tomkins, Corrosion in Molten Salts: An Annotated Bibliography, *Corrosion*, Vol 35, 1979, p 485
102. J. Selman and H. Maru, Physical Chemistry and Electrochemistry of Alkali Carbonate Melts, in *Advances in Molten Salt Chemistry*, Vol 5, G. Mamontov and J. Braunstein, Ed., Plenum Press, 1980, p 160
103. R. Rapp, Materials Selection and Problems in the MCFC, in *Proceedings of the DOE/EPRI Workshop on MCFS*, Oak Ridge National Laboratory, 1979, p 230

Corrosion of Metallic Implants and Prosthetic Devices

Anna C. Fraker, National Bureau of Standards

BIOMEDICAL PROSTHETIC DEVICES are artificial replacements that are used in a biological system, such as the human body, in an effort to provide the function of the original part. Prosthetic devices are made of polymeric, metallic, and ceramic materials or combinations of these materials, depending on the intended use. There are specific degradation processes for all of these materials. Metals are used as surgical implants in the human body primarily for orthopedic purposes. Additional applications include reconstructive surgery, cosmetic surgery, wire leads, heart valve parts, aneurysm clips, and dental uses (see the article "Tarnish and Corrosion of Dental Alloys" in this Volume).

The first requirement for any material to be placed in the body is that it should be biocompatible and not cause any adverse reaction in the body. The material must withstand the body environment and not degrade to the point that it cannot function in the body as intended. For example, metals used in the cardiovascular system must be nonthrombogenic, and in general, the more electronegative the metal with respect to blood, the less thrombogenic the metal will be. Cobalt-chromium alloys of the HS-21 and HS-25 compositions and titanium are used in heart valves. Design also is an important factor in preventing thrombus formation. Corrosion is included in the topic of biocompatibility because it is an important factor in the release of metal ions into the body environment and in the degradation of the implant metal. *In vitro* electrochemical measurements can be conducted in controlled environments, and these techniques provide methods of determining the basic corrosion reactions necessary for predicting the corrosion behavior of materials and for screening and characterizing materials intended for use in surgical applications.

This article will discuss the corrosion of metallic materials. Background information will be given on the metals used in prosthetic devices, biocompatibility, significance of corrosion, and standards. This will be followed by discussions of electrochemistry and corrosion processes, the types of corrosion to be expected, and the test methods used to evaluate the corrosion behavior of surgical implant metals. Information on specific materials and on corrosion processes is provided in other articles in this Volume and in the references cited in this article.

Background

The earliest attempts at repairing the human body probably went unrecorded. There are a number of historical accounts of the development of the use of metals in the human body (Ref 1-5). The first record of metal implantation discusses the repair of a cleft palate with a gold plate by Petronius in 1565 (Ref 1, 3). About 100 years later, Hieronymus Fabricius described the use of gold, iron, and bronze wires for sutures. In 1775, arguments arose between Pujol, a surgeon opposing internal fixation, and Icart, a surgeon who favored internal fixation and used brass wires (Ref 1, 3, 6). Icart encountered problems with infection, but cited the work of two French surgeons, Lapeyode and Sicre, who had successfully used wire to repair bone fractures. In 1886, Hansmann used metal plates for internal fixation (Ref 7). These plates, which were nickel-plated steel, had holes through which screws were inserted into the bone. Some had a bend at one end and protruded through the skin for ease of removal. X-rays were discovered by Roentgen in 1895. The use of x-rays to observe the healing of fractures revealed the advantages of internal fixation and stimulated its use.

It was difficult in early times to determine whether the infection and inflammation were due to the metal or to other factors. The development of aseptic techniques by Baron Joseph Lister in the 1860s made it possible to determine the most suitable metals for use as implants (Ref 1, 8). As the success of surgery increased, it became clear that the metals were an important limiting factor. The metals tested for implant use included platinum, gold, silver, lead, zinc, aluminum, copper, and magnesium (Ref 1); all of these were found to be too malleable. Magnesium was found to be highly reactive in the body. Steel plates coated with gold or nickel came into use. The need for strong and corrosion-resistant metals became apparent. Stainless steels were introduced as implants in 1926, and cobalt-chromium-molybdenum-carbon alloys were first used in 1936 (Ref 2). Titanium was determined to be inert in the body (Ref 9, 10); titanium and titanium alloys were not introduced until the 1960s and came into increased use in the 1970s. Tantalum, which was studied in the early 1950s, does show some tissue reaction (Ref 11, 12).

Observations eventually led to the discovery of electrochemical reactions in the body in the 1930s. Steel screws in a magnesium plate produced the dramatic result of having the magnesium plate disappear before the fracture healed. The combination of copper with zinc was found to be highly reactive, as was the combination of brass screws with an aluminum plate. Surgeons and scientists began to believe that homogeneous metals should be used to prevent these reactions. At this time, many scientists were beginning to notice electrochemical reactions with surgical implants; one of the first actual measurements of an electrode potential difference between uncorroded and corroded areas of screws was made by F. Masmonteil (Ref 13). C.S. Venable, W.P. Stuck, and A. Beach are responsible for bringing electrolytic effects to the attention of surgeons and scientists in the United States (Ref 14). This interest in and discovery of electrochemical effects with surgical implants in the 1930s occurred more than 140 years after Italian physician Luigi Galvani published his electrochemical discoveries (Ref 15). Electrochemistry was already a well-established science. A. Volta showed that the electricity was generated by the contact of dissimilar metals, not from "animal electricity," as stated by Galvani (Ref 15). Galvani did later demonstrate the electrical nature of nerve action.

The latter half of the 20th century has seen marked advances in the successful use of prosthetic devices. The success of the total hip prosthesis increased because of the addition of the acetabular component (hip socket) and the surgical procedures introduced by J. Charnley (Ref 16) and others, who used poly methylmethacrylate bone cement to affix the metal femoral component and ultrahigh molecular weight polyethylene acetabular cup. Biomaterials research was promoted by the founding of the Society for Biomaterials in 1974 and by increased interest among other medical and scientific societies. More emphasis was placed on standards and specifications, and the American Society for Testing and Materials (ASTM) established Committee F-4 on Medical Materials and Devices in 1964 (Ref 17). Congress passed the Medical Device Amendments (Ref 18) and placed responsibility for the implementation of this legislation with the United States Food and Drug Administration.

J.S. Hirschhorn and J.T. Reynolds explored the use of a totally porous cobalt-chromium-molybdenum implant that had an elastic modulus closer to that of bone, but the porosity made the strength insufficient for the load (Ref 19). Metal porous coated implants were introduced, and they provided a porous coating for bony ingrowth attachment and maintained the strength of the solid substrate (Ref 20-24). Hydroxyapatite filler within the porous coating increases bone growth during the first 4 weeks (Ref 24). Electrical stimulation also increases the early stages of bone ingrowth (Ref 25, 26). Corrosion of porous coatings is a concern because of the increased surface

Table 1 Compositions of metals and alloys currently used as surgical implants

Metal or alloy	Composition(a) C	Ti	Cr	Fe	Co	Ni	Mo	Others
AISI type 316 stainless steel	0.08 max	. . .	18.5	rem	. . .	12.0	3.0	0.75Si, 0.03P, 0.03S
Cast cobalt-chromium alloy	0.36 max	. . .	28.5	0.75 max	rem	2.5 max	6.0	1.0 max Si
Wrought cobalt-chromium alloy	0.15 max	. . .	20.0	3.0 max	rem	2.5 max	. . .	15.3W
Unalloyed titanium	0.10	rem	. . .	0.30	. . .	. . .	. . .	0.015H, 0.13O, 0.07N
Ti-6Al-4V	0.08	rem	. . .	0.25	. . .	. . .	. . .	6.0Al, 4.0V, 0.0125H, 0.13O
MP35N	. . .	. . .	20.0	. . .	35.0	35.0	10.0	. . .
Unalloyed tantalum	0.01	0.01	. . .	0.01	. . .	. . .	0.01	rem Ta, 0.001H, 0.015O, 0.01N, 0.005Si, 0.03W, 0.05Nb

(a) Nominal unless otherwise indicated. Source: Ref 17

area and the changed surface morphology. E.P. Lautenschlager (Ref 27) studied the corrosion of porous Ti-6Al-4V, and H.V. Cameron (Ref 28) studied porous cobalt-chromium. Corrosion studies of porous Ti-6Al-4V and cobalt-chromium-molybdenum indicated no changes in corrosion behavior except for increased current due to the increased surface area of the porous materials; however, caution should be exercised regarding possible contamination or crevice corrosion due to porous configurations (Ref 23, 29).

Metals and Alloys

Metals and alloys used as implants undergo an active-passive transition; therefore, corrosion resistance results from the growth of a protective surface film. These metals are in the passive state with a protective surface oxide film when used as implants and are highly corrosion resistant in saline environments. The metals currently used for surgical implants include stainless steel (AISI type 316L), cobalt-chromium-molybdenum-carbon, cobalt-chromium-tungsten-nickel, cobalt-nickel-chromium-molybdenum, titanium, Ti-6Al-4V, and tantalum. The compositions and mechanical properties of these metals and alloys as recommended by ASTM (Ref 17) are given in Tables 1 and 2.

The metals and alloys most frequently used as implant materials will be discussed in terms of metallurgical factors to provide a better understanding of their effect on corrosion resistance. For example, small changes in alloying additions of certain elements can result in significant changes in corrosion behavior. Such microstructural changes as grain size, precipitates, location of precipitates, and the presence of impurities can be important. Additional information on the metallurgical aspects is available in the article "Metallurgically Influenced Corrosion" in this Volume, the article "Failures of Metallic Orthopedic Implants" in Volume 11 of the 9th Edition of *Metals Handbook*, and Ref 5, 30, and 31.

Stainless Steels. Steels and coated steels were used in the early 1900s for surgical purposes (Ref 1-3). The Frenchman Berthier is credited with discovering in 1821 that adding chromium to iron greatly improves corrosion resistance (Ref 32). Stainless steel was introduced in 1912 (Ref 33). In 1926, type 302 and other stainless steels were used in orthopedic surgery, and type 316 stainless steel came into use during World War II (Ref 3).

Type 316L is used most widely in applications in which the implant is temporary, although it is also used for some permanent implants. The composition and mechanical properties are given in Tables 1 and 2, respectively. Type 316L is the least corrosion resistant to body fluids of the implant metals discussed, but its corrosion resistance is adequate for some purposes, and if the material is in the cold-worked condition, the mechanical properties are good. Figure 1 shows type 316L stainless steel in the cold-worked condition. Deformation lines are evident, indicating that the material is work hardened. This material is virtually free of inclusions. Inclusions often contain sulfur, which is detrimental to pitting corrosion resistance.

Table 2 Mechanical properties of metals and alloys currently used as surgical implants

Metal or alloy	Yield strength MPa	Yield strength ksi	Tensile strength MPa	Tensile strength ksi	Elongation, %	Modulus of elasticity (E) GPa	Modulus of elasticity (E) ksi × 10³
316 stainless steel, annealed	207	30	517	75	40	. . .	. . .
316 stainless steel, cold worked	689	100	862	125	12	200	29
Cast cobalt-chromium alloy	450	65	655	95	8	248	36
Wrought cobalt-chromium alloy	379	55	896	130	. . .	242	35
Titanium, grade 4	485	70	550	80	15	110	16
Ti-6Al-4V, annealed	830	120	895	130	10	124	18
Ti-6Al-4V, heat treated	(a)		(a)		. . .	. . .	. . .
Tantalum, annealed	140	20	205	30	. . .	. . .	. . .
Tantalum, cold worked	345	50	480	70	. . .	. . .	. . .
MP35N, annealed	240–655	35–95	795–1000	115–145	. . .	228	33
MP35N, cold worked and aged	1585	230	1790	260	. . .	. . .	. . .

(a) Subject to agreement between purchaser and manufacturer. Source: Ref 17

The effects of alloying are important when variations in steels are under consideration. Iron exists in two different crystal structures (Ref 34), but there are two structure transitions as the structure transforms from the high-temperature body-centered cubic (bcc) δ-ferrite at 1390 °C (2534 °F) to the face-centered cubic (fcc) γ-austenite, which in turn transforms to bcc α-ferrite at 910 °C (1670 °F):

$$\text{bcc } \delta\text{-iron} \xrightarrow{1390\ °C} \text{fcc } \gamma\text{-iron} \xrightarrow{910\ °C} \text{bcc } \alpha\text{-iron}$$

Details of the metallurgy of stainless steels used for surgical implants are available in Ref 32. Chromium, which is the key element in the corrosion resistance of stainless steel, is a ferrite former. Carbon is an austenite stabilizer. Nickel is added to steel to stabilize the austenite phase. Austenitic stainless steels, such as type 316L, contain chromium and nickel. The minimum combined content of these elements is 23%; the minimum chromium content is 16%, and the minimum nickel content is 7%. Other austenitic stainless steels used widely in industry are types 302, 303, and 304. The type 316L stainless steel used for surgical implants contains 17 to 19% Cr and 12 to 14% Ni.

Molybdenum is added in amounts of 2 to 3% to strengthen the protective surface film in saline and acidic environments and to increase resistance to pitting. Molybdenum in amounts above 3% can reduce the corrosion resistance to strongly oxidizing environments and can result in the formation of some ferrite.

Carbon content in the type 316L surgical implant stainless steel should not exceed 0.08%. The greatest corrosion resistance is obtained when the carbon is in solid solution and when there is a homogeneous single-phase structure. A homogeneous austenitic structure can be produced by heat treating type 316L in the range of 1050 to 1100 °C (1920 to 2010 °F) and cooling rapidly. This rapid cooling (water quenching for large pieces) is essential for keeping the carbides in solution. Slower cooling allows chromium carbides to form at the grain boundaries and leaves the steel susceptible to intergranular corrosion. There is a depletion of chromium in areas adjacent to the grain-boundary carbides, and this further enhances corrosion in grain boundaries. The formation of these grain-boundary carbides in stainless steels is known as sensitization. The susceptibility of low-carbon stainless steels to sensitization is less than that of stainless steels with higher carbon contents.

When carbon is added to stainless steels containing little or no nickel and the stainless steel is heat treated, the result is the formation of tetragonal martensite, a nonequilibrium hard needlelike phase. Martensitic stainless steels are part of the type 400 series, which are not used for surgical implants.

Austenitic stainless steels are not ferromagnetic, they work harden easily, and corrosion resistance is acquired by the presence of a surface film. Magnetic alloys should not be used in the body, because they could become dislodged in a magnetic field. Ferrite should not be present in implants, not only from the standpoint of its corrosion resistance but also because it is magnetic. Magnetic-resonance imaging should not be used for an individual with any type of magnetic material in his body, and if possible, it

100 μm

Fig. 1 Microstructure of cold-worked AISI type 316L stainless steel. Etchant: 20 mL HCl, 10 mL HNO_3, and 3 g $FeCl_3$ for 2 to 5 s

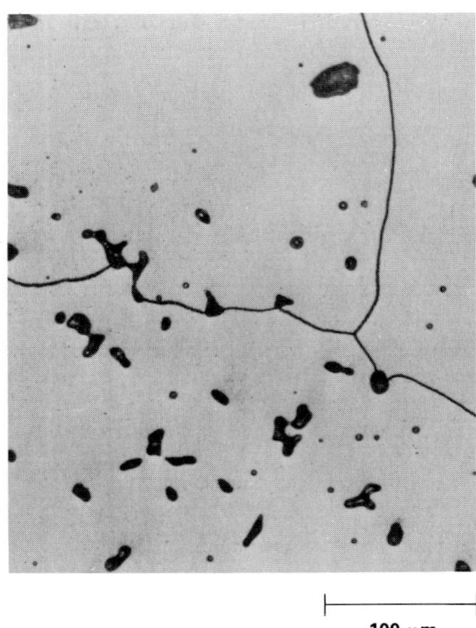

100 μm

Fig. 2 As-cast microstructure of cast cobalt-chromium-molybdenum alloy. Etchant: 20 mL HCl, 10 mL HNO_3, and 3 g $FeCl_3$ for 30 to 60 s

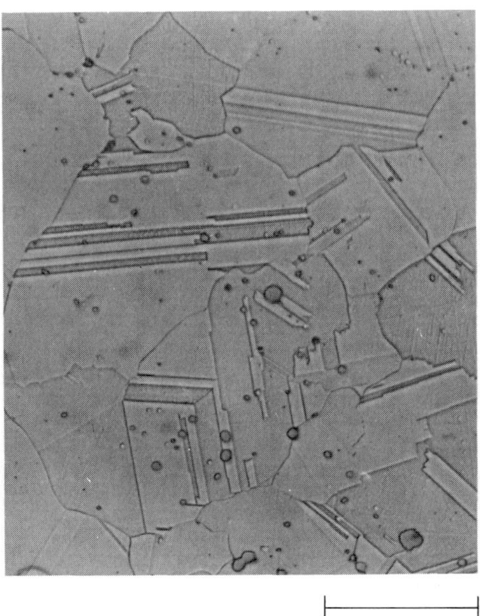

40 μm

Fig. 3 Microstructure of wrought cobalt-chromium-nickel-tungsten alloy. Electrolytic etch: 100 mL HCl and 0.5 mL H_2O_2 at 6 V for 10 s

should be avoided for any patient who has a metallic implant. Two possible problems are the heating of the metal and the distortion of position produced by the metal even though it is not magnetic.

The corrosion resistance of stainless steel can be improved by electropolishing. Electropolishing provides a uniform surface, removes surface defects that could serve as pit sites, and leaves a protective surface film. Another method of improving corrosion resistance is the development of a protective film by passivating the steel in a 20- to 40-vol% nitric acid (HNO_3) solution at 60 °C (140 °F) for 30 min (Ref 35).

Cobalt Alloys. Cobalt-chromium alloys were first studied in 1895 by E. Haynes (Ref 36, 37), who received U.S. Patents in 1907 for a cobalt composition range of 49 to 90% (Ref 38, 39). A report on the alloys appeared in 1913 (Ref 40). Since that time, cobalt-chromium alloys have been further developed and have been used in applications ranging from aircraft engines to surgical implants. The alloys are noted for high strength, good corrosion resistance, and good wear resistance. The influence of refractory metal additions on strengthening, carbide formation, and intermetallic compound formation is discussed in Ref 41. The cobalt-chromium alloys are used for surgical implants in both the cast and wrought forms, which are typified, respectively, by Haynes Stellite-21, a cobalt-chromium-molybdenum-carbon alloy, and Haynes Stellite-25, a cobalt-chromium-tungsten-nickel alloy. The cobalt-chromium alloys have various trade names.

Cobalt undergoes a structure transformation at 450 °C (842 °F) and is fcc above this temperature and hexagonal close-packed (hcp) below it (Ref 34). Mixtures of both structures usually exist at room temperature. The cobalt-chromium alloys are strengthened by solution hardening with refractory metals (elements such as molybdenum or tungsten) and by the addition of carbon to form carbide phases that strengthen the material by dispersion hardening and grain-boundary stabilization (Ref 41). The main carbide found in the alloys is $Cr_{23}C_6$, but other carbides, such as Cr_7C_3 and M_6C, are also present. The role of carbide phases in strengthening the material is discussed in Ref 42 and 43. Agglomeration or large amounts of these phases can reduce fatigue life. The nickel addition to the wrought material stabilizes the fcc phase and increases ductility. Normally, in the cast HS-21 material, both hexagonal (ϵ) and cubic (α) forms of cobalt can be present, along with some σ phase.

Typical microstructures of the cast, wrought, forged, high-strength, and hot isostatically pressed materials are shown in Fig. 2 to 5. The cast material shown in Fig. 2 has large grains with an interdendritic phase consisting of carbides, cobalt, and σ phase. Voids in cast materials can make the materials weaker. The size and shape of the grain and the location and amount of the carbides depend on the manner of casting and the rate of cooling. Thermomechanical treating of the cast material and annealing at 1065 °C (1950 °F) for 30 min will result in some improvement in mechanical properties (Ref 44).

Figure 3 shows the wrought material of the HS-25 composition. This material is strengthened by deformation, twinning, and small grain size. The forged high-strength alloy shown in Fig. 4 is closely related to the cast material (Fig. 2) in terms of composition. This material has been forged in a manner to produce a small, fine grain size and increased mechanical strength. The hot isostatically pressed material shown in Fig. 5 has a small grain size, a fine dispersion of carbides, and improved mechanical properties.

Porous materials are used as prosthetic devices when there is a need for attachment by ingrowth of soft or hard tissue. In one study, a cobalt-chromium alloy was sintered to produce a material with 30% porosity (Ref 19). This lowered the elastic modulus of the material, but the material was too weak. Another researcher sintered cobalt-chromium-molybdenum-carbon spheres to the surface of a substrate of the same material (Ref 21). This technique is used for the production of prosthetic devices. One problem associated with the production of cobalt-chromium-molybdenum-carbon sintered porous coatings that are sintered in the 1150- to 1300-°C (2100- to 2370-°F) range is the formation of carbide phases during sintering and cooling. An example of the sintered spheres on a solid substrate is shown in Fig. 6.

One extensive study of this problem investigated the relationship of microstructure to mechanical properties of the cobalt-chromium-molybdenum-carbon alloy (Ref 45). These studies of heat treating, cooling, microstructures, and mechanical properties showed that there are two methods of reducing the grain-boundary interdendritic and/or carbide phases responsible for decreasing ductility and ultimate tensile strength. The first method—cooling the material from a temperature above 1508 K to a temperature below 1508 K, then quenching—is not very practical, and the second method—reducing the carbon content to levels below 0.17%—results in a large decrease in yield strength. In addition, replacing some carbon by nitriding the alloy was found to increase yield strength (Ref 45, 46).

Alloy MP35M (35Co-35Ni-20Cr-10Mo), a cobalt-nickel alloy, has been used for surgical implants in recent years (Ref 47). Figure 7 shows the microstructure of this alloy in the hardened but aged condition. This material is strengthened by phase transformations induced by deformation, and it has yield strength values ranging from 414 MPa (60 ksi) for annealed material to 2128 MPa (309 ksi) for work-hardened and aged material. Work hardening involves the transformation of an fcc crystal structure to an hcp structure. Aging for 4 h in the 427- to 649-°C (800- to 1200-°F) range results in the precipitation of Co_3Mo and

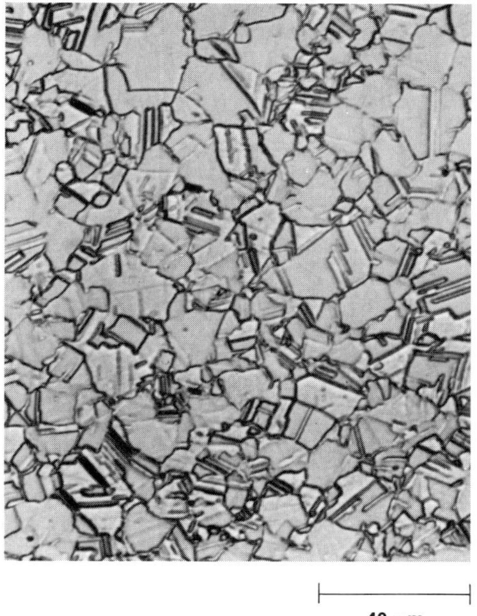

Fig. 4 Microstructure of forged high-strength cobalt-chromium-molybdenum alloy. Etchant: 20 mL HCl, 10 mL HNO_3, and 3 g $FeCl_3$ for 30 to 60 s

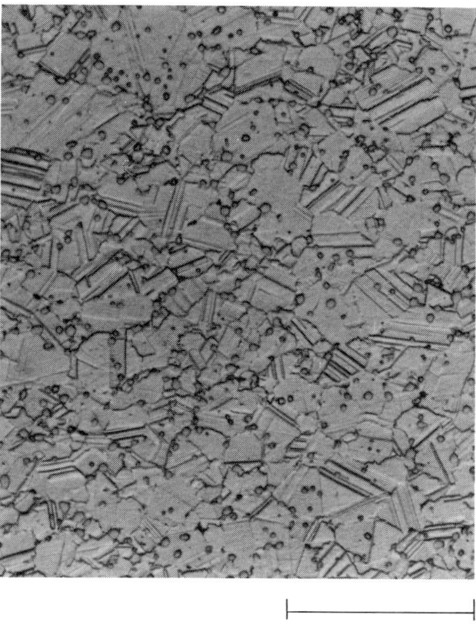

Fig. 5 Microstructure of hot isostatically pressed cobalt-chromium-molybdenum alloy. Etchant: 20 mL HCl, 10 mL HNO_3, and 3 g $FeCl_3$ for 30 to 60 s

Fig. 6 Microstructure of sintered porous coated cobalt-chromium-molybdenum alloy. Etchant: 20 mL HCl, 10 mL HNO_3, and 3 g $FeCl_3$ for 30 to 60 s

further strengthens the alloy; Co_3Mo has a hexagonal crystal structure and is formed by a peritectic reaction at 1020 °C (1868 °F).

Titanium and titanium alloys are relatively new materials compared to steels and cobalt-chromium alloys. Titanium was discovered in 1790 (Ref 48). The ninth most common element in the earth's crust, titanium, occurs as the minerals rutile (TiO_2) and ilmenite ($FeO{\cdot}TiO_2$), with $FeO{\cdot}TiO_2$ being found in larger deposits. Titanium is at least as strong as steel and is about 50% lighter. This high strength-to-weight ratio has made titanium and its alloys attractive materials for use in aircraft, aerospace, and marine applications.

Titanium is easily fabricated, but contamination with such interstitial elements as hydrogen, nitrogen, and oxygen should be avoided, because these elements have an embrittling effect on titanium. The development of the Kroll process in 1936 for extracting the metal and producing titanium sponge made it possible to produce the metal in commercial amounts (Ref 49); by 1948, it was commercially available in the United States. The Kroll process involves the chlorination of the ore to produce titanium tetrachloride ($TiCl_4$), which is reduced with magnesium metal in a sealed reactor with an inert atmosphere. Magnesium chloride ($MgCl_2$) is drained from the reactor vessel, leaving titanium sponge. The sponge is contaminated with ferric chloride ($FeCl_3$) and $MgCl_2$, which are then leached out. Purer titanium can be obtained by decomposing titanium tetraiodide (TiI_4) at high temperatures or by electrolytic reduction.

In the 1950s, titanium alloys were developed to answer the need for materials with high strength, low weight, high melting temperature, and high corrosion resistance for jet aircraft engines and airframe components. It was not until the 1960s that titanium alloys were used as surgical implant materials (Ref 10, 50-52). Use of titanium alloys in surgery has been growing steadily since the mid-1970s and continues to increase. An ASTM symposium on the use of titanium for surgical implants was held in 1981 (Ref 53). Figure 8 shows the microstructures of pure titanium, Ti-6Al-4V, and Beta III alloy (Ti-11.5Mo-6Zr-4.5Sn). Mechanical properties of the α-β alloys and the β alloys depend on such factors as heat treatment, quenching, and aging.

Titanium undergoes an allotropic transformation from an hcp structure (α phase) to a bcc structure (β phase) at 882 °C (1620 °F) (Ref 31, 34). As a result of this structural change, titanium alloys fall into three classes: α alloys, α-β alloys, and β alloys. Selected alloying additions, such as aluminum, oxygen, tin, and zirconium, are α stabilizers, and other elemental additions, such as vanadium, molybdenum, niobium, chromium, iron, and manganese, are β stabilizers. Aluminum acts as an α stabilizer in amounts up to 8%, but aluminum compositions higher than this can result in an embrittling titanium-aluminum phase change. The presence of oxygen as an interstitial element is damaging to toughness.

In one study, porous surface coatings were applied to a solid substrate by using arc plasma spraying (Ref 20). An example is shown in Fig. 9. These coatings and their properties were studied further in another investigation (Ref 54). The pore size of the arc plasma sprayed coating can be varied or can be graduated throughout a given coating. Wire mesh porous coatings are available that are sintered together and then sintered to the substrate. Attempts at sintering spherical powders to titanium alloys have not been very successful until recently, because of the need to sinter at high temperatures. Alloy Ti-6Al-4V should not be sintered at temperatures above the β transus temperature, because mechanical properties, especially fatigue, are adversely affected.

A titanium-nickel alloy known as Nitinol is used in dentistry and has potential use when a high strain recovery is needed, such as for braces

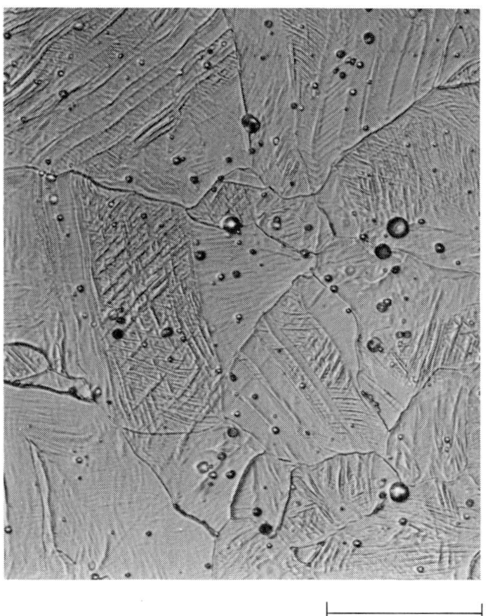

Fig. 7 Microstructure of MP35N cobalt-nickel-chromium-molybdenum alloy. Electrolytic etch: 80 mL $HC_2H_3O_2$ and 20 mL HCl at 6 V for 5 to 60 s

or restraining devices, or when the shape-memory effect of the alloy can be used. According to the shape-memory effect, an alloy wire that is shaped or bent at a given temperature and then reshaped at another temperature will return to the original shape when it is brought back to the shaping temperature. Other alloys, such as gold-cadmium and silver-cadmium, exhibit this effect (Ref 55).

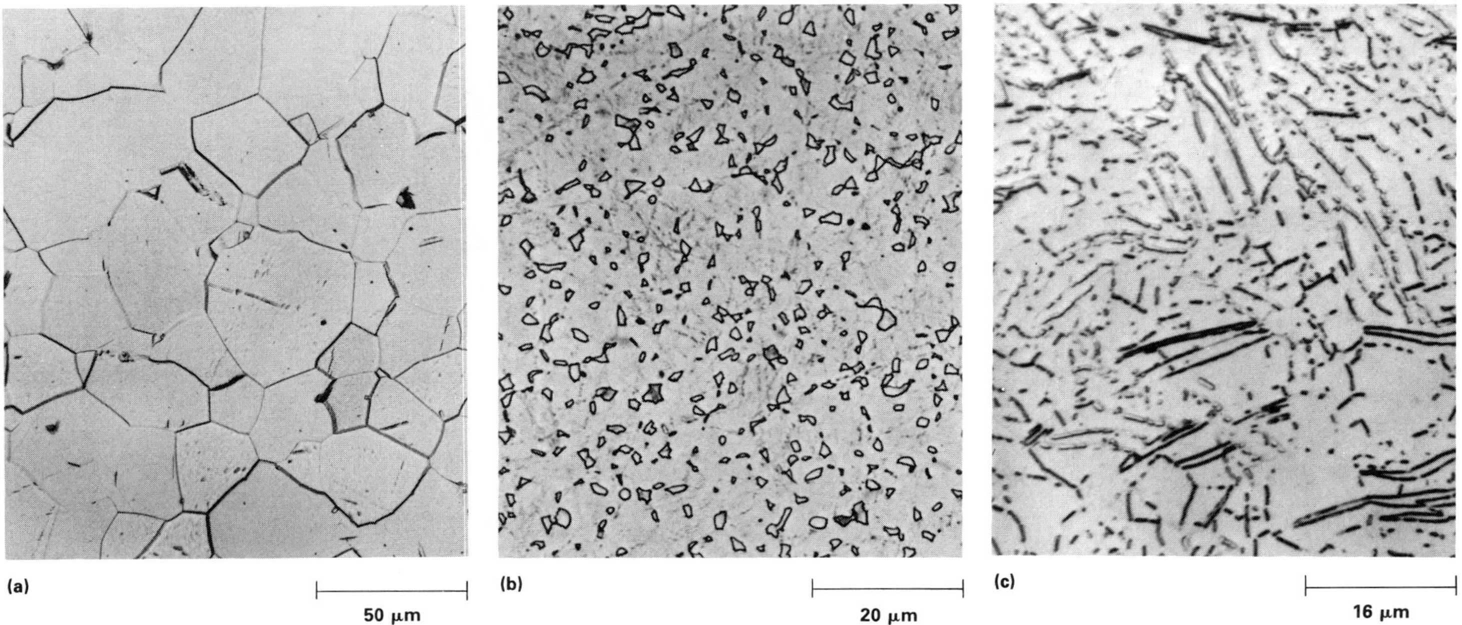

Fig. 8 Microstructures of titanium and titanium alloys. Kroll's etchant: 95 mL H_2O, 3.5 mL HNO_3, and 1.5 mL HF for 20 s. (a) Unalloyed titanium. (b) Ti-6Al-4V. (c) Beta III titanium

The shape-memory effect is associated with a martensitic transformation. Shape memory can also occur when the alloy is already in the martensitic condition. In this case, it probably occurs through a rearrangement of the martensitic plates.

There are two temperatures of concern: the transformation temperature and the reverse transformation (A_s) temperature. The transformation temperature, M_s, below which the material transforms to martensite, for stoichiometric titanium-nickel is approximately 650 °C (1200 °F) (Ref 56). The M_s temperature can be lowered by changing the titanium-nickel ratio or by alloying with other elements, such as cobalt. An M_s temperature more compatible with the body can be achieved by alloying titanium-nickel with titanium-cobalt. The reverse transformation temperature, A_s, for titanium-nickel begins at 165.6 °C (330 °F), and that for titanium-cobalt starts at −237.2 °C (−395 °F). The desired shape (for example, a twist in a wire) that the memory will hold is fixed above the M_s temperature, the shape is changed as needed below the temperature transition range (for example, the wire is straightened), the temperature is raised to the A_s temperature, and the original shape returns.

Beta-titanium alloys are not currently used for surgical implants, but several implant producers are considering these materials. The β-titanium alloys are metastable and are produced by adding alloying elements in sufficient amounts to stabilize the high-temperature bcc structure at room temperature. The mechanical properties of the material depend on the morphology and structure of high-modulus α particles in the low-modulus β matrix. The material can be processed to have a high fracture toughness. There are two types of β stabilizers (Ref 57). The addition of such β stabilizers as molybdenum, vanadium, tantalum, and niobium results in the formation of isomorphous α-titanium from the metastable β. Beta stabilizers such as chromium, manganese, iron, silicon, cobalt, nickel, and copper result in the formation of a eutectoid mixture of α and a compound. The structure and properties of the material can be controlled by alloying and thermomechanical processing. Some β alloys are Ti-13V-11Cr-3Al, Beta III, Ti-3Al-8V-6Cr-4Mo-4Zr, Ti-10V-2Fe-3Al, Ti-15Mo-5Zr-3Al, and Ti-15V-3Al-3Sn-3Cr. Many other possibilities exist for alloying to produce the β structure.

Pure titanium is used in reconstructive surgery and for purposes not subject to high loads. Pure titanium is also used for coatings.

Significance of Corrosion

Corrosion of metal implants is critical because it can adversely affect biocompatibility and mechanical integrity. The material used must not cause any adverse biological reaction in the body, and it must be stable and retain its functional properties. Corrosion and surface film dissolution are two mechanisms for introducing additional ions to the body. Extensive release of metal ions from a prosthesis can result in adverse biological reactions and can lead to mechanical failure of the device.

Metals used in the human body must have a high corrosion resistance and must not be treated or used in a configuration that would degrade the corrosion behavior. Degradation of metals and alloys used as surgical implant orthopedic devices is usually a combination of electrochemical and mechanical effects. Because surgical implants are being placed in younger people and because the older population is living longer, demands on the materials for good long-term durability and corrosion resistance are increasing.

Biocompatibility is the first consideration for materials of any type that are to be used in the body (Ref 58). Various *in vitro* and *in vivo* tests have indicated that selected metals are biocompatible and suitable for use as surgical implants. The long-term effects of metal ions on the body are less well known. Data are being collected, and studies continue in an effort to provide more information. It is desirable to keep metal ion release at a minimum by the use of corrosion-resistant materials. Some effects of incompatible materials include interference with normal tissue growth near the implant, interference with systemic reactions of the body, and transport and deposit of metal ions at selective sites or organs. There is always concern about the carcinogenic effects of foreign materials in the

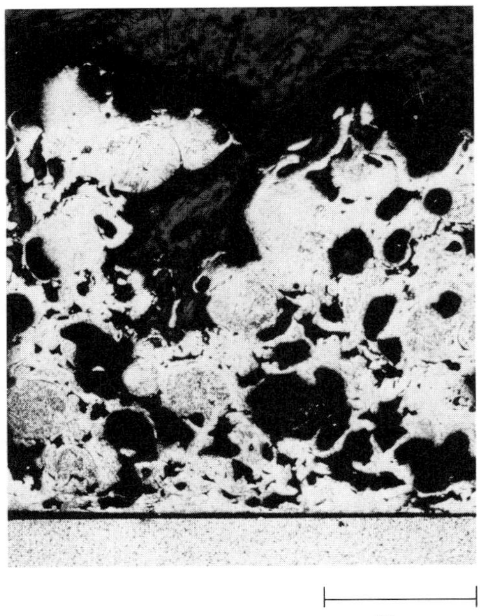

Fig. 9 Microstructure of the arc plasma sprayed coating on porous coated Ti-6Al-4V. Kroll's etchant

body, both short-term and after periods exceeding 20 years.

The first investigation of tissue tolerance to metals was made by studying wires implanted in dogs (Ref 59). Platinum was concluded to be the least irritating of the platinum, gold, silver, and lead tested. Biocompatibility testing was conducted by implanting metals in rabbits (Ref 60). Minimal tissue reaction was reported to cobalt-chromium alloys, type 316L stainless steel, and titanium. Some tissue reaction was credited to the size and shape of the implant. Some individuals are sensitive to metals, and some develop metal sensitivity at a later time after receiving an implant. Stainless steel implants sometimes cause rashes or pain; this is because of nickel ion release. Metal sensitivity is discussed in Ref 61.

Integrity of the Device. Orthopedic devices must maintain mechanical strength. Load-bearing implants at the lower extremities must support three or four times the body weight. Resistance to cyclic loading is important because the metal would probably be subjected to more than 3×10^6 cycles per year. Corrosion can lead to mechanical failure of the device.

Surface Effects and Ion Release

Metal ion release from surgical implants probably results primarily from corrosion. Another source of metal ions is passive films, which are thick and not stable in body fluids. Titanium is highly corrosion resistant, yet ion release occurs. Titanium ion release is discussed in Ref 62. Titanium that is subjected to various surface treatments comes to the same rest potential in saline solution, indicating that changes have taken place in the surface film (Ref 63). The effects of passivation and sterilization have been studied. Films placed by these methods on stainless steel and cobalt-chromium alloys reduced corrosion currents, but the effects were negligible with Ti-6Al-4V (Ref 64). In all cases, surface films exist that must be stable in the body or must change to a stable form. Metal ion release and retention of ions in the body are discussed in Ref 64 to 67.

Standards

The American Society for Testing and Materials instituted the F-4 Committee on Medical Devices and Materials. Task forces associated with this committee have developed 106 approved documents that appear in Ref 17. These documents include specifications for implant metals in terms of composition, mechanical properties, and other factors. Recommended methods for testing corrosion, mechanical properties, and other properties are also available in Ref 17.

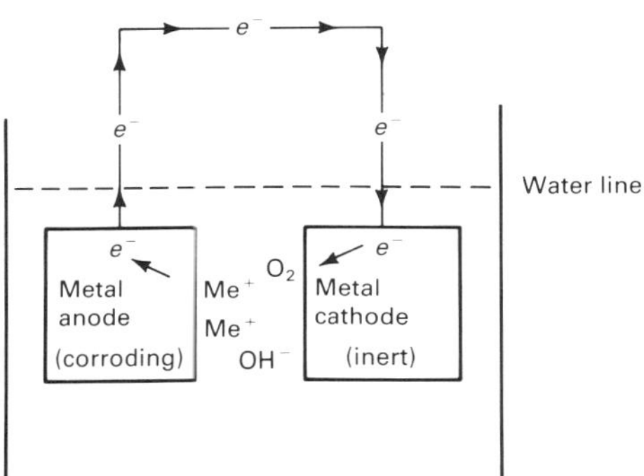

Fig. 10 Schematic of a galvanic-corrosion cell in an aqueous solution

Electrochemistry and Basic Corrosion Processes

Corrosion is the result of an electrochemical reaction of a metal with its environment. Chemical dissolution of the surface films does play a role and should not be ignored; however, the principal concern with most forms of corrosion is electrochemical.

Corrosion Reactions. Electrochemical deterioration of the metal occurs as positive metal ions are released from the reaction site (anode) and electrons are made available to flow to a protected site (cathode). The flow of electricity and material loss during this process obey Faraday's laws, resulting in the release of one equivalent weight of metal ions for every 96 500 C of electricity passed through the electrolytic solution.

The electrochemical reaction cell consists of two conducting and electrically connected electrodes in an electrolytic solution. The two electrodes can be dissimilar metals, or they can result from different surface areas of the same metal, defects, impurities, precipitate phases, concentration differences of gas, solution or metal ions, or other variables. A schematic of a corrosion cell is shown in Fig. 10, and a representative oxidation (anodic) reaction and some typical reduction (cathodic) reactions are:

$$\text{Metal (Me)} \rightarrow Me^{2+} + 2e^-$$

Anodic reaction (loss of electrons)

$$O_2 + 2H_2O + 4e^- \rightarrow 4OH^-$$

$$O_2 + 4H^+ + 4e^- \rightarrow 2H_2O$$

$$2H^+ + 2e^- \rightarrow H_2$$

Cathodic reactions (consumption of electrons)

Table 3 Standard emf series of metals

Metal-metal ion equilibrium (unit activity)	Electrode potential versus SHE at 25 °C (75 °F), V
Noble or cathodic	
$Au\text{-}Au^{3+}$	1.498
$Pt\text{-}Pt^{2+}$	1.2
$Pd\text{-}Pd^{2+}$	0.987
$Ag\text{-}Ag^{+}$	0.799
$Hg\text{-}Hg^{2+}$	0.788
$Cu\text{-}Cu^{2+}$	0.337
$H_2\text{-}H^{+}$	0.000
$Pb\text{-}Pb^{2+}$	−0.126
$Sn\text{-}Sn^{2+}$	−0.136
$Ni\text{-}Ni^{2+}$	−0.250
$Co\text{-}Co^{2+}$	−0.277
$Cd\text{-}Cd^{2+}$	−0.403
$Fe\text{-}Fe^{2+}$	−0.440
$Cr\text{-}Cr^{2+}$	−0.744
$Zn\text{-}Zn^{2+}$	−0.763
$Ti\text{-}Ti^{3+}$	−1.210
$Ti\text{-}Ti^{2+}$	−1.630
$Al\text{-}Al^{2+}$	−1.662
$Mg\text{-}Mg^{2+}$	−2.363
$Na\text{-}Na^{+}$	−2.714
$K\text{-}K^{+}$	−2.925
Active or anodic	

Source: Ref 74

Thermodynamics of Implant Metal Corrosion. Corrosion occurs because the metal oxide or corrosion product is more stable thermodynamically than the metal. There is always a tendency for metals to corrode, with the driving force being the electrode potential difference between the oxidation and reduction reactions. The thermodynamics and kinetics of corrosion of surgical implant metals are discussed in Ref 68 to 70. Corrosion in general is discussed in Ref 68, 69, and 71 to 74. More information on the thermodynamics and kinetics of aqueous corrosion is available in the article "Effects of Environmental Variables on Aqueous Corrosion" in this Volume.

The change in free energy ΔG associated with the electrochemical reaction can be written as:

$$\Delta G = -nFE$$

where n is the number of electrons, F is the Faraday constant (96 500 C), and E is the cell potential. A negative ΔG results in an increased tendency for the electrochemical reaction to occur.

The standard oxidation potentials of metals are measured from a bare metal surface. These potentials can be grouped to indicate active corrosion. This is known as the electromotive force (emf) series (Table 3). The galvanic series (Table 4) is based on reactions of the material with a specific environment. The galvanic series is more practical than the emf series. Table 4 lists the noble or

Table 4 Galvanic series of selected metals and alloys in seawater

Noble or cathodic

- Platinum
- Gold
- Graphite
- Titanium
- Silver
  - Chlorimet 3 (Ni-18Cr-18Mo)
  - Hastelloy C (Ni-17Cr-15Mo)
  - 18–8 stainless steel with molybdenum (passive)
  - 18–8 stainless steel (passive)
  - Chromium stainless steel 11–30% Cr (passive)
  - Inconel (passive)
  - Nickel (passive)
- Silver solder
  - Monel 400
  - Cupronickels (Cu-40Ni to Cu-10Ni)
  - Bronzes (Cu-Sn)
  - Copper
  - Brasses (Cu-Zn)
  - Chlorimet 2 (Ni-32Mo-1Fe)
  - Hastelloy B (Ni-30Mo-6Fe-1Mn)
  - Inconel (active)
  - Nickel (active)
- Tin
- Lead
- Lead-tin solders
  - 18–8 stainless steel with molybdenum (active)
  - 18–8 stainless steel (active)
- Ni-Resist (high-nickel cast iron)
- Chromium stainless steel, 13% Cr (active)
  - Cast iron
  - Steel or iron
- Aluminum alloy 2024
- Cadmium
- Aluminum alloy 1100
- Zinc
- Magnesium and magnesium alloys

Active or anodic

Source: Ref 67

Table 5 Anodic back series of selected metals and alloys in equine serum

Metal or alloy	Potential versus SCE, V
Titanium	3.5
Niobium	1.85
Tantalum	1.65
Platinum	1.45
Palladium	1.35
Rhodium	1.15
Iridium	1.15
Gold	1.0
Chromium-nickel-molybdenum alloy	0.88
Chromium-nickel-molybdenum alloy	0.875
Chromium	0.75
Chromium-cobalt-molybdenum alloy	0.75
Chromium-cobalt-nickel alloy	0.75
Chromium-cobalt-molybdenum alloy	0.65
Chromium-cobalt-molybdenum alloy	0.65
316L stainless steel	0.48
Nickel-chromium-iron alloy	0.35
Nickel-chromium-iron alloy	0.35
Zirconium	0.32
Nickel-chromium-iron alloy	0.25
Nickel	0.20
Nickel-chromium-aluminum-molybdenum alloy	0.16
Tungsten	0.12
Silver	0.11
Molybdenum	−0.020
Copper-nickel alloy	−0.020
Copper	−0.030
Vanadium	−0.070
Aluminum bronze	−0.080
Tin bronze	−0.090
Nickel silver	−0.10
Admiralty brass	−0.10
Brass	−0.11
Tin	−0.20
Antimony	−0.25
Nickel-molybdenum-iron alloy	−0.30
Nickel-molybdenum-iron alloy	−0.33
Cobalt	−0.35
Indium	−0.40
Aluminum-copper alloy	−0.50
Aluminum	−0.60
Cadmium	−0.65
Aluminum-magnesium alloy	−0.65
Zinc	−0.95
Manganese	−1.08
Magnesium	−1.55

Source: Ref 75

active tendencies of metals in seawater. In one study, measurements were conducted in equine serum; the values are given in Table 5.

The thermodynamic consideration of the relationship of electrode potential to corrosion is shown in Pourbaix diagrams (Ref 76), in which the electrode potential is plotted versus pH for a given temperature and other conditions (Ref 69). Most of these diagrams predict corrosion behavior for pure metals in water at 25 °C (75 °F). The pH of a solution describes the hydrogen ion (H^+) activity and is equal to $-\log[H^+]$. Although these diagrams are for pure metals, the information is useful for determining passivation and the reactions of alloys. These diagrams serve as a guide for determining corrosion, immunity, or passivity; however, they should be used with care, and the information should be correlated with experimental data when possible.

Figure 11 shows the Pourbaix diagram for titanium. The known areas of immunity, corrosion, and passivity are indicated. Dashed lines a and b refer to the representative equilibrium

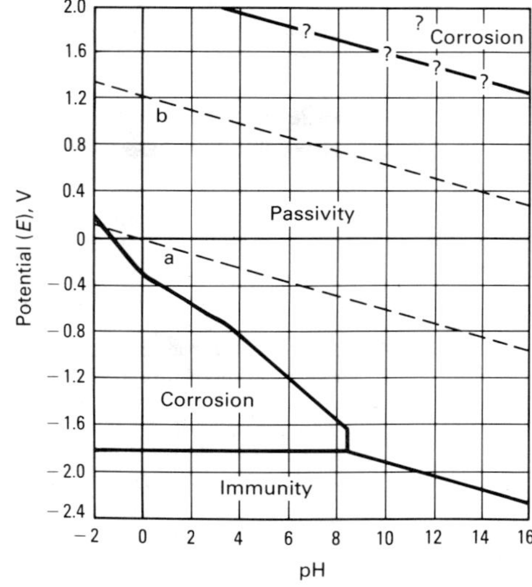

Fig. 11 Pourbaix (potential versus pH) diagram for titanium. Source: Ref 76

reactions $H_2 \rightarrow 2H^+ + 2e^-$ and $2H_2O \rightarrow O_2 + 4H^+ + 4e^-$, respectively. The hydrogen pressure at line a is 1 atm, and the oxygen pressure at line b is 1 atm. Oxygen concentrations in solution will be less below line b, and hydrogen concentration in solution will be less below line a.

Kinetics of Implant Metal Corrosion. The rate at which the corrosion reaction proceeds is related to environmental composition, environmental effects (such as motion or load), and other environmental factors. The rate at which a metal corrodes depends on kinetic factors that are important in determining whether a metal will corrode excessively. For example, there may be a strong driving force or potential difference for a given reaction to proceed, but because of the formation of an impervious surface film or other polarizing action, the reaction rate is slowed or practically stopped. When the polarization occurs mostly at the anode, the system is described as being under anodic control; if the reaction occurs mostly at the cathode, the system is under cathodic control. Figure 12 shows a schematic of an anodic polarization curve for metals that exhibit an active-passive transition.

The physiological environment contains chloride ions (Cl^-) and is controlled at a pH level of 7.4 and a temperature of 37 °C (98.6 °F). Following surgery, the pH can increase to 7.8, decrease to 5.5 (Ref 77), then return to 7.4 within a few weeks. Infection or hematoma can cause variations in the pH from 4 to 9. Physiological solutions are oxygenated and contain organic components in addition to the salts. These solutions are electrical conductors.

Surface films on the metals are dominated by certain elements—for example, chromium for the stainless steels and the cobalt-chromium alloys and titanium for the titanium alloys. Changes beyond a specified amount in chromium content of these materials can result in reduced corrosion resistance. Other metallurgical variables can influence corrosion behavior. The corrosion resistance of a material is specific to a number of factors, including composition, changes in metallurgical heat treatment, microstructural phases present, and surface finish.

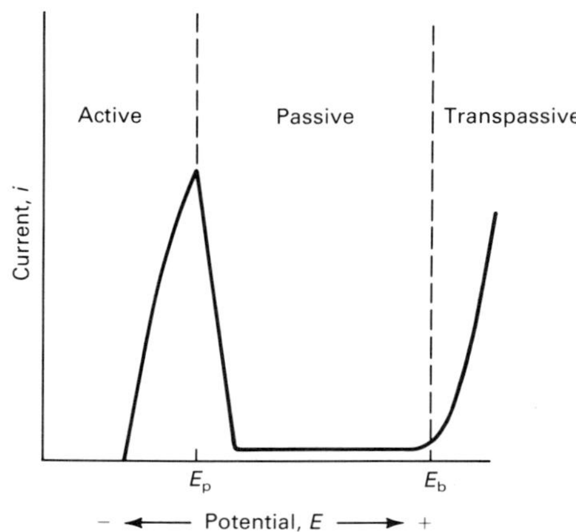

Fig. 12 Schematic of a potentiostatic polarization curve. See text for details

Forms of Corrosion in Implant Materials

The metals and alloys used as surgical implants achieve passivity by the presence of a protective surface film that inhibits corrosion and keeps current flow and the release of corrosion products at a very low level (Ref 69). The types of corrosion that are pertinent to the alloys currently used are pitting, crevice corrosion, corrosion fatigue, stress-corrosion cracking (SCC), fretting, galvanic corrosion, and intergranular corrosion. More information on all of these types of corrosion is available in the Section "Forms of Corrosion" in this Volume.

Pitting is a severe form of localized corrosion attack that results in extensive damage to the part and in the release of significant amounts of metal ions. Pits may be initiated at breaks in the protective film, defects in the material or protective film, inclusions, voids, and dislocations.

Pit initiation occurs when the protective film is broken by some means and exposes the metal to ions (such as Cl^-), body fluids, and water. Once the pit has initiated, metal ions form precipitates at the top of the pit and often form a film covering the pit. The film restricts entry of the solution and oxygen into the pit, and this makes repassivation, which would renew the protection, impossible. The small area of the pit and pit tip are anodic to the rest of the material, which is cathodic; this results in a high corrosion current density at the base of the pit. Movement of metal ions or H^+ ions from the bottom of the pit is restricted by the film covering the top of the pit. As a result, the pH at the bottom of the pit is lowered to the more acid range, and pitting is accelerated. Figure 13 shows an illustration of a pit and gives selected reactions.

Pitting cannot be tolerated in surgical implant metals. The first corrosion test procedure developed by the ASTM F-4/G-1 Joint Committee on Corrosion of Implant Metals was a test to screen passive metals and alloys for resistance to pitting or crevice corrosion (Ref 78). The document specifies an electrode potential stimulation test that is conducted in 0.9% sodium chloride (NaCl) solution at 37 °C (98.6 °F). A Teflon washer is placed on the specimen to create a crevice; therefore, pitting and crevice testing are carried out simultaneously. This test shows that type 316L stainless steel, which was used as a refer-

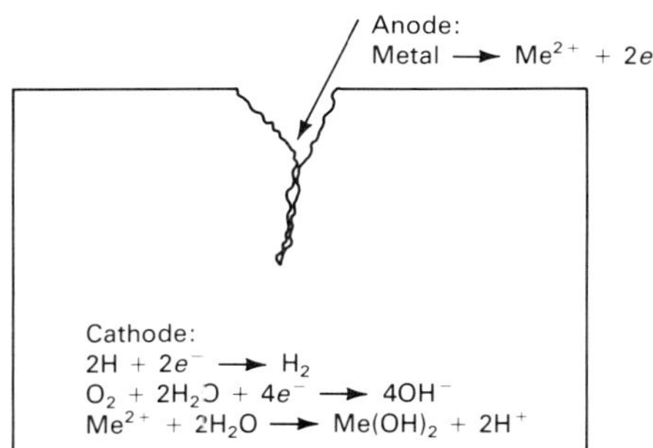

Fig. 13 Schematic of pitting corrosion

ence material, has a pitting potential of 0 ± 50 mV versus a saturated calomel electrode (SCE). Cobalt-chromium-molybdenum-carbon, titanium, and Ti-6Al-4V do not pit or form crevices in this test, indicating that they are not readily subject to pitting or crevice corrosion.

The importance of assessing pitting corrosion resistance by determining pit propagation rate (PPR) curves is discussed in Ref 79 and 80, which also review the usual tests for pitting. These tests involve anodically polarizing the specimen to a breakdown potential, E_b, at which pitting ensues. If the potential is reversed and in the opposite direction, a protection potential, E_p, is reached below which pits repassivate. If the corrosion potential, E_{corr}, is below E_p, pitting does not occur; if E_{corr} is above E_p, pitting occurs immediately upon immersion. Pitting resistance based on determinations of E_b and E_p have the disadvantage that the value of E_p changes, depending on the amount of pit growth and the length of time of pitting.

The PPR curves are an accurate technique for collecting pitting corrosion data. The PPR data are obtained by applying to the specimen a preselected potential between E_b and E_p and holding at this potential for 10 min. No pitting should have occurred, and the recorded current should be from general corrosion. The applied potential is then increased beyond E_b until a nominal current of 10 mA/cm^2 (64.5 mA/in.^{2}) is reached. The next step is to decrease the applied potential to the same preselected potential and hold there for 10 min. The recorded current is a measure of the rate of general corrosion in the nonpitted areas plus the rate of pit growth. The pits can be repassivated by returning the applied potential to the value of the cathodic potentials, E_c, for 5 min. The process can then be repeated. Some assessment of the pitted area can be made microscopically to determine pit growth area and average pit propagation current densities. This technique can also be used to study crevice corrosion. Application of this method to the study of implant metals is reported in Ref 81.

In one study, hysteresis effects were found upon reversing the applied potential in the polarization scanning when testing passivated cast cobalt-chromium-molybdenum alloy (Ref 82). This normally indicates pitting. However, because no pits were observed on the surface and because the hysteresis effect was not present on the unpassivated material, pitting was ruled out, and the hysteresis in the polarization curves for passivated specimens was attributed to a breakdown of the existing surface film, followed by repassivation. Other cyclic polarization measurements on cobalt-chromium-molybdenum alloy did not show hysteresis (Ref 81).

In another study, potentiodynamic scans were conducted on cobalt-chromium materials; a marked hysteresis was observed with cast cobalt-chromium-molybdenum (Ref 83). It was concluded that the material was subject to crevice attack. However, *in vivo* experiments of cobalt-chromium crevice corrosion test specimens in dogs and monkeys conducted over a period of 2 years found no evidence of crevice corrosion (Ref 84). In addition, accelerated corrosion tests of the relative corrosion resistance of cobalt-nickel-chromium-molybdenum-titanium and cobalt-chromium-molybdenum alloys showed that the materials have comparable corrosion behavior (Ref 85).

Crevice corrosion is a local attack. It occurs when a metal surface is partially shielded from the environment. Crevice corrosion can occur on metals that would otherwise be resistant to pitting and other corrosion, and it often occurs at a threaded or other type of junction. Damaging ion species accumulate in the crevice, and an environment similar to that of a pit eventually develops. Crevice corrosion problems can often be eliminated by changing the design of a device.

Corrosion fatigue is a fracture or failure of metal that occurs because of the combined interaction of electrochemical reactions and mechanical damage. Corrosion fatigue resistance is an important factor of consideration for load-bearing surgical implant metals or for metals used in cyclic-motion applications. Many corrosion fatigue failures would not occur without the combined, complementary action of these factors. Normally, a failure would not occur, but cracks can initiate from hidden imperfections, surface damage, minute flaws, chemical attack, and other causes. The corrosive environment may result in local corrosive attack that accentuates the effect of the various imperfections. The corrosive attack will be influenced by solution type, solution pH, oxygen content, and temperature.

Stress cycle frequency does not significantly influence the fatigue resistance of materials, and it is convenient and expedient to test at high frequencies. This is not the case for corrosion fatigue. The effects of corrosion depend on the stress cycle frequency, and they increase with decreasing frequency. Frequency is an important factor in the evaluation of corrosion fatigue resistance, and materials should be tested at frequencies identical to those encountered in the use of the material (Ref 67). Figure 14 shows an *S/N* curve (stress or applied shear strain amplitude/number of cycles to failure). Corrosion is a dominant process for the material represented by curve B, which does not have a corrosion fatigue limit below which failure would not occur.

Fatigue strength measured in aqueous media is usually less than fatigue strength measured in air, as demonstrated in several investigations. In one study, corrosion fatigue tests were conducted on type 316L stainless steel in synthetic physiological solution held at a temperature of 37 °C (98.6 °F) and with a pH of 7.6 ± 0.2 in axial loading with the stress ratio, $R = 0$, at a frequency of 140 Hz (Ref 86). A 10 to 15% decrease in fatigue strength for a given endurance level resulted when testing in the solution instead of in air at 37 °C (98.6 °F) and 25% relative humidity. Fatigue strength was estimated to be 20 to 30% lower when determined at frequency of 1 Hz. Fatigue bending tests were carried out in Ringer's solution (a lactated 0.9% NaCl solution containing about the same concentration of Cl^- ions as body fluids) at a frequency of 1 Hz on type 316L stainless steel and a cobalt-chromium alloy (Ref 87). Results showed a reduction in fatigue strength of both materials when tested in solution:

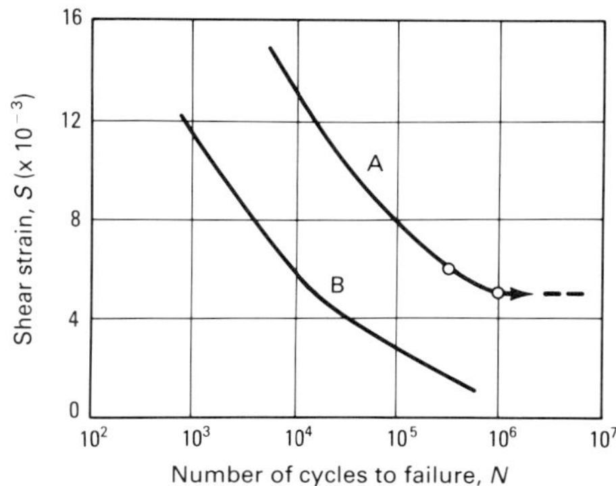

Fig. 14 *S-N* curve. Curve A, the influence of corrosion is small; Curve B, corrosion is occurring.

Alloy	Cycles to failure (40-kg/mm^2, or 57-ksi load) Air	Ringer's solution
Type 316L stainless steel	5 × 10^5	9 × 10^4
Cobalt-chromium	1.5 × 10^7	3.3 × 10^5

Other studies involving the determination of crack growth rates in Ringer's solution showed that cast cobalt-chromium-molybdenum alloys are susceptible to corrosion fatigue in the human body (Ref 88). Crack growth rates increased, and crack growth threshold levels decreased in solution over the values found for air. Failure was attributed to the synergistic effects of corrosion and cyclic stress, but hydrogen embrittlement was not ruled out. Controlled-potential slow strain rate tests conducted in Ringer's solution demonstrated that cobalt-chromium-molybdenum alloys are susceptible to hydrogen embrittlement (Ref 89).

Alloy Ti-6Al-4V was studied by using reversible torsion corrosion fatigue testing with a selected applied shear strain amplitude at a frequency of 1 Hz in Hanks's physiological solution at 37 °C (98.6 °F) and pH of 7.4 (Ref 90). The results were compared with those obtained in air (Ref 91) and showed only a slight decrease in fatigue strength in the solution. The tests in air were conducted at a frequency of 0.2 Hz and at a higher shear strain amplitude (±0.020 compared to ±0.18 in aqueous solution). These torsion tests conducted on type 316L, cobalt-chromium-molybdenum, and Ti-6Al-4V in Hanks's solution showed Ti-6Al-4V to have superior corrosion fatigue resistance over type 316L and cobalt-chromium-molybdenum.

Previous work found this same ranking of the corrosion fatigue resistance in saline solutions (Ref 92). Another study reported that rotating-bending tests at a frequency of 100 Hz in air gave fatigue strengths of 600 to 660 MPa (87 to 96 ksi) for Ti-6Al-4V, 500 to 800 MPa (73 to 116 ksi) for cobalt-chromium-molybdenum, and 400 to 780 MPa (58 to 113 ksi) for cobalt-nickel-chromium-molybdenum (Ref 93). The wide variation in fati ue strength is due to tests on different material of the same composition that were heat treated or processed in different ways and shows a range of strength available for these materials. A minimum rotating-bending fatigue strength needed for a hip prosthesis was estimated at 400 MPa (58 ksi) (Ref 93).

Stress-corrosion cracking is a form of localized corrosion that occurs when a metal is simultaneously subjected to a tensile stress and a corroding medium (Ref 94). A single mechanism to explain the process of SCC has not been found, but several mechanisms have been shown to be important in this complex interaction of electrochemical, mechanical, and material factors. One mechanism is the occurrence of preferred anodic dissolution at the crack tip due to high plastic deformation, which causes rupture of the protective film. Another mechanism involves the adsorption of ion species at strained areas in the crack tip, and this results in weak bonds within the metal and the film. In other cases, cracking appears to propagate as a series of brittle fracture events.

Stress corrosion differs from corrosion fatigue in mode of cracking and in the application of the stress. Stress-corrosion cracks are often branched, but fatigue cracks follow a more direct path with some striations and other microstructural effects. The loading in SCC is static, but the loading in corrosion fatigue is cyclic. Hydrogen embrittlement failure also results from static stress, but in this mechanism of failure, the adsorption of hydrogen and the production of a brittle region at the crack tip are required. Hydrogen can be produced from cathodic reactions. As a result, cathodic protection, which can be used to prevent SCC, cannot be used to stop hydrogen embrittlement.

Stress-corrosion cracking, corrosion fatigue, and hydrogen embrittlement failures can be impeded by a number of methods. The first is to select a material that is not susceptible to this type of failure in the body. Other methods of increasing immunity are to shot peen the surface to produce compressive stresses, to remove surface flaws that could act as crack initiation sites, and to heat treat the material to produce optimum microstructures for resisting mechanical damage.

Fretting occurs when two surfaces are in contact and experience small-amplitude relative oscillatory motion. Fretting corrosion involves wear and corrosion, in which particles removed from the surface form oxides that are abrasive and increase the wear rate. In some cases, the wear debris may be soluble, and increased wear would not occur. Instead, chemical reactions would play a larger role in the deterioration of the material.

Fretting corrosion can have significance for metallic surgical implants because it offers mechanisms of metal ion release in the body and of mechanical failure. Fretting destroys the passive film by which the implant metal achieves its corrosion resistance. Fretting corrosion is extensively discussed and is related to many applications in Ref 95. Fretting corrosion has occurred with plate and screw portions of prosthetic devices, and it could be the cause of fatigue fractures of these appliances (Ref 96). Reference 97 details a test method for measuring fretting corrosion of osteosynthesis plates and screws, and a test method is being developed for measuring fretting corrosion of surgical implant metals.

Galvanic corrosion is different from the various forms of corrosion discussed. It concerns the reactions of dissimilar metals. The electrochemical reaction that results when two dissimilar metals are in contact depends on the difference in potential of the two metals; the less noble metal becomes the anode, and the other the cathode. Table 5 (galvanic series) lists the corrosion potentials of implant metals. This information on the corrosion tendency, coupled with measurements of the corrosion rate and other factors, can help predict reactions of these metals in physiological use. The area ratio of the anode to cathode must be taken into account when making this assessment of corrosion behavior. A large anode coupled to a small cathode could produce a low current density, which would indicate a low corrosion rate. The opposite conclusion could be made if the sizes of the anode and cathode were reversed.

The corrosion behavior of coupled metals is discussed in Ref 98, and corrosion tests of coupled metals are reported in Ref 99. These discussions indicate that in some cases coupled metals can be used for surgical implants. It is possible for the couple to result in enhanced protection; this is the case for titanium with platinum in acid solutions (Ref 67, 69), in which the platinum in the couple causes the potential to be in the passive region of the polarization curve and not active as it was for titanium alone. Expert knowledge of corrosion principles and corrosion testing is required to determine the suitability of coupled metals for surgical implants. If this expertise is not available or if there are any uncertainties, it is best to avoid the use of coupled dissimilar metals.

Intergranular corrosion occurs when the grain boundary becomes anodic or cathodic to the rest of the grain. The change in composition in the grain boundaries may be due to precipitated grain-boundary phases, concentration of impurities, or elemental depletion near the grain-boundary area. The disorder and higher energy of the grain boundaries also provides a means for collecting second phases or contaminating materials that can lead to corrosion.

Corrosion Tests for Evaluating Implant Metals

The experimental apparatus for conducting corrosion measurements can be as simple as that shown in Fig. 15. The components of the apparatus can be altered for making corrosion fatigue, stress corrosion, or other specific measurements. The measurements can be done manually or automatically.

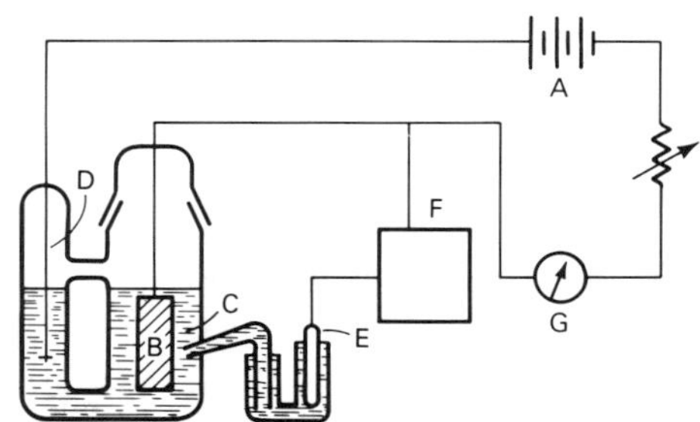

Fig. 15 Schematic of polarization testing apparatus. A, battery or power supply; B, test specimen; C, test environment; D, counter electrode; E, reference electrode; F, high-impedance volt meter; G, ammeter. Source: Ref 69

Corrosion measurements are sensitive to any changes in environment, specimen condition, or electrochemical disturbance. The corrosion process is specific to a given set of conditions. Several factors influence the results of corrosion tests regardless of the type of test being performed:

- Specimens cannot be reused for testing except when the effects of testing can be removed
- Specimens that have been in solution and then exposed to the air should have the surfaces prepared again before testing
- Specimens that are accidentally disturbed electrically during the test will not give a reproducible result
- Surface preparation of the specimen should be kept constant for a given set of experiments; a highly polished surface will be more uniform and will provide a more accurate measure of the surface area
- The specimen surface should be clean and free of all contaminants, including fingerprints
- The pH of the testing solution should be monitored at least before and after the test. Solution pH affects the corrosion processes and the stability of the surface films present
- The oxygen content of the solution should be kept constant by aerating or by deaerating
- The temperature of the solution should be monitored and kept constant
- The reference electrode should be handled with care, kept in good condition, and checked for accuracy. Voltages should be given in reference to an electrode, such as the standard hydrogen electrode (SHE) or the SCE
- All leads should be shielded at the air/solution interface. This is also true for the specimen. It must be totally immersed
- The amount of time between each step in potential and the voltage increment of each step should be kept constant to obtain reproducible results. The time required to reach a reasonably steady current will vary for different metals. It is possible to step the potential too rapidly, and active regions or other variations in the polarization curve would not be recognized
- Caution should be exercised and care taken not to touch any electrical leads in all tests and especially in galvanostatic measurements, because the voltage can produce electric shock.

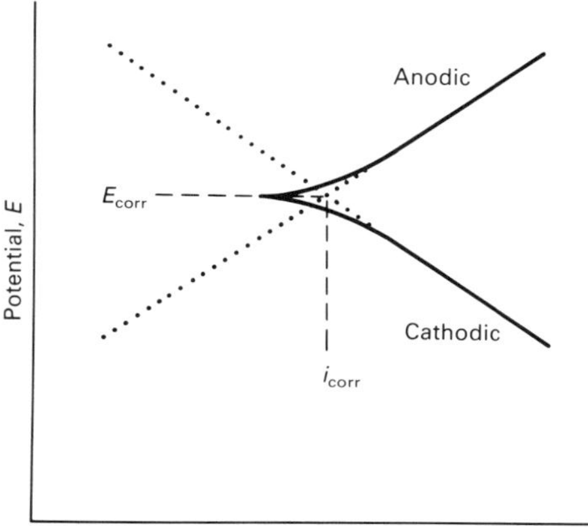

Fig. 16 Schematic of a potential versus log current density curve for anodic and cathodic galvanostatic polarization curves. Source: Ref 69

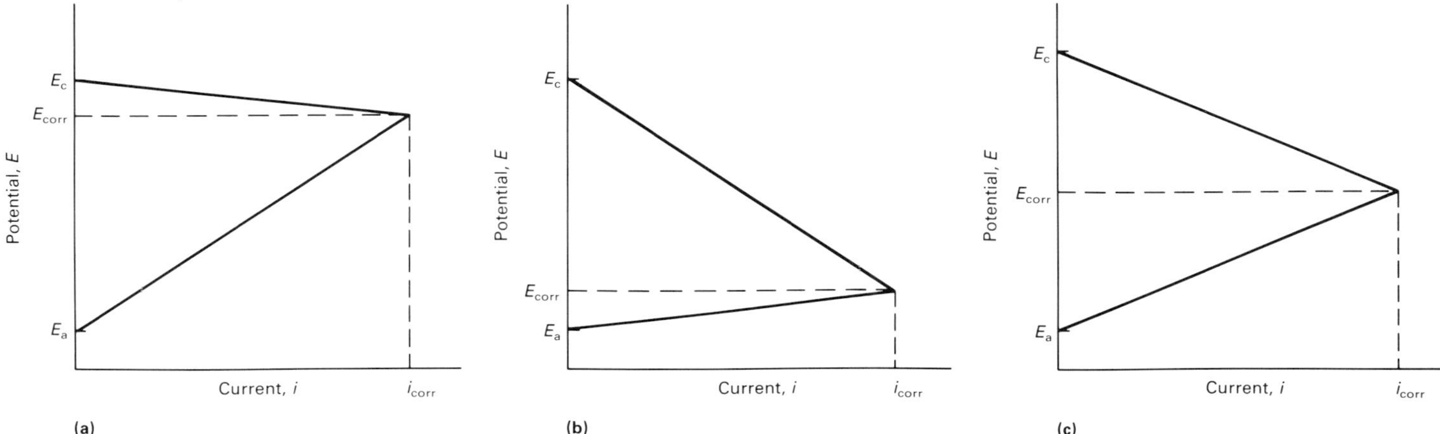

Fig. 17 Evans diagrams showing three types of corrosion rate control. (a) Anodic control (controlled by reactions at the anode). (b) Cathodic control (controlled by reactions at the cathode). (c) Mixed control (controlled by a mixture of both anodic and cathodic reactions). E_{corr}, corrosion potential; E_c, cathodic potential; E_a, anodic potential

The one-hand rule should be observed, and the power always should be turned off before touching the leads and equipment

Galvanostatic polarization measurements are performed by applying a current and measuring the resulting potential. This technique is used for Tafel curve and linear polarization measurements, and its applications include corrosion rate measurements. Tafel curves and Tafel slopes are determined by applying currents ranging from 1 μA to 10 mA in small uniformly spaced increments and recording the steady-state potential after 2 to 5 min. The Tafel slope is evidenced by the straight line portion of this curve. There may be a deviation at the beginning of the curve due to anodic dissolution and a deviation at the latter part of the curve due to concentration polarization and resistance effects between the Luggin probe and the electrode.

The experimental determination and theoretical implications of galvanostatic polarization measurements are discussed in Ref 69. Figure 16 shows a plot of potential versus log current. These curves are obtained by placing the specimen in an aqueous environment, measuring the open-circuit potential after a designated time, and then applying a current. This open-circuit potential is also the corrosion potential E_{corr}. The anodic curve is determined by applying positive currents in small increments and measuring the resulting potential for each increment. The cathodic curve is established in the same manner by applying small negative-current increments and measuring the resulting potential. The linear portions of the curves can be extrapolated and will intersect at E_{corr}.

Further extrapolation of these lines brings them to the potentials for the reversible anodic and cathodic reactions. These reactions determine the signs of the corrosion cell, with the anode being negative and the cathode positive. The extension of the anodic line to the reverse potential brings it to a potential at which the metal is in equilibrium with its ions, and the extension of the cathodic line brings it to the reversible potential for the cathodic reaction. Figure 17 shows plots of these extended lines, known as Evans diagrams.

Polarization resistance measurements are described in Ref 100, which also includes the derivation of an equation to relate polarization resistance and Tafel slopes to corrosion rate. The Stern-Geary equation has been widely used in the determination of corrosion rates. Galvanostatic polarization resistance measurements are described in Ref 101; these measurements do not disturb the system. The use of an ac impedance technique that superimposes a small amplitude overpotential between the test electrode and the reference electrode is discussed in Ref 102 and 103; this superimposition causes a phase shift or perturbation in the current. The ac impedance method is also nondestructive to the specimen and uses the Stern-Geary equation to determine corrosion current density.

Potentiostatic anodic polarization measurements are performed by applying a potential and measuring the resulting current. The applied voltage increment should be small. A recommended rate of applying the potential is 0.006 to 0.012 V/min. This technique is especially suitable for measuring the corrosion behavior of active-passive metals, such as those used for surgical implants. These measurements will show the length of the passive region in terms of applied potential, the breakdown potential, and the magnitude of the current in the passive region. The effects of environment, such as solution pH and organic constituents, can be studied with this method. Figure 12 shows a schematic of an anodic polarization curve.

Another application of the anodic polarization technique involves studies of repassivation kinetics (Ref 104-106). The term repassivation was coined to describe events occurring after surface film rupture in SCC. Repassivation involves the formation of a protective film over the abraded or clean metal surface. Repassivation measurements are made in a device that is designed to permit abrasion of the specimen, apply a preselected potential, and take a measurement (after removal of the abrader) of the anodic current transient by using an oscilloscope or a computer. Curve fitting or analysis of the current decay versus time can be carried out to provide information on the kinetics of the film formation and the stability of the film formed.

Other Corrosion Tests and Chemical Analysis. The electrochemical techniques described in the polarization measurements can be used to apply or measure electrode potentials and currents in corrosion fatigue, stress corrosion, pitting, and other tests. Chemical analysis of corroding solutions can be made with atomic absorption, and surface analysis can be made with several techniques, including x-ray photoelectron spectroscopy and Auger analysis. Chemical analysis can provide data on the amount of metal ions released, the oxidation state of the ions, and the nature of the surface film on the metal after testing.

REFERENCES

1. C.O. Bechtol, A.B. Ferguson, Jr., and P.G. Laing, *Metals and Engineering in Bone and Joint Surgery*, The Williams and Wilkins Company, 1959
2. C.S. Venable and W.G. Stuck, *The Internal Fixation of Fractures*, Charles G. Thomas, 1947
3. D.C. Ludwigson, Requirements for Metallic Surgical Implants and Prosthetic Devices, *Met. Eng. Quart.*, Vol 5 (No. 3), Aug 1965, p 1-6
4. T.P. Hoar and D.C. Mears, Corrosion-Resistant Alloys in Chloride Solutions: Materials for Surgical Implants, *Proc. R. Soc. (London)*, SCR.A294 (1439), 1966, p 486-510
5. D.F. Williams and R. Roaf, *Implants in Surgery*, W.B. Saunders, 1973
6. Icart, Letter in Response to the Memorandum of Mr. Pujol, *J. Med. Chir. et Pharm.*, Roux 44, 1775, p 169
7. H. Hansmann, A New Method of Fixation of Fragments in Complicated Fractures, *Verein Deutsches Gesellschaft fur Chirurgie*, Vol 15, 1886, p 134
8. J. Lister, *Brit. Med. J.*, Vol 2, 1883, p 855
9. F.H. Jergesen, "Studies of Various Factors Influencing Internal Fixation as a Method of Treatment of Fractures of the Long Bones," National Research Council, Dec 1951
10. G.C. Leventhal, Titanium—A Metal for Surgery, *J. Bone Joint Surg.*, Vol 33, 1951, p 473
11. O.T. Bailey, F.D. Ingraham, P.S. Weadon, and A.F. Susen, Tissue Reaction to Powdered Tantalum in the Central Nervous Sys-

tem, *J. Neurosurg.*, Vol 9, 1952, p 83

12. A.M. Mirowsky, L.A. Hazouri, and D.T. Greener, Epidural Granulomata in Presence of Tantalum Plates, *J. Neurosurg.*, Vol 7, 1950, p 485
13. F. Masmonteil, The Tolerance of Bone for Metallic Foreign Bodies, *Pressé Med.*, Vol 43, 1935, p 1915
14. C.S. Venable, W.P. Stuck, and A. Beach, The Effect on Bone of the Presence of Metals Based on Electrolysis, *Ann. Surg.*, Vol 105, 1937, p 917
15. T.M. Brown, Luigi Galvani, in *Encyclopedia Americana*, Vol 12, Grolier, 1984, p 256; B. Dibner, A. Volta in *Encyclopedia Americana*, Vol 28, Grolier, 1984, p 277
16. J. Charnley, *J. Bone Joint Surg.*, Vol 42B, 1960, p 28
17. *Medical Devices*, Vol 13.01, *Annual Book of ASTM Standards*, American Society for Testing and Materials
18. Medical Device Ammendments, Public Law 94-295, United States Food and Drug Administration, Bureau of Medical Devices and Radiological Health, 1976
19. J.S. Hirschhorn and J.T. Reynolds, Powder Metallurgy Fabrication of Cobalt Alloy Surgical Implant Materials, in *Research in Dental and Medical Materials*, Plenum Press, 1969, p 137-150
20. H. Hahn and W.J. Palich, Preliminary Evaluation of Porous Metal Surfaced Titanium for Orthopedic Implants, *J. Biomed. Mater. Res.*, Vol 4, 1970, p 571
21. R.P. Welsh, R.M. Pilliar, and I. MacNab, Surgical Implants: The Role of Surface Porosity in Fixation to Bone and Acrylic, *J. Bone Joint Surg.*, Vol 53-A (No. 5), 1971, p 963-967
22. J. Galante, W. Rostoker, R. Luek, and R.D. Ray, Sintered Fiber Metal Composites as a Basis for Attachment of Implants to Bone, *J. Bone Joint Surg.*, Vol 53-A (No. 1), 1971, p 101-114
23. A.C. Fraker, A.W. Ruff, A.C. Van Orden, H. Hahn, A.J. Bailey, and C.D. Olson, "Studies of Porous Metal Coated Surgical Implants," NBSIR 85-3166, National Bureau of Standards, June 1985
24. P. Ducheyne, L.L. Hench, A. Kagan, M. Martens, A. Burssens, and J.C. Mulier, Effects of Hydroxyapatite Impregnation on Skeletal Bonding of Porous Coated Implants, *J. Biomed. Mater. Res.*, Vol 14, 1980, p 225-237
25. N.N. Salman, "The Effect of Direct Electrical Current Stimulation on the Bone Growth Into Porous Polymeric, Ceramic and Metallic Implants," Ph.D. dissertation, Clemson University, 1980
26. C.T. Brighton, Symposium on Electrically Induced Osteogenesis, *Orth. Clinics N. Am.*, Vol 15 (No. 1), 1984
27. E.P. Lautenschlager, N.K. Sarker, A. Acharya, J.O. Galante, and W. Rostoker, Anodic Polarization of Porous Fiber Metal, *J. Biomed. Mater. Res.*, Vol 8, 1974, p 189
28. H.V. Cameron, R.M. Pilliar, and I. MacNab, Porous Vitallium in Implant Surgery, *J. Biomed. Mater. Res.*, Vol 8, 1974, p 283
29. L.C. Lucas, J.E. Lemons, J. Lee, and P. Dale, In Vitro Corrosion Characteristics of Co-Cr-Mo/Ti-6Al-4V/Ti Alloys, in *Quantitative Characterization and Performance of Porous Implants for Hard Tissue Applications*, STP 953, American Society for Testing and Materials, 1987
30. D.F. Williams, *Biocompatibility of Clinical Implant Materials*, Vol 1, CRC Press, 1981
31. M. Hanson and K. Anderko, *Constitution of Binary Alloys*, McGraw-Hill, 1958
32. E.J. Sutow and S.R. Pollack, The Biocompatibility of Certain Stainless Steels, in *Biocompatibility of Clinical Implant Materials*, Vol 1, D.F. Williams, Ed., CRC Press, 1981, p 45-98
33. J.W.W. Sullivan, Steel, in *Encyclopedia Americana*, Vol 25, Grolier, 1984, p 662-663
34. W.B. Pearson, *A Handbook of Lattice Spacings and Structures of Metals and Alloys*, Pergamon Press, 1958
35. "Standard Recommended Practice for Surface Preparation and Marking of Metallic Surgical Implants," F 86, *Annual Book of ASTM Standards*, American Society for Testing and Materials
36. A.T. Kuhn, Corrosion of Co-Cr Alloys in Aqueous Environments—A Review, *Biomaterials*, Vol 2, 1981, p 68-77
37. R. Earnshaw, *Br. Dent. J.*, Vol 8, 1956, p 67
38. E. Haynes, U.S. Patent 873745, 1907
39. D.F. Williams, The Properties and Clinical Uses of Cobalt-Chromium Alloys, in *Biocompatibility of Clinical Implant Materials*, Vol 1, D.F. Williams, Ed., CRC Press, 1981, p 99-127
40. E. Haynes, Alloys of Cobalt With Chromium and Other Metals, *Trans. AIME*, Vol 44, 1913, p 573
41. M.F. Rothman, R.D. Zordan, and D.R. Muzyka, Role of Refractory Elements in Cobalt-Base Alloys, in *Refractory Alloying Elements in Superalloys*, J.K. Tien and S. Reichman, Ed., American Society for Metals, 1984, p 101-115
42. J.B. Vander Sande, J.R. Coke, and J. Wulff, A Transmission Electron Microscopy Study of the Mechanisms of Strengthening in the Heat-Treated Co-Cr-Mo-C Alloys, *Metall. Trans. A*, Vol 7A, 1976, p 389-397
43. T. Kilner, R.M. Pilliar, G.C. Weatherly, and C. Allibert, Phase Identification and Incipient Melting in a Cast Co-Cr Surgical Implant Alloy, *J. Biomed. Mater. Res.*, Vol 16 (No. 1), 1982, p 63-79
44. T.M. Devine and J. Wulff, Cast vs. Wrought-Cobalt-Chromium Surgical Implant Alloys, *J. Biomed. Mater. Res.*, Vol 9 (No. 2), 1975, p 151-167
45. T. Kilner, "The Relationship of Microstructure to the Mechanical Properties of a Cobalt-Chromium-Molybdenum Alloy Used for Prosthetic Devices," Ph.D. thesis, University of Toronto, 1984
46. T. Kilner, *Trans. Soc. for Biomater.*, Vol VIII, 1985
47. C.N. Younkin, Multiphase MP35N Alloy for Medical Implants, *J. Biomed. Mater. Res.*, No. 5, Part 1, 1974, p 219-226
48. J.C. Van Loon, Titanium, in *Encyclopedia Americana*, Vol 26, Grolier, 1984, p 785
49. J.C. Van Loon, Titanium (Kroll Process), in *Encyclopedia Americana*, Vol 26, Grolier, 1984, p 786
50. D.F. Williams, Titanium and Titanium Alloys, in *Biocompatibility of Clinical Implant Materials*, Vol 1, D.F. Williams, Ed., CRC Press, 1981, p 9-44
51. J.A. McMaster, "Titanium for Prosthetic Devices," Paper presented at the Dental-Medical Committee Meeting, Cleveland, OH, American Institute of Mining, Metallurgical and Petroleum Engineers, Oct 1970
52. G.H. Hille, Titanium for Surgical Implants, *J. Met.*, Vol 1 (No. 2), 1966, p 373-383
53. H.A. Luckey and F. Kubli, Jr., Ed., *Titanium Alloys in Surgical Implants*, STP 796, American Society for Testing and Materials, 1983
54. H. Hahn, P.J. Lare, R.H. Rowe, Jr., A.C. Fraker, and F. Ordway, Mechanical Properties and Structure of Ti-6Al-4V With Graded-Porosity Coatings Applied by Plasma Spraying for Use in Orthopedic Implants, in *Corrosion and Degradation of Implant Materials*, STP 859, A.C. Fraker and C.D. Griffin, Ed., American Society for Testing and Materials, 1985, p 179-191
55. E.W. Collings, *The Physical Metallurgy of Titanium Alloys*, American Society for Metals, 1984
56. L.S. Castleman and S.M. Motzkin, The Biocompatibility of Nitinol, in *Biocompatibility of Clinical Implant Materials*, Vol 1, D.F. Williams, Ed., CRC Press, 1981, p 129-154
57. F.H. Froes and H.B. Bomberger, The Beta Titanium Alloys, *J. Met.*, Vol 37 (No. 7), July 1985, p 28-37
58. J.L. Katz, Prosthetic and Restorative Materials for Bone, in *Workshop in Biomaterials*, Battelle Seattle Research Center, Nov 1969
59. Levert, *J., Am. J. M. Sc.*, Vol 4, 1829, p 17
60. P.G. Laing, Compatibility of Biomaterials, *Orth. Clinics N. Am.*, Vol 4 (No. 2), 1973, p 249-275
61. K. Merritt and S.A. Brown, Metal Sensitivity Reactions to Orthopedic Implants, *Int. J. Dermatol.*, Vol 20, March 1981, p 89-94
62. R.J. Solar, Corrosion Resistance of Titanium Surgical Implant Alloys: A Review, *Corrosion and Degradation of Implant Materials*, STP 684, B.C. Syrett and A. Acharya, Ed., American Society for Testing and Materials, 1979, p 259-273
63. A.C. Fraker, A.W. Ruff, P. Sung, A.C. Van Orden, and K.M. Speck, Surface Preparation and Corrosion Behavior of Titanium Alloys for Surgical Implants, *Titanium Alloys in Surgical Implants*, STP 796, H.A. Luckey and F. Kubli, Jr., Ed., American Society for Testing and Materials, 1983, p 206-219
64. R.W. Revie and N.D. Greene, Corrosion Behavior of Surgical Implant Materials: I Effects of Sterilization; II Effects of Surface Preparation, *Corros. Sci.*, Vol 9, 1969, p 755-770
65. J. Black, E.C. Maitin, H. Gelman, and D.M. Morris, Serum Concentrations of Cobalt and Nickel After Total Hip Replacement: A Six Month Study, *Biomaterials*, Vol 4, July 1983, p 160-164
66. J.L. Woodman, J.J. Jacobs, J.O. Galante, and R.M. Urban, "Titanium, Aluminum and Vanadium Release From Titanium Based Prosthetic Segmental Replacements of Long Bones in Baboons: A Long Term Study," St. Lakes' Hospital, 1985
67. M.G. Fontana and N.D. Greene, *Corrosion Engineering*, McGraw-Hill, 1967
68. H.H. Uhlig, *Corrosion and Corrosion Control*, 2nd ed., John Wiley & Sons, 1973
69. J. Kruger, Fundamental Aspects of the Cor-

rosion of Metallic Implants, in *Corrosion and Degradation of Implant Materials*, STP 684, B.C. Syrett and A. Acharya, Ed., American Society for Testing and Materials, 1979, p 107-127

70. M. Pourbaix, Electrochemical Corrosion of Metallic Biomaterials, *Biomaterials*, Vol 5, 1984, p 122-134
71. U.R. Evans, *An Introduction to Metallic Corrosion*, 3rd ed., Edward Arnold and American Society for Metals, 1981
72. J.C. Scully, *The Fundamentals of Corrosion*, Pergamon Press, 1966
73. J.C. Scully, Ed., *Corrosion: Aqueous Processes and Passive Films*, Vol 23, *Treatise on Materials Science and Technology*, 1983
74. N.D. Thomashov, *The Theory of Corrosion and Protection of Metals*, Macmillan, 1966
75. E.G.C. Clarke and J. Hickman, *J. Bone Joint Surg.*, Vol 35B (No. 3), 1977, p 467
76. M. Pourbaix, *Atlas of Electrochemical Equilibria in Aqueous Solutions*, Pergamon Press, 1966
77. P.G. Laing, Compatibility of Biomaterials, *Orth. Clinics N. Am.*, Vol 4 (No. 2), April 1973, p 249-275
78. Standard Test Method for Pitting or Crevice Corrosion of Metallic Surgical Implant Materials, F 746, *Annual Book of ASTM Standards*, ASTM
79. B.C. Syrett, Pit Propagation Rate Curves for Assessing Pitting Resistance, *Corrosion*, Vol 33, 1977, p 221
80. B.C. Syrett, The Application of Electrochemical Technique to the Study of Corrosion of Metallic Implant Materials, in *Electrochemical Techniques for Corrosion*, R. Baboian, Ed., National Association of Corrosion Engineers, 1977
81. B.C. Syrett and S.S. Wing, Pitting Resistance of New and Conventional Orthopedic Implant Materials—Effects of Metallurgical Condition, *Corrosion*, Vol 34 (No. 4), April 1978, p 138-145
82. L.C. Lucas, R.A. Buchanan, J.E. Lemons, and C.D. Griffin, Susceptibility of Surgical Cobalt-Base Alloy to Pitting Corrosion, *J. Biomed. Mater. Res.*, Vol 16 (No. 6), Nov 1982, p 799-810
83. F.G. Hodge, and T.S. Lee III, *Corrosion*, Vol 31, 1975, p 111
84. B.C. Syrett and E.E. Davis, Crevice Corrosion of Implant Alloys—A Comparison of In Vitro and In Vivo Studies, in *Corrosion and Degradation of Implant Materials*, STP 684, B.C. Syrett and A. Acharya, Ed., American Society for Testing and Materials, 1979, p 229-244
85. P. Sury and M. Semlitsch, Corrosion Behavior of Cast and Forged Cobalt-Based Alloys for Double Alloy Joint Prostheses, *J. Biomed. Mater. Res.*, Vol 12, 1978, p 723-741
86. J.R. Cahoon and R.N. Holte, Corrosion Fatigue of Surgical Stainless Steel in Synthetic Physiological Solution, *J. Biomed. Mater. Res.*, Vol 15, 1981, p 137-145
87. O.E.M. Pohler and F. Straumann, Fatigue and Corrosion Fatigue Studies on Stainless-Steel Implant Material, in *Evaluation of Biomaterials*, G.D. Winter, J.L. Leray, and K. de Groot, Ed., John Wiley & Sons, 1980, p 89-113
88. J.D. Bolton, J. Hayden, and M. Humphreys, A Study of Corrosion Fatigue in Cast Cobalt-Chrome-Molybdenum Alloys, *Eng. Med.*, Vol 11 (No. 2), 1982, p 59-68
89. B.J. Edwards, M.R. Louthan, Jr., and R.D. Sisson, Jr., Hydrogen Embrittlement of Zimaloy: A Cobalt-Chromium-Molybdenum Orthopedic Implant Alloy, in *Corrosion and Degradation of Implant Materials*, Second Symposium, STP 859, A.C. Fraker and C.D. Griffin, Ed., American Society for Testing and Materials, 1985, p 11-29
90. M.A. Imam, A.C. Fraker, and C.M. Gilmore, Corrosion Fatigue of 316L Stainless Steel, Co-Cr-Mo Alloy, and ELI Ti-6Al-4V, in *Corrosion and Degradation of Implant Materials*, STP 684, B.C. Syrett and A. Acharya, Ed., American Society for Testing and Materials, 1979, p 128-143
91. M.A. Imam, "Effect of Microstructure on Fatigue Properties in Ti-6Al-4V," Ph.D. thesis, The George Washington University, 1978
92. D.F. Bowers, "Corrosion Fatigue: Type 304 Stainless Steel in Acid Chloride and Implant Metals in Biological Fluids," Ph.D. thesis, Ohio State University, 1975
93. M.F. Semlitsch, B. Panic, H. Weber, and R. Schoen, Comparison of the Fatigue Strength of Femoral Prosthesis Stems Made of Forged Ti-6Al-4V and Cobalt Based Alloys, in *Titanium Alloys in Surgical Implants*, STP 796, H.A. Luckey and F. Kubli, Jr., Ed., American Society for Testing and Materials, 1983, p 120-147
94. E.N. Pugh, Stress Corrosion Cracking, in *Encyclopedia of Materials Science and Engineering*, Pergamon Press, 1982
95. R.B. Waterhouse, *Fretting Corrosion*, Pergamon Press, 1972
96. L.E. Slotter and H.R. Piehler, Corrosion Fatigue Performance of Stainless Steel Hip Nails—Jewett Type, in *Corrosion and Degradation of Implant Materials*, STP 684, B.C. Syrett and A. Adarya, Ed., American Society for Testing and Materials, 1979, p 173-195
97. "Standard Test Method for Measuring Fretting Corrosion of Osteosynthesis Plates and Screws," F 897, *Annual Book of ASTM Standards*, American Society for Testing and Materials
98. D.C. Mears, The Use of Dissimilar Metals in Surgery, *J. Biomed. Mater. Res.*, Vol 9 (No. 4), 1975, p 133-148
99. C.D. Griffin, "An In Vitro Electrochemical Corrosion Study of Surgical Implant Materials," M.S.E. thesis, University of Alabama at Birmingham, 1979
100. M. Stern and A.L. Geary, Electrochemical Polarization: A Theoretical Analysis of the Shape of Polarization Curves, *J. Electrochem. Soc.*, Vol 104, 1957, p 56-63
101. D.A. Jones, The Advantages of Galvanostatic Polarization Resistance Measurements, *Corrosion*, Vol 39 (No. 11), 1983, p 444-448
102. K.J. Bundy and R. Luedemann, Characterization of the Corrosion Behavior of Porous Biomaterials by AC Impedance Techniques, in *Quantitative Characterization and Performance of Porous Implants for Heart Tissue Applications*, STP 953, American Society for Testing and Materials, 1987
103. L. Lemaitre, M. Moors, and A.P. Van Peteghem, AC Impedance Measurements on High Copper Dental Amalgams, *Biomaterials*, Vol 6, Nov 1985, p 425-426
104. J.R. Ambrose, Repassivation Kinetics, in *Corrosion: Aqueous Processes and Passive Film*, Vol 23, *Treatise on Materials Science and Technology*, J.C. Scully, Ed., Academic Press, 1983, p 175-204
105. J.R. Ambrose and J. Kruger, Tribo-Ellipsometry: A New Technique to Study the Relationship of Repassivation Kinetics to Stress Corrosion, *Corrosion*, Vol 28 (No. 1), 1972, p 30-35
106. P. Sung and A.C. Fraker, Repassivation Kinetics of Ti (6Al)-4V, 316L Stainless Steel, Co-Cr (Cast) and Co-Ni-Cr (MP35N), in *Transactions of the Seventh Annual Meeting of the Society for Biomaterials*, Vol IV, 1981, p 28

Tarnish and Corrosion of Dental Alloys

Herbert J. Mueller, Consultant

DENTAL ALLOY DEVICES serve to restore or align lost or misaligned teeth so that normal biting function and aesthetics can prevail. Depending on the application, the particular design can take a number of different forms. The dental specialties of restorative, crown and bridge, prosthodontia, orthodontia, endodontia, implant, pedodontia, periodontia, and geriatric find applications. Alloys are used for direct fillings, crowns, inlays, onlays, bridges, fixed and removable partial dentures, full denture bases, implanted support structures, and wires and brackets for the controlled movement of teeth. In addition to applications calling for cast or wrought alloys, other uses of alloys include soldered assemblies, porcelain fused to metal, and resin bonded to metal restorations.

Dental Alloy Compositions. The compositions of alloys utilized to fulfill the diverse applications germane to dentistry include the following elements: Au, Pd, Pt, Ag, Cu, Co, Cr, Ni, Fe, Mo, W, Ti, Zn, In, Ir, Rh, Sn, Ga, Ru, Si, Mn, Be, B, Al, V, C, Ta, Zr, and others. Figures 1 to 6 show a number of typical restorations and appliances fabricated from alloys containing some of these metals.

Compositions for direct filling restorations usually consist of silver-tin-copper-zinc alloy amalgams. Pure gold in the form of cohesive foil, mat, or powder is used only in very limited applications.

Alloys for all-alloy cast crown and bridge restorations are usually gold-, silver-, or nickel-base compositions, although iron-base and other alloys have also been used. The gold-base alloys contain silver and copper as principal alloying elements, with smaller additions of palladium, platinum, zinc, indium, and other noble metals as grain refiners. The silver-base alloys contain palladium as a major alloying element, with additions of copper, gold, zinc, indium, and grain refiners. The nickel-base alloys are alloyed with chromium, iron, molybdenum, and others.

Alloys for porcelain fused to alloy restorations are gold-, palladium-, nickel-, or cobalt-base compositions. The gold-base alloys are divided into gold-platinum-palladium, gold-palladium-silver, and gold-palladium types. The palladium-base alloys are palladium-silver alloys or palladium-gallium alloys with additions from either copper or cobalt. The nickel- and cobalt-base alloys are alloyed primarily with chromium and with minor additions of molybdenum and other elements. In contrast to alloys for crown and bridge use, alloys fused to porcelain contain low concentrations of oxidizable elements, such as tin; indium; iron; gallium for the noble metal containing alloys; and aluminum, vanadium, and others for the base metal alloys. During the heating cycle, these elements form oxides on the surface of the alloy and combine with the porcelain at the firing temperatures to promote chemical bonding.

Alloys for removable partial dentures are primarily nickel- and cobalt-base compositions and are similar to alloys used for porcelain fused to alloy applications. However, carbon is present in amounts up to 0.3 to 0.4% with the partial denture alloys. Carbon is not added to alloys to be used for porcelain bonding.

Alloys that have found applications for support structures implanted in the lower or upper jaws are composed of cobalt-chromium, nickel-chromium, stainless steel, and titanium and its alloys. Wrought orthodontic wires are composed of stainless steel, cobalt-chromium-nickel, nickel-titanium, and β-titanium alloys. Silver- and gold-alloy solders are used for the joining of components. High-temperature brazing alloys are used for the joining of a number of high fusing temperature alloys. Additional information on noble metals is available in the article "Corrosion of Noble Metals" in this Volume.

Properties. The diversity in available alloys exists so that alloys with specific properties can be used when needed. For example, the mechanical property requirements of alloys used for crown and bridge applications are different from the requirements of alloys used for porcelain fused to alloy restorations. Even though crown and bridge alloys must possess sufficient hardness and rigidity when used in stress-bearing restorations, excessively high strength is a disadvantage for grinding, polishing, and burnishing. Also, excessive wear of the occluding teeth is also likely to occur. Alloys used with porcelain fused to metal restorations are used as substrates for the overlaying porcelain. In this case, the high strength and rigidity of the alloys more closely matches the properties of the porcelain. Also, a higher sag resistance of the alloy at temperatures used for firing the porcelain means less distortion and less retained residual stresses.

Similarly, alloys used for partial denture and implant applications must possess increased mechanical properties for resistances to failures. However, clasps contained within removable partial denture devices are often fabricated from a more ductile alloy, such as a gold-base alloy, than from cobalt-chromium or nickel-chromium al-

Fig. 1 Various types of crowns. Source: Ref 1

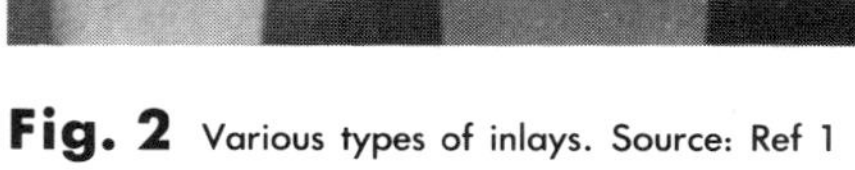

Fig. 2 Various types of inlays. Source: Ref 1

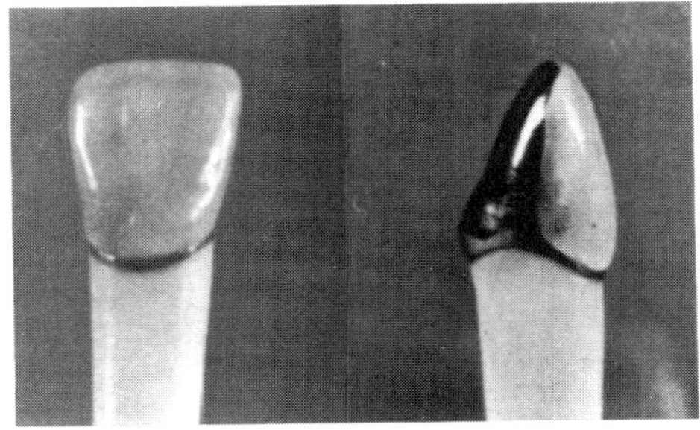

Fig. 3 Porcelain veneer fused to alloy. Source: Ref 1

loys. This ensures that the clasps possess sufficient ductility for adjustments without breakage from brittle fractures.

Other properties required in specific systems include the matching of the thermal expansion coefficients between porcelain and substrate alloy, negligible setting contractions with the direct filling amalgams, and specific modulus to yield strength ratios with orthodontic wires. Due to the use of lower gold content alloys because of the gold price market, alloy color is often a consideration. Lighter and pale-yellow gold alloys, as well as white gold alloys, are currently more prevalent. Tarnish and corrosion of all dental alloy systems have been and will remain of prime importance. Table 1 presents some typical mechanical properties for a number of different alloy systems used in dentistry. Additional information on the compositions, properties, and applications of dental alloys can be found in the section "Classification and Characterization of Dental Alloys" later in this article.

Tarnish and Corrosion Resistance

Dental alloy devices must possess acceptable corrosion resistance because of safety and efficacy. Aesthetics is also a consideration and is discussed in the section "Efficacy" in this article.

Safety

Dental alloys are required to have acceptable corrosion resistance so that biocompatibility is maintained during the time the metallic components are used (Ref 5-7). No harmful ions or corrosion products can be generated such that toxicological conditions result. The effects of the dental alloys on the oral environment have the capabilities for producing local, remote, or systematic changes that may be short-term, long-term, or repetitive (sensitization) in nature (Ref 8). Dental alloy-oral environment interactions have the potential for generating such conditions as metallic taste, discoloration of teeth, galvanic pain, oral lesions, cariogenesis, allergic hypersensitive dermatitis and stomatitis, endodontic failures, dental implant rejection, tumorgenisis, and carcinogenisis. Figure 7 shows a schematic of useful dental anatomy.

Metallic Taste. The symptom of metallic taste has been reported and related to the presence of metallic materials in the mouth (Ref 9). In addition, the release of ions and the formation of products through corrosion, wear and abrasion can occur simultaneously, which can accelerate the process. Therefore, patients with metallic restorations and with an inclination toward bruxism (the unconscious gritting or grinding of the teeth) are likely to be more susceptible to metallic taste. Although this condition is not as prevalent as it once was when metallic materials with lower corrosion resistances were more often used, metallic taste is still known to occur on occasion.

Discoloration of teeth has occurred mainly with amalgam fillings (Ref 10) and with base alloy screwposts (Ref 11). With amalgams, tin and zinc concentrations have been identified in the dentinal tubules of the discolored areas, while with the screwposts, copper and zinc were detected in both the dentin and enamel and the surrounding soft connective tissue. Discoloration is not, however, a definite indicator of the presence of metallic ions.

Galvanic pain results from contacting dissimilar-alloy restorations either continuously or intermittently (Ref 9, 12). An electrochemical circuit occurring between the two dissimilar-alloy restorations is short circuited by the contact. An instantaneous current flows through the external circuit, which is the oral tissues. The placement of dissimilar-alloy restorations in direct contact is ill advised.

Oral lesions resulting from the metallic prosthesis contacting tissue can be due to physical factors alone (Ref 9). An irritation in the opposing tissues of the oral mucosa can be generated because of the shape and location in the mouth of the prosthesis, as well as its metallurgical properties such as surface finish, grain size, and microstructural features. Tarnish and corrosion can change the nature of the alloy surface and be more of an irritant to the opposing tissues. Microgalvanic currents due to chemical differences of microstructural constituents and due to crevices, such as those created by the partial coverage of the alloy surface by the opposing tissues, must also be considered as possible causative factors in traumatizing and damaging tissue. However, no data have related *in vivo* galvanic currents from dental restorations to tissue damage. The released metallic ions from corrosion reactions can interact with the oral tissues to generate redness, swelling, and infection. Oral lesions can then occur. These reactions are discussed in the section "Allergic Hypersensitive Reactions" in this article.

Cariogenesis corresponds to the ability for released metallic ions and formed corrosion products to affect the resistance of either dentin or enamel to decay (caries). The mechanisms involved with caries formation (Ref 13, 14), which include the fermentation of carbohydrate by microorganisms and with the production of acid, are likely to become altered when metallic ions and products from corrosion reactions are included. This may be indicated by the reports that show tin and zinc concentrations originating from amalgam corrosion in softened, demineralized dentin and enamel (Ref 15-17).

Allergic Hypersensitive Reactions. With allergic hypersensitive contact reactions, some people can become sensitized to particular foreign substances, such as ions or products from the corrosion of dental alloys (Ref 9, 18). The metallic ions or products combine with proteins in the skin or mucosa to form complete antigens. Upon first exposure to the foreign substance by the oral mucosa, sensitization of the host may occur in times of up to several weeks and with no adverse reactions. Thereafter, any new exposures to the foreign substance will lead to biological reactions, such as swelling, redness, burning sensation, vesiculation, ulceration, and necrosis. Abstinence from the foreign substance leads to healing. Identification and avoidance are the means for controlling these allergic hypersensitive reactions. Exposure of the oral mucosa to the foreign substance can lead not only to allergic stomatitis reactions (of the oral mucosa) but also to allergic dermatitis reactions (of the skin) at sites well away from the contact site with the oral mucosa. Because the oral mucosa is more resistant to allergic reactions than the skin, the reverse process usually does not occur.

Of the currently used metals contained in dental alloys, nickel, cobalt, chromium, mercury,

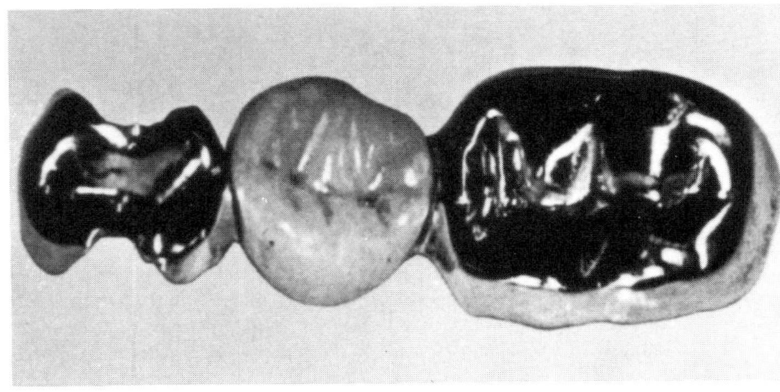

(a)

(b)

Fig. 4 Fixed bridges. (a) Three-unit bridge consisting of inlay (left member), onlay (right member), and porcelain fused to alloy pontic (center member). Source: Ref 2. (b) Five-unit bridge consisting of four porcelain fused to alloy members and one crown. Source: Ref 1

Table 1 Properties of some typical dental alloys

	Proportional limit		Yield strength, 0.1% offset		Modulus of elasticity		Strain,	Ultimate tensile strength(b)		Brinell hardness,
	MPa	Ksi	MPa	Ksi	GPa	psi × 10^6	%	MPa	Ksi	kg/mm^2
Amalgam										
New True Dentalloy	...	...	...	...	21.3	3.1	...	54	7.9 [318 MPa, or 46.1 Ksi]	...
Dispersalloy	...	...	...	...	33.8	4.9	...	48	6.9 [423 MPa, or 61.3 Ksi]	...
Conventional gold-alloys										
Type I, Ney Oro A	69	10	...	...	...	...	29.5	221	32	45
Type II, Ney Oro A-1	190	27.5	...	...	...	...	32	379	55	95
Type III, Ney Oro B-2										
Soft	221	32.0	...	...	...	...	35	421	61	110
Hard	262	38.0	...	...	...	...	34	448	65	120
Type IV, Ney Oro G-3										
Soft	286	41.5	...	...	99.3	14.4	24	469	68	140
Hard	572	83.0	...	...	...	...	6.5	758	110	220
Low-gold-alloys										
40 Au-Ag-Cu (Forticast)										
Soft	...	...	379	55.0	...	...	18	562	81.5	177
Hard	...	...	738	107	...	...	2	889	129	252
10 Au-Ag-Pd (Paliney)										
Soft	438	63.5	...	...	...	...	17	558	81.0	150
Hard	583	84.5	...	...	...	...	7	731	106	205
Ag-Pd (Albacast)										
Soft	...	...	262	38.0	...	...	10	434	63	130
Hard	...	...	324	47.0	...	...	8	469	68	140
Porcelain fused to metal gold alloy										
Ceramco O	...	...	...	...	86.2	12.5	5	...	...	131
Nickel-chromium alloys										
Crown and bridge alloys	...	...	359	52	179.3	26.0	1.1	421	61	330 HV
Partial denture alloys	...	...	710	103	...	...	2.4	807	117	...
Porcelain fused to metal alloys	...	...	...	...	202.7	29.4	16	917	133	270 HV
Cobalt-chromium alloy										
Cast Vitallium	...	...	644	93.4	217.9	31.6	1.5	869	126	...
Wires										
Austenitic stainless steels	...	...	1372	199(a)	200.6	29.1	...	...	...	...
Elgiloy	...	...	1110	161(a)	171.0	24.8	...	...	...	...
β-titanium	...	...	586	85(a)	71.7	10.4	...	...	...	...
Nitinol	...	...	193	28(a)	42.1	6.1	...	...	...	...
Tooth structure										
Enamel	353	51.2	...	...	...	...	...	10	1.5 [384 MPa, or 55.7 ksi]	343 HV
Dentin	167	24.2	...	...	...	...	...	52	7.5 [297 MPa, or 43.1 ksi]	68 HV

(a) 0.05% offset. (b) Bracketed values are ultimate compressive strengths

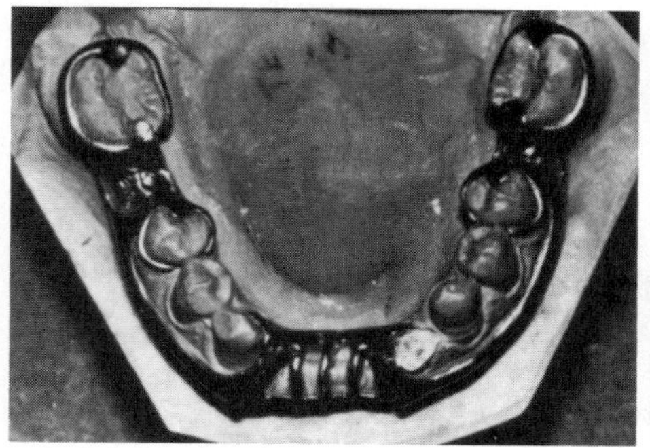

(a)

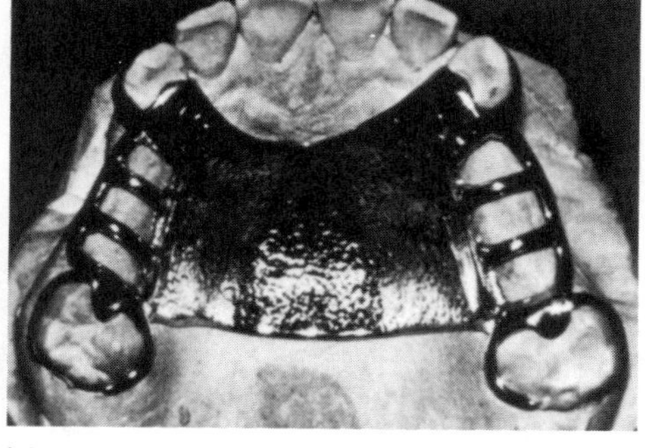

(b)

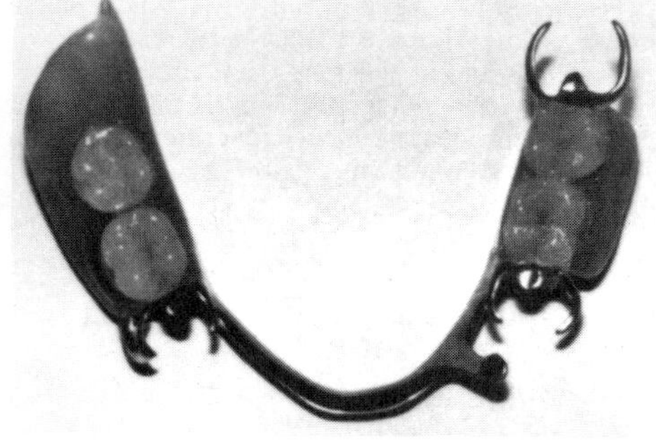

(c)

Fig. 5 Removable partial dentures, lower (a) and upper (b) cobalt-chromium frameworks, and a completed unit in (c). Source: Ref 3

beryllium, and cadmium need to be considered as inducing possible allergic or cytotoxic reactions. Nickel is the primary alloying element in nickel-chromium casting alloys (up to 80%), in nickel-titanium wires (up to 50%) and in lower concentrations in some cobalt-chromium alloys, and in austenitic stainless steels. Nickel from dental alloys is known to react with the oral tissues in some individuals to produce allergic sensitization reactions (Ref 19). About 9% of women and 1% of men are estimated to be allergic to nickel. It is recommended that individuals be screened for possible nickel allergies prior to dental treatments. If an allergy arises from a dental restoration, it is recommended that the individual be tested for allergies to nickel and have the nickel-containing prosthesis replaced with a nickel-free alloy, if so indicated by the testing results.

Cobalt, also a component of some dental alloys, has been known to react with the oral mucosa and cause allergic reactions (Ref 19). However, the occurrences of such allergies are less than 1% of the population and mainly affect women. Testing for cobalt allergies should be

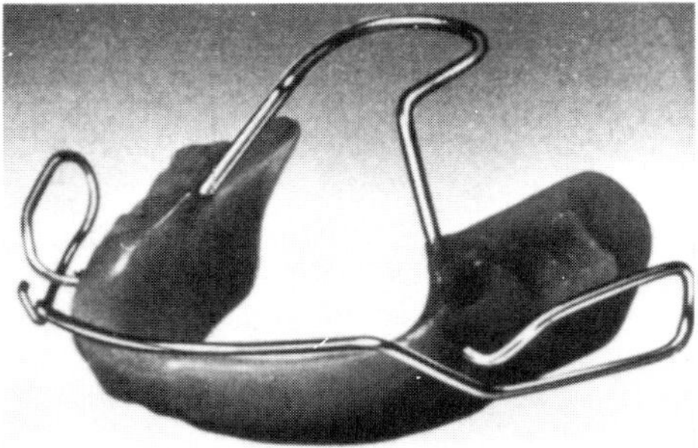

Fig. 6 Removable orthodontic appliance. Source: Ref 4

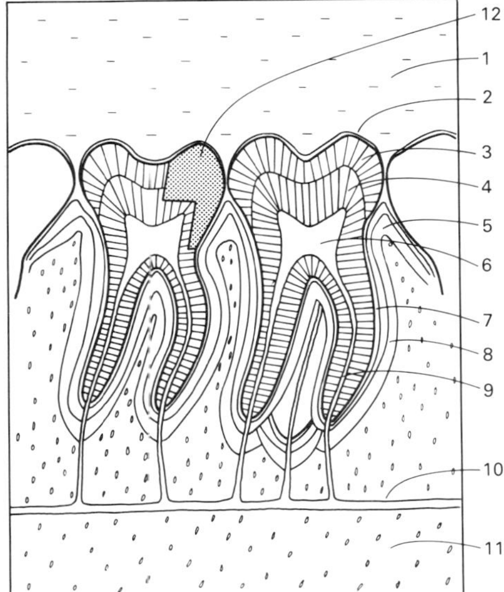

Fig. 7 Useful dental anatomy. 1, saliva; 2, integument; 3, enamel; 4, dentin; 5, gingiva; 6, pulp; 7, cementum; 8, periodontal ligament; 9, root canal; 10, artery; 11, alveolar bone; 12, restoration—amalgam filling

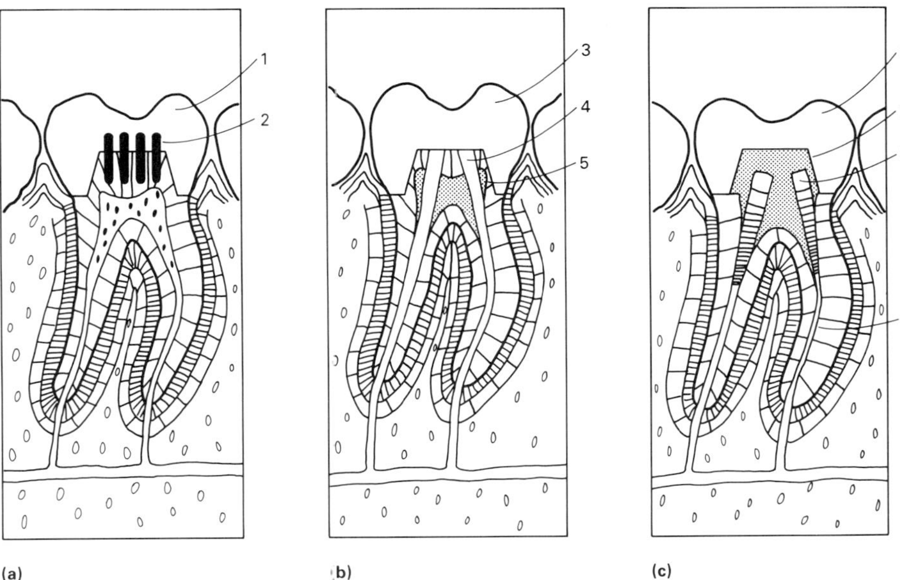

Fig. 8 Typical restored teeth. (a) Pin restored amalgam filling on vital tooth. (b) Cast metal crown restoration on endodontically treated tooth with silver cones and cement to seal root canals. (c) Cast metal crown restoration on endodontically treated tooth with cement core buildup and screwposts. 1, amalgam filling; 2, stainless steel pins; 3, metal crown; 4, silver cones; 5, cement; 6, metal crown; 7, cement core; 8, screw posts; 9, gutta perch or similar sealing material

performed if reactions to cobalt from cobalt-containing materials are suspected. Contact allergic reactions to chromium from dental alloys are also reported (Ref 19) but the occurrences of such reactions are rare.

Mercury is contained in amalgam fillings, which contain microstructural phases composed of silver-mercury and tin-mercury. Mercury ions may be released from microstructural phases through corrosion. However, the concentrations are low and not relatable to toxicological ramifications. Mercury vapors released from amalgam surfaces may also occur. Again, because of the low concentrations emitted, amalgam mercury vapors are not related to toxicity. Allergic reactions to mercury contained in dental amalgams have been reported (Ref 19). If mercury allergic reactions are suspected from the amalgam, it is recommended that testing for mercury allergies be conducted. Mercury vapors pose more of a health risk to dental personnel who routinely handle pure mercury than to individuals having amalgam restorations. Precautions should be followed in handling of pure mercury for amalgams.

Beryllium is contained in some nickel-chromium casting alloys in concentrations up to about 2 wt%. No biological reactions have been related to the beryllium contained in these alloys (Ref 19). More of a health hazard is posed to the dental personnel doing the actual melting and finishing of the alloy than to individuals having a prosthesis made from a beryllium-containing alloy. Precautions for working with beryllium must be followed.

Cadmium is contained in some dental gold and silver solders of up to 15% (Ref 20). No biological reactions have been related to the cadmium contained in these materials. Precautions should be taken in fusing solders containing cadmium. Additional information on the biocompatibility of metals and acceptable exposure limits is available in the article "Toxicity of Metal Powders" in Volume 7 of the 9th Edition of *Metals Handbook*.

Endodontic Failures. Root canals obturated with silver cones have occasionally been associated with corrosion (Ref 21). Development of a fluid-tight seal at the apex of the root canal is the primary objective of endodontic therapy. Corrosion of the silver points is known to lead to failure by allowing the penetration of fluids along the silver cone/root canal interface. Figure 8(b) shows a schematic of a tooth with cones.

Dental Implant Rejection. Dental implants, which are used for permanently attaching bridges, and so on, extend through or up to the maxillary or mandibular bones and must function in both hard and soft tissues as well as within a wide range of applied stresses (Ref 22, 23). Depending on the chemical inertness of the materials used for these devices, the thickness of the tissue connecting implant to bone varies. Released ions can infiltrate thick membranes surrounding loosely held implants, which can lead to an early rejection of the implant through immune response.

Tumorgenisis and Carcinogenisis. Even though dental alloy devices have not been implicated with tumorgenisis and carcinogenisis, their possible formation must never be ruled out and should always be considered as potential biological reactions, especially with new, untried alloys (Ref 23).

Efficacy

The oral environment must not induce changes in physical, mechanical, chemical, optical, and other properties of the dental alloy such that inferior functioning and/or aesthetics result. The effect of the oral environment on the alloy has the potential for altering dimensions, weight, stress versus strain behavior, bonding strengths with other alloys and with nonmetals, appearances, and creating or enhancing crevices. In combination with mechanical forces, the oral environment is capable of generating premature failure through stress corrosion and corrosion fatigue and of generating increased surface deterioration by fretting, abrasion, and wear.

Dimensions, Weight, Mechanical Properties, and Crevices. At least in theory, corrosion of precision castings and attachments, which rely on accurate and close tolerance for proper fit and functioning, can alter their dimensions, thus changing the fit and functionality of the restorations. Similarly, corrosion of margins on crowns and other cast restorations can lead to decreased dimensions and to enhanced crevice conditions. Increased seepage of oral secretions into the crevices created between restoration and tooth, microorganism invasion, generation of acidic conditions, and the operation of differential aeration cells can occur. Under these conditions, the bonding of the restoration to the dentinal walls through the underlying cement is likely to become weakened. In combination with the biting stresses, microcrack formation along the interface is likely to occur; this will cause the penetration of the crevice even further beneath the restoration. Eventually, the loosening of the entire restoration may occur.

With amalgam restorations, however, a slight amount of corrosion on these surfaces adjacent to

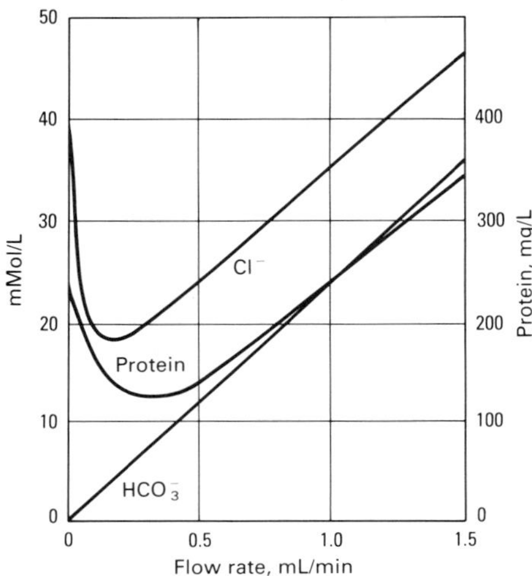

Fig. 9 Variations in the concentrations of Cl^-, HCO_3^-, and protein in human saliva as a function of the flow rate of saliva. Source: Ref 30

the cavity walls may actually be beneficial, because corrosion product buildup increases dimensions and adaptability. The crevice between the amalgam and the cavity is reduced in width, which leads to a decreased seepage of fluids. On the other hand, corrosion of the amalgam deteriorates its subsurface structure; this is likely to lead to an increased occurrence of marginal fracture, a known problem with amalgams, through corrosion fatigue mechanisms with stresses generated from biting (Ref 24).

The loss of sufficient substance from any dental alloy through corrosion can lead to a reduction in mechanical strength, thus enhancing failure or reducing rigidity so that unacceptable strains occur. For silver-soldered wires, corrosion of the solder leads to a weakening of the entire joint (Ref 25). Loosening of crowns and bridges because of corrosion-induced fractures of posts and pins is also known to occur (Ref 26), as shown in Fig. 8(a) and (c). Still other possibilities include the reduction through corrosion in the bond strengths of metal brackets bonded to teeth, as well as the degradation of porcelain fused to metal restorations because of corrosion.

Appearance. Because of the various optical properties of corrosion products, the appearances of tarnished and corroded surfaces can become unacceptable. A degradation in surface appearance without a loss in the properties of the appliance can be taken to be either acceptable or unacceptable, depending on individual preferences. If, however, the tarnished surface promotes additional consequences, such as the attachment of plaque and bacteria or a greater irritation to opposing tissue, then tarnishing must be deemed unacceptable.

Interstitial Versus Oral Fluid Environments and Artificial Solutions

In order to select and/or develop dental alloys, an understanding of the environment to which these materials will be exposed is imperative. This section will define and compare interstitial fluid and oral fluid environments. In addition, artificial solutions developed for testing and evaluation of dental materials will also be discussed.

Table 2 Mean blood plasma composition

Compound	mg/100 mL
Inorganic	
Na^+	327
K^+	13
Ca^{2+}	10
Mg^{2+}	2
Cl^-	372
HCO_3^-	165
PO_4^-	20
SO_4^-	10
Nonprotein organic	
Urea	25
Uric acid	4
Carbohydrates	260
Organic acids	19
Lipids	530
Fatty acids	325
Amino acids	50
Major proteins	
Albumin	3650
Globulins	3250
Fibrinogen	300

Source: Ref 27

Table 3 Mean saliva composition

Compound	mg/100 mL
Inorganic	
Na^+	30
K^+	78
Ca^{2+}	6
Mg^{2+}	1
Cl^-	53
HCO_3^-	31
PO_4^{3-}	48
SCN^-	15
Nonprotein organic	
Urea	4
Uric acid	5
Amino acids	4
Citrate & lactate	5
Ammonia	0.6
Sugars	4
Carbohydrates	73
Lipids	2
Protein	
Glycoproteins	45
Amylase	42
Lysozyme	14
Albumin	2
Gamma-globulin	5

Source: Ref 30, 31

Interstitial Fluid

Applications of metallic materials to oral rehabilitation are confronted with a number of environmental conditions that differentiate most dental uses from other biomedical uses (Ref 27). The one major exception is dental implants, because interstitial fluids (the fluids in direct contact with tissue cells) are encountered by both dental and other types of surgical implants (see, for example, the article "Corrosion of Metallic Implants and Prosthetic Devices" in this Volume). As discussed later in this article, other exceptions occur, because restorations in teeth have their interior surfaces in direct contact with the dentinal and bone fluids, which are more similar to interstitial fluids in composition than to saliva.

Other types of extracellular fluids, such as lymph and blood plasma, contain similar inorganic contents and are also likely to come into contact with dental implants, particularly with plasma during and shortly after surgery. Table 2 presents a composition of blood plasma. The inorganic content is similar to the inorganic content of interstitial and other types of extracellular fluids, while the protein concentration for plasma is higher than for other biofluids. For plasma, the major proteins are albumin, globulins, and fibrinogen. For all extracellular fluids, the inorganic contents are characterized by high sodium (Na) and chloride (Cl^-) and moderate bicarbonate (HCO_3^-) contents. Considerable variations in pH, pO_2, and pCO_2 can occur in the vicinity of an implant. In crevices formed between plates and screws, some extreme values ranging between 5 to 7 in pH, and $<$8 to 110 and $<$10 to 300 mm Hg, respectively, have been determined (Ref 28). Similar corrosive conditions are expected regardless of the extracellular fluid, provided the effects of the protein and cellular contents are minimal.

Tissue cells and other types of cellular matter can also directly contact implant material, with the possibility of intracellular fluid permeating through the cell membrane and effecting corrosion of the alloy. Separation by shearing of biological cells from alloy surfaces almost always generates cohesive failures through the cell instead of adhesive failures along the alloy/cell interface (Ref 29). In these situations, intracellular fluids can gain direct access to the surface of the alloy. In contrast to extracellular fluids, intracellular fluids contain high potassium and organic anion contents. The sodium is replaced by potassium and Cl^- by orthophosphate (HPO_4^{2-}). The effectiveness of intracellular fluids in corroding implant surfaces will be governed by the ability of the larger organic anions to pass through cell membranes, which are usually very restricted. Extracellular fluids are therefore the fluids interacting with the implant in most cases, although the possible effects from intracellular fluids must not be dismissed.

Oral Fluids

Whole mixed saliva is produced by the paratid, submandibular, and sublingual glands, together with the minor accessory glands of the cheeks, lips, tongue, and hard and soft palates from the oral mucosa. Gingival or crevicular fluid is also produced, as well as fluid transport between the hard tissues of the teeth and saliva. The composition of the secretion from each gland is different and varies with flow rate and with the intensity and duration of the stimulus. Saliva composition varies from individual to individual and in the same individual under different circumstances, such as time of day and emotional state.

Although about 1 L of saliva is produced per day in response to stimulation accompanying chewing and eating; for the greater part of the day, the flow rate is at very low levels (0.03 to 0.05 mL/min). During sleep, there is virtually no flow from the major glands. At low flow rates, the concentrations of sodium, Cl^-, and HCO_3^- are reduced notably; the concentration of calcium is elevated slightly; and the concentrations of magnesium, phosphate (PO_4^{3-}), and urea are elevated decidedly when compared with stimulated flow rates (Ref 30). It is therefore impossible to define specific compositions and concentrations that are universally applicable. However, compilations of data encompassing large statistical populations have been made by a number of researchers. One typical analysis for the composition of human saliva that utilized the results from many investigators is shown in Table 3.

The inorganic ions readily detectable in saliva are Na^+, K^+, Ca^{2+}, Mg^{2+}, Cl^-, PO_4^{3-}, HCO_3^-, thiocyanate (SCN^-), and sulfate ($SO_4^=$). Minute traces of F^-, I^-, Br^-, Fe^{2+}, Sn^{2+}, and nitrite (NO_2^-) are also found, and on occasion, Zn^{2+}, Pb^{2+}, Cu^{2+}, and Cr^{3+} are found in trace quantities. Figure 9 shows the Cl^- and HCO_3^- variations in concentration as saliva is stimulated to flow at a rate of 1.5 mL/min. The O_2 and N_2 contents of saliva are 0.18 to 0.25 and 0.9 vol%, respectively. The carbon dioxide (CO_2) content varies greatly with flow rate, being about 20 vol% when unstimulated and up to about 150 vol% when vigorously stimulated. The buffering capacity is chiefly due to CO_2/HCO_3^- system, with that of the PO_4^{3-} system only having a small, limited part. The redox potential of saliva indicates it to possess reducing properties, which is likely due to bacteria reductions, carbohydrate split-offs from glycoproteins, and nitrites. The normal pH of unstimulated saliva is in the 6 to 7 range and increases with flow rate.

The clearance of saliva involves its movement toward the back of the mouth and its eventual introduction into the stomach. Saliva is continually being secreted and replenished, especially during active times. A volume of about 1 L/d is considered average for saliva production. Chemical analysis of human mouth air showed hydrogen sulfide (H_2S), methyl mercaptan, and dimethyl sulfide to be some of the most important constituents (Ref 30).

Organic. Human saliva is composed of nonprotein organic and protein contents, as shown in Table 3. The largest contributions from the nonprotein ingredients are from the carbohydrates, while smaller amounts are from urea, organic acids, amino acids, ammonia, sugars, lipids, blood group substances, water-soluble vitamins, and others. Some of the lipids include the fatty acids, glycerides, and cholesterol. At least eighteen amino acids have been identified, with glycine being the main constituent. Many of these species are produced directly by the salivary glands, while others, such as some carbohydrates and amino acids, are the result of the dissociations of glycoproteins and proteins by bacterial enzymes. Still others are derived from blood plasma. The protein content of human saliva is primarily of salivary gland origin, with a very small amount derived from blood plasma. The protein content may vary from less than 1 to more than 6 g/L. Detailed information on protein and glycoproteins that have been identified to be in saliva can be found in Ref 32 to 34.

Chemicals in food, drink, and atmospheric air. All of the ingredients found in food and drink are capable of becoming incorporated into saliva. However, most of the foods are injested before the breakdown into basic chemicals occurs. Some foods and beverages, though, contain chemicals that are reactive by themselves without any reductions and may become dissolved in saliva and affect the tarnish and corrosion of metallic materials. Some of these include various organic acids, such as lactic, tartaric, oleic, ascorbic, fumaric, maleic, and succinic, as well as sulfates, chlorides, nitrates, sulfides, acetates, bichromates, formaldehyde, sulfoxylates, urea, and the nutrients themselves of lipids, carbohydrates, proteins, vitamins, and minerals (Ref 35).

The components found in atmospheric air and pollutants, coupled with the human respiratory function, have the potential of exposing the oral environment to additional aggressive chemical species. Some of the species known to be in atmospheric air and pollutants are O_2, CO_2, NO_2, carbon monoxide (CO), sulfur dioxide (SO_2), Cl_2, hydrogen chloride (HCl), hydrogen sulfide (H_2S), ammonia (NH_3), formaldehyde, formic acid, acetic acid, Cl^- salts, ammonium salts of sulphate and nitrate, and dust (Ref 36).

Because the volume of lung ventilation is of the order of 8.5 L/min, the amount of potentially hazardous and corrosive material possibly coming into contact with the oral environment is significant. In approximately 2 h, 1 m^3 of air for a mouth breather will have been used during respiration with the potential uptake of the normal urban SO_2 amount of 0.11 to 2.3 mg. Sulfur dioxide can be involved in many interactions, accelerating the tarnish and corrosion of metals. Hydrogen sulfide is another reactive gas. Ammonium salts are known for lowering of the surface tension of water and salt solutions. Dust particles may vary from organic to inorganic components of the earth's surface to industrial pollutants. The proteins in saliva combine with most of the aggressive external stimuli coming into contact with saliva. Therefore, most of these hazardous species are rendered inactive before they can cause tarnish and corrosion. However, the pathway from the atmosphere to the surfaces of dental alloys are certainly potential sources for introducing corrosive species.

A comparison between interstitial and oral fluids shows differences in both inorganic and organic contents. One important difference is the approximately sevenfold higher Cl^- concentration in interstitial fluid. Even though interstitial fluids do undergo variations in pH and pO_2, especially at the site of the implant, saliva is more susceptible to variations in composition. This comes about because the composition of saliva depends to a large degree on flow rate, which in turn depends on a number of physical and emotional factors. Saliva is also subjected to exposures from chemicals contained in the air, food, drink, pharmaceuticals, as well as temperature variations of 0 to 60 °C (32 to 140 °F) and microbiological involvement with the production of acid and plaque.

The operation of crevice conditions and mechanical-environment interactions are common to both saliva and interstitial fluid. Differential aeration cells and the generation of acidic conditions accompany crevices, while the biting stresses or the stresses generated in surgical implants, such as from walking, can develop creep, fatigue, wear, and abrasion process. Stress-corrosion cracking (SCC) and corrosion fatigue are additional potential mechanisms.

Artificial Solutions. Numerous solutions simulating human saliva have been formulated and used for testing the tarnish and corrosion susceptibility of dental alloys (Ref 37-42). Modifications to these solutions have also been made and used (Ref 43-48). Some of the solutions contain only inorganics (Ref 40-42, 46-48), while others include the addition of an organic component consisting mostly of mucin (Ref 37-39, 43-45). Some solutions also purge a $CO_2/O_2/N_2$ gas mixture through the solution to simulate pH control and buffering capacity controlled by the CO_2/HCO_3^- redox reaction. All compositions contain mostly chlorides (Na, K, and Ca) and various forms (mono-, di-, or tri-basic, pyro) of phosphates in smaller amounts. Additional ingredients include bicarbonate, thiocyanate, sulfide, carbonate, organic acids, citrate, hydroxide, and urea.

Table 4 Composition of artificial solutions

Compound	Composition, mg/100 mL Artificial saliva	Ringer's solution
NaCl	40	82–90
KCl	40	2.5–3.5
$CaCl_2 \cdot 2H_2O$	79.5	3.0–3.6
$NaH_2PO_4 \cdot H_2O$	69	. . .
$Na_2S \cdot 9H_2O$	0.5	. . .
Urea	100	. . .

Source: Ref 41, 49

Table 4 presents the composition for an artificial saliva that corresponds very well to human saliva with regard to the anodic polarization of dental alloys. Ringer's physiological saline solution used to simulate interstitial fluid is also included in Table 4. Both solutions are entirely inorganic. The Cl^- concentration of Ringer's is about seven times higher than that of the saliva. The anionic content of Ringer's is entirely chloride, but that of saliva also contains phosphate and sulfide. Urea is also a part of the saliva. Sodium, potassium, and calcium constitute the cationic content of both solutions.

A number of additional artificial physiological solutions, some of which are named Hanks, Tyrod, Locke, and Krebs, appear in the literature and have been used to simulate the interstitial fluids. Basically, these solutions contain small additions of modifying ingredients, such as magnesium chloride, glucose, lactate, amino acids, and organic anions. The Ringer's solution presented in this article, after the National Formulary Designation, does not contain sodium bicarbonate ($NaHCO_3^-$). Some solutions, however, referred to in the literature as Ringer's, do contain bicarbonate.

Effect of Saliva Composition on Alloy Tarnish and Corrosion

Chloride/Orthophosphate/Bicarbonate/Thiocyanate. The interactions of the various salts contained in saliva are complex. The effects from the combined saliva solutions are not simply the additive effects from the isolated individual salts. This synergistic behavior is discussed for the corrosion of an amalgam in the $Cl^-/HPO_4^{2-}/HCO_3^-/SCN^-$ system in Ref 42. Chloride alone produces a powdery, finely crystalline corrosion product in heaps around the sites of attack, such as porosities and pits. The addition of HPO_4^{2-}, which by itself produced very little effect, caused the corrosion products to become organized in conical structures, the bases being over the sites of corrosion. The addition of HCO_3^- to the Cl^-/HPO_4^{2-} system generated increased microstructural corrosion. On the contrary, addition of SCN^- to the Cl^-/HPO_4^{2-} system suppressed the microstructural corrosion. By adding all four salts together, an even more corrosion-resistant system was obtained. Corrosion was much reduced and more localized.

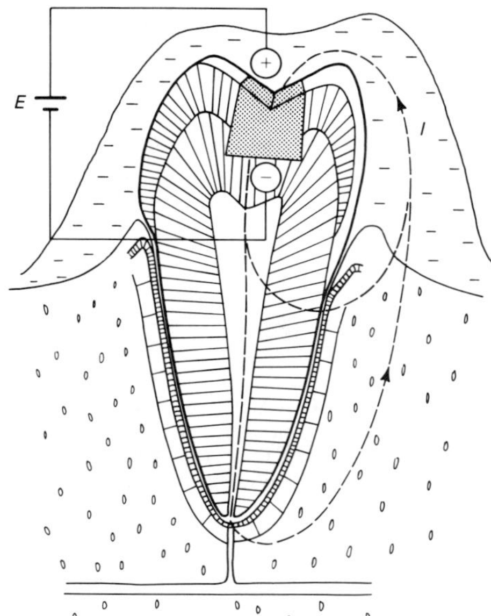

Fig. 10 Schematic of a single metallic restoration showing two possible current (*I*) pathways between external surface exposed to saliva and interior surface exposed to dentinal fluids. Because the dentinal fluids contain a higher Cl^- concentration than saliva, it is assumed the electrode potential of interior surface exposed to dentinal fluids is more active and is therefore given a negative sign (−). The potential difference between the two surfaces is represented by *E*.

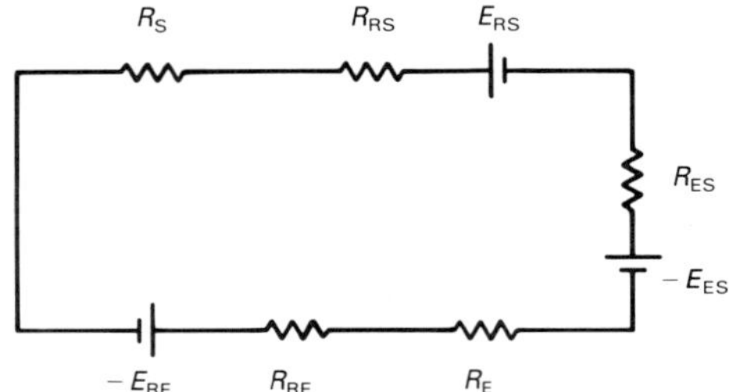

Fig. 11 Electrical schematic representing the equivalent circuit pathway shown in Fig. 10. Terminology is defined in text.

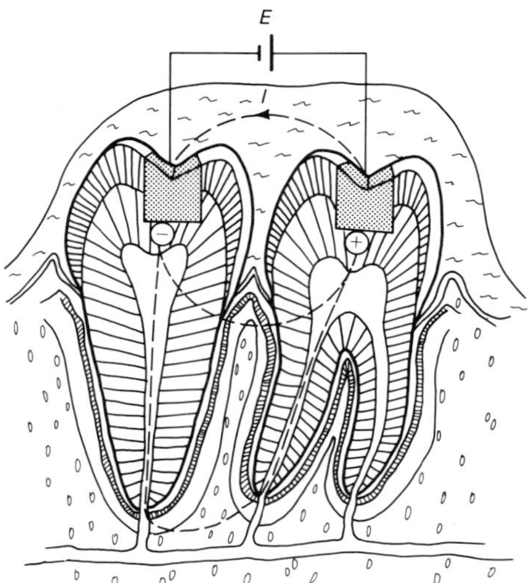

Fig. 12 Schematic of two nonisolated, noncontacting restorations. The alloy restoration on the left, which is an amalgam, is more active than the restoration on right, which is a gold-based alloy

Artificial Salivas. The effect on alloy corrosion from different artificial saliva solutions has been studied (Ref 41). The polarization behavior of a number of dental alloys, including gold-base alloys, nickel-chromium, and cobalt-chromium, in artificial salivas without HCO_3^- and SCN^- but with protein provided the best correlation with the behavior observed with human saliva in both aerated and deaerated conditions. The artificial salivas containing HCO_3^- and SCN^- but no proteins constantly shortened the passivation range of the alloys. The specific contributions from Cl^- and SCN^- shortened the passivation range of the gold-base alloy, but phosphate increased the passivation range of all alloys.

Lowering the pH shifted the amalgam polarization curve to increased currents and potentials, while buffering capacity, which was increased by protein content, influenced corrosion behavior under localized corrosion conditions (Ref 50). In sulfide solution, the polarization curves of amalgams indicated increased corrosion (Ref 39, 51). Dissolved O_2 generated both inhibition and acceleration, as reflected by the formation of anodic films and the consumption of electrons by cathodic depolarization. The particular alloy-environment combination determines whether corrosion is inhibited or accelerated.

Chloride and Organic Content. Anodic polarization of amalgams in human saliva compared to Ringer's solution was shown to be shifted by up to several orders of magnitude to lower currents at constant potentials, depending on the amalgam system (Ref 52). These differences were related to the Cl^- concentration of the solutions. The effect of Cl^- on amalgam polarization is well documented (Ref 47, 53, 54). Pretreatment of gold-base alloys in human saliva prior to galvanic coupling with amalgams in a protein-free artificial saliva reduced the corrosion on some of the amalgams studied (Ref 55). Pretreatment of the amalgams had little effect.

Significant reductions in the weight gains of amalgams stored in artificial saliva with mucin as compared to mucin-free saliva have been reported (Ref 39). Anodic polarization of amalgams in artificial saliva or diluted Ringer's solution with and without additions of mucin or albumin was, however, shown to be very similar (Ref 39, 53). Proteins in artificial saliva on silver-palladium and nickel-chromium alloy polarizations were also reported to have little effect (Ref 50). For a copper-aluminum crown and bridge alloy, anodic polarization differences were detected in an artificial saliva with and without additions of a human salivary dialysate (Ref 56). The total accumulated anodic charge passed from corrosion potentials to +0.3 V versus saturated calomel electrode (SCE) was significantly reduced in protein-containing saliva. Similarly, the polarization resistance of the alloy was more than doubled by progressively adding up to 1.6 mg dialysate/mL to saliva initially free of proteins.

Microorganisms. The tarnishing of dental alloys by three microorganisms likely to be found in the mouth has been reported (Ref 57). Some specificity between the degree of tarnish and the type of microorganism was obtained. A likely tarnishing mechanism was due to the organic acids generated by the fermentation of carbohydrate by the bacteria. The section "Oral Corrosion Processes" in this article discusses in greater detail the effect of microorganisms on accelerating corrosion.

Oral Corrosion Pathways and Electrochemical Properties

The electrochemical properties of dental alloy restorations vary widely. Electrochemical potentials, current pathways, and resistances depend on whether there is no contact, intermittent contact, or continuous contact between alloy restorations. This section will examine the effects of restoration contact on electrochemical parameters and will review concentration cells developed by dental alloy-environment electrochemical reactions.

Noncontacting Alloy Restorations

Isolated. The total liquid environment of a restoration includes, in addition to saliva, fluids contained within the interior of dentin and enamel, which are more like extracellular fluids in composition than saliva. Figure 10 shows a schematic of a likely current path for a single metallic restoration. The current path encompasses a route that includes the restoration, enamel, dentin, membranes such as the periodontal ligament, soft tissues, and saliva (see Fig. 7). The conduction of current through hard tissues, including enamel, dentin, and bone, occurs through the extracellular fluids, which are compositionally similar in all hard and soft tissues. However, the current through these different hard tissues will take pathways of least resistances. For example, the resistance of dentin in a direction parallel to the tubules is about 18 times lower than in a perpendicular direction due to the calcification of the tubule walls. Structural details, including imperfections, orientations, and so on, control the actual resistances for particular hard tissue structures.

The restoration (R) develops electrochemical potentials with the extracellular fluids, E_{RE}, and with saliva, E_{RS}, while a liquid junction potential occurs between extracellular fluids and saliva, E_{ES}. Contact resistances occur between restoration and extracellular fluids, R_{RE}, and between restoration and saliva, R_{RS}. Resistances of the extracellular fluids, R_E, extracellular fluid-saliva junction, R_{ES}, and of saliva, R_S, also occur. Figure 11 shows an electrical schematic for this system. Summing electromotive forces (emf) in one direction and equating to zero yields for the current *I*:

$$I = \frac{E_{RE} + E_{ES} - E_{RS}}{R_{RE} + R_{RS} + R_E + R_S + R_{ES}} \quad \text{(Eq 1)}$$

Taken together, E_{ES} and R_S have negligible effect on current. The extracellular resistance, R_E is usually in the range between 10^4 and 10^6 Ω because of variations in particular hard tissue structures, and to possible variations in membrane/hard tissue interfacial characteristics. The potentials E_{RE} and E_{RS} are characteristic of the metal-electrolyte combinations, and the resistances R_{RE} and R_{RS} are dependent on the polarization characteristics for the particular combinations.

Polarization is related to the corrosion products that form. For soluble or loosely adhered

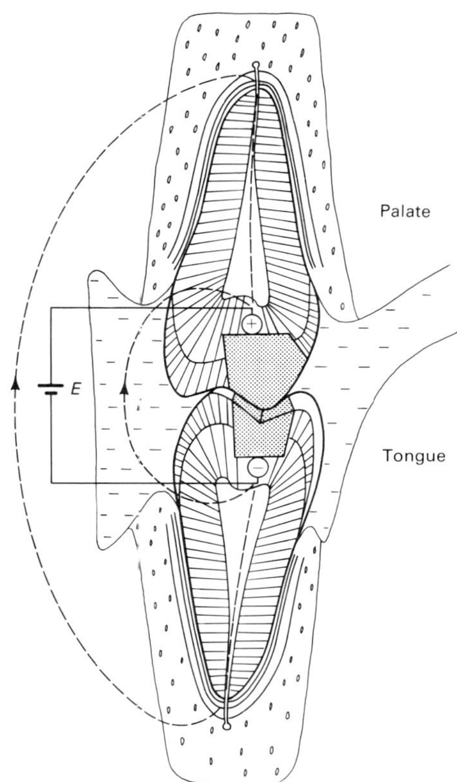

Fig. 13 Schematic of two restorations making intermittent contact due to biting. The restoration in the lower arch, which is an amalgam, is more active than the restoration in the upper arch, which is a gold-based alloy. Two possible current pathways are shown. An additional pathway very likely to occur would be directly through saliva between the two restorations.

products, the contact resistances will not be changed significantly. However, for tenaciously adhering products with semiconducting or insulating electrical characteristics, the contact resistances will be largely affected. These resistances are the primary parameters affecting the magnitude of the generated current. This reasoning is directly in line with the mixed-potential theory for electrochemical corrosion (Ref 58). The corrosion current, I_{corr}, without ohmic resistance control is:

$$I_{corr} = \frac{\beta_a \beta_c}{(\beta_a + \beta_c)\, R_p} \quad \text{(Eq 2)}$$

where β a and β c are the Tafel slopes from the anodic and cathodic polarization curves, and R_p is the polarization resistance or the linear slope of the $\Delta E/\Delta I$ curve within ±10 mV of the corrosion potential

Nonisolated. For two restorations not in contact (Fig. 12), the extracellular fluid-saliva resistance, R_{ES}, determines the extent to which the current will be short circuited through the saliva/extracellular fluid interface. If R_{ES} is high, there is maximum interaction between the separated restorations (the currents are small, of the order of 1 to 10 × 10^{-9} A/cm² between an amalgam and a gold alloy restorations). As R_{RS} decrease, the current through the interface between saliva and extracellular fluids increases. The interaction between the separated restorations will then be minimized. Each restoration, though, will still generate its own current path loop (Ref 59).

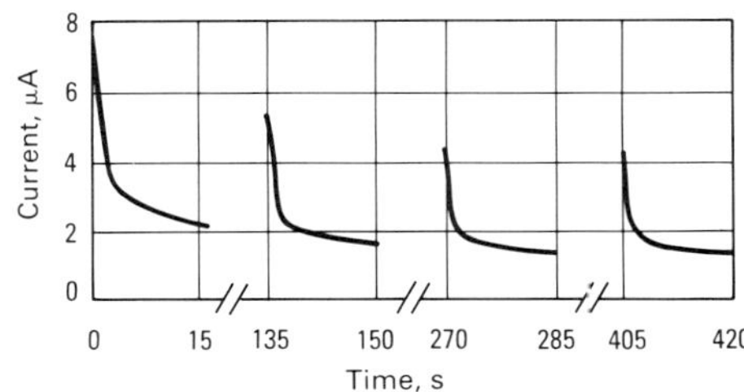

Fig. 14 Current-time responses between gold alloy and amalgam of the same cross-sectional areas. Short circuiting occurred for 15 s, followed by a 2 min delay before recontacting. Source: Ref 62

Intraoral Electrochemical Properties. In a study comprising 115 people, the corrosion potentials from 243 restorations ranged between −0.55 and +0.4 V versus SCE. Amalgam restorations were the most active, followed by cobalt-chromium alloys and gold-base alloys. Variations in potential on different surfaces of the same restoration occurred routinely. This was likely due to the effects from abrasion on the occlusal surfaces and from the accumulation of plaque and debris on nonocclusal surfaces (Ref 60).

For noncontacting amalgam and gold alloy restorations (78 fillings in 66 people), the average currents flowing through the restorations due to saliva-bone fluid liquid junction cells were calculated from measured intraoral potential and resistance data to be 0.48 and 0.26 μA, respectively (Ref 12).

Utilizing constant current pulses (1 to 10 μA) and measuring the corresponding potential changes, the intraoral polarization resistances for noncontacting amalgam restorations ranged between 50 and 300 × 10^3 Ω (Ref 61). With the use of linear polarization theory, corrosion currents are calculated to be 0.2 to 1.0 μA.

Restorations Making Intermittent or Continuous Contact

Intermittent Contact. A situation can occur in the mouth in which two alloy restorations, one in the upper arch and the other in the lower arch, come into contact intermittently by biting (Fig. 13). When the two restorations are in direct contact, a galvanic cell is generated with an associated galvanic current short circuited between the two restorations. The external current path can take a number of directions, with the least resistance path controlling. Figure 13 shows two possible pathways, one entirely through extracellular fluids and the other partly through extracellular fluids and partly through saliva.

The current-time transients have been measured and are presented in Fig. 14. Upon first making contact, currents of the order of 10 μA and more occur and decrease rapidly within a matter of minutes. If, however, the restorations are open circuited for a time interval and then again closed, the current level will again increase but not to the same magnitude as from the previous closure. The amount of recovery will increase as the time lapse between closure increases. This phenomenon is explained by the formation of protective surface films on the electrodes due to the passage of current. Upon making contact on succeeding occasions, the film offers additional resistance to the flow of current, even though the two restorations appear to be in direct intimate contact. The films dissipate with time, thus increasing the level of the initial current upon recontacting restorations.

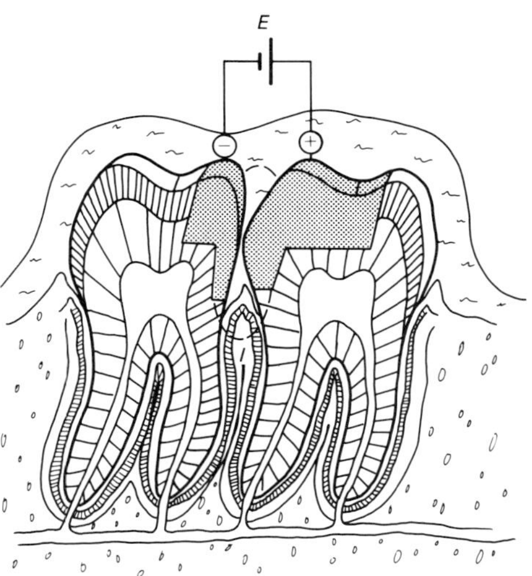

Fig. 15 Schematic of two adjacent restorations in continuous contact. Two possible current pathways shown.

A similar situation can occur because of an alloy restoration contacting, for example, eating utensils or dental instruments during dental treatment. Again a short-circuited galvanic current is generated. The external circuit will be partly through saliva and partly through extracellular fluids.

Continuous Contact. The fourth situation in which metallic restorations in the mouth are capable of generating galvanic currents involves two dissimilar metallic restorations in continuous contact, as shown in Fig. 15. Most attention has been given to the combination of amalgam-gold alloy couples (Ref 44). Other situations occur, for example, between two amalgam restorations (Ref 63)—one a conventional amalgam and the other a high-copper amalgam—and between two gold alloys with differences in noble metal content. Other situations have already been discussed. These include a stainless steel reinforced amalgam (Fig. 8a), an endodontically restored tooth with silver cones making contact with a gold crown (Fig. 8b), and an endontically restored tooth with steel screwposts making contact with a gold crown (Fig. 8c). Soldered appliances are also examples of dissimilar metals making continuous contact. Any multiphase microstructures are situations for galvanic corrosion to occur. Multiphase microstructures occur extensively with dental alloys.

For the amalgam-gold alloy couple making direct contact, the amalgam is the anode and suffers corrosive attack; the gold alloy is the cathode. As with galvanic couples making intermittent contact, large galvanic currents occur upon first contact and decrease rapidly with time. For silver-tin amalgams, the tin from the tin-mercury phase suffers corrosive attack. The freed mercury combines with the gold of the gold alloy to form a gold amalgam that is capable of producing surface discolorations on the gold alloy. In addition to becoming corroded, the amalgam is capable of being degraded in strength by the corrosion generated by the galvanic currents (Ref 63).

Currents calculated from polarization resistance and potential differences of various contacting dissimilar metallic restorations indicate most couples to pass 1 to 5 μA upon first contact (Ref 60). However, amalgam-gold alloy couples indicate a greater percentage of currents in the ranges of 6 to 10 and 11 to 15 μA. All couples initially show a sharp decrease in current with time, followed by a gradual leveling off as zero current is approached. However, disruption of surface protective films is reason for considering possible increases in current at later times.

Concentration Cells

Interior-Exterior Surfaces. Because the interior surfaces of restorations adjacent to the cavity walls are exposed to extracellular fluids and higher concentrations of Cl^- than the exterior surfaces exposed to saliva, the interior instead of the exterior surfaces are more susceptible to anodic attack from Cl^-. However, if the electrons generated by the anodic oxidations are not consumed by reductions, the oxidation reactions will cease. Because the extracellular fluids have low concentrations of dissolved oxygen, corrosion of the interior surfaces would likely cease if it were not for the accessibility of electrons to the exterior surfaces exposed to saliva having a supply of dissolved oxygen from contact with the atmosphere.

Corrosion that is perpetuated by electrochemical reactions occurring on adjacent or opposite surfaces of the same restoration constitutes an important pathway for the tarnishing and corrosion of dental alloys. This pathway is germane to amalgams as well as all types of restorations, including crowns and inlays that are cemented into the cavity preparation. In the mouth, cements are likely to become electrical conductors because the absorption of oral fluids permits the passage of ions.

Marginal Crevices. A second pathway can occur because of the seepage of salivary fluids into crevices or marginal openings formed between the restoration (especially with amalgams) and the cavity walls. The pathway is distinguished from the first in that the conditions developed in the crevice are due to diffusion and charge balances resulting from the salivary fluids instead of the extracellular fluids. Because of a lack of diffusion of the large O_2 molecule into the crevice, low O_2 concentrations result within the crevice. With time, the acidity within the crevice increases because of the accumulation of H^+ ions from the oral environment and from corrosive reactions occurring within the crevice. Chloride and other anion concentrations will also tend to increase within the crevice over time because of charge equalization. Therefore, this pathway results in conditions that are similar to the interior-exterior pathway.

Alloy Surface Characteristics. Porosities, differences in surface finish, pits, weak microstructural phases, and the deposition of organic matter can initiate corrosion by concentration cell effects. For example, gold-base alloys are known to become tarnished more easily when containing porosities and inhomogeneities (Ref 64). Rougher surface finishes of restorations generate increased corrosive conditions (Ref 65). Similarly, the pitting of base metal dental alloys of the stainless steel and nickel-chromium varieties occurs by concentration cell corrosion. Basically, the advancing pit front is free of O_2, but the surfaces of the alloy outside the pit have an ample supply of O_2 from the air. Because the anode-to-cathode surface area is very small, the corrosion occurring at the bottom of the pit is concentrated to a very small area, thus increasing intensity of the attack. Removal of only a small amount of metal has a large effect on advancing the pit front.

Amalgam γ_2 Phase. Deterioration of the weak, corrosion-prone tin-mercury phase (γ-2) in silver-tin amalgams has also been proposed to occur by concentration cell corrosion (Ref 59). In this model, partial removal of the γ-2 phase initially occurs by abrasion resulting from biting and chewing. After removal of the γ-2 phase has progressed to a sufficient depth, an occluded cell is formed between the bottom of the depression and the unabraded surface. Mass transport is restricted from and into the cell. The condition will approach conditions occurring in other types of concentration cells. In the present example, however, Sn^{2+} will be slowly released from the passivated γ-2 regions. The concentration of Sn^{2+} will slowly increase within the occluded cell and will be neutralized by an equivalent amount of Cl^- by migration from the bulk electrolyte.

Consumption of O_2 within the occluded cell will take place by its utilization in the consumption of electrons by cathodic depolarization. Replenishment of O_2 will be restricted, and the concentration of O_2 within the cell will become reduced. When the solubility product of stannous oxide (SnO) is exceeded, SnO precipitates and the H^+ concentration increases. At this point, activation of the γ-2 phase occurs. Dissolution of tin occurs freely. The Cl^- concentration within the cell continuously increases to maintain electrical neutrality. Galvanic coupling of the occluded cell to the external surface generates a galvanic cell by which the cathodic reduction of O_2 occurs. Corrosion of γ-2 tin within the cell continues. Under conditions of high acidity and high concentration of Cl^-, the formation of insoluble tin chloride hydroxide ($Sn(OH)Cl{\cdot}H_2O$) becomes thermodynamically possible.

Oral Corrosion Processes

Whether corrosion is occurring between microstructural phases of a single restoration, between components having different environmental concentrations, or between individual restorations of different compositions and making intermittent or continuous contact, the corrosion processes involved consist of oxidation and reduction. The dissolution of ions is involved with the anodic reaction, and the consumption of electrons is involved with the cathodic reaction. The slowest step in the complete chain of events controls the overall corrosion rate. Corrosion of alloys in the mouth can be viewed as being the result of corrosive and inhibiting factors (Ref 66). Some corrosive factors consist of Cl^- (in most instances), H^+, S^- (at times), O_2, microorganisms, and the clearance rate of corrosion products from the mouth, while some inhibiting factors consist of proteins and glycoproteins (in most instances), CO_2/HCO_3^- buffering system, PO_4^-/PO_4^{3-} buffering system, and salivary flow rate.

Corrosive Factors

Chloride. The effect of Cl^- on the deterioration of passivated surface films on stainless steel, nickel-chromium, and cobalt-chromium alloys is well known. The susceptibility to pitting attack is increased. Increased Cl^- content also increases the attack of corrosion-prone phases in amalgam, other base metal alloys, and the low noble metal content alloys. Because the Cl^- concentration in saliva is about seven times lower than that in the extracellular fluids, the corrosiveness of Cl^- in saliva is usually less. Figure 16 illustrates the effect of Cl^- concentration on the polarization of amalgam by comparing the cyclic voltammetry in deaerated artificial saliva to that in Ringer's. Increases in Cl^- concentrations are also likely to occur in crevices, such as the interfaces between cavity walls and adjacent surfaces of restoration. The Cl^- concentration within crevices is expected to increase to preserve electrical neutrality from the increase in Sn^+ concentration resulting from the γ-2 tin and γ-1 corrosion (Ref 59).

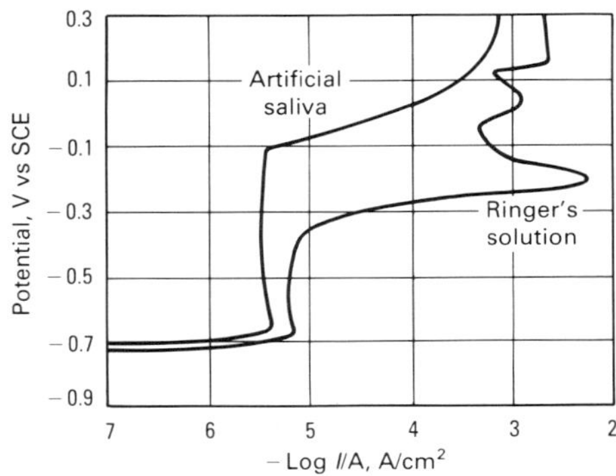

Fig. 16 Anodic polarization at 0.03 V/min of low-copper amalgam (Microalloy) in artificial saliva and Ringer's solution. Source: Ref 67

Chloride is capable of generating numerous compounds as products of corrosion. Chloride combines with zinc, tin, copper, silver, and others contained in dental alloys. Some of the products formed include zinc chloride (ZnCl), stannous chloride ($SnCl_2$), stannic chloride ($SnCl_4$), SnCl compounds such as hydrated $SnOHCl{\cdot}H_2O$ and $Sn_4(OH)_6Cl_2$, copper chloride (CuCl), cupric chloride ($CuCl_2$), complex hydrated cupric chloride ($CuCl_2{\cdot}3Cu(OH)_2$), and silver chloride (AgCl). The solubilities are high for all compounds, except CuCl, AgCl, and the basic tin and copper chlorides. Many additional compounds are to be considered for a complete listing of all potential corrosion products that form from dental alloys. Certainly, the chlorides of indium, gallium, beryllium, iron, nickel, chromium, cobalt, and molybdenum should be included.

Hydrogen Ion. The pH in the mouth can vary from about 4.5 and lower to about 8. In addition to the normal variations in pH of saliva due to human factors (see the section 'Oral Fluids' earlier in this article), increased acidity can also result from a number of additional factors, such as the operation of crevice corrosion conditions, the production of plaque, and the effects of food, drink, and atmospheric conditions. The operation of crevice conditions in amalgams can increase acidity to well below a pH of 4. For amalgams, this acidity is mostly the result of the oxidation of γ-2 and γ-1 tin in aqueous solution. Under these conditions, the freed H^+ will become the cathodic depolarizers. With this increased acidity, dissolution of the tooth structure is also likely to occur. Calcium and phosphorus are likely to be

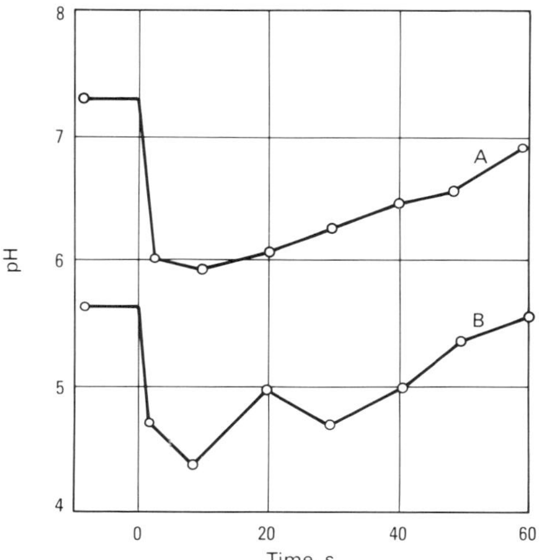

Fig. 17 pH versus time responses (Stephan curves) of plaque from caries-active (B) and caries-free (A) groups following a sugar challenge. Source: Ref 13

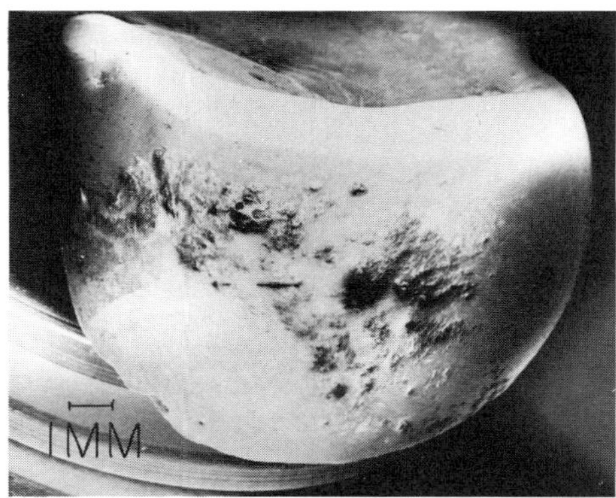

Fig. 18 Alloy restoration after intraoral usage showing the severity of plaque buildup that can occur. Source: Ref 68

dissolved from enamel and dentin. Plaque is produced by the fermentation of carbohydrate by microorganisms (Ref 13, 14). Most of the fermentable carbohydrate responsible for acid production comes from the diet in the form of sugars or starchy foodstuffs.

Figure 17 shows a schematic Stephan pH test curve of plaque. Stephan showed that the pH for all plaques decrease in value following a sugar challenge (Ref 13). This means that the production of acid by fermentable carbohydrate is greater than the rate at which acid can be removed. As time proceeds, the pH again rises. For caries-free and caries-active individuals, the qualitative shapes of Stephan pH curves are similar; however, the relative position of the curve for caries-active individuals is shifted to lower pH values. Values in pH of 4.5 and lower occur. Depending on the source of the sugar challenge, the pH minimum on the Stephan pH curves have been shown to remain for a number of hours. Even though no data are available to show the effect of plaque pH value on the tarnishing and corrosion of dental alloys, it follows from first principles that the reduction in pH will adversely affect tarnishing and corrosion resistance. Metallic restorations can become severely deposited with plaque and organic matter, as shown in Fig. 18.

Sulfide compounds, such as silver sulfide (Ag_2S), cuprous sulfide (Cu_2S), and cupric sulfide (CuS), have very low solubility product constants and often constitute the tarnished films on dental alloys. Mercury and tin sulfides may also be considered when amalgams are considered. The formation of thin insoluble films occurs with very small amounts of formed corrosion products, especially on the higher noble metal content alloys. In spite of even microgram quantities of tarnishing products at times, surface discolorations can still occur and elicit unsatisfactory personal responses. Tarnishing products under these conditions almost always maintain biocompatibility with the alloy system. With the lower noble metal content alloys, however, increased quantities of corrosion products can form, and tarnishing and corrosion can become more involved.

The corrosion potentials for many dental alloys in sulfide-containing solutions are often lower than the standard reduction potentials for the formation of the metal sulfides—an indication that the metal sulfides are thermodynamically stable. In some instances, particularly with amalgams, dissolution rates increase with S^- concentrations. This is probably due to the increased solubility for some of the sulfides (for example, Sn_2S_3 with amalgams) to form complexes with other species. Dietary factors are the main source for increasing S^- levels in saliva. Some foods, such as eggs and fish, as well as some drinking waters, are high in sulfur. Smokers have higher SCN^- saliva concentrations than nonsmokers (Ref 69). Sulfate-reducing bacteria may also generate S^- in the mouth. Hydrogen sulfide that is produced in the crevicular fluid and periodontal pockets can be easily dissolved in oral fluids. Atmospheric pollutants often contain high levels of SO_2 and H_2S and may influence the concentrations of S^- in the oral fluids.

Dissolved oxygen participates in corrosion reactions by either depolarizing cathodic reactions or by reoxidizing disruptive passivated surface films on base metal alloys. The first case increases or perpetuates corrosion, while the second case reduces or inhibits corrosion. Oxygen depolarization occurs by the customary electrochemical redox reactions. Electrons from the anodic process are consumed by the depolarization process. For a typical restoration, the exterior surfaces are exposed to higher O_2 concentrations. Differential aeration conditions can become operative, with the outer surfaces cathodic to the anodic interior surfaces. In other situations, differential aeration cells are set up between the bottoms of pits and the surrounding surfaces. Other types of pores and porosities are also likely to generate concentration cells. In near-neutral solution that corresponds to saliva, the reaction $O_2 + 2H_2O + 4e^- \rightarrow 4OH^-$ occurs most prevalently. If a driving force exists for metal oxidation, dissolution will be perpetuated on surfaces exposed to the lower O_2 concentration.

Oxygen is involved with numerous corrosion products formed on dental alloys. The tin from the γ-1 and γ-2 phases from silver-tin amalgams generates tin oxide products. These products passivate the amalgam at potentials less negative than about −0.7 V versus SCE, as indicated by the passive regions on the anodic polarization curves shown in Fig. 16. At more noble potentials, basic tin chlorides of the type $SnOHCl \cdot H_2O$ are formed, as indicated by the large current increases on the polarization curves. For copper-containing amalgams, basic copper chlorides of the type $CuCl_2 \cdot 3Cu(OH)_2$ form. Some additional products containing oxygen that are likely to occur with the corrosion of dental alloys include SnO_2, $Sn_4(OH)_6Cl_2$, Cu_2O, CuO, ZnO, $Zn(OH)_2$, and the oxides of chromium, nickel, cobalt, molybdenum, iron, titanium, and so on.

Oxygen is usually excluded from solutions during polarization testing with alloys. Dissolved O_2 interferes with the anodic processes. The generated anodic polarization curves obtained in O_2-containing solutions are usually cut off within the negative potential regions. For this reason, deaerated solutions are usually used to obtain entire anodic polarization curves. Passive breakdown potentials were observed to vary depending upon whether aerated, dearated, or air-exposed solutions were used (Ref 70). Within the O_2 concentration range likely to occur for surgical implants, the anodic polarization of type 316L stainless steel in Ringer's solution was independent of oxygen concentration (Ref 71).

Microorganisms. Two types of organisms—sulfate-reducing (*Bacteriodes corrodens*) and acid-producing (*Streptococcus mutans*) bacteria—have been discussed with the corrosion of dental alloys in the mouth (Ref 66). With regard to sulfate-reducing bacteria, depolarization of cathodic sites is thought to occur by removing H^+ from the metal surface. The hydrogen is utilized by the bacteria for the reduction of sulfate to sulfide, such as by the reaction $SO_4^- + 8H^+ \rightarrow S_2^- + 4H_2O$. In the case of acid-producing bacteria, the adsorbed microorganisms on the surface establish differential aeration conditions. As the dissolution of the metal occurs underneath the deposited microorganisms, the released acidic metabolic products, which include organic acids such as lactic, pyruvic, acetic, proprionic, and butyric, increase corrosion of the already-formed anodic sites. Because anodic areas are relatively small compared to the larger cathodic areas, corrosion can be severe.

The fermentation of carbohydrates by microorganisms generates plaque. The effects of plaque on the tarnishing and corrosion of dental alloys are probably more significant than the effect on alloy corrosion of only the microorganisms themselves. The effects of plaque on corrosion have been considered in the section "Hydrogen Ion" in this article.

Clearance Rate. The clearance of corrosion products from the mouth by the movement of saliva toward the back of the mouth and eventually by swallowing and replenishment affects the concentration of products in equilibrium with the metallic restorations. Therefore, a driving force for the continuation of the corrosion processes is maintained. Products of corrosion, like chemical species introduced through the diet, are cleared from the mouth by binding the exterior surfaces of the oral mucosa to the salivary glycoproteins and mucopolysaccharides lining. Detailed information on the binding ability of corroded metallic ions to proteins in human saliva can be found in Ref 32, 33, and 72.

Alloy Factors. Although the effects of alloy selection on tarnish and corrosion behavior are considered in more detail in the section "Tarnish and Corrosion Under Simulated or Accelerated Conditions" later in this article, some of the important factors will be mentioned here. Alloy composition and microstructure are probably the two most important factors. The corrosion resistance of dental alloys is the result of nobility in composition or the protectiveness of oxide films formed on base metal alloys. Multiphase microstructures are capable of exhibiting increased tarnish and corrosion because of the galvanic coupling of the individual components. The heat treatment state of cast alloys has an important influence on corrosion resistance (Ref 73). Surface state or finish also influence corrosion. Rougher surfaces are more prone to corrosion because of increased tendencies for galvanic coupling. Cast restorations with burnished margins are more susceptible to corrosion because of differences in surface cold-worked states. Rougher surfaces are prone to attachment of microorganisms and plaque, which usually increase corrosion (Ref 74).

Inhibiting Factors

Organics in the form of microorganisms and plaque usually have an accelerating effect on the tarnishing and corrosion of dental alloys (see the sections "Effect of Saliva Composition on Alloy Tarnish" and "Corrosion and Oral Corrosion Processes" earlier in this article). Organics in the form of amino acids, proteins, and glycoproteins have received mixed reports. For the amino acids, the building blocks of proteins, the passivation of copper was shown to be improved in Ringer's solution with added cysteine, while nickel became more corrosion prone (Ref 75). Alanine had little effect. For Ti-6Al-4V, the amino acids proline, glycine, tyrosine, and others that constitute many salivary proteins were again shown to have very little effect. For the plasma proteins, which simulate the organic content in blood and which simulate dental and surgical implant applications more closely, additional evidence can be found implicating the effect of proteins on corrosion behavior.

For example, the corrosion rates of cobalt and copper powders increased significantly when exposed to saline solutions with albumin and fibrinogen (Ref 76); however, for chromium and nickel powders only slight increases occurred, and for molybdenum, decreases occurred. Corrosion of stainless steel by applied external currents was shown to be increased when conducted in saline with added calf's serum (Ref 77). When conducted under fretting corrosion conditions, however, the degradation of stainless steel was shown to increase in saline without the added serum (Ref 78). For a copper-zinc alloy, the cyclic voltammetry was reported to be altered by addition of plasma proteins and at plasma concentrations to a phosphated physiological saline solution (Ref 79). Albumin and γ-globulin generated increased passivation currents, while fibrinogen generated decreased critical current densities. The anodic polarizations prior to the onset of critical current densities were also shifted to more active behavior in the protein solutions. Finally, the pitting potential for aluminum increased slightly in human plasma, and current-time transients were shifted to lower values in plasma (Ref 80).

Carbon Dioxide/Bicarbonate Buffering System. The major buffering system in saliva is the CO_2/HCO_3^- system, which has been found to inhibit corrosion processes on dental alloys. Inhibition results from the deposition of such elements as copper, zinc, and calcium as carbonate films. Carbon dioxide, above all other gases, is contained most abundantly in saliva. Up to about 150 vol% (~3000 ppm) is contained in vigorously stimulated saliva. The equilibrium concentration of HCO_3^- in saliva is identified by the redox reaction:

$$CO_2(g) + H_2O \rightarrow H^+ + HCO_3^- \quad \text{(Eq 3)}$$

and with its equilibrium constant pK equal to (Ref 46):

$$pK = 7.9 = -\log \frac{[(H^+)(HCO_3^-)]}{p_{CO_2}} \quad \text{(Eq 4)}$$

At a pH = 7 and rearranging terms yields:

$$\frac{p_{CO_2}}{HCO_3^-} = 7.9 \quad \text{(Eq 5)}$$

Equation 5 states that the partial pressure, p, of CO_2 in units of atmospheres is 7.9 times larger than the HCO_3^- concentration in mol/L. Therefore, for p_{CO_2} of the order of 0.07 atm, HCO_3^- concentrations of the order of 0.009 mol/L are formed. This shows that relatively large concentrations of HCO_3^- can be made available in saliva to form carbonates with cations released from corrosion reactions on dental alloys and with other cations found in the mouth, such as calcium.

Many of the different carbonates likely to form are insoluble in aqueous solution. The calcium carbonates are known for making waters hard. Compounds of this type being deposited as thin-film tarnish and corrosion products on dental alloys are very likely to interfere with the corrosion activity. Deposition over cathodic sites effectively increases corrosion resistance by increasing resistance to depolarization reactions. Because the films of carbonates are also likely to increase the contact resistances between electrodes and saliva, galvanic-corrosion processes are likely to change from purely corrosion control to at least partial ohmic control. Under these conditions, local anodes and cathodes may change in order to maintain lower-resistance paths for both ionic and electronic conduction. The *in vivo* tarnishing of several silver-palladium alloys was shown to be due to the galvanic coupling between microstructural phases located very close to each other on the alloy surface (Ref 81). This was in contrast to laboratory tests in high-conductivity solutions indicating larger distances between microelectrodes.

Phosphate Buffering System. A secondary buffering system in saliva, the PO_4^-/PO_4^{3-} system, has also inhibited the corrosion of dental alloys. The progressive inhibition of chloride (10 millimolar NaCl) amalgam corrosion activity was shown to occur with increasing added phosphate concentrations (Ref 66). A 15 millimolar phosphate addition retarded the anodic polarization almost entirely, while concentrations of 10, 7, 5, and 1 millimolar generated anodic current peaks of about 2.5, 3.0, 3.5, and 6.0 $\mu A/mm^2$, respectively. The 10 millimolar NaCl solution without phosphate generated a continuous increase in current to much larger values. No passivation occurred within the potential range used.

For tin in neutral phosphate solutions, a passive film forms by precipitation or by a nucleation and growth processes (Ref 82). Tin phosphate, basic tin phosphate complexes, and tin hydroxides are formed.

Salivary Flow Rate. Increasing the salivary flow rate increases the concentration of most species in saliva. This tends to inhibit corrosion. The organic content, the CO_2/HCO_3^- content, and the PO_4^-/PO_3^- content, pH, and the Ca^{2+} content all increase with flow rate. Only the increases in Cl^- concentration promote corrosion. Figure 9 shows the effect of flow rate on the concentration of a number of species.

Overview. Saliva acts as an ocean of anions, cations, nonelectrolytes, amino acids, proteins, carbohydrates, and lipids flowing in waves against and into dental surfaces with a diurnal tide and varying degrees of intensity (Ref 83). Whether or not tarnish and/or corrosion of dental metallic materials will occur cannot be categorically stated. It has been discussed that the degree to which dental alloy corrosion occurs in the mouth is dependent on the oral environmental conditions for each person. In addition to effects from the dental alloy itself, competition between corrosive and inhibitory factors of the oral environment will dictate whether or not corrosion will occur and to what extent. In addition to the factors listed above, still others have been isolated and should be included for a more complete assessment of the overall corrosiveness or protectiveness of the oral environment (Ref 66).

Nature of the Intraoral Surface

The composition and characterization of biofilms, corrosion products, and other debris that deposit on dental material surfaces will be discussed in this section. As will be shown, the nature of these deposits is dependent on the substrate material (enamel, alloy, porcelain, and so on).

Acquired Pellicles

Characteristics. Most surfaces that come into contact with saliva, including enamel and metallic, polymeric, and ceramic dental materials, interact almost instantaneously with the proteins and glycoproteins to form a bacteria-free biofilm of the order of several nanometers in thickness (Ref 84-86). This most intimate layer of organic matter adsorbed to the substrate material is called the acquired pellicle. A fourier transform infrared spectroscopy (FT-IR) spectra of the surface of a low-gold dental crown and bridge alloy after *in vivo* exposure is shown in Fig. 19. Detection for protein, carbohydrate, and lipid is indicated. (Information on the principles and applications of FT-IR can be found in the article "Infrared Spectroscopy" in Volume 10 of the 9th Edition of *Metals Handbook*.) Thicknesses of the films increase only slightly with longer exposure times. The pellicles, in contrast to enamel and most dental alloys, are acid insoluble, although an acid-soluble fraction also occurs. The films are diffusion barriers against acids, thus reducing the acid solubility of enamel and metallic materials and inhibiting or at least reducing the adherence of organisms.

Composition. Chemical analysis of 2-h pellicles formed on enamel indicated abundant amounts of glycine, glutamic acid, and serine (Ref 87). Carbohydrate contents of similar pellicles formed on enamel were found to contain about 70% glucose, with a number of other sugars and small molecules. Acidic proline-rich phosphoproteins have also been identified from *in vivo* enamel pellicles (Ref 88). The proline-rich proteins constitute as much as about 37% of the total proteins in new pellicles within the first hour. However, there is a gradual degradation beginning after about 24 h that is reflected by the

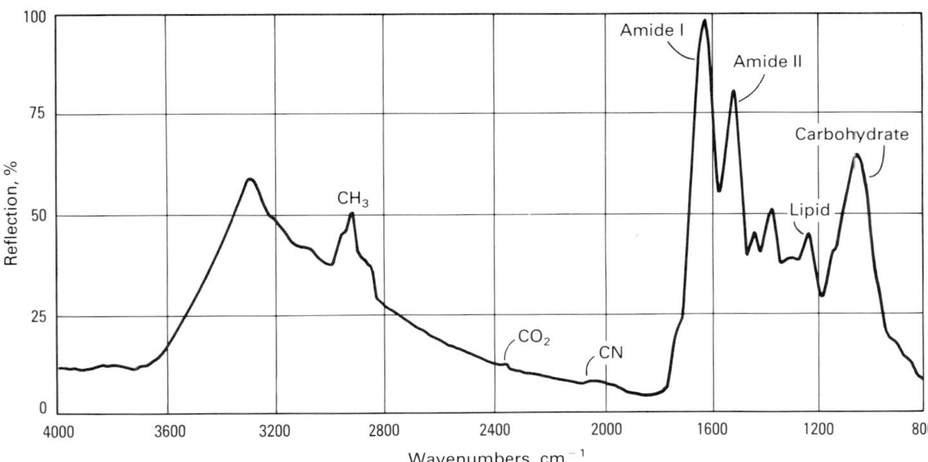

Fig. 19 Fourier transform infrared spectroscopy spectra from surface of a crown and bridge alloy (Midas) after several weeks of intraoral usage. Amide I and II are protein. Additional smaller peaks at 1375 and 1425 cm^{-1} are also protein.

fact that the proline-rich protein content in aged pellicles is less than 0.1%.

Substrate Effects on Pellicle Composition. Chemical analysis of the pellicles formed on several plastics and glass showed that the amino acid content varied and was different from that formed on enamel (Ref 89). It was concluded that the chemical composition of the substrate has an important influence on the type of proteins that become adsorbed. For the pellicle formed on dentures, it was concluded that a specific mechanism was controlling the deposition of protein and that specific proteins seemed to be precursors in forming the film (Ref 90). Isoelectric focusing of the extracted proteins adsorbed from a human saliva preparation onto a number of different powder substrate compositions, including palladium, silver, copper, silver-copper, tin, silver-tin-copper alloy, bismuth, polymethyl methacrylate, porcelain, hydroxyapatite, and enamel, indicated that the same three to four proteins appeared to be involved with the adsorption process on all substrates regardless of composition (Ref 33). Therefore, from this study, substrate composition appeared not to affect the type of proteins becoming adsorbed.

Binding Mechanisms. The binding of salivary macromolecules to surfaces has been proposed to consist of electrostatic interactions between the charged groups in the molecule and the surface charges on the substrate (Ref 91). For enamel, only the hydroxyapatite, and not the organic matrix, contributes a surface charge for binding. Because the negatively charged phosphate group comprises about 90% of the surface area of hydroxyapatite, the phosphate group rather than the calcium ions will be the primary binding sites. The hydration layer contains soluble calcium and phosphate ions as well as soluble cations and anions. Because the salivary molecules adsorbed to enamel are mainly acidic, binding to the negatively charged phosphate group appears to occur through a divalent cation, such as calcium. Phosphorylated and sulfated acidic proteins show a high affinity for hydroxyapatite. A direct replacement of the protein phosphate group and the phosphate in hydroxyapatite is also likely (Ref 92).

Direct binding to the calcium surface ions in enamel will also occur, but will be limited because of the relatively small surface area fraction occupied by the calcium ions. Adsorption of salivary proteins to metals may again occur through a divalent cation. The negative charges in the acidic proteins are likely to be bound to the negative anodic surface sites on the metal surface by the bridging cation. Additional information on protein binding and analyses of variations in protein binding to a metal surface through differential scanning calorimetry can be found in Ref 93.

Plaque, Corrosion Products, and Other Debris

Integument. In addition to the thin acquired biofilms, aged pellicles contain microorganisms, plaque, mineralized products, corrosion products, and other debris. Plaque, which is a by-product of the reaction between microorganisms and carbohydrates, may form in abundance in some environments. Plaque does not form directly onto teeth or other materials. It is deposited or adsorbed onto the acquired pellicle. The combined surface coating, including the adsorbed pellicle and plaque, which includes organic matter and any released ions or corrosion products generated by the substrate, is often referred to as the integument.

Substrate Effects on Integument Characteristics. A study was made of the effect of the restorative material type on plaque composition (Ref 94). The carbohydrate/nitrogen ratios (CHO/N) were similar for amalgam, gold inlay, gold foil, and resin. Plaque analyzed from freshly placed restorations had CHO/N = 1; this value increased to 1.3 and 1.2 at 3 and 6 months, respectively, and decreased to 0.5 at 1 year and for old restorations. It was proposed that the variation in plaque carbohydrate content with the age of the restoration was due to corrosion or to the absorption of impurities into surface porosities and pits. These mechanisms are supported by the data generated with silicate restorations. This was the only material to show significant differences in CHO/N. The CHO/N was 1.0 at 1 year. This suggests that the carbohydrate is metabolized less efficiently by the silicate. It is known that silicate restorations leach fluoride with time. Therefore, the fluoride acts as an enzyme inhibitor.

The thicknesses of the integuments formed in the mouth vary and may depend on the substrate material. For example, sputtering times in Auger electron spectroscopy depth profiling required only 0.3 min to reach the amalgam substrate, while 2.4 min was required to reach the gold alloy substrate (Ref 86). Carbon, nitrogen, and oxygen were distributed in much the same manner as films formed on different substrates. The main difference between the integuments formed on the amalgam and on the gold alloy was the presence of tin ions with the amalgam and the presence of copper ions with the gold alloy. The release of substrate ions is likely to interact with the attachment of microorganisms and therefore with the metabolism of plaque.

Substrate Corrosion. Corrosion reactions involve diffusion of ions—whether cations from oxidations or dissolved O_2 and H^+ for reductions—through the formed integument. The surface coating has the ability to act as a diffusion barrier to the movement of ions. Released ions are likely to become complexed, or bound, to the proteins and glycoproteins constituting the integument and free native proteins in the bulk saliva, provided diffusion is not restricted by the integument. Insoluble corrosion products of the oxides, chlorides, sulfides, carbonates, phosphates, and so on, have the capability of being deposited at the alloy/film interface or becoming an integral part of the integument. Soluble products, in addition, may be released into the bulk saliva.

For one dental restorative alloy, it was shown that the polarization resistance of the alloy increased with protein concentration, while at the same time, the concentration of soluble species in solution also increased (Ref 56). This situation was explained by the increased effect of proteins in solubilizing corrosion products. Energy-dispersive spectroscopy (EDS) spectra of the corroded surfaces showed reduced peak intensities for chlorine and sulfur on surfaces exposed to the proteins. Therefore, even though the severity of corrosion is less in protein-containing solutions, increased levels of soluble products are still generated.

***In Vivo* Tarnished Film Compositions.** Auger thin-film analysis of the surfaces of dental alloys with varying compositions and after functioning in the mouth indicated that the tarnished films were due to chemical reactions between alloy and inorganic species and to the adsorption and deposition of organic matter (Ref 35). Carbon was the dominant nonalloying element by about six times, followed by oxygen, calcium, nitrogen, chlorine, sulfur, magnesium, silicon, phosphorous, aluminum, sodium, and tin. Of the elements from the alloy itself, copper was dominant. In a microprobe analysis of *in vivo* discolorations on gold alloys, both silver sulfides and copper sulfides were detected, depending on the composition of the alloy. Sulfur was found isolated and carbon was present in greatest quantities (Ref 95).

Intraoral (*In Vivo*) Versus Simulated (*In Vitro*) Exposures

Need for Laboratory Testing. The tarnish and corrosion behavior of dental alloys under actual oral environmental conditions is required. However, except for selected clinical trials, the initial testing of new and improved alloys for tarnish and corrosion resistance is usually carried out under laboratory conditions in either simulated or accelerated tests. This is so because of:

- The possible human exposure to harmful species
- The variability in the oral environmental conditions from person to person and even with the same person from location to location and with time
- As a result of the variability in the oral environment, the inability to follow the effects on tarnish and corrosion from changes in parameters in alloys and in solution

Most laboratory tests utilize an artificial saliva or a physiological saline solution, such as Ringer's solution (Table 4), diluted Ringer's solution, various concentrations of NaCl, and various concentrations of Na_2S. The main deficiencies with these solutions is that the nonelectrolytes, including the proteins, glyco-proteins, and microorganisms, are not included. This fails to produce the pellicle and integuments on laboratory samples that otherwise would have formed on all intraoral surfaces.

In spite of these shortcomings, for the most part, the inorganic salt solutions have become indicators for the aggressiveness of the oral environment. However, the inability to correlate *in vivo* to *in vitro* behaviors in some instances is likely because of the failure to account for the shortcomings (Ref 81).

The use of solutions with higher-than-normal concentrations accelerates the tarnish and corrosion processes. For example, 3200 immersions of 15 s/min duration in a 5% Na_2S solution with a Tuccillo and Nielsen tarnishing apparatus (Ref 96) is estimated to simulate 12 months of actual in-service use (Ref 97). Ringer's and 1% NaCl solutions, which contain about seven times the Cl^- concentration of human saliva, are used in anodic polarization tests to amplify peaks in current behavior (Ref 53, 54). Corrosion of conventional amalgams in Ringer's or 1% NaCl generates products that are morphologically similar to those from retrieved amalgams after intraoral use (Ref 98, 99).

A comparison of the tarnishing of three gold alloys, both *in vivo* and *in vitro*, indicated that the cyclic immersions in a 5% Na_2S-air environment predicted with considerable reliability the relative susceptibility for the alloys to tarnish (Ref 100). The tarnishing of 81 gold-silver-copper-palladium alloys also indicated accelerated laboratory exposures in Na_2S solution simulated *in vivo* use (Ref 101). *In vivo* and *in vitro* (Na_2S solutions) tarnishing of gold alloys in Na_2S solutions has shown the same microstructural constituents to be attacked (Ref 102, 103). Silver- and copper-rich lamellae were the constituents exhibiting sulfide deposits. Utilizing linear polarization, good agreement with calculated current densities was obtained between measurements whether performed *in vivo* (baboons) or *in vitro* with an artificial saliva for times up to 45 days (Ref 104). Differences that occurred between *in vivo* and *in vitro* (0.1% NaCl) tarnish measurements as determined by colorimetry were attributed to the effects of abrasion and buildup of plaque that occurs on *in vivo* surfaces (Ref 105). Figure 18 shows the potential plaque buildup that can occur on a nonocclusal surface of a restoration.

Artificial Solutions in Corrosion and Tarnish Testing. As already indicated, the interior surfaces of restorations are exposed to the interstitial fluids and the exterior surfaces to the salivary fluids. A physiological saline solution, such as Ringer's, which contains a Cl^- concentration of about seven times larger than artificial saliva, is therefore more appropriate for simulating *in vivo* interior surfaces in laboratory testing methodologies. The O_2 content should be reduced to simulate *in vivo* levels in dentin. The use of Ringer's and even higher Cl^- concentrations is appropriate for the testing of corrosion that may occur within marginal crevices of restorations, because crevices can become chloride-rich and acidified. However, applying these results to the corrosion occurring on exterior surfaces of restorations may not be appropriate, even when considering that the increased Cl^- corrosion with Ringer's solution would be an even more stringent test and that the results would correspond to maximum corrosion conditions.

An artificial saliva is more appropriate for testing the corrosion of the exterior surfaces of restorations. The artificial saliva should take into account most of the species contained in saliva and not just a selected few that have been known to affect alloy corrosion. The artificial saliva should include the capabilities for generating organic films on the surfaces, even though their effects in isolated tests may prove unimportant. In order to simulate oral environmental conditions for the tarnishing of the exterior surfaces of restorations, an artificial saliva incorporating sulfide is appropriate. Even though the normal sulfide concentrations contained in saliva are within low ranges, accumulations of sulfide can occur along and within crevices to justify the use of higher than normal concentrations. However, the sulfur peak intensities detected with secondary ion mass spectroscopy (SIMS) on alloy surfaces exposed to low levels of sulfide solutions were similar to those from solutions containing higher sulfide concentrations. However, the alloy surface color changes responded more to higher sulfide concentrations.

Classification and Characterization of Dental Alloys

As indicated in the introduction to this article, a wide range of dental alloys exists. This section will review the following types of alloys available for dental applications:

- Direct filling alloys
- Crown and bridge alloys
- Partial denture alloys
- Porcelain fused to metal alloys
- Wrought wire alloys
- Soldering alloys
- Implant alloys

The effects of composition and microstructure on the corrosion of each alloy group will be discussed. Additional information on tarnishing and corrosion behavior of these alloys is discussed in the section "Tarnishing and Corrosion Under Simulated or Accelerated Conditions" in this article.

Direct Filling Alloys

Amalgams. Two types of amalgams are used: low copper (referred to as conventional) and high copper. The alloy particles of the low-copper type are all of the single-particle variety, whereas the high copper type can also be of the dispersed particle variety.

Amalgams are produced by combining mercury with alloy particles by a process referred to as trituration. About 42 to 50% Hg is initially triturated with the high-copper types, while increased quantities of mercury are used with the low-copper types. High-speed mechanical amalgamators achieve mixing in a matter of seconds. The plastic amalgam mass after trituration is inserted into the cavity by a process of condensation. This is accomplished by pressing small amalgam increments together until the entire filling is formed. For amalgams using excess mercury during trituration, the excess mercury is condensed to the top of the setting amalgam mass and scraped away. Dental amalgam alloys can become certified by complying with the requirements of American National Standards Institute (ANSI)/American Dental Association (ADA) Specification No. 1, which covers alloys for dental amalgams (Ref 106, 107). Table 5 presents compositions for a number of different amalgam alloys.

Low-Copper Conventional Amalgams. The alloy particles with the low-copper type are basically Ag_3Sn, the γ phase of the Ag-Sn system, even though smaller amounts of the β phase, a phase richer in silver, may also be present. Copper can be added in amounts up to about 5 wt% and zinc up to 1 to 2%. About 2 to 4% Cu is soluble in Ag_3Sn, while the additional copper

Table 5 Compositions of some dental amalgam alloys

Alloy	Composition, wt%			
	Ag	Sn	Cu	Zn
Low-copper				
Cresilver	75.0	24.6	0.1	0.3
Pure Lab	72.0	26.0	1.0	1.0
New True Dentalloy	72.8	26.2	2.4	1.0
Microalloy	69.0	26.6	3.5	0.9
Lustralloy	68.0	26.0	5.1	0.9
High-copper-dispersed				
Optalloy II	69.8	20.0	9.7	0.5
Dispersalloy	69.3	18.1	11.6	1.0
Cluster	69.8	16.2	13.5	0.5
Phasealloy	62.0	18.5	18.5	1.0
Cupralloy	63.5	16.9	19.5	0.2
High-copper-single				
Tytin	59.5	27.6	12.2	0
Indiloy	60.0	22.0	13.0	(5.0 In)
Valiant	49.5	30.0	20.0	(0.5 Pd)
Cupralloy ESP	41.0	32.5	26.5	0
Sybraloy	40.0	31.2	28.8	0

Source: Ref 67

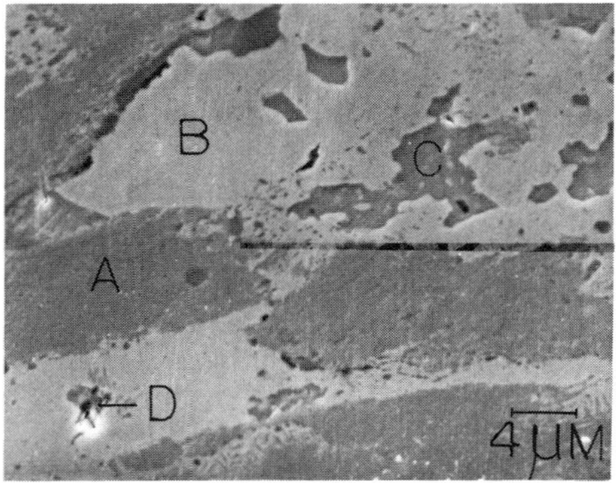

Fig. 20 SEM micrograph of polished, etched, and partially repolished low copper amalgam (Minimax). A, γ; B, γ_1; C, γ_2; D, porosity.

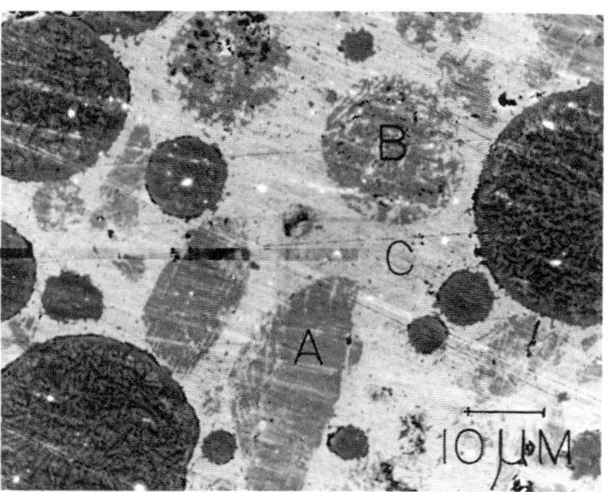

Fig. 21 SEM micrograph of polished, etched, and partially repolished high-copper dispersed-phase amalgam (Cupralloy). A, γ; B, Ag-Cu; C, γ_1. See also Fig. 22.

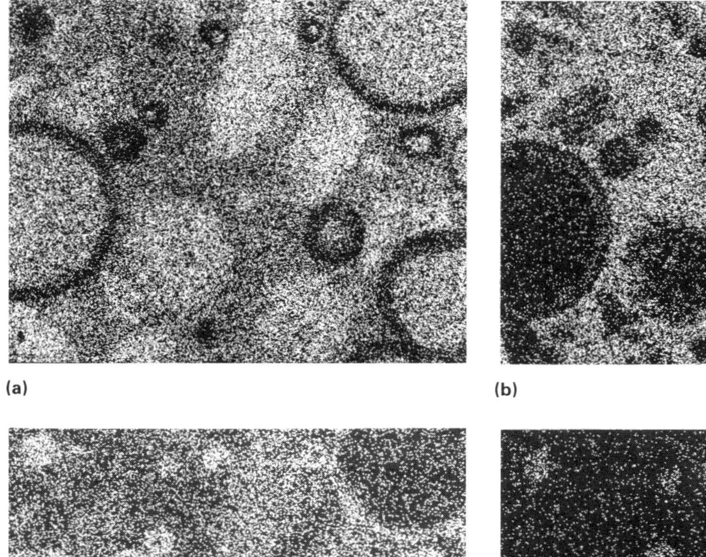

(a)

(b)

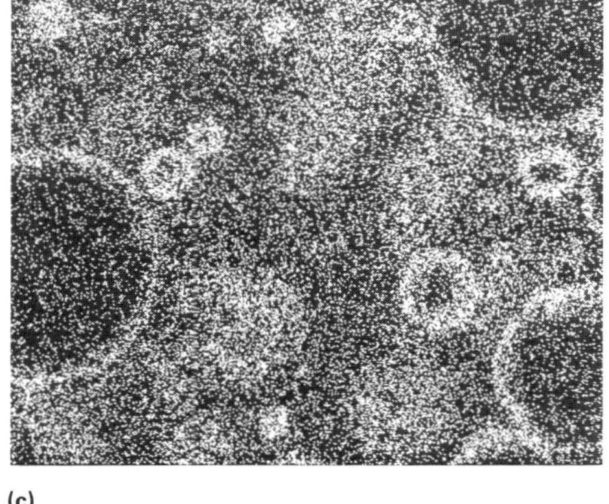

(c)

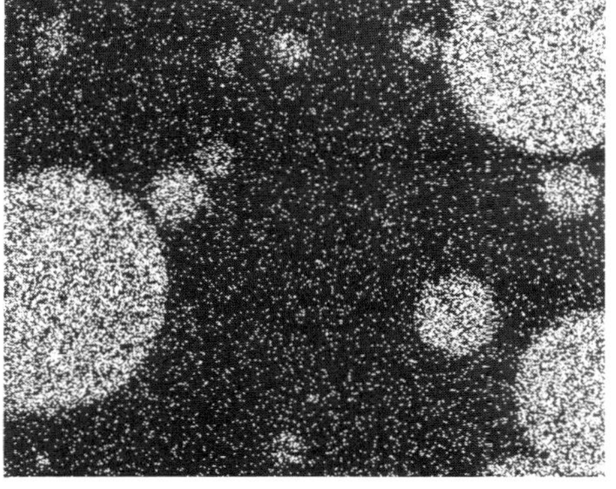

(d)

Fig. 22 Elemental maps obtained by energy-dispersive spectroscopy of the high-copper amalgam shown in Fig. 21. (a) EDS mapping for silver. (b) EDS mapping for mercury. (c) EDS mapping for tin. (d) EDS mapping for copper

usually precipitates as Cu_3Sn, the ϵ phase of the Cu-Sn system, although amounts of Cu_6Sn_5, the η' phase, may also occur. The low-copper particles are mostly lathe cut irregular, although spherical atomized particles are also used. The amalgamation reaction for a low-copper amalgam is:

$$Ag_3Sn\text{-}Cu_3Sn\text{-}Zn + Hg \rightarrow Ag_{22}SnHg_{27} + Sn_8Hg + Ag_3Sn\text{-}Cu_3Sn\text{-}Zn \text{ (unreacted)} \quad \text{(Eq 6)}$$

In Eq 6, Ag_3Sn with Cu_3Sn and zinc react with mercury to form two major reaction products of $Ag_{22}SnSn_{27}$ (the γ phase of the Ag-Hg system with dissolved tin and referred to as the γ_1 amalgam phase) and Sn_8Hg (the γ phase of the Sn-Hg system and referred to as the γ_2 amalgam phase) (Ref 108). Unreacted Ag_3Sn with Cu_3Sn and zinc particles are held together in a γ_1 matrix with γ_2 interspersed within the matrix. Typical distributions of the phases range up to about 30 wt% for γ, 60 to 80% for γ_1, 5 to 30% for γ_2, and up to about 3% for ϵ (Ref 109). Very minimal η' may also form. Zinc is generally distributed uniformly throughout material. Porosities are in all amalgam structures. As high as 6 to 7 vol% occur with some systems (Ref 110). Interconnection of the γ_2 phase throughout the bulk may also occur (Ref 111). Transformation of the γAg-Hg (γ_1 amalgam phase) to the βAg-Hg phase (β_1 amalgam phase) can also occur with aging (Ref 112). However, because of the dissolved tin in the γ_1 structure, stability is increased. Figure 20 presents the microstructure of a polished low-copper amalgam.

High-Copper Amalgams. The alloy particles with the high-copper dispersed-phase type are blends of conventional particles with basically spherical silver-copper eutectic particles in the proportion of about 3 to 1, respectively. The dispersed particles can be composed of a variety of silver-copper compositions, with other alloying elements, and combined in varying proportions with the conventional particles.

The alloy particles with the single-particle high copper are compositions that can contain up to 30% Cu and more. The particles are mostly atomized into spherical shape.

The setting reaction for high-copper dispersed-phase amalgam is:

$$Ag_3Sn\text{-}Cu_3Sn\text{-}Zn + Ag\text{-}Sn + Hg \rightarrow Ag_{22}SnHg_{27} + CU_6Sn_5 + Ag_3Sn\text{-}Cu_3Sn\text{-}Zn \text{ (unreacted)} + Ag\text{-}Cu \text{ (unreacted)} \quad \text{(Eq 7)}$$

and for high-copper single-particle amalgam is:

$$Ag_3Sn\text{-}Cu_3Sn\text{-}Zn \rightarrow Ag_{22}SnHg_{27} + Cu_6Sn_5 + Ag_3Sn\text{-}Cu_3Sn\text{-}Zn \text{ (unreacted)} \quad \text{(Eq 8)}$$

For dispersed-phase amalgam, γ initially reacts with mercury to form γ_1 and γ_2 phases, as with the low-copper amalgam. However, an additional reaction occurs between γ_2 and the silver-copper particles to form η' and additional γ_1. The η' phase forms reaction rings around the dispersed particles as well as islands of reaction phase within γ_1 matrix. Figures 21 and 22 present the microstructure of a dispersed-phase amalgam and EDS x-ray mapping for silver, mercury, tin, and copper, respectively.

For single-particle high-copper amalgam, reaction of the initially formed γ_2 phase occurs with the ϵ phase of the original alloy particles instead of with a dispersed particle to form the η' phase again. The γ_1 phase is likely to become tin enriched. Reaction zones around the original alloy particles, as well as products within the matrix, occur.

Figure 23 shows the microstructure of a polished, etched, and slightly repolished high-copper single-particle amalgam that shows primarily the distribution of the η' phase. The elimination of γ_2 phase and the subsequent formation of η' phase are time dependent and are dependent on the amalgam system (Ref 113). For fast reacting amalgams, formation of η' may be complete within hours, while for slower reacting systems, η' may continue to form for months. The single-particle high-copper amalgams contain higher percentages of the η' phase and lower percentages of the γ_2 phase, although all high-copper amalgams

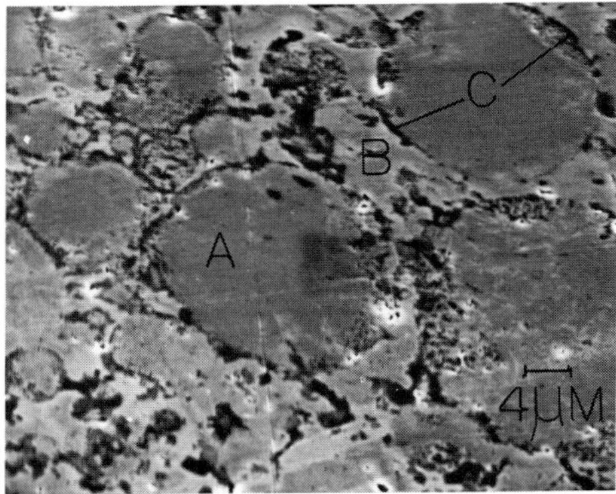

Fig. 23 SEM micrograph of polished, etched, and partially repolished high-copper amalgam (Sybraloy). A, alloy particles; B, γ_1; C, η'

contain minimal γ_2 relative to conventional amalgams. Porosities with the high-copper amalgams can be up to approximately 5 vol % and with a smaller size distribution than with the conventional type (Ref 110). The γ_1 to β_1 transformation can also occur to a limited extent. With the high-copper amalgams, both indium (5%) and palladium (0.5%) containing amalgams add specific characteristics to the amalgams and have gained limited use.

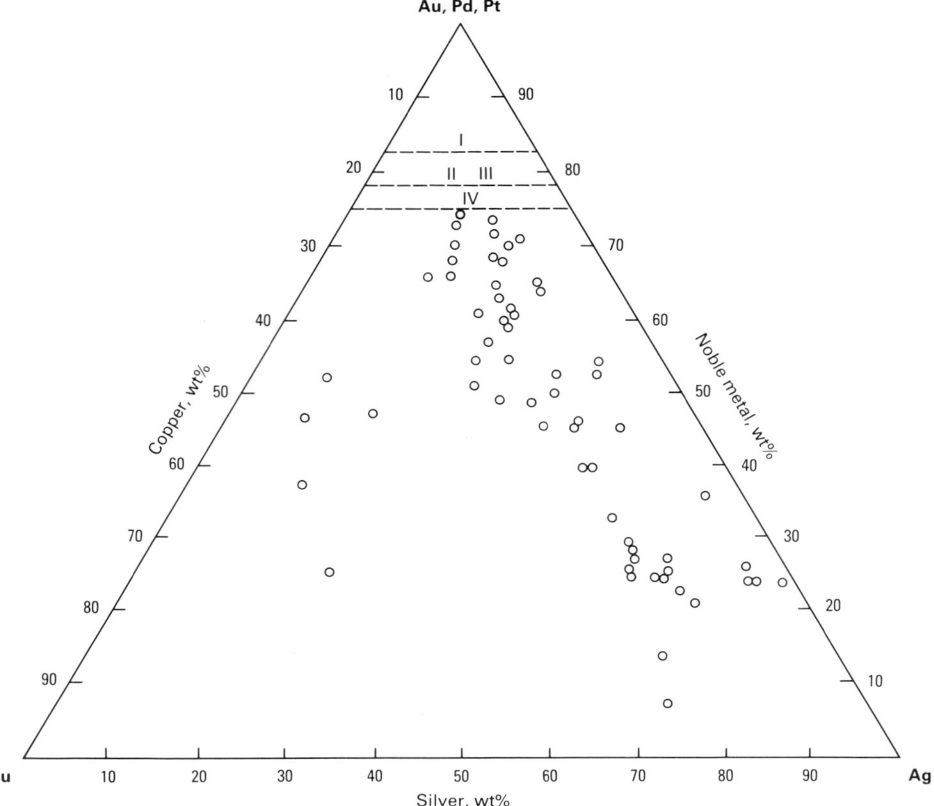

Fig. 24 Compositions of some commercially available low gold content crown and bridge alloys projected onto a pseudo noble metal (Au, Pt, Pd)-Ag-Cu ternary phase diagram. Also shown near the noble metal apex are minimum noble metal contents for Types I, II & III, and IV conventional alloys required by ANSI/ADA Specification No. 5.

Crown and Bridge and Partial Denture Alloys

High Nobility Alloys. Gold-base alloys for cast appliances have traditionally been based on the chemical compositional requirements of ANSI/ADA Specification No. 5, which covers gold dental casting alloys (Ref 114). This specification requires a minimum of 75 wt% Au plus platinum group metals. Alloys meeting the requirements of the specification became known as the conventional alloys, of which there were Types I to IV (Table 6). This requirement has served to reject alloys that would have tarnished in the mouth. Some alloys that have complied with the compositional requirement were, however, found to tarnish in the mouth, while others with lower nobility were developed with very good tarnish resistance. As a result, Specification No. 5 may be subject to revisions.

Composition of High Nobility Alloys. The primary alloys elements of high nobility alloys are gold, silver, and copper, along with additions of palladium, platinum, and zinc, as well as grain-refining elements, such as rubidium and iridium. Figure 24 shows the compositions for about 75 currently available dental casting alloys projected onto a pseudo (Au, Pd, Pt)-Ag-Cu ternary diagram. Actually, all of the compositions shown are referred to as low gold content alloys, which are covered in the following section. Also shown in Fig. 24, near the gold + platinum group metal apex, are dashed lines referring to the minimum nobility content of the four types (I to IV) of conventional alloys from the existing ANSI/ADA Specification No. 5. High, medium, and low nobility alloys are currently used. High and medium nobility alloys refer to compositions with gold plus platinum group metal contents greater than 75 wt%, while the low nobility alloys refer to compositions with less than 75 wt% Au plus platinum-group metals. Table 6 includes compositions for a number of high and low gold content alloys.

For the most part, properties of the high nobility cast dental alloys follow similar patterns as those shown by the gold-silver-copper ternary alloys, although the additional alloying elements in the dental alloys have significant effects on properties. Because the high nobility alloys have relatively small liquidus-solidus gaps, casting segregations, inhomogeneities, and coring effects are not major problems. Microstructurally, single-phase structures predominate because compositions fall within the single-phase region of the Au-Ag-Cu system. Grain refinement by the noble metal additions of ruthenium and iridium decreases grain sizes to 20 to 50 µm and increases strengths and elongations by about 30 and 15%, respectively (Ref 116).

Hardening Mechanisms of High Nobility Alloys. Type III and IV cast alloys, which are used for restorations subjected to high biting stresses, can be hardened by heat treatment. The primary hardening mechanism in gold dental alloys is by disorder-order superlattice transformations of the Au-Cu system. The ordered domains of the Au-Cu binary system also extend into the ternary phase Au-Ag-Cu regions. Typical heat treatment times and temperatures are 15 to 30 min at 350 to 375 °C (660 to 705 °F). Often, it is adequate to bench cool the casting in the investment after casting to gain hardness. Because of the complexity of the dental gold alloy compositions, the exact hardening mechanisms depend on the particular composition of the alloys. Table 7 presents a schematic representation of the age-hardening mechanisms and related microstructures occurring in gold dental alloys. Included are representations for high-gold alloys (HG), low-gold alloys (LG), gold-silver-palladium base alloys (GSP), and 18 karat and 14 karat gold alloys. Five types of phase transformations are found (Ref 117):

- The formation of the AuCu I ordered platelets and twinning characterized by a stair-step fashion
- The formation of the AuCu II superlattice with periodic antiphase domain structure
- The precipitation of the PdCu superlattice with face-centered tetragonal (fct) structure analogous to the AuCu I
- Spinodal decomposition giving rise to a modulated structure
- The formation of the lamellae structure developed from grain boundaries by discontinuous precipitation

Low Nobility Alloys. Ever since the increases in the gold prices have taken place, new economy golds with lower gold contents have been introduced and have become popular. Figure 25 shows the trends in sales of dental casting alloys in the United States from 1976 to 1981. In a more recent survey of 488 dental laboratories, 46.5% of the responses indicated that only 0 to

Table 6 Compositions of noble metal containing crown and bridge alloys

Alloy	Type	Nobility at.%	Nobility wt%	Au	Pt	Pd	Ag	Cu+Zn
22K	I	83.1	91.7	91.7	...	—	5.6	2.7
NeyOro A-A	I	73.9	85.0	81.0	...	4.0	12.0	3.0
NeyOro A-1	II	64.3	80.0	78.0	...	2.0	13.0	7.0
NeyOro B-2	III	60.6	78.0	74.0	...	4.0	12.0	10.0
NeyOro G-3	IV	57.3	76.0	69.0	3.0	4.0	12.0	12.0
NeyOro #5	IV	48.2	68.0	64.0	2.0	2.0	19.0	13.0
NeyOro B-20	IV	47.3	65.0	62.0	...	3.0	26.0	9.0
Densilay	III	46.6	64.0	60.0	...	4.0	27.0	9.0
Sterngold 20	III	45.1	63.5	59.5	...	4.0	25.0	11.5
NeyOro CB	III	43.8	63.0	59.0	...	4.0	23.0	14.0
Midigold	III	34.6	52.0	48.5	...	3.5	35.0	13.0
Tiffany	IV	34.1	54.0	50.0	...	4.0	25.0	21.0
Pentron 20	III	32.2	40.0	20.0	...	20.0	40.0	20.0In+Zn
Sunrise	III	30.8	46.0	39.0	1.0	6.0	41.0	13.0
Paliney CB	III	30.5	39.0	15.0	1.0	23.0	44.0	17.0
Duallor	III	30.4	45.9	40.0	...	5.9	40.5	13.6
Neycast III	III	29.9	52.0	42.0	2.0	8.0	9.0	39.0
Albacast	III	24.4	25.0	...	...	25.0	70.0	5.0
Miracast	III	24.1	46.0	41.0	1.0	4.0	9.0	45.0
Ney 76	III	22.7	25.0	...	...	25.0	59.0	16.0
Econocast	III	18.3	36.0	26.0	...	10.0	...	64.0
Salivan	III	8.2	8.0	...	...	8.0	70.0	22.0In

Source: Ref 115

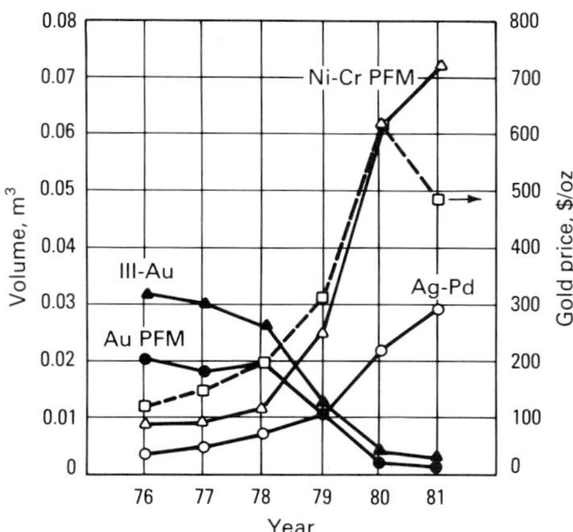

Fig. 25 Trends in alloy use along with the market price of gold for the years 1976 to 1981. Source: Ref 118

25% of the crown and bridge alloys that they processed contained gold. Only 27.9% responded that 51% or more of their alloys contained gold (Ref 119).

Compositions of Low Nobility Alloys. Low noble metal alloys comprise a wide variety of compositions (Table 6). Gold-, palladium-, silver-, and copper-base alloys are used. Platinum and zinc contents are usually held to several weight percent maximum, if present. Microstructurally, the low gold content casting alloys are complex, and examination of phase diagrams of either the Au-Ag-Cu or Ag-Pd-Cu ternary systems indicates that the liquidus-solidus gaps between the various phases can be large (Ref 116). Therefore, coring and casting segregations occur upon solidification during casting. The alloys are characterized by dendritic structures combined with additional phases located within interdendritic positions. Both silver- and copper-rich segregations occur. The presence of gold, platinum, zinc, and other alloying elements further complicates the structures.

The addition of palladium and zinc to the Au-Ag-Cu alloy systems makes heat treatments to single-phase structures difficult. In silver-rich phases, the solubility limit for palladium and zinc is only about 1 to 2% at 500 °C (930 °F), while for copper-rich phases, solubilities are much higher—of the order of 10%. Therefore, precipitation of palladium- and zinc-rich phases occurs. Differences also occur in the gold contents between the phases with the copper-rich phases having the higher contents (Ref 120).

Silver palladium base alloys with additions of copper, gold, and zinc are also complex and contain multiple phases. Microstructurally these alloys are characterized by silver-rich matrices interspersed with Pd-Cu-Zn enriched compounds. As many as three different Pd-Cu-Zn compounds have been detected in one alloy system. The Pd-Cu-Zn compounds are actually composed of two face-centered cubic (fcc) phases in addition to a body-centered cubic (bcc) $PdCu_xZn_{x-1}$ phase. In addition, the silver-rich matrixes are normally cored with silver-enriched dendritic arms and copper segregations in the interdendritic areas (Ref 121).

The hardening mechanisms associated with the low nobility alloys are shown in Table 7. In addition to the gold-copper disorder-order transformations, ordering due to the palladium-copper superlattices is also usually involved because of replacement of some of the gold by palladium.

Silver-indium alloys with small additions of palladium or gold have also been used for crown and bridge applications. Figure 26 presents microstructures for a number of low-gold alloys.

Base Metal Alloys. Most of the base metal alloys for all metal cast crowns and bridges are composed of nickel-chromium alloys, although stainless steels are also used, particularly outside of the United States. Nickel-chromium alloys are also used for the construction of partial dentures. However, cobalt-chromium alloys far exceed any other alloy for use in this application. Titanium alloys are also being developed for dental applications. Table 8 presents the compositions for a number of base metal alloys intended for crown and bridge, partial denture, porcelain fused to metal, and implant applications.

Nickel-Chromium Alloys. The primary alloying element with the nickel-base alloys is chromium between about 10 and 20 wt%. Molybdenum up to about 10%; manganese and aluminum up to about 4% each; and silicon, beryllium, copper, and iron up to several percent each can also be added. The carbon contents range between about 0.05 and 0.4%. Elements such as gallium, titanium, niobium, tin, and cobalt can also be added. Because of differences in properties required for crown and bridge applications versus partial denture applications, minor modifications in compositions occur between nickel-chromium alloys intended for the two applications. This is reflected by the fact that crown and bridge nickel-chromium alloys contain higher percentages of iron, very minimal or no aluminum and carbon, and copper additions (Ref 131).

Chromium and molybdenum are added for corrosion resistance. In order to be effective, these elements must not be concentrated along grain boundaries. Chromium contents below about 10% deplete the interior of the grains leading to corrosion. Molybdenum protects against concentration cell corrosion, such as pitting and crevice corrosion, and is also a solid-solution hardener. Manganese and silicon are reducing agents, while aluminum also improves corrosion resistance and improves strength through its formation of intermetallic compounds with nickel. Silicon, like beryllium and gallium, lowers the melting temperature. Beryllium is also a solid-solution hardener and improves castability. Niobium, like molybdenum and iron, affects the coefficient of thermal expansion. Gallium is a stabilizer and improves corrosion resistance. The oxide-forming elements and the elements promoting good bond strength to porcelain will be discussed in the section "Porcelain Fused To Metal Alloys" in this article.

Although the compositions for many of these alloys are similar, differences occur with regard to microstructure and properties. In comparing the microstructures of four different commercial nickel-chromium alloys, it was noted that one of the alloys had thin, elongated carbides and intragranular precipitates adjacent to grain boundaries. A second alloy showed discontinuous spherical carbides, the third showed a continuous grain-boundary phase and cored dendrites, and the fourth showed a dendritic precipitate (Ref 132). In another alloy, the dark interdendritic carbides changed to a lamellar pattern at certain sites and were surrounded by dispersed carbide particles and the δ′ phase of Ni_3Al (Ref 133). Heat treatments were shown to have drastic effects on the microstructures of several as-cast nickel-chromium alloys (Ref 134).

Cobalt-Chromium Alloys. The primary alloying element with the cobalt-base alloys is chromium between about 20 to 30%. Molybdenum is also present in amounts up to about 10%; small additions of elements such as silicon, manganese, iron, and nickel are sometimes also present. The carbon contents usually range between about 0.05 and 0.4% (Ref 131). This composition forms the basis for two additional generalized compositions. The first includes a group of alloys that has been developed from the above basic composition, but with each modified by the addition of one or more elements in order to obtain a particular range of properties (Ref 135). Some of these modifying elements include gallium, zirconium, boron, tungsten, niobium, tantalum, and titanium.

Table 7 Hardening mechanisms for some dental alloys

The hatched areas represent the hardness peaks on aging. (9/5) °C + 32 °F.

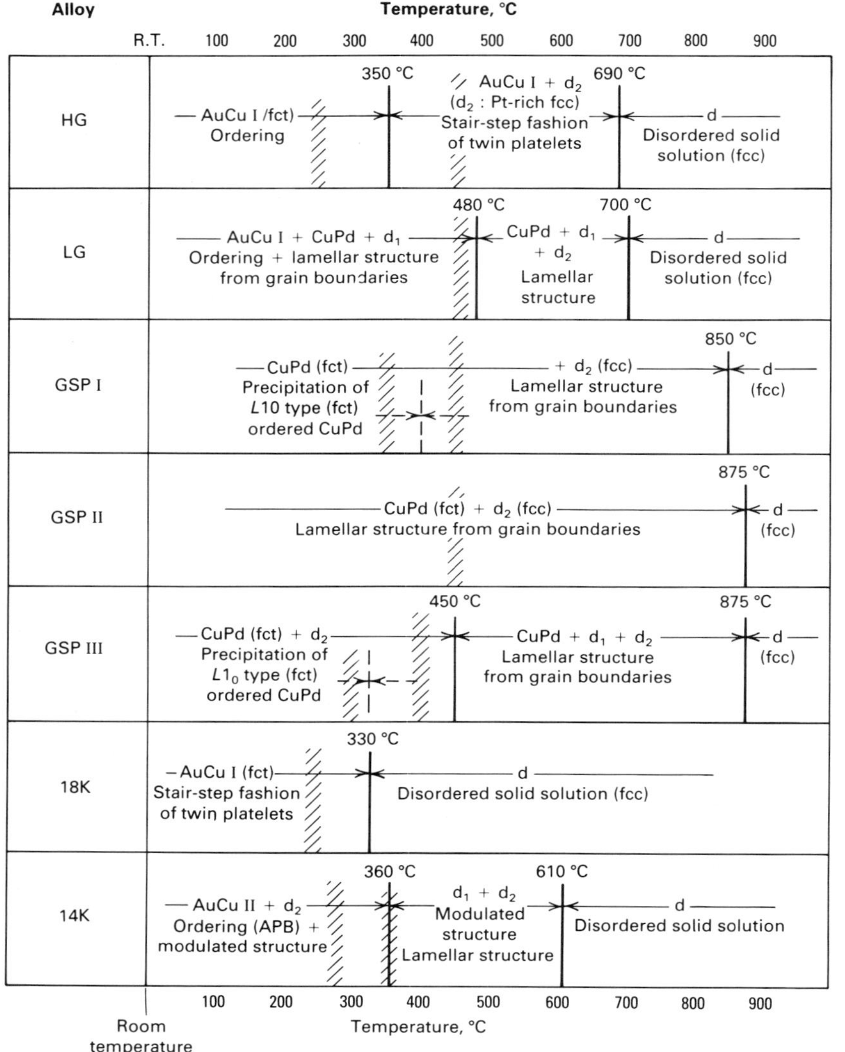

Alloy	Composition, wt% Au	Pt	Pd	Ag	Cu
HG	68	11	6	6	9
LG	30	...	22	29	18
GSP-I	20.0	...	25.2	44.9	9.9
GSP-II	12.0	...	28.0	48.8	11.2
GSP-III	10.0	...	25.4	50.0	12.8
18K	75.0	...	...	8.7	16.3
14K	58.3	...	...	14.6	27.1

Source: Ref 117

The second generalized composition includes replacement-type alloys, with a major portion of the cobalt replaced by nickel and/or iron. The resulting composition is a cross between cobalt-base alloys and stainless steels. The effects of the individual elements on the properties of the alloys are similar to those already discussed for the nickel-chromium alloys. The alloys are hardened primarily by carbide formation. Therefore, the carbon content is of primary importance.

A number of carbides, including MC, M_6C, M_7C_3, and $M_{23}C_6$, have been detected in dental cobalt-chromium alloys (Ref 136). Solid-solution strengthening also occurs, as well as intermetallic compound formation strengthening, with a number of the modified alloy types. Certified alloys for partial dentures must contain a minimum cobalt-chromium-nickel content of 85% with a minimum of 20% Cr as required by ANSI/ADA Specification No. 14 for dental base metal casting alloys (Ref 137).

Microstructurally, the cobalt-chromium alloys are typified by a cored austenitic solid-solution matrix interspersed with isolated carbides, as shown in Fig. 27. However, depending on compositions, the microstructures can vary considerably. In comparing the microstructures of a number of different cobalt-chromium alloys, the unmodified alloys with the basic composition given above exhibited large grains, with both large and small carbides being, respectively, either randomly dispersed or precipitated along grain boundaries (Ref 135). For the modified alloys containing additional alloying elements, the grain sizes were smaller than for the unmodified materials, with the carbides dispersed within the grains rather than being precipitated along grain boundaries.

The microstructures for modified cobalt-chromium alloys were analyzed in greater detail (Ref 138). For one typical alloy, the underlying matrix had the highest cobalt concentration with lowest chromium and molybdenum concentrations. The numerous continuous precipitates located along grain boundaries were moderately high in cobalt and low in chromium and molybdenum. The dark areas within the grain interiors contained high carbon, moderately high chromium and molybdenum, and low cobalt. Dispersed precipitates located within the matrix had high chromium and molybdenum contents and low cobalt.

An extensive project was undertaken regarding the development of a new dental superalloy system (Ref 139). The outcome of the project indicated that a cobalt-base alloy with desirable properties for dental applications could be made from a 40Co-30Ni-30Cr matrix. The alloy is strengthened by the precipitation of coherent intermetallic compounds of tantalum. Alloys that are stronger and more ductile than conventional dental alloys were obtainable. The results were reinvestigated, and it was again concluded that tantalum does improve the properties somewhat (Ref 140).

Stainless Steel Alloys. At least one alloy (Dentillium) is commercially available for crown and bridge applications (Table 8). This alloy, as well as its modified version for partial denture applications, is an iron-base composition with 24 to 28% Cr, 6 to 8% Co, 4 to 6% Ni, and 2.5% Mo. Microstructurally, two phases exist (Ref 138). The matrix is high in chromium, molybdenum, iron, and cobalt and is low in nickel, but the precipitates are high in nickel and reduced in chromium, molybdenum, iron, and cobalt. Stainless steel preformed crowns are also occasionally used for temporary restorations.

Titanium-Base Alloys. Cast titanium alloys have only within about the past decade been shown to possess potential for crown and bridge and partial denture applications, as well as for porcelain bonding. Because titanium alloys have high melting temperatures, casting of these alloys has not been possible until recently in the dental laboratory. The introduction of a lower-melting titanium alloy has made this requirement achievable (Ref 141). The first alloy to be successfully cast had a composition of 82Ti-13Cu-4.5Ni with a melting temperature of 1330 °C (2426 °F). The introduction of an argon/electric arc vertical centrifugal casting machine and a vacuum-argon electric arc pressure casting machine made the casting of the higher melting point titanium alloys also achievable. Pure titanium and Ti-6Al-4V have been successfully cast by these latter methods (Ref 142). Better investment materials are currently being developed.

Copper-Aluminum Alloys. The copper-aluminum restorative alloys have elicited renewed interest because of their similarity in appearance to the yellow gold alloys and because of the volatility of gold prices. The compositions include copper base, with about 10 to 20% Al, and up to approximately 10% iron-manganese nickel (Table 8). The as-cast etched structures are dendritic.

Alloy Color. Color is a surface characteristic that is related to alloy composition. Although color by itself has no affect on biocompatibility and the properties of the alloy, it has become equated with quality for many people, including some professionals. Alloy color remains an important issue. The white golds are held by many to be inferior to the yellow golds. The significance of yellow color as an indicator of alloy quality is so strong that a number of nonnoble yellow alloys, such as copper-aluminum, that resemble gold alloys in appearance and lack tarnish resistance have become available.

Yellowness in dental gold alloys is imparted by the gold and copper. Absence of yellowness is no assurance that the alloy is lacking gold or copper content. Color can be a misleading indicator of composition. For example, some dental gold alloys for porcelain fused to metal restorations contain more than 80% Au, yet yellowness is absent because of the strong whitening effect of palladium and platinum. Detailed information on alloy color and the characterization of dental alloys by colorimetry can be found in Ref 105, 143, and 144 to 149.

Factors Related to Casting. Dental alloys must be cast into thin sections and intricate shapes in order to produce the margins and curvatures on restorations and to provide an accurately fitting casting. The castability of dental alloys is evaluated by the ability to be cast into shapes, such as wedges, spirals, rods of various shapes, and mesh grids of various sizes. The casting accuracy is usually evaluated by the degree of fit a finished casting possesses with a master die.

Many factors determine the castability of alloys. Some of these include the casting temperature and surface tension of the molten alloy as well as many variables associated with the casting technique, some of which include wax pattern preparation, position of the pattern in the casting ring, techniques used in alloy heating, and centrifugal casting force. Some factors affecting casting accuracy include the thermal contraction of the alloy as a result of going from liquid to room temperature, the effectiveness of investment material to compensate for the thermal contraction of the alloy, anisotropic contractions, and the roughness of the casting.

There are indications that the base metal castings produce inferior fits for crown and bridge uses as compared to castings made from conventional gold alloys (Ref 150). Low-gold alloys with about 50% Au generated fits that were satisfactory. Part of the problem with the base metal alloys may be the use of casting techniques that were developed for the gold alloys.

Casting porosity is another important factor in the casting process. Although casting technique variables affect porosity contents, alloy composition can also affect porosity. One way in which composition affects porosity is the generation of internal shrinkage pores between microstructural phases in complex multiphase alloys. This microporosity weakens alloys, makes finishing and polishing more difficult, and is a prime factor in tarnishing. Palladium in alloys is susceptible to occluding gases from the melt. Therefore, palladium-containing alloys have the potential for becoming affected mechanically through embrittlement.

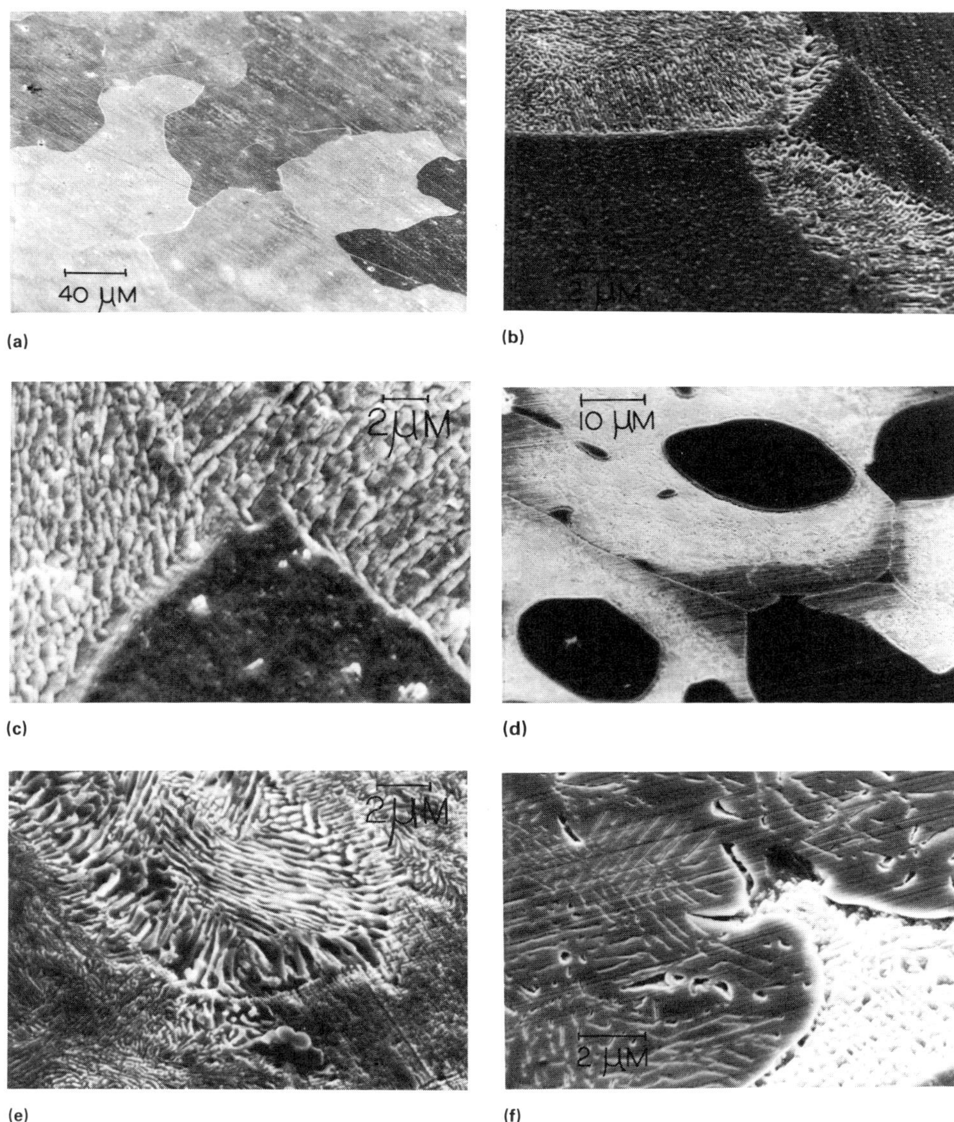

Fig. 26 SEM micrographs of etched as-cast low gold content crown and bridge alloys. (a) Sterngold 20. (b) Tiffany. (c) Midas. (d) Pentron 20. (e) Sunrise. and (f) Econocast. See Table 6 for chemical compositions of these alloys.

Porcelain Fused to Metal Alloys

Stringent demands are placed on the alloy system meant to be used as a substrate for the baking on or firing of a porcelain veneer. The thermal expansion coefficients of alloy and porcelain must be matched so that the porcelain will not crack and break away from the alloy as the temperature is cooled from firing temperature to room temperature. Thermal expansion coefficients of porcelains are in the range of 14×10^{-6} to 15×10^{-6} in./in. °C. Selection of an alloy with a slightly larger coefficient by about 0.05% is recommended so that the alloy will be under slight compression.

The alloy must be high melting so that it can withstand the firing temperatures involved with the porcelain. However, the temperature must not be excessively high so that conventional dental equipment can still be used. A temperature of 1300 to 1350 °C (2370 to 2460 °F) is about maximum. The porcelain firing procedures require an alloy with high hardness, strength, and modulus so that thin sections of the alloy substrate can support the porcelain, especially at the firing temperatures.

High mechanical properties are also required for resisting sag of long span bridge unit assemblies during firing. The alloy should also have the ability to absorb the thermal contraction stresses due to any mismatch in expansion coefficients as well as occlusal stresses without plastic deformation. Therefore, too high of a modulus for a material with an insufficient yield strength is contraindicated, although too high of an elastic deformation is also contraindicated. High bond

Table 8 Compositions of base metal wires, crown and bridge, porcelain fused to metal, partial denture, and implant alloys

Name	Composition, wt%											Ref
	Co	Ni	Cr	Mo	Fe	Si	Mn	Cu	Al	C	Other	
Wires												
18-8 ss	...	8–10	17–19	...	bal	...	2	...	...	0.08–0.2	...	122
Elgiloy	40	15	20	7	bal	...	2	...	...	0.15	...	122
Nitinol	1.4	48.6	...	...	...	...	...	...	...	...	50 Ti	123
β-titanium	...	...	...	11.5	...	...	...	...	...	...	bal Ti	124
Crown and bridge												
Howmedica III	0.3	67.5	19.7	4.2	0.1	3.0	1.2	1.8	...	0.1	2 Sn	125
Gemini II	...	79.9	12.4	2.0	0.1	0.1	0.1	0.1	2.8	0.2	2.1 Be	125
Dentillium	8.0	6.0	28.5	2.5	46.0	...	...	...	...	...	...	126
MS	...	4.7	...	...	3.0	...	0.7	71.8	19.8	...	...	68
Porcelain fused to metal												
Ceramalloy	0.01	69.9	19.9	5.6	0.1	1.0	0.01	...	...	0.2	2.9 B	125
Wiron S	0.02	70.6	15.7	4.5	0.2	1.5	3.2	...	3.8	0.1	...	125
Ultratek	+	81.0	11.4	2.0	...	...	0.02	+	2.2	...	Be	127
Partial denture												
Vitallium	62.5	...	30.0	5.0	1.0	0.5	0.5	...	...	0.5	...	126
Platinore	60.7	2.7	26.7	5.8	2.6	0.6	0.5	...	...	0.3	0.3W,0.1Pt	126
Nobillium	62.0	...	32.0	5.0	...	0.35	...	0.04	...	0.35	0.05 Ga	128
Implant												
Vitallium	61.1	0.3	31.6	4.4	0.6	0.6	0.7	...	...	...	...	129
Stellite 25	49.8	10.0	20.0	...	3.0	...	...	...	...	0.15	15 W	130
MP35N	35.0	35.0	20.0	10.0	...	...	...	...	...	...	...	130
Ticonium	15.4	54.3	24.6	4.3	0.7	0.5	0.03	0.03	0.02	...	...	129

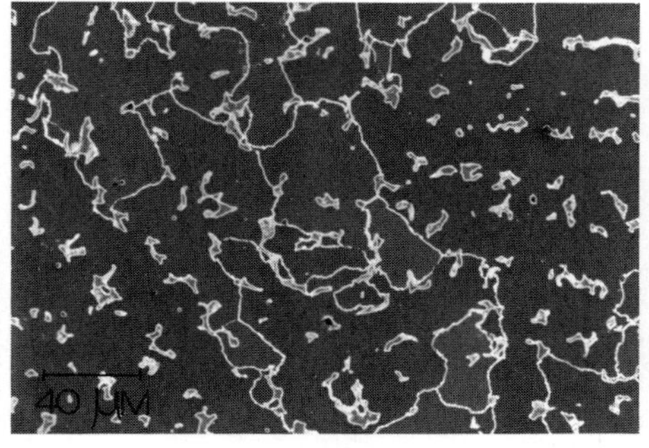

(a)

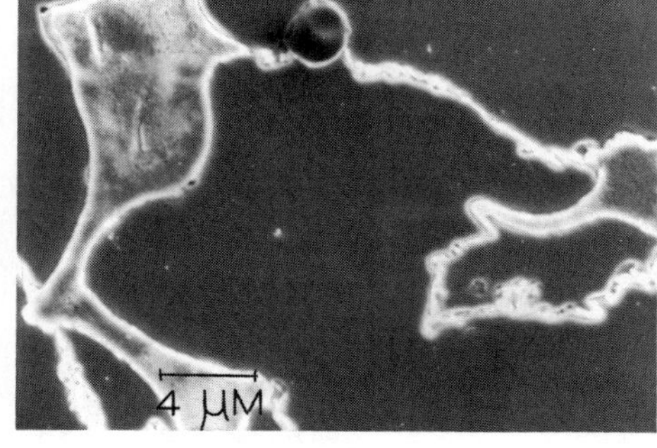

(b)

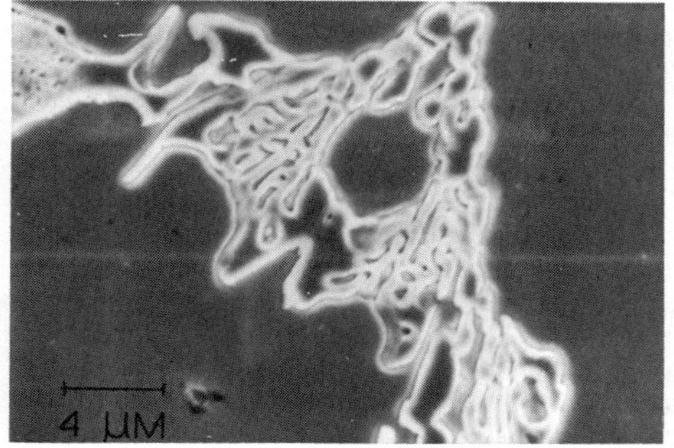

(c)

Fig. 27 SEM micrographs of etched cobalt-chromium partial denture alloys. (a) and (b) Neoloy. (c) Nobillium.

strengths are required so that the porcelain veneers remain attached to the alloy. The alloy system must also be chemically compatible with the porcelain. Alloying elements must not discolor porcelain, yet must have tarnish and corrosion resistance to the fluids in the oral environment. The compositions for a number of porcelain fused to metal (PFM) alloys are presented in Table 9.

In order to promote and form high bond strengths between porcelain and alloy, the alloy must have the ability to form soluble oxides that are compatible with the porcelain. At the firing temperature, the porcelain should spread or wet the surface of the alloy. Therefore, both mechanical and chemical interactions are involved in this process. In order to promote chemical interactions, specific oxide-forming elements, such as tin, indium, or gallium, are added to the alloy in low concentrations. The addition of iron, nickel, cobalt, copper, and zinc provide the means for hardening.

Alloy color with the PFM alloys is not as demanding as with the crown and bridge alloys because the alloy is masked by the porcelain veneer. However, the margins between metal and porcelain can still be seen. This consideration still makes the yellow alloys more aesthetically pleasing, especially for anterior restorations. White alloys have the effect of producing a grayish color.

Table 9 Compositions of noble metal porcelain fused to metal alloys

Alloy	Composition, wt%					Ref
	Au	Pt	Pd	Ag	Other	
Ceramco	87.5	4.2	6.7	0.9	0.3Fe, 0.4Sn	151
Degudent	84.8	7.9	4.6	1.3	1.3In, 0.1Ir	152
Vivostar	54.2	...	25.4	15.7	4.6Sn	153
Cameo	51.4	...	29.5	12.1	6.8In	153
P-D	59.4	...	36.4	...	4.0Ga	154
P-G	19.9	0.9	39.0	35.9	3Ni, 1.2Ga	154
JP92	...	...	60.5	32.0	7.5In	155
A 36	1.8	...	77.8	...	10.4Ga, 10Cu	156
Orion Star	...	...	78.0	...	In,Sn,Cu,Ga,Co	...

The fabrication of the PFM restoration consists of a complex set of processes. After casting, the alloy substrate is subjected to a preoxidation heat treatment to achieve the optimum surface oxides that are important for porcelain bonding. As many as three or more different porcelain firings follow. These include a thin opaque layer adjacent to the alloy, followed by body porcelain layers, including both dentin and enamel porcelain buildups. The opaque layer should mask the color of the alloy from interfering with the appearance of the porcelain. The body porcelain layers build up the restoration to the desired occlusion. The alloy-porcelain systems are slowly cooled from firing temperatures to accommodate dimensional changes occurring in both alloy and porcelain. Therefore, the alloy substrates are subjected to a number of temperature cycles during processing that affect their microstructure and properties. The slow cooling cycles also permit the formation of the compounds important for hardening of alloys.

Noble Metal PFM Alloys. The evolution of alloys for the PFM restoration has generated at least four different noble alloy systems. These are classified as gold-platinum-palladium, gold-palladium with and without silver, silver-palladium, and high-palladium content alloys (Ref 116, 140, 157, 158).

Gold-Platinum-Palladium PFM alloys. The compositions of alloys included within this group are in the range of 80 to 90% Au, 5 to 15% Pt, 0 to 10% Pd, and 0 to 5% Ag, along with about 1% each of tin and indium. Other additions may include up to about 1% each of iron, cobalt, zinc, and copper. Platinum and palladium additions increase melting temperature and decrease thermal expansion coefficients, with platinum having the added effect of hardening the alloy. Iron is the principal hardening agent. Iron promotes the formation of an ordered iron-platinum type intermetallic phase that forms between 850 and 1050 °C (1560 and 1920 °F) upon cooling from the firing temperature (Ref 116). The ordered phase is finely dispersed throughout matrix. Iron, along with tin and indium, promotes bonding to porcelain by diffusing into the porcelain up to about 60 μm at the firing temperature. Tin and indium also promote solid-solution strengthening. Alloys within this group range in color from light yellow to yellow.

Even though these alloys have many advantages, their high cost and low sag resistances have necessitated the development of additional alloy systems. Unless they are used in thick sections, for example, 3 × 3 mm (0.12 × 0.12 in.), plastic deformation of long spans will occur during firing.

Gold-Palladium and Gold-Palladium-Silver PFM Alloys. Gold-palladium with and without silver alloys were developed as alternatives to the costly gold-platinum-palladium alloys. The Au-Pd-Ag system was one of the first alternative systems. Up to 15% Ag and 30% Pd replaced all of the platinum and a large fraction of the gold from the Au-Pt-Pd system, which resulted in substantial cost savings. The gold-palladium-silver alloys possessed better mechanical properties for the PFM restoration. Because gold-palladium, gold-silver, and palladium-silver are all solid-solution alloys, the ternary Au-Pd-Ag system also forms a series of solid-solution alloys over the entire compositional ranges. Therefore, the matrices of the gold-palladium-silver dental alloys are single phase. Up to 5% Sn is added, which hardens the alloy by forming compounds with palladium that are dispersed throughout matrix. Tin is also an oxide former and bonding agent with porcelain. Because platinum was avoided, there was no need to incorporate iron for hardening. Their shortcoming was the ability of silver from the alloy to vaporize, diffuse, and combine with the porcelain at the firing temperature, thus inducing color changes, mostly greenish, along the alloy-porcelain margin. Sodium-containing porcelains were more susceptible to this color change.

The development of the silver-free gold-palladium alloys eliminated the discoloration of the porcelain. Their compositions cover a wide range: 50 to 85% Au, 10 to 40% Pd, 0 to 5% Sn, and 0 to 5% In, along with possible additions from zinc, gallium, and other elements. The alloy matrix is based on the Au-Pd binary system, which is of the solid-solution type. Hardening is due to Pd-(In, Sn, Ga, Zn) complexes that disperse throughout the matrix. Microstructurally, a fine network of gold-rich regions are entwined by second-phase particles of Pd-(Sn, In, Ga, and Zn). Their color is only a pale yellow, unlike some of the deeper yellow gold-platinum-palladium alloys. However, about a 30 to 40% cost savings is obtained. Their mechanical properties are superior, which means good sag resistance at the firing temperatures. The only disadvantage with these alloys is their lower thermal expansion coefficients when used with some of the higher-expanding porcelains.

Some newer gold-palladium alloys have up to 5% Ag, which is much lower than the 15% contents used with the original gold-palladium-silver alloys. This results in better thermal expansion matches with porcelain and avoids the discoloration problem with porcelain because of the lower silver concentrations. The gold-palladium alloys have gained a large percentage of the PFM alloy market.

Palladium-Silver PFM Alloys. These alloys were developed out of the need to reduce the cost of the PFM restoration even more than from those fabricated from gold-palladium alloys. Compositions range from 50 to 60% Pd, 25 to 35% Ag, 5 to 10% Sn, 0 to 5% In, and up to 2% Zn. The silver and palladium contents are just about reversed in magnitude as compared to those compositions used with the silver-palladium crown and bridge alloys discussed earlier in this article.

The microstructures of the alloys are based on the Pd-Ag solid-solution system. Instead of using copper to harden the alloys, hardening occurs through compounds formed between palladium and tin, indium, zinc, and others. Hardening rates are high, which indicates nondiffusional reactions. It is likely that hardening occurs by ordering processes that form by spinodal decomposition (Ref 116). Oxides are imparted to the alloy surface because of the oxide-forming ability of indium, tin, and zinc alloying additions. This promotes high bond strengths. The mechanical properties of the palladium-silver alloys, along with the high-palladium alloys discussed below, are superior to those of any other system, excluding the nickel-chromium alloys. As with the gold-palladium-silver alloys, the chief disadvantage is the ability of silver to discolor porcelains during firing. In order to overcome this problem, various methods have been used, including coupling agents composed of porcelains or colloidal gold.

High-palladium PFM. These alloys represent the most recent developments in alloys for PFM restorations (Ref 159-164). Their compositions are 75 to 85% Pd with 0 to 15% Cu, 0 to 10% Ga, 0 to 8% In, 0 to 5% Co, 0 to 5% Sn, and 0 to 2% Au. The alloys are based on either the Pd-Cu-Ga or the Pd-Co-Ga ternary systems. Regardless of the high copper contents, these alloys do not induce porcelain discoloration and bonding problems. Many of these alloys have better workability than other types of PFM alloys, while retaining high hardnesses. The hardness is dependent on the formation of intermetallic compounds with palladium upon cooling from the firing temperature. The alloy forms strong bonds with porcelain because of the alloy oxidizer content. Oxides form with palladium and the alloying additives. However, palladium oxide (PdO) forms only during heating and cooling because of the relatively low decomposition temperature for the oxide.

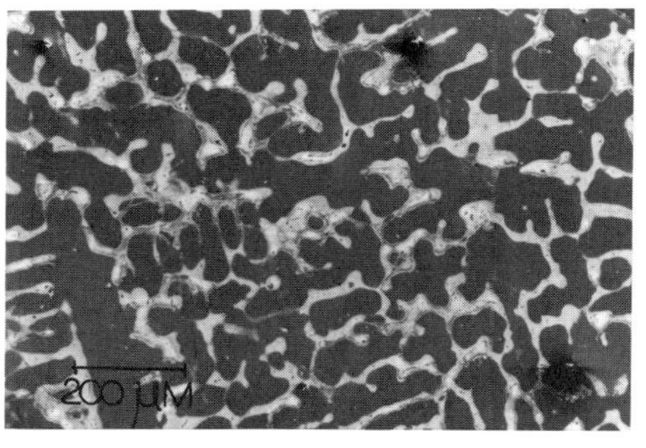

(a)

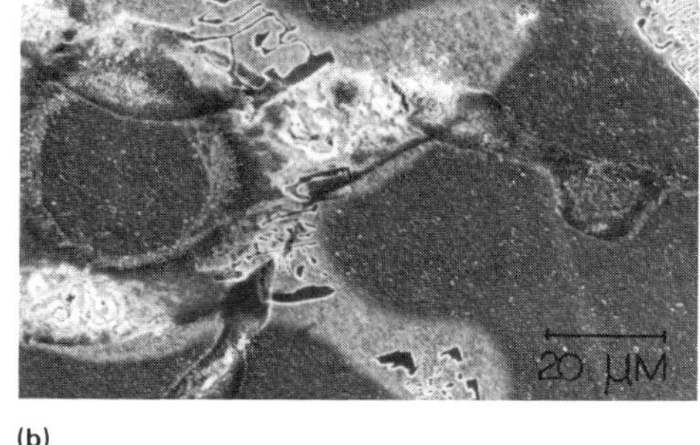

(b)

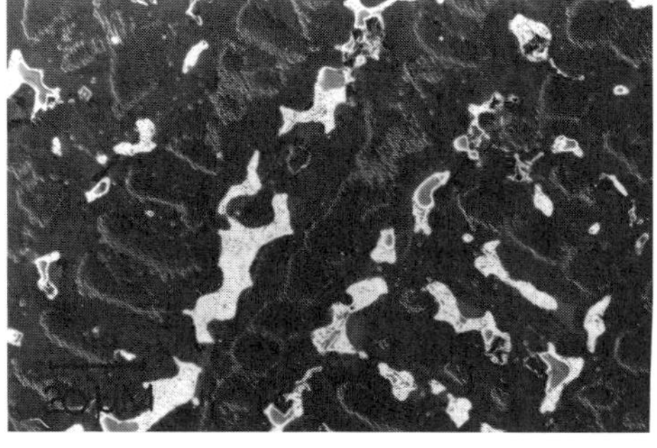
(c)

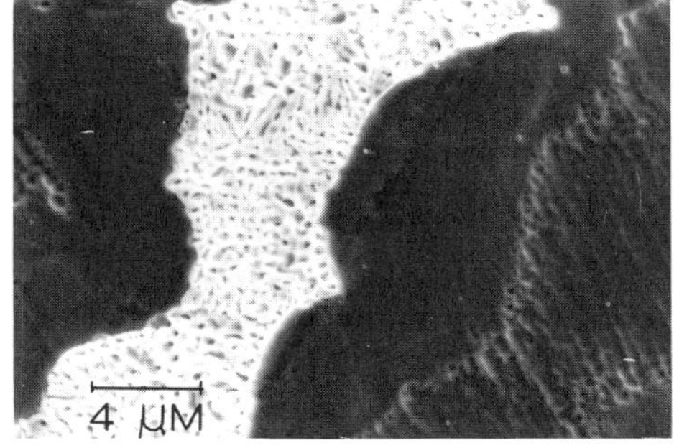

(d)

Fig. 28 SEM micrographs of etched nickel-chromium alloys for fusing with porcelain. (a) and (b) Ceramalloy. (c) and (d) Biobond

The oxide-forming ability of the added oxidizers is dependent on alloy composition, temperature, and time. Indium, gallium, and cobalt oxidize preferentially, while copper and tin show nonpreferential oxidation. Cobalt suppressed the oxidation for copper and tin in one alloy system (Ref 160).

The palladium-gallium eutectic dominates the alloy systems. Additions of both indium and copper reduce the solid solubility of gallium so that the eutectic forms at lower gallium contents (Ref 154). Microstructurally, as-cast structures are dendritic, with moderate compositional variations occurring due to coring. The interdendritic regions contain higher amounts of alloying elements. For a palladium-copper-gallium alloy, the interdendritic precipitates are fcc with the composition of $Pd_3Ga_xCu_{1-x}$ (Ref 164). Upon cooling below 500 °C (930 °F), an ordered fct structure forms rapidly. This lattice straining produces the high hardnesses and yield strengths characteristic of these materials.

Base metal PFM alloys are primarily composed of the nickel-chromium alloys. Cobalt-chromium alloys are also used, but they constitute only a very small percentage of base metal use for PFM. Titanium-base alloys are currently being developed for PFM usage.

The nickel-chromium PFM alloys are very similar, if not the same, as the compositions of the nickel-chromium alloys used for partial dentures, which were discussed earlier in this article. One distinction in the compositions, however, is the absence of carbon with the PFM compositions (Ref 131). The microstructures for two alloy systems are presented in Fig. 28.

The mechanical properties of the nickel-chromium alloys are excellent for the PFM restoration. The high strengths, moduli, yield strengths, and hardnesses are used to advantage with PFM and partial dentures. Thinner alloy sections can be made from nickel-chromium alloys than from the noble metal alloys. The flexibilities of long span partial denture frameworks are only one-half those for the high gold-content alloys. Additionally, the sag resistances of the nickel-chromium alloys at the porcelain firing temperatures are superior to all of the noble metal alloys.

The bond strengths are seriously impaired by nonadherent or loosely attached oxides. A properly attached oxide is characterized by minute protrusions on the underside of the oxide layer at the alloy/oxide interface that extend into the alloy. For alloys containing additional microstructural phases, larger peg-shaped protrusions also occur on the underside of the oxide layer, improving oxide adherence (Ref 165). The oxide layer on nickel-chromium alloys contains nickel oxide (NiO) on the exterior of the oxide scale, chromium oxide (Cr_2O_3) at the interior covering the alloy, and nickel-chromium oxide ($NiCr_2O_3$) in between. The relative amounts of the oxides depend on the chromium concentration and alloying elements in the alloy, as well as temperature, time of oxidation, and pO_2 in atmosphere.

The bond strengths between alloy and porcelain are significantly affected by minor alloying elements. The additional alloying elements of molybdenum, aluminum, silicon, boron, titanium, beryllium, and manganese also form oxides of their own. An aluminum content of 5% is necessary for aluminum oxide (Al_2O_3) to form, while 3% Si is required to form silicon dioxide (SiO_2), which increases in concentration as the alloy/oxide interface is approached. Manganese forms manganese oxide (MnO) and manganese chromite ($MnCr_2O_4$), and these are mainly concentrated at the outermost part of the oxide. Even though molybdenum oxide (MoO) volatizes above 600 °C (1110 °F), molybdenum is still found in the oxide layer close to the alloy/oxide interface with alloys containing more than 3% Mo. Beryllium, which improves the adherence of the oxide layer to the alloy, is also found concentrated near the alloy/oxide interface. Similarly, niobium is found in the oxide layer close to alloy. Tin does not diffuse throughout oxide (Ref 166).

The difficulties experienced with nickel-chromium alloys for dental applications have been at least partly related to their processing due to the use of casting techniques that had been well established for the lower-melting gold-base alloys. Because the nickel-chromium alloys melt between 1200 and 1400 °C (2190 and 2550 °F), as compared to the 850- to 1050-°C (1560- to 1922-°F) melting range for the gold-base alloys, additional casting shrinkages must be compensated for with the nickel-chromium alloys. This involves the use of phosphate- or silicate-bonded investment materials as compared to the gypsum-bonded investments used with the gold-base alloys. Technique variables with these higher-temperature investments are still being resolved. Also, the use of the gas-air torch for alloy melting, which is very popular in the dental field, cannot be used to melt nickel-chromium alloys.

Another disadvantage often cited with the harder nickel-chromium alloys as compared to the softer gold alloys is their increased finishing costs. This factor, though, should be more than offset by the decreased cost of the alloy. Even though casting accuracy may be improved with improvements in casting techniques, there may still be differences in castabilities between the different alloys. The ability to cast thin sections is a prime requirement with the dental technique.

Other PFM Alloy Systems. The compositions of cobalt-chromium alloys for porcelain fused to metal usually have nickel, tungsten, and molybdenum as major alloying elements. Tungsten and molybdenum are high-temperature strengtheners, and therefore increase sag resistances. Tantalum and ruthenium can also be added in minor amounts. The carbon is also reduced or eliminated with PFM alloys (Ref 131). The carbon monoxide gases generated during firing of porcelain are likely to cause porosities in the interface and in porcelain. Carbon along the interface also interferes with porcelain wetting alloy during firing.

Alloy oxidation and alloy-porcelain compatibility must be better controlled in order for success to be achieved with porcelain fused to titanium alloy restorations. Titanium oxidations occurring at the currently used firing temperatures of about 1000 °C (1830 °F) must be decreased in order to achieve satisfactory alloy-porcelain bonding. The use of low firing temperatures has reduced alloy oxidations and porcelain-alloy concentrations. Newly formulated porcelains have been used with only about one-half the coefficient of thermal expansion and with firing temperatures as low as 550 °C (1020 °F) (Ref 167). Pure titanium and Ti-6Al-4V are candidate metal substrates for PFM restorations.

Wrought Alloys for Wires

Property Requirements With Orthodontic Biomechanics. Orthodontic wires (frequently used round sizes are 0.3 to 0.7 mm, or 0.012 to 0.028 in., in diameter) constitute a large percentage of the wrought alloys used in dentistry. The wires currently in use include stainless steels, cobalt-chromium-nickel Elgiloy-type alloys, nickel-titanium Nitinol-type alloys, and β-titanium. The stainless steel and Elgiloy wires have been used extensively with conventional orthodontic biomechanics. That is, the ability to move teeth was based to a large extent on the stiffnesses of the wire appliances. Materials with high yield strength to modulus of elasticity ratios were required. The energy released from the wires during springback corresponded to areas on stress-strain curves with large slopes.

With the newer orthodontic biomechanics methodologies, however, materials with lower yield strength to modulus of elasticity ratios are desired. Here the nickel-titanium and β-titanium wires are used. The released energy under stress-strain curves with lower slopes can be large because of the greater deflections involved. The designs associated with the newer orthodontic approaches also permit greater energies to be released with time. Certified orthodontic wires comply with the property requirements of ANSI/ADA Specification No. 32, which covers base wires for orthodontics (Ref 168).

Gold-base wires are currently used on a very limited basis; in the past, gold-base wires were used more extensively. One application, however, still relies on gold-base wires. This is for clasps on partial denture frameworks and bridges. Because of their greater ductility, clasps fabricated from gold-base alloys permit adjustments to be made with less possibility for clasp breakage. One disadvantage with this application is the need to subject the soldered bridge assembly to a heat treatment operation. The soldering leaves the gold wires in a softened condition. For this reason, base metal wires with better ductility than the cobalt-chromium frameworks but requiring no follow-up heat treatments have been used (Ref 169). Except for cast cobalt-chromium-molybdenum implant alloys, all additional implant alloys, including cobalt-chromium-tungsten, modified cobalt-chromium, and titanium and its alloys, are used in the wrought form and will be covered in the section "Implant Alloys" in this article.

Stainless Steel and Elgiloy Wires. The stainless steels used are usually the austenitic 18-8 type, although precipitation-hardening type steels have also been used. The springback of the 18-8 wires can be improved by a stress-relief heat treatment. A 400-°C (750-°F) treatment for 10 min of the as-received drawn wires generates significant improvements in springback (Ref 170). The Elgiloy wires are of the modified cobalt-chromium type. In addition to cobalt and chromium, other major alloying elements include nickel and iron. Aluminum, silicon, gallium, and copper are not added to the Elgiloy wires, as with the PFM cobalt-chromium alloys, because bonding agents with porcelain are not needed. Mechanical properties are controlled primarily through the carbon additions, which affect carbide formation. Although the operator has some control over the mechanical properties of the cobalt-chromium wires through heat treatments, the Elgiloy wires are supplied in different temper designations from soft to semiresilient to resilient.

Nitinol and β-Titanium Wires. Although Nitinol alloys have attracted interest because of their shape memory effects, this property has been used very little in orthodontics. The elasticity effect

Table 10 Compositions of some dental solders

Type	Composition, wt% Au	Pd	Ag	Cu	Sn	In	Zn
Silver	...	...	52.6	22.2	7.1	...	14.1
Gold	45.0	...	20.6	28.4	4.3	...	2.9
Gold	63.0	2.7	19.0	8.6	...	6.5	...

of Nitinol is its most important characteristic. Nitinol wires can almost be bent back on themselves without taking a permanent set. Even greater deformations by as much as 1.6 times can be achieved with newer superelastic nickel-titanium alloys (Ref 171). The composition of Nitinol is primarily the intermetallic compound of NiTi. The alloy is tough, resilient, and has a low modulus of elasticity. The alloy does not respond to heat treatments except for the normal homogenizing softening treatments of as-cast alloys. Cobalt is added in order to obtain critical temperatures that are useful for the shape memory effect.

Alpha-titanium (hexagonal close-packed structure) is the stable form of titanium at room temperature. By adding alloying elements to the high-temperature form of β-titanium (bcc structure), the β phase can also exist at room temperature, but in the metastable condition. The β stabilizers include molybdenum, vanadium, cobalt, tantalum, manganese, iron, chromium, nickel, cobalt, and copper. Beta-titanium is strengthened by cold working or by precipitating the phase. A variety of heat treatments can be used to alter the properties of the wires (Ref 124).

Permanent magnets are used for the fixation of dentures, bridgework, and for corrective therapy by controlling the movement of teeth. Either the attractive or repulsive forces between magnets with opposite or the same polarities have been used to advantage. For example, the repulsive forces have been used to correct crossbite and to realign teeth. This area of dentistry is in its infancy.

The new generation of rare-earth magnets provides large magnetic forces per unit volume of material, unlike the Alnico and other previously used magnets. Disk-shaped magnets 5 to 10 mm (0.2 to 0.4 in.) in diameter and 2 mm (0.08 in.) thick that are embedded in the properly designed appliance will often provide the forces necessary in dental biomechanic treatments. Some compositions that have proved useful in dental treatment include samarium cobalt ($SmCo_5$, Sm_2Co_5) and neodymium-iron-boron ($Nd_2Fe_{17}B$).

Soldering Alloys

Composition and Applications. Gold-base and silver-base solder alloys are used for the joining of separate alloy components (Table 10). High fusing temperature base alloy solders are also used for the joining of nickel-chromium and other alloys. In many cases, the term brazing would be more appropriate, but the term is seldom used in dentistry. The gold-containing solders are used almost exclusively in bridgework because of their superior tarnish and corrosion resistance. The use of nonnoble metal containing silver-base solders is mainly limited to the joining of stainless steel and cobalt-chromium wires in orthodontic appliances because of the impermanence of the appliances.

The joining by soldering of small units to form a large one-piece partial denture is employed in some processing techniques. This is done to prevent framework distortions that may occur with large one-piece castings. The salvaging of large, poorly fitting castings by sectioning, repositioning, and soldering the pieces together also takes place. Both pre- and postsoldering techniques are used with PFM restorations. These imply that the soldering is carried out either prior to or after the porcelain has been baked onto the alloy substrate. Therefore, presoldering uses high fusing temperature solders, while postsoldering uses lower fusing temperature solders.

Gold-base solders are most often rated according to their fineness, that is, the gold content in weight percent related to a proportional number of units contained in 1000. Conventional gold-base crown and bridge alloys are seldom soldered together with solders having less than a 600 to 650 fineness. The soldering of gold-base wire clasps to cobalt-chromium partial dentures is another application for the higher-fineness solders. However, with the use of the low gold content crown and bridge alloys, lower-fineness solders are used. The lower-fineness solders are also occasionally used to solder the cobalt-chromium Elgiloy wires. The gold-base solders usually do not contain platinum or palladium, which increases melting temperatures. The important requirement to be satisfied during soldering is that the solders melt and flow at temperatures below the melting ranges of the parts to be joined.

The compositions of gold-base solders are largely gold-silver-copper alloys to which amounts of zinc, tin, indium, and others have been added to control melting temperatures and flow during melting. The silver-base solders are basically silver-copper-zinc alloys to which smaller amounts of tin have been added. The higher-fusing solders to be used with the high-fusing alloys are usually specially formulated for a particular alloy composition, because not all alloys have good soldering characteristics.

Microstructure of Solder-Alloy Joints. The microstructural appearance of the gold alloy-solder joints provides information as to their quality. A thin, distinct, continuous demarcation between the solder alloy and the casting alloy should exist, indicating that the solder has flown freely over the surface and that no mutual diffusion between the alloys has occurred. The junction region should be free of isolated and demarcated domains, indicative of the formation of new alloy phases. Obviously, porosity is to be avoided. However, microporosity among the phases in solder may be unavoidable. As with the cooling that occurs with all alloys, differences in the thermal expansion coefficient among phases can generate microporosity. The presence of a distinct layer of columnar dendrites within the solder starting at the solder/alloy interface and projecting into the solder bulk is assurance that there has been no tendency of the alloy surface to melt (Ref 173). The solidified solder tends to match the grain size of the parent alloy by epitaxial nucleation of the solder by the casting alloy. The microstructural characteristics of low gold content casting alloys interfaced by soldering affect the microstructural characteristics of the solidified solder (Ref 173).

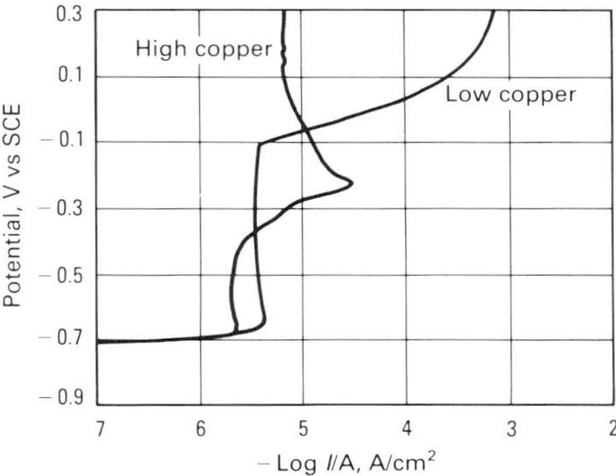

Fig. 29 Anodic polarization at 0.03 V/min of both low (Microalloy) and high (Sybraloy) copper amalgams in artificial saliva. Source: Ref 67

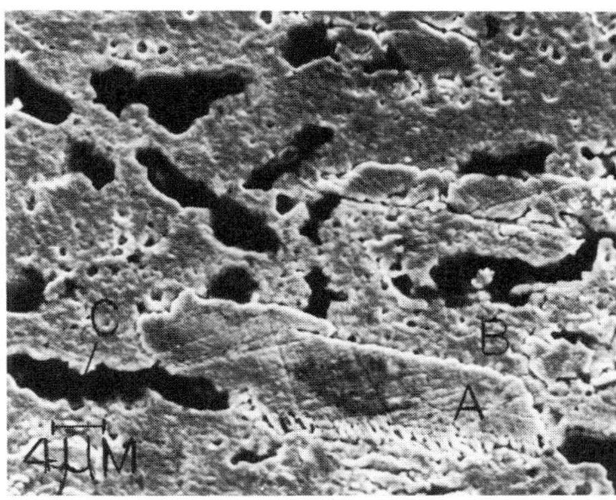

Fig. 30 SEM micrograph of corroded (10 μA/cm²) low-copper amalgam (New True Dentalloy) in 0.2% NaCl after removal of corrosion products by ultrasonics. A, alloy particles; B, matrix; C, regions formerly occupied by γ_2 phase

Microstructurally, a silver-base solder is multiphasal. Both silver and copper-zinc rich areas occur, which is in contrast to some of the higher-fineness gold-base solders that are single phase.

Implant Alloys

Applications and Compositions. Dental implants, which are used for supporting and attaching crowns, bridges, and partial and full dentures, can be of the endosseous and subperiosteal types. The endosseous implants pass into or through the mandibular or maxillary arch bones, while the subperiosteal implants are positioned directly on top of or below the mandibular or maxillary bones, respectively. Endosseous implants are usually selected for size and type from implants already made, while the subperiosteal implants are usually custom made for the particular case. Therefore, both cast and wrought forms of implants are used. The alloys used for dental implants are similar in composition to the alloys mentioned previously for use with crowns and bridges and partial dentures. These include stainless steel, cobalt-chromium, and titanium and its alloys (Ref 22, 131).

Cast Versus Wrought Cobalt-Chromium Alloys. Castable cobalt-chromium alloys are similar to Vitallium and Haynes Stellite-21 alloys, while the wrought forms are similar to surgical

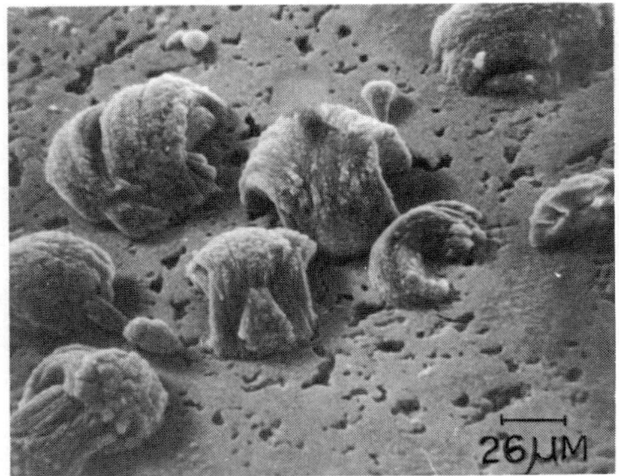

Fig. 31 SEM micrograph of low-copper amalgam (New True Dentalloy) after immersion in artificial saliva. The clumps of corrosion products contain tin. Source: Ref 175

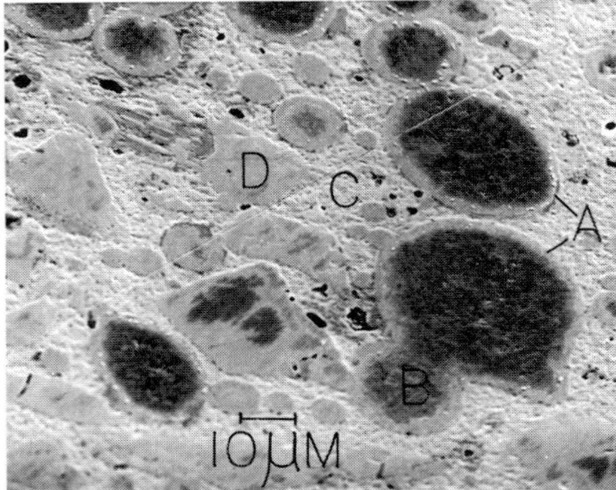

Fig. 32 SEM micrograph of corroded (5 μA/cm²/d) high-copper amalgam (cluster) in 0.2% NaCl solution. Note the definition of the η′ rings (A), Ag-Cu particles (B), matrix (C), and η′ particles (D).

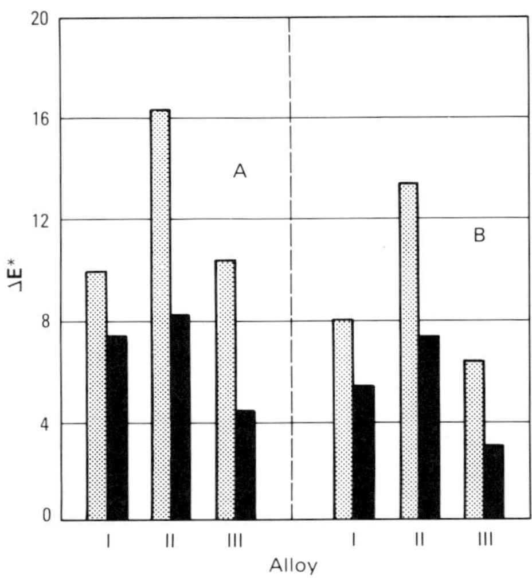

Fig. 33 Color change vector ΔE* for three low-gold alloys (I, Miracast; II, Sunrise; III, Tiffany) in both the as-cast (left bar for each alloy) and solutionized at 750 °C (1380 °F) (right bar for each alloy) conditions after exposure for 3 days to artificial saliva (A) or 0.5% Na_2S solution (B). Source: Ref 146

Vitallium and Haynes Stellite-25 alloys. Table 8 includes compositions for some of these implant alloys. Two important distinctions exist between the cast and wrought forms (Ref 174). The casting alloy contains 5% Mo, while the wrought form contains 14 to 16% W. Molybdenum is added as a hardening agent, an oxide former, and to increase crevice and pitting corrosion resistance. Tungsten is added as a hardener and as an oxide former.

Because the wrought form can be either cold or hot worked, the alloy needs additional protection against deterioration during working. Carbides provide strengthening only up to 880 °C (1615 °F). Tungsten overcomes much of the loss in strength at higher temperatures by being a source of solid-solution strengthening above 1100 °C (2010 °F). The carbon content in the castable alloy is also higher than in the wrought form (0.35% to 0.05 to 0.15%). Carbide precipitations occur in the cobalt-chromium-molybdenum castable alloy, which presents a high resistance to wear. It is one of the materials of choice for applications involving moving parts.

Microstructurally, extreme differences exist between the cast and worked forms. The as-cast and annealed structures consist of distinct grains and grain boundaries interspersed with precipitated phases, similar to the micrographs shown in Fig. 27 for partial denture alloys. The cold-rolled, annealed, and hot-worked structures show a very fine grain structure oriented in the direction of working. For both structures, however, the matrix was fcc and contained carbide particles. The large chemical inhomogeneities across the surface of the alloys were indicated by the large changes in the x-ray intensities for cobalt, chromium, and molybdenum in traversing across the surfaces.

Modified cobalt-chromium alloys contain higher nickel contents. One wrought alloy included in this category is MP35N (35Co-35Ni-20Cr-10Mo).

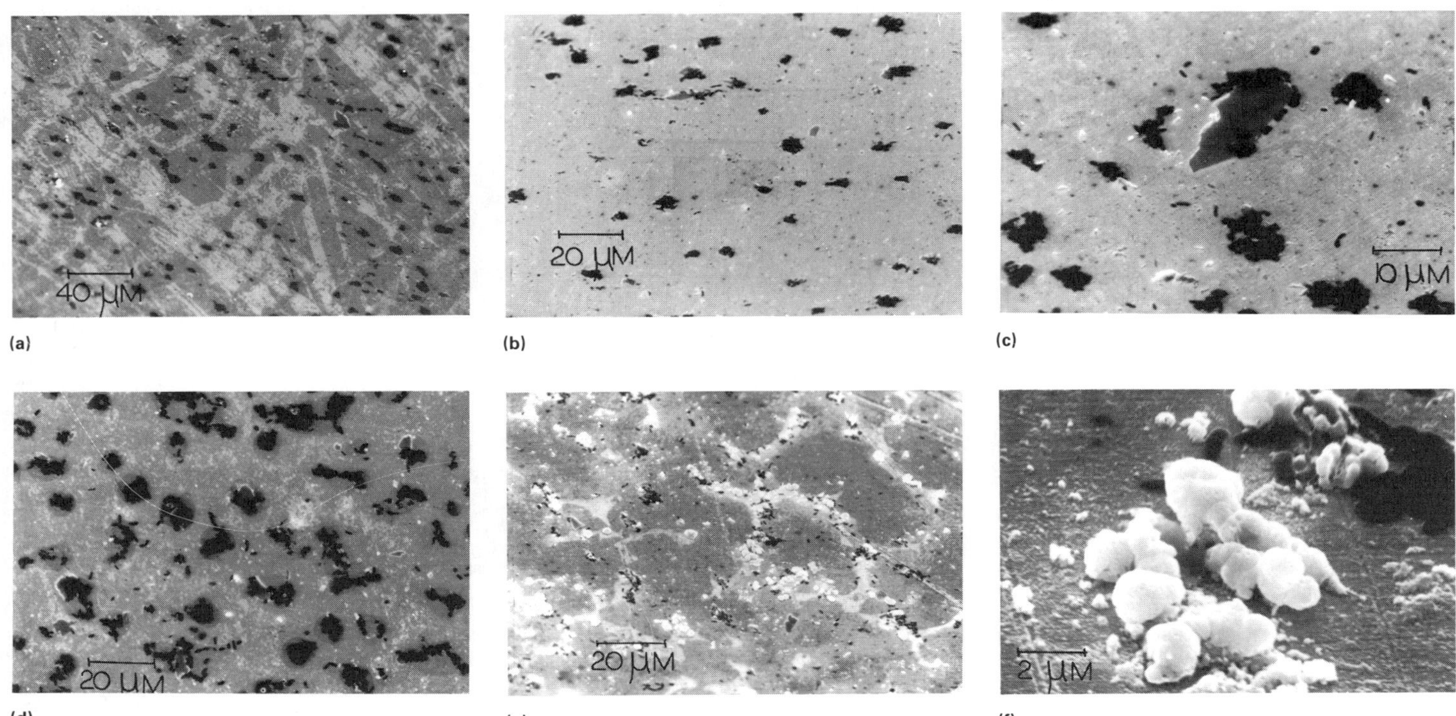

Fig. 34 SEM micrographs of low-gold alloys after 72 h of alternate-immersion tarnishing in artificial saliva containing 0.016% Na_2S. (a) Sterngold 20. (b) Tiffany. (c) Pentron 20. (d) Sunrise. (e) and (f) Econocast. See Table 6 for chemical compositions. Source: Ref 176

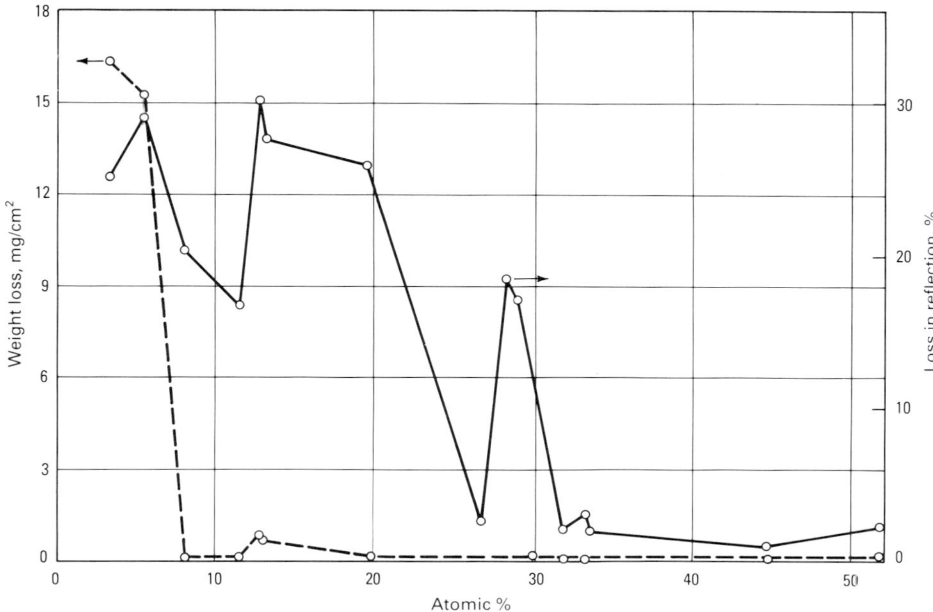

Fig. 35 Effect of nobility (in atomic percent) on tarnish (percent loss in reflection after 3 days in 0.1% Na_2S) and corrosion (weight loss after 7 days in aerated 0.1 *M* lactic acid plus 0.1 M NaCl at 37 °C or 99 °F) of 15 gold alloys. Source: Ref 179

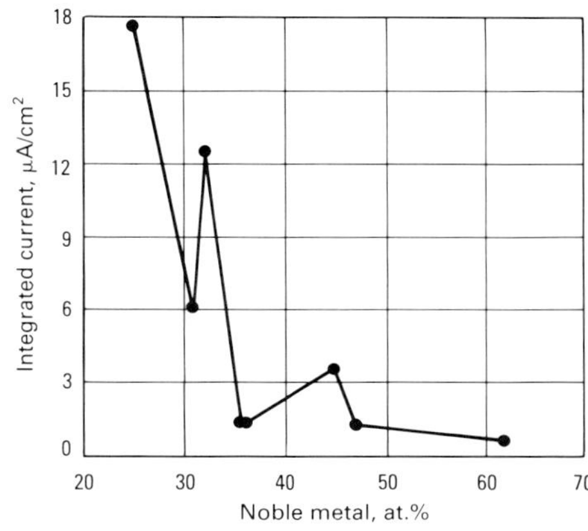

Fig. 36 Integrated anodic currents between −0.3 V versus SCE and +0.3 V (at 0.06 V/min) for eight gold alloys in deaerated 1% NaCl plotted against the atomic nobility. Source: Ref 146

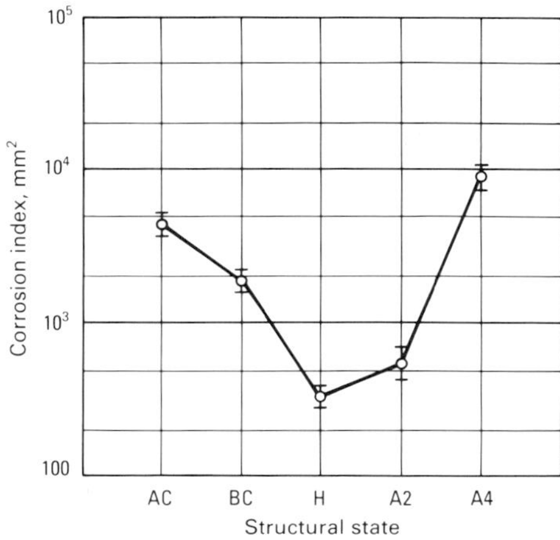

Fig. 37 Corrosion index (area in mm² under polarization curve between −1.0 V versus SCE and +0.4 V at 0.6 V/min) versus the heat treatment state of Midigold, a low gold content alloy. AC, as-cast; BC, bench cool; H, homogenized; A2, aged for 2 h at 350 °C (660 °F); A4, aged for 4 h at 350 °C (660 °F). Source: Ref 73

Microstructurally, the alloy takes the character of cobalt-chromium alloys. However, no carbides are formed, because carbon has not been added to the alloy.

Porous Surfaces. Alloy powders of the same composition as the implants have been sintered onto the surfaces of the implants for generating bone ingrowth to obtain better retention between the implant and the bone.

Tarnish and Corrosion Under Simulated or Accelerated Conditions

Low-Copper Amalgams. The Sn_8Hg (γ_2) phase shown in Fig. 20 is electrochemically the most active phase in conventional amalgam. Upon exposure to an electrolyte, the tin oxide (SnO)/tin couple becomes operative. The formed SnO may or may not protect the γ_2 phase from further corrosion. Depending on environmental conditions, the tin from the γ_2 phase will either be protected by a film of SnO or consumed by additional corrosion reactions.

In the case of an artificial saliva, the γ_2-tin becomes protected. This is shown in Fig. 29 on the anodic polarization curve as the current peak at about −0.7 V versus SCE, which relates in potential to the SnO/Sn couple. With increasing potential, the film protects the amalgam, as shown by the presence of a limiting or passivating current. The film will remain passivating until the potential of another redox reaction is reached that is controlling and nonpassivating. This is the situation that occurs in Cl^- solution when the corrosion potential for the amalgam approaches and surpasses the redox potential for a reaction that produces $SnOCl{\cdot}H_2O$.

If this condition is satisfied, the SnO passivating film breaks down, exposing freely corroding γ_2-tin to the electrolyte. Nonprotective products of the form $SnOHCl{\cdot}H_2O$ precipitate. This is represented on the polarization curve in Fig. 29 as the sharp increase in current at about −0.1 V. Because the interconnection of the γ_2 phase, the interior γ_2 can also become corroded. Figure 30 shows the devastating effect that corrosion of the γ_2 phase has on the microstructure of a conventional amalgam, while Fig. 31 shows typical tin-containing products that precipitate on the surface.

High-Copper Amalgams. Corrosion of high-copper amalgams by γ_2-phase corrosion will not occur, because of its almost complete absence from the microstructure. Although possessing better corrosion resistance than the γ_2-phase, the Cu_6Sn_5 (η') phase will be the least resistant phase in the microstructure (Fig. 21, 22). Upon exposure to solution, any corridable tin within the material first forms a protective SnO film indicated on the anodic polarization curve in Fig. 29 as the small current peak at about −0.7 V. Upon attaining a steady-state corrosion potential in chloride solution, high-copper amalgam is likely to surpass redox potentials for couples of $CuCl_2$/Cu, $Cu(OH)_2$, Cu_2O/Cu, and $CuCl_2{\cdot}3Cu(OH)_2$/Cu. Under these conditions, both soluble and insoluble corrosion products will form. This is indicated on the polarization curve as a small anodic current peak at about −0.25 V.

Microstructurally, if the copper from the η' phase becomes exhausted by corrosion, copper corrosion from the silver-copper and γ particles may also follow. Freed by copper corrosion, tin also becomes corroded. Corrosion of the γ_1-tin decreases the stability of the γ_1 phase, which is likely to be transformed into the β_1 phase. Unlike low-copper amalgam, the interior of high-copper amalgam is not likely to become affected by corrosion, because of the noninterconnection of any of the phases. Figure 32 shows a corroded high-copper dispersed-phase amalgam, emphasizing the reaction zones of the η' phase, the interior of the silver-copper particles, the γ particles, and the matrix.

Tarnish of Gold Alloys. Because tarnish is by definition the surface discoloration of a metallic material by the formation of a thin film of oxide or corrosion product, the quantification of dental alloy tarnish by assessing color changes on surfaces is most appropriate. By determining the color of an alloy before and after exposure to a test solution, the degree of discoloration can be obtained by quantitative colorimetry techniques, which are described in Ref 105 and 144 to 149. The use of quantitative colorimetry in conjunction with SIMS for determining the effects of alloy nobility (in atomic percent) and sulfide concentration on color changes in gold crown and bridge alloys is detailed in Ref 176.

Effect of Microstructure on Tarnishing Behavior. The devastating effect of microgalvanic coupling on tarnishing behavior is shown by a comparison of solid-solution annealed and as-cast structures (Fig. 33). The as-cast alloys, composed of two-phase structures, consistently showed greater color change or tarnish (Ref 146).

Single-phase as-cast gold-silver-copper alloys with gold contents between 50 and 84 wt% were observed microstructurally to tarnish in Na_2S

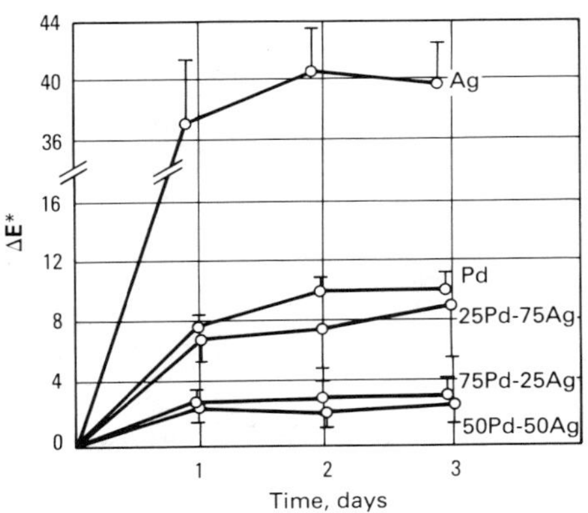

Fig. 38 Color change vector ΔE* for pure silver, palladium, and three Ag-Pd binary compositions after exposure to Na_2S solutions. Source: Ref 187

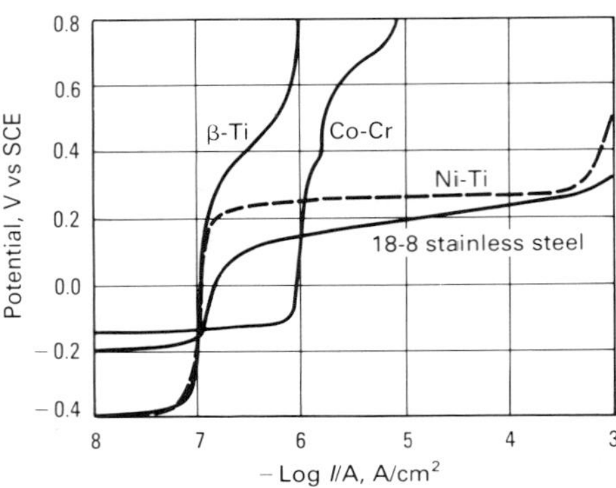

Fig. 39 Anodic polarization at 0.03 V/min of four orthodontic wires in artificial saliva. Source: Ref 194

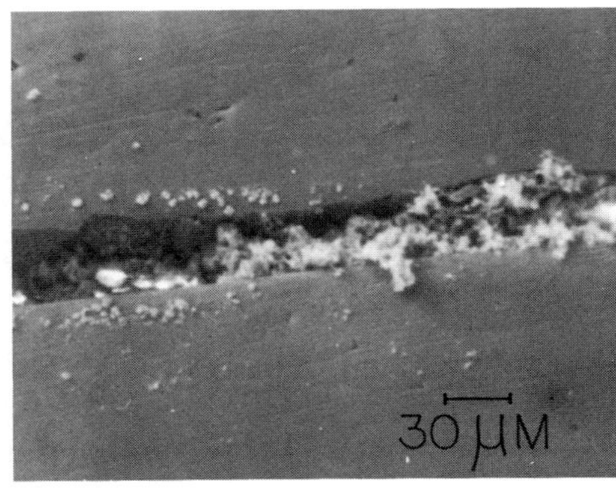

Fig. 40 SEM micrograph of a corroded stainless steel-silver soldered joint after immersion in a 1% H_2O_2 solution. Source: Ref 196

solutions by localized microgalvanic cells. The characteristics of the various tarnished surfaces included a uniformly speckled appearance, dendritic attack, matrix attack, grain-boundary dependent attack, and grain-boundary attack. Silver-rich areas discolored preferentially because of the operation of the silver-rich areas as anodes and the surrounding copper-rich areas as cathodes. The uniformly speckled appearance occurred with high-silver low-copper contents, while the grain orientation dependent appearance occurred with low-silver high-copper contents. The dendritic and matrix attack occurred with alloys containing intermediate silver and copper contents (Ref 96).

For gold-silver-copper-palladium alloys with gold contents between 35 and 73 wt%, the tarnishability in oxygenated 2% Na_2S is shown to be affected by altering the microstructure through heat treatment. Tarnishing occurred on multiphase structures annealed at 500 °C (930 °F), but did not occur on single-phase structures annealed at 700 °C (1290 °F). Silver, copper, and palladium-rich phases were precipitated. Some alloys, though, showed only silver- and copper-rich phases. In these cases, the palladium tended to follow the copper-rich phase. Splitting of the matrix into thin lamellae of alternating silver and copper enrichments occurred. The silver-rich phases in all materials were attacked by the sulfide and were responsible for the tarnish. Age hardening by AuCu(I)-ordered precipitates increased the tendency of the silver-rich lamellae to tarnish (Ref 177).

In sulfide solutions, silver sulfide (Ag_2S) is the principle product of tarnish, although copper sulfides (Cu_2S and CuS) also form. These products are produced by the operation of microgalvanic cells set up between silver-rich and copper-rich lamellae. The addition of palladium to gold-silver-copper alloys considerably reduces the rate of tarnishing by slowing down the formation of a layer of silver and copper sulfides on the surface. This has been shown to be due to the enrichment of palladium and gold on the surface of the alloy when exposed to the atmosphere prior to sulfide exposure (Ref 178). The rate of diffusion from the bulk to the surface is hindered by the palladium enrichment. The active sites on the alloy surface for the sulfidation reaction are selectively blocked by the palladium atoms (Ref 178).

Figure 26 presents chemically etched microstructures for a number of different casting alloys. All structures are multiphase, except for Sterngold 20 and Midas. After the surfaces were repolished and exposed for 3 days to a 0.016% Na_2S solution with a rotating tarnish tester (15 s exposure/min), the surfaces were again examined by scanning electron microscopy (SEM). Some of the results are presented in Fig. 34, which shows that tarnishing for all alloys took the form of dark patches over the alloy surfaces.

Effect of the Silver/Copper Ratio. The silver/copper ratio is an important aspect in affecting the tarnish and corrosion resistance of gold dental alloys. A comparison of Midacast to Neycast III and Miracast (see Table 6) shows that all three alloys have about the same nobility but vastly different Ag/Cu ratios (based on weight percent), ranging from 41/7.4 for Midacast to 10.3/37.9 and 9.8/37.7 for Neycast and Miracast, respectively (Ref 177). A comparison of their polarization behavior in a sulfide solution shows the Midacast exhibits increases in current density up to 10 μA/cm² at −0.3 V, while Neycast III and Miracast exhibit a current density of ~1 μA/cm² extending to positive potentials. Therefore, high silver contents relative to low copper contents in low gold alloys can have detrimental effects upon tarnishing and corrosion. For Forticast (the composition also given in Table 6) unacceptable levels of tarnish would occur if the silver/copper ratio were changed (Ref 158). For some low-gold alloys, the best resistance to tarnishing has been obtained by using ratios between 1.2 and 1.4 and a palladium content of 9 wt%.

Effect of the Palladium/Gold Ratio. Increasing the palladium content in gold alloys increases the tarnish resistance. However, in gold-silver-copper-palladium alloys, this effect is greater. The palladium/gold ratio is just as important as the silver/copper ratio. In gold-silver-copper alloys without palladium, the degree of tarnish (subjective test: 0 = least and 8 = most) was evaluated to be between 6.5 and 8 for all silver/copper ratios (1:3, 1:2, 2:3, 1:1, 3:2, 2:1 3:1) (Ref 101). However, in alloys having palladium/gold ratios of 1:12, the degree of tarnish diminished to between 2 and 3.

Tarnishing and Corrosion Compared. Figure 35 shows reflection loss versus nobility and weight loss versus nobility for the same 15 gold alloys. Tarnishing was by immersion for 3 days in 0.1 *M* Na_2S, while corrosion was by immersion in 0.1 *M* lactic acid. As is evident, a number of alloys that appear not to have been affected by corrosion are, however, largely affected by tarnishing (Ref 179).

Corrosion of Gold Alloys. Electrochemical polarization has been applied to the corrosion evaluation of gold dental alloys. High nobility alloys exhibited low current densities over a wide range in potentials in Ringer's solution, but low nobility alloys exhibited increased current densities and decreased breakdown potentials (Ref 180). Very small current peaks on the anodic polarization curves for some gold alloys in artificial saliva were detected and interpreted to be due to the dissolution of alloying components (Ref 181).

A comparison of the anodic polarization of noble alloys in artificial saliva with and without sulfide indicated that without sulfide the electrochemistry is governed mainly by chloride ions. The alloys passivate in a state with very low current densities, which makes detection of differences among the alloys difficult. With sulfide added to the artificial saliva, a preferential sulfidation of the less noble alloy component is induced. The sulfidation is characterized by a critical potential and limiting current density, both of which may be dependent on composition (Ref 182).

The corrosion susceptibilities for silver and copper in various gold alloys were quantified by an analysis of both forward and reverse polarization scans (Ref 183). Both silver and copper demonstrated characteristic current peaks during either oxidation or reduction. The heights of the current peaks were taken to be a measure of the amount of corrodible silver and copper in the alloys. In a similar technique, the integrated current from the polarization curves within a potential range of −0.3 V versus SCE to +0.3 V was taken to be a measure for the corrodible species (Ref 184). Figure 36 presents a rank ordering of eight as-cast gold alloys in regard to nobility plotted versus their integrated anodic currents, while Fig. 37 shows the effect of heat treatment state on the polarization corrosion index for a low gold content alloy.

Silver-Palladium Alloys. Silver is prone to tarnishing by sulfur and is prone to corrosion by chloride. The addition of palladium to silver generates alloys with much better resistance to tarnishing and silver corrosion. In 1/7 diluted Ringer's solution and 0.1% NaCl, alloys with more than 40% Pd showed passive anodic polarization behavior (Ref 185). In sulfur-saturated air, the amount of sulfur deposited onto the surfaces

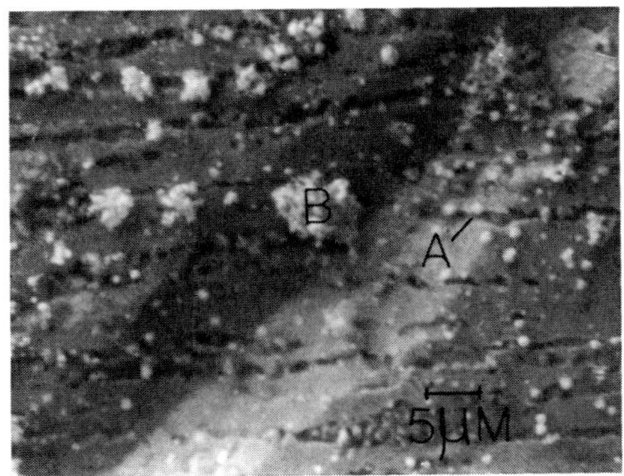

Fig. 41 SEM micrograph of a corroded silver solder in 1% NaCl (held at −0.05 V versus SCE) showing the destruction of the copper-zinc-rich phase (A) and the accumulation of products (B) that contain copper, zinc, and chlorine. Source: Ref 196

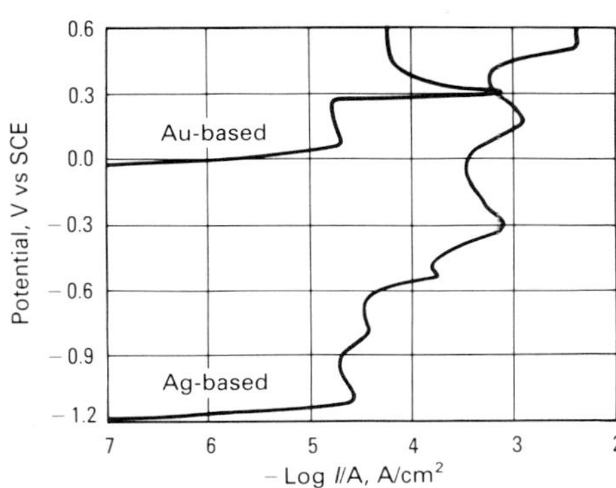

Fig. 42 Anodic polarization at 0.03 V/min of silver and gold (450 fine) solders in 1% NaCl solution. Source: Ref 196

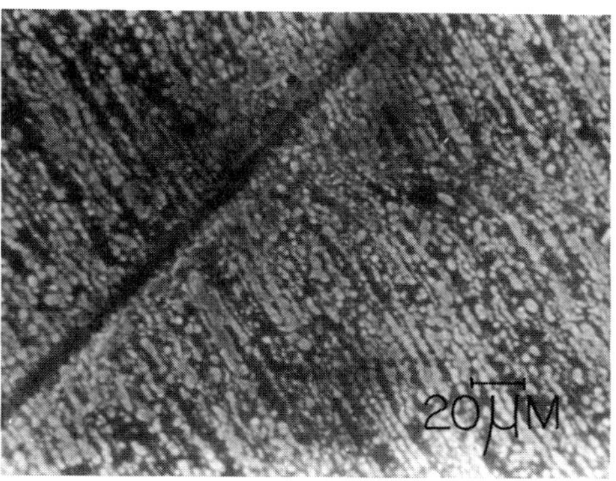

Fig. 43 SEM micrograph of a corroded (polarized to +0.5 V versus SCE) gold solder (450 solder) in 1% NaCl. The light areas contain chlorine. Source: Ref 196

of silver-palladium alloys was minimal for compositions with ≥40 wt% Pd (Ref 182). In artificial saliva, two transitions in the corrosion currents occurred with palladium content. The first occurred at about 22% Pd, where the current decreased from 6 to 1 μA. The second transition occurred at about 29% Pd, where the current decreased to about 0.4 μA and then remained fairly constant throughout the rest of the compositional range (Ref 186). Figure 38 shows the color change vectors for the pure metals and alloys from the Ag-Pd system after tarnishing in artificial saliva with 0.5% Na_2S. Compositions 50Pd-50Ag and 75Pd-25Ag showed the best tarnish resistance (Ref 187).

Corrosion behavior and tarnishing behavior usually must be viewed independently. That is, corrosion is not an indicator for tarnishing, and vice versa. Alloy nobility dominates corrosion behavior, while alloy nobility, composition, and microstructure, in conjunction with environment, influence tarnishing behavior.

Microstructurally, the silver-palladium alloys tarnish by chlorides and/or sulfides becoming deposited over the silver-rich matrix, while the palladium-rich precipitates display resistance to chlorides and sulfides. With forward and reverse scan polarization, the amounts of corrodible silver and copper in silver-palladium alloys were characterized by comparing the relative current magnitudes for reduction peaks. Microstructurally, the alloys were composed of a corrosion-resistant copper- and palladium-rich phase and a nonresistant silver-rich phase. Increased tarnish and corrosion of the silver-rich phase component occurred by microgalvanic coupling (Ref 188). Manipulation of the microstructural features through heat treatments produced structures with varying proportions of the tarnish-resistant and tarnish-prone phases. Age hardening increased the proportion of the tarnish- and corrosion-prone phases (Ref 189, 190).

High-Palladium Ceramic Alloys. Alloys with up to 80% Pd and additions of copper, gallium, tin, indium, gold, and others were shown to exhibit good saline corrosion resistance in the potential range and Cl^- ion concentration associated with oral use. Anodic polarization showed passive behavior until breakdown occurred, which was well above potential magnitudes occurring intraorally (Ref 163). For similar compositions and palladium-base compositions containing cobalt, spontaneous passivation or active-passive behavior occurred in chloride solutions. Differences among the polarization profiles for the various palladium-base alloys were detected, but any differences corresponded to high potentials generally above the biodental range (Ref 161).

The effects of adding copper and gallium to palladium were determined by modeling alloy composition and corresponding electrochemical behavior to a linear regression and equation. The resultant coefficients accounted for the individual elemental effects on corrosion potential and anodic Tafel constants. Elements that increased the nobility of the corrosion potential also increased the rate of corrosion. This was attributed to anodic protection occurring at the surface of the alloys (Ref 162).

Nickel Chromium Alloys. The tarnish resistance and corrosion resistance of these alloys result from balancing the composition with regard to the passivating elements chromium, molybdenum, manganese, and silicon. Alloys containing increased amounts of molybdenum and manganese exhibited increased passivation. Increasing the chromium content too much (above 20%) can precipitate an additional phase and alter the corrosion resistance. By using polarization methods, 3 different behaviors were observed with 12 nickel-chromium alloys with varying compositions in deaerated and aerated artificial saliva (Ref 191). Some alloys were constantly passive, others were either active/passive or passive according to the aeration condition of the electrolyte, and still others (<16% Cr without molybdenum) were constantly active and corroding.

Corrosion potentials for nickel-chromium alloys in artificial saliva were low, ranging between about −0.2 V and −0.8 V versus SCE. Breakdown potentials varied, depending on composition. For alloys with less than 16% Cr and no molybdenum, breakdown potentials as low as −0.2 V occurred. For compositions with higher chromium contents and with molybdenum and various manganese contents, breakdown potentials as high as +0.6 V also occurred.

Pitting attack occurs with these alloys because they rely on protective surface oxide films for imparting protection. From electrochemical and immersion tests, the resistance of nickel-chromium alloys to pitting attack was found to be good in solutions with Cl^- concentrations equivalent to that found in saliva. Only at higher Cl^- concentrations are pitting and tarnishing likely to occur (Ref 192).

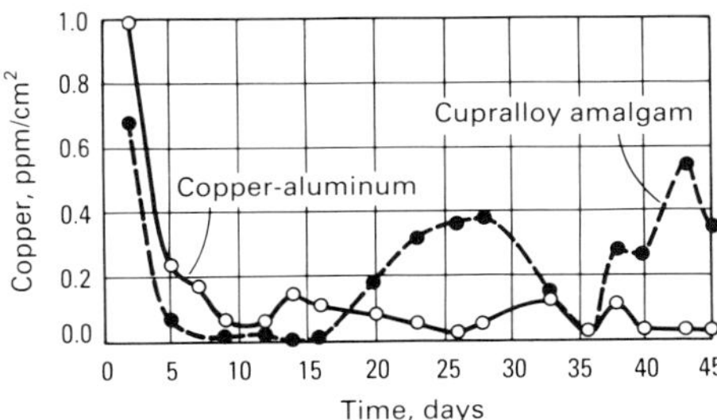

Fig. 44 Released copper into human saliva from a copper-aluminum crown and bridge alloy (MS) and a high-copper amalgam (Cupralloy) plotted against time for up to 45 days. Source: Ref 68

Cobalt-Chromium Alloys. Compared to the nickel-chromium alloys, the cobalt-chromium alloys for partial denture and implant prosthesis exhibit superior tarnish and corrosion resistance. Cobalt-chromium alloys exhibit passive behavior to potentials of at least +0.5 V (Ref 70, 193), but the nickel-chromium alloys exhibit much greater variations in behavior. For some of the nickel-chromium alloys, breakdown potentials can be as low as −0.2 V (Ref 191, 193). In spite of the excellent corrosion resistance of cobalt-chromium alloys, allergic reactions to cobalt, chromium, and nickel contained in appliances made from these alloys are known to have occurred (see the section "Allergic Hypersensitive Reactions" in this article).

Titanium Alloys. Titanium and titanium alloys, like the cobalt-chromium alloys, have proved to be resistant to tarnish and corrosion. The anodic polarization for pure titanium and its alloys indicates passivities over at least several volts in overvoltage (Ref 70, 193). This demonstrates the tenacity and protectiveness of the titanium oxide films formed on these materials.

With regard to the newer casting titanium alloys for crown and bridgework and partial dentures, good corrosion resistance is still preserved (Ref 141). Concerns must be raised because the casting alloys contain higher percentages of alloying elements, with the potential for elucidating diminished chemical stabilities.

One wrought titanium alloy, Nitinol, containing about 50% Ni and 50% Ti, does not exhibit the same high corrosion resistance as alloys with 80 to 90% Ti, and higher. Nitinol is covered in the following section.

(a)

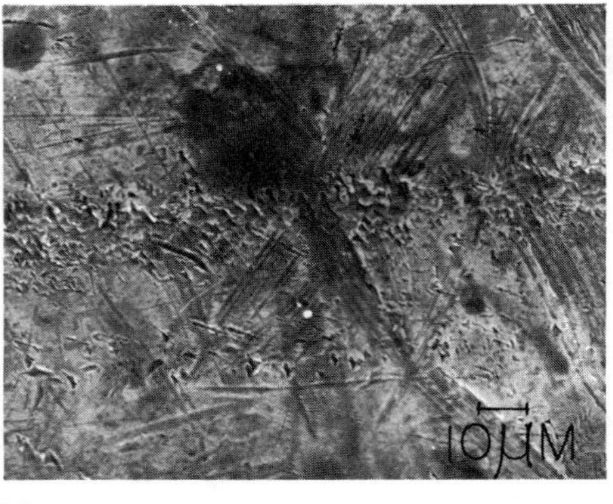

(b)

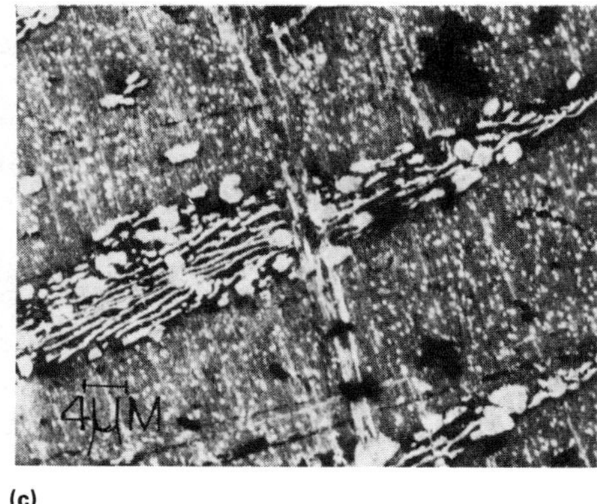

(c)

Fig. 45 SEM micrographs of the copper-aluminum restoration shown in Fig. 18 at higher magnifications. In (a), both accumulated plaque (A) and corrosion products (B) occur. (b) Higher-magnification view of the areas identified by (B) contain copper. The light-appearing areas are probably copper oxides. (c) Still higher magnification of the products shown in (b). Here the copper oxides are deposited over the dark copper-rich microstructural phase. Source: Ref 68

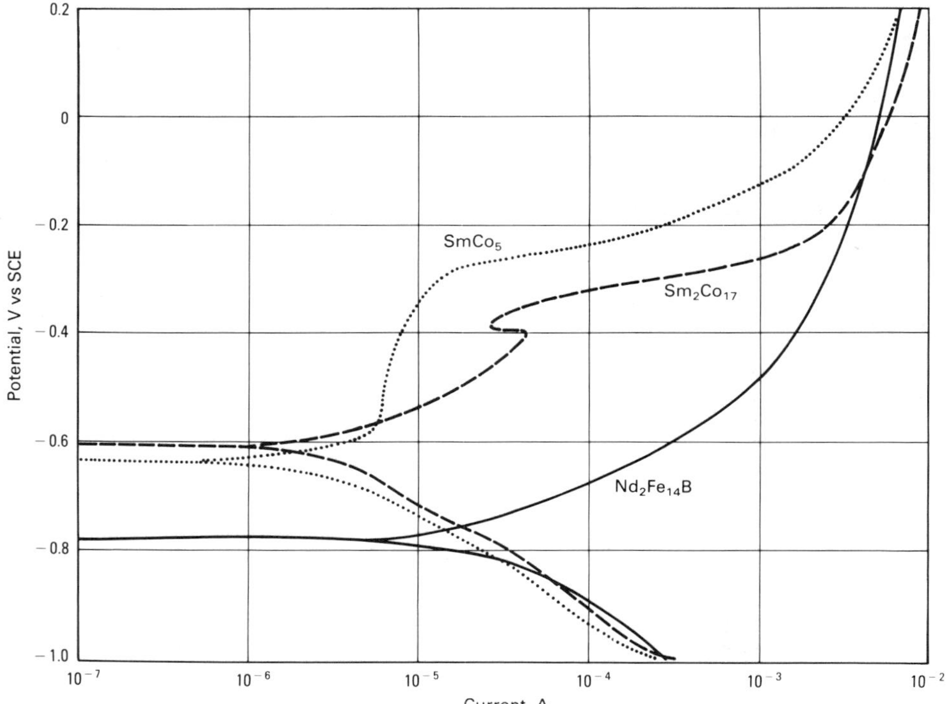

Fig. 46 Anodic polarization at 0.03 V/min of permanent rare-earth magnets in artificial saliva. Source: Ref 198

Fig. 47 Photograph of an orthodontic appliance containing a $Nd_2Fe_{14}B$ magnet after intraoral use for several weeks. Iron was identified in the corrosion products. Source: Ref 198

Wrought Orthodontic Wires. A comparison of their anodic polarization curves (Fig. 39) shows that both β-titanium and Elgiloy exhibit resistance to corrosion in artificial saliva. No breakdown in passivity occurred within the potential ranges employed (+0.8 V). With Nitinol and stainless steel wires, breakdown occurred at +0.2 and 0.05 V, respectively. Nitinol and stainless steel also exhibited current increases upon potential reversals at +0.8 V, an indication of the susceptibility to pitting corrosion. Breakdown potentials differed by as much as 0.6 V between different brands of stainless steel wires. Variations in the polarization characteristics of stainless steel have been related to microstructure (Ref 70) and to surface preparation and finish (Ref 71). Microstructurally, the Nitinol wires were observed to suffer pitting attack after polarization tests (Ref 195).

Silver and Gold Solders. A corroded silver soldered-stainless steel joint is shown in Fig. 40. Microstructurally, silver solders are composed of two phases: silver- and copper-zinc rich segregations (Ref 196). The copper-zinc regions are the least resistant to corrosion. These solders corrode by microgalvanic coupling, either by cells set up between the two microstructural phases or between solder and the parts they join. Figure 41 shows the copper-zinc phase of a silver solder attacked by corrosion. In addition to the release of copper and zinc into the surrounding environment, metallic ions from the wires themselves were also shown to be leached into the wires (Ref 25). The leaching of nickel and chromium from stainless steel wires occurred with greater intensity than from the wires of the cobalt-chromium type.

Figure 42 shows the polarization curves for both silver and gold (450 fine) solders. The silver solder is characterized by active behavior, as indicated by the low corrosion potential of −1.2 V versus SCE in 0.16 *M* NaCl on the anodic polarization traverse and the numerous current peaks related to redox reactions with the elements constituting the solder. Therefore, zinc, tin, copper, and even silver products are likely to be precipitated or become dissolved in solution.

The polarization curve for gold solder indicates intense activity by the sharp current density peak at +0.25 V. Because the solder contains silver, copper, and zinc, in addition to gold, this peak is probably due to the corrosion of one of these elements. Figure 43 shows the gold solder after the polarization test. Corrosion has delineated the basic microstructure of the solder alloy. Chlorine was detected with the white appearing phase.

Silver-Indium Alloys. These alloys rely on the unusual properties of indium oxide for providing tarnish and corrosion control. Small amounts of noble metals, such as palladium, may also be added in an attempt to improve corrosion resistance. Anodic polarization of a silver-indium alloy in artificial saliva indicated only a very narrow potential range of about 0.1 V of reduced current densities. The tarnish resistance of these alloys appears to be acceptable, but the long-term corrosion resistance has yet to be established (Ref 197).

Copper-Aluminum Alloys. A comparison of the released copper in human saliva from a dental copper-aluminum alloy to that from a high-copper amalgam is shown in Fig. 44. The amalgam released more copper over a 45-day interval. No aluminum, iron, manganese, or nickel was detected. Figures 45(a) to (c) show micrographs from an *in vivo* restoration at various magnifications. Large amounts of organic matter were adsorbed onto the surface, as well as light powdery corrosion products composed of copper oxides (Ref 68).

Rare-Earth Permanent Magnets. The anodic polarization of $SmCo_5$, Sm_2Co_{17}, and $Nd_2Fe_{14}B$ in artificial saliva indicates active behavior (Fig. 46). Additional *in vitro* and *in vivo* tests have indicated their poor chemical stability. Scanning electron microscopy, energy dispersive spectroscopy, and secondary ion mass spectroscopy identified some of the products to be oxides and chlorides of samarium, in the case of samarium-cobalt materials, and of iron and neodymium with the neodymium-iron-boron materials. Figure 47 shows an $Nd_2Fe_{17}B$ orthodontic bracket after several weeks of *in vivo* use. Abundant amounts of corrosion products occurred. In their present state, these materials are unacceptable for routine bioapplications. However, because of their extreme usefulness in providing relatively large magnetic forces from small volumes of material, special efforts are being made to apply surface coating technologies to the rare-earth magnets (Ref 198).

REFERENCES

1. *Dental Technician, Prosthetic*, Navpers 10685 c, U.S. Naval Dental School, Bureau of Naval Personnel, 1965, p 256
2. G. Ravasini, *Clinical Procedures for Partial Crowns, Inlays, Onlays, and Pontics, An Atlas*, Quintessence Publishing, 1985, p 136
3. K.L. Stewart, K.D. Rudd, and W.A. Kuebker, *Clinical Removable Partial Dentures*, C.V. Mosby, 1983, p 31, 230, 494
4. T.M. Graber and B.F. Swain, *Orthodontics: Current Principles and Techniques*, C.V. Mosby, 1985, p 385
5. D.C. Smith and D.F. Williams, *Biocompatibility of Dental Materials*, Vol 1-4, CRC Press, 1982
6. *Workshop on Biocompatibility of Metals in Dentristry*, Conference Proceedings, American Dental Association, 1984
7. *An International Workshop: Biocompatibility, Toxicity and Hypersensitivity to Alloy Systems Used in Dentistry, Proceedings*, The University of Michigan School of Dentistry, 1985
8. D.C. Smith. The Biocompatibility of Dental Materials, in *Biocompatibility of Dental Materials*, Vol 1, D.C. Smith and D.F. Williams, Ed., CRC Press, 1982, p 11
9. D.C. Smith, Tissue Reaction to Noble and Base Metal Alloys, in *Biocompatibility of Dental Materials*, Vol IV, D.C. Smith and D.F. Williams, Ed., CRC Press, 1982, p 55
10. A. Halse, Metal in Dentinal Tubles Beneath Amalgam Fillings in Human Teeth, *Arch. Oral Biol.*, Vol 20, 1975, p 87-88
11. K. Arvidson and R. Wroblewski, Migration of Metallic Ions From Screwposts Into Dentin and Surrounding Tissues, *Scand. J. Dent. Res.*, Vol 86, 1978, p 200-203
12. W. Schriever and L.E. Diamond, Electromotive Forces and Electric Currents Caused by Metallic Dental Fillings, *J. Dent. Res.*, Vol 31, 1952, p 205-229
13. I. Kleinberg, Etiology of Dental Caries, *J. Can. Dent. Assn.*, Vol 12, 1979, p 661-668
14. I.D. Mandel, Dental Caries, *Am. Sci.*, Vol 67, 1979, p 680-688
15. E. Hals and A. Halse, Electron Probe Microanalysis of Secondary Carious Lesions Associated With Silver Amalgam Fillings, *Acta Odontol. Scand.*, Vol 33, 1975, p 149-160
16. N. Kurosahi and T. Fusayama, Penetration of Elements from Amalgam Into Dentin, *J. Dent. Res.*, Vol 52, 1973, p 309-317
17. L.W.J. van der Linden and J. van Aken, The Origin of Localized Increased Radiopacity in the Dentin, *Oral Surg.*, Vol 35, 1973, p 862-871
18. B.L. Dahl, Hypersensitivity to Dental Materials, in *Biocompatibility of Dental Materials*, Vol 1, D.C. Smith and D.F. Williams, Ed., CRC Press, 1982, p 177-185
19. E.W. Mitchell, Summary and Recommendations to the Workshop, in *Workshop on Biocompatibility of Metals in Dentistry*, Conference Proceedings, American Dental Association, 1984
20. M. Bergan and O. Ginstrup, Dissolution Rate of Cadmium from Dental Gold Solder Alloys, *Acta Odontol. Scand.*, Vol 33, 1975, p 199-210
21. D.R. Zielke, J.M. Brady, and C.E. del Rio, Corrosion of Silver Cones in Bone: A Scanning Electron Microscope and Microprobe Analysis, *J. Endo.*, Vol 1, 1975, p 356-360
22. R.A. James, Host Response to Dental Implant Devices, in *Bicompatibility of Dental Materials*, Vol IV, D.C. Smith and D.F. Williams, Ed., CRC Press, 1982, p 163-195
23. J.R. Natiella, Local Tissue Reaction/ Carcinogenesis, in *International Workshop on Bicompatibility, Toxicity, and Hypersensitivity to Alloy Systems Used in Dentistry*, Section 6, Conference Proceedings, University of Michigan School of Dentistry, 1985
24. R.S. Mateer and C.D. Reitz, Corrosion of Amalgam Restorations, *J. Dent. Res.*, Vol 49, 1970, p 399-407
25. M. Berge, N.R. Gjerdet, and E.S. Erichsen, Corrosion of Silver Soldered Orthodontic Wires, *Acta Odontol. Scand.*, Vol 40, 1982, p 75-79
26. B. Angmar-Mansson, K.-A. Omnell, and J. Rud, Root Fractures Due to Corrosion, *Odontol. Revy.*, Vol 20, 1969, p 245-265
27. D.C. Mears, Metals in Medicine and Surgery, *Int. Met. Rev.*, Vol 22 (No. 218), 1977, p 119-155
28. J.R. Cahoon and L.D. Hill, Evaluation of a Precipitation Hardened Wrought Cobalt-Nickel-Chromium-Titanium Alloy for Surgical Implants, *J. Biomed. Mater. Res.*, Vol 12, 1978, p 805-821
29. T.C. Ruck and J.F. Fulton, *Medical Physiology and Biophysics.*, 18th ed., W.B. Saunders, 1960
30. H.C. McCann, Inorganic Components of Salivary Secretions, in *Art and Science of Dental Caries Research*, R.S. Harris, Ed., Academic Press, 1968, p 55-73
31. D.B. Ferguson, Salivary Glands and Saliva, in *Applied Physiology of the Mouth*, C.L.B. Lavell, Ed., John Wright, 1975, p 145-179
32. H.J. Mueller, Binding of Corroded Ions to Human Saliva, *Biomater.*, Vol 6, 1985, p 146-149
33. H.J. Mueller, Characterization of the Acquired Biofilms on Materials Exposed to Human Saliva, in *Proteins at Interfaces*, T.A. Horbett and J. Brash, Ed., Advances in Chemistry Series, American Chemical Society, 1987
34. S.A. Ellison, The Identification of Salivary Components, in *Saliva and Dental Caries*, I. Kleinberg, S.A. Ellison, and I.D. Mandell, Ed., *Sp. Supp. Microbiol. Abst.*, 1979, p 13-29
35. C.E. Ingersoll, Characterization of Tarnish, *J. Dent. Res.*, Vol 55, 1976, IADR No. 144
36. K. Barton, chapter 2, in *Protection Against Atmospheric Corrosion*, John Wiley & Sons, 1973
37. I.C. Schoonover and W. Souder, Corrosion of Dental Alloys, *J. Am. Dent. Assoc.*, Vol 28, 1941, p 1278-1291
38. J.C. Muhler and H.M. Swenson, Preparation of Synthetic Saliva From Direct Analysis of Human Saliva, *J. Dent. Res.*, Vol 26, 1947, p 474
39. D.A. Carter, T.K. Ross, and D.C. Smith, Some Corrosion Studies on Silver-Tin Amalgams, *Br. Corros. J.*, Vol 2, 1967, p 199-205
40. G. Tani and F. Zucci, Electrochemical Evaluation of the Corrosion Resistance of the Commonly Used Metals in Dental Prosthesis, *Minerva. Stomat.*, Vol 16, 1967, p 710-713
41. J.M. Meyer and J.N. Nally, Influence of Artificial Salivas on the Corrosion of Dental Alloys, *J. Dent. Res.*, Vol 54, 1975, IADR No. 76
42. B.W. Darvell, The Development of an Artificial Saliva for In-Vitro Amalgam Corrosion Studies, *J. Oral Rehab.*, Vol 5, 1978, p 41-49
43. M.L. Swartz, R.W. Phillips, and M.D. El Tannir, Tarnish of Certain Dental Alloys, *J. Dent. Res.*, Vol 37, 1958, p 837-847
44. F. Fusayama, T. Katayori, and S. Nomoto, Corrosion of Gold and Amalgam Placed in Contact With Each Other, *J. Dent. Res.*, Vol 42, 1963, p 1183-1197
45. C.E. Guthrow, L.B. Johnson, and K.R. Lawless, Corrosion of Dental Amalgam and Its Component Phases, *J. Dent. Res.*, Vol 46, 1967, p 1372-1381
46. F.V. Wald and F.H. Cocks, Investigation of Copper-Manganese-Nickel Alloys for Dental Purposes, *J. Dent. Res.*, Vol 50, 1971, p 44-59
47. M. Marek and R.F. Hockman, Corrosion Behavior of Amalgam Electrode in Artificial Saliva, *J. Dent. Res.*, Vol 51, 1972, IADR No. 63
48. J. Brugirard, R. Bargain, J.C. Dupuy, H. Mazille, and G. Monnier, Study of the Electrochemical Behavior of Gold Dental Alloys, *J. Dent. Res.*, Vol 52, 1973, p 828-836
49. *The National Foundry*, American Pharmaceutical Association, 1970, p 624
50. M. Marek and E. Topfl, Electrolytes for Corrosion Testing of Dental Alloys, *J. Dent. Res.*, Vol 65, 1986, IADR No. 1192
51. N.K. Sarkar and E.H. Greener, In Vitro Corrosion of Dental Amalgam, *J. Dent. Res.*, Vol 50, 1971, IADR No. 13
52. G.F. Finkelstein and E.H. Greener, In Vitro Polarization of Dental Amalgam in Hu-

man Saliva, *J. Oral. Rehab.*, Vol 4, p 355-368

53. G.F. Finkelstein and E.H. Greener, Role of Mucin and Albumin in Saline Polarization of Dental Amalgam, *J. Oral Rehab.*, Vol 5, 1978, p 95-110
54. H. Do Duc, P. Tissot, and J.-M. Meyer, Potential Sweep and Intensiostatic Pulse Studies of Sn, Sn_8Hg, and Dental Amalgam in Chloride Solution, *J. Oral Rehab.*, Vol 6, 1979, p 189-197
55. R.I. Holland, Effect of Pellicle on Galvanic Corrosion of Amalgam, *Scand. J. Dent. Res.*, Vol 92, 1984, p 93-96
56. H.J. Mueller, The Effects of a Human Salivary Dialysate Upon Ionic Release and Electrochemical Corrosion of a Cu-Al Alloy, *J. Electrochem. Soc.*, Vol 134, 1987, p 555–580
57. A. Schulman, H.A.B. Linke, T.K. Vaidyanathan, Tarnish of Dental Alloys by Oral Microorganisms, *J. Dent. Res.*, Vol 63, 1984, IADR No. 55
58. M. Stern and E.D. Weisert, Experimental Observation on the Relation Between Polarization Resistance and Corrosion Rate, *Proc. ASTM*, Vol 59, 1959, p 1280-1291
59. M. Marek, The Corrosion of Dental Materials, in *Corrosion: Aqueous Processes and Passive Films*, Vol 23, *Treatise on Materials Science*, J.C. Scully, Ed., Academic Press, 1983, p 331-394
60. M. Bergman, O. Ginstrup, and B. Nilsson, Potentials of and Currents Between Dental Metallic Restorations, *Scand. J. Dent. Res.*, Vol 90, 1982, p 404-408
61. K. Nilner, P.-O. Glantz, B. Zoger, On Intraoral Potential and Polarization Measurements of Metallic Restorations, *Acta Odontol. Scand.*, Vol 40, 1982, p 275-281
62. J.M. Mumford, Electrolytic Action in the Mouth and Its Relationship to Pain, *J. Dent. Res.*, Vol 36, 1957, p 632-640
63. C.P. Wang Chen and E.H. Greener, A Galvanic Study of Different Amalgams, *J. Oral Rehab.*, Vol 4, 1977, p 23-27
64. R. Soremark, G. Freedman, J. Goldin, and L. Gettleman, Structure and Microdistribution of Gold Alloys, *J. Dent. Res.*, Vol 45, 1966, p 1723-1735
65. D.B. Boyer, K. Chan, C.W. Svare, The Effect of Finishing on the Anodic Polarization of High-Copper Amalgams, *J. Oral Rehab.*, Vol 5, 1978, p 223-228
66. G. Palaghias, Oral Corrosion and Corrosion Inhibition Processes, *Swed. Dent. J.*, Supp 30, 1985
67. H.J. Mueller and A. Edahl, The Effect of Exposure Conditions Upon the Release of Soluble Copper and Tin From Dental Amalgams, *Biomater.*, Vol 5, 1984, p 194-200
68. H.J. Mueller and R.M. Barrie, Intraoral Corrosion of Copper-Aluminum Alloys, *J. Dent. Res.*, Vol 64, 1985, IADR No. 1753
69. G.N. Jenkins, *The Physiology and Biochemistry of the Mouth*, 4th ed., Blackwell, 1978, p 284-359
70. H.J. Mueller and E.H. Greener, Polarization Resistance of Surgical Materials in Ringer's Solution, *J. Biomed. Mater. Res.*, Vol 4, 1970, p 29-41
71. E.J. Sutow, S.R. Pollack, and E. Korostoff, An In Vitro Investigation of the Anodic Polarization and Capacitance Behavior of 316-L Stainless Steel, *J. Biomed. Mater. Res.*, Vol 10, 1976, p 671-693
72. H.J. Mueller, The Binding of Corroded Metallic Ions to Salivary-Type Proteins, *Biomater.*, Vol 4, 1983, p 66-72
73. J.R. Strub, C. Eyer, N.K. Sarkar, Microstructure and Corrosion of a Low-Gold Casting Alloy, *J. Dent. Res.*, Vol 63, 1984, IADR No. 793
74. M.P. Keenan, Effects of Gold Finishing on Plaque Retention, *J. Dent. Res.*, Vol 56, 1977, IADR No. 121(B)
75. C.W. Svare, G. Belton, and E. Korostoff, The Role of Organics in Metallic Passivation, *J. Biomed. Mater. Res.*, Vol 4, 1970, p 457-467
76. G.C.F. Clark and D.F. Williams, The Effects of Proteins on Metallic Passivation, *J. Biomed. Mater. Res.*, Vol 16, 1982, p 125-134
77. S.A. Brown and K. Merritt, Electrochemical Corrosion in Saline and Serum, *J. Biomed. Mater. Res.*, Vol 14, 1980, p 173-175
78. S.A. Brown and K. Merritt, Fretting Corrosion in Saline and Serum, *J. Biomed. Mater. Res.*, Vol 15, p 479-488
79. H.J. Mueller, The Effect of Electrical Signals Upon the Adsorption of Plasma Proteins to a High Cu Alloy, in *Biomaterials: Interfacial Phenomena and Applications*, S.L. Cooper and N.A. Peppas, Ed., ACS monograph series 199, American Chemical Society, 1982
80. R.C. Salvarezza, M.E.L. de Mele, H.H. Videla, and F.R. Goni, Electrochemical Behavior of Aluminum in Human Plasma, *J. Biomed. Mater. Res.*, Vol 19, 1985, p 1073-1084
81. H. Hero and L. Niemi, Tarnishing In Vivo of Ag-Pd-Cu-Zn, *J. Dent. Res.*, Vol 65, 1986, p 1303-1307
82. H. Do Duc and P. Tissot, Rotating Disc and Ring Disc Electrode Studies of Tin in Neutral Phosphate Solution, *Corros. Sci.*, Vol 19, 1979, p 191-197
83. I.D. Mandel, Relation of Saliva and Plaque to Caries, *J. Dent. Res.*, Vol 53, 1974, p 246
84. T. Ericson, K.M. Pruitt, H. Arwin, and I. Lunstrom, Ellipsometric Studies of Film Formation on Tooth Enamel and Hydrophilic Silicon Surfaces, *Acta Odontol. Scand.*, Vol 40, 1982, p 197-201
85. R.E. Baier and P.-O. Glantz, Characterization of Oral In Vivo Films Formed on Different Types of Solid Surfaces, *Acta Odontol. Scand.*, Vol 36, 1978, p 289-301
86. K. Skjorland, Auger Analysis of Integuments Formed on Different Dental Filling Materials In Vivo, *Acta Odontol. Scand.*, Vol 40, 1982, p 129-134
87. K. Hannesson Eggen and G. Rolla, Gel Filtration, Ion Exchange Chromatography and Chemical Analysis of Macromolecules Present in Acquired Enamel Pellicle (2-hr), *Scand. J. Dent. Res.*, Vol 90, 1982, p 182-188
88. A. Bennick, G. Chau, R. Goodlin, S. Abrams, D. Tustian, and G. Mandapallimattam, The Role of Human Salivary Acidic Proline-Rich Proteins in the Formation of Acquired Dental Pellicle In Vivo and Their Fate After Adsorption to the Human Enamel Surface, *Arch. Oral Biol.*, Vol 28, 1983, p 19-27
89. T. Sonju and P.-O. Glantz, Chemical Composition of Salivary Integuments Formed In Vitro on Solids with Some Estalished Surface Characteristics, *Arch. Oral Biol.*, Vol 20, 1975, p 687-691
90. D.I. Hay, The Adsorption of Salivary Proteins by Hydroxyapatite and Enamel, *Arch. Oral Biol.*, Vol 12, 1967, p 937-946
91. G. Rolla, Formation of Dental Integuments—Basic Chemical Considerations, *Swed. Dent. J.*, Vol 1, 1977, p 241-251
92. A.C. Juriaanse, M. Booij, J. Arends, and J.J. Ten Bosch, The Adsorption In Vivo of Purified Salivary Proteins on Bovine Dental Enamel. *Arch. Oral Biol.*, Vol 26, 1981, p 91-96
93. H.J. Mueller, Differential Scanning Calorimetry of Adsorbed Protein Films, in *Transactions of the 13th Annual Meeting Society of the Biomaterials*, 1987
94. R D. Norman, R.V. Mehra, and M.L. Schwartz, The Effects of Restorative Materials on Plaque Composition, *J. Dent. Res.*, Vol 50, 1971, IADR No. 162
95. J.J. Tuccillo and J.P. Nielsen, Microprobe Analysis of an In Vivo Discoloration, *J. Prosthet. Dent.*, Vol 31, 1974, p 285-289
96. J.J. Tuccillo and J.P. Nielson, Observation of Onset of Sulfide Tarnish on Gold-Base Alloys, *J. Prosthet. Dent.*, Vol 25, 1971, p 629-637
97. R.P. Lubovich, R.E. Kovarik, and D.L. Kinser, A Quantitative and Subjective Characterization of Tarnishing in Low-Gold Alloys, *J. Prosthet. Dent.*, Vol 42, 1979, p 534-538
98. G.W. Marshall, N.K. Sarkar, and E.H. Greener, Detection of Oxygen in Corrosion Products of Dental Amalgam, *J. Dent. Res.*, Vol 54, 1975, p 904
99. H. Otani, W.A. Jesser, and H.G.F. Wilsdorf, The In Vivo and the In Vitro Corrosion Products of Dental Amalgam, *J. Biomed. Mater. Res.*, Vol 7, 1973, p 523-539
100. A.B. Burse, M.L. Swartz, R.W. Phillips, and R.W. Oykema, Comparison of the In Vivo and In Vitro Tarnish of Three Gold Alloys, *J. Biomed. Mater. Res.*, Vol 6, 1972, p 267-277
101. B.R. Laing, S.H. Bernier, Z. Giday, and K. Asgar, Tarnish and Corrosion of Noble Metal Alloys, *J. Prosthet. Dent.*, Vol 48, 1982, p 245-252
102. H. Hero and J. Valderhaug, Tarnishing In Vivo and In Vitro of a Low-Gold Alloy Related to Its Structure, *J. Dent. Res.*, Vol 64, 1985, p 139-143
103. H. Hero and R.B. Jorgensen, Tarnishing of a Low-Gold Alloy in Different Structural States, *J. Dent. Res.*, Vol 62, 1983 p 371-376
104. L. Gettlemen, R.F. Cocks, L.A. Darmiento, P.A. Levine, S. Wright, and D. Nathanson, Measurement of In Vivo Corrosion Rates in Baboons and Correlation With In Vivo Tests. *J. Dent. Res.*, Vol 59, 1980, p 689-707
105. L. Gettleman, C. Amman, and N.K. Sarkar, Quantitative In Vivo and In Vitro Measurement of Tarnish. *J. Dent. Res.*, Vol 58, 1979, IADR No. 969
106. Revised American Dental Association Specification No. 1 for Alloy for Amalgam, *J. Am. Dent. Assoc.*, Vol 95, 1977, p 614-617
107. Addendum to ANSI/ADA Specification No. 1 for Alloy for Amalgam, *J. Am. Dent. Assoc.*, Vol 100, 1980, p 246

108. D.B. Mahler and J.D. Adey, Microprobe Analysis of Three High Copper Amalgams, *J. Dent. Res.*, Vol 63, 1984, p 921-925
109. J.W. Edie, D.B. Boyer, and K.C. Chjan, Estimation of the Phase Distribution in Dental Amalgams With Electron Microprobe, *J. Dent. Res.*, Vol 57, 1978, p 277–282
110. J. Leitao, Surface Roughness and Porosity of Dental Amalgam, *Acta Odontol. Scand.*, Vol 40, 1982, p 9-16
111. R.W. Bryant, Gamma-2 Phase in Conventional Amalgam-Discrete Clumps or Continuous Network—A Review, *Aust. Dent. J.*, Vol 29, 1984, p 163-167
112. L.B. Johnson, X-Ray Diffraction Evidence for the Presence of β(Ag-Hg) in Dental Amalgam, *J. Biomed. Mater. Res.*, Vol 1, 1967, p 285-297
113. S.J. Marshall and G.W. Marshall, Jr., Time-Dependent Phase Changes in Cu-Rich Amalgams, *J. Biomed. Mater. Res.*, Vol 13, 1979, p 395-406
114. Revised ANSI/ADA Specification No. 5 For Dental Casting Gold Alloy, *J. Am. Dent. Assoc.*, Vol 104, 1981, p 70
115. J.P. Moffa, Alternative Dental Casting Alloys, *Dent. Clin. N. Am.*, Vol 27, 1983, p 194-200
116. R.M. German, Precious-Metal Dental Casting Alloys, *Int. Met. Rev.*, Vol 27, 1982, p 260-288
117. K. Yasuda and K. Hisatsune, The Development of Dental Alloys Conserving Precious Metals: Improving Corrosion Resistance by Controlled Aging, *Int. Dent. J.*, Vol 33, 1983
118. D.L. Smith, Dental Casting Alloys, Technical and Economic Considerations in the USA, *Int. Dent. J.*, Vol 33, 1983, p 25-34
119. S.A. Aquilino and T.D. Taylor, Prosthodontic Laboratory Survey, *J. Prosthet. Dent.*, Vol 53, 1984, p 879-885
120. H. Hero, Tarnishing and Structures of Some Annealed Dental Low-Gold Alloys, *J. Dent. Res.*, Vol 63, 1984, p 926-931
121. L. Niemi and H. Hero, Structure, Corrosion, and Tarnishing of Ag-Pd-Cu Alloys, *J. Dent. Res.*, Vol 64, 1985, p 1163-1169
122. R.C. Craig, H.J. Skesnick, and F.A. Peyton, Application of 17-7 Precipitation Hardenable Stainless Steel in Dentistry, *J. Dent. Res.*, Vol 44, 1965, p 587-595
123. S. Civjan, E.F. Huget, and L.B. de Simon, Effects of Laboratory Procedures on 55-Nitinol, *J. Dent. Res.*, Vol 52, 1973, IADR No. 51
124. A.J. Goldberg and C.J. Burstone, an Evaluation of Beta-Stabilized Titanium Alloys for Use in Orthodontic Appliances, *J. Dent. Res.*, Vol 57, 1978, p 593-600
125. E.F. Huget and S.G. Vermilyea, Base Metal Dental and Surgical Alloys, in *Biocompatibility of Dental Materials*, Vol IV, D.C. Smith and D.F. Williams, Ed., CRC Press, 1982, p 37-49
126. H.F. Morris and K. Asgar, Physical Properties and Microstructure of Four New Paertial Denture Alloys, *J. Dent. Res.*, Vol 57, 1978, IADR No. 218
127. A.T. Kuhn, The Corrosion of Metals and Alloys Used in Dentistry, in *Restoration of the Partially Dentate Mouth*, J.F. Bates, D.J. Neill, and H.W. Preiskel, Ed., Quintessence Publishing, 1984, p 160-175
128. K. Asgar and F.C. Allan, Microstructure and Physical Properties of Alloys for Partial Denture Castings, *J. Dent. Res.*, Vol 47, 1968, p 189-197
129. S. Civijan, E.F. Huget, W.L. Erhard, and G.J. Vaccaro, Characterization of Surgical Casting Alloys, *J. Dent. Res.*, Vol 50, 1971, IADR No. 584
130. T.M. Devine and J. Wulff, Cast vs Wrought Cobalt-Chromium Surgical Implant Alloys, *J. Biomed. Mater. Res.*, Vol 9, 1975, p 151-167
131. R.G. Craig (Chm), Section One Report, in *International Workshop on Biocompatibility, Toxicity, and Hypersensitivity to Alloy Systems Used in Dentistry*, Conference Proceedings, University of Michigan School of Dentistry, 1985
132. E.F. Huget and S.G. Vermilyea, Base Metal Dental and Surgical Alloys, in *Biocompatibility of Dental Materials*, Vol IV, D.C. Smith and D.F. Williams, Ed., CRC Press, 1982, p 34-49
133. T.G. Goodall, The Metallography of Heat Treatment Effects in a Nickel-Base Casting Alloy, *Aust. Dent. J.*, Vol 24, 1879, p 235-237
134. S. Winkler, H.F. Morris, and J.M. Monteiro, Changes in Mechanical Properties and Microstructure Following Heat Treatment of a Nickel-Chromium Alloy, *J. Prosthet. Dent.*, Vol 52, 1984, p 821-827
135. K. Asgar and F.C. Allan, Microstructure and Physical Properties of Alloys of Partial Denture Castings, *J. Dent. Res.*, Vol 47, 1968, p 189-197
136. K. Asgar and F.A. Peyton, Effect of Microstructure on the Physical Properties of Cobalt-Base Alloys, *J. Dent. Res.*, Vol 40, 1961, p 63-72
137. Revised ANSI/ADA Specification No. 14, Dental Base Metal Casting Alloys, *J. Am. Dent. Assoc.*, Vol 105, 1982, p 686–687
138. H.F. Morris and K. Asgar, Physical Properties and Microstructure of Four New Commercial Partial Denture Alloys, *J. Prosthet. Dent.*, Vol 33, 1975, p 36-46
139. H. Mohammed and K. Asgar, A New Dental Superalloy System, *J. Dent. Res.*, Vol 53, 1973, p 7-14
140. J.F. Bates and A.G. Knapton, Metal and Alloys in Dentistry, *Int. Met. Rev.*, Vol 22 (No. 215), 1977, p 39–60
141. R.M. Waterstrat, N.W. Rupp, and O. Franklin, Production of a Cast Titanium-Base Partial Denture, *J. Dent. Res.*, Vol 57A, 1978, IADR No. 717
142. M. Taira, J.B. Moser, and E.H. Greener, Mechanical Properties of Cast Ti Alloys for Dental Uses, *J. Dent. Res.*, Vol 65, 1986, IADR No. 603
143. E.F.I. Roberts and K.M. Clarke, The Colour Characteristics of Gold Alloys, *Gold Bull.*, Vol 9, 1979, p 9-19
144. R.M. German, M.M. Guzowski, and D.C. Wright, Color and Color Stability as Alloy Design Criterion, *J. Met.*, Vol 32, 1980, p 20-27
145. D.J.L. Treacy and R.M. German, Chemical Stability of Gold Dental Alloys, *Gold Bull.*, Vol 17, 1984, p 46-54
146. P.P. Coroso, Jr., R.M. German, and H.D. Simmons, Jr., Tarnish Evaluation of Gold-Based Dental Alloys, *J. Dent. Res.*, Vol 64, 1965
147. R.M. German, The Role of Microstructure in the Tarnish of Low Gold alloys, *Metallography*, Vol 14, 1981, p 253-266
148. R.M. German, D.C. Wright, and R.F. Gallant, In Vitro Tarnish Measurements on Fixed Prosthodontic Alloys, *J. Prosthet. Dent.*, Vol 47, 1982, p 399-406
149. D.C. Wright and R.M. German, Quantification of Color and Tarnish Resistance of Dental Alloys, *J. Dent. Res.*, Vol 58A, 1979, IADR No. 975
150. D.A. Nitkin and K. Asgar, Evaluation of Alternative Alloys to Type III Gold for Use in Fixed Prosthodontics, *J. Am. Dent. Assoc.*, Vol 93, 1976, p 622-629
151. S. Civjan, E.F. Huget, and J. Marsden, Characterization of Two High-Fusing Gold Alloys, *J. Dent. Res.*, Vol 51, 1972, IADR No. 222
152. J.F. Bates and A.G. Knapton, Metal and Alloys in Dentistry, *Int. Met. Rev.*, Vol 22, 1982, p 39-60
153. S. Civjan, E.F. Huget, N.N. Dvivedi, and H.E. Cosner, Jr., Characterization of Two Au-Pd-Ag Alloys, *J. Dent. Res.*, Vol 52, 1973, IADR No. 46
154. E.F. Huget, S.G. Vermilyea, and J.M. Vilca, Studies on White Crown-and-Bridge Alloys, *J. Dent. Res.*, Vol 57, 1978, IADR No. 722
155. P.F. Mezger, M.M.A. Vrijhoef, and E.H. Greener, Corrosion Resistance of Three High Palladium Alloys, *Dent. Mater.*, Vol 1, 1985, p 177-179
156. M.M.A. Vrijhoef and J.M. van der Zel, Oxidation of Two High-Palladium PFM Alloys, *Dent. Mater.*, Vol 1, 1985, p 214-218
157. R.L. Bertolotti, Selection of Alloys for Today's Crown and Fixed Partial Denture Restorations, *J. Am. Dent. Assoc.*, Vol 108, 1984, p 959-966
158. J.J. Tuccillo, Compositional and Functional Characteristics of Precious Metal Alloys for Dental Restorations, in *Alternatives to Gold Alloys in Dentistry*, T.M. Valega, Ed., Conference Proceedings, DHEW Publication (NIH) 77-1227, Department of Health, Education, and Welfare, 1977
159. P.J. Cascone, Phase Relations of the Palladium-Base, Copper, Gallium, Indium Alloy System, *J. Dent. Res.*, Vol 63, 1984, IADR No. 563
160. M.M.A. Vrijhoef, Oxidation of Two High-Palladium PFM Alloys, *Dent. Mater.*, Vol 1, 1985, p 214-18
161. N. Sumithra, T.K. Vaidyanathan, S. Sastri, and A. Prasad, Chloride Corrosion of Recent Commercial Pd-Based Alloys, *J. Dent. Res.*, Vol 62, 1983, IADR No. 346
162. S.M. Paradiso, Corrosion Evaluation of Pd-Cu-Ga, *J. Dent. Res.*, Vol 43, 1984, IADR No. 43
163. P.R. Mezger, M.M.A. Vrijhoef, and E.H. Greener, Corrosion Resistance of Three High-Palladium Alloys, *Dent. Mater.*, Vol 1, 1985, p 177-180
164. A. Oden and H. Hero, The Relationship Between Hardness and Structure of Pd-Cu-Ga Alloys, *J. Dent. Res.*, Vol 65, 1986, p 75-79
165. J.R. Mackert, Jr., E.E. Parry, and C.W. Fairhurst, Oxide Metal Interface Morphology Related to Oxide Adherence, *J. Dent. Res.*, Vol 63, 1984, IADR No. 405
166. G. Baron, Auger Chemical Analysis of Ox-

ides on Ni-Cr Alloys. *J. Dent. Res.*, Vol 63, 1984, p 76-80
167. D.L. Menis, J.B. Moser, and E.H. Greener, Experimental Porcelain Compositions for Application to Cast Titanium, *J. Dent. Res.*, Vol 65, 1986, IADR No. 1565
168. ANSI/ADA Specification No. 32, New American Dental Association Specification No. 32 for Orthodontic Wires Not Containing Precious Metals, *J. Am. Dent. Assoc.*, Vol 95, 1977, p 1169-71
169. P.J. Brockhurst, Base Metal Wires for Gold Alloy Soldering to Cast Cobalt-Chromium Alloy Partial Dentures, *Aust. Dent. J.*, Vol 15, 1970, p 499-506
170. M.R. Marcotte, Optimum Time and Temperature for Stress Relief Heat Treatment of Stainless Steel Wire, *J. Dent. Res.*, Vol 52, 1973, p 1171-1175
171. C.J. Burstone and J.Y. Morton, Chinese NiTi Wire—A New Orthodontic Wire, *Am. J. Ortho.*, Vol 87, 1985, p 445-452
172. M. Bergman, Combinations of Gold Alloys in Soldered Joints, *Swed. Dent. J.*, Vol 1, 1977, p 99-106
173. C.E. Janus, D.F. Taylor, and G.A. Holland, A Microstructural Study of Soldered Connectors of Low-Gold Casting Alloys, *J. Prosthet. Dent.*, Vol 50, 1983, p 657-663
174. T.M. Devine and J. Wulff, Cast vs Wrought Cobalt-Chromium Surgical Implant Alloys, *J. Biomed. Mater. Res.*, Vol 9, 1975, p 151-167
175. H.J. Mueller and B.C. Marker, Effect of PO_4^{3-} and Cl^- Upon Product Deposition on NTD and Cupralloy, *J. Dent. Res.*, Vol 59, IADR No. 279, 1980
176. H.J. Mueller, SIMS and Colorimetry of In-Vitro Sulfided Crown and Bridge Alloys, in *Fifth International Symposium on New Spectroscopic Methods for Biomedical Research*, Battelle Laboratories and University of Washington, 1986
177. H. Hero, Tarnishing and Structures of Some Annealed Dental Low-Gold Alloys, *J. Dent. Res.*, Vol 63, 1984, p 926-931
178. E. Suoninen and H. Hero, Effect of Palladium on Sulfide Tarnishing of Noble Metal Alloys, *J. Biomed. Mater. Res.*, Vol 19, 1985, p 917-934
179. R. Kropp, Application of Corrosion and Tarnish Tests to Different Dental Alloys, *J. Dent. Res.*, Vol 65, 1986, IADR No. 197
180. T.K. Vaidyanathan and A. Prasad, In Vitro Corrosion and Tarnish Characteristics of Typical Dental Gold Compositions, *J. Biomed. Mater. Res.*, Vol 15, 1981, p 191-201
181. J. Brugirard, Baigain, J.C. Dupuy, H. Mazille, and G. Monnier, Study of the Electrochemical Behavior of Gold Dental Alloys, *J. Dent. Res.*, 1973, p 838-836
182. W. Popp, H. Kaiser, H. Kaesche, W. Bramer, and F. Sperner, Electrochemical Behavior of Noble Metal Dental Alloys in Different Artificial Saliva Solutions, in *Proceedings of the 8th International Congress of Metallic Corrosion*, Vol 1, DECHEMA, 1981, p 76-81
183. N.K. Sarkar, R.A. Fuys, and J.W. Stanford, The Chloride Corrosion Behavior of Silver-Base Casting Alloys, *J. Dent. Res.*, Vol 58, 1979, p 1572-1577
184. D.C. Wright, R.M. German, and R.F. Gallant, Copper and Silver Corrosion Activity in Crown and Bridge Alloys, *J. Dent. Res.*, Vol 60, 1981, p 809-814
185. T.K. Vaidyanathan and A. Prasad, In Vitro Corrosion and Tarnish Analysis of Ag-Pd Binary System, *J. Dent. Res.*, Vol 60, 1981, p 707-715
186. N. Ishizaki, Corrosion Resistance of Ag-Pd Alloy System in Artificial Saliva: An Electrochemical Study, *J. Osaka Dent. Univ.*, Vol 3, 1969, p 121-133
187. L.A. O'Brien and R.M. German, Compositional Effects on Pd-Ag Dental Alloys, *J. Dent. Res.*, Vol 63, 1984, IADR No. 44
188. N.K. Sarkar, R.A. Fuys, and J.W. Stanford, The Chloride Behavior of Silver-Base Casting Alloys, *J. Dent. Res.*, Vol 58, 1979, p 1572-1577
189. L. Niemi and R.I. Holland, Tarnish and Corrosion of a Commercial Dental Ag-Pd-Cu-Au Casting Alloy, *J. Dent. Res.*, Vol 63, 1984, p 1014-1018
190. L. Niemi and H. Hero, Structure, Corrosion, and Tarnishing of Ag-Pd-Cu Alloys, *J. Dent. Res.*, Vol 64, 1985, p 1163-1169
191. J.M. Meyer, Corrosion Resistance of Ni-Cr Dental Casting Alloys, *Corros. Sci.*, Vol 17, 1977, p 971-982
192. R.J. Hodges, The Corrosion Resistance of Gold and Base Metal Alloys, in *Alternatives to Gold Alloys in Dentistry*, T.M. Valega, Ed., DHEW Publication (NIH) 77-1227, Department of Health, Education, and Welfare, 1977
193. N.K. Sarkar and E.H. Greener, In Vitro Corrosion Resistance of New Dental Alloys, *Biomater. Med. Dev. Art. Org.*, Vol 1, 1973, p 121-129
194. H.J. Mueller and C.P. Chen, Properties of a Fe-Cr-Mo Wire, *J. Dent.*, Vol 11, 1983, p 71-79
195. N.K. Sarkar, W. Redmond, B. Schwaninger, and A.J. Goldberg, The Chloride Corrosion Behavior of Four Orthodontic Wires, *J. Oral Rehab.*, Vol 10, 1983, p 121-128
196. H.J. Mueller, Silver and Gold Solders—Analysis Due to Corrosion, *Quint. Int.*, Vol 37, 1981, p 327-337
197. D.L. Johnson, V.W. Rinne, and L.L. Bleich, Polarization-Corrosion Behavior of Commercial Gold- and Silver-Base Casting Alloys in Fusayama Solution, *J. Dent. Res.*, Vol 62, 1983, p 1221-1225
198. A.D. Vardimon and H.J. Mueller, In Vitro and In Vivo Corrosion of Permanent Magnets in Orthodontic Therapy, *J. Dent. Res.*, Vol 64, 1985, IADR No. 89

Corrosion of Emission-Control Equipment

William J. Gilbert and Robert John Chironna, Croll-Reynolds Company, Inc.

CORROSION PROBLEMS and material selection for emission-control equipment can be difficult because of the varied corrosive compounds present and the severe environments encountered. Therefore, a number of the more common emission-control applications will be discussed. More detailed information on the applications is available in the references cited at the end of this article.

Flue Gas Desulfurization

By far the most common cleaning application for flue gases is flue gas desulfurization (FGD). This section will discuss the selection of materials of construction for FGD systems. More information on corrosion in FGD systems is available in the section "Corrosion of Flue Gas Desulfurization Systems" of the article "Corrosion in Fossil Fuel Power Plants" in this Volume.

These systems came into being in the late 1960s and early 1970s because of the tightening of restrictions on the release of sulfur emissions. The oil shortage of the mid-1970s and subsequent oil price increases led to the reuse of coal in new and renovated power plants. In virtually all cases, this meant the potential for increased sulfur emissions. Many more FGD systems were needed.

Flue gas desulfurization systems typically use wet scrubbing units with lime or limestone slurries for sulfur dioxide (SO_2) absorption. Initially, it was thought that the relatively mild pH and temperature conditions found within most of these systems would not present a significant corrosion problem. This was soon found not to be the case. The fact that the FGD system could constitute up to 25% of the total capital and operating expenses of the power plant made it imperative to determine the reasons behind the failure of the material.

Environment. The gases encountered by the FGD system are hot and contain SO_2 at significant levels, some sulfur trioxide (SO_3) as a result of the oxidation of SO_2 at high temperatures, and fly ash. Initially, these gases may be sent to a dry-dust collector, such as an electrostatic precipitator or fabric filter baghouse, for fly ash removal. The gases typically enter a wet scrubber (venturi with separator) and are quenched as SO_2 is absorbed. The components that often have the severest problems, however, are the outlet duct and stack. Here the condensates are more acidic, the gases are highly oxygenated, and the presence of chlorides and fluorides can cause serious corrosion problems. Nevertheless, throughout the entire system, corrosion can occur to various degrees and because of various factors.

Corrosion Factors. Four basic factors affect the severity and type of corrosion that occurs. They are discussed below.

pH. The result of the reactions that take place within the scrubber is a slurry with a typical pH of 4 to 5. This is desirable, because it allows for good absorption of SO_2 and is acidic enough to reduce scale formation. Local pH values as low as 1 may exist from the concentration of chlorides entering the makeup liquid with contributions from fluorides. The low-pH conditions with the presence of chlorides and fluorides limit the use of carbon steels, stainless steels, and a number of higher-nickel alloys (Fig. 1).

Gas Saturation. The dry flue gas is not severely corrosive. However, when the gas reaches its dew point, sulfuric (H_2SO_4) and sulfurous (H_2SO_3) acids can form. In addition, hydrochloric acid (HCl) is produced because of the presence of hydrogen chloride (formed at the elevated temperatures of combustion) plus the condensing water vapor. Again, significant problems arise from the use of carbon or stainless steels.

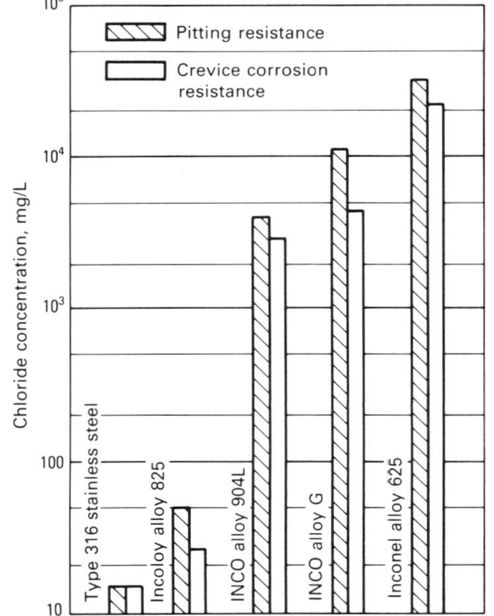

Fig. 1 Minimum levels of chloride that cause pitting and crevice corrosion in 30 days in SO_2-saturated chloride solutions at 80 °C (175 °F). Source: Ref 1

Temperature. The problems caused by temperature excursions are primarily related to the lessening of the corrosion-resistant properties of synthetic coatings, fiberglass-reinforced plastics (FRP), and thermoplastics, possibly to the point of complete destruction at high enough temperatures. This affects metals to a lesser extent, but can make a borderline problem a serious one.

Erosion generally occurs as a result of fly ash within the gas impacting on a surface in a relatively dry area of the system or the liquid slurry impinging upon a wetted surface. In either case, areas susceptible to corrosion attack are produced.

General Materials Selection. An easily overlooked but critical aspect of materials selection is the ability of the manufacturer to construct the equipment properly with correct fabrication techniques. In particular, with regard to the use of high-nickel alloys, the welding recommendations of alloy producers should be precisely followed to maintain the corrosion resistance of the materials (Ref 2). This is of course true for any type of fabrication. The most careful materials selection process can be negated by poor fabrication practices.

Metals. Where pH is neutral or higher, austenitic stainless steels (AISI types 304, 316, and 317, L grades preferred) perform well even at elevated temperatures. If pH is as low as 4 and chloride content is low (less than 100 ppm) but temperatures are above approximately 65 °C (150 °F), then Incoloy 825, Inconel 625, Hastelloy G-3, and alloy 904L (UNS N08904) or their equivalents are usually acceptable. Table 1 lists compositions of alloys commonly used in FGD systems.

When chloride content is up to 0.1% and pH approaches 2, only Hastelloys C-276, G, and G-3, and Inconel 625 can be successfully used. The other alloys mentioned above would be subjected to pitting and crevice corrosion. If a region is encountered with pH as low as 1 and chloride content above 0.1%, one of the only successful alloys acceptable is reported to be Hastelloy C-276 or its equivalent. In terms of metals selection, the higher the molybdenum content in an alloy, the more severe the corrosive environment it can withstand in the FGD system (Ref 3).

Nonmetals. Fiberglass-reinforced plastics can be used in almost any application in which temperatures do not exceed 120 °C (250 °F) (preferably 95 °C, or 205 °F), regardless of whether there are high chlorides or low pHs. The best choices

Table 1 Compositions of some alloys used in FGD systems

Alloy	Composition, %(a) C	Fe	Ni	Cr	Mo	Mn	Others
Type 304L	0.03 max	bal(b)	10.0	19.0	. . .	2.0 max	0.045 max P, 0.03 max S, and 1.00 max Si
Type 316L	0.03 max	bal	12.0	17.0	2.5	2.0 max	1.00 max Si, 0.045 max P, and 0.03 max S
Type 317L	0.03 max	bal	13.0	19.0	3.5	2.0 max	1.00 max Si, 0.045 max P, and 0.03 max S
Inconel alloy 625	0.10 max	5.0 max	bal	21.5	9.0	0.50 max	0.40 max Al, 0.40 max Ti, 3.65 Nb, 0.015 max P, 0.015 max S, and 0.50 max Si
Incoloy alloy 825	0.05 max	bal	42.0	21.5	3.0	1.0 max	0.8 Ti, 0.5 max Si, 0.2 max Al, 2.25 Cu, and 0.03 max S
INCO alloy G	0.05 max	19.5	bal	22.25	6.5	1.5	1.0 max Si, 2.125 Nb, 2.5 max Co, 2.0 Cu, 1.0 max W, and 0.04 max P
INCO alloy G-3	0.15 max	19.5	bal	22.25	7.0	1.0 max	5.0 Co, 2.0 Cu, 0.04 max P, 1.0 max Si, 0.03 max S, 1.5 max W, and 0.50 max Nb + Ta
INCO alloy C-276	0.02 max	5.5	bal	15.5	16.0	1.0 max	2.5 max Co, 0.03 max P, 0.03 max S, 0.08 max Si, and 0.35 max V
INCO alloy 904L	0.02 max	bal	25.5	21.0	4.5	2.0 max	1.5 Cu, 1.0 max Si, 0.045 max P, and 0.035 max S

(a) Nominal composition unless otherwise specified. (b) bal, balance

would be premium grades of vinyl-ester and polyester resins. Polypropylene (PP), chlorinated polyvinyl chloride (CPVC), and other thermoplastics can be used in such applications as mist elimination, in which temperatures are suitably low, for example, 80 °C (175 °F) for PP. Rubber linings can also be used where temperatures are suitable and mechanical damage can be avoided.

Waste Incineration

In a number of ways, the problems associated with materials for incinerator off-gas treatment equipment are similar to those used for FGD systems. Depending on the wastes being burned, however, significantly higher gas temperatures as well as more varied and more highly corrosive compounds may be encountered. Materials selection for waste incineration parallels that for FGD systems to some extent, but can often be more demanding.

The importance of incineration for the treatment of domestic and industrial wastes has increased as the availability of sanitary landfills has lessened and their costs have escalated. At the same time, environmental safety regulations have limited the use of deep below-ground and sea-disposal sites for untreated wastes.

Incineration provides a viable, although not inexpensive, alternative that produces scrubbable gaseous and particulate contaminants from a myriad of waste products. Incinerators are used to burn municipal solid wastes, industrial chemical wastes, and sewage sludge. In general, the off-gases can be classified according to their corrosiveness in descending order as follows: industrial chemical, municipal solid, and sewage sludge.

Industrial Chemical. These gases are characterized by extremely high temperatures (1000 °C, or 1830 °F, is not uncommon) and the presence of halogenated compounds. In many cases, chlorinated hydrocarbons and plastics are burned, producing HCl, chlorine, hydrogen fluoride (HF), and possibly hydrogen bromide. Some sulfur and phosphorus compounds may also be produced.

The typical treatment system uses a gas quench to saturate and cool the gases, a wet venturi scrubber (if particulates pose a problem), a packed tower absorber, exhaust fan, ducting, liquid piping, and liquid recirculation pumps. Figure 2 shows a standard system arrangement.

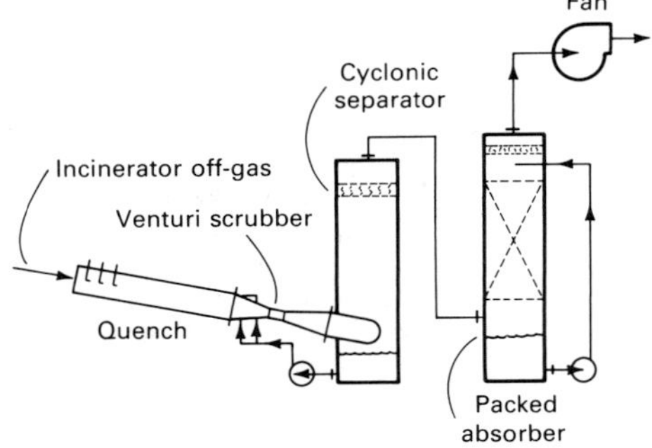

Fig. 2 Schematic of a general scrubber system arrangement

Because of high temperatures, the presence of chlorides, and the fact that the gas becomes saturated with water vapor within the quench, very few materials can be successfully used for the quench construction. The major problem is not uniform attack but local pitting and crevice corrosion of many metals. In particular, chloride stress-corrosion cracking severely affects austenitic stainless steels.

The materials that have been found to perform very well are such high-nickel alloys as Hastelloy C-276, Inconel 625, and titanium for the highest-temperature cases and Hastelloy G and G-3 for slightly less severe cases. These materials have been used in other critical areas of the treatment system, such as fan wheels, dampers, liquid spray nozzles, and piping. Multiple-year service life histories have been reported with these alloys (Ref 4).

Refractory linings for the quench have also been used with some success. This can sometimes prove to be a more economical alternative to the use of high-nickel alloys. Problems do occur, however, because of attack on the binding substances employed and on the carbon steel base material, if exposed.

Following the quench, where temperatures are typically less than 95 °C (205 °F), the major equipment (venturis, tower shells, sump tanks, fan housings, and pump bodies) can be constructed of FRP. A premium polyester or vinyl-ester resin can withstand even the most severe corrosive atmospheres at these milder temperatures. Even the presence of glass-attacking fluorides would not preclude the use of FRP, given the availability of synthetic veils used to replace glass veils within the resin layers closest to the internally exposed surfaces.

The recirculating fluids, often alkaline because of the need to scrub acidic gases, can often be handled satisfactorily by FRP or such thermoplastics as CPVC and PP. In this case, the alkalinity is not the problem. Free chlorides and fluorides may be present even in the most carefully operated and maintained systems.

Fiberglass-reinforced plastic ductwork is used to transport the gases in the milder-temperature areas of the system. Because PP exhibits good resistance to most of the corrosives usually encountered, it is used for tower packing, mist eliminators, and spray nozzles. It is a particularly good choice for environments having the potential for severe fluoride attack. The use of rubber-lined components can be successful, but the emergence of sound FRP construction has limited its popularity.

Caution must be exercised when using plastics in the system following the quench. If the quench loses its liquid and there are no safeguards, a major part of the downstream equipment may be destroyed. Typically, temperatures are monitored so that an emergency cooling liquid source, possibly city water, is injected into the quench to prevent disastrous temperature excursions if the normal liquid source is lost.

A more conservative approach that is implemented in many system designs would also use high-nickel alloy construction for the equipment directly downstream of the quench. In any case, this question must be addressed during the design phase of any incineration project.

Municipal Solid Waste. The by-products of solid municipal wastes can be similar to those found in chemical incineration. The levels of the worst contaminants—chlorides, for example—are usually lower. The nature of the requirements for burning these wastes, which contain large portions of cellulose, result in lower off-gas temperatures than those for chemical incineration.

Nevertheless, corrosion problems are severe, and materials selection is not very different from that for industrial chemicals incineration. Reference 5 provides a ranking of metals with respect to corrosion resistance on the basis of corrosion tests in this service. In addition, Ref 6 shows the results of corrosion tests for a very wide range of alloys in six distinct system zones.

Sewage Sludge. The burning of sewage sludge presents the least corrosive discharge of the three types under discussion. This can be attributed to limited halogen compounds in the gas and somewhat lower temperatures (typically 315 to 650 °C, or 600 to 1200 °F).

Type 304 and 316 stainless steels are suitable for construction in most areas of the system, including the quenching area, whether as a separate quench or part of the wet scrubber. Again, FRP, thermoplastics, and lined carbon steel can be used in the cooler regions.

The predominant contaminants in the environment are odorous sulfur compounds, both organic (mercaptans) and inorganic (hydrogen sulfide, H_2S), and particulate. Chlorides can exist, but they normally originate from the water used for makeup. Their presence sometimes requires the use of high-nickel alloys for such components as fan wheels and pump impellers.

Erosion can be a significant problem in any of these systems. It can wear down critical moving mechanical components and equipment walls at points of liquid and/or gas impingement and, perhaps more importantly, it can contribute to corrosion attack.

The overall effect is not as severe as that found with FGD treatment equipment, but there are a number of areas of concern. The venturi throat and spray nozzles can suffer some abrasion. Fan wheels and pump impellers, however, are usually the most critical areas in these systems with respect to potential problems.

The use of high-nickel alloys at these points has been noted above. Rubber lining can also be used, although generally not on fan wheels. Fiberglass-reinforced plastic can also be fabricated with silicon carbide impregnation for increased abrasion resistance for the internal surfaces.

Bulk Solids

Bulk solids processes include many different industries. The one thing they have in common is the need to handle dust collection for air pollution control.

Examples of this type of process include grain handling, foundries, coal handling, pneumatic conveying systems, and spray-drying systems. In every case, fine particles of dust can become entrained in the exhaust air and must be removed prior to discharge of the air. The three most common types of dust collection equipment are fabric filters, electrostatic precipitators, and wet scrubbers.

The selection of materials of construction does vary with the industry, but the dust handled is generally not severely corrosive. Carbon steel is the most common material of construction.

In selecting a dust collector, the most common construction for the vessel itself is steel, but the fabric used in the collection varies greatly. The manufacturers of fabric filters will have the greatest experience with selection for a specific application. Their expertise should be used in evaluating the relative initial maintenance cost for alternate fabrics. Typically, most bulk-handling applications can be managed with the use of PP for the filter bags. Currently, the cost of PP filter bags is close to that of cotton filter bags. The low initial cost of PP makes it a versatile material of construction for this application. It does not rot when it becomes wet and offers relatively good corrosion resistance. The primary limitation is temperature.

Fans and stacks located downstream from a fabric filter are normally constructed of carbon steel. Because most of the dust has already been collected before it reaches this point, abrasion is not a major concern in the design of the downstream components.

Where extremely high performance is required, an electrostatic precipitator can be applied to a bulk solids application. This normally occurs in relatively dry services with inert particles. As such, the materials of construction are typically carbon steel.

Wet scrubbers are often used where the solids being handled are more reactive. For example, if there is a concern over the potential for an explosive mixture of the dust with air, the wet scrubber eliminates this problem. Wet scrubbers are also versatile and can simultaneously remove dust and gas.

Where scrubbers are applied, their low initial cost is partially offset by the need to recirculate a water-base solution. Care must be taken to ensure that this solution does not become corrosive or, if it does, to select the proper materials of construction for this specific case.

Commonly, general nuisance dust that is collected by a scrubber does not cause a direct corrosion problem. Instead, the problems arise because of the need to minimize the wastewater from the scrubber. For example, if the inlet air is at ambient conditions, the scrubber will evaporate 122 L/h (31.8 gal/h) for every 10 000 m^3/h (5889 ft^3/min). If the solids quantity requires only 10% of the evaporation rate as a liquid bleed rate, then the dissolved solids in the water will be concentrated by a factor of ten. Thus, 200 ppm of chloride would suddenly become 2000 ppm of chloride. This would be sufficient to cause corrosion problems.

In most cases, fiberglass has been considered as a material of construction where abrasion is not particularly severe. In other cases, carbon steel is used, particularly for coal handling or other applications in which abrasion is definitely present.

Spray-drying applications typically require stainless steel. Either type 304 or 316 is used, depending on the particular compound being collected. The use of stainless steel arises from the need for product purity. Because the slurry is usually returned to the spray dryer, care must be taken to avoid any potential corrosion.

Chemical and Related Processing Plants

Chemical process and related industries experience a wide variety of potential corrosion problems. Many of the compounds used have severe effects on many materials of construction. For air pollution control, the quantities of these compounds can be greatly reduced, but the same corrosion problems may still be encountered. Obviously, it is important to rely on the experience of the plant with its process equipment in selecting air pollution control equipment for exhaust ventilation and process vents.

There are some specific differences, and the most important is the difference in operating pressure. Typically, the ventilation systems of chemical reactors will be at atmospheric pressure. By comparison, the reactor itself may be at several atmospheres of pressure. This is important because more economical materials of construction can often be selected for the ventilation system, but they would not have the mechanical strength necessary to handle the pressures in the reactor.

Because fiberglass can be used for atmospheric conditions, many of the clean-up systems used in chemical and process plants are fabricated from fiberglass. The primary reason for using FRP is its low initial cost and good corrosion resistance in a wide variety of services. The corrosion resistance of FRP is a function of both the resin content and the specific resin used in the laminate.

Chloro-Alkali Plants. The production of chlorine results in severe corrosion problems. Quantities of chlorine in the effluent gas are normally scrubbed using dilute caustic solutions. The most common material of construction is fiberglass. Although fiberglass can be used in chlorine service, specific types of resins must be employed for this very difficult application. Vinyl-ester resins are most commonly used. Numerous resins of the vinyl-ester type are available that can handle chlorine and chlorides. In addition to using a high-performance resin, the inner glass reinforcement is usually replaced with a synthetic veil to provide additional protection and to avoid any attack by hypochlorite or caustic on the inner liner.

Fiberglass-reinforced plastic is typically used for the scrubbers handling chlorine removal, ductwork, fans, and stacks. Even recycled pumps are manufactured of this material.

Where heat exchanges are used in the recycled solution, fiberglass is obviously not a practical material because of its low heat transfer coefficient. For dilute hypochlorite solutions, such alloys as Hastelloy C-276 or Inconel 625 have been used. Graphite can be used if the proper binder is selected. With the high heat transfer coefficients of plate heat exchangers, constructions of Hastelloy C-276 can be economical.

Polyvinyl chloride or high temperature PVC (CPVC) is another material of construction that performs well in this service. These materials are sometimes selected for small units or for ductwork construction.

Nitric Acid Plants. In nitric acid (HNO_3) manufacture, stainless steel is the most common material of construction. Concentrated HNO_3 will affect many of the nonmetallic materials of construction; therefore, FRP is not as common or as easily accepted.

For many ventilation systems, either type 304 stainless steel or CPVC could be used. Fiberglass-reinforced plastic could also be used when the acid is being neutralized. Relative costs are shown in Figure 3. Because the cost of stainless steel is still relatively high compared to that of FRP, this alternative should be considered where very dilute acid concentrations are involved. Where concentration and return to the process are involved, stainless steel remains the best solution.

In many cases, HNO_3 manufacture also produces oxides of nitrogen. Nitric oxide (NO) and nitrogen dioxide (NO_2) would be handled by the same materials of construction. Most commonly, these are removed from the air by using scrubbing systems. The recycled solution becomes a dilute HNO_3 solution. A special wet-phase catalyst has been developed for use on this service. This material has properties very similar to those of stainless steel.

In other HNO_3 facilities, thermal reduction has been used to eliminate the residual oxides of nitrogen from the air. Where thermal reduction is

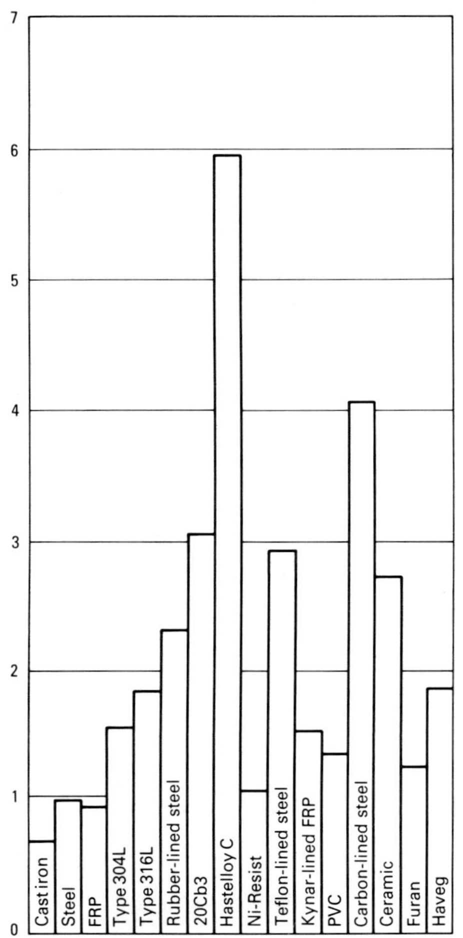

Fig. 3 Relative costs of scrubber materials

used, there is no wet surface. However, the possibility of condensation remains should the system shut down. Therefore, the holders in such units are typically stainless steel. The catalyst itself is normally a ceramic material with a vanadium oxide or similar catalyst applied to the surface. These materials are selected by the manufacturers and would be compatible with stainless steel components selected for ductwork, fans, and other auxiliaries.

Sulfuric Acid Service. Sulfuric acid mist is collected by using fiber bed mist eliminators. Such units employ a glass mat held inside of a vessel operating at low velocities to remove submicron mists.

Fiber bed unit shells for H_2SO_4 are either type 316 stainless steel or alloy 20Cb3. The relative economics suggest the use of type 316 stainless steel, although it may suffer a small amount of attack.

Alloy 20Cb3 can be used for most ranges of acid that would be encountered in air pollution control systems. If the solution is weak enough, type 316 stainless steel can be used. Also, if the temperatures are low enough, FRP should be considered because of its low initial cost. It is best to obtain coupons of the materials and to conduct some initial testing at dilute conditions before making a final decision. The current industry practice is to use more and more FRP on these inorganic acid applications because of low initial cost.

Sulfur Dioxide Service. Sulfur dioxide has similar requirements even though the initial solution formed is usually a neutralized salt; that is, SO_2 is normally absorbed using an alkali solution, such as lime or caustic. This solution is a sodium or calcium sulfite/sulfate mixture. It can be handled at low temperatures in fiberglass and at higher temperatures in type 316L stainless steel.

Because most of the gas is removed in the air pollution control equipment, the downstream equipment can often be handled using liners rather than expensive alloys. This is particularly true for the fan because epoxy coatings can be applied to the fan housing. The wheel itself is recommended in solid alloy construction because of the high speeds involved. The combination of an epoxy liner and a stainless steel wheel can cost as much as 25% less than a solid stainless steel fan.

Discharge stacks are often treated on the same basis. Where the stacks are large enough, a coating can be applied to a steel stack. Of course, the first selection might be an FRP stack if the temperature is low enough because of its elimination of maintenance. More information on materials of construction for the chemical-processing industry is available in the article "Corrosion in the Chemical-Processing Industry" in this Volume.

The Fertilizer Industry. Several severe problems can occur in the manufacture of fertilizers. Trace quantities of fluorides in phosphates result in the formation of HF and silicon tetrafluoride in the gas. Although these can be scrubbed out, the resulting solution is extremely corrosive to most metals and fiberglass.

The most common solution is the use of FRP, but with the substitution of synthetic veils for the inner glass lining. The FRP resin itself is not affected by the HF, but the internal glass could be. A small pinhole leading to the glass would result in a catastrophic attack. This is avoided by substituting a synthetic veil for the glass. Fiberglass is used throughout the industry as a standard. Polypropylene, PVC, and similar thermoplastics are also used. Nitrates and urea products are typically handled by using type 304 stainless steel. As noted above, concentrated HNO_3 could attack FRP-type materials.

Lime Kiln and Similar Kiln Operations. Lime kilns are found in several applications, including the pulp and paper industry. Lime and other kiln applications result in a hot gas that contains dust.

Most kilns use a pollution control system. When possible, a dry collection system is used, because it allows the material to be collected in a form that can be returned directly to the kiln. Some products are simply too reactive for this technique, or the temperature of the kiln is too high. Wet scrubbers are then used.

Most of the scrubbers on kilns are manufactured of carbon steel. The problem of corrosion resistance is usually minimal because the solutions tend to be alkaline or at least neutral. The primary problem is usually abrasion resistance. The basic collectors are manufactured of carbon steel, and high-wear areas are often made of stainless steel. Some units use heat-treated stainless steel, which is hardened for the wear-resistant areas. Another technique is to install a liner to protect areas of greatest wear. Fiberglass-reinforced plastic is typically not used in this application.

Pulp and Paper Industry. Most of the pollution control problems in the pulp and paper industry consist of either the organic sulfur compounds produced from digesting the pulp or the chlorine-related oxidizing agents produced from bleaching the pulp. The reduced sulfur compounds are generally handled in FRP construction. Temperature limitations are not normally a factor, because most of these applications are at temperatures of 80 °C (175 °F) or less. Chlorine or chlorine dioxide applications can be handled by materials of construction similar to those discussed in the section "Chloro-Alkali Plants." More information on materials of construction for the pulp and paper industry is available in the article "Corrosion in the Pulp and Paper Industry" in this Volume.

REFERENCES

1. J.R. Crum, E.L. Hibner, and R.W. Ross, Jr., "Corrosion Resistance of High-Nickel Alloys In Simulated SO_2-Scrubber Environments," Huntington Alloys, Inc.
2. F.G. Hodge, High Performance Alloys...Make Wet Scrubbers Work, *Chem. Eng. Prog.*, Vol 74 (No. 10), 1978, p 84-88
3. R.W. Kirchner, Materials of Construction for Flue-Gas-Desulfurization Systems, *Chem. Eng.*, 19 Sept 1983, p 81-86
4. D.C. Agarwal and F.G. Hodge, "Material Selection Processes and Case Histories Associated with the Hazardous Industrial and Municipal Waste Treatment Industries," Cabot Corporation
5. R.W. Kirchner, Corrosion of Pollution Control Equipment, *Chem. Eng. Prog.*, Vol 71 (No. 3), 1975, p 58-63
6. H.D. Rice, Jr. and R.A. Burford, "Corrosion of Gas-Scrubbing Equipment in Municipal Refuse Incinerators," Paper presented at the International Corrosion Forum, National Association of Corrosion Engineers, 19-23 March 1973

SELECTED REFERENCES

- G.L. Crow and H.R. Horsman, Corrosion in Lime/Limestone Slurry Scrubbers for Coal-Fired Boiler Flue Gases, *Mater. Perform.*, July 1981, p 35-45
- T.G. Gleason, How to Avoid Scrubber Corrosion, *Chem. Eng. Prog.*, Vol 71 (No. 3), 1975, p 43-47
- E.C. Hoxie and G.W. Tuffnell, A Summary of INCO Corrosion Tests in Power Plant Flue Gas Scrubbing Processes, in *Resolving Corrosion Problems in Air Pollution Control Equipment*, National Association of Corrosion Engineers, 1976, p 65-71
- T.S. Lee and R.O. Lewis, Evaluation of Corrosion Behavior of Materials in a Model SO_2 Scrubber System, *Mater. Perform.*, May 1985, p 25-32
- T.S. Lee and B.S. Phull, "Use of a Model Limestone SO_2 Scrubber to Evaluate Slurry Chloride Level Effects on Corrosion Behavior," Paper presented at the APCA/IGCI/NACE Symposium on Solving Problems in Air Pollution Control Equipment, Orlando, FL, Dec 1984
- B.S. Phull and T.S. Lee, "The Effect of Fly Ash and Fluoride on Corrosion Behavior in a Model SO_2 Scrubber," Paper presented at the International Corrosion Forum, National Association of Corrosion Engineers, 25-29 March 1985
- S.L. Sakol and R.A. Schwartz, Construction Materials for Wet Scrubbers, *Chem. Eng. Prog.*, Vol 70 (No. 8), 1974, p 63-68

Metric Conversion Guide

This Section is intended as a guide for expressing weights and measures in the Système International d'Unités (SI). The purpose of SI units, developed and maintained by the General Conference of Weights and Measures, is to provide a basis for world-wide standardization of units and measure. For more information on metric conversions, the reader should consult the following references:

- "Standard for Metric Practice," E 380, *Annual Book of ASTM Standards*, American Society for Testing and Materials, 1916 Race Street, Philadelphia, PA 19103
- "Metric Practice," ANSI/IEEE 268–1982, American National Standards Institute, 1430 Broadway, New York, NY 10018
- *Metric Practice Guide—Units and Conversion Factors for the Steel Industry*, 1978, American Iron and Steel Institute, 1000 16th Street NW, Washington, DC 20036
- *The International System of Units*, SP 330, 1986, National Bureau of Standards. Order from Superintendent of Documents, U.S. Government Printing Office, Washington, DC 20402-9325
- *Metric Editorial Guide*, 4th ed. (revised), 1985, American National Metric Council, 1010 Vermont Avenue NW, Suite 320, Washington, DC 20005–4960
- *ASME Orientation and Guide for Use of SI (Metric) Units*, ASME Guide SI 1, 9th ed., 1982, The American Society of Mechanical Engineers, 345 East 47th Street, New York, NY 10017

Base, supplementary, and derived SI units

Measure	Unit	Symbol
Base units		
Amount of substance	mole	mol
Electric current	ampere	A
Length	meter	m
Luminous intensity	candela	cd
Mass	kilogram	kg
Thermodynamic temperature	kelvin	K
Time	second	s
Supplementary units		
Plane angle	radian	rad
Solid angle	steradian	sr
Derived units		
Absorbed dose	gray	Gy
Acceleration	meter per second squared	m/s^2
Activity (of radionuclides)	becquerel	Bq
Angular acceleration	radian per second squared	rad/s^2
Angular velocity	radian per second	rad/s
Area	square meter	m^2
Capacitance	farad	F
Concentration (of amount of substance)	mole per cubic meter	mol/m^3
Conductance	siemens	S
Current density	ampere per square meter	A/m^2
Density, mass	kilogram per cubic meter	kg/m^3
Electric charge density	coulomb per cubic meter	C/m^3
Electric field strength	volt per meter	V/m
Electric flux density	coulomb per square meter	C/m^2
Electric potential, potential difference, electromotive force	volt	V
Electric resistance	ohm	Ω
Energy, work, quantity of heat	joule	J
Energy density	joule per cubic meter	J/m^3
Entropy	joule per kelvin	J/K
Force	newton	N
Frequency	hertz	Hz
Heat capacity	joule per kelvin	J/K
Heat flux density	watt per square meter	W/m^2
Illuminance	lux	lx
Inductance	henry	H
Irradiance	watt per square meter	W/m^2
Luminance	candela per square meter	cd/m^2
Luminous flux	lumen	lm
Magnetic field strength	ampere per meter	A/m
Magnetic flux	weber	Wb
Magnetic flux density	tesla	T
Molar energy	joule per mole	J/mol
Molar entropy	joule per mole kelvin	J/mol · K
Molar heat capacity	joule per mole kelvin	J/mol · K
Moment of force	newton meter	N · m
Permeability	henry per meter	H/m
Permittivity	farad per meter	F/m
Power, radiant flux	watt	W
Pressure, stress	pascal	Pa
Quantity of electricity, electric charge	coulomb	C
Radiance	watt per square meter steradian	$W/m^2 \cdot sr$
Radiant intensity	watt per steradian	W/sr
Specific heat capacity	joule per kilogram kelvin	J/kg · K
Specific energy	joule per kilogram	J/kg
Specific entropy	joule per kilogram kelvin	J/kg · K
Specific volume	cubic meter per kilogram	m^3/kg
Surface tension	newton per meter	N/m
Thermal conductivity	watt per meter kelvin	W/m · K
Velocity	meter per second	m/s
Viscosity, dynamic	pascal second	Pa · s
Viscosity, kinematic	square meter per second	m^2/s
Volume	cubic meter	m^3
Wavenumber	1 per meter	1/m

Conversion factors

Angle

To convert from	to	multiply by
degree	rad	1.745 329 E − 02

Area

To convert from	to	multiply by
in.^{2}	mm^2	6.451 600 E + 02
in.^{2}	cm^2	6.451 600 E + 00
in.^{2}	m^2	6.451 600 E − 04
ft^2	m^2	9.290 304 E − 02

Bending moment or torque

To convert from	to	multiply by
lbf · in.	N · m	1.129 848 E − 01
lbf · ft	N · m	1.355 818 E + 00
kgf · m	N · m	9.806 650 E + 00
ozf · in.	N · m	7.061 552 E − 03

Bending moment or torque per unit length

To convert from	to	multiply by
lbf · in./in.	N · m/m	4.448 222 E + 00
lbf · ft/in.	N · m/m	5.337 866 E + 01

Corrosion rate

To convert from	to	multiply by
mils/yr	mm/yr	2.540 000 E − 02
mils/yr	μm/yr	2.540 000 E + 01

Current density

To convert from	to	multiply by
A/in.^{2}	A/cm^2	1.550 003 E − 01
A/in.^{2}	A/mm^2	1.550 003 E − 03
A/ft^2	A/m^2	1.076 400 E + 01

Electricity and magnetism

To convert from	to	multiply by
gauss	T	1.000 000 E − 04
maxwell	μWb	1.000 000 E − 02
mho	S	1.000 000 E + 00
Oersted	A/m	7.957 700 E + 01
Ω · cm	Ω · m	1.000 000 E − 02
Ω circular-mil/ft	μΩ · m	1.662 426 E − 03

Energy (impact, other)

To convert from	to	multiply by
ft · lbf	J	1.355 818 E + 00
Btu (thermochemical)	J	1.054 350 E + 03
cal (thermochemical)	J	4.184 000 E + 00
kW · h	J	3.600 000 E + 06
W · h	J	3.600 000 E + 03

Flow rate

To convert from	to	multiply by
ft^3/h	L/min	4.719 475 E − 01
ft^3/min	L/min	2.831 000 E + 01
gal/h	L/min	6.309 020 E − 02
gal/min	L/min	3.785 412 E + 00

Force

To convert from	to	multiply by
lbf	N	4.448 222 E + 00
kip (1000 lbf)	N	4.448 222 E + 03
tonf	kN	8.896 443 E + 00
kgf	N	9.806 650 E + 00

Force per unit length

To convert from	to	multiply by
lbf/ft	N/m	1.459 390 E + 01
lbf/in.	N/m	1.751 268 E + 02

Fracture toughness

To convert from	to	multiply by
ksi $\sqrt{in.}$	MPa $\sqrt{m}$	1.098 800 E + 00

Heat content

To convert from	to	multiply by
Btu/lb	kJ/kg	2.326 000 E + 00
cal/g	kJ/kg	4.186 800 E + 00

Heat input

To convert from	to	multiply by
J/in.	J/m	3.937 008 E + 01
kJ/in.	kJ/m	3.937 008 E + 01

Length

To convert from	to	multiply by
Å	nm	1.000 000 E − 01
μin.	μm	2.540 000 E − 02
mil	μm	2.540 000 E + 01
in.	mm	2.540 000 E + 01
in.	cm	2.540 000 E + 00
ft	m	3.048 000 E − 01
yd	m	9.144 000 E − 01
mile	km	1.609 300 E + 00

Mass

To convert from	to	multiply by
oz	kg	2.834 952 E − 02
lb	kg	4.535 924 E − 01
ton (short, 2000 lb)	kg	9.071 847 E + 02
ton (short, 2000 lb)	kg × 10^3(a)	9.071 847 E − 01
ton (long, 2240 lb)	kg	1.016 047 E + 03

Mass per unit area

To convert from	to	multiply by
oz/in.^{2}	kg/m^2	4.395 000 E + 01
oz/ft^2	kg/m^2	3.051 517 E − 01
oz/yd^2	kg/m^2	3.390 575 E − 02
lb/ft^2	kg/m^2	4.882 428 E + 00

Mass per unit length

To convert from	to	multiply by
lb/ft	kg/m	1.488 164 E + 00
lb/in.	kg/m	1.785 797 E + 01

Mass per unit time

To convert from	to	multiply by
lb/h	kg/s	1.259 979 E − 04
lb/min	kg/s	7.559 873 E − 03
lb/s	kg/s	4.535 924 E − 01

Mass per unit volume (includes density)

To convert from	to	multiply by
g/cm^3	kg/m^3	1.000 000 E + 03
lb/ft^3	g/cm^3	1.601 846 E − 02
lb/ft^3	kg/m^3	1.601 846 E + 01
lb/in.^{3}	g/cm^3	2.767 990 E + 01
lb/in.^{3}	kg/m^3	2.767 990 E + 04

Power

To convert from	to	multiply by
Btu/s	kW	1.055 056 E + 00
Btu/min	kW	1.758 426 E − 02
Btu/h	W	2.928 751 E − 01
erg/s	W	1.000 000 E − 07
ft · lbf/s	W	1.355 818 E + 00
ft · lbf/min	W	2.259 697 E − 02
ft · lbf/h	W	3.766 161 E − 04
hp (550 ft · lbf/s)	kW	7.456 999 E − 01
hp (electric)	kW	7.460 000 E − 01

Power density

To convert from	to	multiply by
W/in.^{2}	W/m^2	1.550 003 E + 03

Pressure (fluid)

To convert from	to	multiply by
atm (standard)	Pa	1.013 250 E + 05
bar	Pa	1.000 000 E + 05
in. Hg (32 °F)	Pa	3.386 380 E + 03
in. Hg (60 °F)	Pa	3.376 850 E + 03
lbf/in.^{2} (psi)	Pa	6.894 757 E + 03
torr (mm Hg, 0 °C)	Pa	1.333 220 E + 02

Specific heat

To convert from	to	multiply by
Btu/lb · °F	J/kg · K	4.186 800 E + 03
cal/g · °C	J/kg · K	4.186 800 E + 03

Stress (force per unit area)

To convert from	to	multiply by
tonf/in.^{2} (tsi)	MPa	1.378 951 E + 01
kgf/mm^2	MPa	9.806 650 E + 00
ksi	MPa	6.894 757 E + 00
lbf/in.^{2} (psi)	MPa	6.894 757 E − 03
MN/m^2	MPa	1.000 000 E + 00

Temperature

To convert from	to	multiply by
°F	°C	5/9 · (°F − 32)
°R	°K	5/9

Temperature interval

To convert from	to	multiply by
°F	°C	5/9

Thermal conductivity

To convert from	to	multiply by
Btu · in./s · ft^2 · °F	W/m · K	5.192 204 E + 02
Btu/ft · h · °F	W/m · K	1.730 735 E + 00
Btu · in./h · ft^2 · °F	W/m · K	1.442 279 E − 01
cal/cm · s · °C	W/m · K	4.184 000 E + 02

Thermal expansion

To convert from	to	multiply by
in./in. · °C	m/m · K	1.000 000 E + 00
in./in. · °F	m/m · K	1.800 000 E + 00

Velocity

To convert from	to	multiply by
ft/h	m/s	8.466 667 E − 05
ft/min	m/s	5.080 000 E − 03
ft/s	m/s	3.048 000 E − 01
in./s	m/s	2.540 000 E − 02
km/h	m/s	2.777 778 E − 01
mph	km/h	1.609 344 E + 00

Velocity of rotation

To convert from	to	multiply by
rev/min (rpm)	rad/s	1.047 164 E − 01
rev/s	rad/s	6.283 185 E + 00

Viscosity

To convert from	to	multiply by
poise	Pa · s	1.000 000 E − 01
stokes	m^2/s	1.000 000 E − 04
ft^2/s	m^2/s	9.290 304 E − 02
in.^{2}/s	mm^2/s	6.451 600 E + 02

Volume

To convert from	to	multiply by
in.^{3}	m^3	1.638 706 E − 05
ft^3	m^3	2.831 685 E − 02
fluid oz	m^3	2.957 353 E − 05
gal (U.S. liquid)	m^3	3.785 412 E − 03

Volume per unit time

To convert from	to	multiply by
ft^3/min	m^3/s	4.719 474 E − 04
ft^3/s	m^3/s	2.831 685 E − 02
in.^{3}/min	m^3/s	2.731 177 E − 07

Wavelength

To convert from	to	multiply by
Å	nm	1.000 000 E − 01

(a) kg × 10^3 = 1 metric ton

SI prefixes—names and symbols

Exponential expression	Multiplication factor	Prefix	Symbol
10^{18}	1 000 000 000 000 000 000	exa	E
10^{15}	1 000 000 000 000 000	peta	P
10^{12}	1 000 000 000 000	tera	T
10^{9}	1 000 000 000	giga	G
10^{6}	1 000 000	mega	M
10^{3}	1 000	kilo	k
10^{2}	100	hecto(a)	h
10^{1}	10	deka(a)	da
10^{0}	1	BASE UNIT	
10^{-1}	0.1	deci(a)	d
10^{-2}	0.01	centi(a)	c
10^{-3}	0.001	milli	m
10^{-6}	0.000 001	micro	μ
10^{-9}	0.000 000 001	nano	n
10^{-12}	0.000 000 000 001	pico	p
10^{-15}	0.000 000 000 000 001	femto	f
10^{-18}	0.000 000 000 000 000 001	atto	a

(a) Nonpreferred. Prefixes should be selected in steps of 10^3 so that the resultant number before the prefix is between 0.1 and 1000. These prefixes should not be used for units of linear measurement, but may be used for higher order units. For example, the linear measurement, decimeter, is nonpreferred, but square decimeter is acceptable.

Relationships among some of the units commonly used for corrosion rates

d is metal density in grams per cubic centimeter (g/cm^3)

Unit	Factor for conversion to— mdd	$g/m^2/d$	μm/yr	mm/yr	mils/yr	in./yr
Milligrams per square decimeter per day (mdd)	1	0.1	36.5/*d*	0.0365/*d*	1.144/*d*	0.00144/*d*
Grams per square meter per day ($g/m^2/d$)	10	1	365/*d*	0.365/*d*	14.4/*d*	0.0144/*d*
Microns per year (μm/yr)	0.0274*d*	0.00274*d*	1	0.001	0.0394	0.0000394
Millimeters per year (mm/yr)	27.4*d*	2.74*d*	1000	1	39.4	0.0394
Mils per year (mils/yr)	0.696*d*	0.0696*d*	25.4	0.0254	1	0.001
Inches per year (in./yr)	696*d*	69.6*d*	25 400	25.4	1000	1

Source: G. Wranglén, *An Introduction to Corrosion and Protection of Metals*, Chapman and Hall, 1985, p 238

Nomograph for conversion of corrosion rates

The example given is for type 304 stainless steel (density 7.87 g/cm^3) and a corrosion rate of 30 mils/yr.

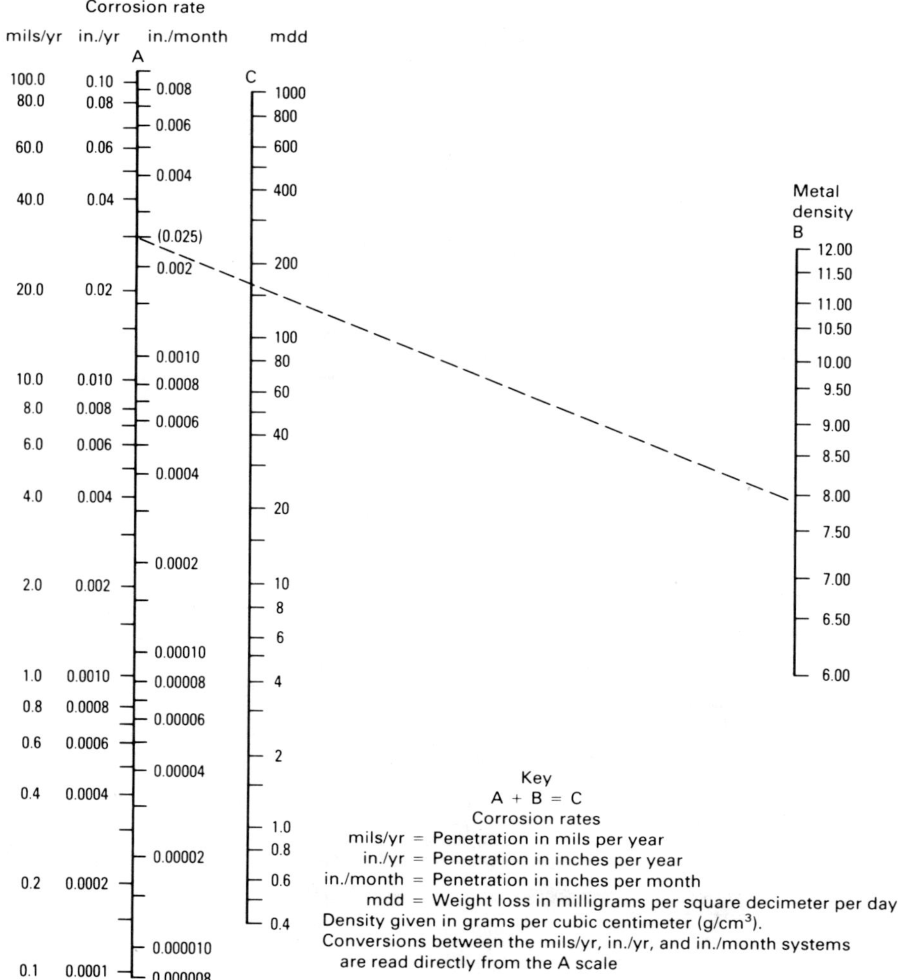

Source: M.G. Fontana, *Corrosion Engineering*, 3rd ed., McGraw-Hill, 1986, p 217

Abbreviations and Symbols

a chemical activity; crack length; crystal lattice length along the *a* axis

A area

A ampere

Å angstrom

A_a anodic area

AA Aluminum Association

ABS acrylonitrile-butadiene-styrene

A_c cathodic area

ac alternating current

ACI Alloy Casting Institute

a_{cr} critical flaw (crack) size

adt air dried ton

AECL Atomic Energy of Canada Limited

AHF anhydrous hydrogen fluoride

AISI American Iron and Steel Institute

AMS Aerospace Material Specification (of SAE)

ANSI American National Standards Institute

API American Petroleum Institute

APU auxiliary power unit (space shuttle)

ASME American Society of Mechanical Engineers

ASTM American Society for Testing and Materials

at.% atomic percent

atm atmospheres (pressure)

ATP acceptance testing procedures

AVT all-volatile water treatment

AWS American Welding Society

b crystal lattice length along the *b* axis; Tafel coefficient

B absolute mobility

bal balance or remainder

bcc body-centered cubic

BCMP bisulfite chemi-mechanical pulping

bct body-centered tetragonal

Bé Baumé hydrometer scale

BFPD barrels of fluid produced per day (oil and gas production)

BWR boiling water reactor(s)

c crystal lattice length along the *c* axis

C concentration

C coulomb

CASS copper-accelerated acetic acid-salt spray (test)

CCI crevice corrosion index

CCT critical crevice temperature

CDA Copper Development Association

CFC corrosion-fatigue cracking

cm centimeter

cpm cycles per minute

cps cycles per second

CPVC chlorinated polyvinyl chloride

CRT collet retainer tube

CTMP chemi-thermomechanical pulping

CVD chemical vapor deposition

d density; used in mathematical expressions involving a derivative (denotes rate of change)

D diffusion coefficient

d day

da/dN crack growth rate per cycle

da/dt crack growth rate per unit time

dc direct current

DCB double cantilever beam

diam diameter

DIP dual in-line package (electronic component)

DNB departure from nucleate (boiling)

e^- electron

e natural log base, 2.71828 . . .

EC electrolytic corrosion (test)

E_{corr} corrosion potential

ECTFE ethylene-chlorotrifluoroethylene

EDTA ethylenediamine tetraacetic acid

E_e equilibrium potential

EPA Environmental Protection Agency

E_{pass} passivation potential

EPDM ethylene propylene diene monomer (rubber)

E_{pr} passive potential range

EPRI Electric Power Research Institute

EPS electrical power system (space shuttle)

Eq equation

f flow rate

F Faraday constant

F farad

fcc face-centered cubic

Fig. figure

FACT Ford anodized aluminum corrosion test

FEP fluorinated ethylene propylene

FGD flue gas desulfurization

FRP fiber-reinforced plastic

ft foot

G Gibbs energy

g gas; gram

gal gallon

GMA gas metal arc

GOR gas-oil ratio (in petroleum production)

GOX gaseous oxygen

GTA gas tungsten arc

h hour

h˙ electron hole

HAZ heat-affected zone

HB Brinell hardness

hcp hexagonal close-packed

HIP hot isostatic pressing

HK Knoop hardness

HLW high-level waste

HR Rockwell hardness; requires scale designation, such as HRC for Rockwell C hardness

HSLA high-strength low-alloy (steel)

HV Vickers (diamond pyramid) hardness

HWR heavy water reactor(s)

Hz hertz

i* or *I current

I* or *i current density

IASCC irradiation-assisted stress-corrosion cracking

IC integrated circuit

i_{corr} corrosion current

i_{crit} critical current for passivation

ID inside diameter

IGA intergranular attack

IGP intergranular penetration

IGSCC intergranular stress-corrosion cracking

ILZRO International Lead Zinc Research Organization

I/M ingot metallurgy

in. inch

INPO Institute for Nuclear Power Operations

i_o exchange current

i_{pass} passive current

IR infrared

ISO International Organization for Standardization

I_{trans} current due to transpassive dissolution

IVD ion vapor deposited

J flux or mass; stress-intensity factor in elastic-plastic fracture mechanics

J joule

K stress-intensity factor in linear elastic fracture mechanics

K Kelvin

k_c potential-dependent rate constant
K_{Ic} plane-strain fracture toughness
K_{IHE} threshold stress intensity for hydrogen embrittlement
K_{ISCC} threshold stress intensity for stress-corrosion cracking
kg kilogram
k_L linear oxidation rate constant
k_p parabolic oxidation rate constant
kPa kilopascal
ksi kips per square inch (1000 pounds per square inch)
K_t stress concentration factor
K_{th} threshold stress-intensity factor
kW kilowatt
L length
L liter
l liquid
lb pound
LME liquid metal embrittlement
LMP Larsen-Miller parameter
ln natural logarithm (base e)
log common logarithm (base 10)
LOX liquid oxygen
LSI Langelier saturation index
LWR light water reactor(s)
m mass; molar (solution)
M molecular weight; molar solution
M metal
mA milliampere
MAPP methyl-acetylene-propadiene (gas)
max maximum
MBT mercaptobenzothiazole
m_c mass transport coefficient
MCA multiple crevice assembly
MDC microdiscontinuous chromium (plating)
mg milligram
Mg megagram
MIL military
MIL-STD military standard
min minimum; minute
mL milliliter
mm millimeter
MMH monomethyl hydrazine
MPa megapascal
MPIF Metal Power Industries Foundation
MPS main propulsion system (space shuttle)
MTI Materials Technology Institute (of the Chemical Process Industry)
MTTC mils per two thousand cooks (pulp and paper industry)
mV millivolt
MW megawatt
MWD measurement while drilling (housings or instruments in oil and gas production)
N mole fraction; normal (solution); number of cycles in corrosion-fatigue testing
N newton
NACE National Assocation of Corrosion Engineers
NASA National Aeronautics and Space Administration
NG nuclear grade
NGPA Natural Gas Producers Association
nm nanometer
No. number
NRC Nuclear Regulatory Commission
NVT neutrons per unit volume × time
NWTI National Wood Tank Institute
O oxidation, oxidizing reagent
OD outside diameter
ODS oxide dispersion strengthened
OMS orbital maneuvering system (space shuttle)
OSHA Occupational Safety and Health Administration
OTSG once-through steam generator
oz ounce
p partial pressure
p page
Pa pascal
pH negative logarithm of hydrogen ion activity
PH precipitation-hardenable
p_{O_2} partial pressure of oxygen
PCB printed circuit board
P/M powder metallurgy
ppb parts per billion
PPB prior particle boundary
ppm parts per million
PP polypropylene
PPRIC Pulp and Paper Research Institute of Canada
psi pounds per square inch
psia pounds per square inch (absolute)
psig pounds per square inch (gage)
PTB pounds per thousand barrels (crude oil)
PTFE polytetrafluoroethylene
PVA polyvinyl alcohol
PVC polyvinyl chloride
PVD physical vapor deposition
PVDC polyvinylidene chloride
PVDF polyvinylidene fluoride
PWHT post-weld heat treatment
PWR pressurized water reactor(s)
Q activation energy for diffusion
r corrosion rate
R radius; ratio of minimum stress to maximum stress; gas constant
R Rankine; reduction; reduced species
RCS reaction control system (space shuttle)
Re Reynold's number
Ref reference
rem roentgen equivalent man; remainder or balance
RHR root height reading
RMP refiner mechanical pulping
rms root mean square
RSG recirculating steam generator
RSI Ryzner stability index
RTV room-temperature vulcanizing (rubber)
s solid
S_a stress amplitude
SAE Society of Automotive Engineers
Sc Schmidt number
SCC stress-corrosion cracking
SCE saturated calomel electrode
SEM scanning electron microscopy
S_f fatigue (endurance) limit
Sh Sherwood number
S_m mean stress
S_r stress range
SHE standard hydrogen electrode
SI Système International d'Unités
SIMS secondary ion mass spectroscopy
SLC sustained-load cracking
SMIE solid metal induced embrittlement
SPS service propulsion system (space shuttle)
SRB sulfate reducing bacteria; solid rocket booster (space shuttle)
SSPC Steel Structures Painting Council
t thickness; time
T absolute temperature
TAPPI Technical Association of the Pulp and Paper Industry
TEM transmission electron microscopy
TFE tetrafluoroethylene
TMP thermomechanical pulping
TPS thermal protection system (tiles on the space shuttle)
TRS total reduced sulfur
TYS tensile yield strength
UNS Unified Numbering System
UTS ultimate tensile strength
UV ultraviolet
V volume
V volt
V″ doubly charged cation vacancy
vol% volume percent
w weight or mass
W watt
wt% weight percent
XPS x-ray photoelectron spectroscopy
yr year
° degree (angular measure)
°C temperature, degrees Celsius (centigrade)
°F temperature, degrees Fahrenheit
$\rightleftharpoons$ direction of a chemical reaction
÷ divided by
= equals
≈ approximately equals
≠ not equal to
> greater than
≫ much greater than

$\geq$ greater than or equal to
$\int$ integral of
$<$ less than
$\ll$ much less than
$\leq$ less than or equal to
$\pm$ maximum deviation
$-$ minus; negative ion charge
$\times$ multiplied by; diameters (magnification)
$\cdot$ multiplied by
‰ parts per thousand
/ per
% percent
$+$ plus; in addition to; positive ion charge
$\sqrt{}$ square root of
$\sim$ similar to; approximately
$\propto$ varies as; is proportional to
β Tafel coefficient
Δ change in quantity, an increment, a range
$\Delta\epsilon_e$ elastic strain range
ϵ_p plastic strain range
ϵ_t total strain range
ΔE_{therm} thermodynamic driving force
ΔG° standard Gibbs energy change
ΔH enthalpy of reaction
ΔK stress-intensity factor range
ΔS entropy change
ϵ applied strain
η overpotential
μ chemical potential
μA microamp
μin. micro-inch
μm micron (micrometer)
μs microsecond
μV microvolt
ν kinematic viscosity
π pi (3.14159 . . .)
ρ density
σ oxide conductivity; stress
Σ summation of
Ω ohm

Greek Alphabet

A, α alpha
B, β beta
Γ, γ gamma
Δ, δ delta
E, ϵ epsilon
Z, ζ zeta
H, η eta
Θ, θ theta
I, ι iota
K, κ kappa
Λ, λ lambda
M, μ mu
N, ν nu
Ξ, ξ xi
O, o omicron
Π, π pi
P, ρ rho
Σ, σ sigma
T, τ tau
Υ, υ upsilon
Φ, φ phi
X, χ chi
Ψ, ψ psi
Ω, ω omega

Index

A

B

C

D

E

F

G

I

J

K

L

M

N

O

P

Q

R

S

U

V

W

X

Y

Z